MASTER ATLAS OF GREATER LONDON

Key to Map Pages	Overleaf
Map Pages @ approx. 3⅓" to 1 mile	2-211
Congestion Charging Zone Map	212
Large Scale Map Pages @ 9" to 1 mile	213-231
West End Cinemas & Theatres Maps	232
Road Map Pages @ 3 miles to 1"	233-241
London's Rail & Tube Services Map	242-243
Underground Railway Map	244-245
Index to Streets, Places, Areas, Hospitals, Stations, and selected Places of Interest	246-431
Emission Zones Map	432-433

REFERENCE

- Motorway M1
- A Road A2
- Tunnel
- B Road B408
- Dual Carriageway
- One-way Street
  Traffic flow on A Roads is also indicated by a heavy line on the driver's left.
- Road Under Construction
  Opening dates are correct at the time of publication.
- Proposed Road
- Junction Name MARBLE ARCH
- Restricted Access
- Pedestrianized Road
- Track / Footpath
- Residential Walkway
- Congestion Charging Zone — Zone edge. Within Zone.
  See page 212 for more information.
- Ultra Low Emission Zone — Zone edge. Within Zone.
  See pages 432-433 for more information.
- Low Emission Zone — Zone edge. Within Zone.
  See pages 432-433 for more information.
- Railway — Tunnel, Level Crossing
- Stations:
  - National Rail Network
  - Crossrail / Elizabeth Line
    See www.tfl.gov.uk for up to date information
  - Docklands Light Railway DLR
  - Overground
  - Underground
- London Tramlink — Tunnel, Stop
  The boarding of Tramlink trams at stops may be limited to a single direction, indicated by the arrow.
- Built-up Area MILL STREET
- Local Authority Boundary
- Postcode Boundary

- Map Continuation 56 — Large Scale Map Pages 214
- Airport
- Car Park (selected) P
- Church or Chapel †
- Fire Station ■
- Hospital H
- House Numbers (A & B Roads only) 37 44
- Information Centre
- National Grid Reference 530
- Park and Ride — Windsor (Home Park) P+R
- Police Station ▲
- Post Office ★
- River Bus Stop R
- Toilet
- Educational Establishment
- Hospital or Healthcare Building
- Industrial Building
- Leisure or Recreational Facility
- Place of Interest
- Public Building
- Shopping Centre or Market
- Other Selected Buildings

SCALE

Map pages 2-211
1:19,000
3⅓ inches (8.47 cm) to 1 mile
5.26 cm to 1 km

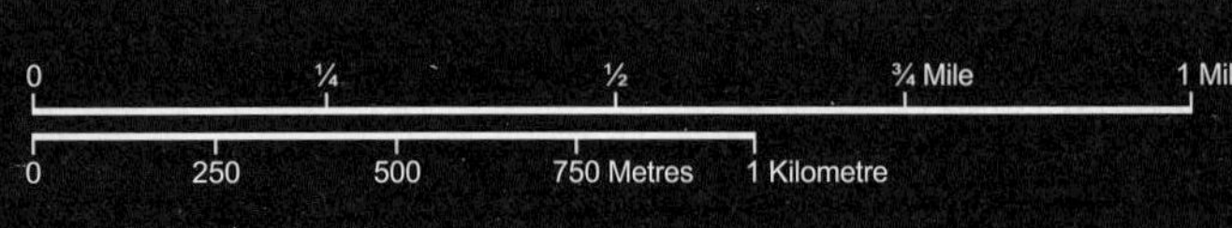

0
¼
½
¾ Mile
1 Mile
0
250
500
750 Metres
1 Kilometre

EDITION 18 2022

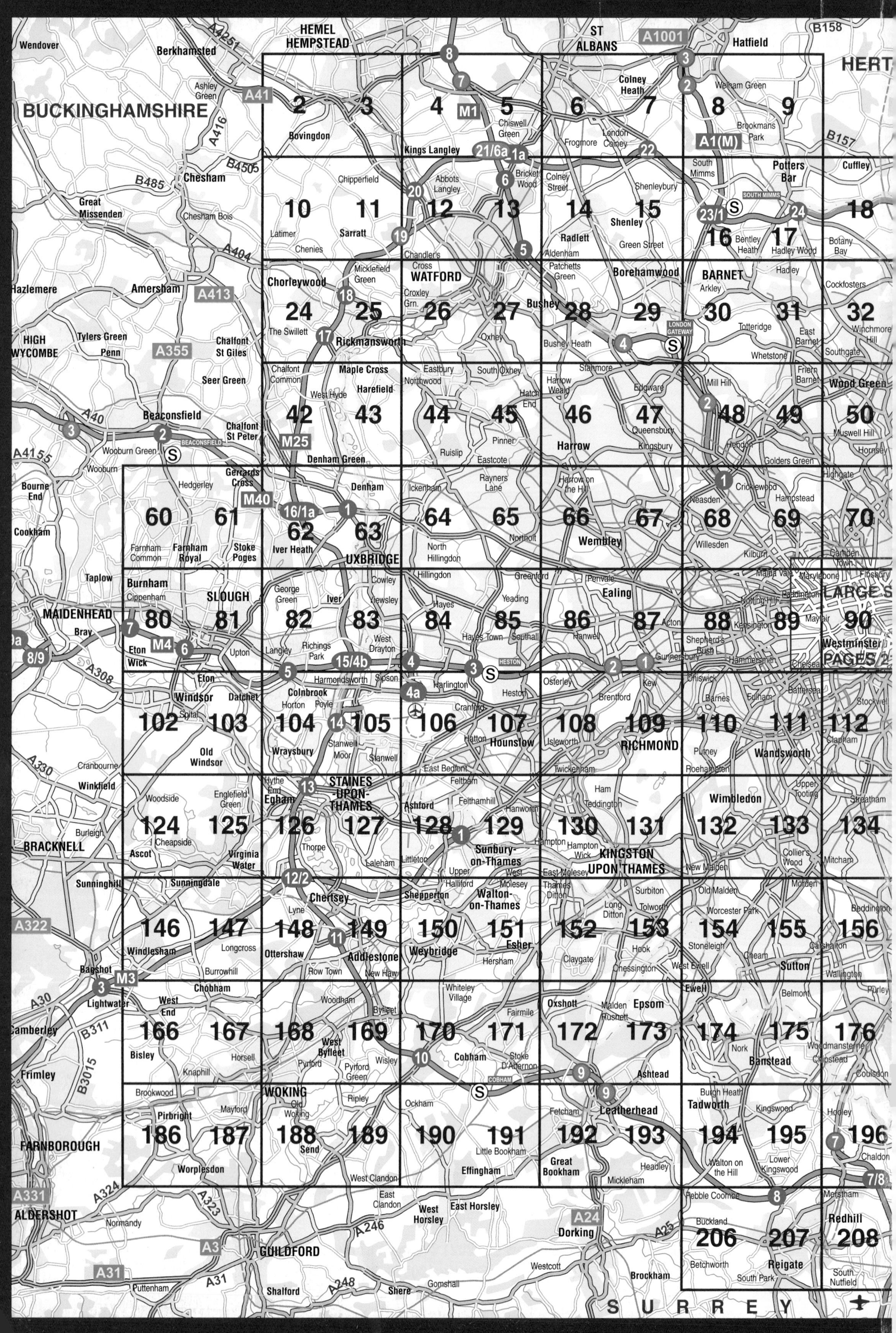

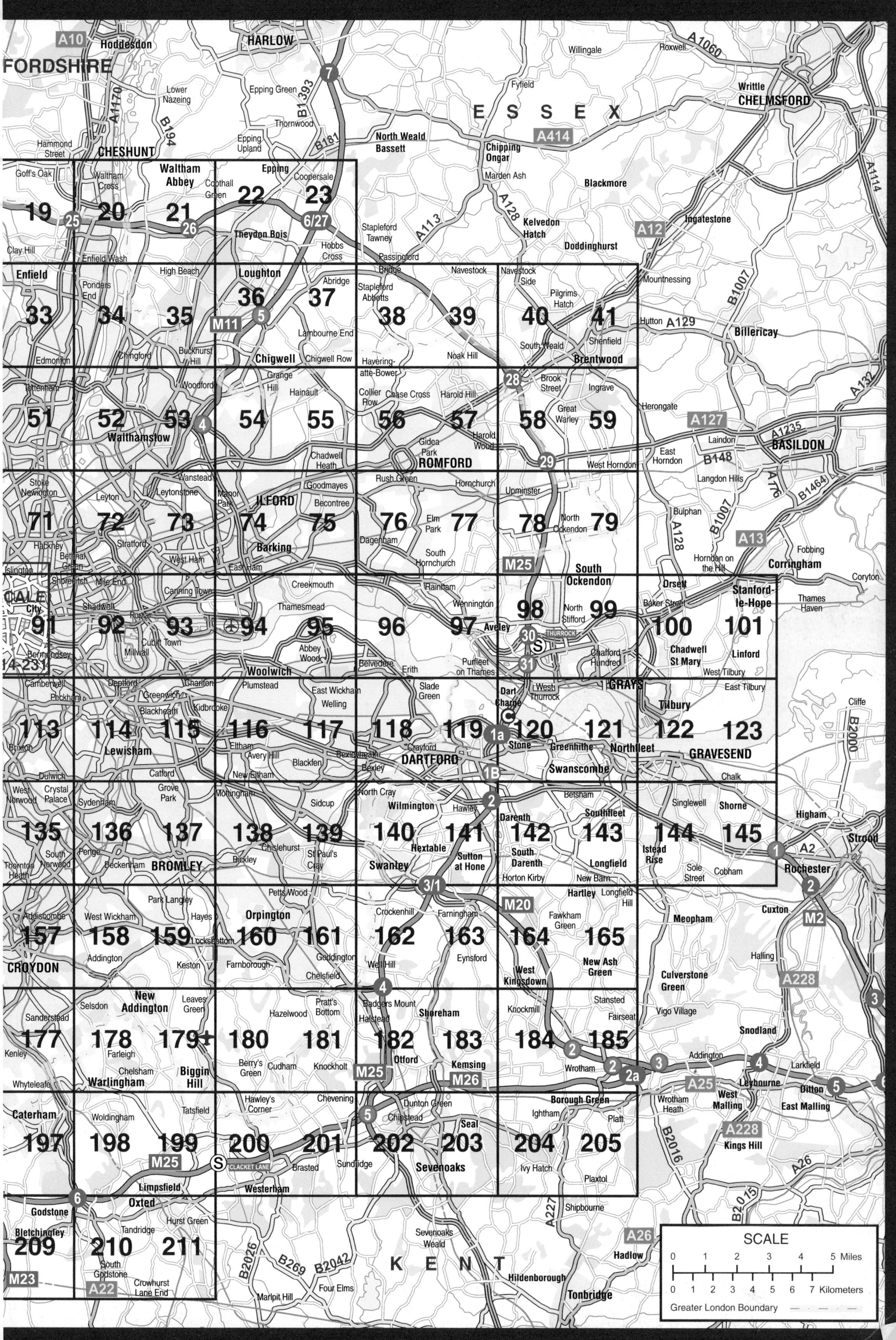
ESSEX
KENT
HARLOW
CHELMSFORD
CHESHUNT
ROMFORD
ILFORD
DARTFORD
GRAVESEND
BROMLEY
CROYDON
BASILDON
GRAYS
Sevenoaks
Rochester
SCALE
0 1 2 3 4 5 Miles
0 1 2 3 4 5 6 7 Kilometers
Greater London Boundary

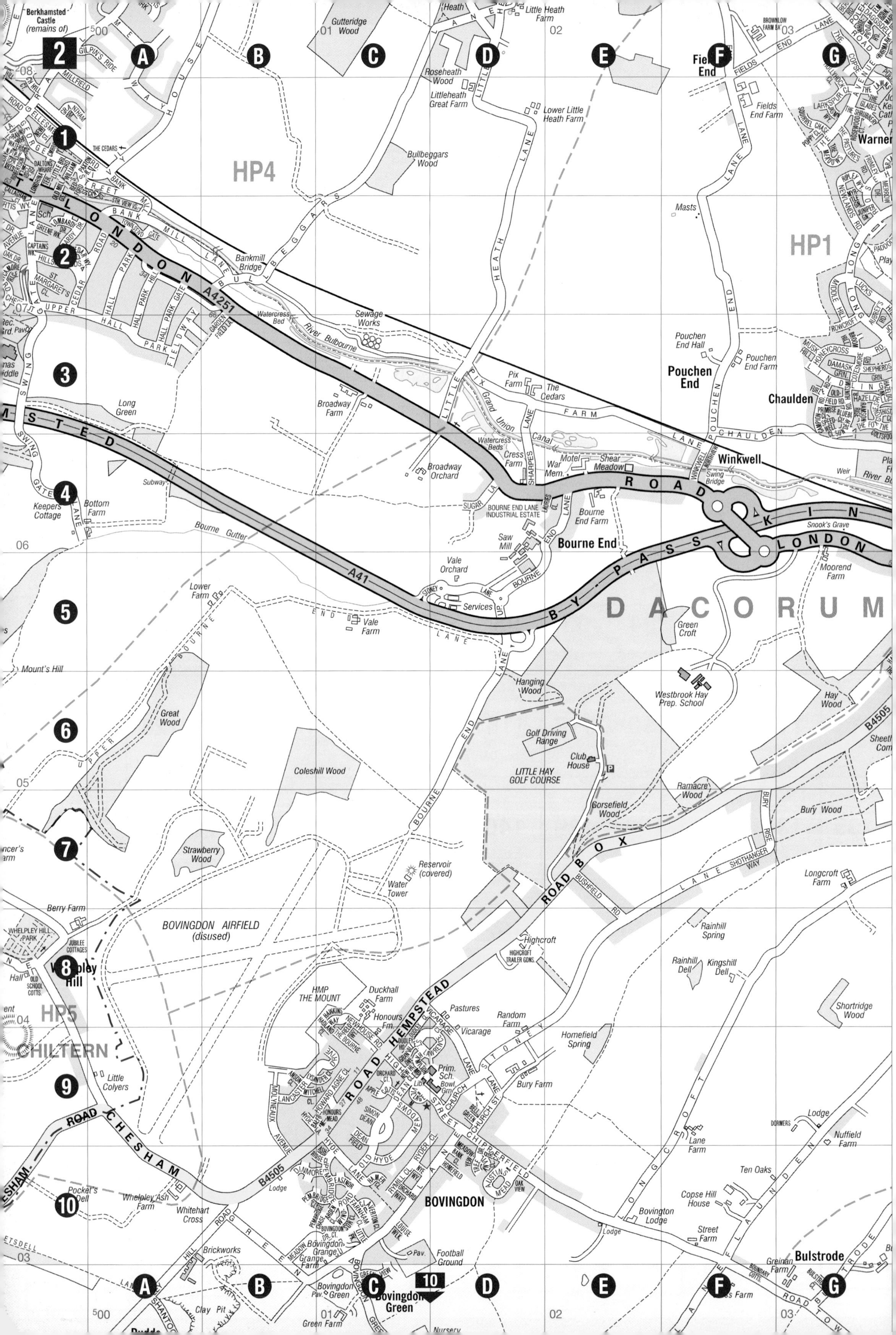

2
Berkhamsted Castle (remains of)
Gutteridge Wood
Little Heath Farm
Roseheath Wood
Littleheath Great Farm
Lower Little Heath Farm
Fields End
Fields End Farm
Warner
HP4
Bullbeggars Wood
HP1
Masts
LONDON ROAD
A4251
Bankmill Bridge
Watercress Bed
River Bulbourne
Sewage Works
Pouchen End Hall
Pouchen End Farm
Pouchen End
Chaulden
Long Green
Broadway Farm
Pix Farm
The Cedars
Grand Union Canal
Watercress Beds
Cress Farm
Motel
War Mem.
Shear Meadow
Winkwell
Swing Bridge
Weir
Broadway Orchard
Subway
Keepers Cottage
Bottom Farm
Bourne Gutter
BOURNE END LANE INDUSTRIAL ESTATE
Bourne End Farm
Saw Mill
Bourne End
Snook's Grave
Moorend Farm
A41
Vale Orchard
Services
Lower Farm
Vale Farm
BY-PASS
DACORUM
Green Croft
Mount's Hill
Hanging Wood
Westbrook Hay Prep. School
Hay Wood
B4505
Great Wood
Golf Driving Range
Club House
LITTLE HAY GOLF COURSE
Coleshill Wood
Ramacre Wood
Gorsefield Wood
Bury Wood
Strawberry Wood
BOX LANE
Reservoir (covered)
Water Tower
Longcroft Farm
Berry Farm
BOVINGDON AIRFIELD (disused)
Highcroft
HIGHCROFT TRAILER GDNS.
Rainhill Spring
Rainhill Dell
Kingshill Dell
Whelpley Hill
HMP THE MOUNT
Duckhall Farm
Honours Fm.
Pastures
Random Farm
Vicarage
HEMPSTEAD ROAD
Homefield Spring
Shortridge Wood
HP5
CHILTERN
Little Colyers
Prim. Sch.
Bowl Grn
Bury Farm
CHESHAM ROAD
Lodge
Nuffield Farm
Lane Farm
Ten Oaks
Pocket's Dell
Whelpley Ash Farm
Whitehart Cross
B4505
Copse Hill House
BOVINGDON
Bovington Lodge
Street Farm
Brickworks
Bovingdon Grange
Grange Farm
Football Ground
Bulstrode
Greinan Farm
Bovingdon Green
Clay Pit
Green Farm
10

3
H
J
K
L
M
N
P
Gadebridge
Highfield
Playing Fields
Galley Hill Prim. Sch.
Randall Park
Gadebridge Park
Skatepark
Weir
Putting Green
St. Cuthbert Mayne Catholic Sch.
Gade Valley Jun. Mixed & Inf. Sch.
Tennis Cts.
Playing Field
The Cavendish Sch.
The Old Town Hall Arts Cen.
Playing Field
Hoblets Manor Schs.
QUEENSWAY
B487
HP2
A4146
BUZZARD
ROAD
West Herts College
Dacorum Wy.
Civic Cen.
Lib.
Wks.
Adeyfield
Adventure Playground
Broadfield Prim. Sch.
Counters End
Hammerfield
Gravelhill Spring
Northridge Park
Shrub Hill Common
Playing Fields
Tennis Courts
The Drey
Lockers Pk. Sch.
Lockers Pk.
Cemy.
Nursy. Sch.
Paradise
HEMEL HEMPSTEAD GENERAL HOSPITAL
Quasar
Paradise Industrial Estate
Superstore
Cineworld Cin.
Planet Ice
The Gym
Jarman Square
A414
ST. ALBANS
Green End
Prim. Sch.
St. Rose's Catholic Inf. Sch.
Social Cen.
Convent
The Hemel Hempstead School
Playing Field
Heath Barn
Sportspace
Sportspace Athletics Track
The Snow Centre
Tudor Prim. Sch.
Lime Wlk. Prim. Sch.
Logandene EMI Unit
Playground
Rugby Ground
Watercress Beds
Cricket Grd.
Boxmoor
STATION ROAD
A4146
Thtre.
Cricket Grd.
Heath Park
Superstore
HEMEL HEMPSTEAD
Corner Hall
Bennetts End
Belswains Playing Field
A41
A4251
ROAD
LONDON
Hemel Hempstead
Roughdown Common
Nursery
Two Waters
A414
TWO WATERS
WAY
Depot
Roughdown Common
River Bulbourne
Grand Union Canal
River Gade
Frogmore End
Vis. Cen.
Roman Fields Sch.
Mast
Felden
Threefields
Belswains Prim. Sch.
Comm. Cen.
Trading Est.
Superstore
Apsley Lock
Apsley Mills Retail Park
Aspley
Prim. Sch.
Howe's Retreat
Felden Lodge
Felden Farm
Hyde Farm Bungalows
Tower Farm
Felden Grange
PRIORY HOSPITAL HEMEL HEMPSTEAD
Felden Barns
Felden Manor
HP3
Felden Orchard
Felden Ash
Felden Ridge
Valley Farm
New Farm
Apsley
SHENDISH MANOR GOLF COURSE
Cock's Head Wood
Apsley Manor Farm
Club House
Shendish Manor
Bowl. Green
Tennis Courts
Great Wood
Pavilion
Tennis Court
The Bungalow
Works
Nash Mills
Nash Mills Recreation Centre
Shafford Cottages
The Lodge
RUCKLERS LANE
Phasels Wood
Stony Dene
Old Dean
High Oak
Rucklers Wood
Playground
The Hangings
Merceys Wood
Barnes Lodge
Dark Wood
The Nucket
HP4
Kielder
Fairydell Farm
Badgerdell
Badgerdell Wood
Barnes Farm
Playing Field
Tennis Cts.
Kings Langley School
Play. Fld.
Kings Langley Prim. Sch.
Kings Langley Common
Cricket Grd.
Hill Farm
Scatterdells Wood
The Grove
Ulstrode Farm
11
HEMPSTEAD ROAD
Grand Union Canal
Langley Wharf
04
05
06
07

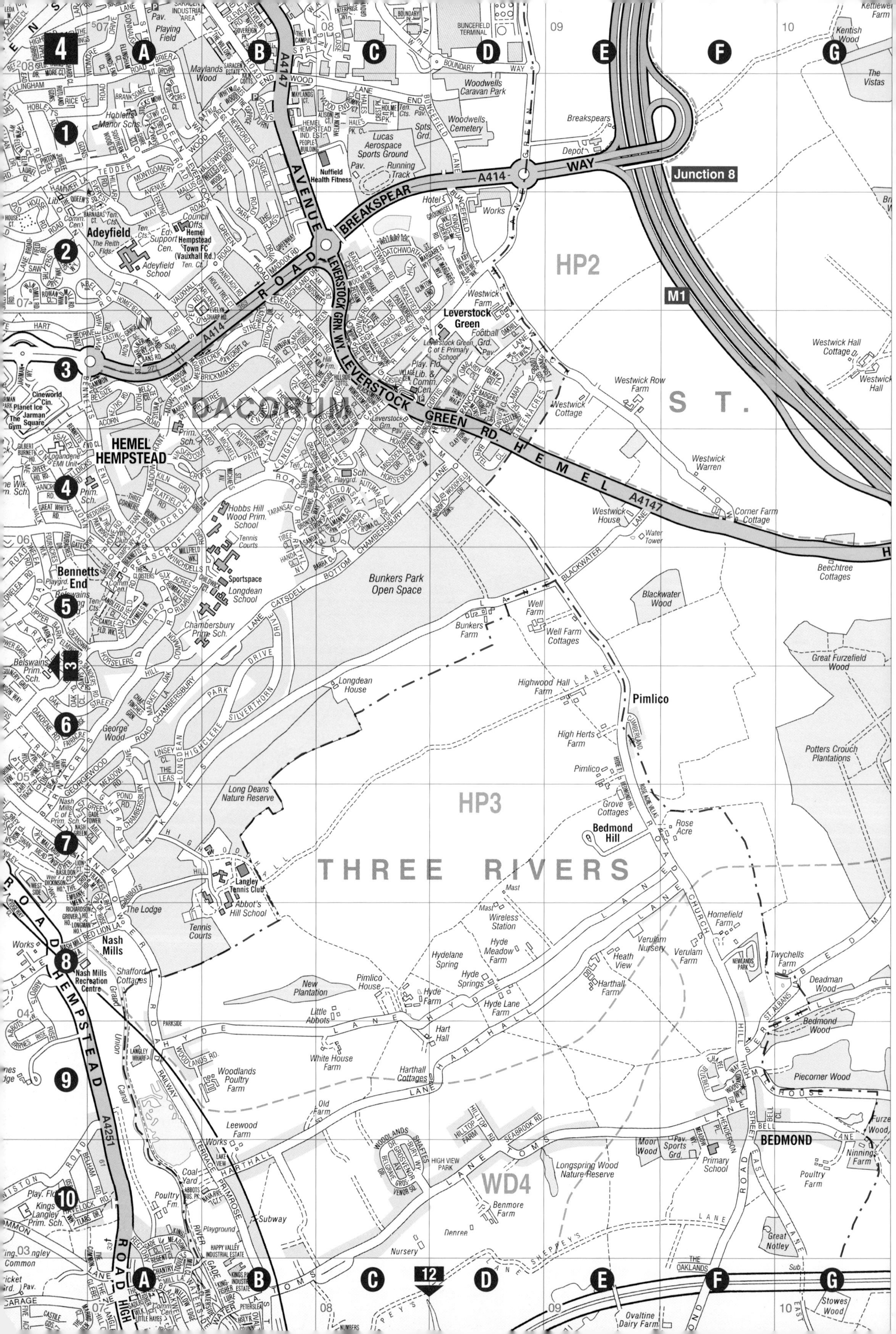

HEMEL HEMPSTEAD
DACORUM
THREE RIVERS
HP2
HP3
WD4
Adeyfield
Leverstock Green
Bennetts End
Nash Mills
Pimlico
Bedmond Hill
BEDMOND
Junction 8
M1
A414
A4147
A4251
BREAKSPEAR WAY
LEVERSTOCK GREEN RD.
Bunkers Park Open Space
Long Deans Nature Reserve
Longspring Wood Nature Reserve
Blackwater Wood
Great Furzefield Wood
Potters Crouch Plantations
Piecorner Wood
Deadman Wood
Bedmond Wood
Kentish Wood
The Vistas
Woodwells Caravan Park
Woodwells Cemetery
Lucas Aerospace Sports Ground
Nuffield Health Fitness
Buncefield Terminal
Maylands Wood
Hemel Hempstead Town FC (Vauxhall Rd.)
Hobletts Manor Schs.
Adeyfield School
Cineworld Cin.
Planet Ice
Jarman Square
Hobbs Hill Wood Prim. School
Sportspace
Longdean School
Chambersbury Prim. Sch.
Langley Tennis Club
Abbot's Hill School
Nash Mills Recreation Centre
Kings Langley Prim. Sch.
Westwick Farm
Westwick Row Farm
Westwick Hall
Westwick Warren
Corner Farm Cottage
Beechtree Cottages
Well Farm
Bunkers Farm
Highwood Hall Farm
High Herts Farm
Grove Cottages
Rose Acre
Homefield Farm
Verulam Nursery
Verulam Farm
Twychells Farm
Wireless Station
Hyde Meadow Farm
Hyde Lane Farm
Hartall Farm
White House Farm
Woodlands Poultry Farm
Leewood Farm
Benmore Farm
Poultry Farm
Ninnings Farm
Great Notley
Stowes Wood
Ovaltine Dairy Farm
12
3
4

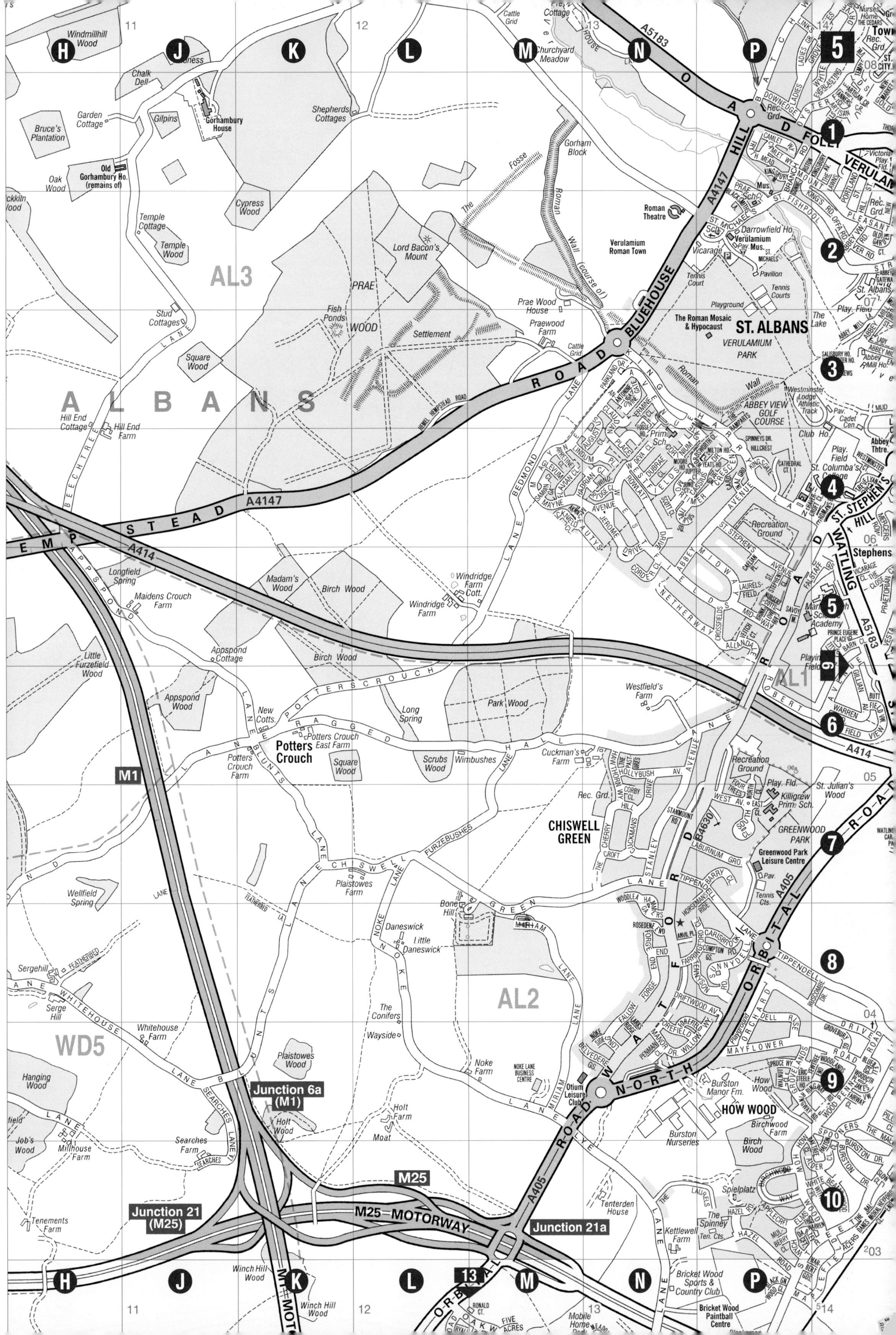

5
11
12
13
14
H
J
K
L
M
N
P
Windmillhill Wood
Chalk Dell
Garden Cottage
Bruce's Plantation
Gilpins
Gorhambury House
Shepherds Cottages
Cattle Grid
Churchyard Meadow
A5183
Old Gorhambury Ho. (remains of)
Oak Wood
Cypress Wood
Temple Cottage
Temple Wood
Gorham Block
The Fosse
Roman Wall (course of)
Roman Theatre
Verulamium Roman Town
Lord Bacon's Mount
AL3
PRAE WOOD
Fish Ponds
Settlement
Stud Cottages
Square Wood
Prae Wood House
Praewood Farm
Cattle Grid
BLUEHOUSE HILL
A4147
ROAD
HEMEL HEMPSTEAD ROAD
Darrowfield Ho.
Verulamium Mus.
Vicarage
Tennis Court
Tennis Courts
Pavilion
Playground
The Roman Mosaic & Hypocaust
ST. ALBANS
VERULAMIUM PARK
The Lake
Roman Wall
ABBEY VIEW GOLF COURSE
Westminster Lodge Athletic Track
Club Ho.
Play. Field
St. Columba's College
Abbey Thtre.
A L B A N S
Hill End Cottage
Hill End Farm
BEECHTREE LANE
HEMPSTEAD
A4147
A414
APPSPOND
Longfield Spring
Maidens Crouch Farm
Madam's Wood
Birch Wood
Windridge Farm Cott.
Windridge Farm
BEDMOND LANE
Recreation Ground
ST. STEPHEN'S HILL
WATLING
Stephens
A5183
Little Furzefield Wood
Appspond Cottage
Birch Wood
POTTERSCROUCH
Appspond Wood
New Cotts.
Long Spring
Park Wood
Westfield's Farm
AL1
ROBERT
WARREN
6
A414
M1
RAGGED HALL LANE
Potters Crouch East Farm
Potters Crouch
Potters Crouch Farm
Square Wood
Scrubs Wood
Wimbushes
Cuckman's Farm
HOLLYBUSH AV.
Recreation Ground
Play. Fld.
Killigrew Prim. Sch.
St. Julian's Wood
05
Rec. Grd.
CHISWELL GREEN
GREENWOOD PARK
Greenwood Park Leisure Centre
7
FURZEBUSHES
CHISWELL
Plaistowes Farm
Wellfield Spring
FEATHERBED LANE
Bone Hill
GREEN
MIRIAM
NOKE LANE
Daneswick
Little Daneswick
A405
8
TIPPENDELL
Sergehill
WHITEHOUSE
Serge Hill
AL2
The Conifers
Wayside
04
MAYFLOWER
WD5
Whitehouse Farm
Plaistowes Wood
Hanging Wood
Noke Farm
NOKE LANE BUSINESS CENTRE
Otium Leisure Club
NORTH ORBITAL ROAD
Burston Manor Fm.
How Wood
HOW WOOD
9
Junction 6a (M1)
Holt Wood
Holt Farm
Moat
Birchwood Farm
Birch Wood
Burston Nurseries
Job's Wood
Millhouse Farm
Searches Farm
SEARCHES LANE
M25
Tenterden House
Spielplatz
10
Junction 21 (M25)
Tenements Farm
M25 MOTORWAY
Junction 21a
The Spinney Ten. Cts.
Kettlewell Farm
03
Winch Hill Wood
13
Bricket Wood Sports & Country Club
Bricket Wood Paintball Centre
Winch Hill Wood
Mobile Home
FIVE ACRES

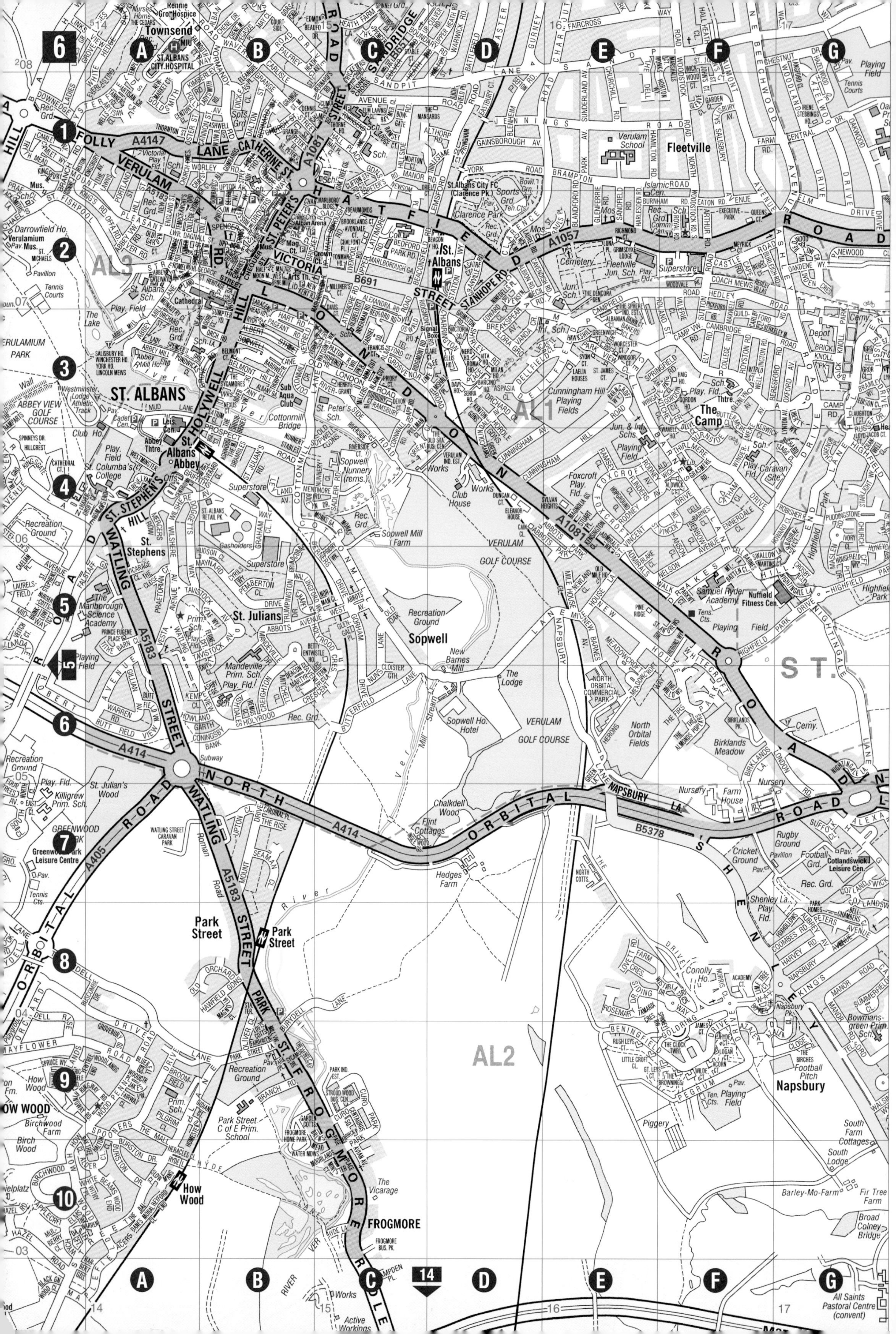
A
B
C
D
E
F
G
14
15
16
17
08
07
06
05
04
03
1
2
3
4
5
6
7
8
9
10
Townsend
ST.ALBANS CITY HOSPITAL
Rennie Gro. Hospice
SANDRIDGE
FAIRCROSS
SANDPIT LANE
JENNINGS ROAD
Verulam School
Fleetville
GAINSBOROUGH AV.
FOLLY LANE
A4147
VERULAM
CATHERINE ST.
A1081
ST. PETER'S
St.Albans City FC (Clarence Pk.)
Clarence Park
HATFIELD ROAD
A1057
Verulamium Mus.
Darrowfield Ho.
AL3
AL1
AL2
VICTORIA
St. Albans
STATION
STANHOPE RD.
Superstore
Cemetery
Fleetville Jun. Sch.
Cathedral
HOLYWELL HILL
LONDON ROAD
B691
Signal Box
VERULAMIUM PARK
The Lake
ST. ALBANS
ABBEY VIEW GOLF COURSE
St. Albans Abbey
Abbey Thtre.
St. Columba's College
Sopwell Nunnery (rems.)
Cunningham Hill Playing Fields
The Camp
Cottonmill Bridge
Sopwell Mill Farm
VERULAM GOLF COURSE
Foxcroft Play. Fld.
Samuel Ryder Academy
Nuffield Fitness Cen.
Highfield Park
ST.
St. Stephens
St. Julians
Gasholders
Superstore
WATLING STREET
A5183
Recreation Ground
Sopwell
New Barnes Mill
The Lodge
Marlborough Science Academy
Mandeville Prim. Sch.
Sopwell Ho. Hotel
North Orbital Commercial Park
North Orbital Fields
Birklands Meadow
NAPSBURY LA.
A414
NORTH ORBITAL ROAD
St. Julian's Wood
Killigrew Prim. Sch.
Greenwood Park Leisure Centre
Chalkdell Wood
Flint Cottages
Hedges Farm
B5378
Rugby Ground
Cricket Ground
Cotlandswick Leisure Cen.
SHENLEY LANE
A405
Park Street
Watling Street Caravan Park
Roman Road
River Ver
Connolly Ho.
Napsbury Pk.
Bowmansgreen Prim. Sch.
Recreation Ground
Park Street C of E Prim. School
How Wood
Birchwood Farm
Birch Wood
How Wood
FROGMORE
The Vicarage
Piggery
Napsbury
South Farm Cottages
South Lodge
Barley-Mo-Farm
Fir Tree Farm
Broad Colney Bridge
All Saints Pastoral Centre (convent)
Active Workings
Works
RADLETT ROAD
14
5

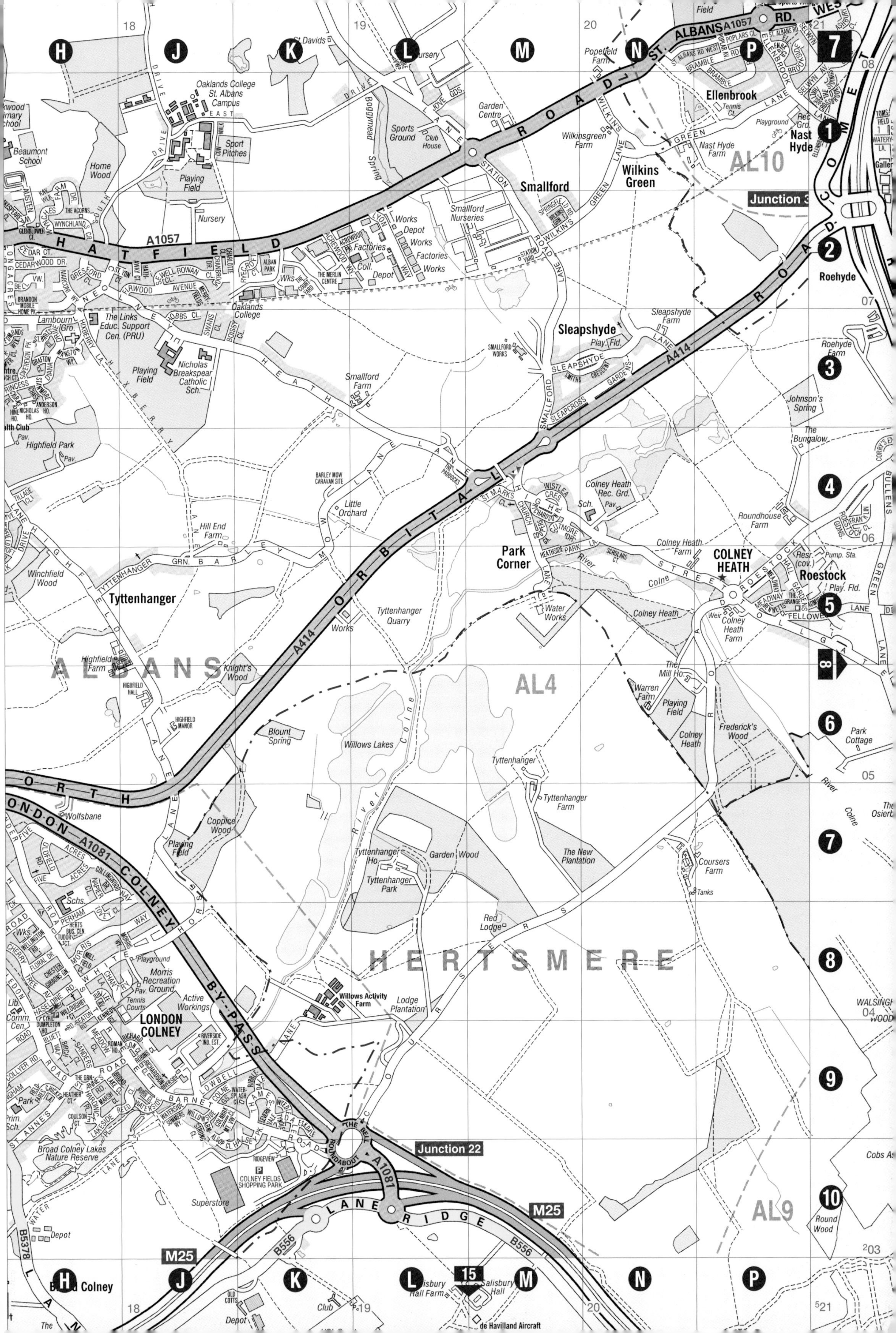

7
H
J
K
L
M
N
P
1
2
3
4
5
6
7
8
9
10
18
19
20
21
08
07
06
05
04
03
ST. ALBANS A1057 RD.
Ellenbrook
Nast Hyde
AL10
Junction 3
Roehyde
Wilkins Green
Smallford
Oaklands College St. Albans Campus
Sport Pitches
Playing Field
Home Wood
Beaumont School
Sports Ground
Club House
Garden Centre
Smallford Nurseries
HATFIELD ROAD
A1057
Oaklands College
The Links Educ. Support Cen. (PRU)
Nicholas Breakspear Catholic Sch.
Highfield Park
Sleapshyde
Sleapshyde Farm
Roehyde Farm
Johnson's Spring
The Bungalow
Smallford Farm
A414
NORTH ORBITAL ROAD
Little Orchard
Hill End Farm
Colney Heath Rec. Grd.
Park Corner
COLNEY HEATH
Roestock
Roundhouse Farm
Colney Heath Farm
Winchfield Wood
Tyttenhanger
Tyttenhanger Quarry
Works
Water Works
Colney Heath
Colney Heath Farm
River Colne
ALBANS
AL4
Knight's Wood
Highfield Farm
The Mill Ho.
Warren Farm
Playing Field
Colney Heath
Frederick's Wood
Park Cottage
8
Blount Spring
Willows Lakes
Tyttenhanger Farm
Wolfsbane
Coppice Wood
LONDON COLNEY A1081 BY-PASS
Tyttenhanger Ho.
Tyttenhanger Park
Garden Wood
The New Plantation
Coursers Farm
Red Lodge
HERTSMERE
Morris Recreation Ground
Active Workings
LONDON COLNEY
Willows Activity Farm
Lodge Plantation
Junction 22
Broad Colney Lakes Nature Reserve
COLNEY FIELDS SHOPPING PARK
Superstore
M25
B556
RIDGE LANE
AL9
Round Wood
Cobs Ash
WALSINGHAM WOOD
15
Salisbury Hall
Salisbury Hall Farm
de Havilland Aircraft
Broad Colney
B5378
Depot

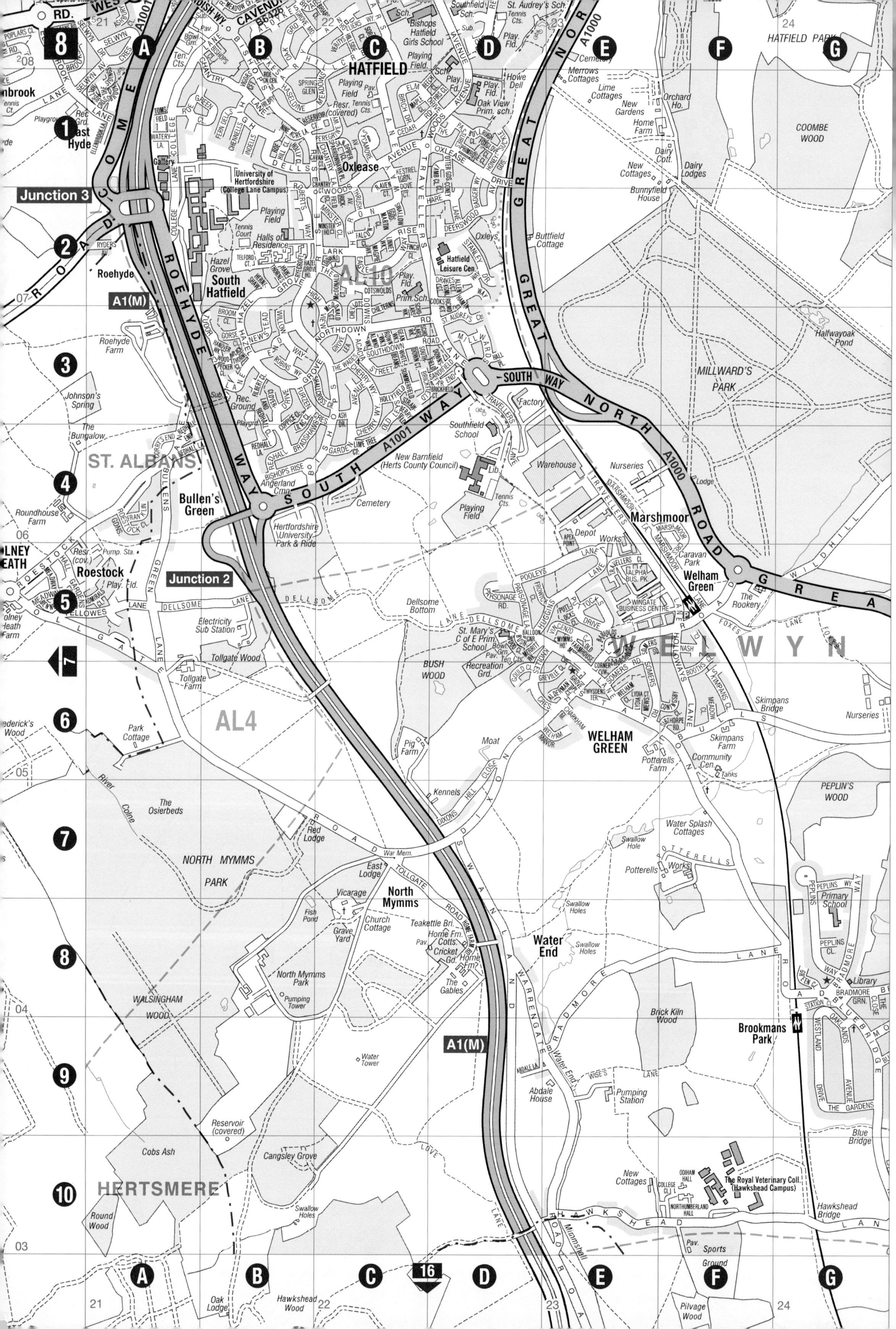

8
HATFIELD
Oxlease
South Hatfield
AL10
University of Hertfordshire (College Lane Campus)
Junction 3
Junction 2
A1(M)
ROEHYDE WAY
SOUTH WAY
A1001
GREAT NORTH ROAD
A1000
ST. ALBANS
AL4
Roestock
Bullen's Green
Marshmoor
Welham Green
WELHAM GREEN
WELWYN
North Mymms
NORTH MYMMS PARK
Water End
Brookmans Park
HERTSMERE
COOMBE WOOD
MILLWARD'S PARK
PEPLIN'S WOOD
WALSINGHAM WOOD
The Royal Veterinary Coll. (Hawkshead Campus)
HATFIELD PARK
7
16

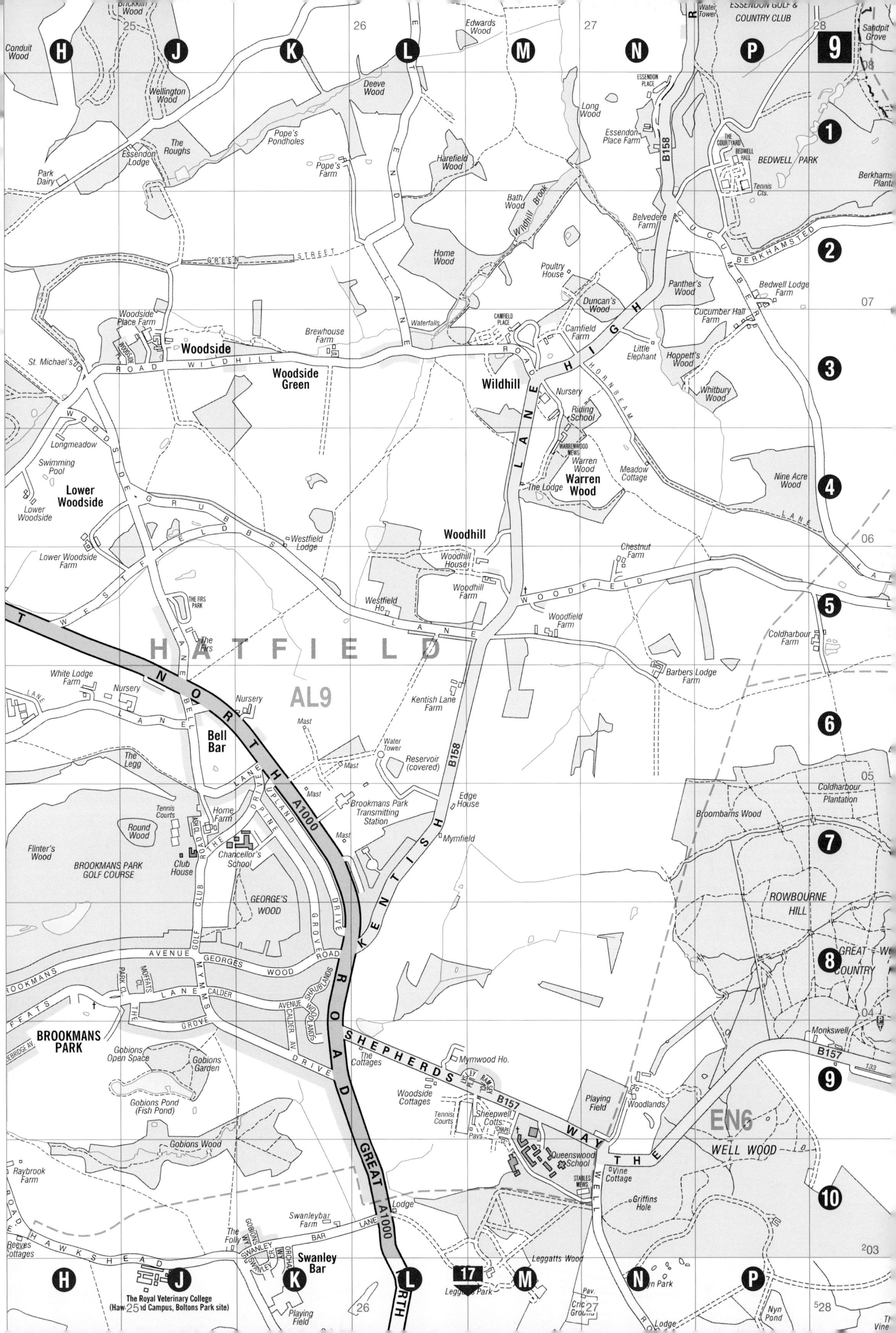
9
HATFIELD
AL9
EN6
Woodside
Woodside Green
Wildhill
Lower Woodside
Woodhill
Warren Wood
Bell Bar
BROOKMANS PARK
Swanley Bar
BEDWELL PARK
ESSENDON GOLF & COUNTRY CLUB
BROOKMANS PARK GOLF COURSE
ROWBOURNE HILL
WELL WOOD
GREAT NORTH ROAD
A1000
B158
B157
HIGH LANE
KENTISH LANE
SHEPHERDS WAY
WILDHILL ROAD
WOODFIELD LANE
GREEN STREET
BERKHAMSTED
CUCUMBER LANE
HORNBEAM
WESTFIELD LANE
GEORGES WOOD ROAD
MYMMS DRIVE
HAWKSHEAD
Conduit Wood
Wellington Wood
The Roughs
Essendon Lodge
Park Dairy
Pope's Pondholes
Pope's Farm
Deeve Wood
Edwards Wood
Long Wood
Essendon Place Farm
Harefield Wood
Bath Wood
Whitehill Brook
Home Wood
Belvedere Farm
Poultry House
Panther's Wood
Bedwell Lodge Farm
Cucumber Hall Farm
Duncan's Wood
Camfield Farm
Little Elephant
Hoppett's Wood
Whitbury Wood
Brewhouse Farm
Woodside Place Farm
St. Michael's
Waterfalls
Nursery
Riding School
Warren Wood
The Lodge
Meadow Cottage
Nine Acre Wood
Longmeadow
Swimming Pool
Lower Woodside Farm
Westfield Lodge
Woodhill House
Woodhill Farm
Westfield Ho.
Chestnut Farm
Woodfield Farm
Coldharbour Farm
Barbers Lodge Farm
The Firs
White Lodge Farm
Kentish Lane Farm
The Legg
Water Tower
Reservoir (covered)
Mast
Brookmans Park Transmitting Station
Edge House
Mymfield
Home Farm
Tennis Courts
Round Wood
Flinter's Wood
Club House
Chancellor's School
GEORGE'S WOOD
Coldharbour Plantation
Broombarns Wood
Monkswell
Gobions Open Space
Gobions Garden
Gobions Pond (Fish Pond)
Gobions Wood
The Cottages
Mymwood Ho.
Woodside Cottages
Sheepwell Cotts.
Playing Field
Woodlands
Queenswood School
Vine Cottage
Griffins Hole
Raybrook Farm
Reeves Cottages
Swanleybar Farm
The Folly
Lodge
Leggatts Wood
The Royal Veterinary College (Hawkshead Campus, Boltons Park site)
Nyn Park
Nyn Pond
17
Sandpit Grove
Berkhamsted Plantation

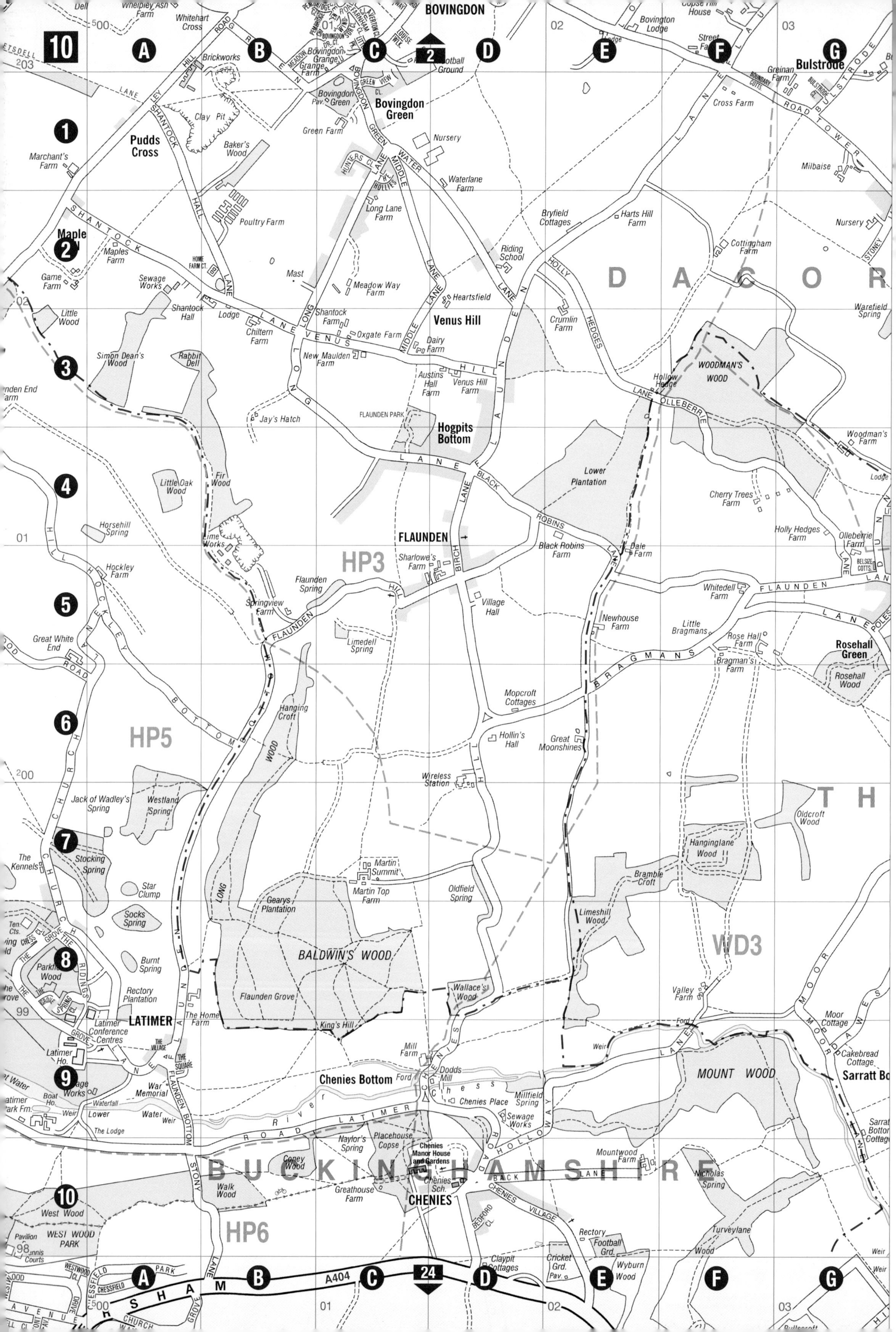
10
BOVINGDON
2
24
Bovingdon Green
Pudds Cross
Maple
Venus Hill
Hogpits Bottom
FLAUNDEN
HP3
HP5
HP6
WD3
LATIMER
Chenies Bottom
CHENIES
BUCKINGHAMSHIRE
Bulstrode
Rosehall Green
DACOR
TH
Sarratt Bo
MOUNT WOOD
BALDWIN'S WOOD
WOODMAN'S WOOD
Lower Plantation
River Chess
A404
Latimer Conference Centres
Chenies Manor House and Gardens

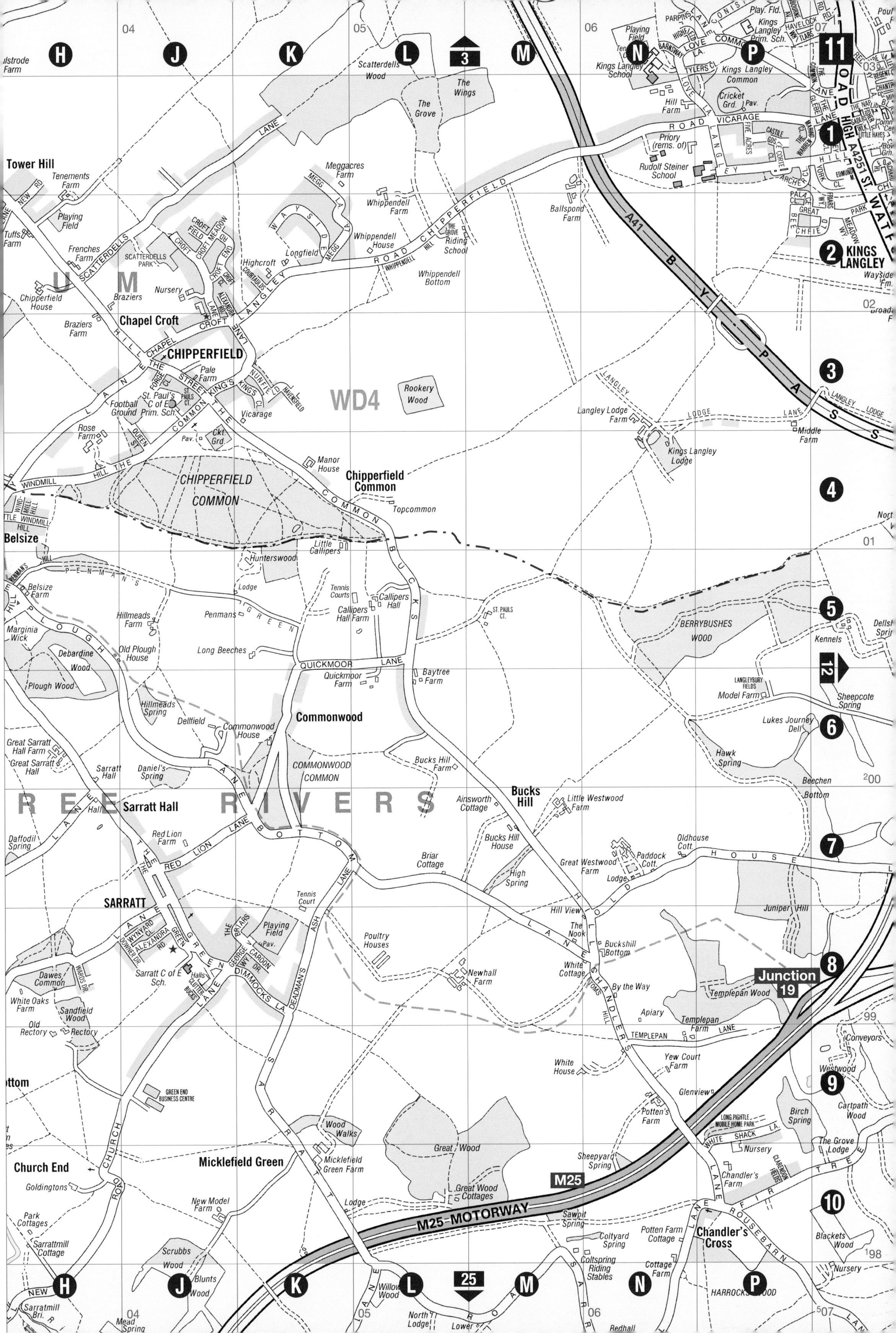

3
11
12
25
Tower Hill
Chapel Croft
CHIPPERFIELD
WD4
Chipperfield Common
CHIPPERFIELD COMMON
Belsize
Commonwood
COMMONWOOD COMMON
THREE RIVERS
Sarratt Hall
SARRATT
Bucks Hill
Micklefield Green
Church End
Chandler's Cross
KINGS LANGLEY
Junction 19
M25
M25 MOTORWAY
A41 BYPASS
Kings Langley Common
Rudolf Steiner School
Scatterdells Wood
The Grove
The Wings
Meggacres Farm
Whippendell Farm
Whippendell House
Whippendell Bottom
Rookery Wood
Ballspond Farm
Langley Lodge Farm
Kings Langley Lodge
Middle Farm
Berrybushes Wood
Model Farm
Sheepcote Spring
Lukes Journey Dell
Hawk Spring
Beechen Bottom
Juniper Hill
Templepan Wood
Templepan Farm
Conveyors
Westwood
Cartpath Wood
Birch Spring
The Grove Lodge
Blackets Wood
Harrocks Wood
Great Wood
Poultry Houses
Newhall Farm
Little Westwood Farm
Great Westwood Farm
Paddock Cott.
Oldhouse Cott.
Bucks Hill Farm
Ainsworth Cottage
Quickmoor Farm
Baytree Farm
Callipers Hall
Callipers Hall Farm
Little Callipers
Hunterswood
Penmans Green
Long Beeches
Old Plough House
Hillmeads Farm
Plough Wood
Debardine Wood
Marginia Wick
Belsize Farm
Great Sarratt Hall Farm
Sarratt Hall
Daniel's Spring
Red Lion Farm
Daffodil Spring
Dawes Common
Sandfield Wood
White Oaks Farm
Old Rectory
Rectory
Goldingtons
Park Cottages
Sarrattmill Cottage
Sarrattmill Bri.
Scrubbs Wood
Blunts Wood
New Model Farm
Micklefield Green Farm
Wood Walks
Great Wood Cottages
Sawpit Spring
Coltyard Spring
Coltspring Riding Stables
Potten Farm Cottage
Cottage Farm
Chandler's Farm
Potten's Farm
Sheepyard Spring
Glenview
Yew Court Farm
White House
Apiary
By the Way
Buckshill Bottom
White Cottage
The Nook
Hill View
High Spring
Bucks Hill House
Briar Cottage
Tenements Farm
Playing Field
Tuffs Farm
Frenches Farm
Chipperfield House
Braziers Farm
Braziers
Nursery
Highcroft
Longfield
Riding School
Pale Farm
Vicarage
Manor House
Topcommon
Football Ground
St. Paul's C of E Prim. Sch.
Rose Farm
Sarratt C of E Sch.
GREEN END BUSINESS CENTRE
Kennels
Nursery
LONG PIGHTLE MOBILE HOME PARK
Kings Langley Prim. Sch.
Kings Langley School
Priory (rems. of)
Hill Farm
Lower
North Lodge
Willow Wood
Redhall
Mead Spring

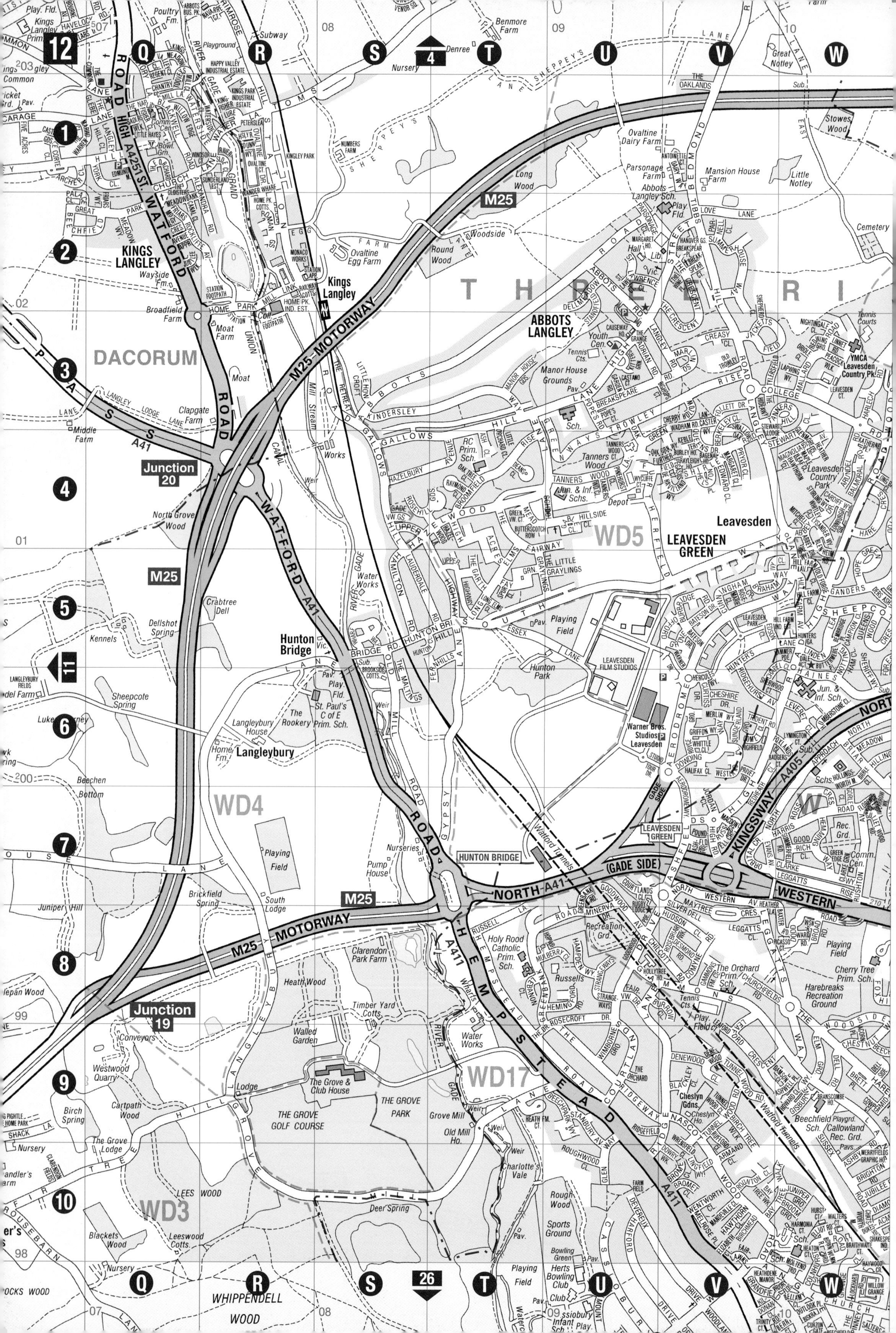

12
Q
R
S
T
U
V
W
4
26
11
1
2
3
4
5
6
7
8
9
10
07
08
09
10
02
01
200
99
98
203
KINGS LANGLEY
Kings Langley
ABBOTS LANGLEY
DACORUM
THREE RIVERS
Leavesden
LEAVESDEN GREEN
WD4
WD5
WD17
WD3
Hunton Bridge
Langleybury
M25
M25 MOTORWAY
Junction 20
Junction 19
WATFORD ROAD
A41
A411
HEMPSTEAD
NORTH A41
WESTERN
KINGSWAY A405
HIGH ST. A4251
HUNTON BRIDGE
LEAVESDEN GREEN
Benmore Farm
Great Notley
Little Notley
Stowes Wood
Ovaltine Dairy Farm
Parsonage Farm
Mansion House Farm
Long Wood
Round Wood
Ovaltine Egg Farm
Woodside
Cemetery
Broadfield Farm
Moat Farm
Moat
Clapgate Farm
Middle Farm
North Grove Wood
Crabtree Dell
Dellshot Spring
Kennels
Sheepcote Spring
Langleybury House
Home Fm.
Beechen Bottom
The Rookery
St. Paul's C of E Prim. Sch.
Water Works
Playing Field
Hunton Park
LEAVESDEN FILM STUDIOS
Warner Bros. Studios Leavesden
Manor House Grounds
Tanners Wood
Leavesden Country Park
YMCA Leavesden Country Pk.
Juniper Hill
Brickfield Spring
South Lodge
Pump House
Nurseries
Clarendon Park Farm
Heath Wood
Timber Yard Cotts.
Walled Garden
The Grove & Club House
THE GROVE GOLF COURSE
THE GROVE PARK
Grove Mill
Old Mill Ho.
Charlotte's Vale
Deer Spring
LEES WOOD
Leeswood Cotts.
Blackets Wood
Cartpath Wood
Birch Spring
Westwood Quarry
Conveyors
The Grove Lodge
WHIPPENDELL WOOD
Rough Wood
Sports Ground
Bowling Green
Herts Bowling Club
Holy Rood Catholic Prim. Sch.
Russells
Recreation Grd.
Cheslyn Gdns.
Beechfield Sch.
Callowland Rec. Grd.
Harebreaks Recreation Ground
Cherry Tree Prim. Sch.
The Orchard Prim. Sch.
Watford Tunnels

13
14
M25
Junction 21 (M25)
Junction 21a
M1 MOTORWAY
Junction 6
Junction 5
BRICKET WOOD
ST. ALBANS
BRICKET WOOD COMMON
HERTSMERE
WD25
WD24
WOODSIDE
GARSTON
MERIDEN
NORTH WATFORD
ALDENHAM
Kingswood
Waterdale
Smug Oak
Garston
Watford North
27

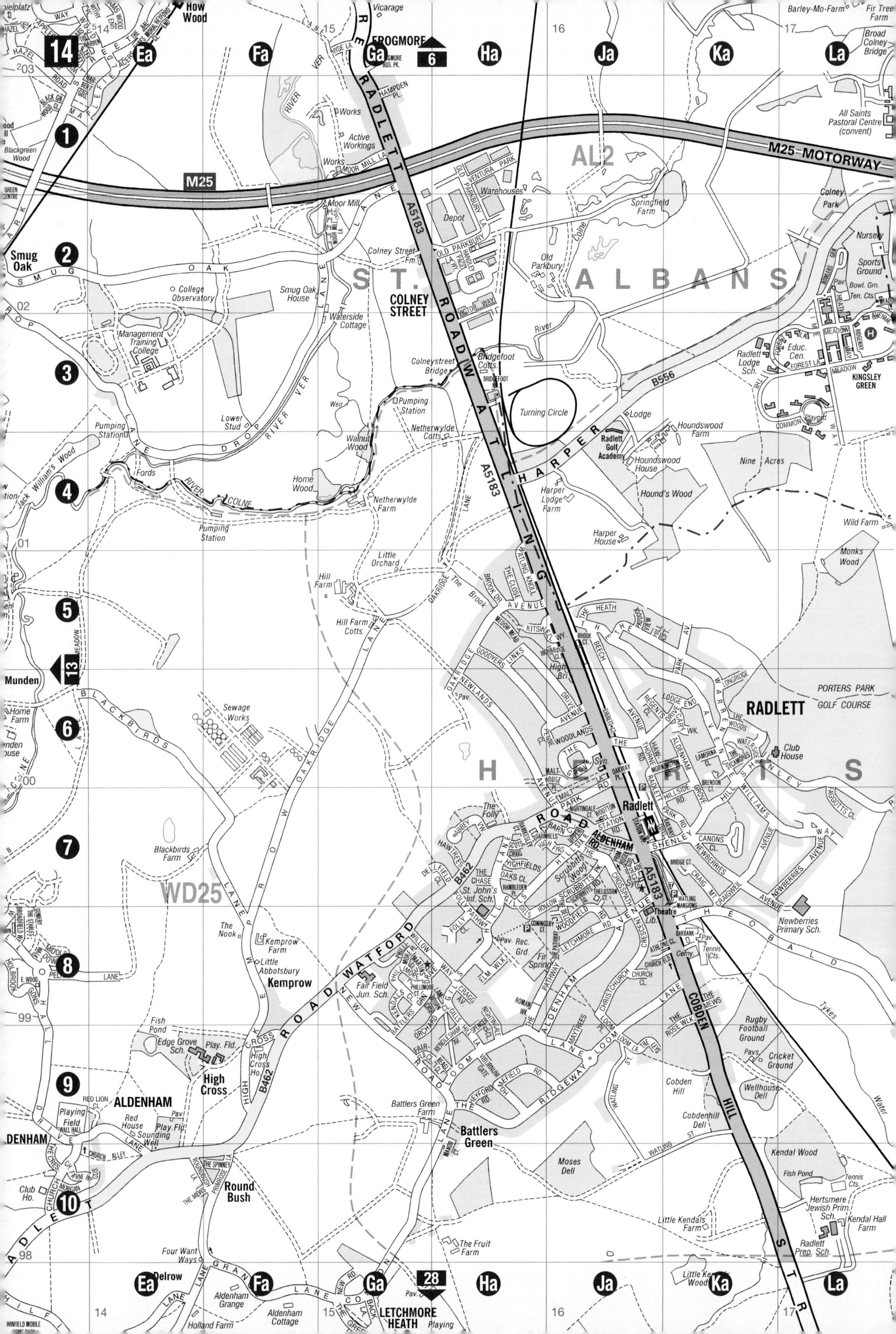
14
Ea
Fa
Ga
Ha
Ja
Ka
La
FROGMORE
6
How Wood
Vicarage
Barley-Mo-Farm
Fir Tree Farm
Broad Colney Bridge
All Saints Pastoral Centre (convent)
M25 MOTORWAY
M25
AL2
Smug Oak
College Observatory
Management Training College
Smug Oak House
Colney Street Fm.
COLNEY STREET
ST. ALBANS
Moor Mill
Works
Active Workings
Depot
Warehouses
Springfield Farm
Old Parkbury
Colney Park
Nursery
Sports Ground
Waterside Cottage
Colneystreet Bridge
Bridgefoot Cotts.
Turning Circle
Lodge
B556
HARPER
Radlett Golf Academy
Houndswood Farm
Houndswood House
Hound's Wood
Nine Acres
Harper Lodge Farm
Harper House
Wild Farm
Monks Wood
Educ. Cen.
Radlett Lodge Sch.
KINGSLEY GREEN
Pumping Station
Lower Stud
Walnut Wood
Netherwylde Cotts.
Home Wood
Netherwylde Farm
Fords
RIVER COLNE
Jack William's Wood
Little Orchard
Hill Farm
Hill Farm Cotts.
Munden
13
Home Farm
Sewage Works
BLACKBIRDS LANE
OAKRIDGE
WATLING STREET
A5183
RADLETT
PORTERS PARK GOLF COURSE
Club House
HERTS
Radlett
WD25
Blackbirds Farm
The Nook
Kemprow Farm
Little Abbotsbury
Kemprow
WATFORD ROAD
B462
St. John's Inf. Sch.
Fair Field Jun. Sch.
Theatre
Newberries Primary Sch.
THEOBALD STREET
Rugby Football Ground
Cricket Ground
Wellhouse Dell
Cobden Hill
Cobdenhill Dell
COBDEN HILL
Fish Pond
Edge Grove Sch.
High Cross
ALDENHAM
Red House
Sounding Well
Playing Field
Battlers Green Farm
Battlers Green
Moses Dell
Kendal Wood
Hertsmere Jewish Prim. Sch.
Kendal Hall Farm
Radlett Prep. Sch.
Little Kendals Farm
The Fruit Farm
Round Bush
Club Ho.
Four Want Ways
Delrow
Aldenham Grange
Aldenham Cottage
Holland Farm
28
LETCHMORE HEATH
Little Kendals Wood
1
2
3
4
5
6
7
8
9
10
14
15
16
17

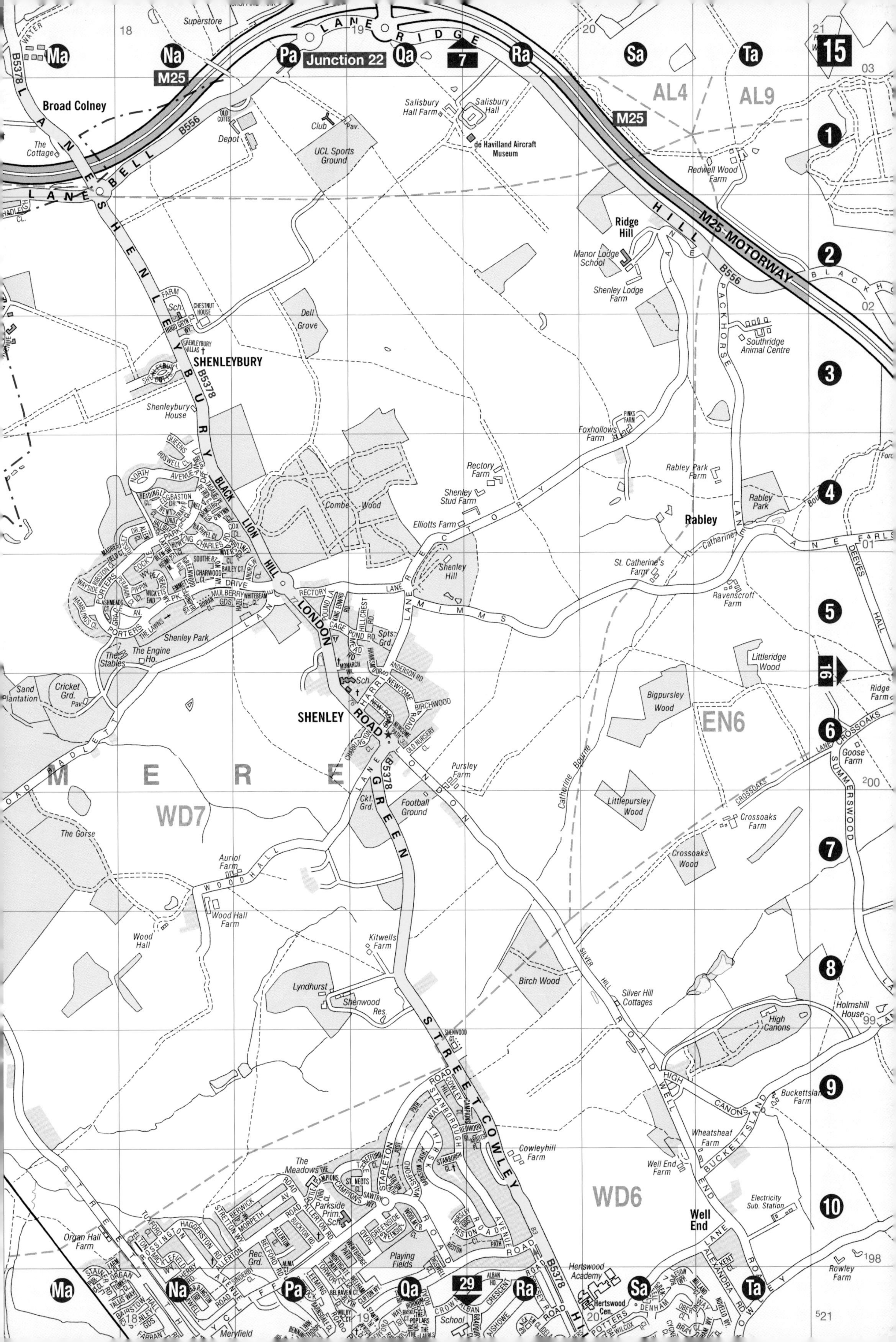

Junction 22
M25
Broad Colney
Shenleybury
Shenley
Ridge Hill
Rabley
Well End
Salisbury Hall
de Havilland Aircraft Museum
UCL Sports Ground
Combe Wood
Shenley Park
Bigpursley Wood
Littlepursley Wood
Crossoaks Wood
Birch Wood
High Canons
M25 MOTORWAY
AL4
AL9
EN6
WD7
WD6
MERE

Round Wood
Swallow Holes
Hawkshead Campus
Northumberland Hall
Hawkshead Bridge
Hawkshead Lane
Mimmshall Road
Pav.
Sports Ground
AL9
Oak Lodge
Hawkshead Wood
Flint Cottage
Mymmshall Wood
Pilvage Wood
Redwell Wood
Brook
The Grange
Woodhill Farm
Motorway
Blackhorse Lane
Warrengate Farm
Furzefield Wood
King George V Playing Field
Furzefield Centre
Tennis Courts
The Enterprise Centre
Cranborne Road
Cranborne Ind. Est.
Southridge Animal Centre
B556
St. Albans Road
A1(M) Motorway
Warrengate Road
Mimms Hall
Mutton Lane
Windmore Hall
Windmore Av.
Wroxham Gardens
Deepdene
The Wroxham School
Town Farm
Cecil Road
Ormesby Dr.
Crescent
Cambridge Drive
Suffolk Rd.
Auckland Road
Way
EN6
Ford
Bourne
Rabley Park
M25
A1(M)
Earls Lane
Deeves Hall
South Mimms
St. Giles C of E Prim. Sch.
Greyhound Rd.
Frowyke Cres.
Blanche Lane
Earls Farm
Bridgefoot Farm
Bridgefoot Ho.
Bridgefoot Lane
Shillitoe Avenue
Kimptons Cl.
Dugdale Hill
Herts
Ludlow's Lake
Dame Alice Owen's Sch.
Rydal Mount
Littleridge Wood
15
The Hawthorns
Church Fm.
Ridge Farm
Ridge
The Meadows
Junction 1 A1(M)
Bignell's Corner
Swanland Road
Depot
Watery Lane
Imperial Cancer Research Centre
Crossoaks Lane
Goose Farm
Blanche Lane Farm
South Mimms Service Area
Playing Field
Sawyers Lane
Bentley
Crossoaks Farm
Summerswood
St. Albans Road
A1081
Junction 23 (M25)
Dancers Lane
Dancershill Farm
Norfolk Lodge Farm
Baker
Summerswood Farm
WD6
Dyrham
Dyrham Park Farm
Laurel House
Laurel Lodge
Dancers Hill
Kitts Lane
Holmshill House
High Canons
Holmshill Lane
Bottom
EN5
Packhorse Lane
Trotters
Dyrham Park Golf Course
Lower Kitt's End Farm
Kitt's End
Dyrham Park Country Club & Club House
Knightsland Farm
Electricity Sub. Station
Valentine's Farm
Dyrham Park
Kitt's End Farm
Galley Lane
Arkley
Strangeways Farm
Little Dyrham
Barnet
By-Pass
A1 Road
Rowley Farm
Rowley Lane
The Old Fold Manor Golf Course
30
8
Ua
Va
Wa
Xa
Ya
Za
Ab

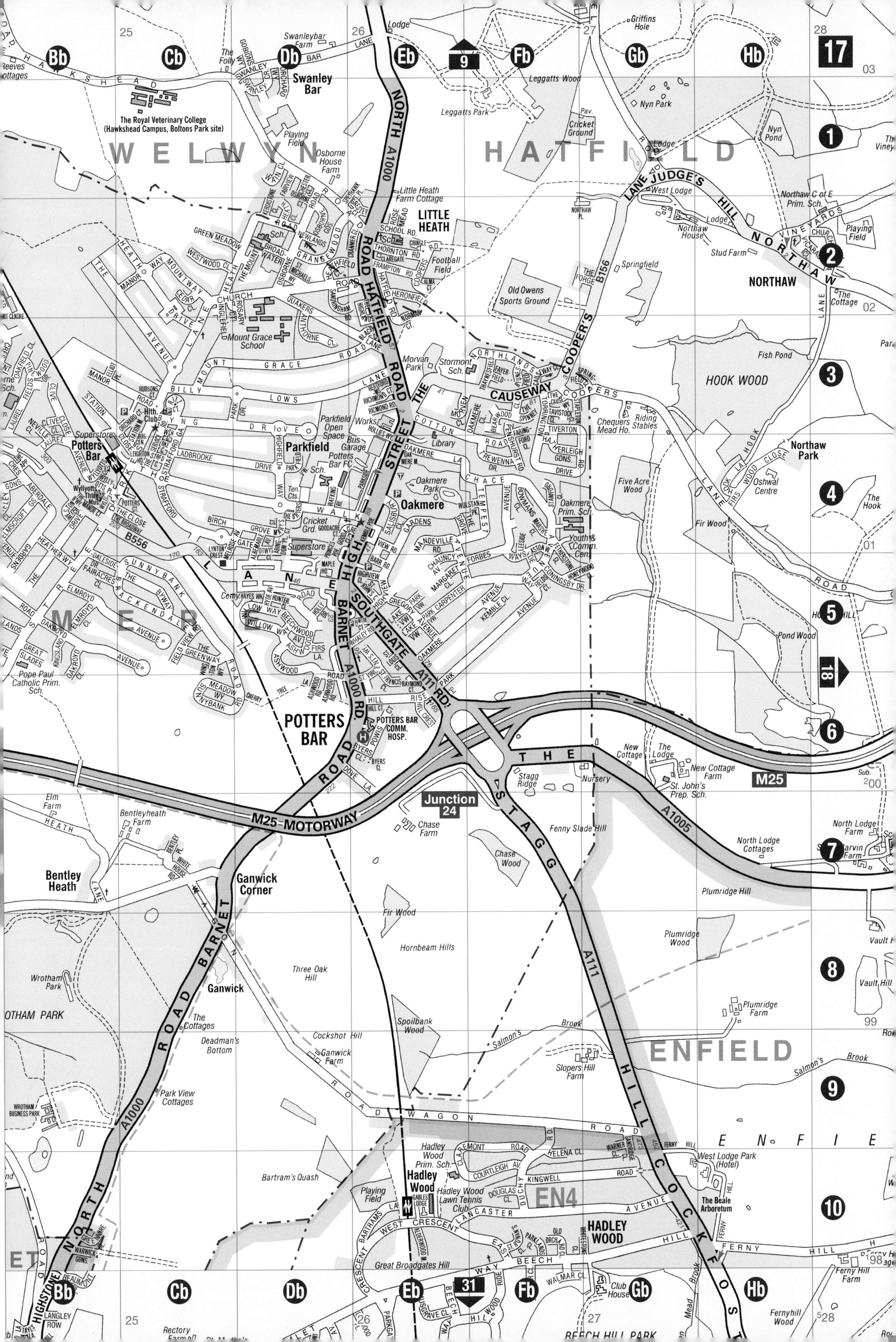
WELWYN HATFIELD
ENFIELD
Swanley Bar
The Royal Veterinary College
(Hawkshead Campus, Boltons Park site)
Little Heath
Northaw
Leggatts Wood
Leggatts Park
Nyn Park
Nyn Pond
Griffins Hole
Cricket Ground
Judge's Hill
Northaw C of E Prim. Sch.
Playing Field
Springfield
Old Owens Sports Ground
Hook Wood
Fish Pond
Northaw Park
Oshwal Centre
The Hook
Fir Wood
Pond Wood
Five Acre Wood
Chequers Mead Ho.
Riding Stables
Mount Grace School
Potters Bar
Parkfield
Parkfield Open Space
Potters Bar FC
Oakmere
Oakmere Park
Oakmere Prim. Sch.
Youth & Comm. Cen.
Library
Causeway
Superstore
Cricket Grd.
Wyllyotts Theatre & Museum
Pope Paul Catholic Prim. Sch.
Potters Bar Comm. Hosp.
Junction 24
M25 Motorway
M25
A1000
A111
A1005
B156
B556
The Stagg
Stagg Ridge
Nursery
New Cottage
The Lodge
New Cottage Farm
St. John's Prep. Sch.
North Lodge Farm
North Lodge Cottages
Plumridge Hill
Plumridge Wood
Plumridge Farm
Vault Hill
Elm Farm
Bentleyheath Farm
Bentley Heath
Ganwick Corner
Ganwick
Chase Farm
Chase Wood
Fenny Slade Hill
Fir Wood
Hornbeam Hills
Three Oak Hill
Wrotham Park
Spoilbank Wood
Cockshot Hill
Ganwick Farm
Deadman's Bottom
The Cottages
Salmon's Brook
Slopers Hill Farm
Park View Cottages
Wrotham Business Park
Wagon Road
Bartram's Quash
Hadley Wood Prim. Sch.
Hadley Wood
Hadley Wood Lawn Tennis Club
Playing Field
EN4
West Lodge Park (Hotel)
The Beale Arboretum
Cockfosters Hill
Ferny Hill
Ferny Hill Farm
Fernyhill Wood
Great Broadgates Hill
Club House
Beech Hill Park
Langley Row
Rectory Farm
9
18
31

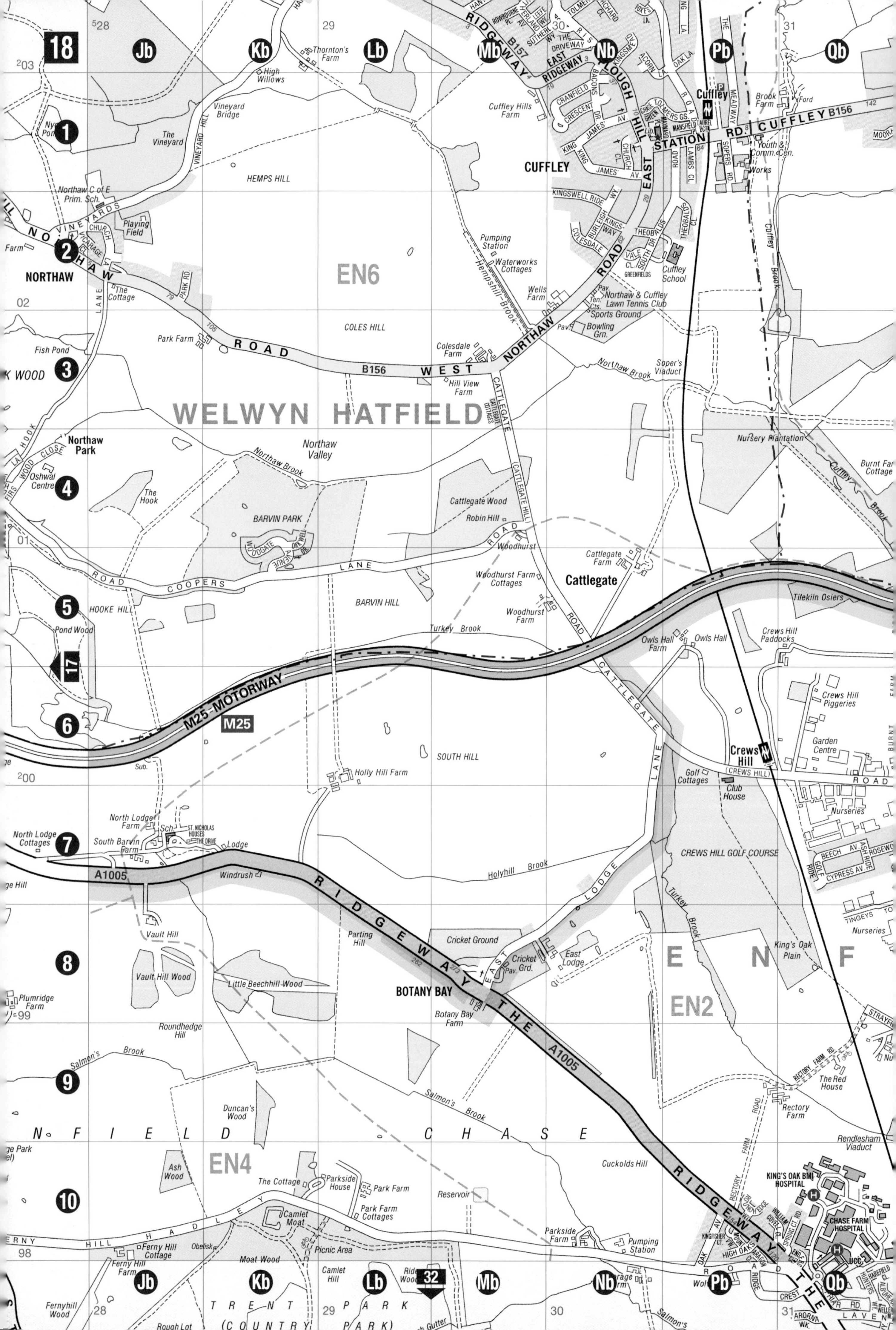
18
Jb
Kb
Lb
Mb
Nb
Pb
Qb
1
2
3
4
5
6
7
8
9
10
17
32
Thornton's Farm
High Willows
Vineyard Bridge
The Vineyard
VINEYARD HILL
HEMPS HILL
Cuffley Hills Farm
CUFFLEY
Cuffley
Brook Farm
Ford
CUFFLEY
B156
STATION RD.
Youth & Comm. Cen.
Works
RIDGEWAY
B157
EAST RIDGEWAY
THE DRIVEWAY
PLOUGH HILL
EAST ROAD
KINGSWELL RIDE
KING JAMES' AV.
CHURCH CL.
THEOBALDS
Cuffley School
Cuffley Brook
Northaw C of E Prim. Sch.
VINEYARDS
Playing Field
CHURCH
VICARAGE CL.
Farm
NORTHAW
NORTHAW
The Cottage
LANE
PARK RD
Pumping Station
Waterworks Cottages
Hempshill Brook
EN6
Wells Farm
Northaw & Cuffley Lawn Tennis Club
Sports Ground
Bowling Grn.
Park Farm
ROAD
COLES HILL
Colesdale Farm
NORTHAW
WEST
B156
Hill View Farm
Northaw Brook
Soper's Viaduct
Fish Pond
WOOD
WELWYN HATFIELD
CATTLEGATE
Northaw Park
Northaw Valley
Nursery Plantation
Burnt Farm Cottage
Oshwal Centre
The Hook
Northaw Brook
BARVIN PARK
Cattlegate Wood
Robin Hill
(CATTLEGATE HILL)
Woodhurst
Cattlegate Farm
Woodhurst Farm Cottages
Cattlegate
ROAD
COOPERS
LANE
BARVIN HILL
Tilekiln Osiers
HOOKE HILL
Pond Wood
Woodhurst Farm
Turkey Brook
Owls Hall Farm
Owls Hall
Crews Hill Paddocks
M25 MOTORWAY
M25
Crews Hill Piggeries
Garden Centre
Crews Hill
SOUTH HILL
Holly Hill Farm
Golf Cottages
Club House
(CREWS HILL)
ROAD
Nurseries
North Lodge Farm
North Lodge Cottages
South Barvin Farm
Sch.
ST. NICHOLAS HOUSES
THE DRIVE
Lodge
A1005
Windrush
CREWS HILL GOLF COURSE
BEECH AV.
CYPRESS AV.
Holyhill Brook
LODGE
Vault Hill
Parting Hill
RIDGEWAY
Cricket Ground
Cricket Grd.
Pav.
East Lodge
Turkey Brook
King's Oak Plain
Nurseries
Vault Hill Wood
Little Beechhill Wood
BOTANY BAY
Botany Bay Farm
EN2
Plumridge Farm
Roundhedge Hill
THE
A1005
Salmon's Brook
The Red House
Rectory Farm
Duncan's Wood
Salmon's Brook
ENFIELD CHASE
Rendlesham Viaduct
Cuckolds Hill
RIDGEWAY
Ash Wood
EN4
The Cottage
Parkside House
Park Farm
Park Farm Cottages
Reservoir
KING'S OAK BMI HOSPITAL
CHASE FARM HOSPITAL
Camlet Moat
HADLEY
Ferny Hill Cottage
Obelisk
Picnic Area
Moat Wood
Ferny Hill Farm
Camlet Hill
Parkside Farm
Pumping Station
Fernyhill Wood
TRENT PARK
(COUNTRY PARK)
Rough Lot
Salmon's
28
29
30
31
02
01
200
99
98
203
ROAD
HIGH OAKS
CREST
LAVENDER

19
Rb
Sb
Tb
Ub
Vb
Wb
Xb
GOFF'S OAK
GOFF'S HILL
GOFF'S LANE
B156
Whitehouse Farm
Rosedale Sports Ground
Jungle Adventure
Woodside Primary Sch.
Poyndon Farm
Theobalds Brook
SILVER STREET
Colesgrove Manor
South Lodge
Halstead Hill House
Nursery
HALSTEAD HILL
CHURCHGATE
Goff's Sports & Arts Cen.
Bonneygrove Prim. Sch.
Council Offs.
Focus Sch.
Cheshunt School
EN7
Cemetery
Woodgreen Farm
BARROW LA.
Broadfield Bungalow
Broadfield Farm
B R O X B O U R N E
Burnt Farm
LIEUTENANT ELLIS WAY
B198
BURY GREEN
St. Mary's C of E High Sch.
Burygreen Plantation
Broom Hills
OLD PARK RIDE
Dysons Osiers
Catlins
Home Wood
Hanging Plantation
THEOBALDS LANE
Rush Lodge
Home Plantation
Spring Farm
Theobalds Manor
Theobald Stud
Caravan & Camping Site
Sports Ground
Theobalds Pk. Conference Centre
EN8
M25-MOTORWAY
M25
Cinder Ash
Square Planting
South Osiers
Littlefield
Bullscross Farm
Theobalds Park Riding School
Theobalds Park Farm
20
Theobalds Camp Site
Western Cemetery (Jewish)
Junction 25
County Brook Plantation
Glasgow Stud
Stud Cottages
Sander's Corner
Sloeman's Farm
Sloeman's Cottage
Whitewebbs Farm
WHITEWEBBS ROAD
Whitewebbs Cottage
Whitewebbs Museum of Transport
CREWS HILL
North Lodge
Guy Lodge
Meux Cottages
Aqueduct
Fish Pond
WHITEWEBBS WOOD
White Webbs
The Limes
Keeper's Cottage
Clydesdale Stud
Capel Manor College
BULLSMOOR
Water Garden Centre
Wildwoods
WHITEWEBBS PARK
WHITEWEBBS GOLF COURSE
Training Centre
Playing Field
Myddelton Farm
Lee Valley Park Information Centre
BULLS CROSS
Myddelton House Gardens
The Lodge
Turkey Brook
New River (Old Course)
Brayside Farm
Club House
Forty Hall Country Park
Maiden's Bri.
St. Ignatius College
Turkey Street
Garden of Remembrance
Enfield Crematorium
Queenswood Farm
Kingswood
CLAY HILL
BEGGARS HOLLOW
Forty Hill C of E Prim. Sch.
EN1
Forty Hall Farm
Forty Hall
HILLY FIELDS PARK
Clovelly
CEMETERY
PHIPPS HATCH LA.
FORTY HILL
Worcesters Prim. Sch.
Lavender Prim. Sch.
A10
GREAT CAMBRIDGE ROAD
Russet Ho. Sch.
Durants Sch.
Aylands Sch.
Gordon Hill
33
1
2
3
4
5
6
7
8
9
10

20
19
34
Yb
Zb
Ac
Bc
Cc
Dc
Ec
35
36
37
38
1
2
3
4
5
6
7
8
9
10
CHESHUNT
BROXBOURNE
WALTHAM CROSS
WALTHAM ABBEY
ENFIELD
FREEZY WATER
ENFIELD WASH
ENFIELD LOCK
Enfield Island Village
Turkey Street
Bullsmoor
Holdbrook
Fishers Green
Meridian Park
RAMMEY MARSH
GUNPOWDER PARK
LEE VALLEY REGIONAL PARK
ROYAL GUNPOWDER MILLS
Royal Gunpowder Mills
Lee Valley White Water Centre
Theobalds Grove
Cheshunt
Waltham Cross
Enfield Lock
Turkey St.
Junction 25
HOLMESDALE TUNNEL
M25 MOTORWAY
M25
GREAT CAMBRIDGE ROAD
A10
A121
A1055
A112
B198
B176
B194
WINSTON CHURCHILL WAY
ELEANOR CROSS RD.
STATION RD.
HIGHBRIDGE ST.
MERIDIAN WAY
COLLEGE RD.
THEOBALDS LANE
BULLSMOOR LANE
MOLLISON AVENUE
HERTFORD ROAD
HIGH ST.
CROOKED MILE
HOLYFIELD
ABBEY VIEW
ORDNANCE ROAD
RIVER LEA NAVIGATION
RIVER LEE NAVIGATION
Seventy Acres
Sailing Club
Cheshunt Lake
Hooks Marsh
Turnershill Marsh
Herts Young Mariners Base
Thistly Marsh
Cheshunt Junction
Bowyer's Water
Hall Marsh
Waltham Marsh
Cheshunt Marsh
Marsh Bridge
Waltham Lock
Hazlemere Marina
Town Mead
Abbey Gdns.
Harold's Bri.
Abbey Fish Ponds
Arboretum
Tree Nursery
Nurseries
Cheshunt School
Cheshunt FC (The Stadium)
Holy Trinity C of E Prim. Sch.
Albury Farm
Rifle Range
Grundy Park
Laura Trott Leis. Cen.
Downfield Prim. Sch.
Wolsey Hall
Buzz Bingo
Bus Sta.
Sports Grd.
PARK PLAZA
Warehouses
Warehouse
MERIDIAN BUSINESS PARK
Keys Meadow Prim. Sch.
Oasis Academy Enfield
Prince of Wales Prim. Sch.
Albany Leisure Cen.
Oasis Academy Hadley
ALBANY PARK
Enfield Crematorium
Garden of Remembrance
Cemy.
Sewardstone Hall
Kingsfield Nurseries
Barnfield Riding Stables
Pumping Station
Overflow Channel
Rammey Marsh Flood Relief Channel
Cattlegate Flood Relief Channel
Small River Lee or Lea
Turkey Brook
Cobbins Brook
EN8
EN3
EN1

21
HOLYFIELD
Holyfield Cotts
North Lodge Farm
Homefield Wood
Stable Lodge
Kennel Wood
Galley Wood House
Aimes Green
Broadgate Springs
Brayshill Spring
Monkhams Cottage
Aimesgreen Farm
Stock Grove
Longcroft Grove
Scatterbushes Wood
Maynards Farm
Cobbinsend Farm
Maple Springs
Playground
BREACH BARNS MOBILE HOME PARK
Playing Field
Fernhall Farm
Fernhall Wood
Poultry Farm
Dallance Farm
Nursery
Brookmeadow Wood
Willows
Galley Hill
EN9
Magic Lakes
Cobbin Pond
Obelisk
Nicholls Farm
Home Farm
Obelisk Farm
WARLIES PARK
Nurseries
Oaklands Farm
Warlies Gardens
Warlies Park House
The Temple
Burgess Farm
West Hill
Temple Hill
Nursing Home
Pickhill Farm
EPPING FOREST
Cricket Ground
Warlie's Lodge
UPSHIRE
Upshire Prim. Sch.
Sergeants Green
Warlies Park Farm
COPTHALL GREEN
Oxleys Wood
Jun. & Inf. Schs.
King Harold Academy
Spts Cen.
Tennis Courts
Playing Field
Hazelwood Children's Cen.
Prim. Sch.
Waltham Abbey Swim. Pool
Recreation Ground
Superstore
Football Ground
Cemetery
Rec. Grd.
The Leverton Infant & Junior Sch.
Honeylands
King George's Field
Waltham Abbey Swim. Pool
Quinton Hill Farm
Woodgreen Farm
Potkiln Wood
Brambly Shaw
Woodgreen Potteries
Southend Farm
Upshire Hall
WOOD GREEN
M25
Woodredon House
Conybury Wood
Woodredon Farm
Sudbury Farm
Junction 26
Skillet Hill Farm
HONEY LANE
WOODRIDDEN
DOWDING WAY
A121
Inner Lodge
Lord Padgets Wood
Poplar Shaw
Prayer Halls
Cemetery
Honey Lane Plain
Round Hills
BEECH HILL PARK
Tile Hill Farm
Honey Lane Quarters
Sunshine Plain
The Downing Nursery
Beechview Nursery
Felicia Nursery
Kennels
Portulaca
Providence Nursery
Normandy Nursery
Aveylane Farm
Riding School
HIGH BEECH GOLF COURSE
The Rookery
Wake Valley Pond
Wake Valley
High Beech Nurseries
Kennels
Piggery
Sewardstone Nursery
Chandlers Farm
Avey Cottages
Pynest Green
Riding School
Club House
Rushey Plain
Great Monk Wood
Bowling Green
Youth Hostel
IG10
HIGH BEECH
Mount Pleasant
A104
Mott Street Nursery
Thompson's Wood
E4
Beechside Nursery
Manor Farm
High Beech C of E Primary Sch
Manor House
35
Epping Forest Field Centre
Little Monk Wood
Oak Farm
Wallsgrove House
Paul's Nursery

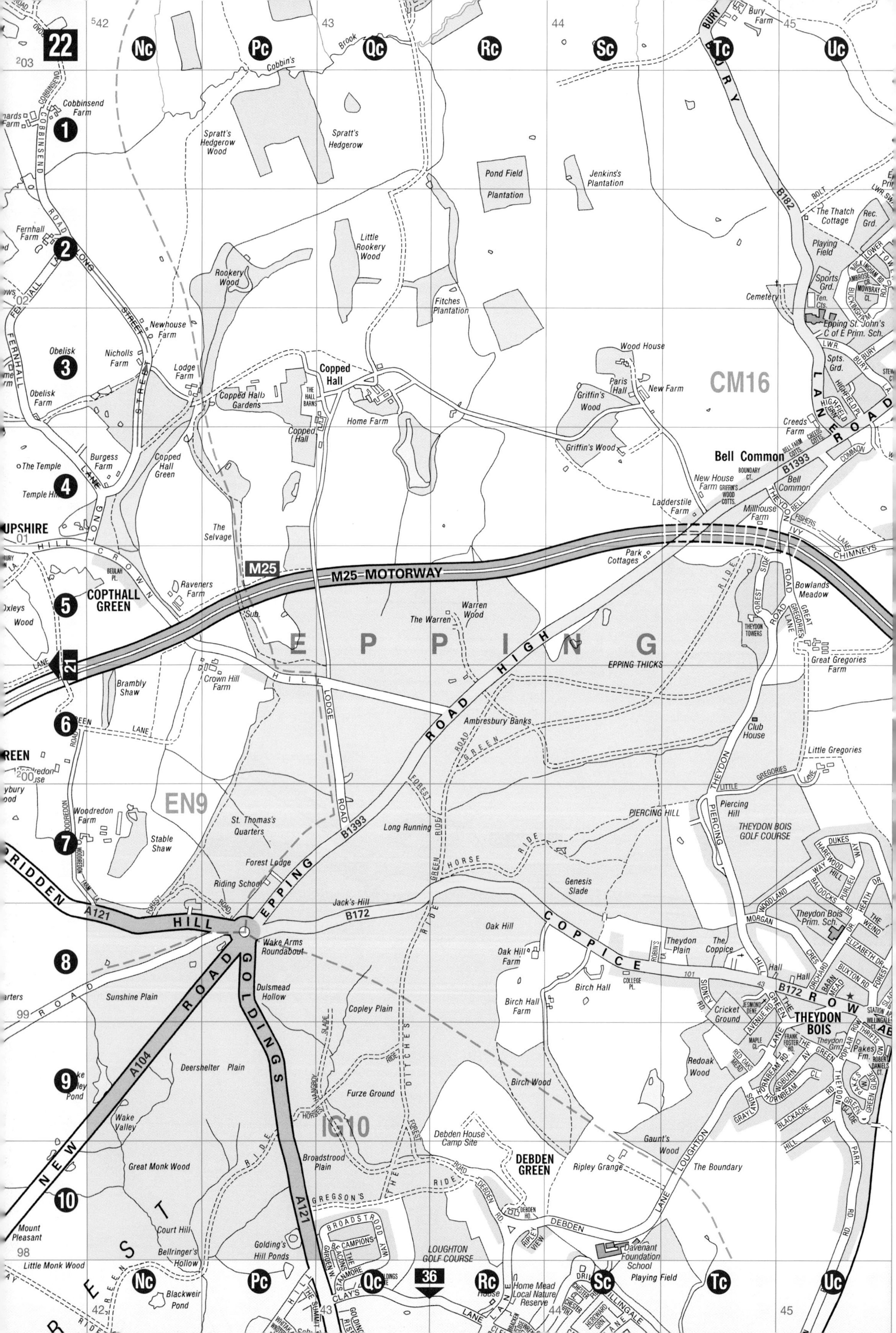
Cobbinsend Farm
Spratt's Hedgerow Wood
Spratt's Hedgerow
Pond Field Plantation
Jenkins's Plantation
Bury Farm
The Thatch Cottage
Fernhall Farm
Little Rookery Wood
Rookery Wood
Fitches Plantation
Newhouse Farm
Nicholls Farm
Obelisk
Obelisk Farm
Lodge Farm
Copped Hall
Copped Hall Gardens
Home Farm
Wood House
Paris Hall
New Farm
Griffin's Wood
CM16
Bell Common
The Temple
Burgess Farm
Copped Hall Green
Temple Hill
New House Farm
Ladderstile Farm
Millhouse Farm
UPSHIRE
The Selvage
M25
M25 MOTORWAY
Park Cottages
Raveners Farm
COPTHALL GREEN
Warren Wood
The Warren
Bowlands Meadow
EPPING
EPPING THICKS
Great Gregories Farm
Brambly Shaw
Crown Hill Farm
Ambresbury Banks
Club House
Little Gregories
EN9
Woodredon Farm
St. Thomas's Quarters
Stable Shaw
Long Running
PIERCING HILL
Piercing Hill
THEYDON BOIS GOLF COURSE
Forest Lodge
Riding School
Genesis Slade
Jack's Hill
B172
Oak Hill
Wake Arms Roundabout
Oak Hill Farm
Theydon Plain
The Coppice
Theydon Bois Prim. Sch.
Sunshine Plain
Dulsmead Hollow
Birch Hall
Birch Hall Farm
Copley Plain
Cricket Ground
THEYDON BOIS
Deershelter Plain
Redoak Wood
Furze Ground
Birch Wood
Wake Valley
IG10
Debden House Camp Site
Gaunt's Wood
Great Monk Wood
Broadstrood Plain
DEBDEN GREEN
Ripley Grange
The Boundary
Court Hill
Mount Pleasant
Golding's Hill Ponds
LOUGHTON GOLF COURSE
Davenant Foundation School
Bellringer's Hollow
Little Monk Wood
Blackweir Pond
Home Mead Local Nature Reserve
Playing Field
36
21

EPPING
COOPERSALE
COOPERSALE STREET
FIDDLERS HAMLET
Steward's Green
Ivy Chimneys
Theydon Garnon
Hobbs Cross
Theydon Bois
Mount End
EPPING GOLF COURSE
HOBBS CROSS GOLF COURSE
ABRIDGE GOLF COURSE
M11 MOTORWAY
M25 MOTORWAY
Junction 27 (M25)
Junction 6 (M11)
RM4

37

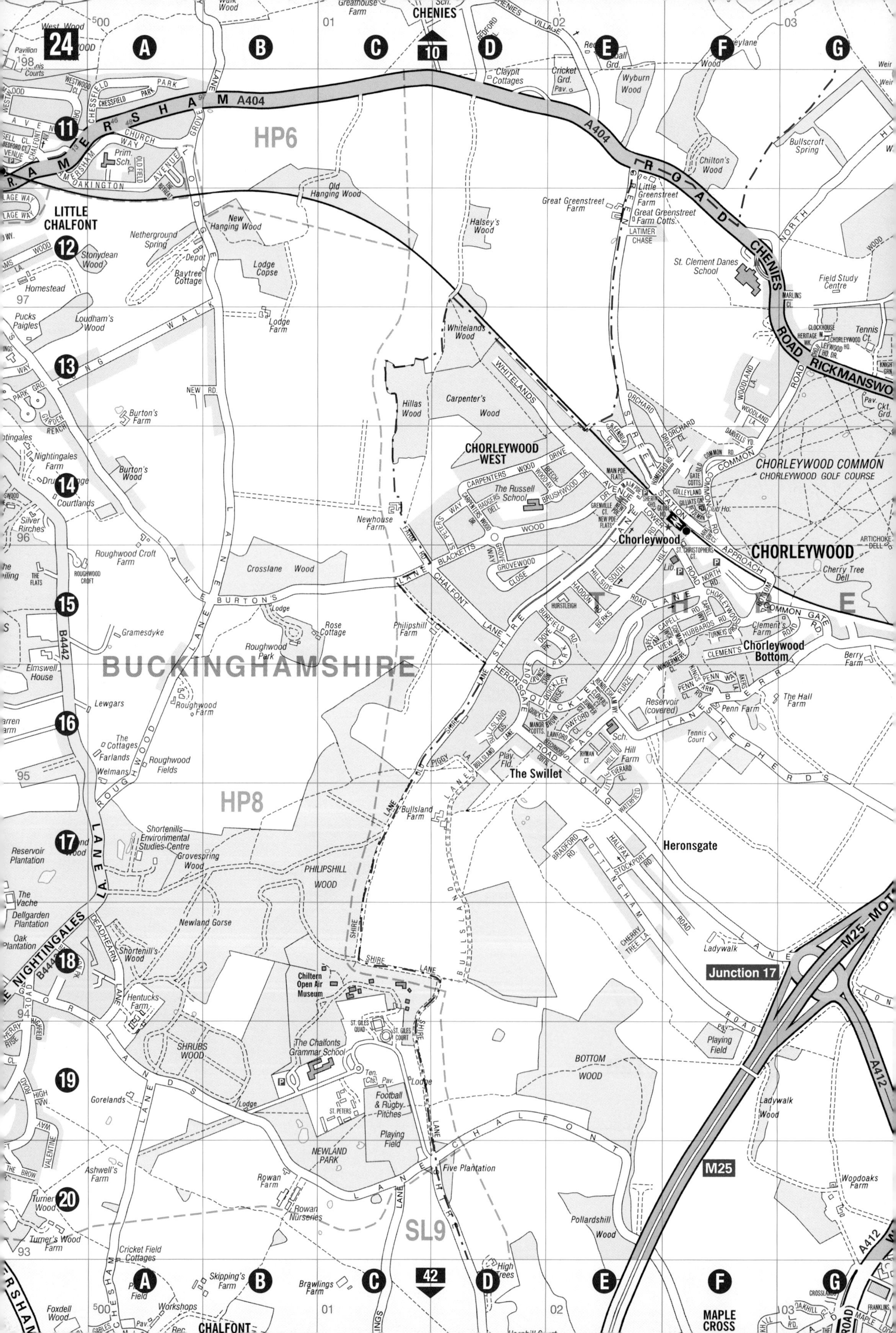
A
B
C
D
E
F
G
10
CHENIES
Greathouse Farm
Walk Wood
West Wood
Pavilion
A404
AMERSHAM ROAD
CHESSFIELD PARK
WESTWOOD CL.
CHURCH WAY
Prim. Sch.
OAKINGTON
OLD FIELD CL.
AVENUE
LODGE LANE
GROVE LANE
HP6
Claypit Cottages
Cricket Grd. Pav.
Wyburn Wood
Chilton's Wood
Bullscroft Spring
CHENIES VILLAGE
BEDFORD AV.
LITTLE CHALFONT
Old Hanging Wood
New Hanging Wood
Netherground Spring
Stonydean Wood
Depot
Baytree Cottage
Lodge Copse
Homestead
Halsey's Wood
Great Greenstreet Farm
Little Greenstreet Farm
Great Greenstreet Farm Cotts.
LATIMER CHASE
GREEN STREET
St. Clement Danes School
CHENIES ROAD
NORTH HILL
Field Study Centre
MARLINS CL.
CLOCKHOUSE
HERITAGE M.
Tennis Ct.
RICKMANSWORTH ROAD
Pucks Paigles
Loudham's Wood
LONG WALK
Lodge Farm
Whitelands Wood
WHITELANDS AV.
NEW RD.
Burton's Farm
PARK GRO.
GARDEN REACH
Hillas Wood
Carpenter's Wood
CHORLEYWOOD WEST
ORCHARD DRIVE
ORCHARD CL.
WOODLAND LA.
DARVELLS YD.
OLD COMMON RD.
COMMON GATE COTTS.
CHORLEYWOOD COMMON
CHORLEYWOOD GOLF COURSE
Pav.
Ckt. Grd.
Nightingales Farm
Burton's Wood
Courtlands
Silver Birches
CARPENTERS WOOD DRIVE
The Russell School
BRUSHWOOD DR.
BADGERS DELL
ST. PETERS WAY
WOOD AV.
Newhouse Farm
MAIN PDE. FLATS
NEW PDE. FLATS
GRENVILLE CT.
STATION APPROACH
Chorleywood
Club Ho.
ARTICHOKE DELL
Roughwood Croft Farm
ROUGHWOOD CROFT
THE FLATS
Crosslane Wood
BLACKETTS WOOD
GROVE WAY
GROVEWOOD CLOSE
CHALFONT LANE
FARM RD.
CHORLEYWOOD
Cherry Tree Dell
Lib.
ST. CHRISTOPHERS CT.
NORTH RD.
BURTON'S LANE
Lodge
Rose Cottage
Gramesdyke
Roughwood Park
Philipshill Farm
HILLSIDE
SOUTH RD.
HADDON RD.
BERKS
HURSTLEIGH
BURFIELD RD.
DOVE PARK
THE
COMMON GATE RD.
Clement's Farm
CLEMENT'S RD.
Chorleywood Bottom
Berry Farm
B4442
Elmswell House
BUCKINGHAMSHIRE
HUBBARDS RD.
CAPELL RD.
WINDERMERE
TURNEYS ORCH.
PENN WAY
PENN CL.
PENN FARM
Lewgars
Roughwood Farm
HERONSGATE
QUICKLEY LANE
QUICKLEY RISE
FURZE VIEW
Reservoir (covered)
Penn Farm
The Hall Farm
SHEPHERDS LANE
Tennis Court
The Cottages
Farlands
Roughwood Fields
Welmans
ROUGHWOOD LANE
BULLSLAND GDS.
BULLSLAND LA.
PIGGY LA.
Play. Fld.
The Swillet
LONG LANE
Sch.
Hill Farm
EVERARD CL.
RYMAN CT.
WATERFIELD
HP8
Bullsland Farm
BULLSLAND LANE
SHIRE LANE
Shortenills Environmental Studies Centre
Grovespring Wood
Reservoir Plantation
The Vache
Dellgarden Plantation
Oak Plantation
PHILIPSHILL WOOD
BRADFORD RD.
NOTTINGHAM ROAD
HALIFAX RD.
STOCKPORT RD.
Heronsgate
CHERRY TREE LA.
Newland Gorse
NIGHTINGALES LANE
DEADHEARN LANE
Shortenill's Wood
Ladywalk
Junction 17
M25
SHIRE LANE
Chiltern Open Air Museum
Hentucks Farm
ORCHEHILL AV.
ST. GILES QUAD
ST. GILES COURT
SHRUBS WOOD
The Chalfonts Grammar School
Ten. Cts.
Pav.
Lodge
BOTTOM WOOD
Playing Field
A412
Gorelands
GORELANDS LANE
Football & Rugby Pitches
ST. PETERS
Lodge
Playing Field
CHALFONT LANE
Ladywalk Wood
VALENTINE WAY
THE BROW
NEWLAND PARK
Five Plantation
Ashwell's Farm
Rowan Farm
Rowan Nurseries
Woodoaks Farm
Turner Wood
Pollardshill Wood
SL9
Turner's Wood Farm
Cricket Field Cottages
High Trees
Skipping's Farm
Brawlings Farm
42
Play. Field
Workshops
Foxdell Wood
Pav.
Rec.
CHALFONT
MAPLE CROSS
CROSSLANDS
FRANKLINS
OAKHILL RD.
CHESHAM LANE
GABLES

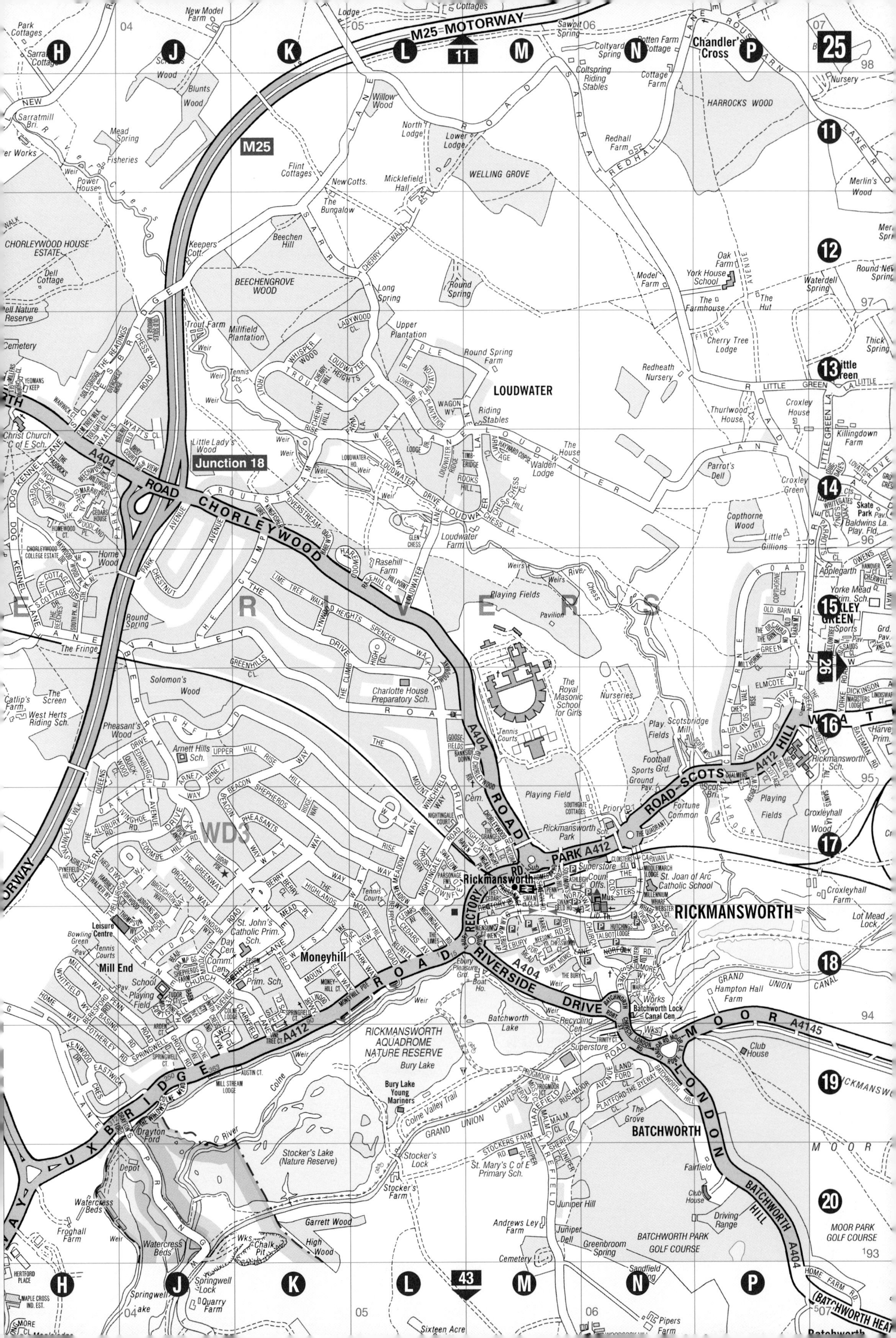

25
M25 MOTORWAY
11
26
43
M25
Junction 18
H
J
K
L
M
N
P
04
05
06
07
98
97
96
95
94
93
11
12
13
14
15
16
17
18
19
20
Park Cottages
New Model Farm
Lodge
Cottages
Sawpit Spring
Coltyard Spring
Potten Farm Cottage
Chandler's Cross
Nursery
Scrubs Wood
Blunts Wood
Willow Wood
Coltspring Riding Stables
Cottage Farm
HARROCKS WOOD
Sarratmill Bri.
Mead Spring
Fisheries
North Lodge
Lower Lodge
Redhall Farm
Power House
Weir
Flint Cottages
New Cotts.
Micklefield Hall
WELLING GROVE
Merlin's Wood
The Bungalow
CHORLEYWOOD HOUSE ESTATE
Dell Cottage
Keepers Cott.
Beechen Hill
BEECHENGROVE WOOD
Oak Farm
York House School
Model Farm
Waterdell Spring
Round Spring
Long Spring
The Farmhouse
The Hut
Cemetery
Trout Farm
Millfield Plantation
Upper Plantation
Cherry Tree Lodge
Thick Spring
Round Spring Farm
Redheath Nursery
LOUDWATER
Riding Stables
Thurlwood House
Croxley House
Killingdown Farm
Christ Church C of E Sch.
Little Lady's Wood
The House
Walden Lodge
Parrot's Dell
Croxley Green
Skate Park
Baldwins La. Play. Fld.
Copthorne Wood
Little Gillions
Loudwater Farm
Rasehill Farm
Home Wood
Applegarth
Yorke Mead Prim. Sch.
Playing Fields
Pavilion
Round Spring
The Fringe
Sports
Solomon's Wood
Charlotte House Preparatory Sch.
The Royal Masonic School for Girls
Nurseries
The Screen
Catlip's Farm
West Herts Riding Sch.
Pheasant's Wood
Arnett Hills Sch.
Tennis Courts
Play Fields
Scotsbridge Mill
Football Grd.
Sports Ground
Rickmansworth Sch.
Playing Field
Fortune Common
Playing Fields
Croxleyhall Wood
WD3
Rickmansworth Park
Rickmansworth
St. Joan of Arc Catholic School
Croxleyhall Farm
RICKMANSWORTH
Lot Mead Lock
Leisure Centre
Bowling Green
St. John's Catholic Prim. Sch.
Moneyhill
Mill End
Ebury Pleasure Grd.
Boat Ho.
Hampton Hall Farm
Batchworth Lake
Batchworth Lock Canal Cen.
Recycling Cen.
RICKMANSWORTH AQUADROME NATURE RESERVE
Bury Lake
Superstore
Club House
Bury Lake Young Mariners
Colne Valley Trail
GRAND UNION CANAL
The Grove
BATCHWORTH
Drayton Ford
River Colne
Stocker's Lake (Nature Reserve)
Stocker's Lock
St. Mary's C of E Primary Sch.
Fairfield
Depot
Stocker's Farm
Watercress Beds
Froghall Farm
Juniper Hill
Club House
Driving Range
MOOR PARK GOLF COURSE
Garrett Wood
High Wood
Andrews Ley Farm
Juniper Dell
Greenbroom Spring
BATCHWORTH PARK GOLF COURSE
Watercress Beds
Chalk Pit
Cemetery
Sandfield
Springwell Lock
Quarry Farm
Springwell Lake
Sixteen Acre
Pipers Farm
MAPLE CROSS IND. EST.
HERTFORD PLACE
A404
A412
A4145
CHORLEYWOOD ROAD
RIVERSIDE DRIVE
UXBRIDGE ROAD
PARK ROAD
SCOTS HILL
LONDON ROAD
BATCHWORTH HILL
MOOR LANE
RECTORY ROAD
LOUDWATER LANE
LITTLE GREEN LANE
R I V E R S

Q
R
S
T
U
V
W
12
44
25
11
12
13
14
15
16
17
18
19
20
Blackets Wood
Leeswood Cotts.
Deer Spring
Nursery
ROCKS WOOD
WHIPPENDELL WOOD
JACOTTS HILL
Merlin's Wood
Newland's Spring
Merlin's Spring
WEST HERTS GOLF COURSE
Round Newland's Spring
Waterdell Spring
Dell Wood
The Hut
Long Newland's Spring
Thick Spring
Little Green
Iron Bridge Lock
GRAND UNION CANAL
Watercress Beds
Boating Pond
CASSIOBURY PARK
Playing Field
Herts Bowling Club
Cassiobury Infant Sch.
Cassiobury Junior Sch.
WD17
WATFORD
Watford
Watford Grammar Sch. for Boys
West Herts College
PROSPECT HO.
Watford Colosseum
Hospice
Croxley House
Killingdown Farm
Little Green Jun. Sch.
Playing Fields
WD3
KENILWORTH DRIVE
Skate Park
Baldwins La. Play. Fld.
Copthorne Wood
Little Gillions
Applegarth
Yorke Mead Prim. Sch.
CROXLEY GREEN
Youth Centre
Inf. Sch.
Cassio Bridge Lock
Fish Pond
Marina
RICKMANSWORTH ROAD
A412
WD18
WEST WATFORD
Mail Centre
Westfield Community Sports Cen.
St. Anthony's Catholic Prim. Sch.
Westfield Academy
Croxley Centre
Croxley
HOLYWELL
WATFORD ROAD
Harvey Rd. Prim. Sch.
Rickmansworth Sch.
Croxleyhall Wood
Long Valley Wood
Croxley Common Moor Local Nature Reserve
Common Moor
Moor Park Crossing
TOLPITS LANE
A4145
King George V Playing Fields
Electricity Transmission Station
Harwoods Rec. Grd.
Laurance Haines Prim. Sch.
Bowling Green
Sports Grd.
Brightwells Farm Cotts.
Brightwells Farm
Croxleyhall Farm
RICKMANSWORTH
Lot Mead Lock
Tolpits Ho.
RIVER COLNE
Brightwells Spring
Oxhey Hall
Moat
THREE RIVERS
Running Track
Merchant Taylors' School
Playing Fields
Pumping Station
Tolpits Plantation
The Withey Beds Local Nature Reserve
Northwood Prep. Sch.
Manor of the Rose
HAMPERMILL
Hampermill Wood
Big Wood
Club House
RICKMANSWORTH GOLF COURSE
MOOR PARK
Club House
The Mansion
Tennis Courts
Moor Park
HA6
SANDY LODGE GOLF COURSE
Farm Cottage
SANDY LODGE LANE
Hampermill Spring
Round Spring
Old Furze Field
The Roughs
A4125
Young Hangings
MOOR PARK GOLF COURSE
BATCHWORTH HILL
A404
Home Farm
EASTBURY
Old Hangings
Oxhey Woods Local Nature Reserve
Bathend

27
Junction 5
13
WD25
WD24
WD23
WD19
HERTSMERE
Watford Junction
Watford High St.
Bushey
BUSHEY
OXHEY
Watford Heath
Merry Hill
CARPENDERS PARK
Carpenders Park
BUSHEY HALL GOLF COURSE
HARTSBOURNE GOLF COURSE
Scotts Wood
Icedale Wood
Bushey Hall Farm
Highwood Prim. Sch.
David Lloyd Leisure
The Purcell School
Queens' School (North Site)
Queens' School (South Site)
Metropolitan Police Sports Ground
Powerleague Soccer Centre (Watford)
Bushey Jewish Cemetery
Farmyard Funworld
The Lincolnsfield Centre
Bushey Grove Leisure Centre
Royal Connaught Park
Bushey Manor Junior School
The Bushey Academy
Sacred Heart Catholic Prim. Sch.
Bournehall Prim. Sch.
Moat Field
Bushey Meads School
Ashfield Jun. Sch.
St. Margaret's Sch.
St. Margaret's Sports Centre
Merry Hill House
Little Hartsbourne Wood
Hartsbourne Country Club (Club House)
Levels Wood
Grim's Ditch
Grim's Dyke
Weald Common
Valley View Farm
Mutton Wood
Carpenders Park Cemetery
Carpenders Park Farm
Woodwalks
St. Meryl Sch.
St. Joseph's Catholic Primary School
Little Furze Field
Beck Mason's Wood
Pond Wood
Sandpit Wood
South Oxhey Playing Fields
Ox Pasture Spring
Warren Dell
Skatepark
Sherwoods Wood
Margeholes Wood
Oxhey Grange
Front Lodge
Woodlands
Brick Works
Riding School
Nursery
Reservoir
Bromet Prim. Sch.
The Rookery
Riverside Park
Oxhey Park
Watford FC (Vicarage Road Stadium)
Watford Health Campus
Vicarage Road Cem.
Hertfordshire Fire Mus.
Waterfields Retail Park
Colne Valley Retail Park
Hollywood Bowl
OTTERSPOOL WAY
ELTON WAY
A41 (NORTH WESTERN AV.)
M1
HARTSPRING LANE
STEPHENSON WAY
WATERFIELDS WAY
LONDON ROAD
HIGH STREET
CHALK HILL
PINNER ROAD
OXHEY ROAD
OXHEY LANE
PRESTWICK ROAD
EASTBURY ROAD
WIGGENHALL ROAD
THOMAS SAWYER WAY
BROOKDENE AV.
SPARROWS HERNE
A4008
A411
A412
A4125
A4178
A4145
B462
B4542
X Y Z Aa Ba Ca Da
11 12 13 14 15 16 17 18 19 20
28
45

28
Ea
Fa
Ga
14
Ha
Ja
Ka
La
Round Bush
Four Want Ways
Delrow
Aldenham Grange
Aldenham Cottage
Holland Farm House
Patchetts Green
Bhaktivedanta Manor School
LETCHMORE HEATH
Playing Field
Aldenham School
Playing Field
Boydens Hill
Little Kendals Farm
Little Kendals Wood
Hertsmere Jewish Prim. Sch.
Kendal Hall Farm
Radlett Prep. Sch.
Medburn House
Play. Fld.
A5183
WATLING STREET
WINFIELD MOBILE HOME PARK
Hartspring Cotts.
Delrow Ho.
Holland Farm
WD25
Electricity Transforming Station
Kennels
Gullimore Farm
SANDY LA. CARAVAN SITE
Bushey Jewish Cemetery
(NORTH WESTERN AV.)
TYLERS WAY
A41
Cemetery
Slades Farm
Letchmore Lodge
Moats
Playing Field
Playing Field
The Haberdashers' Aske's Boys' Sch.
Medburn Bri.
ELSTREE AERODROME
ALDENHAM PARK
Hilfield Farm
Ham Farm
M1
Flying School
Flying Club
The Haberdashers' Aske's School for Girls
Tykes Water
Tykes Water Lake
Boathouse
Lodge
Hilfield Castle
Boat House
Page's Farm
Dam
HILFIELD PARK
Sewage Works
Home Farm
Pumping Station
RESERVOIR
Tyler's Farm
Bushey Meads School
Meadow Wood Sch.
Works
Hart's Farm
Laboratories
Pumping Station
H E R T S
Dam
ALDENHAM COUNTRY PARK
Aldenham Country Pk. Rare Breeds Farm
ALDENHAM RESERVOIR
Aldenham Sailing Club
Pav.
Sports Ground
M1 MOTORWAY
WAY
ROAD
WATFORD
A411
27
Caravan Site
WD23
Caldecote Farm
Tennis Courts
BUSHEY
King George Rec. Grd.
Tennis Cts.
Immanuel College
Rosary Priory
ELSTREE ROAD
Caldecote Hill
LISMIRRANE INDUSTRIAL PARK
THE WATERFRONT
Reservoir
Elstree Hill Open Space
Comm. Cen.
CENTENNIAL PARK
CENTENNIAL AVENUE
A41
BY-PASS
A409
Heath House
SPIRE BUSHEY HOSP.
HERNE
HIGH ROAD
A4140
Sch.
Rec. Grd.
Works
Margaret's Sports Centre
BUSHEY HEATH
HEATHBOURNE ROAD
Belswood Cottage
H A R R O W
Brockley Hill Farm
Sports Ground
A5
BROCKLEY HILL
ROYAL NATIONAL ORTHOPAEDIC HOSPITAL
Brockley Hill Ho.
ASPIRE
Stanmore Common Local Nature Reserve
HA7
Little Hartsbourne Wood
Hartsbourne Country Club (Club House)
Hartsbourne Prim. Sch.
Kestrel Grove (Offices)
Tennis Courts
The Cedars
A4140
COMMON
Lodge
Limes House
Pear Wood Cottages
Pear Wood
Fish Pond
Cricket Ground
Fish Pond
Grove Field Pavilion
Tennis Cts.
Cloisters Wood
HARTSBOURNE GOLF COURSE
HA3
Lodge
Bentley Priory Mus.
Bentley Priory
Heriots
Stanmore Hall
HARROW WEALD
Levels Wood
COMMON
The Princess Alexandra Home
Deer Park
STANMORE COMMON (Local Nature Reserve)
Bentley Priory Local Nature Reserve
46
Grim's Dyke
Grimsdyke Open Space
Harrow Weald Common
Nurseries
Old Barn
Heriot's Wood
STANMORE
11
12
13
14
15
16
17
18
19
20
98
97
96
95
94
93
15
16
17

29
15
30
47
11
12
13
14
15
16
17
18
19
20
BOREHAMWOOD
ELSTREE
Elstree & Borehamwood
BBC Elstree
Elstree Studios
Borehamwood Shopping Park
Hertswood Academy
Meryfield Prim. Sch.
Cowley Hill Sch.
Meadow Park
Boreham Wood FC (Meadow Pk.)
Aberford Park
Sports Ground
Radlett Park Golf Course
Golf Driving Range
Club House
Parkfields Open Space
North Medburn Farm
South Medburn Farm
Organ Hall Farm
Well End
Electricity Sub. Station
Rowley Farm
Rowley Lane Sports Grd.
King George's Field
Kenilworth Prim. Sch.
Borehamwood Ind. Park
Hertsmere Ind. Park
Elstree Distribution Park
Ripon Park
Playing Field
Yavneh Coll.
Supertstore
Recycling Centre
Cemetery
St. Nicholas Elstree C of E Prim. Sch.
Deacons Hill
The Chantry
Deacons Hill Open Space
Woodcock Farm
Elstree Manor
Penniwells Farm
The Leys
Hill House
Edgwarebury Ho. Farm
Elstree Open Space
The Dower House
Scratch Wood
Froghall Wood
Scratchwood Open Space
Boys Hill Wood
Thistle Wood
Hyver Hall
Elstree Tunnel
Junction 4
M1
London Gateway Service Area
BARNET
MERE
NW7
HA8
Mill Hill Golf Course
Clump of Trees Wood
Bury Farm
Edgware Bury
Gate Cottages
Nursery Cottage
Edgwarebury Cemetery
Sports Grd.
Broadfields Prim. Sch.
Edgwarebury Park
Hemmings Wood
Stoney Wood Lake
Wood Fairway Prim. Sch.
Northway Sch.
Nut Wood
Target Wood
Tanglewood
Mill Hill High Sch.
Newlands
Driving Range
Stanmore and Edgware Golf Centre
London Academy
St Mary's Convent
A5183
A411
A5135
B5378
A1
A41
A410
WATFORD BY-PASS
WATLING STREET
ELSTREE WAY
SHENLEY ROAD
ALLUM LANE
BARNET LANE
DEACONS HILL
WOODCOCK HILL
EDGWARE WAY
BARNET BY-PASS
ROWLEY LANE
ELSTREE HILL NTH.
ELSTREE HILL STH.
STIRLING CORNER
Ma
Na
Pa
Qa
Ra
Sa
Ta
18
19
20
21
93
94
95
96
97
98

30
16
48
29
HERTSMERE
WD6
EN5
NW7
DYRHAM PARK
THE SHIRE LONDON GOLF COURSE
OLD FOLD MANOR GOLF COURSE
Saffron Green
Rowley Green
HIGH BARNET
CHIPPING BARNET
ARKLEY
Barnet Gate
Highwood Hill
TOTTERIDGE
BARNET BY-PASS
A1
A411
A5109
A1081
A5135
B552
WOOD STREET
TOTTERIDGE COMMON
HIGHWOOD HILL
MARSH LANE
BARNET ROAD
Moat Mount Open Space
Arkley Golf Course
Rowley Green Common Nature Reserve
Barnet Hospital
Queen Elizabeth's School
Whitings Hill Open Space
Ducks Island
Totteridge Fields Local Nature Reserve
Hendon Wood Lane Open Space
Folly Brook & Darland's Lake Nature Reserve
Mill Hill High Sch.
Dollis Brook
Folly Brook

31
17
32
49
Bb
Cb
Db
Eb
Fb
Gb
Hb
11
12
13
14
15
16
17
18
19
20
25
26
27
28
98
97
96
95
94
93
HADLEY WOOD
The Beale Arboretum
Ferny Hill Farm
Fernyhill Wood
Great Broadgates Hill
Club House
BEECH HILL PARK
HADLEY WOOD GOLF COURSE
ENFIELD
Rectory Farm
St. Martha's Senior School
Monken Hadley C of E Prim. Sch.
Bournwell Hill
MONKEN HADLEY COMMON
Hadley Wood Nature Reserve
Sewits Hill
Fish Pond
Seedfield Spinney
Water Tower
Oak Wood
Reservoir (covered)
Go Ape
MONKEN HADLEY
Priddeon's Hill
Newmans Hill
Beech Hill Lake
Jewish Comm. Sec. Sch.
Playing Field
Artificial Pitches
Play. Fld.
Livingstone Prim. Sch.
Tudor Park Footgolf
Tudor Sports Ground
Club House
HADLEY
King George's Field
BARNET
Cromer Road Prim. Sch.
Playing Fields
Cemetery
Cockfosters
COCKFOSTERS
Queen Elizabeth's Girls' Sch.
Leisure Cen.
High Barnet
Hospice
Depot
Recreation Ground
NEW BARNET
PureGym
New Barnet
Highlands Gds.
Barnet Vale
Lyonsdown
EAST BARNET
Pleasure Ground
Underhill Playing Fields
Cinema
The Rainbow Cen.
Gravesend Avenue Inf. Sch.
KING GEORGE PLAYING FIELDS
Sports Cen.
Tennis Courts
N20
Wyatts Farm Open Space
Recreation Ground
SOUTH HERTS GOLF COURSE
Brook Farm Open Space
Pavilion
Oakleigh Park
East Barnet Sch.
Oakwood Sch.
Oak Hill Nature Reserve
Bandstand
Playground
OAK HILL PARK
Miniature Golf Course
Monkfrith Prim. Sch.
Bowling Green
Oak Hill Campus
Focus School
Warehouses
Pavilion Study Centre
Pupil Referral Unit
Sports Grd.
Club House
Works
Totteridge and Whetstone
OAKLEIGH PARK
South Barnet
Waverley
Windyridge
Dell House
The Croft
Cricket Ground
WHETSTONE
Swan Lane Open Space
Tennis Club
Sports Ground
N11
BRUNSWICK PARK
Laurel Farmhouse
N12
Woodridge Prim. Sch.
North London Hospice
High Sch.
NORTH MIDDLESEX GOLF COURSE
Club House
Bethune Recreation Ground
Sports Ground
Woodside Park Club
Tennis Courts
Friary
GREAT NORTH ROAD
HIGH STREET
BARNET HILL
GREAT NORTH ROAD
STATION ROAD
EAST BARNET RD.
BROOKHILL RD.
CAT HILL
LONGMORE AV.
LYONSDOWN RD.
HIGH ROAD
TOTTERIDGE VILLAGE
TOTTERIDGE LANE
OAKLEIGH ROAD
RUSSELL ROAD
OAKLEIGH ROAD NORTH
FRIERN BARNET LANE
HIGH ROAD
COCKFOSTERS ROAD
CAMLET WAY
HADLEY ROAD
HADLEY COMMON
A1000
A110
A111
A109
A5109
B193
B550
B1453

32
18
31
50
Jb
Kb
Lb
Mb
Nb
Pb
Qb
11
12
13
14
15
16
17
18
19
20
Reservoir
Park Farm Cottages
Camlet Moat
Ferny Hill Cottage
Ferny Hill Farm
Obelisk
Moat Wood
Picnic Area
Parkside Farm
Pumping Station
Chase Farm Hospital
Camlet Hill
Ride Wood
Vicarage Farm
Wolverton
TRENT PARK (COUNTRY PARK)
Rough Lot
Leeging Beech Gutter
Fish Pond
Icehouse Wood
Seedfield Spinney
Mansion House
Williams Wood
Hog Hill
Salmon's Brook
Noddins Well
Hall
College Ho.
Oak Wood
Trent Park
Shaws Wood Cotts.
Shaws Wood
Reservoir (covered)
Lodge
The Rookery
Rookery Cotts.
Go Ape
Southgate Hockey Cen.
Hockey Pitches
Sports Ground
Church Wood
Miniature Rifle Range
EN2
Merryhills Brook
TRENT PARK GOLF COURSE
SLADES
A110 ROAD
Cemetery
Triangular Wood
EN4
Cockfosters
COCKFOSTERS
Trent Park Stables
ENFIELD ROAD
World's End
Driving Range
Jungle Falls Adventure Golf
Club Ho.
Boxer's Lake
Highlands School
Playing Fields
Playing Field
Grange Park Prim. Sch.
Moat
Oakwood
Play. Fld.
Southgate Sch. & Sports Cen.
BRAMLEY ROAD
A110
Highlands Village
Superstore
Nursing Home
Sports Ground
Chickenshed Theatre
De Bohun Prim. Sch.
Bramley Sports Grd
Saracens Amateur Rugby Football Club
Addison Ho. (PRU)
East Barnet Sch.
Oaktree Sch.
Oak Hill Theological College
Oak Hill Woods Nature Reserve
CHASE SIDE
A111
OAKWOOD PARK
OAKWOOD
Yacht Pond
CHASEVILLE PARK RD.
Bandstand Playground
OAK HILL PARK
Pavilion
Miniature Golf Course
Monkfrith Prim. Sch.
N14
Superstore
Bowling Green
Tennis Courts
GROVELANDS PARK
Leisure Cen.
PureGym
Osidge Prim. Sch.
South Barnet
Southgate
NORTH LONDON PRIORY HOSPITAL
Boating Lake
OSIDGE
B1453
Brunswick Park Prim. Sch.
BARNET
Ashmole Academy
Playing Fields
SOUTHGATE
Barnet & Southgate Coll.
HIGH STREET
A1004
BOURNE HILL
Pitch & Putt Course
Sports Grd.
N11
NEW SOUTHGATE CEMETERY
New Southgate Crematorium
War Mem.
BRUNSWICK PARK
Southgate Cricket Club Ground
Cemetery
A1003 ROAD
CANNON HILL
ALDERMANS HILL
PALMERS GREEN
Palmers Green
Recreation Ground
OAKLEIGH ROAD
A109
Cricket Ground
WATERFALL ROAD
ARNOS PARK
Putting Green
Bowling Greens
Tennis Courts
The Avenue
Broomfield Park
Sports Ground
Recreation Ground
Bethune Park

33
19
34
51
Rb
Sb
Tb
Ub
Vb
Wb
Xb
11
12
13
14
15
16
17
18
19
20
ENFIELD
ENFIELD TOWN
EN1
N21
N13
N9
Chase Side
Gordon Hill
Enfield Chase
Enfield Town
Bush Hill Park
Southbury
Grange Park
Winchmore Hill
Edmonton Green
LOWER EDMONTON
EDMONTON
GRANGE PARK
BUSH HILL PARK
WINDMILL A110 HILL
CHURCH ST.
SOUTHBURY RD.
A110
LONDON ROAD
A105
PARK AV.
VILLAGE RD.
CHURCH STREET
B154
GREAT CAMBRIDGE ROAD
A10
HERTFORD ROAD
A1010
FORE ST.
GREEN LANES
RIDGE AV.
BUSH HILL PARK GOLF COURSE
Town Park
Playing Fields
JUBILEE PARK
EDMONTON CEMETERY
Firs Farm Playing Fields
PYMMES PARK
Cemetery
Tottenham Park Cemetery
Federation Cemetery (Jewish)
Enfield Town FC (Queen Elizabeth II Stadium)
David Lloyd Leisure
Southbury Leisure Centre
Cineworld Cinema
MARTIN DALE INDUSTRIAL ESTATE
ENFIELD RETAIL PK.
MARTINBRIDGE TRADING ESTATE
Bishop Stopford's School
Suffolks Primary School
Heron Hall Academy
The Latymer School
Aylward Academy
Hazelwood Sports Club
The Paulin Ground
Sports Cen.
City Amenity Site
Recreation Ground
Bury Lodge Park
Hollywood Gardens
Tatem Rec. Grd.
Salmon's Brook
King George's Field
Miniature Golf Course

34
20
52
33
EN3
N9
ENFIELD HIGHWAY
PONDERS END
BRIMSDOWN
CHINGFORD
SEWARDSTONE
KING GEORGE'S
RESERVOIR
WILLIAM
GIRLING
RESERVOIR
ENFIELD
LEE VALLEY GOLF COURSE
Lee Valley Athletics Centre
Odeon Cinema
Athletics Track
Sewage Works
Refuse Incineration Plant
Teamsport Karting
Brimsdown
Ponders End
Power Station
Pumping Station
Overflow Channel
River Lee Navigation
River Lee Diversion
King George Sailing Club
Chingford Golf Range
Mansfield Park
Chingford Mount Cemetery
Chingford Foundation Sch.
Ridgeway Putting Park
Yardley Hill
Hawkswood PRU
Yardley Prim. Sch.
Pole Hill Obelisk
The Hawk Wood
Sewardstone Lodge
Caravan Park
Picks Farm
Amesbury Mead Farm
Luthers Farm
Woodlands Nursery
Santa Maria Nursery
Kingsfield Nursery
Barnfield Riding Stables
Hannah Nursery
Fouracres Nursery
Northfield Nursery
Northcroft Nursery
Marshfield Nursery
Chapelfield Nursery
Playing Field
Brimsdown Sports & Social Club
Brimsdown Prim. Sch.
Durants Park
Durants Sch.
The College of Haringey, Enfield & NE London
Kingfisher Hall Prim. Acad.
Heron Hall Academy
Woodpecker Hall Prim. Acad.
Nightingale Academy
Montagu Rec. Grd.
Tottenham Cemetery
Federation Cemetery (Jewish)
Oasis Academy Hadley
Boundary Ditch
Brimsdown Industrial Estate
Watermill Business Centre
Centenary Trading Estate
Mill River Trading Estate
Ponders End Industrial Estate
Waterways Business Centre
Sovereign Business Centre
Southbury
HERTFORD ROAD
HIGH STREET
NAGS HEAD RD.
LEA VALLEY ROAD
MOLLISON AVENUE
MERIDIAN WAY
KINGS HEAD HILL
SEWARDSTONE ROAD
MANSFIELD HILL
WALTHAM WAY
OLD CHURCH ROAD
NEW ROAD
HALL LANE
NIGHTINGALE ROAD
MONTAGU ROAD
RIDGEWAY
A1010
A1055
A110
A112
A1037
A1009
B137
B169
Yb
Zb
Ac
Bc
Cc
Dc
Ec
11
12
13
14
15
16
17
18
19
20
35
36
37
38
93
94
95
96
97
98
Suffield Hatch
Nuffield Health Fitness Cen. Chingford
Chingford Mount

35
Fc
Gc
Hc
Jc
Kc
Lc
Mc
39
40
41
42
98
97
96
95
94
193
542
21
36
53
11
12
13
14
15
16
17
18
19
20
Mott Street Nursery
Thompson's Wood
Beechside Nursery
Manor Farm
High Beech C of E Primary School
HIGH BEECH
Manor House
Epping Forest Field Centre
Little Monk Wood
Oak Farm
MOTT STREET
Wallsgrove House
Cricket Ground
Pav.
Paul's Nursery
Aldergrove Wood
The Vicarage
Dairy Farm
Loughton Camp
Barn Hill
Scout Camp Site
Kate's Cellar
Shelleys Hill
Day's Farm
Riding School
Equestrian Centre
EPPING FOREST
Fernhill Wood
Field Study Centre
Hill Wood
Debden Slade
Woodbury Hollow
Police Training Camp
Three Horseshoes Farm
Lippittshill Lodge
Whitehouse Plain
Carrolls Farm
Fairmead Pond
Strawberry Hill Ponds
Round Thicket
North Long Hills
GILWELL PARK
Gilwell Park
Club Ho.
WEST ESSEX GOLF COURSE
Ludgate Water
Caravan Site
Ludgate House
Fairmead Bottom
Reservoir (covered)
Almshouse Plain
IG10
Bury Wood
Long Hills
Nursery
Tennis Courts
Ludgate Plain
Bury Farm
Sewardstonebury
Peartree Plain
Palmer's Bridge
Roding Valley High Sch.
Superstore
Woodman's Glade
Cuckoo Pits
Grimston's Oak
The Hawk Wood
Riding School
Oaklands Sch.
Buttonseed Corner
Warren Hill
Magpie Hill
Connaught Water
Warren Hill Ho.
ROYAL EPPING FOREST GOLF COURSE
CHINGFORD GOLF COURSE
Chingford Plain
Recreation Ground
Powell's Forest
Queen Elizabeth's Hunting Lodge
Albany Ho.
Oak Hall
Hill Fm.
RANGER'S ROAD
A1069
A104
A121
A110
NEW ROAD
EPPING NEW ROAD
HIGH ROAD
The Warren
The Connaught Club
Tennis Courts
Bowl. Grn.
Warren Pond
Barn Hoppet
Linder's Field Local Nature Reserve
Chingford
Bus Sta.
STATION ROAD
Cricket Ground
Club
St.John's C of E Prim. School
CHINGFORD GREEN
Whitehall Plain
PALMERSTON RD.
Buckhurst Hill
Buckhurst Hill Prim. Sch.
Rec. Grd.
WHITEHALL RD.
Play. Fld.
Bluehouse Grove
Playing Field
Pimp Hall Nature Reserve
Pimp Hall Park
Friday Hill
Hatch Forest
REDBRIDGE
Reed's Forest
THE HOLLY PRIVATE HOSP.
BUCKHURST HILL
Playground
LORD'S BUSHES
Hatch Grove
Bancroft's Sch. & Prep. Sch.
Knighton Wood
Football Ground
WOODFORD GOLF COURSE
IG8
IG9
WOODFORD WELLS
Roding Valley
Rugby Football Ground
The Lops
HATCH LA.
LARKSHALL RD.
B160
B170
B146
RODING RD.

36
22
54
35
LOUGHTON
DEBDEN
CHIGWELL
Buckhurst Hill
Arewater Green
IG10
IG9
IG7
EPPING
M11 MOTORWAY
M11
Junction 5
A121
A1168
A113
A123
B170
CHURCH HILL
RECTORY LANE
CHIGWELL LANE
RODING LANE
ABRIDGE ROAD
CHIGWELL RISE
HAINAULT
Loughton
Debden
Chigwell
Roding Valley Meadows Nature Reserve
WOOLSTON MANOR GOLF COURSE
Loughton Golf Course
Debden Park High School
Roding Valley High Sch.
David Lloyd Leisure
Epping Forest Coll.
Loughton Leisure Cen.
Loughton Resource Centre
Rolls Park
Grange Farm (Riding for the Disabled)
Rugby Football Ground
Old Loughtonians Hockey Club
Home Mead Local Nature Reserve
Davenant Foundation School
Blackweir Pond
Blackweir Hill
Baldwins Hill
Baldwins Pond
Loughton Camp
Debden Slade
Woodbury Hollow
Little Monk Wood
Mount Pleasant
Court Hill
Bellringer's Hollow
Hillyfields Open Space
Broadfield Shaw
Long Shaw
High Wood
Patch Park
Broad Oaks
Chigwell Prim. Acad.
River Roding
Recreation Ground
Playing Fields
Nc
Pc
Qc
Rc
Sc
Tc
Uc
11
12
13
14
15
16
17
18
19
20
42
43
44
45
93
94
95
96
97
98

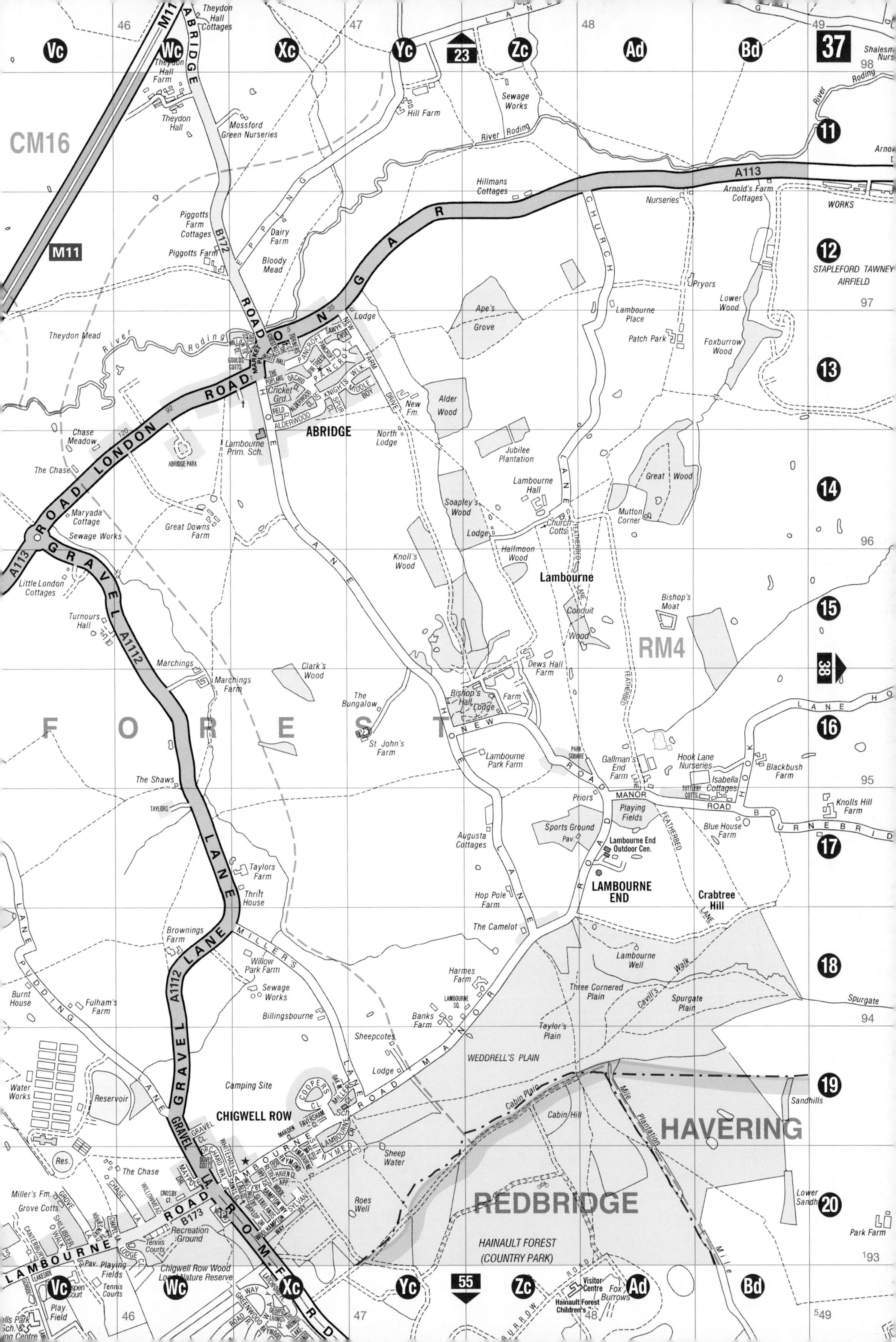
ABRIDGE
Lambourne
LAMBOURNE END
CHIGWELL ROW
HAVERING
REDBRIDGE
HAINAULT FOREST (COUNTRY PARK)
RM4
CM16
FOREST
ROAD LONDON ROAD
ONGAR ROAD
GRAVEL LANE
MANOR ROAD
Theydon Hall
Hill Farm
Sewage Works
River Roding
Hillmans Cottages
Arnold's Farm Cottages
Nurseries
STAPLEFORD TAWNEY AIRFIELD
Lambourne Place
Patch Park
Foxburrow Wood
Lower Wood
Ape's Grove
Alder Wood
Jubilee Plantation
Great Wood
Mutton Corner
Lambourne Hall
Church Cotts
Halfmoon Wood
Soapley's Wood
Knoll's Wood
Bishop's Moat
Conduit Wood
Dews Hall Farm
Bishop's Hall
Clark's Wood
The Bungalow
St. John's Farm
Lambourne Park Farm
Gallman's End Farm
Hook Lane Nurseries
Isabella Cottages
Blackbush Farm
Knolls Hill Farm
Playing Fields
Sports Ground
Lambourne End Outdoor Cen.
Blue House Farm
Crabtree Hill
Hop Pole Farm
The Camelot
Augusta Cottages
Lambourne Well
Three Cornered Plain
Spurgate Plain
Taylor's Plain
WEDDRELL'S PLAIN
Harmes Farm
Banks Farm
Sheepcotes
Cabin Hill
Sheep Water
Roes Well
Visitor Centre
Fox Burrows
Sandhills
Park Farm
Piggotts Farm Cottages
Piggotts Farm
Dairy Farm
Bloody Mead
Mossford Green Nurseries
Theydon Mead
Chase Meadow
The Chase
Maryada Cottage
Sewage Works
Great Downs Farm
Little London Cottages
Turnours Hall
Marchings
Marchings Farm
The Shaws
Taylors Farm
Thrift House
Brownings Farm
Willow Park Farm
Billingsbourne
Camping Site
Reservoir
Water Works
Burnt House
Fulham's Farm
Miller's Fm.
Grove Cotts.
Recreation Ground
Tennis Courts
Chigwell Row Wood Local Nature Reserve
Lambourne Prim. Sch.
Cricket Grd.
North Lodge
New Fm.
A113
A1112
B172
B173
M11
23
38
55
11
12
13
14
15
16
17
18
19
20

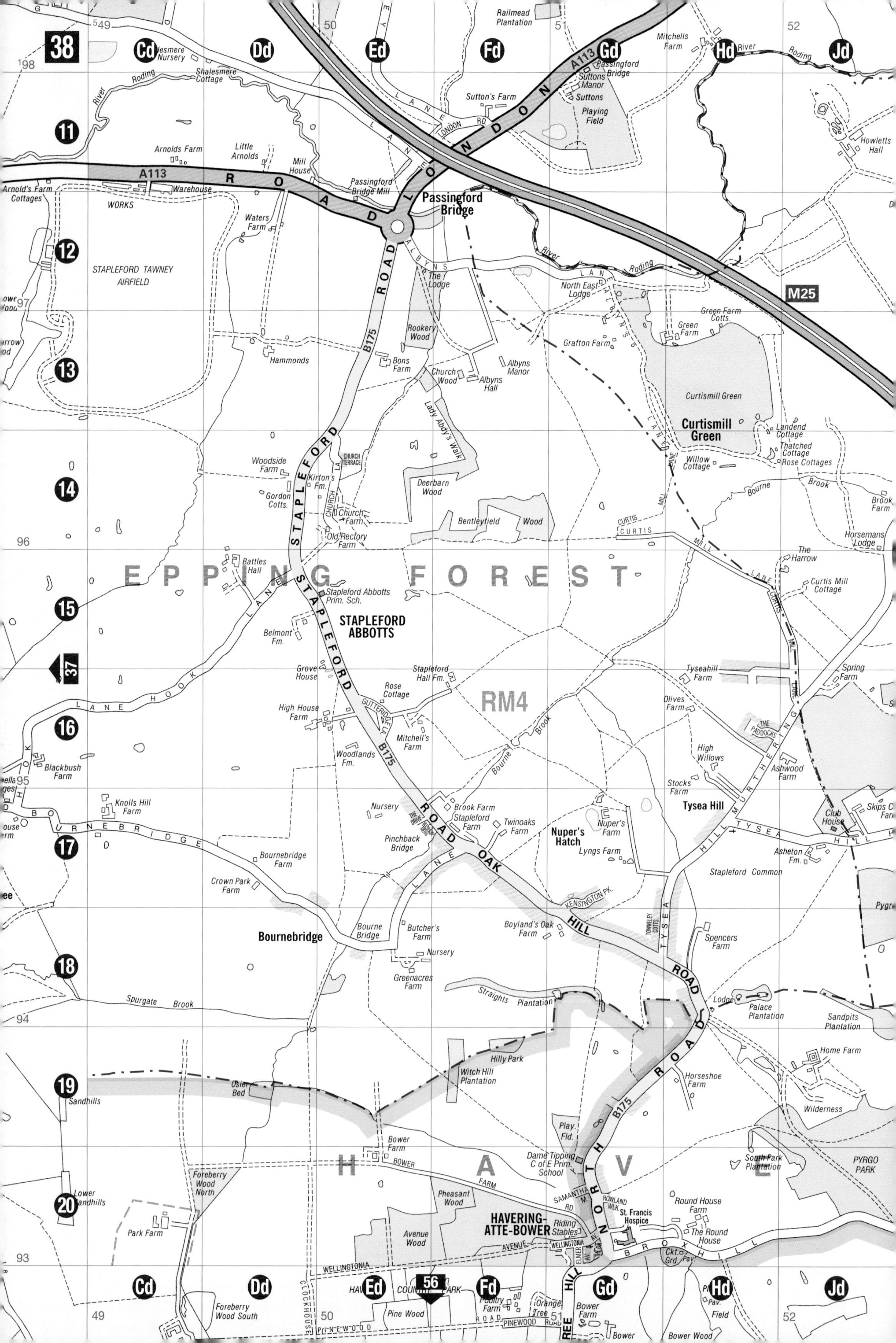
49
50
51
52
Cd
Dd
Ed
Fd
Gd
Hd
Jd
98
97
96
95
94
93
11
12
13
14
15
16
17
18
19
20
Railmead Plantation
Mitchells Farm
River Roding
Hesmere Nursery
Shalesmere Cottage
Passingford Bridge
Suttons Manor
Suttons
Playing Field
Sutton's Farm
LONDON RD
LANE
Howletts Hall
Arnolds Farm
Little Arnolds
Mill House
A113
ROAD
LONDON
Arnold's Farm Cottages
Warehouse
WORKS
Passingford Bridge Mill
Passingford Bridge
Waters Farm
STAPLEFORD TAWNEY AIRFIELD
ALBYNS
The Lodge
River Roding
North East Lodge
ALBYNS LANE
M25
Green Farm Cotts.
Green Farm
Rookery Wood
Grafton Farm
B175
Hammonds
Bons Farm
Church Wood
Albyns Manor
Albyns Hall
Curtismill Green
Curtismill Green
Lady Abdy's Walk
Landend Cottage
Thatched Cottage
Rose Cottages
Willow Cottage
CHURCH TERRACE
Woodside Farm
Kirton's Fm.
Deerbarn Wood
Gordon Cotts.
Church Farm
Bourne
Brook
Brook Farm
Bentleyfield Wood
CURTIS
CURTIS MILL LANE
Horsemans Lodge
STAPLEFORD
Old Rectory Farm
The Harrow
Battles Hall
EPPING FOREST
Curtis Mill Cottage
Stapleford Abbotts Prim. Sch.
STAPLEFORD ABBOTTS
Belmont Fm.
37
Grove House
Stapleford Hall Fm.
Tyseahill Farm
Spring Farm
HOOK LANE
Rose Cottage
RM4
High House Farm
GUTTERIDGE LA
Olives Farm
Mitchell's Farm
THE PADDOCKS
B175
Brook
Bourne
Blackbush Farm
Woodlands Fm.
High Willows
Ashwood Farm
Stocks Farm
Tysea Hill
Knolls Hill Farm
Nursery
Brook Farm
Stapleford Farm
Twinoaks Farm
Nuper's Hatch
Nuper's Farm
Club House
Skips C
BOURNEBRIDGE
Pinchback Bridge
ROAD
OAK
Lyngs Farm
TYSEA HILL
Asheton Fm.
Bournebridge Farm
LANE
Stapleford Common
Crown Park Farm
KENSINGTON PK.
Pygr
HILL
Boyland's Oak Farm
TOWNELEY COTTS.
TYSEA
Bournebridge
Bourne Bridge
Butcher's Farm
Nursery
Spencers Farm
Greenacres Farm
ROAD
Spurgate Brook
Straights Plantation
Lodge
Palace Plantation
Sandpits Plantation
Hilly Park
Home Farm
Witch Hill Plantation
Osier Bed
Sandhills
Horseshoe Farm
B175
Wilderness
Play. Fld.
Bower Farm
BOWER
FARM
Dame Tipping C of E Prim. School
NORTH
South Park Plantation
PYRGO PARK
Foreberry Wood North
SAMANTHA M.
RD
ROWLAND WLK.
St. Francis Hospice
Round House Farm
Lower Sandhills
Pheasant Wood
HAVERING-ATTE-BOWER
Park Farm
Avenue Wood
Riding Stables
AVENUE
WELLINGTONIA
The Round House
BROXHILL
Ckt. Grd.
Pav.
WELLINGTONIA
CLOCKHOUSE LA.
56
COUNTRY PARK
Poultry Farm
Orange Tree
Bower Farm
Foreberry Wood South
Pine Wood
ROAD
PINEWOOD
PINEWOOD
Bower
Bower Wood
Pav.
Field
HAVERING

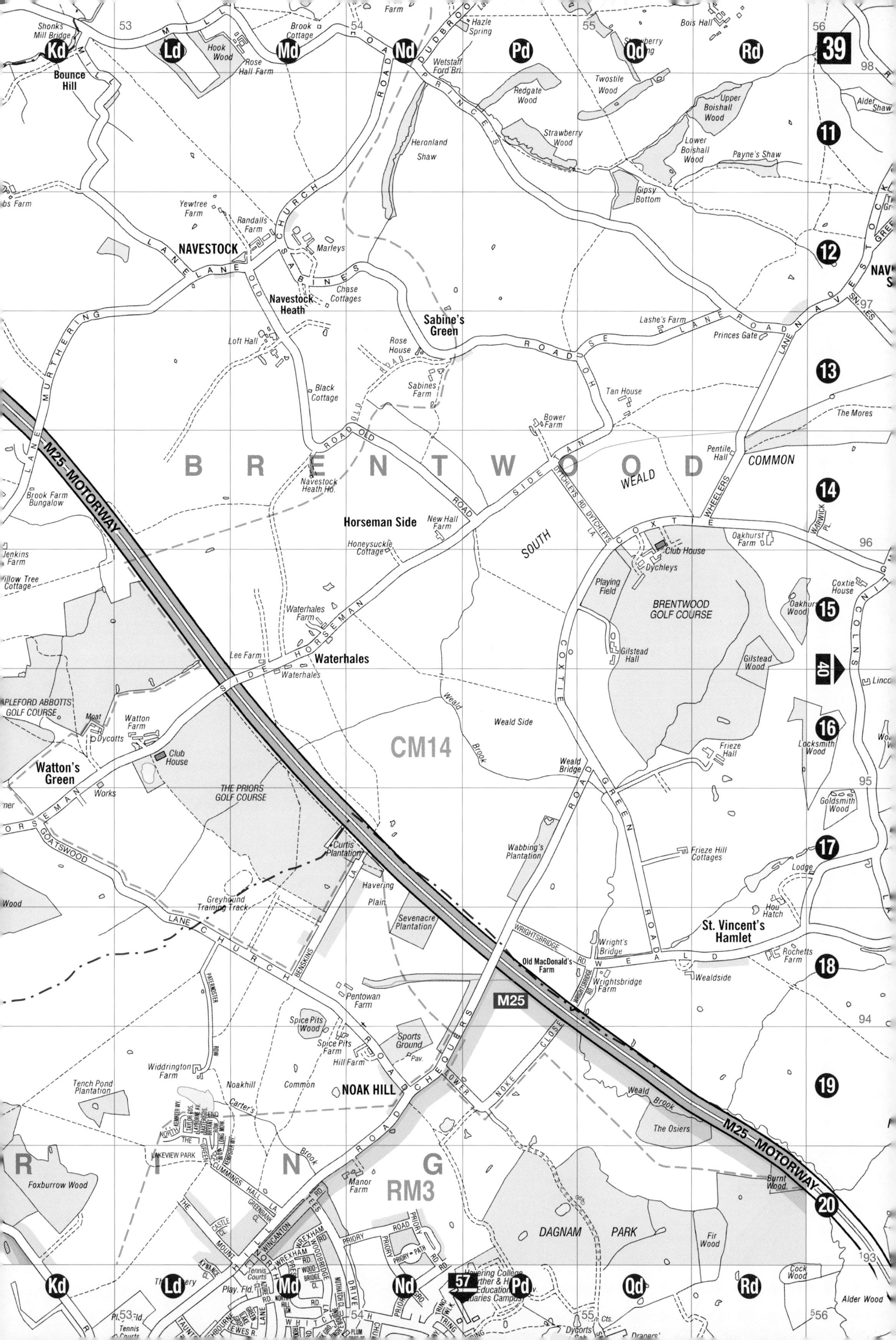
39
40
57
BRENTWOOD
CM14
RM3
NAVESTOCK
Navestock Heath
Sabine's Green
Horseman Side
Waterhales
Watton's Green
NOAK HILL
St. Vincent's Hamlet
WEALD COMMON
SOUTH WEALD
BRENTWOOD GOLF COURSE
THE PRIORS GOLF COURSE
DAGNAM PARK
M25 MOTORWAY
Bounce Hill
Shonks Mill Bridge
Hook Wood
Rose Hall Farm
Brook Cottage
Hazle Spring
Wetstaff Ford Bri.
Heronland Shaw
Redgate Wood
Strawberry Wood
Twostile Wood
Bois Hall
Upper Boishall Wood
Lower Boishall Wood
Payne's Shaw
Alder Shaw
Gipsy Bottom
Yewtree Farm
Randalls Farm
Marleys
Chase Cottages
Loft Hall
Rose House
Sabines Farm
Black Cottage
Lashe's Farm
Princes Gate
Tan House
Bower Farm
The Mores
Pentile Hall
Navestock Heath Ho.
Brook Farm Bungalow
Jenkins Farm
Willow Tree Cottage
New Hall Farm
Honeysuckle Cottage
Oakhurst Farm
Club House
Dychleys
Playing Field
Coxtie House
Waterhales Farm
Lee Farm
Gilstead Hall
Gilstead Wood
Weald Side
Weald Brook
Frieze Hall
Locksmith Wood
Goldsmith Wood
Weald Bridge
Moat
Watton Farm
Dycotts
Works
Curtis Plantation
Havering Plain
Sevenacre Plantation
Wabbing's Plantation
Frieze Hill Cottages
Lodge
Hou Hatch
Greyhound Training Track
Wright's Bridge
Old MacDonald's Farm
Wrightsbridge Farm
Wealdside
Rochetts Farm
Pentowan Farm
Spice Pits Wood
Spice Pits Farm
Hill Farm
Sports Ground
Widdrington Farm
Tench Pond Plantation
Noakhill
Common
The Osiers
Burnt Wood
Foxburrow Wood
Manor Farm
Fir Wood
Cock Wood
Alder Wood
Havering College
Play. Fld.
Tennis Courts
Dycorts
LAKEVIEW PARK
CUMMINGS HALL LA.
WINCANTON RD.
WREXHAM RD.
WOODBRIDGE CL.
PRIORY RD.
PRIORY PATH
TRING WLK.
NUTHATCH CL.
DRIVE
WRIGHTSBRIDGE RD.
NOKE CLOSE
LOWER
CHEQUERS ROAD
BENSKINS LA.
CHURCH LANE
GOATSWOOD
HORSEMAN SIDE
MURTHERING LANE
MILL LANE
CHURCH ROAD
DUDBROOK
PRINCES ROAD
SABINES
OLD ROAD
HOE LANE
DYTCHLEYS RD
COXTIE GREEN ROAD
WHEELERS
WEALD ROAD
WARWICK PL.
PATERNOSTER ROW
Kd Ld Md Nd Pd Qd Rd
11 12 13 14 15 16 17 18 19 20
53 54 55 56
98 97 96 95 94 93

Kelvedon Common
Fox Hatch
Navestock Side
Bentley
Bentley Golf Course
Crow Green
Coxtie Green
Pilgrims Hatch
CM14
Weald Country Park
South Weald
Brentwood
St. Faith's Country Park
Front Park
RM3
Ongar Road
A128
A12
Brentwood By-Pass
M25 Motorway
A1023
High Street
London Road
Weald Country Park Visitors Centre
Bentley Crematorium
Peniel College of Higher Education
Larchwood Prim. Sch.
The Brentwood Centre
Hopefield Animal Sanctuary
Calcott Hall Farm
Langton's Wood
Shepherds Spinney
Fox Wood
Broom Wood
Locksmith Wood
Goldsmith Wood
Alders Wood
Brentwood Camping & Caravaning Site
Ashwells Farm
Wishfields Farm
Days Farm
Clay Hall
Brick House Wood
Halfway House Farm
Bishops Hall Park
St. Peter's C of E Prim. Sch.
St. Paul's C of E Prim. School
Burnt Wood
Cock Wood
Alder Wood
Lower Vicarage Wood
Holly Trees Prim. Sch.

39
58

41
MOUNTNESSING
HUTTON
SHENFIELD
HUTTON MOUNT
CM15
CM13
Junction 12
MARYLANDS INTERCHANGE
MOUNTNESSING A12 BY PASS
SHENFIELD ROAD A1023
CHELMSFORD ROAD
RAYLEIGH ROAD A129
INGRAVE ROAD A128
MERRYMEADE COUNTRY PARK
Shenfield High Sch. & Sports Centre
BRENTWOOD NUFFIELD HEALTH HOSPITAL
BRENTWOOD COMMUNITY HOSPITAL
Thriftwood International Scout Activity Centre
St. Martin's School
Long Ridings Prim. Sch.
Hall Wood
Long Wood
Home Wood
Ingrave Common
Ingrave Hall
59

42
24
62
A
B
C
D
E
F
G
CHALFONT COMMON
THE NATIONAL SOCIETY FOR EPILEPSY
Horn Hill
MAPLE CROSS
Beechen Wood
Franklin's Spring
Chalk Cliff
Maple Cross Mixed Inf. & Jun. Schs.
Lynsters
Recreation Ground
A412 (NORTH ORBITAL ROAD)
M25
M25 MOTORWAY
THREE
WEST HYDE
Pynesfield Manor
Community Centre
CHALFONT LANE
HORNHILL
ROBERTS LANE
RICKMANSWORTH LANE
Tubb's Cottages
Kiln Wood
Village Hall
Springview Farm
Robert's Farm
Robert's Wood
Brawlings Farm
Skipping's Farm
Rowan Nurseries
Playing Field
Workshops
Cricket Field Cottages
Turner's Wood
Pollardshill
High Trees
Hornhill Court
ROBERTSWOOD LODGE
Gravel Hill
Round Rocket Plantation
Beech Ho.
Gravel Pit
THE ORCHARDS CARAVAN SITE
Robertswood School
Playing Field
Cemetery
Bloom Wood
Middle Wood
CHALFONT ST. PETER
Willow Tree Farm
Warren Farm Cottages
Winkers Farm
Warren Farm
SL9
Chalfont Heights Scout Camp
HOGTROUGH WOOD
GERRARDS CROSS GOLF COURSE
Mopes Farm
BUCKINGHAMSHIRE
Chalfont Lodge
Club House
Research Laboratory
Chalfont Park
Cricket Ground
Square Copse
Round Copse
Coldharbour Farm
Coldharbour Farm Cotts.
Meadow Grange
OAKEND WOOD
Birch Boughs
River Misbourne
NOCKHILL WOOD
JUNIPER WOOD
Quarters Wood
Quarters Corner
Boundary Lodge
The Studio
THE MARISH
Inwood
Hillery
Lit. Wotton
Tellisford
The Arches
Tirainia
Denham Park Farm
Marish Farm
The Bungalow
Halings House
Mid. Halings
Pumping Ho.
Great Halings Wood
Little Halings Wood
Little Halings
Lesser Halings
Costers Cott.
Tile House
Lodge
De Vere Leisure Club
TILEHOUSE LANE
DENHAM AERODROME
Flying Training School
Club
White Plains
Shepherds Holt
The Mirrie
The Little House
Owls Oak
Denham Marsh Wood
Isle Of Wight Farm
Lynwood
Pantiles
Beverley
Little Gordons
The Cottage
Nursy
Slade Oak
Oakridge
Mill House
Silver Springs
Greenwoods
Greenacre
DENHAM GOLF COURSE
Collsels Wood
Club House
Bailey Hill
Denham Golf Club
Sherwood House
Cuckoo Oak
The Lodge
Grove Cott.
The Grove
Newst Wood
Works
A413
GRAVEL HILL
CHALFONT ST. PETER BY-PASS
AMERSHAM ROAD
KINGSWAY
PACKHORSE ROAD
B416
AUSTENWOOD
Chiltern Home
St. Mary's Sch.
Gerrards Cross
Superstore
GERRARDS CROSS
St. Michael's Convent
Mem. Cen.
Gerrards Cross Common
The Gerrards Cross C of E School
OXFORD ROAD
A40
Recreation Grd.
Library
Tennis Courts
Cricket Grd.
Hollytree Fm.
Hotel
River Misbourne
21
22
23
24
25
26
27
28
29
30
92
91
90
89
88
500
501
01
02
03
193

43
25
44
63
THREE RIVERS
HILLINGDON
WD3
UB9
Batchworth Heath
Woodcock Hill
Hill End
Mount Pleasant
HAREFIELD
HAREFIELD HOSPITAL
SOUTH HAREFIELD
BROADWATER
Denham Garden Village
DENHAM GREEN
Newyears Green
PARK WOOD
Old Park Wood Nature Reserve
BAYHURST WOOD COUNTRY PARK
Rickmansworth Sailing Club
The Harefield Academy
Harefield United FC (Preston Park)
Harefield Jun. & Inf. Schs.
Hillingdon Outdoor Activities Centre
Dews Farm Sand Pits Nature Reserve
Northmoor Hill Wood Nature Reserve
GRAND UNION CANAL
RIVER COLNE
A412 DENHAM WAY (NORTH ORBITAL ROAD)
A404
Denham

44
26
43
64
Q
R
S
T
U
V
W
07
08
09
10
93
92
91
90
89
88
21
22
23
24
25
26
27
28
29
30
MOOR PARK
MOOR PARK GOLF COURSE
SANDY LODGE GOLF COURSE
Tennis Courts
The Roughs
Old Furze Field
Young Hanging
Old Hanging
Oxhey Wood Local Nature Reserve
The Plantation
Oxhey Wood Lodge
Oxhey Wood House
HMS Warrior
Pretty Corner
THREE RIVERS
EASTBURY
Sch.
Home Farm
HOME FARM RD.
(BATCHWORTH HEATH HILL)
A404
Batchworth Heath
Howard Spring
Batchworth Heath Farm
Whipping Spring
Bathend Clump
Park Wood
WD3
White Hill
BISHOPS WOOD BMI HOSP.
MIU
MOUNT VERNON HOSPITAL
Hospice
Poorfield
Bishop's Wood
Lockwell Wood
Lockwell
Reservoir (covered)
Harefield Reservoir (covered)
French Grove
Battlers Wells Farm
Battlers Wells Wood
David Lloyd Leisure
Tennis Courts
Denville Hall
The Gravel Pits
Holy Trinity C of E Sch.
Park Farm
Playing Field
St. Martin's Sch.
London Sch. of Theology
Club House
RICKMANSWORTH ROAD
GREEN LANE
NORTHWOOD
St. Helen's School
St. Helen's Sports Cen.
Northwood
Northwood College
Frithwood Prim. Sch.
B469
WATFORD ROAD
A4125
HIGH ST.
HA6
PINNER ROAD
NORTHWOOD GOLF COURSE
DUCK'S HILL ROAD
A4180
Farm End
Duck's Hill Farm
Deadman's Grove
Furzefield
Ducks Hill Grange
Scarlet Spring
Duck's Hill
Raveners Spring
Ashby Cottages
Badminton Shooting Ground
Dell
HILLINGDON
COPSE WOOD
Ashby Farm
Miniature Rifle Range
HASTE HILL GOLF COURSE
Club House
The Northwood Club
NORTHWOOD HILLS
Northwood Hills
Northwood FC
Cemetery
Rec. Grd.
RUISLIP WOODS (National Nature Reserve)
Youngwood Farm
Poor's Field
Haste Hill
Haydon School
Joel Street Farm
Playing Field
YOUNG WOOD
MAD BESS WOOD
NORTH RIDING WOOD
Warren Farm
BREAKSPEAR ROAD
Ruislip Lido Railway
Waters Edge
RUISLIP LIDO
Ruislip Lido Woodlands Centre
Nurseries
Mad Bess Wood Cottage
RUISLIP COMMON
Breakspear Crematorium (Ruislip)
Woody Bay
PARK WOOD
Grangewood School
Coteford Junior Sch.
HURST WOOD COUNTRY PARK
Willow Tree Centre
Willow Tree Farm
UB9
Football Ground
Playing Field
Cannons Bridge Farm
Hillingdon Athletic Club
RUISLIP WOODS (National Nature Reserve)
BURY ROAD
FINE BUSH LA.
Oak Cottage
HA4
Kings College Playing Fields
King's Coll. Play. Flds. Track
Eastcote Hockey & Badminton Club
St. Leonard's Farm
GREEN LANE
Pylon Farm
Elm Tree Farm
Whiteheath Jun. & Inf. Schs.
Newyears Green
High View Farm
Braemar Farm
Crows Nest Farm
Old Clack Farm
Winston Churchill Hall
Manor Farm
Prim. Sch.
Youth Cen.
Lib.
Bowl. Grn.
River Pinn
Highgrove Pool & Fitness Cen.
Bishop Ramsey C of E Sch.
Warrender Prim. Sch.
JOEL STREET
B466
B472
FIELD END ROAD

45
27
46
65
SOUTH OXHEY
Carpenders Park
St. Joseph's Catholic Prim. School
Little Furze Field
Beck Mason's Wood
OXHEY WOODS
Sandpit Wood
Woodhall Prim. Sch.
Sir James Altham Pool
Nature Reserve
Nanscot Wood
Birchanger
Reservoir (covered)
Carpenders Park Cemetery
Mutton Wood
Valley View Farm
GRIM'S DYKE GOLF COURSE
Club House
Oxheylane Farm
Grimsdyke Sch.
Grim's Ditch
Grim's Dyke
Weald Wood
Stony Wood
Grimswood Lodge
Grimswood House
Highcroft
The Lodge
HA3
Saddlers Mead
Hatch End Lawn Ten. Club
Playing Fields
Bannister Sports Centre
Pinnerwood House
Pinnerwood Farm
Pinnerwood Cottage
Pinner Wood
PINNER HILL GOLF COURSE
Pinnerwood Park
Shaftesbury Playing Fields
Cricket Club
Hatch End
HATCH END
Harrow Arts Cen.
Hatch End Swim. Pool
Hatch End Play. Flds.
Hatch End High Sch.
Superstore
Rugby Football Ground
Bedford Coll. Sports Grd.
Headstone Lane
PINNER PARK
Pinner Park Farm
Headstone Lane Sports Grd.
HARROW
Pinner Wood Sch.
PINNER GREEN
Montesole Play. Flds.
Pinner Lawn Tennis Club
Paine's Br.
Little Common
Cemetery
East End Farm
Bloomsbury Court
Chiswick Court
Pinner Pk. Schs.
Open Air Theatre
Headstone Manor
Harrow Museum
Headstone Manor Recreation Ground
Yeading Brook
Harlyn Primary Sch.
White Cottage
Cuckoo Hill
HA5
Heath Robinson Mus.
Pinner Village Hall
Pinner Memorial Park
West Lodge Prim. Sch.
Pinner
PINNER
Reddiford Sch.
Nower Hill
Pinner New Cemetery
Nower Hill High Sch.
Woodlands Open Space
Mosque
HEADSTONE
North Harrow
NORTH HARROW
Sports Grd.
Pinner Village Gardens
Yeading Brook Open Space
The Croft
HA2
Longfield Prim. Sch.
EASTCOTE VILLAGE
Eastcote Ho. Gs
Cricket Grd.
Cannon Lane Prim. Sch.
Heathfield Sch. for Girls
Streamside Rec. Grd.
EASTCOTE
Vaughan Prim. Sch.
West Harrow
OXHEY LANE
A4008
UXBRIDGE ROAD
A410
A404
RICKMANSWORTH RD.
GEORGE V AVENUE
PINNER ROAD
A4090 DRIVE
B4542 OXHEY LANE
B466
X
Y
Z
Aa
Ba
Ca
Da
21
22
23
24
25
26
27
28
29
30
93
92
91
90
89
88

Ea
Fa
Ga
Ha
Ja
Ka
La
28
Bentley Priory Mus.
Bentley Priory
The Princess Alexandra Home
Bentley Priory Local Nature Reserve
Deer Park
Heriots
Stanmore Hall
Cloisters Wood
STANMORE COMMON (Local Nature Reserve)
Levels Wood
HARROW WEALD COMMON
Grim's Ditch
Grim's Dyke
Grimsdyke Open Space
Harrow Weald Common
Nurseries
The Old Barn
Heriot's Wood
Weald Wood
The City
Priory House
Scout Hall
Lodge
Harrow Weald Common
Hertfordshire & Middlesex Country Club
Copse Farm
Lower Priory Farm
Nursery
Bentley Wood High Sch.
Hermitage Cottages
Hermitage Gate
Woodland Hall
Stony Wood
Grimswood House
Highcroft
Harrow Weald Cemetery
Harrow Coll. (Harrow Weald Campus)
Brockhurst Cnr.
Boot Pond
Temple Pond
STANMORE
Stanmore College
Stanmore Rec. Grd.
Bowling Green
THE BROADWAY
Club House
STANMORE GOLF COURSE
Sports Ground
Stanburn Schs.
Bannister Sports Centre
Playing Fields
Play Fld.
Grave Yard
The Cedars
Hatch End High Sch.
Cedars Manor Sch.
HARROW WEALD
Superstore
Bus Depot
Recreation Ground
Harrow Weald Lawn Tennis Club
Weald Schs.
HARROW
BELMONT
Belmont Sch.
NHS Walk-In Sch.
Headstone Lane
Headstone Lane Sports Grd.
Pinner Pk. Schs.
Salvatorian Coll.
Sch.
Schs.
HA3
Sports Ground
Priestmead Rec. Grd.
Play Field
The Cliveden Cen.
Works
HA2
Open Air Theatre
Headstone Manor
Harrow Museum
Harrow Crown Court
Harrow & Wealdstone
Headstone Manor Recreation Ground
Yeading Brook
Byron Recreation Ground
WEALDSTONE
Cemetery
Harrow Leisure Centre
Harrow Driving, Cycling and Road Safety Centre
Civic Centre
HA1
HEADSTONE
Ind. Est.
Elmgrove Prim. Sch.
Kenton Recreation Ground
KENTON
Bowling Green
Tennis Courts
Old Lyonian Sports Club
Sports Grd.
Tennis Cts.
Harrow Recreation Ground
GREENHILL
Superstore
Harrow Cemetery
Rec. Grd.
Playgrd.
HARROW
Vaughan Prim. Sch.
West Harrow
Harrow-on-the-Hill
Bus Sta.
Sports Cen.
Superstore
Kenton
Halls of Residence
University of Westminster (Harrow Campus)
Harrow Sports Hall
North Park
HARROW HILL GOLF COURSE
66
The Grove
A410
A409
A4140
A4008
A404
A4005
A4006
UXBRIDGE ROAD
BROOKSHILL
HIGH ROAD
ST. GEORGE GANGE WAY
STATION ROAD
SHEEPCOTE ROAD
KENTON ROAD
CHURCH ROAD
STANMORE HILL
MARSH LANE
LONDON ROAD
REDDING LANE
HARROW VIEW
HEADSTONE LANE
GREENHILL WAY
LOWLANDS RD.
KENTON LANE
PINNER ROAD

47
29
67
48
Stanmore
Stonegrove
EDGWARE WAY (WATFORD BY-PASS)
A41
A410
A5
A5100
A5109
A5150
A4006
A4140
M1
M1 MOTORWAY
Edgware
EDGWARE
Edgware Bus Sta.
Bus Depot
Canons Park
CANONS PARK
WHITCHURCH
LITTLE STANMORE
Chandos Recreation Grd.
North London Collegiate School
King George V Memorial Gdns.
Barnet FC & London Broncos RLFC (The Hive)
The Hive Football & Fitness Centre
Prince Edward Playing Field
BURNT OAK
EDGWARE COMM. HOSP.
Burnt Oak
Watling
WATLING PARK
Mill Hill Broadway
Mill Hill Old Railway Nature Reserve
The Hale
QUEENSBURY
Queensbury
Canons High Sch.
Montrose Recreation Ground
Roe Green
Roe Green Park
Kingsbury Manor
Kingsbury
KINGSBURY
Kingsbury Green
St. Luke's Hospice
Kingsbury High Sch.
Eton Grove Open Space
Queensbury Park
Kingswood Centre
JFS School
St Robert Southwell RC Prim. Sch.
FRYENT COUNTRY PARK (Local Nature Reserve)
EDGWARE ROAD
KINGSBURY ROAD
KENTON ROAD
HONEYPOT LANE
STONEGROVE
HIGH STREET
STATION ROAD
DEANSBROOK RD.
BURNT OAK BROADWAY
FRYENT WAY
CHURCH LANE
HALE LANE
SELVAGE LANE
COLINDEEP LANE
WATFORD WAY
HA8
NW9
BARNET
BRENT
HARROW

48
30
47
68
Ua
Va
Wa
Xa
Ya
Za
Ab
21
22
23
24
25
26
27
28
29
30
Folly Farm
Burtonhole Lane
Farm Estate
Cricket Club
Sports Ground
Nursery
Cemetery
Burtonhole Farm
Sports Ground
Mill Hill Sub. Sta.
London Equestrian Centre
Partingdale Manor
Holcombe Ho.
Riding School
Belmont Farm
The Mill Field
Lawrence Green
St. Paul's C of E Prim. Sch.
National Centre for Medical Research
Oakfield House
Wentworth Hall
St. Vincent's Catholic Prim. Sch.
Convent
Drivers Hill
The Mount Sch.
APEX CORNER
WATFORD WAY
MILL HILL CIRCUS
MILL HILL
Playing Field
DAWS LANE
THE BROADWAY
M1 MOTORWAY
Mill Hill Broadway
Lyndhurst Park
Featherstone Hill
Arrandene Open Space
School Park
Burton Bank
Garden Centre
Mill Hill Cemetery
NW7
Watch Tower House
Bittacy Hill Park
Mill Hill East
RIDGEWAY
The Hale
BARNET BY-PASS
A1
MILL HILL PARK
Copthall Sch. & Sports Cen.
Chase Lodge Playing Fields
Powerleague Soccer Cen.
METRO PLAYGOLF
Metro Golf Centre
Barnet Copthall Leisure Centre
Saracens RFC (Allianz Park)
Shaftesbury Barnet Harriers
Club House
Hendon Crematorium
HENDON CEMETERY
HENDON GOLF COURSE
BARNET COPTHALL SPORTS CENTRE
Cricket Ground
Rugby Club
Junction 2
FIVEWAYS CORNER
The Orion Prim. Sch.
Woodcroft Park
Grahame Park
RAF Museum London
Battle of Britain Hall
NW9
Colindale
COLINDALE
Colindale Park
Peel Centre (Metropolitan Police Training Establishment)
National Blood Service
Montrose Recreation Ground
Elysian
GREAT NORTH WAY
SUNNY HILL PARK
Holders Hill
Windsor Open Space
THE GARDEN BMI HOSP
NW4
Middlesex University
HENDON
Hendon Central
Nuffield Health
HENDON PARK
COLINDEEP LANE
WATFORD WAY
THE HYDE
Silk Stream
Superstore
Hendon
WEST HENDON PLAYING FIELDS
Silver Jubilee Park
Kingsbury Town FC
Kingsbury Green
Cycle Track
Pavilion
Cool Oak Bridge
Youth Sailing Base
BRENT CROSS INTERCHANGE
Brent Cross Shopping Centre
Brent Cross
NORTH CIRCULAR
BRENT CROSS FLYOVER
Superstores
The Decoy
Bowl Grn.
Princes Park
FINCHLEY LANE
CHURCH RD.
QUEENS ROAD
HENDON WAY
STATION ROAD
GOLDERS GREEN
A406
A41
A502
A504
A5150
B552
B551
A4006
A5109
A5100

49
31
50
69
Bb
Cb
Db
Eb
Fb
Gb
Hb
21
22
23
24
25
26
27
28
29
30
25
26
27
28
N12
N20
N11
N3
N2
N10
NW11
N6
FINCHLEY
NORTH FINCHLEY
FRIERN BARNET
COLNEY HATCH
EAST FINCHLEY
HAMPSTEAD GARDEN SUBURB
HIGHGATE
GOLDERS GREEN
HAMPSTEAD
FORTIS GREEN
Woodside Park
West Finchley
Finchley Central
East Finchley
Fallow Corner
HENLYS CORNER
NORTH MIDDLESEX GOLF COURSE
FINCHLEY GOLF COURSE
HIGHGATE GOLF COURSE
HAMPSTEAD GOLF COURSE
Friary Park
Bethune Park
Coppetts Wood & Glebelands Local Nature Reserve
COLDFALL WOOD
HIGHGATE WOOD
Big Wood
Little Wood
Bunkers Hill
Victoria Park
Cherry Tree Wood
ST. PANCRAS & ISLINGTON CEM.
EAST FINCHLEY CEMETERY
St. Marylebone Crematorium
Islington Crematorium
Jewish Cemetery
Golders Green Crematorium
FINCHLEY MEM. HOSP.
HIGHGATE HOSPITAL
North London Hospice
Frith Manor Prim. Sch.
Woodside Park Club
GREAT NORTH LEISURE PARK
NORTH CIRCULAR ROAD
A406
A1000
HIGH ROAD
BALLARDS LANE
A598
REGENTS PARK ROAD
FINCHLEY ROAD
A504
EAST END ROAD
FORTIS GREEN
GREAT NORTH RD.
ARCHWAY ROAD
AYLMER ROAD
LYTTELTON ROAD
FALLODEN WAY
A1
WOODHOUSE RD.
A1003
FRIERN BARNET LANE
COLNEY HATCH LANE
B550
BOBBY MOORE

50
32
49
70
Jb
Kb
Lb
Mb
Nb
Pb
Qb
21
22
23
24
25
26
27
28
29
30
N14
N11
N13
N22
N10
N8
N6
PALMERS GREEN
Palmers Green
ARNOS PARK
Arnos Grove
NEW SOUTHGATE
New Southgate
FRIERN BARNET
BARNET
Bowes Park
Bounds Green
MUSWELL HILL GOLF COURSE
Alexandra Park School
Alexandra Palace
ALEXANDRA PARK
WOOD GREEN
Wood Green
MUSWELL HILL
FORTIS GREEN
Cranley Gardens
HIGHGATE WOOD
QUEEN'S WOOD
Highgate
HIGHGATE
HIGHGATE HOSPITAL
CROUCH END
HORNSEY
Hornsey
HORNSEY VALE
STROUD GREEN
Harringay
PRIORY PARK
Hollickwood Park
Broomfield Park
Clockhouse Junction
Crematorium
ALDERMANS HILL
NORTH CIRCULAR ROAD
BOUNDS GREEN ROAD
BOWES ROAD
TELFORD WAY
PINKHAM WAY
COLNEY HATCH LANE
OAKLEIGH ROAD SOUTH
STATION ROAD
BROWNLOW RD.
DURNSFORD RD.
HIGH ROAD
LORDSHIP LANE
TURNPIKE LA.
MUSWELL HILL
PARK ROAD
PRIORY RD.
ARCHWAY ROAD
MUSWELL HILL RD.
TOTTENHAM LA.
QUEENS AV.
FORTIS GREEN
ALEXANDRA PARK RD.
WOOD GREEN

51
33
EDMONTON
UPPER EDMONTON
PYMMES PARK
Silver Street
NORTH MIDDLESEX UNIVERSITY HOSPITAL
GREAT CAMBRIDGE JUNCTION
NORTH CIRCULAR ROAD
A406
STERLING WAY
ANGEL EDMONTON
FORE ST.
GREAT CAMBRIDGE ROAD
A10
CAMBRIDGE ROAD
TOTTENHAM
TOTTENHAM CEMETERY
White Hart Lane
Tottenham Hotspur FC
ROUNDWAY
BRUCE CASTLE PARK
Bruce Castle Mus.
LORDSHIP LANE
A109
LORDSHIP RECREATION GROUND
The Moselle
DOWNHILLS PARK
WEST GREEN
WEST GREEN ROAD
A504
Bruce Grove
BRUCE GROVE
TOTTENHAM HALE
Tottenham Hale
MONUMENT WAY
FERRY LANE
A503
WATERMEAD WAY
Northumberland Park
Meridian Water
Tottenham Marshes
LOCKWOOD RESERVOIR
LOW MAYNARD RESERVOIR
WARWICK RESERVOIR (EAST)
WARWICK RESERVOIR (WEST)
RESERVOIR No.1
RESERVOIR No.2
RESERVOIR No.3
Lee Valley Park
Seven Sisters
SEVEN SISTERS
SOUTH TOTTENHAM
South Tottenham
ST. ANN'S ROAD
ST. ANN'S HOSPITAL
Harringay Green Lanes
GRAND PARADE
HARRINGAY
Markfield Beam Engine & Museum
Markfield Recreation Ground
Stamford Hill
HIGH ROAD
BROAD LANE
CLAPTON
Walthamstow Marsh Nature Reserve
Copper Mills Junction
Coppermill Bridge
N17
N15
N9
N18
E17
71
52

52
34
72
51
ENFIELD
HARINGEY
WALTHAMSTOW
SOUTH CHINGFORD
CHINGFORD MOUNT
Suffield Hatch
HIGHAM HILL
HIGHAMS PARK
Chapel End
N18
N17
E4
E17
E10
BANBURY RESERVOIR
LOCKWOOD RESERVOIR
LOW MAYNARD RESERVOIR
HIGH MAYNARD RESERVOIR
RESERVOIR No.1
RESERVOIR No.2
RESERVOIR No.3
RESERVOIR No.4
RESERVOIR No.5
WARWICK RESERVOIR
LEE VALLEY PARK
Lee Valley Park
LEA VALLEY VIADUCT
NORTH CIRCULAR ROAD
A406
A1055
A1009
A112
A503
A1006
A104
B179
B160
A1037
SOUTHEND ROAD
CHINGFORD ROAD
CHINGFORD MOUNT RD.
FOREST ROAD
FOREST LANE
HOE STREET
HIGH STREET
LEA BRIDGE RD.
BLACKHORSE LANE
BLACKHORSE ROAD
MARKHOUSE ROAD
ST. JAMES'S STREET
MERIDIAN WAY
MONTAGU ROAD
HALL LANE
NEW ROAD
OLD CHURCH ROAD
LARKSHALL ROAD
WINCHESTER ROAD
FULBOURNE ROAD
WOOD STREET
LEYTON GREEN RD.
BILLET ROAD
HIGHAM HILL ROAD
Northumberland Park
Meridian Water
Blackhorse Rd.
Walthamstow Central
Walthamstow Queens Rd.
St. James St.
Walthamstow Marsh Nature Reserve
Coppermill Bridge
Lockwood Reservoir
Refuse Incineration Plant
Pumping Station
Teamsport Karting
Chase Lane Park
Larkswood Playing Fields
Larkswood Park
Ainslie Wood Nature Reserve
Rolls Sports Grd.
Foundation School Playing Field
Peter May Sports Cen.
Walthamstow Academy
William Morris Sch.
Lloyd Park
Waltham Forest Feel Good Centre
Chestnuts Sports Ground
Town Hall
Thorpe Coombe Hosp
Walthamstow Leisure Cen.
Queens Road Cemetery
Douglas Eyre Playing Fields
Douglas Eyre Sports Centre
St. James's Park
Filter Beds
Water Works
Golf Driving Range
Britannia Sports Grd.
Crooked Billet
The Bell
Stonebridge Locks
Chalk Bridge
River Lee Navigation
River Lee Diversion
Pymme's Brook
Lee Flood Relief Channel
Suffield Hatch
Nuffield Health Fitness Cen.
Chingford Leisure Centre
Powerleague Soccer Cen.
Hillyfield Prim. Sch.
Higham Hill Park
Whittingham Prim. Acad.

WOODFORD WELLS
Knighton Wood
WOODFORD GOLF COURSE
Football Ground
HATCH LA.
CHINGFORD LANE A1009
Roding Valley
Rugby Football Ground
Chigwell Sewage Works
M11
Woodford Side
Highams Park
The Highams Park
Boating Lake
WOODFORD GREEN
IG8
CHINGFORD HATCH
HIGHAMS PARK
Highams Pk.
HALE END
Ashton Playing Fields
Woodford Green Athletic Cen.
James Leal Cen.
Ray Lodge Prim. Sch.
Sports Ground
BROADMEAD
A1009
WOODFORD
Woodford
Woodbridge High Sch.
Playing Field
Wynn Bridge
EPPING FOREST
Hotel
St. Aubyn's Sch.
WATERWORKS CORNER
SOUTHEND ROAD A406
Junction 4
Walthamstow Forest
Reservoir (covered)
Reservoir (cov.)
Hyland Ho. School
Gilbert's Slade
A503
FOREST ROAD
NEW ROAD A104
CHARLIE BROWN'S ROUNDABOUT
NORTH CIRCULAR ROAD
E18
South Woodford
SOUTH WOODFORD
Abbeydale
Boating Pond
Community Hall
Wood St. Walthamstow
UPPER WALTHAMSTOW
Sylvestrian Leisure Cen.
Forest Sch.
Sports Ground
RODING SPIRE HOSPITAL
Hatton School
PDSA (Veterinary Centre)
REDBRIDGE
SNARESBROOK
Snaresbrook
Snaresbrook Crown Court
Eagle Pond
Hollow Pond (Boating Lake)
WHIPPS CROSS UNIV. HOSP.
WOODBURY UNIT
Margaret Centre
Epping Forest
E11
IG4
REDBRIDGE ROUNDABOUT
Redbridge
EASTERN AVENUE A12
Wanstead
WANSTEAD
Leisure Cen.
Wanstead High School
Golf Course
Pumping Station
SOUTH WOODFORD TO BARKING RELIEF ROAD
The Heronry
Lincoln Island
Superstore
35
73
54

54
36
53
74
Junction 4
M11
EPPING FOREST
CHIGWELL
Chigwell
Library
Old Loughtonians Hockey Club
Rugby Football Ground
Anderson School
Chigwell Sewage Works
Playing Fields
West Hatch High School
Club House
CHIGWELL GOLF COURSE
Grange Hill
GRANGE HILL
Limes Farm Inf. & Jun. Schools
Cemetery
The Lodge
Depot
Hainault
IG7
IG8
IG6
IG5
IG4
IG1
IG2
WOODFORD BRIDGE
Repton Park
Hospital Hill Wood
Cocked Hat Plantation
Egg Clump
Ash Plantation
Roding Prim. Sch.
Woodford Green Athletic Cen. (Running Track)
Ashton Playing Fields
Sports Ground
Daisy Lake
Cuthbert Lake
Cemetery
FULLWELL CROSS
Fairlop
FAIRLOP
Forest Farm
Redbridge Sports & Leisure Cen.
Oakfield Playing Field
Ilford & Redbridge Hockey Club
The Jean Brown Indoor Arena
Fairlop Lodge
FAIRLOP WATERS GOLF COURSE
King Solomon High Sch.
Ilford Jewish Prim. Sch.
Ilford High Sch.
CLAYHALL
CLAYHALL PARK
Garden of Rest
BARKINGSIDE
Barkingside
Redbridge FC (Oakside Stadium)
Avanti Court Primary School
Aldborough Ho. Farm
NEWBURY PARK
Newbury Park
Oaks Park High Sch.
GANTS HILL
Gants Hill
Valentines High Sch.
Valentines Mansion & Gardens
Jack Carter Centre
REDBRIDGE
Redbridge
Cranbrook Castle Ten. Club
Beal High Sch.
PDSA (Veterinary Centre)
Recreation Ground
Tennis Courts
EASTERN AVENUE
WOODFORD AVENUE
CRANBROOK RD.
MANOR ROAD
HIGH ROAD
FENCEPIECE ROAD
HAINAULT ROAD
TOMSWOOD ROAD
A12
A123
A113
A1400
B173
B192
M11 MOTORWAY
ROAD TO BARKING RELIEF ROAD
42
43
44
45

55
37
56
75
Vc
Wc
Xc
Yc
Zc
Ad
Bd
HAINAULT FOREST
(COUNTRY PARK)
Visitor Centre
Hainault Forest Children's Zoo
Fox Burrows
Chigwell Row Wood Local Nature Reserve
Recreation Ground
Playing Fields
Tennis Courts
Aspen Court
Play Field
Fairview
Coppice Prim. Sch.
Manford Prim. Sch.
HAINAULT
Forest Academy
Ten. Cts.
Youth Cen.
Old Hainault Oak
Dog Kennel Hill
HAINAULT FOREST GOLF COURSE
Reservoirs (covered)
Cemetery
Forest Park Crematorium
Hainault Business Park
Gardens of Peace (Muslim Cemetery)
Elmbridge Sports Ground
Club House
Skate Park
Forest Road Playing Field
Hainault Lodge Local Nature Reserve
Hog Hill
Redbridge Cycling Centre
Club Ho.
Golf Driving Range
Oaklands Hamlet
Havering Park
Willoughby's Hill
Lower Park Farm
Sandhills
Park Farm
Community Centre
Havering Lodge
Forest Farm Equestrian Centre
Cold Blow Farm
ROMFORD B174 ROAD
HOG HILL RD.
Gobions Farm
RM5
Marks Gate
Football Grd.
Rollerbowl Laserquest
The City Pavilion
Whites Farm
Furze Cottage
Furze House Farm
Hainault Recreation Ground
Fairlop Oak Recreation Ground
Powerleague Soccer Cen.
Fairlop Outdoor Activity Centre
Fairlop Lake
FAIRLOP WATERS COUNTRY PARK
Lagoon
Aldborough Hall Equestrian Centre
Aldborough Hall Farm
ALDBOROUGH HATCH
PAINTERS ROAD
Farm Cottages
Hainault Works
Hainault Farm
Seven Kings Water
BARKING
RM6
RM7
Red House Farm
Hainault House
Willow Farm
Playing Field
Little Heath Sch.
Sports Ground
Chadwell Heath Cemetery
Jun. Sch.
Infant Sch.
Moat
Warren Farm
MOBY DICK
A12
EASTERN AVENUE
A1112
WHALEBONE LANE
KING GEORGE HOSPITAL
North Residence
South Residence
UCC
GOODMAYES HOSPITAL
LITTLE HEATH
Redbridge College
Newbridge Sch.
Grove Prim. Sch.
Moby Golf
Driving Range
Golf Kingdom
Warren Sports Centre
Warren Jun. Sch.
Golf Course
Crown Farm
ST. CHAD'S PARK
Chadwell Millennium Green
CHADWELL HEATH
A118
HIGH ROAD
SEVEN KINGS PARK
IG3
Bowling Green
Sports Ground
Club House
Playing Field
Games Court
Tennis Courts
Bandstand
Farnham Green Prim. Sch.
Brookside
St. Edward's C of E Sch.
West Ham FC Training Grd.
Whalebone Bri.
Westlands Playing Fields

56
HAVERING-ATTE-BOWER
38
Havering Country Park
Pine Wood
Churchpath Plantation North
Kiln Wood
Poultry Farm
Orange Tree Inn
Bower Farm
Bower House
Bower Wood
Round House Farm
St. Francis Hospice
The Round House
Playing Field
RM4
Foreberry Wood South
Lower Park Farm
Willoughby's Hill
Havering Park
Upper Bedfords Farm
Bedfords
Deer Park
Larch Wood
BEDFORDS PARK
Nursery
Bower Park School
CHASE CROSS
Community Centre
Clockhouse Primary School
RM5
Havering Lodge
COLLIER ROW
Sunnyside Farm
Club House
RISEBRIDGE GOLF COURSE
Gobions Farm
Lawns Rec. Grd.
RISE PARK
HAVERING
Works
Football Grd.
55
Crownfield Jun. & Inf. Schs.
Whites Farm
Marshalls Park School
RAPHAEL PARK
Fish Ponds
Mawney
RM7
Warehouse
King Georges Field
MAWNEY PARK
RM1
Romford County Ct.
Town Hall
Romford Mag. Ct.
LODGE FARM PARK
RM6
Golf Course
Crown Farm
Sports Ground
Cottons Park
Crown Cottages
St. Edward's C of E Sch.
Romford Greyhound Stadium
ROMFORD
Nuffield Health Fitness Cen.
Westlands Playing Field
West Ham FC Training Grd.
Crowlands
76
Gas Works
Queen's Hospital
Superstore
HEATH PARK
WESTERN AVENUE A12
EASTERN AVENUE A12
LONDON ROAD A118
MAIN ROAD A118
COLLIER ROW ROAD B174
CHASE CROSS ROAD B1459
HAVERING ROAD B175
ORANGE TREE HILL
LOWER BEDFORDS ROAD
NORTH STREET A125
MERCURY GARDENS
THURLOE GARDENS A125
WATERLOO ROAD
ROM VALLEY WAY A125
ST. EDWARDS WAY
Cd
Dd
Ed
Fd
Gd
Hd
Jd
21
22
23
24
25
26
27
28
29
30

57
39
77
58
Kd
Ld
Md
Nd
Pd
Qd
Rd
53
54
55
56
93
92
91
90
89
88
M25
21
22
23
24
25
26
27
28
29
30
DAGNAM PARK
Fir Wood
Cock Wood
Alder Wood
The Nursery
Havering College of Further & Higher Education (Quarles Campus)
Dycorts Sch.
Drapers' Academy
Hatters Wood
MAYLANDS GOLF COURSE
Duck Wood Community Nature Reserve
Pyrgo Priory Sch.
Club House
Maylands
Maylands Cottage
Putwell Bridge
Central Park Leisure Centre
Central Park
HAROLD HILL
Long Wood
Sage Wood
Min. Golf Course
St. Ursula's Catholic Infant School
St. Ursula's Catholic Junior School
Broxhill Common
Hilldene Prim. Sch.
Mead Prim. Sch.
Maylands Field
HAROLD PARK
Harold Court Prim. Sch.
Harold Court Woods
The Chase
Garden Centre
Sports Ground
HAROLD WOOD
Harold Wood
Cissbury Lodge Farm
Cockabourne Bridge
Miniature Golf Course
Pages Farm
PAGES WOOD
RM14
Montrose
Four Wantz
Emanuel Cottage
Gallows Corner
GALLOWS CORNER
Sports Ground
Superstore
Warehouse
Factory
The Grange
GIDEA PARK
RM2
Gidea Park
Squirrel's Heath
David Lloyd Gidea Park
Redden Court Sch.
Harold Wood Park
Harold Wood Prim. Sch.
Prospect Farm
Ingrebourne Farm
Sunset Play. Fld.
The Campion School
Playing Field
Hotel
Mount Pleasant Farm
The Rosary
Grasslands
Cornlands Farm
College
ARDLEIGH GREEN
Miniature Golf Course
HAYNES PARK
Drill Roundabout
RM11
Lillyputts Equestrian Centre
EMERSON PARK
Goodrington Sch.
HYLANDS PARK
Emerson Pk.
Emerson Park Sch.
Upminster Court
UPMINSTER
EAST ROAD
COLCHESTER
SOUTHEND ARTERIAL
A12
A127
A118
HAROLD WOOD INDUSTRIAL ESTATE
ELMS INDUSTRIAL ESTATE
BATES INDUSTRIAL ESTATE
THE OLD BRICKWORKS INDUSTRIAL ESTATE
DAGNAM PARK DRIVE
GOOSHAYS DRIVE
PETERSFIELD AVENUE
FARINGDON AVENUE
SQUIRRELS HEATH LANE
SHEPHERDS HILL
HUBBARDS CHASE
SYLVAN AVENUE
PARKSTONE AVENUE
WYCH ELM RD.
ERNEST ROAD
HARROW DRIVE
ELM GROVE
NELMES WAY
ROMFORD ROAD

Junction 11 (A12)
Junction 28 (M25)
Junction 29
BROOK STREET INTERCHANGE
BROOK STREET
WARLEY
GREAT WARLEY
HAVERING
TYLERS COMMON
CM14
RM14
M25
MOTORWAY
A12
A1023
A127
B186
MAYLANDS GOLF COURSE
Warley Country Park
Brentwood Mus.
The Duchess of Kent Social Centre
HARTSWOOD SPIRE HOSP.
Brentwood Park Ski & Snowboarding Centre
Brentwood Karting
Fitness First (Brentwood)
Holly Trees Prim. Sch.
Ursuline Prep. Sch.
Warley Place Nature Reserve
Folkes Lane Woodland
Sewage Disposal Works
Oak Royal Nurseries
Emerson Park School
Upminster Hall Playing Fields
UPMINSTER

40
57
78

INGRAVE
HERONGATE
LITTLE WARLEY
WEST HORNDON
Childerditch
Ingrave Common
THORNDON COUNTRY PARK
HARTSWOOD GOLF COURSE
WARLEY PARK GOLF COURSE
THORNDON PARK GOLF COURSE
CM13
BRENTWOOD ROAD
SOUTHEND ARTERIAL ROAD
A128
A127
41
79

A
B
C
D
E
F
G
1
2
3
4
5
6
7
8
9
10
80
HP10
HP9
SL1
Wooburn Common
Littleworth Common
EGYPT
EGYPT WOODS
BURNHAM BEECHES
Burnham Beeches
BURNHAM BEECHES NATIONAL NATURE RESERVE
EAST BURNHAM COMMON
FARNHAM COMMON
EAST BURNHAM
FARNHAM ROYAL
BOWER WOOD
NORTH DORNEY HILL SOUTH
COLLINSWOOD A355
BEACONSFIELD ROAD
A355
Dipple Wood
Dipple Lodge
Woodlands Farm
Hollybush Wood
Hangings Wood
Downey Bottom
Jennings Wood
Jennings Farm
Broomhill Coppice
Hicknaham Plantation
Hicknaham Farm
Ashen Coppice
Bristles Wood
Homer Wood
Boveney Wood
Boveney Wood Farm
Little Jennings Farm
Akroyd Cottages
Akroyd Place
Abbey Park Farm
Abbey Park Cottage
Healy's Gorse
Water Tower
Summerlins Wood
Little Burnt Coppice
Heathfield Wood
Staplefurze Wood
Dorney Bottom
Pennlands Wood
Pennlands Farm
Little Wood
Hillmotts
Furze
Mount Farm
Moll Hayes Wood
Ponds
Hollybush Corner
School Wood
Hollybush Wood
Dropmore Inf. Sch.
Dropmore Park
Sewage Works
Dropmore
Abbey Wood
Brook End Farm
Reservoir (covered)
Twelve Acre Wood
Littleworth Corner
Lower Brook End
Kilnwood
Dorney Wood
Nine Acre Wood
Moat
Whitespark Wood
GREAT BURNT COPPICE
Crabtree Heath
Tower Wood
Fleet Wood
Scout Camp
Towerwood
Sports Ground
Tennis Courts
Jun. Sch.
Playing Field
Inf. Sch.
Club House
Lambournes Wood
Dorney Wood
Home Close
Pumpkin Hill
Juniper Grove
Seven Ways Plain
Upper Pond
Middle Pond
Kiln Wood
Nursery
Lower Lambournes Wood
Snowball Farm
Council Depot
Archery Range
BURNHAM BEECHES GOLF COURSE
Hunts Wood Fm.
Hunt's Wood
Woodview
Swilley Fm.
Swilly Pond
Thompkin's Wood
East Burnham Ho.
Playing Field
Caldicott Sch.
Club House
Burnham Grove
Greenfields Farm
Cant's Hill
Nursery
Grove Wood
Hotel
Bottom Waltons
Lock's Bottom
Allerds Farm
Nursery
Dair House Sch.
SLOUGH COMMUNITY Club
Elm Cottages
Deepwood
The Spinney
Cherry Orchard
Recreation Ground
Sports Field
Beechwood & Arbour Vale Schools
Britwell House
Youth Cen.
Day Cen.
COMMON
HAREHATCH LANE
PARISH LANE
DUKES DRIVE
MCAULIFFE DRIVE
LORD MAYORS DRIVE
HAWTHORN LANE
ALLERDS ROAD
THOMPKINS LANE
CROWN LANE
DROPMORE ROAD
PUMPKIN HILL
HARVEY

61
62
81
H
J
K
L
M
N
P
1
2
3
4
5
6
7
8
9
10
97
98
99
500
88
87
86
85
84
183
Slade Farm
Hedgerley Green
Moat Farm Cotts.
Bulstrode Cotts.
Manor Farm
Coppathorne
BULSTRODE PARK
Menagerie Wood
Grove Plantation
GERRARDS CROSS
Crab Hill
FORT
Sutton's Wood
Sherley Close
Bloomfield Cottage
Leith Grove
Ponders
M40
SL9
HEDGERLEY
Church Wood Reserve
Subway
M40 MOTORWAY
High Meadows
Court Farm
Hanging Wood
Hedgerley Hill
HEDGERLEY PARK
Mount Hill Farm
Mounthill Wood
Duke's Orchard
Riding Stables
Black Grove
Park House
Tara
Parkside Cemetery
Pickeridge Wood
One Pin Farm
Reservoir (Covered)
Hedgerley Park Farm
Low Farm
Fish Pond
Hay's Wood
SL3
FULMER
TIMBER WOOD
Reservoir (covered)
Stoke Wood
Pickeridge Farm
Kestrel Farm
Fulmer Hall
Fulmer House
Fulmer Infant Sch.
Stoke Common
Templemead
Church Farm
LARCHMOOR PARK
STOKE COMMON
Penn
Duffield Fm.
Ingram's Copse
Fernacres Farm
Frame Wood
King George's Field
Black Firs
Oakwood House
Brockhurst Wood
Mill House Fm.
Framewood Manor
Fulmer Court
Parson's Wood
Home Wood
Eight Acres Fm.
Teikyo Sch.
Eight Acres
Woodlands
Play Fld.
Stoke Park
Stoke Court
WEXHAM SPRINGS
Old Orchard Wood
Old Orchard Ho.
Little Orchard
Upton Lake
Upton Farm
Khalsa Sec. Academy
Sch.
Tennis Court
Rec. Grd.
Pav.
Sports Grd.
Tennis Courts
Dogkennel Field
Burners Cottage
Upton Wood
THE SOUTH BUCKINGHAMSHIRE GOLF COURSE
Yew Place
Hockley Hole
SL3
Twin Trees Farm
STOKE POGES
THAMES VALLEY SPIRE HOSPITAL
WEXHAM STREET
Sandhole Wood
FARNHAM PARK GOLF COURSE
Club Ho.
Tithe Fm.
WEXHAM PARK GOLF COURSE
Club House
Driving Range
WEXHAM PARK GOLF COURSE
Rowley Wood
The Clock Ho.
Nuffield Health Fitness & Wellbeing Cen.
Gallions Wood
Rowley Slip
Berry Farm
Stoke Park Fm.
Manor House
Grotes Wood
STOKE POGES GOLF COURSE
Gray's Field
Monument
Appletons
Pinewood Nurseries
Spring Wood
Hollybush
HAMSHIRE
B416
A40
WINDSOR ROAD
STOKE COMMON ROAD
GERRARDS CROSS ROAD
BELLS HILL
GRAYS PARK
HEDGERLEY LANE
COLLUM GREEN ROAD
WOOD LANE
DUFFIELD LANE
WINDMILL ROAD
FULMER ROAD
FRAMEWOOD ROAD
ROWLEY LANE
HOLLYBUSH HILL
WEST END LANE
CHURCH LANE
GREEN LANE
FARTHING GREEN LANE
SNOWCRETE

62
GERRARDS CROSS
A
B
C
D
E
F
G
42
61
82
Gerrards Cross Common
The Gerrards Cross C of E School
St. Michael's Convent
Mill Wood
Gallows Wood
Woodhill
Little Gallows Wood
M25
A40
OXFORD ROAD
The Rancho
Gas Pipeline Station
SL9
St. Huberts
Felbrigg
St. Hubert's Home Farm
Prestwick Place
Little Prestwick
Old Prestwick
Prestwick Cottage
Naboth
Grayleigh
Fulmerfields
Alderbourne Manor
The Barn
Fish Pond
M40
The Oldhouse Farm
Alderbrook
Hay's Wood
Hawk's Wood
Hawkswood
Furzeney
Furzeney Wood
FULMER
Fulmer House
Fulmer Infant Sch.
Fulmer House Farm
Church Farm
Orchard Cottage
Alderbourne Arches
Junction 1a (M40)
Junction 16 (M25)
M25 MOTORWAY
M40 MOTORWAY
Gladwins Wood
Broadspring Wood
Denham Golf Club
Collsels
Club House
The Lodge
Grove Cott
The Grove
Bailey Hill
Higher Denham
Tom William's Wood
Works
El Sub Sta.
Sewage Works
Gasholder Station
Comm. Cen.
Bradbury's Island
Moorhouse Farm
Caps Wood
A413 ROAD
Council Offs
Capswood Business Centre
Baker's Wood
Hedgerows Poultry Farm
Tatling End
Hollybush Cottages
Hollybush Farm
Mount Farm
RED HILL
Huntwyk
Redhill Farm
Council Depot
Denham Mount
Maltmas Green
Holly Bush
Rush Green
Gossams Wood
Rush Hill Farm
Smallhold
Durham House
Oldhouse Wood
Watersplash Farm
Alder Bourne
SL3
BUCKINGHAMSHIRE
Penn Wood
Fernacres Farm
West Hill
Fulmer Rise Manor
Home Wood
The Thatched Cott.
The Spinney
Glen Erica
Parkspring Wood
Brown's Wood
Field End Farm
Ways Farm
Alderbourne Farm
Springfield Cottages
Saul's Farm
Home Farm
Cherry Orchard Farm
Blanchards Farm
Caravan Park
Kennels
Belle Farm
Long Coppice
Sevenhills Farm
Taidswood
Dromenagh
Playing Field
Dromenagh Farm
Black Firs
Oakwood House
Mount Fidget
Woodland Court
Mount Fidget Wood
Fulmer Common Ho.
Upton Ho.
Fernacres Cottage
Langley Corner
Convent
Fulmer Common
Langley Common
Strawberry Wood
Pinewood Film Studios
The Clump
M25
Blacketts
Heathfield Plantation
Upton Plantation
Upton Lake
Upton Farm
Bowlers Plantation
Upton Wood
Spruce Wood
Heatherden Hall
SL0
Broom Warren
Dellfield Fm.
Sycamore Wood
Park Lodge Farm
Brackenwood
Ensbys
Sandhole Wood
Cuckoo Pen
BLACK PARK COUNTRY PARK
Heathlands
Park Lodge
Rec. Grd.
Club Ho.
Rowley Wood
Diamond Wood
Go Ape
Rowleyfield Larches
Rowley Slip
Walk Wood
Gallions Wood
Blackpark Lake
Warren Lodge
Warren House
Chandler's Hill
Ancoats Lodge
Black Park Countryside Cen.
Club
Warren Ho. Fm.
Spring Wood
Hollybush Wood
Rowley Farm
Five Points
IVER HEATH
South Lodge
A412
CHURCH ROAD
DENHAM ROAD
SLOUGH ROAD
BLACK PARK ROAD
PINEWOOD ROAD
SEVEN HILLS ROAD
FULMER ROAD
ALDERBOURNE LANE
HAWKSWOOD LANE
BLACKSMITHS LANE
FROGGY LANE

43
63
64
83
H
J
K
L
M
N
P
Denham
Savay Farm
Gravel Pit
The Alders
Dewes Farm Cottages
Highway Farm
The Bungalow
Newyears Green
Braemar Farm
Hillingdon Outdoor Activities Centre
Dewes Farm
Newyears Green Covert
Sports Ground Club
Altcar Covert
Widows Cruise Covert
Piggery
BUCKINGHAMSHIRE GOLF COURSE
DENHAM COUNTRY PARK
Pumping Station
Denham Place
War Meml.
Research Farm
Brackenbury Barn
Copthall Covert
HILLINGDON
Frays Valley Local Nature Reserve
Works
Depot
DENHAM
Court Farm
Denham Mill
Driving Range
Cowmoor Covert
Club House
Harvil Farm
Copthall Farm
UXBRIDGE GOLF COURSE
Priory Covert
Colne Valley Park Visitor Centre
John's Covert
Harefield Place
UB9
Playing Field
River Misbourne
Denham Lock
Smallholdings
Lea Cottage
The Lea
Junction 1
DENHAM ROUNDABOUT
New House Farm
Rusholt House
WESTERN AVENUE
A40
SWAKELEYS ROUNDABOUT
Common Plantation
Vyners School
Weatherlys Covert
Fray's Farm
The Alderglade Nature Reserve
WILLOWBANK
Ivy House Farm
Oxford Road Nursery
Brickfield Farm
NEW DENHAM
Southlands Manor
Quarry
Osborne Farm
Clare House
Uxbridge Common
Sports Ground
UB10
Hillingdon Athletics Stadium
Hillingdon Sports & Leisure Complex
Playing Fields
Cricket Ground
Uxbridge Lido
Uxbridge Coll. (Uxbridge Campus)
Field Cottage
Warehouse
Uxbridge Moor Nature Reserve
Fassnidge Park
Uxbridge
Odeon Cin.
Civic Cen.
The John Locke Acad.
HILLINGDON COURT PARK (Recreation Ground)
ACS Hillingdon International Sch.
Watergate Fm.
Electricity Sub Station
Nursery
Mansfield Farm
UXBRIDGE MOOR
Recreation Ground
UXBRIDGE
Rec. Grd.
RAF Uxbridge, Battle of Britain Bunker
White Cottage
A4007
SLOUGH ROAD
West Middlesex College
HILLINGDON GOLF COURSE
Uxbridge High Sch.
Stratford Bri.
Depot
Woodlands
Stables
Halls of Residence
Hillingdon Cem.
Works
A4020
A408
COWLEY ROAD
HILLINGDON ROAD
A412
DENHAM ROAD
OXFORD ROAD
A4020
Grand Union Canal
Shire Ditch
Fray's River

64
Q
R
S
44
T
U
V
W
07
08
09
10
88
87
86
85
84
83
31
32
33
34
35
36
37
38
39
40
63
84
St. Leonard's Farm
Pylon Farm
Elm Tree Farm
High View Farm
Newyears Green
Braemar Farm
Crows Nest Farm
Old Clack Farm
UB9
The Corner
Tile Kiln
Grays Cottages
Research Farm
Dunster Cottage
Gatemead Farm
Brackenbury Barn
Moat
White Jun. & Inf. Schs.
King's Coll. Play. Flds. Track
Eastcote Hockey & Badminton Club
Highgrove Pool & Fitness Ce
Bishop C of
Warrender Prim. Sch.
Winston Churchill Hall
Manor Farm
Lib.
Prim. Sch.
Youth Cen.
Play. Fld.
Church Field Gdns.
Pav.
EASTCOTE B466
A4180
HIGH STREET
B466
ROAD
RUISLIP GOLF COURSE
Playground
WEST RUISLIP
Golf Driving Range
Club Ho.
ICKENHAM ROAD
West Ruislip
RUISLIP
Super store
Ruislip
Ruislip Manor
RUISLIP MANOR
Sports Grd.
Wealdstone.F.C (Grosvenor Vale Stadium)
Ruislip Rugby Club
Sacred Heart Catholic Primary School
WEST END RD.
A4180
Works
RUISLIP GARDENS
Ruislip Gardens
Ruislip Gardens Primary School
Playing Field
Ruislip High Sch.
Playing Field
Cricket Ground
Schs.
Moat
Copthall Farm
ICKENHAM
B467
HIGH ROAD
Community Cen.
Lib.
Compass Theatre
Ickenham
Play. Fld.
Ten. Cts.
Bowl. Grn.
Harrington House
Swakeleys
Sports Ground
The Clump
Vyners School
Adult Learning Centre
Sch.
Glebe Primary School
Tennis Courts
Ickenham Manor
Play. Flds.
Long Lane Farm
UB10
HILLINGDON
Ickenham Marsh
Weir
Brook
Park Wood
A40
WESTERN AVENUE
Hillingdon
HILLINGDON CIRCUS
Playing Fields
Playing Field
Freezeland Covert
RAF NORTHOLT AERODROME
Canal Feeder
Yeading Brook
Rec. Grd.
Cutthroat Wood
Gutteridge Wood
HILLINGDON COURT PARK (Recreation Ground)
Bowling Green
Tennis Courts
Playground
Coll.
Play. Fld.
Schs.
A437
LONG LANE
NORTH HILLINGDON
Playing Field
Ryefield Primary Sch.
TACS Hillingdon International Sch.
Convent
Sports Grd
Pav.
Liby.
Prim. Sch.
Playing Fld.
Home Covert
Tennis Courts
Recreation Ground
Rugby Ground
Hare Plantation
The Gorse
Recreation Ground
Ten Acres
Westways Fm.
Lane Covert
Swakeleys School for Girls
Abbotsfield School
Playing Field
Comm. Cen.
Charville Prim. Sch.
Hedgewood School
Grosvenor Playing Field
UB4
Club Ho.
Coney Green
Play. Fld.
Prim. Sch.
HILLINGDON
UXBRIDGE
HILL

65
45
85
99
X
Y
Z
Aa
Ba
Ca
Da
31
32
33
34
35
36
37
38
39
40
11
12
13
14
EASTCOTE
Eastcote
HA5
HA4
HA2
UB5
UB6
WEST HARROW
RAYNERS LANE
Rayners Lane
SOUTH RUISLIP
South Ruislip
SOUTH HARROW
South Harrow
Northolt Pk.
Northolt
NORTHOLT
West End
Wood End
ROXE
GREENFORD
Cavendish Recreation Ground
Bessingby Playing Fields
Roxborough Rough
Roxbourne Park
Miniature Railway
Recreation Ground
Field End Jun. & Inf. Schs.
Deane Park
Queensmead Sports Cen.
Cineworld Cinema
Superstore
Goals Soccer Centre
Solid Waste Transfer Station
Newton Park West
Newton Ecology Park
Roxbourne Complex
Alexandra Park
Hendon FC, Harrow Borough FC (Earlsmead Stadium)
Northolt Park
Northolt High Sch.
Lord Halsbury Memorial Playing Flds.
Islip Manor Park
Belvue Park
TARGET ROUNDABOUT
Alec Reed Academy
John Chilton Sch.
WEST LONDON GOLF COURSE
C & L GOLF COURSE
POLISH WAR MEMORIAL
Dinosaur Escape
Lime Tree Park
NORTHALA FIELDS
NORTHOLT GOLF COURSE
Gifford Primary Sch.
RECTORY PARK (Recreation Ground)
West London Shooting Grounds
Downe Barns Fm.
WHITE HART ROUNDABOUT
PureGym (Northolt)
A40 WESTERN AVENUE
A4090 ALEXANDRA AVENUE
A4180 WEST END ROAD
A4180 RUISLIP ROAD
A312 MANDEVILLE ROAD
A312 CHURCH ROAD
A312 ROAD
NORTHOLT WHITTON HILL
B455 RUISLIP ROAD
IMPERIAL
HARROW
EALING

66
West Harrow
Harrow-on-the-Hill
Kenton
Northwick Park
HARROW
WEST HARROW
University of Westminster (Harrow Campus)
Harrow Sports Hall
NORTHWICK PARK HOSPITAL
ST. MARK'S HOSP.
HARROW ON THE HILL
HA2
HA1
ROXETH
South Harrow
SOUTH HARROW
NORTH WEMBLEY
South Kenton
THE CLEMENTINE CHURCHILL BMI HOSPITAL
HARROW CYGNET HOSPITAL
Sudbury Hill Harrow
Sudbury Hill
SUDBURY
Sudbury & Harrow Road
Sudbury Town
Northolt
Wood End
Greenford Green
Greenford
GREENFORD
GREENFORD ROUNDABOUT
South Greenford
PERIVALE
Perivale
UB5
UB6
HA0
BARHAM PARK
HORSENDEN HILL GOLF COURSE
SUDBURY GOLF COURSE
BERKELEY FIELDS
GRAND UNION CANAL
WESTERN AVENUE
WHITTON AVENUE WEST
WHITTON AVENUE EAST
EALING GOLF COURSE
Perivale Park Athletics Track

67
Ma
Na
Pa
Qa
Ra
Sa
Ta
47
87
KINGSBURY
Kingsbury Green
FRYENT COUNTRY PARK (Local Nature Reserve)
Barn Hill Open Space
Barn Hill
PRESTON
Preston Road
HA3
HA9
NW9
NW10
North Wembley
WEMBLEY PARK
Wembley Park
Wembley Stadium
WEMBLEY
Wembley Central
TOKYNGTON
ALPERTON
Alperton
STONEBRIDGE
Stonebridge Park
Harlesden
Lower Place
PARK ROYAL
PARK ROYAL JUNCTION
Brent Junction
FORTY AVENUE
FORTY LANE
SALMON STREET
BLACKBIRD HILL
NEASDEN LA.
GREAT CENTRAL WAY
HARROW ROAD
HIGH ROAD
EALING ROAD
HANGER LA.
NORTH CIRCULAR ROAD
HILLSIDE
CRAVEN PK.
CHURCH ROAD
CHURCH END
A4140
A4088
A4089
A479
A404
A406
A4005
B4557
B4492
B454
B456
Wealdstone Brook
River Brent
GRAND UNION CANAL (Paddington Branch)
Carriage Shed
Royal Mail Depot
Railway Works Neasden Depot
Wembley Brook Depot
Sherrans Farm Open Space
King Edward VII Park
Preston Park Recreation Ground
John Billam Playing Fields
One Tree Hill Rec. Grd.
Stonebridge Rec. Grd.
St. Gibbons Recreation Ground
Wembley Sailing Club
Seahorse Sailing Club
Twyford Abbey

48
88
67
Ua
Va
Wa
Xa
Ya
Za
Ab
31
32
33
34
35
36
37
38
39
40
21
22
23
24
88
87
86
85
84
83
NW9
NW2
NW10
WEST HENDON
WEST HENDON PLAYING FIELDS
Kingsbury Town FC Grd.
Cool Oak Bridge
Youth Sailing Base
Woodfield Sch.
Field Study Cen.
BRENT RESERVOIR
Welsh Harp (Brent Reservoir) Nature Reserve
Wembley Sailing Club
Seahorse Sailing Club
Neasden Recreation Ground
Sports Ground
Rifle Range
Nursery
WOODFIELD PARK
Brent Cross
Brent Cross Shopping Centre
Brent Cross Bus Sta.
INTERCHANGE
Junc. 1
M1
MOTORWAY
NORTH CIRCULAR ROAD
A406
A41
A5
A407
A404
A4088
B453
B455
B451
B4492
STAPLES CORNER
Staples Corner Retail Park
Clitterhouse Recreation Ground
BRENT
CRICKLEWOOD
Cricklewood
CRICKLEWOOD LANE
CRICKLEWOOD BROADWAY
EDGWARE ROAD
SHOOT-UP HILL
HENDON WAY
Brent Sidings
Superstores
Depot
NEASDEN
Neasden
NEASDEN JUNCTION
Neasden Depot
DOLLIS HILL
Dollis Hill
Dollis Hill House
GLADSTONE PARK
Arts Centre
DUDDEN HILL
DUDDEN HILL LANE
CHICHELE RD.
WILLESDEN
Willesden Green
WILLESDEN GREEN
WILLESDEN NEW CEMETERY
Jewish Cemy.
CHURCH END
HARLESDEN
Harlesden
BRONDESBURY PARK
Brondesbury Park
KENSAL RISE
Kensal Rise
KENSAL GREEN
Kensal Green
QUEENS PARK
Queens Park
ROUNDWOOD PARK
King Edward VII Recreation Ground
Willesden Sports Stadium
Willesden Sports Cen.
Willesden Cen. for Health & Care
BRENT OLDER PEOPLE DAY HOSP.
Capital City Academy
Queens Park Comm. Sch.
Willesden Junction
Freightliner Depot
Power Station
HIGH ROAD
HIGH STREET
WALM LANE
WILLESDEN LANE
CHURCH ROAD
CRAVEN PK.
HARROW ROAD
COLLEGE PARK
MORTIMER
Hindu Temple
Mosque
Syn.

HIGHGATE
GOLDERS GREEN
Golders Green
Golders Green Crematorium
NW11
HAMPSTEAD GOLF COURSE
HAMPSTEAD
HEATH
Bunkers Hill
Club House
Turners Wood
Spts. Grd.
North Wood
Kenwood House
Caen Wood
Nursery
Highgate School Playing Field
Athlone House
N6
N2
Mallinson Sports Centre
Sports Ground
Yth. Hostel
The Spaniards
Sandy Heath
Mount Tyndal
Open Air Stage
KEN WOOD
Bowling Green
GOLDERS HILL PARK
Childs Hill Park
Tennis Courts
Bowling Grn.
The Hill
North End
Leg of Mutton Pond
West Heath
HAMPSTEAD HEATH
Ladies Bathing Pond
Highgate Ponds
VALE OF HEALTH
East Heath
Jack Straw's Castle
Whitestone Pond
Sports Ground
Men's Bathing Pond
CHILD'S HILL
Bathing Pond
PARLIAMENT HILL
Parliament Hill Fields
Hampstead Heath Educ. Centre
HAMPSTEAD
Hampstead Ponds
NW5
Parliament Hill Fields Athletics Track
Parliament Hill School
Lido
Gospel Oak
Kings Coll. (Hampstead Campus)
Burgh Ho.
Hampstead
2 Willow Rd.
Keats Ho.
Hampstead Heath
Sports Ground
Pavilion
Fortune Green
Hampstead Cemetery
CAMDEN
University College School
ROYAL FREE HOSPITAL
Res. (cov.)
Camden Arts Cen.
WEST HAMPSTEAD
Finchley Rd. & Frognal
Hospice
Belsize Pk.
Kingsgate Prim. Sch.
West Hampstead Thameslink
Finchley Rd.
Superstore
Freud Mus.
Sigmund Freud Statue
West Hampstead
SOUTH HAMPSTEAD
Chalk Farm
Swiss Cottage
South Hampstead
Theatre
Kilburn Grange Park
UCS Academy
PRIMROSE HILL
PRIMROSE HILL
Reservoir (covered)
Barrow Hill
Playground
NW6
NW8
Kilburn High Road
KILBURN
Barracks
Kilburn Park
Queens Park
St. John's Wood
NW1
London Zoo
Macclesfield Bridge
The Hub
REGENT'S PARK
WESTMINSTER
Tennis Courts
Paddington Rec. Grd.
Cricket Ground
Maida Vale
ST. JOHN'S WOOD
Recording Studio
Middlesex CCC (Lord's)
London Central Mosque
Winfield House
St. John's Lodge
Open Air Theatre
WEST KILBURN
A598
A502
A41
B519
B510
B520
B509
B518
B511
A4003
B451
B413
B507
B525
A5205
49
70
214
215
89

70
HIGHGATE
CROUCH END
50
Jb
Kb
Lb
Mb
Nb
Pb
Qb
STROUD GREEN
Harringay
31
Highgate School Playing Field
Crouch Hill
32
HIGHGATE CEMETERY
Lauderdale House
Waterlow Park
WATERLOW PARK
Whittington Hospital
Archway
33
Upper Holloway
UPPER HOLLOWAY
DARTMOUTH PARK
FINSBURY PARK
Finsbury Park
34
Parliament Hill Fields
Hampstead Heath Educ. Centre
HOLLOWAY
Islington Ecology Cen.
Arsenal
Ashburton Grove
Arsenal FC (Emirates Stadium)
Tufnell Pk.
TUFNELL PARK
35
Gospel Oak
GOSPEL OAK
Drayton Pk.
Holloway Rd.
69
36
KENTISH TOWN
Kentish Town
LOWER HOLLOWAY
CAMDEN
Caledonian Rd.
37
Kentish Town W.
HMP PENTONVILLE
Chalk Farm
BARNSBURY
Caledonian Rd & Barnsbury
38
Camden Rd.
CAMDEN TOWN
Camden Town
PRIMROSE HILL
39
COPENHAGEN
St. Pancras Hospital
215
216
217
London Zoo
Mornington Cres
PENTONVILLE
40
REGENT'S PARK
King's Cross
KING'S CROSS
St. Pancras International
SOMERS TOWN
British Library
90
EUSTON
Euston
FINSBURY

71
51
91
218
72
Rb
Sb
Tb
Ub
Vb
Wb
Xb
31
32
33
34
35
36
37
38
39
40
Harringay Green Lanes
Stamford Hill
FINSBURY PARK
Manor House
EAST RESERVOIR
WEST RESERVOIR
West Reservoir Centre
The Castle Climbing Cen.
BROWNSWOOD PARK
Abney Park Cemetery (Local Nature Reserve)
STOKE NEWINGTON
Stoke Newington
STAMFORD HILL
UPPER CLAPTON
SPRINGFIELD PARK
Walthamstow Marsh Nature Reserve
Coppermill Bridge
Horse Shoe Point
Springfield Marina
CLISSOLD PARK
Clissold Leisure Cen.
Clapton
North Millfields Recreation Ground
Rectory Road
E5
E17
N16
N5
E8
E2
HACKNEY
HACKNEY DOWNS
LOWER CLAPTON
SHACKLEWELL
HIGHBURY
Highbury Fields
Canonbury
CANONBURY
Dalston Kingsland
Dalston Junc.
Hackney Downs
Hackney Central
KINGSLAND
DALSTON
DE BEAUVOIR TOWN
Haggerston
HAGGERSTON
Hackney City Farm
Essex Rd.
ISLINGTON
HOXTON
Hoxton
London Fields
Cambridge Heath
Bethnal
SHOREDITCH
Angel
Highbury & Islington
SEVEN SISTERS ROAD
AMHURST PARK
STAMFORD HILL
CLAPTON COMMON
UPPER CLAPTON ROAD
LOWER CLAPTON ROAD
STOKE NEWINGTON HIGH ST.
KINGSLAND HIGH ST.
KINGSLAND ROAD
BALLS POND ROAD
DALSTON LA.
GRAHAM ROAD
MARE STREET
NEW NORTH ROAD
ESSEX ROAD
SOUTHGATE ROAD
GREEN LANES
HIGHBURY PARK
ST. PAUL'S ROAD
NORTHWOLD ROAD
AMHURST ROAD
HACKNEY ROAD

72
52
71
92
Yb
Zb
Ac
Bc
Cc
Dc
Ec
31
32
33
34
35
36
37
38
39
40
WALTHAMSTOW
Walthamstow Marsh Nature Reserve
Copper Mills Junction
Low Hall Sports Ground
Walthamstow Leisure Cen.
Horse Shoe Point
Horse Shoe Bri.
Clapton Junction
MARSHES
Leyton Marshes
Lea Valley Riding Centre
Playing Field
Lea Bridge
Lee Valley Ice Centre
Waterworks Centre
Waterworks Nature Reserve
Middlesex Filter Beds Nature Reserve
RIVER LEA or LEE
North Millfields Recreation Ground
UPPER CLAPTON
Clapton
LEA BRIDGE
SOUTH MILL FIELDS
South Millfields Rec. Grd.
E5
E10
E9
E20
E3
E17
LEYTON
Leyton Stadium
Leyton Jubilee Pk.
The Lammas Sch. & Sports College
Marsh Lane Cottage
Playing Fields
Dagenham Brook
Leyton Midland Rd.
Leyton Orient FC
The Score Complex
Waste Management Centre
Football Pitch
HACKNEY MARSH
RECREATION GROUND
Cow Bridge
Changing Rooms
Playing Fields
NEW SPITALFIELDS MARKET
MARKET PAVILION
Temple Mills
Eton Manor
Lee Valley Hockey & Tennis Centre
Mountain Bike Trail
Hackney Marshes Centre
Temple Mills Bri.
CLAPTON PARK
HACKNEY
HOMERTON
HOMERTON UNIVERSITY HOSP.
Homerton
Recreation Ground
Mabley Green
Football Ground
Marshgate Bri.
Wick Woodland
LEA INTERCHANGE
EAST CROSS ROUTE
RUCKHOLT ROAD
Lee Valley VeloPark
Velodrome
Chobham Manor
Chobham Academy
East Village
Hopkin's Fields
Alfred's Meadow
Lane's Walk
Timber Lodge
Eastcross Bridge
Copper Box Arena
HACKNEY WICK
Hackney Wick
Stratford International
The International Quarter
STRATFORD CITY SHOPPING CENTRE
Stratford
QUEEN ELIZABETH OLYMPIC PARK
London Stadium West Ham Utd. FC
ArcelorMittal Orbit
The Slide
London Aquatics Centre
London Marathon Community Track
Bobby Moore Academy (Secondary)
Bobby Moore Academy (Prim.)
GREENWAY
SOUTH HACKNEY
WELL STREET COMMON
VICTORIA PARK
Victoria Park Athletics Arena
Victoria Fountain
Bandstand
Three Colt Bridge
Royal Gate East
Crown Gate West
Skew Bridge
OLD FORD
OLD FORD
Pudding Mill Lane
BOW INTERCHANGE
A12 BLACKWALL TUNNEL
TOWER HAMLETS
BOW
Bow Church
Bow Rd.
GLOBE TOWN
Meath Gdns.
Arts Pav.
Bethnal
DALSTON
Hackney Cen.
London Fields Lido
Cambridge Heath
York Hall
MARE STREET
HOMERTON HIGH ST.
WICK ROAD
VICTORIA PARK ROAD
GROVE ROAD
OLD FORD ROAD
ROMAN ROAD
LEA BRIDGE ROAD
CHURCH ROAD
LEYTON HIGH ROAD
A104
A106
A107
A12
A102
A1006
A1205
A118
B112
B113
B118
B127
B135
B161
DLR

73
53
74
93
WANSTEAD
REDBRIDGE
Wanstead
Wanstead High School
Golf Course
Pumping Station
The Heronry
Lincoln Island
Rook Island
WANSTEAD PARK
The Temple
Keepers Lodge
Heronry Pond
Perch Pond
Shoulder of Mutton Pond
Reservoir Wood
WANSTEAD GOLF COURSE
The Basin (Fish Pond)
Sports Ground
Bush Wood
GREEN MAN ROUNDABOUT
BUSH A114 ROAD
BLAKE A1008 HALL A114 ROAD
CAMBRIDGE A106 PARK
GAINSBOROUGH RD.
Leytonstone
LEYTONSTONE
Leytonstone High Rd.
E11
E12
E7
E15
E13
E6
ALDERSBROOK
ALDERSBROOK A116
WANSTEAD FLATS
CENTRE A114 ROAD WOODFORD
Horse Ride
Playing Field
DAMES ROAD
HARROW B161 ROAD
St. Patrick's RC Cemetery
Langthorne Park
Cann Hall
West Ham Cemetery
THE MANOR PARK CEMETERY
Manor Park Crem.
Manor Park
MANOR PARK
CITY OF LONDON
City of London Crematorium
Wanstead Pk.
Forest Gate
FOREST GATE
Woodgrange Park
Woodgrange Park Cemetery
ROMFORD A118 ROAD
GREEN STREET
KATHERINE ROAD
Maryland
Stratford New Town
Stratford
Stratford High St.
Jubilee Line Depot
Abbey Road
Gas Depot
Pumping Station
Mill Meads
Prescott Channel
Channelsea
West Ham
WEST HAM
WEST HAM PARK
Public Gardens Bowl Grn.
PORTWAY
VICARAGE B164
UPTON
UPTON LANE
PLASHET ROAD
Upton Park
UPTON PARK
PLASHET PARK
Plaistow
PLAISTOW
NEW PLAISTOW RD. PLAISTOW RD. HIGH ST.
BARKING ROAD A124
GREEN STREET
CENTRAL PARK RD.
MANOR A1011 RD
East London Crematorium
EAST LONDON CEMETERY
Memorial Recreation Ground
Pavilion
Flip Out East Ham
Fc Gc Hc Jc Kc Lc Mc
31 32 33 34 35 36 37 38 39 40
39 40 41 42
83 84 85 86 87 88

74
54
73
94
Nc
Pc
Qc
Rc
Sc
Tc
Uc
42
43
44
45
31
32
33
34
35
36
37
38
39
40
88
87
86
85
84
83
REDBRIDGE
CRANBROOK
VALENTINES PARK
Valentines Mansion & Gardens
IG1
IG2
ILFORD
Ilford
Seven Kings
CITY OF LONDON CEMETERY
City of London Crematorium
ILFORD GOLF COURSE
River Roding
MANOR PARK
Manor Park
LITTLE ILFORD
NEWHAM
E12
E6
PLASHET
Woodgrange Park
LOXFORD
BARKING PARK
FAIR CROSS
Barking
BARKING
Wallend
EAST HAM
East Ham
CENTRAL PARK
SOUTH PARK
Showcase Cinema
The Discovery Centre
Refuse & Sewage Works
Hand Trough Creek
Cuckold's Haven
A406
A123
A118
A116
A117
A124
A13
A12
A1083
B165
B167
NORTH CIRCULAR ROAD
SOUTH WOODFORD TO BARKING RELIEF ROAD
HIGH STREET
HIGH STREET NORTH
ILFORD LANE
CRANBROOK ROAD
LONGBRIDGE ROAD
RIPPLE ROAD
FANSHAWE AV.
NORTHERN RELIEF RD.
ABBEY RD.
LONDON RD.
HIGHBRIDGE RD.
GREEN LANE
WINSTON WY.
ROMFORD ROAD
BARKING ROAD
EAST HAM & BARKING BY-PASS
MOVERS LANE

75
55
95
76
SEVEN KINGS PARK
CHADWELL HEATH
Chadwell Heath
Goodmayes
GOODMAYES
GOODMAYES PARK
Mayesbrook Park Arena
MAYESBROOK PARK
Boating Lake
PARSLOES PARK
DAGENHAM
Dagenham Heathway
Becontree
BECONTREE
VALENCE PARK
Valence Ho. Mus.
The Sydney Russell Sch.
Leisure Centre
Barking Abbey Sch.
Barking FC
Upney
LODGE AVENUE FLYOVER JUNC.
Cemetery
Eastbury Manor Ho.
Rippleside
CASTLE GREEN
Castle Green Leisure Cen.
Goals Soccer Cen.
Jo Richardson Comm. Sch.
Mecca Bingo
Vue Cinema
Hollywood Bowl
Goresbrook Park
GORESBROOK INTERCHANGE
HEATHWAY
Scrattons Farm Eco-Park
High Speed 1
Ripple Lane Sidings
Newlands Park
Thames View Jun. Sch.
Dagenham Dock
West Ham FC Training Grd.
Robert Clack Sch.
BEACONTREE HEATH
Whalebone Bri.
Warehouse
Superstore
Depot
Works
RM6
RM8
RM9
IG3
A118
A1083
A124
A1153
A1112
A1240
A13
B177
B191
HIGH ROAD
GREEN LANE
LONGBRIDGE ROAD
WOOD LANE
RIPPLE ROAD
LODGE AVENUE
PORTERS AVENUE
GOODMAYES ROAD
GOODMAYES LANE
WHALEBONE LANE NORTH
WHALEBONE LANE SOUTH
HEATHWAY
CASTLE LANE
BENNETT'S
BARLEY LANE
THAMES AVENUE
Vc
Wc
Xc
Yc
Zc
Ad
Bd
31
32
33
34
35
36
37
38
39
40
46
47
48
49

76
56
75
96
Cd
Dd
Ed
Fd
Gd
Hd
Jd
31
32
33
34
35
36
37
38
39
40
49
50
51
52
83
84
85
86
87
88
ROMFORD
Westlands Playing Field
West Ham FC Training Grd.
Crowlands
Jutsums Rec. Grd.
Gas Works
Romford Cemetery
Crowlands Sports Ground
QUEEN'S HOSPITAL
Oldchurch Park
RUSH GREEN
RUSH GREEN ROAD
A124
HORNCHURCH ROAD
HORNCHURCH
RM8
RM7
RM9
RM10
RM13
BEACONTREE HEATH
CROWLANDS HEATH GOLF COURSE
Robert Clack Leisure Cen.
Fords Sports Ground
Miniature Golf Course
Dagenham Civic Centre
CENTRAL PARK
Barking & Dagenham Coll.
Nursing Home
Fels Farm
Eastbrookend Ho.
EASTBROOKEND COUNTRY PARK
Eastbrook End Cemetery
White Hart Lakes
Hook's Hall Farm
THE CHASE
The Chase Local Nature Reserve
The Millennium Centre
Eastbrook Grove
Playing Field
Eastbrook Comp. Sch.
Sports Ground
Dagenham & Redbridge FC (Victoria Road)
Pondfield Park
Dagenham East
Gas Transmission Station
Beam River
HARROW
Elm Park
Bretons Outdoor Recreation Centre
Scargill Jun. & Inf. Schools
The Brittons Academy
Whybridge Inf. Sch.
Whybridge Jun. Sch.
NHS Walk-in
DAGENHAM
Dagenham Heathway
HEATHWAY
A1240
A1112
A125
A1306
A13
B178
RAINHAM ROAD
SOUTH DAGENHAM RD.
WOOD LANE
UP. RAINHAM RD.
Dagenham Beam Bri.
Newtons Corner
SOUTH HORNCHURCH
Dagenham Park Leisure Cen.
OLD DAGENHAM PARK
BMX Track
THE LEYS
Wantz Stream
BEAM VALLEY COUNTRY PARK
Adventure Playground
Beam Prim. Sch.
NEW ROAD
Works
Dagenham Dock
BEAM PARK DEVELOPMENT
Warehouse
THAMES GATEWAY

77
EMERSON PARK
57
Kd
Ld
Md
Nd
Pd
Qd
Rd
53
54
55
56
88
87
86
85
84
83
Emerson Park School
Playing Field
Upminster Court
UPMINSTER GOLF COURSE
31
32
33
34
35
36
37
38
39
40
RM11
A124
HIGH ROAD
St. Leonards Hamlet
Harrow Lodge
Hornchurch Sports Centre
Towers Inf. Sch.
Towers Jun. Sch.
Sports Ground
Langton Gds.
Queens Theatre
Emerson Pk.
St. Andrew's Park
HIGH STREET
HORNCHURCH
UPMINSTER ROAD
ST. MARY'S LANE
Upminster Bridge
Upminster Windmill
UPMINSTER PARK
Upminster
The Albany College
Ravens Bridge
Hornchurch
Abbs Cross Academy & Arts Coll.
Everyone Active, Abbs Cross
Hacton Prim. Sch.
Branfil Primary School
AFC Hornchurch (Hornchurch Stadium)
Playing Field
The Sanders Draper Sch.
Suttons Prim. Sch.
HAVERING
ELM PARK
RM12
Hacyon Bri.
River Ingrebourne
Suttons Parkway
Gaynes Parkway
CORBETS TEY
Corbets Tey School
Parklands Open Space
Gaynes Language
78
HACTON
Park Corner Farm
Lodge Farm
Rainham Lodge Farm
RM14
Central Cottages
Central Farm
Bonnetts Wood
New House
Refuse Transfer Station
Llantrisant Oaks
Greenacres
Gerpins Farm
Damyn's Hall
Oakfields Montessori Sch.
The Lodge
West Lodge
West Lodge
Playing Field
Runway Wood
Airfield Wood
Flightway Wood
HORNCHURCH COUNTRY PARK
Farmhouse Wood
Nason Waters Centre (Adult Training)
Albyns Farm
Berwick Glades
Berwick Wood
Abbey Wood
Berwick Pond
Berwick Ponds Farm
Berwick Manor
Berwick Manor Farm
Berwick Manor Cottages
Landfill Site
White Post Corner
Aylett's Cotts.
Bedbrick Cotts.
Sand & Gravel Pits
97
Warwick
White Post Wood
CELY WOOD
Prim. Sch.
HILL

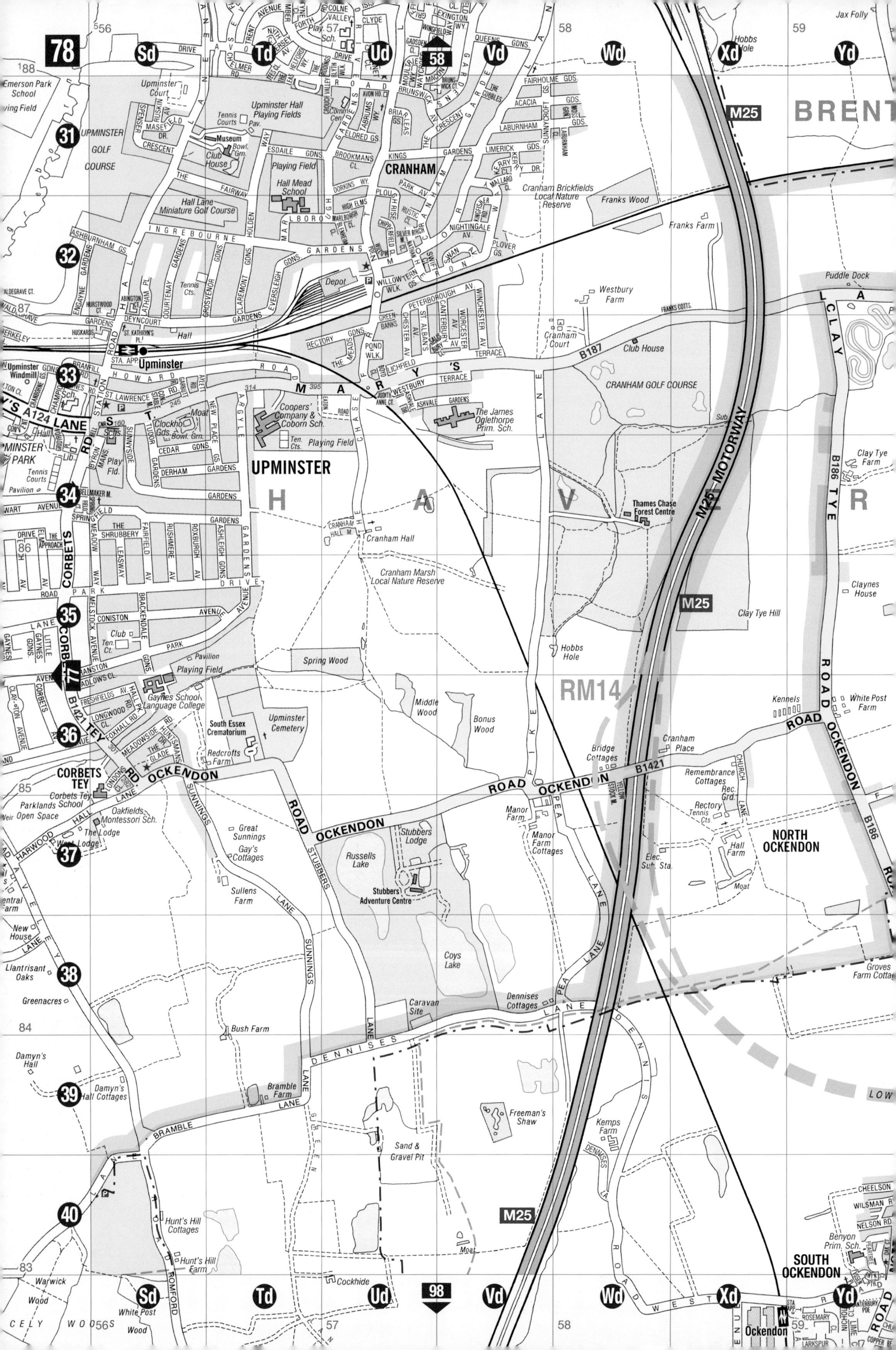

UPMINSTER
CRANHAM
Upminster
Upminster Hall Playing Fields
Upminster Golf Course
Hall Lane Miniature Golf Course
Hall Mead School
Museum
Cranham Brickfields Local Nature Reserve
Franks Wood
Franks Farm
Westbury Farm
Cranham Court
Club House
CRANHAM GOLF COURSE
The James Oglethorpe Prim. Sch.
Coopers' Company & Coborn Sch.
Thames Chase Forest Centre
M25 MOTORWAY
Cranham Hall
Cranham Marsh Local Nature Reserve
Spring Wood
Middle Wood
Bonus Wood
Hobbs Hole
RM14
Clay Tye Hill
Clay Tye Farm
Claynes House
White Post Farm
Kennels
Cranham Place
Bridge Cottages
Remembrance Cottages
NORTH OCKENDON
Hall Farm
Gaynes School Language College
South Essex Crematorium
Upminster Cemetery
Redcrofts Farm
CORBETS TEY
Corbets Tey School
Parklands Open Space
Oakfields Montessori Sch.
The Lodge
Great Sunnings
Gay's Cottages
Sullens Farm
Russells Lake
Stubbers Lodge
Stubbers Adventure Centre
Coys Lake
Manor Farm
Manor Farm Cottages
Dennises Cottages
Caravan Site
Bush Farm
Bramble Farm
Damyn's Hall
Damyn's Hall Cottages
Llantrisant Oaks
Greenacres
New House
Freeman's Shaw
Kemps Farm
Sand & Gravel Pit
Hunt's Hill Cottages
Hunt's Hill Farm
Cockhide
Warwick Wood
White Post Wood
SOUTH OCKENDON
Benyon Prim. Sch.
Ockendon
Puddle Dock
Jax Folly
Hobbs Hole
Groves Farm Cotts.
BRENT
OCKENDON ROAD
DENNISES LANE
BRAMBLE LANE
SUNNINGS LANE
ST MARY'S LANE
CORBETS TEY RD.
B187
B1421
B186
A124

58
98

79
59
99
WEST HORNDON
West Horndon
Lower Shaw
Engineering Works
Horndon Industrial Park
Smallholding
Abattoir
Orchard Farm
WOOD
ST. MARY'S LANE
WARLEY HALL LANE
DUNNINGS LANE
Monks Farm
Old England
Little Tillingham Hall
Tillingham Hall
Moat
Field House
CM13
Clay Tye Wood
Sewage Works
Bury Farm
I N G
Fairplay Farm
Bullens & Herds
Blankets Farm
Mar Dyke
Caylock's Farm
Stone Hall
Hatch Farm
CHINA LANE
Home Farm
Corner Farm
FEN LANE
HARROW LANE
The Orchard
Club House
TOP MEADOW GOLF COURSE
Fen Farm
Judds Farm
Bulphan Fen
The Dowries
Redcrofts
Groves Cottages
Groves Barns
The Wilderness
THAMES CROSSING (PROPOSED)
ROAD
RM15
ORSETT FEN
South Ockendon Hall
HALL LANE
Moat
Reservoir
Cemetery
Bowling Green
Tennis Courts
Rec. Grd.
60
61
62
63
88
87
86
85
84
83
31
32
33
34
35
36
37
38
39
40
Zd
Ae
Be
Ce
De
Ee
Fe

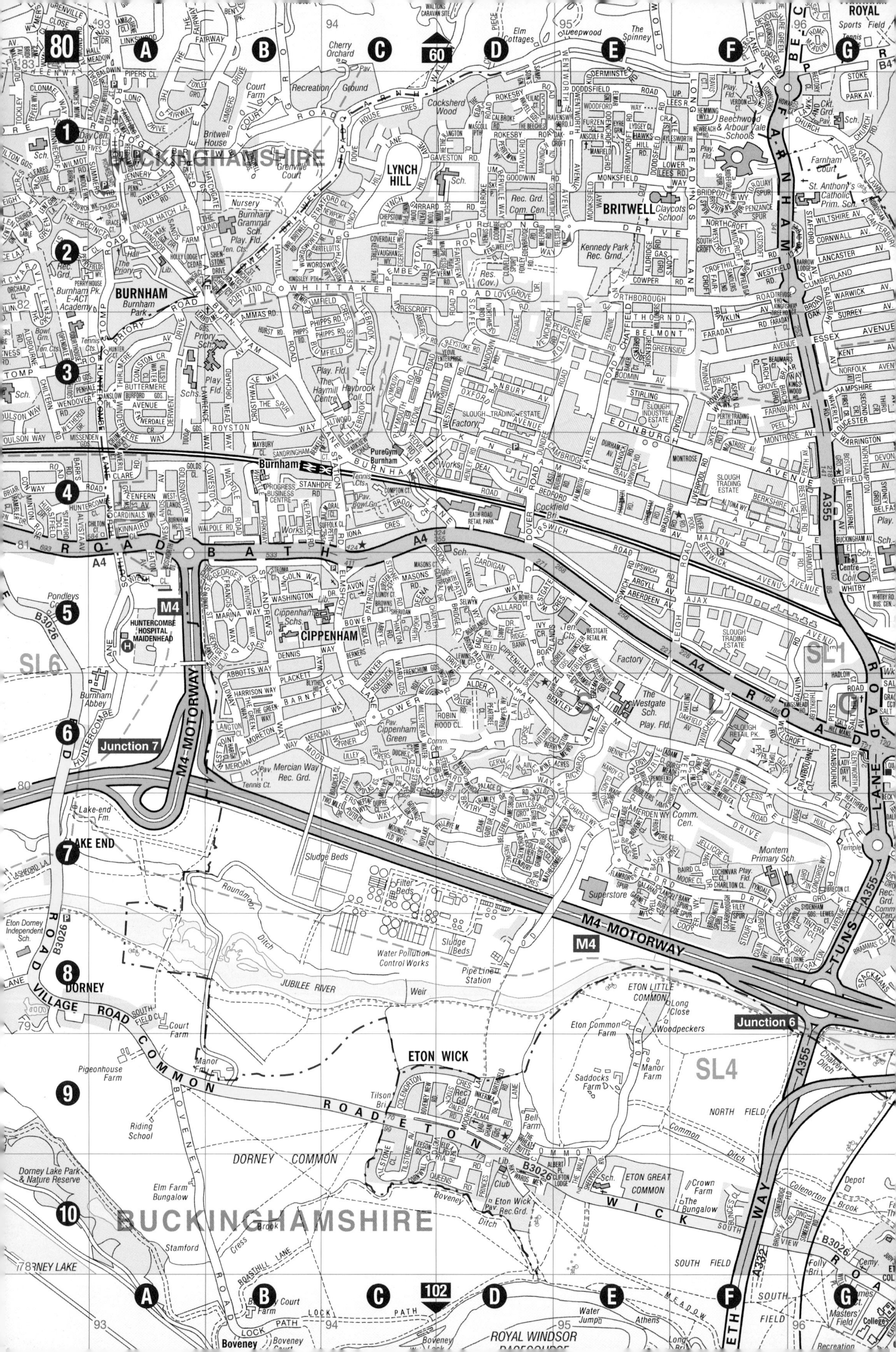
80
60
102
A
B
C
D
E
F
G
1
2
3
4
5
6
7
8
9
10
BUCKINGHAMSHIRE
BURNHAM
Burnham Park
LYNCH HILL
BRITWELL
Claycots School
CIPPENHAM
Cippenham Schs.
Cippenham Green
Burnham
PureGym Burnham
SLOUGH TRADING ESTATE
SLOUGH INDUSTRIAL ESTATE
SLOUGH RETAIL PK.
BATH ROAD RETAIL PARK
WESTGATE RETAIL PK.
The Westgate Sch.
Kennedy Park Rec. Grnd.
Burnham Grammar Sch.
Burnham Priory E-ACT Academy
Britwell House
Grenville Court
Court Farm
Recreation Ground
Cherry Orchard
Elm Cottages
The Spinney
Cocksherd Wood
Beechwood & Arbour Vale Schools
Farnham Court
St. Anthony's Catholic Prim. Sch.
Sports Field
ROYAL
HUNTERCOMBE HOSPITAL MAIDENHEAD
Burnham Abbey
Pondleys Copse
Junction 7
Junction 6
M4
M4 MOTORWAY
A4
A355
A332
B3026
SL6
SL1
SL4
LAKE END
Lake-end Fm.
Eton Dorney Independent Sch.
DORNEY
Court Farm
Manor Farm
Pigeonhouse Farm
Riding School
DORNEY COMMON
Dorney Lake Park & Nature Reserve
Elm Farm Bungalow
Stamford
DORNEY LAKE
Boveney Court Farm
Boveney
Boveney Court
ROYAL WINDSOR RACECOURSE
Water Jump
Athens
Sludge Beds
Filter Beds
Water Pollution Control Works
Pipe Line Station
JUBILEE RIVER
Weir
Roundmoor Ditch
ETON WICK
Eton Wick Rec.Grd.
Bell Farm
Tilson Bri.
ETON LITTLE COMMON
Long Close
Woodpeckers
Eton Common Farm
Manor Farm
Saddocks Farm
ETON GREAT COMMON
Crown Farm
The Bungalow
NORTH FIELD
SOUTH FIELD
Common Ditch
Chalvey Ditch
Colenorton Brook
Folly Bri.
Depot
Cemy.
Masters Field
College
Recreation
Superstore
Montem Primary Sch.
Mercian Way Rec. Grd.
ROAD
BATH ROAD
ETON WICK ROAD
COMMON ROAD
VILLAGE ROAD
FARNHAM ROAD
TUNS LANE
M4 MOTORWAY
ETON WICK
ROYAL WINDSOR RACECOURSE

81
61
103
BUCKINGHAMSHIRE
WINDSOR & MAIDENHEAD
SLOUGH
SL2
SL3
STOKE GREEN
Wexham
Wexham Court
MANOR PARK
GEORGE GREEN
MIDDLE GREEN
Salt Hill
CHALVEY
UPTON
Upton Park
Myrke
DITTON PARK
UPTON COURT PARK
STOKE POGES GOLF COURSE
WEXHAM PARK GOLF COURSE
Stoke Park Hotel
WEXHAM PARK HOSPITAL
THE SOUTH BUCKINGHAMSHIRE ACADEMY
Slough Crematorium
Wexham Park Stadium
The Park Driving Range
Slough Town FC (Arbour Pk.)
GRAND UNION CANAL
WELLINGTON ST.
A4
A412
A332
M4 MOTORWAY
M4
STOKE POGES LANE
UXBRIDGE ROAD
LONDON ROAD
BATH ROAD
SLOUGH ROAD
ETON LANE
DATCHET ROAD
WINDSOR ROAD
Agars Plough (Playing Fields)
Thames Valley Athletics Cen.
Queensmere Shop. Cen.
Jubilee River
Granville Playing Field
Godolphin Playing Field
Lascelles Playing Fields
Recreation Ground
THE HOME PARK

82
62
81
104
A
B
C
D
E
F
G
41
42
43
44
45
46
47
48
49
50
Rowleyfield Larches
Gallions Wood
Rowley Slip
Walk Wood
Black Park
Black Park Countryside Cen.
Sawmill Cotts.
Spring Wood
Hollybush Wood
Rowley Farm
Jubilee Plantation
Rowley Lake
Deodar Wood
Warren Lodge
Warren House
Club
Warren Ho. Fm.
Pav.
Five Points
A412
A4007
SLOUGH ROAD
Play. Sch. Fld.
Lib.
Ancoats Lodge
South Lodge
IVER HEATH
Home Cottage
Fernleigh
Tennis Courts
Sports Ground
Column Larches
Temple Grange
Leslie Lodge
Tudors
Homefield
Nursery
Norwood Grange
Bangors Park Farm
Bangors Park
Clovelly
NORWOOD LANE
Barry's Plantation
LANGLEY PARK COUNTRY PARK
Stone's Wood
Langley Furze
Roughground Wood
Evreham Adult Learning Cen.
Evreham Sports Centre
Tennis Cts.
Nursery
Ankerdine
St. John's
Wood Lane Fm.
Ashen Cross
Treal Farm
Langley Park
BELLSWOOD LANE
Calves Lane Farm
Riding School
Heath Lodge
Stables
Love Green Farm
Love Green
The Orchards
Play. Fld.
Cedar House
UXBRIDGE ROAD
CORONATION
BUCKINGHAMSHIRE
HIGH ROAD
Sch.
SHREDING GREEN
Iver Grove
Masons Fm.
GEORGE GREEN
Rec. Grd.
Park Stile
Love Hill House Fm.
Shreding Green Farm
Recreation Ground
Playgrd.
Larkfields
SL3
Grange Farm
Home Farm
Convent of Our Lady Queen of Apostles
B470 PARK
Poultry Farm & Piggery
Nursery
Devon Ct.
Warehouses
SL0
The Manor House
St. Paul's House
CROMWELLS COURT
Moat Ho.
IVER GOLF COURSE
Nursery
Southend Farm
MIDDLE GREEN
Wildcroft
Driving Range
Club House
DUDLEY WHARF CARAVANS
Moat
Sawyers Green Farm
GRAND UNION CANAL
(SLOUGH ARM)
ORCHARDS RESIDENTIAL PARK
Langley Hall Prim. Academy
MANSION CARAVAN SITE
THORNEY BUSINESS PARK
Langley
LANGLEY BUSINESS PARK
Chequer Bridge
Withy Bridge
LANGLEY BUS. CEN.
E. Berkshire College (Langley Campus)
Langley Comm. Acad.
RICHINGS PARK
Play. Fld.
Sports Field
The Langley Academy
SLOUGH
Memorial Grd.
Academy
LANGLEY
Sch.
Play. Fld.
Kedermister Park
Playground
School
Playing Field
Langley Leisure Cen.
Driving Range
Percy Lodge
Weir
Club Ho.
Home Wood
Harvey Pk. (Rec. Grd.)
AXIS PARK
Sutton Court Farm
RICHINGS PARK GOLF COURSE
Sutton
Mid Sutton
Little Sutton
Sutton End Farm
Oak Plantation
Junction 5
LANGLEY ROUNDABOUT
Marriott
M4 MOTORWAY
M4
Old Wood
Evans Cott.
Sports Grd.
Mayes Cott.
Creak Cott.
DITTON PARK
Canal
King's Oak
Offices
Langley Gate
Ditton Fm.
LONDON RD.
A4
B470
500
01
02
03
183
82
81
80
79
78

83
63
84
105
H
J
K
L
M
N
P
41
42
43
44
45
46
47
48
49
50
M25-MOTORWAY
M25
M4-MOTORWAY
M4
Junction 4b
(M4)
Junction 15
(M25)
A4020
A408
A3044
B470
HILLINGDON
HILLINGDON HILL
UXBRIDGE
HARLINGTON
UB8
UB7
COWLEY
COWLEY PEACHEY
PIELD HEATH
COLHAM GREEN
YIEWSLEY
WEST DRAYTON
THORNEY
HARMONDSWORTH
Little Britain
Cowley Lake
Woodlands Park
Moorwards Cottage
Palmer's Moor Farm
Delaford Manor
Coppins Fm.
Lower Delaford
North Lodge
Iver Lodge
HUNTSMOOR PARK
Huntsmoor Park Fm.
Delaford Park
Yacht Club
Farlows Lake
Sewage Works
Airport Parking
Iver Water Treatment Works
THORNEY PARK GOLF COURSE
Mercers Farm
THORNEY COUNTRY PARK
Thorney Pool
Frays Island Mabey's Meadow Nature Reserve
Old Slade Lake
New Wood
Arboretum
Home Farm
Picnic Area
Manor Court
Heathrow Primary School
Nursery
Playing Field
Cricket Grd.
Laurel Lane Prim. Sch.
Cemetery
St. Georges Meadow
Drayton Hall
West Drayton Prim. Sch.
Brunel University Uxbridge Campus
Halls of Residence
Athletics Centre
Uxbridge High Sch.
Hillingdon Cem.
Uxbridge Cem.
Brunel Uni. Sports Park
The Grove Nature Reserve
Bishopshalt School
Pield Heath House RC Sch.
Garden Centre
HILLINGDON HOSPITAL
Meadow High Sch.
Colham Manor Prim. Sch.
Moorcroft Sch.
Stockley Academy
STOCKLEY COUNTRY PARK
Uxbridge FC
Stockley Park
Recreation Ground
Cherry La. Prim. Sch.
West Drayton
Iver
GOLF COURSE
Stratford Bri.
Aqueduct
Grand Union Canal
River Colne
River Pinn
Frays River

64
83
106
HILLINGDON
HILLINGDON HEATH
COLHAM GREEN
HAYES END
WOOD END GREEN
WOOD END
HAYES
HAYES TOWN
Hayes & Harlington
WEST DRAYTON
Goulds Green
Stockley Park
STOCKLEY COUNTRY PARK
STOCKLEY PARK GOLF COURSE
HILLINGDON HOSPITAL
UB10
UB8
UB4
UB11
UB3
UB7
UXBRIDGE ROAD
HARLINGTON ROAD
DAWLEY ROAD
STOCKLEY ROAD
STATION ROAD
SIPSON ROAD
M4 MOTORWAY
Junction 4
Cherry Lane Roundabout
Yeading Brook Meadows Nature Reserve
Hayes Park
Lake Farm Country Park
Bourne Farm Playing Field
Pinkwell Park
Harlington School
Cranford Park Academy
Harlington Hospice
Little Harlington Field
Sports Ground
Crane Meadows
Gravel Works
CRANFORD PARK
Heathrow Primary School
Goals Soccer Centre Heathrow
Nuffield Health Fitness & Wellbeing Gym
Rosedale College
Uxbridge Coll. (Hayes Campus)
Botwell Green Spts. Cen.
Town Hall Park
Barra Hall
Swakeleys School for Girls
Bishopshalt School
Hedgewood School
Grange Park Jun. & Inf. Schs.
Hillingdon Manor Sch.
Colham Manor Prim. Sch.
Pinkwell Prim. Sch.
Premier Inn
Novotel
Holiday Inn
Crowne Plaza
Airport Parking
Uxbridge FC
Recreation Ground
Tudor Cott.
The Grove Nature Reserve
Grosvenor Playing Field
Mellow Lane Farm
Home Farm
The Paddocks Farm
Frogs Ditch Farm
Wall Garden Farm
Burial Ground
Sam Philp Rec. Grd.
Sipson Meadow
Dog Kennel Covert

85
65
107
98
X
Y
Z
Aa
Ba
Ca
Da
12
13
14
RECTORY PARK
(Recreation Ground)
UB5
UB6
UB1
UB2
TW5
RUISLIP ROAD
B455
GREENFORD
RAVENOR PARK
London Motorcycle Museum (Ravenor Farm)
YEADING
Belmore Playing Fields
Barnhill High Sch.
Belmore Prim. Sch.
THE PARKWAY
A312
Public Open Space
King George's Field
Durdans Park Primary Sch.
Willowtree Marina
Jubilee Park
GREENFORD PARK CEMETERY
GRAND UNION CANAL (PADDINGTON BRANCH)
OSSIE GARVIN ROUNDABOUT
UXBRIDGE ROAD
THE BROADWAY
A4020
HIGH ST.
SOUTHALL
EALING
DORMER'S WELLS
Dormers Wells Leisure Cen.
WEST MIDDLESEX GOLF COURSE
GREENFORD ROAD
Hillingdon Cycle Circuit
Goals Soccer Centre (Hayes)
Guru Nanak Academy
MINET COUNTRY PARK
Hayes & Yeading Utd. FC (The Warren)
Spikes Bridge Track
SOUTH ROAD
A3005
Southall
SOUTHALL PARK
Glade Lane Canalside Park
GREAT WESTERN INDUSTRIAL PARK
WINDMILL LANE
A4127
TENTELOW LANE
NORWOOD ROAD
MERRICK ROAD
NORWOOD GREEN
Havelock Cemetery
Southall Grn.
Featherstone Sports Cen.
HOUNSLOW
Junction 3
David Lloyd Leisure
Golf Driving Range
AIRLINKS GOLF COURSE
NORTH HYDE LANE
RINGWAY
Heston Service Area
M4 MOTORWAY
M4
HESTON PARK
WESTERN INTERNATIONAL MARKET
Cranford Community College Sports Centre
Lower Park Farm
Osterley Park Farm
Sports Ground
41
42
43
44
45
46
47
48
49
50

86
66
85
108
Ea
Fa
Ga
Ha
Ja
Ka
La
41
42
43
44
45
46
47
48
49
50
GREENFORD
GREENFORD ROUNDABOUT
South Greenford
Perivale Park Athletics Track
PERIVALE PARK
PERIVALE PARK GOLF COURSE
UB6
RAVENOR PARK
London Motorcycle Museum (Ravenor Farm)
TEIGNMOUTH GDNS.
A40
EALING GOLF COURSE
Gurnell Leisure Cen.
Perivale Bri. Playing Fields
Pitshanger Park
Brentham Club
RUISLIP ROAD
B455
THE BROADWAY
GREENFORD ROAD
A4127
Greenford Bridge
Cardinal Wiseman RC Sch.
Elthorne Heights
BRENT VALLEY GOLF COURSE
Recreation Ground
Cricket Ground
Brent Lodge Park
Millennium Maze
River Brent
UB1
DORMER'S WELLS
WEST MIDDLESEX GOLF COURSE
Castle Bar Pk.
SCOTCH COMMON
CASTLEBAR HILL
CASTLEBAR
Hanwell Community Centre
W7
Drayton Green
W13
West Ealing
EALING
Hanwell
Kensington Cemetery
UXBRIDGE ROAD
BROADWAY
A4020
HANWELL
City of Westminster Cemetery
EALING HOSP
ST. BERNARD'S HOSPITAL
Meadow Ho. Hospice
Hanwell Locks
WEST EALING
NORTHFIELD
WALPOLE PARK
The Questors Theatre
BOSTON ROAD
A3002
Northfields
Northfields Depot
Blondin Pk. & Nature Area
LAMMAS PARK
Glade Lane Canalside Park
GRAND UNION CANAL
WINDMILL LANE
QPR Training Academy & Community Sports Complex
Osterley Sports Club
Sports Ground
UB2
The Aviary Farm
B454
Wincott Farm
Osterley Lock
Elthorne Park
Elthorne Park High Sch.
Evershed Spts. Grnd.
M4
M4 MOTORWAY
Boston Manor
Boston Manor Park
Boston Manor House
HOUNSLOW
Trumpers Wood
OSTERLEY LANE
Osterley Park Farm
Osterley Park & House
TW7
WYKE GREEN GOLF COURSE
Grand Union Canal Warehouse
Goals Soccer Centre
Gunnersbury Catholic Sch.
Brentford
OSTERLEY
PARK

87
67
109
Ma
Na
Pa
Qa
Ra
Sa
Ta
41
42
43
44
45
46
47
48
49
50
88
BRENT
HOUNSLOW
PARK ROYAL
Park Royal
NORTH ACTON
North Acton
WEST ACTON
West Acton
ACTON
EAST ACTON
SOUTH ACTON
Acton Town
Acton Central
Acton Main Line
South Acton
ACTON GREEN
BEDFORD PARK
SOUTHFIELD
TURNHAM GREEN
GUNNERSBURY
Gunnersbury
GUNNERSBURY PARK
Hanger Hill
Hanger La.
Nth Ealing
Ealing Broadway
Ealing Common
EALING COMMON
Little Ealing
CHISWICK
Kew Bridge
Strand on the Green
W3
W5
NW10
WESTERN AVENUE
HANGER LANE
GUNNERSBURY AVENUE
UXBRIDGE ROAD
HIGH STREET
THE VALE
M4 MOTORWAY
CHISWICK HIGH ROAD
A406
A40
A4000
A4020
A315
A4
Junc. 1
Junc. 2
Central Middlesex Hospital
Gunnersbury Park Mus.
London Transport Mus. Depot
Pitch & Putt Course
Fox Wood Nature Reserve
Grand Union Canal (Paddington Branch)
Twyford Abbey
Metropolitan Police Traffic Unit
Alperton Sports Grd.
Kensington Cemetery
Brentford FC Comm. Stadium
Mus. of Water & Steam

88
68
110
87
Ua
Va
Wa
Xa
Ya
Za
Ab
21
22
23
24
183
82
81
80
79
78
41
42
43
44
45
46
47
48
49
50
NW10
W10
W12
W6
W4
W14
W11
KENSAL RISE
Kensal Rise
HARVIST
B451
HARLESDEN
Willesden Junction
Freightliner Depot
Grand Union Canal
COLLEGE PARK
MORTIMER
KENSAL GREEN
Kensal Green
KENSAL GREEN CEMETERY
ST. MARY'S R.C. CEMETERY
West London Crematorium
Gas Works
Superstore
A404
HARROW ROAD
LADBROKE GROVE
SCRUBS LANE
OLD OAK LANE
OLD OAK COMMON
Old Oak Common Sidings
Carriage Shed
North Pole International Depot
Mitre Bridge
West London Junc.
Little Wormwood Scrubs Rec. Grd.
WORMWOOD SCRUBS PARK
(Local Nature Reserve)
Linford Christie Stadium
HAMMERSMITH HOSP. QUEEN CHARLOTTE'S & CHELSEA HOSP. (Imperial College London)
HMP WORMWOOD SCRUBS
Burlington Danes Academy
Wormwood Scrubs Pony Cen.
NORTH KENSINGTON
ST. CHARLES HOSPITAL
Carmelite Monastery
ST. QUINTIN AV.
B412
B450
CAMBRIDGE GS.
Ladbroke Gro
NORTH ACTON
North Acton
VICTORIA ROAD
A4000
PORTAL WY.
WALES FARM RD.
East Acton
EAST ACTON
WESTWAY
A40
WESTERN AVENUE
SAVOY CIRCUS
WHITE CITY
White City
A219
WOOD LANE
Wood La.
BBC Studios
Wormholt Park
QPR FC (Loftus Rd.)
Latimer Rd.
WEST CROSS ROUTE
A3220
SHEPHERD'S BUSH
Shepherd's Bush Mkt.
Shepherd's Bush
HOLLAND PARK
HOLLAND ROAD
UXBRIDGE ROAD
A4020
UXBRIDGE
HAMMERSMITH
ACTON PARK ESTATE
Acton Cen.
ACTON PARK
Southfields Rec. Grd.
Goldhawk Rd.
GOLDHAWK ROAD
A402
B408
PADDENSWICK RD.
Ravenscourt Park
Stamford Brook
Turnham Green
BEDFORD PARK
B409
BATH RD.
CHISWICK HIGH ROAD
A316
KING ST.
A315
Hammersmith
HAMMERSMITH FLYOVER
TALGARTH RD.
A4
GREAT WEST RD.
Brook Green
Kensington Olympia
Olympia
WEST KENSINGTON
Barons Court
Hammersmith Cemetery
Hammersmith Pier
Riverside Studios
CHARING CROSS HOSP.
FULHAM PALACE RD.
A306
Chiswick Lodge
St. Pauls School
Playing Fields
Brewery
Chiswick Eyot
Homefield Recreation Ground
TURNHAM GREEN
CHISWICK
HOGARTH
DUNSTANS

89
69
Queens Park
Bb
Cb
Db
Eb
Fb
Gb
Hb
CARLTON VALE
MAIDA VALE
John's Wood
WELLINGTON RD.
PRINCE
REGENT'S PARK
NW6
NW1
NW8
W9
W2
W8
SW7
SW10
SW3
SW1
WEST KILBURN
KILBURN PARK
SHIRLAND
MAIDA HILL
Maida Vale
ST. JOHN'S WOOD
St. John's Wood Rd.
Middlesex CCC (Lord's)
The Wellington Hosp
London Central Mosque
Winfield House
QUEEN MARY'S GARDENS
Boating Lake
Regent's Univ. London
Paddington Sports Club
Paddington Rec. Grd.
Cricket Ground
Warwick Av.
CLIFTON GS.
LISSON GROVE
Lisson Green
MARYLEBONE
Baker St.
Madame Tussaud's
BAKER STREET
GLOUCESTER PLACE
MARYLEBONE RD.
Edgware Rd.
EDGWARE ROAD
MARYLEBONE FLYOVER
WESTBOURNE GREEN
Westbourne Park
WESTWAY
A40
HARROW ROAD
Royal Oak
BISHOP'S BRIDGE RD
PADDINGTON
Paddington
Paddington Basin
St. Mary's Hosp
PRAED ST.
SUSSEX GDNS.
BAYSWATER
Bayswater
Queensway
QUEENSWAY
WESTBOURNE GRO.
CHEPSTOW RD
PEMBRIDGE
BOURNE
NOTTING HILL
Notting Hill Gate
Lancaster Gate
BAYSWATER RD
Marble Arch
Speakers' Corner
WESTMINSTER
HYDE PARK
KENSINGTON GARDENS
The Serpentine
THE SERPENTINE
Serpentine Sackler Gallery
Serpentine Gallery
Diana, Princess of Wales Memorial Fountain
Kensington Palace
Round Pond
Peter Pan Statue
The Long Water
Albert Memorial
Royal Albert Hall
ROTTEN ROW
SCOTCH HOUSE
Hyde Park Corner
KNIGHTSBRIDGE
Knightsbridge
Harrods
KENSINGTON GORE
KENSINGTON ROAD
High St. Kensington
KENSINGTON
HOLLAND PARK
Holland Park
KENSINGTON and CHELSEA
Design Mus.
Imperial Coll.
Science Museum
Natural History Mus.
Victoria & Albert Museum
Brompton Oratory
CROMWELL ROAD
Gloucester Rd.
Sth. Kensington
SOUTH KENSINGTON
BROMPTON
BROMPTON ROAD
SLOANE ST.
Sloane Square
EARL'S COURT
Earl's Court
West Kensington
WEST BROMPTON
West Brompton
Brompton Cemetery
FULHAM ROAD
CHELSEA
KINGS RD.
The Royal Marsden Hosp
Royal Brompton Hosp
Saatchi Gallery
Royal Hospital Chelsea
CHELSEA BRIDGE RD
PIMLICO
OLD BROMPTON ROAD
REDCLIFFE GDNS.
214
220
226
227
90
111
41
42
43
44
45
46
47
48
49
50
25
26
27
28

90
70
Jb
Kb
Lb
Mb
Nb
Pb
Qb
41
42
43
44
45
46
47
48
49
50
89
112
227
228
229
REGENT'S PARK
PENTONVILLE
PENTONVILLE RD.
KING'S CROSS
ST. PANCRAS INTERNATIONAL
SOMERS TOWN
British Library
EUSTON
FINSBURY
Queen Mary's Gardens
Open Air Theatre
MARYLEBONE
Regent's Pk.
Baker St.
Madame Tussaud's
Gt. Portland St.
Warren St.
Euston Sq.
University College London
ST. PANCRAS
ST. CLERKENWELL
CLERKENWELL
Russell Sq.
BLOOMSBURY
British Museum
Goodge St.
Holborn
HOLBORN
Royal Courts of Justice
PADDINGTON
Wallace Collection
Oxford Circus
Tottenham Ct. R.
Bond St.
Marble Arch
SOHO
Covent Gdn.
Leicester Sq.
Temple
WESTMINSTER
MAYFAIR
Piccadilly Circus
National Gallery
Charing Cross
Embankment
CHARING CROSS
Green Park
ST. JAMES'S
GREEN PARK
ST. JAMES'S PARK
Buckingham Palace
Hyde Park Corner
Knightsbridge
Westminster
Houses of Parliament
London Eye
WATERLOO
Waterloo East
Lambeth Nth.
St. James's Park
Victoria
VICTORIA
BELGRAVIA
WESTMINSTER
Tate Britain
LAMBETH
Imperial War Mus.
Sloane Square
PIMLICO
Pimlico
Vauxhall
KENNINGTON
RIVER THAMES
Battersea Power Sta.

91
Rb
Sb
Tb
Ub
Vb
Wb
Xb
71
219
225
231
230
113
92
41
42
43
44
45
46
47
48
49
50
SHOREDITCH
HOXTON
BETHNAL GREEN
ST. LUKE'S
SPITALFIELDS
CITY
LONDON
WHITECHAPEL
SHADWELL
WAPPING
BERMONDSEY
SOUTHWARK
THE BOROUGH
NEWINGTON
WALWORTH
Old Street
Shoreditch High St.
Barbican
Moorgate
LIVERPOOL ST.
Aldgate East
Aldgate
Tower Hill
Tower Gateway
Bank
Mansion Ho.
Monument
CANNON STREET
BLACKFRIARS
City Thameslink
St. Paul's
LONDON BRIDGE
Borough
Elephant & Castle
Bermondsey
Whitechapel
Bethnal Grn.
Shadwell
Wapping
Kennington
UPPER POOL
LOWER POOL
THE TOWER OF LONDON
TOWER BRIDGE

92
72
114
91
Yb
Zb
Ac
Bc
Cc
Dc
Ec
41
42
43
44
45
46
47
48
49
50
GLOBE TOWN
MILE END
BOW
BOW COMMON
BROMLEY
STEPNEY
RATCLIFF
LIMEHOUSE
TOWER HAMLETS
POPLAR
SHADWELL
Wapping
Rotherhithe
ROTHERHITHE
SOUTHWARK
Canada Water
Surrey Quays
South Bermondsey
New Bermondsey
LEWISHAM
Canary Wharf
Heron Quays
South Quay
Crossharbour
Mudchute
Island Gds.
MILLWALL
CUBITT TOWN
DEPTFORD
ISLE OF DOGS
WEST INDIA DOCKS
Westferry
West India Quay
Poplar
All Saints
Devons Rd.
Bow Church
Bow Rd.
Bromley-by-Bow
Langdon Pk.
Mile End
Stepney Green
Bethnal Grn.
Whitechapel
Limehouse
Cutty Sark for Maritime Greenwich
E1
LOWER POOL
LIMEHOUSE REACH
GREENWICH REACH
TOWER HAMLETS CEMETERY PARK
MUDCHUTE PARK
MILLWALL PARK
SOUTHWARK PARK
MILLWALL OUTER DOCK
COMMERCIAL ROAD
EAST INDIA DOCK ROAD
THE HIGHWAY
MILE END ROAD
BOW ROAD
BURDETT ROAD
WESTFERRY ROAD
MANCHESTER ROAD
EVELYN STREET
LOWER ROAD
ROTHERHITHE TUNNEL
NORTHERN APPROACH
ASPEN WAY

93
73
115
94
Fc
Gc
Hc
Jc
Kc
Lc
Mc
41
42
43
44
45
46
47
48
49
50
West Ham
PLAISTOW
Plaistow
East London Crematorium
EAST LONDON CEMETERY
Memorial Recreation Ground
E13
E15
E16
E6
CENTRAL PARK
NEWHAM UNIVERSITY HOSPITAL
Newham College
North Beckton
Star La
BOW CREEK
Canning Town
CANNING TOWN
NEWHAM WAY
A13
BARKING ROAD
PRINCE REGENT LANE
CUSTOM HOUSE
West Beckton
Royal Victoria
Custom House for ExCeL
Prince Regent
ExCeL
ROYAL VICTORIA DOCK
LOWER LEA CROSSING
SILVERTOWN VIADUCT
SILVERTOWN WY.
East India
LEAMOUTH
Blackwall
BLACKWALL TUNNEL
The O₂
North Greenwich
North Greenwich Pier
Emirates Air Line Cable Car
Emirates Greenwich Peninsula
West Silvertown
NORTH WOOLWICH ROAD
A1020
Pontoon Dock
SILVERTOWN
Thames Barrier Park
London Regatta Centre
Royal Wharf Pier
RIVER
WOOLWICH REACH
THAMES BARRIER
Thames Barrier Yacht Club Marina
SE10
GREENWICH
Greenwich Millennium Village
Greenwich Yacht Club
New Charlton
Charlton
Charlton Athletic FC (The Valley)
SE7
Maryon Park
MARYON WILSON PARK
WOOLWICH ROAD
A206
BLACKWALL TUNNEL SOUTHERN APPROACH
A102
A2203
Westcombe Park
Maze Hill
TRAFALGAR RD.
CHARLTON
CHARLTON PARK

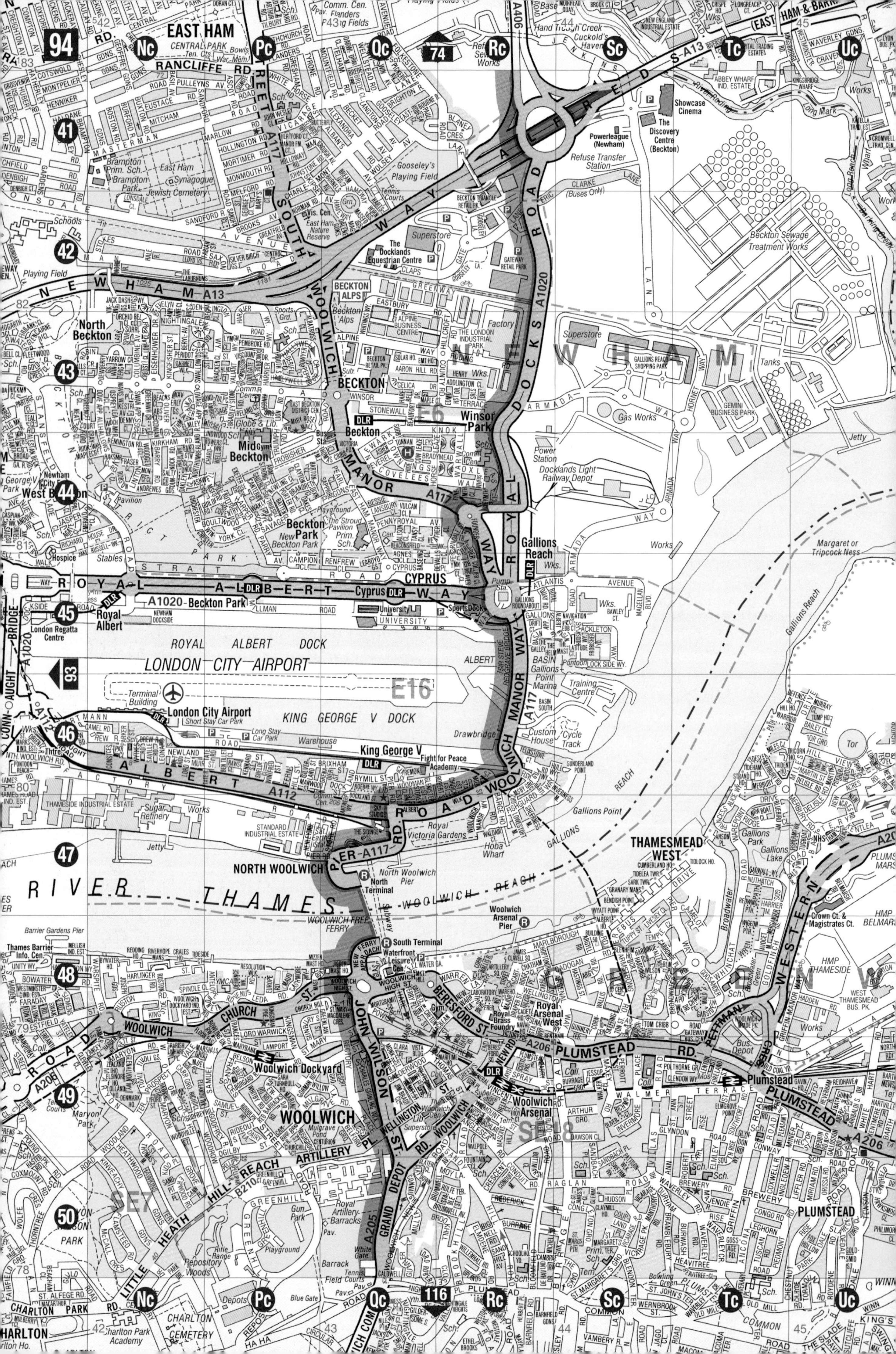

94
EAST HAM
CENTRAL PARK
RANCLIFFE RD.
74
Showcase Cinema
The Discovery Centre (Beckton)
Powerleague (Newham)
Refuse Transfer Station
Beckton Sewage Treatment Works
East Ham Jewish Cemetery
Gooseley's Playing Field
The Docklands Equestrian Centre
Superstore
NEWHAM WAY A13
ALFRED'S WAY A13
WOOLWICH MANOR WAY
ROYAL DOCKS ROAD A1020
BECKTON ALPS
North Beckton
BECKTON
Mid Beckton
West Beckton
Beckton Park
Winsor Park
Gallions Reach
GALLIONS REACH SHOPPING PARK
GEMINI BUSINESS PARK
Gas Works
Power Station
Docklands Light Railway Depot
Margaret or Tripcock Ness
Gallions Reach
ROYAL ALBERT WAY
Cyprus
CYPRUS
A1020 Beckton Park
Royal Albert
London Regatta Centre
University
ROYAL ALBERT DOCK
LONDON CITY AIRPORT
93
E16
E6
Terminal Building
London City Airport
KING GEORGE V DOCK
King George V
Fight for Peace Academy
ALBERT ROAD A112
THAMESIDE INDUSTRIAL ESTATE
Sugar Refinery
STANDARD INDUSTRIAL ESTATE
Royal Victoria Gardens
Hoba Wharf
Gallions Point
Gallions Point Marina
Training Centre
Cycle Track
Custom House
NORTH WOOLWICH
North Woolwich Pier
North Terminal
PIER RD A117
RIVER THAMES
WOOLWICH REACH
WOOLWICH FREE FERRY
South Terminal
Waterfront Leisure Cen.
Woolwich Arsenal Pier
THAMESMEAD WEST
Gallions Park
Gallions Lake
Broadwater
Crown Ct. & Magistrates Ct.
HMP THAMESIDE
WEST THAMESMEAD BUS. PK.
HMP BELMARSH
WESTERN WAY
Thames Barrier Info. Cen.
Barrier Gardens Pier
WOOLWICH CHURCH ST.
WOOLWICH ROAD A206
Woolwich Dockyard
WOOLWICH
JOHN WILSON ST.
BERESFORD ST.
A206 PLUMSTEAD RD.
Royal Arsenal West
Royal Brass Foundry
Woolwich Arsenal
SE18
SE7
Plumstead
PLUMSTEAD
PLUMSTEAD HIGH ST.
ARTILLERY PL.
GRAND DEPOT RD. A205
Royal Artillery Barracks
Gun Park
HILL REACH B210
Repository Woods
Maryon Park
Maryon Wilson Park
CHARLTON PARK RD.
CHARLTON CEMETERY
Charlton Park Academy
CHARLTON
116
Nc
Pc
Qc
Rc
Sc
Tc
Uc
41
42
43
44
45
46
47
48
49
50

95
75
117
96
Vc
Wc
Xc
Yc
Zc
Ad
Bd
41
42
43
44
45
46
47
48
49
50
Barking Riverside
Barking Riverside Rail Extension
Estimated completion: Autumn 2022
The Ripple Nature Reserve
Riverside Bridge Sch.
Power Station (disused)
CREEKMOUTH
IG11
RM9
Horse Shoe Corner
Barking Riverside Resource Centre
Barking Riverside Recycling Park
Dagenham Sunday Market
Scrattons Farm Eco-Park
Thames Gateway Park
Oil Storage Depot
Power Station
Dagenham Dock
RIVER THAMES
BARKING REACH
Barking or False Point
Barking Creek Flood Barrier
Cross Ness
Abbey Sluice
Thamesmead North
The Crossness Pumping Station
SE28
Thamesmere Leisure Cen.
Twin Tumps
THAMESMEAD CENTRAL
Bishop John Robinson Prim. Sch.
Woolwich Polytechnic School
Tump 53 Natural Park
Butts Wood
THAMESMEAD
Birchmere
Birchmere Park
St. Margaret Clitherow RC Prim. Sch.
THAMESMEAD SOUTH
Lakeside Centre
Southmere Lake
Southmere Boating Centre
DA18
BEXLEY
ERITH MARSHES
THAMESMEAD EAST
The Business Academy, Bexley
Thamesmead South West
White Hart Triangle Business Park
West Thamesmead Business Pk.
Abbey Wood Park
Abbey Wood Pk.
Abbey Wood
ABBEY WOOD
St. Paul's Academy
Southern Outfall Sewers
Recreation Ground
Plumstead Gdns.
Lesnes Abbey (Ruins)
Lesnes Abbey Pk.
DA17
LESNES ABBEY WOODS (Nature Reserve)
Cooperative Woods
Greenwich & Bexley Community Hospice
St. Luke's Hosp. (Bostall Ho.)
BOSTALL HEATH
BOSTALL WOODS
Bleak Hill
Clam Field Recreation Ground
Heathview Day Centre
WEST HEATH
Bedonwell Jun. & Inf. Schs.
Sports Grd.
A13
A2016
A2041
A206
B213
WOOLWICH ROAD
BOSTALL HILL
WICKHAM STREET
McLEOD RD.
HARROW MANORWAY
KNEE HILL
EASTERN WAY
CENTRAL WAY
CARLYLE RD.
YARNTON WAY
ABBEY ROAD
46
47
48
49
83
82
81
80
79
78

96
76
95
118
Dagenham Dock
BARKING
Dagenham Breach
Power Station
GATEWAY
A13
A1306
HORNCHURCH MARSHES
RM9
RM13
Car Compound
Beam River
Hornchurch Shoot
Thames Gateway College (CEME)
Sewage Works
Havering Coll. (Rainham Campus)
Rainham Steel Mills
DOVERS CORNER
RAINHAM
Rainham
Rainham Hall
Frog Island
Old Man's Head
RIVER THAMES
ERITH REACH
Jenningtree Point
Coldharbour Point
SEWAGE WORKS
Crossness
Nature Reserve
THAMESMEAD EAST
BEXLEY
A2016
BRONZE AGE WAY
Belvedere
BELVEDERE
HILLSIDE
DA17
FRANK'S PARK
LESSNESS HEATH
DA8
ERITH
Erith
A206
FRASER ROAD
Landing Stage
Chalk Farm Wharf
Garden Wharf
Concrete Works

97
77
119
98
RM15
RM19
THURROCK
HAVERING
A1306
A13
A1090
B1335
NEW ROAD
A13 THAMES GATEWAY
LONDON ROAD
SANDY LANE
WENNINGTON ROAD
PURFLEET ROAD
ARTERIAL ROAD
UPMINSTER ROAD SOUTH
WARWICK LANE
LAUNDERS LANE
NORTH ROAD
RIVER THAMES
Berwick Manor Farm
Berwick Manor Cottages
White Post Corner
Landfill Site
Ayletts Cotts.
Redbrick Cotts.
Sand & Gravel Pits
Jewish Cemetery
Warwick Wood
CELY WOODS
White Post Wood
Common Watercourse
Tennis Courts
Spring Farm Park Pavilion
Prim. Sch.
CHANDLERS CORNER
Playing Field
Harris Academy
Chafford Sports Complex
Brady Prim. Sch.
South Hall Farm
Southall Bridge
East Hall Farm
Landthorpe House
Halldare Cottages
Wennington Hall
WENNINGTON
The Willows
Bretts Cottages
Bretts Farm
Works
Moat
Kenningtons
The Orchard
Kennington's Cottages
Moor Hall
BELHUS WOODS COUNTRY PARK
Whitehouse Cottage
Sandy Lane Farm
NOAKES INDUSTRIAL SITE
Willow Cottage
PURFLEET INDUSTRIAL PARK
THURROCK COMMERCIAL CENTRE
Warehouses
Easter Farm
Ponds Farm
Bowl. Grn.
Sports Ground
Tennis Courts
Fanns Farm
Playing Field
WENNINGTON MARSHES
AVELEY MARSHES
Rainham Marshes Nature Reserve
Salt Lagoons
COLDHARBOUR LANE
Jetty
Vehicle Testing Sta.
Visitor Centre
Purfleet Heritage & Military Centre
MILEHAMS IND. EST.
ENSIGN ESTATE
PURFLEET
Purfleet
Depots
BEACON HILL INDUSTRIAL ESTATE
Beacon Hill
High Speed 1
HARRISONS WHARF
Crayford Ness
The Saltings
Slipway
Yacht Club
DARENT INDUSTRIAL PARK
THAMESIDE IND. EST.
Flood Barrier
Pier
Kd
Ld
Md
Nd
Pd
Qd
Rd
41
42
43
44
45
46
47
48
49
50
53
54
55
56
79
80
81
82
83

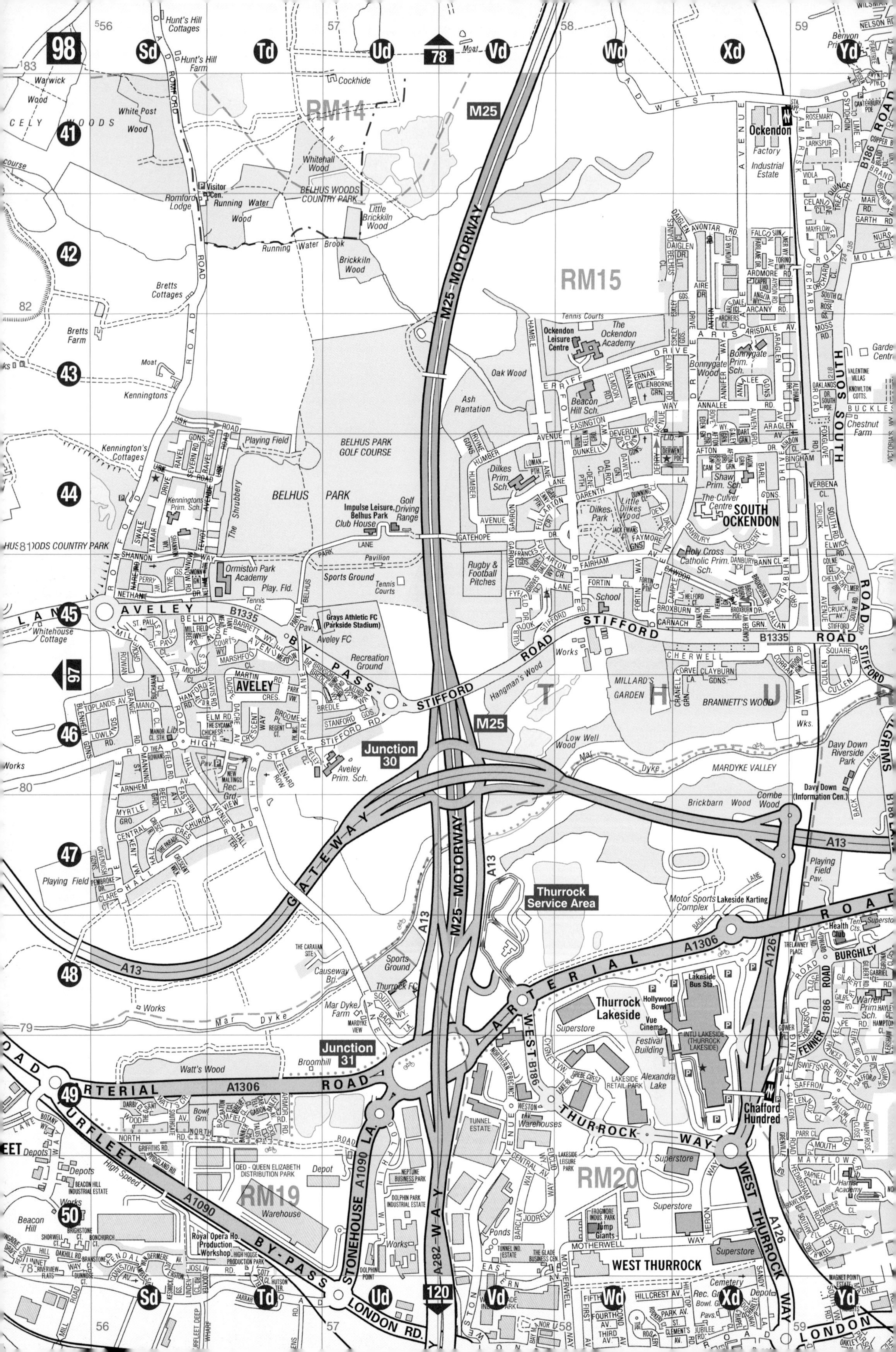

98
78
97
120
Sd
Td
Ud
Vd
Wd
Xd
Yd
41
42
43
44
45
46
47
48
49
50
RM14
RM15
RM19
RM20
M25
M25 MOTORWAY
Junction 30
Junction 31
Thurrock Service Area
AVELEY
SOUTH OCKENDON
WEST THURROCK
Thurrock Lakeside
Chafford Hundred
Ockendon
BELHUS PARK
BELHUS PARK GOLF COURSE
BELHUS WOODS COUNTRY PARK
Running Water Wood
Running Water Brook
Whitehall Wood
Little Brickkiln Wood
Brickkiln Wood
Hunt's Hill Cottages
Hunt's Hill Farm
Cockhide
Warwick Wood
CELY WOODS
White Post Wood
Romford Lodge
Visitor Cen.
Bretts Cottages
Bretts Farm
Kenningtons
Kennington's Cottages
Whitehouse Cottage
Ormiston Park Academy
Impulse Leisure, Belhus Park Club House
Golf Driving Range
Grays Athletic FC (Parkside Stadium)
Aveley FC
Recreation Ground
Sports Ground
Aveley Prim. Sch.
Kenningtons Prim. Sch.
Ockendon Leisure Centre
The Ockendon Academy
Oak Wood
Ash Plantation
Beacon Hill Sch.
Dilkes Prim. Sch.
Dilkes Park
Little Dilkes Wood
Bonnygate Wood
Bonnygate Prim. Sch.
Shaw Prim. Sch.
The Culver Centre
Holy Cross Catholic Prim. Sch.
Rugby & Football Pitches
Ockendon Factory Industrial Estate
Chestnut Farm
Hangman's Wood
MILLARD'S GARDEN
BRANNETT'S WOOD
Low Well Wood
MARDYKE VALLEY
Mar Dyke
Davy Down Riverside Park
Davy Down (Information Cen.)
Brickbarn Wood
Combe Wood
Motor Sports Complex
Lakeside Karting
Playing Field
Causeway Bri.
Thurrock FC
Mar Dyke Farm
Watt's Wood
Broomhill
GATEWAY
A13
A1306
ARTERIAL ROAD
STIFFORD ROAD
B1335
AVELEY BY-PASS
A1090
STONEHOUSE LA.
A282 WAY
A126
WEST THURROCK WAY
THURROCK WAY
B186
SOUTH ROAD
London Rd.
High Speed 1
Lakeside Bus Sta.
Hollywood Bowl
Vue Cinema
Festival Building
INTU LAKESIDE (THURROCK LAKESIDE)
Alexandra Lake
LAKESIDE RETAIL PARK
Superstore
TUNNEL ESTATE
Warehouses
QED - QUEEN ELIZABETH DISTRIBUTION PARK
Royal Opera Ho. Production Workshop
Beacon Hill
BEACON HILL INDUSTRIAL ESTATE
Depots
Depot
NEPTUNE BUSINESS PARK
DOLPHIN PARK INDUSTRIAL ESTATE
FROGMORE INDUS. PARK
Jump Giants
Ponds
Harris Academy
Warren Prim. Sch.
Health Club
BURGHLEY
Cemetery
Rec. Grd.
56
57
58
59
78
79
80
81
82
83

99
79
100
121
Zd
Ae
Be
Ce
De
Ee
Fe
41
42
43
44
45
46
47
48
49
50
60
61
62
63
ORSETT FEN
Hall
Moat
Cemetery
Bowling Green
Tennis Courts
Rec. Grd.
Comm. Cen.
Little Mollands Farm
Lake Alexandria
Great Mollands
Grange Farm
Works
Depot
Hobletts
Reservoir
Old Rectory
Whitfields
Orsett Smock Mill
Grangewaters Outdoor Education Centre
BUCKLES LANE
Little Palmer's Shaw
Great Palmer's Shaw
Stifford Clays
MARDYKE VALLEY GOLF COURSE
Hill Fm.
Club Ho.
Hotel
Mar Dyke
STIFFORD HILL
STIFFORD CLAYS ROAD
STIFFORD ROAD
CLAYS ROAD
A13
A1089
RM16
Bloomfields Farm
William Edwards School
Playing Field
Little Wellhouse Farm
Greygoose Farm Cottage
Greygoose Farm
NORTH STIFFORD
Rec. Grd.
Stifford Clays Jun. & Inf. Schs.
THURROCK COMMUNITY HOSPITAL
Deneholm Prim. Sch.
Thurrock Rugby Club
Playing Fields
Tennis Courts
Thurrock Athletics Stadium
King George's Field
Civic Hall
Impulse Leisure, Blackshots
Hangmans Wood
A1306
ARTERIAL ROAD
Caravan Park
Chafford Gorges Visitor Centre
Warren Gorge
CHAFFORD HUNDRED
Chafford Gorges Nature Park
Essex Fire Mus.
ELIZABETH RD.
A1012
HOGG LANE
A126
LONG LANE
The Grays Sch. Media Arts College
GRAYS
A1013
SOUTHEND ROAD
LODGE LA.
B149
STANFORD ROAD
Little Thurrock Prim. Sch.
St. Thomas of Canterbury Catholic Primary Sch.
Playing Field
RM17
Grays Convent High Sch.
Quarry Hill Academy
CHADWELL ROAD
LITTLE THURROCK
Grays New Cemetery
Palmer's Sports & Fitness Centre
Oil Depot
Lion Gorge
DEVONSHIRE ROAD
Stifford Prim. Sch.
AR Centre
TITAN INDUSTRIAL ESTATE
ORSETT RD.
PALMERS AV.
CLARENCE R.
Grays STATION
E. THURROCK RD.
DOCK RD.
MARSHFOOT ROAD
Rec. Grd.

100
A
B
C
D
E
F
G
1
2
3
4
5
6
7
8
9
10
99
122
Reservoir
Hobletts
Snake Spinney
New Covert
Sticking Hill Covert
Long Covert
Poultry Houses
Home Farm
Fox Holes
Well Wood
ORSETT PARK
Cherry Orchard Farm
The Decoy
Old Hall Farm
Orsett Hall
Linsteads Farm
Cholley's Farm
Orsett Fruit Farm
Poplars Farm
Orsett House
ORSETT
Old Rectory
Loft Hall Farm
Larch Plantation
Barringtons Farm
Whitfields
Baker Street
Mill House
Orsett C of E Prim. Sch.
ORSETT HOSP.-MIU
Orsett Smock Mill
Foxhound Riding School
Southfields
Old Kennels Farm
Club Ho.
Mucking Heath
ORSETT GOLF COURSE
Walnut Tree Cottages
Five Chimney Cottages
Whitecroft's
Heath Place
RM16
T H U R
Thurrock Rugby Club
Playing Fields
Brook Farm
Old House Wood
Rainbow Wood
Ashen Shaw
ORSETT HEATH
Treetops School
Thurrock Athletic Stadium
King George's Field
Civic Hall
Impulse Leisure, Blackshots
DANEHOLES ROUNDABOUT
Hangmans Wood
Woodside Prim. Sch.
Recreation Ground
Tennis Courts
Rec. Grd.
Cemetery
CHADWELL ST. MARY
Herringham Primary Sch.
High House
Becksland
LITTLE THURROCK
Playing Field
Palmer's Sports & Fitness Cen.
Palmer's College
RM17
Chadwell St. Mary Prim. Sch.
Works
Mill House
Chadwell Place
Rookery Hill
Bretts Farm
Hutts Hill
Poultry Farm
Shrove Hill Wood
Holford Farm
The Gateway Academy & Prim. Free Sch.
Hobhill Farm
WEST TILBURY
West Tilbury Hall
Gunhill Farm
Biggin Farm
A13
A128
A1089
A1013
A126
B188
B149
STANFORD ROAD
ORSETT ROAD
PRINCE CHARLES AVENUE
RECTORY RD.
HIGH ROAD
BAKER STREET
FEN LANE
HORNSBY LANE
BRENTWOOD ROAD
LOWER THAMES CROSSING
HEATH ROAD
CHADWELL RD.
MARSHFOOT ROAD
CHADWELL B149 BY-PASS
ASDA APPROACH
LINFORD ROAD
MUCKINGFORD ROAD
TURNPIKE LANE
BLUE ANCHOR LANE
RECTORY RD.
SANDY LANE
BIGGIN LANE
RIVER VIEW
WOODVIEW
PARKSIDE

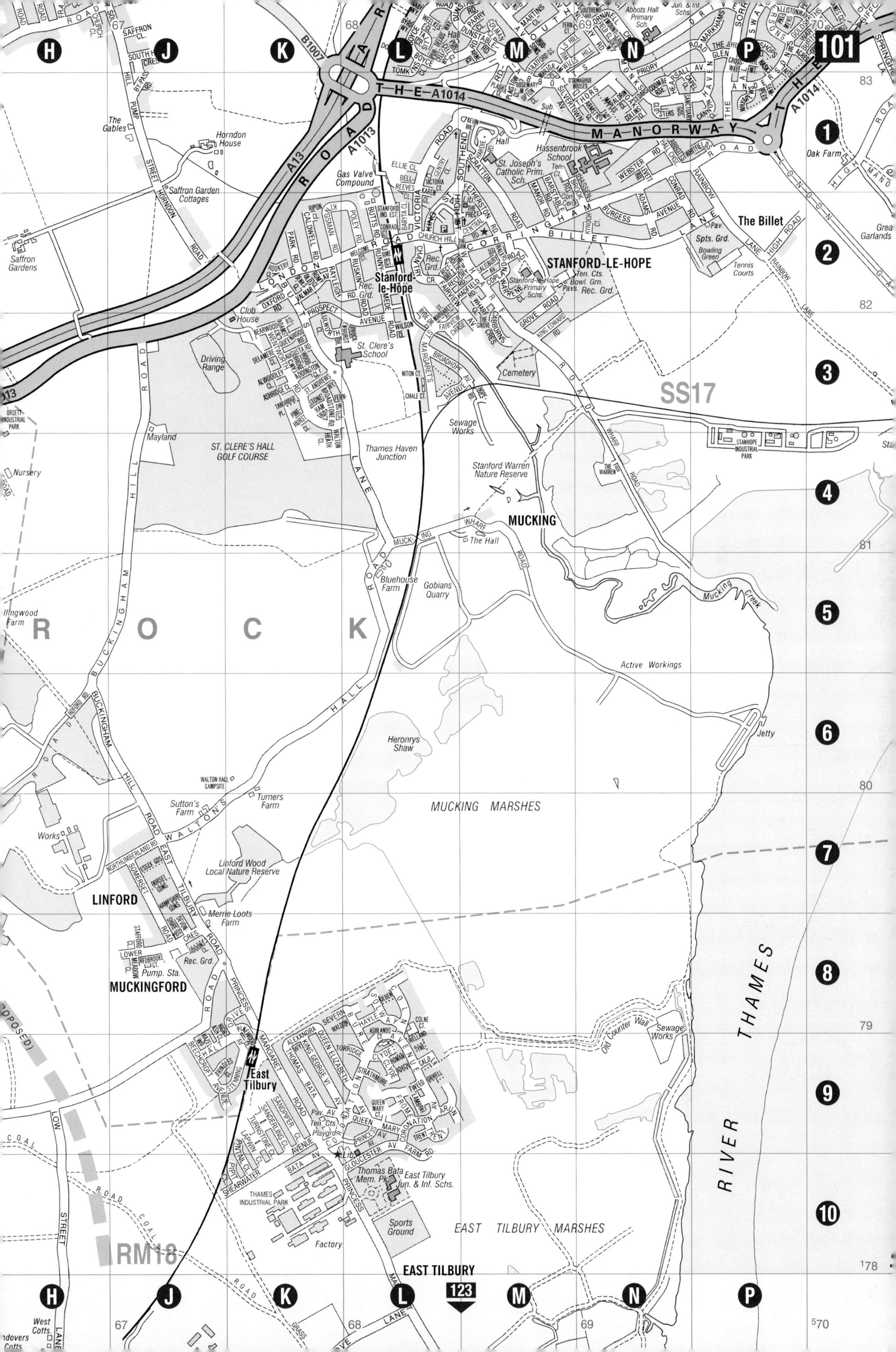

101
123
STANFORD-LE-HOPE
Stanford-le-Hope
MUCKING
LINFORD
MUCKINGFORD
EAST TILBURY
East Tilbury
ROCK
SS17
RM18
RIVER THAMES
MANORWAY
A13
A1013
A1014
B1007
Horndon House
The Gables
Saffron Garden Cottages
Saffron Gardens
Gas Valve Compound
Club House
Driving Range
St. Clere's School
ST. CLERE'S HALL GOLF COURSE
Mayland
Nursery
Hassenbrook School
St. Joseph's Catholic Prim. Sch.
The Billet
Oak Farm
Tennis Courts
Bowling Green
Cemetery
Sewage Works
Thames Haven Junction
Stanford Warren Nature Reserve
Stanhope Industrial Park
The Hall
Bluehouse Farm
Gobians Quarry
Mucking Creek
Active Workings
Heronrys Shaw
Jetty
MUCKING MARSHES
Walton Hall Campsite
Sutton's Farm
Turners Farm
Works
Linford Wood Local Nature Reserve
Merrie Loots Farm
Pump. Sta.
Rec. Grd.
Old Counter Wall
Thomas Bata Mem. Pk.
East Tilbury Jun. & Inf. Schs.
Thames Industrial Park
Sports Ground
Factory
EAST TILBURY MARSHES
West Cotts.
Stanford-le-Hope Primary Schs.
Orsett Industrial Park
Abbots Hall Primary Sch.

80
124
BUCKINGHAMSHIRE
DORNEY LAKE
Stamford
Boveney
Boveney Court
Boveney Court Farm
Eton College Rowing Centre
RIVER THAMES
Windsor Marina
Bullock's Hatch Bri.
ROYAL WINDSOR RACECOURSE
Racecourse Marina
Grandstand
Stables
Mill Stream
Works
Clewer St. Andrew
Clewer Park
CLEWER VILLAGE
Clewer St. Stephen
Recreation Ground
ETON
Masters' Field
Games Ct.
The Brocas
Bird Sanctuary
Leisure Cen.
Windsor & Eton Central
MAIDENHEAD ROAD
A308
B3383
GREEN ROAD
THE HAND CLINIC
Bishops Farmhouse
Fair Acres Farm
OAKLEY GREEN
DEDWORTH
DEDWORTH ROAD
B3024
Playing Field
Schools
Superstore
Dedworth Green
Alexander First Sch.
Gale House Farm
Tarbay Farm
Clewer Hill
Clewer Green
Clewer New Town
SL4
Thames Hospicecare
Windsor Girls' Sch.
IMPERIAL ROAD B3173
CLARENCE RD.
GOSLAR WAY
Clewer Within
QUEEN ELIZABETH A332
ST. LEONARD'S ROAD
B3022
CEMETERY
Windsor FC (Stag Meadow)
Spital
Moat
Queen Mary's Plantation
Miniature Rifle Range
Cavalry Exercise Grounds
Nightingale Corner
St. Leonardshill
Reservoirs (covered)
Water Tower
St.Leonard's Farm
LEGOLAND
Main Entrance
Government Lodge
Windsor (Legoland)
Crown Cotts.
Withy Cover
Ash Plantation
Diningroom Close
Spital Meadow
WINDSOR and
Paddockhill Copse
Palatine Gate
New Cover
Queen Victoria's Tree
High Standinghill Woods
BRACKNELL FOREST
High Standing Hill
Badgers Bridge
WINDSOR FOREST
School Allotment
Northstreet Bridge
Fish Pond
Forest Park
Flemish Farm
Bromley Hill
Star Clump
PRINCE CONSORT'S DRIVE
Cranbourne Cover
Buck Paddocks
CRANBOURNE WOOD
CRANBOURNE CHASE
Devil's Gap
CRANBOURNE PARK
Brook's Corner
Pickleherring Pond
Reservoir (covered)
Cranbourne Tower
WINKFIELD ROAD
Ranger's Gate
Ranger's Lodge
Beehive Plantation
Beehive Hill
Sir Andrew Cope's Clump
Sandy Arch
Seymours Plantation
Fiddle Covert
Russel's Pond
Russel's Plantation
STREET
A332
MOUNTS
B383

81
125
104
WINDSOR
DATCHET
OLD WINDSOR
NEW WINDSOR
MAIDENHEAD
SL3
TW19
Windsor Castle
THE HOME PARK
(Public Recreation Ground)
HOME PARK
ROYAL HOUSEHOLD GOLF COURSE
DATCHET GOLF COURSE
Frogmore
Royal Mausoleum
Frogmore House
ROYAL GARDENS
Southlea
Ham Island
Friary Island
Manor Farm Estate
Runnymede
Sunnymeads
Datchet Common
Windsor & Eton Riverside
Windsor & Eton Central
KING EDWARD VII AV.
ALBERT ROAD
STRAIGHT ROAD
SOUTHLEA ROAD
HORTON ROAD
MAJORS ROAD
LONDON ROAD
DITTON ROAD
WINDSOR ROAD
Albert Bridge
Victoria Bri.
Old Windsor Weir
Old Windsor Lock
Old Windsor Wood
Long Walk
Shaw Farm
Crimp Hill
Beaumont House
Runnymede House
Priest Hill House
John F. Kennedy Memorial
Magna Carta Island
Equestrian Statue George III

Junction 5
82
M4
M4 MOTORWAY
LANGLEY ROUNDABOUT
LONDON RD.
COLNBROOK A4
SLOUGH
Brands Hill
Datchet Sailing Club
COLNBROOK
THE QUEEN MOTHER RESERVOIR
WINDSOR and MAIDENHEAD
SL3
POYLE
HORTON
Sunnymeads
Gravel Pit Lake
Wraysbury Dive Cen.
WRAYSBURY
Wraysbury
WRAYSBURY RESERVOIR
TW19
M25 MOTORWAY
M25
Junction 13
HERON LAKE
British Disabled Water-Ski Association
Ankerwycke Farm
Magna Carta Island
Staines Nature Reserve
The Moor
103
126
Horton Lodge
Berkyn Manor Farm
Arthur Jacob Nature Reserve
Colnbrook C of E Primary School
Colnbrook Bridge
Manor Farm
Gravel Pit
Observation Platform
Old Wood
Tanhouse Farm
Mildridge Farm
Horton Cedars
Ashgood Farm
Welly Farm
Airport Parking
Remenham House
Remenham Park
Tithe Fm.
Whitehall Nurseries
Langley Gate
King's Oak
Fish Pond
Ditton Fm.
Sutton End Farm
A B C D E F G
51 52 53 54 55 56 57 58 59 60
500 01 02 03
73 74 75 76

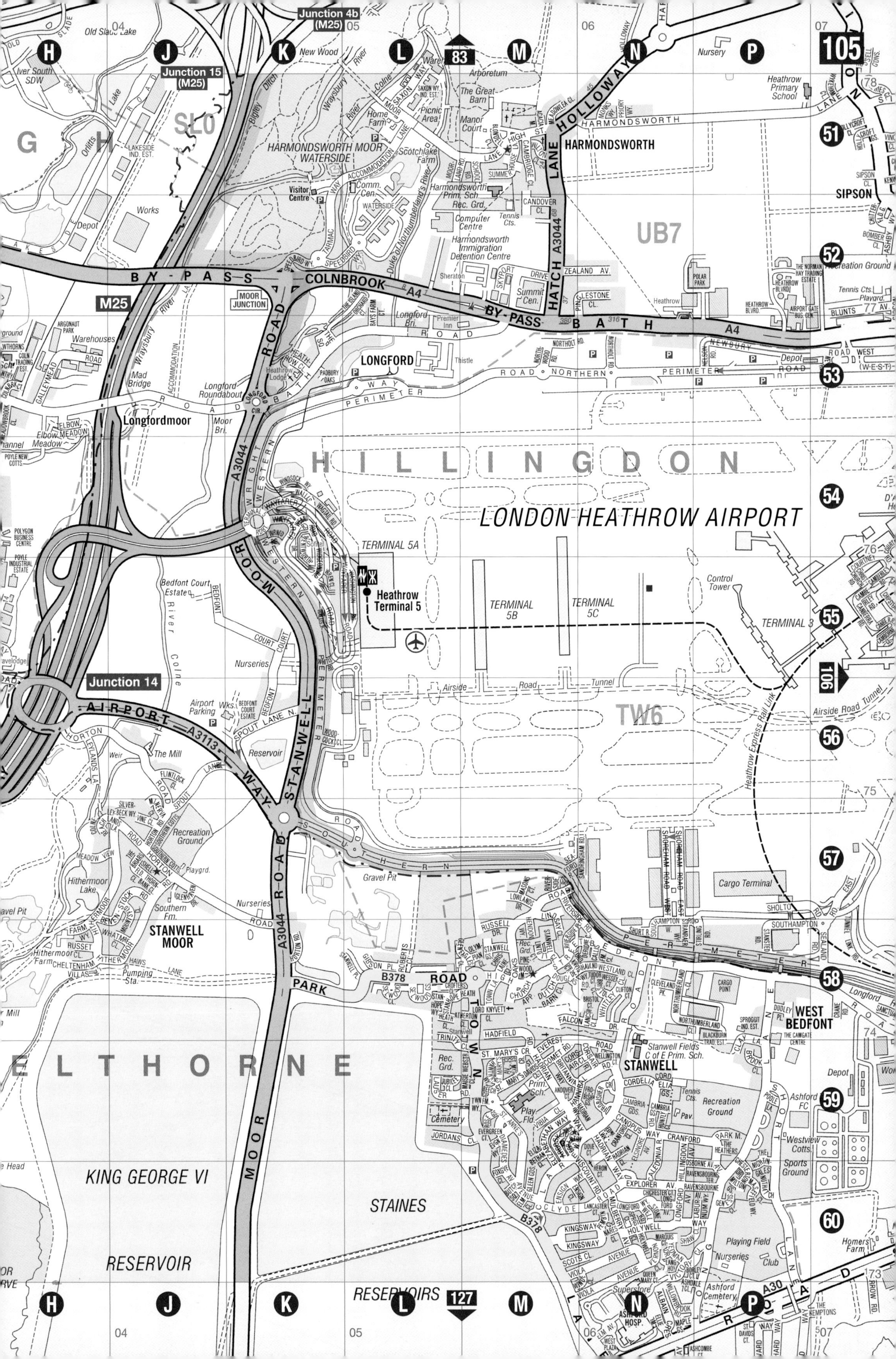
Junction 4b (M25)
Junction 15 (M25)
Junction 14
83
106
127
H
J
K
L
M
N
P
04
05
06
07
SL0
UB7
TW6
HARMONDSWORTH MOOR WATERSIDE
HARMONDSWORTH
SIPSON
LONGFORD
Longfordmoor
HILLINGDON
LONDON HEATHROW AIRPORT
TERMINAL 5A
Heathrow Terminal 5
TERMINAL 5B
TERMINAL 5C
TERMINAL 3
Control Tower
Cargo Terminal
STANWELL MOOR
STANWELL
WEST BEDFORD
ELTHORNE
KING GEORGE VI RESERVOIR
STAINES RESERVOIRS
M25
A4
A3044
A3113
B378
A30
COLNBROOK BY-PASS
BATH ROAD
BATH BY-PASS
HATCH LANE
HOLLOWAY LANE
HARMONDSWORTH LANE
NORTHERN PERIMETER ROAD
PERIMETER WAY
WRIGHT WAY
WESTERN PERIMETER ROAD
MOOR LANE
STANWELL MOOR ROAD
AIRPORT WAY
SOUTHERN PERIMETER ROAD
PARK ROAD
TOWN LANE
Airside Road Tunnel
Heathrow Express Rail Link
Longford Roundabout
Moor Junction
Bedfont Court Estate
River Colne
Nurseries
Reservoir
Recreation Ground
Hithermoor Lake
Gravel Pit
Stanwell Fields C of E Prim. Sch.
Harmondsworth Immigration Detention Centre
Heathrow Primary School
Ashford Hosp.
Ashford Cemetery
Playing Field
Sports Ground
Ashford FC
Westview Cotts.
Homers Farm
51
52
53
54
55
56
57
58
59
60
73
74
75
76
77
78

106
84
128
105
Q
R
S
T
U
V
W
51
52
53
54
55
56
57
58
59
60
SIPSON
HARLINGTON
CRANFORD
CRANFORD PARK
HILLINGDON
LONDON HEATHROW AIRPORT
Heathrow Terminals 2 & 3
TERMINAL 2B
TERMINAL 2A
TERMINAL 3
TERMINAL 4
Heathrow Terminal 4
Hatton Cross
HATTON
WEST BEDFONT
EAST BEDFONT
NORTH FELTHAM
UB7
UB3
TW6
TW14
M4 MOTORWAY
M4
Junction 4a
A408
A437
A4
BATH ROAD
A30
GREAT SOUTH WEST ROAD
A315
STAINES ROAD
B3003
A312
THE PARKWAY
CLOCKHOUSE ROUNDABOUT
Vehicle Tunnel
Cargo Tunnel
Airside Road Tunnel
Heathrow Express Rail Link
Compass Base
Maintenance Area
Airport Car Park
Harlington Hospice
Heathrow Primary School
Recreation Ground
Sports Ground
Playing Field
Cranebank (Nature Reserve)
Hounslow Urban Farm
Hatton Cemetery
Bedfont Town FC
Marjory Kinnon Sch.
Bedfont Primary School
Duke of Northumberland's River
Longford River
Blenheim Park
Rivers Academy
Fairholme Primary Sch.
Bedfont Cemetery
BEDFONT LAKES COUNTRY PARK
Edward Pauling Prim. Sch.
Grosvenor Park
West London Family Court

M4-MOTORWAY
Heston Service Area
AIRLINKS GOLF COURSE
Cranford Community College Sports Centre
Cranford Community College
Sports Ground
Concorde Centre
PARKWAY TRADING EST.
Thorncliffe House (Hotel)
Berkeley Prim. Sch.
HESTON PARK
Heston Pool
Heston Prim. Sch.
Heston Comm. Sports Hall
Heston Comm. Sch.
HESTON
Playing Field
Pavilion
Rectory Farm
Rectory Farm (Proposed Park)
TW5
WAGGONERS ROUNDABOUT
HENLYS ROUNDABOUT
BATH RD.
GREAT WEST ROAD
A4
A30
A312
A3006
A3063
A3005
A315
A314
A244
B363
B361
B358
B3377
A316
Sutton Playing Field
Playground
Lampton Academy
LAMPTON
LAMPTON PARK
Civic Centre
Cemetery
Hounslow West
Hounslow Central
Hounslow East
Bowls Club
Sports Grd.
Beaversfield Park
HOUNSLOW WEST
Cavalry Barracks
Inf. Jun. Sch. Sch.
Alexandra Prim. Sch.
Kingsley Academy
Treaty Centre
St. Mark's Cath. Sch.
HOUNSLOW
Maswell Park
Hounslow
H O U N S L O W
TW4
Rifle Range
Sports Ground
Reservoir
Warehouse
PROLOGIS PARK
CENTRAL PARK ESTATE
NORTH FELTHAM TRADING ESTATE
Works
Heath Nurseries
Baber Bridge
Weir
HOUNSLOW HEATH
Hounslow Heath Nature Reserve
The Heathland School
Mosque
Rec. Grd.
Sports Ground
Comm. Cen.
Hounslow Cemy.
River Crane
HEATHROW GATEWAY
Bridge Farm Nursery
Cemetery
Recreation Grd.
RICHMOND upon THAMES
TW2
TW13
WHITTON
South West Middlesex Crematorium
De Brome Building
Playing Field
Twickenham Cemetery
Feltham Park
Feltham
FELTHAM
Buzz Bingo
Cineworld
Springwest Academy
Crane Park Island Nature Reserve
Sports Ground
HANWORTH RD.
HARLINGTON RD. EAST
HARLINGTON ROAD WEST
UXBRIDGE A312
STAINES ROAD
WELLINGTON ROAD SOUTH
NELSON ROAD
HOSPITAL BRIDGE
CHERTSEY
X
Y
Z
Aa
Ba
Ca
Da
85
129
108
51
52
53
54
55
56
57
58
59
60

86
OSTERLEY
Osterley Park & House
PARK
Osterley
GREAT WEST ROAD
A4
BRENTFORD
BRENTFORD END
Brentford
Syon Lane
SYON PARK
Syon House
Syon Garden Centre
DUKE OF NORTHUMBERLAND ESTATE
LAMPTON
SPRING GROVE
Hounslow East
Isleworth
WOODLANDS
HOUNSLOW
ISLEWORTH
Old Isleworth
King's Observatory
Isleworth Ait Nature Reserve
WEST MIDDLESEX UNIVERSITY HOSP.
LONDON ROAD
A315
TWICKENHAM ROAD
A310
RICHMOND RD.
A3004
ST. MARGARETS
Richmond Lock
Old Deer Park Recreation Ground
Mogden Sewage Treatment Works
REDLEES PARK
Inwood Park
Maswell Park
Hounslow
Murray Park
Royal Military Sch. of Music (Kneller Hall)
KNELLER ROAD
WHITTON ROAD
CHERTSEY ROAD
A316
Twickenham Stadium
World Rugby Museum
Cole Park
Moor Mead Gd.
Richmond upon Thames College
Harlequins RUFC (Twickenham Stoop)
Whitton
WHITTON
Kneller Gdns.
Twickenham Cemetery
Twickenham
RICHMOND ROAD
A305
MARBLE HILL
Marble Hill House
Orleans Ho. Gallery
RIVER THAMES
EEL PIE ISLAND
TWICKENHAM
HEATH ROAD
Ham House & Garden
Miniature Rifle Range
Riverside Lands
107
130
TW2
TW7

109
Junc. 2
Junc. 1
Kew Bridge
WELLESLEY
87
TURNHAM GREEN
CEDARS RD.
ELLESMERE RD. A4 HOGARTH
CHISWICK
HOUNSLOW
Chiswick House & Gardens
BURLINGTON LANE
Chiswick
W4
GROVE PARK
University of Westminster Sports Ground
Strand on the Green
Oliver's Island
The National Archives
Kew Pier
Kew Green
KEW GREEN
KEW
Kew Palace
Botanical Library
Orangery
Princess of Wales Conservatory
Palm House
Plants & People Exhibition
The Pond
Rose Garden
Visitor Centre
Victoria Gate
Temple of Bellona
King William's Temple
ROYAL BOTANIC GARDENS
KEW GARDENS
Syon Vista
The Lake
Arboretum
Pagoda Vista
Isleworth Ferry Gate (Closed)
Waterlilly Pond
Evolution House
Temperate House
Marianne North Gallery
Cedar Vista
Conservation Area
Queen Charlotte's Cottage
Japanese Gateway & Landscape
Pagoda
Pavilion Restaurant
DEER PARK
MID-SURREY GOLF COURSE
London Welsh RFC (Old Deer Park)
Richmond Cricket Grnd.
Richmond Athletic Ground
Richmond and London Scottish RUFC
Richmond Olympus Gym & Squash Club
Pools on the Park
Old Brentford
Brentford Gate
Griffin Pk
The Musical Mus.
Waterman's Arts Centre
Brentford Ait
County Court
Marina
A315
A307
A205
A316
A3003
A305
B353
B351
B321
Kew Gds
NORTH SHEEN
MORTLAKE
Mortlake
Mortlake Crematorium
Hammersmith Cemetery
FULHAM (NORTH SHEEN) CEMETERY
CHALKER'S CORNER
Riverside Lands
Playing Fields
DUKE'S MEADOWS
Duke's Meadow Golf Course
Skieasy
Driving Range
Dukes Meadow Golf & Ten. Cen.
Virgin Active Health Club
Barnes B
RIVER THAMES
Westminster Boathouse
Yacht Basin
Stadium
Sports Ground
CHISWICK BRI.
GREAT CHERTSEY RD.
LOWER RICHMOND RD.
MORTLAKE HIGH ST.
UPPER RICHMOND ROAD
SHEEN LANE
CLIFFORD AV.
MANOR CIRCUS
North Sheen
Holy Trinity C of E Prim. Sch.
RICHMOND CIRCUS
Richmond
RICHMOND
Richmond Green
EAST SHEEN
SW14
East Sheen Prim. Sch.
Sheen Mt. Prim. Sch.
Pesthouse Common
Christ's C of E Sec. Sch.
East Sheen Cemetery
Richmond Cemetery
King's House Sch.
SHEEN COMMON
Football Ground
Bowling Green
Tennis Courts
Palewell Common
Palewell Fields
Playing Fields
East Sheen Gate
Adam's Pond
Bog Gate
Sheen Wood
Teck Plantation
The Bog
Riding Ring
Sheen Cross Wood
Holly Lodge
Polo Field
Playing Field
Barn Wood Pond
Conduit Wood
Bishop's Pond
Richmond Gate
Cambrian Gate
Roehampton Gate
SAWYERS HILL
(Open during daylight hours only)
RICHMOND PARK
Saw Pit Plantation
Kidney Wood
Jubilee Plantation
Queen's Ride
Duchess' Wood
Deer Pen
White Lodge Plantation
Queen Mother's Copse
Victory Plantation
The Royal Ballet Lower School
SIDMOUTH WOOD
King Henry VIII Mound
Pen Ponds
Leg of Mutton Pond
Queen Elizabeth's Plantation
Lawn Plantation
SPANKERS HILL WOOD
Treebox Wood
Oak Lodge
Pembroke Lodge (Cafeteria)
PETERSHAM PARK
King George V
KINGSTON upon THAMES
Petersham Meadows
Petersham Farm
Glover's Island
Terrace Gardens
The American Int. Uni. in London
Royal Star & Garter Home
Petersham Common
Petersham Gate
PETERSHAM
The German Sch. London
The Russell Prim. Sch.
Strathmore Sch.
The Copse
Buccleugh Gdns.
Richmond Bri.
PETERSHAM ROAD
STAR & GARTER HILL
RICHMOND HILL
QUEENS ROAD
FRIARS STILE
Terrace Fld.
Beverley Brook
Record Office
Sports Ground
Pitch and Putt Course
BARNES HOSP.
Academy
131
110
Ma
Na
Pa
Qa
Ra
Sa
Ta
51
52
53
54
55
56
57
58
59
60

110
88
109
132
TURNHAM GREEN
CHISWICK
HOUNSLOW
W4
RICHMOND upon THAMES
SW13
SW15
BARNES
FULHAM
PUTNEY
ROEHAMPTON
PUTNEY HEATH
Castelnau
Hammersmith Pier
Riverside Studios
Chiswick House & Gardens
Chiswick Eyot
Chiswick Pier
Lifeboat Sta.
Leg O'Mutton (Lonsdale Rd.) Reservoir Local Nature Reserve
LONDON WETLAND CENTRE
Visitor Centre
Barn Elms
Barn Elms Playing Fields
Barn Elms Athletics Track
Barn Elms Schools Sports Centre
Boat House
Fulham FC (Craven Cottage)
Fulham Palace
Hurlingham Yacht Club
Putney Pier
Barnes Bridge
Barnes Common Nature Reserve
BARNES COMMON
Old Barnes Cemetery
Putney Cemetery
Putney Lwr. Comn.
Football Ground
Cricket Ground
Barnes
Rosslyn Pk. RUFC
PRIORY HOSPITAL ROEHAMPTON
ROEHAMPTON GOLF COURSE
Roehampton University (Southlands College)
Roehampton University (Digby Stuart College)
Roehampton University (Froebel College)
QUEEN MARY'S HOSPITAL ROEHAMPTON
Roehampton Sport & Fitness Cen.
Maryfield Convent
Club College Halls Ho. (Mount Clare)
Roehampton Gate
RICHMOND PARK GOLF COURSE
Kingsfarm Plantation
Victory Plantation
Killcat Corner
White Lodge Plantation
Queen Mothers Copse
The Royal Ballet Lower School
Deer Pen
Polo Field
Adam's Pond
Horse Ride
TW10
TIBBET'S CORNER
ROEHAMPTON LANE
Putney Vale Crematorium
Scio Pond
King's Mere
Seven Post Pond
ROYAL HOSP. FOR NEURO-DISABILITY
East Putney
HIGH ST.
UPPER RICHMOND RD.
LOWER RICHMOND RD.
KINGSTON ROAD
CASTELNAU
ROCKS LANE
CHURCH RD.
MILL HILL RD.
QUEENS RIDE
ROEHAMPTON LANE
TIBBET'S RIDE
PUTNEY HILL
PUTNEY HIGH ST.
FULHAM PALACE ROAD
HAMMERSMITH BRI.
BURLINGTON LA.
HOGARTH LA.
A4
A205
A219
A306
A3
A3003
B306
B349
B350
Ua Va Wa Xa Ya Za Ab
21 22 23 24
73 74 75 76 77 78

Bb
Cb
Db
Eb
89
Fb
Gb
Hb
25
26
27
28
CHELSEA
West Brompton
WEST BROMPTON
Brompton Cemetery
KENSINGTON
Chelsea FC (Stamford Bridge)
Fulham Broadway
FULHAM BROADWAY
A3218
A3219
A3220
A3212
A304
A308
A217
A3205
A3209
A3036
A3031
A205
A214
A218
A3
B317
B305
B229
B234
B237
L304
EMBANKMENT
RIVER
THAMES
CHELSEA REACH
Peace Pagoda
Children's Zoo
Battersea Evolution
BATTERSEA PARK
General Games Ground
Go Ape Battersea Pk.
Cadogan Pier
Cremorne Riverside Centre
Cremorne Gardens
Chelsea Wharf
Cremorne Wharf
SANDS END
Gas Works
Imperial Wharf
FULHAM
WALHAM GREEN
Eel Brook Common
Parsons Grn.
PARSONS GRN.
Chelsea Harbour Pier
Bascule Bri.
THE LONDON HELIPORT
SOUTH PARK
HURLINGHAM PARK
Running Track
Hurlingham
The Hurlingham Club
Hurlingham Wharf
Trinidad Wharf
Fulham Wharf
Plantation Wharf Pier
Wandsworth Riverside Quarter Pier
Jetties
Solid Waste Transfer Station
Wandsworth Town
WANDSWORTH GYRATORY
CLAPHAM JUNCTION
BATTERSEA
Falcon Park
CLAPHAM COMMON
NORTH SIDE
Spencer Park
Emanuel School
St. Mary's Cemy.
SW18
SW11
SW10
SW12
West Hill
WANDSWORTH
King George's Park
SOUTHFIELDS
Southfields
Southfields Academy
HMP WANDSWORTH
WANDSWORTH COMMON
Wandsworth Common
WANDSWORTH CEMETERY
The Chelsea Group of Children
EARLSFIELD
Earlsfield
Beatrix Potter Prim. Sch.
Nightingale Academy
The Joan Bicknell Centre
Cricket Ground
Playing Field
Pavilion
51
52
53
54
55
56
57
58
59
60
112
133
75
74
76
77
78

90
134
111
Jb
Kb
Lb
Mb
Nb
Pb
Qb
51
52
53
54
55
56
57
58
59
60
GROSVENOR ROAD
EMBANKMENT
A3212
RIVER THAMES
CHELSEA REACH
Grosvenor Bridge
Vauxhall
KENNINGTON
Surrey CCC (The Oval)
Oval
VAUXHALL
NINE ELMS
NEW COVENT GARDEN MARKET
Fruit & Vegetable Market
Battersea Power Sta.
BATTERSEA PARK
Pump House Gallery
Boating Lake
Battersea Park
Queenstown Rd. (Battersea)
BATTERSEA
SOUTH LAMBETH
Stockwell
STOCKWELL
Angell Town
Larkhall Park
Wandsworth Rd.
Clapham High St.
Clapham North
LAMBETH HOSPITAL
CLAPHAM
Clapham Comm.
CLAPHAM COMMON
Trinity Hospice
Brixton
BRIXTON
COLDHARBOUR
Lambeth Academy
CLAPHAM PARK
HMP BRIXTON
Clapham Sth.
Balham
BALHAM
Wandsworth Common
STREATHAM HILL
CHRISTCHURCH
A205
A23
A24
A3
SW4
SW2
SW8

91
CAMBERWELL
CAMBERWELL GREEN
PECKHAM
PECKHAM ROAD
PECKHAM HIGH ST.
QUEEN'S ROAD
SOUTHAMPTON WAY
ALBANY ROAD
Burgess Park
SOUTHWARK
Denmark Hill
KINGS COLLEGE HOSPITAL
THE MAUDSLEY HOSP
Loughborough Junc.
COLDHARBOUR LA.
Peckham Rye
EAST DULWICH
East Dulwich
EAST DULWICH ROAD
GROVE VALE
LORDSHIP LANE
DOG KENNEL HILL
PECKHAM RYE COMMON
PECKHAM RYE PARK
RUSKIN PARK
HERNE HILL
Herne Hill
HALF MOON LA.
VILLAGE WAY
North Dulwich
DULWICH VILLAGE
Dulwich Picture Gallery
DULWICH PARK
SE5
SE15
SE22
SE24
SE21
SE23
Brockwell Park Lido
TULSE HILL
Tulse Hill
West Dulwich
DULWICH COMMON
THE GROVE
Horniman Museum
LONDON ROAD
NORWOOD ROAD
THURLOW PARK ROAD
CROXTED ROAD
Nunhead
NUNHEAD
Camberwell Old Cemetery
DULWICH & SYDENHAM HILL GOLF COURSE
114
135

114
Bermondsey
New Bermondsey
Millwall FC (The Den)
Lewisham Lions Cen.
Deptford Park
92
Convoy's Wharf
Greenwich Reach
Deptford
Cutty Sark for Maritime Greenwich
Greenwich Pier
Creek Road
A200
Evelyn Street
Surrey Canal Road
Deptford Church St.
Greenwich
DLR
Greenwich High Rd.
New Cross
New Cross Gate
New Cross Road
A2
Queens Rd. (Peckham)
Queen's Road
A202
Goldsmiths College
Lewisham Way
Deptford Bri.
Deptford Broadway
Blackheath Road
Blackheath Hill
Brookmill Road
Lewisham Road
A20
Loampit Hill
Loampit Vale
St. Johns
Elverson Rd
Lewisham
Telegraph Hill Park
Nunhead
Drakefell Road
B2142
Evelina Road
Brockley
Brockley Road
B218
Hilly Fields
SE4
SE14
SE15
SE23
SE6
Nunhead Cemetery
Nunhead Cemetery Nature Reserve
Southwark
Reservoirs (covered)
Peckham Rye Park
Harris Sports Cen.
Aquarius Golf Course
Camberwell New Cemetery
Honor Oak Crematorium
One Tree Hill Local Nature Reserve
Honor Oak
Honor Oak Park
Honor Oak Road
B238
Forest Hill
Horniman Museum
Horniman Gardens
Camberwell Old Cemetery
London Road
Stanstead Road
A205
Waldram Pk. Rd.
Brockley Cemetery
Ladywell Cemetery
Crofton Park
Adelaide Av.
B236
Ladywell
Ladywell Arena (Running Track)
Ladywell Flds.
University Hospital Lewisham
Lewisham High Street
A21
Molesworth St.
Lewisham Park
Rushey Grn.
Catford
Catford Bridge
Catford Road
Catford Hill
A212
Brownhill Road
Blythe Hill
Blythe Hill Fields
Guy's Hosp. Athletic Grd.
St. Dunstan's Coll.
Bromley Road
136
Yb Zb Ac Bc Cc Dc Ec
51 52 53 54 55 56 57 58 59 60
113
35 36 37 38
73 74 75 76 77 178

115
93
116
137
Fc
Gc
Hc
Jc
Kc
Lc
Mc
51
52
53
54
55
56
57
58
59
60
GREENWICH
GREENWICH PARK
National Maritime Museum
Royal Observatory Greenwich
Maze Hill
Westcombe Park
CHARLTON
CHARLTON PARK
SE7
BLACK HEATH
Blackheath Vale
BLACKHEATH
Blackheath
GREENWICH
KIDBROOKE
Kidbrooke
SE3
BLACKHEATH PARK
SE13
LEE
Lee
Hither Green
HITHER GREEN
SE12
ELTHAM
SE9
BROMLEY
Horn Park
Sutcliffe Park Athletics Track
HITHER GREEN CEMETERY
ROCHESTER WAY RELIEF ROAD
SHOOTERS HILL
DOVER ROAD
ELTHAM ROAD
WESTHORNE AVENUE
BURNT ASH HILL
BARING ROAD
ST. MILDREDS ROAD
SIDCUP ROAD
KIDBROOKE PARK ROAD
LEE HIGH ROAD
A2
A20
A102
A205
A207
A210
A212
A2213
B210
B211
B212
B220

116
94
115
138
Nc
Pc
Qc
Rc
Sc
Tc
Uc
51
52
53
54
55
56
57
58
59
60
SE7
SE3
SE18
SE9
CHARLTON
CHARLTON PARK
Charlton Park Academy
CHARLTON CEMETERY
Repository Woods
QUEEN ELIZABETH HOSPITAL
Hornfair Park
Charlton Lido
WOOLWICH COMMON
Barrack Field
Rifle Range
Depots
Blue Gate
NIGHTINGALE PL.
PLUMSTEAD
PLUMSTEAD COMMON
Plumstead Common
SHREWSBURY PARK
Shooters Hill Post 16 Campus
Reservoir (covered)
Greenwich Free School
SHOOTERS HILL
SHOOTERS HILL GOLF COURSE
Eaglesfield Rec. Grd.
Club House
Clothworkers Wood
Willow Dene School
Hillview Cemetery
Woodlands Farm
Garden Centre
MEMORIAL HOSPITAL
ELTHAM COMMON
Severndroog Castle
CASTLE WOOD
Rose Cottage
JACK WOOD
Jackwood Ho.
Oxleas Meadows
OXLEAS WOOD
Eastcote Prim. Sch.
BELLEGROVE
GREENWICH
Recreation Ground
Corelli College
CEMETERY
Sports Ground
Sam Montague Boy's Club
WELL HALL ROUNDABOUT
Playing Field
SHEPHERDLEAS WOOD
ELTHAM PARK NORTH
ELTHAM PARK
ELTHAM PARK SOUTH
Falconwood Field
Falconwood
FALCONWOOD
Falconwood Academy
Eltham Cemetery
Eltham Crematorium
ROCHESTER WAY
RELIEF ROAD
A2
A20
A205
A207
A208
A210
A211
B210
B2214
B263
Eltham
ELTHAM WARREN GOLF COURSE
Nature Study Centre
Stationers Crown Woods Academy
University of Greenwich (Avery Hill Campus)
AVERY HILL
AVERY HILL PARK
Alderwood Prim. Sch.
Halls of Residence
Sparrow's Farm Leisure Centre
All Weather Pitches
BEXLEY ROAD
ELTHAM HIGH STREET
ELTHAM COMM. HOSP
Hospice
Eltham Hill Sch.
Harris Academy Greenwich
Queenscroft Park
Eltham Palace & Gardens
KING JOHN'S WALK
ROYAL BLACKHEATH GOLF COURSE
Superstore
Bowling Green
Bird Sanctuary
The Tarn
Community Centre
Mottingham
MOTTINGHAM
MOTTINGHAM RD.
COURT RD.
SIDCUP ROAD
Eric Liddell Spts. Cen.
Eltham Coll.
Coldharbour Leisure Centre
Fairy Hill
New Eltham
Dulverton Sch.
Wyncham Stream
River Shuttle

117
95
118
139
Vc
Wc
Xc
Yc
Zc
Ad
Bd
BOSTALL HILL
WOOLWICH
A206
Greenwich & Bexley Community Hospice
BOSTALL WOODS
SE2
West Heath
Clam Field Recreation Ground
Heathview Day Centre
Turpins Cave
Woolwich Cemetery
Plumstead Cemetery
Hill Grove
East Wickham Open Space
EAST WICKHAM
WICKHAM LANE
UPPER WICKHAM LA.
A209
Stevens Park
Welling
WELLING
WELLING HIGH ST.
PARK VIEW RD.
A207
ROAD
Welling Corner
Welling United FC (Park View Road)
Bexleyheath Sports Club
DA16
DA7
DA6
DA5
DA15
BEXLEY
Bexley Grammar Sch.
Danson House
Danson Park Watersports Centre
DANSON PARK
THE LAKE
DANSON ROAD
A221
Crook Log Leisure Cen.
CROOK LOG
BROADWAY
BEXLEYHEATH
Bexleyheath
Brampton Prim. Acad.
Bexleyheath Academy
Goals Soccer Cen.
Upton
UPTON CENTRE
Red House
BEXLEYHEATH GOLF COURSE
A2
ROCHESTER WAY
EAST
DANSON INTERCHANGE
BLENDON RD.
B2210
Blendon
BRIDGEN RD.
PARKHILL
Bridgen
Bexley Park Wood
Blackfen Sch. for Girls
Sherwood Park Prim. Sch.
BLACKFEN
Hollyoak Wood Park
PENHILL PARK
PENHILL RD.
SIDCUP GOLF COURSE
Marlborough Park
Rose Theatre
Rose Bruford College
Lamorbey
Hurstmere School
Hurst Pl. (Com. Cen.)
HURST ROAD
A222
Old Bexley C of E Prim. Sch
Cricket Ground
Upper College Farm
Albany Park
Sidcup Leisure Centre
Abbeyhill Park
Sidcup
A210
A2 ROCHESTER WAY
ROAD
51
52
53
54
55
56
57
58
59
60
46
47
48
49

118
Cd
Dd
Ed
Fd
Gd
Hd
Jd
96
117
140
51
52
53
54
55
56
57
58
59
60
LESSNESS HEATH
Erith
NORTHUMBERLAND HEATH
North End
Slade Green
SLADE GREEN
BEXLEYHEATH
Barnehurst
BARNEHURST
BARNEHURST GOLF COURSE
BEXLEY
CRAYFORD
Crayford
OLD BEXLEY
Bexley
Bowmans
Maypole
Coldblow
DARTFORD HEATH
DA5
DA6
DA7
DA8
A2
A206
A207
A220
A221
A223
A2000
A2018
A222
A570
B2186
B2174
B2210
Bursted Wood
Hall Place Gardens
BLACK PRINCE INTERCHANGE
Crayford Stadium (Greyhound)
Recreation Ground
Playing Field
Erith Leisure Centre
Erith Stadium
SPORTS FIELD
The Old Orchard
Churchfield Wood
Perry Street Farm
Upper College Farm
Cricket Ground
Sands Spinney

97
RIVER THAMES
LONG REACH
THURROCK
RM19
DARTFORD SALT MARSHES
Dartford Clay Shooting Club
CRAYFORD MARSHES
RIVER DARENT
RIVER CRAY
The Saltings
Erith Yacht Club
Slipway
Flood Barrier
DARENT INDUSTRIAL PARK
THAMESIDE IND. EST.
Works
Lower Fm.
Playground
Playing Field
Howbury Farm
Moat
Carriage Depot
The Grange
Crayford Creek Junction
Joyce Green Farm
Sewage Works
LITTLEBROOK BUSINESS CENTRE
The Bridge
Training Centre
A206
DUNN
BOB
DARTFORD FRESH MARSHES
DA1
Distribution Centre
RIVERSIDE INDUSTRIAL ESTATE
JOYCE GREEN
Temple Hill
BARNES CRAY
Stanham Farm
Holy Trinity C of E Prim. Sch.
A2026
BURNHAM ROAD
VICTORIA RD.
Dartford
Dartford County Ct.
New Town
A226 (WATLING ST.)
WEST HILL
EAST HILL
DARTFORD
CENTRAL PARK
Sports Ground
Fairfield Pool & Leisure Cen.
Central Park Arena
Dartford Grammar Sch. for Girls
Westgate Prim. Sch.
Dartford Science & Technology College
Playing Fields
Beckets Sports Cen.
Postal Delivery Office
A2018
B2174
PRINCES ROAD
A225
Superstore
Brooklands Lake
David Lloyd Leisure
Dartford FC (Princes Pk.)
The Leigh Academy
Goals Soccer Cen.
DARTFORD GOLF COURSE
Brickhill Wood
PRINCES PARK GOLF COURSE
141
Oakfield Prim. Sch.
Miskin Theatre
North West Kent College
DA2
Junction 1a
LITTLEBROOK INTERCHANGE
Junc.1b
PRINCES RD. INTERCHANGE
A282
DARTFORD TUNNEL APPROACH RD.
CROSSWAYS
Littlebrook Hosp.
GREENACRES
Recreation Ground
Dartford Judo Cl.
B2500
B2228
B260
Hesketh Park
Fleetdown Primary Sch.
120

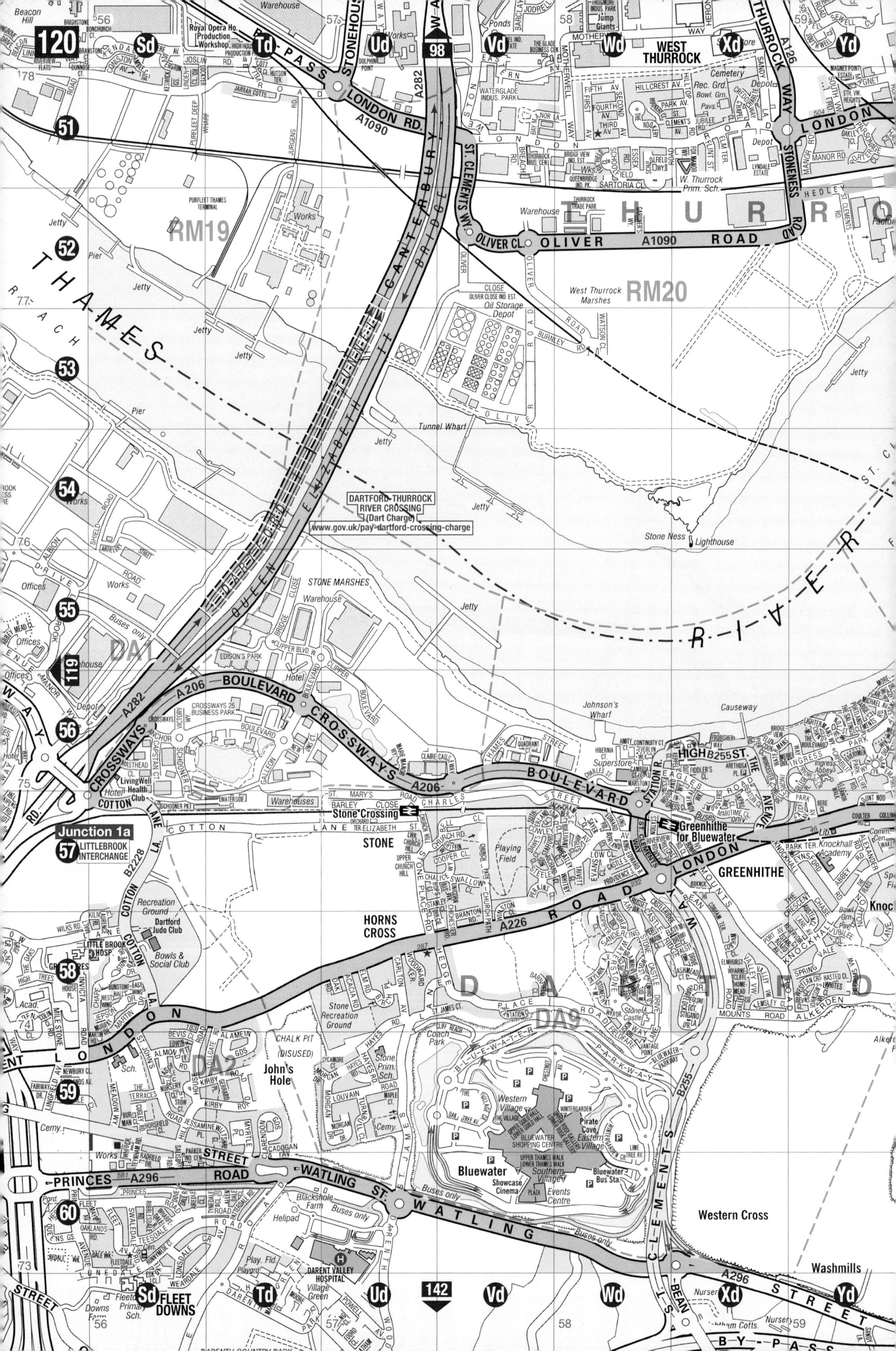

Beacon Hill
Royal Opera Ho. Production Workshop
Warehouse
98
WEST THURROCK
Jump Giants
Cemetery
Rec. Grd.
LONDON RD.
A1090
THURROCK WAY
A126
LONDON
STONENESS ROAD
W. Thurrock Prim. Sch.
THURROCK TRADE PARK
OLIVER ROAD
OLIVER CL.
A1090
RM19
PURFLEET THAMES TERMINAL
Works
Jetty
Pier
THAMES
RM20
West Thurrock Marshes
Oil Storage Depot
OLIVER CLOSE IND. EST.
BURNLEY RD.
Tunnel Wharf
QUEEN ELIZABETH II BRIDGE
DARTFORD TUNNELS
A282
DARTFORD-THURROCK RIVER CROSSING (Dart Charge)
www.gov.uk/pay-dartford-crossing-charge
Stone Ness
Lighthouse
RIVER
STONE MARSHES
Warehouse
Hotel
DA1
119
Offices
Depot
Buses only
CROSSWAYS BOULEVARD
A206
CROSSWAYS 25 BUSINESS PARK
Johnson's Wharf
Causeway
Superstore
HIGH ST.
B255
Stone Crossing
STONE
Greenhithe for Bluewater
GREENHITHE
LivingWell Health Club
Hotel
COTTON LANE
Junction 1a
LITTLEBROOK INTERCHANGE
B2228
Recreation Ground
Dartford Judo Club
Bowls & Social Club
Playing Field
HORNS CROSS
LONDON ROAD
A226
Knockhall Academy
DARTFORD
DA9
DA2
Stone Recreation Ground
CHALK PIT (DISUSED)
John's Hole
Stone Castle
Coach Park
BLUEWATER PARKWAY
Western Village
Pirate Cove
Eastern Village
BLUEWATER SHOPPING CENTRE
Bluewater
Showcase Cinema
Southern Village
Events Centre
Bluewater Bus Sta.
B255
ST. CLEMENTS WAY
Western Cross
Washmills
PRINCES ROAD
A296
WATLING STREET
Blackshole Farm
Buses only
Helipad
DARENT VALLEY HOSPITAL
Play. Fld.
Village Green
FLEET DOWNS
Fleetdown Primary Sch.
Downs Farm
Cemy.
Works
A296
STREET
BY-PASS
Sd Td Ud Vd Wd Xd Yd
51 52 53 54 55 56 57 58 59 60
142

121
Zd
Ae
Be
Ce
De
Ee
Fe
99
60
61
62
63
A126 ROAD
SOUTH STIFFORD
Works (Proposed)
Stifford Prim. Sch.
LAWRENCE TRADING ESTATE
Grays
STATION
CROWN ROAD
CLARENCE R.
DERBY RD. BRI.
B189
THURROCK RD.
DOCK
A126 ROAD
S. Essex College (Thurrock Campus)
ARGENT STREET
THAMES RD.
Thurrock Yacht Club
Grays Beach Riverside Park
Boating Pool
Thameside Prim. Sch.
Playing Field
CURZON DR.
51
52
53
54
55
56
57
58
59
60
DOCK-A1089-ROAD
ST. ANDREW
Jetty
THAMES
REACH
Lighthouse
NORTHFLEET HOPE
Broadness Salt Marshes
Causeway
Mills
Grain Terminal
Rail Container Terminal
Tilbury North Junction
Warehouse
TILBURY DOCKS
No. 1 Gate Main Entrance
Warehouses
Works
RM18
Northfleet Hope Container Terminal
Rolling Lift Bridge
Tilbury Ness
122
Pier
Bell Wharf
Swanscombe Marshes
Sewage Works
Depot
Wharf
Botany Marshes
High Speed 1
MANOR WAY
TOWER WHARF
Ingress Park
NORTHFLEET INDUSTRIAL ESTATE
Warehouses
Ebbsfleet Utd. FC (Stonebridge Rd.) (Kuflink Stadium)
A226 ROAD GALLEY HILL ROAD
Swanscombe
Recreation Ground
SWANSCOMBE
Swanscombe Heritage Park
Swanscombe Skull Site National Nature Reserve
Swanscombe Leisure Cen.
STONEBRIDGE RD.
THAMES WAY
Northfleet
HIGH STREET
B2175
GRAVESHAM
DA11
THE HILL
LONDON ROAD
Mills
Warehouse
Ebbsfleet International
EBBSFLEET
The Ebbsfleet Academy
Manor Primary Sch.
Alkerden
DA10
WHITECLIFF
EBBSFLEET A2260 GATEWAY
A226
Sewage Works
WAY THAMES
Recreation Ground
Northfleet Urban Country Park
DOVER RD.
PERRY STREET
Castle Hill
The Observatory
Cherry Orchard Prim. Acad.
Fastrack Bus Route
Ebbsfleet Valley
Weldon
SOUTHFLEET RD.
CRESSFIELD
143
SPRINGHEAD ENTERPRISE PARK
B262 ROAD
Cemy.
Recreation Ground
Northfleet Technology College
Wombwell
Play. Field
Cygnet Leis. Cen.
Wombwell Park
WATLING STREET

122
100
121
144
A
B
C
D
E
F
G
1
2
3
4
5
6
7
8
9
10
63
64
65
66
73
74
75
76
77
78
DOCK ROAD
MARSHFOOT ROAD
A126
BY-PASS ROAD
CHADWELL
Brett's Farm
Rookery
Hutts Hill
THAMES VIEW
The Gateway Academy & Prim. Free Sch.
Hobhill Farm
HOB HILL
BIGGIN LANE
SANDY LANE
Biggin Farm
Biggin
Shrove Hill Wood
Poultry Farm
Holford Farm
WEST TILBURY
RECTORY RD.
GUN LANE
Manor Farm
West Tilbury Hall
CHURCH ROAD
Gunhill Farm
COOPERS SHAW ROAD
COOPERS LA.
Hallhill Common
Fort Road Common
Parsonage Common
RM16
RM17
RM18
TILBURY MARSHES
DOCK-A1089-ROAD
WINDRUSH
Superstore
THURROCK PARK WAY
CAPSTAN CENTRE
CLIPPER PARK
Playing Field
LONDON DISTRIBUTION PARK
WINCHELSEA RD.
Tilbury FC Ground
ST. CHADS RD.
THURROCK
Tilbury Manor Jun. Sch. & Manor Inf. Sch.
King George's Field
Warehouse
Warehouses
No. 1 Gate Main Entrance
Works
ST. ANDREW'S ROAD
A1089
Tilbury Town
CALCUTTA RD.
MONTREAL RD.
DOCK RD.
TILBURY
Walton Common
West Tilbury Marshes
FORT ROAD
Aggregates Terminal
Cement Terminal
Bulk Metal Terminal
FERRY ROAD
TILBURY FREE PORT
TILBURY 2 (Proposed)
Sewage Works
RIVERSIDE RAIL FREIGHT TERMINAL
FORTRESS DISTRIBUTION PARK
RIVERSIDE APPROACH
LONDON INTERNATIONAL CRUISE TERMINAL
Tilbury Landing Stage
Tilbury Fort
Wharf
Quay
Causeway
Bill Meroy Creek
Jetty
Pier
Swing Bridge
Foot Ferry
RIVER THAMES
Northfleet
GRAVESEND
Royal Terrace Pier
TOWN PIER
Lifeboat Sta.
ROSHERVILLE
Northfleet Nursy. Sch.
CRETE HALL ROAD
OVERCLIFFE ROAD
THAMES WAY
Superstores
Superstore
LONDON ROAD
B2175
Gravesend Sailing Club
Milton Chantry Heritage Cen.
Fort Gdns.
Riverside Leis. Area
Canal Basin
MILTON
CANAL INDUSTRIAL PARK
Timber Yard
Slipways
DENTON
MILTON ROAD
A226
MILTON RD.
Gravesend
Bronte School
A226
DOVER RD.
Northfleet Urban Country Park
UCC
Copperfield Academy
PERRY STREET
EAST OLD ROAD
B261
DA11
Cemetery
WEST B261
OLD ROAD
ROCHESTER ROAD
St. John's Catholic Prim. Sch.
St. John's Catholic Comp. Sch.
Sports Ground
Playing Field
Northfleet Technology College
Wombwell Park
Cold Harbour
Bunker Gravesend
Woodlands Park
Convent
Prep. Sch.
Play. Fld.
Sports Grd.

EAST TILBURY MARSHES
101
EAST TILBURY
Sports Ground
Factory
West Cotts.
Industrial Estate
Barvills Farm
Goshem's Farm
Buckland
Bowaters Farm
Mariner Cotts.
Church Green
Coalhouse Fort & Thameside Aviation Museum
East Tilbury Marshes
Valve Compound
Water Tower
Coalhouse Point
Jetty
Wharf
RIVER THAMES
THE LOWER HOPE
GRAVESEND REACH
Higham Saltings
ME3
Causeway
Shornmead Fort
SHORNE MARSHES
Jetty
London Port Health Authority
North West Kent College
Warehouse
Milton Rifle Range
EASTCOURT MARSHES
Sewage Works
Thames & Medway Canal (disused)
LION BUSINESS PARK
Westcourt Marshes
Great Clane Lane Marshes
DALEFIELD INDUSTRIAL ESTATE
LOWER THAMES CROSSING (PROPOSED)
FILBOROUGH MARSHES
APEX BUSINESS PARK
NORALITE IND. CEN.
Hoo Junction
Fish Pond
QUEEN'S FARM
Queen's Farm Cotts.
DA12
Sports Ground
Eastcourt Cotts.
Filborough Farm
Piggery
Court Manor
King's Farm
145
CHALK
WESTCOURT
A226
Polperro

124
102
146
CRANBOURNE
CRANBOURNE PARK
Pickleherring Pond
Ranger's Lodge
Beehive Plantation
Beehive Hill
Seymours Plantation
Fiddle Covert
Russel's Pond
Russel's Plantation
Russel's Field
Cranbourne Grange
Kingsmead
Fernhill Farm
Fernhill Park Gardens
Fernhill Park
Forest Farm
The Grove
Cranbourne Court
Fernhill Cottages
Springhill
The Vale
Woodside
Crispin Cotts.
Woodside Ho.
Oak Tree Cott.
Woodend Cotts.
Wood End
Quelmans Head
Forest Gate Cottages
Forest Gate
Forest Lodge
Cranbourne Gate
Prince Consort's Workshop
Isle of Wight Pond
Queen Elizabeth II Coronation Grove
The Village
Rec. Grd.
Bowling Green
Watchoak Plantation
Richardson's Lawn
Timberlodge Hill
Albert Plantation
WINDSOR
Poets Lawn
Duncum Plantation
Sandpit Gate
HOLLY WALK
SL4
BRACKNELL FOREST
South Forest Cottage
SOUTH FOREST
Primrose Hill
Jubilee Statue
Dark Wood
Henhouse Covert
Primrosehill Covert
Prince Christian Plantation
Gravel Hill
The Royal First School
Queen's Diamond Jubilee Plantation 1897
Woodcock Pen Covert
Hilton's Covert
Mezel Hill
GREAT WINDSOR PARK
Ivy Bridge
Hawkshill Wood
Square Covert
Shooting Platforms
Oxshed Covert
Great Meadow
Menzies Plantation
Slans Hill
Shooting Platform
North Lodge
Home Farm
Home Farm Cottages
Sunninghill Park
Paddock Wood
Fountainhill Wood
Long Wood
Leiper Hill
Mill Covert
Birch Copse
Fountain Hill
Prince Consort's Gate
Red Cross Grove of Oaks
Norfolk Plantation
Norfolk Farm
Rosy Bottom
Princess Beatrice Plantation
Round Oak
Norfolk Farm Cottages
The Dawrey
Great Pond
The Rookery
Otley Cottage
The Garden House
Ascot Gate
Horse Shoe Clump
Brewer's Pond
Sunninghill Park
Pond Head
Ribblesdale Park
Buckhurst Park
Manor Hill
Dark Plantation
Long Plantation
Johnson's Pond
ROYAL ASCOT GOLF COURSE
Lower Farm
Lower Farm Bungalows
Ash Covert
Platt's Firs
CHEAPSIDE
School
Buckhurst Hill
Manorhill Cottages
Club House
Penslade Bottom
Ascot Farm
Fireball Hill
HILLTOP CLOSE
Heronsbrook Cottage
Emmet's Wood
Millhill Plantation
American Clump
Silwood Farm
Nash's Copse
Mill Pond
The Old Mill
China Island
Straight Course
NEW MILE ROAD
Nursery
The New Mile Course (Disused)
Silwood Lake
Man's Copse
Virginia Water Lodge
Harewood
Titness House
High Bridge
Tetworth Hall
SL5
Halls of Residence
Larklands
Gunne's Bridge
Deer Park
Imperial College London (Silwood Park Campus)
SILWOOD PARK
The Cedars
The Rookery
Queen's Hill Lodge
Blacknest Gate
Lodge
Blacknest
Nutfield
St. George's School
Beech Grove
Science Pk.
Swan Lake Wild Life Sanctuary
Mount Pleasant
Tittenhurst
Victory Field
Play. Fld.
Beggar's Bush
Coworth Park House
Coworth
SUNNINGHILL
Coombe Meadows
The Marist School (Prep. Sch.)
The Marist Schs. (Senior Sch.)
Fish Pond
Coworth Park Farm
Cemetery
Bowling Green
Novello Th.
Play. Flds.
The Gardens
Market Garden
SUNNINGDALE
A332
B383
B3034
A329
B3020
LONDON ROAD
SHEET ST
SUNNINGHILL ROAD
BUCKHURST ROAD
SILWOOD ROAD
MOUNTS HILL

103
147
126
Runnymede
RUNNYMEDE
COOPER'S HILL SLOPES
COOPER'S HILL
Air Forces Memorial (Runnymede Memorial)
John F. Kennedy Memorial
Magna Carta Memorial
Magna Carta Island
Priest Hill Farm
Oak Lane Cottage
Halls of Residence
Englefield Green
ENGLEFIELD GREEN
Bishops Gate
Savill Court Hotel (Closed for refurbishment)
DELL PARK
Bishopsgate School
St. John's Beaumont School
Beaumont House
Cricket Ground
Playing Field
TW20
Royal Holloway (University of London)
Ripley Springs
Egham Wick
Forest Estate
Sandylands
Castlewood House
Egham Wick Plantations
Cheeseman's Gate
RUNNYMEDE
Queenswood
Ulverscroft
ACS Egham International School
Glenwood
Huntersdale
Callow Hill
Great Wood
The Dell
Stroude Farm
Stroude Crossing
VIRGINIA WATER
GU25
Christ Church C of E Inf. Sch.
The Clockcase
Virginia Water Plantations
Broad Ride
Wick Pond
Totem Pole
Larch Bridge
Eton Bridge
VIRGINIA WATER
Botany Bay Point
FROSTFARM PLANTATION
BELVEDERE WOOD
Belvedere Fort
Worldsend Gate
Worldsend Cottage
'Roman Temple'
THE VALLEY GARDENS
Winter Garden
High Flyer's Hill
Heather Garden
Smith's Lawn (Polo & Equestrian Grounds)
Guards Polo Club
Smith's Lawn Cottages
Breakheart Hill
Flying Barn
Scarletoak Plantation
Obelisk Pond
Cumberland Obelisk
Alder Cover
Fuzzy Clump
The Savill Building
The Savill Garden
Cumberland Gate
Wilderness
Shooting Platform
Prince Consort's Statue
Harewarren Clump
Chapel Wood
Plainfields
The Royal First Sch.
King George VI Coronation Grove
Cow Pond
Royal Lodge
Chapel Lodge
Conservatory Covert
Greenbroom Covert
Main Gate Lodges
Bishopsgate Cott.
Park Place
Cookes Hill
Gore Plantation
Queen Victoria's Plantation
Spring Hill
Prince of Wales Plantation
Snow Hill
Equestrian Statue George III (Copper Horse)
Duke of York Plantation
Sultan Plantation
Prince Consort's Plantation
Liddell Plantation
Old Covert
Wild Boar Enclosure
The Tilery
Woodside
A30
A328
A308
A329
B389
LONDON ROAD
CHRISTCHURCH ROAD
BLACKNEST ROAD
ST. JUDES ROAD
PRIEST HILL
BISHOPSGATE ROAD
WICK LANE
WICK ROAD
EGHAM HILL
PRUNE HILL
BAKEHAM LANE
CALLOW HILL
STATION APP
Virginia Water
Wentworth Club
WENTWORTH GOLF COURSE (WEST COURSE)
H J K L M N P
1 2 3 4 5 6 7 8 9 10
97 98 99 500
73 72 71 70 69 68

104
WINDSOR and MAIDENHEAD
RUNNYMEDE
TW19
TW20
GU25
A
B
C
D
E
F
G
61
62
63
64
65
66
67
68
69
70
Ankerwycke Farm
Magna Carta Island
Longfield Farm
Gravel Pits
Watts Hill Cottages
B376 ROAD
WRAYSBURY
Hythe End Bridge
Gravel Pit
HERON LAKE
British Disabled Water Ski Association
Junction 13
M25
M25 MOTORWAY
The Moor
STAINES ROAD
A30
Aqueduct
Church Lammas Quarry
Church Lammas
WRAYSBURY ROAD
Queensmead Lake
B376
The Lammas Rec. Grd.
Hollybock Island
London Stone
Holm Island
Water Works
Gas Works
THE CAUSEWAY
A308
Superstore
EGHAM ROUNDABOUT
CHERTSEY
PINE TREES BUSINESS PK.
EGHAM HYTHE
Hythe Prim. Sch.
Magna Carta Sch.
Playing Field
COOPER'S HILL SLOPES
COOPER'S HILL
Air Forces Memorial (Runnymede Memorial)
Cooper's Hill Coppice
Langham's Pond
RUNNYMEDE
Runnymede Pleasure Ground
Warehouse
RIVER THAMES
RIVERSIDE
A308
ROAD
HYTHE END
Depot
Dearsley Island
Weir
Wraysbury Hall
Bellweir Lock
Runnymede Bridge
RUNNYMEDE ROUNDABOUT
BY-PASS
A30
THE GLANTY
Glanty
Langham Farm
EGHAM
Strode's Coll.
HIGH ST.
RUNNYMEDE PARK
EGHAM HILL
Englefield Lodge
Sports Ground
Comm. Cen.
Egham
Egham Orbit
Garden Cen.
Pooley Green
POOLEY GREEN RD.
VICARAGE ROAD
Sports Fld.
Glenville Farm
Egham Town F.C. (Runnymede Stadium)
THORPE LEA ROAD
B3376
Ripley Springs
Royal Holloway (University of London)
Halls of Residence
Sports Centre
Sports Ground
Playing Field
Manorcroft Prim. Sch.
Bowling Club
Research Laboratories
Rusham Crossing
WHITEHALL LA.
NEW WICKHAM LANE
THORPE LEA
Milton Park Cottages
Crown Cottages
Five Acre Farm
Great Fosters
Moat
Lake
Black Lake Farm
Nursery
Crabtree Corner
Works
EGHAM BUSINESS PARK
THORPE IND. PARK
THORPE BY-PASS
Thorpe Hay Meadow Nature Reserve
Whitehall Farm
Great Wood
Luddington House
Stroude
STROUDE ROAD
Cemetery
Eastly End
NORLANDS LANE
Offices
The Dell
Stroude Farm
Stroude Crossing
Nurseries
Spring Farm
Inf. Sch.
Sch.
Hall
COLDHARBOUR LANE
Manor Lake
The Fleet
Virginia Water
Longside Lake
Frank Muir Memorial Field
THORPE
Playing Field
Water Gardens
Woodcock Farm
Stroude Hill
ST. ANN'S HEATH
Hurst Farm
Tennis Courts
Playing Field
Thorpe Green
B389 ROAD
MILL HOUSE B388 LANE
St. Ann's Lake
Britannia Arena
Thorpe Lakes Wakeboard, Waterski & Wakesurf Centre
Thorpe Park
Home Farm
St. Ann's Park
Junction 12 (M25)
148
M3 MOTORWAY
M3
Junction 2 (M3)
St. Ann's Hill
Anchor Copse
SANDHILLS
LANE GREEN
THORPE
THORPE B388 BY-PASS
125
00
01
02
03
500
68
69
70
72
73

127
105
149
128
KING GEORGE VI RESERVOIR
STAINES RESERVOIR
QUEEN MARY RESERVOIR
ASHFORD
STAINES-UPON-THAMES
KNOWLE GREEN
LALEHAM
Birch Green
Ashford Park
Laleham Reach
Penton Hook
LALEHAM BURWAY
SPELTHORNE
TW15
TW18
TW17
KT16
LONDON ROAD
STAINES BYPASS
KINGSTON ROAD
A30
A308
A3044
A320
B376
B377
B378
CROOKED BILLET ROUNDABOUT
FORDBRIDGE ROUNDABOUT
HMP BRONZEFIELD
Ashford Hosp.
Thomas Knyvett College
Brookland College
St. James Independent Schools (Senior Boys)
The Matthew Arnold Sch.
Buckland Primary School
Laleham C of E Primary School
Staines Town FC (Wheatsheaf Pk.)
Penton Hook Lock
Penton Hook Island
Penton Hook Marina
Penton Hook Yacht Club
Shepperton Film Studios
ASHFORD MANOR GOLF COURSE
Laleham Park
Laleham Abbey
Laleham Nurseries
Shortwood Common
Shortwood Farm
Hengrove Farm
River Ash
Intake Channel
Round Copse
Manor Farm
Mountbatten Pavilion
Beomonds Farm
Abbey Lake
Pumping Station

128
EAST BEDFONT
106
127
150
Q
R
S
T
U
V
W
61
62
63
64
65
66
67
68
69
70
TW14
TW15
TW16
TW17
BEDFONT LAKES COUNTRY PARK
HMYOI FELTHAM
LOWER FELTHAM
CHATTERN HILL
ASHFORD
ASHFORD COMMON
SUNBURY COMMON
FELTHAMHILL
Littleton Common
QUEEN MARY RESERVOIR
ASHFORD MANOR GOLF COURSE
SCHOOL ROAD JUNCTION
Staines Reservoirs Aqueduct
Filter Beds
Water Works
Pumping Station
Upper Halliford
UPPER HALLIFORD
CHARLTON
LITTLETON
SHEPPERTON GREEN
Shepperton Film Studios
SUNBURY GOLF COURSE
Refuse Transfer Station
Driving Range
M3 MOTORWAY
M3
Junc.1
SUNBURY CROSS
Sunbury
Grosvenor Park
Rugby Football Ground
London Irish RFC (Hazelwood)
Vicarage Farm
Grange Farm
Manor Farm
Staines Road Farm
STAINES ROAD
FELTHAM ROAD
ASHFORD ROAD
CHERTSEY ROAD
CONVENT RD.
CLOCKHOUSE LANE
WINDMILL ROAD
UPPER HALLIFORD ROAD
UPPER HALLIFORD BY-PASS
A30
A308
A244
B3003
B377
B378
B376

129
107
130
151
FELTHAM
HANWORTH
HANWORTH PARK
Hanworth Park House
Hanworth Air Park Leisure Centre
Springwest Academy
De Brome Building
Middlesex Crematorium
Crane Park Island Nature Reserve
Twickenham Cemetery
TW2
TW13
TW12
KT8
UXBRIDGE ROAD
HAMPTON RD. WEST
GREAT CHERTSEY ROAD
STAINES ROAD
TWICKENHAM RD.
HAMPTON RD. EAST
HOUNSLOW ROAD
COUNTRY WAY
HANWORTH RD.
APEX CORNER
HOSPITAL BRIDGE ROUNDABOUT
Fulwell Park
HAMPTON GOLF COURSE
FULWELL GOLF COURSE
David Lloyd Centre
Hampton High
Hampton School
The Lady Eleanor Holles Sch.
Rectory Meadow
Kempton Nature Reserve
Staines Reservoirs Aqueduct
Lindon Bennett Sch.
RICHMOND upon THAMES
Marling Park
Hatherop Park
Carlisle Park
HAMPTON HILL
HAMPTON
Hampton
Hampton & Richmond Borough FC (Beveree Stadium)
Kempton Park
KEMPTON PARK RACECOURSE
Reservoir
Half Moon Covert
Sports Ground
Football Ground
Hydes Field
Portlane Bridge
UPPER SUNBURY ROAD
Stain Hill West Reservoir
Stain Hill East Reservoir
Grand Junction Reservoir
Kenton Court Meadow
Sunbury Court
Water Works
Filter Beds
LOWER SUNBURY ROAD
Platt's Eyot
Sunnyside Reservoir
RIVER THAMES
Grand Junction Isle
SUNBURY
Sunbury Park
Rivermead Island
Sunbury Locks
MOLESEY RESERVOIRS
ELMBRIDGE
HURST PARK
Hurst Park Open Space
Garrick's Ait
Hurst Pool
Molesey Hurst Rec. Grd.
East & West Molesey Cemy.
WALTON ROAD
BESSBOROUGH RESERVOIR
KNIGHTS RESERVOIR
Apps Court Farm
The Cottage
WEST MOLESEY
Molesey Hosp.
X
Y
Z
Aa
Ba
Ca
Da
11
12
13
14
61
62
63
64
65
66
67
68
69
70
72
73

TWICKENHAM
STRAWBERRY HILL
TEDDINGTON
HAMPTON HILL
HAMPTON
HAMPTON WICK
HAM
HURST PARK
EAST MOLESEY
HAMPTON COURT
BUSHY PARK
HAMPTON COURT PARK
FULWELL GOLF COURSE
STRAWBERRY HILL GOLF COURSE
HAMPTON COURT PALACE GOLF COURSE
Ham Lands Nature Reserve
Ham Riverside Lands
Fulwell Park
Fulwell
Strawberry Hill
Teddington
Hampton Court
St. Mary's University College
National Phy. Lab
Hampton Court Palace
The Long Water
The Royal Paddocks
Diana Fountain
Hampton Court Gate
Stud House
Teddington Lock
Teddington Weir
Canal Plantation
Round Plantation
Willow Plantation
Warren Plantation
Oval Plantation
Waterhouse Woodland Garden
Heron Pond
Deer Pen
Hampton Open Air Pool
Hampton & Richmond Borough FC (Beveree Stadium)
Molesey Hurst Rec. Grd.
Cigarette Island Park
RIVER THAMES
HAMPTON ROAD
HAMPTON COURT ROAD
WALDEGRAVE ROAD
STRAWBERRY VALE
KINGSTON ROAD
CHESTNUT AVENUE
HIGH STREET
TW12
TW1
TW11
ELMBRIDGE
KINGSTON UPON THAMES
A305 A309 A310 A311 A312 A313 A316 A308 A3050 B358 B369
Ea Fa Ga Ha Ja Ka La
61 62 63 64 65 66 67 68 69 70
108 129 152

109
132
153
Ma
Na
Pa
Qa
Ra
Sa
Ta
18
19
20
21
61
62
63
64
65
66
67
68
69
70
71
72
73
PETERSHAM
PETERSHAM PARK
SIDMOUTH WOOD
King Henry VIII Mound
Leg of Mutton Pond
Queen Elizabeth's Plantation
The Royal Ballet Lower School
Mothers Copse
Pen Ponds
Lawn Plantation
Treebox Wood
SPANKERS HILL WOOD
DEER PARK
King George V Jubilee Plantation
Pond Plantation
Oak Lodge
Pembroke Lodge (Cafeteria)
White Ash Lodge
Sudbrook Park Club House
The Wilderness
SUDBROOK PARK
RICHMOND GOLF COURSE
P A R K
(National Nature Reserve)
Tercentenary Plantation
Horse Ride
(Open during daylight hours only)
Robin Hood Gate
Pond Slade
Prince Charles's Spinney
Ham Cross Plantation
Isabella Plantation Garden
Gibbet Wood
Peg's Pond
Thomson's Pond
Broomfield Hill
Broomfield Hill Wood
ISABELLA PLANTATION
Ham Cross
Ham Gate
HAM GATE AVENUE
B352
HAM COMMON (Nature Reserve)
TW10
Ham Bottom
Ham Dip
High Wood
Still Pond
Deer Fold
Coronation Plantation
The Flat
Dann's Pond
Gallows Pond
King's Clump
Ladderstile Gate
Kingston Gate
THAMES
UPPER HAM RD.
PETERSHAM RD.
RICHMOND ROAD
A307
CASSEL HOSP.
ST. GEORGE'S INDUSTRIAL ESTATE
YMCA Sports Grd.
The Tiffin Girls' Sch.
Tiffin Girls Comm. Spts. Cen.
The Kingston Academy
Fern Hill Prim. Sch.
KT2
Latchmere Sch.
RC Prim. Sch.
Sports Ground
Recreation Ground
Steven's Eyot
Canbury Gardens
Nuffield Health Superstore
Kingston Mos.
KINGSTON HILL
A308
Kingston University (Kingston Hill Campus)
Walkden Hall
Robin Hood Prim. Sch.
KINGSTON BY-PASS
TW15
Canbury Sch.
Warren House
COOMBE HILL GOLF COURSE
COOMBE WOOD GOLF COURSE
COOMBE
Holy Cross Preparatory RC Sch.
Marymount International School
Club House
Park Hill Sch.
KINGSTON HOSPITAL
COOMBE LANE
A238
COOMBE ROAD
Coombe Hill Jun. & Inf. Schools
Coombe Girls' Sch.
TRAPS LANE
COOMBE LANE
Norbiton
NORBITON
Kingston
CROMWELL R.
The Rotunda
Bentall Centre
LONDON ROAD
CAMBRIDGE ROAD
FAIRFIELD
The Fairfield Recreation Ground
Tiffin Sch.
Super-store
KINGSTON UPON THAMES
Kingston Univ.
County Hall
Kingston Cemetery
Kingston Crematorium
Athletics Centre
Indoor Bowls Club
AFC Wimbledon (Kingsmeadow)
Recreation Ground
Sewage Sludge Beds
Hogsmill River
Waste Disposal Centre
Sewage Works
KINGSTON ROAD
A2043
Retail Park
Corpus Christi Catholic Prim. School
KT3
Islamic Cen.
Surbiton Cemetery
Berrylands
KT5
PENRHYN RD.
A240
HIGH ST.
SURBITON ROAD
SURBITON HILL
Hampton Wick
Thames Ditton
Recreation Ground
Pavilion

Ua
Va
Wa
Xa
Ya
Za
Ab
110
131
154
WANDSWORTH
PUTNEY HEATH
Putney Vale Crematorium
PUTNEY VALE CEMETERY
PUTNEY VALE
Wimbledon Windmill Museum
LONDON SCOTTISH GOLF COURSE
Queen's Mere
WIMBLEDON
COMMON
WIMBLEDON COMMON GOLF COURSE
ROYAL WIMBLEDON GOLF COURSE
Caesar's Camp
Cannizaro Park
Beverley Meads & Fishponds Wood Nature Reserve
KINGSTON VALE
Kingston University (Roehampton Vale Centre)
ROBIN HOOD
ROBIN HOOD WAY (KINGSTON BY-PASS)
KINGSTON BY-PASS
KINGSTON VALE
ROEHAMPTON VALE
STAG LANE
SW15
SW19
SW20
TW10
KT2
KT3
PARKSIDE HOSP
All England Lawn Tennis & Croquet Club
Museum
Buddhist Temple
WIMBLEDON
Wimbledon Chase
COPSE HILL
Cottenham Park
COOMBE LANE
NEW VICTORIA HOSPITAL
COOMBE GOLF COURSE
KINGSTON upon THAMES
Raynes Park
RAYNES PARK
Bushey Mead
David Lloyd Leisure, Raynes Park
Prince George's Playing Field
CANNON HILL COMMON (Recreation Ground)
Joseph Hood Recreation Ground
New Malden
NEW MALDEN
SHANNON CORNER
WEST BARNES
Motspur Pk.
A3
A238
A298
A308
A219
B281
B282
B283
B279
A2043
B235

Earlsfield
111
SW18
WANDSWORTH
Springfield Park (Est. Comp. 2023)
SPRINGFIELD UNIVERSITY HOSPITAL
Burntwood School
Garratt Green
STREATHAM CEMETERY
UPPER TOOTING
Tooting Bec
Wimbledon Park
WIMBLEDON PARK
Wimbledon Park Athletics Track
SUMMERSTOWN
AFC Wimbledon Plough La. Stadium (Opens 2020)
Lambeth Crematorium
Wimbledon Cemetery
LAMBETH CEMETERY
SW17
Tooting Broadway
St. George's Uni. of London
ST. GEORGE'S HOSPITAL
TOOTING
Haydons Road
Wandle Meadow Nature Park
TOOTING GRAVENEY
Tooting
134
Wimbledon
SOUTH WIMBLEDON
THE BROADWAY
Colliers Wood
COLLIER'S WOOD
MERTON HIGH STREET
South Wimbledon
Superstore
Merton Park
MERANTUN WAY
WESTERN ROAD
Lavender Park
MERTON
Deen City Farm
MERTON PARK
CR4
South Merton
Morden
MORDEN HALL PARK
Phipps Bridge
Morden Hall
Morden Lodge
Belgrave Wlk
Ravensbury Park
Mitcham
MITCHAM
Beehive Bridge
Canons Leisure Centre
Blue Houses Bridge
Cranmer Bri.
The Wilson Hospital
CARSHALTON
SM4
MORDEN
Morden South
Morden Park School Sports Centre
Morden Leisure Centre
155
Watermeads
Bb
Cb
Db
Eb
Fb
Gb
Hb
25
26
27
28
61
62
63
64
65
66
67
68
69
70

BALHAM
Balham
Jb
Kb
Lb
112
Mb
Nb
Pb
Qb
STREATHAM HILL
Streatham Hill
SW12
SW2
TOOTING BEC
Tooting Bec
Tooting Bec
TOOTING
COMMON
WANDSWORTH
Tooting Graveney Common
Tooting Bec Lido
BEC ROAD A214
Tooting Bec Athletic Track
Tooting Graveney Common
STREATHAM PARK
SW17
FURZEDOWN
STREATHAM
Streatham
Streatham Hub
Streatham Common
STREATHAM COMMON
STREATHAM COMMON NORTH
SW16
GREYHOUND LA.
MITCHAM ROAD A216
STREATHAM HIGH ROAD A23
Tooting
133
Streatham Common
STREATHAM VALE
Streatham Vale Park
Lonesome
MERTON
NORBURY
Norbury
NORBURY PARK
Norwood Grove Recreation Ground
South London Crematorium
Mitcham Eastfields
Streatham Park Cemetery
Jewish Cemetery
CR4
Harris Academy Merton
William Morris Primary Sch.
LONDON ROAD
MITCHAM
Canons Leisure Centre
Cranmer Bri.
The Wilson Hospital
CROYDON ROAD A236
MITCHAM
Sherwood Park
Playing Fields
156
Gonville Academy
Keith Tuckey Stadium
MITCHAM GOLF COURSE
Playing Fields
Croydon

135
113
136
157
Rb
Sb
Tb
Ub
Vb
Wb
Xb
61
62
63
64
65
66
67
68
69
70
TULSE HILL
Tulse Hill
THURLOW PARK
A205
NORWOOD RD.
A215
West Dulwich
DULWICH
SOUTHWARK
Dulwich College
DULWICH & SYDENHAM HILL GOLF COURSE
WEST DULWICH
SE21
West Norwood Crematorium
West Norwood
WEST NORWOOD
CROXTED RD.
A2199
Kingsdale Foundation Sch.
Sydenham Hill
Sydenham Hill Wood & Cox's Walk Nature Reserve
Dulwich Wood
UPPER SYDENHAM
LEWISHAM
SE26
DARTMOUTH
A2216
KIRKDALE
SYDENHAM
Sydenham
WESTWOOD HILL
A212
CRYSTAL PALACE PARK RD.
A234
Gipsy Hill
NORWOOD PARK
KNIGHTS HILL
SE27
CROWN DALE
CENTRAL HILL
A214
NORWOOD NEW TOWN
St. Joseph's RC Jun. & Inf. Schools
CRYSTAL PALACE PARK
Crystal Palace National Sports Centre
CRYSTAL PALACE
Crystal Palace Museum
Crystal Palace Athletics Stadium
Crystal Palace
Crystal Palace Dinosaurs
Penge West
Penge East
BEULAH HILL
A215
WESTOW HILL
NORWOOD
Recreation Ground
Biggin Wood
Downsview Prim. Sch.
UPPER NORWOOD
CHURCH RD.
A212
ANERLEY HILL
ANERLEY RD.
BROMLEY
SE20
Anerley
ANERLEY
PENGE
Harris City Academy
Phil Edwards PRU
David Livingstone Prim. Acad.
Sports Ground
PARCHMORE RD.
CROYDON
South Norwood Lake
Croydon Sailing Club
Orchard School Sports Centre
Spurgeon College
Grangewood Park
Birkbeck
Beckenham Crematorium
ELMERS END RD.
A213
CROYDON RD.
SOUTH NORWOOD
Harris Academy
NORWOOD RD.
A215
SOUTH NORWOOD HILL
SE25
WHITEHORSE RD.
Crystal Palace FC (Selhurst Park)
HIGH ST.
PORTLAND RD.
Norwood Junc.
Thornton Heath
THORNTON HEATH
B266
B273
Blind Corner
Pitch & Putt Course
SOUTH NORWOOD COUNTRY PARK
Croydon Sports Arena
Croydon FC Grd.
Harrington Rd.
Heavers Farm Prim. Sch.
St. Chad's Cath. Prim. Sch.

136
114
135
158
Yb
Zb
Ac
Bc
Cc
Dc
Ec
35
36
37
38
61
62
63
64
65
66
67
68
69
70
FOREST HILL
SE23
CATFORD
SE6
BELLINGHAM
BELL GREEN
LOWER SYDENHAM
SYDENHAM
SE26
NEW BECKENHAM
PENGE
SE20
ANERLEY
BECKENHAM
BR3
ELMERS END
EDEN PARK
PARK LANGLEY
SE25
Forest Hill
Sydenham
Lwr. Sydenham
Bellingham
Beckenham Hill
Penge East
Penge West
New Beckenham
Kent House
Beckenham Road
Beckenham Junc.
Clock Ho.
Avenue Road
Birkbeck
Elmers End
Harrington Rd.
BECKENHAM PLACE PARK
CATOR PARK
Mayow Park
Alexandra Rec. Grd.
Beckenham Crematorium
SOUTH NORWOOD COUNTRY PARK
Kelsey Park
BECKENHAM BEACON
DARTMOUTH ROAD
KIRKDALE
SYDENHAM ROAD
SOUTHEND LANE
BECKENHAM HILL ROAD
SOUTHEND ROAD
BROMLEY ROAD
HIGH STREET
CROYDON ROAD
ELMERS END ROAD
BECKENHAM ROAD
HIGH STREET
PARISH LANE
NEWLANDS PARK
PERRY RISE
PERRY HILL
CATFORD HILL
BELL GRN
A212
A2216
A2218
A2015
A213
A234
A222
A214
A21
B227
B230

137
115
138
159
Grove Park Nature Reserve
HITHER GREEN CEMETERY
Hither Green Crematorium
Mottingham Farm
Eltham College Jun. Sch.
Eltham Coll.
Eric Liddell Spts. Cen.
Playing Fields
Sports Ground
Bannatyne Health Club
Library
SE9
SE12
GROVE PARK
Depot
Comm. Cen.
CHINBROOK ROAD
Grove Park
Chinbrook Meadows
Drumbeat Sch.
Academy
FORSTER MEMORIAL PARK
Playing Field
SOUTHEND
Churchdown Christian Centre
Downham Fields
Downham Health & Leisure Centre
Lower Marvels Wood
Marvels Wood
Grove Park Cemetery
Castlecombe Prim. Sch.
Youth Cen.
ELMSTEAD WOOD
BR7
ELMSTEAD
DOWNHAM
Coun. Depot
Playing Fields
Shaftesbury Park
Sports Grd.
Bromley Hill Cemy.
BR1
King's Meadow Playing Grd.
Hall's Farm
Cemetery
Garden Wood
SUNDRIDGE PARK GOLF COURSE
SUNDRIDGE
Botany Bay
Elmstead Woods
Sports Ground
PLAISTOW
Playing Field
Golf Course
Sundridge Park
Scotts Park Prim. Sch.
Convent
SHORTLANDS GOLF COURSE
BROMLEY PARK
THE SLOANE BMI HOSP
Bromley Nth.
Theatre
Widmore Green
Cricket Ground
Bishop Challoner Sch.
Shortlands
Martin's Hill
Queen's Mead
Queens Gdn.
Churchill Theatre
THE GLADES
Leisure Cen.
BROMLEY
Bromley Civic Centre
Bickley Prim. Sch.
WIDMORE
Superstore
Harris Prim. Acad.
BR2
SHORTLANDS
Bromley South
Recreation Ground
Virgin Active Health Club
Whitehall Pleasure Ground
Highfield Inf. Sch.
Highfield Jun. Sch.
The Ravensbourne Sch.
Bromley Trust Academy
Playing Field
South Hill Wood
Nuffield Health Fitness & Wellbeing Centre
A21
A222
A2212
B226
B228
B251
B264
B265
B2212
Fc
Gc
Hc
Jc
Kc
Lc
Mc
61
62
63
64
65
66
67
68
69
70
39
40
41
42

Mottingham
MOTTINGHAM
Eric Liddell Spts. Cen.
Fairy Hill
116
New Eltham
NEW ELTHAM
Cricket Ground
Sports Ground
St. Olave's Prep. School
Coldharbour Leisure Centre
Playing Fields
Netball Courts
SIDCUP ROAD A20
MOTTINGHAM COURT RD.
B263 LANE GREEN
GREENWICH
SE9
A208
A211
Dulverton Sch.
Sidcup Family Golf
Cray Wanderers FC Est. Comp. 2021
LONGLANDS
Pitch & Putt Course
Kemnal Park Cemetery
Flamingo Park
Ashen Grove
Sports Ground
Playing Field
Rec. Grd.
Lower Marvels Wood
Marvels Wood
Castlecombe Prim. Sch.
Youth Cen.
ELMSTEAD WOOD
Sports Ground
ELMSTEAD
Edgebury Prim. Sch.
Partridge Shaw
Nuffield Health & Fitness Centre
Foxbury
Playing Field
Old Elthamians Sports Club
Chislehurst Sch. for Girls
Athletic Ground
Darul Uloom London
Mead Rd. Inf. Sch.
Red Hill Prim. Sch.
CHISLEHURST WEST
Whyte's Woodland
WHITE HORSE HILL
RED HILL
HIGH ST.
BR7
Elmstead Woods
Botany Bay
137
Rockpit Wood
Kyd Brook
CHISLEHURST
CAMDEN PARK
CHISLEHURST GOLF COURSE
CHISLEHURST COMMON
Camden Place Club House
The Cedars
Hangman's Cnr.
War Mem.
BROMLEY A222 LANE PERRY
BROMLEY RD.
ROYAL PDE.
ST. PAUL'S CRAY ROAD
Farringtons School
Icehouse Wood
PARK WOOD
Hadresham
The Fantails
Holbrook Wood Lodge
The Larches
St. Peter & St. Paul RC Prim. Sch.
Bullers Wood Sch.
Bullers Wood School for Boys
Cricket Grd.
Caves
Chalkpit Wood
Coopers Sch.
Marjorie McClure School
Bansome Wood
Rifle Range
Hawkwood Nursery
Birch Wood
Gosshill Wood
Avalon
Pond Wood
Hawkwood Estate
Hawkwood Farm
Tongs Farm
Walk Wood
The Gorse Children's Home
St. Paul's Cray Common
BROMLEY
Chislehurst
SUMMER HILL
OLD HILL
WATTS LA.
MANOR PARK RD.
B264
CHISLEHURST ROAD
BICKLEY ROAD
BICKLEY
Hadlow
BICKLEY A222 PARK
PAGE HEATH LA.
B265
Bickley
Bickley Junc.
Chislehurst Junction
Great Thrift
PETTS WOOD
Edlmann Memorial Wood
Willett Memorial Wood
St. Mary Cray Junction
ORPINGTON RD.
Virgin Active Health Club
Depot
BR1
Blackbrook Wood
JUBILEE COUNTRY PARK
Bromley High School
Petts Wood Junction
BR2
Whitehall Pleasure Ground
Thornet Wood Country Nature Res.
Sydney Shaw
160
SOUTHBOROUGH
PETTS WOOD
Nc
Pc
Qc
Rc
Sc
Tc
Uc
42
43
44
45
61
62
63
64
65
66
67
68
69
70

139
Lamorbey
Rose Theatre
Lamorbey Park
Rose Bruford College
All Weather Pitch
Hurst Pl. (Com. Cen.)
Cricket Ground
The Glade
Chislehurst & Sidcup Grammar Sch.
Sidcup Leisure Centre
HURST ROAD
A222
Abbeyhill Park
Albany Park
Rutland Shaw
Sidcup Cemy.
Sidcup
DA15
Playground
King George's Rec. Grd.
Marlowe Ho. Offs.
STATION ROAD
Birkbeck Prim. Sch.
WARING PARK
Royal Park Prim. Sch.
Playing Field
The Alders
Woollett Hall Farm
Crayzee Barn
Works
Sports Ground
Loring Hall
Nursery
Bunkers Hill
Pond Farm
Home Wood
Gattons Wood
Little Mana Farm
DA14
Cleeve Pk. Sch.
North Cray Woods
Playing Field
Stable Meadow
Five Arch Bri.
FOOTS CRAY MEADOWS (Nature Reserve)
Foots Cray Place
Dower House
NORTH CRAY
Nurseries
Manor Cotts
Manor Farm
North Cray Poultry Farm
SIDCUP
Sports Ground
A211
ROAD
SIDCUP HIGH ST.
THE GRN.
A2001
Sidcup Green
Sidcup Place
Sidcup Place Park
Christ The King (St Mary's) Sixth Form College
Playing Fields
David Lloyd Leisure
Cray Youth & Community Centre
Catholic Prim. Sch.
Warehouse
FIVE ARCHES BUSINESS PARK
Hope Comm. Sch.
SIDCUP TECHNOLOGY CENTRE
Pitch & Putt
QUEEN MARY'S HOSPITAL
UCC
Garden Centre
Orchard Prim. Sch.
Football Grd.
FOOTS CRAY
MAIDSTONE RD.
NORTH CRAY RD.
FROGNAL CORNER
Frognal Ho.
Chestnut Ho.
Spring Shaw
The Lodge
Black Clump
Bank's Shaw
Little Wood
Square Clump
SCADBURY PARK (Nature Reserve)
SIDCUP BY-PASS
Superstores
EDGINGTON WY.
A223
B2173
Ruxley Corner
RUXLEY
RUXLEY WOOD
Ruxley Manor Garden Centre
Superstore
A20
Kemnal Tech. College
CRITTALLS CORNER
Irish Plantation
Midfield Prim. Sch.
Moat
Archway
Youth Club
Playing Field
SEVENOAKS WAY
A224
Grays Farm Cottages
Bannatyne Health Club (Orpington)
RUXLEY PARK GOLF COURSE
Driving Range
ORPINGTON GOLF CENTRE
Club House
CRAY VALLEY GOLF COURSE
A20 BY PASS
Caravan Site
Upper
Gray's Farm Prim. Acad.
Home Farm Cotts.
Home Farm
Riverside Sch.
Recreation Ground
Garland Hill
The Lodge
Pilgrim Manor
Holly Cottage
The Gardens
Chestnut Cottage
Barnfield Bank
Lower Hockenden
Simon's Field Shaw
The Small House
PAULS CRAY HILL PARK
Highlands Wood
HOCKENDEN WOOD
Fairmile Ho.
ST. PAUL'S CRAY
Recreation Ground
Leesons Prim. Sch.
St. Paul's Cray C of E Prim. Sch.
BR5
Warehouse
St. Mary Cray
Cemetery
The Warren
Sheepcote Farm
POVEREST
Poverest Prim. Sch.
ST. MARY CRAY
B258
BLACKSMITH'S LANE
Manor Oak Primary Sch.
Child's Wood
117
140
161
168
Vc Wc Xc Yc Zc Ad Bd
61 62 63 64 65 66 67 68 69 70

Coldblow
Maypole
OLD BEXLEY LA.
DARTFORD
Leyton Cross
Heath Side
THE BRACTON CENTRE
Woollett Hall Farm
Crayzee Barn
Joydens Wood
JOYDENS WOOD (Woodland Trust)
DA5
Hook Green
BEXLEY
Bunkers Hill
Gattons Plantation
Gattons Wood
Manor Farm
Puddledock
CHALK WOOD
DA14
STONEHILL WOODS PARK
BIRCHWOOD PARK GOLF COURSE
Stonehill Green
Upper Ruxley
HEXTABLE
Hextable Park
Hextable Heritage Centre & Gardens
The Howard Venue
Sports Ground
Miniature Railway
SWANLEY PARK
White Oak Leis. Cen.
Horizon Primary Academy
Burnt House Farm
Hockenden Lodge
Hockenden
Hockenden House
BROMLEY
HOCKENDEN WOOD
BOURNE WOOD
Windmill Hill Wood
Orchards Academy
Swanley Adult Educ. Centre
SWANLEY
St. Mary's C of E Prim. Sch.
Swanley
Downsview Primary Sch.
High Firs Prim. Sch.
Hook Spring
Tile Kilns Wood
Marwood Ho. Caravan & Camping Site
Sheepcote Farm
Child's Wood
BR8
SEVENOAKS
BY PASS
SWANLEY BY-PASS
LONDON ROAD
B2173
A20
A2018
A223
B258
A2
118
139
162
49
50
51
52
68
69
70
71
72
73

119
142
163
Kd
Ld
Md
Nd
Pd
Qd
Rd
53
54
55
56
68
69
70
71
72
73
DARTFORD GOLF COURSE
Brickhill Wood
Club Ho.
Sports Hall
Miskin Theatre
North West Kent College of Technology
Oakfield Prim. Sch.
Pav. Oakfield Park
Comm. Cen.
Superstore
Brooklands Lake
David Lloyd Leisure
Ten. Cts.
Dartford FC (Princes Pk.)
Buses only
The Leigh Academy
Football Pitch
Goals Soccer Cen.
Playing Field
Pitches
PRINCES PARK GOLF COURSE
ORBITAL ONE INDUSTRIAL ESTATE
Fleetdown Primary Sch.
Downs Farm
DARTFORD TRADE PARK
QUESTOR
Weirs
Blackdale Farm Cottages
Blackdale Farm
Trolling Down Hill
DARENTH INTERCHANGE
DA1
DA2
DA4
Wilmington Prim. Sch.
Rec. Grd.
Claremont Cotts.
Holland's Farm
Hawley Manor
Day's Farm
A2
A225
A282
B258
B260
M25
Junction 2
Junction 3 (M25)
DARTFORD BY-PASS
ST. HAWLEY ROAD
CHURCH HILL
OAKFIELD LANE
HIGH ROAD
WILMINGTON
Bybow Farm
Shirehall Farm Cottage
Lower Barnend
Upper Barnend
LANE END
HIGH STREET
SHIREHALL ROAD
BURNT HOUSE LANE
HAWLEY
King George's Field
Hawley Garden Centre
Yewtree Farm Nursery
Works
Nursing Home
Peter Gidney Centre
Darenth Grange
DARENTH
Darenth Court Farm
Depot
Old Mill Farm
The Alders
Summit Farm
Broad Oak
Portway Nursery
PARSONAGE LANE
M25 MOTORWAY
MAIN ROAD
Rec. Ground
Pav.
Cemy.
Sutton-at-Hone C of E Prim. Sch.
SUTTON AT HONE
Lib.
St. John's Jerusalem
RIVER DARENT
SOUTH DARENTH
Chapel Field Cottages
BARN ROAD
B258 ROAD
HIGHFIELD COTTAGES
Playing Field
Wiffen's Farm
Sutton's Cotts.
Clement Street
Clement Ho.
Nursery
Old Mill Equestrian Centre
North View Cotts.
CLEMENT STREET
ARNOLD'S LANE
CHURCH ROAD
Swanley Bottom
Holt's Farm
Gildenhill Farm
GILDENHILL ROAD
St. Paul's C of E Prim. Sch.
SWANLEY VILLAGE
The Orchard
THE OLD STABLE YARD
THE PADDOCKS
SHIP LANE
SWANLEY VILLAGE RD.
BUTTON STREET BUSINESS CENTRE
Old Place Farm Nurseries
Tweed Hill
Franks Park
Rams Wood
Canada Heights Motorcycle Circuit
Homefield Cotts.
Homefield Farm
HOMEFIELD FARM ROAD
The Old Bungalow
The New Bungalow
Farningham Road
STATION
Viaduct
HORTON KIRBY TRAD. EST.
Highcroft
Creswick Nurseries
New Cotts.
Parkwood Hall Co-Operative Academy
Bowling Green
The Olympic Golf Driving Range
Broom Hill
FARNINGHAM WOOD
BUTTON STREET
CALFSTOCK LANE
The Leas
Franks Farm
Franks Hall
Little Franks
Court Lodge
Westminster Field
Grave-yard
OLD PARSONAGE YARD
Reynolds Pl.
RUSSELL TERRACE
HORTON ROAD
FORGE
Hill
61
62
63
64
65
66
67
68
69
70

142
120
164
141
Western Cross
Washmills
WATLING STREET
A296
A2
B255
B260
B262
BEAN BY-PASS
DARENT VALLEY HOSPITAL
Village Green
FLEET DOWNS
Fleetdown Primary Sch.
Downs Farm
DARENTH COUNTRY PARK
Cemetery
Cemy.
Arrow Riding Centre
DARENTH WOOD
Trolling Down Hill
Blackdale Farm
DARENTH INTERCHANGE
Junction 2
Badgers Mount
JJ's Clay Shooting Club
DA2
DA4
Bean Farm
Bean House
Hope Cotts.
Nurseries
Nursery
The Thrift
BEAN
Bean Prim. Sch.
Shellbank
BEACON WOOD COUNTRY PARK
LADIES WOOD
LORDS WOOD
Waller Park
Lane End
Peter Gidney Centre
Darenth Grange
DARENTH
Darenth Court Farm
Depot
The Alders
Swyres Farm
Darenth Hall
Darenth Prim. Sch.
Playing Field
Manor Farm
Warrigal Farm
SANDBANKS
GREEN STREET GREEN
GREEN STREET GREEN ROAD
Rec. Ground
St. John's Jerusalem
St. Margarets
St. Margaret's Farm
Malt House Farm
Meadowcroft
Walnut Tree Farm
Grubb Street
Ryecrofts Farm
Ryecrofts Wood
Gill's Farm
SOUTH DARENTH
RIVER DARENT
ROMAN VIA ROAD
Works
Pinden
Pinden End Farm
Sports Ground
East Lodge
Rabbits Farm
Viaduct
Horton Kirby Environmental Studies Centre
Sports Ground
Dean Bottom
Whitefield
Court Lodge
Westminster Field
HORTON KIRBY
Horton Kirby Environmental Education Centre
Reynolds Place Farm
Beeches Farm
Hill Barn Farm
Little Fran
SEVENOA
DART
A225
ROAD
56
57
58
59
68
69
70
71
72
173

143
121
165
144
Weldon
CRESSFIELD
DA10
WATLING STREET
A2
Wombwell Park
Northfleet Technology College
SPRINGHEAD ROAD
DA11
NEW HOUSE
Pepper Hill
Superstore
Millbrook Garden Centre
Refuse Tip
STATION ROAD
B262
BETSHAM
Northend
Warwick House
Betsham House
Highfield Farm
Manor House Farm
Orchard House
FORD
GRAVESHAM
Northfleet Green
Northfleet Green Farm
SOUTHFLEET
DA13
Village Hall
Court Hill Cottages
Chapter Farm
Broadditch Farm
Redstreet
Madam Wood
Hook Green Farm
Hook Green
Hook Place Fm.
Westwood Farm
Westwood
WESTWOOD
HOOK GREEN
Thomas's Cottages
Brakefield Cottage
BROAD DOWNS
DITCH HILL ROAD
ISTEAD
Willerby Farm
Court Wood
Elmlands Shaw
Rose Farm
Bushfield Shaw
WHITEHILL ROAD
B255
Mile End Green
LONGFIELD
B260
MAIN ROAD
Longfield
Longfield Academy
Rowhill School
Langafel C of E Prim. Sch.
DA3
NEW BARN
Thirty Acre Shaw
Walnut Wood
WALNUT TREE AVENUE
Reservoirs (covered)
Nurstead Hall
LONGFIELD HILL
Gorse Wood
Cobham Valleys
HARTLEY
Steephill School
Hartley Green
HARTLEY WOOD
Cricket Ground
Manor Farm
Skinners Hill
Dean Bottom Farm
Bramblefield Wood
Playing Field
Zd
Ae
Be
Ce
De
Ee
Fe
60
61
62
63
64
65
66
67
68
69
70
71
72
73

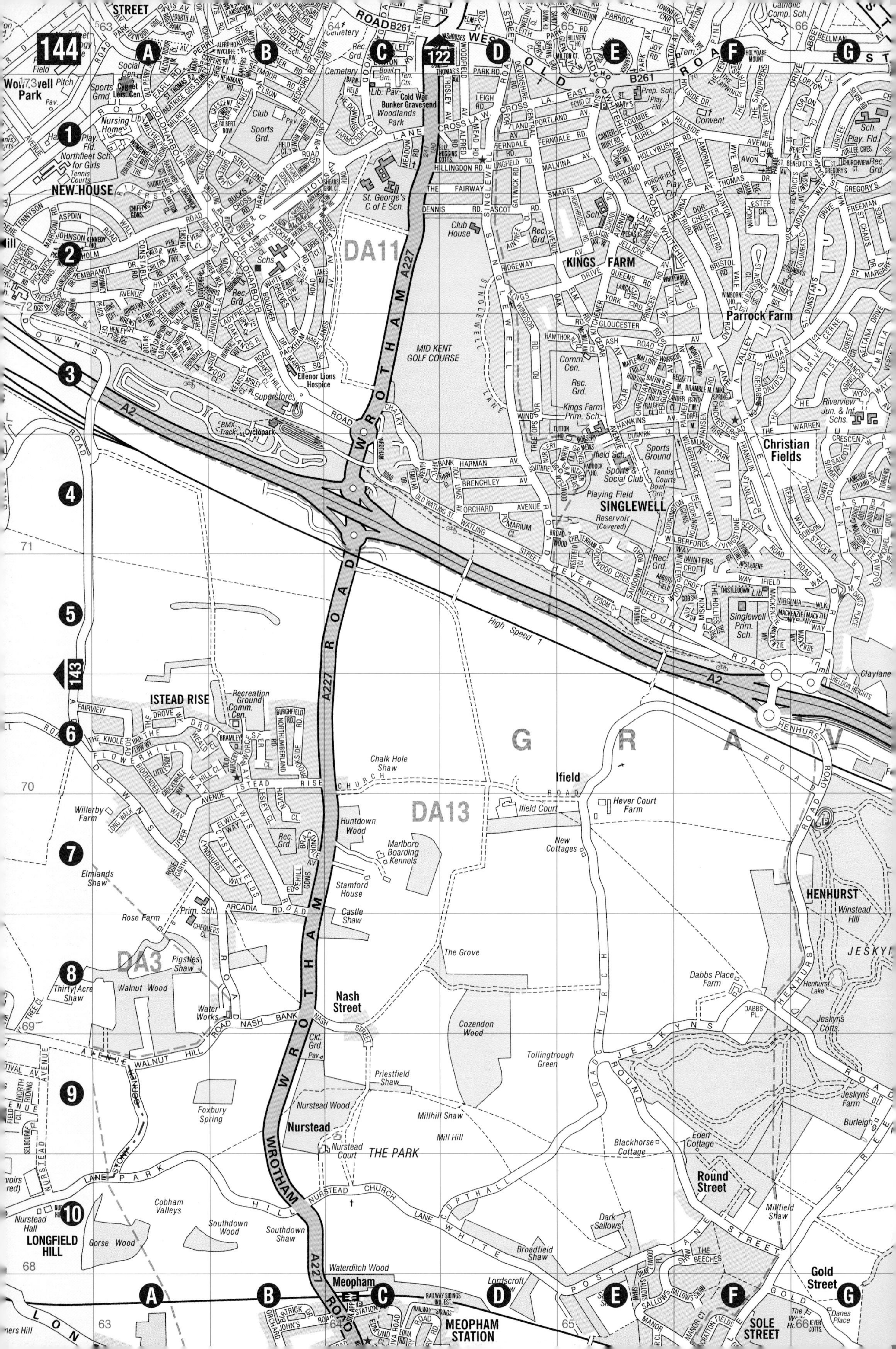
122
143
A
B
C
D
E
F
G
1
2
3
4
5
6
7
8
9
10
63
64
65
66
68
69
70
71
72
DA11
DA13
DA3
WROTHAM ROAD
A227
A2
High Speed 1
B261
WEST
OLD ROAD
EAST
NEW HOUSE
KINGS FARM
SINGLEWELL
Parrock Farm
Christian Fields
Woodlands Park
Cold War Bunker Gravesend
MID KENT GOLF COURSE
Ellenor Lions Hospice
Superstore
Cyclopark
BMX Track
St. George's C of E Sch.
Northfleet Sch. for Girls
Nursing Home
Sports Grd.
Club House
Kings Farm Prim. Sch.
Ifield Sch.
Sports & Social Club
Playing Field
Reservoir (Covered)
Singlewell Prim. Sch.
Riverview Jun. & Inf. Schs.
Claylane
SHELDON HEIGHTS
HENHURST ROAD
G R A V
ISTEAD RISE
Recreation Ground
Comm. Cen.
Ifield
Ifield Court
Hever Court Farm
New Cottages
Chalk Hole Shaw
Huntdown Wood
Marlboro Boarding Kennels
Stamford House
Castle Shaw
Willerby Farm
Elmlands Shaw
Rose Farm
Prim. Sch.
Pigstles Shaw
Thirty Acre Shaw
Walnut Wood
Water Works
Nash Street
NASH BANK
WALNUT HILL
Ckt. Grd.
Priestfield Shaw
Nurstead Wood
Nurstead
Nurstead Court
THE PARK
Foxbury Spring
Millhill Shaw
Mill Hill
The Grove
Cozendon Wood
Tollingtrough Green
Blackhorse Cottage
Dark Sallows
Broadfield Shaw
Lordscroft
HENHURST
Winstead Hill
Henhurst Lake
Jeskyns Cotts.
Jeskyns Farm
Burleigh
Dabbs Place Farm
DABBS PL.
JESKYNS ROAD
Eden Cottage
Round Street
Millfield Shaw
Cobham Valleys
Southdown Wood
Southdown Shaw
Gorse Wood
Nurstead Hall
LONGFIELD HILL
PARK LANE
NURSTEAD CHURCH HILL
COPTHALL
WHITE POST LANE
Waterditch Wood
Meopham
MEOPHAM STATION
RAILWAY SIDINGS IND. EST.
Gold Street
SOLE STREET
Danes Place
THE BEECHES
SALLOWS SHAW
Woodlands Park
Wombwell Park
Cemetery
Rec. Grd.
Convent
Catholic Comp. Sch.
HOLYOAKE MOUNT

HIGHAM
CHALK
WESTCOURT
A226
GRAVESEND ROAD
GRAVESEND GOLF CENTRE
SOUTHERN VALLEY GOLF COURSE
Cascades Leisure Centre
DA12
RIVERVIEW PARK
Lower Shorne
SHORNE
SHORNE RIDGEWAY
THONG
RANDALL WOOD
SHORNE WOODS COUNTRY PARK
Shorne Woods Country Park Visitor Cen.
BREWERS WOOD
A2
Junction 1
High Speed 1
Ashenbank Wood (Ancient Woodland)
ROCHESTER AND COBHAM PARK GOLF COURSE
Cobham Hall
COBHAM PARK
ME2
Darnley Mausoleum (Cobham Mausoleum)
MEDWAY
Battle Street
COBHAM
New College of Cobham
123

124
166
A
B
C
D
E
F
G
1
2
3
4
5
6
7
8
9
10
93
94
95
96
168
67
66
65
64
63
SUNNINGHILL
Coombe Meadows
Beggar's Bush
Pleasant
Coworth Park House
Coworth Park Farm
The Marist Schs. (Senior Sch.)
Novello Th.
Play. Flds.
Market Garden
SUNNINGDALE PARK
SUNNINGDALE
Farm Cottage
Cemetery
Bowling Green
Works
Resr. (covered)
Lynwood Court
SL5
Prim. Sch.
The Devil's Highway
Broomhall Farm
Broomhall
Broom Hall
Wardour Lodge
Rec. Grd.
Ten. Cts.
Sunningdale
Shrubs Hill
Comboni Missionaries
Playing Fields
Sunningdale Sch.
Walter's Cott.
Playing Field
Play. Fld.
Heathermount
Charters School
Charters Leisure Centre
WINDSOR and MAIDENHEAD
Hotel
Lowood House
Forest Ridge
King's Beeches
Cedar House
West Lodge
Broadlands
Clock House
Broadlands Farm
Skears
Old Windsor Bog
The Belt
Keeper's Cottage
Nursery Court
Thankerton Court
Richmond Nurseries
Garden Centre
Kingswood Lodge
Windsor Manor
King's Hill
Hills End
Middleton
Kingsmoor
Park Hill
Club House
Club Ho.
Hostel
Laversholt Hall
Laversholt Court
Windlesham Cottage
Windlemere
Little Court
Sunning Ho.
SUNNINGDALE LADIES GOLF COURSE
SUNNINGDALE GOLF COURSE
Wood Hall
Lodge
Windlesham Hall
Windlesham Hall Fm.
Braids
Green Lanes
Merrywood
Warren Lodge
Mount Cott.
Lennoxwood
Teesdale
High Chimneys
Westlands
Ruddocks Ridge
Long Down
SURREY HEATH
Windlesham Ct.
Snow's Ride Farm
Nursery
Upper Ribsden
Pears Hill
Allerton Hill
Westwood Lodge
Westwoods
Ribsden Holt
Updown Hill House
Brick Hill
Birch Hill
Valley End Farm
Coworth-Flexlands School
Playing Fld.
Hatton Hill
Camus
GU20
Adbury House
Offices
Leech Pond
Playing Field
Woodcote House School
Oakwood
Windlesham House
WINDLESHAM
Heathpark Wood
Valley End
Westley Green
The Old Vicarage
Valley Wood Place
Shergolds Stud Fm.
Sparrow Row
Highams Fm.
Valley End C of E Infant Sch.
Pembroke House
Russell Hill
Oak Wood
Birch Hall
Ashleigh Farm
Cricket Grnd.
Pavilion
Windlesham Park
Woodlands Farm
Park Wood
White Bridge
Westcroft Park
Westcroft Farm
Twelve Oaks
Clappers Brook
Steep Acre Farm
Nursy.
GU18
M3
M3 MOTORWAY
A30
A330
B3020
B383
B386
LONDON ROAD
DEVENISH ROAD
HIGH ROAD
BROOMHALL LANE
STATION ROAD
CHOBHAM ROAD
WESTWOOD ROAD
HATTON HILL
KENNEL ROAD
UPDOWN HILL
CHERTSEY ROAD
WOODLANDS LANE
WINDLESHAM ROAD
BEDFORD LANE
CHURCH ROAD
RIDGEMOUNT ROAD
Chobham Pl

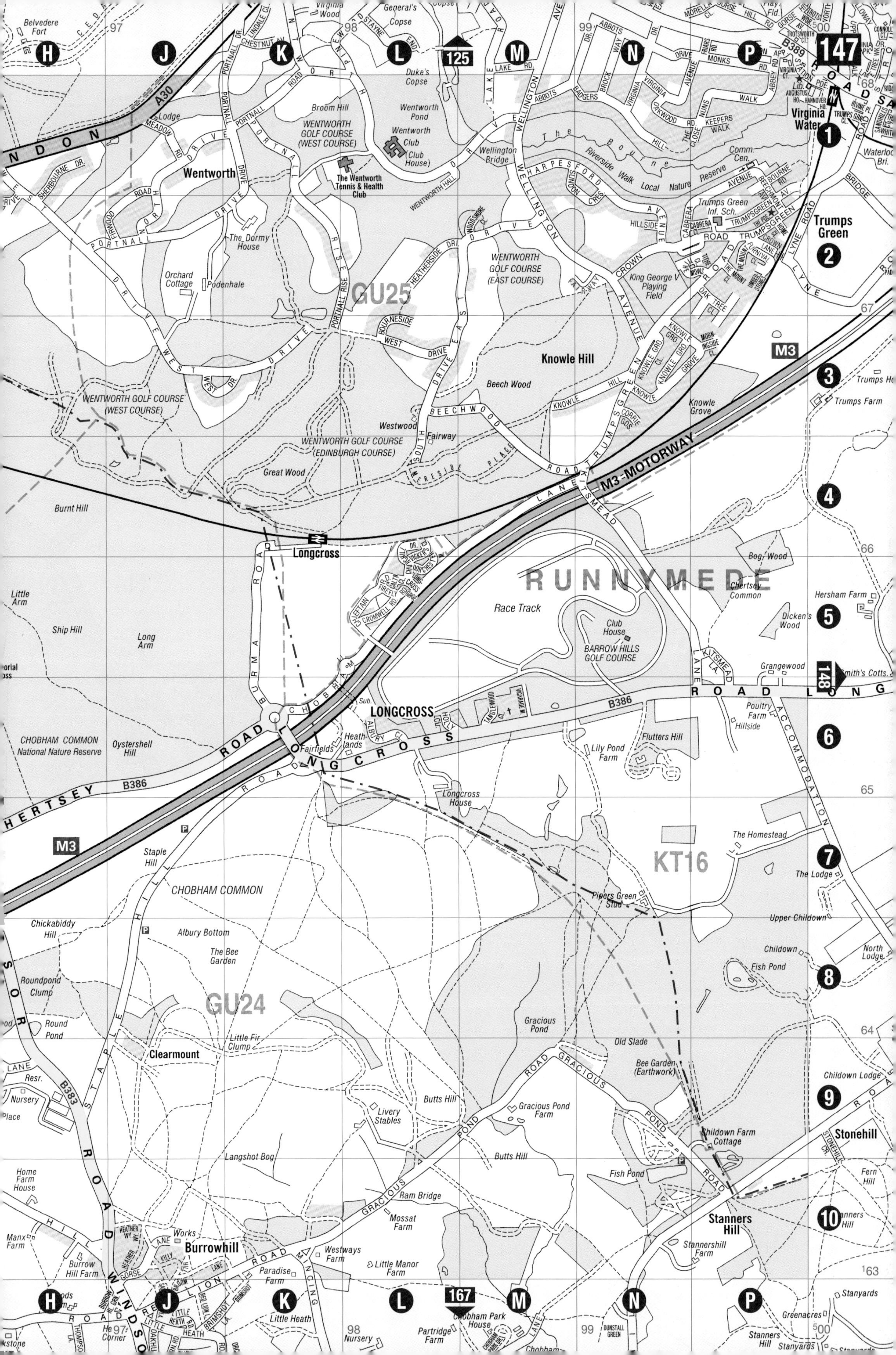

147
125
167
148
H
J
K
L
M
N
P
1
2
3
4
5
6
7
8
9
10
97
98
99
500
68
67
66
65
64
63
Belvedere Fort
Virginia Wood
General's Copse
Duke's Copse
Broom Hill
Wentworth Pond
WENTWORTH GOLF COURSE (WEST COURSE)
Wentworth Club (Club House)
The Wentworth Tennis & Health Club
Wentworth
Lodge
Wellington Bridge
Virginia Water
Waterloo Bri.
Trumps Green
Trumps Green Inf. Sch.
Comm. Cen.
Riverside Walk Local Nature Reserve
The Bourne
The Dormy House
Orchard Cottage
Podenhale
GU25
WENTWORTH GOLF COURSE (EAST COURSE)
King George V Playing Field
Knowle Hill
Beech Wood
Knowle Grove
Trumps Farm
Westwood
Fairway
WENTWORTH GOLF COURSE (EDINBURGH COURSE)
Great Wood
WENTWORTH GOLF COURSE (WEST COURSE)
Burnt Hill
Longcross
M3 MOTORWAY
M3
Bog Wood
RUNNYMEDE
Chertsey Common
Race Track
Club House
BARROW HILLS GOLF COURSE
Hersham Farm
Dicken's Wood
Grangewood
Smith's Cotts.
Little Arm
Ship Hill
Long Arm
CHOBHAM COMMON National Nature Reserve
Oystershell Hill
LONGCROSS
Heathlands
Fairfields
Poultry Farm
Hillside
Flutters Hill
Lily Pond Farm
Longcross House
B386
CHERTSEY ROAD
LONGCROSS ROAD
CHOBHAM ROAD
BURMA ROAD
KITSMEAD LANE
ACCOMMODATION ROAD
The Homestead
KT16
The Lodge
Staple Hill
CHOBHAM COMMON
Chickabiddy Hill
Albury Bottom
The Bee Garden
Pipers Green Stud
Upper Childown
Childown
Fish Pond
North Lodge
Roundpond Clump
Round Pond
GU24
Gracious Pond
Old Slade
Bee Garden (Earthwork)
Little Fir Clump
Clearmount
Childown Lodge
Stonehill
Resr.
Nursery
B383
Butts Hill
Livery Stables
Gracious Pond Farm
GRACIOUS POND ROAD
Childown Farm Cottage
Fern Hill
Home Farm House
Langshot Bog
Butts Hill
Fish Pond
Ram Bridge
Mossat Farm
Stanners Hill
Stannershill Farm
Manx Farm
Burrow Hill Farm
Works
Burrowhill
Westways Farm
Little Manor Farm
Paradise Farm
WINDSOR ROAD
STAPLE HILL
Little Heath
Chobham Park House
Partridge Farm
Nursery
Stanyards
Greenacres
Stanners Hill

148
126
147
168
A
B
C
D
E
F
G
HEATH
Thorpe Green
Home Farm
St. Ann's Park
Virginia Water
Hurst Farm
Playing Field
MILL HOUSE LANE
St. Ann's Lake
Thorpe Lakes Wakeboard, Waterski & Wakesurf Centre
Thorpe Park
M3 MOTORWAY
M3
Junction 2 (M3)
Junction 12 (M25)
THORPE
SANDHILLS ROAD
B389
LANE GREEN ROAD
St. Ann's Heath Jun. Sch.
Waterloo Bri.
Trumps Green
GU25
TRUMPSGREEN
Sewage Disposal Works
Community Recycling Centre
Anchor Copse
St. Ann's Hill
St. Ann's Hill Park
The Dingle
Hill Cottages
Ruxbury Farm
St. Ann's Court
Monk's Grove
Druidswood
B388
ROAD
Redlands
Lyne Farm
LYNE CROSSING
LYNE LANE
Lyne Grove Cotts.
Lyne Hill House
Nursery
ELM FARM CARAVAN PARK
RUXBURY ROAD
The Grange
Chilseygreen Farm
Chertsey Nursy. Sch.
M25
Trumps House
Trumps Farm
Lyne Place Manor
The Rough
Merrywood Farm
Almner Priory
Almner's Farm
Piggery
Lyne Copse
Village Hall
Pantiles Farm
Rec. Grd.
LYNE
ALMNERS
KT16
M25 MOTORWAY
Hersham Copse
Great Cockcrow Railway
Cockcrow Hill
Fan Grove
FANGROVE PARK
Playing Field
Trys Hill Farm
Lynwood Farm
Bog Wood
Chertsey Common
Hersham Farm
Pavilion
HARDWICK LANE
Hardwick Cott.
Hardwick Court Farm
R U N
Lyne & Long Cross C of E Inf. Sch.
Angel Farm
Lyne Farm
Arbon Grove
Silverlands
B386
Trys Hill
LONGCROSS ROAD
Smith's Cotts.
Poultry Farm
Hillside
ABRAHAM COWLEY UNIT
B386 HILL
HOLLOWAY
White Lodge
RUNNYMEDE BMI HOSP.
ST. PETER'S HOSPITAL
France Farmhouse
Silverlands Farm
Salesian Sch.
Chertsey South
FOXHILLS GOLF COURSE
GUILDFORD ROAD
ACCOMMODATION ROAD
The Homestead
Club House
Fox Hills
Foxhills Country Club
Botleys Mansion
Homewood Park
Home Wood
Offices
Parklands Manor
A320
ST. PETER'S
Church Farm
The Lodge
Foxhills Lodge
Kitchenride Corner
Upper Childown
Dower House
Budds Cottage
North Lodge
Childown
Fish Pond
Kitchenride Cotts.
Penny Farthing Cottage
Ether Hill
Playground
Memorial Fields
Ten Acres
Tennis Courts
STONEHILL ROAD
Stonehill Cottage
Stone Hill
Childown Lodge
Merlewood
QUEENWOOD GOLF COURSE
Coach House
The Paddock
Queenwood House
OTTERSHAW
B3121
MURRAY ROAD
Great Grove Farm
Stonehill
Club House
A319
CHOBHAM ROAD
Chaworth Copse
Timber Hill
The Common
Ottershaw Chase
Fern Hill
GU24
Stanners Hill
SURREY HEATH
Hawks Hill
Hawks Hill Cottage
Timber Hill
South Lodge
OTTERSHAW PARK
Playing Field
Meath School
Bousley Farm
Ottershaw C of E Inf. & Jun. Schs.
Piggery
Otterhill Nursery
Southern Wood Farm
Brox
Hill Crest
Tennis Courts
The Mansion
Ottershaw Park
Miniature Rifle Range
Long Copse
Timber Hill Cotts.
Greenacres
Stanyards
Bramshill
Brox Wood
Bourne Rise Farm
Rodwell Farm
71
72
73
74
75
76
77
78
79
80
67
66
65
64
63
500
01
02
03

H
J
K
L
M
N
P
127
LALEHAM
BURWAY
Reservoir
Laleham Park
Pumping Station
Pavilion
M3 MOTORWAY
M3
TW18
SPELTHORNE
Gravel Pit
Works
Pool End
Sheepwalk Lake
ABBEY MEAD
Abbey Bridge
Abbeyfields
Abbey Chase
Abbey Chase Farm
Abbey River
RIVER THAMES
Orchard Cen.
Moat
ST. ANN'S ROAD
LONDON ST.
BRIDGE RD.
GUILDFORD ST.
Cricket Grd.
Chertsey Town FC Grd.
Gogmore Farm Park
CHERTSEY
River Bourne Health Club
Play. Fld.
EASTWORTH
Eastworth
Chertsey Lock
CHERTSEY BRIDGE RD.
Chertsey Bridge
CHERTSEY ROAD
B375
TW17
Riversleigh Farm
Mead Farm
The Range
DOCKETT EDDY LANE
Dumsey Eyot
Chertsey Meads Local Nature Reserve
Chertsey Meads
Dockett Moorings
Marina
Fordwater Trading Estate
FORDWATER RD.
B3387
A317
Salesian Sch.
Tennis Courts
Playing Field
Philip Southcote Sch.
Conference Centre
Woburn Park Farm
Woburn Park
Dockett Point
Pharaoh's Island
Shepperton
Hamhaugh Island
Weir
Bourne
HAMM COURT
Hamm Court Farm
KT13
150
Pannells Farm
M25
Addlestone Moor
ABBEY MOOR GOLF COURSE
Addlestone Cemy.
Club Ho.
WOBURN HILL
Woburn Hill
Hatch Farm
A318
CHERTSEY RD.
WOBURN PARK
Athletic Grd.
Running Track
St. George's College
Tennis Courts
Cricket Ground
Woburn Bri.
Meadowlands Park
Junc. 11
Brook Ho.
Victory Park
Rec. Grd.
Aviator Park
Addlestone
WEYBRIDGE ROAD
Bourne Business Park
Ham Moor
Weybridge Lock
Black Boy Bri
St. Augustine's Nurs. Home
Superstore
B3121
STATION ROAD
HIGH ST
St. Paul's C of E Prim. School
Achieve POWER
Jubilee High Sch.
CHURCH RD
BRIGHTON RD.
ADDLESTONE
Crockford Bridge
Crockford Bridge Farm
Mill Pond
Coxes Lock
RIVER WEY NAVIGATION
Waterside Trading Estate
WEY MEADOWS
Riverside Park
Cemetery
Heathside School
Playing Fields
Brooklands College (Weybridge Campus)
Weybridge
HEATH
Rowhill
Walton Leigh Rec. Ground
KT15
Sayes Court Open Space
Garden Centre
Canal Bank
Addlestone Junction
ELMBRIDGE
NEW HAW ROAD
Coombelands Farm
Moated Farm
TopGolf Surrey
Driving Range
Golf Course
Canal Bridge
Wey Manor Farm
Nine Arches Bridge
Works
Seven Arches Bridge
Conduit Hill
Weybridge Junc.
The London Bus Museum
ROW TOWN
M25 MOTORWAY
NEW HAW
Hall's Farm
169
BYFLEET
B385
Central Veterinary Laboratory
Mercedes-Benz World
Brooklands Museum
04
05
06
07
63
64
65
66
67
68
71
72
73
74
75
76
77
78
79
80

SHEPPERTON GREEN
128
UPPER HALLIFORD
Q
R
S
T
U
V
W
Shepperton
Pool End
Sheepwalk Lake
M3
Funky Footprints Nature Reserve
SHEPPERTON
SHEPPERTON
Lower Halliford
Halliford Mere Lake
Cemetery
War Close
Las Palmas Estate
Playing Field
Desborough Sailing Club
DESBOROUGH ISLAND
TW17
TW16
Gaston Bridge
River Ash
Nursery
Watersplash Farm
The Swan Sanctuary
Gravel Pit
Marina
River Ash Estate
Riverhouse Gdns.
Riverhouse Barn
Thames Meadow
Cowey Sale
Shepperton Lock
Sailing Club
Tennis Club
Hamhaugh Point
Hamm Court Farm
149
Greenlands Farm
Broadwater Farm
PORTMORE PARK
Golf Course
St. James C of E Prim. Sch.
WALTON-ON-THAMES
Ashley Park
Cricket Ground
OATLANDS PARK
Oatlands Park
Cleves School
The Bull Dogs
WEYBRIDGE
Churchfields Rec. Grd.
Weybridge Lock
Cemetery
Heathside School
Brooklands College (Weybridge Campus)
The Heath
Weybridge
Conduit Hill
Weybridge Junc.
KT13
Burwood Park
Walton Common
Upper Common
Rabbit Lane House
Burvale Farm
Tennis Courts
Warren Pond
Club
Sports Ground
Recreation Ground
BURHILL GOLF COURSE
The London Bus Museum
Mercedes-Benz World
Brooklands Museum
170
A317
A3050
A244
B376
B375
B3366
B365
B373
B374
B372
QUEENS ROAD
ST. GEORGE'S AVENUE
HANGER HILL
OATLANDS DRIVE
ASHLEY ROAD
BURWOOD ROAD
HILLS ROAD
RUSSELL ROAD
WALTON LANE
WALTON BRIDGE
HIGH ST.
MONUMENT HILL
HEATH ROAD
NEW ZEALAND AV.
CHURCH STREET
FORDBRIDGE RD.
RENFREE WAY
GASTON BRIDGE ROAD
UPPER HALLIFORD BY-PASS

151
129
152
171
QUEEN ELIZABETH II
RESERVOIR
KNIGHTS
RESERVOIR
ISLAND BARN
RESERVOIR
WEST MOLESEY
KT8
KT12
KT10
HERSHAM
ESHER
Rydens
Fieldcommon
Lower Green
WEST END
Hersham Green
Burhill
Claremont Park
HERSHAM GOLF COURSE
MOORE PLACE GOLF COURSE
WEST END COMMON (Local Nature Reserve)
Molesey Heath Local Nature Reserve
Esher Water Pollution Control Works
Hersham Water Pollution Control Works
Sewage Works
Claremont Landscape Garden
Claremont Lake
Claremont Fan Court School
Esher C of E High School
Three Rivers Academy
Cardinal Newman Catholic Primary School
Esher RUFC (Molesey Rd.)
The Princess Alice Hospice
Sandown Sports Club
Winterhouse Farm
Southwood Manor Farm
Riverside Farm
Thames Field Farm
A3050
A244
A307
B369

Hampton Court
130
HAMPTON COURT PARK
HAMPTON COURT PALACE GOLF COURSE
KT8
KT1
RICHMOND upon THAMES
HOME PARK
Seething Wells
THAMES DITTON
ISLAND BARN RESERVOIR
Nielson Playing Field
Sports Ground
Grist Memorial Sports Ground
Thames Ditton
Giggshill
KT7
WESTON GREEN
Esher College
Metropolitan Police FC (Imber Ct.)
Imber Court
EMBER COURT RD.
STATION RD.
PORTSMOUTH ROAD
A307
A309
B364
B3379
Long Ditton Recreation Ground
LONG DITTON
Stokes Field (Local Nature Reserve)
Cemetery
Esher GOLF COURSE
SANDOWN PARK GOLF COURSE
SANDOWN PARK RACECOURSE
Daytona Sandown Park
Lower Green
151
SCILLY ISLES
KINGSTON BY-PASS
Hinchley Wood
HINCHLEY WOOD
Hinchley Wood Sch. & Sixth Form Centre
Littleworth Common
ESHER
ELMBRIDGE
SURBITON GOLF COURSE
KT10
Slough Farm
Wingham Court
Claygate
CLAYGATE
CLAYGATE COMMON (Local Nature Reserve)
Claygate Prim. Sch.
Rowan Prep. Sch.
Milbourne Lodge Prep. School
Loseberry Farm
Manor Farm
Chessington Equestrian
Five Acres Farm
Lovelace Prim. Sch.
ESHER BY-PASS
A3
A244
COPSEM LANE
Arbrook Common
172
Horridon
Barwell Farm
Winey Hill
Ea
Fa
Ga
Ha
Ja
Ka
La
71
72
73
74
75
76
77
78
79
80

153
131
Berrylands
KT5
KT3
KT6
KT4
KT9
KT19
BERRYLANDS
KINGSTON upon THAMES
SURBITON
Surbiton
SOUTHBOROUGH
TOLWORTH
Tolworth
TOLWORTH JUNCTION
Hollywood Bowl
Alexandra Millennium Green
Recreation Ground
Grand Avenue Prim. Sch.
Richard Challoner School
Malden Manor
Malden Manor Prim. Sch.
Knollmead Prim. Sch.
Maori Sports Ground
Surbiton Raceway
154
Linden Bridge School
KINGSTON BY-PASS
HOOK
HOOK JUNC.
Tolworth Rec. Cen.
King George's Field Playing Field
Kingstonian FC (King George's Field)
Corinthian Casuals FC
Jubilee Way Training Track
Tolworth Court Bridge
Hogsmill Local Nature Reserve
Castle Hill Prim. Sch.
Chessington North
CHESSINGTON
Chessington South
Bonesgate Open Space Local Nature Reserve
Castle Hill Local Nature Reserve
Butcher's Grove
HORTON PARK GOLF COURSE
COUNTRY PARK
Hook Road Arena
EPSOM
Epsom & Ewell High Sch.
Riverview C of E Prim. Sch.
West Ewell Inf. Sch.
WEST EWELL
The Harrier Cen.
St. Ebba's Farm Cottage
Horton Country Park Local Nature Reserve
173
A3
A240
A243
A3210
B3364
B284
B2200
Ma
Na
Pa
Qa
Ra
Sa
Ta
71
72
73
74
75
76
77
78
79
80
18
19
20
21

MALDEN
132
WEST BARNES
Motspur Pk.
Raynes Park Playing Fields
SW20
KINGSTON BY-PASS
A3
A2043
University of London Athletic Ground
KT3
MOTSPUR PARK
BBC Sports Ground
The Sir Joseph Hood Memorial Playing Fields
North East Surrey Crematorium
MORDEN CEMETERY
MORDEN PARK
Old Blues Rugby Football Grd.
Sports Ground
KINGSTON upon THAMES
Playing Fields
Merton & Sutton Joint Cemetery
Green Lane Prim. Sch.
Aragon Prim. Sch.
King George's Field
B279
MANOR PARK
Malden Manor
Malden Green
Worcester Park
Mayflower Park
Refuse Destructor
Richard Challoner School
OLD MALDEN
B284
Playing Field
Maori Sports Ground
Dorchester Prim. Sch.
KT4
St. Anthony's Hosp.
St. Raphael's Hospice
CENTRAL ROAD
CHEAM COMMON ROAD
WORCESTER PARK
Shadbolt Park
153
Linden Bridge School
Wandgas Athletics Ground
Cuddington Prim. Sch.
King George's Field
Auriol Park
Cuddington Recreation Ground
Beverley Brook
LONDON ROAD
A24
NORTH CHEAM
QUEEN VICTORIA
Cheam Leisure Cen.
SM3
MALDEN ROAD
A2043
A240
Sports Direct Fitness Sports Ground
Meadow Prim. Sch.
STONELEIGH
Stoneleigh
Oak Plantation
London Road Plantation
Cheam Recreation Ground
Hogsmill Local Nature Reserve
KT19
KT17
BEGGAR'S HILL
EPSOM
NONSUCH PARK
Mansion House
CHEAM
A232
EWELL RD.
WEST EWELL
Chamber Mead
Hogsmill River
David Lloyd Leisure Cen.
Nonsuch High Sch.
Harefield Bri. Playing Field
ORGAN CROSSROADS
A24 BY-PASS
Playing Fields
174
Warren Fm.
Ewell West
Bourne Hall Mus.
Ewell Castle Senior Sch.
CHESSINGTON
Ua
Va
Wa
Xa
Ya
Za
Ab
21
22
23
24
63
64
65
66
67
68
71
72
73
74
75
76
77
78
79
80

MORDEN
Morden South
Morden Leisure Centre
South Thames College (Merton College)
Morden Park School Sports Centre
School Playing Fields
St. Helier
ST. HELIER
ROSEHILL
Rosehill Park East
Rosehill Recreation Ground
ROSE HILL ROUNDABOUT
St. Helier Open Space
David Weir Leisure Centre
ST. HELIER HOSPITAL
QUEEN MARY'S HOSPITAL FOR CHILDREN
Greenshaw Wood
Sutton Sports Village
Sutton Tennis Academy
Sutton Common
Sutton Cemetery
Superstore
BENHILTON
Royston Park
THE WRYTHE
Wrythe Rec. Grd.
Carshalton Athletic FC (War Memorial Sports Ground)
Carshalton
CARSHALTON
Carshalton Beeches
CARSHALTON BEECHES
CARSHALTON ON THE HILL
SUTTON
Sutton
West Sutton
Sutton United FC (The Knights Comm. Stadium)
Collingwood Road Recreation Ground
Perrett's Field
Overton Grange Sch.
Overton Park
Harris Academy Morden
Poulter Park
Tooting & Mitcham United FC (Imperial Fields)
Watermeads
River Wandle
Mitcham
The Wilson Hospital
Westcroft Leisure Cen.
The Frying Pan
CARSHALTON PARK
Hackbridge
Wilderness Island
Mill Green
Nature Reserves
A24
A217
A232
A237
A239
A297
B278
B279
B2230
LONDON ROAD
BISHOPSFORD ROAD
CENTRAL ROAD
REIGATE AVENUE
CARSHALTON ROAD
CHEAM ROAD
WRYTHE LANE
NORTH STREET
OLDFIELDS
ST. NICHOLAS WAY
HIGH STREET
133
156
175
SM1
SM2
SM4
SM5
CR4

156
134
155
176
Jb
Kb
Lb
Mb
Nb
Pb
Qb
MITCHAM
MERTON
COMMON
MITCHAM GOLF COURSE
Club House
Mitcham Junction
CROYDON ROAD
A236
MITCHAM ROAD
Playing Fields
Keith Tuckey Stadium
Croydon Crematorium
CROYDON CEMETERY
Gonville Academy
The Archbishop Lanfranc Academy
THORNTON HEATH POND
Beddington Lane
Traq Motor Racing
Hundred Acre Bridge
Sludge Beds
Transfer Station
BEDDINGTON CORNER
CR4
Irrigation Bridge
Therapia Lane
LOMBARD ROUNDABOUT
Ampere Way
Beddington Sewage Treatment Works
Mile Road Bridge
Hackbridge
HACKBRIDGE
SUTTON
Superstores
Vue Cinema
Tenpin
Waddon Marsh
WADDON
BEDDINGTON PARK
Carew Manor & Dovecote
Carew Academy
Electricity Switching Station
BEDDINGTON
River Wandle
The Grange
Wilderness Island
Wandle Bri.
Boating Lake
Manor Gdn.
CROYDON ROAD
A232
Westcroft Leisure Cen.
The Grove
WALLINGTON GRN.
Waddon
FIVEWAYS CORNER
SM6
Bandon Hill Cemetery
Bandonhill
WALLINGTON
Wallington
The Frying Pan
CARSHALTON PARK
Mellows Park
SM5
SOUTH BEDDINGTON
Powerleague Soccer Centre (Croydon)
CROYDON AIRPORT
David Lloyd Leisure
Kidspace
Nuffield Health
Roundshaw
Roundshaw Park (Playing Fields)
Amy Johnson Prim. Sch.
Sports Direct Fitness (Croydon)
Roundshaw Downs Local Nature Reserve
Roundshaw Park
Wallington Cen.
PURLEY WAY
A23
A237
LONDON ROAD
WOODCOTE ROAD
MANOR ROAD
B271
B272
SANDY LA.
STAFFORD ROAD
PLOUGH LANE
71
72
73
74
75
76
77
78
79
80
28
29
30
31
63
64
65
66
67
168

157
Rb
Sb
Tb
Ub
Vb
Wb
Xb
135
177
158
SHIRLEY
THORNTON HEATH
SELHURST
Selhurst
SE25
WOODSIDE
Norwood Junc.
Broad Green
CROYDON
West Croydon
East Croydon
ADDISCOMBE
Addiscombe
CR0
CR2
SOUTH CROYDON
South Croydon
LLOYD PARK
Lloyd Park
COOMBE PARK
CROHAM HURST GOLF COURSE
Sanderstead
71
72
73
74
75
76
77
78
79
80
32
33
34
35
68
67
66
65
64
163

158
SOUTH NORWOOD COUNTRY PARK
Yb
Zb
Ac
Bc
Cc
Dc
Ec
ELMERS END
136
Sports Grd.
David Lloyd Leisure
Eden Park High Sch.
EDEN PARK
Playing Field
Pavilion
Langley Court (Research Labs)
PARK LANGLEY
Croydon Sports Arena
Croydon FC Grd.
Oasis Academy Arena
Arena
SE25
World of Golf Centre
Monks Orchard Prim. Sch.
ELMERS END A214 ROAD
UPPER ELMERS END
Crouch Oak Wood
Eden Park
Unicorn Prim. Sch.
The Beck
Langley Park School for Boys
Sports Ground
Langley Park School for Girls
Langley Park Sports Centre
Langley Park Prim. Sch.
LANGLEY PARK GOLF COURSE
BR3
The Rookery
LANGLEY
LINKS WAY
SOUTH EDEN PARK RD.
SOUTH EDEN B230
LONG LANE A222
Long Lane Wood
Bird Sanctuary
71
72
73
74
75
76
77
78
79
80
Woodside
Ashburton Park
Ashburton Playing Fields
Oasis Academy Shirley Park (Seniors)
SHIRLEY OAKS BMI HOSPITAL
Shirley Oaks
Glade Wood
MONKS ORCHARD
Edenham High Sch.
Bowling Green
Sports Ground
Cold Shaw
High Broom Wood
Bethlem & Maudsley Hospital Sch.
BETHLEM ROYAL HOSPITAL
Whitley House
Denis Hill Unit
Chelsham House
River House
Parkfields Recreation Ground
Playing Field
Wickham Park House
Caravan Park
West Wickham
Oak Lodge Prim. Sch.
Blake Recreation Ground
WEST WICKHAM
BECKENHAM ROAD
STATION ROAD A214
HIGH STREET
Playing Field
A215
SHIRLEY ROAD
A232
WICKHAM ROAD
Superstore
Syn. Hall
Liby.
Trinity Sch.
Sports Cen.
Playing Field
Nursery
SHIRLEY
157
CR0
Spring Park
Miller's Pond
Spring Park Wood
SPRING PARK
Forest Acad.
Harris Prim. Academy, Benson
Foxes Wood
Hall Grange
Colma Convent Girls' Sch.
Shirley High School Performing Arts College
SHIRLEY PARK GOLF COURSE
Oaks Farm
Shirley Windmill
Shirley Court
Lacrosse Pitch
Pinewoods
Wychwood
SPRING PARK
UPPER SHIRLEY
CROYDON
Kennel Wood
Club House
The Bungalow
THE ADDINGTON GOLF COURSE
Addington Hills
Threehalfpenny Wood
Pumping Station
Rugby Ground
Foxhill Shaw
BIRCH WOOD
ADDINGTON VILLAGE RD.
Reservoir (covered)
Coombe Lane
Bardy Mount
Sports Grd.
Royal Russell School
Ballards Cotta
ADDINGTON
Vicarage
ADDINGTON PALACE GOLF COURSE
Addington Palace
Rowdown Fields
Time Bridge Community Centre
A2022 WAY
KENT GATE
Addington Village
Bus Sta.
ADDINGTON PARK
Pavilion
Applegarth Academy
Castle Hill Academy
Prim. Sch.
GRAVEL HILL
A212
Heathfield (Train. Cen.)
Heathfield Farm
Bramley Bank Nature Reserve
CR2
Playing Field
Gilbert Scott Primary School & Red Gates School
The Quest Academy
John Ruskin College
Gravel Hill
Fieldway
ADDINGTON COURT GOLF COURSE
Water Tower
PARK A2022
178
Addington Well Pumping Sta.
Play Fld.
Selsdon Prim. Sch.
North Down

159
137
179
160
HAYES
HAYES COMMON
BROMLEY COMMON
NORMAN PARK
CONEY HALL
KESTON
KESTON MARK
NASH
NEW ADDINGTON
BROMLEY
BR2
BR4
CROYDON ROAD
A232
A233
A21
A2022
B265
B251
B2212
WESTERHAM ROAD
HAYES LANE
BASTON ROAD
PICKHURST LANE
WEST WICKHAM COMMON
HAYES GROVE PRIORY HOSP.
Bromley College of Further & Higher Education
Bromley FC & Cray Wanderers FC (Courage Stadium)
Norman Park Athletics Track
Nuffield Health Fitness & Wellbeing Centre
Bromley Trust Academy
Pickhurst Jun. & Inf. Schs.
Hayes Prim. Sch.
Hayes Sch.
Ravens Wood Sch.
Keston C of E Prim. Sch.
Hawes Down Jun. & Inf. Schs.
Glebe Sch.
Nash College
The Emmaus Centre
Metropolitan Police Dog Training Establishment
Electricity Sub Station
Cupola Wood
Scrogginhall Wood
Lord's Wood
Brook Wood
Mazzards Wood
Woodcock Grove
Bolds Wood
Barnet Wood
Padmall Wood
Colyers Wood
Gorsewood
Well Wood
Cooper's Wood
Long Shaw
James's Wood
Fuller's Wood
Bradmanshill Wood
Broomfieldbank Shaw
Rushfield Shaw
Baldwin Shaw
Jackass Shaw
Leafy Grove
Ravensbourne
Keston Windmill
Keston Common
Heathfield House
Hast Hill
Baston Manor
Keston Court Riding School
Rouse Farm
Arlington Orchard
Elm Farm
Fox Hill Nursery
Nash Farm
North Pole Farm
Five Oaks
Waytes
Wickham Court Farm
The Cottages
The Bungalows
Castlehill Ruffs
Rowdown Wood
Knightons
War Bank
The Temple (Roman Mausoleum)
Keston Court Farm
Milchden
Wilberforce Oak
Caesars Camp
Holwood
Westerham Lodge
Oakley House
Hayes Street Farm
Sparrow's Den Playing Fields
Fc Gc Hc Jc Kc Lc Mc
39 40 41 42

138
180
159
BR1
BR2
SOUTHBOROUGH
PETTS WOOD
Petts Wood
BROMLEY COMMON
ORPINGTON
CROFTON
LOCKSBOTTOM
KESTON MARK
FARNBOROUGH
GREEN STREET GREEN
JUBILEE COUNTRY PARK
Jubilee Country Park Sydney Shaw Local Nature Reserve
Bromley High School
Petts Wood Junction
Southborough Prim. Sch.
St. James' RC Prim. Sch.
Southborough Open Space (Rec. Grd.)
Parkfield Recreation Ground
BROMLEY GOLF COURSE
Bishop Justus C of E School
Princes Plain Prim. Sch.
Prince's Plain
Sparrow Wood
Petts Wood Recreation Ground
Crofton Jun. Sch.
Crofton Inf. Sch.
Sparrow Common
Roundabout Wood
Crofton Heath
Scrubs Farm
Clay Wood
Nobody's Wood
Pokeridden Wood
Knowlehill Wood
Knowlehill Farm
Hollydale Recreation Grounds
Gumping Common
Newstead Wood
Newstead Wood School
Bromley Tennis Centre
Tubbenden Prim. Sch.
Darrick Wood Swimming Pool
Darrick Wood School
Darrick Wood Sports Cen.
Darrick Wood Jun. & Inf. Schs.
Farnborough Recreation Ground
PRINCESS ROYAL UNIVERSITY HOSP.
THE FANTAIL
Ravens Wood Sch.
Oakley House
Oakley Farm
Depot Sports Grd.
Bus Depot
Ninhams Wood
SOUTH PARK
Lake Wood
Caesars Camp
Broom Bank
Woodlands Poultry Farm
The Larches
HOLWOOD ESTATE
Holwood
Sevenoaks Lodge
Lower Hook Farm
LOWER HOOK BUSINESS PARK
North End Farm
North End Farm Cotts.
Blacklands
HIGH ELMS COUNTRY PARK (Local Nature Reserve)
High Elms Clockhouse
HIGH ELMS GOLF COURSE
Bromley Environmental Education Centre at High Elms (BEECHE)
Cocksett Plantation
Ramus Wood
Whitelands Wood
Church Field Plantation
The Manse
Mill Hill
Farnborough Prim. Sch.
Westerham Lodge
Holwood Farm
GLEBE FARM BUSINESS PARK
PADDOCK MOBILE HOME PARK
Farthing Street
Farthings
A21
A232
A233
B2158
COMMON ROAD
FARNBOROUGH COMMON
HIGH STREET FARNBOROUGH
CROFTON ROAD
OAKLEY ROAD
HASTINGS ROAD
FARNBOROUGH HILL
SHIRE LANE
WESTERHAM RD.
Nc
Pc
Qc
Rc
Sc
Tc
Uc
42
43
44
45
63
64
65
66
67
71
72
73
74
75
76
77
78
79
80

St. Mary Cray
Cemetery
POVEREST
Poverest Prim. Sch.
ST. MARY CRAY
BLACKSMITH'S LANE
SWEEPS
CROCKENHILL
B258
Manor Oak Primary Sch.
Kent House Hospital
Kent Ho. Hosp. Sch.
Kevington Manor
KEVINGTON
The Warren
Sheepcote Farm
Child's Wood
Little Kevington
Crouch Farm
Waldens Farm
Kynaston Wood
Derry Downs
Walden Manor
Waldens Bungalow
Lark Rise
Heather's Holding
Bleak House Farm
Green Cedar
East Hall Cottage
Eastwood
Griff's Wood Cottage
Griff's Wood
Piggery
Orchard View
Lone Barn Farm
Brambles
Byways Nursery
Sage Wents
Cookham Hill Farm
Cookham Farm
Feathercot
Micklewood Farm
BR5
Covet Wood
Recreation Ground
Bowling Green
Perry Hall Prim. Sch.
PERRY HALL RD.
A208
Educ. Cen.
Broom Hill
Priory Gardens
Grave Yard
Mus.
Grassmeade Recreation Ground
Hillside Primary School
Grigg's Cross
RAMSDEN
Bowls Cen.
Harris Academy
Harris Academy Sports Cen.
Playing Field
Leisure Cen.
Bowl. Alley
Super store
STATION ROAD
SPUR ROAD
A232 ROAD
A224
ORPINGTON BY-PASS
COURT ROAD
CRAY AVENUE
Blenheim Prim. Sch. & Nursery
Burwood School
Play. Fld.
Sports Ground
Piggeries
GODDINGTON
Goddington Park
The Lodge
Rounds Smallams
Club
Lilly's Wood
Black Bush Wood
Playing Field
Tennis Courts
St. Olave's & St. Saviour's Grammar Sch.
Ontario Centre
A223
BR6
Hospice
ORPINGTON HOSPITAL
Prim. Sch.
Play. Fld.
Orpington Sports Club
The Highway Prim. Sch.
CHELSFIELD
Chelsfield
Chelsfield Village
BMI CHELSFIELD PARK HOSPITAL
Brown's School
Bucks Cross Farm
Court Lodge Farm
Court Lodge
Glentrammon Rec. Grd.
High Beeches
Sports Ground
Pecks Cotts.
Chelsfield Riding School
Club House
Driving Range
Julian Brimstone
CHELSFIELD LAKES GOLF COURSE
Eleven Oaks
Chelsfield Hill Wood
Rounds Wood
Chelsfield Hall
Maypole
Hewitts
HEWITTS ROUNDABOUT
A21
M25
SEVENOAKS ROAD
Broke Lodge
139
162
181
163
Vc Wc Xc Yc Zc Ad Bd
71 72 73 74 75 76 77 78 79 80
46 47 48 49
64 65 66 67 68

162
140
161
182
Cd
Dd
Ed
Fd
Gd
Hd
Jd
49
50
51
52
168
67
66
65
64
63
71
72
73
74
75
76
77
78
79
80
Quarry
Hook Spring
Tile Kilns Wood
Marwood Ho. Caravan & Camping Site
Marwood Ho.
Football Ground
Club Ho.
Sheepcote Farm
Nursery
War Mem.
Rec. Grd.
Pav.
GOLDSEL
GREEN COURT RD.
STONES CROSS RD.
Fleetfoot Fm.
Prim. Sch.
Peartrees
CROCKENHILL
Cemetery
Cemy.
Cherry Orchard
Gosenhill Farm
BR8
Waldens Farm
Crouch Farm
Crouch Farm House
Felma House
Crouch Farm Cottages
ROAD
CRAY
B258
EYNSFORD
BROADWAY
GREEN CT.
The Mount
Hall
Middle Farm Nursery
Willow Farm
GREEN RD.
HARVEST WY.
NEWPORTS
Walden Manor
Heathers Holding
Bleak House Farm
Green Cedar
East Hall Cottage
BROMLEY
Piggery
Orchard View
The Oaks
Lone Barn Farm
Brambles
GORSE
Byways Nursery
Highcroft Hall
Heathfield Shaw
Pollard
Horncross Shaw
M25 MOTORWAY
Football Ground
Westerd Farm Cotts.
Little Wested House
Wested Meadows
Cacketshill Wood
Wested Farm
Industrial Estate
FARM COTTAGES
PETHAM CT.
BY PASS
A20
B258
B2173
High Firs Prim. Sch.
Rec. Grd.
DISCOVERY
Hulberry
Baldwins Cottages
Hulberry Farm
Fairfield
Crown Wood
Crown Wood Fm.
BR5
Cookham Farm
Feathercot
Woodside
SKEET HILL
Micklewood Farm
Playing Field
Skeet Hill Farm
DALTONS
Davidson House
LANE
S
E
Roughet
Snow White Cottages
PARKGATE
M25
Park Gate House
Club House
LULLINGSTONE PARK GOLF COURSE
LULLINGSTONE PARK
Driving Range
LULLINGSTONE COUNTRY PARK
UPPER BEECHEN WOOD
LOWER BEECHEN WOOD
Doctor's Wood
OWEN WOOD
BR6
FIRMINGERS
WELL HILL
HOME FARM
Brown School
HAWSTEAD
Bucks Cross Farm
ROCK
WELL HILL LANE
JUBILEE
HOLLYBUSH
SHOREHAM ROAD
Maypole
Reservoir
Goat Hill
Westwood Cottage
REDMANS HILL
REDMANS LANE
Great Cockerhurst
HOME WOOD
COCKERHURST
Ridge Farm
Homewood Farm
TN14
Hewitts
Pascalls Wood
CHELSFIELD
Westwood Farm
WEST WOOD
Brent Farm
Dalhanna
Coneyearth Wood
HOLLOWS WOOD
Junction 4
HEWITTS ROUNDABOUT
ORPINGTON
WHEATSHEAF HILL
Goss Bushes
Keepers Cottage
Coombe Hollow Cottage
FARM
ROAD

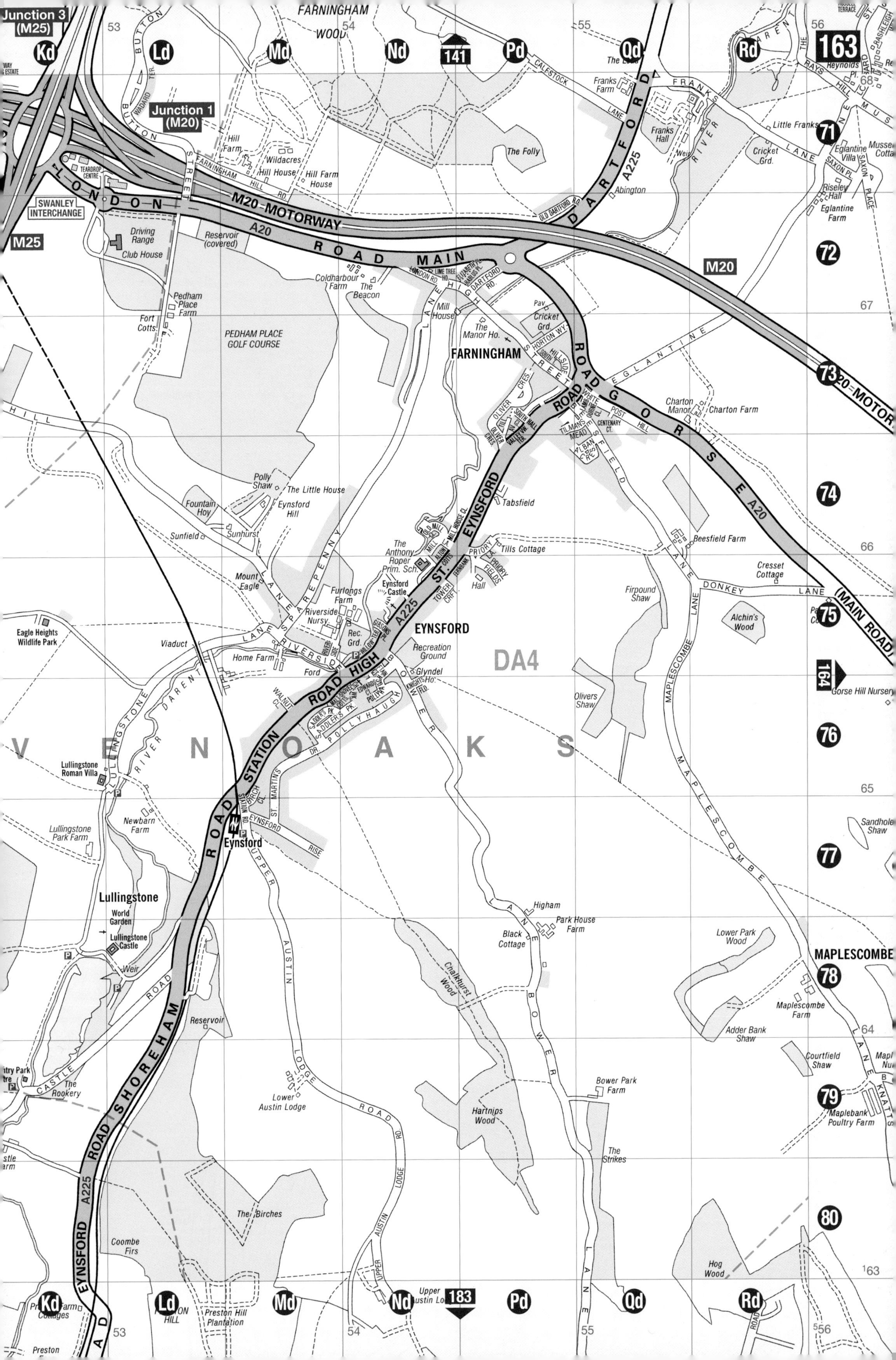
163
Junction 3 (M25)
Junction 1 (M20)
SWANLEY INTERCHANGE
M25
M20
FARNINGHAM WOOD
141
183
164
Kd
Ld
Md
Nd
Pd
Qd
Rd
53
54
55
56
63
64
65
66
67
68
71
72
73
74
75
76
77
78
79
80
M20 MOTORWAY
A20
ROAD MAIN
LONDON
A225
DARTFORD
ROAD GORSE
EYNSFORD ST.
HIGH ROAD STATION
SHOREHAM ROAD
EYNSFORD ROAD
DA4
SEVENOAKS
FARNINGHAM
EYNSFORD
Eynsford
Lullingstone
MAPLESCOMBE
PEDHAM PLACE GOLF COURSE
Driving Range
Club House
Reservoir (covered)
Pedham Place Farm
Fort Cotts.
Hill Farm
Wildacres
Hill House
Hill Farm House
The Folly
Franks Farm
Franks Hall
Abington
Little Franks
Cricket Grd.
Eglantine Villa
Riseley Hall
Eglantine Farm
Coldharbour Farm
The Beacon
Mill House
The Manor Ho.
Cricket Grd.
Charton Manor
Charton Farm
Polly Shaw
The Little House
Eynsford Hill
Fountain Hoy
Sunfield
Sunhurst
Mount Eagle
Tabsfield
Tills Cottage
Beesfield Farm
Cresset Cottage
The Anthony Roper Prim. Sch.
Eynsford Castle
Furlongs Farm
Riverside Nursy.
Rec. Grd.
Priory Hall
Firpound Shaw
Alchin's Wood
Eagle Heights Wildlife Park
Viaduct
Home Farm
Ford
Recreation Ground
Glyndel
Olivers Shaw
Gorse Hill Nursery
Lullingstone Roman Villa
Newbarn Farm
Lullingstone Park Farm
World Garden
Lullingstone Castle
Weir
Reservoir
The Rookery
Lower Austin Lodge
Higham
Park House Farm
Black Cottage
Chalkhurst Wood
Lower Park Wood
Maplescombe Farm
Adder Bank Shaw
Courtfield Shaw
Sandhole Shaw
Bower Park Farm
Hartnips Wood
Maplebank Poultry Farm
The Strikes
The Birches
Coombe Firs
Hog Wood
Preston Hill Plantation
Upper Austin Lodge
Preston
CALFSTOCK LANE
FRANKS LANE
RIVER DARENT
DONKEY LANE
MAPLESCOMBE LANE
EGLANTINE LANE
POST HILL
BUTTON STREET
FARNINGHAM HILL RD.
LULLINGSTONE LANE
SPAREPENNY LANE
RIVERSIDE
POLLYHAUGH
TOWER RD.
UPPER AUSTIN LODGE ROAD
BOWER LANE
CASTLE ROAD
KNATTS LANE
EYNSFORD RISE
ST. MARTINS DR.
HORTON WY.
HILLSIDE
OLIVER CRES.
PRIORY LA.
PRIORY FIELDS

164
HORTON KIRBY
Horton Kirby Environmental Education Centre
Reynolds Place Farm
Franks
Cricket Grd.
Eglantine Villa
Mussenden Cottage
Mussenden Farm
Blue Cottages
Riseley Hall
Eglantine Farm
Beeches Farm
Canada Farm
Hill Barn Farm
Lone Oak Farm
DA4
M20 MOTORWAY
Liberty Farm
THE GROVE CRAFT WORKSHOPS
Grove Farm
Fawkham
Wickhams
The Grange
HORTON WOOD
Fawkham C of E Prim. Sch.
Parkfield Wood
FAWKHAM MANOR BMI HOSPITAL
Hopkins Spring Wood
Hatchfield Wood
Fawkham Manor Farm
Cresset Cottage
Speedgate Farm
Speedgate House
White House Farm
Speedgate
Alchin's Wood
Paygate Cottage
Gabrielspring Wood
Choaks Farm
Choaks Wood
FAWKHAM GREEN
163
Gorse Hill Nursery
Hazlewood
Birchgrove
Greenlands
Cherrywood
The Haven
Springfield
Rowanhurst
Ashen Grove
BRANDS HATCH PK.
Brands Hatch Place
Saxten's Wood
Sandhole Shaw
Shetley Shaws
Grandstand
Cage's Wood
Vineyard Shaw
Hotel
Offices
Woodlands
Rogers Wood
Eastwood
Gallows Wood
Little Haven
BRANDS HATCH MOTOR RACING CIRCUIT
Billet Wood
MAPLESCOMBE
Kingsdown Farm
Cowless Shaw
Battersea at Brands Hatch
Maplescombe Farm
Grove Wood
BLUE CHALET INDUSTRIAL PARK
Courtfield Shaw
Maplebank Nursery
Maplebank Poultry Farm
To the Woods Caravan & Camping Site
FLORENCE FARM MOBILE HOME PK.
Westfield Cotts.
THE LONDON GOLF COURSE
Broomfield Wood
Mace Wood
CHURCH WOOD
Birtley House School
Springham's Cotts.
Crowhurst Farm
Sidehilly Wood
Rec. Grd.
WEST KINGSDOWN
Club House
Ashen Shaw
Sports Ground
WEST KINGSDOWN INDUSTRIAL EST.
184
Richardson's Farm
Inges Shaw
THE LONDON GOLF COURSE

143
185
Zd
Ae
Be
Ce
De
Ee
Fe
HARTLEY
Hartley Green
LONGFIELD HILL
Nurstead Hall
NURSTEAD HILL FARM
Steephill School
CASTLE HILL
OLD DOWNS
Cricket Ground
Manor Farm
Skinners Hill
B260
LONGFIELD ROAD
HARTLEY WOOD
Primary Sch.
Spts. Grd.
Bowling Green
Tennis Courts
Hartley Prim. Acad.
Liby.
Football Ground
Corinthian Sports Club
Pennis Farm
CORINTHIAN GOLF COURSE
Hartley Manor
MANOR DRIVE
MANOR LANE
Middleton Cottages
Middleton Farm
Orange Hill
Broom Wood
GRAVESHAM
DA3
Hartley Dell
Foxbrough Wood
North Lands Shaw
Redlibbets Wood
Chapel Wood
GRANGE ROAD
Pumping Station
Hartley Hill
Goldsmiths Shaw
Redsteadle Wood
Dell Wood
Millfield Shaw
Shipley Hills
Milestone Academy
Hartley Bottom Farm
Bridelands Wood
Northdown Farm
Southdown Farm
Idleigh Farm
Idleigh Court
Sunnymead
Sunnydale
Buzy Bees
Oliver's Farm
NEW ASH GREEN
Nine Horse Wood
Turner's Oak Shaw
West Yoke
BEECHCROFT FARM INDUSTRIES
Delsrop
Hill View
Larksfield
Cuckolds Corner
The Warren
Court View
DA13
Selbyfield Shaw
Ten Acre Shaw
Valley's Shaw
Feathercot
Grubb Wood
Elbows Wood
Recreation Ground
Landway Farm
HEAVER TRADING ESTATE
Swan Meadows Farm
Bowes Wood
White Ash Wood
Middle Wood
Eldridge Shaw
Eight Acre Wood
Ashplat Wood
Ash Place Farm
White Ash Wood
Swan Farm
ASH
Tennis Courts
Ckt. Ground
The Old Rectory
Rectory
Dunmery Grove
Viney Wood
Oxens Shaw
New Street
Eldridge Cottage
New Street Farm
New Street Farm Cottages
Orchard Cottage
Hilly Field Shaw
The Old Rectory
Glebe House
Ridley Court Farm
RIDLEY
Sheeplands Shaw
TN15
Galsmead
Goss Wood
Ridley Wood
Pease Hill Cottages
Rumney Farm Cottages
Berry's Maple
Pilgrim Corner
Ridley Farm
Pond Farm
Long Bottom Wood
Rumney Farm
The Roughets
Haven Wood
Newlands Shaw
Meadowcroft
Jub's Hill Wood
Haven Manor
TONBRIDGE and MALLING
Baker's Shaw
Seafield
Holywell Park Farm
HODSALL STREET
Lion Wood
Pettings Wood
Pettings Court Farm
Martin Hill Spring
Woodlands
Piggery
Little Pettings
Pettings
Dairy House Farm
Holywell Farm
Greenfields
Baker's Wood
The Old Malt House
Pumping Station
Pettings Cotts
Pettings Court
Pettings Park
Martinhill Wood
60
61
62
63
64
65
66
67
68
71
72
73
74
75
76
77
78
79
80

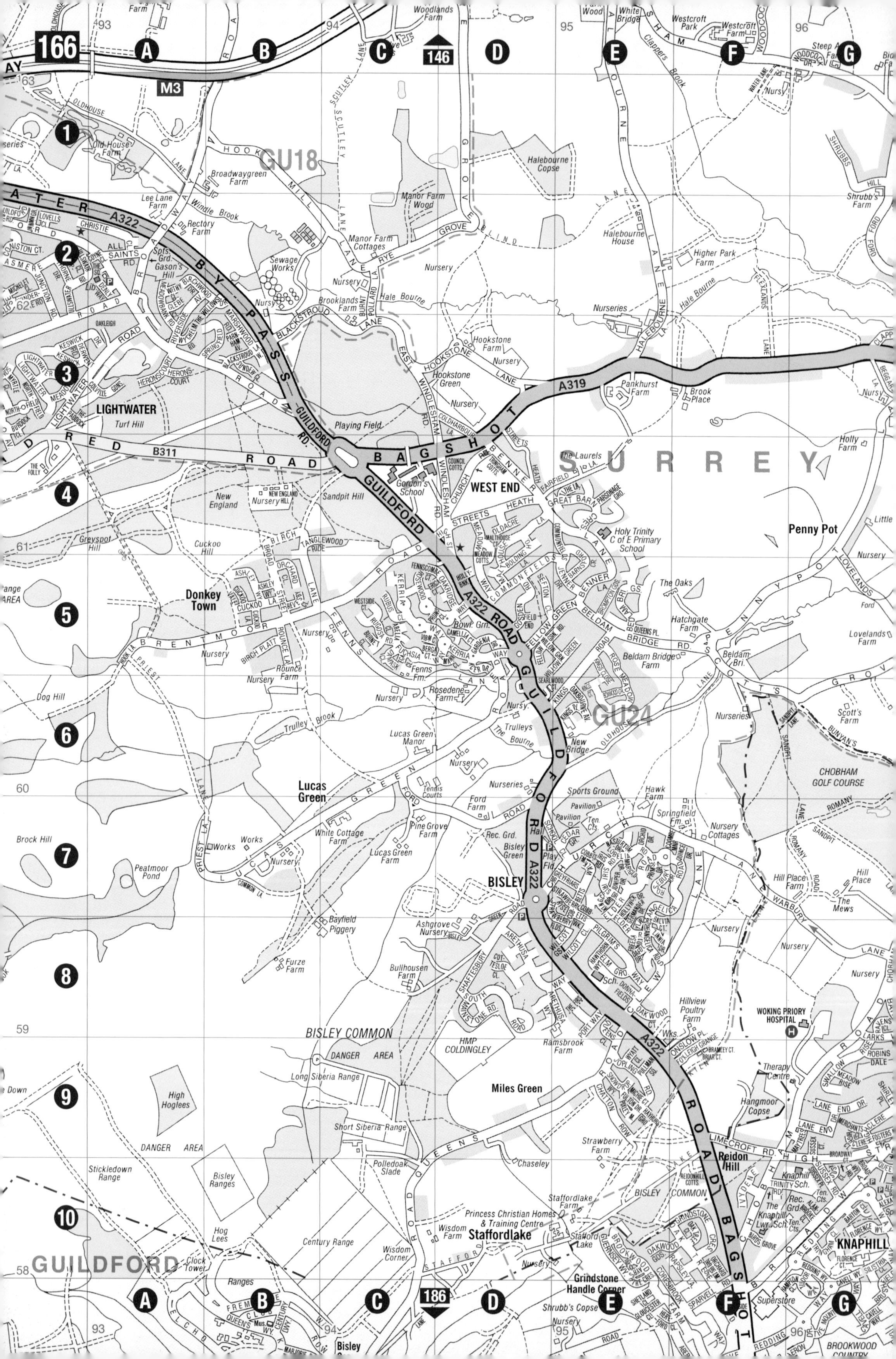

166
146
186
A
B
C
D
E
F
G
1
2
3
4
5
6
7
8
9
10
M3
A322
A319
B311
GU18
GU24
SURREY
LIGHTWATER
WEST END
BISLEY
KNAPHILL
GUILDFORD
Donkey Town
Lucas Green
Penny Pot
Staffordlake
Reidon Hill
Miles Green
BISLEY COMMON
CHOBHAM GOLF COURSE
WOKING PRIORY HOSPITAL
HMP COLDINGLEY
Halebourne Copse
Manor Farm Wood
Holy Trinity C of E Primary School
Gordon's School
Princess Christian Homes & Training Centre
Grindstone Handle Corner
DANGER AREA
Long Siberia Range
Short Siberia Range
Century Range
Stickledown Range
Bisley Ranges
High Hoglees
Brock Hill
Dog Hill
Peatmoor Pond
Turf Hill
New England
Sandpit Hill
Cuckoo Hill
Greyspot Hill
Hookstone Green
Beldam Bridge Farm
Lovelands Farm
Hangmoor Copse
Hillview Poultry Farm
Strawberry Farm
Superstore

147
168
187
Burrowhill
CHOBHAM
HEATH
CASTLE GREEN
MIMBRIDGE
WOKING
LITTLEWICK
GOLDSWORTH PARK
HORSELL
Goldsworth
WOKING
ST. JOHNS
MOUNT HERMON
Stanners Hill
Horsell Common
Chobham Park House
Littlewick Common
Woking Crematorium
GU21

SURREY HEATH
FAIROAKS AIRPORT
GU24
KT16
HORSELL COMMON
SHEERWATER
WOKING
MAYBURY
Woking
Horsell Common
Goldsworth
MOUNT HERMON
Brox
Anningsley Park
NEW ZEALAND GOLF COURSE
WEST BYFLEET GOLF COURSE
HOEBRIDGE GOLF COURSE
THE ROUGH
Pyrford Common
McCLAREN TECHNOLOGY CENTRE
Ottershaw Park
Basingstoke Canal
GUILDFORD ROAD
CHERTSEY ROAD
WOODHAM LANE
SHORES ROAD
CHOBHAM ROAD
VICTORIA WAY
A319
A320
A245
A3046
A324
B382
B385
148
167
188

169
149
170
189
H
J
K
L
M
N
P
ROW TOWN
NEW HAW
Hall's Farm
Central Veterinary Laboratory
Grange Farm
Fallow Farm
New Haw Jun. & The Grange Inf. Schs.
Woodham & New Cen.
Wey Manor Farm
Golf Course
Driving Range
Electricity Switching Station
Byfleet & New Haw
Byfleet Junc.
Mercedes-Benz World
The London Bus Museum
Brooklands Museum
BROOKLANDS
ELMBRIDGE
KT15
KT13
KT14
Brooklands Business Park
Superstores
Heathervale Caravan Park
Heathervale Recreation Grd.
Bottom Lock
Camphill Industrial Estate
Scotland Bri. Lock
Fullbrook Sch.
Sports Cen.
Woodham Lock
West Byfleet
West Byfleet Jun. & Inf. Schs.
Recreation Ground
Dartnell Park
Byfleet Industrial Estate
Byfleet Prim. Sch.
Nursing Home Sports Ground
Ckt. Grd.
Oldwood Pond
Old Wood
Tins Wood
Broadoaks
The Moat
WEST BYFLEET
West Hall Farm
Dodd's Wood
Murray's Bri.
Dodd's Bri.
Nuffield Health Fitness & Wellbeing Centre
Byfleet Lawn Tennis Club
TWISTED STONE GOLF COURSE
Club House
RIDGWAY
Old House
Nurseries
Peatmoor Wood
Pyrford C of E Prim. Sch.
PYRFORD
GU22
Pyrford Marina
Lock Copse
Church Farm
The Bungalow
Wisley Bri.
Sewage Works
Royal Horticultural Society Cotts.
WISLEY
Sports Ground
GU23
Recreation Ground
Village Hall
St. Mary's C of E Prim. Sch.
Sanway
Nursery
BYFLEET
Bluegates Hole
Manor House
The Island
Common Meadows
M25
M25 MOTORWAY
A245
A318
B385
B367
B374
A3
Wisley Common
Woolger's Wood
WISLEY COMMON
Pyrford Place Farm
Pyrford Stone
Stone Farm
Lodge
PYRFORD GREEN
Elveden Cottages
PYRFORD GOLF COURSE
WISLEY GOLF COURSE
Townslow Meadow
Wren's Nest
Pinetum
River Wey
River Wey Navigation
Greenhouses
Aberconway House
The Stable Courtyard
Garden Cen.
Rock Garden
RHS Garden Wisley
Battleston Hill
GUILDFORD
Church Farm
Lady Place
The Hanger
Vicarage
PYRFORD VILLAGE
The Prae
Wheelers Farm
Home Farm
Pyrford Place
Abbey Stream Bri.
Warren Farm Mobile Home Park
Walsham Lock
Plantation
Ockham Mill
Nursery
Plough Bridge Farm
Horizon Business Village
Weymede
David Lloyd Leisure
THE HEIGHTS
Seven Arches Bridge
Arches Bridge
04
05
06
07
58
59
60
61
62
63

150
Q
R
S
T
U
V
W
The London Bus Museum
Mercedes-Benz World
Brooklands Museum
BROOKLANDS
Handling Track
KINGSWAY TER.
ST. GEORGE'S BUSINESS PK.
THE HEIGHTS
B374
(CAENSWOOD HILL)
WHITE KNIGHTS RD.
NORTHFIELD
RIDGEWOOD
GOLF COURSE
LINDEN RD
YAFFLE
BOWATER RIDGE
ABBOTSWOOD
ST. GEORGE'S HILL
David Lloyd Leisure
Club House
KT13
ST. GEORGE'S HILL GOLF COURSE
RAVENSCROFT RD.
RODONA
Blue Barn Farm
BLUE BARN LANE
FAIRWAY
HORIZON BUSINESS VILLAGE
WEYMEDE
BYFLEET
Plough Bridge
Plough Bridge Farm
Foxholm Covert
GOLF COURSE
Bluegates Hole
SILVERMERE GOLF COURSE
Silver Mere
Silvermere
Club House
Driving Range
Home Farm
Silvermere Farm Estate
A245
SILVERMERE GOLF COURSE
Thirty Acre Wood
KT14
Byfleet Mill
169
The Hangers
Queen Anne's Cottage
Queen Anne's Hills
Manor House
The Island
Park Barn Farm
Nursery
Long Orchard Farm
Bell Wood
Highlands Farm
Foxwarren Park
Red Hill
Gothic Tower
Tower Hill
Waterwheel
Wheel Island
Buxton Wood
Clearmount
M25
M25 MOTORWAY
WISLEY INTERCHANGE
Wisley Common
Woolger's Wood
Junction 10
Pond Farm
Cockcrow Hill
Sandpit Hill
Redhill Bottom
Chatley Wood
Oak Pollard
Red Hill
Chatley Farm
RIVER MOLE
Chatley Cott.
GUILDFORD
WISLEY COMMON
GU23
Hut Hill Cottage
Hut Hill
Currie's Clump
OCKHAM COMMON
Chatley Heath
Breach Hill Wood
Telegraph Hill
Chatley Heath Semaphore Tower
The Bogs
M25 MOTORWAY
Bolder Mere
Wisley Common, Ockham & Chatley Heath Nature Reserve
Hatchford Wood
Woodside Cottage
Elm Cottage
Hatchford Park
Oakham Village Green
Fellside Cottages
ELM CORNER
Wilderness Farm
Hatchford End
190
Seven Hills Cl.
B365
SEVEN HILLS ROAD
Foxoak Hill
Recreation Ground
COOMBE LANE
Burhill Golf Course
OCTAGON
CIRCLE
WEST AV.
ELIZA PALMER HUB
HUNTLEY HO.
WHITELEY VILLAGE
Strawberry Hill
KT12
BURHILL GOLF COURSE
Burhill Park
Club House
Dick Turpin's Cottage
KINGS DRIVE
Beards Hill
CONVENT LANE
Notre Dame Preparatory & Senior Schools
Burwood Park
Seven Hills Estate
Spts. Grd.
Pavilion
RIVER MOLE
Chestnut Lodge Farm
E L M B
Walton Firs Scout Camp
Manor Pond
PAINSHILL
ESHER ROAD
Superstore
Play. Fld.
St. George's Nursing Home
Feltonfleet Sch.
Lodge
Hotel
A3 ROAD
PORTSMOUTH ROAD
Cobham Bridge
Pains Hill
PAINS HILL HO.
Education Cen.
Heyswood Girl Guide Camp
The Broom
PAINSHILL PARK
Wood Hill
Gothic Temple
Chinese Bridge
Grotto
The Lake
Lodge Cottages
Cobham Court Farm
Halfpenny Cross
Wrens Nest Cottage
Bramble Wood
Bramble Cottage
Lodge Copse
M25
Poynters Farm
Pointers Green
Brickfield Copse
Pondtail Farm
Little Brickfield Copse
Highfield Farm
Nurseries
Cold Norton Farm
NORTON LA.
Tennis Courts
HATCHFORD MANOR
07
08
09
10
63
62
61
60
59
58
81
82
83
84
85
86
87
88
89
90

171
151
172
191
X
Y
Z
Aa
Ba
Ca
Da
11
12
13
14
63
62
61
60
59
58
81
82
83
84
85
86
87
88
89
90
Claremont Park
Claremont Landscape Gard
Claremont Lake
Tennis Courts
WEST END COMMON (Local Nature Reserve)
Southwood Manor Farm
Riverside Farm
Horseshoe Clump
ALBANY
MEADWAY
BLACKHILLS
Gate
Winterdown Wood
RIVER MOLE
The Homewood
BURHILL GOLF COURSE
KT10
ESHER COMMON (Local Nature Reserve)
Black Pond
Round Hill
A307
A3
A244
ESHER BY-PASS
Winterdown
ACS Cobham International School
Upper Court
ELMEDON
Hollybush Cottages
LAKEWOOD
FAIRMILE COMMON
Middle Pond
BY-PASS
The Fairmile
Spts. Grd.
Pav.
Hook's Corner
Chippings Farm
RIDGE
Old Common (Local Nature Reserve)
Randalls Cotts.
Old Common
Prim. Sch.
Reed's Sch.
Cricket Ground
Fairmile
SANDY LANE
WARREN LANE
OXSHOTT HEATH
Oxshott
Sports
Rec. Grd
Tennis Courts
STEEL'S LANE
OXSHOTT WAY
FAIRMILE LANE
COPSE RD.
LITTLEHEATH ROAD
Littleheath Common
Knowle Hill
Little Heath
Polyapes Scout Camp
The Furze
BRIDLE LANE
KT11
STREETS
HIGH ST.
MILL RD.
STOKE ROAD
A245
Cobham Mill
COBHAM
Pyports
Rose Lodge Wood
Lower Cobham Tilt
Elm Farm
Upper Cobham Tilt
Cemetery
Ash Ford
RIVER MOLE
Recreation Ground
STOKE D'ABERNON
Lower Farm
Cobham Park Nursery
Park Wood
COBHAM PARK
Big Roundabout
Cobham Stud
Cobham & Stoke D'Abernon
Tennis Courts
Chelsea Football Academy
Little Roundabout
Manor Farm
Nursery
Manor Farm
Pine Trees
Woodlands Farm
Woodlands
WOODLANDS
Old Parks Cottage
Old Parks
Little Park
Cossins Farm
Cobham Lodge
Parkside Sch.
South Lodge
Old Parks Copse
M25
MOLE VALLEY
M25 MOTORWAY
Downside Farm Cottage
Downside Saw Mill
Downside Farm
Downside Common
Downside
Stoke D'Abernon Bri.
Weir
Heathfield Copse
Rye Salters
Wrens Hill Wood
Slyfield House
Little Slyfield
Yehudi Menuhin Hall
COBHAM ROAD
DOWNSIDE ROAD

172
152
171
192
KT10
KT22
KT11
CLAYGATE
ELMBRIDGE
OXSHOTT
KINGSTON
MOLE
MALDEN RUSHETT
LEATHERHEAD COMMON
Arbrook Common
Arbrook Farm
Round Hill
Vale Farm
Horridon Farm
CLAYGATE COMMON (Local Nature Reserve)
Oak Cottage
Rowan Prep. Sch.
Barwell Farm
Winey Hill
Sports Ground
Chessington World of Adventures
Birch Wood
Holroyd's Plantation
Great Oaks
Sixty Acre Wood
Milbournehold Plantation
Limekiln Wood
Stokesheath Farm Cottages
Old Carpenters Plantation
The Heckets
Brickyard Plantation
The Milbournehold Wood
NEW WOOD
Stoke Wood
Jessop's Well
PRINCE'S COVERTS
Adecroft Furze Field
Eleven Acres
D'Abernon Chase Cottage
Horns Hill
Oxshott Flat
LEATHERHEAD GOLF COURSE
Fairways
Pachesham Lake
LEATHERHEAD COMMON
Little Brook Park
Crab Apple Court
Pachesham Park
Club House
Byhurst Farm
Bunkers Hill
Rushett Farm
Telegraph Hill
Reservoir (covered)
D'Abernon Ho.
Ashley Court
Hillthorpe
The Oaks
Paddock Lodge Stables
Trevona
Junction 9
Electricity Sub-Station
Jubilee Wood
Garden Centre
SILVERGLADE BUSINESS PARK
OXSHOTT HEATH
Oxshott
Pinewoods Ho.
Prim. Sch.
Danes Hill School
Sports Ground
Sports Club
Tennis Courts
Knott Park
Clouds Hill Farm
Oxshott Mead
Tyrwhitt House
Woodlands Farm
Manor Farm
Pine Trees
Woodlands Park
Old Parks Cottage
Little Park
East Lodge
Queen Elizabeth's Foundation
Dorincourt
Teazle Wood
Golf Course
Club Ho.
Driving Range
PACHESHAM GOLF COURSE
Superstore
Rowhurst Forge
Heathfield Copse
West Orchard
Brook Willow Farm
Sewage Works
Pachesham Farm
The Mounts
Research Laboratories
West Hill Sch.
Works
Barnett Wood
A3
A244
A243
A245
B280
B2430
M25
BY-PASS
ESHER COMMON
Subway
WARREN LANE
LEATHERHEAD ROAD
OXSHOTT ROAD
KINGSTON ROAD
WOODLANDS LANE
MALDEN RUSHETT LANE
FAIROAK LANE
PRINCES ROAD
PRINCES DRIVE
BIRDS HILL
THE AVENUE
MILBOURNEHOLD NEW ROAD
HIGH ST.
Ea Fa Ga Ha Ja Ka La
14 15 16 17
63 62 61 60 59 58
81 82 83 84 85 86 87 88 89 90

173
153
193
174
HORTON PARK GOLF COURSE
Driving Range
St. Ebba's Farm Cottage
Playing Field
Cemetery
ST. EBBA'S
CHESSINGTON ROAD
B2200
Chessington & Hook United FC
Horton Country Park Local Nature Reserve
HORTON COUNTRY
Four Acre Wood
KT19
HORTON
Horton Farm
B284
Hobbledown Adventure Farm Park & Zoo
David Lloyd Leisure
Polo Ground
Epsom Polo Club & Equistrian Centre
West Park Farm
Park Farm
Pond Wood
upon THAMES
Acre Hill Farmhouse
KT9
NEW EPSOM & EWELL COTTAGE HOSPITAL
Rec. Grd.
THE MEADOWS WEST PK. HOSP.
Laundry Cott.
Southfield Park Prim. Sch.
Sports Ground
Sports Club
Court Recreation Grd.
Sewage Wks.
Youth Cen.
Comm. Recycling Cen.
LONGMEAD BUSINESS CEN.
EPSOM TRADE PARK
Depot
Water Works
CHRIST CHURCH ROAD
Glanmire Farm
Hollywood Lodge
EPSOM
Stamford Green Prim. Sch.
Stamford Green
ROAD WEST B280
Kingswood Ho. Sch.
Stew Pond
Great Stew Pond
Epsom
Ashley Centre
HIGH ST.
A24
Playhouse
Epsom Common Local Nature Reserve
EPSOM COMMON
Newton Wood
THE FOREST
The Wells
St. Josephs Catholic Prim. Sch.
EPSOM
Rosebery Sch.
Rosebery Park
EPSOM GENERAL HOSPITAL
Sports Club
Tennis Courts
Woodcote Grove
The Durdans
The Grove
DORKING ROAD
Newton Wood
The Rye
Wood Field
Ashtead
KT18
WOODCOTE
Epsom Common
Woodcote Stud Farm
Halfmoon Clump
Crockingham Wood
Fish Pond
Woodcote Park
Woodcote Park (Club House)
Ridgecourt
Racing Stables
RAC GOLF COURSE
Greenside
Nursery
Ashtead House
KT21
ASHTEAD
Tennis Courts
Rec. Grd.
The Cottage in the Park
ASHTEAD PARK (Local Nature Reserve)
Island Pond
ASHTEAD PARK
Playing Field
City of London Freemen's Sch.
William III Gate
Langley Bottom Wood
Fairways
Spindles
Little Copse
Spinney's
Meridan
EPSOM DOWNS
THE STREET
EPSOM ROAD

174
Ua
Va
Wa
Xa
Ya
Za
Ab
154
194
173
KT19
KT17
KT18
KT20
EWELL
Ewell West
Ewell East
EAST EWELL
E P S O M
EPSOM DOWNS
EPSOM DOWNS RACECOURSE
E P S O M   D O W N S
THE HILL
NORK
NORK PARK
North Looe
North Looe Farm
Priest Hill
Priesthill
Greenacres
Howell Hill Nature Reserve
GOLF COURSE
CUDDINGTON GOLF COURSE
Nonsuch High Sch.
David Lloyd Leisure Cen.
Bourne Hall Mus.
Ewell Castle Jun. Sch.
Ewell Castle Senior Sch.
Hatch Furlong
Gibraltar Rec.Grd.
Sewage Wks.
Youth Cen.
Works Comm. Recycling Cen.
Glyn School
Supermarket
Playing Field
Playing Fields
Nescot Sports Centre
NESCOT (North East Surrey College of Technology)
DW Fitness (Ewell)
Sports Ground
Rainbow Leisure Centre
Wallace Fields Jun. Sch.
Alexandra Rec. Grd.
The Downs Farm
Reservoir (covered)
The Looe
DRIFT BRIDGE
Epsom College
Epsom College Fitness Cen.
Rifle Range
Club House
Littlecourt Cottage
Racing Stables
South Hatch
Warren Mead Jun. & Inf. Schs.
Round Wood
Park Wood
Nork Park Farm
Great Burgh
Tattenham Way Recreation Ground
The Beacon Sch.
Tattenham Corner
TATTENHAM CORNER
Queen's Stand
Grandstand
Buckle's Gap
The Paddock
Greenside
Downs House
Langley Vale
Crockingham Wood
The Durdans
Cemetery
Lonesome Ponds
BURGH HEATH
Superstore
Rugby Football Grd.
Warren Fm.
Cricket Grd.
Howell Hill
CHESSINGTON
A24
A232
A240
A2022
B2200
B284
B288
B289
B290
B291
B2221
A24 RD. EPSOM
EAST ST.
UP. HIGH ST.
ALEXANDRA
ROAD
COLLEGE ROAD
REIGATE ROAD
FIR TREE
YEW TREE BOTTOM RD.
TATTENHAM CRES.
GREAT TATTENHAM
TATTENHAM
TATTENHAMS
CHEAM ROAD
NONSUCH WALK
NORTHEY
KINGSWAY
BURGH HEATH
DOWNS RD.
CHURCH STREET
ASHLEY
HIGH ST.
GRANDSTAND
ROSEBERY
GROSVENOR
81
82
83
84
85
86
87
88
89
90
21
22
23
24
58
59
60
61
62
63

175
155
176
195
BELMONT
SUTTON
CARSHALTON BEECHES
CARSHALTON ON THE HILL
SM2
SM5
SM7
CR5
BANSTEAD DOWNS
BANSTEAD DOWNS GOLF COURSE
HMP DOWNVIEW
HMP HIGH DOWN
SUTTON HOSPITAL
THE ROYAL MARSDEN HOSP. (SUTTON)
THE OAKS GOLF COURSE
OAKS PARK
LITTLE WOODCOTE
BANSTEAD
WOODMANSTERNE
PARK DOWNS
CHIPSTEAD
CHIPSTEAD GOLF COURSE
RUFFETT WOOD
Scratch Wood
Great Solom's Wood
Little Solom's Wood
BRIGHTON ROAD
BURDON LANE
COTSWOLD RD.
SUTTON LANE
CROYDON ROAD
WINKWORTH ROAD
WOODMANSTERNE LANE
WOODMANSTERNE ROAD
CARSHALTON ROAD
HIGH STREET
HOLLY LANE
GARRATTS LANE
PARK ROAD
FAIRLAWN ROAD
A217
A2022
B2230
B2218
B2219
B2217
B278
B2032
Overton Grange Sch.
Overton Park
Stanley Park High Sch.
Banstead Mobility Cen.
Focus Sch. (Carshalton Campus)
Diamond Riding Centre
Drapers Centre (PRU)
Seaton House Sch.
White Lodge
Fairlawn
Hunters Field Farm
Fairlawn Farm
South Lodge
The Oaks Farm
April Cott.
Driving Range
Craggy Island
Club Ho.
Hylands Nurseries
Rose Cott.
Hillside Cotts.
Woodmansterne Primary Sch.
Barn Grove
Sports Ground
Ruffet Wood
Big Wood
Manor Cottage Farm
The Manor Ho.
The Old School
Recreation Ground
Rectory
Hengest Farm
Christmas Farm
Boundary Farm
Nicola Farm
Weston Acres
Yewlands Meadow
Courtlands Stables
The Shrubbery
Apsley
Lady Neville Rec. Grd.
Cricket Grd.
Banstead Place Brain Injury Cen.
Place Fm.
Mint Farm
The Bungalow
Courtlands Farm
Farmside
Piggery
Vyne House
New Place
Haughdell House
Woodend
Walwood
Cramond
Soloms Court
Park Farm
Pages Acre
Lambert's Shaw
Aberdour School
Canons Wood
Canons Farm Cottages
Old Lodge Farm
Caravan Park
White Hill
HOW HILLS
Club House
Chipstead
West Ridge
How Green
Gally Hills Tumuli
Rifle Range
The Cutting
Resr. (Covered)
Club House
Nursing Home
Recreation Ground
Tel. Exch.
Pickstead Shaw
Beckens Shaw
Bb
Cb
Db
Eb
Fb
Gb
Hb
81
82
83
84
85
86
87
88
89
90
25
26
27
28
63
62
61
60
59
58
B A N S T E A D

176
156
175
196
Roundshaw
Roundshaw Downs Local Nature Reserve
CR0
CR5
CR8
SM5
SM6
SUTTON
WOODCOTE GREEN
LITTLE WOODCOTE
WOODCOTE
PURLEY
PURLEY CROSS
Reedham
CLOCK HOUSE
COULSDON
Coulsdon Town
Coulsdon South
Woodmansterne
Woodcote Grove
REIGATE
CHIPSTEAD
Chipstead
WOODCOTE PARK GOLF COURSE
COULSDON COURT GOLF COURSE
CHIPSTEAD GOLF COURSE
A237
A2022
A23
B2030
B2032
B272
B276
B278
WOODMANSTERNE LA.
LITTLE WOODCOTE LA.
FOXLEY LANE
SMITHAM BOTTOM LA.
WOODCOTE GROVE ROAD
BRIGHTON ROAD
FARTHING WAY
MARLPIT LA.
CHIPSTEAD VALLEY ROAD
COULSDON ROAD
SMITHAM DOWNS RD.

177
Rb
Sb
Tb
Ub
Vb
Wb
Xb
157
81
82
83
84
85
86
87
88
89
90
178
197
Sanderstead
Purley Oaks
Riddlesdown
Kenley
Whyteleafe
Whyteleafe South
Upper Warlingham
SANDERSTEAD
KENLEY
WHYTELEAFE
HAMSEY GREEN
TANDRIDGE
CR2
CR3
CR6
CROHAM HURST GOLF COURSE
Thrift Wood
Breakneck Hill
PURLEY BEECHES
PURLEY DOWNS GOLF COURSE
SELSDON PARK HOTEL GOLF COURSE
Sanderstead Plantation
MITCHLEY WOOD
Ragged Grove
Barnfield Shaw
Ansley Berry Shaw
Dipsley's Shaw
KINGS WOOD
LITTLEHEATH WOOD
Coombes Wood
RIDDLESDOWN
KENLEY COMMON
KENLEY AERODROME
Betts Mead Recreation Ground
Dollypers Hill Nature Reserve
Elm Grove Farm
Sunnycroft Farm
Kenley House
Riddlesdown Collegiate
Warlingham Sch.
Hamsey Green Prim. Sch.
Gresham Prim. Sch.
Atwood Prim. Academy
Kenley Prim. Sch.
Ridgeway Prim. Sch.
Croydon High School
Croydon High Sports Club
Kingsdown Secondary School
Halls Wettern Tree Gdn.
Warehouse
Gasholder Sta.
Tithepit Shaw
Warlingham Court Farm
Batts Farm
Whyteleafe FC Grd.
Joysons Hill
Foxes Wood
SELSDON ROAD
ADDINGTON ROAD
LIMPSFIELD ROAD
RECTORY PARK
MITCHLEY AVENUE
A2022 COURT ROAD
GODSTONE ROAD
SANDERSTEAD ROAD
HILLBURY ROAD
A22
A235
A2022
B269
B270
B275

178
158
177
198
SELSDON
FORESTDALE
CROYDON
CR2
CR6
CR0
TANDRIDGE
FARLEIGH
Farleigh Common
HAMSEY GREEN
CHELSHAM
Chelsham Common
WARLINGHAM
KINGS WOOD
ADDINGTON COURT GOLF COURSE
FARLEIGH COURT GOLF COURSE
SELSDON PARK HOTEL GOLF COURSE
Selsdon Wood Nature Reserve
FRITH WOOD
PUPLET WOOD
GREATPARK WOOD
HOLT WOOD
FRYLANDS WOOD (Scout Camp)
Bears Wood (Scout Camp)
Dumpsy Derry
Crab Wood
Henley Wood
Crewe's Wood
Kingswood Shaw
Newlands Wood
LITTLEHEATH WOOD
SELSDON ROAD
SELSDON PARK ROAD A2022
LIMPSFIELD ROAD
B269
B270
FARLEIGH ROAD
OLD FARLEIGH ROAD
FARLEIGH COURT ROAD
HARROW ROAD
LEDGERS ROAD
CHURCH LANE
KING HENRY'S DR.
New Addington
Fieldway

NEW ADDINGTON
FICKLESHOLE
BR2
LEAVES GREEN
LONDON BIGGIN HILL AIRPORT
B R O M L E Y
TN16
BIGGIN HILL
Metropolitan Police Dog Training Establishment
Layham's Farm Cottage
Layham's Farm
Furze Bottom
Hoppershatch Shaw
Greyhound Training Kennels
West Wickham Res. (covered)
Highams Hill
Higham's Hill
Jewels Wood
Saltbox Hill Nature Reserve
Babington Place
Fairchildes Cotts.
Blackmans Cotts.
Skid Hill House
Skid Hill Farm
Mollards Wood
Owls Wood
Jerry Reddins Shaw
Broom Lodge Farm
Chelsham Court Farm
Cony Crook
White Bank
Cowyard Shaw
Beddlestead House
Round Wood
Norheads Farm
Norheads Farm Cotts.
Long Coppice
Oaklands Prim. Sch.
Biggin Hill Memorial Pool
Recreation Ground
Cemetery
South London Aero Club
Cudham Lodge Woods
Airport Industrial Estate
Biggin Hill Business Park
Concorde Business Pk.
Costains Farm
Farley Kennels
Parsons Shaw
Isle Shaw
Kingscote
Gorsey
Club House
Terminal
Milking Lane Farm
High House Farm
Ashmore Farm
Bradmoor
Greenacres Riding Sch.
Vale Farm
Longlands Nurseries
Kimberley Business Park
Glebe Farm Business Park
Westerham Lodge
Keston Court Farm
The Temple (Roman Mausoleum)
Milchder
Scout Camp
Broomfieldbank Shaw
Rushfield Shaw
Electricity Sub Station
Pitch & Putt Golf Course
Rowdown Wood
Rowdown Primary School
Vulcan Business Centre
Addington Business Centre
Meridian Centre
Warehouses
Sports Ground
Bowling Green
Milne Park
Fairchildes Prim. Sch.
Meridian High Sch.
Tennis Courts
Playing Field
Gushybank Shaw
Chapel Hill
Bograms Wood
North Lodge
South Lodge
Fairchildes Farm
Coxs Cottages
Fickleshole Farm
Fickleshole Farm Cotts.
Greathill Shaw
Midgley Shaws
Honeyoak Wood
Church Cotts.
Chelsham Court
Milbury Cott.
A233
MAIN ROAD
LEAVES GREEN ROAD
WESTERHAM RD.
SHEEPBARN LANE
SALTBOX HILL
JEWELS HILL
OAKLANDS LANE
ASH LANE
CROWN ROAD
HESIERS HILL
HESIERS ROAD
CHELSHAM COURT ROAD
BEDDLESTEAD LANE
NORHEADS LANE
LAYHAMS ROAD
HIGHAMS HILL
FAIRCHILDES ROAD
FAIRCHILDES AVENUE
BLACKMAN'S LA.
PARK ROAD
RAILPIT LANE
KINGS ROAD
SUNNINGVALE AV.
STOCK HILL
DOWNE ROAD
CHURCH LANE
159
180
199

160
179
200
Nc
Pc
Qc
Rc
Sc
Tc
Uc
81
82
83
84
85
86
87
88
89
90
42
43
44
45
62
61
60
59
58
BR2
TN16
B R O M L E Y
LOWER HOOK BUSINESS PARK
North End Farm
HIGH ELMS COUNTRY PARK (Local Nature Reserve)
Whitehillands Wood
Playgrd.
Club House
BOGEY
Westerham Lodge
GLEBE FARM BUSINESS PARK
DOWNE ROAD
Holwood Farm
PADDOCK MOBILE HOME PARK
NEW ROAD HILL
Farthing Street Farm
Farthing Street
FARTHING BARN LANE
ORANGE COURT LANE
Farthings
Orange Court
Blacklands
High Elms Clockhouse
The Rookery
HIGH ELMS GOLF COURSE
Rec. Grd.
Downe Court Riding Centre
Green Acres
Meadow Cott.
ROOKERY ROAD
CROWHILL
Little Crowhill
Hazels
ELMS
Cuckoo Lodge
CUCKOO WOOD
Great Molloms Wood
ROOKERY VILLAS
Rookery Cotts.
MILL
BELLE VUE RD.
SNOWDEN CL.
STAUSFIELD PL.
STANDARD RD.
HIGH VW.
Gorringes
LANE
NORTH END
CUDHAM PK. RD.
The Shack
Burnt Gorse
White Earth
Long Acre
Play Fld.
Cemy.
Rec. Grd.
Sch.
BUTCHERS YD.
WELLER PL.
Avondale Farm
Endsleigh
Snag Farm
SNAG LANE
Southfields
LEAVES GREEN
DOWNE
HIGH ST.
Downe Hall
SPINNEY WY.
Milking Lane Farm
Petleys Farm
Petleys
CUDHAM ROAD
Downe Hall Farm
Christmas Tree Farm
HAZELWOOD
DOWNE AV.
HAZELWOOD RD.
Hang Grove Farm
Rose Cott.
Club House
MILKING LANE
LUXTED
Downe Lodge
Thompson Wood
Square Shaw
Hazel Wood
The Oaks
The Vicarage
Green Hill
Down House
Downe Court
Oak Lodge
Snag Lane Fm.
Millfield Wood
GREEN HILL
WEST
BROWNE GDS.
HANGROVE ROAD
LONDON BIGGIN HILL AIRPORT
Lordfield Shaw
Downe Bank Nature Reserve
Mace Farm
Mace Farm Cotts.
WEST KENT GOLF COURSE
Cricket Grd.
Pav.
The Shaws (Girl Guides' Camp)
Overshaws
Hostye Lodge
Hostye Fm.
Hostye Bungalow
Hang Grove
Shaws Cott.
SOW WOOD
Beechwood
The Boundary
DOWNE HILL
Fairways
MACE
Granada Ho.
Holmleigh
Flora Cottage
Spencers Grove
Luxted
Brownsea Hut (Boy Scouts)
Cudham Lodge Woods
Snotsdale Wood
Greenacres
Church Hill Farm
Foxburr Wood
South London Aero Club
Ladies Wood
Spencers Grove
Luxted Fm.
Twenty Acre Shaw
Downe Camp
Angas Home (Site)
Hook Wood
Leasons Wood
Clock Twr.
Bird House Wood
Great Southfield Shaw
SINGLE STREET
Single Street
CHURCH APP.
Cudham Ct. Farm
Rec. Grd.
Pav.
Bottom Barn Farm
CHURCH HILL
CUDHAM
BIRDHOUSE
Bromley Croft
BIGGIN HILL
CONCORDE BUSINESS PK.
BIGGIN HILL BUSINESS PARK
St. Julians Farm
Cudham Hall
The Lodge
Cacket's Cotts.
CACKET'S
LORD DARBY MEWS
Playing Fld.
Darwin Sports Centre
Strakes Shaw
Costains Farm
Bertrey Cotts.
STREET
Charles Darwin School
Barn Farm
Primrose Farm
Parkfield Fm.
Cudham C of E Prim. Sch.
BERRYS LANE
BERRY'S
Hall
Vicarage
MAGNOLIA
DRIVE
LUNAR CL.
ACER
SPRUCE
Cemetery
RESTAVON CARAVAN SITE
Homeleigh Farm
BERRY'S GREEN
Blackbush Shaw
Parsonage Fm.
Westbrook Farm
Maple Farm
Biggin Hill Prim. Sch.
Grubb Grounds Farm
Recreation Ground
Golden Acres
Littlewood Farm
Brickyard Farm
Underhill Farm
Corkers Fm.
Rosehaugh Farm
Biggin Hill Memorial Pool
Piggeries
CHURCH ROAD
APERFIELD
JUNIPER CL.
CHERRY LODGE GOLF COURSE
Oak Farm
Warren Farm
Nurseries
Foal Farm Animal Rescue Centre
Club House
Valley View
GREEN
Cedar Cott.
NELSON CL.
The Grove
New Barn Farm
Aperfield
Pimlico Wood
Laurel Farm
HORNS GREEN
Thrift Farm
THRIFT LA.
A233
MAIN ROAD
ROAD NEW
Barn End
EDWARD RD.
MOSELLE RD.
FOXEARTH CL.
CLARENCE RD.
Park Farm
Silver Beech Fm.
AVENUE
ST. WINIFRED'S
BELVEDERE
WOODBURY
GROVE
Farley Kennels
Buckhurst Farm
Cudham Frith
Baverstock Farm
CUDHAM FRITH
Withins Wood
PAYNESFIELD RD.
HILL
Gorsey
Piggeries
TUDOR CT.
EAGLES DR.

SEVENOAKS ROAD
A21
PRATT'S BOTTOM
Knockholt
HEWITTS ROUNDABOUT
ORPINGTON A224
HALSTEAD
KNOCKHOLT POUND
KNOCKHOLT
LETT'S GREEN
CHEVENING
SEVENOAKS
BR6
TN14
CHELSFIELD LAKES GOLF COURSE
Eleven Oaks
Rounds Wood
Chelsfield Hill Wood
Yew Tree Cottage
Charmwood Farm
Little Molloms Wood
Stonehouse Farm
Rock-Field Shaw
Watercrofts
Charm Wood
Pratts Bottom Wood
Pratt's Grove
Stud Farm
Homefield Spring
Hook Wood
Rushmore Hill Wood
Birthday Wood
Widmore Farm
Norsted Manor Farm
Fairtrough Farm
Hillside Farm
Works
Sumachs
Beech Cott.
Lower Brooms Wood
Upper Brooms Wood
Lattice Coppice
Garden Cen.
Chalkhurst Wood
Broom Wood
High Wood
Halstead Primary Sch.
Kangles Wood
Lower Randle's Wood
Deerleap Wood
The Washneys
Perry Wood
Closeheath Wood
Warren Court Farm
Park Farm
Woodvale Cottage
Hayman's Wood
Piece Wood
Curry Wood Farm
NEWYEARS WOOD
Newlands Wood
New Stables Farm
Curry Farm
Long Bottoms Shaw
Hawfield Wood
Birches Croft
Chapel Croft
Jockey's Wood
Blueberry Farm
The Spinney Nurseries
Aldwyn Lodge
Cophall Wood
Garden Centre
Knockholt Village Centre
Play. Fields
Ivy Farm
Lees Wood
Shelleys
Homefield
Lodge
Burlings
Baston Wood
Cooper's Wood
The Keyhole
Ashfield Wood
Wood Farm
St. Katharine's Knockholt C of E Prim. Sch.
Ash Platt
Chatham's Ride
Minny Wood
Crittleshaw
Knockholt Wood
Knockholt Beeches
Knockholt Lodge
Lord
Bowman's Orchard
Highfield House
Masts
Sand Banks
Park House
High Land
Burwood
Keepers Cottage
Mountross Farm
Summerhouse Wood
Coalhearth Cottage
The Mount
Minquintin
PARK WOOD
The Spinney
Sundridge Hill Farm
Pilgrims' Way (course of)
Hazlet Wood Farm
Roughfields Wood
Pilgrims Ho.
Withlerfield
Stoneings Cottage
Scott's Lodge
Hagbush Wood
Chevening House
161
201
182
158

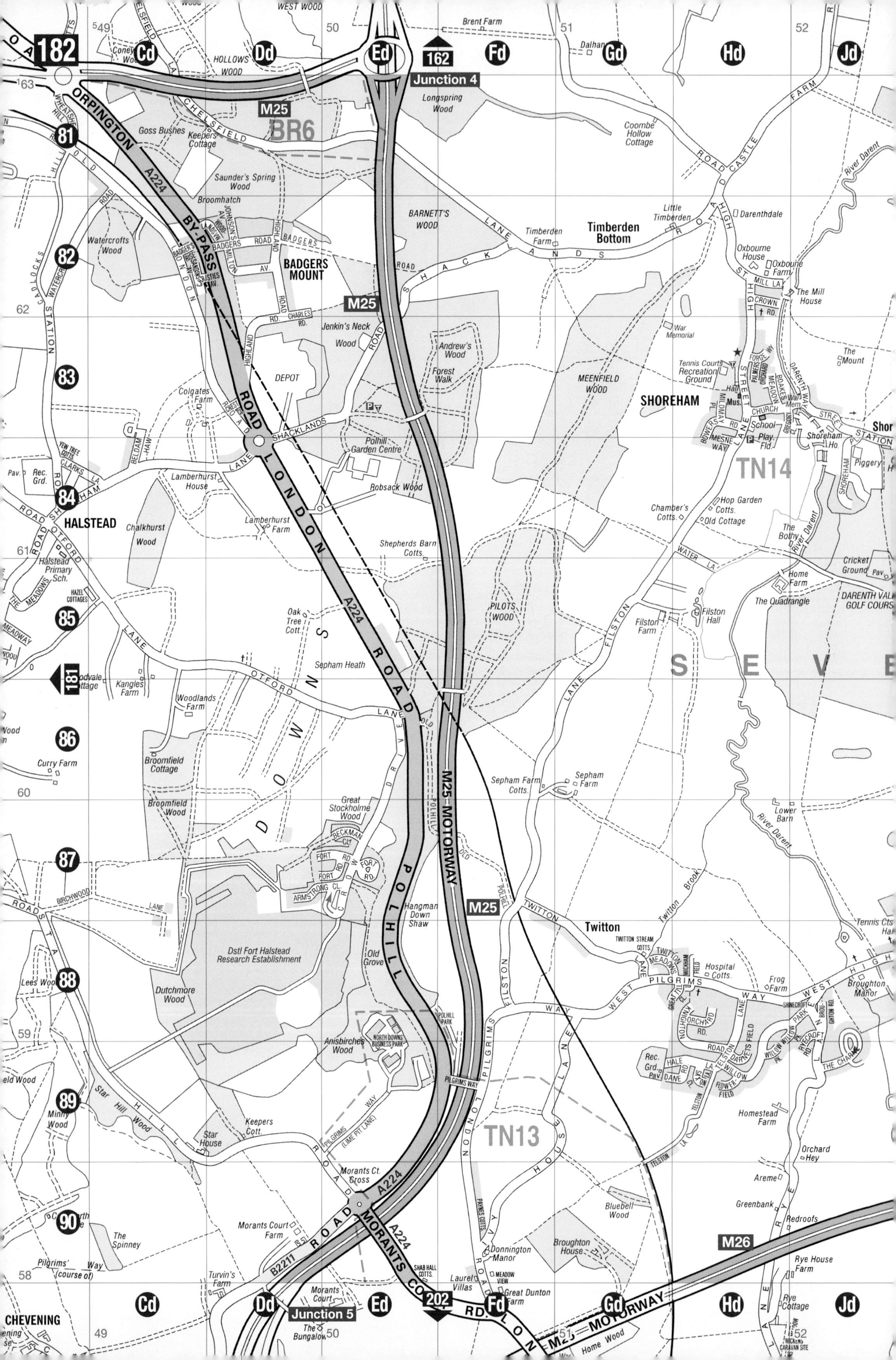
182
162
181
202
Junction 4
Junction 5
M25
M26
M25 MOTORWAY
BR6
TN14
TN13
Cd
Dd
Ed
Fd
Gd
Hd
Jd
49
50
51
52
63
62
61
60
59
58
81
82
83
84
85
86
87
88
89
90
WEST WOOD
Brent Farm
Dalham
HOLLOWS WOOD
Longspring Wood
ORPINGTON BY-PASS
A224
CHELSFIELD
Goss Bushes
Keepers Cottage
Coombe Hollow Cottage
CASTLE FARM ROAD
Saunder's Spring Wood
Broomhatch
BARNETT'S WOOD
Little Timberden
Darenthdale
River Darent
Timberden Farm
Timberden Bottom
SHACKLANDS ROAD
Watercrofts Wood
BADGERS MOUNT
Oxbourne House
Oxbourne Farm
MILL LA.
The Mill House
HIGH ST.
CROWN RD.
CHARLES RD.
HIGHLAND RD.
M25
Jenkin's Neck Wood
War Memorial
Andrew's Wood
Forest Walk
MEENFIELD WOOD
Tennis Courts
Recreation Ground
The Mount
DEPOT
SHOREHAM
Colgates Farm
CHURCH ST.
School
Play. Fld.
Shoreham Ho.
STATION
Polhill Garden Centre
SHACKLANDS
BELDAM HAW
Lamberhurst House
Pav.
Rec. Grd.
Piggery
Robsack Wood
HALSTEAD
Chalkhurst Wood
Lamberhurst Farm
Hop Garden Cotts.
Old Cottage
Chamber's Cotts.
The Bothy
Shepherds Barn Cotts.
WATER LA.
Halstead Primary Sch.
Cricket Ground
Home Farm
The Quadrangle
DARENTH VALLEY GOLF COURSE
THE MEADOWS
Oak Tree Cott.
PILOTS WOOD
Filston Hall
Filston Farm
MEADWAY
A224 ROAD
LONDON ROAD
Sepham Heath
OTFORD LANE
Kangles Farm
Woodlands Farm
FILSTON LANE
Curry Farm
Broomfield Cottage
Broomfield Wood
Sepham Farm Cotts.
Sepham Farm
Great Stockholme Wood
BECKMAN CL.
Lower Barn
FORT RD.
ARMSTRONG CL.
POLHILL
BIRCHWOOD
Hangman Down Shaw
Twitton
Tennis Cts
Dstl Fort Halstead Research Establishment
Old Grove
TWITTON STREAM COTTS.
TWITTON MEADOWS
Hospital Cotts.
Frog Farm
Broughton Manor
Lees Wood
Dutchmore Wood
PILGRIMS WAY WEST
Anisbirches Wood
NORTH DOWNS BUSINESS PARK
POLHILL PARK
Rec. Grd.
Pav.
THE CHART
Star Hill Wood
Minny Wood
PILGRIMS WAY
Keepers Cott.
Star House
Homestead Farm
Orchard Hey
(LIME PIT LANE)
TELSTON LA.
Morants Ct. Cross
Areme
Greenbank
Redroofs
Bluebell Wood
The Spinney
Morants Court Farm
Donnington Manor
Broughton House
Rye House Farm
Pilgrims' Way (course of)
Turvin's Farm
B2211
MORANTS COURT RD.
Morants Court
Laurel Villas
MEADOW VIEW
Great Dunton Farm
Rye Cottage
CHEVENING
The Bungalow
Home Wood
M26 MOTORWAY
WICKENS CARAVAN SITE

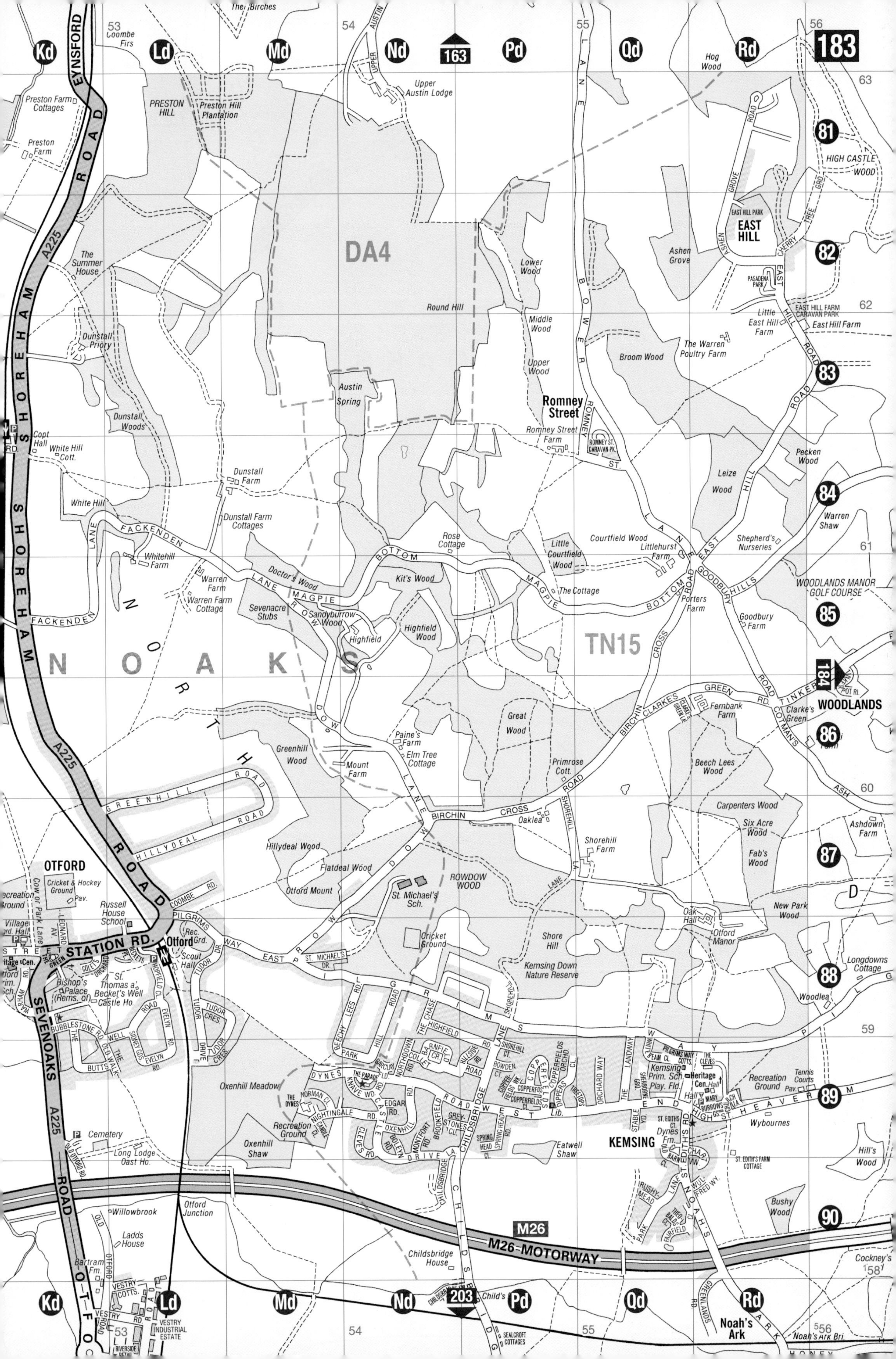

Kd
Ld
Md
Nd
163
Pd
Qd
Rd
EYNSFORD
SHOREHAM ROAD
A225
Preston Farm Cottages
Preston Farm
Coombe Firs
The Birches
PRESTON HILL
Preston Hill Plantation
Upper Austin Lodge
Hog Wood
High Castle Wood
East Hill Park
EAST HILL
Ashen Grove
DA4
The Summer House
Lower Wood
Round Hill
Middle Wood
Upper Wood
Pasadena Park
East Hill Farm Caravan Park
East Hill Farm
Little East Hill Farm
The Warren Poultry Farm
Broom Wood
Dunstall Priory
Austin Spring
Romney Street
Romney Street Farm
Romney St. Caravan Pk.
Dunstall Woods
Copt Hall
White Hill Cott.
Dunstall Farm
Pecken Wood
Leize Wood
White Hill
Dunstall Farm Cottages
Rose Cottage
Little Courtfield Wood
Courtfield Wood
Littlehurst Farm
Shepherd's Nurseries
Warren Shaw
Whitehill Farm
Warren Farm
Warren Farm Cottage
Doctor's Wood
Kit's Wood
The Cottage
Porters Farm
Woodlands Manor Golf Course
Sevenacre Stubs
Sandyburrow Wood
Highfield
Highfield Wood
Goodbury Farm
TN15
FACKENDEN LANE
MAGPIE BOTTOM
SEVENOAKS
NORTH
Fernbank Farm
Clarke's Green
WOODLANDS
184
Greenhill Wood
Paine's Farm
Elm Tree Cottage
Great Wood
Mount Farm
Primrose Cott.
Beech Lees Wood
Greenhill Road
Hillydeal Road
Oaklea
Carpenters Wood
Six Acre Wood
Ashdown Farm
Shorehill Farm
Fab's Wood
Hillydeal Wood
Flatdeal Wood
Otford Mount
Rowdow Wood
St. Michael's Sch.
OTFORD
Cricket & Hockey Ground
Pav.
Russell House School
Recreation Ground
Village Hall
Cricket Ground
Shore Hill
Oak Hall
Otford Manor
New Park Wood
Station Rd.
Otford
Rec. Grd.
Scout Hall
Pilgrims Way
Kemsing Down Nature Reserve
Longdowns Cottage
Woodlea
Bishop's Palace (Rems. of)
St. Thomas a' Becket's Well
Castle Ho.
SEVENOAKS ROAD
A225
Oxenhill Meadow
The Dynes
Recreation Ground
Oxenhill Shaw
Kemsing Prim. Sch. Play. Fld.
Heritage Cen.
Hall
Recreation Ground
Tennis Courts
Pav.
Wybournes
KEMSING
Dynes Fm.
Eatwell Shaw
St. Edith's Farm Cottage
Hill's Wood
Cemetery
Long Lodge Oast Ho.
M26
M26 MOTORWAY
Willowbrook
Otford Junction
Ladds House
Bushy Wood
Childsbridge House
Bartram Fm.
Cockney's
Vestry Cotts.
Vestry Industrial Estate
Riverside
Child's
203
Sealcroft Cottages
Noah's Ark
Noah's Ark Bri.
53
54
55
56
57
58
59
60
61
62
63
81
82
83
84
85
86
87
88
89
90

184
Sd
Td
Ud
Vd
Wd
Xd
Yd
164
Sidehilly Wood
CLEARWAYS CARAVAN PK.
KINGSWOOD CT.
CHURCH
CHANCEL CL.
KAYSLAND PK.
KINGSINGFIELD
SOUTHFIELDS
Crowhurst Farm
Cotts.
Club House
HIGH CASTLE WOOD
Blooming Wood
EAST HILL PARK
EAST HILL
PASADENA PARK
EAST HILL FARM CARAVAN PARK
Little East Hill Farm
East Hill Farm
KNATTS VALLEY
SEVENOAKS
Kingsingfield Shaw
KNATTS VALLEY CARAVAN PARK
Pav.
Lib.
Sports Ground
Ten. Cts.
Hall
Skate Pk.
WEST KINGSDOWN INDUSTRIAL EST.
WEST KINGSDOWN
ST. EDMUND'S
Prim. Academy
Richardson's Farm
Ashen Shaw
M20
Southfield Shaw
Inges Shaw
THE LONDON GOLF COURSE
Little Boutshole Shaw
Hatch Wood
Bouts Corner
West Kingsdown Smockmill
MILLVIEW PARK
Mill Farm
THE GROVE
LONDON ROAD A20
STANSTED
Rose Cottage
Abbott's Wood
Noah's Rough
Stacklands Wood
HOLLYWOOD LA.
Pells Farm
Knatts Farm
Subway
Depot
Kent Firefighting Museum
Garden Cen.
Peckham Wood Corner
Pecken Wood
MANOR ROAD
Water Wood
Knockmill
ST. CLERE
SCHOOL LANE
Gravelpit Wood
PECKHAM WOOD
Warren Shaw
Shepherd's Nurseries
Knockmill Wood
HILL
Barn Gate
WOODLANDS MANOR GOLF COURSE
Tinkerpot Shaw
M20 MOTORWAY
Terry's Lodge Farm
Goodbury Farm
KNATTS
POT LANE
TINKER
Club House
WOODLANDS
183
KNOCK MILL LANE
BIRCHES WOOD
Drane Farm
Clarke's Green
Bali Hai Farm
Whitehill Wood
Arboretum
TERRY'S LODGE ROAD
EXEDOWN
ASH LANE
COTMAN'S
Cotman's Ash
Ashdown Farm
Carpenters Wood
Six Acre Wood
Fab's Wood
Summeryards Wood
Terry's Shoulder of Mutton
Chalk Pit Wood
DOWNS
White Hill
New Park Wood
Kester
Cotmans Ash
Pilgrims' Way
KEMSING ROAD
Downlands
Longdowns Cottage
Woodlea
The Wilderness
ST. CLERE
The Bothy
EXEDOWN ROAD
Coney Shaw
ST. EDITH COTTS.
Recreation Ground
Tennis Courts
Crowdleham House
Crowdleham
HEAVERHAM ROAD
Heaverham
Manor House
Chequers
Lower St. Clere
West Yaldham Farm
YALDHAM MANOR DR.
Yaldham Manor
Wybournes
ST. EDITH'S FARM COTTAGE
Hill's Wood
Martin Spring Wood
FEN POND ROAD
Broughton Cottages
Bushy Wood
M26
Cockney's Wood
Spindles
FEN POND COTTS.
FEN MEADOW
CHAUCER BUSINESS PARK
Hook Wood
Rumleys Wood
204
Noah's Ark
Noah's Ark Bri.
Kemsing
POT LANE
Osier Bed
81
82
83
84
85
86
87
88
89
90
56
57
58
59
60
61
62
63

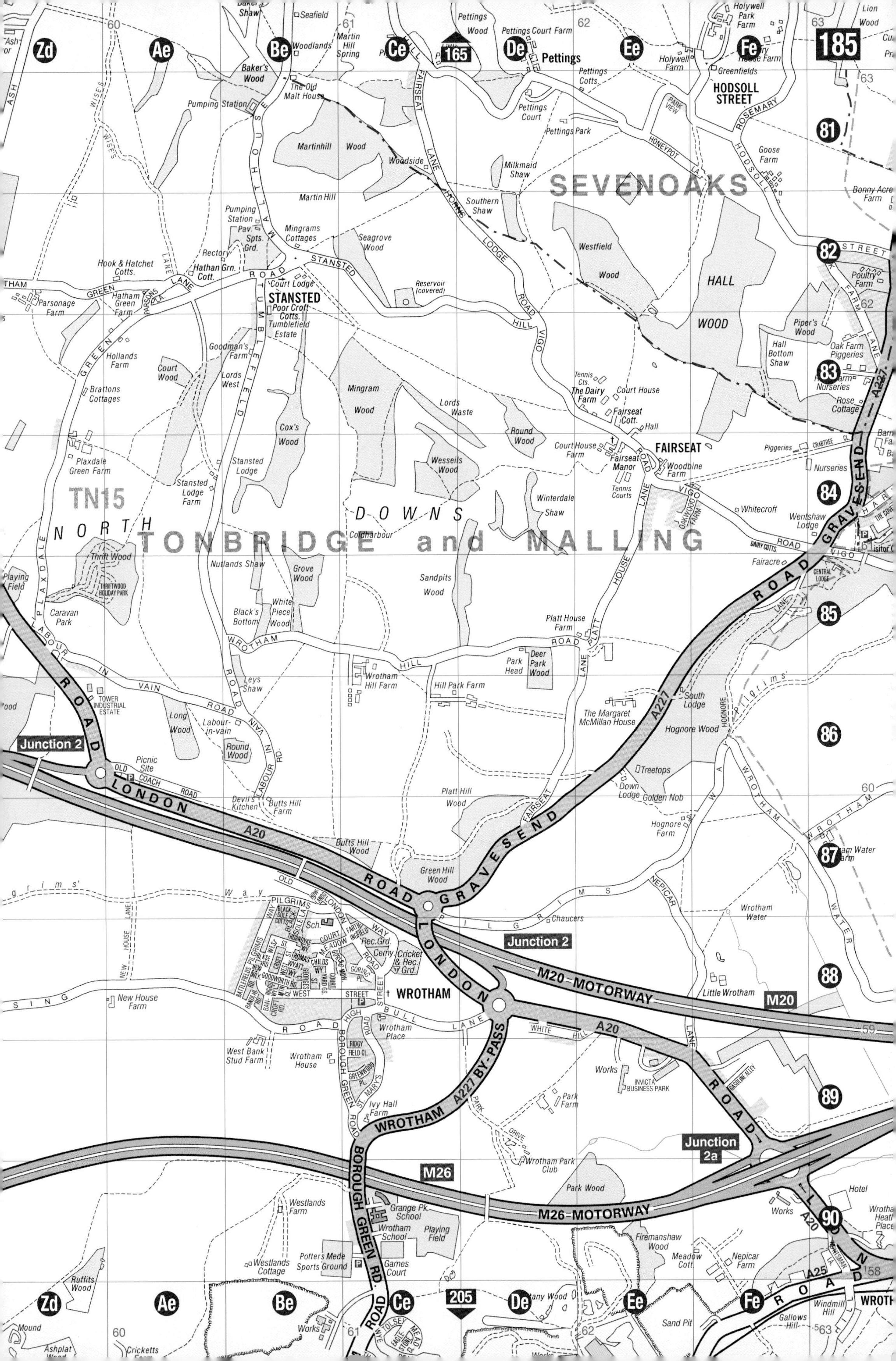
165
205
SEVENOAKS
TONBRIDGE and MALLING
NORTH DOWNS
TN15
STANSTED
FAIRSEAT
WROTHAM
HODSOLL STREET
Pettings
Junction 2
Junction 2a
M20 MOTORWAY
M26 MOTORWAY
A20
A227
A25
LONDON ROAD
GRAVESEND ROAD
WROTHAM BY-PASS
BOROUGH GREEN RD.
Baker's Wood
Martinhill Wood
Mingram Wood
Westfield Wood
HALL WOOD
Cox's Wood
Court Wood
Thrift Wood
Grove Wood
Round Wood
Wesseils Wood
Hognore Wood
Platt Hill Wood
Green Hill Wood
Park Wood
Fireманshaw Wood
Wrotham Water
Little Wrotham
Wrotham Park Club
Grange Pk. School
Wrotham School
Potters Mede Sports Ground
Court Lodge
Fairseat Manor
The Dairy Farm
Court House
Wrotham Hill Farm
Hill Park Farm
The Margaret McMillan House
Caravan Park
Tower Industrial Estate
Invicta Business Park
Works
Hotel

186
166
A
B
C
D
E
F
G
1
2
3
4
5
6
7
8
9
10
93
94
95
96
58
57
56
55
54
53
Hog Lees
Century Range
Wisdom Corner
Wisdom Farm
Princess Christian Homes & Training Centre
Staffordlake
Stafford Lake
Nursery
Grindstone Handle Corner
Knaphill Lwr. Sch.
KNAPHILL
Superstore
BROOKWOOD COUNTRY PARK
GU21
Clock Tower
Ranges
BISLEY CAMP (National Rifle Association)
Bisley Camp
Mus.
Shrubb's Copse
Nursery
Sheet's Heath
Recreation Ground
Cowshot Common
Lord Roberts Centre
Sports Ground
Running Track
Ten. Cts.
PirbrightCamp
Sports Ground
Parade Gd.
Alexander Barracks
BROOKWOOD
Prim. Sch.
Basingstoke Canal
Cowshot Hill
Victoria Farm
Pirbright Bridge
The Slade
Cowshot Manor Bri.
Brookwood
Club House
Memorial
Military Cemetery
PIRBRIGHT COMMON
The Cottage
Works
Dawney's Hill
New Pond
Canal
Stanley Pool
Manyuema
Furzehill Place
Aruwimi Wood
Mazamboni Farm
Holy Island
Ruwenzori Hill
Causeway Farm
Cook's Green
Pirbright Village Prim. School
Sch.
Wickham's Farm
Burrow Hill
West Heath
PIRBRIGHT
Pav.
Hall
BROOKWOOD
CEMETERY
West Hall Farm
GU24
Grove Farm Nurseries
The Manor House
Mill Ho.
Manor Farm
Cove Bri.
Vines Farm
Hazelacre Hill
Pirbright Lodge
Nursery
White's Farm
Hogleys Farm
Crammond House
Burner's Heath
PIRBRIGHT COMMON
No. 2 Range
Piggery
Rowe Lane Farm
Ardeley
Bowling Hill
Furzefield Copse
Admiral's Walk
Heatherwood
Hodge
G U I L D F O R D
Gorselands
Chair Hill
Track End
Bullswater Common
BULLSWATER COMMON
Bullswater Bridge
Fox Corner
Berrylands Farm
Fox Corner Community Wildlife Area
Kiln Hill
Upper Stanford Bridge
Bakersgate
Stream Farm
Rails Farm
Stanford Farm
Sludge Lagoons
The Bungalow
Stanford Common
Lower Stanford Bridge
Filter Beds
Sewage Disposal Works
Oldfield Wood
Sludge Beds
Peat Moor
STANFORD
COMMON
The Animal Virus Research Institute
Westwind
Peatmoor Pond
Henley Park Ranges
MERRIST WOOD
Henleypark Lake
COBBETTHILL
COMMON
Standinghill Wood
Wireless Station (Disused)
NORMANDY BUSINESS PARK
COBBETT HILL
Guildford College of Further & Higher Education (Merrist Wood Campus)
GU3
Slyfield Wood
Leapingbar Copse
Mercury Lodge
Deyrolles
Park Farm
MERRIST WOOD GOLF COURSE
Guildford College of Further & Higher Education (Merrist Wood Campus-Equine Centre)
The Glen
HENLY BUSINESS PARK
Finavon
Club House
Littlefield Common
Littlefield Cottages
Island Copse
Clasford Common
FAIROAKS CARAVAN PARK
PINE PARK
A324
A322
B3012
B3405
B3032
B380
CONNAUGHT ROAD
BROOKWOOD LYE
BAGSHOT ROAD
GOLE ROAD
STANLEY HILL
GRANGE ROAD
SCHOOL LANE
DAWNEY HILL
PIRBRIGHT GRN.
GUILDFORD RD.
GUILDFORD ROAD
ASH ROAD
ALDERSHOT ROAD
CEMETERY PALES
CHAPEL LANE
CHURCH LANE
MILL LANE
ROWE LANE
MALTHOUSE LANE
QUEENS ROAD
BRUNSWICK ROAD
HENLEY GATE
COBBETT HILL
Newbridge Commom
BULLSWATER COMMON RD.
HEATH MILL LA.
BERRY LA.
STORR'S LANE
LAWFORD'S HILL RD.
THE FAIRWAY
WORPLESDON HEATH

WOKING
HERMITAGE ROAD
A324
HERMITAGE RD.
Woking Crematorium
ST. JOHNS
Cricket Ground
St. John's Lye
Football Ground
Club House
HOOK HEATH
WOKING GOLF COURSE
MOUNT HERMON
WYCH HILL LA.
GUILDFORD ROAD
A320
Barnsbury Primary School
Westfield Prim. Sch.
Westfield Common
Mayford Meadows Local Nature Reserve
MAYFORD
Freemantles School
GU22
W O K I N G
WORPLESDON GOLF COURSE
SMART'S HEATH LANE
B380
Smart's Heath
Prey Heath
PREY HEATH ROAD
Worplesdon
Kemishford Bridge
Burdenshott Hill
BURDENSHOTT ROAD
PYLE HILL
SUTTON GREEN GOLF COURSE
188
Bridley Manor Farm
Bridley Manor
Bridley Copse
Blanketmill Farm
Rickford Mill
Hoe Stream
Rye Farm
Hockley Lands
Gooserye
Whitmoor Common
Whitmoor Pond
Whitmoor Wood
Poor Jack's Wood
JORDAN HILL
Rickford Common
Perryhill Cottage
Norton Farm
WORPLESDON
Whitmoor & Rickford Commons Local Nature Reserve
WHITMOOR COMMON
Mount Pleasant
Maryland
WORPLESDON ROAD
A322
GU4
Britten's Pond
PITCH PLACE
Stringer's Common
JACOBS WELL
SALT BOX ROAD
WOKING ROAD
Frog Lane Farm
Cox's Farm
Sunridge House
Whitmoor House
Elm Nursery
Tangley
Driving Range
Rifle & Pistol Range
97
98
99
500
53
54
55
56
57
58
H
J
K
L
M
N
P
1
2
3
4
5
6
7
8
9
10
167

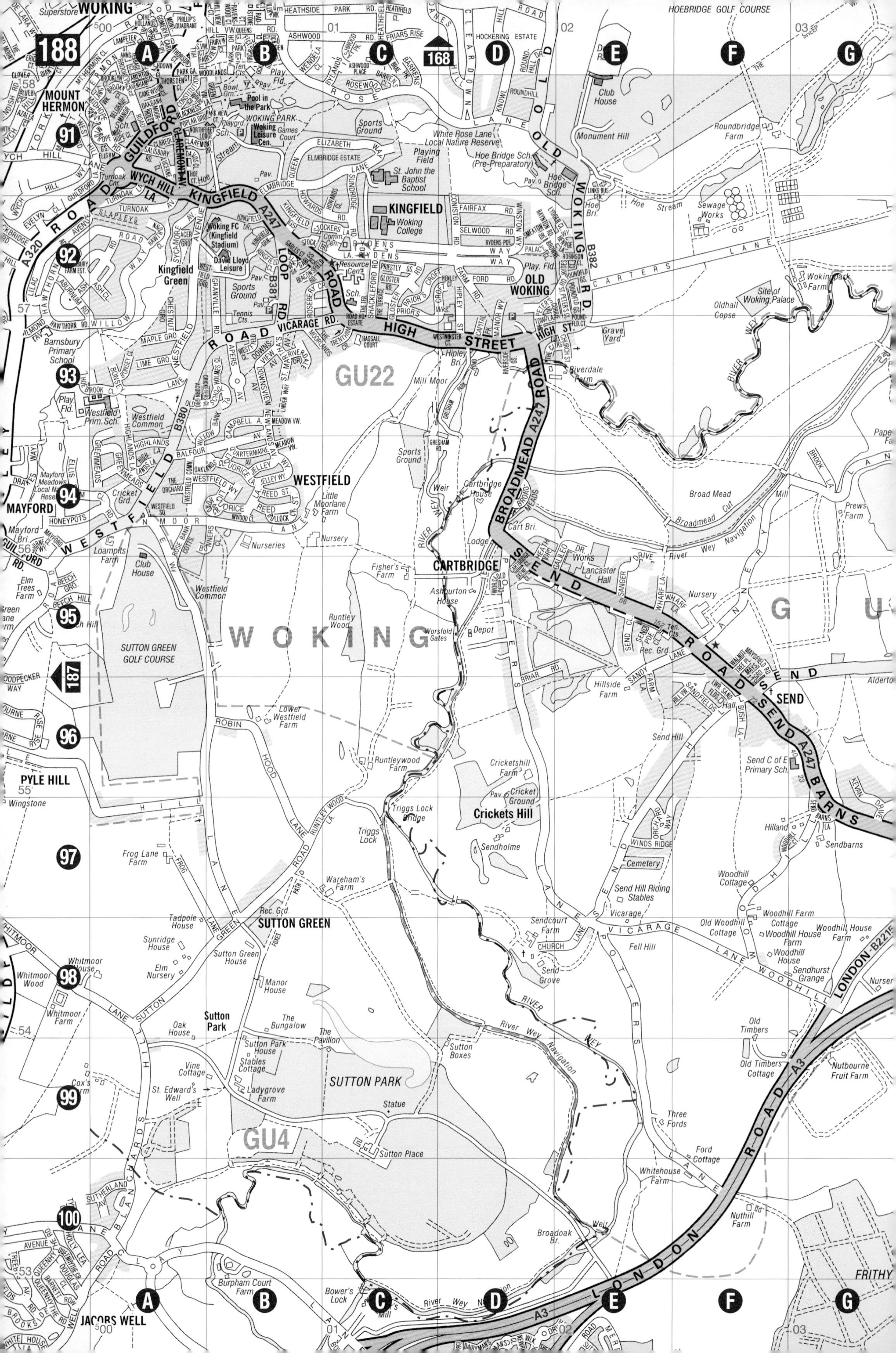

188
WOKING
168
187
HOEBRIDGE GOLF COURSE
MOUNT HERMON
MAYFORD
PYLE HILL
JACOBS WELL
KINGFIELD
Kingfield Green
WESTFIELD
OLD WOKING
CARTBRIDGE
SEND
SUTTON GREEN
Sutton Park
SUTTON PARK
Crickets Hill
FRITHY
W O K I N G
GU22
GU4
SUTTON GREEN GOLF COURSE
Woking Leisure Cen.
Woking FC (Kingfield Stadium)
David Lloyd Leisure
St. John the Baptist School
Woking College
White Rose Lane Local Nature Reserve
Hoe Bridge Sch. (Pre-Preparatory)
Monument Hill
Club House
Roundbridge Farm
Sewage Works
Site of Woking Palace
Oldhall Copse
Woking Park Farm
Riverdale Farm
Broad Mead
Broadmead Cut
Prews Farm
Westfield Common
Westfield Prim. Sch.
Barnsbury Primary School
Mayford Meadows Local Nature Reserve
Loampits Farm
Elm Trees Farm
Little Moorlane Farm
Fisher's Farm
Runtley Wood
Ashburton House
Worsfold Gates
Lancaster Hall
Hillside Farm
Send Hill
Send C of E Primary Sch.
Hilland
Sendbarns
Lower Westfield Farm
Runtleywood Farm
Triggs Lock
Cricketshill Farm
Cricket Ground
Sendholme
Frog Lane Farm
Wareham's Farm
Send Hill Riding Stables
Cemetery
Woodhill Cottage
Woodhill Farm Cottage
Woodhill House Farm
Woodhill House
Sendhurst Grange
Old Woodhill Cottage
Fell Hill
Sendcourt Farm
Send Grove
Tadpole House
Sunridge House
Elm Nursery
Sutton Green House
Manor House
Whitmoor House
Whitmoor Wood
Whitmoor Farm
Oak House
The Bungalow
The Pavilion
Sutton Park House
Stables Cottage
Vine Cottage
St. Edward's Well
Ladygrove Farm
Statue
Sutton Place
Sutton Boxes
River Wey Navigation
Old Timbers
Old Timbers Cottage
Nutbourne Fruit Farm
Three Fords
Ford Cottage
Whitehouse Farm
Nuthill Farm
Broadoak Br.
Burpham Court Farm
Bower's Lock
HIGH STREET
BROADMEAD ROAD A247
SEND ROAD
SEND BARNS
LONDON ROAD A3
WOKING RD. B382
GUILDFORD ROAD A320
KINGFIELD A247
WESTFIELD ROAD
VICARAGE ROAD
LOOP RD. B381
B380
ROBIN HOOD LANE
SUTTON GREEN
BLANCHARDS HILL
POTTERS LANE
VICARAGE LANE
WOODHILL
CARTERS LANE
WYCH HILL LA.
A B C D E F G
91 92 93 94 95 96 97 98 99 100
500 01 02 03
53 54 55 56 57 58

PYRFORD VILLAGE
Vicarage
189
169
190
H
J
K
L
M
N
P
04
05
06
07
The Prae
Wheelers Farm
The Bourne
Abbey Stream Bri
Abbey Stream
Newark Priory (Remains of)
Newark New Bri
Wey Navigation
Newark Lock
Mill House
Mill Stream
Ockham Mill Stream
Irish Hole
RIVER WEY
Walsham Meadow
Walsham Plantation
Walsham Lock
WARREN FARM MOBILE HOME PARK
Sludge Bed
WARREN LA
Ockham Mill
Ockham Mill Farm
Nursery
Rock Garden
Garden Cen.
RHS Garden Wisley
Battleston Hill
58
A3
Stratford Bridge
El. Sub. Sta.
91
92
93
94
95
96
97
98
99
100
Dunsborough Farm
Dunsborough Park
Bridgefoot Farm
Cricket Grd.
Sewage Works
BARATARIA PARK
NEWARK COTTS
Homewood Farm
Lodge
B367
Ripley Green
HALF MOON COTTS.
GREENSIDE COTTS.
PORTSMOUTH ROAD
HIGH STREET
Mus.
Sch.
Hall
Chapel Farm
RIPLEY
Ripley Ct. Sch.
ROSE LANE
Tennis Court
Buckingham Lodge
Tankards
Chimneys
Old Rectory Farm
Church End
Chestnut Farm
B2039
OCKHAM LANE
Ockham Park
OCKHAM PARK
Park Wood
Guileshill Lodge
GUILESHILL LANE
Guileshill Farm
57
56
55
54
53
B2215
RIPLEY BY PASS
PAPERCOURT LANE
Works
Devonia Farm
Hurst Farm
GRANDIS COTTS.
MILESTONE CL.
Papercourt Sailing Club
Industrial Estate
GUILDFORD
B368
MANOR RD.
SEND MARSH
HEATH M.
HEATH RISE
GROVE HEATH
Willow Farm
Grove Heath Farm
Grove Heath
GU23
Loveland's Copse
Tennis Court
Ryde Farm
Roborow
Roborow Wood
Bachelor's Copse
Romer Cottage
The Old Cottage
Russets
BACHELORS LANE
Hangover Ho.
GROVE HEATH CT.
GAMBLES LANE
Hungry Hill Cottage
Hungry Hill
Garlick's Arch Copse
Garlick's Arch
KILN LANE
HAWTHORN RD.
WILLOW DR.
MAPLE
BURNTCOMMON
Oldlands
Oldlands Copse
Three Bridges
The Paddocks
Nursery
Sheppardsgrove Copse
Brambleride Copse
Depot
CLANDON ROAD
Tithe Barns
Tythebarns Farm
TITHEBARNS LANE
Cedarwood
Oak Ride Farm
Sussex Farm
Holride Farm
August Hill
Jury Farm
Brambleride
Blendon
Highcotts Wood
The Oaks
Hazelhurst Cottage
Southernwood
Nursery
HMP SEND
THE SPINNEY
Furze Copse
KINGSTON LANE
A247
GREEN LANE
GU4
RIPLEY LANE
Lower Hammond's Farm
Dedswell Manor Farm House
Dedswell Manor Farm
MALACCA FARM
Humphreys Copse
Nursery
LIME GROVE
CLOSE
Fish Pond
Roam Wells
Martin Furze
Gason Wood
WEST CLANDON
THE ROAD
Birch Farm
Buttinham Copse
CROCKERY
WALK

190
170
189
Q
R
S
T
U
V
W
91
92
93
94
95
96
97
98
99
100
ELM CORNER
Hatchford End
Hatchford
Hatchford Wood
Hatchford Park
Wisley Airfield (disused)
MARTYR'S GREEN
May's Green
Bridge End
OCKHAM
KT11
GU23
Chaffers Copse
Hook Wood
Stumps Grove
Effingham Junction
DRIFT GOLF COURSE
BARNSTHORNS WOOD
Blue Ride
The Forest Nature Reserve
EFFINGHAM COMMON
Effingham Common
Even Wood
G U I L D F O
KT24
Horsley
WEST HORSLEY
EAST HORSLEY
Lollesworth Wood
Frenchlands Copse
Nomad Theatre
Horsley Towers
M25
A3
B2039
OCKHAM ROAD NORTH
OCKHAM ROAD SOUTH
LONG REACH
RIPLEY LANE
EAST LANE
GREEN LANE
FOREST ROAD
THE DRIFT
The Raleigh School
Norton Wood
Pound Farm
Old Oak Common
The New Preserve
Roundabout
Hurst Hill
Brick Kiln Copse
Clamp Rough
Barns Thorns
Gallows Grove
Ridings Wood
Little Ridings
Garden Grove
Gravel Hill Pond
Pennymead Lake
House Pond
Dancing Pit
Jury Farm
Lower Hammond's Farm
Hammonds Farm
Pincott Farm
Dene Place
Barcombe Farm
West Horsley Place
Place Farm
Parkrow Copse
Kingston Meadows Rec. Grd.
Football Ground
Gas Pumping Station
Broom House
Manor Farm
Roundtree Farm
Cricket Ground
Waterloo Farm
Slade Farm
Bramble Cottage
Slade Cottage
Ridings Copse
Rydings Farm
Romer Cottage
The Old Cottage
Russets
Guileshill Lodge
Old Rectory Farm
Chestnut Farm
Appstree Farm
Pavilion Recreation Ground
Alms Heath
Trulliber Copse
Little Upton Cottages
Upton Farm
The Lodge
Tanner's Copse
Bridge End Farm
Blackmoor Farm
Stumpsgrove Cottages
Stumps Grove Farm
Wilderness Farm
Hunt's Copse
Highfield Farm
Cold Norton Farm
Orchard Farm
Hazeldene Farm
Nurseries
Poynters Farm
Brickfield Copse
Pondtail Farm

191
171
192
X
Y
Z
Aa
Ba
Ca
Da
11
12
13
14
53
54
55
56
57
58
91
92
93
94
95
96
97
98
99
100
ELMBRIDGE
Downside
Downside Common
Downside Farm Cottage
Downside Saw Mill
Downside Farm
D'Abernon Bri.
Heathfield
M25 MOTORWAY
M25
Muggeridge Wood
Wrens Hill Wood
Thornets Wood
Spring Meadow Wood
Down Wood
Rye Salters
Bull Riding Farm
Cobham Service Area
Withy Bed
Peaked Rough
Chasemore Wood
Chasemoor Farm
Yehudi Menuhin Hall
Yehudi Menuhin School
Bookham Lodge Stud
Bookham Lodge
Little Slyfield
Slyfield House
Slyfield Farm
Little Wood
Cattle Grid
Slyfield Cotts.
Sheepbell Farm
Bushy Copse
Gardeners Cottage
Gate Lodge
The Birches
Nursery
Godstone Ho.
Ashspring Copse
Barracks Farm
Monksgrove Farm
KT22
New Barn Wood
Hundred Pound Bridge Wood
Hill Farm
Hundred Pound Bridge
Newmarsh Farm
North Surrey Gun Club
Bank's Common
Banks Farm
Little Bookham Common
GREAT BOOKHAM COMMON
The Common
Merritt's Farm
Manor Cottage
Oak Tree Cottage
The White House
Bushy Thicket
Thicket Copse
Little Mornshill Wood
Banks Way House
Mornshill Farm
GREAT MORNSHILL WOOD
Sewage Works
Beehive Farm
Bookham
Slater's Oak
Norwood Farm
Long Maddox Farm
Pikes Pool
Herons Reach
Mallards Mere
Lower Farm
Little Bookham
OAKEN WOOD
KT23
MOLE VALLEY
Indian Farm
Moat
THORNET WOOD
Upper Leewood Farm
GREATLEE WOOD
Littlelee Wood
Great Ridings Plantation
Cook's Coppice
Effingham Lodge Farm
The Vineries
The Grange (Training Centre)
Preston Cross
Rec. Grd.
The Dawnay Sch.
GREAT BOOKHAM
LEATHERHEAD ROAD
Jun. Sch.
Inf. Sch.
Keswick
Bowling Green Pav.
Sports Centre
Howard of Effingham School
Manor House Sch.
Manor Farm
Prim. Sch.
Club
Orestan Farm
Parrott's Coppice
Home Farm
King George's Field
EFFINGHAM
Hall
Rolls Farm
Nursery
The Nook Farm
Works
The Short Wood
Floriana Farm
GUILDFORD ROAD
A246
Club House
Tennis Courts
Polesden Lacey Infant Sch.
Kenilworth Riding Stables
Goldstone Farm
Phoenice Park Cottage
Standard Hill
North Lodge
Connicut

Ea
Fa
Ga
Ha
Ja
Ka
La
172
191
14
15
16
17
53
54
55
56
57
58
91
92
93
94
95
96
97
98
99
100
Heathfield Copse
West Orchard
PACHESHAM GOLF COURSE
Pachesham Farm
Brook Willow Farm
Sewage Works
Refuse Composting Plant
Household Waste Recycling Site
River Mole Local Nature Reserve
Slyfield Farm
Bushy Copse
The Mounts
Research Laboratories
Gutter's Bri.
Bushy Platt
LEATHERHEAD COMMON
West Hill Sch.
Therfield Sch.
Barnett Wood
M25
Junc. 9
Play Field
St. Peter's Prim. Sch.
Randalls Park Crematorium (Leatherhead)
Randalls Park Lawn Cemetery
Randalls Research Park
Leatherhead
Barracks Farm
Ashspring Copse
Monksgreen Farm
Nurseries
Sports Field
Tennis Cts.
KT22
Nature Reserve
Fairfield
Leatherhead Trinity Prim. Sch.
LEATHERHEAD COMMUNITY HOSP.
Woodlands Sch.
LEATHERHEAD
Leatherhead FC (Fetcham Grove)
Thorncroft Manor
Leatherhead Leisure Centre
Rifle Range
Vale Lodge
Downside Farm
Reservoir (covered)
Downside
Naomi Acres
FETCHAM
Fetcham Village Inf. Sch.
Oakfield Jun. Sch.
Rec. Grd.
Hall
Elmer Water Works
Cherry Orchard Farm
Hawk's Hill
River Mole Local Nature Reserve
MOLE
Bocketts Farm Park
YOUNG STREET
GUILDFORD ROAD
BY-PASS
A246
A245
A243
A24
B2122 (HAWK'S HILL)
B2450
B2033
B2122
B2430
B2209
GREAT BOOKHAM
Keswick
The Junipers
The Hazels
The Crown of the Hill
Roaringhouse Farm
Fetcham Downs
Denshire Hill
The Scrubs
Walnut Tree Clump
Updown Wood
Lilac Cottage
Icehouse Combe
Norbury Park Farm
Norbury Park
Pressforward Cotts.
Givons Grove
Oakwood Court
Seven Gables
The Priory
Nursery
Polesden Lacey Infant Sch.
Kenilworth Riding Stables
KT23
Saw Mill
NORBURY PARK
Bookham Wood
Druids Grove
Swanworth Farm
Mole Cott.
Box Hill Sch.
MICKLEHAM
St.Michael C of E Inf.
RH5
Mickleham Hall
Charley Mount
Juniperhill
Lodge Farm
Phoenice Farm
Cowslip Farm
Cowslip Cottage
Fredley Manor
North Lodge
MICKLEHAM BY-PASS
LONDON ROAD
DORKING ROAD
KINGSTON ROAD
RANDALLS ROAD
EPSOM ROAD
HIGHLANDS RD.
COBHAM ROAD

193
ASHTEAD PARK
173
ASHTEAD
EPSOM DOWNS
Langley Bottom
EPSOM
KT21
KT18
ASHTEAD HOSP
City of London Freemen's Sch.
William III Gate
Langley Bottom Wood
Langley Bottom Farm
Larch Field
Tudor Croft
Ermyn Lodge
Chase Farm Stud
Thirty Acres Barn
Gillette Cottages
Nohome Farm
M25 MOTORWAY
M25
Stane Street
Addlestead Wood
Addlestead Farm
North Lodge
Loretta Lodge
Fourfield Close
Little Hurst Wood
Walton Hurst Farm
Highdale
Highlands Farm
Reservoir (covered)
Hambleton Wood
Headley Court Rehabilitation Cen.
Headley Park Farm
Headley Park
Edes Barn Cottages
194
Hook Wood
Tyrrell's Wood
Club House
Round Wood
Pignut Wood
Court Farm
Lee Green
Shepherds Close
REIGATE ROAD
B2033
MILL WAY
VALLEY
TYRRELLS WOOD GOLF COURSE
GREAT HURST
Oyster Hill
NOWER WOOD
Great Hurst Wood
Golf Course
Paddock Cottage
Radio Bungalow
Leatherhead Downs
The Garden Ho.
Tumber House
Costal Wood
Subway
Tunbarr
Tot Hill
Hayes Cotts.
HEADLEY
Rectory
WALTON
Wellbottom Cottage
Cherkley Court
CHERKLEY WOOD
LEECH LA.
Kingsbarns
Manor House
Heath Farm
Mickleham Downs
Garden Cottage
Goodman's Furze
Dean Wood
Cricket Ground
Great Hayes
White Lodge
Cockshoot Wood
LODGE BOTTOM
Cockshot Cottage
White Hill
Warren Farm
Warren Farm Cottages
High Ashurst (Camping Site)
HEADLEY HEATH
HEADLEY COMMON
Headley Grove Farm
Headley Grove
Herring Grove
TYE LANE
Junipers
Whitehill Cottage
Middle Wood
Bullen Wood
Headley Plantation
Headley Reservoir
Pebble Coombe
Fraser Heath
Ma
Na
Pa
Qa
Ra
Sa
Ta
18
19
20
21
58
57
56
55
54
53
91
92
93
94
95
96
97
98
99
100

EPSOM DOWNS
EPSOM DOWNS RACECOURSE
Tattenham Corner
Tattenham Cnr.
174
TATTENHAM CORNER
THE HILL
Downs House
Racing Stables
Langley Vale
Langley Bottom
Trial Course
The Warren
Prim. Sch.
The Warren Rec. Grd.
ROSEBERY
GROSVENOR ROAD
E P S O M
WALTON DOWNS
Nohome Farm
KT18
Downs View Wood
Downs View
Round Wood
Little Hurst Wood
The Hurst
Edes Barn Cottages
193
GREAT HURST WOOD
M25
Great Hurst Wood
Subway
WALTON PARK WOOD
M25 MOTORWAY
Rectory
Manor House
Heath Farm
Great Hayes
White Lodge
Headley Grove Farm
Headley Grove
Herring Grove
Queen's Wood
Frith Park
Frith Park Farm
Hedgecroft
Walton Heath
Walton Oaks
The Hermitage
LITTLE HEATH
Little Heath Cottages
B2033
DORKING ROAD
Pebble Coombe
Headley Reservoir
Fraser Heath
Round Wood
Watson's Wood
206
MOGADOR
Mogador Cottages
Wayside Farm
Natural Gas Pressure Reducing Station
Rookery Cottages
Red Lodge Mobile Home Park
Stables
B A N S T E A D
H E A T H
The Gallops
WALTON HEATH GOLF COURSE
WALTON HEATH
Chussex Plain
Castle House
Dowding Lodge
Heath Farm
Hockey Ground
Recreation Ground
Works
Pav.
PRIORY HOSPITAL STURT GRANGE
B2220
CHEQUERS LANE
B2032
WALTON ON THE HILL
WALTON
Walton Manor
Walton Manor Farm
Bramley Sch.
Recreation Ground
Club House
Ebbisham Farm
SANDHILL WOOD
Mere Pond
Priests' Mere Pond
Hedge Corner
NEW B290 ROAD
MILL B290 ROAD
Sports Ground
Workhouse Pond
Twistwood
Wingfield Stud Farm
Wildwoods Riding Centre
Wildwoods
EBBISHAM LANE
MOTTS HILL
TADWORTH
Tadworth
TADWORTH STREET
B2220
Tadworth Prim. Sch.
THE CHILDREN'S TRUST
Tadworth Park
KT20
BONSOR
BRIGHTON ROAD
A217
Heather Cl.
Workhouse Pond
Garden Farm
Millennia Ho.
KINGSWOOD PARK
TADWORTH ROUNDABOUT
BURGH HEATH
Marbles Pond
Lonesome Ponds
A240
Superstore
Epsom Downs Prim. Sch. & Children's Cen.
Banstead Athletic FC Grd.
Tadworth Leis. & Comm. Centre
Pit Wood
Play. Fld.
B290
EPSOM LANE NORTH
ASHURST B290 ROAD
CROSS RD.
STATION APP
Lothian Wd.
Ua
Va
Wa
Xa
Ya
Za
Ab
21
22
23
24
53
54
55
56
57
58
91
92
93
94
95
96
97
98
99
100

195
175
196
207
BANSTEAD WOOD
CHIPSTEAD
Chipstead Bottom
Kingswood
KINGSWOOD
LOWER KINGSWOOD
MUGSWELL
MARGERY
Upper Gatton
GATTON
UPPER GATTON PARK
GATTON PARK
SM7
CR5
RH2
SURREY DOWNS GOLF COURSE
KINGSWOOD GOLF COURSE
OUTWOOD LANE
WATERHOUSE LANE
HIGH ROAD
MARKEDGE LANE
CHIPSTEAD LANE
BRIGHTON ROAD
A217
B2032
M25
MOTORWAY
GATTON BOTTOM
ROCKY LANE
RECTORY LANE
WHITE HILL LANE
HOGSCROSS LANE
HOGDEN LANE
SMUGGLERS PIT PLANTATION
Canons Farm
Canons Farm Cottages
Reads Rest Cottages
Lunch Wood
Ambrose Cottages
Sports Ground
Perrotts Farm
Fames Rough
Chiphouse Wood
Well Copse
Little Manor
Outwood Shaw
Out Wood
Poorfield Wood
Porters Wood
Shabden Park
Embers Shaw
The Rectory
Ley Cottage
Court Lodge
Church Green
Longshaw
Courtyard Theatre
Elmore
Elmore Pond
Chipstead Meads (Recreation Ground)
Vincents Grn.
Starrock Farm
Noke Farm
The Grove
Fairdene
South Lodge
The Belt
Hog's Cross
The Ruffett
Round Wood
Ruffetts Wood
Fourteen Acre Shaw
Tickners Wood
Northdean Shaw
The Long Plantation
Eyhurst Farm
Eyhurst Farm House
Eyhurst Farm Lodge
Club Ho.
Cornerways
Windlestraw
Longcroft Shaw
Woodcot Cottages
The Lodge
Warren Cotts.
Kingswood Warren
Kingswood House
Training Cen.
Furze Hill
Club House
Pigeonhouse Farm
Top Shaw
Round Wood
Prior's Field
South Manor
Reeves Rest
Homefield Shaws
Northfield Shaw
Field View
Penlan
Wellfield Shaw
May Cottages
Old Farm House
Southerns Farm
April Cottage
Parson's Green
Old Rectory
Orchard Cottage
Marksland
Park Farm
Sopers Cottage
Bugspit Shaw
Boars Green Shaw
Boars Green Farm
Woodfield Place
Mawingo
Merton Vean
Sandy Crest
Sports Ground
Beechen Copse
Millfield Wood
Mill House
Corner Cottage
Lilac Cottage
Furzefield Copse
Keepers Cottage
Long Wood
Little Wood
UPPER GATTON WOOD
Pilgrims
Lodge Fm.
Dell Wood
Furzefield Shaw
High Bank
Grub Wood
Gatwick Wood
Colts Bushes
Home Field Farm
Lovelands Mead
Colts Lodge
Upper Gatton Park
Markedge Shaw
Ashtead Hill
Old Mint House
The Old Forge
Jubilee Plantations
Park Shaw
Ishstar Gate
Council Yard
Tennis Courts
Recreation Ground
Kingswood Prim. Sch.
Kingswood Manor
Kingswood Knoll
The Crossways
Crossways Farm
Majority Plantation
Manships Shaw
Kingswood Hill
Glebe Shaw
Catsbrain East
Gatton Field Shaw
Grove Shaw
Gatlands Shaw
Whitehall Farm
North Lodge
Tower Wood
High Beeches
Catsbrain Shaw
Royal Alexandra & Albert School
East Lodge Farm
Reservoir
Margery Ho.
Margery Hall
REIGATE GOLF CO
Bb
Cb
Db
Eb
Fb
Gb
Hb
25
26
27
28
58
57
56
55
54
53
91
92
93
94
95
96
97
98
99
100

196
176
195
208
CHIPSTEAD
HOOLEY
Netherne-on-the-Hill
REIGATE
CHALDON
MERSTHAM
GATTON
WARWICK WOLD
CR5
RH1
RH2
Junction 7
Junction 7 (M25)
Junction 8 (M23)
M23 MOTORWAY
M25 MOTORWAY
M23
M25
A23
B2031
BRIGHTON ROAD
LONDON ROAD
HAPPY VALLEY
FAIRDEAN DOWNS
Starrock Farm
Starrock Wood
Hooley Farm
Woodplace Farm
Football Ground
Boxers Wood
Three Cornered Shaw
Petersshole Shaw
Netherne Leisure Cen.
Netherne Wood
Sports Ground
Bowling Green
Grasscuts Shaw
Lord's Wood
Dean Farm
Coldroast Shaw
Sparticles Shaw
Sparticles Wood
Mersham Tunnel
Alderstead Heath
Alderstead Farm
Alderstead Fort (Disused)
Furzefield Wood
Cold Blow
Five Pillars
Tollsworth Manor
Tollsworth Cottages
Uplands Farm
Court Farm
Church Green
Glebe House
The Rookery
Figgs Wood
The Gullet
Devilsden Wood
The Devil's Den
Sparklie Wood
Ditches Shaw
Tollers Farm
Coulsdon Coll.
Playing Field
Lacey Green
Parson's Pightle
Piles Wood
Hooley Lodge
Community Centre
The Rectory
Ley Cottage
Embers Shaw
Court Lodge
Church Green
Church Shaw
Noke Farm
Fairdene
Tennis Courts
The Ruffett
Ruffetts Wood
Fourteen Acre Shaw
Chipstead Meads (Recreation Ground)
Vincents Grn.
Prior's Field
Coldroast Shaw
Hedgefield Shaw
Harps Wood
Bugspit Shaw
Sopers Cottage
Woodfield Place
Fox Shaw
Boars Green Shaw
Mawingo
Field End
Harps Oak Shaw
Boars Green Farm
Hoaths Farm House
Furzefield Shaw
Marling Glen Wood
Gatton Bottom
Church Hill
Church Hill Covert
Beech Wood
Wellhead Wood
North Lodge
Ashtead Hill
Grave Yard
Denehole
Ockley Hill
Ockley Wood
Subway
Bedlams Bank
Meadshaw
Heronswood Mere
Withy Shaw
The Glade House
Sunacre
Orchard Cottage
Cricket Ground
REIGATE HILL GOLF COURSE
Club House
Middle Lodge
The Bushetts
Whitehall Farm
East Lodge Farm
Merstham
Rec. Grd.
Playing Field
The Oakley Youth Club
Hawthorns
Brambleside
Rest Harrow
Hilltop Farm
The Beacon
Six Brothers Field
Rocksha House
The Grove
Coppice Lea

197
177
209
198
KENLEY AERODROME
WHYTELEAFE
Whyteleafe South
OLD COULSDON
Oasis Academy Coulsdon
Windmill Farm Kennels
COULSDON COMMON
Sports Ground
Rifle Ranges
Dean Hill
SURREY NATIONAL GOLF COURSE
Club House
Westway Common
Public Open Space
Clifton Hill Sch.
Fryern Farm
Fryern Broom Wood
Rook Hill
Round Wood
Bulstaff Cottage
Five Acres
TANDRIDGE
Green Hollows
Roverdene (Scout Camp Site)
Oakhyrst Grange Sch.
Willey Park Farm
Stanstead Farm
WHITE HILL
Grubs Wood
Whitehill Tower
Arthur's Seat
Quarry Hanger Lodge
Quarry Cottages
Quarry Ho.
Spring Bottom
Rocky Shaw
Quarry Wood
Quarry Hall Farm
Pendell Wood
Little Park Hill
The Mound
War Coppice House
White Rose Farm
Black Bushes
Green Croft
The Yews
OLDPARK WOOD
Ten Acre Shaw
GRAVELLY HILL
WT Station
Gravelly Place
Fosterdown Wood
Pilgrim Fort
Little Moyle
Treetops
The Moyle House
THE DOWNS RESIDENTIAL SITE
Freer Farm
Castle View
GODSTONE INTERCHANGE
Junc. 6
M25 MOTO
CATERHAM
Caterham-on-the-Hill
Caterham School
Caterham Preparatory School
NORTH DOWNS PRIVATE HOSP.
Upwood Gorse
Eight Acre Wood
The Chalet Farm
The Chalet
Valley Sports Ground
St. John's C. of E Prim. Sch.
QUEENS PARK
Playing Field
Beech Hanger
Resr. (covered)
Reservoir (cov.)
Recreation Ground
Tillingdown Farm
Stony Hill
Park Shaw
Paddock Barn
Ockley Wood
The Devils Hole
Dial-bank Wood
Quarry Cott.
Kennel Shaw
Round Shaw
Birchwood House Farm
Playing Field
MANOR PARK
Cloisters
Burntwood Hill
Little Manor
Blize Wood
Bryngwyn Farm
Whyteleafe FC Grd.
Joyson Hill
Foxes Wood
Shepherds Shaw
Rifle Range
Rydon's Wood
St. Francis' Catholic Prim. Sch.
Audley Prim. Sch.
Sunnydown School
de Stafford School
de Stafford Sports Cen.
WASPES LODGE ROUNDABOUT
Windmill Shaw
CR8
CR6
CR3
RH9
A22
B2208
B2030
B2031
CATERHAM BY-PASS
GODSTONE ROAD
BANSTEAD ROAD
CHALDON LANE
WESTWAY
TOWN END
CROYDON ROAD
GODSTONE HILL
WOLDINGHAM ROAD
STITES HILL

198
WARLINGHAM
178
210
197
Reservoir (covered)
LIMPSFIELD ROAD
Highlands Farm
Gard Centre
Worms Heath
Birchen Shaw
WT Station
B269
Slines Green
Slynes Oak
Barnards Lodge
Derwent House
Nore Hill House
Slines Oaks
Owlpen
Nore Hill Pinnacle (Local Nature Reserve)
NORE HILL
Worms Heath Cotts.
HOVINGS HOLE
Green Dragons Airsports
Warren Barn Farm
Plantation
WOLDINGHAM GOLF COURSE
Club Ho.
Cottershill Shaw
Windmill Shaw
WAPSES LODGE
Hogerty Hill
Hogerty Hill Plantation
Dukes Hill House
Halliloo Cotts.
Scoldhill Plantation
Dukes Hill Shaw
Langlands
Butlers Dene
Highways
Playing Field
Viaduct Lodge
WOLDINGHAM GARDEN VILLAGE
Shrublands
Round Shaw
Birchwood House Farm
North Downs Farm
Garden Cen.
Park View
Birchwood House
Stony Hill
BIRCH WOOD
CR3
Woldingham
Station Lodge
Station Cottage
Woodlea Prim. Sch.
The Bushes
Church Road Farm
Sports Grd.
Orchard Farm
Riding Stables
Valleyfields
Greenhill Shaw
Worlds End
Park Shaw
Marden Park Farm
WOLDINGHAM
Marden Park
Beechwood Riding School
The Nash
Marden Hillboxes Farm
Hunters Hollow
Little Church Wood
TANDRIDGE
Whistlers Wood Farm
Whistlers Wood
Bundanoon
The Warren
Thorn Acres
Whistlers Hollow
Templehill Plantation
Club House
NORTH DOWNS GOLF COURSE
Woldingham School
Rookery
Garden Cott.
Great Church Wood
Oxted Tunnel
Heathdown
Chalk Pit
Carr's Croft
Tennis Court
East Lodge
MARDEN PARK
Pilgrims Place
Wood Rising
Quarry Cotts.
Works
Williams Cottages
St. John's C of E Prim. Sch.
Paddock Barn
Paddock Wood
South Hawke
Stubbs Copse
Paddock Barn
Playing Field
Ockley Wood
The Rumps
Whitefield Plantation
South Hawke
Keepers Cott.
Oxted Downs (West)
Armitage Wood
Rye Shaw
Lodge Wood
Hamfield Shaw
Five Acre Shaw
THE DOWNS RESIDENTIAL SITE
The Devils Hole
Dialbank Wood
Wks.
WINDERS HILL
Winders Hill Cotts.
Quarry Cott.
South Lodge
Horse Shaw
Hazel Shaw
Chaldon's Farm
TANDRIDGE HILL
Rye Wood
Quarry Farm
Godstone Vineyards
RH9
Hanging Wood
Kennel Shaw
Freer Farm
Primrose Bank
M25
Robins Grove Wood
Castle View
Flint Hall Farm
Oak Pollard
Nixon's Pit
Tandridge Hill Farm
Barrow Green Court
GODSTONE INTERCHANGE
Generals Grove
Junction 6
M25 MOTORWAY
THE ABBEYS
Barrow Green Farm
North Model
Greenacres
The Firs
Palmer's Wood
Priory Shaw
Barrow Green Farm Cotts.
CATERHAM BY-PASS
A22
GODSTONE ROAD
GODSTONE HILL
WOLDINGHAM ROAD
STATION ROAD
CHURCH ROAD
NORTHDOWN ROAD
LUNGHURST ROAD
SLINES OAK ROAD
UPLAND ROAD
SOUTHFIELDS ROAD
GANGERS LANE
TANDRIDGE HILL LANE
HOGTROUGH LANE
Yb
Zb
Ac
Bc
Cc
Dc
Ec
35
36
37
38
53
54
55
56
57
58
91
92
93
94
95
96
97
98
99
100

Cony Crook
White Bank
Cowyard Shaw
Long Coppice
Milbury Cott.
Round Wood
Beddlestead House
Forge Cott.
Beddlestead Farm
Culver Keys
Beddlestead Farm Cotts.
Lumberdine Wood
CR6
BROMLEY
Kingscote
Isle Shaw
Parsons Shaw
Gorsey Down Farm
Home Wood
Lusted Hall Farm
Cedar Lodge
Cherry Tree Shaw
Shepherds Shaw
Paynesfield Coppice
TATSFIELD
Manor House
Ashen Shaw
TN16
Longlands Shaw
Kitchen Grove
Tatsfield Green
Park Farm
Colegates
Glengarry
Tatsfield Primary School
Sports Field
Furze Corner
Cheverells Farm
Cheverells
PARK WOOD GOLF COURSE
Beacon Shaw
Clarkslane Shaw
PITCHERS WOOD
Gas Installation
Wireless Station
Hill Park (Local Nature Reserve)
Clarks Lane Farm
CLARKS LANE
B2024
Paygate Cott.
Coldharbour Beeches
High Trees
BOTLEY HILL
Botley Hill Farm
TITSEY ROAD
WT Station
Park Lane Cotts.
Botley Hill Cotts.
The Old Rectory
The Old Rectory Lodge
Pilgrims' Farm
Pilgrims Cotts.
RIDGE
WT Station
TITSEY PLANTATION
Titsey Place & Gardens
TITSEY
Lodge
Rectory
Titsey Court
Hell Shaw
Flint House
Whistlers Steep
Pitchfont Farm
Barton Shaw
The Rookery
Limpsfield Lodge Fm.
Lodge
TITSEY PARK
Roughfield Shaw
Church Wood
Square Wood
Clacket Lane Service Area
Oxted Downs (East)
M25
RH8
Park Farm
South Green
TITSEY WOOD
Lodge
M25 MOTORWAY
Chalkpit Wood
Limpsfield Grange Sch.
Broomlands Farm
Ashen Coppice
Moorhouse Sand Pits
Downs Way Sch.
Playing Field
St. Mary's C of E Jun. Sch.
Tennis Courts
Games Ct.
Swimming Pool
Oxted Sch.
Theatre
Hookwood
Hookwood Park
Grubstreet Shaw
Grubstreet Copse
Streatfield's Copse
Sunridge
Wymondley
Kentish Border
Hookwood Bungalows
Grubbe Street Wood
Kennels
Cemetery
OXTED
LIMPSFIELD
Tennis Club
De-Tillens
Scout Hall
Pebble Hill
Pebble Hill Ho.
The Gate House
The Rips
Oxted
Leisure
Tandridge DC Offs.
WESTERHAM ROAD
A25
179
200
211

Nc
Pc
Qc
180
Rc
Sc
Tc
Uc
42
43
44
45
58
57
56
55
54
53
91
92
93
94
95
96
97
98
99
100
Farley Kennels
Park Farm
Silver Beech Fm.
Buckhurst Farm
Cudham Frith
CUDHAM FRITH
Fairmead
Baverstock Farm
Stud Farm
Withins Wood
Horsey Down Farm
Leecroft
SOUTH STREET
A233
HARROW
BOMBERS
Piggeries
Bombers Lodge
Shellem Wood
TATSFIELD
The Hermitage
Little South Street Farm
WESTERHAM HILL
Southwood Farm
CUDHAM RD.
Manor House
Ashen Shaw
Painter's Wood
Great South Street Farm
Southwood
Bombers Farm
Squirrel Wood
Cudham Frith Farm
B R O M L E Y
Buckhurst
Bombers Cott.
Joelands
Woodlands
Cudham Grange
GRAYS
SILVERSTEAD
Sheepshed Cott.
Westerham Riding School
Tatsfield Green
Park Farm
RAG HILL
SILVERSTEAD LANE
Silversted
Gray's Farm
Colegates
Tor Croft
Hawley's Corner
Broomcocks Wood
Westerham Heights
Heights View
Fort Cotts.
Deepdale
Deepdale Cott.
Pilgrim House
Play. Fld.
GOLF COURSE
Park Wood
Club House
Compton Chase
Homestead
Little Betsom's Farm
Round Shaw
Lime Villa
BETSOM'S HILL
Highfield
Whitelands Shaw
Hill Park (Local Nature Reserve)
CHESTNUT AVENUE
Cerne Easter
Whitlands
CLARKS LANE
B2024
Halston
Rysted
Betsoms
Playing Field
Pav.
Hill Park
REDWOOD HO.
PILGRIMS WAY
S E V
The Old Rectory
Tatsfield Court Farm
Betsomhill Farm
The Old Rectory Lodge
LONDON ROAD
CROYDON
Force Green Farm Cottages
199
TANDRIDGE
Gaysham
Green Leas
Force Green Farm
Rowtye Wood
Force Green
M25
Hartley Wood
WESTERHAM WOOD
BEGGARS
Clacket Green
Church Wood
Clacket Wood
Cupids Coppice
Gaysham Cottages
M25-MOTORWAY
Roughfield Shaw
Square Wood
Tennis Court
Churchill C of E Prim. Sch.
Playing Field
Court Lodge
TN16
Wet Wood
Devil of Kent
Clacket Lane Service Area
Green Croft Farm
TITSEY WOOD
Greencroft Cotts.
King George's Field
Farley Common
Squerryes Sand Pit
Westwood Farm
Broomlands Farm
Farley Grange
WESTERHAM
High Cottage
Bower Cottage
Hosey Trees
Ashen Coppice
Works
RH8
Gotley
New Winterton
Covers Farm
Hosey Shaw
Moorhouse Sand Pits
Moorhouse
Park Lodge
Thrift Wood
Squerryes
Spring Shaw
Glebe House
Packway
Red Eaves
Croft Cott.
A25
Moorhouse Farm
Moorhouse Bank
Lodges Wood
Squerryes Home Farm
Hoseyridge
Hosey Hill
WESTERHAM ROAD
HIGH STREET
BRASTED
The Rips
The Birches
Grey Timbers
Moor House
The Mead
Moorhouse Bank Common
Rooks Nest
Chart's Edge
SQUERRYES PARK
Loampit Wood
Watts Hill
Cronklands
TOWER WOOD

Vc
Wc
Xc
Yc
Zc
Ad
Bd
181
202
46
47
48
49
58
57
56
55
54
153
91
92
93
94
95
96
97
98
99
100
The Mount
Minquintin
Hazlet Wood Farm
Stoneings
Stoneings Cottage
Laurel Cottage
Scott's Lodge
Roughfields Wood
Pilgrims Ho.
Hitherfield
Sundridge Hill Farm
PARK WOOD
Hagbush Wood
Shootfield
Shootfield Cottage
Shootfield Wood
Brasted Hill
Brasted Hill Farm
West Lodge
Home Farm
CHEVENING
Chevening House
Boat Ho.
CHEVENING PARK
Home Wood
Pilgrims' Way (course of)
CHEVENING CROSS
Grove House
Claypit
Homewood
Sundridge Lodge
Ovenden Lodge
OVENDEN ROAD
The Nower
Oveny Green Cottages
Oveny Green Farm
Combe Bank Farm
Combe Bank Cottages
North Lodge
Woodlands
Hogtrough Hill
Hogtrough Cott.
Court Lodge Farm
Watercress Beds
Court Lodge Cottages
PILGRIMS WAY
BRASTED HILL
HOGTROUGH HILL
OVENDEN
Combebank Wood
Combe Bank Wood
M25 MOTORWAY
M25
COMBE WOOD
Combe Wood
Ashwood
B2211
TN14
SEVENOAKS
Holywell Shaw
Park Wood
The Cottage
Pumping Station
Glebe Cotts.
ST MARTINS MEADOW
THORNS MEADOW
STATION RD.
COLES LA.
RECTORY LANE
Tanners
Hall
Rectory
The Old Rectory
PYM ORCHARD
Depot
BRASTED
THE OLD YARD
Mill
Weir
Mill Cottage
Mill Farm
Boat Ho.
Radnor House Sch.
Tennis Courts
Pav.
Recreation Ground
Weirs
COMBE BANK DRIVE
Rectory
Little Combe Bank
CHEVENING ROAD
R. Darent
Offices
Pumping Station
Market Gdns.
Bowl. Grn.
War Mem.
A25
MAIN
HIGH STREET
ROAD
SUNDRIDGE
MONDS CTS.
MANOR RD.
CHAPMANS RD.
WOODSIDE RD.
MAPLE T.
CHAPMANS CL.
The Old Rectory
Tennis Ct.
Oast Hill Farm
Meadlands
Sundridge & Brasted C of E Prim. Sch.
ST. MARY'S CHURCH RD.
Applecroft
The Warren
Park Farm
Charmans Farm
Darent
Tennis Courts
Rec. Grd.
Hall
Park Fm. Cott.
River
WILKINS CL.
WATERMILL CL.
ALMS ROW
ELLIOTTS LA.
MILBERRY PL.
THE CARRIAGE WAY
Lodge
Orchard Farm
Brasted Place
The Rookeries
Tennis Cts.
Playing Fld.
Brasted Place Farm
Heverswood Lodge
WEST END
Camrie
Lodge Wood
Quarry Cott.
The Quarry
Netherdowns
Birchwood
The Red Ho.
PARK
The White House
Sundridge Place Farm
Oast Cottages
Long Barn
GREYSTONE
Little Mead
Birchfield Farm
Mullion
WESTERHAM ROAD
LANE
HORIZON CL.
Fountain Cotts.
Twin Cedars
Slaney
Old Merralls
Twelve Acre Cott.
Pav.
Cricket Grd.
Club House
Valence Sch.
Neptune Pond
The Gardens
Heverswood Farm
Colinette Cotts.
Little Cacketts
Greenacre
Spring Shaw
Birchfield Wood
Birchfield Cottage
Lodge
Dunsdale Cott.
Foxwold Lodge
CACKETTS COTTS.
Overdale Meadow
Cacketts Lodge
Cobs Lea
Cacketts
Colinette
Colinette Farm
Vines Gate Shaw
Kennels
Park Shaw
Old Bartholmy Brooms
Lodge Barn Farm
VALENCE WOOD
WESTERHAM GOLF COURSE
PIPERS LANE
Vines Gate
Greenhills
GREAT WOOD
Les Etoiles
BIRCHFIELD
Gowanbank
Hill Water Bottom Wood
WILLOW
Wayside Cott.
Ivy Cott.
Mile End
Foxwold
Vale Cott.
The Chantry
Piper's Farm
Piper's Green
Brasted Chart
Bracken House
Gillhams
Lodge
Cricket Grd.
Pav.
Fieldway
Rec. Grd.
Turtle House
Thatches
PIPER'S GREEN RD.
Manning's Wood
QUORNDEN WOOD
Chapman's Wood
Plateau
Penn Farm
PENN LANE
CHURCH ROAD
IDE HILL ROAD
Phillipines Shaw
Newmans Rise
Brook Place Cotts.
The Shaw
Chartside Cottage
Ebor Wood
LITTLE NORMAN ST.
Little Norman Street Farm
The Bungalow
NORMAN
Brook Place
Stanhope Wood
Little Redwood
ROUND WOOD
Quornden
Starvecrow
North Lodge
Valance End
COMMON
French Street
Cleeve
Pilgroves
Saltwood
Parson's Marsh
Quakers
Greenways
Willowcroft
Roundwood
PHILLIPPINES
Pigeon Croft
Newman's Hill
GREAT
Great Norman
Brooms

Cd
Dd
Ed
Fd
Gd
Hd
Jd
182
49
50
51
52
58
57
56
55
54
53
CHEVENING
91
92
93
94
95
96
97
98
99
100
201
The Spinney
Way (course of)
Morants Court Farm
Turvin's Farm
B2211
Morants Court
The Bungalow
M25
Edward Shaw
Alder Shaw
M26
Laurel Villas
Great Dunton Farm
Broughton House
MORANTS COURT RD.
LONDON
A224
M26 MOTORWAY
Home Wood
War Mem.
DUNTON GREEN
Dunton Green
Rye Wood
Redroofs
Rye House Farm
Rye Cottage
WICKENS CARAVAN SITE
River Darent
Boat Ho.
Home Wood
CHEVENING CROSS
SUNDRIDGE
M25 MOTORWAY
Rakefield Wood
Donnington Hall
Subway
Rec. Grd.
Tennis Cts.
Prim. Sch.
Days Cotts.
Longford
ROAD
Grove House
Claypit Shaw
Homewood
OVENDEN ROAD
Junction 5 (M25)
Windmill Farm
Froghole
Sevenoaks Wildlife Reserve
Jeffery Harrison Visitor Centre
Works
Knole Academy Sports & Leisure Centre
Bradbourne Riding Cen.
Superstore
Hardy's Yard
Sch.
Chevening (St. Botolph's) C of E Prim. Sch.
Froghall Farm
Chipstead Sailing Club
Chipstead Lake Cheshire Home
R. Darent
CHIPSTEAD
RIVERHEAD
BRADBOURNE
A25
Brickfield Cott.
Rec. Grd.
Pav.
Willow Fm.
Works
Moat Farm
HIGH STREET
CHIPSTEAD LANE
Chipstead Ho.
The Rectory
Chipstead Comn.
Amherst Sch.
St. Mary's Dr.
The Stone Ho.
Children's Play Area
Bradbourne Lakes
Works
Council Refuse Transfer Cen.
Council Yard
Garden Centre
WESTERHAM ROAD
Tennis Courts
Pav.
Holmesdale Sports Grd.
Riverhead Infants' School
Sevenoaks Mag. Ct.
A25
Offices
R. Darent
Warren Farm
DRYHILL
SEVENOAKS
Bessels Green
Byways
COLD ARBOR RD.
WORSHIPS HILL
Ant Wood
AMHERST HILL
LONDON
Sevenoaks
RIDGE
Rectory
Wellers Farm
Dryhill Local Nature Reserve
Cold Arbor Wood
MONTREAL PARK
Ash Platt Wood
Brittains Farm
Dryhill
The Warren
Dryhill Farm
TN14
A21
BY-PASS
COLD ARBOR RD.
B2042
Salters Heath
Obelisk Wood
Cods Corner
REDLANDS RD.
TN13
Sundridge Place Farm
Middlings Wood
Little Britain's Wood
Red Grove
Greenlane Farm
Greenlane Farm Cotts.
Kippington
The Coach House
Manor Farm
Lodge Barn Farm
Greenlane Wood
DIBDEN LANE
Great Britain's Wood
Dibden
Dibden Farm
Furzefield Wood
Shrubs Corner
Pollard's Barn
Playing Field
The New Beacon Sch.
Mildridge Wood
Whitley
Cross Keys
Hill Water Wood
WILLOW WOOD
Nursery
Hawks Wood
Mill Bank Wood
Fig Street Farm
Fig Street
Brook Place Cotts.
Mackerels Plain
Nightingale Cott.
B2042
Whitley Forest
Mill Bank Wood
Goldsmith's Bottom
SEVENOAKS
A21
Norman Street Farm
The Bungalow
NORMAN STREET
Whitley Row
Rustwood
Seven Wents
The Woodman
Longspring Wood
Greystones

203
204
183
M26 MOTORWAY
Junction
Willowbrook
Kd
Ld
Md
Nd
Pd
Qd
Rd
53
54
55
56
57
58
Childsbridge House
Child's Bridge
CHILDSBRIDGE LA.
SEALCROFT COTTAGES
Noah's Ark
NOAH'S ARK
HONEY
TN14
Bartram Fm.
OTFORD ROAD
VESTRY COTTS.
VESTRY RD.
VESTRY INDUSTRIAL ESTATE
RIVERSIDE RETAIL PARK
Superstore
The Moors
JUBILEE COTTS.
CONNECTIONS BUS. PK.
Warehouse
Council Depot
CRAMPTONS ROAD
SWANZY RD.
Gas Wks
Retail Park
Works
COOMBE CT.
SEVENOAKS BUS. CEN.
Comm. Cen.
Bat & Ball
A225 ROAD
BRICKWORKS COTTS.
WATERCRESS CL.
WATERCRESS DR.
Wks
Greatness Farm
ORCHARD CL.
Greatness
Greatness Cemy.
Greatness Park
Playground
Playing Fields
THE PINES
GREATNESS LA.
MILL LA.
Millpond Wood
PINEWOOD
SEAL ROAD
HILLINGDON AV.
SEVENOAKS HOSP.
St. John's
ST. JOHN'S HILL
WICKENDEN RD.
HILLINGDON RISE
THE GREEN WAY
LITTLE WOOD
Rec. Grd.
Sevenoaks Prim. Sch.
Sevenoaks Jun. Sch.
Sevenoaks Adult Educ. Cen.
CAMDEN RD.
ST GEORGE'S RD.
ST JAMES'S RD.
QUAKERS HALL LA.
St. John's Prim. Sch.
BAYHAM RD.
HOLMESDALE RD.
Walthamstow Hall
Hollybush Spts. Complex
All Weather Pitch
Quarry Shaw
DARTFORD RD.
A225
AVENUE RD.
VINE CT. RD.
Vine Ckt. Grd.
War Mem.
ST. BOTOLPH'S RD.
THE HILL
A224
Club House
Lodges
Youth Club
Rayley's Pav. Field
Sevenoaks Rugby Club
Knole Paddock
Lady Boswell's C of E Prim. Sch.
Sevenoaks Leisure Centre
SEVENOAKS
Environmental Park
Bus Sta.
Arts Cen.
St. Thomas' Prim. Sch.
Drill Hall
PEMBROKE RD.
SUFFOLK WAY
KNOLE LA.
Offices
Sevenoaks School
Sevenoaks Sch.
Playing Field
Tennis Centre
Running Track
Playing Field
Sch.
TONBRIDGE ROAD
A225
Playing Field
Pav.
SOLEFIELDS ROAD
Sevenoaks Common
The New Sch. at West Heath
WHITE HART CL.
The Sevenoaks
DEER PARK
KNOLE PARK GOLF COURSE
KNOLE PARK
Echo Mount
DUCHESS WALK
Knole
The Wilderness
Birdhouse
Cedarholme
BROAD WALK
CHESTNUT WALK
Badcock Wood
Rumshott Wood
Blackhall Farm
Duchess Plantation
Blackhall
Godden Wood
HOLLOW LA.
B2019
Trinity Sch.
Weald of Kent Grammar Sch. (Sevenoaks Annexe)
Summerhouse Wood
WILDERNESSE AVENUE
Quarry Dean
Angle House
Wildernesse
Scotch Corner
Westwood
Blackhall Barns
WOODLAND RISE
BLACKHALL LA.
Barnett Field
Lone Oak
PARKFIELD
TN15
Blackhall Spinney
Kilnwood
Great Oaks Farm
Seal C of E Prim. Sch.
HIGHLANDS PK.
Works
Rec. Grd.
Pav.
New House
SEAL DRIVE
ZAMBRA WY.
SEAL
Church Fm.
Vicarage
GARDEN TERRACE
THE STABLES
HIGH STREET
Dorton College of Further Education
Dorton House School
Wychden
Red Court
Chance Wood
Club House
CHANCE WOOD
WILDERNESSE GOLF COURSE
PARK LANE
GROVE
Penfield
Tanners Cross
FULLER STREET
Money Shaw
Fullers Hill
Seal Chase
MAIDSTONE ROAD
A25
WATERY
Waterden
Wildwood
Tennis Cts.
Warren
The Grove
Lodge
Oak Bank Hall
Playing Field
HALL HILL ROAD
Chapel Wood
Richardson's Wood
Hall Cott.
Apps's Wood
Hall Place
Old Post House
Godden Green
Hillsmead
Torrington's Wood
Hall Wood
Tavern Cotts.
Merry Mead Fm.
CYGNET HOSPITAL GODDEN GREEN
Knole Development Centre
Stormont Court
Playing Field
Stake Farm Cott.
Sevenoaks Prep. Sch.
Fawke Cott.
Lord's Spring Wood
Perry Hill Quarry
Bow Petts Cotts.
Rambles Wood
Baker's Wood
South Field Shaw
Spinners Wood
RAMBLES WOOD
FAWKE COMMON
Nether Fawke
Mare Reeds
Fawke Fm.
Little Fawke
Lower Fawke Farm
Keepers Cott.
The Kennels
Fawke Common
Fawke Common Cotts.
Fawke House
Fawke Wood
ST. JULIAN ROAD
FAWKE WOOD ROAD
Redlands Wood
Clacketts Wood
Weald Cottage
Chestnut Plantation
Carter's Hill
Carter's Hill Cott.
White Rock
Tudor House
Quarry Shaw
Shepherds Mead
Broadfield Shaw
Bente
91
92
93
94
95
96
97
98
99
100

204
184
203
SEVENOAKS
TONBRIDGE
Kemsing
Noah's Ark
Fullers Hill
Seal Chart
SEAL CHART
Godden Green
Blakes Green
Bitchet Green
Lower Bitchet
STONE STREET
Styants Bottom
Oldbury
OLDBURY HILL
OLDBURY WOOD
IGHTHAM COMMON
IVY HATCH
Redwell
Carter's Hill
Fairlawne
TN11
A25
A227
SEVENOAKS ROAD
IGHTHAM ROAD
TONBRIDGE ROAD
HONEY POT LANE
WATERY LANE
Ightham Mote
Oldbury Hillfort (Remains)
Patchgrove Wood
Rumleys Wood
Hook Wood
Osier Bed Wood
Clay Pits
Stonepitts Farm
Broomsleigh Farm
Chart Farm
Styants Wood
Picnic Site
Crown Point
Fish Ponds Wood
Crown Point Nurseries
Broadhoath Wood
Martins Wood
Burnt Wood
Scathes Wood
East Bank Wood
Shingle Hill
Wilmot Hill
Cinderfield Wood
One Tree Hill
White Rocks
Rooks Hill
South Seers Wood
Plantation Farm
Orchard Shaw
Alders Shaw
Fatting Pen

205
185
M26 MOTORWAY
WROTHAM HEATH
Wrotham School
Playing Field
Potters Mede Sports Ground
Firemanshaw Wood
Meadow Cott.
Ruffits Wood
Mound
Ashplat Wood
Cricketts Farm
West Bank
The Dene
Brick Works
Borough Green & Wrotham
WESTERN RD.
BOROUGH GREEN
SEVENOAKS RD.
BY - PASS
Gracelands Park
Dark Hill Farm
A227
BOROUGH GREEN RD.
WROTHAM ROAD
A25
MAIDSTONE ROAD
Botany Wood
Sand Pit
Works
PLATT INDUSTRIAL ESTATE
HOMLESDALE BUSINESS CENTRE
King George's Field
Platt Mem. Hall
Windmill Hill
Gallows Hill
Valley Wood
Ingleborough Farm
High Haugh
Mays Cottages
Potter's Hole
Platt Wood
Rough Acres
Windmill House
ST. MARYS PLATT
Comp
Great Comp Garden
The Pavilion
Bracken Hill
Rowan Wood
Stone House Farm
Playing Field
Staddles Wood
Pine Close
BOROUGH GREEN
Trycewell Farm
Ash Tree Fm.
Stangate Bri.
Stangate Hill
Basted House
Long Wood
Platt Farm
Paddock Orchard
Potash Farm
Beechin Wood Farm
Comp Shoot Cottages
Green Acres
Beechin Oast
Napps Farm
Birchin Napps Farm
Gowles Farm
The Napps
Reservoir
Haw's Wood
Mill Pond
Butlers Wood
Longbrook Wood
Sewage Works (Disused)
Long Bottom Wood
Kingshill Farm
TN15
Warren Farm
Crowhurst Farm
Basted
Kings Cotts.
Landway Farm
CROUCH
Crouch Cottage
Goddens Cotts.
Claygate Cross
Claygate House
Claygate Cross Farm
Winfield Farm
Hatcham's Wood
and MALLING
Pondbottom Wood
Hatcham House
Hurst Wood Farm
Rye Field Cott.
Sheet Hill
Bourne Farm
Golding Hop Farm
Sheet Hill Farm
Brookside Farm
River Bourne
HURST WOOD
Boot Wood
Bourne Hill Farm
Yopps Green Farm
Lady's Wood
Yopps Green
Cage House
Old Soar Oast
Old Soar Manor
Keeper's Cottage
Dux Farm
Duck's Farm
Bartons Farm
Old Soar Farm
Old Soar Cotts.
Marsh Cottage
Broadfield Shaw
Plaxtol Spoute
Broadfield Farm
Quarry Wood
PLAXTOL
The Wilderness
Brook Farm
Rec. Grd.
Oast House
Spinners Wents
FAIRLAWNE PARK
Plaxtol Primary School
Allen's House
Little Allen's
Old Allen's
Cricket Ground
Rec. Grd.
Allen's Farm
Gover Hill
Roundels
Upper Farm
The White Cott.
Plaxtol Lane Cottages
Roughway Farm
Rats Castle
Home Coverts
Mill Shaw
Ivy Cottage
Roughway
Brakybank Wood
Barton Cotts.
DUNK'S GREEN
Roughway Mills
Roughway Farm Cottage
Well House
Gover Hill
Round Shaw
St Giles Cottage
Zd
Ae
Be
Ce
De
Ee
Fe
91
92
93
94
95
96
97
98
99
100
60
61
62
63
58
57
56
55
54
53

194
LITTLE HEATH
Little Heath Cottages
The Hermitage
Natural Gas Pressure Reducing Station
Mogador Cottages
MOGADOR
Wayside Farm
Dent's Farm
M25
B2033
B2032
PEBBLEHILL ROAD
DORKING ROAD
Pebble Coombe
Headley Reservoir (covered)
Fraser Heath
Round Wood
Watson's Wood
Dewriding Plantation
Lady Hill
Thistle Hill
White Cottage
Buckland Hills
Dawcombe Wood
White's Wood
Harebeating Wood
Pilgrims' Way
Mount Hill
Conybury Hill
Colley Hill
The Saddle Knob
Juniper Hill
Betchworth Hills
Poor's Field
Kemp's Farm
Kemp's Farm Cottages
Wildecroft
Underhill Farm
Northland Shaw
Quarry Orchard
Bushy Shaw
Clifton's Shaw
Colleyland Shaw
Buckland Green
Dowdes Farm
Bushfield Coppice
Claypit Shaw
Glebe House
MOLE VALLEY
Betchworth
STATION ROAD
The Sloughs
Petty's Farm
Warehouses
Works
Sports Ground
Depot
Colley Pit
A25
REIGATE ROAD
BUCKLAND
Tap Wood
Play. Fld.
Offices
Fairhall
Nursery
Crossways Farm
BUCKLAND COURT GARDENS
PARK LAKE DRIVE
Round Alders
Buckland Corner
BUCKLAND ROAD
Shagbrook
Broome Park
BETCHWORTH
Cricket Ground
OLD ROAD
Buckland Lodge
Buckland Park Lake
Shag Brook
REIGATE HEATH GOLF COURSE
Reigate Heath (Local Nature Reserve)
Reigate Heath
Reigate Heath Windmill
Club House
Hartsfield Manor
Sandhills
RH3
Dungate's Farm
Long Alders
Wonham Hill
Wonham Hill Cottages
Wonham Wood Cottage
The Alders
Ivy Cottage
Cemetery
The Old House
More Place
Wonham Manor
Garthlands
Heathfield Farm
SKIMMINGTON COTTAGES
Mill Pond
Wonham Cottage
Wonham Mill
Dungate Manor
The Lithe
Trumpet Hill
Trumpetshill Copse
Heathfield
Betchworth House
Betchworth Bridge
Fryleigh
Fryleigh Shaw
Mill Stream
Wilderness Wood
Agland Copse
June Farm
Little Santon Farm
Littleton Manor Farm
Stepstile Meadow
Littleton
Oldpark
Dendy's Wood
Keepers Cott.
Snowerhill Farm
Ricebridge Farm
Guzzle Copse
Wallace Brook
Gilbert's Farm
Santon House
Holly Cottage
Duffle's Plantation
Nine Acres
Rice Bridge
RIVER MOLE
Clayhall Farm
Gadbrook Farm
John's Copse
Riding's Copse
Knight's Gorse
Greenlane Cottages
Gadbrook Copse
Flanchford Farm
Little Flanchford Farm
Mill Pond
Gadbrook House
Elm Cottages
Little Abbot's Farm
Great Oaks
Flanchford Bridge
The Lodge
Little Flanchford Cottages

195
208
KT20
RH2
RH1
M25
M25 MOTORWAY
Junction 8
REIGATE HILL INTERCHANGE
MARGERY
Margery Ho.
Margery Hall Nursery
Margery Hall Pig Farm
Margery Farm
Reservoir (covered)
Kingswood Knoll
Grove Shaw
Gatlands Shaw
Catsbrain Shaw
Gatton East Field Shaw
GATTON
North Lodge
High Beeches
Jubilee Plantations
Royal Alexandra & Albert School
GATTON PARK
East Lodge Farm
Halfmoon Clump
Hop Garden Pond
Engine Pond
The Burpit
The Serpentine
THE LAKE
Boat Ho.
Little Buck Wood
Great Buck Wood
Wingate Hill
Serpentine Wood
Temple Wood
Temple Lodge
Nut Wood
Greenclump Hill
Piggeries
Two Acre Wood
Blackhorse Wood
Margery Hall
Pilgrims Holt
Water Tower
Four Acre Upper Woods
Rock Farm
Fort (dismantled)
Queen's Park
The Beeches
Reigate Hill
Pilgrims' Way
Broadleas
Colley Wood
REIGATE HILL
GATTON PARK ROAD
A242
A217
A23
A25
B2034
A2044
COLES MEADS
St. Bede's Sch.
Redhill Tennis Club
REDHILL
Redhill
Depot
STATION
Brooklands Sch.
Fire Brigade East Division HQ
Wray Common
Wray Common Prim. Sch.
Indoor Bowls Club
Donyngs Lei. Cen.
Reigate
Superstore
Play. Fld.
Reigate Coll.
Castle Grds.
Town Hall
Moat Hall
Cricket Grd.
Pavilion
REIGATE
Priory Museum
Reigate Priory Playing Fields
Priory Pond
REIGATE PARK
Park Hill
Monument
Reigate Gram. Sch.
Cemetery
Reigate Parish Church Sch.
REIGATE ROAD
CROYDON ROAD
BLACKBOROUGH ROAD
LESBOURNE RD.
BELL STREET
CHURCH ST.
HIGH ST.
LONDON RD.
HATCHLANDS RD.
REDSTONE HILL
BRIGHTON ROAD
THE CUTTING
HORLEY ROAD
COCKSHOT HILL
DOVERS GREEN
WOODHATCH ROAD
MEAD VALE
REDHILL COMMON
Jubilee Plantation
St. John's
Dunottar Day Sch. for Girls
Redstone Hall
EARLSWOOD
Earlswood
YMCA Sports Centre
Playing Field
Sports Ground
Rosebank Cotts.
Woodhatch
WOODHATCH
Reigate Sch.
Club Ho.
REDHILL & REIGATE GOLF COURSE
EARLSWOOD COMMON (Local Nature Reserve)
Cricket Ground
Boat Houses
Boating Lake
New Pond
Earlswood Lakes
Sewage Works
Household Waste Recycling Centre
Redhill Football Club
New Pond Farm
SOUTH PARK
King George's Field
Community Centre
Rushetts Farm
Dovers Green Sch.
Dovers Farm
DOVERSGREEN
Recreation Ground
Felland Copse
Withy Wood
Hartswood Farm Cotts.
Cricket Grnd.
THREE ARCH BUSINESS PARK
Burnt Oak
Earlswood Brook
25
26
27
28
53
52
51
50
49
48
H
J
K
L
M
N
P
G
1
2
3
4
5
6
7
8
9
10

208
196
207
MERSTHAM
Merstham
Club House
The Bushetts
East Lodge Farm
Gatton Park
RH2
THE LAKE
Boat Ho.
Jetty
Gatton Point
Temple Wood
Temple Lodge
LONDON ROAD A23
SOUTH
A242
Junction 8 (M23)
Junction (M25)
M23
Furzefield Wood
Playing Field
The Oakley Youth Club
Coppice Lea
Poultry Houses
Woodfield Sch.
St. Nicholas Sch.
SOUTH MERSTHAM
Mercers West
Spynes Mere Nature Reserve
Sewage Works
Mercers Country Park
Mercers Farm
Warwick Wold Farm
WARWICK WOLD
North Lodge
Residential Caravan Park
Crossways Rest Gdn.
East Surrey Coll. (Gatton Point North)
East Surrey Coll. (Gatton Point S.)
HOLMETHORPE
Holmethorpe Lagoons Nature Reserve
The Moors
The Moors Nature reserve
Ford Bri.
Nutfield Marsh
Ckt. Grd.
Marsh Cotts.
Glebe Lake
Chilmead Farm
RH1
Redhill Tennis Club
REDHILL
Redhill
REIGATE
Sandpit
Play. Fld.
The Warwick Sch.
Kennedys Wood
Works
Patteson Court
Pimlico Hill
Park Wood
Sports Grd.
NUTFIELD
BLETCHINGLEY
HIGH ST.
ROAD
A25
NUTFIELD
HILL
Robert Denholm House
Capenor Nursery
Adlington Cotts.
Fox Hole
North Park
Playing Field
Nutfield Priory Hotel
Furze Hill
Kingfisher House
Nursery Cottage
Steners Hill
Redstone Hall
REDSTONE
Hillsbrow
Byes Wood
Nutfield Lodge
Stockers Hill
Priory Farm
Ride The Hill, Mountainboarding Centre
Nutfield Priory Lake
Bowerhill Wood
Little Cormongers Farm
High Beech
Redstone Hill
Cemetery
ROYAL HILL PARK DEVELOPMENT
Bower Hill
Bower Hill Farm
Bray's Farm
Rec. Grd.
Nutfield Church C of E Primary School
Lyttel Hall
Ckt. Grd.
EARLSWOOD
Playing Field
Sports Ground
Nutfield
SOUTH NUTFIELD
Old Gladstone's House
Maple Farm
King's Mill
Cricket Grd.
Thepps Shaw
Earlswood
YMCA Sports Centre
Old Garston's House
Garston's Wood
Works
Nutfield Park
Ketwin's Wood
Crab Hill House
Theppswood Hse.
Ridge Green
Redhill Football Club
EAST SURREY HOSPITAL
Robin Cooks Farm
Depot
Warehouse
Staplehurst Farm
Hale Farm
Crab Hill
Crab Hill Farm
Nutfield Nurseries
The Rookery
Helicopter Landing Square
Tennis Courts
Eugene Bann Indoor Tennis Centre
White Bushes
Hazelhurst Farm
REDHILL AERODROME AND HELIPORT
Perimeter

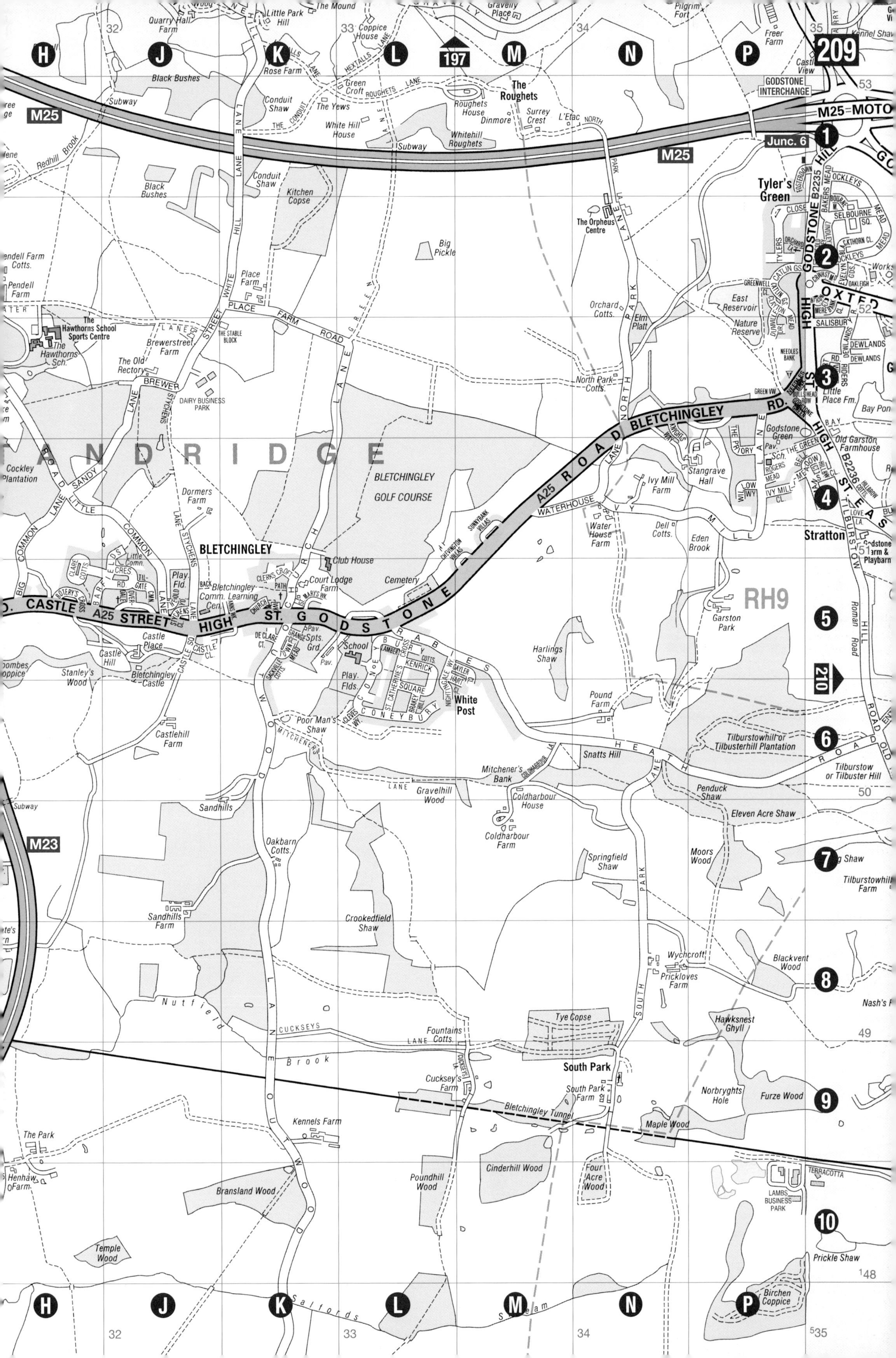

H
J
K
L
M
N
P
197
209
210
32
33
34
35
53
52
51
50
49
48
1
2
3
4
5
6
7
8
9
10
M25
M25
M23
Junc. 6
GODSTONE INTERCHANGE
M25 MOTO
Quarry Hall Farm
Little Park Hill
The Mound
Coppice House
Gravelly Place
Pilgrim Fort
Freer Farm
Castle View
Kennel Shaw
Rose Farm
HEXTALLS LANE
Black Bushes
Green Croft
ROUGHETS LANE
The Roughets
Conduit Shaw
The Yews
Roughets House
Dinmore
Surrey Crest
L'Etac
Subway
THE CONDUIT
White Hill House
Whitehill Roughets
Redhill Brook
Black Bushes
Conduit Shaw
Kitchen Copse
WHITE HILL LANE
NORTH PARK LANE
Tyler's Green
The Orpheus Centre
Big Pickle
GODSTONE HILL B2235
OCKLEYS MEAD
SELBOURNE SQ.
Pendell Farm Cotts.
Pendell Farm
Place Farm
PLACE FARM ROAD
CHURCH LANE
GREEN LANE
Orchard Cotts.
Elm Platt
East Reservoir
Nature Reserve
OXTED
SALISBURY
DEWLANDS
The Hawthorns School Sports Centre
The Hawthorns Sch.
Brewerstreet Farm
THE STABLE BLOCK
The Old Rectory
BREWER STREET
LANE
DAIRY BUSINESS PARK
North Park Cotts.
GREEN VW.
HIGH ST.
Little Place Fm.
Bay Pond
BLETCHINGLEY RD.
Godstone Green
Old Garston Farmhouse
B2236
TANDRIDGE
Cockley Plantation
SANDY LANE
ROAD
BLETCHINGLEY GOLF COURSE
Stangrave Hall
Ivy Mill Farm
THE PRIORY
IVY MILL LANE
Dormers Farm
STYCHENS LANE
WATERHOUSE LANE
Water House Farm
Dell Cotts.
Eden Brook
IVY MILL LANE
Stratton
Godstone Farm & Playbarn
TILBURSTOW HILL
LITTLE COMMON LANE
BIG COMMON LANE
BLETCHINGLEY
Club House
Court Lodge Farm
Cemetery
CHEVINGTON VILLAS
SUNNYBANK VILLAS
A25 ROAD
RH9
Garston Park
CASTLE STREET
A25
HIGH ST.
GODSTONE
Play. Fld.
Bletchingley Comm. Learning Cen.
Castle Place
Castle Hill
Stanley's Wood
Bletchingley Castle
School
Play. Flds.
KENRICK SQUARE
CONEYBURY
White Post
Harlings Shaw
Roman Road
Pound Farm
Poor Man's Shaw
Castlehill Farm
TWOOD LANE
MITCHENERS LANE
HEATH ROAD
Snatts Hill
Tilburstowhill or Tilbusterhill Plantation
Tilburstow or Tilbuster Hill
Mitchener's Bank
COLDHARBOUR LA.
Gravelhill Wood
Coldharbour House
Coldharbour Farm
Penduck Shaw
Eleven Acre Shaw
Subway
Sandhills
Oakbarn Cotts.
Springfield Shaw
SOUTH PARK LANE
Moors Wood
Tilburstowhill Farm
Sandhills Farm
Crookedfield Shaw
Wychcroft
Prickloves Farm
Blackvent Wood
Nash's
Nutfield
CUCKSEYS LANE
Fountains Cotts.
Tye Copse
Hawksnest Ghyll
Brook
South Park
Cucksey's Farm
South Park Farm
Norbryghts Hole
Furze Wood
Bletchingley Tunnel
Kennels Farm
Maple Wood
The Park
Henhaw Farm
Bransland Wood
Poundhill Wood
Cinderhill Wood
Four Acre Wood
TERRACOTTA
LAMBS BUSINESS PARK
Prickle Shaw
Temple Wood
Salfords Stream
Birchen Coppice

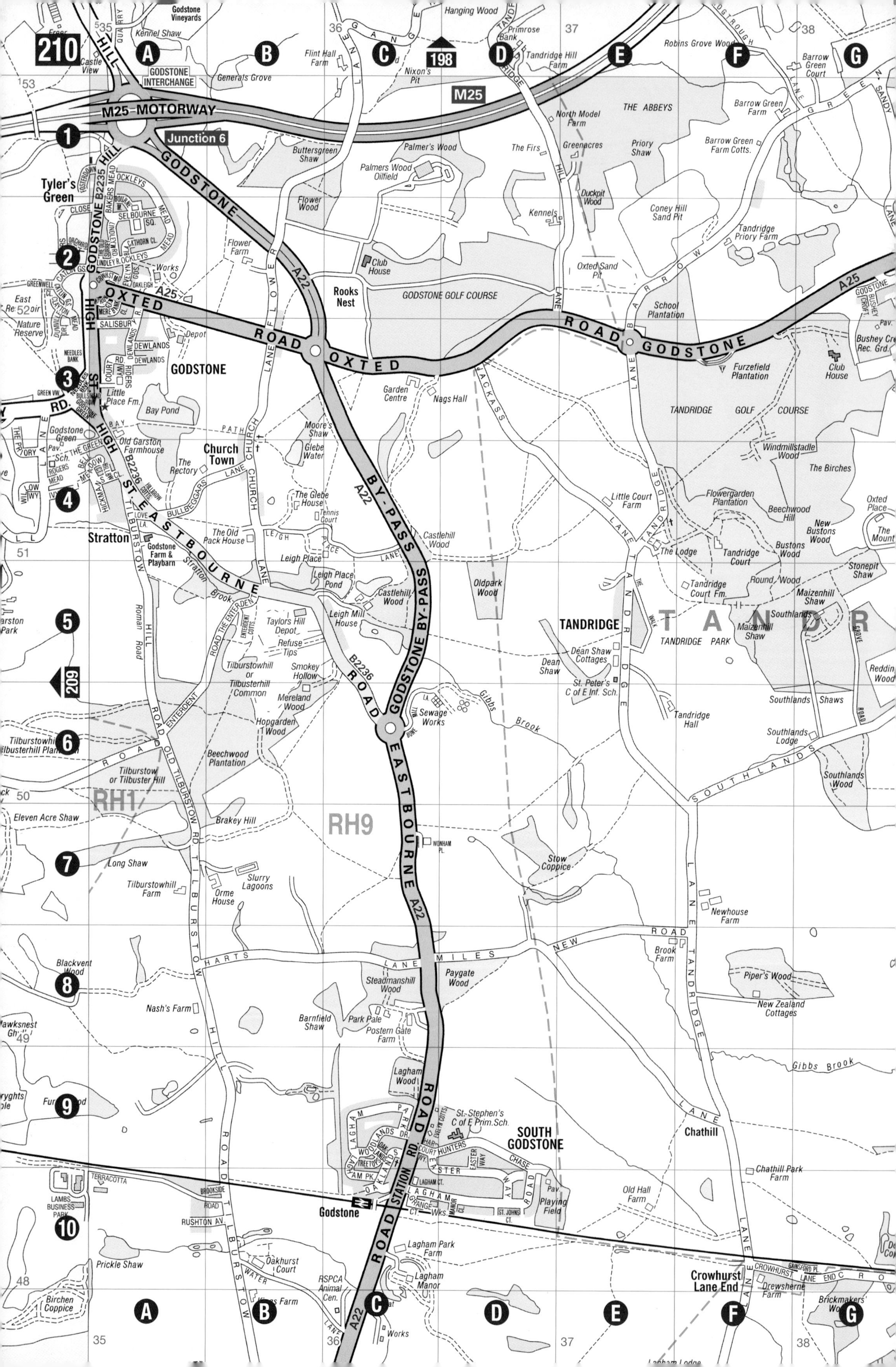
Godstone Vineyards
Kennel Shaw
Castle View
GODSTONE INTERCHANGE
Generals Grove
Flint Hall Farm
Nixon's Pit
198
Hanging Wood
Primrose Bank
Tandridge Hill Farm
Robins Grove Wood
Barrow Green Court
M25 MOTORWAY
M25
Junction 6
North Model Farm
THE ABBEYS
Barrow Green Farm
Barrow Green Farm Cotts.
Buttersgreen Shaw
Palmer's Wood
Palmers Wood Oilfield
The Firs
Greenacres
Priory Shaw
Tyler's Green
GODSTONE HILL
B2235
Flower Wood
Duckpit Wood
Coney Hill Sand Pit
Kennels
Tandridge Priory Farm
Flower Farm
Club House
Works
Oxted Sand Pit
GODSTONE
A25
Rooks Nest
GODSTONE GOLF COURSE
School Plantation
OXTED ROAD
East Reservoir Nature Reserve
GODSTONE ROAD
Depot
Furzefield Plantation
Club House
Bushey Croft Rec. Grd.
TANDRIDGE GOLF COURSE
Garden Centre
Nags Hall
Little Place Fm.
Bay Pond
Moore's Shaw
Glebe Water
Windmillstadle Wood
The Birches
Godstone Green
Old Garston Farmhouse
Church Town
The Rectory
CHURCH LANE
BULLBEGGARS LANE
A22 BY-PASS
Little Court Farm
Flowergarden Plantation
Beechwood Hill
Oxted Place
The Glebe House
Tennis Court
Stratton
The Old Pack House
Leigh Place
Castlehill Wood
Newbustons Wood
The Mount
Godstone Farm & Playbarn
ST. EASTBOURNE
Leigh Place Pond
Castlehill Wood
Oldpark Wood
The Lodge
Tandridge Court
Bustons Wood
Round Wood
Stonepit Shaw
Tandridge Court Fm.
Maizenhill Shaw
Southlands
Leigh Mill House
Taylors Hill Depot
Refuse Tips
TANDRIDGE
TANDRIDGE PARK
Roman Road
Tilburstowhill or Tilbusterhill Common
Smokey Hollow
B2236 ROAD
GODSTONE BY-PASS
Dean Shaw Cottages
Dean Shaw
St. Peter's C of E Inf. Sch.
Mereland Wood
Sewage Works
Gibbs Brook
Tandridge Hall
Southlands Shaws
Reddin Wood
Hopgarden Wood
Southlands Lodge
Beechwood Plantation
Tilburstow or Tilbuster Hill
SOUTHLANDS
Southlands Wood
RH1
Eleven Acre Shaw
Brakey Hill
RH9
EASTBOURNE A22
Wonham Pl.
Long Shaw
Stow Coppice
Tilburstowhill Farm
Slurry Lagoons
Orme House
Newhouse Farm
HARTS LANE
MILES
NEW ROAD
Brook Farm
Blackvent Wood
Steadmanshill Wood
Paygate Wood
Piper's Wood
Nash's Farm
New Zealand Cottages
Barnfield Shaw
Park Pale
Postern Gate Farm
Gibbs Brook
Lagham Wood
St. Stephen's C of E Prim.Sch.
SOUTH GODSTONE
Chathill
Chathill Park Farm
LAMBS BUSINESS PARK
TERRACOTTA
BROOKSIDE ROAD
Godstone
STATION RD
LAGHAM CT.
Pav.
Playing Field
Old Hall Farm
RUSHTON AV.
Lagham Park Farm
Prickle Shaw
Oakhurst Court
WATER
Lagham Manor
RSPCA Animal Cen.
Crowhurst Lane End
Drewsherne Farm
Birchen Coppice
Kings Farm
Works
Brickmakers' Wood
TILBURSTOW HILL ROAD
TANDRIDGE LANE
35
36
37
38
48
49
50
51
52
53
A
B
C
D
E
F
G
1
2
3
4
5
6
7
8
9
10
209

211
OXTED
Oxted
LIMPSFIELD
199
OLD OXTED
West Heath
Hurst Green
HURST GREEN
Broadham Green
HOLLAND
Pains Hill
LIMPSFIELD CHART
Itchingwood Common
Merle Common
RH8
TN8
RH7
WESTERHAM A25
B269
B2025
LIMPSFIELD COMMON
ITCHINGWOOD COMMON
Honesland Wood
Mollstones Wood
Staffhurst Wood Nature Reserve
Foyle Tolt
Pound Wood
Tenchleys Wood
Crabbett Wood
Parishcroft Wood
Langhurst

- The daily charge applies every day, 7.00am to 10.00pm, except Christmas Day (25th December).
- Payment of the daily charge allows you to drive in, around, leave and re-enter the charging zone as many times as required.
- Payment can be made in advance, or on the day of travel, or by midnight of the third day after travel. Payment after the day of travel will incur an increased cost.
- You can pay using Auto Pay (registration required), online, or using the official TfL App.
- Some vehicle types and classes are exempt, and some classes of road users can apply for a discount scheme.
- Penalty charge for non-payment of the daily charge by midnight on the third day after the day of travel.

This information is correct at the time of publication.

Visit www.tfl.gov.uk/modes/driving for more information on London's driving zones.

LARGE SCALE SECTION

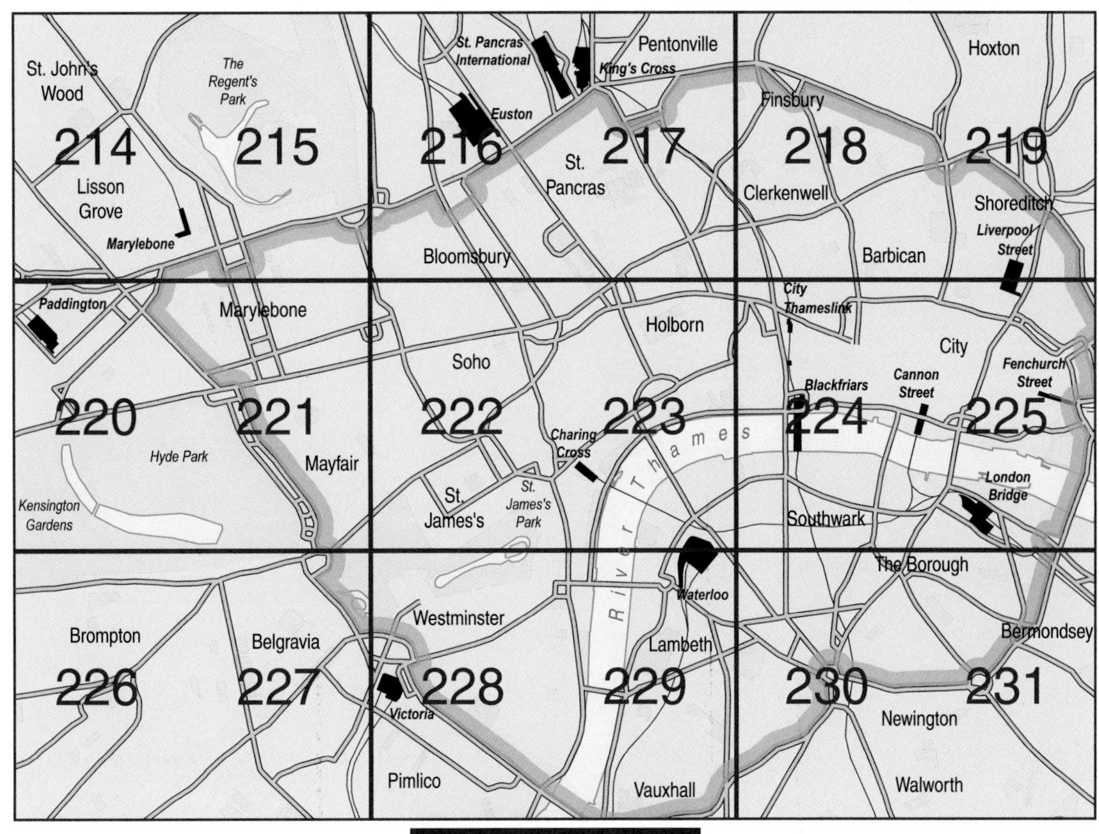

St. John's Wood
The Regent's Park
214
215
Lisson Grove
Marylebone
St. Pancras International
Pentonville
King's Cross
Euston
216
217
St. Pancras
Bloomsbury
Finsbury
Hoxton
218
219
Clerkenwell
Shoreditch
Liverpool Street
Barbican
Paddington
Marylebone
Holborn
City Thameslink
City
Fenchurch Street
Soho
Blackfriars
Cannon Street
220
221
222
223
224
225
Hyde Park
Mayfair
Charing Cross
Thames
Kensington Gardens
St. James's
St. James's Park
London Bridge
Southwark
River
Waterloo
The Borough
Brompton
Belgravia
Westminster
Lambeth
Bermondsey
226
227
228
229
230
231
Victoria
Newington
Pimlico
Vauxhall
Walworth

REFERENCE

A Road	A2
B Road	B408
Dual Carriageway	
One-way Street	
Road Under Construction Opening dates are correct at the time of publication.	
Proposed	
Inner Ring Road	R
Junction Name	MARBLE ARCH
Restricted Access	
Pedestrianized Road	
Congestion Charging Zone See Page 212 for more information	
Railway Station	
Railway Station Entrance	
National Rail Network	
Crossrail / Elizabeth Line See www.tfl.gov.uk for up to date information	
Docklands Light Railway	DLR
Overground	
Underground	

Local Authority Boundary	
Postal Boundary	
Map Continuation	90 · Large Scale Map Pages 222
Car Park (selected)	P
Cinema	
Cycle Hire Docking Station	
Fire Station	
Hospital	H
House Numbers (A & B Roads only)	34 62
Information Centre	i
National Grid Reference	179
Park and Ride	P+R
Police Station	▲
Post Office	★
River Boat Trip	
River Bus Stop	R

Theatre
Toilet:
Education Establishment
Hospital or Healthcare Building
Industrial Building
Leisure or Recreational Facility
Office Building
Place of Interest - Public Access
Place of Interest - no Public Access
Place of Worship
Public Building
Residential Building
Shopping Centre or Market
Other Selected Buildings

SCALE

Pages 214-231
1:7040
9 inches (22.7cm) to 1 mile
14.2cm to 1 km

0 50 100 200 300 Yards ¼ ½ Mile

0 50 100 200 300 400 500 750 Metres

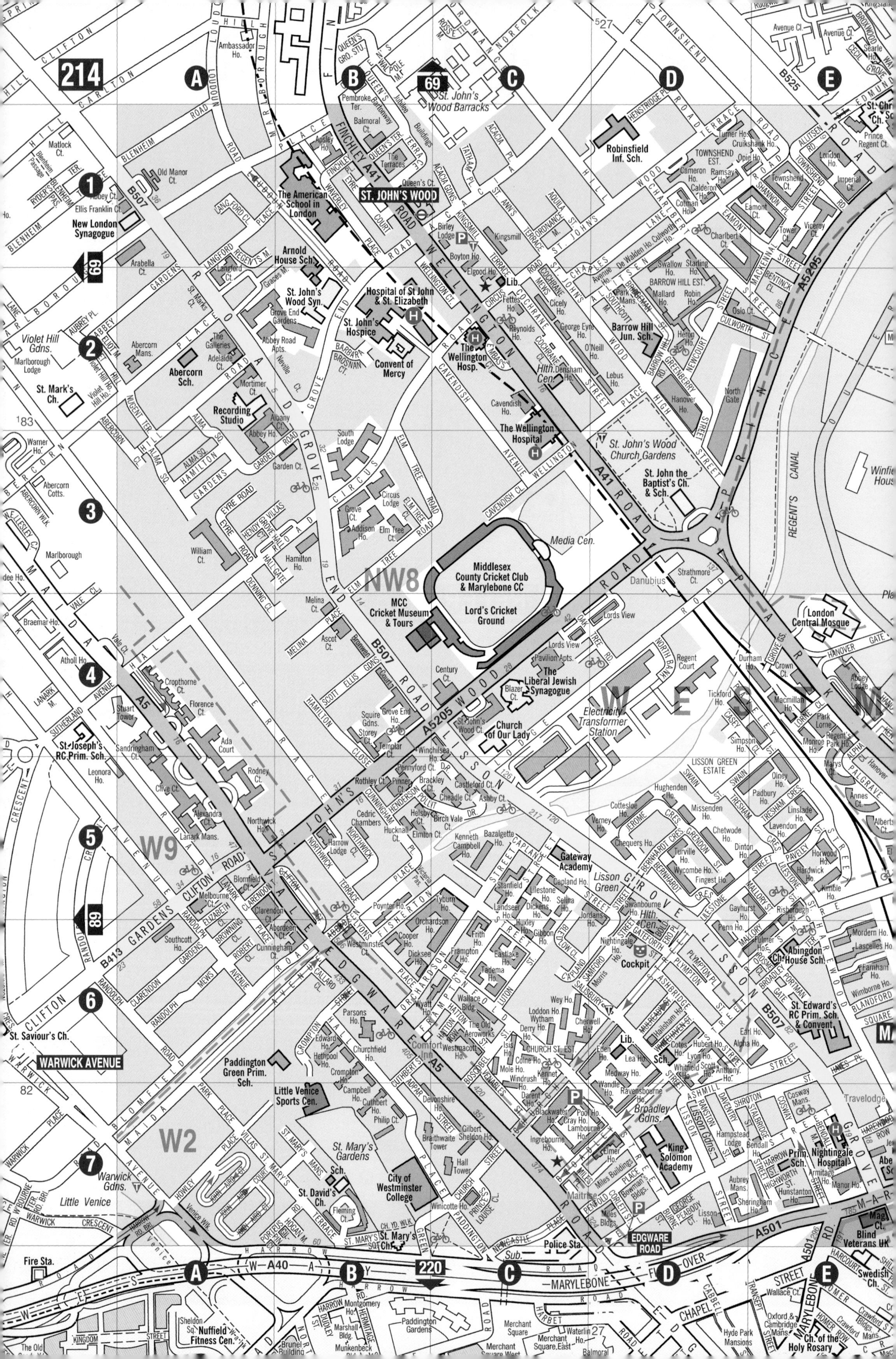

214
A
B
C
D
E
1
2
3
4
5
6
7
69
89
220
NW8
W9
W2
ST. JOHN'S WOOD
WARWICK AVENUE
EDGWARE ROAD
St. John's Wood Barracks
Robinsfield Inf. Sch.
The American School in London
New London Synagogue
Arnold House Sch.
St. John's Wood Syn.
Hospital of St John & St. Elizabeth
St. John's Hospice
Convent of Mercy
The Wellington Hosp.
The Wellington Hospital
Abercorn Sch.
Recording Studio
Barrow Hill Jun. Sch.
St. John's Wood Church Gardens
St. John the Baptist's Ch. & Sch.
Middlesex County Cricket Club & Marylebone CC
Lord's Cricket Ground
MCC Cricket Museum & Tours
Media Cen.
London Central Mosque
The Liberal Jewish Synagogue
Church of Our Lady
Electricity Transformer Station
St. Joseph's RC Prim. Sch.
Gateway Academy
Lisson Green
Cockpit
Abingdon House Sch.
St. Edward's RC Prim. Sch. & Convent
Paddington Green Prim. Sch.
Little Venice Sports Cen.
St. Mary's Gardens
City of Westminster College
St. David's Ch.
St. Mary's Ch.
King Solomon Academy
Nightingale Hospital
Police Sta.
Fire Sta.
Nuffield Fitness Cen.
St. Saviour's Ch.
St. Mark's Ch.
Little Venice
Warwick Gdns.
Regent's Canal
Blind Veterans UK
Swedish Ch.
Ch. of the Holy Rosary
Paddington Gardens
Merchant Square
Travelodge
FINCHLEY ROAD
WELLINGTON ROAD
ST. JOHN'S WOOD ROAD
GROVE END ROAD
MAIDA VALE
EDGWARE ROAD
LISSON GROVE
PRINCE ALBERT ROAD
PARK ROAD
CLIFTON ROAD
HARROW ROAD
MARYLEBONE FLYOVER
A41
A5
A5205
B507
B413
A40
A501
B525

REGENT'S PARK
CAMDEN
NW1
W1
WESTMINSTER
London Zoo
Entrance
Cricket Ground
The Hub
Running Track
Regent's Canal
Primrose Hill Br.
Playground
Cumberland Basin
Car & Coach Park
CLOSED Midnight - 7am
Refreshments
Children's Boating Pond
Boating Lake
The Holme
Bandstand
Queen Mary's Gardens
Regent's Park Open Air
Nursery
Lodge
Chester Road Gates
Jubilee Gates
Inner Circle
Outer Circle
Chester Road
York Bridge
St. John's Lodge
Frazer Lodge
Parks Department
New Lodge
Tennis Courts
Regent's University London
Avenue Gdns.
English Gdns.
Chester Walk Gate
Cumberland Gate
Clarence Gate
Park Square Gardens
Crescent Gardens
Regent's Park Barracks
Royal College of Physicians
Holy Trinity Ch.
Christ Ch. C of E Prim. Sch.
Danish Ch.
Nash Ho.
Silsoe Ho.
Albany Street
A4201
A5205
A503
A501
A41
A4380
B424
B506
Delancey
Gloucester Gate
Park Village West
Cumberland Terrace
Chester Terrace
Cambridge Gate
Park Square East
Park Square West
Park Crescent
Ulster Ter.
Harley House
International Comm. Sch.
Ryl. Academy of Music Mus.
London Business Sch.
Royal Coll. of Obstetricians & Gynaecologists
Francis Holland Sch.
St. Cyprian's Ch.
The Sherlock Holmes Mus.
Baker Street
Baker St.
Lost Property Off.
Madame Tussaud's
Marylebone Road
Marylebone
University of Westminster (Marylebone Campus)
St. Marylebone Ch.
The St. Marylebone Sch.
The Princess Grace Hosp.
The London Clinic
Central London Civil Justice Centre
Central London County Ct.
Regent's Park
The Portland Hosp.
RIBA
Portland Pl. Sch.
Central Synagogue
Franco-Anglaise Sch.
PureGym
BBC Broadcasting House
King Edward VII's Hosp.
The Harley St. Clinic
University Coll. Hosp.
Paddington Street
Paddington Street Gdns.
Dorset Sq.
Town Hall
Lib.
Meth. Ch.
Park Plaza Sherlock Holmes
Everyman Baker St.
St. Mary's Prim. Sch.
St. Mary's Ch.
West London Coll.
BDA Dental Museum
Devonshire Place
Harley Street
Wimpole Street
Weymouth Street
New Cavendish Street
Portland Place
Gloucester Place
Dorset Square
Melcombe Street
York Street
George Street
215
216
221
69
70

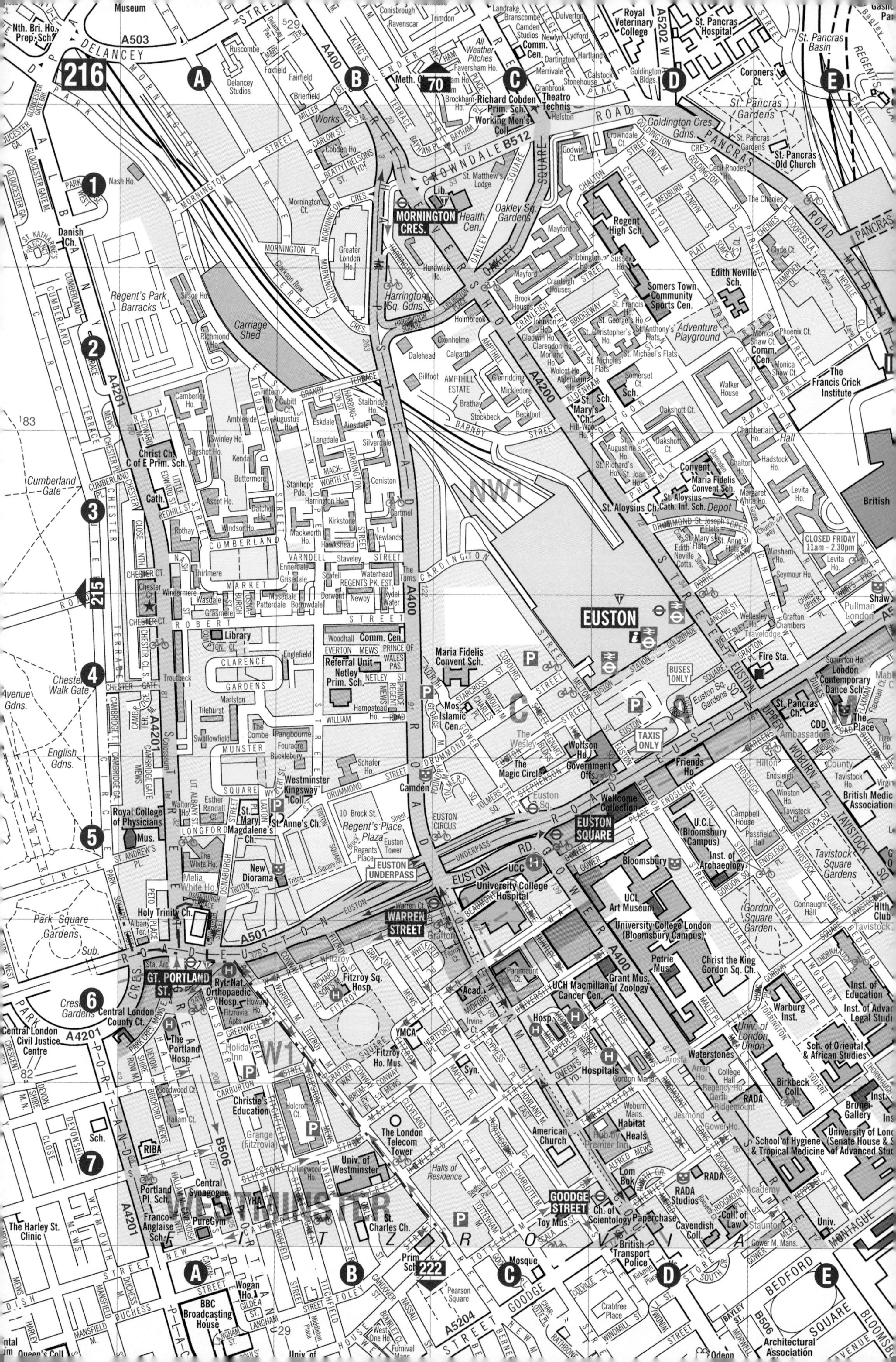

216
70
215
222
A
B
C
D
E
1
2
3
4
5
6
7
Museum
Nth. Bri. Ho. Prep. Sch.
DELANCEY
A503
Delancey Studios
Ruscombe
Royal Veterinary College
St. Pancras Hospital
St. Pancras Basin
Coroners' Ct.
St. Pancras Gardens
St. Pancras Old Church
Meth. Ch.
Richard Cobden Prim. Sch.
Theatre Technis
Working Men's Coll.
All Weather Pitches
CROWNDALE
B512
St. Matthew's Lodge
Oakley Sq. Gardens
Goldington Cres. Gdns.
PANCRAS
Lib.
MORNINGTON CRES.
Health Cen.
Mornington Ct.
Greater London Ho.
Nash Ho.
Danish Ch.
Regent's Park Barracks
Carriage Shed
Harrington Sq. Gdns.
Regent High Sch.
Somers Town Community Sports Cen.
Edith Neville Sch.
Adventure Playground
The Francis Crick Institute
Walker House
St. Mary's Ch.
A4200
A4201
A400
NW1
British
Convent Maria Fidelis Convent Sch.
St. Aloysius Cath. Inf. Sch.
St. Aloysius Ch.
Depot
CLOSED FRIDAY 11am - 2.30pm
Cumberland Gate
Chester Walk Gate
Avenue Gdns.
English Gdns.
Christ Ch. C of E Prim. Sch.
Cath.
Library
Comm. Cen.
Referral Unit
Netley Prim. Sch.
Maria Fidelis Convent Sch.
Mos. Islamic Cen.
EUSTON
BUSES ONLY
TAXIS ONLY
Euston Sq. Gardens
The Wesley
The Magic Circle
Wolfson Ho.
Government Offs.
Friends Ho.
Wellcome Collection
EUSTON SQUARE
St. Pancras Ch.
London Contemporary Dance Sch.
The Place
Shaw
Pullman London
Travelodge
Fire Sta.
Westminster Kingsway Coll.
St. Anne's Ch.
Camden
Regent's Place Plaza
Euston Tower
EUSTON UNDERPASS
EUSTON CIRCUS
Royal College of Physicians
Mus.
St. Mary Magdalene's Ch.
New Diorama
Holy Trinity Ch.
WARREN STREET
A501
UCH
University College Hospital
Bloomsbury
UCL Art Museum
University College London (Bloomsbury Campus)
U.C.L. (Bloomsbury Campus)
Inst. of Archaeology
Tavistock Square Gardens
Gordon Square Garden
Petrie Mus.
Christ the King Gordon Sq. Ch.
Grant Mus. of Zoology
UCH Macmillan Cancer Cen.
British Medical Association
Warburg Inst.
Inst. of Education
Univ. of London Union
Waterstones
Sch. of Oriental & African Studies
Birkbeck Coll.
RADA
Brunei Gallery
School of Hygiene & Tropical Medicine
University of London (Senate House & School of Advanced Studies)
GT. PORTLAND ST.
Ryl. Nat. Orthopaedic Hosp.
Fitzroy Sq. Hosp.
Central London County Ct.
Central London Civil Justice Centre
The Portland Hosp.
Holiday Inn
W1
YMCA
Fitzroy Ho. Mus.
Syn.
Hosp.
Hospitals
Christie's Education
RIBA
The London Telecom Tower
American Church
Habitat
Heals
Hub by Premier Inn
Univ. of Westminster
Halls of Residence
GOODGE STREET
Ch. of Scientology
Paperchase
RADA Studios
Coll. of Law
Cavendish Coll.
Toy Mus.
Central Synagogue
Portland Pl. Sch.
YHA
PureGym
Franco Anglaise Sch.
The Harley St. Clinic
WESTMINSTER
FITZROVIA
St. Charles Ch.
Prim. Sch.
Mosque
British Transport Police
BBC Broadcasting House
Wogan Ho.
A5204
Pearson Square
Crabtree Place
Architectural Association
BEDFORD SQUARE
B506
Odeon
Park Square Gardens
Crescent Gardens
Sub.
EUSTON ROAD
TOTTENHAM COURT ROAD
GOWER STREET
EVERSHOLT STREET
HAMPSTEAD ROAD

217
218
223
70
F
G
H
J
K
1
2
3
4
5
6
7
Western Transit Shed
UAL Central Saint Martins
West Handyside Canopy
Handyside Canopy
Midland Goods Shed
House of Illustration
Granary Square
York Way Ct.
Sch.
Comm. Cen.
Blessed Sacrament Cath. Prim. Sch.
Ch. of the Blessed Sacrament
Copenhagen Prim. Sch.
Playground
Young Actors Theatre Sch.
Regents Canal
Thornhill Bridge
Islington
Kings Place
Ice Wharf
Regents Wharf
Ice Wharf Marina
London Canal Mus.
Marina One
Battlebridge Wharf
Bridge Wharf
Vittoria Prim. Sch.
Superstore
Tolpuddle
Police Sta.
Pancras Sq. Leisure Cen.
Town Hall (Until 2021)
Pancras Square
The Gymnasium
St. Pancras National
King's Cross
King's Cross St. Pancras
St. Pancras Renaissance
Wharfdale Road
Health Cen.
Priory Heights
Priory Green Estate
Elizabeth Garrett Anderson School
St. Silas's Ch.
Chalbury Walk
Baron Close
Utd. Reformed Ch.
Crafts Council
White Lion Cen.
DoubleTree by Hilton
The Gower Sch.
Joseph Grimaldi Park
Winton Prim. Sch.
Mos.
Radius Apartments
Pentonville Road
A501
A5203
A5200
A5202
A201
Reservoir
Angel Building
Claremont
Myddelton
Paul Robeson Ho.
Holford Yard
Holford Gdns.
College of Naturopathic Med.
Sch. of Oriental & African Studies
Baptist Ch.
St. Mark's Ch.
Baptist Chap.
Islington
EC1
Cobden Bldgs.
Derby Lodge
Travelodge
Pioneer Ho.
Nuffield Cen.
Arriva
Swinton Street
Acton Street
Clerkenwell Parochial C of E Prim. Sch.
The Nautilus Bldg.
New River Head
The Hydra Bldg.
The Remus Bldg.
Kings Cross Mission Ch.
Comfort Inn
Crestfield
Megaro
Fairway
European
Alhambra
Argyle Sq. Gdns.
Argyle Prim. Sch.
Town Hall Closed until 2021
Central
Queen Alexandra Mans.
Elmwood
Fleetfield
Clink 261
Clink 78
Point A
RNIB
Comm. Cen.
Holy Cross Ch.
St. Mary's German Ch.
Harrison
St. Peter's Ho.
Westminster Kingsway Coll. (Kings Cross Cen.)
Wells Square
Travelodge
Charles Simmons Ho.
Crowne Plaza
Depot
Lumen Utd. Ref. Ch.
St. George's Gdns.
Thomas Coram Early Childhood Cen.
Goodenough Club
Mecklenburgh Square Gardens
St. Andrews Gdns.
Postmark Development
Mount Pleasant Sorting Office
The Postal Museum
The Urdang
RC Ch.
Ch. of Our Most Holy Redeemer
London Metropolitan Archives
AR Cen.
Brunswick Centre
Sch. of Pharmacy
The Foundling Mus.
The Wolfson Assessment Cen.
Coram's Fields
Nuffield Fitness Cen.
Calthorpe
B502
B504
A401
WC1
Herbrand Estate
The Gym (Bloomsbury)
Holiday Inn
Curzon
Russell Sq.
Royal National
Gov. Offs.
Kimpton Fitzroy
President
Imperial
Russell Square Gardens
The National Hospital for Neurology and Neurosurgery
Gt. Ormond Street Hospital for Children
Royal London Hospital for Integrated Medicine
St. Giles Coll. London Central
Mercure (Bloomsbury)
The Charles Dickens Mus.
Independent Television News
Res. Cen.
St. George the Martyr C of E Prim. Sch.
Buddhist Cen. London
Library
Comm. Cen.
Gray's Inn Gardens
Gray's Inn
Atkin Building
Chapel
South Square
Islamic Cen.
St. Alban's C of E Prim. Sch.
Ch.
St. Peter's Ch.
Herbal Hill Gdns.
St. George the Martyr Ch.
Mary Ward Cen.
Sch.
PureGym
Conway Hall
Premier Inn
The Old School
Doubletree by Hilton
College
Bloomsbury Square Gdns.
Le Cordon Bleu
British Museum Court
Holborn Bars
Greville St.
St. Etheldreda's
Theobald's Road
Holborn
A40
Southampton Row
Russell Square
Bedford
Montague
Portland
Clarendon
Lancaster
Beauchamp
Chancellors Ct.
Holborn
Vesage Ct.

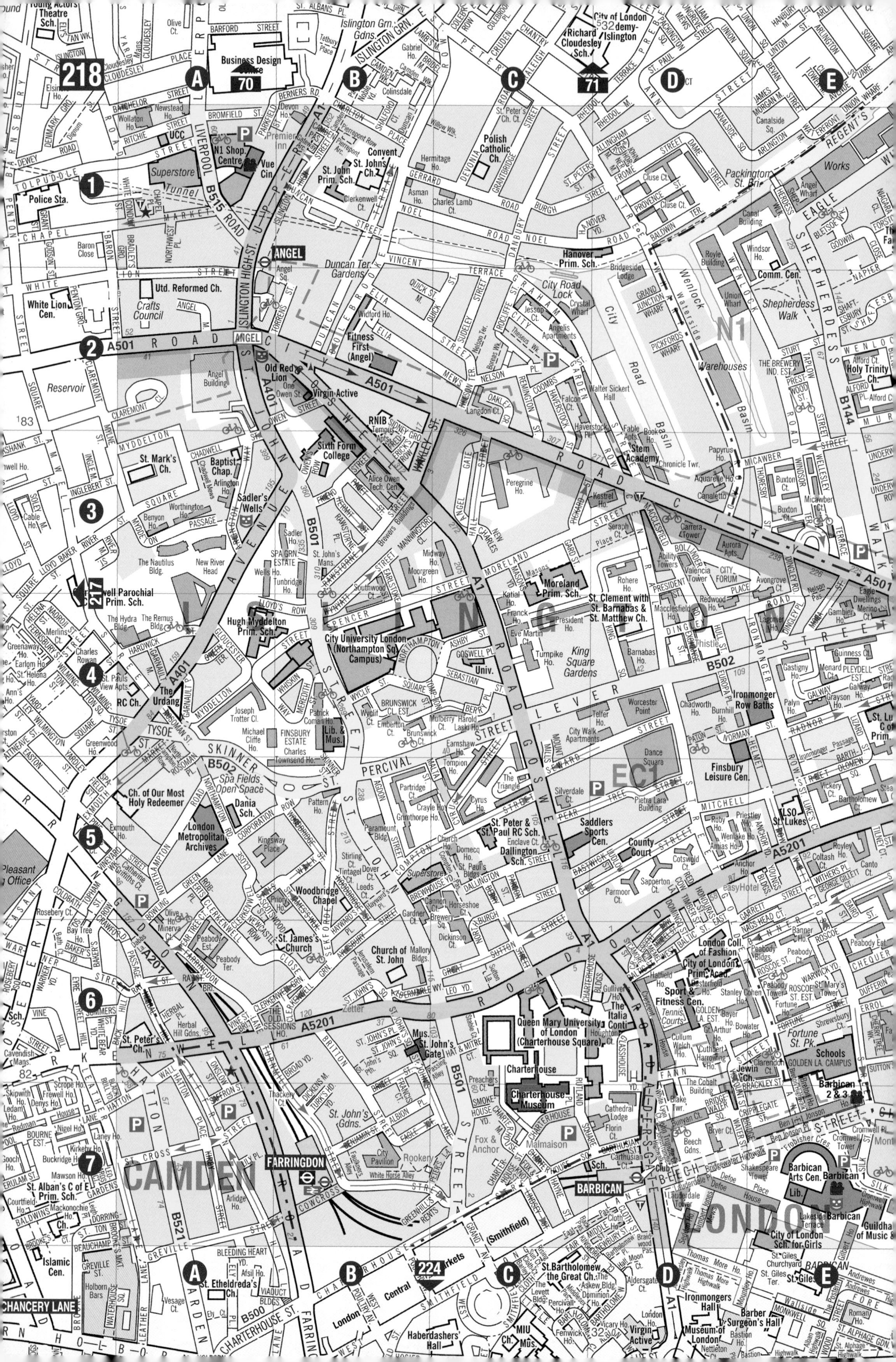

218
70
71
217
224
A
B
C
D
E
1
2
3
4
5
6
7
Business Design Centre
Islington Grn. Gdns.
ISLINGTON GRN.
City of London Academy Islington
Richard Cloudesley Sch.
UCC
N1 Shop. Centre
Vue Cin.
Superstore
Tunnel
Police Sta.
LIVERPOOL ROAD
B515
ISLINGTON HIGH ST.
UPPER ST.
ANGEL
Premier Inn
Convent St. John's Ch.
St. John Prim. Sch.
Polish Catholic Ch.
Packington St. Bri.
Works
REGENT'S
Duncan Ter. Gardens
Utd. Reformed Ch.
Crafts Council
White Lion Cen.
A501
CITY ROAD
Old Red Lion
Virgin Active
Fitness First (Angel)
Angel Building
Reservoir
RNIB
Sixth Form College
Alice Owen Tech. Cen.
Hanover Prim. Sch.
City Road Lock
City Road Basin
Wenlock Basin
Warehouses
N1
Shepherdess Walk
Comm. Cen.
Holy Trinity Ch.
B144
St. Mark's Ch.
Baptist Chap.
Sadler's Wells
A401
ST. JOHN STREET
B501
ISLINGTON
Moreland Prim. Sch.
St. Clement with St. Barnabas & St. Matthew Ch.
King Square Gardens
Hugh Myddelton Prim. Sch.
City University London (Northampton Sq. Campus)
Univ.
B502
Ironmonger Row Baths
Finsbury Leisure Cen.
Clerkenwell Parochial C of E Prim. Sch.
The Urdang
RC Ch.
Lib. & Mus.
Dance Square
EC1
GOSWELL ROAD
A1
Ch. of Our Most Holy Redeemer
Spa Fields Open Space
Dania Sch.
London Metropolitan Archives
Saddlers Sports Cen.
County Court
St. Peter & St. Paul RC Sch.
Dallington Sch.
LSO St. Lukes
A5201
Woodbridge Chapel
St. James's Church
Church of St. John
London Coll. of Fashion
City of London Prim. Acad.
Sport & Fitness Cen.
Fortune St. Pk.
Schools GOLDEN LA. CAMPUS
Barbican 2 & 3
Queen Mary University of London (Charterhouse Square)
Charterhouse
Charterhouse Museum
Mus. St. John's Gate
St. John's Gdns.
Fox & Anchor
Malmaison
Sch.
BARBICAN
Barbican Arts Cen.
Barbican 1
Barbican
Lib.
City of London Sch. for Girls
Guildhall Sch. of Music & Drama
St. Giles
A201
St. Peter's Ch.
CAMDEN
FARRINGDON
COWCROSS
City Pavilion
Rookery
B521
St. Alban's C of E Prim. Sch.
Islamic Cen.
CHANCERY LANE
Holborn Bars
St. Etheldreda's Ch.
B500
Central Markets
(Smithfield)
London Central Markets
Haberdashers' Hall
St. Bartholomew the Great Ch.
Virgin Active
MIU Mus.
Ironmongers Hall
Barber Surgeons' Hall
Museum of London
LONDON
Pleasant Sorting Office

219
Shoreditch Park
Play Park
Tennis Courts
Britannia Leisure Centre
Whitmore Prim. Sch.
Comet Nursery Sch.
Bridge Academy
Mosque
St. Annes with St. Columba Ch.
Burbage Prim. Sch.
Randal Cremer Prim. Sch.
St. Mary's Community Garden
Haggerston Sch.
Geffrye Mus.
Geffrye's Gdns.
HOXTON
Hoxton Hall
Trust HQ & Hlth. Cen.
Hackney Comm. Coll. (Shoreditch Campus)
Univ. Tech. College
St. John the Baptist C of E Prim. Sch.
St. John the Baptist Ch.
True Jesus Ch.
St. Chad's Ch.
HACKNEY
TOWER HAMLETS
Ravenscroft Park
Columbia Mkt. Nursy. Sch.
COLUMBIA ROAD
Mildmay Hosp.
St. Leonard's Ch.
Shoreditch Town Hall
St. Monica's RC Prim. Sch.
St. Monica's RC Church
Aske Gdns.
Courtyard
The Energy Centre
Courthouse
Boundary Gds.
ARNOLD CIRCUS
Electric Cinema
Moorfields Eye Hosp.
New Regent's College
City Coll.
Fire Sta.
OLD STREET
St. Agnes Well
Central Foundation Boys' Sch.
Energie Fitness
Wesley's Chapel & Mus.
Wesley's Ho.
Bunhill Fields Burial Ground
Armoury House
Virgin Active
HAC Ground
The College of Law
Government Offices
The Lyceum Sch.
Army Reserve Cen.
EC2
E1
London Coll. of Fashion
Holywell Cen.
SHOREDITCH HIGH STREET
Dennis Severs House
Old Spitalfields Market
Christ Ch.
Christ Church Prim. Sch.
Mosque
Milton Ct. Concert Hall
Citypoint
MOORGATE
Chapel of the Open Book
St. Mary Moorfields Ch.
BROADGATE CIRCLE
Broadgate Tower
Exchange House
Exchange Square
Bishopsgate Institute & Lib.
London Fruit Exchange
London Metropolitan Univ.
Police
Bowling Green
Bandstand
Finsbury Sq.
71
91
225
A10
A1200
A1202
A1208
A1209
A5201
A501
B100
B101
B118
B122
B144
B1

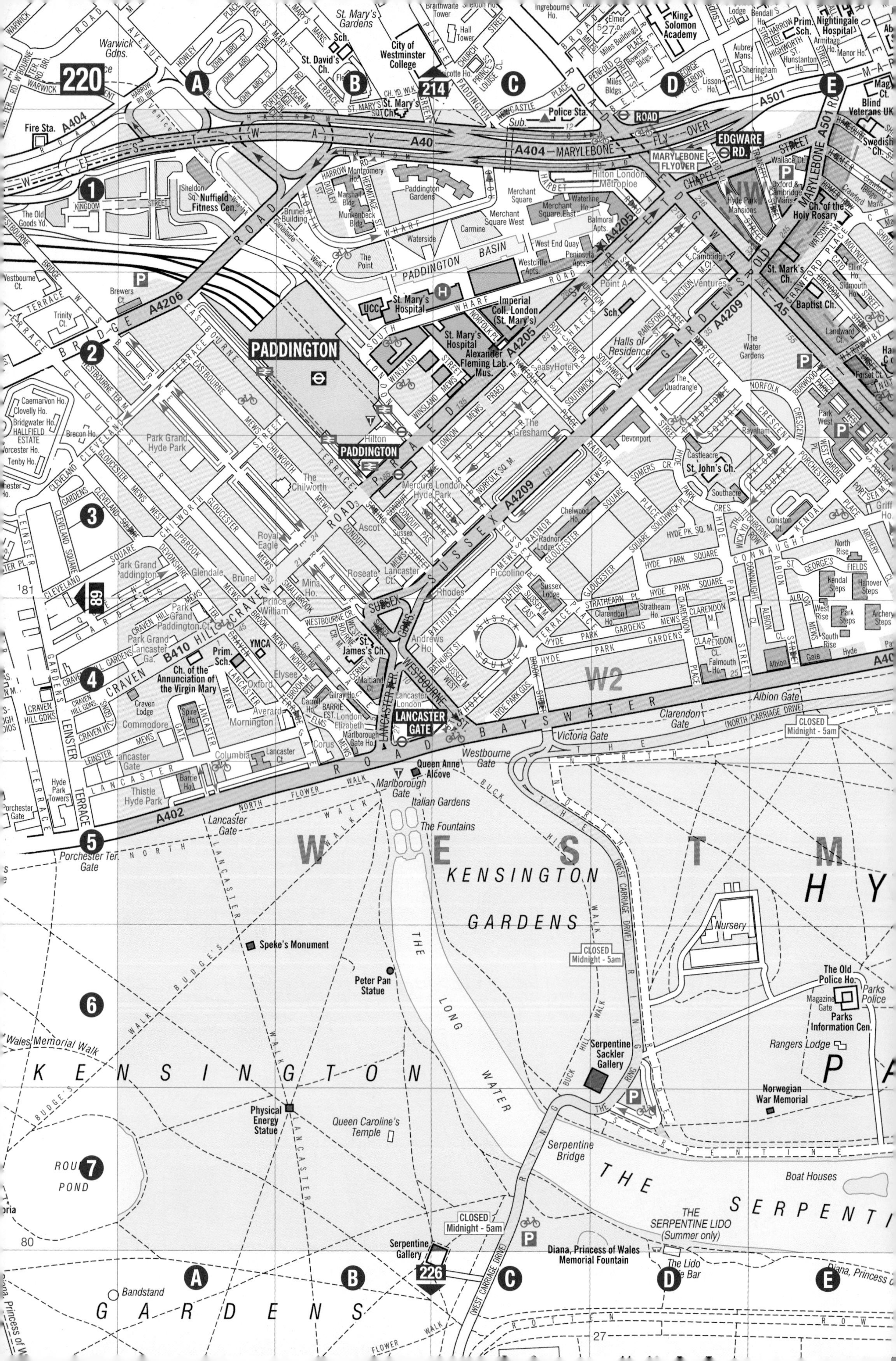

220
214
226
68
A
B
C
D
E
1
2
3
4
5
6
7
PADDINGTON
LANCASTER GATE
EDGWARE RD.
MARYLEBONE FLYOVER
W2
WESTMINSTER
KENSINGTON GARDENS
HYDE PARK
THE SERPENTINE
THE LONG WATER
ROUND POND
St. Mary's Gardens
City of Westminster College
St. David's Ch.
St. Mary's Ch.
Police Sta.
King Solomon Academy
Nightingale Hospital
Blind Veterans UK
Warwick Gdns.
Fire Sta.
Nuffield Fitness Cen.
Paddington Gardens
Merchant Square
PADDINGTON BASIN
St. Mary's Hospital
Imperial Coll. London (St. Mary's)
Alexander Fleming Lab. Mus.
Halls of Residence
Hilton London Metropole
St. Mark's Ch.
Ch. of the Holy Rosary
Baptist Ch.
The Water Gardens
St. John's Ch.
Ch. of the Annunciation of the Virgin Mary
YMCA
St. James's Ch.
Thistle Hyde Park
Queen Anne Alcove
Italian Gardens
The Fountains
Victoria Gate
Albion Gate
Clarendon Gate
CLOSED Midnight - 5am
Speke's Monument
Peter Pan Statue
Physical Energy Statue
Queen Caroline's Temple
Serpentine Sackler Gallery
Serpentine Bridge
Serpentine Gallery
THE SERPENTINE LIDO (Summer only)
Diana, Princess of Wales Memorial Fountain
The Lido Bar
Boat Houses
Norwegian War Memorial
Rangers Lodge
Parks Information Cen.
The Old Police Ho.
Parks Police
Nursery
Bandstand
A40
A404
A4206
A4205
A4209
A5
A501
A402
B410
BAYSWATER ROAD
SUSSEX GARDENS
PRAED STREET
EDGWARE ROAD

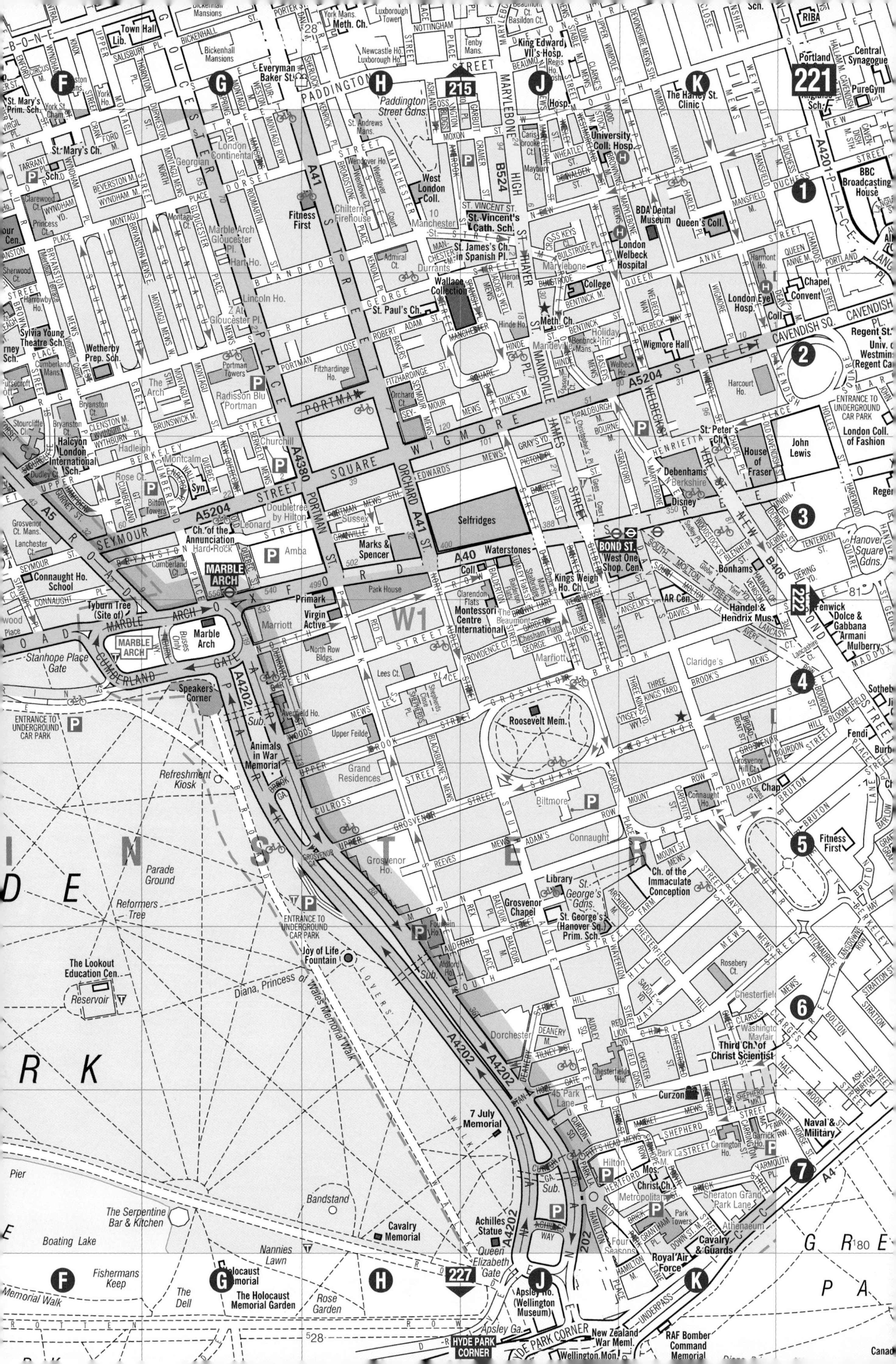
215
221
222
227
F
G
H
J
K
1
2
3
4
5
6
7
Town Hall
Lib.
Bickenhall Mansions
Everyman Baker St.
Meth. Ch.
York Mans.
Luxborough Tower
Newcastle Ho.
Luxborough Ho.
Tenby Mans.
King Edward VII's Hosp.
RIBA
Portland Hosp.
Central Synagogue
PureGym
The Harley St. Clinic
St. Mary's Prim. Sch.
St. Mary's Ch.
Sch.
Paddington Street Gdns.
St. Andrews Mans.
London Continental
Georgian
West London Coll.
University Coll. Hosp.
BBC Broadcasting House
Fitness First
Chiltern Firehouse
Manchester St.
St. Vincent's Cath. Sch.
St. James's Ch. Spanish Pl.
BDA Dental Museum
Queen's Coll.
London Welbeck Hospital
Marble Arch Gloucester Pl.
Hart Ho.
Admiral Ct.
Durrants
Marylebone
Wallace Collection
College
London Eye Hosp.
Chapel
Convent
Coll.
Lincoln Ho.
Z Att Gloucester Pl.
St. Paul's Ch.
Meth. Ch.
Holiday Inn
Wigmore Hall
Regent St.
Univ. of Westminster (Regent Campus)
Sylvia Young Theatre Sch.
Wetherby Prep. Sch.
Portman Towers
Fitzhardinge Ho.
Harcourt Ho.
The Arch
Radisson Blu Portman
ENTRANCE TO UNDERGROUND CAR PARK
London Coll. of Fashion
Halcyon London International Sch.
Bryanston Ct.
Hadleigh
Churchill
John Lewis
St. Peter's Ch.
House of Fraser
Debenhams
Berkshire
Montcalm
Syn.
Rose Ct.
Bilton Towers
Disney
Doubletree by Hilton
Sussex
Selfridges
Ch. of the Annunciation
Hard Rock
Amba
Marks & Spencer
Waterstones
BOND ST.
West One Shop. Cen.
Hanover Square Gdns.
Bonhams
Connaught Ho. School
MARBLE ARCH
Primark
Park House
Coll.
Kings Weigh Ho. Ch.
AR Cen.
Handel & Hendrix Mus.
Fenwick
Dolce & Gabbana
Armani
Mulberry
Tyburn Tree (Site of)
Marble Arch
Marriott
Virgin Active
W1
Clarendon Flats
Montessori Centre International
The Beaumont
Chesham Flats
Marriott
Claridge's
North Row Bldgs.
Lees Ct.
Stanhope Place Gate
Speakers' Corner
Avenfield Ho.
Upper Feilde
Roosevelt Mem.
Sothebys
Fendi
Burberry
ENTRANCE TO UNDERGROUND CAR PARK
Animals in War Memorial
Grand Residences
Biltmore
Connaught Ho.
Chap.
Refreshment Kiosk
Connaught
Fitness First
Grosvenor Ho.
Parade Ground
Library
St. George's Gdns.
Ch. of the Immaculate Conception
Reformers Tree
Grosvenor Chapel
St. George's (Hanover Sq.) Prim. Sch.
ENTRANCE TO UNDERGROUND CAR PARK
Joy of Life Fountain
Fountain Ho.
Aldford Ho.
Rosebery Ct.
The Lookout Education Cen.
Reservoir
Diana, Princess of Wales Memorial Walk
Chesterfield
Washington Mayfair
Third Ch. of Christ Scientist
Dorchester
Chesterfield Ho.
45 Park Lane
Curzon
7 July Memorial
Naval & Military
Pier
Hilton
Mos.
Christ Ch.
Metropolitan
Carrington Ho.
Bandstand
The Serpentine Bar & Kitchen
Boating Lake
Cavalry Memorial
Achilles Statue
Park Towers
Sheraton Grand Park Lane
Athenaeum
Four Seasons
Cavalry & Guards
Royal Air Force
Nannies Lawn
Queen Elizabeth Gate
Fishermans Keep
Holocaust Memorial
The Holocaust Memorial Garden
Rose Garden
The Dell
Memorial Walk
Apsley Ho. (Wellington Museum)
Apsley Ga.
HYDE PARK CORNER
New Zealand War Meml.
Wellington Mon.
RAF Bomber Command Memorial
HYDE PARK
GREEN PARK
A41
A4380
A5204
A5
A40
A4202
A4201
B524
B406
A4
BAKER STREET
PADDINGTON STREET
MARYLEBONE HIGH STREET
GLOUCESTER PLACE
WIGMORE STREET
OXFORD STREET
PORTMAN SQUARE
ORCHARD ST.
SEYMOUR STREET
CUMBERLAND GATE
PARK LANE
UPPER BROOK ST.
GROSVENOR SQUARE
MOUNT STREET
SOUTH AUDLEY STREET
CURZON STREET
CAVENDISH SQ.
NEW BOND STREET
DUKE ST.
ROTTEN ROW
HYDE PARK CORNER

222
A
B
C
D
E
1
2
3
4
5
6
7
216
221
228
29
FITZROVIA
W1
SW1
WEST END
GREEN PARK
ST. JAMES'S
OXFORD CIRCUS
PICCADILLY CIRCUS
TOTTENHAM COURT ROAD
GOODGE STREET
GREEN PARK
RIBA
Portland Pl. Sch.
Central Synagogue
Franco-Anglaise Sch.
YHA
Univ. of Westminster
St. Charles Ch.
Telecom Tower
Halls of Residence
Toy Mus.
Ch. of Scientology
British Transport Police
RADA
RADA Studios
Coll. of Law
Cavendish Coll.
The Harley St. Clinic
BBC Broadcasting House
All Souls' Ch.
Queen's Coll.
Chapel
Convent
London Eye Hosp.
Prim. Sch.
Mosque
Pearson Square
Crabtree Place
Rathbone Place
Charlotte St.
Odeon
Architectural Association
Radisson Edwardian Kenilworth
Central YMCA
Bloomsbury
Dominion
Cartoon Mus.
All Saints Ch.
Sanderson
Temple
Welsh Baptist Chapel
Rathbone Square
Academy
Primark
Regent St.
Univ. of Westminster (Regent Campus)
ENTRANCE TO UNDERGROUND CAR PARK
London Coll. of Fashion
Buses & Taxis only Mon. - Sat. 7am - 7pm
Waterstones
French Protestant Ch.
St. Patrick's Ch.
Centrepoint
St. Giles Square
John Lewis
House of Fraser
Debenhams
Disney
Top Shop
Marks & Spencer
Photographers' Gallery
College of Beauty Therapy
Youth Hostel
Soho
NHS Cen.
Hazlitts
Phoenix
Foyles
Regent Hall
Courthouse
London Palladium
Fitness First
Apple
Hanover Square Gdns.
Bonhams
Hendrix Mus.
Liberty
Marshall St. Leisure Cen.
Prince Edward
Adult Ed. Cen.
Palace
Fenwick
Dolce & Gabbana
Armani
Mulberry
Hamleys
Hackett
Westminster Kingsway Coll.
St. Anne's Ch.
Curzon
Fire Sta.
St. George's Ch.
Sothebys
Jimmy Choo
Lalique
Brooks Brothers
Sanctum Soho
St. James Residences
Queen's
Gielgud
Apollo
Prince Charles
Leicester Sq.
Vue
Cineworld
Fendi
Burberry
Vuitton
Chanel
Dior
Asprey
Royal Institution Mus.
The Faraday Mus.
Tiffany
Lauren
Penhaligon
Ham Yard
Piccadilly
Lyric
Pictureho.
London Trocadero
London Pav.
Prince of Wales
Leicester Fields
Odeon
Fitness First
Gerrard
Royal Academy of Arts (Burlington Ho.)
Yves Saint Laurent
Gucci
Cartier
De Beers
Mappin & Webb
Cafe Royal
Le Meridien Piccadilly
Eros
Criterion
Jermyn St.
Waterstones
Vue (Apollo)
PureGym
Comedy Store
Harold Pinter
Empire
Theatre Royal Haymarket
Cen. Ref. Lib.
Sainsbury Wing
National Gallery
St. James's Ch.
Paxton & Whitfield
Fortnum & Mason
Her Majesty's
Canada Ho.
Trafalgar
The May Fair
Holiday Inn
Ritz
Curzon Plaza
Third Ch. of Christ Scientist
Boodle's
Brooke's
London Library
Christies
HQ of the Inst. of Directors
Admiralty Arch
Royal Marines Meml.
Mall Galleries
ICA
Old Admiralty Offices
Duke of York Column
Travellers'
Athenaeum
Reform
RAC
Naval & Military
Fox Club
Hilton Green Park
Spencer Ho.
Oxford & Cambridge
Marlborough House (Commonwealth Secretariat)
Queen's Chapel
King George VI & Queen Mother Memorials
Chapel Royal
St. James's Palace
Clarence Ho.
Lancaster Ho.
Stable Yard Gate
CLOSED SUNDAY
Refreshments
Duck Island
Guards Mem.
Cavalry & Guards
Royal Air Force
RAF Bomber Command Memorial
Canada Memorial
Park Towers
Curzon
A4201
A5204
A40
A400
A401
A4
B406
B506
OXFORD STREET
REGENT STREET
PICCADILLY
SHAFTESBURY AVENUE
CHARING CROSS ROAD
PALL MALL
THE MALL
TOTTENHAM COURT ROAD
GOODGE STREET
MORTIMER STREET
LANGHAM PLACE
CAVENDISH SQ.
HANOVER SQUARE
BERKELEY SQUARE
ST. JAMES'S SQUARE
LEICESTER SQUARE
SOHO SQUARE
BEDFORD SQUARE

223
217
224
229
F
G
H
J
K
1
2
3
4
5
6
7
WC2
SE1
LAMBETH
HOLBORN
CHANCERY LANE
COVENT GARDEN
TEMPLE
EMBANKMENT
CHARING CROSS
WATERLOO
WATERLOO EAST
RIVER THAMES
Bloomsbury Square Gdns.
Gray's Inn Gardens
Gray's Inn
Lincoln's Inn Fields
Lincoln's Inn
Royal Courts of Justice
London School of Economics
Sir John Soane's Museum
Royal Coll. of Surgeons
The Hunterian Mus. (Closed until 2021)
Old Curiosity Shop
Freemasons' Hall
Royal Opera House
Covent Gdn Piazza
The Market
St. Paul's Ch.
London Transport Mus.
Theatre Royal Drury Lane
Somerset House
Courtauld Gallery
King's Coll. (Strand Campus)
Embankment Galleries
Australia Ho.
India Ho.
Bush Ho. Inland Revenue
St. Mary-le-Strand Ch.
St. Clement Danes Ch.
Temple Bar
Temple
Inner Temple
Mid. Temple Hall
Inner Temple Garden
Victoria Embankment Gardens
HQS Wellington
Temple Pier
Savoy Pier
Cleopatra's Needle
Hungerford Bridge
The Golden Jubilee Bridges
Festival Pier
Embankment Pier
RS Hispaniola
PS Tattershall Castle
Jubilee Gardens
London Eye
London Eye Millennium Pier
RAF Memorial
Whitehall Stairs
Lifeboat Station
BFI Southbank
National Theatre
Royal Festival Hall
Queen Elizabeth Hall
Purcell Room
Hayward Gallery
Southbank Centre
Rambert
King's College London (Waterloo Campus)
IMAX
Old War Office
Ministry of Defence
Banqueting House
Dept. for Energy
Scotland Office
Wales Office
New Scotland Yard
Cenotaph
St. Martin-in-the-Fields Ch.
Charing Cross Underground Shop. Cen.
Adelphi
Savoy
Savoy Chapel
Shell-Mex House
Strand Palace
Waldorf
Aldwych
Bush Ho.
Stamford Wharf
OXO TOWER WHARF
Gabriels Wharf
Bernie Spain Gardens
The Queen's Walk
Kings Reach
Bank South
A40
A4200
A401
A5200
B400
A4
B402
B401
A3211
A301
A3200
B308
B521
HIGH HOLBORN
KINGSWAY
STRAND
FLEET STREET
WATERLOO BRIDGE
WATERLOO ROAD
STAMFORD STREET
NORTHUMBERLAND AV.
WHITEHALL
VICTORIA EMBANKMENT
HORSE GUARDS AVENUE
THEOBALDS ROAD

224
218
223
230
CAMDEN
LAMBETH
EC1
EC4
SOUTHWARK
SE1
CHANCERY LANE
FARRINGDON
BARBICAN
CITY THAMESLINK
ST. PAUL'S
MANSION HO.
BLACKFRIARS
WATERLOO EAST
SOUTHWARK
HOLBORN
HOLBORN CIRCUS
HOLBORN VIADUCT
NEWGATE STREET
FLEET STREET
NEW BRIDGE STREET
FARRINGDON STREET
FETTER LANE
NEW FETTER LANE
QUEEN VICTORIA ST.
UPPER THAMES STREET
CANNON STREET
CHEAPSIDE
KING EDWARD ST.
ST. MARTIN'S LE GRAND
ALDERSGATE STREET
LONDON WALL
BLACKFRIARS BRIDGE
SOUTHWARK BRIDGE
BLACKFRIARS ROAD
SOUTHWARK STREET
STAMFORD STREET
UNION STREET
VICTORIA EMBANKMENT
RIVER THAMES
Smithfield
Central Markets
St. Bartholomew's Hospital
St. Bartholomew the Great Ch.
Museum of London
Barbican Arts Cen.
City of London Sch. for Girls
Central Criminal Ct.
Old Bailey
St. Paul's Cathedral
The Queen's Diamond Jubilee Gdn.
Festival Gdns.
London Stock Exchange
British Telecom Centre
Postmans Park
Goldsmiths Hall
Wax Chandlers Hall
Haberdashers' Hall
Ironmongers Hall
Barber Surgeon's Hall
Pewterer's Hall
Police Mus.
St. Dunstan-in-the-West Ch.
King's College Maughan Lib.
Dr. Johnson's Ho.
Temple Ch.
Inner Temple
Inner Temple Garden
Middle Temple Hall
Hamilton House
Sion Coll.
Unilever House
Former City of London School
St. Bride Foundation
Fleetbank Ho.
Youth Hostel
Salvation Army HQ
City of London Sch.
Millennium Bridge
Blackfriars Millennium Pier
Bankside Pier
Vintners Hall
Shakespeare's Globe & Exhibition
Tate Modern
Bankside Gallery
Rose Playho. (site of)
Financial Times
OXO TOWER WHARF
Sea Containers
Stamford Wharf
Bernie Spain Gardens
The London Nautical Sch.
Christ Church
Colombo Centre
Hilton Bankside
Contemporary Applied Arts
Blue Fin Building
Bankside Mix
Great Guildford Business Square
Menier Chocolate Factory
King's College London
The Young Vic
Lewisham Coll. (Waterloo Cen.)
Jerwood Space Art Gallery
Fire Brigade HQ
Nelson Sq. Gdn.
Blackfriars Crown Court
The Cathedral of St. Saviour & St. Mary Overie
HQS Wellington

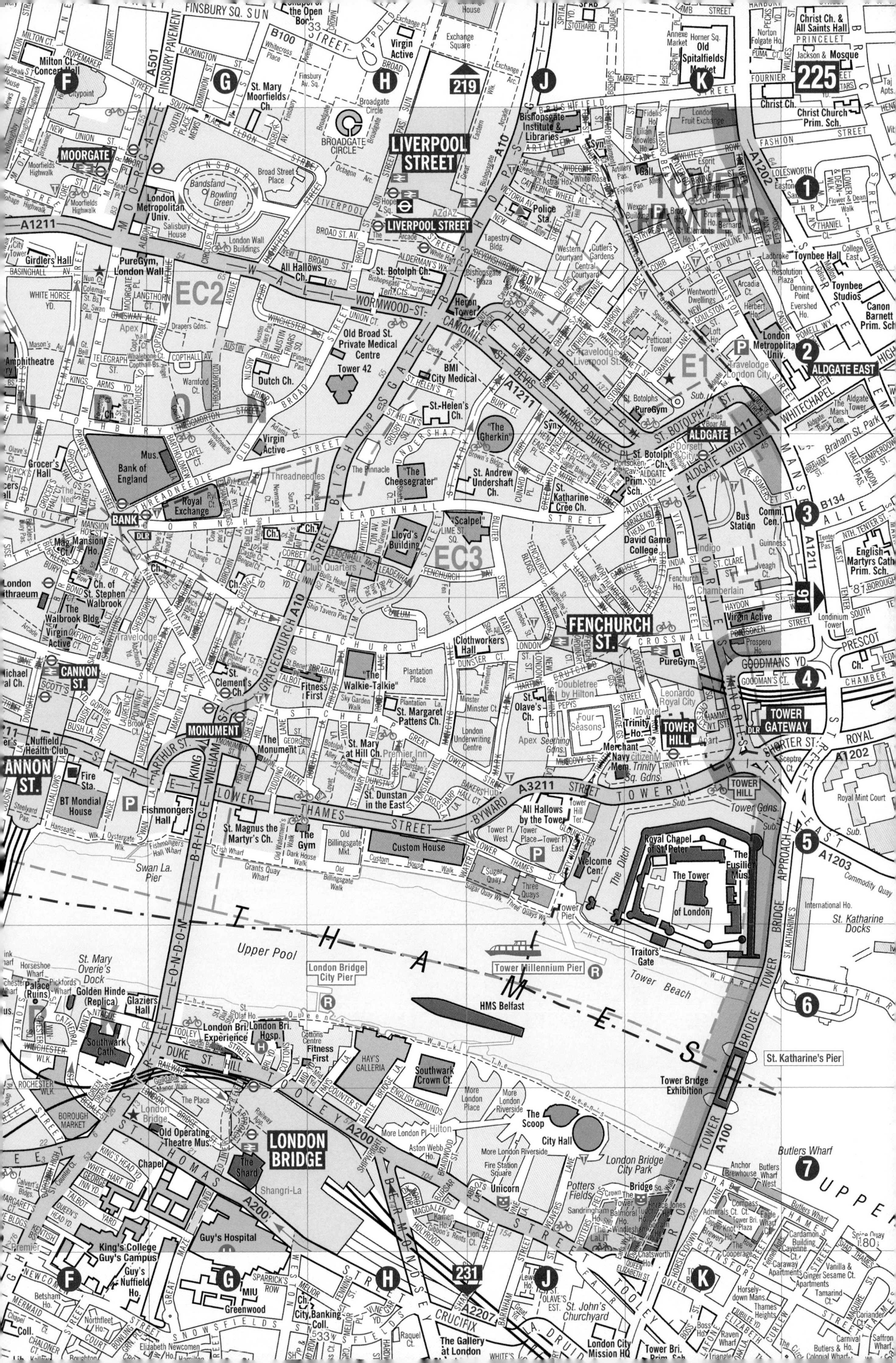

F
G
H
J
K
219
225
1
2
3
4
5
6
7
91
231
FINSBURY SQ.
Milton Ct. Concert Hall
MOORGATE
FINSBURY PAVEMENT
A501
St. Mary Moorfields Ch.
Virgin Active
Exchange Square
Broadgate Circle
BROADGATE CIRCLE
LIVERPOOL STREET
Bishopsgate Institute & Libraries
Old Spitalfields Market
London Fruit Exchange
Christ Ch. & All Saints Hall
Mosque
Christ Ch.
Christ Church Prim. Sch.
TOWER HAMLETS
Bandstand
Bowling Green
FINSBURY CIRCUS
London Metropolitan Univ.
Salisbury House
London Wall Buildings
Broad Street Place
A1211
Girdlers Hall
PureGym, London Wall
EC2
All Hallows Ch.
St. Botolph Ch.
WORMWOOD ST.
Heron Tower
Police Sta.
Tapestry Bldg.
Western Courtyard
Central Courtyard
Cutlers Gardens
Toynbee Hall
Toynbee Studios
Canon Barnett Prim. Sch.
Old Broad St. Private Medical Centre
Tower 42
BMI City Medical
Dutch Ch.
Drapers Gdns.
Apex
Amphitheatre
Travelodge Liverpool St.
Petticoat Tower
E1
London Metropolitan Univ.
ALDGATE EAST
Travelodge London City
St. Helen's Ch.
"The Gherkin"
PureGym
ALDGATE
St. Botolph Ch.
The Pinnacle
"The Cheesegrater"
St. Andrew Undershaft Ch.
St. Katharine Cree Ch.
Bank of England
Mus.
Royal Exchange
BANK
Grocer's Hall
LEADENHALL
"Scalpel"
Lloyd's Building
EC3
Club Quarters
David Game College
Bus Station
Comm. Cen.
English Martyrs Cath. Prim. Sch.
Indigo
Chamberlain
Mansion Ho.
Ch. of St. Stephen Walbrook
The Walbrook Bldg.
Virgin Active
CANNON ST.
Travelodge
FENCHURCH ST.
Clothworkers Hall
Plantation Place
"The Walkie-Talkie"
Fitness First
Virgin Active
Londinium Tower
Prospero Ho.
PureGym
St. Clement's Ch.
St. Margaret Pattens Ch.
Minster Ct.
Doubletree by Hilton
Four Seasons
Leonardo Royal City
GOODMANS YD.
Novotel
Trinity Ho.
TOWER HILL
TOWER GATEWAY
MONUMENT
Nuffield Health Club
CANNON ST.
Fire Sta.
BT Mondial House
Fishmongers Hall
The Monument
St. Mary at Hill Ch.
Premier Inn
London Underwriting Centre
Apex
Seething Gdns.
Merchant Navy Mem.
Trinity Sq. Gdns.
Royal Mint Court
St. Dunstan in the East
LOWER THAMES STREET
A3211
BYWARD STREET
TOWER HILL
All Hallows by the Tower
Tower Place West
Tower Place East
Welcome Cen.
The Ditch
Royal Chapel of St Peter
The Fusilier Mus.
The Tower of London
St. Magnus the Martyr's Ch.
The Gym
Old Billingsgate Mkt.
Custom House
Swan La. Pier
Grants Quay Wharf
Sugar Quay
Three Quays
Tower Pier
A1203
International Ho.
St. Katharine Docks
THAMES
Upper Pool
London Bridge City Pier
Tower Millennium Pier
Traitors' Gate
Tower Beach
HMS Belfast
St. Mary Overie's Dock
Golden Hinde (Replica)
Glaziers Hall
LONDON BRIDGE
London Bri. Experience
London Bri. Hosp.
Fitness First
HAY'S GALLERIA
Southwark Crown Ct.
St. Katharine's Pier
Tower Bridge Exhibition
TOWER BRIDGE
Southwark Cath.
BOROUGH MARKET
Old Operating Theatre Mus.
The Place
More London Place
More London Riverside
The Scoop
City Hall
London Bridge City Park
Butlers Wharf
Anchor Brewhouse
The Shard
Shangri-La
LONDON BRIDGE
A200
Fire Station Square
Unicorn
Potters Fields
TOOLEY STREET
A100
TOWER BRIDGE ROAD
Chapel
King's College Guy's Campus
Guy's Hospital
Guy's Nuffield Ho.
MIU
Greenwood
City Banking Coll.
The Gallery at London
A2207
St. John's Churchyard
London City Mission HQ
Premier
Betsham Ho.
Northfleet Ho.
Elizabeth Newcomen

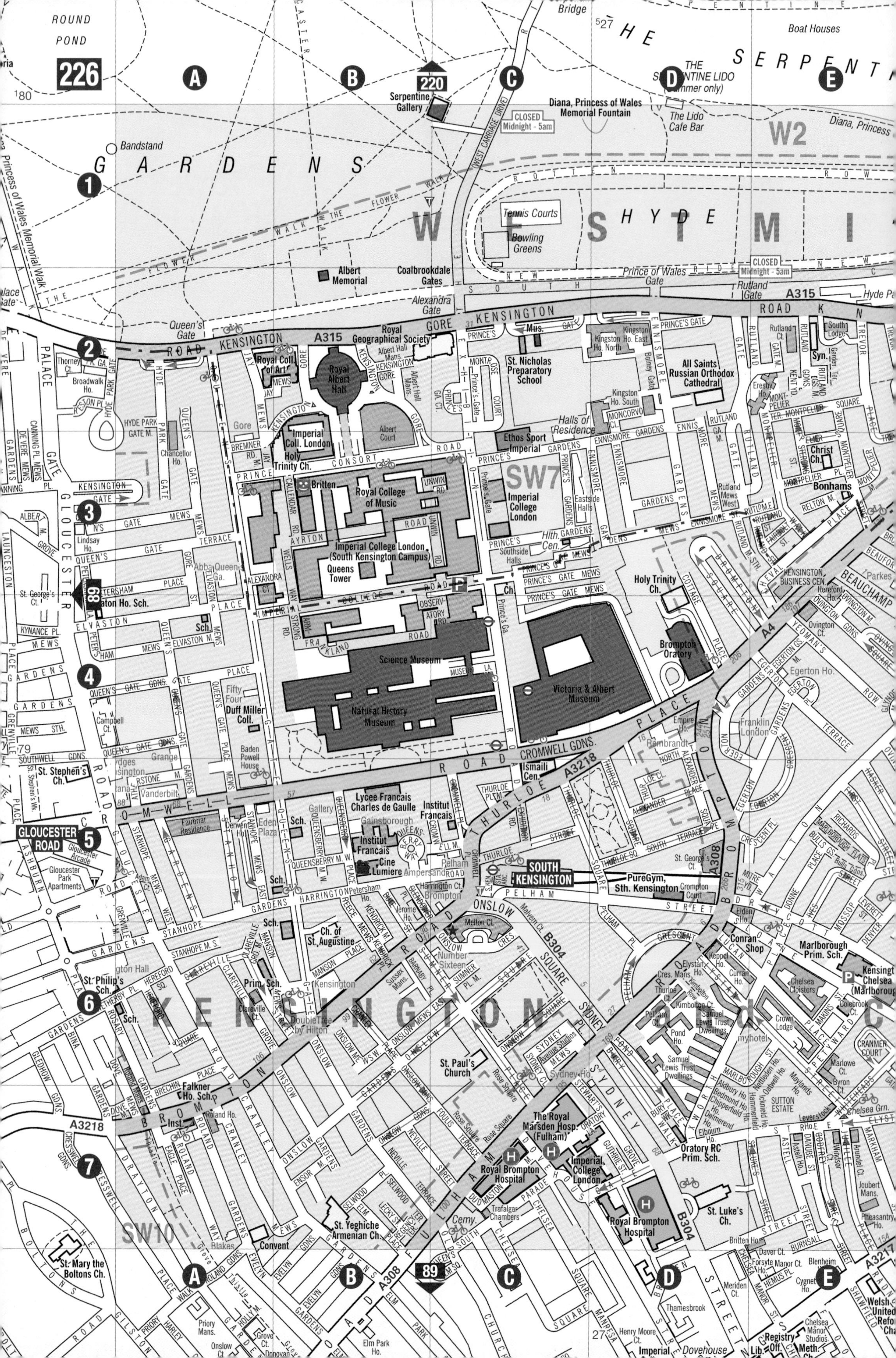
226
220
68
89
ROUND POND
GARDENS
THE SERPENTINE
THE SERPENTINE LIDO (Summer only)
Boat Houses
Serpentine Gallery
Diana, Princess of Wales Memorial Fountain
The Lido Cafe Bar
CLOSED Midnight - 5am
W2
Bandstand
Princess of Wales Memorial Walk
THE FLOWER WALK
WEST CARRIAGE DRIVE
ROTTEN ROW
HYDE
WESTMI
Tennis Courts
Bowling Greens
Albert Memorial
Coalbrookdale Gates
Prince of Wales Gate
Rutland Gate
SOUTH CARRIAGE DRIVE
Alexandra Gate
Queen's Gate
KENSINGTON GORE
KENSINGTON ROAD
A315
Royal Geographical Society
Royal Coll. of Art
Royal Albert Hall
Albert Court
Imperial Coll. London
Holy Trinity Ch.
PRINCE CONSORT ROAD
Britten
Royal College of Music
Imperial College London (South Kensington Campus)
Queens Tower
IMPERIAL COLLEGE ROAD
St. Nicholas Preparatory School
Kingston Ho. North
Kingston Ho. East
Kingston Ho. South
All Saints Russian Orthodox Cathedral
Halls of Residence
Ethos Sport Imperial
SW7
Imperial College London
Southside Halls
Eastside Halls
Hlth. Cen.
ENNISMORE GARDENS
Holy Trinity Ch.
Brompton Oratory
Science Museum
Natural History Museum
Victoria & Albert Museum
EXHIBITION ROAD
CROMWELL ROAD
CROMWELL GDNS.
Ismaili Cen.
THURLOE PLACE
A3218
Lycee Francais Charles de Gaulle
Institut Francais
Cine Lumiere
SOUTH KENSINGTON
PureGym, Sth. Kensington
PELHAM STREET
ONSLOW SQUARE
B304
SYDNEY STREET
A308
FULHAM ROAD
BROMPTON ROAD
A4
Bonhams
Christ Ch.
KENSINGTON BUSINESS CEN.
BEAUCHAMP PLACE
Egerton Ho.
Franklin
Conran Shop
Marlborough Prim. Sch.
Chelsea Cloisters
myhotel
SUTTON ESTATE
St. Paul's Church
The Royal Marsden Hosp. (Fulham)
Royal Brompton Hospital
Imperial College London
Oratory RC Prim. Sch.
St. Luke's Ch.
St. Yeghiche Armenian Ch.
Convent
SW10
St. Mary the Boltons Ch.
THE BOLTONS
Duff Miller Coll.
Baden Powell House
Fairbriar Residence
GLOUCESTER ROAD
Gloucester Arcade
Gloucester Park Apartments
St. Stephen's Ch.
St. Philip's Sch.
Falkner Ho. Sch.
Ch. of St. Augustine
DoubleTree by Hilton
KENSINGTON
Number Sixteen
Melton Ct.
Cerny.
ELM PARK GARDENS
DRAYTON GARDENS
A3217
Welsh United Refo. Ch.
Registry Off.
Lib.
Imperial
Dovehouse

227
228
221
89
90
28
80
79
179
F
G
H
J
K
1
2
3
4
5
6
7
SW1
SW3
HYDE PARK CORNER
KNIGHTSBRIDGE
SLOANE SQ.
VICTORIA
The Serpentine Bar & Kitchen
Boating Lake
Fishermans Keep
Memorial Walk
The Dell
Nannies Lawn
Holocaust Memorial
The Holocaust Memorial Garden
Rose Garden
Bandstand
Cavalry Memorial
Achilles Statue
Queen Elizabeth Gate
Apsley Ho. (Wellington Museum)
Apsley Ga.
CLOSED Midnight - 5am
Wellington Mon.
Wellington Arch
Royal Artillery Mem.
Australian War Meml.
New Zealand War Meml.
RAF Bomber Command Memorial
Royal Air Force
Cavalry & Guards
Inter-Continental
Diana, Princess of Wales Memorial Walk
CLOSED SUNDAY
Commonwealth Memorial Gates
CONSTITUTION
PALACE GARDENS
The Royal Mews
Royal Mews
'Pan' (Epstein)
Mandarin Oriental
One Hyde Park
SCOTCH HOUSE
Harvey Nichols
Burberry
Harrods
Ferragamo
Rossetti
Vuitton
Gucci
Versace
D & G
Dior
Jimmy Choo
Valentino
Yves Saint Laurent
Chanel
Armani
Prada
Millennium
Hugo Ho.
Berkeley
Lanesborough
Wellesley
St. Paul's Ch.
Jumeirah Lowndes
Belgravia Ho.
Lowndes Lodge
Halkin
Montrose Ho.
BELGRAVE SQUARE
St. Peter's Ch.
Diplomat
Inchbald Sch.
St. Michael's Ch.
Lord Milner
Victoria Coach Sta. (Arrivals)
Victoria Coach Sta. (Departures)
Liby.
Victoria Place Shop. Cen.
Fountain Square
Coach Station
National Audit Office
Inchbald Sch. of Design
Ebury Bri. Cen.
St. Peter's Prim. Sch.
Tophams
Eaton Square Gardens
More Ho. Sch.
St. Saviour's Ch.
Claverley Court Apartments
Knightsbridge Sch.
St. Columba's Ch.
St. Simon Zelotes Ch.
Sussex Ho. Sch.
Cadogan Hall
Hackett
Smythson
Cartier
Tiffany
Holy Trinity Ch.
Sloane Ter. Mans.
Prim. Sch.
Royal Ct.
Peter Jones
Hugo Boss
St. Mary's Ch.
St. Thomas More Language Coll. & St. Joseph RC Prim. Sch.
Saatchi Gallery
Duke of York Square
Garden Ho. Sch.
St. Mary's Ch.
Francis Holland School
Carmel Hall
St. Barnabas C of E Prim. Sch.
Church of St. Barnabas
The Old Chapel
KINGS WALK SHOP. CEN.
Royal Hospital Chelsea
Mus.
Great Hall
Chap.
LIGHT HORSE CT.
Burton's Court
Chelsea Affiliated Synagogue
Carriage Shed
A4
A4202
A302
A3216
A3217
A3213
A3214
B310
B312
B313
B319
B302
GROSVENOR PLACE
CHESHAM STREET
PONT STREET
SLOANE STREET
BUCKINGHAM PALACE ROAD
ECCLESTON STREET
CHELSEA BRIDGE ROAD
EBURY BRIDGE ROAD
PIMLICO ROAD
KING'S ROAD

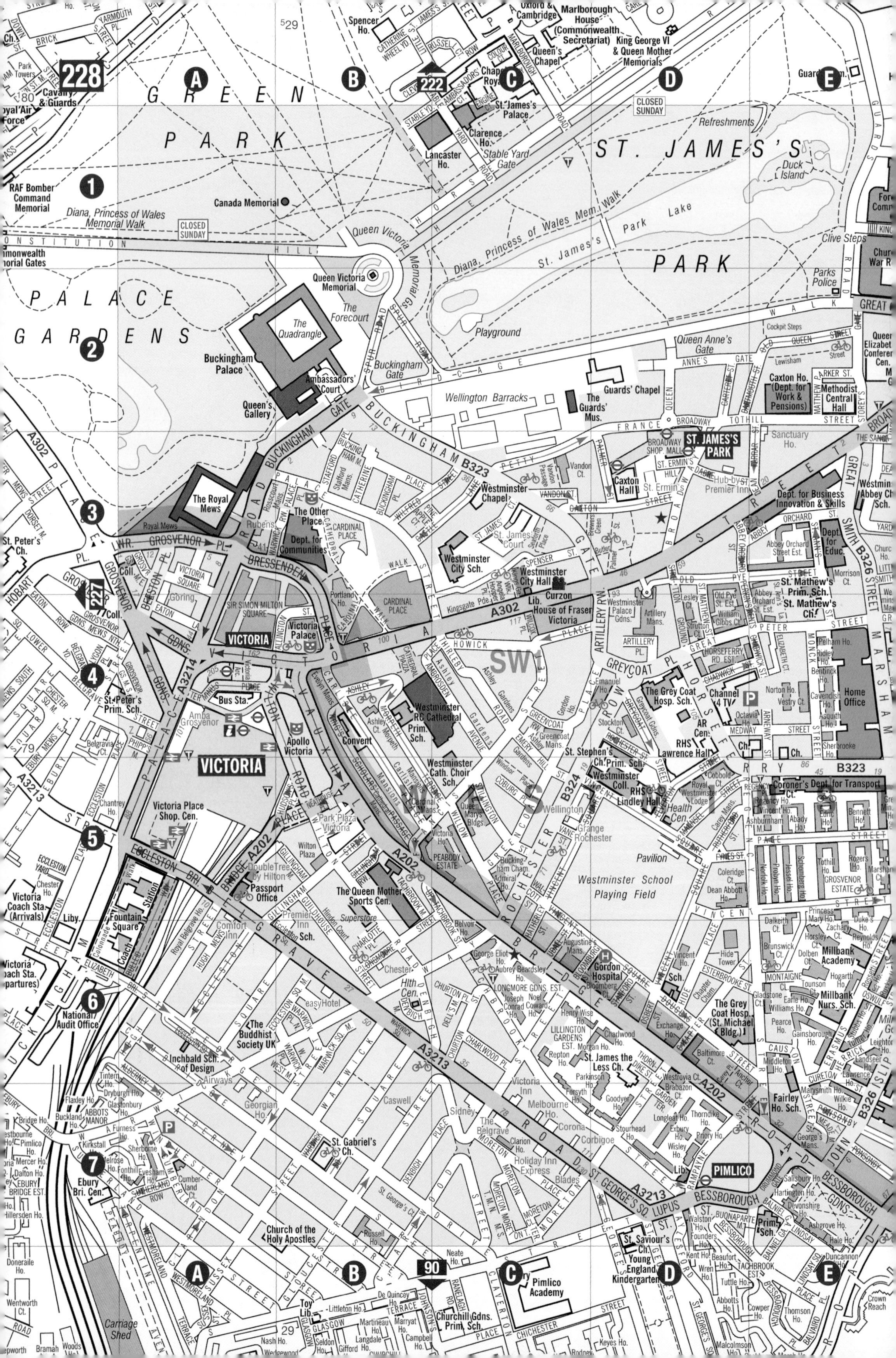

228
529
222
227
90
A
B
C
D
E
1
2
3
4
5
6
7
GREEN PARK
PALACE GARDENS
ST. JAMES'S PARK
SW1
WESTMINSTER
Spencer Ho.
Oxford & Cambridge
Marlborough House (Commonwealth Secretariat)
King George VI & Queen Mother Memorials
Queen's Chapel
Chapel Royal
St. James's Palace
Clarence Ho.
Lancaster Ho.
Stable Yard Gate
CLOSED SUNDAY
Refreshments
Duck Island
St. James's Park Lake
Cavalry & Guards
Royal Air Force
Park Towers
RAF Bomber Command Memorial
Canada Memorial
Diana, Princess of Wales Memorial Walk
Diana, Princess of Wales Mem. Walk
CONSTITUTION HILL
Commonwealth Memorial Gates
Queen Victoria Memorial
The Forecourt
The Quadrangle
Buckingham Palace
Ambassadors' Court
Queen's Gallery
Buckingham Gate
BIRDCAGE WALK
Playground
Parks Police
Clive Steps
Queen Anne's Gate
Cockpit Steps
Wellington Barracks
Guards' Chapel
The Guards' Mus.
Caxton Ho. (Dept. for Work & Pensions)
Methodist Central Hall
Sanctuary Ho.
ST. JAMES'S PARK
Broadway Shop. Mall
Caxton Hall
St. Ermin's
Hub by Premier Inn
Dept. for Business Innovation & Skills
Westminster Abbey Ch. Sch.
Dept. for Educ.
The Royal Mews
Royal Mews
Rubens
The Other Place
Dept. for Communities
Westminster Chapel
Westminster City Sch.
Westminster City Hall
Curzon
House of Fraser Victoria
Cardinal Place
Victoria Palace
VICTORIA
Goring
St. Peter's Ch.
St. Peter's Prim. Sch.
Bus Sta.
Apollo Victoria
Victoria Place Shop. Cen.
Westminster RC Cathedral
Westminster Cathedral Prim. Sch.
Westminster Cath. Choir Sch.
Convent
The Grey Coat Hosp. Sch.
Channel 4 TV
Home Office
St. Stephen's Ch. Prim. Sch.
Westminster Coll.
RHS Lawrence Hall
RHS Lindley Hall
Health Cen.
Coroner's Ct.
Dept. for Transport
Westminster School Playing Field
Pavilion
Park Plaza Victoria
Wilton Plaza
DoubleTree by Hilton
Passport Office
Victoria Coach Sta. (Arrivals)
Victoria Coach Sta. (Departures)
Fountain Square
Coach Station
Liby.
National Audit Office
Premier Inn
The Queen Mother Sports Cen.
Superstore
Gordon Hospital
St. James the Less Ch.
The Grey Coat Hosp. (St. Michael Bldg.)
Millbank Academy
Millbank Nurs. Sch.
Fairley Ho. Sch.
easyHotel
The Buddhist Society UK
Inchbald Sch. of Design
St. Gabriel's Ch.
Holiday Inn Express
Church of the Holy Apostles
Ebury Bri. Cen.
PIMLICO
St. Saviour's Ch.
Young England Kindergarten
Pimlico Academy
Churchill Gdns. Prim. Sch.
Carriage Shed
Toy Lib.
Crown Reach
A302
A3214
A3213
A202
B323
B324
B326
VICTORIA STREET
GROSVENOR PL.
BUCKINGHAM PALACE ROAD
VAUXHALL BRIDGE ROAD
BELGRAVE ROAD
ST. GEORGE'S SQ.
LUPUS ST.
BESSBOROUGH GDNS.
JOHN ISLIP ST.
HORSEFERRY RD.
ROCHESTER ROW
ARTILLERY RW.
GREYCOAT PL.
ECCLESTON BRI.
WARWICK WAY

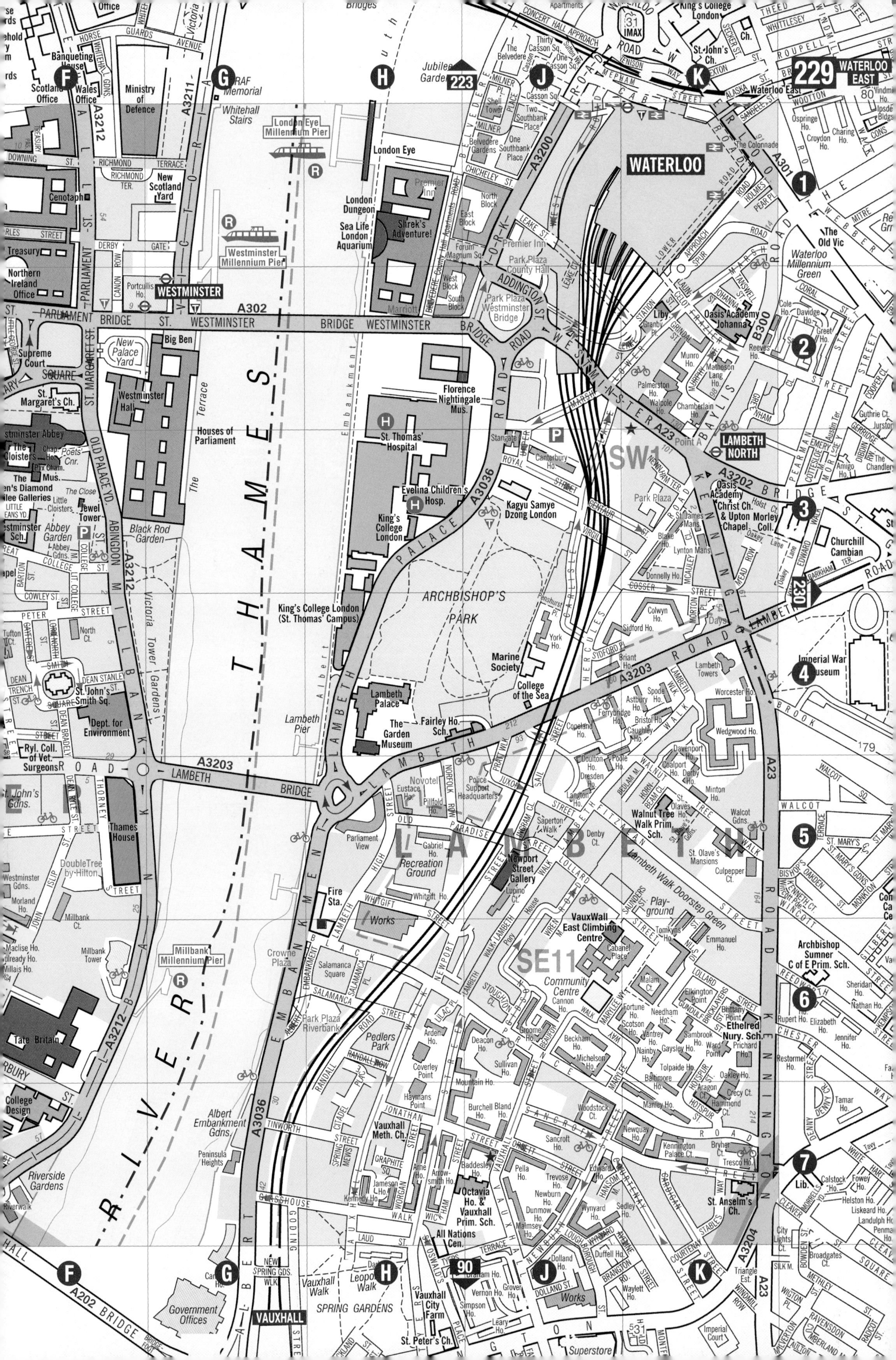

229
WATERLOO EAST
223
230
WATERLOO
WESTMINSTER
LAMBETH NORTH
VAUXHALL
Waterloo East
RIVER THAMES
ARCHBISHOP'S PARK
LAMBETH
SW1
SE11
A3200
A3036
A3203
A302
A3212
A3202
A23
A3204
A301
A202
London Eye
London Eye Millennium Pier
Westminster Millennium Pier
Millbank Millennium Pier
Jubilee Gardens
London Dungeon
Sea Life London Aquarium
Shrek's Adventure!
Houses of Parliament
Big Ben
Westminster Hall
Westminster Abbey
Supreme Court
St. Margaret's Ch.
Jewel Tower
Victoria Tower Gardens
Black Rod Garden
Ministry of Defence
Banqueting House
Scotland Office
Wales Office
New Scotland Yard
Cenotaph
Treasury
Northern Ireland Office
Portcullis Ho.
St. Thomas' Hospital
Evelina Children's Hosp.
King's College London
King's College London (St. Thomas' Campus)
Florence Nightingale Mus.
Lambeth Palace
The Garden Museum
Lambeth Pier
Kagyu Samye Dzong London
Marine Society
College of the Sea
Fairley Ho. Sch.
The Old Vic
Waterloo Millennium Green
Oasis Academy Johanna
Oasis Academy Christ Ch. & Upton Morley Chapel
Morley Coll.
Churchill Cambian
Imperial War Museum
Walnut Tree Walk Prim. Sch.
Newport Street Gallery
VauxWall East Climbing Centre
Community Centre
Ethelred Nury. Sch.
Archbishop Sumner C of E Prim. Sch.
St. Anselm's Ch.
Octavia Ho. & Vauxhall Prim. Sch.
Vauxhall Meth. Ch.
Vauxhall City Farm
All Nations Cen.
St. Peter's Ch.
Pedlers Park
Spring Gardens
Vauxhall Walk
Leopold Walk
Government Offices
Albert Embankment Gdns.
Riverside Gardens
Tate Britain
College Design
Thames House
DoubleTree by Hilton
Dept. for Environment
Ryl. Coll. of Vet. Surgeons
St. John's Smith Sq.
Recreation Ground
Lambeth Walk Doorstep Green
Playground
Crowne Plaza
Park Plaza Riverbank
Park Plaza Westminster Bridge
Park Plaza County Hall
Premier Inn
Novotel
Police Support Headquarters
Fire Sta.
Lib.
Superstore
Works
Westminster Bridge
Lambeth Bridge
Westminster Bridge Road
Lambeth Road
Lambeth Palace Road
Kennington Road
Millbank
Albert Embankment
Belvedere Road
York Road
Baylis Road
Lambeth Walk
Kennington Lane

230
224
229
90
91
A
B
C
D
E
1
2
3
4
5
6
7
WATERLOO
WATERLOO EAST
SOUTHWARK
LAMBETH NORTH
BOROUGH
ELEPHANT & CASTLE
KENNINGTON
SE11
LAMBETH
King's College London
St. John's Ch.
Waterloo East
The Colonnade
The Young Vic
Lewisham Coll. (Waterloo Cen.)
Theatre Vw. Apts.
St. Andrews Ch.
The Old Vic
Waterloo Millennium Green
Hampton by Hilton
Oasis Academy Johanna
Travelodge
Teacher Training Cen.
Nelson Sq. Gdn.
Union
Cervantes
Blackfriars Crown Court
Sports Grd.
Youth & Community Cen.
Friars Primary Foundation Sch.
Jerwood Space Art Gallery
Fire Brigade HQ
Great Guildford Business Square
Menier Chocolate Factory
Precious Blood
The Cathedral of St. Saviour & St. Mary Overie
St. Joseph's Cath. Prim. Sch.
Lit. Dorrit Park
Charles Dickens Prim. Sch.
Theatre Rites
McLaren House
Rec. Grnd.
Deeper Life Bible Ch.
London South Bank University
London Sth. Bank Univ.
London South Bank Univ. Technopark
Court Annexe
The Henry Wood Hall
Inner London Crown Court (Sessions House)
Newington Gardens
Mosque
Southwark Playhouse
Oasis Academy Christ & Upton Chapel
Morley Coll.
Churchill Cambian
Prim. Sch.
St. George's RC Cathedral
Notre Dame School
Siobhan Davis Dance Studios
Imperial War Museum
Geraldine Mary Harmsworth Park
Charlotte Sharman Foundation Prim. Sch.
West Sq. Gdn.
Imperial War Mus. (All Saints Annexe)
Perry Lib.
Metro Central Heights
London College of Communication
Metropolitan Tabernacle
The London Palace Superbowl
The Castle Centre
One The Elephant
St. Mary's Churchyard
Strata
Elephant Park Development Area
Crossway Ch.
Comm. Cen. St. Matthews Ch.
Halls of Residence
Brotherhood of the Cross & Star Church
Comm. Care Cen.
Archbishop Sumner C of E Prim. Sch.
Jamyang Buddhist Centre
Ethelred Nury. Sch.
St. Anselm's Ch.
Pedlars Acre
Guinness Trust Bldgs.
St. Mary's Ch.
Crampton Prim. Sch.
Alberta Estate
Safestay
Eurotraveller
Police Station
Tibetan Buddhist Cen.
Depot
St. John's Ch.
St. John's Prim. Sch.
Lib.
City & Guilds London Art Sch.
AR Cen.
Walworth Gdn. Farm
Keyworth Prim. Sch.
Pasley Park
Tel. Ex.
A3200
A3201
A3202
A3203
A3204
A301
A201
A300
A302
A3
A23
A215
B300
B240
BLACKFRIARS ROAD
BOROUGH ROAD
ST. GEORGE'S ROAD
LAMBETH ROAD
KENNINGTON ROAD
NEWINGTON BUTTS
ELEPHANT & CASTLE
WALWORTH ROAD
NEW KENT ROAD
LONDON ROAD
KENNINGTON LANE
BOROUGH HIGH STREET
MARSHALSEA ROAD
UNION STREET
THE CUT
WESTMINSTER BRIDGE ROAD
NEWINGTON CAUSEWAY
KENNINGTON PARK ROAD
BROOK DRIVE
HARPER ROAD

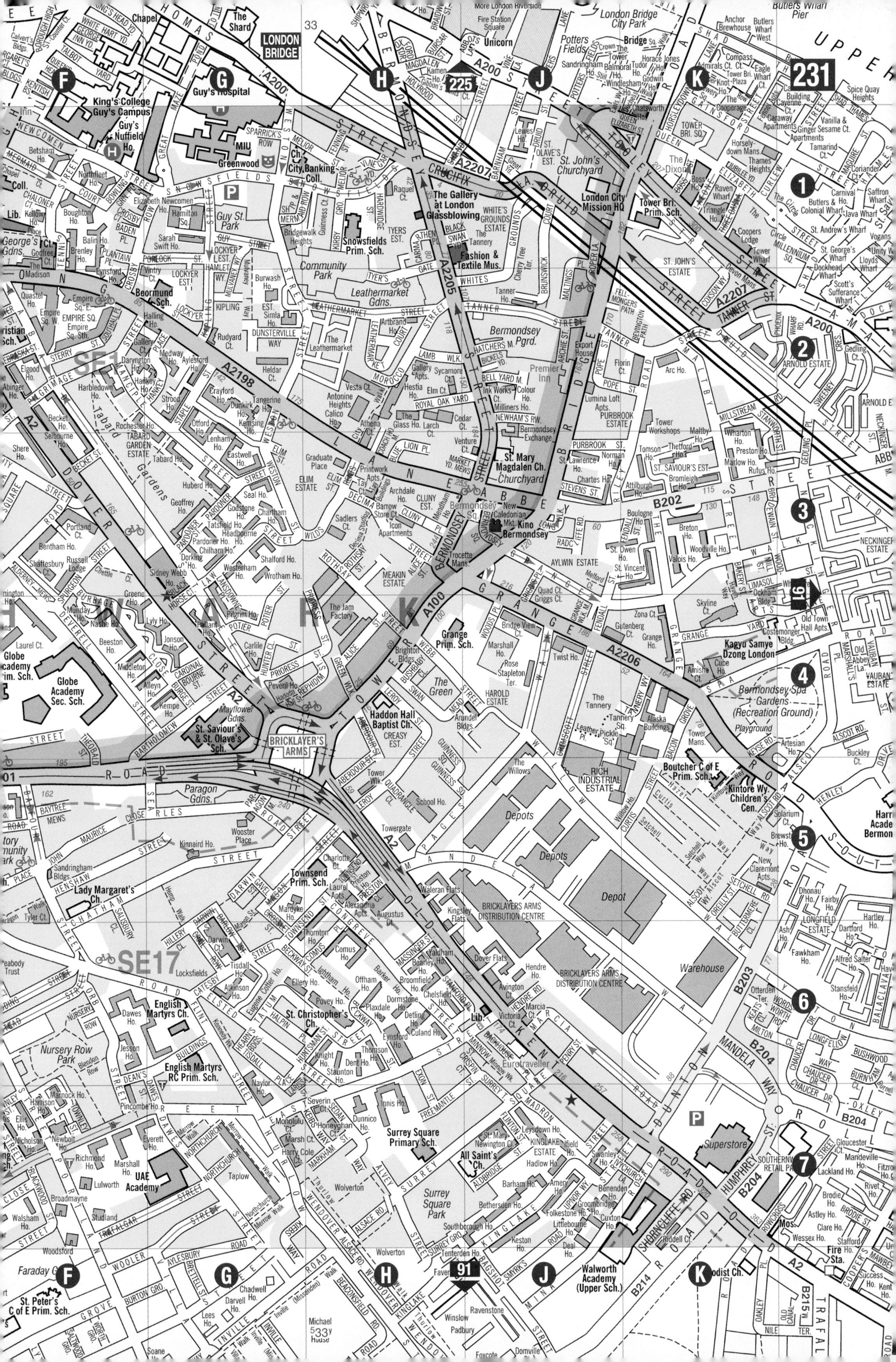

33
231
225
91
F
G
H
J
K
1
2
3
4
5
6
7
SE1
SE17
LONDON BRIDGE
The Shard
Guy's Hospital
King's College Guy's Campus
Guy's Nuffield Ho.
MIU
London Bridge City Park
Potters Fields
Unicorn
Fire Station Square
More London Riverside
Anchor Brewhouse
Butlers Wharf Pier
Tower Bri. Prim. Sch.
London City Mission HQ
St. John's Churchyard
The Gallery at London Glassblowing
Fashion & Textile Mus.
Snowsfields Prim. Sch.
Community Park
Leathermarket Gdns.
Guy St. Park
Beormund Sch.
Bermondsey Pgrd.
St. Mary Magdalen Ch.
Churchyard
Bermondsey Sq.
Kino Bermondsey
Tabard Gardens
TABARD GARDEN ESTATE
Globe Academy Prim. Sch.
Globe Academy Sec. Sch.
Grange Prim. Sch.
The Green
Haddon Hall Baptist Ch.
BRICKLAYER'S ARMS
St. Saviour's & St. Olave's Sch.
Mayflower Gdns.
Paragon Gdns.
Bermondsey Spa Gardens (Recreation Ground)
Kagyu Samye Dzong London
Boutcher C of E Prim. Sch.
Kintore Wy. Children's Cen.
RICH INDUSTRIAL ESTATE
Depots
Depot
Warehouse
BRICKLAYERS ARMS DISTRIBUTION CENTRE
Townsend Prim. Sch.
Lady Margaret's Ch.
English Martyrs Ch.
English Martyrs RC Prim. Sch.
Nursery Row Park
St. Christopher's Ch.
Surrey Square Primary Sch.
All Saint's Ch.
Surrey Square Park
UAE Academy
Walworth Academy (Upper Sch.)
Superstore
SOUTHERNWOOD RETAIL PARK
St. Peter's C of E Prim. Sch.
Eurotraveller
Faraday
Michael Faraday House
A2
A100
A200
A2198
A2205
A2206
A2207
B202
B203
B204
B214
B215
OLD KENT ROAD
TOWER BRIDGE ROAD
GRANGE ROAD
LONG LANE
DOVER STREET
GREAT DOVER STREET
BERMONDSEY STREET
TOOLEY STREET
MANDELA WAY
DUNTON ROAD

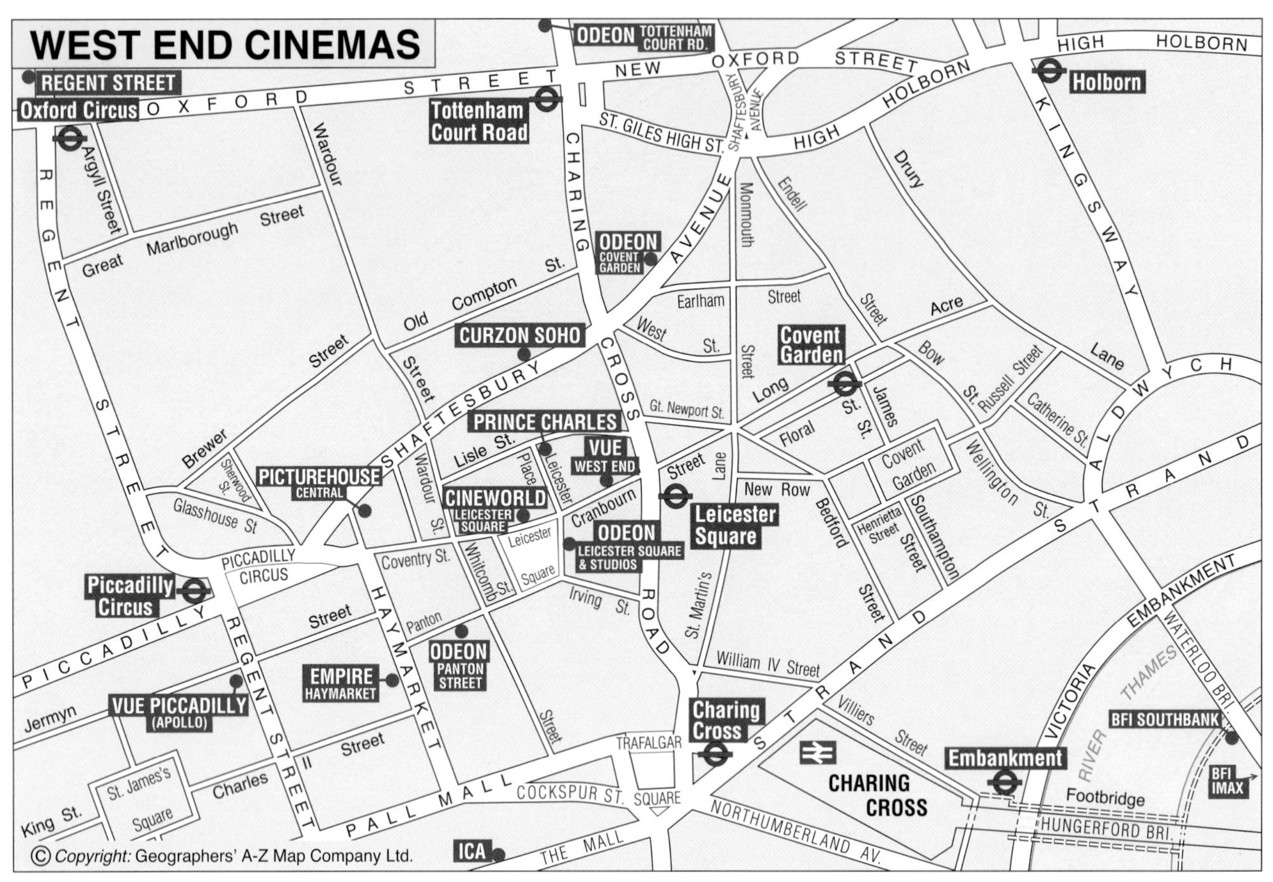
WEST END CINEMAS
ODEON TOTTENHAM COURT RD.
REGENT STREET
Oxford Circus
Tottenham Court Road
Holborn
ODEON COVENT GARDEN
CURZON SOHO
PRINCE CHARLES
VUE WEST END
PICTUREHOUSE CENTRAL
CINEWORLD LEICESTER SQUARE
ODEON LEICESTER SQUARE & STUDIOS
Covent Garden
Leicester Square
Piccadilly Circus
EMPIRE HAYMARKET
ODEON PANTON STREET
VUE PICCADILLY (APOLLO)
Charing Cross
CHARING CROSS
Embankment
BFI SOUTHBANK
BFI IMAX
ICA
OXFORD STREET
NEW OXFORD STREET
HIGH HOLBORN
KINGSWAY
ALDWYCH
STRAND
CHARING CROSS ROAD
SHAFTESBURY AVENUE
REGENT STREET
PICCADILLY
HAYMARKET
PALL MALL
THE MALL
TRAFALGAR SQUARE
NORTHUMBERLAND AV.
VICTORIA EMBANKMENT
RIVER THAMES
WATERLOO BRI.
HUNGERFORD BRI.
Footbridge
© Copyright: Geographers' A-Z Map Company Ltd.

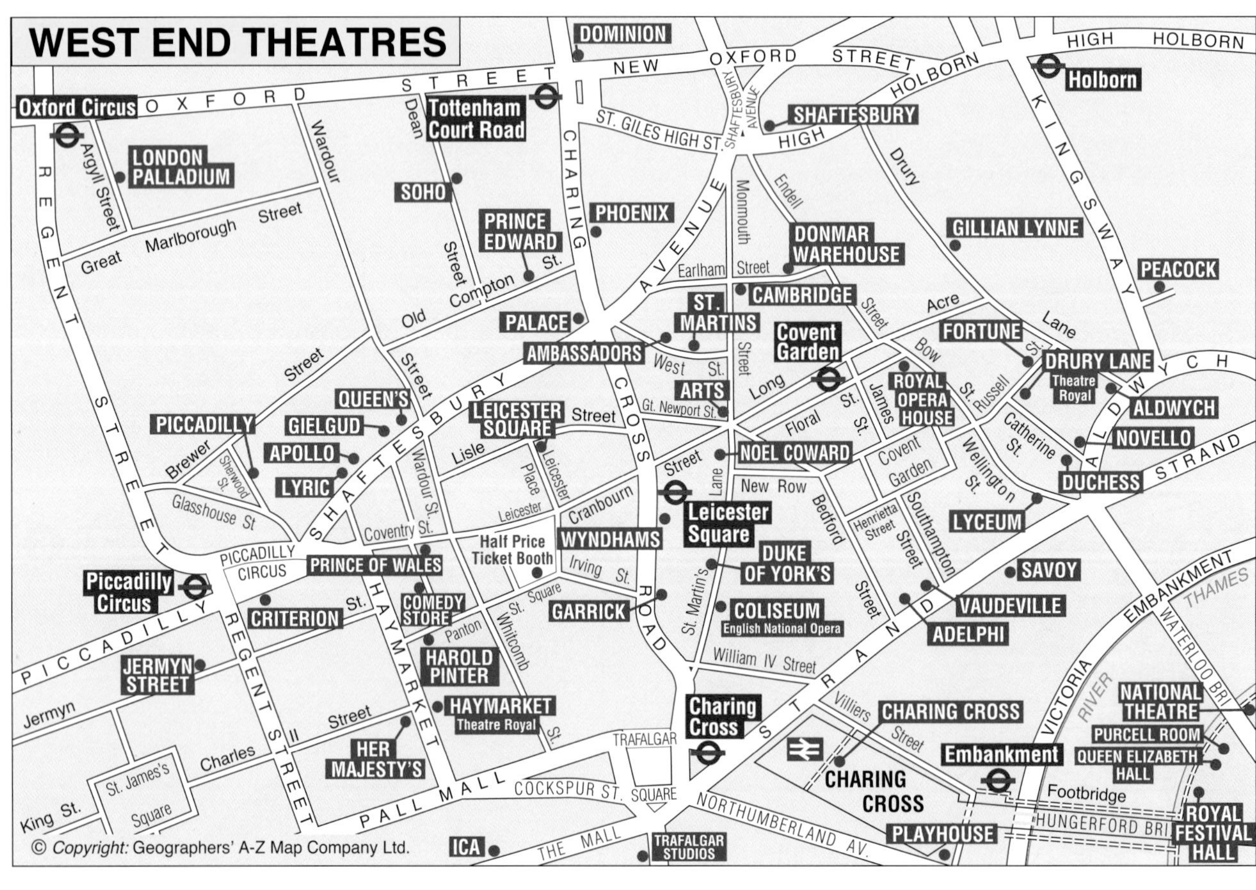
WEST END THEATRES
DOMINION
Oxford Circus
Tottenham Court Road
Holborn
SHAFTESBURY
LONDON PALLADIUM
SOHO
PRINCE EDWARD
PHOENIX
DONMAR WAREHOUSE
GILLIAN LYNNE
PEACOCK
CAMBRIDGE
ST. MARTINS
PALACE
AMBASSADORS
Covent Garden
FORTUNE
DRURY LANE Theatre Royal
ROYAL OPERA HOUSE
ALDWYCH
NOVELLO
ARTS
QUEEN'S
GIELGUD
PICCADILLY
APOLLO
LYRIC
LEICESTER SQUARE
NOEL COWARD
DUCHESS
LYCEUM
Leicester Square
WYNDHAMS
Half Price Ticket Booth
PRINCE OF WALES
DUKE OF YORK'S
SAVOY
Piccadilly Circus
CRITERION
COMEDY STORE
GARRICK
COLISEUM English National Opera
VAUDEVILLE
ADELPHI
HAROLD PINTER
JERMYN STREET
HAYMARKET Theatre Royal
Charing Cross
CHARING CROSS
NATIONAL THEATRE
PURCELL ROOM
QUEEN ELIZABETH HALL
HER MAJESTY'S
Embankment
CHARING CROSS
PLAYHOUSE
ROYAL FESTIVAL HALL
ICA
TRAFALGAR STUDIOS
© Copyright: Geographers' A-Z Map Company Ltd.

ROAD MAPS

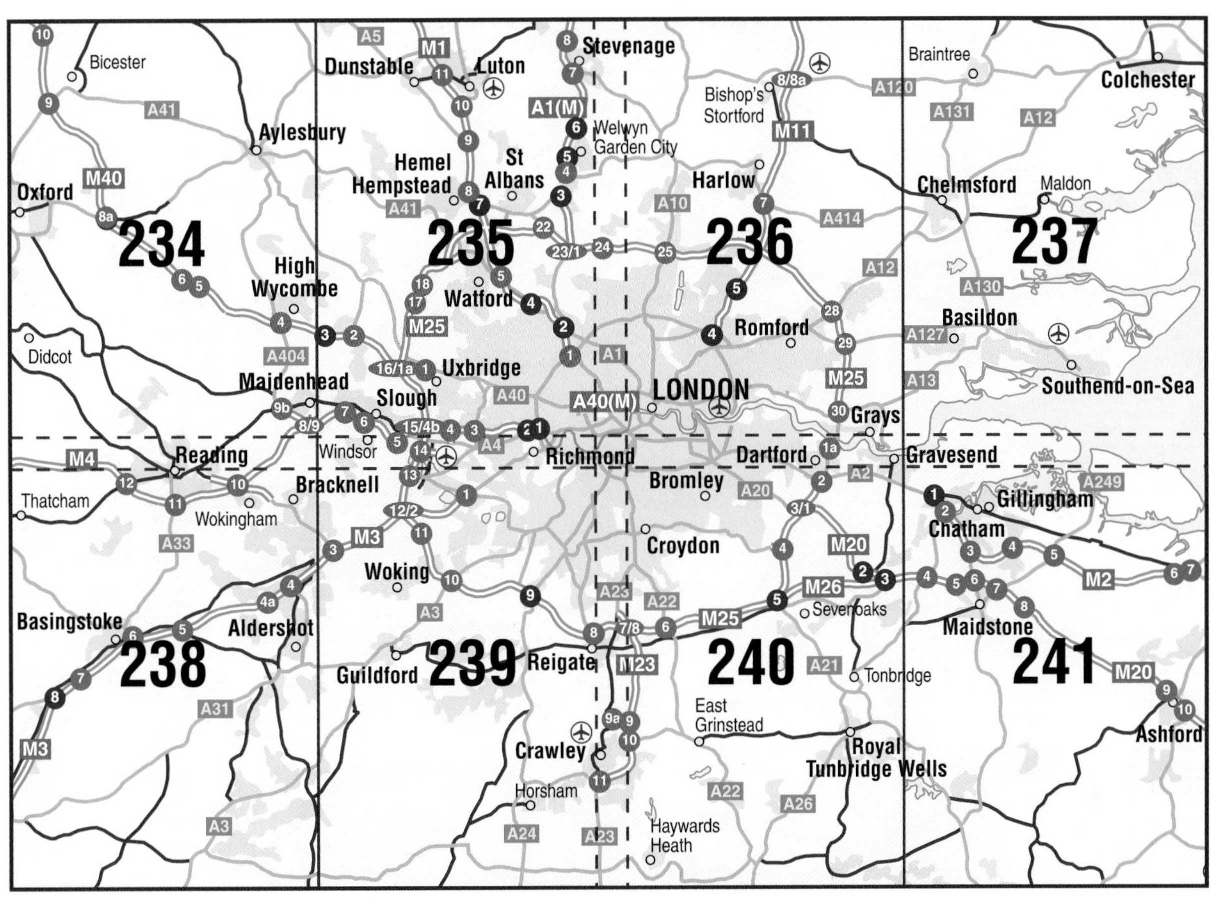

Bicester
Aylesbury
Oxford
M40
234
High Wycombe
Didcot
Maidenhead
Reading
M4
Thatcham
Wokingham
Bracknell
Windsor
Basingstoke
Aldershot
238
M3
A31
A3
A33
A41
A404
Dunstable
Luton
M1
A5
Hemel Hempstead
St Albans
235
Watford
M25
Uxbridge
Slough
A40
Richmond
Woking
Guildford
239
Reigate
Crawley
Horsham
A24
A23
M23
Stevenage
A1(M)
Welwyn Garden City
Bishop's Stortford
M11
Harlow
A10
236
Romford
LONDON
A40(M)
Bromley
Croydon
Dartford
Grays
A414
A12
A120
M20
M26
Sevenoaks
240
East Grinstead
Tonbridge
Royal Tunbridge Wells
Haywards Heath
A22
A21
A26
A20
A2
Braintree
Colchester
A131
Chelmsford
Maldon
237
A130
Basildon
A127
A13
Southend-on-Sea
Gravesend
Gillingham
Chatham
A249
M2
Maidstone
241
Ashford

REFERENCE AND TOURIST INFORMATION

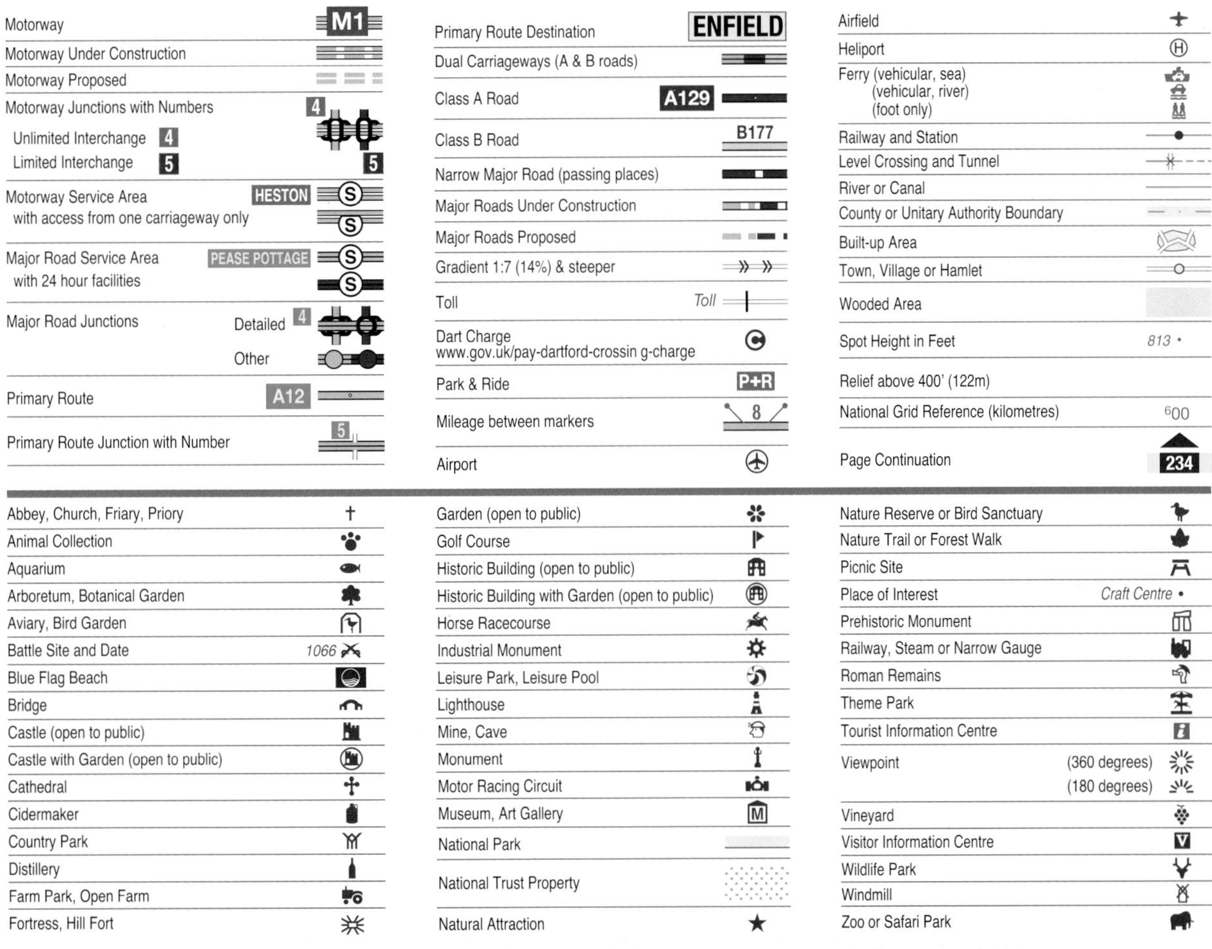

Motorway M1
Motorway Under Construction
Motorway Proposed
Motorway Junctions with Numbers 4
Unlimited Interchange 4
Limited Interchange 5
Motorway Service Area HESTON
with access from one carriageway only
Major Road Service Area PEASE POTTAGE
with 24 hour facilities
Major Road Junctions Detailed 4
Other
Primary Route A12
Primary Route Junction with Number 5
Primary Route Destination ENFIELD
Dual Carriageways (A & B roads)
Class A Road A129
Class B Road B177
Narrow Major Road (passing places)
Major Roads Under Construction
Major Roads Proposed
Gradient 1:7 (14%) & steeper
Toll Toll
Dart Charge
www.gov.uk/pay-dartford-crossin g-charge
Park & Ride P+R
Mileage between markers 8
Airport
Airfield
Heliport
Ferry (vehicular, sea)
(vehicular, river)
(foot only)
Railway and Station
Level Crossing and Tunnel
River or Canal
County or Unitary Authority Boundary
Built-up Area
Town, Village or Hamlet
Wooded Area
Spot Height in Feet 813
Relief above 400' (122m)
National Grid Reference (kilometres) 600
Page Continuation 234
Abbey, Church, Friary, Priory
Animal Collection
Aquarium
Arboretum, Botanical Garden
Aviary, Bird Garden
Battle Site and Date 1066
Blue Flag Beach
Bridge
Castle (open to public)
Castle with Garden (open to public)
Cathedral
Cidermaker
Country Park
Distillery
Farm Park, Open Farm
Fortress, Hill Fort
Garden (open to public)
Golf Course
Historic Building (open to public)
Historic Building with Garden (open to public)
Horse Racecourse
Industrial Monument
Leisure Park, Leisure Pool
Lighthouse
Mine, Cave
Monument
Motor Racing Circuit
Museum, Art Gallery
National Park
National Trust Property
Natural Attraction
Nature Reserve or Bird Sanctuary
Nature Trail or Forest Walk
Picnic Site
Place of Interest Craft Centre
Prehistoric Monument
Railway, Steam or Narrow Gauge
Roman Remains
Theme Park
Tourist Information Centre
Viewpoint (360 degrees)
(180 degrees)
Vineyard
Visitor Information Centre
Wildlife Park
Windmill
Zoo or Safari Park

SCALE 1:200,000

0 1 2 3 4 5 10 miles

0 1 2 3 4 5 10 15 kilometres

Map Pages 234-241
3.156 miles to 1 inch
2kms 20 1 cm

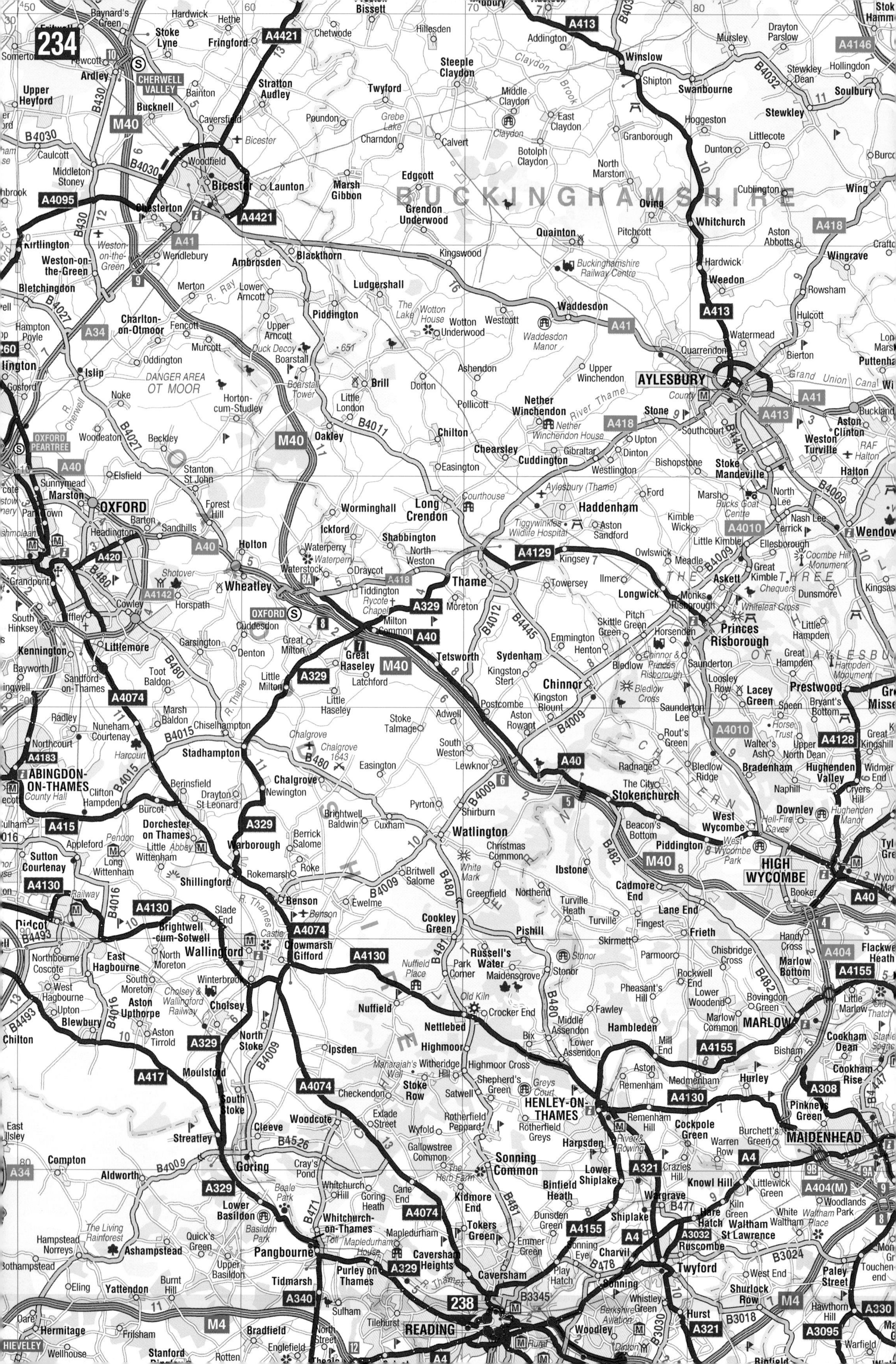
BUCKINGHAMSHIRE
Fritwell
Baynard's Green
Hardwick
Hethe
Stoke Lyne
Fringford
Somerton
Fewcott
Ardley
CHERWELL VALLEY
Bainton
Upper Heyford
Bucknell
Caversfield
B4030
Caulcott
Middleton Stoney
Bicester
Woodfield
Launton
Chesterton
A4095
A4421
Kirtlington
Weston-on-the-Green
Wendlebury
Ambrosden
Blackthorn
Bletchingdon
Merton
Lower Arncott
Upper Arncott
R. Ray
Charlton-on-Otmoor
Fencott
Murcott
Hampton Poyle
Islip
Oddington
Gosford
DANGER AREA
OT MOOR
Noke
Horton-cum-Studley
Beckley
Woodeaton
OXFORD PEARTREE
Stanton St John
Elsfield
Sunnymead
Marston
OXFORD
Forest Hill
Barton
Headington
Sandhills
Holton
Wheatley
Shotover
Horspath
Cowley
Grandpont
Iffley
South Hinksey
Littlemore
Kennington
Garsington
Cuddesdon
Denton
Great Milton
Little Milton
Bayworth
Sandford-on-Thames
Toot Baldon
Marsh Baldon
Radley
Nuneham Courtenay
Chiselhampton
Stadhampton
Northcourt
ABINGDON-ON-THAMES
County Hall
Clifton Hampden
Burcot
Berinsfield
Drayton St Leonard
Dorchester on Thames
Abbey
Little Wittenham
Long Wittenham
Appleford
Pendon
Sutton Courtenay
Shillingford
Warborough
Rokemarsh
Benson
Didcot
Railway
Brightwell-cum-Sotwell
Slade End
Wallingford
Castle
Crowmarsh Gifford
Northbourne
Coscote
East Hagbourne
West Hagbourne
Upton
North Moreton
South Moreton
Aston Upthorpe
Aston Tirrold
Cholsey
Cholsey & Wallingford Railway
Winterbrook
Blewbury
Chilton
North Stoke
Moulsford
South Stoke
Streatley
Goring
Compton
Aldworth
East Ilsley
Lower Basildon
Basildon Park
Beale Park
Pangbourne
Hampstead Norreys
The Living Rainforest
Ashampstead
Quick's Green
Upper Basildon
Bothampstead
Eling
Yattendon
Burnt Hill
Tidmarsh
Sulham
Hermitage
Frilsham
Bradfield
Englefield
Stanford Dingley
North Street
Wellhouse
Rotten Row
HIEVELEY
Tilehurst
READING
Bissett
Chetwode
Hillesden
Stratton Audley
Twyford
Steeple Claydon
Poundon
Grebe Lake
Charndon
Calvert
Middle Claydon
Claydon
East Claydon
Botolph Claydon
Claydon Brook
Edgcott
Marsh Gibbon
Grendon Underwood
Kingswood
Ludgershall
Piddington
Duck Decoy
Boarstall
Boarstall Tower
Wotton House
The Lake
Wotton Underwood
Westcott
Waddesdon
Waddesdon Manor
Brill
Little London
Dorton
Ashendon
Pollicott
Oakley
Chilton
B4011
Easington
Chearsley
Long Crendon
Courthouse
Worminghall
Ickford
Shabbington
Waterperry
Waterstock
Draycot
North Weston
Tiddington
Rycote Chapel
Thame
Moreton
Milton Common
Great Haseley
Latchford
Little Haseley
Tetsworth
Stoke Talmage
Adwell
South Weston
Lewknor
Chalgrove
Chalgrove 1643
Easington
Newington
Berrick Salome
Brightwell Baldwin
Cuxham
Pyrton
Shirburn
Watlington
Roke
Britwell Salome
Ewelme
White Mark
Christmas Common
Cookley Green
Greenfield
Northend
Pishill
Nuffield Place
Park Corner
Russell's Water
Maidensgrove
Nuffield
Old Kiln
Crocker End
Nettlebed
Highmoor
Witheridge Hill
Highmoor Cross
Maharajah's Well
Ipsden
Checkendon
Stoke Row
Shepherd's Green
Satwell
Exlade Street
Woodcote
Cleeve
Greys Court
Rotherfield Peppard
Rotherfield Greys
Wyfold
Gallowstree Common
Cray's Pond
Whitchurch Hill
Goring Heath
Cane End
The Herb Farm
Kidmore End
Sonning Common
Whitchurch-on-Thames
Mapledurham
Mapledurham House
Toll
Caversham Heights
Purley on Thames
Tokers Green
Emmer Green
Caversham
Addington
Winslow
Shipton
Swanbourne
Mursley
Drayton Parslow
Stewkley Dean
Hollingdon
Soulbury
Stewkley
Hoggeston
Granborough
Dunton
Littlecote
North Marston
Oving
Cublington
Wing
Whitchurch
Quainton
Pitchcott
Aston Abbotts
Wingrave
Buckinghamshire Railway Centre
Hardwick
Weedon
Rowsham
Hulcott
Watermead
Quarrendon
Bierton
Upper Winchendon
AYLESBURY
County
Nether Winchendon
Nether Winchendon House
River Thame
Stone
Southcourt
Grand Union Canal
Aston Clinton
Buckland
Weston Turville
RAF Halton
Halton
Upton
Dinton
Gibraltar
Westlington
Bishopstone
Stoke Mandeville
Cuddington
Aylesbury (Thame)
Haddenham
Ford
Marsh
Bucks Goat Centre
North Lee
Nash Lee
Terrick
Wendover
Tiggywinkles Wildlife Hospital
Aston Sandford
Kimble Wick
Little Kimble
Ellesborough
Coombe Hill Monument
Kingsey
Owlswick
Meadle
Great Kimble
Chequers
Towersey
Ilmer
Longwick
Askett
Monks Risborough
Whiteleaf Cross
Dunsmore
Pitch Green
Skittle Green
Princes Risborough
Horsenden
Little Hampden
Emmington
Henton
Sydenham
Kingston Stert
Chinnor
Chinnor & Princes Risborough Railway
Bledlow
Bledlow Cross
Saunderton
Great Hampden
Hampden Monument
Loosley Row
Lacey Green
Prestwood
Speen
Bryant's Bottom
Great Missenden
Kingston Blount
Postcombe
Aston Rowant
Saunderton Lee
Horse Trust
Upper North Dean
Rout's Green
Walter's Ash
Radnage
Bledlow Ridge
Bradenham
Hughenden Valley
Great Kingshill
Widmer End
The City
Stokenchurch
Naphill
Cryers Hill
Downley
Hughenden Manor
Hell-Fire Caves
West Wycombe
West Wycombe Park
Beacon's Bottom
Piddington
HIGH WYCOMBE
Ibstone
Cadmore End
Booker
Turville Heath
Turville
Lane End
Fingest
Frieth
Skirmett
Parmoor
Chisbridge Cross
Handy Cross
Marlow Bottom
Flackwell Heath
Stonor
Rockwell End
Lower Woodend
Pheasant's Hill
Bovingdon Green
Little Marlow
Fawley
Middle Assendon
Lower Assendon
Hambleden
Marlow Common
MARLOW
Mill End
Bisham
Cookham Dean
Cookham Rise
Bix
Aston
Remenham
Medmenham
Hurley
HENLEY-ON-THAMES
Remenham Hill
Cockpole Green
Pinkneys Green
Harpsden
River & Rowing
Burchett's Green
MAIDENHEAD
Warren Row
Lower Shiplake
Crazies Hill
Knowl Hill
Littlewick Green
Binfield Heath
Wargrave
Kiln Green
Woodlands Park
Dunsden Green
Shiplake
Hare Hatch
White Waltham
Waltham Place
Waltham St Lawrence
Sonning Eye
Charvil
Ruscombe
Twyford
Play Hatch
Sonning
Whistley Green
West End
Shurlock Row
Paley Street
Berkshire Aviation
Woodley
Hurst
Hawthorn Hill
Dinton
Warfield
M40
M4
A34
A40
A41
A413
A418
A420
A4010
A4074
A4095
A4129
A4130
A4142
A4146
A4155
A4183
A415
A417
A4128
A4421
A404
A404(M)
A308
A321
A329
A3032
A3095
A330
A340
A4
B4009
B4011
B4012
B4015
B4016
B4027
B4030
B4032
B430
B4445
B4443
B4493
B4526
B471
B477
B478
B480
B481
B482
B3018
B3024
B3030
B3345
238
260

235
Potsgrove
Harlington
Pegsdon
Westmill
William
Weston
TODDINGTON
Heath and Reach
Battlesden
Toddington
Watergate Farm
Streatley
Upper Sundon
Sundon Hills
Lower Sundon
Barton Hills
HITCHIN
Great Wymondley
Little Wymondley
Graveley
Fancott
Tebworth
Hockliffe
Chalton
A6
A505
Great Offley
Offley Hoo
Charlton
Gosmore
St Ippolyts
Walkern
Leighton Buzzard
Eggington
Tilsworth
Wingfield
A4012
A5
Houghton Regis
Sundon Park
Waulud's Bank
Lilley
Ley Green
Preston
A1(M)
A1072
STEVENAGE
Stanbridge
A505
Leagrave
Limbury
Wardown House
Cockernhoe
King's Walden
Rush Green
Aston End
Benington Lordship
Billington
Sewell
DUNSTABLE
A5505
Stopsley
LUTON
Breachwood Green
St Paul's Walden
Langley
Shephall
Aston
Little Billington
Mead Open Farm
Honeywick
Totternhoe
Church End
Chaul End
Priory Ho.
New Town
LONDON LUTON
Bendish
Knebworth House
Bragbury End
Hebing End
Eaton Bray
Slapton
Northall
Edlesborough
A5183
M1
Caddington
Stockwood Discovery Cen.
Someries Castle
Peters Green
Whitwell
Old Knebworth
Knebworth
Hook's Cross
A602
Datchworth
Chilterns Gateway
Tree Cathedral
Kensworth
Whipsnade
Aley Green
Slip End
Woodside Farm
Chiltern Green
Kimpton
Rableyheath
Codicote
Datchworth Green
Walton at Stone
Horton
Ivinghoe Aston
Cheddington
B4540
Kensworth Common
A1081
East Hyde
Blackmore End
Ayot St Lawrence
Roman Baths
Woolmer Green
Burnham Green
Bull's Green
Stapleford
Ford End Watermill
Ivinghoe Beacon
Ivinghoe
Pitstone
Pitstone Green
Whipsnade
White Lion
Dagnall
Studham
Markyate
FLAMSTEAD
Kinsbourne Green
Mackerye End
Shaw's Corner
Welwyn
Harmer Green
Digswell
Bramfield
Tewin
Waterford
Ringshall
Flamstead
Jockey End
Trowley Bottom
Hertfordshire Showground
Hatching Green
HARPENDEN
Wheathampstead
The Folly
Ayot St Peter
Ayot Green
Bridgewater Monument
Bulbourne
New Mill
Upper Dunsley
Tring
Aldbury
Little Gaddesden
St Margaret's
Great Gaddesden
Redbourn
HERTFORDSHIRE
Amwell
Coleman Green
Lemsford
Stanborough
WELWYN GARDEN CITY
Hatfield Hyde
Cole Green
Letty Green
Toll
Holtsmere End
Water End
Childwick Green
Sandridge
Nettleden
Redbournbury Watermill
Wigginton
Natural History
Frithsden
Piccotts End
A4147
Hatfield Garden Village
New Greens
Marshalswick
Mill Green
Essendon
Little Berkhamsted
Hatfield House
A41
Northchurch
Berkhamsted
Potten End
Warners End
Old Gorhambury House
Roman Theatre
Verulamium Roman Town
Townsend
Ellenbrook
A1057
HATFIELD
West End
Cholesbury Camp
BERKHAMSTED
A4251
Chaulden
A414
Fleetville
ST ALBANS
Galleries
Woodside
Cholesbury
Bourne End
Boxmoor
Colney Heath
A1000
Epping Green
Buckland Common
Ashley Green
HEMEL HEMPSTEAD
Potters Crouch
Park Street
London Colney
Welham Green
Brookmans Park
Newgate Street
Chiswell Green
A416
Bellingdon
Whelpley Hill
Felden
Rucklers Lane
Bedmond
Frogmore
Water End
Great Wood
Chartridge
Lye Green
Kings Langley
Bovingdon
Ballinger Common
Botley
CHESHAM
Chipperfield
Abbots Langley
Bricket Wood
Colney Street
Shenleybury
de Havilland Aircraft
South Mimms
POTTERS BAR
Northaw
South Heath
B485
Ley Hill
Flaunden
Belsize
Bucks Hill
Langleybury
A405
RADLETT
Shenley
Ridge
SOUTH MIMMS
Waterside
Hyde Heath
Sarratt
Hunton Bridge
Crews Hill
A413
Chesham Bois
Latimer
Chenies Manor
Chenies
Micklefield Green
Chandler's Cross
Aldenham
Letchmore Heath
Green Street
Bentley Heath
Botany Bay
Little Missenden
Little Chalfont
A411
WATFORD
Patchetts Green
BOREHAMWOOD
High Barnet
Monken Hadley
Hadley Wood
Trent Park
Cockfosters
Beamond End
AMERSHAM
A404
CHORLEYWOOD
Three Rivers
Croxley Green
BUSHEY
Elstree
Arkley
BARNET
EAST BARNET
Penn Street
Coleshill
Winchmore Hill
Heronsgate
RICKMANSWORTH
A4145
Oxhey
Bushey Heath
LONDON GATEWAY
A5109
Totteridge
Whetstone
Chalfont St Giles
Batchworth
South Oxhey
A4008
A409
Chiltern Open Air
Penn
Knotty Green
Seer Green
Milton's Cottage
Maple Cross
NORTHWOOD
Eastbury
A4125
Hatch End
STANMORE
Mill Hill
FINCHLEY
FRIERN BARNET
A598
Forty Green
Bekonscot Model Village
Chalfont Common
Jordans
Layter's Green
West Hyde
Harefield
A4180
A4140
EDGWARE
Muswell Hill
WOOD GREEN
HORNSEY
BEACONSFIELD
Chalfont St Peter
A40
M25
Bayhurst Wood
Ruislip Common
PINNER
Harrow Weald
Wealdstone
A4006
HARROW
Kingsbury
HENDON
Wooburn Green
M40
Denham
Denham Green
Newyears Green
Eastcote
KENTON
Fryent
Golders Green
Highgate
Wooburn
Hedgerley
Gerrards Cross
Colne Valley
RUISLIP
A4090
Harrow on the Hill
A4088
Brent Resr.
Brent Cross
LONDON
Odds Farm Park
A355
Tatling End
Denham
Ickenham
Rayners Lane
WEMBLEY
Neasden
Egypt
Farnham Common
Fulmer
B467
Northolt
NORTHOLT
Stadium
Cricklewood
HAMPSTEAD
East Burnham
Stoke Poges
Wexham Street
New Denham
UXBRIDGE
Black Park
Iver Heath
A4090
WILLESDEN
Kilburn
CAMDEN TOWN
Burnham
Farnham Royal
A412
Hillingdon
Langley Park
B455
Brent
MARYLEBONE
George Green
Shreding Green
Cowley
A312
Yeading
GREENFORD
A4127
London Zoo
A4
SLOUGH
B470
Stockley
A437
HAYES
A4020
EALING
ACTON
PADDINGTON
WEST END
Cippenham
Richings Park
Yiewsley
West Drayton
SOUTHALL
A406
Shepherd's Bush
KENSINGTON
Hyde Park
Eton Wick
Dorney Court
Langley
Upton
Sutton
Thorney
Motor
Hanwell
WESTMINSTER
Windsor Castle
Boveney
Eton College
HESTON
M4
CHISWICK
HAMMERSMITH
CHELSEA
A308
Harmondsworth
Sipson
Harlington
Heston
Osterley
Brentford
Kew
FULHAM
Oakley Green
WINDSOR
Datchet
Colnbrook
Poyle
Longford
Cranford
Syon Pk.
North Sheen
Chiswick Ho.
Palace
BARNES
BATTERSEA
Windsor Frogmore House
Horton
LONDON HEATHROW
Hatton
HOUNSLOW
ISLEWORTH
Rugby Grd.
A205
PUTNEY
CLAPHAM
Legoland
Wraysbury
Stanwell Moor
Stanwell
239
RICHMOND
Roehampton
WANDSWORTH
B3022
A332
Old Windsor
Hythe End
A3044
A30
FELTHAM
East Bedfont
Marble Hill
Orleans
Richmond Park
Ham Ho.
A316
TWICKENHAM
Lawn Tennis
A218
A3
Cranbourne
WINDSOR GREAT PARK
A308
ASHFORD
A244
STREATHAM
236

236
Weston
Buntingford
Cottered
A507
Aspenden
Luffenhall
Cromer
Ardeley
Walkern
Bassus Green
Westmill
Hare Street
Great Hormead
Little Hormead
Dassels
Hay Street
Braughing
Furneux Pelham
Stocking Pelham
Berden
Rickling Green
Ugley
Ugley Green
M11
Henham
Woodend Green
Manuden
East End
Mallows Green
Farnham Green
Farnham
Stansted Mountfitchet
Patmore Heath
Clapgate
Albury
Upwick Green
Elsenham
Pledgdon Green
Broxted
Molehill Green
Great Easton
Little Easton
Tye Green
Mountfitchet Castle
Easton Lodge
London Stansted
Bamber's Green
Great Dunmow
B184
B1051
B1038
B1368
B1037
B1383
STEVENAGE
A1155
Benington Lordship
Benington
Aston
Aston End
Shephall
Clay End
Wood End
Great Munden
Nasty
Haultwick
Levens Green
Green End
Puckeridge
Standon
A10
A120
Hebing End
Bragbury End
A602
HERTFORDSHIRE
Knebworth
Hook's Cross
Dane End
Colliers End
Standon Green End
Wellpond Green
Little Hadham
Hadham Ford
Bromley
Latchford
BISHOP'S STORTFORD
Birchanger
BIRCHANGER GREEN
Bury Green
Rhodes
Thorley Street
Thorley
Little Hallingbury
Great Hallingbury
Start Hill
B1256
Takeley
Takeley Street
Smith's Green
Little Canfield
Bedlar's Green
Hatfield Forest
Puttock's End
Great Canfield
Bacon End
Datchworth
Datchworth Green
Walton at Stone
Sacombe
High Cross
Barwick
Hopleys
Forge
Much Hadham
B1004
Hadham Cross
Perry Green
Green Tye
Spellbrook
Wright's Green
Hatfield Broad Oak
Taverners Green
THE RODINGS
High Roding
Bishop's Green
Woolmer Green
Bull's Green
Burnham Green
Harmer Green
Digswell
Tewin
Bramfield
Stapleford
Chapmore End
A119
Waterford
Wadesmill
Tonwell
Thundridge
Babbs Green
Wareside
Widford
Allen's Green
A1184
Sawbridgeworth
High Wych
Hatfield Heath
Sheering
A1060
Roundbush Green
Leaden Roding
High Easter
B158
WARE
Hunsdon
Hunsdonbury
B1000
WELWYN GARDEN CITY
HERTFORD
Great Amwell
Stanstead Abbotts
A414
St Margarets
Eastwick
Pye Corner
Gibberd
Junc. 7a under const.
Matching
Matching Green
White Roding
Nether Street
Margaret Roding
Good Easter
Hatfield Hyde
B195
Cole Green
Hertingfordbury
Hertford Heath
Scott's Grotto
Letty Green
HARLOW
Roydon
Rye House Gatehouse
Mark Hall
A1025
Church Langley
Matching Tye
Abbess Roding
Beauchamp Roding
Chalk End
Essendon
Bayford
Little Berkhamsted
HODDESDON
Lowewood
A1170
Great Parndon
Hare Street
Potter Street
Foster Street
High Laver
Little Laver
Birds Green
Shellow Bowells
West End
Hatfield House
Broxbourne Wood
Broxbourne
Roydon Hamlet
Redwings
A1169
Magdalen Laver
Moreton
Fyfield
Willingale
Woodside
Epping Green
Paradise
Lower Nazeing
Nazeing
Broadley Common
B1393
Hastingwood
Tyler's Green
Bovinger
Bobbingworth
Brookmans Park
Newgate Street
Hammond Street
Wormley
B194
Epping Green
B181
Thornwood Common
North Weald Airfield
North Weald Bassett
Epping Ongar Railway
Norton Mandeville
Norton Heath
High Ongar
Great Wood
Goff's Oak
Cuffley
Northaw
Turnford
Holyfield Hall Farm
Hayes Hill Farm
Lee Valley Pk.
Holyfield
Bumble's Green
Epping Upland
Coopersale
Greensted
Greensted Green
Saxon Wooden Church
Chipping Ongar
Nine Ashes
Elkins Green
POTTERS BAR
235
B157
B156
CHESHUNT
Gunpowder Mills
Harold's Bridge
Waltham Cross
Epping Forest
Upshire
EPPING
Coopersale Street
Toot Hill
Stanford Rivers
A113
A128
Stondon Massey
Blackmore
Beggar Hill
M25
Fiddlers Hamlet
Tunnel
B198
Crews Hill
Capel Manor
Transport
WALTHAM ABBEY
Copthall Green
Theydon Bois
Little End
Stapleford Tawney
Secret Nuclear Bunker
Wyatt's Green
Fryerning
Bentley Heath
A111
A1005
Enfield Wash
A1055
Sewardstone
A121
A104
B172
Hobbs Cross
Fox Hatch
Kelvedon Hatch
Doddinghurst
Mountnessing
Botany Bay
Forty Hill
High Beech
M11
Navestock
Monken Hadley
Hadley Wood
Trent Park
ENFIELD
A1010
A112
Debden Green
Stapleford
Abridge
Passingford Bridge
Navestock Side
Crow Green
Pilgrims Hatch
A12
Cockfosters
A110
Winchmore Hill
A105
Queen's Lodge
LOUGHTON
Lambourne End
Stapleford Abbotts
Old MacDonald's Farm
Weald
South Weald
Hutton
BARNET
EAST BARNET
Ponders End
Buckhurst Hill
CHIGWELL
Hainault Forest
Noak Hill
Shenfield
BRENTWOOD
Totteridge
Whetstone
SOUTHGATE
EDMONTON
B137
CHINGFORD
B170
Grange Hill
A1112
Chigwell Row
Havering-Atte-Bower
A128
Ingrave
FRIERN BARNET
A598
A406
Woodford Green
Hainault
Havering
Collier Row
Harold Hill
Brook Street
Thorndon
Great Warley
Little Warley
Herongate
FINCHLEY
Bruce Castle
TOTTENHAM
WOODFORD
A123
B173
B174
Harold Wood
B186
Muswell Hill
WOOD GREEN
HORNSEY
Vestry
A503
A1400
Gants Hill
Barkingside
Gidea Park
A127
Cranham
West Horndon
Golders Green
Highgate
Beam
WALTHAMSTOW
Chadwell Heath
ROMFORD
Nostalgia
UPMINSTER
Kenwood
Cemy.
STOKE NEWINGTON
LEYTON
WANSTEAD
A118
Goodmayes
A124
Eastbrookend
HORNCHURCH
B187
Thames Chase
Crow Green
North Ockendon
Bulphan
LONDON
A400
Leytonstone
Manor Park
ILFORD
Valence
Becontree
Elm Park
HAMPSTEAD
ISLINGTON
HACKNEY
Olympic Park
STRATFORD
BARKING
A1240
A125
Hornchurch
B1421
B186
A5
A41
CAMDEN TOWN
FINSBURY
BETHNAL GREEN
DAGENHAM
South Hornchurch
Belhus Woods
MARYLEBONE
SHOREDITCH
A11
WEST HAM
EAST HAM
Eastbury Ho.
Creekmouth
A1306
Rainham
South Ockendon
Orsett
PADDINGTON
WEST END
CITY
STEPNEY
POPLAR
A13
A1020
Beckton
City Farm
Thamesmead
Wennington
B1335
North Stifford
Baker Street
KENSINGTON
Hyde Park
WESTMINSTER
LAMBETH
BERMONDSEY
A102
Blackwall Tunnel
Barrier
LONDON CITY
A2016
Aveley
THURROCK
Little Thurrock
CHELSEA
DEPTFORD
A206
Abbey Wood
Belvedere
Thames
Chafford Hundred
South Stifford
CAMBERWELL
A2
GREENWICH
Charlton
WOOLWICH
Plumstead
ERITH
Slade Green
Purfleet
Dartford-Thurrock River Crossing
Dart Charge
West Thurrock
GRAYS
A23
BATTERSEA
A20
Blackheath
Kidbrooke
A205
A207
Welling
A209
Bexleyheath
A282
A1089
CLAPHAM
BRIXTON
LEWISHAM
Lee
Danson Pk.
Red Ho.
Crayford
DARTFORD
Greenhithe
Stone
Northfleet
WANDSWORTH
Horniman
Palace
ELTHAM
240
Blackfen
Bexley
Hall Place
A226
Swanscombe
Ebbsfleet
A218
STREATHAM
Dulwich
CATFORD
A21
Grove Park
Mottingham
A211
SIDCUP
Old Bexley
Wilmington
Bluewater
Bean
Betsham
Perry Street

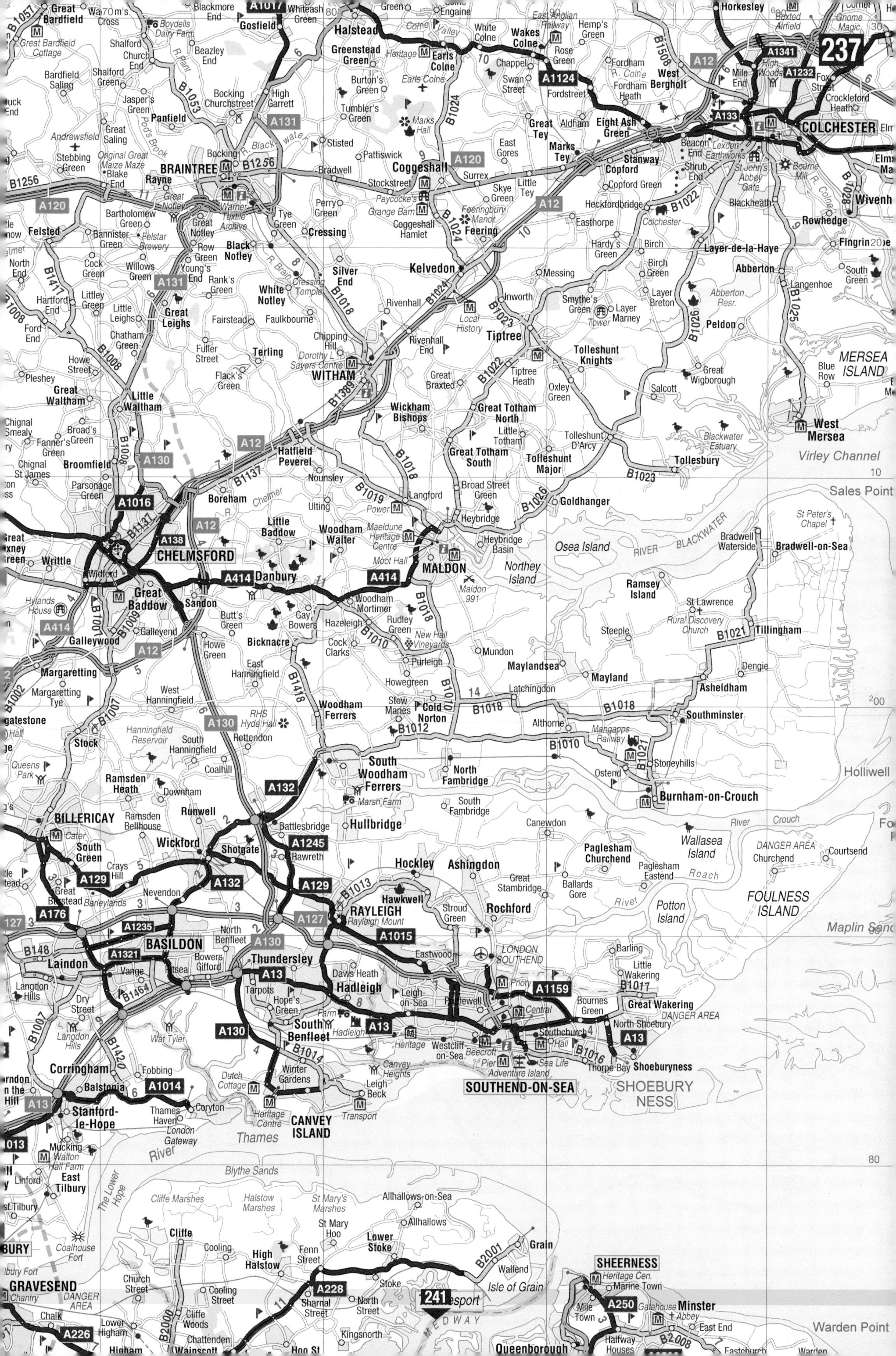

237
241
Great Bardfield
Halstead
Earls Colne
Wakes Colne
West Bergholt
COLCHESTER
Wivenhoe
Rowhedge
Layer-de-la-Haye
Abberton
Peldon
MERSEA ISLAND
West Mersea
Virley Channel
Tollesbury
Tolleshunt D'Arcy
Tolleshunt Major
Tolleshunt Knights
Tiptree
Goldhanger
Great Totham North
Great Totham South
Heybridge
MALDON
Northey Island
Osea Island
RIVER BLACKWATER
Bradwell-on-Sea
Bradwell Waterside
Ramsey Island
St Lawrence
Tillingham
Mayland
Maylandsea
Asheldham
Southminster
Burnham-on-Crouch
Wallasea Island
FOULNESS ISLAND
Maplin Sands
Potton Island
Great Wakering
Shoeburyness
SHOEBURY NESS
SOUTHEND-ON-SEA
Rochford
Hockley
Ashingdon
Hullbridge
RAYLEIGH
Hadleigh
Leigh-on-Sea
Westcliff-on-Sea
Thundersley
South Benfleet
CANVEY ISLAND
BASILDON
Laindon
Wickford
BILLERICAY
Stanford-le-Hope
Corringham
East Tilbury
GRAVESEND
Cliffe
High Halstow
Allhallows-on-Sea
Isle of Grain
Grain
SHEERNESS
Minster
Queenborough
Warden Point
MEDWAY
Thames
South Woodham Ferrers
North Fambridge
Woodham Ferrers
Danbury
CHELMSFORD
Great Baddow
Writtle
Broomfield
Great Waltham
Little Waltham
Boreham
Hatfield Peverel
WITHAM
Terling
Kelvedon
Feering
Coggeshall
BRAINTREE
Rayne
Black Notley
White Notley
Great Leighs
Silver End
Cressing
Felsted
Stock
Margaretting
Galleywood
Sandon
Bicknacre
Langford
Woodham Walter
Wickham Bishops
Great Braxted
Copford
Stanway
Marks Tey
Great Tey
Eight Ash Green
A12
A120
A130
A414
A13
A127
A131
A1124
A133
A132
A129
A1245
A1015
A1159
A228
A250
A226
A1014
A176
A1016
A138
B1018
B1010
B1021
B1023
B1026
B1022
B1024

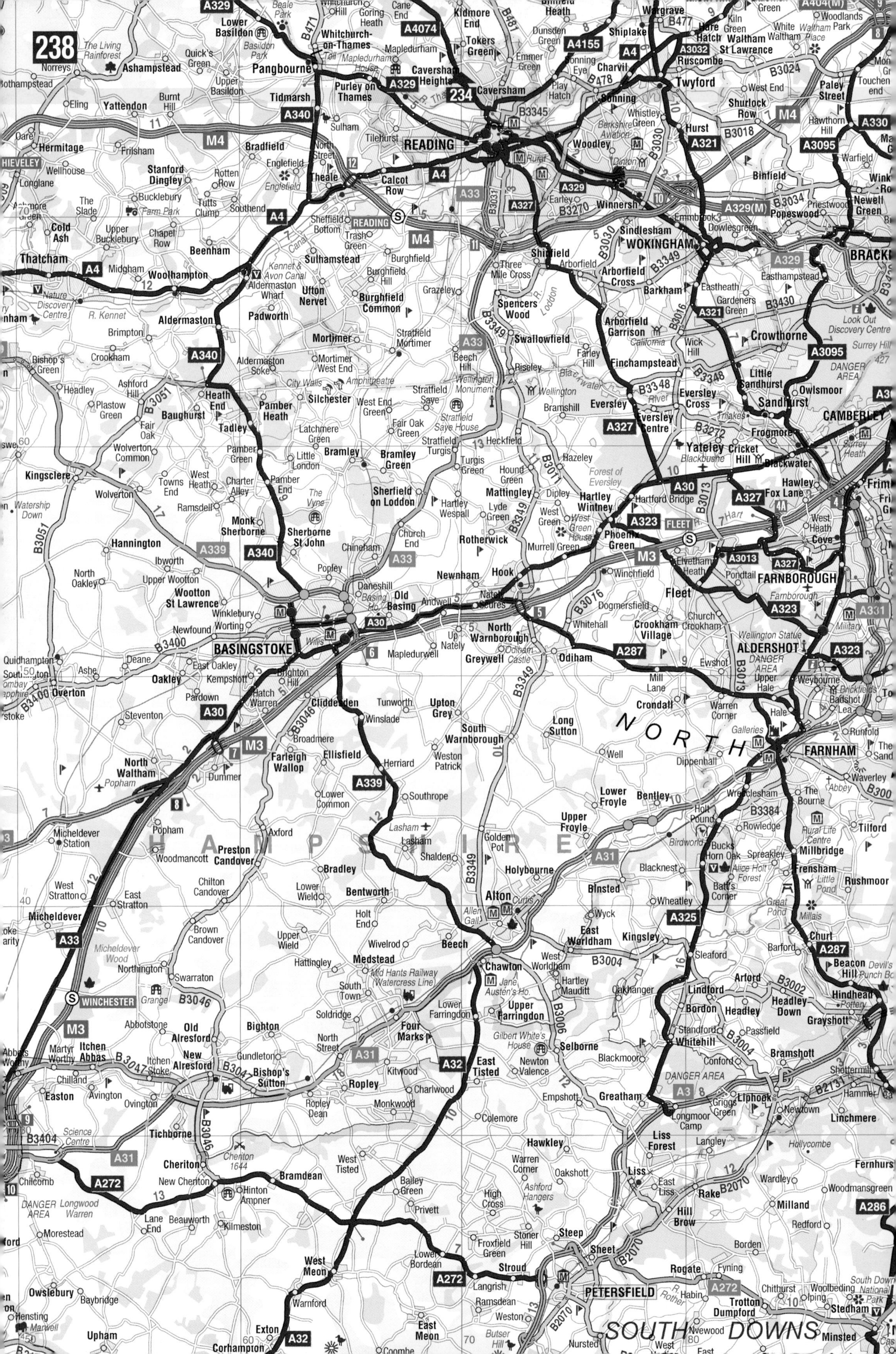

A329
Lower Basildon
Beale Park
The Living Rainforest
Ashampstead
Norreys
Quick's Green
Upper Basildon
Basildon Park
Pangbourne
Whitchurch-on-Thames
Goring Heath
Cane End
A4074
Kidmore End
Binfield Heath
Wargrave
Shiplake
Hare Hatch
Waltham St Lawrence
White Waltham
Woodlands Park
Tokers Green
Dunsden Green
A4155
A4
A3032
Ruscombe
Mapledurham
Mapledurham House
Caversham Heights
A329
Caversham
234
Emmer Green
Sonning Eye
Charvil
B478
Twyford
B3024
West End
Paley Street
Touchen end
Purley on Thames
Tidmarsh
Eling
Yattendon
Burnt Hill
A340
Sulham
Tilehurst
B3345
Play Hatch
Sonning
Whistley Green
Berkshire Aviation
Hurst
Shurlock Row
M4
Hawthorn Hill
A330
Oare
Hermitage
Frilsham
M4
Bradfield
North Street
READING
Woodley
Dinton
B3030
A321
B3018
A3095
Warfield
HIEVELEY
Wellhouse
Longlane
Stanford Dingley
Rotten Row
Englefield
Theale
12
Calcot Row
A4
Binfield
Bucklebury
Farm Park
The Slade
Tutts Clump
Southend
A4
A33
B3031
A327
A329
Earley
B3270
Winnersh
10
A329(M)
B3034
Priestwood
Newell Green
Popeswood
Cold Ash
Upper Bucklebury
Chapel Row
Beenham
Sheffield Bottom
READING
Trash Green
M4
11
Sindlesham
WOKINGHAM
Emmbrook
Dowlesgreen
Thatcham
A4
Midgham
Woolhampton
Kennet & Avon Canal
Aldermaston Wharf
Sulhamstead
Burghfield
Burghfield Hill
Three Mile Cross
Shinfield
Arborfield
Arborfield Cross
B3349
Barkham
Eastheath
A329
Easthampstead
BRACKNELL
Nature Discovery Centre
R. Kennet
Ufton Nervet
Burghfield Common
Grazeley
Spencers Wood
Gardeners Green
B3430
Aldermaston
Padworth
Brimpton
Crookham
Mortimer
Stratfield Mortimer
A33
Swallowfield
Arborfield Garrison
California
Wick Hill
A321
B3016
Crowthorne
Look Out Discovery Centre
A340
Bishop's Green
Aldermaston Soke
Mortimer West End
Beech Hill
Farley Hill
Finchampstead
A3095
Surrey Hill
Headley
Ashford Hill
City Walls
Amphitheatre
Wellington Monument
Riseley
Blackwater
B3348
Little Sandhurst
Owlsmoor
Plastow Green
B3051
Heath End
Pamber Heath
Silchester
West End Green
Stratfield Saye
Stratfield Saye House
Wellington
Bramshill
Eversley
Eversley Cross
Sandhurst
Baughurst
Tadley
Latchmere Green
Fair Oak Green
Eversley Centre
Trilakes
CAMBERLEY
Fair Oak
Wolverton Common
Pamber Green
Little London
Bramley
Bramley Green
Stratfield Turgis
Heckfield
A327
B3272
Frogmore
Yateley
Cricket Hill
Blackbushe
Blackwater
Surrey Heath
Kingsclere
Wolverton
Towns End
West Heath
Charter Alley
Pamber End
Turgis Green
Hound Green
B3011
Hazeley
Forest of Eversley
10
A30
Hawley
Fox Lane
Frimley
Watership Down
Ramsdell
The Vyne
Sherfield on Loddon
Mattingley
Dipley
Hartley Wintney
Hartford Bridge
B3013
A327
Lyde Green
Hartley Wespall
West Green
West Green House
A323
FLEET
Hart
West Heath
Monk Sherborne
Sherborne St John
Church End
Chineham
B3349
Rotherwick
Murrell Green
Phoenix Green
M3
Elvetham Heath
A3013
Cove
B3051
Hannington
A339
A340
A33
A327
Ibworth
North Oakley
Upper Wootton
Wootton St Lawrence
Popley
Daneshill
Newnham
Hook
Winchfield
Pondtail
FARNBOROUGH
Basing Ho.
Old Basing
Andwell
Nately Scures
B3076
Dogmersfield
Fleet
Farnborough
A323
A331
Winklebury
Worting
A30
Up Nately
North Warnborough
Whitehall
Crookham Village
Church Crookham
Military
Newfound
B3400
BASINGSTOKE
Willis
Mapledurwell
Odiham Castle
A287
Wellington Statue
ALDERSHOT
A323
Quidhampton
Deane
Ashe
East Oakley
Oakley
Kempshott
Brighton Hill
Greywell
Odiham
Mill Lane
Ewshot
DANGER AREA
Upper Hale
Weybourne
Bourne
Badshot Lea
B3400
Overton
Pardown
Hatch Warren
Cliddesden
Tunworth
Upton Grey
B3349
Crondall
Warren Corner
Hale
Steventon
A30
B3046
Winslade
South Warnborough
Long Sutton
NORTH
Galleries
FARNHAM
Runfold
The Sands
Broadmere
Farleigh Wallop
Ellisfield
Weston Patrick
Well
Dippenhall
North Waltham
M3
Dummer
Popham
Herriard
Lower Froyle
Bentley
Wrecclesham
Waverley Abbey
B3001
A339
Lower Common
Southrope
The Bourne
B3384
Popham
Axford
Lasham
Upper Froyle
Holt Pound
Rowledge
Tilford
Micheldever Station
HAMPSHIRE
Woodmancott
Preston Candover
Shalden
Golden Pot
Birdworld
Bucks Horn Oak
Spreakley
Millbridge
Rural Life Centre
A31
Alice Holt Forest
Blacknest
Bradley
Holybourne
Frensham
Little Pond
Rushmoor
Chilton Candover
Lower Wield
Bentworth
Binsted
Batt's Corner
Wheatley
Great Pond
West Stratton
East Stratton
Holt End
Alton
Curtis
Allen Gall.
Wyck
Millais
Micheldever
Upper Wield
Wivelrod
Beech
East Worldham
Kingsley
A325
Churt
A33
Brown Candover
Micheldever Wood
Hattingley
Medstead
West Worldham
B3004
Sleaford
Barford
A287
Northington
Chawton
Jane Austen's Ho.
Hartley Mauditt
Oakhanger
Arford
Beacon Hill
Devil's Punch Bowl
Swarraton
Grange
Mid Hants Railway (Watercress Line)
South Town
Upper Farringdon
B3006
Lindford
B3002
Hindhead
WINCHESTER
B3046
Soldridge
Lower Farringdon
Bordon
Headley
Headley Down
Grayshott
M3
Abbotstone
Old Alresford
Bighton
Four Marks
Gilbert White's House
Standford
Passfield
Martyr Worthy
Itchen Abbas
New Alresford
Gundleton
North Street
Whitehill
B3004
Bramshott
Abbots Worthy
B3047
Itchen Stoke
Bishop's Sutton
A31
A32
East Tisted
Newton Valence
Selborne
Blackmoor
Conford
Chilland
Avington
Kitwood
DANGER AREA
Shottermill
Easton
Ovington
Ropley
Charlwood
Empshott
Greatham
A3
Griggs Green
Liphook
B2131
Hammer
Ropley Dean
Monkwood
Longmoor Camp
Newtown
Linchmere
B3404
Science Centre
Tichborne
Colemore
Liss Forest
Langley
Hollycombe
A31
Cheriton
Cheriton 1644
West Tisted
Hawkley
Warren Corner
Oakshott
Fernhurst
Chilcomb
A272
New Cheriton
Bramdean
Hinton Ampner
Bailey Green
Ashford Hangers
Liss
East Liss
Rake
B2070
Wardley
Woodmansgreen
DANGER AREA
Longwood Warren
Lane End
Beauworth
Kilmeston
Privett
High Cross
Hill Brow
Milland
A286
Morestead
Froxfield Green
Stoner Hill
Steep
Redford
Borden
Lower Bordean
Sheet
B2070
West Meon
A272
Stroud
Rogate
Fyning
South Downs National Park
Owslebury
Langrish
PETERSFIELD
A272
Chithurst
Woolbeding
Trotton
Iping
Baybridge
Warnford
Ramsdean
R. Rother
Habin
Dumpford
Stedham
Hensting
Marwell
Exton
Weston
B2070
SOUTH DOWNS
Nyewood
Minsted
Upham
A32
East Meon
Butser Hill
Nursted
Corhampton
Coombe
West

Dorney
Eton Wick
Upton
Sutton
Harlington
SOUTHALL
HESTON
Brentford
CHISWICK
HAMMERSMITH
CHELSEA
239
Oakley Green
WINDSOR
Eton
Boveney
Windsor
Datchet
Colnbrook
Poyle
Harmondsworth
Sipson
Longford
Cranford
Heston
Osterley
Kew
North Sheen
FULHAM
BARNES
BATTERSEA
Legoland
Old Windsor
Horton
Wraysbury
Stanwell Moor
Stanwell
LONDON HEATHROW
Hatton
235
HOUNSLOW
ISLEWORTH
RICHMOND
PUTNEY
CLAPHAM
WANDSWORTH
Cranbourne
WINDSOR GREAT PARK
Hythe End
East Bedfont
FELTHAM
TWICKENHAM
Roehampton
STREATHAM
Woodside
Englefield Green
EGHAM
STAINES-UPON-THAMES
ASHFORD
Hanworth
TEDDINGTON
WIMBLEDON
Upper Tooting
North Ascot
Ascot
Cheapside
Virginia Water
Thorpe
Laleham
Littleton
SUNBURY
Hampton
Hampton Wick
KINGSTON UPON THAMES
NEW MALDEN
MERTON
MITCHAM
Sunninghill
Sunningdale
Upper Halliford
West Molesey
East Molesey
Thames Ditton
MORDEN
South Ascot
Broomhall
Trumps Green
Shepperton
WALTON-ON-THAMES
SURBITON
Tolworth
Long Ditton
Worcester Park
CARSHALTON
Beddington
Lyne
CHERTSEY
WEYBRIDGE
Windlesham
Longcross
Ottershaw
Addlestone
Row Town
New Haw
Hersham
ESHER
Hook
Chessington
Claygate
EWELL
SUTTON
WALLINGTON
Valley End
Burrowhill
Chobham
Woodham
Whiteley Village
St George's Hill
Fairmile
Malden Rushett
EPSOM
Lightwater
West End
Mimbridge
Sheerwater
West Byfleet
Byfleet
Cobham
Oxshott
Woodmansterne
Donkey Town
Bisley
Castle Green
Horsell
Pyrford
Stoke D'Abernon
BANSTEAD
Knaphill
WOKING
Pyrford Village
Wisley
Downside
Ashtead
Epsom Downs
Burgh Heath
Chipstead
Brookwood
Old Woking
Ripley
Mayford
Send Marsh
Ockham
Martyr's Green
Effingham Common
LEATHERHEAD
Tadworth
Kingswood
Hooley
Pirbright
Fox Corner
Cartbridge
Burntcommon
Little Bookham
Fetcham
Headley
Walton on the Hill
Mugswell
Lower Kingswood
Pyle Hill
Send
Sutton Green
West Horsley
East Horsley
Great Bookham
Worplesdon
Jacobs Well
Pitch Place
West Clandon
East Clandon
Effingham
Mickleham
Margery
Gatton
Merstham
Willey Green
Fairlands
Stoughton
Burpham
Westhumble
Polesden Lacey
Ranmore Common
Box Hill
B2032
Buckland
REDHILL
Wyke
Normandy
Wood Street Village
Merrow
Hatchlands Park
Denbies
Betchworth
Flexford
Wanborough
Onslow Village
GUILDFORD
Newlands Corner
Silent Pool
Sherbourne Farm
DORKING
Brockham
REIGATE
240
Hog's Back
Artington
Albury
Gomshall
Shere
Wotton
Westcott
Strood Green
Gadbrook
Dawesgreen
Doversgreen
South Nutfield
Puttenham
Littleton
Shalford
Chilworth
Abinger Hammer
Sutton Abinger
Friday Street
North Holmwood
Stonebridge
Leigh
Sidlow
Redhill
Compton
Farncombe
Peasmarsh
Blackheath
Brook
Blackbrook
Salfords
Shackleford
Hurtmore
Wonersh
Bramley
Peaslake
Abinger Common
South Holmwood
Nalderswood
Eashing
Catteshall
GODALMING
Shamley Green
Farley Green
Holmbury St Mary
Leith Hill
Coldharbour
Parkgate
Norwood Hill
Meath Green
Peper Harow
Hurtwood Common
Beare Green
HORLEY
Busbridge
Thorncombe Street
Grafham
Milford
Tuesley
Leith Hill Tower
Newdigate
Cudworth
Hookwood
Wheelerstreet
Winkworth Arboretum
Rowly
Ewhurst
Forest Green
Capel
Charlwood
LONDON GATWICK
Witley
Hydestile
Hascombe
Cranleigh
Mayes Green
Ockley
Clark's Green
Lowfield Heath
Russ Hill
Fernhill
Bowlhead Green
Wormley
Hambledon
Loxhill
Ewhurst Green
Ifieldwood
Langley Green
Tinsley Green
Brook
Sandhills
Walliswood
Oakwoodhill
Rusper
Ifield
CRAWLEY
Three Bridges
Chiddingfold
Dunsfold
Vachery Pond
Ellen's Green
Kingsfold
Lambs Green
Gossops Green
Worth
Alfold Crossways
Cox Green
Faygate
Maidenbower
Grayswood
Highstreet Green
Alfold
Rudgwick
Rowhook
Warnham
Broadfield
Haslemere
Ramsnest Common
Alfold Bars
Tisman's Common
Bucks Green
Strood Green
Broadbridge Heath
Roffey
Colgate
Pease Pottage
Ansteadbrook
Fisherstreet
Loxwood
Ifold
Park Street
The Haven
Slinfold
HORSHAM
ST LEONARDS FOREST
Tilgate Forest Row
Roundhurst
Plaistow
Tower Hill
Doomsday Green
Handcross
Black Down
The Lake
Northchapel
Roundstreet Common
Five Oaks
Itchingfield
Mannings Heath
Plummers Plain
Slaugham
Staplefield
Ebernoe
Southwater Street
Monk's Gate
Ashfold Crossways
Dial Green
Lurgashall
Kirdford
Newpound Common
Wisborough Green
Barns Green
Southwater
Lower Beeding
Slough Green
Lickfold
Balls Cross
Billingshurst
Brooks Green
Nuthurst
Warninglid
Bexleyhill
Strood Green
Parbrook
Coneyhurst
Copsale
Maplehurst
Crabtree
Cuckfield
WEST SUSSEX
Lodsworth
River
Upperton
Adversane
Dragon's Green
Coolham
Cowfold
Bolney
Ansty
Tillington
Petworth
Bedham
Broadford Bridge
Shipley
Kent Street
Gay Street
Whitehall
Little Bognor
Selham
Codmore Hill
North Heath
Broomer's Corner
Knepp Castle
West Grinstead
Littleworth
Wineham
Hickstead
SURREY
NORTH DOWNS

240
WESTMINSTER
LAMBETH
BERMONDSEY
CHELSEA
CAMBERWELL
DEPTFORD
GREENWICH
A206
Charlton
WOOLWICH
Plumstead
Abbey Wood
ERITH
Purfleet
Slade Green
Dartford-Thurrock River Crossing
Dart Charge
West Thurrock
GRAYS
A1089
BATTERSEA
A23
A2
A20
A205
Blackheath
Kidbrooke
A207
Welling
A209
A206
Bexleyheath
Red Ho.
Crayford
DARTFORD
A282
Greenhithe
Stone
Swanscombe
Northfleet
Ebbsfleet
WANDSWORTH
CLAPHAM
BRIXTON
LEWISHAM
Lee
236
Danson Pk.
Bexley
Hall Place
A226
Bluewater
Horniman
Dulwich
CATFORD
Palace
ELTHAM
Blackfen
New Eltham
A211
SIDCUP
Old Bexley
Wilmington
A2
Hawley
Darenth
Green
Betsham
Perry Street
A218
STREATHAM
A214
West Norwood
A21
Grove Park
Mottingham
A20
Foots Cray
North Cray
B2173
B258
Sutton at Hone
St John's Jerusalem
B260
St. Green
Southfleet
Upper Tooting
Crystal Palace
Sydenham
CHISLEHURST
BROMLEY
St Paul's Cray
Hextable
Swanley
M25
A225
South Darenth
Hook Green
Istead Rise
MERTON
A216
Thornton Heath
Penge
South Norwood
BECKENHAM
A222
A208
Caves
St Mary Cray
A20
New Barn
Longfield
MITCHAM
Deen City
A23
Petts Wood
B258
Horton Kirby
Hartley
Longfield Hill
MORDEN
A237
A236
Addiscombe
B251
West Wickham
B2212
A21
ORPINGTON
Crockenhill
Farningham
M20
Fawkham Green
New Ash Green
Meopham
A217
Carew
Beddington
A232
Hayes
A232
Bromley
Eagle Heights
Villa
Eynsford
Maplescombe
Lullingstone Castle
Meopham Green
CARSHALTON
CROYDON
Addington
A2022
Keston
Farnborough
A224
Chelsfield
Green St. Green
Well Hill
West Kingsdown
Brands Hatch
Ash
Ridley
A227
SUTTON
WALLINGTON
A212
Selsdon
New Addington
A233
High Elms
A21
Pratt's Bottom
Badgers Mount
Lullingstone
Aircraft
A225
Shoreham
DANGER AREA
Knatts Valley
A20
Hodsoll Street
Firefighting
Stansted
Fairseat
A237
B278
B275
PURLEY
Sanderstead
Selsdon Wood
Leaves Green
Downe
Down House
Halstead
A224
M25
S
Woodlands
A2022
Woodmansterne
Farleigh
London Biggin Hill
Cudham
Knockholt Pound
Otford
Kemsing
Heaverham
Wrotham
Trottiscliffe
BANSTEAD
Kenley
COULSDON
A22
Hamsey Green
Chelsham
Biggin Hill
Berry's Green
Knockholt
N
B2219
Chipstead
Whyteleafe
WARLINGHAM
A23
Old Coulsdon
B2030
B269
Tatsfield
W
A233
Chevening
Chevening Park
Dunton Green
Seal
M26
Ightham
Borough Green
B2032
Kingswood
Hooley
Kenley
East Surrey
Woldingham
O
B2221
M25
5
Chipstead
A25
St Marys Platt
Great Comp
Mugswell
Chaldon
D
B2024
Titsey Place
CLACKET LANE
Sundridge
Riverhead
B2019
Crouch
Claygate Cross
Ivy Hatch
7
Lower Kingswood
H
B2031
CATERHAM
Titsey
Brasted
A25
SEVENOAKS
Stone Street
Knole
Mereworth Woods
8
M25
6
Limpsfield
Quebec Ho.
Brasted Chart
Emmetts
Ightham Mote
Plaxtol
Old Soar Manor
Gatton
Merstham
Warwick Wold
Godstone
Oxted
A25
Moorhouse
Westerham
Squerryes
Limpsfield Chart
B2026
The Chart
Ide Hill
Goathurst Common
A225
Riverhill
Underriver
Shipbourne
Dunk's Green
Mercers
REDHILL
Nutfield
Bletchingley
A25
Hurst Green
Tandridge
B269
Chartwell
Toy's Hill
B2042
Crockham Hill
A21
Sevenoaks Weald
B245
Fletcher's Green
A227
Coldharbour
Broadview Hadlow
Pittswood
239
A2044
South Nutfield
M23
Godstone Farm
South Godstone
B269
B2026
Pootings
Cooper's Corner
Bough Beech Reservoir
Hildenborough
Higham Wood
A26
Golden Green
S U R R E Y
A23
Redhill
A22
Crowhurst Lane End
Crowhurst
Marlpit Hill
Four Elms
B2027
Chiddingstone Causeway
Leigh
TONBRIDGE
K
Sidlow
Salfords
Outwood
Blindley Heath
River Eden
Edenbridge
Eden Valley
How Green
Bough Beech
Hever Castle
Chiddingstone
Lower Haysden
A21
Tudeley
Five Oak Green
A217
Meath Green
Wasp Green
Horne
B2029
Haxted Mill
Marsh Green
Hever
Chiddingstone Castle
B2176
Penshurst Place
Penshurst
Bidborough
Crockhurst Street
B2017
HORLEY
B2036
Smallfield
British Wildlife Centre
Newchapel
Lingfield
B2028
B2026
Chiddingstone Hoath
A26
Southborough
Hookwood
9A
9
Burstow
B2028
Felcourt
Lingfield Park
Dormansland
Markbeech
Smart's Hill
B2176
Speldhurst
Salomons
High Brooms
Pembury
LONDON GATWICK
Fernhill
B2037
Cowden
Fordcombe
Rusthall
Lowfield Heath
A23
Shipley Bridge
Dormans Park
Hammerwood
Blackham
B2188
Langton Green
A21
Kippings Cross
Langley Green
Tinsley Green
Copthorne
A264
Felbridge
Hammerwood Park
Holtye
A264
Ashurst
Stone Cross
A264
The Pantiles
ROYAL TUNBRIDGE WELLS
Lamberhurst Quarter
Three Bridges
Pound Hill
A2220
Crawley Down
EAST GRINSTEAD
Groombridge Place
High Rocks
Groombridge
Bells Yew Green
Sunnyside
Ashurst Wood
Ball's Green
Spa Valley Railway
A267
B2169
Worth
M23
Standen
B2110
Saint Hill
Forest Row
River Medway
Hartfield
Withyham
Park Corner
A26
Frant
Maidenbower
Tilgate
Go Ape!
10A
Turners Hill
Manor
Forest Way
Upper Hartfield
B2110
Pooh Bridge
B2188
Mott's Mill
Harrison's Rocks
Eridge Green
The Platt
B2099
Broadfield
Selsfield Common
Priest House
Weir Wood Resr.
Lye Green
W E A L D
Pease Pottage
PEASE POTTAGE
B2028
H I G H
Sharpthorne
Wych Cross
Forest
Coleman's Hatch
Chuck Hatch
Friar's Gate
Boarshead
Sparrow's Green
Tilgate Forest Row
B2110
Balcombe Lane
Wakehurst Place
West Hoathly
Llama Park
ASHDOWN FOREST
Steel Cross
Durgates
A23
High Beeches
Wings
Balcombe
South of England Centre
Highbrook
Chelwood Gate
National Cat Centre
A22
CROWBOROUGH
Town Row
Mark Cross
Best Beech Hill
Riseden
Handcross
Nymans
B2036
West Hill
Ardingly Resr.
Ardingly
Horsted Keynes
A275
Chelwood Common
Nutley
Marlpits
B2026
Poundgate
A26
Poundfield
Jarvis Brook
B2100
Rotherfield
B2101
Tidebrook
A267
Slaugham
Staplefield
B2114
WEST SUSSEX
Viaduct
Bluebell Railway
Danehill
Duddleswell
Heron's Ghyll
Argos Hill
Coggins Mill
Brook Street
Borde Hill
Slough Green
HAYWARDS HEATH
Heaven Farm
Furner's Green
Fairwarp
High Hurstwood
E A S T
Butcher's Cross
Mayfield
Warninglid
Whitemans Green
B2028
Lindfield
Sheffield Green
Five Ash Down
River Rother
Cuckfield
Scaynes Hill
Fletching
Maresfield
A272
Buxted
Hadlow Down
Five Ashes
Paine's Corner
Bolney
Ansty
A272
A272
Sheffield Park
A272
Pound Green
Wilderness Wood
Broad Oak
A23
B2036
A273
B2112
North Common
Rural Life
Newick
Grisling Common
Piltdown
Uckfield
Pounsley
B2102
Heathfield
Punnett's Town
Hickstead
World's End
Wivelsfield
North
Shortbridge
Framfield
Roser's Cross
Cross in Hand
Cade Street

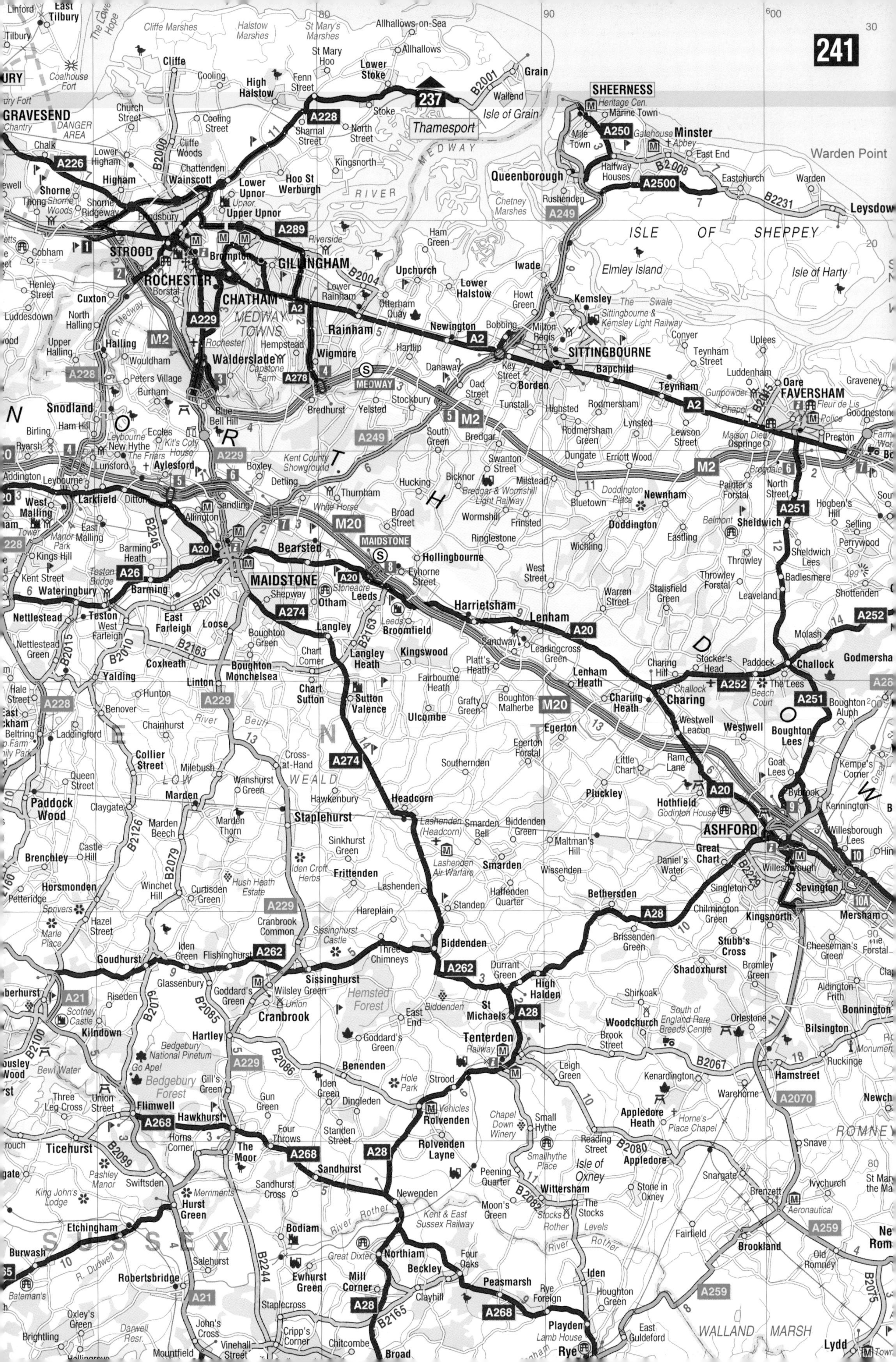

241
East Tilbury
Linford
Tilbury
Coalhouse Fort
Cliffe Marshes
Halstow Marshes
St Mary's Marshes
Allhallows-on-Sea
Allhallows
St Mary Hoo
Lower Stoke
Grain
Cliffe
Cooling
High Halstow
Fenn Street
Stoke
237
Wallend
Isle of Grain
Thamesport
SHEERNESS
Heritage Cen.
Marine Town
Minster
Gatehouse
Abbey
East End
Warden Point
GRAVESEND
Chantry
Church Street
Cooling Street
DANGER AREA
A228
Sharnal Street
North Street
Mile Town
A250
Chalk
A226
Lower Higham
Higham
Cliffe Woods
Chattenden
Wainscott
Kingsnorth
Hoo St Werburgh
MEDWAY
RIVER
Queenborough
Halfway Houses
A2500
B2008
Eastchurch
Warden
Shorne
Thong
Shorne Woods
Shorne Ridgeway
Lower Upnor
Upnor
Upper Upnor
Frindsbury
Chetney Marshes
Rushenden
A249
B2231
Leysdown
Cobham
STROOD
A289
Riverside
Brompton
GILLINGHAM
Ham Green
ISLE OF SHEPPEY
Elmley Island
Isle of Harty
Iwade
Henley Street
ROCHESTER
Borstal
CHATHAM
Lower Rainham
B2004
Otterham Quay
Upchurch
Lower Halstow
Howt Green
Kemsley
The Swale
Sittingbourne & Kemsley Light Railway
Cuxton
Luddesdown
North Halling
A229
MEDWAY TOWNS
Rainham
Newington
A2
Bobbing
Milton Regis
Conyer
Upper Halling
Halling
M2
Rochester
Hempstead
Wigmore
Hartlip
SITTINGBOURNE
Teynham Street
Uplees
Wouldham
Walderslade
Capstone Farm
A278
MEDWAY
Danaway
Key Street
Borden
Bapchild
Luddenham
Oare
Graveney
Peters Village
Burham
A228
Snodland
Blue Bell Hill
Bredhurst
Yelsted
Stockbury
Oad Street
Tunstall
Highsted
Rodmersham
Teynham
FAVERSHAM
Gunpowder Chapel
Fleur de Lis
Goodnestone
Birling
Ham Hill
Eccles
Kit's Coty House
South Green
Bredgar
Rodmersham Green
Lynsted
Lewson Street
Maison Dieu
Ospringe
Preston
Ryarsh
Leybourne
New Hythe
The Friars
Aylesford
Kent County Showground
Swanton Street
Dungate
Erriott Wood
Brogdale
Addington
Lunsford
Boxley
Detling
Hucking
Bicknor
Milstead
Doddington Place
Newnham
Painter's Forstal
North Street
West Malling
Larkfield
Ditton
Sandling
Thurnham
Bredgar & Wormshill Light Railway
Bluetown
Hogben's Hill
Allington
White Horse
M20
Broad Street
Wormshill
Frinsted
Doddington
Belmont
Sheldwich
Selling
East Malling
Manor Park
Tower
Barming Heath
B2246
A20
MAIDSTONE
Ringlestone
Eastling
Perrywood
Kings Hill
Bearsted
Hollingbourne
Wichling
Throwley
Sheldwich Lees
Kent Street
Teston Bridge
A26
Barming
Eyhorne Street
West Street
Throwley Forstal
Badlesmere
Wateringbury
Shepway
Stoneacre
Otham
Leeds
Warren Street
Stalisfield Green
Leaveland
Shottenden
Nettlestead
Teston
West Farleigh
East Farleigh
Loose
B2010
A274
Leeds Broomfield
Harrietsham
Lenham
Molash
A252
Nettlestead Green
B2015
B2163
Boughton Green
Langley
Sandway
Leadingcross Green
Chart Corner
Langley Heath
Kingswood
Platt's Heath
Charing Hill
Stocker's Head
Paddock
Challock
Godmersham
Yalding
Coxheath
Boughton Monchelsea
Fairbourne Heath
Lenham Heath
The Lees
Beech Court
Hale Street
Linton
Chart Sutton
Sutton Valence
Grafty Green
Boughton Malherbe
Charing Heath
Charing
A251
Boughton Aluph
Hunton
Benover
Ulcombe
Westwell Leacon
Westwell
Boughton Lees
Beltring
Laddingford
Chainhurst
River Beult
Egerton
Egerton Forstal
Collier Street
Cross-at-Hand
Southernden
Little Chart
Ram Lane
Goat Lees
Kempe's Corner
Queen Street
Milebush
LOW WEALD
Wanshurst Green
Marden
Hawkenbury
Headcorn
Pluckley
Hothfield
Godinton House
Bybrook
Kennington
Paddock Wood
Claygate
Staplehurst
Lashenden (Headcorn)
Smarden Bell
Biddenden Green
Maltman's Hill
ASHFORD
Willesborough Lees
Marden Beech
Marden Thorn
Sinkhurst Green
Lashenden Air Warfare
Smarden
Great Chart
Castle Hill
B2126
B2079
Iden Croft Herbs
Frittenden
Daniel's Water
Willesborough
Brenchley
Hush Heath Estate
Lashenden
Haffenden Quarter
Wissenden
Bethersden
B2229
Singleton
Sevington
Horsmonden
Winchet Hill
Curtisden Green
Standen
Chilmington Green
Kingsnorth
Mersham
Petteridge
Sprivers
Hareplain
A28
Hazel Street
Marle Place
Cranbrook Common
Sissinghurst Castle
Brissenden Green
Stubb's Cross
Cheeseman's Green
Forstal
Iden Green
Three Chimneys
Biddenden
Goudhurst
Flishinghurst
A262
Durrant Green
Shadoxhurst
Bromley Green
Glassenbury
Sissinghurst
High Halden
Aldington Frith
Lamberhurst
Riseden
A21
Goddard's Green
Wilsley Green
Union
Hemsted Forest
Biddenden
Shirkoak
Bonnington
Scotney Castle
B2085
Cranbrook
East End
St Michaels
Woodchurch
South of England Rare Breeds Centre
Orlestone
Kilndown
Hartley
Goddard's Green
Tenterden
Brook Street
Bilsington
Bedgebury National Pinetum
B2086
Railway
B2067
Ruckinge
Bewl Water
Go Ape!
Benenden
Hole Park
Leigh Green
Kenardington
Hamstreet
Bedgebury Forest
Gill's Green
Iden Green
Strood
Warehorne
A2070
Three Leg Cross
Union Street
Gun Green
Dingleden
Vehicles
Rolvenden
Chapel Down Winery
Small Hythe
Appledore Heath
Horne's Place Chapel
Newchurch
Flimwell
A268
Hawkhurst
Four Throws
Standen Street
Rolvenden Layne
Smallhythe Place
Reading Street
B2080
Appledore
ROMNEY
Snave
Ticehurst
Horns Corner
The Moor
Isle of Oxney
Snargate
Ivychurch
St Mary in the Marsh
B2099
Pashley Manor
Swiftsden
Sandhurst
Peening Quarter
Wittersham
Stone in Oxney
Brenzett
Aeronautical
King John's Lodge
Merriments
Hurst Green
Sandhurst Cross
Newenden
Moon's Green
B2082
Stocks
The Stocks
Rother Levels
Etchingham
Bodiam
River Rother
Kent & East Sussex Railway
Fairfield
A259
SUSSEX
Burwash
Salehurst
B2244
Great Dixter
Northiam
Four Oaks
River Rother
Brookland
Old Romney
R. Dudwell
Robertsbridge
Ewhurst Green
Mill Corner
Beckley
Peasmarsh
Iden
Bateman's
A21
Clayhill
Rye Foreign
Houghton Green
B2075
Oxley's Green
Staplecross
B2165
Darwell Resr.
John's Cross
Cripp's Corner
Playden
Lamb House
East Guldeford
WALLAND MARSH
Brightling
Vinehall Street
Chitcombe
Mountfield
Broad
Rye
Lydd
Town
KENT
NORTH DOWNS

London's Rail & Tube services

Key to lines and symbols

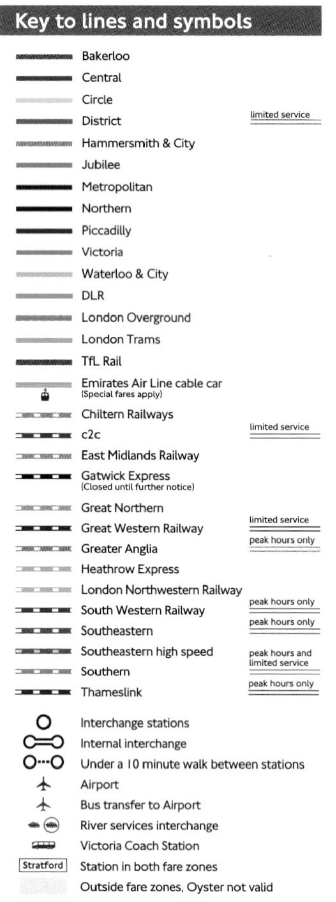

Bakerloo
Central
Circle
District
limited service
Hammersmith & City
Jubilee
Metropolitan
Northern
Piccadilly
Victoria
Waterloo & City
DLR
London Overground
London Trams
TfL Rail
Emirates Air Line cable car (Special fares apply)
Chiltern Railways
c2c
limited service
East Midlands Railway
Gatwick Express (Closed until further notice)
Great Northern
limited service
Great Western Railway
peak hours only
Greater Anglia
Heathrow Express
London Northwestern Railway
peak hours only
South Western Railway
peak hours only
Southeastern
Southeastern high speed
peak hours and limited service
Southern
peak hours only
Thameslink
Interchange stations
Internal interchange
Under a 10 minute walk between stations
Airport
Bus transfer to Airport
River services interchange
Victoria Coach Station
Stratford
Station in both fare zones
Outside fare zones, Oyster not valid

Check before you travel

oyster

Pay as you go with contactless (card or device) or Oyster is valid at all stations within the pay as you go area, except where shown. Pay as you go fares vary according to the services used, the time of travel and the zones travelled through.
To check your fare, please visit tfl.gov.uk/tube-rail-fares

Pay as you go area
Outside fare zones. Oyster not valid

London Trams fare zone. National Rail tickets not valid unless specifically stated.

Travelcards

Travelcards are valid within the zones covered, except on:

- Emirates Air Line
- Heathrow Express
- Southeastern high speed

Travelcards valid in zones 3, 4, 5 or 6 can be used on all tram services.

You may be liable for a penalty fare if you cannot show a validated smartcard (or device), a valid ticket or other authority to travel.

Due to coronavirus, some stations have amended opening times. Some services may be suspended or stations closed. Please check before you travel by visiting nationalrail.co.uk or tfl.gov.uk/plan-a-journey

 September 2021 Version G TfL Reg. User No. 21/S/3563/P

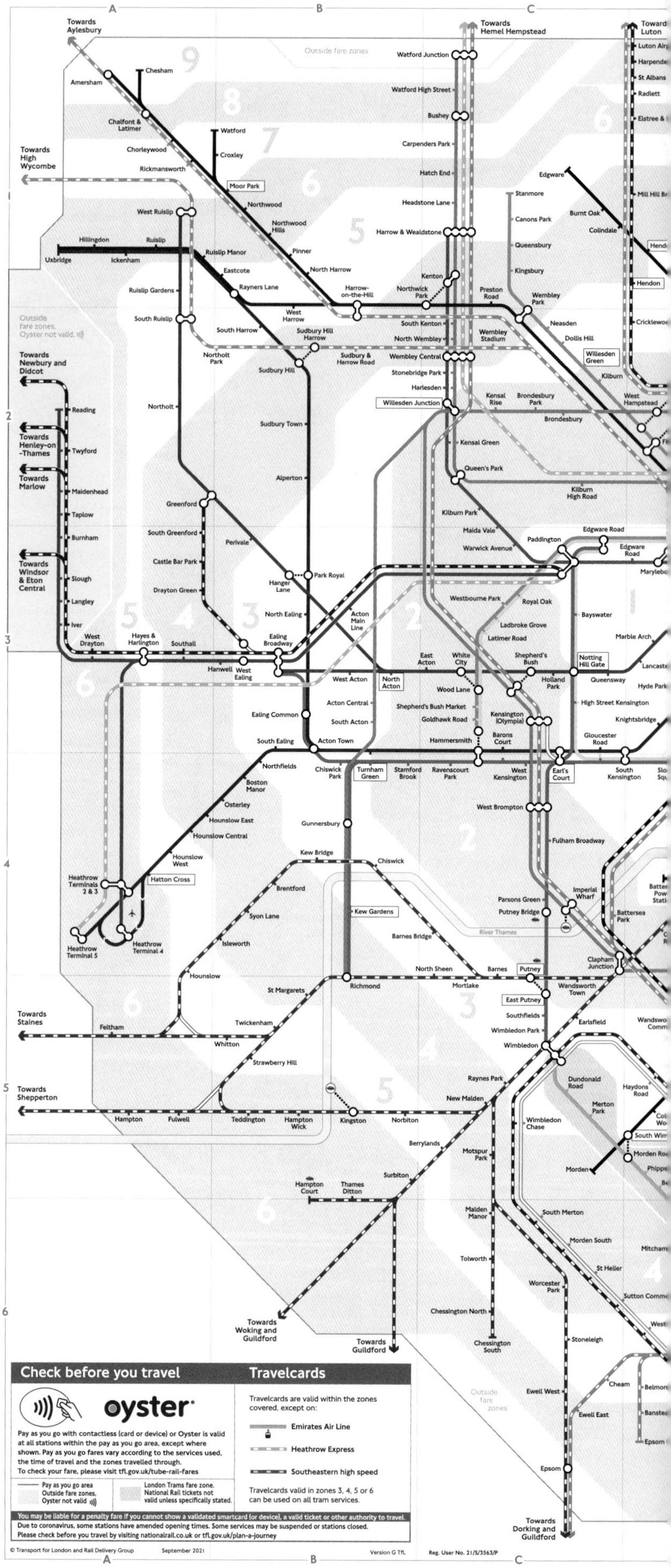

tfl.gov.uk
nationalrail.co.uk

Notes
The National Rail services on this map are shown as a guide to the routes and the train operators that serve these routes.
The map does not necessarily show service patterns or guarantee direct trains between the stations shown and some peak only or limited services are omitted.
Please check nationalrail.co.uk for details.
Correct at time of going to print.
For the latest information see www.tfl.gov.uk
National Rail
TRANSPORT FOR LONDON
EVERY JOURNEY MATTERS
Correct at time of going to print

Watford Junction
Towards St Albans City and Luton Airport Parkw
Chesham
Chalfont & Latimer
Watford
Amersham
Chorleywood
Croxley
Rickmansworth
Moor Park
Watford High Street
Bushey
Carpenders Park
Hatch End
Headstone Lane
Harrow & Wealdstone
Elstree & Borehamwood
Totteridge &
Edgware
Burnt Oak
Colindale
Hendon
Mill Hill Broadway
Hendon Central
Brent Cross
Golders Green
Hampstead
Cricklewood
West Ruislip
Northwood
Northwood Hills
Pinner
North Harrow
Harrow-on-the-Hill
Kenton
Preston Road
Stanmore
Canons Park
Queensbury
Kingsbury
Neasden
Dollis Hill
Willesden Green
Kilburn
West Hampstead
West Hampstead Thameslink
Gospel Oak
Belsize Park
Hampstead Heath
Chalk Farm
Camden Town
Finchley Road & Frognal
Finchley Road
South Hampstead
Mornington Crescent
Swiss Cottage
St John's Wood
Euston
Hillingdon
Uxbridge
Ickenham
Ruislip
Ruislip Manor
Eastcote
Rayners Lane
West Harrow
Northwick Park
Ruislip Gardens
South Ruislip
Northolt
South Harrow
Sudbury Hill
Sudbury Town
Greenford
Alperton
South Kenton
North Wembley
Wembley Central
Wembley Park
Stonebridge Park
Harlesden
Willesden Junction
Kensal Rise
Brondesbury Park
Brondesbury
Kensal Green
Queen's Park
Kilburn High Road
Kilburn Park
Maida Vale
Warwick Avenue
Royal Oak
Paddington
Edgware Road
Marylebone
Baker Street
Great Portland Street
Euston Square
Warren Street
Regent's Park
Goodge Street
Reading
Twyford
Maidenhead
Taplow
Burnham
Slough
Langley
Iver
West Drayton
Hayes & Harlington
Southall
Hanwell
West Ealing
Ealing Broadway
Perivale
Hanger Lane
Park Royal
Acton Main Line
North Ealing
Westbourne Park
Ladbroke Grove
Latimer Road
Bayswater
East Acton
White City
Shepherd's Bush
Notting Hill Gate
Lancaster Gate
Bond Street
Oxford Circus
Tottenham Court Road
West Acton
North Acton
Wood Lane
Holland Park
Queensway
Marble Arch
High Street Kensington
Green Park
Piccadilly Circus
Leicester Square
Charing Cross
Ealing Common
Acton Central
South Acton
Shepherd's Bush Market
Goldhawk Road
Kensington (Olympia)
Hyde Park Corner
Knightsbridge
Acton Town
Hammersmith
Barons Court
Gloucester Road
Sloane Square
St James's Park
Westminster
South Ealing
Northfields
Boston Manor
Chiswick Park
Turnham Green
Stamford Brook
Ravenscourt Park
West Kensington
Earl's Court
South Kensington
Victoria
Osterley
Hounslow East
Hounslow Central
Hounslow West
West Brompton
Waterloo
Gunnersbury
Fulham Broadway
Parsons Green
Putney Bridge
Pimlico
Heathrow Terminals 2 & 3
Hatton Cross
Kew Gardens
Imperial Wharf
River Thames
Richmond
Nine Elms
Kennington
East Putney
Southfields
Battersea Power Station
Vauxhall
Clapham Junction
Wimbledon Park
Wandsworth Road
Clapham High Street
Oval
Stockwell
Heathrow Terminal 5
Heathrow Terminal 4
Wimbledon
Haydons Road
Clapham North
Clapham Common
Clapham South
Brixton
Wimbledon Chase
Dundonald Road
Tooting Bec
Balham
Merton Park
South Wimbledon
Tooting Broadway
Tooting
South Merton
Colliers Wood
Morden South
Morden Road
Phipps Bridge
Belgrave Walk
Mitcham
St Helier
Sutton Common
Morden
West Sutton
Sutton
Carshalton
Bakerloo
Central
Circle
District
Hammersmith & City
Jubilee
Metropolitan
Northern
Piccadilly
Victoria
Waterloo & City
DLR
Emirates Air Line cable car (Special fares apply)
London Overground
TfL Rail
Thameslink - Not operated by TfL. If you need assistance, please speak to staff at your origin station so this can be arranged at the destination station or check assisted travel at www.thameslinkrailway.co.uk
London Trams
District open at weekends and on some public holidays
Interchange stations
Internal interchange
Under a 10 minute walk between stations
Step-free from train to street
Step-free from platform to street
National Rail interchange
River services interchange
Victoria Coach Station
Airport
Emirates Air Line cable car
† Services or access at these stations are subject to variation. To check before you travel, visit tfl.gov.uk/plan-a-journey
† Heathrow - TfL Rail customers should change at Terminals 2&3 for free rail transfer to Terminal 5.
† Hounslow West - Step-free access for manual wheelchairs only.
Correct at time of going to print. For th
MAYOR OF LONDON
tfl.gov.uk
24 hour travel info
0343 222 1
*Network charges may apply. See tfl.gov.uk/terms for details.
© Transport for London
Reg. user No. 21/S/3563/P
Improvement works may affect your

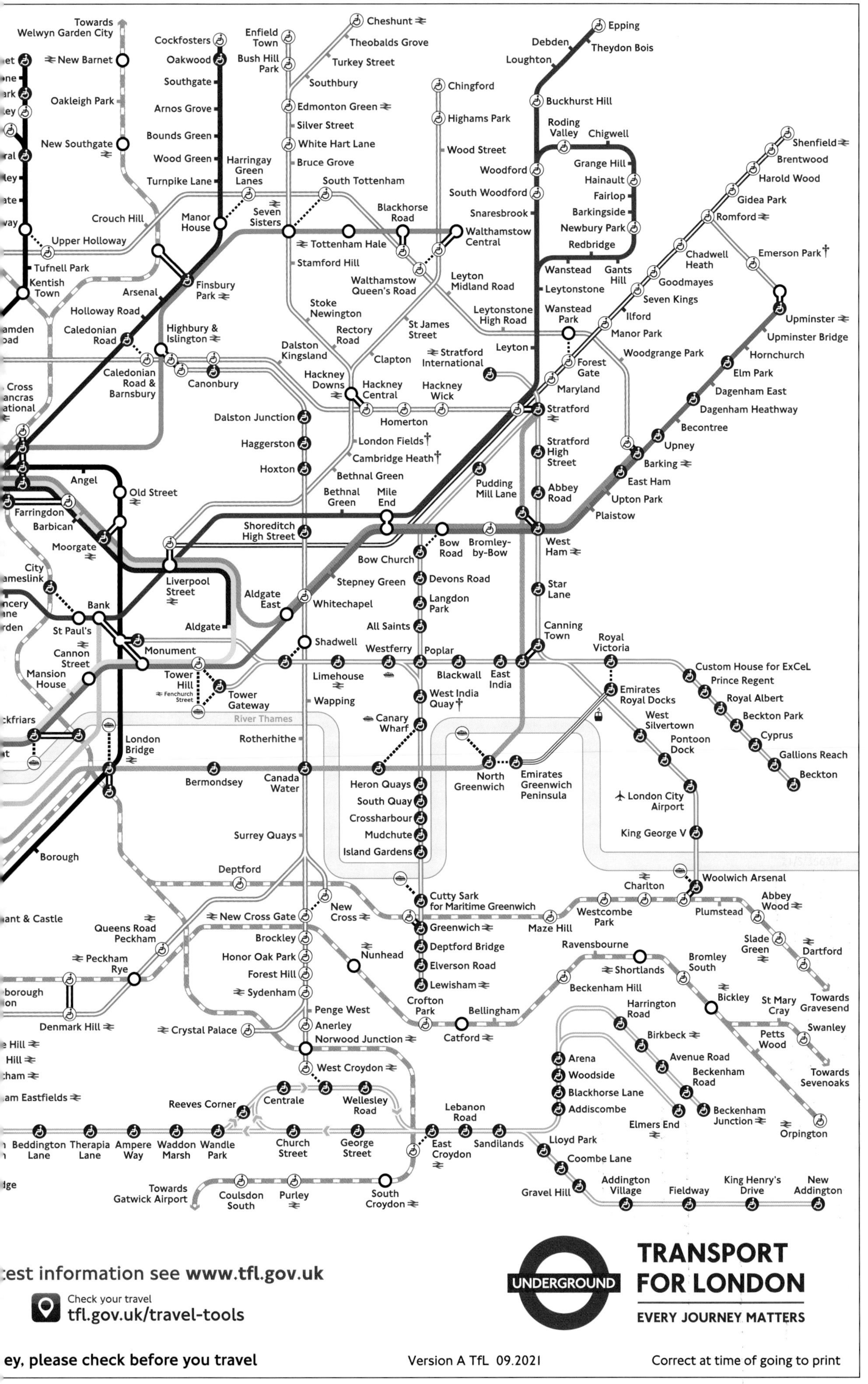
Towards Welwyn Garden City
New Barnet
Oakleigh Park
New Southgate
Cockfosters
Oakwood
Southgate
Arnos Grove
Bounds Green
Wood Green
Turnpike Lane
Harringay Green Lanes
Enfield Town
Bush Hill Park
Cheshunt
Theobalds Grove
Turkey Street
Southbury
Edmonton Green
Silver Street
White Hart Lane
Bruce Grove
South Tottenham
Seven Sisters
Tottenham Hale
Blackhorse Road
Walthamstow Central
Chingford
Highams Park
Wood Street
Epping
Theydon Bois
Debden
Loughton
Buckhurst Hill
Roding Valley
Chigwell
Grange Hill
Hainault
Fairlop
Barkingside
Newbury Park
Redbridge
Woodford
South Woodford
Snaresbrook
Wanstead
Gants Hill
Leytonstone
Shenfield
Brentwood
Harold Wood
Gidea Park
Romford
Chadwell Heath
Emerson Park
Goodmayes
Seven Kings
Ilford
Manor Park
Upminster
Upminster Bridge
Hornchurch
Elm Park
Dagenham East
Dagenham Heathway
Becontree
Upney
Barking
East Ham
Upton Park
Plaistow
Crouch Hill
Manor House
Upper Holloway
Tufnell Park
Kentish Town
Arsenal
Finsbury Park
Holloway Road
Caledonian Road
Highbury & Islington
Caledonian Road & Barnsbury
Canonbury
Stamford Hill
Walthamstow Queen's Road
Leyton Midland Road
Stoke Newington
Rectory Road
St James Street
Leytonstone High Road
Wanstead Park
Dalston Kingsland
Clapton
Stratford International
Leyton
Forest Gate
Woodgrange Park
Maryland
Hackney Downs
Hackney Central
Hackney Wick
Homerton
Stratford
Stratford High Street
Dalston Junction
Haggerston
Hoxton
London Fields
Cambridge Heath
Bethnal Green
Pudding Mill Lane
Abbey Road
Angel
Old Street
Farringdon
Barbican
Moorgate
Shoreditch High Street
Bethnal Green
Mile End
Bow Road
Bromley-by-Bow
West Ham
Bow Church
Devons Road
Stepney Green
Star Lane
City Thameslink
Liverpool Street
Aldgate East
Whitechapel
Langdon Park
Bank
St Paul's
Aldgate
All Saints
Canning Town
Royal Victoria
Cannon Street
Monument
Shadwell
Westferry
Poplar
Mansion House
Tower Hill
Fenchurch Street
Tower Gateway
Limehouse
Blackwall
East India
Custom House for ExCeL
Prince Regent
Royal Albert
Beckton Park
Cyprus
Gallions Reach
Beckton
Emirates Royal Docks
West Silvertown
Pontoon Dock
Wapping
West India Quay
Canary Wharf
River Thames
London Bridge
Rotherhithe
North Greenwich
Emirates Greenwich Peninsula
Bermondsey
Canada Water
Heron Quays
South Quay
Crossharbour
Mudchute
Island Gardens
London City Airport
King George V
Surrey Quays
Borough
Deptford
Woolwich Arsenal
Charlton
Abbey Wood
Plumstead
Westcombe Park
Maze Hill
Cutty Sark for Maritime Greenwich
Greenwich
New Cross Gate
New Cross
Queens Road Peckham
Brockley
Honor Oak Park
Forest Hill
Sydenham
Nunhead
Deptford Bridge
Elverson Road
Lewisham
Peckham Rye
Denmark Hill
Crofton Park
Bellingham
Catford
Penge West
Anerley
Norwood Junction
Crystal Palace
Ravensbourne
Shortlands
Beckenham Hill
Bromley South
Slade Green
Dartford
Bickley
St Mary Cray
Towards Gravesend
Swanley
Petts Wood
Harrington Road
Birkbeck
Avenue Road
Beckenham Road
Beckenham Junction
Arena
Woodside
Blackhorse Lane
Addiscombe
Elmers End
Towards Sevenoaks
Orpington
West Croydon
Reeves Corner
Centrale
Wellesley Road
Lebanon Road
Beddington Lane
Therapia Lane
Ampere Way
Waddon Marsh
Wandle Park
Church Street
George Street
East Croydon
Sandilands
Lloyd Park
Coombe Lane
Gravel Hill
Addington Village
Fieldway
King Henry's Drive
New Addington
Towards Gatwick Airport
Coulsdon South
Purley
South Croydon
www.tfl.gov.uk
Check your travel
tfl.gov.uk/travel-tools
UNDERGROUND
TRANSPORT FOR LONDON
EVERY JOURNEY MATTERS
Version A TfL 09.2021
Correct at time of going to print

INDEX

Including Streets, Places & Areas, Hospitals etc., Industrial Estates,
Selected Flats & Walkways, Service Areas, Stations and Selected Places of Interest.

HOW TO USE THIS INDEX

1. Each street name is followed by its Postcode District (or, if outside the London Postcodes, by its Locality Abbreviation(s)) and then by its map reference; e.g. **Abbey Av.** HA0: Wemb.......40Na **67** is in the HA0 Postcode District and the Wembley Locality and is to be found in square 40Na on page **67**. The page number being shown in bold type.
2. A strict alphabetical order is followed in which Av., Rd., St., etc. (though abbreviated) are read in full and as part of the street name; e.g. **Alder M.** appears after **Aldermary Rd.** but before **Aldermoor Rd.**
3. Streets and a selection of flats and walkways that cannot be shown on the mapping, appear in the index with the thoroughfare to which they are connected shown in brackets; e.g. **Abbey Ct.** AL1: St A......3B **6** *(off Holywell Hill)*
4. Addresses that are in more than one part are referred to as not continuous.
5. Places and areas are shown in the index in BLUE TYPE and the map reference is to the actual map square in which the town centre or area is located and not to the place name shown on the map; e.g. ABBEY WOOD.......49Yc 95
6. An example of a selected place of interest is **Barnet Mus.**......**14Ab 30**
7. Junction names and Service Areas are shown in the index in BOLD CAPITAL TYPE; e.g. **ANGEL EDMONTON**.......**22Wb 51**
8. Map references for entries that appear on large scale pages **214-231** are shown first, with small scale map references shown in brackets; e.g. **Abbey Gdns.** NW8 . . . .2A **214** (40Eb **69**)

GENERAL ABBREVIATIONS

All. : Alley
Apts. : Apartments
App. : Approach
Arc. : Arcade
Av. : Avenue
Bk. : Back
Blvd. : Boulevard
Bri. : Bridge
B'way. : Broadway
Bldg. : Building
Bldgs. : Buildings
Bungs. : Bungalows
Bus. : Business
Cvn. : Caravan
C'way. : Causeway
Cen. : Centre
Chu. : Church
Circ. : Circle
Cir. : Circus
Cl. : Close
Coll. : College
Comn. : Common
Cnr. : Corner
Cott. : Cottage
Cotts. : Cottages
Ct. : Court
Ctyd. : Courtyard
Cres. : Crescent
Cft. : Croft
Dpt. : Depot
Dr. : Drive
E. : East
Emb. : Embankment
Ent. : Enterprise
Est. : Estate
Fld. : Field
Flds. : Fields
Gdn. : Garden
Gdns. : Gardens
Gth. : Garth
Ga. : Gate
Gt. : Great
Grn. : Green
Gro. : Grove
Hgts. : Heights
Ho. : House
Ho's. : Houses
Ind. : Industrial
Info. : Information
Intl. : International
Junc. : Junction
La. : Lane
Lit. : Little
Lwr. : Lower
Mnr. : Manor
Mans. : Mansions
Mkt. : Market
Mdw. : Meadow
Mdws. : Meadows
M. : Mews
Mt. : Mount
Mus. : Museum
Nth. : North
No. : Number
Pal. : Palace
Pde. : Parade
Pk. : Park
Pas. : Passage
Pav. : Pavilion
Pl. : Place
Pct. : Precinct
Prom. : Promenade
Quad. : Quadrant
Res. : Residential
Ri. : Rise
Rd. : Road
Rdbt. : Roundabout
Shop. : Shopping
Sth. : South
Sq. : Square
Sta. : Station
St. : Street
Ter. : Terrace
Twr. : Tower
Trad. : Trading
Up. : Upper
Va. : Vale
Vw. : View
Vs. : Villas
Vis. : Visitors
Wlk. : Walk
W. : West
Yd. : Yard

LOCALITY ABBREVIATIONS

Abbots Langley: WD5 ... Ab L
Abridge: CM16, IG7, RM4 ... Abr
Addington: BR4:, CR0 ... Addtn
Addlestone: KT15 ... Add
Aldenham: WD25 ... A'ham
Arkley: EN5 ... Ark
Ascot: SL5 ... Asc
Ash: TN15 ... Ash
Ashford: TW15 ... Ashf
Ashtead: KT21, KT22 ... Asht
Aveley: RM14, RM15 ... Avel
Aveley: RM19 ... Avel
Badger's Mount: TN14 ... Bad M
Banstead: CR5, KT17, SM2, SM7 ... Bans
Barking: IG11, IG3, RM8, RM9 ... Bark
Barnet: EN4, EN5 ... Barn
Beaconsfield: HP9 ... Beac
Bean: DA2 ... Bean
Beckenham: BR3 ... Beck
Beddington: CR0, SM6 ... Bedd
Bedfont: TW14 ... Bedf
Bedmond: WD5 ... Bedm
Belvedere: DA17, DA18, DA7 ... Belv
Berkhamsted: HP4 ... Berk
Bessels Green: TN13 ... Bes G
Betchworth: RH3 ... B'wth
Bexley: DA15, DA1, DA5 ... Bexl
Bexleyheath: DA5, DA6, DA7 ... Bex
Biggin Hill: TN16 ... Big H
Bisley: GU24 ... Bisl
Bletchingley: CR3, RH1 ... Blet
Bluewater: DA2, DA9 ... Bluew
Bookham: KT23 ... Bookh
Borehamwood: WD6 ... Bore
Borough Green: TN15 ... Bor G
Bovingdon: HP3 ... Bov
Brasted Chart: TN16 ... B Char
Brasted: TN16 ... Bras
Brentford: TW7, TW8 ... Bford
Brentwood: CM13, CM14, CM15 ... B'wood
Bricket Wood: AL2 ... Brick W
Brimsdown: EN3 ... Brim
Bromley: BR1, BR2 ... Broml
Brookmans Park: AL9 ... Brk P
Brookwood: GU21, GU24 ... Brkwd
Buckhurst Hill: IG8, IG9 ... Buck H
Buckland: RH3 ... Bkld
Bucks Hill: WD3, WD4 ... Bucks
Bulphan: RM14 ... Bulp
Burnham: SL1, SL2 ... Burn
Burpham: GU4 ... Burp
Bushey: WD23, WD24 ... Bush
Bushy Heath: WD23 ... B Hea
Byfleet: KT14 ... Byfl
Carshalton: CR4, SM4, SM5 ... Cars
Caterham: CR3 ... Cat'm
Chadwell Heath: IG7, RM6, RM7 ... Chad H
Chafford Hundred: RM16, RM20 ... Chaf H
Chalfont St Giles: HP8 ... Chal G
Chalfont St Peter: SL9, WD3 ... Chal P
Chandler's Cross ... WD3 Chan C
Cheam: KT17, Cheam
Cheam: SM3 ... Cheam
Chelsfield: BR6 ... Chels
Chenies: WD3 ... Chen
Chertsey: KT16 ... Chert
Cheshunt: EN7, EN8 ... Chesh
Chessington: KT9 ... Chess
Chevening: TN14 ... Chev
Chigwell: IG6, IG7 ... Chig
Chipperfield: WD3, WD4 ... Chfd
Chipstead: CR5, TN13:, TN14 ... Chips
Chislehurst: BR7 ... Chst
Chiswell Green: AL2 ... Chis G
Chobham: GU24 ... Chob
Chorleywood: WD3 ... Chor
Claygate: KT10 ... Clay
Cobham: DA12, DA13, KT11 ... Cobh
Cobham: KT12 ... Cobh
Cockfosters: EN4, EN5 ... Cockf
Collier Row: RM5, RM6, RM7 ... Col R
Colnbrook: SL3 ... Coln
Colney Heath: AL4 ... Col H
Colney Street: AL2 ... Col S
Coopersale: CM16 ... Coop
Corringham: SS17 ... Corr
Coryton: SS17 ... Cory
Coulsdon: CR3, CR5, RH1 ... Coul
Cowley: UB8 ... Cowl
Cranford: TW4, TW5, TW6, UB3 ... Cran
Crayford: DA1 ... Cray
Crews Hill: EN2 ... Crew H
Crockenhill: BR8 ... Crock
Crockham Hill: RH8 ... Crock H
Crouch: TN15 ... Crou
Crowhurst: RH7, RH8 ... C'rst
Croxley Green: WD3 ... Crox G
Croydon: CR0 ... C'don
Cudham: TN14, TN16 ... Cud
Cuffley: EN6, EN7 ... Cuff
Dagenham: IG11, RM10, RM6, RM8, RM9 ... Dag
Darenth: DA2 ... Daren
Dartford: DA1, DA2, DA3, DA5, DA9 ... Dart
Datchet: SL3 ... Dat
Denham: SL0, UB9 ... Den
Doddinghurst: CM15 ... Dodd
Dorney: SL4 ... Dor
Downe: BR6 ... Downe
Downside: KT11 ... D'side
Dunk's Green: TN11 ... Dun G
Dunton Green: TN13, TN14 ... Dun G
Dunton: CM13 ... Dun
East Barnet: EN4 ... E Barn
East Clandon: GU4 ... E Clan
East Horsley: KT24 ... E Hor
East Molesey: KT1, KT8, TW12 ... E Mos
East Tilbury: RM18, SS17 ... E Til
Eastcote: HA4, Eastc
Ebbsfleet: DA10 ... Ebbs
Edenbridge: RH8, TN8 ... Eden
Edgware: HA8 ... Edg
Effingham Junction: KT24 ... Eff J
Effingham: KT24 ... Eff
Egham: TW20 ... Egh
Elstree: WD6 ... E'tree
Enfield Highway: EN3 ... Enf H
Enfield Lock: EN3 ... Enf L
Enfield Wash: EN3 ... Enf W
Enfield: EN1, EN2 ... Enf
Englefield Green: TW20 ... Eng G
Epping: CM16, EN9 ... Epp
Epsom Downs: KT17:, KT18 ... Eps D
Epsom: KT17, KT18, KT19 ... Eps
Erith: DA17, DA18, DA1, DA7, DA8 ... Erith
Esher: KT10 ... Esh
Essendon: AL9 ... Ess
Eton Wick: SL4 ... Eton W
Eton: SL4 ... Eton
Ewell: KT17, KT19, SM2 ... Ewe
Eynsford: BR8, DA4 ... Eyns
Eynsford: TN14 ... Eyns
Fairseat: TN15 ... Fair
Farnborough: BR5, BR6 ... Farnb
Farnham Common: SL1, SL2 ... Farn C
Farnham Royal: SL2 ... Farn R
Farningham: BR8, DA4 ... Farni
Fawkham: DA3 ... Fawk
Feltham: TW13, TW14 ... Felt
Fetcham: KT22, KT23 ... Fet
Fiddlers Hamlet: CM16 ... Fidd H
Flaunden: HP3, HP5 ... Flau
Frogmore: AL2 ... F'mre
Fulmer: SL3 ... Ful
George Green: SL3 ... Geor G
Gerrards Cross: SL2, SL9 ... Ger X
Godden Green: TN15 ... God G
Godstone: RH8, RH9 ... G'stone
Goff's Oak: EN7 ... G Oak
Gravesend: DA11, DA12 ... Grav'nd
Grays: RM16, RM17, RM18, RM20 ... Grays
Great Warley: CM13, CM14, RM14 ... Gt War
Green Street Green: DA2 ... G St G
Greenford: UB6 ... G'frd
Greenhithe: DA10, DA2, DA9 ... Ghithe
Guildford: GU2, GU3, GU4 ... Guild
Hadley Wood: EN4 ... Had W
Halstead: TN14 ... Hals
Ham: TW10 ... Ham
Hampton Hill: TW12 ... Hamp H
Hampton Wick: KT1, TW11 ... Hamp W
Hampton: TW12 ... Hamp
Hanworth: TW13 ... Hanw
Harefield: UB9, WD3 ... Hare
Harlington: UB3 ... Harl
Harlington: UB7 ... Harl
Harmondsworth: UB7 ... Harm
Harold Wood: RM11, RM3 ... Hrld W
Harrow Weald: HA3 ... Hrw W
Harrow: HA1, HA2, HA3 ... Harr
Hartley: DA3, TN15 ... Hartl
Hatch End: HA5 ... Hat E
Hatfield: AL10, AL9 ... Hat
Havering-Atte-Bower: RM1, RM4 ... Have B
Hawley: DA2 ... Hawl
Hayes: BR2, BR4, UB3, UB4 ... Hayes
Headley: KT18 ... Head
Hedgerley: SL2 ... Hedg
Hemel Hempstead: HP1, HP2, HP3 ... Hem H
Herongate: CM13 ... Heron
Heronsgate: WD3 ... Herons
Hersham: KT10, KT12 ... Hers
Heston: TW5 ... Hest
Hextable: BR8 ... Hext
High Beech: EN9, IG10 ... H Beech
Higham: DA12, ME2, ME3 ... High'm
Hillingdon: UB10, UB4, UB8 ... Hil
Hinchley Wood: KT10 ... Hin W
Hodsoll Street: TN15 ... Hod S
Hornchurch: RM11, RM12, RM13, RM2 ... Horn
Horndon-on-the-Hill: SS17 ... Horn H
Horton Kirby: DA3, DA4 ... Hort K
Horton: SL3 ... Hort
Hounslow: TW1, TW3, TW4, TW7 ... Houn
Hunton Cross: WD4 ... Hunt C
Hutton: CM13, CM15 ... Hut
Ickenham: UB10 ... Ick
Ide Hill: TN14 ... Ide H
Ightham: TN15 ... Igh
Ilford: IG1, IG2, IG3, IG4, IG5, IG6, IG7, IG8, RM6 ... Ilf
Ingrave: CM13 ... Ingve
Isleworth: TW1, TW3, TW5, TW7 ... Isle
Istead Rise: DA11, DA13, DA3 ... Ist R
Iver Heath: SL0 ... Iver H
Iver: SL0 ... Iver
Ivy Hatch: TN15 ... Ivy H
Jacobs Well: GU4 ... Jac W
Kelvedon Common: CM14 ... Kel C
Kelvedon Hatch: CM14, CM15 ... Kel H
Kemsing: TN15 ... Kems'g
Kenley: CR, CR8 ... Kenley
Kenton: HA3, HA7, HA9 ... Kenton
Keston: BR2 ... Kes
Kew: TW9 ... Kew
Kings Langley: WD4 ... K Lan
Kingston upon Thames: KT1, KT2 ... King T
Kingswood: CR5, KT20 ... Kgswd
Knaphill: GU21 ... Knap
Knatts Valley: TN15 ... Knat
Knockholt: TN14 ... Knock
Laleham: TW18 ... Lale
Langley: SL3 ... L'ly
Langleybury: WD4 ... Lang
Latimer: HP5 ... Lat
Leatherhead: KT11, KT18, KT22, T9 ... Lea
Letchmore Heath: WD25 ... Let H
Ley Hill: HP5 ... Ley H
Lightwater: GU18 ... Light
Limpsfield: RH8, TN8 ... Limp
Linford: SS17 ... Linf
Little Chalfont: HP6 ... L Chal
Little Warley: CM13 ... L War
London Colney: AL2 ... Lon C
London Heathrow Airport: TW6 ... H'row A
Longcross: KT16 ... Longc
Longfield Hill: DA3 ... Long H
Longfield: DA13, DA3 ... Lfield
Longford: TW6, UB7 ... Lford
Loudwater: WD3 ... Loud
Loughton: CM16, EN9, IG10, IG7, G9 ... Lough
Lower Kingswood: KT20 ... Lwr K
Luddesdown: DA12 ... Lud'n
Lyne: KT16 ... Lyne
Maple Cross: HP8, WD3 ... Map C
Mawney: RM7 ... Mawney
Meopham: DA13 ... Meop
Merstham: RH1 ... Mers
Mickleham: RH5 ... Mick
Mitcham: CR0, CR4 ... Mitc
Morden: SM4 ... Mord
Mountnessing: CM13, CM15 ... Mount
Navestock: CM14, RM4 ... Nave
Navestockside: CM14, RM4 ... N'side
New Addington: CR0 ... New Ad
New Ash Green: DA3, TN15 ... Nw A G
New Barnet: EN5 ... New Bar
New Haw: KT15 ... New H
New Malden: KT3 ... N Mald
Newgate Street: SG13 ... New S
Noak Hill: RM4 ... Noak H
Normandy: GU3 ... Norm
North Mymms: AL9 ... N Mym
North Ockendon: RM14 ... N Ock
North Stifford: RM15, RM16 ... N Stif
North Weald: CM16 ... N Weald
Northaw: EN6 ... N'thaw
Northfleet Green: DA13 ... Nflt G
Northfleet: DA10, DA11 ... Nflt
Northolt: UB4, UB5 ... N'olt
Northwood: HA6 ... Nwood
Nutfield: RH1 ... Nutf
Oakley Green: SL4 ... Oak G
Ockham: GU23 ... Ock
Old Windsor: SL4, TW20 ... Old Win
Orpington: BR5, BR6 ... Orp
Orsett: RM16, SS17 ... Ors
Otford: TN13, TN14 ... Otf
Ottershaw: KT16 ... Ott
Oxshott: KT10, KT11, KT22 ... Oxs
Oxted: RH8 ... Oxt
Park Street: AL2 ... Park
Petts Wood: BR5, BR6 ... Pet W
Pilgrims Hatch: CM14, CM15 ... Pil H
Pinner: HA5, HA6, WD19 ... Pinn
Pirbright: GU24 ... Pirb
Platt: TN15 ... Plat
Plaxtol: TN11, TN15 ... Plax
Ponders End: EN3 ... Pond E
Potten End: HP4 ... Pott E
Potters Bar: EN5, EN6 ... Pot B
Potters Crouch: AL2 ... Pot C
Poyle: SL3 ... Poyle
Pratts Bottom: BR6, TN14 ... Prat B
Purfleet: RM15, RM19 ... Purf
Purley: CR5, CR8 ... Purl
Pyrford: GU22 ... Pyr
Radlett: WD25, WD7 ... R'lett
Rainham: RM12, RM13, RM14, RM9 ... Rain
Ranmore Common: KT24 ... Ran C
Redhill: RH1 ... Redh
Reigate: KT20, RH2, RH3 ... Reig
Richings Park: SL0, SL3 ... Rich P
Richmond: TW10, TW9 ... Rich
Rickmansworth: UB9, WD3 ... Rick
Ridge: EN6 ... Ridge
Ripley: GU23 ... Rip
Riverhead: TN13 ... Riv
Romford: CM14, RM11, RM1, RM2, RM3, RM5, RM7 ... Rom
Roughway: TN11 ... Roug
Ruislip: HA4 ... Ruis
Rush Green: RM10, RM1, RM7 ... Rush G
Salfords: RH1 ... Salf
Sanderstead: CR2 ... Sande
Sarratt: WD3 ... Sarr
Seal: TN15 ... Seal
Selsdon: CR0, CR2 ... Sels
Send: GU23 ... Send
Sevenoaks: TN13, TN14, TN15, TN16 ... S'oaks
Shenfield: CM15 ... Shenf
Shenley: WD7 ... Shenl
Shepperton: TW17, TW18 ... Shep
Shipbourne: TN11, TN15 ... S'brne
Shoreham: TN14 ... S'ham
Shorne: DA12 ... Shorne
Sidcup: DA14, DA15, SE9 ... Sidc
Sidlow: RH2 ... Sid
Sipson: UB7 ... Sip
Slough: SL1, SL2, SL3 ... Slou
Smallford: AL4 ... S'ford
Sole Street: DA12, DA13 ... Sole S
South Croydon: CR2 ... S Croy
South Darenth: DA4 ... S Dar
South Godstone: RH9 ... S God
South Mimms: EN6 ... S Mim
South Nutfield: RH1 ... S Nut
South Ockendon: RM15 ... S Ock
South Weald: CM14 ... S Weald
Southall: UB1, UB2 ... S'hall
Southfleet: DA13, DA2 ... Sflt
St Albans: AL1, AL2, AL3, AL4 ... St A
St Mary Cray: BR5, BR6 ... St M Cry
St Pauls Cray: BR5 ... St P
Staines: TW18, TW19 ... Staines
Stanford-le-Hope: SS17 ... Stan H
Stanmore: HA3, HA7 ... Stan
Stansted: TN15 ... Stans
Stanwell Moor: TW19 ... Stanw M
Stanwell: TW19, TW6 ... Stanw
Stapleford Abbotts: RM4 ... Stap A
Stapleford Tawney: RM4 ... Stap T
Stapleford: RM4 ... Stfrd
Stockley Park: UB11 ... Stock P
Stoke D'Abernon: KT11 ... Stoke D
Stoke Poges: SL2, SL3 ... Stoke P
Strood: DA12, ME2 ... Strood
Sunbury: TW16 ... Sun
Sundridge: TN14 ... Sund
Sunningdale: SL5 ... S'dale
Sunninghill: SL5 ... S'hill
Surbiton: KT10, KT1, KT5, KT6 ... Surb
Sutton Green: GU4 ... Sut G
Sutton at Hone: DA2, DA4 ... Sut H
Sutton: SM1, SM2, SM3 ... Sutt
Swanley: BR5, BR8 ... Swan
Swanley: DA14, DA4 ... Swan
Swanscombe: DA10 ... Swans
Tadworth: KT18, KT20 ... Tad
Tandridge: RH8 ... Tand
Taplow: SL6 ... Tap
Tatsfield: TN16 ... Tats
Tattenham Corner: KT18 ... Tatt C
Teddington: TW11, TW1 ... Tedd
Thames Ditton: KT10, KT7 ... T Ditt
Theydon Bois: CM16, EN9 ... They B
Theydon Garnon: CM16 ... They G
Theydon Mount: CM16 ... They M
Thorney: SL0 ... Thorn
Thornton Heath: CR7 ... Thor H
Thorpe: TW20 ... Thorpe
Tilbury: RM18 ... Tilb
Titsey: RH8, TN16 ... T'sey
Toy's Hill: TN16 ... Toy H
Trottiscliffe: ME19 ... Tros
Turnford: EN8 ... Turn
Twickenham: TW13, TW1, TW2, TW7 ... Twick
Underriver: TN15 ... Under
Upminster: CM13, CM14, RM11, RM12, RM13, RM14, RM15 ... Upm
Uxbridge: UB10, UB8 ... Uxb
Virginia Water: GU25, KT16, SL5 ... Vir W
Waddon: CR0 ... Wadd
Wallington: SM5, SM6 ... W'gton
Waltham Abbey: CM16, EN9 ... Walt A
Waltham Cross: EN2, EN7, EN8, EN9 ... Walt C
Walton on the Hill: KT18, KT20 ... Walt H
Walton-on-Thames: KT12 ... Walt T
Warley: CM13, CM14 ... W'ley
Warlingham: CR3, CR6 ... W'ham
Water Oakley: SL4 ... Wat O
Watford: WD17, WD18, WD19, WD24, WD25, WD3 ... Wat
Wealdstone: HA3 ... W'stone
Welham Green: AL9 ... Wel G
Well Hill: BR6 ... Well H
Welling: DA16 ... Well
Wembley: HA0, HA9, NW10 ... Wemb
Wennington: RM13, RM15 ... Wenn
West Byfleet: KT14 ... W Byf
West Clandon: GU4 ... W Cla
West Drayton: UB7 ... W Dray
West End: GU24 ... W End
West Horndon: CM13 ... W H'dn
West Horsley: KT24 ... W Hor
West Hyde: WD3 ... W Hyd
West Kingsdown: TN15 ... W King
West Molesey: KT8 ... W Mole
West Thurrock: RM19, RM20 ... W Thur
West Tilbury: RM16, RM18 ... W Til
West Wickham: BR4 ... W W'ck
Westerham: TN16 ... Westrm
Westhumble: KT23 ... Westh
Wexham: SL2, SL3 ... Wex
Weybridge: KT13 ... Weyb
Whelpley Hill: HP5 ... Whel H
Whiteley Village: KT12 ... W Vill
Whitton: TW2 ... Whitt
Whyteleafe: CR3 ... Whyt
Wilmington: DA2, DA5 ... Wilm
Windlesham: GU20 ... W'sham
Windsor: SL4 ... Wind
Winkfield: SL4 ... Wink
Wisley: GU23 ... Wis
Woking: GU21, GU22 ... Wok
Woking: GU24 ... Wok
Woldingham: CR3, CR6 ... Wold
Woodford Green: IG4, IG8, IG9 ... Wfd G
Woodham: KT15 ... Wdhm
Worcester Park: KT4, SM3 ... Wor Pk
Worplesdon: GU3 ... Worp
Wraysbury: TW19 ... Wray
Wrotham Heath: TN15 ... Wro H
Wrotham: TN15 ... Wro
Yeading: UB4 ... Yead
Yiewsley: UB7, UB8 ... Yiew

1

2 Temple Place 4K 223 (45Qb 90) (off Temple Pl.)
2 Willow Rd. 35Gb 69
7 July Memorial 7J 221 (46Jb 90)
10 Brock St. NW1 5B 216 (42Lb 90) (off Triton Sq.)
18 Stafford Terrace 48Cb 89 (off Stafford Ter.)
60 St Martins La. WC2 4F 223 (45Nb 90) (off St Martin's La.)
198 Contemporary Arts & Learning 58Rb 113 (off Railton Rd.)
201 Bishopsgate EC2 7J 219 (43Ub 91) (off Bishopsgate)

A

A1 Activity Cen. 13Ua 30
Aaron Hill Rd. E6 43Qc 94
Aashiana Ct. WD18: Wat 16W 26
Abady Ho. SW1 5E 228 (49Mb 90) (off Page St.)
Abberley M. SW4 55Kb 112
Abberton Wlk. RM13: Rain 39Hd 76
Abbess Cl. E6 43Nc 94
Abbess Cl. SW2 60Rb 113
Abbess Ct. IG10: Lough 13Rc 36
Abbess Ter. IG10: Lough 13Qc 36
Abbeville M. SW4 56Mb 112
Abbeville Rd. N8 29Mb 50
Abbeville Rd. SW4 58Lb 112
Abbey Av. AL3: St A 5N 5
Abbey Av. HA0: Wemb 40Na 67
Abbey Chase KT16: Chert 73K 149
Abbey Churchyard EN9: Walt A 5Ec 20
Abbey Cl. BR6: Chels 77Xc 161
Abbey Cl. E5 35Wb 71
Abbey Cl. GU22: Pyr 88G 168
Abbey Cl. HA5: Pinn 27X 45
Abbey Cl. RM1: Rom 30Jd 56
Abbey Cl. SL1: Slou 5C 80
Abbey Cl. SW8 53Mb 112
Abbey Cl. UB3: Hayes 46X 85
Abbey Cl. UB5: N'olt 41Ba 85
Abbey Ct. AL1: St A 3B 6 (off Holywell Hill)
Abbey Ct. EN9: Walt A 6Dc 20
Abbey Ct. KT16: Chert 73K 149
Abbey Ct. NW8 1A 214 (40Eb 69) (off Abbey Rd.)
Abbey Ct. SE17 50Sb 91 (off Macleod St.)
Abbey Ct. SE6 58Cc 114
Abbey Ct. TW12: Hamp 66Ca 129
Abbey Ct. TW18: Lale 70L 127
Abbey Cres. DA17: Belv 49Cd 96
Abbeydale Rd. HA0: Wemb 39Pa 67
Abbey Dr. DA2: Wilm 61Gd 140
Abbey Dr. SW17 64Jb 134
Abbey Dr. TW18: Lale 69L 127
Abbey Dr. WD5: Ab L 4W 12
Abbey Est. NW8 39Db 69
Abbeyfield Cl. CR4: Mitc 68Gb 133
Abbeyfield Est. SE16 49Yb 92
Abbeyfield Rd. SE16 49Yb 92 (not continuous)
Abbeyfields KT16: Chert 73M 149
Abbeyfields Cl. NW10 41Qa 87
Abbey Gdns. BR7: Chst 67Qc 138
Abbey Gdns. EN9: Walt A 5Ec 20
Abbey Gdns. KT16: Chert 72J 149
Abbey Gdns. NW8 2A 214 (40Eb 69)
Abbey Gdns. SE16 49Wb 91
Abbey Gdns. SW1 3F 229 (48Nb 90) (off Great College St.)
Abbey Gdns. TW15: Ashf 64R 128
Abbey Gdns. W6 51Ab 110
Abbey Gateway AL3: St A 2A 6
Abbey Grn. KT16: Chert 72J 149
Abbey Gro. SE2 49Xc 95
Abbeyhill Rd. DA15: Sidc 61Yc 139
Abbey Ho. E15 40Gc 73 (off Baker's Row)
Abbey Ho. NW8 3A 214 (41Eb 89) (off Garden Rd.)
Abbey Ind. Est. CR4: Mitc 71Hb 155
Abbey Ind. Est. HA0: Wemb 39Pa 67
Abbey La. BR3: Beck 66Cc 136
Abbey La. E15 40Ec 72 (not continuous)
Abbey La. Commercial Est. E15 40Gc 73
Abbey Leisure Cen. Barking 39Sc 74
Abbey Life Ct. E16 43Kc 93
Abbey Lodge NW8 4E 214 (41Gb 89) (off Park Rd.)
Abbey Mansion M. SE24 57Rb 113
Abbey Mead Cl. DA1: Dart 55Rd 119
Abbey Mead Ind. Est. EN9: Walt A 7Ec 20
Abbey Mead Ind. Pk. EN9: Walt A 6Dc 20
Abbey Mdws. KT16: Chert 73L 149
Abbey M. AL1: St A 3B 6 (off Holywell Hill)
Abbey M. E17 29Cc 52
Abbey M. TW18: Lale 70L 127
Abbey M. TW7: Isle 53Ka 108
Abbey Mill End AL3: St A 3A 6
Abbey Mill La. AL3: St A 3A 6
Abbey Moor Golf Course 76J 149
Abbey Mt. DA17: Belv 50Bd 95
Abbey Orchard St. SW1 3D 228 (48Mb 90)
Abbey Orchard St. Est. SW1 3E 228 (48Mb 90) (not continuous)
Abbey Pde. SW19 66Eb 133 (off Merton High St.)
Abbey Pde. W5 41Pa 87
Abbey Pk. BR3: Beck 66Cc 136
Abbey Pk. Ind. Est. IG11: Bark 40Sc 74
Abbey Pk. La. SL1: Burn 3C 60
Abbey Pl. DA1: Dart 57Md 119
Abbey Pl. KT16: Chert 69J 127
Abbey Retail Pk. Barking 38Rc 74
Abbey Rd. CR0: C'don 76Rb 157
Abbey Rd. CR2: Sels 82Zb 178
Abbey Rd. DA12: Grav'nd 10G 122
Abbey Rd. DA17: Belv 49Zc 95
Abbey Rd. DA7: Bex 56Ad 117
Abbey Rd. DA9: Ghithe 57Yd 120
Abbey Rd. E15 40Fc 73
Abbey Rd. EN1: Enf 15Ub 33
Abbey Rd. EN8: Walt C 6Ac 20
Abbey Rd. GU21: Wok 9N 167
Abbey Rd. GU25: Vir W 1P 147
Abbey Rd. IG11: Bark 39Rc 74
Abbey Rd. IG2: Ilf 29Tc 54
Abbey Rd. KT16: Chert 73K 149
Abbey Rd. NW10 39Ra 67
Abbey Rd. NW6 38Db 69
Abbey Rd. NW8 39Eb 69
Abbey Rd. SE2 49Zc 95
Abbey Rd. SW19 66Eb 133
Abbey Rd. TW17: Shep 74Q 150
Abbey Rd. Apts. NW8 2A 214 (40Eb 69) (off Abbey Rd.)
Abbey Road Station (DLR) 40Gc 73
Abbey Sports Cen. Barking 39Sc 74
Abbey St. E13 42Jc 93
Abbey St. SE1 3J 231 (48Ub 91)
Abbey St. SE16 48Wb 91
Abbey Ter. SE2 49Yc 95
Abbey Theatre St Albans 4A 6
Abbey Trad. Est. SE26 64Bc 136
Abbey Vw. AL1: St A 4B 6
Abbey Vw. EN9: Walt A 5Dc 20
Abbey Vw. NW7 20Va 30
Abbey Vw. WD25: Wat 8Z 13
Abbey Vw. WD7: R'lett 7Ha 14
Abbey View Golf Course 4A 6
Abbey Vw. Rd. AL3: St A 2A 6
Abbey Wlk. KT8: W Mole 69Da 129
Abbey Wharf Ind. Est. IG11: Bark 41Tc 94
ABBEY WOOD 49Yc 95
Abbey Wood SL5: S'dale 3E 146
Abbey Wood Cvn. Club Site SE2 49Yc 95
Abbey Wood La. RM13: Rain 40Md 77
Abbey Wood Rd. SE2 49Xc 95
Abbey Wood Station (Rail & Crossrail) 48Yc 95
Abbot Cl. HA4: Ruis 34Z 65
Abbot Cl. KT14: Byfl 82M 169
Abbot Cl. TW18: Staines 66M 127
Abbot Ct. SW8 52Nb 112 (off Hartington Rd.)
Abbot Ho. E14 45Dc 92 (off Smythe St.)
Abbots Av. AL1: St A 5C 6
Abbots Av. KT19: Eps 83Qa 173
Abbots Av. W. AL1: St A 5B 6
Abbotsbury NW1 38Mb 70 (off Camley St.)
Abbotsbury Cl. E15 40Ec 72
Abbotsbury Cl. W14 47Ab 88
Abbotsbury Ct. WD25: Wat 4X 13
Abbotsbury Gdns. HA5: Eastc 31Y 65
Abbotsbury Ho. W14 47Ab 88
Abbotsbury M. SE15 55Yb 114
Abbotsbury Rd. BR2: Hayes 75Hc 159
Abbotsbury Rd. SM4: Mord 71Db 155
Abbotsbury Rd. W14 47Ab 88
Abbots Bus. Pk. WD4: K Lan 10A 4
Abbots Cl. BR5: Farnb 74Sc 160
Abbots Cl. CM15: Shenf 18Ce 41
Abbots Cl. RM13: Rain 40Ld 77
Abbots Cl. RM3: Hrld W 25Pd 57 (off Queen's Pk. Rd.)
Abbots Ct. W8 47Db 89 (off Thackeray St.)
Abbots Dr. GU25: Vir W 1M 147
Abbots Dr. HA2: Harr 33Ca 65
Abbots Fld. DA12: Grav'nd 5E 144
Abbotsford Av. N15 28Sb 51
Abbotsford Cl. GU22: Wok 89C 168
Abbotsford Gdns. IG8: Wfd G 24Jc 53
Abbotsford Rd. IG3: Ilf 33Wc 75
Abbots Gdns. N2 28Fb 49
Abbots Grn. CR0: Addtn 79Zb 158
Abbotshade Rd. SE16 46Zb 92
Abbotshall Av. N14 20Lb 32
Abbotshall Rd. SE6 60Fc 115
Abbots Hill HP3: Hem H 7A 4
Abbot's Ho. W14 48Bb 89 (off St Mary Abbot's Ter.)
Abbots La. CR8: Kenley 88Sb 177
Abbots La. SE1 7J 225 (46Ub 91)
ABBOTS LANGLEY 3U 12
Abbotsleigh Cl. SM2: Sutt 80Db 155
Abbotsleigh Rd. SW16 63Lb 134
Abbots Mnr. SW1 6K 227 (49Kb 90) (not continuous)
Abbots Pk. AL1: St A 4E 6
Abbots Pk. SW2 60Qb 112
Abbot's Pl. NW6 39Db 69
Abbots Pl. WD6: Bore 9Ra 15
Abbots Ri. RH1: Redh 4A 208
Abbots Ri. WD4: K Lan 8P 3
Abbot's Rd. E6 39Mc 73
Abbots Rd. HA8: Edg 24Sa 47
Abbots Rd. WD5: Ab L 3S 12
Abbots Ter. N8 30Nb 50
Abbotstone Rd. SW15 55Ya 110
Abbot St. E8 37Vb 71
Abbots Vw. WD4: K Lan 9P 3
Abbots Wlk. SL4: Wind 4C 102
Abbots Wlk. W8 48Db 89
Abbots Way BR3: Beck 71Ac 158
Abbots Way KT16: Chert 73H 149
Abbotswell Rd. SE4 57Bc 114
Abbotswood KT13: Weyb 76V 150
Abbotswood Cl. DA17: Belv 48Ad 95
Abbotswood Dr. KT13: Weyb 82T 170
Abbotswood Gdns. IG5: Ilf 27Pc 54
Abbotswood Rd. SE22 56Ub 113
Abbotswood Rd. SW16 62Mb 134
Abbotswood Way UB3: Hayes 46X 85
Abbott Av. SW20 67Za 132
Abbott Cl. TW12: Hamp 65Aa 129
Abbott Cl. UB5: N'olt 37Ba 65
Abbott Rd. E14 43Ec 92 (not continuous)
Abbott Rd. TN15: Bor G 92Be 205
Abbott's Cl. UB8: Cowl 43M 83
Abbotts Cl. BR8: Swan 70Jd 140
Abbotts Cl. N1 37Sb 71
Abbotts Cl. RM7: Mawney 27Dd 56
Abbotts Cl. SE28 45Yc 95
Abbotts Cres. E4 21Fc 53
Abbotts Cres. EN2: Enf 12Rb 33
Abbotts Dr. EN9: Walt A 5Jc 21
Abbotts Dr. HA0: Wemb 33Ka 66
Abbotts Dr. SS17: Stan H 1M 101
Abbotts Hall Chase SS17: Stan H 1N 101
Abbotts Ho. SW1 7D 228 (50Mb 90) (off Aylesford St.)
Abbotts Mead TW10: Ham 63Ma 131
Abbottsmede Cl. TW1: Twick 61Ha 130
Abbotts Pk. Rd. E10 31Ec 72
Abbotts Rd. CR4: Mitc 70Lb 134
Abbotts Rd. EN5: New Bar 14Db 31
Abbotts Rd. SM3: Cheam 77Ab 154
Abbotts Rd. UB1: S'hall 46Aa 85
Abbott's Tilt KT12: Hers 76Aa 151
Abbott's Wlk. DA7: Bex 52Zc 117
Abbotts Wlk. CR3: Cat'm 94Xb 197
Abbotts Way SL1: Slou 6B 80
Abbott's Wharf E14 44Cc 92 (off Stainsby Pl.)
Abbotts Wharf Moorings E14 44Cc 92 (off Stainsby Rd.)
Abbs Cross RM12: Horn 32Ld 77
Abbs Cross Gdns. RM12: Horn 32Ld 77
Abbs Cross La. RM12: Horn 34Ld 77
Abchurch La. EC4 4G 225 (45Tb 91) (not continuous)
Abchurch Yd. EC4 4F 225 (45Tb 91)
Abdale La. AL9: N Mym 9D 8
Abdale Rd. W12 46Xa 88
Abelard Pl. W5 48Ma 87
Abel Cl. HP2: Hem H 2A 4
Abel Ho. SE11 51Qb 112 (off Kennington Rd.)
Abelia Cl. GU24: W End 5C 166
Abell Ct. KT15: Add 78K 149
Abenberg Way CM13: Hut 19De 41
Abenglen Ind. Est. UB3: Hayes 47T 84
Aberavon Rd. E3 41Ac 92
Abercairn Rd. SW16 66Lb 134
Aberconway Rd. SM4: Mord 70Db 133
Abercorn Cl. CR2: Sels 85Zb 178
Abercorn Cl. NW7 24Ab 48
Abercorn Cl. NW8 3A 214 (41Eb 89)
Abercorn Commercial Cen. HA0: Wemb 39Ma 67
Abercorn Cotts. NW8 3A 214 (41Eb 89) (off Abercorn Pl.)
Abercorn Cres. HA2: Harr 32Da 65
Abercorn Dell WD23: B Hea 19Ea 28
Abercorn Gdns. HA3: Kenton 31Ma 67
Abercorn Gdns. RM6: Chad H 30Xc 55
Abercorn Gro. HA4: Ruis 28T 44
Abercorn Mans. NW8 2A 214 (40Eb 69) (off Abercorn Pl.)
Abercorn M. TW10: Rich 57Pa 109
Abercorn Pl. NW8 3A 214 (41Eb 89)
Abercorn Rd. HA7: Stan 24La 46
Abercorn Rd. NW7 24Ab 48
Abercorn Wlk. NW8 3A 214 (41Eb 89)
Abercorn Way GU21: Wok 10L 167
Abercorn Way SE1 50Wb 91
Abercrombie Dr. EN1: Enf 11Wb 33
Abercrombie Rd. E20 36Dc 72
Abercrombie St. SW11 54Gb 111
Aberdale Ct. SE16 47Zb 92 (off Garter Way)
Aberdale Gdns. EN6: Pot B 4Bb 17
Aberdare Cl. BR4: W W'ck 75Ec 158
Aberdare Gdns. NW6 38Db 69
Aberdare Gdns. NW7 24Za 48
Aberdare Rd. EN3: Pond E 14Yb 34
Aberdeen Av. SL1: Slou 5E 80
Aberdeen Cotts. HA7: Stan 24La 46
Aberdeen Ct. W9 6B 214 (42Fb 89) (off Maida Vale)
Aberdeen La. N5 36Sb 71
Aberdeen Mans. WC1 5F 217 (42Nb 90) (off Kenton St.)
Aberdeen Pde. N18 22Xb 51 (off Aberdeen Rd.)
Aberdeen Pk. N5 36Sb 71
Aberdeen Pl. NW8 6B 214 (42Fb 89)
Aberdeen Rd. CR0: C'don 77Sb 157
Aberdeen Rd. HA3: W'stone 26Ha 46
Aberdeen Rd. N18 22Wb 51 (not continuous)
Aberdeen Rd. N5 35Sb 71
Aberdeen Rd. NW10 36Va 68
Aberdeen Sq. E14 46Bc 92
Aberdeen Ter. SE3 54Fc 115
Aberdeen Way GU21: Knap 1F 186
Aberdeen Wharf E1 46Xb 91 (off Wapping High St.)
Aberdour Rd. IG3: Ilf 34Xc 75
Aberdour St. SE1 5H 231 (49Ub 91)
Aberfeldy Ho. SE5 52Rb 113 (not continuous)
Aberfeldy St. E14 43Ec 92 (not continuous)
Aberford Gdns. SE18 53Nc 116
Aberford Rd. WD6: Bore 12Qa 29
Aberfoyle Rd. SW16 65Mb 134
Abergeldie Rd. SE12 58Kc 115
Abernethy Ho. EC1 1D 224 (43Sb 91) (off Bartholomew Cl.)
Abernethy Rd. SE13 56Gc 115
Abersham Rd. E8 36Vb 71
Abery St. SE18 49Uc 94
Abid M. SE15 54Wb 113
Abigail M. RM3: Hrld W 26Pd 57
Ability Pl. E14 47Dc 92
Ability Plaza E8 38Vb 71 (off Arbutus St.)
Ability Towers EC1 3D 218 (41Sb 91) (off Macclesfield Rd.)
Abingdon W14 49Bb 89 (off Kensington Village)
Abingdon Cl. GU21: Wok 10N 167
Abingdon Cl. KT4: Wor Pk 76Xa 154
Abingdon Cl. NW1 37Mb 70
Abingdon Cl. SE1 49Vb 91 (off Bushwood Dr.)
Abingdon Cl. SW19 65Eb 133
Abingdon Cl. UB10: Hil 39P 63
Abingdon Ct. EN8: Walt C 5Ac 20 (off High St.)
Abingdon Ct. GU22: Wok 90B 168
Abingdon Ct. W8 48Cb 89 (off Abingdon Vs.)
Abingdon Gdns. W8 48Cb 89
Abingdon Ho. BR1: Broml 66Kc 137
Abingdon Ho. E2 5K 219 (42Vb 91) (off Boundary St.)
Abingdon Lodge BR2: Broml 68Hc 137 (off Beckenham La.)
Abingdon Lodge W8 48Cb 89
Abingdon Mans. W8 48Cb 89 (off Pater St.)
Abingdon Pl. EN6: Pot B 4Db 17
Abingdon Rd. N3 26Eb 49
Abingdon Rd. SW16 68Nb 134
Abingdon Rd. W8 48Cb 89
Abingdon St. SW1 3F 229 (48Nb 90)
Abingdon Vs. W8 48Cb 89
Abingdon Way BR6: Chels 77Xc 161
Abinger Av. SM2: Cheam 81Ya 174
Abinger Cl. BR1: Broml 69Nc 138
Abinger Cl. CR0: New Ad 79Ec 158
Abinger Cl. IG11: Bark 35Wc 75
Abinger Cl. SM6: W'gton 78Nb 156
Abinger Ct. SM6: W'gton 78Nb 156
Abinger Ct. W5 45La 86
Abinger Dr. RH1: Redh 8N 207
Abinger Gdns. TW7: Isle 55Ga 108
Abinger Gro. SE8 51Bc 114
Abinger Ho. SE1 2F 231 (47Tb 91) (off Gt. Dover St.)
Abinger M. W9 42Cb 89
Abinger Rd. W4 48Ua 88
Abington Ct. RM14: Upm 32Sd 78
Ablett St. SE16 50Yb 92
Abney Gdns. N16 33Vb 71
Abney Pk. Cemetery Local Nature Reserve 33Ub 71
Abney Pk. Ter. N16 33Vb 71 (off Cazenove Rd.)
Aborfield NW5 36Lb 70
Aboyne Dr. SW20 68Wa 132
Aboyne Rd. NW10 34Ua 68
Aboyne Rd. SW17 62Fb 133
Abraham Cl. WD19: Wat 21X 45
Abraham Ct. RM14: Upm 33Qd 77
ABRAHAM COWLEY UNIT 76F 148
Abraham Fisher Ho. E12 36Qc 74
ABRIDGE 13Xc 37
Abridge Cl. EN8: Walt C 7Zb 20
Abridge Gdns. RM5: Col R 23Cd 56
Abridge Golf Course 9Bd 23
Abridge Pk. RM4: Abr 14Wc 37
Abridge Rd. CM16: Abr 9Vc 23
Abridge Rd. CM16: Lough 9Vc 23
Abridge Rd. CM16: They B 9Vc 23
Abridge Rd. IG7: Abr 16Tc 36
Abridge Rd. IG7: Chig 16Tc 36
Abridge Rd. RM4: Abr 11Wc 37
Abyssinia Cl. SW11 56Gb 111
Abyssinia Ct. N8 29Pb 50
Abyssinia Rd. SW11 56Gb 111
Acacia Av. GU22: Wok 2P 187
Acacia Av. HA4: Ruis 32W 64
Acacia Av. HA9: Wemb 36Na 67
Acacia Av. N17 24Tb 51
Acacia Av. RM12: Horn 33Hd 76
Acacia Av. TW17: Shep 71Q 150
Acacia Av. TW19: Wray 5A 104
Acacia Av. TW8: Bford 52Ka 108
Acacia Av. UB3: Hayes 44V 84
Acacia Av. UB7: Yiew 45P 83
Acacia Bus. Cen. E11 34Gc 73
Acacia Cl. BR5: Pet W 71Tc 160
Acacia Cl. HA7: Stan 23Ga 46
Acacia Cl. KT15: Wdhm 82H 169
Acacia Cl. SE20 68Wb 135
Acacia Cl. SE8 49Ac 92
Acacia Ct. DA11: Grav'nd 9C 122
Acacia Ct. EN9: Walt A 6Jc 21 (off Lamplighters Cl.)
Acacia Ct. HA1: Harr 29Da 45
Acacia Dr. KT15: Wdhm 81H 169
Acacia Dr. RM14: Upm 35Qd 77
Acacia Dr. SM3: Sutt 74Bb 155
Acacia Dr. SM7: Bans 86Za 174
Acacia Gdns. BR4: W W'ck 75Ec 158
Acacia Gdns. NW8 1C 214 (40Fb 69)
Acacia Gdns. RM14: Upm 31Vd 78
Acacia Gro. KT3: N Mald 69Ta 131
Acacia Gro. SE21 61Tb 135
Acacia Hall 59Nd 119
Acacia Ho. N22 25Qb 50 (off Douglas Rd.)
Acacia Ho. SL9: Chal P 25A 42
Acacia M. UB7: Harm 51M 105
Acacia Pl. NW8 1C 214 (40Fb 69)
Acacia Rd. BR3: Beck 69Bc 136
Acacia Rd. CR4: Mitc 69Jb 134
Acacia Rd. DA1: Dart 60Md 119
Acacia Rd. DA9: Ghithe 58Ud 120
Acacia Rd. E11 33Gc 73
Acacia Rd. E17 30Ac 52
Acacia Rd. EN2: Enf 11Tb 33
Acacia Rd. N22 25Qb 50
Acacia Rd. NW8 1C 214 (40Fb 69)
Acacia Rd. SW16 67Nb 134
Acacia Rd. TW12: Hamp 65Ca 129
Acacia Rd. TW18: Staines 64K 127
Acacia Rd. W3 45Sa 87
Acacias, The EN4: E Barn 15Fb 31
Acacia St. AL10: Hat 3C 8
Acacia Wlk. BR8: Swan 68Fd 140
Acacia Wlk. SW10 52Eb 111 (off Tadema Rd.)
Acacia Way DA15: Sidc 60Vc 117
Academia Way N17 23Ub 51
Academy, Middlesex County Cricket Club, The 26Db 49
Academy, The SW8 51Pb 112
Academy Apts. E8 36Xb 71 (off Dalston La.)
Academy Bldgs. N1 3H 219 (41Ub 91) (off Fanshaw St.)
Academy Ct. AL2: Lon C 8F 6
Academy Ct. DA5: Bexl 61Gd 140 (off Beaconsfield Rd.)
Academy Ct. E2 41Yb 92 (off Kirkwall Pl.)
Academy Ct. NW6 39Cb 69
Academy Ct. RM8: Dag 35Wc 75
Academy Ct. WD6: Bore 14Qa 29
Academy Flds. Cl. RM2: Rom 29Jd 56
Academy Flds. Rd. RM2: Rom 29Kd 57
Academy Gdns. CR0: C'don 74Vb 157
Academy Gdns. UB5: N'olt 41Aa 85
Academy Gdns. W8 47Cb 89
Academy Ho. E3 43Dc 92 (off Violet Rd.)
Academy Ho. WD6: Bore 14Qa 29 (off Academy Ct.)
Academy Pl. SE18 53Pc 116
Academy Pl. TW7: Isle 53Ga 108
Academy Rd. SE18 53Pc 116
Academy Way E17 25Cc 52
Academy Way RM8: Dag 35Wc 75
Acanthus Dr. SE1 50Wb 91
Acanthus Rd. SW11 55Jb 112
Access Bus. Pk. KT14: Byfl 83M 169
Accommodation La. UB7: Harm 51L 105
Accommodation La. UB7: Lford 53J 105
Accommodation Rd. E4 20Fc 35 (off Ashwood Rd.)
Accommodation Rd. KT16: Longc 6P 147
Accommodation Rd. KT17: Ewe 78Wa 154
Accommodation Rd. NW11 32Bb 69
AC Ct. KT7: T Ditt 72Ja 152
Accrington Ho. RM3: Rom 22Md 57 (off Montgomery Cres.)
Ace Pde. KT9: Chess 76Na 153
Acer Av. RM13: Rain 41Md 97
Acer Av. UB4: Yead 43Aa 85
Acer Cl. IG8: Wfd G 23Nc 54
Acer Cl. KT19: Eps 81Sa 173
Acer Cl. SM7: Bans 87Za 174
Acer Ct. EN3: Enf H 13Ac 34 (off Enstone Rd.)
Acer Dr. GU24: W End 5D 166
Acer Gro. GU22: Wok 92A 188
Acer Hgts. CR8: Purl 83Sb 177
Acer Rd. E8 38Vb 71
Acer Rd. TN16: Big H 88Mc 179
Acers AL2: Park 10A 6
Acers BR7: Chst 66Nc 138
Aces Ct. TW3: Houn 54Ea 108
Ace Way SW11 51Mb 112
Acfold Rd. SW6 53Db 111
Achieve POWER 77J 149
Achieve ZONE Addlestone 77L 149
Achilles Cl. HP2: Hem H 1P 3
Achilles Cl. SE1 50Wb 91
Achilles Ho. E2 40Xb 71 (off Old Bethnal Grn. Rd.)
Achilles Pl. GU21: Wok 9N 167
Achilles Rd. NW6 36Cb 69
Achilles Statue 7J 221 (46Jb 90)
Achilles St. SE14 52Ac 114
Achilles Way W1 7J 221 (46Jb 90)
Ackers Dr. DA10: Swans 60Be 121
Acklam Rd. W10 43Ab 88 (not continuous)
Acklington Dr. NW9 25Ua 48
Ackmar Rd. SW6 53Cb 111
Ackroyd Dr. E3 43Bc 92
Ackroyd Rd. SE23 59Zb 114
Acland Cl. SE18 52Tc 116
Acland Cres. SE5 55Tb 113
Acland Ho. SW9 53Pb 112
Acland Rd. NW2 37Xa 68
Acle Cl. IG6: Ilf 24Rc 54
Acme Rd. WD24: Wat 10W 12
Acme Studios E14: Gillender St. 43Ec 92
Acme Studios E14: Leven Rd. 43Ec 92 (off Leven Rd.)
Acock Gro. UB5: N'olt 35Da 65
Acol Ct. NW6 38Cb 69
Acol Cres. HA4: Ruis 36X 65
Acol Rd. NW6 38Cb 69
Aconbury Rd. RM9: Dag 39Xc 75
Acorn, Cen., The IG6: Ilf 23Xc 55
Acorn Cl. BR7: Chst 64Sc 138
Acorn Cl. E4 22Dc 52
Acorn Cl. EN2: Enf 11Rb 33
Acorn Cl. HA7: Stan 24Ka 46
Acorn Cl. RM1: Rom 26Gd 56
Acorn Cl. SL3: L'ly 50D 82
Acorn Cl. SM7: Bans 87Ab 174
Acorn Cl. TW12: Hamp 65Da 129
Acorn Ct. AL2: Lon C 9F 6
Acorn Ct. E3 40Cc 72 (off Morville St.)
Acorn Ct. E6 38Nc 74
Acorn Ct. EN8: Walt C 5Zb 20
Acorn Ct. IG2: Ilf 30Uc 54
Acorn Gdns. RH8: Oxt 99Fc 199
Acorn Gdns. SE19 67Vb 135
Acorn Gdns. W3 43Ta 87
Acorn Gro. GU22: Wok 93A 188
Acorn Gro. HA4: Ruis 35V 64
Acorn Gro. KT20: Kgswd 96Bb 195
Acorn Gro. UB3: Harl 52V 106
Acorn Ho. BR8: Swan 68Hd 140 (off Squirrels Cl.)
Acorn Ind. Pk. DA1: Cray 57Jd 118
Acorn La. EN6: Cuff 1Nb 18
Acorn Pde. SE15 52Xb 113
Acorn Pl. WD24: Wat 9W 12
Acorn Production Cen. N7 38Nb 70
Acorn Rd. DA1: Cray 57Hd 118
Acorn Rd. HP3: Hem H 3A 4
Acorns, The AL4: St A 2H 7
Acorns, The IG7: Chig 21Uc 54
Acorns, The TN13: S'oaks 95Jd 202
Acorns Way KT10: Esh 78Ea 152
Acorn Trad. Est. RM20: Grays 51Zd 121
Acorn Wlk. SE16 46Ac 92
Acorn Way BR3: Beck 71Ec 158
Acorn Way BR6: Farnb 77Rc 160
Acorn Way SE23 62Zb 136
Acqua Ho. TW9: Kew 52Ra 109
Acre Dr. SE22 56Wb 113
Acrefield Ho. NW4 28Za 48 (off Belle Vue Est.)
Acrefield Rd. SL9: Chal P 27A 42
Acre La. SM5: Cars 77Jb 156
Acre La. SM6: W'gton 77Kb 156
Acre La. SW2 56Nb 112
Acre Pas. SL4: Wind 3H 103
Acre Path UB5: N'olt 37Aa 65 (off Arnold Rd.)
Acre Rd. KT2: King T 67Na 131
Acre Rd. RM10: Dag 38Dd 76
Acre Rd. SW19 65Fb 133
Acres, The SS17: Stan H 1P 101
Acres Gdns. KT20: Tad 91Za 194
Acre Vw. RM11: Horn 28Nd 57
Acre Way HA6: Nwood 25V 44
Acrewood HP2: Hem H 3N 3
Acrewood Pk. AL4: St A 2K 7
Acrewood Way AL4: St A 2K 7
Acris St. SW18 57Eb 111
Acropolis Ho. KT1: King T 69Pa 131 (off Winery La.)
Actaeon M. SE18 51Rc 116
ACTON 46Sa 87
Acton Apts. N1 39Tb 71 (off Branch Pl.)
Acton Central Ind. Est. W3 46Ra 87
Acton Central Station (Overground) 46Ta 87
Acton Cl. EN8: Chesh 3Ac 20
Acton Cl. N9 19Wb 33
ACTON GREEN 48Sa 87
Acton Hill M. W3 46Ra 87
Acton Ho. E8 39Vb 71 (off Lee St.)
Acton Ho. W3 44Sa 87
Acton La. NW10 41Sa 87
Acton La. W3 47Sa 87
Acton La. W4 49Sa 87 (not continuous)
Acton Main Line Station (Rail & Crossrail) 44Sa 87
Acton M. E8 39Ub 71
Acton Pk. Est. W3 47Ta 87
Acton St. WC1 4H 217 (41Pb 90)
Acton Swimming Baths 46Sa 87 (off Salisbury St.)
Acton Town Station (Underground) 47Qa 87
Acton Va. Ind. Pk. W3 47Va 88
Acton Wlk. N20 18Eb 31
Acuba Rd. SW18 61Db 133
Acworth Cl. N9 17Yb 34
Acworth Ho. SE18 52Rc 116 (off Barnfield Rd.)
Acworth Pl. DA1: Dart 58Ld 119
Ada Cl. N11 20Hb 31
Ada Ct. N1 39Sb 71 (off Packington St.)
Ada Ct. W9 4A 214 (41Eb 89)
Ada Gdns. E14 44Fc 93

Ada Gdns. E15 39Hc 73
Adagio Point SE8 51Dc 114
(off Copperas St.)
Ada Ho. E2 39Wb 71
(off Ada Pl.)
Adair Cl. SE25 69Xb 135
Adair Gdns. CR3: Cat'm 93Sb 197
Adair Ho. SW3 51Gb 111
(off Oakley St.)
Adair Rd. W10 42Ab 88
Adair Twr. W10 42Ab 88
(off Appleford Rd.)
Adair Wlk. GU24: Brkwd 3A 186
Ada Kennedy Ct. SE10 52Ec 114
(off Greenwich Sth. St.)
Ada Lewis Ho. HA9: Wemb 35Pa 67
Adam & Eve Ct. W1 2C 222 (44Lb 90)
(off Oxford St.)
Adam & Eve M. W8 48Cb 89
Ada Maria Ct. E1 44Xb 91
(off James Voller Way)
Adam Cl. NW7 23Za 48
Adam Cl. SE6 63Bc 136
Adam Cl. SL1: Slou 6E 80
Adam Ct. DA2: Wilm 63Ld 141
Adam Ct. SE11 6B 230 (49Rb 91)
(off Opal St.)
Adam Ct. SW7 5A 226 (49Eb 89)
(off Gloucester Rd.)
Adamfields NW3 38Fb 69
(off Adamson Rd.)
Adam Rd. E4 23Bc 52
Adams Bri. Bus. Cen. HA9: Wemb 36Ra 67
Adams Cl. KT5: Surb 72Pa 153
Adams Cl. N3 24Cb 49
Adams Cl. NW9 33Ra 67
Adams Cl. RM5: Col R 25Ed 56
Adams Ct. E17 30Ac 52
Adams Ct. EC2 2H 225 (44Ub 91)
Adams Ct. WD25: Wat 8Z 13
Adams Cft. GU24: Brkwd 2A 186
Adams Gdns. Est. SE16 47Yb 92
Adams Ho. E14 44Fc 93
(off Aberfeldy St.)
Adamsley Rd. SS17: Stan H 3K 101
Adams M. N22 24Pb 50
Adams M. SW17 61Hb 133
Adamson Ct. N2 27Gb 49
Adamson Rd. E16 44Jc 93
Adamson Rd. NW3 38Fb 69
Adamson Way BR3: Beck 71Ec 158
Adams Pl. E14 46Dc 92
(off The Nth. Colonnade)
Adams Pl. N7 36Pb 70
Adams Quarter TW8: Bford 52La 108
Adamsrill Cl. EN1: Enf 16Tb 33
Adamsrill Rd. SE26 63Zb 136
Adams Rd. BR3: Beck 71Ac 158
Adams Rd. N17 26Tb 51
Adams Rd. SS17: Stan H 2N 101
Adam's Row W1 5J 221 (45Jb 90)
Adams Sq. DA6: Bex 55Ad 117
Adams Ter. E3 41Cc 92
(off Rainhill Way)
Adam St. WC2 5G 223 (45Nb 90)
Adams Wlk. KT1: King T 68Na 131
Adams Way CR0: C'don 72Vb 157
Adams Way SE25 71Xb 157
Adam Wlk. SW6 52Ya 110
Adana SE13 54Ec 114
Ada Pl. E2 39Wb 71
Adare Wlk. SW16 62Pb 134
Ada Rd. HA0: Wemb 34Ma 67
Ada Rd. SE5 52Ub 113
Adastral Ho. WC1 7H 217 (43Pb 90)
(off Harpur St.)
Adastra Way SM6: W'gton 79Nb 156
Ada St. E8 39Xb 71
Adcock Wlk. BR6: Orp 77Vc 161
Adderley Gdns. SE9 63Qc 138
Adderley Gro. SW11 57Jb 112
Adderley Rd. HA3: W'stone 25Ha 46
Adderley St. E14 44Ec 92
Addey Ho. SE8 52Bc 114
ADDINGTON 78Cc 158
Addington Bus. Cen. CR0: New Ad 82Gc 179
Addington Bus Station 79Cc 158
Addington Cl. SL4: Wind 5E 102
Addington Cl. SS17: Stan H 3K 101
Addington Cl. UB2: S'hall 46Fa 86
Addington Ct. SW14 55Ta 109
Addington Court Falconwood Course & Driving Range 81Cc 178
Addington Court Golf Course 81Cc 178
Addington Dr. N12 23Fb 49
Addington Golf Course, The 77Bc 158
Addington Gro. SE26 63Ac 136
Addington Hgts. CR0: New Ad 83Ec 178
Addington Hill Viewpoint 78Yb 158
Addington Ho. SW9 54Pb 112
(off Stockwell Rd.)
Addington Lofts SE5 52Sb 113
(off Bethwin Rd.)
Addington Palace Golf Course 79Ac 158
Addington Rd. BR4: W W'ck 77Ec 158
Addington Rd. CR0: C'don 74Qb 156
Addington Rd. CR2: Sande 83Wb 177
Addington Rd. CR2: Sels 83Wb 177
Addington Rd. E16 42Gc 93
Addington Rd. E3 41Cc 92
Addington Rd. N4 30Qb 50
Addington Sq. SE5 51Tb 113
(not continuous)
Addington St. SE1 2J 229 (47Pb 90)
Addington Village Rd. CR0: Addtn 79Bc 158
(not continuous)
Addington Village Stop (London Tramlink) 79Cc 158
Addis Cl. EN3: Enf H 11Zb 34
ADDISCOMBE 74Wb 157
Addiscombe Av. CR0: C'don 74Wb 157
Addiscombe Cl. HA3: Kenton 29La 46
Addiscombe Ct. Rd. CR0: C'don 74Ub 157
Addiscombe Gro. CR0: C'don 75Ub 157
Addiscombe Rd. CR0: C'don 75Tb 157
(not continuous)
Addiscombe Rd. WD18: Wat 14X 27
Addiscombe Stop (London Tramlink) 74Wb 157
Addis Ho. E1 43Yb 92
(off Lindley St.)
Addisland Ct. W14 47Ab 88
(off Holland Vs. Rd.)
Addison Av. N14 16Kb 32
Addison Av. TW3: Houn 53Ea 108
Addison Av. W11 46Ab 88
Addison Bri. Pl. W14 49Bb 89
Addison Cl. BR5: Pet W 72Sc 160
Addison Cl. CR3: Cat'm 94Tb 197
Addison Cl. HA6: Nwood 25W 44
Addison Cl. SL0: Iver 45G 82
Addison Ct. CM16: Epp 3Wc 23
Addison Ct. NW6 39Cb 69
(off Brondesbury Rd.)
Addison Cres. W14 48Ab 88
(not continuous)
Addison Dr. SE12 57Kc 115
Addison Gdns. KT5: Surb 70Pa 131
Addison Gdns. RM17: Grays 49Ee 99
Addison Gdns. W14 48Za 88
Addison Gro. W4 48Ua 88
Addison Ho. NW8 3B 214 (41Fb 89)
(off Grove End Rd.)
Addison Pk. Mans. W14 48Za 88
(off Richmond Way)
Addison Pl. SE25 70Wb 135
Addison Pl. UB1: S'hall 45Ca 85
Addison Pl. W11 46Ab 88
Addison Rd. BR2: Broml 71Lc 159
Addison Rd. CR3: Cat'm 93Tb 197
Addison Rd. E11 30Jc 53
Addison Rd. E17 29Dc 52
Addison Rd. EN3: Enf H 11Yb 34
Addison Rd. GU21: Wok 89B 168
Addison Rd. IG6: Ilf 25Sc 54
Addison Rd. SE25 70Wb 135
Addison Rd. TW11: Tedd 65Ka 130
Addison Rd. W14 47Bb 88
Addisons Cl. CR0: C'don 75Bc 158
Addison Ter. W4 49Sa 87
(off Chiswick Rd.)
Addison Way HA6: Nwood 25V 44
Addison Way NW11 28Bb 49
Addison Way UB3: Hayes 44W 84
Addle Hill EC4 3C 224 (44Rb 91)
ADDLESTONE 77L 149
Addlestone Ho. KT15: Add 76K 149
Addlestone Ho. W10 43Ya 88
(off Sutton Way)
Addlestone Moor KT15: Add 75L 149
ADDLESTONE MOOR 75K 149
Addlestone ONE Shop. Cen. 77L 149
(off Market St.)
Addlestone Pk. KT15: Add 78K 149
Addlestone Rd. KT13: Weyb 77P 149
Addlestone Rd. KT15: Add 77N 149
Addlestone Station (Rail) 77M 149
Addle St. EC2 2E 224 (44Sb 91)
Addy Ho. SE16 49Yb 92
Adecroft Way KT8: W Mole 69Ea 130
Adela Av. KT3: N Mald 71Xa 154
Adela Ho. W6 50Ya 88
(off Queen Caroline St.)
Adelaide Av. SE4 56Bc 114
Adelaide Cl. EN1: Enf 10Ub 19
Adelaide Cl. HA7: Stan 21Ja 46
Adelaide Cl. SL1: Slou 7E 80
Adelaide Cl. SW9 56Qb 112
Adelaide Community Gdn. 38Hb 69
(off Adelaide Rd.)
Adelaide Ct. BR3: Beck 66Cc 136
Adelaide Ct. E9 36Ac 72
(off Kenworthy Rd.)
Adelaide Ct. NW8 2A 214 (40Eb 69)
(off Abbey Rd.)
Adelaide Ct. W7 47Ha 86
Adelaide Gdns. RM6: Chad H 29Ad 55
Adelaide Gro. W12 46Wa 88
Adelaide Ho. E15 40Hc 73
Adelaide Ho. E17 26Bc 52
Adelaide Ho. SE5 54Ub 113
Adelaide Ho. W11 44Bb 89
(off Portobello Rd.)
Adelaide Pl. KT13: Weyb 77T 150
Adelaide Rd. BR7: Chst 64Rc 138
Adelaide Rd. E10 34Dc 72
Adelaide Rd. IG1: Ilf 33Rc 74
Adelaide Rd. KT12: Walt T 76W 150
Adelaide Rd. KT6: Surb 71Na 153
Adelaide Rd. NW3 38Fb 69
Adelaide Rd. RM18: Tilb 3B 122
Adelaide Rd. SL4: Wind 3K 103
Adelaide Rd. SW18 57Cb 111
Adelaide Rd. TW11: Tedd 65Ha 130
Adelaide Rd. TW15: Ashf 64M 127
Adelaide Rd. TW5: Hest 53Aa 107
Adelaide Rd. TW9: Rich 56Pa 109
Adelaide Rd. UB2: S'hall 49Aa 85
Adelaide Rd. W13 46Ja 86
Adelaide Sq. SL4: Wind 4H 103
Adelaide St. AL3: St A 1B 6
Adelaide St. WC2 5F 223 (45Nb 90)
(not continuous)
Adelaide Ter. TW8: Bford 50Ma 87
Adela St. W10 42Ab 88
Adelina Gro. E1 43Yb 92
Adelina M. SW12 60Mb 112
Adelina Yd. E1 43Yb 92
(off Adelina Gro.)
Adeline Pl. WC1 1E 222 (43Mb 90)
Adeliza Cl. IG11: Bark 38Sc 74
Adelphi Ct. E8 38Vb 71
(off Celandine Dr.)
Adelphi Ct. SE16 47Zb 92
(off Garter Way)
Adelphi Ct. W4 51Ta 109
Adelphi Cres. RM12: Horn 33Jd 76
Adelphi Cres. UB4: Hayes 41U 84
Adelphi Gdns. SL1: Slou 7J 81
Adelphi Rd. KT17: Eps 85Ta 173
Adelphi Ter. WC2 5G 223 (45Nb 90)
Adelphi Theatre 5G 223 (45Nb 90)
(off Strand)
Adelphi Way UB4: Hayes 41V 84
Adeney Cl. W6 51Za 110
Aden Gro. N16 35Tb 71
Adenmore Rd. SE6 59Cc 114
Aden Rd. EN3: Brim 14Ac 34
Aden Rd. IG1: Ilf 31Sc 74
Aden Ter. N16 35Tb 71
ADEYFIELD 2A 4
Adeyfield Gdns. HP2: Hem H 1P 3
Adeyfield Ho. EC1 4G 219 (41Tb 91)
(off Cranwood St.)
Adeyfield Rd. HP2: Hem H 2N 3
Adhara Rd. HA6: Nwood 22W 44
Adie Rd. W6 48Ya 88
Adine Rd. E13 42Kc 93
Adler Ind. Est. UB3: Hayes 47T 84
Adler St. E1 44Wb 91
Adley St. E5 36Ac 72
Adlington Cl. N18 22Tb 51
Admark Ho. KT18: Eps 87Ra 173
Admaston Rd. SE18 52Sc 116
Admiral Cl. BR5: St P 70Zc 139
Admiral Cl. KT13: Weyb 75U 150
Admiral Ct. IG11: Bark 40Xc 75
Admiral Ct. SE5 53Ub 113
(off Havil St.)
Admiral Ct. SM5: Cars 74Gb 155
Admiral Ct. SW10 53Eb 111
(off Admiral Sq.)
Admiral Ct. W1 1H 221 (43Jb 90)
(off Blandford St.)
Admiral Hood Ho. SL9: Chal P 21A 42
Admiral Ho. SW1 5C 228 (49Lb 90)
(off Willow Pl.)
Admiral Ho. TW11: Tedd 63Ja 130
Admiral Hyson Ind. Est. SE16 50Xb 91
Admiral M. SW19 66Eb 133
Admiral M. W10 42Za 88
Admiral Pl. N8 28Rb 51
Admiral Pl. SE16 46Ac 92
Admirals Cl. AL4: Col H 5A 8
Admirals Cl. E18 28Kc 53
Admirals Ct. E6 44Rc 94
(off Trader Rd.)
Admirals Ct. HA6: Nwood 22V 44
Admirals Ct. SE1 7K 225 (46Vb 91)
(off Horselydown La.)
Admiral Seymour Rd. SE9 56Pc 116
Admiral's Ga. SE10 53Dc 114
Admirals Lodge RM1: Rom 28Hd 56
Admiral Sq. SW10 53Eb 111
Admirals Rd. KT22: Fet 98Fa 192
Admirals Rd. KT23: Bookh 100Ea 192
Admiral Stirling Ct. KT13: Weyb 77P 149
Admiral's Twr. SE10 51Dc 114
(off Dowells St.)
Admiral St. SE8 54Cc 114
Admirals Wlk. AL1: St A 4E 6
Admirals Wlk. CR5: Coul 92Pb 196
Admirals Wlk. DA9: Ghithe 57Xd 120
Admirals Wlk. NW3 34Eb 69
Admirals Way DA12: Grav'nd 8F 122
Admirals Way E14 47Cc 92
Admiralty & Commercial Court 2K 223 (44Qb 90)
Admiralty Arch London 6E 222 (46Mb 90)
Admiralty Av. E16 47Kc 93
Admiralty Bldg. KT2: King T 67Ma 131
(off Down Hall Rd.)
Admiralty Cl. SE8 52Cc 114
Admiralty Cl. UB7: W Dray 47N 83
Admiralty Ho. E1 45Wb 91
(off Vaughan Way)
Admiralty Rd. TW11: Tedd 65Ha 130
Admiralty Way TW11: Tedd 65Ha 130
Admiral Wlk. W9 43Cb 89
Adnams Wlk. RM13: Rain 37Jd 76
Adolf St. SE6 63Dc 136
Adolphus Rd. N4 33Rb 71
Adolphus St. SE8 52Bc 114
Adomar Rd. RM8: Dag 34Ad 75
Adpar St. W2 7B 214 (43Fb 89)
Adrian Av. NW2 32Xa 68
Adrian Boult Ho. E2 41Xb 91
(off Mansford St.)
Adrian Cl. EN5: Barn 16Za 30
Adrian Cl. HP1: Hem H 3K 3
Adrian Cl. UB9: Hare 25M 43
Adrian Ho. E15 38Fc 73
(off Jupp Rd.)
Adrian Ho. N1 1J 217 (39Pb 70)
(off Barnsbury Est.)
Adrian Ho. SW8 52Nb 112
(off Wyvil Rd.)
Adrian M. SW10 51Db 111
Adrian Rd. WD5: Ab L 3U 12
Adrians Wlk. SL2: Slou 6K 81
Adriatic Apts. E16 45Jc 93
(off Western Gateway)
Adriatic Bldg. E14 45Ac 92
(off Horseferry Rd.)
Adriatic Ho. E1 42Zb 92
(off Ernest St.)
Adrienne Av. UB1: S'hall 42Ba 85
Adrienne Bus. Cen. UB1: S'hall 41Ba 85
Adron Ho. SE16 49Yb 92
(off Millender Wlk.)
Adstock Ho. N1 38Rb 71
(off The Sutton Est.)
Adstock M. SL9: Chal P 25A 42
Adstock Way RM17: Grays 49Ce 99
Advance Rd. SE27 63Sb 135
Adventure Kingdom 68Kc 137
(off Stockwell Cl.)
Adventurers Ct. E14 45Fc 93
(off Newport Av.)
Advent Way N18 22Yb 52
Advice Av. RM16: Grays 47Ce 99
Adys Lawn NW2 37Xa 68
Ady's Rd. SE15 55Vb 113
Aegean Apts. E16 45Jc 93
(off Western Gateway)
Aegon Ho. E14 48Dc 92
(off Lanark Sq.)
Aerodrome Rd. NW9 26Va 48
Aerodrome Way TW5: Hest 51Y 107
Aerodrome Way WD25: Wat 6V 12
Aeroville NW9 26Ua 48
AFC Hornchurch 33Od 77
AFC Wimbledon 69Qa 131
AFC Wimbledon Plough La. Stadium 63Eb 133
Affleck St. N1 2J 217 (40Pb 70)
Afghan Rd. SW11 54Gb 111
Afsil Ho. EC1 1A 224 (43Qb 90)
(off Viaduct Bldgs.)
Aftab Ter. E1 42Xb 91
(off Tent St.)
Afton Dr. RM15: S Ock 44Xd 98
Agamemnon Rd. NW6 35Bb 69
Agar Cl. KT6: Surb 75Pa 153
Agar Gro. NW1 38Lb 70
Agar Gro. Est. NW1 38Mb 70
Agar Ho. KT1: King T 69Na 131
(off Denmark Rd.)
Agar Pl. NW1 38Lb 70
Agars Pl. SL3: Dat 1L 103
Agar St. WC2 5F 223 (45Nb 90)
Agate Cl. E16 44Mc 93
Agate Cl. NW10 41Qa 87
Agate Rd. W6 48Ya 88
Agates La. KT21: Asht 90Ma 173
Agatha Cl. E1 46Xb 91
Agaton Path SE9 61Sc 138
Agaton Rd. SE9 61Sc 138
Agave Rd. NW2 35Ya 68
Agdon St. EC1 5B 218 (42Rb 91)
Age Exchange 55Hc 115
(off Blackheath Village)
Ager Av. RM8: Dag 32Zc 75
Agincourt SL5: Asc 9A 124
Agincourt Rd. NW3 35Hb 69
Agister Rd. IG7: Chig 22Wc 55
Agnes Av. IG1: Ilf 35Qc 74
Agnes Cl. E6 45Qc 94
Agnesfield Cl. N12 23Gb 49
Agnes Gdns. RM8: Dag 35Zc 75
Agnes George Wlk. E16 46Mc 93
Agnes Ho. W11 45Za 88
(off St Ann's Rd.)
Agnes Rd. W3 46Va 88
Agnes Scott Ct. KT13: Weyb 76R 150
(off Palace Dr.)
Agnes St. E14 44Bc 92
Agnew Rd. SE23 59Zb 114
Agricola Ct. E3 39Bc 72
(off Parnell Rd.)
Agricola Pl. EN1: Enf 15Vb 33
Agua Ho. KT16: Chert 73L 149
Ahoy Cen., The 50Cc 92
(off Stretton Mans.)
Aida M. RM13: Rain 42Ld 97
Aidan Cl. RM8: Dag 35Ad 75
Aigburth Mans. SW9 52Qb 112
(off Mowll St.)
Ailantus Ct. HA8: Edg 22Pa 47
Aileen Wlk. E15 38Hc 73
Ailsa Av. TW1: Twick 57Ja 108
Ailsa Ho. E16 45Qc 94
(off University Way)
Ailsa Rd. TW1: Twick 57Ka 108
Ailsa St. E14 43Ec 92
AIMES GREEN 1Hc 21
Ainger M. NW3 38Hb 69
(off Ainger Rd.)
Ainger Rd. NW3 38Hb 69
Ainsdale NW1 2B 216 (40Lb 70)
(off Harrington St.)
Ainsdale Cl. BR6: Orp 74Tc 160
Ainsdale Cres. HA5: Pinn 27Ca 45
Ainsdale Dr. SE1 50Wb 91
Ainsdale Rd. W5 42Ma 87
Ainsdale Rd. WD19: Wat 20Y 27
Ainsdale Way GU21: Wok 10L 167
Ainsley Av. RM7: Rom 30Dd 56
Ainsley Cl. N9 18Ub 33
Ainsley St. E2 41Xb 91
Ainslie Ct. HA0: Wemb 40Na 67
Ainslie Wlk. SW12 59Kb 112
Ainslie Wood Cres. E4 22Dc 52
Ainslie Wood Gdns. E4 21Dc 52
Ainslie Wood Local Nature Reserve 22Dc 52
Ainslie Wood Rd. E4 22Cc 52
Ainsty Est. SE16 47Zb 92
Ainsty St. SE16 47Yb 92
Ainsworth Cl. N20 19Fb 31
Ainsworth Cl. NW2 34Wa 68
Ainsworth Cl. SE15 54Ub 113
Ainsworth Ct. NW10 41Xa 88
(off Plough Cl.)
Ainsworth Ho. NW8 39Db 69
(off Ainsworth Way)
Ainsworth Ho. W10 41Ab 88
(off Kilburn La.)
Ainsworth Rd. CR0: C'don 74Rb 157
Ainsworth Rd. E9 38Yb 72
Ainsworth Way NW8 39Eb 69
Aintree Av. E6 39Nc 74
Aintree Cl. DA12: Grav'nd 2D 144
Aintree Cl. SL3: Poyle 53G 104
Aintree Cl. UB8: Hil 44R 84
Aintree Cres. IG6: Ilf 26Sc 54
Aintree Est. SW6 52Ab 110
Aintree Gro. RM14: Upm 34Pd 77
Aintree Rd. UB6: G'frd 40Ka 66
Aintree St. SW6 52Ab 110
Airbourne Ho. SM6: W'gton 77Lb 156
(off Maldon Rd.)
Air Call Bus. Cen. NW9 27Ta 47
Airco Cl. NW9 27Ta 47
Aird Ho. SE1 4D 230 (48Sb 91)
(off Rockingham St.)
Aird Point E16 45Sc 94
(off Lock Side Way)
Airdrie Cl. N1 38Pb 70
Airdrie Cl. UB4: Yead 43Aa 85
Airedale Av. W4 49Va 88
Airedale Av. Sth. W4 50Va 88
Airedale Cl. DA2: Dart 60Sd 120
Airedale Rd. SW12 59Hb 111
Airedale Rd. W5 48La 86
Aire Dr. RM15: S Ock 42Xd 98
Airey Neave Ct. RM17: Grays 47Ce 99
Airfield Pathway RM12: Horn 38Ld 77
Airfield Way RM12: Horn 37Kd 77
Air Forces Memorial 3P 125
Airlie Gdns. IG1: Ilf 32Rc 74
Airlie Gdns. W8 46Cb 89
Airlinks Golf Course 50Y 85
Air Links Ind. Est. TW13: Hanw 62Aa 129
Airlinks Ind. Est. TW5: Cran 50Y 85
Air Pk. Way TW13: Felt 61X 129
Airport Bowl 53U 106
Airport Ga. Bus. Cen. UB7: Sip 52P 105
Airport Ind. Est. TN16: Big H 86Mc 179
Airport Way TW19: Stanw M 56H 105
Air Sea M. TW2: Twick 61Fa 130
Air St. W1 5C 222 (45Lb 90)
Airthrie Rd. IG3: Ilf 33Xc 75
Aisgill Av. W14 50Bb 89
(not continuous)
Aisher Rd. SE28 45Yc 95
Aisher Way TN13: Riv 93Gd 202
Aislibie Rd. SE12 56Gc 115
Aissele Pl. KT10: Esh 77Da 151
Aiten Pl. W6 49Wa 88
Aithan Ho. E14 44Bc 92
(off Copenhagen Pl.)
Aitken Cl. CR4: Mitc 73Hb 155
Aitken Cl. E8 39Wb 71
Aitken Cl. HA4: Eastc 30W 44
Aitken Rd. EN5: Barn 15Ya 30
Aitken Rd. SE6 61Dc 136
Aitman Dr. TW8: Bford 50Qa 87
Aitons Ho. TW8: Bford 50Na 87
Aits Vw. KT8: W Mole 69Da 129
Ajax Av. NW9 27Ua 48
Ajax Av. SL1: Slou 5F 80
Ajax Ho. E2 40Xb 71
(off Old Bethnal Grn. Rd.)
Ajax Rd. NW6 35Bb 69
Akabusi Cl. CR0: C'don 72Wb 157
Akbar Ho. E14 49Dc 92
(off Cahir St.)
Akehurst La. TN13: S'oaks 97Ld 203
Akehurst St. SW15 58Wa 110
Akeman Cl. AL3: St A 4M 5
Akenside Rd. NW3 36Fb 69
Akerman Cl. SL3: L'ly 9N 81
Akerman Rd. KT6: Surb 72La 152
Akerman Rd. SW9 54Rb 113
Akers Ct. EN8: Walt C 4Ac 20
Akers La. WD3: Chor 16F 24
Akintaro Ho. SE8 51Bc 114
(off Alverton St.)
Alabama St. SE18 52Tc 116
Alacia Ct. W3 48Sa 87
(off Bassington Rd.)
Alacross Rd. W5 47La 86
Alamaro Lodge SE10 48Hc 93
(off Teal St.)
Alameda Pl. E3 40Dc 72
Alamein Gdns. DA2: Dart 59Td 120
Alamein Rd. DA10: Swans 58Zd 121
Alana Hgts. E4 17Dc 34
Alanbrooke DA12: Grav'nd 9E 122
Alanbrooke Cl. GU21: Knap 10G 166
Alan Cl. DA1: Dart 56Ld 119
Alan Coren Cl. NW2 35Ya 68
Alandale Dr. HA5: Pinn 25X 45
Aland Ct. SE16 48Ac 92
Alander M. E17 28Ec 52
Alan Dr. EN5: Barn 16Ab 30
Alan Gdns. RM7: Rush G 31Cd 76
Alan Hilton Ct. KT16: Ott 79F 148
(off Cheshire Cl.)
Alan Hocken Way E15 40Gc 73
Alan Preece Ct. NW6 38Za 68
Alan Rd. SW19 64Ab 132
Alanthus Cl. SE12 58Jc 115
Alan Way SL3: Geor G 44A 82
Alaska Apts. E16 45Jc 93
(off Western Gateway)
Alaska Bldg. SE13 53Dc 114
(off Deal's Gateway)
Alaska Bldgs. SE1 4K 231 (48Vb 91)
Alaska St. SE1 7K 223 (46Qb 90)
Alastor Ho. E14 48Ec 92
(off Strattondale Ho.)
Alba Cl. UB4: Yead 42Z 85
Albacore Cres. SE13 58Dc 114
Albacore Way UB3: Hayes 45V 84
Alba Gdns. NW11 30Ab 48
Albain Cres. TW15: Ashf 61N 127
Alba M. SW18 61Cb 133
Alban Arena 2B 6
Alban Av. AL3: St A 1B 6
Alban Ct. AL1: St A 2F 6
(off Burleigh Rd.)
Alban Cres. DA4: Farni 74Qd 163
Alban Cres. WD6: Bore 11Ra 29
Alban Highwalk EC2 1E 224 (43Sb 91)
(off Wood St.)
Alban Ho. WD6: Bore 11Ra 29
Albanian Ct. AL1: St A 3E 6
Alban Pk. AL4: St A 2K 7
Albans Vw. WD25: Wat 5X 13
Albanwood WD25: Wat 5X 13
Albany N12 23Db 49
Albany W1 5B 222 (45Lb 90)
Albany, The IG8: Wfd G 21Hc 53
Albany, The 52Cc 114
Albany Cl. DA5: Bexl 59Yc 117
Albany Cl. KT10: Esh 81Ca 171
Albany Cl. N15 28Rb 51
Albany Cl. RH2: Reig 3J 207
Albany Cl. SW14 56Ra 109
Albany Cl. UB10: Ick 36Q 64
Albany Cl. WD23: Bush 16Fa 28
Albany Ct. E4: Chelwood Cl. 16Cc 34
Albany Ct. KT13: Weyb Hillcrest 77R 150
Albany Ct. KT13: Weyb Oakhill Gdns. 75U 150
Albany Ct. E4: Westward Rd. 22Bc 52
Albany Ct. CM16: Epp 2Vc 23
Albany Ct. E1 44Wb 91
(off Plumber's Row)
Albany Ct. E10 31Cc 72
Albany Ct. HA8: Edg 25Ta 47
Albany Ct. NW10 41Xa 88
(off Trenmar Gdns.)
Albany Ct. NW8 2B 214 (40Fb 69)
(off Abbey Rd.)
Albany Ct. TW15: Ashf 66S 128
Albany Ct. TW20: Egh 64D 126
Albany Ctyd. W1 5B 222 (45Lb 90)
Albany Cres. HA8: Edg 24Qa 47
Albany Cres. KT10: Clay 79Ga 152
Albany Ga. AL1: St A 3B 6
Albany Hgts. RM17: Grays 50Ce 99
(off Hogg La.)
Albany Leisure Cen. 10Zb 20
Albany Mans. SW11 52Gb 111
Albany M. AL2: Chis G 9N 5
(off Nth. Orbital Rd.)
Albany M. BR1: Broml 65Jc 137
Albany M. KT2: King T 65Ma 131
Albany M. N1 38Qb 70
Albany M. SE5 51Sb 113
Albany M. SM1: Sutt 78Db 155
Albany Pde. TW8: Bford 51Na 109
Albany Pk. SL3: Coln 53F 104
Albany Pk. Av. EN3: Enf W 11Yb 34
Albany Pk. Rd. KT2: King T 65Ma 131
Albany Pk. Rd. KT22: Lea 91Ja 192
Albany Park Station (Rail) 61Zc 139
Albany Pas. TW10: Rich 57Pa 109
Albany Pl. TW20: Egh 63D 126
Albany Pl. TW8: Bford 51Ma 109
Albany Reach KT7: T Ditt 71Ha 152
Albany Rd. BR7: Chst 64Rc 138
Albany Rd. CM15: Pil H 16Xd 40
Albany Rd. DA17: Belv 51Bd 117
Albany Rd. DA5: Bexl 59Yc 117
Albany Rd. E10 31Cc 72
Albany Rd. E12 35Mc 73
Albany Rd. E17 30Ac 52
Albany Rd. EN3: Enf W 9Zb 20
Albany Rd. KT12: Hers 77Z 151
Albany Rd. KT3: N Mald 70Ta 131
Albany Rd. N18 22Yb 52
Albany Rd. N4 30Qb 50
Albany Rd. RM12: Horn 32Jd 76
Albany Rd. RM18: Tilb 3C 122
Albany Rd. RM6: Chad H 30Bd 55
Albany Rd. SE5 51Sb 113
Albany Rd. SL4: Old Win 7L 103
Albany Rd. SL4: Wind 4H 103
Albany Rd. SW19 64Db 133
Albany Rd. TW10: Rich 57Pa 109
Albany Rd. TW8: Bford 51Ma 109
Albany Rd. W13 45Ka 86
Albany St. NW1 1K 215 (40Kb 70)
Albany Ter. NW1 6A 216 (42Kb 90)
(off Marylebone Rd.)
Albany Ter. TW10: Rich 57Pa 109
(off Albany Pas.)
Albany Vw. IG9: Buck H 18Jc 35
Albany Way TW18: Staines 65M 127
Albany Works E3 39Ac 72
(off Gunmakers La.)
Alba Pl. W11 44Bb 89
Albatross NW9 26Va 48

Albatross Cl. E6 43Pc 94
Albatross Gdns. CR2: Sels 83Zb 178
Albatross St. SE18 52Uc 116
Albatross Way SE16 47Zb 92
Albemarle SW19 61Za 132
Albemarle App. IG2: Ilf 30Rc 54
Albemarle Av. EN6: Pot B 5Db 17
Albemarle Av. TW2: Whitt 60Ba 107
Albemarle Cl. RM17: Grays 47Ce 99
Albemarle Ct. N17 27Xb 51
(off Perkyn Sq.)
Albemarle Gdns. IG2: Ilf 30Rc 54
Albemarle Gdns. KT3: N Mald 70Ta 131
Albemarle Ho. SE8 49Bc 92
(off Foreshore)
Albemarle Pk. BR3: Beck 67Dc 136
Albemarle Pk. HA7: Stan 22La 46
Albemarle Rd. BR3: Beck 67Dc 136
Albemarle Rd. EN4: E Barn 17Gb 31
Albemarle St. W1 5A 222 (45Kb 90)
Albemarle Wlk. SW9 55Qb 112
Albemarle Way EC1 6B 218 (42Rb 91)
Alberon Gdns. NW11 28Bb 49
Alberta Av. SM1: Sutt 77Ab 154
Alberta Est. SE17 7C 230 (50Rb 91)
(off Alberta St.)
Alberta Ho. UB4: Yead 41X 85
(off Ayles Rd.)
Alberta Rd. DA8: Erith 53Ed 118
Alberta Rd. EN1: Enf 16Vb 33
Alberta St. SE17 7B 230 (50Rb 91)
Albert Av. E4 21Cc 52
Albert Av. KT16: Chert 69J 127
Albert Av. SW8 52Pb 112
Albert Barnes Ho. SE1 4D 230 (48Sb 91)
(off New Kent Rd.)
Albert Basin 45Rc 94
Albert Basin Way E16 45Sc 94
Albert Bigg Point E15 40Ec 72
(off Godfrey St.)
Albert Bri. 51Gb 111
Albert Bri. Rd. SW11 52Gb 111
Albert Broccoli Rd. UB9: Den 29J 43
Albert Carr Gdns. SW16 64Nb 134
Albert Cl. E9 39Xb 71
Albert Cl. N22 25Mb 50
Albert Cl. RM16: Grays 48Ee 99
Albert Cl. SL1: Slou 8K 81
Albert Cotts. E1 43Wb 91
(off Deal St.)
Albert Ct. E7 35Jc 73
Albert Ct. EN8: Walt C 6Bc 20
(off Holdbrook Sth.)
Albert Ct. SW7 3B 226 (48Fb 89)
Albert Ct. Ga. SW1 2F 227 (47Hb 89)
(off Knightsbridge)
Albert Cres. E4 21Cc 52
Albert Dane Cen. UB2: S'hall 48Aa 85
Albert Dr. GU21: Wok 87D 168
Albert Dr. SW19 61Ab 132
Albert Dr. TW18: Staines 64H 127
Albert Emb. SE1: Kennington La. 7G 229 (50Nb 90)
Albert Emb. SE1:
Lambeth Pal. Rd. 4H 229 (48Pb 90)
Albert Gdns. E1 44Zb 92
Albert Ga. SW1 1G 227 (47Hb 89)
Albert Gray Ho. SW10 52Fb 111
(off Worlds End Est.)
Albert Gro. SW20 67Za 132
Albert Hall Mans. SW7 2B 226 (47Fb 89)
Albert Ho. E18 27Kc 53
(off Albert Rd.)
Albert Ho. SE28 48Sc 94
Albertine Cl. KT17: Eps D 88Xa 174
Albertine Gr. BR4: W W'ck 77Fc 159
Albert Mans. CR0: C'don 74Tb 157
(off Lansdowne Rd.)
Albert Mans. SW11 53Hb 111
(off Albert Bri. Rd.)
Albert Memorial London 2B 226 (47Fb 89)
Albert M. E14 45Ac 92
(off Northey St.)
Albert M. N4 32Pb 70
Albert M. RH1: Redh 9A 208
Albert M. SE4 56Ac 114
Albert M. UB9: Den 29H 43
Albert M. W8 3A 226 (48Eb 89)
Albert Murray Cl. DA12: Grav'nd 9E 122
Albert Pal. Mans. SW11 53Kb 112
(off Lurline Gdns.)
Albert Pl. N17 27Vb 51
Albert Pl. N3 25Cb 49
Albert Pl. SL4: Eton W 10E 80
Albert Pl. W8 48Db 89
Albert Rd. BR2: Broml 71Mc 159
Albert Rd. BR5: St M Cry 72Xc 161
Albert Rd. BR6: Chels 78Wc 161
Albert Rd. CR4: Mitc 69Hb 133
Albert Rd. CR6: W'ham 89Bc 178
Albert Rd. DA10: Swans 58Be 121
Albert Rd. DA17: Belv 50Bd 95
Albert Rd. DA2: Wilm 62Ld 141
Albert Rd. DA5: Bexl 58Cd 118
Albert Rd. E10 33Ec 72
Albert Rd. E16 46Nc 94
Albert Rd. E17 29Cc 52
Albert Rd. E18 27Kc 53
Albert Rd. EN4: E Barn 14Eb 31
Albert Rd. HA2: Harr 27Ea 46
Albert Rd. IG1: Ilf 34Rc 74
Albert Rd. IG9: Buck H 19Mc 35
Albert Rd. KT1: King T 68Pa 131
Albert Rd. KT15: Add 76M 149
Albert Rd. KT17: Eps 85Va 174
Albert Rd. KT21: Asht 90Pa 173
Albert Rd. KT3: N Mald 70Va 132
Albert Rd. N15 30Ub 51
Albert Rd. N22 25Lb 50
Albert Rd. N4 32Pb 70
Albert Rd. NW4 28Za 48
Albert Rd. NW6 40Bb 69
Albert Rd. NW7 22Va 48
Albert Rd. RH1: Mers 1C 208
Albert Rd. RM1: Rom 29Hd 56
Albert Rd. RM8: Dag 32Cd 76
Albert Rd. SE20 65Zb 136
Albert Rd. SE25 70Wb 135
Albert Rd. SE9 62Nc 138
Albert Rd. SL4: Old Win 5H 103
Albert Rd. SL4: Wind 5H 103
Albert Rd. SM1: Sutt 78Fb 155
Albert Rd. TW1: Twick 60Ha 108
Albert Rd. TW10: Rich 57Na 109
Albert Rd. TW11: Tedd 65Ha 130
Albert Rd. TW12: Hamp H 64Ea 130
Albert Rd. TW15: Ashf 64P 127
Albert Rd. TW20: Eng G 5P 125
Albert Rd. TW3: Houn 56Ca 107
Albert Rd. UB2: S'hall 48Z 85
Albert Rd. UB3: Hayes 48U 84
Albert Rd. UB7: Yiew 46N 83
Albert Rd. W5 42Ka 86
Albert Rd. Est. DA17: Belv 50Bd 95
Albert Rd. Nth. RH2: Reig 5H 207
Albert Rd. Nth. WD17: Wat 13X 27
Albert Rd. Sth. WD17: Wat 13X 27
Alberts Ct. NW1 5E 214 (42Gb 89)
(off Palgrave Gdns.)
Albert Sleet Ct. N9 20Xb 33
(off Colthurst Dr.)
Albert Sq. E15 36Gc 73
Albert Sq. SW8 52Pb 112
Albert Starr Ho. SE8 49Zb 92
(off Haddonfield)
Albert St. AL1: St A 3B 6
Albert St. CM14: W'ley 22Yd 58
Albert St. N12 22Eb 49
Albert St. NW1 39Kb 70
Albert St. SL1: Slou 8K 81
(not continuous)
Albert St. SL4: Wind 3F 102
Albert Studios SW11 53Hb 111
Albert Ter. IG9: Buck H 19Nc 36
Albert Ter. NW1 39Jb 70
Albert Ter. NW10 39Sa 67
Albert Ter. W5 42Ka 86
Albert Ter. W6 50Wa 88
(off Beavor La.)
Albert Ter. M. NW1 39Jb 70
Albert Victoria Ho. N22 25Qb 50
Albert Wlk. E16 47Qc 94
Albert Way SE15 52Xb 113
Albert Westcott Ho. SE17 7C 230 (50Rb 91)
Albert Whicher Ho. E17 28Ec 52
Albert Yd. SE19 65Vb 135
Albery Ct. E8 38Vb 71
(off Middleton Rd.)
Albion Av. N10 25Jb 50
Albion Av. SW8 54Mb 112
Albion Bldgs. N1 2G 217 (40Nb 70)
(off Albion Yd.)
Albion Cl. RM7: Rom 30Fd 56
Albion Cl. SL2: Slou 6L 81
Albion Cl. W2 4E 220 (45Gb 89)
Albion Ct. SE10 49Gc 93
(off Azof St.)
Albion Ct. SM2: Sutt 80Fb 155
Albion Ct. W6 49Xa 88
(off Albion Pl.)
Albion Dr. E8 38Vb 71
Albion Est. SE16 47Zb 92
Albion Gdns. W6 49Xa 88
Albion Ga. W2 4E 220 (45Gb 89)
(off Albion St.)
Albion Gro. N16 35Ub 71
Albion Hill HP2: Hem H 3M 3
Albion Hill IG10: Lough 15Lc 35
Albion Ho. E16 46Rc 94
(off Church St.)
Albion Ho. GU21: Wok 89B 168
Albion Ho. SE8 52Cc 114
(off Watsons St.)
Albion M. N1 39Qb 70
Albion M. NW6 38Bb 69
Albion M. W2 4E 220 (45Gb 89)
Albion M. W6 49Xa 88
Albion Pde. DA12: Grav'nd 8F 122
Albion Pde. N16 35Tb 71
Albion Pk. IG10: Lough 15Mc 35
Albion Pl. EC1 7B 218 (43Rb 91)
Albion Pl. EC2 1G 225 (43Tb 91)
Albion Pl. SL4: Wind 4E 102
Albion Pl. W6 49Xa 88
Albion Riverside Bldg. SW11 52Gb 111
Albion Rd. AL1: St A 2D 6
Albion Rd. DA1: Dart 55Rd 119
Albion Rd. DA12: Grav'nd 9E 122
Albion Rd. DA6: Bex 56Bd 117
Albion Rd. E17 27Ec 52
Albion Rd. KT2: King T 67Sa 131
Albion Rd. N16 35Tb 71
Albion Rd. N17 26Wb 51
Albion Rd. RH2: Reig 7L 207
Albion Rd. SM2: Sutt 79Fb 155
Albion Rd. TW2: Twick 60Ga 108
Albion Rd. TW3: Houn 56Ca 107
Albion Rd. UB3: Hayes 44U 84
Albion Sq. E8 38Vb 71
(not continuous)
Albion St. CR0: C'don 74Rb 157
Albion St. SE16 47Yb 92
Albion St. W2 3E 220 (44Gb 89)
Albion Ter. DA12: Grav'nd 8E 122
Albion Ter. E4 14Dc 34
Albion Ter. E8 38Vb 71
Albion Vs. Rd. SE26 62Yb 136
Albion Wlk. N1 2G 217 (40Nb 70)
(off York Way)
Albion Way E6 39Mc 73
Albion Way EC1 1D 224 (43Sb 91)
Albion Way HA9: Wemb 34Qa 67
Albion Way SE13 56Ec 114
Albion Yd. E1 43Xb 91
Albion Yd. N1 2G 217 (40Nb 70)
Albon Ho. SW18 58Db 111
(off Neville Gill Cl.)
Albright Ind. Est. RM13: Rain 42Hd 96
Albrighton Rd. SE22 55Ub 113
Albuhera Cl. EN2: Enf 11Qb 32
Albuhera M. NW7 22Za 48
Albury Av. DA7: Bex 54Ad 117
Albury Av. SM2: Cheam 81Ya 174
Albury Av. TW7: Isle 52Ha 108
Albury Cl. KT16: Longc 6L 147
Albury Cl. KT19: Eps 81Ra 173
Albury Cl. TW12: Hamp 65Da 129
Albury Ct. CR0: C'don 75Sb 157
(off Tanfield Rd.)
Albury Ct. CR4: Mitc 68Fb 133
Albury Ct. SE8 51Cc 114
(off Albury St.)
Albury Ct. SM1: Sutt 77Eb 155
Albury Ct. UB5: N'olt 41Y 85
(off Canberra Dr.)
Albury Dr. HA5: Pinn 25Y 45
Albury Gro. Rd. EN8: Chesh 2Zb 20
Albury Ho. SE1 2C 230 (47Rb 91)
(off Boyfield St.)
Albury M. E12 33Lc 73
Albury Pl. KT10: Clay 79Ga 152
Albury Pl. RH1: Mers 1C 208
Albury Ride EN8: Chesh 3Zb 20
Albury Rd. KT12: Hers 79U 150
Albury Rd. KT9: Chess 78Na 153
Albury Rd. RH1: Mers 1C 208
Albury St. SE8 51Cc 114
Albury Wlk. EN8: Chesh 2Yb 20
(not continuous)
Albyfield BR1: Broml 70Pc 138
Albyn Ho. HP2: Hem H 2M 3
Albyn Rd. SE8 53Cc 114
Albyns Cl. RM13: Rain 38Jd 76
Albyns La. RM4: Nave 12Ed 38
Albyns La. RM4: Stap T 12Ed 38
Alcester Ct. SM6: W'gton 77Kb 156
Alcester Cres. E5 33Xb 71
Alcester Ho. RM3: Rom 22Md 57
(off Northallerton Way)
Alcester Rd. SM6: W'gton 77Kb 156
Alcock Cl. SM6: W'gton 80Mb 156
Alcock Cres. DA1: Cray 57Jd 118
Alcock Rd. TW5: Hest 52Z 107
Alcocks Cl. KT20: Tad 92Ab 194
Alcocks La. KT20: Kgswd 93Ab 194
Alcocks La. KT20: Tad 93Ab 194
Alconbury DA6: Bex 57Dd 118
Alconbury Cl. WD6: Bore 11Pa 29
Alconbury Rd. E5 33Wb 71
Alcorn Cl. SM3: Sutt 75Cb 155
Alcott Cl. TW14: Felt 60V 106
Alcott Cl. W7 43Ha 86
Alcuin Ct. HA7: Stan 24La 46
Aldam Pl. N16 33Vb 71
Aldborough Ct. IG2: Ilf 29Vc 55
(off Aldborough Rd. Nth.)
Aldborough Hall Equestrian Cen. 27Vc 55
ALDBOROUGH HATCH 28Vc 55
Aldborough Rd. RM10: Dag 37Ed 76
Aldborough Rd. RM14: Upm 33Pd 77
Aldborough Rd. Nth. IG2: Ilf 29Vc 55
Aldborough Rd. Sth. IG3: Ilf 32Uc 74
Aldborough Spur SL1: Slou 4J 81
Aldbourne Rd. SL1: Burn 3A 80
Aldbourne Rd. W12 46Va 88
Aldbridge St. SE17 7J 231 (50Ub 91)
Aldburgh M. W1 2J 221 (44Jb 90)
Aldbury Av. HA9: Wemb 38Ra 67
Aldbury Cl. WD25: Wat 8Z 13
Aldbury Ho. SW3 6D 226 (49Gb 89)
(off Cale St.)
Aldbury M. N9 17Tb 33
Aldbury Rd. WD3: Rick 17H 25
Aldebert Ter. SW8 52Nb 112
Aldeburgh Cl. E5 33Xb 71
Aldeburgh Pl. IG8: Wfd G 21Jc 53
Aldeburgh Pl. SE10 49Jc 93
(off Aldeburgh St.)
Aldeburgh St. SE10 50Jc 93
Alden Av. E15 41Hc 93
Alden Ct. CR0: C'don 76Ub 157
ALDENHAM 10Da 13
Aldenham Av. WD7: R'lett 8Ja 14
Aldenham Cl. SL3: L'ly 8P 81
Aldenham Country Pk. 15Ka 28
Aldenham Country Pk. Rare Breeds
Farm 15Ka 28
Aldenham Dr. UB8: Hil 42R 84
Aldenham Golf Course 10Da 13
Aldenham Gro. WD7: R'lett 6Ka 14
Aldenham Ho. NW1 2C 216 (40Lb 70)
(off Aldenham St.)
Aldenham Rd. WD19: Wat 16Aa 27
Aldenham Rd. WD23: Bush 16Aa 27
Aldenham Rd. WD25: Let H 11Ga 28
Aldenham Rd. WD6: Elstree 13Ja 28
Aldenham Rd. WD7: R'lett 7Ja 14
Aldenham Sailing Club 15La 28
Aldenham St. NW1 2C 216 (40Lb 70)
Aldenholme KT13: Weyb 79U 150
Alden Ho. E8 39Xb 71
(off Duncan Rd.)
Alden Mead HA5: Hat E 23Ca 45
(off The Avenue)
Aldensley Rd. W6 48Xa 88
Alden Vw. SL4: Wind 3B 102
Alder Av. RM14: Upm 35Pd 77
Alderbourne La. SL0: Iver H 36C 62
Alderbourne La. SL3: Ful 35A 62
Alderbrook Rd. SW12 58Kb 112
Alderbury Rd. SL3: L'ly 47B 82
Alderbury Rd. SW13 51Wa 110
Alderbury Rd. W. SL3: L'ly 47B 82
Alder Cl. AL2: Park 10P 5
Alder Cl. DA18: Erith 47Bd 95
Alder Cl. SE15 51Vb 113
Alder Cl. SL1: Slou 6D 80
Alder Cl. TW20: Eng G 64A 126
Aldercombe La. CR3: Cat'm 99Ub 197
Alder Ct. E7 36Jc 73
Alder Ct. N11 23Lb 50
Alder Cft. CR5: Coul 88Pb 176
Alder Dr. RM15: S Ock 42Yd 98
Alderglade Nature Reserve, The 36M 63
Alder Gro. NW2 33Wa 68
Aldergrove Gdns. TW3: Houn 54Aa 107
Aldergrove Wlk. RM12: Horn 37Ld 77
Alder Ho. E3 39Bc 72
(off Hornbeam Sq.)
Alder Ho. NW3 37Hb 69
Alder Ho. SE15 51Vb 113
(off Alder Cl.)
Alder Ho. SE4 55Cc 114
Alderley Gdns. SL2: Farn C 5G 60
Alder Lodge SW6 53Ya 110
Alderman Av. IG11: Bark 41Wc 95
Aldermanbury EC2 2E 224 (44Sb 91)
Aldermanbury Sq. EC2 1E 224 (43Sb 91)
Alderman Cl. AL9: Wel G 6E 8
Alderman Cl. DA1: Cray 59Gd 118
Alderman Judge Mall KT1: King T 68Na 131
(off Eden St.)
Aldermans Hill N13 21Nb 50
Aldermans Ho. E9 36Ac 72
(off Ward La.)
Aldermans Wlk. EC2 1H 225 (43Ub 91)
Aldermary Rd. BR1: Broml 67Jc 137
Alder M. N19 33Lb 70
Aldermoor Rd. SE6 62Bc 136
Alderney Av. TW5: Hest 52Da 107
Alderney Av. TW5: Isle 52Da 107
Alderney Cl. NW9 27Wa 48
Alderney Ct. SE10 51Fc 115
(off Trafalgar Rd.)
Alderney Gdns. UB5: N'olt 38Ba 65
Alderney Ho. EN3: Enf W 10Zb 20
Alderney Ho. N1 37Sb 71
(off Arran Wlk.)
Alderney Ho. WD18: Wat 16V 26
Alderney M. SE1 3F 231 (48Tb 91)
Alderney Rd. DA8: Erith 52Jd 118
Alderney Rd. E1 42Zb 92
Alderney St. SW1 6A 228 (49Kb 90)
Alder Pl. DA1: Dart 57Pd 119
Alder Rd. DA14: Sidc 62Vc 139
Alder Rd. SL0: Iver H 40F 62
Alder Rd. SW14 55Ta 109
Alder Rd. UB9: Den 37L 63
Alders, The BR4: W W'ck 74Dc 158
Alders, The KT14: W Byf 84L 169
Alders, The N21 16Qb 32
Alders, The SW16 63Lb 134
Alders, The TW13: Hanw 63Aa 129
Alders, The TW5: Hest 51Ba 107
Alders, The UB9: Den 37L 63
Alders Av. IG8: Wfd G 23Gc 53
ALDERSBROOK 33Kc 73
Aldersbrook Av. EN1: Enf 12Ub 33
Aldersbrook Dr. KT2: King T 65Pa 131
Aldersbrook La. E12 34Pc 74
Aldersbrook Rd. E11 33Kc 73
Aldersbrook Rd. E12 33Kc 73
Alders Cl. E11 33Kc 73
Alders Cl. HA8: Edg 22Sa 47
Alders Cl. W5 48Ma 87
Alders Ct. DA11: Nflt 2B 144
Aldersey Gdns. IG11: Bark 37Tc 74
Aldersford Cl. SE4 57Zb 114
Aldersgate Ct. EC1 1D 224 (43Sb 91)
(off Bartholomew Cl.)
Aldersgate St. EC1 7D 218 (43Sb 91)
Alders Gro. CR3: Cat'm 92Ub 197
Alders Gro. KT8: E Mos 71Fa 152
Aldersgrove EN9: Walt A 6Gc 21
Aldersgrove Av. SE9 62Mc 137
Aldershot Rd. GU24: Pirb 9B 186
Aldershot Rd. NW6 39Bb 69
Aldershot Ter. SE18 52Qc 116
Alderside Wlk. TW20: Eng G 64A 126
Aldersmead Av. CR0: C'don 72Zb 158
Aldersmead Rd. BR3: Beck 66Ac 136
Alderson Gro. KT12: Hers 76Y 151
Alderson Pl. UB2: S'hall 46Ea 86
Alderson St. W10 42Ab 88
Alders Rd. HA8: Edg 22Sa 47
Alders Rd. RH2: Reig 4K 207
ALDERSTEAD HEATH 97Mb 196
Alderstead Heath Cvn. Club Site
RH1: Mers 96Mb 196
Alderstead La. RH1: Mers 97Mb 196
Alderton Cl. CM15: Pil H 15Xd 40
Alderton Cl. IG10: Lough 14Qc 36
Alderton Cl. NW10 34Ta 67
Alderton Ct. KT8: W Mole 70Ba 129
(off Dunstable Rd.)
Alderton Cres. NW4 29Xa 48
Alderton Hall La. IG10: Lough 14Qc 36
Alderton Hill IG10: Lough 15Nc 36
Alderton M. IG10: Lough 14Qc 36
Alderton Ri. IG10: Lough 14Qc 36
Alderton Rd. CR0: C'don 73Vb 157
Alderton Rd. RM16: Ors 4G 100
Alderton Rd. SE24 55Sb 113
Alderton Way IG10: Lough 15Pc 36
Alderton Way NW4 29Xa 48
Alderville Rd. SW6 54Bb 111
Alder Wlk. IG1: Ilf 36Sc 74
Alder Wlk. WD25: Wat 7X 13
Alder Way BR8: Swan 68Fd 140
Alderwick Ct. N7 37Pb 70
(off Cornelia St.)
Alderwick Dr. TW3: Houn 55Fa 108
Alderwood Cl. CR3: Cat'm 97Ub 197
Alderwood Cl. RM4: Abr 13Xc 37
Alderwood Dr. RM4: Abr 13Xc 37
Alderwood Ho. WD19: Wat 20Y 27
Alderwood M. EN4: Had W 10Eb 17
Alderwood Rd. SE9 58Tc 116
Alderwood Ter. IG7: Chig 22Ad 55
Aldford Ho. W1 6H 221 (46Jb 90)
(off Park St.)
Aldford St. W1 6H 221 (46Jb 90)
Aldgate E1 2K 225 (44Vb 91)
(off Whitechapel High St.)
ALDGATE 3K 225 (44Vb 91)
(off Aldgate High St.)
Aldgate EC3 3K 225 (44Vb 91)
Aldgate Av. E1 2K 225 (44Vb 91)
Aldgate Barrs E1 2K 225 (44Vb 91)
(off Whitechapel High St.)
Aldgate Bus Station 3K 225 (44Vb 91)
Aldgate East Station
(Underground) 2K 225 (44Vb 91)
Aldgate High St. EC3 3K 225 (44Vb 91)
Aldgate Pl. E1 44Vb 91
Aldgate Sq. EC3 3K 225 (44Vb 91)
Aldgate Station
(Underground) 3K 225 (44Vb 91)
Aldgate Twr. E1 2K 225 (44Vb 91)
Aldham Dr. RM15: S Ock 43Yd 98
Aldham Ho. SE4 54Bc 114
(off Malpas Rd.)
Aldin Av. Nth. SL1: Slou 7L 81
Aldin Av. Sth. SL1: Slou 7L 81
Aldine Ct. W12 47Ya 88
(off Aldine St.)
Aldine Pl. W12 47Ya 88
Aldine St. W12 47Ya 88
Aldingham Ct. RM12: Horn 36Kd 77
(off Easedale Dr.)
Aldingham Gdns. RM12: Horn 36Jd 76
Aldington Cl. RM8: Dag 32Yc 75
Aldington Ct. E8 38Wb 71
(off London Flds. W. Side)
Aldington Rd. SE18 48Mc 93
Aldis M. EN3: Enf L 9Cc 20
Aldis M. SW17 64Gb 133
Aldis St. SW17 64Gb 133
Aldred Rd. NW6 36Cb 69
Aldren Rd. SW17 62Eb 133
Aldrich Cres. CR0: New Ad 81Ec 178
Aldriche Way E4 23Ec 52
Aldrich Gdns. SM3: Cheam 76Bb 155
Aldrich Ter. SW18 61Eb 133
Aldrick Ho. N1 1J 217 (39Pb 70)
(off Barnsbury Est.)
Aldridge Av. EN3: Enf L 10Cc 20
Aldridge Av. HA4: Ruis 33Y 65
Aldridge Av. HA7: Stan 25Na 47
Aldridge Av. HA8: Edg 20Ra 29
Aldridge Ct. W11 43Bb 89
(off Aldridge Rd. Vs.)
Aldridge Pl. SL2: Stoke P 8K 61
Aldridge Ri. KT3: N Mald 73Ua 154
Aldridge Rd. SL2: Slou 2E 80
Aldridge Rd. Vs. W11 43Bb 89
Aldridge Wlk. N14 17Nb 32
Aldrin Cl. SS17: Stan H 1N 101
Aldrington Rd. SW16 64Lb 134
Aldsworth Cl. W9 42Db 89
Aldwick AL1: St A 4F 6
Aldwick Cl. SE9 62Tc 138
Aldwick Ct. AL1: St A 4F 6
Aldwick Rd. CR0: Bedd 76Pb 156
Aldworth Gro. SE13 58Ec 114
Aldworth Rd. E15 38Gc 73
Aldwych WC2 3H 223 (44Pb 90)
Aldwych Av. IG6: Ilf 28Sc 54
Aldwych Bldgs. WC2 2G 223 (44Nb 90)
(off Parker M.)
Aldwych Cl. RM12: Horn 33Jd 76
Aldwych Ct. E8 38Vb 71
(off Middleton Rd.)
Aldwych Theatre 3H 223 (44Pb 90)
(off Aldwych)
Aldwyck Ct. HP1: Hem H 1L 3
Aldwyn Ho. SW8 52Nb 112
(off Davidson Gdns.)
Aldwyn Pl. TW20: Eng G 5M 125
Aldykes AL10: Hat 1B 8
Alers Rd. DA6: Bex 57Zc 117
Alesia Cl. N22 24Nb 50
Alestan Beck Rd. E16 44Mc 93
Alexa Ct. SM2: Sutt 79Cb 155
Alexa Ct. W8 49Cb 89
Alexander Av. NW10 38Xa 68
Alexander Cl. BR2: Hayes 74Jc 159
Alexander Cl. DA15: Sidc 58Uc 116
Alexander Cl. EN4: E Barn 14Fb 31
Alexander Cl. TW2: Twick 61Ga 130
Alexander Cl. UB2: S'hall 46Ea 86
Alexander Ct. BR3: Beck 67Fc 137
Alexander Ct. EN8: Chesh 2Zb 20
Alexander Ct. HA7: Stan 27Pa 47
Alexander Ct. TW16: Sun 65V 128
Alexander Cres. CR3: Cat'm 93Sb 197
Alexander Evans M. SE23 61Zb 136
Alexander Fleming Laboratory
Mus. 2C 220 (44Fb 89)
Alexander Godley Cl. KT21: Asht 91Pa 193
Alexander Ho. E14 48Cc 92
(off Tiller Rd.)
Alexander Ho. KT2: King T 67Na 131
(off Seven Kings Way)
Alexander Ho. RM14: Upm 30Ud 58
Alexander Ho. SE15 54Xb 113
(off Godman Rd.)
Alexander La. CM13: Hut 16De 41
Alexander La. CM15: Shenf 15Ce 41
Alexander M. SW16 64Lb 134
Alexander M. W2 44Db 89
Alexander Pl. RH8: Oxt 100Gc 199
Alexander Pl. SW7 5D 226 (49Gb 89)
Alexander Raby Mill KT15: Add 78N 149
(off Bourneside Rd.)
Alexander Rd. AL2: Lon C 7G 6
Alexander Rd. BR7: Chst 65Rc 138
Alexander Rd. CR5: Coul 87Kb 176
Alexander Rd. DA7: Bex 54Zc 117
Alexander Rd. DA9: Ghithe 57Yd 120
Alexander Rd. N19 34Nb 70
Alexander Rd. RH2: Reig 9J 207
Alexander Rd. TW20: Egh 64E 126
Alexander Sq. SW3 5D 226 (49Gb 89)
Alexander St. W2 44Cb 89
Alexander Studios SW11 56Fb 111
(off Haydon Way)
Alexanders Wlk. CR3: Cat'm 98Vb 197
Alexander Ter. SE2 50Xc 95
Alexandra Av. CR6: W'ham 89Bc 178
Alexandra Av. HA2: Harr 32Ba 65
Alexandra Av. N22 25Mb 50
Alexandra Av. SM1: Sutt 76Cb 155
Alexandra Av. SW11 53Jb 112
Alexandra Av. UB1: S'hall 45Ba 85
Alexandra Av. W4 52Ta 109
Alexandra Cl. BR8: Swan 68Gd 140
Alexandra Cl. HA2: Harr 34Ca 65
Alexandra Cl. KT12: Walt T 75W 150
Alexandra Cl. RM16: Grays 7D 100
Alexandra Cl. SE8 51Bc 114
Alexandra Cl. TW15: Ashf 66T 128
Alexandra Cl. TW18: Staines 65M 127
Alexandra Cotts. SE14 53Bc 114
Alexandra Ct. HA9: Wemb 35Pa 67
Alexandra Ct. N14 15Lb 32
Alexandra Ct. SE5 51Sb 113
(off Urlwin St.)
Alexandra Ct. SL4: Wind 4H 103
(off Alexandra Rd.)
Alexandra Ct. SW7 3A 226 (48Eb 89)
(off Queen's Ga.)
Alexandra Ct. TW15: Ashf 65T 128
Alexandra Ct. TW3: Houn 54Da 107
Alexandra Ct. UB6: G'frd 40Da 65
Alexandra Ct. W2 45Db 89
(off Moscow Rd.)
Alexandra Ct. W9 5A 214 (42Eb 89)
(off Maida Vale)
Alexandra Ct. WD24: Wat 12Y 27
Alexandra Cres. BR1: Broml 65Hc 137
Alexandra Dr. KT5: Surb 73Qa 153
Alexandra Dr. SE19 64Ub 135
Alexandra Gdns. GU21: Knap 10G 166
Alexandra Gdns. N10 28Kb 50
Alexandra Gdns. SM5: Cars 80Jb 156
Alexandra Gdns. TW3: Houn 54Da 107
Alexandra Gdns. W4 52Ua 110
Alexandra Gro. N12 22Db 49
Alexandra Gro. N4 32Rb 71
Alexandra Ho. E16 46Kc 93
(off Wesley Av.)
Alexandra Ho. IG8: Wfd G 24Qc 54
Alexandra Ho. W6 50Ya 88
(off Queen Caroline St.)
Alexandra Lodge KT13: Weyb 77R 150
(off Monument Hill)
Alexandra Mans. KT17: Eps 85Va 174
(off Alexandra Rd.)
Alexandra Mans. SW3 51Fb 111
(off King's Rd.)
Alexandra Mans. W12 46Ya 88
(off Stanlake Rd.)
Alexandra M. N2 27Hb 49
Alexandra M. N4 33Rb 71
Alexandra M. SW19 65Bb 133
Alexandra M. WD17: Wat 12W 26
Alexandra Palace 26Mb 50
Alexandra Palace Ice Rink 26Mb 50
Alexandra Palace Station (Rail) 26Nb 50
Alexandra Palace Theatre 26Mb 50
Alexandra Pal. Way N22 28Lb 50
Alexandra Pal. Way N8 28Lb 50
Alexandra Pde. HA2: Harr 35Da 65
Alexandra Pk. Rd. N10 26Kb 50
Alexandra Pk. Rd. N22 25Lb 50
Alexandra Pl. CR0: C'don 74Ub 157
Alexandra Pl. NW8 39Eb 69
Alexandra Pl. SE25 71Tb 157
Alexandra Plaza SL1: Slou 7H 81
(off Chalvey Rd. W.)
Alexandra Rd. AL1: St A 2C 6
Alexandra Rd. CM14: B'wood 20Yd 40
Alexandra Rd. CR0: C'don 74Ub 157
Alexandra Rd. CR4: Mitc 66Gb 133

Alexandra Rd. CR6: W'ham 89Bc **178**
Alexandra Rd. DA12: Grav'nd 9G **122**
Alexandra Rd. DA8: Erith 51Hd **118**
Alexandra Rd. E10 34Ec **72**
Alexandra Rd. E17 30Bc **52**
Alexandra Rd. E18 27Kc **53**
Alexandra Rd. E6 41Qc **94**
Alexandra Rd. EN3: Pond E 14Zb **34**
Alexandra Rd. HP2: Hem H 1M **3**
Alexandra Rd. KT15: Add 77M **149**
(not continuous)
Alexandra Rd. KT17: Eps 85Va **174**
Alexandra Rd. KT2: King T 66Qa **131**
Alexandra Rd. KT7: T Ditt 71Ha **152**
Alexandra Rd. N10 24Kb **50**
Alexandra Rd. N15 29Tb **51**
Alexandra Rd. N8 27Qb **50**
Alexandra Rd. N9 17Xb **33**
Alexandra Rd. NW4 28Za **48**
Alexandra Rd. NW8 39Eb **69**
Alexandra Rd. RM1: Rom 30Hd **56**
Alexandra Rd. RM13: Rain 39Hd **76**
Alexandra Rd. RM18: Tilb 4B **122**
Alexandra Rd. RM6: Chad H 30Ad **55**
Alexandra Rd. SE26 65Zb **136**
Alexandra Rd. SL1: Slou 8H **81**
Alexandra Rd. SL4: Wind 4H **103**
Alexandra Rd. SW14 55Ta **109**
Alexandra Rd. SW19 65Bb **133**
Alexandra Rd. TN16: Big H 91Kc **199**
Alexandra Rd. TW1: Twick 58La **108**
Alexandra Rd. TW15: Ashf 66T **128**
Alexandra Rd. TW20: Eng G 5N **125**
Alexandra Rd. TW3: Houn 54Da **107**
Alexandra Rd. TW8: Bford 51Ma **109**
Alexandra Rd. TW9: Kew 54Pa **109**
Alexandra Rd. UB8: Uxb 40M **63**
Alexandra Rd. W4 47Ta **87**
Alexandra Rd. WD17: Wat 12W **26**
Alexandra Rd. WD3: Sarr 8J **11**
Alexandra Rd. WD4: Chfd 2J **11**
Alexandra Rd. WD4: K Lan 1Q **12**
Alexandra Rd. WD6: Bore 10Ta **15**
Alexandra Rd. Ind. Est. EN3: Pond E 14Zb **34**
Alexandra Sq. SM4: Mord 71Cb **155**
Alexandra St. E16 43Jc **93**
Alexandra St. SE14 52Ac **114**
Alexandra Ter. DA1: Dart 58Md **119**
Alexandra Ter. E14 50Dc **92**
(off Westferry Rd.)
Alexandra Wlk. DA4: S Dar 68Ud **142**
Alexandra Wlk. SE19 64Ub **135**
Alexandra Way EN8: Walt C 6Bc **20**
Alexandra Way KT19: Eps 83Qa **173**
Alexandra Way RM18: E Til 9K **101**
Alexandra Wharf E2 39Xb **71**
(off Darwen Pl.)
Alexandra Yd. E9 39Zb **72**
Alexandria Apts. SE17 5H **231** (49Ub **91**)
(off Townsend St.)
Alexandria Rd. W13 45Ja **86**
Alex Ct. HP2: Hem H 1M **3**
Alex Guy Gdns. RM8: Dag 32Dd **76**
Alexia Sq. E14 48Dc **92**
Alexis St. SE16 49Wb **91**
Alfan La. DA2: Wilm 64Fd **140**
Alfearn Rd. E5 35Yb **72**
Alford Ct. N1 2E **218** (40Sb **71**)
(off Shepherdess Wlk.)
Alford Grn. CR0: New Ad 79Fc **159**
Alford Ho. N6 30Lb **50**
Alford Pl. N1 2E **218** (40Sb **71**)
Alford Rd. DA8: Erith 50Ed **96**
Alfoxton Av. N15 28Rb **51**
Alfreda St. SW11 53Kb **112**
Alfred Cl. W4 49Ta **87**
Alfred Ct. SE16 49Xb **91**
(off Bombay St.)
Alfred Dickens Ho. E16 44Hc **93**
(off Hallsville Rd.)
Alfred Finlay Ho. N22 26Rb **51**
Alfred Gdns. UB1: S'hall 45Aa **85**
Alfred Ho. DA11: Nflt 10B **122**
Alfred Ho. E12 38Nc **74**
(off Tennyson Av.)
Alfred Ho. E9 36Ac **72**
(off Homerton Rd.)
Alfred M. W1 7D **216** (43Mb **90**)
Alfred Nunn Ho. NW10 39Va **68**
Alfred Pl. DA11: Nflt 10B **122**
Alfred Pl. WC1 7D **216** (43Mb **90**)
Alfred Prior Ho. E12 35Qc **74**
Alfred Rd. CM14: B'wood 19Zd **41**
Alfred Rd. DA11: Grav'nd 1D **144**
Alfred Rd. DA17: Belv 50Bd **95**
Alfred Rd. DA2: Hawl 63Nd **141**
Alfred Rd. E15 36Hc **73**
Alfred Rd. IG9: Buck H 19Mc **35**
Alfred Rd. KT1: King T 69Na **131**
Alfred Rd. RM15: Avel 46Sd **98**
Alfred Rd. SE25 71Wb **157**
Alfred Rd. SM1: Sutt 78Eb **155**
Alfred Rd. TW13: Felt 61Y **129**
Alfred Rd. W2 43Cb **89**
Alfred Rd. W3 46Sa **87**
Alfred Salter Ho. SE1 6K **231** (49Vb **91**)
(off Fort Rd.)
Alfred's Gdns. IG11: Bark 40Uc **74**
Alfred St. E3 41Bc **92**
Alfred St. RM17: Grays 51Ee **121**
Alfreds Way IG11: Bark 41Rc **94**
Alfreds Way Ind. Est. IG11: Bark 40Wc **75**
Alfred Vs. E17 28Ec **52**
Alfreton Cl. SW19 62Za **132**
Alfriston KT5: Surb 72Pa **153**
Alfriston Av. CR0: C'don 73Nb **156**
Alfriston Av. HA2: Harr 30Ca **45**
Alfriston Cl. DA1: Cray 58Gd **118**
Alfriston Cl. KT5: Surb 71Pa **153**
Alfriston Rd. SW11 57Hb **111**
Algar Cl. HA7: Stan 22Ha **46**
Algar Cl. TW7: Isle 55Ja **108**
Algar Ho. SE1 2B **230** (47Rb **91**)
(off Webber Row)
Algar Rd. TW7: Isle 55Ja **108**
Algarve Rd. SW18 60Db **111**
Algernon Rd. NW4 30Wa **48**
Algernon Rd. NW6 39Cb **69**
Algernon Rd. SE13 56Dc **114**
Algers Cl. IG10: Lough 15Mc **35**
Algers Mead IG10: Lough 15Mc **35**
Algers Rd. IG10: Lough 15Mc **35**
Algiers Rd. SE13 56Cc **114**
Alibon Gdns. RM10: Dag 36Cd **76**
Alibon Rd. RM10: Dag 36Cd **76**
Alibon Rd. RM9: Dag 36Bd **76**
Alice Cl. EN5: New Bar 14Eb **31**
(off Station App.)
Alice Gilliatt Ct. W14 51Bb **111**
(off Star Rd.)
Alice La. E3 39Bc **72**
Alice M. TW11: Tedd 64Ha **130**
Alice Owen Technology Cen. EC1 3B **218** (41Rb **91**)
(off Goswell Rd.)
Alice Ruston Pl. GU22: Wok 1N **187**
Alice Shepherd Ho. E14 47Ec **92**
(off Manchester Rd.)
Alice St. SE1 4H **231** (48Ub **91**)
(not continuous)
Alice Thompson Cl. SE12 61Lc **137**
Alice Walker Cl. SE24 56Rb **113**
Alice Way TW3: Houn 56Da **107**
Alicia Av. HA3: Kenton 28Ka **46**
Alicia Cl. HA3: Kenton 28La **46**
Alicia Gdns. HA3: Kenton 28Ka **46**
Alicia Ho. DA16: Well 53Xc **117**
Alie St. E1 3K **225** (44Vb **91**)
Alington Cres. NW9 31Sa **67**
Alington Gro. SM6: W'gton 81Lb **176**
Alison Cl. CR0: C'don 74Zb **158**
Alison Cl. E6 44Qc **94**
Alison Cl. GU21: Wok 87A **168**
Alison Cl. HA5: Eastc 30X **45**
Alison Ct. HP2: Hem H 1C **4**
Alissa Dr. EN5: New Bar 15Eb **31**
Aliwal M. SW11 56Gb **111**
Aliwal Rd. SW11 56Gb **111**
ALKERDEN 59Zd **121**
Alkerden La. DA10: Swans 58Zd **121**
Alkerden La. DA9: Ghithe 58Yd **120**
Alkerden Rd. W4 50Ua **88**
Alkham Rd. N16 33Vb **71**
Allan Barclay Cl. N15 30Vb **51**
Allan Cl. KT3: N Mald 71Ta **153**
Allandale AL3: St A 5P **5**
Allandale HP2: Hem H 1M **3**
Allandale Av. N3 27Ab **48**
Allandale Cres. EN6: Pot B 4Ab **16**
Allandale Pl. BR6: Chels 76Zc **161**
Allandale Rd. EN3: Enf W 8Zb **20**
Allandale Rd. RM11: Horn 31Hd **76**
Allan Ho. CR8: Purl 83Pb **176**
Allanson Ct. E10 33Cc **72**
(off Leyton Grange Est.)
Allans Way WD25: Wat 2X **13**
Allan Way W3 43Sa **87**
Allard Cl. BR5: Orp 73Yc **161**
Allard Cres. WD23: B Hea 18Ea **28**
Allard Gdns. SW4 57Mb **112**
Allard Ho. NW9 26Va **48**
(off Boulevard Dr.)
Allardyce St. SW4 56Pb **112**
Allbrook Cl. TW11: Tedd 64Ga **130**
Allcroft Rd. NW5 36Jb **70**
Allder Way CR2: S Croy 80Rb **157**
Alldicks Rd. HP3: Hem H 4P **3**
Allenby Av. CR2: S Croy 81Sb **177**
Allenby Cl. UB6: G'frd 41Ca **85**
Allenby Cres. RM17: Grays 50Ee **99**
Allenby Dr. RM11: Horn 32Nd **77**
Allenby Rd. SE23 62Ac **136**
Allenby Rd. SE28 48Sc **94**
Allenby Rd. TN16: Big H 89Nc **180**
Allenby Rd. UB1: S'hall 41Ca **85**
Allen Cl. CR4: Mitc 67Kb **134**
Allen Cl. TW16: Sun 67X **129**
Allen Cl. WD7: Shenl 4Na **15**
Allen Ct. AL10: Hat 2D **8**
Allen Ct. E17 30Cc **52**
(off Yunus Khan Cl.)
Allendale Av. UB1: S'hall 44Ca **85**
Allendale Cl. DA2: Dart 60Td **120**
Allendale Cl. SE26 64Zb **136**
Allendale Cl. SE5 54Tb **113**
Allendale Rd. HA0: Wemb 37Ka **66**
Allendale Rd. UB6: G'frd 37Ka **66**
Allen Edwards Dr. SW8 53Nb **112**
Allenford Ho. SW15 58Va **110**
(off Tunworth Cres.)
Allen Ho. W8 48Cb **89**
(off Allen St.)
Allen Ho. Pk. GU22: Wok 2N **187**
Allen Mans. W8 48Cb **89**
(off Allen St.)
Allen Rd. BR3: Beck 68Zb **136**
Allen Rd. CR0: C'don 74Qb **156**
Allen Rd. E3 40Bc **72**
Allen Rd. KT23: Bookh 98Da **191**
Allen Rd. N16 35Ub **71**
Allen Rd. RM13: Rain 40Ld **77**
Allen Rd. TW16: Sun 67X **129**
Allensbury Pl. NW1 38Mb **70**
Allens La. TN15: Plax 100Ce **205**
Allens Mead DA12: Grav'nd 10H **123**
Allens Rd. EN3: Pond E 15Yb **34**
Allen St. W8 48Cb **89**
Allensway SS17: Stan H 1P **101**
Allenswood SW19 60Ab **110**
Allenswood Rd. SE9 55Nc **116**
Allen Way SL3: Dat 3N **103**
Allerds Rd. SL2: Farn R 9D **60**
Allerds Way SL2: Farn R 9F **60**
Allerford Ct. HA2: Harr 29Ea **46**
Allerford Rd. SE6 62Dc **136**
Allerton Cl. WD6: Bore 10Pa **15**
Allerton Gro. IG7: Chig 23Ad **55**
Allerton Ho. N1 3G **219** (41Tb **91**)
(off Provost St.)
Allerton Rd. N16 33Sb **71**
Allerton Rd. WD6: Bore 10Na **15**
Allerton St. N1 3F **219** (41Tb **91**)
Allerton Wlk. N7 33Pb **70**
Allestree Rd. SW6 52Ab **110**
Alleyn Cres. SE21 61Tb **135**
Alleyndale Rd. RM8: Dag 33Yc **75**
Alleyn Ho. SE1 4G **231** (48Tb **91**)
(off Burbage Cl.)
Alleyn Pk. SE21 61Tb **135**
Alleyn Pk. UB2: S'hall 50Ca **85**
Alleyn Rd. SE21 62Tb **135**
Alleys, The HP2: Hem H 1M **3**
Alley Way UB8: Uxb 38M **63**
Allfarthing La. SW18 58Db **111**
Allgood Cl. SM4: Mord 72Za **154**
Allgood St. E2 40Vb **71**
Allhallows La. EC4 5F **225** (45Tb **91**)
All Hallows Rd. N17 25Ub **51**
Allhallows Rd. E6 43Nc **94**
Allhusen Gdns. SL3: Ful 35A **62**
Allhusen Pl. SL2: Stoke P 8L **61**
Alliance Cl. HA0: Wemb 35Ma **67**
Alliance Cl. TW4: Houn 57Ba **107**
Alliance Ct. TW15: Ashf 63S **128**
Alliance Ct. W3 43Ra **87**
Alliance Rd. E13 43Lc **93**
Alliance Rd. SE18 51Wc **117**
Alliance Rd. W3 42Ra **87**
Allianz Pk. 24Ya **48**
Allied Ct. N1 38Ub **71**
(off Enfield Rd.)
Allied Ind. Est. W3 47Ua **88**
Allied Way W3 47Ua **88**
Allingham Cl. W7 45Ha **86**
Allingham Ct. BR2: Broml 70Hc **137**
Allingham M. N1 1D **218** (40Sb **71**)
(off Allingham St.)
Allingham Rd. RH2: Reig 9J **207**
Allingham Rd. SW4 58Mb **112**
Allingham St. N1 1D **218** (40Sb **71**)
Allington Av. N17 23Ub **51**
Allington Av. TW17: Shep 69U **128**
Allington Cl. DA12: Grav'nd 10H **123**
Allington Cl. SW19 64Za **132**
Allington Cl. UB6: G'frd 38Ea **66**
Allington Ct. CR0: C'don 72Yb **158**
(off Chart Cl.)
Allington Ct. EN3: Pond E 15Zb **34**
Allington Ct. SL2: Slou 4K **81**
Allington Ct. SW8 54Lb **112**
Allington Rd. BR6: Orp 75Tc **160**
Allington Rd. HA2: Harr 29Ea **46**
Allington Rd. NW4 29Xa **48**
Allington Rd. W10 41Ab **88**
Allingtons RH2: Reig 3J **207**
Allington St. SW1 4A **228** (48Kb **90**)
Allington Way BR8: Swan 70Hd **140**
Allison Cl. EN9: Walt A 4Jc **21**
Allison Cl. SE10 53Ec **114**
Allison Gro. SE21 60Ub **113**
Allison Rd. N8 29Qb **50**
Allison Rd. W3 44Sa **87**
Alliston Ho. E2 41Vb **91**
(off Gibraltar Wlk.)
Allistonway SS17: Stan H 1P **101**
Allitsen Rd. NW8 2D **214** (40Gb **69**)
(not continuous)
Allium Ri. DA1: Dart 56Md **119**
Allkins Ct. SL4: Wind 4H **103**
All Nations Ho. E8 38Xb **71**
(off Martello St.)
Allnutts Rd. CM16: Epp 5Wc **23**
Allnutt Way SW4 57Mb **112**
Alloa Rd. IG3: Ilf 33Wc **75**
Alloa Rd. SE8 50Zb **92**
Allom Ho. W11 45Ab **88**
(off Clarendon Rd.)
Allonby Dr. HA4: Ruis 31R **64**
Allonby Gdns. HA9: Wemb 32La **66**
Allotment La. TN13: S'oaks 94Ld **203**
Allotment Way NW2 34Za **68**
Alloway Cl. GU21: Wok 10M **167**
Alloway Rd. E3 41Ac **92**
Alloy Ho. SE14 51Bc **114**
(off Moulding La.)
Allport Ho. SE5 55Tb **113**
(off Champion Pk.)
Allport M. E1 42Yb **92**
(off Hayfield Pas.)
All Saints Cl. DA10: Swans 57Be **121**
All Saints Cl. IG7: Chig 20Xc **37**
All Saints Cl. N9 19Wb **33**
All Saints Cl. SW8 53Nb **112**
All Saint's Ct. TW5: Hest 53Z **107**
(off Springwell Rd.)
All Saints Ct. E1 45Yb **92**
(off Johnson St.)
All Saints Ct. SW11 52Kb **112**
(off Prince of Wales Dr.)
All Saints Cres. WD25: Wat 5Z **13**
All Saints Dr. CR2: Sande 84Vb **177**
All Saints Dr. SE3 54Gc **115**
(not continuous)
All Saints Ho. W11 43Bb **89**
(off All Saints Rd.)
All Saints La. WD3: Crox G 16Q **26**
All Saints M. HA3: Hrw W 23Ga **46**
All Saints Pas. SW18 57Cb **111**
All Saints Rd. DA11: Nflt 10B **122**
All Saints Rd. GU18: Light 2A **166**
All Saints Rd. SM1: Sutt 76Db **155**
All Saints Rd. SW19 66Eb **133**
(not continuous)
All Saints Rd. W11 43Bb **89**
All Saints Rd. W3 48Sa **87**
(off Ashbridge St.)
All Saints Russian Orthodox Cathedral 2D **226** (47Gb **89**)
All Saints Station (DLR) 45Dc **92**
All Saints St. N1 1H **217** (40Pb **70**)
All Saints Wlk. SE15 52Vb **113**
Allsop Pl. NW1 6G **215** (42Hb **89**)
All Souls Av. NW10 40Xa **68**
All Souls' Pl. W1 1A **222** (43Kb **90**)
Allum Gro. KT20: Tad 93Xa **194**
Allum La. WD6: E'tree 15Ma **29**
Allum Way N20 18Eb **31**
Alluvium Ct. SE1 3H **231** (48Ub **91**)
(off Long La.)
Allwood Cl. SE26 63Zb **136**
Allyn Cl. TW18: Staines 65H **127**
Alma, The DA12: Grav'nd 4H **145**
Alma Av. E4 24Ec **52**
Alma Av. RM12: Horn 35Nd **77**
Alma Barn M. BR6: Orp 75Zc **161**
Alma Birk Ho. NW6 38Ab **68**
Almack Rd. E5 35Yb **72**
Alma Cl. GU21: Knap 9J **167**
Alma Cl. N10 25Kb **50**
Alma Cl. CR3: Cat'm 93Sb **197**
(off Coulsdon Rd.)
Alma Ct. EN6: Pot B 2Eb **17**
Alma Ct. HA2: Harr 33Fa **66**
Alma Ct. SL1: Burn 1A **80**
Alma Ct. WD6: Bore 10Pa **15**
Alma Cres. SM1: Sutt 78Ab **154**
Alma Cut AL1: St A 3C **6**
Alma Gro. SE1 49Vb **91**
Alma Ho. N9 21Wb **51**
Alma Ho. TW8: Bford 51Na **109**
Almanza Pl. IG11: Bark 40Xc **75**
Alma Pl. CR7: Thor H 71Qb **156**
Alma Pl. NW10 41Xa **88**
Alma Pl. SE19 66Vb **135**
Alma Pl. WD25: Wat 8Aa **13**
Alma Rd. AL1: St A 3C **6**
Alma Rd. BR5: Orp 75Zc **161**
Alma Rd. DA10: Swans 57Be **121**
Alma Rd. DA14: Sidc 62Wc **139**
Alma Rd. EN3: Enf H 15Ac **34**
Alma Rd. EN3: Pond E 15Ac **34**
Alma Rd. KT10: Esh 74Ga **152**
Alma Rd. N10 24Kb **50**
Alma Rd. RH2: Reig 5K **207**
Alma Rd. SL4: Eton W 9D **80**
Alma Rd. SL4: Wind 4G **102**
Alma Rd. SM5: Cars 78Gb **155**
Alma Rd. SW18 56Eb **111**
Alma Rd. UB1: S'hall 45Aa **85**
Alma Rd. Ind. Est. EN3: Pond E 14Zb **34**
Alma Row HA3: Hrw W 25Fa **46**
Alma Sq. NW8 2A **214** (40Eb **69**)
Alma St. E15 37Fc **73**
Alma St. NW5 37Kb **70**
Alma Ter. E3 39Bc **72**
(off Beale Rd.)
Alma Ter. SW18 59Fb **111**
Alma Ter. W8 48Cb **89**
Almeida St. N1 39Rb **71**
Almeida Theatre 39Rb **71**
(off Almeida St.)
Almeric Rd. SW11 56Hb **111**
Almer Rd. SW20 66Wa **132**
Almington St. N4 32Pb **70**
Almners Rd. KT16: Lyne 74C **148**
(not continuous)
Almond Av. GU22: Wok 3P **187**
Almond Av. SM5: Cars 75Hb **155**
Almond Av. UB10: Ick 34R **64**
Almond Av. UB7: W Dray 48Q **84**
Almond Av. W5 48Ma **87**
Almond Cl. BR2: Broml 73Qc **160**
Almond Cl. E17 28Ac **52**
Almond Cl. HA4: Ruis 34V **64**
Almond Cl. RM16: Grays 8C **100**
Almond Cl. SE15 54Wb **113**
Almond Cl. SL4: Wind 4F **102**
Almond Cl. TW13: Felt 60W **106**
Almond Cl. TW17: Shep 68S **128**
Almond Cl. TW20: Eng G 5M **125**
Almond Cl. UB3: Hayes 45U **84**
Almond Dr. BR8: Swan 68Fd **140**
Almond Gro. TW8: Bford 52Ka **108**
Almond Ho. E15 41Gc **93**
(off Teasel Way)
Almond Rd. DA2: Dart 59Sd **120**
Almond Rd. KT19: Eps 83Ta **173**
Almond Rd. N17 24Wb **51**
Almond Rd. SE16 49Xb **91**
Almond Rd. SL1: Burn 1A **80**
Almonds, The AL1: St A 6F **6**
Almonds Av. IG9: Buck H 19Jc **35**
Almond Wlk. AL10: Hat 3C **8**
Almond Way BR2: Broml 73Qc **160**
Almond Way CR4: Mitc 71Mb **156**
Almond Way HA2: Harr 26Da **45**
Almond Way WD6: Bore 14Ra **29**
Almons Way SL2: Slou 3M **81**
Almorah Rd. N1 38Tb **71**
Almorah Rd. TW5: Hest 53Z **107**
Alms Heath GU23: Ock 93R **190**
Almshouse La. EN1: Enf 9Xb **19**
Almshouse La. KT9: Chess 81La **172**
Almshouses IG10: Lough 11Pc **36**
Almshouses KT16: Chert 73J **149**
Alms Ho's., The IG11: Bark 37Sc **74**
Almshouses, The EN8: Chesh 2Zb **20**
(off Turner's Hill)
Alms Row TN16: Bras 96Xc **201**
Alnmouth Ct. UB1: S'hall 44Ea **86**
(off Fleming Rd.)
Alnwick N17 24Xb **51**
Alnwick Ct. DA2: Dart 58Rd **119**
(off Osbourne Rd.)
Alnwick Gro. SM4: Mord 70Db **133**
Alnwick Rd. E16 44Lc **93**
Alnwick Rd. SE12 58Kc **115**
ALPERTON 40Na **67**
Alperton La. HA0: Wemb 41Ma **87**
Alperton La. UB6: G'frd 41La **86**
Alperton Station (Underground) 39Ma **67**
Alperton St. W10 42Bb **89**
Alphabet Gdns. SM5: Cars 72Fb **155**
Alphabet M. SW9 53Qb **112**
Alphabet Sq. E3 43Cc **92**
Alpha Bus. Pk. AL9: Wel G 5E **8**
Alpha Cl. NW1 5E **214** (42Gb **89**)
Alpha Ct. CR3: Whyt 90Wb **177**
Alpha Ct. WD17: Wat 13Y **27**
(off Grosvenor Rd.)
Alpha Est. UB3: Hayes 47U **84**
Alpha Gro. E14 47Cc **92**
Alpha Ho. NW6 40Cb **69**
Alpha Ho. NW8 6D **214** (42Gb **89**)
(off Ashbridge St.)
Alpha Ho. SW4 56Pb **112**
Alpha Pl. NW6 40Cb **69**
Alpha Pl. SM4: Mord 74Za **154**
Alpha Pl. SW3 51Gb **111**
Alpha Rd. CM13: Hut 16Fe **41**
Alpha Rd. CR0: C'don 74Ub **157**
Alpha Rd. E4 20Cc **34**
Alpha Rd. EN3: Pond E 14Ac **34**
Alpha Rd. GU22: Wok 88D **168**
Alpha Rd. GU24: Chob 2K **167**
Alpha Rd. KT5: Surb 72Pa **153**
Alpha Rd. N18 23Wb **51**
Alpha Rd. SE14 53Bc **114**
Alpha Rd. TW11: Tedd 64Fa **130**
Alpha Rd. UB10: Hil 42R **84**
Alpha St. SE15 54Wb **113**
Alpha St. Nth. SL1: Slou 7L **81**
Alpha St. Sth. SL1: Slou 8K **81**
Alpha Way TW20: Thorpe 67E **126**
Alphea Cl. SW19 66Gb **133**
Alpine Av. KT5: Surb 75Sa **153**
Alpine Bus. Cen. E6 43Qc **94**
Alpine Cl. CR0: C'don 76Ub **157**
Alpine Cl. KT19: Ewe 78Sa **153**
Alpine Cl. RM3: Rom 21Nd **57**
Alpine Cl. SL5: S'hill 2B **146**
Alpine Copse BR1: Broml 68Qc **138**
Alpine Gro. E9 38Yb **72**
Alpine Rd. E10 33Dc **72**
Alpine Rd. KT12: Walt T 73W **150**
Alpine Rd. NW9 28Qa **47**
Alpine Rd. RH1: Redh 3A **208**
Alpine Rd. SE16 50Zb **92**
Alpine Vw. SM5: Cars 78Gb **155**
Alpine Wlk. HA7: Stan 19Ga **28**
Alpine Way E6 43Qc **94**
Alric Av. KT3: N Mald 69Ua **132**
Alric Av. NW10 38Ta **67**
Alroy Rd. N4 31Qb **70**
Alsace Rd. SE17 7H **231** (50Ub **91**)
Alsager KT22: Lea 91Ja **192**
(off Clements Mead)
Alscot Rd. SE1 5K **231** (49Vb **91**)
Alscot Way SE1 5K **231** (49Vb **91**)
Alsford Pl. SL3: L'ly 9P **81**
Alsike Rd. DA18: Erith 48Zc **95**
Alsike Rd. SE2 48Zc **95**
Alsom Av. KT4: Wor Pk 77Wa **154**
Alsop Cl. AL2: Lon C 9J **7**
Alston Cl. KT6: Surb 73Ka **152**
Alstonfield KT10: Esh 78Ea **152**
Alston Rd. EN5: Barn 13Ab **30**
Alston Rd. HP1: Hem H 3J **3**
Alston Rd. N18 22Xb **51**
Alston Rd. SW17 63Fb **133**
Alston Works EN5: Barn 12Ab **30**
Altab Ali Pk. 44Wb **91**
(off Adler St.)
Altair Cl. N17 23Vb **51**
Altair Way HA6: Nwood 21V **44**
Altamont CR6: W'ham 91Xb **197**
Altash Way SE9 61Pc **138**
Altenburg Av. W13 48Ka **86**
Altenburg Gdns. SW11 56Hb **111**
Alterton Cl. GU21: Wok 9L **167**
Alt Gro. SW19 66Bb **133**
Altham Ct. HA2: Harr 25Da **45**
Altham Gdns. WD19: Wat 21Z **45**
Altham Rd. HA5: Pinn 24Aa **45**
Altham Way WD19: Wat 22Aa **45**
Althea St. SW6 54Db **111**
Althorne Gdns. E18 28Hc **53**
Althorne Rd. RH1: Redh 8A **208**
Althorne Way RM10: Dag 33Cd **76**
Althorp Cl. EN5: Ark 17Wa **30**
Althorpe M. SW11 53Fb **111**
Althorpe Rd. HA1: Harr 29Ea **46**
Althorp Rd. AL1: St A 1D **6**
Althorp Rd. SW17 60Hb **111**
Altima Ct. SE22 56Wb **113**
(off E. Dulwich Rd.)
Altior Ct. N6 30Lb **50**
Altissima Ho. SW11 52Kb **112**
Altitude Apts. CR0: C'don 76Tb **157**
(off Altyre Rd.)
Altitude Point E1 44Wb **91**
(off Alie St.)
Altius Apts. E3 40Cc **72**
(off Wick La.)
Altius Ct. E4 23Ec **52**
Altius Wlk. E20 37Ec **72**
Altmore Av. E6 38Pc **74**
Alton Av. HA7: Stan 24Ha **46**
Altona Way SL1: Slou 4F **80**
Alton Cl. DA5: Bexl 60Ad **117**
Alton Cl. TW7: Isle 54Ha **108**
Alton Cotts. DA4: Eyns 74Nd **163**
Alton Ct. TW18: Staines 67G **126**
Alton Gdns. BR3: Beck 66Cc **136**
Alton Gdns. TW2: Whitt 59Fa **108**
Alton Ho. E3 41Dc **92**
(off Bromley High St.)
Alton Ho. RH1: Redh 4A **208**
Alton Rd. CR0: Wadd 76Qb **156**
Alton Rd. N17 27Tb **51**
Alton Rd. SW15 60Wa **110**
Alton Rd. TW9: Rich 56Na **109**
Alton St. E14 43Dc **92**
Altura Twr. SW11 54Fb **111**
Altus Ho. SE6 63Ec **136**
Altwood Cl. SL1: Slou 3C **80**
Altyre Cl. BR3: Beck 71Bc **158**
Altyre Rd. CR0: C'don 75Tb **157**
Altyre Way BR3: Beck 71Bc **158**
Aluna Ct. SE15 55Yb **114**
Aluric Cl. RM16: Grays 9D **100**
Alvanley Gdns. NW6 36Db **69**
Alva Way WD19: Wat 19Z **27**
Alverstoke Rd. RM3: Rom 24Nd **57**
Alverstone Av. EN4: E Barn 17Gb **31**
Alverstone Av. SW19 61Cb **133**
Alverstone Gdns. SE9 60Sc **116**
Alverstone Ho. SE11 51Qb **112**
Alverstone Rd. E12 35Qc **74**
Alverstone Rd. HA9: Wemb 32Pa **67**
Alverstone Rd. KT3: N Mald 70Va **132**
Alverstone Rd. NW2 38Ya **68**
Alverston Gdns. SE25 71Ub **157**
Alverton St. SE8 50Bc **92**
(not continuous)
Alveston Av. HA3: Kenton 27Ka **46**
Alveston Sq. E18 26Jc **53**
Alvey St. SE17 7H **231** (50Ub **91**)
Alvia Gdns. SM1: Sutt 77Eb **155**
Alvington Cres. E8 36Vb **71**
Alvista Av. SL6: Tap 4A **80**
Alway Av. KT19: Ewe 78Ta **153**
Alwen Gro. RM15: S Ock 43Xd **98**
Alwin Pl. WD18: Wat 14U **26**
Alwold Cres. SE12 58Kc **115**
Alwoodley Cl. SS17: Stan H 3K **101**
Alwyn Av. W4 50Ta **87**
Alwyn Cl. CR0: New Ad 80Dc **158**
Alwyn Cl. WD6: E'tree 16Pa **29**
Alwyne Av. CM15: Shenf 16Ce **41**
Alwyne Ct. GU21: Wok 88A **168**
Alwyne La. N1 38Rb **71**
Alwyne Pl. N1 37Sb **71**
Alwyne Rd. N1 38Sb **71**
Alwyne Rd. SW19 65Bb **133**
Alwyne Rd. W7 45Ga **86**
Alwyne Sq. N1 37Sb **71**
Alwyne Vs. N1 38Rb **71**
Alwyn Gdns. NW4 28Wa **48**
Alwyn Gdns. W3 44Ra **87**
Alwyns Cl. KT16: Chert 72J **149**
Alwyns La. KT16: Chert 72H **149**
Alyth Gdns. NW11 30Cb **49**
Alzette Ho. E2 40Zb **72**
(off Mace St.)
Amalgamated Dr. TW8: Bford 51Ka **108**
Amanda Cl. IG7: Chig 23Tc **54**
Amanda Ct. SL3: L'ly 8P **81**
Amanda Ct. TW15: Ashf 61P **127**
(off Edward Way)
Amanda M. RM7: Rom 29Ed **56**
Amar Ct. SE18 49Vc **95**
Amar Deep Ct. SE18 50Vc **95**
Amarelle Apts. CR0: C'don 74Tb **157**
(off Cherry Orchard Rd.)
Amazon Bldg. N8 28Pb **50**
Amazon St. E1 44Wb **91**
Ambassador, The SL5: S'dale 3F **146**
Ambassador Bldg. SW11 51Mb **112**
Ambassador Cl. TW3: Houn 54Aa **107**
Ambassador Gdns. E6 43Pc **94**
Ambassador Ho. CR7 70Sb **135**
(off Brigstock Rd.)
Ambassador Ho. CR7: Thor H 70Sb **135**
(off Brigstock Rd.)
Ambassador Ho. NW8 1A **214** (39Eb **69**)
Ambassadors Cinema Woking 89A **168**
(off Victoria Way)
Ambassador's Ct. SW1 7C **222** (46Lb **90**)
(off St James's Pal.)
Ambassadors Ct. E8 38Vb **71**
(off Holly St.)
Ambassador Sq. E14 49Dc **92**
Ambassadors Theatre 4E **222** (45Mb **90**)
(off West St.)

Amber Av. E17 25Ac 52
Amber Cl. EN5: New Bar 16Db 31
Amber Cl. KT17: Eps D 87Xa 174
Amber Ct. CR0: C'don 74Ub 157
Amber Ct. E15 39Ec 72
(off Warton Rd.)
Amber Ct. KT5: Surb 73Pa 153
Amber Ct. N7 37Qb 70
(off Bride St.)
Amber Ct. SW17 63Jb 134
(off Brudenell Rd.)
Amber Ct. TW18: Staines 64H 127
(off Laleham Rd.)
Ambercroft Way CR5: Coul 91Rb 197
Amberden Av. N3 27Cb 49
Ambergate St. SE17 7B 230 (50Rb 91)
Amber Gro. NW2 32Za 68
Amber Ho. E1 44Zb 92
(off Aylward St.)
Amber La. IG6: Ilf 24Rc 54
Amberley Cl. BR6: Chels 78Vc 161
Amberley Cl. GU23: Send 97H 189
Amberley Cl. HA5: Pinn 27Ba 45
Amberley Ct. BR3: Beck 66Bc 136
Amberley Ct. DA14: Sidc 64Yc 139
Amberley Dr. KT15: Wdhm 82H 169
Amberley Gdns. EN1: Enf 17Ub 33
Amberley Gdns. KT19: Ewe 77Va 154
Amberley Gro. CR0: C'don 73Vb 157
Amberley Gro. SE26 64Xb 135
Amberley Pl. SL4: Wind 3H 103
(off Peascod St.)
Amberley Rd. E10 31Cc 72
Amberley Rd. EN1: Enf 17Vb 33
Amberley Rd. IG9: Buck H 18Lc 35
Amberley Rd. N13 19Pb 32
Amberley Rd. SE2 51Zc 117
Amberley Rd. SL2: Slou 3C 80
Amberley Rd. W9 43Cb 89
Amberley Ter. WD19: Wat 16Aa 27
(off Villiers Rd.)
Amberley Way RM7: Mawney 28Dd 56
Amberley Way SM4: Mord 73Bb 155
Amberley Way TW4: Houn 57Y 107
Amberley Way UB10: Uxb 40N 63
Amberlith Ho. CR7: Thor H 71Qb 156
(off Thornton Rd.)
Amber M. N22 27Qb 50
(off High Rd.)
Amberside Cl. TW7: Isle 58Fa 108
Amberside Ct. HP3: Hem H 5K 3
Amber Way W3 47Ua 88
Amber Wharf E2 39Vb 71
(off Nursery La.)
Amberwood Cl. SM6: W'gton 78Nb 156
Amberwood Ri. KT3: N Mald 72Ua 154
Amblecote KT11: Cobh 83Aa 171
Amblecote Cl. SE12 62Kc 137
Amblecote Mdws. SE12 62Kc 137
Amblecote Rd. SE12 62Kc 137
Ambler Rd. N4 34Rb 71
Ambleside BR1: Broml 65Fc 137
Ambleside CM16: Epp 3Wc 23
Ambleside NW1 2A 216 (40Kb 70)
(off Augustus St.)
Ambleside RM19: Purf 50Sd 98
Ambleside SW19 60Ab 110
Ambleside Av. BR3: Beck 71Ac 158
Ambleside Av. KT12: Walt T 74Y 151
Ambleside Av. RM12: Horn 36Kd 77
Ambleside Av. SW16 63Mb 134
Ambleside Cl. E10 31Dc 72
Ambleside Cl. E9 36Yb 72
Ambleside Cl. N17 27Vb 51
Ambleside Cl. RH1: Redh 10B 208
Ambleside Cres. EN3: Enf H 13Zb 34
Ambleside Dr. TW14: Felt 60V 106
Ambleside Gdns. CR2: Sels 81Zb 178
Ambleside Gdns. HA9: Wemb 32Ma 67
Ambleside Gdns. IG4: Ilf 28Nc 54
Ambleside Gdns. SM2: Sutt 79Eb 155
Ambleside Gdns. SW16 64Mb 134
Ambleside Point SE15 52Yb 114
(off Tustin Est.)
Ambleside Rd. DA7: Bex 54Cd 118
Ambleside Rd. NW10 38Va 68
Ambleside Wlk. UB8: Uxb 39M 63
Ambleside Way TW20: Egh 66D 126
Ambrey Way SM6: W'gton 81Mb 176
Ambridge Ho. CR3: Cat'm 95Tb 197
Ambrook Rd. DA17: Belv 48Cd 96
Ambrosden Av. SW1 4B 228 (48Lb 90)
Ambrose Av. NW11 31Ab 68
Ambrose Cl. BR6: Orp 76Vc 161
Ambrose Cl. DA1: Cray 56Hd 118
Ambrose Cl. E6 43Pc 94
Ambrose Ct. N18 23Vb 51
(off Cannon Rd.)
Ambrose Cres. CM16: Epp 2Uc 22
Ambrose Ho. E14 43Cc 92
(off Selsey St.)
Ambrose M. SW11 54Hb 111
Ambrose St. SE16 49Xb 91
Ambrose Wlk. E3 40Cc 72
Ambulance Rd. E11 29Fc 53
AMC Bus. Cen. NW10 41Ra 87
Amelia Cl. TW4: Houn 60Aa 107
Amelia Cl. W3 46Ra 87
Amelia Ct. CR8: Purl 82Pb 176
Amelia Ho. E14 44Gc 93
(off Lyell St.)
Amelia Ho. NW9 26Va 48
(off Boulevard Dr.)
Amelia Ho. TW9: Kew 52Ra 109
Amelia Ho. W6 50Ya 88
(off Queen Caroline St.)
Amelia Mans. E20 37Dc 72
(off Olympic Pk. Av.)
Amelia St. SE17 7D 230 (50Sb 91)
Amelle Gdns. RM3: Hrld W 23Rd 57
Amen Cnr. EC4 3C 224 (44Rb 91)
Amen Cnr. SW17 65Hb 133
Amen Ct. EC4 2C 224 (44Rb 91)
Amenity Way SM4: Mord 73Ya 154
Amerara Ct. DA3: Lfield 69Be 143
(off Harrison Av.)
Amerden Way SL1: Slou 7E 80
American International University in London, The Kensington Campus, Ansdell Street 48Db 89
(off Ansdell St.)
American International University in London, The Kensington Campus, St Albans Grove 48Db 89
American International University in London, The Kensington Campus, Young Street 47Db 89
American International University in London, The Richmond Hill Campus 58Na 109
American University of London, The 34Pb 70
America Sq. EC3 4K 225 (45Vb 91)
America St. SE1 7D 224 (46Sb 91)
Amerland Rd. SW18 57Bb 111
Amersham Av. N18 23Tb 51
Amersham Cl. RM3: Rom 23Pd 57
Amersham Dr. RM3: Rom 23Nd 57
Amersham Gro. SE14 52Bc 114
Amersham Ho. WD18: Wat 17U 26
(off Chenies Way)
Amersham Rd. CR0: C'don 72Sb 157
Amersham Rd. HP6: L Chal 11A 24
Amersham Rd. RM3: Rom 23Nd 57
Amersham Rd. SE14 53Bc 114
Amersham Rd. SL9: Ger X 30C 42
Amersham Rd. UB9: Den 32E 62
Amersham Rd. WD3: Chen 11C 24
Amersham Va. SE14 52Bc 114
Amersham Wlk. RM3: Rom 23Pd 57
Amersham Way HP6: L Chal 11A 24
Amery Gdns. NW10 39Ya 68
Amery Gdns. RM2: Rom 27Md 57
Amery Ho. SE17 7J 231 (50Ub 91)
(off Kinglake St.)
Amery Rd. HA1: Harr 33Ja 66
Amesbury EN9: Walt A 4Jc 21
Amesbury Av. SW2 61Nb 134
Amesbury Cl. CM16: Epp 3Vc 23
Amesbury Cl. KT4: Wor Pk 74Ya 154
Amesbury Ct. EN2: Enf 12Qb 32
Amesbury Dr. E4 16Dc 34
Amesbury Rd. BR1: Broml 69Mc 137
Amesbury Rd. CM16: Epp 3Vc 23
Amesbury Rd. RM9: Dag 38Zc 75
Amesbury Rd. SL1: Slou 7D 80
Amesbury Rd. TW13: Felt 61Z 129
Amesbury Twr. SW8 54Lb 112
Ames Cotts. E14 43Ac 92
(off Maroon St.)
Ames Ho. E2 40Zb 72
(off Mace St.)
Ames Rd. DA10: Swans 58Ae 121
Amethyst Cl. EN5: Ark 15Va 30
Amethyst Cl. N11 24Mb 50
Amethyst Ct. BR6: Chels 78Uc 160
(off Farnborough Hill)
Amethyst Ct. DA8: Erith 52Jd 118
Amethyst Ct. EN3: Enf H 13Ac 34
(off Enstone Rd.)
Amethyst Rd. E15 35Fc 73
Amey Dr. KT23: Bookh 96Ea 192
Amherst Av. W13 44La 86
Amherst Cl. BR5: St M Cry 70Xc 139
Amherst Dr. BR5: St M Cry 70Vc 139
Amherst Gdns. W13 44La 86
(off Amherst Rd.)
Amherst Hill TN13: Riv 95Hd 202
Amherst Ho. SE16 47Zb 92
(off Wolfe Cres.)
Amherst Pl. TN13: Riv 94Hd 202
Amherst Rd. TN13: S'oaks 94Kd 203
Amherst Rd. W13 44La 86
Amhurst Ct. EN3: Enf W 9Zb 20
Amhurst Gdns. TW7: Isle 54Ja 108
Amhurst Pde. N16 31Vb 71
(off Amhurst Pk.)
Amhurst Pk. N16 31Tb 71
Amhurst Pas. E8 35Wb 71
Amhurst Rd. E8 36Xb 71
Amhurst Rd. N16 35Vb 71
Amhurst Ter. E8 35Wb 71
Amhurst Wlk. SE28 46Wc 95
Amias Dr. HA8: Edg 21Na 47
Amias Ho. EC1 5D 218 (42Sb 91)
(off Central St.)
Amicia Gdns. SL2: Stoke P 9K 61
Amidas Gdns. RM8: Dag 35Xc 75
Amiel St. E1 42Yb 92
Amies St. SW11 55Hb 111
Amigo Ho. SE1 3A 230 (48Qb 90)
(off Morley St.)
Amina Way SE16 48Wb 91
Amiot Ho. NW9 26Va 48
(off Heritage Av.)
Amis Av. KT15: New H 82J 169
Amis Av. KT19: Ewe 79Ra 153
Amisha Ct. SE1 4K 231 (48Vb 91)
(off Grange Rd.)
Amis Rd. GU21: Wok 1J 187
Amity Ct. DA9: Ghithe 56Wd 120
Amity Gro. SW20 67Xa 132
Amity Rd. E15 38Hc 73
Ammanford Grn. NW9 30Ua 48
Ammonite Ho. E15 38Hc 73
Amner Rd. SW11 58Jb 112
Amor Rd. W6 48Ya 88
Amory Ho. N1 1J 217 (39Pb 70)
(off Barnsbury Est.)
Amott Rd. SE15 55Wb 113
Amoy Pl. E14 45Cc 92
(not continuous)
Ampere Way CR0: Wadd 73Nb 156
Ampere Way Stop (London Tramlink) 74Pb 156
Amphlett Cl. DA13: Sflt 65Ce 143
Ampleforth Cl. BR6: Chels 77Yc 161
Ampleforth Rd. SE2 47Xc 95
Amport Pl. NW7 23Ab 48
Ampthill Est. NW1 2B 216 (40Lb 70)
Ampthill Ho. RM3: Rom 22Md 57
(off Montgomery Cres.)
Ampthill Sq. NW1 2C 216 (40Lb 70)
Ampton Pl. WC1 4H 217 (41Pb 70)
Ampton St. WC1 4H 217 (41Pb 70)
Amroth Cl. SE23 60Xb 113
Amroth Grn. NW9 30Ua 48
Amstel Ct. SE15 52Vb 113
Amstel Way GU21: Wok 10K 167
Amsterdam Rd. E14 48Ec 92
Amundsen Ct. E14 50Cc 92
(off Napier Av.)
Amundsen Ho. NW10 38Ta 67
(off Stonebridge Pk.)
Amwell Cl. EN2: Enf 15Tb 33
Amwell Cl. WD25: Wat 7Aa 13
Amwell Ct. EN9: Walt A 5Hc 21
Amwell Ct. Est. N4 33Sb 71
Amwell Ho. WC1 3K 217 (41Qb 70)
(off Cruikshank St.)
Amwell St. EC1 3K 217 (41Qb 70)
Amwell Vw. IG6: Chig 22Xc 55
Amwell Vw. IG6: Ilf 22Xc 55
Amyand Cotts. TW1: Twick 58Ka 108
Amyand La. TW1: Twick 59Ka 108
Amyand Pk. Gdns. TW1: Twick 59Ka 108
Amyand Pk. Rd. TW1: Twick 59Ja 108
Amy Cl. SM6: W'gton 80Nb 156
Amy Ct. CR0: C'don 78Rb 157
Amy Johnson Ct. HA8: Edg 26Ra 47
Amy Rd. RH8: Oxt 1J 211
Amyruth Rd. SE4 57Cc 114
Amy Warne Cl. E6 42Nc 94
Anarth Ct. KT13: Weyb 74U 150
Anastasia M. N12 22Db 49
Anatola Rd. N19 33Lb 70
Anayah Apts. SE8 50Zb 92
(off Trundleys Rd.)
Ancaster Cres. KT3: N Mald 72Wa 154
Ancaster M. BR3: Beck 69Zb 136
Ancaster Rd. BR3: Beck 69Zb 136
Ancaster St. SE18 52Uc 116
Anchor SW18 56Db 111
Anchorage Cl. SW19 64Cb 133
Anchorage Ho. E14 45Fc 93
(off Clove Cres.)
Anchorage Point E14 47Bc 92
(off Cuba St.)
Anchorage Point Ind. Est. SE7 48Lc 93
Anchor & Hope La. SE7 48Kc 93
Anchor Bay Ind. Est. DA8: Erith 51Jd 118
Anchor Blvd. DA2: Dart 56Sd 120
Anchor Brewhouse SE1 7K 225 (46Vb 91)
Anchor Bus. Cen. CR0: Bedd 76Nb 156
Anchor Cl. IG11: Bark 41Xc 95
Anchor Ct. DA8: Erith 52Hd 118
Anchor Ct. EN1: Enf 15Ub 33
Anchor Ct. RM17: Grays 52De 121
Anchor Ct. SW1 6D 228 (49Mb 90)
(off Vauxhall Bri. Rd.)
Anchor Cres. GU21: Knap 9H 167
Anchor Dr. N15 28Ub 51
Anchor Dr. RM13: Rain 41Kd 97
Anchor Hill GU21: Knap 9H 167
Anchor Ho. E16: Barking Rd. 43Hc 93
(off Barking Rd.)
Anchor Ho. E16: Prince Regent La. 44Lc 93
(off Prince Regent La.)
Anchor Ho. EC1 5D 218 (42Sb 91)
(off Old St.)
Anchor Ho. SW10 51Fb 111
(off Cremorne Est.)
Anchor Iron Wharf SE10 50Fc 93
Anchor La. HP1: Hem H 4K 3
(not continuous)
Anchor M. N1 37Ub 71
Anchor M. SW12 58Kb 112
Anchor Retail Pk. Stepney Green 42Yb 92
Anchor St. SE16 49Xb 91
Anchor Ter. E1 42Yb 92
Anchor Ter. SE1 6E 224 (46Sb 91)
(off Southwark Bri. Rd.)
Anchor Wharf E3 43Dc 92
(off Yeo St.)
Anchor Yd. EC1 5D 218 (42Sb 91)
Ancient Almshouses EN8: Chesh 2Zb 20
(off Turner's Hill)
Ancill Cl. W6 51Ab 110
Ancona Rd. NW10 40Wa 68
Ancona Rd. SE18 50Tc 94
Andace Pk. Gdns. BR1: Broml 68Lc 137
Andalus Rd. SW9 55Nb 112
Ander Cl. HA0: Wemb 35Ma 67
Andermans SL4: Wind 3B 102
Anderson Cl. KT19: Eps 84Ra 173
Anderson Cl. N21 15Pb 32
Anderson Cl. SM3: Sutt 74Cb 155
Anderson Cl. UB9: Hare 25J 43
Anderson Cl. W3 44Ta 87
Anderson Ct. NW2 32Ya 68
Anderson Ct. RH1: Redh 9A 208
Anderson Dr. TW15: Ashf 63S 128
Anderson Hgts. SW16 68Pb 134
Anderson Ho. AL4: St A 3H 7
Anderson Ho. E14 45Ec 92
(off Woolmore St.)
Anderson Ho. IG11: Bark 39Tc 74
Anderson Ho. SW17 64Fb 133
Anderson Pl. TW3: Houn 56Da 107
Anderson Rd. E9 37Zb 72
Anderson Rd. IG8: Wfd G 27Mc 53
Anderson Rd. KT13: Weyb 76T 150
Anderson Rd. SE3 56Kc 115
Anderson Rd. WD7: Shenl 5Qa 15
Anderson Sq. N1 39Rb 71
(off Gaskin St.)
Anderson St. SW3 7F 227 (50Hb 89)
Anderson Way DA17: Belv 47Dd 96
Anderton Cl. SE5 55Tb 113
Anderton Ct. N22 26Mb 50
Andora Ct. NW6 38Ab 68
(off Brondesbury Pk.)
Andora Ho. E10 32Ac 72
(off Colin Cl.)
Andorra Ct. BR1: Broml 67Lc 137
Andover Av. E16 44Mc 93
Andover Cl. KT19: Eps 83Ta 173
Andover Cl. TW14: Felt 60V 106
Andover Cl. UB6: G'frd 42Da 85
Andover Cl. UB8: Uxb 40K 63
Andover Ct. E2 42Zb 92
(off Thee Colts La.)
Andover Ct. TW19: Stanw 59M 105
Andover Pl. NW6 40Db 69
Andover Rd. BR6: Orp 74Tc 160
Andover Rd. N7 33Pb 70
Andover Rd. TW2: Twick 60Fa 108
Andoversford Ct. SE15 51Ub 113
(off Bibury Cl.)
Andover Ter. W6 49Xa 88
(off Raynham Rd.)
Andrea Av. RM16: Grays 47Ce 99
Andreck Ct. BR3: Beck 68Ec 136
(off Crescent Rd.)
Andre St. E8 36Wb 71
Andrew Cl. DA1: Cray 57Fd 118
Andrew Cl. IG6: Ilf 23Tc 54
Andrew Cl. WD7: Shenl 5Pa 15
Andrew Ct. SE23 61Zb 136
Andrewes Gdns. E6 44Nc 94
Andrewes Highwalk EC2 1E 224 (43Sb 91)
(off Fore St.)
Andrewes Ho. EC2 1E 224 (43Sb 91)
(off Fore St.)
Andrewes Ho. SM1: Sutt 77Cb 155
Andrew Gibb Memorial, The 53Hc 115
Andrew Hill La. SL2: Hedg 3G 60
Andrew Pl. SW8 52Mb 112
Andrew Reed Ct. WD24: Wat 12Y 27
(off Keele Cl.)
Andrew Reed Ho. SW18 59Ab 110
(off Linstead Way)
Andrew's Cl. BR5: St P 68Zc 139
Andrew's Cl. KT17: Eps 86Va 174
Andrews Cl. HA1: Harr 31Fa 66
Andrews Cl. HP2: Hem H 1M 3
Andrews Cl. IG9: Buck H 19Lc 35
Andrews Cl. KT4: Wor Pk 75Ya 154
Andrews Crosse WC2 3K 223 (44Qb 90)
(off Chancery La.)
Andrews Ga. TW17: Shep 68S 128
Andrew's Ho. CR2: S Croy 79Sb 157
Andrews Ho. NW3 38Hb 69
(off Fellows Rd.)
Andrews Pl. DA2: Wilm 61Gd 140
Andrews Pl. SE9 58Rc 116
Andrew's Rd. E8 39Xb 71
Andrew St. E14 44Ec 92
Andrews Wlk. SE17 51Rb 113
Andringham Lodge BR1: Broml 67Kc 137
(off Palace Gro.)
Andrula Ct. N22 25Rb 51
Andwell Cl. SE2 47Xc 95
Anelle Ri. HP3: Hem H 6P 3
ANERLEY 67Xb 135
Anerley Gro. SE19 66Vb 135
Anerley Hill SE19 65Vb 135
Anerley Pk. SE20 66Wb 135
Anerley Pk. Rd. SE20 66Xb 135
Anerley Rd. SE19 66Wb 135
Anerley Rd. SE20 66Wb 135
Anerley Station (Rail & Overground) 67Xb 135
Anerley Sta. Rd. SE20 67Xb 135
Anerley St. SW11 54Hb 111
Anerley Va. SE19 66Vb 135
Aneurin Bevan Ct. NW2 33Xa 68
Aneurin Bevan Ho. N11 24Mb 50
Anfield Cl. SW12 59Lb 112
Angas Ct. KT13: Weyb 78S 150
ANGEL 2A 218 (40Qb 70)
Angela Carter Cl. SW9 55Qb 112
Angela Davies Ind. Est. SE24 56Rb 113
Angela Hooper Pl. SW1 3C 228 (48Lb 90)
(off Victoria St.)
Angel Bldg. N1 2A 218 (40Qb 70)
Angel Cl. N18 22Vb 51
Angel Cl. TW12: Hamp H 64Ea 130
Angel Cnr. Pde. N18 21Wb 51
Angel Ct. E15 37Fc 73
Angel Ct. EC2 2G 225 (44Tb 91)
Angel Ct. SW1 7C 222 (46Lb 90)
ANGEL EDMONTON 22Wb 51
Angelfield TW3: Houn 56Da 107
Angel Ga. EC1 3C 218 (41Rb 91)
(not continuous)
Angel Hill SM1: Sutt 76Db 155
Angel Hill Dr. SM1: Sutt 76Db 155
Angel Ho. E3 41Cc 92
(off Campbell Rd.)
Angelica Cl. UB7: Yiew 44N 83
Angelica Dr. E6 43Qc 94
Angelica Gdns. CR0: C'don 74Zb 158
Angelica Ho. E3 39Bc 72
(off Sycamore Av.)
Angelica Rd. GU24: Bisl 7E 166
Angelina Ho. SE15 53Wb 113
(off Goldsmith Rd.)
Angelis Apts. N1 2C 218 (40Rb 71)
(off Graham St.)
Angel La. E15 37Fc 73
Angel La. EC4 5F 225 (45Tb 91)
Angel La. UB3: Hayes 43T 84
Angell Pk. Gdns. SW9 55Qb 112
Angell Rd. SW9 55Qb 112
ANGELL TOWN 53Qb 112
Angell Town Est. SW9 54Qb 112
Angel M. E1 45Xb 91
Angel M. N1 2A 218 (40Qb 70)
Angel M. SW15 59Wa 110
Angelo M. SW16 69Pb 134
Angelo's SL4: Eton 10H 81
(off Common La.)
Angel Pl. N18 21Wb 51
Angel Pl. RH2: Reig 9K 207
Angel Pl. SE1 1F 231 (47Tb 91)
Angel Rd. HA1: Harr 30Ga 46
Angel Rd. KT7: T Ditt 73Ja 152
Angel Rd. N18 22Wb 51
Angel Rd. Works N18 22Yb 52
Angel Sq. EC1 2B 218 (40Rb 71)
Angel Station (Underground) 1A 218 (40Qb 70)
Angel St. EC1 2D 224 (44Sb 91)
Angel Wlk. W6 49Ya 88
Angel Way RM1: Rom 29Gd 56
Angel Wharf N1 1E 218 (40Sb 71)
Angel Yd. N6 32Jb 70
Angerstein Bus. Pk. SE10 49Jc 93
Angerstein La. SE3 53Hc 115
Angie M. DA1: Dart 54Pd 119
Anglais M. NW9 28Ua 48
Anglebury W2 44Cb 89
(off Talbot Rd.)
Angle Cl. UB10: Hil 39Q 64
Angle Grn. RM8: Dag 32Yc 75
Angle Rd. RM20: Grays 51Zd 121
Anglers, The KT1: King T 69Ma 131
(off High St.)
Anglers Cl. TW10: Ham 63La 130
Angler's La. NW5 37Kb 70
Anglers Reach KT6: Surb 71Ma 153
Anglesea Av. SE18 49Rc 94
Anglesea Ho. KT1: King T 70Ma 131
(off Anglesea Rd.)
Anglesea M. SE18 49Rc 94
Anglesea Pl. DA11: Grav'nd 8D 122
(off Clive Rd.)
Anglesea Rd. BR5: St M Cry 72Yc 161
Anglesea Rd. KT1: King T 70Ma 131
Anglesea Rd. SE18 49Rc 94
Anglesea Ter. W6 48Xa 88
(off Wellesley Av.)
Anglesey Cl. TW15: Ashf 62Q 128
Anglesey Ct. Rd. SM5: Cars 79Jb 156
Anglesey Dr. RM13: Rain 42Jd 96
Anglesey Gdns. SM5: Cars 79Jb 156
Anglesey Ho. E14 44Cc 92
(off Lindfield St.)
Anglesey Rd. EN3: Pond E 14Xb 33
Anglesey Rd. WD19: Wat 22Y 45
Anglesmede Cres. HA5: Pinn 27Ca 45
Anglesmede Way HA5: Pinn 27Ca 45
Angles Rd. SW16 63Nb 134
Anglia Cl. N17 24Xb 51
Anglia Ct. RM8: Dag 32Zc 75
(off Spring Cl.)
Anglia Ho. E14 44Ac 92
(off Salmon La.)
Anglian Cl. WD24: Wat 12Y 27
Anglian Ind. Est. IG11: Bark 42Vc 95
Anglian Rd. E11 34Fc 73
Anglia Wlk. E6 39Qc 74
(off Napier Rd.)
Anglia Way RM15: S Ock 42Xd 98
Anglo Rd. E3 40Bc 72
Angora Cl. SM6: W'gton 75Jb 156
Angrave Ct. E8 39Vb 71
(off Scriven St.)
Angrave Pas. E8 39Vb 71
Angus Cl. KT9: Chess 78Qa 153
Angus Dr. HA4: Ruis 35Y 65
Angus Gdns. NW9 25Ta 47
Angus Ho. SW2 59Mb 112
Angus Rd. E13 41Lc 93
Angus St. SE14 52Ac 114
Anhalt Rd. SW11 52Gb 111
Anise Ct. DA11: Nflt 59Fe 121
Ankerdine Cres. SE18 52Rc 116
Anlaby Rd. TW11: Tedd 64Ga 130
Anley Rd. W14 47Za 88
Anmer Cl. KT20: Tad 92Xa 194
Anmersh Gro. HA7: Stan 25Ma 47
Annabel Cl. E14 44Dc 92
Annabell Av. RM16: Ors 4F 100
Annabels M. W5 42Ma 87
Annables Ct. AL1: St A 2C 6
(off Victoria St.)
Anna Cl. E8 39Vb 71
Annadale Ct. RH1: Redh 5P 207
(off Warwick Rd.)
Annalee Gdns. RM15: S Ock 43Xd 98
Annalee Rd. RM15: S Ock 43Xd 98
Annaleigh Pl. KT12: Hers 77Z 151
Annandale Gro. UB10: Ick 34S 64
Annandale Rd. CR0: C'don 75Wb 157
Annandale Rd. DA15: Sidc 59Uc 116
Annandale Rd. SE10 51Hc 115
Annandale Rd. W4 50Ua 88
Annan Dr. SM5: Cars 81Jb 176
Anna Neagle Cl. E7 35Jc 73
Annan Way RM1: Rom 25Gd 56
Anne Boleyn Ct. SE9 58Sc 116
Anne Boleyn's Wlk. KT2: King T 64Na 131
Anne Boleyn's Wlk. SM3: Cheam 80Za 154
Anne Case M. KT3: N Mald 69Ta 131
Anne Compton M. SE12 59Hc 115
Anne Goodman Ho. E1 44Yb 92
(off Jubilee St.)
Anne Heart Cl. RM16: Chaf H 49Zd 99
Anne Matthews Ct. E14 43Cc 92
(off Selsey St.)
Anne M. IG11: Bark 38Sc 74
Anne Nastri Ct. RM2: Rom 29Kd 57
(off Heath Pk. Rd.)
Anne of Cleeves Ct. SE9 58Tc 116
Anne of Cleves Rd. DA1: Dart 57Md 119
Anners Cl. TW20: Thorpe 69E 126
Annes Ct. NW1 5E 214 (42Gb 89)
(off Palgrave Gdns.)
Annesley Apts. E3 43Cc 92
(off Gresham Pl.)
Annesley Av. NW9 27Ta 47
Annesley Cl. NW10 34Ua 68
Annesley Dr. CR0: C'don 76Bc 158
Annesley Ho. SW9 53Qb 112
Annesley Pl. BR2: Broml 72Nc 160
Annesley Rd. SE3 53Kc 115
Annesley Wlk. N19 33Lb 70
Annesmere Gdns. SE3 55Mc 115
Anne St. E13 42Jc 93
Anne Sutherland Ho. BR3: Beck 66Ac 136
Anne's Wlk. CR3: Cat'm 92Ub 197
Annett Cl. TW17: Shep 70U 128
Annette Cl. HA3: W'stone 26Ga 46
Annette Cres. N1 38Sb 71
Annette Rd. N7 34Pb 70
(not continuous)
Annett Rd. KT12: Walt T 73W 150
Annetts Hall TN15: Bor G 91Ce 205
Anne Way IG6: Ilf 23Sc 54
Anne Way KT8: W Mole 70Da 129
Annexe Mkt. E1 7K 219 (43Vb 91)
(off Spital Sq.)
Annie Besant Cl. E3 39Bc 72
Annie Brookes Cl. TW18: Staines 62F 126
Annie Taylor Ho. E12 35Qc 74
(off Walton Rd.)
Annifer Way RM15: S Ock 43Xd 98
Anningsley Pk. KT16: Ott 82D 168
ANNINGSLEY PARK 82E 168
Anning St. EC2 5J 219 (42Ub 91)
Annington Rd. N2 27Hb 49
Annis Rd. E9 37Ac 72
Ann La. SW10 51Fb 111
Ann Moss Way SE16 48Yb 92
Ann's Cl. SW1 2G 227 (47Hb 89)
(off Kinnerton St.)
Ann's Pl. E1 1K 225 (43Vb 91)
(off Wentworth St.)
Ann St. N1 1D 218 (39Sb 71)
Ann St. SE18 50Sc 94
(not continuous)
Ann Stroud Ct. SE12 57Jc 115
Anns Wlk. RH1: Blet 5K 209
Annsworthy Av. CR7: Thor H 69Tb 135
Annsworthy Cres. SE25 68Tb 135
Ansar Gdns. E17 29Bc 52
Ansculf Rd. SL2: Slou 1E 80
Ansdell Rd. SE15 54Yb 114
Ansdell St. W8 48Db 89
Ansdell Ter. W8 48Db 89
Ansell Gro. SM5: Cars 74Jb 156
Ansell Ho. E1 43Yb 92
(off Mile End Rd.)
Ansell Rd. SW17 62Gb 133
Anselm Cl. CR0: C'don 76Vb 157
Anselm Rd. HA5: Hat E 24Ba 45
Anselm Rd. SW6 51Cb 111
Ansford Rd. BR1: Broml 64Ec 136
Ansleigh Pl. W11 45Za 88
Anslie Cl. CR2: Sande 86Xb 177
Anslow Gdns. SL0: Iver H 40F 62
Anslow Pl. SL1: Slou 3A 80
Anson Cl. AL1: St A 4F 6
Anson Cl. HP3: Bov 9B 2
Anson Cl. RM7: Mawney 26Dd 56
Anson Ho. E1 42Ac 92
(off Shandy St.)
Anson Ho. SW1 51Lb 112
(off Churchill Gdns.)
Anson M. SW19 66Cb 133
Anson Pl. SE28 47Tc 94
Anson Rd. N7 35Lb 70
Anson Rd. NW2 35Xa 68
Anson Ter. UB5: N'olt 37Da 65
Anson Wlk. HA6: Nwood 21S 44
Anstead Dr. RM13: Rain 40Jd 76
Anstey Ct. W3 47Ra 87
Anstey Ho. E9 39Yb 72
(off Templecombe Rd.)
Anstey Rd. SE15 55Wb 113
Anstey Wlk. N15 28Rb 51
Anstice Cl. W4 52Ua 110
Anstridge Path SE9 58Tc 116
Anstridge Rd. SE9 58Tc 116
Antelope Av. RM16: Grays 48Ce 99
Antelope Rd. SE18 48Pc 94
Antelope Wlk. KT6: Surb 71Ma 153

Antenor Ho. E2 40Xb 71
(off Old Bethnal Grn. Rd.)
Anthems Way E20 37Dc 72
Anthony Cl. NW7 21Ua 48
Anthony Cl. TN13: Dun G 92Gd 202
Anthony Cl. WD19: Wat 18Y 27
Anthony Cope Ct. N1 3G 219 (41Tb 91)
(off Chart St.)
Anthony Ct. W3 47Ua 88
Anthony Ho. CR5: Coul 94Mb 196
Anthony Ho. NW8 6D 214 (42Gb 89)
(off Ashbridge St.)
Anthony La. BR8: Swan 67Jd 140
Anthony M. RM13: Rain 40Ld 77
Anthony Rd. DA16: Well 53Wc 117
Anthony Rd. SE25 72Wb 157
Anthony Rd. UB6: G'frd 41Ga 86
Anthony Rd. WD6: Bore 12Pa 29
ANTHONYS 84C 168
Anthony St. E1 44Xb 91
Anthony Way N18 23Zb 52
Anthony Way SL1: Slou 5B 80
Anthorne Cl. EN6: Pot B 3Db 17
Anthus M. HA6: Nwood 24U 44
Antigua M. E13 41Kc 93
Antigua Wlk. SE19 64Tb 135
Antila Ct. E1 5K 219 (42Vb 91)
(off Sclater St.)
Antilles Bay E14 47Ec 92
Antill Rd. E3 41Ac 92
Antill Rd. N15 28Wb 51
Antill Ter. E1 44Zb 92
Antlers Hill E4 15Dc 34
Antoinette Ct. WD5: Ab L 1V 12
Antoinette M. NW2 37Wa 68
Anton Cres. SM1: Sutt 76Cb 155
Antoneys Cl. HA5: Pinn 26Z 45
Antonine Ga. AL3: St A 3N 5
Antonine Hgts. SE1 2H 231 (47Ub 91)
(off City Wlk.)
Anton Pl. HA9: Wemb 34Ra 67
Anton Rd. RM15: S Ock 43Xd 98
Anton St. E8 36Wb 71
Antony Ho. SE16 49Yb 92
(off Raymouth Rd.)
Antrim Gro. NW3 37Hb 69
Antrim Rd. NW3 37Hb 69
Antrobus Cl. SM1: Sutt 78Bb 155
Antrobus Rd. W4 49Sa 87
Antwerp Way E16 47Qc 94
Anvil Cl. HP3: Bov 10D 2
Anvil Cl. SW16 66Lb 134
Anvil Ct. SL3: L'ly 49C 82
Anvil La. KT11: Cobh 86W 170
Anvil M. RM13: Rain 39Hd 76
Anvil M. WD25: A'ham 12Ea 28
Anvil Pl. AL2: Chis G 8N 5
Anvil Rd. TW16: Sun 69W 128
Anvil Ter. DA2: Wilm 61Gd 140
Anworth Cl. IG8: Wfd G 23Kc 53
Anyards Rd. KT11: Cobh 85X 171
Anytime Fitness Woking 89B 168
(off Chertsey Rd.)
Anzio Gdns. CR3: Cat'm 93Sb 197
Apeldoorn Dr. SM6: W'gton 81Nb 176
Apelles St. SE18 53Nc 116
(off Tellson Av.)
Aperdele Rd. KT22: Lea 90Ja 172
APERFIELD 89Nc 180
Aperfield Rd. DA8: Erith 51Hd 118
Aperfield Rd. TN16: Big H 89Nc 180
Aperfields TN16: Big H 89Nc 180
Apers Av. GU22: Wok 93B 188
Apex Bus. Pk. DA12: Grav'nd 9P 123
Apex Cl. BR3: Beck 67Dc 136
APEX CORNER Feltham 62Ba 129
APEX CORNER Mill Hill 21Ua 48
Apex Ct. KT14: W Byf 84K 169
Apex Ct. W13 45Ja 86
Apex Ho. BR5: St M Cry 70Yc 139
(off Ridge Pl.)
Apex Ho. E1 5K 219 (42Vb 91)
(off Bacon St.)
Apex Ind. Est. NW10 42Va 88
Apex Pde. NW7 21Ta 47
(off Selvage La.)
Apex Point AL9: Wel G 4E 8
Apex Retail Pk. 62Ba 129
Aphrodite Ct. E14 49Cc 92
(off Homer Dr.)
Aphrodite Ct. E15 39Ec 72
(off Warton Rd.)
Apley Rd. RH2: Reig 9J 207
Aplin Way TW7: Isle 53Ga 108
Apollo Av. BR1: Broml 67Kc 137
Apollo Av. HA6: Nwood 22W 44
Apollo Bldg. E14 49Cc 92
Apollo Bus. Cen. SE8 50Zb 92
Apollo Cl. RM12: Horn 33Kd 77
Apollo Ct. E1 45Wb 91
(off Thomas More St.)
Apollo Ct. E15 39Ec 72
(off High St.)
Apollo Ct. SW9 53Qb 112
(off Southey Rd.)
Apollo Ho. E2 40Xb 71
(off St Jude's Rd.)
Apollo Ho. E3 40Cc 72
(off Garrison Rd.)
Apollo Ho. N6 31Hb 69
Apollo Ho. SW10 52Fb 111
(off Milman's St.)
Apollo Pl. E11 34Gc 73
Apollo Pl. GU21: Wok 1L 187
Apollo Pl. SW10 52Fb 111
Apollo Theatre Piccadilly 4D 222 (45Mb 90)
(off Shaftesbury Av.)
Apollo Victoria Theatre 4B 228 (48Lb 90)
(off Wilton Rd.)
Apollo Way DA8: Erith 49Fd 96
Apollo Way SE28 48Tc 94
Apostle Ct. SE20 66Zb 136
Apostles, The SE23 61Zb 136
(off Church Ri.)
Apostle Way CR7: Thor H 68Rb 135
Apothecary St. EC4 3B 224 (44Rb 91)
Appach Rd. SW2 57Qb 112
Appian Ct. E3 40Bc 72
(off Parnell Rd.)
Apple Barn Ct. TN15: Crou 94Ee 205
Apple Blossom Ct. SW8 52Mb 112
(off Pascal St.)
Appleby Cl. BR5: Pet W 73Uc 160
Appleby Cl. E4 23Ec 52
Appleby Cl. N15 29Tb 51
Appleby Cl. TW2: Twick 61Fa 130
Appleby Cl. UB8: Hil 44S 84
Appleby Ct. SE6 58Cc 114
Appleby Ct. W3 47Sa 87
(off Newport Rd.)
Appleby Dr. RM3: Rom 22Ld 57
Appleby Dr. WD3: Crox G 14R 26
Appleby Gdns. TW14: Felt 60V 106
Appleby Grn. RM3: Rom 22Ld 57
Appleby Ho. KT19: Eps 83Ta 173
Appleby Rd. E16 44Hc 93
Appleby Rd. E8 38Wb 71
Appleby St. E2 1K 219 (40Vb 71)
Apple Cotts. HP3: Bov 9C 2
Apple Ct. KT21: Asht 92Ma 193
Applecroft AL2: Park 10P 5
Appledore Av. DA7: Bex 53Ed 118
Appledore Av. HA4: Ruis 34X 65
Appledore Cl. BR2: Broml 71Hc 159
Appledore Cl. HA8: Edg 25Qa 47
Appledore Cl. RM3: Rom 25Ld 57
Appledore Cl. SW17 61Hb 133
Appledore Cres. DA14: Sidc 62Uc 138
Appledore Way NW7 24Za 48
Appledown Ri. CR5: Coul 87Lb 176
Appleford Ho. W10 42Ab 88
(off Bosworth Rd.)
Appleford Rd. W10 42Ab 88
Apple Gth. TW8: Bford 49Ma 87
Applegarth CR0: New Ad 80Dc 158
(not continuous)
Applegarth KT10: Clay 78Ha 152
Applegarth Dr. DA1: Dart 61Nd 141
Applegarth Dr. IG2: Ilf 28Vc 55
Applegarth Ho. DA8: Erith 54Hd 118
Applegarth Ho. SE1 1C 230 (47Rb 91)
(off Nelson Sq.)
Applegarth Ho. SE15 52Wb 113
(off Bird in Bush Rd.)
Applegarth Rd. SE28 46Xc 95
Applegarth Rd. W14 48Za 88
Apple Ga. CM14: Pil H 15Vd 40
Applegate Ho. E20 36Ec 72
(off Victory Pde.)
Apple Gro. EN1: Enf 13Ub 33
Apple Gro. HA2: Harr 32Ca 65
Apple Gro. KT9: Chess 77Na 153
Apple Gro. TW1: Twick 58Ja 108
Apple Mkt. KT1: King T 68Ma 131
(off Market Pl.)
Apple Orchard BR8: Swan 70Fd 140
Apple Orchard, The HP2: Hem H 1P 3
Apple Rd. E11 34Gc 73
Appleshaw Cl. DA11: Grav'nd 4C 144
Appleshaw Ho. SE5 55Ub 113
Appleton Cl. DA7: Bex 54Ed 118
Appleton Dr. DA2: Wilm 62Kd 141
Appleton Gdns. KT3: N Mald 72Wa 154
Appleton Rd. IG10: Lough 13Rc 36
Appleton Rd. SE9 55Nc 116
Appleton Sq. CR4: Mitc 67Gb 133
Appleton Way RM12: Horn 32Md 77
Apple Tree Av. UB7: Yiew 44P 83
Apple Tree Av. UB8: Hil 43P 83
Apple Tree Cl. KT22: Fet 96Ea 192
Appletree Cl. SE20 67Xb 135
Appletree Gdns. EN4: E Barn 14Gb 31
Apple Tree La. RM13: Rain 40Fd 76
Appletree La. SL3: L'ly 8N 81
APPLE TREE RDBT. 44P 83
Apple Trees Pl. GU22: Wok 1N 187
Appletree Wlk. WD25: Wat 6X 13
Apple Tree Yd. E1 41Zb 92
(off Longnor Rd.)
Apple Tree Yd. SW1 6C 222 (46Lb 90)
Applewood Cl. N20 18Gb 31
Applewood Cl. NW2 34Xa 68
Applewood Cl. UB10: Ick 35N 63
Applewood Dr. E13 42Kc 93
Apple Yd. SE20 67Wb 135
Appleyard Ter. EN3: Enf W 9Yb 20
Appold St. DA8: Erith 51Hd 118
Appold St. EC2 7H 219 (43Ub 91)
Apprentice Gdns. UB5: N'olt 41Ba 85
Apprentice Way E5 35Xb 71
Approach, The BR6: Orp 75Vc 161
Approach, The EN1: Enf 12Xb 33
Approach, The EN6: Pot B 4Bb 17
Approach, The KT23: Bookh 95Aa 191
Approach, The NW4 29Za 48
Approach, The RM14: Upm 34Rd 77
Approach, The W3 44Ta 87
Approach Cl. N16 35Ub 71
Approach Rd. AL1: St A 3C 6
Approach Rd. CR8: Purl 84Qb 176
Approach Rd. E2 40Yb 72
Approach Rd. EN4: E Barn 14Fb 31
Approach Rd. HA8: Edg 23Qa 47
Approach Rd. KT8: W Mole 71Ca 151
Approach Rd. SW20 68Ya 132
Approach Rd. TW15: Ashf 65S 128
Apps Mdw. Cl. KT8: W Mole 70Ba 129
Appspond La. AL2: Pot C 5J 5
Appspond La. HP3: Hem H 5H 5
Aprey Gdns. NW4 28Ya 48
April Cl. BR6: Chels 78Vc 161
April Cl. KT21: Asht 90Pa 173
April Cl. TW13: Felt 62W 128
April Cl. W7 45Ga 86
April Ct. E2 40Wb 71
(off Teale St.)
April Glen SE23 62Zb 136
April St. E8 35Vb 71
Aprilwood Cl. KT15: Wdhm 83H 169
Apsledene DA12: Grav'nd 5F 144
APSLEY 6N 3
Apsley Cen., The NW2 33Wa 68
Apsley Cl. HA2: Harr 29Ea 46
Apsley Ct. BR5: St M Cry 72Xc 161
Apsley Grange HP3: Hem H 7N 3
Apsley Ho. E1 43Yb 92
(off Stepney Way)
Apsley Ho. NW8 1B 214 (40Fb 69)
(off Finchley Rd.)
Apsley Ho. SL1: Slou 7L 81
Apsley Ho. SW15 58Wa 110
(off Holford Way)
Apsley Ho. TW4: Houn 56Ba 107
Apsley House 1J 227 (47Jb 90)
Apsley Ind. Est. HP3: Hem H 6M 3
Apsley Lock HP3: Hem H 7N 3
Apsley Mills Retail Pk. 6N 3
Apsley Rd. KT3: N Mald 69Sa 132
Apsley Rd. SE25 70Xb 135
Apsley Station (Rail) 7N 3
Apsley Wlk. SL0: Rich P 48H 83
Apsley Way NW2 33Wa 68
Apsley Way W1 1J 227 (47Jb 90)
(not continuous)
Aqua Ho. NW10 41Qa 87
Aqua Ho. UB8: Cowl 44L 83
Aquarelle Ho. EC1 3D 218 (41Sb 91)
Aquarius TW1: Twick 60Ka 108
Aquarius Bus. Pk. NW2 32Wa 68
(off Priestley Way)
Aquarius Ct. HA8: Edg 24Qa 47
Aquarius Golf Course 58Yb 114
Aquarius Way HA6: Nwood 22W 44
Aqua Vista Sq. E3 43Cc 92
(off Bow Comn. La.)
Aquila Cl. KT22: Lea 93Na 193
Aquila St. NW8 1C 214 (40Fb 69)
Aquinas St. SE1 7A 224 (46Qb 90)
Arabella Ct. NW8 1A 214 (40Eb 69)
(off Marlborough Pl.)
Arabella Dr. SW15 56Ua 110
Arabella St. SE16 48Wb 91
Arabia Cl. E4 17Fc 35
Arabian Ho. E1 42Ac 92
(off Ernest St.)
Arabin Rd. SE4 56Ac 114
Arado Ho. NW9 26Va 48
(off Boulevard Dr.)
Araglen Av. RM15: S Ock 43Xd 98
Aragon Av. KT17: Ewe 82Xa 174
Aragon Av. KT7: T Ditt 71Ha 152
Aragon Cl. BR2: Broml 74Pc 160
Aragon Cl. CR0: New Ad 82Gc 179
Aragon Cl. EN2: Enf 10Pb 18
Aragon Cl. IG10: Lough 16Nc 36
Aragon Cl. RM5: Col R 23Dd 56
Aragon Cl. TW16: Sun 65V 128
Aragon Ct. GU21: Knap 9H 167
Aragon Ct. IG6: Ilf 24Sc 54
Aragon Ct. KT8: E Mos 70Ea 130
Aragon Ct. SE11 7K 229 (50Qb 90)
(off Hotspur St.)
Aragon Dr. HA4: Ruis 32Z 65
Aragon Dr. IG6: Ilf 24Sc 54
Aragon Ho. E16 46Jc 93
(off Capulet M.)
Aragon M. CM16: Epp 1Yc 23
Aragon Pl. SM4: Mord 73Ab 154
Aragon Rd. KT2: King T 64Na 131
Aragon Rd. SM4: Mord 72Za 154
Aragon Twr. SE8 49Bc 92
Aragon Wlk. KT14: Byfl 85P 169
Aral Ho. E1 42Zb 92
(off Ernest St.)
Aram Ct. GU22: Wok 87E 168
Aran Ct. KT13: Weyb 75T 150
Arandora Cres. RM6: Chad H 31Xc 75
Aran Dr. HA7: Stan 21La 46
Aran Lodge NW6 38Cb 69
(off Woodchurch Rd.)
Aran M. N7 38Qb 70
(off St Clements St.)
Arapiles Ho. E14 44Fc 93
(off Blair St.)
Arbery Rd. E3 41Ac 92
Arbon Ct. N1 39Sb 71
(off Linton St.)
Arbor Cl. BR3: Beck 68Dc 136
Arbor Ct. N16 33Tb 71
Arboretum Ct. N1 37Tb 71
(off Dove Rd.)
Arboretum Pl. IG11: Bark 38Sc 74
(off Clockhouse Av.)
Arborfield Cl. SL1: Slou 8J 81
Arborfield Cl. SW2 60Pb 112
Arborfield Ho. E14 45Cc 92
(off E. India Dock Rd.)
Arbor Ho. BR6: Orp 75Wc 161
(off Orchard Gro.)
Arbor Ho. SE14 51Bc 114
Arbor Ho. TW8: Bford 52La 108
Arbor Rd. E4 20Fc 35
Arbot Ho. SE10 49Gc 93
(off Manilla Wlk.)
Arbour Cl. CM14: W'ley 22Yd 58
Arbour Cl. KT22: Fet 95Ha 192
Arbour Ho. E1 44Zb 92
(off Arbour Sq.)
Arbour Pk. 4L 81
Arbour Rd. EN3: Pond E 13Zb 34
Arbour Sq. E1 44Zb 92
Arbour Way RM12: Horn 36Kd 77
Arbroath Grn. WD19: Wat 20W 26
Arbroath Rd. SE9 55Nc 116
Arbrook Chase KT10: Esh 79Ea 152
Arbrook Cl. BR5: St P 69Wc 139
Arbrook Hall KT10: Clay 79Ha 152
Arbrook La. KT10: Esh 79Ea 152
Arbury Ter. SE26 62Wb 135
Arbus Cres. HA2: Harr 27Ea 46
Arbuthnot La. DA5: Bexl 58Ad 117
Arbuthnot Rd. SE14 54Zb 114
Arbuthnot Rd. SM4: Mord 70Eb 133
Arbutus Cl. RH1: Redh 8L 207
Arbutus Rd. RH1: Redh 9L 207
Arbutus St. E8 39Vb 71
Arc, The Caterham 94Sb 197
Arcade CR0: C'don 75Sb 157
Arcade, The CR0: C'don 76Sb 157
(off High St.)
Arcade, The E20 37Ec 72
(within Westfield Shop. Cen.)
Arcade, The EC2 1H 225 (43Ub 91)
(off Liverpool St.)
Arcade, The N7 35Nb 70
(off Macready Pl.)
Arcade, The RH1: Redh 5P 207
(off Station Rd.)
Arcade, The RM3: Rom 22Md 57
(off Farnham Rd.)
Arcade, The SE9 58Qc 116
(off High St.)
Arcade Chambers SE9 58Qc 116
Arcade Pde. KT9: Chess 78Ma 153
Arcade Pl. RM1: Rom 29Gd 56
Arcadia Av. N3 26Cb 49
Arcadia Caravans TW18: Staines 67K 127
Arcadia Cen., The 45Ma 87
Arcadia Cl. SM5: Cars 77Jb 156
Arcadia Ct. E1 2K 225 (44Vb 91)
(off Old Castle St.)
Arcadia M. KT3: N Mald 69Ua 132
Arcadian Av. DA5: Bexl 58Ad 117
Arcadian Cl. DA5: Bexl 58Ad 117
Arcadian Gdns. N22 24Pb 50
Arcadian Pl. SW18 59Bb 111
Arcadian Rd. DA5: Bexl 58Ad 117
Arcadia Rd. DA13: Ist R 7B 144
Arcadia St. E14 44Cc 92
Arcany Rd. RM15: S Ock 42Xd 98
Arc Ct. N11 22Kb 50
Arc Ct. RM7: Rush G 30Gd 56
ArcelorMittal Orbit 38Dc 72
AR Cen. 41Ac 92
(off Rhondda Gro.)
AR Cen. 6H 219 (42Ub 91)
(off Worship St.)
Archangel St. SE16 47Zb 92
Archates Av. RM16: Grays 48Ce 99
Archbishop Lanfranc School Sports Cen. 72Nb 156
Archbishop's Pl. SW2 59Pb 112
Archdale Bus. Cen. HA2: Harr 33Ea 66
Archdale Ct. W12 46Xa 88
Archdale Ho. SE1 3H 231 (48Ub 91)
(off Long La.)
Archdale Pl. KT3: N Mald 69Ra 131
Archdale Rd. SE22 57Vb 113
Archel Rd. W14 51Bb 111
Archer Apts. N1 1H 219 (40Ub 71)
(off Fern Cl.)
Archer Cl. CM16: Coop 1Zc 23
Archer Cl. EN5: Barn 16Bb 31
Archer Cl. KT2: King T 66Na 131
Archer Cl. WD4: K Lan 1P 11
Archer Ho. N1 1J 219 (39Ub 71)
(off Whitmore Est.)
Archer Ho. SE14 53Ac 114
Archer Ho. SW11 53Fb 111
Archer Ho. W11 45Bb 89
(off Westbourne Gro.)
Archer Ho. W13 46Ka 86
(off Sherwood Cl.)
Archer M. SW9 55Nb 112
Archer M. TW12: Hamp H 65Ea 130
Archer Rd. BR5: St M Cry 71Wc 161
Archer Rd. SE25 70Xb 135
Archers Cl. EN6: Pot B 4Ab 16
Archers Ct. CR2: S Croy 78Sb 157
(off Nottingham Rd.)
Archers Ct. RH1: Redh 7P 207
(off Brighton Rd.)
Archers Ct. RM15: S Ock 43Xd 98
Archers Dr. EN3: Enf H 12Yb 34
Archers Lodge SE16 50Wb 91
(off Culloden Cl.)
Archers Pl. KT22: Oxs 87Ga 172
Archer Sq. SE14 51Ac 114
Archer St. W1 4D 222 (45Mb 90)
Archer Ter. UB7: Yiew 45N 83
Archer Way BR8: Swan 68Hd 140
Archery Cl. HA3: W'stone 27Ha 46
Archery Cl. W2 3E 220 (44Gb 89)
Archery Flds. Ho. WC1 3K 217 (41Qb 90)
(off Wharton St.)
Archery La. BR2: Broml 72Mc 159
Archery Rd. SE9 57Pc 116
Archery Steps W2 4E 220 (45Gb 89)
(off St George's Flds.)
Arches SW8 51Nb 112
Arches, The E16 42Gc 93
Arches, The HA2: Harr 33Da 65
Arches, The NW1 38Kb 70
Arches, The SE8 51Ac 114
Arches, The SL4: Wind 3G 102
(off Alma Rd.)
Arches, The SW8 52Mb 112
Arches, The WC2 6G 223 (46Nb 90)
(off Villiers St.)
Arches Bus. Cen., The UB2: S'hall 47Ba 85
(off Merrick Rd.)
Arches La. SW11 51Kb 112
Archgate Bus. Cen. N12 22Eb 49
Archibald Cl. EN3: Enf W 8Zb 20
Archibald M. W1 5K 221 (45Kb 90)
Archibald Rd. N7 35Mb 70
Archibald Rd. RM3: Hrld W 25Qd 57
Archibald St. E3 41Cc 92
Archie Cl. UB7: W Dray 47Q 84
Archie St. SE1 2J 231 (47Ub 91)
Archington Ct. AL1: St A 3C 6
(off Oswald Rd.)
Arc Ho. SE1 2K 231 (47Vb 91)
(off Tanner St.)
Arch Rd. KT12: Hers 76Z 151
Arch St. SE1 4D 230 (48Sb 91)
ARCHWAY 33Lb 70
Archway RM3: Rom 23Kd 57
Archway Bus. Cen. N19 34Mb 70
Archway Bus Station 33Lb 70
Archway Cl. SM6: Bedd 76Mb 156
Archway Cl. SW19 62Db 133
Archway Cl. W10 43Za 88
Archway Ct. HP1: Hem H 1M 3
(off Chapel St.)
Archway Leisure Cen. 33Lb 70
Archway Mall N19 33Lb 70
Archway M. SW15 56Ab 110
(off Putney Bri. Rd.)
Archway Rd. N19 32Lb 70
Archway Rd. N6 30Jb 50
Archway Station (Underground) 33Lb 70
Archway St. SW13 55Ua 110
Arcola St. E8 36Vb 71
Arcola Theatre 36Vb 71
Arcon Dr. UB5: N'olt 42Aa 85
Arcon Ter. N9 17Wb 33
Arctic Ho. NW9 26Va 48
(off Heritage Av.)
Arctic St. NW5 36Kb 70
Arcus Rd. BR1: Broml 65Gc 137
Ardbeg Rd. SE24 57Tb 113
Arden Cl. HA1: Harr 34Fa 66
Arden Cl. HP3: Bov 10C 2
Arden Cl. RH2: Reig 10K 207
Arden Cl. SE28 44Zc 95
Arden Cl. TW2: Whitt 59Ba 107
Arden Cl. UB4: Yead 42X 85
Arden Cl. WD23: B Hea 17Ha 28
Arden Ct. WD17: Wat 11W 26
(off Lockhart Rd.)
Arden Ct. WD3: Rick 19J 25
Arden Ct. Gdns. N2 30Fb 49
Arden Cres. E14 49Cc 92
Arden Cres. RM9: Dag 38Yc 75
Arden Est. N1 2H 219 (40Ub 71)
Arden Grange N12 21Eb 49
Arden Gro. BR6: Farnb 77Rc 160
Arden Ho. N1 2H 219 (40Ub 71)
(off Arden Est.)
Arden Ho. SE11 6H 229 (49Pb 90)
(off Black Prince Rd.)
Arden Ho. SE13 55Dc 114
(off Thurston Rd.)
Arden Ho. SW9 54Nb 112
(off Grantham Rd.)
Arden M. E17 29Dc 52
Arden Mhor HA5: Eastc 28X 45
Arden Rd. N3 27Bb 49
Arden Rd. W13 45La 86
Ardent Cl. SE25 69Ub 135
Ardent Ho. E3 40Ac 72
(off Roman Rd.)
Ardentinny AL1: St A 3C 6
(off Grosvenor Rd.)
Ardesley Wood KT13: Weyb 77U 150
Ardfern Av. SW16 69Qb 134
Ardfillan Rd. SE6 60Fc 115
Ardgowan Rd. SE6 59Gc 115
Ardilaun Rd. N5 35Sb 71
Ardingly Cl. CR0: C'don 76Zb 158
Ardingly Ct. KT18: Eps 86Ta 173
Ardleigh Ct. RM11: Horn 27Md 57
Ardleigh Ct. BR1: Broml 66Hc 137
(off London Rd.)
Ardleigh Ct. CM15: Shenf 17Be 41
Ardleigh Gdns. CM13: Hut 16Fe 41
Ardleigh Gdns. SM3: Sutt 73Cb 155
ARDLEIGH GREEN 28Ld 57
Ardleigh Grn. Rd. RM11: Horn 29Md 57
Ardleigh Ho. IG11: Bark 39Sc 74
(off Cooke St.)
Ardleigh M. IG1: Ilf 34Rc 74
Ardleigh Rd. E17 25Bc 52
Ardleigh Rd. N1 37Ub 71
Ardleigh Ter. E17 25Bc 52
Ardley Cl. HA4: Ruis 31S 64
Ardley Cl. NW10 34Ua 68
Ardley Cl. SE6 62Ac 136
Ardley Cl. TN14: Dun G 92Hd 202
(off Campion Sq.)
Ardlui Rd. SE27 61Sb 135
Ardmay Gdns. KT6: Surb 71Na 153
Ardmere Rd. SE13 58Fc 115
Ardmore La. IG9: Buck H 17Kc 35
Ardmore Pl. IG9: Buck H 17Kc 35
Ardmore Rd. RM15: S Ock 42Xd 98
Ardoch Rd. SE6 61Fc 137
Ardra Rd. N9 20Zb 34
Ardrossan Cl. SL2: Slou 2G 80
Ardrossan Gdns. KT4: Wor Pk 76Wa 154
Ardross Av. HA6: Nwood 22U 44
Ardshiel Cl. SW15 55Za 110
Ardshiel Dr. RH1: Redh 8N 207
Ardwell Av. IG6: Ilf 29Sc 54
Ardwell Rd. SW2 61Nb 134
Ardwick Rd. NW2 35Cb 69
Arena, The EN3: Enf L 10Bc 20
Arena Bus. Cen. N4 30Sb 51
Arena Ho. E3 40Cc 72
(off Lefevre Wlk.)
Arena Shop. Pk. 30Rb 51
Arena Sq. HA9: Wemb 35Qa 67
Arena Stop (London Tramlink) 71Yb 158
Arena Twr. E14 47Dc 92
Ares Ct. E14 49Cc 92
(off Homer Dr.)
Arethusa Ho. E14 49Cc 92
(off Cahir St.)
Arethusa Pl. DA9: Ghithe 56Xd 120
Arethusa Way GU24: Bisl 8D 166
AREWATER GREEN 12Qc 36
Argali Ho. DA18: Erith 48Ad 95
(off Kale Rd.)
Argall Av. E10 31Zb 72
Argall Way E10 32Zb 72
Argan Cl. EN5: Barn 13Bb 31
Argenta Way NW10 38Ra 67
Argent Cen., The UB3: Hayes 47W 84
Argent Cl. TW20: Egh 65E 126
Argent Ct. E14 43Cc 92
(off Thomas Rd.)
Argent Ct. EN5: New Bar 14Eb 31
(off Leicester Rd.)
Argent Ct. KT6: Surb 76Qa 153
Argent Ct. RM17: Grays 52Ce 121
Argent Ho. IG11: Bark 42Wc 95
Argenton Twr. SW18 58Db 111
(off Mapleton Cres.)
Argent St. RM17: Grays 51Ae 121
Argles Cl. DA9: Ghithe 57Wd 120
Argo Apts. E16 44Hc 93
(off Sylvia Pankhurst St.)
Argo Bus. Cen. NW6 41Cb 89
Argonaut Pk. SL3: Poyle 53H 105
Argon M. SW6 52Cb 111
Argon Rd. N18 22Yb 52
Argos Ct. SW9 53Qb 112
(off Caldwell St.)
Argos Ho. E2 40Xb 71
(off Old Bethnal Grn. Rd.)
Argosy Gdns. TW18: Staines 65H 127
Argosy Ho. SE8 49Ac 92
Argosy La. TW19: Stanw 59M 105
Argus Cl. RM7: Mawney 25Dd 56
Argus Way UB5: N'olt 41Aa 85
Argyle Av. TW3: Houn 58Ca 107
Argyle Cl. W13 42Ja 86
Argyle Cl. WD18: Wat 14V 26
Argyle Gdns. RM14: Upm 33Td 78
Argyle Ho. E14 48Ec 92
Argyle Pas. N17 25Vb 51
Argyle Pl. W6 49Xa 88
Argyle Rd. E1 42Zb 92
Argyle Rd. E15 35Gc 73
Argyle Rd. E16 44Kc 93
Argyle Rd. EN5: Barn 14Ya 30
Argyle Rd. HA2: Harr 30Da 45
Argyle Rd. IG1: Ilf 33Qc 74
Argyle Rd. N12 22Db 49
Argyle Rd. N17 25Wb 51
Argyle Rd. N18 21Wb 51
Argyle Rd. TN13: S'oaks 97Kd 203
Argyle Rd. TW3: Houn 57Da 107
Argyle Rd. UB6: G'frd 41Ha 86
Argyle Rd. W13 43Ja 86
Argyle Sq. WC1 3G 217 (41Nb 90)
(not continuous)
Argyle St. WC1 3F 217 (41Nb 90)
Argyle Wlk. WC1 4G 217 (41Nb 90)
Argyle Way SE16 50Wb 91
Argyll Av. SL1: Slou 5E 80
Argyll Av. UB1: S'hall 46Da 85
Argyll Cl. SW9 55Pb 112
Argyll Ct. SW2 59Nb 112
(off New Pk. Rd.)
Argyll Gdns. HA8: Edg 26Ra 47
Argyll Mans. SW3 51Fb 111
Argyll Mans. W14 49Ab 88
(off Hammersmith Rd.)
Argyll Rd. RM17: Grays 50Ce 99
Argyll Rd. SE18 48Sc 94
Argyll Rd. W8 47Cb 89
Argyll St. W1 3B 222 (44Lb 90)
Argyll Ter. SW19 59Za 110
Aria Ct. IG2: Ilf 30Sc 54
Arica Ho. SE16 48Xb 91
(off Slippers Pl.)
Arica Rd. SE4 56Ac 114
Ariel Apts. E16 44Jc 93
Ariel Cl. DA12: Grav'nd 3H 145
Ariel Ct. SE11 6B 230 (49Rb 91)

Ariel Ho. E1 45Wb 91
(off Vaughan Way)
Ariel Rd. NW6 37Cb 69
Ariel Way TW4: Houn 55X 107
Ariel Way W12 46Ya 88
Arisdale Av. RM15: S Ock 43Xd 98
Arisdale Cl. RM15: S Ock 42Xd 98
Arista Ct. TW20: Eng G 4P 125
Aristotle Rd. SW4 55Mb 112
Arizona Bldg. SE13 53Dc 114
(off Deal's Gateway)
Ark, The W6 50Za 88
(off Talgarth Rd.)
Ark Apts. CR2: S Croy 81Vb 177
Ark Av. RM16: Grays 48Ce 99
Arkell Gro. SE19 66Rb 135
Arkindale Rd. SE6 62Ec 136
Arklay Cl. UB8: Hil 42P 83
ARKLEY 15Wa 30
Arkley Cres. E17 29Bc 52
Arkley Dr. EN5: Ark 14Wa 30
Arkley Golf Course 15Va 30
Arkley La. EN5: Ark 10Va 16
Arkley Pk. EN5: Ark 17Ta 29
Arkley Rd. E17 29Bc 52
Arkley Vw. EN5: Ark 14Xa 30
Arklow Ct. WD3: Chor 14F 24
Arklow Ho. SE17 51Tb 113
Arklow M. KT6: Surb 75Na 153
Arklow Rd. SE14 51Bc 114
Arkwright Rd. CR2: Sande 82Vb 177
Arkwright Rd. NW3 36Eb 69
Arkwright Rd. RM18: Tilb 4C 122
Arkwright Rd. SL3: Poyle 54G 104
Arla Pl. HA4: Ruis 35Y 65
Arlesey Cl. SW15 57Ab 110
Arlesford Rd. SW9 55Nb 112
Arlidge Ho. EC1 7A 218 (43Qb 90)
(off Kirby St.)
Arlingford Rd. SW2 57Qb 112
Arlingham M. EN9: Walt A 5Ec 20
(off Sun St.)
Arlington N12 20Cb 31
Arlington Av. N1 1E 218 (40Sb 71)
Arlington Bldg. E3 40Cc 72
Arlington Cl. DA15: Sidc 59Uc 116
Arlington Cl. SE13 57Fc 115
Arlington Cl. SM1: Sutt 75Cb 155
Arlington Cl. TW1: Twick 58La 108
Arlington Ct. RH2: Reig 4K 207
Arlington Ct. W3 47Ra 87
(off Mill Hill Rd.)
Arlington Cres. EN8: Walt C 6Ac 20
Arlington Dr. HA4: Ruis 30T 44
Arlington Dr. SM5: Cars 75Hb 155
Arlington Gdns. IG1: Ilf 32Qc 74
Arlington Gdns. RM3: Hrld W 25Nd 57
Arlington Gdns. W4 50Sa 87
Arlington Grn. NW7 24Za 48
Arlington Ho. EC1 3A 218 (41Qb 90)
(off Arlington Way)
Arlington Ho. SE8 51Bc 114
(off Evelyn St.)
Arlington Ho. SW1 6B 222 (46Lb 90)
Arlington Ho. TW9: Kew 52Ra 109
Arlington Ho. UB7: W Dray 47P 83
Arlington Ho. W12 46Xa 88
(off Tunis Rd.)
Arlington Lodge KT13: Weyb 77R 150
Arlington Lodge SW2 56Pb 112
Arlington M. SE13 57Fc 115
Arlington Pk. Mans. W4 50Sa 87
(off Sutton La. Nth.)
Arlington Pas. TW11: Tedd 63Ha 130
Arlington Pl. SE10 52Ec 114
Arlington Rd. IG8: Wfd G 25Jc 53
Arlington Rd. KT6: Surb 72Ma 153
Arlington Rd. N14 19Kb 32
Arlington Rd. NW1 39Kb 70
Arlington Rd. TW1: Twick 58La 108
Arlington Rd. TW10: Ham 61Ma 131
Arlington Rd. TW11: Tedd 63Ha 130
Arlington Rd. TW15: Ashf 64P 127
Arlington Rd. W13 44Ka 86
Arlington Sq. N1 39Sb 71
Arlington St. SW1 6B 222 (46Lb 90)
Arlington Way EC1 3A 218 (41Qb 90)
Arliss Ho. HA1: Harr 29Ha 46
Arliss Way UB5: N'olt 39Y 65
Arlow Rd. N21 18Qb 32
Armada Ct. RM16: Grays 48Ce 99
Armada Ct. SE8 51Cc 114
Armadale Cl. N17 28Xb 51
Armadale Rd. GU21: Wok 9L 167
Armadale Rd. SW6 52Cb 111
Armadale Rd. TW14: Felt 57W 106
Armada St. SE8 51Cc 114
Armada Way E6 43Rc 94
Armagh Rd. E3 39Bc 72
Armand Cl. WD17: Wat 10V 12
Arments Ct. SE5 51Tb 113
(off Albany Rd.)
Armfield Cl. KT8: W Mole 71Ba 151
Armfield Cres. CR4: Mitc 68Hb 133
Armfield Rd. EN2: Enf 11Tb 33
Arminger Rd. W12 46Xa 88
Armistice Gdns. SE25 69Wb 135
Armitage Cl. WD3: Loud 14M 25
Armitage Ct. SL5: S'hill 2A 146
Armitage Ho. NW1 7E 214 (43Gb 89)
(off Lisson Gro.)
Armitage Rd. NW11 32Ab 68
Armitage Rd. SE10 50Hc 93
Armor Rd. RM19: Purf 49Td 98
Armour Cl. N7 37Pb 70
Armoury, The 35Gb 69
(off Pond St.)
Armoury Dr. DA12: Grav'nd 9E 122
Armoury Ho. E3 39Ac 72
(off Gunmakers La.)
Armoury Rd. SE8 54Dc 114
Armoury Way SW18 57Cb 111
Armsby Ho. E1 43Yb 92
(off Stepney Way)
Armstead Wlk. RM10: Dag 38Cd 76
Armstrong Av. IG8: Wfd G 23Gc 53
Armstrong Cl. AL2: Lon C 9J 7
Armstrong Cl. BR1: Broml 69Nc 138
Armstrong Cl. E6 44Pc 94
Armstrong Cl. HA5: Eastc 30W 44
Armstrong Cl. KT12: Walt T 72W 150
Armstrong Cl. RM8: Dag 31Zc 75
Armstrong Cl. SE3 56Kc 115
Armstrong Cl. SS17: Stan H 1N 101
Armstrong Cl. TN14: Hals 87Dd 182
Armstrong Cl. WD6: Bore 13Sa 29
Armstrong Cres. EN4: Cockf 13Fb 31
Armstrong Gdns. WD7: Shenl 4Na 15
Armstrong Ho. E14 44Ac 92
(off Commercial Rd.)
Armstrong Ho. UB8: Uxb 38L 63
(off High St.)
Armstrong Pl. HP1: Hem H 1M 3
(off High St.)
Armstrong Rd. NW10 38Ua 68
Armstrong Rd. SE18 48Sc 94
Armstrong Rd. SW7 4B 226 (48Fb 89)
Armstrong Rd. TW13: Hanw 64Aa 129
Armstrong Rd. TW20: Eng G 5N 125
Armstrong Rd. W3 46Va 88
Armstrong Way UB2: S'hall 47Da 85
Armytage Rd. TW5: Hest 52Z 107
Arnal Cres. SW18 59Ab 110
Arncliffe NW6 40Db 69
Arncliffe Cl. N11 23Jb 50
Arncroft Ct. IG11: Bark 41Xc 95
Arndale Wlk. SW18 57Db 111
Arndale Way TW20: Egh 64C 126
Arne Cl. SS17: Stan H 1M 101
Arne Gro. BR6: Orp 76Vc 161
Arne Ho. SE11 7H 229 (50Pb 90)
(off Tyers St.)
Arne St. WC2 3G 223 (44Nb 90)
Arnett Cl. WD3: Rick 16J 25
Arnett Sq. E4 23Bc 52
Arnett Way WD3: Rick 16J 25
Arne Wlk. SE3 56Hc 115
Arneways Av. RM6: Chad H 27Zc 55
Arneway St. SW1 4E 228 (48Mb 90)
Arnewood Cl. KT22: Oxs 86Da 171
Arnewood Cl. SW15 60Wa 110
Arneys La. CR4: Mitc 72Jb 156
Arngask Rd. SE6 59Fc 115
Arnhem Av. RM15: Avel 46Sd 98
Arnhem Dr. CR0: New Ad 83Fc 179
Arnhem Pl. E14 48Cc 92
Arnhem Way SE22 57Ub 113
Arnhem Wharf E14 48Bc 92
Arnison Rd. KT8: E Mos 70Fa 130
Arnold Av. E. EN3: Enf L 10Cc 20
Arnold Av. W. EN3: Enf L 10Bc 20
Arnold Bennett Way N8 27Qb 50
Arnold Cir. E2 4K 219 (41Vb 91)
(off Hunton St.)
Arnold Cl. HA3: Kenton 31Pa 67
Arnold Ct. N22 24Nb 50
Arnold Cres. TW7: Isle 57Fa 108
Arnold Dr. KT9: Chess 79Ma 153
Arnold Est. SE1 2K 231 (47Vb 91)
(not continuous)
Arnold Gdns. N13 22Rb 51
Arnold Ho. SE17 50Rb 91
(off Doddington Gro.)
Arnold Ho. SE3 52Lc 115
(off Shooters Hill Rd.)
Arnold Mans. W14 51Bb 111
(off Queen's Club Gdns.)
Arnold Pl. RM18: Tilb 3E 122
Arnold Rd. DA12: Grav'nd 1F 144
Arnold Rd. E3 41Cc 92
Arnold Rd. EN9: Walt A 7Ec 20
Arnold Rd. GU21: Wok 87D 168
Arnold Rd. N15 27Vb 51
Arnold Rd. RM10: Dag 38Cd 76
Arnold Rd. RM9: Dag 38Bd 75
Arnold Rd. SW17 66Hb 133
Arnold Rd. TW18: Staines 66L 127
Arnold Rd. UB5: N'olt 37Z 65
Arnolds Av. CM13: Hut 15Ee 41
Arnolds Cl. CM13: Hut 15Ee 41
Arnolds Farm La. CM13: Mount 13Fe 41
Arnold's La. DA4: Sut H 65Pd 141
Arnold Ter. HA7: Stan 22Ha 46
Arnold Way KT19: Eps 84Pa 173
Arnos Gro. N14 21Mb 50
Arnos Gro. Ct. N11 22Lb 50
(off Palmer's Rd.)
Arnos Grove Station (Underground) 22Lb 50
Arnos Rd. N11 22Lb 50
Arnos Swimming Pool 22Mb 50
Arnot Ho. SE5 52Sb 113
(off Comber Gro.)
Arnott Cl. SE28 46Yc 95
Arnott Cl. W4 49Ta 87
Arnould Av. SE5 56Tb 113
Arnsberg Way DA6: Bex 56Cd 118
Arnsberg Way DA7: Bex 56Cd 118
Arnside Gdns. HA9: Wemb 32Ma 67
Arnside Ho. SE17 51Tb 113
(off Arnside St.)
Arnside Rd. DA7: Bex 53Cd 118
Arnside St. SE17 51Tb 113
Arnulf St. SE6 63Dc 136
Arnulls Rd. SW16 65Rb 135
Arnwil Dr. RM3: Rom 22Ld 57
Arodene Rd. SW2 58Pb 112
Arona Ho. BR3: Beck 68Ec 136
Arora Twr. SE10 46Fc 93
Arosa Rd. TW1: Twick 58Ma 109
Arpley Sq. SE20 66Yb 136
(off High St.)
Arragon Gdns. BR4: W W'ck 76Dc 158
Arragon Gdns. SW16 66Nb 134
Arragon Rd. E6 39Mc 73
Arragon Rd. SW18 60Cb 111
Arragon Rd. TW1: Twick 59Ja 108
Arran Cl. DA8: Erith 51Fd 118
Arran Cl. HP3: Hem H 4C 4
Arran Cl. SM6: W'gton 77Kb 156
Arran Ct. NW10 34Ta 67
Arran Ct. NW9 26Va 48
Arran Dr. E12 32Mc 73
Arran Grn. WD19: Wat 21Z 45
Arran Ho. E14 46Ec 92
(off Raleana Rd.)
Arran Ho. WD18: Wat 16W 26
Arran M. W5 46Pa 87
Arranmore Ct. WD23: Bush 14Aa 27
Arran Rd. SE6 61Dc 136
Arran Wlk. N1 38Sb 71
Arran Way KT10: Esh 75Da 151
Arras Av. SM4: Mord 71Eb 155
Arretine Cl. AL3: St A 4M 5
Arreton Mead GU21: Wok 86B 168
Arrival Sq. E1 45Wb 91
Arrol Ho. SE1 4E 230 (48Sb 91)
Arrol Rd. BR3: Beck 69Yb 136
Arrow Ct. SW5 49Cb 89
(off W. Cromwell Rd.)
Arrowhead Quay E14 47Cc 92
Arrow Ho. N1 1J 219 (39Ub 71)
(off Wilmer Gdns.)
Arrow Rd. E3 41Dc 92
Arrowscout Wlk. UB5: N'olt 41Aa 85
(off Argus Way)
Arrows Ho. SE15 52Yb 114
(off Clifton Way)
Arrowsmith Cl. IG7: Chig 22Vc 55
Arrowsmith Ho. SE11 7H 229 (50Pb 90)
(off Tyers St.)
Arrowsmith Path IG7: Chig 22Vc 55
Arrowsmith Rd. IG7: Chig 22Uc 54
Arsenal FC 35Qb 70
Arsenal Rd. SE9 54Pc 116
Arsenal Station (Underground) 34Qb 70
Arsenal Way SE18 48Sc 94
Arta Ho. E1 44Yb 92
(off Devonport St.)
Artbrand Ho. SE1 2H 231 (47Ub 91)
(off Leathermarket St.)
Artemis Cl. DA12: Grav'nd 9G 122
Artemis Ct. E14 49Cc 92
(off Homer Dr.)
Artemis Pl. SW18 59Bb 111
Arterberry Rd. SW20 66Ya 132
Arterial Av. RM13: Rain 42Kd 97
Arterial Rd. RM16: N Stif 47Zd 99
Arterial Rd. RM19: Purf 49Rd 97
Arterial Rd. RM19: W Thur 49Rd 97
Arterial Rd. RM20: W Thur 48Vd 98
Arterial Rd. SS17: Stan H 1L 101
Artesian Cl. NW10 38Ta 67
Artesian Cl. RM11: Horn 30Hd 56
Artesian Cl. RM11: Rom 30Hd 56
Artesian Gro. EN5: New Bar 14Eb 31
Artesian Ho. SE1 4K 231 (48Vb 91)
(off Grange Rd.)
Artesian Rd. W2 44Cb 89
Artesian Wlk. E11 34Gc 73
Arthaus Apts. E8 37Xb 71
(off Richmond Rd.)
Arthingworth St. E15 39Gc 73
Arthouse Crouch End 29Nb 50
Arthur Barnes Ct. RM16: Grays 8E 100
Arthur Ct. CR0: C'don 76Ub 157
(off Fairfield Path)
Arthur Ct. SW11 53Jb 112
Arthur Ct. W10 44Za 88
(off Silchester Rd.)
Arthur Ct. W2 44Db 89
(off Queensway)
Arthur Deakin Ho. E1 43Wb 91
(off Hunton St.)
Arthurdon Rd. SE4 57Cc 114
Arthur Gro. SE18 49Sc 94
Arthur Henderson Ho. SW6 54Bb 111
(off Fulham Rd.)
Arthur Horsley Wlk. E7 36Hc 73
(off Tower Hamlets Rd.)
Arthur Ho. N1 39Ub 71
(off Halcomb St.)
Arthur Jacob Nature Reserve 55E 104
Arthur Lovell Ct. E14 44Bc 92
(off Lovat Cl.)
Arthur Newton Ho. SW11 55Fb 111
(off Winstanley Est.)
Arthur Rd. AL1: St A 2F 6
Arthur Rd. E6 40Pc 74
Arthur Rd. KT2: King T 66Qa 131
Arthur Rd. KT3: N Mald 71Xa 154
Arthur Rd. N7 35Pb 70
Arthur Rd. N9 19Vb 33
Arthur Rd. RM6: Chad H 30Yc 55
Arthur Rd. SL1: Slou 7H 81
Arthur Rd. SL4: Wind 3G 102
Arthur Rd. SW19 64Bb 133
Arthur Rd. TN16: Big H 87Lc 179
Arthur's Bri. Rd. GU21: Wok 9N 167
Arthur's Bri. Wharf GU21: Wok 9P 167
Arthur St. DA11: Grav'nd 9C 122
Arthur St. DA8: Erith 52Hd 118
Arthur St. EC4 4G 225 (45Tb 91)
Arthur St. RM17: Grays 51Ee 121
Arthur St. WD23: Bush 14Z 27
Arthur St. W. DA11: Grav'nd 9C 122
Arthur Toft Ho. RM17: Grays 51De 121
(off New Rd.)
Arthur Wade Ho. E2 3K 219 (41Vb 91)
(off Baroness Rd.)
Arthur Wallis Ho. E12 34Qc 74
(off Grantham Rd.)
Artichoke Dell WD3: Chor 14G 24
Artichoke Hill E1 45Xb 91
Artichoke M. SE5 53Tb 113
(off Artichoke Pl.)
Artichoke Pl. SE5 53Tb 113
Artichoke Wlk. TW9: Rich 57Ma 109
(off Red Lion St.)
Artillery Bldg., The E1 1J 225 (43Ub 91)
(off Artillery La.)
Artillery Cl. IG2: Ilf 30Sc 54
Artillery Ho. E15 37Gc 73
Artillery Ho. E3 39Ac 72
(off Barge La.)
Artillery Ho. SE18 50Qc 94
(off Connaught M.)
Artillery La. E1 1J 225 (43Ub 91)
Artillery La. W12 44Wa 88
Artillery Mans. SW1 4D 228 (48Mb 90)
(off Victoria St.)
Artillery Pas. E1 1J 225 (43Ub 91)
(off Artillery La.)
Artillery Pl. HA3: Hrw W 24Ea 46
Artillery Pl. SE18 49Pc 94
Artillery Pl. SW1 4D 228 (48Mb 90)
Artillery Row DA12: Grav'nd 9E 122
Artillery Row SW1 4C 228 (48Mb 90)
Artillery Sq. SE18 48Rc 94
Artillery St. DA1: Dart 54Sd 120
Artington Cl. BR6: Farnb 77Sc 160
Artisan Cl. E6 44Rc 94
Artisan Ct. E8 37Wb 71
Artisan Cres. AL3: St A 1A 6
(not continuous)
Artisan M. NW10 41Za 88
(off Warfield Rd.)
Artisan Pl. HA3: W'stone 26Ga 46
Artisan Quarter NW10 41Za 88
(off Wellington Rd.)
Artizan St. E1 2K 225 (44Vb 91)
(off Harrow Pl.)
Art School Yd. AL1: St A 2B 6
(off Victoria St.)
Arts Depot 22Eb 49
Arts Development Unit & Weir Hall Library 21Tb 51
Arts La. SE16 4K 231 (48Vb 91)
Arts Sq. E1 42Ac 92
Arts Theatre Covent Garden 4F 223 (45Nb 90)
(off Gt. Newport St.)
Arun RM18: E Til 9L 101
Arun Ct. SE25 71Wb 157
Arundale KT1: King T 70Ma 131
(off Anglesea Rd.)
Arundel Av. CR2: Sande 82Wb 177
Arundel Av. KT17: Ewe 82Xa 174
Arundel Av. SM4: Mord 70Bb 133
Arundel Bldgs. SE1 4J 231 (48Ub 91)
Arundel Cl. CR0: Wadd 76Rb 157
Arundel Cl. DA5: Bexl 58Bd 117
Arundel Cl. E15 35Gc 73
Arundel Cl. EN8: Chesh 1Yb 20
Arundel Cl. HP2: Hem H 1B 4
Arundel Cl. SW11 57Gb 111
Arundel Cl. TW12: Hamp H 64Da 129
Arundel Ct. BR2: Broml 68Gc 137
Arundel Ct. HA2: Harr 35Ca 65
Arundel Ct. N12 23Gb 49
Arundel Ct. N17 25Wb 51
Arundel Ct. SE16 50Xb 91
(off Verney Rd.)
Arundel Ct. SL3: L'ly 9P 81
Arundel Ct. SW13 51Xa 110
(off Arundel Ter.)
Arundel Ct. SW3 7E 226 (50Gb 89)
(off Jubilee Pl.)
Arundel Ct. W11 45Bb 89
(off Arundel Gdns.)
Arundel Dr. BR6: Chels 78Xc 161
Arundel Dr. HA2: Harr 35Ba 65
Arundel Dr. IG8: Wfd G 24Jc 53
Arundel Dr. WD6: Bore 14Sa 29
Arundel Gdns. HA8: Edg 24Ta 47
Arundel Gdns. IG3: Ilf 33Wc 75
Arundel Gdns. N21 18Qb 32
Arundel Gdns. W11 45Bb 89
Arundel Gt. Ct. WC2 4J 223 (45Pb 90)
Arundel Gro. N16 36Ub 71
Arundel Ho. CR0: C'don 78Tb 157
(off Heathfield Rd.)
Arundel Ho. E17 25Bc 52
Arundel Ho. UB8: Cowl 42L 83
Arundel Ho. W3 47Ra 87
(off Park Rd. Nth.)
Arundel Ho. WD6: Bore 14Sa 29
(off Arundel Dr.)
Arundel Mans. SW6 53Bb 111
(off Kelvedon Rd.)
Arundel Pl. N1 37Qb 70
Arundel Rd. CR0: C'don 72Tb 157
Arundel Rd. DA1: Dart 56Ld 119
Arundel Rd. EN4: Cockf 13Gb 31
Arundel Rd. KT1: King T 68Ra 131
Arundel Rd. RM3: Hrld W 24Pd 57
Arundel Rd. SM2: Cheam 80Bb 155
(not continuous)
Arundel Rd. SM2: Sutt 80Bb 155
(not continuous)
Arundel Rd. TW4: Houn 55Y 107
Arundel Rd. UB8: Uxb 40K 63
Arundel Rd. WD5: Ab L 4W 12
Arundel Sq. N7 37Qb 70
Arundel St. WC2 4J 223 (45Pb 90)
Arundel Ter. SW13 51Xa 110
Arun Ho. KT2: King T 67Ma 131
Arvon Rd. N5 36Qb 70
(not continuous)
Asa Ct. UB3: Harl 48V 84
Asbaston Ter. IG11: Ilf 37Sc 74
Asbridge Ct. W6 48Xa 88
(off Dalling Rd.)
Asbury Ct. N21 15Nb 32
(off Pennington Dr.)
Ascalon Ho. SW8 52Lb 112
(off Thessaly Rd.)
Ascalon St. SW8 52Lb 112
Ascension Rd. RM5: Col R 23Ed 56
Ascensis Twr. SW18 56Eb 111
Ascent Ho. KT13: Weyb 78U 150
(off Ellesmere Rd.)
Ascent Ho. NW9 26Va 48
(off Boulevard Dr.)
Ascham Dr. E4 24Dc 52
Ascham End E17 25Ac 52
Ascham St. NW5 36Lb 70
Aschurch Rd. CR0: C'don 73Vb 157
Ascot Cl. IG6: Ilf 23Uc 54
Ascot Cl. TN15: Bor G 92De 205
Ascot Cl. UB5: N'olt 36Ca 65
Ascot Cl. WD6: E'tree 15Qa 29
Ascot Ct. DA5: Bexl 59Bd 117
Ascot Ct. NW8 4B 214 (41Fb 89)
(off Grove End Rd.)
Ascot Ct. WD3: Crox G 16R 26
Ascot Gdns. EN3: Enf W 9Yb 20
Ascot Gdns. RM12: Horn 35Nd 77
Ascot Gdns. UB1: S'hall 42Ba 85
Ascot Ho. NW1 3A 216 (41Kb 90)
(off Redhill St.)
Ascot Ho. SL4: Wind 3D 102
(off Paddock Cl.)
Ascot Ho. W9 42Cb 89
(off Harrow Rd.)
Ascot Lodge NW6 39Db 69
Ascot M. SM6: W'gton 81Lb 176
Ascot Pl. HA7: Stan 22La 46
Ascot Rd. BR5: St M Cry 70Vc 139
Ascot Rd. DA12: Grav'nd 2D 144
Ascot Rd. E6 41Pc 94
Ascot Rd. N15 29Tb 51
Ascot Rd. N18 21Wb 51
Ascot Rd. SW17 65Jb 134
Ascot Rd. TW14: Bedf 60Q 106
Ascot Rd. WD18: Wat 15U 26
Ascott Av. W5 47Na 87
Ascott Cl. HA5: Eastc 28W 44
ASH 78Ae 165
Ashanti M. E8 36Yb 72
Ash Av. SE17 5D 230 (49Sb 91)
Ashbeam Cl. CM13: Gt War 23Yd 58
Ashbee Ho. E2 41Yb 92
(off Portman Pl.)
Ashbourne AL2: Brick W 3Ba 13
Ashbourne Av. DA7: Bex 52Ad 117
Ashbourne Av. E18 28Kc 53
Ashbourne Av. HA2: Harr 33Fa 66
Ashbourne Av. N20 19Hb 31
Ashbourne Av. NW11 29Bb 49
Ashbourne Cl. CR5: Coul 90Lb 176
Ashbourne Cl. N12 21Db 49
Ashbourne Cl. W5 43Qa 87
Ashbourne Ct. AL4: St A 5G 6
Ashbourne Ct. E5 35Ac 72
Ashbourne Ct. N12 21Db 49
(off Ashbourne Cl.)
Ashbourne Gro. NW7 22Ta 47
Ashbourne Gro. SE22 56Vb 113
Ashbourne Gro. W4 50Ua 88
Ashbourne Ho. SL1: Slou 7J 81
Ashbourne Pde. NW11 28Bb 49
Ashbourne Pde. W5 42Pa 87
Ashbourne Ri. BR6: Orp 77Uc 160
Ashbourne Rd. CR4: Mitc 66Jb 134
Ashbourne Rd. RM3: Rom 21Ld 57
Ashbourne Rd. W5 42Pa 87
Ashbourne Sq. HA6: Nwood 23U 44
Ashbourne Ter. SW19 66Bb 133
Ashbourne Way NW11 28Bb 49
Ashbridge Rd. E11 31Gc 73
Ashbridge St. NW8 6D 214 (42Gb 89)
Ashbrook HA8: Edg 23Pa 47
Ashbrook Rd. N19 32Mb 70
Ashbrook Rd. RM10: Dag 34Dd 76
Ashbrook Rd. SL4: Old Win 9M 103
Ashburn Gdns. SW7 49Eb 89
Ashburnham Av. HA1: Harr 30Ha 46
Ashburnham Cl. N2 27Fb 49
Ashburnham Cl. TN13: S'oaks 99Ld 203
Ashburnham Cl. WD19: Wat 20W 26
Ashburnham Ct. BR3: Beck 68Ec 136
Ashburnham Dr. WD19: Wat 20W 26
Ashburnham Gdns. HA1: Harr 30Ha 46
Ashburnham Gdns. RM14: Upm 32Rd 77
Ashburnham Gro. SE10 52Dc 114
Ashburnham Mans. SW10 52Eb 111
(off Ashburnham Rd.)
Ashburnham M. SW1 5E 228 (49Mb 90)
(off Regency St.)
Ashburnham Pk. KT10: Esh 77Ea 152
Ashburnham Pl. SE10 52Dc 114
Ashburnham Retreat SE10 52Dc 114
Ashburnham Rd. DA17: Belv 49Ed 96
Ashburnham Rd. NW10 41Ya 88
Ashburnham Rd. SW10 52Eb 111
Ashburnham Rd. TW10: Ham 62Ka 130
Ashburnham Twr. SW10 52Fb 111
(off Worlds End Est.)
Ashburn Pl. SW7 5A 226 (49Eb 89)
Ashburton Av. CR0: C'don 74Xb 157
Ashburton Av. IG3: Ilf 36Uc 74
Ashburton Cl. CR0: C'don 74Wb 157
Ashburton Ent. Cen. SW15 58Ya 110
Ashburton Gdns. CR0: C'don 75Wb 157
ASHBURTON GROVE 35Qb 70
Ashburton Ho. W9 42Bb 89
(off Fernhead Rd.)
Ashburton Memorial Homes CR0: C'don 73Xb 157
Ashburton Pl. W1 6A 222 (46Kb 90)
Ashburton Rd. CR0: C'don 75Wb 157
Ashburton Rd. E16 44Jc 93
Ashburton Rd. HA4: Ruis 33W 64
Ashburton Ter. E13 40Jc 73
Ashburton Triangle N5 35Qb 70
Ashbury Cl. AL10: Hat 1A 8
Ashbury Dr. UB10: Ick 34R 64
Ashbury Gdns. RM6: Chad H 29Zc 55
Ashbury Pl. SW19 65Eb 133
Ashbury Rd. SW11 55Hb 111
Ashby Av. KT9: Chess 79Qa 153
Ashby Cl. BR4: W W'ck 76Fc 159
Ashby Cl. RM11: Horn 32Qd 77
Ashby Cl. RM16: Ors 4G 100
Ashby Ct. NW8 5C 214 (42Fb 89)
(off Pollitt Dr.)
Ashby Gdns. AL1: St A 6B 6
Ashby Gro. N1 38Sb 71
(not continuous)
Ashby Ho. N1 38Sb 71
(off Essex Rd.)
Ashby Ho. SW9 54Rb 113
Ashby Ho. UB5: N'olt 42Ba 85
(off Waxlow Way)
Ashby M. SE4 54Bc 114
Ashby M. SW2 57Nb 112
(off Prague Pl.)
Ashby Rd. N15 29Wb 51
Ashby Rd. SE4 54Bc 114
Ashby Rd. WD24: Wat 10W 12
Ashbys Ct. E3 40Bc 72
(off Centurion La.)
Ashby St. EC1 4C 218 (41Rb 91)
Ashby Wlk. CR0: C'don 72Sb 157
Ashby Way UB7: Sip 52Q 106
Ashchurch Gro. W12 48Wa 88
Ashchurch Pk. Vs. W12 48Wa 88
Ashchurch Ter. W12 48Wa 88
Ash Cl. AL9: Brk P 7J 9
Ash Cl. BR5: Pet W 71Tc 160
Ash Cl. BR8: Swan 68Ed 140
Ash Cl. CM15: Pil H 15Vd 40
Ash Cl. DA14: Sidc 62Xc 139
Ash Cl. GU22: Pyr 87J 169
Ash Cl. GU22: Wok 92A 188
Ash Cl. HA7: Stan 23Ja 46
Ash Cl. HA8: Edg 21Sa 47
Ash Cl. KT3: N Mald 68Ta 131
Ash Cl. RH1: Mers 2C 208
Ash Cl. RM5: Col R 24Dd 56
Ash Cl. SE20 68Yb 136
Ash Cl. SL3: L'ly 48D 82
Ash Cl. SM5: Cars 75Hb 155
Ash Cl. SM7: Bans 87Ab 174
Ash Cl. TW7: Isle 53Ha 108
Ash Cl. UB9: Hare 25M 43
Ash Cl. WD25: Wat 7X 13
Ash Cl. WD5: Ab L 4T 12
Ashcombe Av. KT6: Surb 73Ma 153
Ashcombe Cl. TW15: Ashf 62N 127
Ashcombe Ct. TW15: Ashf 61P 127
Ashcombe Gdns. HA8: Edg 21Qa 47
Ashcombe Ho. E3 41Dc 92
(off Bruce Rd.)
Ashcombe Ho. EN3: Pond E 13Zb 34
Ashcombe Pde. GU22: Wok 92C 188
(off Kingfield Rd.)
Ashcombe Pk. NW2 34Ua 68
Ashcombe Rd. RH1: Mers 99Lb 196
Ashcombe Rd. SM5: Cars 79Jb 156
Ashcombe Rd. SW19 64Cb 133
Ashcombe Sq. KT3: N Mald 69Sa 131
Ashcombe St. SW6 54Db 111
Ashcombe Ter. KT20: Tad 92Xa 194
Ash Copse AL2: Brick W 3Ba 13
Ash Ct. KT15: Add 78K 149
Ash Ct. KT19: Ewe 77Sa 153
Ash Ct. KT22: Lea 92Ha 192
Ash Ct. N11 23Lb 50
Ash Ct. SW19 66Ab 132
Ashcroft HA5: Hat E 23Ca 45
Ashcroft N14 19Mb 32
Ashcroft Av. DA15: Sidc 58Wc 117
Ash Cft. Ct. DA3: Nw A G 76Ae 165
Ashcroft Ct. DA1: Dart 59Qd 119
Ashcroft Ct. N20 19Fb 31
Ashcroft Ct. SL1: Burn 10A 60
Ashcroft Cres. DA15: Sidc 58Wc 117
Ashcroft Dr. UB9: Den 30H 43
Ashcroft Ho. SW8 53Lb 112
(off Wadhurst Rd.)
Ashcroft Pk. KT11: Cobh 84Aa 171
Ashcroft Pl. KT22: Lea 93La 192
Ashcroft Ri. CR5: Coul 88Nb 176
Ashcroft Rd. KT9: Chess 76Pa 153
Ashcroft Sq. W6 49Ya 88

Ashcroft Theatre Croydon 76Tb 157
(within Fairfield Halls)
Ashdale KT23: Bookh 98Ea 192
Ashdale Cl. TW19: Stanw 61N 127
Ashdale Cl. TW2: Whitt 59Ea 108
Ashdale Gro. HA7: Stan 23Ha 46
Ashdale Ho. N4 31Tb 71
Ashdale Rd. SE12 60Kc 115
Ashdales AL1: St A 6B 6
Ashdale Way TW2: Whitt 59Da 107
Ashdene HA5: Pinn 27Y 45
Ashdene SE15 52Xb 113
Ashdene Cl. TW15: Ashf 66S 128
Ashdene Ho. TW20: Eng G 5N 125
Ashdon Cl. CM13: Hut 16Ee 41
Ashdon Cl. IG8: Wfd G 23Kc 53
Ashdon Cl. RM15: S Ock 44Xd 98
Ashdon Rd. NW10 39Va 68
Ashdon Rd. WD23: Bush 13Z 27
Ashdown W13 43Ka 86
(off Clivedon Ct.)
Ashdown Cl. BR3: Beck 68Dc 136
Ashdown Cl. DA5: Bexl 59Ed 118
Ashdown Cl. GU22: Wok 90A 168
Ashdown Cl. RH2: Reig 10K 207
Ashdown Ct. E17 26Ec 52
Ashdown Ct. IG11: Bark 37Rc 74
Ashdown Ct. SM2: Sutt 79Eb 155
Ashdown Cres. EN8: Chesh 1Ac 20
Ashdown Cres. NW5 36Jb 70
Ashdown Dr. WD6: Bore 12Pa 29
Ashdowne Ct. N17 25Wb 51
Ashdown Gdns. CR2: Sande 87Xb 177
Ashdown Pl. KT17: Ewe 80Va 154
Ashdown Pl. KT7: T Ditt 73Ja 152
Ashdown Rd. EN3: Enf H 12Yb 34
Ashdown Rd. KT1: King T 68Na 131
Ashdown Rd. KT17: Eps 85Va 174
Ashdown Rd. RH2: Reig 10K 207
Ashdown Rd. UB10: Hil 40Q 64
Ashdown Wlk. E14 49Cc 92
Ashdown Wlk. RM7: Mawney 25Dd 56
Ashdown Way SW17 61Jb 134
Ash Dr. AL10: Hat 3C 8
Ash Dr. RH1: Redh 8A 208
Ashe Ho. TW1: Twick 58Ma 109
Ashen E6 44Qc 94
Ashenbank Wood 8J 145
Ashenden Rd. E5 36Ac 72
Ashenden Wlk. SL2: Farn C 5H 61
Ashen Dr. DA1: Dart 58Jd 118
Ashen Gro. SW19 62Cb 133
Ashen Gro. Rd. TN15: Knat 82Rd 183
Ashentree Ct. EC4 3A 224 (44Qb 90)
(off Whitefriars St.)
Ashen Va. CR2: Sels 81Zb 178
Asher Loftus Way N11 23Hb 49
Asher Way E1 45Wb 91
Ashfield Av. TW13: Felt 60X 107
Ashfield Av. WD23: Bush 16Da 27
Ashfield Cl. BR3: Beck 66Cc 136
Ashfield Cl. KT21: Asht 91Na 193
Ashfield Cl. TW10: Ham 60Na 109
Ashfield Ct. SW9 54Nb 112
(off Clapham Rd.)
Ashfield Ho. W14 50Bb 89
(off W. Cromwell Rd.)
Ashfield La. BR7: Chst 65Rc 138
(not continuous)
Ashfield Pde. N14 18Mb 32
Ashfield Rd. N14 20Lb 32
Ashfield Rd. N4 30Sb 51
Ashfield Rd. W3 46Va 88
Ashfields IG10: Lough 12Pc 36
Ashfields WD25: Wat 7V 12
Ashfields Ct. RH2: Reig 4K 207
Ashfield St. E1 43Xb 91
(not continuous)
Ashfield Yd. E1 43Yb 92
Ashflower Dr. RM3: Hrld W 25Nd 57
ASHFORD 63P 127
Ashford Av. CM14: B'wood 20Xd 40
Ashford Av. N8 28Nb 50
Ashford Av. TW15: Ashf 65R 128
Ashford Av. UB4: Yead 44Z 85
Ashford Bus. Complex T
W15: Ashf 64S 128
Ashford Cl. TW15: Ashf West Cl 63N 127
Ashford Cl. E17 30Bc 52
ASHFORD COMMON 66T 128
Ashford Ct. HA8: Edg 20Ra 29
Ashford Ct. KT19: Eps 84Pa 173
Ashford Ct. NW2 35Za 68
Ashford Ct. RM17: Grays 50Fe 99
Ashford Cres. EN3: Enf H 12Yb 34
Ashford Cres. TW15: Ashf 62N 127
Ashford Gdns. KT11: Cobh 88Z 171
Ashford Grn. WD19: Wat 22Z 45
ASHFORD HOSPITAL 61N 127
Ashford Ho. SE8 51Bc 114
Ashford Ho. SW9 56Rb 113
Ashford Ind. Est. TW15: Ashf 63S 128
Ashford La. SL4: Dor 7A 80
Ashford Manor Golf Course 65P 127
Ashford M. N17 25Wb 51
ASHFORD PARK 63M 127
Ashford Pas. NW2 35Za 68
Ashford Rd. E18 26Kc 53
Ashford Rd. E6 38Qc 74
Ashford Rd. NW2 35Za 68
Ashford Rd. SL0: Iver H 39E 62
Ashford Rd. TW13: Felt 63T 128
Ashford Rd. TW15: Ashf 66S 128
Ashford Rd. TW18: Lale 68L 127
Ashford Rd. TW18: Staines 68L 127
Ashford Station (Rail) 63P 127
Ashford St. N1 3H 219 (41Ub 91)
Ashford Tennis Club 63M 127
Ash Grn. UB9: Den 37K 63
Ash Gro. BR4: W W'ck 75Ec 158
Ash Gro. E8 39Xb 71
(not continuous)
Ash Gro. EN1: Enf 17Ub 33
Ash Gro. HA0: Wemb 35Ja 66
Ash Gro. HP3: Hem H 6P 3
Ash Gro. N10 28Kb 50
Ash Gro. N13 20Sb 33
Ash Gro. NW2 35Za 68
Ash Gro. SE12 60Jc 115
Ash Gro. SE20 68Yb 136
Ash Gro. SL2: Stoke P 8K 61
Ash Gro. TW14: Felt 60U 106
Ash Gro. TW18: Staines 65L 127
Ash Gro. TW5: Hest 53Z 107
Ash Gro. UB1: S'hall 43Ca 85
Ash Gro. UB3: Hayes 45T 84
Ash Gro. UB7: Yiew 45P 83
Ash Gro. UB9: Hare 25M 43
Ash Gro. W5 47Na 87
Ashgrove TN14: Knock 87Ad 181
Ashgrove Ct. W9 43Cb 89
(off Elmfield Way)
Ashgrove Ho. SW1 7E 228 (50Mb 90)
(off Lindsay Sq.)
Ashgrove Rd. BR1: Broml 65Fc 137
Ashgrove Rd. IG3: Ilf 32Vc 75
Ashgrove Rd. TN13: S'oaks 99Jd 202
Ashgrove Rd. TW15: Ashf 64S 128
Ash Hill Cl. WD23: Bush 18Da 27
Ash Hill Dr. HA5: Pinn 27Y 45
Ash Ho. DA3: Nw A G 75Ae 165
Ash Ho. E14 47Ec 92
(off E. Ferry Rd.)
Ash Ho. SE1 6K 231 (49Vb 91)
(off Longfield Est.)
Ash Ho. TW18: Staines 63H 127
Ash Ho. W10 42Ab 88
(off Heather Wlk.)
Ashingdon Cl. E4 20Ec 34
Ashington Ho. E1 42Xb 91
(off Barnsley St.)
Ashington Rd. SW6 54Bb 111
Ash Island KT8: E Mos 69Fa 130
Ashlake Rd. SW16 63Nb 134
Ashland Pl. W1 7H 215 (43Jb 90)
Ashlands Ct. RM18: E Til 8L 101
Ash La. RM1: Rom 23Jd 56
Ash La. RM11: Horn 28Qd 57
Ash La. SL4: Wind 4B 102
Ash La. TN15: Ash 84Yd 184
Ash La. TN15: W King 84Yd 184
Ashlar Pl. SE18 49Rc 94
Ashlea Ct. CR6: W'ham 90Wb 177
Ashlea Rd. SL9: Chal P 26A 42
Ashleigh Av. TW20: Egh 66E 126
Ashleigh Commercial Est. SE7 48Lc 93
Ashleigh Ct. EN9: Walt A 6Jc 21
Ashleigh Ct. N14 17Lb 32
Ashleigh Ct. W5 49Ma 87
(off Murray Rd.)
Ashleigh Ct. WD17: Wat 13Y 27
(off Loates La.)
Ashleigh Ct. WD3: Rick 17M 25
Ashleigh Gdns. RM14: Upm 34Td 78
Ashleigh Gdns. SM1: Sutt 75Db 155
Ashleigh M. SE15 55Vb 113
(off Oglander Rd.)
Ashleigh Point SE23 62Zb 136
Ashleigh Rd. SE20 69Xb 135
Ashleigh Rd. SW14 55Ua 110
Ashley Av. IG6: Ilf 26Rc 54
Ashley Av. KT18: Eps 85Ta 173
Ashley Av. SM4: Mord 71Cb 155
Ashley Cen. 85Ta 173
Ashley Cl. HA5: Pinn 26X 45
Ashley Cl. HP3: Hem H 4P 3
Ashley Cl. KT12: Walt T 74V 150
Ashley Cl. KT23: Bookh 97Ba 191
Ashley Cl. NW4 26Ya 48
Ashley Cl. TN13: S'oaks 96Kd 203
Ashley Ct. E3 41Ec 92
(off Bolinder Way)
Ashley Ct. EN5: New Bar 15Eb 31
Ashley Ct. GU21: Wok 10K 167
Ashley Ct. KT18: Eps 85Ta 173
Ashley Ct. NW4 26Ya 48
Ashley Ct. SW1 4B 228 (48Lb 90)
(off Morpeth Ter.)
Ashley Ct. UB5: N'olt 39Aa 65
Ashley Cres. N22 26Qb 50
Ashley Cres. SW11 55Jb 112
Ashley Dr. KT12: Walt T 76W 150
Ashley Dr. SM7: Bans 86Cb 175
Ashley Dr. TW2: Whitt 59Da 107
Ashley Dr. TW7: Isle 51Ga 108
Ashley Dr. WD6: Bore 15Sa 29
Ashley Gdns. BR6: Orp 78Uc 160
Ashley Gdns. HA9: Wemb 33Na 67
Ashley Gdns. N13 21Sb 51
Ashley Gdns. RM16: Grays 46Ee 99
Ashley Gdns. SW1 4C 228 (48Lb 90)
(not continuous)
Ashley Gdns. TW10: Ham 61Ma 131
Ashley Gro. IG10: Lough 13Nc 36
Ashley La. CR0: Wadd 77Rb 157
Ashley La. NW4 26Ya 48
ASHLEY PARK 76W 150
Ashley Pk. Av. KT12: Walt T 75V 150
Ashley Pk. Cres. KT12: Walt T 74W 150
Ashley Pk. Rd. KT12: Walt T 75W 150
Ashley Pl. KT12: Walt T 75W 150
(off Ashley Rd.)
Ashley Pl. SW1 4B 228 (48Lb 90)
(not continuous)
Ashley Ri. KT12: Walt T 77W 150
Ashley Rd. AL1: St A 2G 6
Ashley Rd. CR7: Thor H 70Pb 134
Ashley Rd. E4 23Cc 52
Ashley Rd. E7 38Lc 73
Ashley Rd. EN3: Enf H 12Yb 34
Ashley Rd. GU21: Wok 10K 167
Ashley Rd. KT12: Walt T 77V 150
Ashley Rd. KT18: Eps 85Ta 173
Ashley Rd. KT18: Eps D 85Ta 173
Ashley Rd. KT7: T Ditt 72Ha 152
Ashley Rd. N17 27Wb 51
Ashley Rd. N19 32Nb 70
Ashley Rd. SW19 65Db 133
Ashley Rd. TN13: S'oaks 96Kd 203
Ashley Rd. TW12: Hamp 67Ca 129
Ashley Rd. TW9: Rich 55Na 109
Ashley Rd. UB8: Uxb 40K 63
Ashleys WD3: Rick 17H 25
Ashleys All. N15 28Sb 51
Ashley Sq. KT18: Eps 85Ta 173
(off South St.)
Ashley Wlk. NW7 24Ya 48
Ashley Way GU24: W End 5B 166
Ashling Rd. CR0: C'don 74Wb 157
Ashlin Rd. E15 35Fc 73
Ash Lodge KT12: Walt T 74W 150
Ash Lodge TW16: Sun 66V 128
(off Forest Dr.)
Ashlone Rd. SW15 55Ya 110
Ashlyn Cl. WD23: Bush 14Aa 27
Ashlyn Ct. WD23: Bush 14Aa 27
Ashlyn Gro. RM11: Horn 27Md 57
Ashlyns Pk. KT11: Cobh 85Aa 171
Ashlyns Rd. CM16: Epp 2Vc 23
Ashlyns Way KT9: Chess 79Ma 153
Ashmead N14 15Lb 32
Ashmead Bus. Cen. E16 42Fc 93
Ashmead Cl. TW15: Ashf Oxford Cl 66S 128
Ashmead Ct. DA9: Ghithe 58Xd 120
Ashmead Dr. UB9: Den 33J 63
Ashmead Ga. BR1: Broml 67Lc 137
Ashmead Ho. E9 36Ac 72
(off Homerton Rd.)
Ashmead Ho. W13 46Ja 86
(off Tewkesbury Rd.)
Ashmead La. UB9: Den 33J 63
Ashmead M. SE8 54Cc 114
Ashmead Rd. SE8 54Cc 114
Ashmead Rd. TW14: Felt 60W 106
Ashmeads IG10: Lough 13Pc 36
Ashmeads Ct. WD7: Shenl 5Ma 15
Ashmere Av. BR3: Beck 68Fc 137
Ashmere Cl. SM3: Cheam 78Za 154
Ashmere Gro. SW2 56Nb 112
Ash M. CR3: Cat'm 95Tb 197
Ash M. KT18: Eps 85Ua 174
Ash M. NW5 36Lb 70
Ashmill St. NW1 7D 214 (43Gb 89)
Ashmole Pl. SW8 51Pb 112
Ashmole St. SW8 51Pb 112
Ashmore NW1 38Mb 70
(off Agar Gro.)
Ashmore Cl. DA1: Dart 55Nd 119
Ashmore Cl. SE15 52Vb 113
Ashmore Ct. N11 23Hb 49
Ashmore Ct. TW5: Hest 51Ca 107
Ashmore Gdns. DA11: Nflt 62Fe 143
Ashmore Gdns. HP3: Hem H 3B 4
Ashmore Gro. DA16: Well 55Tc 116
Ashmore Ho. W14 48Ab 88
(off Russell Rd.)
Ashmore La. BR2: Kes 83Lc 179
Ashmore Rd. SE18 52Pc 116
Ashmore Rd. W9 40Bb 69
Ashmount Cres. SL1: Slou 7E 80
Ashmount Est. N19 31Mb 70
Ashmount Rd. N15 29Vb 51
Ashmount Rd. N19 31Lb 70
Ashmount Ter. W5 49Ma 87
Ashmour Gdns. RM1: Rom 26Fd 56
Ashneal Gdns. HA1: Harr 34Fa 66
Ashness Gdns. UB6: G'frd 37Ka 66
Ashness Rd. SW11 57Hb 111
Ashpark Ho. E14 44Bc 92
(off Norbiton Rd.)
Ash Platt, The TN15: Seal 92Nd 203
Ash Platt Rd. TN15: Seal 93Nd 203
Ash Ride EN2: Crew H 7Qb 18
Ashridge Cl. HA3: Kenton 30La 46
Ashridge Cl. HP3: Bov 10B 2
Ashridge Cl. N3 27Cb 49
Ashridge Cl. SS17: Stan H 3K 101
Ashridge Ct. N14 15Lb 32
Ashridge Ct. UB1: S'hall 44Ea 86
(off Redcroft Rd.)
Ashridge Cres. SE18 52Sc 116
Ashridge Dr. AL2: Brick W 2Aa 13
Ashridge Dr. WD19: Wat 22Y 45
Ashridge Gdns. HA5: Pinn 28Aa 45
Ashridge Gdns. N13 22Mb 50
Ashridge Ho. WD18: Wat 17U 26
(off Chenies Way)
Ashridge Way SM4: Mord 69Bb 133
Ashridge Way TW16: Sun 65W 128
Ash Rd. BR6: Chels 80Vc 161
Ash Rd. CR0: C'don 75Cc 158
Ash Rd. DA1: Dart 60Md 119
Ash Rd. DA12: Grav'nd 3E 144
Ash Rd. DA2: Hawl 63Pd 141
Ash Rd. DA3: Hartl 69Ae 143
Ash Rd. DA3: Lfield 69Ae 143
Ash Rd. DA3: Nw A G 69Ae 143
Ash Rd. E15 36Gc 73
Ash Rd. GU22: Wok 2P 187
Ash Rd. GU24: Pirb 8D 186
Ash Rd. IG7: Chig 21Tc 54
Ash Rd. SM3: Sutt 73Ab 154
Ash Rd. TN15: Ash 77Zd 165
Ash Rd. TN15: Nw A G 77Zd 165
Ash Rd. TN16: Westrm 97Tc 200
Ash Rd. TW17: Shep 70Q 128
Ash Row BR2: Broml 73Qc 160
ASHTEAD 90Pa 173
Ashtead Common 87La 172
Ashtead Gap KT22: Lea 88Ka 172
Ashtead Pk. 89Qa 173
ASHTEAD PARK 90Qa 173
ASHTEAD PRIVATE
HOSPITAL 91Na 193
Ashtead Rd. E5 31Wb 71
Ashtead Station (Rail) 89Na 173
Ashtead Woods Rd. KT21: Asht 89La 172
Ashton Cl. KT12: Hers 79X 151
Ashton Cl. SM1: Sutt 77Cb 155
Ashton Ct. E4 20Gc 35
Ashton Ct. HA1: Harr 34Ha 66
Ashton Gdns. RM6: Chad H 30Ad 55
Ashton Gdns. TW4: Houn 56Ba 107
Ashton Ga. RM3: Rom 24Md 57
Ashton Hgts. SE23 60Yb 114
Ashton Ho. SE11 7B 230 (50Rb 91)
(off Cottington St.)
Ashton Ho. SW9 52Qb 112
Ashton Pl. KT10: Clay 80Ha 152
Ashton Reach SE16 49Ac 92
Ashton Rd. E15 36Fc 73
Ashton Rd. EN3: Enf W 8Ac 20
Ashton Rd. GU21: Wok 9K 167
Ashton Rd. RM3: Rom 24Md 57
Ashton St. E14 45Ec 92
Ashtree Av. CR4: Mitc 68Fb 133
Ash Tree Cl. BR6: Farnb 77Rc 160
Ash Tree Cl. CR0: C'don 72Ac 158
Ash Tree Cl. KT6: Surb 75Na 153
Ash Tree Cl. TN15: W King 81Vd 184
Ash Tree Ct. TN15: W King 81Vd 184
Ash Tree Ct. TW15: Ashf 64R 128
(off Feltham Hill Rd.)
Ashtree Ct. AL1: St A 2D 6
Ashtree Ct. EN9: Walt A 6Jc 21
(off Horseshoe Cl.)
Ash Tree Dell NW9 29Sa 47
Ash Tree Dr. TN15: W King 80Vd 164
Ash Tree Ho. SE5 52Sb 113
(off Pitman St.)
Ash Tree Rd. WD24: Wat 8X 13
Ashtree Wlk. IG7: Chig 22Ad 55
Ash Tree Way CR0: C'don 71Zb 158
Ashtree Way HP1: Hem H 3J 3
Ashurst KT18: Eps 86Ta 173
Ashurst Cl. CR8: Kenley 87Tb 177
Ashurst Cl. DA1: Cray 55Hd 118
Ashurst Cl. HA6: Nwood 24U 44
Ashurst Cl. KT22: Lea 93Ja 192
Ashurst Cl. SE20 67Xb 135
Ashurst Dr. IG2: Ilf 30Rc 54
Ashurst Dr. IG6: Ilf 29Sc 54
Ashurst Dr. TW17: Shep 71N 149
Ashurst Gdns. SW2 60Qb 112
Ashurst Pk. SL5: S'hill 9C 124
Ashurst Rd. EN4: Cockf 15Hb 31
Ashurst Rd. KT20: Tad 93Xa 194
Ashurst Rd. N12 22Gb 49
Ashurst Wlk. CR0: C'don 75Xb 157
Ash Va. WD3: Map C 22F 42
Ashvale Ct. E3 40Cc 72
(off Matilda Gdns.)
Ashvale Dr. RM14: Upm 33Ud 78
Ashvale Gdns. RM14: Upm 33Ud 78
Ashvale Gdns. RM5: Col R 22Fd 56
Ashvale Rd. SW17 64Hb 133
Ashview Apts. N4 31Sb 71
(off Katherine Cl.)
Ashview Cl. TW15: Ashf 64N 127
Ashview Gdns. TW15: Ashf 64N 127
Ashville Rd. E11 33Fc 73
Ash Wlk. HA0: Wemb 35La 66
Ash Wlk. RM15: S Ock 41Zd 99
Ashwater Rd. SE12 60Jc 115
Ash Way IG8: Wfd G 26Mc 53
Ashway Cen., The KT2: King T 67Na 131
Ashwell Cl. E6 44Nc 94
Ashwell Ct. TW15: Ashf 61N 127
Ashwell Pl. WD24: Wat 9W 12
Ashwells Rd. CM15: Pil H 13Td 40
Ashwell St. AL3: St A 1B 6
Ashwick Cl. CR3: Cat'm 96Wb 197
Ashwindham Ct. GU21: Wok 10K 167
Ashwin St. E8 37Vb 71
Ashwood CR6: W'ham 92Yb 198
Ashwood Av. RM13: Rain 42Kd 97
Ashwood Av. UB8: Hil 44Q 84
Ashwood Gdns. CR0: New Ad 79Ec 158
Ashwood Gdns. UB3: Harl 49V 84
Ashwood Ho. HA5: Hat E 23Ca 45
(off The Avenue)
Ashwood Ho. NW4 28Ya 48
(off Belle Vue Est.)
Ashwood M. AL1: St A 4B 6
Ashwood Pk. GU22: Wok 90C 168
Ashwood Pk. KT22: Fet 95Ea 192
Ashwood Pl. DA2: Bean 62Xd 142
Ashwood Pl. GU22: Wok 90C 168
Ashwood Pl. SL5: S'dale 3C 146
Ashwood Rd. E4 20Fc 35
Ashwood Rd. EN6: Pot B 5Db 17
Ashwood Rd. GU22: Wok 90B 168
Ashwood Rd. TW20: Eng G 5M 125
Ashworth Av. TN15: W King 80Td 164
(off London Rd.)
Ashworth Cl. SE5 54Tb 113
Ashworth Est. CR0: Bedd 74Nb 156
Ashworth Mans. W9 41Db 89
(off Elgin Av.)
Ashworth Rd. W9 41Db 89
Aske Ho. N1 3H 219 (41Ub 91)
(off Fanshaw St.)
Asker Ho. N7 35Nb 70
Askern Cl. DA6: Bex 56Zc 117
Aske St. N1 3H 219 (41Ub 91)
Askew Bldg. The EC1 1D 224 (43Sb 91)
(off Bartholomew Cl.)
Askew Cres. W12 47Va 88
Askew Rd. HA6: Nwood 19T 26
Askew Rd. W12 47Va 88
Askews Farm La. RM17: Grays 50Ae 99
Askham Ct. W12 46Wa 88
Askham Rd. W12 46Wa 88
Askill Dr. SW15 57Ab 110
Askwith Rd. RM13: Rain 41Fd 96
Asland Rd. E15 39Gc 73
Aslett St. SW18 59Db 111
Aslin Ct. AL1: St A 2C 6
(off Hatfield Rd.)
Asman Ho. N1 1B 218 (40Rb 71)
(off Colebrooke Rd.)
Asmara Rd. NW2 36Ab 68
Asmar Cl. CR5: Coul 87Nb 176
Asmuns Hill NW11 29Cb 49
Asmuns Pl. NW11 29Bb 49
Asolando Dr. SE17 6E 230 (49Sb 91)
Aspasia Cl. AL1: St A 3D 6
Aspdin Rd. DA11: Nflt 62Fe 143
Aspect Ct. E14 47Ec 92
(off Manchester Rd.)
Aspect Ct. SW6 54Eb 111
Aspects SM1: Sutt 78Db 155
Aspects Ct. SL1: Slou 7J 81
Aspen Cl. AL2: Brick W 2Aa 13
Aspen Cl. BR6: Chels 78Wc 161
Aspen Cl. BR8: Swan 67Fd 140
Aspen Cl. KT1: Hamp W 67La 130
Aspen Cl. KT11: Stoke D 88Aa 171
Aspen Cl. KT19: Eps 81Ta 173
Aspen Cl. N19 33Lb 70
Aspen Cl. SL2: Slou 3F 80
Aspen Cl. TW18: Staines 62H 127
Aspen Cl. UB7: Yiew 46P 83
Aspen Cl. W5 47Pa 87
Aspen Copse BR1: Broml 68Pc 138
Aspen Ct. CM13: B'wood 20Ce 41
Aspen Ct. DA1: Dart 58Qd 119
Aspen Ct. GU25: Vir W 70A 126
Aspen Ct. IG7: Chig 20Vc 37
Aspen Ct. NW4 26Ab 48
Aspen Dr. HA0: Wemb 34Ja 66
Aspen Gdns. CR4: Mitc 71Jb 156
Aspen Gdns. TW15: Ashf 64S 128
Aspen Gdns. W6 50Xa 88
Aspen Grn. DA18: Erith 48Bd 95
Aspen Gro. HA5: Eastc 27V 44
Aspen Gro. RM14: Upm 35Qd 77
Aspen Ho. CR6: W'ham 87Dc 178
Aspen Ho. DA15: Sidc 61Wc 139
Aspen Ho. E15 41Gc 93
(off Teasel Way)
Aspen Ho. SE15 51Yb 114
(off Sharratt St.)
Aspen La. UB5: N'olt 41Aa 85
Aspenlea Rd. W6 51Za 110
Aspen Lodge W8 48Db 89
(off Abbots Wlk.)
Aspen M. SE20 65Yb 136
Aspen Pk. Dr. WD25: Wat 7X 13
Aspen Pl. WD23: B Hea 17Ga 28
Aspens, The EN9: Walt A 7Lc 21
(within Woodbine Cl. Caravan Pk.)
Aspens Pl. HP1: Hem H 5H 3
Aspen Sq. KT13: Weyb 76T 150
Aspen Va. CR3: Whyt 89Vb 177
Aspen Way E14 45Dc 92
Aspen Way EN3: Enf W 7Zb 20
Aspen Way RM15: S Ock 41Zd 99
Aspen Way SM7: Bans 86Za 174
Aspen Way TW13: Felt 62X 129
Aspern Gro. NW3 36Gb 69
Aspinall Rd. SE4 55Zb 114
(not continuous)
Aspinden Rd. SE16 49Xb 91
ASPIRE 19Ka 28
ASPIRE Nat. Training Cen. 19Ka 28
Aspire Sport & Fitness Cen. 18Ub 33
Aspland Gro. E8 37Xb 71
(off Amhurst Rd.)
Aspley Rd. SW18 57Db 111
Asplins Rd. N17 25Wb 51
Asprey Gro. CR3: Cat'm 96Wb 197
Asprey La. HA6: Nwood 23U 44
Asprey M. BR3: Beck 71Bc 158
Asprey Pl. BR1: Broml 68Nc 138
Asquith Cl. RM8: Dag 32Yc 75
Asquith Ho. SM7: Bans 87Bb 175
(off Dunnymans Rd.)
Asquith Ho. SW1 4E 228 (48Mb 90)
(off Monck St.)
Assam St. E1 44Wb 91
(off White Church La.)
Assata M. N1 37Rb 71
Assembley Ho. SE14 51Bc 114
(off Arklow Rd.)
Assembly Apts. SE15 53Yb 114
(off York Gro.)
Assembly Pas. E1 43Yb 92
Assembly Wlk. SM5: Cars 73Gb 155
Assher Rd. KT12: Hers 76Aa 151
Ass Ho. La. HA3: Hrw W 21Da 45
Astall Cl. HA3: Hrw W 25Ga 46
Astbury Bus. Pk. SE15 53Yb 114
Astbury Ho. SE11 4K 229 (48Qb 90)
(off Lambeth Wlk.)
Astbury Rd. SE15 53Yb 114
Astede Pl. KT21: Asht 90Pa 173
Astell Ho. E14 44Gc 93
(off Lyell St.)
Astell Ho. SW3 7E 226 (50Gb 89)
(off Astell St.)
Astell Rd. SE3 56Lc 115
Astell St. SW3 7E 226 (50Gb 89)
Asten Way RM7: Mawney 26Dd 56
Aster Ct. E5 33Yb 72
(off Woodmill Rd.)
Asterid Hgts. E20 36Ec 72
(off Liberty Bri. Rd.)
Aster Pl. E9 38Yb 72
(off Frampton Pk. Rd.)
Aster Rd. IG7: Chig 22Ad 55
Asters, The EN7: G Oak 1Tb 19
Asters, The SL5: S'dale 4C 146
Aste St. E14 47Ec 92
Astey's Row N1 38Sb 71
Asthall Gdns. IG6: Ilf 28Sc 54
Astins Ho. E17 28Dc 52
Astleham Rd. TW17: Shep 69N 127
Astle St. SW11 54Jb 112
Astley RM17: Grays 51Be 121
Astley Av. NW2 36Ya 68
Astley Ho. SE1 7K 231 (50Vb 91)
(off Rowcross St.)
Astley Ho. SW13 51Xa 110
(off Wyatt Dr.)
Astley Ho. W2 43Cb 89
(off Alfred Rd.)
Astley Rd. HP1: Hem H 2L 3
Aston Av. HA3: Kenton 31La 66
Aston Cl. DA14: Sidc 62Wc 139
Aston Cl. HP3: Hem H 6M 3
Aston Cl. KT21: Asht 90La 172
Aston Cl. WD23: Bush 16Ea 28
Aston Cl. WD24: Wat 12Y 27
Aston Ct. IG8: Wfd G 23Jc 53
Aston Grn. TW4: Cran 54Y 107
Aston Ho. EC4 1K 223 (43Qb 90)
(off Furnival St.)
Aston Ho. RM8: Dag 35Wc 75
Aston Ho. SW8 53Mb 112
Aston Ho. W11 45Bb 89
(off Westbourne Gro.)
Aston Mead SL4: Wind 3C 102
Aston M. RM6: Chad H 31Yc 75
Aston M. W10 41Za 88
Aston Pl. SW16 65Rb 135
Aston Rd. KT10: Clay 78Ga 152
Aston Rd. SW20 68Ya 132
Aston Rd. W5 44Ma 87
Astons Rd. HA6: Nwood 20S 26
Aston St. E14 43Ac 92
Aston Ter. SW12 58Kb 112
Astonville St. SW18 60Cb 111
Aston Way EN6: Pot B 4Fb 17
Aston Way KT18: Eps 87Va 174
Aston Webb Ho. SE1 7H 225 (46Ub 91)
(off Tooley St.)
Astor Av. RM7: Rom 30Ed 56
Astor Cl. KT15: Add 77M 149
Astor Cl. KT2: King T 65Ra 131
Astor Coll. W1 7C 216 (43Lb 90)
(off Charlotte St.)
Astor Ct. E16 44Lc 93
(off Ripley Rd.)
Astor Ct. SW6 52Eb 111
(off Maynard Cl.)
Astoria Ct. CR8: Purl 83Rb 177
(off High St.)
Astoria Ct. E8 38Vb 71
(off Queensbridge Rd.)
Astoria Ho. NW9 26Va 48
(off Boulevard Dr.)
Astoria Mans. SW16 62Nb 134
Astoria Wlk. SW9 55Qb 112
Astor Rd. TN15: W King 79Ud 164
Astra Cl. RM12: Horn 37Kd 77
Astra Ct. WD18: Wat 15V 26
Astra Dr. DA12: Grav'nd 4G 144
Astra Ho. E3 41Bc 92
(off Alfred St.)
Astra Ho. SE14 51Bc 114
(off Arklow Rd.)
Astral Ho. E1 1J 225 (43Ub 91)
(off Middlesex Rd.)
Astral Ho. SE6 63Ec 136
Astrid Ho. TW13: Felt 61Y 129
Astrop M. W6 48Ya 88
Astrop Ter. W6 47Ya 88
Astwood Cl. RM11: Horn 31Pd 77
Astwood Dr. HA7: Stan 19Ga 28
Astwood M. SW7 49Eb 89
Asylum Arch Rd. RH1: Redh 9P 207
Asylum Rd. SE15 52Xb 113
Atalanta Cl. CR8: Purl 82Qb 176
Atalanta St. SW6 52Za 110
Atbara Rd. TW11: Tedd 65Ka 130
Atcham Rd. TW3: Houn 56Ea 108
Atcost Rd. IG11: Bark 43Wc 95
Atcraft Cen. HA0: Wemb 39Na 67
Atelier Ct. SE8 52Cc 114
(off Watson's St.)
Atelier Ct. Central E14 43Ec 92
(off Leven Rd.)
Atelier Ct. Nth. E14 43Ec 92
(off Leven Rd.)

Atelier Ct. Sth. E14 43Ec 92
(off Leven Rd.)
Atfield Gro. GU20: W'sham 9B 146
Athboy Rd. DA11: Grav'nd 8D 122
Atheldene Rd. SW18 60Db 111
Athelney St. SE6 62Cc 136
Athelstan Cl. RM3: Hrld W 26Pd 57
Athelstane Gro. E3 40Bc 72
Athelstane M. N4 32Qb 70
Athelstan Gdns. NW6 38Ab 68
Athelstan Ho. E9 36Bc 72
(off Homerton Rd.)
Athelstan Ho. KT1: King T 70Pa 131
(off Athelstan Rd.)
Athelstan Pl. TW2: Twick 60Ga 108
Athelstan Recreation Ground 69Pa 131
(off Chapel Mill Rd.)
Athelstan Rd. HP3: Hem H 5P 3
Athelstan Rd. KT1: King T 70Pa 131
Athelstan Rd. RM3: Hrld W 25Pd 57
Athelstan Way BR5: St P 67Wc 139
Athelstone Rd. HA3: W'stone 26Fa 46
Athena Cl. HA2: Harr 33Fa 66
Athena Cl. KT1: King T 69Pa 131
Athena Ct. SE1 3H 231 (48Ub 91)
(off City Wlk.)
Athenaeum Ct. N5 35Sb 71
Athenaeum Lawn Tennis Club 30Vc 55
Athenaeum Pl. N10 27Kb 50
Athenaeum Rd. N20 18Eb 31
Athena Pl. HA6: Nwood 25V 44
Athene Pl. EC4 2A 224 (44Qb 90)
(off Thavie's Inn)
Athenia Cl. EN7: G Oak 1Rb 19
Athenia Ho. E14 44Fc 93
(off Blair St.)
Athenlay Rd. SE15 57Zb 114
Athens Gdns. W9 42Cb 88
(off Harrow Rd.)
Atherden Rd. E5 35Yb 72
Atherfield Ho. RH2: Reig 9L 207
(off Atherfield Rd.)
Atherfield Rd. RH2: Reig 9L 207
Atherfold Rd. SW9 55Nb 112
Atherley Way TW4: Houn 59Ba 107
Atherstone Ct. W2 43Db 89
(off Delamere Ter.)
Atherstone M. SW7 5A 226 (49Eb 89)
Atherton Cl. TW19: Stanw 58M 105
Atherton Ct. SL4: Eton 2H 103
Atherton Dr. SW19 63Za 132
Atherton Gdns. RM16: Grays 9E 100
Atherton Hgts. HA0: Wemb 38La 66
Atherton Ho. RM3: Rom 24Nd 57
(off Leyburn Cres.)
Atherton Leisure Cen. 37Hc 73
Atherton M. E7 37Hc 73
Atherton Pl. HA2: Harr 27Fa 46
Atherton Pl. UB1: S'hall 45Ca 85
Atherton Rd. E7 37Hc 73
Atherton Rd. IG5: Ilf 26Nc 54
Atherton Rd. SW13 52Wa 110
Atherton St. SW11 54Gb 111
Athill Ct. TN13: S'oaks 94Ld 203
Athlone KT10: Clay 79Ga 152
Athlone Cl. E5 36Xb 71
Athlone Cl. WD7: R'lett 8Ja 14
Athlone Ct. E17 27Fc 53
Athlone Ho. E1 44Yb 92
(off Sidney St.)
Athlone Pl. W10 43Ab 88
Athlone Rd. SW2 59Pb 112
Athlone Sq. SL4: Wind 3G 102
Athlone St. NW5 37Jb 70
Athlon Ind. Est. HA0: Wemb 39Ma 67
Athlon Rd. HA0: Wemb 40Ma 67
Athol Cl. HA5: Pinn 25X 45
Athole Gdns. EN1: Enf 15Ub 33
Athol Gdns. HA5: Pinn 25X 45
Atholl Ho. W9 4A 214 (41Eb 89)
(off Maida Vale)
Atholl Rd. IG3: Ilf 31Wc 75
Athol Rd. DA8: Erith 50Ed 96
Athol Sq. E14 44Ec 92
Athol Way UB10: Hil 41Q 84
Atkin Bldg. WC1 7J 217 (43Pb 90)
(off Raymond Bldgs.)
Atkins Cl. GU21: Wok 10L 167
Atkins Cl. TN16: Big H 84Lc 179
Atkins Ct. E3 39Bc 72
(off Willow Tree Cl.)
Atkins Dr. BR4: W W'ck 75Fc 159
Atkins Lodge BR6: Orp 73Wc 161
(off High St.)
Atkins Lodge W8 47Cb 89
(off Thornwood Gdns.)
Atkinson Cl. BR6: Chels 78Wc 161
Atkinson Cl. SW20 66Wa 132
Atkinson Cl. WD23: Bush 17Ga 28
Atkinson Ct. E10 31Dc 72
(off Kings Cl.)
Atkinson Ho. E13 42Hc 93
(off Sutton Rd.)
Atkinson Ho. E2 40Wb 71
(off Pritchards Rd.)
Atkinson Ho. SE17 6G 231 (49Tb 91)
(off Catesby St.)
Atkinson Ho. SW11 53Jb 112
(off Austin Rd.)
Atkinson Morley Av. SW17 62Fb 133
Atkinson Rd. E16 43Lc 93
Atkins Rd. E10 30Dc 52
Atkins Rd. SW12 59Lb 112
Atkins Sq. E8 36Xb 71
Atlanta Blvd. RM1: Rom 30Gd 56
Atlanta Bldg. SE13 53Dc 114
(off Deal's Gateway)
Atlanta Ct. CR7: Thor H 69Sb 135
Atlanta Ho. SE16 48Ac 92
(off Brunswick Quay)
Atlantic Apts. E16 45Jc 93
(off Seagull La.)
Atlantic Bldg. E15 36Fc 73
(off Property Row)
Atlantic Cl. DA10: Swans 57Ae 121
Atlantic Ct. E14 45Fc 93
(off Jamestown Way)
Atlantic Ct. SW3 7F 227 (50Hb 89)
Atlantic Rd. SW9 56Qb 112
Atlantic Wharf E1 45Zb 92
Atlantis Av. E16 45Rc 94
Atlantis Cl. IG11: Bark 41Xc 95
Atlas Bus. Cen. NW2 32Xa 68
Atlas Cres. HA8: Edg 19Ra 29
Atlas Gdns. SE7 49Lc 93
Atlas M. E8 37Vb 71
Atlas M. N7 37Pb 70
Atlas M. SE13 56Fc 115
Atlas Rd. DA1: Dart 55Pd 119
Atlas Rd. E13 40Jc 73
Atlas Rd. HA9: Wemb 35Sa 67
Atlas Rd. N11 24Jb 50
Atlas Rd. NW10 41Ua 88
Atlas Trade Pk. DA8: Erith 50Fd 96
Atlas Wharf E9 37Cc 72
Atlip Rd. HA0: Wemb 39Na 67
Atney Rd. SW15 56Ab 110
Atria Rd. HA6: Nwood 22W 44
Atrium, The IG9: Buck H 19Mc 35
Atrium, The W12 46Za 88
Atrium Apts. N1 39Tb 71
(off Felton St.)
Atrium Hgts. SE8 51Dc 114
(off Creekside)
Atrium Ho. SE8 52Bc 114
Attel Cl. KT15: Add 80J 149
Attenborough Cl. WD19: Wat 20Aa 27
Atterbury Cl. TN16: Westrm 98Tc 200
Atterbury Rd. N4 30Qb 50
Atterbury St. SW1 6F 229 (49Nb 90)
Attewood Av. NW10 34Ua 68
Attewood Rd. UB5: N'olt 37Aa 65
Attfield Cl. N20 19Fb 31
Attfield Cl. KT1: King T 68Pa 131
(off Albert Rd.)
Attilburgh Ho. SE1 3K 231 (48Vb 91)
(off St Saviour's Est.)
Attleborough Ct. SE23 61Wb 135
Attle Cl. UB10: Hil 40Q 64
Attlee Cl. CR7: Thor H 71Sb 157
Attlee Cl. UB4: Yead 41X 85
Attlee Cl. UB5: N'olt 41X 85
Attlee Ct. RM17: Grays 48Ce 99
Attlee Dr. DA1: Dart 57Qd 119
Attlee Rd. SE28 45Xc 95
Attlee Rd. UB4: Yead 41W 84
Attlee Ter. E17 28Dc 52
Attneave St. WC1 4K 217 (41Qb 90)
Attock M. E17 29Dc 52
Attwood Cl. CR2: Sande 86Xb 177
Attwood La. SE10 49Gc 93
Attwood Pl. Yd. TN15: Ash 78Zd 165
Atunbi Ct. NW1 38Lb 70
(off Farrier St.)
Atwater Cl. SW2 60Qb 112
Atwell Cl. E10 30Dc 52
Atwell Pl. KT7: T Ditt 74Ha 152
Atwell Rd. SE15 54Wb 113
Atwood KT23: Bookh 96Aa 191
Atwood Av. TW9: Kew 54Qa 109
Atwood Ho. W14 49Bb 89
(off Beckford Cl.)
Atwood Rd. W6 49Xa 88
Atwoods All. TW9: Kew 53Qa 109
Aube Ho. SE6 63Ec 136
Aubers Ridge Ct. E3 40Bc 72
(off Festubert Pl.)
Aubert Ct. N5 35Rb 71
Aubert Pk. N5 35Rb 71
Aubert Rd. N5 35Rb 71
Aubrey Av. AL2: Lon C 8G 6
Aubrey Beardsley Ho. SW1 6C 228 (49Lb 90)
(off Vauxhall Bri. Rd.)
Aubrey Mans. NW1 7D 214 (43Gb 89)
(off Lisson St.)
Aubrey Moore Point E15 40Ec 72
(off Abbey La.)
Aubrey Pl. NW8 2A 214 (40Eb 69)
Aubrey Rd. E17 27Cc 52
Aubrey Rd. N8 29Nb 50
Aubrey Rd. W8 46Bb 89
Aubrey's Rd. HP1: Hem H 3G 2
Aubrey Wlk. W8 46Bb 89
Aubrey Way E17 28Dc 52
Aubrietia Cl. RM3: Hrld W 25Nd 57
Auburn Cl. SE14 52Ac 114
Aubyn Hill SE27 63Sb 135
Aubyn Sq. SW15 57Wa 110
Auckland Av. RM13: Rain 41Hd 96
Auckland Cl. EN1: Enf 9Xb 19
Auckland Cl. RM18: Tilb 4C 122
Auckland Cl. SE19 67Vb 135
Auckland Ct. UB4: Yead 42Y 85
Auckland Gdns. SE19 67Ub 135
Auckland Hill SE27 63Sb 135
Auckland Ho. KT12: Walt T 74W 150
Auckland Ho. W12 45Xa 88
(off White City Est.)
Auckland Ri. SE19 67Ub 135
Auckland Rd. CR3: Cat'm 94Ub 197
Auckland Rd. E10 34Dc 72
Auckland Rd. EN6: Pot B 4Ab 16
Auckland Rd. IG1: Ilf 32Rc 74
Auckland Rd. KT1: King T 70Pa 131
Auckland Rd. SE19 67Vb 135
Auckland Rd. SW11 56Gb 111
Auckland St. SE11 50Pb 90
Audax NW9 26Va 48
Auden Dr. WD6: Bore 15Qa 29
Auden Pl. NW1 39Jb 70
Auden Pl. SM3: Cheam 77Ya 154
Audleigh Pl. IG7: Chig 23Qc 54
Audley Cl. KT15: Add 78K 149
Audley Cl. N10 24Kb 50
Audley Cl. SW11 55Jb 112
Audley Cl. WD6: Bore 13Qa 29
Audley Ct. E18 28Hc 53
Audley Ct. HA5: Pinn 26Y 45
Audley Ct. TW2: Twick 62Fa 130
Audley Ct. UB5: N'olt 41Y 85
Audley Dr. CR6: W'ham 87Yb 178
Audley Dr. E16 46Kc 93
Audley Firs KT12: Hers 77Y 151
Audley Gdns. EN9: Walt A 6Ec 20
Audley Gdns. IG10: Lough 12Sc 36
(not continuous)
Audley Gdns. IG3: Ilf 33Vc 75
Audley Ho. KT15: Add 78K 149
Audley Pl. SM2: Sutt 80Db 155
Audley Rd. EN2: Enf 12Rb 33
Audley Rd. NW4 29Wa 48
Audley Rd. TW10: Rich 57Pa 109
Audley Rd. W5 43Pa 87
Audley Sq. W1 6J 221 (46Jb 90)
Audley Wlk. BR5: St M Cry 72Yc 161
Audrey Cl. BR3: Beck 72Dc 158
Audrey Gdns. HA0: Wemb 33Ka 66
Audrey Rd. IG1: Ilf 34Rc 74
Audrey St. E2 40Wb 71
Audric Cl. KT2: King T 67Qa 131
Audwick Cl. EN8: Chesh 1Ac 20
Augurs La. E13 41Kc 93
Augur Cl. TW18: Staines 64H 127
Augusta Cl. KT8: W Mole 69Ba 129
Augusta Rd. SS17: Stan H 3K 101
Augusta Rd. TW2: Twick 61Ea 130
Augustas La. N1 38Qb 70
Augusta St. E14 44Dc 92
Augusta Wlk. W5 43Ma 87
August End SL3: Geor G 44A 82
Augustine Bell Twr. E3 40Dc 72
(off Pancras Way)
Augustine Cl. SL3: Poyle 55G 104
Augustine Ct. EN9: Walt A 5Dc 20
Augustine Rd. BR5: St P 69Zc 139
Augustine Rd. DA12: Grav'nd 9E 122
(not continuous)
Augustine Rd. HA3: Hrw W 25Da 45
Augustine Rd. W14 48Za 88
Augustus Bldg. E1 44Xb 91
(off Tarling St.)
Augustus Cl. AL3: St A 4N 5
Augustus Cl. HA7: Stan 20Ma 29
Augustus Cl. TW8: Bford 52La 108
Augustus Cl. W12 47Xa 88
Augustus Ct. SE1 5H 231 (49Ub 91)
(off Old Kent Rd.)
Augustus Ct. SW16 61Mb 134
Augustus Ct. TW13: Hanw 63Ba 129
Augustus Dr. KT17: Ewe 83Wa 174
Augustus Ho. GU25: Vir W 1P 147
Augustus Ho. NW1 2A 216 (40Kb 70)
(off Augustus St.)
Augustus La. BR6: Orp 75Wc 161
Augustus Rd. SW19 60Za 110
Augustus St. NW1 2A 216 (40Kb 70)
Aulay Ho. SE16 48Vb 91
Aultone Way SM1: Sutt 75Db 155
Aultone Way SM5: Cars 76Hb 155
Aultone Yd. Ind. Est. SM5: Cars 76Hb 155
Aulton Pl. SE11 50Qb 90
Aumonier M. N2 28Gb 49
Aura Ct. SE15 56Xb 113
Aura Ho. TW9: Kew 53Ra 109
Aurelia Gdns. CR0: C'don 71Pb 156
Aurelia Ho. E20 36Ec 72
(off Sunrise Cl.)
Aurelia Rd. CR0: C'don 72Nb 156
Auriel Av. RM10: Dag 37Fd 76
Auriga M. N1 36Tb 71
Auriol Cl. KT4: Wor Pk 76Ua 154
Auriol Dr. UB10: Hil 37Q 64
Auriol Dr. UB6: G'frd 38Fa 66
Auriol Ho. W12 46Xa 88
(off Ellerslie Rd.)
Auriol Mans. W14 49Ab 88
(off Edith Rd.)
Auriol Pk. Rd. KT4: Wor Pk 76Ua 154
Auriol Rd. W14 49Ab 88
Aurora Apts. EC1 3D 218 (41Sb 91)
(off Bollinder Pl.)
Aurora Apts. SW18 57Cb 111
(off Buckhold Rd.)
Aurora Bldg. E14 46Ec 92
(off Blackwall Way)
Aurora Bldg., The N1 3G 219 (41Tb 91)
(off East Rd.)
Aurora Cl. WD25: Wat 6Y 13
Aurora Ct. DA12: Grav'nd 8E 122
(off Romulus Rd.)
Aurora Gdns. 51Kb 112
Aurora Ho. E14 44Dc 92
(off Kerbey St.)
Aurora Ho. SE6 63Ec 136
Aurora Point SE8 49Bc 92
(off Grove St.)
Ausden Pl. WD17: Wat 15Y 27
(off Pumphouse Cres.)
Austell Gdns. NW7 20Ua 30
Austell Hgts. NW7 20Ua 30
(off Austell Gdns.)
Austen Apts. SE20 68Xb 135
(off Devas St.)
Austen Cl. DA9: Ghithe 58Yd 120
Austen Cl. IG10: Lough 13Tc 36
Austen Cl. RM18: Tilb 4E 122
Austen Cl. SE28 46Xc 95
Austen Ct. KT22: Lea 93Ja 192
(off Highbury Dr.)
Austen Gdns. DA1: Dart 56Pd 119
Austen Ho. NW6 41Cb 89
(off Cambridge Rd.)
Austen Ho. SW17 62Fb 133
(off St George's Gro.)
Austen Rd. DA8: Erith 52Dd 118
Austen Rd. HA2: Harr 33Da 65
Austen Vw. SL3: L'ly 51B 104
Austen Way AL4: St A 2H 7
Austen Way SL3: L'ly 51B 104
Austenway SL9: Chal P 27A 42
Austin Av. BR2: Broml 71Nc 160
Austin Cl. CR5: Coul 90Rb 177
Austin Cl. SE23 59Ac 114
Austin Cl. TW1: Twick 57La 108
Austin Ct. E6 39Lc 73
Austin Ct. EN1: Enf 15Ub 33
Austin Ct. SE15 55Wb 113
(off Peckham Rye)
Austin Ct. WD3: Rick 19K 25
Austin Friars EC2 2G 225 (44Tb 91)
Austin Friars Sq. EC2 2G 225 (44Tb 91)
(off Austin Friars)
Austin Ho. SE14 52Bc 114
(off Achilles St.)
Austin Pl. KT13: Weyb 75U 150
Austin Rd. BR5: St M Cry 72Wc 161
Austin Rd. DA11: Nflt 10B 122
Austin Rd. SW11 53Jb 112
Austin Rd. UB3: Hayes 47V 84
Austin's La. HA4: Ruis 36T 64
Austin's La. UB10: Ick 34S 64
Austins Mead HP3: Bov 10D 2
Austins Pl. HP2: Hem H 1M 3
Austin St. E2 4K 219 (41Vb 91)
Austin Ter. SE1 3A 230 (48Qb 90)
(off Morley St.)
Austin Vs. WD25: Wat 3X 13
Austin Waye UB8: Uxb 39L 63
Austinwood La. WD25: Wat 3X 13
Austral Cl. DA15: Sidc 62Vc 139
Austral Dr. RM11: Horn 31Md 77
Australian War Memorial 1J 227 (47Jb 90)
Australia Rd. SL1: Slou 7M 81
Australia Rd. W12 45Xa 88
Austral St. SE11 5B 230 (49Rb 91)
Austyn Gdns. KT5: Surb 74Ra 153
Austyns Pl. KT17: Ewe 81Wa 174
Autumn Cl. EN1: Enf 11Wb 33
Autumn Cl. SL1: Slou 6D 80
Autumn Cl. SW19 65Eb 133
Autumn Ct. RM7: Rom 30Ed 56
Autumn Dr. SM2: Sutt 81Db 175
Autumn Glades HP3: Hem H 4C 4
Autumn Gro. BR1: Broml 65Kc 137
Autumn Lodge CR2: S Croy 77Ub 157
(off South Pk. Hill Rd.)
Autumn St. E3 39Cc 72
Autumn Way UB7: W Dray 47P 83
Avalon Cl. BR6: Chels 76Zc 161
Avalon Cl. EN2: Enf 12Qb 32
Avalon Cl. SE6 61Gc 137
Avalon Cl. SW20 68Ab 132
Avalon Cl. W13 43Ja 86
Avalon Cl. WD25: Wat 4Aa 13
Avalon Ct. CR0: C'don 73Vb 157
Avalon Ct. WD25: A'ham 11Ca 27
Avalon Rd. BR6: Chels 75Yc 161
Avalon Rd. SW6 53Db 111
Avalon Rd. W13 42Ja 86
Avante KT1: King T 69Ma 131
Avantgarde Pl. E1 5K 219 (42Vb 91)
(off Sclater St.)
Avantgarde Twr. E1 5K 219 (42Vb 91)
(off Sclater St.)
Avard Gdns. BR6: Farnb 77Sc 160
Avarn Rd. SW17 65Hb 133
Avebury SL1: Slou 5E 80
Avebury Ct. N1 1F 219 (39Tb 71)
(off Imber St.)
Avebury Ct. SE16 49Yb 92
(off Debnams Rd.)
Avebury Pk. KT6: Surb 73Ma 153
Avebury Rd. BR6: Orp 76Tc 160
Avebury Rd. E11 32Fc 73
Avebury Rd. SW19 67Bb 133
Avebury St. N1 39Tb 71
Avedon Cl. HA2: Harr 26Ea 46
AVELEY 46Td 98
Aveley By-Pass RM15: Avel 45Sd 98
Aveley Cl. DA8: Erith 51Hd 118
Aveley Cl. RM15: Avel 46Td 98
Aveley Mans. IG11: Bark 38Rc 74
(off Whiting Av.)
Aveley Rd. RM1: Rom 28Fd 56
Aveley Rd. RM14: Avel 37Rd 77
Aveley Rd. RM14: Upm 37Rd 77
Aveline St. SE11 7K 229 (50Qb 90)
Aveling Cl. CR8: Purl 85Pb 176
Aveling Pk. Rd. E17 26Cc 52
Avelon Rd. RM13: Rain 39Jd 76
Avelon Rd. RM5: Col R 23Fd 56
Ave Maria La. EC4 3C 224 (44Rb 91)
Avenell Mans. N5 35Rb 71
Avenell Rd. N5 34Rb 71
Avenfield Ho. W1 4G 221 (45Hb 89)
(off Park La.)
Avening Rd. SW18 59Cb 111
Avening Ter. SW18 59Cb 111
Avenir Ho. E15 37Fc 73
(off Forrester Way)
Avenons Rd. E13 42Jc 93
Aventine Av. CR4: Mitc 69Kb 134
Aventine Ct. AL1: St A 3B 6
(off Holywell Hill)
Avenue, The BR1: Broml 69Mc 137
Avenue, The BR2: Kes 77Mc 159
Avenue, The BR3: Beck 67Dc 136
Avenue, The BR4: W W'ck 73Ec 158
Avenue, The BR5: St P 66Xc 139
Avenue, The BR6: Orp 75Vc 161
Avenue, The CM13: B'wood 23Ae 59
Avenue, The CM15: Kel H 11Ud 40
Avenue, The CR0: C'don 76Ub 157
Avenue, The CR3: Whyt 91Wb 197
Avenue, The CR5: Coul 87Mb 176
Avenue, The DA11: Grav'nd 10C 122
Avenue, The DA12: Cobh 10K 145
Avenue, The DA5: Bexl 59Zc 117
Avenue, The DA9: Ghithe 56Xd 120
Avenue, The E11 30Kc 53
Avenue, The E3 42Dc 92
Avenue, The E4 23Fc 53
Avenue, The EC2 2J 225 (44Ub 91)
Avenue, The EN5: Barn 13Ab 30
Avenue, The EN6: Pot B 2Bb 17
Avenue, The GU24: Chob 1L 167
Avenue, The GU3: Worp 9J 187
Avenue, The HA3: Hrw W 25Ha 46
Avenue, The HA5: Hat E 23Ba 45
Avenue, The HA5: Pinn 30Ba 45
Avenue, The HA6: Nwood 23S 44
Avenue, The HA9: Wemb 32Na 67
Avenue, The HP1: Hem H 1G 2
Avenue, The IG10: Lough 16Mc 35
Avenue, The IG9: Buck H 19Lc 35
Avenue, The KT10: Clay 79Ga 152
Avenue, The KT15: New H 82J 169
Avenue, The KT17: Ewe 80Xa 154
Avenue, The KT20: Tad 94Xa 194
Avenue, The KT22: Oxs 83Ha 172
Avenue, The KT4: Wor Pk 75Va 154
Avenue, The KT5: Surb 72Pa 153
Avenue, The N10 26Lb 50
Avenue, The N11 22Kb 50
Avenue, The N17 27Tb 51
Avenue, The N3 26Cb 49
Avenue, The N8 27Qb 50
Avenue, The NW6 39Za 68
Avenue, The RH1: S Nut 9E 208
Avenue, The RM1: Rom 28Fd 56
Avenue, The RM12: Horn 33Ld 77
Avenue, The SE10 52Fc 115
Avenue, The SL2: Farn C 5F 60
Avenue, The SL3: Dat 3M 103
Avenue, The SL4: Old Win 7M 103
Avenue, The SM2: Cheam 81Bb 175
Avenue, The SM5: Cars 80Jb 156
Avenue, The SW18 59Gb 111
Avenue, The SW4 56Jb 112
Avenue, The TN15: Bor G 91Ce 205
Avenue, The TN16: Tats 94Pc 200
Avenue, The TN16: Westrm 94Pc 200
Avenue, The TW1: Twick 57Ka 108
Avenue, The TW12: Hamp 65Ba 129
Avenue, The TW16: Sun 67X 129
Avenue, The TW18: Staines 67K 127
Avenue, The TW19: Wray 5P 103
Avenue, The TW20: Egh 63D 126
Avenue, The TW3: Houn 57Da 107
Avenue, The TW5: Cran 52W 106
Avenue, The TW9: Kew 54Pa 109
Avenue, The UB10: Ick 35Q 64
Avenue, The UB8: Cowl 42M 83
Avenue, The W13 44Ka 86
Avenue, The W4 48Ua 88
Avenue, The WD17: Wat 12W 26
Avenue, The WD23: Bush 14Ba 27
Avenue, The WD7: R'lett 5Ja 14
Avenue App. WD4: K Lan 2Q 12
Avenue C KT15: Add 76N 149
Avenue Cl. KT20: Tad 94Xa 194
Avenue Cl. N14 16Lb 32
Avenue Cl. NW8 1E 214 (39Gb 69)
(not continuous)
Avenue Cl. RM3: Hrld W 24Pd 57
Avenue Cl. TW5: Cran 53X 107
Avenue Cl. UB7: W Dray 48M 83
Avenue Ct. IG5: Ilf 27Nc 54
Avenue Ct. KT20: Tad 95Xa 194
Avenue Ct. N14 16Lb 32
Avenue Ct. NW2 34Bb 69
Avenue Ct. SW3 6F 227 (49Hb 89)
(off Draycott Av.)
Avenue Cres. TW5: Cran 53X 107
Avenue Cres. W3 47Ra 87
Avenue de Cagny GU24: Pirb 4D 186
Avenue Elmers KT6: Surb 71Na 153
Avenue Gdns. SE25 68Wb 135
Avenue Gdns. SW14 55Ua 110
Avenue Gdns. TW11: Tedd 66Ha 130
Avenue Gdns. TW5: Cran 52X 107
Avenue Gdns. W3 47Ra 87
Avenue Ga. IG10: Lough 16Lc 35
Avenue Ho. NW10 40Xa 68
(off All Souls Av.)
Avenue Ho. NW6 38Ab 68
(off The Avenue)
Avenue Ho. NW8 2D 214 (40Gb 69)
(off Allitsen Rd.)
Avenue Ind. Est. E4 23Cc 52
Avenue Ind. Est. RM3: Hrld W 26Md 57
Avenue Lodge NW8 38Fb 69
(off Avenue Rd.)
Avenue Lodge RM17: Grays 50Ee 99
Avenue Mans. NW3 36Db 69
(off Finchley Rd.)
Avenue M. N10 27Kb 50
Avenue One KT15: Add 77N 149
Avenue Pde. N21 17Tb 33
Avenue Pde. TW16: Sun 69X 129
Avenue Pk. Rd. SE27 61Rb 135
Avenue Ri. WD23: Bush 15Ca 27
Avenue Rd. AL1: St A 1C 6
Avenue Rd. BR3: Beck 68Zb 136
Avenue Rd. CM14: W'ley 21Yd 58
Avenue Rd. CM16: They B 9Tc 22
Avenue Rd. CR3: Cat'm 94Tb 197
Avenue Rd. DA17: Belv 49Ed 96
Avenue Rd. DA17: Erith 49Ed 96
Avenue Rd. DA7: Bex 55Ad 117
Avenue Rd. DA8: Erith 52Ed 118
Avenue Rd. E7 35Kc 73
Avenue Rd. HA5: Pinn 27Aa 45
Avenue Rd. IG8: Wfd G 23Lc 53
Avenue Rd. KT1: King T 69Na 131
Avenue Rd. KT11: Cobh 88Z 171
Avenue Rd. KT18: Eps 86Ta 173
Avenue Rd. KT3: N Mald 70Ua 132
Avenue Rd. N12 21Eb 49
Avenue Rd. N14 17Lb 32
Avenue Rd. N15 29Tb 51
Avenue Rd. N6 31Lb 70
Avenue Rd. NW10 40Va 68
Avenue Rd. NW3 38Fb 69
Avenue Rd. NW8 38Fb 69
Avenue Rd. RM3: Hrld W 24Pd 57
Avenue Rd. RM6: Chad H 31Xc 75
Avenue Rd. SE20 67Yb 136
Avenue Rd. SE25 68Vb 135
Avenue Rd. SM2: Sutt 82Cb 175
Avenue Rd. SM6: W'gton 80Lb 156
Avenue Rd. SM7: Bans 87Db 175
Avenue Rd. SW16 68Mb 134
Avenue Rd. SW20 68Xa 132
Avenue Rd. TN13: S'oaks 96Ld 203
Avenue Rd. TN16: Tats 92Nc 200
Avenue Rd. TW11: Tedd 66Ja 130
Avenue Rd. TW12: Hamp 67Da 129
Avenue Rd. TW13: Felt 62V 128
Avenue Rd. TW18: Staines 64F 126
Avenue Rd. TW7: Isle 53Ha 108
Avenue Rd. TW8: Bford 50La 86
Avenue Rd. UB1: S'hall 46Ba 85
Avenue Rd. W3 47Ra 87
Avenue Road Stop (London Tramlink) 68Zb 136
Avenue Sth. KT5: Surb 73Qa 153
Avenue Studios SW3 6C 226 (49Fb 89)
(off Sydney Cl.)
Avenue Ter. KT3: N Mald 69Sa 131
Avenue Ter. WD19: Wat 16Aa 27
Avenue Three KT15: Add 76N 149
Avenue Two KT15: Add 77N 149
Avenue Vs. RH1: Mers 1C 208
Averil Ct. SL6: Tap 4A 80
Averil Gro. SW16 65Rb 135
Averill St. W6 51Za 110
Avern Gdns. KT8: W Mole 70Da 129
Avern Rd. KT8: W Mole 70Da 129
Avershaw Ho. SW15 57Za 110
Avery Cl. BR3: Beck 65Bc 136
Avery Ct. NW9 27Sa 47
Avery Farm Row SW1 6J 227 (49Jb 90)
Avery Gdns. IG2: Ilf 29Pc 54
AVERY HILL 58Tc 116
Avery Hill Rd. SE9 58Tc 116
Avery Row W1 4K 221 (45Kb 90)
Avery Wlk. SW11 55Jb 112
Avery Way DA1: Dart 61Pd 141
Avey La. EN9: Lough 8Fc 21
Avey La. EN9: Walt A 8Fc 21
Avey La. IG10: H Beech 10Jc 21
Avey La. IG10: Lough 10Jc 21
Avia Cl. HP3: Hem H 6M 3
Avian Av. AL2: F'mre 10C 6
Aviary Cl. E16 43Hc 93
Aviary Rd. GU22: Pyr 88J 169
Aviation Dr. NW9 26Wa 48
Aviator Pk. KT15: Add 77M 149
Aviemore Cl. BR3: Beck 71Bc 158
Aviemore Way BR3: Beck 71Ac 158
Avigdor M. N16 33Tb 71
Avignon Rd. SE4 55Zb 114
Avingdor Ct. W3 46Sa 87
(off Horn La.)
Avington Ct. SE1 6J 231 (49Ub 91)
(off Old Kent Rd.)
Avington Gro. SE20 66Yb 136
Avion Cres. NW9 25Wa 48
Avior Dr. HA6: Nwood 21V 44
Avis Sq. E1 44Zb 92
Avoca Rd. SW17 63Jb 134
Avocet Cl. SE1 50Wb 91
Avocet M. SE28 48Tc 94
Avocet Rd. HP3: Hem H 7L 3
Avon Cl. DA12: Grav'nd 1F 144
Avon Cl. KT15: Add 79J 149
Avon Cl. KT4: Wor Pk 75Wa 154
Avon Cl. SL1: Slou 5C 80
Avon Cl. SM1: Sutt 77Eb 155
Avon Cl. UB4: Yead 42Y 85
Avon Cl. WD25: Wat 6Y 13
Avon Ct. E4 18Ec 34
Avon Ct. HA5: Hat E 24Ca 45
(off The Avenue)

Avon Ct. IG9: Buck H 18Kc 35
Avon Ct. N12 22Db 49
Avon Ct. SW15 57Ab 110
Avon Ct. UB6: G'frd 42Da 85
Avon Ct. W9 43Cb 89
(off Elmfield Way)
Avondale Av. EN4: E Barn 18Hb 31
Avondale Av. KT10: Hin W 76Ja 152
Avondale Av. KT4: Wor Pk 74Va 154
Avondale Av. N12 22Db 49
Avondale Av. NW2 34Ua 68
Avondale Av. TW18: Staines 66H 127
Avondale Cl. IG10: Lough 17Pc 36
Avondale Cl. KT12: Hers 78Y 151
Avondale Ct. AL1: St A 2C 6
Avondale Ct. E11 32Gc 73
Avondale Ct. E16 43Gc 93
Avondale Ct. E18 25Kc 53
Avondale Ct. SM2: Sutt 80Eb 155
(off Brighton Rd.)
Avondale Cres. EN3: Enf H 13Ac 34
Avondale Cres. IG4: Ilf 29Mc 53
Avondale Dr. IG10: Lough 17Pc 36
Avondale Dr. UB3: Hayes 46W 84
Avondale Gdns. TW4: Houn 57Ba 107
Avondale High CR3: Cat'm 93Xb 197
Avondale Ho. SE1 50Wb 91
(off Avondale Sq.)
Avondale Mans. SW6 53Bb 111
(off Rostrevor Rd.)
Avondale Pk. Gdns. W11 45Ab 88
Avondale Pk. Rd. W11 45Ab 88
Avondale Pavement SE1 50Wb 91
Avondale Ri. SE15 55Vb 113
Avondale Rd. BR1: Broml 65Gc 137
Avondale Rd. CR2: S Croy 79Sb 157
Avondale Rd. DA16: Well 54Yc 117
Avondale Rd. E16 43Gc 93
Avondale Rd. E17 31Cc 72
Avondale Rd. HA3: W'stone 27Ha 46
Avondale Rd. N13 19Qb 32
Avondale Rd. N15 29Rb 51
Avondale Rd. N3 25Eb 49
Avondale Rd. SE9 61Nc 138
Avondale Rd. SW14 55Ua 110
Avondale Rd. SW19 64Db 133
Avondale Rd. TW15: Ashf 62M 127
Avondale Sq. SE1 50Wb 91
Avonfield Ct. E17 27Fc 53
Avon Grn. RM15: S Ock 44Xd 98
Avongrove Ct. EC1 3E 218 (41Sb 91)
(off Bollinder Pl.)
Avon Ho. KT2: King T 67Ma 131
Avon Ho. RM14: Upm 31Ud 78
Avon Ho. W14 49Bb 89
(off Kensington Village)
Avon Ho. W8 48Cb 89
(off Allen St.)
Avonhurst Ho. NW2 38Ab 68
Avonley Rd. SE14 52Yb 114
Avonmead GU21: Wok 10N 167
Avon M. HA5: Hat E 24Ba 45
Avonmore Gdns. W14 49Bb 89
Avonmore Mans. W14 49Ab 88
(off Avonmore Rd.)
Avonmore Pl. W14 49Ab 88
Avonmore Rd. W14 49Ab 88
Avonmor M. GU23: Rip 94K 189
Avonmouth Apts. SW11 56Gb 111
(off Monarch Sq.)
Avonmouth Rd. DA1: Dart 57Md 119
Avonmouth St. SE1 3D 230 (48Sb 91)
Avon Path CR2: S Croy 79Sb 157
Avon Pl. SE1 2E 230 (47Sb 91)
Avon Rd. E17 27Fc 53
Avon Rd. RM14: Upm 30Td 58
Avon Rd. SE4 55Cc 114
Avon Rd. TW16: Sun 66V 128
Avon Rd. UB6: G'frd 42Ca 85
Avonstowe Cl. BR6: Farnb 76Sc 160
Avontar Ct. RM15: S Ock 42Xd 98
Avontar Rd. RM15: S Ock 42Xd 98
Avon Ter. IG10: Lough 16Pc 36
Avon Way E18 27Jc 53
Avonwick Rd. TW3: Houn 54Da 107
Avril Way E4 22Ec 52
Avro Ct. E9 36Ac 72
(off Mabley St.)
Avro Ho. NW9 26Va 48
(off Boulevard Dr.)
Avro Ho. SW8 52Kb 112
(off Havelock Ter.)
Avro Pl. TW5: Hest 52Y 107
Avro Way KT13: Weyb 82N 169
Avro Way SM6: W'gton 80Nb 156
Awberry Ct. WD18: Wat 16T 26
Awlfield Av. N17 25Tb 51
Awliscombe Rd. DA16: Well 54Vc 117
Axe St. IG11: Bark 39Sc 74
(not continuous)
Axholme Av. HA8: Edg 25Qa 47
Axiom Apts. BR2: Broml 70Kc 137
(off Masons Hill)
Axiom Apts. RM1: Rom 28Hd 56
(off Mercury Gdns.)
Axio Way E3 43Cc 92
Axis Apts. E1 5K 219 (42Vb 91)
(off Sclater St.)
Axis Ct. SE10 51Gc 115
(off Woodland Cres.)
Axis Ct. SE16 47Wb 91
(off East La.)
Axis Ho. SE13 56Ec 114
(off Lewisham High St.)
Axis Pk. SL3: L'ly 50D 82
Axminster Cres. DA16: Well 53Yc 117
Axminster Rd. N7 34Nb 70
Axon Pl. IG1: Ilf 33Sc 74
Axtaine Rd. BR5: St M Cry 73Zc 161
Axtane DA13: Sflt 66Be 143
Axtane Cl. DA4: S Dar 68Sd 142
Axwood KT18: Eps 87Sa 173
Aybrook St. W1 1H 221 (43Jb 90)
Aycliffe Cl. BR1: Broml 70Pc 138
Aycliffe Ho. SE17 51Tb 113
(off Portland St.)
Aycliffe Rd. W12 46Wa 88
Aycliffe Rd. WD6: Bore 11Na 29
Ayebridges Av. TW20: Egh 66E 126
Ayelands DA3: Nw A G 75Ae 165
Ayelands La. DA3: Nw A G 76Ae 165
Ayerst Ct. E10 31Ec 72
Aylands Cl. HA9: Wemb 33Na 67
Aylands Rd. EN3: Enf W 8Yb 20
Aylesbury Cl. E7 37Hc 73
Aylesbury Ct. SM1: Sutt 76Eb 155
Aylesbury Cres. SL1: Slou 4H 81
Aylesbury Ho. HA0: Wemb 39Na 67
(off Hatton Rd.)
Aylesbury Ho. SE15 51Wb 113
(off Friary Est.)
Aylesbury Rd. BR2: Broml 69Jc 137
Aylesbury Rd. SE17 7G 231 (50Tb 91)
Aylesbury St. EC1 6B 218 (42Rb 91)
Aylesbury St. NW10 34Ta 67
Aylesford Av. BR3: Beck 71Ac 158
Aylesford Ho. SE1 2G 231 (47Tb 91)
(off Long La.)
Aylesford St. SW1 7D 228 (50Mb 90)
Aylesham Cen. 53Wb 113
Aylesham Cl. NW7 24Wa 48
Aylesham Rd. BR6: Orp 73Vc 161
Ayles Rd. UB4: Yead 41X 85
Aylestone Av. NW6 38Za 68
Aylesworth Av. SL2: Slou 1E 80
Aylesworth Spur SL4: Old Win 9M 103
Aylett Rd. RM14: Upm 33Sd 78
Aylett Rd. SE25 70Xb 135
Aylett Rd. TW7: Isle 54Ga 108
Ayley Cft. EN1: Enf 15Wb 33
Ayliffe Cl. KT1: King T 68Qa 131
Aylmer Cl. HA7: Stan 21Ja 46
Aylmer Ct. N2 29Hb 49
Aylmer Dr. HA7: Stan 21Ja 46
Aylmer Ho. SE10 50Fc 93
Aylmer Pde. N2 29Hb 49
Aylmer Rd. E11 32Hc 73
Aylmer Rd. N2 29Gb 49
Aylmer Rd. RM8: Dag 34Ad 75
Aylmer Rd. W12 47Va 88
Ayloffe Rd. RM9: Dag 37Bd 75
Ayloffs Cl. RM11: Horn 28Md 57
Ayloffs Wlk. RM11: Horn 29Md 57
Aylsham Dr. UB10: Ick 33S 64
Aylsham La. RM3: Rom 21Ld 57
Aylton Est. SE16 47Yb 92
Aylward Rd. SE23 61Zb 136
Aylward Rd. SW20 68Bb 133
Aylwards Ri. HA7: Stan 21Ja 46
Aylward St. E1: Jamaica St. 44Yb 92
Aylward St. E1: Jubilee St. 44Yb 92
Aylwin Est. SE1 3J 231 (48Ub 91)
Aymer Cl. TW18: Staines 67G 126
Aymer Dr. TW18: Staines 67G 126
Aynhoe Mans. W14 49Za 88
(off Aynhoe Rd.)
Aynhoe Rd. W14 49Za 88
Aynho St. WD18: Wat 15X 27
Aynscombe Angle BR6: Orp 73Wc 161
Aynscombe Path SW14 54Sa 109
Ayot Path WD6: Bore 9Qa 15
Ayr Ct. W3 43Qa 87
Ayres Cl. E13 41Jc 93
Ayres St. SE1 1E 230 (47Sb 91)
Ayr Grn. RM1: Rom 25Gd 56
Ayron Rd. RM15: S Ock 42Xd 98
Ayrshire Cres. GU21: Brkwd 10E 166
Ayrsome Rd. N16 34Ub 71
Ayrton Gould Ho. E2 41Zb 92
(off Roman Rd.)
Ayrton Rd. SW7 3B 226 (48Fb 89)
Ayr Way RM1: Rom 25Gd 56
Aysgarth Ct. SM1: Sutt 76Db 155
Aysgarth Pl. SL0: Iver H 39F 62
Aysgarth Rd. SE21 59Ub 113
Ayston Ho. SE16 49Zb 92
(off Plough Way)
Aytoun Pl. SW9 54Pb 112
Aytoun Rd. SW9 54Pb 112
Azalea Cl. AL2: Lon C 9F 6
Azalea Cl. IG1: Ilf 36Rc 74
Azalea Cl. W7 46Ha 86
Azalea Ct. GU22: Wok 1P 187
Azalea Ct. IG8: Wfd G 23Gc 53
Azalea Ct. W7 46Ha 86
Azalea Dr. BR8: Swan 70Fd 140
Azalea Ho. SE14 52Bc 114
(off Achilles St.)
Azalea Ho. TW13: Felt 60X 107
Azalea Wlk. HA5: Eastc 29X 45
Azalea Wlk. SL3: Geor G 44A 82
Azania M. NW5 37Kb 70
Azenby Rd. SE15 54Vb 113
Azof St. SE10 49Gc 93
Azor Pl. KT18: Tatt C 91Xa 194
Azov Ho. E1 42Ac 92
(off Commodore St.)
Aztec Ho. IG1: Ilf 33Tc 74
Aztec Ho. IG6: Ilf 25Sc 54
Azura Ct. E15 39Ec 72
(off Warton Rd.)
Azure Bldg. E15 38Fc 73
(off Gt. Eastern Rd.)
Azure Cl. RM3: Rom 21Md 57
Azure Ct. NW9 29Qa 47
Azure Ho. E2 41Wb 91
(off Buckfast St.)
Azure Pl. TW3: Houn 56Da 107

B

Baalbec Rd. N5 36Rb 71
Babbacombe Cl. KT9: Chess 78Ma 153
Babbacombe Gdns. IG4: Ilf 28Nc 54
Babbacombe Ho. BR1: Broml 67Jc 137
(off Babbacombe Rd.)
Babbacombe Rd. BR1: Broml 67Jc 137
Babbage Ct. SE17 51Rb 113
(off Cook's Rd.)
Babbage Point SE10 51Dc 114
(off Norman Rd.)
Babell Ho. N1 37Rb 71
(off Canonbury Rd.)
Baber Bri. Cvn. Site TW14: Felt 57Y 107
Baber Bri. Pde. TW14: Felt 58Y 107
Baber Dr. TW14: Felt 58Y 107
Babington Ct. WC1 7G 217 (43Nb 90)
(off Orde Hall St.)
Babington Ho. SE1 1E 230 (47Sb 91)
(off Disney St.)
Babington Ri. HA9: Wemb 37Qa 67
Babington Rd. NW4 28Xa 48
Babington Rd. RM12: Horn 32Kd 77
Babington Rd. RM8: Dag 36Yc 75
Babington Rd. SW16 64Mb 134
Babmaes St. SW1 5D 222 (45Mb 90)
Babylon La. KT20: Lwr K 99Cb 195
Bachelors Acre SL4: Wind 3H 103
Bachelors La. GU23: Ock 96P 189
Bache's St. N1 3G 219 (41Tb 91)
Back All. EC3 3J 225 (44Ub 91)
(off Lloyd's Av.)
Back Church La. E1 44Wb 91
Back Grn. KT12: Hers 79Y 151
Back Hill EC1 6A 218 (42Qb 90)
Backhouse Pl. SE17 6J 231 (49Ub 91)
Back La. DA5: Bexl 59Cd 118
Back La. HA8: Edg 25Sa 47
Back La. IG9: Buck H 19Mc 35
Back La. N8 29Nb 50
Back La. NW3 35Eb 69
Back La. NW9 26Ta 47
Back La. RH2: Reig 1M 207
Back La. RM16: N Stif 47Yd 98
Back La. RM19: Purf 48Ud 98
Back La. RM20: W Thur 48Xd 98
Back La. RM6: Chad H 31Zc 75
Back La. TN13: Bes G 96Fd 202
Back La. TN15: God G 96Qd 203
Back La. TN15: Igh 96Yd 204
Back La. TW10: Ham 62La 130
Back La. TW8: Bford 51Ma 109
Back La. WD25: Let H 11Ga 28
Back La. WD3: Chen 10D 10
Backley Gdns. SE25 72Wb 157
Back of High St. GU24: Chob 3J 167
Back Path RH1: Blet 5J 209
Back Rd. DA14: Sidc 63Wc 139
Back Rd. E11 30Gc 53
Back Rd. E17 28Ec 52
Back Rd. TW11: Tedd 66Ga 130
Bacon Gro. SE1 4K 231 (48Vb 91)
Bacon La. HA8: Edg 25Qa 47
Bacon La. NW9 28Ra 47
(not continuous)
Bacon Link RM5: Col R 23Dd 56
Bacon's College Sports Cen. 46Ac 92
Bacons Dr. EN6: Cuff 1Nb 18
Bacons La. N6 32Jb 70
Baconsmead UB9: Den 33J 63
Bacon St. E1 5K 219 (42Vb 91)
Bacon St. E2 5K 219 (42Vb 91)
Bacon Ter. RM8: Dag 36Xc 75
Bacton NW5 36Jb 70
Bacton St. E2 41Yb 92
Badburgham Ct. EN9: Walt A 5Hc 21
Baddeley Cl. EN3: Enf L 9Cc 20
Baddeley Ho. KT8: W Mole 71Ca 151
(off Down St.)
Baddesley Ho. SE11 7J 229 (50Pb 90)
(off Jonathan St.)
Baddow Cl. IG8: Wfd G 23Lc 53
Baddow Cl. RM10: Dag 39Cd 76
Baden Cl. TW18: Staines 66K 127
Baden Dr. E4 14Dc 34
Baden Pl. SE1 1F 231 (47Tb 91)
Baden Powell Cl. KT6: Surb 75Pa 153
Baden Powell Cl. RM9: Dag 39Ad 75
Baden Powell Ho. DA17: Belv 48Cd 96
(off Ambrooke Rd.)
Baden Powell Ho. 5A 226 (49Eb 89)
(off Queen's Ga.)
Baden Powell Rd. TN13: Riv 93Gd 202
Baden Rd. IG1: Ilf 36Rc 74
Baden Rd. N8 28Mb 50
Bader Cl. CR8: Kenley 87Tb 177
Bader Ct. NW9 26Va 48
(off Runway Cl.)
Bader Gdns. SL1: Slou 7E 80
Bader Wlk. DA11: Nflt 2B 144
Bader Way RM13: Rain 37Jd 76
Bader Way SW15 58Wa 110
Bader Way UB10: Uxb 38N 63
Badgemore Path SE18 50Tc 94
(off Tuscan Rd.)
Badger Cl. IG2: Ilf 30Sc 54
Badger Cl. TW13: Felt 62X 129
Badger Cl. TW4: Houn 55Y 107
Badger Ct. NW2 34Ya 68
Badgersbridge Ride SL4: Wink 9A 102
Badgers Cl. EN2: Enf 13Rb 33
Badgers Cl. GU21: Wok 10N 167
Badgers Cl. HA1: Harr 30Fa 46
Badgers Cl. TW15: Ashf 64P 127
Badgers Cl. UB3: Hayes 45U 84
Badgers Cl. WD6: Bore 12Pa 29
Badgers Copse BR6: Orp 75Vc 161
Badgers Copse KT4: Wor Pk 75Va 154
Badger's Ct. KT17: Eps 85Ua 174
Badgers Ct. WD25: Wat 6V 12
Badgers Cft. HP2: Hem H 3D 4
Badgers Cft. N20 17Ab 30
Badgers Cft. SE9 62Qc 138
Badgers Dell WD3: Chor 14D 24
Badgers Hill GU25: Vir W 1N 147
Badgers Hole CR0: C'don 77Zb 158
Badgers La. CR6: W'ham 92Yb 198
Badger's Lodge KT17: Eps 85Ua 174
BADGERS MOUNT 82Dd 182
Badgers Mt. RM16: Ors 7B 100
Badger's Ri. TN14: Bad M 82Dd 182
Badgers Rd. TN14: Bad M 82Dd 182
Badgers Rd. TN14: S'ham 82Dd 182
Badgers Wlk. CR3: Whyt 90Vb 177
Badgers Wlk. CR8: Purl 83Lb 176
Badgers Wlk. KT3: N Mald 68Ua 132
Badgers Wlk. WD3: Chor 14H 25
Badgers Wood CR3: Cat'm 97Tb 197
Badgers Wood SL2: Farn C 6G 60
Badger Wlk. GU3: Norm 10A 186
Badgers Way AL10: Hat 2D 8
Badingham Dr. KT22: Fet 95Ga 192
Badlis Rd. E17 27Cc 52
Badlow Cl. DA8: Erith 52Gd 118
Badma Cl. N9 20Yb 34
Badminton Cl. HA1: Harr 28Ga 46
Badminton Cl. UB5: N'olt 37Ca 65
Badminton Cl. WD6: Bore 12Qa 29
Badminton Ct. N4 32Rb 71
Badminton Ho. WD24: Wat 12Y 27
(off Anglian Cl.)
Badminton M. E16 46Jc 93
Badminton Rd. SW12 58Jb 112
Badric Ct. SW11 54Fb 111
Badsworth Rd. SE5 53Sb 113
Baffin Rd. DA12: Grav'nd 3E 144
Baffin Way E14 46Ec 92
Baffin Ga. BR2: Hayes 74Kc 159
Bagenal Ho. WD5: Ab L 4V 12
Bagley Cl. UB7: W Dray 47N 83
Bagley's La. SW6 53Db 111
Bagleys Spring RM6: Chad H 28Ad 55
Bagley Wlk. N1 1E 216 (39Mb 70)
Bagnigge Ho. WC1 4K 217 (41Qb 90)
(off Margery St.)
Bagot Cl. KT21: Asht 88Pa 173
Bagshot Ct. SE18 53Qc 116
Bagshot Ho. NW1 3A 216 (41Kb 90)
(off Redhill St.)
Bagshot Rd. EN1: Enf 17Vb 33
Bagshot Rd. GU21: Knap 10F 166
Bagshot Rd. GU22: Wok 4G 186
Bagshot Rd. GU24: Brkwd 3F 186
Bagshot Rd. GU24: Chob 4C 166
Bagshot Rd. GU24: W End 4C 166
Bagshot Rd. GU24: Wok 3F 186
Bagshot Rd. GU3: Worp 6G 186
Bagshot Rd. TW20: Eng G 6N 125
Bagshot St. SE17 7J 231 (50Ub 91)
Bahram Ct. E2 42Xb 91
(off Three Colts La.)
Bahram Rd. KT19: Eps 82Ta 173
Baigents La. GU20: W'sham 9B 146
Baildon E2 40Yb 72
(off Cyprus St.)
Baildon St. SE8 52Bc 114
Bailes Pl. BR3: Beck 67Ac 136
Bailey Cl. E4 21Ec 52
Bailey Cl. N11 24Mb 50
Bailey Cl. RM19: Purf 49Td 98
Bailey Cl. SE28 46Uc 94
Bailey Cl. SL4: Wind 4E 102
Bailey Cotts. E14 43Ac 92
(off Maroon St.)
Bailey Ct. NW9 27Ua 48
(off Lingard Av.)
Bailey Cres. KT9: Chess 80Ma 153
Bailey Dr. DA10: Swans 59Ae 121
Bailey Ho. E3 41Dc 92
(off Talwin St.)
Bailey Ho. SW10 52Db 111
(off Coleridge Gdns.)
Bailey M. SW2 57Qb 112
Bailey M. W4 51Ra 109
Bailey Pl. N16 36Ub 71
Bailey Pl. SE26 65Zb 136
Baileys Ho. SW11 52Mb 112
(off Charles Clowes Wlk.)
Baileys M. HP1: Hem H 1M 3
(off High St.)
Bailey St. SE8 49Ac 92
Bailey Twr. E1 45Xb 91
Baillie Cl. RM13: Rain 42Kd 97
Baillie M. KT16: Ott 79F 148
Baillies Wlk. W5 47Ma 87
Baily Gdns. RH2: Reig 6M 207
(off Wray Comn. Rd.)
Bainbridge Cl. TW10: Ham 64Na 131
Bainbridge Ct. SE10 48Jc 93
(off Rennie Street)
Bainbridge Rd. RM9: Dag 35Bd 75
Bainbridge St. WC1 2E 222 (44Mb 90)
Baines Cl. CR2: S Croy 78Tb 157
Bainton Mead GU21: Wok 9L 167
Baird Av. UB1: S'hall 45Da 85
Baird Cl. E10 32Cc 72
Baird Cl. NW9 30Sa 47
Baird Cl. SL1: Slou 7F 80
Baird Cl. WD23: Bush 16Da 27
Baird Gdns. SE19 63Ub 135
Baird Ho. W12 45Xa 88
(off White City Est.)
Baird Memorial Cotts. N14 19Mb 32
(off Balaams La.)
Baird Rd. EN1: Enf 13Xb 33
Baird St. EC1 5E 218 (42Sb 91)
Bairny Wood App. IG8: Wfd G 23Kc 53
Bairstow Cl. WD6: Bore 11Na 29
Baizdon Rd. SE3 54Gc 115
Bakeham La. TW20: Eng G 6P 125
Bakehouse M. TW12: Hamp 66Ca 129
Baker Beal Ct. DA7: Bex 55Dd 118
Baker Boy La. CR0: Sels 85Ac 178
Baker Ct. WD6: Bore 12Ra 29
Baker Cres. DA1: Dart 59Ld 119
Baker Hill Cl. DA11: Nflt 3B 144
Baker Ho. E3 41Dc 92
(off Bromley High St.)
Baker Ho. W7 46Ha 86
Baker Ho. WC1 6G 217 (42Nb 90)
(off Colonnade)
Baker La. CR4: Mitc 68Jb 134
Baker Pas. NW10 39Ua 68
Baker Pl. KT19: Ewe 79Sa 153
Baker Rd. NW10 39Ua 68
Baker Rd. SE18 52Nc 116
Bakers Av. E17 30Dc 52
Bakers Av. TN15: W King 80Ud 164
Bakers Cl. AL1: St A 3E 6
Bakers Cl. CR8: Kenley 86Sb 177
Bakers Ct. CM14: B'wood 20Yd 40
Bakers Ct. RH1: Redh 7P 207
Bakers Ct. SE25 69Ub 135
Bakers Ct. UB8: Uxb 38M 63
Bakers End SW20 68Ab 132
Baker's Fld. N7 35Mb 70
Bakers Field Cl. KT19: Ewe 81Ta 173
Bakers Gdns. SM5: Cars 75Gb 155
Bakersgate Ctyd. GU24: Pirb 8E 186
Bakersgate Gdns. GU24: Pirb 7E 186
Bakers Hall Ct. EC3 5J 225 (45Ub 91)
(off Harp La.)
Bakers Hill E5 32Yb 72
Bakers Hill EN5: New Bar 12Db 31
Bakers Ho. W5 46Ma 87
(off The Grove)
Bakers La. CM16: Epp 2Vc 23
Bakers La. N6 30Hb 49
Bakers Mead RH9: G'stone 2A 210
Baker's M. W1 2H 221 (44Jb 90)
Bakers M. BR6: Chels 79Vc 161
Bakers Pas. NW3 35Eb 69
(off Heath St.)
Baker's Rents E2 4K 219 (41Vb 91)
Bakers Rd. EN7: Chesh 2Xb 19
Bakers Rd. UB8: Uxb 38M 63
Baker's Row E15 40Gc 73
Baker's Row EC1 6K 217 (42Qb 90)
Baker St. EN1: Enf 13Tb 33
Baker St. EN6: Pot B 7Ab 16
Baker St. KT13: Weyb 77Q 150
Baker St. NW1 6G 215 (42Hb 89)
Baker St. RM16: Ors 4A 100
Baker St. W1 7G 215 (43Hb 89)
BAKER STREET 7G 215 (43Hb 89)
BAKER STREET 4A 100
Baker Street Station (Underground) 6G 215 (42Hb 89)
Bakers Vs., The CM16: Epp 2Vc 23
BAKER'S WOOD 32F 62
Bakers Wood UB9: Den 33F 62
Baker's Yd. EC1 6K 217 (42Qb 90)
(off Bakers Rd.)
Bakers Yd. UB8: Uxb 38M 63
Bakery Cl. RM6: Chad H 27Ad 55
Bakery Cl. SW9 52Pb 112
Bakery M. KT6: Surb 74Qa 153
Bakery Path HA8: Edg 22Ra 47
(off St Margaret's Rd.)
Bakery Pl. SW11 56Hb 111
Bakery St. SE16 3K 231 (48Vb 91)
Bakewell Way KT3: N Mald 68Ua 132
Balaam Ho. SM1: Sutt 77Cb 155
Balaam Leisure Cen. 42Jc 93
Balaams La. N14 19Mb 32
Balaam St. E13 42Jc 93
Balaclava Rd. KT6: Surb 73La 152
Balaclava Rd. SE1 6K 231 (49Vb 91)
Bala Grn. NW9 30Ua 48
(off Ruthin Cl.)
Balboa Ct. E14 43Ac 92
(off Pechora Way)
Balcaskie Rd. SE9 57Pc 116
Balchen Rd. SE3 54Mc 115
Balchier Rd. SE22 58Xb 113
Balcombe Cl. DA6: Bex 56Zc 117
Balcombe Ho. NW1 5E 214 (42Gb 89)
(off Taunton Pl.)
Balcombe St. NW1 6F 215 (42Hb 89)
Balcon Ct. W5 44Pa 87
Balcon Way WD6: Bore 11Sa 29
Balcony, The W12 46Za 88
Balcorne St. E9 38Yb 72
Balder Ri. SE12 61Kc 137
Balderton Flats W1 3J 221 (44Jb 90)
(off Balderton St.)
Balderton St. W1 3J 221 (44Jb 90)
Baldewyne Ct. N17 25Wb 51
Baldocks Rd. CM16: They B 7Uc 22
Baldock St. E3 40Dc 72
Baldock Way WD6: Bore 11Pa 29
Baldrey Ho. SE10 50Hc 93
(off Blackwall La.)
Baldry Gdns. SW16 65Nb 134
Baldwin Cres. SE5 53Sb 113
Baldwin Gdns. TW3: Houn 53Ea 108
Baldwin Ho. SW2 60Qb 112
Baldwin Rd. SL1: Burn 1A 80
Baldwin Rd. SW11 58Jb 112
Baldwin Rd. WD17: Wat 10W 12
Baldwin's Bec SL4: Eton 1H 103
(off Baldwin's Shore)
Baldwins Gdns. EC1 7K 217 (43Qb 90)
Baldwins Hill IG10: Lough 12Pc 36
Baldwin's La. WD3: Crox G 14Q 26
Baldwin's Shore SL4: Eton 1H 103
Baldwin St. EC1 4F 219 (41Tb 91)
Baldwin Ter. N1 1D 218 (40Sb 71)
Baldwyn Gdns. W3 45Ta 87
Baldwyn's Pk. DA5: Bexl 61Fd 140
Baldwyns Rd. DA5: Bexl 61Fd 140
Balearic Apts. E16 45Jc 93
(off Western Gateway)
Bale Rd. E1 43Ac 92
Bales Ter. N9 20Vb 33
Balfern Gro. W4 50Ua 88
Balfern St. SW11 54Gb 111
Balfe St. N1 1G 217 (40Nb 70)
Balfont Cl. CR2: Sande 85Wb 177
Balfour Av. GU22: Wok 94A 188
Balfour Av. W7 46Ha 86
Balfour Bus. Cen. UB2: S'hall 48Y 85
Balfour Ct. RM17: Grays 49Fe 99
Balfour Gro. N20 20Hb 31
Balfour Ho. KT13: Weyb 77Q 150
(off Balfour Rd.)
Balfour Ho. SW11 53Jb 112
(off Forfar Rd.)
Balfour Ho. W10 43Za 88
(off St Charles Sq.)
Balfour M. HP3: Bov 9C 2
Balfour M. N9 20Wb 33
Balfour M. W1 6J 221 (46Jb 90)
Balfour Pl. SW15 56Xa 110
Balfour Pl. W1 5J 221 (45Jb 90)
Balfour Rd. BR2: Broml 71Mc 159
Balfour Rd. HA1: Harr 29Fa 46
Balfour Rd. IG1: Ilf 33Rc 74
Balfour Rd. KT13: Weyb 77Q 150
Balfour Rd. N5 35Sb 71
Balfour Rd. RM17: Grays 49Ee 99
Balfour Rd. SE25 71Wb 157
Balfour Rd. SM5: Cars 80Hb 155
Balfour Rd. SW19 66Db 133
Balfour Rd. TW3: Houn 55Da 107
Balfour Rd. UB2: S'hall 48Z 85
Balfour Rd. W13 47Ja 86
Balfour Rd. W3 43Sa 87
Balfour St. SE17 5F 231 (49Tb 91)
Balfour Ter. N3 26Db 49
Balfron Twr. E14 44Ec 92
Balgonie Rd. E4 18Fc 35
Balgores Cres. RM2: Rom 27Kd 57
Balgores La. RM2: Rom 27Kd 57
Balgores Sq. RM2: Rom 28Kd 57
Balgove Ct. NW10 37Xa 68
(off Eden Gro.)
Balgowan Cl. KT3: N Mald 71Ua 154
Balgowan Rd. BR3: Beck 69Ac 136
Balgowan St. SE18 49Vc 95
BALHAM 60Jb 112
Balham Gro. SW12 59Jb 112
Balham High Rd. SW12 60Jb 112
Balham High Rd. SW17 62Jb 134
Balham Hill SW12 59Kb 112
Balham Leisure Cen. 61Kb 134
Balham New Rd. SW12 59Kb 112
Balham Pk. Rd. SW12 60Hb 111
Balham Rd. N9 20Wb 33
Balham Station (Rail & Underground) 60Kb 112
Balham Sta. Rd. SW12 60Kb 112
Balin Ho. SE1 1F 231 (47Tb 91)
(off Long La.)
Balladier Wlk. E14 43Dc 92
Ballamore Rd. BR1: Broml 62Jc 137
Ballance Rd. E9 37Zb 72
Ballands Nth., The KT22: Fet 94Ga 192
Ballands Sth., The KT22: Fet 95Ga 192
Ballantine St. SW18 56Eb 111
Ballantrae Ho. NW2 35Bb 69
Ballantyne Cl. SE9 63Nc 138
Ballantyne Dr. KT20: Kgswd 93Bb 195
Ballard Cl. KT2: King T 66Ta 131
Ballard Grn. SL4: Wind 2C 102
Ballard Ho. SE10 51Dc 114
(off Thames St.)
Ballards Cl. RM10: Dag 39Dd 76
Ballards Farm Rd. CR0: C'don 79Xb 157
Ballards Farm Rd. CR2: S Croy 79Wb 157
Ballards Grn. KT20: Tad 91Ab 194
Ballards La. N12 23Db 49
Ballards La. N3 25Cb 49
Ballards La. RH8: Limp 1N 211
Ballards M. HA8: Edg 23Qa 47
Ballards Ri. CR2: Sels 79Wb 157
Ballards Rd. NW2 33Wa 68
Ballards Rd. RM10: Dag 40Dd 76
Ballards Way CR0: C'don 79Xb 157
Ballards Way CR2: Sels 79Wb 157
Ballast Quay SE10 50Fc 93
Ballater Cl. WD19: Wat 21Y 45
Ballater Rd. CR2: S Croy 78Vb 157
Ballater Rd. SW2 56Nb 112
Ball Ct. EC3 3G 225 (44Tb 91)
(off Birchin La.)

Ballencrieff Rd. SL5: S'dale 3D 146
Balletica Apts. WC2 3G 223 (44Nb 90)
(off Long Acre)
Ball Ho. NW9 27Va 48
(off Aerodrome Rd.)
Ballie Apts. E16 45Sc 94
(off Lock Side Way)
Ballina St. SE23 59Zb 114
Ballin Ct. E14 47Ec 92
(off Stewart St.)
Ballingdon Rd. SW11 58Jb 112
Ballinger Ct. WD18: Wat 13X 27
Ballinger Point E3 41Dc 92
(off Bromley High St.)
Ballinger Way UB5: N'olt 42Aa 85
Balliol Av. E4 21Gc 53
Balliol Rd. DA16: Well 54Xc 117
Balliol Rd. N17 25Ub 51
Balliol Rd. W10 44Ya 88
Balloch Rd. SE6 60Fc 115
Ballogie Av. NW10 35Ua 68
Balloon Cnr. AL9: Wel G 5D 8
Ballow Cl. SE5 52Ub 113
Balls Pond Pl. N1 37Tb 71
Balls Pond Rd. N1 37Tb 71
Balmain Cl. W5 46Ma 87
Balmain Ct. TW3: Houn 53Da 107
Balmain Lodge KT5: Surb 70Na 131
(off Cranes Pk. Av.)
Balman Ho. SE16 49Zb 92
(off Rotherhithe New Rd.)
Balmer Rd. E3 40Bc 72
Balmes Rd. N1 39Tb 71
Balmoral Apts. W2 1D 220 (43Gb 89)
(off Praed St.)
Balmoral Av. BR3: Beck 70Ac 136
Balmoral Av. N11 23Jb 50
Balmoral Cl. AL2: Park 10A 6
Balmoral Cl. SL1: Slou 4C 80
Balmoral Cl. SW15 58Za 110
Balmoral Ct. BR3: Beck 67Ec 136
(off The Avenue)
Balmoral Ct. HA9: Wemb 34Pa 67
Balmoral Ct. KT4: Wor Pk 75Xa 154
Balmoral Ct. NW8 1B 214 (40Fb 69)
(off Queen's Ter.)
Balmoral Ct. SE12 63Kc 137
Balmoral Ct. SE16 46Zb 92
(off King & Queen Wharf)
Balmoral Ct. SE17 50Tb 91
(off Merrow St.)
Balmoral Ct. SE27 63Sb 135
Balmoral Ct. SM2: Sutt 80Cb 155
Balmoral Cres. KT8: W Mole 69Ca 129
Balmoral Dr. GU22: Wok 88E 168
Balmoral Dr. UB1: S'hall 42Ba 85
Balmoral Dr. UB4: Hayes 42U 84
Balmoral Dr. WD6: Bore 15Ta 29
Balmoral Gdns. CR2: Sande 82Tb 177
Balmoral Gdns. DA5: Bexl 59Bd 117
Balmoral Gdns. IG3: Ilf 32Vc 75
Balmoral Gdns. SL4: Wind 5H 103
Balmoral Gdns. W13 48Ja 86
Balmoral Ho. E14 48Dc 92
(off Lanark Sq.)
Balmoral Ho. E16 46Kc 93
(off Keats Av.)
Balmoral Ho. SE1 7K 225 (46Vb 91)
(off Duchess Wlk.)
Balmoral Ho. W14 49Ab 88
(off Windsor Way)
Balmoral M. W12 48Va 88
Balmoral Rd. CM15: Pil H 16Xd 40
Balmoral Rd. DA4: Sut H 66Rd 141
Balmoral Rd. E10 33Dc 72
Balmoral Rd. E7 35Lc 73
Balmoral Rd. EN3: Enf W 8Zb 20
Balmoral Rd. HA2: Harr 35Ca 65
Balmoral Rd. KT1: King T 70Pa 131
Balmoral Rd. KT4: Wor Pk 76Xa 154
Balmoral Rd. NW2 37Xa 68
Balmoral Rd. RM12: Horn 34Md 77
Balmoral Rd. RM2: Rom 29Kd 57
Balmoral Rd. WD24: Wat 10Y 13
Balmoral Rd. WD5: Ab L 4W 12
Balmoral Trad. Est. IG11: Bark 43Vc 95
Balmoral Way SM2: Sutt 82Cb 175
Balmore Cl. E14 44Ec 92
Balmore Cres. EN4: Cockf 15Jb 32
Balmore St. N19 33Kb 70
Balmuir Gdns. SW15 56Ya 110
Balnacraig Av. NW10 35Ua 68
Balniel Ga. SW1 7E 228 (50Mb 90)
Balquhain Cl. KT21: Asht 89Ma 173
Balsam Ho. E14 45Dc 92
(off E. India Dock Rd.)
Baltic Apts. E16 45Jc 93
(off Western Gateway)
Baltic Av. TW8: Bford 50Ma 87
Baltic Cl. SW19 66Fb 133
Baltic Ct. E1 46Yb 92
(off Clave St.)
Baltic Ct. SE16 47Zb 92
Baltic Ho. SE5 54Sb 113
Baltic Pl. N1 39Ub 71
Baltic St. E. EC1 6D 218 (42Sb 91)
Baltic St. W. EC1 6D 218 (42Sb 91)
Baltic Wharf DA11: Grav'nd 7C 122
Baltimore Cl. DA17: Belv 47Dd 96
Baltimore Ct. SW1 6D 228 (49Mb 90)
(off Vauxhall Bri. Rd.)
Baltimore Ho. SE11 6K 229 (49Qb 90)
(off Hotspur St.)
Baltimore Ho. SW18 55Eb 111
Baltimore Pl. DA16: Well 54Vc 117
Baltimore Wharf E14 48Dc 92
Balvaird Pl. SW1 50Mb 90
Balvernie Gro. SW18 59Bb 111
Balvernie M. SW18 59Cb 111
Bamber Ho. IG11: Bark 39Sc 74
Bamber Rd. SE15 53Vb 113
Bamboo Ct. E5 33Yb 72
(off Woodmill Rd.)
Bamborough Gdns. W12 47Ya 88
Bamburgh N17 24Xb 51
Bamford Av. HA0: Wemb 39Pa 67
Bamford Rd. BR1: Broml 64Ec 136
Bamford Rd. IG11: Bark 37Sc 74
Bamford Way RM5: Col R 22Dd 56
Bampfylde Cl. SM6: W'gton 76Lb 156
Bampton Ct. W5 44Ma 87
Bampton Dr. NW7 24Wa 48
Bampton Rd. RM3: Rom 25Nd 57
Bampton Rd. SE23 62Zb 136
Bampton Way GU21: Wok 10L 167
Banavie Gdns. BR3: Beck 67Ec 136
Banbury Av. SL1: Slou 3D 80
Banbury Cl. EN2: Enf 11Rb 33
Banbury Ct. SM2: Sutt 80Cb 155
Banbury Ct. WC2 4F 223 (45Nb 90)
(off Long Acre)
Banbury Ho. E9 38Zb 72
Banbury Rd. E17 24Zb 52
Banbury Rd. E9 38Zb 72
Banbury St. SW11 54Gb 111
Banbury St. WD18: Wat 15W 26
Banbury Vs. DA13: Sflt 65Be 143
Banbury Wlk. UB5: N'olt 40Ca 65
(off Brabazon Rd.)
Banchory Rd. SE3 52Kc 115
Banckside DA3: Hartl 70Ae 143
Bancroft Av. IG9: Buck H 19Jc 35
Bancroft Av. N2 29Gb 49
Bancroft Chase RM12: Horn 33Hd 76
Bancroft Cl. TW15: Ashf 64Q 128
Bancroft Ct. RH2: Reig 6K 207
Bancroft Ct. SW8 52Nb 112
(off Allen Edwards Dr.)
Bancroft Ct. UB5: N'olt 39Y 65
Bancroft Gdns. BR6: Orp 74Vc 161
Bancroft Gdns. HA3: Hrw W 25Ea 46
Bancroft Ho. E1 42Yb 92
(off Cephas St.)
Bancroft Rd. E1 41Yb 92
Bancroft Rd. HA3: Hrw W 26Ea 46
Bancroft Rd. RH2: Reig 6J 207
Bancroft Rd. TN15: Wro 88Be 185
Band La. TW20: Egh 64B 126
Bandon Cl. UB10: Uxb 40P 63
BANDONHILL 78Mb 156
Bandon Ri. SM6: W'gton 78Mb 156
Banfield Rd. SE15 55Xb 113
Banfor Ct. SM6: W'gton 78Lb 156
Bangalore St. SW15 55Ya 110
Bangays Way TN15: Bor G 93Ae 205
Bangla Ho. E8 39Vb 71
(off Clarissa St.)
Bangor Cl. UB5: N'olt 36Da 65
Bangors Cl. SL0: Iver 44G 82
Bangors Pk. SL0: Iver H 42G 82
Bangors Rd. Nth. SL0: Iver H 39F 62
Bangors Rd. Sth. SL0: Iver 41G 82
Bangors Rd. Sth. SL0: Iver H 41G 82
Banim St. W6 49Xa 88
Banister Ho. E9 36Zb 72
Banister Ho. SW8 53Lb 112
(off Wadhurst Rd.)
Banister Ho. W10 41Ab 88
(off Bruckner St.)
Banister M. NW6 38Db 69
Banister Rd. W10 41Za 88
Bank, The N6 32Kb 70
Bank Av. CR4: Mitc 68Fb 133
Bank Bldgs. E4 23Fc 53
(off The Avenue)
Bank Ct. DA1: Dart 58Nd 119
Bank Ct. E17 28Ec 52
Bank Ct. HP1: Hem H 3L 3
Bank End SE1 6E 224 (46Sb 91)
Bankfoot RM17: Grays 50Be 99
Bankfoot Rd. BR1: Broml 63Gc 137
Bank Ho. KT15: Add 77K 149
Bankhurst Rd. SE6 59Bc 114
Bank La. KT2: King T 66Na 131
Bank La. SW15 57Ua 110
Bank M. SM1: Sutt 79Eb 155
Bank Mill HP4: Berk 2B 2
Bank Mill La. HP4: Berk 2A 2
Bank of England Mus. 3G 225 (44Tb 91)
(off Bartholomew La.)
Bank of England Sports Cen. 57Ua 110
Bank Pl. CM14: B'wood 19Yd 40
Banks Ho. SE1 4D 230 (48Sb 91)
(off Rockingham St.)
Banksian Wlk. TW7: Isle 53Ga 108
Banksia Rd. N18 22Zb 52
Bankside CR2: S Croy 79Vb 157
Bankside DA11: Nflt 58Ee 121
Bankside EN2: Enf 11Rb 33
Bankside GU21: Wok 10M 167
(not continuous)
Bankside KT17: Eps D 88Ya 174
Bankside SE1 5D 224 (45Sb 91)
(not continuous)
Bankside TN13: Dun G 93Fd 202
Bankside UB1: S'hall 46Z 85
Bankside Av. SE13 55Ec 114
Bankside Av. UB5: N'olt 40W 64
Bankside Cl. DA5: Bexl 63Fd 140
Bankside Cl. SM5: Cars 79Gb 155
Bankside Cl. TN16: Big H 90Lc 179
Bankside Cl. TW7: Isle 56Ha 108
Bankside Cl. UB9: Hare 23J 43
Bankside Down WD3: Rick 16L 25
Bankside Dr. KT7: T Ditt 74Ka 152
Bankside Gallery 5C 224 (45Rb 91)
Bankside Lofts SE1 6C 224 (46Rb 91)
(off Hopton St.)
Bankside Mix 6D 224 (46Sb 91)
Bankside Pk. IG11: Bark 41Wc 95
Bankside Pl. N4 30Sb 51
Bankside Rd. IG1: Ilf 36Sc 74
Bankside Way SE19 65Ub 135
Bank's La. KT24: Eff J 95W 190
Banks La. CM16: Fidd H 5Ad 23
Banks La. CM16: They M 5Ad 23
Banks La. DA6: Bex 56Bd 117
Bank Spur SL1: Slou 7F 80
Banks Rd. WD6: Bore 12Sa 29
Bank Station (Underground & DLR) 3F 225 (44Tb 91)
Bank St. DA12: Grav'nd 8D 122
Bank St. E14 46Dc 92
Bank St. TN13: S'oaks 97Ld 203
Banks Way E12 35Qc 74
Banks Yd. TW5: Hest 51Ba 107
Bankton Rd. SW2 56Qb 112
Bankwell Rd. SE13 56Gc 115
Bannatyne Health Club Chafford Hundred 48Yd 98
Bannatyne Health Club Chingford 23Cc 52
Bannatyne Health Club Grove Park 61Kc 137
Bannatyne Health Club Maida Vale 40Db 69
(off Greville Rd.)
Bannatyne Health Club Orpington 67Zc 139
Bannatyne Health Club Russell Square 5E 216 (42Mb 90)
(off Woburn Pl.)
Bann Cl. RM15: S Ock 45Xd 98
Banner Cl. RM19: Purf 49Td 98
Banner Ct. SE16 49Yb 92
(off Rotherhithe New Rd.)
Banner Ho. EC1 6E 218 (42Sb 91)
(off Roscoe St.)
Banner La. RM8: Dag 32Ad 75
Bannerman Ho. SW8 51Pb 112
Banner St. EC1 6E 218 (42Sb 91)
Banning St. SE10 50Gc 93
Bannister Cl. SL3: L'ly 47A 82
Bannister Cl. SW2 60Qb 112
Bannister Cl. UB6: G'frd 36Fa 66
Bannister Dr. CM13: Hut 16Ee 41
Bannister Gdns. BR5: St P 69Yc 139
Bannister Ho. HA3: W'stone 27Ga 46
(off Headstone Dr.)
Bannister Ho. SE14 51Zb 114
(off John Williams Cl.)
Bannister Sports Cen. 23Ea 46
Bannockburn Rd. SE18 49Uc 94
Bannon Ct. SW6 53Db 111
(off Michael Rd.)
Bannow Cl. KT19: Ewe 77Ua 154
Banqueting House 7F 223 (46Nb 90)
BANSTEAD 87Db 175
Banstead Ct. W12 45Va 88
Banstead Downs Golf Course 83Cb 175
Banstead Gdns. N9 20Ub 33
BANSTEAD MOBILITY CEN. 82Hb 175
BANSTEAD PLACE BRAIN INJURY CEN. 88Eb 175
Banstead Rd. CR3: Cat'm 93Tb 197
Banstead Rd. CR8: Purl 83Qb 176
Banstead Rd. KT17: Ewe 82Xa 174
Banstead Rd. SM5: Cars 81Fb 175
Banstead Rd. SM7: Bans 84Za 174
Banstead Rd. Sth. SM2: Sutt 83Eb 175
Banstead Station (Rail) 86Bb 175
Banstead St. SE15 55Yb 114
Banstead Way SM6: W'gton 78Nb 156
Banstead Wood SM7: Bans 90Eb 175
Banstock Rd. HA8: Edg 23Ra 47
Bantam Ho. NW9 26Va 48
(off Heritage Av.)
Banting Dr. N21 15Pb 32
Banting Ho. NW2 34Wa 68
Bantock Ho. W10 41Ab 88
(off Third Av.)
Banton Cl. EN1: Enf 12Xb 33
Bantry Ho. E1 42Zb 92
(off Ernest St.)
Bantry Rd. SL1: Slou 7D 80
Bantry St. SE5 52Tb 113
Banwell Rd. DA5: Bexl 58Zc 117
Banyan Ct. E16 46Kc 93
(off Regalia Close)
Banyard Rd. SE16 48Xb 91
Banyards RM11: Horn 28Nd 57
Baptist Gdns. NW5 37Jb 70
Baquba SE13 54Dc 114
Barandon Rd. W11 45Za 88
(off Grenfell Rd.)
Barandon Wlk. W11 45Za 88
Barataria Pk. GU23: Rip 93H 189
Barbanel Ho. E1 42Yb 92
(off Cephas St.)
Barbara Brosnan Ct. NW8 2B 214 (40Fb 69)
Barbara Castle Cl. SW6 51Bb 111
Barbara Cl. TW17: Shep 71R 150
Barbara Hucklesby Cl. N22 26Rb 51
Barbauld Rd. N16 34Ub 71
Barbel Cl. EN8: Walt C 6Cc 20
Barber Beaumont Ho. E1 41Zb 92
(off Bancroft Rd.)
Barber Cl. N21 17Qb 32
Barberry Cl. RM3: Rom 24Ld 57
Barberry Ct. E15 37Gc 73
Barberry Rd. HP1: Hem H 2J 3
Barbers All. E13 41Kc 93
Barbers Rd. E15 40Dc 72
Barbican EC2 7E 218 (43Sb 91)
(off Silk St.)
Barbican Art Gallery 7E 218 (43Sb 91)
Barbican Arts Cen. 7E 218 (43Sb 91)
Barbican Cinema 1 Silk St. 7E 218 (43Sb 91)
(within Arts Cen.)
Barbican Cinema 2 & 3 Whitecross St. 7E 218 (43Sb 91)
(off Whitecross St.)
Barbican Rd. UB6: G'frd 44Da 85
Barbican Station (Underground) 7D 218 (43Sb 91)
Barbican Theatre London 7E 218 (43Sb 91)
(within Arts Cen.)
Barb M. W6 48Ya 88
Barbon Cl. WC1 7H 217 (43Pb 90)
Barbot Cl. N9 20Wb 33
Barchard St. SW18 57Db 111
Barchester Cl. UB8: Cowl 42L 83
Barchester Cl. W7 46Ha 86
Barchester Rd. HA3: Hrw W 26Fa 46
Barchester Rd. SL3: L'ly 47B 82
Barchester St. E14 43Dc 92
Barcino Ho. AL1: St A 3D 6
Barclay Cl. KT22: Fet 95Da 191
Barclay Cl. SW6 52Cb 111
Barclay Cl. WD18: Wat 16W 26
Barclay Fld. TN15: Kems'g 89Nd 183
Barclay Ho. E9 38Yb 72
(off Well St.)
Barclay Oval IG8: Wfd G 21Jc 53
Barclay Path E17 29Ec 52
Barclay Rd. CR0: C'don 76Tb 157
Barclay Rd. E11 32Hc 73
Barclay Rd. E13 42Lc 93
Barclay Rd. E17 29Ec 52
Barclay Rd. N18 23Tb 51
Barclay Rd. SW6 52Cb 111
Barclay Way RM20: W Thur 50Vd 98
Barcombe Av. SW2 61Nb 134
Barcombe Cl. BR5: St P 69Vc 139
Bardell Ho. SE1 47Wb 91
(off Parkers Row)
Barden Cl. UB9: Hare 24L 43
Barden St. SE18 52Uc 116
Bardeswell Cl. CM14: B'wood 19Yd 40
Bardfield Av. RM6: Chad H 27Zc 55
Bardney Rd. SM4: Mord 70Db 133
Bardolph Av. CR0: Sels 81Ac 178
Bardolph Rd. N7 35Nb 70
Bardolph Rd. TW9: Rich 55Pa 109
Bardon Wlk. GU21: Wok 9M 167
Bard Rd. W10 45Za 88
Bards Cnr. HP1: Hem H 1K 3
Bards Ct. RM3: Rom 24Kd 57
Bardsey Pl. E1 42Yb 92
Bardsey Wlk. N1 37Sb 71
(off Douglas Rd. Nth.)
Bardsley Cl. CR0: C'don 76Vb 157
Bardsley Ho. SE10 51Ec 114
(off Bardsley La.)
Bardsley La. SE10 51Ec 114
Bardwell Ct. AL1: St A 3B 6
(off Bardwell Rd.)
Bardwell Rd. AL1: St A 3B 6
Barents Ho. E1 42Zb 92
(off White Horse La.)
Barfett St. W10 42Bb 89
Barfield DA4: Sut H 67Rd 141
Barfield Av. N20 19Hb 31
Barfield Ct. RH1: Redh 4A 208
Barfield Rd. BR1: Broml 69Qc 138
Barfield Rd. E11 32Hc 73
Barfields IG10: Lough 14Qc 36
Barfields RH1: Blet 5H 209
Barfields Gdns. IG10: Lough 14Qc 36
Barfields Path IG10: Lough 14Qc 36
Barfleur La. SE8 49Bc 92
Barfolds AL9: Wel G 5E 8
Barford Cl. NW4 26Wa 48
Barford Ho. E3 40Bc 72
(off Tredegar Rd.)
Barford St. N1 1A 218 (39Qb 70)
Barforth Rd. SE15 55Xb 113
Barfreston Way SE20 67Xb 135
Bargate Cl. KT3: N Mald 73Wa 154
Bargate Cl. SE18 50Vc 95
Barge Ct. DA9: Ghithe 56Yd 120
Barge Dr. UB2: S'hall 48Da 85
Barge Ho. HP3: Hem H 6P 3
Barge Ho. Rd. E16 47Rc 94
Barge Ho. St. SE1 6A 224 (46Qb 90)
Barge La. E3 39Ac 72
Bargery Rd. SE6 60Dc 114
Barge Wlk. KT8: E Mos Boyle Farm Island 72Ja 152
Barge Wlk. KT8: E Mos Hampton Ct. Cres. 69Fa 130
Barge Wlk. KT1: Hamp W 69Ma 131
Barge Wlk. KT2: King T 67Ma 131
Barge Wlk. SE10 48Hc 93
Bargrove Av. HP1: Hem H 3J 3
Bargrove Cl. SE20 66Wb 135
Bargrove Cres. SE6 61Bc 136
Barham Av. WD6: E'tree 13Pa 29
Barham Cl. BR2: Broml 74Nc 160
Barham Cl. BR7: Chst 64Rc 138
Barham Cl. DA12: Grav'nd 10H 123
Barham Cl. HA0: Wemb 37Ka 66
Barham Cl. KT13: Weyb 77S 150
Barham Cl. RM7: Mawney 26Dd 56
Barham Ct. CR2: S Croy 77Sb 157
(off Barham Rd.)
Barham Ho. SE17 7J 231 (50Ub 91)
(off Kinglake Est.)
Barham Rd. BR7: Chst 64Rc 138
Barham Rd. CR2: S Croy 77Sb 157
Barham Rd. DA1: Dart 59Od 119
Barham Rd. SW20 66Wa 132
Baring Cl. SE12 61Jc 137
Baring Ct. N1 39Tb 71
(off Baring St.)
Baring Ho. E14 44Cc 92
(off Canton St.)
Baring Rd. CR0: C'don 74Wb 157
Baring Rd. EN4: Cockf 14Fb 31
Baring Rd. SE12 59Jc 115
Baring St. N1 39Tb 71
Baritone Ct. E15 39Hc 73
(off Church St.)
Bark Burr Rd. RM16: Chaf H 47Be 99
Barker Cl. HA6: Nwood 24V 44
Barker Cl. KT16: Chert 73G 148
Barker Cl. KT3: N Mald 70Ra 131
Barker Cl. TW9: Kew 54Ra 109
Barker Dr. NW1 38Lb 70
Barker Flds. DA13: Sflt 65Ce 143
Barker Ho. SE17 6H 231 (49Ub 91)
(off Congreve St.)
Barker M. SW4 56Kb 112
Barker Rd. KT16: Chert 73G 148
Barker's Arc. 47Db 89
Barker St. SW10 51Eb 111
Barker Wlk. SW16 62Mb 134
Barkham Rd. N17 24Tb 51
Barkham Ter. SE1 3A 230 (48Qb 90)
(off Lambeth Rd.)
Bark Hart Rd. BR6: Orp 74Xc 161
BARKING 38Sc 74
Barking Abbey School Leisure Cen. 37Wc 75
Barking Bus. Cen. IG11: Bark 41Wc 95
BARKING HOSPITAL 38Wc 75
Barking Ind. Pk. IG11: Bark 39Vc 75
Barking Northern Relief Rd. IG11: Bark 38Rc 74
Barking Pk. Miniature Railway 37Tc 74
BARKING RIVERSIDE 41Xc 95
Barking Rd. E13 42Jc 93
Barking Rd. E16 43Gc 93
Barking Rd. E6 40Mc 73
BARKINGSIDE 27Sc 54
Barkingside Station (Underground) 27Tc 54
Barking Splash Pk. 36Tc 74
Barking Station (Rail, Underground & Overground) 38Sc 74
Barkis Ho. W11 46Za 88
Bark Pl. W2 45Db 89
Barkston Gdns. SW5 49Db 89
Barkston Path WD6: Bore 10Qa 15
Barkway Ct. N4 33Sb 71
Barkway Dr. BR6: Farnb 77Qc 160
Barkwith Ho. SE14 51Zb 114
(off Cold Blow La.)
Barkwood Cl. RM7: Rom 29Ed 56
Barkworth Rd. SE16 50Xb 91
Barlborough St. SE14 52Zb 114
Barlby Gdns. W10 42Za 88
Barlby Rd. W10 43Ya 88
Barlee Cres. UB8: Cowl 43L 83
Barle Gdns. RM15: S Ock 44Xd 98
Barley Brow WD25: Wat 3X 13
Barley Cl. DA9: Ghithe 57Ud 120
Barley Cl. HA0: Wemb 35Ma 67
Barley Cl. WD23: Bush 15Da 27
Barleycorn Way E14 45Bc 92
(not continuous)
Barleycorn Way RM11: Horn 30Pd 57
Barley Ct. E5 33Yb 72
Barley Ct. RM13: Rain 40Fd 76
(off Lwr Mardyke Av.)
Barley Cft. HP2: Hem H 2C 4
Barleyfields Cl. RM6: Chad H 30Xc 55
Barley La. IG3: Ilf 31Wc 75
Barley La. RM6: Chad H 28Xc 55
Barley Mow Cvn. Site AL4: St A 4K 7
Barley Mow Cl. GU21: Knap 9H 167
Barley Mow La. AL4: St A 5J 7
Barley Mow La. GU21: Knap 8G 166
Barley Mow Pas. W4 50Ta 87
Barley Mow Rd. TW20: Eng G 4N 125
Barley Mow Way TW17: Shep 70Q 128
Barley Shotts Bus. Pk. W10 43Bb 89
Barley Wlk. SW18 57Db 111
Barling NW1 37Kb 70
(off Castlehaven Rd.)
Barlow Cl. SM6: W'gton 79Nb 156
Barlow Dr. SE18 53Nc 116
Barlow Ho. N1 2F 219 (40Tb 71)
(off Provost Est.)
Barlow Ho. SE16 49Xb 91
(off Rennie Est.)
Barlow Ho. W11 45Ab 88
(off Walmer Rd.)
Barlow Pl. W1 5A 222 (45Kb 90)
Barlow Rd. NW6 37Bb 69
Barlow Rd. TW12: Hamp 66Ca 129
Barlow Rd. W3 46Ra 87
Barlow St. SE17 5G 231 (49Tb 91)
Barlow Way RM13: Rain 43Fd 96
Barmeston Rd. SE6 61Dc 136
Barmor Cl. HA2: Harr 26Da 45
Barmouth Av. UB6: G'frd 40Ha 66
Barmouth Rd. CR0: C'don 75Zb 158
Barmouth Rd. SW18 58Eb 111
Barn, The RM17: Grays 49De 99
Barnabas Ct. EN2: Enf 14Qb 32
Barnabas Ho. EC1 4D 218 (41Sb 91)
(off King Sq.)
Barnabas Lodge SW8 53Nb 112
(off Guildford Rd.)
Barnabas Rd. E9 36Zb 72
Barnaby Cl. HA2: Harr 33Ea 66
Barnaby Ct. NW9 27Ua 48
Barnaby Ct. SE16 47Wb 91
(off Scott Lidgett Cres.)
Barnaby Ho. SE15 51Yb 114
Barnaby Pl. SW7 6B 226 (49Fb 89)
(off Old Brompton Rd.)
Barnaby Way IG7: Chig 20Rc 36
Barnacre Cl. UB8: Cowl 44M 83
Barnacres Rd. HP3: Hem H 7P 3
Barnard Cl. BR7: Chst 67Tc 138
Barnard Cl. SE18 48Qc 94
Barnard Cl. SM6: W'gton 80Mb 156
Barnard Cl. TW16: Sun 66X 129
Barnard Ct. DA2: Dart 58Rd 119
(off Osbourne Rd.)
Barnard Ct. GU21: Wok 10J 167
Barnard Gdns. KT3: N Mald 70Wa 132
Barnard Gdns. UB4: Yead 42X 85
Barnard Gro. E15 38Hc 73
Barnard Hill N10 25Kb 50
Barnard Ho. E2 41Xb 91
(off Ellsworth St.)
Barnard Lodge EN5: New Bar 14Eb 31
Barnard Lodge W9 43Cb 89
(off Admiral Wlk.)
Barnard M. SW11 56Gb 111
Barnardo Dr. IG6: Ilf 28Sc 54
Barnardo Gdns. E1 45Zb 92
Barnardo St. E1 44Zb 92
Barnardo Village Wlk. IG6: Ilf 27Sc 54
Barnard Rd. CR4: Mitc 69Jb 134
Barnard Rd. CR6: W'ham 91Dc 198
Barnard Rd. EN1: Enf 12Xb 33
Barnard Rd. SW11 56Gb 111
Barnards Ho. SE16 47Bc 92
(off Wyatt Cl.)
Barnard's Inn EC1 2A 224 (44Qb 90)
(off Fetter La.)
Barnards Pl. CR2: S Croy 81Rb 177
Barnard Way HP3: Hem H 3N 3
Barnato Cl. KT14: Byfl 84N 169
Barnbrough NW1 39Lb 70
(off Camden St.)
Barnby Cl. KT21: Asht 89La 172
Barnby Rd. GU21: Knap 9H 167
Barnby Sq. E15 39Gc 73
Barnby St. E15 39Gc 73
Barnby St. NW1 2C 216 (40Lb 70)
Barn Cl. HP3: Hem H 5P 3
Barn Cl. KT18: Eps 87Sa 173
Barn Cl. NW5 36Mb 70
(off Torriano Av.)
Barn Cl. SL2: Farn C 5F 60
Barn Cl. SM7: Bans 87Fb 175
Barn Cl. TW15: Ashf 64R 128
Barn Cl. UB5: N'olt 40Y 65
Barn Cl. WD7: R'lett 7Ja 14
Barn Cres. CR8: Purl 85Tb 177
Barn Cres. HA7: Stan 23La 46
Barncroft Cl. IG10: Lough 15Qc 36
Barncroft Cl. UB8: Hil 43R 84
Barncroft Grn. IG10: Lough 15Qc 36
Barncroft Rd. IG10: Lough 15Qc 36
Barncroft Way AL1: St A 3E 6
Barndale Ct. DA12: Shorne 4N 145
Barneby Cl. TW2: Twick 60Ga 108
BARNEHURST 55Ed 118
Barnehurst Av. DA7: Bex 53Ed 118
Barnehurst Av. DA8: Erith 53Ed 118
Barnehurst Cl. DA8: Erith 53Ed 118
Barnehurst Golf Course 55Fd 118
Barnehurst Rd. DA7: Bex 54Ed 118
Barnehurst Station (Rail) 54Ed 118
Barn Elms Athletics Track 54Xa 110
Barn Elms Cl. KT4: Wor Pk 76Va 154
Barn Elms Pk. SW15 55Ya 110
Barn End Dr. DA2: Wilm 63Ld 141
Barn End La. DA2: Wilm 64Ld 141
BARNES 54Va 110
Barnes All. TW12: Hamp 68Ea 130
Barnes Av. SW13 52Wa 110
Barnes Av. UB2: S'hall 49Ba 85
Barnes Bri. 54Ua 110
Barnes Bridge Station (Rail) 54Ua 110
Barnes Cl. E12 35Mc 73
Barnes Cl. HA8: Edg 21Pa 47
Barnes Common Nature Reserve 55Wa 110
Barnes Ct. CR7: Thor H 69Sb 135
Barnes Ct. E16 43Lc 93
Barnes Ct. EN5: New Bar 14Db 31
Barnes Ct. IG8: Buck H 22Mc 53
Barnes Ct. IG8: Wfd G 22Mc 53
Barnes Ct. N1 38Qb 70
BARNES CRAY 56Jd 118
Barnes Cray Rd. DA1: Cray 56Jd 118
Barnesdale Cres. BR5: St M Cry 72Wc 161
Barnes End KT3: N Mald 71Wa 154
Barnes High St. SW13 54Va 110
BARNES HOSPITAL 55Ua 110
Barnes Ho. E2 40Yb 72
(off Wadeson St.)
Barnes Ho. NW1 38Kb 70
(off Camden Rd.)
Barnes Ho. SE14 51Zb 114
(off John Williams Cl.)
Barnes La. WD4: K Lan 9K 3
Barnes Pikle W5 45Ma 87
Barnes Ri. WD4: K Lan 9P 3
Barnes Rd. IG1: Ilf 36Sc 74
Barnes Rd. N18 21Yb 52
Barnes Station (Rail) 55Wa 110
Barnes St. E14 44Ac 92
Barnes Ter. SE8 50Bc 92
Barnes Wallis Cl. KT24: Eff 99Z 191

Barnes Wallis Ct. HA9: Wemb 34Sa **67**
Barnes Wallis Dr. KT13: Weyb 83N **169**
Barnes Wallis Way AL2: Brick W 3Aa **13**
Barnes Way EN9: Walt A 4Jc **21**
Barnes Way SL0: Iver 45H **83**
BARNET 13Ab **30**
Barnet (Arkley) Windmill 16Va **30**
Barnet Burnt Oak Leisure Cen. 25Ta **47**
Barnet Bus. Cen. EN5: Barn 13Ab **30**
Barnet By-Pass NW7 23Va **48**
Barnet By-Pass Rd. EN5: Ark 13Ua **30**
Barnet By-Pass Rd. EN5: Barn 13Ua **30**
Barnet By-Pass Rd. WD6: Bore 16Ta **29**
Barnet Copthall Leisure Cen. 24Xa **48**
Barnet Copthall Sports Cen. 25Xa **48**
Barnet Dr. BR2: Broml 75Nc **160**
Barnet FC 25Na **47**
BARNET GATE 16Va **30**
Barnet Ga. La. EN5: Ark 16Va **30**
Barnet Gro. E2 41Wb **91**
Barnet Hill EN5: Barn 14Bb **31**
BARNET HOSPITAL 14Za **30**
Barnet Ho. N20 19Eb **31**
Barnet La. EN5: Barn 16Bb **31**
Barnet La. N20 18Bb **31**
Barnet La. WD6: Bore 16Ma **29**
(not continuous)
Barnet La. WD6: E'tree 16Ma **29**
(not continuous)
Barnet Mus. 14Ab **30**
Barnet Rd. AL2: Lon C 9J **7**
Barnet Rd. EN5: Ark 16Ta **29**
Barnet Rd. EN5: Barn 8Cb **17**
Barnet Rd. EN5: Pot B 8Cb **17**
Barnet Rd. EN6: Pot B 5Db **17**
Barnett Cl. DA8: Erith 54Hd **118**
Barnett Cl. KT22: Lea 91Ka **192**
Barnett Row GU4: Jac W 10P **187**
Barnetts Ct. HA2: Harr 34Da **65**
Barnett's Shaw RH8: Oxt 99Fc **199**
Barnett St. E1 44Xb **91**
Barnetts Way RH8: Oxt 99Fc **199**
Barnett Wood La. KT21: Asht 90La **172**
Barnett Wood La. KT22: Lea 92Ka **192**
BARNET VALE 15Db **31**
Barnet Way NW7 20Ta **29**
Barnet Wood Rd. BR2: Broml 75Lc **159**
Barney Cl. SE7 50Lc **93**
Barn Fld. NW3 36Hb **69**
Barnfield CM16: Epp 1Wc **23**
Barnfield DA11: Grav'nd 1C **144**
Barnfield HP3: Hem H 5P **3**
Barnfield KT3: N Mald 72Ua **154**
Barnfield SL0: Iver 44G **82**
Barnfield SL1: Slou 6B **80**
Barnfield SM7: Bans 86Db **175**
Barnfield Av. CR0: C'don 75Yb **158**
Barnfield Av. CR4: Mitc 70Kb **134**
Barnfield Av. KT2: King T 63Ma **131**
Barnfield Cl. BR8: Crock 73Ed **162**
Barnfield Cl. CR5: Coul 91Sb **197**
Barnfield Cl. DA3: Lfield 69Fe **143**
Barnfield Cl. DA9: Ghithe 58Vd **120**
Barnfield Cl. N4 31Nb **70**
Barnfield Cl. SW17 62Fb **133**
Barnfield Cres. TN15: Kems'g 89Nd **183**
Barnfield Gdns. KT2: King T 63Na **131**
Barnfield Gdns. SE18 51Rc **116**
Barnfield Pk. TN15: Ash 78Zd **165**
Barnfield Pl. E14 49Cc **92**
Barnfield Rd. BR5: St P 69Zc **139**
Barnfield Rd. CR2: Sande 81Ub **177**
Barnfield Rd. DA17: Belv 51Bd **117**
Barnfield Rd. HA8: Edg 25Sa **47**
Barnfield Rd. SE18 51Rc **116**
(not continuous)
Barnfield Rd. TN13: Riv 95Gd **202**
Barnfield Rd. TN16: Tats 92Mc **199**
Barnfield Rd. W5 42La **86**
Barnfield Wlk. RM2: Rom 28Ld **57**
Barnfield Way RH8: Oxt 5L **211**
Barnfield Wood Cl. BR3: Beck 72Fc **159**
Barnfield Wood Rd. BR3: Beck 72Fc **159**
Barnham Dr. SE28 46Vc **95**
(not continuous)
Barnham Rd. UB6: G'frd 41Ea **86**
Barnham St. SE1 1J **231** (47Ub **91**)
Barn Hill HA9: Wemb 32Qa **67**
Barnhill HA5: Eastc 29Y **45**
Barnhill Av. BR2: Broml 71Hc **159**
Barnhill La. UB4: Yead 41X **85**
Barnhill Rd. HA9: Wemb 34Sa **67**
Barnhill Rd. UB4: Yead 41X **85**
Barnhurst Path WD19: Wat 22Y **45**
Barningham Way NW9 30Ta **47**
Barn Lea WD3: Rick 18J **25**
Barnlea Cl. TW13: Hanw 61Aa **129**
Barn Mead CM16: They B 8Uc **22**
Barnmead GU24: Chob 2K **167**
Barnmead Ct. RM9: Dag 36Bd **75**
Barnmead Gdns. RM9: Dag 36Bd **75**
Barnmead Mdw. RM16: Grays 8B **100**
Barn Mdw. La. KT23: Bookh 96Ba **191**
Barnmead Rd. BR3: Beck 67Zb **136**
Barnmead Rd. RM9: Dag 36Bd **75**
Barn M. HA2: Harr 34Ca **65**
Barnock Cl. DA1: Cray 59Gd **118**
Barn Ri. HA9: Wemb 32Qa **67**
BARNSBURY 38Pb **70**
Barnsbury Cl. KT3: N Mald 70Sa **131**
Barnsbury Cres. KT5: Surb 74Sa **153**
Barnsbury Est. N1 1J **217** (39Pb **70**)
(not continuous)
Barnsbury Farm Est. GU22: Wok 2P **187**
Barnsbury Gro. N7 38Pb **70**
Barnsbury Ho. SW4 58Mb **112**
Barnsbury La. KT5: Surb 75Ra **153**
Barnsbury Pk. N1 38Qb **70**
Barnsbury Rd. N1 1K **217** (40Qb **70**)
Barnsbury Sq. N1 38Qb **70**
Barnsbury St. N1 38Qb **70**
Barnsbury Ter. N1 38Pb **70**
Barnscroft SW20 69Xa **132**
Barnsdale Av. E14 49Dc **92**
Barnsdale Cl. WD6: Bore 11Pa **29**
Barnsdale Rd. W9 42Bb **89**
Barnsfield Pl. UB8: Uxb 39L **63**
Barnsford Cres. GU24: W End 5E **166**
Barnsley Rd. RM3: Rom 24Pd **57**
Barnsley St. E1 42Xb **91**
Barnstaple Ho. SE10 52Dc **114**
(off Devonshire Dr.)
Barnstaple Ho. SE12 57Hc **115**
(off Taunton Rd.)
Barnstaple La. SE13 56Ec **114**
Barnstaple Path RM3: Rom 22Ld **57**
Barnstaple Rd. HA4: Ruis 34Y **65**
Barnstaple Rd. RM3: Rom 22Ld **57**
Barnston Wlk. N1 39Sb **71**
(off Popham St.)
Barnston Way CM13: Hut 15Ee **41**
Barn St. N16 33Ub **71**
Barnsway WD4: K Lan 10N **3**
Barn Theatre Oxted, The 100Gc **199**
Barn Theatre Sidcup, The 60Wc **117**
Barn Theatre West Molesey, The 70Ca **129**
Barn Way HA9: Wemb 32Qa **67**
Barnway TW20: Eng G 4N **125**
Barnwell Cl. HA8: Edg 21Pa **47**
Barnwell Ho. SE5 53Ub **113**
(off St Giles Rd.)
Barnwell Rd. DA1: Dart 55Pd **119**
Barnwell Rd. SW2 57Qb **112**
Barnwood Cl. HA4: Ruis 33T **64**
Barnwood Cl. N20 18Bb **31**
Barnwood Cl. W9 42Db **89**
Barnyard, The KT20: Walt H 96Wa **194**
Baron Cl. N1 1K **217** (40Qb **70**)
Baron Cl. N11 22Jb **50**
Baron Cl. SM2: Sutt 82Db **175**
Baroness Rd. E2 3K **219** (41Vb **91**)
Baronet Gro. N17 25Wb **51**
Baronet Rd. N17 25Wb **51**
Baron Gdns. IG6: Ilf 27Sc **54**
Baron Gro. CR4: Mitc 70Gb **133**
Baron Ho. SW19 67Fb **133**
Baron Rd. RM8: Dag 32Zc **75**
Barons, The TW1: Twick 58Ka **108**
Barons, The TW20: Eng G 2N **125**
Baronsclere Ct. N6 31Lb **70**
BARONS COURT 50Ab **88**
Barons Ct. IG1: Ilf 33Tc **74**
Barons Ct. NW9 30Ta **47**
Barons Ct. SM6: Bedd 76Mb **156**
Baron's Ct. Rd. W14 50Ab **88**
Barons Court Station (Underground) 50Ab **88**
Barons Court Theatre 50Ab **88**
(off Comeragh Rd.)
Baronsfield Rd. TW1: Twick 58Ka **108**
Barons Ga. EN4: E Barn 16Gb **31**
Barons Ga. W4 48Sa **87**
Baron's Hurst KT18: Eps 88Sa **173**
Barons Keep W14 50Ab **88**
Barons Lodge E14 49Fc **93**
(off Manchester Rd.)
Baronsmead HA1: Harr 28Ga **46**
Baronsmead Rd. SW13 53Wa **110**
Baronsmede W5 47Pa **87**
Baronsmere Ct. EN5: Barn 14Ab **30**
Baronsmere Rd. N2 28Gb **49**
Baron's Pl. SE1 2A **230** (47Qb **90**)
Baron St. N1 1K **217** (40Qb **70**)
Baron's Wlk. CR0: C'don 72Ac **158**
Baron's Way RH2: Reig 10J **207**
Barons Way TW20: Egh 65F **126**
Barons Wood TW20: Eng G 4P **125**
Baron Wlk. E16: Canning Town 43Hc **93**
Baron Wlk. CR4: Mitc Mitcham 70Gb **133**
Baroque Ct. TW3: Houn 55Da **107**
Baroque Gdns. SE8 49Ac **92**
(off Grand Canal Av.)
Barque M. SE8 51Cc **114**
Barquentine Hgts. SE10 48Jc **93**
Barrack La. SL4: Wind 3H **103**
Barrack Path GU21: Wok 10J **167**
(not continuous)
Barrack Rd. TW4: Houn 56Z **107**
Barrack Row DA11: Grav'nd 8D **122**
Barracks La. EN5: Barn 13Ab **30**
Barra Cl. HP3: Hem H 5C **4**
Barracouta Ho. SE18 51Vc **117**
Barra Hall Cir. UB3: Hayes 45U **84**
Barra Hall Rd. UB3: Hayes 45U **84**
Barra Ho. WD18: Wat 16V **26**
(off Scammell Way)
Barrass Cl. EN3: Enf L 9Cc **20**
Barratt Av. N22 26Pb **50**
Barratt Ho. N1 38Rb **71**
(off Sable St.)
Barratt Ind. Est. UB1: S'hall 47Ca **85**
Barratt Ind. Pk. E3 42Ec **92**
Barratt Way HA3: W'stone 26Fa **46**
Barra Wood Cl. UB3: Hayes 44U **84**
Barrenger Rd. N10 25Hb **49**
Barrens Brae GU22: Wok 90C **168**
Barrens Cl. GU22: Wok 91C **188**
Barrens Pk. GU22: Wok 90C **168**
Barret Ho. NW6 39Cb **69**
Barret Ho. SW9 55Pb **112**
(off Benedict Rd.)
Barrett Cl. RM3: Rom 24Kd **57**
Barrett Ct. SE5 52Tb **113**
(off Dobson Wlk.)
Barrett Ho. SE17 7E **230** (50Sb **91**)
(off Browning St.)
Barrett Pl. UB10: Uxb 39N **63**
Barrett Rd. E17 28Ec **52**
Barrett Rd. KT22: Fet 96Fa **192**
Barrett's Grn. Rd. NW10 41Sa **87**
Barrett's Gro. N16 36Ub **71**
Barretts Rd. TN13: Dun G 92Fd **202**
Barrett St. W1 3J **221** (44Jb **90**)
Barrett Way RM15: Avel 45Td **98**
Barrhill Rd. SW2 61Nb **134**
Barricane GU21: Wok 1M **187**
Barrie Cl. CR5: Coul 88Lb **176**
Barrie Ct. EN5: New Bar 15Eb **31**
(off Lyonsdown Rd.)
Barriedale SE14 54Ac **114**
Barrie Est. W2 4B **220** (45Fb **89**)
Barrie Ho. KT15: Add 80J **149**
Barrie Ho. NW8 1E **214** (39Gb **69**)
(off St Edmund's Ter.)
Barrie Ho. W2 5A **220** (45Eb **89**)
(off Lancaster Ga.)
Barrier App. SE7 48Mc **93**
Barrier Point Rd. E16 46Lc **93**
Barringers Ct. HA4: Ruis 31T **64**
Barringer Sq. SW17 63Jb **134**
Barrington Cl. IG10: Lough 13Sc **36**
Barrington Cl. IG5: Ilf 25Pc **54**
Barrington Cl. NW5 36Jb **70**
Barrington Ct. CM13: Hut 17Ee **41**
Barrington Ct. N10 26Jb **50**
Barrington Ct. RH1: Redh 4A **208**
Barrington Ct. SW4 54Nb **112**
Barrington Ct. TW18: Staines 65H **127**
(off Thameside)
Barrington Ct. W3 47Ra **87**
(off Cheltenham Pl.)
Barrington Dr. KT22: Fet 97Fa **192**
Barrington Dr. UB9: Hare 24J **43**
Barrington Grn. IG10: Lough 14Sc **36**
Barrington Lodge KT13: Weyb 78S **150**
Barrington Pk. Gdns. HP8: Chal G 18A **24**
Barrington Rd. CR8: Purl 84Lb **176**
Barrington Rd. DA7: Bex 54Zc **117**
Barrington Rd. E12 37Qc **74**
Barrington Rd. IG10: Lough 14Sc **36**
Barrington Rd. N8 29Mb **50**
Barrington Rd. SM3: Sutt 75Cb **155**
Barrington Rd. SW9 55Rb **113**
Barrington Vs. SE18 53Qc **116**
Barrington Wlk. SE19 65Ub **135**
Barrons Chase TW10: Ham 64Pa **131**
Barrow Av. SM5: Cars 80Hb **155**
Barrow Cl. N21 20Rb **33**
Barrow Ct. SE6 60Hc **115**
(off Cumberland Pl.)
Barrowdene Cl. HA5: Pinn 26Aa **45**
Barrowell Grn. N21 19Rb **33**
Barrowfield Cl. N9 20Xb **33**
Barrow Gdns. RH1: Redh 4B **208**
Barrowgate Rd. W4 50Sa **87**
Barrow Grn. Rd. RH8: Oxt 2E **210**
Barrow Hall Ho. RM16: Grays 8B **100**
Barrow Hedges Cl. SM5: Cars 80Gb **155**
Barrow Hedges Way SM5: Cars 80Gb **155**
Barrow Hill KT4: Wor Pk 75Ua **154**
Barrow Hill Cl. KT4: Wor Pk 75Ua **154**
Barrow Hill Est. NW8 2D **214** (40Gb **69**)
(off Barrow Hill Rd.)
Barrow Hill Rd. NW8 2D **214** (40Gb **69**)
Barrow Hills Golf Course 5N **147**
Barrow La. EN7: Chesh 2Vb **19**
(not continuous)
Barrow La. EN7: G Oak 2Vb **19**
(not continuous)
Barrow Lodge SL2: Slou 2G **80**
Barrow Point Av. HA5: Pinn 26Aa **45**
Barrow Point La. HA5: Pinn 26Aa **45**
Barrow Rd. CR0: Wadd 78Qb **156**
Barrow Rd. SW16 65Mb **134**
Barrowsfield CR2: Sande 84Vb **177**
Barrow Store Ct. SE1 3H **231** (48Ub **91**)
(off Decima St.)
Barrow Wlk. TW8: Bford 51La **108**
Barr Rd. DA12: Grav'nd 1H **145**
Barr Rd. EN6: Pot B 5Eb **17**
Barrsbrook Farm Rd. KT16: Chert 74G **148**
Barrsbrook Hall KT16: Chert 74G **148**
Barr's La. GU21: Knap 8H **167**
(not continuous)
Barr's Rd. SL6: Tap 4A **80**
Barrs Rd. NW10 38Ta **67**
Barry Av. DA7: Bex 52Ad **117**
Barry Av. N15 30Vb **51**
Barry Av. SL4: Wind 2G **102**
Barry Blandford Way E3 42Dc **92**
Barry Cl. AL2: Chis G 7P **5**
Barry Cl. BR6: Orp 76Uc **160**
Barry Cl. RM16: Grays 8C **100**
Barry Ct. RM5: Col R 22Fd **56**
Barry Ct. WD18: Wat 15Y **27**
(off Cardiff Rd.)
Barrydene N20 18Fb **31**
Barry Ho. SE16 49Xb **91**
(off Rennie Est.)
Barry Pde. SE22 57Wb **113**
Barry Rd. E6 44Nc **94**
Barry Rd. NW10 38Sa **67**
Barry Rd. SE22 58Wb **113**
Barry Ter. TW15: Ashf 61P **127**
(off Orchard Way)
Barset Rd. SE15 55Yb **114**
(not continuous)
Barson Cl. SE20 66Yb **136**
Barstable Rd. SS17: Stan H 1M **101**
Barston Rd. SE27 62Sb **135**
Barstow Cres. SW2 60Pb **112**
Bartel Cl. HP3: Hem H 4D **4**
Bartelotts Rd. SL2: Slou 2B **80**
Barter St. WC1 1G **223** (43Nb **90**)
Barters Wlk. HA5: Pinn 27Aa **45**
Barth M. SE18 49Uc **94**
Bartholomew Cl. EC1 1D **224** (43Sb **91**)
(not continuous)
Bartholomew Cl. SW18 56Eb **111**
Bartholomew Ct. E14 45Fc **93**
(off Newport Av.)
Bartholomew Ct. EC1 5E **218** (42Sb **91**)
(off Old St.)
Bartholomew Ct. HA8: Edg 24Ma **47**
Bartholomew Dr. RM3: Hrld W 26Md **57**
Bartholomew Ho. EN3: Enf W 9Ac **20**
Bartholomew Ho. IG8: Wfd G 24Rc **54**
Bartholomew Ho. W10 42Ab **88**
(off Appleford Rd.)
Bartholomew La. EC2 3G **225** (44Tb **91**)
Bartholomew Pl. EC1 1D **224** (43Sb **91**)
(off Bartholomew Cl.)
Bartholomew Rd. NW5 37Lb **70**
Bartholomew Sq. E1 42Xb **91**
Bartholomew Sq. EC1 4E **218** (41Sb **91**)
Bartholomew St. SE1 4F **231** (48Tb **91**)
Bartholomew Vs. NW5 37Lb **70**
Bartholomew Way BR8: Swan 69Gd **140**
Barth Rd. SE18 49Uc **94**
Bartle Av. E6 40Nc **74**
Bartle Rd. W11 44Ab **88**
Bartlett Cl. E14 44Cc **92**
Bartlett Ct. EC4 2A **224** (44Qb **90**)
Bartlett Ho. KT4: Wor Pk 75Va **154**
(off The Avenue)
Bartlett Ho's. RM10: Dag 38Dd **76**
(off Vicarage Rd.)
Bartlett M. E14 50Dc **92**
Bartlett Rd. DA11: Grav'nd 10C **122**
Bartlett Rd. TN16: Westrm 98Sc **200**
Bartletts Hillside Cl. SL9: Chal P 24A **42**
Bartletts Pas. EC4 2A **224** (44Qb **90**)
(off Fetter La.)
Bartlett St. CR2: S Croy 78Tb **157**
Bartlow Gdns. RM5: Col R 25Fd **56**
Bartok Ho. W11 46Bb **89**
(off Lansdowne Wlk.)
Barton, The KT11: Cobh 84Z **171**
Barton Av. RM7: Rush G 32Dd **76**
Barton Cl. DA6: Bex 57Ad **117**
Barton Cl. E6 44Pc **94**
Barton Cl. E9 36Yb **72**
Barton Cl. GU21: Knap 10G **166**
Barton Cl. IG7: Chig 19Sc **36**
Barton Cl. KT15: Add 79J **149**
Barton Cl. NW4 29Wa **48**
Barton Cl. SE15 55Xb **113**
Barton Cl. TW17: Shep 72R **150**
Barton Ct. CR3: W'ham 91Wb **197**
(off Godstone Rd.)
Barton Ct. W14 50Ab **88**
(off Baron's Ct. Rd.)
Barton Friars IG7: Chig 19Sc **36**
Barton Grn. KT3: N Mald 68Ta **131**
Barton Ho. E3 41Dc **92**
(off Bow Rd.)
Barton Ho. N1 38Rb **71**
(off Sable St.)
Barton Ho. SW6 55Db **111**
(off Wandsworth Bri. Rd.)
Barton Mdws. IG6: Ilf 28Rc **54**
Barton M. E14 47Dc **92**
Barton M. SW19 65Eb **133**
Barton Rd. DA14: Sidc 65Ad **139**
Barton Rd. DA4: Sut H 67Rd **141**
Barton Rd. RM12: Horn 32Jd **76**
Barton Rd. SL3: L'ly 47B **82**
Barton Rd. W14 50Ab **88**
Bartons, The WD6: E'tree 16Ma **29**
Barton St. SW1 3F **229** (48Nb **90**)
Barton Way WD3: Crox G 15R **26**
Barton Way WD6: Bore 12Qa **29**
Bartonway NW8 1B **214** (39Fb **69**)
(off Queen's Ter.)
Bartram Cl. UB8: Hil 42R **84**
Bartram Rd. SE4 57Ac **114**
Bartrams La. EN4: Had W 10Eb **17**
Bartrip St. E9 37Bc **72**
Barts & London School of Medicine & Dentistry, The Whitechapel Campus 43Xb **91**
(off Turner St.)
Barts Cl. BR3: Beck 71Cc **158**
Barville Cl. SE4 56Ac **114**
Barwell Bus. Pk. KT9: Chess 80Ma **153**
Barwell Ct. KT9: Chess 80Ka **152**
Barwell Cres. TN16: Big H 84Lc **179**
Barwell Ho. E2 42Wb **91**
(off Menotti St.)
Barwell La. KT9: Chess 80La **152**
Barwick Dr. UB8: Hil 43R **84**
Barwick Ho. W3 47Sa **87**
(off Strafford Rd.)
Barwick Rd. E7 35Kc **73**
Barwood Av. BR4: W W'ck 74Dc **158**
Baryta Cl. SS17: Stan H 2L **101**
Bascombe Gro. DA1: Bexl 59Gd **118**
Bascombe Gro. DA1: Cray 59Gd **118**
Bascombe St. SW2 58Qb **112**
Basden Gro. TW13: Hanw 61Ca **129**
Basden Ho. TW13: Hanw 61Ca **129**
Basedale Rd. RM9: Dag 38Xc **75**
Baseing Cl. E6 45Qc **94**
Baseline Bus. Studios W11 45Za **88**
(off Barandon Wlk.)
Basepoint Bus. Cen. RM13: Rain 42Fd **96**
Basevi Way SE8 51Dc **114**
Basford Way SL4: Wind 5B **102**
Bashley Rd. NW10 42Ta **87**
Basi Cl. RM11: Horn 28Pd **57**
Basil Av. E6 41Nc **94**
Basildene Rd. TW4: Houn 55Z **107**
Basildon Av. IG5: Ilf 25Qc **54**
Basildon Cl. SM2: Sutt 81Db **175**
Basildon Cl. WD18: Wat 16S **26**
Basildon Ct. HP3: Hem H 7P **3**
Basildon Ct. W1 7J **215** (43Jb **90**)
(off Devonshire Pl.)
Basildon Rd. SE2 50Wc **95**
Basil Gdns. CR0: C'don 74Zb **158**
Basil Gdns. SE27 64Sb **135**
Basil Ho. E1 44Wb **91**
(off Henriques St.)
Basil Ho. SW8 52Nb **112**
(off Wyvil Rd.)
Basilica M. SW12 58Jb **112**
Basilica Pl. E3 40Bc **72**
Basil Mans. SW3 2G **227** (47Hb **89**)
(off Basil St.)
Basilon Rd. DA7: Bex 54Ad **117**
Basil Spence Ho. N22 25Pb **50**
Basil St. SW3 3F **227** (48Hb **89**)
Basin App. E14 44Ac **92**
Basin App. E16 45Rc **94**
Basing Cl. KT7: T Ditt 73Ha **152**
Basing Ct. SE15 53Vb **113**
Basingdon Way SE5 56Tb **113**
Basing Dr. DA5: Bexl 58Bd **117**
Basingfield Rd. KT7: T Ditt 73Ha **152**
Basinghall Av. EC2 2F **225** (44Tb **91**)
Basinghall Gdns. SM2: Sutt 81Db **175**
Basinghall St. EC2 2F **225** (44Tb **91**)
Basing Hill HA9: Wemb 33Pa **67**
Basing Hill NW11 32Bb **69**
Basing Ho. Yd. E2 3J **219** (41Ub **91**)
(off Kingsland Rd.)
Basing Pl. E2 3J **219** (41Ub **91**)
Basing Rd. SM7: Bans 86Bb **175**
Basing Rd. WD3: Rick 18H **25**
Basing St. W11 44Bb **89**
Basing Way KT7: T Ditt 73Ha **152**
Basing Way N3 27Cb **49**
Basin Mill Apts. E2 1K **219** (39Vb **71**)
(off Laburnum St.)
Basin Sth. E16 46Rc **94**
Basire St. N1 39Sb **71**
Baskerville Gdns. NW10 35Ua **68**
Baskerville Rd. SW18 59Gb **111**
Basket Gdns. SE9 57Nc **116**
Baslow Cl. HA3: Hrw W 25Fa **46**
Baslow Wlk. E5 35Zb **72**
Basnett Rd. SW11 55Jb **112**
Basque Ct. SE16 47Zb **92**
(off Garter Way)
Bassano St. SE22 57Vb **113**
Bassant Rd. SE18 51Vc **117**
Bass Ct. E15 39Hc **73**
(off Plaistow Rd.)
Bassein Pk. Rd. W12 47Va **88**
Basset Cl. KT15: New H 82K **169**
Bassett Cl. SM2: Sutt 81Db **175**
Bassett Dr. RH2: Reig 5J **207**
Bassett Gdns. TW7: Isle 52Ea **108**
Bassett Ho. SW19 64Db **133**
Bassett Rd. GU22: Wok 88E **168**
Bassett Rd. UB8: Uxb 38L **63**
Bassett Rd. W10 44Za **88**
Bassetts TN16: Tats 93Lc **199**
Bassett's Cl. BR6: Farnb 77Rc **160**
Bassetts Ho. BR6: Farnb 77Rc **160**
Bassett St. NW5 37Jb **70**
Bassett's Way BR6: Farnb 77Rc **160**
Bassett Way SL2: Slou 2C **80**
Bassett Way UB6: G'frd 44Da **85**
Bassil Rd. HP2: Hem H 3M **3**
Bassingbourn Ho. N1 38Qb **70**
(off The Sutton Est.)
Bassingham Rd. HA0: Wemb 37Ma **67**
Bassingham Rd. SW18 59Eb **111**
Bassington Rd. W3 48Sa **87**
Bassishaw Highwalk EC2 1F **225** (43Tb **91**)
(off London Wall)
Bass M. SE22 56Wb **113**
Basswood Cl. SE15 55Xb **113**
Bastable Av. IG11: Bark 40Uc **74**
BASTED 95Be **205**
Basted La. TN15: Crou 95Ce **205**
Basted Mill TN15: Bor G 93Ae **205**
Basterfield Ho. EC1 6D **218** (42Sb **91**)
(off Golden La. Est.)
Bastion Highwalk EC2 1D **224** (43Sb **91**)
(off London Wall)
Bastion Ho. EC2 1E **224** (43Sb **91**)
(off London Wall)
Bastion Rd. SE2 50Wc **95**
Baston Mnr. Rd. BR2: Hayes 76Kc **159**
Baston Mnr. Rd. BR2: Kes 76Kc **159**
Baston Rd. BR2: Hayes 75Kc **159**
Bastwick St. EC1 5D **218** (42Sb **91**)
Basuto Rd. SW6 53Cb **111**
Bata Av. RM18: E Til 10K **101**
Bataleur Ct. RM11: Horn 30Md **57**
BAT & BALL 93Ld **203**
Bat & Ball Ent. Cen. TN14: S'oaks 93Ld **203**
Bat & Ball Rd. TN14: S'oaks 93Ld **203**
Bat & Ball Station (Rail) 93Ld **203**
Batavia Cl. TW16: Sun 67X **129**
Batavia Ho. SE14 52Ac **114**
(off Batavia Rd.)
Batavia M. SE14 52Ac **114**
Batavia Rd. SE14 52Ac **114**
Batavia Rd. TW16: Sun 67X **129**
Batchelor St. N1 1A **218** (39Qb **70**)
Batchwood Dr. AL3: St A 1P **5**
Batchwood Grn. BR5: St P 69Wc **139**
BATCHWORTH 19N **25**
Batchworth Heath WD3: Rick 21R **44**
BATCHWORTH HEATH 21Q **44**
Batchworth Heath Hill WD3: Rick 21Q **44**
Batchworth Hill WD3: Rick 19N **25**
(not continuous)
Batchworth La. HA6: Nwood 22S **44**
Batchworth Lock Canal Cen. 18N **25**
Batchworth Locks 18N **25**
(off Church St.)
Batchworth Pk. Golf Course 20N **25**
BATCHWORTH RDBT. 18N **25**
Bateman Cl. IG11: Bark 37Sc **74**
Bateman Ho. SE17 51Rb **113**
(off Otto St.)
Bateman M. SW4 58Mb **112**
Bateman Rd. E4 23Cc **52**
Bateman Rd. WD3: Crox G 16Q **26**
Bateman's Bldgs. W1 3D **222** (44Mb **90**)
(off Bateman St.)
Batemans M. CM14: W'ley 21Xd **58**
Bateman's Row EC2 5J **219** (42Ub **91**)
Bateman St. W1 3D **222** (44Mb **90**)
Bates Bus. Cen. RM3: Hrld W 24Qd **57**
Bates Cl. SL3: Geor G 44A **82**
Bates Cres. CR0: Wadd 78Qb **156**
Bates Cres. SW16 66Lb **134**
Bates Hill TN15: Igh 94Yd **204**
Bates Ind. Est. RM3: Hrld W 24Qd **57**
Bateson Dr. WD25: Wat 5V **12**
Bateson St. SE18 49Uc **94**
Bateson Way GU21: Wok 86E **168**
Bates Rd. RM3: Hrld W 24Qd **57**
Bate St. E14 45Bc **92**
Bates Wlk. KT15: Add 79L **149**
Bat Gdns. KT2: King T 65Pa **131**
Bath Cl. SE15 52Xb **113**
Bath Ct. EC1: St Luke's Est. 4F **219** (41Tb **91**)
(off St Luke's Est.)
Bath Ct. EC1: Warner St. 6K **217** (42Qb **90**)
(off Warner St.)
Bath Ct. SE26 62Wb **135**
(off Droitwich Cl.)
Bathgate Ho. SW9 53Rb **113**
(off Lothian Rd.)
Bathgate Rd. SW19 62Za **132**
Bath Gro. E2 40Wb **71**
(off Horatio St.)
Bath Ho. E2 42Wb **91**
(off Ramsey St.)
Bath Ho. IG11: Bark 38Sc **74**
Bath Ho. SE1 3D **230** (48Sb **91**)
(off Bath Ter.)
Bath Ho. Rd. CR0: Bedd 74Nb **156**
Bath Pas. KT1: King T 68Ma **131**
Bath Pl. EC2 4H **219** (41Ub **91**)
Bath Pl. EN5: Barn 13Bb **31**
Bath Pl. W6 50Ya **88**
(off Peabody Est.)
Bath Rd. DA1: Dart 59Kd **119**
Bath Rd. E7 37Mc **73**
Bath Rd. N9 19Xb **33**
Bath Rd. RM6: Chad H 30Ad **55**
Bath Rd. SL1: Slou 5B **80**
Bath Rd. SL3: Coln 52E **104**
Bath Rd. SL3: Coln 53G **104**
Bath Rd. SL3: Poyle 53G **104**
Bath Rd. TW3: Houn 55Da **107**
Bath Rd. TW4: Houn 54Z **107**
Bath Rd. TW5: Cran 53Y **107**
Bath Rd. TW6: H'row A 53U **106**
Bath Rd. UB3: Harl 53U **106**
Bath Rd. UB7: Harm 53K **105**
Bath Rd. UB7: Lford 53K **105**
Bath Rd. UB7: Sip 53K **105**
Bath Rd. W4 49Ua **88**
Bath Rd. Retail Pk. 4D **80**
Baths Ct. W12 47Xa **88**
Baths Rd. BR2: Broml 70Mc **137**
Bath St. DA11: Grav'nd 8D **122**
Bath St. EC1 4E **218** (41Sb **91**)
Bath Ter. SE1 4D **230** (48Sb **91**)
Bathurst Av. SW19 67Db **133**
Bathurst Cl. SL0: Rich P 47H **83**
Bathurst Gdns. NW10 40Xa **68**
Bathurst Ho. W12 45Xa **88**
(off White City Est.)
Bathurst M. W2 3C **220** (44Fb **89**)
Bathurst Rd. IG1: Ilf 32Rc **74**
Bathurst Sq. N15 28Ub **51**
Bathurst St. W2 4C **220** (45Fb **89**)
Bathurst Wlk. SL0: Rich P 47G **82**
Bathway SE18 49Qc **94**
Batley Cl. CR4: Mitc 73Hb **155**
Batley Pl. N16 34Vb **71**
Batley Rd. EN2: Enf 11Sb **33**
Batley Rd. N16 34Vb **71**
Batman Cl. W12 46Xa **88**
Baton Cl. RM19: Purf 49Td **98**
Batoum Gdns. W6 48Ya **88**
Batsford Ho. SW19 63Db **133**
(off Durnsford Rd.)
Batson Ho. E1 44Wb **91**
(off Fairclough St.)
Batson St. W12 47Wa **88**
Batsworth Rd. CR4: Mitc 69Fb **133**
Battalion Ho. NW9 26Va **48**
(off Heritage Av.)
Batten Av. GU21: Wok 1J **187**
Battenberg Wlk. SE19 65Ub **135**

Batten Cl. E6 44Pc 94
Batten Cotts. E14 43Ac 92
(off Maroon St.)
Batten Ho. SW4 57Lb 112
Batten Ho. W10 41Ab 88
(off Third Av.)
Batten St. SW11 55Gb 111
Battersby Rd. SE6 61Fc 137
BATTERSEA 53Jb 112
Battersea Arts Cen. 55Hb 111
(off Lavender Hill)
Battersea at Brands Hatch 78Xd 164
Battersea Bri. 52Fb 111
Battersea Bri. Rd. SW11 52Gb 111
Battersea Bus. Cen. SW11 55Jb 112
Battersea Bus. Pk. SW8 53Lb 112
Battersea Church Rd. SW11 53Fb 111
Battersea Dogs' Home 52Kb 112
Battersea Evolution 52Jb 112
Battersea High St. SW11 53Fb 111
(not continuous)
Battersea Pk. 52Hb 111
BATTERSEA PARK 52Kb 112
Battersea Pk. Children's Zoo 52Jb 112
Battersea Pk. Millennium Arena 52Jb 112
Battersea Pk. Rd. SW11 54Gb 111
Battersea Pk. Rd. SW8 53Kb 112
Battersea Park Station (Rail) 52Kb 112
Battersea Power Sta. Development SW11 51Kb 112
Battersea Ri. SW11 57Gb 111
Battersea Roof Gdns. SW11 52Kb 112
Battersea Sports Cen. 55Fb 111
Battersea Sq. SW11 53Fb 111
Battery Rd. SE28 47Uc 94
Battis, The RM1: Rom 30Gd 56
Battishill St. N1 38Rb 71
Battlebridge Ct. N1 1G 217 (40Nb 70)
(off Wharfdale Rd.)
Battle Bri. La. SE1 7H 225 (46Ub 91)
Battlebridge La. RH1: Mers 2B 208
Battlebridge La. RH1: Redh 2B 208
Battle Cl. SW19 65Eb 133
Battledean Rd. N5 36Rb 71
Battlefield Rd. AL1: St A 1D 6
Battlefields Rd. TN15: Wro 88Be 185
Battle Ho. SE15 51Wb 113
(off Haymerle Rd.)
Battle Rd. DA17: Belv 49Ed 96
Battle Rd. DA17: Erith 49Ed 96
Battle Rd. DA8: Erith 49Ed 96
BATTLERS GREEN 8Ga 14
Battlers Grn. Dr. WD7: R'lett 9Ga 14
Battle St. DA12: Cobh 9H 145
BATTLE STREET 9H 145
Batts Hill RH1: Redh 4N 207
Batts Hill RH2: Reig 4M 207
Batty St. E1 44Wb 91
Batwa Ho. SE16 50Xb 91
Baudwin Rd. SE6 61Gc 137
Baugh Rd. DA14: Sidc 64Yc 139
Baulk, The SW18 59Cb 111
Bavant Rd. SW16 68Nb 134
Bavaria Rd. N19 33Nb 70
Bavdene M. NW4 28Xa 48
(off The Burroughs)
Bavent Rd. SE5 54Sb 113
Bawdale Rd. SE22 57Vb 113
Bawdsey Av. IG2: Ilf 28Vc 55
Bawley Ct. E16 45Sc 94
Bawley Ter. E15 39Fc 73
(off Rick Roberts Way)
Bawtree Cl. SM2: Sutt 82Eb 175
Bawtree Rd. SE14 52Ac 114
Bawtree Rd. UB8: Uxb 37M 63
Bawtry Rd. N20 20Hb 31
Baxendale N20 19Eb 31
Baxendale St. E2 41Wb 91
Baxter Av. RH1: Redh 6P 207
Baxter Cl. BR1: Broml 69Rc 138
Baxter Cl. SL1: Slou 8J 81
Baxter Cl. UB10: Hil 41R 84
Baxter Cl. UB2: S'hall 48Da 85
Baxter Gdns. RM3: Rom 11Jd 39
Baxter Ho. E3 41Dc 92
(off Bromley High St.)
Baxter Rd. E16 44Lc 93
Baxter Rd. IG1: Ilf 36Rc 74
Baxter Rd. N1 37Tb 71
Baxter Rd. N18 21Xb 51
Baxter Rd. WD24: Wat 8W 12
Baxter Wlk. SW16 61Mb 134
Bayard Ct. DA6: Bex 56Dd 118
Bayards CR6: W'ham 90Yb 178
Bay Ct. E1 42Zb 92
(off Frimley Way)
Bay Ct. W5 48Na 87
Baycroft Cl. HA5: Eastc 27Y 45
Baydon Ct. BR2: Broml 69Hc 137
Bayer Ho. EC1 6D 218 (42Sb 91)
(off Golden La. Est.)
Bayes Cl. SE26 64Yb 136
Bayes Ct. NW3 38Hb 69
(off Primrose Hill Rd.)
Bayes Ho. N1 38Qb 70
(off Augustas La.)
Bayeux KT20: Tad 94Za 194
Bayfield Ho. SE4 56Zb 114
(off Coston Wlk.)
Bayfield Pl. BR6: Orp 75Zc 161
Bayfield Rd. SE9 56Mc 115
Bayford M. E8 38Xb 71
(off Bayford St.)
Bayford Rd. NW10 41Za 88
Bayford St. E8 38Xb 71
Bayford St. Bus. Cen. E8 38Xb 71
(off Sidworth St.)
Baygrove M. KT1: Hamp W 67La 130
Bayham Pl. NW1 1B 216 (39Lb 70)
Bayham Rd. SM4: Mord 70Db 133
Bayham Rd. TN13: S'oaks 95Ld 203
Bayham Rd. W13 45Ka 86
Bayham Rd. W4 48Ta 87
Bayham St. NW1 39Lb 70
Bay Ho. SE16 48Ac 92
Bayhurst Dr. HA6: Nwood 23V 44
Bayhurst Wood Country Pk. 29P 43
Bayleaf Cl. TW12: Hamp H 64Fa 130
Bayley Mead HP1: Hem H 4K 3
Bayleys Mead CM13: Hut 19Ee 41
Bayley St. WC1 1D 222 (43Mb 90)
Bayley Wlk. SE2 51Ad 117
Baylie Ct. HP2: Hem H 1N 3
Baylie La. HP2: Hem H 1N 3
Baylis Bus. Cen. SL1: Slou 5H 81
Baylis Hgts. SE10 49Jc 93
(off Horn Link Way)
Baylis M. TW1: Twick 59Ja 108
Baylis Pde. SL1: Slou 4J 81
Baylis Pl. BR1: Broml 69Nc 138
Baylis Rd. SE1 2K 229 (47Qb 90)
Baylis Rd. SL1: Slou 5H 81
Bayliss Av. SE28 45Zc 95
Bayliss Cl. N21 15Nb 32
Bayliss Cl. UB1: S'hall 44Da 85
(off Whitecote Rd.)
Bayly Rd. DA1: Dart 58Qd 119
Bay Mnr. La. RM20: W Thur 51Vd 120
Baymans Wood CM15: Shenf 19Be 41
Baynard Ho. EC4 4C 224 (45Rb 91)
(off Queen Victoria St.)
Bayne Cl. E6 44Pc 94
Baynes Cl. EN1: Enf 11Wb 33
Baynes Cres. RM10: Dag 38Ed 76
Baynes M. NW3 37Fb 69
Baynes St. NW1 38Lb 70
Baynham Cl. DA5: Bexl 58Bd 117
Baynton Rd. GU22: Wok 92D 188
Bayonne Rd. W6 51Ab 110
Bay Path RH9: G'stone 3A 210
Bays Ct. HA8: Edg 22Ra 47
Bays Farm Ct. UB7: Lford 53L 105
Bayshill Ri. UB5: N'olt 37Da 65
Baysixty6 Skate Pk. 43Bb 89
(off Acklam Rd.)
Bayston Rd. N16 34Vb 71
BAYSWATER 4A 220 (45Eb 89)
Bayswater Cl. N13 21Rb 51
Bayswater Rd. W2 45Db 89
Bayswater Station (Underground) 45Db 89
Baythorne Ho. E16 44Hc 93
(off Turner St.)
Baythorne St. E3 43Bc 92
Bayton Ct. E8 38Wb 71
(off Lansdowne Dr.)
Bay Tree Av. KT22: Lea 92Ja 192
Bay Tree Cl. AL2: Park 10A 6
Bay Tree Cl. BR1: Broml 67Mc 137
Bay Tree Cl. IG6: Ilf 24Rc 54
Baytree Cl. DA15: Sidc 60Vc 117
Bay Tree Ct. SL1: Burn 1A 80
Baytree Ct. SW2 56Pb 112
Bay Tree Ho. EC1 6K 217 (42Qb 90)
(off Baker's Row)
Baytree Ho. E4 17Dc 34
Baytree M. SE17 5F 231 (49Tb 91)
Baytree Rd. SW2 56Pb 112
Bay Trees RH8: Oxt 5M 211
Baytrees SL9: Ger X 28B 42
Baytree Shop. Cen., The 19Yd 40
Bay Tree Wlk. WD17: Wat 10V 12
Baywillow Av. SM5: Cars 74Hb 155
Baywood Sq. IG7: Chig 21Wc 55
Bazalgette Cl. KT3: N Mald 71Ta 153
Bazalgette Gdns. KT3: N Mald 71Ta 153
Bazalgette Ho. NW8 5C 214 (42Fb 89)
(off Orchardson St.)
Bazalgette Wlk. EC4 4A 224 (45Qb 90)
Bazalgette Way SE2 46Yc 95
Bazeley Ho. SE1 2B 230 (47Rb 91)
(off Library St.)
Bazely St. E14 45Ec 92
Bazes Shaw DA3: Nw A G 75Be 165
(not continuous)
Bazile Rd. N21 16Qb 32
BBC Broadcasting House 1A 222 (43Kb 90)
BBC Elstree 13Qa 29
BBC Maida Vale Studios 42Db 89
(off Delaware Rd.)
BBC Studios 46Ya 88
BBC Worldwide 45Ya 88
BDA Dental Mus. 1K 221 (43Kb 90)
Beacham Cl. SE7 50Mc 93
Beachborough Rd. BR1: Broml 63Ec 136
Beach Ct. SE9 58Nc 116
Beachcroft Av. UB1: S'hall 46Aa 85
Beachcroft Rd. E11 34Gc 73
Beachcroft Way N19 32Mb 70
Beach Gro. TW13: Hanw 61Ca 129
Beach Ho. SW5 50Cb 89
(off Philbeach Gdns.)
Beach Ho. TW13: Hanw 61Ca 129
Beach Ho. Apts. IG11: Bark 42Wc 95
Beach's Ho. TW18: Staines 64J 127
Beachy Rd. E3 38Cc 72
Beacon Bingo Cricklewood 35Za 68
Beacon Cl. SL9: Chal P 24A 42
Beacon Cl. SM7: Bans 88Za 174
Beacon Cl. UB8: Uxb 36M 63
Beacon Ct. SL3: Coln 52E 104
Beacon Dr. DA2: Bean 62Xd 142
Beaconfield Av. CM16: Epp 1Vc 23
Beaconfield Rd. CM16: Epp 1Vc 23
Beaconfields TN13: S'oaks 98Hd 202
Beaconfield Way CM16: Epp 1Vc 23
Beacon Ga. SE14 55Zb 114
Beacon Gro. SM5: Cars 77Jb 156
Beacon Hill GU21: Wok 1N 187
Beacon Hill N7 36Nb 70
Beacon Hill RM19: Purf 50Rd 97
Beacon Hill Ind. Est. RM19: Purf 50Rd 97
Beacon Ho. AL1: St A 2D 6
Beacon Ho. E14 50Dc 92
(off Burrells Wharf Sq.)
Beacon Ho. SE5 52Ub 113
(off Southampton Way)
Beacon Pl. CR0: Bedd 76Nb 156
Beacon Point SE10 51Dc 114
(off Dowells St.)
Beacon Ri. TN13: S'oaks 98Jd 202
Beacon Rd. DA8: Erith 52Kd 119
Beacon Rd. SE13 58Fc 115
Beacon Rd. TW6: H'row A 58Q 106
Beacon Row TW15: W King 81Vd 184
Beacons, The IG10: Lough 10Qc 22
Beacons Cl. E6 43Nc 94
Beaconsfield WC1 1H 223 (43Pb 90)
(off Red Lion St.)
Beaconsfield Cl. N11 22Jb 50
Beaconsfield Cl. SE3 51Jc 115
Beaconsfield Cl. W4 50Sa 87
Beaconsfield Comn. La. HP9: Beac 1G 60
Beaconsfield Ct. WD25: Wat 4X 13
(off Horseshoe La.)
Beaconsfield Gdns. KT10: Clay 80Ga 152
Beaconsfield Pde. SE9 63Nc 138
Beaconsfield Pl. KT17: Eps 84Ua 174
Beaconsfield Rd. AL1: St A 2C 6
Beaconsfield Rd. BR1: Broml 69Mc 137
Beaconsfield Rd. CR0: C'don 72Tb 157
Beaconsfield Rd. DA5: Bexl 61Gd 140
Beaconsfield Rd. E10 33Ec 72
Beaconsfield Rd. E16 42Hc 93
Beaconsfield Rd. E17 30Bc 52
Beaconsfield Rd. EN3: Enf W 9Zb 20
Beaconsfield Rd. GU22: Wok 92B 188
Beaconsfield Rd. KT10: Clay 80Ga 152
Beaconsfield Rd. KT18: Eps D 91Ta 193
Beaconsfield Rd. KT3: N Mald 68Ta 131
Beaconsfield Rd. KT5: Surb 73Pa 153
Beaconsfield Rd. N11 20Jb 32
Beaconsfield Rd. N15 28Ub 51
Beaconsfield Rd. N9 20Wb 33
Beaconsfield Rd. NW10 37Va 68
Beaconsfield Rd. SE17 50Tb 91
Beaconsfield Rd. SE3 52Hc 115
Beaconsfield Rd. SE9 61Nc 138
Beaconsfield Rd. SL2: Farn C 10G 60
Beaconsfield Rd. SL2: Farn R 10G 60
Beaconsfield Rd. TW1: Twick 58Ka 108
Beaconsfield Rd. UB1: S'hall 46Z 85
Beaconsfield Rd. UB4: Yead 46Y 85
Beaconsfield Rd. W4 48Ta 87
Beaconsfield Rd. W5 47La 86
Beaconsfield St. N1 39Nb 70
Beaconsfield Ter. RM6: Chad H 30Zc 55
Beaconsfield Ter. Rd. W14 48Ab 88
Beaconsfield Wlk. E6: Beckton 44Qc 94
Beaconsfield Wlk. SW6: Parsons Green 53Bb 111
Beacontree Av. E17 25Fc 53
BEACONTREE HEATH 32Cd 76
Beacontree Rd. E11 32Hc 73
Beacon Way SM7: Bans 88Za 174
Beacon Way WD3: Rick 17J 25
Beacon Wood Country Pk. 63Xd 142
Beadle Pl. DA8: Erith 52Hd 118
Beadles La. RH8: Oxt 2H 211
Beadle's Pde. RM10: Dag 37Ed 76
Beadlow Cl. SM5: Cars 72Fb 155
Beadman Pl. SE27 63Rb 135
Beadman Rd. EN8: Chesh 2Yb 20
Beadman St. SE27 63Rb 135
Beadnell Ct. E1 45Wb 91
(off Cable St.)
Beadnell Rd. SE23 60Zb 114
Beadon Rd. BR2: Broml 70Jc 137
Beadon Rd. W6 49Ya 88
Beads Hall La. CM15: Pil H 14Xd 40
Beaford Gro. SW20 69Ab 132
Beagle Cl. TW13: Felt 63X 129
Beagle Cl. WD7: R'lett 9Ha 14
Beagles Cl. BR5: Orp 75Zc 161
Beak St. W1 4B 222 (45Lb 90)
Beal Cl. DA16: Well 53Wc 117
Beale Arboretum, The 10Hb 17
Beale Cl. N13 22Rb 51
Beale Pl. E3 40Bc 72
Beale Rd. E3 39Bc 72
Beales La. KT13: Weyb 76R 150
Beales Rd. KT23: Bookh 99Da 191
Beal Rd. IG1: Ilf 33Qc 74
Beam Av. RM10: Dag 39Dd 76
Beames Rd. NW10 39Ta 67
Beaminster Gdns. IG6: Ilf 26Rc 54
Beaminster Ho. SW8 52Pb 112
(off Dorset Rd.)
Beamish Dr. WD23: B Hea 18Ea 28
Beamish Ho. SE16 49Xb 91
(off Rennie Est.)
Beamish Rd. BR5: Orp 73Yc 161
Beamish Rd. N9 18Wb 33
Beam Pk. Development RM9: Dag 41Dd 96
Beam Reach Bus. Pk. RM13: Rain 41Fd 96
Beam Valley Country Pk. 39Ed 76
Beamway RM10: Dag 38Fd 76
BEAN 62Xd 142
Beanacre Cl. E9 37Bc 72
Beane Cft. DA12: Grav'nd 10H 123
Bean Hill Cotts. DA2: Bean 63Yd 142
(not continuous)
Bean La. DA2: Bean 61Xd 142
Bean Rd. DA6: Bex 56Zc 117
Bean Rd. DA9: Ghithe 57Xd 120
Beanshaw SE9 63Qc 138
Beansland Gro. RM6: Chad H 27Ad 55
Bear All. EC4 2B 224 (44Rb 91)
Bear Cl. RM7: Rom 30Dd 56
Beardell St. SE19 65Vb 135
Beardow Gro. N14 16Lb 32
Beard Rd. KT2: King T 64Pa 131
Beardsfield E13 40Jc 73
Beard's Hill TW12: Hamp 67Ca 129
Beard's Hill Cl. TW12: Hamp 67Ca 129
Beardsley Ter. RM8: Dag 36Xc 75
(off Stonard Rd.)
Beardsley Way W3 47Ta 87
Beard's Rd. TW15: Ashf 65U 128
Bearfield Rd. KT2: King T 66Na 131
Bear Gdns. SE1 6D 224 (46Sb 91)
Bearing Cl. IG7: Chig 21Wc 55
Bearing Way IG7: Chig 21Wc 55
Bear La. SE1 7C 224 (46Rb 91)
Bear Pit Apts. SE1 6D 224 (46Sb 91)
(off New Globe Wlk.)
Bear Rd. TW13: Hanw 63Z 129
Bears Den KT20: Kgswd 94Bb 195
Bearsden Ct. SL5: S'dale 3D 146
Bears Rails Pk. SL4: Old Win 9K 103
Bearstead Ri. SE4 57Bc 114
Bearsted Ter. BR3: Beck 67Cc 136
Bear St. WC2 4E 222 (45Mb 90)
Bearwood Cl. EN6: Pot B 3Fb 17
Bearwood Cl. KT15: Add 79J 149
Bearwood Rd. SS17: Stan H 3K 101
Beasley's Ait TW16: Sun 72V 150
Beasley's Ait La. TW16: Sun 72V 150
Beaton Cl. DA9: Ghithe 57Xd 120
Beaton Cl. SE15 53Vb 113
Beatrice Av. HA9: Wemb 36Na 67
Beatrice Av. SW16 69Pb 134
Beatrice Cl. E13 42Jc 93
Beatrice Cl. HA5: Eastc 28W 44
Beatrice Ct. IG9: Buck H 19Mc 35
Beatrice Gdns. DA11: Nflt 1A 144
Beatrice Ho. W6 50Ya 88
(off Queen Caroline St.)
Beatrice Pl. SW19 59Za 110
Beatrice Pl. W8 48Db 89
Beatrice Rd. E17 29Cc 52
Beatrice Rd. N4 31Qb 70
Beatrice Rd. N9 17Yb 34
Beatrice Rd. RH8: Oxt 1J 211
Beatrice Rd. SE1 49Wb 91
Beatrice Rd. TW10: Rich 57Pa 109
Beatrice Rd. UB1: S'hall 46Ba 85
Beatrice Webb Ho. E3 40Ac 72
(off Chisenhale Rd.)
Beatrice Wilson Flats TN13: S'oaks 97Kd 203
Beatrix Apts. E3 42Bc 92
(off English St.)
Beatrix Ho. SW5 50Db 89
(off Old Brompton Rd.)
Beatson Wlk. SE16 46Ac 92
(not continuous)
Beattie Cl. KT23: Bookh 96Ba 191
Beattie Cl. TW14: Felt 59V 106
Beattie Ho. SW8 53Lb 112
Beattock Ri. N10 28Kb 50
Beatty Ho. E14 47Cc 92
(off Admirals Way)
Beatty Ho. SW1 50Lb 90
(off Dolphin Sq.)
Beatty Rd. EN8: Walt C 6Bc 20
Beatty Rd. HA7: Stan 23La 46
Beatty Rd. N16 35Ub 71
Beatty St. NW1 1B 216 (40Lb 70)
Beattyville Gdns. IG6: Ilf 28Qc 54
Beauchamp Cl. W4 48Sa 87
Beauchamp Ct. EN5: Barn 14Bb 31
(off Victors Way)
Beauchamp Ct. HA7: Stan 22La 46
Beauchamp Gdns. WD3: Rick 18J 25
Beauchamp Pl. SW3 3E 226 (48Gb 89)
Beauchamp Rd. E7 38Kc 73
Beauchamp Rd. KT8: E Mos 71Da 151
Beauchamp Rd. KT8: W Mole 71Da 151
Beauchamp Rd. SE19 67Tb 135
Beauchamp Rd. SM1: Sutt 77Cb 155
Beauchamp Rd. SW11 56Gb 111
Beauchamp Rd. TW1: Twick 59Ja 108
Beauchamp St. EC1 1K 223 (43Qb 90)
Beauchamp Ter. SW15 55Xa 110
Beauclare Cl. KT22: Lea 92Ma 193
Beauclerc Ct. TW16: Sun 68Y 129
Beauclerc Rd. W6 48Xa 88
Beauclere Ho. SM2: Sutt 79Eb 155
Beauclerk Cl. TW13: Felt 60X 107
Beauclerk Ho. SW16 62Nb 134
Beau Ct. HA7: Stan 24Ma 47
(off Hitchin La.)
Beaudesert M. UB7: W Dray 47N 83
Beaufort E6 43Qc 94
Beaufort Av. HA3: Kenton 28Ja 46
Beaufort Cl. E4 23Dc 52
Beaufort Cl. GU22: Wok 88E 168
Beaufort Cl. RH2: Reig 5H 207
Beaufort Cl. RM16: Chaf H 48Be 99
Beaufort Cl. RM7: Mawney 28Ed 56
Beaufort Cl. SW15 59Xa 110
Beaufort Cl. W5 43Pa 87
Beaufort Ct. E14 47Cc 92
(off Admirals Way)
Beaufort Ct. EN5: New Bar 15Eb 31
Beaufort Ct. N11 22Kb 50
(off The Limes Av.)
Beaufort Ct. SW6 51Cb 111
Beaufort Ct. TW10: Ham 63La 130
Beaufort Dr. NW11 28Cb 49
Beaufort Gdns. E1 43Zb 92
Beaufort Gdns. IG1: Ilf 32Qc 74
Beaufort Gdns. NW4 30Ya 48
Beaufort Gdns. SW16 66Pb 134
Beaufort Gdns. SW3 3E 226 (48Gb 89)
Beaufort Gdns. TW5: Hest 53Aa 107
Beaufort Ho. E16 46Kc 93
(off Fairfax M.)
Beaufort Ho. SW1 7D 228 (50Mb 90)
(off Aylesford St.)
Beaufort Ho. SW3 51Fb 111
(off Beaufort St.)
Beaufort Mans. SW3 51Fb 111
Beaufort M. GU21: Wok 10P 167
Beaufort M. SW6 51Bb 111
Beaufort Pk. NW11 28Cb 49
Beaufort Pl. BR5: St P 68Yc 139
Beaufort Rd. GU22: Wok 88E 168
Beaufort Rd. HA4: Ruis 33T 64
Beaufort Rd. KT1: King T 70Na 131
Beaufort Rd. RH2: Reig 5H 207
Beaufort Rd. TW1: Twick 59La 108
Beaufort Rd. TW10: Ham 63La 130
Beaufort Rd. W5 43Pa 87
Beauforts TW20: Eng G 4N 125
Beaufort Sq. NW9 26Wa 48
Beaufort St. SW3 51Fb 111
Beaufort Ter. E14 50Ec 92
(off Ferry St.)
Beaufort Way KT17: Ewe 80Wa 154
Beaufoy Ho. SE27 62Rb 135
Beaufoy Ho. SW8 52Pb 112
(off Rita Rd.)
Beaufoy Rd. N17 24Ub 51
(not continuous)
Beaufoy Wlk. SE11 6J 229 (49Pb 90)
Beaulieu Av. E16 46Kc 93
Beaulieu Av. SE26 63Xb 135
Beaulieu Cl. CR4: Mitc 67Jb 134
Beaulieu Cl. NW9 28Ua 48
Beaulieu Cl. SE5 55Tb 113
Beaulieu Cl. SL3: Dat 3M 103
Beaulieu Cl. TW1: Twick 58Ma 109
Beaulieu Cl. TW4: Houn 57Ba 107
Beaulieu Cl. WD19: Wat 18Y 27
Beaulieu Ct. W5 43Na 87
Beaulieu Dr. EN9: Walt A 5Dc 20
Beaulieu Dr. HA5: Pinn 30Z 45
Beaulieu Gdns. N21 17Sb 33
Beaulieu Hgts. SE25 67Ub 135
Beaulieu Lodge E14 48Fc 93
(off Schooner Cl.)
Beaulieu Pl. W4 48Sa 87
Beauly Way RM1: Rom 25Gd 56
Beaumanor Gdns. SE9 63Qc 138
Beaumanor Mans. W2 45Db 89
(off Queensway)
Beaumans Dr. E17 25Bc 52
Beaumaris Ct. SL2: Slou 3F 80
Beaumaris Dr. IG8: Wfd G 24Mc 53
Beaumaris Gdns. SE19 66Sb 135
Beaumaris Grn. NW9 30Ua 48
Beaumaris Twr. W3 47Ra 87
(off Park Rd. Nth.)
Beaumayes Cl. HP1: Hem H 3K 3
Beaumonds AL1: St A 2C 6
Beaumont W14 49Bb 89
(off Kensington Village)
Beaumont Av. AL1: St A 1F 6
Beaumont Av. HA0: Wemb 36La 66
Beaumont Av. HA2: Harr 30Da 45
Beaumont Av. TW9: Rich 55Pa 109
Beaumont Av. W14 50Bb 89
Beaumont Bldgs. WC2 3G 223 (44Nb 90)
(off Martlett Ct.)
Beaumont Cen. EN8: Chesh 2Zb 20
Beaumont Cl. KT2: King T 66Qa 131
Beaumont Cl. N2 28Gb 49
Beaumont Cl. RM2: Rom 26Ld 57
Beaumont Ct. E1 41Zb 92
Beaumont Ct. E5 34Xb 71
Beaumont Ct. HA0: Wemb 36La 66
Beaumont Ct. NW1 39Mb 70
Beaumont Ct. NW9 26Va 48
(off Cherry Cl.)
Beaumont Ct. W1 7J 215 (43Jb 90)
(off Beaumont St.)
Beaumont Ct. W4 50Sa 87
Beaumont Cres. RM13: Rain 37Jd 76
Beaumont Cres. W14 50Bb 89
Beaumont Dr. DA11: Nflt 9A 122
Beaumont Dr. KT4: Wor Pk 73Xa 154
Beaumont Dr. TW15: Ashf 64T 128
Beaumont Gdns. CM13: Hut 16Ee 41
Beaumont Gdns. NW3 34Cb 69
Beaumont Ga. WD7: R'lett 7Ja 14
(off Shenley Hill)
Beaumont Gro. E1 42Zb 92
Beaumont Ho. E10 31Dc 72
(off Skelton's La.)
Beaumont Ho. W9 41Bb 89
(off Fernhead Rd.)
Beaumont Lodge E8 37Wb 71
(off Greenwood Rd.)
Beaumont M. HA5: Pinn 27Aa 45
Beaumont M. NW5 36Mb 70
(off Charlton King's Rd.)
Beaumont M. W1 7J 215 (43Jb 90)
Beaumont Pl. EN5: Barn 11Bb 31
Beaumont Pl. TW7: Isle 57Ha 108
Beaumont Pl. UB10: Ick 36Q 64
Beaumont Pl. W1 5C 216 (42Lb 90)
Beaumont Pl. WD18: Wat 15W 26
Beaumont Ri. N19 32Mb 70
Beaumont Rd. BR5: Pet W 72Tc 160
Beaumont Rd. CR8: Purl 85Qb 176
Beaumont Rd. E10 31Dc 72
(not continuous)
Beaumont Rd. E13 41Kc 93
Beaumont Rd. SE19 65Sb 135
Beaumont Rd. SL2: Slou 2H 81
Beaumont Rd. SL4: Wind 4G 102
Beaumont Rd. SW19 59Ab 110
Beaumont Rd. W4 48Sa 87
Beaumont Sq. E1 43Zb 92
Beaumont St. W1 7J 215 (43Jb 90)
Beaumont Ter. SE13 59Gc 115
(off Wellmeadow Rd.)
Beaumont Wlk. NW3 38Hb 69
Beaumont Works AL1: St A 2F 6
(off Hedley Rd.)
Beaumore Pl. SL9: Ger X 28B 42
Beauvais Ter. UB5: N'olt 41Z 85
Beauvale NW1 38Jb 70
(off Ferdinand St.)
Beauval Rd. SE22 58Vb 113
Beaux Arts Bldg., The N7 34Nb 70
Beaverbank Rd. SE9 60Tc 116
BEAVERBROOK RDBT. 95Ma 193
Beaver Cl. SE20 66Wb 135
Beaver Cl. SM4: Mord 73Ya 154
Beaver Cl. TW12: Hamp 67Da 129
Beaver Ct. BR3: Beck 66Dc 136
Beaver Gro. UB5: N'olt 41Aa 85
Beaver Ind. Est. UB2: S'hall 48Y 85
Beaver Rd. IG6: Ilf 22Yc 55
Beavers Cres. TW4: Houn 56Y 107
Beavers La. TW4: Houn 54Y 107
Beavers La. Campsite TW4: Houn 56Z 107
Beavers Lodge DA14: Sidc 63Vc 139
Beaverwood Rd. BR7: Chst 65Uc 138
Beavor Gro. W6 50Wa 88
(off Beavor La.)
Beavor La. W6 50Wa 88
Bebbington Rd. SE18 49Uc 94
Beblets Cl. BR6: Chels 78Vc 161
Beccles Dr. IG11: Bark 37Uc 74
Beccles St. E14 44Bc 92
Bec Cl. HA4: Ruis 34Z 65
Bechervaise Ct. E10 32Dc 72
(off Leyton Grange Est.)
Bechtel Ho. W6 49Za 88
(off Hammersmith Rd.)
Beck Cl. SE13 53Dc 114
Beck Ct. BR3: Beck 69Zb 136
BECKENHAM 68Cc 136
BECKENHAM BEACON 68Bc 136
Beckenham Bus. Cen. BR3: Beck 65Ac 136
Beckenham Civic Halls 67Dc 136
Beckenham Crematorium 69Yb 136
Beckenham Gdns. N9 20Ub 33
Beckenham Gro. BR2: Broml 68Fc 137
Beckenham Hill Est. BR3: Beck 64Dc 136
Beckenham Hill Rd. BR3: Beck 65Dc 136
Beckenham Hill Rd. SE6 64Ec 136
Beckenham Hill Station (Rail) 64Ec 136
Beckenham Junction Station (Rail & London Tramlink) 67Cc 136
Beckenham La. BR2: Broml 68Gc 137
Beckenham Pl. Pk. BR3: Beck 66Dc 136
Beckenham Place Pk. 65Ec 136
Beckenham Rd. BR3: Beck 67Zb 136
Beckenham Rd. BR4: W W'ck 73Dc 158
Beckenham Road Stop (London Tramlink) 67Ac 136
Beckenham Theatre Cen., The 68Dc 136
Beckenshaw Gdns. SM7: Bans 87Gb 175
Beckers, The N16 35Wb 71
Becket Av. E6 41Qc 94
Becket Cl. CM13: Gt War 23Yd 58
Becket Cl. IG8: Wfd G 25Nc 54
Becket Cl. SE25 72Wb 157
Becket Cl. SW19 67Db 133
(off High Path)
Becket Fold HA1: Harr 29Ha 46
Becket Ho. E16 46Kc 93
(off Constable Av.)
Becket Ho. SE1 2F 231 (47Tb 91)
(off Tabard St.)
Becket Ho. WC1 4H 217 (41Pb 90)
(off Westking Pl.)
Becket Rd. N18 21Yb 52
Beckets Pl. TN14: Otf 88Ld 183
Beckets Sports Cen. 58Ld 119
Becket St. SE1 3F 231 (48Tb 91)
Beckett Av. CR8: Kenley 87Rb 177
Beckett Chase SL3: L'ly 50B 82
Beckett Cl. DA17: Belv 48Bd 95
Beckett Cl. NW10 37Ua 68
Beckett Cl. SW16 61Mb 134
Beckett Ho. E1 43Yb 92
(off Jubilee St.)
Beckett Ho. SW9 54Nb 112
Beckett M. DA12: Grav'nd 3F 144
Beckett Rd. CR5: Coul 94Mb 196
Becketts Cl. BR6: Orp 76Vc 161
Becketts Cl. DA5: Bexl 60Ed 118
Becketts Cl. TW14: Felt 58X 107
Becketts Ho. IG1: Ilf 34Qc 74
Becketts Pl. KT1: Hamp W 67Ma 131
Becketts Wharf KT1: Hamp W 67Ma 131
(off Lwr. Teddington Rd.)
Beckett Wlk. BR3: Beck 65Ac 136
Beckfoot NW1 2C 216 (40Lb 70)
(off Ampthill Est.)
Beckford Cl. W14 49Bb 89
Beckford Dr. BR5: Orp 73Tc 160

Beckford Ho. N16 ....36Ub 71
Beckford Pl. SE17 ....7E 230 (50Sb 91)
Beckford Rd. CR0: C'don 72Vb 157
Beckham Ho. SE11 ....6J 229 (49Pb 90)
Beckhaven Ho. SE11 ....6A 230 (49Qb 90)
(off Gilbert Rd.)
Beck Ho. N18 ....22Xb 51
(off Upton Rd.)
Beckingham Metro TW20: Egh 64C 126
(off Station Rd.)
Beck La. BR3: Beck 69Zb 136
Beckley Cl. DA12: Grav'nd 1K 145
Beckley Ho. E3 ....42Bc 92
(off Hamlets Way)
Becklow Gdns. W12 ....47Wa 88
(off Becklow Rd.)
Becklow M. W12 ....47Wa 88
(off Becklow Rd.)
Becklow Rd. W12 ....47Va 88
Beckman Cl. TN14: Hals 87Ed 182
Beck River Pk. BR3: Beck 67Bc 136
Beck Rd. CR4: Mitc 72Hb 155
Beck Rd. E8 ....39Xb 71
Beck Sq. E10 ....32Ac 72
Becks Rd. DA14: Sidc 62Wc 139
Beck Theatre, The 44V 84
BECKTON 43Qc 94
BECKTON ALPS 42Pc 94
Beckton Globe & Library 43Pc 94
BECKTON PARK 44Pc 94
Beckton Park Station (DLR) 45Pc 94
Beckton Retail Pk. 43Qc 94
Beckton Rd. E16 ....43Hc 93
Beckton Station (DLR) 43Qc 94
Beckton Triangle Retail Pk. 42Rc 94
Beck Way BR3: Beck 69Bc 136
Beckway Rd. SW16 ....68Mb 134
Beckway St. SE17 ....6H 231 (49Ub 91)
(not continuous)
Beckwell Rd. SL1: Slou 7G 80
Beckwith Cl. EN2: Enf 11Rb 33
Beckwith Ho. E2 ....40Xb 71
(off Wadeson St.)
Beckwith Rd. SE24 ....57Tb 113
Beclands Rd. SW17 ....65Jb 134
Becmead Av. HA3: Kenton 29Ka 46
Becmead Av. SW16 ....63Mb 134
Becondale Rd. SE19 ....64Ub 135
BECONTREE 35Zc 75
Becontree Av. RM8: Dag 35Xc 75
Becontree Heath Leisure Cen. 33Cd 76
Becontree Station (Underground) 37Zc 75
Becquerel Ct. SE10 ....48Hc 93
(off West Parkside)
Bective Pl. SW15 ....56Bb 111
Bective Rd. E7 ....35Jc 73
Bective Rd. SW15 ....56Bb 111
Becton Pl. DA8: Erith 53Dd 118
Bedale Rd. EN2: Enf 10Sb 19
Bedale Rd. RM3: Rom 22Qd 57
Bedale St. SE1 ....7F 225 (46Tb 91)
Bedale Wlk. DA2: Dart 60Rd 119
Beddalls Farm Ct. E6 ....43Mc 93
BEDDINGTON 77Nb 156
BEDDINGTON CORNER 73Jb 156
Beddington Cross CR0: Bedd 73Mb 156
Beddington Farm Rd. CR0: Bedd 73Nb 156
Beddington Farm Rd. CR0: Wadd 73Nb 156
Beddington Gdns. SM5: Cars 79Jb 156
Beddington Gdns. SM6: W'gton 79Kb 156
Beddington Grn. BR5: St P 67Vc 139
Beddington Gro. SM6: W'gton 78Mb 156
Beddington La. CR0: Bedd 71Lb 156
Beddington La. CR0: C'don 71Lb 156
Beddington Lane Stop (London Tramlink) 72Lb 156
Beddington Pk. 76Kb 156
Beddington Pk. Cotts. SM6: Bedd 76Mb 156
Beddington Path BR5: St P 67Vc 139
Beddington Rd. BR5: St P 68Uc 138
Beddington Rd. IG3: Ilf 31Vc 75
Beddington Ter. CR0: C'don 73Pb 156
Beddington Trad. Est. CR0: Bedd 74Nb 156
Beddlestead La. CR6: W'ham 89Hc 179
Bede Cl. HA5: Pinn 25Z 45
Bedefield WC1 ....4G 217 (41Nb 90)
Bede Ho. SE14 ....53Bc 114
(off Clare Rd.)
Bedens Rd. DA14: Sidc 65Ad 139
Bede Rd. RM6: Chad H 30Yc 55
Bede Sq. E3 ....42Bc 92
(off Joseph St.)
Bedevere Rd. N9 ....20Wb 33
Bedfont Cl. CR4: Mitc 68Jb 134
Bedfont Cl. TW14: Bedf 58S 106
Bedfont Ct. TW19: Stanw M 55J 105
Bedfont Ct. Est. TW19: Stanw M 56K 105
Bedfont Grn. Cl. TW14: Bedf 60S 106
Bedfont Ind. Pk. TW15: Ashf 62S 128
Bedfont Ind. Pk. Nth. TW15: Ashf 62S 128
Bedfont Lakes Country Pk. 61S 128
Bedfont Lakes Country Pk. Vis. Cen. 62R 128
Bedfont La. TW13: Felt 60X 107
Bedfont La. TW14: Felt 59U 106
Bedfont Rd. TW13: Felt 61U 128
Bedfont Rd. TW14: Bedf 60S 106
Bedfont Rd. TW19: Stanw 58N 105
Bedfont Trad. Est. TW14: Bedf 61T 128
Bedford Av. EN5: Barn 15Bb 31
Bedford Av. SL1: Slou 4D 80
Bedford Av. UB4: Yead 44X 85
Bedford Av. WC1 ....1E 222 (43Mb 90)
Bedfordbury WC2 ....4F 223 (45Nb 90)
(not continuous)
Bedford Cl. GU21: Wok 7N 167
Bedford Cl. N10 ....24Jb 50
Bedford Cl. W4 ....51Ua 110
Bedford Cl. WD3: Chen 10D 10
Bedford Cnr. W4 ....49Ua 88
(off South Pde.)
Bedford Ct. CR0: C'don 74Tb 157
(off Tavistock Rd.)
Bedford Ct. WC2 ....5F 223 (45Nb 90)
Bedford Ct. Mans. WC1 ....1E 222 (43Mb 90)
(off Bedford Av.)
Bedford Cres. EN3: Enf W 7Ac 20
Bedford Dr. SL2: Farn C 6F 60
Bedford Gdns. RM12: Horn 33Ld 77
Bedford Gdns. W8 ....46Cb 89
Bedford Gdns. Ho. W8 ....46Cb 89
(off Bedford Gdns.)
Bedford Hill SW12 ....60Kb 112
Bedford Hill SW16 ....62Lb 134
Bedford Ho. CR0: C'don 73Rb 157
Bedford Ho. SW4 ....56Nb 112
(off Solon New Rd. Est.)
Bedford La. SL5: S'dale 1F 146
Bedford M. N2 ....27Gb 49
Bedford M. SE6 ....61Dc 136
Bedford Pk. CR0: C'don 74Sb 157
BEDFORD PARK 48Ta 87
Bedford Pk. Cnr. W4 ....49Ua 88
Bedford Pk. Mans. W4 ....49Ta 87
Bedford Pk. Rd. AL1: St A 2C 6
Bedford Pas. SW6 ....52Ab 110
(off Dawes Rd.)
Bedford Pas. W1 ....7C 216 (43Lb 90)
Bedford Pl. CR0: C'don 74Tb 157
Bedford Pl. WC1 ....7F 217 (43Nb 90)
Bedford Rd. AL1: St A 3C 6
Bedford Rd. BR6: Orp 75Xc 161
Bedford Rd. DA1: Dart 59Qd 119
Bedford Rd. DA11: Nflt 1B 144
Bedford Rd. DA15: Sidc 62Uc 138
Bedford Rd. E17 ....26Cc 52
Bedford Rd. E18 ....26Jc 53
Bedford Rd. E6 ....39Qc 74
Bedford Rd. HA1: Harr 30Ea 46
Bedford Rd. HA4: Ruis 35V 64
Bedford Rd. HA6: Nwood 20S 26
Bedford Rd. IG1: Ilf 34Rc 74
Bedford Rd. KT4: Wor Pk 75Ya 154
Bedford Rd. N15 ....28Ub 51
Bedford Rd. N2 ....27Gb 49
Bedford Rd. N22 ....26Nb 50
Bedford Rd. N8 ....30Mb 50
Bedford Rd. N9 ....17Xb 33
Bedford Rd. NW7 ....19Ua 30
Bedford Rd. RM17: Grays 50De 99
Bedford Rd. SW4 ....56Nb 112
Bedford Rd. TW2: Twick 62Fa 130
Bedford Rd. W13 ....45Ka 86
Bedford Rd. W4 ....48Ta 87
Bedford Row SE1 ....3E 230 (48Sb 91)
Bedford Row WC1 ....7J 217 (43Pb 90)
Bedford Sq. DA3: Lfield 69Zd 143
Bedford Sq. WC1 ....1E 222 (43Mb 90)
Bedford St. HP4: Berk 1A 2
Bedford St. WC2 ....4F 223 (45Nb 90)
Bedford St. WD24: Wat 11X 27
Bedford Ter. SM2: Sutt 79Eb 155
Bedford Ter. SW2 ....57Nb 112
Bedford Way WC1 ....6E 216 (42Mb 90)
Bedgebury Ct. E17 ....26Ec 52
Bedgebury Gdns. SW19 ....61Ab 132
Bedgebury Rd. SE9 ....56Mc 115
Bedivere Rd. BR1: Broml 62Jc 137
Bedlam Bri. Gdns. GU24: W End 5E 166
Bedlam M. SE11 ....5K 229 (49Qb 90)
(off Walnut Tree Wlk.)
Bedlow Way CR0: Bedd 77Pb 156
BEDMOND 9F 4
Bedmond Hill HP3: Hem H 7E 4
BEDMOND HILL 7E 4
Bedmond Ho. SW3 ....7D 226 (50Gb 89)
(off Cale St.)
Bedmond La. AL2: Pot C 6J 5
Bedmond La. AL3: St A 4M 5
Bedmond La. WD5: Bedm 8G 4
Bedmond Rd. HP3: Hem H 3C 4
Bedmond Rd. WD5: Ab L 1V 12
Bedmond Rd. WD5: Bedm 1V 12
Bedonwell Rd. DA17: Belv 51Ad 117
Bedonwell Rd. DA7: Belv 51Bd 117
Bedonwell Rd. DA7: Bex 51Bd 117
Bedonwell Rd. DA7: Erith 51Bd 117
Bedonwell Rd. SE2 ....51Ad 117
Bedser Cl. CR7: Thor H 69Sb 135
Bedser Cl. GU21: Wok 88C 168
Bedser Cl. SE11 ....51Pb 112
Bedser Dr. UB6: G'frd 36Fa 66
Bedster Gdns. KT8: W Mole 68Da 129
Bedwardine Rd. SE19 ....66Ub 135
Bedwell Cl. CR0: C'don 71Tb 157
Bedwell Ct. RM6: Chad H 31Zc 75
(off Chapel La.)
Bedwell Gdns. UB3: Harl 50U 84
(not continuous)
Bedwell Hall AL9: Ess 1P 9
Bedwell Ho. SW9 ....54Qb 112
Bedwell Pk. 1P 9
Bedwell Rd. DA17: Belv 50Cd 96
Bedwell Rd. N17 ....25Ub 51
Beeby Rd. E16 ....43Kc 93
Beech Av. BR8: Swan 70Hd 140
Beech Av. CM13: B'wood 20Be 41
Beech Av. CR2: Sande 83Tb 177
Beech Av. DA15: Sidc 59Wc 117
Beech Av. EN2: Crew H 7Qb 18
Beech Av. HA4: Ruis 32X 65
Beech Av. IG9: Buck H 19Kc 35
Beech Av. KT24: Eff 100Z 191
Beech Av. N20 ....18Gb 31
Beech Av. RM14: Upm 34Rd 77
Beech Av. TN16: Tats 91Mc 199
Beech Av. TW8: Bford 52Ka 108
Beech Av. W3 ....46Ua 88
Beech Av. WD7: R'lett 5Ja 14
Beech Cl. AL10: Hat 1C 8
Beech Cl. IG10: Lough 13Rc 36
Beech Cl. KT11: Cobh 84Ca 171
Beech Cl. KT12: Hers 77Y 151
Beech Cl. KT14: Byfl 84N 169
Beech Cl. KT22: Fet 97Fa 192
Beech Cl. KT24: Eff 100Z 191
Beech Cl. N9 ....16Wb 33
Beech Cl. RM12: Horn 34Kd 77
Beech Cl. SE8 ....51Cc 114
Beech Cl. SM5: Cars 75Hb 155
Beech Cl. SW15 ....59Wa 110
Beech Cl. SW19 ....65Ya 132
Beech Cl. TW15: Ashf 64T 128
Beech Cl. TW16: Sun 68Z 129
Beech Cl. TW19: Stanw 59M 105
Beech Cl. UB7: W Dray 48Q 84
Beech Cl. Ct. KT11: Cobh 83Ba 171
Beech Copse BR1: Broml 67Pc 138
Beech Copse CR2: S Croy 78Ub 157
Beech Ct. BR1: Broml 67Hc 137
(off Blyth Rd.)
Beech Ct. BR3: Beck 66Bc 136
Beech Ct. DA1: Dart 58Qd 119
Beech Ct. E17 ....27Fc 53
Beech Ct. IG1: Ilf 34Qc 74
(off Riverdene Rd.)
Beech Ct. KT6: Surb 73Ma 153
Beech Ct. UB5: N'olt 39Aa 65
Beech Ct. W9 ....43Cb 89
(off Elmfield Way)
Beech Cres. Ct. N5 ....35Rb 71
Beechcroft BR7: Chst 66Qc 138
Beechcroft KT21: Asht 91Pa 193
Beechcroft Av. CR8: Kenley 87Tb 177
Beechcroft Av. DA7: Bex 53Fd 118
Beechcroft Av. HA2: Harr 31Ca 65
Beechcroft Av. KT3: N Mald 67Sa 131
Beechcroft Av. NW11 ....31Bb 69
Beechcroft Av. SS17: Linf 9J 101
Beechcroft Av. WD3: Crox G 16S 26
Beechcroft Cl. BR6: Orp 77Tc 160
Beechcroft Cl. SL5: S'hill 10B 124
Beechcroft Cl. SW16 ....64Pb 134
Beechcroft Cl. TW5: Hest 52Aa 107
Beechcroft Ct. N12 ....21Db 49
Beechcroft Ct. NW11 ....31Bb 69
(off Beechcroft Av.)
Beechcroft Farm Industries TN15: Ash 76Zd 165
Beechcroft Gdns. HA9: Wemb 34Pa 67
Beechcroft Ho. W5 ....43Na 87
Beechcroft Lodge SM2: Sutt 80Eb 155
Beechcroft Mnr. KT13: Weyb 76T 150
Beechcroft Pl. HA6: Nwood 22V 44
(off Eastbury Av.)
Beechcroft Rd. BR6: Orp 77Tc 160
Beechcroft Rd. E18 ....26Kc 53
Beechcroft Rd. KT9: Chess 76Pa 153
Beechcroft Rd. SW14 ....55Sa 109
Beechcroft Rd. SW17 ....61Gb 133
Beechcroft Rd. WD23: Bush 15Aa 27
Beechdale N21 ....19Pb 32
Beechdale Rd. SW2 ....58Pb 112
Beech Dell BR2: Kes 77Pc 160
Beechdene KT20: Tad 94Xa 194
Beechdene SE15 ....53Xb 113
(off Carlton Gro.)
Beech Dr. GU23: Rip 96J 189
Beech Dr. KT20: Kgswd 94Bb 195
Beech Dr. N2 ....26Hb 49
Beech Dr. RH2: Reig 6M 207
Beech Dr. WD6: Bore 12Pa 29
Beechen Cliff Way TW7: Isle 54Ha 108
Beechen Gro. HA5: Pinn 27Ba 45
Beechen Gro. WD17: Wat 13X 27
Beechen La. KT20: Kgswd 97Bb 195
Beechenlea La. BR8: Swan 70Jd 140
Beechen Pl. SE23 ....61Zb 136
Beechen Wood WD3: Map C 22F 42
Beechers Cft. TN16: Big H 89Lc 179
Beeches, The AL2: Park 9B 6
Beeches, The BR8: Hext 66Hd 140
Beeches, The CM14: B'wood 20Xd 40
Beeches, The CR2: S Croy 78Tb 157
(off Blunt Rd.)
Beeches, The DA13: Sole S 10F 144
Beeches, The DA3: Lfield 68Ee 143
Beeches, The E12 ....38Nc 74
Beeches, The EN9: Walt A 7Lc 21
(within Woodbine Cl. Caravan Pk.)
Beeches, The KT22: Fet 96Ga 192
Beeches, The RM18: Tilb 4D 122
Beeches, The SL2: Slou 1D 80
Beeches, The SM7: Bans 88Cb 175
Beeches, The TW18: Staines 64J 127
Beeches, The TW3: Houn 53Da 107
Beeches, The WD18: Wat 13X 27
(off Halsey Rd.)
Beeches, The WD3: Chor 15H 25
Beeches Av. SM5: Cars 80Gb 155
Beeches Cl. KT20: Kgswd 95Cb 195
Beeches Cl. SE20 ....67Yb 136
Beeches Dr. KT23: Bookh 97Da 191
Beeches Dr. SL2: Farn C 6F 60
Beeches Rd. SL2: Farn C 6F 60
Beeches Rd. SM3: Sutt 74Ab 154
Beeches Rd. SW17 ....62Gb 133
Beeches Wlk. SM5: Cars 81Fb 175
Beeches Way IG7: Chig 22Ad 55
Beeches Wood KT20: Kgswd 94Cb 195
Beechey Ho. E1 ....46Xb 91
(off Watts St.)
Beech Farm Rd. CR6: W'ham 92Ec 198
Beechfield KT13: Weyb 77R 150
Beechfield SM7: Bans 85Db 175
Beechfield WD4: K Lan 2P 11
(off Station Av.)
Beechfield Cl. WD6: Bore 12Na 29
Beechfield Cotts. BR1: Broml 67Lc 137
Beechfield Ct. CR2: S Croy 77Sb 157
(off Bramley Hill)
Beechfield Ct. WD17: Wat 11W 26
Beechfield Gdns. RM7: Rush G 31Ed 76
Beechfield Rd. BR1: Broml 68Lc 137
Beechfield Rd. DA8: Erith 52Gd 118
Beechfield Rd. HP1: Hem H 3K 3
Beechfield Rd. N4 ....30Sb 51
Beechfield Rd. SE6 ....60Bc 114
Beechfield Wlk. EN9: Walt A 7Fc 21
Beech Gdns. EC2 ....7D 218 (43Sb 91)
(off Beech St.)
Beech Gdns. GU21: Wok 87A 168
Beech Gdns. RM10: Dag 38Ed 76
Beech Gdns. W5 ....47Na 87
Beech Gro. CR3: Cat'm 98Ub 197
Beech Gro. CR4: Mitc 71Mb 156
Beech Gro. GU22: Wok 5P 187
Beech Gro. GU24: Brkwd 2A 186
(not continuous)
Beech Gro. IG6: Ilf 23Uc 54
Beech Gro. KT15: Add 77K 149
Beech Gro. KT18: Tatt C 89Xa 174
Beech Gro. KT23: Bookh 99Ca 191
Beech Gro. KT3: N Mald 69Ta 131
Beech Gro. RM15: Avel 47Sd 98
Beech Hall KT16: Ott 80E 148
Beech Hall Cres. E4 ....24Fc 53
Beech Hall Rd. E4 ....24Ec 52
Beech Haven Ct. DA1: Cray 57Fd 118
(off London Rd.)
Beech Hill EN4: Had W 10Fb 17
Beech Hill GU22: Wok 5P 187
Beech Hill Av. EN4: Had W 11Eb 31
Beech Hill Ct. HP4: Berk 1A 2
Beech Hill Gdns. EN9: Walt A 9Kc 21
Beech Hill Rd. SL5: S'dale 2D 146
Beechhill Rd. SE9 ....57Qc 116
Beech Holt KT22: Lea 94La 192
Beech Ho. CM13: Hut 16Ee 41
Beech Ho. CR0: New Ad 79Dc 158
Beech Ho. KT15: Add 77M 149
(off Victory Pk. Rd.)
Beech Ho. SE16 ....47Yb 92
(off Ainsty Est.)
Beech Ho. Rd. CR0: C'don 76Tb 157
Beech Hurst Cl. BR7: Chst 67Sc 138
Beeching Ct. W3 ....48Sa 87
(off Bollo Bri. Rd.)
Beechin Wood La. TN15: Plat 94Ee 205
Beechlands Cl. DA3: Hartl 71Ce 165
Beech La. IG9: Buck H 19Kc 35
Beech Lawns N12 ....22Fb 49
Beechlee SM6: W'gton 82Lb 176
Beech Lodge TW18: Staines 64G 126
Beechmeads KT11: Cobh 85Z 171
Beechmont Av. GU25: Vir W 1P 147
Beechmont Cl. BR1: Broml 64Gc 137
Beechmore Gdns. SM3: Cheam 75Za 154
Beechmore Rd. SW11 ....53Hb 111
Beechmount Av. W7 ....43Fa 86
Beecholme N12 ....22Db 49
Beecholme SM7: Bans 86Ab 174
Beecholme Av. CR4: Mitc 67Kb 134
Beecholme Est. E5 ....34Xb 71
Beecholm M. EN8: Chesh 1Ac 20
(off The Colonnade)
Beechpark Way WD17: Wat 9U 12
Beech Pl. CM16: Epp 3Vc 23
Beech Rd. BR6: Chels 80Wc 161
Beech Rd. DA1: Dart 60Md 119
Beech Rd. KT13: Weyb 77T 150
Beech Rd. KT17: Eps 87Va 174
Beech Rd. N11 ....23Nb 50
Beech Rd. RH1: Mers 98Lb 196
Beech Rd. RH2: Reig 3J 207
Beech Rd. SL3: L'ly 47A 82
Beech Rd. SW16 ....68Nb 134
Beech Rd. TN13: S'oaks 97Kd 203
Beech Rd. TN16: Big H 91Kc 199
Beech Rd. TW14: Bedf 59U 106
Beech Rd. WD24: Wat 9W 12
Beechrow TW10: Ham 63Na 131
Beech St. EC2 ....7D 218 (43Sb 91)
Beech St. RM7: Rom 28Ed 56
Beech Ter. IG10: Lough 14Nc 36
Beechtree Av. TW20: Eng G 5M 125
Beech Tree Cl. HA7: Stan 22La 46
Beech Tree Cl. KT23: Fet 97Fa 192
Beech Tree Cl. N1 ....38Qb 70
Beech Tree Glade E4 ....18Hc 35
Beech Tree La. TW18: Lale 68K 127
Beechtree La. AL3: St A 4H 5
Beech Tree Pl. SM1: Sutt 78Db 155
Beechvale GU22: Wok 90B 168
(off Hill Vw. Rd.)
Beechvale Cl. N12 ....22Gb 49
Beech Wlk. DA1: Cray 56Jd 118
Beech Wlk. GU20: W'sham 9B 146
Beech Wlk. KT17: Ewe 83Wa 174
Beech Wlk. N17 ....27Vb 51
Beech Wlk. NW7 ....23Ua 48
Beech Way CR2: Sels 85Zb 178
Beech Way KT17: Eps 87Va 174
Beech Way NW10 ....38Ta 67
Beech Way TW2: Twick 62Ca 129
Beechway DA5: Bexl 58Zc 117
Beech Waye SL9: Ger X 31B 62
Beechwood CR3: Cat'm 96Wb 197
Beechwood Av. AL1: St A 1F 6
Beechwood Av. BR6: Chels 78Uc 160
Beechwood Av. CR5: Coul 87Kb 176
Beechwood Av. CR7: Thor H 70Rb 135
Beechwood Av. EN6: Pot B 5Db 17
Beechwood Av. HA2: Harr 34Da 65
Beechwood Av. HA4: Ruis 33V 64
Beechwood Av. KT13: Weyb 77U 150
Beechwood Av. KT20: Kgswd 93Cb 195
Beechwood Av. N3 ....27Bb 49
Beechwood Av. TW16: Sun 65W 128
Beechwood Av. TW18: Staines 65K 127
Beechwood Av. TW9: Kew 53Qa 109
Beechwood Av. UB3: Hayes 45T 84
Beechwood Av. UB6: G'frd 41Da 85
Beechwood Av. UB8: Hil 44Q 84
Beechwood Av. WD3: Chor 14D 24
Beechwood Cen., The BR2: Broml 74Pc 160
(off Lwr. Gravel Rd.)
Beechwood Circ. HA2: Harr 34Da 65
Beechwood Cl. GU21: Knap 9J 167
Beechwood Cl. KT13: Weyb 77U 150
Beechwood Cl. KT6: Surb 73La 152
Beechwood Cl. N2 ....27Hb 49
Beechwood Cl. NW7 ....22Ua 48
Beechwood Cotts. WD3: Chor 16E 24
Beechwood Ct. KT12: Walt T 76W 150
(off Station Av.)
Beechwood Ct. SM5: Cars 77Hb 155
Beechwood Ct. TW16: Sun 65W 128
Beechwood Ct. W4 ....51Ta 109
Beechwood Cres. DA7: Bex 55Zc 117
Beechwood Dr. BR2: Kes 77Mc 159
Beechwood Dr. IG8: Wfd G 22Hc 53
Beechwood Dr. KT11: Cobh 83Ca 171
Beechwood Gdns. CR3: Cat'm 94Wb 197
Beechwood Gdns. HA2: Harr 34Da 65
Beechwood Gdns. IG5: Ilf 29Pc 54
Beechwood Gdns. NW10 ....41Pa 87
Beechwood Gdns. RM13: Rain 43Kd 97
Beechwood Gdns. SL1: Slou 7J 81
Beechwood Gro. KT6: Surb 73La 152
Beechwood Gro. W3 ....45Ua 88
Beechwood Hall KT20: Kgswd 95Eb 195
Beechwood Hall N3 ....27Bb 49
Beechwood Ho. E2 ....40Wb 71
(off Teale St.)
Beechwood La. CR6: W'ham 91Zb 198
Beechwood Mnr. KT13: Weyb 77U 150
Beechwood M. N9 ....19Wb 33
Beechwood Pk. E18 ....27Jc 53
Beechwood Pk. HP3: Hem H 5H 3
Beechwood Pk. KT22: Lea 94La 192
Beechwood Pk. WD3: Chor 14H 25
Beechwood Pl. SE10 ....53Ec 114
Beechwood Ri. BR7: Chst 63Rc 138
Beechwood Ri. WD24: Wat 8X 13
Beechwood Rd. CR2: Sande 82Ub 177
Beechwood Rd. CR3: Cat'm 94Wb 197
Beechwood Rd. E8 ....37Vb 71
Beechwood Rd. GU21: Knap 9J 167
Beechwood Rd. GU25: Vir W 3L 147
Beechwood Rd. N8 ....28Mb 50
Beechwood Rd. SL2: Slou 3H 81
Beechwoods Ct. SE19 ....64Vb 135
Beechworth NW6 ....38Ab 68
Beechworth Cl. NW3 ....33Cb 69
Beechy Lees Rd. TN14: Otf 89Md 183
Beecot La. KT12: Walt T 75Y 151
Beecroft La. SE4 ....57Ac 114
Beecroft M. SE4 ....57Ac 114
Beecroft Rd. SE4 ....57Ac 114
Beefeater Distillery 50Qb 90
Beehive Cl. E8 ....38Vb 71
Beehive Cl. UB10: Uxb 38P 63
Beehive Cl. WD6: E'tree 16Ma 29
Beehive Ct. HA8: Edg 22Ra 47
Beehive Ct. IG1: Ilf 30Pc 54
Beehive Ct. RM3: Hrld W 24Pd 57
Beehive La. IG1: Ilf 30Pc 54
Beehive La. IG4: Ilf 29Pc 54
Beehive Pl. SW9 ....55Qb 112
Beehive Rd. TW18: Staines 64H 127
Beehive Way RH2: Reig 10K 207
Beeken Dene BR6: Farnb 77Sc 160
Beeleigh Rd. SM4: Mord 70Db 133
Beemans Row SW18 ....61Eb 133
Beesfield La. DA4: Farni 73Qd 163
Beesons Yd. WD3: Rick 18M 25
Beeston Cl. E8 ....36Wb 71
Beeston Cl. WD19: Wat 21Z 45
Beeston Ct. DA2: Dart 58Rd 119
(off Hardwick Cres.)
Beeston Ho. SE1 ....4F 231 (48Tb 91)
(off Burbage Cl.)
Beeston Pl. SW1 ....3A 228 (48Kb 90)
Beeston Rd. EN4: E Barn 16Fb 31
Beeston Way TW14: Felt 58Y 107
Beethoven Rd. WD6: E'tree 16La 28
Beethoven St. W10 ....41Ab 88
Beeton Cl. HA5: Hat E 24Ca 45
Beeton Way SE27 ....63Tb 135
Begbie Rd. SE3 ....53Lc 115
BEGGAR'S BUSH 10D 124
Beggars Bush La. WD18: Wat 15T 26
Beggar's Hill KT17: Ewe 80Va 154
BEGGAR'S HILL 79Va 154
Beggars Hollow EN2: Enf 9Tb 19
Beggars La. GU24: Chob 3G 166
Beggars La. TN16: Westrm 97Tc 200
Beggars Roost La. SM1: Sutt 79Cb 155
Begonia Cl. E6 ....43Pc 94
Begonia Pl. TW12: Hamp 65Ca 129
Begonia Wlk. W12 ....44Va 88
Beirach Moshe Sq. E5 ....33Xb 71
Beira St. SW12 ....59Kb 112
Bejun Ct. EN5: New Bar 14Eb 31
Beken Ct. WD25: Wat 7Y 13
Bekesbourne St. E14 ....44Ac 92
Bekesbourne Twr. BR5: Orp 74Zc 161
(off Wichling Cl.)
Belcroft Cl. BR1: Broml 66Hc 137
Beldam Bri. Rd. GU24: Chob 5E 166
Beldam Bri. Rd. GU24: W End 5E 166
Beldam Haw TN14: Hals 84Cd 182
Beldam Way TW3: Houn 55Ba 107
Beldanes Lodge NW10 ....38Wa 68
Beldham Gdns. KT8: W Mole 68Da 129
Belfairs Dr. RM6: Chad H 31Yc 75
Belfairs Grn. WD19: Wat 22Z 45
Belfast Av. SL1: Slou 4G 80
Belfast Rd. N16 ....33Vb 71
Belfast Rd. SE25 ....70Xb 135
Belfield Rd. KT19: Ewe 81Ta 173
Belfont Wlk. N7 ....35Nb 70
(not continuous)
Belford Gro. SE18 ....49Qc 94
Belford Ho. E8 ....39Vb 71
Belford Rd. WD6: Bore 10Pa 15
Belfort Rd. SE15 ....54Yb 114
Belfry, The RH1: Redh 5P 207
Belfry, The WD24: Wat 11Y 27
Belfry Av. UB9: Hare 24J 43
Belfry Cl. BR1: Broml 70Rc 138
Belfry Cl. SE16 ....50Xb 91
Belfry La. WD3: Rick 18L 25
Belgrade Rd. N16 ....35Ub 71
Belgrade Rd. TW12: Hamp 67Da 129
Belgrave Av. RM2: Rom 27Ld 57
Belgrave Av. WD18: Wat 15V 26
Belgrave Cl. BR5: St M Cry 70Yc 139
Belgrave Cl. KT12: Hers 77X 151
Belgrave Cl. N14 ....15Lb 32
Belgrave Cl. NW7 ....22Ta 47
Belgrave Cl. W3 ....47Ra 87
Belgrave Ct. E13 ....42Lc 93
Belgrave Ct. E14 ....45Bc 92
(off Westferry Cir.)
Belgrave Ct. E2 ....40Xb 71
(off Temple St.)
Belgrave Ct. SW8 ....52Lb 112
(off Ascalon St.)
Belgrave Ct. W4 ....50Sa 87
Belgrave Cres. TW16: Sun 67X 129
Belgrave Dr. WD4: K Lan 10C 4
Belgrave Gdns. HA7: Stan 22La 46
Belgrave Gdns. N14 ....15Mb 32
Belgrave Gdns. NW8 ....39Db 69
Belgrave Hgts. E11 ....32Jc 73
Belgrave Ho. SW9 ....52Qb 112
Belgrave Mnr. GU22: Wok 91A 188
Belgrave Mans. NW8 ....39Db 69
(off Belgrave Gdns.)
Belgrave M. UB8: Cowl 42M 83
Belgrave M. Nth. SW1 ....2H 227 (47Jb 90)
Belgrave M. Sth. SW1 ....3J 227 (48Jb 90)
Belgrave M. W. SW1 ....3H 227 (48Jb 90)
Belgrave Pde. SL1: Slou 7L 81
(off Stoke Poges La.)
Belgrave Pl. SL1: Slou 7L 81
Belgrave Pl. SW1 ....3J 227 (48Jb 90)
Belgrave Rd. CR4: Mitc 69Fb 133
Belgrave Rd. E10 ....32Ec 72
Belgrave Rd. E11 ....33Jc 73
Belgrave Rd. E13 ....42Lc 93
Belgrave Rd. E17 ....29Cc 52
Belgrave Rd. IG1: Ilf 31Qc 74
Belgrave Rd. SE25 ....70Vb 135
Belgrave Rd. SL1: Slou 5J 81
Belgrave Rd. SW1 ....5A 228 (49Kb 90)
Belgrave Rd. SW13 ....52Va 110
Belgrave Rd. TW16: Sun 67X 129
Belgrave Rd. TW4: Houn 55Ba 107
Belgrave Sq. SW1 ....3H 227 (48Jb 90)
Belgrave St. E1 ....43Zb 92
Belgrave Ter. IG8: Wfd G 20Jc 35
Belgrave Wlk. CR4: Mitc 69Fb 133
Belgrave Walk Stop (London Tramlink) 70Fb 133
Belgrave Yd. SW1 ....4K 227 (48Kb 90)
(off Lwr. Belgrave St.)
BELGRAVIA 4J 227 (48Jb 90)
Belgravia Cl. EN5: Barn 13Bb 31
Belgravia Ct. SW1 ....4K 227 (48Kb 90)
(off Ebury St.)
Belgravia Gdns. BR1: Broml 65Gc 137
Belgravia Ho. SW1 ....3H 227 (48Jb 90)
(off Halkin Pl.)
Belgravia Ho. SW4 ....58Mb 112
Belgravia M. KT1: King T 70Ma 131
Belgravia Workshops N19 ....33Nb 70
(off Marlborough Rd.)
Belgrove St. WC1 ....3G 217 (41Nb 90)
Belham Rd. WD4: K Lan 10P 3
Belham Wlk. SE5 ....53Tb 113
Belhaven Ct. WD6: Bore 11Pa 29
Belhouse Av. RM15: Avel 45Td 98
Belhus Pk. 44Td 98
Belhus Pk. Golf Course 44Ud 98
Belhus Pk. La. RM15: Avel 45Td 98
Belhus Vw. RM15: Avel 46Ud 98
Belhus Woods Country Pk. Romford Road 41Td 98
Belhus Woods Country Pk. Sandy Lane 44Rd 97
Belhus Woods Country Pk. Vis. Cen. 41Td 98
Belinda Rd. SW9 ....55Rb 113
Belitha Vs. N1 ....38Pb 70
BELL, THE 27Cc 52

Bella Best Ho. SW1 7A 228 (50Kb 90)
(off Westmoreland Ter.)
Bellamy Cl. E14 47Cc 92
Bellamy Cl. HA8: Edg 19Sa 29
Bellamy Cl. UB10: Ick 34Q 64
Bellamy Cl. W14 50Bb 89
Bellamy Cl. WD17: Wat 11W 26
Bellamy Ct. HA7: Stan 25Ka 46
Bellamy Dr. HA7: Stan 25Ka 46
Bellamy Ho. SW17 63Fb 133
Bellamy Ho. TW5: Hest 51Ca 107
Bellamy Rd. E4 23Dc 52
Bellamy Rd. EN2: Enf 12Tb 33
Bellamy Rd. EN8: Chesh 1Ac 20
Bellamy's Ct. SE16 46Zb 92
(off Abbotshade Rd.)
Bellamy St. SW12 59Kb 112
Bel La. TW13: Hanw 62Aa 129
Bellarmine Cl. SE28 46Vc 95
Bellasis Av. SW2 61Nb 134
Bell Av. RM3: Rom 25Kd 57
Bell Av. UB7: W Dray 49P 83
BELL BAR 6J 9
Bell Bri. Rd. KT16: Chert 74H 149
Bell Brook Ri. N11 21Kb 50
Bellchambers Cl. AL2: Lon C 8G 6
Bell Cl. DA9: Ghithe 57Vd 120
Bell Cl. HA4: Ruis 34V 64
Bell Cl. HA5: Pinn 26Y 45
Bell Cl. SL2: Slou 3M 81
Bell Cl. WD5: Bedm 9F 4
Bellclose Rd. UB7: W Dray 47N 83
BELL COMMON 4Tc 22
Bell Comn. CM16: Epp 4Uc 22
Bell Cnr. KT16: Chert 73H 149
Bell Cnr. RM14: Upm 33Sd 78
Bell Ct. NW4 28Ya 48
Bell Cres. CR5: Coul 93Kb 196
Bell Dr. SW18 59Ab 110
Bellefield SL1: Slou 7D 80
Bellefield Rd. BR5: St M Cry 71Xc 161
Bellefields Rd. SW9 55Pb 112
Bellegrove Cl. DA16: Well 54Vc 117
Bellegrove Pde. DA16: Well 55Vc 117
Bellegrove Rd. DA16: Well 54Tc 116
Bellenden Rd. SE15 53Vb 113
Bellenden Rd. Retail Pk. 53Wb 113
Bellestaines Pleasaunce E4 19Cc 34
Belleville Ho. SE10 52Dc 114
(off Norman Rd.)
Belleville Rd. SW11 57Gb 111
Belle Vue UB6: G'frd 39Fa 66
Belle Vue Cl. TW18: Staines 67J 127
Bellevue Ct. TW3: Houn 56Ca 107
Belle Vue Est. NW4 28Za 48
Belle Vue La. WD23: B Hea 18Fa 28
Bellevue M. N11 22Jb 50
Bellevue Pde. SW17 60Hb 111
Belle Vue Pk. CR7: Thor H 69Sb 135
Bellevue Pl. E1 42Yb 92
Belle Vue Rd. BR6: Downe 82Qc 180
Belle Vue Rd. E17 26Fc 53
Belle Vue Rd. NW4 28Ya 48
Belle Vue Rd. RM5: Col R 23Ed 56
Bellevue Rd. DA6: Bex 57Bd 117
Bellevue Rd. KT1: King T 69Na 131
(not continuous)
Bellevue Rd. N11 21Jb 50
Bellevue Rd. RM11: Horn 32Pd 77
Bellevue Rd. SW13 54Wa 110
Bellevue Rd. SW17 60Gb 111
Bellevue Rd. W13 42Ka 86
Bellevue Ter. UB9: Hare 24J 43
Bellew St. SW17 62Eb 133
Bell Farm Av. RM10: Dag 34Ed 76
Bell Farm Cotts. CM16: Epp 4Uc 22
Bell Farm Way KT12: Hers 77Y 151
Bellfield CR0: Sels 81Ac 178
Bellfield Av. HA3: Hrw W 23Fa 46
Bellfield Cl. SE3 52Kc 115
Bellflower Cl. E6 43Nc 94
Bellflower Path RM3: Rom 24Ld 57
Bell Foundry Cl. CR0: C'don 72Tb 157
Bell Gdns. BR5: St M Cry 71Yc 161
Bell Gdns. E10 32Cc 72
(off Church Rd.)
Bellgate M. NW5 35Kb 70
BELL GREEN 63Bc 136
Bell Grn. HP3: Bov 9D 2
Bell Grn. SE26 63Bc 136
Bell Grn. La. SE26 64Bc 136
Bell Grn. Retail Pk. 62Bc 136
Bell Grn. Trade City SE6 62Bc 136
Bellhaven E15 37Fc 73
Bell Hill CR0: C'don 75Sb 157
Bell Ho. HA9: Wemb 34Na 67
Bell Ho. RM17: Grays 51Be 121
Bell Ho. SE10 51Ec 114
(off Haddo St.)
Bellhouse Cotts. UB3: Hayes 45U 84
Bellhouse La. CM14: Pil H 15Ud 40
Bell Ho. Rd. RM7: Rush G 32Ed 76
Bellina M. NW5 35Kb 70
Bell Ind. Est. W4 49Sa 87
Belling Cres. EN3: Pond E 15Yb 34
Bellingdon WD19: Wat 20Aa 27
Bellingham N17 24Xb 51
(off Park La.)
BELLINGHAM 62Dc 136
Bellingham Ct. IG11: Bark 41Xc 95
Bellingham Dr. RH2: Reig 6H 207
Bellingham Grn. SE6 62Cc 136
Bellingham Grn. Pk. 62Cc 136
(off Bellingham Green)
Bellingham Leisure & Lifestyle Cen. 62Dc 136
Bellingham Rd. SE6 62Dc 136
Bellingham Station (Rail) 62Dc 136
Bellingham Trad. Est. SE6 62Dc 136
Bell Inn Yd. EC3 3G 225 (44Tb 91)
Bell La. AL2: Lon C 1Na 15
Bell La. AL9: Brk P 6J 9
Bell La. E1 1K 225 (43Vb 91)
Bell La. E16 46Hc 93
Bell La. EN3: Enf H 10Zb 20
Bell La. EN3: Enf W 10Zb 20
Bell La. HA9: Wemb 33Ma 67
Bell La. KT22: Fet 95Fa 192
Bell La. NW4 28Za 48
Bell La. SL4: Eton W 9D 80
Bell La. TW1: Twick 60Ja 108
Bell La. TW14: Bedf 59T 106
Bell La. WD5: Bedm 9F 4
Bell La. Cl. KT22: Fet 95Fa 192
Bellmaker Ct. E3 43Cc 92
Bellmaker M. RM14: Upm 34Rd 77
Bellman Av. DA12: Grav'nd 10G 122
Bellman Cl. SL3: Slou 7M 81
Bellmarsh Rd. KT15: Add 77K 149
Bell Mdw. RH9: G'stone 4P 209
Bell Mdw. SE19 64Ub 135
Bell Moor NW3 34Eb 69
Bellmount Wood Av. WD17: Wat 11U 26
Bello Cl. SE24 59Rb 113
Bellot Gdns. SE10 50Gc 93
(off Bellot St.)
Bellot St. SE10 50Gc 93
Bellows La. TN15: Bor G 92Ae 205
Bell Pde. BR4: W W'ck 75Ec 158
Bell Pde. SL4: Wind 4D 102
Bell-Reeves Cl. SS17: Stan H 1L 101
Bellring Cl. DA17: Belv 51Cd 118
Bell Rd. EN1: Enf 11Tb 33
Bell Rd. KT8: E Mos 71Fa 152
Bell Rd. TW3: Houn 55Da 107
BELL RDBT., THE 9L 7
Bells All. SW6 54Cb 111
Bellsfield Ct. SL4: Eton W 9D 80
(off Bell La.)
Bells Hill EN5: Barn 15Za 30
Bells Hill SL2: Stoke P 9L 61
Bells Hill Grn. SL2: Stoke P 8L 61
Bellsize Ct. NW3 36Fb 69
Bells La. SL3: Hort 55D 104
Bell St. NW1 7D 214 (43Gb 89)
Bell St. RH2: Reig 6J 207
Bell St. SE18 53Nc 116
Bellswood La. SL0: Iver 43D 82
Bell Ter. DA1: Dart 55Pd 119
Belltrees Gro. SW16 64Pb 134
Bell Vw. AL4: St A 2H 7
Bell Vw. SL4: Wind 5D 102
Bell Vw. Cl. SL4: Wind 4D 102
Bellview Ct. TW3: Houn 56Da 107
Bell Vw. Mnr. HA4: Ruis 31T 64
Bell Vue Pl. SL1: Slou 8K 81
Bell Water Ga. SE18 48Qc 94
Bellway Ho. RH1: Mers 100Lb 196
Bell Weir Cl. TW19: Staines 61D 126
Bellwether La. SW18 57Db 111
(off Ryland Blvd.)
Bell Wharf La. EC4 5E 224 (45Sb 91)
Bellwood Rd. SE15 56Zb 114
Bell Yd. WC2 3K 223 (44Qb 90)
Bell Yd. M. SE1 2J 231 (47Ub 91)
Belmarsh Rd. SE28 47Uc 94
BELMONT 26Ja 46
Belmont KT13: Weyb 79S 150
Belmont SL2: Slou 3E 80
BELMONT 82Cb 175
Belmont Av. DA16: Well 55Uc 116
Belmont Av. EN4: Cockf 15Hb 31
Belmont Av. HA0: Wemb 39Pa 67
Belmont Av. KT3: N Mald 71Wa 154
Belmont Av. N13 22Pb 50
Belmont Av. N17 27Sb 51
Belmont Av. N9 18Wb 33
Belmont Av. RM14: Upm 33Pd 77
Belmont Av. UB2: S'hall 48Aa 85
Belmont Circ. HA3: Kenton 25Ka 46
Belmont Cl. E4 22Fc 53
Belmont Cl. EN4: Cockf 14Hb 31
Belmont Cl. IG8: Wfd G 21Kc 53
Belmont Cl. N20 18Db 31
Belmont Cl. SW4 55Lb 112
Belmont Cl. UB8: Uxb 37M 63
Belmont Cotts. SL3: Coln 52E 104
(off High St.)
Belmont Ct. AL1: St A 3B 6
Belmont Ct. KT23: Bookh 97Ea 192
Belmont Ct. N5 35Sb 71
Belmont Ct. NW11 29Bb 49
Belmont Gro. SE13 55Fc 115
Belmont Gro. W4 49Ta 87
Belmont Hall Ct. SE13 55Fc 115
Belmont Hill AL1: St A 3B 6
Belmont Hill SE13 55Fc 115
Belmont La. BR7: Chst 64Sc 138
Belmont La. HA7: Stan 25La 46
Belmont Lodge HA3: Hrw W 24Fa 46
Belmont M. SW19 61Za 132
Belmont Pde. BR7: Chst 64Sc 138
Belmont Pde. NW11 29Bb 49
Belmont Pk. SE13 56Fc 115
Belmont Pk. Cl. SE13 56Gc 115
Belmont Pk. Rd. E10 30Dc 52
Belmont Ri. SM2: Sutt 80Bb 155
Belmont Rd. BR3: Beck 68Ac 136
Belmont Rd. BR7: Chst 64Rc 138
Belmont Rd. DA8: Erith 52Cd 118
Belmont Rd. HA3: W'stone 27Ha 46
Belmont Rd. HP3: Hem H 5N 3
Belmont Rd. IG1: Ilf 34Sc 74
Belmont Rd. KT22: Lea 94Ja 192
Belmont Rd. N15 28Sb 51
Belmont Rd. N17 28Sb 51
Belmont Rd. RH2: Reig 7L 207
Belmont Rd. RM12: Horn 34Md 77
Belmont Rd. RM17: Grays 51Be 121
Belmont Rd. SE25 71Xb 157
Belmont Rd. SM2: Sutt 82Cb 175
Belmont Rd. SM6: W'gton 78Kb 156
Belmont Rd. SW4 55Lb 112
Belmont Rd. TW2: Twick 61Fa 130
Belmont Rd. UB8: Uxb 38M 63
Belmont Rd. W4 49Ta 87
Belmont Rd. WD23: Bush 15Aa 27
Belmont Station (Rail) 82Db 175
Belmont St. NW1 38Jb 70
Belmont Ter. W4 49Ta 87
Belmor WD6: E'tree 16Qa 29
Belmore Av. GU22: Pyr 88F 168
Belmore Av. UB4: Hayes 44W 84
Belmore Ho. N7 36Mb 70
Belmore La. N7 36Mb 70
Belmore St. SW8 53Mb 112
Beloe Cl. SW15 56Wa 110
Belsham St. E9 37Yb 72
BELSIZE 4H 11
Belsize Av. N13 23Pb 50
Belsize Av. NW3 37Fb 69
Belsize Av. W13 48Ka 86
Belsize Cl. HP3: Hem H 3A 4
Belsize Cotts. WD3: Chfd 5G 10
Belsize Ct. SM1: Sutt 77Db 155
Belsize Ct. Garages NW3 36Fb 69
(off Belsize La.)
Belsize Cres. NW3 36Fb 69
Belsize Gdns. SM1: Sutt 77Db 155
Belsize Grange KT16: Chert 73L 149
Belsize Gro. NW3 37Gb 69
Belsize La. NW3 37Fb 69
Belsize M. NW3 37Fb 69
Belsize Pk. NW3 37Fb 69
Belsize Pk. Gdns. NW3 37Fb 69
Belsize Pk. M. NW3 37Fb 69
Belsize Park Station (Underground) 36Gb 69
Belsize Pl. NW3 36Fb 69
Belsize Rd. HA3: Hrw W 24Fa 46
Belsize Rd. HP3: Hem H 3A 4
Belsize Rd. NW6 39Db 69
Belsize Sq. NW3 37Fb 69
Belsize Ter. NW3 37Fb 69
Belson Rd. SE18 49Pc 94
Belswains Grn. HP3: Hem H 5N 3
Belswains La. HP3: Hem H 5N 3
Beltana Dr. DA12: Grav'nd 3G 144
Beltane Dr. SW19 62Za 132
Belthorn Cres. SW12 59Lb 112
Beltinge Rd. RM3: Hrld W 26Pd 57
Belton Rd. DA14: Sidc 63Wc 139
Belton Rd. E11 35Gc 73
Belton Rd. E7 38Kc 73
Belton Rd. N17 27Ub 51
Belton Rd. NW2 37Wa 68
Belton Way E3 43Cc 92
Beltran Rd. SW6 54Db 111
Beltwood Rd. DA17: Belv 49Ed 96
BELVEDERE 49Cd 96
Belvedere, The SE1 7J 223 (46Pb 90)
Belvedere, The SW10 53Eb 111
(off Chelsea Harbour)
Belvedere Av. IG5: Ilf 26Rc 54
Belvedere Av. SW19 64Ab 132
Belvedere Bldgs. SE1 2C 230 (47Rb 91)
Belvedere Bus. Pk. DA17: Belv 47Dd 96
Belvedere Cl. DA12: Grav'nd 10E 122
Belvedere Cl. KT10: Esh 78Da 151
Belvedere Cl. KT13: Weyb 78Q 150
Belvedere Cl. TW11: Tedd 64Ga 130
Belvedere Ct. DA17: Belv 48Bd 95
Belvedere Ct. N1 39Ub 71
(off De Beauvoir Cres.)
Belvedere Ct. N2 29Fb 49
Belvedere Ct. NW2 37Za 68
(off Willesden La.)
Belvedere Ct. RH1: Redh 2A 208
Belvedere Ct. SW15 56Ya 110
Belvedere Ct. WD19: Wat 16Aa 27
Belvedere Dr. SW19 64Ab 132
Belvedere Gdns. AL2: Chis G 9N 5
Belvedere Gdns. KT8: W Mole 71Ba 151
Belvedere Gdns. SE1 1J 229 (47Pb 90)
Belvedere Grange SL5: S'dale 4E 146
Belvedere Gro. SW19 64Ab 132
Belvedere Ho. KT13: Weyb 78R 150
Belvedere Ho. TW13: Felt 60W 106
(off Lemon Gro.)
Belvedere Ind. Est. DA17: Belv 46Ed 96
Belvedere Link Bus. Pk. DA8: Erith 48Ed 96
Belvedere Mans. SL1: Slou 7H 81
Belvedere M. SE15 55Yb 114
Belvedere M. SE3 52Kc 115
Belvedere Pl. SE1 2C 230 (47Rb 91)
Belvedere Pl. SW2 56Pb 112
Belvedere Rd. CM14: B'wood 20Vd 40
Belvedere Rd. DA7: Bex 55Bd 117
Belvedere Rd. E10 32Ac 72
Belvedere Rd. SE1 2H 229 (47Pb 90)
Belvedere Rd. SE19 66Vb 135
Belvedere Rd. TN16: Big H 90Pc 180
Belvedere Rd. W7 48Ha 86
Belvedere Row Apts. W12 45Ya 88
(off Fountain Park Way)
Belvederes, The RH2: Reig 8K 207
Belvedere Sq. SW19 64Ab 132
Belvedere Station (Rail) 48Dd 96
Belvedere Strand NW9 26Va 48
Belvedere Way HA3: Kenton 30Na 47
Belvoir Cl. SE9 62Nc 138
Belvoir Ho. SW1 6C 228 (49Lb 90)
Belvoir Rd. SE22 59Wb 113
Belvue Bus. Cen. UB5: N'olt 38Da 65
Belvue Cl. UB5: N'olt 38Ca 65
Belvue Rd. UB5: N'olt 38Ca 65
Belz Dr. N15 28Ub 51
Belz Ter. E5 31Wb 71
Bembridge Cl. NW6 38Ab 68
Bembridge Ct. SL1: Slou 7K 81
Bembridge Gdns. HA4: Ruis 33T 64
Bembridge Ho. KT2: King T 68Qa 131
(off Coombe Rd.)
Bembridge Ho. SE8 49Bc 92
(off Longshore)
Bembridge Ho. SW18 58Db 111
(off Iron Mill Rd.)
Bembridge Pl. WD25: Wat 5W 12
Bemersyde Point E13 41Kc 93
(off Dongola Rd. W.)
Bemerton Est. N1 38Nb 70
Bemerton St. N1 39Pb 70
Bemish Rd. SW15 55Za 110
Bempton Dr. HA4: Ruis 33X 65
Bemsted Rd. E17 27Bc 52
Benares Rd. SE18 49Vc 95
Benbow Cl. AL1: St A 4F 6
Benbow Ct. W6 48Ya 88
(off Benbow Rd.)
Benbow Ho. SE8 51Cc 114
(off Benbow St.)
Benbow M. E3 41Bc 92
(off Tredegar Sq.)
Benbow Moorings UB8: Cowl 43L 83
Benbow Rd. W6 48Xa 88
Benbow St. SE8 51Cc 114
Benbow Waye UB8: Cowl 43L 83
Benbury Cl. BR1: Broml 64Ec 136
Bence, The TW20: Thorpe 69D 126
Bence Ho. SE8 49Ac 92
(off Rainsborough Av.)
Bench, The TW10: Ham 62La 130
Bench Fld. CR2: S Croy 79Vb 157
Benchleys Rd. HP1: Hem H 4H 3
Bencombe Rd. CR8: Purl 86Qb 176
Bencroft Rd. HP2: Hem H 2N 3
Bencroft Rd. SW16 66Lb 134
Bencurtis Pk. BR4: W W'ck 76Fc 159
Bendall Ho. NW1 7E 214 (43Gb 89)
(off Penfold St.)
Bendall M. NW1 7E 214 (43Gb 89)
(off Bell St.)
Bendemeer Rd. SW15 55Za 110
Benden Ho. SE13 57Ec 114
(off Monument Gdns.)
Bendish Point SE28 47Sc 94
Bendish Rd. E6 38Nc 74
Bendmore Av. SE2 50Wc 95
Bendon Valley SW18 59Db 111
Bendysh Rd. WD23: Bush 13Aa 27
Benedict Cl. BR6: Orp 76Uc 160
Benedict Cl. DA17: Belv 48Ad 95
Benedict Ct. RM6: Chad H 30Bd 55
Benedict Dr. TW14: Bedf 59T 106
Benedict Ho. AL3: St A 1B 6
(off Adelaide St.)
Benedictine Pl. AL1: St A 2B 6
(off London Rd.)
Benedict Rd. CR4: Mitc 69Fb 133
Benedict Rd. SW9 55Pb 112
Benedicts Wharf IG11: Bark 39Rc 74
Benedict Way N2 27Eb 49
Benedict Wharf CR4: Mitc 69Gb 133
Benenden Grn. BR2: Broml 71Jc 159
Benenden Ho. SE17 7J 231 (50Ub 91)
(off Mina Rd.)
Benen-Stock Rd. TW19: Stanw M 57J 105
Benets Rd. RM11: Horn 32Qd 77
Benett Gdns. SW16 68Nb 134
Ben Ezra Ct. SE17 6E 230 (49Sb 91)
(off Asolando Dr.)
Benfleet Cl. KT11: Cobh 84Aa 171
Benfleet Cl. SM1: Sutt 76Eb 155
Benfleet Ct. E8 39Vb 71
Benfleet Way N11 19Jb 32
Bengal Ct. EC3 3G 225 (44Tb 91)
(off Birchin La.)
Bengal Rd. IG1: Ilf 35Rc 74
Bengarth Dr. HA3: Hrw W 26Fa 46
Bengarth Rd. UB5: N'olt 39Aa 65
Bengeo Gdns. RM6: Chad H 30Yc 55
Bengeworth Rd. HA1: Harr 33Ja 66
Bengeworth Rd. SE5 55Sb 113
Ben Hale Cl. HA7: Stan 22Ka 46
Benham Cl. CR5: Coul 90Rb 177
Benham Cl. KT9: Chess 79La 152
Benham Cl. SW11 55Fb 111
Benham Gdns. TW4: Houn 57Ba 107
Benham Ho. SW10 52Db 111
(off Coleridge Gdns.)
Benham Rd. W7 43Ga 86
Benham's Pl. NW3 35Eb 69
Benhill Av. SM1: Sutt 77Db 155
(not continuous)
Benhill Rd. SE5 52Tb 113
Benhill Rd. SM1: Sutt 76Eb 155
Benhill Wood Rd. SM1: Sutt 76Eb 155
BENHILTON 75Db 155
Benhilton Gdns. SM1: Sutt 76Db 155
Benhurst Av. RM12: Horn 35Kd 77
Benhurst Cl. CR2: Sels 82Zb 178
Benhurst Ct. SW16 64Qb 134
Benhurst Gdns. CR2: Sels 82Yb 178
Benhurst La. SW16 64Qb 134
Benina Cl. IG2: Ilf 29Vc 55
Beningfield Dr. AL2: Lon C 9E 6
Benin Ho. WC1 1H 223 (43Pb 90)
(off Procter St.)
Benin St. SE13 59Fc 115
Benison Ct. SL1: Slou 8K 81
Benjafield Cl. N18 21Xb 51
Benjamin Cl. E8 39Wb 71
Benjamin Cl. RM11: Horn 30Jd 56
Benjamin Ct. DA17: Belv 51Bd 117
Benjamin Ct. TW15: Ashf 66S 128
Benjamin Franklin House 6F 223 (46Nb 90)
(off Craven St.)
Benjamin La. SL3: Wex 2M 81
Benjamin M. SW12 59Lb 112
Benjamin St. EC1 7B 218 (43Rb 91)
Benjamin Truman Cl. E1 42Wb 91
Ben Jonson Ct. N1 1J 219 (40Ub 71)
Ben Jonson Ho. EC2 7D 218 (43Sb 91)
(off Beech St.)
Ben Jonson Pl. EC2 7D 218 (43Sb 91)
(off Beech St.)
Ben Jonson Rd. E1 43Zb 92
Benkart M. SW15 58Wa 110
Benledi Rd. E14 44Fc 93
Benlow Works UB3: Hayes 47V 84
(off Silverdale Rd.)
Ben More SL9: Ger X 29B 42
Benn Cl. RH8: Oxt 6L 211
Benneck Ho. WD18: Wat 16U 26
Bennelong Cl. W12 45Xa 88
Benner La. GU24: W End 4D 166
Bennerley Rd. SW11 57Gb 111
Bennet Cl. KT1: Hamp W 67La 130
Bennet M. N19 34Mb 70
(off Wedmore St.)
Bennets Ctyd. SW19 67Eb 133
Bennets Fld. Rd. UB11: Stock P 46R 84
Bennet's Hill EC4 4D 224 (45Sb 91)
Bennets Lodge EN2: Enf 13Rb 33
Bennet St. SW1 6B 222 (46Lb 90)
Bennett Cl. CR5: Coul 89Kb 176
Bennett Cl. DA16: Well 54Wc 117
Bennett Cl. HA6: Nwood 24V 44
Bennett Cl. KT11: Cobh 85W 170
Bennett Cl. TW4: Houn 57Aa 107
Bennett Ct. N7 34Pb 70
Bennett Gro. SE13 53Dc 114
Bennett Ho. DA11: Nflt 2B 144
Bennett Ho. SW1 5E 228 (49Mb 90)
(off Page St.)
Bennett Pk. SE3 55Hc 115
Bennett Pl. DA1: Dart 55Pd 119
Bennett Rd. E13 42Lc 93
Bennett Rd. N16 35Ub 71
Bennett Rd. RM6: Chad H 30Ad 55
Bennett Rd. SW9 54Qb 112
Bennetts Av. CR0: C'don 75Ac 158
Bennetts Av. TN15: W King 84Yd 184
Bennetts Av. UB6: G'frd 39Ga 66
Bennett's Castle La. RM8: Dag 33Yc 75
Bennetts Cl. AL4: Col H 5P 7
Bennetts Cl. CR4: Mitc 67Kb 134
Bennetts Cl. N17 23Vb 51
Bennetts Cl. SL1: Slou 6E 80
Bennetts Copse BR7: Chst 65Nc 138
BENNETTS END 4A 4
Bennetts End Cl. HP3: Hem H 4P 3
Bennetts End Rd. HP3: Hem H 3P 3
Bennetts Farm Pl. KT23: Bookh 97Ba 191
Bennetts Fld. WD23: Bush 14Aa 27
Bennetts Ga. HP3: Hem H 5A 4
Bennett St. W4 51Ua 110
Bennetts Way CR0: C'don 75Ac 158
Bennett's Yd. SW1 4E 228 (48Mb 90)
Bennett's Yd. UB8: Uxb 38L 63
Bennett Way DA2: Daren 63Td 142
Bennett Way GU4: W Cla 100J 189
Benning Cl. SL4: Wind 5B 102
Benning Dr. RM8: Dag 32Ad 75
Benningholme Rd. HA8: Edg 23Ua 48
Bennington Cl. CR7: Thor H 70Sb 135
Bennington Dr. WD6: Bore 11Pa 29
Bennington Rd. IG8: Wfd G 24Gc 53
Bennington Rd. N17 25Ub 51
Bennions Cl. RM12: Horn 37Md 77
Bennison Dr. RM3: Hrld W 26Md 57
Benn's All. TW12: Hamp 68Da 129
Benn St. E9 37Ac 72
Benns Wlk. TW9: Rich 56Na 109
(off Michelsdale Dr.)
Benrek Cl. IG6: Ilf 25Sc 54
Bensbury Cl. SW15 59Xa 110
Bensham Cl. CR7: Thor H 70Sb 135
Bensham Gro. CR7: Thor H 68Sb 135
Bensham La. CR0: C'don 73Rb 157
Bensham La. CR7: Thor H 71Rb 157
Bensham Mnr. Rd. CR7: Thor H 70Sb 135
Bensham Mnr. Rd. Pas. CR7: Thor H 70Sb 135
Benskin Rd. WD18: Wat 15W 26
Benskins La. RM4: Noak H 18Md 39
Bensley Cl. N11 22Hb 49
Ben Smith Way SE16 48Wb 91
Benson Av. E6 40Lc 73
Benson Cl. EN5: Barn 15Bb 31
Benson Cl. SL2: Slou 6L 81
Benson Cl. TW3: Houn 56Ca 107
Benson Cl. UB8: Hil 43N 83
Benson Ct. SW8 53Nb 112
(off Hartington Rd.)
Benson Ho. E2 5K 219 (42Vb 91)
(off Ligonier St.)
Benson Ho. SE1 7A 224 (46Qb 90)
(off Hatfields)
Benson Ho. W14 49Bb 89
(off Radnor Ter.)
Benson M. BR7: Chst 64Qc 138
Benson Quay E1 45Yb 92
Benson Rd. CR0: Wadd 76Qb 156
Benson Rd. RM17: Grays 51Ee 121
Benson Rd. SE23 60Yb 114
Bentall Cen., The 68Ma 131
Bentfield Gdns. SE9 62Mc 137
Bentfield Ho. NW9 27Va 48
(off Heritage Av.)
Benthall Gdns. CR8: Kenley 89Sb 177
Benthal Rd. N16 33Wb 71
Bentham Av. GU21: Wok 87E 168
Bentham Ct. N1 38Sb 71
(off Ecclesbourne Rd.)
Bentham Ho. SE1 3F 231 (48Tb 91)
(off Falmouth Rd.)
Bentham Ho. SE18 48Rc 94
Bentham Rd. E9 37Zb 72
Bentham Rd. SE28 45Xc 95
Bentham Wlk. NW10 36Sa 67
Ben Tillet Cl. E16 46Pc 94
Ben Tillet Cl. IG11: Bark 38Wc 75
Ben Tillet Ho. N15 27Rb 51
Bentinck Cl. NW8 2E 214 (40Gb 69)
Bentinck Ho. SW1 4E 228 (48Mb 90)
(off Monck St.)
Bentinck Ho. W12 45Xa 88
(off White City Est.)
Bentinck Ho. WD17: Wat 15Y 27
(off Pumphouse Cres.)
Bentinck Mans. W1 2J 221 (44Jb 90)
(off Bentinck St.)
Bentinck M. W1 2J 221 (44Jb 90)
Bentinck Rd. UB7: Yiew 46M 83
Bentinck St. W1 2J 221 (44Jb 90)
Bentine La. WD18: Wat 13X 27
BENTLEY 12Td 40
Bentley Cl. DA3: Lfield 69Ee 143
Bentley Cl. SW19 62Cb 133
Bentley Cl. W7 46Ha 86
Bentley Ct. SE13 56Ec 114
(off Whitburn Rd.)
Bentley Crematorium 14Vd 40
Bentley Dr. IG2: Ilf 30Sc 54
Bentley Dr. KT13: Weyb 81Q 170
Bentley Dr. NW2 34Bb 69
Bentley Golf Course 11Ud 40
BENTLEY HEATH 7Bb 17
Bentley Heath La. EN5: Barn 6Ab 16
Bentley Ho. E3 42Cc 92
(off Wellington Way)
Bentley Ho. SE5 53Ub 113
(off Peckham Rd.)
Bentley Lodge WD23: B Hea 19Ga 28
Bentley M. EN1: Enf 16Tb 33
Bentley Pk. SL1: Burn 10B 60
Bentley Pl. EN5: Barn 7Cb 17
Bentley Pl. KT13: Weyb 77R 150
(off Baker St.)
Bentley Priory Local Nature Reserve 21Ga 46
Bentley Priory Mus. 20Ga 28
Bentley Rd. N1 37Ub 71
Bentley Rd. SL1: Slou 6E 80
Bentley's Mdw. TN15: Seal 92Pd 203
Bentley St. DA12: Grav'nd 8E 122
Bentley St. Ind. Est. DA12: Grav'nd 8E 122
Bentley Way EN5: New Bar 14Eb 31
Bentley Way HA7: Stan 22Ja 46
Bentley Way IG8: Buck H 19Jc 35
Bentley Way IG8: Wfd G 19Jc 35
Benton Rd. IG1: Ilf 32Tc 74
Benton Rd. WD19: Wat 22Z 45
Bentons La. SE27 63Sb 135
Benton's Ri. SE27 64Tb 135
Bentry Cl. RM8: Dag 33Ad 75
Bentry Rd. RM8: Dag 33Ad 75
Bentworth Ct. E2 42Wb 91
(off Granby St.)
Bentworth Rd. W12 44Xa 88
Ben Uri Gallery 39Db 69
Benville Ho. SW8 52Pb 112
(off Dorset Rd.)
Benwell Cen. TW16: Sun 67W 128
Benwell Ct. TW16: Sun 67W 128
Benwell Rd. GU24: Brkwd 1D 186
Benwell Rd. N7 35Qb 70
Benwick Cl. SE16 49Xb 91
Benwick M. SE20 67Yb 136
Benwood Ct. SM1: Sutt 76Eb 155
Benworth St. E3 41Bc 92
Benyon Ct. N1 39Ub 71
(off De Beauvoir Est.)
Benyon Ho. EC1 3A 218 (41Qb 90)
(off Myddelton Pas.)
Benyon Path RM15: S Ock 41Yd 98
Benyon Rd. N1 39Tb 71
Benyon Wharf E8 39Ub 71
(off Kingsland Rd.)
Beomonds KT16: Chert 73J 149
Beomonds Row KT16: Chert 73J 149
Berberis Ct. IG1: Ilf 37Rc 74
Berberis Ho. E3 43Cc 92
(off Gale St.)
Berberis Ho. TW13: Felt 61W 128
Berberis Wlk. UB7: W Dray 49N 83
Berber Pde. SE18 53Nc 116
Berber Pl. E14 45Cc 92
Berber Rd. SW11 57Hb 111
Berberry Cl. HA8: Edg 21Sa 47
Berceau Wlk. WD17: Wat 11U 26
Bercta Rd. SE9 61Sc 138
Berebinder Ho. E3 40Bc 72
(off Tredegar Rd.)
Bere Cl. DA9: Ghithe 57Yd 120
Beredens La. CM13: Gt War 27Vd 58
Beregaria Ct. SE11 51Qb 112
(off Kennington Pk. Rd.)

Berengers Ct. RM6: Chad H 31Bd 75
(off Whalebone La. Sth.)
Berengers Pl. RM9: Dag 37Xc 75
Berenger Twr. SW10 52Fb 111
(off Worlds End Est.)
Berenger Wlk. SW10 52Fb 111
(off Worlds End Est.)
Berens Ct. DA14: Sidc 63Vc 139
Berens Rd. BR5: St M Cry 71Zc 161
Berens Rd. NW10 41Za 88
Berens Way BR7: Chst 69Vc 139
Beresford Av. HA0: Wemb 39Pa 67
Beresford Av. KT5: Surb 74Ra 153
Beresford Av. N20 19Hb 31
Beresford Av. SL2: Slou 5N 81
Beresford Av. TW1: Twick 58La 108
Beresford Av. W7 43Fa 86
Beresford Ct. E9 36Ac 72
(off Mabley St.)
Beresford Dr. BR1: Broml 69Nc 138
Beresford Dr. IG8: Wfd G 21Lc 53
Beresford Gdns. EN1: Enf 14Ub 33
Beresford Gdns. RM6: Chad H 29Ad 55
Beresford Gdns. TW4: Houn 57Ba 107
Beresford Rd. AL1: St A 3F 6
Beresford Rd. DA11: Nflt 9A 122
Beresford Rd. E17 25Dc 52
Beresford Rd. E4 18Gc 35
Beresford Rd. HA1: Harr 29Fa 46
Beresford Rd. KT2: King T 67Pa 131
Beresford Rd. KT3: N Mald 70Sa 131
Beresford Rd. N2 27Gb 49
Beresford Rd. N5 36Tb 71
Beresford Rd. N8 29Qb 50
Beresford Rd. SM2: Sutt 80Bb 155
Beresford Rd. UB1: S'hall 46Z 85
Beresford Rd. WD3: Rick 18H 25
Beresford Sq. SE18 49Rc 94
Beresford St. SE18 48Rc 94
Beresford Ter. N5 36Sb 71
Berestede Rd. W6 50Va 88
Bere St. E1 45Zb 92
Bergen Ho. SE5 54Sb 113
(off Carew St.)
Bergenia Ct. GU24: W End 5C 166
Bergenia Ho. TW13: Felt 60W 106
Bergen Sq. SE16 48Ac 92
Berger Cl. BR5: Pet W 72Tc 160
Berger Ct. E3 41Ec 92
(off Bolinder Way)
Berger Rd. E9 37Zb 72
Berghem M. W14 48Za 88
Bergholt Av. IG4: Ilf 29Nc 54
Bergholt Cres. N16 31Ub 71
Bergholt M. NW1 38Mb 70
Berglen Ct. E14 44Ac 92
Bergman Ho. E17 28Cc 52
(off Hoe St.)
Bering Sq. E14 50Cc 92
Bering Wlk. E16 44Mc 93
Berisford M. SW18 58Eb 111
Berkeley Av. DA7: Bex 53Zc 117
Berkeley Av. IG5: Ilf 26Qc 54
Berkeley Av. RM5: Col R 24Ed 56
Berkeley Av. TW4: Cran 53W 106
Berkeley Av. UB6: G'frd 37Fa 66
(not continuous)
Berkeley Cl. BR5: Pet W 73Uc 160
Berkeley Cl. EN6: Pot B 4Ab 16
Berkeley Cl. HA4: Ruis 34W 64
Berkeley Cl. KT2: King T 66Na 131
Berkeley Cl. RM11: Horn 33Rd 77
Berkeley Cl. TW19: Staines 61F 126
Berkeley Cl. WD5: Ab L 4V 12
Berkeley Cl. WD6: E'tree 15Qa 29
Berkeley Ct. BR2: Broml 70Kc 137
Berkeley Ct. BR8: Swan 69Gd 140
Berkeley Ct. CR0: C'don 77Tb 157
(off Coombe Rd.)
Berkeley Ct. KT13: Weyb 75T 150
Berkeley Ct. KT21: Asht 90Pa 173
Berkeley Ct. KT6: Surb 73Ma 153
Berkeley Ct. N14 16Lb 32
Berkeley Ct. N3 25Db 49
Berkeley Ct. NW1 6G 215 (42Hb 89)
(off Marylebone Rd.)
Berkeley Ct. NW10 35Ua 68
Berkeley Ct. NW11 31Bb 69
(off Ravenscroft Av.)
Berkeley Ct. SM6: W'gton 76Lb 156
Berkeley Ct. W5 45La 86
Berkeley Ct. WD3: Crox G 15T 26
Berkeley Cres. DA1: Dart 60Pd 119
Berkeley Cres. EN4: E Barn 15Fb 31
Berkeley Dr. KT8: W Mole 69Ba 129
Berkeley Dr. RM11: Horn 32Qd 77
Berkeley Gdns. KT10: Clay 79Ja 152
Berkeley Gdns. KT12: Walt T 73V 150
Berkeley Gdns. KT14: W Byf 86H 169
Berkeley Gdns. N21 17Tb 33
Berkeley Gdns. W8 46Cb 89
Berkeley Ho. E3 41Cc 92
(off Wellington Way)
Berkeley Ho. SE8 50Bc 92
(off Grove St.)
Berkeley Ho. TW8: Bford 51Ma 109
(off Albany Rd.)
Berkeley M. AL1: St A 3F 6
Berkeley M. SL1: Slou 4B 80
Berkeley M. TW16: Sun 69Y 129
Berkeley M. W1 2G 221 (44Hb 89)
Berkeley Pl. KT18: Eps 87Ta 173
Berkeley Pl. SW19 65Za 132
Berkeley Rd. E12 36Nc 74
Berkeley Rd. N15 30Tb 51
Berkeley Rd. N8 29Mb 50
Berkeley Rd. NW9 28Qa 47
Berkeley Rd. SW13 53Wa 110
Berkeley Rd. UB10: Hil 38S 64
Berkeleys, The KT22: Fet 96Ga 192
Berkeleys, The SE25 70Wb 135
Berkeley Sq. W1 5A 222 (45Kb 90)
Berkeley St. W1 5A 222 (45Kb 90)
Berkeley Ter. RM18: Tilb 2C 122
Berkeley Twr. E14 46Bc 92
(off Westferry Cir.)
Berkeley Wlk. N7 33Pb 70
(off Durham Rd.)
Berkeley Waye TW5: Hest 51Z 107
Berkhampstead Rd. DA17: Belv 50Cd 96
Berkhamsted Av. HA9: Wemb 37Pa 67
Berkhamsted La. AL9: Ess 2P 9
Berkley Av. EN8: Walt C 6Zb 20
Berkley Cl. TW2: Twick 62Ga 130
(off Wellesley Rd.)
Berkley Cres. DA12: Grav'nd 8E 122
Berkley Gro. NW1 38Jb 70
Berkley Pl. EN8: Walt C 6Zb 20
Berkley Rd. DA12: Grav'nd 8D 122
Berkley Rd. NW1 38Hb 69
Berks Hill WD3: Chor 15E 24
Berkshire Av. SL1: Slou 4F 80
Berkshire Cl. CR3: Cat'm 94Tb 197
Berkshire Ct. W7 42Ha 86
(off Copley Cl.)
Berkshire Gdns. N13 23Qb 50
Berkshire Gdns. N18 22Xb 51
Berkshire Ho. SE6 63Cc 136
Berkshire Rd. E9 37Bc 72
Berkshire Way CR4: Mitc 70Nb 134
Berkshire Way RM11: Horn 29Od 57
Berkshire Yeomanry Mus. 6H 103
Berley Rd. E17 25Bc 52
Berlin Ter. E15 36Gc 73
Berman's Cl. CM13: Hut 19De 41
Bermans Way NW10 35Ua 68
Bermer Rd. WD24: Wat 11Y 27
BERMONDSEY 47Wb 91
Bermondsey Exchange SE1 3J 231 (48Ub 91)
(off Bermondsey St.)
Bermondsey Spa Gdns. 4K 231 (48Vb 91)
Bermondsey Sq. SE1 3J 231 (48Ub 91)
Bermondsey Station (Underground) 48Wb 91
Bermondsey St. SE1 7H 225 (46Ub 91)
Bermondsey Trad. Est. SE16 50Yb 92
Bermondsey Wall E. SE16 47Wb 91
Bermondsey Wall W. SE16 47Wb 91
Bermuda Rd. RM18: Tilb 4C 122
Bermuda Way E1 43Ac 92
(off Dongola Rd.)
Bernal Cl. SE28 45Zc 95
Bernard Angell Ho. SE10 51Fc 115
(off Trafalgar Rd.)
Bernard Ashley Dr. SE7 50Kc 93
Bernard Av. W13 48Ka 86
Bernard Cassidy St. E16 43Hc 93
Bernard Gdns. SW19 64Bb 133
Bernard Gro. EN9: Walt A 5Dc 20
Bernard Hegarty Lodge E8 38Wb 71
(off Lansdowne Dr.)
Bernard Ho. E1 1K 225 (43Vb 91)
(off Toynbee St.)
Bernard Mans. WC1 6F 217 (42Nb 90)
(off Bernard St.)
Bernard Myers Ho. SE5 52Ub 113
(off Havil St.)
Bernard Pl. KT17: Ewe 82Ya 174
Bernard Rd. N15 29Vb 51
Bernard Rd. RM7: Rush G 31Ed 76
Bernard Rd. SM6: W'gton 77Kb 156
Bernards Cl. IG6: Ilf 24Tc 54
Bernard Shaw Ct. NW1 38Lb 70
(off St Pancras Way)
Bernard Shaw Ho. NW10 39Ta 67
(off Knatchbull Rd.)
Bernard St. AL3: St A 1B 6
Bernard St. DA12: Grav'nd 8D 122
Bernard St. WC1 6F 217 (42Nb 90)
Bernard Sunley Ho. SW9 52Qb 112
(off Sth. Island Pl.)
Bernays Cl. HA7: Stan 23La 46
Bernays Gro. SW9 56Pb 112
Bernel Dr. CR0: C'don 76Bc 158
Berne Rd. CR7: Thor H 71Sb 157
Berners Cl. SL1: Slou 5C 80
Berners Dr. AL1: St A 5C 6
Berners Dr. W13 45Ja 86
Berners Ho. N1 1K 217 (40Qb 70)
(off Barnsbury Est.)
Berners M. W1 1C 222 (43Lb 90)
Berners Pl. W1 2C 222 (44Lb 90)
Berners Rd. N1 1B 218 (39Rb 71)
Berners Rd. N22 25Qb 50
Berners St. W1 1C 222 (43Lb 90)
Berner Ter. E1 44Wb 91
(off Fairclough St.)
Berney Ho. BR3: Beck 71Ac 158
Berney Rd. CR0: C'don 73Tb 157
Bernhard Baron Ho. E1 44Wb 91
(off Henriques St.)
Bernhardt Cres. NW8 5D 214 (42Gb 89)
Bernice Cl. HA8: Edg 24Sa 47
Bernice Cl. RM13: Rain 42Ld 97
Bernie Grant Arts Cen. 28Vb 51
Bernville Way HA3: Kenton 29Pa 47
Bernwelle Av. RM3: Rom 23Nd 57
Bernwell Rd. E4 20Gc 35
Berridge Grn. HA8: Edg 24Qa 47
Berridge M. NW6 36Cb 69
Berridge Rd. SE19 64Tb 135
Berriman Rd. N7 34Pb 70
Berrington Dr. KT24: E Hor 96V 190
Berrington Ho. W2 45Cb 89
(off Herrington Rd.)
Berrington M. SL1: Slou 6D 80
Berriton Rd. HA2: Harr 32Ba 65
Berry Av. WD24: Wat 8X 13
Berrybank Cl. E4 19Ec 34
Berry Cl. N21 18Rb 33
Berry Cl. RM10: Dag 36Cd 76
Berry Cl. RM12: Horn 36Ld 77
Berry Cl. WD3: Rick 17K 25
Berry Cotts. E14 44Ac 92
(off Maroon St.)
Berry Ct. TW4: Houn 57Ba 107
Berrydale Rd. UB4: Yead 42Aa 85
Berryfield SL2: Slou 4N 81
Berryfield Cl. BR1: Broml 67Nc 138
Berryfield Cl. E17 28Dc 52
Berryfield Rd. SE17 7C 230 (50Rb 91)
BERRYGROVE 10Ba 13
Berry Gro. La. WD25: A'ham
Otterspool La. 10Aa 13
Berry Gro. La. WD25: A'ham
Otterspool Way 11Ca 27
Berry Hill HA7: Stan 21Ma 47
Berryhill SE9 56Rc 116
Berryhill Gdns. SE9 56Rc 116
Berry Ho. E1 42Xb 91
(off Headlam St.)
Berry Ho. SW11 54Hb 111
(off Culvert Rd.)
Berrylands BR6: Chels 76Yc 161
Berrylands DA3: Hartl 72Ce 165
Berrylands KT5: Surb 72Pa 153
BERRYLANDS 72Qa 153
Berrylands SW20 69Ya 132
Berrylands Rd. KT5: Surb 72Pa 153
Berrylands Station (Rail) 70Ra 131
Berry La. GU22: Wok 5J 187
Berry La. GU3: Worp 7G 186
(not continuous)
Berry La. KT12: Hers 78Z 151
(off New Berry La.)
Berry La. SE21 63Tb 135
Berry La. WD3: Chor 16F 24
Berry La. WD3: Rick 16F 24
Berryman Cl. RM8: Dag 34Yc 75
Berryman's La. SE26 63Zb 136
Berrymead HP2: Hem H 1P 3
Berry Meade KT21: Asht 89Pa 173
Berry Meade Cl. KT21: Asht 89Pa 173
Berrymead Gdns. W3 46Sa 87
Berrymede Rd. W4 48Ta 87
Berry Pl. EC1 4C 218 (41Rb 91)
Berrys Ct. KT14: Byfl 83M 169
Berryscroft Ct. TW18: Staines 66L 127
Berryscroft Rd. TW18: Staines 66L 127
BERRY'S GREEN 88Rc 180
Berry's Grn. Rd. TN16: Big H 88Rc 180
Berry's Hill TN16: Big H 88Rc 180
Berryside Apts. N4 32Sb 71
(off Swan La.)
Berry's La. KT14: Byfl 83M 169
Berry St. EC1 5C 218 (42Rb 91)
Berry Wlk. KT21: Asht 91Pa 193
Berry Way W5 48Na 87
Berry Way WD3: Rick 17K 25
Bersham La. RM17: Grays 49Be 99
Bertal Rd. SW17 63Fb 133
Bertelli Pl. TW13: Felt 60X 107
Bertha Hollamby Ct. DA14: Sidc 64Yc 139
(off Sidcup Hill)
Bertha James Ct. BR2: Broml 70Kc 137
Berther Rd. RM11: Horn 31Md 77
Berthold M. EN9: Walt A 5Dc 20
Berthons Gdns. E17 29Fc 53
(off Wood St.)
Berthon St. SE8 52Cc 114
Bertie Rd. NW10 37Wa 68
Bertie Rd. SE26 65Zb 136
Bertram Cotts. SW19 66Cb 133
Bertram Rd. EN1: Enf 14Wb 33
Bertram Rd. KT2: King T 66Qa 131
Bertram Rd. NW4 30Wa 48
Bertram St. N19 33Kb 70
Bertrand Ho. E16 44Kc 93
(off Russell Rd.)
Bertrand Ho. SW16 62Nb 134
(off Leigham Av.)
Bertrand St. SE13 55Dc 114
Bertrand Way SE28 45Xc 95
Bert Rd. CR7: Thor H 71Sb 157
Bert Way EN1: Enf 14Vb 33
Berwick Av. SL1: Slou 5F 80
Berwick Av. UB4: Yead 44Z 85
Berwick Cl. EN8: Walt C 6Cc 20
Berwick Cl. HA7: Stan 23Ha 46
Berwick Cl. TW2: Whitt 60Ca 107
Berwick Ct. SE1 2E 230 (47Sb 91)
(off Swan St.)
Berwick Cres. DA15: Sidc 59Uc 116
Berwick Gdns. SM1: Sutt 76Eb 155
Berwick Ho. BR6: Orp 74Wc 161
Berwick Ho. N2 26Fb 49
Berwick Pond Cl. RM13: Rain 40Md 77
Berwick Pond Rd. RM13: Rain 40Md 77
Berwick Pond Rd. RM14: Rain 38Pd 77
Berwick Pond Rd. RM14: Upm 38Pd 77
Berwick Rd. DA16: Well 53Xc 117
Berwick Rd. E16 44Kc 93
Berwick Rd. N22 25Rb 51
Berwick Rd. RM13: Rain 40Md 77
Berwick Rd. WD6: Bore 10Pa 15
Berwick St. W1 2C 222 (44Lb 90)
Berwick Way BR6: Orp 74Wc 161
Berwick Way TN14: S'oaks 92Kd 203
Berwyn Av. TW3: Houn 53Da 107
Berwyn Rd. SE24 60Rb 113
Berwyn Rd. TW10: Rich 56Ra 109
Beryl Av. E6 43Nc 94
Beryl Ho. SE18 50Vc 95
(off Spinel Cl.)
Beryl Rd. W6 50Za 88
Berystede KT2: King T 66Ra 131
Besant Cl. NW2 34Ab 68
Besant Ct. N1 36Tb 71
Besant Ct. SE28 46Xc 95
(off Titmuss Av.)
Besant Ho. NW8 39Eb 69
(off Boundary Rd.)
Besant Ho. WD24: Wat 12Z 27
Besant Pl. SE22 56Vb 113
Besant Rd. NW2 35Ab 68
Besant Wlk. N7 33Pb 70
Besant Way NW10 36Sa 67
Besford Ho. E2 40Wb 71
(off Pritchard's Rd.)
Besley St. SW16 65Lb 134
Bessant Dr. TW9: Kew 53Ra 109
Bessborough Gdns. SW1 7E 228 (50Mb 90)
Bessborough Ho. DA9: Ghithe 56Yd 120
(off Carmichael Av.)
Bessborough Pl. SW1 7E 228 (50Mb 90)
Bessborough Rd. HA1: Harr 32Fa 66
Bessborough Rd. SW15 60Wa 110
Bessborough St. SW1 7D 228 (50Mb 90)
Bessborough Vw. KT8: W Mole 70Aa 129
BESSELS GREEN 96Fd 202
Bessels Grn. Rd. TN13: Bes G 95Fd 202
Bessels Mdw. TN13: Bes G 96Fd 202
Bessels Way TN13: Bes G 96Fd 202
Bessemer Cl. SL3: L'ly 50B 82
Bessemer Ct. NW1 38Lb 70
(off Rochester Sq.)
Bessemer Pk. Ind. Est. SE24 56Rb 113
Bessemer Pl. SE10 48Hc 93
Bessemer Rd. SE5 54Sb 113
Bessie Lansbury Cl. E6 44Qc 94
Bessingby Rd. HA4: Ruis 33X 65
Bessingham Wlk. SE4 56Zb 114
(off Aldersford Cl.)
Besson St. SE14 53Yb 114
Bessy St. E2 41Yb 92
Bestobell Rd. SL1: Slou 4G 80
Best Ter. BR8: Crock 72Ed 162
Bestwood St. SE8 49Zb 92
Beswick M. NW6 37Db 69
Beta Ct. CR0: C'don 74Tb 157
(off Sydenham Rd.)
Betam Rd. UB3: Hayes 47T 84
Beta Pl. SW4 56Pb 112
Beta Rd. GU22: Wok 88D 168
Beta Rd. GU24: Chob 2K 167
Beta Way TW20: Thorpe 67E 126
BETCHWORTH 6A 206
Betchworth Cl. SM1: Sutt 78Fb 155
Betchworth Rd. IG3: Ilf 33Uc 74
Betchworth Station (Rail) 4A 206
Betchworth Way CR0: New Ad 81Ec 178
Betenson Av. TN13: S'oaks 94Hd 202
Betham Rd. UB6: G'frd 41Fa 86
Bethany Cl. RM12: Horn 33Ld 77
Bethany Pl. GU21: Wok 10P 167
Bethany Waye TW14: Bedf 59U 106
Bethecar Rd. HA1: Harr 29Ga 46
Bethel Cl. NW4 29Za 48
Bethell Av. E16 42Hc 93
Bethell Av. IG1: Ilf 31Pc 74
Bethel Rd. DA16: Well 55Yc 117
Bethel Rd. TN13: S'oaks 95Ld 203
Bethersden Cl. BR3: Beck 66Bc 136
Bethersden Ho. SE17 7J 231 (50Ub 91)
(off Kinglake Est.)
Bethlehem Ct. UB6: G'frd 40La 66
Bethlehem Ho. E14 45Bc 92
(off Limehouse C'way.)
BETHLEM ROYAL HOSPITAL 73Cc 158
BETHNAL GREEN 41Xb 91
Bethnal Green Cen. for Sports & Performing Arts 41Vb 91
Bethnal Grn. Rd. E1 5K 219 (42Vb 91)
Bethnal Grn. Rd. E2 5K 219 (42Vb 91)
Bethnal Green Station (Overground) 42Xb 91
Bethnal Green Station (Underground) 41Yb 92
Bethune Av. N11 21Hb 49
Bethune Rd. N16 31Tb 71
Bethune Rd. NW10 42Ta 87
Bethwin Rd. SE5 52Rb 113
Betjeman Cl. CR5: Coul 89Pb 176
Betjeman Cl. HA5: Pinn 28Ca 45
Betjeman Ct. UB7: Yiew 46M 83
Betjeman Gdns. WD3: Chor 14F 24
Betjeman M. N5 35Sb 71
Betjeman Way HP1: Hem H 1K 3
Betley Ct. KT12: Walt T 76X 151
Betony Cl. CR0: C'don 74Zb 158
Betony Rd. RM3: Rom 23Ld 57
Betoyne Av. E4 21Gc 53
BETSHAM 63Be 143
Betsham Ho. SE1 1F 231 (47Tb 91)
(off Newcomen St.)
Betsham Rd. DA10: Swans 59Ae 121
Betsham Rd. DA13: Sflt 64Zd 143
Betsham Rd. DA8: Erith 52Hd 118
BETSTYLE CIR. 21Kb 50
Betstyle Ho. N10 24Jb 50
Betstyle Rd. N11 21Kb 50
Betterton Cl. BR7: Chst 64Pc 138
Better Gym Brentwood 19Yd 40
(within The Baytree Centre)
Better Gym East Village 37Ec 72
Better Gym Greenwich 47Gc 93
Better Gym Harrow 30Ga 46
(within St George's Shop. & Leisure Cen.)
Better Gym Pinner 27Z 45
Better Gym Woolwich 48Rc 94
Betterton Dr. DA14: Sidc 61Ad 139
Betterton Ho. WC2 3G 223 (44Nb 90)
(off Betterton St.)
Betterton Rd. RM13: Rain 41Gd 96
Betterton St. WC2 3F 223 (44Nb 90)
Bettles Cl. UB8: Uxb 40L 63
Bettons Pk. E15 39Gc 73
Bettridge Rd. SW6 54Bb 111
Betts Cl. BR3: Beck 68Ac 136
Betts Ho. E1 45Xb 91
(off Betts St.)
Betts M. E17 30Bc 52
Betts Rd. E16 45Kc 93
Betts St. E1 45Xb 91
Betts Way KT6: Surb 74Ka 152
Betts Way SE20 67Xb 135
Betty Brooks Ho. E11 34Fc 73
Betty Entwistle Ho. AL1: St A 5B 6
Betty May Gray Ho. E14 49Ec 92
(off Pier St.)
Betty Paterson Ho. HP1: Hem H 2L 3
(off Astley Rd.)
Betula Cl. CR8: Kenley 87Tb 177
Betula Wlk. RM13: Rain 41Md 97
Between Streets KT11: Cobh 86W 170
Beulah Av. CR7: Thor H 68Sb 135
Beulah Cl. HA8: Edg 20Ra 29
Beulah Cres. CR7: Thor H 68Sb 135
Beulah Gro. CR0: C'don 72Sb 157
Beulah Hill SE19 65Rb 135
Beulah Path E17 29Ec 52
Beulah Pl. EN9: Walt A 5Nc 22
Beulah Rd. CM16: Epp 1Wc 23
Beulah Rd. CR7: Thor H 69Sb 135
Beulah Rd. E17 29Dc 52
Beulah Rd. RM12: Horn 34Ld 77
Beulah Rd. SM1: Sutt 77Cb 155
Beulah Rd. SW19 66Bb 133
Beulah Wlk. CR3: Wold 92Ac 198
Beult Rd. DA1: Cray 56Jd 118
Bevan Av. IG11: Bark 38Wc 75
Bevan Cl. HP3: Hem H 4M 3
Bevan Ct. CR0: Wadd 78Qb 156
Bevan Ct. E3 40Cc 72
(off Tredegar Rd.)
Bevan Ct. TW1: Twick 58Ma 109
Bevan Ho. IG11: Bark 38Xc 75
Bevan Ho. N1 39Ub 71
(off Halcomb St.)
Bevan Ho. RM16: Grays 47Fe 99
Bevan Ho. WC1 7G 217 (43Nb 90)
(off Boswell St.)
Bevan Ho. WD24: Wat 12Z 27
Bevan M. W12 47Wa 88
Bevan Pk. KT17: Ewe 82Va 174
Bevan Pl. BR8: Swan 70Hd 140
Bevan Rd. EN4: Cockf 14Hb 31
Bevan Rd. SE2 50Xc 95
Bevans Cl. DA9: Ghithe 58Yd 120
Bevans Ho. SW18 56Eb 111
(off Eltringham St.)
Bevan St. N1 1E 218 (39Sb 71)
Bevan Way RM12: Horn 35Pd 77
Bev Callender Cl. SW8 55Kb 112
Bevenden St. N1 3G 219 (41Tb 91)
Bevercote Wlk. DA17: Belv 51Bd 117
(off Osborne Rd.)
Beveree Stadium 67Da 129
Beveridge Ct. N21 15Nb 32
(off Pennington Dr.)
Beveridge Ct. SE28 45Xc 95
(off Saunders Way)
Beveridge M. E1 43Yb 92
Beveridge Rd. NW10 38Ua 68
Beverley Av. DA15: Sidc 59Vc 117
Beverley Av. SW20 67Va 132
Beverley Av. TW4: Houn 56Ba 107
Beverley Cl. EN1: Enf 14Ub 33
Beverley Cl. KT13: Weyb 75U 150
Beverley Cl. KT15: Add 78M 149
Beverley Cl. KT17: Ewe 83Ya 174
Beverley Cl. KT9: Chess 77La 152
Beverley Cl. N21 18Sb 33
Beverley Cl. RM11: Horn 31Pd 77
Beverley Cl. RM16: Ors 4F 100
Beverley Cl. SW11 56Fb 111
Beverley Cl. SW13 54Wa 110
Beverley Cotts. SW15 62Ua 132
Beverley Ct. HA2: Harr 27Fa 46
Beverley Ct. HA3: Kenton 28La 46
Beverley Ct. N14 17Lb 32
Beverley Ct. N2 28Hb 49
(off Western Rd.)
Beverley Ct. NW6 38Eb 69
(off Fairfax Rd.)
Beverley Ct. SE4 55Bc 114
(not continuous)
Beverley Ct. SL1: Slou 7M 81
Beverley Ct. TW4: Houn 56Ba 107
Beverley Ct. W4 50Sa 87
Beverley Cres. IG8: Wfd G 25Kc 53
Beverley Dr. HA8: Edg 27Qa 47
Beverley Gdns. EN7: Chesh 2Wb 19
Beverley Gdns. HA7: Stan 25Ja 46
Beverley Gdns. HA9: Wemb 32Pa 67
Beverley Gdns. KT4: Wor Pk 74Wa 154
Beverley Gdns. NW11 31Ab 68
Beverley Gdns. RM11: Horn 31Pd 77
Beverley Gdns. SW13 55Va 110
Beverley Hgts. RH2: Reig 4K 207
Beverley Ho. BR1: Broml 64Fc 137
(off Brangbourne Rd.)
Beverley Hyrst CR0: C'don 75Vb 157
Beverley La. KT2: King T 66Ua 132
Beverley La. SW15 62Va 132
Beverley Meads & Fishponds Wood Nature Reserve 65Va 132
Beverley M. E4 23Fc 53
Beverley Path SW13 54Va 110
Beverley Rd. BR2: Broml 75Nc 160
Beverley Rd. CR3: Whyt 88Ub 177
Beverley Rd. CR4: Mitc 70Mb 134
Beverley Rd. DA7: Bex 54Ed 118
Beverley Rd. E4 23Fc 53
Beverley Rd. E6 41Mc 93
Beverley Rd. HA4: Ruis 33W 64
Beverley Rd. KT1: Hamp W 67La 130
Beverley Rd. KT3: N Mald 70Wa 132
Beverley Rd. KT4: Wor Pk 75Ya 154
Beverley Rd. RM9: Dag 35Ad 75
Beverley Rd. SE20 68Xb 135
Beverley Rd. SW13 55Va 110
Beverley Rd. TW16: Sun 67V 128
Beverley Rd. UB2: S'hall 49Aa 85
Beverley Rd. W4 50Va 88
Beverley Trad. Est. SM4: Mord 73Za 154
Beverley Way KT3: N Mald 69Wa 132
Beverley Way SW20 67Va 132
Beversbrook Rd. N19 34Mb 70
Beverstone Rd. CR7: Thor H 70Qb 134
Beverstone Rd. SW2 57Pb 112
Beverston M. W1 1F 221 (43Hb 89)
(off Up. Montagu St.)
Bevile Ho. RM17: Grays 52De 121
Bevill Allen Cl. SW17 64Hb 133
Bevill Cl. SE25 69Wb 135
Bevin Cl. SE16 46Ac 92
Bevin Ct. WC1 3J 217 (41Pb 90)
Bevington Path SE1 2K 231 (47Vb 91)
(off Tanner St.)
Bevington Rd. BR3: Beck 68Dc 136
Bevington Rd. W10 43Ab 88
Bevington St. SE16 47Wb 91
Bevin Ho. E2 41Yb 92
(off Butler St.)
Bevin Ho. E3 41Cc 92
(off Alfred St.)
Bevin Rd. UB4: Yead 41W 84
Bevin Sq. SW17 62Hb 133
Bevin Wlk. SS17: Stan H 1M 101
Bevin Way WC1 2K 217 (40Qb 70)
Bevis Cl. DA2: Dart 59Sd 120
Bevis Marks EC3 2J 225 (44Ub 91)
Bewcastle Gdns. EN2: Enf 14Nb 32
Bew Ct. SE22 59Wb 113
Bewdley Pl. IG7: Chig 23Ad 55
Bewdley St. N1 38Qb 70
Bewick M. SE15 52Xb 113
Bewick St. SW8 54Kb 112
Bewley Cl. EN8: Chesh 3Zb 20
Bewley Ho. E1 45Xb 91
(off Bewley St.)
Bewley La. TN15: Bor G 97Yd 204
Bewley La. TN15: Plax 97Yd 204
Bewley St. E1 45Yb 92
Bewley St. SW19 65Eb 133
Bewlys Rd. SE27 64Rb 135
Bexhill Cl. TW13: Felt 61Aa 129
Bexhill Dr. RM17: Grays 51Ae 121
Bexhill Rd. N11 22Mb 50
Bexhill Rd. SE4 58Bc 114
Bexhill Rd. SW14 55Sa 109
Bexhill Wlk. E15 39Gc 73
BEXLEY 59Cd 118
Bexley Cl. DA1: Cray 57Gd 118
Bexley Cotts. DA4: Hort K 70Rd 141
Bexley Ct. TN14: Dun G 92Hd 202
(off Campion Sq.)
Bexley Gdns. N9 20Tb 33
Bexley Gdns. RM6: Chad H 29Xc 55
BEXLEYHEATH 56Cd 118
Bexleyheath Golf Course 57Ad 117
Bexleyheath Sports Club 56Yc 117
Bexleyheath Station (Rail) 54Ad 117
Bexley High St. DA5: Bexl 59Cd 118
Bexley Ho. SE4 56Ac 114
Bexley La. DA1: Cray 57Gd 118
Bexley La. DA14: Sidc 63Yc 139
Bexley Lawn Tennis, Squash & Racketball Club 59Cd 118
Bexley Local Studies & Archive Cen. 56Cd 118
(off Townley Rd.)
Bexley Mus. Collection, The 58Ed 118
Bexley Music & Dance Cen. 63Wc 139
(off Station Rd.)
Bexley Rd. DA8: Erith 52Ed 118
(not continuous)
Bexley Rd. SE9 57Rc 116
Bexley Station (Rail) 60Cd 118
Bexley St. SL4: Wind 3G 102
Bexley Tourist Info. Cen. 56Bd 117
Beynon Rd. SM5: Cars 78Hb 155
Bezier Apts. EC2 5G 219 (42Tb 91)
BFI Southbank 6J 223 (46Pb 90)
(off Waterloo Rd.)
Bianca Ho. N1 2H 219 (40Ub 71)
(off Crondall St.)
Bianca Rd. SE15 51Wb 113
Bibsworth Rd. N3 26Bb 49
Bibury Cl. SE15 51Ub 113
(not continuous)
Bicester Rd. TW9: Rich 55Qa 109
Bickels Yd. SE1 2J 231 (47Ub 91)
Bickenhall Mans. W1 7G 215 (43Hb 89)
(off Bickenhall St.)
Bickenhall St. W1 7G 215 (43Hb 89)
Bickersteth Rd. SW17 65Hb 133

Bickerton Ho. WD17: Wat 14Y 27
(off Pumphouse Cres.)
Bickerton Rd. N19 33Lb 70
BICKLEY 69Nc 138
Bickley Cres. BR1: Broml 70Nc 138
Bickley Pk. Rd. BR1: Broml 69Nc 138
Bickley Rd. BR1: Broml 68Mc 137
Bickley Rd. E10 31Dc 72
Bickley Station (Rail) 69Nc 138
Bickley St. SW17 64Gb 133
Bicknell Ho. E1 44Wb 91
(off Ellen St.)
Bicknell Rd. SE5 55Sb 113
Bickney Way KT22: Fet 94Ea 192
Bicknoller Cl. SM2: Sutt 82Db 175
Bicknoller Rd. EN1: Enf 11Ub 33
Bicknor Rd. BR6: Orp 73Uc 160
Bicycle M. SW4 55Mb 112
Bidborough Cl. BR2: Broml 71Hc 159
Bidborough St. WC1 4F 217 (41Nb 90)
Biddenden Way DA13: Ist R 6A 144
Biddenden Way SE9 63Qc 138
Biddenham Ho. SE16 49Zb 92
(off Plough Way)
Biddenham Turn WD25: Wat 7Y 13
Bidder St. E16 43Gc 93
(not continuous)
Biddesden Ho. SW3 6F 227 (49Hb 89)
(off Cadogan St.)
Biddestone Rd. N7 35Pb 70
Biddles Cl. SL1: Slou 6C 80
Biddulph Ho. SE18 49Pc 94
Biddulph Mans. W9 41Db 89
(off Elgin Av.)
Biddulph Rd. CR2: S Croy 82Sb 177
Biddulph Rd. W9 41Db 89
Bideford Av. UB6: G'frd 40Ka 66
Bideford Cl. HA8: Edg 25Qa 47
Bideford Cl. RM3: Rom 25Ld 57
Bideford Cl. TW13: Hanw 62Ba 129
Bideford Gdns. EN1: Enf 17Ub 33
Bideford Rd. BR1: Broml 62Hc 137
Bideford Rd. DA16: Well 52Xc 117
Bideford Rd. EN3: Enf L 10Bc 20
Bideford Rd. HA4: Ruis 34X 65
Bideford Spur SL2: Slou 1F 80
Bidhams Cres. KT20: Tad 93Ya 194
Bidwell Gdns. N11 24Lb 50
Bidwell St. SE15 53Xb 113
Bield, The RH2: Reig 8J 207
Big Apple Woking, The 89B 168
Big Ben 2G 229 (47Nb 90)
Bigbury Cl. N17 24Tb 51
Big Comn. La. RH1: Blet 5H 209
Biggerstaff Rd. E15 39Ec 72
Biggerstaff St. N4 33Qb 70
BIGGIN 1D 122
Biggin Av. CR4: Mitc 67Hb 133
Biggin Hill SE19 66Rb 135
BIGGIN HILL 89Mc 179
Biggin Hill Airport 84Mc 179
Biggin Hill Bus. Pk.
TN16: Big H 87Mc 179
Biggin Hill Cl. KT2: King T 64La 130
Biggin Hill Memorial Pool 88Mc 179
Biggin La. RM16: Grays 1D 122
Biggin Way SE19 66Rb 135
Bigginwood Rd. SW16 66Rb 135
Biggs Ct. NW9 26Ua 48
(off Harvey Cl.)
Biggs Row SW15 55Za 110
Biggs Sq. E9 37Bc 72
Big Hill E5 32Xb 71
Bigland St. E1 44Xb 91
Bignell Rd. SE18 50Rc 94
BIGNELL'S CORNER 5Xa 16
BIGNELL'S CORNER 6Wa 16
Bignells Cnr. EN6: S Mim 6Xa 16
Bignold Rd. E7 35Jc 73
Bigsworth Ct. SE20 67Xb 135
Bigwood Ct. NW11 29Db 49
Bigwood Rd. NW11 29Db 49
Biko Cl. UB8: Cowl 44L 83
Bilberry Ho. E3 43Cc 92
(off Watts Gro.)
Billericay Rd. CM13: Heron 24Fe 59
BILLET THE 2P 101
Billet Cl. RM6: Chad H 27Zc 55
Billet Hill TN15: Ash 77Yd 164
Billet La. RM11: Horn 32Md 77
Billet La. SL0: Iver H 41D 82
Billet La. SL3: L'ly 45D 82
Billet La. SS17: Stan H 2M 101
Billet Rd. E17 25Zb 52
Billet Rd. RM6: Chad H 27Xc 55
Billet Rd. TW18: Staines 62J 127
Billets Hart Cl. W7 47Ga 86
Bill Hamling Cl. SE9 61Pc 138
Billingford Cl. SE4 56Zb 114
Billing Ho. E1 44Zb 92
(off Bower St.)
Billinghurst Way SE10 48Jc 93
Billingley NW1 39Lb 70
(off Pratt St.)
Billing Pl. SW10 52Db 111
Billing Rd. SW10 52Db 111
Billings Cl. RM9: Dag 38Yc 75
Billingsgate Market 46Dc 92
Billings Hill Shaw DA3: Hartl 72Be 165
Billing St. SW10 52Db 111
Billington M. W3 46Ra 87
(off High St.)
Billington Rd. SE14 52Zb 114
Billinton Hill CR0: C'don 75Tb 157
Billiter St. EC3 3J 225 (44Ub 91)
Bill Nicholson Way N17 24Vb 51
(off High Rd.)
Billockby Cl. KT9: Chess 79Pa 153
Billson St. E14 49Ec 92
Bill Voisey Ct. E14 44Ac 92
(off Repton St.)
Billy Lows La. EN6: Pot B 3Cb 17
Bilsby Gro. SE9 63Mc 137
Bilsby Lodge HA9: Wemb 34Sa 67
(off Chalklands)
Bilton Cen. KT22: Lea 91Ha 192
Bilton Cen., The UB6: G'frd 39Ka 66
Bilton Cl. SL3: Poyle 54G 104
Bilton Rd. DA8: Erith 52Jd 118
Bilton Rd. UB6: G'frd 39Ja 66
Bilton Towers W1 3G 221 (44Hb 89)
(off Gt. Cumberland Pl.)
Bilton Way EN3: Enf L 11Ac 34
Bilton Way UB3: Hayes 47X 85
Bina Gdns. SW5 6A 226 (49Eb 89)
Binbrook Ho. W10 43Ya 88
(off Sutton Way)
Bincote Rd. EN2: Enf 13Pb 32
Binden Rd. W12 48Va 88
Bindon Grn. SM4: Mord 70Db 133
Binfield Cl. KT14: Byfl 84P 169
Binfield Rd. CR2: S Croy 78Vb 157
Binfield Rd. KT14: Byfl 84N 169
Binfield Rd. SW4 53Nb 112
Bingfield St. N1 39Nb 70
(not continuous)
Bingham Cl. RM15: S Ock 44Yd 98
Bingham Ct. N1 38Rb 71
(off Halton Rd.)
Bingham Dr. GU21: Wok 10K 167
Bingham Dr. TW18: Staines 66M 127
Bingham Pl. W1 7H 215 (43Jb 90)
Bingham Point SE18 49Rc 94
(off Wilmount St.)
Bingham Rd. CR0: C'don 74Wb 157
Bingham St. N1 37Tb 71
Bingley Rd. E16 44Lc 93
Bingley Rd. TW16: Sun 66W 128
Bingley Rd. UB6: G'frd 42Ea 86
Binley Ho. SW15 58Wa 110
Binnacle Ho. E1 46Xb 91
(off Cobblestone Sq.)
Binney St. W1 3J 221 (44Jb 90)
Binnie Ho. SE1 4D 230 (48Sb 91)
(off Bath Ter.)
Binnie Rd. DA1: Dart 54Nd 119
Binnington Twr. BR2: Broml 72Nc 160
Binns Rd. W4 50Ua 88
Binns Ter. W4 50Ua 88
Binsey Wlk. SE2 47Yc 95
(not continuous)
Binstead Cl. UB4: Yead 44Aa 85
Binyon Cres. HA7: Stan 22Ha 46
Bioko Ct. E1 43Xc 92
(off Ocean Est.)
Biraj Ho. E6 39Pc 74
Birbetts Rd. SE9 61Pc 138
Bircham Path SE4 56Zb 114
(off Aldersford Cl.)
Birchanger Rd. SE25 71Wb 157
Birch Av. CR3: Cat'm 96Tb 197
Birch Av. KT22: Lea 92Ha 192
Birch Av. N13 20Sb 33
Birch Av. UB7: Yiew 44P 83
Birch Cl. DA3: Lfield 68Ee 143
Birch Cl. DA4: Eyns 76Md 163
Birch Cl. E16 43Gc 93
Birch Cl. GU21: Wok 1N 187
Birch Cl. GU23: Send 97H 189
Birch Cl. IG9: Buck H 20Mc 35
Birch Cl. KT15: New H 81M 169
Birch Cl. N19 33Lb 70
Birch Cl. RM15: S Ock 42Zd 99
Birch Cl. RM7: Mawney 27Dd 56
Birch Cl. SE15 54Wb 113
Birch Cl. SL0: Iver H 40F 62
Birch Cl. SM7: Bans 86Ab 174
Birch Cl. TN13: S'oaks 95Kd 203
Birch Cl. TW11: Tedd 64Ja 130
Birch Cl. TW17: Shep 68U 128
Birch Cl. TW3: Houn 54Fa 108
Birch Cl. TW8: Bford 52Ka 108
Birch Copse AL2: Brick W 2Aa 13
Birch Ct. HA6: Nwood 23S 44
Birch Ct. KT22: Lea 92Ha 192
Birch Ct. N12 21Db 49
Birch Ct. RM6: Chad H 30Yc 55
Birch Ct. SM1: Sutt 77Eb 155
Birch Ct. SM6: W'gton 77Kb 156
Birch Cres. RM11: Horn 28Nd 57
Birch Cres. RM15: S Ock 41Zd 99
Birch Cres. UB10: Uxb 39P 63
Birchcroft Cl. CR3: Cat'm 97Sb 197
Birchdale SL9: Ger X 2P 61
Birchdale Cl. KT14: W Byf 83L 169
Birchdale Gdns. RM6: Chad H 31Zc 75
Birchdale Rd. E7 36Lc 73
Birchdene Dr. SE28 46Wc 95
Birchdown Ho. E3 41Dc 92
(off Rainhill Way)
Birch Dr. AL10: Hat 1C 8
Birch Dr. WD3: Map C 22F 42
Birchen Cl. NW9 33Ta 67
Birchend Cl. CR2: S Croy 79Tb 157
Birchen Gro. NW9 33Ta 67
Birches, The AL2: Lon C 9G 6
Birches, The BR2: Broml 70Hc 137
(off Durham Rd.)
Birches, The BR6: Farnb 77Qc 160
Birches, The BR8: Swan 68Gd 140
Birches, The CM13: B'wood 20Ae 41
Birches, The E12 35Nc 74
Birches, The EN9: Walt A 6Hc 21
Birches, The GU22: Wok 90B 168
Birches, The HP3: Hem H 5H 3
Birches, The KT24: E Hor 98U 190
Birches, The N21 16Pb 32
Birches, The SE5 54Ub 113
Birches, The SE7 51Kc 115
Birches, The TW4: Houn 59Ba 107
Birches, The WD23: Bush 15Ea 28
Birches Cl. CR4: Mitc 69Hb 133
Birches Cl. HA5: Pinn 29Aa 45
Birches Cl. KT18: Eps 87Ua 174
Birches Cl. N17 24Wb 51
Birchfield RM16: N Stif 46Zd 99
Birchfield TN14: Sund 99Ad 201
Birchfield Cl. CR5: Coul 88Pb 176
Birchfield Cl. KT15: Add 77K 149
Birchfield Ct. KT12: Walt T 73X 151
(off Grove Cres.)
Birchfield Gro. KT17: Ewe 82Ya 174
Birchfield Ho. E14 45Cc 92
(off Birchfield St.)
Birchfield Rd. EN8: Chesh 1Xb 19
Birchfield St. E14 45Cc 92
Birch Gdns. RM10: Dag 34Ed 76
BIRCH GREEN 63J 127
Birch Grn. HP1: Hem H 1H 3
Birch Grn. NW9 24Ua 48
Birch Grn. TW18: Staines 63H 127
Birch Gro. DA16: Well 56Wc 117
Birch Gro. E11 35Gc 73
Birch Gro. EN6: Pot B 4Cb 17
Birch Gro. GU22: Pyr 87F 168
Birch Gro. KT11: Cobh 86Y 171
Birch Gro. KT20: Kgswd 96Ab 194
Birch Gro. SE12 59Hc 115
Birch Gro. SL2: Slou 3F 80
Birch Gro. SL4: Wind 3B 102
Birch Gro. TW17: Shep 68U 128
Birch Gro. W3 46Qa 87
Birchgrove Ho. TW9: Kew 52Ra 109
Birch Hill CR0: C'don 78Zb 158
Birch Ho. N22 25Qb 50
(off Acacia Rd.)
Birch Ho. SE14 53Bc 114
Birch Ho. SW2 58Qb 112
Birch Ho. UB7: W Dray 47P 83
(off Park Lodge Av.)
Birch Ho. W10 42Ab 88
(off Droop St.)
Birchin Cross Rd. TN15: Knat 87Nd 183
Birchington Cl. BR5: Orp 74Yc 161
Birchington Cl. DA7: Bex 53Dd 118
Birchington Ct. NW6 39Db 69
(off West End La.)
Birchington Ho. E5 36Xb 71
Birchington Rd. KT5: Surb 73Pa 153
Birchington Rd. N8 30Mb 50
Birchington Rd. NW6 39Cb 69
Birchington Rd. SL4: Wind 4E 102
Birchin La. EC3 3G 225 (44Tb 91)
Birchlands Av. SW12 59Hb 111
Birch La. CR8: Purl 83Nb 176
Birch La. GU24: W End 4B 166
Birch La. HP3: Flau 5D 10
Birch Mead BR6: Farnb 75Qc 160
Birchmead WD17: Wat 10V 12
Birchmead Av. HA5: Pinn 27Y 45
Birchmere Bus. Pk. SE28 47Wc 95
Birchmere Lodge SE16 50Xb 91
(off Sherwood Gdns.)
Birchmere Row SE3 54Hc 115
Birchmore Hall N5 34Sb 71
Birchmore Wlk. N5 34Sb 71
(not continuous)
Birch Pk. HA3: Hrw W 24Ea 46
Birch Pl. DA9: Ghithe 58Ud 120
Birch Pl. TN13: S'oaks 96Jd 202
Birch Platt GU24: W End 5B 166
Birch Rd. GU20: W'sham 9C 146
Birch Rd. RM7: Mawney 27Dd 56
Birch Rd. TW13: Hanw 64Z 129
Birch Row BR2: Broml 73Qc 160
Birchside Apts. NW6 40Bb 69
Birch Tree Av. BR4: W W'ck 78Hc 159
Birch Tree Ho. SE7 51Kc 115
Birch Tree Way CR0: C'don 75Xb 157
Birch Va. KT11: Cobh 84Ca 171
Birch Va. Ct. NW8 5C 214 (42Fb 89)
(off Pollitt Dr.)
Birch Vw. CM16: Epp 1Xc 23
Birch Vw. HA1: Harr 29Fa 46
Birchville Ct. WD23: B Hea 18Ga 28
Birch Wlk. CR4: Mitc 67Kb 134
Birch Wlk. DA8: Erith 51Ed 118
Birch Wlk. IG3: Ilf 35Uc 74
(off Loxford La.)
Birch Wlk. KT14: W Byf 84J 169
(not continuous)
Birch Wlk. WD6: Bore 11Qa 29
Birch Way AL2: Lon C 9H 7
Birch Way CR6: W'ham 90Ac 178
Birch Way RH1: Redh 8B 208
Birchway TN15: W King 81Vd 184
Birchway UB3: Hayes 46W 84
Birchwood EN9: Walt A 6Gc 21
Birchwood WD7: Shenl 6Qa 15
Birchwood Apts. N4 31Sb 71
(off Woodberry Gro.)
Birchwood Av. BR3: Beck 70Bc 136
Birchwood Av. DA14: Sidc 61Xc 139
Birchwood Av. N10 27Jb 50
Birchwood Av. SM6: W'gton 76Jb 156
Birchwood Cl. CM13: Gt War 23Yd 58
Birchwood Cl. SM4: Mord 70Db 133
Birchwood Ct. HA8: Edg 26Sa 47
Birchwood Ct. KT13: Weyb 78S 150
Birchwood Ct. N13 22Rb 51
Birchwood Dr. DA2: Wilm 63Gd 140
Birchwood Dr. GU18: Light 2A 166
Birchwood Dr. KT14: W Byf 84J 169
Birchwood Dr. NW3 34Db 69
Birchwood Gro. TW12: Hamp 65Ca 129
Birchwood La. CR3: Cat'm 97Rb 197
Birchwood La. KT10: Esh 81Fa 172
Birchwood La. KT22: Oxs 82Fa 172
Birchwood La. TN14: Dun G 87Bd 181
Birchwood Pde. DA2: Wilm 63Gd 140
Birchwood Pk. Av. BR8: Swan 69Gd 140
Birchwood Pk. Golf Course 66Ed 140
Birchwood Rd. BR5: Pet W 70Tc 138
Birchwood Rd. BR8: Swan 67Ed 140
Birchwood Rd. DA2: Wilm 64Gd 140
Birchwood Rd. KT14: W Byf 84J 169
Birchwood Rd. SW17 64Kb 134
Birchwood Way AL2: Park 10P 5
Birdbrook Cl. CM13: Hut 16De 41
Birdbrook Cl. RM10: Dag 38Ed 76
Birdbrook Ho. N1 38Sb 71
(off Popham Rd.)
Birdbrook Rd. SE3 56Lc 115
Birdcage Wlk. SW1 2B 228 (47Lb 90)
Birdham Cl. BR1: Broml 71Nc 160
Birdhouse La. BR6: Downe 87Qc 180
Birdhurst Av. CR2: S Croy 77Tb 157
Birdhurst Ct. SM6: W'gton 80Lb 156
(off Woodcote Av.)
Birdhurst Gdns. CR2: S Croy 77Tb 157
Birdhurst Ri. CR2: S Croy 78Ub 157
Birdhurst Rd. CR2: S Croy 78Ub 157
Birdhurst Rd. SW18 57Eb 111
Birdhurst Rd. SW19 65Gb 133
Bird in Bush BMX Track 52Xb 113
(off Bird in Bush Rd.)
Bird in Bush Rd. SE15 52Wb 113
Bird in Hand La. BR1: Broml 68Mc 137
Bird in Hand M. SE23 61Yb 136
(off Bird-in-Hand Pas.)
Bird in Hand Pas. SE23 61Yb 136
Bird in Hand Path CR0: C'don 73Tb 157
(off Sydenham Rd.)
Bird in Hand Yd. NW3 35Eb 69
Bird La. CM13: Gt War 27Yd 58
Bird La. CM13: L War 27Yd 58
Bird La. RM14: Upm 29Td 58
Bird La. UB9: Hare 26L 43
Birdsall Ho. SE5 55Ub 113
Birds Farm Av. RM5: Col R 24Dd 56
Birdsfield La. E3 39Bc 72
Birds Gro. GU21: Knap 10F 166
Birds Hill Dr. KT22: Oxs 85Fa 172
Birds Hill Ri. KT22: Oxs 85Fa 172
Birds Hill Rd. KT22: Oxs 84Fa 172
Birdsmouth Ct. N15 28Ub 51
(off Bathurst Sq.)
Bird St. W1 3J 221 (44Jb 90)
Birdswood Dr. GU21: Wok 1J 187
Bird Wlk. TW2: Whitt 60Ba 107
Birdwood Av. DA1: Dart 54Pd 119
Birdwood Av. SE13 58Fc 115
Birdwood Cl. CR2: Sels 83Yb 178
Birdwood Cl. TW11: Tedd 63Ga 130
Birdwood Sq. DA1: Dart 54Qd 119
Birkbeck Av. UB6: G'frd 39Ea 66
Birkbeck Av. W3 45Sa 87
Birkbeck Ct. W3 46Ta 87
Birkbeck Gdns. IG8: Wfd G 19Jc 35
Birkbeck Gro. W3 47Ta 87
Birkbeck Hill SE21 60Rb 113
Birkbeck M. E8 36Vb 71
Birkbeck M. W3 46Ta 87
Birkbeck Pl. SE21 61Sb 135
Birkbeck Rd. BR3: Beck 68Yb 136
Birkbeck Rd. CM13: Hut 16Fe 41
Birkbeck Rd. DA14: Sidc 62Wc 139
Birkbeck Rd. E8 36Vb 71
Birkbeck Rd. EN2: Enf 11Tb 33
Birkbeck Rd. IG2: Ilf 29Tc 54
Birkbeck Rd. N12 22Eb 49
Birkbeck Rd. N17 25Vb 51
Birkbeck Rd. N8 28Nb 50
Birkbeck Rd. NW7 22Va 48
Birkbeck Rd. RM7: Rush G 32Fd 76
Birkbeck Rd. SW19 64Db 133
Birkbeck Rd. W3 46Ta 87
Birkbeck Rd. W5 49La 86
Birkbeck Station (Rail & London
Tramlink) 69Yb 136
Birkbeck St. E2 41Xb 91
Birkbeck Way UB6: G'frd 39Fa 66
Birkdale Av. HA5: Pinn 27Ca 45
Birkdale Av. RM3: Hrld W 24Pd 57
Birkdale Cl. BR6: Orp 73Tc 160
Birkdale Cl. SE16 50Xb 91
Birkdale Cl. SE28 44Zc 95
Birkdale Ct. UB1: S'hall 44Ea 86
(off Redcroft Rd.)
Birkdale Gdns. CR0: C'don 77Zb 158
Birkdale Gdns. WD19: Wat 20Z 27
Birkdale Ho. E14 44Bc 92
(off Keymer Pl.)
Birkdale Rd. SE2 49Wc 95
Birkdale Rd. W5 42Na 87
Birkenhead Av. KT2: King T 68Pa 131
Birkenhead St. WC1 3G 217 (41Nb 90)
Birken M. HA6: Nwood 22R 44
Birkhall Rd. SE6 60Fc 115
Birkheads Rd. RH2: Reig 5J 207
Birkin Ct. KT14: Byfl 83M 169
Birklands La. AL1: St A 6F 6
Birklands Pk. AL1: St A 6F 6
Birkwood Cl. SW12 59Mb 112
Birley Lodge NW8 1C 214 (40Fb 69)
(off Acacia Rd.)
Birley Rd. N20 19Eb 31
Birley Rd. SL1: Slou 4H 81
Birley St. SW11 54Jb 112
Birling Rd. DA8: Erith 52Fd 118
Birnam Cl. GU23: Rip 96J 189
Birnam Cres. IG7: Chig 22Ad 55
Birnam Rd. N4 33Pb 70
Birnbeck Ct. EN5: Barn 14Za 30
Birnbeck Ct. NW11 29Bb 49
Birrell Ho. SW9 54Pb 112
(off Stockwell Rd.)
Birse Cres. NW10 34Ua 68
Birstal Grn. WD19: Wat 21Z 45
Birstall Rd. N15 29Ub 51
Birtrick Dr. DA13: Meop 10B 144
(not continuous)
Birtwhistle Ho. E3 39Bc 72
(off Parnell Rd.)
Biscay Ho. E1 42Zb 92
(off Mile End Rd.)
Biscayne Av. E14 46Fc 93
Biscay Rd. W6 50Za 88
Biscoe Cl. TW5: Hest 51Ca 107
Biscoe Way SE13 55Fc 115
Biscott Ho. E3 42Dc 92
Bisenden Rd. CR0: C'don 75Ub 157
Bisham Cl. SM5: Cars 74Hb 155
Bisham Ct. SL1: Slou 7K 81
(off Park St.)
Bisham Gdns. N6 32Jb 70
Bishams Ct. CR3: Cat'm 96Vb 197
Bishop Butt Cl. BR6: Orp 76Vc 161
Bishop Ct. TW9: Rich 55Na 109
Bishop Duppas Pk. TW17: Shep 73U 150
Bishop Fox Way KT8: W Mole 70Ba 129
Bishop Ken Rd. HA3: W'stone 26Ha 46
Bishop King's Rd. W14 49Ab 88
Bishop Ramsey Cl. HA4: Ruis 31V 64
Bishop Rd. N14 17Kb 32
Bishop's Av. E13 39Kc 73
Bishop's Av. SW6 54Za 110
Bishops Av. BR1: Broml 68Lc 137
Bishops Av. HA6: Nwood 21U 44
Bishops Av. RM6: Chad H 30Yc 55
Bishops Av. WD6: E'tree 15Pa 29
Bishops Av., The N2 31Fb 69
Bishop's Bri. Rd. W2 44Db 89
Bishop's Cl. CR5: Coul 90Qb 176
Bishop's Cl. N19 34Lb 70
Bishop's Cl. SE9 61Sc 138
Bishop's Cl. SM1: Sutt 76Cb 155
Bishops Cl. AL10: Hat 1B 8
Bishops Cl. E17 28Dc 52
Bishops Cl. EN1: Enf 12Xb 33
Bishops Cl. EN5: Barn 16Za 30
Bishops Cl. TW10: Ham 62Ma 131
Bishops Cl. UB10: Hil 40Q 64
Bishops Cl. W4 50Sa 87
Bishop's Ct. EC4 2C 224 (44Rb 91)
(off Old Bailey)
Bishop's Ct. WC2 2K 223 (44Qb 90)
(off Star Yd.)
Bishops Ct. CR0: C'don 75Vb 157
Bishops Ct. DA9: Ghithe 57Vd 120
Bishops Ct. EN8: Chesh 2Yb 20
Bishops Ct. HA0: Wemb 35Ka 66
Bishops Ct. N2 29Gb 49
Bishops Ct. RM16: Ors 3C 100
Bishops Ct. W2 44Db 89
(off Bishop's Bri. Rd.)
Bishops Ct. WD5: Ab L 3V 12
Bishopsdale Ho. NW6 39Cb 69
(off Kilburn Vale)
Bishop's Dr. TW14: Bedf 58T 106
Bishops Dr. UB5: N'olt 39Aa 65
Bishops Farm Cl. SL4: Oak G 4A 102
Bishopsford Ho. SM5: Cars 72Gb 155
Bishopsford Rd. SM4: Mord 73Eb 155
BISHOPS GATE 2L 125
Bishopsgate EC2 3H 225 (44Ub 91)
Bishopsgate Arc. EC2 1J 225 (43Ub 91)
(off Bishopsgate)
Bishopsgate Churchyard EC2 1H 225 (43Ub 91)
Bishopsgate Institute &
Libraries 1J 225 (43Ub 91)
(off Bishopsgate)
Bishopsgate Plaza EC3 2J 225 (44Ub 91)
(off Bishopsgate)
Bishopsgate Rd. TW20: Eng G 2K 125
Bishops Grn. BR1: Broml 67Lc 137
Bishop's Gro. TW12: Hamp 63Ba 129
Bishops Gro. GU20: W'sham 9A 146
Bishops Gro. KT20: Tad 94Xa 194
Bishops Gro. N2 30Gb 49
Bishops Gro. Cvn. Site TW12: Hamp 63Ca 129
Bishop's Hall KT1: King T 68Ma 131
Bishop's Hall Rd. CM15: Pil H 16Xd 40
Bishops Hill KT12: Walt T 73W 150
Bishops Ho. SW8 52Nb 112
(off Sth. Lambeth Rd.)
Bishop's Mans. SW6 54Za 110
Bishops Mead HP1: Hem H 4K 3
Bishops Mead SE5 52Sb 113
(off Camberwell Rd.)
Bishopsmead Cl. KT19: Ewe 82Ta 173
Bishopsmead Cl. KT24: E Hor 100U 190
Bishopsmead Ct. KT19: Ewe 82Ua 174
Bishopsmead Dr. KT24: E Hor 100V 190
Bishopsmead Pde. KT24: E Hor 100U 190
Bishops Orchard SL2: Farn R 1F 80
Bishop's Pk. Rd. SW6 54Za 110
Bishops Pk. Rd. SW16 67Nb 134
Bishops Pl. SM1: Sutt 78Eb 155
Bishops Ri. AL10: Hat 1B 8
Bishop's Rd. CR0: C'don 73Rb 157
Bishop's Rd. SW11 52Gb 111
Bishop's Rd. UB3: Hayes 44S 84
Bishops Rd. N6 30Jb 50
Bishops Rd. SL1: Slou 7L 81
Bishops Rd. SS17: Stan H 1P 101
Bishops Rd. SW6 53Ab 110
Bishops Rd. W7 47Ga 86
Bishops Sq. E1 7J 219 (43Ub 91)
Bishop's Ter. SE11 5A 230 (49Qb 90)
Bishopsthorpe Rd. SE26 63Zb 136
Bishop St. N1 39Sb 71
Bishops Vw. Ct. N10 28Kb 50
Bishops Wlk. BR7: Chst 67Sc 138
Bishops Wlk. CR0: Addtn 78Zb 158
Bishops Wlk. HA5: Pinn 27Aa 45
Bishop's Way E2 40Xb 71
Bishops Way TW20: Egh 65F 126
Bishops Wharf Ho. SW11 52Gb 111
(off Parkgate Rd.)
Bishops Wood GU21: Wok 9K 167
Bishops Wood Almshouses E5 35Xb 71
(off Lwr. Clapton Rd.)
BISHOPS WOOD BMI HOSPITAL 23R 44
Bishopswood Rd. N6 31Hb 69
Bishop Wlk. CM15: Shenf 19Be 41
Bishop Way NW10 38Ua 68
Bishop Wilfred Wood Cl. SE15 54Wb 113
Bishop Wilfred Wood Ct. E13 40Lc 73
(off Pragel St.)
Biskra WD17: Wat 11W 26
BISLEY 7D 166
BISLEY CAMP 1B 186
Bisley Cl. EN8: Walt C 5Zb 20
Bisley Cl. KT4: Wor Pk 74Ya 154
Bisley Grn. GU24: Bisl 8D 166
Bisley Pl. TW3: Houn 54Da 107
Bison Ct. TW14: Felt 59X 107
Bispham Rd. NW10 41Pa 87
Bissagos Ct. E1 43Ac 92
(off Ocean Est.)
Bissextile Ho. SE13 54Dc 114
Bisson Rd. E15 40Ec 72
Bisterne Av. E17 27Fc 53
BITCHET GREEN 98Td 204
Bittacy Bus. Cen. NW7 24Ab 48
Bittacy Cl. NW7 23Za 48
Bittacy Ct. NW7 24Ab 48
Bittacy Hill NW7 23Za 48
Bittacy Pk. Av. NW7 22Za 48
Bittacy Ri. NW7 23Ya 48
Bittacy Rd. NW7 23Za 48
Bittams La. KT16: Chert 77F 148
Bittern Cl. HP3: Hem H 7P 3
Bittern Cl. UB4: Yead 43Z 85
Bittern Ct. NW9 26Ua 48
Bittern Ct. SE8 51Cc 114
Bitterne Dr. GU21: Wok 9K 167
Bittern Ho. SE1 2D 230 (47Sb 91)
(off Gt. Suffolk St.)
Bittern Ho. UB7: Yiew 45M 83
(off Wraysbury Dr.)
Bittern Pl. N22 26Pb 50
Bittern St. SE1 2D 230 (47Sb 91)
Bittoms, The KT1: King T 69Ma 131
Bittoms Ct. KT1: King T 69Ma 131
Bixley Cl. UB2: S'hall 49Ba 85
Blackacre Rd. CM16: They B 9Uc 22
Blackall St. EC2 5H 219 (42Ub 91)
Blackberry Cl. E17 27Dc 52
Blackberry Cl. TW17: Shep 70U 128
Blackberry Ct. HA3: Kenton 31Na 67
Blackberry Farm Cl. TW5: Hest 52Aa 107
Blackberry Fld. BR5: St P 67Wc 139
Blackberry Way HP2: Hem H 2D 4
Blackbird Ct. NW9 34Ta 67
Blackbird Hill NW9 33Sa 67
Blackbirds La. WD25: A'ham 6Ea 14
Blackbird Yd. E2 41Vb 91
Blackborne Rd. RM10: Dag 37Cd 76
Blackborough Cl. RH2: Reig 6L 207
Blackborough Ho. IG9: Buck H 19Mc 35
(off Beatrice Ct.)
Blackborough Rd. RH2: Reig 7L 207
Black Boy La. N15 29Sb 51
Black Boy Wood AL2: Brick W 2Ca 13
Blackbridge Rd. GU22: Wok 1P 187
Blackbrook La. BR1: Broml 70Qc 138
Blackbrook La. BR2: Broml 71Qc 160
Black Bull Yd. EC1 7A 218 (43Qb 90)
(off Hatton Wall)
Blackburn NW9 26Va 48
Blackburn, The KT23: Bookh 96Ba 191
Blackburne's M. W1 4H 221 (45Jb 90)
Blackburn Rd. NW6 37Db 69
Blackburn Trad. Est. TW19: Stanw 58P 105
Blackburn Way TW4: Houn 57Aa 107
Blackbury Cl. EN6: Pot B 3Eb 17
Blackbush Av. RM6: Chad H 29Zc 55
Blackbush Cl. SM2: Sutt 80Db 155
Black Cut AL1: St A 3C 6
Blackdown Av. GU22: Pyr 87G 168
Blackdown Cl. GU22: Pyr 88E 168
Blackdown Cl. N2 26Eb 49
Blackdown Ter. SE18 53Pc 116
Black Eagle Cl. TN16: Westrm 99Sc 200
Black Eagle Dr. DA11: Nflt 57Ce 121
Black Eagle Sq. TN16: Westrm 99Sc 200
Blackett Apts. E3 42Bc 92
(off Hamlets Way)
Blackett Cl. TW18: Staines 68G 126
Blackett St. SW15 55Za 110
Blacketts Wood Dr. WD3: Chor 15D 24
Black Fan Cl. EN2: Enf 11Sb 33
BLACKFEN 58Wc 117

Blackfen Pde. DA15: Sidc 58Wc 117
Blackfen Rd. DA15: Sidc 57Uc 116
Blackford Cl. CR2: S Croy 81Rb 177
Blackford Rd. WD19: Wat 22Z 45
Blackford's Path SW15 59Wa 110
Blackfriars Bri. EC4 5B 224 (45Rb 91)
Blackfriars Ct. AL1: St A 1C 6
(off Newsom Pl.)
Blackfriars Ct. EC4 4B 224 (45Rb 91)
(off New Bri. St.)
Black Friars La. EC4 3B 224 (44Rb 91)
(not continuous)
Blackfriars Pas. EC4 4B 224 (45Rb 91)
Blackfriars Rd. SE1 2B 230 (47Rb 91)
Blackfriars Station (Rail & Underground) 4B 224 (45Rb 91)
Blackfriars Underpass EC4 4A 224 (45Qb 90)
Black Gates HA5: Pinn 27Ba 45
Black Grn. Wood Cl. AL2: Park 1Da 13
Blackhall La. TN15: God G 95Md 203
Blackhall La. TN15: S'oaks 95Md 203
BLACKHEATH 54Hc 115
Blackheath Av. SE10 52Fc 115
BLACKHEATH BMI HOSPITAL, THE 55Hc 115
BLACKHEATH BMI HOSPITAL (OUTPATIENT DEPARTMENT), THE 55Hc 115
Blackheath Bus. Cen. SE10 53Ec 114
Blackheath Bus. Est. SE10 53Ec 114
(off Blackheath Hill)
Blackheath Concert Halls 55Hc 115
Blackheath Gro. SE3 54Hc 115
Blackheath Hill SE10 53Ec 114
Blackheath Pk. SE3 55Hc 115
BLACKHEATH PARK 55Jc 115
Blackheath Ri. SE13 54Ec 114
(not continuous)
Blackheath Rd. SE10 53Dc 114
Blackheath RUFC 52Kc 115
Blackheath Station (Rail) 55Hc 115
Blackheath Va. SE3 54Gc 115
BLACKHEATH VALE 54Hc 115
Blackheath Village SE3 54Hc 115
Blackhills KT10: Esh 81Ba 171
Black Horse Cl. SL4: Wind 4B 102
Black Horse Ct. SE1 3G 231 (48Tb 91)
Blackhorse La. CR0: C'don 73Wb 157
Blackhorse La. E17 26Zb 52
Blackhorse La. EN6: S Mim 2Ua 16
Blackhorse La. KT20: Lwr K 1K 207
Blackhorse Lane Stop (London Tramlink) 73Wb 157
Black Horse M. TN15: Bor G 92Ce 205
Blackhorse M. E17 27Zb 52
Black Horse Pde. HA5: Eastc 29X 45
Black Horse Pl. UB8: Uxb 39K 63
Blackhorse Rd. DA14: Sidc 63Wc 139
Blackhorse Rd. E17 28Zb 52
Blackhorse Rd. GU22: Wok 2H 187
Blackhorse Rd. SE8 51Ac 114
BLACKHORSE ROAD 28Zb 52
Blackhorse Road Station (Underground & Overground) 28Zb 52
Black Horse Yd. SL4: Wind 3H 103
Black Horse Yd. UB8: Uxb 39K 63
Black Lake Cl. TW20: Egh 67C 126
Blacklands Dr. UB4: Hayes 42S 84
Blacklands Mdw. RH1: Nutf 5E 208
Blacklands Rd. SE6 63Ec 136
Blacklands Ter. SW3 6F 227 (49Hb 89)
Blackley Cl. WD17: Wat 9V 12
Black Lion Hill WD7: Shenl 4Na 15
Black Lion La. W6 49Wa 88
Black Lion M. W6 49Wa 88
Blackmans Cl. DA1: Dart 60Ld 119
Blackman's La. CR6: W'ham 86Gc 179
Blackmans Yd. E2 42Wb 91
Blackmead TN13: Riv 93Gd 202
Blackmoor La. WD18: Wat 15T 26
Blackmore Av. UB1: S'hall 46Fa 86
Blackmore Cl. RM17: Grays 50De 99
Blackmore Ct. EN9: Walt A 5Jc 21
Blackmore Cres. GU21: Wok 87E 168
Blackmore Dr. NW10 38Ra 67
Blackmore Ho. N1 39Pb 70
(off Barnsbury Est.)
Blackmore Rd. IG9: Buck H 17Nc 36
Blackmore's Gro. TW11: Tedd 65Ja 130
Blackmore Way UB8: Uxb 37M 63
Blackness La. BR2: Kes 81Mc 179
Blackness La. GU22: Wok 91A 188
BLACKNEST 9F 124
Blacknest Ga. Rd. SL5: S'hill 9F 124
Blacknest Rd. GU25: Vir W 9J 125
Blacknest Rd. SL5: S'hill 9H 125
Black Pk. Country Pk. 39C 62
Black Pk. Rd. SL3: Ful 40A 62
Black Pk. Rd. SL3: Wex 40A 62
Black Path E10 31Ac 72
Blackpond La. SL2: Farn C 7F 60
Blackpond La. SL2: Farn R 7F 60
Blackpool Gdns. UB4: Hayes 42U 84
Blackpool Rd. SE15 54Xb 113
Black Prince Cl. KT14: Byfl 86P 169
BLACK PRINCE INTERCHANGE 58Dd 118
Black Prince Rd. SE1 6H 229 (49Pb 90)
Black Prince Rd. SE11 6H 229 (49Pb 90)
Black Prince St. SE18 51Wc 117
Black Robins La. HP3: Flau 4D 10
Black Rod Cl. UB3: Hayes 48V 84
Blackshaw Rd. SW17 63Eb 133
Blackshots La. RM16: Grays 45Ee 99
Blacksmith Cl. KT21: Asht 91Pa 193
Blacksmith Cl. RM6: Chad H 30Yc 55
Blacksmith Cl. TW16: Sun 69W 128
Blacksmith Row HP3: Hem H 3C 4
Blacksmith Row SL3: L'ly 49C 82
Blacksmiths Hill CR2: Sande 85Wb 177
Blacksmiths Ho. E14 44Fc 93
(off Valencia Cl.)
Blacksmiths Ho. E17 28Cc 52
(off Gillards M.)
Blacksmith's La. BR5: St M Cry 71Yc 161
Blacksmith's La. RM13: Rain 39Hd 76
Blacksmiths La. AL3: St A 2P 5
Blacksmiths La. KT16: Chert 73J 149
Blacksmiths La. TW18: Lale 69K 127
Blacksmiths La. UB9: Den 33E 62
Blacksole Cotts. TN15: Wro 87Be 185
Blacksole La. TN15: Wro 88Be 185
Blacksole Rd. TN15: Wro 88Be 185
Blacks Rd. W6 50Ya 88
Blackstock M. N4 33Rb 71
Blackstock Rd. N4 33Rb 71
Blackstock Rd. N5 34Rb 71
Blackstone Cl. RH1: Redh 7N 207
Blackstone Est. E8 38Wb 71
Blackstone Hill RH1: Redh 7M 207
Blackstone Ho. SW1 50Lb 90
(off Churchill Gdns.)
Blackstone Rd. NW2 36Ya 68
Blackstroud La. E. GU18: Light 3B 166
Blackstroud La. W. GU18: Light 3B 166
Black Swan Yd. SE1 1J 231 (47Ub 91)
Black's Yd. TN13: S'oaks 95Jd 203
(off Bank St.)
Blackthorn Av. IG7: Chig 22Ad 55
Blackthorn Av. N7 37Qb 70
Blackthorn Av. UB7: W Dray 49Q 84
Blackthorn Cl. KT17: Eps D 88Xa 174
Blackthorn Cl. RH2: Reig 8L 207
Blackthorn Cl. RH9: G'stone 2A 210
Blackthorn Cl. TN15: W King 81Vd 184
Blackthorn Cl. WD25: Wat 4X 13
Blackthorn Ct. E15 35Fc 73
(off Hall Rd.)
(not continuous)
Blackthorn Ct. TW5: Hest 52Aa 107
Blackthorne Av. CR0: C'don 74Yb 158
Blackthorne Cl. AL10: Hat 3B 8
Blackthorne Ct. SE15 52Vb 113
(off Cator St.)
Blackthorne Ct. TW15: Ashf 66S 128
Blackthorne Ct. UB1: S'hall 46Da 85
(off Dormer's Wells La.)
Blackthorne Cres. SL3: Poyle 54G 104
Blackthorne Dell SL3: L'ly 8N 81
Blackthorne Dr. E4 21Fc 53
Blackthorne Ind. Est. SL3: Poyle 55G 104
Blackthorne Rd. KT23: Bookh 98Ea 192
Blackthorne Rd. SL3: Poyle 55G 104
Blackthorn Gro. BR5: St M Cry 72Xc 161
Blackthorn Gro. DA7: Bex 55Ad 117
Blackthorn Ho. SE16 47Ac 92
(off Blondin Way)
Blackthorn Rd. CR3: Cat'm 95Tb 197
Blackthorn Rd. IG1: Ilf 36Tc 74
Blackthorn Rd. RH2: Reig 8L 207
Blackthorn Rd. RM16: Grays 46De 99
Blackthorn Rd. TN16: Big H 88Mc 179
Blackthorn St. E3 42Cc 92
Blackthorn Way CM14: W'ley 22Zd 59
Blacktree M. SW9 55Qb 112
BLACKWALL 46Ec 92
Blackwall Basin Moorings 46Ec 92
Blackwall La. SE10 50Gc 93
Blackwall Station (DLR) 45Ec 92
Blackwall Trad. Est. E14 43Fc 93
Blackwall Tunnel 46Fc 93
Blackwall Tunnel App. E14 45Ec 92
Blackwall Tunnel Northern App. E3 40Cc 72
Blackwall Tunnel Southern App. SE10 48Gc 93
Blackwall Way E14 46Ec 92
Blackwater CM14: B'wood 18Xd 40
Blackwater Cl. E7 35Hc 73
Blackwater Cl. RM13: Rain 43Fd 96
Blackwater Ho. NW8 7C 214 (43Fb 89)
(off Church St.)
Blackwater La. HP3: Hem H 5E 4
Blackwater St. SE22 57Vb 113
Blackwell Cl. E5 35Zb 72
Blackwell Cl. HA3: Hrw W 24Fa 46
Blackwell Cl. N21 15Nb 32
Blackwell Ct. SL3: L'ly 9N 81
Blackwell Dr. WD19: Wat 16Y 27
Blackwell Gdns. HA8: Edg 21Qa 47
Blackwell Ho. HP3: Hem H 7P 3
(off The Embankment)
Blackwell Ho. SW4 58Mb 112
Blackwell Rd. WD4: K Lan 1Q 12
Blackwood Cl. KT14: W Byf 84L 169
Blackwood Ho. E1 42Xb 91
(off Collingwood St.)
Blackwood St. SE17 7F 231 (50Tb 91)
Blade Ct. RM7: Rush G 30Gd 56
Blade M. SW15 56Bb 111
Bladen Cl. KT13: Weyb 79T 150
Bladen Ho. E1 44Zb 92
(off Dunelm St.)
Blades Cl. KT22: Lea 92Ma 193
Blades Ct. SW15 56Bb 111
Blades Ct. W6 50Xa 88
(off Lower Mall)
Blades Ho. SE11 51Qb 112
(off Kennington Oval)
Bladindon Dr. DA5: Bexl 59Yc 117
Bladon Ct. SW16 65Nb 134
Bladon Gdns. HA2: Harr 30Da 45
Blagdens Cl. N14 19Mb 32
Blagdens La. N14 19Lb 32
Blagdon Ct. W7 45Ga 86
Blagdon Rd. KT3: N Mald 70Va 132
Blagdon Rd. SE13 58Dc 114
Blagdon Wlk. TW11: Tedd 65La 130
Blagrove Cres. HA4: Eastc 30X 45
Blagrove Rd. TW11: Hamp W 66Ka 130
Blagrove Rd. W10 43Ab 88
Blair Av. KT10: Esh 75Ea 152
Blair Av. NW9 31Ua 68
Blair Cl. DA15: Sidc 57Uc 116
Blair Cl. N1 37Sb 71
Blair Cl. UB3: Harl 49W 84
Blair Ct. BR3: Beck 67Dc 136
Blair Ct. NW8 39Fb 69
Blair Ct. SE6 60Hc 115
Blairderry Rd. SW2 61Nb 134
Blair Dr. TN13: S'oaks 95Kd 203
Blairgowrie Ct. E14 44Fc 93
(off Blair St.)
Blairhead Dr. WD19: Wat 20X 27
Blair Ho. SW9 54Pb 112
Blair Rd. SL1: Slou 6J 81
Blair St. E14 44Ec 92
Blake Av. IG11: Bark 39Uc 74
Blakeborough Dr. RM3: Hrld W 26Nd 57
Blake Bldg. N8 27Pb 50
Blake Cl. AL1: St A 5E 6
Blake Cl. DA16: Well 53Uc 116
Blake Cl. RM13: Rain 39Hd 76
Blake Cl. SM5: Cars 74Gb 155
Blake Cl. UB4: Hayes 40T 64
Blake Cl. W10 43Ya 88
Blake Ct. KT19: Eps 84Pa 173
Blake Ct. N21 15Pb 32
Blake Ct. NW6 41Cb 89
(off Malvern Rd.)
Blake Ct. SE16 50Xb 91
(off Stubbs Dr.)
Blakeden Dr. KT10: Clay 79Ha 152
Blakefield Gdns. CR5: Coul 90Pb 176
Blake Gallery, The (off Woodville Pl.)
Blake Gdns. DA1: Dart 56Pd 119
Blake Gdns. SW6 53Db 111
Blake Hall Cres. E11 32Jc 73
Blake Hall Rd. E11 31Jc 73
Blakehall Rd. SM5: Cars 79Hb 155
Blake Ho. E14 47Cc 92
(off Admirals Way)
Blake Ho. SE1 3K 229 (48Qb 90)
Blake Ho. SE8 51Cc 114
(off New King St.)
Blake M. TW9: Kew 53Qa 109
Blakemore Gdns. SW13 51Xa 110
Blakemore Rd. CR7: Thor H 71Pb 156
Blakemore Rd. SW16 62Nb 134
Blakemore Way DA17: Belv 48Ad 95
Blakeney Av. BR3: Beck 67Bc 136
Blakeney Cl. E8 36Wb 71
Blakeney Cl. KT19: Eps 83Ta 173
Blakeney Cl. N20 18Eb 31
Blakeney Cl. NW1 38Mb 70
Blakeney Rd. BR3: Beck 66Bc 136
Blakenham Rd. SW17 63Hb 133
Blaker Ct. SE7 52Lc 115
Blake Rd. CR0: C'don 75Ub 157
Blake Rd. CR4: Mitc 69Gb 133
Blake Rd. E16 42Hc 93
Blake Rd. N11 24Lb 50
Blaker Rd. E15 40Ec 72
Blakes Av. KT3: N Mald 71Va 154
Blakes Ct. KT16: Chert 74J 149
Blake's Grn. BR4: W W'ck 74Ec 158
BLAKES GREEN 97Td 204
Blakes La. KT3: N Mald 71Va 154
Blakesley Av. W5 44La 86
Blakesley Ho. E12 34Qc 74
(off Grantham Rd.)
Blakesley Wlk. SW20 68Bb 133
Blake's Rd. SE15 52Ub 113
Blakes Ter. KT3: N Mald 71Wa 154
Blakesware Gdns. N9 17Tb 33
Blake Twr. EC2 7D 218 (43Sb 91)
(off Fann St.)
Blake Way RM18: Tilb 4E 122
(off Coleridge Rd.)
Blakewood Cl. TW13: Hanw 63Y 129
Blakewood Ct. SE20 66Xb 135
(off Anerley Pk.)
Blanchard Cl. SE9 62Nc 138
Blanchard Dr. WD18: Wat 14U 26
Blanchard Gro. EN3: Enf L 10Dc 20
Blanchard Ho. TW1: Twick 58Ma 109
(off Clevedon Rd.)
Blanchard M. RM3: Hrld W 24Pd 57
Blanchards Hill GU4: Jac W 100A 188
Blanchards Hill GU4: Sut G 100A 188
Blanchard Way E8 37Wb 71
Blanch Cl. SE15 52Yb 114
Blanchedowne SE5 56Tb 113
Blanche La. EN6: S Mim 4Wa 16
Blanche St. E16 42Hc 93
Blanchland Rd. SM4: Mord 71Db 155
Blanchman's Rd. CR6: W'ham 90Ac 178
Blandfield Rd. SW12 59Jb 112
Blandford Av. BR3: Beck 68Ac 136
Blandford Av. TW2: Whitt 60Da 107
Blandford Cl. CR0: Bedd 76Nb 156
Blandford Cl. GU22: Wok 89D 168
Blandford Cl. N2 28Eb 49
Blandford Cl. RM7: Mawney 28Dd 56
Blandford Cl. SL3: L'ly 8P 81
Blandford Ct. N1 38Ub 71
(off St Peter's Way)
Blandford Ct. NW6 38Za 68
Blandford Ct. SL3: L'ly 8P 81
Blandford Cres. E4 17Ec 34
Blandford Ho. SW8 52Pb 112
(off Richborne Ter.)
Blandford Rd. AL1: St A 2E 6
Blandford Rd. BR3: Beck 68Yb 136
Blandford Rd. TW11: Tedd 64Fa 130
Blandford Rd. UB2: S'hall 49Ca 85
Blandford Rd. W4 48Ua 88
Blandford Rd. W5 47Ma 87
Blandford Rd. Nth. SL3: L'ly 8P 81
Blandford Rd. Sth. SL3: L'ly 8P 81
Blandford Sq. NW1 6E 214 (42Gb 89)
Blandford St. W1 2G 221 (44Hb 89)
Blandford Waye UB4: Yead 44Y 85
Bland Ho. SE11 7J 229 (50Pb 90)
(off Vauxhall St.)
Bland St. SE9 56Mc 115
Blaney Cres. E6 41Rc 94
Blanford M. RH2: Reig 6M 207
Blanford Rd. RH2: Reig 7L 207
Blanmerle Rd. SE9 60Rc 116
Blann Cl. SE9 58Mc 115
Blantyre St. SW10 52Fb 111
Blantyre Twr. SW10 52Fb 111
(off Worlds End Est.)
Blantyre Wlk. SW10 52Fb 111
(off Worlds End Est.)
Blashford NW3 38Hb 69
(off Adelaide Rd.)
Blashford St. SE13 59Fc 115
Blashill Ct. E14 45Ec 92
(off Bullivant St.)
Blasker Wlk. E14 50Dc 92
Blatchford Ct. KT12: Walt T 75W 150
Blattner Cl. WD6: E'tree 14Na 29
Blaven Path E16 42Hc 93
Blawith Rd. HA1: Harr 28Ga 46
Blaxland Ho. W12 45Xa 88
(off White City Est.)
Blaydon Cl. HA4: Ruis 31U 64
Blaydon Cl. N17 24Xb 51
Blaydon Ct. UB5: N'olt 37Ca 65
Blaydon Wlk. N17 24Xb 51
Blays Cl. TW20: Eng G 5N 125
Blay's La. TW20: Eng G 6M 125
Blazer Ct. NW8 4C 214 (41Fb 89)
(off St John's Wood Rd.)
Bleak Hill La. SE18 51Vc 117
Bleak Ho. La. W4 50Ta 87
(off Chiswick High Rd.)
Blean Gro. SE20 66Yb 136
Bleasdale Av. UB6: G'frd 40Ja 66
Blechynden Ho. W10 44Za 88
(off Kingsdown Cl.)
Blechynden St. W10 45Za 88
Bledlow Cl. NW8 6C 214 (42Fb 89)
Bledlow Cl. SE28 45Yc 95
Bledlow Ri. UB6: G'frd 40Ea 66
Bleeding Heart Yd. EC1 1A 224 (43Qb 90)
(off Greville St.)
Blegborough Rd. SW16 65Lb 134
Blemundsbury WC1 7H 217 (43Pb 90)
(off Dombey St.)
Blencarn Cl. GU21: Wok 8K 167
BLENDON 58Zc 117
Blendon Dr. DA5: Bexl 58Zc 117
Blendon Path BR1: Broml 66Hc 137
Blendon Rd. DA5: Bexl 58Zc 117
Blendon Row SE17 6F 231 (49Tb 91)
(off Orb St.)
Blendon Ter. SE18 50Sc 94
Blendworth Point SW15 60Xa 110
Blenheim Av. IG2: Ilf 30Qc 54
Blenheim Bus. Cen. CR4: Mitc 68Hb 133
(off London Rd.)
Blenheim Cen., The 55Da 107
Blenheim Cl. DA1: Dart 58Ld 119
Blenheim Cl. KT14: W Byf 85H 169
Blenheim Cl. N21 18Sb 33
Blenheim Cl. RM14: Upm 32Ud 78
Blenheim Cl. RM7: Mawney 28Ed 56
Blenheim Cl. SE12 60Kc 115
Blenheim Cl. SL3: L'ly 46B 82
Blenheim Cl. SM6: W'gton 80Lb 156
Blenheim Cl. SW20 69Ya 132
Blenheim Cl. UB6: G'frd 40Fa 66
Blenheim Cl. WD19: Wat 17Y 27
Blenheim Ct. BR2: Broml 70Hc 137
Blenheim Ct. DA14: Sidc 62Tc 138
Blenheim Ct. HA3: Kenton 30Ja 46
Blenheim Ct. IG8: Wfd G 24Kc 53
Blenheim Ct. N19 33Nb 70
Blenheim Ct. N7 37Nb 70
Blenheim Ct. RM12: Horn 36Ld 77
Blenheim Ct. RM13: Rain 40Fd 76
(off Lowen Rd.)
Blenheim Ct. SE10 50Jc 93
(off Denham St.)
Blenheim Ct. SE16 46Zb 92
(off King & Queen Wharf)
Blenheim Ct. SM2: Sutt 79Eb 155
Blenheim Ct. TW18: Staines 63F 126
Blenheim Cres. CR2: S Croy 80Sb 157
Blenheim Cres. HA4: Ruis 33T 64
Blenheim Cres. W11 45Ab 88
Blenheim Dr. DA16: Well 53Vc 117
Blenheim Gdns. CR2: Sande 84Wb 177
Blenheim Gdns. GU22: Wok 1M 187
Blenheim Gdns. HA9: Wemb 34Na 67
Blenheim Gdns. KT2: King T 66Ra 131
Blenheim Gdns. NW2 37Ya 68
Blenheim Gdns. RM15: Avel 46Rd 97
Blenheim Gdns. SM6: W'gton 79Lb 156
Blenheim Gdns. SW2 58Pb 112
Blenheim Gro. DA12: Grav'nd 9E 122
Blenheim Gro. SE15 54Wb 113
Blenheim Ho. E16 46Kc 93
(off Constable Av.)
Blenheim Ho. SE18 48Sc 94
Blenheim Ho. SW3 7E 226 (50Gb 89)
(off Kings Rd.)
Blenheim Ho. TW3: Houn 55Ca 107
Blenheim M. RM4: Stfrd 17Fd 38
Blenheim M. WD7: Shenl 5Na 15
Blenheim Pde. UB10: Hil 42R 84
Blenheim Pk. Rd. CR2: S Croy 81Sb 177
Blenheim Pas. NW8 1A 214 (40Eb 69)
Blenheim Pl. TW11: Tedd 64Ha 130
Blenheim Ri. N15 28Vb 51
Blenheim Rd. AL1: St A 1D 6
Blenheim Rd. BR1: Broml 70Nc 138
Blenheim Rd. BR6: Orp 75Yc 161
Blenheim Rd. CM15: Pil H 16Wd 40
Blenheim Rd. DA1: Dart 58Ld 119
Blenheim Rd. DA15: Sidc 60Yc 117
Blenheim Rd. E15 35Gc 73
Blenheim Rd. E17 27Zb 52
Blenheim Rd. E6 41Mc 93
Blenheim Rd. EN5: Barn 13Za 30
Blenheim Rd. HA2: Harr 30Da 45
Blenheim Rd. KT19: Eps 83Ta 173
Blenheim Rd. NW8 1A 214 (40Eb 69)
Blenheim Rd. SE20 66Yb 136
Blenheim Rd. SL3: L'ly 9P 81
Blenheim Rd. SM1: Sutt 76Cb 155
Blenheim Rd. SW20 69Ya 132
Blenheim Rd. UB5: N'olt 37Da 65
Blenheim Rd. W4 48Ua 88
Blenheim Rd. WD5: Ab L 4W 12
Blenheim Shop. Cen. 66Yb 136
Blenheim St. W1 3K 221 (44Kb 90)
Blenheim Ter. NW8 1A 214 (40Eb 69)
Blenheim Twr. SE14 52Ac 114
(off Batavia Rd.)
Blenheim Way TW7: Isle 53Ja 108
Blenkarne Rd. SW11 58Hb 111
Bleriot Av. KT15: Add 80J 149
Bleriot Rd. TW5: Hest 52Y 107
Blessbury Rd. HA8: Edg 25Sa 47
Blessington Cl. SE13 55Fc 115
Blessington Rd. SE13 55Fc 115
Blessing Way IG11: Bark 41Yc 95
BLETCHINGLEY 5J 209
Bletchingley Castle 5J 209
Bletchingley Cl. CR7: Thor H 70Rb 135
Bletchingley Cl. RH1: Mers 1C 208
Bletchingley Golf Course 5K 209
Bletchingley Rd. RH1: Mers 1C 208
Bletchingley Rd. RH1: Nutf 5G 208
Bletchingley Rd. RH9: G'stone 3N 209
Bletchley Ct. HA7: Stan 24Na 47
(off Hitchin Way)
Bletchley Ct. N1 2F 219 (40Tb 71)
(off Bletchley St.)
Bletchley St. N1 2F 219 (40Tb 71)
Bletchmore Cl. UB3: Harl 50T 84
Bletsoe Wlk. N1 1E 218 (40Sb 71)
Blewbury Ho. SE2 47Yc 95
(not continuous)
Blewitts Cotts. RM13: Rain 41Hd 96
(off Dunedin Rd.)
Blick Ho. SE16 48Yb 92
(off Neptune St.)
Bligh Rd. DA11: Grav'nd 8C 122
Bligh's Ct. TN13: S'oaks 97Ld 203
(off Bligh's Rd.)
Bligh's Mdw. TN13: S'oaks 97Ld 203
(off High St.)
Bligh's Rd. TN13: S'oaks 97Ld 203
Bligh's Wlk. TN13: S'oaks 97Kd 203
(off Bligh's Rd.)
Blincoe Cl. SW19 61Za 132
Blinco La. SL3: Geor G 44A 82
Blind La. EN9: Walt A 5Lc 21
Blind La. GU24: Chob 2D 166
Blind La. IG10: Lough 12Gc 35
Blind La. RH8: Oxt 2J 211
Blind La. SM7: Bans 87Gb 175
Blindman's La. EN8: Chesh 2Zb 20
Bliss Cres. SE13 54Dc 114
Blissett St. SE10 53Ec 114
Bliss Ho. EN1: Enf 10Wb 19
Bliss M. W10 41Ab 88
Blisworth Cl. UB4: Yead 42Aa 85
Blisworth Ho. E2 39Wb 71
(off Whiston Rd.)
Blithbury Rd. RM9: Dag 37Xc 75
Blithdale Rd. SE2 49Wc 95
Blithehale Ct. E2 41Xb 91
(off Witan St.)
Blithfield St. W8 48Db 89
Blockhouse Rd. RM17: Grays 51Ee 121
Blockley Rd. HA0: Wemb 33Ka 66
Block Wharf E14 47Cc 92
(off Cuba St.)
Bloemfontein Av. W12 46Xa 88
Bloemfontein Rd. W12 45Xa 88
Bloemfontein Way W12 46Xa 88
Blomfield Ct. W9 5A 214 (42Eb 89)
(off Maida Vale)
Blomfield Mans. W12 46Ya 88
(off Stanlake Rd.)
Blomfield Rd. W9 43Db 89
Blomfield St. EC2 1G 225 (43Tb 91)
Blomfield Vs. W2 43Db 89
Blomville Rd. RM8: Dag 34Ad 75
Blondell Cl. UB7: Harm 51M 105
Blondel St. SW11 54Jb 112
Blondin Av. W5 49La 86
Blondin Pk. & Nature Area 49Ka 86
Blondin St. E3 40Cc 72
Blondin Way SE16 47Ac 92
Blomberg Arc. EC4 3F 225 (44Tb 91)
Bloomberg Ct. SW1 6D 228 (49Mb 90)
(off Vauxhall Bri. Rd.)
Bloomburg St. SW1 6D 228 (49Mb 90)
Bloomfield Cl. GU21: Knap 9J 167
Bloomfield Ct. E10 34Dc 72
(off Brisbane Rd.)
Bloomfield Ct. N6 30Jb 50
Bloomfield Cres. IG2: Ilf 30Rc 54
Bloomfield Ho. E1 43Wb 91
(off Old Montague St.)
Bloomfield Pl. W1 4A 222 (45Kb 90)
(off Bourdon St.)
Bloomfield Rd. BR2: Broml 71Mc 159
Bloomfield Rd. KT1: King T 70Na 131
Bloomfield Rd. N6 30Jb 50
Bloomfield Rd. SE18 51Rc 116
Bloomfield Ter. SW1 7J 227 (50Jb 90)
Bloomfield Ter. TN16: Westrm 97Tc 200
Bloomfield Wlk. RM16: Ors 3C 100
Bloom Gro. SE27 62Rb 135
Bloomhall Rd. SE19 64Tb 135
Bloom Ho. E3 40Dc 72
(off Alameda Pl.)
Bloom Pk. Rd. SW6 52Bb 111
BLOOMSBURY 7F 217 (43Nb 90)
Bloomsbury Cl. KT19: Eps 82Ta 173
Bloomsbury Cl. NW7 24Wa 48
Bloomsbury Cl. W5 45Pa 87
Bloomsbury Ct. HA5: Pinn 27Ba 45
Bloomsbury Ct. TW5: Cran 53X 107
Bloomsbury Ct. WC1 1G 223 (43Nb 90)
(off Barter St.)
Bloomsbury Ho. SW4 58Mb 112
Bloomsbury Mans. BR1: Broml 67Kc 137
(off Widmore Rd.)
Bloomsbury M. IG8: Wfd G 23Nc 54
Bloomsbury Pl. SW18 57Eb 111
Bloomsbury Pl. WC1 1G 223 (43Nb 90)
Bloomsbury Sq. WC1 1G 223 (43Nb 90)
Bloomsbury St. WC1 1E 222 (43Mb 90)
Bloomsbury Theatre 5D 216 (42Mb 90)
(off Gordon St.)
Bloomsbury Way WC1 1F 223 (43Nb 90)
Blore Cl. SW8 53Mb 112
Blore Ct. W1 3D 222 (44Mb 90)
(off Berwick St.)
Blore Ho. SW10 52Db 111
(off Coleridge Gdns.)
Blossom Av. HA2: Harr 33Da 65
Blossom Cl. CR2: S Croy 78Vb 157
Blossom Cl. RM9: Dag 39Bd 75
Blossom Cl. W5 47Na 87
Blossom Ct. SE15 52Vb 113
(off All Saints Walk)
Blossom Dr. BR6: Orp 75Vc 161
Blossom La. EN2: Enf 11Sb 33
Blossom Pl. SE28 48Sc 94
Blossom St. E1 6J 219 (42Ub 91)
Blossom Way UB10: Hil 38P 63
Blossom Way UB7: W Dray 49Q 84
Blossom Waye TW5: Hest 51Aa 107
Blount M. UB10: Uxb 40N 63
Blount St. E14 44Ac 92
Bloxam Gdns. SE9 57Nc 116
Bloxhall Rd. E10 32Bc 72
Bloxham Cres. TW12: Hamp 66Ba 129
Bloxworth Cl. SM6: W'gton 76Lb 156
Blucher Rd. SE5 52Sb 113
Blue Anchor All. TW9: Rich 56Na 109
Blue Anchor La. RM18: W Til 9F 100
Blue Anchor La. SE16 49Wb 91
Blue Anchor Yd. E1 45Wb 91
Blue Ball La. TW20: Egh 64B 126
Blue Ball Yd. SW1 7B 222 (46Lb 90)
Blue Barn La. KT13: Weyb 83Q 170
Bluebell Apts. N4 32Sb 71
(off Swan La.)
Bluebell Av. E12 36Mc 73
Bluebell Cl. AL2: Park 9A 6
Bluebell Cl. BR6: Farnb 75Sc 160
Bluebell Cl. E9 39Yb 72
Bluebell Cl. HP1: Hem H 3G 2
Bluebell Cl. RM7: Rush G 33Gd 76
Bluebell Cl. SE26 63Vb 135
Bluebell Cl. SM6: W'gton 74Kb 156
Bluebell Cl. UB5: N'olt 37Ba 65
Bluebell Ct. GU22: Wok 1P 187
Bluebell Ct. NW9 25Ua 48
(off Heybourne Cres.)
Bluebell Dr. EN7: G Oak 1Tb 19
Bluebell Dr. WD5: Bedm 9F 4
Bluebell Ho. SE16 47Ac 92
(off Bondin Way)
Blue Bell La. KT11: Stoke D 89Ca 171
Bluebell La. KT24: E Hor 100U 190
Bluebell M. RM17: Grays 50De 99
Bluebell Ter. UB7: W Dray 47P 83
Bluebell Wlk. HP2: Hem H 2D 4
Bluebell Way IG1: Ilf 37Rc 74
Blueberry Cl. IG8: Wfd G 23Jc 53
Blueberry Cl. WD19: Wat 17Y 27
Blueberry Gdns. CR5: Coul 88Pb 176
Blueberry La. TN14: Knock 88Yc 181
Bluebird Cl. SW20 68Ya 132
Bluebird Ho. IG11: Bark 41Vc 95
Bluebird La. RM10: Dag 38Cd 76
Bluebird Way AL2: Brick W 2Ba 13
Bluebird Way SE28 47Tc 94
Blue Boar All. EC3 2K 225 (44Vb 91)
(off Aldgate High St.)
Bluebridge Av. AL9: Brk P 9G 8
Bluebridge Rd. AL9: Brk P 8G 8
Blue Bldg. SE10 50Hc 93
(off Glenforth St.)
Blue Cedars SM7: Bans 86Za 174
Blue Cedars Pl. KT11: Cobh 84Z 171
Blue Chalet Ind. Pk. TN15: W King 78Td 164

Blue Ct. N1 .... 39Tb 71
(off Sherborne St.)
Blue Elephant Theatre 52Sb 113
(off Bethwin Rd.)
Bluefield Cl. TW12: Hamp 64Ca 129
Blue Fin Bldg. SE1 .... 6C 224 (46Rb 91)
(off Summer St.)
Bluegate M. E1 .... 45Xb 91
Bluegate Pk. CM14: B'wood 20Xd 40
Bluegates KT17: Ewe 80Wa 154
Blue Ho. Cotts. DA2: Bean 62Zd 143
Bluehouse Gdns. RH8: Oxt 100Jc 199
Bluehouse Hill AL3: St A 3N 5
Bluehouse La. RH8: Limp 100Gc 199
Bluehouse La. RH8: Oxt 100Gc 199
Bluehouse Rd. E4 .... 19Gc 35
Blue Leaves Av. CR5: Coul 93Mb 196
Blue Lion Pl. SE1 .... 3H 231 (48Ub 91)
Blueprint Apts. SW12 .... 59Kb 112
(off Balham Gro.)
Blue Riband Ind. Est. CR0: C'don 75Rb 157
Blues St. E8 .... 37Vb 71
Bluett Rd. AL2: Lon C 9H 7
Blue Water SW18 .... 56Db 111
BLUEWATER 60Vd 120
Bluewater Bus Station 60Wd 120
Bluewater Parkway DA9: Bluew 59Vd 120
Bluewater Parkway DA9: Ghithe 59Vd 120
Bluewater Shop. Cen. 59Vd 120
Blumenthal Cl. TW7: Isle 52Fa 108
Blumfield Ct. SL1: Slou 2B 80
Blumfield Cres. SL1: Slou 2B 80
Blundel La. KT11: Stoke D 88Ba 171
Blundell Cl. BR5: St M Cry 72Yc 161
Blundell Cl. E8 .... 36Wb 71
Blundell Rd. HA8: Edg 25Ta 47
Blundell St. N7 .... 38Nb 70
Blunden Cl. RM8: Dag 32Yc 75
Blunden Ct. SW6 .... 52Cb 111
(off Farm La.)
Blunden Dr. SL3: L'ly 49D 82
Blunesfield EN6: Pot B 3Fb 17
Blunt Rd. CR2: S Croy 78Tb 157
Blunts Av. UB7: Sip 52Q 106
Blunts La. AL2: Pot C 6K 5
Blunts La. WD5: Bedm 9K 5
Blunts Rd. SE9 .... 57Qc 116
Blurton Rd. E5 .... 35Yb 72
Blydon Ct. N21 .... 15Pb 32
(off Chaseville Pk. Rd.)
Blyth Cl. E14 .... 49Fc 93
Blyth Cl. TW1: Twick 58Ha 108
Blyth Cl. WD6: Bore 11Pa 29
Blyth Ct. BR1: Broml 67Hc 137
(off Blyth Rd.)
Blythe Cl. SE6 .... 59Bc 114
Blythe Cl. SL0: Iver 44H 83
Blythe Hill BR5: St P 67Vc 139
Blythe Hill SE6 .... 59Bc 114
BLYTHE HILL 59Bc 114
Blythe Hill La. SE6 .... 59Bc 114
Blythe Hill Pl. SE23 .... 59Ac 114
Blythe Ho. SE11 .... 51Qb 112
Blythe Ho. SL1: Slou 6B 80
Blythe M. W14 .... 48Za 88
Blythendale Ho. E2 .... 40Wb 71
(off Mansford St.)
Blythe Rd. W14 .... 48Za 88
Blythe St. E2 .... 41Xb 91
Blytheswood HP3: Hem H 5J 3
Blytheswood Pl. SW16 .... 63Pb 134
Blythe Va. SE6 .... 60Bc 114
Blyth Hill Pl. SE23 .... 59Ac 114
(off Brockley Pk.)
Blyth Ho. DA8: Erith 50Gd 96
Blyth Rd. BR1: Broml 67Hc 137
Blyth Rd. E17 .... 31Bc 72
Blyth Rd. SE28 .... 45Yc 95
Blyth Rd. UB3: Hayes 47U 84
Blyth's Wharf E14 .... 45Ac 92
Blythswood Rd. IG3: Ilf 32Wc 75
Blyth Wlk. RM14: Upm 30Ud 58
Blyth Wood Pk. BR1: Broml 67Hc 137
Blythwood Rd. HA5: Pinn 25Z 45
Blythwood Rd. N4 .... 31Nb 70
BMI CAVELL HOSPITAL 12Qb 32
BMI CHELSFIELD PARK HOSPITAL 78Bd 161
BMI CITY MEDICAL 2H 225 (44Ub 91)
(off St. Helen's Pl.)
BMI URGENT CARE CEN. 34Ha 66
BMX Track London, The 51Tb 113
Boades M. NW3 .... 35Fb 69
Boadicea Cl. SL1: Slou 6C 80
Boadicea St. N1 .... 1H 217 (39Pb 70)
Boakes Cl. NW9 .... 28Sa 47
Boakes Mdw. TN14: S'ham 83Hd 182
Boar Cl. IG7: Chig 22Wc 55
Boardman Av. E4 .... 15Dc 34
Boardman Cl. EN5: Barn 15Ab 30
Boardman Pl. CM14: B'wood 20Xd 40
Board School Rd. GU21: Wok 88B 168
Boardwalk Pl. E14 .... 46Ec 92
Boarlands Cl. SL1: Slou 5D 80
Boarlands Path SL1: Slou 5D 80
Boarley Ho. SE17 .... 6H 231 (49Ub 91)
(off Massinger St.)
Boars Head Yd. TW8: Bford 52Ma 109
Boatemah Wlk. SW9 .... 54Qb 112
(off Peckford Pl.)
Boaters Av. TW8: Bford 52La 108
Boathouse, The E14 .... 44Cc 92
Boathouse Cen., The W10 .... 42Za 88
(off Canal Cl.)
Boathouse Wlk. SE15 .... 52Vb 113
(not continuous)
Boat La. E2 .... 1K 219 (39Vb 71)
Boat Lifter Way SE16 .... 49Ac 92
Boat Quay E16 .... 45Lc 93
Boatyard Apts. E14 .... 50Dc 92
Bob Anker Cl. E13 .... 41Jc 93
Bobbin Cl. SM6: W'gton 75Jb 156
Bobbin Cl. SW4 .... 55Lb 112
Bobby Moore Way IG11: Bark 39Sc 74
Bobby Moore Way N10 .... 24Hb 49
Bob Dunn Way DA1: Dart 56Ld 119
Bob Hope Theatre, The 58Pc 116
Bob Marley Way SE24 .... 56Qb 112
Bobs La. RM1: Rom 24Jd 56
Bocketts Farm Pk. 97Ha 192
Bocketts La. KT22: Fet 96Ha 192
Bockhampton Rd. KT2: King T 66Pa 131
Bocking St. E8 .... 39Xb 71
Boddicott Cl. SW19 .... 61Ab 132
Boddington Gdns. W3 .... 47Qa 87
Boddington Ho. SE14 .... 53Yb 114
(off Pomeroy St.)
Boddington Ho. SW13 .... 51Xa 110
(off Wyatt Dr.)
Bodell Cl. RM16: Grays 48De 99
Bodeney Ho. SE5 .... 53Ub 113
(off Peckham Rd.)
Boden Ho. E1 .... 43Wb 91
(off Woodseer St.)
Bodiam Cl. EN1: Enf 12Ub 33
Bodiam Rd. SW16 .... 66Mb 134
Bodiam Way NW10 .... 41Pa 87
Bodicea M. TW4: Houn 58Ba 107
Bodington Ct. W12 .... 47Za 88
Bodium Ct. E17 .... 25Bc 52
(off Thornbury Way)
Bodle Av. DA10: Swans 59Ae 121
Bodleian Ho. SE20 .... 67Wb 135
Bodley Cl. CM16: Epp 2Vc 23
Bodley Cl. KT3: N Mald 71Ua 154
Bodley Ho. SL0: Iver H 40F 62
Bodley Mnr. Way SW2 .... 59Qb 112
Bodley Rd. KT3: N Mald 72Ta 153
Bodley Way SE17 .... 5E 230 (49Sb 91)
Bodmin Av. SL2: Slou 3E 80
Bodmin Cl. BR5: Orp 74Yc 161
Bodmin Cl. HA2: Harr 34Ba 65
Bodmin Ct. RM3: Rom 25Md 57
Bodmin Gro. SM4: Mord 71Db 155
Bodmin St. SW18 .... 60Cb 111
Bodnant Gdns. SW20 .... 69Wa 132
Bodney Rd. E8 .... 36Xb 71
Bodwell Cl. HP1: Hem H 1J 3
Boeing Way UB2: S'hall 48X 85
Boevey Path DA17: Belv 50Bd 95
Bogart Ct. E14 .... 45Cc 92
(off Premiere Pl.)
Bogey La. BR6: Downe 80Qc 160
Bognor Gdns. WD19: Wat 22Y 45
Bognor Rd. DA16: Well 53Zc 117
Bohemia HP2: Hem H 1N 3
Bohemia Pl. E8 .... 37Yb 72
Bohn Rd. E1 .... 43Ac 92
Bohun Gro. EN4: E Barn 16Gb 31
Boileau Pde. W5 .... 44Pa 87
(off Boileau Rd.)
Boileau Rd. SW13 .... 52Wa 110
Boileau Rd. W5 .... 44Pa 87
Boiler Ho., The UB3: Hayes 47U 84
(off Material Wlk.)
Bois Hall Rd. KT15: Add 78M 149
Boisseau Ho. E1 .... 43Yb 92
(off Stepney Way)
Boissy Cl. AL4: St A 3J 7
Bolanachi Bldg. SE16 .... 3K 231 (48Vb 91)
Bolander Gro. SW6 .... 51Cb 111
Bolberry Rd. RM5: Col R 22Fd 56
Bolden St. SE8 .... 54Dc 114
Boldero Pl. NW8 .... 6D 214 (42Gb 89)
(off Gateforth St.)
Bolderwood Way BR4: W W'ck 75Dc 158
Bolding Ho. La. GU24: W End 5D 166
Boldmere Rd. HA5: Eastc 31Y 65
Bold's Ct. SL2: Stoke P 8L 61
Bolen M. KT16: Chert 75J 149
Boleyn Av. EN1: Enf 11Xb 33
Boleyn Av. KT17: Ewe 82Xa 174
Boleyn Cl. E17 .... 28Cc 52
Boleyn Cl. IG10: Lough 16Nc 36
Boleyn Cl. RM16: Chaf H 48Be 99
Boleyn Cl. TW18: Staines 64G 126
Boleyn Ct. GU21: Knap 10G 166
(off Tudor Way)
Boleyn Ct. IG9: Buck H 18Jc 35
Boleyn Ct. KT8: E Mos 70Fa 130
(off Bridge Rd.)
Boleyn Ct. RH1: Redh 5A 208
(off St Anne's Ri.)
Boleyn Dr. AL1: St A 4B 6
Boleyn Dr. HA4: Ruis 33Z 65
Boleyn Dr. KT8: W Mole 69Ba 129
Boleyn Gdns. BR4: W W'ck 75Dc 158
Boleyn Gdns. CM13: B'wood 20Ce 41
Boleyn Gdns. RM10: Dag 38Ed 76
Boleyn Gro. BR4: W W'ck 75Ec 158
Boleyn Ho. E16 .... 46Jc 93
(off Southey M.)
Boleyn Rd. E6 .... 40Mc 73
Boleyn Rd. E7 .... 38Jc 73
Boleyn Rd. N16 .... 36Ub 71
Boleyn Rd. TN15: Kems'g 89Nd 183
Boleyn Row CM16: Epp 1Yc 23
Boleyn Wlk. KT22: Lea 92Ha 192
Boleyn Way DA10: Swans 59Ae 121
Boleyn Way EN5: New Bar 13Eb 31
Boleyn Way IG6: Ilf 23Sc 54
Bolina Rd. SE16 .... 50Yb 92
Bolinder Way E3 .... 41Ec 92
Bolingbroke Cl. EN4: Cockf 13Hb 31
Bolingbroke Gro. SW11 .... 56Gb 111
Bolingbroke Rd. W14 .... 48Za 88
Bolingbroke Wlk. SW11 .... 53Fb 111
Bolingbroke Way UB3: Hayes 46T 84
Bolliger Ct. NW10 .... 42Sa 87
Bollinder Pl. EC1 .... 3D 218 (41Sb 91)
Bollo Bri. Rd. W3 .... 48Ra 87
Bollo La. W3 .... 47Ra 87
Bollo La. W4 .... 49Sa 87
Bolney Ga. SW7 .... 2D 226 (47Gb 89)
Bolney St. SW8 .... 52Pb 112
Bolney Way TW13: Hanw 62Aa 129
Bolsover Gro. RH1: Mers 1E 208
Bolsover St. W1 .... 6A 216 (42Kb 90)
Bolstead Rd. CR4: Mitc 67Kb 134
Bolster Gro. N22 .... 24Mb 50
Bolt Cellar La. CM16: Epp 2Uc 22
Bolt Ct. EC4 .... 3A 224 (44Qb 90)
Bolters La. SM7: Bans 86Bb 175
Bolt Ho. N1 .... 1J 219 (39Ub 71)
(off Phillipp St.)
Boltmore Cl. NW4 .... 27Za 48
Bolton Av. SL4: Wind 5G 102
Bolton Cl. KT9: Chess 79Ma 153
Bolton Cl. SE20 .... 68Wb 135
Bolton Cres. SE11 .... 52Qb 112
Bolton Cres. SE5 .... 52Rb 113
Bolton Cres. SL4: Wind 5G 102
Bolton Dr. SM4: Mord 73Eb 155
Bolton Gdns. BR1: Broml 65Hc 137
Bolton Gdns. NW10 .... 40Za 68
Bolton Gdns. SW5 .... 50Db 89
Bolton Gdns. TW11: Tedd 65Ja 130
Bolton Gdns. M. SW10 .... 50Eb 89
Bolton Ho. SE10 .... 50Gc 93
(off Trafalgar Rd.)
Bolton Ho. SE11 .... 5B 230 (49Rb 91)
(off George Mathers Rd.)
Bolton Pl. NW8 .... 39Db 69
(off Bolton Rd.)
Bolton Rd. E15 .... 37Hc 73
Bolton Rd. HA1: Harr 28Ea 46
Bolton Rd. KT9: Chess 79Ma 153
Bolton Rd. N18 .... 22Vb 51
Bolton Rd. NW10 .... 39Ua 68
Bolton Rd. NW8 .... 39Db 69
Bolton Rd. SL4: Wind 5G 102
Bolton Rd. W4 .... 52Sa 109
Boltons, The HA0: Wemb 35Ha 66
Boltons, The IG8: Wfd G 21Jc 53
Boltons, The SW10 .... 7A 226 (50Eb 89)
Boltons Cl. GU22: Pyr 88J 169
Boltons Ct. SW5 .... 50Db 89
(off Old Brompton Rd.)
Bolton's La. UB3: Harl 53S 106
Boltons La. GU22: Pyr 88J 169
Boltons Pl. SW5 .... 50Eb 89
Bolton St. W1 .... 6A 222 (46Kb 90)
Bolton Studios SW10 .... 50Eb 89
Bolton Wlk. N7 .... 33Pb 70
(off Durham Rd.)
Bombay Ct. SE16 .... 47Yb 92
(off St Marychurch St.)
Bombay St. SE16 .... 49Xb 91
Bombers La. TN16: Westrm 91Tc 200
Bomer Cl. UB7: Sip 52Q 106
Bomore Rd. W11 .... 45Ab 88
Bonaly Ho. RH8: Oxt 3G 210
Bonar Pl. BR7: Chst 66Nc 138
Bonar Rd. SE15 .... 52Wb 113
Bonaventure Ct. DA12: Grav'nd 3H 145
Bonchester Cl. BR7: Chst 66Qc 138
Bonchurch Cl. SM2: Sutt 80Db 155
Bonchurch Ct. RM19: Purf 50Sd 98
Bonchurch Rd. W10 .... 43Ab 88
Bonchurch Rd. W13 .... 46Ka 86
Bond Cl. DA16: Well 54Uc 116
Bond Cl. SL0: Iver H 38D 62
Bond Cl. TN14: Knock 87Zc 181
Bond Cl. UB7: Yiew 45M 83
Bond Ct. EC4 .... 4F 225 (45Tb 91)
Bondfield Av. UB4: Yead 41W 84
Bondfield Rd. E6 .... 43Pc 94
Bondfield Wlk. DA1: Dart 56Pd 119
Bond Gdns. SM6: W'gton 77Lb 156
Bond Ho. NW6 .... 40Bb 69
(off Rupert Rd.)
Bond Ho. SE14 .... 52Ac 114
(off Goodwood Rd.)
Bond Ho. TW8: Bford 50Ma 87
Bonding Yd. Wlk. SE16 .... 48Ac 92
Bond Rd. CR4: Mitc 68Gb 133
Bond Rd. CR6: W'ham 90Zb 178
Bond Rd. KT6: Surb 75Pa 153
Bond St. E15 .... 36Gc 73
Bond St. RM17: Grays 51Ee 121
Bond St. TN14: Knock 87Ad 181
Bond St. TW20: Eng G 4M 125
Bond St. W4 .... 49Ta 87
Bond St. W5 .... 45Ma 87
Bond Street Station (Underground & Crossrail) 3K 221 (44Kb 90)
Bondway SW8 .... 51Nb 112
Boneashe La. TN15: Plat 93Fe 205
Bone Mill La. RH9: G'stone 6C 210
Bonesgate Open Space Local Nature Reserve 79Qa 153
Boneta Rd. SE18 .... 48Pc 94
Bonfield Rd. SE13 .... 56Ec 114
Bonham Cl. DA17: Belv 50Bd 95
Bonham Dr. RM16: Ors 3C 100
Bonham Gdns. RM8: Dag 33Zc 75
Bonham Ga. KT12: Walt T 74V 150
(off New Zealand Av.)
Bonham Ho. W11 .... 46Bb 89
(off Boyne Ter. M.)
Bonham Rd. RM8: Dag 33Zc 75
Bonham Rd. SW2 .... 57Pb 112
Bonham Way DA11: Nflt 60Ee 121
Bonheur Rd. W4 .... 47Ta 87
Bonhill St. EC2 .... 6G 219 (42Tb 91)
Boniface Gdns. HA3: Hrw W 24Da 45
Boniface Rd. UB10: Ick 34R 64
Boniface Wlk. HA3: Hrw W 24Da 45
Bonington Ho. EN1: Enf 15Wb 33
Bonington Rd. RM12: Horn 36Md 77
Bonita M. SE4 .... 55Zb 114
Bon Marche M. SE27 .... 63Ub 135
Bonner Hill Rd. KT1: King T 68Pa 131
Bonner Rd. E2 .... 40Yb 72
Bonners Cl. GU22: Wok 94B 188
Bonnersfield Cl. HA1: Harr 30Ha 46
Bonnersfield La. HA1: Harr 30Ha 46
(not continuous)
Bonner St. E2 .... 40Yb 72
Bonner Wlk. RM16: Chaf H 48Be 99
Bonnet St. E16 .... 47Kc 93
Bonnett M. RM11: Horn 32Nd 77
Bonneville Gdns. SW4 .... 58Lb 112
Bonney Gro. EN7: Chesh 2Wb 19
Bonney Way BR8: Swan 68Gd 140
Bonnington Ct. UB5: N'olt 40Aa 65
(off Gallery Gdns.)
Bonnington Ho. N1 .... 2H 217 (40Pb 70)
Bonningtons CM13: B'wood 20De 41
Bonnington Sq. SW8 .... 51Pb 112
Bonnys Rd. RH2: Reig 7F 206
Bonny St. NW1 .... 38Lb 70
Bonser Rd. TW1: Twick 61Ha 130
Bonsey Cl. GU22: Wok 93A 188
Bonsey La. GU22: Wok 93A 188
Bonseys La. GU24: Chob 81B 168
Bonsey's Yd. UB8: Uxb 38M 63
Bonsor Dr. KT20: Kgswd 94Ab 194
Bonsor Ho. SW8 .... 53Lb 112
Bonsor St. SE5 .... 52Ub 113
Bonville Gdns. NW4 .... 28Xa 48
Bonville Rd. BR1: Broml 64Hc 137
Bonwick Ct. KT8: E Mos 71Fa 152
Bookbinders Cott. Homes N20 .... 20Hb 31
Bookbinders Ct. E1 .... 42Xb 91
(off Cudworth St.)
Booker Cl. E14 .... 43Bc 92
Booker Rd. N18 .... 22Wb 51
Bookham Comn. Rd. KT23: Bookh 93Aa 191
Bookham Ct. CR4: Mitc 68Gb 133
Bookham Ct. KT23: Bookh 95Ba 191
Bookham Gro. KT23: Bookh 98Da 191
Bookham Ind. Est. KT23: Bookh 96Ba 191
Bookham Rd. KT11: D'side 91Y 191
Bookham Station (Rail) 95Ba 191
Book Ho. N1 .... 2D 218 (40Sb 71)
Book M. WC2 .... 3E 222 (44Mb 90)
Boomes Ind. Est. RM13: Rain 42Hd 96
Boone Ct. N9 .... 20Yb 34
Boones Rd. SE13 .... 56Gc 115
Boone St. SE13 .... 56Gc 115
Boord St. SE10 .... 48Gc 93
Boot All. AL3: St A 2B 6
(off Chequer St.)
Boothby Ct. E4 .... 20Ec 34
Boothby Rd. N19 .... 33Mb 70
Booth Cl. E9 .... 39Xb 71
Booth Cl. SE28 .... 46Xc 95
Booth Ct. SE13 .... 55Dc 114
Booth Dr. TW18: Staines 65M 127
Booth Ho. TW8: Bford 52La 108
(off High St.)
Booth La. EC4 .... 4D 224 (45Sb 91)
(off Baynard St.)
Boothman Ho. HA3: Kenton 27Ma 47
Booth Rd. CR0: C'don 75Rb 157
Booth Rd. E16 .... 47Lc 93
Booth Rd. NW9 .... 26Ta 47
Booths Cl. AL9: Wel G 6F 8
Booth's Ct. CM13: Hut 16Ee 41
Booth's Pl. W1 .... 1C 222 (43Lb 90)
Boot Pde. HA8: Edg 23Qa 47
(off High St.)
Boot St. N1 .... 4H 219 (41Ub 91)
Bordars Rd. W7 .... 43Ga 86
Bordars Wlk. W7 .... 43Ga 86
Bordeaux Ho. E15 .... 36Gc 73
(off Luxembourg M.)
Borden Av. EN1: Enf 16Tb 33
Border Cres. SE26 .... 64Xb 135
Border Gdns. CR0: C'don 77Dc 158
Bordergate CR4: Mitc 67Hb 133
Border Rd. SE26 .... 64Xb 135
Borders Cres. IG10: Lough 14Rc 36
(off Border's La.)
Borderside SL2: Slou 4L 81
Border's La. IG10: Lough 14Qc 36
Borders Wlk. IG10: Lough 14Qc 36
(off Border's La.)
Bordesley Rd. SM4: Mord 71Db 155
Bordeston Ct. TW8: Bford 52La 108
(off The Ham)
Bordon Wlk. SW15 .... 59Wa 110
Boreas Wlk. N1 .... 2C 218 (40Rb 71)
(off Nelson Pl.)
Boreham Av. E16 .... 44Jc 93
Boreham Cl. E11 .... 32Ec 72
Boreham Holt WD6: E'tree 14Pa 29
Boreham Rd. N22 .... 26Sb 51
BOREHAMWOOD 13Qa 29
Borehamwood Ent. Cen. WD6: Bore 13Pa 29
Boreham Wood FC 12Ra 29
Borehamwood Ind. Pk. WD6: Bore 12Ta 29
Borehamwood Shop. Pk. 13Qa 29
Borehamwood Tourist Info. Point 12Ra 29
Boreman Ho. SE10 .... 51Ec 114
(off Thames St.)
Borgard Rd. SE18 .... 49Pc 94
Borkwood Pk. BR6: Orp 77Vc 161
Borkwood Way BR6: Orp 77Uc 160
Borland Cl. DA9: Ghithe 57Wd 120
Borland Rd. SE15 .... 56Yb 114
Borland Rd. TW11: Tedd 66Ka 130
Borley Ct. RM16: Ors 4F 100
Borley Ct. TW19: Stanw 60N 105
Bornedene EN6: Pot B 3Ab 16
Borneo St. SW15 .... 55Ya 110
BOROUGH, THE 2G 231 (47Tb 91)
Borough Grange CR2: Sande 84Wb 177
BOROUGH GREEN 92Be 205
Borough Green & Wrotham Station (Rail) 92Be 205
Borough Grn. Rd. TN15: Bor G 93Zd 205
(not continuous)
Borough Grn. Rd. TN15: Igh 93Zd 205
(not continuous)
Borough Grn. Rd. TN15: Wro 89Ce 185
Borough High St. SE1 .... 2E 230 (47Sb 91)
Borough Hill CR0: Wadd 76Rb 157
Borough Mkt. SE1 .... 7F 225 (46Tb 91)
(off Borough High St.)
Borough Rd. CR4: Mitc 68Gb 133
Borough Rd. KT2: King T 67Qa 131
Borough Rd. SE1 .... 3B 230 (48Rb 91)
Borough Rd. TN16: Tats 93Mc 199
Borough Rd. TW7: Isle 53Ga 108
Borough Sq. SE1 .... 2D 230 (47Sb 91)
(off McCoid Way)
Borough Station (Underground) 2E 230 (47Sb 91)
Borough Way EN6: Pot B 4Ab 16
Borrett Cl. SE17 .... 7D 230 (50Sb 91)
Borrodaile Rd. SW18 .... 58Db 111
Borromeo Way CM14: B'wood 18Xd 40
Borrowdale NW1 .... 4B 216 (41Lb 90)
(off Robert St.)
Borrowdale Av. HA3: W'stone 26Ja 46
Borrowdale Cl. CR2: Sande 85Vb 177
Borrowdale Cl. IG4: Ilf 28Nc 54
Borrowdale Cl. N2 .... 26Eb 49
Borrowdale Cl. TW20: Egh 66D 126
Borrowdale Ct. EN2: Enf 11Sb 33
Borrowdale Dr. CR2: Sande 84Vb 177
Borthwick M. E15 .... 35Gc 73
Borthwick Rd. E15 .... 35Gc 73
Borthwick Rd. NW9 .... 30Va 48
Borthwick St. SE8 .... 50Cc 92
Borwick Av. E17 .... 27Bc 52
Bosanquet Cl. UB8: Cowl 42M 83
Bosbury Rd. SE6 .... 62Ec 136
Boscastle Rd. NW5 .... 34Kb 70
Boscobel Cl. BR1: Broml 68Pc 138
Boscobel Ho. E8 .... 37Xb 71
Boscobel Pl. SW1 .... 5J 227 (49Jb 90)
Boscobel St. NW8 .... 6C 214 (42Fb 89)
Bosco Cl. BR6: Orp 77Vc 161
Boscombe Av. E10 .... 31Fc 73
Boscombe Av. RM11: Horn 31Md 77
Boscombe Av. RM17: Grays 49Fe 99
Boscombe Cir. NW9 .... 25Ta 47
Boscombe Cl. E5 .... 36Ac 72
Boscombe Cl. TW20: Egh 67E 126
Boscombe Gdns. SW16 .... 65Nb 134
Boscombe Ho. CR0: C'don 74Tb 157
(off Sydenham Rd.)
Boscombe Rd. KT4: Wor Pk 74Ya 154
Boscombe Rd. SW17 .... 65Jb 134
Boscombe Rd. SW19 .... 67Db 133
Boscombe Rd. W12 .... 46Wa 88
Bose Cl. N3 .... 25Ab 48
Bosgrove E4 .... 19Ec 34
Boshers Gdns. TW20: Egh 65B 126
Bosman Dr. GU20: W'sham 6A 146
Boss Ho. SE1 .... 1K 231 (47Vb 91)
(off Boss St.)
Boss St. SE1 .... 1K 231 (47Vb 91)
Bostall Heath SE2 .... 50Yc 95
Bostall Hill SE2 .... 50Wc 95
Bostall La. SE2 .... 49Xc 95
Bostall Mnr. Way SE2 .... 49Xc 95
Bostall Pk. Av. DA7: Bex 52Ad 117
Bostall Rd. BR5: St P 66Xc 139
Bostock Ho. TW5: Hest 51Ca 107
Boston Bus. Pk. W7 .... 48Ga 86
Boston Ct. SE25 .... 70Vb 135
Boston Ct. SM2: Sutt 80Eb 155
Boston Gdns. TW8: Bford 49Ja 86
Boston Gdns. W4 .... 51Ua 110
Boston Gdns. W7 .... 49Ja 86
Boston Gro. HA4: Ruis 30S 44
Boston Gro. SL1: Slou 4G 80
Boston Ho. SW5 .... 49Db 89
(off Collingham Rd.)
BOSTON MANOR 49Ja 86
Boston Manor House 50Ka 86
Boston Mnr. Rd. TW8: Bford 49Ka 86
Boston Manor Station (Underground) 49Ja 86
Boston Pde. W7 .... 48Ja 86
Boston Pk. Rd. TW8: Bford 50La 86
Boston Pl. NW1 .... 6F 215 (42Hb 89)
Boston Rd. CR0: C'don 72Pb 156
Boston Rd. E17 .... 30Cc 52
Boston Rd. E6 .... 41Nc 94
Boston Rd. HA8: Edg 24Sa 47
Boston Rd. W7 .... 46Ga 86
Bostonthorpe Rd. W7 .... 47Ga 86
Boston Va. W7 .... 49Ja 86
Bosun Cl. E14 .... 47Cc 92
Bosville Av. TN13: S'oaks 95Jd 202
Bosville Dr. TN13: S'oaks 95Jd 202
Bosville Rd. TN13: S'oaks 95Jd 202
Boswell Cl. BR5: Orp 72Yc 161
Boswell Cl. WD7: Shenl 4Na 15
Boswell Ct. KT2: King T 67Pa 131
(off Clifton Rd.)
Boswell Ct. NW9 .... 26Ua 48
(off Charcot Rd.)
Boswell Ct. W14 .... 48Za 88
(off Blythe Rd.)
Boswell Ct. WC1 .... 7G 217 (43Nb 90)
Boswell Ho. WC1 .... 7G 217 (43Nb 90)
(off Boswell St.)
Boswell Path UB3: Harl 49V 84
Boswell Rd. CR7: Thor H 70Sb 135
Boswell Row CR3: Cat'm 94Wb 197
Boswell St. WC1 .... 7G 217 (43Nb 90)
Boswood Ct. TW3: Houn 55Ba 107
Bosworth Cl. E17 .... 25Bc 52
Bosworth Ct. SL1: Slou 5A 80
Bosworth Cres. RM3: Rom 23Ld 57
Bosworth Ho. DA8: Erith 50Gd 96
(off Saltford Cl.)
Bosworth Ho. W10 .... 42Ab 88
(off Bosworth Rd.)
Bosworth Rd. EN5: New Bar 13Cb 31
Bosworth Rd. N11 .... 23Mb 50
Bosworth Rd. RM10: Dag 34Cd 76
Bosworth Rd. W10 .... 42Ab 88
Botanic Sq. E14 .... 44Gc 93
BOTANY BAY 8Mb 18
Botany Bay La. BR7: Chst 69Sc 138
Botany Cl. EN4: E Barn 14Gb 31
Botany Cotts. RM19: Purf 50Qd 97
Botany Rd. DA11: Nflt 56Ce 121
Botany Ter. RM19: Purf 50Qd 97
Botany Way RM19: Purf 49Rd 97
Boteley Cl. E4 .... 19Fc 35
Botery's Cross RH1: Blet 5H 209
Botham Cl. HA8: Edg 24Sa 47
Botham Dr. SL1: Slou 8J 81
Botha Rd. E13 .... 43Kc 93
Bothwell Cl. E16 .... 43Hc 93
Bothwell Rd. CR0: New Ad 82Ec 178
Bothwell St. W6 .... 51Za 110
Bothy, The GU22: Pyr 89H 169
Bothy, The SM7: Bans 90Eb 175
Botolph All. EC3 .... 4H 225 (45Ub 91)
(off Botolph La.)
Botolph La. EC3 .... 5H 225 (45Ub 91)
Botsford Rd. SW20 .... 68Ab 132
Botsom La. TN15: W King 79Sd 164
Bottle Cotts. TN13: S'oaks 94Jd 202
Bottom La. WD3: Bucks 7K 11
Bottom La. WD4: Bucks 7L 11
Bottom Waltons Cvn. Site SL2: Farn R 10C 60
Bott Rd. DA2: Hawl 63Pd 141
Botts M. W2 .... 44Cb 89
Botwell Comn. Rd. UB3: Hayes 45T 84
Botwell Cres. UB3: Hayes 44U 84
Botwell Green Sports & Leisure Cen. 46V 84
Botwell La. UB3: Hayes 45U 84
Boucher Cl. TW11: Tedd 64Ha 130
Boucher Dr. DA11: Nflt 2B 144
Bouchier Ho. N2 .... 26Fb 49
(off The Grange)
Bouchier Wlk. RM13: Rain 37Jd 76
Bough Beech Ct. EN3: Enf W 9Zb 20
Boughton Av. BR2: Hayes 73Hc 159
Boughton Hall Av. GU23: Send 96H 189
Boughton Ho. SE1 .... 1F 231 (47Tb 91)
(off Tennis St.)
Boughton Rd. SE28 .... 48Uc 94
Boulcott St. E1 .... 44Zb 92
Boulevard, The DA9: Ghithe 56Yd 120
Boulevard, The IG8: Wfd G 23Qc 54
Boulevard, The SW17 .... 61Jb 134
Boulevard, The SW18 .... 56Db 111
Boulevard, The SW6 .... 53Eb 111
Boulevard, The WD18: Wat 15T 26
Boulevard 25 WD6: Bore 13Qa 29
Boulevard Dr. NW9 .... 26Va 48
Boulevard Walkway E1 .... 44Wb 91
Boulmer Rd. UB8: Cowl 41L 83
Boulogne Ho. SE1 .... 3K 231 (48Vb 91)
(off St Saviour's Est.)
Boulogne Rd. CR0: C'don 72Sb 157
Boulter Cl. BR1: Broml 69Qc 138
Boulter Gdns. RM13: Rain 37Jd 76
Boulter Ho. SE14 .... 53Yb 114
(off Kender St.)
Boulters Cl. SL1: Slou 6E 80
Boulthurst Way RH8: Oxt 4M 211
Boulton Ho. TW8: Bford 50Na 87
Boulton Rd. RM8: Dag 33Ad 75
Boultwood Rd. E6 .... 44Pc 94
BOUNCE HILL 11Kd 39
Bounces La. N9 .... 19Xb 33
Bounces Rd. N9 .... 19Xb 33
Boundaries Rd. SW12 .... 61Hb 133
Boundaries Rd. TW13: Felt 60Y 107
Boundary Av. E17 .... 31Bc 72
Boundary Bus. Ct. CR4: Mitc 69Fb 133
Boundary Cl. EN5: Barn 11Bb 31
Boundary Cl. IG3: Ilf 35Uc 74
Boundary Cl. KT1: King T 69Ra 131
Boundary Cl. SE20 .... 68Wb 135
Boundary Cl. UB2: S'hall 50Ca 85
Boundary Cotts. HP3: Bov 1F 10
Boundary Ct. CM16: Epp 4Tc 22
Boundary Ct. N18 .... 23Vb 51
(off Snells Pk.)
Boundary Dr. CM13: Hut 17Fe 41
Boundary Dr. SL2: Wex 3M 81
Boundary Ho. DA11: Nflt 10B 122
(off Victoria Rd.)

Boundary Ho. EN8: Chesh 4Zb **20**
Boundary Ho. SE5 52Sb **113**
Boundary Ho. W11 46Za **88**
(off Queensdale Cres.)
Boundary La. E13 41Mc **93**
Boundary La. SE17 51Sb **113**
Boundary Pk. HP2: Hem H 1C **4**
Boundary Pk. KT13: Weyb 76U **150**
Boundary Pas. E2 5K **219** (42Vb **91**)
Boundary Rd. IG11: Bark
King Edwards Rd. 39Tc **74**
Boundary Rd. IG11: Bark The Clarksons 40Sc **74**
Boundary Rd. AL1: St A 1C **6**
Boundary Rd. DA15: Sidc 57Uc **116**
Boundary Rd. E13 40Lc **73**
Boundary Rd. E17 31Bc **72**
Boundary Rd. GU21: Wok 88C **168**
Boundary Rd. HA5: Eastc 31Z **65**
Boundary Rd. HA9: Wemb 34Na **67**
Boundary Rd. N2 25Fb **49**
Boundary Rd. N22 27Rb **51**
Boundary Rd. N9 16Yb **34**
Boundary Rd. NW8 39Db **69**
Boundary Rd. RM1: Rom 30Jd **56**
Boundary Rd. RM14: Upm 34Qd **77**
Boundary Rd. SM5: Cars 79Kb **156**
Boundary Rd. SM5: W'gton 79Kb **156**
Boundary Rd. SM6: W'gton 79Kb **156**
Boundary Rd. SW19 65Fb **133**
Boundary Rd. TW15: Ashf 64L **127**
Boundary Row SE1 1B **230** (47Rb **91**)
Boundary St. DA8: Erith 52Hd **118**
Boundary St. E2 4K **219** (41Vb **91**)
Boundary Way CR0: Addtn 78Cc **158**
Boundary Way GU21: Wok 87C **168**
Boundary Way WD25: Wat 4X **13**
Boundfield Rd. SE6 62Gc **137**
BOUNDS GREEN 23Mb **50**
Bounds Grn. Ct. N11 23Mb **50**
(off Bounds Grn. Rd.)
Bounds Grn. Ind. Est. N11 23Lb **50**
Bounds Grn. Rd. N11 23Lb **50**
Bounds Grn. Rd. N22 24Nb **50**
Bounds Green Station
(Underground) 23Mb **50**
Bourbon Ho. SE6 64Ec **136**
Bourbon La. W12 46Za **88**
Bourbon Rd. SW9 53Qb **112**
Bourchier Cl. TN13: S'oaks 98Kd **203**
Bourchier St. W1 4D **222** (45Mb **90**)
Bourdon Pl. W1 4A **222** (45Kb **90**)
(off Bourdon St.)
Bourdon Rd. SE20 68Yb **136**
Bourdon St. W1 5K **221** (45Kb **90**)
Bourke Cl. NW10 37Ua **68**
Bourke Cl. SW4 58Nb **112**
Bourke Hill CR5: Chips 90Hb **175**
Bourlet Cl. W1 1B **222** (43Lb **90**)
Bourn Av. EN4: E Barn 15Fb **31**
Bourn Av. N15 28Tb **51**
Bourn Av. UB8: Hil 42Q **84**
Bournbrook Rd. SE3 55Mc **115**
Bourne, The HP3: Bov 9C **2**
Bourne, The KT16: Chert 73J **149**
(off Guildford St.)
Bourne, The N14 18Mb **32**
Bourne Av. HA4: Ruis 36Y **65**
Bourne Av. KT16: Chert 69J **127**
Bourne Av. N14 19Nb **32**
Bourne Av. SL4: Wind 5G **102**
Bourne Av. UB3: Harl 48S **84**
BOURNEBRIDGE 18Dd **38**
Bournebridge Cl. CM13: Hut 17Fe **41**
Bournebridge La. RM4: Stap A 17Bd **37**
Bournebrook Gro. RM7: Rush G 30Fd **56**
Bourne Bus. Pk. KT15: Add 77N **149**
Bourne Cir. UB3: Harl 48S **84**
Bourne Cl. KT14: W Byf 85K **169**
Bourne Cl. KT7: T Ditt 75Ha **152**
Bourne Cl. TW7: Isle 55Ga **108**
Bourne Ct. CR3: Cat'm 95Wb **197**
Bourne Ct. E11 29Jc **53**
Bourne Ct. HA4: Ruis 36X **65**
Bourne Ct. IG8: Wfd G 27Mc **53**
Bourne Ct. W4 51Sa **109**
Bourne Dr. CR4: Mitc 68Fb **133**
BOURNE END 4E **2**
Bourne End RM11: Horn 31Qd **77**
Bourne End La. HP1: Hem H 7C **2**
Bourne End La. Ind. Est. HP1: Hem H 4D **2**
Bourne End Rd. HA6: Nwood 21U **44**
Bourne Ent. Cen. TN15: Bor G 91Ce **205**
Bourne Est. EC1 7K **217** (43Qb **90**)
Bournefield Rd. CR3: Whyt 90Wb **177**
Bourne Gdns. E4 21Dc **52**
Bourne Gro. KT21: Asht 91Ma **193**
Bournehall Av. WD23: Bush 16Ca **27**
Bournehall Av. WD23: Bush 15Ca **27**
Bournehall La. WD23: Bush 16Ca **27**
Bourne Hall Mus. 81Va **174**
Bournehall Rd. WD23: Bush 16Ca **27**
Bourne Hill N13 19Nb **32**
Bourne Hill Cl. N13 19Pb **32**
Bourne Ho. IG9: Buck H 20Mc **35**
Bourne Ho. TW15: Ashf 64Q **128**
Bourne Ind. Pk. DA1: Cray 57Gd **118**
Bourne La. CR3: Cat'm 93Tb **197**
Bourne La. TN15: Plax 98Be **205**
Bourne Mead DA5: Bexl 57Fd **118**
Bournemead WD23: Bush 16Da **27**
Bournemead Av. UB5: N'olt 40W **64**
Bournemead Cl. UB5: N'olt 41W **84**
Bourne Mdw. TW20: Thorpe 70D **126**
Bournemead Way UB5: N'olt 40X **65**
Bourne M. RH9: G'stone 2A **210**
Bourne M. W1 3J **221** (44Jb **90**)
Bournemouth Cl. SE15 54Wb **113**
Bournemouth Rd. SE15 54Wb **113**
Bournemouth Rd. SW19 67Cb **133**
Bourne Pde. DA5: Bexl 59Dd **118**
Bourne Pk. Cl. CR8: Kenley 88Ub **177**
Bourne Pl. KT16: Chert 74K **149**
Bourne Pl. W4 50Ta **87**
Bourne Rd. BR2: Broml 70Mc **137**
Bourne Rd. DA1: Cray 57Fd **118**
Bourne Rd. DA12: Grav'nd 1H **145**
Bourne Rd. DA5: Bexl 59Dd **118**
Bourne Rd. DA5: Dart 59Dd **118**
Bourne Rd. E7 34Hc **73**
Bourne Rd. GU25: Vir W 1P **147**
Bourne Rd. N8 30Nb **50**
Bourne Rd. RH1: Mers 2C **208**
Bourne Rd. SL1: Slou 7G **80**
Bourne Rd. WD23: Bush 15Ca **27**
Bournes Ho. N15 30Ub **51**
(off Chisley Rd.)
Bourneside GU25: Vir W 3L **147**
Bourneside Cres. N14 18Mb **32**
Bourneside Gdns. SE6 64Ec **136**
Bourneside Rd. KT15: Add 77M **149**
Bourne St. CR0: C'don 75Rb **157**
Bourne St. SW1 6H **227** (49Jb **90**)
Bourne Ter. W2 43Db **89**
Bourne Va. BR2: Hayes 74Hc **159**
Bourne Va. TN15: Plax 99Ce **205**
Bournevale Rd. SW16 63Nb **134**
Bourne Vw. CR8: Kenley 87Tb **177**
Bourne Vw. UB6: G'frd 37Ha **66**
Bourne Way BR2: Hayes 75Hc **159**
Bourne Way BR8: Swan 69Ed **140**
Bourne Way GU22: Wok 4P **187**
Bourne Way KT15: Add 78L **149**
Bourne Way KT19: Ewe 77Sa **153**
Bourne Way SM1: Sutt 78Bb **155**
Bournewood Gro. CR6: W'ham 91Yb **198**
Bournewood Rd. BR5: Orp 73Yc **161**
Bournewood Rd. SE18 52Wc **117**
Bournville Rd. SE6 59Cc **114**
Bournwell Cl. EN4: Cockf 13Hb **31**
Bourton Cl. UB3: Hayes 46W **84**
Bousfield Rd. SE14 54Zb **114**
Bousley Ri. KT16: Ott 79F **148**
Boutflower Rd. SW11 56Gb **111**
Boutique Hall SE13 56Ec **114**
Bouton Pl. N1 38Rb **71**
(off Waterloo Ter.)
Bouverie Gdns. CR8: Purl 86Pb **176**
Bouverie Gdns. HA3: Kenton 30Ma **47**
Bouverie M. N16 33Ub **71**
Bouverie Pl. W2 2C **220** (44Fb **89**)
Bouverie Rd. CR5: Chips 90Jb **176**
Bouverie Rd. HA1: Harr 30Ea **46**
Bouverie Rd. N16 33Ub **71**
Bouverie St. EC4 3A **224** (44Qb **90**)
Bouverie Way SL3: L'ly 50A **82**
Bouvier Rd. EN3: Enf W 10Yb **20**
BOVENEY 1B **102**
Boveney Cl. SE23 58Zb **114**
Boveney Cl. SL1: Slou 7E **80**
Boveney New Rd. SL4: Eton W 9C **80**
Boveney Rd. SE23 59Zb **114**
Boveney Rd. SL4: Dor 9A **80**
Boveney Wood La. SL1: Burn 4B **60**
Bovet Ct. E1 43Ac **92**
(off Ocean Est.)
Bovey Way RM15: S Ock 43Xd **98**
Bovill Rd. SE23 59Zb **114**
BOVINGDON 9C **2**
Bovingdon Av. HA9: Wemb 37Qa **67**
Bovingdon Cl. N19 33Lb **70**
Bovingdon Ct. HP3: Bov 10C **2**
Bovingdon Ct. WD23: Bush 15Da **27**
(off Farrington Av.)
Bovingdon Cres. WD25: Wat 6Z **13**
BOVINGDON GREEN 1C **10**
Bovingdon Grn. HP3: Bov 1C **10**
Bovingdon La. NW9 25Ua **48**
Bovingdon Rd. SW6 53Db **111**
Bovril Ct. SW6 52Db **111**
(off Fulham Rd.)
BOW 41Bc **92**
Bow Arrow La. DA1: Dart 58Qd **119**
Bow Arrow La. DA2: Dart 58Rd **119**
Bowater Cl. NW9 29Ta **47**
Bowater Cl. SW2 58Nb **112**
Bowater Gdns. TW16: Sun 68Y **129**
Bowater Ho. EC1 6D **218** (42Sb **91**)
(off Golden La. Est.)
Bowater Pl. SE3 52Kc **115**
Bowater Ridge KT13: Weyb 82T **170**
Bowater Rd. HA9: Wemb 34Ra **67**
Bowater Rd. SE18 48Mc **93**
Bow Bell Twr. E3 39Cc **72**
(off Pancras Way)
Bow Bri. Est. E3 41Dc **92**
Bow Brook, The E2 40Zb **72**
(off Mace St.)
Bow Church Station (DLR) 41Cc **92**
Bow Churchyard EC4 3E **224** (44Sb **91**)
(off Bow La.)
BOW COMMON 43Cc **92**
Bow Comn. La. E3 42Bc **92**
Bow Creek Ecology Pk. 44Gc **93**
Bowden Cl. TW14: Bedf 60U **106**
Bowden Ct. TN15: Kems'g 89Pd **183**
Bowden Dr. RM11: Horn 32Nd **77**
Bowden Ho. E3 41Dc **92**
(off Rainhill Way)
Bowden Rd. SL5: S'hill 1A **146**
Bowden St. SE11 7A **230** (50Qb **90**)
Bowditch SE8 49Bc **92**
Bowdon Rd. E17 31Cc **72**
Bowen Ct. SE16 49Yb **92**
(off Debnams Rd.)
Bowen Dr. SE21 62Ub **135**
Bowen Rd. HA1: Harr 31Ea **66**
Bowen St. E14 44Dc **92**
Bowens Wood CR0: Sels 81Bc **178**
Bowen Way CR5: Coul 94Mb **196**
Bower Av. SE10 53Gc **115**
Bower Cl. RM5: Col R 24Fd **56**
Bower Cl. UB5: N'olt 40Y **65**
Bower Ct. CM16: Epp 4Wc **23**
Bower Ct. E4 18Ec **34**
(off The Ridgeway)
Bower Ct. GU22: Wok 88D **168**
Bower Ct. SL1: Slou 5D **80**
Bowerdean St. SW6 53Db **111**
Bowerden Ct. NW10 40Xa **68**
Bower Farm Rd. RM4: Have B 20Ed **38**
Bower Hill CM16: Epp 3Wc **23**
Bower Hill Cl. RH1: S Nut 9E **208**
Bower Hill Ind. Est. CM16: Epp 4Wc **23**
Bower Hill La. RH1: S Nut 7D **208**
Bower La. DA4: Eyns 75Nd **163**
Bower La. TN15: Knat 82Qd **183**
Bowerman Av. SE14 51Ac **114**
Bowerman Ct. N19 33Mb **70**
(off St John's Way)
Bowerman Rd. RM16: Grays 9C **100**
Bower Rd. BR8: Hext 66Jd **140**
Bowers RM11: Horn 29Md **57**
Bowers Av. DA11: Nflt 3B **144**
Bowers Rd. TN14: S'ham 83Hd **182**
Bower St. E1 44Zb **92**
Bowers Wlk. E6 44Pc **94**
Bower Ter. CM16: Epp 4Wc **23**
Bower Va. CM16: Epp 4Wc **23**
Bower Way SL1: Slou 5C **80**
Bowery Apts. W12 45Ya **88**
(off Fountain Park Way)
Bowery Ct. RM10: Dag 37Dd **76**
Bowes Cl. DA15: Sidc 58Xc **117**
Bowes Ct. DA2: Dart 58Rd **119**
(off Osbourne Rd.)
Bowesden La. DA12: Shorne 6N **145**
Bowesden La. DA12: Strood 6N **145**
Bowe's Ho. IG11: Bark 38Rc **74**
Bowes Lyon Cl. SL4: Wind 3G **102**
(off Mountbatten Sq.)
Bowes-Lyon Hall E16 46Jc **93**
(off Wesley Av.)
Bowes Lyon M. AL3: St A 2B **6**
BOWES PARK 23Nb **50**
Bowes Park Station (Rail) 23Nb **50**
Bowes Rd. KT12: Walt T 75X **151**
Bowes Rd. N11 22Lb **50**
Bowes Rd. N13 22Nb **50**
Bowes Rd. RM8: Dag 35Yc **75**
Bowes Rd. TW18: Staines 64G **126**
Bowes Rd. W3 45Ua **88**
Bowes Wood DA3: Nw A G 76Be **165**
Bow Exchange E3 43Dc **92**
(off Yeo St.)
Bow Fair E3 40Cc **72**
(off Fairfield Rd.)
Bowfell Rd. W6 51Ya **110**
Bowford Av. DA7: Bex 53Ad **117**
Bowgate AL1: St A 1C **6**
Bowhay CM13: Hut 19De **41**
Bowhill Cl. SW9 52Qb **112**
Bow Ho. N1 1H **219** (39Ub **71**)
(off Wilmer Gdns.)
Bowie Cl. SW4 59Mb **112**
BOW INTERCHANGE 40Dc **72**
Bowland Rd. IG8: Wfd G 23Lc **53**
Bowland Rd. SW4 56Mb **112**
Bowland Yd. SW1 2G **227** (47Hb **89**)
(off Kinnerton St.)
Bow La. EC4 3E **224** (44Sb **91**)
Bow La. N12 24Eb **49**
Bow La. SM4: Mord 72Ab **154**
Bowlby Ho. SE4 56Zb **114**
(off Frendsbury Rd.)
Bowl Ct. EC2 6J **219** (42Ub **91**)
Bowlers Grn. WD7: Shenl 2La **14**
Bowles Ct. N12 24Gb **49**
Bowles Grn. EN1: Enf 8Xb **19**
Bowles Rd. SE1 51Wb **113**
Bowley Cl. SE19 65Vb **135**
Bowley Ho. SE16 48Wb **91**
Bowley La. SE19 64Vb **135**
Bowline Ct. SE10 49Gc **93**
Bowline Ct. TW8: Bford 52La **108**
(off Durham Wharf Dr.)
Bowling, The KT12: Walt T 73W **150**
Bowling Cl. UB10: Uxb 39P **63**
Bowling Cl. WD18: Wat 14W **26**
Bowling Grn. Cl. SW15 59Xa **110**
Bowling Grn. Ct. HA9: Wemb 33Pa **67**
Bowling Grn. Ho. SW10 52Fb **111**
(off Riley St.)
Bowling Grn. La. EC1 5A **218** (42Qb **90**)
Bowling Grn. Pl. SE1 1F **231** (47Tb **91**)
Bowling Grn. Rd. GU24: Chob 1J **167**
Bowling Grn. Row SE18 48Pc **94**
Bowling Grn. St. SE11 51Qb **112**
Bowling Grn. Wlk. N1 3H **219** (41Ub **91**)
Bow Locks E3 42Ec **92**
Bowls, The IG7: Chig 20Uc **36**
Bowls Cl. HA7: Stan 22Ka **46**
Bowman Av. E16 45Hc **93**
Bowman Ho. N1 1J **219** (39Ub **71**)
(off Nuttall St.)
Bowman M. SW18 60Bb **111**
BOWMANS 59Hd **118**
Bowman's Bldgs. NW1 7D **214** (43Gb **89**)
(off Penfold Pl.)
Bowmans Cl. EN6: Pot B 4Fb **17**
Bowmans Cl. SL1: Burn 9A **60**
Bowmans Cl. W13 46Ka **86**
Bowmans Grn. WD25: Wat 8Aa **13**
Bowmans Lea SE23 59Yb **114**
Bowmans Mdw. SM6: W'gton 76Kb **156**
Bowman's M. E1 44Wb **91**
Bowman's M. N7 34Nb **70**
Bowman's Pl. N7 34Nb **70**
Bowman's Rd. DA1: Dart 59Hd **118**
Bowman Trad. Est. NW9 27Qa **47**
Bowmead SE9 61Pc **138**
Bowmont Cl. CM13: Hut 16De **41**
Bowmore Wlk. NW1 38Mb **70**
Bown Cl. RM18: Tilb 5D **122**
Bowness Cl. E8 37Vb **71**
(off Beechwood Rd.)
Bowness Cres. SW15 64Ua **132**
Bowness Dr. TW4: Houn 56Aa **107**
Bowness Ho. SE15 52Yb **114**
(off Hillbeck Cl.)
Bowness Rd. DA7: Bex 54Dd **118**
Bowness Rd. SE6 59Dc **114**
Bowness Way RM12: Horn 36Jd **76**
Bowood Rd. EN3: Enf H 12Zb **34**
Bowood Rd. SW11 57Jb **112**
Bowring Grn. WD19: Wat 22Y **45**
Bow River Village E3 40Ec **72**
(off Global App.)
Bow Rd. E3 41Bc **92**
Bow Road Station (Underground) 41Cc **92**
Bowrons Av. HA0: Wemb 38Ma **67**
Bowry Dr. TW19: Wray 58B **104**
Bowry Ho. E14 43Bc **92**
(off Wallwood St.)
Bowsley Ct. TW13: Felt 61W **128**
Bowspirit Apts. SE8 52Dc **114**
Bowsprit, The KT11: Cobh 87Y **171**
Bowsprit Point E14 48Cc **92**
(off Westferry Rd.)
Bow St. E15 36Gc **73**
Bow St. WC2 3G **223** (44Nb **90**)
Bowstring Plaza E3 41Bc **92**
(off St Clements Av.)
Bowthorpe Cl. RM11: Horn 31Pd **77**
Bow Triangle Bus. Cen. E3 42Cc **92**
(not continuous)
Bowyer Cl. E6 43Pc **94**
Bowyer Ct. EN8: Walt C 5Bc **20**
Bowyer Cres. UB9: Den 30H **43**
Bowyer Dr. SL1: Slou 6C **80**
Bowyer Ho. N1 1J **219** (39Ub **71**)
(off Whitmore Est.)
Bowyer Pl. SE5 52Sb **113**
Bowyer Plaza E3 41Bc **92**
(off St Clements Av.)
Bowyers HP2: Hem H 1M **3**
Bowyers Cl. KT21: Asht 90Pa **173**
Bowyers Ct. TW1: Isle 56Ka **108**
Bowyer St. SE5 52Sb **113**
Boxall Rd. SE21 58Ub **113**
Boxall Way SL3: L'ly 10N **81**
Boxelder Cl. HA8: Edg 22Sa **47**
Boxford Cl. CR2: Sels 84Zb **178**
Boxgrove Rd. SE2 48Yc **95**
Box La. HP3: Hem H 7E **2**
Box La. IG11: Bark 40Xc **75**
Boxley Rd. SM4: Mord 70Eb **133**
Boxley St. E16 46Kc **93**
BOXMOOR 4K **3**
Boxmoor Ho. E2 39Wb **71**
(off Whiston Rd.)
Boxmoor Ho. W11 46Za **88**
(off Queensdale Cres.)
Boxmoor Playhouse Theatre 4L **3**
Boxmoor Rd. HA3: Kenton 28Ka **46**
Boxmoor Rd. RM5: Col R 22Ed **56**
Boxoll Rd. RM9: Dag 35Bd **75**
Box Pk. 75Tb **157**
Box Ridge Av. CR8: Purl 84Pb **176**
Boxted Cl. IG9: Buck H 18Nc **36**
Boxted Rd. HP1: Hem H 1H **3**
Box Tree Ho. SE8 51Ac **114**
Boxtree La. HA3: Hrw W 25Ea **46**
Boxtree Rd. HA3: Hrw W 24Fa **46**
Box Tree Wlk. BR5: Orp 74Zc **161**
Box Tree Wlk. RH1: Redh 9L **207**
Boxwood Cl. UB7: W Dray 47P **83**
Boxwood Gro. GU24: W End 6E **166**
Boxwood Way CR6: W'ham 89Zb **178**
Boxworth Cl. N12 22Fb **49**
Boxworth Gro. N1 39Pb **70**
Boyce Cl. WD6: Bore 11Na **29**
Boyce Ct. WD17: Wat 11W **26**
(off Lockhart Rd.)
Boyce Ho. SW16 64Lb **134**
Boyce Ho. W10 41Bb **89**
(off Bruckner St.)
Boyce Rd. SS17: Stan H 1L **101**
Boyce Way E13 42Jc **93**
Boycroft Av. NW9 30Sa **47**
Boyd Av. UB1: S'hall 46Ba **85**
Boyd Bldg. E16 45Sc **94**
(off Frobisher Yd.)
Boyd Cl. KT2: King T 66Qa **131**
Boyd Cl. RM14: Upm 33Ud **78**
Boydell Ct. NW8 38Fb **69**
(not continuous)
Boyden Ho. E17 27Ec **52**
Boyd Rd. SW19 65Fb **133**
Boyd St. E1 44Wb **91**
Boyd Way SE3 56Lc **115**
Boyes Cres. AL2: Lon C 8F **6**
Boyfield St. SE1 2C **230** (47Rb **91**)
Boyland Rd. BR1: Broml 64Hc **137**
Boyle Av. HA7: Stan 23Ja **46**
Boyle Cl. UB10: Uxb 40P **63**
Boyle Farm Island KT7: T Ditt 72Ja **152**
Boyle Farm Rd. KT7: T Ditt 72Ja **152**
Boyle St. W1 4B **222** (45Lb **90**)
Boyne Av. NW4 28Za **48**
Boyne Rd. RM10: Dag 34Cd **76**
Boyne Rd. SE13 55Ec **114**
Boyne Ter. M. W11 46Bb **89**
Boyseland Ct. HA8: Edg 19Sa **29**
Boyson Rd. SE17 51Sb **113**
(not continuous)
Boyson Wlk. SE17 51Tb **113**
Boyton Cl. E1 42Yb **92**
Boyton Cl. N8 27Nb **50**
Boyton Ho. NW8 1C **214** (40Fb **69**)
(off Wellington Rd.)
Boyton Rd. N8 27Nb **50**
Brabant Ct. EC3 4H **225** (45Ub **91**)
(off Philpot La.)
Brabant Rd. N22 26Pb **50**
Brabazon Av. SM6: W'gton 80Nb **156**
Brabazon Ct. SW1 7D **228** (50Mb **90**)
(off Moreton St.)
Brabazon Rd. TW5: Hest 52Y **107**
Brabazon Rd. UB5: N'olt 40Ca **65**
Brabazon St. E14 44Dc **92**
Brabiner Gdns. CR0: New Ad 82Fc **179**
Brabner Ho. E2 41Wb **91**
(off Wellington Row)
Brabourne Cl. SE19 64Ub **135**
Brabourne Cres. DA7: Bex 51Bd **117**
Brabourne Hgts. NW7 20Ua **30**
Brabourne Ri. BR3: Beck 71Ec **158**
Brabourn Gro. SE15 54Yb **114**
Brabrook Ct. SM6: W'gton 77Kb **156**
Brabstone Ho. UB6: G'frd 40Ha **66**
Bracer Ho. N1 1J **219** (40Ub **71**)
(off Whitmore Est.)
Bracewell Av. UB6: G'frd 36Ha **66**
Bracewell Rd. W10 43Ya **88**
Bracewood Gdns. CR0: C'don 76Vb **157**
Bracey M. N4 33Nb **70**
Bracey St. N4 33Nb **70**
Bracken, The E4 19Ec **34**
Bracken Av. CR0: C'don 76Dc **158**
Bracken Av. SW12 58Jb **112**
Brackenbridge Dr. HA4: Ruis 34Z **65**
Brackenbridge Ho. HA4: Ruis 35Aa **65**
(off Brackenhill)
Brackenbury N4 32Qb **70**
(off Osborne Rd.)
Brackenbury Gdns. W6 48Xa **88**
Brackenbury Rd. N2 27Eb **49**
Brackenbury Rd. W6 48Xa **88**
Bracken Cl. E6 43Pc **94**
Bracken Cl. GU22: Wok 90B **168**
Bracken Cl. IG10: Lough 11Rc **36**
Bracken Cl. KT23: Bookh 96Ba **191**
Bracken Cl. SL2: Farn C 5H **61**
Bracken Cl. TW16: Sun 65V **128**
Bracken Cl. TW2: Whitt 59Ca **107**
Bracken Cl. WD6: Bore 11Ra **29**
Bracken Ct. IG6: Ilf 23Vc **55**
Brackendale EN6: Pot B 5Cb **17**
Brackendale N21 19Pb **32**
Brackendale Cl. TW20: Eng G 5N **125**
Brackendale Cl. TW3: Houn 53Da **107**
Brackendale Gdns. RM14: Upm 35Sd **78**
Brackendene AL2: Brick W 2Ba **13**
Brackendene DA2: Wilm 63Gd **140**
Brackendene Cl. GU21: Wok 87C **168**
Bracken Dr. IG7: Chig 23Rc **54**
Bracken End TW7: Isle 57Fa **108**
Brackenfield Cl. E5 34Xb **71**
Brackenforde SL3: L'ly 7N **81**
Bracken Gdns. SW13 54Wa **110**
Brackenhill HA4: Ruis 35Aa **65**
Brackenhill KT11: Cobh 84Da **171**
Bracken Hill Cl. BR1: Broml 67Hc **137**
Bracken Hill Cl. HA6: Nwood 22W **44**
Bracken Hill La. BR1: Broml 67Hc **137**
Bracken Ho. E3 43Cc **92**
(off Devons Rd.)
Bracken Ind. Est. IG6: Ilf 24Uc **54**
Bracken M. E4 18Ec **34**
Bracken M. RM7: Rom 30Dd **56**
Bracken Path KT18: Eps 85Ra **173**
Brackens BR3: Beck 66Cc **136**
Brackens, The BR6: Chels 78Wc **161**
Brackens, The EN1: Enf 17Ub **33**
Brackens, The HP2: Hem H 1M **3**
Brackens, The TN13: S'oaks 95Ld **203**
Brackens Dr. CM14: W'ley 22Yd **58**
Bracken Way GU24: Chob 2K **167**
Brackenwood TW16: Sun 67W **128**
Brackenwood Lodge EN5:
New Bar 14Cb **31**
(off Prospect Rd.)
Brackenwood Rd. GU21: Wok 1H **187**
Brackley KT13: Weyb 78T **150**
Brackley Av. SE15 55Yb **114**
Brackley Cl. SM6: W'gton 80Nb **156**
Brackley Ct. NW8 5B **214** (42Fb **89**)
(off Pollitt Dr.)
Brackley Rd. BR3: Beck 66Bc **136**
Brackley Rd. W4 50Ua **88**
Brackley Sq. IG8: Wfd G 24Mc **53**
Brackley St. EC1 6E **218** (42Sb **91**)
Brackley Ter. W4 50Ua **88**
Brackley Wlk. HA8: Edg 24Ra **47**
Bracklyn Ct. N1 1F **219** (40Tb **71**)
(not continuous)
Bracklyn St. N1 1F **219** (40Tb **71**)
Bracknell Cl. N22 25Qb **50**
Bracknell Gdns. NW3 35Db **69**
Bracknell Ga. NW3 36Db **69**
Bracknell Way NW3 35Db **69**
Bracondale KT10: Esh 78Ea **152**
Bracondale Av. DA13: Ist R 7B **144**
Bracondale Rd. SE2 49Wc **95**
BRACTON CEN. 61Hd **140**
Bracton La. DA2: Wilm 61Hd **140**
Bradbeer Ho. E2 41Yb **92**
(off Cornwall Av.)
Bradbery WD3: Map C 22F **42**
Bradbourne Ct. TN13: S'oaks 94Kd **203**
Bradbourne Pk. Rd. TN13: S'oaks 95Jd **202**
Bradbourne Rd. DA5: Bexl 59Cd **118**
Bradbourne Rd. RM17: Grays 51De **121**
Bradbourne Rd. TN13: S'oaks 94Kd **203**
Bradbourne St. SW6 54Cb **111**
Bradbourne Va. Rd. TN13: S'oaks 94Hd **202**
Bradbrook Dr. DA3: Lfield 69Be **143**
Bradbury Cl. UB2: S'hall 49Ba **85**
Bradbury Cl. WD6: Bore 11Ra **29**
Bradbury Ct. DA11: Nflt 10B **122**
Bradbury Gdns. SL3: Ful 5P **61**
Bradbury M. N16 36Ub **71**
Bradbury St. N16 36Ub **71**
Bradby Ho. NW8 40Db **69**
(off Hamilton Ter.)
Bradby's HA1: Harr 32Ga **66**
(off High St.)
Bradcaster Gro. KT22: Lea 93Na **193**
Bradd Cl. RM15: S Ock 41Yd **98**
Braddock Cl. RM5: Col R 23Ed **56**
Braddock Cl. TW7: Isle 54Ha **108**
Braddon Ct. EN5: Barn 13Ab **30**
Braddon Rd. TW9: Rich 55Pa **109**
Braddyll St. SE10 50Gc **93**
Bradenham Av. DA16: Well 56Wc **117**
Bradenham Cl. SE17 51Tb **113**
Bradenham Rd. HA3: Kenton 28Ka **46**
Bradenham Rd. UB4: Hayes 41U **84**
Bradenhurst Cl. CR3: Cat'm 98Vb **197**
Braden St. W9 42Db **89**
Bradfield Cl. GU22: Wok 90A **168**
Bradfield Ct. NW1 38Kb **70**
(off Hawley Rd.)
Bradfield Ct. RM19: Purf 50Rd **97**
(off Linnet Way)
Bradfield Dr. IG11: Bark 36Wc **75**
Bradfield Ho. IG8: Wfd G 23Qc **54**
Bradfield Rd. E16 47Jc **93**
Bradfield Rd. HA4: Ruis 36Aa **65**
Bradford Cl. BR2: Broml 74Pc **160**
Bradford Cl. N17 23Vb **51**
Bradford Cl. SE26 63Xb **135**
Bradford Dr. KT19: Ewe 79Ua **154**
Bradford Ho. W14 48Za **88**
(off Spring Va. Ter.)
Bradford Rd. IG1: Ilf 32Tc **74**
Bradford Rd. SL1: Slou 4E **80**
Bradford Rd. W3 47Ua **88**
Bradford Rd. WD3: Herons 17E **24**
Bradfords Cl. IG9: Buck H 21Mc **53**
Bradgate Rd. SE6 58Dc **114**
Brading Cres. E11 33Kc **73**
Brading Rd. CR0: C'don 72Pb **156**
Brading Rd. SW2 59Pb **112**
Brading Ter. W12 48Wa **88**
Bradiston Rd. W9 41Bb **89**
Bradleigh Av. RM17: Grays 50Ee **99**
Bradley Cl. N7 37Nb **70**
Bradley Cl. SM2: Sutt 82Cb **175**
Bradley Ct. EN3: Enf L 10Ac **20**
(off Bradley Rd.)
Bradley Gdns. W13 44Ka **86**
Bradley Ho. E3 41Dc **92**
(off Bromley High St.)
Bradley Ho. IG8: Wfd G 24Jc **53**
Bradley Ho. SE16 49Yb **92**
(off Raymouth Rd.)
Bradley M. SW17 60Hb **111**
Bradley Rd. EN3: Enf L 10Ac **20**
Bradley Rd. EN9: Walt A 7Ec **20**
Bradley Rd. N22 26Pb **50**
Bradley Rd. SE19 65Sb **135**
Bradley Rd. SL1: Slou 5H **81**
Bradley's Cl. N1 1A **218** (40Qb **70**)
Bradley Stone Rd. E6 43Pc **94**
Bradman Ho. NW8 41Eb **89**
(off Abercorn Pl.)
Bradman Row HA8: Edg 24Sa **47**
Bradmead SW8 52Kb **112**
Bradmore Ct. EN3: Enf H 13Ac **34**
(off Enstone Rd.)
BRADMORE GREEN 90Pb **176**
Bradmore Grn. AL9: Brk P 8G **8**
Bradmore Grn. CR5: Coul 90Pb **176**
Bradmore La. AL9: N Mym 9E **8**
Bradmore Pk. Rd. W6 49Xa **88**
Bradmore Way AL9: Brk P 8G **8**
Bradmore Way CR5: Coul 89Nb **176**
Bradshaw Cl. SL4: Wind 3C **102**
Bradshaw Cl. SW19 65Cb **133**
Bradshaw Cotts. E14 44Ac **92**
(off Repton St.)
Bradshaw Dr. NW7 24Za **48**
Bradshawe Rd. RM16: Grays 46Ce **99**
Bradshawe Waye UB8: Hil 43P **83**
Bradshaw Rd. WD24: Wat 11Y **27**
Bradshaws AL10: Hat 4B **8**
Bradshaws Cl. SE25 69Wb **135**
Bradstock Ho. E9 38Zb **72**
Bradstock Rd. E9 37Zb **72**
Bradstock Rd. KT17: Ewe 78Wa **154**
Brad St. SE1 7A **224** (46Qb **90**)
Bradwell Av. RM10: Dag 33Cd **76**
Bradwell Cl. E18 28Hc **53**
Bradwell Cl. RM12: Horn 37Kd **77**

Bradwell Ct. CM13: Hut 16Ee 41
(off Bradwell Grn.)
Bradwell Ct. CR3: Whyt 91Wb 197
(off Godstone Rd.)
Bradwell Grn. CM13: Hut 16Ee 41
Bradwell Ho. NW6 39Db 69
(off Mortimer Cres.)
Bradwell M. N18 21Wb 51
Bradwell Rd. IG9: Buck H 17Nc 36
Bradwell St. E1 41Zb 92
Brady Av. IG10: Lough 12Sc 36
Brady Ct. RM8: Dag 32Zc 75
Brady Dr. BR1: Broml 69Qc 138
Brady Ho. SW8 53Lb 112
(off Corunna Rd.)
Bradymead E6 44Qc 94
Brady St. E1 42Xb 91
Braeburn Cl. BR8: Swan 70Hd 140
Braeburn Ct. BR6: Orp 75Vc 161
(off Blossom Dr.)
Braeburn Ct. EN4: E Barn 14Fb 31
Braeburn Ct. RM13: Rain 40Fd 76
(off Broadis Way)
Brae Ct. KT2: King T 67Qa 131
Braemar SW15 58Za 110
Braemar Av. CR2: S Croy 82Sb 177
Braemar Av. CR7: Thor H 69Qb 134
Braemar Av. DA7: Bex 56Ed 118
Braemar Av. HA0: Wemb 38Ma 67
Braemar Av. N22 25Nb 50
Braemar Av. NW10 34Ta 67
Braemar Av. SW18 61Cb 133
Braemar Av. SW19 61Cb 133
Braemar Cl. SE16 50Xb 91
(off Masters Dr.)
Braemar Ct. SE6 60Hc 115
(off Cumberland Pl.)
Braemar Ct. WD23: Bush 16Ca 27
Braemar Gdns. BR4: W W'ck 74Ec 158
Braemar Gdns. DA15: Sidc 62Tc 138
Braemar Gdns. NW9 25Ta 47
Braemar Gdns. RM11: Horn 30Qd 57
Braemar Gdns. SL1: Slou 7E 80
Braemar Ho. W9 4A 214 (41Eb 89)
(off Maida Vale)
Braemar Mans. SW7 48Db 89
(off Cornwall Gdns.)
Braemar Rd. E13 42Hc 93
Braemar Rd. KT4: Wor Pk 76Xa 154
Braemar Rd. N15 29Ub 51
Braemar Rd. TW8: Bford 51Ma 109
Braeside BR3: Beck 64Cc 136
Braeside KT15: New H 83K 169
Braeside Av. SW19 67Ab 132
Braeside Av. TN13: S'oaks 96Hd 202
Braeside Cl. HA5: Hat E 24Ca 45
Braeside Cl. TN13: S'oaks 95Hd 202
Braeside Cres. DA7: Bex 56Ed 118
Braeside Rd. SW16 66Lb 134
Braes Mead RH1: S Nut 7E 208
Braes St. N1 38Rb 71
Braesyde Cl. DA17: Belv 49Bd 95
Brafferton Rd. CR0: C'don 77Sb 157
Braganza St. SE17 7B 230 (50Rb 91)
Bragg Cl. RM8: Dag 37Xc 75
Bragmans La. WD3: Sarr 6E 10
Braham Ct. E2 42Xb 91
(off Three Colts La.)
Braham Cres. WD25: Wat 5V 12
Braham Ho. SE11 7J 229 (50Pb 90)
Braham St. E1 3K 225 (44Vb 91)
Braham St. Pk. 3K 225 (44Vb 91)
(off Half Moon Pas.)
Braid Av. W3 44Ua 88
Braid Cl. TW13: Hanw 61Ba 129
Braidwood Pas. EC1 7D 218 (43Sb 91)
(off Aldersgate St.)
Braidwood Rd. SE6 60Fc 115
Braidwood St. SE1 7H 225 (46Ub 91)
Brailey Ho. E10 32Gc 73
Brailsford Cl. CR4: Mitc 66Gb 133
Brailsford Rd. SW2 57Qb 112
Brainton Av. TW14: Felt 59X 107
Braintree Av. IG4: Ilf 28Nc 54
Braintree Ho. E1 42Yb 92
(off Malcolm Rd.)
Braintree Rd. HA4: Ruis 35X 65
Braintree Rd. RM10: Dag 34Cd 76
Braintree St. E2 41Yb 92
Braithwaite Av. RM7: Rush G 31Cd 76
Braithwaite Ct. WD17: Wat 10W 12
Braithwaite Gdns. HA7: Stan 25La 46
Braithwaite Ho. E15 37Fc 73
(off Forrester Way)
Braithwaite Ho. EC1 5F 219 (42Tb 91)
(off Bunhill Row)
Braithwaite Rd. EN3: Brim 13Bc 34
Braithwaite St. E1 6K 219 (42Vb 91)
Braithwaite Twr. W2 7C 214 (43Fb 89)
(off Hall Pl.)
Brakefield Rd. DA13: Sflt 65De 143
Brakes Pl. TN15: W King 79Ud 164
Brakey Hill RH1: Blet 6L 209
Bramah Grn. SW9 53Qb 112
(off Eythorne Rd.)
Bramah Ho. SW1 7K 227 (50Kb 90)
Bramah Rd. SW9 53Qb 112
Bramalea Cl. N6 30Jb 50
Bramall Cl. E15 36Hc 73
Bramall Ct. N7 37Pb 70
(off Watkinson Rd.)
Bramber WC1 4F 217 (41Nb 90)
(off Cromer St.)
Bramber Cl. KT20: Tad 95Xa 194
Bramber Ct. DA2: Dart 58Rd 119
(off Bow Arrow La.)
Bramber Ct. SL1: Slou 6E 80
Bramber Ct. TW8: Bford 49Na 87
Bramber Ct. W14 51Bb 111
(off Bramber Rd.)
Bramber Ho. KT2: King T 67Na 131
(off Seven Kings Way)
Bramber Rd. N12 22Gb 49
Bramber Rd. W14 51Bb 111
Bramber Way CR6: W'ham 88Bc 178
Brambleacres Cl. SM2: Sutt 80Cb 155
Bramble Av. DA2: Bean 62Yd 142
Bramble Banks SM5: Cars 81Jb 176
Bramblebury Rd. SE18 50Sc 94
Bramble Cl. BR3: Beck 71Ec 158
Bramble Cl. CR0: C'don 77Cc 158
Bramble Cl. HA7: Stan 24Ma 47
Bramble Cl. IG7: Chig 18Sc 36
Bramble Cl. N15 28Wb 51
Bramble Cl. RH1: Redh 8A 208
Bramble Cl. RH8: Oxt 5M 211
Bramble Cl. SE19 67Tb 135
Bramble Cl. SL9: Chal P 23A 42
Bramble Cl. TW17: Shep 69T 128
Bramble Cl. UB8: Hil 44P 83
Bramble Cl. WD25: Wat 6W 12
Bramble Ct. IG6: Ilf 23Vc 55
Bramble Cft. DA8: Erith 49Ed 96
Brambledene Cl. GU21: Wok 10N 167
Brambledown DA3: Hartl 70Be 143
Brambledown TW18: Staines 67K 127
Brambledown Cl. BR4: W W'ck 71Gc 159
Brambledown Rd. CR2: Sande 80Ub 157
Brambledown Rd. SM5: Cars 80Jb 156
Brambledown Rd. SM6: W'gton 80Jb 156
Bramblefield Cl. DA3: Lfield 69Zd 143
Bramble Gdns. W12 45Va 88
Bramble Ho. E3 43Cc 92
(off Devons Rd.)
Bramble La. RM14: Upm 39Sd 78
Bramble La. TN13: S'oaks 100Kd 203
Bramble La. TW12: Hamp 65Ba 129
Bramble M. DA12: Grav'nd 3F 144
Bramble Ri. KT11: Cobh 87Y 171
Bramble Rd. AL10: Hat 1P 7
Brambles, The AL1: St A 4B 6
Brambles, The EN8: Chesh 3Zb 20
Brambles, The IG7: Chig 23Sc 54
Brambles, The SM1: Sutt 75Fb 155
Brambles, The SW19 64Bb 133
(off Woodside)
Brambles, The UB7: W Dray 49M 83
Brambles Cl. CR3: Cat'm 94Ub 197
Brambles Cl. TW7: Isle 52Ka 108
Brambles Farm Dr. UB10: Hil 41Q 84
Brambletye Pk. Rd. RH1: Redh 8P 207
Bramble Wlk. KT18: Eps 86Ra 173
Bramble Wlk. RH1: Redh 8A 208
Bramble Way GU23: Rip 96H 189
Bramblewood RH1: Mers 1B 208
Bramblewood Cl. SM5: Cars 74Gb 155
Brambling Cl. DA9: Ghithe 58Wd 120
Brambling Cl. WD23: Bush 14Aa 27
Brambling Ct. SE8 51Bc 114
(off Abinger Gro.)
Bramblings, The E4 21Fc 53
Bramcote Av. CR4: Mitc 70Hb 133
Bramcote Ct. CR4: Mitc 70Hb 133
(off Bramcote Av.)
Bramcote Gro. SE16 50Yb 92
Bramcote Ho. KT13: Weyb 77S 150
Bramcote Rd. SW15 56Xa 110
Bramdean Cres. SE12 60Jc 115
Bramdean Gdns. SE12 60Jc 115
Bramerton NW6 38Za 68
(off Willesden La.)
Bramerton Rd. BR3: Beck 69Bc 136
Bramerton St. SW3 51Gb 111
Bramfield WD25: Wat 6Aa 13
Bramfield Ct. N4 33Sb 71
(off Queen's Dr.)
Bramfield Rd. SW11 58Gb 111
Bramford Ct. N14 19Mb 32
Bramford Rd. SW18 56Eb 111
Bramham Ct. HA6: Nwood 23U 44
Bramham Gdns. KT9: Chess 77Ma 153
Bramham Gdns. SW5 50Db 89
Bramhope La. SE7 51Kc 115
Bramlands Cl. SW11 55Gb 111
Bramleas WD18: Wat 14V 26
Bramley Av. CR5: Coul 87Lb 176
Bramley Av. TW17: Shep 69U 128
Bramley Bank Local Nature
Reserve 79Yb 158
Bramley Cl. BR6: Farnb 74Rc 160
Bramley Cl. BR8: Swan 70Gd 140
Bramley Cl. CR2: S Croy 78Sb 157
Bramley Cl. DA13: Ist R 6B 144
Bramley Cl. E17 26Ac 52
Bramley Cl. HA5: Eastc 27V 44
Bramley Cl. IG8: Wfd G 24Lc 53
Bramley Cl. KT16: Chert 74K 149
Bramley Cl. N14 15Kb 32
Bramley Cl. NW7 20Ua 30
Bramley Cl. RH1: Redh 8N 207
Bramley Cl. TW18: Staines 65L 127
Bramley Cl. TW2: Whitt 58Ea 108
Bramley Cl. UB3: Hayes 45W 84
Bramley Ct. BR6: Orp 75Vc 161
(off Blossom Dr.)
Bramley Ct. CR4: Mitc 68Fb 133
Bramley Ct. DA16: Well 53Xc 117
Bramley Ct. E4 18Ec 34
(off The Ridgeway)
Bramley Ct. EN4: E Barn 14Gb 31
Bramley Ct. GU24: Bisl 9F 166
Bramley Ct. RH1: Redh 4N 207
Bramley Ct. RM13: Rain 40Fd 76
(off Broadis Way)
Bramley Ct. UB1: S'hall 45Ea 86
(off Haldane Rd.)
Bramley Ct. WD25: Wat 4X 13
Bramley Cres. IG2: Ilf 30Qc 54
Bramley Cres. SW8 52Mb 112
Bramley Gdns. WD19: Wat 23Y 45
Bramley Gro. KT21: Asht 91Na 193
Bramley Hill CR2: S Croy 78Rb 157
Bramley Ho. RH1: Redh 7A 208
Bramley Ho. SW15 58Va 110
(off Tunworth Cres.)
Bramley Ho. TW4: Houn 56Ba 107
Bramley Ho. W10 44Za 88
Bramley Ho. Ct. EN2: Enf 9Tb 19
(off Meadow Rd.)
Bramley Hyrst CR2: S Croy 78Sb 157
Bramley Lodge HA0: Wemb 35Ma 67
Bramley Pde. N14 14Mb 32
Bramley Pl. DA1: Cray 56Jd 118
Bramley Rd. N14 15Kb 32
Bramley Rd. SM1: Sutt 78Fb 155
Bramley Rd. SM2: Cheam 81Za 174
Bramley Rd. W10 44Za 88
Bramley Rd. W5 48La 86
Bramley Shaw EN9: Walt A 5Hc 21
Bramley Sports Ground 15Jb 32
Bramley Way AL4: St A 3G 6
Bramley Way BR4: W W'ck 75Dc 158
Bramley Way KT21: Asht 89Pa 173
Bramley Way TW4: Houn 57Ba 107
Brammas Cl. SL1: Slou 8G 80
Brampton WC1 1H 223 (43Pb 90)
(off Red Lion Sq.)
Brampton Cl. E5 33Xb 71
Brampton Cl. EN7: Chesh 1Wb 19
Brampton Ct. NW4 28Xa 48
Brampton Ct. RM7: Rush G 30Fd 56
(off Union Rd.)
Brampton Gdns. KT12: Hers 78Y 151
Brampton Gro. HA3: Kenton 28Ja 46
Brampton Gro. HA9: Wemb 32Qa 67
Brampton Gro. NW4 28Xa 48
Brampton Ho. SE16 47Yb 92
(off Albatross Way)
Brampton La. NW4 28Ya 48
Brampton Pk. Rd. N22 27Qb 50
Brampton Rd. AL1: St A 1E 6
Brampton Rd. CR0: C'don 73Vb 157
Brampton Rd. DA7: Bex 55Zc 117
Brampton Rd. E6 41Mc 93
Brampton Rd. N15 29Sb 51
Brampton Rd. NW9 28Qa 47
Brampton Rd. SE2 51Yc 117
Brampton Rd. UB10: Hil 40R 64
Brampton Rd. WD19: Wat 20W 26
Brampton Ter. WD6: Bore 10Qa 15
(off Stapleton Rd.)
Bramshaw Gdns. WD19: Wat 22Z 45
Bramshaw Ri. KT3: N Mald 72Ua 154
Bramshaw Rd. E9 37Zb 72
Bramshill Cl. IG7: Chig 22Uc 54
Bramshill Gdns. NW5 34Kb 70
Bramshill Rd. NW10 40Va 68
Bramshot Av. SE7 51Jc 115
Bramshot Way WD19: Wat 19W 26
Bramshurst NW8 39Db 69
(off Abbey Rd.)
Bramston Cl. IG6: Ilf 23Vc 55
Bramston Rd. NW10 40Wa 68
Bramston Rd. SW17 62Eb 133
Bramwell Cl. TW16: Sun 68Z 129
Bramwell Ho. SE1 4E 230 (48Sb 91)
Bramwell Ho. SW1 50Lb 90
(off Churchill Gdns.)
Bramwell M. N1 39Pb 70
Bramwell Way E16 46Lc 93
Brancaster Dr. NW7 24Wa 48
Brancaster Ho. E1 41Zb 92
(off Moody St.)
Brancaster La. CR8: Purl 82Sb 177
Brancaster Pl. IG10: Lough 13Pc 36
Brancaster Rd. E12 35Pc 74
Brancaster Rd. IG2: Ilf 30Uc 54
Brancaster Rd. SW16 62Nb 134
Brancepeth Gdns. IG9: Buck H 19Jc 35
Branch Hill NW3 34Eb 69
Branch Hill Ho. NW3 34Db 69
Branch Pl. N1 39Tb 71
Branch Rd. AL2: Park 9B 6
Branch Rd. AL3: St A 1P 5
Branch Rd. E14 45Ac 92
Branch Rd. IG6: Ilf 22Xc 55
Branch St. SE15 52Ub 113
Brancker Rd. HA3: Kenton 27Ma 47
Brancroft Way EN3: Brim 11Ac 34
Brand Av. UB10: Uxb 40N 63
Brand Cl. N4 32Rb 71
Brandesbury Sq. IG8: Wfd G 24Qc 54
Brandlehow Rd. SW15 56Bb 111
Brandon Cl. RM16: Chaf H 47Be 99
Brandon Ct. E16 42Hc 93
Brandon Est. SE17 51Rb 113
Brandon Groves Av. RM15: S Ock 41Yd 98
Brandon Ho. BR3: Beck 64Dc 136
(off Beckenham Hill Rd.)
Brandon Mans. W14 51Ab 110
(off Queen's Club Gdns.)
Brandon M. EC2 1F 225 (43Tb 91)
(off Moor La.)
Brandon M. SE17 6E 230 (49Sb 91)
Brandon Mobile Home Pk. AL4: St A 2H 7
Brandon Rd. DA1: Dart 59Qd 119
Brandon Rd. E17 28Ec 52
Brandon Rd. N7 38Nb 70
Brandon Rd. SM1: Sutt 77Db 155
Brandon Rd. UB2: S'hall 50Ba 85
Brandon St. DA11: Grav'nd 9D 122
Brandon St. SE17 6E 230 (49Sb 91)
(not continuous)
Brandram M. SE13 56Gc 115
(off Brandram Rd.)
Brandram Rd. SE13 55Gc 115
Brandrams Wharf SE16 47Yb 92
Brandreth Ct. HA1: Harr 30Ha 46
Brandreth Rd. E6 44Pc 94
Brandreth Rd. SW17 61Kb 134
Brandries, The SM6: Bedd 76Mb 156
Brands Hatch 78Ud 164
Brands Hatch Pk. DA3: Fawk 77Vd 164
Brands Hatch Rd. DA3: Fawk 76Wd 164
BRANDS HILL 51D 104
Brands Ho. NW6 39Bb 69
(off Lincoln M.)
Brandsland RH2: Reig 10K 207
Brands Rd. SL3: L'ly 51D 104
Brand St. SE10 52Ec 114
Brandville Gdns. IG6: Ilf 28Rc 54
Brandville Rd. UB7: W Dray 47N 83
Brandy Way SM2: Sutt 80Cb 155
Branfill Rd. RM14: Upm 33Rd 77
Brangbourne Rd. BR1: Broml 64Ec 136
Brangton Rd. SE11 7J 229 (50Pb 90)
Brangwyn Ct. W14 48Ab 88
(off Blythe Rd.)
Brangwyn Cres. SW19 67Eb 133
Branksea St. SW6 52Ab 110
Branksome KT13: Weyb 79T 150
(off Gower Rd.)
Branksome Av. N18 23Vb 51
Branksome Cl. HP2: Hem H 1A 4
Branksome Cl. KT12: Walt T 75Z 151
Branksome Cl. TW11: Tedd 63Fa 130
Branksome Ho. SW8 52Pb 112
(off Meadow Rd.)
Branksome Rd. SW19 67Cb 133
Branksome Rd. SW2 57Nb 112
Branksome Way HA3: Kenton 30Pa 47
Branksome Way KT3: N Mald 67Sa 131
Branksome Ct. N2 27Eb 49
Brannigan Way HA8: Edg 21Na 47
Bransby Rd. KT9: Chess 79Na 153
Branscombe NW1 1C 216 (39Lb 70)
(off Plender St.)
Branscombe Ct. BR2: Broml 71Hc 159
Branscombe Gdns. N21 17Qb 32
Branscombe Ho. WD24: Wat 9W 12
Branscombe St. SE13 55Dc 114
Bransdale Cl. NW6 39Cb 69
Bransell Cl. BR8: Crock 72Ed 162
Bransgrove Rd. HA8: Edg 25Pa 47
Branston Cl. WD19: Wat 17Y 27
Branston Cres. BR5: Pet W 74Tc 160
Branstone Ct. RM19: Purf 50Rd 97
Branstone Rd. TW9: Kew 53Pa 109
Brants Rd. DA9: Ghithe 58Vd 120
Brants Wlk. W7 42Ga 86
Brantwood Av. DA8: Erith 52Ed 118
Brantwood Av. TW7: Isle 56Ja 108
Brantwood Cl. E17 27Dc 52
Brantwood Cl. KT14: W Byf 85J 169
Brantwood Ct. KT14: W Byf 85H 169
(off Brantwood Dr.)
Brantwood Dr. KT14: W Byf 85H 169
Brantwood Gdns. EN2: Enf 14Nb 32
Brantwood Gdns. IG4: Ilf 28Nc 54
Brantwood Gdns. KT14: W Byf 85H 169
Brantwood Ho. SE5 52Sb 113
(off Wyndam Est.)
Brantwood Rd. CR2: S Croy 81Sb 177
Brantwood Rd. DA7: Bex 54Dd 118
Brantwood Rd. N17 23Wb 51
Brantwood Rd. SE24 57Sb 113
Brantwood Way BR5: St P 69Yc 139
Branxholme Ct. BR1: Broml 67Hc 137
(off Highland Rd.)
Braque Bldg. SE1 7D 224 (46Sb 91)
(off Union St.)
Brasenose Dr. SW13 51Ya 110
Brasher Cl. UB6: G'frd 36Fa 66
Brassett Point E15 39Gc 73
(off Abbey Rd.)
Brassey Cl. RH8: Oxt 1L 211
Brassey Cl. TW14: Felt 60W 106
Brassey Hill RH8: Oxt 2L 211
Brassey Ho. E14 49Dc 92
(off Cahir St.)
Brassey Rd. NW6 37Bb 69
Brassey Rd. RH8: Oxt 2K 211
Brassey Sq. SW11 55Jb 112
Brassie Av. W3 44Ua 88
Brass Talley All. SE16 47Zb 92
BRASTED 96Yc 201
BRASTED CHART 100Xc 201
Brasted Cl. BR6: Orp 75Wc 161
Brasted Cl. DA6: Bex 57Zc 117
Brasted Cl. SE26 63Yb 136
Brasted Cl. SM2: Sutt 82Cb 175
Brasted Hill TN14: Knock 91Wc 201
Brasted Hill TN14: Sund 91Wc 201
Brasted Hill Rd. TN16: Bras 93Xc 201
Brasted La. TN14: Knock 91Wc 201
Brasted Lodge BR3: Beck 66Cc 136
Brasted Rd. DA8: Erith 52Gd 118
Brasted Rd. TN16: Westrm 98Uc 200
Brathay NW1 2C 216 (40Lb 70)
(off Ampthill Est.)
Brathway Rd. SW18 59Cb 111
Bratley St. E1 42Wb 91
Bratten Ct. CR0: C'don 72Tb 157
Braund Av. UB6: G'frd 42Da 85
Braundton Av. DA15: Sidc 60Vc 117
Braunston Dr. UB4: Yead 42Aa 85
Braunston Ho. HA0: Wemb 39Na 67
Bravington Cl. TW17: Shep 71P 149
Bravington Pl. W9 42Bb 89
Bravington Rd. W9 40Bb 69
Bravingtons Wlk. N1 2G 217 (40Nb 70)
(off York Way)
Brawlings La. SL9: Chal P 21C 42
Brawne Ho. SE17 51Rb 113
(off Brandon Est.)
Braxfield Rd. SE4 56Ac 114
Braxted Pk. SW16 65Pb 134
Bray NW3 38Gb 69
Brayards Rd. SE15 54Xb 113
Brayards Rd. Est. SE15 54Xb 113
(off Caulfield Rd.)
Braybourne Cl. UB8: Uxb 37L 63
Braybrooke Dr. TW7: Isle 52Ha 108
Braybrooke Gdns. SE19 66Ub 135
Braybrook St. W12 43Va 88
Brayburne Av. SW4 54Lb 112
Bray Cl. WD6: Bore 11Sa 29
Bray Ct. E2 41Zb 92
(off Meath Cres.)
Bray Ct. SW16 64Nb 134
Braycourt Av. KT12: Walt T 73X 151
Bray Cres. SE16 47Zb 92
Braydon Rd. N16 32Wb 71
Bray Dr. E16 45Hc 93
Brayfield Ter. N1 38Qb 70
Brayford Sq. E1 44Yb 92
Bray Gdns. GU22: Pyr 88G 168
Bray Pas. E16 45Jc 93
Bray Pl. SW3 6F 227 (49Hb 89)
Bray Rd. KT11: Stoke D 88Aa 171
Bray Rd. NW7 23Za 48
Brays Gdns. SE13 58Ec 114
Brays Gdns. SE6 59Dc 114
Bray Springs EN9: Walt A 6Gc 21
Brayton Gdns. EN2: Enf 14Mb 32
Braywood Av. TW20: Egh 65B 126
Braywood Rd. SE9 56Tc 116
Brazier Cres. UB5: N'olt 42Ba 85
Brazil Cl. CR0: Bedd 73Nb 156
Breach Barns La. EN9: Walt A 2Hc 21
Breach Barns Mobile Home Pk.
EN9: Walt A 2Kc 21
Breacher Ho. Apts. IG11: Bark 42Wc 95
Breach La. RM9: Dag 41Cd 96
Breach Rd. RM20: W Thur 51Vd 120
Bread St. EC4 3E 224 (44Sb 91)
(not continuous)
Breakfield CR5: Coul 88Nb 176
Breakneck Hill DA9: Ghithe 57Xd 120
Breakspear Av. AL1: St A 3D 6
Breakspear Ct. WD5: Ab L 2V 12
Breakspear Crematorium 29S 44
Breakspeare Cl. WD24: Wat 10X 13
Breakspeare Rd. WD5: Ab L 3U 12
Breakspear Ho. UB9: Hare 27N 43
Breakspear M. UB9: Hare 27M 43
Breakspear Path UB9: Hare 27M 43
Breakspear Pl. WD5: Ab L 2V 12
Breakspear Rd. HA4: Ruis 31R 64
Breakspear Rd. Nth. UB9: Hare 25L 43
Breakspear Rd. Sth. UB10: Ick 34P 63
Breakspear Rd. Sth. UB9: Hare 32Q 64
Breakspears Dr. BR5: St P 67Wc 139
Breakspears M. SE4 54Bc 114
Breakspears Rd. SE4 56Bc 114
(not continuous)
Breakspear Way HP2: Hem H 2C 4
Breakwell Ct. W10 42Ab 88
(off Wornington Rd.)
Bream Cl. N17 28Xb 51
Bream Gdns. E6 41Qc 94
Breamore Cl. SW15 60Wa 110
Breamore Ct. IG3: Ilf 33Wc 75
Breamore Ho. SE15 52Wb 113
(off Friary Est.)
Breamore Rd. IG3: Ilf 33Vc 75
Bream's Bldgs. EC4 2K 223 (44Qb 90)
Bream St. E3 38Cc 72
Breamwater Gdns. TW10: Ham 62Ka 130
Brearley Cl. HA8: Edg 24Sa 47
Brearley Cl. UB8: Uxb 37N 63
Breasley Cl. SW15 56Xa 110
Breasy Pl. NW4 28Xa 48
(off Burroughs Gdns.)
Brechin Pl. SW7 6A 226 (49Eb 89)
Brecknock Rd. N19 35Lb 70
Brecknock Rd. N7 36Mb 70
Brecknock Rd. Est. N19 35Lb 70
Breckonmead BR1: Broml 68Lc 137
Brecon Cl. CR4: Mitc 69Nb 134
Brecon Cl. KT4: Wor Pk 75Ya 154
Brecon Ct. SL1: Slou 7G 80
Brecon Grn. NW9 30Ua 48
Brecon Ho. E3 40Bc 72
(off Ordell Rd.)
Brecon Ho. UB5: N'olt 41Ba 85
(off Taywood Rd.)
Brecon Ho. W2 2A 220 (44Eb 89)
(off Hallfield Est.)
Brecon Lodge UB7: W Dray 47P 83
Brecon M. N7 36Mb 70
Brecon Rd. EN3: Pond E 14Yb 34
Brecon Rd. W6 51Ab 110
Brecons, The KT13: Weyb 77T 150
Brede Cl. E6 41Qc 94
Bredel Ho. E14 43Cc 92
(off St Paul's Way)
Brede M. E6 41Qc 94
Bredgar SE13 57Dc 114
Bredgar Rd. N19 33Lb 70
Bredhurst Cl. SE20 65Yb 136
Bredinghurst SE22 59Wb 113
Bredin Ho. SW10 52Db 111
(off Coleridge Gdns.)
Bredle Way RM15: Avel 46Ud 98
Bredon Rd. CR0: C'don 73Vb 157
Bredune CR8: Kenley 87Tb 177
Breech La. KT20: Walt H 96Wa 194
Bree Ct. NW9 27Sa 47
Breer St. SW6 55Db 111
Breething Rd. TN14: Dun G 92Hd 202
Breezers Ct. E1 45Wb 91
(off The Highway)
Breezer's Hill E1 45Wb 91
Bremer Rd. HA2: Harr 33Ea 66
Bremer M. E17 28Dc 52
Bremer Rd. TW18: Staines 62J 127
Bremner Cl. BR8: Swan 70Jd 140
Bremner Rd. SW7 3A 226 (48Eb 89)
Brenchley Av. DA11: Grav'nd 4D 144
Brenchley Cl. BR2: Broml 72Hc 159
Brenchley Cl. BR7: Chst 67Qc 138
Brenchley Gdns. SE23 58Yb 114
Brenchley Rd. BR5: St P 68Vc 139
Bren Ct. EN3: Enf L 9Cc 20
(off Colgate Pl.)
Brendans Cl. RM11: Horn 32Nd 77
Brenda Rd. SW17 61Hb 133
Brenda Ter. DA10: Swans 59Ae 121
Brende Gdns. KT8: W Mole 70Da 129
Brendon Av. NW10 35Ua 68
Brendon Cl. DA8: Erith 53Gd 118
Brendon Cl. KT10: Esh 79Ea 152
Brendon Cl. UB3: Harl 52S 106
Brendon Ct. UB2: S'hall 49Da 85
Brendon Ct. WD7: R'lett 6Ka 14
Brendon Dr. KT10: Esh 79Ea 152
Brendon Gdns. HA2: Harr 35Da 65
Brendon Gdns. IG2: Ilf 29Uc 54
Brendon Gro. N2 26Eb 49
Brendon Rd. RM8: Dag 32Bd 75
Brendon Rd. SE9 61Tc 138
Brendon St. W1 2E 220 (44Gb 89)
Brendon Vs. N21 18Sb 33
Brendon Way EN1: Enf 17Ub 33
Brenley Cl. CR4: Mitc 69Jb 134
Brenley Gdns. SE9 56Mc 115
Brenley Ho. SE1 1F 231 (47Tb 91)
(off Tennis St.)
Brennand Ct. N19 34Lb 70
Brennan Rd. RM18: Tilb 4D 122
Brent, The DA1: Dart 59Qd 119
Brent, The DA2: Dart 59Rd 119
Brent Cl. DA2: Dart 58Rd 119
Brent Cl. DA5: Bexl 60Ad 117
Brentcot Cl. W13 42Ka 86
Brent Ct. NW11 31Za 68
Brent Ct. W7 45Fa 86
Brent Cres. NW10 40Pa 67
BRENT CROSS 31Ya 68
Brent Cross Bus Station 31Ya 68
Brent Cross Fly-Over NW4 31Za 68
Brent Cross Gdns. NW4 30Za 48
BRENT CROSS INTERCHANGE 30Ya 48
Brent Cross Shop. Cen. 31Ya 68
Brent Cross Station (Underground) 31Za 68
Brentfield NW10 38Ra 67
Brentfield Cl. NW10 37Ta 67
Brentfield Gdns. NW2 31Za 68
Brentfield Ho. NW10 38Ta 67
Brentfield Rd. DA1: Dart 58Qd 119
Brentfield Rd. NW10 37Ta 67
BRENTFORD 51Ma 109
Brentford Bus. Cen. TW8: Bford 52La 108
Brentford Cl. UB4: Yead 42Z 85
Brentford Community Stadium 50Pa 87
BRENTFORD END 52Ka 108
Brentford FC 50Pa 87
Brentford Fountain Leisure Cen. 50Qa 87
Brentford Ho. TW1: Twick 58Ka 108
Brentford Station (Rail) 51La 108
Brent Grn. NW4 29Ya 48
Brent Grn. Wlk. HA9: Wemb 34Sa 67
Brentham Club 41La 86
Brentham Way W5 42Ma 87
Brent Ho. E9 37Yb 72
(off Brenthouse Rd.)
Brenthouse Rd. E9 38Yb 72
Brenthurst Rd. NW10 37Va 68
Brentlands Dr. DA1: Dart 60Qd 119
Brent La. DA1: Dart 59Pd 119
Brent Lea TW8: Bford 52La 108
Brentleigh Ct. CM14: B'wood 20Wd 40
Brentmead Cl. W7 45Ga 86
Brentmead Gdns. NW10 40Pa 67
Brentmead Pl. NW11 30Za 48
Brentmoor Rd. GU24: W End 5A 166
Brent Mus. 37Xa 68
Brent New Ent. Cen. NW10 37Va 68
BRENT OLDER PEOPLE DAY
HOSPITAL 39Wa 68
Brenton Ct. E9 36Ac 72
(off Mabley St.)
Brenton St. E14 44Ac 92
Brent Pk. Ind. Est. UB2: S'hall 48X 85
Brent Pk. Rd. NW4 31Xa 68
Brent Pk. Rd. NW9 31Xa 68
Brent Pl. EN5: Barn 15Bb 31
Brent Reservoir 32Va 68
Brent Rd. CR2: Sels 81Xb 177
Brent Rd. E16 44Jc 93
Brent Rd. SE18 52Rc 116
Brent Rd. TW8: Bford 51La 108
Brent Rd. UB2: S'hall 48Y 85
Brent Side TW8: Bford 51La 108
Brentside Cl. W13 42Ja 86

Brentside Executive Cen. TW8: Bford ...51Ka 108
Brent Sth. Shop. Pk. ...32Ya 68
Brent St. NW4 ...28Ya 48
Brent Ter. NW2 ...32Ya 68
(not continuous)
Brent Trad. Cen. NW10 ...36Ua 68
Brentvale Av. HA0: Wemb ...39Pa 67
Brentvale Av. UB1: S'hall ...46Fa 86
Brent Valley Golf Course ...45Ga 86
Brent Vw. Rd. NW9 ...30Wa 48
Brentwaters Bus. Pk. TW8: Bford ...52La 108
Brent Way DA2: Dart ...58Rd 119
Brent Way HA9: Wemb ...37Ra 67
Brent Way N3 ...23Cb 49
Brent Way TW8: Bford ...52Ma 109
Brentwick Gdns. TW8: Bford ...49Na 87
BRENTWOOD ...19Zd 41
Brentwood By-Pass CM14: B'wood ...21Td 58
Brentwood Camping & Caravaning Site CM15: Dodd ...12Vd 40
Brentwood Cen., The ...16Yd 40
Brentwood Cl. SE9 ...60Sc 116
BRENTWOOD COMMUNITY HOSPITAL ...18Ae 41
Brentwood Ct. KT15: Add ...77K 149
Brentwood Golf Course ...15Qd 39
Brentwood Ho. SE18 ...52Mc 115
(off Portway Gdns.)
Brentwood Karting ...24Yd 58
Brentwood Leisure Pk. ...24Yd 58
Brentwood Lodge NW4 ...29Za 48
(off Holmdale Gdns.)
Brentwood Mus. ...21Yd 58
BRENTWOOD NUFFIELD HEALTH HOSPITAL ...18Ae 41
Brentwood Pk. Ski & Snowboarding Cen. ...24Xd 58
Brentwood Pl. CM15: B'wood ...18Zd 41
Brentwood RC Cathedral ...19Zd 41
Brentwood Rd. CM13: Heron ...21Ce 59
Brentwood Rd. CM13: Ingve ...21Ce 59
Brentwood Rd. CM13: W H'dn ...21Ce 59
Brentwood Rd. RM1: Rom ...30Hd 56
Brentwood Rd. RM16: Grays ...9D 100
Brentwood Rd. RM16: Ors ...9D 100
Brentwood Rd. RM2: Rom ...30Kd 57
Brentwood School Sports Cen. ...19Zd 41
Brentwood Station (Rail & Crossrail) ...20Yd 40
Brentwood Theatre ...18Zd 41
Brereton Ct. HP3: Hem H ...4N 3
Brereton Rd. N17 ...24Vb 51
Bressay Dr. NW7 ...24Wa 48
Bressenden Pl. SW1 ...3A 228 (48Kb 90)
Bressey Av. EN1: Enf ...11Wb 33
Bressey Gro. E18 ...26Hc 53
Bresslaw Ct. E3 ...43Bc 92
(off Wager St.)
Bretlands Rd. KT16: Chert ...75G 148
Breton Highwalk EC2 ...7E 218 (43Sb 91)
(off Golden La.)
Breton Ho. EC2 ...7E 218 (43Sb 91)
(off Golden La.)
Breton Ho. SE1 ...3K 231 (48Vb 91)
(off St Saviour's Est.)
Bretons Outdoor Recreation Cen. ...37Hd 76
Brett Cl. N16 ...33Ub 71
Brett Cl. UB5: N'olt ...41Z 85
Brett Ct. N9 ...19Yb 34
Brettell St. SE17 ...7G 231 (50Tb 91)
Brettenham Av. E17 ...25Cc 52
Brettenham Rd. E17 ...26Cc 52
Brettenham Rd. N18 ...21Wb 51
Brett Gdns. RM9: Dag ...38Ad 75
Brettgrave KT19: Eps ...82Sa 173
Brett Ho. EN8: Chesh ...1Zb 20
Brett Ho. Cl. SW15 ...59Za 110
Brettinghurst SE1 ...50Wb 91
(off Avondale Sq.)
Brett Pas. E8 ...36Xb 71
Brett Pl. WD24: Wat ...9W 12
Brett Rd. E8 ...36Xb 71
Brett Rd. EN5: Barn ...15Ya 30
Brevet Cl. RM19: Purf ...49Td 98
Brewers Bldgs. EC1 ...3B 218 (41Rb 91)
(off Rawstorne St.)
Brewers Ct. W2 ...1A 220 (43Eb 89)
Brewer's Fld. DA2: Wilm ...63Ld 141
Brewer's Grn. SW1 ...3D 228 (48Mb 90)
(off Buckingham Ga.)
Brewer's Hall Gdn. EC2 ...1E 224 (43Sb 91)
(off Aldermanbury Sq.)
Brewers La. E20 ...36Dc 72
Brewers La. TW9: Rich ...57Ma 109
Brewers Rd. DA12: Shorne ...7L 145
Brewer St. RH1: Blet ...3J 209
Brewer St. W1 ...5C 222 (45Lb 90)
Brewery, The EC2 ...7E 218 (43Sb 91)
Brewery, The RM1: Rom ...29Gd 56
Brewery Cl. HA0: Wemb ...36Ja 66
Brewery Ind. Est., The N1 ...2E 218 (40Sb 71)
(off Wenlock Rd.)
Brewery La. KT14: Byfl ...85N 169
Brewery La. TN13: S'oaks ...97Kd 203
Brewery La. TW1: Twick ...59Ha 108
Brewery M. WD17: Wat ...11X 27
Brewery M. Cen. TW7: Isle ...55Ja 108
Brewery Rd. BR2: Broml ...74Nc 160
Brewery Rd. GU21: Wok ...9P 167
Brewery Rd. N7 ...38Nb 70
Brewery Rd. SE18 ...50Tc 94
Brewery Sq. EC1 ...6C 218 (42Rb 91)
Brewery Sq. SE1 ...1K 231 (47Vb 91)
(off Horselydown La.)
Brewery Wlk. RM1: Rom ...29Gd 56
Brewhouse La. E1 ...46Xb 91
Brewhouse La. SW15 ...55Ab 110
Brewhouse Rd. SE18 ...49Pc 94
Brewhouse Wlk. SE16 ...46Ac 92
Brewhouse Yd. DA12: Grav'nd ...8D 122
Brewhouse Yd. EC1 ...5B 218 (42Rb 91)
Brewin Ter. UB4: Yead ...43Y 85
(off Larch Cres.)
Brewood Rd. RM8: Dag ...37Xc 75
Brewster Ct. CM14: W'ley ...22Xd 58
Brewster Gdns. W10 ...43Ya 88
Brewster Ho. E14 ...45Bc 92
(off Three Colt St.)
Brewster Ho. SE1 ...5K 231 (49Vb 91)
(off Dunton Rd.)
Brewster Pl. KT1: King T ...68Sa 131
Brewster Rd. E10 ...32Dc 72
Breyer Group Stadium, The ...34Dc 72
(shown as Leyton Orient FC)
Brian Av. CR2: Sande ...84Ub 177
Brian Cl. RM12: Horn ...35Kd 77
Briane Rd. KT19: Eps ...82Sa 173
Brian Rd. RM6: Chad H ...29Yc 55
Briant Ho. SE1 ...4J 229 (48Pb 90)
(off Hercules Rd.)
Briants Cl. HA5: Pinn ...26Ba 45
Briant St. SE14 ...53Zb 114
Briar Av. SW16 ...66Pb 134
Briarbank Rd. W13 ...44Ja 86
Briar Banks SM5: Cars ...81Jb 176
Briarcliff HP1: Hem H ...1G 2
Briar Cl. CR6: W'ham ...88Cc 178
Briar Cl. EN8: Chesh ...1Yb 20
Briar Cl. IG9: Buck H ...19Mc 35
Briar Cl. KT14: W Byf ...83K 169
Briar Cl. N13 ...20Sb 33
Briar Cl. N2 ...27Db 49
Briar Cl. TW12: Hamp ...64Ba 129
Briar Cl. TW7: Isle ...57Ha 108
Briar Ct. E3 ...40Cc 72
(off Morville St.)
Briar Ct. GU24: Bisl ...9F 166
Briar Ct. SM3: Cheam ...77Ya 154
Briar Ct. SW15 ...56Xa 110
Briar Cres. UB5: N'olt ...37Da 65
Briardale HA8: Edg ...21Ta 47
Briardale Gdns. NW3 ...34Cb 69
Briarfield Av. N3 ...26Db 49
(not continuous)
Briarfield Cl. DA7: Bex ...54Cd 118
Briar Gdns. BR2: Hayes ...74Hc 159
Briar Gro. CR2: Sande ...85Wb 177
Briar Hill CR8: Purl ...83Nb 176
Briaris Cl. N17 ...24Xb 51
Briar La. BR4: Addtn ...77Dc 158
Briar La. SM5: Cars ...81Jb 176
Briarleas Gdns. RM14: Upm ...31Td 78
Briar Rd. DA5: Bexl ...62Fd 140
Briar Rd. GU23: Send ...96D 188
Briar Rd. HA3: Kenton ...29La 46
Briar Rd. NW2 ...35Ya 68
Briar Rd. RM3: Rom ...24Ld 57
Briar Rd. SW16 ...69Nb 134
Briar Rd. TW17: Shep ...71P 149
Briar Rd. TW2: Twick ...60Ga 108
Briar Rd. WD25: Wat ...6W 12
Briars, The CM15: Kel H ...11Ud 40
Briars, The EN8: Chesh ...3Ac 20
Briars, The SL3: L'ly ...50B 82
Briars, The TN15: W King ...79Td 164
Briars, The TW19: Stanw M ...57J 105
Briars, The WD23: B Hea ...17Ga 28
Briars, The WD3: Sarr ...8J 11
Briars Ct. KT22: Oxs ...86Fa 172
Briars Wlk. RM3: Hrld W ...26Pd 57
Briars Way DA3: Hartl ...71Ce 165
Briarswood EN7: G Oak ...1Tb 19
Briarswood Way BR6: Chels ...78Vc 161
Briar Wlk. HA8: Edg ...24Sa 47
Briar Wlk. KT14: W Byf ...84J 169
Briar Wlk. SW15 ...56Xa 110
Briar Wlk. W10 ...42Ab 88
Briar Way SL2: Slou ...3F 80
Briar Way UB7: W Dray ...47Q 84
Briar Wood Cl. BR2: Broml ...75Nc 160
Briarwood Cl. NW9 ...30Sa 47
Briarwood Cl. TW13: Felt ...63U 128
Briarwood Ct. KT4: Wor Pk ...74Wa 154
(off The Avenue)
Briarwood Dr. HA6: Nwood ...26W 44
Briarwood Rd. GU21: Wok ...1H 187
Briarwood Rd. KT17: Ewe ...79Wa 154
Briarwood Rd. SW4 ...57Mb 112
Briary Cl. NW3 ...38Gb 69
Briary Ct. DA14: Sidc ...64Xc 139
Briary Ct. E16 ...44Hc 93
Briary Gdns. BR1: Broml ...64Kc 137
Briary Gro. HA8: Edg ...26Ra 47
Briary La. N9 ...20Vb 33
Briary Lodge BR3: Beck ...67Ec 136
Briavels Ct. KT18: Eps ...87Ua 174
Brickbarn Cl. SW10 ...52Eb 111
(off King's Barn)
Brickbat All. KT22: Lea ...93Ka 192
Brick Ct. EC4 ...3K 223 (44Qb 90)
Brick Ct. RM17: Grays ...51Ce 121
(off Columbia Wharf Rd.)
Brickenden Ct. EN9: Walt A ...5Hc 21
Bricket Rd. AL1: St A ...2B 6
Brickett Cl. HA4: Ruis ...29S 44
BRICKET WOOD ...2Ba 13
Bricket Wood Common ...4Ca 13
Bricket Wood Paintball Cen. ...1Da 13
Bricket Wood Station (Rail) ...2Ca 13
Brick Farm Cl. TW9: Kew ...53Ra 109
Brickfield AL10: Hat ...3C 8
Brickfield Av. HP3: Hem H ...3B 4
Brickfield Cl. E9 ...37Yb 72
Brickfield Cl. TN14: Dun G ...92Hd 202
Brickfield Cl. TW8: Bford ...52La 108
Brickfield Cotts. BR7: Chst ...64Qc 138
Brickfield Cotts. SE18 ...51Vc 117
Brickfield Ct. AL10: Hat ...3D 8
Brickfield Farm DA3: Lfield ...69Ce 143
Brickfield Farm Cl. DA3: Lfield ...69Be 143
Brickfield Farm Gdns. BR6: Farnb ...77Sc 160
Brickfield Ho. N1 ...39Ub 71
(off Hertford Rd.)
Brickfield La. EN5: Ark ...16Va 30
Brickfield La. UB3: Harl ...51T 106
Brickfield M. WD19: Wat ...16Aa 27
Brickfield Rd. CM16: Coop ...1Zc 23
Brickfield Rd. CR4: Mitc ...69Gb 133
Brickfield Rd. CR7: Thor H ...67Rb 135
Brickfield Rd. E3 ...42Dc 92
Brickfield Rd. SW19 ...63Db 133
Brickfield Rd. SW4 ...58Mb 112
(off Parkfield Rd.)
Brickfields HA2: Harr ...33Fa 66
(not continuous)
Brickfields Cotts. WD6: Bore ...13Pa 29
Brickfields Way UB7: W Dray ...48P 83
BRICK HILL ...7G 146
Brick Kiln Cl. WD19: Wat ...16Aa 27
Brick Kiln La. RH8: Limp ...2N 211
Brick Kiln One SE13 ...55Ec 114
(off Station Rd.)
Brick Kiln Rd. RM3: Rom ...21Ld 57
Brick Kiln Two SE13 ...55Ec 114
(off Station Rd.)
Brick Knoll Pk. AL1: St A ...3G 6
Brick La. E1 ...5K 219 (42Vb 91)
Brick La. E2 ...4K 219 (41Vb 91)
Brick La. EN1: Enf ...12Xb 33
Brick La. EN3: Enf H ...12Yb 34
Brick La. HA7: Stan ...24Ma 47
Brick La. UB5: N'olt ...41Ba 85
Brick Lane Music Hall ...46Mc 93
BRICKLAYER'S ARMS ...5G 231 (49Tb 91)
Bricklayers Arms Distribution Cen. SE1 ...6J 231 (49Ub 91)
(not continuous)
Bricklayers St. SE11 ...6K 229 (49Qb 90)
Brickmakers La. HP3: Hem H ...3B 4
Brickmakers Mdws. TN15: Plat ...92Ee 205
Brickstock Furze CM15: Shenf ...18Ce 41
Brick St. W1 ...7K 221 (46Kb 90)
Brickwall La. HA4: Ruis ...32U 64
Brickwood Cl. SE26 ...62Xb 135
Brickwood Rd. CR0: C'don ...75Ub 157
Brickworks Cotts. TN14: S'oaks ...92Md 203
Brideale Cl. SE15 ...51Vb 113
Bride Ct. EC4 ...3B 224 (44Rb 91)
(off Bride La.)
Bride La. EC4 ...3B 224 (44Rb 91)
Bridel M. N1 ...1B 218 (39Rb 71)
(off Colebrook Row)
Brides M. N7 ...37Pb 70
Brides Pl. N1 ...38Ub 71
Bride St. N7 ...37Pb 70
Bridewain St. SE1 ...3K 231 (48Vb 91)
Bridewell Pl. E1 ...46Xb 91
Bridewell Pl. EC4 ...3B 224 (44Rb 91)
Bridford M. W1 ...7A 216 (43Kb 90)
Bridge, The HA3: W'stone ...28Ha 46
Bridge, The SW11 ...52Kb 112
BRIDGE, THE ...54Qd 119
Bridge App. NW1 ...38Jb 70
Bridge Av. RM14: Upm ...34Qd 77
Bridge Av. W6 ...49Ya 88
Bridge Av. W7 ...43Fa 86
Bridge Av. Mans. W6 ...50Ya 88
(off Bridge Av.)
Bridge Barn La. GU21: Wok ...10P 167
Bridge Bus. Cen., The UB2: S'hall ...47Ca 85
Bridge Cl. CM13: B'wood ...20Ae 41
Bridge Cl. DA2: Dart ...55Td 120
Bridge Cl. EN1: Enf ...12Xb 33
Bridge Cl. GU21: Wok ...9N 167
Bridge Cl. KT12: Walt T ...73V 150
Bridge Cl. KT14: Byfl ...84P 169
Bridge Cl. RM7: Rush G ...30Gd 56
Bridge Cl. SL1: Slou ...5D 80
Bridge Cl. TW11: Tedd ...63Ha 130
Bridge Cl. TW18: Staines ...63G 126
Bridge Cl. W10 ...44Za 88
Bridge Ct. E10 ...32Bc 72
Bridge Ct. E14 ...45Fc 93
(off Newport Av.)
Bridge Ct. GU21: Wok ...9P 167
Bridge Ct. HA2: Harr ...33Ea 66
Bridge Ct. KT12: Walt T ...74V 150
(off Bridge St.)
Bridge Ct. KT13: Weyb ...77R 150
Bridge Ct. KT22: Lea ...94Ja 192
Bridge Ct. RM17: Grays ...51De 121
(off Bridge Rd.)
Bridge Ct. WD7: R'lett ...7Ka 14
Bridge Dr. N13 ...21Pb 50
Bridge End E17 ...25Ec 52
BRIDGE END ...92R 190
Bridge End Cl. KT2: King T ...67Qa 131
Bridgefield Cl. SM7: Bans ...87Ya 174
Bridgefield Ho. W2 ...44Db 89
(off Queensway)
Bridgefield Rd. SM1: Sutt ...79Cb 155
Bridgefields Cl. RM11: Horn ...32Kd 77
Bridgefoot SE1 ...50Nb 90
Bridgefoot TW16: Sun ...67V 128
Bridgefoot Ho. WD7: R'lett ...3Ha 14
Bridgefoot La. EN6: Pot B ...5Ya 16
Bridgeford Ho. WD18: Wat ...13X 27
Bridge Gdns. KT8: E Mos ...70Fa 130
Bridge Gdns. N16 ...35Tb 71
Bridge Gdns. TW15: Ashf ...66S 128
Bridge Ga. N21 ...17Sb 33
Bridgeham Cl. KT13: Weyb ...78Q 150
Bridge Hill CM16: Epp ...5Vc 23
Bridgehill Cl. HA0: Wemb ...39Ma 67
Bridge Ho. CR0: C'don ...76Sb 157
(off Surrey St.)
Bridge Ho. DA1: Dart ...59Nd 119
Bridge Ho. E9 ...37Zb 72
(off Shepherds La.)
Bridge Ho. KT16: Chert ...73L 149
Bridge Ho. NW10 ...40Za 68
(off Chamberlayne Rd.)
Bridge Ho. NW3 ...38Jb 70
(off Adelaide Rd.)
Bridge Ho. SE4 ...56Bc 114
Bridge Ho. SM2: Sutt ...79Db 155
(off Bridge Rd.)
Bridge Ho. SW1 ...7K 227 (50Kb 90)
(off Ebury Bri.)
Bridge Ho. UB7: W Dray ...46M 83
Bridgehouse Ct. SE1 ...1B 230 (47Rb 91)
(off Blackfriars Rd.)
Bridge Ho. Quay E14 ...46Ec 92
Bridgeland Rd. E16 ...45Jc 93
Bridgelands Cl. BR3: Beck ...66Bc 136
Bridge La. GU25: Vir W ...71A 148
Bridge La. NW11 ...28Ab 48
Bridge La. SW11 ...53Gb 111
Bridge Leisure Cen., The ...63Bc 136
Bridgeman Ct. SL4: Wind ...4E 102
Bridgeman Dr. SL4: Wind ...4E 102
Bridgeman Ho. E9 ...38Yb 72
(off Frampton Pk. Rd.)
Bridgeman Ho. W14 ...49Bb 89
(off Radnor Ter.)
Bridgeman Rd. N1 ...38Pb 70
Bridgeman Rd. TW11: Tedd ...65Ja 130
Bridgeman St. NW8 ...2D 214 (40Gb 69)
Bridge Mead GU24: Pirb ...8D 186
Bridge Mdws. SE14 ...51Zb 114
Bridgemere Cl. RM3: Hrld W ...25Pd 57
Bridge M. GU21: Wok ...9P 167
Bridge M. NW10 ...37Ua 68
Bridgemount M. N4 ...31Pb 70
BRIDGEN ...59Ad 117
Bridgend Rd. EN1: Enf ...7Yb 20
Bridgend Rd. SW18 ...56Eb 111
Bridgenhall Rd. EN1: Enf ...11Vb 33
Bridgen Ho. E1 ...44Xb 91
(off Nelson St.)
Bridgen Rd. DA5: Bexl ...59Ad 117
Bridge Pde. N21 ...17Sb 33
(off Ridge Av.)
Bridgepark SW18 ...57Cb 111
Bridge Pk. Community Leisure Cen. ...38Ra 67
Bridge Pl. CR0: C'don ...74Tb 157
Bridge Pl. SW1 ...5A 228 (49Kb 90)
Bridge Pl. WD17: Wat ...15Z 27
Bridge Point RM11: Horn ...27Nd 57
Bridgepoint Ct. AL2: Brick W ...2Aa 13
(off Old Watford Rd.)
Bridgepoint Lofts E7 ...38Lc 73
Bridgepoint Pl. N6 ...32Lb 70
(off Hornsey La.)
Bridgeport Pl. E1 ...46Wb 91
Bridger Cl. WD25: Wat ...5Z 13
Bridge Rd. BR3: Beck ...66Bc 136
Bridge Rd. BR5: St M Cry ...72Xc 161
Bridge Rd. DA7: Bex ...54Ad 117
Bridge Rd. DA8: Erith ...54Hd 118
Bridge Rd. E11 ...30Fc 53
Bridge Rd. E15 ...38Fc 73
Bridge Rd. E17 ...31Bc 72
Bridge Rd. E6 ...38Pc 74
Bridge Rd. HA9: Wemb ...34Qa 67
Bridge Rd. KT13: Weyb ...77P 149
Bridge Rd. KT16: Chert ...73L 149
Bridge Rd. KT17: Eps ...84Va 174
Bridge Rd. KT8: E Mos ...70Fa 130
Bridge Rd. KT9: Chess ...78Na 153
Bridge Rd. N22 ...25Nb 50
Bridge Rd. N9 ...20Wb 33
Bridge Rd. NW10 ...37Ua 68
Bridge Rd. RM13: Rain ...42Jd 96
Bridge Rd. RM17: Grays ...51De 121
Bridge Rd. SL5: S'hill ...1B 146
Bridge Rd. SM2: Sutt ...79Db 155
Bridge Rd. SM6: W'gton ...78Kb 156
Bridge Rd. TW1: Twick ...58Ka 108
Bridge Rd. TW3: Houn ...55Fa 108
Bridge Rd. TW3: Isle ...55Fa 108
Bridge Rd. TW7: Isle ...55Fa 108
Bridge Rd. UB2: S'hall ...47Ba 85
Bridge Rd. UB8: Uxb ...40L 63
Bridge Rd. WD4: Hunt C ...5S 12
Bridge Rd. Dpt. E15 ...39Gc 73
Bridge Row CR0: C'don ...74Tb 157
Bridges Av. KT8: E Mos ...72Fa 152
Bridges Ct. Rd. SW11 ...55Fb 111
Bridges Dr. DA1: Dart ...57Rd 119
Bridges Ho. SE5 ...52Tb 113
(off Elmington Est.)
Bridgeside Lodge N1 ...2D 218 (40Sb 71)
(off Wharf Rd.)
Bridges La. CR0: Bedd ...77Nb 156
Bridges Pl. SW6 ...53Bb 111
Bridges Rd. HA7: Stan ...22Ha 46
Bridges Rd. SW19 ...65Db 133
Bridges Rd. M. SW19 ...65Db 133
Bridge St. HA5: Pinn ...27Aa 45
Bridge St. HP1: Hem H ...3M 3
Bridge St. KT12: Walt T ...74U 150
Bridge St. KT22: Lea ...94Ja 192
Bridge St. SL3: Coln ...52F 104
Bridge St. SW1 ...2F 229 (47Nb 90)
Bridge St. TW18: Staines ...63G 126
Bridge St. TW9: Rich ...57Ma 109
Bridge St. W4 ...49Ta 87
Bridges Wharf SW11 ...55Fb 111
Bridge Ter. E15 ...38Fc 73
Bridge Theatre ...7K 225 (46Vb 91)
(off Tower Bri. Rd.)
Bridgetown Cl. SE19 ...64Ub 135
Bridge Vw. DA9: Ghithe ...56Xd 120
Bridge Vw. SL5: S'dale ...3F 146
Bridge Vw. W6 ...50Ya 88
Bridgeview Ct. HP3: Hem H ...5N 3
Bridge Vw. Ct. SE1 ...4J 231 (48Ub 91)
(off Grange Rd.)
Bridgeview Ct. IG6: Ilf ...23Tc 54
Bridge Vw. Ind. Est. RM20: W Thur ...51Wd 120
Bridge Wlk. IG1: Ilf ...33Rc 74
(within The Exchange)
Bridge Wlk. SE8 ...51Dc 114
(off Copperas St.)
Bridgewalk Hgts. SE1 ...1G 231 (47Tb 91)
(off Weston St.)
Bridgewater GU21: Wok ...88B 168
Bridgewater Cl. BR7: Chst ...69Uc 138
Bridgewater Ct. SL3: L'ly ...49C 82
Bridgewater Gdns. HA8: Edg ...26Pa 47
Bridgewater Highwalk EC2 ...7D 218 (43Sb 91)
(off Beech St.)
Bridgewater Ho. E14 ...44Gc 93
(off Lookout Lane)
Bridgewater Rd. E15 ...39Ec 72
Bridgewater Rd. HA0: Wemb ...37La 66
Bridgewater Rd. KT13: Weyb ...79T 150
Bridgewater Sq. EC2 ...7D 218 (43Sb 91)
Bridgewater St. EC2 ...7D 218 (43Sb 91)
Bridgewater Ter. SL4: Wind ...3H 103
Bridgewater Way SL4: Wind ...3H 103
Bridgewater Way WD23: Bush ...16Da 27
Bridge Way CR5: Chips ...91Gb 195
Bridge Way KT11: Cobh ...85V 170
Bridge Way N11 ...20Lb 32
Bridge Way NW11 ...29Bb 49
Bridge Way TW2: Whitt ...59Ea 108
Bridge Way UB10: Ick ...36R 64
Bridgeway HA0: Wemb ...38Na 67
Bridgeway IG11: Bark ...38Vc 75
BRIDGEWAYS DAY HOSPITAL ...72Nc 160
Bridgeway St. NW1 ...2C 216 (40Lb 70)
Bridge Wharf E2 ...40Zb 72
Bridge Wharf KT16: Chert ...73L 149
Bridge Wharf N1 ...1H 217 (40Pb 70)
(off Calshot St.)
Bridge Wharf Rd. TW7: Isle ...55Ka 108
Bridgewood Cl. SE20 ...66Xb 135
Bridgewood Rd. KT4: Wor Pk ...77Wa 154
Bridgewood Rd. SW16 ...66Mb 134
Bridge Works UB8: Cowl ...42L 83
Bridge Yd. SE1 ...6G 225 (46Tb 91)
Bridgford St. SW18 ...62Eb 133
Bridgland Rd. RM19: Purf ...50Sd 98
Bridgman Rd. W4 ...48Sa 87
Bridgnorth Ho. SE15 ...51Wb 113
(off Friary Est.)
Bridgwater Cl. RM3: Rom ...22Md 57
Bridgwater Ho. W2 ...2A 220 (44Eb 89)
(off Hallfield Est.)
Bridgwater Rd. HA4: Ruis ...35W 64
Bridgwater Rd. RM3: Rom ...22Ld 57
Bridgwater Wlk. RM3: Rom ...22Md 57
Bridle Cl. EN3: Enf L ...9Bc 20
Bridle Cl. KT1: King T ...70Ma 131
Bridle Cl. KT19: Ewe ...78Ta 153
Bridle Cl. TW16: Sun ...69W 128
Bridle End KT17: Eps ...85Va 174
Bridle La. KT11: Stoke D ...87Da 171
Bridle La. KT22: Oxs ...87Da 171
Bridle La. TW1: Twick ...58Ka 108
Bridle La. W1 ...4C 222 (45Lb 90)
Bridle La. WD3: Loud ...13L 25
Bridle M. E1 ...44Wb 91
Bridle M. EN5: Barn ...14Bb 31
Bridle Path CR0: Bedd ...76Nb 156
Bridle Path WD17: Wat ...12X 27
Bridle Path, The IG8: Wfd G ...24Gc 53
Bridle Path, The KT17: Ewe ...82Ya 174
Bridlepath, The CR3: Cat'm ...95Qb 196
Bridlepath Way TW14: Bedf ...59U 106
Bridle Rd. CR0: C'don ...76Cc 158
Bridle Rd. CR2: Sande ...81Wb 177
Bridle Rd. HA5: Eastc ...30X 45
Bridle Rd. KT10: Clay ...79Ka 152
Bridle Rd. KT17: Eps ...85Va 174
Bridle Rd., The CR8: Purl ...82Nb 176
Bridle Way BR6: Farnb ...77Sc 160
Bridle Way CR0: C'don ...78Cc 158
Bridle Way, The CR0: Sels ...82Ac 178
Bridle Way, The SM6: W'gton ...77Lb 156
Bridleway, The CR5: Coul ...86Kb 176
Bridleway Cl. KT17: Ewe ...82Ya 174
Bridlington Cl. TN16: Big H ...91Kc 199
Bridlington Rd. N9 ...17Xb 33
Bridlington Rd. WD19: Wat ...20Z 27
Bridlington Spur SL1: Slou ...8F 80
Bridport SE17 ...7F 231 (50Tb 91)
(off Cadiz St.)
Bridport Av. RM7: Rom ...30Dd 56
Bridport Ho. N1 ...39Tb 71
(off Bridport Pl.)
Bridport Ho. N18 ...22Vb 51
(off College Gdns.)
Bridport Pl. N1 ...39Tb 71
(not continuous)
Bridport Rd. CR7: Thor H ...69Qb 134
Bridport Rd. N18 ...22Ub 51
Bridport Rd. UB6: G'frd ...39Da 65
Bridport Ter. SW8 ...53Mb 112
(off Deeley Rd.)
Bridport Way SL2: Slou ...2F 80
Bridstow Pl. W2 ...44Cb 89
Brief St. SE5 ...53Rb 113
Brierfield NW1 ...1B 216 (39Lb 70)
(off Arlington Rd.)
Brier Lea KT20: Lwr K ...98Bb 195
Brierley CR0: New Ad ...79Dc 158
(not continuous)
Brierley Av. N9 ...18Yb 34
Brierley Cl. RM11: Horn ...30Ld 57
Brierley Cl. SE25 ...70Wb 135
Brierley Ct. W7 ...45Ga 86
Brierley Rd. E11 ...35Fc 73
Brierley Rd. SW12 ...61Lb 134
Brierly Gdns. E2 ...40Yb 72
Brier Rd. KT20: Tad ...91Xa 194
Briers, The EN6: Pot B ...4Db 17
Briery Ct. WD3: Chor ...14J 25
Briery Fld. WD3: Chor ...14J 25
Briery Way HP2: Hem H ...1A 4
Brigade Cl. HA2: Harr ...33Fa 66
Brigade Pl. CR3: Cat'm ...94Sb 197
Brigade St. SE3 ...54Hc 115
(off Tranquil Va.)
Brigadier Av. EN2: Enf ...10Sb 19
Brigadier Hill EN2: Enf ...10Sb 19
Brigadier Ho. NW9 ...26Va 48
(off Heritage Av.)
Briggeford Cl. E5 ...33Wb 71
Briggs Cl. CR4: Mitc ...67Kb 134
Briggs Ho. E2 ...3K 219 (41Vb 91)
(off Chambord St.)
Brighstone Ct. RM19: Purf ...50Rd 97
Bright Cl. DA17: Belv ...49Zc 95
Brightfield Rd. SE12 ...57Gc 115
Bright Ho. KT1: King T ...69Ma 131
(off Kingston Hall Rd.)
Brightlands DA11: Nflt ...3A 144
Brightlands Rd. RH2: Reig ...4L 207
Brightling Rd. SE4 ...58Bc 114
Brightlingsea Pl. E14 ...45Bc 92
Brightman Rd. SW18 ...60Fb 111
Brighton Av. E17 ...29Bc 52
Brighton Bldgs. SE1 ...4H 231 (48Ub 91)
(off Tower Bri. Rd.)
Brighton Cl. KT15: Add ...78L 149
Brighton Cl. UB10: Hil ...38R 64
Brighton Dr. UB5: N'olt ...37Ca 65
Brighton Gro. SE14 ...53Ac 114
Brighton Ho. SE5 ...53Tb 113
(off Camberwell Grn.)
Brighton Rd. CR2: S Croy ...78Sb 157
Brighton Rd. CR5: Coul ...95Kb 196
Brighton Rd. CR8: Purl ...83Qb 176
Brighton Rd. E6 ...41Qc 94
(not continuous)
Brighton Rd. KT15: Add ...78L 149
Brighton Rd. KT20: Kgswd ...93Ab 194
Brighton Rd. KT20: Lwr K ...93Ab 194
Brighton Rd. KT20: Tad ...93Ab 194
Brighton Rd. KT6: Surb ...72La 152
Brighton Rd. N16 ...35Ub 71
Brighton Rd. N2 ...26Eb 49
Brighton Rd. RH1: Mers ...95Kb 196
Brighton Rd. RH1: Redh ...7P 207
Brighton Rd. SM2: Bans ...83Cb 175
Brighton Rd. SM2: Sutt ...83Cb 175
Brighton Rd. SM7: Bans ...88Bb 175
Brighton Rd. WD24: Wat ...10W 12
Brighton Spur SL2: Slou ...2F 80
Brighton Ter. RH1: Redh ...7P 207
Brighton Ter. SW9 ...56Pb 112
Brights Av. RM13: Rain ...42Kd 97
Brightside, The EN3: Enf H ...11Zb 34
Brightside Av. TW18: Staines ...66L 127
Brightside Rd. SE13 ...58Fc 115
Bright St. E14 ...43Dc 92
Brightview Cl. AL2: Brick W ...1Aa 13
Brightwell Cl. CR0: C'don ...74Qb 156
Brightwell Ct. N7 ...36Pb 70
(off Mackenzie Rd.)
Brightwell Cres. SW17 ...64Hb 133
Brightwell Rd. WD18: Wat ...15W 26
Brightwen Gro. HA7: Stan ...19Ja 28
Brig M. SE8 ...51Cc 114
Brigstock Ho. SE5 ...54Sb 113
Brigstock Rd. CR5: Coul ...87Kb 176
Brigstock Rd. CR7: Thor H ...71Qb 156
Brigstock Rd. DA17: Belv ...49Dd 96
Brill Pl. NW1 ...2E 216 (40Mb 70)
Brimfield Rd. RM19: Purf ...49Td 98
Brim Hill N2 ...28Eb 49
Brimpsfield Cl. SE2 ...48Xc 95
(not continuous)
BRIMSDOWN ...12Bc 34
Brimsdown Av. EN3: Enf H ...12Ac 34
Brimsdown Ho. E3 ...42Dc 92
Brimsdown Ind. Est. EN3: Brim Lockfield Av. ...11Bc 34
Brimsdown Ind. Est. EN3: Brim Stockingswater La. ...12Bc 34
Brimsdown Station (Rail) ...13Ac 34
Brimshot La. GU24: Chob ...1J 167
Brimstone Cl. BR6: Chels ...80Yc 161
Brimstone Ho. E15 ...38Gc 73
(off Victoria St.)
Brind Cotts. GU24: Chob ...2K 167
Brindle Ga. DA15: Sidc ...60Uc 116
Brindles RM11: Horn ...28Nd 57
Brindles, The SM7: Bans ...89Bb 175
Brindles Cl. CM13: Hut ...19Ee 41
Brindles Cl. SS17: Linf ...9J 101
Brindlewick Gdns. BR3: Beck ...65Cc 136
Brindley Cl. DA7: Bex ...55Dd 118

Brindley Cl. HA0: Wemb 39Ma 67
Brindley Ct. HA7: Stan 24Ma 47
Brindley Ho. W2 43Cb 89
(off Alfred Rd.)
Brindley Pl. UB8: Cowl 44L 83
Brindley St. SE14 53Bc 114
Brindley Way BR1: Broml 64Jc 137
Brindley Way HP3: Hem H 7P 3
Brindley Way UB1: S'hall 45Da 85
Brindwood Rd. E4 20Bc 34
Brine Ho. E3 40Ac 72
(off St Stephen's Rd.)
Brinkburn Cl. HA8: Edg 27Ra 47
Brinkburn Cl. SE2 49Wc 95
Brinkburn Gdns. HA8: Edg 27Qa 47
Brinkley KT1: King T 68Qa 131
Brinkley Rd. KT4: Wor Pk 75Xa 154
Brinklow Ct. AL3: St A 5P 5
Brinklow Cres. SE18 52Rc 116
Brinklow Ho. W2 43Db 89
(off Torquay St.)
Brinkworth Pl. SL4: Old Win 9M 103
Brinkworth Rd. IG5: Ilf 27Nc 54
Brinkworth Way E9 37Bc 72
Brinley Cl. EN8: Chesh 3Zb 20
Brinsdale Rd. NW4 27Za 48
Brinsley Ho. E1 44Yb 92
(off Tarling St.)
Brinsley Rd. HA3: Hrw W 26Fa 46
Brinsmead AL2: F'mre 9C 6
Brinsmead Rd. RM3: Hrld W 26Qd 57
Brinson Way RM15: Avel 45Td 98
Brinsworth Cl. TW2: Twick 60Fa 108
Brinsworth Ho. TW2: Twick 61Fa 130
Brinton Wlk. SE1 7B 224 (46Rb 91)
(off Nicholson St.)
Brion Pl. E14 43Ec 92
Brisbane Av. SW19 67Db 133
Brisbane Ho. RM18: Tilb 3C 122
Brisbane Ho. W12 45Xa 88
(off White City Est.)
Brisbane Rd. E10 33Dc 72
Brisbane Rd. IG1: Ilf 31Rc 74
Brisbane Rd. W13 47Ja 86
Brisbane St. SE5 52Tb 113
Briscoe Cl. E11 33Hc 73
Briscoe M. TW2: Twick 61Fa 130
Briscoe Rd. RM13: Rain 40Ld 77
Briscoe Rd. SW19 65Fb 133
Briset Rd. SE9 55Mc 115
Briset St. EC1 6B 218 (42Rb 91)
Briset Way N7 33Pb 70
Brisson Cl. KT10: Esh 79Ba 151
Bristol Av. NW9 25Va 48
Bristol Cl. SM6: W'gton 80Nb 156
Bristol Cl. TW19: Stanw 58N 105
Bristol Cl. TW4: Houn 59Ca 107
Bristol Ct. TW19: Stanw 58N 105
Bristol Gdns. SW15 59Ya 110
Bristol Gdns. W9 42Db 89
Bristol Ho. IG11: Bark 38Wc 75
(off Margaret Bondfield Av.)
Bristol Ho. SE11 4K 229 (48Qb 90)
(off Lambeth Wlk.)
Bristol Ho. SW1 7H 227 (50Jb 90)
(off Lwr. Sloane St.)
Bristol Ho. WC1 7G 217 (43Nb 90)
(off Southampton Row)
Bristol Ho. WD6: Bore 12Qa 29
(off Eldon Av.)
Bristol M. W9 42Db 89
Bristol Pk. Rd. E17 28Ac 52
Bristol Rd. DA12: Grav'nd 2F 144
Bristol Rd. E7 37Lc 73
Bristol Rd. SM4: Mord 71Eb 155
Bristol Rd. UB6: G'frd 39Da 65
Bristol Wlk. NW6 40Cb 69
(off Alpha Pl.)
Bristol Way SL1: Slou 6K 81
Briston Gro. N8 30Nb 50
Briston M. NW7 24Wa 48
Bristow Ct. E8 39Xb 71
(off Triangle Rd.)
Bristowe Cl. SW2 58Qb 112
Bristowe Dr. RM16: Ors 4F 100
Bristow Rd. CR0: Bedd 77Nb 156
Bristow Rd. DA7: Bex 53Ad 117
Bristow Rd. SE19 64Ub 135
Bristow Rd. TW3: Houn 55Ea 108
Britannia Bldg. N1 3F 219 (41Tb 91)
(off Ebenezer St.)
Britannia Bus. Cen. NW2 35Za 68
Britannia Bus. Pk. EN8: Walt C 7Bc 20
Britannia Cen., The IG10: Lough 14Tc 36
Britannia Cl. DA8: Erith 51Hd 118
Britannia Cl. SW4 56Mb 112
Britannia Cl. UB5: N'olt 41Z 85
Britannia Ct. EN8: Walt C 6Bc 20
(off Eleanor Cross Rd.)
Britannia Ct. KT2: King T 67Ma 131
(off Skerne Wlk.)
Britannia Ct. UB7: W Dray 48M 83
Britannia Dr. DA12: Grav'nd 4H 145
Britannia Ga. E16 46Jc 93
Britannia Ind. Est. SL3: Poyle 54G 104
BRITANNIA JUNC. 39Kb 70
Britannia La. TW2: Whitt 59Ea 108
Britannia Leisure Cen. 1G 219 (39Tb 71)
Britannia Rd. CM14: W'ley 22Yd 58
Britannia Rd. E14 49Cc 92
Britannia Rd. EN8: Walt C 6Bc 20
Britannia Rd. IG1: Ilf 34Rc 74
Britannia Rd. KT5: Surb 73Pa 153
Britannia Rd. N12 20Eb 31
Britannia Rd. SW6 52Db 111
Britannia Row N1 39Rb 71
Britannia St. WC1 3H 217 (41Pb 90)
Britannia Wlk. N1 2F 219 (40Tb 71)
(not continuous)
Britannia Way NW10 42Ra 87
Britannia Way SW6 52Db 111
(off Britannia Rd.)
Britannia Way TW19: Stanw 59M 105
Britannic Highwalk EC2 1F 225 (43Tb 91)
(off Moor La.)
British Disabled Water-Ski
Association 60E 104
British Gro. W4 50Va 88
British Gro. Nth. W4 50Va 88
British Gro. Pas. W4 50Va 88
British Gro. Sth. W4 50Va 88
British Legion Rd. E4 19Hc 35
British Library 3E 216 (41Mb 90)
British Mus. 1F 223 (43Nb 90)
British Rail Goods Yd. EN8: Walt C 4Zb 20
British St. E3 41Bc 92
British Telecom Cen. 2D 224 (44Sb 91)
(off Newgate St.)
British Wharf Ind. Est. SE14 50Zb 92
Britley Ho. E14 44Bc 92
(off Copenhagen Pl.)
Briton Cl. CR2: Sande 83Ub 177
Briton Cres. CR2: Sande 83Ub 177
Briton Hill Rd. CR2: Sande 82Ub 177
Brittain Ct. SE9 60Nc 116
Brittain Rd. KT12: Hers 78Z 151
Brittain Rd. RM8: Dag 34Ad 75
Brittains La. TN13: S'oaks 96Hd 202
Brittany Ho. EN2: Enf 10Sb 19
Brittany Point SE11 6K 229 (49Qb 90)
(off Lollard Street)
Britten Cl. NW11 32Db 69
Britten Cl. WD6: E'tree 16Ma 29
Britten Ct. E15 40Fc 73
Brittenden Cl. BR6: Chels 79Vc 161
Brittenden Pde. BR6: Chels 79Vc 161
Britten Dr. UB1: S'hall 44Ca 85
Britten Ho. SW3 7D 226 (50Gb 89)
(off Britten St.)
Brittens Cl. GU2: Guild 10L 187
Britten St. SW3 7D 226 (50Gb 89)
Britten Theatre 3B 226 (48Fb 89)
(off Prince Consort Rd.)
Brittidge Rd. NW10 38Ua 68
Britton Av. AL3: St A 2B 6
Britton Cl. SE6 59Fc 115
Britton St. EC1 6B 218 (42Rb 91)
BRITWELL 1E 80
Britwell Gdns. SL1: Burn 1B 80
Britwell Rd. SL1: Burn 1A 80
Brixham Cres. HA4: Ruis 32W 64
Brixham Gdns. IG3: Ilf 36Uc 74
Brixham Rd. DA16: Well 53Zc 117
Brixham St. E16 46Qc 94
BRIXTON 56Pb 112
Brixton Hill SW2 59Nb 112
Brixton Hill Ct. SW2 57Pb 112
Brixton Hill Pl. SW2 59Nb 112
Brixton Oval SW2 56Qb 112
Brixton Recreation Cen. 55Qb 112
(off Brixton Sta. Rd.)
Brixton Rd. SE11 51Qb 112
Brixton Rd. SW9 56Qb 112
Brixton Rd. WD24: Wat 11X 27
Brixton Station (Rail &
Underground) 56Qb 112
Brixton Sta. Rd. SW9 55Qb 112
Brixton Water La. SW2 57Pb 112
Broad Acre AL2: Brick W 2Aa 13
Broadacre TW18: Staines 64J 127
Broadacre Cl. UB10: Ick 34R 64
Broadash Cl. SW16 62Qb 134
Broadbent Cl. N6 32Kb 70
Broadbent St. W1 4K 221 (45Kb 90)
Broadberry Ct. N18 23Xb 51
Broadbridge Cl. SE3 52Jc 115
Broad Cl. KT12: Hers 76Z 151
BROAD COLNEY 1Ma 15
Broad Colney Lakes Nature
Reserve 10H 7
Broad Comn. Est. N16 32Wb 71
(off Osbaldeston Rd.)
Broadcoombe CR2: Sels 80Yb 158
Broad Ct. WC2 3G 223 (44Nb 90)
Broadcroft Av. HA7: Stan 26Ma 47
Broadcroft Rd. BR5: Pet W 73Tc 160
Broad Ditch Rd. DA13: Nflt G 66Ee 143
Broadeaves Cl. CR2: S Croy 78Ub 157
Broadfield NW6 37Db 69
Broadfield Cl. CR0: Wadd 75Pb 156
Broadfield Cl. KT20: Tad 92Ya 194
Broadfield Cl. NW2 34Ya 68
Broadfield Cl. RM1: Rom 29Hd 56
Broadfield Ct. HA2: Harr 25Da 45
(off Broadfields)
Broadfield Ct. WD23: B Hea 19Ga 28
Broadfield La. NW1 38Nb 70
Broadfield Pde. HA8: Edg 20Ra 29
(off Glengall Rd.)
Broadfield Rd. HP2: Hem H 2P 3
Broadfield Rd. SE6 59Gc 115
Broadfields EN7: G Oak 1Rb 19
Broadfields HA2: Harr 26Da 45
Broadfields KT8: E Mos 72Ga 152
Broadfields Av. HA8: Edg 21Ra 47
Broadfields Av. N21 16Qb 32
Broadfields Hgts. HA8: Edg 21Ra 47
Broadfields La. WD19: Wat 18X 27
Broadfield Sq. EN1: Enf 12Xb 33
Broadfields Way NW10 36Va 68
Broadfield Way IG9: Buck H 20Lc 35
Broadfield Way WD25: A'ham 8Da 13
Broadford Ho. E1 42Ac 92
(off Commodore St.)
Broadford La. GU24: Chob 4J 167
Broadgate EC2 1H 225 (43Ub 91)
(off Broadgate Circ.)
Broadgate EN9: Walt A 5Hc 21
Broadgate Circ. EC2 1H 225 (43Ub 91)
Broadgate Circle 1H 225 (43Ub 91)
Broadgate Plaza EC2 7J 219 (43Ub 91)
Broadgate Rd. E16 44Mc 93
Broadgates Av. EN4: Had W 11Db 31
Broadgates Ct. SE11 7A 230 (50Qb 90)
(off Cleaver St.)
Broadgates Rd. SW18 60Fb 111
Broadgate Twr. EC2 6J 219 (42Ub 91)
BROAD GREEN 73Rb 157
Broad Grn. Av. CR0: C'don 73Rb 157
Broad Hall AL10: Hat 2B 8
(off Bishops Ri.)
BROADHAM GREEN 4H 211
Broadham Grn. Rd. RH8: Oxt 4H 211
Broadham Pl. RH8: Oxt 3H 211
Broadhead Apts. E3 41Bc 92
(off St Clements Av.)
Broadhead Strand NW9 25Va 48
Broadheath Dr. BR7: Chst 64Pc 138
Broad Highway KT11: Cobh 86Z 171
Broadhinton Rd. SW4 55Kb 112
Broadhoath TN15: Ivy H 98Ud 204
Broadhope Av. SS17: Stan H 3L 101
Broadhurst KT21: Asht 88Na 173
Broadhurst Av. HA8: Edg 21Ra 47
Broadhurst Av. IG3: Ilf 35Vc 75
Broadhurst Cl. NW6 37Eb 69
Broadhurst Cl. TW10: Rich 57Pa 109
Broadhurst Gdns. HA4: Ruis 33Y 65
Broadhurst Gdns. IG7: Chig 21Sc 54
Broadhurst Gdns. NW6 37Db 69
Broadhurst Gdns. RH2: Reig 9K 207
Broadhurst Wlk. RM13: Rain 37Jd 76
Broadis Way RM13: Rain 40Fd 76
Broadlake Cl. AL2: Lon C 9H 7
Broadlands E17 27Ac 52
Broadlands RM17: Grays 50Be 99
Broadlands TW13: Hanw 62Ca 129
Broadlands Av. EN3: Enf H 13Xb 33
Broadlands Av. SW16 61Nb 134
Broadlands Av. TW17: Shep 72S 150
Broadlands Cl. EN3: Enf H 13Yb 34
Broadlands Cl. EN8: Walt C 6Yb 20
Broadlands Cl. N6 31Jb 70
Broadlands Cl. SW16 61Nb 134
Broadlands Ct. TW9: Kew 52Qa 109
(off Kew Gdns. Rd.)
Broadlands Dr. CR6: W'ham 91Yb 198
Broadlands Dr. SL5: Asc 3A 146
Broadlands Dr. SL5: S'hill 3A 146
Broadlands Lodge N6 31Hb 69
Broadlands Rd. BR1: Broml 63Kc 137
Broadlands Rd. N6 31Hb 69
Broadlands Way KT3: N Mald 72Va 154
Broad La. DA2: Wilm 63Jd 140
Broad La. EC2 7H 219 (43Ub 91)
(not continuous)
Broad La. N15 28Vb 51
Broad La. N17 28Wb 51
Broad La. N8 29Pb 50
Broad La. TW12: Hamp 66Ba 129
Broad Lawn SE9 61Qc 138
Broadlawns Ct. HA3: Hrw W 25Ha 46
Broadley Gdns. WD7: Shenl 4Na 15
Broadley Grn. GU20: W'sham 10B 146
Broadleys SL4: Wind 2D 102
Broadley St. NW8 7C 214 (43Fb 89)
Broadley Ter. NW1 6E 214 (42Gb 89)
Broadmark Rd. SL2: Slou 5M 81
Broadmayne SE17 7F 231 (50Tb 91)
(off Portland St.)
Broadmead KT21: Asht 89Pa 173
Broadmead RH1: Mers 100Lb 196
(off Station Rd.)
Broadmead SE6 62Cc 136
Broadmead W14 49Ab 88
Broadmead Av. KT4: Wor Pk 73Wa 154
Broadmead Cen. IG8: Wfd G 24Lc 53
(off Navestock Cres.)
Broadmead Cl. HA5: Hat E 24Aa 45
Broadmead Cl. TW12: Hamp 65Ca 129
Broadmead Ct. IG8: Wfd G 23Jc 53
Broadmead Rd. GU22: Wok 93D 188
Broadmead Rd. GU23: Send 94D 188
Broadmead Rd. IG8: Wfd G 23Jc 53
(not continuous)
Broadmead Rd. UB4: Yead 42Aa 85
Broadmead Rd. UB5: N'olt 42Aa 85
Broadmeads GU23: Send 94D 188
Broad Oak IG8: Wfd G 22Kc 53
Broad Oak SL2: Slou 2G 80
Broad Oak TW16: Sun 65V 128
Broadoak Av. EN3: Enf W 7Zb 20
Broad Oak Cl. BR5: St P 68Wc 139
Broad Oak Cl. E4 22Cc 52
Broadoak Cl. DA4: Sut H 65Qd 141
Broad Oak Ct. SL2: Slou 2G 80
Broadoak Ct. SW9 55Qb 112
(off Gresham Rd.)
Broadoak Ho. NW6 39Db 69
(off Mortimer Cres.)
Broadoak Rd. DA8: Erith 52Fd 118
Broadoaks CM16: Epp 3Vc 23
Broadoaks KT6: Surb 75Ra 153
Broadoaks Cres. KT14: W Byf 85K 169
Broadoaks Way BR2: Broml 71Hc 159
Broad Pas. W3 46Va 88
Broad Platts SL3: L'ly 8P 81
Broad Rd. DA10: Swans 58Ae 121
Broad Rd. WD24: Wat 11Y 27
Broad Sanctuary SW1 2E 228 (47Mb 90)
Broadstone NW1 38Mb 70
(off Agar Gro.)
Broadstone Ho. SW8 52Pb 112
(off Dorset Rd.)
Broadstone Pl. W1 1H 221 (43Jb 90)
Broadstone Rd. RM12: Horn 33Jd 76
Broadstone Rd. SS17: Stan H 3K 101
Broad St. GU24: W End 5B 166
Broad St. HP2: Hem H 1M 3
Broad St. RM10: Dag 38Cd 76
Broad St. TW11: Tedd 65Ha 130
Broad St. Av. EC2 1H 225 (43Ub 91)
Broad St. Mkt. RM10: Dag 38Cd 76
Broad St. Pl. EC2 1G 225 (43Tb 91)
(off Blomfield St.)
Broadstrood IG10: Lough 10Qc 22
Broadview NW9 30Qa 47
Broadview Av. RM16: Grays 47Fe 99
Broadview Pl. E5 32Yb 72
Broadview Rd. SW16 66Mb 134
Broad Wlk. W1: Hyde Pk. 5G 221 (45Hb 89)
Broad Wlk. NW1: Regent's Pk. 1J 215 (39Jb 70)
Broad Wlk. BR6: Chels 76Zc 161
Broad Wlk. CR3: Cat'm 94Vb 197
Broad Wlk. CR5: Coul 95Jb 196
Broad Wlk. KT18: Tatt C 91Za 194
Broad Wlk. N21 19Pb 32
Broad Wlk. SE3 54Lc 115
Broad Wlk. TN15: S'oaks 100Nd 203
Broad Wlk. TW5: Hest 53Z 107
Broad Wlk. TW9: Kew 52Pa 109
Broadwalk E18 27Hc 53
Broadwalk HA2: Harr 29Ca 45
Broad Wlk., The KT8: E Mos 70Ha 130
Broad Wlk., The W8 46Db 89
Broadwalk, The HA6: Nwood 26S 44
Broadwalk Ct. W8 46Cb 89
(off Palace Gdns. Ter.)
Broadwalk Ho. EC2 6H 219 (42Ub 91)
(off Appold St.)
Broadwalk Ho. SW7 2A 226 (47Eb 89)
(off Hyde Pk. Ga.)
Broad Wlk. La. NW11 31Bb 69
Broad Wlk. (North), The CM13:
B'wood 20Ce 41
Broadwalk Shop. Cen. Edgware 23Ra 47
Broad Wlk. Sth., The CM13: B'wood 21Ce 59
Broadwall SE1 6A 224 (46Qb 90)
Broadwater EN6: Pot B 2Db 17
Broadwater Cl. GU21: Wok 84F 168
Broadwater Cl. KT12: Hers 78W 150
Broadwater Cl. TW19: Wray 59B 104
Broadwater Farm Est. N17 26Tb 51
Broadwater Gdns. BR6: Farnb 77Rc 160
Broadwater Gdns. UB9: Hare 28L 43
Broadwater La. UB9: Hare 28K 43
Broadwater Pk. UB9: Den 30J 43
Broadwater Pl. KT13: Weyb 75U 150
Broadwater Rd. N17 25Ub 51
Broadwater Rd. SE28 48Tc 94
Broadwater Rd. SW17 63Gb 133
Broadwater Rd. Nth. KT12: Hers 78V 150
Broadwater Rd. Sth. KT12: Hers 78V 150
Broadway BR8: Crock 72Ed 162
Broadway DA6: Bex 56Ad 117
(not continuous)
Broadway DA7: Bex 56Dd 118
Broadway E15 38Fc 73
Broadway GU21: Knap 10F 166
Broadway IG11: Bark 39Sc 74
Broadway RM13: Rain 42Jd 96
Broadway RM17: Grays 51Ee 121
Broadway RM18: Tilb 4B 122
Broadway RM2: Rom 26Jd 56
Broadway SW1 3D 228 (48Mb 90)
Broadway TW18: Staines 64K 127
Broadway UB9: Den 30J 43
Broadway W13 46Ja 86
Broadway W7 46Ga 86
Broadway, The AL1: St A 2B 6
Broadway, The CR0: Bedd 77Nb 156
Broadway, The E13 40Kc 73
Broadway, The E4 23Fc 53
Broadway, The EN6: Pot B 4Bb 17
(not continuous)
Broadway, The GU21: Wok 89B 168
Broadway, The HA2: Harr 33Fa 66
Broadway, The HA3: W'stone 26Ga 46
Broadway, The HA6: Nwood 26W 44
Broadway, The HA7: Stan 22La 46
Broadway, The HA9: Wemb 34Na 67
Broadway, The IG10: Lough 14Sc 36
Broadway, The IG8: Wfd G 23Kc 53
Broadway, The KT15: New H 82J 169
Broadway, The KT7: T Ditt 74Ga 152
Broadway, The N11 22Hb 49
Broadway, The N14 18Mb 32
(off The Bourne)
Broadway, The N22 26Qb 50
Broadway, The N8 30Nb 50
Broadway, The N9 20Wb 33
Broadway, The NW7 22Ua 48
Broadway, The RM12: Horn 35Kd 77
Broadway, The RM8: Dag 32Bd 75
Broadway, The SL2: Farn C 7G 60
Broadway, The SL9: Chal P 25A 42
(off Market Pl.)
Broadway, The SM1: Sutt 78Eb 155
Broadway, The SM3: Cheam 79Ab 154
Broadway, The SW13 54Ua 110
Broadway, The SW19 65Bb 133
Broadway, The TW18: Lale 69L 127
Broadway, The UB1: S'hall 45Z 85
Broadway, The UB6: G'frd 42Ea 86
Broadway, The W3 47Qa 87
(off Ridgeway Dr.)
Broadway, The W5 45Ma 87
Broadway, The WD17: Wat 13Y 27
Broadway Arc. W6 49Ya 88
(off Hammersmith B'way.)
Broadway Av. CR0: C'don 71Tb 157
Broadway Av. TW1: Twick 58Ka 108
Broadway Cen., The 49Ya 88
B'way. Chambers W6 49Ya 88
(off Hammersmith B'way.)
Broadway Cl. CR2: Sande 86Xb 177
Broadway Cl. IG8: Wfd G 23Kc 53
Broadway Ct. BR3: Beck 69Ec 136
Broadway Ct. GU21: Knap 9G 166
Broadway Ct. SW19 65Cb 133
Broadway Gdns. CR4: Mitc 70Gb 133
Broadway Gdns. IG8: Wfd G 23Kc 53
Broadway Ho. BR1: Broml High St. 69Jc 137
Broadway Ho. BR1: Broml 64Fc 137
(off Bromley Rd.)
Broadway Ho. E8 39Xb 71
(off Ada St.)
Broadway Ho. GU21: Knap 10G 166
Broadway Mans. SW6 52Cb 111
(off Fulham Rd.)
Broadway Mkt. E8 39Xb 71
(not continuous)
Broadway Mkt. IG6: Ilf 25Tc 54
(not continuous)
Broadway Mkt. SW17 63Hb 133
Broadway Mkt. M. E8 39Wb 71
Broadway M. E5 31Vb 71
Broadway M. N13 22Pb 50
Broadway M. N21 18Rb 33
Broadway Pde. E4 23Ec 52
(off The Broadway)
Broadway Pde. HA2: Harr 29Da 45
Broadway Pde. N8 30Nb 50
Broadway Pde. RM12: Horn 35Kd 77
(off The Broadway)
Broadway Pde. UB3: Hayes 46W 84
Broadway Pl. SW19 65Bb 133
Broadway Retail Pk. 35Za 68
B'way. Rd. GU18: Light 2A 166
B'way. Rd. GU20: W'sham 1B 166
Broadway Shop. Cen. Bexleyheath 56Cd 118
Broadway Shop. Mall St James's
Park 3D 228 (48Mb 90)
Broadway Sq. DA6: Bex 56Cd 118
Broadway Theatre Barking, The 39Sc 74
Broadway Theatre Catford, The 59Dc 114
Broadway Wlk. E14 47Cc 92
Broadwell Ct. TW5: Hest 53Z 107
(off Springwell Rd.)
Broadwell Pde. NW6 37Db 69
(off Broadhurst Gdns.)
Broadwick St. W1 4C 222 (45Lb 90)
Broadwood DA11: Grav'nd 4D 144
Broadwood Av. HA4: Ruis 30T 44
Broadwood Rd. CR5: Coul 93Mb 196
Broadwood Ter. W8 49Bb 89
(off Pembroke Rd.)
Broad Yd. EC1 6B 218 (42Rb 91)
Brocade Cl. SM6: W'gton 75Kb 156
Brocas Cl. NW3 38Gb 69
Brocas St. SL4: Eton 2H 103
Brocas Ter. SL4: Eton 2H 103
Brockbridge Ho. SW15 58Ua 110
Brockdene Dr. BR2: Kes 77Mc 159
Brockdish Av. IG11: Bark 36Vc 75
Brockenhurst KT8: W Mole 71Ba 151
Brockenhurst Av. KT4: Wor Pk 74Ua 154
Brockenhurst Cl. GU21: Wok 86B 168
Brockenhurst Dr. SS17: Stan H 3L 101
Brockenhurst Gdns. IG1: Ilf 36Sc 74
Brockenhurst Gdns. NW7 22Ua 48
Brockenhurst M. N18 21Wb 51
Brockenhurst Rd. CR0: C'don 73Xb 157
Brockenhurst Way SW16 68Mb 134
Brocket Cl. IG7: Chig 21Vc 55
Brocket Ho. SW8 54Mb 112
Brockett Rd. RM16: Grays 8C 100
Brocket Way IG7: Chig 22Uc 54
Brock Grn. RM15: S Ock 44Xd 98
Brockham Cl. SW19 64Bb 133
Brockham Ct. CR2: S Croy 78Sb 157
Brockham Ct. RH1: Redh 4B 208
(off Goodworth Rd.)
Brockham Cres. CR0: New Ad 80Fc 159
Brockham Dr. IG2: Ilf 30Rc 54
Brockham Dr. SW2 59Pb 112
Brockham Ho. NW1 1C 216 (39Lb 70)
(off Bayham Pl.)
Brockham Ho. SW2 59Pb 112
(off Brockham Dr.)
Brockham St. SE1 3E 230 (48Sb 91)
Brockhill GU21: Wok 9L 167
Brockhurst Cl. HA7: Stan 23Ha 46
Brockill Cres. SE4 56Ac 114
Brocklebank Ct. CR3: Whyt 90Wb 177
Brocklebank Ho. E16 46Qc 94
(off Glenister St.)
Brocklebank Retail Pk. 49Kc 93
Brocklebank Rd. SE7 49Kc 93
Brocklebank Rd. SW18 59Eb 111
Brocklehurst St. SE14 52Zb 114
Brocklesbury Cl. WD24: Wat 13Z 27
Brocklesby Rd. SE25 70Xb 135
BROCKLEY 56Zb 114
Brockley Av. HA7: Stan 20Na 29
Brockley Cl. HA7: Stan 21Na 47
Brockley Combe KT13: Weyb 77T 150
Brockley Cres. RM5: Col R 24Ed 56
Brockley Cross SE4 55Ac 114
Brockley Cross Bus. Cen. SE4 55Ac 114
Brockley Footpath SE15 55Yb 114
(not continuous)
Brockley Gdns. SE4 54Bc 114
Brockley Gro. CM13: Hut 18Ce 41
Brockley Gro. SE4 57Bc 114
Brockley Hall Rd. SE4 57Ac 114
Brockley Hill HA7: Stan 18La 28
Brockley Jack Theatre 57Ac 114
Brockley M. SE4 57Ac 114
Brockley Pk. SE23 59Ac 114
Brockley Ri. SE23 60Ac 114
Brockley Rd. SE4 55Bc 114
Brockleyside HA7: Stan 21Na 47
Brockley Station (Rail &
Overground) 55Ac 114
Brockley Vw. SE23 59Ac 114
Brockley Way SE4 57Zb 114
Brockman Ri. BR1: Broml 63Fc 137
Brockmer Ho. E1 45Xb 91
(off Crowder St.)
Brock Pl. E3 42Dc 92
Brock Rd. E13 43Kc 93
Brocks Dr. SM3: Cheam 76Ab 154
Brockshot Cl. TW8: Bford 50Ma 87
Brocksparkwood CM13: B'wood 20De 41
Brock St. NW1 5B 216 (42Lb 90)
(off Triton Sq.)
Brock St. SE15 55Yb 114
Brockton Cl. RM1: Rom 28Hd 56
Brock Way GU25: Vir W 1N 147
Brockway TN15: Bor G 92Ce 205
Brockway Cl. E11 33Gc 73
Brockweir E2 40Yb 72
(off Cyprus St.)
Brockwell Av. BR3: Beck 71Dc 158
Brockwell Cl. BR5: St M Cry 71Vc 161
Brockwell Ct. SW2 57Qb 112
Brockwell Ho. SE11 51Pb 112
(off Vauxhall St.)
Brockwell Pk. 58Rb 113
Brockwell Pk. Gdns. SE24 59Qb 112
Brockwell Pk. Lido 58Rb 113
Brockwell Pk. Row SW2 59Qb 112
Brockwell Pas. SE24 58Rb 113
Brodewater Rd. WD6: Bore 12Ra 29
Brodia Rd. N16 34Ub 71
Brodick Ho. E3 40Bc 72
(off Saxon Rd.)
Brodie Ho. SE1 7K 231 (50Vb 91)
(off Cooper's Rd.)
Brodie Rd. E4 18Ec 34
Brodie Rd. EN2: Enf 10Sb 19
Brodie St. SE1 7K 231 (50Vb 91)
Brodlove La. E1 45Zb 92
Brodrick Gro. KT23: Bookh 98Ca 191
Brodrick Gro. SE2 49Xc 95
Brodrick Rd. SW17 61Gb 133
Brody Ho. E1 1K 225 (43Vb 91)
(off Strype St.)
Brograve Gdns. BR3: Beck 68Dc 136
Broke Farm Dr. BR6: Prat B 81Yc 181
Broken Furlong SL4: Eton 10F 80
Broken Ga. La. UB9: Den 32E 62
Broken Wharf EC4 4D 224 (45Sb 91)
Brokes Cres. RH2: Reig 4J 207
Brokesley St. E3 41Bc 92
Brokes Rd. RH2: Reig 4J 207
Broke Wlk. E8 39Vb 71
Bromar Rd. SE5 55Ub 113
Bromborough Grn. WD19: Wat 22Y 45
Bromefield HA7: Stan 25La 46
Bromefield Ct. EN9: Walt A 5Jc 21
Bromehead St. E1 44Yb 92
Bromell's Rd. SW4 56Lb 112
Brome Rd. SE9 55Pc 116
Bromet Cl. WD17: Wat 10V 12
(not continuous)
Bromfelde Rd. SW4 55Mb 112
Bromfelde Wlk. SW4 54Mb 112
Bromfield Ct. SE16 48Wb 91
(off Ben Smith Way)
Bromfield St. N1 1A 218 (39Qb 70)
Bromford Cl. RH8: Oxt 5L 211
Bromhall Rd. RM8: Dag 37Xc 75
Bromhall Rd. RM9: Dag 37Xc 75
Bromhead Rd. E1 44Yb 92
(off Jubilee St.)
Bromhedge SE9 62Pc 138
Bromholm Rd. SE2 48Xc 95
Bromleigh Ct. SE23 61Wb 135
Bromleigh Ho. SE1 3K 231 (48Vb 91)
(off St Saviour's Est.)
BROMLEY 68Jc 137
BROMLEY 41Dc 92
Bromley RM17: Grays 51Ce 121
Bromley Av. BR1: Broml 66Gc 137
Bromley-by-Bow Station
(Underground) 41Dc 92
Bromley Coll. BR1: Broml 67Jc 137
BROMLEY COMMON 73Nc 160
Bromley Comn. BR2: Broml 70Lc 137
Bromley Cres. BR2: Broml 69Hc 137
Bromley Cres. HA4: Ruis 35V 64
Bromley FC 71Kc 159
Bromley Gdns. BR2: Broml 69Hc 137
Bromley Golf Course 73Nc 160
Bromley Gro. BR2: Broml 68Fc 137
Bromley Hall Rd. E14 43Ec 92
Bromley High St. E3 41Dc 92
Bromley Hill BR1: Broml 65Gc 137
Bromley Ho. BR1: Broml 67Jc 137
(off North St.)
Bromley Indoor Bowls Cen. 74Yc 161
Bromley Ind. Cen. BR1: Broml 69Mc 137
Bromley La. BR7: Chst 66Sc 138
Bromley Little Theatre 67Jc 137
Bromley Mus. 73Xc 161

Bromley North Station (Rail) 67Jc 137
Bromley Pk. BR1: Broml 67Hc 137
BROMLEY PARK 67Gc 137
Bromley Pl. W1 7B 216 (43Lb 90)
Bromley Rd. BR1: Broml 63Fc 137
Bromley Rd. BR2: Broml 68Fc 137
Bromley Rd. BR3: Beck 67Dc 136
Bromley Rd. BR7: Chst 67Rc 138
Bromley Rd. E10 30Dc 52
Bromley Rd. E17 27Cc 52
Bromley Rd. N17 25Vb 51
Bromley Rd. N18 20Tb 33
Bromley Rd. SE6 60Dc 114
Bromley South Station (Rail) 69Jc 137
Bromley St. E1 43Zb 92
Bromley Tennis Cen. 76Tc 160
Bromley Valley Gymnastics Cen. 68Wc 139
BROMPTON 4E 226 (48Gb 89)
Brompton Arc. SW3 2F 227 (47Hb 89)
(off Brompton Rd.)
Brompton Cemetery 51Db 111
Brompton Cl. SE20 68Wb 135
Brompton Cl. TW4: Houn 57Ba 107
Brompton Cotts. SW10 51Eb 111
(off Hollywood Rd.)
Brompton Ct. BR1: Broml 67Jc 137
(off Tweedy Rd.)
Brompton Dr. DA8: Erith 52Kd 119
Brompton Gdns. GU24: W End 5E 166
Brompton Gro. N2 28Gb 49
Brompton M. N12 23Eb 49
Brompton Pk. Cres. SW6 51Db 111
Brompton Pl. SW3 3E 226 (48Gb 89)
Brompton Rd. SW1 2F 227 (47Hb 89)
Brompton Rd. SW3 5D 226 (49Gb 89)
Brompton Rd. SW7 4D 226 (48Gb 89)
Brompton Sq. SW3 3D 226 (48Gb 89)
Brompton Ter. SE18 53Pc 116
Brompton Vs. SW6 51Cb 111
(off Lillie Rd.)
Bromwich Av. N6 33Jb 70
Bromyard Av. W3 45Ua 88
Bromyard Ho. SE15 52Xb 113
(off Commercial Way)
Bromycroft Rd. SL2: Slou 1E 80
Bromyward Ho. W3 46Va 88
Bron Ct. NW6 39Cb 69
BRONDESBURY 38Bb 69
Brondesbury Ct. NW2 37Za 68
Brondesbury M. NW6 38Cb 69
BRONDESBURY PARK 39Ya 68
Brondesbury Pk. NW2 37Ya 68
Brondesbury Pk. NW6 38Ya 68
Brondesbury Pk. Mans. NW6 39Ab 68
(off Salusbury Rd.)
Brondesbury Park Station (Overground) 39Ab 68
Brondesbury Rd. NW6 40Bb 69
Brondesbury Station (Overground) 38Bb 69
Brondesbury Vs. NW6 40Bb 69
Bronhill Ter. N17 25Wb 51
Bronsart Rd. SW6 52Ab 110
Bronsdon Way UB9: Den 33H 63
Bronson Rd. SW20 68Za 132
Bronte Cl. DA8: Erith 52Dd 118
Bronte Cl. E7 35Jc 73
Bronte Cl. IG2: Ilf 28Qc 54
Bronte Cl. RM18: Tilb 4E 122
Bronte Cl. SL1: Slou 7J 81
Bronte Ct. RH1: Redh 5A 208
Bronte Ct. W14 48Za 88
(off Girdler's Rd.)
Bronte Ct. W3 47Qa 87
Bronte Gro. DA1: Dart 56Pd 119
Bronte Ho. N16 36Ub 71
Bronte Ho. SW17 62Fb 133
(off Grosvenor Way)
Bronte Ho. SW4 59Lb 112
Bronte Vw. DA12: Grav'nd 10E 122
Bronti Cl. SE17 7E 230 (50Sb 91)
Bronwen Ct. NW8 4B 214 (41Fb 89)
(off Grove End Rd.)
Bronze Age Way DA17: Belv 47Dd 96
Bronze Age Way DA8: Erith 49Fd 96
Bronze St. SE8 52Cc 114
Bronze Wlk. W12 45Za 88
Brook Av. HA8: Edg 23Ra 47
Brook Av. HA9: Wemb 33Pa 67
Brook Av. RM10: Dag 38Dd 76
Brook Bank EN1: Enf 9Xb 19
Brookbank Av. W7 43Fa 86
Brookbank Rd. SE13 55Cc 114
Brook Bus. Cen. UB8: Uxb 40K 63
Brook Cl. HA4: Ruis 31U 64
Brook Cl. KT19: Ewe 81Ua 174
Brook Cl. NW7 24Ab 48
Brook Cl. RM2: Rom 25Hd 56
Brook Cl. SW17 61Jb 134
Brook Cl. SW20 69Xa 132
Brook Cl. TW19: Stanw 59P 105
Brook Cl. W3 46Qa 87
Brook Cl. WD6: Bore 12Ra 29
Brook Ct. IG11: Bark Sebastian Ct. 39Vc 75
Brook Ct. IG11: Bark Spring Pl. 40Sc 74
Brook Ct. BR3: Beck 67Bc 136
Brook Ct. E11 34Gc 73
Brook Ct. E17 27Ac 52
Brook Ct. EC4 4F 225 (45Tb 91)
(off Laurence Pountney La.)
Brook Ct. HA8: Edg 22Ra 47
Brook Ct. SE12 62Lc 137
Brook Ct. WD7: R'lett 5Ja 14
Brook Cres. E4 21Cc 52
Brook Cres. N9 21Xb 51
Brook Cres. SL1: Slou 4C 80
Brookdale N11 21Lb 50
Brookdale Av. RM14: Upm 34Qd 77
Brookdale Cl. RM14: Upm 33Rd 77
Brookdale Rd. SE6: Catford B'way 59Dc 114
Brookdale Rd. SE6: Medusa Rd 58Dc 114
Brookdale Rd. DA5: Bexl 58Ad 117
Brookdale Rd. E17 27Cc 52
Brookdales NW11 28Ab 48
Brookdene Av. WD19: Wat 17X 27
Brookdene Dr. HA6: Nwood 24V 44
Brookdene Rd. SE18 49Uc 94
Brook Dr. HA1: Harr 28Ea 46
Brook Dr. HA4: Ruis 31U 64
Brook Dr. SE11 4A 230 (48Qb 90)
Brook Dr. WD7: R'lett 5Ha 14
Brooke Av. HA2: Harr 34Ea 66
Brooke Cl. WD23: Bush 17Ea 28
Brooke Ct. W10 40Ab 68
(off Kilburn La.)
Brooke Dr. DA12: Grav'nd 10K 123
Brooke End RH8: Oxt 5L 211
Brooke Ho. SE14 53Ac 114
Brooke Ho. WD23: Bush 17Ea 28
Brookehowse Rd. SE6 61Cc 136
Brookend Rd. DA15: Sidc 60Uc 116
Brooke Rd. E17 28Ec 52
Brooke Rd. E5 34Wb 71
Brooke Rd. N16 34Vb 71
Brooke Rd. RM17: Grays 50Ce 99
Brooker Rd. EN9: Walt A 6Ec 20
Brookers Cl. KT21: Asht 89La 172
Brooke's Ct. EC1 1K 223 (43Qb 90)
Brooke's Mkt. EC1 7K 217 (43Qb 90)
(off Dorrington St.)
Brooke St. EC1 1K 223 (43Qb 90)
Brooke Trad. Est. RM1: Rom 31Hd 76
Brooke Way WD23: Bush 17Ea 28
Brook Farm Rd. KT11: Cobh 87Z 171
Brookfield GU21: Wok 8M 167
Brookfield N6 34Jb 70
Brookfield TN15: Kems'g 89Nd 183
Brookfield Av. E17 28Ec 52
Brookfield Av. NW7 23Xa 48
Brookfield Av. SM1: Sutt 77Fb 155
Brookfield Av. W5 42Ma 87
Brookfield Cl. CM13: Hut 16Ee 41
Brookfield Cl. KT16: Ott 79F 148
Brookfield Cl. KT21: Asht 92Na 193
Brookfield Cl. NW7 23Xa 48
Brookfield Ct. N12 21Db 49
Brookfield Ct. UB6: G'frd 41Ea 86
Brookfield Cres. HA3: Kenton 29Na 47
Brookfield Cres. NW7 23Xa 48
Brookfield Gdns. KT10: Clay 79Ha 152
Brookfield Ho. HP2: Hem H 3M 3
(off Selden Hill)
Brookfield Pk. NW5 34Kb 70
Brookfield Path IG8: Wfd G 23Gc 53
Brookfield Pl. KT11: Cobh 87Aa 171
Brookfield Rd. E9 37Ac 72
Brookfield Rd. N9 20Wb 33
Brookfield Rd. W4 47Ta 87
Brookfields EN3: Pond E 14Zb 34
Brookfields Av. CR4: Mitc 71Gb 155
Brook Gdns. E4 21Dc 52
Brook Gdns. KT2: King T 67Sa 131
Brook Gdns. SW13 55Va 110
Brook Ga. W1 5G 221 (45Hb 89)
BROOK GREEN 49Za 88
Brook Grn. GU24: Chob 2K 167
(off Chertsey Rd.)
Brook Grn. W6 48Za 88
Brook Grn. Flats W14 48Za 88
(off Dunsany Rd.)
Brook Hill RH8: Oxt 2G 210
Brookhill Cl. EN4: E Barn 15Gb 31
Brookhill Cl. SE18 50Rc 94
Brookhill Ct. EN4: E Barn 15Gb 31
Brookhill Rd. EN4: E Barn 15Gb 31
Brookhill Rd. SE18 51Rc 116
Brook Ho. E1 45Wb 91
(off Fletcher St.)
Brook Ho. SL1: Slou 8H 81
Brook Ho. W6 49Ya 88
(off Shepherd's Bush Rd.)
Brookhouse Gdns. E4 21Gc 53
Brook Ho's. NW1 2C 216 (40Lb 70)
(off Cranleigh St.)
Brookhurst Rd. KT15: Add 79K 149
Brook Ind. Est. UB4: Yead 46Z 85
Brook Ind. Pk. BR5: St M Cry 70Yc 139
Brooking CM14: B'wood 20Xd 40
Brooking Cl. RM8: Dag 34Yc 75
Brooking Rd. E7 36Jc 73
Brookland Cl. NW11 28Cb 49
Brookland Ct. RH2: Reig 4K 207
Brookland Dr. EN9: Walt A 4Jc 21
Brookland Gth. NW11 28Cb 49
Brookland Hill NW11 28Db 49
Brookland Ri. NW11 28Cb 49
Brooklands DA1: Dart 60Nd 119
BROOKLANDS 82P 169
Brooklands, The TW7: Isle 53Fa 108
Brooklands App. RM1: Rom 28Fd 56
Brooklands Av. DA15: Sidc 61Tc 138
Brooklands Av. SW19 61Db 133
Brooklands Bus. Pk. KT13: Weyb 83N 169
Brooklands Cl. KT11: Cobh 87Aa 171
Brooklands Cl. RM7: Rom 28Fd 56
Brooklands Cl. TW16: Sun 67U 128
Brooklands Ct. AL1: St A 2C 6
Brooklands Ct. CR4: Mitc 68Fb 133
Brooklands Ct. KT1: King T 70Ma 131
(off Surbiton Rd.)
Brooklands Ct. KT15: New H 82M 169
Brooklands Ct. N21 15Tb 33
Brooklands Ct. NW6 38Bb 69
Brooklands Dr. KT13: Weyb 82P 169
Brooklands Dr. UB6: G'frd 39La 66
Brooklands Gdns. EN6: Pot B 4Ab 16
Brooklands Gdns. KT10: Esh 75Da 151
Brooklands Gdns. RM11: Horn 30Ld 57
Brooklands Ho. KT15: Add 77L 149
(off Market St.)
Brooklands Ind. Est. KT13: Weyb 82N 169
Brooklands La. KT13: Weyb 79P 149
Brooklands La. RM7: Rom 28Fd 56
(not continuous)
Brooklands Mus. 81P 169
Brooklands Pk. SE3 55Jc 115
Brooklands Pas. SW8 53Mb 112
Brooklands Pl. TW12: Hamp H 64Da 129
Brooklands Rd. KT13: Weyb 84Q 170
Brooklands Rd. KT7: T Ditt 74Ha 152
Brooklands Rd. RM7: Rom 28Fd 56
Brooklands Ter. TW16: Sun 70W 128
Brooklands Way RH1: Redh 4N 207
Brook La. BR1: Broml 65Jc 137
Brook La. DA5: Bex 58Zc 117
Brook La. DA5: Bexl 58Zc 117
Brook La. GU23: Send 94G 188
Brook La. GU24: Chob 3H 167
Brook La. SE3 54Kc 115
Brook La. TN15: Plax 99Ce 205
Brook La. Bus. Cen. TW8: Bford 50Ma 87
Brook La. Nth. TW8: Bford 50Ma 87
(not continuous)
Brooklea Cl. NW9 25Ua 48
Brookleys GU24: Chob 2K 167
Brooklime Path RM3: Rom 24Ld 57
Brook Lodge NW11 29Za 48
(off Nth. Circular Rd.)
Brook Lodge RM7: Rom 28Fd 56
(off Brooklands Rd.)
Brooklyn SE20 66Wb 135
Brooklyn Av. IG10: Lough 14Nc 36
Brooklyn Av. SE25 70Xb 135
Brooklyn Cl. GU22: Wok 91A 188
Brooklyn Cl. SM5: Cars 75Gb 155
Brooklyn Ct. GU22: Wok 91A 188
Brooklyn Ct. IG10: Lough 14Nc 36
Brooklyn Ct. W12 46Ya 88
(off Frithville Gdns.)
Brooklyn Gro. SE25 70Xb 135
Brooklyn Rd. BR2: Broml 71Mc 159
Brooklyn Rd. GU22: Wok 90A 168
Brooklyn Rd. SE25 70Xb 135
Brooklyn Way UB7: W Dray 48M 83
Brookman Ho. E3 40Bc 72
(off Mostyn Gro.)
Brookman's Av. RM16: Grays 46Ee 99
Brookmans Av. AL9: Brk P 8H 9
Brookmans Cl. RM14: Upm 31Ud 78
BROOKMANS PARK 8G 8
Brookmans Pk. Dr. RM14: Upm 29Ud 58
Brookmans Pk. Golf Course 7J 9
Brookmans Park Station (Rail) 9G 8
Brookmarsh Ind. Est. SE10 52Dc 114
Brook Mead KT19: Ewe 79Ua 153
Brookmead CR0: Bedd 72Lb 156
Brookmead Av. BR1: Broml 71Pc 160
Brookmead Cl. BR5: St M Cry 73Xc 161
Brookmead Ind. Est. CR0: Bedd 72Lb 156
Brook Mdw. N12 20Db 31
Brook Mdw. Cl. IG8: Wfd G 23Gc 53
Brookmeadow Way EN9: Walt A 2Kc 21
Brookmead Rd. CR0: C'don 72Lb 156
Brookmead Way BR5: St M Cry 72Xc 161
Brook M. IG7: Chig 20Rc 36
Brook M. N13 22Qb 50
Brook M. Nth. W2 4A 220 (45Eb 89)
Brookmill Cl. WD19: Wat 17X 27
Brookmill Rd. SE8 53Cc 114
Brook Pde. IG7: Chig 20Rc 36
Brook Pk. DA1: Dart 61Qd 141
Brook Pk. Cl. N21 15Rb 33
Brook Path IG10: Lough 14Nc 36
Brook Path SL1: Slou 5D 80
(not continuous)
Brook Pl. EN5: Barn 15Cb 31
Brook Retail Pk. South Ruislip 36Z 65
Brook Ri. IG7: Chig 20Qc 36
Brook Rd. BR8: Swan 69Fd 140
Brook Rd. CM14: B'wood 20Vd 40
Brook Rd. CM16: Epp 5Wc 23
Brook Rd. CR7: Thor H 70Sb 135
Brook Rd. DA11: Nflt 10A 122
Brook Rd. EN8: Walt C 6Bc 20
Brook Rd. IG10: Lough 14Nc 36
Brook Rd. IG2: Ilf 30Uc 54
Brook Rd. IG9: Buck H 19Jc 35
Brook Rd. IG9: Wfd G 19Jc 35
Brook Rd. KT6: Surb 75Na 153
Brook Rd. N22 27Pb 50
Brook Rd. N8 28Nb 50
Brook Rd. NW2 33Va 68
Brook Rd. RH1: Mers 1C 208
Brook Rd. RH1: Redh 7P 207
Brook Rd. RM2: Rom 26Hd 56
Brook Rd. TW1: Twick 58Ja 108
Brook Rd. WD6: Bore 11Qa 29
Brook Rd. Sth. TW8: Bford 51Ma 109
Brooks Apts. E3 43Cc 92
(off Geoff Cade Way)
Brooks Av. E6 42Pc 94
Brooksbank Ho. E9 37Yb 72
(off Retreat Pl.)
Brooksbank St. E9 37Yb 72
Brooksby Ho. N1 38Qb 70
(off Liverpool Rd.)
Brooksby M. N1 38Qb 70
Brooksby St. N1 38Qb 70
Brooksby's Wlk. E9 36Zb 72
Brooks Cl. KT13: Weyb 82Q 170
Brooks Cl. SE9 61Qc 138
Brooks Ct. SW11 52Lb 112
Brookscroft CR0: Sels 82Bc 178
Brookscroft E17 27Dc 52
Brookscroft Rd. E17 25Dc 52
(not continuous)
Brooks Farm 31Dc 72
Brookshill HA3: Hrw W 22Fa 46
Brookshill Av. HA3: Hrw W 22Fa 46
Brookshill Dr. HA3: Hrw W 22Fa 46
Brookshill Ga. HA3: Hrw W 22Fa 46
Brooks Ho. CM14: B'wood 18Yd 40
Brookside AL10: Hat 1P 7
Brookside BR6: Orp 73Vc 161
Brookside EN4: E Barn 16Gb 31
Brookside EN6: S Mim 4Wa 16
Brookside EN9: Walt A 4Gc 21
Brookside GU4: Jac W 10P 187
Brookside IG6: Ilf 23Sc 54
Brookside KT16: Chert 73G 148
Brookside N21 16Pb 32
Brookside RH9: S God 10B 210
Brookside RM11: Horn 29Nd 57
Brookside SL3: Coln 52E 104
Brookside SM5: Cars 78Jb 156
Brookside UB10: Uxb 38P 63
Brookside WD24: Wat 9Z 13
Brookside Av. TW15: Ashf 64L 127
Brookside Av. TW19: Wray 55A 104
Brookside Caravans WD19: Wat 17X 27
Brookside Cl. EN5: Barn 16Ab 30
Brookside Cl. HA2: Harr 35Aa 65
Brookside Cl. HA3: Kenton 29Ma 47
Brookside Cl. TW13: Felt 62W 128
Brookside Cotts. WD4: Hunt C 6S 12
Brookside Cres. KT4: Wor Pk 74Wa 154
Brookside Gdns. EN1: Enf 9Yb 20
Brookside Rd. DA13: Ist R 6B 144
Brookside Rd. N19 33Lb 70
Brookside Rd. N9 21Xb 51
(not continuous)
Brookside Rd. NW11 30Ab 48
Brookside Rd. UB4: Yead 45Y 85
Brookside Rd. WD19: Wat 17X 27
Brookside Sth. EN4: E Barn 17Jb 32
Brookside Wlk. N3 27Ab 48
Brookside Way CR0: C'don 72Zb 158
Brooks La. W4 51Qa 109
Brooks Lodge N1 1J 219 (40Ub 71)
Brook's M. W1 4K 221 (45Kb 90)
Brook Sq. SE18 53Nc 116
Brooks Rd. E13 39Jc 73
Brooks Rd. W4 50Qa 87
Brook St. CM14: B'wood 22Td 58
Brook St. DA17: Belv 50Dd 96
Brook St. DA17: Erith 50Dd 96
Brook St. DA8: Erith 50Dd 96
Brook St. KT1: King T 68Na 131
Brook St. N17 26Vb 51
Brook St. SL4: Wind 4H 103
Brook St. W1 4K 221 (45Kb 90)
Brook St. W2 4C 220 (45Fb 89)
BROOK STREET 21Vd 58
BROOK STREET INTERCHANGE 22Td 58
Brooksville Av. NW6 39Ab 68
Brooks Way BR5: St P 68Yc 139
Brooks Way RM3: Rom 22Md 57
Brook Va. DA8: Erith 53Dd 118
Brook Valley Gdns. EN5: Barn 15Bb 31
Brookview Ct. EN1: Enf 15Ub 33
Brookview Rd. SW16 64Lb 134
Brookville Rd. SW6 52Bb 111
Brook Wlk. HA8: Edg 23Ta 47
Brook Wlk. N2 25Fb 49
Brook Way IG7: Chig 20Qc 36
Brook Way KT22: Lea 90Ja 172
Brookway RM13: Rain 43Kd 97
Brookway SE3 55Jc 115
Brookwell Ho. E17 28Zb 52
BROOKWOOD 2E 186
Brookwood Av. SW13 54Va 110
Brookwood Cl. BR2: Broml 70Hc 137
Brookwood Country Pk. 1G 186
Brookwood Farm Dr. GU21: Knap 10E 166
Brookwood Ho. SE1 2C 230 (47Rb 91)
(off Webber St.)
Brookwood Lye Rd. GU24: Brkwd 2F 186
Brookwood Rd. SW18 60Bb 111
Brookwood Rd. TW3: Houn 54Da 107
Brookwood Station (Rail) 3E 186
Broom Av. BR5: St P 68Xc 139
Broom Bank CR6: W'ham 91Ec 198
Broom Cl. AL10: Hat 3B 8
Broom Cl. BR2: Broml 72Nc 160
Broom Cl. KT10: Esh 78Da 151
Broom Cl. TW11: Tedd 66Ma 131
Broomcroft Av. UB5: N'olt 41Y 85
Broomcroft Cl. GU22: Pyr 88F 168
Broomcroft Dr. GU22: Pyr 87F 168
Broome Cl. KT18: Head 98Sa 193
Broome Ct. KT20: Tad 91Ab 194
Broome Lodge TW18: Staines 64J 127
(off Kingston Rd.)
Broome Pl. RM15: Avel 46Td 98
Broome Rd. TW12: Hamp 66Ba 129
Broomer Pl. EN8: Chesh 1Yb 20
Broome Way SE5 52Tb 113
Broom Farm Est. SL4: Wind 4A 102
Broomfield AL2: Park 9A 6
Broomfield E17 31Bc 72
Broomfield NW1 38Jb 70
(off Ferdinand St.)
Broomfield TW16: Sun 67W 128
Broomfield TW18: Staines 65J 127
Broomfield Av. IG10: Lough 16Pc 36
Broomfield Av. N13 22Pb 50
Broomfield Cl. RM5: Col R 24Fd 56
Broomfield Cl. SL5: S'dale 3F 146
Broomfield Ct. KT13: Weyb 79R 150
Broomfield Ct. N2 28Gb 49
Broomfield Ga. SL2: Slou 2F 80
Broomfield Ho. HA7: Stan 20Ja 28
(off Stanmore Hill)
Broomfield Ho. SE17 6H 231 (49Ub 91)
(off Massinger St.)
Broomfield La. N13 21Nb 50
Broomfield Pk. SL5: S'dale 2F 146
Broomfield Pl. W13 46Ka 86
Broomfield Ride KT22: Oxs 84Fa 172
Broomfield Ri. WD5: Ab L 4T 12
Broomfield Rd. BR3: Beck 69Ac 136
Broomfield Rd. DA10: Swans 57Ae 121
Broomfield Rd. DA6: Bex 57Cd 118
Broomfield Rd. KT15: New H 83K 169
Broomfield Rd. KT5: Surb 74Pa 153
Broomfield Rd. N13 22Nb 50
Broomfield Rd. RM6: Chad H 31Zc 75
Broomfield Rd. TN13: S'oaks 94Hd 202
Broomfield Rd. TW11: Tedd 65La 130
Broomfield Rd. TW9: Kew 53Pa 109
Broomfield Rd. W13 46Ka 86
Broomfields DA3: Hartl 71Ae 165
Broomfields KT10: Esh 78Ea 152
Broomfield St. E14 43Cc 92
Broom Gdns. CR0: C'don 76Cc 158
Broom Gro. WD17: Wat 10W 12
Broomgrove Gdns. HA8: Edg 25Qa 47
Broomgrove Rd. SW9 54Pb 112
Broom Hall KT22: Oxs 86Fa 172
BROOMHALL 2E 146
Broomhall End GU21: Wok 88A 168
(off Broomhall La.)
Broomhall La. GU21: Wok 88A 168
Broomhall La. SL5: S'dale 2E 146
Broomhall Rd. CR2: Sande 81Tb 177
Broomhall Rd. GU21: Wok 88A 168
BROOM HILL 73Vc 161
Broom Hill HP1: Hem H 3G 2
Broom Hill SL2: Stoke P 8L 61
Broomhill Ct. IG8: Wfd G 23Jc 53
Broomhill Ri. DA6: Bex 57Cd 118
Broomhill Rd. BR6: Orp 73Wc 161
Broomhill Rd. DA1: Dart 58Kd 119
Broomhill Rd. IG3: Ilf 33Wc 75
Broomhill Rd. IG8: Wfd G 23Jc 53
(not continuous)
Broomhill Rd. SW18 57Cb 111
Broomhills DA13: Sflt 63Ae 143
Broomhill Wlk. IG8: Wfd G 23Hc 53
Broom Ho. SL3: L'ly 49B 82
Broomhouse La. SW6 54Cb 111
Broomhouse Rd. SW6 54Cb 111
Broomlands La. RH8: Limp 98Mc 199
Broomlands La. RH8: T'sey 98Mc 199
Broom La. GU24: Chob 1J 167
Broomloan La. SM1: Sutt 75Cb 155
Broom Lock TW11: Tedd 65La 130
Broom Mead DA6: Bex 58Cd 118
Broom Pk. TW11: Tedd 66Ma 131
Broom Rd. CR0: C'don 76Cc 158
Broom Rd. TW11: Tedd 64Ka 130
Broomsleigh Bus. Pk. SE26 64Bc 136
Broomsleigh St. NW6 36Bb 69
Broomstick Hall Rd. EN9: Walt A 5Gc 21
Broom Water TW11: Tedd 65La 130
Broom Water W. TW11: Tedd 64La 130
Broom Way KT13: Weyb 77U 150
Broomwood Cl. CR0: C'don 71Zb 158
Broomwood Cl. DA5: Bexl 61Fd 140
Broomwood Gdns. CM15: Pil H 16Wd 40
Broomwood Rd. BR5: St P 68Xc 139
Broomwood Rd. SW11 58Hb 111
Broseley Gdns. RM3: Rom 21Nd 57
Broseley Gro. SE26 64Ac 136
Broseley Rd. RM3: Rom 21Nd 57
Brosse Way BR2: Broml 72Nc 160
Broster Gdns. SE25 69Vb 135
Brotherstone Wlk. TW9: Kew 53Ra 109
Brouard Ct. BR1: Broml 69Jc 137
Brougham Ct. DA2: Dart 58Rd 119
(off Hardwick Cres.)
Brougham Rd. E8 39Wb 71
Brougham Rd. W3 44Sa 87
Brougham St. SW11 54Hb 111
Brough Cl. KT2: King T 64Ma 131
Brough Cl. SW8 52Nb 112
Broughinge Rd. WD6: Bore 12Ra 29
Broughton Av. N3 27Ab 48
Broughton Av. TW10: Ham 62Ka 130
Broughton Ct. W13 45Ka 86
Broughton Dr. SW9 56Qb 112
Broughton Gdns. N6 30Lb 50
Broughton Pl. E17 25Bc 52
Broughton Rd. BR6: Orp 75Tc 160
Broughton Rd. CR7: Thor H 72Qb 156
Broughton Rd. SW6 54Db 111
Broughton Rd. TN14: Otf 88Jd 182
Broughton Rd. W13 45Ka 86
Broughton Rd. App. SW6 54Db 111
Broughton St. SW8 54Jb 112
Broughton St. Ind. Est. SW11 54Jb 112
Broughton Way WD3: Rick 17J 25
Brouncker Rd. W3 47Sa 87
Brow, The RH1: Redh 10A 208
Brow, The WD25: Wat 5X 13
Brow Cl. BR5: Orp 73Zc 161
Brow Cres. BR5: Orp 74Yc 161
Browells La. TW13: Felt 61X 129
(not continuous)
Brown Bear Ct. TW13: Hanw 63Z 129
Brown Cl. SM6: W'gton 80Nb 156
Browne Cl. CM14: B'wood 18Xd 40
Browne Cl. GU22: Wok 92D 188
Browne Cl. RM5: Col R 22Dd 56
Browne Ho. SE8 52Cc 114
(off Deptford Chu. St.)
Brownell Pl. W7 47Ha 86
Brownfield Area E14 44Dc 92
Brownfield St. E14 44Dc 92
Browngraves Rd. UB3: Harl 52S 106
Brown Hart Gdns. W1 4J 221 (45Jb 90)
Brownhill Rd. SE6 59Dc 114
Browning Apts. E3 42Bc 92
(off Hamlets Way)
Browning Av. KT4: Wor Pk 74Xa 154
Browning Av. SM1: Sutt 77Gb 155
Browning Av. W7 44Ha 86
Browning Cl. DA16: Well 53Uc 116
Browning Cl. E17 28Ec 52
Browning Cl. RM5: Col R 24Bd 55
Browning Cl. TW12: Hamp 63Ba 129
Browning Cl. W9 6A 214 (42Eb 89)
Browning Ct. W14 51Bb 111
(off Turneville Rd.)
Browning Ho. N16 35Ub 71
(off Shakspeare Wlk.)
Browning Ho. SE14 53Ac 114
(off Loring Rd.)
Browning Ho. W12 44Ya 88
(off Wood La.)
Browning M. W1 1K 221 (43Kb 90)
Browning Pl. CR5: Coul 89Lb 176
Browning Rd. DA1: Dart 56Pd 119
Browning Rd. E11 31Hc 73
Browning Rd. E12 36Pc 74
Browning Rd. EN2: Enf 9Tb 19
Browning Rd. KT22: Fet 97Fa 192
Brownings, The AL2: Lon C 9F 6
Browning St. SE17 7E 230 (50Sb 91)
Browning Wlk. RM18: Tilb 4E 122
(off Coleridge Rd.)
Browning Way TW5: Hest 53Z 107
Brownlea Gdns. IG3: Ilf 33Wc 75
Brownlow Cl. EN4: E Barn 15Fb 31
Brownlow Ct. N11 23Nb 50
(off Brownlow Rd.)
Brownlow Ct. N2 29Eb 49
Brownlow Farm Barns HP1: Hem H 1F 2
Brownlow Ho. SE16 47Wb 91
(off George Row)
Brownlow M. WC1 6J 217 (42Pb 90)
Brownlow Rd. CR0: C'don 77Ub 157
Brownlow Rd. E7 35Jc 73
Brownlow Rd. E8 39Vb 71
Brownlow Rd. N11 23Nb 50
Brownlow Rd. N3 24Db 49
Brownlow Rd. NW10 38Ua 68
Brownlow Rd. RH1: Redh 6N 207
Brownlow Rd. W13 46Ja 86
Brownlow Rd. WD6: Bore 14Qa 29
Brownlow St. WC1 1J 223 (43Pb 90)
Brownrigg Rd. TW15: Ashf 63Q 128
Brown Rd. DA12: Grav'nd 10G 122
Brown's Bldgs. EC3 3J 225 (44Ub 91)
Browns Ct. SL1: Slou 5C 80
Brownsea Wlk. NW7 23Za 48
Browns La. KT24: Eff 99Z 191
Browns La. NW5 36Kb 70
Brownspring Dr. SE9 63Rc 138
Brown's Rd. KT5: Surb 73Pa 153
Browns Rd. E17 27Cc 52
Brown St. W1 2F 221 (44Hb 89)
Brownswell Rd. N2 26Fb 49
BROWNSWOOD PARK 33Rb 71
Brownswood Rd. N4 34Rb 71
BROX 80E 148
Broxash Rd. SW11 58Jb 112
Broxbourne Av. E18 28Kc 53
Broxbourne Ho. E3 42Dc 92
(off Empson St.)
Broxbourne Rd. BR6: Orp 74Vc 161
Broxbourne Rd. E7 34Jc 73
Broxburn Ct. RM15: S Ock 45Xd 98
Broxburn Dr. RM15: S Ock 45Wd 98
Broxburn Pde. RM15: S Ock 45Xd 98
Broxhill Cen. RM4: Have B 21Kd 57
Broxhill Rd. RM4: Have B 20Gd 38
Broxholme Cl. SE25 70Tb 135
Broxholme Ho. SW6 53Db 111
(off Harwood Rd.)
Broxholm Rd. SE27 62Qb 134
Brox La. KT15: Add 81G 168
Brox La. KT16: Ott 80E 148
Brox M. KT16: Ott 79E 148
Brox Rd. KT16: Ott 79E 148
Broxted M. CM13: Hut 16Ee 41
Broxted Rd. SE6 61Bc 136
Broxwood Cl. DA9: Ghithe 58Xd 120
Broxwood Way NW8 1E 214 (39Gb 69)
Bruce Av. RM12: Horn 33Ld 77
Bruce Av. TW17: Shep 72S 150
Bruce Castle Ct. N17 25Vb 51
(off Lordship La.)
Bruce Castle Mus. 25Vb 51
Bruce Castle Rd. N17 25Vb 51
Bruce Cl. DA16: Well 53Xc 117
Bruce Cl. KT14: Byfl 85M 169
Bruce Cl. KT15: Add 76L 149
Bruce Cl. SL1: Slou 6E 80
Bruce Cl. W10 43Za 88
Bruce Ct. DA15: Sidc 63Vc 139
Bruce Dr. CR2: Sels 81Zb 178
Bruce Gdns. N20 20Hb 31
Bruce Gro. BR6: Orp 74Wc 161
Bruce Gro. N17 25Ub 51
Bruce Gro. WD24: Wat 10Y 13
Bruce Grove Station (Overground) 26Vb 51

Bruce Hall M. SW17 63Jb 134
Bruce Ho. W10 43Za 88
Bruce Rd. CR4: Mitc 66Jb 134
Bruce Rd. E3 41Dc 92
Bruce Rd. EN5: Barn 13Ab 30
Bruce Rd. HA3: W'stone 26Ga 46
Bruce Rd. NW10 38Ta 67
Bruce Rd. SE25 70Tb 135
Bruces Wharf Rd. RM17: Grays 51Ce 121
Bruce Wlk. SL4: Wind 4B 102
Bruce Way EN8: Walt C 5Zb 20
Bruckner St. W10 41Ab 88
Brudenell SL4: Wind 5D 102
Brudenell Rd. SW17 62Hb 133
Bruffs Mdw. UB5: N'olt 37Aa 65
Bruford Ct. SE8 51Cc 114
Bruges Pl. NW1 38Lb 70
(off Randolph St.)
Brumana Cl. KT13: Weyb 79R 150
Brumfield Rd. KT19: Ewe 78Sa 153
Brummel Cl. DA7: Bex 55Ed 118
Brumwell Av. SE18 50Qc 94
Brune Ho. E1 1K 225 (43Vb 91)
(off Bell La.)
Brunei Gallery 7E 216 (43Mb 90)
Brunel Bldg. W2 1B 220 (43Fb 89)
(off Nth. Wharf Rd.)
Brunel Cl. RM1: Rom 28Gd 56
Brunel Cl. RM18: Tilb 5D 122
Brunel Cl. SE19 65Vb 135
Brunel Cl. TW5: Cran 52X 107
Brunel Cl. UB5: N'olt 41Ba 85
Brunel Ct. AL1: St A 1C 6
(off Newsom Pl.)
Brunel Ct. HA8: Edg 21Pa 47
Brunel Ct. HP3: Hem H 4M 3
Brunel Ct. SE16 47Yb 92
(off Canon Beck Rd.)
Brunel Ct. SW13 54Va 110
(off Westfields Av.)
Brunel Est. W2 43Cb 89
Brunel Ho. BR2: Broml 72Nc 160
(off Wells Vw. Dr.)
Brunel Ho. CM14: B'wood 20Yd 40
Brunel Ho. DA2: Dart 58Sd 120
(off Stone Ho. La.)
Brunel Ho. E14 50Dc 92
(off Ship Yd.)
Brunel Ho. RM8: Dag 35Wc 75
Brunel Ho. SW10 52Fb 111
(off Cheyne Rd.)
Brunel M. W10 41Za 88
Brunel Mus. 47Yb 92
Brunel Pl. UB1: S'hall 44Da 85
Brunel Rd. E17 30Ac 52
Brunel Rd. IG8: Wfd G 22Pc 54
Brunel Rd. SE16 47Yb 92
Brunel Rd. W3 43Ua 88
Brunel Science Pk. UB8: Cowl 41N 83
Brunel St. E16 44Hc 93
Brunel University Indoor Athletics
Cen. 41N 83
Brunel University Sports Pk. 42P 83
Brunel University Uxbridge Campus 41M 83
Brunel Wlk. N15 29Ub 51
Brunel Wlk. TW2: Whitt 59Ca 107
Brunel Way DA1: Dart 55Pd 119
Brunel Way SL1: Slou 6K 81
Brune St. E1 1K 225 (43Vb 91)
Brunlees Ho. SE1 4D 230 (48Sb 91)
(off Bath Ter.)
Brunner Cl. NW11 29Db 49
Brunner Ct. KT16: Ott 78E 148
Brunner Ho. SE6 63Ec 136
Brunner Rd. E17 29Ac 52
Brunner Rd. W5 42Ma 87
Bruno Pl. NW9 33Sa 67
Brunswick Av. N11 20Jb 32
Brunswick Av. RM14: Upm 31Ud 78
Brunswick Cen. 5F 217 (42Nb 90)
Brunswick Cl. DA6: Bex 56Zc 117
Brunswick Cl. HA5: Pinn 30Aa 45
Brunswick Cl. KT12: Walt T 75Y 151
Brunswick Cl. KT7: T Ditt 74Ha 152
Brunswick Cl. TW2: Twick 62Fa 130
Brunswick Cl. Est. EC1 4B 218 (41Rb 91)
Brunswick Ct. CM14: W'ley 22Xd 58
Brunswick Ct. EC1 4B 218 (41Rb 91)
(off Tompion St.)
Brunswick Ct. EN4: E Barn 15Fb 31
Brunswick Ct. RM14: Upm 31Vd 78
Brunswick Ct. SE1 2J 231 (47Ub 91)
Brunswick Ct. SM1: Sutt 77Db 155
Brunswick Ct. SW1 6E 228 (49Mb 90)
(off Regency St.)
Brunswick Cres. N11 20Jb 32
Brunswick Dr. GU24: Brkwd 2B 186
Brunswick Flats W11 44Cb 89
(off Westbourne Gro.)
Brunswick Gdns. IG6: Ilf 24Sc 54
Brunswick Gdns. W5 42Na 87
Brunswick Gdns. W8 46Cb 89
Brunswick Gro. KT11: Cobh 85Y 171
Brunswick Gro. N11 20Jb 32
Brunswick Ho. E2 1K 219 (40Vb 71)
(off Thurtle Rd.)
Brunswick Ho. N3 25Bb 49
Brunswick Ho. SE16 48Ac 92
(off Brunswick Quay)
Brunswick Ho. W6 50Ya 88
(off Parrs Way)
Brunswick Ind. Pk. N11 21Kb 50
Brunswick Mans. WC1 5G 217 (42Nb 90)
(off Handel St.)
Brunswick M. SW16 65Mb 134
Brunswick M. W1 2G 221 (44Hb 89)
BRUNSWICK PARK 20Hb 31
Brunswick Pk. SE5 53Ub 113
Brunswick Pk. Gdns. N11 19Jb 32
Brunswick Pk. Rd. N11 19Jb 32
Brunswick Pl. N1 4G 219 (41Tb 91)
Brunswick Pl. NW1 6J 215 (42Jb 90)
(not continuous)
Brunswick Pl. SE19 66Wb 135
Brunswick Quay SE16 48Zb 92
Brunswick Rd. DA6: Bex 56Zc 117
Brunswick Rd. E10 32Ec 72
Brunswick Rd. E14 44Ec 92
Brunswick Rd. EN3: Enf L 10Cc 20
Brunswick Rd. KT2: King T 67Qa 131
Brunswick Rd. N15 29Ub 51
Brunswick Rd. SM1: Sutt 77Db 155
Brunswick Rd. W5 42Ma 87
Brunswick Sq. N17 23Ub 51
Brunswick Sq. WC1 5G 217 (42Nb 90)
(not continuous)
Brunswick St. E17 29Ec 52
Brunswick Ter. BR3: Beck 67Dc 136
Brunswick Vs. SE5 53Ub 113
Brunswick Wlk. DA12: Grav'nd 9F 122
Brunswick Way N11 21Kb 50
Brunton Pl. E14 44Ac 92
Brushfield St. E1 7J 219 (43Ub 91)
Brushfield Way GU21: Knap 1G 186
Brushrise WD24: Wat 8X 13
Brushwood Cl. E14 43Dc 92
Brushwood Dr. WD3: Chor 14E 24
Brussels Rd. SW11 56Fb 111
Bruton Cl. BR7: Chst 66Pc 138
Bruton La. W1 5A 222 (45Kb 90)
Bruton Pl. W1 5A 222 (45Kb 90)
Bruton Rd. SM4: Mord 70Eb 133
Bruton St. W1 5A 222 (45Kb 90)
Bruton Way W13 43Ja 86
Brutus Ct. SE11 6B 230 (49Rb 91)
(off Kennington La.)
Bryan Av. NW10 38Xa 68
Bryan Cl. TW16: Sun 66W 128
Bryan Ho. NW10 38Xa 68
Bryan Ho. SE16 47Bc 92
Bryan Rd. SE16 47Bc 92
Bryan's All. SW6 54Db 111
Bryanston Av. TW2: Whitt 60Da 107
Bryanston Cl. UB2: S'hall 49Ba 85
Bryanston Ct. HP2: Hem H 3M 3
Bryanston Ct. W1 2F 221 (44Hb 89)
(off Seymour Pl.)
Bryanstone Ct. SM1: Sutt 76Eb 155
Bryanstone Rd. EN8: Walt C 6Bc 20
Bryanstone Rd. N8 29Mb 50
Bryanston Mans. W1 7F 215 (43Hb 89)
(off York St.)
Bryanston M. E. W1 1F 221 (43Hb 89)
Bryanston M. W. W1 1F 221 (43Hb 89)
Bryanston Pl. W1 1F 221 (43Hb 89)
Bryanston Rd. RM18: Tilb 4E 122
Bryanston Sq. W1 2F 221 (44Hb 89)
Bryanston St. W1 3G 221 (44Hb 89)
Bryan St. N1 1J 217 (39Pb 70)
Bryant Av. RM10: Dag 38Ed 76
Bryant Av. RM3: Hrld W 25Md 57
Bryant Av. SL2: Slou 3H 81
Bryant Cl. EN5: Barn 15Bb 31
Bryant Ct. E2 1K 219 (39Vb 71)
(off Whiston Rd.)
Bryant Ct. W3 46Ta 87
Bryant Ho. E3 40Cc 72
(off Thomas Fyre Dr.)
Bryant Rd. UB5: N'olt 41Y 85
Bryant Row RM3: Rom 19Ld 39
Bryant St. E15 38Fc 73
Bryant St. E2 1K 219 (39Vb 71)
Bryantwood Rd. N7 36Qb 70
Brycedale Cres. N14 21Mb 50
Bryce Ho. SE14 51Zb 114
(off John Williams Cl.)
Bryce Rd. RM8: Dag 35Yc 75
Brydale Ho. SE16 49Zb 92
(off Rotherhithe New Rd.)
Bryden Cl. SE26 64Ac 136
Brydges Pl. WC2 5F 223 (45Nb 90)
Brydges Rd. E15 36Fc 73
Brydon Wlk. N1 39Nb 70
Bryer Ct. EC2 7D 218 (43Sb 91)
(off Bridgewater Sq.)
Bryer Pl. SL4: Wind 5B 102
Bryett Rd. N7 34Nb 70
Bryher Ct. SE11 7K 229 (50Qb 90)
(off Sancroft St.)
Brymay Cl. E3 40Cc 72
Brymcourt W9 40Db 69
Brynford Cl. GU21: Wok 87A 168
Brynmaer Rd. SW11 53Hb 111
Bryn-y-mawr Rd. EN1: Enf 14Vb 33
Bryony Cl. IG10: Lough 14Rc 36
Bryony Cl. UB8: Hil 43P 83
Bryony Rd. W12 45Wa 88
Bryony Way TW16: Sun 65W 128
Bubblestone Rd. TN14: Otf 88Kd 183
Bubbling Well Sq. SW18 57Db 111
(off Ryland Blvd.)
Buccleuch Rd. SL3: Dat 2L 103
Buccleugh Ho. E5 31Wb 71
Buchanan Cl. N21 15Pb 32
Buchanan Cl. RM15: Avel 46Sd 98
Buchanan Cl. SE16 49Zb 92
(off Worgan St.)
Buchanan Ct. WD6: Bore 12Sa 29
Buchanan Gdns. NW10 40Xa 68
Buchanan Pl. KT17: Ewe 82Wa 174
Buchan Cl. UB8: Cowl 41L 83
Buchan Ho. W3 47Ra 87
(off Hanbury Rd.)
Buchan Rd. SE15 55Yb 114
Bucharest Rd. SW18 59Eb 111
Buckbean Path RM3: Rom 24Ld 57
Buckden Cl. N2 28Hb 49
Buckden Cl. SE12 58Jc 115
Buckettsland La. WD6: Bore 10Ta 15
Buckfast Ct. W13 45Ja 86
(off Romsey Rd.)
Buckfast Ho. N14 15Lb 32
Buckfast Rd. SM4: Mord 70Db 133
Buckfast St. E2 41Wb 91
Buckfield Ct. SL0: Rich P 47H 83
Buckham Thorns Rd. TN16: Westrm 98Sc 200
Buck Hill Wlk. W2 5C 220 (45Fb 89)
Buckhold Rd. SW18 58Cb 111
Buckhurst Av. SM5: Cars 74Gb 155
Buckhurst Av. TN13: S'oaks 97Ld 203
Buckhurst Cl. RH1: Redh 4N 207
Buckhurst Ct. IG9: Buck H 18Mc 35
BUCKHURST HILL 19Mc 35
BUCKHURST HILL 7D 124
Buckhurst Hill Ho. IG9: Buck H 19Kc 35
Buckhurst Hill Station
(Underground) 19Mc 35
Buckhurst Ho. N7 36Mb 70
Buckhurst La. SL5: S'hill 9D 124
Buckhurst La. TN13: S'oaks 97Ld 203
Buckhurst Rd. SL5: Asc 7D 124
Buckhurst Rd. SL5: S'hill 7D 124
Buckhurst Rd. TN16: Westrm 93Qc 200
Buckhurst St. E1 42Xb 91
Buckhurst St. E2 42Xb 91
Buckhurst Way IG9: Buck H 21Mc 53
Buckingham Arc. WC2 5G 223 (45Nb 90)
(off Strand)
Buckingham Av. CR7: Thor H 67Qb 134
Buckingham Av. DA16: Well 56Uc 116
Buckingham Av. KT8: W Mole 68Da 129
Buckingham Av. N20 17Eb 31
Buckingham Av. SL1: Slou 4C 80
Buckingham Av. TW14: Felt 58X 107
Buckingham Av. UB6: G'frd 39Ja 66
Buckingham Av. E. SL1: Slou 4G 80
Buckingham Chambers SW1... 5C 228 (49Lb 90)
(off Greencoat Pl.)
Buckingham Cl. BR5: Pet W 73Uc 160
Buckingham Cl. EN1: Enf 12Ub 33
Buckingham Cl. RM11: Horn 30Md 57
Buckingham Cl. TW12: Hamp 64Ba 129
Buckingham Cl. W5 43La 86
Buckingham Ct. AL1: St A 1D 6
(off Lemsford Rd.)
Buckingham Ct. NW4 27Wa 48
Buckingham Ct. SM2: Sutt 81Cb 175
Buckingham Ct. TW18: Staines 63J 127
(off Kingston Rd.)
Buckingham Ct. UB5: N'olt 40Aa 65
Buckingham Ct. W11 45Cb 89
(off Kensington Pk. Rd.)
Buckingham Ct. W7 42Ha 86
(off Copley Cl.)
Buckingham Dr. BR7: Chst 63Sc 138
Buckingham Gdns. CR7: Thor H 68Qb 134
Buckingham Gdns. HA8: Edg 24Na 47
Buckingham Gdns. KT8: W Mole 68Da 129
Buckingham Gdns. SL1: Slou 7K 81
Buckingham Ga. SW1 3B 228 (48Lb 90)
Buckingham Gro. UB10: Hil 40Q 64
Buckingham Gro. WD6: Bore 14Ta 29
Buckingham Hill Rd. SS17: Ors 5H 101
Buckingham Hill Rd. SS17: Stan H 5H 101
Buckingham La. SE23 59Ac 114
Buckingham Mans. NW6 36Db 69
(off West End La.)
Buckingham M. N1 37Ub 71
Buckingham M. NW10 40Va 68
Buckingham M. SW1 3B 228 (48Lb 90)
(off Stafford Pl.)
Buckingham Palace 2A 228 (47Kb 90)
Buckingham Pal. Rd. SW1 6K 227 (49Kb 90)
Buckingham Pde. HA7: Stan 22La 46
Buckingham Pde. SL9: Chal P 25A 42
(off Market Pl.)
Buckingham Pl. SW1 3B 228 (48Lb 90)
Buckingham Rd. CM16: Epp 2Uc 22
Buckingham Rd. CR4: Mitc 70Nb 134
Buckingham Rd. DA11: Nflt 59Fe 121
Buckingham Rd. E10 34Dc 72
Buckingham Rd. E11 29Lc 53
Buckingham Rd. E15 36Hc 73
Buckingham Rd. E18 25Hc 53
Buckingham Rd. HA1: Harr 29Fa 46
Buckingham Rd. HA8: Edg 24Pa 47
Buckingham Rd. IG1: Ilf 33Tc 74
Buckingham Rd. KT1: King T 70Pa 131
Buckingham Rd. N1 37Ub 71
Buckingham Rd. N22 25Nb 50
Buckingham Rd. NW10 40Va 68
Buckingham Rd. TW10: Ham 61Ma 131
Buckingham Rd. TW12: Hamp 63Ba 129
Buckingham Rd. WD24: Wat 9Y 13
Buckingham Rd. WD6: Bore 14Ta 29
Buckinghamshire Golf Course 33L 63
Buckinghamshire New University
Uxbridge Campus 37L 63
Buckingham St. WC2 5G 223 (45Nb 90)
Buckingham Way SM6: W'gton 81Lb 176
BUCKLAND 5C 206
Buckland Av. SL3: Slou 9M 81
Buckland Cl. NW7 21Wa 48
Buckland Ct. N1 1H 219 (40Ub 71)
(off St John's Est.)
Buckland Ct. RH1: Redh 4B 208
Buckland Ct. UB10: Ick 33S 64
Buckland Ct. Gdns. RH3: Bkld 5C 206
Buckland Cres. NW3 38Fb 69
Buckland Cres. SL4: Wind 3D 102
Buckland Ga. SL3: Wex 1M 81
Buckland Ho. SW1 7K 227 (50Kb 90)
(part of Abbots Mnr.)
Buckland La. KT20: Walt H 100Wa 194
Buckland La. RH3: Bkld 2C 206
Buckland Park Lake RH3: Reig 5D 206
Buckland Ri. HA5: Pinn 25Y 45
Buckland Rd. BR6: Orp 77Uc 160
Buckland Rd. E10 33Ec 72
Buckland Rd. KT20: Lwr K 1H 207
Buckland Rd. KT9: Chess 78Pa 153
Buckland Rd. RH2: Reig 5F 206
Buckland Rd. SM2: Cheam 82Ya 174
Bucklands WD19: Wat 20Z 27
Bucklands, The WD3: Rick 17J 25
Bucklands Rd. TW11: Tedd 65La 130
Buckland St. N1 2G 219 (40Tb 71)
Buckland's Wharf KT1: King T 68Ma 131
Buckland Wlk. SM4: Mord 70Eb 133
Buckland Wlk. W3 47Sa 87
Buckland Way KT4: Wor Pk 74Ya 154
Buckland Way RM13: Rain 40Md 77
Buck La. NW9 29Ta 47
Bucklebury NW1 5B 216 (42Lb 90)
(off Stanhope St.)
Buckleigh Av. SW20 69Ab 132
Buckleigh Rd. SW16 65Mb 134
Buckleigh Way SE19 66Vb 135
Buckler Ct. N7 36Pb 70
Buckler Gdns. SE9 62Pc 138
Bucklers All. SW6 51Bb 111
Bucklersbury EC4 3F 225 (44Tb 91)
Bucklersbury Pas. EC4 3F 225 (44Tb 91)
Bucklers Ct. CM14: W'ley 22Yd 58
Buckler's Way SM5: Cars 76Hb 155
Buckles Ct. DA17: Belv 49Zc 95
Buckles La. RM15: S Ock 43Yd 98
Buckle St. E1 44Vb 91
Buckles Way SM7: Bans 88Ab 174
Buckley Cl. DA1: Cray 54Hd 118
Buckley Cl. SE23 59Xb 113
Buckley Ct. NW6 38Bb 69
Buckley Ct. SE1 4K 231 (48Vb 91)
Buckley Ho. W14 47Ab 88
(off Holland Pk. Av.)
Buckley Rd. NW6 38Bb 69
Buckmaster Cl. SW9 55Qb 112
(off Stockwell Pk. Rd.)
Buckmaster Ho. N7 35Pb 70
Buckmaster Rd. SW11 56Gb 111
Bucknall Pl. WD19: Wat 20Z 27
Bucknalls Cl. WD25: Wat 4Aa 13
Bucknalls Dr. AL2: Brick W 3Ba 13
Bucknalls La. WD25: Wat 4Z 13
Bucknall St. WC2 2F 223 (44Mb 90)
Bucknall Way BR3: Beck 70Dc 136
Bucknell Cl. SW2 56Pb 112
Buckner Rd. SW2 56Pb 112
Bucknill Ho. SW1 7K 227 (50Kb 90)
(off Ebury Bri. Rd.)
Bucknills Cl. KT18: Eps 86Sa 173
Buckrell Rd. E4 19Fc 35
Buckridge Ho. EC1 7K 217 (43Qb 90)
(off Portpool La.)
Buck's Av. WD19: Wat 17Aa 27
Bucks Cl. KT14: W Byf 86K 169
Bucks Cross Rd. BR6: Chels 78Ad 161
Bucks Cross Rd. DA11: Nflt 2B 144
Buckshead Ho. W2 43Cb 89
(off Gt. Western Rd.)
Bucks Hill WD4: Bucks 5L 11
BUCKS HILL 7M 11
Buckston Browne Gdns. BR6: Downe 84Pc 180
Buckstone Cl. SE23 58Yb 114
Buckstone Rd. N18 22Wb 51
Buck St. NW1 38Kb 70
Buckters Rents SE16 46Ac 92
Buckthorne Rd. SE4 57Ac 114
Buckthorn Ho. DA15: Sidc 62Vc 139
(off Longlands Rd.)
Buckthorn Ho. E15 41Gc 93
(off Manor Rd.)
Buckton Rd. WD6: Bore 10Pa 15
Buck Wlk. E17 28Fc 53
Buckwell Pl. TN13: S'oaks 100Ld 203
Buckwheat Ct. DA18: Erith 48Zc 95
Budd Cl. N12 21Db 49
Buddings Circ. HA9: Wemb 34Sa 67
Buddleia Ho. TW13: Felt 60W 106
Budd's All. TW1: Twick 57La 108
Budebury Rd. TW18: Staines 64J 127
Bude Cl. E17 29Bc 52
Budge La. CR4: Mitc 73Hb 155
Budgen Dr. RH1: Redh 3A 208
Budgin's Hill BR6: Prat B 84Yc 181
Budleigh Cres. DA16: Well 53Yc 117
Budleigh Ho. SE15 52Wb 113
(off Bird in Bush Rd.)
Budoch Ct. IG3: Ilf 33Wc 75
Budoch Dr. IG3: Ilf 33Wc 75
Buer Rd. SW6 54Ab 110
Buff Av. SM7: Bans 86Db 175
Buffers La. KT22: Lea 91Ja 192
Bug Hill CR3: Wold 92Zb 198
Bug Hill CR6: W'ham 92Zb 198
Bug Hill CR6: Wold 92Zb 198
Bugsby's Way SE10 49Hc 93
Bugsby's Way SE7 49Jc 93
Buick Ho. E3 42Cc 92
(off Wellington Way)
Buick Ho. KT2: King T 68Pa 131
Building 50 SE18 48Sc 94
Bulbarrow NW8 39Db 69
(off Abbey Rd.)
Bulbourne Cl. HP1: Hem H 3J 3
Bulbourne Ho. HP1: Hem H 4L 3
(off Cotterells)
Bulganak Rd. CR7: Thor H 70Sb 135
Bulinga St. SW1 6F 229 (49Nb 90)
(off John Islip St.)
Bulkeley Av. SL4: Wind 5F 102
Bulkeley Cl. TW20: Eng G 4N 125
Bullace Cl. HP1: Hem H 1J 3
Bullace La. DA1: Dart 58Nd 119
Bullace Row SE5 53Tb 113
Bull All. DA16: Well 55Xc 117
Bullard's Pl. E2 41Zb 92
Bullbanks Rd. DA17: Belv 49Ed 96
Bullbeggars La. GU21: Wok 8M 167
Bullbeggars La. HP4: Berk 2B 2
Bullbeggars La. HP4: Pott E 2B 2
Bullbeggars La. RH9: G'stone 4A 210
Bull Cl. RM16: Chaf H 47Be 99
Bulleid Way SW1 6A 228 (49Kb 90)
Bullen Ho. E1 42Xb 91
(off Collingwood St.)
BULLEN'S GREEN 4B 8
Bullens Grn. La. AL4: Col H 4A 8
Bullen St. SW11 54Gb 111
Buller Cl. SE15 52Wb 113
Buller Rd. CR7: Thor H 68Tb 135
Buller Rd. IG11: Bark 38Uc 74
Buller Rd. N17 26Wb 51
Buller Rd. N22 26Qb 50
Buller Rd. NW10 41Za 88
Bullers Cl. DA14: Sidc 64Ad 139
Bullers Wood Dr. BR7: Chst 66Pc 138
Bullescroft Rd. HA8: Edg 20Qa 29
Bullfinch Cl. TN13: Riv 94Fd 202
Bullfinch Dene TN13: Riv 94Fd 202
Bullfinch Ho. NW9 30Va 48
(off Perryfield Way)
Bullfinch La. TN13: Riv 94Fd 202
Bullfinch Rd. CR2: Sels 82Zb 178
Bullhead Rd. WD6: Bore 15Sa 29
Bull Hill DA4: Hort K 70Sd 142
Bull Hill KT22: Lea 93Ja 192
Bullingham Mans. W8 47Cb 89
(off Pitt St.)
Bull Inn Ct. WC2 5G 223 (45Nb 90)
(off Strand)
Bullivant Cl. DA9: Ghithe 57Wd 120
Bullivant St. E14 45Ec 92
Bull La. BR7: Chst 66Tc 138
Bull La. N18 22Ub 51
Bull La. RM10: Dag 34Dd 76
Bull La. TN15: Wro 88Ce 185
Bullman Cl. DA7: Bex 55Dd 118
Bullock Cres. GU22: Wok 94B 188
Bull Rd. E15 40Hc 73
Bullrush Cl. AL10: Hat 1D 8
Bullrush Cl. CR0: C'don 72Ub 157
Bullrush Cl. SM5: Cars 75Gb 155
Bullrush Gro. UB8: Cowl 42L 83
Bull's All. SW14 54Ta 109
Bulls Bri. Cen. UB3: Hayes 48W 84
Bullsbridge Ind. Est. UB2: S'hall 49X 85
Bulls Bri. Rd. UB2: S'hall 49Y 85
Bulls Bri. Rd. UB3: Hayes 48X 85
Bullsbrook Rd. UB4: Yead 46Y 85
Bulls Cross EN2: Enf 7Wb 19
BULLS CROSS 7Wb 19
Bulls Cross Ride EN7: Walt C 5Wb 19
Bulls Gdns. SW3 5E 226 (49Gb 89)
Bulls Head Pas. EC3 3H 225 (44Ub 91)
(off Lime St. Pas.)
Bulls Head Row RH9: G'stone 3P 209
Bulls Head Yd. DA1: Dart 58Nd 119
(off High St.)
Bullsland Gdns. WD3: Chor 16D 24
Bullsland La. WD3: Chor 18D 24
Bulls La. AL9: Brk P 6F 8
Bulls La. AL9: Wel G 6F 8
BULLSMOOR 7Yb 20
Bullsmoor Cl. EN8: Walt C 7Yb 20
Bullsmoor Gdns. EN8: Walt C 7Xb 19
Bullsmoor La. EN1: Enf 7Wb 19
Bullsmoor La. EN3: Enf W 7Yb 20
Bullsmoor La. EN7: Walt C 7Wb 19
Bullsmoor Ride EN8: Walt C 7Yb 20
Bullsmoor Way EN8: Walt C 7Yb 20
BULLSWATER COMMON 8E 186
Bullswater Comn. Rd. GU24: Pirb 8E 186
Bullswater La. GU24: Pirb 8E 186
Bull Theatre, The 14Bb 31
Bullwell Cres. EN8: Chesh 1Ac 20
Bull Yd. DA12: Grav'nd 8D 122
(off Horn Yd.)
Bull Yd. SE15 53Wb 113
Bulmer Gdns. HA3: Kenton 31Ma 67
Bulmer M. W11 45Cb 89
Bulmer Pl. W11 46Cb 89
Bulmer Wlk. RM13: Rain 40Ld 77
BULSTRODE 1G 10
Bulstrode Av. TW3: Houn 54Ba 107
Bulstrode Cl. WD4: Chfd 1G 10
Bulstrode Ct. SL9: Ger X 1P 61
Bulstrode Gdns. TW3: Houn 55Ca 107
Bulstrode La. HP3: Hem H 7J 3
Bulstrode La. WD4: Chfd 1G 10
Bulstrode La. WD4: K Lan 1G 10
Bulstrode Pl. SL1: Slou 8K 81
Bulstrode Pl. W1 1J 221 (43Jb 90)
Bulstrode Rd. TW3: Houn 55Ca 107
Bulstrode St. W1 2J 221 (44Jb 90)
Bulwark Cl. E14 49Ec 92
(off Parkside Sq.)
Bulwer Ct. E11 32Fc 73
Bulwer Ct. Rd. E11 32Fc 73
Bulwer Gdns. EN5: New Bar 14Eb 31
Bulwer Rd. E11 31Fc 73
Bulwer Rd. EN5: New Bar 14Db 31
Bulwer Rd. N18 21Ub 51
Bulwer St. W12 46Ya 88
Bumpstead Mead RM15: Avel 46Ud 98
Bunbury Ho. SE15 52Wb 113
(off Fenham Rd.)
Bunbury Way KT17: Eps D 88Xa 174
Bunby Rd. SL2: Stoke P 8K 61
Bunce Dr. CR3: Cat'm 95Tb 197
Buncefield Terminal HP2: Hem H 1D 4
Bunce's Cl. SL4: Eton W 10F 80
Bunce's La. IG8: Wfd G 24Hc 53
Bundy's Way TW18: Staines 65H 127
Bungalow Rd. SE25 70Ub 135
Bungalows, The E10 30Ec 52
Bungalows, The HA2: Harr 35Ba 65
Bungalows, The IG6: Ilf 25Uc 54
Bungalows, The RM20: Grays 51Zd 121
Bungalows, The SM6: W'gton 78Kb 156
Bungalows, The SW16 66Kb 134
Bungalows, The UB4: Yead 42Z 85
Bunhill Row EC1 5F 219 (42Tb 91)
Bunhouse Pl. SW1 7H 227 (50Jb 90)
Bunkers Hill DA14: Sidc 62Bd 139
Bunkers Hill DA17: Belv 49Cd 96
Bunkers Hill NW11 31Eb 69
Bunkers Hill TN15: Ash 79De 165
Bunkers La. HP3: Hem H 7A 4
Bunning Way N7 38Nb 70
Bunns La. NW7 23Ua 48
(not continuous)
Bunny Hill DA12: Shorne 5N 145
Bunsen Ho. E3 40Ac 72
(off Grove Rd.)
Bunsen St. E3 40Ac 72
Bunstone Hall DA2: Dart 58Sd 120
Bunten Meade SL1: Slou 6F 80
Buntingbridge Rd. IG2: Ilf 29Tc 54
Bunting Cl. CR4: Mitc 71Hb 155
Bunting Cl. HP3: Hem H 7L 3
Bunting Cl. N9 18Zb 34
Bunting Ct. NW9 26Ua 48
Bunting Ho. UB10: Ick 33S 64
(off Coyle Dr.)
Bunton St. SE18 48Qc 94
Bunwell Ho. E3 42Bc 92
(off William Whiffin Sq.)
Bunyan Ct. EC2 7D 218 (43Sb 91)
(off Fann St.)
Bunyan Rd. E17 27Ac 52
Bunyan's La. GU24: Chob 6G 166
Bunyard Dr. GU21: Wok 86E 168
Buonaparte M. SW1 7D 228 (50Mb 90)
Burbage Cl. EN8: Chesh 3Dc 20
Burbage Cl. SE1 4F 231 (48Tb 91)
Burbage Cl. UB3: Hayes 44T 84
Burbage Ho. N1 39Tb 71
(off Poole St.)
Burbage Ho. SE14 51Zb 114
(off Samuel Cl.)
Burbage Rd. SE21 59Tb 113
Burbage Rd. SE24 58Sb 113
Burberry Cl. KT3: N Mald 68Ua 132
Burbery Cl. UB9: Hare 26M 43
Burbidge Rd. TW17: Shep 70Q 128
Burbridge Gdns. UB10: Uxb 40N 63
Burbridge Rd. WD25: Wat 5V 12
Burbridge Way N17 26Wb 51
Burcham Cl. TW12: Hamp 66Ca 129
Burcham St. E14 44Dc 92
Burcharbro Rd. SE2 51Zc 117
Burchell Ct. WD23: Bush 17Ea 28
Burchell Ho. SE11 7J 229 (50Pb 90)
(off Jonathan St.)
Burchell Rd. E10 32Dc 72
Burchell Rd. SE15 53Xb 113
Burcher Gale Gro. SE15 52Vb 113
Burchetts Way TW17: Shep 72R 150
Burchett Way RM6: Chad H 30Ed 56
Burch Rd. DA11: Nflt 8B 122
Burchwall Cl. RM5: Col R 24Ed 56
Burcote KT13: Weyb 79T 150
Burcote Rd. SW18 59Fb 111
Burcott Gdns. KT15: Add 79L 149
Burcott Rd. CR8: Purl 86Qb 176
Burden Cl. TW8: Bford 50La 86
Burden Ho. SW8 52Nb 112
(off Thorncroft St.)
Burdenshott Hill GU3: Worp 7L 187
Burdenshott Av. TW10: Rich 56Ra 109
Burdenshott Rd. GU3: Worp 7L 187
Burden Way E11 33Kc 73
Burder Cl. N1 37Ub 71
Burder Rd. N1 37Ub 71
Burdett Av. DA12: Shorne 3N 145
Burdett Av. SW20 67Wa 132
Burdett Cl. DA14: Sidc 64Ad 139
Burdett Cl. W7 46Ha 86
Burdett M. NW3 37Fb 69
Burdett M. W2 44Db 89
Burdett Rd. CR0: C'don 72Tb 157
Burdett Rd. E14 43Bc 92
Burdett Rd. E3 42Ac 92
Burdett Rd. TW9: Rich 54Pa 109
Burdetts Rd. RM9: Dag 39Bd 75
Burdock Cl. CR0: C'don 74Zb 158
Burdock Cl. GU18: Light 3A 166
Burdock Rd. N17 27Wb 51
Burdon La. SM2: Cheam 80Ab 154
Burdon Pk. SM2: Cheam 81Bb 175
Bure RM18: E Til 8L 101
Bure Ct. EN5: New Bar 15Eb 31
Burfield Cl. SW17 63Fb 133
Burfield Dr. CR6: W'ham 91Yb 198

Burfield Rd. SL4: Old Win 8L 103
Burfield Rd. WD3: Chor 15E 24
Burford Cl. IG6: Ilf 28Sc 54
Burford Cl. RM8: Dag 34Yc 75
Burford Cl. UB10: Ick 35N 63
Burford Gdns. N13 20Pb 32
Burford Gdns. SL1: Slou 3A 80
Burford Ho. KT17: Ewe 83Ya 174
Burford Ho. TW8: Bford 50Ma 87
Burford La. KT17: Ewe 83Ya 174
Burford Rd. BR1: Broml 70Nc 138
Burford Rd. E15 39Fc 73
Burford Rd. E6 41Nc 94
Burford Rd. KT4: Wor Pk 73Va 154
Burford Rd. SE6 61Bc 136
Burford Rd. SM1: Sutt 75Cb 155
Burford Rd. TW8: Bford 50Na 87
Burford Wlk. SW6 52Eb 111
Burford Way CR0: New Ad 79Ec 158
Burford Wharf Apts. E15 39Fc 73
(off Cam Rd.)
Burgate Cl. DA1: Cray 55Hd 118
Burges Cl. RM11: Horn 30Pd 57
Burges Gro. SW13 52Xa 110
Burges Rd. E6 38Nc 74
Burgess Av. NW9 30Ta 47
Burgess Av. SS17: Stan H 2N 101
Burgess Bus. Pk. SE5 52Tb 113
Burgess Cl. TW13: Hanw 63Aa 129
Burgess Ct. CM15: B'wood 18Zd 41
Burgess Ct. E6 38Qc 74
Burgess Ct. SE6 59Cc 114
Burgess Ct. UB1: S'hall 44Da 85
(off Fleming Rd.)
Burgess Ct. WD6: Bore 10Pa 15
(off Aycliffe Rd.)
Burgess Hill NW2 35Cb 69
Burgess Ho. SE5 52Sb 113
(off Bethwin Rd.)
Burgess Lofts SE5 52Sb 113
(off Bethwin Rd.)
Burgess M. SW19 65Db 133
Burgess Pk. 51Tb 113
Burgess Rd. E15 35Gc 73
Burgess Rd. E6 38Qc 74
Burgess Rd. SM1: Sutt 77Db 155
Burgess St. E14 43Cc 92
Burge St. SE1 4G 231 (48Tb 91)
Burges Way TW18: Staines 64J 127
Burgett Rd. SL1: Slou 8F 80
Burgh Cft. KT17: Eps 87Va 174
Burghfield KT17: Eps 87Va 174
Burghfield Rd. DA13: Ist R 6B 144
BURGH HEATH 91Ab 194
Burgh Heath Rd. KT17: Eps 86Va 174
Burgh Heath Rd. KT17: Eps D 86Va 174
Burgh House 35Fb 69
Burghill Rd. SE26 63Ac 136
Burghley Av. KT3: N Mald 67Ta 131
Burghley Av. WD6: Bore 15Sa 29
Burghley Hall Cl. SW19 60Ab 110
Burghley Ho. SW19 62Ab 132
Burghley Pas. E11 32Gc 73
(off Burghley Rd.)
Burghley Pl. CR4: Mitc 71Hb 155
Burghley Rd. E11 32Gc 73
Burghley Rd. N8 27Qb 50
Burghley Rd. NW5 35Kb 70
Burghley Rd. RM16: Chaf H 48Yd 98
Burghley Rd. SW19 63Za 132
Burghley Twr. W3 45Va 88
Burgh Mt. SM7: Bans 87Bb 175
Burgh St. N1 1C 218 (40Rb 71)
Burgh Wood SM7: Bans 87Ab 174
Burgoine Quay KT1: Hamp W 67Ma 131
Burgon St. EC4 3C 224 (44Rb 91)
Burgos Cl. CR0: Wadd 79Qb 156
Burgos Gro. SE10 53Dc 114
Burgoyne Ho. TW8: Bford 50Ma 87
(off Ealing Rd.)
Burgoyne Rd. N4 30Rb 51
Burgoyne Rd. SE25 70Vb 135
Burgoyne Rd. SW9 55Pb 112
Burgoyne Rd. TW16: Sun 65V 128
Burgundy Ct. HA4: Ruis 35Y 65
Burgundy Ho. E20 36Ec 72
(off Liberty Bri. Rd.)
Burgundy Ho. EN2: Enf 10Sb 19
(off Bedale Rd.)
Burgundy Pl. W12 46Za 88
Burham Cl. SE20 66Yb 136
BURHILL 81X 171
Burhill Golf Course 81W 170
Burhill Gro. HA5: Pinn 26Aa 45
Burhill Rd. KT12: Hers 81X 171
Burke Cl. SW15 56Ua 110
Burke Lodge E13 41Kc 93
Burke St. E16 43Hc 93
(not continuous)
Burket Cl. UB2: S'hall 49Aa 85
Burland Rd. CM15: B'wood 18Zd 41
Burland Rd. RM5: Col R 23Ed 56
Burland Rd. SW11 57Hb 111
Burlea Cl. KT12: Hers 78X 151
Burleigh Av. DA15: Sidc 57Vc 117
Burleigh Av. SM6: W'gton 76Jb 156
Burleigh Cl. KT15: Add 78K 149
Burleigh Cl. RM7: Mawney 28Dd 56
Burleigh Ct. KT22: Lea 94Ja 192
Burleigh Gdns. GU21: Wok 89B 168
Burleigh Gdns. N14 18Lb 32
Burleigh Gdns. TW15: Ashf 64S 128
Burleigh Ho. SW3 51Fb 111
(off Beaufort St.)
Burleigh Ho. W10 43Ab 88
(off St Charles Sq.)
Burleigh Ho. WC1 4H 217 (41Pb 90)
(off Westking Pl.)
Burleigh Pde. N14 18Mb 32
Burleigh Pk. KT11: Cobh 84Aa 171
Burleigh Pl. SW15 57Za 110
Burleigh Rd. AL1: St A 2F 6
Burleigh Rd. EN1: Enf 14Ub 33
Burleigh Rd. EN8: Chesh 4Ac 20
Burleigh Rd. HP2: Hem H 3C 4
Burleigh Rd. KT15: Add 78K 149
Burleigh Rd. SM3: Sutt 74Ab 154
Burleigh Rd. UB10: Hil 39R 64
Burleigh St. WC2 4H 223 (45Pb 90)
Burleigh Wlk. SE6 60Ec 114
Burleigh Way EN2: Enf 13Tb 33
Burleigh Way EN6: Cuff 2Nb 18
Burlescombe Ho. RH1: Redh 4A 208
(off Burrage Rd.)
Burley Cl. E4 22Cc 52
Burley Cl. SW16 68Mb 134
Burley Ho. E1 44Zb 92
(off Chudleigh St.)
Burley Ho. WD5: Ab L 4V 12
Burley Orchard KT16: Chert 72J 149
Burley Rd. E16 44Lc 93
BURLINGS 89Vc 181
Burlings La. TN14: Knock 89Vc 181
Burlington Arc. W1 5B 222 (45Lb 90)
Burlington Av. RM7: Rom 30Dd 56
Burlington Av. SL1: Slou 7J 81
Burlington Av. TW9: Kew 53Qa 109
Burlington Cl. BR6: Farnb 75Rc 160
Burlington Cl. E6 44Nc 94
Burlington Cl. HA5: Eastc 27X 45
Burlington Cl. TW14: Bedf 59T 106
Burlington Cl. W9 42Cb 89
Burlington Cnr. NW1 38Lb 70
(off Camden Rd.)
Burlington Ct. E1 45Wb 91
(off Cable St.)
Burlington Ct. RH1: Redh 5P 207
(off Station Rd.)
Burlington Ct. SL1: Slou 7J 81
Burlington Gdns. RM6: Chad H 31Ad 75
Burlington Gdns. SW6 54Ab 110
Burlington Gdns. W1 5B 222 (45Lb 90)
Burlington Gdns. W3 46Sa 87
Burlington Gdns. W4 50Sa 87
Burlington Ho. N15 30Tb 51
(off Tewkesbury Rd.)
Burlington Ho. SE16 47Zb 92
(off Province Dr.)
Burlington Ho. UB7: W Dray 47P 83
(off Park Lodge Av.)
Burlington La. W4 52Sa 109
Burlington M. SW15 57Bb 111
Burlington M. W3 46Sa 87
Burlington Pl. IG8: Wfd G 20Kc 35
Burlington Pl. RH2: Reig 5J 207
Burlington Pl. SW6 54Ab 110
Burlington Pl. TN13: S'oaks 95Jd 202
Burlington Ri. EN4: E Barn 18Gb 31
Burlington Rd. CR7: Thor H 68Sb 135
Burlington Rd. EN2: Enf 11Tb 33
Burlington Rd. KT3: N Mald 70Va 132
Burlington Rd. N10 27Jb 50
Burlington Rd. N17 25Wb 51
Burlington Rd. SL1: Slou 7J 81
Burlington Rd. SW6 54Ab 110
Burlington Rd. TW7: Isle 53Fa 108
Burlington Rd. W4 50Sa 87
Burma M. N16 35Tb 71
Burman Cl. DA2: Dart 59Sd 120
Burma Rd. GU24: Chob 6K 147
Burma Rd. N16 35Tb 71
Burmarsh NW5 37Jb 70
Burmarsh Ct. SE20 67Yb 136
Burma Ter. SE19 64Ub 135
Burmester Rd. SW17 62Eb 133
Burnaby Cres. W4 51Sa 109
Burnaby Gdns. W4 51Ra 109
Burnaby Rd. DA11: Nflt 9A 122
Burnaby St. SW10 52Eb 111
Burnand Ho. W14 48Za 88
(off Redan St.)
Burnbrae Cl. N12 23Db 49
Burnbury Rd. SW12 60Lb 112
Burn Cl. KT15: Add 77M 149
Burn Cl. KT22: Oxs 87Fa 172
Burn Cl. WD25: A'ham 13Fa 28
Burncroft Av. EN3: Enf H 12Yb 34
Burndell Way UB4: Yead 43Z 85
Burne Jones Ho. W14 49Ab 88
Burnell Av. DA16: Well 54Wc 117
Burnell Av. TW10: Ham 64La 130
Burnell Bldg. NW2 33Ya 68
Burnell Ct. KT16: Chert 74J 149
Burnell Gdns. HA7: Stan 26Ma 47
Burnell Ho. E20 36Dc 72
(off Peloton Av.)
Burnelli Bldg. SW11 52Kb 112
(off Sopwith Way)
Burnell Rd. SM1: Sutt 77Db 155
Burnell Wlk. CM13: Gt War 23Yd 58
Burnell Wlk. SE1 50Vb 91
(off Cadet Dr.)
Burnels Av. E6 41Qc 94
Burness Cl. N7 37Pb 70
Burness Cl. UB8: Uxb 40M 63
Burne St. NW1 7D 214 (43Gb 89)
Burnet Cl. GU24: W End 5C 166
Burnet Cl. HP3: Hem H 3N 3
Burnet Gro. KT19: Eps 85Sa 173
Burnett Cl. E9 36Yb 72
Burnett Ho. SE13 54Ec 114
(off Lewisham Hill)
Burnett Rd. DA8: Erith 51Md 119
Burnett Rd. IG6: Ilf 24Rc 54
Burnetts Rd. SL4: Wind 3C 102
Burney Av. KT5: Surb 71Pa 153
Burney Cl. KT22: Fet 97Ea 192
Burney Dr. IG10: Lough 12Rc 36
Burney Ho. KT22: Lea 93Ja 192
(off Highbury Dr.)
Burney St. SE10 52Ec 114
Burnfoot Av. SW6 53Ab 110
Burnham NW3 38Gb 69
BURNHAM 1A 80
Burnham Av. UB10: Ick 35S 64
BURNHAM BEECHES 6F 60
Burnham Beeches Golf Course 9B 60
Burnham Beeches Nat. Nature Reserve 7D 60
Burnham Cl. EN1: Enf 10Ub 19
Burnham Cl. GU21: Knap 10H 167
Burnham Cl. HA3: W'stone 28Ja 46
Burnham Cl. NW7 24Wa 48
Burnham Cl. SE1 6K 231 (49Vb 91)
Burnham Cl. SL4: Wind 4B 102
Burnham Ct. NW4 28Ya 48
(off Brent St.)
Burnham Ct. NW6 38Eb 69
(off Fairhazel Gdns.)
Burnham Ct. W2 45Db 89
(off Moscow Rd.)
Burnham Cres. DA1: Dart 56Ld 119
Burnham Cres. E11 28Lc 53
Burnham Dr. KT4: Wor Pk 75Za 154
Burnham Dr. RH2: Reig 5J 207
Burnham Est. E2 41Yb 92
(off Burnham St.)
Burnham Gdns. CR0: C'don 73Vb 157
Burnham Gdns. TW4: Cran 53X 107
Burnham Gdns. UB3: Harl 48T 84
Burnham Hgts. SL1: Slou 4A 80
Burnham La. SL1: Slou 3B 80
Burnham Rd. AL1: St A 2E 6
Burnham Rd. DA1: Dart 56Ld 119
Burnham Rd. DA14: Sidc 61Ad 139
Burnham Rd. E4 22Bc 52
Burnham Rd. GU21: Knap 10H 167
Burnham Rd. RM7: Rom 27Fd 56
Burnham Rd. RM9: Dag 38Xc 75
Burnham Rd. SM4: Mord 70Db 133
Burnhams Gro. KT19: Eps 83Ra 173
Burnhams Rd. KT23: Bookh 96Aa 191
Burnham Station (Rail & Crossrail) 4B 80
Burnham St. E2 41Yb 92
Burnham St. KT2: King T 67Qa 131
Burnham Ter. DA1: Dart 57Md 119
Burnham Trad. Est. DA1: Dart 56Md 119
Burnham Way SE26 64Bc 136
Burnham Way W13 49Ka 86
Burnhill Cl. SE15 52Xb 113
Burnhill Ho. EC1 4D 218 (41Sb 91)
(off Norman St.)
Burnhill Rd. BR3: Beck 68Cc 136
Burnley Cl. WD19: Wat 22Y 45
Burnley Rd. NW10 36Va 68
Burnley Rd. RM20: W Thur 53Vd 120
Burnley Rd. SW9 54Pb 112
Burnsall St. SW3 7E 226 (50Gb 89)
Burns Av. DA15: Sidc 58Xc 117
Burns Av. RM6: Chad H 31Yc 75
Burns Av. TW14: Felt 58W 106
Burns Av. UB1: S'hall 45Ca 85
Burns Cl. DA16: Well 53Vc 117
Burns Cl. DA8: Erith 53Hd 118
Burns Cl. E17 28Ec 52
Burns Cl. SM5: Cars 81Jb 176
Burns Cl. SW19 65Fb 133
Burns Cl. UB4: Hayes 43V 84
Burns Dr. SM7: Bans 86Ab 174
Burns Ho. E2 41Yb 92
(off Cornwall Av.)
Burns Ho. SE17 50Rb 91
(off Doddington Gro.)
Burnside AL1: St A 4F 6
Burnside KT21: Asht 90Pa 173
Burnside Av. E4 23Bc 52
Burnside Cl. EN5: New Bar 13Cb 31
Burnside Cl. SE16 46Zb 92
Burnside Cl. TW1: Twick 58Ja 108
Burnside Ct. SM5: Cars 76Jb 156
Burnside Cres. HA0: Wemb 39Ma 67
Burnside Ind. Est. IG6: Ilf 22Xc 55
Burnside Rd. RM8: Dag 33Yc 75
Burns Pl. RM18: Tilb 3D 122
Burns Rd. HA0: Wemb 40Na 67
Burns Rd. NW10 39Va 68
Burns Rd. SW11 54Hb 111
Burns Rd. W13 47Ka 86
Burns Way CM13: Hut 17Fe 41
Burns Way TW5: Hest 54Z 107
Burnt Ash Hgts. BR1: Broml 64Kc 137
Burnt Ash Hill SE12 58Hc 115
Burnt Ash La. BR1: Broml 66Jc 137
Burnt Ash Rd. SE12 57Hc 115
BURNTCOMMON 97H 189
Burnt Comn. Cl. GU23: Rip 97H 189
Burnt Comn. La. GU23: Rip 98K 189
Burnt Farm Ride EN2: Crew H 6Qb 18
Burnt Farm Ride EN7: Walt C 5Rb 19
Burnt Ho. La. DA1: Dart 62Pd 141
Burnt Ho. La. DA2: Hawl 63Nd 141
Burnthwaite M. SW6 52Cb 111
(off Burnthwaite Rd.)
Burnthwaite Rd. SW6 52Bb 111
BURNT OAK 25Ra 47
Burnt Oak Apts. E16 44Jc 93
(off Pacific Rd.)
Burnt Oak B'way. HA8: Edg 24Qa 47
Burnt Oak Flds. HA8: Edg 25Sa 47
Burnt Oak La. DA15: Sidc 58Wc 117
(not continuous)
Burnt Oak Station (Underground) 25Sa 47
Burnt Pollard La. GU18: Light 2C 166
Burntwood CM14: B'wood 20Yd 40
Burntwood Av. RM11: Horn 30Md 57
Burntwood Cl. CM13: W H'dn 30Fe 59
Burntwood Cl. CR3: Cat'm 93Wb 197
Burntwood Cl. SW18 60Gb 111
Burntwood Dr. RH8: Oxt 3J 211
Burntwood Grange Rd. SW18 60Fb 111
Burntwood Gro. TN13: S'oaks 99Kd 203
Burntwood La. CR3: Cat'm 94Ub 197
Burntwood La. SW17 62Eb 133
Burntwood Rd. TN13: S'oaks 100Kd 203
Burntwood Vw. SE19 64Vb 135
Burntwood Way CM14: B'wood 20Wd 40
Burn Wlk. SL1: Burn 1A 80
Burnway RM11: Horn 31Nd 77
Buross St. E1 44Xb 91
Burpham Cl. UB4: Yead 43Z 85
Burrage Ct. SE16 49Zb 92
(off Worgan St.)
Burrage Gro. SE18 49Sc 94
Burrage Pl. SE18 50Rc 94
Burrage Rd. RH1: Redh 4B 208
Burrage Rd. SE18 50Sc 94
Burrard Ho. E2 40Yb 72
(off Bishop's Way)
Burrard Rd. E16 44Kc 93
Burrard Rd. NW6 36Cb 69
Burr Bank Ter. DA2: Wilm 63Ld 141
Burr Cl. AL2: Lon C 9J 7
Burr Cl. DA7: Bex 55Bd 117
Burr Cl. E1 46Wb 91
Burreed M. RM13: Rain 40Ed 76
Burrell Cl. CR0: C'don 72Ac 158
Burrell Cl. HA8: Edg 19Ra 29
Burrell Row BR3: Beck 68Cc 136
Burrells, The KT16: Chert 74K 149
Burrell St. SE1 6B 224 (46Rb 91)
Burrells Wharf Sq. E14 50Dc 92
Burrell Towers E10 31Cc 72
Burrfield Dr. BR5: St M Cry 71Zc 161
Burrhill Ct. SE16 48Zb 92
(off Worgan St.)
Burr Hill La. GU24: Chob 1K 167
Burritt Rd. KT1: King T 68Qa 131
Burroughs, The NW4 28Xa 48
Burroughs Club, The 28Xa 48
Burroughs Cotts. E14 43Ac 92
(off Halley St.)
Burroughs Dr. DA1: Dart 57Pd 119
Burroughs Gdns. NW4 28Xa 48
Burroughs Pde. NW4 28Xa 48
Burroway Rd. SL3: L'ly 48D 82
Burrow Cl. IG7: Chig 22Vc 55
Burrow Cl. WD17: Wat 8V 12
Burrow Grn. IG7: Chig 22Vc 55
BURROW HILL 4C 186
BURROWHILL 1J 167
Burrow Hill Grn. GU24: Chob 1H 167
Burrow Ho. SW9 54Qb 112
(off Stockwell Pk. Rd.)
Burrow Rd. IG7: Chig 22Vc 55
Burrow Rd. SE22 56Ub 113
Burrows Chase EN9: Walt A 7Fc 21
Burrows Cl. KT23: Bookh 96Ba 191
Burrows M. SE1 1B 230 (47Rb 91)
Burrows Rd. NW10 41Ya 88
Burrow Wlk. SE21 59Sb 113
Burr Rd. SW18 60Cb 111
Bursar St. SE1 7H 225 (46Ub 91)
Bursdon Cl. DA15: Sidc 61Vc 139
Burses Way CM13: Hut 17De 41
Bursland Rd. EN3: Pond E 14Zb 34
Burslem Av. IG6: Ilf 23Wc 55
Burslem St. E1 44Wb 91
Burstead Cl. KT11: Cobh 84Z 171
Burstock Rd. SW15 56Ab 110
Burston Dr. AL2: Park 10A 6
Burston Rd. SW15 57Za 110
Burston Vs. SW15 57Za 110
(off St John's Av.)
Burstow Rd. SW20 67Ab 132
Burtenshaw Rd. KT7: T Ditt 73Ja 152
Burtley Cl. N4 32Sb 71
Burton Av. WD18: Wat 14W 26
Burton Bank N1 38Tb 71
(off Yeate St.)
Burton Cl. CR7: Thor H 69Tb 135
Burton Cl. GU20: W'sham 9B 146
Burton Cl. KT9: Chess 80Ma 153
Burton Ct. KT7: T Ditt 72Ja 152
Burton Ct. SE20 68Yb 136
Burton Ct. SW3 7G 227 (50Hb 89)
(off Franklin's Row)
Burton Dr. EN3: Enf L 9Cc 20
Burton Gdns. TW5: Hest 53Ba 107
Burton Gro. SE17 7F 231 (50Tb 91)
Burtonhole Cl. NW7 21Za 48
Burtonhole La. N12 21Ab 48
Burtonhole La. NW7 22Ya 48
Burton Ho. SE16 47Xb 91
(off Cherry Gdn. St.)
Burton La. EN7: G Oak 1Ub 19
Burton La. SW9 54Qb 112
(not continuous)
Burton M. SW1 6J 227 (49Jb 90)
Burton Pl. WC1 4E 216 (41Mb 90)
Burton Ridge Ct. E3 40Bc 72
(off Festubert Pl.)
Burton Rd. SW9: Akerman Rd. 54Rb 113
Burton Rd. SW9: Evesham Wlk. 54Qb 112
Burton Rd. DA12: Grav'nd 3E 144
Burton Rd. E18 27Kc 53
Burton Rd. IG10: Lough 14Sc 36
Burton Rd. KT2: King T 66Na 131
Burton Rd. NW6 38Bb 69
Burtons Ct. E15 38Fc 73
Burton's La. WD3: Chor 15C 24
Burton's Rd. TW12: Hamp H 63Da 129
Burton St. WC1 4E 216 (41Mb 90)
Burton Way SL4: Wind 5C 102
Burtonwood Ho. N4 31Tb 71
Burtop Rd. Est. SW17 62Eb 133
Burt Rd. E16 46Lc 93
Burts Wharf DA17: Belv 45Ed 96
Burtt Ho. N1 3H 219 (41Ub 91)
(off Aske St.)
Burtwell La. SE27 63Tb 135
Burvale Ct. WD18: Wat 13X 27
Burwash Ct. BR5: St M Cry 71Yc 161
Burwash Ho. SE1 2G 231 (47Tb 91)
(off Kipling Est.)
Burwash Rd. SE18 50Tc 94
Burway Cl. CR2: S Croy 79Ub 157
Burway Cres. KT16: Chert 70J 127
Burwell KT1: King T 68Qa 131
(off Excelsior Cl.)
Burwell Av. UB6: G'frd 37Ga 66
Burwell Cl. E1 44Xb 91
Burwell Rd. E10 32Ac 72
Burwell Wlk. E3 42Cc 92
Burwood Av. BR2: Hayes 75Kc 159
Burwood Av. CR8: Kenley 86Rb 177
Burwood Av. HA5: Eastc 29X 45
Burwood Cl. KT12: Hers 79Y 151
Burwood Cl. KT6: Surb 74Qa 153
Burwood Cl. RH2: Reig 6M 207
Burwood Gdns. RM13: Rain 41Hd 96
Burwood Ho. SW9 56Rb 113
Burwood Pde. KT16: Chert 73J 149
(off Guildford St.)
BURWOOD PARK 84W 170
BURWOOD PARK 78W 150
Burwood Pk. Rd. KT12: Hers 77X 151
Burwood Pl. EN4: Had W 11Eb 31
Burwood Pl. W2 2E 220 (44Gb 89)
Burwood Rd. KT12: Hers 80U 150
Bury, The HP1: Hem H 1L 3
Bury, The WD3: Rick 18M 25
Bury Av. HA4: Ruis 30S 44
Bury Av. UB4: Hayes 40U 64
Bury Cl. GU21: Wok 8P 167
Bury Cl. SE16 46Zb 92
Bury Ct. EC3 2J 225 (44Ub 91)
Bury Ct. HP1: Hem H 2L 3
Burydell La. AL2: Park 9B 6
Buryfield Ct. SE8 49Zb 92
(off Lower Rd.)
BURY GREEN 3Wb 19
Bury Grn. HP1: Hem H 1L 3
Bury Grn. Rd. EN7: Chesh 3Wb 19
Bury Grn. Rd. EN7: Walt C 4Wb 19
Bury Gro. SM4: Mord 71Db 155
Bury Hall Vs. N9 17Vb 33
Bury Hill HP1: Hem H 1K 3
Bury Hill Cl. HP1: Hem H 1L 3
Bury Lake Young Mariners 19L 25
Bury La. CM16: Epp 1Tc 22
Bury La. GU21: Wok 8N 167
Bury La. WD3: Rick 18M 25
Bury Mdws. WD3: Rick 18M 25
Bury M. RM1: Rom 30Hd 56
Bury M. WD3: Rick 18M 25
(off Bury La.)
Bury Pl. WC1 1F 223 (43Nb 90)
Bury Ri. HP3: Hem H 7F 2
Bury Rd. CM16: Epp 3Uc 22
Bury Rd. E4 13Fc 35
Bury Rd. HP1: Hem H 1L 3
Bury Rd. N22 26Qb 50
Bury Rd. RM10: Dag 36Dd 76
Buryside Cl. IG2: Ilf 28Vc 55
Bury St. EC3 3J 225 (44Ub 91)
Bury St. HA4: Ruis 29S 44
Bury St. N9 17Vb 33
Bury St. SW1 6B 222 (46Lb 90)
Bury St. W. N9 17Tb 33
Bury Wlk. SW3 6D 226 (49Gb 89)
Busbridge Ho. E14 43Cc 92
(off Brabazon St.)
Busby Ho. SW16 63Lb 134
Busby M. NW5 37Mb 70
Busby Pl. NW5 37Mb 70
Busch Cl. TW7: Isle 53Ka 108
Bushbaby Cl. SE1 4H 231 (48Ub 91)
Bushbarns EN7: Chesh 1Wb 19
Bushberry Rd. E9 37Ac 72
Bush Cl. IG2: Ilf 29Tc 54
Bush Cl. KT15: Add 78L 149
Bush Cotts. SW18 57Cb 111
Bush Ct. N14 18Mb 32
Bush Ct. W12 47Za 88
Bushell Cl. SW2 61Pb 134
Bushell Grn. WD23: B Hea 19Fa 28
Bushell St. E1 46Wb 91
Bushell Way BR7: Chst 64Qc 138
Bush Elms Rd. RM11: Horn 31Jd 76
Bushetts Gro. RH1: Mers 1B 208
BUSHEY 16Ca 27
Bushey Av. BR5: Pet W 73Tc 160
Bushey Av. E18 27Hc 53
Bushey Cl. CR8: Kenley 88Vb 177
Bushey Cl. E4 20Ec 34
Bushey Cl. UB10: Ick 33Q 64
Bushey Ct. DA8: Erith 53Jd 118
Bushey Ct. SW20 69Xa 132
Bushey Cft. RH8: Oxt 2G 210
Bushey Down SW12 61Kb 134
Bushey Grove Leisure Cen. 14Ba 27
Bushey Gro. Rd. WD23: Bush 14Z 27
Bushey Hall Dr. WD23: Bush 14Aa 27
Bushey Hall Golf Course 14Aa 27
Bushey Hall Pk. WD23: Bush 13Aa 27
Bushey Hall Rd. WD23: Bush 14Z 27
BUSHEY HEATH 18Fa 28
Bushey Hill Rd. SE5 53Ub 113
Bushey La. SM1: Sutt 77Cb 155
Bushey Lees DA15: Sidc 58Vc 117
BUSHEY MEAD 68Za 132
Bushey Mill Cres. WD24: Wat 9Y 13
Bushey Mill La. WD23: Bush 10Z 13
Bushey Mill La. WD24: Wat 9Y 13
Bushey Mus. & Art Gallery 16Ca 27
Bushey Pk. WD23: Bush 16Ca 27
Bushey Rd. CR0: C'don 75Cc 158
Bushey Rd. E13 40Lc 73
Bushey Rd. N15 30Ub 51
Bushey Rd. SM1: Sutt 77Cb 155
(not continuous)
Bushey Rd. SW20 69Xa 132
Bushey Rd. UB10: Ick 33Q 64
Bushey Rd. UB3: Harl 49U 84
Bushey Shaw KT21: Asht 89Ka 172
Bushey Station (Rail & Overground) 16Z 27
Bushey Vw. Wlk. WD24: Wat 12Z 27
Bushey Way BR3: Beck 72Fc 159
Bush Fair Ct. N14 16Kb 32
Bushfield Cl. HA8: Edg 19Ra 29
Bushfield Cres. HA8: Edg 19Ra 29
Bushfield Dr. RH1: Redh 10A 208
Bushfield Rd. HP3: Bov 7E 2
Bushfields IG10: Lough 15Qc 36
Bushfield Wlk. DA10: Swans 58Ae 121
Bush Gro. HA7: Stan 25Ma 47
Bush Gro. NW9 31Sa 67
Bushgrove Rd. RM8: Dag 35Zc 75
Bush Hill N21 17Sb 33
Bush Hill Gdns. 17Tb 33
(off Bush Hill Road)
Bush Hill Pde. EN1: Enf 17Tb 33
Bush Hill Pde. N9 17Tb 33
BUSH HILL PARK 16Vb 33
Bush Hill Pk. Golf Course 15Sb 33
Bush Hill Park Station (Overground) 16Vb 33
Bush Hill Rd. HA3: Kenton 30Pa 47
Bush Hill Rd. N21 16Tb 33
Bush Ind. Est. N19 34Lb 70
Bush Ind. Est. NW10 42Ta 87
Bush La. EC4 4F 225 (45Tb 91)
Bush La. GU23: Send 96F 188
Bushmead Cl. N15 28Vb 51
Bushmoor Cres. SE18 52Rc 116
Bushnell Rd. SW17 61Kb 134
Bush Rd. E11 31Hc 73
Bush Rd. E8 39Xb 71
Bush Rd. IG9: Buck H 21Mc 53
Bush Rd. SE8 49Zb 92
Bush Rd. TW17: Shep 71P 149
Bush Rd. TW9: Kew 51Pa 109
Bush Theatre 47Ya 88
Bushway RM8: Dag 35Zc 75
Bushwood E11 31Hc 73
Bushwood Cl. AL9: Wel G 5D 8
Bushwood Dr. SE1 6K 231 (49Vb 91)
Bushwood Rd. TW9: Kew 51Qa 109
Bushy Cl. RM1: Rom 23Fd 56
Bushy Ct. KT1: Hamp W 67La 130
(off Beverley Rd.)
Bushy Pk. 66Fa 130
Bushy Pk. Gdns. TW11: Tedd 64Fa 130
Bushy Pk. Rd. TW11: Tedd 66Ka 130
(not continuous)
Bushy Rd. KT22: Fet 94Da 191
Bushy Rd. TW11: Tedd 65Ha 130
Business Cen., The RM3: Rom 24Md 57
Business Innovation Cen., The EN3: Enf L 8Bc 20
(off Innova Bus. Pk.)
Business Pk. 5 KT22: Lea 92Ha 192
Business Pk. 8 KT22: Lea 91Ka 192
Business Pk. 25 RH1: Redh 4B 208
Business Village, The SL2: Slou 6M 81
Buspace Studios W10 42Ab 88
(off Conlan St.)
Busty La. TN15: Igh 93Zd 205
Butcher Row E14 45Zb 92
Butchers Hill DA12: Shorne 4N 145
Butcher's La. TN15: Ash 75Zd 165
Butcher's La. TN15: Hartl 75Zd 165
Butchers M. UB3: Hayes 45V 84
(off Hemmen La.)
Butchers Rd. E16 44Jc 93
Butchers Yd. BR6: Downe 83Qc 180
Butcher Wlk. DA10: Swans 59Ae 121
Bute Av. TW10: Ham 61Na 131
Bute Ct. SM6: W'gton 78Lb 156
Bute Gdns. SM6: W'gton 78Lb 156
Bute Gdns. TW10: Ham 60Na 109
Bute Gdns. W6 49Za 88
Bute Gdns. W. SM6: W'gton 78Lb 156
Bute M. NW11 29Eb 49
Bute Rd. CR0: C'don 74Qb 156
Bute Rd. IG6: Ilf 29Rc 54
Bute Rd. SM6: W'gton 77Lb 156
Bute St. SW7 5B 226 (49Fb 89)
Bute Wlk. N1 37Tb 71
Butfield Ho. E9 37Yb 72
(off Stevens Av.)
Butler Av. HA1: Harr 31Fa 66
Butler Cl. HA8: Edg 26Ra 47
Butler Ct. HA0: Wemb 35Ja 66
Butler Ct. RM8: Dag 33Cd 76
(off Gosfield Rd.)

Butler Ct. SW11 53Gb 111
(off Hyde La.)
Butler Dr. DA8: Erith 52Gd 118
Butler Farm Cl. TW10: Ham 63Ma 131
Butler Hall AL10: Hat 2B 8
(off Bishops Ri.)
Butler Ho. E14 44Bc 92
(off Burdett St.)
Butler Ho. E2 41Yb 92
(off Bacton St.)
Butler Ho. E3 43Bc 92
(off Geoffrey Chaucer Way)
Butler Ho. RM17: Grays 51De 121
(off Argent St.)
Butler Ho. SW9 53Rb 113
(off Lothian Rd.)
Butler Pl. SW1 3D 228 (48Mb 90)
Butler Rd. HA1: Harr 31Ea 66
Butler Rd. NW10 38Va 68
Butler Rd. RM8: Dag 35Xc 75
Butlers & Colonial Wharf SE1 1K 231 (47Vb 91)
(off Shad Thames)
Butlers Cl. SL4: Wind 3B 102
Butlers Cl. TW4: Houn 55Ba 107
Butlers Ct. EN8: Walt C 4Ac 20
Butlers Dene Rd. CR3: Wold 92Bc 198
Butlers Dr. E4 10Ec 20
Butler's Pl. TN15: Ash 76Ae 165
Butler St. E2 41Yb 92
Butler St. UB10: Hil 42R 84
Butlers Wharf SE1 7K 225 (46Vb 91)
Butlers Wharf W. SE1 7K 225 (46Vb 91)
(off Shad Thames)
Butler Wlk. RM17: Grays 49Fe 99
Butley Ct. E3 40Ac 72
(off Ford St.)
Buttell Cl. RM17: Grays 50Fe 99
Buttercross La. CM16: Epp 2Wc 23
Buttercup Cl. RM3: Hrld W 25Md 57
Buttercup Cl. UB5: N'olt 37Ba 65
Buttercup Sq. TW19: Stanw 60M 105
Butterfield Cl. N17 23Sb 51
Butterfield Cl. SE16 47Xb 91
Butterfield Cl. TW1: Twick 58Ha 108
Butterfield La. AL1: St A 6C 6
Butterfields E17 29Ec 52
Butterfield Sq. E6 44Pc 94
Butterfly Apts. SW11 56Gb 111
(off Comyn Rd.)
Butterfly Ct. E6 41Pc 94
Butterfly Ct. N15 28Ub 51
(off Bathurst Sq.)
Butterfly Ct. NW9 25Ua 48
Butterfly Cres. HP3: Hem H 7A 4
Butterfly La. SE9 58Rc 116
Butterfly La. WD6: E'tree 13Ja 28
Butterfly Wlk. CR6: W'ham 92Yb 198
Butterfly Wlk. CR6: Wold 92Yb 198
Butterfly Wlk. SE5 53Tb 113
(off Denmark Hill)
Butter Hill SM5: Cars 76Jb 156
Butter Hill SM6: W'gton 76Jb 156
Butteridges Cl. RM9: Dag 39Bd 75
Butterley M. RM3: Rom 23Nd 57
Butterly Av. DA1: Dart 61Pd 141
Buttermere NW1 3A 216 (41Kb 90)
(off Augustus St.)
Buttermere Av. SL1: Slou 3A 80
Buttermere Cl. AL1: St A 3F 6
Buttermere Cl. DA1: Dart 55Qd 119
Buttermere Cl. E15 35Fc 73
Buttermere Cl. SE1 6K 231 (49Vb 91)
Buttermere Cl. SM4: Mord 72Za 154
Buttermere Cl. TW14: Felt 60V 106
Buttermere Ct. NW8 39Fb 69
(off Boundary Rd.)
Buttermere Dr. SW15 57Ab 110
Buttermere Gdns. CR8: Purl 85Tb 177
Buttermere Ho. E3 41Bc 92
(off Mile End Rd.)
Buttermere Pl. WD25: Wat 5W 12
Buttermere Rd. BR5: St P 70Zc 139
Buttermere Wlk. E8 37Vb 71
Buttermere Way TW20: Egh 66D 126
Butterscotch Row WD5: Ab L 4T 12
Butterwick W6 49Za 88
Butterwick WD25: Wat 8Aa 13
Butterworth Gdns. IG8: Wfd G 23Jc 53
Butterworth Ter. SE17 50Sb 91
(off Sutherland Wlk.)
Buttery M. N14 20Nb 32
Buttesland St. N1 3G 219 (41Tb 91)
Buttfield Cl. RM10: Dag 37Dd 76
Butt Fld. Vw. AL1: St A 6A 6
Buttlehide WD3: Map C 22F 42
Buttmarsh Cl. SE18 50Rc 94
Button Lodge E17 28Cc 52
Button Rd. RM17: Grays 49Be 99
Buttonscroft Cl. CR7: Thor H 69Sb 135
Button St. BR8: Swan 68Ld 141
Button St. Bus. Cen. BR8: Swan 68Ld 141
Butts, The TN14: Otf 88Kd 183
Butts, The TW16: Sun 69Y 129
Butts, The TW8: Bford 51La 108
Buttsbury Rd. IG1: Ilf 36Sc 74
Butts Cotts. TW13: Hanw 62Aa 129
Butts Cres. TW13: Hanw 62Ca 129
Butts End HP1: Hem H 1J 3
Butts Grn. Rd. RM11: Horn 30Md 57
Butts La. SS17: Stan H 2K 101
Buttsmead HA6: Nwood 24S 44
Butts Piece UB5: N'olt 40X 65
Butts Rd. BR1: Broml 64Gc 137
Butts Rd. GU21: Wok 89A 168
Butts Rd. SS17: Stan H 2L 101
Buxhall Cres. E9 37Bc 72
Buxted Rd. E8 38Vb 71
Buxted Rd. N12 22Gb 49
Buxted Rd. SE22 56Ub 113
Buxton Av. CR3: Cat'm 93Ub 197
Buxton Cl. IG8: Wfd G 23Mc 53
Buxton Cl. KT19: Eps 83Ra 173
Buxton Cl. N9 19Yb 34
Buxton Ct. E11 31Hc 73
Buxton Ct. N1 3E 218 (41Sb 91)
(off Thoresby St.)
Buxton Cres. SM3: Cheam 77Ab 154
Buxton Dr. E11 28Gc 53
Buxton Dr. KT3: N Mald 68Ta 131
Buxton Gdns. W3 45Ra 87
Buxton Ho. E11 28Gc 53
Buxton La. CR3: Cat'm 92Tb 197
Buxton M. SW4 54Mb 112
Buxton Path WD19: Wat 20Y 27
Buxton Pl. CR3: Cat'm 92Tb 197
Buxton Rd. CM16: They B 8Uc 22
Buxton Rd. CR7: Thor H 71Rb 157
Buxton Rd. DA8: Erith 52Fd 118
Buxton Rd. E15 36Gc 73
Buxton Rd. E17 28Ac 52
(not continuous)
Buxton Rd. E4 17Fc 35
Buxton Rd. E6 41Nc 94
Buxton Rd. EN9: Walt A 4Jc 21
Buxton Rd. IG2: Ilf 30Uc 54
Buxton Rd. N19 32Mb 70
Buxton Rd. NW2 37Xa 68
Buxton Rd. RM16: Grays 7A 100
Buxton Rd. SW14 55Ua 110
Buxton Rd. TW15: Ashf 64M 127
Buxton St. E1 6K 219 (42Vb 91)
Buzy Bees 75Zd 165
Buzzard Creek Ind. Est. IG11: Bark 43Wc 95
Buzz Bingo Bexleyheath 56Dd 118
Buzz Bingo Enfield 14Wb 33
Buzz Bingo Feltham 61X 129
Buzz Bingo Slough 6J 81
Buzz Bingo Stratford 39Fc 73
Buzz Bingo Waltham Cross 6Ac 20
Byam St. SW6 54Eb 111
Byards Ct. SE16 49Zb 92
(off Worgan St.)
Byards Cft. SW16 67Mb 134
Byas Ho. E3 41Bc 92
(off Benworth St.)
Byatt Wlk. TW12: Hamp 65Aa 129
Bybend Cl. SL2: Farn R 9F 60
Bychurch End TW11: Tedd 64Ha 130
Bycliffe M. DA11: Grav'nd 9B 122
(off Bycliffe Ter.)
Bycliffe Ter. DA11: Grav'nd 9B 122
Bycroft Rd. UB1: S'hall 42Ca 85
Bycroft St. SE20 66Zb 136
Bycullah Av. EN2: Enf 13Rb 33
Bycullah Rd. EN2: Enf 12Rb 33
Bye, The W3 44Ua 88
Byegrove Rd. SW19 65Fb 133
Byelands Cl. SE16 46Zb 92
Byers Cl. EN6: Pot B 6Eb 17
(not continuous)
Byewaters WD18: Wat 16S 26
Bye Way, The HA3: W'stone 25Ga 46
Byeway, The SW14 55Sa 109
Byeway, The WD3: Rick 19N 25
Byeways TW2: Twick 62Da 129
Byeways, The KT5: Surb 71Qa 153
Byfeld Gdns. SW13 53Wa 110
Byfield Cl. SE16 47Bc 92
Byfield Ct. CM13: W H'dn 30Ee 59
Byfield Rd. TW7: Isle 55Ja 108
BYFLEET 84N 169
Byfleet & New Haw Station (Rail) 82M 169
Byfleet Ind. Est. KT14: Byfl 82M 169
Byfleet Ind. Est. WD18: Wat 18S 26
Byfleet Rd. KT11: Cobh 84Q 170
Byfleet Rd. KT14: Byfl 84Q 170
Byfleet Rd. KT15: New H 80M 149
Byfleet Technical Cen. KT14: Byfl 83M 169
Byford Cl. E15 38Gc 73
Byford Ho. EN5: Barn 14Za 30
Byford Ho. HA2: Harr 32Ca 65
Bygrove CR0: New Ad 79Dc 158
Bygrove St. E14 44Dc 92
(not continuous)
Byland Cl. N21 17Pb 32
Byland Cl. SE2 48Xc 95
Byland Cl. SM4: Mord 73Eb 155
Bylands GU22: Wok 91C 188
Byne Rd. SE26 65Yb 136
Byne Rd. SM5: Cars 75Gb 155
Bynes Rd. CR2: S Croy 80Tb 157
Byng Dr. EN6: Pot B 3Cb 17
Byng Pl. WC1 6D 216 (42Mb 90)
Byng Rd. EN5: Barn 12Za 30
Byng St. E14 47Cc 92
Bynon Av. DA7: Bex 55Bd 117
By-Pass Rd. KT22: Lea 92Ka 192
By-Pass Rd. SS17: Horn H 1J 101
Byrd Way SS17: Stan H 1L 101
Byre Rd. N14 16Kb 32
Byrne Cl. CR0: C'don 72Sb 157
Byrne Rd. SW12 60Kb 112
Byron SL3: L'ly 50D 82
Byron Av. CR5: Coul 87Nb 176
Byron Av. E12 37Nc 74
Byron Av. E18 27Hc 53
Byron Av. KT3: N Mald 71Wa 154
Byron Av. NW9 28Ra 47
Byron Av. SM1: Sutt 77Fb 155
Byron Av. TW4: Cran 54W 106
Byron Av. WD24: Wat 11Z 27
Byron Av. WD6: Bore 15Qa 29
Byron Av. E. SM1: Sutt 77Fb 155
Byron Cl. E8 39Wb 71
Byron Cl. GU21: Knap 9J 167
Byron Cl. KT12: Walt T 74Aa 151
Byron Cl. KT23: Bookh 96Ca 191
Byron Cl. SE20 69Xb 135
Byron Cl. SE26 63Ac 136
Byron Cl. SE28 46Yc 95
Byron Cl. SW16 65Nb 134
Byron Cl. TW12: Hamp 63Ba 129
Byron Ct. E11 28Kc 53
(off Makepeace Rd.)
Byron Ct. EN2: Enf 12Rb 33
Byron Ct. HA1: Harr 30Ga 46
Byron Ct. NW6 38Eb 69
(off Fairfax Rd.)
Byron Ct. SE22 60Wb 113
Byron Ct. SL4: Wind 5E 102
Byron Ct. SW3 6E 226 (49Gb 89)
(off Elystan St.)
Byron Ct. W7 49Ja 86
(off Boston Rd.)
Byron Ct. W9 42Cb 89
(off Lanhill Rd.)
Byron Ct. WC1 5H 217 (42Pb 90)
(off Mecklenburgh Sq.)
Byron Dr. DA8: Erith 52Dd 118
Byron Dr. N2 30Fb 49
Byron Gdns. RM18: Tilb 3E 122
Byron Gdns. SM1: Sutt 77Fb 155
Byron Hill Rd. HA2: Harr 32Fa 66
Byron Ho. DA1: Cray 57Gd 118
Byron Mans. RM14: Upm 34Sd 78
Byron M. NW3 35Gb 69
Byron M. W9 42Cb 89
Byron Pde. UB10: Hil 42S 84
Byron Pl. KT22: Lea 94Ka 192
Byron Rd. CM13: Hut 17Fe 41
Byron Rd. CR2: Sels 82Xb 177
Byron Rd. DA1: Dart 56Rd 119
Byron Rd. E10 32Dc 72
Byron Rd. E17 27Cc 52
Byron Rd. HA0: Wemb 33La 66
Byron Rd. HA1: Harr 30Ga 46
Byron Rd. HA3: W'stone 26Ha 46
Byron Rd. KT15: Add 77N 149
Byron Rd. NW2 33Xa 68
Byron Rd. NW7 22Wa 48
Byron Rd. W5 46Pa 87
Byron St. E14 44Ec 92
Byron Ter. N9 16Yb 34
Byron Ter. SE7 52Lc 115
Byron Way RM3: Rom 25Ld 57
Byron Way UB4: Hayes 42V 84
Byron Way UB5: N'olt 41Aa 85
Byron Way UB7: W Dray 49P 83
Bysouth Cl. IG5: Ilf 25Rc 54
Bysouth Cl. N15 28Tb 51
By the Wood WD19: Wat 19Z 27
Bythorn St. SW9 55Pb 112
Byton Rd. SW17 65Hb 133
Byttom Hill RH5: Mick 98La 192
Byward Av. TW14: Felt 58Y 107
Byward St. EC3 5J 225 (45Ub 91)
Bywater Ho. SE18 48Nc 94
Bywater Pl. SE16 46Ac 92
Bywater St. SW3 7F 227 (50Hb 89)
Byway, The EN6: Pot B 5Cb 17
Byway, The KT19: Ewe 77Va 154
Byway, The SM2: Sutt 81Fb 175
Byways, The KT21: Asht 90Ma 173
Bywell Pl. E16 43Hc 93
Bywell Pl. W1 1B 222 (43Lb 90)
(off Wells St.)
Bywood Av. CR0: C'don 72Yb 158
Bywood Cl. CR8: Kenley 87Rb 177
Bywood Cl. SM7: Bans 89Bb 175
By-Wood End SL9: Chal P 22C 42
Byworth Wlk. N19 32Nb 70

C

Cabanel Pl. SE11 6K 229 (49Qb 90)
Cabbell Pl. KT15: Add 77L 149
Cabbell St. NW1 1D 220 (43Gb 89)
Cabborns Cres. SS17: Stan H 3M 101
Cabelfeigh Cl. RH1: Redh 6M 207
Cabinet Ct. NW2 34Va 68
Cabinet Way E4 23Bc 52
Cable Car 47Hc 93
Cable Ct. SE16 49Ac 92
(off Rope St.)
Cable Ho. WC1 3K 217 (41Qb 90)
(off Gt. Percy St.)
Cable Ho. Ct. GU21: Wok 87A 168
Cable Pl. SE10 53Ec 114
Cable St. E1 45Wb 91
Cable St. E16 47Kc 93
Cable Trade Pk. SE7 49Lc 93
Cable Wlk. SE10 49Gc 93
Cabot Cl. CR0: Wadd 76Qb 156
Cabot Ct. SE16 49Zb 92
(off Worgan St.)
Cabot Hall 46Cc 92
Cabot Sq. E14 46Cc 92
Cabot Way E6 39Mc 73
Cabrail Ct. E14 43Ac 92
(off Aston Street)
Cabrera Av. GU25: Vir W 2N 147
Cabrera Cl. GU25: Vir W 2P 147
Cab Rd. SE1 1K 229 (47Qb 90)
(off West Rd.)
Cabul Rd. SW11 54Gb 111
Caci Ho. W14 49Bb 89
(off Kensington Village)
Cacket's La. TN14: Cud 87Tc 180
Cacketts Cotts. TN16: B Char 98Yc 201
Cactus Cl. SE15 54Ub 113
Cactus Wlk. W12 44Va 88
Cadbury Cl. N20 18Eb 31
Cadbury Cl. TW16: Sun 66U 128
Cadbury Cl. TW7: Isle 53Ja 108
Cadbury Rd. TW16: Sun 66U 128
Cadbury Way SE16 48Vb 91
Caddington Cl. EN4: E Barn 15Gb 31
Caddington Rd. NW2 34Ab 68
Caddis Cl. HA7: Stan 24Ha 46
Caddy Cl. TW20: Egh 64C 126
Cade La. TN13: S'oaks 100Ld 203
Cadell Cl. E2 2K 219 (40Vb 71)
Cade Rd. SE10 53Fc 115
Cader Rd. SW18 58Eb 111
Cadet Dr. SE1 49Vb 91
Cadet Ho. SE18 48Rc 94
Cadiz Ct. RM10: Dag 38Fd 76
Cadiz Rd. RM10: Dag 38Ed 76
Cadiz St. SE17 7E 230 (50Sb 91)
Cadley Ter. SE23 61Yb 136
Cadlocks Hill TN14: Hals 82Bd 181
Cadman Cl. SW9 52Rb 113
Cadman Ct. W4 50Ra 87
(off Chaseley Dr.)
Cadmer Cl. KT3: N Mald 70Ua 132
Cadmium Sq. E2 41Zb 92
(off Palmer's Rd.)
Cadmore Ct. EN8: Chesh 1Zb 20
Cadmore Ho. N1 38Rb 71
(off The Sutton Est.)
Cadmore La. EN8: Chesh 1Zb 20
Cadmore La. EN8: Turn 1Zb 20
Cadmus Cl. SW4 55Mb 112
Cadmus Ct. SE16 49Ac 92
(off Seafarer Way)
Cadmus Ct. SW9 53Qb 112
(off Southey Rd.)
Cadnam Lodge E14 48Ec 92
(off Schooner Cl.)
Cadnam Point SW15 60Xa 110
Cadogan Av. CM13: W H'dn 30Fe 59
Cadogan Av. DA2: Dart 59Td 120
Cadogan Cl. BR3: Beck 67Fc 137
Cadogan Cl. E9 38Bc 72
Cadogan Cl. HA2: Harr 35Da 65
Cadogan Cl. TW11: Tedd 64Ga 130
Cadogan Ct. E9 38Bc 72
(off Cadogan Ter.)
Cadogan Ct. SM2: Sutt 79Db 155
Cadogan Ct. SW3 6F 227 (49Hb 89)
(off Draycott Av.)
Cadogan Ct. Gdns. SW1 5H 227 (49Jb 90)
(off D'Oyley St.)
Cadogan Gdns. E18 27Kc 53
Cadogan Gdns. N21 15Rb 33
Cadogan Gdns. N3 25Db 49
Cadogan Gdns. SW3 5G 227 (49Hb 89)
Cadogan Ga. SW1 5G 227 (49Hb 89)
Cadogan Hall 5H 227 (49Jb 90)
(off Sloane Ter.)
Cadogan Ho. IG8: Wfd G 24Rc 54
Cadogan Ho. SW3 51Fb 111
(off Beaufort St.)
Cadogan La. SW1 4H 227 (48Jb 90)
Cadogan Mans. SW3 6G 227 (49Hb 89)
(off Cadogan Gdns.)
Cadogan Pl. CR8: Kenley 89Sb 177
Cadogan Pl. SW1 3G 227 (48Hb 89)
Cadogan Rd. KT6: Surb 71Ma 153
Cadogan Rd. SE18 48Sc 94
Cadogan Sq. SW1 4G 227 (48Hb 89)
Cadogan St. SW3 6F 227 (49Hb 89)
Cadogan Ter. E9 37Bc 72
Cadoxton Av. N15 30Vb 51
Cadwal Apts. N1 38Nb 70
(off Caledonian Rd.)
Cadwallon Rd. SE9 61Rc 138
Caedmon Rd. N7 35Pb 70
Caelian Pl. AL3: St A 5P 5
Caenshill Ho. KT13: Weyb 80Q 150
Caenshill Pl. KT13: Weyb 80Q 150
Caenshill Rd. KT13: Weyb 80Q 150
Caenswood Hill KT13: Weyb 82Q 170
Caenwood Cl. KT13: Weyb 79Q 150
Caen Wood Rd. KT21: Asht 90La 172
Caerleon Cl. DA14: Sidc 64Yc 139
Caerleon Cl. KT10: Clay 80Ka 152
Caerleon Ter. SE2 49Xc 95
Caernafon Ho. HA7: Stan 22Ja 46
Caernarvon Cl. CR4: Mitc 69Nb 134
Caernarvon Cl. HP2: Hem H 2M 3
Caernarvon Cl. RM11: Horn 32Qd 77
Caernarvon Ct. HP2: Hem H 2M 3
Caernarvon Dr. IG5: Ilf 25Qc 54
Caernarvon Ho. E16 46Kc 93
(off Audley Dr.)
Caernarvon Ho. W2 2A 220 (44Eb 89)
(off Hallfield Est.)
Caesar Ct. E2 40Zb 72
(off Palmer's Rd.)
Caesars Ct. AL3: St A 2B 6
(off Verulam Rd.)
Caesars Wlk. CR4: Mitc 71Hb 155
Caesars Way TW17: Shep 72T 150
Cage Pond Rd. WD7: Shenl 5Pa 15
Cages Wood Dr. SL2: Farn C 5F 60
Cage Yd. RH2: Reig 6J 207
Cagney Cl. TW16: Sun 67W 128
Cagny Ho. SM1: Sutt 77Db 155
Cahill St. EC1 6E 218 (42Sb 91)
Cahir St. E14 49Dc 92
Caillard Rd. KT14: Byfl 83N 169
Cain Cl. AL1: St A 4D 6
Cain Ct. EN8: Chesh 1Zb 20
(off Wycliffe Cl.)
Cain Ct. W5 43La 86
(off Castlebar M.)
Caine Ho. W3 47Ra 87
(off Hanbury Rd.)
Cain's La. TW14: Felt 57U 106
Caird St. W10 41Ab 88
Cairn Av. W5 46Ma 87
Cairn Ct. KT17: Ewe 82Va 174
Cairncross M. N8 30Nb 50
Cairndale Cl. BR1: Broml 66Hc 137
Cairnfield Av. NW2 34Ua 68
Cairngorm Cl. TW11: Tedd 64Ja 130
Cairngorm Pl. SL2: Slou 2H 81
Cairns Av. IG8: Wfd G 23Nc 54
Cairns Av. SW16 69Lb 134
Cairns Cl. AL4: St A 3H 7
Cairns Cl. DA1: Dart 57Md 119
Cairns M. SE18 53Nc 116
Cairns Pl. SW16 68Lb 134
Cairns Rd. SW11 57Gb 111
Cairn Way HA7: Stan 23Ha 46
Cairo New Rd. CR0: C'don 75Rb 157
Cairo Rd. E17 28Cc 52
Caishowe Rd. WD6: Bore 11Ra 29
Caisson Moor Ct. E3 42Ec 92
(off Navigation Rd.)
Caister Cl. HP2: Hem H 3N 3
Caister Ho. N7 37Pb 70
Caistor Ho. E15 39Hc 73
(off Caistor Pk. Rd.)
Caistor M. SW12 59Kb 112
Caistor Pk. Rd. E15 39Hc 73
Caistor Rd. SW12 59Kb 112
Caithness Dr. KT18: Eps 86Ta 173
Caithness Gdns. DA15: Sidc 58Vc 117
Caithness Ho. N1 39Pb 70
(off Twyford St.)
Caithness Rd. CR4: Mitc 66Kb 134
Caithness Rd. W14 48Za 88
Caithness Wlk. CR0: C'don 75Tb 157
Calabria Rd. N5 37Rb 71
Calais Cotts. DA3: Fawk 75Wd 164
Calais Ga. SE5 53Rb 113
Calais St. SE5 53Rb 113
Calbourne Av. RM12: Horn 36Kd 77
Calbourne Rd. SW12 59Hb 111
Calbroke Rd. SL2: Slou 2D 80
Calcott Cl. CM14: B'wood 18Xd 40
Calcott Ct. W14 48Ab 88
(off Blythe Rd.)
Calcott Wlk. SE9 63Nc 138
Calcraft Ho. E2 40Yb 72
(off Bonner Rd.)
Calcroft Av. DA9: Ghithe 57Yd 120
Calcutta Rd. RM18: Tilb 4B 122
Caldbeck EN9: Walt A 6Fc 21
(not continuous)
Caldbeck Av. KT4: Wor Pk 75Wa 154
Caldecot Av. EN7: Chesh 1Vb 19
Caldecote KT1: King T 68Qa 131
(off Excelsior Cl.)
Caldecote Gdns. WD23: Bush 17Ga 28
CALDECOTE HILL 17Ha 28
Caldecote La. WD23: Bush 16Ha 28
Caldecot Rd. SE5 54Sb 113
Caldecott Way E5 34Zb 72
Calder RM18: E Til 9L 101
Calder Av. AL9: Brk P 8J 9
Calder Av. UB6: G'frd 40Ha 66
Calder Cl. EN1: Enf 13Ub 33
Calder Ct. SE16 46Bc 92
Calder Ct. SL3: L'ly 50B 82
Calder Gdns. HA8: Edg 27Qa 47
Calderon Ho. NW8 1D 214 (40Gb 69)
(off Townshend Est.)
Calderon Pl. W10 43Ya 88
Calderon Rd. E11 35Ec 72
Calder Rd. SM4: Mord 71Eb 155
Caldervale Rd. SW4 57Mb 112
Calder Way SL3: Poyle 55G 104
Calderwood DA12: Grav'nd 4G 144
Calderwood Pl. EN4: Had W 11Db 31
Calderwood St. SE18 49Qc 94
Caldew Ct. NW7 24Wa 48
Caldew St. SE5 52Tb 113
Caldicote Grn. NW9 30Ua 48
Caldon Apts. CR0: C'don 75Tb 157
Caldon Ho. UB5: N'olt 42Ba 85
Caldwell Cl. SE18 50Qc 94
Caldwell Ho. SW13 52Ya 110
(off Trinity Chu. Rd.)
Caldwell Rd. GU20: W'sham 8B 146
Caldwell Rd. SS17: Stan H 2K 101
Caldwell Rd. WD19: Wat 21Z 45
Caldwell St. SW9 52Pb 112
Caldy Rd. DA17: Belv 48Dd 96
Caldy Wlk. N1 38Sb 71
Caleb St. SE1 1E 230 (47Sb 91)
Caledonia Ct. IG11: Bark 40Yc 75
(off Keel Cl.)
Caledonia Ho. E14 44Ac 92
(off Salmon La.)
Caledonian Cl. IG3: Ilf 32Xc 75
Caledonian Ct. BR2: Broml 72Nc 160
(off Wells Vw. Dr.)
Caledonian Ct. UB5: N'olt 42Aa 85
Caledonian Ct. WD17: Wat 12X 27
Caledonian Point SE10 51Dc 114
(off Norman Rd.)
Caledonian Rd. N1 2G 217 (40Nb 70)
Caledonian Rd. N7 35Pb 70
Caledonian Road & Barnsbury Station (Overground) 38Pb 70
Caledonian Road Station (Underground) 37Pb 70
Caledonian Sq. NW1 37Mb 70
Caledonian Wharf E14 49Fc 93
Caledonia Rd. TW19: Stanw 60N 105
Caledonia St. N1 2G 217 (40Nb 70)
Caledon Rd. AL2: Lon C 8G 6
Caledon Rd. E6 39Pc 74
Caledon Rd. SM6: W'gton 77Jb 156
Calendar M. KT6: Surb 72Ma 153
Cale St. SW3 7D 226 (50Gb 89)
Caletock Way SE10 50Hc 93
Calfstock La. DA4: Farni 70Pd 141
Calgarth NW1 2C 216 (40Lb 70)
(off Ampthill Est.)
Calgary Ct. RM7: Mawney 28Dd 56
Calgary Ct. SE16 47Yb 92
(off Canada Est.)
Calia Ho. E20 37Ec 72
(off Anthems Way)
Caliban Twr. N1 2H 219 (40Ub 71)
(off Arden Est.)
Calico Av. SM6: W'gton 75Jb 156
Calico Ct. SE16 48Wb 91
(off Marine St.)
Calico Ho. E20 37Ec 72
(off Mirabelle Gdns.)
Calico Ho. EC4 3E 224 (44Sb 91)
(off Well Ct.)
Calico Ho. SE1 2H 231 (47Ub 91)
(off Long La.)
Calico Row SW11 55Eb 111
Calidore Cl. SW2 58Pb 112
California Bldg. SE13 53Dc 114
(off Deal's Gateway)
California Cl. SM2: Sutt 82Cb 175
California Ct. WD23: B Hea 18Fa 28
(off High Rd.)
California La. WD23: B Hea 18Fa 28
California Rd. KT3: N Mald 70Ra 131
Caling Cft. DA3: New A G 74Be 165
Caliph Cl. DA12: Grav'nd 2H 145
Callaby Ter. N1 37Tb 71
Callaghan Cl. SE13 56Gc 115
Callahan Cotts. E1 43Yb 92
(off Lindley St.)
Callander Rd. SE6 61Dc 136
Callanders, The WD23: B Hea 18Ga 28
Callan Gro. RM15: S Ock 45Xd 98
Callard Av. N13 21Rb 51
Callard Cl. W2 6B 214 (42Fb 89)
Callcott Ct. NW6 38Bb 69
Callcott Rd. NW6 38Bb 69
Callcott St. W8 46Cb 89
Callendar Rd. SW7 3B 226 (48Fb 89)
Callender Ct. CR0: C'don 72Sb 157
(off Harry Cl.)
Callender Rd. DA8: Erith 52Gd 118
Callcey Down Cres. CR0: New Ad 82Fc 179
Callingham Cl. E14 43Bc 92
Callington M. TW2: Twick 61Ga 130
Callis Ct. SE18 48Qc 94
Callis Farm Cl. TW19: Stanw 58N 105
Callisons Pl. SE10 50Gc 93
Callis Rd. E17 30Bc 52
Calliston Ct. E16 43Jc 93
(off Hammersley Rd.)
Callonfield E17 28Zb 52
Callow Fld. CR8: Purl 85Qb 176
Callow Hill GU25: Vir W 9N 125
Callowland Pl. WD24: Wat 10X 13
Callowlands WD24: Wat 11X 27
(off Leavesden Rd.)
Callow St. SW3 51Fb 111
Calluna Ct. GU22: Wok 90B 168
Cally Pool & Gym 39Pb 70
Calmont Rd. BR1: Broml 65Fc 137
Calmore Cl. RM12: Horn 36Ld 77
Calne Av. IG5: Ilf 25Rc 54
Calonne Rd. SW19 63Za 132
Calshot Av. RM16: Chaf H 47Be 99
Calshot Ct. DA2: Dart 58Rd 119
(off Osbourne Rd.)
Calshot Ho. N1 1H 217 (40Pb 70)
(off Calshot St.)
Calshot Rd. TW6: H'row A 54Q 106
Calshot St. N1 1H 217 (40Pb 70)
Calshot Way EN2: Enf 13Rb 33
Calshot Way TW6: H'row A 54Q 106
Calstock NW1 1C 216 (39Lb 70)
(off Royal Coll. St.)
Calstock Ho. SE11 7A 230 (50Qb 90)
(off Kennings Way)
Calthorpe Gdns. HA8: Edg 22Na 47
Calthorpe Gdns. SM1: Sutt 76Eb 155
Calthorpe St. WC1 5J 217 (42Pb 90)
Calton Av. SE21 58Ub 113
Calton Cl. KT16: Vir W 5L 147
Calton Rd. EN5: New Bar 16Eb 31
Calverley Cl. BR3: Beck 65Dc 136
Calverley Ct. KT19: Ewe 77Ta 153
Calverley Cres. RM10: Dag 33Cd 76
Calverley Gdns. HA3: Kenton 31Ma 67
Calverley Gro. N19 32Mb 70
Calverley Rd. KT17: Ewe 79Wa 154
Calvert Av. E2 4J 219 (41Ub 91)
Calvert Cl. DA14: Sidc 65Ad 139
Calvert Cl. DA17: Belv 49Cd 96
Calvert Cl. KT19: Eps 82Ra 173
Calvert Ct. TW9: Rich 56Pa 109
Calvert Dr. DA2: Wilm 61Fd 140
Calvert Ho. W12 45Xa 88
(off White City Est.)
Calvert M. RM8: Dag 32Ad 75
Calverton SE5 51Ub 113
(off Albany Rd.)
Calverton Pl. UB6: G'frd 37Ha 66

Calverton Rd. E6 39Qc 74
Calvert Rd. EN5: Barn 12Za 30
Calvert Rd. KT24: Eff 100X 191
Calvert Rd. SE10 50Hc 93
Calvert's Bldgs. SE1 7F 225 (46Tb 91)
Calvert St. NW1 39Jb 70
Calvin Cl. BR5: St P 69Zc 139
Calvin St. E1 6K 219 (42Vb 91)
Calydon Rd. SE7 50Kc 93
Calypso Cres. SE15 52Vb 113
Calypso Way SE16 48Bc 92
Camac Rd. TW2: Twick 60Fa 108
Camarthen Grn. NW9 29Ua 48
Cambalt Rd. SW15 57Za 110
Cambay Ho. E1 42Ac 92
(off Harford St.)
Camber Ho. SE15 51Yb 114
Camberley Av. EN1: Enf 14Ub 33
Camberley Av. SW20 68Xa 132
Camberley Cl. SM3: Cheam 76Za 154
Camberley Ho. NW1 2A 216 (40Kb 70)
(off Redhill St.)
Camberley Rd. TW6: H'row A 55Q 106
Cambert Way SE3 56Kc 115
CAMBERWELL 53Tb 113
Camberwell Bus. Cen. SE5 52Tb 113
Camberwell Chu. St. SE5 53Tb 113
Camberwell Glebe SE5 53Ub 113
CAMBERWELL GREEN 53Sb 113
Camberwell Grn. SE5 53Tb 113
Camberwell Gro. SE5 53Tb 113
Camberwell Leisure Cen. 53Tb 113
Camberwell New Rd. SE5 51Qb 112
Camberwell Pas. SE5 53Sb 113
Camberwell Rd. SE5 51Sb 113
Camberwell Sta. Rd. SE5 53Sb 113
Camberwell Ter. DA1: Dart 56Md 119
(off Priory Rd.)
Camberwell Trad. Est. SE5 53Rb 113
Cambeys Rd. RM10: Dag 36Dd 76
Cambisgate SW19 64Ab 132
Cambium Apts. SW19 59Za 110
(off Beatrice Pl.)
Cambium Apts. W8 48Db 89
(off Beatrice Pl.)
Cambium Ho. HA9: Wemb 35Qa 67
(off Palace Arts Way)
Camborne Av. RM3: Rom 24Nd 57
Camborne Av. W13 47Ka 86
Camborne Cl. TW6: H'row A 55Q 106
Camborne Cres. TW6: H'row A 55Q 106
(off Camborne Rd.)
Camborne M. SW18 59Cb 111
Camborne M. W11 44Ab 88
Camborne Rd. CR0: C'don 73Wb 157
Camborne Rd. DA14: Sidc 62Yc 139
Camborne Rd. DA16: Well 54Vc 117
Camborne Rd. HA8: Edg 21Pa 47
Camborne Rd. SM2: Sutt 80Cb 155
Camborne Rd. SM4: Mord 71Za 154
Camborne Rd. SW18 59Cb 111
Camborne Way RM3: Rom 24Nd 57
Camborne Way TW5: Hest 53Ca 107
Cambourne Av. N9 17Zb 34
Cambourne Wlk. TW10: Rich 58Ma 109
Cambrai Ct. N13 20Nb 32
Cambray Rd. BR6: Orp 73Vc 161
Cambray Rd. SW12 60Lb 112
Cambria Cl. DA15: Sidc 60Tc 116
Cambria Cl. TW3: Houn 56Ca 107
Cambria Ct. DA9: Ghithe 56Wd 120
Cambria Ct. E17 27Ac 52
Cambria Ct. SL3: L'ly 7N 81
Cambria Ct. TW14: Felt 59X 107
Cambria Ct. TW18: Staines 63G 126
Cambria Cres. DA12: Grav'nd 3G 144
Cambria Gdns. TW19: Stanw 59N 105
(not continuous)
Cambria Ho. E14 44Ac 92
(off Salmon La.)
Cambria Ho. SE26 63Wb 135
(off High Level Dr.)
Cambrian Av. IG2: Ilf 29Uc 54
Cambrian Cl. SE27 62Rb 135
Cambrian Grn. NW9 29Ua 48
(off Snowdon Dr.)
Cambrian Gro. DA11: Grav'nd 9C 122
Cambrian Rd. E10 31Cc 72
Cambrian Rd. TW10: Rich 58Pa 109
Cambria Rd. SE5 55Sb 113
Cambria St. SW6 52Db 111
Cambridge Av. DA16: Well 56Vc 117
Cambridge Av. KT3: N Mald 69Ua 132
(not continuous)
Cambridge Av. NW6 40Cb 69
Cambridge Av. RM2: Rom 27Ld 57
Cambridge Av. SL1: Burn 10A 60
Cambridge Av. SL1: Slou 4E 80
Cambridge Av. UB6: G'frd 36Ha 66
Cambridge Barracks Rd. SE18 49Pc 94
Cambridge Cir. WC2 3E 222 (44Mb 90)
Cambridge Cl. E17 30Bc 52
Cambridge Cl. EN4: E Barn 18Jb 32
Cambridge Cl. EN8: Chesh 1Yb 20
Cambridge Cl. GU21: Wok 10K 167
Cambridge Cl. N22 25Qb 50
Cambridge Cl. NW10 34Sa 67
Cambridge Cl. SW20 67Xa 132
Cambridge Cl. TW4: Houn 56Aa 107
Cambridge Cl. UB7: Harm 51M 105
Cambridge Cotts. TW9: Kew 51Qa 109
Cambridge Ct. E2 40Xb 71
(off Cambridge Heath Rd.)
Cambridge Ct. N16 31Ub 71
(off Amhurst Pk.)
Cambridge Ct. NW6 40Cb 69
(not continuous)
Cambridge Ct. W2 1D 220 (43Gb 89)
(off Edgware Rd.)
Cambridge Ct. W6 49Ya 88
(off Shepherd's Bush Rd.)
Cambridge Cres. E2 40Xb 71
Cambridge Cres. TW11: Tedd 64Ja 130
Cambridge Dr. EN6: Pot B 3Za 16
Cambridge Dr. HA4: Ruis 33Y 65
Cambridge Dr. SE12 57Jc 115
Cambridge Gdns. EN1: Enf 12Wb 33
Cambridge Gdns. KT1: King T 68Qa 131
Cambridge Gdns. N10 25Jb 50
Cambridge Gdns. N17 24Tb 51
Cambridge Gdns. N21 17Tb 33
Cambridge Gdns. NW6 40Cb 69
Cambridge Gdns. RM16: Grays 9C 100
Cambridge Gdns. W10 44Za 88
Cambridge Ga. NW1 5K 215 (42Kb 90)
Cambridge Ga. M. NW1 5A 216 (42Kb 90)
Cambridge Grn. SE9 60Rc 116
Cambridge Gro. SE20 67Xb 135
Cambridge Gro. W6 49Xa 88
Cambridge Gro. Rd. KT1: King T 69Qa 131
(not continuous)
Cambridge Heath Rd. E1 43Xb 91
Cambridge Heath Rd. E2 42Yb 92
Cambridge Heath Station (Overground) 40Xb 71
Cambridge Ho. SL4: Wind 3G 102
Cambridge Ho. W13 44Ja 86
Cambridge Ho. W6 49Xa 88
(off Cambridge Gro.)
Cambridge Pde. EN1: Enf 11Wb 33
Cambridge Pk. E11 31Jc 73
Cambridge Pk. TW1: Twick 58La 108
Cambridge Pk. Ct. TW1: Twick 59Ma 109
Cambridge Pk. Rd. E11 31Jc 73
(off Lonsdale Rd.)
Cambridge Pas. E9 38Yb 72
Cambridge Pl. W8 47Db 89
Cambridge Rd. AL1: St A 2E 6
Cambridge Rd. BR1: Broml 66Jc 137
Cambridge Rd. CR4: Mitc 69Lb 134
Cambridge Rd. DA14: Sidc 63Uc 138
Cambridge Rd. E11 30Hc 53
Cambridge Rd. E4 18Fc 35
Cambridge Rd. HA2: Harr 29Ca 45
Cambridge Rd. IG11: Bark 38Sc 74
Cambridge Rd. IG3: Ilf 32Uc 74
Cambridge Rd. KT1: King T 68Pa 131
Cambridge Rd. KT12: Walt T 72X 151
Cambridge Rd. KT2: King T 68Pa 131
Cambridge Rd. KT3: N Mald 70Ta 131
Cambridge Rd. KT8: W Mole 70Ba 129
Cambridge Rd. NW6 40Cb 69
(not continuous)
Cambridge Rd. SE20 69Xb 135
Cambridge Rd. SM5: Cars 79Gb 155
Cambridge Rd. SW11 53Hb 111
Cambridge Rd. SW13 54Va 110
Cambridge Rd. SW20 67Wa 132
Cambridge Rd. TW1: Twick 58Ma 109
Cambridge Rd. TW11: Tedd 63Ha 130
Cambridge Rd. TW12: Hamp 66Ba 129
Cambridge Rd. TW15: Ashf 66S 128
Cambridge Rd. TW4: Houn 56Aa 107
Cambridge Rd. TW9: Kew 52Qa 109
Cambridge Rd. UB1: S'hall 46Ba 85
Cambridge Rd. UB8: Uxb 37M 63
Cambridge Rd. W7 47Ha 86
Cambridge Rd. WD18: Wat 14Y 27
Cambridge Rd. Nth. W4 50Ra 87
Cambridge Rd. Sth. W4 50Ra 87
Cambridge Row SE18 50Rc 94
Cambridge Sq. RH1: Redh 9A 208
Cambridge Sq. W2 2D 220 (44Gb 89)
Cambridge St. SW1 6A 228 (49Kb 90)
Cambridge Ter. HP4: Berk 1A 2
Cambridge Ter. N9 17Ub 33
Cambridge Ter. NW1 4K 215 (41Kb 90)
Cambridge Ter. M. NW1 4A 216 (41Kb 90)
Cambridge Theatre 3F 223 (44Nb 90)
(off Earlham St.)
Cambstone Cl. N11 19Jb 32
Cambus Cl. UB4: Yead 43Aa 85
Cambus Rd. E16 43Jc 93
Cam Ct. SE15 51Vb 113
Camdale Rd. SE18 52Vc 117
Camden Arts Cen. 36Db 69
Camden Av. TW13: Felt 60Y 107
Camden Av. UB4: Yead 45Z 85
Camden Cl. BR7: Chst 67Sc 138
Camden Cl. DA11: Nflt 60Ee 121
Camden Cl. RM16: Grays 9D 100
Camden Cotts. KT13: Weyb 76Q 150
Camden Ct. DA17: Belv 50Cd 96
Camden Ct. NW1 38Lb 70
(off Rousden St.)
Camden Gdns. CR7: Thor H 69Rb 135
Camden Gdns. NW1 38Kb 70
Camden Gdns. SM1: Sutt 78Db 155
Camden Gro. BR7: Chst 65Rc 138
Camden High St. NW1 38Kb 70
Camden Hill Rd. SE19 65Ub 135
Camden Ho. HP1: Hem H 3M 3
Camden Ho. SE8 50Bc 92
Camdenhurst St. E14 44Ac 92
Camden La. N7 36Mb 70
Camden Lock Market 38Kb 70
Camden Lock Pl. NW1 38Kb 70
Camden Markets 38Kb 70
Camden M. NW1 38Lb 70
CAMDEN MEWS DAY HOSPITAL 38Lb 70
Camden Pk. Rd. BR7: Chst 66Pc 138
Camden Pk. Rd. NW1 37Mb 70
Camden Pas. N1 1B 218 (40Rb 71)
Camden People's Theatre 5B 216 (42Lb 90)
(off Hampstead Rd.)
Camden Rd. DA5: Bexl 60Ad 117
Camden Rd. E11 30Kc 53
Camden Rd. E17 30Bc 52
Camden Rd. N7 35Nb 70
Camden Rd. NW1 38Lb 70
Camden Rd. RM16: Chaf H 48Ae 99
Camden Rd. SM1: Sutt 78Db 155
Camden Rd. SM5: Cars 77Hb 155
Camden Rd. TN13: S'oaks 94Kd 203
Camden Road Station (Overground) 38Lb 70
Camden Row HA5: Pinn 27Y 45
Camden Row SE3 54Gc 115
Camden Sq. NW1 38Mb 70
(not continuous)
Camden Sq. SE15 53Vb 113
Camden St. NW1 38Lb 70
Camden Studios NW1 1C 216 (39Lb 70)
(off Camden St.)
Camden Ter. NW1 37Mb 70
Camden Ter. TN15: Seal 93Pd 203
CAMDEN TOWN 39Kb 70
Camden Town Station (Underground) 39Kb 70
Camden Wlk. N1 1B 218 (39Rb 71)
Camden Way BR7: Chst 66Pc 138
Camden Way CR7: Thor H 69Rb 135
Cameford Ct. SW2 59Nb 112
Camelford NW1 39Lb 70
(off Royal Coll. St.)
Camelford Ct. W11 44Ab 88
Camelford Ho. RM3: Rom 21Nd 57
(off Chudleigh Rd.)
Camelford Ho. SE1 7G 229 (50Nb 90)
Camelford Wlk. W11 44Ab 88
Camel Gro. KT2: King T 64Ma 131
Camellia Cl. CM13: W H'dn 30Ee 59
Camellia Cl. E10 33Cc 72
Camellia Cl. RM3: Hrld W 25Nd 57
Camellia Ct. GU24: W End 5D 166
Camellia Ho. SE8 52Bc 114
(off Idonia St.)
Camellia Ho. SW11 52Kb 112
Camellia Ho. TW13: Felt 60W 106
(off Bedfont La.)
Camellia La. KT5: Surb 70Ra 131
Camellia M. TW20: Eng G 6N 125
Camellia Pl. TW2: Whitt 59Da 107
Camellia St. SW8 52Nb 112
(not continuous)
Camelot Cl. SE28 47Tc 94
Camelot Cl. SW19 63Bb 133
Camelot Cl. TN16: Big H 88Lc 179
Camelot Ho. NW1 37Mb 70
Camel Rd. E16 46Nc 94
Camera Pl. SW10 51Fb 111
Camera Press Gallery, The 1K 231 (47Vb 91)
(off Queen Elizabeth St.)
Cameret Ct. W11 47Za 88
(off Holland Rd.)
Cameron Cl. CM14: W'ley 21Zd 59
Cameron Cl. DA5: Bexl 62Gd 140
Cameron Cl. N18 21Xb 51
Cameron Cl. N20 19Fb 31
Cameron Cl. N22 24Pb 50
Cameron Cres. HA8: Edg 25Ra 47
Cameron Dr. DA1: Dart 54Qd 119
Cameron Dr. EN8: Walt C 6Zb 20
Cameron Ho. BR1: Broml 67Hc 137
Cameron Ho. NW8 1D 214 (40Gb 69)
(off St John's Wood Ter.)
Cameron Ho. SE5 52Sb 113
Cameron Pl. E1 44Xb 91
Cameron Pl. SW16 61Qb 134
Cameron Rd. BR2: Broml 71Jc 159
Cameron Rd. CR0: C'don 72Rb 157
Cameron Rd. IG3: Ilf 32Uc 74
Cameron Rd. SE6 61Bc 136
Cameron Sq. CR4: Mitc 67Gb 133
Cameron Ter. SE12 62Kc 137
Camerton Cl. E8 37Vb 71
Camfield Pl. AL9: Ess 3M 9
Camgate Cen., The TW19: Stanw 58P 105
Camgate Mans. SE5 51Sb 113
(off Camberwell Rd.)
Cam Grn. RM15: S Ock 44Xd 98
Camilla Cl. KT23: Bookh 97Da 191
Camilla Cl. TW16: Sun 65V 128
Camilla Rd. SE16 49Xb 91
Camille Cl. SE25 69Wb 135
Camlan Rd. BR1: Broml 63Hc 137
Camlet St. E2 5K 219 (42Vb 91)
Camlet Way AL3: St A 1P 5
Camlet Way EN4: Barn 12Cb 31
Camlet Way EN4: Had W 12Cb 31
Camley St. N1 38Mb 70
Camley Street Natural Pk. 1F 217 (40Nb 70)
Camley Street Natural Pk. Vis. Cen. 1E 216 (40Mb 70)
(off Campsfield Rd.)
Camm Av. SL4: Wind 5C 102
Camm Cl. KT15: Add 76M 149
Camm Gdns. KT1: King T 68Pa 131
Camm Gdns. KT7: T Ditt 73Ha 152
Camms Ter. RM10: Dag 36Ed 76
(off Cambeys Rd.)
Camomile Av. CR4: Mitc 67Hb 133
Camomile Rd. RM7: Rush G 33Fd 76
Camomile St. EC3 2J 225 (44Ub 91)
Camomile Way UB7: Yiew 44N 83
CAMP, THE 4F 6
Campaign Ct. W9 42Bb 89
(off Chantry Cl.)
Campana Rd. SW6 53Cb 111
Campania Bldg. E1 45Zb 92
(off Jardine Rd.)
Campaspe Bus. Pk. TW16: Sun 71V 150
Campbell Av. GU22: Wok 93B 188
Campbell Av. IG6: Ilf 28Rc 54
Campbell Cl. HA4: Ruis 30W 44
Campbell Cl. KT14: Byfl 84M 169
Campbell Cl. RM1: Rom 23Gd 56
Campbell Cl. SE18 53Qc 116
Campbell Cl. SW16 63Mb 134
Campbell Cl. TW2: Twick 60Fa 108
Campbell Ct. N17 25Vb 51
Campbell Ct. NW9 30Sa 47
Campbell Ct. SE22 60Wb 113
Campbell Ct. SW7 4A 226 (48Eb 89)
(off Gloucester Rd.)
Campbell Cft. HA8: Edg 22Qa 47
Campbell Gordon Way NW2 35Xa 68
Campbell Ho. KT13: Weyb 78S 150
Campbell Ho. SW1 7B 228 (50Lb 90)
(off Churchill Gdns.)
Campbell Ho. W12 45Xa 88
(off White City Est.)
Campbell Ho. W2 7B 214 (43Fb 89)
(off Hall Pl.)
Campbell Ho. WC1 5D 216 (42Mb 90)
Campbell Rd. CR0: C'don 73Rb 157
Campbell Rd. CR3: Cat'm 93Tb 197
Campbell Rd. DA11: Grav'nd 10B 122
Campbell Rd. E15 35Hc 73
Campbell Rd. E17 28Bc 52
Campbell Rd. E3 41Cc 92
Campbell Rd. E6 39Nc 74
Campbell Rd. KT13: Weyb 80Q 150
Campbell Rd. KT8: E Mos 69Ga 130
Campbell Rd. N17 25Wb 51
Campbell Rd. TW2: Twick 61Fa 130
Campbell Rd. W7 45Ga 86
Campbell Wlk. N1 39Nb 70
(off Outram Pl.)
Campdale Rd. N7 34Mb 70
Campden Cres. HA0: Wemb 34Ka 66
Campden Cres. RM8: Dag 35Xc 75
Campden Gro. W8 47Cb 89
Campden Hill W8 47Cb 89
Campden Hill Ct. W8 47Cb 89
Campden Hill Gdns. W8 46Cb 89
Campden Hill Ga. W8 47Cb 89
Campden Hill Mans. W8 46Cb 89
(off Edge St.)
Campden Hill Pl. W11 46Bb 89
Campden Hill Rd. W8 46Cb 89
Campden Hill Sq. W8 46Bb 89
Campden Hill Towers W11 46Cb 89
Campden Ho. NW6 38Fb 69
(off Harben Rd.)
Campden Ho. W8 46Cb 89
(off Sheffield Ter.)
Campden Ho. Cl. W8 47Cb 89
Campden Ho's. W8 46Cb 89
Campden Ho. Ter. W8 46Cb 89
(off Kensington Chu. St.)
Campden Mans. W8 46Cb 89
(off Kensington Mall)
Campden Rd. CR2: S Croy 78Ub 157
Campden Rd. UB10: Ick 34P 63
Campden St. W8 46Cb 89
Campden Way RM8: Dag 35Xc 75
Campe Ho. N10 24Jb 50
Campen Cl. SW19 61Ab 132
Camp End Rd. KT13: Weyb 84S 170
Camperdown Ho. SL4: Wind 4G 102
Camperdown St. E1 3K 225 (44Vb 91)
Campfield Rd. AL1: St A 3E 6
Campfield Rd. SE9 59Mc 115
Camphill Ct. KT14: W Byf 84J 169
Camphill Ind. Est. KT14: W Byf 83K 169
Camphill Rd. KT14: W Byf 84J 169
Campion Cl. CR0: C'don 77Ub 157
Campion Cl. DA11: Nflt 3A 144
Campion Cl. E6 45Pc 94
Campion Cl. HA3: Kenton 30Pa 47
Campion Cl. RM7: Rush G 33Fd 76
Campion Cl. UB8: Hil 43P 83
Campion Cl. UB9: Den 34J 63
Campion Cl. WD25: Wat 5W 12
Campion Ct. HA0: Wemb 40Na 67
Campion Ct. RM17: Grays 51Fe 121
Campion Dr. KT20: Tad 92Xa 194
Campion Gdns. IG8: Wfd G 22Jc 53
Campion Gro. RM3: Rom 23Nd 57
Campion Ho. E14 44Bc 92
(off Frances Wharf)
Campion Ho. RH1: Redh 3P 207
Campion Ho. SE16 47Ac 92
(off Blondin Way)
Campion Pl. SE28 46Wc 95
Campion Rd. E10 31Dc 72
Campion Rd. HP1: Hem H 3G 2
Campion Rd. SW15 56Ya 110
Campion Rd. TW7: Isle 53Ha 108
Campions IG10: Lough 10Qc 22
Campions, The WD6: Bore 10Pa 15
Campions Cl. WD6: Bore 9Ra 15
Campion Sq. TN14: Dun G 92Hd 202
Campion Ter. NW2 34Za 68
Campion Way HA8: Edg 21Sa 47
Cample La. RM15: S Ock 45Wd 98
Camplin Rd. HA3: Kenton 29Na 47
Camplin St. SE14 52Zb 114
Camp Rd. AL1: St A 2D 6
Camp Rd. CR3: Wold 92Ac 198
Camp Rd. SL9: Ger X 1N 61
Camp Rd. SW19 64Xa 132
Campsbourne, The N8 28Nb 50
Campsbourne Ho. N8 28Nb 50
(off Pembroke Rd.)
Campsbourne Pde. N8 28Nb 50
(off High St.)
Campsbourne Rd. N8 27Nb 50
(not continuous)
Campsey Gdns. RM9: Dag 38Xc 75
Campsey Rd. RM9: Dag 38Xc 75
Campsfield Ho. N8 27Nb 50
(off Campsfield Rd.)
Campsfield Rd. N8 27Nb 50
Campshill Pl. SE13 57Ec 114
Campshill Rd. SE13 57Ec 114
Campus, The HP2: Hem H 1B 4
Campus, The IG10: Lough 14Rc 36
Campus Av. RM8: Dag 35Wc 75
Campus Ct. IG10: Lough 14Qc 36
Campus Ho. TW7: Isle 52Ga 108
Campus Rd. E17 30Bc 52
Campus Way NW4 27Xa 48
Camp Vw. SW19 64Xa 132
Camp Vw. Rd. AL1: St A 3F 6
Cam Rd. E15 39Fc 73
Camrose Av. DA8: Erith 51Dd 118
Camrose Av. HA8: Edg 26Na 47
Camrose Av. TW13: Felt 63Y 129
Camrose Cl. CR0: C'don 73Ac 158
Camrose Cl. SM4: Mord 70Cb 133
Camrose St. SE2 50Wc 95
Camwal Ct. CR4: Mitc 68Gb 133
Canada Av. N18 23Sb 51
Canada Av. RH1: Redh 10A 208
Canada Cres. W3 43Sa 87
Canada Dr. RH1: Redh 10A 208
Canada Est. SE16 48Yb 92
Canada Farm Rd. DA2: G St G 68Xd 142
Canada Farm Rd. DA3: Fawk 71Xd 164
Canada Farm Rd. DA4: S Dar 70Wd 142
Canada Gdns. SE13 57Ec 114
Canada Heights Motorcycle Circuit 69Md 141
Canada House 6E 222 (46Mb 90)
(off Trafalgar Sq.)
Canada Memorial 1B 228 (47Lb 90)
Canada Pl. E14 46Dc 92
(off Canada Sq.)
Canada Rd. DA8: Erith 52Kd 119
Canada Rd. KT11: Cobh 85Y 171
Canada Rd. KT14: Byfl 83M 169
Canada Rd. SL1: Slou 7M 81
Canada Rd. W3 43Sa 87
Canada Sq. E14 46Dc 92
Canada St. SE16 47Zb 92
Canada Water Bus Station 47Yb 92
(off Surrey Quays)
Canada Water Station (Underground & Overground) 47Yb 92
Canada Way W12 45Xa 88
Canada Wharf SE16 46Bc 92
Canadian Av. SE6 60Dc 114
Canadian Memorial Av. TW20: Eng G 8K 125
Canal App. SE8 50Ac 92
Canal Bank KT15: Add 79M 149
Canal Bank M. GU21: Wok 89A 168
Canal Basin DA12: Grav'nd 8F 122
Canal Blvd. NW1 37Mb 70
Canal Bri. KT15: Add 80M 149
CANAL BRIDGE 51Wb 113
Canal Bldg. N1 1D 218 (40Sb 71)
Canal Cl. E1 42Ac 92
Canal Cl. W10 42Ya 88
Canal Cotts. E3 39Bc 72
(off Parnell Rd.)
Canal Ct. HP4: Berk 1A 2
Canaletto EC1 3D 218 (41Sb 91)
(off City Rd.)
Canal Gro. SE15 51Xb 113
Canal Ind. Pk. DA12: Grav'nd 8F 122
Canal Market 38Kb 70
(off Castlehaven Rd.)
Canal Mill Apts. E2 1K 219 (39Vb 71)
(off Boat La.)
Canal Path E2 39Vb 71
Canal Reach N1 39Mb 70
Canal Rd. DA12: Grav'nd 8E 122
Canal Side UB9: Hare 24J 43
Canalside RH1: Mers 3B 208
Canalside RH1: Redh 3B 208
Canalside Activity Cen. 42Za 88
Canalside Gdns. UB2: S'hall 49Aa 85
Canalside Sq. N1 1D 218 (39Sb 71)
Canal Side Studios NW1 39Mb 70
(off St Pancras Way)
Canalside Studios N1 39Ub 71
(off Orsman Rd.)
Canalside Wlk. W2 1B 220 (43Fb 89)
Canal St. SE5 51Tb 113
Canal Wlk. CR0: C'don 72Ub 157
Canal Wlk. E17 27Zb 52
Canal Wlk. N1 39Tb 71
Canal Wlk. NW10 38Sa 67
(off Westend Cl.)
Canal Wlk. SE26 64Yb 136
Canal Way UB9: Hare 23J 43
Canal Way W10 42Za 88
Canal Wharf E8 39Ub 71
(off Kingsland Rd.)
Canal Wharf SL3: L'ly 47C 82
Canal Wharf UB6: G'frd 39Ja 66
Canal Wharf Ind. Est. SL3: L'ly 47C 82
Canal Yd. UB2: S'hall 49X 85
Canary Vw. SE10 51Dc 114
(off Dowells St.)
Canary Wharf Station (Crossrail) 46Dc 92
Canary Wharf Station (Underground & DLR) 46Cc 92
Canberra Cl. NW4 27Wa 48
Canberra Cl. RM10: Dag 39Fd 76
Canberra Cl. RM12: Horn 35Ld 77
Canberra Cres. RM10: Dag 38Fd 76
Canberra Dr. UB4: N'olt 41Z 85
Canberra Dr. UB5: N'olt 41Y 85
Canberra Ho. AL1: St A 2B 6
(off London Rd.)
Canberra Path E10 31Dc 72
(off Whitney Rd.)
Canberra Pl. TW9: Rich 55Qa 109
Canberra Rd. DA7: Bex 51Zc 117
Canberra Rd. E6 39Pc 74
Canberra Rd. SE7 51Lc 115
Canberra Rd. TW6: H'row A 55Q 106
Canberra Rd. W13 46Ja 86
Canberra Sq. RM18: Tilb 4C 122
Canbury Av. KT2: King T 67Pa 131
Canbury Bus. Cen. KT2: King T 67Na 131
Canbury Bus. Pk. KT2: King T 67Na 131
(off Canbury Pk. Rd.)
Canbury Ct. KT2: King T 66Ma 131
Canbury M. SE26 62Wb 135
Canbury Pk. Rd. KT2: King T 67Na 131
Canbury Pas. KT2: King T 67Ma 131
Canbury Path BR5: St M Cry 70Wc 139
Cancell Rd. SW9 53Qb 112
Candahar Rd. SW11 54Gb 111
Cander Way RM15: S Ock 45Xd 98
Candida Ct. NW1 38Kb 70
Candid Ho. NW10 41Xa 88
(off Trenmar Gdns.)
Candlefield Cl. HP3: Hem H 5A 4
Candlefield Rd. HP3: Hem H 5A 4
Candlefield Wlk. HP3: Hem H 5A 4
Candle Gro. SE15 55Xb 113
Candlelight Ct. E15 37Hc 73
(off Romford Rd.)
Candler M. TW1: Twick 59Ja 108
Candler St. N15 30Tb 51
Candlerush Cl. GU22: Wok 89D 168
Candlestick La. EN7: Chesh 1Wb 19
Candle St. E1 43Ac 92
Candover Cl. UB7: Harm 52M 105
Candover Rd. RM12: Horn 32Kd 77
Candover St. W1 1B 222 (43Lb 90)
Candy Cft. KT23: Bookh 98Da 191
Candy Dene DA10: Swans 60Ce 121
Candy St. E3 39Bc 72
Candy Wharf E3 42Ac 92
Cane Hill RM3: Hrld W 26Md 57
Cane Hill Development CR5: Coul 89Lb 176
Cane Hill Dr. CR5: Coul 89Lb 176
Caneland Ct. EN9: Walt A 6Hc 21
Canewdon Cl. GU22: Wok 91A 188
Caney M. NW2 33Za 68
Canfield Dr. HA4: Ruis 36X 65
Canfield Gdns. NW6 38Db 69
Canfield Ho. N15 30Ub 51
(off Albert Rd.)
Canfield Pl. NW6 37Eb 69
Canfield Rd. IG8: Wfd G 24Nc 54
Canfield Rd. RM13: Rain 39Hd 76
Canford Av. UB5: N'olt 39Ba 65
Canford Cl. EN2: Enf 12Qb 32
Canford Dr. KT15: Add 75K 149
Canford Gdns. KT3: N Mald 72Ua 154
Canford Pl. TW11: Tedd 65La 130
Canford Rd. SW11 57Jb 112
Cangels Cl. HP1: Hem H 4J 3
Canham Gdns. TW4: Houn 59Ba 107
Canham Rd. SE25 69Ub 135
Canham Rd. W3 47Ua 88
Can Hatch KT20: Tad 90Ab 174
Canius Ho. CR0: C'don 76Sb 157
(off Scarbrook Rd.)
Canmore Gdns. SW16 66Lb 134
CANN HALL 35Gc 73
Cann Hall Rd. E11 35Gc 73
Cann Ho. W14 48Ab 88
(off Russell Rd.)
Canning Cres. N22 25Pb 50
Canning Cross SE5 54Ub 113
Canning Ho. W12 45Xa 88
(off Australia Rd.)
Canning Pas. W8 48Eb 89
Canning Pl. W8 48Eb 89
Canning Pl. M. W8 3A 226 (48Eb 89)
(off Canning Pl.)
Canning Rd. CR0: C'don 75Vb 157
Canning Rd. E15 40Gc 73
Canning Rd. E17 28Ac 52
Canning Rd. HA3: W'stone 27Ga 46
Canning Rd. N5 34Rb 71
Cannington NW5 37Kb 70
Cannington Rd. RM9: Dag 37Yc 75
CANNING TOWN 43Gc 93
CANNING TOWN 44Hc 93
Canning Town Bus Station 44Gc 93
Canning Town Station (Underground & DLR) 44Gc 93
Cannizaro Rd. SW19 65Ya 132
Cannock Ct. E17 26Ec 52
Cannock Ho. N4 31Sb 71
Cannonbury Av. HA5: Pinn 30Z 45
Cannon Cl. SS17: Stan H 1P 101
Cannon Cl. SW20 69Ya 132
Cannon Cl. TW12: Hamp 65Da 129
Cannon Ct. EC1 5B 218 (42Rb 91)
(off Brewhouse Yd.)
Cannon Cres. GU24: Chob 3J 167
Cannon Dr. E14 45Cc 92
Cannon Ga. SL2: Slou 5N 81
Cannon Gro. KT22: Fet 94Ga 192
Cannon Hill N14 20Nb 32
Cannon Hill NW6 36Cb 69
Cannon Hill La. SW20 71Za 154
Cannon Hill M. N14 20Nb 32
Cannon Ho. SE11 6J 229 (49Pb 90)
(off Beaufoy Wlk.)

Cannon La. HA5: Pinn 29Aa **45**
Cannon La. NW3 34Fb **69**
Cannon M. EN9: Walt A 5Dc **20**
Cannon M. KT22: Fet 93Ga **192**
Cannon Pl. NW3 34Fb **69**
Cannon Pl. SE7 50Nc **94**
Cannon Retail Pk. 45Wc **95**
Cannon Rd. DA7: Bex 53Ad **117**
Cannon Rd. N14 20Nb **32**
Cannon Rd. N17 23Vb **51**
Cannon Rd. WD18: Wat 15Y **27**
Cannonside KT22: Fet 94Ga **192**
Cannon St. AL3: St A 1B **6**
Cannon St. EC4 3D **224** (44Sb **91**)
Cannon St. Rd. E1 44Xb **91**
Cannon Street Station (Rail & Underground) 4F 225 (45Tb 91)
Cannon Trad. Est. HA9: Wemb 35Ra **67**
Cannon Wlk. DA12: Grav'nd 9E **122**
(off Albert Murray Cl.)
Cannon Way KT22: Fet 93Ga **192**
Cannon Way KT8: W Mole 70Ca **129**
Cannon Wharf Bus. Cen. SE8 49Ac **92**
(off Pell St.)
Cannon Wharf Development SE8 49Ac **92**
(off Yeoman St.)
Cannon Workshops E14 45Cc **92**
(off Cannon Dr.)
Canoe Wlk. E14 44Bc **92**
Canon All. EC4 3D **224** (44Sb **91**)
(off Queen's Head Pas.)
Canon Av. RM6: Chad H 29Yc **55**
Canon Beck Rd. SE16 47Yb **92**
Canonbie Rd. SE23 59Yb **114**
CANONBURY 37Sb 71
Canonbury Bus. Cen. N1 39Sb **71**
Canonbury Cotts. EN1: Enf 11Ub **33**
Canonbury Ct. N1 38Rb **71**
(off Hawes St.)
Canonbury Cres. N1 38Sb **71**
Canonbury Gro. N1 38Sb **71**
Canonbury Hgts. N1 37Tb **71**
(off Dove Rd.)
Canonbury La. N1 38Rb **71**
Canonbury Pk. Nth. N1 37Sb **71**
Canonbury Pk. Sth. N1 37Sb **71**
Canonbury Pl. N1 37Rb **71**
(not continuous)
Canonbury Rd. EN1: Enf 11Ub **33**
Canonbury Rd. N1 37Rb **71**
Canonbury Sq. N1 38Rb **71**
Canonbury Station (Overground) 36Sb 71
Canonbury St. N1 38Sb **71**
Canonbury Vs. N1 38Rb **71**
Canon Ho. W10 41Bb **89**
(off Bruckner St.)
Canon Mohan Cl. N14 16Jb **32**
Canon Rd. BR1: Broml 69Lc **137**
Canon Row SW1 2F **229** (47Nb **90**)
(not continuous)
Canon's Cl. N2 31Fb **69**
Canons Cl. CR4: Mitc 70Hb **133**
Canons Cl. HA8: Edg 23Pa **47**
Canons Cl. RH2: Reig 5H **207**
Canons Cl. WD7: R'lett 7Ka **14**
Canons Cnr. HA8: Edg 21Na **47**
Canons Ct. E15 35Gc **73**
Canons Ct. HA8: Edg 23Pa **47**
Canons Dr. HA8: Edg 23Na **47**
Canon's Hill CR5: Coul 89Rb **177**
Canon's Hill CR5: Purl 89Rb **177**
Canons La. KT20: Tad 90Ab **174**
Canonsleigh Rd. RM9: Dag 38Xc **75**
Canons Leisure Cen. Mitcham 70Hb **133**
Canons Pk. 23Ma **47**
CANONS PARK 24Ma 47
Canons Pk. Cl. HA8: Edg 24Na **47**
Canons Park Station (Underground) 24Na 47
Canons Row HA8: Edg 21Na **47**
Canon St. N1 39Sb **71**
Canon's Wlk. CR0: C'don 76Zb **158**
Canons Way HA8: Edg 21Pa **47**
Canons Yd. RH2: Reig 9K **207**
Canopus Way HA6: Nwood 21W **44**
Canopus Way TW19: Stanw 59N **105**
Canrobert St. E2 40Xb **71**
Cantelowes Rd. NW1 37Mb **70**
Canterbury Av. DA15: Sidc 61Xc **139**
Canterbury Av. IG1: Ilf 31Nc **74**
Canterbury Av. RM14: Upm 32Vd **78**
Canterbury Av. SL2: Slou 2G **80**
Canterbury Cl. BR3: Beck 67Dc **136**
Canterbury Cl. DA1: Dart 59Qd **119**
Canterbury Cl. E6 44Pc **94**
Canterbury Cl. HA6: Nwood 23V **44**
Canterbury Cl. IG7: Chig 20Vc **37**
Canterbury Cl. KT4: Wor Pk 75Za **154**
Canterbury Cl. SE5 54Sb **113**
(off Lilford Rd.)
Canterbury Cl. UB6: G'frd 43Da **85**
Canterbury Ct. AL1: St A 1D **6**
(off Battlefield Rd.)
Canterbury Ct. CM15: Pil H 15Vd **40**
Canterbury Ct. CR2: S Croy 80Sb **157**
(off St Augustine's Av.)
Canterbury Ct. NW6 40Cb **69**
(off Canterbury Rd.)
Canterbury Ct. NW9 26Ua **48**
Canterbury Ct. SE12 62Kc **137**
Canterbury Ct. SE5 52Qb **112**
Canterbury Ct. TW15: Ashf 63P **127**
Canterbury Cres. SW9 55Qb **112**
Canterbury Gro. SE27 63Qb **134**
Canterbury Hall KT4: Wor Pk 73Xa **154**
Canterbury Ho. CR0: C'don 74Tb **157**
(off Sydenham Rd.)
Canterbury Ho. DA8: Erith 52Hd **118**
Canterbury Ho. E3 41Dc **92**
(off Bow Rd.)
Canterbury Ho. IG11: Bark 38Wc **75**
(off Margaret Bondfield Av.)
Canterbury Ho. KT19: Eps 83Qa **173**
(off Queen Alexandra's Way)
Canterbury Ho. RM8: Dag 35Wc **75**
(off Academy Way)
Canterbury Ho. SE1 3J **229** (48Pb **90**)
Canterbury Ho. SE8 50Cc **92**
(off Wharf St.)
Canterbury Ho. WD24: Wat 12Y **27**
(off Anglian Cl.)
Canterbury Ho. WD6: Bore 12Qa **29**
(off Stratfield Rd.)
Canterbury Ind. Pk. SE15 51Yb **114**
Canterbury M. KT22: Oxs 85Ea **172**
Canterbury M. SL4: Wind 4E **102**
Canterbury Pde. RM15: S Ock 41Yd **98**
Canterbury Pl. RM17: Grays 50Fe **99**
Canterbury Pl. SE17 7C **230** (50Rb **91**)
Canterbury Rd. NW6: Carlton Va. 40Bb **69**
Canterbury Rd. NW6: Princess Rd. 40Cb **69**
Canterbury Rd. CR0: C'don 73Pb **156**
Canterbury Rd. DA12: Grav'nd 1E **144**
Canterbury Rd. E10 31Ec **72**
Canterbury Rd. HA1: Harr 29Da **45**
Canterbury Rd. HA2: Harr 29Da **45**
Canterbury Rd. SM4: Mord 73Db **155**
Canterbury Rd. TW13: Hanw 61Aa **129**
Canterbury Rd. WD17: Wat 12X **27**
Canterbury Rd. WD6: Bore 12Qa **29**
Canterbury Ter. NW6 40Cb **69**
Canterbury Way CM13: Gt War 23Yd **58**
Canterbury Way RM19: Purf 51Ud **120**
Canterbury Way RM20: W Thur 52Ud **120**
Canterbury Way WD3: Crox G 13S **26**
Canter Cl. WD25: A'ham 12Da **27**
Canter Way E1 44Wb **91**
Cantium Retail Pk. 51Wb **113**
Cantley Gdns. IG2: Ilf 30Sc **54**
Cantley Gdns. SE19 67Vb **135**
Cantley Rd. W7 48Ja **86**
Canto Ct. EC1 5E **218** (42Sb **91**)
(off Old St.)
Canton Cl. EN7: Chesh 2Vb **19**
Canton St. E14 44Cc **92**
Cantrell Rd. E3 42Bc **92**
Cantwell Rd. SE18 52Rc **116**
Canute Gdns. SE16 49Zb **92**
Canvey St. SE1 6D **224** (46Sb **91**)
Canville Ri. TN16: Westrm 97Tc **200**
Canyon Gdn. E3 41Bc **92**
(off St Clements Av.)
Capability Way DA9: Ghithe 56Yd **120**
Cape Cl. IG11: Bark 38Rc **74**
Cape Henry Ct. E14 45Fc **93**
(off Jamestown Way)
Cape Ho. E16 46Kc **93**
(off Cunningham Av.)
Cape Ho. E8 37Vb **71**
(off Dalston La.)
Capelands DA3: Nw A G 75Ce **165**
Capel Av. SM6: W'gton 78Pb **156**
Capel Cl. BR2: Broml 74Nc **160**
Capel Cl. N20 20Eb **31**
Capel Cl. SS17: Stan H 1N **101**
Capel Ct. EC2 3G **225** (44Tb **91**)
(off Bartholomew La.)
Capel Ct. SE20 67Yb **136**
Capel Cres. HA7: Stan 19Ja **28**
Capel Gdns. HA5: Pinn 28Ba **45**
Capel Gdns. IG3: Bark 35Vc **75**
Capel Gdns. IG3: Ilf 35Vc **75**
Capel Ho. E9 38Yb **72**
(off Loddiges Rd.)
Capel Ho. WD19: Wat 21Z **45**
Capella Rd. HA6: Nwood 21V **44**
Capell Av. WD3: Chor 15E **24**
Capell Rd. WD3: Chor 15F **24**
Capell Way WD3: Chor 15F **24**
Capel Manor Gdns. 7Wb **19**
Capel Pl. DA2: Wilm 63Ld **141**
Capel Rd. E12 35Mc **73**
Capel Rd. E7 35Kc **73**
Capel Rd. EN1: Enf 8Xb **19**
Capel Rd. EN4: E Barn 16Gb **31**
Capel Rd. WD19: Wat 16Aa **27**
Capelvere Wlk. WD17: Wat 11U **26**
Capener's Cl. SW1 2H **227** (47Jb **90**)
(off Kinnerton St.)
Capern Rd. SW18 60Eb **111**
Cape Rd. AL1: St A 2F **6**
Cape Rd. N17 27Wb **51**
Cape Yd. E1 46Wb **91**
Capital Bus. Cen. CR2: S Croy 80Tb **157**
Capital Bus. Cen. HA0: Wemb 40Ma **67**
Capital Bus. Cen. WD24: Wat 8Z **13**
Capital Bus. Pk. WD6: Bore 13Sa **29**
Capital E. Apts. E16 45Jc **93**
(off Western Gateway)
Capital Ho. SW15 57Ab **110**
(off Plaza Gdns.)
Capital Ind. Est. CR4: Mitc 71Hb **155**
Capital Ind. Est. DA17: Belv 48Dd **96**
Capital Interchange Way TW8: Bford 50Qa **87**
Capital Mill Apts. E2 1K **219** (39Vb **71**)
(off Whiston Rd.)
Capital Pk. GU22: Wok 93D **188**
Capital Trad. Est. IG11: Bark 40Tc **74**
Capital Wharf E1 46Wb **91**
Capitol Bldg. SW11 51Mb **112**
(off New Union Sq.)
Capitol Ind. Pk. NW9 27Sa **47**
Capitol Sq. KT17: Eps 85Ua **174**
Capitol Wlk. SE23 61Yb **136**
(off London Rd.)
Capitol Way NW9 27Sa **47**
Capland Ho. NW8 5C **214** (42Fb **89**)
(off Capland St.)
Capland St. NW8 5C **214** (42Fb **89**)
Caple Ho. SW10 52Eb **111**
(off King's Rd.)
Caple Rd. NW10 40Va **68**
Capon Cl. CM14: B'wood 18Xd **40**
Capper St. WC1 6C **216** (42Lb **90**)
Caprea Cl. UB4: Yead 43Z **85**
Capricorn Cen. RM8: Dag 31Bd **75**
Capricorn Ct. HA8: Edg 24Qa **47**
(off Zodiac Cl.)
Capri Ho. E17 26Bc **52**
Capri Ho. NW9 27Va **48**
(off Caversham Rd.)
Capri Ho. RM15: S Ock 42Xd **98**
Capri Rd. CR0: C'don 74Vb **157**
Capstan Cen. RM18: Tilb 52Fe **121**
Capstan Cl. RM6: Chad H 30Xc **55**
Capstan Ct. DA2: Dart 56Sd **120**
Capstan Ct. E1 45Yb **92**
(off Wapping Wall)
Capstan Dr. RM13: Rain 42Jd **96**
Capstan Ho. E14: Clove Cres. 45Fc **93**
(off Clove Cres.)
Capstan Ho. E14: Stebondale St. 49Ec **92**
(off Stebondale St.)
Capstan M. DA11: Nflt 9A **122**
Capstan Ride EN2: Enf 12Qb **32**
Capstan Rd. SE8 49Bc **92**
Capstan Sq. E14 47Ec **92**
Capstans Wharf GU21: Wok 10K **167**
Capstan Way SE16 46Ac **92**
Capstone Rd. BR1: Broml 63Hc **137**
Capswood Bus. Cen. UB9: Den 32E **62**
Captain St. SE18 53Nc **116**
(off Tellson Av.)
Captains Wlk. HP4: Berk 2A **2**
Capthorne Av. HA2: Harr 32Aa **65**
Capuchin Cl. HA7: Stan 23Ka **46**
Capulet M. E16 46Jc **93**
Capulet Sq. E3 41Dc **92**
(off Talwin St.)
Capworth St. E10 32Cc **72**
Caractacus Cott. Vw. WD18: Wat 17W **26**
Caractacus Grn. WD18: Wat 16V **26**
Caradoc Cl. W2 44Cb **89**
Caradoc Evans Cl. N11 22Kb **50**
(off Springfield Rd.)
Caradoc St. SE10 50Gc **93**
Caradon Cl. E11 32Gc **73**
Caradon Cl. GU21: Wok 10M **167**
Caradon Way N15 28Tb **51**
Caragh M. UB2: S'hall 49Aa **85**
Cara Ho. N1 38Qb **70**
(off Liverpool Rd.)
Cara Ho. NW9 27Sa **47**
Caramel Ct. E3 40Dc **72**
(off Taylor Pl.)
Caranday Vs. W11 46Za **88**
(off Norland Rd.)
Carat Ho. E14 43Cc **92**
(off Ursula Gould Way)
Caravan La. WD3: Rick 17N **25**
Caravan Site, The RM15: Avel Aveley 48Td **98**
Caravel Cl. E14 48Cc **92**
Caravel Cl. RM16: Chaf H 48Be **99**
Caravel Ho. E16 47Kc **93**
(off Regalia Close)
Caravelle Gdns. UB5: N'olt 41Z **85**
Caravel M. SE8 51Cc **114**
Caraway Apts. SE1 1K **231** (47Vb **91**)
(off Cayenne Ct.)
Caraway Cl. E13 43Kc **93**
Caraway Hgts. E14 45Ec **92**
(off Poplar High St.)
Caraway Pl. SM6: W'gton 76Kb **156**
Carberry Rd. SE19 65Ub **135**
Carbery Av. W3 47Pa **87**
Carbery La. SL5: Asc 9A **124**
Carbis Cl. E4 18Fc **35**
Carbis Rd. E14 44Bc **92**
Carbrooke Ho. E9 39Yb **72**
(off Templecombe Rd.)
Carbuncle Pas. N17 26Wb **51**
Carburton St. W1 7A **216** (43Kb **90**)
Carbury Cl. RM12: Horn 37Ld **77**
Cardale St. E14 47Ec **92**
Cardamon Bldg. SE1 1K **231** (47Vb **91**)
(off Shad Thames)
Carden Ct. KT8: W Mole 70Da **129**
Carden Rd. SE15 55Xb **113**
Cardiff Cl. RM5: Col R 24Fd **56**
Cardiff Ho. SE15 51Wb **113**
(off Friary Est.)
Cardiff Rd. EN3: Pond E 14Xb **33**
Cardiff Rd. W7 48Ja **86**
Cardiff Rd. WD18: Wat 15X **27**
Cardiff St. SE18 52Uc **116**
Cardiff Way WD5: Ab L 4W **12**
Cardigan Cl. GU21: Wok 10J **167**
Cardigan Cl. SL1: Slou 5D **80**
Cardigan Ct. W7 42Ha **86**
(off Copley Cl.)
Cardigan Gdns. IG3: Ilf 33Wc **75**
Cardigan Ho. RM3: Rom 22Md **57**
(off Bridgwater Wlk.)
Cardigan Pl. SE3 54Fc **115**
Cardigan Rd. E3 40Bc **72**
Cardigan Rd. SW13 54Wa **110**
Cardigan Rd. SW19 65Eb **133**
Cardigan Rd. TW10: Rich 58Na **109**
Cardigan St. SE11 7K **229** (50Qb **90**)
Cardigan Wlk. N1 38Sb **71**
(off Ashby Gro.)
Cardinal Av. KT2: King T 64Na **131**
Cardinal Av. SM4: Mord 72Ab **154**
Cardinal Av. WD6: Bore 13Ra **29**
Cardinal Bourne St. SE1 4G **231** (48Tb **91**)
Cardinal Cap All. SE1 6D **224** (46Sb **91**)
Cardinal Cl. BR7: Chst 67Uc **138**
Cardinal Cl. CR2: Sande 85Wb **177**
Cardinal Cl. HA8: Edg 24Sa **47**
Cardinal Cl. KT4: Wor Pk 77Wa **154**
Cardinal Cl. SM4: Mord 72Ab **154**
Cardinal Ct. E1 45Wb **91**
(off Thomas More St.)
Cardinal Cres. KT3: N Mald 68Sa **131**
Cardinal Dr. IG6: Ilf 23Sc **54**
Cardinal Dr. KT12: Walt T 74Z **151**
Cardinal Gro. AL3: St A 4P **5**
Cardinal Hinsley Cl. NW10 40Wa **68**
Cardinal Mans. SW1 5B **228** (49Lb **90**)
(off Carlisle Pl.)
Cardinal Pl. AL2: Park 7B **6**
Cardinal Pl. GU22: Wok 90A **168**
Cardinal Pl. SW1 3B **228** (48Lb **90**)
(not continuous)
Cardinal Pl. SW15 56Za **110**
Cardinal Rd. HA4: Ruis 32Z **65**
Cardinal Rd. RM16: Chaf H 48Ae **99**
Cardinal Rd. TW13: Felt 60X **107**
Cardinals Wlk. SL6: Tap 4A **80**
Cardinals Wlk. TW12: Hamp 66Ea **130**
Cardinals Wlk. TW16: Sun 65U **128**
Cardinals Way N19 32Mb **70**
Cardinal Wlk. SW1 4B **228** (48Lb **90**)
(off Victoria St.)
Cardinal Way HA3: W'stone 27Ga **46**
Cardinal Way RM13: Rain 40Md **77**
Cardine M. SE15 52Xb **113**
Cardingham GU21: Wok 9L **167**
Cardington Sq. TW4: Houn 56Z **107**
Cardington St. NW1 3C **216** (41Lb **90**)
Cardinham Rd. BR6: Chels 77Vc **161**
Cardozo Rd. N7 36Nb **70**
Cardrew Av. N12 22Fb **49**
Cardrew Cl. N12 22Gb **49**
Cardrew Ct. N12 22Fb **49**
Cardross Ho. W6 48Xa **88**
(off Cardross St.)
Cardross St. W6 48Xa **88**
Cardwell Cres. SL5: S'hill 1A **146**
Cardwell Rd. N7 35Nb **70**
Cardwell Ter. N7 35Nb **70**
(off Cardwell Rd.)
Cardy Rd. HP1: Hem H 3K **3**
Career Ct. SE16 47Zb **92**
(off Christopher Cl.)
Carew Cl. CR5: Coul 91Rb **197**
Carew Cl. N7 33Pb **70**
Carew Ct. RM16: Chaf H 48Ae **99**
Carew Ct. RM6: Chad H 30Xc **55**
(off Quarles Pk. Rd.)
Carew Ct. SE14 51Zb **114**
(off Samuel Cl.)
Carew Ct. SM2: Sutt 81Db **175**
Carew Manor & Dovecote 76Mb **156**
Carew Mnr. Cotts. SM6: Bedd 76Mb **156**
Carew Rd. CR4: Mitc 68Jb **134**
Carew Rd. CR7: Thor H 70Rb **135**
Carew Rd. HA6: Nwood 23U **44**
Carew Rd. N17 26Wb **51**
Carew Rd. SM6: W'gton 79Lb **156**
Carew Rd. TW15: Ashf 65S **128**
Carew Rd. W13 47La **86**
Carew St. SE5 54Sb **113**
Carew Way BR5: Orp 74Yc **161**
Carew Way WD19: Wat 20Ba **27**
Carey Cl. SL4: Wind 5F **102**
Carey Ct. DA6: Bex 57Dd **118**
Carey Ct. SE5 52Sb **113**
Carey Gdns. SW8 53Lb **112**
Carey La. EC2 2D **224** (44Sb **91**)
Carey Mans. SW1 5D **228** (49Mb **90**)
(off Rutherford St.)
Carey Pl. SW1 6D **228** (49Mb **90**)
Carey Pl. WD17: Wat 14Y **27**
Carey Rd. RM9: Dag 35Ad **75**
Carey's Fld. TN13: Dun G 92Gd **202**
Carey St. WC2 3J **223** (44Pb **90**)
Carey Way HA9: Wemb 35Ra **67**
Carfax Ho. SE20 67Wb **135**
Carfax Pl. SW4 56Mb **112**
Carfax Rd. RM12: Horn 35Hd **76**
Carfax Rd. UB3: Harl 50V **84**
Carfree Cl. N1 38Qb **70**
Cargill Rd. SW18 60Db **111**
Cargo Point TW19: Stanw 58P **105**
Cargreen Pl. SE25 70Vb **135**
(off Cargreen Rd.)
Cargreen Rd. SE25 70Vb **135**
Cargrey Ho. HA7: Stan 22La **46**
Carholme Rd. SE23 60Bc **114**
Carillon Ct. E1 43Wb **91**
(off Greatorex St.)
Carillon Ct. W5 45Ma **87**
Carina Ho. E20 36Ec **72**
(off Cheering La.)
Carinthia Ct. SE16 49Ac **92**
(off Plough Way)
Carisbrook N10 26Kb **50**
Carisbrook Cl. CM16: Epp 3Wc **23**
Carisbrook Cl. EN1: Enf 11Vb **33**
Carisbrooke Av. DA5: Bexl 60Zc **117**
Carisbrooke Av. WD24: Wat 11Z **27**
Carisbrooke Cl. HA7: Stan 26Ma **47**
Carisbrooke Cl. RM11: Horn 32Qd **77**
Carisbrooke Cl. TW4: Houn 59Aa **107**
Carisbrooke Ct. DA2: Dart 58Rd **119**
(off Osbourne Rd.)
Carisbrooke Ct. SL1: Slou 5K **81**
Carisbrooke Ct. SM2: Cheam 80Bb **155**
Carisbrooke Ct. UB5: N'olt 39Ba **65**
(off Eskdale Av.)
Carisbrooke Ct. W1 1J **221** (43Jb **90**)
(off Weymouth St.)
Carisbrooke Ct. W3 47Sa **87**
(off Brouncker Rd.)
Carisbrooke Gdns. SE15 52Vb **113**
Carisbrooke Ho. HA6: Nwood 22V **44**
Carisbrooke Ho. KT2: King T 67Na **131**
(off Seven Kings Way)
Carisbrooke Ho. TW10: Rich 57Qa **109**
Carisbrooke Ho. UB7: W Dray 47P **83**
(off Park Lodge Av.)
Carisbrooke Rd. BR2: Broml 70Lc **137**
Carisbrooke Rd. CR4: Mitc 70Mb **134**
Carisbrooke Rd. E17 28Ac **52**
Carisbrook Rd. AL2: Chis G 8P **5**
Carisbrook Rd. CM15: Pil H 16Xd **40**
Carker's La. NW5 36Kb **70**
Carlbury Cl. AL1: St A 3F **6**
Carlcott Cl. KT12: Walt T 73X **151**
Carl Ekman Ho. DA11: Nflt 59Fe **121**
Carleton Av. KT8: E Mos 72Fa **152**
Carleton Av. SM6: W'gton 81Mb **176**
Carleton Cl. KT10: Esh 74Fa **152**
Carleton Gdns. N19 36Lb **70**
Carleton Ho. NW9 27Va **48**
(off Boulevard Dr.)
Carleton Pl. DA4: Hort K 70Sd **142**
Carleton Rd. DA1: Dart 59Qd **119**
Carleton Rd. N19 36Mb **70**
Carleton Rd. N7 36Mb **70**
Carleton Vs. NW5 36Lb **70**
Carlile Cl. E3 40Bc **72**
Carlile Ho. SE1 4G **231** (48Tb **91**)
(off Tabard St.)
Carlile Pl. TW10: Rich 58Pa **109**
Carlina Gdns. IG8: Wfd G 22Kc **53**
Carlingford Gdns. CR4: Mitc 66Hb **133**
Carlingford Rd. N15 27Rb **51**
Carlingford Rd. NW3 35Fb **69**
Carlingford Rd. SM4: Mord 72Za **154**
Carlisle Av. AL1: St A 1C **6**
Carlisle Av. AL3: St A 1B **6**
Carlisle Av. EC3 3J **225** (44Ub **91**)
Carlisle Av. W3 44Ua **88**
Carlisle Cl. HA5: Pinn 31Aa **65**
Carlisle Cl. KT2: King T 67Qa **131**
Carlisle Gdns. HA3: Kenton 31Ma **67**
Carlisle Gdns. IG1: Ilf 30Nc **54**
Carlisle Ho. IG1: Ilf 30Nc **54**
Carlisle Ho. WD6: Bore 12Qa **29**
Carlisle La. SE1 4J **229** (48Pb **90**)
Carlisle Mans. SW1 5B **228** (49Lb **90**)
(off Carlisle Pl.)
Carlisle M. KT2: King T 67Qa **131**
Carlisle Pl. N11 21Kb **50**
Carlisle Pl. SW1 4B **228** (48Lb **90**)
Carlisle Rd. DA1: Dart 58Qd **119**
Carlisle Rd. E10 32Cc **72**
Carlisle Rd. N4 31Qb **70**
Carlisle Rd. NW6 39Ab **68**
Carlisle Rd. NW9 27Sa **47**
Carlisle Rd. RM1: Rom 29Jd **56**
Carlisle Rd. SL1: Slou 5H **81**
Carlisle Rd. SM1: Sutt 79Bb **155**
Carlisle Rd. TW12: Hamp 66Ca **129**
Carlisle St. W1 3D **222** (44Mb **90**)
Carlisle Wlk. E8 37Vb **71**
Carlisle Way SW17 64Jb **134**
Carlos Pl. W1 5J **221** (45Jb **90**)
Carlow St. NW1 1B **216** (40Lb **70**)
Carlson Ct. SW15 56Bb **111**
Carlton Av. CR2: S Croy 80Ub **157**
Carlton Av. DA9: Ghithe 58Ud **120**
Carlton Av. HA3: Kenton 29Ka **46**
Carlton Av. N14 15Mb **32**
Carlton Av. TW14: Felt 58Y **107**
Carlton Av. UB3: Harl 49U **84**
Carlton Av. E. HA9: Wemb 33Ma **67**
Carlton Av. W. HA0: Wemb 33Ka **66**
Carlton Cl. GU21: Wok 86B **168**
Carlton Cl. HA8: Edg 22Qa **47**
Carlton Cl. KT9: Chess 79Ma **153**
Carlton Cl. NW3 33Cb **69**
Carlton Cl. RM14: Upm 33Rd **77**
Carlton Cl. UB5: N'olt 36Ea **66**
Carlton Cl. WD6: Bore 14Ta **29**
Carlton Ct. IG6: Ilf 27Tc **54**
Carlton Ct. N3 24Cb **49**
Carlton Ct. SE20 67Xb **135**
Carlton Ct. TW18: Staines 64J **127**
Carlton Ct. UB8: Cowl 43M **83**
Carlton Ct. W9 40Db **69**
(off Maida Vale)
Carlton Ct. WD19: Wat 16Y **27**
Carlton Cres. SM3: Cheam 77Ab **154**
Carlton Dr. IG6: Ilf 27Tc **54**
Carlton Dr. SW15 57Za **110**
Carlton Gdns. SW1 7D **222** (46Mb **90**)
Carlton Gdns. W5 44La **86**
Carlton Grn. DA14: Sidc 63Vc **139**
Carlton Grn. RH1: Redh 3N **207**
Carlton Gro. SE15 53Xb **113**
Carlton Hill NW8 40Db **69**
Carlton Ho. IG10: Lough 15Mc **35**
Carlton Ho. NW6 40Cb **69**
(off Canterbury Ter.)
Carlton Ho. SE16 47Zb **92**
(off Wolfe Cres.)
Carlton Ho. TW14: Felt 58V **106**
Carlton Ho. TW3: Houn 58Ca **107**
Carlton Ho. Ter. SW1 7D **222** (46Mb **90**)
Carlton Lodge N4 31Qb **70**
(off Carlton Rd.)
Carlton Mans. N16 32Vb **71**
Carlton Mans. NW6 38Cb **69**
(off West End La.)
Carlton Mans. W14 47Ab **88**
(off Holland Pk. Gdns.)
Carlton Mans. W9 41Db **89**
Carlton M. IG10: Lough 13Hc **35**
Carlton M. NW6 36Cb **69**
(off West Cotts.)
Carlton Pde. BR6: St M Cry 73Xc **161**
Carlton Pde. HA9: Wemb 33Na **67**
Carlton Pde. TN13: S'oaks 94Ld **203**
Carlton Pk. Av. SW20 68Za **132**
Carlton Pl. HA6: Nwood 22R **44**
Carlton Pl. KT13: Weyb 77R **150**
(off Castle Vw. Rd.)
Carlton Rd. CR2: S Croy 79Tb **157**
Carlton Rd. DA14: Sidc 64Vc **139**
Carlton Rd. DA16: Well 55Xc **117**
Carlton Rd. DA8: Erith 51Dd **118**
Carlton Rd. E11 32Hc **73**
Carlton Rd. E12 35Mc **73**
Carlton Rd. E17 25Ac **52**
Carlton Rd. GU21: Wok 86C **168**
Carlton Rd. KT12: Walt T 73X **151**
Carlton Rd. KT3: N Mald 68Ua **132**
Carlton Rd. N11 22Jb **50**
Carlton Rd. N4 31Qb **70**
Carlton Rd. RH1: Redh 3N **207**
Carlton Rd. RH2: Reig 4M **207**
Carlton Rd. RM16: Grays 7B **100**
Carlton Rd. RM2: Rom 29Hd **56**
Carlton Rd. SL2: Slou 5M **81**
Carlton Rd. SW14 55Sa **109**
Carlton Rd. TW16: Sun 66V **128**
Carlton Rd. W4 47Ta **87**
Carlton Rd. W5 45La **86**
Carlton Sq. E1 42Zb **92**
(not continuous)
Carlton St. SW1 5D **222** (45Mb **90**)
Carlton Ter. E11 29Kc **53**
(not continuous)
Carlton Ter. E7 38Lc **73**
Carlton Ter. N18 20Tb **33**
Carlton Ter. SE26 62Yb **136**
Carlton Twr. Pl. SW1 3G **227** (48Hb **89**)
Carlton Towers SM5: Cars 76Hb **155**
Carlton Va. NW6 40Bb **69**
Carlton Vs. SW15 57Ab **110**
Carlton Works, The SE15 52Xb **113**
(off Asylum Rd.)
Carlwell St. SW17 64Gb **133**
Carlyle Av. BR1: Broml 69Mc **137**
Carlyle Av. UB1: S'hall 45Ba **85**
Carlyle Cl. KT8: W Mole 68Da **129**
Carlyle Cl. N2 30Eb **49**
Carlyle Ct. SW10 53Eb **111**
(off Chelsea Harbour Dr.)
Carlyle Ct. SW6 53Db **111**
(off Imperial Rd.)
Carlyle Gdns. UB1: S'hall 45Ba **85**
Carlyle Ho. KT8: W Mole 71Ca **151**
(off Down St.)
Carlyle Ho. N16 34Ub **71**
Carlyle Ho. SE5 52Sb **113**
(off Bethwin Rd.)
Carlyle Ho. SW3 51Fb **111**
(off Old Church St.)
Carlyle Mans. SW3 51Gb **111**
(off Cheyne Wlk.)
Carlyle Mans. W8 46Cb **89**
(off Kensington Mall)
Carlyle M. E1 42Zb **92**
Carlyle Pl. SW15 56Za **110**
Carlyle Rd. CR0: C'don 75Wb **157**
Carlyle Rd. E12 35Nc **74**
Carlyle Rd. NW10 39Ta **67**
Carlyle Rd. SE28 45Xc **95**
Carlyle Rd. TW18: Staines 66J **127**
Carlyle Rd. W5 49La **86**
Carlyle's House 51Gb **111**
(off Cheyne Row)
Carlyle Sq. SW3 50Fb **111**
Carly M. E2 41Wb **91**
(off Barnet Gro.)
Carlyon Av. HA2: Harr 35Ba **65**
Carlyon Cl. HA0: Wemb 39Na **67**
Carlyon Rd. HA0: Wemb 40Na **67**
Carlyon Rd. UB4: Yead 43Y **85**
(not continuous)
Carlys Cl. BR3: Beck 68Zb **136**
Carmalt Gdns. KT12: Hers 78Y **151**
Carmalt Gdns. SW15 56Ya **110**
Carmarthen Ct. W7 42Ha **86**
(off Copley Cl.)
Carmarthen Pl. SE1 1H **231** (47Ub **91**)
Carmarthen Rd. SL1: Slou 5J **81**
Carmel Cl. GU22: Wok 90A **168**
Carmel Ct. HA9: Wemb 33Ra **67**
Carmel Ct. W8 47Db **89**
(off Holland St.)
Carmelite Cl. HA3: Hrw W 25Ea **46**
Carmelite Rd. HA3: Hrw W 25Ea **46**
Carmelite St. EC4 4A **224** (45Qb **90**)
Carmelite Wlk. HA3: Hrw W 25Ea **46**
Carmelite Way DA3: Hartl 71Be **165**
Carmelite Way HA3: Hrw W 26Ea **46**
Carmel Lodge SW6 51Cb **111**
(off Lillie Rd.)
Carmelo M. E1 43Zb **92**
(off Maria Ter.)
Carmel Way TW9: Rich 54Ra **109**

Carmen Ct. WD6: Bore 10Pa 15
(off Aycliffe Rd.)
Carmen St. E14 44Dc 92
Carmichael Av. DA9: Ghithe 56Yd 120
Carmichael Cl. HA4: Ruis 35W 64
Carmichael Cl. SW11 55Fb 111
Carmichael Ct. SW13 54Va 110
(off Grove Rd.)
Carmichael Ho. E14 45Ec 92
(off Poplar High St.)
Carmichael M. SW18 59Fb 111
Carmichael Rd. SE25 71Vb 157
Carmine W2 1C 220 (43Fb 89)
(off Nth. Wharf Rd.)
Carmine Ct. BR1: Broml 66Hc 137
Carmine Wharf E14 44Bc 92
Carminia Rd. SW17 61Kb 134
Carnaby St. W1 3B 222 (44Lb 90)
Carnach Grn. RM15: S Ock 45Wd 98
Carnac St. SE27 63Tb 135
Carnanton Rd. E17 25Fc 53
Carnarvon Av. EN1: Enf 13Vb 33
Carnarvon Dr. UB3: Harl 48S 84
Carnarvon Rd. E10 29Ec 52
Carnarvon Rd. E15 37Hc 73
Carnarvon Rd. E18 25Hc 53
Carnarvon Rd. EN5: Barn 13Ab 30
Carnation Cl. RM7: Rush G 33Gd 76
Carnation St. SE2 50Xc 95
Carnbrook M. SE3 55Mc 115
Carnbrook Rd. SE3 55Mc 115
Carnecke Gdns. SE9 57Nc 116
Carnegie Cl. EN3: Enf L 10Dc 20
Carnegie Cl. KT6: Surb 75Pa 153
Carnegie Cl. SL2: Farn C 7F 60
Carnegie Pl. SW19 62Za 132
Carnegie Rd. HA1: Harr 31Ha 66
Carnegie St. N1 1H 217 (39Pb 70)
Carnell Apts. E14 44Bc 92
(off St Anne's Row)
Carnet Cl. DA1: Cray 59Gd 118
Carney Pl. SW9 56Qb 112
Carnforth Cl. KT19: Ewe 79Ra 153
Carnforth Gdns. RM12: Horn 36Hd 76
Carnforth Rd. SW16 66Mb 134
(not continuous)
Carnie Lodge SW17 62Kb 134
Carnival Ho. SE1 1K 231 (47Vb 91)
(off Gainsford St.)
Carnoustie Cl. SE28 44Zc 95
Carnoustie Dr. N1 38Pb 70
(not continuous)
Carnwath Rd. SW6 55Cb 111
Caroe Ct. N9 18Xb 33
Caro La. HP3: Hem H 4B 4
Carol Cl. NW4 28Za 48
Carolean Cres. SE8 49Ac 92
Carole Ho. NW1 39Hb 69
(off Regent's Pk. Rd.)
Carolina Cl. E15 36Gc 73
Carolina Rd. CR7: Thor H 68Rb 135
Caroline Cl. CR0: C'don 77Ub 157
Caroline Cl. N10 26Kb 50
Caroline Cl. SW16 62Pb 134
Caroline Cl. TW7: Isle 52Fa 108
Caroline Cl. UB7: W Dray 47M 83
Caroline Cl. W2 45Db 89
Caroline Ct. HA7: Stan 23Ja 46
Caroline Ct. SE6 63Fc 137
Caroline Ct. TW15: Ashf 65M 128
Caroline Gdns. E2 3J 219 (41Ub 91)
Caroline Gdns. SE15 52Xb 113
(not continuous)
Caroline Ho. W2 45Db 89
(off Bayswater Rd.)
Caroline Ho. W6 50Ya 88
(off Queen Caroline St.)
Caroline Pl. SW11 54Jb 112
Caroline Pl. UB3: Harl 52U 106
Caroline Pl. W2 45Db 89
Caroline Pl. WD19: Wat 16Aa 27
Caroline Pl. M. W2 45Db 89
Caroline Rd. SW19 66Bb 133
Caroline St. E1 44Zb 92
Caroline Ter. SW1 6H 227 (49Jb 90)
Caroline Wlk. W6 51Ab 110
(off Lillie Rd.)
Carol St. NW1 39Lb 70
Carolyn Cl. GU21: Wok 1K 187
Carolyn Dr. BR6: Chels 76Wc 161
Caronia Ct. SE16 49Ac 92
(off Plough Way)
Caroon Dr. WD3: Sarr 8K 11
Caro Pl. KT3: N Mald 70Va 132
Carpenders Av. WD19: Wat 20Aa 27
CARPENDERS PARK 20Z 27
Carpenders Park Station (Overground) 20Z 27
Carpenter Cl. KT17: Ewe 81Va 174
Carpenter Gdns. N21 19Rb 33
Carpenter Ho. E1 43Zb 92
(off Trafalgar Gdns.)
Carpenter Ho. E14 43Cc 92
(off Burgess St.)
Carpenter Ho. NW11 30Eb 49
Carpenter Path CM13: Hut 15Fe 41
Carpenters Arms Apts. SE1 49Wb 91
(off Welsford St.)
Carpenters Arms Path SE9 58Qc 116
(off Eltham High St.)
Carpenters Cl. EN5: New Bar 16Db 31
Carpenters Ct. BR1: Broml 68Mc 137
Carpenters Ct. NW1 39Lb 70
(off Pratt St.)
Carpenters Ct. TW2: Twick 61Ga 130
Carpenters M. N7 36Nb 70
Carpenters Pl. SW4 56Mb 112
Carpenter's Rd. E15 38Ec 72
Carpenter's Rd. E20 37Cc 72
Carpenters Rd. EN1: Enf 8Yb 20
Carpenter St. W1 5K 221 (45Kb 90)
Carpenters Wood Dr. WD3: Chor 14D 24
Carpenter Way EN6: Pot B 5Eb 17
Carp Ho. E3 39Bc 72
(off Old Ford Rd.)
Carrack Ho. DA8: Erith 50Gd 96
(off Saltford Cl.)
Carradale Ho. E14 44Ec 92
(off St Leonard's Rd.)
Carraige Wlk. RH1: Redh 8P 207
Carrara Cl. SE24 56Qb 112
Carrara Cl. SW9 56Rb 113
Carrara M. E8 37Wb 71
(off Dalston La.)
Carrara Wharf SW6 55Ab 110
Carr Cl. HA7: Stan 23Ja 46
Carre M. SE5 53Rb 113
(off Calais St.)
Carrera Twr. EC1 3D 218 (41Sb 91)
(off Bollinder Pl.)
Carr Gro. SE18 49Nc 94
Carr Ho. DA1: Cray 57Gd 118
Carriage Dr. E. SW11 52Jb 112
Carriage Dr. Nth. SW11: Carriage Dr. E. 51Jb 112
Carriage Dr. Nth. SW11: The Parade 52Hb 111
Carriage Dr. Sth. SW11 53Hb 111
(not continuous)
Carriage Dr. W. SW11 52Hb 111
Carriage Ho. E15 37Fc 73
(off Leyton Rd.)
Carriage Ho. N4 33Qb 70
Carriage M. IG1: Ilf 33Sc 74
Carriage Pl. N16 34Tb 71
Carriage Pl. SW16 64Lb 134
Carriages Apts. CR8: Purl 83Rb 177
(off Brighton Rd.)
Carriage St. SE18 48Rc 94
Carriage Way SE8 52Cc 114
(off Deptford High St.)
Carriageway, The TN16: Bras 96Zc 201
Carrick Cl. TW7: Isle 55Ja 108
Carrick Ct. E3 41Ec 92
(off Bolinder Way)
Carrick Dr. IG6: Ilf 25Sc 54
Carrick Dr. TN13: S'oaks 95Kd 203
Carrick Gdns. N17 24Ub 51
Carrick Ga. KT10: Esh 76Ea 152
Carrick Ho. N7 37Pb 70
(off Caledonian Rd.)
Carrick Ho. SE11 7A 230 (50Qb 90)
Carrick M. SE8 51Cc 114
Carrick Sq. TW8: Bford 52La 108
(off Narrowboat Av.)
Carrigshaun KT13: Weyb 78T 150
Carrill Way DA17: Belv 48Zc 95
Carrington Av. TW3: Houn 57Da 107
Carrington Av. WD6: Bore 15Ra 29
Carrington Cl. CR0: C'don 73Ac 158
Carrington Cl. EN5: Ark 15Wa 30
Carrington Cl. KT2: King T 64Sa 131
Carrington Cl. RH1: Redh 5P 207
Carrington Cl. WD6: Bore 15Sa 29
Carrington Ct. SW11 56Gb 111
(off Barnard Rd.)
Carrington Gdns. E7 35Jc 73
Carrington Ho. W1 7K 221 (46Kb 90)
(off Carrington St.)
Carrington Pl. KT10: Esh 77Da 151
Carrington Rd. DA1: Dart 58Pd 119
Carrington Rd. SL1: Slou 5J 81
Carrington Rd. TW10: Rich 56Qa 109
Carrington Sq. HA3: Hrw W 23Ea 46
Carrington St. W1 7K 221 (46Kb 90)
Carrock Ct. RM7: Rush G 30Fd 56
(off Union Rd.)
Carrol Cl. NW5 35Kb 70
Carroll Cl. E15 36Hc 73
Carroll Hill IG10: Lough 13Pc 36
Carroll Ho. W2 4B 220 (45Fb 89)
(off Craven Ter.)
Carrolls Way RH8: Oxt 5L 211
Carronade Ct. N7 36Pb 70
Carronade Pl. SE28 48Sc 94
Carron Cl. E14 44Dc 92
Carroun Rd. SW8 52Pb 112
Carroway La. UB6: G'frd 41Fa 86
Carrow Rd. KT12: Walt T 76Z 151
Carrow Rd. RM9: Dag 38Xc 75
Carr Rd. E17 26Bc 52
Carr Rd. UB5: N'olt 37Ca 65
Carrs La. N21 15Sb 33
Carr St. E14 43Ac 92
CARSHALTON 77Jb 156
Carshalton Athletic FC 77Gb 155
CARSHALTON BEECHES 81Gb 175
Carshalton Beeches Station (Rail) 79Hb 155
Carshalton Boys Sports Coll. Cen. 75Gb 155
Carshalton Gro. SM1: Sutt 77Hb 155
Carshalton Lodge KT13: Weyb 76T 150
(off Oatlands Dr.)
CARSHALTON ON THE HILL 80Jb 156
Carshalton Pk. Rd. SM5: Cars 78Hb 155
Carshalton Pl. SM5: Cars 78Jb 156
Carshalton Rd. CR4: Mitc 70Jb 134
Carshalton Rd. SM1: Sutt 78Eb 155
Carshalton Rd. SM5: Cars 78Gb 155
Carshalton Rd. SM7: Bans 86Hb 175
Carshalton Station (Rail) 77Hb 155
Carsington Gdns. DA1: Dart 61Md 141
Carslake Rd. SW15 58Ya 110
Carson Rd. E16 42Jc 93
Carson Rd. EN4: Cockf 14Hb 31
Carson Rd. SE21 61Tb 135
Carson Ter. W11 46Ab 88
(off Princes Pl.)
Carstairs Rd. SE6 62Ec 136
Carston Cl. SE12 57Hc 115
Carswell Cl. CM13: Hut 16Fe 41
Carswell Cl. IG4: Ilf 28Mc 53
Carswell Rd. SE6 59Ec 114
CARTBRIDGE 94D 188
Cartbridge Cl. GU23: Send 95D 188
Cartel Cl. RM19: Purf 49Td 98
Carter Av. TW15: W King 80Td 164
Carter Cl. EN5: Barn 15Ab 30
Carter Cl. NW9 30Ta 47
Carter Cl. RM5: Col R 24Cd 56
Carter Cl. SL4: Wind 4E 102
Carter Ct. EC4 3C 224 (44Rb 91)
(off Carter La.)
Carter Dr. RM5: Col R 22Dd 56
Carteret Ho. W12 45Xa 88
(off White City Est.)
Carteret St. SW1 2D 228 (47Mb 90)
Carteret Way SE8 49Ac 92
Carterhatch La. EN1: Enf 10Vb 19
Carterhatch Rd. EN3: Enf H 11Yb 34
Carter Ho. E1 1K 225 (43Vb 91)
(off Brune St.)
Carter La. EC4 3C 224 (44Rb 91)
Carter Pl. SE17 7E 230 (50Sb 91)
Carter Rd. E13 39Kc 73
Carter Rd. SW19 65Fb 133
Carters Cl. KT4: Wor Pk 75Za 154
Carters Cl. NW5 36Mb 70
(off Torriano Av.)
Carter's Cotts. RH1: Redh 8N 207
Cartersfield Rd. EN9: Walt A 6Ec 20
CARTER'S HILL 100Rd 203
Carters Hill Cl. SE9 60Lc 115
Carters La. GU22: Wok 92E 188
Carters La. SE23 61Ac 136
Carter Sq. E14 44Dc 92
(off Bowen St.)
Carters Rd. KT17: Eps 87Va 174
Carters Row DA11: Nflt 10B 122
Carter St. SE17 51Sb 113
Carter's Yd. SW18 57Cb 111
Carthew Rd. W6 48Xa 88
Carthew Vs. W6 48Xa 88
Carthouse La. GU21: Wok 6J 167
Carthusian Ct. EC1 7D 218 (43Sb 91)
(off Carthusian St.)
Carthusian St. EC1 7D 218 (43Sb 91)
Cartier Circ. E14 46Dc 92
Carting La. WC2 5G 223 (45Nb 90)
Cart La. E4 18Gc 35
Cart La. RM17: Grays 50De 99
Cart Lodge M. CR0: C'don 74Ub 157
Cartmel NW1 3B 216 (41Lb 90)
(off Harrington St.)
Cartmel Cl. N17 24Xb 51
Cartmel Cl. RH2: Reig 4N 207
Cartmel Ct. UB5: N'olt 37Aa 65
Cartmel Gdns. SM4: Mord 71Eb 155
Cartmel Rd. DA7: Bex 53Cd 118
Carton Ho. W11 45Za 88
(off St Ann's Rd.)
Cartoon Mus. 2B 222 (44Lb 90)
Cart Path WD25: Wat 5Y 13
Cartridge Pl. SE18 48Rc 94
Cart Track, The HP3: Hem H 7P 3
Cartwright Gdns. WC1 4F 217 (41Nb 90)
Cartwright Ho. SE1 4E 230 (48Sb 91)
(off County St.)
Cartwright Pl. CR5: Coul 88Mb 176
Cartwright Rd. RM9: Dag 38Bd 75
Cartwright St. E1 45Vb 91
Cartwright Way SW13 52Xa 110
Carvel Ho. E14 50Ec 92
(off Manchester Rd.)
Carvell Ho. NW9 27Va 48
(off Aerodrome Rd.)
Carver Cl. W4 48Sa 87
Carver Rd. SE24 58Sb 113
Carville Cres. TW8: Bford 49Na 87
Carville Hall Pk. TW8: Bford 50Na 87
(off Clayponds La.)
Carville St. N4 33Qb 70
Cary Av. SE16 49Ac 92
Cary Rd. E11 35Gc 73
Carysfort Rd. N16 34Tb 71
Carysfort Rd. N8 29Mb 50
Cary Wlk. WD7: R'lett 6Ka 14
Casbeard St. N4 33Sb 71
Casby Ho. SE16 48Wb 91
(off Marine St.)
Cascade Av. N10 28Lb 50
Cascade Cl. BR5: St P 69Yc 139
Cascade Cl. IG9: Buck H 19Mc 35
Cascade Ct. SW11 52Kb 112
Cascade Rd. IG9: Buck H 19Mc 35
Cascades CR0: Sels 82Bc 178
Cascades Ct. SW19 66Bb 133
Cascades Leisure Cen. Gravesend 3J 145
Cascades Twr. E14 46Bc 92
Casel Ct. HA7: Stan 19Ja 28
(off Brightwen Gro.)
Caselden Cl. KT15: Add 78L 149
Casella Rd. SE14 52Zb 114
Casewick Rd. SE27 64Qb 134
Casey Av. UB5: N'olt 40Ba 65
Casey Cl. NW8 4D 214 (41Gb 89)
Casey Ct. SE14 53Zb 114
(off Besson St.)
Cashmere Ho. E8 39Vb 71
(off Pamela St.)
Casia Point E20 38Ec 72
Casimir Rd. E5 33Yb 72
Casings Way E3 38Cc 72
Casino Av. SE24 57Sb 113
Caspian Cl. RM19: Purf 49Qd 97
Caspian St. SE5 52Tb 113
Caspian Wlk. E16 44Mc 93
Caspian Way DA10: Swans 57Ae 121
Caspian Way RM19: Purf 50Qd 97
Caspian Wharf E3 43Dc 92
(off Violet Rd.)
Cassander Pl. HA5: Pinn 25Aa 45
(off Holly Gro.)
Cassandra Cl. RM13: Rain 39Hd 76
Cassandra Cl. UB5: N'olt 35Fa 66
Casselden Rd. NW10 38Ta 67
CASSEL HOSPITAL 63Ma 131
Cassell Cl. RM16: Ors 3C 100
Cassell Ho. SW9 54Pb 112
(off Stockwell Gdns. Est.)
Cass Ho. E9 37Zb 72
(off Harrowgate Rd.)
Cassia Ho. E1 44Wb 91
(off Piazza Wlk.)
Cassidy Rd. SW6 52Cb 111
(not continuous)
Cassilda Rd. SE2 49Wc 95
Cassilis Rd. E14 47Cc 92
Cassilis Rd. TW1: Twick 57Ka 108
Cassini Apts. E16 44Jc 93
Cassio Apts. WD17: Wat 12X 27
(off Malden Rd.)
Cassiobridge WD18: Wat 14T 26
Cassiobridge Rd. WD18: Wat 14U 26
Cassiobridge Ter. WD3: Wat 15T 26
Cassiobury Av. TW14: Felt 59V 106
Cassiobury Ct. WD17: Wat 12U 26
Cassiobury Dr. WD17: Wat 10U 12
Cassiobury Pk. 13U 26
Cassiobury Pk. Av. WD18: Wat 13U 26
Cassiobury Pk. Locks 10T 12
(off Grove Mill La.)
Cassiobury Rd. E17 29Ac 52
Cassio Ho. WD18: Wat 14V 26
(off Manhattan Av.)
Cassio Pl. WD18: Wat 14U 26
Cassio Rd. WD18: Wat 13X 27
Cassio Wharf WD18: Wat 15T 26
Cassis Ct. IG10: Lough 14Sc 36
Cassius Dr. AL3: St A 2F 6
Cassland Rd. CR7: Thor H 70Tb 135
Cassland Rd. E9 38Zb 72
Casslee Rd. SE6 59Bc 114
Cassocks Sq. TW17: Shep 73T 150
Casson Apts. E14 44Cc 92
(off Upper Nth. St.)
Casson Ho. E1 43Wb 91
(off Hanbury St.)
Casson Sq. SE1 7J 223 (46Pb 90)
Casson St. E1 43Wb 91
Casstine Cl. BR8: Hext 66Hd 140
Castalia Ct. DA1: Dart 55Pd 119
Castalia Sq. E14 47Ec 92
Castano Ct. WD5: Ab L 3U 12
Castellain Mans. W9 42Db 89
(off Castellain Rd.)
Castellain Rd. W9 42Db 89
Castellan Av. RM2: Rom 27Kd 57
Castellane Cl. HA7: Stan 24Ha 46
Castell Ho. SE8 52Cc 114
Castello Av. SW15 57Ya 110
Castell Rd. IG10: Lough 11Sc 36
Castelnau SW13 53Wa 110
CASTELNAU 51Xa 110
Castelnau Gdns. SW13 51Xa 110
Castelnau Mans. SW13 51Xa 110
(off Castelnau)
Castelnau Row SW13 51Xa 110
Casterbridge NW6 39Db 69
(off Abbey Rd.)
Casterbridge W11 44Bb 89
(off Dartmouth Cl.)
Casterbridge Rd. SE3 55Jc 115
Casterton St. E8 37Xb 71
Castile Ct. EN8: Walt C 6Cc 20
Castile Gdns. WD4: K Lan 1P 11
Castile Rd. SE18 49Qc 94
Castillon Rd. SE6 61Gc 137
Casting Ho. SE14 51Bc 114
Castlands Rd. SE6 61Bc 136
Castleacre W2 3D 220 (44Gb 89)
(off Hyde Pk. Cres.)
Castle Av. E4 22Fc 53
Castle Av. KT17: Ewe 81Wa 174
Castle Av. RM13: Rain 38Gd 76
Castle Av. SL3: Dat 1L 103
Castle Av. UB7: Yiew 45N 83
Castlebar Ct. W5 43La 86
Castlebar Hill W5 43La 86
Castlebar M. W5 43La 86
Castlebar Pk. W5 43Ka 86
Castle Bar Park Station (Rail) 43Ha 86
Castlebar Rd. W5 43La 86
Castle Baynard St. EC4 4C 224 (45Rb 91)
Castlebrook Cl. SE11 5B 230 (49Rb 91)
Castle Bus. Cen. TW12: Hamp 67Da 129
(off Castle M.)
Castle Cen., The 5C 230 (49Rb 91)
Castle Cinema, The 36Zb 72
Castle Climbing Cen., The 33Sb 71
Castle Cl. BR2: Broml 69Gc 137
Castle Cl. E9 36Ac 72
Castle Cl. RH1: Blet 5J 209
Castle Cl. RH2: Reig 10K 207
Castle Cl. RM3: Rom 20Ld 39
Castle Cl. SW19 62Za 132
Castle Cl. TW16: Sun 66U 128
Castle Cl. W3 47Ra 87
Castle Cl. WD23: Bush 16Da 27
Castlecombe Dr. SW19 59Za 110
Castlecombe Rd. SE9 63Nc 138
Castle Ct. EC3 3G 225 (44Tb 91)
(off Birchin La.)
Castle Ct. SE26 63Ac 136
Castle Ct. SW15 55Ab 110
(off Brewhouse La.)
Castleden Ho. NW3 38Fb 69
(off Hilgrove Rd.)
Castledine Rd. SE20 66Xb 135
Castle Dr. IG4: Ilf 30Nc 54
Castle Dr. RH2: Reig 10J 207
Castle Dr. TN15: Kems'g 89Nd 183
Castle Farm Cvn. Site SL4: Wind 4B 102
Castle Farm Rd. TN14: S'ham 81Hd 182
Castlefield Ct. RH2: Reig 6K 207
Castlefield Rd. RH2: Reig 6J 207
Castlefields DA13: Ist R 7B 144
Castleford Av. SE9 60Rc 116
Castleford Cl. N17 23Vb 51
Castleford Cl. WD6: Bore 10Pa 15
Castleford Ct. NW8 5C 214 (42Fb 89)
(off Henderson Dr.)
Castlegate TW9: Rich 55Pa 109
CASTLE GREEN 5H 167
Castle Grn. KT13: Weyb 76U 150
Castle Green Leisure Cen. 39Zc 75
Castle Gro. Rd. GU24: Chob 5H 167
Castlehaven Rd. NW1 38Kb 70
Castle Hgts. RM9: Dag 39Xc 75
CASTLE HILL 59Be 121
Castle Hill DA3: Fawk 70Zd 143
Castle Hill DA3: Hartl 70Zd 143
Castle Hill SL4: Wind 3H 103
Castle Hill Av. CR0: New Ad 81Dc 178
Castle Hill Dr. DA10: Swans 60Ae 121
Castle Hill Local Nature Reserve
Chessington 79Qa 153
Castle Hill Pde. W13 45Ka 86
(off The Avenue)
Castle Hill Rd. TW20: Eng G 3M 125
Castle Ho. SM2: Sutt 79Cb 155
Castle Ho. SW8 52Nb 112
(off Sth. Lambeth Rd.)
Castle La. DA12: Grav'nd 1K 145
Castle La. SW1 3B 228 (48Lb 90)
Castleleigh Ct. EN2: Enf 15Tb 33
Castlemaine SW11 54Hb 111
Castlemaine Av. CR2: S Croy 78Vb 157
Castlemaine Av. KT17: Ewe 81Xa 174
Castlemain St. E1 43Xb 91
Castle Mead HP1: Hem H 4K 3
Castle Mead SE5 52Sb 113
Castle M. KT13: Weyb 76U 150
Castle M. N12 22Eb 49
Castle M. NW1 37Kb 70
Castle M. SW17 63Gb 133
Castle M. TW12: Hamp 67Da 129
(not continuous)
Castle Pde. KT17: Ewe 80Wa 154
Castle Pl. NW1 37Kb 70
Castle Pl. W4 49Ua 88
Castle Point E13 40Lc 73
(off Boundary Rd.)
Castlereagh Ho. HA7: Stan 23Ka 46
Castlereagh St. W1 2F 221 (44Hb 89)
Castleridge Dr. DA9: Ghithe 57Wd 120
Castle Rd. AL1: St A 2F 6
Castle Rd. CR5: Chips 92Gb 195
Castle Rd. DA10: Swans 58Be 121
Castle Rd. DA4: Eyns 79Kd 163
Castle Rd. EN3: Enf H 11Ac 34
Castle Rd. GU21: Wok 86B 168
Castle Rd. KT13: Weyb 76T 150
Castle Rd. KT18: Eps 87Ra 173
Castle Rd. N12 22Eb 49
Castle Rd. NW1 37Kb 70
Castle Rd. RM17: Grays 51Be 121
Castle Rd. RM9: Dag 39Xc 75
Castle Rd. TW7: Isle 54Ha 108
Castle Rd. UB2: S'hall 48Ba 85
Castle Rd. UB5: N'olt 37Da 65
Castle Row W4 50Ta 87
Castle Sq. RH1: Blet 5J 209
Castle Sq. SE17 5D 230 (49Sb 91)
Castle St. DA10: Swans 58Be 121
Castle St. DA9: Ghithe 57Wd 120
Castle St. E13 40Lc 73
Castle St. E6 40Lc 73
Castle St. KT1: King T 68Na 131
Castle St. RH1: Blet 5H 209
Castle St. SL1: Slou 8K 81
Castleton Av. DA7: Bex 53Fd 118
Castleton Av. HA9: Wemb 35Na 67
Castleton Cl. CR0: C'don 72Ac 158
Castleton Cl. SM7: Bans 87Cb 175
Castleton Dr. SM7: Bans 87Cb 175
Castleton Gdns. HA9: Wemb 34Na 67
Castleton Ho. E14 49Ec 92
(off Pier St.)
Castleton Rd. CR4: Mitc 70Mb 134
Castleton Rd. E17 26Fc 53
Castleton Rd. HA4: Ruis 32Z 65
Castleton Rd. IG3: Ilf 32Wc 75
Castleton Rd. SE9 63Mc 137
Castletown Rd. W14 50Ab 88
Castle Vw. KT18: Eps 86Ra 173
Castleview Cl. N4 33Sb 71
Castleview Gdns. IG1: Ilf 30Nc 54
Castleview Pde. SL3: L'ly 9P 81
Castleview Rd. KT13: Weyb 77R 150
Castleview Rd. SL3: L'ly 9N 81
Castle Wlk. RH2: Reig 6J 207
Castle Wlk. TW16: Sun 69Y 129
Castle Way KT17: Ewe 82Wa 174
Castle Way SW19 62Za 132
Castle Way TW13: Hanw 63Y 129
Castle Wharf E14 45Gc 93
(off Orchard Pl.)
Castlewood Dr. SE9 54Pc 116
Castlewood Rd. EN4: Cockf 13Fb 31
Castlewood Rd. N15 30Wb 51
Castlewood Rd. N16 30Wb 51
Castle Yd. N6 31Jb 70
Castle Yd. SE1 6C 224 (46Rb 91)
Castle Yd. TW10: Rich 57Ma 109
Castor La. E14 45Dc 92
Catalina Av. RM16: Chaf H 47Be 99
Catalina Ct. AL1: St A 2C 6
(off Beaconsfield Rd.)
Catalina Ho. E1 44Wb 91
(off Piazza Wlk.)
Catalina Rd. TW6: H'row A 54R 106
Catalin Ct. EN9: Walt A 5Fc 21
(off Howard Cl.)
Catalonia Apts. WD18: Wat 14V 26
(off Metropolitan Sta. App.)
Catalpa Ct. SE13 58Fc 115
Caterfield La. RH8: Oxt 10L 211
CATERHAM 96Wb 197
Caterham Av. IG5: Ilf 26Pc 54
Caterham By-Pass CR3: Cat'm 92Xb 197
Caterham Cl. GU24: Pirb 3C 186
Caterham Ct. CR3: Cat'm 96Ub 197
Caterham Ct. EN9: Walt A 6Hc 21
CATERHAM DENE HOSPITAL 95Vb 197
Caterham Dr. CR5: Coul 90Rb 177
Caterham High School Fitness Cen. 26Pc 54
CATERHAM-ON-THE-HILL 94Ub 197
Caterham Rd. SE13 55Fc 115
Caterham Station (Rail) 96Wb 197
Catesby Ho. E9 38Yb 72
(off Frampton Pk. Rd.)
Catesby St. SE17 6G 231 (49Tb 91)
CATFORD 59Dc 114
Catford Bridge Station (Rail) 59Cc 114
Catford B'way. SE6 59Dc 114
CATFORD GYRATORY 59Dc 114
Catford Hill SE6 60Bc 114
Catford Island SE6 59Dc 114
Catford M. SE6 59Dc 114
Catford Rd. SE6 59Cc 114
Catford Station (Rail) 59Cc 114
Catford Trad. Est. SE6 61Dc 136
Cathall Rd. E11 33Fc 73
Catham Cl. AL1: St A 4F 6
Catharine Cl. RM16: Chaf H 47Be 99
Catharine Ho. WD19: Wat 20X 27
(off Brabazon Rd.)
Cathay Ho. SE16 47Xb 91
Cathay St. SE16 47Xb 91
Cathay Wlk. UB5: N'olt 40Ca 65
(off Brabazon Rd.)
Cathcart Dr. BR6: Orp 75Uc 160
Cathcart Hill N19 34Lb 70
Cathcart Ho. SW10 50Eb 89
(off Cathcart Rd.)
Cathcart Rd. SW10 51Db 111
Cathcart St. NW5 37Kb 70
Cathedral Ct. AL3: St A 4P 5
Cathedral Lodge EC1 7D 218 (43Sb 91)
(off Aldersgate St.)
Cathedral Mans. SW1 5B 228 (49Lb 90)
(off Vauxhall Bri. Rd.)
Cathedral Piazza SW1 4B 228 (48Lb 90)
Cathedral Pl. CM14: B'wood 19Zd 41
Cathedral St. SE1 6F 225 (46Tb 91)
Cathedral Vw. AL3: St A 2B 6
(off High St.)
Cathedral Wlk. SW1 3B 228 (48Lb 90)
Catherall Rd. N5 34Sb 71
Catherine Cl. CM15: Pil H 15Wd 40
Catherine Cl. IG10: Lough 16Pc 36
Catherine Cl. KT14: Byfl 86N 169
Catherine Ct. NW4 28Xa 48
Catherine Ct. IG2: Ilf 29Sc 54
Catherine Ct. N14 15Lb 32
Catherine Ct. SW19 64Bb 133
Catherine Ct. SW3 51Fb 111
(off Callow St.)
Catherine Dr. TW16: Sun 65V 128
Catherine Dr. TW9: Rich 56Na 109
Catherine Gdns. TW3: Houn 56Fa 108
Catherine Griffiths Ct. EC1 5A 218 (42Qb 90)
(off Northampton Rd.)
Catherine Gro. SE10 53Dc 114
Catherine Ho. E3 40Cc 72
(off Thomas Fyre Dr.)
Catherine Ho. N1 1H 219 (39Ub 71)
(off Whitmore Est.)
Catherine Howard Ct. KT13: Weyb 76R 150
(off Old Palace Rd.)
Catherine Howard Ct. SE9 58Tc 116
Catherine of Aragon Ct. SE9 58Sc 116
Catherine Parr Ct. SE9 58Tc 116
Catherine Pl. HA1: Harr 29Ha 46
Catherine Pl. SW1 3B 228 (48Lb 90)
Catherine Rd. EN3: Enf W 9Ac 20
Catherine Rd. KT6: Surb 71Ma 153
Catherine Rd. RM2: Rom 29Kd 57
Catherines Cl. UB7: W Dray 47M 83
Catherine St. AL3: St A 1B 6
Catherine St. WC2 4H 223 (45Pb 90)
Catherine Wheel All. E1 1J 225 (43Ub 91)
Catherine Wheel Rd. TW8: Bford 52Ma 109
Catherine Wheel Yd. SW1 7B 222 (46Lb 90)
(off Lit. St James's St.)
Catherwood Ct. N1 2F 219 (40Tb 71)
(off Murray Gro.)
Cat Hill EN4: E Barn 16Gb 31
Cathles Rd. SW12 58Kb 112

Cathnor Rd. W12 .... 47Xa 88
Catisfield Rd. EN3: Enf W 9Ac 20
Catkin Cl. E14 .... 44Fc 93
Catkin Cl. HP1: Hem H 1K 3
Catkin Ho. RM3: Hrld W 26Nd 57
Catlin Cres. TW17: Shep 71T 150
Catlin Gdns. RH9: G'stone 2P 209
Catling Cl. SE23 .... 62Yb 136
Catlin's La. HA5: Eastc 27X 45
Catlin St. HP3: Hem H 5K 3
Catlin St. SE16 .... 50Wb 91
Cator Cl. CR0: New Ad 83Gc 179
Cator Cres. CR0: New Ad 83Gc 179
Cator La. BR3: Beck 67Bc 136
Cato Rd. SW4 .... 55Mb 112
Cator Rd. SE26 .... 65Zb 136
Cator Rd. SM5: Cars 78Hb 155
Cator St. SE15: Commercial Way 52Vb 113
Cator St. SE15: Ebley Cl 51Vb 113
Cato's Hill KT10: Esh 77Da 151
Cato St. W1 .... 1E 220 (43Gb 89)
Catsdell Bottom HP3: Hem H 5B 4
Catsey La. WD23: Bush 17Ea 28
Catsey Woods WD23: Bush 17Ea 28
Catterick Cl. N11 .... 23Jb 50
Catterick Way WD6: Bore 11Pa 29
Cattistock Rd. SE9 .... 64Nc 138
CATTLEGATE 5Nb 18
Cattlegate Cotts. EN6: N'thaw 3Mb 18
Cattlegate Hill EN6: Cuff 4Mb 18
Cattlegate Hill EN6: N'thaw 4Mb 18
Cattlegate Rd. EN2: Crew H 6Nb 18
Cattlegate Rd. EN6: Cuff 3Mb 18
Cattlegate Rd. EN6: N'thaw 3Mb 18
Cattley Cl. EN5: Barn 14Ab 30
Cattley Pl. CR5: Coul 89Mb 176
Cattlins Cl. EN7: Chesh 1Ub 19
Catton St. WC1 .... 1H 223 (43Pb 90)
Cattsdell HP2: Hem H 1N 3
Catwalk Pl. N15 .... 30Tb 51
Caudwell Ter. SW18 .... 58Fb 111
Caughley Ho. SE11 .... 4K 229 (48Qb 90)
(off Lambeth Wlk.)
Caulfield Ct. NW1 .... 38Lb 70
(off Baynes St.)
Caulfield Gdns. HA5: Pinn 26Y 45
Caulfield Rd. E6 .... 39Nc 74
Caulfield Rd. SE15 .... 54Xb 113
Caulfield Rd. W3 .... 48Sa 87
Causeway, The EN6: Pot B 3Eb 17
(not continuous)
Causeway, The KT10: Clay 80Ha 152
Causeway, The KT9: Chess 77Na 153
Causeway, The N18 .... 23Yb 52
Causeway, The N2 .... 28Gb 49
Causeway, The SM2: Sutt 81Eb 175
Causeway, The SM5: Cars 76Jb 156
Causeway, The SW18 .... 57Db 111
Causeway, The SW19 .... 64Ya 132
Causeway, The TW11: Tedd 65Ha 130
Causeway, The TW14: Felt 56W 106
Causeway, The TW14: Houn 56W 106
Causeway, The TW18: Staines 63E 126
Causeway Cl. EN6: Pot B 3Fb 17
Causeway Corporate Cen. TW18: Staines 63E 126
Causeway Ct. GU21: Wok 10K 167
Causeway Ho. BR6: Orp 74Wc 161
Causeway Ho. WD5: Ab L 3U 12
Causeyware Rd. N9 .... 17Yb 34
Causton Cotts. E14 .... 44Ac 92
(off Galsworthy Av.)
Causton Ho. SE5 .... 51Sb 113
Causton Ho. SW9 .... 53Pb 112
Causton Rd. N6 .... 31Kb 70
Causton Sq. RM10: Dag 38Cd 76
Causton St. SW1 .... 6E 228 (49Mb 90)
Cautley Av. SW4 .... 57Lb 112
Cavalier Cl. RM6: Chad H 28Zc 55
Cavalier Cl. SM6: W'gton 79Lb 156
Cavalier Ct. KT5: Surb 72Pa 153
Cavalier Gdns. UB3: Hayes 44T 84
Cavalier Ter. SE7 .... 51Lc 115
Cavalli Apts. WD18: Wat 14U 26
(off Moderna M.)
Cavalry Cres. SL4: Wind 5G 102
Cavalry Cres. TW4: Houn 56Z 107
Cavalry Gdns. SW15 .... 57Bb 111
Cavalry Pl. E17 .... 26Zb 52
Cavalry Sq. SW3 .... 7G 227 (50Hb 89)
Cavan Ct. AL10: Hat 1C 8
Cavan Pl. HA5: Hat E 25Ba 45
Cavatina Point SE8 .... 51Dc 114
(off Copperas St.)
Cavaye Ho. SW10 .... 51Eb 111
(off Cavaye Pl.)
Cavaye Pl. SW10 .... 50Eb 89
Cavell Cres. DA1: Dart 56Qd 119
Cavell Cres. RM3: Hrld W 26Nd 57
Cavell Dr. EN2: Enf 12Qb 32
Cavell Ho. N1 .... 39Ub 71
(off Colville Est.)
Cavell Rd. N17 .... 24Tb 51
Cavell St. E1 .... 43Xb 91
Cavell Way GU21: Knap 1G 186
Cavell Way KT19: Eps 83Qa 173
Cavendish Av. DA15: Sidc 59Wc 117
Cavendish Av. DA16: Well 55Vc 117
Cavendish Av. DA8: Erith 51Ed 118
Cavendish Av. HA1: Harr 35Fa 66
Cavendish Av. HA4: Ruis 36X 65
Cavendish Av. IG8: Wfd G 25Kc 53
Cavendish Av. KT3: N Mald 71Wa 154
Cavendish Av. N3 .... 26Cb 49
Cavendish Av. NW8 .... 2C 214 (40Fb 69)
Cavendish Av. RM12: Horn 37Kd 77
Cavendish Av. TN13: S'oaks 94Jd 202
Cavendish Av. W13 .... 43Ja 86
Cavendish Cl. N18 .... 22Xb 51
Cavendish Cl. NW6 .... 37Bb 69
Cavendish Cl. NW8 .... 3C 214 (41Fb 89)
Cavendish Cl. TW16: Sun 65V 128
Cavendish Cl. UB4: Hayes 43U 84
Cavendish Ct. EC3 .... 2J 225 (44Ub 91)
(off Devonshire Sq.)
Cavendish Ct. HP3: Hem H 7N 3
Cavendish Ct. KT13: Weyb 79S 150
Cavendish Ct. KT16: Chert 74J 149
(off Victory Rd.)
Cavendish Ct. SE6 .... 60Dc 114
(off Bromley Rd.)
Cavendish Ct. SL3: Poyle 53G 104
Cavendish Ct. SM6: W'gton 79Kb 156
Cavendish Ct. TW16: Sun 65V 128
Cavendish Ct. WD3: Crox G 15T 26
Cavendish Cres. RM12: Horn 37Kd 77
Cavendish Cres. WD6: E'tree 14Qa 29
Cavendish Dr. E11 .... 32Fc 73
Cavendish Dr. HA8: Edg 23Pa 47
Cavendish Dr. KT10: Clay 78Ga 152
Cavendish Gdns. IG1: Ilf 32Qc 74
Cavendish Gdns. IG11: Bark 36Uc 74
Cavendish Gdns. RH1: Redh 5A 208
Cavendish Gdns. RM15: Avel 47Sd 98
Cavendish Gdns. RM6: Chad H 29Ad 55
Cavendish Gdns. SW4 .... 58Lb 112
Cavendish Ho. CR0: C'don 74Tb 157
(off Tavistock Rd.)
Cavendish Ho. KT12: Hers 78Y 151
Cavendish Ho. NW8 .... 2C 214 (40Fb 69)
(off Wellington Rd.)
Cavendish Ho. NW9 .... 27Va 48
Cavendish Ho. SW1 .... 4E 228 (48Mb 90)
(off Monck St.)
Cavendish Ho. UB7: W Dray 47P 83
(off Park Lodge Av.)
Cavendish Ho. W1 .... 1B 222 (43Lb 90)
(off Mortimer St.)
Cavendish Mans. EC1 .... 6K 217 (42Qb 90)
(off Rosebery Av.)
Cavendish Mans. NW6 .... 36Cb 69
Cavendish Meads SL5: S'hill 2B 146
Cavendish M. Nth. W1 .... 7A 216 (43Kb 90)
Cavendish M. Sth. W1 .... 1A 222 (43Kb 90)
Cavendish Pde. SW4 .... 58Kb 112
(off Clapham Comn. Sth. Side)
Cavendish Pde. TW4: Houn 54Aa 107
Cavendish Pl. AL10: Hat 1B 8
(off Aldykes)
Cavendish Pl. BR1: Broml 70Pc 138
Cavendish Pl. NW2 .... 37Za 68
Cavendish Pl. SW4 .... 57Mb 112
Cavendish Pl. W1 .... 2A 222 (44Kb 90)
Cavendish Rd. AL1: St A 2D 6
Cavendish Rd. CR0: C'don 74Rb 157
Cavendish Rd. E4 .... 23Ec 52
Cavendish Rd. EN5: Barn 13Ya 30
Cavendish Rd. GU22: Wok 1P 187
Cavendish Rd. KT13: Weyb 81R 170
Cavendish Rd. KT3: N Mald 70Va 132
Cavendish Rd. N18 .... 22Xb 51
Cavendish Rd. N4 .... 30Rb 51
Cavendish Rd. NW6 .... 38Ab 68
Cavendish Rd. RH1: Redh 6A 208
Cavendish Rd. SM2: Sutt 80Eb 155
Cavendish Rd. SW12 .... 58Kb 112
Cavendish Rd. SW19 .... 66Fb 133
Cavendish Rd. TW16: Sun 65V 128
Cavendish Rd. W4 .... 53Sa 109
Cavendish Sq. DA3: Lfield 69Zd 143
Cavendish Sq. W1 .... 2K 221 (44Kb 90)
Cavendish St. N1 .... 2F 219 (40Tb 71)
Cavendish Ter. E3 .... 41Bc 92
Cavendish Ter. TW13: Felt 61W 128
Cavendish Wlk. KT19: Eps 83Ra 173
Cavendish Way BR4: W W'ck 74Dc 158
Cavenham Cl. GU22: Wok 91A 188
Cavenham Gdns. IG1: Ilf 34Tc 74
Cavenham Gdns. RM11: Horn 29Ld 57
Caverleigh Pl. BR1: Broml 68Lc 137
Caverleigh Way KT4: Wor Pk 74Wa 154
Cave Rd. E13 .... 41Kc 93
Cave Rd. TW10: Ham 63La 130
Caversham Av. N13 .... 20Qb 32
Caversham Av. SM3: Cheam 75Ab 154
Caversham Ct. N11 .... 20Jb 32
(off Brunswick Pk. Rd.)
Caversham Ho. KT1: King T 68Na 131
(off Lady Booth Rd.)
Caversham Ho. N15 .... 28Sb 51
(off Caversham Rd.)
Caversham Ho. SE15 .... 51Wb 113
(off Haymerle Rd.)
Caversham M. SW3 .... 51Hb 111
(off Caversham St.)
Caversham Rd. KT1: King T 68Pa 131
Caversham Rd. N15 .... 28Sb 51
Caversham Rd. NW5 .... 37Lb 70
Caversham Rd. NW9 .... 27Wa 48
Caversham St. SW3 .... 51Hb 111
Caverswall St. W12 .... 44Ya 88
Caveside Cl. BR7: Chst 67Qc 138
Cavesson Ho. E20 .... 36Ec 72
(off Ribbons Wlk.)
Cavour Ho. SE17 .... 7C 230 (50Rb 91)
(off Alberta Est.)
Cawcott Dr. SL4: Wind 3C 102
Cawdor Av. RM15: S Ock 45Xd 98
Cawdor Cres. W7 .... 49Ja 86
Cawdor Ho. CM14: W'ley 21Zd 59
Cawdor Wlk. E14 .... 44Ec 92
Cawnpore St. SE19 .... 64Ub 135
Cawsey Way GU21: Wok 89A 168
Cawston Ct. BR1: Broml 66Hc 137
Cawston M. SW16 .... 64Rb 135
Caxton Av. KT15: Add 79J 149
Caxton Cl. DA3: Hartl 69Be 143
Caxton Ct. EN8: Walt C 7Ac 20
Caxton Ct. SW11 .... 54Gb 111
Caxton Dr. UB8: Uxb 40M 63
Caxton Gro. E3 .... 41Cc 92
Caxton Hall SW1 .... 3D 228 (48Mb 90)
(off Caxton St.)
Caxton Ho. 2E 228 (47Mb 90)
(off Tothill St.)
Caxton La. RH8: Limp 3P 211
Caxton M. TW8: Bford 51Ma 109
Caxton Pk. DA11: Nflt 61Ee 143
Caxton Pl. IG1: Ilf 34Qc 74
Caxton Ri. RH1: Redh 5A 208
Caxton Rd. N22 .... 26Pb 50
Caxton Rd. SW19 .... 64Eb 133
Caxton Rd. UB2: S'hall 48Z 85
Caxton Rd. W12 .... 47Za 88
Caxtons, The SW9 .... 52Rb 113
(off Langton Rd.)
Caxton St. SW1 .... 3C 228 (48Lb 90)
Caxton St. Nth. E16 .... 44Hc 93
Caxton Trad. Est. UB3: Hayes 47U 84
Caxton Wlk. WC2 .... 3E 222 (44Mb 90)
Caxton Way RM1: Rom 28Gd 56
Caxton Way WD18: Wat 17T 26
Cayenne Ct. SE1 .... 1K 231 (47Vb 91)
Caygill Cl. BR2: Broml 70Hc 137
Cayley Rd. TW6: H'row A 55R 106
Cayley Rd. UB2: S'hall 48Da 85
Cayton Pl. EC1 .... 4F 219 (41Tb 91)
(off Cayton St.)
Cayton Rd. CR5: Coul 94Lb 196
Cayton Rd. UB6: G'frd 40Ga 66
Cayton St. EC1 .... 4F 219 (41Tb 91)
Cazenove Rd. E17 .... 25Cc 52
Cazenove Rd. N16 .... 33Vb 71
Cearns Ho. E6 .... 39Mc 73
Cearn Way CR5: Coul 87Pb 176
Cecil Av. EN1: Enf 14Vb 33
Cecil Av. HA9: Wemb 36Pa 67
Cecil Av. IG11: Bark 38Tc 74
Cecil Av. RM11: Horn 27Nd 57
Cecil Av. RM16: Chaf H 47Be 99
Cecil Cl. KT9: Chess 77Ma 153
Cecil Cl. TW15: Ashf 66S 128
Cecil Cl. W5 .... 43Ma 87
Cecil Ct. CR0: C'don 75Vb 157
Cecil Ct. EN2: Enf 14Tb 33
Cecil Ct. EN5: Barn 13Za 30
Cecil Ct. EN8: Chesh 4Ac 20
Cecil Ct. NW6 .... 38Db 69
Cecil Ct. SW10 .... 51Eb 111
(off Fawcett St.)
Cecil Ct. WC2 .... 4F 223 (45Nb 90)
Cecile Pk. N8 .... 30Nb 50
Cecil Gro. NW8 .... 1E 214 (39Gb 69)
Cecil Hepworth Playhouse, The 74V 150
(off Hepworth Way)
Cecil Ho. E17 .... 25Cc 52
Cecilia Cl. N2 .... 27Eb 49
Cecilia Rd. E8 .... 36Vb 71
Cecil Manning Cl. UB6: G'frd 39Ja 66
Cecil Mans. SW17 .... 61Jb 134
Cecil Pk. HA5: Pinn 28Aa 45
Cecil Pl. CR4: Mitc 71Hb 155
Cecil Rhodes Ho. NW1 .... 1D 216 (40Mb 70)
(off Goldington St.)
Cecil Rd. AL1: St A 2D 6
Cecil Rd. CR0: C'don 72Nb 156
Cecil Rd. DA11: Grav'nd 10B 122
Cecil Rd. E11 .... 34Hc 73
Cecil Rd. E13 .... 39Jc 73
Cecil Rd. E17 .... 25Cc 52
Cecil Rd. EN2: Enf 13Sb 33
Cecil Rd. EN6: S Mim 4Wa 16
Cecil Rd. EN8: Chesh 4Ac 20
Cecil Rd. HA3: W'stone 27Ga 46
Cecil Rd. IG1: Ilf 35Rc 74
Cecil Rd. N10 .... 26Kb 50
Cecil Rd. N14 .... 18Lb 32
Cecil Rd. NW10 .... 39Ua 68
Cecil Rd. NW9 .... 27Ua 48
Cecil Rd. RM6: Chad H 31Zc 75
Cecil Rd. SL0: Iver 44G 82
Cecil Rd. SM1: Sutt 79Bb 155
Cecil Rd. SW19 .... 66Db 133
Cecil Rd. TW15: Ashf 66S 128
Cecil Rd. TW3: Houn 54Ea 108
Cecil Rd. W3 .... 43Sa 87
Cecil Rosen Ct. HA0: Wemb 34Ka 66
Cecil Rosen Ct. WD23: B Hea 18Ga 28
Cecil Sharp House 39Jb 70
(off Gloucester Av.)
Cecil St. WD24: Wat 10X 13
Cecil Way BR2: Hayes 74Jc 159
Cecil Way SL2: Slou 2D 80
Cedar Av. DA12: Grav'nd 3E 144
Cedar Av. DA15: Sidc 59Wc 117
Cedar Av. EN3: Enf H 12Yb 34
Cedar Av. EN4: E Barn 17Gb 31
Cedar Av. EN8: Walt C 5Zb 20
Cedar Av. HA4: Ruis 36Y 65
Cedar Av. KT11: Cobh 87Y 171
Cedar Av. RM14: Upm 35Qd 77
Cedar Av. RM6: Chad H 29Ad 55
Cedar Av. TW2: Whitt 58Da 107
Cedar Av. UB3: Hayes 44W 84
Cedar Av. UB7: Yiew 45P 83
Cedar Cl. BR2: Broml 76Nc 160
Cedar Cl. BR8: Swan 68Ed 140
Cedar Cl. CM13: Hut 17Fe 41
Cedar Cl. CR6: W'ham 91Ac 198
Cedar Cl. E3 .... 39Bc 72
Cedar Cl. EN6: Pot B 2Cb 17
Cedar Cl. IG1: Ilf 36Tc 74
Cedar Cl. IG9: Buck H 19Mc 35
Cedar Cl. KT10: Esh 79Ba 151
Cedar Cl. KT17: Eps 86Va 174
Cedar Cl. KT8: E Mos 70Ga 130
Cedar Cl. RH2: Reig 8L 207
Cedar Cl. RM7: Rom 28Ed 56
Cedar Cl. SE21 .... 60Sb 113
Cedar Cl. SL0: Iver H 39E 62
Cedar Cl. SL1: Burn 2A 80
Cedar Cl. SM5: Cars 79Hb 155
Cedar Cl. SW15 .... 63Ta 131
Cedar Cl. TW18: Lale 69L 127
Cedar Copse BR1: Broml 68Pc 138
Cedar Ct. AL4: St A 2H 7
Cedar Ct. CM16: Epp 3Wc 23
Cedar Ct. E11 .... 29Kc 53
Cedar Ct. E18 .... 25Jc 53
Cedar Ct. KT15: Add 77L 149
Cedar Ct. KT16: Ott 78E 148
Cedar Ct. KT22: Fet 94Ja 192
Cedar Ct. N1 .... 38Sb 71
Cedar Ct. N10 .... 26Jb 50
Cedar Ct. N11 .... 22Lb 50
Cedar Ct. N20 .... 18Fb 31
Cedar Ct. SE1 .... 2H 231 (47Ub 91)
(off Royal Oak Yd.)
Cedar Ct. SE13 .... 56Fc 115
Cedar Ct. SE7 .... 51Lc 115
Cedar Ct. SE9 .... 58Nc 116
Cedar Ct. SL4: Wind 4E 102
Cedar Ct. SM2: Sutt 79Eb 155
Cedar Ct. SW19 .... 62Za 132
Cedar Ct. TW20: Egh 63C 126
Cedar Ct. TW8: Bford 51La 108
Cedar Ct. WD25: Wat 5Z 13
(off Lych Ga.)
Cedar Cres. BR2: Broml 76Nc 160
Cedar Cres. WD23: Bush 13Z 27
Cedarcroft Rd. KT9: Chess 77Pa 153
Cedar Dr. DA4: Sut H 68Rd 141
Cedar Dr. HA5: Hat E 23Ca 45
Cedar Dr. IG10: Lough 13Rc 36
Cedar Dr. KT22: Fet 95Ga 192
Cedar Dr. N2 .... 28Gb 49
Cedar Dr. SL5: S'dale 3E 146
Cedar Dr. SL5: S'hill 10E 125
Cedar Gdns. GU21: Wok 10M 167
Cedar Gdns. GU24: Chob 2K 167
Cedar Gdns. RM14: Upm 34Sd 78
Cedar Gdns. SM2: Sutt 79Eb 155
Cedar Grange EN1: Enf 15Ub 33
Cedar Gro. DA5: Bexl 58Zc 117
Cedar Gro. GU24: Bisl 7E 166
Cedar Gro. KT13: Weyb 77S 150
Cedar Gro. UB1: S'hall 43Ca 85
Cedar Gro. W5 .... 48Na 87
Cedar Hgts. NW2 .... 37Bb 69
Cedar Hgts. TW10: Ham 60Na 109
Cedar Hill KT18: Eps 88Sa 173
Cedar Ho. CR0: New Ad 79Dc 158
Cedar Ho. CR8: Purl 84Nb 176
Cedar Ho. E14 .... 47Ec 92
(off Manchester Rd.)
Cedar Ho. E2 .... 40Xb 71
(off Mowlem St.)
Cedar Ho. HA9: Wemb 35Qa 67
(off Engineers Way)
Cedar Ho. KT22: Lea 91Ha 192
Cedar Ho. N22 .... 25Qb 50
(off Acacia Rd.)
Cedar Ho. SE14 .... 53Zb 114
Cedar Ho. SE16 .... 47Zb 92
(off Woodland Cres.)
Cedar Ho. SW6 .... 54Eb 111
(off Lensbury Av.)
Cedar Ho. TW16: Sun 66V 128
(off Spelthorne Gro.)
Cedar Ho. TW9: Kew 53Ra 109
Cedar Ho. UB4: Yead 42Y 85
Cedar Ho. W8 .... 48Db 89
(off Marloes Rd.)
Cedarhurst BR1: Broml 66Gc 137
Cedarhurst Cotts. DA5: Bexl 59Cd 118
Cedarhurst Dr. SE9 .... 57Lc 115
Cedarland Ter. SW20 .... 66Xa 132
Cedar Lawn Av. EN5: Barn 15Ab 30
Cedar Lodge SL5: S'dale 2C 146
Cedar M. SE4 .... 55Bc 114
Cedar M. SW15 .... 57Za 110
Cedar Mt. SE9 .... 60Mc 115
Cedarne Rd. SW6 .... 52Db 111
Cedar Pk. CR3: Cat'm 93Ub 197
Cedar Pk. IG7: Chig 21Qc 54
Cedar Pk. Gdns. RM6: Chad H 31Zc 75
Cedar Pk. Gdns. SW19 .... 64Xa 132
Cedar Pk. Rd. EN2: Enf 10Sb 19
Cedar Pl. HA6: Nwood 23S 44
Cedar Pl. SE7 .... 50Lc 93
Cedar Ri. N14 .... 17Jb 32
Cedar Ri. RM15: S Ock 42Yd 98
Cedar Rd. AL10: Hat 1C 8
Cedar Rd. BR1: Broml 68Lc 137
Cedar Rd. CM13: Hut 16Fe 41
Cedar Rd. CR0: C'don 75Tb 157
Cedar Rd. DA1: Dart 60Md 119
Cedar Rd. DA8: Erith 53Jd 118
Cedar Rd. EN2: Enf 10Rb 19
Cedar Rd. GU22: Wok 2M 187
Cedar Rd. HP4: Berk 2A 2
Cedar Rd. KT11: Cobh 86X 171
Cedar Rd. KT13: Weyb 77Q 150
Cedar Rd. KT8: E Mos 70Ga 130
Cedar Rd. N17 .... 25Vb 51
Cedar Rd. NW2 .... 35Ya 68
Cedar Rd. RM12: Horn 34Ld 77
Cedar Rd. RM16: Grays 8C 100
Cedar Rd. RM7: Rom 28Ed 56
Cedar Rd. SM2: Sutt 79Eb 155
Cedar Rd. TW11: Tedd 64Ja 130
Cedar Rd. TW14: Bedf 60T 106
Cedar Rd. TW4: Cran 54Y 107
Cedar Rd. WD19: Wat 16Y 27
Cedars SM7: Bans 86Hb 175
Cedars SS17: Stan H 1N 101
Cedars, The AL3: St A 1A 6
Cedars, The E15 .... 38Hc 73
Cedars, The E9 .... 38Zb 72
(off Banbury Rd.)
Cedars, The EN9: Walt A 7Lc 21
(within Woodbine Cl. Caravan Pk.)
Cedars, The GU24: Pirb 4B 186
Cedars, The HP4: Berk 1A 2
Cedars, The IG9: Buck H 18Jc 35
Cedars, The KT14: Byfl 84P 169
Cedars, The KT22: Lea 93Ma 193
Cedars, The KT23: Bookh 98Ea 192
Cedars, The RH2: Reig 6M 207
Cedars, The SL2: Slou 1D 80
Cedars, The SL3: Dat 3N 103
Cedars, The SM6: W'gton 77Lb 156
Cedars, The TW11: Tedd 65Ha 130
Cedars, The W13 .... 44La 86
Cedars, The WD3: Rick 17M 25
Cedars Av. CR4: Mitc 70Jb 134
Cedars Av. E17 .... 29Cc 52
Cedars Av. WD3: Rick 18L 25
Cedars Cl. NW4 .... 27Za 48
Cedars Cl. SE13 .... 55Fc 115
Cedars Cl. SL9: Chal P 22A 42
Cedars Cl. WD6: Bore 14Ra 29
Cedars Ct. N9 .... 19Ub 33
Cedars Dr. UB10: Hil 40P 63
Cedars Ho. E17 .... 27Dc 52
Cedars Ho. WD3: Chor 14H 25
Cedarside Apts. NW6 .... 40Bb 69
Cedars M. SW4 .... 56Kb 112
(not continuous)
Cedars Rd. BR3: Beck 68Ac 136
Cedars Rd. CR0: Bedd 76Nb 156
Cedars Rd. E15 .... 37Gc 73
Cedars Rd. KT1: Hamp W 67La 130
Cedars Rd. N21 .... 19Rb 33
Cedars Rd. N9 .... 19Wb 33
Cedars Rd. SM4: Mord 70Cb 133
Cedars Rd. SW13 .... 54Wa 110
Cedars Rd. SW4 .... 55Kb 112
Cedars Rd. W4 .... 50Sa 87
Cedars Wlk. WD3: Chor 14H 25
Cedar Ter. TW9: Rich 56Na 109
Cedar Ter. Rd. TN13: S'oaks 95Ld 203
Cedar Tree Gro. SE27 .... 64Rb 135
Cedar Vw. KT1: King T 69Ma 131
(off Milner Rd.)
Cedar Vw. Cl. CR5: Coul 91Rb 197
Cedarville Gdns. SW16 .... 65Pb 134
Cedar Wlk. CR8: Kenley 88Sb 177
Cedar Wlk. EN9: Walt A 6Fc 21
Cedar Wlk. HP3: Hem H 4M 3
Cedar Wlk. KT10: Clay 79Ha 152
Cedar Wlk. KT20: Tad 92Ab 194
Cedar Way HP4: Berk 2A 2
Cedar Way N1 .... 38Mb 70
Cedar Way SL3: L'ly 50A 82
Cedar Way TW16: Sun 66U 128
Cedar Way Ind. Est. N1 .... 38Mb 70
Cedar Wood Dr. WD25: Wat 7X 13
Cedarwood Dr. AL4: St A 2H 7
Cedarwood Pl. DA14: Sidc 61Zc 139
Cedra Ct. N16 .... 32Wb 71
Cedric Av. RM1: Rom 27Gd 56
Cedric Chambers NW8 .... 5B 214 (42Fb 89)
(off Northwick Cl.)
Cedric Rd. SE9 .... 62Sc 138
Celadon Cl. EN3: Enf H 13Ac 34
Celandine Cl. E14 .... 43Cc 92
Celandine Cl. RM15: S Ock 42Yd 98
Celandine Ct. E4 .... 20Dc 34
Celandine Dr. E8 .... 38Vb 71
Celandine Dr. SE28 .... 46Xc 95
Celandine Gro. N14 .... 15Lb 32
Celandine Rd. KT12: Hers 77Aa 151
Celandine Way E15 .... 41Gc 93
Celbridge M. W2 .... 43Db 89
Celebration Av. E20 .... 36Ec 72
Celebration Way E4 .... 23Ec 52
Celedon Cl. RM16: Chaf H 48Ae 99
Celestial Gdns. SE13 .... 56Fc 115
Celia Av. TN15: W King 80Ud 164
(off London Rd.)
Celia Cres. TW15: Ashf 65M 127
Celia Ho. N1 .... 1H 219 (40Ub 71)
(off Arden Est.)
Celia Johnson Cl. UB9: Den 29J 43
Celia Johnson Ho. WD6: Bore 11Sa 29
Celia Rd. N19 .... 35Lb 70
Cell Barnes Cl. AL1: St A 4F 6
Cell Barnes La. AL1: St A 3E 6
(not continuous)
Cell Farm Av. SL4: Old Win 7M 103
Cellini St. SW8 .... 52Mb 112
Celtic Av. BR2: Broml 69Gc 137
Celtic Farm Rd. RM13: Rain 42Jd 96
Celtic Rd. KT14: Byfl 86N 169
Celtic St. E14 .... 43Dc 92
Cement Block Cotts. RM17: Grays 51Ee 121
Cemetery Hill HP1: Hem H 3L 3
Cemetery La. SE7 .... 51Nc 116
Cemetery La. TW17: Shep 73R 150
Cemetery Pales GU24: Brkwd 4D 186
Cemetery Rd. E7 .... 36Hc 73
Cemetery Rd. N17 .... 24Ub 51
Cemetery Rd. SE2 .... 52Xc 117
Cemetery Way E4 .... 20Cc 34
Cemmaes Ct. Rd. HP1: Hem H 2L 3
Cemmaes Mdw. HP1: Hem H 2L 3
Cenacle Cl. NW3 .... 34Cb 69
Cena Ho. CR8: Kenley 72Tb 157
Cenotaph 1F 229 (47Nb 90)
Centaur Ct. TW8: Bford 50Na 87
Centaurs Bus. Pk. TW7: Isle 51Ja 108
Centaur St. SE1 .... 3J 229 (48Pb 90)
Centaurus Sq. AL2: F'mre 9C 6
Centenary Cl. TN13: Dun G 92Gd 202
Centenary Ct. DA4: Farni 73Qd 163
Centenary Ct. RH1: Redh 5P 207
(off Warwick Rd.)
Centenary Pl. WD6: Bore 13Qa 29
Centenary Rd. EN3: Brim 14Bc 34
Centenary Trad. Est. EN3: Brim 13Bc 34
Centennial Av. WD6: E'tree 17Ka 28
Centennial Ct. WD6: E'tree 17La 28
Centennial Pk. WD6: E'tree 17La 28
Central Apts. HA9: Wemb 36Na 67
Central Arc. TW15: Ashf 63P 127
(off Woodthorpe Rd.)
Central Av. DA12: Grav'nd 1D 144
Central Av. DA16: Well 54Vc 117
Central Av. E11 .... 33Fc 73
Central Av. EN1: Enf 12Xb 33
Central Av. EN8: Walt C 5Ac 20
Central Av. HA5: Pinn 30Ba 45
Central Av. KT8: W Mole 70Ba 129
Central Av. N2 .... 26Fb 49
Central Av. N9 .... 20Ub 33
Central Av. RM15: Avel 47Sd 98
Central Av. RM18: Tilb 3C 122
Central Av. RM20: W Thur 50Vd 98
Central Av. SM6: W'gton 78Nb 156
Central Av. SW11 .... 52Hb 111
Central Av. SW6 .... 55Eb 111
Central Av. TW3: Houn 56Ea 108
Central Av. UB3: Hayes 46V 84
Central Bus. Cen. NW10 .... 36Ua 68
Central Cir. NW4 .... 29Xa 48
Central Ct. KT15: Add 77L 149
Central Ctyd. EC2 .... 2J 225 (44Ub 91)
(off Cutlers Gdns.)
Central Criminal Court
Old Bailey 2C 224 (44Rb 91)
Central Cross Apts. CR0: C'don 77Sb 157
(off South End)
Central Dr. AL4: St A 1G 6
Central Dr. RM12: Horn 34Nd 77
Central Dr. SL1: Slou 5D 80
Centrale Shop. Cen. 75Sb 157
Centrale Stop (London Tramlink) 75Sb 157
Central Gallery IG1: Ilf 33Rc 74
(within The Exchange)
Central Gdns. SM4: Mord 71Db 155
Central Hgts. WD18: Wat 14V 26
(off Manhattan Av.)
Central Hill SE19 .... 64Tb 135
Central Ho. E15 .... 40Dc 72
Central Ho. IG11: Bark 38Sc 74
Central Lawn RM8: Dag 32Ad 75
(off Ager Av.)
Central Lodge TN15: Wro 85Fe 185
Central Mall SW18 .... 58Db 111
(within Southside Shop. Cen.)
Central Mans. NW4 .... 29Xa 48
(off Watford Way)
CENTRAL MIDDLESEX HOSPITAL 41Sa 87
Central Mill Apts. E8 .... 39Vb 71
(off Samuel St.)
Central Pde. CR0: New Ad 82Ec 178
Central Pde. DA15: Sidc 62Wc 139
Central Pde. E17 .... 28Cc 52
Central Pde. EN3: Enf H 12Yb 34
Central Pde. HA1: Harr 29Ha 46
Central Pde. IG2: Ilf 30Tc 54
Central Pde. KT6: Surb 72Na 153
Central Pde. KT8: W Mole 70Ba 129
Central Pde. RH1: Redh 5P 207
Central Pde. SE20 .... 66Zb 136
(off High St.)
Central Pde. TW14: Felt 59Y 107
Central Pde. TW5: Hest 52Ba 107
Central Pde. UB6: G'frd 41Ja 86
Central Pde. W3 .... 47Ra 87
Central Pk. NW10 .... 41Ua 88
Central Pk. Arena 60Nd 119
Central Pk. Av. RM10: Dag 34Dd 76
Central Pk. Est. TW4: Houn 57Z 107
Central Pk. Leisure Cen. 22Nd 57
Central Pk. Rd. E6 .... 40Mc 73
Central Pl. SE25 .... 71Wb 157
Central Rd. DA1: Dart 57Nd 119
Central Rd. HA0: Wemb 36Ka 66
Central Rd. KT4: Wor Pk 74Wa 154
Central Rd. SM4: Mord 72Cb 155
Central Rd. SS17: Stan H 2M 101
Central Rd. Nth. DA1: Dart 55Nd 119
Central St Giles Piazza WC2 .... 2F 223 (44Nb 90)
(off St Giles High St.)
Central St Martins College of Art & Design Back Hill Site 6A 218 (42Qb 90)
(off Back Hill)
Central St Martins College of Art & Design Byam Shaw Campus 33Mb 70
Central School of Speech & Drama, The 38Fb 69
Central School Path SW14 .... 55Sa 109
Central Sq. HA9: Wemb 36Na 67
(off High Rd.)

Central Sq. KT8: W Mole 71Ba 151
Central Sq. N18 23Yb 52
Central Sq. NW11 30Db 49
Central Stores IG10: Lough 16Pc 36
Central St. EC1 3D 218 (41Sb 91)
Central Ter. BR3: Beck 69Zb 136
Central Wlk. KT19: Eps 85Ta 173
Central Walkway N19 35Mb 70
(off Pleshey Rd.)
Central Way HA6: Nwood 24U 44
Central Way NW10 41Sa 87
Central Way RH8: Oxt 99Fc 199
Central Way SE28 46Wc 95
Central Way SM5: Cars 80Gb 155
Central Way TW14: Felt 57W 106
Central W. UB6: G'frd 42Ea 86
Central Carpenders Pk., The 20Y 27
Central Slough, The 5G 80
Centre, The KT12: Walt T 74W 150
Centre, The TW13: Felt 60X 107
Centre, The TW3: Houn 55Da 107
Centre Av. CM16: Epp 4Vc 23
Centre Av. W3 46Ta 87
Centre Cl. CM16: Epp 4Vc 23
Centre Comn. Rd. BR7: Chst 65Sc 138
Centre Ct. Shop. Cen. 65Bb 133
Centre Dr. CM16: Epp 4Vc 23
Central for Wildlife Gardening Vis. Cen. 55Vb 113
Centre Grn. CM16: Epp 4Vc 23
Centre Hgts. NW3 38Fb 69
(off Finchley Rd.)
Central Point SE1 50Wb 91
Centrepoint WC2 2E 222 (44Mb 90)
(off St Giles High St.)
Central Point Ho. WC2 2E 222 (44Mb 90)
(off St Giles High St.)
Centre Rd. DA3: Nw A G 76Ae 165
Centre Rd. E11 33Jc 73
Centre Rd. E7 33Jc 73
Centre Rd. RM10: Dag 40Dd 76
Centre Rd. SL4: Wind 2A 102
Centre Sq. KT1: King T 68Na 131
(off Eden Walk Shop. Cen.)
Centre St. E2 40Xb 71
Centre Vw. Apts. CR0: C'don 76Sb 157
(off Whitgift St.)
Centre Way E17 24Ec 52
Centre Way N9 19Yb 34
Centreway IG1: Ilf 33Sc 74
(off High Rd.)
Centreway Apts. IG1: Ilf 33Sc 74
(off Axon Pl.)
Centric Cl. NW1 39Jb 70
Centrillion Point CR0: C'don 77Sb 157
(off Mason's Av.)
Centrium AL1: St A 4A 6
Centrium GU22: Wok 89B 168
Centro Ct. E6 42Pc 94
Centurian Sq. SE18 53Nc 116
Centurion Bldg. SW11 51Kb 112
Centurion Cl. N7 38Pb 70
Centurion Ct. AL1: St A 3E 6
(off Camp Rd.)
Centurion Ct. SE18 49Qc 94
Centurion Ct. SM6: W'gton 75Kb 156
Centurion Ho. UB3: Hayes 44T 84
Centurion La. E3 39Bc 72
Centurion Way DA18: Erith 48Bd 95
Centurion Way RM19: Purf 49Pd 97
(not continuous)
Centuryan Pl. DA1: Cray 56Kd 119
Century Cl. AL3: St A 1A 6
Century Cl. NW4 29Za 48
Century Ct. GU21: Wok 88B 168
Century Ct. NW8 4C 214 (41Fb 89)
(off Grove End Rd.)
Century Ct. WD18: Wat 17S 26
Century Gdns. CR2: Sande 85Wb 177
Century Ho. HA9: Wemb 33Pa 67
Century Ho. SM7: Bans 87Db 175
Century Ho. SW15 56Za 110
Century La. SL2: Wex 2M 81
Century M. E5 35Yb 72
Century M. N5 34Rb 71
(off Conewood St.)
Century Pk. WD17: Wat 15Y 27
Century Plaza HA8: Edg 23Qa 47
(off Station Rd.)
Century Rd. E17 27Ac 52
Century Rd. TW18: Staines 64E 126
Century Way BR3: Beck 65Bc 136
Century Way GU24: Brkwd 1B 186
Century Yd. SE23 61Yb 136
(not continuous)
Cephas Av. E1 42Yb 92
Cephas Ho. E1 42Yb 92
(off Doveton St.)
Cephas St. E1 42Yb 92
Ceres Cres. KT17: Ewe 82Wa 174
Ceres Rd. SE18 49Vc 95
Ceres Vw. TN16: Big H 86Lc 179
Cerise Apts. E3 39Ac 72
(off Gunmaker's La.)
Cerise Rd. SE15 53Wb 113
Cerne Cl. UB4: Yead 45Y 85
Cerne Rd. DA12: Grav'nd 3G 144
Cerne Rd. SM4: Mord 72Eb 155
Cerney M. W2 4B 220 (45Fb 89)
(off Gloucester Ter.)
Cerotus Pl. KT16: Chert 73H 149
Cervantes Ct. HA6: Nwood 24V 44
Cervantes Ct. W11 44Ab 88
(off Rushton M.)
Cervantes Ct. W2 44Db 89
Cervantes Theatre 1C 230 (47Rb 91)
Cervia Way DA12: Grav'nd 2H 145
Cester St. E2 39Wb 71
Ceylon Ho. E1 44Wb 91
(off Alie St.)
Ceylon Rd. W14 48Za 88
Ceylon Wharf Apts. SE16 47Yb 92
(off St Marychurch St.)
Cezanne Rd. WD25: Wat 8Z 13
CFGS Community Performing Arts & Sports Cen. 41Bc 92
Chabot Dr. SE15 55Xb 113
Chace Av. EN6: Pot B 4Fb 17
Chadacre Av. IG5: Ilf 27Pc 54
Chadacre Ct. E15 39Jc 73
(off Vicars Cl.)
Chadacre Ho. SW9 56Rb 113
(off Loughborough Pk.)
Chadacre Rd. KT17: Ewe 79Xa 154
Chadbourn St. E14 43Dc 92
Chadbury Ct. NW7 25Wa 48
Chad Cres. N9 20Yb 34
Chadd Dr. BR1: Broml 69Nc 138
Chadd Grn. E13 39Jc 73
(not continuous)
Chadfields RM18: Tilb 2C 122
Chadston Ho. N1 38Rb 71
(off Halton Rd.)
Chadswell WC1 4G 217 (41Nb 90)
(off Cromer St.)
Chadview Ct. RM6: Chad H 31Zc 75
Chadville Gdns. RM6: Chad H 29Zc 55
Chadway RM8: Dag 32Yc 75
Chadwell Av. EN8: Chesh 1Yb 20
Chadwell Av. RM6: Chad H 31Xc 75
Chadwell By-Pass RM16: Grays 10B 100
CHADWELL HEATH 31Zc 75
Chadwell Heath Ind. Pk. RM8: Dag 32Zc 75
Chadwell Heath La. RM6: Chad H 28Xc 55
Chadwell Heath Station (Rail & Crossrail) 31Zc 75
Chadwell Hill RM16: Grays 10D 100
Chadwell Ho. SE17 7G 231 (50Tb 91)
(off Inville Rd.)
Chadwell La. N8 27Pb 50
Chadwell Rd. RM17: Grays 49Ee 99
CHADWELL ST MARY 8D 100
Chadwell St. EC1 3A 218 (41Qb 90)
Chadwick Av. E4 21Fc 53
Chadwick Av. N21 15Pb 32
Chadwick Av. SW19 65Cb 133
Chadwick Cl. DA11: Nflt 1A 144
Chadwick Cl. SW15 59Va 110
Chadwick Cl. TW11: Tedd 65Ja 130
Chadwick Cl. W7 43Ha 86
Chadwick Ct. E14 44Cc 92
(off Jonzen Wlk.)
Chadwick Dr. RM3: Hrld W 26Md 57
Chadwick Gdns. UB8: Uxb 38N 63
Chadwick M. W4 51Ra 109
Chadwick Pl. KT6: Surb 73La 152
Chadwick Rd. E11 30Gc 53
Chadwick Rd. IG1: Ilf 34Rc 74
Chadwick Rd. NW10 39Va 68
Chadwick Rd. SE15 54Vb 113
Chadwick Rd. SL3: L'ly 7P 81
Chadwick St. SW1 4D 228 (48Mb 90)
Chadwick Way SE28 45Zc 95
Chadwin Rd. E13 43Kc 93
Chadworth Ho. EC1 4D 218 (41Sb 91)
(off Lever St.)
Chadworth Ho. N4 32Sb 71
Chadworth Way KT10: Clay 78Fa 152
Chaffers Mead KT21: Asht 88Pa 173
Chaffinch Av. CR0: C'don 72Zb 158
Chaffinch Bus. Pk. BR3: Beck 70Zb 136
Chaffinch Cl. CR0: C'don 71Zb 158
Chaffinch Cl. KT6: Surb 76Qa 153
Chaffinch Cl. N9 18Zb 34
Chaffinches Grn. HP3: Hem H 6A 4
Chaffinch La. WD18: Wat 17V 26
Chaffinch Rd. BR3: Beck 67Ac 136
Chafford CM14: B'wood 18Xd 40
Chafford Gdns. CM13: W H'dn 30Fe 59
Chafford Gorges Nature Pk. 48Zd 99
Chafford Gorges Vis. Cen. 48Zd 99
CHAFFORD HUNDRED 48Be 99
Chafford Hundred Station (Rail) 49Xd 98
Chafford Sports Complex 43Ld 97
Chafford Wlk. RM13: Rain 40Ld 77
Chafford Way RM16: Grays 46Ce 99
Chafford Way RM6: Chad H 28Yc 55
Chagford Ho. E3 41Dc 92
(off Talwin St.)
Chagford St. NW1 6F 215 (42Hb 89)
Chailey Av. EN1: Enf 12Vb 33
Chailey Cl. TW5: Hest 53Z 107
Chailey Ind. Est. UB3: Hayes 47W 84
Chailey Pl. KT12: Hers 77Aa 151
Chailey St. E5 34Yb 72
Chainmakers Ho. E14 44Fc 93
(off Blair St.)
Chairman's Wlk. UB9: Den 29H 43
Chalbury Wlk. N1 1J 217 (40Pb 70)
Chalcombe Rd. SE2 48Xc 95
Chalcot Cl. SM2: Sutt 80Cb 155
Chalcot Cres. NW1 39Hb 69
Chalcot Gdns. NW3 37Hb 69
Chalcot M. SW16 62Nb 134
Chalcot Rd. NW1 38Jb 70
Chalcot Sq. NW1 38Jb 70
(not continuous)
Chalcott SL1: Slou 8J 81
Chalcott Gdns. KT6: Surb 74La 152
Chalcraft Ct. KT16: Chert 74J 149
Chalcroft Rd. SE13 57Gc 115
CHALDON 96Qb 196
Chaldon Cl. RH1: Redh 8N 207
Chaldon Comn. Rd. CR3: Cat'm 96Sb 197
Chaldon Ct. SE19 68Tb 135
Chaldon Path CR7: Thor H 70Rb 135
Chaldon Rd. CR3: Cat'm 96Tb 197
Chaldon Rd. SW6 52Ab 110
Chaldon Way CR5: Coul 89Nb 176
Chale Ct. SS17: Stan H 3L 101
Chale Rd. SW2 58Nb 112
Chalet Cl. DA5: Bexl 63Fd 140
Chalet Cl. TW15: Ashf 65T 128
Chalet Ct. CR7: Thor H 71Sb 157
Chalet Est. NW7 21Wa 48
Chale Wlk. SM2: Sutt 81Db 175
Chalfont Av. HA9: Wemb 37Ra 67
Chalfont Av. HP6: L Chal 11A 24
Chalfont Cen. SL9: Chal P 22B 42
CHALFONT COMMON 22B 42
Chalfont Ct. HA1: Harr 30Ha 46
(off Northwick Pk. Rd.)
Chalfont Ct. NW1 6G 215 (42Hb 89)
(off Baker St.)
Chalfont Ct. NW9 27Va 48
Chalfont Dene SL9: Chal P 22B 42
Chalfont Grn. N9 20Ub 33
Chalfont Ho. SE16 48Xb 91
(off Keetons Rd.)
Chalfont Ho. SL9: Chal P 22B 42
Chalfont Ho. WD18: Wat 16U 26
Chalfont La. WD3: Chor 15D 24
Chalfont La. WD3: W Hyd 23E 42
Chalfont M. SW19 60Bb 111
Chalfont M. UB10: Hil 38R 64
Chalfont Pk. SL9: Chal P 26A 42
Chalfont Pk. 27B 42
Chalfont Pl. AL1: St A 2C 6
Chalfont Rd. HP8: Chal G 20D 24
Chalfont Rd. HP8: Map C 20D 24
Chalfont Rd. N9 20Ub 33
Chalfont Rd. SE25 69Vb 135
Chalfont Rd. UB3: Hayes 47W 84
Chalfont Rd. WD3: Map C 19D 24
CHALFONT ST PETER 25A 42
Chalfont St Peter By-Pass SL9: Chal P 25A 42
CHALFONTS & GERRARDS CROSS HOSPITAL 25A 42
Chalfont Wlk. HA5: Pinn 26Y 45
Chalfont Way W13 48Ka 86
Chalford NW6 37Eb 69
(off Finchley Rd.)
Chalford Cl. KT8: W Mole 70Ca 129
Chalforde Gdns. RM2: Rom 28Kd 57
Chalford Rd. SE21 63Tb 135
Chalford Wlk. IG8: Wfd G 25Mc 53
Chalgrove Av. SM4: Mord 71Cb 155
Chalgrove Cres. IG5: Ilf 26Nc 54
Chalgrove Gdns. N3 27Ab 48
Chalgrove Rd. N17 25Xb 51
Chalgrove Rd. SM2: Sutt 80Fb 155
Chalice Cl. SM6: W'gton 79Mb 156
Chalice Ct. N2 28Gb 49
Chalice Ct. SE19 65Tb 135
Chalice Way DA9: Ghithe 57Ud 120
CHALK 10K 123
Chalk Cl. DA1: Dart 60Pd 119
Chalk Ct. RM17: Grays 51Ce 121
(off Argent St.)
Chalkdell Hill HP2: Hem H 2N 3
Chalkdell Ho. WD25: Wat 5Y 13
Chalk Dene DA10: Swans 60Ae 121
Chalkenden Cl. SE20 66Xb 135
CHALKER'S CORNER 55Ra 109
CHALK FARM 38Jb 70
Chalk Farm Pde. NW3 38Jb 70
(off Adelaide Rd.)
Chalk Farm Rd. NW1 38Jb 70
Chalk Farm Station (Underground) 38Jb 70
Chalk Hill WD19: Wat 16Aa 27
Chalkhill Apts. CR8: Purl 82Pb 176
Chalk Hill Rd. W6 49Za 88
Chalkhill Rd. HA9: Wemb 34Ra 67
Chalklands HA9: Wemb 34Sa 67
Chalk La. EN4: Cockf 13Hb 31
Chalk La. KT18: Eps 87Ta 173
Chalk La. KT18: Eps D 87Ta 173
Chalk La. KT21: Asht 91Pa 193
Chalkley Cl. CR4: Mitc 68Hb 133
Chalkmead RH1: Mers 2C 208
Chalkmill Dr. EN1: Enf 13Xb 33
Chalk Paddock KT18: Eps 87Ta 173
Chalk Pit Av. BR5: St P 69Yc 139
Chalk Pit La. KT23: Bookh 99Ba 191
Chalkpit La. CR3: Wold 97Fc 199
Chalkpit La. RH8: Oxt 97Ec 198
Chalk Pit Rd. KT18: Eps D 91Sa 193
Chalk Pit Rd. SM7: Bans 89Cb 175
Chalk Pit Way SM1: Sutt 78Eb 155
Chalkpit Wood RH8: Oxt 99Fc 199
Chalk Rd. DA12: Grav'nd 10J 123
Chalk Rd. E13 43Kc 93
Chalkstone Cl. DA16: Well 53Wc 117
Chalkwell Ho. E1 44Zb 92
(off Pitsea St.)
Chalkwell Pk. Av. EN1: Enf 14Ub 33
Chalky Bank DA11: Grav'nd 3C 144
Chalky La. KT9: Chess 82Ma 173
Challacombe Cl. CM13: Hut 18De 41
(off Orde Hall St.)
Challenge Cl. DA12: Grav'nd 3H 145
Challenge Cl. NW10 39Ua 68
Challenge Ct. KT22: Lea 91Ka 192
Challenge Ct. TW2: Twick 59Ga 108
Challenger Ho. E14 45Ac 92
(off Victory Pl.)
Challenge Rd. TW15: Ashf 62T 128
Challice Way SW2 60Pb 112
Challin St. SE20 67Yb 136
Challis Ho. SW11 54Hb 111
Challis Rd. TW8: Bford 50Ma 87
Challock Cl. TN16: Big H 88Lc 179
Challoner Cl. N2 26Fb 49
Challoner Ct. BR2: Broml 68Fc 137
Challoner Ct. W14 50Bb 89
(off Challoner St.)
Challoner Cres. W14 50Bb 89
Challoner Mans. W14 50Bb 89
(off Challoner St.)
Challoners Cl. KT8: E Mos 70Fa 130
Challoner St. W14 50Bb 89
Challoner Wlk. E1 44Wb 91
(off Christian St.)
Chalmers Ct. WD3: Crox G 16P 25
Chalmers Ho. E17 29Dc 52
Chalmers Rd. SM7: Bans 87Fb 175
Chalmers Rd. TW15: Ashf 64R 128
Chalmers Rd. E. TW15: Ashf 63R 128
Chalmers Wlk. SE17 51Rb 113
(off Hillingdon St.)
Chalmers Way TW1: Isle 56Ka 108
Chalmers Way TW14: Felt 57X 107
Chaloner Ct. SE1 1F 231 (47Tb 91)
(off Tennis St.)
Chalsey Rd. SE4 56Bc 114
Chalton Dr. N2 30Fb 49
Chalton Ho. NW1 3D 216 (41Mb 90)
(off Chalton St.)
Chalton St. NW1 1C 216 (40Lb 70)
(not continuous)
CHALVEY 8H 81
Chalvey Gdns. SL1: Slou 7J 81
Chalvey Gro. SL1: Slou 8F 80
Chalvey Pk. SL1: Slou 7J 81
Chalvey Rd. E. SL1: Slou 7J 81
Chalvey Rd. W. SL1: Slou 7H 81
Chamberlain Cl. IG1: Ilf 34Sc 74
Chamberlain Cl. KT19: Eps 82Ta 173
Chamberlain Cl. SE28 48Tc 94
Chamberlain Cl. UB3: Hayes 45V 84
Chamberlain Cotts. SE5 53Tb 113
Chamberlain Ct. SE16 49Yb 92
(off Silwood St.)
Chamberlain Cres. BR4: W W'ck 74Dc 158
Chamberlain Gdns. TW3: Houn 53Ea 108
Chamberlain Ho. E1 45Yb 92
(off Cable St.)
Chamberlain Ho. EC2 5H 219 (42Ub 91)
(off Blackall St.)
Chamberlain Ho. NW1 3E 216 (41Mb 90)
(off Ossulston St.)
Chamberlain Ho. SE1 2K 229 (47Qb 90)
(off Westminster Bri. Rd.)
Chamberlain La. HA5: Eastc 27W 44
Chamberlain Pl. E17 27Ac 52
Chamberlain Rd. N2 26Eb 49
Chamberlain Rd. W13 47Ja 86
Chamberlain St. NW1 38Hb 69
Chamberlain Wlk. TW13: Hanw 63Aa 129
Chamberlain Way HA5: Eastc 27X 45
Chamberlain Way KT6: Surb 73Na 153
Chamberlayne Av. HA9: Wemb 34Na 67
Chamberlayne Mans. NW10 41Za 88
(off Chamberlayne Rd.)
Chamberlayne Rd. NW10 39Ya 68
Chamberlens Garages W6 49Xa 88
(off Dalling Rd.)
Chambers, The SW10 53Eb 111
(off Chelsea Harbour Dr.)
Chambers Av. DA14: Sidc 65Ad 139
Chambersbury La. HP3: Hem H 7A 4
(not continuous)
Chambers Bus. Pk. UB7: Sip 51Q 106
Chambers Cl. DA9: Ghithe 57Wd 120
Chambers Gdns. N2 25Fb 49
Chambers La. NW10 38Xa 68
Chambers Pk. Hill SW20 66Xa 132
Chambers Pl. CR2: S Croy 80Tb 157
Chambers Rd. N7 35Nb 70
Chambers St. SE16 47Wb 91
Chamber St. E1 4K 225 (45Vb 91)
Chambers Wlk. HA7: Stan 22Ka 46
Chambers Wharf SE16 47Wb 91
Chambon Pl. W6 49Wa 88
Chambord Ho. E2 4K 219 (41Vb 91)
(off Chambord St.)
Chambord St. E2 4K 219 (41Vb 91)
Chamomile Ct. E17 30Cc 52
(off Yunus Khan Cl.)
Champa Cl. N17 26Vb 51
Champion Cres. SE26 63Ac 136
Champion Down KT24: Eff 100Aa 191
Champion Gro. SE5 55Tb 113
Champion Hill SE5 55Tb 113
Champion Hill Est. SE5 55Ub 113
Champion Hill Stadium 56Ub 113
Champion Ho. SE7 51Lc 115
(off Charlton Rd.)
Champion Pk. SE5 54Tb 113
Champion Rd. RM14: Upm 33Rd 77
Champion Rd. SE26 63Ac 136
Champions Wlk. E20 36Ec 72
Champions Way NW4 25Xa 48
Champions Way NW7 25Xa 48
Champlain Ho. W12 45Xa 88
(off White City Est.)
Champness Cl. E17 27Zb 52
Champness Cl. SE27 63Tb 135
Champness Rd. IG11: Bark 37Vc 75
Champney Cl. SL3: Hort 55C 104
Champneys WD19: Wat 19Aa 27
Champneys Cl. SM2: Cheam 80Bb 155
Chance Cl. RM16: Chaf H 48Be 99
Chancel Cl. TN15: W King 80Ud 164
Chancel Cl. UB8: Uxb 40K 63
Chancel Ct. W1 4D 222 (45Mb 90)
(off Old Compton St.)
Chancel Ind. Est. NW10 36Va 68
Chancellor Gdns. CR2: S Croy 81Rb 177
Chancellor Gro. SE21 61Sb 135
Chancellor Ho. E1 46Xb 91
(off Green Bank)
Chancellor Ho. SW7 3A 226 (48Eb 89)
(off Hyde Pk. Ga.)
Chancellor Pas. E14 46Cc 92
Chancellor Pl. NW9 26Va 48
Chancellors Cl. BR3: Beck 73Cc 158
Chancellors Ct. WC1 7H 217 (43Pb 90)
(off Orde Hall St.)
Chancellor's Rd. W6 50Ya 88
Chancellor's St. W6 50Ya 88
Chancellors Wharf W6 50Ya 88
Chancellor Way RM8: Dag 35Wc 75
Chancellor Way TN13: S'oaks 94Jd 202
Chancelot Rd. SE2 49Xc 95
Chancel St. SE1 7B 224 (46Rb 91)
Chance Mead KT15: New H 82L 169
Chancery Bldg. SW11 52Mb 112
Chancery Bldgs. E1 45Xb 91
(off Lowood St.)
Chancery Ct. DA1: Dart 59Qd 119
Chancerygate UB7: Yiew 46Q 84
Chancerygate Bus. Cen. HA4: Ruis 35Aa 65
Chancerygate Bus. Cen. SL3: L'ly 47A 82
Chancery Ga. Bus. Pk. KT6: Surb 75Pa 153
Chancerygate Ind. Pk. DA14: Sidc 66Zc 139
Chancery La. BR3: Beck 68Dc 136
Chancery La. WC2 2K 223 (44Qb 90)
Chancery Lane Station (Underground) 1K 223 (43Qb 90)
Chancery M. SW17 61Gb 133
Chance St. E1 5K 219 (42Vb 91)
Chance St. E2 5K 219 (42Vb 91)
Chanctonbury Chase RH1: Redh 6A 208
Chanctonbury Cl. SE9 62Rc 138
Chanctonbury Dr. SL5: S'dale 3C 146
Chanctonbury Gdns. SM2: Sutt 80Db 155
Chanctonbury Way N12 21Bb 49
Chandaria Ct. CR0: C'don 76Sb 157
(off Church Rd.)
Chandler Av. E16 43Jc 93
Chandler Cl. TW12: Hamp 67Ca 129
Chandler Ct. TW14: Felt 58W 106
Chandler Ho. NW6 39Bb 69
(off Willesden La.)
Chandler Ho. WC1 6G 217 (42Nb 90)
(off Colonnade)
Chandler Rd. IG10: Lough 11Rc 36
Chandlers Av. SE10 47Hc 93
Chandlers Cl. GU21: Wok 10K 167
Chandlers Cl. KT8: W Mole 71Da 151
Chandlers Cl. TW14: Felt 59V 106
Chandlers Cnr. RM13: Rain 41Ld 97
CHANDLERS CORNER 41Ld 97
Chandlers Ct. SE12 60Kc 115
CHANDLER'S CROSS 10P 11
Chandlers Dr. DA8: Erith 49Fd 96
Chandler's La. WD3: Chan C 8N 11
Chandlers M. DA9: Ghithe 56Yd 120
Chandlers M. E14 47Cc 92
Chandler St. E1 46Xb 91
Chandlers Way RM1: Rom 29Gd 56
Chandlers Way SW2 59Qb 112
Chandler Way SE15 51Ub 113
Chandlery, The SE1 3A 230 (48Qb 90)
(off Gerridge St.)
Chandlery Ho. E1 44Wb 91
(off Gower's Wlk.)
Chandon Lodge SM2: Sutt 80Eb 155
Chandos Av. E17 26Cc 52
Chandos Av. N14 20Lb 32
Chandos Av. N20 18Eb 31
Chandos Av. W5 49La 86
Chandos Cl. IG9: Buck H 19Kc 35
Chandos Ct. HA7: Stan 23Ka 46
Chandos Ct. HA8: Edg 24Pa 47
Chandos Ct. N14 19Mb 32
Chandos Cres. HA8: Edg 24Pa 47
Chandos Gdns. CR5: Coul 91Rb 197
Chandos Mall SL1: Slou 7K 81
(within Queensmere Shop. Cen.)
Chandos Pde. HA8: Edg 24Pa 47
Chandos Pl. WC2 5F 223 (45Nb 90)
Chandos Rd. E15 36Fc 73
Chandos Rd. HA1: Harr 29Ea 46
Chandos Rd. HA5: Eastc 31Z 65
Chandos Rd. N17 26Ub 51
Chandos Rd. N2 26Fb 49
Chandos Rd. NW10 42Ua 88
Chandos Rd. NW2 36Ya 68
Chandos Rd. TW18: Staines 64F 126
Chandos Rd. WD6: Bore 12Pa 29
Chandos St. W1 1A 222 (43Kb 90)
Chandos Way NW11 32Db 69
Change All. EC3 3G 225 (44Tb 91)
Chanin M. NW2 36Ya 68
Chanlock Path RM15: S Ock 45Xd 98
Channel 4 TV 4D 228 (48Mb 90)
(off Horseferry Rd.)
Channel Cl. TW5: Hest 53Ca 107
Channel Ga. Rd. NW10 41Ua 88
Channel Ho. SE16 47Zb 92
(off Water Gdns. Sq.)
Channel Islands Est. N1 37Sb 71
Channelsea Ho. E15 40Fc 73
Channelsea Path E15 39Fc 73
Channelsea Rd. E15 39Fc 73
Channing Cl. RM11: Horn 31Pd 77
Channings GU21: Wok 87A 168
TW14: Felt
Channon Ct. KT6: Surb 71Na 153
(off Maple Rd.)
Chantilly Way KT19: Eps 82Ra 173
Chanton Dr. KT17: Cheam 82Ya 174
Chanton Dr. SM2: Cheam 82Ya 174
Chanton Dr. SM2: Ewe 82Ya 174
Chantrelle Ct. SE8 49Ac 92
(off Yeoman St.)
Chantress Cl. RM10: Dag 39Ed 76
Chantrey Ho. SW1 5K 227 (49Kb 90)
(off Eccleston St.)
Chantrey Rd. SW9 55Pb 112
Chantreywood CM13: B'wood 20Ce 41
Chantry, The E4 18Ec 34
Chantry, The UB8: Hil 41P 83
Chantry Av. DA3: Hartl 72Ae 165
Chantry Cl. DA14: Sidc 64Ad 139
Chantry Cl. EN2: Enf 10Sb 19
Chantry Cl. HA3: Kenton 29Pa 47
Chantry Cl. KT21: Asht 91La 192
Chantry Cl. NW7 16Va 30
Chantry Cl. SE2 48Yc 95
Chantry Cl. SL4: Wind 3E 102
Chantry Cl. TW16: Sun 66W 128
Chantry Cl. UB7: Yiew 45M 83
Chantry Cl. W9 42Cb 89
Chantry Cl. WD4: K Lan 1Q 12
Chantry Ct. AL10: Hat 1C 8
Chantry Ct. DA12: Grav'nd 8E 122
Chantry Ct. SM5: Cars 76Gb 155
Chantry Cres. NW10 37Va 68
Chantry Cres. SS17: Stan H 2L 101
Chantry Ho. KT1: King T 70Na 131
Chantry Hurst KT18: Eps 87Ta 173
Chantry La. AL10: Hat College La. 1B 8
Chantry La. AL10: Hat Sparrowhawk Pl. 1C 8
Chantry La. AL2: Lon C 8H 7
Chantry La. BR2: Broml 71Mc 159
Chantry Pl. HA3: Hrw W 25Da 45
Chantry Rd. HA3: Hrw W 25Da 45
Chantry Rd. KT16: Chert 73L 149
Chantry Rd. KT9: Chess 78Pa 153
Chantry Sq. W8 48Db 89
Chantry St. N1 1C 218 (39Rb 71)
Chantry Way CR4: Mitc 69Fb 133
Chant Sq. E15 38Fc 73
Chant St. E15 38Fc 73
(not continuous)
Chapel, The CM14: W'ley 22Xd 58
(off The Galleries)
Chapel, The SW15 58Ab 110
Chapel Av. KT15: Add 77K 149
Chapel Cl. AL9: Brk P 9M 9
Chapel Cl. DA1: Cray 57Gd 118
Chapel Cl. NW10 36Va 68
Chapel Cl. RM20: W Thur 51Xd 120
Chapel Cl. WD25: Wat 6V 12
Chapel Cotts. HP2: Hem H 1M 3
Chapel Ct. DA10: Swans 58Ae 121
Chapel Ct. E10 33Dc 72
(off Rosedene Ter.)
Chapel Ct. N2 27Gb 49
Chapel Ct. RM7: Rush G 30Fd 56
(off Bournebrook Gro.)
Chapel Ct. SE1 1F 231 (47Tb 91)
Chapel Ct. SE18 51Wc 117
Chapel Ct. UB3: Hayes 45V 84
Chapel Cft. WD4: Chfd 3J 11
CHAPEL CROFT 3J 11
Chapel Dr. DA2: Dart 58Sd 120
CHAPEL END 25Dc 52
Chapel Farm Rd. SE9 62Pc 138
Chapel Ga. M. SW4 55Nb 112
(off Bedford Rd.)
Chapel Ga. Pl. BR7: Chst 65Rc 138
Chapel Grn. CR8: Purl 85Qb 176
Chapel Gro. KT15: Add 77Kk 149
Chapel Gro. KT18: Tatt C 91Ya 194
Chapel High CM14: B'wood 19Yd 40
(off High St.)
Chapel Hill DA1: Cray 57Gd 118
Chapel Hill KT24: Eff 99Z 191
Chapel Ho. St. E14 50Dc 92
Chapelier Ho. SW18 56Cb 111
Chapel La. GU24: Pirb 4E 186
Chapel La. HA5: Pinn 27Z 45
Chapel La. IG7: Chig 20Vc 37
Chapel La. KT23: Bookh 100Ea 192
Chapel La. KT23: Westh 100Ea 192
Chapel La. RM6: Chad H 31Zc 75
Chapel La. SL2: Stoke P 8M 61
Chapel La. UB8: Hil 44Q 84
Chapel Lodge RM13: Rain 41Jd 96
Chapel Mkt. N1 1K 217 (40Qb 70)
Chapel M. IG8: Wfd G 23Oc 54
Chapel M. RM11: Horn 31Ld 77
Chapel Mill Rd. KT1: King T 69Pa 131
Chapelmount Rd. IG8: Wfd G 23Pc 54
Chapel Pk. Rd. KT15: Add 77K 149
Chapel Path E11 30Kc 53
(off Woodbine Pl.)
Chapel Pl. AL1: St A 5B 6
Chapel Pl. EC2 4H 219 (41Ub 91)
Chapel Pl. N1 1A 218 (40Qb 70)
Chapel Pl. N17 24Vb 51
Chapel Pl. W1 3K 221 (44Kb 90)
Chapel Rd. CM16: Epp 2Vc 23
Chapel Rd. CR6: W'ham 90Zb 178
Chapel Rd. DA7: Bex 56Cd 118
Chapel Rd. IG1: Ilf 34Qc 74
Chapel Rd. KT20: Tad 95Ya 194
Chapel Rd. RH1: Redh 6P 207
Chapel Rd. RH8: Limp 2N 211
Chapel Rd. RM18: Tilb 3C 122
Chapel Rd. SE27 63Rb 135
Chapel Rd. TW1: Twick 59Ka 108

Chapel Rd. TW3: Houn 55Da 107
Chapel Rd. W13 46Ka 86
Chapel Row TN15: Igh 93Yd 204
Chapel Row UB9: Hare 25L 43
Chapels Cl. SL1: Slou 6C 80
Chapel Side W2 45Db 89
Chapel Sq. GU25: Vir W 70A 126
Chapel Stones N17 25Vb 51
Chapel St. EN2: Enf 13Sb 33
Chapel St. GU21: Wok 89B 168
Chapel St. HP2: Hem H 1M 3
Chapel St. NW1 1D 220 (43Gb 89)
Chapel St. SL1: Slou 7K 81
Chapel St. SW1 3J 227 (48Jb 90)
Chapel St. UB8: Uxb 39L 63
Chapel Ter. IG10: Lough 14Nc 36
Chapel Vw. CR2: Sels 79Yb 158
Chapel Vw. TN15: Igh 93Yd 204
Chapel Wlk. CR0: C'don 75Sb 157
Chapel Wlk. CR5: Coul 94Mb 196
Chapel Wlk. DA2: Wilm 61Gd 140
Chapel Wlk. DA5: Wilm 61Gd 140
Chapel Wlk. NW4 28Xa 48
(not continuous)
Chapel Way KT18: Tatt C 91Ya 194
Chapel Way N7 34Pb 70
Chapel Way WD5: Bedm 9F 4
Chapel Wood DA3: Nw A G 74Ae 165
(not continuous)
Chapelwood Pl. DA13: Sole S 10E 144
Chapel Wood Rd. DA3: Hartl 75Ae 165
Chapel Wood Rd. TN15: Ash 76Ae 165
Chapel Yd. SW18 57Db 111
(off Wandsworth High St.)
Chaplaincy Gdns. RM11: Horn 32Nd 77
Chaplin Cl. HA0: Wemb 37Ma 67
Chaplin Cl. SE1 1A 230 (47Qb 90)
Chaplin Ct. DA4: Sut H 65Qd 141
Chaplin Ct. E3 42Bc 92
(off Joseph St.)
Chaplin Ct. SE14 53Zb 114
(off Besson St.)
Chaplin Ct. SE17 51Rb 113
(off Royal Rd.)
Chaplin Cres. TW16: Sun 65U 128
Chaplin Ho. DA14: Sidc 63Wc 139
(off Sidcup High St.)
Chaplin Ho. E17 28Cc 52
(off Hoe St.)
Chaplin Ho. N1 39Tb 71
(off Shepperton Rd.)
Chaplin Ho. W3 48Sa 87
(off All Saints Rd.)
Chaplin M. SL3: L'ly 50B 82
Chaplin Pl. CR5: Coul 89Lb 176
Chaplin Rd. E15 40Hc 73
Chaplin Rd. HA0: Wemb 37La 66
Chaplin Rd. N17 27Vb 51
Chaplin Rd. NW2 37Wa 68
Chaplin Rd. RM9: Dag 38Ad 75
Chaplin Sq. N12 24Fb 49
Chapman Cl. UB7: W Dray 48P 83
Chapman Ct. DA1: Dart 54Qd 119
Chapman Ct. EN8: Chesh 2Zb 20
Chapman Cres. HA3: Kenton 30Na 47
Chapman Grn. N22 25Qb 50
Chapman Hall AL10: Hat 2B 8
(off Bishops Ri.)
Chapman Ho. E1 44Xb 91
(off Bigland St.)
Chapman Ho. NW9 27Va 48
(off Aerodrome Rd.)
Chapman Pl. N4 33Rb 71
Chapman Rd. CR0: C'don 74Qb 156
Chapman Rd. DA17: Belv 50Cd 96
Chapman Rd. E9 37Bc 72
Chapmans Cl. TN14: Sund 96Ad 201
Chapman's Hill DA13: Meop 79Fe 165
Chapman's La. BR5: St P 68Zc 139
Chapman's La. BR5: Swan 68Zc 139
Chapman's La. DA17: Belv 49Zc 95
Chapman's La. SE2 49Yc 95
Chapmans Pk. Ind. Est. NW10 37Va 68
Chapman Sq. SW19 61Za 132
Chapmans Rd. TN14: Sund 96Ad 201
Chapman's Ter. N22 25Rb 51
Chapman St. E1 45Xb 91
Chapone Pl. W1 3D 222 (44Mb 90)
Chapter Chambers SW1 6D 228 (49Mb 90)
(off Chapter St.)
Chapter Cl. UB10: Hil 38P 63
Chapter Cl. W4 48Sa 87
Chapter Ho. E2 42Wb 91
(off Dunbridge St.)
Chapter House 3C 224 (44Rb 91)
(off St Paul's Chyd.)
Chapter M. SL4: Wind 2H 103
Chapter Rd. NW2 36Wa 68
Chapter Rd. SE17 50Rb 91
Chapter St. SW1 6D 228 (49Mb 90)
Chapter Way SW19 67Fb 133
Chapter Way TW12: Hamp 63Ca 129
Chara Pl. W4 51Ta 109
Charcot Ho. SW15 58Va 110
Charcot Rd. NW9 26Ua 48
Charcroft Ct. W14 47Za 88
(off Minford Gdns.)
Charcroft Gdns. EN3: Pond E 14Zb 34
Chardin Ho. SW9 53Qb 112
(off Gosling Way)
Chardin Rd. W4 49Ua 88
Chardins Cl. HP1: Hem H 1H 3
Chardmore Rd. N16 32Wb 71
Chard Rd. TW6: H'row A 54R 106
Chardwell Cl. E6 44Pc 94
Charecroft Way W14 47Za 88
Charfield Ct. W9 42Db 89
(off Shirland Rd.)
Charford Rd. E16 43Jc 93
Chargate Cl. KT12: Hers 79V 150
Chargeable La. E13 42Hc 93
Chargeable St. E16 42Hc 93
Chargrove Cl. SE16 47Zb 92
Charing Cl. BR6: Orp 77Vc 161
Charing Ct. BR2: Broml 68Gc 137
Charing Cross SW1 6F 223 (46Nb 90)
(off Whitehall)
CHARING CROSS HOSPITAL 51Za 110
Charing Cross Rd. WC2 3E 222 (44Mb 90)
Charing Cross Sports Club 51Za 110
Charing Cross Station (Rail & Underground) London 6G 223 (46Nb 90)
Charing Cross Theatre 6G 223 (46Nb 90)
(off Villiers St.)
Charing Cross Underground Shop. Cen. 5F 223 (45Nb 90)
(off The Strand)
Charing Ho. SE1 1A 230 (47Qb 90)
(off Windmill Wlk.)
Chariot Cl. E3 39Cc 72
Chariotts Pl. SL4: Wind 3H 103
Charis Ho. E3 41Dc 92
(off Grace St.)
Charkham M. AL9: Wel G 6E 8
Charlbert Ct. NW8 1D 214 (40Gb 69)
(off Charlbert St.)
Charlbert St. NW8 1D 214 (40Gb 69)
Charlbury Av. HA7: Stan 22Ma 47
Charlbury Cl. RM3: Rom 23Ld 57
Charlbury Cres. RM3: Rom 23Ld 57
Charlbury Gdns. IG3: Ilf 33Vc 75
Charlbury Gro. W5 44La 86
Charlbury Rd. UB10: Ick 34P 63
Charldane Rd. SE9 62Rc 138
Charlecombe Ct. TW18: Staines 64K 127
Charlecote Gro. SE26 62Xb 135
Charlecote Rd. RM8: Dag 34Ad 75
Charlemont Rd. E6 42Pc 94
Charles II Pl. SW3 7F 227 (50Hb 89)
Charles II St. SW1 6D 222 (46Mb 90)
Charles Auffray Ho. E1 43Yb 92
(off Smithy St.)
Charles Babbage Cl. KT9: Chess 80La 152
Charles Baker Pl. SW17 60Gb 111
Charles Barry Cl. SW4 55Lb 112
Charles Bradlaugh Ho. N17 24Xb 51
(off Haynes Cl.)
Charles Burton Ct. E5 36Ac 72
(off Ashenden Rd.)
Charles Chu. Wlk. IG1: Ilf 30Pc 54
Charles Cl. DA14: Sidc 63Xc 139
Charles Clowes Wlk. SW11 52Mb 112
Charles Cobb Gdns. CR0: Wadd 78Qb 156
Charles Ct. DA8: Erith 51Gd 118
Charles Coveney Rd. SE15 53Vb 113
Charles Cres. HA1: Harr 31Fa 66
(not continuous)
Charles Cryer Studio Theatre, The 77Jb 156
Charles Darwin Ho. E16 44Hc 93
(off Minnie Baldock St.)
Charles Darwin Ho. E2 41Xb 91
(off Canrobert St.)
Charles Dickens Ho. E2 41Wb 91
(off Mansford St.)
Charles Dickens Mus., The 6J 217 (42Pb 90)
(off Doughty St.)
Charlesfield SE9 62Lc 137
Charles Flemwell M. E16 46Jc 93
Charles Gdns. SL2: Slou 4M 81
Charles Gardner Ct. N1 3G 219 (41Tb 91)
(off Haberdasher St.)
Charles Grinling Wlk. SE18 49Qc 94
Charles Gro. N14 18Lb 32
Charles Haller St. SW2 59Qb 112
Charles Harrod Ct. SW13 51Ya 110
(off Somerville Av.)
Charles Hocking Ho. W3 47Sa 87
(off Bollo Bri. Rd.)
Charles Ho. KT16: Chert 74H 149
(off Sth. Guildford St.)
Charles Ho. N17 24Vb 51
(off Love La.)
Charles Ho. SL4: Wind 3G 102
Charles Ho. UB2: S'hall 47Ca 85
Charles Ho. W14 49Bb 89
Charles Howell Dr. CR5: Coul 89Kb 176
Charles Lamb Ct. N1 1C 218 (40Rb 71)
(off Gerrard Rd.)
Charles La. NW8 1D 214 (40Gb 69)
Charles Lesser Ho. KT9: Chess 78Ma 153
Charles Mackenzie Ho. SE16 49Wb 91
(off Linsey St.)
Charlesmere Gdns. SE28 47Uc 94
(off Thames Reach)
Charles Nex M. SE21 61Sb 135
Charles Pl. E4 17Fc 35
Charles Pl. NW1 4C 216 (41Lb 90)
Charles Rd. E7 38Lc 73
Charles Rd. RM10: Dag 37Fd 76
Charles Rd. RM6: Chad H 30Zc 55
Charles Rd. SW19 67Cb 133
Charles Rd. TN14: Bad M 82Dd 182
Charles Rd. TW18: Staines 65M 127
Charles Rd. W13 44Ja 86
Charles Rowan Ho. WC1 4K 217 (41Qb 90)
(off Margery St.)
Charles Sevright Way NW7 22Za 48
Charles Simmons Ho. WC1 4J 217 (41Pb 90)
(off Margery St.)
Charles Sq. N1 4G 219 (41Tb 91)
Charles Sq. Est. N1 4G 219 (41Tb 91)
(off Pitfield St.)
Charles St. CM16: Epp 4Wc 23
Charles St. CR0: C'don 76Sb 157
Charles St. DA9: Ghithe 57Ud 120
(not continuous)
Charles St. E16 46Lc 93
Charles St. EN1: Enf 15Vb 33
Charles St. HP1: Hem H 3L 3
Charles St. KT16: Chert 74H 149
Charles St. N19 32Nb 70
Charles St. RM17: Grays 51De 121
Charles St. SL4: Wind 3G 102
Charles St. SW13 54Ua 110
Charles St. TW3: Houn 54Ba 107
Charles St. UB10: Hil 42R 84
Charles St. W1 6K 221 (46Kb 90)
Charles Talbot M. SE22 60Xb 113
Charleston Cl. TW13: Felt 62W 128
Charleston St. SE17 6E 230 (49Sb 91)
Charles Townsend Ho. EC1 4B 218 (41Rb 91)
(off Skinner St.)
Charles Uton Ct. E8 35Wb 71
Charles Whincup Rd. E16 46Kc 93
Charlesworth Cl. HP3: Hem H 4M 3
Charlesworth Ho. E14 44Cc 92
(off Dod St.)
Charlesworth Pl. SW13 55Ua 110
Charleville Cir. SE26 64Wb 135
Charleville Ct. W14 50Bb 89
(off Charleville Rd.)
Charleville Mans. W14 50Ab 88
(off Charleville Rd.)
Charleville M. TW7: Isle 56Ka 108
Charleville Rd. W14 50Ab 88
CHARLIE BROWN'S RDBT. 26Lc 53
Charlie Chaplin Adventure Playground 52Rb 113
Charlie Chaplin Wlk. SE1 7J 223 (46Pb 90)
Charlieville Rd. DA8: Erith 52Ed 118
Charlmont Rd. SW17 65Gb 133
Charlock Cl. RM3: Hrld W 25Nd 57
Charlock Way WD18: Wat 16V 26
Charlotte Av. SL2: Slou 5K 81
Charlotte Cl. AL4: St A 2J 7
Charlotte Cl. DA6: Bex 57Ad 117
Charlotte Cl. IG6: Ilf 25Sc 54
Charlotte Cl. KT21: Asht 90Na 173
Charlotte Cl. WD19: Wat 17Y 27
Charlotte Ct. IG2: Ilf 30Qc 54
Charlotte Ct. KT10: Esh 78Ea 152
Charlotte Ct. N8 30Mb 50
Charlotte Ct. RM11: Horn 31Md 77
Charlotte Ct. SE1 5H 231 (49Ub 91)
(off Old Kent Rd.)
Charlotte Ct. W6 49Wa 88
Charlotte Despard Av. SW11 53Jb 112
Charlotte Gdns. RM5: Col R 23Dd 56
Charlotte Ho. E16 46Kc 93
(off Fairfax M.)
Charlotte Ho. W6 50Ya 88
(off Queen Caroline St.)
Charlotte M. KT10: Esh 77Da 151
(off Heather Pl.)
Charlotte M. RM13: Rain 40Md 77
Charlotte M. W1 7C 216 (43Lb 90)
Charlotte M. W10 44Za 88
Charlotte M. W14 49Ab 88
Charlotte Pk. Av. BR1: Broml 69Nc 138
Charlotte Pl. NW9 29Sa 47
Charlotte Pl. RM20: W Thur 51Xd 120
Charlotte Pl. SW1 6B 228 (49Lb 90)
Charlotte Pl. W1 1C 222 (43Lb 90)
Charlotte Rd. EC2 4H 219 (41Ub 91)
Charlotte Rd. RM10: Dag 37Dd 76
Charlotte Rd. SM6: W'gton 79Lb 156
Charlotte Rd. SW13 53Va 110
Charlotte Row SW4 55Lb 112
Charlotte Sq. TW10: Rich 58Pa 109
Charlotte St. W1 7C 216 (43Lb 90)
Charlotte Ter. KT10: Esh 79Ea 152
(off Princess Sq.)
Charlotte Ter. N1 1J 217 (39Pb 70)
Charlow Cl. SW6 54Eb 111
CHARLTON 52Mc 115
Charlton SL4: Wind 4A 102
CHARLTON 69S 128
Charlton Athletic FC 50Lc 93
Charlton Av. KT12: Hers 77X 151
Charlton Chu. La. SE7 50Lc 93
Charlton Cl. SL1: Slou 7F 80
Charlton Cl. UB10: Ick 33R 64
Charlton Ct. E2 1K 219 (39Vb 71)
Charlton Ct. NW5 36Mb 70
Charlton Cres. IG11: Bark 40Vc 75
Charlton Dene SE7 52Lc 115
Charlton Dr. TN16: Big H 89Mc 179
Charlton Gdns. CR5: Coul 90Lb 176
Charlton Ga. Bus. Pk. SE7 49Lc 93
Charlton Ho. TW8: Bford 51Na 109
Charlton Kings KT13: Weyb 76U 150
Charlton King's Rd. NW5 36Mb 70
Charlton La. SE7 49Mc 93
Charlton La. TW17: Shep 69S 128
(not continuous)
Charlton Lido 52Mc 115
Charlton Pk. La. SE7 52Mc 115
Charlton Pk. Rd. SE7 51Mc 115
Charlton Pl. N1 1B 218 (40Rb 71)
Charlton Pl. SL4: Wind 4A 102
(off Charlton)
Charlton Riverside Pl. SE7 49Kc 93
Charlton Rd. HA3: Kenton 28Ma 47
Charlton Rd. HA9: Wemb 32Pa 67
Charlton Rd. N9 18Zb 34
Charlton Rd. NW10 39Ua 68
Charlton Rd. SE3 52Jc 115
Charlton Rd. SE7 51Kc 115
Charlton Rd. TW17: Shep 69S 128
Charlton Row SL4: Wind 4A 102
Charlton Sq. SL4: Wind 4A 102
(off Charlton)
Charlton Station (Rail) 50Lc 93
Charlton St. RM20: Grays 51Zd 121
Charlton Ter. SE11 51Qb 112
Charlton Wlk. SL4: Wind 4A 102
Charlton Way SE3 53Gc 115
Charlton Way SL4: Wind 4A 102
Charlwood CR0: Sels 81Bc 178
Charlwood Cl. HA3: Hrw W 23Ga 46
Charlwood Cl. KT23: Bookh 96Da 191
Charlwood Ct. KT22: Oxs 87Fa 172
Charlwood Ho. SW1 6D 228 (49Mb 90)
(off Vauxhall Bri. Rd.)
Charlwood Ho. TW9: Kew 52Ra 109
Charlwood Ho's. WC1 4G 217 (41Nb 90)
(off Midhope St.)
Charlwood Pl. RH2: Reig 6H 207
Charlwood Pl. SW1 6C 228 (49Lb 90)
Charlwood Rd. SW15 56Za 110
Charlwood St. SW1 7B 228 (50Lb 90)
(not continuous)
Charlwood Ter. SW15 56Za 110
Charman Rd. RH1: Redh 6N 207
Charmans Ho. SW8 52Nb 112
(off Wandsworth Rd.)
Charmeuse Ct. E2 40Xb 71
(off Silk Weaver Way)
Charmian Av. HA7: Stan 27Ma 47
Charmian Ho. N1 2H 219 (40Ub 71)
(off Crondall St.)
Charmille Av. NW9 25Va 48
Charminster Av. SW19 68Cb 133
Charminster Ct. KT6: Surb 73Ma 153
Charminster Rd. KT4: Wor Pk 74Za 154
Charminster Rd. SE9 63Mc 137
Charmouth Ct. TW10: Rich 57Pa 109
Charmouth Ho. SW8 52Pb 112
Charmouth Rd. AL1: St A 1E 6
Charmouth Rd. DA16: Well 53Yc 117
Charmwood La. BR6: Prat B 81Xc 181
Charne, The TN14: Otf 89Jd 182
Charnock BR8: Swan 70Gd 140
Charnock Ho. W12 45Xa 88
(off White City Est.)
Charnock Rd. E5 34Xb 71
Charnwood SL5: S'dale 2D 146
Charnwood Av. SW19 68Cb 133
Charnwood Cl. KT3: N Mald 70Ua 132
Charnwood Dr. E18 27Kc 53
Charnwood Gdns. E14 49Cc 92
Charnwood Pl. N20 20Eb 31
Charnwood Rd. EN1: Enf 8Xb 19
Charnwood Rd. SE25 71Tb 157
Charnwood Rd. UB10: Hil 40Q 64
Charnwood St. E5 33Xb 71
Charrington Cl. WD7: Shenl 6Qa 15
Charrington Ct. RM1: Rush G 30Gd 56
Charrington Pl. AL1: St A 3D 6
Charrington Rd. CR0: C'don 75Sb 157
Charrington St. NW1 1D 216 (40Mb 70)
(not continuous)
Charsley Rd. SE6 61Dc 136
Charta Rd. TW20: Egh 64E 126
Chart Cl. BR2: Broml 67Gc 137
Chart Cl. CR0: C'don 72Yb 158
Chart Cl. CR4: Mitc 70Hb 133
Charter Av. IG2: Ilf 32Tc 74
Charter Bldgs. SE10 53Dc 114
(off Catherine Gro.)
Charter Cl. AL1: St A 2B 6
Charter Cl. SL1: Slou 8K 81
Charter Ct. HP2: Hem H 2M 3
Charter Ct. KT3: N Mald 69Ua 132
Charter Ct. N22 25Mb 50
Charter Ct. N4 32Qb 70
Charter Ct. UB1: S'hall 46Ca 85
Charter Cres. TW4: Houn 56Aa 107
Charter Dr. DA5: Bexl 59Ad 117
Charter Ho. SM2: Sutt 79Db 155
(off Mulgrave Rd.)
Charter Ho. WC2 3G 223 (44Nb 90)
(off Crown Ct.)
Charterhouse 6C 218 (42Rb 91)
Charterhouse Apts. SW18 56Eb 111
Charterhouse Av. HA0: Wemb 35La 66
Charterhouse Bldgs. EC1 6D 218 (42Sb 91)
Charterhouse Dr. TN13: S'oaks 95Jd 202
Charterhouse M. EC1 7C 218 (43Rb 91)
Charterhouse Mus. 7C 218 (43Rb 91)
(within Charterhouse)
Charterhouse Rd. BR6: Chels 76Wc 161
Charterhouse Rd. E8 35Wb 71
Charterhouse Sq. EC1 7C 218 (43Rb 91)
Charterhouse St. EC1 1A 224 (43Qb 90)
Charteris Community Sports Cen. 39Cb 69
Charteris Rd. IG8: Wfd G 24Kc 53
Charteris Rd. N4 32Qb 70
Charteris Rd. NW6 39Bb 69
Charter Pl. TW18: Staines 65J 127
Charter Pl. TW20: Eng G 2P 125
Charter Pl. UB8: Uxb 38M 63
Charter Pl. WD17: Wat 14Y 27
Charter Quay KT1: King T 68Ma 131
(off Wadbrook St.)
Charter Rd. KT1: King T 69Ra 131
Charter Rd. SL1: Slou 5C 80
Charter Rd., The IG8: Wfd G 23Gc 53
Charters Cl. SE19 64Ub 135
Charters Cl. SL5: S'hill 1B 146
Charters Ct. SL5: S'hill 2B 146
Charters Gdn. Ho. SL5: S'hill 2C 146
Charters Health Club 3G 6
Charters La. SL5: S'hill 1B 146
Charters Leisure Cen. 3B 146
Charter Sq. KT1: King T 68Ra 131
Charters Rd. SL5: S'dale 3B 146
Charter St. E14 46Ec 92
Charters Way SL5: S'dale 3D 146
Charter Way N14 16Lb 32
Charter Way N3 28Bb 49
Chartes Ho. SE1 3J 231 (48Ub 91)
(off Stevens St.)
Chartfield Av. SW15 57Xa 110
Chartfield Pl. KT13: Weyb 78R 150
Chartfield Rd. RH2: Reig 7L 207
Chartfield Sq. SW15 57Za 110
Chartham Ct. SW9 55Qb 112
(off Canterbury Cres.)
Chartham Gro. SE27 62Rb 135
Chartham Ho. SE1 3G 231 (48Tb 91)
(off Weston St.)
Chartham Rd. SE25 69Xb 135
Chart Hills Cl. SE28 44Ad 95
Chart Ho. CR4: Mitc 68Hb 133
Chart Ho. E14 50Dc 92
(off Burrells Wharf Sq.)
Chart La. RH2: Reig 6K 207
Chart La. TN16: B Char 100Xc 201
Chart La. TN16: Bras 100Xc 201
Chart La. TN16: Toy H 100Xc 201
Chartley Av. HA7: Stan 23Ha 46
Chartley Av. NW2 34Ua 68
Charton Cl. DA17: Belv 51Bd 117
Chartres Ct. UB6: G'frd 40Fa 66
Chartridge SE17 51Tb 113
(off Westmoreland Rd.)
Chartridge WD19: Wat 19Z 27
Chartridge Cl. EN5: Ark 15Wa 30
Chartridge Cl. WD23: Bush 16Ea 28
Chartridge Way HP2: Hem H 2C 4
Chart St. N1 3G 219 (41Tb 91)
Chart Vw. TN15: Kems'g 89Rd 183
Chartway RH2: Reig 5K 207
Chartway TN13: S'oaks 96Ld 203
Chartwell GU22: Wok 90A 168
(off Mt. Hermon Rd.)
Chartwell Bus. Cen. BR1: Broml 69Mc 137
Chartwell Cl. CR0: C'don 74Tb 157
Chartwell Cl. EN9: Walt A 5Gc 21
Chartwell Cl. SE9 61Tc 138
Chartwell Cl. UB6: G'frd 39Da 65
Chartwell Ct. EN5: Barn 14Ab 30
Chartwell Ct. IG8: Wfd G 24Hc 53
Chartwell Ct. NW2 34Wa 68
Chartwell Ct. UB3: Hayes 45V 84
Chartwell Dr. BR6: Farnb 78Tc 160
Chartwell Gdns. SM3: Cheam 77Ab 154
Chartwell Ho. SW10 53Eb 111
Chartwell Ho. W11 46Bb 89
(off Ladbroke Rd.)
Chartwell La. DA3: Lfield 69Ae 143
Chartwell Lodge BR3: Beck 66Cc 136
Chartwell Pl. HA2: Harr 33Fa 66
Chartwell Pl. KT18: Eps 86Ua 174
Chartwell Pl. SM3: Cheam 77Ab 154
Chartwell Rd. HA6: Nwood 23V 44
Chartwell Way SE20 67Xb 135
Charville Ct. HA1: Harr 30Ha 46
(off Gayton Rd.)
Charville Ct. SE10 51Fc 115
(off Trafalgar Gro.)
Charville La. UB4: Hayes 41S 84
Charville La. W. UB10: Hil 41R 84
(not continuous)
Charwood SW16 63Qb 134
Charwood Cl. WD7: Shenl 5Na 15
Chase, The CM14: W'ley Cromwell Rd. 21Xd 58
Chase, The CM14: W'ley Nelson Cl. 22Zd 59
Chase, The UB10: Ick The Grove 36Q 64
Chase, The BR1: Broml 69Kc 137
Chase, The BR6: Prat B 84Wc 181
Chase, The CM13: Ingve 22Ee 59
Chase, The CM14: B'wood 20Zd 41
Chase, The CR5: Coul 86Lb 176
Chase, The DA3: Lfield 69Be 143
Chase, The DA7: Bex 55Dd 118
Chase, The E12 35Mc 73
Chase, The EN7: G Oak 1Rb 19
Chase, The HA5: Eastc 30Y 45
Chase, The HA5: Pinn 28Ba 45
Chase, The HA7: Stan 23Ja 46
Chase, The HA8: Edg 25Ra 47
Chase, The HP2: Hem H 3N 3
Chase, The IG10: Lough 17Mc 35
Chase, The IG7: Chig 21Sc 54
Chase, The KT20: Kgswd 94Cb 195
Chase, The KT21: Asht 90La 172
Chase, The KT22: Oxs 87Ea 172
Chase, The KT24: E Hor 98V 190
Chase, The RH2: Reig 7M 207
Chase, The RM1: Rom 27Gd 56
Chase, The RM13: Rain 39Kd 77
Chase, The RM14: Upm 34Ud 78
Chase, The RM20: Grays 51Zd 121
Chase, The RM6: Chad H 30Ad 55
Chase, The RM7: Rush G 34Fd 76
Chase, The SM6: W'gton 78Nb 156
Chase, The SW16 66Pb 134
Chase, The SW20 67Ab 132
Chase, The SW4 55Kb 112
Chase, The TN15: Kems'g 88Nd 183
Chase, The TW16: Sun 67X 129
Chase, The WD18: Wat 14U 26
Chase, The WD7: R'lett 7Ha 14
Chase Bank Ct. N14 16Lb 32
(off Avenue Rd.)
Chase Cen., The NW10 41Ta 87
Chase Ct. SW20 68Ab 132
Chase Ct. SW3 4F 227 (48Hb 89)
(off Beaufort Gdns.)
Chase Ct. TW7: Isle 54Ja 108
Chase Ct. Gdns. EN2: Enf 13Sb 33
CHASE CROSS 23Gd 56
Chase Cross Rd. RM5: Col R 24Ed 56
Chase End KT19: Eps 84Ta 173
CHASE FARM HOSPITAL 10Qb 18
Chasefield Rd. SW17 63Hb 133
Chase Gdns. E4 21Cc 52
Chase Gdns. TW2: Whitt 59Fa 108
Chase Grn. EN2: Enf 13Sb 33
Chase Grn. EN6: Cuff 1Nb 18
Chase Grn. Av. EN2: Enf 12Rb 33
Chase Hill EN2: Enf 13Sb 33
Chase Ho. NW6 41Cb 89
(off Hansel Rd.)
Chase Ho. Gdns. RM11: Horn 29Pd 57
Chase La. IG2: Ilf 29Tc 54
Chase La. IG6: Ilf 29Tc 54
Chase La. IG7: Chig 20Vc 37
Chaseley Ct. KT13: Weyb 74U 150
Chaseley Dr. CR2: Sande 82Tb 177
Chaseley Dr. W4 50Ra 87
Chaseley St. E14 44Ac 92
Chasemore Cl. CR4: Mitc 73Hb 155
Chasemore Gdns. CR0: Wadd 78Qb 156
Chasemore Ho. SW6 52Ab 110
(off Williams Cl.)
Chase Nature Reserve, The 34Hd 76
Chase Ridings EN2: Enf 12Qb 32
Chase Rd. CM14: B'wood 20Yd 40
Chase Rd. KT19: Eps 84Ta 173
Chase Rd. N14 15Lb 32
Chase Rd. NW10 42Ta 87
Chase Rd. Trad. Est. NW10 42Ta 87
Chase Side EN2: Enf 13Sb 33
CHASE SIDE 11Tb 33
Chase Side N14 16Jb 32
Chase Side Av. EN2: Enf 12Sb 33
Chase Side Av. SW20 67Ab 132
Chaseside Cl. RM1: Rom 23Gd 56
Chase Side Cres. EN2: Enf 11Sb 33
Chaseside Gdns. KT16: Chert 73K 149
Chase Side Pl. EN2: Enf 12Sb 33
Chase Sq. DA11: Grav'nd 8D 122
Chaseville Pde. N21 15Pb 32
Chaseville Pk. Rd. N21 15Nb 32
Chase Way N14 19Kb 32
Chase Way RM20: Grays 51Zd 121
Chaseway Lodge E16 44Jc 93
(off Butchers Rd.)
Chaseways Vs. RM5: Col R 25Bd 55
Chasewood Av. EN2: Enf 12Rb 33
Chasewood Ct. NW7 22Ta 47
Chasewood Pk. HA1: Harr 34Ha 66
Chastilian Rd. DA1: Dart 59Hd 118
Chaston Pl. NW5 36Jb 70
(off Grafton Ter.)
Chater Ho. E2 41Zb 92
(off Roman Rd.)
Chatfield SL2: Slou 3E 80
Chatfield Ct. CR3: Cat'm 94Tb 197
Chatfield Rd. CR0: C'don 74Rb 157
Chatfield Rd. SW11 55Eb 111
Chatham Av. BR2: Hayes 73Hc 159
Chatham Cl. NW11 29Cb 49
Chatham Cl. SE18 48Rc 94
Chatham Cl. SM3: Sutt 73Bb 155
Chatham Ct. SL1: Slou 8L 81
(off Grove Cl.)
Chatham Hill Rd. TN14: S'oaks 93Ld 203
Chatham Ho. SM6: W'gton 78Kb 156
(off Melbourne Rd.)
Chatham Pl. E9 37Yb 72
Chatham Rd. E17 27Ac 52
Chatham Rd. E18 26Hc 53
Chatham Rd. KT1: King T 68Qa 131
Chatham Rd. SW11 58Hb 111
Chatham St. SE17 5F 231 (49Tb 91)
Chatham Way CM14: B'wood 19Yd 40
CHATHILL 9F 210
Chatley Heath Semaphore Tower 90T 170
Chatsfield KT17: Ewe 82Wa 174
Chatsfield Pl. W5 44Na 87
Chats Palace Arts Cen. 36Zb 72
Chatswood M. DA14: Sidc 63Vc 139
Chatsworth Av. BR1: Broml 63Kc 137
Chatsworth Av. DA15: Sidc 60Wc 117
Chatsworth Av. HA9: Wemb 36Pa 67
Chatsworth Av. NW4 26Ya 48
Chatsworth Av. SW20 67Ab 132
Chatsworth Cl. BR4: W W'ck 75Hc 159
Chatsworth Cl. NW4 26Ya 48
Chatsworth Cl. W4 51Sa 109
Chatsworth Cl. WD6: Bore 13Qa 29
Chatsworth Ct. AL1: St A 2D 6
(off Stanhope Rd.)
Chatsworth Ct. HA7: Stan 22La 46
Chatsworth Ct. SW16 69Pb 134
Chatsworth Ct. W8 49Cb 89
(off Pembroke Rd.)
Chatsworth Cres. TW3: Houn 56Fa 108
Chatsworth Dr. EN1: Enf 17Wb 33
Chatsworth Est. E5 35Zb 72
Chatsworth Gdns. HA2: Harr 32Da 65
Chatsworth Gdns. KT3: N Mald 71Va 154
Chatsworth Gdns. W3 46Ra 87
Chatsworth Ho. BR2: Broml 70Jc 137
(off Westmoreland Rd.)
Chatsworth Ho. E16 46Kc 93
(off Wesley Av.)
Chatsworth Ho. SE1 7K 225 (46Vb 91)
(off Duchess Wlk.)

Chatsworth Lodge W4 .....50Ta **87**
(off Bourne Pl.)
Chatsworth M. WD24: Wat 10W **12**
Chatsworth Pde. BR5: Pet W 71Sc **160**
Chatsworth Pk. SM7: Bans 89Db **175**
Chatsworth Pl. CR4: Mitc 69Hb **133**
Chatsworth Pl. KT22: Oxs 85Fa **172**
Chatsworth Pl. TW11: Tedd 63Ja **130**
Chatsworth Ri. W5 .....42Pa **87**
Chatsworth Rd. CR0: C'don 77Tb **157**
Chatsworth Rd. DA1: Dart 57Ld **119**
Chatsworth Rd. E15 .....36Hc **73**
Chatsworth Rd. E5 .....34Yb **72**
Chatsworth Rd. NW2 .....37Ya **68**
Chatsworth Rd. SM3: Cheam 78Za **154**
Chatsworth Rd. UB4: Yead 42X **85**
Chatsworth Rd. W4 .....51Sa **109**
Chatsworth Rd. W5 .....42Pa **87**
Chatsworth Way SE27 .....62Rb **135**
Chatteris Av. RM3: Rom 23Ld **57**
Chattern Hill TW15: Ashf 63R **128**
CHATTERN HILL 63R **128**
Chattern Rd. TW15: Ashf 63S **128**
Chatterton Ct. TW9: Kew 54Pa **109**
Chatterton M. N4 .....34Rb **71**
(off Chatterton Rd.)
Chatterton Rd. BR2: Broml 70Mc **137**
Chatterton Rd. N4 .....34Rb **71**
Chatton Row GU24: Bisl 9E **166**
Chatto Rd. SW11 .....57Hb **111**
Chaucer Av. KT13: Weyb 80Q **150**
Chaucer Av. TW4: Cran 54X **107**
Chaucer Av. TW9: Rich 55Qa **109**
Chaucer Av. UB4: Hayes 43W **84**
Chaucer Bus. Pk. TN15: Kems'g 90Td **184**
Chaucer Cl. N11 .....22Lb **50**
Chaucer Cl. RM18: Tilb 4E **122**
Chaucer Cl. SL4: Wind 5H **103**
Chaucer Cl. SM7: Bans 86Ab **174**
Chaucer Ct. BR2: Broml 71Lc **159**
Chaucer Ct. EN5: New Bar 15Db **31**
Chaucer Ct. N16 .....35Ub **71**
Chaucer Ct. RH1: Redh 3A **208**
Chaucer Ct. SW17 .....62Fb **133**
(off Lanesborough Way)
Chaucer Dr. SE1 .....6K **231** (49Vb **91**)
Chaucer Gdns. CR5: Coul 89Lb **176**
Chaucer Gdns. E1 .....44Wb **91**
(off Piazza Wlk.)
Chaucer Gdns. SM1: Sutt 76Cb **155**
Chaucer Grn. CR0: C'don 73Xb **157**
Chaucer Gro. WD6: Bore 14Qa **29**
Chaucer Ho. EN5: Barn 14Za **30**
Chaucer Ho. SM1: Sutt 76Cb **155**
(off Chaucer Gdns.)
Chaucer Ho. SW1 .....50Lb **90**
(off Churchill Gdns.)
Chaucer Mans. W14 .....51Ab **110**
(off Queen's Club Gdns.)
Chaucer Pk. DA1: Dart 59Pd **119**
Chaucer Rd. DA11: Nflt 62Fe **143**
Chaucer Rd. DA15: Sidc 60Yc **117**
Chaucer Rd. DA16: Well 53Uc **116**
Chaucer Rd. E11 .....30Jc **53**
Chaucer Rd. E17 .....26Ec **52**
Chaucer Rd. E7 .....37Jc **73**
Chaucer Rd. RM3: Rom 24Kd **57**
Chaucer Rd. SE24 .....57Qb **112**
Chaucer Rd. SM1: Sutt 77Cb **155**
Chaucer Rd. TW15: Ashf 63N **127**
Chaucer Rd. W3 .....46Sa **87**
Chaucer Way DA1: Dart 56Qd **119**
(not continuous)
Chaucer Way KT15: Add 79J **149**
Chaucer Way SL1: Slou 6K **81**
Chaucer Way SW19 .....65Fb **133**
CHAULDEN 3H **3**
Chaulden Ho. EC1 .....4G **219** (41Tb **91**)
(off Cranwood St.)
Chaulden Ho. Gdns. HP1: Hem H 4H **3**
Chaulden La. HP1: Hem H 4F **2**
Chaulden Ter. HP1: Hem H 3H **3**
Chauncey Cl. N9 .....20Wb **33**
Chauncey Ho. WD18: Wat 16U **26**
Chauncy Av. EN6: Pot B 5Eb **17**
Chaundrye Cl. SE9 .....58Pc **116**
Chauntler Cl. E16 .....44Kc **93**
Chave Cft. KT18: Tatt C 91Ya **194**
Chavecroft Ter. KT18: Tatt C 91Ya **194**
Chave Rd. DA2: Wilm 62Nd **141**
Chaville Ct. N11 .....21Jb **50**
Chaville Way N3 .....25Cb **49**
Chaworth Cl. KT16: Ott 79E **148**
Chaworth Rd. KT16: Ott 79E **148**
Cheadle Ct. NW8 .....5C **214** (42Fb **89**)
(off Henderson Dr.)
Cheadle Ho. E14 .....44Bc **92**
(off Copenhagen Pl.)
CHEAM 79Ab **154**
Cheam Cl. KT20: Tad 93Xa **194**
Cheam Comn. Rd. KT4: Wor Pk 75Xa **154**
Cheam Leisure Cen. 77Za **154**
Cheam Mans. SM3: Cheam 80Ab **154**
Cheam Pk. Way SM3: Cheam 79Ab **154**
Cheam Rd. KT17: Ewe 82Wa **174**
Cheam Rd. SM1: Sutt 79Bb **155**
Cheam Rd. SM2: Cheam 82Ya **174**
Cheam Station (Rail) 80Ab **154**
Cheam St. SE15 .....55Yb **114**
CHEAM VILLAGE 79Ab **154**
Cheapside EC2 .....2D **224** (44Sb **91**)
Cheapside GU21: Wok 6P **167**
Cheapside N13 .....21Tb **51**
Cheapside N22 .....27Qb **50**
CHEAPSIDE 7C **124**
Cheapside La. UB9: Den 33H **63**
Cheapside Pas. EC2 .....3D **224** (44Sb **91**)
(off One New Change)
Cheapside Rd. SL5: Asc 9A **124**
Cheddar Cl. N11 .....23Hb **49**
Cheddar Waye UB4: Yead 44X **85**
Cheddington Ho. E2 .....39Wb **71**
(off Whiston Rd.)
Cheddington Rd. N18 .....20Ub **33**
Chedworth Ho. N15 .....28Tb **51**
(off West Grn. Rd.)
Cheeky Monkeys Jungle Adventure 1Wb **19**
Cheelson Rd. RM15: S Ock 40Yd **78**
Cheena Ho. SL9: Chal P 24B **42**
Cheering La. E20 .....36Ec **72**
Cheeseman Cl. TW12: Hamp 65Aa **129**
Cheesemans Ter. W14 .....50Bb **89**
(not continuous)
Cheffery Ct. TW15: Ashf 65R **128**
Cheldon Av. NW7 .....24Za **48**
Chelford Rd. BR1: Broml 64Fc **137**
Chelmer Cres. IG11: Bark 40Xc **75**
Chelmer Dr. CM13: Hut 16Fe **41**
Chelmer Dr. RM15: S Ock 45Yd **98**
Chelmer Ho. RM16: Grays 10C **100**
(off River Vw.)
Chelmer Rd. E9 .....36Zb **72**
Chelmer Rd. RM14: Upm 30Td **58**
Chelmer Rd. RM16: Grays 10C **100**
Chelmsford Av. RM5: Col R 24Fd **56**
Chelmsford Cl. E6 .....44Pc **94**
Chelmsford Cl. SM2: Sutt 81Cb **175**
Chelmsford Cl. W6 .....51Za **110**
Chelmsford Ct. N14 .....17Mb **32**
(off Chelmsford Rd.)
Chelmsford Dr. RM14: Upm 34Pd **77**
Chelmsford Gdns. IG1: Ilf 31Nc **74**
Chelmsford Ho. N7 .....35Pb **70**
(off Holloway Rd.)
Chelmsford Rd. CM15: Shenf 16Be **41**
Chelmsford Rd. E11 .....32Fc **73**
Chelmsford Rd. E17 .....30Cc **52**
Chelmsford Rd. E18 .....25Hc **53**
Chelmsford Rd. N14 .....17Lb **32**
Chelmsford Sq. NW10 .....39Ya **68**
Chelmsine Ct. HA4: Ruis 29S **44**
CHELSEA 7D **226** (50Gb **89**)
CHELSEA & WESTMINSTER HOSPITAL 51Eb **111**
Chelsea Bri. 51Kb **112**
Chelsea Bri. Rd. SW1 .....7H **227** (50Jb **90**)
Chelsea Bri. Wharf SW11 .....51Kb **112**
Chelsea Cloisters SW3 .....6E **226** (49Gb **89**)
Chelsea Cl. HA8: Edg 26Qa **47**
Chelsea Cl. KT4: Wor Pk 73Wa **154**
Chelsea Cl. NW10 .....39Ta **67**
Chelsea Cl. TW12: Hamp H 64Ea **130**
Chelsea Ct. BR1: Broml 69Nc **138**
(off Holmdene Ct.)
Chelsea Ct. KT18: Eps 85Ta **173**
(off Ashley Rd.)
Chelsea Ct. SW3 .....51Hb **111**
(off Embankment Gdns.)
Chelsea Cres. NW2 .....37Bb **69**
Chelsea Cres. SW10 .....53Eb **111**
Chelsea Emb. SW3 .....51Gb **111**
Chelsea Farm Ho. Studios SW10 .....51Fb **111**
(off Cremorne Est.)
Chelsea FC 52Db **111**
Chelsea Flds. SW19 .....67Fb **133**
Chelsea Football Academy 89Aa **171**
Chelsea Gdns. SM3: Cheam 77Ab **154**
Chelsea Gdns. SW1 .....50Jb **90**
Chelsea Gdns. W13 .....43Ha **86**
Chelsea Ga. SW1 .....7J **227** (50Jb **90**)
(off Ebury Bri. Rd.)
Chelsea Harbour SW10 .....53Eb **111**
Chelsea Harbour Design Cen. SW10 .....53Eb **111**
(off Chelsea Harbour Dr.)
Chelsea Harbour Dr. SW10 .....53Eb **111**
Chelsea Lodge SW3 .....51Hb **111**
(off Tite St.)
Chelsea Mnr. Ct. SW3 .....51Gb **111**
Chelsea Mnr. Gdns. SW3 .....50Gb **89**
Chelsea Mnr. St. SW3 .....7D **226** (50Gb **89**)
Chelsea Mnr. Studios SW3 .....7E **226** (50Gb **89**)
(off Flood St.)
Chelsea M. RM11: Horn 32Kd **77**
Chelsea Pk. Gdns. SW3 .....51Fb **111**
Chelsea Physic Gdn. 51Hb **111**
Chelsea Reach Twr. SW10 .....52Fb **111**
(off Worlds End Est.)
Chelsea Sports Cen. 50Gb **89**
Chelsea Sq. SW3 .....7C **226** (50Fb **89**)
Chelsea Studios SW6 .....52Db **111**
(off Fulham Rd.)
Chelsea Theatre, The 52Eb **111**
Chelsea Towers SW3 .....51Gb **111**
(off Chelsea Mnr. Gdns.)
Chelsea Village SW6 .....52Db **111**
(off Fulham Rd.)
Chelsea Vista SW6 .....53Eb **111**
Chelsea Way CM14: B'wood 18Yd **40**
Chelsea Wharf SW10 .....52Fb **111**
(off Lots Rd.)
CHELSFIELD 78Xc **161**
Chelsfield Av. N9 .....17Zb **34**
Chelsfield Gdns. SE26 .....62Yb **136**
Chelsfield Grn. N9 .....17Zb **34**
(not continuous)
Chelsfield Hill BR6: Chels 81Yc **181**
Chelsfield Ho. SE17 .....6H **231** (49Ub **91**)
(off Massinger St.)
Chelsfield Lakes Golf Course 80Ad **161**
Chelsfield La. BR5: Orp 73Zc **161**
Chelsfield La. BR6: Chels 75Zc **161**
Chelsfield La. BR6: Orp 75Zc **161**
Chelsfield La. BR6: Well H 80Cd **162**
Chelsfield La. TN14: S'ham 81Ed **182**
Chelsfield Point E9 .....38Zb **72**
(off Penshurst Rd.)
Chelsfield Rd. BR5: St M Cry 72Yc **161**
Chelsfield Station (Rail) 78Xc **161**
CHELSFIELD VILLAGE 78Ad **161**
CHELSHAM 89Cc **178**
Chelsham Cl. CR6: W'ham 90Ac **178**
CHELSHAM COMMON 88Cc **178**
Chelsham Comn. Rd. CR6: W'ham 89Cc **178**
Chelsham Ct. Rd. CR6: W'ham 90Fc **179**
Chelsham Rd. CR2: S Croy 80Tb **157**
Chelsham Rd. CR6: W'ham 90Bc **178**
Chelsham Rd. SW4 .....55Mb **112**
Chelsham Ter. CR6: W'ham 90Bc **178**
Chelsing Ri. HP2: Hem H 3C **4**
Chelsiter Ct. DA14: Sidc 63Vc **139**
Chelston App. HA4: Ruis 33W **64**
Chelston Ct. E11 .....29Kc **53**
Chelston Rd. HA4: Ruis 32W **64**
Chelsworth Cl. RM3: Hrld W 24Pd **57**
Chelsworth Dr. RM3: Hrld W 25Nd **57**
Chelsworth Dr. SE18 .....51Tc **116**
Cheltenham Av. TW1: Twick 59Ja **108**
Cheltenham Cl. DA12: Grav'nd 4E **144**
Cheltenham Cl. KT3: N Mald 69Sa **131**
Cheltenham Cl. UB5: N'olt 37Da **65**
Cheltenham Ct. AL1: St A 3E **6**
(off Dexter Cl.)
Cheltenham Ct. HA7: Stan 22La **46**
(off Marsh La.)
Cheltenham Gdns. E6 .....40Nc **74**
Cheltenham Gdns. IG10: Lough 16Nc **36**
Cheltenham Ho. IG8: Wfd G 23Qc **54**
Cheltenham Ho. UB3: Harl 48S **84**
(off Skipton Dr.)
Cheltenham Ho. WD24: Wat 12Y **27**
(off Exeter Cl.)
Cheltenham Pl. HA3: Kenton 28Na **47**
Cheltenham Pl. W3 .....46Ra **87**
Cheltenham Rd. BR6: Chels 76Wc **161**
Cheltenham Rd. E10 .....30Ec **52**
Cheltenham Rd. SE15 .....56Yb **114**
Cheltenham Ter. SW3 .....7G **227** (50Hb **89**)
Cheltenham Vs. TW19: Stanw M 58H **105**
Chelverton Rd. SW15 .....56Za **110**
Chelwood N20 .....19Fb **31**
Chelwood Cl. CR5: Coul 91Lb **196**
Chelwood Cl. E4 .....16Dc **34**
Chelwood Cl. HA6: Nwood 24S **44**
Chelwood Cl. KT17: Eps 84Va **174**
Chelwood Ct. CR2: S Croy 78Sb **157**
Chelwood Ct. SW11 .....53Fb **111**
(off Westbridge Rd.)
Chelwood Gdns. TW9: Kew 54Qa **109**
Chelwood Gdns. Pas. TW9: Kew 54Qa **109**
Chelwood Ho. W2 .....3C **220** (44Fb **89**)
(off Gloucester Sq.)
Chelwood Wlk. SE4 .....56Ac **114**
Chenappa Cl. E13 .....41Jc **93**
Chenduit Way HA7: Stan 22Ha **46**
Chene Dr. AL3: St A 1B **6**
Chene M. AL3: St A 1B **6**
Cheney Ct. SE23 .....60Zb **114**
Cheney Row E17 .....25Bc **52**
Cheneys Rd. E11 .....34Gc **73**
Cheney St. HA5: Eastc 28Y **45**
Chenies HA6: Nwood 22V **44**
CHENIES 10D **10**
Chenies, The BR6: Pet W 72Uc **160**
Chenies, The DA2: Wilm 63Gd **140**
Chenies, The NW1 .....1E **216** (40Mb **70**)
(off Pancras Rd.)
CHENIES BOTTOM 9C **10**
Chenies Hill HP3: Flau 7D **10**
Chenies Hill HP5: Flau 8D **10**
Chenies Hill WD3: Chen 9D **10**
Chenies Ho. W2 .....45Db **89**
(off Moscow Rd.)
Chenies Ho. W4 .....52Va **110**
(off Corney Reach Way)
Chenies Manor House & Gdns. 10C **10**
Chenies M. WC1 .....6D **216** (42Mb **90**)
Chenies Pl. EN5: Ark 15Wa **30**
Chenies Pl. NW1 .....1E **216** (40Mb **70**)
Chenies Rd. WD3: Chen 12F **24**
Chenies Rd. WD3: Chor 12F **24**
Chenies St. WC1 .....7D **216** (43Mb **90**)
Chenies Village WD3: Chen 10D **10**
Chenies Way WD18: Wat 17U **26**
Cheniston Cl. KT14: W Byf 85J **169**
Cheniston Ct. SL5: S'dale 3E **146**
Cheniston Gdns. W8 .....48Db **89**
Chenla SE13 .....54Dc **114**
Chennells AL10: Hat 1B **8**
Chenston Cl. TW16: Sun 67W **128**
Chepping Ho. W10 .....45Za **88**
(off Shalfleet Dr.)
Chepstow Av. RM12: Horn 34Nd **77**
Chepstow Cl. SW15 .....57Ab **110**
Chepstow Cnr. W2 .....44Cb **89**
(off Chepstow Pl.)
Chepstow Ct. SL2: Slou 2C **80**
Chepstow Ct. W11 .....45Cb **89**
(off Chepstow Vs.)
Chepstow Cres. IG3: Ilf 30Uc **54**
Chepstow Cres. W11 .....45Cb **89**
Chepstow Gdns. UB1: S'hall 44Ba **85**
Chepstow Ho. RM3: Rom 22Qd **57**
(off Leamington Rd.)
Chepstow Pl. W2 .....44Cb **89**
Chepstow Ri. CR0: C'don 76Ub **157**
Chepstow Rd. CR0: C'don 76Ub **157**
Chepstow Rd. W2 .....44Cb **89**
Chepstow Rd. W7 .....48Ja **86**
Chepstow Vs. W11 .....45Bb **89**
Chequers IG9: Buck H 18Kc **35**
Chequers, The HA5: Pinn 27Z **45**
Chequers Cl. BR5: St P 70Vc **139**
Chequers Cl. DA13: Ist R 8A **144**
Chequers Cl. KT20: Walt H 97Wa **194**
Chequers Cl. NW9 .....27Ua **48**
Chequers Ct. DA17: Belv 49Cd **96**
Chequers Ct. EC1 .....6F **219** (42Tb **91**)
(off Chequer St.)
Chequers Ho. NW8 .....5D **214** (42Gb **89**)
(off Jerome Cres.)
Chequers La. KT20: Walt H 97Wa **194**
Chequers La. RM9: Dag 42Bd **95**
(not continuous)
Chequers La. WD25: Wat 2X **13**
Chequers Orchard SL0: Iver 44H **83**
Chequers Pde. N13 .....22Sb **51**
Chequers Pde. RM9: Dag 39Bd **75**
Chequers Pde. SE9 .....58Pc **116**
(off Eltham High St.)
Chequers Rd. CM14: N'side 19Nd **39**
Chequers Rd. CM14: Rom 19Nd **39**
Chequers Rd. IG10: Lough 15Qc **36**
Chequers Rd. RM3: Rom 19Nd **39**
Chequers Sq. UB8: Uxb 38L **63**
Chequer St. AL1: St A 2B **6**
Chequer St. EC1 .....6E **218** (42Sb **91**)
(not continuous)
Chequers Wlk. EN9: Walt A 5Hc **21**
Chequers Way N13 .....22Rb **51**
Chequer Tree Cl. GU21: Knap 8J **167**
Cherbury Cl. SE28 .....44Zc **95**
Cherbury Ct. N1 .....2G **219** (40Tb **71**)
(off St John's Est.)
Cherbury St. N1 .....2G **219** (40Tb **71**)
Cherchefelle M. HA7: Stan 22Ka **46**
Cherimoya Gdns. KT8: W Mole 69Da **129**
Cherington Rd. W7 .....46Ga **86**
Cheriton Av. BR2: Broml 71Hc **159**
Cheriton Av. IG5: Ilf 26Pc **54**
Cheriton Cl. EN4: Cockf 13Hb **31**
Cheriton Cl. W5 .....43La **86**
Cheriton Ct. KT12: Walt T 74Y **151**
Cheriton Ct. SE12 .....59Jc **115**
Cheriton Ct. SE25 .....71Ub **157**
Cheriton Dr. SE18 .....52Tc **116**
Cheriton Lodge HA4: Ruis 32V **64**
Cheriton Sq. SW17 .....61Jb **134**
Cherkley Hill KT22: Lea 98La **192**
Cherries, The SL2: Slou 4M **81**
Cherry Acre SL9: Chal P 21A **42**
Cherry Av. BR8: Swan 70Fd **140**
Cherry Av. CM13: B'wood 20Be **41**
Cherry Av. SL3: L'ly 7P **81**
Cherry Av. UB1: S'hall 46Z **85**
Cherry Bank HP2: Hem H 1M **3**
(off Chapel St.)
Cherry Blossom Cl. N13 .....22Rb **51**
Cherry Bounce HP1: Hem H 1M **3**
Cherry Cl. E17 .....29Dc **52**
Cherry Cl. HA0: Wemb 34Ma **67**
Cherry Cl. HA4: Ruis 34V **64**
Cherry Cl. NW9 .....26Ua **48**
Cherry Cl. SL0: Iver 45H **83**
Cherry Cl. SM4: Mord 70Ab **132**
Cherry Cl. SM5: Cars 75Hb **155**
Cherry Cl. SM7: Bans 86Za **174**
Cherry Cl. SW2 .....59Qb **112**
Cherry Cl. W5 .....48Ma **87**
Cherrycot Hill BR6: Farnb 77Sc **160**
Cherrycot Ri. BR6: Farnb 77Sc **160**
Cherry Cotts. KT20: Walt H 96Xa **194**
Cherry Ct. HA5: Pinn 26Z **45**
Cherry Ct. IG6: Ilf 27Rc **54**
Cherry Ct. W3 .....46Ua **88**
Cherry Cres. TW8: Bford 52Ka **108**
Cherry Cft. WD3: Crox G 16R **26**
Cherrycroft Gdns. HA5: Hat E 24Ba **45**
Cherrydale WD18: Wat 14V **26**
Cherrydown RM16: Grays 46Fe **99**
Cherrydown Av. E4 .....20Bc **34**
Cherrydown Cl. E4 .....20Cc **34**
Cherrydown Rd. DA14: Sidc 61Zc **139**
Cherrydown Wlk. RM7: Mawney 26Dd **56**
Cherry Gdn. Ho. SE16 .....47Xb **91**
(off Cherry Gdn. St.)
Cherry Gdns. RM9: Dag 36Bd **75**
Cherry Gdns. UB5: N'olt 38Da **65**
Cherry Gdn. St. SE16 .....47Xb **91**
Cherry Gth. TW8: Bford 50Ma **87**
Cherry Grn. Cl. RH1: Redh 8B **208**
Cherry Gro. UB3: Hayes 46X **85**
Cherry Gro. UB8: Hil 43R **84**
Cherry Hill AL2: Chis G 7N **5**
Cherry Hill EN5: New Bar 16Db **31**
Cherry Hill HA3: Hrw W 23Ha **46**
Cherry Hill WD3: Loud 13K **25**
(not continuous)
Cherry Hill Gdns. CR0: Wadd 77Pb **156**
Cherry Hills WD19: Wat 22Aa **45**
Cherry Hollow WD5: Ab L 3V **12**
Cherrylands Cl. NW9 .....33Sa **67**
Cherry La. UB7: W Dray 49P **83**
Cherry Laurel Wlk. SW2 .....58Pb **112**
Cherry Lodge Golf Course 89Qc **180**
Cherry Orchard DA10: Swans 60Ae **121**
Cherry Orchard KT21: Asht 90Ra **173**
Cherry Orchard SE7 .....51Lc **115**
Cherry Orchard SL2: Stoke P 8M **61**
Cherry Orchard TW18: Staines 64J **127**
Cherry Orchard UB7: W Dray 47N **83**
Cherry Orchard Cl. BR5: St M Cry 71Yc **161**
Cherry Orchard Gdns. CR0: C'don 74Tb **157**
Cherry Orchard Gdns. KT8: W Mole 69Ba **129**
Cherry Orchard Rd. BR2: Broml 75Nc **160**
Cherry Orchard Rd. CR0: C'don 75Tb **157**
Cherry Orchard Rd. KT8: W Mole 69Ca **129**
Cherry Pk. La. E20 .....38Ec **72**
Cherry Red Records Stadium 69Ra **131**
Cherry Rd. EN3: Enf W 10Yb **20**
Cherry St. GU21: Wok 90A **168**
Cherry St. RM7: Rom 29Fd **56**
Cherry Tree Av. AL2: Lon C 8H **7**
Cherry Tree Av. TW18: Staines 65K **127**
Cherry Tree Av. UB7: Yiew 44P **83**
Cherry Tree Cl. E9 .....39Yb **72**
Cherry Tree Cl. HA0: Wemb 35Ja **66**
Cherry Tree Cl. RM13: Rain 40Jd **76**
Cherry Tree Cl. RM17: Grays 51Fe **121**
Cherry Tree Ct. CR5: Coul 90Pb **176**
Cherry Tree Ct. KT22: Lea 92Ha **192**
(off Park Vw. Rd.)
Cherry Tree Ct. NW1 .....38Lb **70**
(off Camden Rd.)
Cherry Tree Ct. NW9 .....28Sa **47**
Cherry Tree Ct. SE7 .....51Lc **115**
Cherry Tree Dr. RM15: S Ock 42Zd **99**
Cherry Tree Dr. SW16 .....62Nb **134**
Cherry Tree Grn. CR2: Sande 86Xb **177**
Cherry Tree Gro. TN15: Knat 82Rd **183**
Cherry Tree Hill N2 .....29Gb **49**
Cherry Tree Ho. N22 .....24Nb **50**
Cherry Tree La. DA2: Wilm 62Hd **140**
Cherry Tree La. EN6: Pot B 6Db **17**
Cherry Tree La. RM13: Rain 41Gd **96**
Cherry Tree La. SL0: Iver H 39J **63**
Cherry Tree La. SL3: Ful 37B **62**
Cherry Tree La. WD3: Herons 18E **24**
Cherry Tree Ri. IG9: Buck H 21Lc **53**
Cherry Tree Rd. E15 .....36Gc **73**
Cherry Tree Rd. N2 .....28Hb **49**
Cherry Tree Rd. SL2: Farn R 8G **60**
Cherry Tree Rd. WD24: Wat 8X **13**
Cherry Trees DA3: Hartl 71Be **165**
Cherrytrees CR5: Coul 93Mb **196**
Cherry Tree Ter. SE1 .....2J **231** (47Ub **91**)
(off Whites Grounds)
Cherry Tree Wlk. BR3: Beck 70Bc **136**
Cherry Tree Wlk. BR4: W W'ck 77Hc **159**
Cherry Tree Wlk. EC1 .....6E **218** (42Sb **91**)
Cherry Tree Way E13 .....42Mc **93**
Cherry Tree Way HA7: Stan 23Ka **46**
Cherry Wlk. BR2: Hayes 74Jc **159**
Cherry Wlk. RM13: Rain 40Jd **76**
Cherry Wlk. RM16: Grays 8C **100**
Cherry Wlk. WD3: Sarr 12L **25**
Cherry Way AL10: Hat 3C **8**
Cherry Way KT19: Ewe 79Ta **153**
Cherry Way SL3: Hort 55E **104**
Cherry Way TW17: Shep 70T **128**
Cherrywood Av. TW20: Eng G 6M **125**
Cherrywood Cl. E3 .....41Ac **92**
Cherrywood Cl. KT2: King T 66Qa **131**
Cherrywood Ct. TW11: Tedd 64Ja **130**
Cherrywood Dr. DA11: Nflt 3A **144**
Cherrywood Dr. SW15 .....57Za **110**
Cherrywood La. SM4: Mord 70Ab **132**
Cherrywood Lodge SE13 .....58Fc **115**
(off Birdwood Av.)
Cherry Wood Way W5 .....43Qa **87**
Cherston Gdns. IG10: Lough 14Qc **36**
Cherston Rd. IG10: Lough 14Qc **36**
CHERTSEY 73J **149**
Chertsey Blvd. KT16: Chert 74H **149**
Chertsey Bri. Rd. KT16: Chert 73M **149**
Chertsey Camping & Caravanning Club Site KT16: Chert 73L **149**
Chertsey Cemetery 74J **149**
Chertsey Cl. CR8: Kenley 87Rb **177**
Chertsey Ct. SW14 .....55Ra **109**
Chertsey Cres. CR0: New Ad 82Ec **178**
Chertsey Dr. SM3: Cheam 75Ab **154**
Chertsey Ho. E2 .....4K **219** (41Vb **91**)
(off Arnold Cir.)
Chertsey Ho. KT16: Chert 74L **149**
Chertsey La. KT19: Eps 84Qa **173**
Chertsey La. TW18: Staines 64G **126**
CHERTSEY LOCK 73L **149**
Chertsey Meads KT16: Chert 74M **149**
CHERTSEY MEADS 74N **149**
Chertsey Meads Local Nature Reserve 74N **149**
Chertsey Mus. 72J **149**
Chertsey Rd. GU24: Chob Chobham Rd. 2K **167**
Chertsey Rd. GU24: Chob Windsor Rd. 7H **147**
Chertsey Rd. E11 .....33Fc **73**
Chertsey Rd. GU20: W'sham 9B **146**
Chertsey Rd. GU21: Wok 89B **168**
Chertsey Rd. IG1: Ilf 35Tc **74**
Chertsey Rd. KT14: Byfl 83M **169**
Chertsey Rd. KT15: Add 75K **149**
Chertsey Rd. TW1: Twick 58Ha **108**
Chertsey Rd. TW13: Felt 62U **128**
Chertsey Rd. TW15: Ashf 66T **128**
Chertsey Rd. TW16: Sun 65U **128**
Chertsey Rd. TW17: Shep 73N **149**
Chertsey Rd. TW2: Twick 61Da **129**
CHERTSEY SOUTH 76G **148**
Chertsey Station (Rail) 74H **149**
Chertsey St. SW17 .....64Jb **134**
Chertsey Wlk. KT16: Chert 73J **149**
Chervil Cl. TW13: Felt 62W **128**
Chervil Ct. DA11: Nflt 9A **122**
(off Watercress Way)
Chervil M. SE28 .....46Xc **95**
Chervil Rd. HP2: Hem H 2D **4**
Cherwell Cl. SL3: L'ly 51D **104**
Cherwell Cl. WD3: Crox G 15Q **26**
Cherwell Ct. KT19: Ewe 77Sa **153**
Cherwell Gro. RM15: S Ock 45Xd **98**
Cherwell Ho. NW8 .....6C **214** (42Fb **89**)
(off Church St. Est.)
Cherwell M. SW11 .....56Gb **111**
Cherwell Way HA4: Ruis 30S **44**
Cheryls Cl. SW6 .....53Db **111**
Cheseman St. SE26 .....62Xb **135**
Chesfield Rd. KT2: King T 66Na **131**
Chesham Apts. E4 .....24Dc **52**
Chesham Av. BR5: Pet W 72Rc **160**
Chesham Cl. NW7 .....21Ua **48**
Chesham Cl. RM7: Rom 28Fd **56**
Chesham Cl. SM2: Cheam 82Ab **174**
Chesham Cl. SW1 .....4H **227** (48Jb **90**)
(off Lyall St.)
Chesham Ct. HA6: Nwood 23V **44**
Chesham Cres. SE20 .....67Yb **136**
Chesham Flats W1 .....4J **221** (45Jb **90**)
(off Brown Hart Gdns.)
Chesham Hgts. KT20: Kgswd 93Bb **195**
Chesham Ho. KT18: Eps 85Ta **173**
(off South St.)
Chesham Ho. RM3: Rom 23Nd **57**
(off Leyburn Cres.)
Chesham Ho. SE8 .....53Cc **114**
(off Brookmill Rd.)
Chesham La. HP8: Chal G 20A **24**
Chesham La. SL9: Chal P 21A **42**
Chesham M. SW1 .....3H **227** (48Jb **90**)
(off Belgrave M. W.)
Chesham Pl. SW1 .....4H **227** (48Jb **90**)
(not continuous)
Chesham Rd. HP3: Bov 10B **2**
Chesham Rd. KT1: King T 68Qa **131**
Chesham Rd. SE20 .....68Yb **136**
Chesham Rd. SW19 .....64Fb **133**
Chesham St. NW10 .....34Ta **67**
Chesham St. SW1 .....4H **227** (48Jb **90**)
Chesham Ter. W13 .....47Ka **86**
Chesham Way WD18: Wat 16U **26**
Cheshire Cl. CR4: Mitc 69Nb **134**
Cheshire Cl. E17 .....25Dc **52**
Cheshire Cl. KT16: Ott 79F **148**
Cheshire Cl. RM11: Horn 29Qd **57**
Cheshire Cl. SE4 .....54Bc **114**
Cheshire Ct. EC4 .....3A **224** (44Qb **90**)
(off Fleet St.)
Cheshire Ct. SL1: Slou 7M **81**
Cheshire Dr. WD25: Wat 6V **12**
Cheshire Gdns. KT9: Chess 79Ma **153**
Cheshire Ho. KT16: Ott 79F **148**
(off Crawshaw Rd.)
Cheshire Ho. N18 .....21Xb **51**
Cheshire Ho. SM4: Mord 73Db **155**
Cheshire Rd. N22 .....24Pb **50**
Cheshire St. E2 .....42Vb **91**
Cheshir Ho. NW4 .....28Ya **48**
Chesholm Rd. N16 .....34Ub **71**
CHESHUNT 1Zb **20**
Cheshunt Cen., The 1Zb **20**
CHESHUNT COMMUNITY HOSPITAL 3Ac **20**
Cheshunt Ho. NW6 .....39Db **69**
(off Mortimer Cres.)
Cheshunt Rd. DA17: Belv 50Cd **96**
Cheshunt Rd. E7 .....37Kc **73**
Cheshunt Station (Rail) 2Bc **20**
Chesil Ct. E2 .....40Yb **72**
Chesil Ct. SW3 .....51Gb **111**
Chesilton Rd. SW6 .....53Bb **111**
Chesil Way UB4: Hayes 41V **84**
Chesley Gdns. E6 .....40Mc **73**
Cheslyn Gdns. 9V **12**
Chesney Ct. W9 .....42Cb **89**
(off Shirland Rd.)
Chesney Cres. CR0: New Ad 80Ec **158**
Chesney Ho. SE13 .....56Fc **115**
(off Mercator Rd.)
Chesney St. SW11 .....53Jb **112**
Chesnut Gro. N17 .....27Vb **51**
Chesnut Rd. N17 .....27Vb **51**
Chesnut Row N3 .....24Cb **49**
(off Nether St.)
Chess Cl. HP5: Lat 8A **10**
Chess Cl. WD3: Loud 14M **25**
Chessell Cl. CR7: Thor H 70Rb **135**
Chessfield Pk. HP6: L Chal 11A **24**
Chess Hill WD3: Loud 14M **25**
Chessholme Ct. TW16: Sun 66U **128**
(off Scotts Av.)
Chessholme Rd. TW15: Ashf 65S **128**
Chessing Ct. N2 .....27Hb **49**
(off Fortis Grn.)
CHESSINGTON 78Na **153**
Chessington Av. DA7: Bex 52Ad **117**
Chessington Av. N3 .....27Ab **48**
Chessington Cl. KT19: Ewe 79Sa **153**
Chessington Ct. HA5: Pinn 28Ba **45**
Chessington Ct. N3 .....27Bb **49**
(off Charter Way)
Chessington Hall Gdns. KT9: Chess 80Ma **153**
Chessington Hill Pk. KT9: Chess 78Qa **153**
Chessington Ho. KT17: Ewe 81Va **174**
(off Spring St.)
Chessington Ho. SW8 .....54Mb **112**
Chessington Lodge N3 .....27Bb **49**
Chessington Mans. E10 .....31Cc **72**
Chessington Mans. E11 .....31Gc **73**
Chessington North Station (Rail) 78Na **153**
Chessington Pde. KT9: Chess 79Ma **153**
Chessington Pk. KT9: Chess 77Qa **153**
Chessington Rd. KT17: Ewe 81Ua **174**
Chessington Rd. KT19: Ewe 79Qa **153**
Chessington South Station (Rail) 80Ma **153**
Chessington Sports Cen. 80Ma **153**
Chessington Trade Pk. KT9: Chess 77Qa **153**
Chessington Way BR4: W W'ck 75Dc **158**
Chessington World of Adventures 82La **172**

Chess La. WD3: Loud ... 14M 25
Chesson Rd. W14 ... 51Bb 111
Chess Va. Ri. WD3: Crox G ... 16P 25
Chess Way WD3: Chor ... 13J 25
Chesswood Ct. WD3: Rick ... 18M 25
Chesswood Way HA5: Pinn ... 26Z 45
Chestbrook Ct. EN1: Enf ... 15Ub 33
... (off Forsyth Pl.)
Chester Av. RM14: Upm ... 33Ud 78
Chester Av. TW10: Rich ... 58Pa 109
Chester Av. TW2: Whitt ... 60Ba 107
Chester Cl. EN6: Pot B ... 1Db 17
Chester Cl. IG10: Lough ... 11Sc 36
Chester Cl. RM16: Chaf H ... 47Ae 99
Chester Cl. SM1: Sutt ... 75Cb 155
Chester Cl. SW1 ... 2K 227 (47Kb 90)
Chester Cl. SW13 ... 55Xa 110
Chester Cl. TW10: Rich ... 58Pa 109
Chester Cl. TW15: Ashf ... 64T 128
Chester Cl. UB8: Hil ... 44R 84
Chester Cl. Nth. NW1 ... 3K 215 (41Kb 90)
Chester Cl. Sth. NW1 ... 4A 216 (41Kb 90)
Chester Cotts. SW1 ... 6H 227 (49Jb 90)
... (off Bourne St.)
Chester Ct. BR2: Broml ... 70Hc 137
... (off Durham Rd.)
Chester Ct. NW1 ... 3A 216 (41Kb 90)
... (not continuous)
Chester Ct. SE5 ... 52Tb 113
... (off Lomond Gro.)
Chester Ct. SE8 ... 50Zb 92
Chester Ct. W6 ... 49Za 88
... (off Wolverton Gdns.)
Chester Cres. E8 ... 36Vb 71
Chester Dr. HA2: Harr ... 30Ba 45
Chesterfield Cl. BR5: St M Cry ... 70Ad 139
Chesterfield Cl. SE13 ... 54Fc 115
Chesterfield Ct. KT5: Surb ... 71Na 153
... (off Cranes Pk.)
Chesterfield Dr. DA1: Dart ... 57Kd 119
Chesterfield Dr. KT10: Hin W ... 75Ja 152
Chesterfield Dr. TN13: Riv ... 93Fd 202
Chesterfield Flats EN5: Barn ... 15Za 30
... (off Bells Hill)
Chesterfield Gdns. N4 ... 29Rb 51
Chesterfield Gdns. SE10 ... 52Fc 115
Chesterfield Gdns. W1 ... 6K 221 (46Kb 90)
Chesterfield Gro. SE22 ... 57Vb 113
Chesterfield Hill W1 ... 6K 221 (46Kb 90)
Chesterfield Ho. W1 ... 6J 221 (46Jb 90)
... (off Chesterfield Gdns.)
Chesterfield Lodge N21 ... 17Pb 32
... (off Church Hill)
Chesterfield M. N4 ... 29Rb 51
Chesterfield M. TW15: Ashf ... 63N 127
Chesterfield Rd. E10 ... 30Ec 52
Chesterfield Rd. EN3: Enf W ... 9Ac 20
Chesterfield Rd. EN5: Barn ... 15Za 30
Chesterfield Rd. KT19: Ewe ... 80Ta 153
Chesterfield Rd. N3 ... 23Cb 49
Chesterfield Rd. TW15: Ashf ... 63N 127
Chesterfield Rd. W4 ... 51Sa 109
Chesterfield St. W1 ... 6K 221 (46Kb 90)
Chesterfield Wlk. SE10 ... 53Fc 115
Chesterfield Way SE15 ... 52Yb 114
Chesterfield Way UB3: Hayes ... 47W 84
Chesterford Gdns. NW3 ... 35Db 69
Chesterford Ho. SE18 ... 53Mc 115
... (off Tellson Av.)
Chesterford Rd. E12 ... 36Pc 74
Chester Gdns. EN3: Pond E ... 16Xb 33
Chester Gdns. SM4: Mord ... 72Eb 155
Chester Gdns. W13 ... 44Ka 86
Chester Ga. NW1: Regent's Pk ... 4K 215 (41Kb 90)
Chester Gibbons Grn. AL2: Lon C ... 8H 7
Chester Grn. IG10: Lough ... 11Sc 36
Chester Ho. N10 ... 26Kb 50
Chester Ho. SE8 ... 51Bc 114
Chester Ho. SW1 ... 5K 227 (49Kb 90)
... (off Eccleston Pl.)
Chester Ho. SW9 ... 52Qb 112
... (off Cranmer Rd.)
Chester Ho. UB8: Cowl ... 42L 83
Chesterman Ct. W4 ... 52Ua 110
... (off Corney Reach Way)
Chester M. E17 ... 26Cc 52
Chester M. SW1 ... 3K 227 (48Kb 90)
Chester Path IG10: Lough ... 11Sc 36
Chester Pl. HA6: Nwood ... 24U 44
... (off Green La.)
Chester Pl. NW1 ... 3K 215 (41Kb 90)
Chester Rd. DA15: Sidc ... 57Uc 116
Chester Rd. E11 ... 30Kc 53
Chester Rd. E16 ... 42Gc 93
Chester Rd. E17 ... 29Zb 52
Chester Rd. E7 ... 38Mc 73
Chester Rd. HA6: Nwood ... 24U 44
Chester Rd. IG10: Lough ... 12Rc 36
Chester Rd. IG3: Ilf ... 32Vc 75
Chester Rd. IG7: Chig ... 20Qc 36
Chester Rd. KT24: Eff ... 100X 191
Chester Rd. N17 ... 27Tb 51
Chester Rd. N19 ... 33Kb 70
Chester Rd. N9 ... 18Xb 33
Chester Rd. NW1 ... 4J 215 (41Jb 90)
Chester Rd. SL1: Slou ... 4H 81
Chester Rd. SW19 ... 65Ya 132
Chester Rd. TW4: Houn ... 55X 107
Chester Rd. TW6: H'row A ... 55Q 106
Chester Rd. WD18: Wat ... 15W 26
Chester Rd. WD6: Bore ... 13Sa 29
Chester Row SW1 ... 6H 227 (49Jb 90)
Chesters, The KT3: N Mald ... 67Ua 132
Chester Sq. SW1 ... 5J 227 (49Jb 90)
Chester Sq. M. SW1 ... 4K 227 (48Kb 90)
... (off Chester Sq.)
Chester St. E2 ... 42Wb 91
Chester St. SW1 ... 3J 227 (48Jb 90)
Chester Ter. IG11: Bark ... 37Tc 74
Chester Ter. NW1 ... 3K 215 (41Kb 90)
... (not continuous)
Chesterton Cl. SW18 ... 57Cb 111
Chesterton Cl. UB6: G'frd ... 40Da 65
Chesterton Ct. W5 ... 43Ma 87
Chesterton Dr. RH1: Mers ... 100Nb 196
Chesterton Dr. TW19: Stanw ... 60P 105
Chesterton Ho. CR0: C'don ... 77Tb 157
... (off Heathfield Rd.)
Chesterton Ho. SW11 ... 55Fb 111
... (off Ingrave St.)
Chesterton Ho. W10 ... 43Ab 88
... (off Portobello Rd.)
Chesterton Rd. E13 ... 41Jc 93
Chesterton Rd. W10 ... 43Za 88
Chesterton Sq. W8 ... 49Cb 89
Chesterton Ter. E13 ... 41Jc 93
Chesterton Ter. KT1: King T ... 68Qa 131
Chesterton Way RM18: Tilb ... 4E 122
... (off Brennan Rd.)
Chester Way SE11 ... 6A 230 (49Qb 90)
Chesthunte Rd. N17 ... 25Sb 51
Chestlands Ct. UB10: Hil ... 37Q 64
Chestnut All. SW6 ... 51Bb 111
Chestnut Apts. E3 ... 40Dc 72
... (off Alameda Pl.)
Chestnut Av. BR4: W W'ck ... 78Gc 159
Chestnut Av. CM14: S Weald ... 17Ud 40
Chestnut Av. DA9: Bluew ... 59Vd 120
Chestnut Av. E7 ... 35Kc 73
Chestnut Av. GU25: Vir W ... 10K 125
Chestnut Av. HA0: Wemb ... 36Ka 66
Chestnut Av. HA6: Nwood ... 26V 44
Chestnut Av. HA8: Edg ... 23Na 47
Chestnut Av. IG9: Buck H ... 20Mc 35
Chestnut Av. KT10: Esh ... 73Fa 152
Chestnut Av. KT12: W Vill ... 81U 170
Chestnut Av. KT13: Weyb ... 80S 150
Chestnut Av. KT19: Ewe ... 77Ua 154
Chestnut Av. KT8: E Mos ... 69Ha 130
Chestnut Av. N8 ... 29Nb 50
Chestnut Av. RM12: Horn ... 33Hd 76
Chestnut Av. RM16: Grays ... 47De 99
Chestnut Av. SL3: L'ly ... 47A 82
Chestnut Av. SW14 ... 55Ta 109
Chestnut Av. TN16: Westrm ... 94Mc 199
Chestnut Av. TW11: Tedd ... 68Ha 130
Chestnut Av. TW12: Hamp ... 66Ca 129
Chestnut Av. TW8: Bford ... 49Ma 87
Chestnut Av. UB7: Yiew ... 45P 83
Chestnut Av. WD3: Rick ... 15J 25
Chestnut Av. Nth. E17 ... 28Fc 53
Chestnut Av. Sth. E17 ... 29Ec 52
Chestnut Cl. BR6: Chels ... 78Wc 161
Chestnut Cl. DA11: Nflt ... 8B 122
Chestnut Cl. DA15: Sidc ... 60Wc 117
Chestnut Cl. GU23: Rip ... 97H 189
Chestnut Cl. IG9: Buck H ... 20Mc 35
Chestnut Cl. KT15: Add ... 78M 149
Chestnut Cl. KT20: Kgswd ... 95Cb 195
Chestnut Cl. N14 ... 15Lb 32
Chestnut Cl. N16 ... 33Tb 71
Chestnut Cl. RH1: Redh ... 8B 208
Chestnut Cl. RM12: Horn ... 35Ld 77
Chestnut Cl. SE14 ... 53Bc 114
Chestnut Cl. SE6 ... 64Ec 136
Chestnut Cl. SL9: Chal P ... 25B 42
Chestnut Cl. SM5: Cars ... 74Hb 155
Chestnut Cl. SW16 ... 63Qb 134
Chestnut Cl. TW15: Ashf ... 63R 128
Chestnut Cl. TW16: Sun ... 65V 128
Chestnut Cl. TW20: Eng G ... 5M 125
Chestnut Cl. UB3: Hayes ... 45U 84
Chestnut Cl. UB7: Harl ... 52R 106
Chestnut Cl. UB7: Sip ... 52R 106
Chestnut Copse RH8: Oxt ... 4M 211
Chestnut Ct. CR2: S Croy ... 77Sb 157
... (off Bramley Hill)
Chestnut Ct. KT22: Lea ... 92Ha 192
Chestnut Ct. N8 ... 29Nb 50
Chestnut Ct. RH1: Redh ... 8P 207
Chestnut Ct. SW6 ... 51Bb 111
Chestnut Ct. TW13: Hanw ... 64Z 129
Chestnut Ct. TW19: Stanw ... 60N 105
... (off Mulberry Av.)
Chestnut Ct. W8 ... 48Db 89
... (off Abbots Wlk.)
Chestnut Ct. WD18: Wat ... 14U 26
Chestnut Cres. KT12: W Vill ... 81U 170
Chestnut Dr. AL4: St A ... 1F 6
Chestnut Dr. DA7: Bex ... 55Zc 117
Chestnut Dr. E11 ... 30Jc 53
Chestnut Dr. HA3: Hrw W ... 24Ha 46
Chestnut Dr. HA5: Pinn ... 30Z 45
Chestnut Dr. HP4: Berk ... 2A 2
Chestnut Dr. SL4: Wind ... 6C 102
Chestnut Dr. TW20: Egh ... 5P 125
Chestnut Glen RM12: Horn ... 33Hd 76
Chestnut Gro. CM14: B'wood ... 19Yd 40
Chestnut Gro. CR2: Sels ... 80Xb 157
Chestnut Gro. CR4: Mitc ... 71Mb 156
Chestnut Gro. DA2: Wilm ... 63Fd 140
Chestnut Gro. EN4: E Barn ... 15Hb 31
Chestnut Gro. GU22: Wok ... 92A 188
Chestnut Gro. HA0: Wemb ... 36Ka 66
Chestnut Gro. IG6: Ilf ... 23Uc 54
Chestnut Gro. KT3: N Mald ... 69Ta 131
Chestnut Gro. SE20 ... 66Xb 135
Chestnut Gro. SW12 ... 59Jb 112
Chestnut Gro. TW18: Staines ... 65L 127
Chestnut Gro. TW7: Isle ... 56Ja 108
Chestnut Gro. W5 ... 48Ma 87
Chestnut Ho. BR8: Swan ... 69Fd 140
... (off Squirrels Cl.)
Chestnut Ho. E3 ... 39Bc 72
... (off Sycamore Av.)
Chestnut Ho. SW15 ... 56Va 110
Chestnut Ho. W4 ... 49Ua 88
... (off The Orchard)
Chestnut Ho. WD7: Shenl ... 2Na 15
Chestnut La. GU24: Chob ... 8G 146
Chestnut La. KT13: Weyb ... 78R 150
Chestnut La. N20 ... 18Ab 30
Chestnut La. TN13: S'oaks ... 96Kd 203
Chestnut Mnr. Cl. TW18: Staines ... 64K 127
Chestnut Mead RH1: Redh ... 5N 207
Chestnut M. SW14 ... 56Sa 109
Chestnut Pl. KT13: Weyb ... 78R 150
... (off Pine Gro.)
Chestnut Pl. KT17: Ewe ... 83Wa 174
Chestnut Pl. KT21: Asht ... 91Na 193
Chestnut Pl. SE26 ... 63Vb 135
Chestnut Plaza E20 ... 37Ec 72
... (within Westfield Shop. Cen.)
Chestnut Ri. SE18 ... 51Tc 116
Chestnut Ri. WD23: Bush ... 17Da 27
Chestnut Rd. DA1: Dart ... 60Md 119
Chestnut Rd. EN3: Enf W ... 8Ac 20
Chestnut Rd. KT2: King T ... 66Na 131
Chestnut Rd. SE27 ... 62Rb 135
Chestnut Rd. SW20 ... 68Za 132
Chestnut Rd. TW15: Ashf ... 63R 128
Chestnut Rd. TW2: Twick ... 61Ga 130
Chestnuts CM13: Hut ... 18De 41
Chestnuts, The BR3: Beck ... 69Zb 136
Chestnuts, The HA5: Hat E ... 24Ba 45
Chestnuts, The HP3: Hem H ... 6H 3
Chestnuts, The IG10: Lough ... 15Mc 35
Chestnuts, The KT12: Walt T ... 75W 150
Chestnuts, The N5 ... 35Sb 71
... (off Highbury Grange)
Chestnuts, The RM4: Abr ... 13Xc 37
Chestnuts, The UB10: Uxb ... 38N 63
Chestnuts Ho. E17 ... 30Cc 52
... (off Hoe St.)
Chestnut Ter. SM1: Sutt ... 77Db 155
Chestnut Wlk. IG8: Wfd G ... 22Jc 53
Chestnut Wlk. KT12: W Vill ... 81U 170
Chestnut Wlk. KT14: Byfl ... 84N 169
Chestnut Wlk. SL9: Chal P ... 24A 42
Chestnut Wlk. TN15: S'oaks ... 100Nd 203
Chestnut Wlk. TW17: Shep ... 70U 128
Chestnut Wlk. WD24: Wat ... 9W 12
Chestnut Way KT17: Eps D ... 88Xa 174
Chestnut Way TW13: Felt ... 62X 129
Cheston Av. CR0: C'don ... 75Ac 158
Chestwood Gro. UB10: Hil ... 38P 63
Cheswick Cl. DA1: Cray ... 56Hd 118
Chesworth Cl. DA8: Erith ... 54Gd 118
Chesworth Ct. E1 ... 43Yb 92
... (off Fulneck Pl.)
Chettle Cl. SE1 ... 3F 231 (48Tb 91)
... (off Spurgeon St.)
Chettle Ct. N8 ... 30Qb 50
Chetwode Dr. KT18: Tatt C ... 90Za 174
Chetwode Ho. NW8 ... 5D 214 (42Gb 89)
... (off Grendon St.)
Chetwode Rd. KT20: Tad ... 91Ya 194
Chetwode Rd. SW17 ... 62Hb 133
Chetwood Wlk. E6 ... 43Nc 94
... (off Greenwich Cres.)
Chetwynd Av. EN4: E Barn ... 18Hb 31
Chetwynd Dr. UB10: Hil ... 40P 63
Chetwynd Rd. NW5 ... 35Kb 70
Chetwynd Vs. NW5 ... 35Kb 70
... (off Chetwynd Rd.)
Chevalier Cl. HA7: Stan ... 21Na 47
Cheval Pl. SW7 ... 3E 226 (48Gb 89)
Cheval St. E14 ... 48Cc 92
Cheveley Cl. RM3: Hrld W ... 25Nd 57
Cheveley Gdns. SL1: Burn ... 10A 60
Chevely Cl. CM16: Coop ... 1Zc 23
Cheveney Wlk. BR2: Broml ... 69Jc 137
CHEVENING ... 91Bd 201
Chevening Cross TN14: Chev ... 92Cd 202
Chevening La. TN14: Knock ... 87Ad 181
Chevening Rd. NW6 ... 40Za 68
Chevening Rd. SE10 ... 50Hc 93
Chevening Rd. SE19 ... 65Tb 135
Chevening Rd. TN13: Chips ... 93Dd 202
Chevening Rd. TN14: Chev ... 91Bd 201
Chevening Rd. TN14: Chips ... 91Bd 201
Chevening Rd. TN14: Sund ... 91Bd 201
Chevening Rd. TN14: Sund ... 95Ad 201
Chevenings, The DA14: Sidc ... 62Yc 139
Cheverell Ho. E2 ... 40Wb 71
... (off Pritchard's Rd.)
CHEVERELLS ... 94Gc 199
Cheverton Rd. N19 ... 32Mb 70
Chevet St. E9 ... 36Ac 72
Chevington NW2 ... 37Bb 69
Chevington Pl. RM12: Horn ... 36Md 77
Chevington Vs. RH1: Blet ... 4L 209
Chevington Way RM12: Horn ... 35Md 77
Cheviot N17 ... 24Xb 51
... (off Northumberland Gro.)
Cheviot Cl. DA7: Bex ... 54Gd 118
Cheviot Cl. EN1: Enf ... 12Tb 33
Cheviot Cl. SM2: Sutt ... 81Fb 175
Cheviot Cl. SM7: Bans ... 87Db 175
Cheviot Cl. UB3: Harl ... 52T 106
Cheviot Cl. WD23: Bush ... 16Ea 28
Cheviot Ct. SE14 ... 51Yb 114
... (off Avonley Rd.)
Cheviot Ct. UB2: S'hall ... 49Da 85
Cheviot Gdns. NW2 ... 33Za 68
Cheviot Gdns. SE27 ... 63Rb 135
Cheviot Ga. NW2 ... 33Ab 68
Cheviot Ho. DA11: Nflt ... 58Fe 121
... (off Laburnum Gro.)
Cheviot Ho. E1 ... 44Xb 91
... (off Commercial Rd.)
Cheviot Rd. RM11: Horn ... 31Jd 76
Cheviot Rd. SE27 ... 64Qb 134
Cheviot Rd. SL3: L'ly ... 50C 82
Cheviots AL10: Hat ... 3C 8
Cheviot Way IG2: Ilf ... 28Uc 54
Chevron Cl. E16 ... 44Jc 93
Chevron Ho. RM17: Grays ... 52De 121
Chevy Rd. UB2: S'hall ... 47Ea 86
Chewker La. GU20: W'sham ... 7A 146
Chewton Rd. E17 ... 28Ac 52
Cheyham Gdns. SM2: Cheam ... 82Za 174
Cheyham Way SM2: Cheam ... 82Ab 174
Cheylesmore Ho. SW1 ... 7K 227 (50Kb 90)
... (off Ebury Bri. Rd.)
Cheyne Av. E18 ... 27Hc 53
Cheyne Av. TW2: Whitt ... 60Ba 107
Cheyne Cl. BR2: Broml ... 76Nc 160
Cheyne Cl. NW4 ... 29Ya 48
Cheyne Cl. SL9: Ger X ... 32A 62
Cheyne Ct. SM7: Bans ... 87Db 175
Cheyne Ct. SW3 ... 51Hb 111
Cheyne Ct. WD23: Bush ... 14Aa 27
Cheyne Gdns. SW3 ... 51Gb 111
Cheyne Hill KT5: Surb ... 70Pa 131
Cheyne Ho. SW3 ... 51Hb 111
... (off Chelsea Emb.)
Cheyne M. SW3 ... 51Gb 111
Cheyne Pk. Dr. BR4: W W'ck ... 76Ec 158
Cheyne Path W7 ... 43Ha 86
Cheyne Pl. SW3 ... 51Hb 111
Cheyne Rd. TW15: Ashf ... 66T 128
Cheyne Row SW3 ... 51Gb 111
Cheyne Wlk. CR0: C'don ... 75Wb 157
Cheyne Wlk. DA3: Lfield ... 69Zd 143
Cheyne Wlk. N21 ... 15Rb 33
Cheyne Wlk. NW4 ... 30Ya 48
Cheyne Wlk. SW10 ... 52Fb 111
Cheyne Wlk. SW3 ... 51Gb 111
... (not continuous)
Cheyneys Av. HA8: Edg ... 23Ma 47
Chichele Gdns. CR0: C'don ... 77Ub 157
Chichele Rd. NW2 ... 36Za 68
Chichele Rd. RH8: Oxt ... 100Gc 199
Chicheley Gdns. HA3: Hrw W ... 24Ea 46
... (not continuous)
Chicheley Rd. HA3: Hrw W ... 24Ea 46
Chicheley St. SE1 ... 1J 229 (47Pb 90)
Chichester Av. HA4: Ruis ... 33T 64
Chichester Cl. E6 ... 44Nc 94
Chichester Cl. RM15: Avel ... 46Td 98
Chichester Cl. RM16: Chaf H ... 49Zd 99
Chichester Cl. SE3 ... 52Lc 115
Chichester Cl. TW12: Hamp ... 65Ba 129
Chichester Ct. HA7: Stan ... 27Na 47
Chichester Ct. HA8: Edg ... 23Qa 47
... (off Whitchurch La.)
Chichester Ct. KT17: Ewe ... 81Va 174
Chichester Ct. NW1 ... 38Lb 70
... (off Royal Coll. St.)
Chichester Ct. SL1: Slou ... 8M 81
Chichester Ct. TW19: Stanw ... 60N 105
Chichester Ct. UB5: N'olt ... 39Ba 65
Chichester Dr. CR8: Purl ... 84Pb 176
Chichester Dr. TN13: S'oaks ... 97Hd 202
Chichester Gdns. IG1: Ilf ... 31Nc 74
Chichester Ho. CM14: B'wood ... 19Yd 40
... (off Sir Francis Way)
Chichester Ho. KT19: Eps ... 84Qa 173
Chichester Ho. NW6 ... 40Cb 69
Chichester Ho. SW9 ... 52Qb 112
... (off Cranmer Rd.)
Chichester Lodge SE10 ... 49Jc 93
... (off Peartree Way)
Chichester M. SE27 ... 63Qb 134
Chichester Rents WC2 ... 2K 223 (44Qb 90)
... (off Chancery La.)
Chichester Ri. DA12: Grav'nd ... 3F 144
Chichester Rd. CR0: C'don ... 76Ub 157
Chichester Rd. DA9: Ghithe ... 58Vd 120
Chichester Rd. E11 ... 34Gc 73
Chichester Rd. N9 ... 18Wb 33
Chichester Rd. NW6 ... 40Cb 69
Chichester Rd. W2 ... 43Db 89
Chichester St. SW1 ... 7C 228 (50Lb 90)
Chichester Way E14 ... 49Fc 93
Chichester Way TW14: Felt ... 59Y 107
Chichester Way WD25: Wat ... 5Aa 13
Chichester Wharf DA8: Erith ... 50Gd 96
Chicken La. AL2: Lon C ... 9H 7
Chickenshed Theatre ... 15Jb 32
Chicksand Ho. E1 ... 43Wb 91
... (off Chicksand St.)
Chicksand St. E1 ... 43Vb 91
... (not continuous)
Chidbrook Ho. WD18: Wat ... 16U 26
Chiddingfold N12 ... 20Cb 31
Chiddingstone SE13 ... 57Ec 114
Chiddingstone Av. DA7: Bex ... 52Bd 117
Chiddingstone Cl. SM2: Sutt ... 82Cb 175
Chiddingstone St. SW6 ... 54Cb 111
Chieftain Rd. KT16: Vir W ... 5L 147
Chieftan Dr. RM19: Purf ... 49Qd 97
Chieveley Pde. DA7: Bex ... 55Dd 118
Chieveley Rd. DA7: Bex ... 56Dd 118
Chiffinch Gdns. DA11: Nflt ... 2A 144
Chignell Pl. W13 ... 46Ja 86
CHIGWELL ... 20Rc 36
Chigwell Ct. E9 ... 37Ac 72
... (off Ballance Rd.)
Chigwell Golf Course ... 21Qc 54
Chigwell Grange IG7: Chig ... 18Sc 36
Chigwell Hill E1 ... 45Xb 91
Chigwell Hurst Ct. HA5: Pinn ... 27Z 45
Chigwell La. IG10: Lough ... 15Sc 36
Chigwell La. IG7: Chig ... 15Sc 36
Chigwell La. IG7: Lough ... 15Sc 36
Chigwell Pk. IG7: Chig ... 21Rc 54
Chigwell Pk. Dr. IG7: Chig ... 21Qc 54
Chigwell Ri. IG7: Chig ... 19Qc 36
Chigwell Rd. E18 ... 27Kc 53
Chigwell Rd. IG8: Wfd G ... 23Pc 54
CHIGWELL ROW ... 20Xc 37
Chigwell Row Wood Local Nature Reserve ... 21Wc 55
Chigwell Station (Underground) ... 20Rc 36
Chigwell Vw. RM5: Col R ... 23Cd 56
Chilberton Dr. RH1: Mers ... 2C 208
Chilbolton TW20: Egh ... 64A 126
Chilbrook Rd. KT11: D'side ... 90W 170
Chilcombe Ho. SW15 ... 59Wa 110
... (off Fontley Way)
Chilcot Cl. E14 ... 44Dc 92
Chilcott Cl. HA0: Wemb ... 35La 66
Chilcott Rd. WD24: Wat ... 8U 12
Childebert Rd. SW17 ... 61Kb 134
CHILDERDITCH ... 27Ce 59
Childerditch Hall Dr. CM13: L War ... 27Ce 59
Childerditch Ind. Pk. CM13: L War ... 27Be 59
Childerditch La. CM13: L War ... 24Ae 59
Childerditch La. CM13: W H'dn ... 24Ae 59
Childerditch St. CM13: L War ... 26Ce 59
Childeric Rd. SE14 ... 52Ac 114
Childerley KT1: King T ... 69Qa 131
... (off Burritt Rd.)
Childerley St. SW6 ... 53Ab 110
Childers, The IG8: Wfd G ... ??Pc 54
Childers St. SE8 ... 51Ac 114
Child La. SE10 ... 48Hc 93
CHILDREN'S HOSPITAL (LEWISHAM), THE ... 57Dc 114
... (within Lewisham University Hospital)
CHILDREN'S TRUST, THE ... 93Za 194
Childs Av. UB9: Hare ... 26L 43
Childsbridge Farm Pl. TN15: Seal ... 91Nd 203
Childsbridge La. TN15: Kems'g ... 90Nd 183
Childsbridge La. TN15: Seal ... 90Nd 183
Childsbridge Way TN15: Seal ... 92Pd 203
Childs Cl. RM11: Horn ... 30Ld 57
Childs Ct. UB3: Hayes ... 45W 84
Childs Cres. DA10: Swans ... 58Zd 121
Childs Hall Cl. KT23: Bookh ... 97Ba 191
Childs Hall Dr. KT23: Bookh ... 97Ba 191
Childs Hall Rd. KT23: Bookh ... 97Ba 191
CHILD'S HILL ... 34Bb 69
Childs Hill Wlk. NW2 ... 34Bb 69
... (off Cricklewood La.)
Child's La. SE19 ... 65Ub 135
Child's M. SW5 ... 49Cb 89
... (off Child's Pl.)
Child's Pl. SW5 ... 49Cb 89
Child's St. SW5 ... 49Cb 89
Child's Wlk. SW5 ... 49Cb 89
... (off Child's St.)
Childs Way NW11 ... 29Bb 49
Childs Way TN15: Wro ... 88Be 185
Childwick Ct. HP3: Hem H ... 5B 4
Chilham Cl. DA5: Bexl ... 59Bd 117
Chilham Cl. HP2: Hem H ... 3N 3
Chilham Cl. UB6: G'frd ... 40Ja 66
Chilham Ho. SE1 ... 3G 231 (48Tb 91)
Chilham Ho. SE15 ... 51Yb 114
Chilham Rd. SE9 ... 63Nc 138
Chilham Way BR2: Hayes ... 73Jc 159
Chilianwallah Memorial ... 51Jb 112
Chillerton Rd. SW17 ... 64Jb 134
Chillingford Ho. SW17 ... 63Eb 133
Chillington Dr. SW11 ... 56Fb 111
Chillingworth Gdns. TW1: Twick ... 62Ha 130
Chillingworth Rd. N7 ... 36Qb 70
Chill La. N1 ... 39Mb 70
Chilmans Dr. KT23: Bookh ... 97Da 191
Chilmark Gdns. KT3: N Mald ... 72Wa 154
Chilmark Gdns. RH1: Mers ... 1E 208
Chilmark Rd. SW16 ... 68Mb 134
Chilmead RH1: Redh ... 5P 207
Chilmead La. RH1: Nutf ... 4D 208
Chilsey Grn. Rd. KT16: Chert ... 72G 148
Chiltern Av. TW2: Whitt ... 60Ca 107
Chiltern Av. WD23: Bush ... 16Ea 28
Chiltern Bus. Village UB8: Uxb ... 40K 63
Chiltern Cl. CR0: C'don ... 76Ub 157
Chiltern Cl. DA7: Bex ... 53Gd 118
Chiltern Cl. GU22: Wok ... 4N 187
Chiltern Cl. KT4: Wor Pk ... 74Ya 154
Chiltern Cl. TW18: Staines ... 64J 127
Chiltern Cl. UB10: Ick ... 33Q 64
Chiltern Cl. WD18: Wat ... 14V 26
Chiltern Cl. WD23: Bush ... 16Da 27
Chiltern Cl. WD6: Bore ... 12Pa 29
Chiltern Ct. BR2: Broml ... 75Nc 160
... (off Gravel Rd.)
Chiltern Ct. EN5: New Bar ... 15Eb 31
Chiltern Ct. HA1: Harr ... 29Fa 46
Chiltern Ct. N10 ... 26Jb 50
Chiltern Ct. NW1 ... 6G 215 (42Hb 89)
... (off Baker St.)
Chiltern Ct. SE14 ... 52Yb 114
... (off Avonley Rd.)
Chiltern Ct. SL4: Wind ... 3F 102
... (off Fawcett Rd.)
Chiltern Ct. UB8: Hil ... 42R 84
Chiltern Ct. M. SL4: Wind ... 3F 102
... (off Fawcett Rd.)
Chiltern Dene EN2: Enf ... 14Pb 32
Chiltern Dr. KT5: Surb ... 72Qa 153
Chiltern Dr. WD3: Rick ... 17H 25
Chiltern Est. WD24: Wat ... 11Y 27
Chiltern Gdns. BR2: Broml ... 70Hc 137
Chiltern Gdns. NW2 ... 34Za 68
Chiltern Gdns. RM12: Horn ... 34Ld 77
Chiltern Hgts. N1 ... 38Pb 70
... (off Caledonian Rd.)
Chiltern Hill SL9: Chal P ... 25A 42
Chiltern Ho. N9 ... 20Wb 33
Chiltern Ho. SE17 ... 51Tb 113
... (off Portland St.)
Chiltern Ho. W10 ... 43Ab 88
... (off Telford Rd.)
Chiltern Ho. W5 ... 43Na 87
Chiltern Ho. WD18: Wat ... 14Y 27
... (off King St.)
Chiltern Open Air Mus. ... 18B 24
Chiltern Rd. DA11: Nflt ... 2A 144
Chiltern Rd. E3 ... 42Cc 92
Chiltern Rd. HA5: Eastc ... 29Y 45
Chiltern Rd. IG2: Ilf ... 29Uc 54
Chiltern Rd. SL1: Burn ... 3A 80
Chiltern Rd. SM2: Sutt ... 81Db 175
Chilterns AL10: Hat ... 3C 8
Chilterns, The BR1: Broml ... 68Kc 137
... (off Murray Av.)
Chilterns, The SM2: Sutt ... 81Db 175
Chiltern St. W1 ... 7H 215 (43Jb 90)
Chiltern Vw. Rd. UB8: Uxb ... 40L 63
Chiltern Way IG8: Wfd G ... 20Jc 35
Chilthorne Cl. SE6 ... 59Bc 114
Chilton Av. W5 ... 49Ma 87
Chilton Ct. KT12: Walt T ... 77W 150
Chilton Ct. N22 ... 24Nb 50
... (off Truro Rd.)
Chilton Ct. SL6: Tap ... 4A 80
Chilton Gro. SE8 ... 49Zb 92
Chiltonian Ind. Est. SE12 ... 58Hc 115
Chiltonian M. SE13 ... 57Fc 115
Chilton Rd. HA8: Edg ... 23Qa 47
Chilton Rd. RM16: Grays ... 8C 100
Chilton Rd. TW9: Rich ... 55Qa 109
Chiltons, The E18 ... 26Jc 53
Chiltons Cl. SM7: Bans ... 87Db 175
Chilton St. E2 ... 42Vb 91
Chilvers Cl. TW2: Twick ... 61Ga 130
Chilver St. SE10 ... 50Hc 93
Chilvers Way DA11: Nflt ... 61Ee 143
Chilwell Gdns. WD19: Wat ... 21Y 45
Chilwick Rd. SL2: Slou ... 2D 80
Chilworth Ct. SW19 ... 60Za 110
Chilworth Gdns. SM1: Sutt ... 76Eb 155
Chilworth M. W2 ... 3B 220 (44Fb 89)
Chilworth Pl. IG11: Bark ... 42Wc 95
Chilworth St. W2 ... 3A 220 (44Eb 89)
Chimes Av. N13 ... 22Qb 50
Chimes Ho. BR3: Beck ... 67Ac 136
Chime Sq. AL3: St A ... 1C 6
Chimes Shop. Cen., The ... 38M 63
Chimes Ter. N8 ... 29Nb 50
Chimney Ct. E1 ... 46Xb 91
... (off Brewhouse La.)
Chimneys, The WD23: Bush ... 15Ba 27
China Ct. E1 ... 46Xb 91
... (off Asher Way)
China Hall M. SE16 ... 48Yb 92
China La. RM14: Bulp ... 34Fe 79
China M. SW2 ... 59Pb 112
China Wharf SE1 ... 47Wb 91
Chinbrook Cres. SE12 ... 62Kc 137
Chinbrook Rd. SE12 ... 62Kc 137
Chinchilla Dr. TW4: Houn ... 54Y 107
Chindits La. CM14: W'ley ... 22Yd 58
Chine, The HA0: Wemb ... 36Ka 66
Chine, The N10 ... 28Lb 50
Chine, The N21 ... 16Rb 33
Chine Farm Pl. TN14: Knock ... 88Zc 181
Ching Ct. WC2 ... 3F 223 (44Nb 90)
... (off Monmouth St.)
Chingdale Rd. E4 ... 20Gc 35
CHINGFORD ... 18Ec 34
Chingford Av. E4 ... 20Cc 34
Chingford Bus Station ... 17Gc 35
Chingford Golf Course ... 17Gc 35
Chingford Golf Range ... 18Cc 34
CHINGFORD GREEN ... 18Fc 35
CHINGFORD HATCH ... 21Fc 53
CHINGFORD HEALTH CEN. ... 21Bc 52
... (off York Rd.)
Chingford Ind. Cen. E4 ... 22Ac 52
Chingford La. IG8: Wfd G ... 21Gc 53
Chingford Leisure Cen. ... 21Ec 52
CHINGFORD MOUNT ... 21Cc 52
Chingford Mt. Rd. E4 ... 21Cc 52
Chingford Rd. E17 ... 25Dc 52
Chingford Rd. E4 ... 23Cc 52
Chingford Station (Overground) ... 17Gc 35
Chingley Cl. BR1: Broml ... 65Gc 137
Chingstone Ter. E4 ... 18Ec 34
Ching Way E4 ... 23Bc 52
... (not continuous)
Chinnery Cl. EN1: Enf ... 11Vb 33
Chinnock's Wharf E14 ... 45Ac 92
... (off Narrow St.)
Chinnor Cres. UB6: G'frd ... 40Da 65
Chinthurst M. CR5: Coul ... 88Jb 176
Chipka St. E14 ... 47Ec 92
... (not continuous)
Chipley St. SE14 ... 51Ac 114
Chipmunk Gro. UB5: N'olt ... 41Aa 85
Chippendale All. UB8: Uxb ... 38M 63
Chippendale Ho. SW1 ... 50Kb 90
... (off Churchill Gdns.)
Chippendale St. E5 ... 34Zb 72
Chippendale Waye UB8: Uxb ... 38M 63
Chippenham KT1: King T ... 68Pa 131
... (off Excelsior Cl.)
Chippenham Av. HA9: Wemb ... 36Ra 67

Chippenham Cl. HA5: Eastc....28V **44**
Chippenham Cl. RM3: Rom....22Md **57**
Chippenham Gdns. NW6....41Cb **89**
Chippenham M. W9....42Cb **89**
Chippenham Rd. RM3: Rom....23Md **57**
Chippenham Rd. W9....42Cb **89**
Chippenham Wlk. RM3: Rom....23Md **57**
CHIPPERFIELD....3J **11**
Chipperfield Cl. RM14: Upm....32Ud **78**
CHIPPERFIELD COMMON....4K **11**
Chipperfield Ho. SW3....7D **226** (50Gb **89**)
....*(off Cale St.)*
Chipperfield Rd. BR5: St P....67Wc **139**
....(not continuous)
Chipperfield Rd. HP3: Bov....9D **2**
Chipperfield Rd. HP3: Hem H....6L **3**
Chipperfield Rd. WD4: Chfd....1G **10**
Chipperfield Rd. WD4: K Lan....2L **11**
CHIPPING BARNET....14Ab **30**
Chipping Cl. EN5: Barn....13Ab **30**
CHIPSTEAD....90Hb **175**
CHIPSTEAD....94Ed **202**
Chipstead Av. CR7: Thor H....70Rb **135**
CHIPSTEAD BOTTOM....93Fb **195**
Chipstead Cl. CR5: Coul....88Jb **176**
Chipstead Cl. RH1: Redh....7P **207**
Chipstead Cl. SE19....66Vb **135**
Chipstead Cl. SM2: Sutt....81Db **175**
Chipstead Ct. GU21: Knap....9J **167**
Chipstead Gdns. NW2....33Xa **68**
Chipstead Golf Course....**90Hb 175**
Chipstead La. CR5: Chips....96Eb **195**
Chipstead La. CR5: Coul....96Eb **195**
Chipstead La. KT20: Kgswd....97Bb **195**
Chipstead La. TN13: Chips....94Ed **202**
Chipstead La. TN13: Riv....94Ed **202**
Chipstead Pk. TN13: Chips....94Fd **202**
Chipstead Pk. Cl. TN13: Chips....94Ed **202**
Chipstead Pl. Gdns. TN13: Chips....94Ed **202**
Chipstead Rd. DA8: Erith....52Gd **118**
Chipstead Rd. SM7: Bans....89Bb **175**
Chipstead Rd. TW6: H'row A....55Q **106**
Chipstead Sailing Club....**94Ed 202**
Chipstead Sq. TN13: Chips....94Ed **202**
Chipstead Sta. Pde. CR5: Chips....90Hb **175**
Chipstead Station (Rail)....90Hb **175**
Chipstead St. SW6....53Cb **111**
Chipstead Valley Rd. CR5: Coul....88Jb **176**
Chipstead Way SM7: Bans....88Hb **175**
Chip St. SW4....55Mb **112**
Chirdland Ho. WD18: Wat....16U **26**
Chirk Cl. UB4: Yead....42Aa **85**
Chirton Wlk. GU21: Wok....10L **167**
Chisenhale Rd. E3....40Ac **72**
Chisholm Ct. W6....50Wa **88**
Chisholm Rd. CR0: C'don....75Ub **157**
Chisholm Rd. TW10: Rich....58Pa **109**
Chisledon Wlk. E9....37Bc **72**
....*(off Osborne Rd.)*
CHISLEHURST....65Rc **138**
Chislehurst Av. N12....24Eb **49**
Chislehurst Caves....**67Qc 138**
Chislehurst Golf Course....**66Rc 138**
Chislehurst Rd. BR1: Broml....68Mc **137**
Chislehurst Rd. BR5: Pet W....70Uc **138**
Chislehurst Rd. BR6: Orp....72Uc **160**
Chislehurst Rd. BR6: Pet W....72Uc **160**
Chislehurst Rd. BR6: St M Cry....72Uc **160**
Chislehurst Rd. BR7: Chst....67Nc **138**
Chislehurst Rd. DA14: Sidc....64Wc **139**
Chislehurst Rd. TW10: Rich....57Na **109**
Chislehurst Station (Rail)....68Qc **138**
CHISLEHURST WEST....65Qc **138**
Chislet Cl. BR3: Beck....66Cc **136**
Chisley Rd. N15....30Ub **51**
Chiswell Ct. WD24: Wat....10Y **13**
CHISWELL GREEN....7N **5**
Chiswell Grn. La. AL2: Chis G....7K **5**
Chiswell Grn. La. AL2: Pot C....7K **5**
Chiswell Sq. SE3....54Kc **115**
Chiswell St. EC1....7E **218** (43Sb **91**)
Chiswell St. SE5....52Tb **113**
....*(off Edmund St.)*
CHISWICK....50Ta **87**
Chiswick Bri.....54Sa **109**
Chiswick Cl. CR0: Bedd....76Pb **156**
Chiswick Comn. Rd. W4....49Ta **87**
Chiswick Community Sports Hall....**52Ta 109**
Chiswick Ct. HA5: Pinn....27Ba **45**
Chiswick Ct. W4....49Ra **87**
Chiswick High Rd. TW8: Bford....50Qa **87**
Chiswick High Rd. W4....50Qa **87**
Chiswick House & Gdns.....**51Ta 109**
Chiswick Ho. Grounds W4....51Ua **110**
Chiswick La. W4....50Ua **88**
Chiswick La. Sth. W4....51Va **110**
Chiswick Mall W4....51Va **110**
Chiswick Mall W6....50Wa **88**
Chiswick Pk. W4....49Ra **87**
Chiswick Park Station (Underground)....49Sa **87**
Chiswick Pier....**52Va 110**
Chiswick Plaza W4....51Sa **109**
Chiswick Quay W4....53Sa **109**
Chiswick Rd. N9....19Wb **33**
Chiswick Rd. W4....49Sa **87**
CHISWICK RDBT.....50Qa **87**
Chiswick Sq. W4....51Ua **110**
Chiswick Staithe W4....53Sa **109**
Chiswick Station (Rail)....52Sa **109**
Chiswick Ter. W4....49Sa **87**
....*(off Chiswick High Rd.)*
Chiswick Village W4....51Qa **109**
Chiswick Wharf W4....51Va **110**
Chittenden Cotts. GU23: Wis....88N **169**
Chitterfield Ga. UB7: Sip....52Q **106**
Chitty's La. RM8: Dag....33Zc **75**
Chitty St. W1....7C **216** (43Lb **90**)
Chivalry Rd. SW11....57Gb **111**
Chivelston SW19....60Za **110**
Chivenor Gro. KT2: King T....64Ma **131**
Chivenor Pl. AL4: St A....4G **6**
Chivers Pas. SW18....57Db **111**
Chivers Rd. E4....20Dc **34**
Choats Mnr. Way RM9: Dag....40Ad **75**
Choats Rd. IG11: Bark....40Yc **75**
Choats Rd. RM9: Dag....41Zc **95**
CHOBHAM....3J **167**
Chobham Academy Sports Cen.....**36Ec 72**
Chobham Bus. Cen. GU24: Chob....2P **167**
Chobham Cl. KT16: Ott....79D **148**
Chobham Common Memorial Cross....**5H 147**
Chobham Common Nat. Nature Reserve....**6H 147**
Chobham Gdns. SW19....61Za **132**
Chobham Golf Course....**7H 167**
Chobham La. KT16: Longc....6K **147**
CHOBHAM MANOR....36Dc **72**
Chobham Mus.....**3J 167**
Chobham Pk. Dr. GU24: Chob....1M **167**
Chobham Pk. La. GU24: Chob....2L **167**
Chobham Rd. GU21: Wok Horsell Comn. Rd.....5N **167**
Chobham Rd. GU21: Wok Wheatsheaf Cl.....88A **168**
....(not continuous)
Chobham Rd. E15....36Fc **73**
Chobham Rd. GU21: Knap....10F **166**
Chobham Rd. GU21: Wok....10F **166**
Chobham Rd. GU24: Chob....5G **146**
Chobham Rd. KT16: Ott....80C **148**
Chobham Rd. SL5: S'dale....3F **146**
Chocolate Factory 1, The N22....26Pb **50**
....*(off Clarendon Rd.)*
Chocolate Factory 2, The N22....26Pb **50**
....*(off Coburg Rd.)*
Chocolate Studios N1....3F **219** (41Tb **91**)
....*(off Shepherdess Pl.)*
Choice Vw. IG1: Ilf....33Sc **74**
....*(off Axon Pl.)*
Choir Grn. GU21: Knap....9J **167**
Cholesbury WD19: Wat....19Aa **27**
Cholmeley Cl. N6....31Kb **70**
Cholmeley Cres. N6....31Kb **70**
Cholmeley Lodge N6....32Kb **70**
Cholmeley Pk. N6....32Kb **70**
Cholmley Gdns. NW6....36Cb **69**
Cholmley Rd. KT7: T Ditt....72Ka **152**
Cholmley Ter. KT7: T Ditt....73Ka **152**
....*(off Portsmouth Rd.)*
Cholmley Vs. KT7: T Ditt....72Ka **152**
....*(off Portsmouth Rd.)*
Cholmondeley Av. NW10....40Wa **68**
Cholmondeley Wlk. TW9: Rich....57La **108**
Choppin's Ct. E1....46Xb **91**
Chopwell Cl. E15....38Fc **73**
CHORLEYWOOD....14F **24**
Chorleywood Bottom WD3: Chor....15F **24**
CHORLEYWOOD BOTTOM....15F **24**
Chorleywood Cl. WD3: Rick....17M **25**
Chorleywood Coll. Est. WD3: Chor....14H **25**
Chorleywood Cres. BR5: St P....68Vc **139**
Chorleywood Golf Course....**14F 24**
Chorleywood Ho. WD3: Chor....13G **24**
Chorleywood Ho. Dr. WD3: Chor....13G **24**
Chorleywood House Estate....**12H 25**
Chorleywood Lodge La. WD3: Chor....13H **25**
Chorleywood Rd. WD3: Rick....14J **25**
Chorleywood Station (Rail & Underground)....14F **24**
CHORLEYWOOD WEST....14D **24**
Choudhury Mans. N1....38Nb **70**
....*(off Pembroke St.)*
Choumert Gro. SE15....54Wb **113**
Choumert M. SE15....54Wb **113**
Choumert Rd. SE15....55Vb **113**
Choumert Sq. SE15....54Wb **113**
Chown Ct. KT24: E Hor....100V **190**
Chow Sq. E8....36Vb **71**
Chrislaine Cl. TW19: Stanw....58M **105**
Chrislea Cl. TW5: Hest....51Ba **107**
Chrisp Ho. SE10....51Gc **115**
....*(off Maze Hill)*
Chrisp St. E14....43Dc **92**
....(not continuous)
Chris Pullen Way N7....37Nb **70**
Christabel Cl. TW7: Isle....55Ga **108**
Christabel Pankhurst Ct. SE5....52Tb **113**
....*(off Brisbane St.)*
Christchurch Av. DA8: Erith....51Fd **118**
Christchurch Av. HA0: Wemb....37Na **67**
Christchurch Av. HA3: Kenton....28Ha **46**
Christchurch Av. HA3: W'stone....28Ha **46**
Christchurch Av. N12....23Eb **49**
Christchurch Av. NW6....39Za **68**
Christchurch Av. RM13: Rain....40Hd **76**
Christchurch Av. TW11: Tedd....64Ja **130**
Christchurch Cl. AL3: St A....1A **6**
Christchurch Cl. EN2: Enf....12Sb **33**
Christchurch Cl. N12....24Fb **49**
Christchurch Cl. SW19....66Fb **133**
Christ Church Ct. NW10....39Ua **68**
Christchurch Ct. EC4....2C **224** (44Rb **91**)
....*(off Warwick La.)*
Christchurch Ct. NW6....38Ab **68**
....*(off Willesden La.)*
Christchurch Ct. UB4: Yead....42Y **85**
....*(off Dunedin Way)*
Christchurch Cres. DA12: Grav'nd....9E **122**
Christchurch Cres. WD7: R'lett....8Ja **14**
Christchurch Flats TW9: Rich....55Na **109**
Christchurch Gdns. HA3: W'stone....28Ja **46**
Christchurch Gdns. KT19: Eps....83Ra **173**
Christchurch Grn. HA0: Wemb....37Na **67**
Christchurch Hill NW3....34Fb **69**
Christchurch Ho. RM8: Dag....36Xc **75**
Christchurch Ho. SW2....60Pb **112**
....*(off Christchurch Rd.)*
Christchurch La. EN5: Barn....12Ab **30**
Christchurch Lodge EN4: Cockf....14Hb **31**
Christ Chu. Mt. KT19: Eps....84Ra **173**
....(not continuous)
Christchurch Pk. SM2: Sutt....80Eb **155**
Christchurch Pas. NW3: Hampstead....34Eb **69**
Christchurch Pas. EN5: Barn High Barnet....12Ab **30**
Christchurch Path UB3: Harl....48S **84**
Christchurch Pl. KT19: Eps....83Ra **173**
Christchurch Pl. SW8....54Mb **112**
Christ Chu. Rd. BR3: Beck....68Cc **136**
Christ Chu. Rd. KT19: Eps....84Na **173**
Christ Chu. Rd. KT5: Surb....72Pa **153**
Christchurch Rd. CR8: Purl....83Rb **177**
Christchurch Rd. DA1: Dart....59Ld **119**
Christchurch Rd. DA12: Grav'nd....10E **122**
Christchurch Rd. DA15: Sidc....63Vc **139**
Christchurch Rd. GU25: Vir W....9L **125**
Christchurch Rd. HP2: Hem H....1M **3**
Christchurch Rd. IG1: Ilf....32Rc **74**
Christchurch Rd. N8....30Nb **50**
Christchurch Rd. RM18: Tilb....3C **122**
Christchurch Rd. SW14....57Ra **109**
Christchurch Rd. SW19....66Fb **133**
Christchurch Rd. SW2....60Pb **112**
Christchurch Rd. TW6: H'row A....55Q **106**
Christchurch Sq. E9....39Yb **72**
Christchurch St. SW3....51Hb **111**
Christchurch Ter. SW3....51Hb **111**
....*(off Christchurch St.)*
Christchurch Way GU21: Wok....89B **168**
Christchurch Way SE10....50Gc **93**
Christian Ct. SE16....46Bc **92**
CHRISTIAN FIELDS....3F **144**
Christian Flds. SW16....66Qb **134**
Christian Flds. Av. DA12: Grav'nd....3E **144**
Christian Pl. E1....44Wb **91**
....*(off Burslem St.)*
Christian Sq. SL4: Wind....3G **102**
Christian St. E1....44Wb **91**
Christie Cl. GU18: Light....2A **166**
Christie Cl. KT23: Bookh....97Ba **191**
Christie Ct. CM14: B'wood....20Yd **40**
Christie Ct. N19....33Nb **70**
Christie Ct. WD18: Wat....15W **26**
Christie Dr. CR0: C'don....71Wb **157**
Christie Gdns. RM6: Chad H....30Xc **55**
Christie Ho. E16....47Jc **93**
....*(off Hammersley Rd.)*
Christie Ho. SE10....50Hc **93**
....*(off Blackwall La.)*
Christie Ho. W12....44Xa **88**
....*(off Du Cane Rd.)*
Christie Rd. E9....37Ac **72**
Christie Rd. EN9: Walt A....7Dc **20**
Christies Av. TN14: Bad M....82Cd **182**
Christie Wlk. CR3: Cat'm....94Tb **197**
Christina Sq. N4....32Rb **71**
Christina St. EC2....5H **219** (42Ub **91**)
Christine Ct. RM13: Rain....43Jd **96**
Christine Worsley Cl. N21....18Rb **33**
Christmas La. SL2: Farn C....4G **60**
Christmas Tree Farm....**84Qc 180**
Christopher Av. W7....48Ja **86**
Christopher Bell Twr. E3....40Cc **72**
....*(off Pancras Way)*
Christopher Boones Ct. SE13....56Fc **115**
....*(off Bessington Rd.)*
Christopher Cl. DA15: Sidc....57Vc **117**
Christopher Cl. RM12: Horn....35Md **77**
Christopher Cl. SE16....47Zb **92**
Christopher Cl. DA15: Sidc....61Wc **139**
....*(off Station Rd.)*
Christopher Ct. E1....44Wb **91**
....*(off Leman St.)*
Christopher Ct. HP3: Hem H....5M **3**
Christopher Ct. KT20: Tad....95Ya **194**
Christopher Ct. TW15: Ashf....64N **127**
Christopher Gdns. RM9: Dag....36Zc **75**
Christopher Pl. AL3: St A....2B **6**
....*(off Verulam Rd.)*
Christopher Pl. N22....24Nb **50**
....*(off Myddleton Rd.)*
Christopher Pl. NW1....4E **216** (41Mb **90**)
Christopher Rd. UB2: S'hall....49X **85**
Christophers M. W11....46Ab **88**
Christopher St. EC2....6G **219** (42Tb **91**)
Christy Rd. TN16: Big H....87Lc **179**
Chroma Mans. E20....37Ec **72**
....*(off Penny Brookes St.)*
Chrome Rd. DA8: Erith....52Jd **118**
Chronicle Av. NW9....27Ua **48**
Chronicle Twr. N1....3D **218** (41Sb **91**)
Chryssell Rd. SW9....52Qb **112**
Chrystie La. KT23: Bookh....98Da **191**
Chubworthy St. SE14....51Ac **114**
Chucks La. KT20: Walt H....96Xa **194**
Chudleigh DA14: Sidc....63Xc **139**
Chudleigh Cres. IG3: Ilf....35Uc **74**
Chudleigh Gdns. SM1: Sutt....76Eb **155**
Chudleigh Rd. NW6....38Za **68**
Chudleigh Rd. RM3: Rom....21Nd **57**
Chudleigh Rd. SE4....57Bc **114**
Chudleigh Rd. TW2: Twick....58Ga **108**
Chudleigh St. E1....44Zb **92**
Chudleigh Way HA4: Ruis....32W **64**
Chulsa Rd. SE26....64Xb **135**
Chumleigh Gdns. SE5....51Ub **113**
....*(off Chumleigh St.)*
Chumleigh St. SE5....51Ub **113**
Chumleigh Wlk. KT5: Surb....70Pa **131**
Church All. CR0: C'don....74Qb **156**
Church All. DA11: Grav'nd....8D **122**
....*(off High St.)*
Church All. WD25: A'ham....10Ea **14**
Church App. SE21....62Tb **135**
Church App. TN14: Cud....87Sc **180**
Church App. TW19: Stanw....58M **105**
Church App. TW20: Thorpe....69E **126**
Church Av. BR3: Beck....67Cc **136**
Church Av. DA14: Sidc....64Wc **139**
Church Av. E4....23Fc **53**
....(not continuous)
Church Av. HA4: Ruis....32T **64**
Church Av. HA5: Pinn....30Aa **45**
Church Av. NW1....37Kb **70**
Church Av. SW14....55Ta **109**
Church Av. UB2: S'hall....48Aa **85**
Church Av. UB5: N'olt....38Ba **65**
Churchbank E17....28Cc **52**
....*(off Eastfield Rd.)*
Churchbury Cl. EN1: Enf....12Ub **33**
Churchbury Ho. EN8: Walt C....4Zb **20**
....*(off High St.)*
Churchbury La. EN1: Enf....13Tb **33**
Churchbury Rd. EN1: Enf....12Ub **33**
Churchbury Rd. SE9....59Mc **115**
Church Cloisters EC3....5H **225** (45Ub **91**)
....*(off Lovat La.)*
Church Cl. CM15: Mount....11Fe **41**
Church Cl. EN6: Cuff....1Nb **18**
Church Cl. GU21: Wok....8P **167**
Church Cl. GU24: Brkwd....3D **186**
Church Cl. HA6: Nwood....24V **44**
Church Cl. HA8: Edg....22Sa **47**
Church Cl. IG10: Lough....12Pc **36**
Church Cl. KT15: Add....77K **149**
Church Cl. KT17: Eps....85Ua **174**
Church Cl. KT20: Lwr K....99Bb **195**
Church Cl. KT22: Fet....96Fa **192**
Church Cl. N20....20Gb **31**
Church Cl. SL4: Eton....1H **103**
Church Cl. SS17: Horn H....1H **101**
Church Cl. TN18: Lale....69L **127**
Church Cl. TW3: Houn....54Aa **107**
Church Cl. UB4: Hayes....43T **84**
Church Cl. UB7: W Dray....48N **83**
Church Cl. UB8: Uxb....40K **63**
Church Cl. W8....47Db **89**
Church Cl. WD7: R'lett....8Ja **14**
Church Cotts. KT15: Add....76N **149**
Church Ct. EC4....3K **223** (44Qb **90**)
Church Ct. KT10: Esh....79Ea **152**
....*(off Princess Sq.)*
Church Ct. RH2: Reig....6K **207**
Church Ct. SE16....47Ac **92**
....*(off Rotherhithe St.)*
Church Ct. TW9: Rich....57Ma **109**
Church Cres. AL3: St A....1A **6**
Church Cres. CM15: Mount....11Fe **41**
Church Cres. E9....38Zb **72**
Church Cres. N10....28Kb **50**
Church Cres. N20....20Gb **31**
Church Cres. N3....25Bb **49**
Church Cres. RM15: S Ock....41Yd **98**
Church Cft. AL4: St A....4G **6**
Churchcroft Cl. SW12....59Jb **112**
Churchdown BR1: Broml....63Gc **137**
Church Dr. BR4: W W'ck....76Gc **159**
Church Dr. HA2: Harr....30Ca **45**
Church Dr. NW9....32Ta **67**
Church Elm La. RM10: Dag....37Cd **76**
Church End E17....28Dc **52**
CHURCH END....93P **189**
CHURCH END....25Bb **49**
CHURCH END....38Ua **68**
Church End NW4....27Xa **48**
CHURCH END....10H **11**
Church Entry EC4....3C **224** (44Rb **91**)
....*(off Carter La.)*
Church Est. Almshouses TW9: Rich....56Pa **109**
....*(off Sheen Rd.)*
Church Farm Cl. BR8: Crock....72Ed **162**
Church Farm La. SM3: Cheam....79Ab **154**
Church Farm Leisure Cen.....**17Hb 31**
Church Farm Way WD25: A'ham....10Da **13**
Church Fld. CM16: Epp....1Wc **23**
Church Fld. DA2: Wilm....61Md **141**
Church Fld. TN13: Riv....94Hd **202**
Church Fld. WD7: R'lett....8Ja **14**
Churchfield Av. N12....23Fb **49**
Churchfield Cl. HA2: Harr....28Ea **46**
Churchfield Cl. UB3: Hayes....45V **84**
Church Fld. Cotts. TN15: Seal....92Pd **203**
Churchfield Ct. RH2: Reig....6K **207**
Churchfield Ho. KT11: Cobh....86X **171**
....*(off Lushington Dr.)*
Churchfield Ho. KT12: Walt T....74W **150**
Churchfield Ho. W2....6B **214** (42Fb **89**)
....*(off Hall Pl.)*
Churchfield Mans. SW6....54Bb **111**
....*(off New Kings Rd.)*
Churchfield M. SL2: Slou....4L **81**
Churchfield Path EN8: Chesh....1Yb **20**
....(not continuous)
Churchfield Pl. KT13: Weyb....77Q **150**
Churchfield Pl. TW17: Shep....73R **150**
Churchfield Rd. DA16: Well....55Wc **117**
Churchfield Rd. KT12: Walt T....74W **150**
Churchfield Rd. KT13: Weyb....77Q **150**
Churchfield Rd. RH2: Reig....5H **207**
Churchfield Rd. SL9: Chal P....25A **42**
Churchfield Rd. W13....46Ka **86**
Churchfield Rd. W3....46Sa **87**
Churchfield Rd. W7....47Ga **86**
Churchfields E18....25Jc **53**
Churchfields GU21: Wok....88A **168**
Churchfields IG10: Lough....14Nc **36**
Churchfields KT8: W Mole....69Ca **129**
Churchfields SE10....51Ec **114**
Churchfields Av. KT13: Weyb....77R **150**
Churchfields Av. TW13: Hanw....62Ba **129**
Churchfields Rd. BR3: Beck....68Zb **136**
Churchfields Rd. WD24: Wat....8V **12**
Churchfield Way N12....23Eb **49**
Church Gdns. HA0: Wemb....35Ja **66**
Church Gdns. KT22: Lea....92Ka **192**
Church Gdns. W5....47Ma **87**
Church Gth. N19....33Mb **70**
....*(off St John's Gro.)*
Church Ga. SW6....55Ab **110**
Churchgate EN8: Chesh....1Xb **19**
CHURCHGATE....1Xb **19**
Churchgate Rd. EN8: Chesh....1Xb **19**
Church Grn. KT12: Hers....79Y **151**
Church Grn. KT20: Walt H....96Wa **194**
Church Grn. SW9....53Qb **112**
Church Grn. UB3: Hayes....44V **84**
Church Gro. HP6: L Chal....11A **24**
Church Gro. KT1: Hamp W....67La **130**
Church Gro. SE13....57Dc **114**
Church Gro. SL3: Wex....3N **81**
Church Hill BR6: Orp....73Wc **161**
Church Hill CM16: Epp....1Wc **23**
Church Hill CR3: Cat'm....96Vb **197**
Church Hill CR8: Purl....82Nb **176**
Church Hill DA1: Cray....56Gd **118**
Church Hill DA2: Wilm....61Md **141**
Church Hill DA9: Ghithe....57Ud **120**
Church Hill E17....28Cc **52**
Church Hill GU21: Wok....8P **167**
Church Hill GU22: Pyr....89H **169**
Church Hill HA1: Harr....32Ga **66**
Church Hill IG10: Lough....13Nc **36**
Church Hill N21....17Pb **32**
Church Hill RH1: Mers....98Kb **196**
Church Hill RH1: Nutf....5F **208**
Church Hill SE18....48Pc **94**
Church Hill SM5: Cars....78Hb **155**
Church Hill SS17: Stan H....2L **101**
Church Hill SW19....64Bb **133**
Church Hill TN14: Cud....87Sc **180**
Church Hill TN15: Plax....99Ae **205**
Church Hill TN16: Tats....94Mc **199**
Church Hill UB9: Hare....27L **43**
Church Hill WD5: Bedm....7F **4**
Church Hill Rd. E17....28Dc **52**
Church Hill Rd. EN4: E Barn....16Gb **31**
Church Hill Rd. KT6: Surb....71Na **153**
Church Hill Rd. SM3: Cheam....76Za **154**
Church Hill Wood BR5: St M Cry....71Vc **161**
Church Hollow RM19: Purf....50Qd **97**
Church Ho. EC1....5C **218** (42Rb **91**)
....*(off Compton St.)*
Church Ho. SW1....3E **228** (48Mb **90**)
....*(off Gt. Smith St.)*
Church Hyde SE18....51Uc **116**
Churchill Av. HA3: Kenton....30Ka **46**
Churchill Av. UB10: Hil....41R **84**
Churchill Bus. Pk. TN16: Westrm....98Tc **200**
CHURCHILL CAMBIAN HOSPITAL....**3A 230 (48Qb 90)**
Churchill Cl. CR6: W'ham....89Yb **178**
Churchill Cl. DA1: Dart....60Rd **119**
Churchill Cl. KT22: Fet....95Ga **192**
Churchill Cl. RH1: Redh....5P **207**
Churchill Cl. TW14: Felt....60V **106**
Churchill Cl. UB10: Hil....41R **84**
Churchill Ct. HA2: Harr Eastcote Av.....33Da **65**
Churchill Ct. HA2: Harr Montrose Ct.....29Da **45**
Churchill Ct. BR6: Farnb....78Sc **160**
Churchill Ct. HA5: Hat E....25Aa **45**
Churchill Ct. HA6: Nwood....23T **44**
Churchill Ct. N4....31Qb **70**
Churchill Ct. N9....17Bb **33**
Churchill Ct. SE18....49Pc **94**
Churchill Ct. TN16: Westrm....98Tc **200**
Churchill Ct. TW18: Staines....65K **127**
Churchill Ct. UB5: N'olt....36Ca **65**
....*(off Newmarket Av.)*
Churchill Ct. W5....42Pa **87**
Churchill Cres. AL9: Wel G....6E **8**
Churchill Dr. KT13: Weyb....77S **150**
Churchill Dr. KT16: Vir W....5L **147**
Churchill Gdns. RH8: Oxt....99Fc **199**
Churchill Gdns. SW1....50Lb **90**
....*(off Churchill Gdns. Rd.)*
Churchill Gdns. W3....44Qa **87**
Churchill Gdns. Rd. SW1....50Kb **90**
Churchill Ho. SM7: Bans....86Bb **175**
....*(off Dunnymans Rd.)*
Churchill Ct. TN14: Dun G....92Hd **202**
Churchill Lodge IG6: Ilf....28Sc **54**
Churchill Pk. DA1: Dart....57Qd **119**
Churchill Pl. E14....46Dc **92**
Churchill Pl. HA1: Harr....28Ga **46**
Churchill Rd. AL1: St A....1E **6**
Churchill Rd. CR2: S Croy....81Sb **177**
Churchill Rd. DA11: Grav'nd....10B **122**
Churchill Rd. DA4: Hort K....70Sd **142**
Churchill Rd. E16....44Lc **93**
Churchill Rd. HA8: Edg....23Pa **47**
Churchill Rd. KT19: Eps....83Qa **173**
Churchill Rd. NW2....37Xa **68**
Churchill Rd. NW5....35Kb **70**
Churchill Rd. RM17: Grays....51Fe **121**
Churchill Rd. SL3: L'ly....49B **82**
Churchill Rd. UB10: Uxb....40N **63**
Churchills M. IG8: Wfd G....23Hc **53**
Churchill Ter. E4....21Cc **52**
Churchill Theatre....**68Jc 137**
Churchill Wlk. E9....36Yb **72**
Churchill War Rooms....**1E 228 (47Mb 90)**
Churchill Way BR1: Broml....68Jc **137**
Churchill Way TN16: Big H....87Mc **179**
Churchill Way TW16: Sun....64W **128**
Church Island TW18: Staines....63F **126**
CHURCH LAMMAS....63G **126**
Churchlands Way KT4: Wor Pk....75Za **154**
Church La. CR6: W'ham Church Rd.....89Zb **178**
Church La. CR6: W'ham Ledgers Rd.....88Dc **178**
Church La. AL4: Col H....4M **7**
Church La. BR2: Broml....74Nc **160**
Church La. BR7: Chst....67Sc **138**
Church La. CM13: Gt War....30Zd **59**
Church La. CR3: Cat'm....96Qb **196**
Church La. CR5: Coul....94Jb **196**
Church La. DA12: Grav'nd....2L **145**
Church La. E11....32Gc **73**
Church La. E17....28Dc **52**
Church La. EN1: Enf....13Tb **33**
Church La. EN6: N'thaw....2Jb **18**
Church La. EN8: Chesh....1Xb **19**
Church La. GU23: Send....98D **188**
Church La. GU24: Bisl....7E **166**
Church La. GU24: Pirb....4B **186**
Church La. GU3: Worp....9J **187**
Church La. HA3: W'stone....25Ha **46**
Church La. HA5: Pinn....27Aa **45**
Church La. HP3: Bov....9D **2**
Church La. HP5: Lat....7A **10**
Church La. IG10: Lough....13Pc **36**
Church La. KT13: Weyb....77Q **150**
Church La. KT18: Head....96Sa **193**
Church La. KT18: Tad....89Za **174**
Church La. KT7: T Ditt....72Ha **152**
Church La. KT9: Chess....79Pa **153**
Church La. N17....25Ub **51**
Church La. N2....27Fb **49**
Church La. N8....28Pb **50**
Church La. N9....19Wb **33**
Church La. NW9....30Sa **47**
Church La. RH1: Blet....5K **209**
Church La. RH8: Oxt....2H **211**
Church La. RH9: G'stone....4B **210**
Church La. RM1: Rom....28Gd **56**
Church La. RM10: Dag....38Ed **76**
Church La. RM13: Wenn....44Md **97**
Church La. RM14: N Ock....36Xd **78**
Church La. RM19: Purf....50Qd **97**
Church La. RM4: Abr....12Ad **37**
Church La. RM4: Stap A....14Ed **38**
Church La. SL2: Stoke P....2K **81**
Church La. SL3: Wex....2M **81**
Church La. SL4: Wind....3H **103**
Church La. SL5: S'dale....1F **146**
Church La. SL5: S'hill....10B **124**
Church La. SL9: Chal P....25A **42**
Church La. SM6: Bedd....76Mb **156**
....(not continuous)
Church La. SW17....64Hb **133**
Church La. SW19....67Bb **133**
Church La. TN15: Kems'g....89Rd **183**
Church La. TW1: Twick....60Ja **108**
Church La. TW10: Ham....60Na **109**
Church La. TW11: Tedd....64Ha **130**
Church La. UB8: Uxb....40K **63**
Church La. W5....47La **86**
Church La. WD25: A'ham....10Da **13**
Church La. WD3: Rick....18J **25**
Church La. WD3: Sarr....10H **11**
Church La. WD4: K Lan....1Q **12**
Church La. Av. CR5: Coul....94Kb **196**
Church La. Dr. CR5: Coul....94Kb **196**
Churchley Rd. SE26....63Xb **135**
Church Manorway DA17: Belv....47Ed **96**
Church Manorway DA8: Erith....49Fd **96**
Church Manorway SE2....50Wc **95**
Church Mead SE5....52Sb **113**
....*(off Camberwell Rd.)*
Churchmead Cl. EN4: E Barn....16Gb **31**
Church Mdw. KT6: Surb....75La **152**
Churchmead Rd. NW10....37Wa **68**
Church M. KT15: Add....77L **149**
Churchmore Rd. SW16....67Lb **134**
Church Mt. N2....29Fb **49**
Church Paddock Ct. SM6: Bedd....76Mb **156**
Church Pde. TW15: Ashf....63P **127**
Church Pas. EN5: Barn Barnet....13Ab **30**
Church Pas. KT6: Surb....71Na **153**
Church Pas. TW1: Twick....60Ka **108**
Church Path SM6: Bedd Beddington....76Mb **156**
Church Path CR0: C'don Croydon....75Sb **157**
Church Path RM17: Grays Grays....51Ce **121**
Church Path DA9: Ghithe Greenhithe....57Vd **120**
Church Path N5: Highbury....36Rb **71**
Church Path N8: Hornsey....29Pb **50**
....*(off Tottenham La.)*
Church Path SW14: Mortlake....55Ta **109**
....(not continuous)
Church Path DA11: Nflt Northfleet....58Ee **121**
Church Path UB1: S'hall Southall....46Ca **85**
Church Path UB2: S'hall Southall Grn.....48Ba **85**
....(not continuous)
Church Path W3: Sth. Acton....47Sa **87**
....(not continuous)
Church Path BR8: Swan Swanley....67Kd **141**
Church Path E17: Walthamstow....28Dc **52**
Church Path NW10: Willesden....38Ua **68**
Church Path N12: Woodside La.....20Eb **31**
Church Path N12: Woodside Pk. Rd.....21Eb **49**
Church Path CR4: Mitc....69Gb **133**
....(not continuous)
Church Path CR5: Coul....90Qb **176**
Church Path E11....29Jc **53**
Church Path GU21: Wok....89B **168**
Church Path KT11: Cobh....86X **171**
Church Path N17....24Ub **51**

Church Path RH1: Mers 99Kb 196
Church Path RM1: Rom 29Gd 56
(off Market Pl.)
Church Path SL5: S'hill 9C 124
Church Path SW19 68Bb 133
Church Path W4 48Sa 87
Church Pl. CR4: Mitc 69Gb 133
Church Pl. SW1 5C 222 (45Lb 90)
Church Pl. UB10: Ick 34S 64
Church Pl. W5 47Ma 87
Church Ri. KT9: Chess 79Pa 153
Church Ri. SE23 61Zb 136
Church Rd. GU21: Wok Beech Gdns. 87A 168
Church Rd. SW19: Courthorpe Rd. 64Ab 132
Church Rd. BR2: Broml Edison Rd. 68Jc 137
Church Rd. BR2: Broml Hazelwood Ho's. 69Gc 137
Church Rd. SW19: Reynolds Cl. 67Fb 133
Church Rd. GU21: Wok St John's Hill Rd. 1L 187
Church Rd. BR2: Kes 80Mc 159
Church Rd. BR6: Chels 80Yc 161
Church Rd. BR6: Farnb 78Sc 160
Church Rd. BR8: Crock 73Fd 162
Church Rd. BR8: Swan 67Md 141
Church Rd. CM15: Mount 11Fe 41
Church Rd. CR0: C'don 76Sb 157
(not continuous)
Church Rd. CR3: Cat'm 95Vb 197
Church Rd. CR3: Whyt 90Vb 177
Church Rd. CR3: Wold 94Zb 198
Church Rd. CR4: Mitc 68Fb 133
Church Rd. CR6: W'ham 89Zb 178
Church Rd. CR8: Kenley 87Tb 177
Church Rd. CR8: Purl 82Nb 176
Church Rd. DA10: Swans 58Be 121
Church Rd. DA12: Grav'nd 5E 144
Church Rd. DA13: Cobh 6C 144
Church Rd. DA13: Ist R 6C 144
Church Rd. DA14: Sidc 63Wc 139
Church Rd. DA16: Well 54Xc 117
Church Rd. DA3: Hartl 70Be 143
Church Rd. DA3: Nw A G 70Be 143
Church Rd. DA4: Sut H 66Nd 141
Church Rd. DA7: Bex 54Bd 117
Church Rd. DA8: Erith 50Fd 96
Church Rd. DA9: Ghithe 57Vd 120
Church Rd. E10 32Cc 72
Church Rd. E12 36Nc 74
Church Rd. E17 26Ac 52
Church Rd. EN3: Pond E 16Yb 34
Church Rd. EN6: Pot B 2Cb 17
Church Rd. GU20: W'sham 9A 146
Church Rd. GU24: W End 4D 166
Church Rd. HA6: Nwood 24V 44
Church Rd. HA7: Stan 22Ka 46
Church Rd. HP3: Hem H 4C 4
Church Rd. IG10: H Beech 13Jc 35
Church Rd. IG10: Lough 13Jc 35
Church Rd. IG11: Bark 37Sc 74
Church Rd. IG2: Ilf 30Uc 54
Church Rd. IG9: Buck H 18Kc 35
Church Rd. KT1: King T 68Pa 131
Church Rd. KT10: Clay 79Ha 152
Church Rd. KT14: Byfl 86N 169
Church Rd. KT15: Add 78J 149
Church Rd. KT17: Eps 84Ua 174
Church Rd. KT19: Ewe 80Ta 153
Church Rd. KT21: Asht 90Ma 173
Church Rd. KT22: Lea 94Ka 192
Church Rd. KT23: Bookh 95Ba 191
Church Rd. KT4: Wor Pk 74Ua 154
Church Rd. KT6: Surb 74La 152
Church Rd. KT8: E Mos 70Fa 130
Church Rd. N1 37Sb 71
Church Rd. N17 25Ub 51
(not continuous)
Church Rd. N6 30Jb 50
Church Rd. NW10 38Ua 68
Church Rd. NW4 28Xa 48
Church Rd. RH1: Redh 8N 207
Church Rd. RH2: Reig 8J 207
Church Rd. RM18: Tilb 3B 122
Church Rd. RM18: W Til 1G 122
Church Rd. RM3: Hrld W 25Qd 57
Church Rd. RM4: Nave 12Md 39
Church Rd. RM4: Noak H 18Ld 39
Church Rd. SE19 67Ub 135
Church Rd. SL0: Iver H 41E 82
Church Rd. SL2: Farn R 1G 80
Church Rd. SL4: Old Win 7M 103
Church Rd. SL5: S'dale 2E 146
Church Rd. SM3: Cheam 79Ab 154
Church Rd. SM6: Bedd 76Mb 156
Church Rd. SW13 54Va 110
Church Rd. TN14: Hals 83Ad 181
Church Rd. TN14: Sund 99Ad 201
Church Rd. TN15: Ash 78Ae 165
Church Rd. TN15: Ivy H 95Ud 204
Church Rd. TN15: Seal 93Pd 203
Church Rd. TN15: Seal 95Ud 204
Church Rd. TN15: W King 80Ud 164
Church Rd. TN16: Big H 89Mc 179
Church Rd. TN16: Bras 96Xc 201
Church Rd. TW10: Ham 63Ma 131
Church Rd. TW10: Rich 57Na 109
Church Rd. TW11: Tedd 63Ga 130
Church Rd. TW13: Hanw 64Z 129
Church Rd. TW15: Ashf 62P 127
Church Rd. TW17: Shep 73R 150
Church Rd. TW20: Egh 64B 126
Church Rd. TW5: Cran 50X 85
Church Rd. TW5: Hest 52Ca 107
Church Rd. TW7: Isle 53Fa 108
Church Rd. TW9: Rich 56Na 109
Church Rd. UB2: S'hall 48Ba 85
Church Rd. UB3: Hayes 46V 84
Church Rd. UB5: N'olt 40Z 65
Church Rd. UB7: W Dray 48M 83
Church Rd. UB8: Cowl 42M 83
Church Rd. UB9: Hare 27L 43
Church Rd. W3 46Sa 87
Church Rd. W7 45Fa 86
Church Rd. WD17: Wat 11W 26
Church Rd. Almshouses E10 33Dc 72
(off Church Rd.)
Church Rd. Ind. Est. E10 32Cc 72
Church Row BR7: Chst 67Sc 138
Church Row NW3 35Eb 69
Church Row RM16: Ors 2C 100
(off Malting La.)
Church Row SW18 57Db 111
Church Row SW6 52Db 111
(off Moore Pk. Rd.)
Church Row M. BR7: Chst 66Sc 138
Church Side KT18: Eps 85Ra 173
Churchside Cl. TN16: Big H 89Lc 179
Church Sq. TW17: Shep 73R 150
Church St. SL1: Slou Damson Gro. 7G 80
Church St. SL1: Slou Osborne St. 7K 81
Church St. AL3: St A 1B 6
Church St. CR0: C'don 76Rb 157
Church St. DA11: Grav'nd 8D 122
Church St. DA13: Sflt 64Ce 143
Church St. E15 39Gc 73
Church St. E16 46Rc 94
Church St. EN2: Enf 13Sb 33
Church St. EN9: Walt A 5Ec 20
Church St. GU22: Wok 93E 188
Church St. HP2: Hem H 1M 3
Church St. HP3: Bov 9D 2
Church St. KT1: King T 68Ma 131
Church St. KT10: Esh 77Da 151
Church St. KT11: Cobh 87X 171
Church St. KT12: Walt T 74W 150
Church St. KT13: Weyb 77Q 150
Church St. KT17: Eps 85Ua 174
Church St. KT17: Ewe 81Wa 174
Church St. KT22: Lea 94Ka 192
Church St. KT24: Eff 99Z 191
Church St. N9 17Tb 33
Church St. NW8 7C 214 (43Fb 89)
Church St. RH2: Reig 6J 207
Church St. RH3: B'wth 7A 206
Church St. RM10: Dag 37Dd 76
Church St. RM17: Grays 51Ee 121
Church St. SL1: Burn 2A 80
Church St. SL4: Wind 3H 103
Church St. SM1: Sutt 78Db 155
Church St. TN14: S'ham 83Hd 182
Church St. TN15: Seal 93Qd 203
Church St. TW1: Twick 60Ja 108
Church St. TW12: Hamp 67Ea 130
Church St. TW16: Sun 69X 129
Church St. TW18: Staines 63F 126
Church St. TW7: Isle 55Ka 108
Church St. W2 7C 214 (43Fb 89)
(not continuous)
Church St. W4 51Va 110
Church St. WD18: Wat 14Y 27
Church St. WD3: Rick 18N 25
Church St. E. GU21: Wok 89B 168
Church St. Est. NW8 6C 214 (42Fb 89)
Church St. Nth. E15 39Gc 73
Church St. Pas. E15 39Gc 73
(off Church St.)
Church Street Stop (London Tramlink) 75Sb 157
Church St. W. GU21: Wok 89A 168
Church Stretton Rd. TW3: Houn 57Ea 108
Church Ter. NW4 27Xa 48
Church Ter. RM4: Stap A 14Ed 38
Church Ter. SE13 55Gc 115
Church Ter. SL4: Wind 4C 102
Church Ter. TW10: Rich 57Ma 109
CHURCH TOWN 4B 210
Church Trad. Est. DA8: Erith 52Jd 118
Church Va. N2 27Hb 49
Church Va. SE23 61Zb 136
Church Vw. BR8: Swan 69Fd 140
Church Vw. RM14: Upm 33Rd 77
Church Vw. RM15: Avel 47Sd 98
Church Vw. TW10: Rich 57Na 109
Churchview Cl. CR3: Cat'm 96Wb 197
Churchview Ct. DA12: Grav'nd 1G 144
Church Vw. Gro. SE26 65Zb 136
Churchview Rd. TW2: Twick 60Fa 108
Church Vs. TN13: Riv 94Gd 202
Church Wlk. SW13: Barnes 53Wa 110
Church Wlk. N6: Dartmouth Pk. 34Jb 70
Church Wlk. EN2: Enf Enfield 13Tb 33
Church Wlk. NW4: Hendon 27Ya 48
Church Wlk. SW15: Putney 57Xa 110
Church Wlk. TW9: Rich Richmond 57Ma 109
Church Wlk. NW9: Wembley 33Ta 67
Church Wlk. CM15: B'wood 17Xd 40
Church Wlk. CR3: Cat'm 96Wb 197
Church Wlk. DA12: Grav'nd 10F 122
Church Wlk. DA2: Wilm 62Md 141
Church Wlk. DA4: Eyns 76Nd 163
Church Wlk. KT12: Walt T 74W 150
(not continuous)
Church Wlk. KT13: Weyb 76Q 150
Church Wlk. KT16: Chert 72J 149
Church Wlk. KT22: Lea 94Ka 192
Church Wlk. KT7: T Ditt 72Ha 152
Church Wlk. N16 34Tb 71
(not continuous)
Church Wlk. NW2 34Bb 69
Church Wlk. RH1: Blet 5K 209
(not continuous)
Church Wlk. RH2: Reig 6K 207
(not continuous)
Church Wlk. SL1: Burn 2A 80
(not continuous)
Church Wlk. SW16 68Lb 134
Church Wlk. SW20 69Ya 132
Church Wlk. TW8: Bford 51La 108
(not continuous)
Church Wlk. UB3: Hayes 44U 84
Church Wlk. WD23: Bush 16Ca 27
Churchward Ho. SE17 51Rb 113
(off Lorrimore Sq.)
Churchward Ho. W14 50Bb 89
(off Ivatt Pl.)
Church Way CR2: Sande 82Vb 177
Church Way EN4: Cockf 14Hb 31
Church Way HA8: Edg 23Qa 47
Church Way N20 20Gb 31
Church Way RH8: Oxt 4K 211
Churchway NW1 3E 216 (41Mb 90)
(not continuous)
Churchwell Path E9 36Yb 72
Churchwood Gdns. IG8: Wfd G 21Jc 53
Churchwood Gdns. SE23 59Yb 114
Church Wood Nature Reserve 2K 61
Churchyard Pas. SE5 54Tb 113
Churchyard Row SE11 5C 230 (49Rb 91)
Churston Av. E13 39Kc 73
Churston Cl. SW2 60Qb 112
Churston Dr. SM4: Mord 71Za 154
Churston Gdns. N11 23Lb 50
Churston Mans. WC1 6J 217 (42Pb 90)
(off Gray's Inn Rd.)
Churton Pl. SW1 6C 228 (49Lb 90)
Churton Pl. W4 51Ra 109
(off Chiswick Village)
Churton St. SW1 6C 228 (49Lb 90)
Chusan Pl. E14 44Bc 92
Chute Ho. SW9 54Qb 112
(off Stockwell Pk. Rd.)
Chuter Ede Ho. SW6 51Bb 111
(off Clem Attlee Ct.)
Chuters Cl. KT14: Byfl 84N 169
Chuters Gro. KT17: Eps 84Va 174
Chyne, The SL9: Ger X 29B 42
Chyngton Cl. DA15: Sidc 62Vc 139
Chynham Pl. CR2: Sande 82Ub 177
Cibber Rd. SE23 61Zb 136
Cicada Rd. SW18 58Eb 111
Cicely Ct. CR0: Wadd 78Qb 156
Cicely Ho. NW8 2C 214 (40Fb 69)
(off Cochrane St.)
Cicely Rd. SE15 53Wb 113
Cimba Wood DA12: Grav'nd 3G 144
Cinderella Path NW11 32Db 69
Cinderford Way BR1: Broml 63Gc 137
Cinder Path GU22: Wok 1N 187
Ciné Lumière 5B 226 (49Fb 89)
(off Queensberry Pl.)
Cineworld Cinema Bexleyheath 56Dd 118
Cineworld Cinema Chelsea, Fulham Rd. 50Eb 89
Cineworld Cinema Enfield 14Wb 33
Cineworld Cinema Feltham 61X 129
Cineworld Cinema Hemel Hempstead 3P 3
Cineworld Cinema Ilford 34Rc 74
(off Clements Rd.)
Cineworld Cinema Leicester Sq. 4E 222 (45Mb 90)
(off Leicester Sq.)
Cineworld Cinema South Ruislip 35X 65
Cineworld Cinema O2, The 46Gc 93
(within The O2)
Cineworld Cinema Wandsworth 57Db 111
Cineworld Cinema Wembley 35Qa 67
Cineworld Cinema West India Quay 45Cc 92
Cineworld Cinema Wood Green 26Qb 50
(within Wood Green Shop. City)
Cinnabar Wharf Central E1 46Wb 91
(off Wapping High St.)
Cinnabar Wharf E. E1 46Wb 91
(off Wapping High St.)
Cinnabar Wharf W. E1 46Wb 91
(off Wapping High St.)
Cinnamon Cl. CR0: C'don 73Nb 156
Cinnamon Cl. SE15 52Vb 113
Cinnamon Cl. SL4: Wind 3D 102
Cinnamon M. N13 19Qb 32
Cinnamon Row SW11 55Eb 111
Cinnamon St. E1 46Xb 91
Cinnamon Wharf SE1 47Vb 91
(off Shad Thames)
Cintra Pk. SE19 66Vb 135
Cipher Ct. NW2 34Wa 68
CIPPENHAM 5C 80
Cippenham Cl. SL1: Slou 5D 80
Cippenham La. SL1: Slou 5D 80
Circa Apts. NW1 38Jb 70
Circle, The NW2 34Ua 68
Circle, The NW7 23Ta 47
Circle, The RM18: Tilb 3C 122
Circle, The SE1 1K 231 (47Vb 91)
(off Queen Elizabeth St.)
Circle Gdns. KT14: Byfl 85P 169
Circle Gdns. SW19 68Cb 133
Circle Rd. KT12: W Vill 81U 170
Circuit Cen., The KT13: Weyb 83N 169
Circuits, The HA5: Pinn 28Y 45
Circular Rd. N17 27Vb 51
Circular Way SE18 51Pc 116
Circus, The KT22: Lea 92Ka 192
(off By-Pass Rd.)
Circus Lodge NW8 3B 214 (41Fb 89)
(off Circus Rd.)
Circus M. W1 7F 215 (43Hb 89)
(off Enford St.)
Circus Pl. EC2 1G 225 (43Tb 91)
Circus Rd. NW8 3B 214 (41Fb 89)
Circus Rd. E. SW11 52Kb 112
Circus Rd. W. SW11 52Kb 112
Circus St. SE10 52Ec 114
Circus W. SW11 51Kb 112
Cirencester St. W2 43Db 89
Cirrus Apts. E1 5K 219 (42Vb 91)
(off Bacon St.)
Cirrus Cl. SM6: W'gton 80Nb 156
Cirrus Cres. DA12: Grav'nd 4G 144
Cissbury Ho. SE26 62Wb 135
Cissbury Ring Nth. N12 22Bb 49
Cissbury Ring Sth. N12 22Bb 49
Cissbury Rd. N15 29Tb 51
Citadel Pl. SE11 7H 229 (50Pb 90)
Citius Apts. E3 40Cc 72
(off Tredegar La.)
Citius Ct. E4 23Ec 52
Citius Wlk. E20 37Ec 72
Citizen Ho. N7 35Qb 70
Citizen Rd. N7 35Qb 70
Citrine Apts. E3 39Ac 72
(off Gunmaker's La.)
Citrus Ho. SE8 50Bc 92
(off Alverton St.)
CITY AIRPORT 46Nc 94
CITY & HACKNEY CENTRE FOR MENTAL HEALTH 36Zb 72
City Apts. E1 44Wb 91
(off White Church La.)
City Bus. Cen. SE16 48Yb 92
City Bus. Library 1E 224 (43Sb 91)
City Ct. CR0: C'don 73Rb 157
City Cross Bus. Pk. SE10 49Gc 93
City E. Bldg. E1 45Xb 91
(off Cable St.)
City Forum EC1 3D 218 (41Sb 91)
City Gdn. Row EC1 3D 218 (41Sb 91)
City Gdn. Row N1 2C 218 (40Rb 71)
City Ga. Ho. IG2: Ilf 30Qc 54
(off Eastern Av.)
City Gateway E1 45Wb 91
(off Ensign St.)
City Hall Southwark 7J 225 (46Ub 91)
City Harbour E14 48Dc 92
(off Selsdon Way)
City Hgts. E8 39Vb 71
(off Kingsland Rd.)
City Ho. BR2: Broml 72Nc 160
City Island E14 44Gc 93
City Island Way E14 44Gc 93
City Lights Ct. SE11 7A 230 (50Qb 90)
(off Bowden St.)
City Mill Apts. E8 39Vb 71
(off Lovelace St.)
City Mill River Path E15 39Ec 72
City Nth. E. Twr. N4 33Qb 70
(off City Nth. Pl.)
City Nth. Pl. N4 33Qb 70
City Nth. W. Twr. N4 33Qb 70
CITY OF LONDON 2G 225 (44Tb 91)
City of London Almshouses SW9 56Pb 112
City of London Crematorium 34Nc 74
City of London Distillery 3B 224 (44Rb 91)
City of London Point N7 37Mb 70
(off York Way)
City of London Police Mus. 2E 224 (44Sb 91)
City of London Tourist Info. Cen. 3D 224 (44Sb 91)
City Pav. EC1 7B 218 (43Rb 91)
(off Britton St.)
City Pavilion Marks Gate, The 26Bd 55
City Pl. Ho. EC2 1F 225 (43Tb 91)
(off Basinghall St.)
Citypoint EC2 7F 219 (43Tb 91)
(off Ropemaker St.)
City Pride Development E14 46Cc 92
City Rd. EC1 2B 218 (40Rb 71)
City Thameslink Station (Rail) 2B 224 (44Rb 91)
City Twr. EC2 1F 225 (43Tb 91)
(off Basinghall St.)
City Twr. SW8 51Nb 112
City University London Goswell Pl. 4C 218 (41Rb 91)
City University London Northampton Square Campus 4B 218 (41Rb 91)
City University London Saddlers Sports Cen., The 5C 218 (42Rb 91)
City Vw. IG1: Ilf 33Sc 74
(off Axon Pl.)
Cityview SE7 51Lc 115
City Vw. Apts. N1 38Sb 71
(off Essex Rd.)
City Vw. Apts. N4 31Tb 71
(off Devan Gro.)
City Vw. Ct. SE22 59Wb 113
City Wlk. SE1 2H 231 (47Ub 91)
City Wlk. Apts. EC1 4D 218 (41Sb 91)
(off Seward St.)
City Wharf Ho. KT7: T Ditt 72Ka 152
Civic Cl. AL1: St A 2B 6
Civic Sq. RM18: Tilb 4C 122
Civic Way HA4: Ruis 36Z 65
Civic Way IG6: Ilf 28Sc 54
Civil & Family Court Barnet 25Cb 49
Civil Justice Cen. Central London 6K 215 (42Kb 90)
(off Park Cres.)
Clabon M. SW1 4F 227 (48Hb 89)
Clacket La. TN16: Westrm 96Nc 200
CLACKET LANE SERVICE AREA 98Nc 200
Clack La. HA4: Ruis 32S 64
Clack St. SE16 47Yb 92
Clacton Rd. E17 30Ac 52
Clacton Rd. E6 41Mc 93
Clacton Rd. N17 26Vb 51
Claigmar Gdns. N3 25Db 49
Claire C'way. DA2: Dart 56Ud 120
Claire Cl. CM13: B'wood 21Be 59
Claire Ct. EN8: Chesh 4Zb 20
Claire Ct. HA5: Hat E 24Ba 45
Claire Ct. N12 20Eb 31
Claire Ct. NW2 37Ab 68
Claire Ct. WD23: B Hea 18Fa 28
Claire Gdns. HA7: Stan 22La 46
Claire Ho. IG1: Ilf 35Rc 74
Claire Pl. E14 48Cc 92
Claireville Ct. RH2: Reig 6M 207
Clairvale RM11: Horn 31Nd 77
Clairvale Rd. TW5: Hest 53Z 107
Clairview Rd. SW16 64Kb 134
Clairville Gdns. W7 46Ga 86
Clairville Point SE23 62Zb 136
(off Dacres Rd.)
Clammas Way UB8: Cowl 43L 83
Clamp Hill HA7: Stan 21Fa 46
Clancarty Rd. SW6 54Cb 111
Clandon Av. TW20: Egh 66E 126
Clandon Cl. KT17: Ewe 79Va 154
Clandon Cl. W3 47Ra 87
Clandon Gdns. N3 27Cb 49
Clandon Ho. SE1 2C 230 (47Rb 91)
(off Webber St.)
Clandon Rd. GU23: Send 97H 189
Clandon Rd. GU4: W Cla 98J 189
Clandon Rd. IG3: Ilf 33Uc 74
Clandon St. SE8 54Cc 114
Clandon Ter. SW20 68Za 132
Clanricarde Gdns. W2 45Cb 89
Clapgate Rd. WD23: Bush 16Da 27
CLAPHAM 56Lb 112
CLAPHAM COMMON 56Mb 112
Clapham Comn. Northside SW4 56Lb 112
(off Rookery Rd.)
Clapham Comn. Nth. Side SW4 56Hb 111
Clapham Comn. Sth. Side SW4 58Kb 112
Clapham Common Station (Underground) 56Lb 112
Clapham Comn. W. Side SW4 56Hb 111
(not continuous)
Clapham Cres. SW4 56Mb 112
Clapham High St. SW4 56Mb 112
Clapham High Street Station (Overground) 55Mb 112
CLAPHAM JUNCTION 55Gb 111
Clapham Junction Station (Rail & Overground) 55Gb 111
Clapham Leisure Cen. 55Mb 112
Clapham Mnr. Ct. SW4 55Lb 112
Clapham Mnr. St. SW4 55Lb 112
Clapham North Station (Underground) 55Nb 112
CLAPHAM PARK 57Mb 112
Clapham Pk. Est. SW4 58Mb 112
Clapham Pk. Rd. SW4 56Lb 112
Clapham Pk. Ter. SW2 57Nb 112
(off Lyham Rd.)
Clapham Picturehouse 56Lb 112
Clapham Rd. SW9 55Nb 112
Clapham Rd. Est. SW4 55Nb 112
Clapham South Station (Underground) 58Kb 112
Clap La. RM10: Dag 33Dd 76
Clap La. RM10: Rush G 33Dd 76
Clapperknapper DA10: Swans 60Ae 121
Clappers La. GU24: Chob 3G 166
Claps Ga. La. E6 42Qc 94
Clapton Comn. E5 31Vb 71
(not continuous)
CLAPTON PARK 35Zb 72
Clapton Pk. Est. E5 35Zb 72
Clapton Pas. E5 36Yb 72
Clapton Sq. E5 36Yb 72
Clapton Station (Overground) 33Xb 71
Clapton Ter. E5 32Wb 71
Clapton Way E5 35Wb 71
Clara Grant Ho. E14 48Cc 92
(off Mellish St.)
Clara Nehab Ho. NW11 29Bb 49
(off Leeside Cres.)
Clara Pl. SE18 49Qc 94
Clare Cl. KT14: W Byf 85J 169
Clare Cl. N2 27Eb 49
Clare Cl. WD6: E'tree 16Pa 29
Clare Cnr. SE9 59Rc 116
Clare Cotts. RH1: Blet 5H 209
Clare Ct. AL1: St A 3D 6
Clare Ct. CR3: Wold 95Cc 198
Clare Ct. EN3: Enf W 7Ac 20
Clare Ct. HA6: Nwood 22U 44
Clare Ct. RM15: Avel 47Sd 98
Clare Ct. W11 45Ab 88
(off Clarendon Rd.)
Clare Ct. WC1 4G 217 (41Nb 90)
(off Judd St.)
Clare Cres. KT22: Lea 90Ja 172
Claredale GU22: Wok 91A 188
Claredale Ho. E2 40Xb 71
(off Claredale St.)
Claredale St. E2 40Wb 71
Clare Dr. SL2: Farn C 5F 60
Clarefield Ct. SL5: S'dale 3E 146
Clare Gdns. E7 35Jc 73
Clare Gdns. IG11: Bark 37Vc 75
Clare Gdns. TW20: Egh 64C 126
Clare Gdns. W11 44Ab 88
Claregate EN6: Pot B 2Eb 17
Clare Hill KT10: Esh 78Da 151
Clare Ho. E16 45Qc 94
(off University Way)
Clare Ho. E3 39Bc 72
Clare Ho. HA8: Edg 26Sa 47
(off Burnt Oak B'way.)
Clare Ho. SE1 7K 231 (50Vb 91)
(off Cooper's Rd.)
Clare La. N1 38Sb 71
Clare Lawn Av. SW14 57Ta 109
Clare Mkt. WC2 3J 223 (44Pb 90)
Clare M. SW6 52Db 111
Claremont AL2: Brick W 3Ca 13
Claremont EN7: Chesh 1Vb 19
Claremont TW17: Shep 72R 150
(off Laleham Rd.)
Claremont Av. GU22: Wok 91A 188
Claremont Av. HA3: Kenton 29Na 47
Claremont Av. KT10: Esh 79Ba 151
Claremont Av. KT12: Hers 77Z 151
Claremont Av. KT3: N Mald 71Wa 154
Claremont Av. TW16: Sun 67X 129
Claremont Cl. BR6: Farnb 77Qc 160
Claremont Cl. CR2: Sande 87Xb 177
Claremont Cl. E16 46Qc 94
Claremont Cl. KT12: Hers 78Y 151
Claremont Cl. N1 2A 218 (40Qb 70)
Claremont Cl. RM16: Grays 48Ee 99
Claremont Cl. SW2 60Nb 112
Claremont Ct. E2: Cambridge Heath Rd. 40Xb 71
(off Cambridge Heath Rd.)
Claremont Ct. E2: Claredale St. 40Wb 71
(off Claredale St.)
Claremont Ct. W2 44Db 89
(off Queensway)
Claremont Ct. W9 40Bb 69
(off Claremont Rd.)
Claremont Cres. DA1: Cray 56Gd 118
Claremont Cres. WD3: Crox G 15S 26
Claremont Dr. GU22: Wok 91A 188
Claremont Dr. KT10: Esh 81Ca 171
Claremont Dr. TW17: Shep 72R 150
Claremont End KT10: Esh 79Da 151
Claremont Gdns. IG3: Ilf 33Uc 74
Claremont Gdns. KT6: Surb 71Na 153
Claremont Gdns. RM14: Upm 32Td 78
Claremont Gro. IG8: Wfd G 23Lc 53
Claremont Gro. W4 52Ua 110
Claremont Ho. NW9 27Va 48
Claremont Ho. SE16 47Ac 92
Claremont Ho. SM2: Sutt 80Db 155
Claremont Ho. WD18: Wat 16T 26
Claremont Landscape Gdn. 80Ba 151
Claremont La. KT10: Esh 78Da 151
Claremont M. DA1: Dart 55Qd 119
CLAREMONT PARK 80Da 151
Claremont Pk. N3 25Ab 48
Claremont Pk. Rd. KT10: Esh 79Da 151
Claremont Pl. DA11: Grav'nd 9D 122
(off Arthur St.)
Claremont Pl. IG7: Chig 20Rc 36
Claremont Pl. KT10: Clay 79Ha 152
Claremont Rd. BR1: Broml 70Nc 138
Claremont Rd. BR8: Hext 66Gd 140
Claremont Rd. CR0: C'don 74Wb 157
Claremont Rd. E11 34Fc 73
Claremont Rd. E17 26Ac 52
Claremont Rd. E7 36Kc 73
Claremont Rd. EN4: Had W 10Eb 17
Claremont Rd. HA3: W'stone 26Ga 46
Claremont Rd. KT10: Clay 80Ga 152
Claremont Rd. KT14: W Byf 84J 169
Claremont Rd. KT6: Surb 71Na 153
Claremont Rd. N6 31Lb 70
Claremont Rd. NW2 31Za 68
Claremont Rd. RH1: Redh 3A 208
Claremont Rd. RM11: Horn 30Jd 56
Claremont Rd. SL4: Wind 4G 102
Claremont Rd. TW1: Twick 58Ka 108
Claremont Rd. TW11: Tedd 64Ha 130
Claremont Rd. TW18: Staines 64F 126
Claremont Rd. W13 43Ja 86
Claremont Rd. W9 40Ab 68
Claremont Sq. N1 2K 217 (40Qb 70)
Claremont St. E16 47Qc 94
Claremont St. N18 23Wb 51
Claremont St. SE10 51Dc 114
Claremont Ter. KT7: T Ditt 73Ka 152
Claremont Vs. SE5 52Tb 113
(off Southampton Way)
Claremont Way NW2 32Ya 68
(not continuous)
Claremont Way Ind. Est. NW2 32Ya 68
Claremount Cl. KT18: Tatt C 89Ya 174
Claremount Gdns. KT18: Tatt C 89Ya 174
Clarence Av. BR1: Broml 70Nc 138
Clarence Av. IG2: Ilf 30Qc 54
Clarence Av. KT3: N Mald 68Sa 131
Clarence Av. RM14: Upm 33Qd 77
Clarence Av. SW4 59Mb 112
Clarence Cl. EN4: E Barn 15Fb 31
Clarence Cl. KT12: Hers 77X 151
Clarence Cl. WD23: B Hea 17Ha 28
Clarence Ct. NW7 22Ua 48
Clarence Ct. RM17: Grays 51De 121
(off Clarence Rd.)
Clarence Ct. SL3: L'ly 51D 104
Clarence Ct. SL4: Wind 3F 102
Clarence Ct. TW20: Egh 64B 126
(off Clarence St.)
Clarence Ct. W6 49Xa 88
(off Cambridge Gro.)
Clarence Cres. DA14: Sidc 62Xc 139
Clarence Cres. SL4: Wind 3G 102
Clarence Cres. SW4 58Mb 112
Clarence Dr. TW20: Eng G 3N 125
Clarence Gdns. NW1 4A 216 (41Kb 90)
Clarence Ga. IG8: Ilf 23Qc 54
Clarence Ga. IG8: Wfd G 23Qc 54
Clarence Ga. Gdns. NW1 6G 215 (42Hb 89)
(off Glentworth St.)

Clarence Ho. KT12: Hers 78X 151
(off Queens Rd.)
Clarence Ho. SE17 51Tb 113
(off Merrow St.)
Clarence House 1C 228 (47Lb 90)
(off St James's Pal.)
Clarence La. SW15 58Ua 110
Clarence M. E5 36Xb 71
Clarence M. SE16 46Zb 92
Clarence M. SW12 59Kb 112
Clarence Pk. 2D 6
Clarence Pk. Cres. HA7: Stan 20Ga 28
Clarence Pk. M. AL1: St A 2D 6
Clarence Pl. DA12: Grav'nd 9D 122
Clarence Pl. E5 36Xb 71
Clarence Pl. KT17: Ewe 82Wa 174
Clarence Rd. AL1: St A 2D 6
Clarence Rd. BR1: Broml 69Mc 137
Clarence Rd. CM15: Pil H 16Xd 40
Clarence Rd. CR0: C'don 73Tb 157
Clarence Rd. DA14: Sidc 62Xc 139
Clarence Rd. DA6: Bex 56Ad 117
Clarence Rd. E12 35Lc 73
Clarence Rd. E16 42Gc 93
Clarence Rd. E17 26Zb 52
Clarence Rd. E5 35Xb 71
Clarence Rd. EN3: Pond E 15Xb 33
Clarence Rd. KT12: Hers 77X 151
Clarence Rd. N15 29Sb 51
Clarence Rd. N22 24Nb 50
Clarence Rd. NW6 38Bb 69
Clarence Rd. RH1: Redh 9M 207
Clarence Rd. RM17: Grays 51Ce 121
Clarence Rd. SE8 51Dc 114
Clarence Rd. SE9 61Nc 138
Clarence Rd. SL4: Wind 3E 102
Clarence Rd. SM1: Sutt 78Db 155
Clarence Rd. SM6: W'gton 78Kb 156
Clarence Rd. SW19 65Db 133
Clarence Rd. TN16: Big H 90Pc 180
Clarence Rd. TW11: Tedd 65Ha 130
Clarence Rd. TW9: Kew 53Pa 109
Clarence Rd. W4 50Qa 87
Clarence Row DA12: Grav'nd 9D 122
Clarence St. KT1: King T 68Ma 131
Clarence St. TW18: Staines 63G 126
Clarence St. TW20: Egh 65B 126
Clarence St. TW9: Rich 56Na 109
Clarence St. UB2: S'hall 48Z 85
Clarence Ter. NW1 5G 215 (42Hb 89)
Clarence Ter. TW3: Houn 56Da 107
Clarence Wlk. RH1: Redh 9M 207
Clarence Wlk. SW4 54Nb 112
Clarence Way NW1 38Kb 70
Clarence Way RM15: S Ock 44Zd 99
Clarenden Pl. DA2: Wilm 64Gd 140
Clarendon Cl. BR5: St P 69Wc 139
Clarendon Cl. E9 38Yb 72
Clarendon Cl. HP2: Hem H 1M 3
Clarendon Cl. W2 4D 220 (45Gb 89)
Clarendon Ct. BR3: Beck 67Dc 136
(off Albemarle Rd.)
Clarendon Ct. EC1 6E 218 (42Sb 91)
(off Brackley St.)
Clarendon Ct. HA6: Nwood 22V 44
Clarendon Ct. NW11 28Bb 49
Clarendon Ct. NW2 38Ya 68
Clarendon Ct. SL2: Slou 5M 81
Clarendon Ct. SL4: Wind 3F 102
Clarendon Ct. TW5: Cran 53W 106
Clarendon Ct. TW9: Kew 53Pa 109
Clarendon Ct. W9 5A 214 (42Eb 89)
(off Maida Va.)
Clarendon Cres. TW2: Twick 62Fa 130
Clarendon Cross W11 45Ab 88
Clarendon Dr. SW15 56Ya 110
Clarendon Flds. WD3: Chan C 10P 11
Clarendon Flats W1 3J 221 (44Jb 90)
(off Balderton St.)
Clarendon Gdns. DA2: Dart 59Td 120
Clarendon Gdns. HA9: Wemb 34Ma 67
Clarendon Gdns. IG1: Ilf 31Pc 74
Clarendon Gdns. NW4 27Wa 48
Clarendon Gdns. W9 6A 214 (42Eb 89)
Clarendon Ga. KT16: Ott 79F 148
Clarendon Grn. BR5: St P 70Wc 139
Clarendon Gro. BR5: St P 70Wc 139
Clarendon Gro. CR4: Mitc 69Hb 133
Clarendon Gro. NW1 3D 216 (41Mb 90)
Clarendon Ho. KT2: King T 67Na 131
(off Cowleaze Rd.)
Clarendon Ho. NW1 2C 216 (40Lb 70)
(off Werrington St.)
Clarendon Ho. W2 4D 220 (45Gb 89)
(off Strathearn Pl.)
Clarendon Lodge W11 45Ab 88
(off Clarendon Rd.)
Clarendon Lofts WD17: Wat 13X 27
(off Clarendon Rd.)
Clarendon M. DA5: Bexl 60Dd 118
Clarendon M. KT21: Asht 91Pa 193
(off Rectory La.)
Clarendon M. W2 4D 220 (45Gb 89)
Clarendon M. WD6: Bore 13Qa 29
Clarendon Pde. EN8: Chesh 1Zb 20
Clarendon Path BR5: St P 70Wc 139
(not continuous)
Clarendon Pl. TN13: S'oaks 97Jd 202
Clarendon Pl. W2 4D 220 (45Gb 89)
Clarendon Ri. SE13 56Ec 114
Clarendon Rd. CR0: C'don 75Rb 157
Clarendon Rd. DA12: Grav'nd 8E 122
Clarendon Rd. E11 32Fc 73
Clarendon Rd. E17 30Dc 52
Clarendon Rd. E18 27Jc 53
Clarendon Rd. EN8: Chesh 1Zb 20
Clarendon Rd. HA1: Harr 30Ga 46
Clarendon Rd. N15 28Sb 51
Clarendon Rd. N18 23Wb 51
Clarendon Rd. N22 26Pb 50
Clarendon Rd. N8 27Pb 50
Clarendon Rd. RH1: Redh 5P 207
Clarendon Rd. SM6: W'gton 79Lb 156
Clarendon Rd. SW19 66Gb 133
Clarendon Rd. TN13: S'oaks 96Jd 202
Clarendon Rd. TW15: Ashf 63P 127
Clarendon Rd. UB3: Hayes 47V 84
Clarendon Rd. W11 45Ab 88
Clarendon Rd. W5 41Na 87
Clarendon Rd. WD17: Wat 12X 27
Clarendon Rd. WD6: Bore 13Qa 29
Clarendon St. SW1 7A 228 (50Kb 90)
Clarendon Ter. W9 5A 214 (42Eb 89)
Clarendon Wlk. W11 44Ab 88
Clarendon Way BR5: St P 69Wc 139
Clarendon Way BR7: Chst 69Vc 139
Clarendon Way N21 16Sb 33
Clarens St. SE6 61Bc 114
Clare Pl. SW15 59Va 110
Clare Point NW2 32Za 68
(off Whitefield Av.)
Clare Rd. E11 30Fc 53
Clare Rd. NW10 38Wa 68
Clare Rd. SE14 53Bc 114
Clare Rd. SL6: Tap 4A 80
Clare Rd. TW19: Stanw 60M 105
Clare Rd. TW4: Houn 55Ba 107
Clare Rd. UB6: G'frd 37Fa 66
Clares, The CR3: Cat'm 96Wb 197
Clare St. E2 40Xb 71
Claret Gdns. SE25 69Ub 135
Claret Ho. WD17: Wat 12X 27
Clareville Ct. SW7 6A 226 (49Eb 89)
(off Clareville St.)
Clareville Gro. SW7 6A 226 (49Eb 89)
Clareville Gro. M. SW7 6A 226 (49Eb 89)
Clareville Rd. BR5: Farnb 75Sc 160
Clareville Rd. CR3: Cat'm 96Wb 197
Clareville St. SW7 6A 226 (49Eb 89)
Clare Way DA7: Bex 53Ad 117
Clare Way TN13: S'oaks 100Ld 203
Clare Wood KT22: Lea 90Ka 172
Clarewood Ct. W1 1F 221 (43Hb 89)
(off Seymour Pl.)
Clarewood Wlk. SW9 56Qb 112
Clarges M. W1 6K 221 (46Kb 90)
Clarges St. W1 6A 222 (46Kb 90)
Claribel Rd. SW9 54Rb 113
Clarice Way SM6: W'gton 81Nb 176
Claridge Ct. SW6 54Bb 111
Claridge Rd. RM8: Dag 32Zc 75
Clarinda Ho. DA9: Ghithe 56Yd 120
(off Clovelly Pl.)
Clarinet Ct. HA8: Edg 24Ra 47
Clarion Ho. E3 40Ac 72
(off Roman Rd.)
Clarion Ho. SW1 7C 228 (50Lb 90)
(off Moreton Pl.)
Clarion Ho. W1 3D 222 (44Mb 90)
(off St Anne's Ct.)
Clarissa Ho. E14 44Dc 92
(off Cordela St.)
Clarissa Rd. RM6: Chad H 31Zc 75
Clarissa St. E8 39Vb 71
Clark Cl. DA8: Erith 53Jd 118
Clarke Apts. E3 42Bc 92
(off Heath Pl.)
Clarkebourne Dr. RM17: Grays 51Fe 121
Clarke Cl. CR0: C'don 72Sb 157
Clarke Grn. WD25: Wat 7W 12
Clarke Mans. IG11: Bark 38Vc 75
(off Upney La.)
Clarke M. N9 20Xb 33
Clarke Path N16 32Wb 71
Clarkes Av. KT4: Wor Pk 74Za 154
Clarkes Dr. UB8: Hil 43N 83
Clarke's Grn. La. TN15: Knat 86Qd 183
Clarke's Grn. Rd. TN15: Knat 86Qd 183
Clarke's M. W1 7J 215 (43Jb 90)
Clarke Way WD25: Wat 7W 12
Clarkfield WD3: Rick 18K 25
Clark Gro. IG3: Ilf 35Uc 74
Clark Ho. SW10 52Eb 111
(off Coleridge Gdns.)
Clarks La. CM16: Epp 3Vc 23
Clarks La. RH8: T'sey 95Jc 199
Clarks La. TN14: Hals 84Bd 181
Clarks La. TN16: Tats 95Mc 199
Clarks La. TN16: Westrm 95Mc 199
Clarks Mead WD23: Bush 17Ea 28
Clarks M. CM16: Epp 2Wc 23
Clarkson Rd. E16 44Hc 93
Clarkson Row NW1 2B 216 (40Lb 70)
(off Mornington Ter.)
Clarksons, The IG11: Bark 40Sc 74
Clarkson St. E2 41Xb 91
Clarks Rd. IG1: Ilf 33Tc 74
Clark St. E1 43Xb 91
(not continuous)
Clark Way TW5: Hest 52Z 107
Clarnico Rd. E20 37Cc 72
Clarson Ho. SE5 52Rb 113
(off Midnight Av.)
Clarson Ho. SE8 51Ac 114
Classic Mans. E9 38Xb 71
(off Wells St.)
Classon Cl. UB7: W Dray 47N 83
Claston Cl. DA1: Cray 56Gd 118
Claude Av. NW9 25Va 48
Claude Rd. E10 33Ec 72
Claude Rd. E13 39Kc 73
Claude Rd. SE15 54Xb 113
Claude St. E14 49Cc 92
Claudia Jones Ho. N17 25Sb 51
Claudia Jones Way SW2 58Nb 112
Claudian Pl. AL3: St A 3N 5
Claudian Way RM16: Grays 8D 100
Claudia Pl. SW19 60Ab 110
Claudius Cl. HA7: Stan 20Ma 29
Claudius Ct. AL4: St A 3G 6
Claughton Rd. E13 40Lc 73
Claughton Way CM13: Hut 16Fe 41
Clauson Av. UB5: N'olt 36Da 65
Clavell St. SE10 51Ec 114
Claverdale Rd. SW2 59Pb 112
Claver Dr. SL5: S'hill 10B 124
Claverhambury Rd. EN9: Walt A 2Hc 21
Clavering Av. SW13 51Xa 110
Clavering Cl. TW1: Twick 63Ja 130
Clavering Gdns. CM13: W H'dn 30Fe 59
Clavering Ho. SE13 56Fc 115
(off Blessington Rd.)
Clavering Pl. SW12 58Jb 112
Clavering Rd. E12 32Mc 73
Claverings Ind. Est. N9 19Yb 34
(off Centre Way)
Clavering Way CM13: Hut 16Ee 41
Claverley Gro. N3 25Db 49
Claverley Vs. N3 24Db 49
Claverton KT21: Asht 89Na 173
Claverton Cl. HP3: Bov 10C 2
Claverton St. SW1 7C 228 (50Lb 90)
Clave St. E1 46Yb 92
Claxton Gro. W6 50Za 88
Claxton Path SE4 56Zb 114
(off Coston Wlk.)
Clay Av. CR4: Mitc 68Kb 134
Claybank Gro. SE13 55Dc 114
Claybourne M. SE19 66Ub 135
Claybridge Rd. SE12 63Lc 137
Claybrook Cl. N2 27Fb 49
Claybrook Rd. W6 51Za 110
Clayburn Gdns. RM15: S Ock 45Xd 98
Claybury WD23: Bush 17Da 27
Claybury B'way. IG5: Ilf 27Nc 54
Claybury Hall IG8: Wfd G 24Pc 54
Claybury M. IG5: Ilf 25Pc 54
Claybury Rd. IG8: Wfd G 24Nc 54
Clay Cl. KT15: Add 78K 149
(off Monks Cres.)
Claycorn Ct. KT16: Clay 79Ga 152
Clay Cnr. KT16: Chert 74K 149
Clay Ct. E17 27Fc 53
Clay Ct. SE1 3H 231 (48Ub 91)
(off Long La.)
Claydon Ct. TW18: Staines 63J 127
(off Kingston Rd.)
Claydon Dr. CR0: Bedd 77Nb 156
Claydon End SL9: Chal P 27A 42
Claydon Ho. NW4 26Za 48
(off Holders Hill Rd.)
Claydon Ho. SW10 53Eb 111
Claydon La. SL9: Chal P 27A 42
Claydon Rd. GU21: Wok 8L 167
Claydown M. SE18 50Qc 94
Clayfarm Rd. SE9 61Sc 138
Clayfield Cl. RM14: Upm 30Vd 58
CLAYGATE 79Ha 152
Claygate Cl. RM12: Horn 35Jd 76
Claygate Common 80Ha 152
Claygate Cres. CR0: New Ad 79Ec 158
CLAYGATE CROSS 96Ce 205
CLAYGATE DAY CEN. 79Ha 152
Claygate La. EN9: Walt A 2Ec 20
Claygate La. KT10: Clay 75Ja 152
Claygate La. KT10: Hin W 75Ja 152
Claygate La. KT7: T Ditt 74Ja 152
Claygate Lodge Cl. KT10: Clay 80Ga 152
Claygate Rd. W13 48Ka 86
Claygate Station (Rail) 79Ga 152
CLAYHALL 27Nc 54
Clayhall Av. IG5: Ilf 27Nc 54
Clayhall Ct. E3 40Bc 72
(off St Stephen's Rd.)
Clayhall Ho. RH2: Reig 5J 207
(off Somers Cl.)
Clayhall La. RH2: Reig 10F 206
Clayhall La. SL4: Old Win 7K 103
Clay Hill EN2: Enf 9Sb 19
CLAY HILL 9Sb 19
Clayhill KT5: Surb 71Qa 153
Clayhill Cres. SE9 63Mc 137
Claylands Pl. SW8 52Qb 112
Claylands Rd. SW8 51Pb 112
Clay La. GU4: Burp 10P 187
Clay La. GU4: Jac W 10P 187
Clay La. HA3: Kenton 28Ma 47
Clay La. HA8: Edg 19Qa 29
Clay La. KT18: Head 96Ra 193
Clay La. RH1: S Nut 7C 208
Clay La. TW19: Stanw 59P 105
Clay La. WD23: B Hea 17Ga 28
Claymill Ho. SE18 50Sc 94
Claymills M. HP3: Hem H 5A 4
Claymore Cl. SM4: Mord 73Cb 155
Clay Path E17 26Cc 52
Claypit Hill EN9: H Beech 8Lc 21
Claypit Hill EN9: Lough 8Lc 21
Claypit Hill EN9: Walt A 8Lc 21
Claypole Ct. E17 29Cc 52
(off Yunus Khan Cl.)
Claypole Dr. TW5: Hest 53Aa 107
Claypole Rd. E15 40Ec 72
Clayponds Av. TW8: Bford 49Ma 87
Clayponds Gdns. W5 49Ma 87
(not continuous)
CLAYPONDS HOSPITAL 49Na 87
Clayponds La. TW8: Bford 50Na 87
(not continuous)
Clay Ride IG10: Lough 11Mc 35
Clayside IG7: Chig 22Sc 54
Clay's La. IG10: Lough 11Qc 36
Clay St. W1 1G 221 (43Hb 89)
Clayton Av. HA0: Wemb 38Na 67
Clayton Av. RM14: Upm 36Rd 77
Clayton Bus. Cen. UB3: Hayes 47U 84
Clayton Cl. E6 44Pc 94
Clayton Ct. SL3: L'ly 48C 82
Clayton Cres. N1 39Nb 70
Clayton Cres. TW8: Bford 50Ma 87
Clayton Cft. Rd. DA2: Wilm 61Jd 140
Clayton Dr. HP3: Hem H 4D 4
Clayton Dr. SE8 50Ac 92
Clayton Fld. NW9 24Ua 48
Clayton Ho. E9 38Yb 72
(off Frampton Pk. Rd.)
Clayton Ho. KT7: T Ditt 74Ka 152
Clayton Ho. SW13 52Ya 110
(off Trinity Chu. Rd.)
Clayton Mead RH9: G'stone 2P 209
Clayton M. SE10 53Fc 115
Clayton Pde. EN8: Chesh 2Zb 20
Clayton Rd. KT17: Eps 84Ua 174
Clayton Rd. KT9: Chess 77La 152
Clayton Rd. RM7: Rush G 32Ed 76
Clayton Rd. SE15 53Wb 113
Clayton Rd. TW7: Isle 55Ga 108
Clayton Rd. UB3: Hayes 47U 84
Clayton St. SE11 51Qb 112
Clayton Ter. UB4: Yead 43Aa 85
Claytonville Ter. DA17: Belv 47Ed 96
Clayton Way UB8: Cowl 42M 83
Clay Tye Rd. RM14: Upm 33Yd 78
Clay Wood Cl. BR6: Orp 73Uc 160
Claywood La. DA2: Bean 62Zd 143
Clayworth Cl. DA15: Sidc 58Xc 117
Cleall Av. EN9: Walt A 6Ec 20
Cleanthus Cl. SE18 53Rc 116
Cleanthus Rd. SE18 54Rc 116
(not continuous)
Clearbrook Way E1 44Yb 92
Cleardown GU22: Wok 90D 168
Cleares Pasture SL1: Burn 1A 80
CLEARMOUNT 9J 147
Clears, The RH2: Reig 4G 206
Clears Cotts. RH2: Reig 4G 206
Clearwater La. DA1: Dart 61Qd 141
Clearwater Pl. KT6: Surb 72La 152
Clearwater Ter. W11 47Za 88
Clearwater Yd. NW1 39Kb 70
(off Inverness St.)
Clearway Ct. CR3: Cat'm 94Wb 197
Clearways Bus. Est. TN15: W King 80Ud 164
Clearways Cvn. Pk. TN15: W King 80Td 164
Clearwell Dr. W9 42Db 89
Cleave Av. BR6: Chels 79Uc 160
Cleave Av. UB3: Harl 49U 84
Cleaveland Rd. KT6: Surb 71Ma 153
Cleave Prior CR5: Chips 91Gb 195
Cleaverholme Cl. SE25 72Xb 157
Cleaver Ho. NW3 38Hb 69
(off Adelaide Rd.)
Cleaver Sq. SE11 7A 230 (50Qb 90)
Cleaver St. SE11 7A 230 (50Qb 90)
Cleaves Almshouses KT2: King T 68Na 131
(off London Rd.)
Cleeve Ct. TW14: Bedf 60U 106
Cleeve Hill SE23 60Xb 113
Cleeve Ho. E2 4J 219 (41Ub 91)
(off Calvert Av.)
Cleeve Pk. Gdns. DA14: Sidc 61Xc 139
Cleeve Rd. KT22: Lea 92Ha 192
Cleeves Ct. RH1: Redh 5A 208
(off St Anne's Mt.)
Cleeves Vw. DA1: Dart 58Md 119
(off Priory Pl.)
Cleeve Way SM1: Sutt 74Db 155
Cleeve Way SW15 59Va 110
Cleeve Workshops E2 4J 219 (41Ub 91)
(off Boundary Rd.)
Clegg Ho. SE16 48Yb 92
(off Moodkee St.)
Clegg St. E1 46Xb 91
Clegg St. E13 40Jc 73
Cleland Ho. E2 40Yb 72
(off Sewardstone Rd.)
Cleland Path IG10: Lough 11Rc 36
Cleland Rd. SL9: Chal P 26A 42
Clematis Apts. E3 41Bc 92
(off Merchant St.)
Clematis Cl. RM3: Rom 24Ld 57
Clematis Gdns. IG8: Wfd G 22Jc 53
Clematis St. W12 45Wa 88
Clem Attlee Ct. SW6 51Bb 111
Clem Attlee Pde. SW6 51Bb 111
(off North End Rd.)
Clemence Rd. RM10: Dag 39Ed 76
Clemence St. E14 43Bc 92
Clement Av. SW4 56Mb 112
Clement Cl. CR8: Purl 88Rb 177
Clement Cl. NW6 38Ya 68
Clement Cl. W4 49Ta 87
Clement Danes Ho. W12 44Xa 88
Clement Gdns. UB3: Harl 49U 84
Clementhorpe Rd. RM9: Dag 37Yc 75
Clement Ho. SE8 49Ac 92
Clement Ho. W10 43Ya 88
(off Dalgarno Gdns.)
Clementina Ct. E3 42Ac 92
(off Copperfield Rd.)
Clementina Rd. E10 32Bc 72
CLEMENTINE CHURCHILL BMI HOSPITAL 33Ha 66
Clementine Cl. W13 47Ka 86
Clementine Wlk. IG8: Wfd G 24Jc 53
Clementine Way HP1: Hem H 4K 3
Clement Rd. BR3: Beck 68Zb 136
Clement Rd. SW19 64Ab 132
Clement's Av. E16 45Jc 93
Clements Cl. N12 21Db 49
Clements Cl. SL1: Slou 7M 81
Clements Ct. IG1: Ilf 34Rc 74
Clements Ct. TW4: Houn 56Z 107
Clements Ct. WD25: Wat 7Y 13
Clements Ho. KT22: Lea 91Ja 192
Clement's Inn WC2 3J 223 (44Pb 90)
Clements La. EC4 4G 225 (45Tb 91)
Clements La. IG1: Ilf 34Rc 74
Clements Mead KT22: Lea 91Ja 192
Clements Pl. TW8: Bford 50Ma 87
Clement's Rd. WD3: Chor 15F 24
Clements Rd. E6 38Nc 74
Clements Rd. IG1: Ilf 34Rc 74
Clements Rd. KT12: Walt T 75X 151
Clements Rd. SE16 48Wb 91
Clement St. BR8: Swan 65Ld 141
Clement St. DA4: Swan 65Pd 141
CLEMENT STREET 65Md 141
Clement Way RM14: Upm 34Pd 77
Clemson Ho. E8 39Vb 71
Clemson M. KT17: Eps 84Va 174
Clenches Farm TN13: S'oaks 99Jd 202
Clenches Farm La. TN13: S'oaks 99Jd 202
Clenches Farm Rd. TN13: S'oaks 98Jd 202
Clendon Way SE18 49Tc 94
Clennam St. SE1 1E 230 (47Sb 91)
Clensham Ct. SM1: Sutt 75Cb 155
Clensham La. SM1: Sutt 75Cb 155
Clenston M. W1 2F 221 (44Hb 89)
Cleopatra Cl. HA7: Stan 20Ma 29
Cleopatra's Needle 5H 223 (45Pb 90)
Clephane Rd. N1 37Sb 71
Clephane Rd. Nth. N1 37Sb 71
Clephane Rd. Sth. N1 37Sb 71
Clere Pl. EC2 5G 219 (42Tb 91)
Clere St. EC2 5G 219 (42Tb 91)
Clerics Wlk. TW17: Shep 73T 150
CLERKENWELL 5A 218 (42Qb 90)
Clerkenwell Cl. EC1 5A 218 (42Qb 90)
(not continuous)
Clerkenwell Ct. N1 1B 218 (40Rb 71)
(off Duncan St.)
Clerkenwell Grn. EC1 6A 218 (42Qb 90)
Clerkenwell Rd. EC1 6K 217 (42Qb 90)
Clerks Cft. RH1: Blet 5K 209
Clerk's Piece IG10: Lough 13Pc 36
Clerks Pl. EC3 2H 225 (44Ub 91)
Clermont Pl. RM1: Rom 29Jd 56
Clermont Rd. E9 39Yb 72
Clevedon KT13: Weyb 78T 150
Clevedon Cl. N16 34Vb 71
Clevedon Ct. CR2: S Croy 78Ub 157
Clevedon Ct. SW11 53Gb 111
(off Bolingbroke Wlk.)
Clevedon Gdns. TW5: Cran 53X 107
Clevedon Gdns. UB3: Harl 48T 84
Clevedon Ho. SM1: Sutt 77Eb 155
Clevedon Mans. NW5 35Jb 70
Clevedon Pas. N16 33Vb 71
Clevedon Rd. KT1: King T 68Qa 131
Clevedon Rd. SE20 67Zb 136
Clevedon Rd. TW1: Twick 58Ma 109
Cleve Ho. NW6 38Db 69
Clevehurst Cl. SL2: Stoke P 7L 61
Cleveland Av. SW20 68Bb 133
Cleveland Av. TW12: Hamp 66Ba 129
Cleveland Av. W4 49Va 88
Cleveland Cl. KT12: Walt T 76X 151
Cleveland Ct. W13 43Ka 86
Cleveland Cres. WD6: Bore 15Sa 29
Cleveland Dr. TW18: Staines 68K 127
Cleveland Gdns. KT4: Wor Pk 75Ua 154
Cleveland Gdns. N4 29Sb 51
Cleveland Gdns. NW2 33Za 68
Cleveland Gdns. SW13 54Va 110
Cleveland Gdns. W2 3A 220 (44Eb 89)
Cleveland Gro. E1 42Yb 92
Cleveland Ho. DA11: Nflt 58Fe 121
Cleveland Ho. N2 26Fb 49
(off The Grange)
Cleveland Mans. NW6 38Bb 69
(off Willesden La.)
Cleveland Mans. SW9 52Qb 112
(off Mowll St.)
Cleveland Mans. W9 42Cb 89
Cleveland M. W1 7B 216 (43Lb 90)
Cleveland Pk. TW19: Stanw 58N 105
Cleveland Pk. Av. E17 28Cc 52
Cleveland Pk. Cres. E17 28Cc 52
Cleveland Pl. SW1 6C 222 (46Lb 90)
Cleveland Ri. SM4: Mord 73Za 154
Cleveland Rd. DA16: Well 54Vc 117
Cleveland Rd. E18 27Jc 53
Cleveland Rd. HP2: Hem H 1B 4
Cleveland Rd. IG1: Ilf 34Rc 74
Cleveland Rd. KT3: N Mald 70Ua 132
Cleveland Rd. KT4: Wor Pk 75Ua 154
Cleveland Rd. N1 38Tb 71
Cleveland Rd. N9 17Xb 33
Cleveland Rd. SW13 54Va 110
Cleveland Rd. TW7: Isle 56Ja 108
Cleveland Rd. UB8: Cowl 42M 83
Cleveland Rd. UB8: Uxb 42M 83
Cleveland Rd. W13 43Ja 86
Cleveland Rd. W4 48Sa 87
Cleveland Row SW1 7B 222 (46Lb 90)
Cleveland Sq. W2 3A 220 (44Eb 89)
Cleveland St. W1 6A 216 (42Kb 90)
Cleveland Ter. W2 3A 220 (44Eb 89)
Cleveland Way E1 42Yb 92
Cleveley Cl. SE7 49Mc 93
Cleveley Ct. SE16 49Ac 92
(off Ashton Reach)
Cleveley Cres. W5 40Na 67
Cleveleys Rd. E5 34Xb 71
Cleve Pl. KT13: Weyb 78T 150
Cleverly Est. W12 46Wa 88
Cleve Rd. DA14: Sidc 62Zc 139
Cleve Rd. NW6 38Cb 69
Cleves, The TN15: Kems'g 89Rd 183
Cleves Av. CM14: B'wood 18Xd 40
Cleves Av. KT17: Ewe 81Xa 174
Cleves Cl. IG10: Lough 16Nc 36
Cleves Cl. KT11: Cobh 86X 171
Cleves Ct. DA1: Dart 59Nd 119
Cleves Ct. KT17: Eps 84Va 174
Cleves Ct. SL4: Wind 5D 102
Cleves Cres. CR0: New Ad 83Ec 178
Cleves Ho. E16 46Jc 93
(off Southey M.)
Cleves Rd. E6 39Mc 73
Cleves Rd. TN15: Kems'g 89Nd 183
Cleves Rd. TW10: Ham 62La 130
Cleves Wlk. IG6: Ilf 24Sc 54
Cleves Way HA4: Ruis 32Z 65
Cleves Way RH1: Blet 6L 209
Cleves Way TW12: Hamp 66Ba 129
Cleves Way TW16: Sun 65V 128
Cleves Wood KT13: Weyb 77U 150
Clewer Av. SL4: Wind 4E 102
Clewer Ct. E10 32Cc 72
(off Leyton Grange Est.)
Clewer Ct. Rd. SL4: Wind 2F 102
Clewer Cres. HA3: Hrw W 25Fa 46
Clewer Flds. SL4: Wind 3G 102
CLEWER GREEN 4D 102
CLEWER HILL 5C 102
Clewer Hill Rd. SL4: Wind 4C 102
Clewer Ho. SE2 47Zc 95
(off Wolvercote Rd.)
Clewer New Town SL4: Wind 4E 102
CLEWER NEW TOWN 4E 102
Clewer Pk. SL4: Wind 2E 102
CLEWER ST ANDREW 2F 102
CLEWER ST STEPHEN 2F 102
CLEWER VILLAGE 2E 102
CLEWER WITHIN 3G 102
Clew's La. GU24: Bisl 8E 166
Cley Ho. SE4 56Zb 114
C&L Golf Course 38X 65
Clichy Est. E1 43Yb 92
Clichy Ho. E1 43Yb 92
(off Stepney Way)
Clifden M. E5 35Zb 72
Clifden Rd. E5 36Yb 72
Clifden Rd. TW1: Twick 60Ha 108
Clifden Rd. TW8: Bford 51Ma 109
Cliffe Ho. SE10 50Hc 93
(off Blackwall La.)
Cliff End CR8: Purl 84Rb 177
Cliffe Rd. CR2: S Croy 78Tb 157
Cliffe Wlk. SM1: Sutt 78Eb 155
(off Greyhound Rd.)
Clifford Av. BR7: Chst 65Pc 138
Clifford Av. IG5: Ilf 25Rc 54
Clifford Av. SM6: W'gton 77Lb 156
Clifford Av. SW14 55Ra 109
Clifford Cl. UB5: N'olt 39Aa 65
Clifford Ct. W2 43Db 89
(off Westbourne Pk. Vs.)
Clifford Dr. SW9 56Rb 113
Clifford Gdns. NW10 40Ya 68
Clifford Gdns. UB3: Harl 49U 84
Clifford Gro. TW15: Ashf 63Q 128
Clifford Haigh Ho. SW6 52Za 110
Clifford Ho. BR3: Beck 65Dc 136
(off Calverley Cl.)
Clifford Ho. W14 49Bb 89
(off Edith Vs.)
Clifford Rd. E16 42Hc 93
Clifford Rd. E17 26Ec 52
Clifford Rd. EN5: New Bar 13Db 31
Clifford Rd. HA0: Wemb 38Ma 67
Clifford Rd. N1 39Ub 71
Clifford Rd. N9 16Yb 34
Clifford Rd. RM16: Chaf H 47Be 99
Clifford Rd. SE25 70Wb 135
Clifford Rd. TW10: Ham 61Ma 131
Clifford Rd. TW4: Houn 55Z 107
Clifford's Inn EC4 3K 223 (44Qb 90)
(off Fetter La.)
Clifford's Inn Pas. EC4 3K 223 (44Qb 90)
Clifford St. W1 5B 222 (45Lb 90)
Clifford Way NW10 35Va 68
Cliff Pl. RM15: S Ock 41Zd 99
Cliff Reach DA9: Bluew 58Vd 120
Cliff Rd. NW1 37Mb 70
Cliffsend Ho. SW9 53Qb 112
(off Cowley Rd.)
Cliff Ter. SE8 54Cc 114
Cliffview Rd. SE13 55Cc 114
Cliff Vs. NW1 37Mb 70
Cliff Wlk. E16 43Hc 93
Clifton Av. E17 27Zb 52
Clifton Av. HA7: Stan 26Ka 46
Clifton Av. HA9: Wemb 37Pa 67
Clifton Av. N3 25Bb 49
Clifton Av. SM2: Sutt 83Db 175
Clifton Av. TW13: Felt 62Y 129
Clifton Av. W12 46Va 88
Clifton Cl. BR6: Farnb 78Sc 160
Clifton Cl. CR3: Cat'm 95Tb 197
Clifton Cl. EN8: Chesh 1Ac 20
Clifton Cl. KT15: Add 75K 149
Clifton Ct. BR3: Beck 67Dc 136
Clifton Ct. HP3: Hem H 4M 3

Clifton Ct. IG8: Wfd G 23Jc 53
Clifton Ct. KT19: Eps 84Pa 173
Clifton Ct. KT5: Surb 73Pa 153
Clifton Ct. N4 33Qb 70
Clifton Ct. NW8 5B 214 (42Fb 89)
(off Maida Vale)
Clifton Ct. SE15 52Xb 113
Clifton Ct. TW19: Stanw 58N 105
Clifton Cres. SE15 52Xb 113
Clifton Est. SE15 53Xb 113
Clifton Gdns. EN2: Enf 14Nb 32
Clifton Gdns. N15 30Vb 51
Clifton Gdns. NW11 30Bb 49
Clifton Gdns. UB10: Hil 40R 64
Clifton Gdns. W4 49Ta 87
(not continuous)
Clifton Gdns. W9 6A 214 (42Eb 89)
Clifton Ga. SW10 51Eb 111
Clifton Gro. DA11: Grav'nd 9D 122
Clifton Gro. E8 37Wb 71
Clifton Hill NW8 40Db 69
Clifton Hill Studios NW8 40Eb 69
Clifton Ho. E11 33Gc 73
Clifton Ho. E2 5K 219 (42Vb 91)
(off Club Row)
Clifton Lodge SL4: Eton W 10E 80
Clifton Marine Pde. DA11: Grav'nd 8B 122
(not continuous)
Clifton M. SE25 70Ub 135
Clifton Pde. TW13: Felt 62Y 129
Clifton Pk. Av. SW20 68Ya 132
Clifton Pl. SE16 47Yb 92
Clifton Pl. SM7: Bans 87Cb 175
Clifton Pl. SW10 51Eb 111
(off Hollywood Rd.)
Clifton Pl. W2 3C 220 (44Fb 89)
Clifton Ri. SE14 52Ac 114
(not continuous)
Clifton Ri. SL4: Wind 3B 102
Clifton Rd. CR5: Coul 87Kb 176
Clifton Rd. DA11: Grav'nd 8C 122
Clifton Rd. DA14: Sidc 63Uc 138
Clifton Rd. DA16: Well 55Yc 117
Clifton Rd. E16 43Gc 93
Clifton Rd. E7 37Mc 73
Clifton Rd. HA3: Kenton 28Pa 47
Clifton Rd. IG10: Lough 14Nc 36
Clifton Rd. IG2: Ilf 30Tc 54
Clifton Rd. KT2: King T 66Pa 131
Clifton Rd. N1 37Sb 71
Clifton Rd. N22 25Lb 50
Clifton Rd. N3 25Eb 49
Clifton Rd. N8 30Mb 50
Clifton Rd. NW10 40Wa 68
Clifton Rd. RM11: Horn 30Jd 56
Clifton Rd. SE25 70Ub 135
Clifton Rd. SL1: Slou 7M 81
Clifton Rd. SM6: W'gton 78Kb 156
Clifton Rd. SW19 65Za 132
Clifton Rd. TW11: Tedd 63Ga 130
Clifton Rd. TW7: Isle 54Ga 108
Clifton Rd. UB2: S'hall 49Aa 85
Clifton Rd. UB6: G'frd 42Ea 86
Clifton Rd. W9 5A 214 (42Eb 89)
Clifton Rd. WD18: Wat 15X 27
Clifton's La. RH2: Reig 4F 206
Clifton St. AL1: St A 1C 6
Clifton St. EC2 6H 219 (42Ub 91)
Clifton Ter. N4 33Qb 70
Clifton Vs. W9 43Eb 89
Cliftonville Ct. SE12 60Jc 115
Clifton Wlk. DA2: Dart 58Rd 119
Clifton Wlk. W6 49Xa 88
(off King St.)
Clifton Way CM13: Hut 18Fe 41
Clifton Way GU21: Wok 9K 167
Clifton Way HA0: Wemb 39Na 67
Clifton Way SE15 52Xb 113
Clifton Way WD6: Bore 11Qa 29
Climb, The WD3: Rick 16K 25
Climsland Ho. SE1 6A 224 (46Qb 90)
Cline Rd. N11 23Lb 50
Clinger Ct. N1 1H 219 (39Ub 71)
(off Hobbs Pl. Est.)
Clink Prison Mus. 6F 225 (46Tb 91)
(off Clink St.)
Clink St. SE1 6F 225 (46Tb 91)
Clink Wharf SE1 6F 225 (46Tb 91)
(off Clink St.)
Clinton Av. DA16: Well 56Wc 117
Clinton Av. KT8: E Mos 70Ea 130
Clinton Cl. GU21: Knap 10H 167
Clinton Cl. KT13: Weyb 75R 150
Clinton Cres. IG6: Ilf 23Uc 54
Clinton End HP2: Hem H 2C 4
Clinton Ho. KT6: Surb 73Ma 153
(off Lovelace Gdns)
Clinton Rd. E3 41Ac 92
Clinton Rd. E7 35Jc 73
Clinton Rd. KT22: Lea 95La 192
Clinton Rd. N15 28Tb 51
Clinton Ter. SE8 51Cc 114
(off Watergate St.)
Clinton Ter. SM1: Sutt 77Eb 155
Clipper Apts. SE10 51Ec 114
(off Welland St.)
Clipper Blvd. DA2: Dart 55Ud 120
Clipper Blvd. W. DA2: Dart 55Td 120
Clipper Cl. SE16 47Zb 92
Clipper Cres. DA12: Grav'nd 3H 145
Clipper Ho. E14 50Ec 92
(off Manchester Rd.)
Clipper Pk. RM18: Tilb 52Fe 121
Clipper Way SE13 56Ec 114
Clippesby Cl. KT9: Chess 79Pa 153
Clipstone M. W1 7B 216 (43Lb 90)
Clipstone Rd. TW3: Houn 55Ca 107
Clipstone St. W1 7A 216 (43Kb 90)
Clissold Cl. N2 27Hb 49
Clissold Ct. N4 33Sb 71
Clissold Cres. N16 34Tb 71
Clissold Leisure Cen. 34Tb 71
Clissold Rd. N16 34Tb 71
Clitheroe Av. HA2: Harr 32Ca 65
Clitheroe Gdns. WD19: Wat 20Z 27
Clitheroe Rd. RM5: Col R 22Ed 56
Clitheroe Rd. SW9 54Nb 112
Clitherow Av. W7 48Ja 86
Clitherow Ct. TW8: Bford 50La 86
Clitherow Pas. TW8: Bford 50La 86
Clitherow Rd. TW8: Bford 50Ka 86
Clitterhouse Cres. NW2 32Ya 68
Clitterhouse Rd. NW2 32Ya 68
Clive Av. DA1: Cray 58Hd 118
Clive Av. N18 23Wb 51
Clive Cl. EN6: Pot B 3Bb 17
Clive Ct. SL1: Slou 7H 81
Clive Ct. W9 5A 214 (42Eb 89)
Cliveden Cl. CM15: Shenf 17Be 41
Cliveden Cl. N12 21Eb 49
Cliveden Ho. E16 46Jc 93
(off Fitzwilliam M.)
Cliveden Ho. SW1 5H 227 (49Jb 90)
(off Cliveden Pl.)
Cliveden Pl. SW1 5H 227 (49Jb 90)
Cliveden Pl. TW17: Shep 72S 150
Cliveden Rd. SW19 67Bb 133
Clivedon Ct. W13 43Ka 86
Clivedon Rd. E4 22Gc 53
Clive Ho. SE10 51Ec 114
(off Haddo St.)
Clive Lloyd Ho. N15 29Sb 51
(off Woodlands Pk. Rd.)
Clive Lodge NW4 30Za 48
Clive Pde. HA6: Nwood 24U 44
Clive Pas. SE21 62Tb 135
Clive Rd. CM13: Gt War 24Yd 58
Clive Rd. DA11: Grav'nd 8D 122
Clive Rd. DA17: Belv 49Cd 96
Clive Rd. EN1: Enf 14Wb 33
Clive Rd. KT10: Esh 77Da 151
Clive Rd. RM2: Rom 29Kd 57
Clive Rd. SE21 62Tb 135
Clive Rd. SW19 65Gb 133
Clive Rd. TW1: Twick 63Ha 130
Clive Rd. TW14: Felt 58W 106
Clivesdale Dr. UB3: Hayes 46X 85
Clive Way EN1: Enf 14Wb 33
Clive Way WD24: Wat 11Y 27
Cloak La. EC4 4E 224 (45Sb 91)
Clochar Ct. NW10 39Va 68
Clock Ct. E11 28Kc 53
Clock Ho. E17 28Fc 53
(off Wood St.)
Clock Ho. E3 41Ec 92
CLOCK HOUSE 86Kb 176
Clockhouse, The SW19 62Ya 132
Clockhouse Av. IG11: Bark 39Sc 74
Clock Ho. Cl. KT14: Byfl 84P 169
Clockhouse Cl. SW19 61Ya 132
Clockhouse Ct. BR3: Beck 68Ac 136
CLOCKHOUSE JUNC. 22Pb 50
Clock Ho. La. TN13: S'oaks 95Jd 202
Clockhouse La. RM16: Chaf H 46Zd 99
(not continuous)
Clockhouse La. RM16: N Stif 46Zd 99
(not continuous)
Clockhouse La. RM4: Have B 21Dd 56
Clockhouse La. RM5: Col R 24Dd 56
Clockhouse La. TW14: Bedf 62R 128
Clockhouse La. TW15: Ashf 63Q 128
Clockhouse La. E. TW20: Egh 66D 126
Clockhouse La. W. TW20: Egh 66C 126
Clock Ho. Mead KT22: Oxs 86Da 171
Clockhouse M. WD3: Chor 13G 24
Clock Ho. Pde. E11 30Kc 53
Clockhouse Pde. N13 22Qb 50
Clockhouse Pl. SW15 58Ab 110
Clock Ho. Rd. BR3: Beck 69Ac 136
CLOCKHOUSE RDBT. East Bedfont 60R 106
Clockhouse Stables DA2:
G St G 65Wd 142
Clock House Station (Rail) 67Ac 136
Clock Pde. EN2: Enf 15Tb 33
Clock Pl. SE1 5C 230 (49Rb 91)
(off Newington Butts)
Clock Twr. CM14: W'ley 22Xd 58
Clock Twr., The AL2: Lon C 9F 6
Clock Twr. Ind. Est. TW7: Isle 55Ha 108
Clock Twr. M. N1 1E 218 (39Sb 71)
Clock Twr. M. SE28 45Xc 95
Clock Twr. M. W7 46Ga 86
Clock Twr. Rd. TW7: Isle 55Ha 108
Clock Vw. Cres. N7 37Nb 70
(not continuous)
Clockwork Apts. EC4 3C 224 (44Rb 91)
(off Ludgate Sq.)
Clockwork M. E5 36Zb 72
Clodhouse Hill GU22: Wok 4H 187
Cloister Cl. RM13: Rain 42Kd 97
Cloister Cl. TW11: Tedd 64Ka 130
Cloister Gdns. HA8: Edg 22Sa 47
Cloister Gdns. SE25 72Xb 157
Cloister Gth. AL1: St A 6C 6
Cloister Rd. NW2 34Bb 69
Cloister Rd. W3 43Sa 87
Cloisters SS17: Stan H 1N 101
Cloisters, The DA1: Dart 58Nd 119
(off Orchard St.)
Cloisters, The E1 7K 219 (43Vb 91)
(off Commercial St.)
Cloisters, The GU22: Wok 93D 188
Cloisters, The HP3: Hem H 5A 4
Cloisters, The SL1: Slou 7H 81
(off Henry Rd.)
Cloisters, The SL4: Wind 4E 102
Cloisters, The 3F 229 (48Nb 90)
(within Westminster Abbey)
Cloisters, The SW9 53Qb 112
Cloisters, The WD18: Wat 14Y 27
Cloisters, The WD23: Bush 16Da 27
Cloisters, The WD3: Rick 17N 25
Cloisters, The WD4: K Lan 1Q 12
Cloisters Av. BR2: Broml 71Pc 160
Cloisters Bus. Cen. SW8 52Kb 112
(off Battersea Pk. Rd.)
Cloisters Ct. DA7: Bex 55Dd 118
Cloisters Ct. WD3: Rick 17N 25
Cloisters Mall KT1: King T 68Na 131
Clonard Way HA5: Hat E 23Ca 45
Clonbrock Rd. N16 35Ub 71
Cloncurry St. SW6 54Za 110
Clonmel Cl. HA2: Harr 33Fa 66
Clonmell Rd. N17 27Tb 51
Clonmel Rd. SW6 52Bb 111
Clonmel Rd. TW11: Tedd 63Fa 130
Clonmel Way SL1: Burn 1A 80
Clonmore St. SW18 60Bb 111
Cloonmore Av. BR6: Chels 77Vc 161
Clorane Gdns. NW3 34Cb 69
Close, The CR8: Purl Russell Hill 82Pb 176
Close, The CR8: Purl Wyvern Rd. 82Rb 177
Close, The AL1: St A 5A 6
Close, The AL9: Brk P 8G 8
Close, The BR3: Beck 70Ac 136
Close, The BR5: Pet W 72Uc 160
Close, The CM14: B'wood 20Zd 41
Close, The CR4: Mitc 70Hb 133
Close, The DA14: Sidc 63Xc 139
Close, The DA2: Wilm 62Ld 141
Close, The DA3: Lfield 68De 143
Close, The DA5: Bexl 58Cd 118
Close, The E4 24Ec 52
Close, The EN4: E Barn 16Hb 31
Close, The EN6: Pot B 4Cb 17
Close, The GU25: Vir W 1P 147
Close, The HA0: Wemb 37Na 67
Close, The HA2: Harr 26Ea 46
Close, The HA5: Eastc 31Y 65
Close, The HA5: Pinn 31Ba 65
Close, The HA9: Wemb 34Sa 67
Close, The IG2: Ilf 30Uc 54
Close, The KT10: Esh 79Da 151
Close, The KT12: Hers 78W 150
Close, The KT14: W Byf 85J 169
Close, The KT3: N Mald 68Sa 131
Close, The KT6: Surb 72Na 153
Close, The N10 26Kb 50
Close, The N14 19Mb 32
Close, The N20 19Bb 31
Close, The RH2: Reig 7K 207
Close, The RM16: Grays 47Ee 99
Close, The RM6: Chad H 30Ad 55
Close, The SE25 72Wb 157
Close, The SE3 54Fc 115
Close, The SL0: Iver H 41E 82
Close, The SL1: Slou 5B 80
Close, The SM3: Sutt 73Bb 155
Close, The SM5: Cars 81Gb 175
Close, The TN13: S'oaks 96Gd 202
Close, The TN15: Bor G 91Ce 205
Close, The TN15: Igh 93Zd 205
Close, The TN16: Big H 88Rc 180
Close, The TW7: Isle 54Fa 108
Close, The TW9: Rich 55Ra 109
Close, The UB10: Hil 39Q 64
Close, The WD23: Bush 16Da 27
Close, The WD3: Rick 18K 25
Close, The WD7: R'lett 5Ha 14
Closemead Cl. HA6: Nwood 23S 44
Cloth Ct. EC1 1C 224 (43Rb 91)
(off Cloth Fair)
Cloth Fair EC1 1C 224 (43Rb 91)
Cloth Ho. EC1 7D 218 (43Sb 91)
(off Cloth St.)
Clothier St. E1 2J 225 (44Ub 91)
Cloth St. EC1 7D 218 (43Sb 91)
Clothworkers Rd. SE18 52Tc 116
Cloudberry Rd. RM3: Rom 23Md 57
Cloud Cl. DA1: Dart 54Pd 119
Cloudesdale Rd. SW17 61Kb 134
Cloudeseley Cl. DA14: Sidc 63Vc 139
Cloudesley Ho. N1 39Qb 70
(off Cloudesley St.)
Cloudesley Mans. N1 1A 218 (39Qb 70)
(off Cloudesley Pl.)
Cloudesley Pl. N1 1K 217 (39Qb 70)
Cloudesley Rd. DA7: Bex 53Bd 117
Cloudesley Rd. DA8: Erith 53Hd 118
Cloudesley Rd. N1 39Qb 70
(not continuous)
Cloudesley Sq. N1 39Qb 70
Cloudesley St. N1 39Qb 70
Cloud Way TW6: H'row A 55R 106
Clouston Cl. SM6: W'gton 78Nb 156
Clova Rd. E7 37Hc 73
Clove Cres. E14 45Ec 92
Clove Hitch Quay SW11 55Eb 111
Clovelly Av. CR6: W'ham 91Xb 197
Clovelly Av. NW9 28Va 48
Clovelly Av. UB10: Ick 35S 64
Clovelly Cl. HA5: Eastc 27X 45
Clovelly Cl. UB10: Ick 35S 64
Clovelly Ct. KT17: Eps 85Va 174
(off Alexandra Rd.)
Clovelly Ct. NW2 36Ya 68
Clovelly Ct. RM11: Horn 33Qd 77
Clovelly Gdns. EN1: Enf 17Ub 33
Clovelly Gdns. RM7: Mawney 25Dd 56
Clovelly Gdns. SE19 67Vb 135
Clovelly Ho. W2 2A 220 (44Eb 89)
(off Hallfield Est.)
Clovelly Pl. DA9: Ghithe 56Yd 120
Clovelly Rd. DA7: Bex 51Ad 117
Clovelly Rd. N8 28Mb 50
Clovelly Rd. TW3: Houn 54Ca 107
Clovelly Rd. W4 47Ta 87
Clovelly Rd. W5 47La 86
Clovelly Spur SL2: Slou 2D 80
Clovelly Way BR6: St M Cry 72Vc 161
Clovelly Way E1 44Yb 92
Clovelly Way HA2: Harr 33Ba 65
Clover Cl. E11 33Fc 73
Clover Ct. E14 48Dc 92
(off Watergate Walk)
Clover Ct. GU22: Wok 10P 167
Clover Ct. RM17: Grays 51Fe 121
Cloverdale Ct. SM6: W'gton 79Kb 156
Cloverdale Gdns. DA15: Sidc 58Vc 117
Clover Fld., The WD23: Bush 16Ba 27
Clover Hill CR5: Coul 93Kb 196
Cloverland AL10: Hat 3B 8
Clover Leas CM16: Epp 1Vc 23
Cloverleys IG10: Lough 15Mc 35
Clover M. SW3 51Hb 111
Clovers, The DA11: Nflt 3A 144
Clovers Ct. WD3: Chor 16E 24
Clover Way HP1: Hem H 1K 3
Clover Way SM6: W'gton 74Jb 156
Clove St. E13 42Jc 93
Clove St. RM9: Dag 41Ad 95
Clovis Cl. IG10: Lough 12Hc 35
Clowders Rd. SE6 62Bc 136
Clowser Cl. SM1: Sutt 78Eb 155
Cloysters Grn. E1 46Wb 91
Cloyster Wood HA8: Edg 24Ma 47
Club Gdns. Rd. BR2: Hayes 73Jc 159
Clubhouse, The E14 44Cc 92
Clubhouse La. UB10: Ick 35Q 64
Club Row E1 5K 219 (42Vb 91)
Club Row E2 5K 219 (42Vb 91)
Club Row GU24: Brkwd 1B 186
Cluff Ct. CM14: W'ley 21Yd 58
Clump, The WD3: Rick 15J 25
Clumps, The TW15: Ashf 63T 128
Clunas Gdns. RM2: Rom 27Md 57
Clunbury Av. UB2: S'hall 50Ba 85
Clunbury St. N1 2G 219 (40Tb 71)
Clune Ct. CM15: Shenf 17Ce 41
Clunie Ho. SW1 3G 227 (48Hb 89)
(off Hans Pl.)
Cluny Est. SE1 3H 231 (48Ub 91)
Cluny M. SW5 49Cb 89
Cluny Pl. SE1 3H 231 (48Ub 91)
Cluse Ct. N1 1D 218 (40Sb 71)
(off St Peters St.)
Cluster Ho. W3 46Va 88
Clutterbucks WD3: Sarr 8J 11
Clutton St. E14 43Dc 92
Clydach Rd. EN1: Enf 14Vb 33
Clyde RM18: E Til 9L 101
Clyde Av. CR2: Sande 87Xb 177
Clyde Cir. N15 28Ub 51
Clyde Cl. RH1: Redh 5A 208
Clyde Ct. NW1 1E 216 (40Mb 70)
(off Hampden Cl.)
Clyde Cres. RM14: Upm 30Ud 58
Clyde Flats SW6 52Bb 111
(off Rylston Rd.)
Clyde Ho. KT2: King T 67Ma 131
Clyde Ho. SW18 56Cb 111
(off Enterprise Way)
Clyde Pl. E10 31Dc 72
Clyde Rd. CR0: C'don 75Vb 157
Clyde Rd. N15 28Ub 51
(not continuous)
Clyde Rd. N22 25Mb 50
Clyde Rd. SM1: Sutt 78Cb 155
Clyde Rd. SM6: W'gton 79Lb 156
Clyde Rd. TW19: Stanw 60M 105
Clydesdale EN3: Pond E 14Zb 34
Clydesdale Av. HA7: Stan 27Ma 47
Clydesdale Cl. TW7: Isle 55Ha 108
Clydesdale Cl. WD6: Bore 15Ta 29
Clydesdale Ct. N20 18Fb 31
Clydesdale Gdns. TW10: Rich 56Ra 109
Clydesdale Ho. DA18: Erith 47Ad 95
(off Kale Rd.)
Clydesdale Ho. W11 44Bb 89
(off Clydesdale Rd.)
Clydesdale Path WD6: Bore 15Ta 29
(off Clydesdale Cl.)
Clydesdale Rd. RM11: Horn 31Hd 76
Clydesdale Rd. W11 44Bb 89
Clydesdale Way DA17: Belv 47Dd 96
Clyde Sq. E14 44Bc 92
Clyde St. SE8 51Bc 114
Clyde Ter. SE23 61Yb 136
Clyde Va. SE23 61Yb 136
Clyde Way RM1: Rom 24Gd 56
Clyde Wharf E16 46Jc 93
Clydon Cl. DA8: Erith 51Gd 118
Clyfford Rd. HA4: Ruis 35V 64
Clymping Dene TW14: Felt 59X 107
Clynes Ho. E2 41Zb 92
(off Knottisford St.)
Clynes Ho. RM10: Dag 34Cd 76
(off Uvedale Rd.)
Clyston Rd. WD18: Wat 16V 26
Clyston St. SW8 54Lb 112
Clyve Way TW18: Staines 67G 126
Coach & Horses Yd. W1 4B 222 (45Lb 90)
Coach Ho. La. N5 35Rb 71
Coach Ho. La. SW19 63Za 132
Coach Ho. M. KT12: W Vill 81U 170
Coach Ho. M. RH1: Redh 7P 207
Coach Ho. M. SE1 3H 231 (48Ub 91)
(off Long La.)
Coach Ho. M. SE14 54Zb 114
Coach Ho. M. SE23 58Zb 114
Coach Ho. M. SM2: Sutt 79Db 155
Coachhouse M. SE20 66Xb 135
Coach Ho. Yd. NW3 35Eb 69
(off Heath St.)
Coach Ho. Yd. SW18 56Db 111
Coachmaker M. SW4 55Nb 112
(off Bedford Rd.)
Coachmaker M. W4 48Ta 87
(off Berrymede Rd.)
Coachmans Lodge SL4: Wind 4H 103
(off Frances Rd.)
Coach M. AL1: St A 2F 6
Coach Rd. KT16: Ott 79E 148
Coach Rd. TN15: Igh 96Wd 204
Coach Rd. TN15: Ivy H 96Wd 204
Coach Yd. M. N19 32Nb 70
Coalbrookdale Gates 2C 226 (47Fb 89)
Coal Ct. RM17: Grays 52Ce 121
Coaldale Wlk. SE21 59Sb 113
Coal Drops Yard 1F 217 (39Nb 70)
Coalecroft Rd. SW15 56Ya 110
Coales Hall AL10: Hat 2B 8
(off Bishops Ri.)
Coalhouse Fort 3M 123
Coal La. SW9 56Qb 112
Coalmakers Wharf E14 44Cc 92
Coalport Ho. SE11 5K 229 (49Qb 90)
(off Walnut Tree Wlk.)
Coal Post Cl. BR6: Chels 79Vc 161
Coal Rd. RM18: E Til 9H 101
Coal Rd. RM18: W Til 9H 101
Coalstore Ct. E1 43Ac 92
Coatbridge Ho. N1 38Pb 70
(off Carnoustie Dr.)
Coates Av. SW18 58Gb 111
Coates Cl. CR7: Thor H 69Sb 135
Coates Dell WD25: Wat 5Aa 13
Coates Hill Rd. BR1: Broml 68Qc 138
Coates Rd. WD6: E'tree 17Ma 29
Coate St. E2 40Wb 71
Coates Wlk. TW8: Bford 51Na 109
Coates Way WD25: Wat 5Y 13
Cobalt SE28 47Wc 95
Cobalt Bldg., The EC2 7D 218 (43Sb 91)
(off Bridgewater Sq.)
Cobalt Cl. BR3: Beck 70Zb 136
Cobalt Cl. NW9 25Va 48
Cobalt Ct. HA4: Ruis 35Y 65
Cobalt Pl. SW11 53Fb 111
Cobalt Rd. DA8: Erith 52Jd 118
Cobalt Sq. SW8 51Pb 112
(off Sth. Lambeth Rd.)
Cobalt Twr. SE14 51Ac 114
Cobb Cl. SL3: Dat 3P 103
Cobb Cl. WD6: Bore 15Ta 29
Cobbett Cl. EN3: Enf W 8Yb 20
Cobbett Hill Rd. GU3: Norm 10C 186
Cobbett Rd. SE9 55Nc 116
Cobbett Rd. TW2: Whitt 60Ca 107
Cobbetts Av. IG4: Ilf 29Mc 53
Cobbetts Cl. GU21: Wok 9M 167
Cobbetts Cl. GU3: Norm 10D 186
Cobbetts Farm GU24: Bisl 7E 166
Cobbetts Hill KT13: Weyb 79R 150
Cobbett St. SW8 52Pb 112
Cobbetts Wlk. GU24: Bisl 7E 166
Cobb Grn. WD25: Wat 4X 13
Cobbins, The EN9: Walt A 5Gc 21
Cobbinsbank EN9: Walt A 5Fc 21
Cobbinsend Rd. EN9: Walt A 1Mc 21
Cobble La. N1 38Rb 71
Cobble M. N5 34Sb 71
Cobble M. N6 32Jb 70
(off Highgate West Hill)
Cobble Path E17 29Cc 52
(off Cedars Av.)
Cobblers Cl. SL2: Farn R 10F 60
Cobbles, The CM15: B'wood 19Ae 41
Cobbles, The RM14: Upm 31Vd 78
Cobblestone Pl. CR0: C'don 74Sb 157
Cobblestone Sq. E1 46Xb 91
Cobbold Ct. SW1 5D 228 (49Mb 90)
(off Elverton St.)
Cobbold Ind. Est. NW10 37Va 68
Cobbold M. W12 47Va 88
Cobbold Rd. E11 34Hc 73
Cobbold Rd. NW10 37Ua 68
Cobbold Rd. W12 47Ua 88
Cobb's Ct. EC4 3C 224 (44Rb 91)
(off Carter La.)
Cobb's Hall SW6 51Za 110
(off Fulham Pal. Rd.)
Cobb's Rd. TW4: Houn 56Ba 107
Cobbsthorpe Vs. SE26 63Zb 136
Cobb St. E1 1K 225 (43Vb 91)
Cobden Bldgs. WC1 3H 217 (41Pb 90)
(off King's Cross Rd.)
Cobden Cl. UB8: Uxb 39L 63
Cobden Ct. BR2: Broml 70Lc 137
Cobden Hill WD7: R'lett 8Ka 14
Cobden Ho. E2 41Wb 91
(off Nelson Gdns.)
Cobden Ho. NW1 1B 216 (40Lb 70)
(off Arlington Rd.)
Cobden M. SE26 64Xb 135
Cobden Rd. BR6: Farnb 77Tc 160
Cobden Rd. E11 34Gc 73
Cobden Rd. SE25 71Wb 157
Cobden Rd. TN13: S'oaks 95Ld 203
Cobdens IG7: Chig 23Tc 54
Cob Dr. DA12: Shorne 4N 145
COBHAM 10J 145
COBHAM 86X 171
Cobham RM16: Grays 47De 99
Cobham & Stoke D'Abernon Station
(Rail) 89Aa 171
Cobham Av. KT3: N Mald 71Wa 154
Cobham Bowling Club 86X 171
Cobhambury Rd. DA12: Cobh 10J 145
Cobham Cl. BR2: Broml 73Nc 160
Cobham Cl. DA15: Sidc 58Xc 117
Cobham Cl. DA9: Ghithe 58Xd 120
Cobham Cl. EN1: Enf 13Wb 33
Cobham Cl. HA8: Edg 26Ra 47
Cobham Cl. SL1: Slou 7D 80
Cobham Cl. SM6: W'gton 79Nb 156
Cobham Cl. SW11 58Gb 111
Cobham Ct. CR4: Mitc 68Fb 133
Cobham Ga. KT11: Cobh 86X 171
Cobham Grange KT11: Cobh 86X 171
(off Between Streets)
Cobham Ho. DA8: Erith 52Hd 118
Cobham Mausoleum 10N 145
Cobham M. NW1 38Mb 70
Cobham Mill 87Y 171
Cobham Pk. KT11: D'side 88X 171
Cobham Pk. Rd. KT11: D'side 89X 171
Cobham Pl. DA6: Bex 57Ad 117
Cobham Rd. E17 25Ec 52
Cobham Rd. IG3: Ilf 35Uc 74
Cobham Rd. KT1: King T 68Qa 131
Cobham Rd. KT11: Stoke D 90Ca 171
Cobham Rd. KT22: Fet 93Ca 191
Cobham Rd. N22 27Rb 51
Cobham Rd. TW5: Hest 52Y 107
COBHAM SERVICE AREA 92Y 191
Cobham St. DA11: Grav'nd 9C 122
Cobham Ter. DA11: Nflt 10B 122
(off Southfleet Rd.)
Cobham Ter. DA9: Ghithe 57Xd 120
Cobham Way KT24: E Hor 98U 190
Cobham Youth Cen. 86X 171
(off High St.)
Cobill Cl. RM11: Horn 28Ld 57
Cobland Rd. SE12 63Lc 137
Cobmead Gro. EN9: Walt A 6Hc 21
COBORN CENTRE FOR ADOLESCENT
MENTAL HEALTH, THE 42Mc 93
Coborn M. E3 41Bc 92
(off Coborn St.)
Coborn Rd. E3 40Bc 72
Coborn St. E3 41Bc 92
Cobourg Rd. SE5 51Vb 113
Cobourg St. NW1 4C 216 (41Lb 90)
Cobs Cl. TN15: Igh 93Yd 204
Cobsdene DA12: Grav'nd 5F 144
Cobs Way KT15: New H 82L 169
Cobtree Ct. UB1: S'hall 44Ea 86
(off Fleming Rd.)
Coburg Cl. SW1 5C 228 (49Lb 90)
Coburg Cres. SW2 60Pb 112
Coburg Dwellings E1 44Yb 92
(off Hardinge St.)
Coburg Gdns. IG5: Ilf 26Mc 53
Coburg Rd. N22 27Pb 50
Cochrane Cl. NW8 2C 214 (40Fb 69)
(off Cochrane St.)
Cochrane Ct. E10 32Cc 72
(off Leyton Grange Est.)
Cochrane Dr. DA1: Dart 58Md 119
Cochrane Ho. E14 47Cc 92
(off Admirals Way)
Cochrane Ho. UB8: Uxb 39L 63
Cochrane M. NW8 2C 214 (40Fb 69)
Cochrane Pl. GU20: W'sham 8B 146
Cochrane Rd. SW19 66Ab 132
Cochrane St. NW8 2C 214 (40Fb 69)
Cockabourne Ct. RM3: Hrld W 26Qd 57
(off Archibald Rd.)
Cockburn Ho. SW1 50Mb 90
(off Aylesford St.)
COCKCROW HILL 74Ma 153
Cockerell Rd. E17 31Ac 72
Cockerhurst Rd. TN14: S'ham 79Fd 162
Cocker Rd. EN1: Enf 8Xb 19
Cockett Rd. SL3: L'ly 48A 82
COCKFOSTERS 14Hb 31
Cockfosters Pde. EN4: Cockf 14Jb 32
Cockfosters Rd. EN4: Cockf 10Gb 17
Cockfosters Rd. EN4: Had W 10Gb 17
Cockfosters Station (Underground) 14Jb 32
Cock Hill E1 1J 225 (43Ub 91)
Cock La. EC1 1B 224 (43Rb 91)
Cock La. KT22: Fet 94Ea 192
Cockle Way WD7: Shenl 5Na 15
Cockmannings La. BR5: St M Cry 74Zc 161
Cockmannings Rd. BR5: St M Cry 73Zc 161
Cockpit Steps SW1 2E 228 (47Mb 90)
(off Birdcage Wlk.)
Cockpit Theatre 6D 214 (42Gb 89)
(off Gateforth St.)
Cockpit Yd. WC1 7J 217 (43Pb 90)
Cockram Rd. WD23: Bush 15Z 27
Cocks Cres. KT3: N Mald 70Ua 132
Cocksett Av. BR6: Chels 79Uc 160
Cockshot Hill RH2: Reig 7K 207
Cockshot Rd. RH2: Reig 7K 207
Cockspur Ct. SW1 6E 222 (46Mb 90)
Cockspur St. SW1 6E 222 (46Mb 90)
Cocksure La. DA14: Sidc 62Cd 140
Cock's Yd. UB8: Uxb 38M 63
(off High St.)
Coda Cen., The SW6 52Ab 110
Coda Residences SW11 55Fb 111
(of York Place)

Code Ct. NW2 34Wa **68**
Code St. E1 42Vb **91**
Codham Hall La. CM13: Gt War 29Xd **58**
Codicote Dr. WD25: Wat 6Z **13**
Codicote Ho. SE8 49Zb **92**
(off Chilton Gro.)
Codling Cl. E1 46Wb **91**
Codling Way HA0: Wemb 35Ma **67**
Codrington Ct. E1 42Xb **91**
Codrington Ct. GU21: Wok 10K **167**
Codrington Ct. SE16 45Ac **92**
Codrington Cres. DA12: Grav'nd 4E **144**
Codrington Gdns. DA12: Grav'nd 4F **144**
(not continuous)
Codrington Hill SE23 59Ac **114**
Codrington M. W11 44Ab **88**
Cody Cl. HA3: Kenton 27Ma **47**
Cody Cl. SM6: W'gton 80Mb **156**
Cody Rd. E16 42Fc **93**
Coe Av. SE25 72Wb **157**
Coe's All. EN5: Barn 14Ab **30**
Coe Spur SL1: Slou 7F **80**
Cofferdam Way SE8 52Dc **114**
Coffey St. SE8 52Cc **114**
Coftards SL2: Slou 4N **81**
Cogan Av. E17 25Ac **52**
Cohen Cl. EN8: Chesh 3Ac **20**
Coin St. SE1 6K **223** (46Qb **90**)
(not continuous)
Coity Rd. NW5 37Jb **70**
Cokers La. SE21 60Tb **113**
Coke St. E1 44Wb **91**
Colas M. NW6 39Cb **69**
Colbeck M. SE6 59Fc **115**
Colbeck M. SW7 49Db **89**
Colbeck Rd. HA1: Harr 31Ea **66**
Colberg Pl. N16 31Vb **71**
Colbert SE5 53Ub **113**
(off Sceaux Gdns.)
Colborne Cl. KT17: Ewe 79Va **154**
Colborne Cl. SL0: Iver H 42E **82**
Colborne Ho. E14 45Cc **92**
(off E. India Dock Rd.)
Colborne Ho. WD18: Wat 16U **26**
Colborne Way KT4: Wor Pk 76Ya **154**
Colbrook Av. UB3: Harl 48T **84**
Colbrook Cl. UB3: Harl 48T **84**
Colburn Av. CR3: Cat'm 96Vb **197**
Colburn Av. HA5: Hat E 23Aa **45**
Colburn Way RM16: Grays 8A **100**
Colburn Way SM1: Sutt 76Fb **155**
Colby M. DA11: Nflt 61Ee **143**
Colby M. SE19 64Ub **135**
Colby Rd. KT12: Walt T 74W **150**
Colby Rd. SE19 64Ub **135**
Colchester Av. E12 35Pc **74**
Colchester Dr. HA5: Pinn 29Z **45**
Colchester Ho. E3 39Bc **72**
(off Parnell Rd.)
Colchester Rd. CM14: B'wood 22Sd **58**
Colchester Rd. E10 31Ec **72**
Colchester Rd. E17 30Cc **52**
Colchester Rd. HA6: Nwood 26W **44**
Colchester Rd. HA8: Edg 24Sa **47**
Colchester Rd. RM3: Hrld W 25Md **57**
Colchester Rd. RM3: Rom 25Md **57**
Colclough Ct. CR0: C'don 72Sb **157**
(off Simpson Cl.)
Colcokes Rd. SM7: Bans 88Cb **175**
Cold Arbor Rd. TN13: Bes G 96Fd **202**
Coldart Bus. Cen. DA1: Dart 58Md **119**
Coldbath Sq. EC1 5K **217** (42Qb **90**)
Coldbath St. SE13 53Dc **114**
COLDBLOW 60Ed 118
Cold Blow Cres. DA5: Bexl 60Fd **118**
Cold Blow La. SE14 52Zb **114**
(not continuous)
Cold Blows CR4: Mitc 69Hb **133**
Coldershaw Rd. W13 46Ja **86**
Coldershaw Rd. W7 46Ja **86**
Coldfall Av. N10 26Jb **50**
Coldham Ct. N22 25Rb **51**
Coldham Gro. EN3: Enf W 9Ac **20**
Coldharbour E14 47Ec **92**
Coldharbour Cl. TW20: Thorpe 69E **126**
Coldharbour Crest SE9 62Qc **138**
Coldharbour Ho. WD25: Wat 8Aa **13**
Coldharbour Ind. Est. SE5 54Sb **113**
Coldharbour La. CR8: Purl 83Qb **176**
Coldharbour La. GU22: Pyr 87H **169**
Coldharbour La. GU24: W End 3D **166**
Coldharbour La. RH1: Blet 6M **209**
Coldharbour La. RM13: Rain 44Gd **96**
Coldharbour La. RM19: Purf 49Kd **97**
Coldharbour La. SE5 55Sb **113**
Coldharbour La. SW9 56Qb **112**
Coldharbour La. TW20: Thorpe 69E **126**
Coldharbour La. UB3: Hayes 46W **84**
Coldharbour La. WD23: Bush 16Da **27**
Coldharbour Leisure Cen. **61Pc 138**
Coldharbour Pl. SE5 54Sb **113**
Coldharbour Rd. CR0: Wadd 78Qb **156**
Coldharbour Rd. DA11: Nflt 1A **144**
Coldharbour Rd. GU22: Pyr 87H **169**
Coldharbour Rd. KT14: W Byf 86H **169**
Coldharbour Way CR0: Wadd 78Qb **156**
Coldshott RH8: Oxt 5L **211**
Coldstream Gdns. SW18 58Bb **111**
Coldstream Rd. CR3: Cat'm 93Sb **197**
Cold War Bunker
Gravesend **1C 144**
Cole Av. RM16: Grays 9E **100**
Colebeck M. N1 37Rb **71**
Colebert Av. E1 42Yb **92**
Colebert Ho. E1 42Yb **92**
(off Colebert Av.)
Colebrook KT16: Ott 79F **148**
Colebrook Cl. NW7 23Za **48**
Colebrook Cl. SW15 59Za **110**
Colebrook Ct. SW3 6E **226** (49Gb **89**)
(off Makins St.)
Colebrooke Av. W13 44Ka **86**
Colebrooke Ct. DA14: Sidc 62Xc **139**
Colebrooke Dr. E11 31Lc **73**
Colebrooke Pl. KT16: Ott 80D **148**
Colebrooke Pl. N1 1C **218** (39Rb **71**)
Colebrooke Ri. BR2: Broml 68Gc **137**
Colebrooke Rd. RH1: Redh 4N **207**
Colebrooke Row N1 2B **218** (40Rb **71**)
Colebrook Gdns. IG10: Lough 12Rc **36**
Colebrook Ho. E14 44Dc **92**
(off Ellesmere St.)
Colebrook Ho. SE18 52Qc **116**
Colebrook La. IG10: Lough 12Rc **36**
Colebrook Path IG10: Lough 12Rc **36**
Colebrook Rd. SW16 67Nb **134**
Colebrook St. DA8: Erith 51Hd **118**
Colebrook Way N11 22Kb **50**
Coleby Path SE5 52Tb **113**
Colechurch Ho. SE1 50Wb **91**
(off Avondale Sq.)
Cole Cl. SE28 46Xc **95**
Cole Ct. RM3: Rom 22Nd **57**
Cole Ct. TW1: Twick 59Ja **108**
Coledale Dr. HA7: Stan 25La **46**
Colefax Bldgs. E1 44Wb **91**
(off Plumber's Row)
Coleford Ho. RM3: Rom 23Nd **57**
(off Kingsbridge Rd.)
Coleford Rd. SW18 57Eb **111**
Cole Gdns. TW5: Cran 52W **106**
Colegrave Rd. E15 36Fc **73**
Colegrove Rd. SE15 51Vb **113**
Coleherne Ct. SW5 50Db **89**
Coleherne Mans. SW5 50Db **89**
(off Old Brompton Rd.)
Coleherne M. SW10 50Db **89**
Coleherne Rd. SW10 50Db **89**
Colehill Gdns. SW6 54Ab **110**
Colehill La. SW6 53Ab **110**
Cole Ho. SE1 2A **230** (47Qb **90**)
(off Baylis Rd.)
Coleman Cl. SE25 68Wb **135**
Coleman Ct. SW18 59Cb **111**
Coleman Flds. N1 39Sb **71**
Coleman Mans. N8 31Nb **70**
Coleman Rd. DA17: Belv 49Cd **96**
Coleman Rd. RM9: Dag 37Ad **75**
Coleman Rd. SE5 52Ub **113**
Colemans Heath SE9 62Qc **138**
Coleman St. EC2 2F **225** (44Tb **91**)
Coleman St. Bldgs. EC2 2F **225** (44Tb **91**)
(off Coleman St.)
Colenorton Cres. SL4: Eton W 9C **80**
Colenso Dr. NW7 24Wa **48**
Colenso Rd. E5 35Yb **72**
Colenso Rd. IG2: Ilf 32Uc **74**
COLE PARK 58Ja 108
Cole Pk. Gdns. TW1: Twick 57Ja **108**
Cole Pk. Rd. TW1: Twick 58Ja **108**
Cole Pk. Vw. TW1: Twick 58Ja **108**
Colepits Wood Rd. SE9 57Tc **116**
Coleraine Rd. N8 27Qb **50**
Coleraine Rd. SE3 51Hc **115**
Coleridge Av. E12 37Nc **74**
Coleridge Av. SM1: Sutt 77Gb **155**
Coleridge Cl. SW8 54Kb **112**
Coleridge Ct. EN5: New Bar 15Db **31**
(off Station Rd.)
Coleridge Ct. KT22: Lea 91Ja **192**
(off Kingston Rd.)
Coleridge Ct. N1 39Sb **71**
(off Dibden St.)
Coleridge Ct. SW1 5D **228** (49Mb **90**)
(off Regency St.)
Coleridge Ct. W14 48Za **88**
(off Blythe Rd.)
Coleridge Cres. SL3: Poyle 53G **104**
Coleridge Dr. HA4: Ruis 30X **45**
Coleridge Gdns. NW6 38Eb **69**
Coleridge Gdns. SW10 52Db **111**
Coleridge Ho. SE17 7E **230** (50Sb **91**)
(off Browning St.)
Coleridge Ho. SW1 50Lb **90**
(off Churchill Gdns.)
Coleridge La. N8 30Nb **50**
Coleridge Rd. CR0: C'don 73Yb **158**
Coleridge Rd. DA1: Dart 56Rd **119**
Coleridge Rd. E17 28Bc **52**
Coleridge Rd. N12 22Eb **49**
Coleridge Rd. N4 33Qb **70**
Coleridge Rd. N8 30Mb **50**
Coleridge Rd. RM18: Tilb 4E **122**
Coleridge Rd. RM3: Rom 24Kd **57**
Coleridge Rd. TW15: Ashf 63N **127**
Coleridge Sq. SW10 52Eb **111**
(off Coleridge Gdns.)
Coleridge Sq. W13 44Ja **86**
Coleridge Wlk. CM13: Hut 17Ee **41**
Coleridge Wlk. NW11 28Cb **49**
Coleridge Way BR6: St M Cry 72Wc **161**
Coleridge Way UB4: Hayes 44W **84**
Coleridge Way UB7: W Dray 49N **83**
Coleridge Way WD6: Bore 14Qa **29**
Cole Rd. TW1: Twick 58Ja **108**
Cole Rd. WD17: Wat 11X **27**
Colesburg Rd. BR3: Beck 69Bc **136**
Coles Cres. HA2: Harr 33Da **65**
Colescroft Hill CR8: Purl 87Qb **176**
Colesdale EN6: Cuff 2Nb **18**
Coles Grn. IG10: Lough 11Qc **36**
Coles Grn. WD23: B Hea 18Ea **28**
Coles Grn. Ct. NW2 33Wa **68**
Coles Grn. Rd. NW2 32Wa **68**
Coles Hill HP1: Hem H 1J **3**
Coleshill Flats SW1 6J **227** (49Jb **90**)
(off Pimlico Rd.)
Coleshill Rd. TW11: Tedd 65Ga **130**
Coles La. TN16: Bras 95Yc **201**
Colesmead Rd. RH1: Redh 3P **207**
COLES MEADS 3P 207
Colestown St. SW11 54Gb **111**
Cole St. SE1 2E **230** (47Sb **91**)
Colesworth Ho. HA8: Edg 26Sa **47**
(off Burnt Oak B'way.)
Colet Cl. N13 23Rb **51**
Colet Ct. W6 49Za **88**
(off Hammersmith Rd.)
Colet Flats E1 44Ac **92**
(off Troon St.)
Colet Gdns. W14 49Za **88**
Colet Ho. SE17 50Rb **91**
(off Doddington Gro.)
Colet Rd. CM13: Hut 15Ee **41**
Colets Orchard TN14: Otf 88Kd **183**
Colette Ct. SE16 47Zb **92**
(off Eleanor Cl.)
Coley Av. GU22: Wok 90C **168**
Coley St. WC1 6J **217** (42Pb **90**)
Colfe & Hatcliffe Glebe SE13 57Dc **114**
(off Lewisham High St.)
Colfe Rd. SE23 60Ac **114**
Colfes Leisure Cen. **58Jc 115**
Colgate Ct. EN5: Barn 15Ab **30**
(off Leecroft Rd.)
Colgate Pl. EN3: Enf L 9Cc **20**
Colham Av. UB7: Yiew 46N **83**
COLHAM GREEN 43Q 84
Colham Grn. Rd. UB8: Hil 43Q **84**
Colham Mill Rd. UB7: W Dray 47M **83**
Colham Rd. UB8: Hil 42P **83**
COLHAM RDBT. 44Q **84**
Colina M. N15 28Rb **51**
Colina Rd. N15 29Rb **51**
Colin Chapman Way DA3: Fawk 77Ud **164**
Colin Cl. BR4: W W'ck 76Hc **159**
Colin Cl. CR0: C'don 76Bc **158**
Colin Cl. DA2: Dart 58Rd **119**
Colin Cl. NW9 28Ua **48**
Colin Ct. SE6 59Bc **114**
Colin Cres. NW9 28Va **48**
COLINDALE 27Ta 47
Colindale Av. AL1: St A 4D **6**
Colindale Av. NW9 27Ta **47**
Colindale Bus. Pk. NW9 27Sa **47**
Colindale Gdns. NW9 27Va **48**
Colindale Retail Pk. 28Ta **47**
Colindale Station (Underground) 27Ua 48
Colindeep Gdns. NW4 28Wa **48**
Colindeep La. NW4 28Wa **48**
Colindeep La. NW9 27Ua **48**
Colin Dr. NW9 29Va **48**
Colinette Rd. SW15 56Ya **110**
Colin Gdns. NW9 28Va **48**
Colin Pde. NW9 28Ua **48**
Colin Pk. Rd. NW9 28Ua **48**
Colin Pond Ct. RM6: Chad H 28Zc **55**
Colin Rd. CR3: Cat'm 95Wb **197**
Colin Rd. NW10 37Wa **68**
Colinsdale N1 1B **218** (39Rb **71**)
(off Camden Wlk.)
Colinswood SL2: Farn C 4G **60**
Colinton Rd. IG3: Ilf 33Xc **75**
Colin Way SL1: Slou 8F **80**
Colin Winter Ho. E1 42Yb **92**
(off Nicholas Rd.)
Coliseum Theatre London **5F 223 (45Nb 90)**
(off St Martin's La.)
Coliston Pas. SW18 59Cb **111**
(off Coliston Rd.)
Coliston Rd. SW18 59Cb **111**
Collamore Av. SW18 60Gb **111**
Collapit Cl. HA1: Harr 30Da **45**
Collard Av. IG10: Lough 12Sc **36**
Collard Cl. CR8: Kenley 92Tb **197**
Collard Grn. IG10: Lough 12Sc **36**
Collard Pl. NW1 38Kb **70**
Collards Almshouses E17 29Ec **52**
(off Maynard Rd.)
Collection Pl. NW8 39Db **69**
(off Bolton Pl.)
College SL4: Eton 1H **103**
(off Westons Yd.)
College, The KT17: Eps 87Wa **174**
College App. SE10 51Ec **114**
College Av. HA3: Hrw W 25Ga **46**
College Av. KT17: Eps 86Va **174**
College Av. RM17: Grays 49De **99**
College Av. SL1: Slou 8J **81**
College Av. TW20: Egh 65D **126**
College Cl. AL9: N Mym 10F **8**
College Cl. E9 36Yb **72**
College Cl. HA3: Hrw W 24Ga **46**
College Cl. IG10: Lough 14Rc **36**
College Cl. N18 22Vb **51**
College Cl. RM17: Grays 49Ee **99**
College Cl. TW2: Twick 60Fa **108**
College Ct. EN3: Pond E 15Yb **34**
College Ct. EN8: Chesh 2Yb **20**
College Ct. NW3 37Fb **69**
(off College Cres.)
College Ct. RM2: Rom 29Kd **57**
(off Scholars Way)
College Ct. SW3 50Hb **89**
(off West Rd.)
College Ct. W5 45Na **87**
College Ct. W6 50Ya **88**
(off Queen Caroline St.)
College Ct. Rd. EN3: Pond E 15Yb **34**
College Cres. NW3 37Eb **69**
College Cres. RH1: Redh 3A **208**
College Cres. SL4: Wind 4F **102**
College Cross N1 38Qb **70**
College Dr. HA4: Ruis 31W **64**
College Dr. KT7: T Ditt 73Ga **152**
College E. E1 1K **225** (43Vb **91**)
College Flds. Bus. Cen. SW19 67Gb **133**
College Gdns. E4 17Dc **34**
College Gdns. EN2: Enf 11Tb **33**
College Gdns. IG4: Ilf 29Nc **54**
College Gdns. KT3: N Mald 71Va **154**
College Gdns. N18 22Wb **51**
College Gdns. SE21 60Ub **113**
College Gdns. SW17 61Gb **133**
College Grn. SE19 66Ub **135**
College Gro. NW1 39Lb **70**
College Hall WC1 6D **216** (42Mb **90**)
College Hill EC4 4E **224** (45Sb **91**)
College Hill Rd. HA3: Hrw W 24Ga **46**
College Ho. SW15 57Za **110**
College Ho. TW7: Isle 52Ga **108**
College La. AL10: Hat 2A **8**
(not continuous)
College La. GU22: Wok 1N **187**
College La. NW5 35Kb **70**
College Mans. NW6 39Ab **68**
(off Salusbury Rd.)
College M. N1 38Qb **70**
(off College Cross)
College M. SW1 3F **229** (48Nb **90**)
(off Gt. College St.)
College M. SW18 57Db **111**
College Pde. NW6 39Ab **68**
(off Salusbury Rd.)
COLLEGE PARK 41Xa 88
College Pk. Cl. SE13 56Fc **115**
College Pk. Rd. N17 23Vb **51**
College Pl. AL3: St A 2A **6**
College Pl. CM16: They B 8Sc **22**
College Pl. DA9: Ghithe 56Yd **120**
College Pl. E17 28Gc **53**
College Pl. NW1 39Lb **70**
College Pl. SW10 52Eb **111**
College Point E15 37Hc **73**
College Rd. AL1: St A 3F **6**
College Rd. BR1: Broml 67Jc **137**
College Rd. BR8: Hext 67Gd **140**
College Rd. CR0: C'don 75Tb **157**
College Rd. DA11: Nflt 57De **121**
College Rd. E17 29Ec **52**
College Rd. EN2: Enf 12Tb **33**
College Rd. EN8: Chesh 2Yb **20**
College Rd. GU22: Wok 88D **168**
College Rd. HA1: Harr 30Ga **46**
College Rd. HA3: Hrw W 25Ga **46**
College Rd. HA9: Wemb 32Ma **67**
College Rd. KT17: Eps 86Va **174**
College Rd. N17 23Vb **51**
College Rd. N21 19Qb **32**
College Rd. NW10 40Ya **68**
College Rd. RM17: Grays 49Ee **99**
College Rd. SE19 64Vb **135**
College Rd. SE21 59Ub **113**
College Rd. SL1: Slou 6D **80**
College Rd. SW19 65Fb **133**
College Rd. TW7: Isle 53Ha **108**
College Rd. W13 44Ka **86**
College Rd. WD5: Ab L 3V **12**
COLLEGE RDBT. Kingston upon Thames 69Na **131**
College Row E9 36Zb **72**
(off Homerton Gro.)
College Slip BR1: Broml 67Jc **137**
College St. AL3: St A 2B **6**
College St. EC4 4E **224** (45Sb **91**)
College Ter. E3 41Bc **92**
College Ter. N3 26Bb **49**
College Vw. KT17: Eps 87Va **174**
College Vw. SE9 60Mc **115**
College Wlk. KT1: King T 69Na **131**
College Way HA6: Nwood 23T **44**
College Way RM16: Grays 8A **100**
College Way RM8: Dag 35Wc **75**
College Way TW15: Ashf 63P **127**
College Way UB3: Hayes 45W **84**
College Yd. AL3: St A 2B **6**
(off Lwr. Dagnall St.)
College Yd. NW5 35Kb **70**
College Yd. NW6 39Ab **68**
College Yd. WD24: Wat 10X **13**
Collendale Rd. E17 27Zb **52**
Collens Fld. GU24: Pirb 6D **186**
Collent St. E9 37Yb **72**
Coller Cres. DA2: Daren 64Ud **142**
Collerne St. RM3: Rom 22Nd **57**
Collerston Ho. SE10 50Hc **93**
(off Armitage Rd.)
Colless Rd. N15 29Vb **51**
Collet Rd. TN15: Kems'g 89Nd **183**
Collett Rd. HP1: Hem H 2L **3**
Collett Rd. SE16 48Wb **91**
Collett Way UB2: S'hall 47Da **85**
Colley Hill La. SL2: Hedg 4K **61**
Colley Hill Viewpoint **2H 207**
Colley Ho. UB8: Uxb 39M **63**
Colleyland WD3: Chor 14F **24**
Colley La. RH2: Reig 5G **206**
Colley Mnr. Dr. RH2: Reig 5F **206**
Colley Way RH2: Reig 3G **206**
Collier Cl. E6 45Rc **94**
Collier Cl. KT19: Ewe 79Qa **153**
Collier Cl. SL1: Slou 7D **80**
Collier Ct. RM16: Grays 46De **99**
Collier Dr. HA8: Edg 26Qa **47**
COLLIER ROW 24Dd 56
Collier Row La. RM5: Col R 24Dd **56**
Collier Row Rd. RM5: Col R 25Bd **55**
Colliers CR3: Cat'm 97Wb **197**
Colliers Cl. GU21: Wok 9M **167**
Colliers Ct. CR0: C'don 77Tb **157**
(off St Peter's Rd.)
Colliers Shaw BR2: Kes 77Mc **159**
Collier St. N1 2H **217** (40Pb **70**)
Colliers Water La. CR7: Thor H 71Qb **156**
COLLIERS WOOD 66Fb 133
COLLIERS WOOD 66Fb **133**
Colliers Wood Station (Underground) 66Fb 133
Colliford Ct. HA8: Edg 21Pa **47**
(off King's Dr.)
Collindale Av. DA15: Sidc 60Wc **117**
Collindale Av. DA8: Erith 52Dd **118**
Collingbourne Rd. W12 46Xa **88**
Collingham Gdns. SW5 49Db **89**
Collingham Pl. SW5 49Db **89**
Collingham Rd. SW5 49Db **89**
Collingridge Way KT17: Ewe 82Wa **174**
Collingsbourne KT15: Add 77L **149**
(off High St.)
Collings Cl. N22 23Pb **50**
Collington Cl. DA11: Nflt 9A **122**
Collington St. SE10 50Fc **93**
Collingtree Rd. SE26 63Yb **136**
Collingwood Av. KT5: Surb 74Sa **153**
Collingwood Av. N10 27Jb **50**
Collingwood Cl. SE20 67Xb **135**
Collingwood Cl. TW2: Whitt 59Ca **107**
Collingwood Ct. EN5: New Bar 15Db **31**
Collingwood Ct. W5 43Pa **87**
Collingwood Dr. AL2: Lon C 7H **7**
Collingwood Ho. DA9: Ghithe 57Yd **120**
Collingwood Ho. E1 42Xb **91**
(off Darling Row)
Collingwood Ho. SE16 47Xb **91**
(off Cherry Gdn. St.)
Collingwood Ho. SW1 50Mb **90**
(off Dolphin Sq.)
Collingwood Ho. W1 7B **216** (43Lb **90**)
(off Clipstone St.)
Collingwood Pl. KT12: Walt T 76W **150**
Collingwood Rd. CR4: Mitc 69Gb **133**
Collingwood Rd. E17 30Cc **52**
Collingwood Rd. N15 28Ub **51**
Collingwood Rd. RM13: Rain 40Hd **76**
Collingwood Rd. SM1: Sutt 76Cb **155**
Collingwood Rd. UB8: Hil 42R **84**
Collingwood St. E1 42Xb **91**
Collins Av. HA7: Stan 26Na **47**
Collins Bldg. NW2 33Ya **68**
Collins Cl. SS17: Stan H 1N **101**
Collins Ct. E8 37Wb **71**
Collins Ct. WD23: Bush 15Da **27**
(off Lea Cl.)
Collins Dr. HA4: Ruis 33Y **65**
Collins Ho. E14 45Ec **92**
(off Newby Pl.)
Collins Ho. SE10 50Hc **93**
(off Armitage Rd.)
Collinson Ct. SE1 2D **230** (47Sb **91**)
(off Gt. Suffolk St.)
Collinson Ho. SE15 52Wb **113**
(off Peckham Pk. Rd.)
Collinson Ho. SE3 56Kc **115**
(off Wallace Ct.)
Collinson St. SE1 2D **230** (47Sb **91**)
Collinson Wlk. SE1 2D **230** (47Sb **91**)
Collins Path TW12: Hamp 65Ba **129**
Collins Rd. N5 35Sb **71**
Collins Sq. SE3 54Hc **115**
Collins St. SE3 54Gc **115**
(not continuous)
Collins Way DA8: Erith 52Hd **118**
Collinswood Rd. SL2: Farn C 2E **60**
Collin's Yd. N1 39Rb **71**
Collinwood Av. EN3: Enf H 13Yb **34**
Collinwood Gdns. IG5: Ilf 29Pc **54**
Collis All. TW2: Twick 60Ga **108**
Collison Av. EN5: Barn 14Ya **30**
Collison Pl. N16 32Ub **71**
Coll's Rd. SE15 53Yb **114**
Collum Grn. Rd. SL2: Farn C 4H **61**
Collum Grn. Rd. SL2: Hedg 4H **61**
Collum Grn. Rd. SL2: Stoke P 4H **61**
Collyer Av. CR0: Bedd 77Nb **156**
Collyer Pl. SE15 53Wb **113**
Collyer Rd. AL2: Lon C 9H **7**
Collyer Rd. CR0: Bedd 77Nb **156**
Colman Cl. KT18: Tatt C 89Ya **174**
Colman Cl. SS17: Stan H 1M **101**
Colman Ct. HA7: Stan 23Ka **46**
Colman Ct. N12 23Eb **49**
Colman Ho. RH1: Redh 4P **207**
Colman Pde. EN1: Enf 13Ub **33**
Colman Rd. E16 43Lc **93**
Colmans Wharf E14 43Dc **92**
(off Morris Rd.)
Colman Way RH1: Redh 4N **207**
Colmar Cl. E1 42Zb **92**
Colmer Pl. HA3: Hrw W 24Fa **46**
Colmer Rd. SW16 67Nb **134**
Colmore M. SE15 53Xb **113**
Colmore Rd. EN3: Pond E 14Yb **34**
COLNBROOK 52F 104
Colnbrook By-Pass SL3: L'ly 51E **104**
Colnbrook Cl. AL2: Lon C 9J **7**
Colnbrook Ct. SL3: Poyle 53H **105**
Colnbrook St. SE1 4B **230** (48Rb **91**)
Colndale Rd. SL3: Poyle 54G **104**
Colne RM18: E Til 8L **101**
Colne Av. UB7: W Dray 47L **83**
Colne Av. WD19: Wat 16X **27**
Colne Av. WD3: Rick 19J **25**
Colne Bank SL3: Hort 55E **104**
Colnebridge Cl. TW18: Staines 63G **126**
Colne Bri. Retail Pk. 16Z **27**
Colne Cl. RM15: S Ock 45Yd **98**
Colne Ct. KT19: Ewe 77Sa **153**
Colne Ct. RM18: E Til 8L **101**
Colne Ct. W7 44Fa **86**
(off High La.)
Colnedale Rd. UB8: Uxb 36M **63**
Colne Dr. KT12: Walt T 76Z **151**
Colne Dr. RM3: Rom 23Pd **57**
Colne Gdns. AL2: Lon C 9J **7**
Colne Ho. IG11: Bark 38Rc **74**
Colne Ho. NW8 6C **214** (42Fb **89**)
(off Penfold St.)
Colne Lodge WD23: Bush 14Z **27**
Colne Lodge WD3: Rick 18J **25**
Colne Mead WD3: Rick 19J **25**
Colne Orchard SL0: Iver 44H **83**
Colne Pk. Cvn. Site UB7: W Dray 49L **83**
Colne Pl. UB9: Den 31J **63**
Colne Reach TW19: Stanw M 57H **105**
Colne Rd. E5 35Ac **72**
Colne Rd. N21 17Tb **33**
Colne Rd. TW1: Twick 60Ha **108**
Colne Rd. TW2: Twick 60Ga **108**
Colne St. E13 41Jc **93**
Colne Valley RM14: Upm 30Ud **58**
Colne Valley Pk. Visitor Cen. **34K 63**
Colne Valley Retail Pk. 15Z **27**
Colne Way TW19: Staines 61D **126**
Colne Way WD24: Wat 8Z **13**
Colne Way WD25: Wat 8Y **13**
Colne Way Ct. WD24: Wat 9Z **13**
Colne Way Ind. Est. WD25: Wat 8Z **13**
Colney Flds. Shop. Pk. 10K **7**
COLNEY HATCH 23Hb 49
Colney Hatch La. N10 24Jb **50**
Colney Hatch La. N11 23Hb **49**
COLNEY HEATH 5P 7
Colney Heath La. AL4: St A 2H **7**
Colney Rd. DA1: Dart 58Pd **119**
COLNEY STREET 2Ga 14
Colnhurst Rd. WD17: Wat 11W **26**
Colnmore Ct. E2 41Zb **92**
(off Meath Cres.)
Coln Trad. Est. SL3: Poyle 53H **105**
Cologne Rd. SW11 56Fb **111**
Coloma Ct. BR4: W W'ck 77Gc **159**
Colombo Cen. **7B 224 (46Rb 91)**
(off Colombo St.)
Colombo Rd. IG1: Ilf 32Sc **74**
Colombo St. SE1 7B **224** (46Rb **91**)
Colomb St. SE10 50Gc **93**
Colonel's La. KT16: Chert 72J **149**
Colonel's Wlk. EN2: Enf 13Rb **33**
Colonial Av. TW2: Whitt 58Ea **108**
Colonial Bus. Pk. WD24: Wat 11Y **27**
Colonial Ct. N7 34Pb **70**
Colonial Dr. W4 49Sa **87**
Colonial Rd. SL1: Slou 7L **81**
Colonial Rd. TW14: Felt 59U **106**
Colonial Way WD24: Wat 11Y **27**
Colonnade WC1 6F **217** (42Nb **90**)
Colonnade, The AL3: St A 2B **6**
(off Verulam Rd.)
Colonnade, The EN8: Chesh 1Zb **20**
Colonnade, The SE1 1K **229** (47Qb **90**)
(off Waterloo Rd.)
Colonnade, The SE8 49Bc **92**
Colonnade Gdns. W3 46Va **88**
Colonnades, The W2 44Db **89**
Colonnades Leisure Pk., The **79Qb 156**
Colonnade Wlk. SW1 6K **227** (49Kb **90**)
Colonsay HP3: Hem H 4C **4**
Colony Mans. SW5 50Db **89**
(off Earl's Ct. Rd.)
Colony M. N1 36Tb **71**
(off Mildmay Gro. Nth.)
Colorado Apts. N8 27Pb **50**
(off Gt. Amwell La.)
Colorado Bldg. SE13 53Dc **114**
(off Deal's Gateway)
Colosseum Apts. E2 40Zb **72**
(off Palmers Rd.)
Colosseum Ter. NW1 4A **216** (41Kb **90**)
(off Albany St.)
Colour Ct. SW1 7C **222** (46Lb **90**)
(off Marlborough Rd.)
Colour Ho. SE1 2J **231** (47Ub **91**)
(off Bell Yd M.)
Colour House Theatre Merton **67Eb 133**
Colroy Ct. NW11 29Ab **48**
Colson Gdns. IG10: Lough 14Rc **36**
Colson Grn. IG10: Lough 15Rc **36**
Colson Path IG10: Lough 14Qc **36**
Colson Rd. CR0: C'don 75Ub **157**
Colson Rd. IG10: Lough 14Rc **36**
Colson Way SW16 63Lb **134**
Colstead Ho. E1 44Xb **91**
(off Watney Mkt.)
Colsterworth Rd. N15 28Vb **51**
(not continuous)
Colston Av. SM5: Cars 77Gb **155**
Colston Ct. SL9: Ger X 30A **42**
Colston Ct. SM5: Cars 77Hb **155**
Colston Rd. E7 37Mc **73**
Colston Rd. SW14 56Sa **109**
Coltash Ct. EC1 5E **218** (42Sb **91**)
(off Whitecross St.)
Colthurst Cres. N4 33Rb **71**
Colthurst Dr. N9 20Xb **33**
Coltishall Rd. RM12: Horn 37Ld **77**
Coltman Ho. SE10 51Ec **114**
(off Welland St.)

Coltman St. E14 43Ac **92**
Colt M. EN3: Enf L 9Cc **20**
Coltness Cres. SE2 50Xc **95**
Colton Gdns. N17 27Sb **51**
Colton Rd. HA1: Harr 29Ga **46**
Coltsfoot, The HP1: Hem H 3G **2**
Coltsfoot Ct. RM17: Grays 51Fe **121**
Coltsfoot Dr. UB7: Yiew 44N **83**
Coltsfoot La. RH8: Oxt 5K **211**
Coltsfoot Path RM3: Rom 24Ld **57**
Coltstead DA3: Nw A G 75Ae **165**
Columbas Dr. NW3 32Fb **69**
Columbia Av. HA4: Ruis 32X **65**
Columbia Av. HA8: Edg 25Ra **47**
Columbia Av. KT4: Wor Pk 73Va **154**
Columbia Gdns. SW6 51Cb **111**
(off One Lillie Sq.)
Columbia Gdns. Nth. SW6 51Cb **111**
(off Rickett St.)
Columbia Gdns. Sth. SW6 51Cb **111**
(off Rickett St.)
Columbia Ho. E3 42Bc **92**
(off Hamlets Way)
Columbia Point SE16 48Yb **92**
(off Canada Est.)
Columbia Rd. E13 42Hc **93**
Columbia Rd. E2 3K **219** (41Vb **91**)
Columbia Rd. Flower Market 3K **219** (41Vb **91**)
(off Columbia Rd.)
Columbia Sq. SW14 56Sa **109**
Columbia Wharf EN3: Pond E 16Ac **34**
Columbia Wharf Rd. RM17: Grays 51Ce **121**
Columbine Av. CR2: S Croy 80Rb **157**
Columbine Av. E6 43Nc **94**
Columbine Way RM3: Hrld W 25Nd **57**
Columbine Way SE13 54Ec **114**
Columbus Ct. DA8: Erith 52Hd **118**
Columbus Ct. SE16 46Yb **92**
(off Rotherhithe St.)
Columbus Ctyd. E14 46Cc **92**
Columbus Gdns. HA6: Nwood 25W **44**
Columbus Sq. DA8: Erith 51Hd **118**
Colva Wlk. N19 33Kb **70**
Colvern Ho. RM7: Rom 29Ed **56**
Colverson Ho. E1 43Yb **92**
(off Lindley St.)
Colvestone Cres. E8 36Vb **71**
Colview Ct. SE9 60Mc **115**
Colville Est. N1 39Ub **71**
Colville Est. W. E2 41Vb **91**
(off Turin St.)
Colville Gdns. GU18: Light 3A **166**
Colville Gdns. W11 44Bb **89**
(not continuous)
Colville Ho. E2 40Yb **72**
(off Waterloo Gdns.)
Colville Ho's. W11 44Bb **89**
Colville Mans. E20 36Dc **72**
(off Victory Pde.)
Colville M. W11 44Bb **89**
Colville Pl. W1 1D **222** (43Mb **90**)
Colville Rd. E11 34Ec **72**
Colville Rd. E17 26Ac **52**
Colville Rd. N9 18Xb **33**
Colville Rd. W11 44Bb **89**
Colville Rd. W3 48Ra **87**
Colville Sq. W11 44Bb **89**
Colville Ter. W11 44Bb **89**
Colvin Cl. SE26 64Yb **136**
Colvin Gdns. E11 28Kc **53**
Colvin Gdns. E4 20Ec **34**
Colvin Gdns. EN8: Walt C 7Zb **20**
Colvin Gdns. IG6: Ilf 25Sc **54**
Colvin Ho. W10 44Za **88**
(off Kingsdown Cl.)
Colvin Rd. CR7: Thor H 71Qb **156**
Colvin Rd. E6 38Nc **74**
Colwall Gdns. IG8: Wfd G 22Jc **53**
Colwell Cres. EN3: Pond E 15Yb **34**
Colwell Ho. KT12: Walt T 74V **150**
(off Hepworth Way)
Colwell Rd. SE22 57Vb **113**
Colwick Cl. N6 31Mb **70**
Colwith Rd. W6 51Ya **110**
Colwood Gdns. SW19 66Fb **133**
Colworth Gro. SE17 6E **230** (49Sb **91**)
Colworth Rd. CR0: C'don 74Wb **157**
Colworth Rd. E11 30Gc **53**
Colwyn Av. UB6: G'frd 40Ha **66**
Colwyn Cl. SW16 64Lb **134**
Colwyn Cres. TW3: Houn 53Ea **108**
Colwyn Grn. NW9 30Ua **48**
(off Snowdon Dr.)
Colwyn Ho. SE1 4K **229** (48Qb **90**)
Colwyn Rd. NW2 34Xa **68**
Colyer Cl. SE9 61Rc **138**
Colyer Rd. DA11: Nflt 61Ee **143**
Colyers Cl. DA8: Erith 53Fd **118**
Colyers La. DA8: Erith 53Ed **118**
Colyers Wlk. DA8: Erith 53Gd **118**
Colyton Cl. DA16: Well 53Zc **117**
Colyton Cl. GU21: Wok 10N **167**
Colyton Cl. HA0: Wemb 37La **66**
Colyton La. SW16 64Qb **134**
Colyton Rd. SE22 57Xb **113**
Colyton Way N18 22Wb **51**
Combe, The NW1 4A **216** (41Kb **90**)
Combe Av. SE3 52Hc **115**
Combe Bank Dr. TN14: Sund 94Ad **201**
Combedale Rd. SE10 50Jc **93**
Combe Ho. W2 43Cb **89**
(off Gt. Western Rd.)
Combe Ho. WD18: Wat 16U **26**
Combemartin Rd. SW18 59Ab **110**
Combe M. SE3 52Hc **115**
Combe Pl. KT11: Cobh 86X **171**
Comber Cl. NW2 34Xa **68**
Comber Gro. SE5 52Sb **113**
Comber Ho. SE5 52Sb **113**
Combermere Cl. SL4: Wind 4F **102**
Combermere Rd. SM4: Mord 72Db **155**
Combermere Rd. SW9 55Pb **112**
Combe Rd. WD18: Wat 16V **26**
Comberton KT1: King T 68Qa **131**
(off Eureka Rd.)
Comberton Rd. E5 33Xb **71**
Combeside SE18 52Vc **117**
Combe St. HP1: Hem H 2L **3**
Combewood WD25: Wat 4Y **13**
Combwell Cres. SE2 48Wc **95**
Comedy Store 5D **222** (45Mb **90**)
(off Oxendon St.)
Comely Bank Rd. E17 29Ec **52**
Comeragh Cl. GU22: Wok 2L **187**
Comeragh M. W14 50Ab **88**
Comeragh Rd. W14 50Ab **88**
Comer Cres. UB2: S'hall 47Ea **86**
(off Windmill Av.)
Comerford Rd. SE4 56Ac **114**

Comer Ho. EN5: New Bar 14Eb **31**
Comet Cl. E12 35Mc **73**
Comet Cl. RM19: Purf 49Qd **97**
Comet Cl. WD25: Wat 6V **12**
Comet Ho. UB3: Harl 52S **106**
Comet Pl. SE8 52Cc **114**
(not continuous)
Comet Rd. TW19: Stanw 59M **105**
Comet St. SE8 52Cc **114**
Comet Way AL10: Hat 1A **8**
UB3: Harl
Comforts Farm Av. RH8: Oxt 5K **211**
Comfort St. SE15 51Ub **113**
Comfrey Ct. RM17: Grays 51Fe **121**
Commander Av. NW9 27Wa **48**
Commerce Pk. CR0: Wadd 75Pb **156**
Commerce Rd. N22 25Pb **50**
Commerce Rd. TW8: Bford 51La **108**
Commerce Way CR0: Wadd 75Pb **156**
Commercial Pl. DA12: Grav'nd 8E **122**
Commercial Rd. E1 44Wb **91**
Commercial Rd. E14 44Ac **92**
Commercial Rd. N18 22Ub **51**
Commercial Rd. TW18: Staines 65J **127**
Commercial Rd. Ind. Est. N18 23Vb **51**
Commercial St. E1 6K **219** (42Vb **91**)
Commercial Way GU21: Wok 89A **168**
Commercial Way NW10 40Ra **67**
Commercial Way SE10 49Hc **93**
Commercial Way SE15 52Vb **113**
Commercial Wharf E8 39Ub **71**
(off Kingsland Rd.)
Commerell Pl. SE10 50Hc **93**
Commerell St. SE10 50Gc **93**
Commodity Quay E1 5K **225** (45Vb **91**)
Commodore Ct. SE8 53Cc **114**
(off Albyn Rd.)
Commodore Ho. E14 45Ec **92**
(off Poplar High St.)
Commodore Ho. E16 47Kc **93**
(off Royal Crest Av.)
Commodore Ho. SW18 55Eb **111**
Commodore St. E1 42Ac **92**
Common, The E15 37Gc **73**
Common, The HA7: Stan 19Ga **28**
Common, The KT21: Asht 88Ma **173**
COMMON, THE 94Ba **191**
Common, The UB2: S'hall 49Y **85**
Common, The UB7: W Dray 49L **83**
Common, The W5 45Na **87**
(not continuous)
Common, The WD4: Chfd 4J **11**
Common, The WD4: K Lan 10A **4**
Common, The WD7: Shenl 3Ka **14**
Common Cl. GU21: Wok 6P **167**
Commondale SW15 54Ya **110**
Commonfield La. SW17 64Gb **133**
Commonfield Rd. SM7: Bans 86Cb **175**
Commonfields GU24: W End 4E **166**
Common Ga. Rd. WD3: Chor 15F **24**
Common La. DA2: Wilm 61Jd **140**
Common La. GU24: W End 7B **166**
Common La. KT10: Clay 80Ja **152**
Common La. KT15: New H 81L **169**
Common La. SL1: Burn 4B **60**
Common La. SL4: Eton 10H **81**
Common La. WD25: Let H 11Ga **28**
Common La. WD4: K Lan 10P **3**
Common La. WD7: R'lett 10Ga **14**
Common La. Ho. SL4: Eton 10H **81**
(off Common La.)
Commonmeadow La. WD25: A'ham 6Da **13**
Common Mile Cl. SW4 57Mb **112**
Common Moor Lock 16S **26**
(off Mill La.)
Common Rd. CM13: Ingve 22Ee **59**
Common Rd. HA7: Stan 21Fa **46**
Common Rd. KT10: Clay 79Ja **152**
Common Rd. RH1: Redh 8P **207**
Common Rd. SL3: L'ly 49C **82**
Common Rd. SL4: Dor 9A **80**
Common Rd. SL4: Eton W 9A **80**
Common Rd. SL4: Eton W 9D **80**
Common Rd. SW13 55Xa **110**
Common Rd. TN15: Igh 95Xd **204**
Common Rd. WD3: Chor 14F **24**
Common Side KT18: Eps 87Qa **173**
Commonside KT23: Kes 77Lc **159**
Commonside KT23: Bookh 94Ca **191**
(not continuous)
Commonside Cl. CR5: Coul 92Rb **197**
Commonside Cl. SM2: Sutt 83Db **175**
Commonside E. CR4: Mitc 69Jb **134**
Commonside W. CR4: Mitc 69Hb **133**
Commons La. HP2: Hem H 1N **3**
Commonwealth Av. UB3: Hayes 44T **84**
Commonwealth Av. W12 45Xa **88**
(not continuous)
Commonwealth Ho. RM18: Tilb 4C **122**
(off Montreal Rd.)
Commonwealth Memorial Gates 1K **227** (47Kb **90**)
Commonwealth Rd. CR3: Cat'm 95Wb **197**
Commonwealth Rd. N17 24Wb **51**
Commonwealth Way SE2 50Xc **95**
Common Wood SL2: Farn C 6G **60**
COMMONWOOD 6K **11**
Community Cl. TW5: Cran 53X **107**
Community Cl. UB10: Ick 34R **64**
Community La. N7 36Mb **70**
Community Rd. E15 36Fc **73**
Community Rd. UB6: G'frd 39Ea **66**
Community Wlk. KT10: Esh 77Ea **152**
Community Way WD3: Crox G 15Q **26**
Como Rd. SE23 61Ac **136**
Como St. RM7: Rom 29Fd **56**
COMP 93Fe **205**
Compass Bus. Pk. KT9: Chess 77Qa **153**
Compass Cl. HA8: Edg 21Pa **47**
Compass Cl. TW15: Ashf 66S **128**
Compass Ct. SE1 7K **225** (46Vb **91**)
(off Shad Thames)
Compass Hill TW10: Rich 58Ma **109**
Compass Ho. E1 46Xb **91**
(off Raine St.)
Compass Ho. SW18 56Db **111**
Compass La. BR1: Broml 67Jc **137**
(off North St.)
Compass Point E14 45Bc **92**
(off Grenade St.)
Compass Theatre 34S **64**
Compayne Gdns. NW6 38Db **69**
Compayne Mans. NW6 37Db **69**
(off Fairhazel Gdns.)
Comp La. TN15: Plat 93Ee **205**
Comp La. TN15: Wro H 93Ee **205**
Comport Grn. CR0: New Ad 84Gc **179**
Compter Pas. EC2 3E **224** (44Sb **91**)
(off Wood St.)

Compton Av. CM13: Hut 18Ee **41**
Compton Av. E6 40Mc **73**
Compton Av. HA0: Wemb 35La **66**
Compton Av. N1 37Rb **71**
Compton Av. N6 31Gb **69**
Compton Av. RM2: Rom 27Ld **57**
Compton Cl. E3 43Cc **92**
Compton Cl. HA8: Edg 24Sa **47**
Compton Cl. KT10: Esh 79Fa **152**
Compton Cl. NW1 4A **216** (41Kb **90**)
(off Robert St.)
Compton Cl. NW11 34Za **68**
Compton Cl. SE15 52Wb **113**
Compton Cl. W13 44Ja **86**
Compton Ct. SE19 65Ub **135**
Compton Ct. SL1: Slou 4C **80**
Compton Ct. SM1: Sutt 77Eb **155**
Compton Cres. KT9: Chess 78Na **153**
Compton Cres. N17 24Sb **51**
Compton Cres. UB5: N'olt 39Z **65**
Compton Cres. W4 51Sa **109**
Compton Gdns. AL2: Chis G 8P **5**
Compton Gdns. KT15: Add 78K **149**
(off Monks Cres.)
Compton Ho. E20 36Dc **72**
(off Peloton Av.)
Compton Ho. SW10 53Eb **111**
Compton Ho. SW11 53Gb **111**
Compton Leisure Cen. 23Gb **49**
Compton Pas. EC1 5C **218** (42Rb **91**)
Compton Pl. DA8: Erith 51Hd **118**
Compton Pl. WC1 5F **217** (42Nb **90**)
Compton Pl. WD19: Wat 20Aa **27**
Compton Ri. HA5: Pinn 29Aa **45**
Compton Rd. CR0: C'don 74Xb **157**
Compton Rd. N1 37Rb **71**
Compton Rd. N21 18Rb **33**
Compton Rd. NW10 41Za **88**
Compton Rd. SW19 65Bb **133**
Compton Rd. UB3: Hayes 45U **84**
Compton St. EC1 5B **218** (42Rb **91**)
Compton Ter. N1 37Rb **71**
Compton Ter. N21 18Qb **32**
Compton Ter. N4 30Sb **51**
Comreddy Cl. EN2: Enf 11Rb **33**
Comus Ho. SE17 6H **231** (49Ub **91**)
(off Comus Pl.)
Comus Pl. SE17 6H **231** (49Ub **91**)
Comyne Rd. WD24: Wat 8V **12**
Comyn Rd. SW11 56Gb **111**
Comyns, The WD23: B Hea 18Ea **28**
Comyns Cl. E16 43Hc **93**
Comyns Rd. RM9: Dag 38Cd **76**
Conant Ho. SE11 51Rb **113**
(off St Agnes Pl.)
Conant M. E1 45Wb **91**
Conaways Cl. KT17: Ewe 82Wa **174**
Concanon Rd. SW2 56Pb **112**
Concert Hall App. SE1 7J **223** (46Pb **90**)
Concord Bus. Cen. W3 42Ra **87**
Concord Cl. UB5: N'olt 41Z **85**
Concord Ct. KT1: King T 69Pa **131**
(off Winery La.)
Concorde Bus. Pk. TN16: Big H 87Mc **179**
Concorde Cl. TW3: Houn 54Da **107**
Concorde Cl. UB10: Uxb 40N **63**
Concorde Ct. SL4: Wind 4E **102**
Concorde Dr. E6 43Pc **94**
Concorde Dr. HP2: Hem H 2M **3**
Concorde Ho. RM12: Horn 37Kd **77**
(off Cavendish Av.)
Concorde Way SE16 49Zb **92**
Concorde Way SL1: Slou 7G **80**
Concord Ho. CR0: C'don 72Rb **157**
Concord Ho. KT3: N Mald 69Ua **132**
Concord Ho. N17 24Vb **51**
(off Park La.)
Concordia Wharf E14 46Ec **92**
(off Coldharbour)
Concord Rd. EN3: Pond E 15Xb **33**
Concord Rd. W3 42Ra **87**
Concord Ter. HA2: Harr 33Da **65**
(off Coles Cres.)
Concourse, The N9 19Wb **33**
(within Edmonton Grn. Shop. Cen.)
Concourse, The NW9 25Va **48**
(off Quakers Course)
Condell Rd. SW8 53Lb **112**
Conder St. E14 44Ac **92**
Condor Ct. WD24: Wat 10X **13**
Condor Ho. E13 39Jc **73**
(off Brooks Rd.)
Condor Path UB5: N'olt 40Ca **65**
(off Union Rd.)
Condor Rd. TW18: Lale 69L **127**
Condor Wlk. RM12: Horn 38Kd **77**
Condor Way TW6: H'row A 55Q **106**
Condover Cres. SE18 52Rc **116**
Condray Pl. SW11 52Gb **111**
Conduit, The RH1: Blet 1K **209**
Conduit Av. SE10 53Fc **115**
Conduit Ct. WC2 4F **223** (45Nb **90**)
(off Floral St.)
Conduit La. CR0: C'don 78Wb **157**
Conduit La. CR2: S Croy 78Wb **157**
Conduit La. EN3: Pond E 16Ac **34**
Conduit La. N18 22Yb **52**
Conduit La. SL3: L'ly 50A **82**
Conduit M. SE18 50Rc **94**
Conduit M. W2 3B **220** (44Fb **89**)
Conduit Pas. W2 3B **220** (44Fb **89**)
(off Conduit Pl.)
Conduit Pl. W2 3B **220** (44Fb **89**)
Conduit Rd. SE18 50Rc **94**
Conduit St. W1 4A **222** (45Kb **90**)
Conduit Way NW10 38Sa **67**
Conegar Ct. SL1: Slou 6J **81**
Conewood St. N5 34Rb **71**
Coney Acre SE21 60Sb **113**
Coneybear Point SE10 49Jc **93**
Coneyberry RH2: Reig 10L **207**
Coney Burrows E4 19Gc **35**
Coneybury RH1: Blet 6L **209**
Coneybury Cl. CR6: W'ham 91Xb **197**
Coney Cl. AL10: Hat 2D **8**
Coney Gro. UB8: Hil 41Q **84**
Coneygrove Path UB5: N'olt 37Aa **65**
(off Arnold Rd.)
CONEY HALL 76Gc **159**
Coney Hall Pde. BR4: W W'ck 76Gc **159**
Coney Hill Rd. BR4: W W'ck 75Gc **159**
Coney Way SW8 51Pb **112**
Conference Cl. E4 19Ec **34**
Conference Rd. SE2 49Yc **95**
Confluence Plaza SE13 55Ec **114**
(off Station Rd.)
Congers Ho. SE8 52Cc **114**
Congleton Gro. SE18 50Sc **94**
Congo Dr. N9 20Yb **34**

Congo Rd. SE18 50Tc **94**
Congress Rd. SE2 49Yc **95**
Congreve Ho. N16 36Ub **71**
Congreve Rd. EN9: Walt A 5Gc **21**
Congreve Rd. SE9 55Pc **116**
Congreve St. SE17 5H **231** (49Ub **91**)
Congreve Wlk. E16 43Mc **93**
(off Fulmer Rd.)
Conical Cnr. EN2: Enf 12Sb **33**
Conifer Av. DA3: Hartl 72Ae **165**
Conifer Av. RM5: Col R 22Dd **56**
Conifer Cl. BR6: Orp 77Tc **160**
Conifer Cl. EN7: Chesh 1Vb **19**
Conifer Cl. RH2: Reig 4J **207**
Conifer Ct. TW15: Ashf 64P **127**
(off The Crescent)
Conifer Dr. CM14: W'ley 22Zd **59**
Conifer Gdns. EN1: Enf 16Ub **33**
Conifer Gdns. SM1: Sutt 75Db **155**
Conifer Gdns. SW16 62Pb **134**
Conifer Ho. SE4 56Bc **114**
(off Brockley Rd.)
Conifer La. TW20: Egh 64E **126**
Conifer Pk. KT17: Eps 83Ua **174**
Conifers KT13: Weyb 77U **150**
Conifers, The HP3: Hem H 5H **3**
Conifers, The WD25: Wat 7Y **13**
Conifers Cl. TW11: Tedd 66Ka **130**
Conifer Wlk. SL4: Wind 2A **102**
Conifer Way BR8: Swan 67Ed **140**
Conifer Way HA0: Wemb 34La **66**
Conifer Way UB3: Hayes 45W **84**
Coniger Rd. SW6 54Cb **111**
Coningsby Dr. WD17: Wat 11U **26**
Coningham Ct. SW10 52Eb **111**
(off King's Rd.)
Coningham M. W12 46Wa **88**
Coningham Rd. W12 47Xa **88**
Coningsby Av. NW9 26Ua **48**
Coningsby Bank AL1: St A 6B **6**
Coningsby Cl. AL9: Wel G 6F **8**
Coningsby Cotts. W5 47Ma **87**
Coningsby Ct. CR4: Mitc 68Jb **134**
Coningsby Ct. WD7: R'lett 8Ha **14**
Coningsby Dr. EN6: Pot B 5Fb **17**
Coningsby Gdns. E4 23Dc **52**
Coningsby Rd. CR2: S Croy 81Sb **177**
Coningsby Rd. N4 31Rb **71**
Coningsby Rd. W5 47Ma **87**
Conington Rd. SE13 54Dc **114**
Conisbee Ct. N14 15Lb **32**
Conisborough Ct. DA2: Dart 58Rd **119**
(off Osbourne Rd.)
Conisborough Cres. SE6 62Ec **136**
Conisbrough NW1 1B **216** (39Lb **70**)
(off Bayham St.)
Coniscliffe Cl. BR7: Chst 67Qc **138**
Coniscliffe Rd. N13 20Sb **33**
Conista Ct. GU21: Wok 8K **167**
Coniston NW1 3B **216** (41Lb **90**)
(off Harrington St.)
Coniston Av. DA16: Well 55Uc **116**
Coniston Av. IG11: Bark 38Uc **74**
Coniston Av. RM14: Upm 35Sd **78**
Coniston Av. RM19: Purf 50Sd **98**
Coniston Av. UB6: G'frd 41Ka **86**
Coniston Cl. DA1: Dart 60Kd **119**
Coniston Cl. DA7: Bex 53Ed **118**
Coniston Cl. DA8: Erith 52Gd **118**
Coniston Cl. HP3: Hem H 3C **4**
Coniston Cl. IG11: Bark 38Uc **74**
Coniston Cl. N20 20Eb **31**
Coniston Cl. SW13 52Va **110**
Coniston Cl. SW20 72Za **154**
Coniston Cl. W4 52Sa **109**
Coniston Ct. CM16: Epp 3Wc **23**
Coniston Ct. KT13: Weyb 79R **150**
Coniston Ct. NW7 24Ab **48**
(off Langstone Way)
Coniston Ct. SE16 47Zb **92**
(off Eleanor Cl.)
Coniston Ct. SM6: W'gton 77Kb **156**
Coniston Ct. TW15: Ashf 62M **127**
Coniston Ct. W2 3E **220** (44Gb **89**)
(off Kendal St.)
Coniston Ctes. SL1: Slou 3A **80**
Conistone Way N7 38Nb **70**
Coniston Gdns. HA5: Eastc 28W **44**
Coniston Gdns. HA9: Wemb 32La **66**
Coniston Gdns. IG4: Ilf 28Nc **54**
Coniston Gdns. N9 18Yb **34**
Coniston Gdns. NW9 29Ta **47**
Coniston Gdns. SM2: Sutt 79Fb **155**
Coniston Ho. E3 42Bc **92**
(off Southern Gro.)
Coniston Ho. SE5 52Sb **113**
(off Wyndham Rd.)
Coniston Lodge WD17: Wat 12W **26**
Coniston Rd. BR1: Broml 65Gc **137**
Coniston Rd. CR0: C'don 73Wb **157**
Coniston Rd. CR5: Coul 88Lb **176**
Coniston Rd. DA7: Bex 53Ed **118**
Coniston Rd. GU22: Wok 92D **188**
Coniston Rd. N10 26Kb **50**
Coniston Rd. N17 23Wb **51**
Coniston Rd. TW2: Whitt 58Da **107**
Coniston Rd. WD4: K Lan 10P **3**
Coniston Wlk. E9 36Yb **72**
Coniston Way KT9: Chess 76Na **153**
Coniston Way RH2: Reig 5N **207**
Coniston Way RM12: Horn 36Jd **76**
Coniston Way TW20: Egh 66D **126**
Conlan St. W10 42Ab **88**
Conley Rd. NW10 37Ua **68**
Conley St. SE10 50Gc **93**
Connaught Av. E4 17Fc **35**
Connaught Av. EN1: Enf 12Ub **33**
Connaught Av. EN4: E Barn 18Hb **31**
Connaught Av. IG10: Lough 14Mc **35**
Connaught Av. RM16: Grays 47De **99**
Connaught Av. SW14 55Sa **109**
Connaught Av. TW15: Ashf 63N **127**
Connaught Av. TW4: Houn 56Aa **107**
Connaught Bri. E16 46Mc **93**
Connaught Bus. Cen. CR0: Wadd 79Pb **156**
Connaught Bus. Cen. CR4: Mitc 71Hb **155**
Connaught Bus. Cen. NW9 29Va **48**
Connaught Cl. E10 33Ac **72**
Connaught Cl. EN1: Enf 12Ub **33**
Connaught Cl. HP2: Hem H 1A **4**
Connaught Cl. SM1: Sutt 75Fb **155**
Connaught Cl. UB8: Hil 42S **84**
Connaught Cl. W2 3E **220** (44Gb **89**)
Connaught Club, The 17Hc **35**
Connaught Ct. E17 28Dc **52**
(off Orford Rd.)
Connaught Ct. W2 3F **221** (44Hb **89**)
(off Connaught St.)
Connaught Cres. GU24: Brkwd 2D **186**

CONNAUGHT DAY CEN. 30Gc **53**
(off James La.)
Connaught Dr. KT13: Weyb 83Q **170**
Connaught Dr. NW11 28Cb **49**
Connaught Gdns. N10 29Kb **50**
Connaught Gdns. N13 21Rb **51**
Connaught Gdns. SM4: Mord 70Eb **133**
Connaught Hall WC1 5E **216** (42Mb **90**)
Connaught Hgts. E16 46Mc **93**
(off Agnes George Wlk.)
Connaught Hgts. UB10: Hil 42S **84**
(off Uxbridge Rd.)
Connaught Hill IG10: Lough 14Mc **35**
Connaught Ho. NW10 41Xa **88**
(off Trenmar Gdns.)
Connaught Ho. W1 5K **221** (45Kb **90**)
(off Davies St.)
Connaught Ho. WD23: Bush 14Ba **27**
(off Royal Connaught Dr.)
Connaught La. IG1: Ilf 33Sc **74**
Connaught Lodge N4 31Qb **70**
(off Connaught Rd.)
Connaught M. NW3 35Gb **69**
Connaught M. SE18 50Qc **94**
Connaught M. SW6 53Ab **110**
Connaught Pl. IG10: Lough 14Nc **36**
Connaught Pl. W2 4F **221** (45Hb **89**)
Connaught Rd. E11 32Fc **73**
Connaught Rd. E16 46Mc **93**
Connaught Rd. E17 29Cc **52**
Connaught Rd. E4 17Gc **35**
Connaught Rd. EN5: Barn 16Za **30**
Connaught Rd. GU24: Brkwd 3C **186**
Connaught Rd. HA3: W'stone 25Ha **46**
Connaught Rd. IG1: Ilf 33Tc **74**
Connaught Rd. KT3: N Mald 70Ua **132**
Connaught Rd. N4 31Qb **70**
Connaught Rd. NW10 39Ua **68**
Connaught Rd. RM12: Horn 34Md **77**
Connaught Rd. SL1: Slou 7M **81**
Connaught Rd. SM1: Sutt 75Fb **155**
Connaught Rd. TW10: Rich 57Pa **109**
Connaught Rd. TW11: Tedd 64Fa **130**
Connaught Rd. W13 45Ka **86**
CONNAUGHT RDBT. 45Mc **93**
(off Victoria Dock Rd.)
Connaught Sq. W2 3F **221** (44Hb **89**)
Connaught St. W2 3D **220** (44Gb **89**)
Connaught Way N13 21Rb **51**
Connaught Works E3 39Ac **72**
(off Old Ford Rd.)
Connections Bus. Pk. TN14: S'oaks 91Ld **203**
Connect La. IG6: Ilf 26Sc **54**
Connell Ct. SE14 51Zb **114**
(off Myers La.)
Connell Cres. W5 42Pa **87**
Connemara Cl. WD6: Bore 16Ta **29**
Connersville Way CR0: Wadd 76Qb **156**
Connicut La. KT23: Bookh 100Da **191**
Conniffe Ct. SE9 57Rc **116**
Conningham Ct. SE9 57Lc **115**
Connington Cres. E4 20Fc **35**
Connolly Ct. GU25: Vir W 70A **126**
Connolly Ct. RM7: Rush G 30Gd **56**
(off Union Rd.)
Connop Rd. EN3: Enf W 10Zb **20**
Connor Cl. E11 31Gc **73**
Connor Cl. IG6: Ilf 25Sc **54**
Connor Ct. SW11 53Kb **112**
Connor Rd. RM9: Dag 35Bd **75**
Connor St. E9 39Zb **72**
Conolly Dell W7 46Ga **86**
(off Conolly Rd.)
Conolly Rd. W7 46Ga **86**
Conqueror Ct. RM3: Rom 24Md **57**
Conquest Rd. KT15: Add 78J **149**
Conrad Cl. RM16: Grays 47De **99**
Conrad Ct. NW9 26Ua **48**
(off Needleman Cl.)
Conrad Ct. SE16 49Ac **92**
(off Cary Av.)
Conrad Ct. SS17: Stan H 2L **101**
Conrad Dr. KT4: Wor Pk 74Ya **154**
Conrad Gdns. RM16: Grays 47Ce **99**
Conrad Ho. E14 45Ac **92**
(off Victory Pl.)
Conrad Ho. E16 46Kc **93**
(off Wesley Av.)
Conrad Ho. E8 37Wb **71**
Conrad Ho. N16 36Ub **71**
(off Matthias Rd.)
Conrad Ho. SW8 52Nb **112**
(off Wyvil Rd.)
Conrad M. DA11: Nflt 61De **143**
Conrad Rd. SS17: Stan H 1N **101**
Consfield Av. KT3: N Mald 70Wa **132**
Consort Cl. CM14: W'ley 22Yd **58**
Consort Cl. GU22: Wok 90A **168**
Consort Ct. W8 48Db **89**
(off Wright's La.)
Consort Ho. E14 50Dc **92**
(off St Davids Sq.)
Consort Ho. SW6 54Eb **111**
(off Lensbury Av.)
Consort Ho. W2 45Db **89**
(off Queensway)
Consort Lodge NW8 1F **215** (39Hb **69**)
(off Prince Albert Rd.)
Consort M. TW7: Isle 57Fa **108**
Consort Pk. 54Xb **113**
(off Gordon Road)
Consort Rd. SE15 53Xb **113**
Cons St. SE1 1A **230** (47Qb **90**)
Constable Av. E16 46Kc **93**
Constable Cl. KT17: Ewe 82Wa **174**
Constable Cl. N11 22Hb **49**
Constable Cl. NW11 30Db **49**
Constable Cl. UB4: Hayes 40S **64**
Constable Cl. SE16 50Xb **91**
(off Stubbs Dr.)
Constable Ct. W4 50Ra **87**
(off Chaseley Dr.)
Constable Cres. N15 29Wb **51**
Constable Gdns. HA8: Edg 25Qa **47**
Constable Gdns. TW7: Isle 57Fa **108**
Constable Ho. E14 47Cc **92**
(off Cassilis Rd.)
Constable Ho. NW3 38Hb **69**
Constable Ho. UB5: N'olt 40Z **65**
(off Gallery Gdns.)
Constable M. BR1: Broml 68Kc **137**
Constable M. RM14: Upm 33Sd **78**
Constable M. RM8: Dag 35Xc **75**
Constable Rd. DA11: Nflt 2A **144**
Constable Wlk. SE21 62Ub **135**
Constabulary Cl. UB7: W Dray 48N **83**
Constance Allen Ho. W10 44Za **88**
(off Bridge Cl.)
Constance Cl. SW15 63Ta **131**

Constance Cres. BR2: Hayes 73Hc 159
Constance Gro. DA1: Dart 58Md 119
Constance Rd. CR0: C'don 73Rb 157
Constance Rd. EN1: Enf 16Ub 33
Constance Rd. SM1: Sutt 77Eb 155
Constance Rd. TW2: Whitt 59Da 107
Constance St. E16 46Nc 94
Constant Ho. E14 45Dc 92
(off Harrow La.)
Constantine Ct. E1 44Wb 91
(off Fairclough St.)
Constantine Ho. NW9 27Va 48
(off Boulevard Dr.)
Constantine Ho. UB3: Hayes 44T 84
Constantine Pl. UB10: Hil 39P 63
Constantine Rd. NW3 35Gb 69
Constant Mdw. TN13: S'oaks 97Ld 203
Constellation Way TW6: H'row A 55R 106
Constitution Cres. DA12: Grav'nd 10E 122
(off Constitution Hill)
Constitution Hill DA12: Grav'nd 10E 122
Constitution Hill GU22: Wok 91A 188
Constitution Hill SW1 1K 227 (47Kb 90)
Constitution Ri. SE18 53Qc 116
Consul Av. RM13: Rain 41Fd 96
Consul Av. RM9: Dag 41Ed 96
Consul Av. RM9: Rain 41Ed 96
Consul Gdns. BR8: Hext 66Hd 140
Consul Ho. E3 42Cc 92
(off Wellington Way)
Container City 1 E14 45Gc 93
Container City 2 E14 45Gc 93
Contemporary Applied Arts 7C 224 (46Rb 91)
(off Southwark St.)
Content St. SE17 6F 231 (49Tb 91)
Contessa Cl. BR6: Farnb 78Uc 160
Continuity Ct. DA9: Ghithe 56Wd 120
Contrail Way TW6: H'row A 55R 106
Convair Wlk. UB5: N'olt 41Z 85
Convent Cl. BR3: Beck 66Ec 136
Convent Cl. EN5: Barn 12Bb 31
Convent Cl. GU22: Wok 89D 168
Convent Cl. RM14: Upm 33Rd 77
Convent Gdns. W11 44Ab 88
Convent Gdns. W5 49La 86
Convent Hill SE19 65Sb 135
Convent La. KT11: Cobh 83U 170
Convent Rd. SL4: Wind 4D 102
Convent Rd. TW15: Ashf 64R 128
Convent Way UB2: S'hall 49Y 85
Conway Cl. BR3: Beck 67Ac 136
Conway Cl. HA7: Stan 23Ja 46
Conway Cl. RM13: Rain 38Jd 76
Conway Ct. SL9: Ger X 32D 62
Conway Cres. RM6: Chad H 30Yc 55
Conway Cres. UB6: G'frd 40Ga 66
Conway Dr. SM2: Sutt 79Db 155
Conway Dr. TW15: Ashf 65S 128
Conway Dr. UB3: Harl 48S 84
Conway Gdns. CR4: Mitc 70Nb 134
Conway Gdns. EN2: Enf 10Ub 19
Conway Gdns. HA9: Wemb 31La 66
Conway Gdns. RM17: Grays 52De 121
Conway Gro. W3 43Ta 87
Conway Ho. E14 49Cc 92
(off Cahir St.)
Conway Ho. SW3 50Hb 89
(off Ormonde Ga.)
Conway Ho. WD6: Bore 14Sa 29
Conway M. W1 6B 216 (42Lb 90)
(off Conway St.)
Conway Rd. N14 20Nb 32
Conway Rd. N15 29Rb 51
Conway Rd. NW2 33Ya 68
Conway Rd. SE18 49Tc 94
Conway Rd. SL6: Tap 4A 80
Conway Rd. SW20 67Ya 132
Conway Rd. TW13: Hanw 64Z 129
Conway Rd. TW4: Houn 59Ba 107
Conway St. W1 6B 216 (42Lb 90)
(not continuous)
Conway Wlk. TW12: Hamp 65Ba 129
Conybeare NW3 38Gb 69
Conybury Cl. EN9: Walt A 4Jc 21
Conyerd Rd. TN15: Bor G 92Be 205
Conyers Cl. IG8: Wfd G 23Gc 53
Conyers Cl. KT12: Hers 78Z 151
Conyer's Rd. SW16 64Mb 134
Conyer St. E3 40Ac 72
Conyers Way IG10: Lough 13Rc 36
Cooden Cl. BR1: Broml 66Kc 137
Cook Ct. DA8: Erith 52Hd 118
Cook Ct. SE16 46Yb 92
(off Rotherhithe St.)
Cook Ct. SE8 50Ac 92
(off Evelyn St.)
Cooke Cl. RM16: Chaf H 48Ae 99
Cooke Cl. SE2 48Yc 95
Cookes Cl. E11 33Hc 73
Cookes La. SM3: Cheam 79Ab 154
Cooke St. IG11: Bark 39Sc 74
(not continuous)
Cookham Cl. UB2: S'hall 48Da 85
Cookham Cres. SE16 47Zb 92
Cookham Dene Cl. BR7: Chst 67Tc 138
Cookham Hill BR5: Orp 76Cd 162
Cookham Ho. E2 5K 219 (42Vb 91)
(off Montclare St.)
Cookham Rd. BR8: Swan 67Cd 140
Cookhill Rd. SE2 47Xc 95
Cook Rd. RM9: Dag 39Ad 75
Cook's Cl. RM5: Col R 25Ed 56
Cooks Cl. E14 46Cc 92
(off Cabot Sq.)
Cooks Cl. SL9: Chal P 23A 42
Cooks Hole Rd. EN2: Enf 10Rb 19
Cooks Mead WD23: Bush 16Da 27
Cookson Gro. DA8: Erith 52Dd 118
Cook Sq. DA8: Erith 52Hd 118
Cook's Rd. E15 40Dc 72
Cooks Rd. SE17 51Rb 113
Cooks Way AL10: Hat 2D 8
Coolfin Rd. E16 44Jc 93
Coolgardie Av. E4 22Fc 53
Coolgardie Av. IG7: Chig 20Qc 36
Coolgardie Rd. TW15: Ashf 64S 128
Coolhurst Rd. N8 30Mb 50
Coolhurst Tennis & Squash Club 30Mb 50
Cool Oak La. NW9 31Ua 68
Coomassie Rd. W9 42Bb 89
COOMBE 66Sa 131
Coombe, The RH3: B'wth 3A 206
Coombe Av. CR0: C'don 77Ub 157
Coombe Av. TN14: S'oaks 92Kd 203
Coombe Bank KT2: King T 67Ua 132
Coombe Cl. HA8: Edg 26Pa 47
Coombe Cl. SL2: Slou 2D 80
Coombe Cl. TW3: Houn 56Ca 107
Coombe Cnr. N21 18Rb 33
Coombe Ct. CR0: C'don 77Tb 157
(off Coombe Rd.)
Coombe Ct. KT20: Tad 95Ya 194
Coombe Ct. TN14: S'oaks 92Kd 203
Coombe Cres. TW12: Hamp 66Ba 129
Coombe Dene BR2: Broml 70Hc 137
(off Cumberland Rd.)
Coombe Dr. HA4: Ruis 32X 65
Coombe Dr. KT15: Add 79H 149
Coombe End KT2: King T 66Ta 131
Coombefield Cl. KT3: N Mald 71Ua 154
Coombe Gdns. KT3: N Mald 70Va 132
Coombe Gdns. SW20 68Wa 132
Coombe Hill Ct. SL4: Wind 6B 102
Coombe Hill Glade KT2: King T 66Ua 132
Coombe Hill Golf Course 66Ta 131
Coombe Hill Rd. KT2: King T 66Ua 132
Coombe Hill Rd. WD3: Rick 17J 25
Coombe Ho. N7 36Mb 70
Coombe Ho. Chase KT3: N Mald 67Ta 131
Coombehurst Cl. EN4: Cockf 12Hb 31
Coombelands La. KT15: Add 79J 149
Coombe La. CR0: C'don 78Xb 157
Coombe La. GU3: Worp 10G 186
(not continuous)
Coombe La. KT12: W Vill 81V 170
Coombe La. SL5: S'hill 10A 124
Coombe La. SW20 67Va 132
COOMBE LANE 67Va 132
Coombe La. Flyover SW20 67Va 132
Coombe Lane Stop (London Tramlink) 78Yb 158
Coombe La. W. KT2: King T 67Ra 131
Coombe Lea BR1: Broml 69Nc 138
Coombe Lodge SE7 51Lc 115
Coombe Mnr. GU24: Bisl 7E 166
Coombe Neville KT2: King T 66Ta 131
Coombe Pk. KT2: King T 64Sa 131
Coombe Pl. KT2: King T 64Sa 131
Coomber Ho. SW6 55Db 111
(off Wandsworth Bri. Rd.)
Coombe Ridings KT2: King T 64Sa 131
Coombe Ri. CM15: Shenf 18Be 41
Coombe Ri. KT2: King T 67Sa 131
Coombe Ri. SS17: Stan H 1N 101
Coombe Rd. CR0: C'don 77Tb 157
Coombe Rd. DA12: Grav'nd 1E 144
Coombe Rd. KT2: King T 67Qa 131
Coombe Rd. KT3: N Mald 68Ua 132
Coombe Rd. N22 26Qb 50
Coombe Rd. NW10 34Ta 67
Coombe Rd. RM3: Hrld W 27Pd 57
Coombe Rd. SE26 63Xb 135
Coombe Rd. TN14: Otf 87Ld 183
Coombe Rd. TW12: Hamp 65Ba 129
Coombe Rd. W13 48Ka 86
Coombe Rd. W4 50Ua 88
Coombe Rd. WD23: Bush 17Fa 28
Coomber Way CR0: Bedd 73Mb 156
Coombes Ho. CR8: Purl 84Sb 177
Coombes Rd. AL2: Lon C 8G 6
Coombes Rd. RM9: Dag 39Bd 75
Coombe Va. SL9: Ger X 32A 62
Coombe Wlk. SM1: Sutt 76Db 155
Coombe Way KT14: Byfl 84P 169
Coombewood Dr. RM6: Chad H 30Bd 55
Coombe Wood Golf Course 66Ra 131
Coombe Wood Hill CR8: Purl 85Sb 177
Coombe Wood Local Nature
Reserve 66Va 132
Coombe Wood Rd. KT2: King T 64Sa 131
Coombfield Dr. DA2: Daren 63Td 142
Coombrook Ct. SE16 47Ac 92
(off Elgar St.)
Coombs St. N1 2C 218 (40Rb 71)
Coomer M. SW6 51Bb 111
Coomer Pl. SW6 51Bb 111
Coomer Rd. SW6 51Bb 111
Cooms Wlk. HA8: Edg 25Sa 47
Coope Ct. RM7: Rush G 30Gd 56
(off Union Rd.)
Cooperage, The SE1 1K 231 (47Vb 91)
(off Gainsford St.)
Cooperage, The SW8 52Pb 112
(off Regent's Bri. Gdns.)
Cooperage Cl. N17 23Vb 51
Cooperage Yd. E15 40Ec 72
Co-operative Ho. SE15 55Wb 113
Cooper Av. E17 25Ac 52
Cooper Cl. DA9: Ghithe 57Vd 120
Cooper Cl. SE1 2A 230 (47Qb 90)
Cooper Ct. SE18 51Rc 116
Cooper Cres. SM5: Cars 76Hb 155
Cooper Ho. NW8 6B 214 (42Fb 89)
(off Lyons Pl.)
Cooper Ho. RM16: Grays 8B 100
Cooper Ho. SE4 57Zb 114
(off St Norbert Rd.)
Cooper Ho. TW4: Houn 55Ba 107
Cooper Rd. CR0: Wadd 77Rb 157
Cooper Rd. GU20: W'sham 9B 146
Cooper Rd. NW10 36Wa 68
Cooper Rd. NW4 30Za 48
Coopers, The TN14: Dun G 92Hd 202
COOPERSALE 1Zc 23
Coopersale Cl. IG8: Wfd G 24Lc 53
Coopersale La. CM16: They B 9Wc 23
Coopersale La. CM16: They G 9Wc 23
Coopersale Rd. E9 36Zb 72
Coopersale St. CM16: Epp 4Yc 23
Coopersale St. CM16: Fidd H 4Yc 23
COOPERSALE STREET 3Yc 23
Coopers Cl. DA4: S Dar 67Td 142
Coopers Cl. E1 42Yb 92
Coopers Cl. IG7: Chig 19Xc 37
Coopers Cl. RM10: Dag 37Dd 76
Coopers Cl. TW18: Staines 64G 126
Coopers Ct. E3 42Bc 92
(off Eric St.)
Coopers Ct. TW7: Isle 54Ha 108
(off Woodlands Rd.)
Coopers Ct. W3 46Sa 87
(off Church Rd.)
Coopers Cres. WD6: Bore 11Sa 29
Coopers Dr. DA2: Wilm 61Gd 140
Coopers Ga. AL4: Col H 4M 7
Coopers Hill Dr. GU24: Brkwd 2A 186
Cooper's Hill Ho. TW20: Eng G 2N 125
Coopers Hill La. TW20: Egh 2N 125
Coopers Hill La. TW20: Eng G 2N 125
Cooper's Hill Rd. RH1: Nutf 6G 208
Cooper's Hill Rd. RH1: S Nut 6G 208
Cooper's La. EN6: N'thaw 3Fb 17
Cooper's La. EN6: Pot B 3Fb 17
Cooper's La. SE12 61Kc 137
Coopers La. E10 32Dc 72
Coopers La. E20 36Dc 72
Coopers La. NW1 1E 216 (40Mb 70)
Coopers La. RM18: W Til 1G 122
Coopers La. TW18: Staines 63J 127
Coopers La. Rd. EN6: N'thaw 3Fb 17
Coopers La. Rd. EN6: Pot B 3Fb 17
Coopers Lodge SE1 1K 231 (47Vb 91)
(off Tooley St.)
Coopers M. BR3: Beck 68Cc 136
Coopers M. WD25: Wat 3Y 13
Cooper's Rd. SE1 7K 231 (50Vb 91)
Coopers Rd. DA10: Swans 59Be 121
Coopers Rd. DA11: Nflt 10B 122
Coopers Rd. EN6: Pot B 2Eb 17
Coopers Row EC3 4K 225 (45Vb 91)
Coopers Row SL0: Iver H 42E 82
Coopers Shaw Rd. RM18: W Til 2F 122
Cooper St. E16 43Hc 93
Coopers Wlk. E15 36Gc 73
Coopers Wlk. EN8: Chesh 1Zb 20
Cooper's Yd. SE19 65Ub 135
Coopers Yd. N1 38Rb 71
(off Upper St.)
Cooper Way HP4: Berk 1A 2
Cooper Way SL1: Slou 8F 80
Coote Gdns. RM8: Dag 34Bd 75
Coote Rd. DA7: Bex 53Bd 117
Coote Rd. RM8: Dag 34Bd 75
Cope Ho. EC1 4E 218 (41Sb 91)
(off Bath St.)
Copeland Dr. E14 49Cc 92
Copeland Ho. SE11 4K 229 (48Qb 90)
(off Lambeth Wlk.)
Copeland Ho. SW17 63Fb 133
Copeland Rd. E17 30Dc 52
Copeland Rd. SE15 54Wb 113
Copeman Cl. SE26 64Yb 136
Copeman Rd. CM13: Hut 17Fe 41
Copenhagen Ct. SE8 49Ac 92
(off Pell St.)
Copenhagen Gdns. W4 47Ta 87
Copenhagen Ho. N1 1J 217 (39Pb 70)
(off Barnsbury Est.)
Copenhagen Pl. E14 44Bc 92
(not continuous)
Copenhagen St. N1 39Nb 70
Copenhagen Way KT12: Walt T 76X 151
Cope Pl. W8 48Cb 89
Copers Cope Rd. BR3: Beck 66Bc 136
Cope St. SE16 49Zb 92
Copford Cl. IG8: Wfd G 23Nc 54
Copford Wlk. N1 39Sb 71
(off Popham St.)
Copgate Path SW16 65Pb 134
Copinger Wlk. HA8: Edg 25Ra 47
Copland Av. HA0: Wemb 36Ma 67
Copland Cl. HA0: Wemb 36La 66
Copland M. HA0: Wemb 37Na 67
Copland Rd. HA0: Wemb 37Na 67
Copland Rd. SS17: Stan H 2M 101
Copleigh Dr. KT20: Tad 92Ab 194
Copleston M. SE15 54Vb 113
Copleston Pas. SE5 54Vb 113
Copleston Rd. SE15 55Vb 113
Copley Cl. GU21: Wok 1J 187
Copley Cl. RH1: Redh 4N 207
Copley Cl. SE17 51Sb 113
Copley Cl. W7 42Ha 86
Copley Dene BR1: Broml 67Mc 137
Copley Pk. SW16 65Pb 134
Copley Rd. HA7: Stan 22La 46
Copley St. E1 43Zb 92
Copley Way KT20: Tad 92Za 194
Copmans Wick WD3: Chor 15F 24
Coppard Gdns. KT9: Chess 79La 152
Copped Hall 3Qc 22
COPPED HALL 3Qc 22
Coppelia Rd. SE3 56Hc 115
Coppen Rd. RM8: Dag 31Bd 75
Copperas St. SE8 51Dc 114
Copper Beech Cl. BR5: St M Cry 71Yc 161
Copper Beech Cl. DA12: Grav'nd 9F 122
Copper Beech Cl. GU22: Wok 3M 187
Copper Beech Cl. HP3: Hem H 5H 3
Copper Beech Cl. IG5: Ilf 25Qc 54
Copper Beech Cl. SL4: Wind 3B 102
Copperbeech Cl. NW3 36Fb 69
Copper Beech Ct. IG10: Lough 11Qc 36
Copper Beeches Ct. TW7: Isle 53Fa 108
Copper Beech Ho. GU22: Wok 89B 168
Copper Beech M. TW20: Eng G 6M 125
Copper Beech Rd. RM15: S Ock 41Yd 98
Copper Box Arena 37Cc 72
Copper Cl. N17 24Xb 51
Copper Cl. SE19 66Vb 135
Copper Ct. E5 33Yb 72
Copperdale Ct. WD18: Wat 15U 26
Copperdale Rd. UB3: Hayes 47W 84
Copperfield IG7: Chig 22Tc 54
Copperfield Av. UB8: Hil 43Q 84
Copperfield Cl. CR2: Sande 83Sb 177
Copperfield Cl. DA12: Grav'nd 10J 123
Copperfield Ct. KT22: Lea 93Ja 192
Copperfield Dr. N15 28Vb 51
Copperfield Gdns. CM14: B'wood 18Xd 40
Copperfield Ho. SE1 47Wb 91
(off Wolseley St.)
Copperfield Ho. W1 7J 215 (43Jb 90)
(off Marylebone High St.)
Copperfield Ho. W11 46Za 88
(off St Ann's Rd.)
Copperfield M. E2 40Wb 71
(off Claredale St.)
Copperfield M. N18 21Ub 51
Copperfield Ri. KT15: Add 78H 149
Copperfield Rd. E3 42Ac 92
Copperfield Rd. SE28 44Yc 95
Copperfields BR3: Beck 67Ec 136
Copperfields DA1: Dart 58Nd 119
Copperfields HA1: Harr 31Ga 66
Copperfields KT22: Fet 94Ea 192
Copperfields TN15: Kems'g 89Pd 183
Copperfields TW16: Sun 65V 128
Copperfields Cl. TN15: Kems'g 89Pd 183
Copperfields Ct. W3 47Qa 87
Copperfields Orchard TN15: Kems'g 89Pd 183
Copperfields Shop. Cen. 58Nd 119
(off Spital St.)
Copperfield St. SE1 1C 230 (47Rb 91)
Copperfields Wlk. TN15: Kems'g 89Pd 183
Copperfield Way BR7: Chst 65Sc 138
Copperfield Way HA5: Pinn 28Ba 45
Coppergate Cl. BR1: Broml 67Kc 137
Coppergate Ct. EN9: Walt A 6Jc 21
(off Farthingale La.)
Copper Horse Ct. SL4: Wind 4E 102
Copper La. N16 35Tb 71
Copperlight Apts. SW18 57Cb 111
(off Buckhold Rd.)
Coppermead Cl. NW2 34Ya 68
Copper M. W4 48Sa 87
Coppermill Ct. WD3: W Hyd 24H 43
Copper Mill Dr. TW7: Isle 54Ha 108
Coppermill Hgts. N17 27Xb 51
(off Daneland Wlk.)
Copper Mill La. SW17 63Eb 133
Coppermill La. E17 30Yb 52
Coppermill La. UB9: Hare 24J 43
Coppermill La. WD3: Hare 24G 42
Coppermill La. WD3: W Hyd 24G 42
Copper Mill Lock UB9: Hare 24J 43
Coppermill Rd. TW19: Wray 58C 104
Copper Ridge SL9: Chal P 22B 42
Copper Row SE1 7K 225 (46Vb 91)
(off Horselydown La.)
Copperwood Pl. SE10 53Ec 114
Copperworks, The N1 2G 217 (40Nb 70)
(off Railway St.)
Coppetts Cen. 24Hb 49
Coppetts Cl. N12 24Gb 49
Coppetts Rd. N10 24Hb 49
Coppetts Wood & Glebelands Local
Nature Reserve 23Hb 49
Coppice, The DA5: Bexl 62Fd 140
Coppice, The EN2: Enf 14Rb 33
Coppice, The EN5: New Bar 16Db 31
(off Great Nth. Rd.)
Coppice, The HP2: Hem H 1B 4
Coppice, The TW15: Ashf 65R 128
Coppice, The UB7: Yiew 44N 83
Coppice, The WD19: Wat 16Y 27
Coppice Av. KT11: Cobh 86Ba 171
Coppice Cl. AL10: Hat 3B 8
Coppice Cl. BR3: Beck 70Dc 136
Coppice Cl. HA4: Ruis 30T 44
Coppice Cl. HA7: Stan 23Ha 46
Coppice Cl. SW20 69Ya 132
Coppice Dr. SW15 58Xa 110
Coppice Dr. TW19: Wray 9P 103
Coppice End GU22: Pyr 88G 168
Coppice La. RH2: Reig 4H 207
Coppice Path IG7: Chig 21Xc 55
Coppice Rd. RH2: Reig 5G 206
Coppice Row CM16: They B 8Sc 22
Coppice Wlk. N20 20Cb 31
Coppice Way E18 28Hc 53
Coppice Way SL2: Hedg 3H 61
Coppies Gro. N11 21Kb 50
Copping Cl. CR0: C'don 77Ub 157
Coppins, The CR0: New Ad 79Dc 158
Coppins, The HA3: Hrw W 23Ga 46
Coppins La. SL0: Iver 43H 83
Coppock Cl. SW11 54Gb 111
Coppsfield KT8: W Mole 69Ca 129
Copse, The CR3: Cat'm 98Wb 197
Copse, The CR6: W'ham 89Ac 178
Copse, The E4 18Hc 35
Copse, The GU23: Send 96H 189
Copse, The HP1: Hem H 1G 2
Copse, The KT22: Fet 95Da 191
Copse, The N2 27Hb 49
Copse, The RH1: S Nut 8E 208
Copse, The TN16: Tats 92Lc 199
Copse, The WD23: Bush 13Z 27
Copse Av. BR4: W W'ck 75Dc 158
Copse Av. RM3: Hrld W 25Nd 57
Copse Bank TN15: Seal 92Pd 203
Copse Cl. HA6: Nwood 26S 44
Copse Cl. SE7 51Kc 115
Copse Cl. SL1: Slou 6D 80
Copse Cl. UB7: W Dray 48M 83
Copse Edge Av. KT17: Eps 85Va 174
Copse Glade KT6: Surb 73Ma 153
Copse Hill CR8: Purl 85Nb 176
Copse Hill SM2: Sutt 80Db 155
Copse Hill SW20 67Wa 132
COPSE HILL 66Wa 132
Copsem Dr. KT10: Esh 79Da 151
Copse M. KT13: Weyb 77T 150
Copsem La. KT10: Esh 79Ea 152
Copsem La. KT10: Oxs 79Ea 152
Copsem La. KT22: Oxs 83Ea 172
Copsem Way KT10: Esh 79Ea 152
Copsen Wood KT22: Oxs 83Ea 172
Copse Rd. GU21: Wok 10K 167
Copse Rd. KT11: Cobh 85X 171
Copse Rd. RH1: Redh 8L 207
Copse Side DA3: Hartl 69Ae 143
Copse Vw. CR2: Sels 81Zb 178
Copse Wood SL0: Iver H 39F 62
Copsewood Cl. DA15: Sidc 58Uc 116
Copse Wood Ct. RH2: Reig 4N 207
Copsewood Rd. WD24: Wat 11X 27
Copse Wood Way HA6: Nwood 25R 44
Coptain Ho. SW18 56Cb 111
Coptefield Dr. DA17: Belv 48Zc 95
Coptfold Rd. CM14: B'wood 19Yd 40
Copthall Av. EC2 2G 225 (44Tb 91)
(not continuous)
Copthall Bldgs. EC2 2G 225 (44Tb 91)
(off Copthall Av.)
Copthall Cl. EC2 2F 225 (44Tb 91)
Copthall Cl. SL9: Chal P 24B 42
Copthall Cnr. SL9: Chal P 24A 42
Copthall Dr. NW7 24Wa 48
Copthall Gdns. NW7 24Wa 48
Copthall Gdns. TW1: Twick 60Ha 108
COPTHALL GREEN 5Nc 22
Copthall La. SL9: Chal P 24A 42
Copt Hall Rd. TN15: Igh 95Wd 204
Copthall Rd. DA13: Meop 10D 144
Copthall Rd. DA13: Sole S 10D 144
Copthall Rd. E. UB10: Ick 33Q 64
Copthall Rd. W. UB10: Ick 33Q 64
Copthall Way KT15: New H 82H 169
Copt Hill La. KT20: Tad 92Ab 194
Copthorne Av. BR2: Broml 75Pc 160
Copthorne Av. IG6: Ilf 23Rc 54
Copthorne Av. SW12 59Mb 112
Copthorne Chase TW15: Ashf 63P 127
Copthorne Cl. TW17: Shep 72S 150
Copthorne Cl. WD3: Crox G 15P 25
Copthorne Ct. KT22: Lea 94Ja 192
Copthorne Dr. GU18: Light 2A 166
Copthorne Gdns. RM11: Horn 29Qd 57
Copthorne M. UB3: Harl 49U 84
Copthorne Ri. CR2: Sande 85Tb 177
Copthorne Rd. KT22: Lea 92Ka 192
Copthorne Rd. WD3: Crox G 16P 25
Coptic St. WC1 1F 223 (43Nb 90)
Copt Pl. NW7 23Ab 48
Copwood Cl. N12 21Fb 49
Coral Apts. E16 45Jc 93
(off Western Gateway)
Coral Cl. RM6: Chad H 27Zc 55
Coral Gdns. HP2: Hem H 1P 3
Coral Ho. E1 42Ac 92
(off Harford St.)
Coral Ho. NW10 41Qa 87
Coraline Cl. UB1: S'hall 41Ba 85
Coralline Wlk. SE2 48Yc 95
Coral Mans. NW6 39Cb 69
(off Kilburn High Rd.)
Coral Row SW11 55Eb 111
Coral St. SE1 2A 230 (47Qb 90)
Coram Grn. CM13: Hut 16Fe 41
Coram Ho. W4 50Ua 88
(off Wood St.)
Coram Ho. WC1 5F 217 (42Nb 90)
(off Herbrand St.)
Coram Mans. WC1 6H 217 (42Pb 90)
(off Millman St.)
Coram St. WC1 6F 217 (42Nb 90)
Coran Cl. N9 17Zb 34
Corban Rd. TW3: Houn 55Ca 107
Corbar Cl. EN4: Had W 11Fb 31
Corbden Cl. SE15 53Vb 113
Corben M. SW8 54Lb 112
Corbet Cl. SM6: W'gton 75Jb 156
Corbet Ct. EC3 3G 225 (44Tb 91)
Corbet Ho. N1 1K 217 (40Qb 70)
(off Barnsbury Est.)
Corbet Ho. SE5 52Sb 113
(off Wyndham Rd.)
Corbet Pl. E1 7K 219 (43Vb 91)
Corbet Rd. KT17: Ewe 82Ua 174
Corbets Av. RM14: Upm 36Rd 77
CORBETS TEY 36Sd 78
Corbets Tey Rd. RM14: Upm 35Rd 77
Corbett Av. KT8: E Mos 72Fa 152
Corbett Cl. CR0: New Ad 84Fc 179
Corbett Ct. SE26 63Bc 136
Corbett Gro. N22 24Nb 50
Corbett Ho. SW10 51Eb 111
(off Cathcart Rd.)
Corbett Rd. E11 30Lc 53
Corbett Rd. E17 27Ec 52
Corbetts La. SE16 49Yb 92
(not continuous)
Corbetts Pas. SE16 49Yb 92
(off Corbetts La.)
Corbetts Wharf SE16 47Xb 91
(off Bermondsey Wall E.)
Corbett Theatre 13Rc 36
Corbicum E11 31Gc 73
Corbidge Ct. SE8 51Dc 114
Corbiere Ct. SW19 65Za 132
Corbiere Ho. N1 39Tb 71
(off De Beauvoir Est.)
Corbin Ho. E3 41Dc 92
(off Bromley High St.)
Corbins La. HA2: Harr 34Da 65
Corbould Cl. SM5: Cars 79Hb 155
Corbridge N17 24Xb 51
Corbridge Cres. E2 40Xb 71
Corbridge M. RM1: Rom 29Hd 56
Corby Cl. AL2: Chis G 7N 5
Corby Cl. TW20: Eng G 5N 125
Corby Cres. EN2: Enf 14Nb 32
Corby Dr. TW20: Eng G 5M 125
Corbylands Rd. DA15: Sidc 59Uc 116
Corbyn St. N4 32Nb 70
Corby Rd. NW10 40Ta 67
Corby Way E3 42Cc 92
Corcorans CM15: Pil H 16Yd 40
Cordage Ho. E1 46Xb 91
(off Cobblestone Sq.)
Cordelia Cl. SE24 56Rb 113
Cordelia Gdns. TW19: Stanw 59N 105
Cordelia Ho. N1 1J 219 (40Ub 71)
(off Arden Est.)
Cordelia Rd. TW19: Stanw 59N 105
Cordelia St. E14 44Dc 92
Cordell Ho. N15 29Vb 51
(off Newton Rd.)
Corder Cl. AL3: St A 5N 5
Corderoy Pl. KT16: Chert 72G 148
Cordingley Rd. HA4: Ruis 33T 64
Cording St. E14 43Dc 92
Cordons Cl. SL9: Chal P 25A 42
Cordrey Gdns. CR5: Coul 87Nb 176
(not continuous)
Cordrey Ho. KT15: Add 75J 149
Cordwainer Ho. E8 39Xb 71
Cordwainers Ct. E9 38Yb 72
(off St Thomas's Sq.)
Cordwainers Wlk. E13 40Jc 73
Cord Way E14 48Cc 92
Cordwell Rd. SE13 57Gc 115
Corefield Cl. N11 19Jb 32
Corelli Ct. SE1 49Xb 91
Corelli Ct. SW5 49Cb 89
(off W. Cromwell Rd.)
Corelli Rd. SE3 54Nc 116
Coresbrook Way GU21: Knap 10E 166
Corfe Av. HA2: Harr 35Ca 65
Corfe Cl. HP2: Hem H 3N 3
Corfe Cl. KT21: Asht 90La 172
Corfe Cl. TW4: Houn 60Aa 107
Corfe Cl. UB4: Yead 44Y 85
Corfe Cl. WD6: Bore 13Ta 29
Corfe Gdns. SL1: Slou 5E 80
Corfe Ho. SW8 52Pb 112
(off Dorset Rd.)
Corfe Twr. W3 47Ra 87
Corfield Rd. N21 15Pb 32
Corfield St. E2 41Xb 91
Corfton Lodge W5 43Na 87
Corfton Rd. W5 44Na 87
Coriander Av. E14 44Fc 93
Coriander Ct. SE1 1K 231 (47Vb 91)
(off Gainsford St.)
Cories Cl. RM8: Dag 33Zc 75
Corinium Cl. HA9: Wemb 35Pa 67
Corinium Ga. AL3: St A 4N 5
Corinne Rd. N19 35Lb 70
Corinthian Golf Course 72Zd 165
Corinthian Manorway DA8: Erith 49Fd 96
Corinthian Rd. DA8: Erith 49Fd 96
Corinthian Sports Club 72Zd 165
Corinthian Way TW19: Stanw 59M 105
Corkers Path IG1: Ilf 33Sc 74
Corker Wlk. N7 33Pb 70
Cork Ho. SW19 63Db 133
Corkran Rd. KT6: Surb 73Ma 153
Corkscrew Hill BR4: W W'ck 75Ec 158
Cork Sq. E1 46Xb 91
Cork St. W1 5B 222 (45Lb 90)
Cork St. M. W1 5B 222 (45Lb 90)
(off Cork St.)
Cork Tree Ho. SE27 64Rb 135
(off Lakeview Rd.)
Cork Tree Retail Pk. 22Ac 52
Cork Tree Way E4 22Ac 52
Corlett St. NW1 7D 214 (43Gb 89)
Cormongers La. RH1: Nutf 3D 208
Cormont Rd. SE5 53Rb 113
Cormorant Cl. E17 24Ac 52

Cormorant Ct. SE8 51Bc 114
(off Pilot Cl.)
Cormorant Ho. EN3: Pond E 15Zb 34
Cormorant Lodge E1 46Wb 91
(off Thomas More St.)
Cormorant Pl. SM1: Sutt 78Bb 155
Cormorant Rd. E7 36Hc 73
Cormorant Wlk. RM12: Horn 37Kd 77
Cornburgh Wlk. RM3: Rom 22Md 57
(off Quilter Way)
Cornbury Ho. SE8 51Bc 114
(off Evelyn St.)
Cornbury Rd. HA8: Edg 24Ma 47
Corncrake Gro. HP3: Hem H 7L 3
Cornel Ho. DA15: Sidc 62Wc 139
Cornel Ho. SL4: Wind 5H 103
Cornelia Dr. UB4: Yead 42Y 85
Cornelia Ho. TW1: Twick 58Ma 109
(off Denton Rd.)
Cornelia Pl. DA8: Erith 51Gd 118
Cornelia St. N7 37Pb 70
Cornelius Ho. WD18: Wat 14V 26
(off Chiltern Cl.)
Cornell Bldg. E1 44Wb 91
(off Coke St.)
Cornell Cl. DA14: Sidc 65Ad 139
Cornell Ct. EN3: Enf H 13Ac 34
Cornell Gdns. EN4: E Barn 15Jb 32
Cornell Ho. HA2: Harr 34Ba 65
Cornell Sq. SW8 53Mb 112
Cornell Way RM5: Col R 22Cd 56
Corner, The KT14: W Byf 85J 169
Corner, The W5 46Na 87
Corner Ct. E2 42Xb 91
(off Three Colts La.)
Cornercroft SM3: Cheam 78Za 154
(off Wickham Av.)
Corner Farm Cl. KT20: Tad 94Ya 194
Corner Fielde SW2 60Pb 112
Corner Grn. SE3 54Jc 115
CORNER HALL 4M 3
Corner Hall HP3: Hem H 4L 3
(not continuous)
Corner Hall Av. HP3: Hem H 4M 3
Corner Ho. NW6 40Db 69
(off Oxford Rd.)
Corner HOUSE Arts Cen., The 74Pa 153
Corner Ho. St. WC2 6F 223 (46Nb 90)
(off Northumberland St.)
Corner Mead NW9 24Va 48
Cornerside TW15: Ashf 66S 128
Cornerstone Ho. CR0: C'don 73Sb 157
Corner Vw. AL9: Wel G 6E 8
Corney Reach Way W4 52Ua 110
Corney Rd. W4 51Ua 110
Cornfield Cl. UB8: Uxb 40M 63
Cornfield Rd. RH2: Reig 7L 207
Cornfield Rd. WD23: Bush 14Da 27
Cornfields, The HP1: Hem H 3K 3
Cornflower La. CR0: C'don 74Zb 158
Cornflower Ter. SE22 58Xb 113
Cornflower Way RM3: Hrld W 25Nd 57
Cornford Cl. BR2: Broml 71Jc 159
Cornford Gro. SW12 61Kb 134
Cornhill EC3 3G 225 (44Tb 91)
Cornhill Cl. KT15: Add 75K 149
Cornhill Dr. EN3: Enf W 9Ac 20
Cornick Ho. SE16 48Xb 91
(off Slippers Pl.)
Cornish Ct. N9 17Xb 33
Cornish Gro. SE20 67Xb 135
Cornish Ho. SE17 51Rb 113
(off Brandon Est.)
Cornish Ho. TW8: Bford 50Pa 87
Cornmill EN9: Walt A 5Dc 20
Corn Mill Dr. BR6: Orp 73Wc 161
Cornmill Ho. RM7: Rom 30Ed 56
Cornmill Ho. SE8 50Cc 92
(off Wharf St.)
Cornmill La. SE13 55Ec 114
Cornmow Dr. NW10 36Va 68
Cornshaw Rd. RM8: Dag 32Zc 75
Cornsland CM14: B'wood 20Zd 41
Cornsland Cl. RM14: Upm 27Sd 58
Cornsland Ct. CM14: B'wood 20Yd 40
Cornthwaite Rd. E5 34Yb 72
Cornwall Av. DA16: Well 55Uc 116
Cornwall Av. E2 41Yb 92
Cornwall Av. KT10: Clay 80Ha 152
Cornwall Av. KT14: Byfl 86P 169
Cornwall Av. N22 25Nb 50
Cornwall Av. N3 24Cb 49
Cornwall Av. SL2: Slou 2G 80
Cornwall Av. UB1: S'hall 43Ba 85
Cornwall Cl. EN8: Walt C 5Ac 20
Cornwall Cl. IG11: Bark 37Vc 75
Cornwall Cl. RM11: Horn 28Qd 57
Cornwall Cl. SL4: Eton W 10C 80
Cornwall Ct. HA5: Hat E 24Ba 45
Cornwall Ct. TW18: Staines 65G 126
(off Cornwall Way)
Cornwall Ct. W7 42Ha 86
(off Copley Cl.)
Cornwall Cres. W11 44Ab 88
Cornwall Dr. BR5: St P 66Yc 139
Cornwall Gdns. NW10 37Xa 68
Cornwall Gdns. SE25 70Vb 135
Cornwall Gdns. SW7 48Db 89
Cornwall Gdns. Wlk. SW7 48Db 89
Cornwall Ga. RM19: Purf 49Qd 97
Cornwall Gro. W4 50Ua 88
Cornwall Ho. SW7 48Db 89
(off Cornwall Gdns.)
Cornwallis Av. N9 19Xb 33
Cornwallis Av. SE9 61Tc 138
Cornwallis Cl. CR3: Cat'm 94Sb 197
Cornwallis Cl. DA8: Erith 51Hd 118
Cornwallis Ct. SW8 53Nb 112
(off Lansdowne Grn.)
Cornwallis Gro. N9 19Xb 33
Cornwallis Ho. SE16 47Xb 91
(off Cherry Gdn. St.)
Cornwallis Ho. W12 45Xa 88
(off India Way)
Cornwallis Rd. SE18: Gunnery Ter. 48Sc 94
Cornwallis Rd. SE18: Warren La. 48Rc 94
Cornwallis Rd. E17 28Zb 52
Cornwallis Rd. N19 33Nb 70
Cornwallis Rd. N9 19Xb 33
Cornwallis Rd. RM9: Dag 35Zc 75
Cornwallis Sq. N19 33Nb 70
Cornwallis Wlk. SE9 55Pc 116
Cornwall Mans. SW10 52Eb 111
(off Cremorne Rd.)
Cornwall Mans. W14 48Za 88
(off Blythe Rd.)
Cornwall Mans. W8 47Db 89
(off Kensington Ct.)
Cornwall M. Sth. SW7 4A 226 (48Eb 89)
Cornwall M. W. SW7 48Db 89
Cornwall Pl. E4 14Dc 34
Cornwall Rd. AL1: St A 4C 6
Cornwall Rd. CM15: Pil H 15Xd 40
Cornwall Rd. CR0: C'don 75Rb 157
Cornwall Rd. DA1: Dart 55Pd 119
Cornwall Rd. HA1: Harr 30Ea 46
Cornwall Rd. HA4: Ruis 34V 64
Cornwall Rd. HA5: Hat E 24Ba 45
Cornwall Rd. N15 29Tb 51
Cornwall Rd. N18 22Wb 51
Cornwall Rd. N4 31Qb 70
Cornwall Rd. SE1 6K 223 (46Qb 90)
Cornwall Rd. SM2: Sutt 80Bb 155
Cornwall Rd. TW1: Twick 59Ja 108
Cornwall Rd. UB8: Uxb 37M 63
Cornwall Sq. SE11 7B 230 (50Rb 91)
(off Seaton Cl.)
Cornwall St. E1 45Xb 91
Cornwall Ter. NW1 6G 215 (42Hb 89)
Cornwall Ter. M. NW1 6G 215 (42Hb 89)
(off Allsop Pl.)
Cornwall Way TW18: Staines 65G 126
Corn Way E11 34Fc 73
Cornwell Av. DA12: Grav'nd 2E 144
Cornwell Cres. SS17: Stan H 1N 101
Cornwell Gdns. E10 31Cc 72
Cornwell Rd. SL4: Old Win 8L 103
Cornwood Cl. N2 29Fb 49
Cornwood Dr. E1 44Yb 92
Cornworthy Rd. RM8: Dag 36Yc 75
Corona Bldg. E14 46Ec 92
(off Blackwall Way)
Corona Rd. SE12 59Jc 115
Coronation Av. N16 35Vb 71
Coronation Av. RM18: E Til 9K 101
Coronation Av. SL3: Geor G 43A 82
Coronation Av. SL4: Wind 4L 103
Coronation Cl. DA5: Bexl 58Zc 117
Coronation Cl. IG6: Ilf 28Sc 54
Coronation Cl. DA8: Erith 52Fd 118
Coronation Ct. E15 37Hc 73
Coronation Ct. KT1: King T 70Na 131
(off Surbiton Rd.)
Coronation Ct. RM18: E Til 9L 101
(off Coronation Av.)
Coronation Ct. W10 43Ya 88
(off Brewster Gdns.)
Coronation Dr. RM12: Horn 36Kd 77
Coronation Hill CM16: Epp 2Vc 23
Coronation Rd. E13 41Lc 93
Coronation Rd. NW10 41Qa 87
Coronation Rd. UB3: Harl 49V 84
Coronation Vs. NW10 42Ra 87
Coronation Wlk. TW2: Whitt 60Ca 107
Coroner's Court City of London 4F 225 (45Tb 91)
Coroner's Court North London 14Bb 31
Coroner's Court Poplar 45Dc 92
(off Poplar High St.)
Coroner's Court South London 76Tb 157
(off Barclay Rd.)
Coroner's Court Southwark 1F 231 (47Tb 91)
Coroner's Court St Pancras 1E 216 (39Mb 70)
Coroner's Court West London 53Eb 111
Coroner's Court Westminster 5E 228 (49Mb 90)
Coronet Pde. HA0: Wemb 37Na 67
Coronet St. N1 4H 219 (41Ub 91)
Corporate Dr. TW13: Felt 62X 129
Corporate Ho. HA3: Hrw W 25Fa 46
Corporation Av. TW4: Houn 56Aa 107
Corporation Row EC1 5A 218 (42Qb 90)
Corporation St. E15 40Gc 73
Corporation St. N7 36Nb 70
Corrance Rd. SW2 56Nb 112
Corran Way RM15: S Ock 45Xd 98
Corri Av. N14 21Mb 50
Corrib Ct. N13 20Pb 32
Corrib Dr. SM1: Sutt 78Gb 155
Corrie Gdns. GU25: Vir W 3N 147
Corrie Rd. GU22: Wok 92D 188
Corrie Rd. KT15: Add 77M 149
Corrigan Av. CR5: Coul 87Jb 176
Corrigan Cl. NW4 27Ya 48
Corringham Ct. AL1: St A 3F 6
Corringham Ct. NW11 31Cb 69
Corringham Ho. E1 44Zb 92
(off Pitsea St.)
Corringham Rd. SS17: Stan H
Central Rd. 2M 101
Corringham Rd. SS17: Stan H
Warburtons 1P 101
Corringham Rd. HA9: Wemb 33Qa 67
Corringham Rd. NW11 31Cb 69
Corringway NW11 31Db 69
Corringway W5 42Qa 87
Corris Grn. NW9 29Ua 48
Corry Dr. SW9 56Rb 113
Corry Ho. E14 45Dc 92
(off Wade's Pl.)
Corry's End AL4: Col H 4A 8
Corsair Cl. TW19: Stanw 59M 105
Corsair Ho. E16 47Lc 93
(off Starboard Way)
Corsair Rd. TW19: Stanw 59N 105
Corscombe Cl. KT2: King T 64Sa 131
Corsehill St. SW16 65Lb 134
Corsellis Sq. TW1: Isle 56Ka 108
(off Varley Dr.)
Corsham St. N1 4G 219 (41Tb 91)
Corsica St. N5 37Rb 71
Corsley Way E9 37Bc 72
Corston Hollow RH1: Redh 7P 207
(off Woodlands Rd.)
Cortayne Ct. TW2: Twick 61Ga 130
Cortayne Rd. SW6 54Bb 111
Cortis Rd. SW15 58Xa 110
Cortis Ter. SW15 58Xa 110
Cortland Cl. DA1: Cray 58Gd 118
Cortland Cl. IG8: Wfd G 25Lc 53
Corunna Rd. SW8 53Lb 112
Corunna Ter. SW8 53Lb 112
Corve La. RM15: S Ock 45Xd 98
Corvette Sq. SE10 51Fc 115
Corwell Gdns. UB8: Hil 44S 84
Corwell La. UB8: Hil 44S 84
Cory Dr. CM13: Hut 17De 41
Coryton Path W9 42Bb 89
(off Ashmore Rd.)
Cosbycote Av. SE24 57Sb 113
Cosdach Av. SM6: W'gton 80Mb 156
Cosedge Cres. CR0: Wadd 78Qb 156
Cosgrove Cl. N21 19Sb 33
Cosgrove Cl. UB4: Yead 42Z 85
Cosgrove Ho. E2 39Wb 71
(off Whiston Rd.)
Cosgrove Ho. HA0: Wemb 39Na 67
(off Hatton Rd.)
Cosmia Ct. WD23: Bush 15Aa 27
(off Vale Rd.)
Cosmo Pl. WC1 7G 217 (43Nb 90)
Cosmopolitan Ct. EN1: Enf 15Wb 33
Cosmopolitan Way TW6: H'row A 55R 106
Cosmur Cl. W12 48Va 88
Cossall Wlk. SE15 54Xb 113
Cossar M. SW2 57Qb 112
Cosser St. SE1 4K 229 (48Qb 90)
Costa St. SE15 54Wb 113
Costead Mnr. Rd. CM14: B'wood 18Xd 40
Costells Mdw. TN16: Westrm 98Tc 200
Costemonger Bldg. SE16 4K 231 (48Vb 91)
(off Arts La.)
Coster Av. N4 32Sb 71
Costins Wlk. HP4: Berk 1A 2
(off Robertson Rd.)
Costons Av. UB6: G'frd 41Fa 86
Costons La. UB6: G'frd 41Fa 86
(not continuous)
Coston Wlk. SE4 56Zb 114
Cosway Mans. NW1 7E 214 (43Gb 89)
(off Shroton St.)
Cosway St. NW1 7E 214 (43Gb 89)
Cotall St. E14 43Cc 92
Coteford Cl. HA5: Eastc 29W 44
Coteford Cl. IG10: Lough 12Rc 36
Coteford St. SW17 63Hb 133
Cotelands CR0: C'don 76Ub 157
Cotesbach Rd. E5 34Yb 72
Cotes Ho. NW8 6D 214 (42Gb 89)
(off Broadley St.)
Cotesmore Gdns. RM8: Dag 35Yc 75
Cotesmore Rd. HP1: Hem H 3G 2
Cotford Rd. CR7: Thor H 70Sb 135
Cotham St. SE17 6E 230 (49Sb 91)
Cotherstone KT19: Ewe 82Ta 173
Cotherstone Ct. E2 42Xb 91
(off Three Colts La.)
Cotherstone Rd. SW2 60Pb 112
Cotland Acres RH1: Redh 8M 207
Cotlandswick AL2: Lon C 7G 6
Cotlandswick Leisure Cen. 7G 6
Cotleigh Av. DA5: Bexl 61Zc 139
Cotleigh Rd. NW6 38Cb 69
Cotleigh Rd. RM7: Rom 30Ed 56
Cotman Cl. NW11 30Eb 49
Cotman Cl. SW15 58Za 110
Cotmandene Cres. BR5: St P 68Wc 139
Cotman Gdns. HA8: Edg 26Qa 47
Cotman Ho. NW8 1D 214 (40Gb 69)
(off Townshend Est.)
Cotman Ho. UB5: N'olt 40Z 65
(off Academy Gdns.)
Cotman M. RM8: Dag 36Yc 75
(off Highgrove Rd.)
COTMAN'S ASH 87Sd 184
Cotman's Ash La. TN15: Kems'g 86Rd 183
Cotmans Cl. UB3: Hayes 46W 84
Coton Dr. UB10: Ick 34S 64
Coton Rd. DA16: Well 55Wc 117
Cotsford Av. KT3: N Mald 71Sa 153
Cotsmoor AL1: St A 2D 6
(off Granville Rd.)
Cotswold Av. WD23: Bush 16Ea 28
Cotswold Cl. DA7: Bex 54Gd 118
Cotswold Cl. KT10: Hin W 75Ha 152
Cotswold Cl. KT2: King T 65Sa 131
Cotswold Cl. N11 21Jb 50
Cotswold Cl. SL1: Slou 8G 80
Cotswold Cl. TW18: Staines 64J 127
Cotswold Cl. UB8: Uxb 39L 63
Cotswold Ct. EC1 5D 218 (42Sb 91)
(off Gee St.)
Cotswold Ct. UB6: G'frd 40Ha 66
(off Hodder Dr.)
Cotswold Gdns. CM13: Hut 17Fe 41
Cotswold Gdns. E6 41Mc 93
Cotswold Gdns. IG2: Ilf 31Tc 74
Cotswold Gdns. NW2 33Za 68
Cotswold Ga. NW2 32Ab 68
Cotswold Grn. EN2: Enf 14Pb 32
Cotswold M. SW11 53Fb 111
Cotswold Ri. BR6: St M Cry 72Vc 161
Cotswold Rd. DA11: Nflt 2A 144
Cotswold Rd. RM3: Hrld W 26Pd 57
Cotswold Rd. SM2: Sutt 82Db 175
Cotswold Rd. TW12: Hamp 64Ca 129
Cotswolds AL10: Hat 2C 8
(off Park Cres.)
Cotswold St. SE27 63Rb 135
Cotswold Way EN2: Enf 13Pb 32
Cotswold Way KT4: Wor Pk 75Ya 154
Cottage Av. BR2: Broml 74Nc 160
Cottage Cl. E1 42Yb 92
(off Mile End Rd.)
Cottage Cl. HA2: Harr 33Fa 66
Cottage Cl. HA4: Ruis 32T 64
Cottage Cl. KT16: Ott 79E 148
Cottage Cl. WD17: Wat 12V 26
Cottage Cl. WD3: Crox G 16P 25
Cottage Farm Way TW20: Thorpe 69E 126
Cottage Fld. Cl. DA14: Sidc 60Yc 117
Cottage Gdns. EN8: Chesh 1Zb 20
Cottage Grn. SE5 52Tb 113
Cottage Gro. KT6: Surb 72Ma 153
Cottage Gro. SW9 55Nb 112
Cottage M. RM11: Horn 28Ld 57
Cottage Pk. Rd. SL2: Hedg 3H 61
Cottage Pl. SW3 4D 226 (48Gb 89)
Cottage Rd. KT19: Ewe 80Ta 153
Cottage Rd. N7 36Pb 70
(not continuous)
Cottages, The UB10: Ick 33N 63
Cottage St. E14 45Dc 92
Cottage Wlk. N16 34Vb 71
Cottenham Dr. NW9 27Va 48
Cottenham Dr. SW20 66Xa 132
Cottenham Pde. SW20 68Xa 132
COTTENHAM PARK 67Xa 132
Cottenham Pk. Rd. SW20 67Wa 132
(not continuous)
Cottenham Pl. SW20 66Xa 132
Cottenham Rd. E17 28Bc 52
Cotterells HP1: Hem H 2L 3
Cotterells Hill HP1: Hem H 2L 3
Cotterill Rd. KT6: Surb 75Na 153
Cottesbrooke Cl. SL3: Coln 53F 104
Cottesbrook St. SE14 52Ac 114
Cottesloe Cl. GU24: Bisl 8D 166
Cottesloe Ho. NW8 5D 214 (42Gb 89)
(off Jerome Cres.)
Cottesloe M. SE1 3A 230 (48Qb 90)
(off Emery St.)
Cottesmore Av. IG5: Ilf 26Qc 54
Cottesmore Ct. W8 48Db 89
(off Stanford Rd.)
Cottesmore Gdns. W8 48Db 89
Cottesmore Ho. UB10: Ick 33S 64
Cottimore Av. KT12: Walt T 74X 151
Cottimore Cres. KT12: Walt T 73X 151
Cottimore La. KT12: Walt T 73X 151
Cottimore Ter. KT12: Walt T 73X 151
Cottingham Chase HA4: Ruis 34W 64
Cottingham Rd. SE20 66Zb 136
Cottingham Rd. SW8 52Pb 112
Cottington Rd. TW13: Hanw 64Z 129
Cottington St. SE11 7A 230 (50Qb 90)
Cottis La. CM16: Epp 2Vc 23
Cottle Way SE16 47Xb 91
(off Paradise St.)
Cotton Apts. E1 43Zb 92
(off Killick Way)
Cotton Av. W3 44Ta 87
Cotton Cl. CR4: Mitc 69Gb 133
Cotton Cl. E11 33Gc 73
Cotton Cl. RM9: Dag 38Yc 75
Cottongrass Cl. CR0: C'don 74Zb 158
Cotton Hall Ho. SL4: Eton 1G 102
(off Eton Wick Rd.)
Cottonham Cl. N12 22Fb 49
Cotton Hill BR1: Broml 63Ec 136
Cotton Ho. SW2 59Nb 112
Cotton La. DA2: Dart 57Sd 120
Cotton La. DA2: Ghithe 57Sd 120
Cotton La. DA9: Ghithe 57Sd 120
Cottonmill Cres. AL1: St A 3B 6
Cottonmill La. AL1: St A 4B 6
Cotton Rd. EN6: Pot B 3Eb 17
Cotton Row SW11 55Eb 111
Cottons App. RM7: Rom 29Fd 56
Cottons Cen. SE1 6H 225 (46Ub 91)
Cottons Ct. RM7: Rom 29Fd 56
Cotton's Gdns. E2 3J 219 (41Ub 91)
Cottons La. SE1 6G 225 (46Tb 91)
Cotton St. E14 45Ec 92
Cotton Way SM6: W'gton 75Jb 156
Cottonwood Cl. BR6: Farnb 77Rc 160
Cottonworks Ho. N7 34Pb 70
(off Seven Sisters Rd.)
Cottrell Ct. SE10 49Hc 93
(off Hop St.)
Cottrill Gdns. E8 37Xb 71
Cotts Cl. W7 43Ha 86
Couchmore Av. IG5: Ilf 26Pc 54
Couchmore Av. KT10: Hin W 75Ga 152
Coulgate St. SE4 55Ac 114
COULSDON 88Mb 176
Coulsdon Common 93Sb 197
Coulsdon Court Golf Course 88Pb 176
Coulsdon Ct. Rd. CR5: Coul 88Pb 176
Coulsdon La. CR5: Chips 91Hb 195
Coulsdon Nth. Ind. Est. CR5: Coul 88Mb 176
Coulsdon Pl. CR3: Cat'm 94Tb 197
Coulsdon Ri. CR5: Coul 89Nb 176
Coulsdon Rd. CR3: Cat'm 94Tb 197
Coulsdon Rd. CR3: Coul 94Tb 197
Coulsdon Rd. CR5: Coul 87Pb 176
Coulsdon South Station (Rail) 88Mb 176
Coulsdon Town Station (Rail) 87Nb 176
Coulson Cl. RM8: Dag 32Yc 75
Coulson Ct. AL2: Lon C 9H 7
Coulson St. SW3 7F 227 (50Hb 89)
Coulter Cl. UB4: Yead 42Aa 85
Coulter Ho. DA9: Ghithe 57Yd 120
Coulter Rd. W6 48Xa 88
Coulthurst Ct. SW16 66Nb 134
(off Heybridge Av.)
Coulton Av. DA11: Nflt 9A 122
Council Av. DA11: Nflt 58Ee 121
Council Cotts. GU23: Wis 87M 169
Council Cotts. GU24: W End 4D 166
Councillor St. SE5 52Sb 113
Counter Ct. SE1 7F 225 (46Tb 91)
(off Borough High St.)
Counter Ho. E1 45Wb 91
Counters Ct. HP1: Hem H 2J 3
Counters Ct. W14 48Ab 88
(off Holland Rd.)
COUNTERS END 2J 3
Counter St. SE1 6H 225 (46Ub 91)
Countess Cl. UB9: Hare 26L 43
Countess Rd. NW5 36Lb 70
Countisbury Av. EN1: Enf 17Vb 33
Countisbury Gdns. KT15: Add 78K 149
Country Way TW13: Hanw 65X 129
County Court Brentford 51Ma 109
County Court Bromley 67Jc 137
County Court Central London 6A 216 (42Kb 90)
County Court Clerkenwell &
Shoreditch 5D 218 (42Sb 91)
County Court Croydon 75Tb 157
County Court Dartford 58Nd 119
County Court Edmonton 23Wb 51
County Court Kingston upon Thames 68Ma 131
County Court Romford 28Hd 56
County Court Slough 7J 81
County Court Staines upon Thames 64J 127
County Court Uxbridge 43V 84
County Court Wandsworth 57Ab 110
County Court Watford 12X 27
County Court West London 50Za 88
(off Talgarth Rd.)
County Court Willesden 40Va 68
County Gdns. TW7: Isle 56Fa 108
County Ga. EN5: New Bar 16Db 31
County Ga. SE9 62Sc 138
County Ground Beckenham,
The 65Cc 136
County Gro. SE5 53Sb 113
County Hall Apts. SE1 2H 229 (47Pb 90)
(off Westminster Bri. Rd.)
County Hall (Former) 1H 229 (47Pb 90)
(off Westminster Bri. Rd.)
County Ho. BR3: Beck 67Ac 136
County Ho. SW9 53Qb 112
(off Brixton Rd.)
County Pde. TW8: Bford 52Ma 109
County Rd. CR7: Thor H 68Rb 135
County Rd. E6 43Rc 94
County St. SE1 4E 230 (48Sb 91)
Couper Ho. WD18: Wat 16U 26
Coupland Pl. SE18 50Sc 94
Courage Cl. RM11: Horn 30Ld 57
Courage Ct. CM13: Hut 16Ee 41
Courage Stadium 71Kc 159
Courage Wlk. CM13: Hut 16Fe 41
Courcy Rd. N8 27Qb 50
Courier Rd. RM9: Dag 42Ed 96
Courland Gro. SW8 53Mb 112
Courland Rd. KT15: Add 76K 149
Courland St. SW8 53Mb 112
Course, The SE9 62Qc 138
Coursers Rd. AL4: Col H 9L 7
Court, The CR6: W'ham 90Ac 178
Court, The HA4: Ruis 35Aa 65
Court Annexe 2E 230 (47Sb 91)
Courtauld Cl. SE28 46Wc 95
Courtauld Gallery 4H 223 (45Pb 90)
(off Strand)
Courtauld Ho. E2 39Wb 71
(off Goldsmiths Row)
Courtauld Institute of Art, The
4H 223 (45Pb 90)
(off Strand)
Courtauld Rd. N19 32Nb 70
Courtaulds WD4: Chfd 2K 11
Court Av. CR5: Coul 90Qb 176
Court Av. DA17: Belv 50Bd 95
Court Av. RM3: Hrld W 24Qd 57
Court Bushes Rd. CR3: W'ham 91Wb 197
Court Cl. HA3: Kenton 27Na 47
Court Cl. NW8 38Fb 69
(off Boydell Ct.)
Court Cl. SM6: W'gton 80Mb 156
Court Cl. TW2: Twick 62Da 129
Court Cl. Av. TW2: Twick 62Da 129
Court Cres. BR8: Swan 70Gd 140
Court Cres. KT9: Chess 78Ma 153
Court Cres. SL1: Slou 4H 81
Court Downs Rd. BR3: Beck 68Dc 136
Court Dr. CR0: Wadd 77Pb 156
Court Dr. HA7: Stan 21Na 47
Court Dr. SM1: Sutt 77Gb 155
Court Dr. UB10: Hil 39P 63
Courtenay Av. HA3: Hrw W 24Ea 46
Courtenay Av. N6 31Gb 69
Courtenay Av. SM2: Sutt 81Cb 175
Courtenay Dr. BR3: Beck 68Fc 137
Courtenay Dr. RM16: Chaf H 48Be 99
Courtenay Gdns. HA3: Hrw W 26Ea 46
Courtenay Gdns. RM14: Upm 32Sd 78
Courtenay Ho. CR0: C'don 73Sb 157
(off Oakfield Rd.)
Courtenay M. E17 29Ac 52
Courtenay M. GU21: Wok 88C 168
Courtenay Pl. E17 29Ac 52
Courtenay Rd. E11 34Hc 73
Courtenay Rd. E17 28Zb 52
Courtenay Rd. GU21: Wok 88C 168
Courtenay Rd. HA9: Wemb 34Ma 67
Courtenay Rd. KT4: Wor Pk 76Ya 154
Courtenay Rd. SE20 65Zb 136
Courtenay Sq. SE11 7K 229 (50Qb 90)
Courtenay St. SE11 7K 229 (50Qb 90)
Courtens M. HA7: Stan 24La 46
Court Farm Av. KT19: Ewe 78Ta 153
Court Farm Cl. SL1: Slou 6F 80
Court Farm Gdns. KT19: Eps 83Sa 173
Court Farm La. RH8: Oxt 100Gc 199
Court Farm La. UB5: N'olt 38Ca 65
Court Farm Pk. CR6: W'ham 88Wb 177
Court Farm Rd. CR6: W'ham 90Wb 177
Court Farm Rd. SE9 61Mc 137
Court Farm Rd. UB5: N'olt 38Ca 65
Courtfield W5 43La 86
Courtfield Av. HA1: Harr 29Ha 46
Courtfield Cres. HA1: Harr 29Ha 46
Courtfield Gdns. HA4: Ruis 33V 64
Courtfield Gdns. SW5 49Db 89
Courtfield Gdns. UB9: Den 34J 63
Courtfield Gdns. W13 44Ja 86
Courtfield Ho. EC1 7K 217 (43Qb 90)
(off Baldwins Gdns.)
Courtfield Ho. UB8: Uxb 37M 63
(off Fairfield Rd.)
Courtfield M. SW5 49Eb 89
Courtfield Ri. BR4: W W'ck 76Fc 159
Courtfield Rd. SW7 49Eb 89
Courtfield Rd. TW15: Ashf 65R 128
Court Gdns. N1 37Qb 70
Court Gdns. N7 37Qb 70
Court Gdns. RM3: Hrld W 23Qd 57
Courtgate Cl. NW7 23Va 48
Court Grn. Hgts. GU22: Wok 2N 187
Court Haw SM7: Bans 87Gb 175
Court Hill CR2: Sande 84Ub 177
Court Hill CR5: Chips 90Gb 175
Courthill Rd. SE13 56Ec 114
Courthope Ho. SE16 48Yb 92
(off Lower Rd.)
Courthope Ho. SW8 52Nb 112
(off Hartington Rd.)
Courthope Rd. NW3 35Hb 69
Courthope Rd. SW19 64Ab 132
Courthope Rd. UB6: G'frd 40Fa 66
Courthope Vs. SW19 66Ab 132
Courthouse, The SW1 4F 229 (48Nb 90)
(off Horseferry Rd.)
Courthouse Gdns. N3 23Cb 49
Court Ho. La. N16 35Vb 71
Court Ho. Mans. KT19: Eps 84Ta 173
Courthouse Rd. N12 23Db 49
Courthouse Ter. SL1: Burn 3A 80
Courthouse Way SW18 57Db 111
Courtland Av. E4 19Hc 35
Courtland Av. IG1: Ilf 33Pc 74
Courtland Av. NW7 20Ta 29
Courtland Av. SW16 66Pb 134
Courtland Dr. IG7: Chig 20Rc 36
Courtland Gro. SE28 44Zc 95
Courtland Ho. SE20 67Wb 135
Courtland Rd. E6 39Nc 74
Courtlands KT12: Walt T 73W 150
Courtlands TW10: Rich 57Qa 109
Courtlands Av. BR2: Hayes 74Gc 159
Courtlands Av. KT10: Esh 79Ba 151
Courtlands Av. SE12 57Kc 115
Courtlands Av. SL3: L'ly 9P 81
Courtlands Av. TW12: Hamp 65Ba 129
Courtlands Av. TW9: Kew 54Ra 109
Courtlands Cl. CR2: Sande 82Vb 177
Courtlands Cl. HA4: Ruis 31V 64
Courtlands Cl. WD24: Wat 7U 12
Courtlands Cres. SM7: Bans 87Cb 175
Courtlands Dr. KT19: Ewe 79Ua 154
Courtlands Dr. WD17: Wat 9U 12
Courtlands Dr. WD24: Wat 8U 12
Courtlands Rd. KT5: Surb 73Qa 153
Court La. KT19: Eps 85Sa 173
Court La. SE21 58Ub 113
Court La. SL0: Iver 46H 83
(not continuous)
Court La. SL1: Burn 1B 80
Court La. Gdns. SE21 59Ub 113
Courtleas KT11: Cobh 85Ca 171
Courtleet Dr. DA8: Erith 53Dd 118
Courtleigh NW11 29Bb 49
Courtleigh Av. EN4: Had W 10Fb 17
Courtleigh Gdns. NW11 28Ab 48
Court Lodge DA12: Shorne 5N 145
Court Lodge DA17: Belv 50Cd 96
Court Lodge SW1 6H 227 (49Jb 90)
(off Sloane Sq.)
Courtman Rd. N17 24Sb 51
Court Mead UB5: N'olt 41Ba 85
Courtmead Cl. SE24 58Sb 113
Court Mdw. TN15: Wro 88Be 185
Court M. SE13 58Gc 115

Courtnell St. W2 44Cb 89
Courtney Cl. SE19 65Ub 135
Courtney Ct. N7 36Qb 70
Courtney Cres. SM5: Cars 80Hb 155
Courtney Ho. NW4 27Ya 48
(off Mulberry Cl.)
Courtney Ho. W14 48Ab 88
(off Russell Rd.)
Courtney Pl. CR0: Wadd 76Qb 156
Courtney Pl. KT11: Cobh 84Ba 171
Courtney Rd. CR0: Wadd 76Qb 156
Courtney Rd. N7 36Qb 70
Courtney Rd. RM16: Grays 7E 100
Courtney Rd. SW19 66Gb 133
Courtney Rd. TW6: H'row A 55Q 106
Courtney Way TW6: H'row A 54Q 106
Court Pde. HA0: Wemb 34Ka 66
Courtrai Rd. SE23 58Ac 114
Court Rd. BR6: Chels 73Xc 161
Court Rd. BR6: Orp 73Xc 161
Court Rd. CR3: Cat'm 95Tb 197
Court Rd. DA2: Daren 64Ud 142
Court Rd. RH9: G'stone 3A 210
Court Rd. SE25 68Vb 135
Court Rd. SE9 58Pc 116
Court Rd. SM7: Bans 88Cb 175
Court Rd. UB10: Ick 36R 64
Court Rd. UB2: S'hall 49Ba 85
Court Royal SW15 57Ab 110
Courtside AL3: St A 1B 6
Courtside N8 30Mb 50
Courtside SE26 62Xb 135
Court St. BR1: Broml 68Jc 137
Court St. E1 43Xb 91
Courts Way RM15: Avel 45Td 98
Courtville Ho. W10 41Ab 88
(off Third Av.)
Court Way NW9 28Ua 48
Court Way RM3: Hrld W 26Nd 57
Court Way TW2: Twick 59Ha 108
Court Way W3 43Sa 87
Courtway IG6: Ilf 27Sc 54
Courtway IG8: Wfd G 22Lc 53
Courtway, The WD19: Wat 19Aa 27
Courtwood Dr. TN13: S'oaks 96Jd 202
Courtwood La. CR0: Sels 83Bc 178
Court Yd. SE9 58Pc 116
Courtyard SW3 7F 227 (50Hb 89)
(off Smith St.)
Courtyard, The AL4: St A 2K 7
Courtyard, The AL9: Ess 1P 9
Courtyard, The BR2: Kes 79Nc 160
Courtyard, The CM15: B'wood 17Xd 40
Courtyard, The CR3: Whyt 90Vb 177
Courtyard, The E2 41Vb 91
(off Ezra St.)
Courtyard, The EC3 3G 225 (44Tb 91)
(within Royal Exchange)
Courtyard, The HP3: Hem H 8M 3
Courtyard, The KT14: W Byf 84J 169
Courtyard, The KT20: Kgswd 95Eb 195
Courtyard, The N1 38Pb 70
Courtyard, The NW1 38Jb 70
Courtyard, The SE14 53Zb 114
(off Besson St.)
Courtyard, The SL3: L'ly 47C 82
Courtyard, The SW3 51Fb 111
(off Trident Pl.)
Courtyard, The TN16: Westrm 99Tc 200
Courtyard, The WD3: Crox G 13R 26
Courtyard Apts. E1 5K 219 (42Vb 91)
(off Sclater St.)
Courtyard Gdns. TN15: Wro 88Be 185
Courtyard Ho. SW6 54Eb 111
(off Lensbury Av.)
Courtyard M. BR5: St P 66Wc 139
Courtyard M. DA9: Ghithe 58Wd 120
Courtyard M. RM13: Rain 39Hd 76
Courtyards, The WD18: Wat 17T 26
Courtyard Theatre Chipstead 92Hb 195
Courtyard Theatre Hoxton 4H 219 (41Ub 91)
Cousin La. EC4 5F 225 (45Tb 91)
Cousins Cl. UB7: Yiew 45N 83
Couthurst Rd. SE3 51Kc 115
Coutts Av. DA12: Shorne 3N 145
Coutts Av. KT9: Chess 78Na 153
Coutt's Cres. NW5 34Jb 70
Couzens Ho. E3 43Bc 92
(off Weatherley Cl.)
Couzins Wlk. DA1: Dart 54Qd 119
Coval Gdns. SW14 56Ra 109
Coval La. SW14 56Ra 109
Coval Pas. SW14 56Sa 109
Coval Rd. SW14 56Ra 109
Coveham Cres. KT11: Cobh 85W 170
Covelees Wall E6 44Qc 94
Covell Ct. EN2: Enf 10Pb 18
(off The Ridgeway)
Covell Ct. SE8 52Cc 114
Covell Ho. KT19: Eps 82Ra 173
Covenbrook CM13: B'wood 20De 41
COVENT GARDEN 4G 223 (45Nb 90)
Covent Gdn. WC2 4G 223 (45Nb 90)
Covent Garden Piazza WC2 4G 223 (45Nb 90)
(off Covent Garden)
Covent Garden Station (Underground) 4F 223 (45Nb 90)
Coventry Cl. E6 44Pc 94
Coventry Cl. NW6 40Cb 69
Coventry Hall SW16 64Nb 134
Coventry Rd. E1 42Xb 91
Coventry Rd. E2 42Xb 91
Coventry Rd. IG1: Ilf 33Rc 74
Coventry Rd. SE25 70Wb 135
Coventry St. W1 5D 222 (45Mb 90)
Coverack Cl. CR0: C'don 73Ac 158
Coverack Cl. N14 16Lb 32
Coverdale Cl. HA7: Stan 22Ka 46
Coverdale Ct. EN3: Enf W 9Ac 20
Coverdale Gdns. CR0: C'don 76Vb 157
Coverdale Rd. N11 23Jb 50
Coverdale Rd. NW2 38Za 68
Coverdale Rd. W12 47Xa 88
Coverdales, The IG11: Bark 40Tc 74
Coverdale Way SL2: Slou 2C 80
Coverham Ho. SE4 56Zb 114
(off Billingford Cl.)
Coverley Cl. CM13: Gt War 23Yd 58
Coverley Cl. E1 43Wb 91
Coverley Point SE11 6H 229 (49Pb 90)
(off Tyers St.)
Covert, The BR6: Pet W 72Uc 160
Covert, The HA6: Nwood 25S 44
Covert, The SE19 66Vb 135
(off Fox Hill)
Covert, The SL5: Asc 3A 146
Coverton Rd. SW17 64Gb 133
Covert Rd. IG6: Ilf 22Vc 55
Coverts, The CM13: Hut 18Ce 41
Coverts Rd. KT10: Clay 80Ha 152
Covert Way EN4: Had W 12Eb 31
Covesfield DA11: Grav'nd 9B 122
Covet Wood Cl. BR5: St M Cry 72Vc 161
Covey Cl. SW19 68Db 133
Covey Rd. KT4: Wor Pk 75Za 154
Covington Gdns. SW16 66Rb 135
Covington Way SW16 65Pb 134
(not continuous)
Cowan Cl. E6 43Nc 94
Coward Ind. Est. RM16: Grays 10D 100
Cowbridge La. IG11: Bark 38Rc 74
Cowbridge Mdw. GU24: Pirb 5D 186
Cowbridge Rd. HA3: Kenton 28Pa 47
Cowcross St. EC1 7B 218 (43Rb 91)
Cowdenbeath Path N1 39Pb 70
Cowden Rd. BR6: Orp 73Vc 161
Cowden St. SE6 63Cc 136
Cowdray Rd. UB10: Hil 39S 64
Cowdray Way RM12: Horn 35Jd 76
Cowdray Way WD3: Crox G 14R 26
Cowdrey Cl. EN1: Enf 12Ub 33
Cowdrey Ct. DA1: Dart 59Kd 119
Cowdrey M. SE6 63Cc 136
Cowdrey Rd. SW19 64Db 133
Cowen Av. HA2: Harr 33Fa 66
Cowgate Rd. UB6: G'frd 41Fa 86
Cowick Rd. SW17 63Hb 133
Cowings Mead UB5: N'olt 37Aa 65
Cowland Av. EN3: Pond E 14Yb 34
Cow La. UB6: G'frd 40Fa 66
Cow La. WD23: Bush 16Ca 27
Cow La. WD25: Wat 8Y 13
Cow Leaze E6 44Qc 94
Cowleaze Rd. KT2: King T 67Na 131
COWLEY 42M 83
Cowley Av. DA9: Ghithe 57Vd 120
Cowley Av. KT16: Chert 73H 149
Cowley Bus. Pk. UB8: Cowl 41L 83
Cowley Cl. CR2: Sels 81Yb 178
Cowley Cres. KT12: Hers 77Y 151
Cowley Cres. UB8: Cowl 43L 83
Cowley Hill WD6: Bore 9Qa 15
(not continuous)
Cowley La. E11 34Gc 73
Cowley La. KT16: Chert 73H 149
Cowley Lodge KT16: Chert 73H 149
Cowley Mill Rd. UB8: Uxb 40K 63
Cowley Mill Trad. Est. UB8: Uxb 40K 63
COWLEY PEACHEY 44M 83
Cowley Pl. NW4 29Ya 48
Cowley Retail Pk. 45M 83
Cowley Rd. E11 29Kc 53
Cowley Rd. IG1: Ilf 31Pc 74
Cowley Rd. RM3: Rom 24Kd 57
Cowley Rd. SW14 55Ua 110
Cowley Rd. SW9 53Qb 112
(not continuous)
Cowley Rd. UB8: Uxb 40L 63
Cowley Rd. W3 46Va 88
Cowley St. SW1 3F 229 (48Nb 90)
Cowling Cl. W11 46Ab 88
Coworth Cl. SL5: S'dale 1F 146
Coworth Pk. 10G 124
Coworth Rd. SL5: S'dale 1E 146
Cowper Av. E6 38Nc 74
Cowper Av. RM18: Tilb 3D 122
Cowper Av. SM1: Sutt 77Fb 155
Cowper Cl. BR2: Broml 70Mc 137
Cowper Cl. DA16: Well 57Wc 117
Cowper Cl. KT16: Chert 72H 149
Cowper Ct. WD24: Wat 9W 12
Cowper Gdns. N14 16Kb 32
Cowper Gdns. SM6: W'gton 79Lb 156
Cowper Ho. SE17 7E 230 (50Sb 91)
(off Browning St.)
Cowper Ho. SW1 7E 228 (50Mb 90)
(off Aylesford St.)
Cowper Rd. BR2: Broml 70Mc 137
Cowper Rd. DA17: Belv 49Cd 96
Cowper Rd. HP1: Hem H 4K 3
Cowper Rd. KT2: King T 64Pa 131
Cowper Rd. N14 18Kb 32
Cowper Rd. N16 36Ub 71
Cowper Rd. N18 22Wb 51
Cowper Rd. RM13: Rain 42Jd 96
Cowper Rd. SL2: Slou 2E 80
Cowper Rd. SW19 65Eb 133
Cowper Rd. W3 46Ta 87
Cowper Rd. W7 45Ha 86
Cowper's Ct. EC3 3G 225 (44Tb 91)
(off Birchin La.)
Cowper St. EC2 5G 219 (42Tb 91)
Cowper Ter. W10 43Za 88
COWSHOT COMMON 2C 186
Cowshot Cres. GU24: Brkwd 2B 186
Cowslip Cl. UB10: Uxb 38N 63
Cowslip La. GU21: Wok 7M 167
Cowslip La. RH5: Mick 100Ja 192
(not continuous)
Cowslip Rd. E18 26Kc 53
Cowthorpe Rd. SW8 53Mb 112
Cow Wlk. AL4: St A 1J 7
Cox Cl. WD7: Shenl 4Pa 15
Cox Ct. EN4: E Barn 14Gb 31
Coxdean KT18: Tatt C 91Ya 194
Coxe Pl. HA3: W'stone 28Ja 46
Coxfield Cl. HP2: Hem H 3N 3
Cox Ho. W6 51Ab 110
(off Field Rd.)
Cox La. KT19: Ewe 78Ra 153
(not continuous)
Cox La. KT9: Chess 77Pa 153
Coxley Ri. CR8: Purl 85Sb 177
Coxmount Rd. SE7 50Mc 93
Coxon Dr. RM16: Chaf H 48Ae 99
Coxs Av. TW17: Shep 69U 128
Coxson Way SE1 2K 231 (47Vb 91)
Cox's Wlk. SE21 60Wb 113
COXTIE GREEN 15Sd 40
Coxtie Grn. Rd. CM14: Pil H 14Qd 39
Coxtie Grn. Rd. CM14: S Weald 14Qd 39
Coxwell Blvd. NW9 24Va 48
Coxwell Rd. SE18 50Tc 94
Coxwell Rd. SE19 66Ub 135
Coxwold Path KT9: Chess 80Na 153
Coyle Dr. UB10: Ick 33S 64
Coyne Cl. GU18: Light 2A 166
Crabbs Cft. Cl. BR6: Farnb 78Sc 160
Crab Hill BR3: Beck 66Fc 137
Crab Hill La. RH1: S Nut 10F 208
Crab La. WD25: A'ham 7Da 13
Crabtree Av. HA0: Wemb 40Na 67
Crabtree Av. RM6: Chad H 28Zc 55
Crabtree Cl. E2 1K 219 (40Vb 71)
Crabtree Cl. HP3: Hem H 4M 3
Crabtree Cl. KT23: Bookh 98Ea 192
Crabtree Cl. TN15: Fair 84Fe 185
Crabtree Cl. WD23: Bush 15Da 27
Crabtree Ct. EN5: New Bar 14Db 31
Crabtree Ct. HP3: Hem H 4N 3
Crabtree Dr. KT22: Lea 96La 192
Crabtree Hall SW6 52Ya 110
(off Crabtree La.)
CRABTREE HILL 17Bd 37
Crabtree La. HP3: Hem H 4M 3
Crabtree La. KT18: Head 98Sa 193
Crabtree La. KT23: Bookh 98Ea 192
Crabtree La. SW6 52Ya 110
(not continuous)
Crabtree Manorway Nth. DA17: Belv 47Ed 96
Crabtree Manorway Nth. DA17: Erith 47Ed 96
Crabtree Manorway Sth. DA17: Belv 48Ed 96
Crabtree Office Village TW20: Thorpe 68E 126
Crabtree Pl. W1 1C 222 (43Lb 90)
Crabtree Rd. TW20: Thorpe 68E 126
Crabtree Wlk. CR0: C'don 74Wb 157
Crabwood RH8: Oxt 100Gc 199
Crace St. NW1 3D 216 (41Mb 90)
Cracknell Cl. EN1: Enf 8Xb 19
Craddock Rd. EN1: Enf 13Vb 33
Craddocks Av. KT21: Asht 89Na 173
Craddocks Cl. KT21: Asht 88Qa 173
Craddocks Pde. KT21: Asht 89Na 173
Craddock St. NW5 37Jb 70
Cradford Ho. Nth. E2 41Zb 92
Cradford Ho. Sth. E2 41Zb 92
Cradley Rd. SE9 60Tc 116
Crafts Council & Gallery 2A 218 (40Qb 70)
Craftsman Cl. RM16: Grays 47De 99
Cragg Av. WD7: R'lett 8Ha 14
Craggy Island 83Gb 175
Cragie Ho. SE1 49Vb 91
(off Balaclava Rd.)
Craigdale Rd. RM11: Horn 30Hd 56
Craig Dr. UB8: Hil 44R 84
Craigen Av. CR0: C'don 74Xb 157
Craigen Gdns. IG3: Ilf 35Uc 74
Craigerne Rd. SE3 52Kc 115
Craig Gdns. E18 26Hc 53
Craigholm SE18 54Qc 116
Craig Ho. E17 28Cc 52
(off High St.)
Craigie Ct. DA1: Dart 59Qd 119
Craigmore Ct. HA6: Nwood 24U 44
Craigmore Twr. GU22: Wok 91A 188
(off Constitution Hill)
Craig Mt. WD7: R'lett 7Ka 14
Craigmuir Pk. HA0: Wemb 39Pa 67
Craignair Rd. SW2 59Qb 112
Craignish Av. SW16 68Pb 134
Craig Pk. Rd. N18 21Xb 51
Craig Rd. TW10: Ham 63La 130
Craig's Ct. SW1 6F 223 (46Nb 90)
Craig Rd. SE9 56Pc 116
Craigweil Av. WD7: R'lett 7Ka 14
Craigweil Cl. HA7: Stan 22Ma 47
Craigweil Dr. HA7: Stan 22Ma 47
Craigwell Av. TW13: Felt 62W 128
Craigwell Cl. TW18: Staines 67G 126
Craik Ct. NW6 40Bb 69
(off Carlton Vale)
Crail Row SE17 6G 231 (49Tb 91)
Crakell Rd. RH2: Reig 7L 207
Crakers Mead WD18: Wat 13X 27
Crales Ho. SE18 48Nc 94
Cramer St. W1 1J 221 (43Jb 90)
Crammavill St. RM16: Grays 46Ce 99
Crammavill St. RM16: Grays 45Ce 99
Crammerville Wlk. RM13: Rain 42Kd 97
Crammond Cl. W6 51Ab 110
Cramond Ct. TW14: Bedf 60U 106
Cramonde Ct. DA16: Well 54Wc 117
Crampshaw La. KT21: Asht 91Pa 193
Crampton Ho. SW8 53Lb 112
Crampton Rd. SE20 65Yb 136
Cramptons Rd. TN14: S'oaks 92Kd 203
Crampton St. SE17 6D 230 (49Sb 91)
Cranberry Cl. NW7 23Va 48
Cranberry Cl. UB5: N'olt 40Z 65
Cranberry Ent. Pk. N17 24Vb 51
(off White Hart La.)
Cranberry La. E16 42Gc 93
Cranborne Av. EN6: Pot B 2Ab 16
Cranborne Av. KT6: Surb 76Qa 153
Cranborne Av. UB2: S'hall 49Ca 85
Cranborne Cl. EN6: Pot B 3Ab 16
Cranborne Ct. EN3: Enf W 8Zb 20
Cranborne Cres. EN6: Pot B 2Ab 16
Cranborne Gdns. RM14: Upm 33Rd 77
Cranborne Ind. Est. EN6: Pot B 2Ab 16
(not continuous)
Cranborne Pde. EN6: Pot B 3Za 16
Cranborne Rd. EN6: Pot B 2Ab 16
Cranborne Rd. EN8: Chesh 4Zb 20
Cranborne Rd. IG11: Bark 39Tc 74
Cranborne Waye UB4: Yead 44X 85
(not continuous)
Cranbourn All. WC2 4E 222 (45Mb 90)
(off Cranbourn St.)
Cranbourne NW1 38Mb 70
(off Agar Gro.)
CRANBOURNE 10A 102
Cranbourne Av. E11 28Kc 53
Cranbourne Av. SL4: Wind 4D 102
Cranbourne Cl. KT12: Hers 79Y 151
Cranbourne Cl. SL1: Slou 6G 80
Cranbourne Cl. SW16 69Nb 134
Cranbourne Cotts. SL4: Wink 2A 124
Cranbourne Ct. SW11 52Gb 111
(off Albert Bri. Rd.)
Cranbourne Dr. HA5: Pinn 29Z 45
Cranbourne Gdns. IG6: Ilf 27Sc 54
Cranbourne Gdns. NW11 29Ab 48
Cranbourne Hall Cotts. SL4: Wink 10A 102
(off Squirrel La.)
Cranbourne Rd. E12 36Nc 74
Cranbourne Rd. E15 35Ec 72
Cranbourne Rd. HA6: Nwood 27V 44
Cranbourne Rd. N10 26Kb 50
Cranbourne Rd. SL1: Slou 6G 80
Cranbourn Ho. SE16 47Xb 91
(off Marigold St.)
Cranbourn Pas. SE16 47Xb 91
Cranbourn St. WC2 4E 222 (45Mb 90)
CRANBROOK 32Pc 74
Cranbrook NW1 1C 216 (39Lb 70)
(off Camden St.)
Cranbrook Castle Tennis Club 31Pc 74
Cranbrook Cl. BR2: Hayes 72Jc 159
Cranbrook Ct. CR2: S Croy 78Ub 157
Cranbrook Ct. TW8: Bford 51La 108
Cranbrook Dr. AL4: St A 2J 7
Cranbrook Dr. KT10: Esh 74Ea 152
Cranbrook Dr. RM2: Rom 28Ld 57
Cranbrook Dr. TW2: Whitt 60Da 107
Cranbrook Est. E2 40Zb 72
Cranbrook Ho. DA8: Erith 52Hd 118
(off Boundary St.)
Cranbrook La. N11 21Kb 50
Cranbrook M. E17 29Bc 52
Cranbrook Pk. N22 25Qb 50
Cranbrook Ri. IG1: Ilf 30Pc 54
Cranbrook Rd. CR7: Thor H 68Sb 135
Cranbrook Rd. DA7: Bex 53Bd 117
Cranbrook Rd. EN4: E Barn 16Fb 31
Cranbrook Rd. IG1: Ilf 31Qc 74
Cranbrook Rd. IG2: Ilf 29Qc 54
Cranbrook Rd. IG6: Ilf 29Rc 54
Cranbrook Rd. SE8 53Cc 114
Cranbrook Rd. SW19 66Ab 132
Cranbrook Rd. TW4: Houn 56Ba 107
Cranbrook Rd. W4 50Ua 88
Cranbrook St. E2 40Zb 72
Cranbury Rd. SW6 54Db 111
Crandley Ct. SE8 49Ac 92
(not continuous)
Crandon Wlk. DA4: S Dar 68Ud 142
Crane Av. TW7: Isle 57Ja 108
Crane Av. W3 45Sa 87
Cranebank 54W 106
Cranebank M. TW1: Twick 56Ja 108
Cranebrook TW2: Twick 61Ea 130
Crane Cl. CR3: Cat'm 94Wb 197
Crane Cl. HA2: Harr 34Ea 66
Crane Cl. RM10: Dag 37Cd 76
Crane Ct. EC4 3A 224 (44Qb 90)
Crane Ct. KT19: Ewe 77Sa 153
Crane Ct. SW14 56Sa 109
Cranefield Dr. WD25: Wat 4Aa 13
Craneford Cl. TW2: Twick 59Ha 108
Craneford Way TW2: Twick 59Ga 108
Crane Gdns. UB3: Harl 49V 84
Crane Gro. N7 37Qb 70
Crane Hgts. N17 27Xb 51
(off Waterside Way)
Crane Ho. E3 40Ac 72
(off Roman Rd.)
Crane Ho. SE15 53Vb 113
Crane Ho. TW13: Hanw 62Ca 129
Cranell Grn. RM15: S Ock 46Xd 98
Crane Lodge Rd. TW5: Cran 51X 107
Crane Mead SE16 50Yb 92
Crane Mead Ct. TW1: Twick 59Ha 108
Crane Pk. Island Nature Reserve 61Ba 129
Crane Pk. Rd. TW2: Whitt 61Da 129
Crane Rd. TW19: Stanw 58Q 106
Crane Rd. TW2: Twick 60Ga 108
Cranesbill Cl. NW9 27Ta 47
Cranesbill Cl. SW16 68Mb 134
Cranes Dr. KT5: Surb 70Na 131
Cranes Pk. KT5: Surb 70Na 131
Cranes Pk. Av. KT5: Surb 70Na 131
Cranes Pk. Cres. KT5: Surb 70Pa 131
Crane St. SE10 50Fc 93
Crane St. SE15 53Vb 113
Craneswater UB3: Harl 52V 106
Craneswater Pk. UB2: S'hall 50Ba 85
Cranes Way WD6: Bore 15Sa 29
Crane Way TW2: Whitt 59Ea 108
Cranfield Cl. SE27 62Sb 135
Cranfield Ct. GU21: Wok 10L 167
Cranfield Ct. W1 1E 220 (43Gb 89)
(off Homer St.)
Cranfield Cres. EN6: Cuff 1Nb 18
Cranfield Dr. NW9 24Ua 48
Cranfield Ho. WC1 7F 217 (43Nb 90)
(off Southampton Row)
Cranfield Rd. SE4 55Bc 114
Cranfield Rd. E. SM5: Cars 81Jb 176
Cranfield Rd. W. SM5: Cars 81Jb 176
Cranfield Row SE1 3A 230 (48Qb 90)
(off Gerridge St.)
Cranfield Wlk. SE3 55Kc 115
CRANFORD 52W 106
Cranford Av. N13 22Nb 50
Cranford Av. TW19: Stanw 59N 105
Cranford Cl. CR8: Purl 85Sb 177
Cranford Cl. SW20 66Xa 132
Cranford Cl. TW19: Stanw 59N 105
Cranford Community College Sports Cen. 51X 107
Cranford Cotts. E1 45Zb 92
(off Cranford St.)
Cranford Dr. SL1: Slou 7D 80
Cranford Dr. UB3: Harl 49V 84
Cranford La. TW6: H'row A Bath Rd. 53V 106
Cranford La. TW6: H'row A Elmdon Rd. 55V 106
Cranford La. TW5: Cran 52X 107
Cranford La. TW5: Hest 52X 107
Cranford La. UB3: Cran 51T 106
Cranford La. UB3: Harl 51T 106
Cranford M. BR2: Broml 71Nc 160
Cranford Pk. Rd. UB3: Harl 49V 84
Cranford Ri. KT10: Esh 78Ea 152
Cranford Rd. DA1: Dart 60Nd 119
Cranford St. E1 45Zb 92
Cranford Way N8 28Pb 50
CRANHAM 31Ud 78
Cranham Brickfields Local Nature Reserve 31Vd 78
Cranham Gdns. RM14: Upm 32Ud 78
Cranham Golf Course 33Wd 78
Cranham Hall M. RM14: Upm 34Ud 78
Cranham Marsh Nature Reserve 35Ud 78
Cranham Rd. RM11: Horn 30Kd 57
Cranhurst Rd. NW2 36Ya 68
Cranleigh W11 46Bb 89
(off Ladbroke Rd.)
Cranleigh Cl. BR6: Chels 76Wc 161
Cranleigh Cl. CR2: Sande 84Wb 177
Cranleigh Cl. DA5: Bexl 58Dd 118
Cranleigh Cl. EN7: Chesh 1Wb 19
Cranleigh Cl. SE20 68Xb 135
Cranleigh Ct. CR4: Mitc 69Fb 133
Cranleigh Ct. TW9: Rich 55Qa 109
Cranleigh Ct. UB1: S'hall 44Ba 85
Cranleigh Dr. BR8: Swan 70Gd 140
Cranleigh Gdns. CR2: Sande 84Wb 177
Cranleigh Gdns. HA3: Kenton 29Na 47
Cranleigh Gdns. IG10: Lough 16Pc 36
Cranleigh Gdns. IG11: Bark 38Tc 74
Cranleigh Gdns. KT2: King T 65Pa 131
Cranleigh Gdns. N21 15Qb 32
Cranleigh Gdns. SE25 69Ub 135
Cranleigh Gdns. SM1: Sutt 75Db 155
Cranleigh Gdns. UB1: S'hall 44Ba 85
Cranleigh Gdns. Ind. Est. UB1: S'hall 44Ba 85
Cranleigh Ho's. NW1 2C 216 (40Lb 70)
(off Cranleigh St.)
Cranleigh M. SW11 54Gb 111
Cranleigh Rd. KT10: Esh 74Ea 152
Cranleigh Rd. N15 29Sb 51
Cranleigh Rd. SW19 69Cb 133
Cranleigh Rd. TW13: Felt 63V 128
Cranleigh St. NW1 2C 216 (40Lb 70)
Cranley Dene Ct. N10 28Kb 50
Cranley Dr. HA4: Ruis 33V 64
Cranley Dr. IG2: Ilf 31Sc 74
CRANLEY GARDENS 28Kb 50
Cranley Gdns. N10 28Kb 50
Cranley Gdns. N13 20Pb 32
Cranley Gdns. SM6: W'gton 80Lb 156
Cranley Gdns. SW7 7A 226 (50Eb 89)
Cranley M. SW7 7A 226 (50Eb 89)
Cranley Pde. SE9 63Nc 138
(off Beaconsfield Rd.)
Cranley Pl. GU21: Knap 10H 167
Cranley Pl. SW7 6B 226 (49Fb 89)
Cranley Rd. E13 43Kc 93
Cranley Rd. IG2: Ilf 30Sc 54
Cranley Rd. KT12: Hers 78Y 150
Cranmer Av. W13 48Ka 86
Cranmer Cl. CR6: W'ham 89Ac 178
Cranmer Cl. EN6: Pot B 2Eb 17
Cranmer Cl. HA4: Ruis 32Z 65
Cranmer Cl. HA7: Stan 24La 46
Cranmer Cl. KT13: Weyb 80Q 150
Cranmer Cl. SM4: Mord 72Za 154
Cranmer Ct. GU21: Knap 10G 166
(off Hampton Cl.)
Cranmer Ct. N3 26Ab 48
Cranmer Ct. SW3 6E 226 (49Gb 89)
Cranmer Ct. SW4 55Mb 112
Cranmer Ct. TW12: Hamp H 64Da 129
Cranmere Ct. EN2: Enf 12Qb 32
Cranmer Farm Cl. CR4: Mitc 70Hb 133
Cranmer Gdns. CR6: W'ham 89Ac 178
Cranmer Gdns. RM10: Dag 35Ed 76
Cranmer Ho. SW11 53Gb 111
(off Surrey La. Est.)
Cranmer Ho. SW9 52Qb 112
(off Cranmer Rd.)
Cranmer Rd. CR0: C'don 76Rb 157
Cranmer Rd. CR4: Mitc 70Hb 133
Cranmer Rd. E7 35Kc 73
Cranmer Rd. HA8: Edg 20Ra 29
Cranmer Rd. KT2: King T 64Na 131
Cranmer Rd. SW9 52Qb 112
Cranmer Rd. TN13: Riv 95Gd 202
Cranmer Rd. TW12: Hamp H 64Da 129
Cranmer Rd. UB3: Hayes 44T 84
Cranmer Ter. SW17 64Fb 133
Cranmore Av. TW7: Isle 52Ea 108
Cranmore Ct. AL1: St A 1D 6
(off Avenue Rd.)
Cranmore La. KT24: W Hor 100R 190
Cranmore Rd. BR1: Broml 62Hc 137
Cranmore Rd. BR7: Chst 64Pc 138
Cranmore Way N10 28Lb 50
Cranston Cl. RH2: Reig 7K 207
Cranston Cl. TW3: Houn 54Aa 107
Cranston Cl. UB10: Ick 33T 64
Cranstone Lodge HP1: Hem H 4L 3
(off Cotterells)
Cranston Est. N1 2G 219 (40Tb 71)
Cranston Gdns. E4 23Dc 52
Cranston Pk. Av. RM14: Upm 35Rd 77
Cranston Rd. SE23 60Ac 114
Cranswick Rd. SE16 50Xb 91
Crantock Rd. SE6 61Dc 136
Cranwell Cl. AL4: St A 4G 6
Cranwell Cl. E3 42Dc 92
Cranwell Gro. TW17: Shep 70P 127
Cranwell Rd. TW6: H'row A 54R 106
Cranwells La. SL2: Farn C 4G 60
Cranwich Av. N21 17Tb 33
Cranwich Rd. N16 31Tb 71
Cranwood Ct. EC1 4G 219 (41Tb 91)
(off Vince St.)
Cranwood St. EC1 4G 219 (41Tb 91)
Cranworth Cres. E4 18Fc 35
Cranworth Gdns. SW9 53Qb 112
Craster Rd. SW2 59Pb 112
Crathie Rd. SE12 58Kc 115
Cravan Av. TW13: Felt 61W 128
Craven Av. UB1: S'hall 43Ba 85
Craven Av. W5 45La 86
Craven Cl. N16 31Wb 71
Craven Cl. UB4: Hayes 44W 84
Craven Cottage 54Za 110
Craven Ct. NW10 39Ua 68
Craven Ct. RM6: Chad H 30Ad 55
Craven Gdns. IG11: Bark 40Uc 74
Craven Gdns. IG6: Ilf 26Tc 54
Craven Gdns. RM3: Hrld W 23Sd 58
Craven Gdns. RM5: Col R 22Cd 56
Craven Gdns. SW19 64Cb 133
Craven Hill W2 4A 220 (45Eb 89)
Craven Hill Gdns. W2 4A 220 (45Eb 89)
(not continuous)
Craven Hill M. W2 4A 220 (45Eb 89)
Craven Ho. N2 26Fb 49
(off High Rd. E. Finchley)
Craven Lodge SW6 53Za 110
(off Harbord St.)
Craven Lodge W2 4A 220 (45Eb 89)
(off Craven Hill)
Craven M. SW11 55Jb 112
Craven Pk. NW10 39Ta 67
Craven Pk. M. NW10 38Ua 68
Craven Pk. Rd. N15 30Vb 51
Craven Pk. Rd. NW10 39Ua 68
Craven Pas. WC2 6F 223 (46Nb 90)
(off Craven St.)
Craven Rd. BR6: Chels 76Zc 161
Craven Rd. CR0: C'don 74Xb 157
Craven Rd. KT2: King T 67Pa 131
Craven Rd. NW10 39Ta 67
Craven Rd. W2 4A 220 (45Eb 89)
Craven Rd. W5 45La 86
Craven St. WC2 6F 223 (46Nb 90)
Craven Ter. W2 4A 220 (45Eb 89)
Craven Wlk. N16 31Wb 71
Crawford Av. DA1: Dart 58Md 119
Crawford Av. HA0: Wemb 36Ma 67
Crawford Av. RM16: Grays 46De 99
Crawford Bldgs. W1 1E 220 (43Gb 89)
(off Homer St.)
Crawford Cl. TW7: Isle 54Ga 108
Crawford Compton Cl. RM12: Horn 37Ld 77
Crawford Ct. NW9 27Ua 48
(off Charcot Rd.)
Crawford Cres. CR5: Coul 89Ld 176
Crawford Est. SE5 54Sb 113
Crawford Gdns. N13 20Rb 33
Crawford Gdns. UB5: N'olt 41Ba 85
Crawford Mans. W1 1E 220 (43Gb 89)
(off Crawford St.)
Crawford M. SW20 66Xa 132
Crawford M. W1 1F 221 (43Hb 89)
Crawford Pas. EC1 6K 217 (42Qb 90)
Crawford Pl. W1 2E 220 (44Gb 89)
Crawford Rd. SE5 53Sb 113
Crawfords BR8: Hext 66Gd 140

Crawford St. NW10 38Ta 67
Crawford St. W1 1E 220 (43Gb 89)
Crawley Ct. DA11: Grav'nd 7D 122
Crawley Rd. E10 32Dc 72
Crawley Rd. EN1: Enf 17Ub 33
Crawley Rd. N22 26Sb 51
Crawshaw Rd. KT16: Ott 79F 148
Crawshay Cl. TN13: S'oaks 95Jd 202
Crawshay Rd. SW9 53Qb 112
Crawthew Gro. SE22 56Vb 113
Cray Av. BR5: St M Cry 72Xc 161
Cray Av. KT21: Asht 88Na 173
Craybrooke Rd. DA14: Sidc 63Xc 139
Crayburne DA13: Sflt 64Be 143
Craybury End SE9 61Sc 138
Cray Cl. DA1: Cray 56Jd 118
Craydene Rd. DA8: Erith 53Hd 118
Crayfields Bus. Pk. BR5: St P 67Yc 139
Crayfields Ind. Pk. BR5: St P 68Yc 139
CRAYFORD 57Hd 118
Crayford Cl. E6 44Nc 94
Crayford Ct. W3 48Sa 87
(off Bollo Bri. Rd.)
Crayford High St. DA1: Cray 57Gd 118
Crayford Ho. SE1 2G 231 (47Tb 91)
(off Long La.)
Crayford Ind. Est. DA1: Cray 57Hd 118
Crayford M. N7 35Nb 70
Crayford Rd. DA1: Cray 57Hd 118
Crayford Rd. N7 35Mb 70
Crayford Stadium (Greyhound) 58Gd 118
Crayford Station (Rail) 58Gd 118
Crayford Way DA1: Cray 57Hd 118
Cray Ho. NW8 7C 214 (43Fb 89)
(off Penfold St.)
Crayke Hill KT9: Chess 80Na 153
Craylands BR5: St P 69Yc 139
Craylands La. DA10: Ghithe 57Zd 121
Craylands La. DA10: Swans 57Zd 121
Craylands Sq. DA10: Swans 57Zd 121
Crayle Ho. EC1 5B 218 (42Rb 91)
(off Malta St.)
Crayleigh Ter. DA14: Sidc 65Yc 139
Crayle St. SL2: Slou 1E 80
Craymill Sq. DA1: Cray 54Hd 118
Crayonne Cl. TW16: Sun 67U 128
Cray Rd. BR8: Crock 72Ed 162
Cray Rd. DA1: Cray 56Jd 118
Cray Rd. DA14: Sidc 65Yc 139
Cray Rd. DA17: Belv 51Cd 118
Crayside Ind. Est. DA1: Cray 56Kd 119
Crayside Leisure Cen. 58Gd 118
Cray's Pde. BR5: St P 68Yc 139
Cray Valley Rd. BR5: St M Cry 71Wc 161
Cray Vw. Cl. BR5: St M Cry 70Yc 139
(off Market Mdw.)
Cray Wanderers FC 71Kc 159
Crayzee Barn 61Bd 139
Crealock Gro. IG8: Wfd G 22Hc 53
Crealock St. SW18 58Db 111
Creasey Cl. RM11: Horn 33Kd 77
Creasy Cl. WD5: Ab L 3V 12
Creasy Ct. EN8: Walt C 6Zb 20
Creasy Est. SE1 4H 231 (48Ub 91)
Creative Ho. SW8 52Kb 112
(off Prince of Wales Dr.)
Creative Rd. SE8 52Dc 114
Crebor St. SE22 58Wb 113
Crecy Ct. SE11 7K 229 (50Qb 90)
(off Hotspur St.)
Credenhall Dr. BR2: Broml 74Pc 160
Credenhill Ho. SE15 52Xb 113
Credenhill St. SW16 65Lb 134
Crediton Hgts. NW10 39Za 68
(off Okehampton Rd.)
Crediton Hill NW6 36Db 69
Crediton Rd. E16 44Jc 93
Crediton Rd. NW10 39Za 68
Crediton Way KT10: Clay 78Ja 152
Credon Rd. E13 40Lc 73
Credon Rd. SE16 50Xb 91
Credo Way RM20: W Thur 51Xd 120
Creechurch La. EC3 3J 225 (44Ub 91)
(not continuous)
Creechurch Pl. EC3 3J 225 (44Ub 91)
(off Creechurch La.)
Creed Ct. E1 41Ac 92
Creed Ct. EC4 3C 224 (44Rb 91)
(off Ludgate Sq.)
Creed La. EC4 3C 224 (44Rb 91)
Creed Pas. SE10 50Gc 93
(off Hoskins St.)
Creeds Cotts. CM16: Epp 4Uc 22
Creek, The DA11: Nflt 57De 121
Creek, The TW16: Sun 71W 150
Creek Cotts. KT8: E Mos 70Ga 130
(off Creek Rd.)
Creek Ho. W14 48Ab 88
(off Russell Rd.)
Creek Mill Way DA1: Dart 56Md 119
CREEKMOUTH 42Vc 95
Creekmouth Ind. Pk. IG11: Bark 42Vc 95
Creek Rd. IG11: Bark 41Vc 95
Creek Rd. KT8: E Mos 70Ga 130
Creek Rd. SE10 51Dc 114
Creek Rd. SE8 51Cc 114
Creekside RM13: Rain 42Gd 96
Creekside SE8 52Dc 114
Creekside Foyer SE8 51Dc 114
(off Stowage)
Creek Way RM13: Rain 43Gd 96
Creeland Gro. SE6 60Bc 114
Cree's Mdw. GU20: W'sham 9A 146
Cree Way KT15: Add 80J 149
Cree Way RM1: Rom 24Gd 56
Crefeld Cl. W6 51Ab 110
Creffield Rd. W3 45Qa 87
Creffield Rd. W5 45Pa 87
Creighton Av. AL1: St A 6B 6
Creighton Av. E6 40Mc 73
Creighton Av. N10 26Hb 49
Creighton Av. N2 27Gb 49
Creighton Cl. W12 45Wa 88
Creighton Rd. N17 24Ub 51
Creighton Rd. NW6 40Za 68
Creighton Rd. W5 48Ma 87
Cremer Bus. Cen. E2 2K 219 (40Vb 71)
(off Cremer St.)
Cremer Ho. SE8 52Cc 114
(off Deptford Chu. St.)
Cremer St. E2 2K 219 (40Vb 71)
Cremorne Est. SW10 51Fb 111
(not continuous)
Cremorne Gdns. KT19: Ewe 82Ta 173
Cremorne Riverside Cen. 52Fb 111
Cremorne Rd. DA11: Nflt 9B 122
Cremorne Rd. SW10 52Eb 111
Creon Ct. SW9 52Qb 112
(off Caldwell St.)
Crescent EC3 4K 225 (45Vb 91)
Crescent, The AL2: Brick W 2Ca 13
Crescent, The BR3: Beck 67Cc 136
Crescent, The BR4: W W'ck 72Gc 159
Crescent, The CM16: Epp 4Vc 23
Crescent, The CR0: C'don 71Tb 157
Crescent, The CR3: Wold 95Cc 198
Crescent, The DA11: Nflt 1B 144
Crescent, The DA14: Sidc 63Vc 139
Crescent, The DA3: Lfield 69Ae 143
Crescent, The DA5: Bexl 59Yc 117
Crescent, The DA9: Ghithe 57Yd 120
Crescent, The E17 30Ac 52
Crescent, The EN5: New Bar 12Db 31
Crescent, The HA0: Wemb 33Ka 66
Crescent, The HA2: Harr 32Ea 66
Crescent, The IG10: Lough 15Mc 35
Crescent, The IG2: Ilf 30Qc 54
Crescent, The KT13: Weyb 76Q 150
Crescent, The KT16: Chert 70J 127
Crescent, The KT18: Eps 86Qa 173
(not continuous)
Crescent, The KT22: Lea 94Ka 192
Crescent, The KT3: N Mald 69Sa 131
Crescent, The KT6: Surb 71Na 153
Crescent, The KT8: W Mole 70Ca 129
Crescent, The N11 21Hb 49
Crescent, The NW2 34Xa 68
Crescent, The RH1: Redh 9M 207
Crescent, The RH2: Reig 6K 207
Crescent, The RM14: Upm 31Ud 78
Crescent, The SE8 53Cc 114
(off Seager Pl.)
Crescent, The SL1: Slou 7J 81
(not continuous)
Crescent, The SM1: Sutt 78Fb 155
Crescent, The SM2: Sutt 83Cb 175
Crescent, The SW13 54Va 110
Crescent, The SW19 62Cb 133
Crescent, The TN13: S'oaks 93Md 203
Crescent, The TN15: Bor G 91Ce 205
Crescent, The TW15: Ashf 64P 127
Crescent, The TW17: Shep 73V 150
Crescent, The TW20: Egh 65A 126
Crescent, The UB1: S'hall 47Ba 85
Crescent, The UB3: Harl 52S 106
Crescent, The W12 46Ya 88
Crescent, The W3 44Ua 88
Crescent, The WD18: Wat 14Y 27
Crescent, The WD25: A'ham 10Da 13
Crescent, The WD3: Crox G 16R 26
Crescent, The WD5: Ab L 2V 12
Crescent Arc. SE10 51Ec 114
(off Creek Rd.)
Crescent Av. RM12: Horn 33Hd 76
Crescent Av. RM17: Grays 50Fe 99
(not continuous)
Crescent Cotts. TN13: Dun G 92Gd 202
Crescent Ct. KT6: Surb 71Ma 153
Crescent Ct. RH1: Redh 4A 208
(off Foxboro Rd.)
Crescent Ct. RM17: Grays 50Fe 99
Crescent Ct. SW4 57Mb 112
(off Park Hill)
Crescent Ct. Bus. Cen. E16 42Fc 93
Crescent Dr. BR5: Pet W 71Rc 160
Crescent Dr. CM15: Shenf 18Ae 41
Crescent E. EN4: Had W 10Eb 17
Crescent Gdns. BR8: Swan 68Ed 140
Crescent Gdns. HA4: Ruis 31X 65
Crescent Gdns. SW19 62Cb 133
Crescent Gro. CR4: Mitc 70Gb 133
Crescent Gro. SW4 56Lb 112
Crescent Ho. EC1 6D 218 (42Sb 91)
(off Golden La. Est.)
Crescent Ho. SE13 54Dc 114
Crescent La. SW4 56Lb 112
Crescent Mans. SW3 6D 226 (49Gb 89)
(off Fulham Rd.)
Crescent Mans. W11 45Ab 88
(off Elgin Cres.)
Crescent M. N22 25Nb 50
Crescent Pde. UB10: Hil 41Q 84
Crescent Pl. SW3 5E 226 (49Gb 89)
Crescent Ri. EN4: E Barn 15Gb 31
Crescent Ri. N22 25Mb 50
Crescent Ri. N3 25Bb 49
Crescent Rd. BR1: Broml 66Jc 137
Crescent Rd. BR3: Beck 68Dc 136
Crescent Rd. CM14: W'ley 21Xd 58
Crescent Rd. CR3: Cat'm 96Wb 197
Crescent Rd. DA15: Sidc 62Vc 139
Crescent Rd. DA8: Erith 51Hd 118
Crescent Rd. E10 33Dc 72
Crescent Rd. E13 39Jc 73
Crescent Rd. E18 25Lc 53
Crescent Rd. E4 17Gc 35
Crescent Rd. E6 39Lc 73
Crescent Rd. EN2: Enf 14Rb 33
Crescent Rd. EN4: E Barn 14Fb 31
Crescent Rd. HP2: Hem H 2M 3
Crescent Rd. KT2: King T 66Qa 131
Crescent Rd. N11 21Hb 49
Crescent Rd. N15 27Rb 51
Crescent Rd. N22 25Mb 50
Crescent Rd. N3 25Bb 49
Crescent Rd. N8 30Mb 50
Crescent Rd. N9 18Wb 33
Crescent Rd. RH1: Blet 5J 209
Crescent Rd. RH2: Reig 8J 207
Crescent Rd. RM10: Dag 34Dd 76
Crescent Rd. RM15: Avel 47Sd 98
Crescent Rd. SE18 50Rc 94
Crescent Rd. SW20 67Za 132
Crescent Rd. TW17: Shep 71S 150
Crescent Row EC1 6D 218 (42Sb 91)
Crescent Stables SW15 57Ab 110
Crescent St. N1 38Pb 70
Crescent Vw. IG10: Lough 16Mc 35
Crescent Wlk. RM15: Avel 47Sd 98
Crescent Way BR6: Orp 78Uc 160
Crescent Way N12 23Gb 49
Crescent Way RM15: Avel 46Td 98
Crescent Way SE4 55Cc 114
Crescent Way SW16 65Pb 134
Crescent W. EN4: Had W 11Eb 31
Crescent Wood Rd. SE26 62Wb 135
Cresford Rd. SW6 53Db 111
Crespigny Rd. NW4 30Xa 48
Cressage Cl. UB1: S'hall 42Ca 85
Cressage Ho. TW8: Bford 51Na 109
(off Ealing Rd.)
Cressal Ho. E14 48Cc 92
(off Tiller Rd.)
Cressall Cl. KT22: Lea 92Ka 192
Cressall Mead KT22: Lea 92Ka 192
Cress Ct. TN14: S'oaks 92Kd 203
Cress End WD3: Rick 18J 25
Cressener Pl. DA1: Dart 57Md 119
Cresset Ho. E9 37Yb 72
Cresset Rd. E9 37Yb 72
Cresset St. SW4 55Mb 112
CRESSFIELD 61Ce 143
Cressfield Cl. NW5 36Jb 70
Cressida Rd. N19 32Lb 70
Cressingham Gro. SM1: Sutt 77Eb 155
Cressingham Rd. HA8: Edg 23Ta 47
Cressingham Rd. SE13 55Ec 114
Cressinghams, The KT18: Eps 85Ta 173
Cressington Cl. N16 36Ub 71
Cress M. BR1: Broml 64Fc 137
Cress Rd. SL1: Slou 7F 80
Cresswell Gdns. SW5 7A 226 (50Eb 89)
Cresswell Ho. HA9: Wemb 34Na 67
Cresswell Ho. TW19: Stanw 58N 105
(off Douglas Rd.)
Cresswell Pk. SE3 55Hc 115
Cresswell Pl. SW10 7A 226 (50Eb 89)
Cresswell Rd. SE25 70Wb 135
Cresswell Rd. TW1: Twick 58Ma 109
Cresswell Rd. TW13: Hanw 62Aa 129
Cresswell Way N21 17Qb 32
Cressy Ct. E1 43Yb 92
Cressy Ct. W6 48Xa 88
Cressy Ho. SW15 55Xa 110
Cressy Ho's. E1 43Yb 92
(off Hannibal Rd.)
Cressy Pl. E1 43Yb 92
Cressy Rd. NW3 36Hb 69
Crest, The KT5: Surb 71Qa 153
Crest, The N13 21Qb 50
Crest, The NW4 29Ya 48
Cresta Ct. W5 42Pa 87
Cresta Dr. KT15: Wdhm 82H 169
Cresta Ho. E3 42Cc 92
(off Dimson Cres.)
Cresta Ho. NW3 38Fb 69
(off Finchley Rd.)
Crest Av. RM17: Grays 52De 121
Crestbrook Av. N13 20Rb 33
Crestbrook Pl. N13 20Rb 33
(off Green Lanes)
Crest Cl. TN14: Bad M 83Dd 182
Crest Ct. NW4 29Ya 48
Crest Dr. EN3: Enf W 10Yb 20
Crested Ct. NW9 31Ua 68
Crestfield St. WC1 3G 217 (41Nb 90)
Crest Gdns. HA4: Ruis 34Y 65
Cresthill Av. RM17: Grays 49Ee 99
Creston Av. GU21: Knap 9J 167
Creston Ho. SW11 55Fb 111
(off Gartons Way)
Creston Way KT4: Wor Pk 74Za 154
Crest Pk. HP2: Hem H 1C 4
Crest Rd. BR2: Hayes 73Hc 159
Crest Rd. CR2: Sels 80Xb 157
Crest Rd. NW2 33Va 68
Crest Vw. DA9: Ghithe 56Wd 120
Crest Vw. HA5: Pinn 28Z 45
Crest Vw. Dr. BR5: Pet W 71Rc 160
Crest Wlk. E18 26Lc 53
Crestway SW15 58Wa 110
Crestwood Way TW4: Houn 57Aa 107
Creswell GU21: Knap 9H 167
Creswell Cnr. GU21: Knap 9H 167
Creswell Dr. BR3: Beck 71Dc 158
Creswick Ct. W3 45Ra 87
Creswick Rd. W3 45Ra 87
Creswick Wlk. E3 41Cc 92
Creswick Wlk. NW11 28Bb 49
Crete Hall Rd. DA11: Nflt 58Fe 121
Creton St. SE18 48Qc 94
Creukhorne Rd. NW10 38Ua 68
Crewdson Rd. SW9 52Qb 112
Crewe Ct. KT20: Tad 94Ya 194
Crewe Pl. NW10 41Va 88
Crewe's Av. CR6: W'ham 88Yb 178
Crewe's Cl. CR6: W'ham 89Yb 178
Crewe's Farm La. CR6: W'ham 88Zb 178
Crewe's La. CR6: W'ham 88Yb 178
Crewkerne Ct. SW11 53Fb 111
(off Bolingbroke Wlk.)
Crews Hill EN2: Crew H 6Pb 18
CREWS HILL 7Rb 19
Crews Hill Golf Course 7Pb 18
Crews Hill Station (Rail) 6Pb 18
Crews St. E14 49Cc 92
Crewys Rd. NW2 33Bb 69
Crewys Rd. SE15 54Xb 113
Crichton Av. SM6: Bedd 78Mb 156
Crichton Ho. DA14: Sidc 65Zc 139
Crichton Rd. SM5: Cars 79Hb 155
Crichton St. SW8 54Lb 112
Crick Ct. IG11: Bark 40Sc 74
(off Spring Pl.)
Cricketers Arms Rd. EN2: Enf 12Sb 33
Cricketers Cl. AL3: St A 1C 6
Cricketers Cl. DA8: Erith 50Gd 96
Cricketers Cl. GU22: Wok 93A 188
Cricketers Cl. KT9: Chess 77Ma 153
Cricketers Cl. N14 17Lb 32
Cricketers Ct. SE11 6B 230 (49Rb 91)
(off Kennington La.)
Cricketers La. CM13: Heron 23Fe 59
Cricketers La. GU20: W'sham 8B 146
Cricketers M. SW18 57Db 111
Cricketers Row CM13: Heron 24Fe 59
Cricketers Ter. SM5: Cars 76Gb 155
Cricketers Wlk. SE26 64Yb 136
Cricket Fld. Rd. UB8: Uxb 39M 63
Cricketfield Rd. E5 35Xb 71
Cricketfield Rd. UB7: W Dray 49L 83
Cricket Grn. CR4: Mitc 69Hb 133
Cricket Ground Rd. BR7: Chst 67Rc 138
Cricket Hill RH1: S Nut 8F 208
Cricket La. BR3: Beck 65Ac 136
Cricket La. TW12: Hamp H 65Ea 130
Cricket Marsh Wlk. DA12: Grav'nd 10J 123
CRICKETS HILL 97D 188
Crickett Ho. SE6 59Dc 114
Cricket Way KT13: Weyb 78R 150
Cricket Way KT13: Weyb 75U 150
Cricklade Av. RM3: Rom 23Md 57
Cricklade Av. SW2 61Nb 134
Cricklefield Pl. IG1: Ilf 33Uc 74
CRICKLEWOOD 35Za 68
Cricklewood B'way. NW2 34Ya 68
Cricklewood La. NW2 35Za 68
Cricklewood Station (Rail) 35Za 68
Cridland St. E15 39Hc 73
Crieff Ct. TW11: Tedd 66La 130
Crieff Rd. SW18 58Eb 111
Criffel Av. SW2 61Mb 134
Crimp Hill SL4: Old Win 9K 103
Crimp Hill TW20: Eng G 2L 125
Crimscott St. SE1 4J 231 (48Ub 91)
Crimson Rd. DA8: Erith 52Jd 118
Crimsworth Rd. SW8 53Mb 112
Crinan St. N1 1G 217 (40Nb 70)
Cringle Ct. EN6: Pot B 2Eb 17
Cringle St. SW11 52Lb 112
Crinoline M. E1 1K 225 (43Vb 91)
Cripplegate St. EC2 7D 218 (43Sb 91)
Cripps Cl. IG6: Ilf 27Sc 54
Cripps Grn. UB4: Yead 42X 85
Crispe Ho. IG11: Bark 40Tc 74
Crispe Ho. N1 1H 217 (39Pb 70)
(off Barnsbury Est.)
Crispen Rd. TW13: Hanw 63Aa 129
Crispian Cl. NW10 35Ua 68
Crispin Cl. CR0: Bedd 75Nb 156
Crispin Cl. KT21: Asht 90Pa 173
Crispin Ct. SE17 6J 231 (49Ub 91)
Crispin Cres. CR0: Bedd 76Mb 156
Crispin Ind. Cen. N18 22Yb 52
Crispin Lodge N11 22Hb 49
Crispin M. NW11 29Bb 49
Crispin Pl. E1 7K 219 (43Vb 91)
Crispin Rd. HA8: Edg 23Sa 47
Crispin St. E1 1K 225 (43Vb 91)
Crispin Way SL2: Farn C 5H 61
Crispin Way UB8: Hil 42P 83
Crisp Rd. W6 50Ya 88
Cristie Ct. E16 42Hc 93
Cristowe Rd. SW6 54Bb 111
Critchley Av. DA1: Dart 58Md 119
Criterion Bldgs. KT7: T Ditt 73Ka 152
(off Portsmouth Rd.)
Criterion Ct. E8 38Vb 71
(off Middleton Rd.)
Criterion M. N19 33Mb 70
Criterion M. SE24 57Rb 113
(off Shakespeare Rd.)
Criterion Theatre London 5D 222 (45Mb 90)
(off Piccadilly Cir.)
Criton Ind. Est. RM16: Ors 4G 100
CRITTALLS CORNER 66Yc 139
Crockenhall Way DA13: Ist R 6A 144
CROCKENHILL 72Fd 162
Crockenhill La. BR8: Crock 73Jd 162
Crockenhill La. BR8: Eyns 73Jd 162
Crockenhill La. BR8: Farni 73Jd 162
Crockenhill La. DA4: Eyns 73Kd 163
Crockenhill La. DA4: Farni 73Kd 163
Crockenhill Rd. BR5: St M Cry 71Zc 161
Crockenhill Rd. BR8: Crock 72Bd 161
Crockerton Rd. SW17 61Hb 133
Crockery La. GU4: E Clan 100M 189
Crockford Cl. KT15: Add 77L 149
Crockford Pk. Rd. KT15: Add 78L 149
Crockham Way SE9 63Qc 138
Crocus Cl. CR0: C'don 74Zb 158
Crocus Fld. EN5: Barn 16Bb 31
Croffets KT20: Tad 93Za 194
Croft, The AL2: Chis G 7N 5
Croft, The BR8: Swan 69Ed 140
Croft, The CR0: C'don 76Vb 157
Croft, The E4 19Gc 35
Croft, The EN5: Barn 14Ab 30
Croft, The HA0: Wemb 36La 66
Croft, The HA4: Ruis 35Y 65
Croft, The HA5: Pinn 31Ba 65
Croft, The HA8: Edg 24Ra 47
Croft, The IG10: Lough 12Qc 36
Croft, The KT17: Eps 86Va 174
Croft, The KT22: Fet 95Ga 192
Croft, The NW10 40Va 68
Croft, The TW5: Hest 51Aa 107
Croft, The W5 43Na 87
Croft Av. BR4: W W'ck 74Ec 158
Croft Cl. BR7: Chst 64Pc 138
Croft Cl. DA17: Belv 50Bd 95
Croft Cl. NW7 20Ua 30
Croft Cl. SL9: Chal P 26A 42
Croft Cl. TN13: S'oaks 100Hd 202
Croft Cl. UB10: Hil 38Q 64
Croft Cl. UB3: Harl 52S 106
Croft Cl. WD4: Chfd 2J 11
Croft Cnr. SL4: Old Win 7M 103
Croftdown Rd. NW5 34Jb 70
Croft End Cl. KT9: Chess 76Pa 153
Croft End Rd. WD4: Chfd 2J 11
Crofters SL4: Old Win 8L 103
Crofters Cl. RH1: Redh 8B 208
Crofters Cl. TW19: Stanw 58L 105
Crofters Cl. TW7: Isle 57Fa 108
Crofters Ct. SE8 49Ac 92
(off Croft St.)
Crofters Mead CR0: Sels 81Bc 178
Crofters Rd. HA6: Nwood 21U 44
Crofters Way NW1 39Mb 70
Croft Fld. WD4: Chfd 2J 11
Croft Gdns. HA4: Ruis 32V 64
Croft Gdns. W7 47Ja 86
Crofthill Rd. SL2: Slou 2F 80
Croft Ho. E17 28Dc 52
Croft Ho. NW9 26Va 48
Croft Ho. W10 41Ab 88
(off Third Av.)
Croft La. WD4: Chfd 2J 11
Croftleigh Av. CR8: Purl 88Qb 176
Croft Lodge Cl. IG8: Wfd G 23Kc 53
Croft Mdw. WD4: Chfd 2J 11
Croft M. N12 20Eb 31
CROFTON 75Sc 160
Crofton KT21: Asht 90Na 173
Crofton Albion Sports Ground 56Jc 115
Crofton Av. BR6: Farnb 75Sc 160
Crofton Av. DA5: Bexl 59Zc 117
Crofton Av. KT12: Walt T 76Y 151
Crofton Av. W4 52Sa 109
Crofton Cl. KT16: Ott 80E 148
Croftongate Way SE4 57Ac 114
Crofton Gro. E4 21Fc 53
Crofton Ho. SW3 51Gb 111
(off Old Church St.)
Crofton La. BR5: Farnb 75Tc 160
Crofton La. BR5: Orp 75Tc 160
Crofton La. BR5: Pet W 75Tc 160
Crofton La. BR6: Pet W 73Tc 160
CROFTON PARK 57Bc 114
Crofton Pk. Rd. SE4 58Bc 114
Crofton Park Station (Rail) 57Bc 114
Crofton Rd. BR6: Farnb 76Qc 160
Crofton Rd. BR6: Orp 76Qc 160
Crofton Rd. E13 42Kc 93
Crofton Rd. RM16: Grays 7A 100
Crofton Rd. SE5 53Ub 113
Crofton Roman Villa 75Uc 160
Crofton Ter. E5 36Ac 72
Crofton Ter. TW9: Rich 56Pa 109
Crofton Way EN2: Enf 12Qb 32
Crofton Way EN5: New Bar 16Db 31
Croft Rd. BR1: Broml 65Jc 137
Croft Rd. CR3: Wold 94Cc 198
Croft Rd. EN3: Enf H 11Ac 34
Croft Rd. SL9: Chal P 26A 42
Croft Rd. SM1: Sutt 78Gb 155
Croft Rd. SW16 67Qb 134
Croft Rd. SW19 66Eb 133
Croft Rd. TN16: Westrm 98Rc 200
Crofts, The HP3: Hem H 3B 4
Crofts, The TW17: Shep 70U 128
Crofts Ho. E2 40Wb 71
(off Teale St.)
Croftside, The SE25 69Wb 135
Crofts La. N22 24Qb 50
Crofts Path HP3: Hem H 4A 4
Crofts Rd. HA1: Harr 30Ja 46
Crofts St. E1 45Wb 91
Croft St. SE8 49Ac 92
Crofts Vs. HA1: Harr 30Ja 46
Croft Way DA15: Sidc 62Uc 138
Croft Way TN13: S'oaks 97Hd 202
Croft Way TW10: Ham 62Ka 130
Croftway NW3 35Cb 69
Crogsland Rd. NW1 38Jb 70
Croham Cl. CR2: S Croy 80Ub 157
Croham Hurst Golf Course 79Vb 157
Croham Mnr. Rd. CR2: S Croy 80Ub 157
Croham Mt. CR2: S Croy 80Ub 157
Croham Pk. Av. CR2: S Croy 78Ub 157
Croham Rd. CR2: S Croy 78Tb 157
Croham Valley Rd. CR2: Sels 79Wb 157
Croindene Rd. SW16 67Nb 134
Crokesley Ho. HA8: Edg 26Sa 47
(off Burnt Oak B'way.)
Cromar Ct. GU21: Wok 8N 167
Cromartie Rd. N19 31Mb 70
Cromarty Ct. SW2 57Pb 112
Cromarty Ho. E1 43Ac 92
(off Ben Jonson Rd.)
Cromarty Rd. HA8: Edg 19Ra 29
Cromberdale Ct. N17 25Wb 51
(off Spencer Rd.)
Crombie Cl. IG4: Ilf 29Pc 54
Crombie M. SW11 54Gb 111
Crombie Rd. DA15: Sidc 60Tc 116
Crome Ho. UB5: N'olt 40Aa 65
(off Parkfield Dr.)
Cromer Cl. UB8: Hil 44S 84
Cromer Ct. SL1: Slou 4J 81
Cromer Hyde SM4: Mord 71Db 155
Crome Rd. NW10 37Ua 68
Cromer Pl. BR6: Orp 74Uc 160
Cromer Rd. E10 31Fc 73
Cromer Rd. EN5: New Bar 14Eb 31
Cromer Rd. IG8: Wfd G 21Jc 53
Cromer Rd. N17 26Wb 51
Cromer Rd. RM11: Horn 31Md 77
Cromer Rd. RM6: Chad H 30Ad 55
Cromer Rd. RM7: Rom 30Ed 56
Cromer Rd. SE25 69Xb 135
Cromer Rd. SW17 65Jb 134
Cromer Rd. TW6: H'row A 54Q 106
Cromer Rd. WD24: Wat 10Y 13
Cromer St. WC1 4F 217 (41Nb 90)
Cromer Ter. E8 36Wb 71
Cromer Ter. RM6: Chad H 29Xc 55
Cromer Vs. Rd. SW18 58Bb 111
Cromford Cl. BR6: Orp 76Uc 160
Cromford Path E5 35Zb 72
Cromford Rd. SW18 57Cb 111
Cromford Way KT3: N Mald 67Ta 131
Cromie Cl. N13 20Qb 32
Cromlix Cl. BR7: Chst 68Rc 138
Crompton Cl. KT21: Asht 89La 172
Crompton Ct. BR2: Broml 69Jc 137
(off St Mark's Sq.)
Crompton Ct. SW3 5D 226 (49Gb 89)
Crompton Hall SL9: Ger X 29B 42
Crompton Ho. SE1 4E 230 (48Sb 91)
(off County St.)
Crompton Ho. W2 6B 214 (42Fb 89)
(off Hall Pl.)
Crompton Pl. EN3: Enf L 10Cc 20
Crompton St. W2 6B 214 (42Fb 89)
Cromwell Av. BR2: Broml 70Kc 137
Cromwell Av. EN7: Chesh 2Xb 19
Cromwell Av. KT3: N Mald 71Va 154
Cromwell Av. N6 32Kb 70
Cromwell Av. W6 50Xa 88
CROMWELL BUPA HOSPITAL 49Db 89
Cromwell Cen. IG11: Bark 41Wc 95
Cromwell Cen. IG6: Ilf 22Xc 55
Cromwell Cen. NW10 41Ta 87
Cromwell Cen., The RM8: Dag 31Bd 75
(off Coppen Rd.)
Cromwell Cl. BR2: Broml 70Kc 137
Cromwell Cl. E1 46Wb 91
Cromwell Cl. KT12: Walt T 74X 151
Cromwell Cl. N2 28Fb 49
Cromwell Cl. TW18: Staines 65L 127
Cromwell Cl. W3 46Sa 87
(not continuous)
Cromwell Cl. W4 50Ra 87
(off Harvard Rd.)
Cromwell Ct. EN3: Pond E 15Zb 34
Cromwell Ct. GU21: Knap 1G 186
(off Tudor Way)
Cromwell Cres. SW5 49Cb 89
Cromwell Dr. SL1: Slou 4J 81
Cromwell Gdns. SW7 4C 226 (48Fb 89)
Cromwell Gro. CR3: Cat'm 93Sb 197
Cromwell Gro. W6 48Ya 88
Cromwell Highwalk EC2 7E 218 (43Sb 91)
(off Silk St.)
Cromwell Ho. CR0: C'don 76Rb 157
Cromwell Ho. SW11 53Jb 112
(off Charlotte Despard Av.)
Cromwell Ind. Est. E10 32Ac 72
Cromwell Lodge DA6: Bex 57Ad 117
Cromwell Lodge E1 42Yb 92
(off Cleveland Gro.)
Cromwell Lodge IG11: Bark 36Uc 74
Cromwell Mans. SW5 49Cb 89
(off Cromwell Rd.)
Cromwell M. SW7 5B 226 (49Fb 89)
Cromwell Pl. EC2 7E 218 (43Sb 91)
(off Silk St.)
Cromwell Pl. N6 32Kb 70
Cromwell Pl. SW14 55Sa 109
Cromwell Pl. SW7 5C 226 (49Fb 89)
Cromwell Rd. BR3: Beck 68Ac 136
Cromwell Rd. CM14: W'ley 21Xd 58
Cromwell Rd. CR0: C'don 73Tb 157
Cromwell Rd. CR3: Cat'm 93Sb 197
Cromwell Rd. E17 29Ec 52
Cromwell Rd. E7 38Lc 73
Cromwell Rd. HA0: Wemb 40Na 67
Cromwell Rd. KT12: Walt T 74X 151
Cromwell Rd. KT16: Vir W 5L 147

Cromwell Rd. KT2: King T 67Na **131**
Cromwell Rd. KT4: Wor Pk 76Ta **153**
Cromwell Rd. N10 ....24Jb **50**
....................(not continuous)
Cromwell Rd. N3 ....25Eb **49**
Cromwell Rd. RH1: Redh 6P **207**
Cromwell Rd. RM17: Grays 49Ce **99**
Cromwell Rd. SW19 ....64Cb **133**
Cromwell Rd. SW5 ....49Cb **89**
Cromwell Rd. SW7 ....49Db **89**
Cromwell Rd. SW9 ....53Qb **112**
Cromwell Rd. TW11: Tedd 65Ja **130**
Cromwell Rd. TW13: Felt 60X **107**
Cromwell Rd. TW3: Houn 56Ca **107**
Cromwell Rd. UB3: Hayes 44T **84**
Cromwell Rd. WD6: Bore 11Na **29**
Cromwell Road Bus Station 67Na 131
Cromwells Ct. SL3: L'ly 46B **82**
Cromwells Mere RM1: Rom 23Fd **56**
Cromwell St. TW3: Houn 56Ca **107**
Cromwell Twr. EC2 ....7E **218** (43Sb **91**)
....................*(off Silk St.)*
Cromwell Trad. Cen. IG11: Bark 41Uc **94**
Cromwell Wlk. RH1: Redh 6P **207**
Crondace Rd. SW6 ....53Cb **111**
Crondall Ct. N1 ....2H **219** (40Ub **71**)
....................*(off St John's Est.)*
Crondall Ho. SW15 ....59Wa **110**
Crondall St. N1 ....2G **219** (40Tb **71**)
Crone Ct. NW6 ....40Bb **69**
....................*(off Denmark Rd.)*
Cronin St. SE15 ....52Vb **113**
Cronks Hill RH1: Redh 8M **207**
Cronks Hill RH2: Reig 8L **207**
Cronks Hill Cl. RH1: Redh 8M **207**
Cronks Hill Rd. RH1: Redh 8M **207**
CROOKED BILLET 25Cc **52**
Crooked Billet SW19 ....65Ya **132**
CROOKED BILLET RDBT. 63J **127**
Crooked Billet Yd. E2 ....3J **219** (41Ub **91**)
....................*(off Kingsland Rd.)*
Crooked La. DA12: Grav'nd 8D **122**
Crooked Mile EN9: Walt A 1Ec **20**
Crooked Usage N3 ....27Ab **48**
Crooke Rd. SE8 ....50Ac **92**
Crookham Rd. SW6 ....53Bb **111**
Crook Log DA6: Bex 55Zc **117**
Crook Log Leisure Cen. **55Zc 117**
Crookston Rd. SE9 ....55Qc **116**
Croombs Rd. E16 ....43Lc **93**
Croom's Hill SE10 ....52Ec **114**
Croom's Hill Gro. SE10 ....52Ec **114**
Cropley Ct. N1 ....1F **219** (40Tb **71**)
....................*(off Cropley St.)*
Cropley St. N1 ....1F **219** (40Tb **71**)
Croppath Rd. RM10: Dag 35Cd **76**
Cropthorne Ct. W9 ....4A **214** (41Eb **89**)
Crosbie Ho. E17 ....27Ec **52**
....................*(off Prospect Hill)*
Crosby Cl. AL4: St A 5G **6**
Crosby Cl. TW13: Hanw 62Aa **129**
Crosby Ct. IG7: Chig 20Wc **37**
Crosby Ct. SE1 ....1F **231** (47Tb **91**)
Crosby Gdns. UB8: Uxb 38N **63**
Crosby Ho. BR1: Broml 68Jc **137**
....................*(off Elmfield Rd.)*
Crosby Ho. E14 ....48Ec **92**
....................*(off Manchester Rd.)*
Crosby Ho. E7 ....37Jc **73**
Crosby Rd. E7 ....37Jc **73**
Crosby Rd. RM10: Dag 40Dd **76**
Crosby Row SE1 ....2F **231** (47Tb **91**)
Crosby Sq. EC3 ....3H **225** (44Ub **91**)
Crosby Wlk. E8 ....37Vb **71**
Crosby Wlk. SW2 ....59Qb **112**
Crosby Way SW2 ....59Qb **112**
Crosfield Ct. WD18: Wat 15Y **27**
....................*(off Lwr. High St.)*
Crosier Cl. SE3 ....53Nc **116**
Crosier Rd. UB10: Ick 35S **64**
Crosier Way HA4: Ruis 34U **64**
Crosland Pl. SW11 ....55Jb **112**
Crossacres GU22: Pyr 88G **168**
Cross Av. SE10 ....51Fc **115**
Crossbones Graveyard **7E 224** (**46Sb 91**)
....................(off Redcross Way)
Crossbow Ho. N1 ....1H **219** (39Ub **71**)
....................*(off Whitmore Est.)*
Crossbow Ho. W13 ....46Ka **86**
....................*(off Sherwood Cl.)*
Crossbow Rd. IG7: Chig 22Vc **55**
Crossbrook AL10: Hat 1A **8**
Crossbrook Ct. EN8: Chesh 3Ac **20**
Crossbrook Rd. SE3 ....54Nc **116**
Crossbrook St. EN8: Chesh 3Zb **20**
Crossby Cl. CM15: Mount 11Fe **41**
Cross Cl. SE15 ....54Xb **113**
Cross Ct. SE28 ....45Xc **95**
....................*(off Titmuss Av.)*
Cross Deep TW1: Twick 61Ha **130**
Cross Deep Gdns. TW1: Twick 61Ha **130**
Crossett Grn. HP3: Hem H 4C **4**
Crossfell Rd. HP3: Hem H 3C **4**
Crossfield Ct. W10 ....44Za **88**
....................*(off Cambridge Gdns.)*
Crossfield Ho. SL9: Ger X 28A **42**
Crossfield Ho. W11 ....45Ab **88**
....................*(off Mary Pl.)*
Crossfield Pl. KT13: Weyb 80R **150**
Crossfield Rd. N17 ....27Sb **51**
Crossfield Rd. NW3 ....37Fb **69**
Crossfields AL3: St A 5P **5**
Crossfields IG10: Lough 15Rc **36**
Crossfield St. SE8 ....52Cc **114**
....................(not continuous)
Crossford St. SW9 ....54Pb **112**
Cross Ga. HA8: Edg 20Qa **29**
Crossgate UB6: G'frd 37Ka **66**
Crossharbour Plaza E14 ....48Dc **92**
Crossharbour Station (Underground & DLR) 48Dc 92
CROSS KEYS 99Jd 202
Cross Keys Cl. N9 ....19Wb **33**
Cross Keys Cl. TN13: S'oaks 99Jd **202**
Cross Keys Cl. W1 ....1J **221** (43Jb **90**)
Cross Keys Sq. EC1 ....1D **224** (43Sb **91**)
....................*(off Little Britain)*
Cross Lances Rd. TW3: Houn 56Da **107**
Crossland Ho. GU25: Vir W 70A **126**
....................*(off Holloway Dr.)*
Crossland Rd. CR7: Thor H 72Rb **157**
Crossland Rd. RH1: Redh 6A **208**
Crosslands KT16: Chert 77G **148**
Crosslands WD3: Map C 21G **42**
Crosslands Av. UB2: S'hall 50Ba **85**
Crosslands Av. W5 ....46Pa **87**
Crosslands Pde. UB2: S'hall 50Ca **85**
Crosslands Rd. KT19: Ewe 79Ta **153**
Cross La. DA5: Bexl 59Bd **117**
Cross La. EC3 ....5H **225** (45Ub **91**)
....................(not continuous)
Cross La. KT16: Ott 79D **148**
Cross La. N8 ....27Pb **50**
....................(not continuous)
Cross La. E. DA12: Grav'nd 1D **144**
Cross Lanes SL9: Chal P 22A **42**
Cross Lanes Cl. SL9: Chal P 22B **42**
Cross La. W. DA11: Grav'nd 1D **144**
Crossleigh Ct. SE14 ....52Bc **114**
....................*(off New Cross Rd.)*
Crosslet St. SE17 ....6G **231** (49Tb **91**)
Crosslet Va. SE10 ....53Dc **114**
Crossley Cl. TN16: Big H 87Mc **179**
Crossley St. N7 ....37Qb **70**
Crossmead SE9 ....60Pc **116**
Crossmead WD19: Wat 16X **27**
Crossmead Av. UB6: G'frd 41Ca **85**
Crossmount Ho. SE5 ....52Sb **113**
....................*(off Bowyer St.)*
Crossness Footpath DA18: Erith 46Bd **95**
Crossness La. SE2 ....45Ad **95**
Crossness Nature Reserve **46Cd 96**
Crossness Pumping Station, The **44Ad 95**
Crossness Rd. IG11: Bark 41Vc **95**
Cross Oak SL4: Wind 4E **102**
Crossoaks La. EN6: Ridge 6Ua **16**
Crossoaks La. EN6: S Mim 6Ua **16**
Crossoaks La. WD6: Bore 7Ta **15**
Crosspath, The WD7: R'lett 7Ja **14**
Crosspoint Ho. SE8 ....52Cc **114**
....................*(off Watson's St.)*
Crossrail Pl. E14 ....46Dc **92**
Crossrail Wlk. E14 ....46Dc **92**
Cross Rd. BR2: Broml 75Nc **160**
Cross Rd. BR5: St M Cry 71Xc **161**
Cross Rd. CR0: C'don 74Tb **157**
Cross Rd. CR8: Purl 85Rb **177**
Cross Rd. DA1: Dart 58Ld **119**
Cross Rd. DA11: Nflt 8B **122**
Cross Rd. DA14: Sidc 63Xc **139**
Cross Rd. DA2: Hawl 63Pd **141**
Cross Rd. E4 ....18Fc **35**
Cross Rd. EN1: Enf 14Ub **33**
Cross Rd. EN8: Walt C 5Ac **20**
Cross Rd. HA1: Harr 28Fa **46**
Cross Rd. HA2: Harr 34Da **65**
Cross Rd. HA3: W'stone 26Ja **46**
Cross Rd. IG8: Wfd G 23Pc **54**
Cross Rd. KT13: Weyb 76T **150**
Cross Rd. KT2: King T 66Pa **131**
Cross Rd. KT20: Tad 94Ya **194**
Cross Rd. N11 ....22Kb **50**
Cross Rd. N22 ....24Qb **50**
Cross Rd. RM6: Chad H 31Yc **75**
Cross Rd. RM7: Mawney 28Cd **56**
Cross Rd. SE5 ....54Ub **113**
Cross Rd. SL5: S'dale 4D **146**
Cross Rd. SM1: Sutt 78Fb **155**
Cross Rd. SM2: Sutt 82Cb **175**
Cross Rd. SW19 ....66Cb **133**
Cross Rd. TW13: Hanw 63Aa **129**
Cross Rd. UB8: Uxb 38L **63**
Cross Rd. WD19: Wat 16Aa **27**
Cross Roads IG10: H Beech 12Kc **35**
Crossroads, The KT24: Eff 100Z **191**
Cross St. AL3: St A 2B **6**
Cross St. DA12: Grav'nd 8D **122**
....................*(off Terrace St.)*
Cross St. DA8: Erith 51Gd **118**
Cross St. N1 ....39Rb **71**
Cross St. N18 ....22Wb **51**
Cross St. SE5 ....55Tb **113**
Cross St. SW13 ....54Ua **110**
Cross St. TW12: Hamp H 64Ea **130**
Cross St. UB8: Uxb 38L **63**
Cross St. WD17: Wat 13Y **27**
Cross Ter. EN9: Walt A 6Gc **21**
....................*(off Stonyshotts)*
Crossthwaite Av. SE5 ....56Tb **113**
Crosstrees Ho. E14 ....48Cc **92**
....................*(off Cassilis Rd.)*
Crosswall EC3 ....4K **225** (45Vb **91**)
Cross Way NW10 ....38Wa **68**
Crossway BR5: Pet W 70Tc **138**
Crossway EN1: Enf 17Ub **33**
Crossway HA4: Ruis 35Y **65**
Crossway HA5: Pinn 26X **45**
Crossway IG8: Wfd G 21Lc **53**
Crossway KT12: Walt T 75X **151**
Crossway N12 ....23Fb **49**
Crossway N16 ....36Ub **71**
Crossway NW9 ....28Va **48**
Crossway RM8: Dag 34Yc **75**
Crossway SE28 ....44Xc **95**
Crossway SS17: Stan H 1P **101**
Crossway SW20 ....70Ya **132**
Crossway UB3: Hayes 46W **84**
Crossway W13 ....42Ja **86**
Cross Way, The HA3: W'stone 26Ga **46**
Crossway, The N22 ....24Rb **51**
Crossway, The SE9 ....61Mc **137**
Crossway, The UB10: Hil 40P **63**
Crossway Ct. SE4 ....54Ac **114**
Crossway Pde. N22 ....24Rb **51**
....................*(off The Crossway)*
Crossways CM15: Shenf 16Ce **41**
Crossways CR2: Sels 80Ac **158**
Crossways DA2: Dart 56Sd **120**
Crossways HP3: Hem H 2B **4**
Crossways IG10: Lough 15Qc **36**
Crossways KT24: Eff 99Z **191**
Crossways N21 ....16Sb **33**
Crossways RM2: Rom 27Kd **57**
Crossways SM2: Sutt 81Fb **175**
Crossways TN16: Tats 92Lc **199**
Crossways TW16: Sun 66V **128**
Crossways TW20: Egh 65F **126**
Crossways, The CR5: Coul 91Pb **196**
Crossways, The HA9: Wemb 33Qa **67**
Crossways, The KT5: Surb 74Ra **153**
Crossways, The RH1: Mers 2C **208**
Crossways, The TW5: Hest 52Ba **107**
Crossways 25 Bus. Pk. DA2: Dart 56Sd **120**
Crossways Blvd. DA2: Dart 56Sd **120**
Crossways Blvd. DA9: Ghithe 56Vd **120**
Crossways Ct. SL4: Wind 4G **102**
....................*(off Osborne Rd.)*
Crossways Ct. TN16: Tats 92Lc **199**
Crossways Ho. KT17: Eps 84Ua **174**
Crossways La. RH2: Reig 100Eb **195**
....................(not continuous)
Crossways Rd. BR3: Beck 70Cc **136**
Crossways Rd. CR4: Mitc 69Kb **134**
Crossways Ter. E5 ....35Yb **72**
Crosswell Cl. TW17: Shep 68S **128**
Crosthwaite Way SL1: Slou 3B **80**
Croston St. E8 ....39Wb **71**
Crothall Cl. N13 ....20Pb **32**
CROUCH 95De 205
Crouch Av. IG11: Bark 40Xc **75**
Crouch Cl. BR3: Beck 65Cc **136**
Crouch Cft. SE9 ....62Qc **138**
CROUCH END 31Mb 70
Crouch End Hill N8 ....31Mb **70**
Croucher Av. DA10: Swans 60Ae **121**
Crouchfield HP1: Hem H 3K **3**
Crouch Hall Ct. N19 ....32Nb **70**
Crouch Hall Rd. N8 ....30Mb **50**
Crouch Hill N4 ....31Nb **70**
Crouch Hill N8 ....30Nb **50**
Crouch Hill Station (Overground) 31Pb 70
Crouch Ind. Est. KT22: Lea 91Ka **192**
Crouch La. KT10: Hin W 76Ha **152**
Crouch La. TN15: Bor G 92Ce **205**
Crouchman Cl. RM16: Grays 47Ee **99**
Crouchman's Cl. SE26 ....62Vb **135**
Crouch Oak La. KT15: Add 77L **149**
Crouch Rd. NW10 ....38Ta **67**
Crouch Rd. RM16: Grays 10C **100**
Crouch Valley RM14: Upm 31Ud **78**
Crowborough Cl. CR6: W'ham 90Ac **178**
Crowborough Dr. CR6: W'ham 90Ac **178**
Crowborough Path WD19: Wat 21Z **45**
Crowborough Rd. SW17 ....65Jb **134**
Crowden Way SE28 ....45Yc **95**
Crowder Cl. N12 ....25Eb **49**
Crowder St. E1 ....45Xb **91**
CROWDLEHAM 89Td 184
Crow Dr. TN14: Hals 87Ed **182**
Crowfield Ho. N5 ....35Sb **71**
Crowfoot Cl. E9 ....36Bc **72**
Crowfoot Cl. SE28 ....46Uc **94**
CROW GREEN 13Vd 40
Crow Grn. La. CM15: Pil H 15Wd **40**
Crow Grn. Rd. CM15: Pil H 15Vd **40**
Crow Hill TN15: Bor G 92Ce **205**
Crowhill BR6: Downe 82Qc **180**
Crow Hill Rd. TN15: Bor G 92Ce **205**
Crowhurst SE25 ....71Vb **157**
Crowhurst Cl. SW9 ....54Qb **112**
Crowhurst Ho. SW9 ....54Pb **112**
....................*(off Aytoun Rd.)*
Crowhurst La. RH7: C'rst 10G **210**
Crowhurst La. RH8: Oxt 10H **211**
Crowhurst La. TN15: Ash 81Wd **184**
Crowhurst La. TN15: Bor G 96Zd **205**
Crowhurst La. TN15: W King 81Wd **184**
Crowhurst La. End RH8: C'rst 10F **210**
Crowhurst La. End RH8: Tand 10F **210**
CROWHURST LANE END 10F 210
Crowhurst Mead RH9: G'stone 2A **210**
Crowhurst Rd. TN15: Bor G 93Be **205**
Crowhurst Way BR5: St M Cry 71Yc **161**
Crowland Av. UB3: Harl 49U **84**
Crowland Gdns. N14 ....17Nb **32**
Crowland Ho. NW8 ....39Eb **69**
....................*(off Springfield Rd.)*
Crowland Rd. CR7: Thor H 70Tb **135**
Crowland Rd. N15 ....29Vb **51**
CROWLANDS 31Dd 76
Crowlands Av. RM7: Rom 30Dd **56**
Crowlands Heath Golf Course **32Dd 76**
Crowland Ter. N1 ....38Tb **71**
Crowland Wlk. SM4: Mord 72Db **155**
Crow La. RM7: Rush G 31Bd **75**
Crowley Cres. CR0: Wadd 78Qb **156**
Crowley M. SW16 ....69Lb **134**
Crowline Wlk. N1 ....37Sb **71**
Crowmarsh Gdns. SE23 ....59Yb **114**
Crown, The TN16: Westrm 98Tc **200**
Crown All. SE9 ....58Pc **116**
....................*(off Court Yd.)*
Crown Apts. HA4: Ruis 32W **64**
Crown Arc. KT1: King T 68Ma **131**
Crown Ash Hill TN16: Big H 86Kc **179**
Crown Ash La. CR6: W'ham 88Jc **179**
Crown Ash La. TN16: Big H 88Jc **179**
Crownbourne Ct. SM1: Sutt 77Db **155**
....................*(off St Nicholas Way)*
Crown Bldgs. E4 ....18Ec **34**
Crown Cl. BR6: Chels 78Wc **161**
Crown Cl. E3 ....39Cc **72**
Crown Cl. IG9: Buck H 18Kc **35**
Crown Cl. KT12: Walt T 73Y **151**
Crown Cl. N22 ....25Qb **50**
Crown Cl. NW6 ....37Db **69**
Crown Cl. NW7 ....19Va **30**
Crown Cl. SL3: Coln 52E **104**
Crown Cl. UB3: Hayes 47V **84**
Crown Cl. Bus. Cen. E3 ....39Cc **72**
....................*(off Crown Cl.)*
Crown Cotts. RM5: Col R 25Cd **56**
Crown Cotts. SL4: Wind 6H **103**
....................*(off Bucklers All.)*
Crown Court Blackfriars 1C **230** (47Rb **91**)
Crown Court Croydon 75Tb **157**
Crown Court Harrow 27Fa **46**
Crown Court Inner London 3D **230** (48Sb **91**)
Crown Court Isleworth 53Ga **108**
Crown Court Kingston upon Thames 69Ma **131**
Crown Court Snaresbrook 29Hc **53**
Crown Court Southwark 6H **225** (46Ub **91**)
Crown Court St Albans 2C **6**
Crown Court Wood Green 25Qb **50**
Crown Court Woolwich 47Uc **94**
Crown Ct. EC2 ....3E **224** (44Sb **91**)
....................*(off Cheapside)*
Crown Ct. N10 ....24Jb **50**
Crown Ct. NW8 ....4E **214** (41Gb **89**)
....................*(off Park Rd.)*
Crown Ct. RM18: Tilb 4C **122**
Crown Ct. SE12 ....58Kc **115**
Crown Ct. WC2 ....3G **223** (44Nb **90**)
Crown Crest Ct. TN13: S'oaks 93Ld **203**
....................*(off Seal Rd.)*
Crown Dale SE19 ....65Rb **135**
....................*(off Crown Dale)*
Crowndale Ct. NW1 ....1D **216** (40Mb **70**)
....................*(off Crowndale Rd.)*
Crowndale Pl. E17 ....26Ec **52**
Crowndale Rd. NW1 ....1C **216** (40Lb **70**)
Crown Dr. RM7: Rush G 30Fd **56**
Crown Dr. SL2: Farn R 9E **60**
E16: Lon
UB7: W Dray
Crownfield Av. IG2: Ilf 30Uc **54**
Crownfield Rd. E15 ....35Fc **73**
Crownfields TN13: S'oaks 97Kd **203**
Crown Gdns. SL2: Farn R 9F **60**
Crowngate Ho. E3 ....40Bc **72**
....................*(off Hereford Rd.)*
Crown Grn. DA12: Shorne 4N **145**
Crown Grn. M. HA9: Wemb 33Na **67**
Crown Hill CR0: C'don 75Sb **157**
Crown Hill EN9: Walt A 5Nc **22**
Crownhill Rd. IG8: Wfd G 24Nc **54**
Crownhill Rd. NW10 ....39Va **68**
Crown Ho. KT3: N Mald 69Sa **131**
Crown Ho. NW10 ....40Qa **67**
Crown La. BR2: Broml 71Mc **159**
Crown La. BR7: Chst 67Sc **138**
Crown La. DA12: Shorne 4N **145**
Crown La. GU25: Vir W 2P **147**
Crown La. N14 ....18Lb **32**
Crown La. SL2: Farn R 9F **60**
Crown La. SM4: Mord 70Cb **133**
Crown La. SW16 ....64Qb **134**
Crown La. Gdns. SW16 ....64Qb **134**
Crown La. Spur BR2: Broml 72Mc **159**
Crown Lodge SW3 ....6E **226** (49Gb **89**)
....................*(off Elystan St.)*
Crown Mdw. SL3: Coln 52D **104**
Crown Mdw. Ct. BR2: Broml 72Nc **160**
Crownmead Way RM7: Mawney 28Dd **56**
Crown M. E1 ....43Zb **92**
....................*(off White Horse La.)*
Crown M. E13 ....39Lc **73**
Crown M. TW13: Felt 60X **107**
Crown M. W6 ....49Wa **88**
Crown Mill CR4: Mitc 71Gb **155**
Crown Office Row EC4 ....4K **223** (45Qb **90**)
Crown Pde. N14 ....18Lb **32**
Crown Pde. SM4: Mord 69Cb **133**
Crown Pas. KT1: King T 68Ma **131**
....................*(off Church St.)*
Crown Pas. SW1 ....7C **222** (46Lb **90**)
Crown Pas. WD18: Wat 14Y **27**
Crown Pl. EC2 ....7H **219** (43Ub **91**)
Crown Pl. NW5 ....37Kb **70**
Crown Pl. SE16 ....50Xb **91**
Crown Point SE19 ....65Rb **135**
Crown Point Pde. SE19 ....65Rb **135**
....................*(off Crown Dale)*
Crown Reach SW1 ....7E **228** (50Mb **90**)
Crown Ri. KT16: Chert 74H **149**
Crown Ri. WD25: Wat 6Y **13**
Crown Rd. CM14: Kel H Royds La 11Td **40**
Crown Rd. BR6: Chels 78Wc **161**
Crown Rd. EN1: Enf 13Wb **33**
Crown Rd. GU25: Vir W 2N **147**
Crown Rd. HA4: Ruis 36Z **65**
....................(not continuous)
Crown Rd. IG6: Ilf 28Tc **54**
Crown Rd. KT3: N Mald 67Sa **131**
Crown Rd. N10 ....24Jb **50**
Crown Rd. RM17: Grays 51Ce **121**
Crown Rd. SM1: Sutt 77Db **155**
Crown Rd. SM4: Mord 70Db **133**
Crown Rd. TN14: S'ham 82Hd **182**
Crown Rd. TW1: Twick 58Ka **108**
Crown Rd. WD6: Bore 11Qa **29**
Crown Sq. GU21: Wok 89B **168**
....................*(off Chertsey Rd.)*
Crown Sq. SE1 ....7K **225** (46Vb **91**)
....................*(off Duchess Wlk.)*
Crownstone Ct. SW2 ....57Qb **112**
Crownstone Rd. SW2 ....57Qb **112**
Crown St. CM14: B'wood 19Yd **40**
Crown St. HA2: Harr 32Fa **66**
Crown St. RM10: Dag 37Ed **76**
....................(not continuous)
Crown St. SE5 ....52Sb **113**
Crown St. TW20: Egh 63C **126**
Crown St. W3 ....46Ra **87**
Crown Ter. N14 ....18Mb **32**
....................*(off Crown La.)*
Crown Ter. TW9: Rich 56Pa **109**
Crown Trad. Cen. UB3: Hayes 47U **84**
Crowntree Cl. TW7: Isle 51Ha **108**
Crown Village Green **72Nc 160**
....................(off Crown Lane)
Crown Wlk. HA9: Wemb 34Pa **67**
Crown Wlk. HP3: Hem H 6N **3**
Crown Wlk. UB8: Uxb 38L **63**
Crown Way UB7: Yiew 46P **83**
Crown Wharf E14 ....46Ec **92**
....................*(off Coldharbour)*
Crown Wharf SE8 ....50Bc **92**
....................*(off Grove St.)*
Crown Woods La. SE18 ....54Rc **116**
Crown Woods Way SE9 ....57Tc **116**
Crown Yd. E2 ....40Xb **71**
Crown Yd. SW6 ....54Cb **111**
Crown Yd. TW3: Houn 55Ea **108**
Crow Piece La. SL2: Farn R 8D **60**
Crowshott Av. HA7: Stan 26La **46**
Crows Rd. CM16: Epp 2Vc **23**
Crows Rd. E3 ....41Fc **93**
Crows Rd. IG11: Bark 37Rc **74**
Crowstone Rd. RM16: Grays 47Ee **99**
Crowther Av. TW8: Bford 49Na **87**
Crowther Cl. SW6 ....51Bb **111**
....................*(off Bucklers All.)*
Crowther Rd. SE25 ....71Wb **157**
Crowthorne Cl. SW18 ....59Bb **111**
Crowthorne Rd. W10 ....44Za **88**
Croxall Ho. KT12: Walt T 72Y **151**
Croxdale Rd. WD6: Bore 12Pa **29**
Croxden Cl. HA8: Edg 27Qa **47**
Croxden Wlk. SM4: Mord 72Eb **155**
Croxford Gdns. N22 ....24Rb **51**
Croxford Way RM7: Rush G 32Fd **76**
CROXLEY CEN. 16T 26
Croxley Cl. BR5: St P 68Xc **139**
Croxley Common Moor Local Nature Reserve **17S 26**
CROXLEY GREEN 16R 26
Croxley Grn. BR5: St P 67Xc **139**
Croxley Green Skate Pk. **14Q 26**
Croxleyhall Wood **17Q 26**
Croxley Rd. HP3: Hem H 7P **3**
Croxley Rd. W9 ....41Bb **89**
Croxley Station (Underground) 16R 26
Croxley Vw. WD18: Wat 16Q **26**
Croxted Cl. SE21 ....59Sb **113**
Croxted M. SE24 ....58Sb **113**
Croxted Rd. SE21 ....59Sb **113**
Croxted Rd. SE24 ....59Sb **113**
Croxteth Ho. SW8 ....54Mb **112**
Croyde Av. UB3: Harl 49U **84**
Croyde Av. UB6: G'frd 41Ea **86**
Croyde Cl. DA15: Sidc 59Tc **116**
CROYDON 75Sb 157
Croydon N17 ....26Tb **51**
....................*(off Gloucester Rd.)*
Croydon Airport Ind. Est. CR0: Wadd 79Pb **156**
Croydon Airport Vis. Cen. **79Qb 156**
Croydon Clocktower **76Sb 157**
....................(off Katharine St.)
Croydon Crematorium **71Pb 156**
Croydon Flyover, The CR0: C'don 77Rb **157**
Croydon Gro. CR0: C'don 74Rb **157**
Croydon High Sports Club **83Yb 178**
Croydon Ho. SE1 ....1A **230** (47Qb **90**)
....................*(off Wootton St.)*
Croydon La. SM7: Bans 86Eb **175**
Croydon La. Sth. SM7: Bans 86Eb **175**
Croydon Rd. BR2: Hayes 76Lc **159**
Croydon Rd. BR2: Kes 76Lc **159**
Croydon Rd. BR3: Beck 70Zb **136**
Croydon Rd. BR4: Hayes 76Gc **159**
Croydon Rd. BR4: W W'ck 76Gc **159**
Croydon Rd. CR0: Bedd 77Mb **156**
Croydon Rd. CR0: C'don 71Lb **156**
Croydon Rd. CR0: Wadd 77Mb **156**
Croydon Rd. CR3: Cat'm 95Wb **197**
Croydon Rd. CR4: Mitc 70Jb **134**
Croydon Rd. E13 ....42Hc **93**
Croydon Rd. RH2: Reig 6K **207**
Croydon Rd. SE20 ....68Xb **135**
Croydon Rd. SM6: Bedd 77Kb **156**
Croydon Rd. SM6: W'gton 77Kb **156**
Croydon Rd. TN16: Westrm 95Pc **200**
Croydon Rd. TW6: H'row A 54R **106**
Croydon Rd. Ind. Est. BR3: Beck 70Zb **136**
Croydon Sailing Club **68Wb 135**
Croydon Sports Arena **71Yb 158**
CROYDON UNIVERSITY HOSPITAL **72Rb 157**
Croydon Valley Trade Pk. CR0: Bedd 73Nb **156**
....................*(off Therapia La.)*
Croyland Rd. N9 ....18Wb **33**
Croylands Dr. KT6: Surb 73Na **153**
Croysdale Av. TW16: Sun 69W **128**
Crozier Dr. CR2: Sels 82Xb **177**
Crozier Ho. SW8 ....52Pb **112**
....................*(off Wilkinson St.)*
Crozier Ter. E9 ....36Zb **72**
....................(not continuous)
Crucible Cl. RM6: Chad H 30Xc **55**
Crucifix La. SE1 ....1H **231** (47Ub **91**)
Cruden Ho. E3 ....40Bc **72**
....................*(off Vernon Rd.)*
Cruden Ho. SE17 ....51Rb **113**
....................*(off Brandon Est.)*
Cruden Pl. CR5: Coul 89Lb **176**
Cruden Rd. DA12: Grav'nd 2H **145**
Cruden St. N1 ....1C **218** (39Rb **71**)
Cruick Av. RM15: S Ock 44Yd **98**
Cruikshank Ho. NW8 ....1D **214** (40Gb **69**)
....................*(off Townshend Rd.)*
Cruikshank Rd. E15 ....35Gc **73**
Cruikshank St. WC1 ....3K **217** (41Qb **90**)
Crummock Cl. SL1: Slou 4A **80**
Crummock Gdns. NW9 ....29Ua **48**
Crumpsall St. SE2 ....49Yc **95**
Crundale Av. NW9 ....29Qa **47**
Crundale Twr. BR5: Orp 74Yc **161**
....................*(off Tintagel Rd.)*
Crunden Rd. CR2: S Croy 80Tb **157**
Crusader Cl. RM19: Purf 49Qd **97**
Crusader Ct. DA1: Dart 57Pd **119**
Crusader Gdns. CR0: C'don 76Ub **157**
Crusader Ind. Est. N4 ....30Sb **51**
Crusader Way WD18: Wat 16V **26**
Crushes Cl. CM13: Hut 16Fe **41**
Crusoe M. N16 ....33Tb **71**
Crusoe Rd. CR4: Mitc 66Hb **133**
Crusoe Rd. DA8: Erith 50Fd **96**
Crutched Friars EC3 ....4J **225** (45Ub **91**)
Crutchfield La. KT12: Walt T 75X **151**
Crutchley Rd. SE6 ....61Gc **137**
Crystal, The 45Jc **93**
Crystal Av. RM12: Horn 35Nd **77**
Crystal Ct. E8 ....39Xb **71**
Crystal Ct. N14 ....15Lb **32**
Crystal Ct. SE19 ....64Vb **135**
....................*(off College Rd.)*
Crystal Ho. SE18 ....50Vc **95**
CRYSTAL PALACE 65Vb 135
Crystal Palace Athletics Stadium **65Wb 135**
Crystal Pal. Cvn. Club Site SE19 ....64Wb **135**
Crystal Palace Dinosaurs **65Wb 135**
Crystal Palace FC **70Ub 135**
Crystal Palace Indoor Bowling Club **67Xb 135**
Crystal Palace Mus. **65Wb 135**
Crystal Palace Nat. Sports Cen. **64Wb 135**
Crystal Palace Pde. SE19 ....65Vb **135**
Crystal Palace Pk. **64Wb 135**
Crystal Palace Pk. Farm **65Wb 135**
Crystal Palace Pk. Rd. SE26 ....64Wb **135**
Crystal Palace Rd. SE22 ....58Vb **113**
Crystal Palace Station (Rail & Overground) 65Wb 135
Crystal Palace Sta. Rd. SE19 ....65Wb **135**
Crystal Pl. KT4: Wor Pk 75Xa **154**
Crystal Ter. SE19 ....65Tb **135**
Crystal Vw. Ct. BR1: Broml 63Fc **137**
Crystal Way HA1: Harr 29Ha **46**
Crystal Way RM8: Dag 32Yc **75**
Crystal Wharf N1 ....2C **218** (40Rb **71**)
Cuba Dr. EN3: Enf H 12Yb **34**
Cuba St. E14 ....47Cc **92**
Cube North East London Gymnastic Club, The **35Tb 71**
Cube Ho. SE16 ....4K **231** (48Vb **91**)
Cubitt Apts. SW11 ....55Eb **111**
....................*(off Chatfield Rd.)*
Cubitt Bldg. SW1 ....50Kb **90**
Cubitt Ct. NW1 ....2B **216** (40Lb **70**)
....................*(off Park Village E.)*
Cubitt Ho. SW4 ....58Lb **112**
Cubitt Sq. UB2: S'hall 46Ea **86**
Cubitt Steps E14 ....46Cc **92**
Cubitt St. WC1 ....4J **217** (41Pb **90**)
Cubitt Ter. SW4 ....55Lb **112**
CUBITT TOWN 49Ec 92
Cubitt Way GU21: Knap 10H **167**
Cuckmans Dr. AL2: Chis G 7N **5**
Cuckmere Way BR5: Orp 74Zc **161**
Cuckoo Av. W7 ....42Ga **86**
Cuckoo Dene W7 ....43Fa **86**
Cuckoo Hall La. N9 ....17Yb **34**
Cuckoo Hall Rd. N9 ....17Yb **34**
Cuckoo Hill HA5: Eastc 27Y **45**
Cuckoo Hill HA5: Pinn 27Y **45**
Cuckoo Hill Dr. HA5: Pinn 27Y **45**
Cuckoo Hill Rd. HA5: Pinn 28Y **45**
Cuckoo La. GU24: W End 5B **166**
Cuckoo La. RM16: N Stif 46Be **99**
....................(not continuous)
Cuckoo La. W7 ....45Ga **86**
Cuckoo Pound TW17: Shep 71U **150**
Cuckoo Va. GU24: W End 5B **166**
Cucksey La. RH1: Blet 8K **209**
Cucumber La. AL9: Ess 2N **9**
Cucumber La. SG13: New S 5P **9**
Cudas Cl. KT19: Ewe 77Va **154**
Cuddington Av. KT4: Wor Pk 76Va **154**
Cuddington Ct. SM2: Cheam 81Za **174**
Cuddington Glade KT19: Eps 84Qa **173**
Cuddington Golf Course **85Ab 174**
Cuddington Pk. Cl. SM7: Bans 85Bb **175**
Cuddington Pl. KT20: Tad 92Za **194**
Cuddington Way SM2: Cheam 84Za **174**
CUDHAM 87Tc 180
Cudham Cl. SM2: Sutt 82Cb **175**

Cudham Dr. CR0: New Ad 82Ec 178
Cudham La. Nth. BR6: Downe 82Uc 180
Cudham La. Nth. TN14: Cud 86Sc 180
Cudham La. Sth. TN14: Cud 87Sc 180
Cudham La. Sth. TN14: Knock 87Sc 180
Cudham Pk. Rd. TN14: Cud 82Uc 180
Cudham Rd. BR6: Downe 83Qc 180
Cudham Rd. TN16: Tats 91Nc 200
Cudham St. SE6 59Ec 114
Cudweed Ct. E14 48Dc 92
(off Watergate Walk)
Cudworth Ho. SW8 53Lb 112
Cudworth St. E1 42Xb 91
Cuff Cres. SE9 58Mc 115
CUFFLEY 1Nb 18
Cuffley Av. WD25: Wat 6Z 13
Cuffley Hill EN7: Cuff 1Pb 18
Cuffley Hill EN7: G Oak 1Pb 18
Cuffley Ho. W10 43Ya 88
(off Sutton Way)
Cuffley Station (Rail) 1Pb 18
Cuff Point E2 3K 219 (41Vb 91)
(off Columbia Rd.)
Cugley Rd. DA2: Dart 59Sd 120
Culand Ho. SE17 6H 231 (49Ub 91)
(off Congreve St.)
Culcroft DA3: Hartl 69Be 143
Culford Gdns. SW3 6G 227 (49Hb 89)
Culford Gro. N1 37Ub 71
Culford Mans. SW3 6G 227 (49Hb 89)
(off Culford Gdns.)
Culford M. N1 37Ub 71
Culford Rd. N1 38Ub 71
Culford Rd. RM16: Grays 47Ee 99
Culford Ter. N1 37Ub 71
(off Balls Pond Rd.)
Culgaith Gdns. EN2: Enf 14Nb 32
Culham Ho. E2 4K 219 (41Vb 91)
(off Palissy St.)
Culham Ho. W2 43Cb 89
(off Gt. Western Rd.)
Cullen Sq. RM15: S Ock 46Yd 98
Cullen Way NW10 42Sa 87
Cullera Cl. HA6: Nwood 23V 44
Cullerne Cl. KT17: Ewe 82Va 174
Cullesden Rd. CR8: Kenley 87Rb 177
Culling Rd. DA17: Belv 48Ed 96
Culling Rd. SE16 48Yb 92
Cullings Ct. EN9: Walt A 5Hc 21
Cullington Cl. HA3: W'stone 28Ja 46
Cullingworth Rd. NW10 36Wa 68
Culloden Cl. SE16 50Wb 91
Culloden Cl. SE7 51Kc 115
Culloden Ho. SE14 52Ac 114
(off Batavia Rd.)
Culloden Rd. EN2: Enf 12Rb 33
Cullum St. EC3 4H 225 (45Ub 91)
Cullum Welch Ct. N1 3G 219 (41Tb 91)
(off Haberdasher St.)
Cullum Welch Ho. EC1 6D 218 (42Sb 91)
(off Golden La. Est.)
Culmington Pde. W13 46La 86
(off Uxbridge Rd.)
Culmington Rd. CR2: S Croy 81Sb 177
Culmington Rd. W13 46La 86
Culmore Rd. SE15 52Xb 113
Culmstock Rd. SW11 57Jb 112
Culpeper Cl. IG6: Ilf 23Rc 54
Culpepper Cl. N18 22Xb 51
Culpepper Ct. SE11 5K 229 (49Qb 90)
(off Kennington Rd.)
Culross Cl. N15 28Sb 51
Culross Ho. W10 44Za 88
(off Bridge Cl.)
Culross St. W1 5H 221 (45Jb 90)
Culsac Rd. KT6: Surb 75Na 153
Culverden Ct. KT13: Weyb 76T 150
(off Oatlands Dr.)
Culverden Rd. SW12 61Lb 134
Culverden Rd. WD19: Wat 20X 27
Culverden Ter. KT13: Weyb 76T 150
Culver Dr. RH8: Oxt 2J 211
Culver Gro. HA7: Stan 26La 46
Culverhay KT21: Asht 88Na 173
Culverhouse WC1 1H 223 (43Pb 90)
(off Red Lion Sq.)
Culverhouse Gdns. SW16 62Pb 134
Culverin Av. RM16: Grays 9B 100
Culverlands Cl. HA7: Stan 21Ka 46
Culverley Rd. SE6 60Dc 114
Culver Rd. AL1: St A 1C 6
Culvers Av. SM5: Cars 75Hb 155
Culvers Ct. DA12: Grav'nd 10H 123
Culvers Retreat SM5: Cars 74Hb 155
Culverstone Cl. BR2: Broml 72Hc 159
Culvers Way SM5: Cars 75Hb 155
Culvert Dr. E3 41Ec 92
Culvert La. UB8: Uxb 40K 63
Culvert Pl. SW11 54Jb 112
Culvert Rd. N15 29Ub 51
(not continuous)
Culvert Rd. SW11 54Hb 111
Culvey Cl. DA3: Hartl 71Ae 165
Culworth Ho. NW8 1D 214 (40Gb 69)
(off Allitsen Rd.)
Culworth St. NW8 2D 214 (40Gb 69)
Culzean Cl. SE27 62Rb 135
Cumberland Av. DA12: Grav'nd 9E 122
Cumberland Av. DA16: Well 55Uc 116
Cumberland Av. NW10 41Ra 87
Cumberland Av. RM12: Horn 34Nd 77
Cumberland Av. SL2: Slou 2G 80
Cumberland Basin Primrose Hill 1H 215 (39Jb 70)
Cumberland Basin NW1 1J 215 (39Jb 70)
Cumberland Bus. Pk. NW10 41Ra 87
Cumberland Cl. E8 37Vb 71
Cumberland Cl. HP3: Hem H 6E 4
Cumberland Cl. IG6: Ilf 25Sc 54
Cumberland Cl. KT19: Ewe 82Ua 174
Cumberland Cl. RM12: Horn 34Nd 77
Cumberland Cl. SW20 66Za 132
Cumberland Cl. TW1: Twick 58Ka 108
Cumberland Ct. AL3: St A 1C 6
Cumberland Ct. CR0: C'don 74Tb 157
Cumberland Ct. DA16: Well 54Uc 116
Cumberland Ct. HA1: Harr 27Ga 46
(off Princes Dr.)
Cumberland Ct. SW1 7A 228 (50Kb 90)
(off Cumberland St.)
Cumberland Ct. TN13: Dun G 92Fd 202
Cumberland Ct. W1 3G 221 (44Hb 89)
(off Gt. Cumberland Pl.)
Cumberland Cres. W14 49Ab 88
(not continuous)
Cumberland Dr. DA1: Dart 59Pd 119
Cumberland Dr. DA7: Bex 52Ad 117
Cumberland Dr. KT10: Hin W 75Ja 152
Cumberland Dr. KT9: Chess 76Na 153
Cumberland Gdns. NW4 26Ab 48
Cumberland Gdns. WC1 3K 217 (41Qb 90)
Cumberland Ga. W1 4F 221 (45Hb 89)
Cumberland Ho. E16 46Jc 93
(off Wesley Av.)
Cumberland Ho. KT2: King T 66Ra 131
Cumberland Ho. N9 18Yb 34
(off Cumberland Rd.)
Cumberland Ho. SE28 47Sc 94
Cumberland Ho. W8 47Db 89
(off Kensington Ct.)
Cumberland Mans. W1 2F 221 (44Hb 89)
(off George St.)
Cumberland Mkt. NW1 3A 216 (41Kb 90)
Cumberland M. SE11 50Qb 90
Cumberland Mills Sq. E14 50Fc 93
Cumberland Obelisk 6B 125
Cumberland Pk. NW10 41Wa 88
Cumberland Pk. W3 45Sa 87
Cumberland Pl. NW1 3K 215 (41Kb 90)
Cumberland Pl. SE6 60Hc 115
Cumberland Pl. TW16: Sun 70W 128
Cumberland Rd. BR2: Broml 70Gc 137
Cumberland Rd. E12 35Mc 73
Cumberland Rd. E13 43Kc 93
Cumberland Rd. E17 26Ac 52
Cumberland Rd. HA1: Harr 29Da 45
Cumberland Rd. HA7: Stan 27Pa 47
Cumberland Rd. N22 26Pb 50
Cumberland Rd. N9 18Yb 34
Cumberland Rd. RM16: Chaf H 47Ae 99
Cumberland Rd. SE25 72Xb 157
Cumberland Rd. SW13 53Va 110
Cumberland Rd. TW15: Ashf 62M 127
Cumberland Rd. TW9: Kew 52Qa 109
Cumberland Rd. W3 45Sa 87
Cumberland Rd. W7 47Ha 86
Cumberlands CR8: Kenley 87Tb 177
Cumberland St. SW1 7A 228 (50Kb 90)
Cumberland St. TW18: Staines 64F 126
Cumberland Ter. NW1 2K 215 (40Kb 70)
Cumberland Ter. M. NW1 2K 215 (40Kb 70)
(off Cumberland Ter.)
Cumberland Vs. W3 45Sa 87
(off Cumberland Rd.)
Cumberland Wharf SE16 47Yb 92
(off Rotherhithe St.)
Cumberlow Av. SE25 69Vb 135
Cumberlow Pl. HP2: Hem H 3C 4
Cumbernauld Gdns. TW16: Sun 64V 128
Cumberton Rd. N17 25Tb 51
Cumbrae Cl. SL2: Slou 6L 81
Cumbrae Gdns. KT6: Surb 75Ma 153
Cumbria Ct. RH2: Reig 5M 207
Cumbrian Av. DA7: Bex 54Gd 118
Cumbrian Gdns. NW2 33Za 68
Cumbrian Way UB8: Uxb 38M 63
Cummings Hall La. RM3: Rom 20Ld 39
Cumming St. N1 2J 217 (40Pb 70)
Cumnor Cl. SW9 54Pb 112
(off Robsart St.)
Cumnor Gdns. KT17: Ewe 79Wa 154
Cumnor Ri. CR8: Kenley 89Sb 177
Cumnor Rd. SM2: Sutt 79Eb 155
Cunard Ct. HA7: Stan 19Ja 28
(off Brightwen Gro.)
Cunard Cres. N21 16Tb 33
Cunard Pl. EC3 3J 225 (44Ub 91)
Cunard Rd. NW10 41Ta 87
Cunard Wlk. SE16 49Zb 92
Cundy Rd. E16 44Lc 93
Cundy St. SW1 6J 227 (49Jb 90)
Cuneo M. NW7 22Ab 48
Cunliffe Cl. KT18: Head 96Ra 193
Cunliffe Pde. KT19: Ewe 77Va 154
Cunliffe Rd. KT19: Ewe 77Va 154
Cunliffe St. SW16 65Lb 134
Cunningham Av. AL1: St A 4D 6
Cunningham Av. E16 47Kc 93
Cunningham Av. EN3: Enf W 8Ac 20
Cunningham Cl. BR4: W W'ck 75Dc 158
Cunningham Cl. RM6: Chad H 29Yc 55
Cunningham Ct. E10 34Dc 72
(off Oliver Rd.)
Cunningham Ct. W9 6A 214 (42Eb 89)
(off Maida Vale)
Cunningham Dr. UB10: Ick 33S 64
Cunningham Hill Rd. AL1: St A 4D 6
Cunningham Ho. SE5 52Tb 113
(off Elmington Est.)
Cunningham Pk. HA1: Harr 29Ea 46
Cunningham Pl. NW8 5B 214 (42Fb 89)
Cunningham Rd. N15 28Wb 51
Cunningham Rd. SM7: Bans 87Fb 175
Cunningham Way WD25: Wat 5V 12
Cunnington St. W4 48Sa 87
Cupar Rd. SW11 53Jb 112
Cupola Cl. BR1: Broml 64Kc 137
Curates Wlk. DA2: Wilm 62Md 141
Curchin Cl. TN16: Big H 84Lc 179
Cureton St. SW1 6E 228 (49Mb 90)
Curfew Bell Rd. KT16: Chert 73H 149
Curfew Ho. IG11: Bark 39Sc 74
Curfew Tower, The 39Sc 74
Curfew Yd. SL4: Wind 2H 103
Curie Ct. HA1: Harr 31Ka 66
Curie Gdns. NW9 26Ua 48
Curlew Cl. CR2: Sels 83Zb 178
Curlew Cl. SE28 45Zc 95
Curlew Ct. KT6: Surb 76Qa 153
Curlew Ct. W13 42Ha 86
Curlew Ho. EN3: Pond E 15Zb 34
Curlew Ho. SE15 53Vb 113
Curlew Ho. SE4 56Ac 114
(off St Norbert Rd.)
Curlews, The DA12: Grav'nd 1F 144
Curlew St. SE1 1K 231 (47Vb 91)
Curlew St. SE18 53Nc 116
(off Tellson Av.)
Curlew Way UB4: Yead 43Z 85
Curling Cl. CR5: Coul 92Pb 196
Curling La. RM17: Grays 50Be 99
Curness St. SE13 56Ec 114
Curnick's La. SE27 63Sb 135
Curo Pk. AL2: Frog 9C 6
Curran Av. DA15: Sidc 57Vc 117
Curran Av. SM6: W'gton 76Jb 156
Curran Cl. UB8: Cowl 42L 83
Curran Ho. SW3 6D 226 (49Gb 89)
(off Lucan Pl.)
Currey Rd. UB6: G'frd 37Fa 66
Curricle St. W3 46Ua 88
Currie Hill Cl. SW19 63Bb 133
Curriers La. SL1: Burn 6B 60
Curry Ri. NW7 23Za 48
Cursitor St. EC4 2K 223 (44Qb 90)
Curtain Pl. EC2 4J 219 (41Ub 91)
Curtain Rd. EC2 6H 219 (42Ub 91)
Curthwaite Gdns. EN2: Enf 14Mb 32
Curtis & Staub Health Club Golders Green 31Bb 69
Curtis Cl. WD3: Rick 18J 25
Curtis Dr. W3 44Ta 87
Curtis Fld. Rd. SW16 63Pb 134
Curtis Ho. SE17 7F 231 (50Tb 91)
(off Morecambe St.)
Curtis La. HA0: Wemb 37Na 67
Curtismill Cl. BR5: St P 69Xc 139
CURTISMILL GREEN 13Hd 38
Curtis Mill La. RM4: Nave 14Gd 38
Curtismill Way BR5: St P 69Xc 139
Curtis Rd. HP3: Hem H 3D 4
Curtis Rd. KT19: Ewe 77Sa 153
Curtis Rd. RM11: Horn 32Pd 77
Curtis Rd. TW4: Houn 59Ba 107
Curtiss Dr. WD25: Wat 6V 12
Curtiss Ho. NW9 27Va 48
Curtis St. SE1 5K 231 (49Vb 91)
Curtis Way SE1 5K 231 (49Vb 91)
Curtis Way SE28 45Xc 95
Curtlington Ho. HA8: Edg 26Sa 47
(off Burnt Oak B'way.)
Curvan Cl. KT17: Ewe 82Va 174
Curve, The W12 45Wa 88
Curwen Av. E7 35Kc 73
Curwen Rd. W12 47Wa 88
Curzon Av. EN3: Pond E 15Zb 34
Curzon Av. HA7: Stan 25Ja 46
Curzon Cinema Bloomsbury 5F 217 (42Nb 90)
Curzon Cinema Mayfair 6K 221 (46Kb 90)
(off Curzon St.)
Curzon Cinema Richmond 57Ma 109
Curzon Cinema Soho 4E 222 (45Mb 90)
(off Shaftesbury Av.)
Curzon Cinema Victoria 3C 228 (48Lb 90)
(off Victoria St.)
Curzon Cl. BR6: Orp 77Tc 160
Curzon Cl. KT13: Weyb 77Q 150
Curzon Ct. SW6 53Eb 111
(off Imperial Rd.)
Curzon Cres. IG11: Bark 40Vc 75
Curzon Cres. NW10 38Ua 68
Curzon Dr. RM17: Grays 52Ee 121
Curzon Ga. W1 7J 221 (46Jb 90)
Curzon Ga. Ct. WD17: Wat 11W 26
Curzon Mall SL1: Slou 7K 81
(within Queensmere Shop. Cen.)
Curzon Pl. HA5: Eastc 29Y 45
Curzon Rd. CR7: Thor H 72Qb 156
Curzon Rd. KT13: Weyb 78Q 150
Curzon Rd. N10 26Kb 50
Curzon Rd. W5 42Ka 86
Curzon Sq. W1 7J 221 (46Jb 90)
Curzon St. W1 7J 221 (46Jb 90)
Cusack Cl. TW1: Tedd 63Ha 130
Cussans Ho. WD18: Wat 16U 26
Cussons Cl. EN7: Chesh 1Wb 19
Custance Ho. N1 2F 219 (40Tb 71)
(off Provost St.)
Custance St. N1 3F 219 (41Tb 91)
Custom Ho. EC3 5H 225 (45Ub 91)
CUSTOM HOUSE 44Lc 93
Custom House for ExCeL Station (DLR & Crossrail) 45Kc 93
(off Cyrus St.)
Custom House Reach SE16 47Bc 92
Custom Ho. Wlk. EC3 5H 225 (45Ub 91)
Cut, The SE1 1A 230 (47Qb 90)
Cut, The SL2: Slou 2E 80
Cutbush Ho. N7 36Mb 70
Cutcombe Rd. SE5 54Sb 113
Cuthberga Cl. IG11: Bark 38Sc 74
Cuthbert Bell Twr. E3 40Cc 72
(off Pancras Way)
Cuthbert Ct. CR3: W'ham 91Wb 197
(off Godstone Rd.)
Cuthbert Gdns. SE25 69Ub 135
Cuthbert Harrowing Ho. EC1 6D 218 (42Sb 91)
(off Golden La. Est.)
Cuthbert Ho. W2 7B 214 (43Fb 89)
(off Hall Pl.)
Cuthbert Rd. CR0: C'don 75Rb 157
Cuthbert Rd. E17 27Ec 52
Cuthbert Rd. N18 22Wb 51
Cuthberts Cl. EN7: Chesh 1Vb 19
Cuthbert St. W2 7B 214 (43Fb 89)
Cuthered M. CR5: Coul 89Lb 176
Cuthill Wlk. SE5 53Tb 113
Cutlers Gdns. EC2 1J 225 (43Ub 91)
Cutlers Gdns. Arc. EC2 2J 225 (44Ub 91)
(off Devonshire Sq.)
Cutlers Sq. E14 49Cc 92
Cutlers Ter. N1 37Ub 71
(off Balls Pond Rd.)
Cutler St. E1 2J 225 (44Ub 91)
Cutmore Dr. AL4: Col H 4M 7
Cutmore St. DA11: Grav'nd 9D 122
Cutter Ho. DA8: Erith 49Gd 96
Cutter Ho. E16 47Kc 93
(off Admiralty Av.)
Cutter La. SE10 47Gc 93
Cutthroat All. TW10: Ham 61La 130
Cuttinge, The RH1: Redh 8P 207
Cuttsfield Ter. HP1: Hem H 3H 3
Cutty Sark 51Ec 114
Cutty Sark for Maritime Greenwich Station (DLR) 51Ec 114
Cutty Sark Gdns. SE10 51Ec 114
(off King William Wlk.)
Cutty Sark Hall SE10 51Ec 114
(off Welland St.)
Cuxton BR5: Pet W 71Sc 160
Cuxton Cl. DA6: Bex 57Ad 117
Cuxton Ho. SE17 7J 231 (50Ub 91)
(off Mina Rd.)
Cyan Apts. E3 40Bc 72
(off Gunmaker's La.)
Cyclamen Cl. TW12: Hamp 65Ca 129
Cyclamen Rd. BR8: Swan 70Fd 140
Cyclamen Way KT19: Ewe 78Sa 153
Cyclopark 3B 144
Cyclops M. E14 49Cc 92
Cygmus Ct. CR8: Purl 83Rb 177
(off Brighton Rd.)
Cygnet Av. TW14: Felt 59Y 107
Cygnet Cl. BR5: St P 69Wc 139
Cygnet Cl. GU21: Wok 8M 167
Cygnet Cl. HA6: Nwood 23S 44
Cygnet Cl. NW10 36Ta 67
Cygnet Cl. WD6: Bore 11Sa 29
Cygnet Gdns. DA11: Nflt 1B 144
CYGNET HOSPITAL, BECKTON 44Qc 94
CYGNET HOSPITAL, BLACKHEATH 53Ec 114
CYGNET HOSPITAL GODDEN GREEN 97Rd 203
CYGNET HOSPITAL, WOKING 10H 167
Cygnet Ho. CR0: C'don 74Tb 157
Cygnet Ho. DA12: Grav'nd 9D 122
(off Windmill St.)
Cygnet Ho. SE15 51Wb 113
Cygnet Ho. SW3 7E 226 (50Gb 89)
(off King's Rd.)
Cygnet Ho. Nth. E14 44Dc 92
(off Chrisp St.)
Cygnet Ho. Sth. E14 44Dc 92
(off Chrisp St.)
Cygnet Leisure Cen. 1A 144
CYGNET LODGE 58Ec 114
Cygnets, The TW13: Hanw 63Aa 129
Cygnets, The TW18: Staines 64H 127
Cygnets Cl. RH1: Redh 4A 208
Cygnet St. E1 5K 219 (42Vb 91)
Cygnet Vw. RM20: W Thur 49Vd 98
Cygnet Way UB4: Yead 43Z 85
Cygnus Bus. Cen. NW10 36Va 68
Cymbeline Ct. AL3: St A 1A 6
(off The Lawns)
Cymbeline Ct. HA1: Harr 30Ha 46
Cynthia St. N1 2J 217 (40Pb 70)
Cyntra Pl. E8 38Xb 71
Cypress Av. EN2: Crew H 7Qb 18
Cypress Av. TW2: Whitt 59Ea 108
Cypress Cl. E5 33Wb 71
Cypress Cl. EN9: Walt A 6Fc 21
Cypress Cl. KT19: Eps 81Ta 173
Cypress Ct. GU25: Vir W 70A 126
Cypress Ct. NW9 28Qa 47
(off Alpine Rd.)
Cypress Ct. SM1: Sutt 78Cb 155
Cypress Gdns. SE4 57Ac 114
Cypress Gro. IG6: Ilf 23Uc 54
Cypress Ho. SE14 53Zb 114
Cypress Ho. SE16 47Zb 92
(off Woodland Cres.)
Cypress Ho. SL3: L'ly 50D 82
Cypress Path RM3: Rom 24Md 57
Cypress Pl. W1 6C 216 (42Lb 90)
Cypress Rd. HA3: Hrw W 26Fa 46
Cypress Rd. SE25 68Ub 135
Cypress Tree Cl. DA15: Sidc 60Vc 117
Cypress Wlk. TW20: Eng G 5M 125
Cypress Wlk. WD25: Wat 7X 13
Cypress Way SM7: Bans 86Za 174
CYPRUS 45Qc 94
Cyprus Av. N3 26Ab 48
Cyprus Cl. N4 30Rb 51
Cyprus Gdns. N3 26Ab 48
Cyprus Pl. E2 40Yb 72
Cyprus Pl. E6 45Qc 94
Cyprus Rd. N3 26Bb 49
Cyprus Rd. N9 19Vb 33
Cyprus Station (DLR) 45Qc 94
Cyprus St. E2 40Yb 72
(not continuous)
Cyrena Rd. SE22 58Vb 113
Cyril Dumpleton Ho. AL2: Lon C 8H 7
Cyril Lodge DA14: Sidc 63Wc 139
Cyril Mans. SW11 53Hb 111
Cyril Rd. BR6: Orp 73Wc 161
Cyril Rd. DA7: Bex 54Ad 117
Cyrils Way AL1: St A 5B 6
Cyrus Fld. St. SE10 49Gc 93
Cyrus Ho. EC1 5C 218 (42Rb 91)
(off Cyrus St.)
Cyrus St. EC1 5C 218 (42Rb 91)
(off Cyrus St.)
Cyrus Ter. UB10: Ick 34S 64
(off Pentland Way)
Czar St. SE8 51Cc 114

D

Dabbling Cl. DA8: Erith 52Kd 119
Dabbs Hill La. UB5: N'olt 37Ba 65
(not continuous)
Dabbs La. EC1 6A 218 (42Qb 90)
(off Farringdon Rd.)
Dabbs Pl. DA13: Cobh 8F 144
D'Abernon Chase KT22: Oxs 86Ja 172
D'Abernon Cl. KT10: Esh 77Ca 151
D'Abernon Dr. KT11: Stoke D 88Aa 171
Dabin Cres. SE10 53Ec 114
Dacca St. SE8 51Bc 114
Dace Rd. E3 39Cc 72
Dacorum Way HP1: Hem H 2L 3
(not continuous)
Dacre Av. IG5: Ilf 26Qc 54
Dacre Av. RM15: Avel 46Td 98
Dacre Cl. CR5: Chips 90Hb 175
Dacre Cl. IG7: Chig 21Sc 54
Dacre Cl. UB6: G'frd 40Da 65
Dacre Cres. RM15: Avel 46Td 98
Dacre Gdns. IG7: Chig 21Sc 54
Dacre Gdns. SE13 56Gc 115
Dacre Gdns. WD6: Bore 15Ta 29
Dacre Ho. SW3 51Fb 111
(off Beaufort St.)
Dacre Ind. Est. EN8: Chesh 1Bc 20
Dacre Pk. SE13 55Gc 115
Dacre Pl. SE13 55Gc 115
Dacre Rd. CR0: C'don 73Nb 156
Dacre Rd. E11 32Hc 73
Dacre Rd. E13 39Kc 73
Dacres Cl. EN7: Chesh 1Vb 19
Dacres Est. SE23 62Zb 136
Dacres Ho. SW4 55Kb 112
Dacres Rd. SE23 61Zb 136
Dacre St. SW1 3D 228 (48Mb 90)
Dade Way UB2: S'hall 50Ba 85
Daerwood Cl. BR2: Broml 74Pc 160
Daffodil Av. CM15: Pil H 15Xd 40
Daffodil Cl. CR0: C'don 74Zb 158
Daffodil Dr. GU24: Bisl 8E 166
Daffodil Gdns. IG1: Ilf 36Rc 74
Daffodil Pl. TW12: Hamp 65Ca 129
Daffodil St. W12 45Va 88
Dafforne Rd. SW17 62Jb 134
Da Gama Pl. E14 50Cc 92
DAGENHAM 37Cd 76
Dagenham & Redbridge FC 36Dd 76
Dagenham Av. RM9: Dag 39Ad 75
(not continuous)
Dagenham Dock Station (Rail) 40Bd 75
Dagenham East Station (Underground) 36Ed 76
Dagenham Heathway Station (Underground) 37Bd 75
Dagenham Pk. Leisure Cen. 38Cd 76
Dagenham Rd. E10 32Bc 72
Dagenham Rd. RM10: Dag 35Ed 76
Dagenham Rd. RM13: Rain 38Fd 76
Dagenham Rd. RM7: Rush G 31Fd 76
Dagger La. WD6: E'tree 16Ja 28
Dagmar Av. HA9: Wemb 35Pa 67
Dagmar Ct. E14 48Ec 92
Dagmar Gdns. NW10 40Za 68
Dagmar M. UB2: S'hall 48Aa 85
(off Dagmar Rd.)
Dagmar Pas. N1 39Rb 71
(off Cross St.)
Dagmar Rd. KT2: King T 67Pa 131
Dagmar Rd. N15 28Tb 51
Dagmar Rd. N22 25Mb 50
Dagmar Rd. N4 31Qb 70
Dagmar Rd. RM10: Dag 38Ed 76
Dagmar Rd. SE25 71Ub 157
Dagmar Rd. SE5 53Ub 113
Dagmar Rd. SL4: Wind 4H 103
Dagmar Rd. UB2: S'hall 48Aa 85
Dagmar Ter. N1 39Rb 71
Dagnall Cres. UB8: Cowl 43L 83
Dagnall Pk. SE25 72Ub 157
Dagnall Rd. SE25 71Ub 157
Dagnall St. SW11 54Hb 111
Dagnam Pk. Cl. RM3: Rom 22Qd 57
Dagnam Pk. Dr. RM3: Rom 22Nd 57
Dagnam Pk. Gdns. RM3: Rom 23Qd 57
Dagnam Pk. Sq. RM3: Rom 23Rd 57
Dagnan Rd. SW12 59Kb 112
Dagobert Ho. E1 43Yb 92
(off Smithy St.)
Dagonet Gdns. BR1: Broml 62Jc 137
Dagonet Rd. BR1: Broml 62Jc 137
Dahlia Dr. BR8: Swan 68Hd 140
Dahlia Gdns. CR4: Mitc 70Mb 134
Dahlia Gdns. IG1: Ilf 37Rc 74
Dahlia Rd. SE2 49Xc 95
Dahomey Rd. SW16 65Lb 134
Daiglen Dr. RM15: S Ock 42Xd 98
Daimler Ho. E3 42Cc 92
(off Wellington Way)
Daimler Way SM6: W'gton 80Nb 156
Dain Ct. W8 49Cb 89
(off Lexham Gdns.)
Daines Cl. E12 34Pc 74
Daines Cl. RM15: S Ock 42Wd 98
Dainford Cl. BR1: Broml 64Fc 137
Dainton Cl. BR1: Broml 67Kc 137
Dainton Ho. W2 43Cb 89
(off Gt. Western Rd.)
Daintry Cl. HA3: W'stone 28Ja 46
Daintry Lodge HA6: Nwood 23V 44
Daintry Way E9 37Bc 72
Dairsie Ct. BR1: Broml 67Lc 137
Dairsie Rd. SE9 55Qc 116
Dairy Bus. Pk. RH1: Blet 3J 209
Dairy Cl. BR1: Broml 66Kc 137
Dairy Cl. CR7: Thor H 68Sb 135
Dairy Cl. DA4: Sut H 66Rd 141
Dairy Cl. EN3: Enf W 9Yb 20
Dairy Cl. NW10 39Wa 68
Dairy Cl. SW6 53Cb 111
Dairy Cl. UB6: G'frd 40Fa 66
Dairy Cotts. TN15: Fair 84Fe 185
Dairy Farm La. UB9: Hare 26L 43
Dairy Farm Pl. SE15 53Yb 114
Dairyglen Av. EN8: Chesh 3Ac 20
Dairy La. SE18 49Pc 94
Dairyman Cl. NW2 34Za 68
Dairy M. GU20: W'sham 9B 146
Dairy M. N2 28Gb 49
Dairy M. RM6: Chad H 31Zc 75
Dairy M. SW9 55Nb 112
Dairy M. WD18: Wat 15W 26
Dairy Wlk. SW19 63Ab 132
Dairy Way WD5: Ab L 1V 12
Daisy Cl. CR0: C'don 74Zb 158
Daisy Cl. NW9 33Sa 67
Daisy Dobbings Wlk. N19 31Nb 70
(off Jessie Blythe La.)
Daisy La. SW6 55Cb 111
Daisy Mdw. TW20: Egh 64C 126
Daisy Rd. E16 42Gc 93
Daisy Rd. E18 26Kc 53
Dakin Pl. E1 43Ac 92
Dakota Bldg. SE13 53Dc 114
(off Deal's Gateway)
Dakota Cl. SM6: W'gton 80Pb 156
Dakota Gdns. E6 42Nc 94
Dakota Gdns. UB5: N'olt 41Aa 85
Dakota Ho. CR7: Thor H 72Rb 157
Dalberg Rd. SW2 56Qb 112
(not continuous)
Dalberg Way SE2 48Zc 95
Dalby Rd. SW18 56Eb 111
Dalbys Cres. N17 23Ub 51
Dalby St. NW5 37Kb 70
Dalcross Rd. TW4: Houn 54Aa 107
Dale, The BR2: Kes 77Mc 159
Dale, The EN9: Walt A 6Gc 21
Dale Av. HA8: Edg 25Pa 47
Dale Av. TW4: Houn 55Aa 107
Dalebury Pl. TW5: Hest 50Ca 85
Dalebury Rd. SW17 61Hb 133
Dale Cl. DA1: Cray 58Hd 118
Dale Cl. E4 20Ec 34
Dale Cl. EN5: New Bar 16Db 31
Dale Cl. HA5: Pinn 25X 44
Dale Cl. KT15: Add 78K 149
Dale Cl. KT23: Bookh 97Ea 192
Dale Cl. RM15: S Ock 44Wd 98
Dale Cl. SE3 55Jc 115
Dale Cl. SL5: S'dale 1E 146
Dale Ct. EN2: Enf 11Sb 33
Dale Ct. KT2: King T 66Pa 131
(off York Rd.)
Dale Ct. SL1: Slou 7G 80
Dale Ct. WD25: Wat 5W 12
Dale Dr. UB4: Hayes 42V 84
Dale End DA1: Cray 58Hd 118
Dalefield IG9: Buck H 18Lc 35
(off Roebuck La.)
Dalefield Ind. Est. DA12: Grav'nd 9H 123
Dalefield Way DA12: Grav'nd 9H 123
Dale Gdns. IG8: Wfd G 21Kc 53
Dalegarth Gdns. CR8: Purl 85Tb 177
Dale Grn. Rd. N11 20Kb 32
Dale Gro. N12 22Eb 49
Daleham Av. TW20: Egh 65C 126
Daleham Dr. UB8: Hil 44R 84
Daleham Gdns. NW3 36Fb 69
Daleham M. NW3 37Fb 69
Dalehead NW1 2B 216 (40Lb 70)
(off Hampstead Rd.)
Dale Ho. N1 39Ub 71
(off Halcomb St.)
Dale Ho. NW8 39Eb 69
(off Boundary Rd.)
Dale Ho. SE4 56Ac 114
Dale Lodge N6 30Lb 50
Dale Lodge Rd. SL5: S'dale 1E 146
Dalemain M. E16 46Jc 93
Dale Pk. Av. SM5: Cars 75Hb 155
Dale Pk. Rd. SE19 67Sb 135
Dale Rd. BR8: Swan 68Ed 140
Dale Rd. CR8: Purl 84Qb 176
Dale Rd. DA1: Cray 58Hd 118

Dale Rd. DA13: Sflt 63Ce 143
Dale Rd. KT12: Walt T 73V 150
Dale Rd. NW5 36Jb 70
Dale Rd. SE17 51Rb 113
Dale Rd. SM1: Sutt 77Bb 155
Dale Rd. TW16: Sun 66V 128
Dale Rd. UB6: G'frd 43Da 85
Dale Row W11 44Ab 88
Dale Side SL9: Ger X 32A 62
Daleside BR6: Chels 78Wc 161
Daleside Cl. BR6: Chels 79Wc 161
Daleside Dr. EN6: Pot B 5Bb 17
Daleside Gdns. IG7: Chig 20Sc 36
Daleside Rd. KT19: Ewe 79Ta 153
Daleside Rd. SW16 64Kb 134
Dales Path WD6: Bore 15Ta 29
Dales Rd. WD6: Bore 15Ta 29
Dalestone M. RM3: Rom 23Kd 57
Dale St. DA1: Dart 57Pd 119
Dale St. W4 50Ua 88
Dale Vw. DA8: Erith 54Hd 118
Dale Vw. GU21: Wok 10M 167
Dale Vw. KT18: Head 95Ra 193
Dale Vw. Av. E4 19Ec 34
Dale Vw. Cres. E4 19Ec 34
Dale Vw. Gdns. E4 20Fc 35
Daleview Rd. N15 30Ub 51
Dale Wlk. DA2: Dart 60Sd 120
Dalewood Cl. RM11: Horn 31Pd 77
Dalewood Gdns. KT4: Wor Pk 75Xa 154
Dale Wood Rd. BR6: Orp 73Uc 160
Daleworth Cl. BR3: Beck 65Cc 136
Daley Ho. W12 44Xa 88
Daley St. E9 37Zb 72
Daley Thompson Way SW8 54Kb 112
Dalgarno Gdns. W10 43Ya 88
Dalgarno Way W10 42Ya 88
Dalgleish St. E14 44Ac 92
Daling Way E3 39Ac 72
Dalkeith Ct. SW1 6E 228 (49Mb 90)
(off Vincent St.)
Dalkeith Gro. HA7: Stan 22Ma 47
Dalkeith Ho. SW9 53Rb 113
(off Lothian Rd.)
Dalkeith Rd. IG1: Ilf 34Sc 74
Dalkeith Rd. SE21 60Sb 113
Dallas Rd. NW4 31Wa 68
Dallas Rd. SE26 62Xb 135
Dallas Rd. SM3: Cheam 79Ab 154
Dallas Rd. W5 43Pa 87
Dallas Ter. UB3: Harl 48V 84
Dallega Cl. UB3: Hayes 45T 84
Dallinger Rd. SE12 58Hc 115
Dalling Rd. W6 49Xa 88
Dallington Cl. KT12: Hers 79Y 151
Dallington Sq. EC1 5C 218 (42Rb 91)
(off Dallington St.)
Dallington St. EC1 5C 218 (42Rb 91)
Dallin Rd. DA6: Bex 56Zc 117
Dallin Rd. SE18 52Rc 116
Dalmain Rd. SE23 60Zb 114
Dalmally Rd. CR0: C'don 73Vb 157
Dalmany Pas. CR0: C'don 73Vb 157
Dalmeny Av. N7 35Mb 70
Dalmeny Av. SW16 68Qb 134
Dalmeny Cl. HA0: Wemb 37La 66
Dalmeny Cres. TW3: Houn 56Fa 108
Dalmeny Rd. DA8: Erith 53Dd 118
Dalmeny Rd. EN5: New Bar 16Eb 31
Dalmeny Rd. KT4: Wor Pk 76Xa 154
Dalmeny Rd. N7 34Mb 70
(not continuous)
Dalmeny Rd. SM5: Cars 80Jb 156
Dalmeny Way KT18: Eps 86Sa 173
Dalmeyer Rd. NW10 37Va 68
Dalmore Av. KT10: Clay 79Ha 152
Dalmore Rd. SE21 61Sb 135
Dalo Lodge E3 43Cc 92
(off Gale St.)
Dalroy Cl. RM15: S Ock 44Wd 98
Dalrymple Cl. N14 17Mb 32
Dalrymple Rd. SE4 56Ac 114
DALSTON 37Vb 71
Dalston Gdns. HA7: Stan 25Na 47
Dalston Junction (Overground) 37Vb 71
Dalston Kingsland Station
(Overground) 36Ub 71
Dalston La. E8 37Vb 71
Dalston Sq. E8 37Vb 71
(not continuous)
Dalton Av. CR4: Mitc 68Gb 133
Dalton Cl. BR6: Orp 76Uc 160
Dalton Cl. CR8: Purl 84Sb 177
Dalton Cl. UB4: Hayes 42T 84
Dalton Grn. SL3: L'ly 51B 104
Dalton Ho. E3 40Ac 72
(off Ford St.)
Dalton Ho. HA7: Stan 22Ja 46
Dalton Ho. SE14 51Zb 114
(off John Williams Cl.)
Dalton Ho. SW1 7K 227 (50Kb 90)
(off Ebury Bri. Rd.)
Dalton Ho. Apts. IG11: Bark 42Wc 95
Dalton Rd. HA3: W'stone 26Fa 46
Daltons Rd. BR6: Well H 76Dd 162
Daltons Rd. BR8: Crock 75Ed 162
Daltons Shaw RM16: Ors 3C 100
Dalton St. AL3: St A 1B 6
Dalton St. SE27 62Rb 135
Daltons Wharf HP4: Berk 1A 2
Dalton Way WD17: Wat 15Z 27
Dalwood St. SE5 53Ub 113
Daly Dr. BR1: Broml 69Qc 138
Dalyell Rd. SW9 55Pb 112
Damascene Wlk. SE21 60Sb 113
Damask Cl. GU24: W End 5C 166
Damask Ct. SM1: Sutt 74Db 155
Damask Cres. E16 42Gc 93
Damask Grn. HP1: Hem H 3G 2
Damer Ter. SW10 52Eb 111
Dames Rd. E7 34Jc 73
Dame St. N1 1D 218 (40Sb 71)
Damien Ct. E1 44Xb 91
(off Damien St.)
Damien St. E1 44Xb 91
Damigos Rd. DA12: Grav'nd 10H 123
Damon Cl. DA14: Sidc 62Xc 139
Damory Ho. SE16 49Xb 91
(off Abbeyfield Est.)
Damsel Ho. SE4 55Ac 114
(off Dragonfly Pl.)
Damsel Wlk. NW9 30Wa 48
(off Perryfield Way)
Damson Cl. WD24: Wat 9W 12
Damson Ct. BR8: Swan 70Fd 140
Damson Dr. UB3: Hayes 45W 84
Damson Gro. SL1: Slou 7G 80
Damson Way SM5: Cars 81Hb 175
Damsonwood Rd. UB2: S'hall 48Ca 85
Danbrook Rd. SW16 67Nb 134
Danbury Cl. CM15: Pil H 15Vd 40
Danbury Cl. RM6: Chad H 27Zc 55
Danbury Cres. RM15: S Ock 44Xd 98
Danbury Mans. IG11: Bark 38Rc 74
(off Whiting Av.)
Danbury M. SM6: W'gton 77Kb 156
Danbury Rd. IG10: Lough 17Nc 36
Danbury Rd. RM13: Rain 39Hd 76
Danbury St. N1 1C 218 (40Rb 71)
Danbury Way IG8: Wfd G 23Lc 53
Danby Ct. EN2: Enf 13Sb 33
Danby Ho. E9 38Yb 72
(off Frampton Pk. Rd.)
Danby Ho. W10 41Ab 88
(off Bruckner St.)
Danby St. SE15 55Vb 113
Dance Ho. SE4 57Zb 114
(off St Norbert Rd.)
Dancer Rd. SW6 53Bb 111
Dancer Rd. TW9: Rich 55Qa 109
DANCERS HILL 8Za 16
Dancers Hill Rd. EN5: Barn 8Ya 16
Dancers La. EN5: Barn 7Ya 16
Dancers Way SE8 51Dc 114
Dance Sq. EC1 4D 218 (41Sb 91)
Dan Ct. NW10 41Qa 87
Dandelion Cl. RM7: Rush G 33Gd 76
Dandelion Ct. E14 48Dc 92
(off Watergate Walk)
Dandelion Pl. RM3: Rom 24Md 57
Dandridge Cl. SE10 50Hc 93
Dandridge Cl. SL3: L'ly 9P 81
Dandridge Ho. E1 7K 219 (43Vb 91)
(off Lamb St.)
Danebury CR0: New Ad 79Ec 158
Danebury Av. SW15 58Ua 110
Daneby Rd. SE6 62Dc 136
Dane Cl. BR6: Farnb 78Tc 160
Dane Cl. DA5: Bexl 59Cd 118
Dane Ct. GU22: Pyr 87H 169
Danecourt Gdns. CR0: C'don 76Vb 157
Danecroft Rd. SE24 57Sb 113
Danehill Wlk. DA14: Sidc 62Wc 139
DANEHOLES RDBT. 48Fe 99
Danehurst TW8: Bford 52La 108
Danehurst Cl. TW20: Egh 65A 126
Danehurst Ct. KT17: Eps 85Va 174
Danehurst Gdns. IG4: Ilf 29Nc 54
Danehurst St. SW6 53Ab 110
Daneland EN4: E Barn 16Hb 31
Daneland Wlk. N17 27Xb 51
Danemead Gro. UB5: N'olt 36Da 65
Danemere St. SW15 55Ya 110
Dane Pl. E3 40Ac 72
Dane Rd. CR6: W'ham 89Zb 178
Dane Rd. IG1: Ilf 36Sc 74
Dane Rd. N18 20Yb 34
Dane Rd. SW19 67Eb 133
Dane Rd. TN14: Otf 89Hd 182
Dane Rd. TW15: Ashf 65S 128
Dane Rd. UB1: S'hall 45Aa 85
Dane Rd. W13 46La 86
Danes, The AL2: Park 10A 6
Danesbury Rd. TW13: Felt 60X 107
Danes Cl. DA11: Nflt 62Ee 143
Danes Cl. KT22: Oxs 86Ea 172
Danescombe SE12 60Jc 115
Danes Ct. HA9: Wemb 34Ra 67
Danes Ct. NW8 1F 215 (39Hb 69)
(off St Edmund's Ter.)
Danescourt Cres. SM1: Sutt 75Eb 155
Danescroft NW4 29Za 48
Danescroft Av. NW4 29Za 48
Danescroft Gdns. NW4 29Za 48
Danesdale Rd. E9 37Ac 72
Danesfield GU23: Rip 95H 189
Danesfield GU23: Send 95H 189
Danesfield SE5 51Ub 113
(off Albany Rd.)
Danesfield Cl. KT12: Walt T 76X 151
Danes Ga. HA1: Harr 27Ga 46
Danes Hill GU22: Wok 90C 168
Daneshill RH1: Redh 5N 207
Daneshill Cl. RH1: Redh 5N 207
Danes Hill School Dr. KT22: Oxs 86Ea 172
Danes Ho. W10 43Ya 88
(off Sutton Way)
Danesmead KT11: Cobh 83Ca 171
Danes Rd. RM7: Rush G 31Ed 76
Dane St. WC1 1H 223 (43Pb 90)
Danes Way CM15: Pil H 15Wd 40
Danes Way KT22: Oxs 86Fa 172
Daneswood Av. SE6 62Ec 136
Daneswood Cl. KT13: Weyb 78R 150
Dane's Yd. E15 40Ec 72
Danethorpe Rd. HA0: Wemb 37Ma 67
Danetree Cl. KT19: Ewe 80Sa 153
Danetree Rd. KT19: Ewe 80Sa 153
Danette Gdns. RM10: Dag 33Cd 76
Daneville Rd. SE5 53Tb 113
Danewood Dr. N2 31Fb 69
Dangan Rd. E11 30Jc 53
Daniel Bolt Cl. E14 43Dc 92
Daniel Cl. N18 21Yb 52
Daniel Cl. RM16: Chaf H 47Ae 99
Daniel Cl. RM16: Grays 8D 100
Daniel Cl. SW17 65Gb 133
Daniel Cl. TW4: Houn 59Ba 107
Daniel Ct. BR3: Beck 66Cc 136
(off Brackley Rd.)
Daniel Ct. NW9 25Ua 48
Daniel Gdns. SE15 52Vb 113
Daniel Lambert Mill KT15: Add 78N 149
(off Bourneside Rd.)
Daniell Ho. N1 1G 219 (40Tb 71)
(off Cranston Est.)
Daniell Way CR0: Wadd 74Nb 156
Daniel Pl. NW4 31Xa 68
Daniel Rd. W5 45Pa 87
Daniels La. CR6: W'ham 88Bc 178
Daniels Rd. SE15 55Yb 114
Daniel Way SM7: Bans 86Db 175
Danleigh Ct. N14 17Mb 32
Dan Leno Wlk. SW6 52Db 111
Dan Mason Dr. W4 54Sa 109
Dannatt Cl. N20 19Fb 31
Danny Fiszman Bri. 35Qb 70
Dansey Pl. W1 4D 222 (45Mb 90)
(off Wardour St.)
Dansington Rd. DA16: Well 56Wc 117
Danson Cres. DA16: Well 55Xc 117
Danson House 56Yc 117
DANSON INTERCHANGE 57Yc 117
Danson La. DA16: Well 56Xc 117
Danson Mead DA16: Well 55Yc 117
Danson Pk. 57Yc 117
Danson Pk. Watersports Cen. 56Yc 117
Danson Rd. DA5: Bex 58Yc 117
Danson Rd. DA5: Bexl 58Yc 117
Danson Rd. DA6: Bex 57Zc 117
Danson Rd. SE17 7C 230 (50Rb 91)
Danson Underpass DA15: Sidc 58Yc 117
Dante Pl. SE11 6C 230 (49Rb 91)
Dante Rd. SE11 5B 230 (49Rb 91)
Danube Apts. N8 28Pb 50
(off Gt. Amwell La.)
Danube Cl. N9 20Yb 34
Danube Ct. SE15 52Vb 113
(off Daniel Gdns.)
Danube St. SW3 7E 226 (50Gb 89)
Danvers Av. SW11 56Gb 111
Danvers Ho. E1 44Wb 91
(off Christian St.)
Danvers Rd. N8 28Mb 50
Danvers St. SW3 51Fb 111
Danvers Way CR3: Cat'm 95Sb 197
Danyon Cl. RM13: Rain 40Ld 77
Danziger Way WD6: Bore 11Sa 29
Dao Ct. E13 39Kc 73
Da Palma Ct. SW6 51Cb 111
(off Anselm Rd.)
Daphne Ct. KT4: Wor Pk 75Ua 154
Daphne Gdns. E4 20Ec 34
Daphne Ho. N22 25Qb 50
(off Acacia Rd.)
Daphne St. SW18 58Eb 111
Daplyn St. E1 43Wb 91
Dara Ho. NW9 27Ta 47
Darbishire Pl. E1 45Wb 91
(off John Fisher St.)
D'Arblay St. W1 3C 222 (44Lb 90)
Darby Cl. CR3: Cat'm 94Sb 197
Darby Cl. RM19: Purf 49Sd 98
Darby Cres. TW16: Sun 68Y 129
Darby Dr. EN9: Walt A 5Ec 20
Darby Gdns. TW16: Sun 68Y 129
Darcies M. N8 30Nb 50
Darcy Av. SM6: W'gton 77Lb 156
Darcy Cl. CM13: Hut 17De 41
Darcy Cl. CR5: Coul 91Rb 197
Darcy Cl. EN8: Chesh 3Ac 20
Darcy Cl. N20 19Fb 31
D'Arcy Dr. HA3: Kenton 28Ma 47
D'Arcy Gdns. HA3: Kenton 28Ma 47
Darcy Gdns. RM9: Dag 39Bd 75
Darcy Ho. E8 39Xb 71
(off London Flds. E. Side)
Darcy Ho. RM9: Dag 39Cd 76
D'Arcy Pl. BR2: Broml 70Jc 137
D'Arcy Pl. KT21: Asht 89Pa 173
D'Arcy Rd. KT21: Asht 89Pa 173
D'Arcy Rd. SM3: Cheam 77Za 154
Darcy Rd. SW16 68Nb 134
Darcy Rd. TW7: Isle 53Ja 108
Dare Gdns. RM8: Dag 34Ad 75
Darell Rd. TW9: Rich 55Qa 109
Darent Cl. TN13: Chips 94Ed 202
DARENTH 64Sd 142
Darenth Country Pk. 61Td 142
Darenth Dr. DA12: Grav'nd 10K 123
Darenth Gdns. TN16: Westrm 98Tc 200
Darenth Hill DA2: Daren 64Sd 142
DARENTH INTERCHANGE 62Rd 141
Darenth La. RM15: S Ock 44Wd 98
Darenth La. TN13: Dun G 93Gd 202
Darenth Mill La. DA2: Daren 64Rd 141
Darent Ho. BR1: Broml 64Fc 137
Darent Ho. NW8 7C 214 (43Fb 89)
(off Church St. Est.)
Darenth Pk. Av. DA2: Dart 61Td 142
Darenth Pl. DA2: Daren 64Td 142
Darenth Rd. DA1: Dart 59Pd 119
Darenth Rd. DA16: Well 53Wc 117
Darenth Rd. N16 31Vb 71
Darenth Rd. Sth. DA2: Daren 63Rd 141
Darenth Valley Golf Course 84Jd 182
Darenth Way TN14: S'ham 83Jd 182
Darenth Wood Rd. DA2: Dart 60Ud 120
(not continuous)
Darent Ind. Pk. DA8: Erith 51Md 119
Darent Mead DA4: Sut H 67Rd 141
DARENT VALLEY HOSPITAL 60Ud 120
Darfield NW1 39Lb 70
(off Bayham St.)
Darfield Rd. SE4 57Bc 114
Darfield Way W10 44Za 88
Darfur St. SW15 55Za 110
Dargate Cl. SE19 66Vb 135
Dariel Cl. SL1: Slou 7D 80
Darien Rd. SW11 55Fb 111
Daring Ho. E3 40Ac 72
(off Roman Rd.)
Darkes La. EN6: Pot B 4Cb 17
Dark Hill Rd. TN15: Bor G 92Ae 205
Dark Ho. Wlk. EC3 5G 225 (45Tb 91)
Dark La. CM14: Gt War 23Vd 58
Dark La. EN7: Chesh 2Wb 19
Darlands Dr. EN5: Barn 15Za 30
Darlan Rd. SW6 52Bb 111
Darlaston Rd. SW19 66Za 132
Darley Cl. CR0: C'don 72Ac 158
Darley Cl. KT15: Add 78L 149
Darley Cft. AL2: Park 10P 5
Darley Dene Ct. KT15: Add 77L 149
Darley Dr. KT3: N Mald 68Ta 131
Darley Gdns. SM4: Mord 72Eb 155
Darley Ho. SE11 7H 229 (50Pb 90)
(off Laud St.)
Darley Rd. N9 18Vb 33
Darley Rd. SW11 58Hb 111
Darling Ho. TW1: Twick 58Ma 109
Darling Rd. SE4 55Cc 114
Darling Row E1 42Xb 91
Darlington Ct. CM15: Pil H 16Wd 40
Darlington Ct. SE6 60Hc 115
Darlington Gdns. RM3: Rom 22Md 57
Darlington Ho. SW8 52Mb 112
(off Hemans St.)
Darlington Path RM3: Rom 22Md 57
Darlington Rd. SE27 64Rb 135
Darlton Cl. DA1: Cray 55Hd 118
Darmaine Cl. CR2: S Croy 80Sb 157
Darnall Ho. SE10 53Ec 114
(off Royal Hill)
Darnaway Pl. E14 43Ec 92
(off Aberfeldy St.)
Darnay Apts. E15 37Fc 73
Darndale Cl. E17 26Bc 52
Darnets Fld. TN14: Otf 89Hd 182
Darnhills WD7: R'lett 7Ha 14
Darnley Cl. DA15: Sidc 57Vc 117
Darnley Ct. DA11: Grav'nd 9C 122
(off Darnley Rd.)
Darnley Ho. E14 44Ac 92
(off Camdenhurst St.)
Darnley Mausoleum 10N 145
Darnley Pk. KT13: Weyb 76R 150
Darnley Rd. DA11: Grav'nd 10C 122
(not continuous)
Darnley Rd. E9 37Yb 72
Darnley Rd. IG8: Wfd G 25Jc 53
Darnley Rd. RM17: Grays 51De 121
Darnley St. DA11: Grav'nd 9C 122
Darnley Ter. W11 46Za 88
Darns Hill BR8: Crock 73Ed 162
Darrell Charles Ct. UB8: Uxb 38N 63
Darrell Cl. SL3: L'ly 49B 82
Darrell Rd. SE22 57Wb 113
Darren Cl. N4 31Pb 70
Darren Ct. N7 35Nb 70
Darrick Wood Rd. BR6: Orp 75Tc 160
Darrick Wood Sports Cen. 76Sc 160
Darrick Wood Swimming Pool 76Sc 160
Darrington Rd. WD6: Bore 11Na 29
Darris Cl. UB4: Yead 42Aa 85
Darsley Dr. SW8 53Mb 112
Dart Cl. RM14: Upm 30Td 58
Dart Cl. SL3: L'ly 50D 82
Dartfields RM3: Rom 23Md 57
DARTFORD 58Nd 119
Dartford Av. N9 16Yb 34
Dartford Borough Mus. 59Nd 119
Dartford Bus. Pk. DA1: Dart 57Md 119
Dartford By-Pass DA5: Bexl 60Gd 118
Dartford Clay Shooting Club 52Nd 119
Dartford Ct. KT19: Eps 84Na 173
Dartford FC 60Qd 119
Dartford Gdns. RM6: Chad H 29Xc 55
Dartford Golf Course 61Kd 141
DARTFORD HEATH 60Hd 118
Dartford Heath Retail Pk. 60Ld 119
Dartford Ho. SE1 6K 231 (49Vb 91)
(off Longfield Est.)
Dartford Judo Club 58Sd 120
Dartford Rd. DA1: Dart 58Jd 118
Dartford Rd. DA4: Farni 72Pd 163
Dartford Rd. DA4: Hort K 72Pd 163
Dartford Rd. DA4: S Dar 72Pd 163
Dartford Rd. DA5: Bexl 60Ed 118
Dartford Rd. TN13: S'oaks 96Ld 203
Dartford Station (Rail) 58Nd 119
Dartford Stone Lodge Bowls &
Social Club 58Sd 120
Dartford St. SE17 51Sb 113
Dartford-Thurrock River Crossing 54Ud 120
Dartford Trade Pk. DA1: Dart 61Nd 141
Dartford Tunnel 55Td 120
Dartford Tunnel App. Rd. DA1: Dart 59Rd 119
Dart Grn. RM15: S Ock 43Xd 98
Dartington NW1 1C 216 (39Lb 70)
(off Plender St.)
Dartington Ho. SW8 54Mb 112
(off Union Gro.)
Dartington Ho. W2 43Db 89
(off Senior St.)
Dartle Ct. SE16 47Wb 91
(off Scott Lidgett Cres.)
Dartmoor Wlk. E14 49Cc 92
(off Severnake Cl.)
Dartmouth Av. GU21: Wok 86E 168
Dartmouth Cl. W11 44Bb 89
Dartmouth Ct. SE10 53Ec 114
Dartmouth Grn. GU21: Wok 86F 168
Dartmouth Gro. SE10 53Ec 114
Dartmouth Hill SE10 53Ec 114
Dartmouth Ho. KT2: King T 67Na 131
(off Seven Kings Way)
Dartmouth Ho. SE10 53Dc 114
(off Catherine Gro.)
DARTMOUTH PARK 34Kb 70
Dartmouth Pk. Av. NW5 34Kb 70
Dartmouth Pk. Hill N19 32Kb 70
Dartmouth Pk. Hill NW5 34Kb 70
Dartmouth Pk. Rd. NW5 35Kb 70
Dartmouth Path GU21: Wok 86F 168
Dartmouth Pl. SE23 61Yb 136
Dartmouth Pl. W4 51Ua 110
Dartmouth Rd. BR2: Hayes 73Jc 159
Dartmouth Rd. HA4: Ruis 34W 64
Dartmouth Rd. NW2 37Za 68
Dartmouth Rd. NW4 30Wa 48
Dartmouth Rd. SE23 62Yb 136
Dartmouth Rd. SE26 62Xb 135
Dartmouth Row SE10 54Ec 114
Dartmouth St. SW1 2D 228 (47Mb 90)
Dartmouth Ter. SE10 53Fc 115
Dartnell Av. KT14: W Byf 84K 169
Dartnell Cl. KT14: W Byf 84K 169
Dartnell Ct. KT14: W Byf 84L 169
Dartnell Cres. KT14: W Byf 84K 169
DARTNELL PARK 84K 169
Dartnell Pk. Rd. KT14: W Byf 84K 169
Dartnell Pl. KT14: W Byf 84K 169
Dartnell Rd. CR0: C'don 73Vb 157
Darton Ct. W3 46Sa 87
Dartrey Twr. SW10 52Eb 111
(off Worlds End Est.)
Dartrey Wlk. SW10 52Eb 111
Dart St. W10 41Ab 88
Dartview Cl. RM17: Grays 9A 100
Darvel Cl. GU21: Wok 8L 167
Darvell Ho. SE17 7G 231 (50Tb 91)
(off Inville Rd.)
Darvells Yd. WD3: Chor 14F 24
Darville Rd. N16 34Vb 71
Darvill's La. SL1: Slou 7H 81
Darwell Cl. E6 40Qc 74
Darwell M. E6 40Qc 74
Darwen Pl. E2 40Xb 71
Darwin Av. DA1: Dart 54Nd 119
Darwin Cl. BR6: Farnb 78Tc 160
Darwin Cl. N11 20Kb 32
Darwin Ct. CR2: S Croy 78Rb 157
(off Warham Rd.)
Darwin Ct. E13 41Kc 93
Darwin Ct. NW1 39Kb 70
(not continuous)
Darwin Ct. SE17 6G 231 (49Tb 91)
(off Barlow St.)
Darwin Dr. UB1: S'hall 44Ds 85
Darwin Gdns. WD19: Wat 22Y 45
Darwin Ho. SE20 67Wb 135
Darwin Ho. SW1 51Lb 112
(off Grosvenor Rd.)
Darwin Ri. DA11: Nflt 60De 121
Darwin Rd. DA16: Well 55Vc 117
Darwin Rd. N22 25Rb 51
Darwin Rd. RM18: Tilb 3B 122
Darwin Rd. SL3: L'ly 47B 82
Darwin Rd. W5 50La 86
Darwin Sports Cen. 87Pc 180
Darwin St. SE17 5G 231 (49Tb 91)
(not continuous)
Darwood Ct. NW6 38Eb 69
(off Belsize Rd.)
Daryngton Dr. UB6: G'frd 40Fa 66
Daryngton Ho. SE1 2F 231 (47Tb 91)
(off Hankey Pl.)
Daryngton Ho. SW8 52Nb 112
(off Hartington Rd.)
Dashwood Cl. DA6: Bex 57Cd 118
Dashwood Cl. KT14: W Byf 84L 169
Dashwood Cl. SL3: L'ly 9N 81
Dashwood Lang Rd. KT15: Add 77M 149
Dashwood Rd. DA11: Grav'nd 10C 122
Dashwood Rd. N8 30Pb 50
Dashwood Studios SE17 6D 230 (49Sb 91)
(off Walworth Rd.)
Dassett Rd. SE27 64Rb 135
Data Point Bus. Cen. E16 42Fc 93
Datchelor Pl. SE5 53Tb 113
DATCHET 3M 103
DATCHET COMMON 3P 103
Datchet Golf Course 2L 103
Datchet Ho. E2 4K 219 (41Vb 91)
(off Virginia Rd.)
Datchet Ho. NW1 3A 216 (41Kb 90)
(off Augustus St.)
Datchet Pl. SL3: Dat 3M 103
Datchet Rd. SE6 61Bc 136
Datchet Rd. SL3: Hort 55B 104
Datchet Rd. SL3: Slou 9K 81
Datchet Rd. SL4: Old Win 6L 103
Datchet Rd. SL4: Wind 2H 103
Datchet Sailing Club 52D 104
Datchet Station (Rail) 3M 103
Datchworth Ct. EN1: Enf 15Ub 33
Datchworth Ho. N1 38Rb 71
(off The Sutton Est.)
Datchworth Turn HP2: Hem H 2C 4
Date St. SE17 7F 231 (50Tb 91)
Daubeney Gdns. N17 24Sb 51
Daubeney Pl. TW12: Hamp 67Ea 130
(off High St.)
Daubeney Rd. E5 35Ac 72
Daubeney Rd. N17 24Sb 51
Daubeney Twr. SE8 50Bc 92
(off Bowditch)
Dault Rd. SW18 58Eb 111
Dauncey Ho. SE1 2B 230 (47Rb 91)
(off Webber Row)
Davall Ho. RM17: Grays 51De 121
(off Argent St.)
Dave Adams Ho. E3 40Bc 72
(off Norman Gro.)
Davema Cl. BR7: Chst 67Qc 138
Davenant Ho. E1 43Wb 91
(off Old Montague St.)
Davenant Pl. SE26 64Xb 135
Davenant Rd. CR0: C'don 77Rb 157
Davenant Rd. N19 33Mb 70
Davenant St. E1 43Wb 91
Davenham Av. HA6: Nwood 22V 44
Davenham Pl. HA6: Nwood 22V 44
Davenport Cen. IG11: Bark 39Xc 75
Davenport Cl. TW11: Tedd 65Ja 130
Davenport Ct. CR0: C'don 73Pb 157
Davenport Ho. SE11 5K 229 (49Qb 90)
(off Walnut Tree Wlk.)
Davenport Ho. UB7: W Dray 47P 83
Davenport Lodge TW5: Hest 52Aa 107
Davenport Rd. DA14: Sidc 61Ad 139
Davenport Rd. SE6 58Dc 114
Daventer Dr. HA7: Stan 24Ha 46
Daventry Av. E17 30Cc 52
Daventry Cl. SL3: Poyle 53H 105
Daventry Gdns. RM3: Rom 22Ld 57
Daventry Grn. RM3: Rom 22Ld 57
Daventry Rd. RM3: Rom 22Ld 57
Daventry St. NW1 7D 214 (43Gb 89)
Daver Ct. SW3 7E 226 (50Gb 89)
Daver Ct. W5 42Ma 87
Davern Cl. SE10 49Hc 93
Davey Cl. N13 22Pb 50
Davey Cl. N7 37Pb 70
Davey Gdns. IG11: Bark 42Wc 95
Davey Rd. E9 38Cc 72
Davey's Ct. WC2 4F 223 (45Nb 90)
(off Bedfordbury)
Davey St. SE15 51Vb 113
David Av. UB6: G'frd 41Ga 86
David Cl. UB3: Harl 52U 106
David Coffer Ct. DA17: Belv 49Dd 96
David Ct. E14 43Dc 92
(off Hillary M.)
David Ct. N20 20Eb 31
David Dr. RM3: Hrld W 23Qd 57
Davidge Ho. SE1 2A 230 (47Qb 90)
(off Coral St.)
Davidge St. SE1 2B 230 (47Rb 91)
David Hewitt Ho. E3 43Dc 92
(off Watts Gro.)
David Ho. DA15: Sidc 62Wc 139
David Ho. SW8 52Nb 112
(off Wyvil Rd.)
David Lean Cinema Croydon 76Tb 157
(within Fairfield Halls)
David Lean Ct. UB9: Den 29H 43
(off Patrons Way W.)
David Lee Point E15 39Gc 73
(off Leather Gdns.)
David Lloyd Leisure Barnet 24Fb 49
David Lloyd Leisure Beckenham 70Bc 136
David Lloyd Leisure Bushey 12Ca 27
David Lloyd Leisure Cheam 80Za 154
David Lloyd Leisure Chigwell 18Pc 36
David Lloyd Leisure Dartford 60Pd 119
David Lloyd Leisure Enfield 12Wb 33
David Lloyd Leisure Epsom 82Qa 173
David Lloyd Leisure Fulham 52Cb 111
(within Fulham Broadway Shop. Cen.)
David Lloyd Leisure Hampton 62Ca 129
David Lloyd Leisure Heston 50Y 85
David Lloyd Leisure Hornchurch 27Md 57
David Lloyd Leisure Kidbrooke 56Kc 115
David Lloyd Leisure Kingston
upon Thames 68Na 131
(within The Rotunda Cen.)
David Lloyd Leisure Northwood 23R 44
David Lloyd Leisure Purley 79Pb 156
David Lloyd Leisure Raynes Park 69Za 132
David Lloyd Leisure Sidcup 64Yc 139
David Lloyd Leisure Sudbury Hill 36Fa 66
David Lloyd Leisure Weybridge 82Q 170
David Lloyd Leisure Woking 92B 188
David M. SE10 52Ec 114
David M. W1 7G 215 (43Hb 89)
David Rd. RM8: Dag 33Ad 75
David Rd. SL3: Poyle 54H 105
Davidson Gdns. SW8 52Nb 112
Davidson Ho. HP2: Hem H 1M 3
Davidson La. HA1: Harr 31Ha 66
Davidson Rd. CR0: C'don 74Ub 157
Davidson Terraces E7 36Kc 73
(off Claremont Rd.)

Davidson Way RM7: Rush G 31Gd 76
David's Rd. SE23 60Yb 114
David St. E15 37Fc 73
David's Way IG6: Ilf 24Uc 54
David Ter. RM3: Hrld W 24Qd 57
David Twigg Cl. KT2: King T 67Na 131
David Weir Leisure Cen. 73Fb 155
David Wildman La. NW7 23Ab 48
Davies Cl. CR0: C'don 72Wb 157
Davies Cl. RM13: Rain 41Ld 97
Davies La. E11 33Gc 73
Davies M. W1 4K 221 (45Kb 90)
Davies St. W1 3K 221 (44Kb 90)
Davies Wlk. TW7: Isle 53Fa 108
Da Vinci Ct. SE16 50Xb 91
(off Rossetti Rd.)
Da Vinci Ct. WD25: Wat 8Z 13
Da Vinci Lodge SE10 48Hc 93
(off W. Parkside)
Da Vinci Torre SE13 55Dc 114
(off Loampit Va.)
Davington Gdns. RM8: Dag 36Xc 75
Davington Rd. RM8: Dag 37Xc 75
Davinia Cl. IG8: Wfd G 23Pc 54
Davis Av. DA11: Nflt 10A 122
Davis Cl. TN13: S'oaks 94Ld 203
Davis Ct. AL1: St A 2C 6
Davis Ho. W12 45Xa 88
(off White City Est.)
Davison Cl. KT19: Eps 83Ra 173
Davison Ct. TN14: Dun G 93Hd 202
Davison Dr. EN8: Chesh 1Zb 20
Davison Rd. SL3: L'ly 50B 82
Davis Rd. KT13: Weyb 82P 169
Davis Rd. KT9: Chess 77Qa 153
Davis Rd. RM15: Avel 46Td 98
Davis Rd. RM16: Chaf H 48Be 99
Davis Rd. W3 46Va 88
Davis Rd. Ind. Pk. KT9: Chess 77Qa 153
Davis St. E13 40Kc 73
Davisville Rd. W12 47Wa 88
Davis Way DA14: Sidc 65Ad 139
Davmor Ct. TW8: Bford 50La 86
Davos Cl. GU22: Wok 91A 188
Davy Down (Info. Cen.) 46Yd 98
Davy Down Riverside Pk. 46Yd 98
Davy Ho. AL1: St A 3D 6
Davy's Pl. DA12: Grav'nd 5G 144
Dawburn Pl. TW5: Hest 52Z 107
Dawell Dr. TN16: Big H 89Lc 179
Dawes Av. RM12: Horn 34Md 77
Dawes Av. TW7: Isle 57Ja 108
Dawes Cl. DA9: Ghithe 57Vd 120
Dawes Cl. UB10: Uxb 40N 63
Dawes Ct. KT10: Esh 77Da 151
Dawes East Rd. SL1: Burn 2A 80
Dawes Ho. SE17 6F 231 (49Tb 91)
(off Orb St.)
Dawes La. WD3: Sarr 9G 10
Dawes Moor Cl. SL2: Slou 4N 81
Dawe's Rd. UB10: Uxb 40N 63
Dawes Rd. SW6 52Ab 110
Dawes St. SE17 7G 231 (50Tb 91)
Dawkins Ct. SE1 4F 231 (48Tb 91)
Dawley Av. UB8: Hil 43S 84
Dawley Grn. RM15: S Ock 44Wd 98
Dawley Pde. UB3: Hayes 45S 84
Dawley Pk. UB3: Hayes 47T 84
Dawley Ride SL3: Poyle 53G 104
Dawley Rd. UB3: Harl 45S 84
Dawley Rd. UB3: Hayes 45S 84
Dawlish Av. N13 21Nb 50
Dawlish Av. SW18 61Db 133
Dawlish Av. UB6: G'frd 40Ja 66
Dawlish Dr. HA4: Ruis 33W 64
Dawlish Dr. HA5: Pinn 29Aa 45
Dawlish Dr. IG3: Ilf 35Uc 74
Dawlish Rd. E10 32Ec 72
Dawlish Rd. N17 27Wb 51
Dawlish Rd. NW2 37Za 68
Dawlish Wlk. RM3: Rom 25Ld 57
Dawnay Gdns. SW18 61Fb 133
Dawnay Rd. KT23: Bookh 98Da 191
Dawnay Rd. SW18 61Eb 133
Dawn Cl. TW4: Houn 55Aa 107
Dawn Ct. AL1: St A 3E 6
Dawn Cres. E15 39Fc 73
Dawney Hill GU24: Pirb 3C 186
Dawneys Rd. GU24: Pirb 4C 186
Dawn Redwood Cl. SL3: Hort 55C 104
Dawpool Rd. NW2 33Va 68
Daws Ct. SL0: Iver 44H 83
Daws Hill E4 12Ec 34
Daws La. NW7 22Va 48
Dawson Av. BR5: St P 68Xc 139
Dawson Av. IG11: Bark 38Vc 75
Dawson Cl. SE18 49Sc 94
Dawson Cl. SL4: Wind 4E 102
Dawson Cl. UB3: Hayes 43T 84
Dawson Ct. W3 48Sa 87
(off Palmerston Rd.)
Dawson Dr. BR8: Hext 66Gd 140
Dawson Dr. RM13: Rain 38Kd 77
Dawson Gdns. IG11: Bark 38Vc 75
Dawson Ho. E2 41Yb 92
(off Sceptre Rd.)
Dawson Pl. W2 45Cb 89
Dawson Rd. KT1: King T 69Pa 131
Dawson Rd. KT14: Byfl 83M 169
Dawson Rd. NW2 36Ya 68
Dawson St. E2 2K 219 (40Vb 71)
Dawson Ter. N9 17Yb 34
Daws Pl. RH1: Mers 3C 208
Dax Ct. TW16: Sun 69Y 129
Daybrook Rd. SW19 68Db 133
Day Dr. RM8: Dag 32Zc 75
Day Ho. SE5 52Sb 113
(off Bethwin Rd.)
Daylesford Av. SW15 56Wa 110
Daylesford Gro. SL1: Slou 7D 80
Daylop Dr. IG7: Chig 20Xc 37
Daymer Gdns. HA5: Eastc 28X 45
Daymerslea Ridge KT22: Lea 93La 192
Daynor Ho. NW6 39Cb 69
(off Quex Ho.)
Days Acre CR2: Sande 82Vb 177
Daysbrook Rd. SW2 60Pb 112
Days La. CM15: Dodd 14Wd 40
Days La. CM15: Pil H 14Wd 40
Days La. DA15: Sidc 59Uc 116
Daytona Sandown Pk. 76Ea 152
Dayton Dr. DA8: Erith 50Md 97
Dayton Gro. SE15 53Yb 114
Deacon Cl. AL1: St A 6B 6
Deacon Cl. CR8: Purl 81Nb 176
Deacon Cl. KT11: D'side 91X 191
Deacon Ct. SL4: Wind 4B 102
Deaconess Ct. N15 28Vb 51
(off Tottenham Grn. E.)
Deacon Est., The E4 23Bc 52
Deacon Ho. SE11 6J 229 (49Pb 90)
(off Black Prince Rd.)
Deacon M. N1 38Tb 71
Deacon Pl. CR3: Cat'm 95Sb 197
Deacon Rd. KT2: King T 67Pa 131
Deacon Rd. NW2 36Wa 68
Deacons Cl. HA5: Pinn 26X 45
Deacons Cl. WD6: E'tree 14Qa 29
Deacons Ct. TW1: Twick 61Ha 130
Deaconsfield Rd. HP3: Hem H 5M 3
Deacons Hgts. WD6: E'tree 16Qa 29
Deacons Hill WD19: Wat 16Y 27
Deacons Hill WD6: E'tree 16Pa 29
Deacon's Hill Rd. WD6: E'tree 14Pa 29
Deacons Leas BR6: Orp 77Tc 160
Deacon's Ri. N2 29Fb 49
Deacons Ter. N1 37Sb 71
(off Harecourt Rd.)
Deacons Wlk. TW12: Hamp 63Ca 129
Deacon Way IG8: Wfd G 24Pc 54
Deadhearn La. HP8: Chal G 18A 24
Deadman's Ash La. WD3: Sarr 8K 11
Deakin Cl. WD18: Wat 17U 26
Deakins Ter. BR6: Orp 73Wc 161
Deal Av. SL1: Slou 4D 80
Deal Ct. NW9 26Va 48
(off Hazel Cl.)
Deal Ct. UB1: S'hall 44Ea 86
(off Haldane Rd.)
Deal Ho. SE15 51Zb 114
(off Lovelinch La.)
Deal Ho. SE17 7J 231 (50Ub 91)
(off Mina Rd.)
Deal M. W5 49Ma 87
Deal Porters Wlk. SE16 47Zb 92
Deal Porters Way SE16 48Yb 92
Deal Rd. SW17 65Jb 134
Deal's Gateway SE10 53Cc 114
Deal's Gateway SE13 53Cc 114
Deal St. E1 43Wb 91
Dealtry Rd. SW15 56Ya 110
Deal Wlk. SW9 52Qb 112
Dean Abbott Ho. SW1 5D 228 (49Mb 90)
(off Vincent St.)
Deanacre Cl. SL9: Chal P 23A 42
DEAN BOTTOM 69Xd 142
Dean Bradley St. SW1 4F 229 (48Nb 90)
Dean Cl. E9 36Yb 72
Dean Cl. GU22: Pyr 87G 168
Dean Cl. SE16 46Zb 92
Dean Cl. SL4: Wind 5B 102
Dean Cl. UB10: Hil 38P 63
Dean Ct. HA0: Wemb 34Ka 66
Dean Ct. HA8: Edg 23Ra 47
Dean Ct. RM7: Rom 29Fd 56
Dean Ct. SW8 52Nb 112
(off Thorncroft St.)
Dean Ct. W3 44Ta 87
Dean Ct. WD25: Wat 5Z 13
Deancroft Rd. SL9: Chal P 23A 42
Deancross St. E1 44Yb 92
Dean Dr. HA7: Stan 26Na 47
Deane Av. HA4: Ruis 36Y 65
Deane Ct. HA6: Nwood 25U 44
Deane Cft. Rd. HA5: Eastc 30Y 45
Deanery Cl. N2 28Gb 49
Deanery M. W1 6J 221 (46Jb 90)
(off Deanery St.)
Deanery Rd. E15 37Gc 73
Deanery St. W1 6J 221 (46Jb 90)
Deane Way HA4: Ruis 30X 45
Dean Farrar St. SW1 3E 228 (48Mb 90)
Dean Fld. HP3: Bov 9C 2
Deanfield Gdns. CR0: C'don 77Tb 157
Dean Gdns. E17 28Fc 53
Deanhill Ct. SW14 56Ra 109
Deanhill Rd. SW14 56Ra 109
Dean Ho. E1 44Yb 92
(off Tarling St.)
Dean Ho. SE14 52Ac 114
(off New Cross Rd.)
Dean La. RH1: Mers 95Kb 196
Dean Moore Cl. AL1: St A 3A 6
Dean Path RM8: Bark 35Wc 75
Dean Rd. CR0: C'don 77Tb 157
Dean Rd. NW2 37Ya 68
Dean Rd. SE28 46Wc 95
Dean Rd. TW12: Hamp 64Ca 129
Dean Rd. TW3: Houn 57Da 107
Dean Ryle St. SW1 5F 229 (49Nb 90)
Dean's Bldgs. SE17 6F 231 (49Tb 91)
Deansbrook Cl. HA8: Edg 24Sa 47
Deansbrook Rd. HA8: Edg 24Ra 47
Deans Cl. CR0: C'don 76Vb 157
Deans Cl. HA8: Edg 23Sa 47
Deans Cl. KT20: Walt H 96Xa 194
Deans Cl. SL2: Stoke P 9M 61
Deans Cl. W4 51Ra 109
Deans Cl. WD5: Ab L 4T 12
Dean's Ct. EC4 3C 224 (44Rb 91)
Deans Ct. GU20: W'sham 10B 146
Deanscroft Av. NW9 32Sa 67
Deans Dr. HA8: Edg 22Ta 47
Deans Dr. N13 23Rb 51
Deans Factory Est. RM13: Rain 42Ld 97
Deansfield CR3: Cat'm 97Vb 197
Deans Ga. Cl. SE23 62Zb 136
Deanshanger Ho. SE8 49Zb 92
(off Chilton Gro.)
Deans La. HA8: Edg 23Sa 47
Deans La. KT20: Walt H 96Xa 194
Deans La. RH1: Nutf 5G 208
Deans La. W4 51Ra 109
(off Deans Cl.)
Dean's M. W1 2A 222 (44Kb 90)
Deans Rd. CM14: W'ley 21Xd 58
Deans Rd. RH1: Mers 2C 208
Deans Rd. SM1: Sutt 76Db 155
Deans Rd. W7 46Ha 86
Dean Stanley St. SW1 4F 229 (48Nb 90)
Deanston Wharf E16 47Kc 93
(not continuous)
Dean St. E7 36Jc 73
Dean St. W1 2D 222 (44Mb 90)
Dean's Wlk. CR5: Coul 90Qb 176
Deans Way HA8: Edg 22Sa 47
Deansway HP3: Hem H 5P 3
Deansway N2 28Fb 49
Deansway N9 20Ub 33
Deanswood N11 23Mb 50
Dean's Yd. SW1 3E 228 (48Mb 90)
Dean Trench St. SW1 4F 229 (48Nb 90)
Dean Wlk. HA8: Edg 23Sa 47
Dean Wlk. KT23: Bookh 98Da 191
Dean Way UB2: S'hall 47Da 85
Dearmer Ho. SL9: Chal P 22A 42
(off Micholls Av.)
Dearne Cl. HA7: Stan 22Ja 46
De'Arn Gdns. CR4: Mitc 69Gb 133
Dearsley Ho. RM13: Rain 40Fd 76
Dearsley Rd. EN1: Enf 13Wb 33
Deauville Cl. E14 44Fc 93
Deauville Ct. SE16 47Zb 92
(off Eleanor Cl.)
Deauville Ct. SW4 58Lb 112
De Barowe M. N5 35Rb 71
Debdale Ho. E2 39Wb 71
(off Whiston Rd.)
DEBDEN 14Sc 36
Debden N17 26Tb 51
(off Gloucester Rd.)
Debden Cl. IG8: Wfd G 24Mc 53
Debden Cl. KT2: King T 64Ma 131
Debden Cl. NW9 25Ua 48
DEBDEN GREEN 10Rc 22
Debden Ho. IG10: Lough 10Rc 22
Debden La. IG10: Lough 10Sc 22
Debden Pl. UB10: Uxb 39N 63
Debden Rd. IG10: Lough 10Rc 22
Debden Station (Underground) 14Sc 36
Debden Wlk. RM12: Horn 37Kd 77
De Beauvoir Ct. N1 38Tb 71
(off Northchurch Rd.)
De Beauvoir Cres. N1 39Ub 71
De Beauvoir Est. N1 39Ub 71
De Beauvoir Pl. N1 37Ub 71
De Beauvoir Rd. N1 39Ub 71
De Beauvoir Sq. N1 38Ub 71
DE BEAUVOIR TOWN 39Ub 71
De Beauvoir Wharf N1 39Ub 71
(off Hertford Rd.)
Deben RM18: E Til 8L 101
Debenham Ct. E8 39Wb 71
(off Pownall Rd.)
Debenham Ct. EN5: Barn 15Ya 30
Debham Ct. NW2 34Ya 68
Deblin Dr. UB10: Uxb 40N 63
Debnams Rd. SE16 49Yb 92
De Bohun Av. N14 16Kb 32
Deborah Cl. TW7: Isle 53Ga 108
Deborah Ct. E18 27Kc 53
(off Victoria Rd.)
Deborah Cres. HA4: Ruis 31T 64
Deborah Lodge HA8: Edg 25Ra 47
Debrabant Cl. DA8: Erith 51Fd 118
De Brome Rd. TW13: Felt 60Y 107
De Bruin Ct. E14 50Ec 92
(off Ferry St.)
De Burgh Gdns. KT20: Tad 91Za 194
De Burgh Pk. SM7: Bans 87Db 175
Deburgh Rd. SW19 66Eb 133
Debussy NW9 26Va 48
Decapod St. E15 36Fc 73
Decies Way SL2: Stoke P 9L 61
Decima St. SE1 3H 231 (48Ub 91)
Decima Studios SE1 3H 231 (48Ub 91)
(off Decima St.)
Decimus Cl. CR7: Thor H 70Tb 135
Deck Cl. SE16 46Zb 92
De Clare Ct. RH1: Blet 5K 209
De Coubertin St. E20 37Ec 72
Decoy Av. NW11 29Ab 48
De Crespigny Pk. SE5 54Tb 113
Dedswell Dr. GU4: W Cla 100J 189
DEDWORTH 3C 102
Dedworth Dr. SL4: Wind 3D 102
DEDWORTH GREEN 4C 102
Dedworth Mnr. SL4: Wind 3D 102
Dedworth Rd. SL4: Wind 4A 102
Dee Cl. RM14: Upm 30Ud 58
Dee Ct. W7 44Fa 86
(off Hobbayne Rd.)
Deedsworth Ho. SL9: Ger X 28A 42
Dee Ho. KT2: King T 67Ma 131
(off May Bate Av.)
Deeley Rd. SW8 53Mb 112
Deena Cl. SL1: Slou 5C 80
Deena Cl. W3 44Pa 87
Deen City Farm 68Eb 133
Deepak Ho. SW17 63Gb 133
Deepdale SW19 63Za 132
Deepdale Av. BR2: Broml 70Hc 137
Deepdale Cl. N11 23Jb 50
Deepdale Ct. CR2: S Croy 78Tb 157
(off Birdhurst Av.)
Deep Dene W5 42Pa 87
Deepdene EN6: Pot B 3Za 16
Deepdene Av. CR0: C'don 76Vb 157
Deepdene Cl. E11 28Jc 53
Deepdene Ct. BR2: Broml 69Gc 137
Deepdene Ct. N21 16Rb 33
Deepdene Gdns. SW2 59Pb 112
Deepdene Mans. SW6 53Bb 111
(off Rostrevor Rd.)
Deepdene Path IG10: Lough 14Qc 36
(not continuous)
Deepdene Point SE23 62Zb 136
Deepdene Rd. DA16: Well 55Wc 117
Deepdene Rd. IG10: Lough 14Qc 36
Deepdene Rd. SE5 56Tb 113
Deep Fld. SL3: Dat 2M 103
Deepfield Way CR5: Coul 88Nb 176
Deep Pool La. GU24: Wok 6M 167
Deepwell Cl. TW7: Isle 53Ja 108
Deepwood La. UB6: G'frd 41Fa 86
Deerbrook Rd. SE24 60Rb 113
Deercote Ct. EN8: Chesh 2Zb 20
Deerdale Rd. SE24 56Sb 113
Deere Av. RM13: Rain 37Jd 76
Deerfield Cl. NW9 29Va 48
Deerfield Cotts. NW9 29Va 48
Deerhurst Cl. DA3: Lfield 69Ee 143
Deerhurst Cl. TW13: Felt 63X 129
Deerhurst Ct. CR0: C'don 74Rb 157
(off Parson's Mead)
Deerhurst Cres. TW12: Hamp H 64Ea 130
Deerhurst Ho. SE15 51Wb 113
(off Haymerle Rd.)
Deerhurst Rd. NW2 37Za 68
Deerhurst Rd. SW16 64Pb 134
Deering Ho. SE3 56Lc 115
Deerings Dr. HA5: Eastc 29W 44
Deerings Rd. RH2: Reig 6K 207
Deerleap Gro. E4 15Dc 34
Deerleap La. TN14: Knock 85Ad 181
Deer Mead Ct. RM1: Rom 29Hd 56
Dee Rd. SL4: Wind 2A 102
Dee Rd. TW9: Rich 56Pa 109
Deer Pk. Cl. KT2: King T 66Ra 131
Deer Pk. Gdns. CR4: Mitc 70Fb 133
Deer Pk. Rd. SW19 68Db 133
Deer Pk. Way BR4: W W'ck 75Hc 159
Deer Pk. Way EN9: Walt A 8Dc 20
Deers Farm Cl. GU23: Wis 88N 169
Deerswood Av. AL10: Hat 2D 8
Deerswood Cl. CR3: Cat'm 96Wb 197
Deeside Rd. SW17 62Fb 133
Dee St. E14 44Ec 92
Deeves Hall La. EN6: Ridge 5Ua 16
Dee Way KT19: Ewe 82Ua 174
Dee Way RM1: Rom 24Gd 56
Defence Cl. SE28 46Uc 94
Defiance Wlk. SE18 48Pc 94
Defiant Way SM6: W'gton 80Nb 156
Defoe Av. TW9: Kew 52Qa 109
Defoe Cl. DA8: Erith 53Gd 118
Defoe Cl. SE16 47Bc 92
Defoe Cl. SW17 65Gb 133
Defoe Ct. KT17: Eps 84Ua 174
Defoe Ho. EC2 7D 218 (43Sb 91)
(off Beech St.)
Defoe Pde. RM16: Grays 8D 100
Defoe Pl. EC2 7D 218 (43Sb 91)
(off Beech St.)
Defoe Pl. SW17 63Hb 133
Defoe Rd. N16 34Ub 71
Defoe Way RM5: Col R 23Cd 56
De Frene Rd. SE26 63Zb 136
Degema Rd. BR7: Chst 64Rc 138
Dehar Cres. NW9 31Va 68
de Havilland Aircraft Mus. 1Ra 15
Dehavilland Cl. UB5: N'olt 41Z 85
De Havilland Ct. N17 27Xb 51
De Havilland Ct. WD7: Shenl 4Na 15
De Havilland Dr. KT13: Weyb 83N 169
De Havilland Dr. SE18 51Rc 116
De Havilland Rd. HA8: Edg 26Qa 47
De Havilland Rd. TW5: Hest 52Y 107
Dehavilland Studios E5 33Yb 72
(off Theydon Rd.)
De Havilland Way TW19: Stanw 58M 105
De Havilland Way WD5: Ab L 4V 12
Deirdre Chapman Ho. DA10: Swans 58Ae 121
(off Craylands La.)
Dekker Cl. RM5: Col R 24Cd 56
Dekker Ho. SE5 52Tb 113
(off Elmington Est.)
Dekker Rd. SE21 58Ub 113
Dekota HA9: Wemb 35Qa 67
(off Engineers Way)
Delabole Rd. RH1: Mers 1E 208
Delacourt Rd. SE3 52Kc 115
Delacy Ct. SM2: Sutt 83Cb 175
Delafield Ho. E1 44Wb 91
(off Christian St.)
Delafield Rd. RM17: Grays 50Fe 99
Delafield Rd. SE7 50Kc 93
Delaford Cl. SL0: Iver 44J 83
Delaford Rd. SE16 50Xb 91
Delaford St. SW6 52Ab 110
Delagarde Rd. TN16: Westrm 98Sc 200
Delahay Ho. SW3 51Hb 111
(off Chelsea Emb.)
Delamare Ct. SE6 62Dc 136
Delamare Cres. CR0: C'don 72Yb 158
Delamare Rd. EN8: Chesh 2Ac 20
Delamere Cl. SS17: Stan H 3K 101
Delamere Ct. E17 26Ec 52
Delamere Gdns. NW7 23Ta 47
Delamere Rd. RH2: Reig 10K 207
Delamere Rd. SW20 67Za 132
Delamere Rd. UB4: Yead 45Z 85
Delamere Rd. W5 47Na 87
Delamere Rd. WD6: Bore 11Ra 29
Delamere St. W2 43Db 89
Delamere Ter. W2 43Db 89
Delancey Pas. NW1 1A 216 (39Kb 70)
(off Delancey St.)
Delancey St. NW1 1K 215 (39Kb 70)
Delancey Studios NW1 1A 216 (39Kb 70)
Delany Ho. SE10 51Ec 114
(off Thames St.)
Delaporte Cl. KT17: Eps 84Ua 174
De Lapre Cl. BR5: St M Cry 73Zc 161
De Lara Way GU21: Wok 10P 167
Delarch Ho. SE1 2B 230 (47Rb 91)
(off Webber Row)
Delargy Cl. RM16: Grays 8D 100
De Laune St. SE17 50Rb 91
Delaware Mans. W9 42Db 89
(off Delaware Rd.)
Delaware Rd. W9 42Db 89
Delawyk Cres. SE24 58Sb 113
Delcombe Av. KT4: Wor Pk 74Ya 154
Delderfield KT22: Lea 92Ma 193
Delderfield Ho. RM1: Rom 26Fd 56
(off Portnoi Cl.)
Delft Ho. KT2: King T 66Pa 131
(off Acre Rd.)
Delft Way SE22 57Ub 113
Delhi Rd. EN1: Enf 17Vb 33
Delhi St. N1 39Nb 70
Delia St. SW18 59Db 111
Delisle Rd. SE28 46Uc 94
Delius Cl. WD6: E'tree 16La 28
Delius Gro. E15 40Fc 73
Delius Way SS17: Stan H 1L 101
Dell, The AL1: St A 1E 6
Dell, The CM13: Gt War 23Xd 58
Dell, The DA5: Bexl 60Gd 118
Dell, The DA9: Ghithe 57Xd 120
Dell, The EN9: Walt A 8Ec 20
Dell, The GU21: Wok 1N 187
Dell, The HA0: Wemb 36Ka 66
Dell, The HA5: Pinn 26Z 45
Dell, The HA6: Nwood 19U 26
Dell, The IG8: Wfd G 20Kc 35
Dell, The KT20: Tad 93Ya 194
Dell, The RH2: Reig 5J 207
Dell, The SE19 67Vb 135
Dell, The SE2 50Wc 95
Dell, The SL9: Chal P 23A 42
Dell, The TW14: Felt 59X 107
Dell, The TW20: Eng G 2L 125
Dell, The TW8: Bford 51La 108
Dell, The WD7: R'lett 8Ja 14
Della Path E5 34Wb 71
Dellbow Rd. TW14: Felt 57X 107
Dell Cl. E15 39Fc 73
Dell Cl. IG8: Wfd G 20Kc 35
Dell Cl. KT22: Fet 95Fa 192
Dell Cl. RH5: Mick 99La 192
Dell Cl. SL2: Farn C 6G 60
Dell Cl. SM6: W'gton 77Lb 156
Dell Ct. HA6: Nwood 24T 44
Dell Ct. RM12: Horn 33Nd 77
Delle Gro. RM3: Rom 23Md 57
Dell Farm Rd. HA4: Ruis 29T 44
Dellfield AL1: St A 3D 6
Dellfield Cl. BR3: Beck 67Ec 136
Dellfield Cl. WD17: Wat 12W 26
Dellfield Cl. WD7: R'lett 7Ha 14
Dellfield Ct. WD17: Wat 12W 26
Dellfield Cres. UB8: Cowl 42M 83
Dellfield Pde. UB8: Cowl 42L 83
Dell La. KT17: Ewe 78Wa 154
Dell Mdw. HP3: Hem H 6N 3
Dellmeadow WD5: Ab L 2U 12
Dell Nature Reserve 12H 25
Dellors Cl. EN5: Barn 15Za 30
Dellow Cl. IG2: Ilf 31Tc 74
Dellow Ho. E1 45Xb 91
(off Dellow St.)
Dellow St. E1 45Xb 91
Dell Ri. AL2: Park 8P 5
Dell Rd. EN3: Enf W 10Yb 20
Dell Rd. KT17: Ewe 79Wa 154
Dell Rd. RM17: Grays 49De 99
Dell Rd. UB7: W Dray 49P 83
Dell Rd. WD24: Wat 9W 12
Dells, The HP3: Hem H 3B 4
Dells Cl. E4 17Dc 34
Dells Cl. TW11: Tedd 65Ha 130
Dell Side WD24: Wat 9W 12
Dellside UB9: Hare 29L 43
Dell's M. SW1 6C 228 (49Lb 90)
(off Churton Pl.)
Dellsome La. AL4: Col H 5A 8
Dellsome La. AL9: Wel G 5B 8
Dell View Rd. DA8: Erith 52Gd 118
Dell Wlk. KT3: N Mald 68Ua 132
Dell Way W13 44La 86
Dellwood WD3: Rick 18K 25
Dellwood Gdns. IG5: Ilf 27Qc 54
Delmar Av. HP2: Hem H 3D 4
Delmare Cl. SW9 56Pb 112
Delme Cres. SE3 54Kc 115
Delmer Ct. WD6: Bore 10Pa 15
(off Aycliffe Rd.)
Delmerend Ho. SW3 7D 226 (50Gb 89)
(off Cale St.)
Delmey Cl. CR0: C'don 76Vb 157
Deloraine Ho. SE8 53Cc 114
Delorme St. W6 51Za 110
Delphina Ho. CM14: B'wood 19Zd 41
DELROW 11Ea 28
Delroy Ct. N20 17Eb 31
Delta Bldg. E14 44Ec 92
(off Ashton St.)
Delta Bldg., The RM7: Rush G 30Gd 56
Delta Cen. HA0: Wemb 39Pa 67
Delta Cl. GU24: Chob 2K 167
Delta Cl. KT4: Wor Pk 76Va 154
Delta Ct. NW2 33Wa 68
Delta Ct. SE8 51Ac 114
(off Trundleys Rd.)
Delta Gain WD19: Wat 19Z 27
Delta Gro. UB5: N'olt 41Z 85
Delta Ho. KT16: Chert 73L 149
Delta Ho. N1 3F 219 (41Tb 91)
(off Nile St.)
Delta M. NW7 22Ab 48
Delta Pk. SW18 56Db 111
Delta Pk. Ind. Est. EN3: Brim 13Bc 34
Delta Point CR0: C'don 74Sb 157
(off Wellesley Rd.)
Delta Point E2 41Wb 91
(off Delta St.)
Delta Rd. CM13: Hut 16Fe 41
Delta Rd. GU21: Wok 88C 168
Delta Rd. GU24: Chob 2K 167
Delta Rd. KT4: Wor Pk 76Ua 154
Delta St. E2 41Wb 91
Delta Way TW20: Thorpe 67E 126
De Luci Rd. DA8: Erith 50Ed 96
De Lucy St. SE2 49Xc 95
Delvan Cl. SE18 52Qc 116
Delvers Mead RM10: Dag 35Ed 76
Delverton Ho. SE17 7C 230 (50Rb 91)
(off Delverton Rd.)
Delverton Rd. SE17 7C 230 (50Rb 91)
Delves KT20: Tad 93Za 194
Delves Cl. CR8: Purl 85Pb 176
Delvino Rd. SW6 53Cb 111
De Mel Cl. KT19: Eps 84Ra 173
DEMELZA HOSPICE CARE FOR CHILDREN 58Pc 116
Demesne Rd. SM6: W'gton 77Mb 156
Demeta Cl. HA9: Wemb 34Sa 67
De Montfort Pde. SW16 62Nb 134
De Montfort Rd. SW16 62Nb 134
De Morgan Rd. SW6 55Db 111
Dempsey Ct. SE6 59Cc 114
Dempster Cl. KT6: Surb 74La 152
Dempster Rd. SW18 57Eb 111
Den, The 50Yb 92
Denbar Pde. RM7: Rom 28Ed 56
Denberry Dr. DA14: Sidc 62Xc 139
Denbigh Cl. BR7: Chst 65Pc 138
Denbigh Cl. HA4: Ruis 33V 64
Denbigh Cl. HP2: Hem H 3N 3
Denbigh Cl. RM11: Horn 28Qd 57
Denbigh Cl. SM1: Sutt 78Bb 155
Denbigh Cl. UB1: S'hall 44Ba 85
Denbigh Cl. W11 45Bb 89
Denbigh Ct. E6 41Mc 93
Denbigh Ct. W7 43Ha 86
(off Copley Cl.)
Denbigh Dr. UB3: Harl 47S 84
Denbigh Gdns. TW10: Rich 57Pa 109
Denbigh Ho. RM3: Rom 23Nd 57
(off Kingsbridge Cir.)
Denbigh Ho. SW1 3G 227 (48Hb 89)
(off Hans Pl.)
Denbigh Ho. W11 45Bb 89
(off Westbourne Gro.)
Denbigh M. SW1 6B 228 (49Lb 90)
(off Denbigh St.)
Denbigh Pl. SW1 7B 228 (50Lb 90)
Denbigh Rd. E6 41Mc 93
Denbigh Rd. TW3: Houn 54Da 107
Denbigh Rd. UB1: S'hall 44Ba 85
Denbigh Rd. W11 45Bb 89
Denbigh Rd. W13 45Ka 86
Denbigh St. SW1 6B 228 (49Lb 90)
(not continuous)
Denbigh Ter. W11 45Bb 89
Denbridge Rd. BR1: Broml 68Pc 138
Denbury Ho. E3 41Dc 92
(off Talwin St.)
Denby Ct. SE11 5J 229 (49Pb 90)
(off Lambeth Wlk.)
Denby Rd. KT11: Cobh 84Y 171
Dence Ho. E2 41Wb 91
(off Turin St.)
Denchworth Ho. SW9 54Qb 112
Dencliffe TW15: Ashf 64Q 128
Den Cl. BR3: Beck 69Fc 137
Dencora Cen., The AL1: St A 2E 6
Dencora Cen., The EN3: Brim 13Ac 34
Dendridge Cl. EN1: Enf 9Xb 19
Dene, The CR0: C'don 77Zb 158
Dene, The HA9: Wemb 35Na 67
Dene, The KT8: W Mole 71Ba 151
Dene, The SM2: Cheam 83Bb 175

Dene, The TN13: S'oaks....98Kd **203**
Dene, The W13....43Ka **86**
Dene Av. DA15: Sidc....59Xc **117**
Dene Av. TW3: Houn....55Ba **107**
Dene Cl. BR2: Hayes....74Hc **159**
Dene Cl. CR5: Chips....91Gb **195**
Dene Cl. DA2: Wilm....63Gd **140**
Dene Cl. E10....33Dc **72**
Dene Cl. KT4: Wor Pk....75Va **154**
Dene Cl. SE4....55Ac **114**
Dene Ct. CR2: S Croy....78Sb **157**
(off Warham Rd.)
Dene Ct. W5....43La **86**
Denecroft Cres. UB10: Hil....39R **64**
Denecroft Gdns. RM17: Grays....48Fe **99**
Dene Dr. BR6: Chels....76Xc **161**
Dene Dr. DA3: Lfield....68De **143**
Denefield Dr. CR8: Kenley....87Tb **177**
Dene Gdns. HA7: Stan....22La **46**
Dene Gdns. KT7: T Ditt....75Ja **152**
Dene Holm Rd. DA11: Nflt....62Fe **143**
Dene Ho. N14....17Mb **32**
Denehurst Gdns. IG8: Wfd G....21Kc **53**
Denehurst Gdns. NW4....30Ya **48**
Denehurst Gdns. TW10: Rich....56Qa **109**
Denehurst Gdns. TW2: Twick....59Fa **108**
Denehurst Gdns. W3....46Ra **87**
Dene Lodge Cl. TN15: Bor G....92Be **205**
Dene Path RM15: S Ock....44Wd **98**
Dene Pl. GU21: Wok....10N **167**
Dene Rd. DA1: Dart....59Pd **119**
Dene Rd. HA6: Nwood....23S **44**
Dene Rd. IG9: Buck H....18Mc **35**
Dene Rd. KT21: Asht....91Pa **193**
Dene Rd. N11....18Hb **31**
Denes, The HP3: Hem H....6P **3**
Denesfield Ct. TN13: Chips....95Dd **202**
Denesmead SE24....57Sb **113**
Dene Wlk. DA3: Lfield....69Ae **143**
Denewood EN5: New Bar....15Eb **31**
Denewood KT17: Eps....85Ua **174**
Denewood Cl. WD17: Wat....9V **12**
Denewood M. WD17: Wat....9V **12**
Denewood Pl. HA6: Nwood....23T **44**
Denewood Rd. N6....30Hb **49**
Denford St. SE10....50Hc **93**
(off Glenforth St.)
Dengie Wlk. N1....39Sb **71**
(off Basire St.)
DENHAM....34H 63
Denham Aerodrome....29G **42**
Denham Av. UB9: Den....33H **63**
Denham Cl. DA16: Well....55Yc **117**
Denham Cl. UB9: Den....34J **63**
Denham Country Pk.....32K **63**
Denham Ct. NW6....38Eb **69**
(off Fairfax Rd.)
Denham Ct. SE26....62Xb **135**
(off Kirkdale)
Denham Ct. UB1: S'hall....45Ea **86**
(off Baird Av.)
Denham Ct. Dr. UB9: Den....35J **63**
Denham Cres. CR4: Mitc....70Hb **133**
Denham Dr. IG2: Ilf....30Sc **54**
DENHAM GARDEN VILLAGE....29H 43
Denham Golf Club Station (Rail)....31F 62
Denham Golf Course....30F **42**
DENHAM GREEN....30H 43
Denham Grn. Cl. UB9: Den....31J **63**
Denham Grn. La. UB9: Den....29G **42**
Denham Ho. UB7: W Dray....47P **83**
(off Park Lodge Av.)
Denham Ho. W12....45Xa **88**
(off White City Est.)
Denham La. SL9: Chal P....23B **42**
Denham Lodge UB9: Den....37L **63**
Denham Pde. UB9: Den....34H **63**
(off Oxford Gdns.)
Denham Pl. UB9: Den....32H **63**
Denham Rd. KT17: Eps....84Va **174**
Denham Rd. N20....20Hb **31**
Denham Rd. SL0: Iver H....39F **62**
Denham Rd. TW14: Felt....59Y **107**
Denham Rd. TW20: Egh....63C **126**
Denham Rd. UB9: Den....36H **63**
DENHAM RDBT.....35J **63**
Denham Station (Rail)....31J 63
Denham St. SE10....50Jc **93**
Denham Wlk. SL9: Chal P....23B **42**
Denham Way UB9: Den Old Mill Rd.....34J **63**
Denham Way UB9: Den Wyatt's Covert....27H **43**
Denham Way IG11: Bark....39Uc **74**
Denham Way WD3: Map C....23G **42**
Denham Way WD3: W Hyd....23G **42**
Denham Way WD6: Bore....11Sa **29**
Denholme Rd. W9....41Bb **89**
Denholme Wlk. RM13: Rain....37Hd **76**
Denison Cl. N2....27Eb **49**
Denison Ho. E14....48Dc **92**
Denison Rd. SW19....65Fb **133**
Denison Rd. TW13: Felt....63Y **128**
Denison Rd. W5....42La **86**
Deniston Av. DA5: Bexl....60Ad **117**
Denis Way SW4....55Mb **112**
Denland Ho. SW8....52Pb **112**
(off Dorset Rd.)
Denleigh Gdns. KT7: T Ditt....72Ga **152**
Denleigh Gdns. N21....18Qb **32**
Denley Sq. UB8: Uxb....38M **63**
Denly Way GU18: Light....2A **166**
Denman Av. UB2: S'hall....46Fa **86**
Denman Dr. KT10: Clay....78Ja **152**
Denman Dr. NW11....29Cb **49**
Denman Dr. TW15: Ashf....65R **128**
Denman Dr. Nth. NW11....29Cb **49**
Denman Dr. Sth. NW11....29Cb **49**
Denman Ho. N16....33Ub **71**
Denman Pl. W1....4D **222** (45Mb **90**)
(off Denman St.)
Denman Rd. SE15....53Vb **113**
Denman St. W1....5D **222** (45Mb **90**)
Denmark Av. SW19....66Ab **132**
Denmark Ct. KT13: Weyb....76R **150**
(off Grotto Rd.)
Denmark Ct. SM4: Mord....72Cb **155**
Denmark Gdns. SM5: Cars....76Hb **155**
Denmark Gro. N1....1K **217** (40Qb **70**)
Denmark Hill SE5....53Tb **113**
DENMARK HILL....55Sb 113
Denmark Hill Dr. NW9....27Wa **48**
Denmark Hill Est. SE5....56Tb **113**
Denmark Hill Station (Rail & Overground)....54Tb 113
Denmark Ho. SE7....49Nc **94**
Denmark Lodge RM3: Hrld W....25Nd **57**
Denmark Mans. SE5....54Sb **113**
(off Coldharbour La.)
Denmark Path SE25....71Xb **157**
Denmark Pl. E3....41Cc **92**
Denmark Pl. WC2....2E **222** (44Mb **90**)
Denmark Rd. BR1: Broml....67Kc **137**
Denmark Rd. KT1: King T....69Na **131**
Denmark Rd. N8....28Qb **50**
Denmark Rd. NW6....40Bb **69**
Denmark Rd. SE25....71Wb **157**
Denmark Rd. SE5....53Sb **113**
Denmark Rd. SM5: Cars....76Hb **155**
Denmark Rd. SW19....65Za **132**
Denmark Rd. TW2: Twick....62Fa **130**
Denmark Rd. W13....45Ka **86**
Denmark St. E11....34Gc **73**
Denmark St. E13....43Kc **93**
Denmark St. N17....25Xb **51**
Denmark St. WC2....3E **222** (44Mb **90**)
Denmark St. WD17: Wat....12X **27**
Denmark Ter. N2....27Hb **49**
Denmark Wlk. SE27....63Sb **135**
Denmead Cl. SL9: Ger X....31A **62**
Denmead Ho. SW15....58Va **110**
Denmead Rd. CR0: C'don....74Rb **157**
Denmead Way SE15....52Vb **113**
Denmore Ct. SM6: W'gton....78Kb **156**
Dennan Rd. KT6: Surb....74Pa **153**
Dennard Way BR6: Farnb....77Rc **160**
Denner Rd. E4....19Cc **34**
Dene Ter. E8....39Vb **71**
Dennett Rd. CR0: C'don....74Qb **156**
Dennett's Gro. SE14....54Zb **114**
Dennett's Rd. SE14....53Yb **114**
Denning Av. CR0: Wadd....77Qb **156**
Denning Cl. NW8....3A **214** (41Eb **89**)
Denning Cl. TW12: Hamp....64Ba **129**
Denning M. SW12....58Jb **112**
Denning Point E1....2K **225** (44Vb **91**)
(off Commercial St.)
Denning Rd. NW3....35Fb **69**
Dennington Cl. E5....33Yb **72**
Dennington Pk. Rd. NW6....37Cb **69**
Denningtons, The KT4: Wor Pk....75Ua **154**
Dennis Av. HA9: Wemb....36Pa **67**
Dennis Cl. RH1: Redh....4N **207**
Dennis Cl. TW15: Ashf....66T **128**
Dennis Ct. AL3: St A....1B **6**
Dennises La. RM14: Upm....39Ud **78**
Dennises La. RM15: S Ock....39Wd **78**
Dennis Gdns. HA7: Stan....22La **46**
Dennis Ho. E3....40Bc **72**
(off Roman Rd.)
Dennis Ho. SM1: Sutt....77Db **155**
Dennis La. HA7: Stan....20Ka **28**
Dennison Point E15....38Ec **72**
Dennis Pde. N14....18Mb **32**
Dennis Pk. Cres. SW20....67Ab **132**
Dennis Reeve Cl. CR4: Mitc....67Hb **133**
Dennis Rd. DA11: Grav'nd....2C **144**
Dennis Rd. KT8: E Mos....70Ea **130**
Dennis Rd. RM15: S Ock....38Wd **78**
Dennis Severs' House....7J **219** (43Ub **91**)
Dennis Way SL1: Slou....5B **80**
Dennis Way SW4....55Mb **112**
Denny Av. EN9: Walt A....6Fc **21**
Denny Cl. E6....43Nc **94**
Denny Ct. DA2: Dart....58Rd **119**
(off Bow Arrow La.)
Denny Cres. SE11....7A **230** (50Qb **90**)
Denny Gdns. RM9: Dag....38Xc **75**
Denny Rd. N9....18Xb **33**
Denny Rd. SL3: L'ly....49B **82**
Denny St. SE11....7A **230** (50Qb **90**)
De Novo Pl. AL1: St A....2D **6**
(off Stanhope Rd.)
Den Rd. BR2: Broml....69Fc **137**
Densham Dr. CR8: Purl....86Qb **176**
Densham Ho. NW8....2C **214** (40Fb **69**)
(off Cochrane St.)
Densham Rd. E15....39Gc **73**
Densole Cl. BR3: Beck....67Ac **136**
Denstone Ho. SE15....51Wb **113**
(off Haymerle Rd.)
Densworth Gro. N9....19Yb **34**
Dent Cl. RM15: S Ock....44Wd **98**
Dent Ho. SE17....6H **231** (49Ub **91**)
(off Peacock St.)
DENTON....9G 122
Denton NW1....37Jb **70**
Denton Cl. EN5: Barn....15Ya **30**
Denton Cl. RH1: Redh....10A **208**
Denton Ct. Rd. DA12: Grav'nd....9G **122**
Denton Gro. KT12: Walt T....75Aa **151**
Denton Ho. N1....38Rb **71**
(off Halton Rd.)
Denton Ho. WD19: Wat....20Y **27**
Denton Rd. DA1: Dart....59Gd **118**
Denton Rd. DA16: Well....52Yc **117**
Denton Rd. DA5: Bexl....61Gd **140**
Denton Rd. N18....21Ub **51**
Denton Rd. N8....29Pb **50**
Denton Rd. TW1: Twick....58Ma **109**
Denton St. DA12: Grav'nd....9G **122**
Denton St. SW18....58Db **111**
Denton Ter. DA5: Bexl....61Gd **140**
Denton Way E5....34Zb **72**
Denton Way GU21: Wok....9K **167**
Denton Way SL3: Slou....7M **81**
Denton Wharf DA12: Grav'nd....8G **122**
Dents Gro. KT20: Lwr K....100Bb **195**
Dents Rd. SW11....58Hb **111**
Denvale Wlk. GU21: Wok....10L **167**
Denver Cl. BR6: Pet W....72Uc **160**
Denver Ind. Est. RM13: Rain....43Hd **96**
Denver Rd. DA1: Dart....59Jd **118**
Denver Rd. N16....31Ub **71**
Denwood SE23....62Zb **136**
Denyer St. SW3....6E **226** (49Gb **89**)
Denys Ho. EC1....7K **217** (43Qb **90**)
(off Bourne Est.)
Denziloe Av. UB10: Hil....41R **84**
Denzil Rd. NW10....36Va **68**
Deodar Rd. SW15....56Ab **110**
Deodora Cl. N20....20Gb **31**
Department for Business, Energy & Industrial Strategy....3E **228** (48Mb **90**)
Department for Communities....3B **228** (48Lb **90**)
(off Bressenden Pl.)
Department for Education....3E **228** (48Mb **90**)
(off Gt. Smith St.)
Department for Energy & Climate Change....7F **223** (46Nb **90**)
(off Whitehall Pl.)
Department for Transport....5E **228** (49Mb **90**)
Department for Work & Pensions....2E **228** (47Mb **90**)
(off Tothill St.)
Department of Environment
Parks Department....4J **215** (41Jb **90**)
De Pass Gdns. IG11: Bark....42Wc **95**
De Paul Way CM14: B'wood....18Xd **40**
Depot App. NW10: Wemb....37Qa **67**
Depot App. NW2....35Za **68**
Depot Rd. KT17: Eps....85Ua **174**
Depot Rd. TW3: Houn....55Fa **108**
Depot Rd. W12....45Ya **88**
Depot St. SE5....51Tb **113**
DEPTFORD....52Cc 114
Deptford Bri. SE8....53Cc **114**
Deptford Bridge Station (DLR)....53Cc 114
Deptford B'way. SE8....53Cc **114**
Deptford Bus. Pk. SE15....51Yb **114**
Deptford Chu. St. SE8....51Cc **114**
Deptford Ferry Rd. E14....49Cc **92**
Deptford Grn. SE8....51Cc **114**
Deptford High St. SE8....51Cc **114**
Deptford Pk. Bus. Cen. SE8....50Ac **92**
Deptford Station (Rail)....52Cc 114
Deptford Strand SE8....49Bc **92**
Deptford Trad. Est. SE8....51Ac **114**
Deptford Wharf SE8....49Bc **92**
De Quincey Ho. SW1....7B **228** (50Lb **90**)
(off Lupus St.)
De Quincey M. E16....46Jc **93**
De Quincey Rd. N17....25Tb **51**
Derby Arms Rd. KT18: Eps D....89Va **174**
Derby Av. HA3: Hrw W....25Fa **46**
Derby Av. N12....22Eb **49**
Derby Av. RM14: Upm....34Pd **77**
Derby Av. RM7: Rom....30Ed **56**
Derby Cl. KT18: Tatt C....91Xa **194**
Derby Ga. SW1....1F **229** (47Nb **90**)
(not continuous)
Derby Hill SE23....61Yb **136**
Derby Hill Cres. SE23....61Yb **136**
Derby Ho. HA5: Pinn....26Z **45**
Derby Ho. SE11....5K **229** (49Qb **90**)
(off Walnut Tree Wlk.)
Derby Ho. WD23: Bush....14Ca **27**
Derby Lodge N3....26Bb **49**
Derby Lodge WC1....3H **217** (41Pb **90**)
(off Britannia St.)
Derby Rd. CR0: C'don....74Rb **157**
Derby Rd. E18....25Hc **53**
Derby Rd. E7....38Mc **73**
Derby Rd. E9....39Zb **72**
Derby Rd. EN3: Pond E....15Xb **33**
Derby Rd. KT5: Surb....74Qa **153**
Derby Rd. N18....22Yb **52**
Derby Rd. RM17: Grays....50De **99**
Derby Rd. SM1: Sutt....79Bb **155**
Derby Rd. SW14....56Ra **109**
Derby Rd. SW19....66Cb **133**
Derby Rd. TW3: Houn....56Da **107**
Derby Rd. UB6: G'frd....39Da **65**
Derby Rd. UB8: Uxb....40L **63**
Derby Rd. WD17: Wat....13Y **27**
(not continuous)
Derby Rd. Bri.....51De **121**
Derby Rd. Ind. Est. TW3: Houn....56Da **107**
Derbyshire St. E2....41Wb **91**
(not continuous)
Derby Sq. KT19: Eps....85Ta **173**
Derby Stables Rd. KT18: Eps D....89Va **174**
Derby St. W1....7J **221** (46Jb **90**)
Dere Cl. SW6....53Ab **110**
Dereham Ho. SE4....56Zb **114**
(off Frendsbury Rd.)
Dereham Pl. EC2....4J **219** (41Ub **91**)
Dereham Pl. RM5: Col R....23Dd **56**
Dereham Rd. IG11: Bark....36Vc **75**
Derek Av. HA9: Wemb....38Ra **67**
Derek Av. KT19: Ewe....79Qa **153**
Derek Av. SM6: W'gton....77Kb **156**
Derek Cl. KT19: Ewe....78Ra **153**
Derek Walcott Cl. SE24....57Rb **113**
Derham Gdns. RM14: Upm....34Sd **78**
Deri Av. RM13: Rain....42Kd **97**
Dericote St. E8....39Xb **71**
Deridene Cl. TW19: Stanw....58N **105**
Derifall Cl. E6....43Pc **94**
Dering Pl. CR0: C'don....77Sb **157**
Dering Rd. CR0: C'don....77Sb **157**
Dering St. W1....3A **222** (44Kb **90**)
Dering Way DA12: Grav'nd....9H **123**
Dering Yd. W1....3A **222** (44Kb **90**)
Derinton Rd. SW17....63Hb **133**
Derisley Cl. KT14: Byfl....84M **169**
Derley Rd. UB2: S'hall....48Y **85**
Dermody Gdns. SE13....57Fc **115**
Dermody Rd. SE13....57Fc **115**
Derny Av. E20....36Dc **72**
Deronda Rd. SE24....60Rb **113**
De Ros Pl. TW20: Egh....65C **126**
Deroy Cl. SM5: Cars....79Hb **155**
Derrick Av. CR2: Sande....82Sb **177**
Derrick Gdns. SE7....48Lc **93**
Derrick Rd. BR3: Beck....69Bc **136**
Derry Av. RM15: S Ock....44Wd **98**
Derrycombe Ho. W2....43Cb **89**
(off Gt. Western Rd.)
Derrydown GU22: Wok....3N **187**
Derry Downs BR5: St M Cry....72Yc **161**
DERRY DOWNS....72Yc 161
Derry Ho. NW8....6C **214** (42Fb **89**)
(off Church St. Est.)
Derry M. N19....33Nb **70**
Derry Rd. CR0: Bedd....76Nb **156**
Derry St. W8....47Db **89**
Dersingham Av. E12....35Pc **74**
Dersingham Rd. NW2....34Ab **68**
Derwent NW1....4B **216** (41Lb **90**)
(off Robert St.)
Derwent Av. EN4: E Barn....18Hb **31**
Derwent Av. HA5: Hat E....23Aa **45**
Derwent Av. N18....22Tb **51**
Derwent Av. NW7....23Ta **47**
Derwent Av. NW9....29Ua **48**
Derwent Av. SW15....63Ua **132**
Derwent Av. UB10: Ick....33Q **64**
Derwent Cl. DA1: Dart....60Kd **119**
Derwent Cl. KT10: Clay....79Ga **152**
Derwent Cl. KT15: Add....78M **149**
Derwent Cl. TW14: Felt....60V **106**
Derwent Cl. WD25: Wat....6Y **13**
Derwent Ct. SE16....47Zb **92**
(off Eleanor Cl.)
Derwent Cres. DA7: Bex....54Cd **118**
Derwent Cres. HA7: Stan....26La **46**
Derwent Cres. N20....20Eb **31**
Derwent Dr. BR5: Pet W....73Tc **160**
Derwent Dr. CR8: Purl....85Tb **177**
Derwent Dr. SL1: Slou....3A **80**
Derwent Dr. UB4: Hayes....43U **84**
Derwent Gdns. HA9: Wemb....31La **66**
Derwent Gdns. IG4: Ilf....28Nc **54**
Derwent Gro. SE22....56Vb **113**
Derwent Ho. E3....42Bc **92**
(off Southern Gro.)
Derwent Ho. KT2: King T....67Ma **131**
(off May Bate Av.)
Derwent Ho. SW7....5A **226** (49Eb **89**)
(off Cromwell Rd.)
Derwent Lodge KT4: Wor Pk....75Xa **154**
Derwent Lodge TW7: Isle....54Fa **108**
Derwent Pde. RM15: S Ock....44Wd **98**
Derwent Point EC1....3B **218** (41Rb **91**)
(off Goswell Rd.)
Derwent Ri. NW9....30Ua **48**
Derwent Rd. GU18: Light....3A **166**
Derwent Rd. HP3: Hem H....3C **4**
Derwent Rd. N13....21Pb **50**
Derwent Rd. SE20....68Wb **135**
Derwent Rd. SW20....72Za **154**
Derwent Rd. TW2: Whitt....58Da **107**
Derwent Rd. TW20: Egh....66D **126**
Derwent Rd. UB1: S'hall....44Ba **85**
Derwent Rd. W5....48La **86**
Derwent St. SE10....50Gc **93**
Derwent Wlk. SM6: W'gton....80Kb **156**
Derwentwater Rd. W3....46Sa **87**
Derwent Way RM12: Horn....36Kd **77**
Derwent Yd. W5....48La **86**
(off Derwent Rd.)
De Salis Rd. UB10: Hil....42S **84**
Des Barres Ct. SE10....48Jc **93**
(off Peartree Way)
Desborough Cl. TW17: Shep....73Q **150**
Desborough Cl. W2....43Db **89**
Desborough Ho. W14....51Bb **111**
(off North End Rd.)
Desborough Sailing Club....73R **150**
Desborough St. W2....43Db **89**
(off Cirencester St.)
Desenfans Rd. SE21....58Ub **113**
Deseronto Trad. Est. SL3: L'ly....47A **82**
Desford Rd. E16....42Gc **93**
Desford Way TW15: Ashf....61P **127**
Design Mus.....48Bb **89**
Deslandes Hgts. CR8: Purl....85Pb **176**
Desmond Ho. EN4: E Barn....16Gb **31**
Desmond Rd. WD24: Wat....8V **12**
Desmond St. SE14....51Ac **114**
Desmond Tutu Dr. SE23....60Bc **114**
Despard Rd. N19....32Lb **70**
de Stafford Sports Cen.....93Vb **197**
Desvignes Dr. SE13....58Fc **115**
De Tany Ct. AL1: St A....3B **6**
Dethick Ct. E3....39Ac **72**
Detillens La. RH8: Limp....1L **211**
Detling Cl. RM12: Horn....36Ld **77**
Detling Ho. SE17....6H **231** (49Ub **91**)
(off Congreve St.)
Detling Rd. BR1: Broml....64Jc **137**
Detling Rd. DA11: Nflt....60Fe **121**
Detling Rd. DA8: Erith....52Fd **118**
Detmold Rd. E5....33Yb **72**
Dettingen Pl. IG11: Bark....40Xc **75**
Deva Cl. AL3: St A....4N **5**
Devalls Cl. E6....45Rc **94**
Devana End SM5: Cars....76Hb **155**
Devane Way SE27....62Rb **135**
Devan Gro. N4....31Tb **71**
Devas Rd. SW20....67Ya **132**
Devas St. E3....42Dc **92**
Devenay Rd. E15....38Hc **73**
Devenish La. SL5: S'dale....4C **146**
Devenish Rd. SE2....47Wc **95**
Devenish Rd. SL5: S'dale....2A **146**
Devenish Rd. SL5: S'hill....2A **146**
Deventer Cres. SE22....57Ub **113**
Deveraux Cl. BR3: Beck....71Ec **158**
De Vere Cl. SM6: W'gton....80Nb **156**
De Vere Cotts. W8....48Eb **89**
(off Canning Pl.)
De Vere Gdns. IG1: Ilf....33Pc **74**
De Vere Gdns. W8....2A **226** (47Eb **89**)
De Vere Leisure Club Denham....27G **42**
Deverell St. SE1....4F **231** (48Tb **91**)
De Vere M. W8....3A **226** (48Eb **89**)
(off Canning Pl.)
Devereux Ct. WC2....3K **223** (44Qb **90**)
(off Essex St.)
Devereux Dr. WD17: Wat....10U **12**
Devereux La. SW13....52Xa **110**
Devereux Rd. RM16: Chaf H....48Be **99**
Devereux Rd. SL4: Wind....4H **103**
Devereux Rd. SW11....58Hb **111**
De Vere Wlk. WD17: Wat....12U **26**
Deverills Way SL3: L'ly....49E **82**
Deveron Gdns. RM15: S Ock....43Wd **98**
Deveron Way RM1: Rom....25Gd **56**
De Vesci Ct. TW20: Eng G....2N **125**
Devey Cl. KT2: King T....66Va **132**
Devil's La. TW18: Staines....66F **126**
Devil's La. TW20: Egh....65E **126**
Devitt Cl. KT21: Asht....88Qa **173**
Devitt Ho. E14....45Dc **92**
(off Wade's Pl.)
Devizes Ho. RM3: Rom....22Md **57**
(off Montgomery Cres.)
Devizes St. N1....39Tb **71**
Devoke Way KT12: Walt T....75Z **151**
Devon Av. SL1: Slou....4G **80**
Devon Av. TW2: Twick....60Ea **108**
Devon Cl. CR8: Kenley....88Vb **177**
Devon Cl. IG9: Buck H....19Kc **35**
Devon Cl. N17....27Vb **51**
Devon Cl. UB6: G'frd....39La **66**
Devon Ct. AL1: St A....3C **6**
Devon Ct. DA4: Sut H....67Rd **141**
Devon Ct. KT18: Eps....86Ua **174**
(off St Martin's Av.)
Devon Ct. TW12: Hamp....66Ca **129**
Devon Ct. W7....43Ha **86**
(off Copley Cl.)
Devon Cres. RH1: Redh....6M **207**
Devoncroft Gdns. TW1: Twick....59Ja **108**
Devon Gdns. N4....30Rb **51**
Devon Ho. CR3: Cat'm....96Vb **197**
Devon Ho. E17....26Bc **52**
Devon Ho. N1....1B **218** (40Rb **71**)
(off Upper St.)
Devonhurst Pl. W4....50Ta **87**
Devonia Gdns. N18....23Sb **51**
Devonia Rd. N1....1C **218** (40Rb **71**)
Devon Mans. HA3: Kenton....29La **46**
(off Woodcock Hill)
Devon Mans. SE1....1K **231** (47Vb **91**)
(off Tooley St.)
Devon Pde. HA3: Kenton....29La **46**
Devonport W2....3D **220** (44Gb **89**)
Devonport Gdns. IG1: Ilf....30Pc **54**
Devonport Ho. W2....43Cb **89**
(off Gt. Western Rd.)
Devonport M. W12....47Xa **88**
Devonport Rd. W12....46Xa **88**
(not continuous)
Devonport St. E1....44Yb **92**
Devon Ri. N2....28Fb **49**
Devon Rd. DA4: S Dar....67Rd **141**
Devon Rd. DA4: Sut H....67Rd **141**
Devon Rd. IG11: Bark....39Uc **74**
Devon Rd. KT12: Hers....77Y **151**
Devon Rd. RH1: Mers....2C **208**
Devon Rd. SM2: Cheam....81Ab **174**
Devon Rd. WD24: Wat....11Z **27**
Devons Est. E3....41Dc **92**
Devonshire Av. DA1: Dart....58Kd **119**
Devonshire Av. GU21: Wok....8E **168**
Devonshire Av. SM2: Sutt....80Eb **155**
Devonshire Bus. Cen. EN6: Pot B....2Ab **16**
Devonshire Bus. Pk. WD6: Bore....13Ta **29**
Devonshire Cl. E15....35Gc **73**
Devonshire Cl. N13....20Qb **32**
Devonshire Cl. SL2: Farn R....10F **60**
Devonshire Cl. W1....7K **215** (43Kb **90**)
Devonshire Ct. E1....41Yb **92**
(off Bancroft Rd.)
Devonshire Ct. HA5: Hat E....25Ba **45**
(off Devonshire Rd.)
Devonshire Ct. N17....23Sb **51**
Devonshire Ct. TW13: Felt....61X **129**
Devonshire Ct. WC1....7G **217** (43Nb **90**)
(off Boswell St.)
Devonshire Cres. NW7....24Za **48**
Devonshire Dr. KT6: Surb....74Ma **153**
Devonshire Dr. SE10....52Dc **114**
Devonshire Gdns. N17....23Sb **51**
Devonshire Gdns. N21....17Sb **33**
Devonshire Gdns. SS17: Linf....8J **101**
Devonshire Gdns. W4....52Sa **109**
Devonshire Grn. SL2: Farn R....10F **60**
Devonshire Gro. SE15....51Xb **113**
Devonshire Hall E9....37Yb **72**
(off Frampton Pk. Rd.)
Devonshire Hill La. N17....23Rb **51**
Devonshire Ho. E14....49Cc **92**
(off Westferry Rd.)
Devonshire Ho. IG8: Wfd G....24Rc **54**
Devonshire Ho. NW6....37Bb **69**
(off Kilburn High Rd.)
Devonshire Ho. SE1....3D **230** (48Sb **91**)
(off Bath Ter.)
Devonshire Ho. SM2: Sutt....80Eb **155**
Devonshire Ho. SW1....7E **228** (50Mb **90**)
(off Lindsay Sq.)
Devonshire Ho. SW15....57Za **110**
Devonshire Ho. W2....7C **214** (43Fb **89**)
(off Adpar St.)
Devonshire Ho. WD23: Bush....14Ba **27**
Devonshire Ho. Bus. Cen. BR2: Broml....70Kc **137**
(off Devonshire Sq.)
Devonshire M. N13....21Qb **50**
Devonshire M. SW10....51Fb **111**
(off Park Wlk.)
Devonshire M. W4....50Ua **88**
Devonshire M. Nth. W1....7K **215** (43Kb **90**)
Devonshire M. Sth. W1....7K **215** (43Kb **90**)
Devonshire M. W. W1....6J **215** (42Jb **90**)
Devonshire Pas. W4....50Ua **88**
Devonshire Pl. NW2....34Cb **69**
Devonshire Pl. W1....6J **215** (42Jb **90**)
Devonshire Pl. W8....48Db **89**
Devonshire Pl. M. W1....7J **215** (43Jb **90**)
Devonshire Point TW15: Ashf....62S **128**
Devonshire Rd. BR6: Orp....73Wc **161**
Devonshire Rd. CR0: C'don....73Tb **157**
Devonshire Rd. DA12: Grav'nd....10D **122**
Devonshire Rd. DA6: Bex....56Ad **117**
Devonshire Rd. E16....44Kc **93**
Devonshire Rd. E17....30Cc **52**
Devonshire Rd. HA1: Harr....30Fa **46**
Devonshire Rd. HA5: Eastc....30Y **45**
Devonshire Rd. HA5: Hat E....25Ba **45**
Devonshire Rd. IG2: Ilf....31Uc **74**
Devonshire Rd. KT13: Weyb....77Q **150**
Devonshire Rd. N13....21Pb **50**
Devonshire Rd. N17....23Sb **51**
Devonshire Rd. N9....18Yb **34**
Devonshire Rd. NW7....24Za **48**
Devonshire Rd. RM12: Horn....33Ld **77**
Devonshire Rd. RM16: Chaf H....50Ae **99**
Devonshire Rd. RM16: Grays....50Ae **99**
Devonshire Rd. SE23....60Yb **114**
Devonshire Rd. SE9....61Nc **138**
Devonshire Rd. SM2: Sutt....80Eb **155**
Devonshire Rd. SM5: Cars....77Jb **156**
Devonshire Rd. SW19....66Gb **133**
Devonshire Rd. TW13: Hanw....62Aa **129**
Devonshire Rd. UB1: S'hall....43Ca **85**
Devonshire Rd. W4....50Ua **88**
Devonshire Rd. W5....48La **86**
Devonshire Road Nature Reserve....59Zb **114**
Devonshire Road Nature Reserve Vis. Cen.....59Zb **114**
Devonshire Row EC2....1J **225** (43Ub **91**)
Devonshire Row M. W1....6A **216** (42Kb **90**)
(off Devonshire St.)
Devonshires, The KT17: Eps....86Va **174**
Devonshire Sq. BR2: Broml....70Kc **137**
Devonshire Sq. EC2....2J **225** (44Ub **91**)
Devonshire St. W1....7J **215** (43Jb **90**)
Devonshire St. W4....50Ua **88**
Devonshire Ter. W2....3A **220** (44Eb **89**)
Devonshire Way CR0: C'don....75Ac **158**
Devonshire Way UB4: Yead....44X **85**
Devons Rd. E3....41Dc **92**
Devons Road Station (DLR)....42Dc 92
Devon St. SE15....51Xb **113**
Devon Way KT19: Ewe....78Ra **153**
Devon Way KT9: Chess....78La **152**
Devon Way UB10: Hil....40P **63**
Devon Waye TW5: Hest....52Ba **107**
De Walden Ho. NW8....1D **214** (40Gb **69**)
(off Allitsen Rd.)
De Walden St. W1....1J **221** (43Jb **90**)
Dewar Spur SL3: L'ly....51B **104**
Dewar St. SE15....55Wb **113**
Dewberry Gdns. E6....43Nc **94**
Dewberry St. E14....43Ec **92**
Dewey Ct. BR1: Broml....69Jc **137**
Dewey La. SW2....58Qb **112**
(off Tulse Hill)
Dewey Path RM12: Horn....37Ld **77**
Dewey Rd. N1....1K **217** (40Qb **70**)
Dewey Rd. RM10: Dag....37Dd **76**
Dewey St. SW17....64Hb **133**
Dewgrass Gro. EN8: Walt C....7Zb **20**
Dewhurst Ct. TW3: Houn....56Ca **107**
Dewhurst Rd. EN8: Chesh....1Xb **19**
Dewhurst Rd. W14....48Za **88**
Dewlands RH9: G'stone....3A **210**
(not continuous)
Dewlands Av. DA2: Dart....59Rd **119**
Dewlands Rd. RH9: G'stone....3A **210**
Dewsbury Cl. HA5: Pinn....30Aa **45**
Dewsbury Cl. RM3: Rom....23Nd **57**
Dewsbury Ct. W4....49Sa **87**

Dewsbury Gdns. KT4: Wor Pk 76Wa 154
Dewsbury Gdns. RM3: Rom 23Md 57
Dewsbury Rd. NW10 36Wa 68
Dewsbury Rd. RM3: Rom 23Md 57
Dewsbury Ter. NW1 39Kb 70
Dews Farm Sand Pits Nature Reserve 30M 43
De Wyndsor Ct. SE16 48Wb 91
(off Jamaica Rd.)
Dexter Apts. SE15 53Xb 113
(off Queen's Rd.)
Dexter Cl. AL1: St A 3E 6
Dexter Cl. RM17: Grays 48Ce 99
Dexter Ho. DA18: Erith 48Ad 95
(off Kale Rd.)
Dexter Rd. EN5: Barn 16Za 30
Dexter Rd. UB9: Hare 26L 43
Deyncourt Gdns. RM14: Upm 33Sd 78
Deyncourt Rd. N17 25Sb 51
Deynecourt Gdns. E11 28Lc 53
D'Eynsford Rd. SE5 53Tb 113
Dharam Marg WD25: A'ham 12Ea 28
Dharam Marg WD25: Let H 12Ea 28
Dhonau Ho. SE1 5K 231 (49Vb 91)
(off Longfield Est.)
Diadem Ct. W1 3D 222 (44Mb 90)
(off Dean St.)
Dial Cl. DA9: Ghithe 57Yd 120
Dialmead EN6: Ridge 6Va 16
Dial Stone Ct. KT13: Weyb 78S 150
Dial Wlk., The W8 47Db 89
Diamedes Av. TW19: Stanw 59M 105
Diameter Rd. BR5: Pet W 73Rc 160
Diamond Cl. RM16: Chaf H 48Be 99
Diamond Cl. RM8: Dag 32Yc 75
Diamond Ct. BR5: St P 67Xc 139
Diamond Ct. RH1: Redh 5A 208
(off St Anne's Mt.)
Diamond Ct. RM11: Horn 32Jd 76
Diamond Est. SW17 62Gb 133
Diamond Gdns. E3 43Bc 92
Diamond Ho. E3 40Ac 72
(off Roman Rd.)
Diamond Jubilee Way SM5: Cars 80Hb 155
Diamond Pl. RH2: Reig 5K 207
Diamond Rd. HA4: Ruis 35Z 65
Diamond Rd. SL1: Slou 7L 81
Diamond Rd. WD24: Wat 10W 12
Diamond St. NW10 38Ta 67
Diamond St. SE15 52Ub 113
Diamond Ter. SE10 53Ec 114
Diamond Way SE8 51Cc 114
Diana Cl. DA14: Sidc 61Ad 139
Diana Cl. E18 25Kc 53
Diana Cl. RM16: Chaf H 48Be 99
Diana Cl. SE8 51Bc 114
Diana Cl. SL3: Geor G 44A 82
Diana Ct. DA8: Erith 51Gd 118
Diana Gdns. KT6: Surb 75Pa 153
Diana Ho. SW13 53Va 110
Diana, Princess of Wales Memorial Fountain 1D 226 (47Gb 89)
Diana, Princess of Wales Memorial Playground 46Db 89
Diana, Princess of Wales Memorial Wlk. Kensington Gdns. 46Db 89
Diana Rd. E17 27Bc 52
Dianne Ct. SE12 60Jc 115
Dianne Way EN4: E Barn 14Gb 31
Dianthus Cl. KT16: Chert 73G 148
Dianthus Cl. SE2 50Xc 95
Dianthus Ct. GU22: Wok 10P 167
Diban Av. RM12: Horn 35Kd 77
Diban Ct. RM12: Horn 35Kd 77
(off The Broadway)
DIBDEN 99Gd 202
Dibden Ho. SE5 52Ub 113
Dibden La. TN14: Ide H 98Gd 202
Dibden La. TN14: S'oaks 98Gd 202
Dibden St. N1 39Sb 71
Dibdin Cl. SM1: Sutt 76Cb 155
Dibdin Ho. W9 40Db 69
Dibdin Rd. SM1: Sutt 76Cb 155
Dibdin Row SE1 3A 230 (48Qb 90)
Diceland Lodge SM7: Bans 88Bb 175
(off Diceland Rd.)
Diceland Rd. SM7: Bans 88Bb 175
Dicey Av. NW2 35Ya 68
Dickens Av. DA1: Dart 56Qd 119
Dickens Av. N3 25Eb 49
Dickens Av. RM18: Tilb 3D 122
Dickens Av. UB8: Hil 44R 84
Dickens Cl. AL3: St A 1B 6
Dickens Cl. DA3: Hartl 71Be 165
Dickens Cl. DA8: Erith 52Dd 118
Dickens Cl. TW10: Ham 61Na 131
Dickens Cl. UB3: Harl 49U 84
Dickens Ct. E11 28Jc 53
(off Makepeace Rd.)
Dickens Ct. SW17 62Fb 133
(off Grosvenor Way)
Dickens Dr. BR7: Chst 65Sc 138
Dickens Dr. CR5: Coul 89Lb 176
Dickens Dr. KT15: Add 79H 149
Dickens Est. SE1 47Wb 91
Dickens Est. SE16 48Wb 91
Dickens Ho. NW6 41Cb 89
(off Malvern Rd.)
Dickens Ho. NW8 5C 214 (42Fb 89)
(off Fisherton St.)
Dickens Ho. SE17 50Rb 91
(off Doddington Gro.)
Dickens Ho. SE19 66Ub 135
Dickens Ho. WC1 5F 217 (42Nb 90)
(off Herbrand St.)
Dickens La. N18 22Ub 51
Dickens M. EC1 7B 218 (43Rb 91)
(off Turnmill St.)
Dickenson Cl. N9 18Wb 33
Dickenson Rd. N8 31Nb 70
Dickenson Rd. TW13: Hanw 64Y 129
Dickenson's La. SE25 72Wb 157
Dickensons La. SE25 71Wb 157
(not continuous)
Dickensons Pl. SE25 72Wb 157
Dickens Pl. SL3: Poyle 53G 104
Dickens Ri. IG7: Chig 20Rc 36
Dickens Rd. DA12: Grav'nd 10G 122
Dickens Rd. E6 40Mc 73
Dickens Sq. SE1 3E 230 (48Sb 91)
Dickens St. SW8 54Kb 112
Dickens Way RM1: Rom 28Gd 56
Dickenswood Cl. SE19 66Rb 135
Dickerage La. KT3: N Mald 69Sa 131
Dickerage Rd. KT1: King T 67Sa 131
Dickerage Rd. KT3: N Mald 68Sa 131
Dickinson Av. WD3: Crox G 16Q 26
Dickinson Ct. EC1 5C 218 (42Rb 91)
(off Brewhouse Yd.)
Dickinson Ho. HP3: Hem H 7P 3
Dickinson Quay HP3: Hem H 7N 3
Dickinson Sq. WD3: Crox G 16R 26
Dicksee Ho. NW8 6B 214 (42Fb 89)
(off Lyons Pl.)
Dickson Fold HA5: Pinn 28Z 45
Dickson Ho. E1 44Xb 91
(off Philpot St.)
Dickson Ho. N1 38Qb 70
(off Drummond Way)
Dickson Rd. SE9 55Nc 116
Dick Turpin Way TW14: Felt 56V 106
Didsbury Cl. E6 39Pc 74
Dieppe Cl. W14 50Bb 89
Digby Bus. Cen. E9 37Zb 72
(off Digby Rd.)
Digby Cres. N4 33Sb 71
Digby Gdns. RM10: Dag 39Cd 76
Digby Mans. W6 50Xa 88
(off Hammersmith Bri. Rd.)
Digby M. RM3: Rom 23Nd 57
Digby Pl. CR0: C'don 76Vb 157
Digby Rd. E9 37Zb 72
Digby Rd. IG11: Bark 38Vc 75
Digby St. E2 41Yb 92
Digby Stuart Coll. Roehampton University 57Wa 110
Digby Wlk. RM12: Horn 37Ld 77
Digby Way KT14: Byfl 84P 169
Digdens Ri. KT18: Eps 87Sa 173
Diggens Ct. IG10: Lough 13Nc 36
Diggon St. E1 43Zb 92
Dighton Ct. SE5 51Sb 113
Dighton Rd. SW18 57Eb 111
Dignum St. N1 1K 217 (40Qb 70)
Digswell Cl. WD6: Bore 10Qa 15
Digswell St. N7 37Qb 70
Dilhorne Cl. SE12 62Kc 137
Dilke St. SW3 51Hb 111
Dillmount Dr. KT12: Walt T 71Y 151
Dillon Cl. KT19: Eps 84Pa 173
Dillon Hgts. BR3: Beck 67Ec 136
Dilloway La. UB2: S'hall 47Aa 85
Dillwyn Cl. SE26 63Ac 136
Dilston Cl. UB5: N'olt 41Y 85
Dilston Gro. SE16 49Yb 92
Dilston Rd. KT22: Lea 91Ja 192
Dilton Gdns. SW15 60Wa 110
Dilwyn Ct. E17 26Ac 52
Dimes Pl. W6 49Xa 88
Dimmock Dr. UB6: G'frd 36Fa 66
Dimmocks La. WD3: Sarr 8K 11
Dimond Cl. E7 35Jc 73
Dimsdale Dr. EN1: Enf 17Wb 33
Dimsdale Dr. NW9 32Sa 67
Dimsdale Dr. SL2: Farn C 6C 60
Dimsdale Hgts. E1 44Xb 91
(off Spencer Way)
Dimsdale Wlk. E13 40Jc 73
Dimson Cres. E3 41Cc 92
Dinerman Ct. NW8 39Eb 69
Dingle, The UB10: Hil 41R 84
Dingle Cl. EN5: Ark 16Va 30
Dingle Gdns. E14 45Cc 92
Dingle Rd. TW15: Ashf 64R 128
Dingles Ct. HA5: Pinn 25Z 45
Dingley La. SW16 61Mb 134
Dingley Pl. EC1 4E 218 (41Sb 91)
Dingley Rd. EC1 4D 218 (41Sb 91)
Dingwall Av. CR0: C'don 75Sb 157
Dingwall Gdns. NW11 30Cb 49
Dingwall Rd. CR0: C'don 74Tb 157
Dingwall Rd. SM5: Cars 81Hb 175
Dingwall Rd. SW18 59Eb 111
Dinmont Est. E2 40Wb 71
Dinmont Ho. E2 40Wb 71
(off Pritchard's Rd.)
Dinmont St. E2 40Xb 71
Dinmore HP3: Bov 10B 2
Dinmore Ho. E9 39Yb 72
(off Templecombe Rd.)
Dinnington Ho. E1 42Xb 91
(off Coventry Rd.)
Dinosaur Escape Adventure Golf 39Y 65
Dinsdale Cl. GU22: Wok 90C 168
Dinsdale Gdns. EN5: New Bar 15Db 31
Dinsdale Gdns. SE25 71Ub 157
Dinsdale Rd. SE3 51Hc 115
Dinsmore Rd. SW12 59Kb 112
Dinton Ho. NW8 5D 214 (42Gb 89)
(off Lilestone St.)
Dinton Rd. KT2: King T 66Pa 131
Dinton Rd. SW19 65Fb 133
Diploma Av. N2 28Gb 49
Diploma Ct. N2 28Gb 49
Dippers Cl. TN15: Kems'g 89Pd 183
Dippingwell Ct. SL2: Farn C 6G 60
Diprose Lodge SW17 63Fb 133
Dirdene Cl. KT17: Eps 84Va 174
Dirdene Gdns. KT17: Eps 84Va 174
Dirdene Gro. KT17: Eps 84Ua 174
Dirleton Rd. E15 39Hc 73
Dirtham La. KT24: Eff 100X 191
Dirty La. SE1 6E 224 (46Sb 91)
Disbrowe Rd. W6 51Ab 110
Discover Stratford 38Fc 73
Discover Greenwich Visitor Cen. 51Ec 114
Discovery Bus. Pk. SE16 48Wb 91
(off St James's Rd.)
Discovery Cen., The Becton 41Sc 94
Discovery Dock Apts. E. E14 47Dc 92
(off Sth. Quay Sq.)
Discovery Dock Apts. W. E14 47Dc 92
(off Sth. Quay Sq.)
Discovery Dr. BR8: Swan 70Gd 140
Discovery Ho. E14 45Ec 92
(off Newby Pl.)
Discovery Twr. E16 44Hc 93
(off Minnie Baldock St.)
Discovery Wlk. E1 46Xb 91
Dishforth La. NW9 24Ua 48
Disley Ct. UB1: S'hall 44Da 85
(off Howard Rd.)
Disney Pl. SE1 1E 230 (47Sb 91)
Disney St. SE1 1E 230 (47Sb 91)
Dison Cl. EN3: Enf H 11Zb 34
Dispensary La. E8 36Xb 71
Disraeli Cl. SE28 46Yc 95
Disraeli Cl. W4 48Ta 87
Disraeli Ct. SL3: L'ly 51D 104
Disraeli Gdns. SW15 56Bb 111
Disraeli Rd. E7 37Jc 73
Disraeli Rd. NW10 40Ta 67
Disraeli Rd. SW15 56Ab 110
Disraeli Rd. W5 46Ma 87
Diss St. E2 3K 219 (41Vb 91)
Distaff La. EC4 4D 224 (45Sb 91)
Distel Apts. SE10 49Gc 93
(off Telegraph Av.)
Distillery La. W6 50Ya 88
Distillery Rd. W6 50Ya 88
Distillery Twr. SE8 53Cc 114
Distillery Wlk. TW8: Bford 51Na 109
Distillery Wharf W6 51Ya 110
Distin St. SE11 6K 229 (49Qb 90)
District Rd. HA0: Wemb 36Ka 66
Ditch All. SE13 53Dc 114
Ditchburn St. E14 45Ec 92
Ditches La. CR3: Cat'm 94Pb 196
Ditches La. CR3: Coul 94Pb 196
Ditches La. CR5: Coul 92Nb 196
Ditches Ride, The IG10: Lough 10Qc 22
Ditchfield Rd. UB4: Yead 42Aa 85
Ditchley Ct. W7 43Ha 86
(off Templeman Rd.)
Ditchling Ct. AL1: St A 2C 6
(off Bricket Rd.)
Dittisham Rd. SE9 63Nc 138
Ditton Cl. KT7: T Ditt 73Ja 152
Dittoncroft Cl. CR0: C'don 77Ub 157
Ditton Grange Cl. KT6: Surb 74Ma 153
Ditton Grange Dr. KT6: Surb 74Ma 153
Ditton Gro. KT10: Esh 76Fa 152
Ditton Hill KT6: Surb 74La 152
Ditton Hill Rd. KT6: Surb 74La 152
Ditton Lawn KT7: T Ditt 74Ja 152
Ditton Pk. 10P 81
Ditton Pk. Rd. SL3: L'ly 51A 104
Ditton Pl. SE20 67Xb 135
Ditton Reach KT7: T Ditt 72Ka 152
Ditton Rd. DA6: Bex 57Zc 117
Ditton Rd. KT6: Surb 75Ma 153
Ditton Rd. SL3: Dat 3P 103
Ditton Rd. SL3: L'ly 50B 82
Ditton Rd. UB2: S'hall 50Ba 85
Diversity Av. RM13: Rain 40Fd 76
Divine Way UB3: Hayes 44T 84
Divis Way SW15 58Xa 110
(off Dover Pk. Dr.)
Dixie Ct. SE6 58Cc 114
Dixon Butler M. W9 42Cb 89
Dixon Clark Ct. N1 37Rb 71
Dixon Cl. E6 44Pc 94
Dixon Dr. KT13: Weyb 82P 169
Dixon Ho. W10 44Za 88
(off Darfield Way)
Dixon Pl. BR4: W W'ck 74Dc 158
Dixon Rd. SE14 53Ac 114
Dixon Rd. SE25 69Ub 135
Dixon's All. SE16 47Xb 91
Dixons Hill Cl. AL9: N Mym 7D 8
Dixons Hill Rd. AL9: N Mym 7D 8
Dixons Hill Rd. AL9: Wel G 7D 8
Dixon Way NW10 38Ua 68
Dobbin Cl. HA3: Kenton 26Ja 46
Dobell Rd. SE9 57Pc 116
Doble Ct. CR2: Sande 84Wb 177
Dobles Cl. CR5: Coul 87Nb 176
Dobree Av. NW10 38Xa 68
Dobson Cl. CR5: Coul 88Mb 176
Dobson Cl. NW6 38Fb 69
Dobson Ho. SE14 51Zb 114
(off John Williams Cl.)
Dobson Rd. DA12: Grav'nd 4G 144
Dobson Wlk. SE5 52Tb 113
Doby Ct. EC4 4E 224 (45Sb 91)
(off Skinners La.)
Dock, The E1 45Xb 91
Dock App. Rd. RM16: Grays 9B 100
Dock App. Rd. RM17: Grays 1A 122
Dock Cotts. E1 45Yb 92
(off The Highway)
Dockers Tanner Rd. E14 48Cc 92
Dockett Eddy KT16: Chert 74N 149
Dockett Eddy La. TW17: Shep 74P 149
Dockett Moorings KT16: Chert 74N 149
Dockhead SE1 47Vb 91
Dockhead Wharf SE1 1K 231 (47Vb 91)
(off Shad Thames)
Dock Hill Av. SE16 46Zb 92
Docklands Ct. E14 44Bc 92
(off Wharf La.)
Docklands Equestrian Cen., The 42Qc 94
Docklands Sailing & Watersports Cen. 48Cc 92
Dockland St. E16 46Qc 94
(not continuous)
Dockley Rd. SE16 48Wb 91
Dockley Rd. Ind. Est. SE16 48Wb 91
(off Dockley Rd.)
Dock Mdw. Reach W7 48Ga 86
Dock Offices SE16 48Yb 92
(off Surrey Quays Rd.)
Dock Rd. E16 45Hc 93
Dock Rd. IG11: Bark 40Sc 74
Dock Rd. RM17: Grays 51Fe 121
(not continuous)
Dock Rd. RM18: Tilb 3A 122
Dock Rd. TW8: Bford 52Ma 109
Dockside Ct. E5 33Yb 72
Dockside Rd. E16 45Mc 93
Dock St. E1 45Wb 91
Dock Wlk. RM3: Rom 22Ld 57
Dockweed Ct. E14 48Dc 92
(off Watergate Walk)
Dockwell Cl. TW14: Felt 56W 106
Dockwell's Ind. Est. TW14: Felt 57X 107
Dockyard La. E14 47Cc 92
Dr Harold Moody Pk. 54Xb 113
Dr Johnson Av. SW17 62Kb 134
Dr Johnson's House 2A 224 (44Qb 90)
(off Pemberton Row)
Doctors Cl. SE26 64Yb 136
Doctors La. CR3: Cat'm 95Qb 196
Doctor's Lock 52Ma 109
Docura Ho. N7 33Pb 70
Docwra's Bldgs. N1 37Ub 71
Dodbrooke Rd. SE27 62Qb 134
Dodd Ho. SE16 49Xb 91
(off Rennie Est.)
Doddington Gro. SE17 51Rb 113
Doddington Pl. SE17 51Rb 113
Dodd Rd. WD24: Wat 8W 12
Dodds Cres. KT14: W Byf 86K 169
Doddsfield Rd. SL2: Slou 1E 80
Dodds Ho. SE10 48Jc 93
(off Peartree Way)
Dodd's La. GU22: Pyr 86J 169
Dodsley Pl. N9 20Yb 34
Dodson St. SE1 2A 230 (47Qb 90)
Dod St. E14 44Bc 92
Doebury Wlk. SE18 52Wc 117
(off Prestwood Cl.)
Doel Cl. SW19 66Eb 133
Dog & Duck Yd. WC1 7J 217 (43Pb 90)
(off Princeton St.)
Doggett Rd. SE6 59Cc 114
Doggetts Cl. EN4: E Barn 15Gb 31
Doggett's Cnr. RM11: Horn 33Pd 77
Doggetts Farm Rd. UB9: Den 31E 62
Doggetts Way AL1: St A 4A 6
Doghurst Av. UB3: Harl 52R 106
Doghurst Dr. UB7: Sip 52R 106
Doghurst La. CR5: Chips 92Hb 195
Dog Kennel Hill SE22 55Ub 113
Dog Kennel Hill Est. SE22 55Ub 113
(off Albrighton Rd.)
Dog Kennel La. WD3: Chor 14H 25
Dog La. NW10 35Ua 68
Dogwood Cl. DA11: Nflt 3B 144
Dokal Ind. Est. UB2: S'hall 47Aa 85
Dolben Ct. SW1 6E 228 (49Mb 90)
(off Montaigne Cl.)
Dolben St. SE1 7B 224 (46Rb 91)
(not continuous)
Dolby Rd. SW6 54Bb 111
Dolland Ho. SE11 7J 229 (50Pb 90)
(off Newburn St.)
Dolland St. SE11 7J 229 (50Pb 90)
Dollar Bay Pl. E14 47Ec 92
Dollary Pde. KT1: King T 69Ra 131
(off Kingston Rd.)
Dolliffe Cl. CR4: Mitc 68Gb 133
Dollis Av. N3 25Bb 49
Dollis Brook Wlk. EN5: Barn 16Ab 30
Dollis Cres. HA4: Ruis 32Y 65
Dolliscroft NW7 24Ab 48
DOLLIS HILL 33Xa 68
Dollis Hill Av. NW2 34Xa 68
Dollis Hill Est. NW2 34Wa 68
Dollis Hill La. NW2 35Va 68
Dollis Hill Station (Underground) 36Wa 68
Dollis M. N3 25Cb 49
Dollis Pk. N3 25Bb 49
Dollis Rd. N3 25Bb 48
Dollis Rd. NW7 24Ab 48
Dollis Valley Dr. EN5: Barn 16Bb 31
Dollis Valley Way EN5: Barn 16Bb 31
Dollypers Hill Nature Reserve 90Rb 177
Dolman Cl. N3 26Eb 49
Dolman Rd. W4 49Ta 87
Dolman St. SW4 56Pb 112
Dolomite Ct. HA4: Ruis 35Y 65
Dolphin App. RM1: Rom 28Hd 56
Dolphin Cl. KT6: Surb 71Ma 153
Dolphin Cl. SE16 47Zb 92
Dolphin Cl. SE28 44Zc 95
Dolphin Ct. IG7: Chig 20Rc 36
Dolphin Ct. NW11 30Ab 48
Dolphin Ct. SL1: Slou 7M 81
Dolphin Ct. TW18: Staines 62J 127
Dolphin Ct. Nth. TW18: Staines 62J 127
Dolphin Est. TW16: Sun 67U 128
Dolphin Ho. SW18 56Db 111
Dolphin Ho. SW6 54Eb 111
(off Lensbury Av.)
Dolphin Ho. TW16: Sun 67U 128
(off Windmill Rd.)
Dolphin La. E14 45Dc 92
Dolphin Pk. Ind. Est. RM19: Purf 50Ud 98
Dolphin Point RM19: Purf 50Ud 98
Dolphin Rd. SL1: Slou 7M 81
Dolphin Rd. TW16: Sun 67U 128
Dolphin Rd. UB5: N'olt 40Ba 65
Dolphin Rd. Nth. TW16: Sun 67U 128
Dolphin Rd. Sth. TW16: Sun 67U 128
Dolphin Rd. W. TW16: Sun 67U 128
Dolphin Sq. SW1 50Lb 90
Dolphin Sq. W4 52Ua 110
Dolphin St. KT1: King T 68Na 131
Dolphin Twr. SE8 51Bc 114
(off Abinger Gro.)
Dolphin Way RM19: Purf 49Ud 98
Dolphin Yd. AL1: St A 3B 6
(off Holywell Hill)
Dombey Ho. SE1 47Wb 91
(off Wolseley St.)
Dombey Ho. W11 46Za 88
(off St Ann's Rd.)
Dombey St. WC1 7H 217 (43Pb 90)
(not continuous)
Dome, The RH1: Redh 5P 207
DOME, THE 8Y 13
Domecq Ho. EC1 5C 218 (42Rb 91)
(off Dallington St.)
Dome Hill CR3: Cat'm 99Ub 197
Dome Hill Pk. SE26 63Vb 135
Dome Hill Peak CR3: Cat'm 98Ub 197
Domelton Ho. SW18 58Db 111
(off Iron Mill Rd.)
Domett Cl. SE5 56Tb 113
Dome Way RH1: Redh 5P 207
Domfe Pl. E5 35Yb 72
Domingo St. EC1 5D 218 (42Sb 91)
Dominica Cl. E13 40Mc 73
Dominic Ct. EN9: Walt A 5Dc 20
Dominion Apts. E17 29Bc 52
Dominion Bus. Pk. N9 19Zb 34
Dominion Cen., The UB2: S'hall 47Aa 85
Dominion Cl. TW3: Houn 54Fa 108
Dominion Ct. E8 38Vb 71
(off Middleton Rd.)
Dominion Dr. RM5: Col R 23Dd 56
Dominion Dr. SE16 47Zb 92
Dominion Ho. E14 50Dc 92
(off St Davids Sq.)
Dominion Ho. EC1 1D 224 (43Sb 91)
(off Bartholomew Cl.)
Dominion Ind. Est. UB2: S'hall 47Aa 85
(off Feather Rd.)
Dominion Pde. HA1: Harr 29Ha 46
Dominion Rd. CR0: C'don 73Vb 157
Dominion Rd. UB2: S'hall 47Aa 85
Dominion St. EC2 7G 219 (43Tb 91)
Dominion Theatre 2E 222 (44Mb 90)
(off Tottenham Ct. Rd.)
Dominion Wlk. E14 45Fc 93
(off Fairmont Av.)
Dominion Way RM13: Rain 41Jd 96
Domonic Dr. SE9 63Rc 138
Domville Cl. N20 19Fb 31
Domville Ct. SE17 7J 231 (50Ub 91)
(off Bagshott St.)
Donald Biggs Dr. DA12: Grav'nd 10E 122
Donald Dr. RM6: Chad H 29Yc 55
Donald Hunter Ho. E7 36Kc 73
(off Woodgrange Rd.)
Donald Rd. CR0: C'don 73Qb 156
Donald Rd. E13 39Kc 73
Donaldson Rd. NW6 39Bb 69
Donaldson Rd. SE18 53Qc 116
Donald Woods Gdns. KT5: Surb 75Ra 153
Donato Dr. SE15 51Ub 113
Doncaster Dr. UB5: N'olt 36Ba 65
Doncaster Gdns. N4 30Sb 51
Doncaster Gdns. UB5: N'olt 36Ba 65
Doncaster Grn. WD19: Wat 22Y 45
Doncaster Rd. N9 17Xb 33
Doncaster Way RM14: Upm 34Pd 77
Doncel Ct. E4 17Fc 35
Doncella Cl. RM16: Chaf H 48Zd 99
Donegal Ho. E1 42Xb 91
(off Cambridge Heath Rd.)
Donegal St. N1 2J 217 (40Pb 70)
Doneraile Ho. SW1 7K 227 (50Kb 90)
(off Ebury Bri. Rd.)
Doneraile St. SW6 54Za 110
Dongola Rd. E1 43Ac 92
Dongola Rd. E13 41Kc 93
Dongola Rd. N17 27Ub 51
Dongola Rd. W. E13 41Kc 93
Don Gratton Ho. E1 43Wb 91
(off Old Montague St.)
Donington Av. IG2: Ilf 29Sc 54
Donington Av. IG6: Ilf 29Sc 54
Donkey All. SE22 59Wb 113
Donkey La. DA4: Farni 75Rd 163
Donkey La. EN1: Enf 12Wb 33
Donkey La. UB7: W Dray 49L 83
DONKEY TOWN 5B 166
Donkin Ho. SE16 49Xb 91
(off Rennie Est.)
Donmar Warehouse Theatre 3F 223 (44Nb 90)
(off Earlham St.)
Donnafields GU24: Bisl 8E 166
Donnatt's Rd. SE14 53Bc 114
Donne Ct. SE24 58Sb 113
Donnefield Av. HA8: Edg 24Na 47
Donne Gdns. GU22: Pyr 87G 168
Donne Ho. E14 44Cc 92
(off Dod St.)
Donne Ho. SE14 51Zb 114
(off Samuel Cl.)
Donnelly Ct. SW6 52Ab 110
(off Dawes Rd.)
Donnelly Ho. SE1 3K 229 (48Qb 90)
(off McAuley Cl.)
Donne Pl. CR4: Mitc 70Kb 134
Donne Pl. SW3 5E 226 (49Gb 89)
Donne Rd. RM8: Dag 33Yc 75
Donnington Ct. DA2: Dart 58Rd 119
(off Osbourne Rd.)
Donnington Ct. NW1 38Kb 70
(off Castlehaven Rd.)
Donnington Ct. NW10 38Xa 68
(off Donnington Rd.)
Donnington Mans. NW10 39Ya 68
(off Donnington Rd.)
Donnington Rd. HA3: Kenton 29Ma 47
Donnington Rd. KT4: Wor Pk 75Wa 154
Donnington Rd. NW10 38Xa 68
Donnington Rd. TN13: Dun G 92Fd 202
Donnybrook Ct. E3 39Bc 72
(off Old Ford Rd.)
Donnybrook Rd. SW16 66Lb 134
Donoghue Bus. Pk. NW2 34Za 68
Donoghue Cotts. E14 43Ac 92
(off Galsworthy Av.)
Donoghue Ct. E3 42Dc 92
(off Barry Blandford Way)
Donovan Av. N10 26Kb 50
Donovan Cl. KT19: Eps 82Ta 173
Donovan Ct. SW10 50Fb 89
(off Drayton Gdns.)
Donovan Ho. E1 45Yb 92
(off Cable St.)
Donovan Pl. N21 15Pb 32
Donovan's Gdn. CM13: Heron 24Fe 59
Don Phelan Cl. SE5 53Tb 113
Dons Ct. BR1: Broml 67Hc 137
(off London Rd.)
Don Way RM1: Rom 24Gd 56
Donyngs Leisure Cen. 5N 207
Doods Pk. Rd. RH2: Reig 5L 207
Doods Pl. RH2: Reig 5M 207
Doods Rd. RH2: Reig 5L 207
Doods Way RH2: Reig 5M 207
Doolittle Mdws. HP3: Hem H 7N 3
Doone Cl. TW11: Tedd 65Ja 130
Doon St. SE1 6K 223 (46Qb 90)
Dorado Gdns. BR6: Chels 76Zc 161
Dora Ho. E14 44Bc 92
(off Rhodeswell Rd.)
Dora Ho. W11 45Za 88
(off St Ann's Rd.)
Doral Way SM5: Cars 78Hb 155
Doran Ct. E6 40Pc 74
Doran Ct. RH1: Redh 6M 207
Dorando Cl. W12 45Xa 88
Doran Dr. RH1: Redh 6M 207
Doran Gdns. RH1: Redh 6M 207
Doran Gro. SE18 52Uc 116
Doran Mnr. N2 29Hb 49
(off Great Nth. Rd.)
Doran Wlk. E15 38Ec 72
Dora Rd. SW19 64Cb 133
Dora St. E14 44Bc 92
Dora Way SW9 54Qb 112
Dorchester Av. DA5: Bexl 60Zc 117
Dorchester Av. HA2: Harr 30Ea 46
Dorchester Av. N13 21Sb 51
Dorchester Cl. BR5: St P 66Xc 139
Dorchester Cl. DA1: Dart 59Pd 119
Dorchester Cl. KT10: Hin W 75Ha 152
Dorchester Cl. UB5: N'olt 36Da 65
Dorchester Ct. AL1: St A 3E 6
(off Dexter Cl.)
Dorchester Ct. E18 25Hc 53
(off Buckingham Rd.)
Dorchester Ct. GU22: Wok 88C 168
Dorchester Ct. N1 38Ub 71
(off Englefield Rd.)
Dorchester Ct. N10 27Kb 50
Dorchester Ct. N14 17Kb 32
Dorchester Ct. NW2 34Za 68
Dorchester Ct. RH2: Reig 5M 207
Dorchester Ct. SE24 57Sb 113
Dorchester Ct. SW1 4G 227 (48Hb 89)
(off Sloane St.)
Dorchester Ct. TW18: Staines 63J 127
Dorchester Ct. WD19: Wat 16Aa 27
(off Chalk Hill)
Dorchester Ct. WD3: Crox G 15S 26
Dorchester Dr. SE24 57Sb 113
Dorchester Dr. TW14: Bedf 58U 106
Dorchester Gdns. E4 21Cc 52
Dorchester Gdns. NW11 28Cb 49
Dorchester Gro. W4 50Ua 88
Dorchester Ho. TW9: Kew 52Ra 109
Dorchester M. KT16: Vir W 5L 147
Dorchester M. KT3: N Mald 70Ta 131
Dorchester M. TW1: Twick 58La 108
Dorchester Rd. DA12: Grav'nd 2F 144
Dorchester Rd. KT13: Weyb 76R 150
Dorchester Rd. KT4: Wor Pk 74Ya 154

Dorchester Rd. SM4: Mord 73Db 155
Dorchester Rd. UB5: N'olt 36Da 65
Dorchester Ter. NW2 34Za 68
(off Needham Ter.)
Dorchester Way HA3: Kenton 30Pa 47
Dorchester Waye UB4: Yead 44X 85
(not continuous)
Dorcis Av. DA7: Bex 54Ad 117
Dordrecht Rd. W3 46Ua 88
Dore Av. E12 36Oc 74
Doreen Av. NW9 32Ta 67
Doreen Capstan Ho. E11 34Gc 73
(off Apollo Pl.)
Dore Gdns. SM4: Mord 73Db 155
Dorell Cl. UB1: S'hall 43Ba 85
Doresa Cl. KT15: Add 78J 149
Dorey Ho. TW8: Bford 52La 108
(off High St.)
Dorfman Theatre 6K 223 (46Qb 90)
(within National Theatre)
Doria Dr. DA12: Grav'nd 2G 144
Dorian Dr. SL5: Asc 7C 124
Dorian Rd. RM12: Horn 32Jd 76
Doria Rd. SW6 54Bb 111
Doric Dr. KT20: Tad 92Bb 195
Doric Ho. E2 40Zb 72
(off Mace St.)
Doric Way NW1 3D 216 (41Mb 90)
Dorie M. N12 21Db 49
(off Ashbourne Cl.)
Dorien Rd. SW20 68Za 132
Dorin Ct. CR6: W'ham 92Xb 197
Dorin Ct. GU22: Pyr 87G 168
Doris Ashby Cl. UB6: G'frd 39Ja 66
Doris Av. DA8: Erith 53Ed 118
Doris Emmerton Ct. SW11 56Eb 111
Doris Rd. E7 38Jc 73
Doris Rd. TW15: Ashf 65T 128
Dorking Cl. KT4: Wor Pk 75Za 154
Dorking Cl. SE8 51Bc 114
Dorking Ct. N17 25Wb 51
(off Hampden La.)
Dorking Gdns. RM3: Rom 22Md 57
Dorking Ho. SE1 3G 231 (48Tb 91)
Dorking Ri. RM3: Rom 21Md 57
Dorking Rd. KT18: Eps 88Qa 173
Dorking Rd. KT20: Tad 1A 206
Dorking Rd. KT20: Walt H 1A 206
Dorking Rd. KT22: Lea 94Ka 192
Dorking Rd. KT23: Bookh 98Da 191
Dorking Rd. RM3: Rom 22Md 57
Dorking Vs. GU21: Knap 9H 167
Dorking Wlk. RM3: Rom 21Md 57
Dorkins Way RM14: Upm 31Ud 78
Dorlcote Rd. SW18 59Gb 111
Dorling Dr. KT17: Eps 84Va 174
Dorly Cl. TW17: Shep 71U 150
Dorman Pl. N9 19Wb 33
Dormans Cl. HA6: Nwood 24T 44
Dorman Wlk. NW10 36Ta 67
Dorman Way NW8 39Fb 69
Dorma Trad. Pk. E10 32Zb 72
Dormay St. SW18 57Db 111
Dormer Cl. E15 37Hc 73
Dormer Cl. EN5: Barn 15Za 30
Dormers HP3: Bov 9G 2
Dormer's Av. UB1: S'hall 44Ca 85
Dormers Ri. UB1: S'hall 45Da 85
DORMER'S WELLS 44Ca 85
Dormer's Wells La. UB1: S'hall 44Ca 85
Dormers Wells Leisure Cen. 44Da 85
Dormstone Ho. SE17 6H 231 (49Ub 91)
(off Congreve St.)
Dormywood HA4: Ruis 29V 44
Dornberg Cl. SE3 52Jc 115
Dornberg Rd. SE3 52Kc 115
Dorncliffe Rd. SW6 54Ab 110
Dornels SL2: Slou 4N 81
Dorney NW3 38Gb 69
DORNEY 8A 80
Dorney Gro. KT13: Weyb 75R 150
Dorney Hill Sth. HP9: Beac 1E 60
Dorney Lake 10A 80
Dorney Lake Pk. & Nature Reserve 10A 80
Dorney Pl. DA1: Dart 55Qd 119
Dorney Ri. BR5: St M Cry 70Vc 139
Dorney Way TW4: Houn 57Aa 107
Dorneywood Cl. SL1: Burn 10A 60
Dorney Wood Rd. SL1: Burn 4A 60
Dornfell St. NW6 36Bb 69
Dornford Gdns. CR5: Coul 91Sb 197
Dornoch Ho. E3 40Bc 72
(off Anglo Rd.)
Dornton Rd. CR2: S Croy 78Tb 157
Dornton Rd. SW12 61Kb 134
Dorothy Av. HA0: Wemb 38Na 67
Dorothy Evans Cl. DA7: Bex 56Dd 118
Dorothy Gdns. RM8: Dag 35Xc 75
Dorothy Pettingell Ho. SM1: Sutt 76Db 155
(off Vermont Rd.)
Dorothy Rd. SW11 55Hb 111
Dorothy Smith La. N17 24Tb 51
Dorrell Pl. SW9 55Qb 112
Dorrien Wlk. SW16 61Mb 134
Dorrington Cl. IG11: Bark 38Wc 75
Dorrington Ct. SE25 68Ub 135
Dorrington Gdns. RM12: Horn 32Md 77
Dorrington Point E3 41Dc 92
(off Bromley High St.)
Dorrington St. EC1 7K 217 (43Qb 90)
Dorrington Way BR3: Beck 71Ec 158
Dorrit Ho. W11 46Za 88
(off St Ann's Rd.)
Dorrit M. N18 22Ub 51
Dorrit St. SE1 1E 230 (47Sb 91)
Dorrit Way BR7: Chst 65Sc 138
Dorrofield Cl. WD3: Crox G 15S 26
Dorryn Ct. SE26 64Zb 136
Dors Cl. NW9 32Ta 67
Dorset Av. DA16: Well 56Vc 117
Dorset Av. RM1: Rom 28Fd 56
Dorset Av. UB2: S'hall 49Ca 85
Dorset Av. UB4: Hayes 41U 84
Dorset Bldgs. EC4 3B 224 (44Rb 91)
Dorset Cl. KT9: Chess 77Ma 153
Dorset Cl. NW1 7F 215 (43Hb 89)
Dorset Cl. UB4: Hayes 41U 84
Dorset Ct. HA6: Nwood 25V 44
Dorset Ct. KT17: Eps 84Va 174
Dorset Ct. N1 38Ub 71
(off Hertford Rd.)
Dorset Ct. UB5: N'olt 41Aa 85
Dorset Ct. W7 43Ha 86
(off Copley Cl.)
Dorset Cres. DA12: Grav'nd 3G 144
Dorset Dr. GU22: Wok 89D 168
Dorset Dr. HA8: Edg 23Pa 47
Dorset Gdns. CR4: Mitc 70Pb 134
Dorset Gdns. HA0: Wemb 36La 66
Dorset Gdns. SS17: Linf 7J 101
Dorset Ho. NW1 6G 215 (42Hb 89)
(off Gloucester Pl.)
Dorset Mans. SW6 51Za 110
(off Lillie Rd.)
Dorset M. N3 25Cb 49
Dorset M. SW1 3K 227 (48Kb 90)
Dorset Pl. E15 37Fc 73
Dorset Ri. EC4 3B 224 (44Rb 91)
Dorset Rd. BR3: Beck 69Zb 136
Dorset Rd. CR4: Mitc 68Gb 133
Dorset Rd. E7 38Lc 73
Dorset Rd. HA1: Harr 30Ea 46
Dorset Rd. N15 28Tb 51
Dorset Rd. N22 25Nb 50
Dorset Rd. SE9 61Nc 138
Dorset Rd. SL4: Wind 3G 102
Dorset Rd. SM2: Sutt 82Cb 175
Dorset Rd. SW19 67Cb 133
Dorset Rd. SW8 52Nb 112
Dorset Rd. TW15: Ashf 62M 127
Dorset Rd. W5 48La 86
Dorset Sq. KT19: Ewe 82Ta 173
Dorset Sq. NW1 6F 215 (42Hb 89)
Dorset St. TN13: S'oaks 97Ld 203
Dorset St. W1 1G 221 (43Hb 89)
Dorset Way KT14: Byfl 82M 169
Dorset Way TW2: Twick 60Fa 108
Dorset Way UB10: Hil 40P 63
Dorset Waye TW5: Hest 52Ba 107
Dorset Wharf W6 52Ya 110
(off Rainville Rd.)
Dorsey Ho. N1 37Rb 71
(off Canonbury Rd.)
Dorton Cl. SE15 52Ub 113
Dorton Dr. TN15: Seal 94Pd 203
(not continuous)
Dorton Vs. UB7: Sip 52Q 106
Dorton Way GU23: Rip 93K 189
Dorville Cres. W6 48Xa 88
Dorville Rd. SE12 57Hc 115
Dothill Rd. SE18 52Sc 116
Douai Gro. TW12: Hamp 67Ea 130
Doubleday Rd. IG10: Lough 13Sc 36
Doughty Ct. E1 46Xb 91
(off Prusom St.)
Doughty Ho. SW10 51Eb 111
(off Netherton Gro.)
Doughty M. WC1 6H 217 (42Pb 90)
Doughty St. WC1 5H 217 (42Pb 90)
Douglas Av. E17 25Bc 52
Douglas Av. HA0: Wemb 38Na 67
Douglas Av. KT3: N Mald 70Xa 132
Douglas Av. RM3: Hrld W 26Nd 57
Douglas Av. WD24: Wat 9Z 13
Douglas Bader Ho. TW7: Isle 55Fa 108
Douglas Cl. EN4: Had W 10Fb 17
Douglas Cl. GU4: Jac W 10P 187
Douglas Cl. HA7: Stan 22Ja 46
Douglas Cl. IG6: Ilf 24Rc 54
Douglas Cl. RM16: Chaf H 48Ae 99
Douglas Cl. SM6: W'gton 79Nb 156
Douglas Ct. CR3: Cat'm 94Sb 197
Douglas Ct. KT1: King T 70Na 131
(off Geneva Rd.)
Douglas Ct. N3 26Db 49
Douglas Ct. NW6 38Cb 69
(off Quex Rd.)
Douglas Ct. TN16: Big H 89Nc 180
Douglas Cres. UB4: Yead 42Y 85
Douglas Dr. CR0: C'don 76Cc 158
Douglas Eyre Sports Cen. 29Zb 52
Douglas Ho. EN8: Chesh 1Zb 20
(off Davison Dr.)
Douglas Ho. KT23: Bookh 96Ca 191
Douglas Ho. KT6: Surb 74Pa 153
Douglas Ho. RH2: Reig 5J 207
Douglas Johnstone Ho. SW6 51Bb 111
(off Clem Attlee Ct.)
Douglas La. TW19: Wray 57B 104
Douglas Mans. TW3: Houn 55Da 107
Douglas M. NW2 34Ab 68
Douglas M. SM7: Bans 88Bb 175
Douglas Path E14 50Ec 92
(off Manchester Rd.)
Douglas Rd. DA16: Well 53Xc 117
Douglas Rd. E16 43Jc 93
Douglas Rd. E4 17Gc 35
Douglas Rd. IG3: Ilf 31Wc 75
Douglas Rd. KT1: King T 68Ra 131
Douglas Rd. KT10: Esh 75Da 151
Douglas Rd. KT15: Add 76K 149
Douglas Rd. KT6: Surb 75Pa 153
Douglas Rd. N1 38Sb 71
Douglas Rd. N22 25Qb 50
Douglas Rd. NW6 39Bb 69
Douglas Rd. RH2: Reig 5J 207
Douglas Rd. RM11: Horn 30Hd 56
Douglas Rd. SL2: Slou 3H 81
Douglas Rd. TW19: Stanw 58M 105
Douglas Rd. TW3: Houn 55Da 107
Douglas Rd. Nth. N1 37Sb 71
Douglas Rd. Sth. N1 37Sb 71
Douglas Robinson Ct. SW16 66Nb 134
(off Streatham High Rd.)
Douglas Sq. SM4: Mord 72Cb 155
Douglas St. SW1 6D 228 (49Mb 90)
Douglas Ter. E17 25Bc 52
Douglas Waite Ho. NW6 38Cb 69
Douglas Way SE8: Stanley St. 52Bc 114
Douglas Way SE8: Watsons St. 52Cc 114
Doug Siddons Ct. RM17: Grays 51De 121
Doulton Ho. SE11 4J 229 (48Pb 90)
(off Lambeth Wlk.)
Doulton M. NW6 37Db 69
Doultons, The TW18: Staines 66J 127
Dounesforth Gdns. SW18 60Db 111
Dounsell Ct. CM15: Pil H 16Wd 40
Douro Pl. W8 48Db 89
Douro St. E3 40Cc 72
Douthwaite Sq. E1 46Wb 91
Dove App. E6 43Nc 94
Dove Cl. CR2: Sels 83Zb 178
Dove Cl. NW7 24Va 48
Dove Cl. RM16: Chaf H 48Ae 99
Dove Cl. SM6: W'gton 80Pb 156
Dove Cl. UB5: N'olt 42Z 85
Dove Commercial Cen. NW5 36Lb 70
Dovecot Cl. HA5: Eastc 29Y 45
Dovecote Av. N22 27Qb 50
Dovecote Cl. KT13: Weyb 76R 150
Dovecote Gdns. SW14 55Ta 109
Dovecote Ho. SE16 47Zb 92
(off Water Gdns. Sq.)
Dovecote M. UB9: Hare 27N 43
Dove Ct. AL10: Hat 1C 8
Dove Ct. TW19: Stanw 59N 105
Dovedale Av. HA3: Kenton 30La 46
Dovedale Av. IG5: Ilf 26Qc 54
Dovedale Bus. Est. SE15 54Wb 113
(off Blenheim Gro.)
Dovedale Cl. DA16: Well 54Wc 117
Dovedale Cl. UB9: Hare 26L 43
Dovedale Ri. CR4: Mitc 66Hb 133
Dovedale Rd. DA2: Dart 60Sd 120
Dovedale Rd. SE22 57Xb 113
Dovedon Cl. N14 19Nb 32
Dovehouse Ct. UB5: N'olt 41Z 85
(off Delta Gro.)
Dove Ho. Cres. SL2: Slou 1C 80
Dove Ho. Gdns. E4 19Cc 34
Dovehouse Grn. KT13: Weyb 76T 150
Dovehouse Mead IG11: Bark 40Tc 74
Dovehouse St. SW3 7C 226 (50Fb 89)
Dove La. EN6: Pot B 6Eb 17
Dove M. SW5 6A 226 (49Eb 89)
Doveney Cl. BR5: St P 69Yc 139
Dove Pk. HA5: Hat E 24Ca 45
Dove Pk. WD3: Chor 16D 24
Dover Cl. NW2 33Za 68
Dover Cl. RM5: Col R 26Ed 56
Dover Ct. EC1 5B 218 (42Rb 91)
(off St John St.)
Dover Ct. N1 38Tb 71
(off Southgate Rd.)
Dovercourt Av. CR7: Thor H 71Qb 156
Dovercourt Est. N1 37Tb 71
Dovercourt Gdns. HA7: Stan 22Na 47
Dovercourt La. SM1: Sutt 76Eb 155
Dovercourt Rd. SE22 58Ub 113
Doverfield EN7: G Oak 1Sb 19
Doverfield Rd. SW2 59Nb 112
Dover Flats SE1 6J 231 (49Ub 91)
Dover Gdns. SM5: Cars 76Hb 155
Dover Ho. N18 22Vb 51
Dover Ho. SE15 51Yb 114
Dover Ho. Rd. SW15 56Wa 110
Doveridge Gdns. N13 21Rb 51
Dove Rd. N1 37Tb 71
Dove Row E2 39Wb 71
Dover Pk. Dr. SW15 58Xa 110
Dover Patrol SE3 54Kc 115
Dover Rd. DA11: Nflt 59Fe 121
Dover Rd. E12 33Lc 73
Dover Rd. N9 19Yb 34
Dover Rd. RM6: Chad H 30Ad 55
Dover Rd. SE19 65Tb 135
Dover Rd. SL1: Slou 4D 80
Dover Rd. E. DA11: Grav'nd 9A 122
DOVERS CORNER 41Jd 96
Dovers Cnr. Ind. Est. RM13: Rain 41Hd 96
DOVERSGREEN 10K 207
Dovers Grn. Rd. RH2: Reig 9K 207
Doversmead GU21: Knap 8J 167
Dover St. W1 5A 222 (45Kb 90)
Dovers Wlk. RH2: Reig 10K 207
Dover Ter. TW9: Rich 54Pa 109
(off Sandycombe Rd.)
Dover Way WD3: Crox G 14S 26
Dover Yd. W1 6B 222 (46Lb 90)
(off Berkeley St.)
Doves Cl. BR2: Broml 75Nc 160
Doves Cotts. IG7: Chig 20Wc 37
Doves Yd. N1 39Qb 70
Dovet Ct. SW9 53Pb 112
Doveton Ho. E1 42Yb 92
(off Doveton St.)
Doveton Rd. CR2: S Croy 78Tb 157
Doveton St. E1 42Yb 92
Dove Tree Cl. KT19: Eps 81Ta 173
Dovetree Ct. RM3: Rom 21Md 57
(off North Hill Dr.)
Dove Wlk. RM12: Horn 37Kd 77
Dove Wlk. SW1 7H 227 (50Jb 90)
Dovey Lodge N1 38Qb 70
(off Bewdley St.)
Dovoll Ct. SE16 48Wb 91
(off Old Jamaica Rd.)
Dowanhill Rd. SE6 60Fc 115
Dowd Cl. N11 19Jb 32
Dowdeswell Cl. SW15 56Ua 110
Dowding Dr. SE9 57Lc 115
Dowding Ho. N6 31Jb 70
(off Hillcrest)
Dowding Pl. HA7: Stan 23Ja 46
Dowding Rd. TN16: Big H 87Mc 179
Dowding Rd. UB10: Uxb 38P 63
Dowding Wlk. DA11: Nflt 2A 144
Dowding Way EN9: Walt A 8Fc 21
Dowding Way RM12: Horn 38Kd 77
Dowding Way WD25: Wat 6V 12
Dowdney Cl. NW5 36Lb 70
Dowe Ho. SE3 55Gc 115
Dowells St. SE10 51Dc 114
Dower Av. SM6: W'gton 81Kb 176
Dower Ct. SE16 49Yb 92
(off Silwood Rd.)
Dower Pk. SL4: Wind 6C 102
Dowes Ho. SW16 62Nb 134
Dowgate Hill EC4 4F 225 (45Tb 91)
Dowland Ho. KT13: Weyb 76Q 150
Dowland Cl. SS17: Stan H 1L 101
Dowland St. W10 41Ab 88
Dowlands Cl. KT23: Bookh 99Ca 191
Dowlans Rd. KT23: Bookh 99Da 191
Dowlas St. SE5 52Ub 113
Dowler Ct. KT2: King T 67Na 131
Dowler Ho. E1 44Wb 91
(off Burslem St.)
Dowlerville Rd. BR6: Chels 79Vc 161
Dowletts Rd. RM8: Dag 32Ad 75
Dowling Ct. HP3: Hem H 5M 3
Dowling Ho. DA17: Belv 48Bd 95
Dowman Cl. SW19 66Db 133
Downage NW4 27Ya 48
(not continuous)
Downage, The DA11: Grav'nd 1C 144
Downalong WD23: B Hea 18Fa 28
Downbank Av. DA7: Bex 53Fd 118
Down Barns Rd. HA4: Ruis 34Z 65
Downbarton Ho. SW9 53Qb 112
(off Gosling Way)
Downbury M. SW18 57Cb 111
Down Cl. UB5: N'olt 40X 65
Downderry Rd. BR1: Broml 62Fc 137
DOWNE 83Qc 180
Downe Av. TN14: Cud 84Tc 180
Downe Bank Nature Reserve 85Rc 180
Downe Cl. DA16: Well 52Yc 117
Downedge AL3: St A 1P 5
Downend SE18 52Rc 116
Downend Ct. SE15 51Ub 113
(off Bibury Cl.)
Downer Dr. WD3: Sarr 8J 11
Downe Rd. BR2: Kes 81Nc 180
Downe Rd. CR4: Mitc 68Hb 133
Downe Rd. TN14: Cud 85Sc 180
Downer's Cott. SW4 56Lb 112
Downesbury NW3 37Hb 69
(off Steele's Rd.)
Downes Cl. TW1: Twick 58Ka 108
Downes Ct. N21 18Qb 32
Downes Ho. CR0: Wadd 77Rb 157
(off Violet La.)
Downe Ter. TW10: Rich 58Na 109
Downey Ho. E1 42Zb 92
(off Globe Rd.)
Downfield KT4: Wor Pk 74Va 154
Downfield Cl. W9 42Db 89
Downfield Rd. EN8: Chesh 3Ac 20
Down Hall Rd. KT2: King T 67Ma 131
DOWNHAM 64Fc 137
Downham Cl. RM5: Col R 24Cd 56
Downham Ct. KT12: Walt T 76Y 151
(off Long Lodge Dr.)
Downham Ct. N1 38Tb 71
(off Downham Rd.)
Downham Ent. Cen. SE6 61Hc 137
Downham Health & Leisure Cen. 63Hc 137
Downham La. BR1: Broml 64Fc 137
Downham Rd. N1 38Tb 71
Downham Way BR1: Broml 64Fc 137
Downham Wharf N1 39Ub 71
(off Downham Rd.)
Downhills Av. N17 27Tb 51
Downhills Pk. Rd. N17 27Sb 51
Downhills Way N17 27Sb 51
Down House 84Qc 180
Downhurst Av. NW7 22Ta 47
Downhurst Ct. NW4 27Ya 48
Downie Wlk. SE18 50Rc 94
(off Brumwell Av.)
Downing Cl. HA2: Harr 27Ea 46
Downing Ct. WC1 6G 217 (42Nb 90)
(off Grenville St.)
Downing Ct. WD6: Bore 11Pa 29
(off Bennington Dr.)
Downing Dr. UB6: G'frd 39Fa 66
Downing Ho. W10 44Za 88
(off Cambridge Gdns.)
Downing Path SL2: Slou 2C 80
Downing Rd. RM9: Dag 38Bd 75
Downings E6 44Qc 94
Downing St. SW1 1F 229 (47Nb 90)
Downings Wood WD3: Map C 22F 42
Downland Cl. KT18: Tatt C 90Xa 174
Downland Cl. N20 18Eb 31
Downland Ct. E11 33Gc 73
Downland Gdns. KT18: Tatt C 90Xa 174
Downlands EN9: Walt A 6Gc 21
Downlands Cl. CR5: Coul 86Kb 176
Downlands Rd. CR8: Purl 85Nb 176
Downland Way KT18: Tatt C 90Xa 174
Downleys Cl. SE9 61Nc 138
Downman Rd. SE9 55Nc 116
Down Pl. W6 49Xa 88
Down Rd. TW11: Tedd 65Ka 130
Downs, The AL10: Hat 2C 8
Downs, The KT22: Lea 97La 192
Downs, The SW20 66Za 132
Downs Av. BR7: Chst 64Pc 138
Downs Av. DA1: Dart 59Qd 119
Downs Av. HA5: Pinn 30Aa 45
Downs Av. KT18: Eps 86Ua 174
Downs Bri. Rd. BR3: Beck 67Fc 137
Downs Ct. RH1: Redh 3A 208
Downs Ct. UB6: G'frd 41Ja 86
Downs Ct. Pde. E8 36Xb 71
(off Amhurst Rd.)
Downs Ct. Rd. CR8: Purl 84Rb 177
Downsell Rd. E15 35Ec 72
Downsfield AL10: Hat 3D 8
Downsfield Rd. E17 30Ac 52
Downshall Av. IG3: Ilf 30Uc 54
Downs Hill BR3: Beck 66Fc 137
Downs Hill DA13: Nflt G 66Ee 143
Downs Hill Rd. KT18: Eps 86Ua 174
Downshire Hill NW3 35Fb 69
Downs Ho. TN13: S'oaks 94Ld 203
Downs Ho. Rd. KT18: Eps D 90Ua 174
Downside HP2: Hem H 1N 3
DOWNSIDE 90X 171
Downside KT16: Chert 74H 149
Downside KT18: Eps 86Ua 174
Downside TW1: Twick 62Ha 130
Downside TW16: Sun 67W 128
Downside Bri. Rd. KT11: Cobh 86X 171
Downside Cl. SW19 65Eb 133
Downside Comn. KT11: D'side 90X 171
Downside Comn. Rd. KT11: D'side 90X 171
Downside Ct. RH1: Mers 1C 208
Downside Cres. NW3 36Gb 69
Downside Cres. W13 42Ja 86
Downside Orchard GU22: Wok 89C 168
Downside Rd. KT11: D'side 88X 171
Downside Rd. SM2: Sutt 79Fb 155
Downside Wlk. TW8: Bford 51Ma 109
(off Windmill Rd.)
Downside Wlk. UB5: N'olt 41Ba 85
Downsland Dr. CM14: B'wood 20Yd 40
Downs La. AL10: Hat 2C 8
Downs La. E5 35Xb 71
Downs La. KT22: Lea 95Ka 192
Downs Lodge Ct. KT17: Eps 86Ua 174
Downs Pk. Rd. E5 36Wb 71
Downs Pk. Rd. E8 36Wb 71
Downs Reach KT17: Eps D 89Ya 174
Downs Res. Site, The CR3: Cat'm 99Xb 197
Downs Rd. BR3: Beck 68Dc 136
Downs Rd. CR5: Coul 90Mb 176
Downs Rd. CR7: Thor H 67Sb 135
Downs Rd. CR8: Purl 83Rb 177
Downs Rd. DA11: Ist R 63Fe 143
Downs Rd. DA11: Nflt 63Fe 143
Downs Rd. DA13: Ist R 6A 144
Downs Rd. DA13: Nflt G 6A 144
Downs Rd. E5 35Wb 71
Downs Rd. EN1: Enf 14Ub 33
Downs Rd. KT18: Eps 87Ua 174
Downs Rd. KT18: Eps D 87Ua 174
Downs Rd. KT18: Eps D 92Sa 193
Downs Rd. RH5: Mick 100La 192
Downs Rd. SL3: L'ly 7P 81
Downs Rd. SM2: Sutt 82Db 175
Downs Side SM2: Cheam 83Bb 175
Down St. KT8: W Mole 71Ca 151
Down St. W1 7K 221 (46Kb 90)
Down St. M. W1 7K 221 (46Kb 90)
Downs Valley DA3: Hartl 70Ae 143
Downs Vw. KT20: Tad 93Xa 194
Downs Vw. TW7: Isle 53Ha 108
Downsview Av. GU22: Wok 93B 188
Downs Vw. Cl. BR6: Prat B 82Yc 181
Downsview Cl. BR8: Swan 69Hd 140
Downsview Cl. KT11: D'side 91X 191
Downsview Gdns. SE19 66Rb 135
Downs Vw. Rd. KT23: Bookh 99Ea 192
Downsview Rd. SE19 66Sb 135
Downsview Rd. TN13: S'oaks 97Hd 202
Downs Way KT18: Eps 88Va 174
Downs Way KT20: Tad 93Xa 194
Downs Way KT23: Bookh 98Ea 192
Downs Way RH8: Oxt 99Gc 199
Downsway BR6: Orp 78Uc 160
Downsway CR2: Sande 83Ub 177
Downsway CR3: Whyt 88Vb 177
Downsway, The SM2: Sutt 81Eb 175
Downsway Cl. KT20: Tad 93Wa 194
Downs Wood KT18: Tatt C 89Xa 174
Downswood RH2: Reig 3M 207
Downton Av. SW2 61Nb 134
Downton M. DA8: Erith 52Gd 118
Downtown Rd. SE16 47Ac 92
Down Way UB5: N'olt 41X 85
Downy Ho. W3 46Ua 88
Dowrey St. N1 39Qb 70
Dowry Wlk. WD17: Wat 10V 12
Dowsett Rd. N17 26Vb 51
Dowson Cl. SE5 56Tb 113
Dowson Ho. E1 44Zb 92
(off Bower St.)
Doyce St. SE1 1D 230 (47Sb 91)
Doyle Cl. DA8: Erith 53Gd 118
Doyle Gdns. NW10 39Wa 68
Doyle Ho. SW13 52Ya 110
(off Trinity Chu. Rd.)
Doyle Rd. SE25 70Wb 135
Doyle Way RM18: Tilb 4E 122
(off Coleridge Rd.)
D'Oyley St. SW1 5H 227 (49Jb 90)
D'Oyly Carte Island KT13: Weyb 74R 150
Doynton St. N19 33Kb 70
Draco Ga. SW15 55Ya 110
Draco St. SE17 51Sb 113
Dragmore St. SW4 58Mb 112
Dragonfly Cl. E13 41Kc 93
Dragonfly Cl. KT5: Surb 74Sa 153
Dragonfly Ct. NW9 25Ua 48
(off Heybourne Cres.)
Dragonfly Pl. SE4 55Ac 114
Dragon La. KT13: Weyb 83Q 170
Dragon Rd. SE15 51Ub 113
Dragons Way EN5: Barn 15Bb 31
Dragon Yd. WC1 2G 223 (44Nb 90)
Dragoon Rd. SE8 50Bc 92
Dragor Rd. NW10 42Sa 87
Drake Av. CR3: Cat'm 94Sb 197
Drake Av. SL3: L'ly 9P 81
Drake Av. TW18: Staines 64H 127
Drake Cl. CM14: W'ley 22Ae 59
Drake Cl. IG11: Bark 42Wc 95
Drake Cl. SE16 47Zb 92
Drake Ct. DA8: Erith 52Hd 118
(off Frobisher Rd.)
Drake Ct. KT5: Surb 70Na 131
(off Cranes Pk. Av.)
Drake Ct. SE1 2E 230 (47Sb 91)
(off Swan St.)
Drake Ct. SE19 64Vb 135
Drake Ct. W12 47Ya 88
(off Scott's Rd.)
Drake Cres. SE28 44Yc 95
Drakefell Rd. SE14 54Zb 114
Drakefell Rd. SE4 55Ac 114
Drakefield Rd. SW17 62Jb 134
Drake Hall E16 46Kc 93
(off Wesley Av.)
Drake Ho. E1 43Yb 92
(off Stepney Way)
Drake Ho. E14 45Ac 92
(off Victory Pl.)
Drake Ho. SW1 51Mb 112
(off Dolphin Sq.)
Drakeland Ho. W9 42Bb 89
(off Fernhead Rd.)
Drakeley Ct. N5 35Rb 71
Drake M. BR2: Broml 70Lc 137
Drake M. DA12: Grav'nd 3F 144
Drake M. RM12: Horn 37Jd 76
Drake Point DA8: Erith 50Gd 96
Drake Rd. CR0: C'don 73Pb 156
Drake Rd. CR4: Mitc 72Jb 156
Drake Rd. HA2: Harr 33Ba 65
Drake Rd. KT9: Chess 78Qa 153
Drake Rd. RM16: Chaf H 48Ae 99
Drake Rd. SE4 55Cc 114
Drakes, The SE8 51Cc 114
Drake's Cl. KT10: Esh 77Ca 151
Drakes Ct. SE23 60Yb 114
Drakes Ctyd. NW6 38Bb 69
Drakes Dr. AL1: St A 5F 6
Drakes Dr. HA6: Nwood 25R 44
Drake St. EN2: Enf 11Tb 33
Drake St. WC1 1H 223 (43Pb 90)
Drakes Wlk. E6 39Pc 74
Drakes Way AL10: Hat 2D 8
Drakes Way GU22: Wok 4P 187
Drakewood Rd. SW16 66Mb 134
Draper Cl. DA17: Belv 49Bd 95
Draper Cl. RM20: Grays 51Zd 121
Draper Cl. TW7: Isle 54Fa 108
Draper Ct. BR1: Broml 70Nc 138
Draper Ho. SE1 5C 230 (49Rb 91)
(off Newington Butts)
Draper Pl. N1 39Rb 71
(off Dagmar Ter.)
Drapers Almshouses E3 41Dc 92
(off Rainhill Way)
Drapers Cott. Homes NW7 21Va 48
(not continuous)
Draper's Ct. SW11 53Jb 112
(off Battersea Pk. Rd.)
Drapers Ct. RM12: Horn 33Nd 77
Drapers Cres. KT12: W Vill 82V 170
Drapers Gdns. EC2 2G 225 (44Tb 91)
Drapers Rd. E15 35Fc 73
Drapers Rd. EN2: Enf 12Rb 33
Drapers Rd. N17 27Wb 51
Drapers Yd. SW18 57Db 111
(off Ryland Blvd.)
Drappers Way SE16 49Wb 91
Draven Cl. BR2: Hayes 73Hc 159
Drawdock Rd. SE10 47Fc 93
Drawell Cl. SE18 50Uc 94
Drax Av. SW20 66Wa 132
Draxmont SW19 65Ab 132
Draycot Rd. E11 30Kc 53
Draycot Rd. KT6: Surb 74Qa 153
Draycott Av. HA3: Kenton 30Ka 46
Draycott Av. SW3 5E 226 (49Gb 89)
Draycott Cl. HA3: Kenton 30Ka 46
Draycott Cl. NW2 34Za 68
Draycott Cl. SE5 52Tb 113
(not continuous)
Draycott Ct. SW11 53Gb 111
(off Westbridge Rd.)

Draycott M. SW6 54Bb **111**
(off Laurel Bank Gdns.)
Draycott Pl. SW3 6F **227** (49Hb **89**)
Draycott Ter. SW3 5F **227** (49Hb **89**)
Dray Ct. HA0: Wemb 36Ja **66**
(off Brewery Cl.)
Drayford Cl. W9 42Bb **89**
Dray Gdns. SW2 57Pb **112**
Draymans M. SE15 54Vb **113**
Draymans Way TW7: Isle 55Ha **108**
Drayside M. UB2: S'hall 47Ba **85**
Drayson Cl. EN9: Walt A 4Gc **21**
Drayson M. W8 47Cb **89**
Drayton Av. BR6: Farnb 74Rc **160**
Drayton Av. EN6: Pot B 4Ab **16**
Drayton Av. IG10: Lough 17Pc **36**
Drayton Av. W13 45Ja **86**
Drayton Bri. Rd. W13 44Ha **86**
Drayton Bri. Rd. W7 45Ha **86**
Drayton Cl. IG1: Ilf 32Tc **74**
Drayton Cl. KT22: Fet 96Ga **192**
Drayton Cl. TW4: Houn 57Ba **107**
Drayton Ct. UB7: W Dray 49P **83**
Drayton Ford WD3: Rick 19J **25**
Drayton Gdns. N21 17Rb **33**
Drayton Gdns. SW10 7A **226** (50Eb **89**)
Drayton Gdns. UB7: W Dray 47N **83**
Drayton Gdns. W13 45Ja **86**
Drayton Grn. W13 45Ja **86**
Drayton Grn. Rd. W13 45Ka **86**
Drayton Green Station (Rail) 44Ha 86
Drayton Gro. W13 45Ja **86**
Drayton Ho. E11 32Fc **73**
Drayton Ho. SE5 52Tb **113**
(off Elmington Rd.)
Drayton Pk. N5 35Qb **70**
Drayton Pk. M. N5 36Qb **70**
Drayton Park Station (Rail) 35Qb 70
Drayton Rd. CR0: C'don 75Rb **157**
Drayton Rd. E11 32Fc **73**
Drayton Rd. N17 26Ub **51**
Drayton Rd. NW10 39Va **68**
Drayton Rd. W13 45Ja **86**
Drayton Rd. WD6: Bore 14Qa **29**
Drayton Waye HA3: Kenton 30Ka **46**
Dray Wlk. E1 6K **219** (42Vb **91**)
Dreadnought Cl. SW19 68Fb **133**
Dreadnought Library 51Ec **114**
Dreadnought St. SE10 48Gc **93**
Dreadnought Wlk. SE10 51Dc **114**
Drenon Sq. UB3: Hayes 45V **84**
Dresden Cl. NW6 37Db **69**
Dresden Ho. SE11 5J **229** (49Pb **90**)
(off Lambeth Wlk.)
Dresden Ho. SW11 54Jb **112**
(off Dagnall St.)
Dresden Rd. N19 32Lb **70**
Dresden Way KT13: Weyb 78S **150**
Dressington Av. SE4 58Cc **114**
Drewery Ct. SE3 55Gc **115**
Drewett Ho. E1 44Wb **91**
(off Christian St.)
Drew Gdns. UB6: G'frd 37Ha **66**
Drew Ho. SE8 50Cc **92**
Drew Ho. SW16 62Nb **134**
Drew Mdw. SL2: Farn C 5G **60**
Drew Pl. CR3: Cat'm 95Tb **197**
Drew Rd. E16 46Mc **93**
(not continuous)
Drewstead La. SW16 61Mb **134**
Drewstead Rd. SW16 61Mb **134**
Drey, The SL9: Chal P 22A **42**
Drey Ct. KT4: Wor Pk 75Wa **154**
(off The Avenue)
Driffield Ct. NW9 25Ua **48**
(off Pageant Av.)
Driffield Rd. E3 40Ac **72**
Drift, The BR2: Broml 76Mc **159**
DRIFT BRIDGE 86Ya **174**
Drift Ct. E16 45Rc **94**
Drift Golf Course 95V **190**
Drift La. KT11: Stoke D 80Ba **171**
Drift Rd. KT24: E Hor 96T **190**
Drift Rd. KT24: Eff J 96T **190**
Drift Way SL3: Coln 53E **104**
Driftway, The CR4: Mitc 67Jb **134**
Driftway, The HP2: Hem H 2P **3**
Driftway, The KT22: Lea 95Ka **192**
(not continuous)
Driftway, The SM7: Bans 87Ya **174**
Driftway Ho. E3 40Bc **72**
(off Stafford Rd.)
Driftwood Av. AL2: Chis G 8N **5**
Driftwood Dr. CR8: Kenley 89Rb **177**
Drill Hall Rd. KT16: Chert 73J **149**
Drinkwater Ho. SE5 52Tb **113**
(off Picton St.)
Drinkwater Rd. HA2: Harr 33Da **65**
Driscoll Way CR3: Cat'm 95Tb **197**
Drive, The EN2: Enf Farr Rd. 11Tb **33**
Drive, The EN2: Enf St Nicholas Ho's. 7Jb **18**
Drive, The AL2: Lon C 7E **6**
Drive, The AL9: Brk P 7J **9**
Drive, The BR3: Beck 68Cc **136**
Drive, The BR4: W W'ck 73Fc **159**
Drive, The BR6: Orp 75Vc **161**
Drive, The BR7: Chst 69Vc **139**
Drive, The CM13: Gt War 22Yd **58**
Drive, The CR5: Coul 86Nb **176**
Drive, The CR7: Thor H 70Tb **135**
Drive, The DA12: Grav'nd 3F **144**
Drive, The DA14: Sidc 62Xc **139**
Drive, The DA3: Lfield 69De **143**
Drive, The DA5: Bexl 58Yc **117**
Drive, The DA8: Erith 52Dd **118**
Drive, The E17 27Dc **52**
Drive, The E18 28Jc **53**
Drive, The E4 17Fc **35**
Drive, The EN5: Barn 13Ab **30**
Drive, The EN5: New Bar 16Eb **31**
Drive, The EN6: Pot B 5Bb **17**
Drive, The EN7: G Oak 1Rb **19**
Drive, The GU22: Wok 2M **187**
Drive, The GU25: Vir W 71B **148**
Drive, The HA2: Harr 31Ca **65**
Drive, The HA6: Nwood 26U **44**
Drive, The HA8: Edg 22Qa **47**
Drive, The HA9: Wemb 33Sa **67**
Drive, The IG1: Ilf 30Nc **54**
Drive, The IG10: Lough 13Nc **36**
Drive, The IG11: Bark 38Vc **75**
Drive, The IG9: Buck H 17Lc **35**
Drive, The KT10: Esh 74Ea **152**
Drive, The KT11: Cobh 86Aa **171**
Drive, The KT18: Head 96Qa **193**
Drive, The KT19: Ewe 79Va **154**
Drive, The KT2: King T 66Sa **131**
Drive, The KT20: Lwr K 98Bb **195**
Drive, The KT22: Fet 94Ga **192**
Drive, The KT22: Lea 95Na **193**
Drive, The KT6: Surb 73Na **153**
Drive, The N11 23Lb **50**
Drive, The N3 24Cb **49**
Drive, The N6 29Hb **50**
Drive, The N7 37Pb **70**
(not continuous)
Drive, The NW10 39Va **68**
Drive, The NW11 31Ab **68**
Drive, The RM3: Hrld W 25Nd **57**
Drive, The RM4: Stap A 17Ed **38**
Drive, The RM5: Col R 25Ed **56**
Drive, The SL3: Dat 3M **103**
Drive, The SL3: L'ly 47A **82**
Drive, The SL4: Wind 3D **102**
Drive, The SL9: Chal P 24A **42**
Drive, The SM2: Cheam 84Bb **175**
Drive, The SM4: Mord 71Eb **155**
Drive, The SM6: W'gton 82Lb **176**
Drive, The SM7: Bans 89Ab **174**
Drive, The SW20 66Ya **132**
Drive, The SW6 54Ab **110**
Drive, The TN13: S'oaks 96Kd **203**
Drive, The TW14: Felt 59Y **107**
Drive, The TW15: Ashf 66T **128**
Drive, The TW19: Wray 7P **103**
Drive, The TW3: Houn 54Fa **108**
Drive, The TW7: Isle 54Fa **108**
Drive, The UB10: Ick 35N **63**
Drive, The W3 44Sa **87**
Drive, The WD17: Wat 9T **12**
Drive, The WD3: Rick 15K **25**
Drive, The WD7: R'lett 6Ja **14**
Drive Ct. HA8: Edg 22Qa **47**
Drive Mans. SW6 54Ab **110**
(off Fulham Rd.)
Drive Mead CR5: Coul 86Nb **176**
Drive Rd. CR5: Coul 92Mb **196**
(not continuous)
Drive Spur KT20: Kgswd 93Db **195**
Driveway, The EN6: Cuff 1Nb **18**
Driveway, The HP1: Hem H 3K **3**
Droitwich Cl. SE26 62Wb **135**
Dromey Gdns. HA3: Hrw W 24Ha **46**
Dromore Rd. SW15 58Ab **110**
Dronfield Gdns. RM8: Dag 36Yc **75**
Dron Ho. E1 43Yb **92**
(off Adelina Gro.)
Droop St. W10 41Za **88**
Drop La. AL2: Brick W 2Da **13**
Dropmore Ho. SL1: Burn 4A **60**
Dropmore Pk. 4A **60**
Dropmore Rd. SL1: Burn 5A **60**
Drovers Ct. KT1: King T 68Na **131**
(off Fairfield E.)
Drovers Mead CM14: W'ley 21Xd **58**
Drovers Pl. SE15 52Yb **114**
Drovers Rd. CR2: S Croy 78Tb **157**
Drovers Way AL3: St A 2B **6**
Drovers Way N7 37Nb **70**
Droveway IG10: Lough 12Rc **36**
Drove Way, The DA13: Ist R 6A **144**
Druce Rd. SE21 58Ub **113**
Drudgeon Way DA2: Bean 62Xd **142**
Druids Cl. KT21: Asht 92Pa **193**
Druid St. SE1 1J **231** (47Ub **91**)
(not continuous)
Druids Way BR2: Broml 70Fc **137**
Drumaline Ridge KT4: Wor Pk 75Ua **154**
Drum Ct. N1 38Nb **70**
(off Gifford St.)
Drummer Stagpole M. NW7 22Ab **48**
Drummond Av. RM7: Rom 28Fd **56**
Drummond Cl. DA8: Erith 53Gd **118**
Drummond Ct. CM15: B'wood 17Yd **40**
Drummond Ct. KT19: Eps 84Pa **173**
Drummond Ct. N12 24Gb **49**
Drummond Ct. W3 48Sa **87**
(off Palmerston Rd.)
Drummond Cres. NW1 3D **216** (41Mb **90**)
Drummond Dr. HA7: Stan 24Ha **46**
Drummond Gdns. KT19: Eps 83Sa **173**
Drummond Ga. SW1 7E **228** (50Mb **90**)
Drummond Ho. E2 40Wb **71**
(off Goldsmiths Row)
Drummond Ho. N2 26Eb **49**
(off Font Hills)
Drummond Ho. SL4: Wind 5H **103**
(off Balmoral Gdns.)
Drummond Pl. TW1: Twick 59Ka **108**
Drummond Rd. CR0: C'don 75Sb **157**
Drummond Rd. E11 30Lc **53**
Drummond Rd. RM7: Rom 28Fd **56**
Drummond Rd. SE16 48Xb **91**
Drummonds, The CM16: Epp 2Wc **23**
Drummonds, The IG9: Buck H 19Kc **35**
Drummonds Pl. TW9: Rich 56Na **109**
Drummond St. NW1 5B **216** (42Lb **90**)
Drummond Way N1 38Qb **70**
Druries HA1: Harr 32Ga **66**
(off High St.)
Drury Cl. SL9: Chal P 22B **42**
Drury Cl. SW15 58Wa **110**
Drury Cres. CR0: Wadd 75Qb **156**
Drury Ho. SW8 53Lb **112**
Drury La. WC2 2G **223** (44Nb **90**)
Drury Lane Theatre Royal 3G **223** (44Nb **90**)
(off Catherine St.)
Drury Rd. HA1: Harr 31Ea **66**
Drury Way NW10 36Ta **67**
Drury Way Ind. Est. NW10 36Sa **67**
Dryad St. SW15 55Za **110**
Dry Arch Rd. SL5: S'dale 2D **146**
Dryburgh Gdns. NW9 27Qa **47**
Dryburgh Ho. SW1 7K **227** (50Kb **90**)
(part of Abbots Mnr.)
Dryburgh Rd. SW15 55Xa **110**
Dryden Av. W7 44Ha **86**
Dryden Bldg. E1 44Wb **91**
(off Commercial Rd.)
Dryden Cl. IG6: Ilf 23Vc **55**
Dryden Cl. SW4 57Mb **112**
Dryden Ct. SE11 6A **230** (49Qb **90**)
Dryden Mans. W14 51Ab **110**
(off Queen's Club Gdns.)
Dryden Pl. RM18: Tilb 3D **122**
Dryden Rd. DA16: Well 53Vc **117**
Dryden Rd. EN1: Enf 16Ub **33**
Dryden Rd. HA3: W'stone 25Ha **46**
Dryden Rd. SW19 65Eb **133**
Dryden St. WC2 3G **223** (44Nb **90**)
Dryden Towers RM3: Rom 24Kd **57**
Dryden Way BR6: Orp 74Wc **161**
Dryfield Cl. NW10 37Sa **67**
Dryfield Rd. HA8: Edg 23Sa **47**
Dryfield Wlk. SE8 51Cc **114**
DRYHILL 96Dd 202
Dryhill La. TN14: Sund 95Dd **202**
Dryhill Local Nature Reserve 96Dd **202**
Dryhill Rd. DA17: Belv 51Bd **117**
Dryland Av. BR6: Orp 77Vc **161**
Dryland Rd. TN15: Bor G 93Be **205**
Drylands Rd. N8 30Nb **50**
Drynham Pk. KT13: Weyb 76U **150**
Drysdale Av. E4 17Dc **34**
Drysdale Cl. HA6: Nwood 24U **44**
Drysdale Dwellings E8 36Vb **71**
(off Dunn St.)
Drysdale Pl. N1 3J **219** (41Ub **91**)
Drysdale St. N1 3J **219** (41Ub **91**)
Duarte Pl. RM16: Chaf H 48Be **99**
Dublin Av. E8 39Wb **71**
Dublin Ct. HA2: Harr 33Fa **66**
(off Northolt Rd.)
Dubrae Cl. AL3: St A 4N **5**
Du Burstow Ter. W7 47Ga **86**
Ducaine Apts. E3 41Bc **92**
(off Merchant St.)
Ducal St. E2 4K **219** (41Vb **91**)
Du Cane Cl. W12 44Ya **88**
Du Cane Ct. SW17 60Jb **112**
Du Cane Rd. W12 44Va **88**
Ducavel Ho. SW2 60Pb **112**
Duchess Cl. N11 22Kb **50**
Duchess Cl. SM1: Sutt 77Eb **155**
Duchess Ct. KT13: Weyb 76T **150**
Duchess Cres. HA7: Stan 20Ha **28**
Duchess Dr. E13 42Kc **93**
Duchess Gro. IG9: Buck H 19Kc **35**
Duchess M. W1 1A **222** (43Kb **90**)
Duchess M. W3 46Qa **87**
Duchess of Bedford Ho. W8 47Cb **89**
(off Duchess of Bedford's Wlk.)
Duchess of Bedford's Wlk. W8 47Cb **89**
Duchess St. SL1: Slou 6C **80**
Duchess St. W1 1A **222** (43Kb **90**)
Duchess Theatre 4H **223** (45Pb **90**)
(off Catherine St.)
Duchess Wlk. SE1 7K **225** (46Vb **91**)
Duchess' Wlk. TN15: S'oaks 98Nd **203**
Duchy Rd. EN4: Had W 10Fb **17**
Duchy St. SE1 6A **224** (46Qb **90**)
(not continuous)
Ducie St. SW4 56Pb **112**
Duckett M. N4 30Rb **51**
Duckett Rd. N4 30Rb **51**
Duckett's Apts. E3 38Bc **72**
(off Wick La.)
Ducketts Rd. DA1: Cray 57Hd **118**
Duckett St. E1 42Zb **92**
Duckham Ct. E14 49Dc **92**
(off Nauticus Wlk.)
Ducking Stool Ct. RM1: Rom 28Gd **56**
Duck La. W1 3D **222** (44Mb **90**)
(off Broadwick St.)
Duck Lees La. EN3: Pond E 14Ac **34**
Duck's Hill Rd. HA4: Ruis 28S **44**
Duck's Hill Rd. HA6: Nwood 25R **44**
DUCKS ISLAND 16Za 30
Ducks Wlk. TW1: Twick 57La **108**
Duck Wood Community Nature Reserve 22Rd **57**
Duckworth Dr. KT22: Asht 92Ma **193**
Du Cros Dr. HA7: Stan 23Ma **47**
Du Cros Rd. W3 46Ua **88**
Dudbrook Rd. CM14: Kel C 11Nd **39**
Dudbrook Rd. CM14: Nave 11Nd **39**
DUDDEN HILL 36Xa 68
Dudden Hill La. NW10 35Va **68**
Dudden Hill Pde. NW10 35Va **68**
Duddington Cl. SE9 63Mc **137**
Dudley Av. EN8: Walt C 4Zb **20**
Dudley Av. HA3: Kenton 27La **46**
Dudley Cl. HP3: Bov 9C **2**
Dudley Cl. KT15: Add 76L **149**
Dudley Cl. RM16: Chaf H 47Ae **99**
Dudley Ct. NW11 28Bb **49**
Dudley Ct. SL1: Slou 8L **81**
Dudley Ct. W1 3F **221** (44Hb **89**)
(off Up. Berkeley St.)
Dudley Ct. WC2 2F **223** (44Nb **90**)
Dudley Dr. HA4: Ruis 36X **65**
Dudley Dr. SM4: Mord 74Ab **154**
Dudley Gdns. HA2: Harr 32Fa **66**
Dudley Gdns. RM3: Rom 23Md **57**
Dudley Gdns. W13 47Ka **86**
Dudley Gro. KT18: Eps 86Sa **173**
Dudley Ho. HP3: Bov 9C **2**
Dudley M. SW2 58Qb **112**
Dudley Pl. TW19: Stanw 58P **105**
Dudley Pl. UB3: Harl 49T **84**
Dudley Rd. DA11: Nflt 9A **122**
Dudley Rd. E17 26Cc **52**
Dudley Rd. HA2: Harr 33Ea **66**
Dudley Rd. IG1: Ilf 35Rc **74**
Dudley Rd. KT1: King T 69Pa **131**
Dudley Rd. KT12: Walt T 72W **150**
Dudley Rd. N3 26Db **49**
Dudley Rd. NW6 40Ab **68**
Dudley Rd. RM3: Rom 23Md **57**
Dudley Rd. SW19 65Cb **133**
Dudley Rd. TW14: Bedf 60S **106**
Dudley Rd. TW15: Ashf 64P **127**
Dudley Rd. TW9: Rich 54Pa **109**
Dudley Rd. UB2: S'hall 47Z **85**
Dudley St. W2 1B **220** (43Fb **89**)
Dudley Wharf SL0: Iver 46E **82**
Dudley Wharf Caravans SL0: Iver 46D **82**
Dudlington Rd. E5 33Yb **72**
Dudmaston M. SW3 7C **226** (50Fb **89**)
(off Fulham Rd.)
Dudrich Cl. N11 23Hb **49**
Dudrich M. EN2: Enf 11Qb **32**
Dudrich M. SE22 57Vb **113**
Dudsbury Rd. DA1: Dart 58Kd **119**
Dudsbury Rd. DA14: Sidc 65Xc **139**
Dudset La. TW5: Cran 53W **106**
Duel Ct. TW20: Eng G 6M **125**
Duett Ct. TW5: Hest 52Aa **107**
Duffell Ho. SE11 7J **229** (50Pb **90**)
(off Loughborough St.)
Dufferin Av. EC1 6F **219** (42Tb **91**)
(off Dufferin St.)
Dufferin Ct. EC1 6F **219** (42Tb **91**)
(off Dufferin St.)
Dufferin St. EC1 6E **218** (42Sb **91**)
Duffield Cl. HA1: Harr 29Ha **46**
Duffield Cl. RM16: Chaf H 48Be **99**
Duffield Dr. N15 28Vb **51**
Duffield La. SL2: Stoke P 7K **61**
Duffield Pk. SL2: Stoke P 1L **81**
Duffield Rd. KT20: Walt H 96Xa **194**
Duffins Orchard KT16: Ott 80E **148**
Duffins Orchard Mobile Homes KT16: Ott 80E **148**
Duff St. E14 44Dc **92**
Dufour's Pl. W1 3C **222** (44Lb **90**)
Dufton Dwellings E15 35Gc **73**
(off High Rd. Leyton)
Dugard Way SE11 5B **230** (49Rb **91**)
Dugdale Cen. Enfield 14Tb **33**
Dugdale Ct. NW10 41Xa **88**
(off Harrow Rd.)
DUGDALE HILL 5Ab 16
Dugdale Hill La. EN6: Pot B 5Ab **16**
Dugdale Ho. TW20: Egh 64E **126**
(off Rowan Av.)
Dugdales WD3: Crox G 14Q **26**
Duggan Dr. BR7: Chst 65Nc **138**
Dugolly Av. HA9: Wemb 34Ra **67**
Dujardin M. EN3: Pond E 16Zb **34**
Duke Ct. TW3: Houn 56Ba **107**
Duke Gdns. IG6: Ilf 28Tc **54**
Duke Humphrey Rd. SE3 53Gc **115**
Duke of Cambridge Cl. TW2: Whitt 58Fa **108**
Duke of Clarence Ct. SE17 7D **230** (50Sb **91**)
(off Manor Pl.)
Duke of Edinburgh Rd. SM1: Sutt 75Fb **155**
Duke of Wellington Av. SE18 48Rc **94**
Duke of Wellington Pl. SW1 1J **227** (47Jb **90**)
Duke of York Column 7E **222** (46Mb **90**)
Duke of York Sq. SW3 7G **227** (50Hb **89**)
Duke of York's Theatre 5F **223** (45Nb **90**)
(off St Martin's La.)
Duke of York St. SW1 6C **222** (46Lb **90**)
Duke Pl. SL1: Slou 5K **81**
(off Montague Rd.)
Duke Rd. IG6: Ilf 28Tc **54**
Duke Rd. W4 50Ta **87**
Duke's Av. HA8: Edg 23Pa **47**
Duke's Av. N10 27Kb **50**
Duke's Av. W4 50Ta **87**
Dukes Av. CM16: They B 7Uc **22**
Dukes Av. HA1: Harr 28Ga **46**
Dukes Av. HA2: Harr 30Ba **45**
Dukes Av. KT2: King T 63Ma **131**
Dukes Av. KT3: N Mald 69Ua **132**
Dukes Av. N3 25Db **49**
Dukes Av. RM17: Grays 48Ce **99**
Dukes Av. TW10: Ham 63La **130**
Dukes Av. TW4: Houn 56Aa **107**
Dukes Av. UB5: N'olt 38Aa **65**
Dukes Cl. SL9: Ger X 2P **61**
Dukes Cl. TW12: Hamp 64Ba **129**
Dukes Cl. TW15: Ashf 63S **128**
Dukes Ct. E6 39Qc **74**
(not continuous)
Dukes Ct. GU21: Wok 89B **168**
Dukes Ct. KT15: Add 77L **149**
Dukes Ct. KT19: Ewe 81Ua **174**
Dukes Ct. SE13 54Ec **114**
Dukes Ct. SW14 54Ta **109**
Dukes Ct. W2 45Db **89**
(off Moscow Rd.)
Dukes Dr. SL2: Farn C 6D **60**
Dukes Ga. W4 49Sa **87**
Dukes Grn. Av. TW14: Felt 57W **106**
Dukes Head Pas. TW12: Hamp 66Ea **130**
Duke's Head Yd. N6 32Kb **70**
Dukes Hill CR3: Wold 92Ac **198**
Dukes Shore Wharf E14 45Bc **92**
Duke's Ho. SW1 5E **228** (49Mb **90**)
(off Vincent St.)
Dukes Kiln Dr. SL9: Ger X 2N **61**
Dukes La. SL4: Wind 5F **124**
Dukes La. SL5: Asc 6E **124**
Dukes La. SL9: Ger X 31A **62**
Dukes La. W8 47Db **89**
Duke's La. Chambers W8 47Db **89**
(off Dukes La.)
Duke's La. Mans. W8 47Db **89**
(off Dukes La.)
Dukes Lodge W8 46Bb **89**
(off Holland Wlk.)
Dukes Meadow Golf & Tennis 54Ta **109**
Duke's Meadow Golf Course 54Ta **109**
Duke's Meadows 54Ta **109**
Duke's M. W1 2J **221** (44Jb **90**)
(off Duke St.)
Dukes M. N10 27Kb **50**
Dukes Orchard DA5: Bexl 60Ed **118**
Duke's Pas. E17 28Ec **52**
Duke's Pl. CM14: B'wood 18Yd **40**
Duke's Pl. EC3 3J **225** (44Ub **91**)
Dukes Point N6 32Kb **70**
(off Dukes Head Yd.)
Dukes Ride SL9: Ger X 32A **62**
Dukes Ride UB10: Ick 35N **63**
Duke's Rd. WC1 4E **216** (41Mb **90**)
Dukes Rd. E6 39Qc **74**
Dukes Rd. KT12: Hers 78Z **151**
Dukes Rd. W3 42Qa **87**
Dukesthorpe Rd. SE26 63Zb **136**
Duke St. GU21: Wok 89B **168**
Duke St. SL4: Wind 2G **102**
Duke St. SM1: Sutt 77Fb **155**
Duke St. TW9: Rich 56Ma **109**
Duke St. W1 2J **221** (44Jb **90**)
Duke St. WD17: Wat 13Y **27**
Duke St. Hill SE1 6G **225** (46Tb **91**)
Duke St. Mans. W1 3J **221** (44Jb **90**)
(off Duke St.)
Duke St. St James's SW1 6C **222** (46Lb **90**)
Dukes Valley SL9: Ger X 3M **61**
Dukes Way BR4: W W'ck 76Gc **159**
Dukes Way HA9: Wemb 36Na **67**
Dukes Way UB8: Uxb 39K **63**
Dukes Wood Av. SL9: Ger X 31A **62**
Dukes Wood Dr. SL9: Ger X 2N **61**
Duke's Yd. W1 4J **221** (45Jb **90**)
Dukes Yd. WD24: Wat 10W **12**
Dulas St. N4 32Pb **70**
Dulcie Cl. DA9: Ghithe 58Ud **120**
Dulford St. W11 45Ab **88**
Dulka Rd. SW11 57Hb **111**
Dullshot Grn. KT17: Eps 85Ua **174**
Dulverton NW1 1C **216** (39Lb **70**)
(off Royal College St.)
Dulverton Mans. WC1 6J **217** (42Pb **90**)
(off Gray's Inn Rd.)
Dulverton Rd. CR2: Sels 82Yb **178**
Dulverton Rd. HA4: Ruis 33W **64**
Dulverton Rd. RM3: Rom 23Md **57**
Dulverton Rd. SE9 61Tc **138**
DULWICH 61Ub 135
Dulwich & Sydenham Hill Golf Course 61Vb **135**
Dulwich Bus. Cen. SE23 60Zb **114**
Dulwich Comn. SE21 60Ub **113**
Dulwich Comn. SE22 60Wb **113**
DULWICH COMMUNITY HOSPITAL 56Ub **113**
Dulwich Hamlet FC 56Ub **113**
Dulwich Lawn Cl. SE22 57Vb **113**
Dulwich Leisure Cen. 56Wb **113**
Dulwich Oaks, The SE21 62Vb **135**
Dulwich Picture Gallery 59Tb **113**
Dulwich Ri. Gdns. SE22 57Vb **113**
Dulwich Rd. SE24 57Qb **112**
Dulwich Upper Wood Nature Pk. 64Vb **135**
Dulwich Village SE21 58Tb **113**
DULWICH VILLAGE 59Ub 113
Dulwich Way WD3: Crox G 15Q **26**
Dulwich Wood Av. SE19 63Ub **135**
Dulwich Wood Pk. SE19 63Ub **135**
Dumain Ct. SE11 6B **230** (49Rb **91**)
(off Opal St.)
Dumas Way WD18: Wat 14U **26**
Dumbarton Av. EN8: Walt C 6Zb **20**
Dumbarton Ct. SW2 58Nb **112**
Dumbarton Rd. SW2 58Nb **112**
Dumbarton Way SL3: L'ly 10N **81**
Dumbleton Cl. KT1: King T 67Ra **131**
Dumbletons, The WD3: Map C 21G **42**
Dumbreck Rd. SE9 56Pc **116**
Dumfries Cl. WD19: Wat 20V **26**
Dumont Rd. N16 34Ub **71**
Dumpton Pl. NW1 38Jb **70**
Dumsey Eyot KT16: Chert 73N **149**
Dumville Dr. RH9: G'stone 3P **209**
Dunally Pk. TW17: Shep 73T **150**
Dunbar Av. BR3: Beck 70Ac **136**
Dunbar Av. RM10: Dag 34Cd **76**
Dunbar Av. SW16 68Qb **134**
Dunbar Cl. SL2: Slou 5L **81**
Dunbar Cl. UB4: Hayes 43X **85**
Dunbar Ct. BR2: Broml 69Hc **137**
(off Durham Rd.)
Dunbar Ct. KT12: Walt T 74Y **151**
Dunbar Ct. SM1: Sutt 78Fb **155**
Dunbar Gdns. RM10: Dag 36Cd **76**
Dunbar Rd. E7 37Jc **73**
Dunbar Rd. KT3: N Mald 70Sa **131**
Dunbar Rd. N22 25Qb **50**
Dunbar St. SE27 62Sb **135**
Dunbar Twr. E8 37Vb **71**
(off Dalston Sq.)
Dunbar Wharf E14 45Bc **92**
(off Narrow St.)
Dunblane Cl. HA8: Edg 19Ra **29**
Dunblane Rd. SE9 55Nc **116**
Dunboe Pl. TW17: Shep 73S **150**
Dunboyne Pl. SL4: Old Win 6L **103**
Dunboyne Rd. NW3 36Hb **69**
Dunbridge Ho. SW15 58Va **110**
Dunbridge St. E2 42Wb **91**
Duncan Cl. EN5: New Bar 14Eb **31**
Duncan Ct. AL1: St A 4D **6**
Duncan Ct. E14 43Ec **92**
(off Teviot St.)
Duncan Ct. N21 18Rb **33**
Duncan Gdns. TW18: Staines 65J **127**
Duncan Gro. W3 44Ua **88**
Duncan Ho. E15 39Fc **73**
Duncan Ho. NW3 38Hb **69**
(off Fellows Rd.)
Duncan Ho. SW1 50Lb **90**
(off Dolphin Sq.)
Duncannon Cres. SL4: Wind 5B **102**
Duncannon Ho. SW1 7E **228** (50Mb **90**)
(off Lindsay Sq.)
Duncannon Pl. DA9: Ghithe 56Yd **120**
Duncannon St. WC2 5F **223** (45Nb **90**)
Duncan Rd. E8 39Xb **71**
Duncan Rd. KT20: Tad 91Ab **194**
Duncan Rd. TW9: Rich 56Na **109**
Duncan St. N1 1B **218** (40Rb **71**)
Duncans Yd. TN16: Westrm 98Tc **200**
Duncan Ter. N1 2B **218** (40Rb **71**)
(not continuous)
Duncan Way WD23: Bush 12Ba **27**
Dunch St. E1 44Xb **91**
Dunchurch Ho. RM10: Dag 38Cd **76**
Duncombe Ct. RM19: Purf 50Rd **97**
(off Wingrove Dr.)
Duncombe Ct. TW18: Staines 66H **127**
Duncombe Hill SE23 59Ac **114**
Duncombe Rd. N19 32Mb **70**
Duncrievie Rd. SE13 58Fc **115**
Duncroft SE18 52Uc **116**
Duncroft SL4: Wind 5D **102**
Duncroft Cl. RH2: Reig 6H **207**
Duncroft Mnr. TW18: Staines 63G **126**
Dundalk Ho. E1 44Yb **92**
(off Clark St.)
Dundalk Rd. SE4 55Ac **114**
Dundas Ct. SE10 51Dc **114**
(off Dowells St.)
Dundas Gdns. KT8: W Mole 69Da **129**
Dundas Ho. E2 40Yb **72**
(off Bishop's Way)
Dundas M. EN3: Enf L 9Cc **20**
Dundas Rd. SE15 54Yb **114**
Dundas Rd. SW9 53Qb **112**
Dundee Ct. E1 46Xb **91**
(off Wapping High St.)
Dundee Ct. SE1 3H **231** (48Ub **91**)
(off Long La.)
Dundee Ho. W9 3A **214** (41Eb **89**)
(off Maida Vale)
Dundee Rd. E13 40Kc **73**
Dundee Rd. SE25 71Xb **157**
Dundee Rd. SL1: Slou 4D **80**
Dundee St. E1 46Xb **91**
Dundee Way EN3: Brim 13Ac **34**
Dundee Wharf E14 45Bc **92**
Dundela Gdns. KT4: Wor Pk 77Xa **154**
Dundonald Cl. E6 44Nc **94**
Dundonald Rd. NW10 39Za **68**
Dundonald Rd. SW19 66Ab **132**
Dundonald Road Stop (London Tramlink) 66Bb 133
Dundrey Cres. RH1: Mers 1E **208**
Dundry Ho. SE26 62Wb **135**
Dunedin Dr. CR3: Cat'm 97Ub **197**
Dunedin Ho. E16 46Pc **94**
(off Manwood St.)
Dunedin M. SW2 60Nb **112**
Dunedin Rd. E10 34Dc **72**
Dunedin Rd. IG1: Ilf 32Sc **74**
Dunedin Rd. RM13: Rain 41Hd **96**
Dunedin Way UB4: Yead 42Y **85**
Dunelm Gro. SE27 62Sb **135**
Dunelm St. E1 44Zb **92**
Dunfee Way KT14: Byfl 84N **169**
Dunfermline Ho. WD19: Wat 20Y **27**
Dunfield Gdns. SE6 64Dc **136**
Dunfield Rd. SE6 64Dc **136**
(not continuous)
Dunford Ct. HA5: Hat E 24Ba **45**
Dunford Rd. N7 35Pb **70**
Dungannon Ho. SW6 52Cb **111**
(off Vanston Pl.)
Dungarvan Av. SW15 56Wa **110**
Dungates La. RH3: Bkld 5C **206**
Dunheved Cl. CR7: Thor H 72Qb **156**
Dunheved Rd. Nth. CR7: Thor H 72Qb **156**
Dunheved Rd. Sth. CR7: Thor H 72Qb **156**

Dunheved Rd. W. CR7: Thor H 72Qb 156
Dunhill Point SW15 60Wa 110
Dunholme Grn. N9 20Vb 33
Dunholme La. N9 20Vb 33
Dunholme Rd. N9 20Vb 33
Dunkeld Rd. RM8: Dag 33Xc 75
Dunkeld Rd. SE25 70Tb 135
Dunkellin Gro. RM15: S Ock 44Wd 98
Dunkellin Way RM15: S Ock 44Wd 98
Dunkery Rd. SE9 63Mc 137
Dunkin Rd. DA1: Dart 56Qd 119
Dunkirk Cl. DA12: Grav'nd 4E 144
Dunkirk Ho. SE1 2G 231 (47Tb 91)
Dunkirk St. SE27 63Sb 135
Dunlace Rd. E5 35Yb 72
Dunleary Cl. TW4: Houn 59Ba 107
Dunley Dr. CR0: New Ad 80Dc 158
Dunlin Ho. SE16 49Zb 92
(off Tawny Way)
Dunloe Av. N17 27Tb 51
Dunloe Ct. E2 2K 219 (40Vb 71)
Dunloe St. E2 2K 219 (40Vb 71)
Dunlop Cl. DA1: Dart 55Nd 119
Dunlop Cl. RM18: Tilb 4B 122
Dunlop Pl. SE16 48Vb 91
Dunlop Rd. RM18: Tilb 3B 122
Dunmail Dr. CR8: Purl 86Ub 177
Dunmore Point E2 3K 219 (41Vb 91)
(off Gascoigne Pl.)
Dunmore Rd. NW6 39Ab 68
Dunmore Rd. SW20 67Ya 132
Dunmow Cl. IG10: Lough 16Nc 36
Dunmow Cl. RM6: Chad H 29Yc 55
Dunmow Cl. TW13: Hanw 62Aa 129
Dunmow Dr. RM13: Rain 39Hd 76
Dunmow Gdns. CM13: W H'dn 30Fe 59
Dunmow Ho. KT14: Byfl 85N 169
Dunmow Ho. SE11 7J 229 (50Pb 90)
(off Newburn St.)
Dunmow Rd. E15 35Fc 73
Dunmow Wlk. N1 39Sb 71
(off Popham St.)
Dunnage Cres. SE16 49Ac 92
(not continuous)
Dunnell Cl. TW16: Sun 67W 128
Dunnets GU21: Knap 9J 167
Dunnett Ho. E3 40Bc 72
(off Vernon Rd.)
Dunnico Ho. SE17 7H 231 (50Ub 91)
(off East St.)
Dunning Cl. RM15: S Ock 44Wd 98
Dunningford Cl. RM12: Horn 36Hd 76
Dunnings La. CM13: W H'dn 31De 79
Dunnings La. RM14: Bulp 35De 79
Dunn Mead NW9 24Va 48
Dunnock Cl. HP3: Hem H 7L 3
Dunnock Cl. N9 18Zb 34
Dunnock Cl. WD6: Bore 14Qa 29
Dunnock Dr. HA7: Stan 22Ha 46
Dunnock Ho. NW9 30Va 48
Dunnock M. E5 34Wb 71
Dunnock Rd. E6 44Nc 94
Dunnose Ct. RM19: Purf 50Rd 97
Dunn St. E8 36Vb 71
Dunny La. WD3: Chfd 5G 10
Dunny La. WD4: Chfd 4G 10
Dunnymans Rd. SM7: Bans 87Bb 175
Dunollie Pl. NW5 36Lb 70
Dunollie Rd. NW5 36Lb 70
Dunoon Gdns. SE23 59Zb 114
Dunoon Ho. N1 39Pb 70
(off Bemerton Est.)
Dunoon Rd. SE23 59Yb 114
Dunoran Home BR1: Broml 67Nc 138
Dunottar Cl. RH1: Redh 8M 207
Dunraven Dr. EN2: Enf 12Qb 32
Dunraven Rd. W12 46Wa 88
Dunraven St. W1 4G 221 (45Hb 89)
Dunsany Rd. W14 48Za 88
DUNSBOROUGH PARK 93L 189
Dunsbury Cl. SM2: Sutt 81Db 175
Dunsfold Cl. SM2: Sutt 80Db 155
(off Blackbush Cl.)
Dunsfold Ri. CR5: Coul 85Mb 176
Dunsfold Way CR0: New Ad 81Dc 178
Dunsford Ho. KT15: Add 77L 149
Dunsford Way SW15 58Xa 110
Dunsmore WD19: Wat 18Z 27
Dunsmore Cl. UB4: Yead 42Z 85
Dunsmore Cl. WD23: Bush 16Fa 28
Dunsmore Rd. KT12: Walt T 72X 151
Dunsmore Way WD23: Bush 16Fa 28
Dunsmure Rd. N16 32Ub 71
Dunspring La. IG5: Ilf 26Rc 54
Dunstable Cl. RM3: Rom 23Md 57
Dunstable M. W1 7J 215 (43Jb 90)
Dunstable Rd. KT8: W Mole 70Ba 129
Dunstable Rd. RM3: Rom 23Md 57
Dunstable Rd. SS17: Stan H 1L 101
Dunstable Rd. TW9: Rich 56Na 109
Dunstall Grn. GU24: Chob 1N 167
Dunstall Rd. SW20 65Xa 132
Dunstall Way KT8: W Mole 69Da 129
Dunstall Welling Est. DA16: Well 54Xc 117
Dunstan Cl. N2 27Eb 49
Dunstan Glade BR5: Pet W 72Tc 160
Dunstan Gro. SE20 66Xb 135
Dunstan Ho's. E1 43Yb 92
(off Stepney Grn.)
Dunstan M. EN1: Enf 13Ub 33
Dunstan Rd. CR5: Coul 89Mb 176
Dunstan Rd. NW11 32Bb 69
Dunstan's Gro. SE22 58Xb 113
Dunstan's Rd. SE22 59Wb 113
Dunster Av. SM4: Mord 74Za 154
Dunster Cl. EN5: Barn 14Za 30
Dunster Cl. RM5: Col R 26Ed 56
Dunster Cl. UB9: Hare 25K 43
Dunster Ct. EC3 4H 225 (45Ub 91)
Dunster Ct. WD6: Bore 13Ta 29
Dunster Cres. RM11: Horn 33Qd 77
Dunster Dr. NW9 32Sa 67
Dunster Gdns. NW6 38Bb 69
Dunster Gdns. SL1: Slou 5E 80
Dunster Ho. SE6 62Ec 136
Dunsterville Way SE1 2G 231 (47Tb 91)
Dunster Way HA2: Harr 34Aa 65
Dunster Way SM6: W'gton 74Jb 156
Dunston Ct. TW18: Staines 63J 127
Dunstone Ct. SE6 59Cc 114
Dunston Rd. E8 39Vb 71
Dunston Rd. SW11 54Jb 112
Dunston St. E8 39Ub 71
Dunton Cl. KT6: Surb 74Na 153
Dunton Ct. SE23 61Xb 135
DUNTON GREEN 92Gd 202
Dunton Green Station (Rail) 91Gd 202
Dunton Rd. E10 31Dc 72
Dunton Rd. RM1: Rom 28Gd 56
Dunton Rd. SE1 7K 231 (50Vb 91)
Duntshill Rd. SW18 60Db 111
Dunvegan Cl. KT8: W Mole 70Ca 129
Dunvegan Ho. RH1: Redh 6P 207
Dunvegan Rd. SE9 56Pc 116
Dunwich Ct. RM6: Chad H 29Xc 55
(off Glandford Way)
Dunwich Rd. DA7: Bex 53Bd 117
Dunworth M. W11 44Bb 89
Duplex Ride SW1 2G 227 (47Hb 89)
Dupont Rd. SW20 68Za 132
Duppas Av. CR0: Wadd 77Rb 157
Duppas Cl. TW17: Shep 71T 150
Duppas Ct. CR0: C'don 76Rb 157
(off Duppas Hill Ter.)
Duppas Hill La. CR0: C'don 77Rb 157
Duppas Hill Rd. CR0: Wadd 77Qb 156
Duppas Hill Ter. CR0: C'don 76Rb 157
Duppas Rd. CR0: Wadd 76Qb 156
Dupre Cl. RM16: Chaf H 48Ae 99
Dupre Cl. SL1: Slou 7C 80
Dupree Rd. SE7 50Kc 93
Dura Den Cl. BR3: Beck 66Dc 136
Durand Gdns. SW9 53Pb 112
Durands Wlk. SE16 47Bc 92
Durand Way NW10 38Sa 67
Durant Rd. BR8: Hext 65Jd 140
Durants Pk. 13Yb 34
Durants Pk. Av. EN3: Pond E 14Zb 34
Durants Rd. EN3: Pond E 14Yb 34
Durant St. E2 40Wb 71
Durban Ct. E7 38Mc 73
Durban Gdns. RM10: Dag 38Ed 76
Durban Ho. W12 45Xa 88
(off White City Est.)
Durban Rd. BR3: Beck 68Bc 136
Durban Rd. E15 41Gc 93
Durban Rd. E17 25Bc 52
Durban Rd. IG2: Ilf 32Uc 74
Durban Rd. N17 23Ub 51
Durban Rd. SE27 63Sb 135
Durban Rd. E. WD18: Wat 14W 26
Durban Rd. W. WD18: Wat 14W 26
Durbin Rd. KT9: Chess 77Na 153
Durdan Cotts. UB1: S'hall 44Ba 85
(off Denbigh Rd.)
Durdans Ho. NW1 38Kb 70
(off Farrier St.)
Durdans Rd. UB1: S'hall 44Ba 85
Durell Gdns. RM9: Dag 36Zc 75
Durell Ho. SE16 47Zb 92
(off Wolfe Cres.)
Durell Rd. RM9: Dag 36Zc 75
Durfey Pl. SE5 52Tb 113
Durfold Dr. RH2: Reig 6L 207
Durford Cres. SW15 60Xa 110
Durham Av. BR2: Broml 70Hc 137
Durham Av. IG8: Buck H 22Mc 53
Durham Av. IG8: Wfd G 22Mc 53
Durham Av. RM2: Rom 28Ld 57
Durham Av. SL1: Slou 4E 80
Durham Av. TW5: Hest 50Ba 85
Durham Cl. SW20 68Xa 132
Durham Ct. KT22: Lea 94Ja 192
Durham Ct. NW6 40Cb 69
(off Kilburn Pk. Rd.)
Durham Ct. TW11: Tedd 63Ga 130
Durham Hill BR1: Broml 63Hc 137
Durham Ho. BR2: Broml 70Gc 137
Durham Ho. IG11: Bark 38Wc 75
(off Margaret Bondfield Av.)
Durham Ho. NW8 4E 214 (41Gb 89)
(off Lorne Cl.)
Durham Ho. RM10: Dag 36Ed 76
Durham Ho. WD6: Bore 12Qa 29
(off Canterbury Rd.)
Durham Ho. St. WC2 5G 223 (45Nb 90)
(off John Adam St.)
Durham La. EN8: Chesh 2Xb 19
Durham Pl. IG1: Ilf 35Sc 74
Durham Pl. SW3 7F 227 (50Hb 89)
Durham Ri. SE18 50Sc 94
Durham Rd. BR2: Broml 69Hc 137
Durham Rd. DA14: Sidc 64Xc 139
Durham Rd. E12 35Mc 73
Durham Rd. E16 42Gc 93
Durham Rd. HA1: Harr 29Da 45
Durham Rd. N2 27Gb 49
Durham Rd. N7 33Pb 70
Durham Rd. N9 19Wb 33
Durham Rd. RM10: Dag 36Ed 76
Durham Rd. SW20 67Xa 132
Durham Rd. TW14: Felt 59Y 107
Durham Rd. W5 48Ma 87
Durham Rd. WD6: Bore 13Sa 29
Durham Row E1 43Ac 92
Durham St. SE11 50Pb 90
Durham Ter. W2 44Db 89
Durham Wharf Dr. TW8: Bford 52La 108
Durham Yd. E2 41Xb 91
Durian Way DA8: Erith 52Kd 119
Durleston Pk. Dr. KT23: Bookh 97Ea 192
Durley Av. HA5: Pinn 31Aa 65
Durley Gdns. BR6: Chels 76Xc 161
Durley Rd. N16 31Ub 71
Durlings Orchard TN15: Igh 93Zd 205
Durlston Rd. E5 33Wb 71
Durlston Rd. KT2: King T 65Na 131
Durndale La. DA11: Nflt 3A 144
Durnell Way IG10: Lough 13Qc 36
Durnford Ho. SE6 62Ec 136
Durnford Ho. SL4: Eton 1H 103
(off Slough Rd.)
Durnford St. N15 29Ub 51
Durnford St. SE10 51Ec 114
Durning Rd. SE19 64Tb 135
Durnsford Av. SW19 61Cb 133
Durnsford Ct. EN3: Enf H 13Ac 34
(off Enstone Rd.)
Durnsford Rd. N11 25Mb 50
Durnsford Rd. SW19 61Cb 133
Durrant Ct. HA3: Hrw W 26Ga 46
Durrant Ho. EC1 7F 219 (43Tb 91)
(off Chiswell St.)
Durrants Cl. RM13: Rain 40Ld 77
Durrants Dr. WD3: Crox G 13S 26
Durrants Hill Rd. HP3: Hem H 5M 3
Durrants Ho. WD3: Crox G 14R 26
Durrant Way BR6: Farnb 78Tc 160
Durrant Way DA10: Swans 59Ae 121
Durrell Dene DA1: Dart 54Pd 119
Durrell Rd. SW6 53Bb 111
Durrell Way TW17: Shep 72T 150
Durrels Ho. W14 49Bb 89
(off Warwick Gdns.)
Durrington Av. SW20 66Ya 132
Durrington Pk. Rd. SW20 67Ya 132
Durrington Rd. E5 35Ac 72
Durrington Twr. SW8 54Lb 112
Durrisdeer Ho. NW2 35Bb 69
(off Lyndale)
Dursley Cl. SE3 54Lc 115
Dursley Gdns. SE3 53Mc 115
Dursley Rd. SE3 54Lc 115
Durward Ho. W8 47Db 89
(off Kensington Ct.)
Durward St. E1 43Xb 91
Durweston M. W1 7G 215 (43Hb 89)
(off York St.)
Durweston St. W1 1G 221 (43Hb 89)
Dury Falls Cl. RM11: Horn 32Qd 77
Dury Falls Ct. RM5: Col R 26Ed 56
Dury Rd. EN5: Barn 11Bb 31
Dutch Barn Cl. TW19: Stanw 58M 105
Dutch Elm Av. SL4: Wind 2K 103
Dutch Gdns. KT2: King T 65Ra 131
Dutch Yd. SW18 57Cb 111
Dutton St. SE10 53Ec 114
Dutton Way SL0: Iver 44G 82
Duval Ho. N19 33Mb 70
(off Ashbrook Rd.)
Duval Ho. SW11 55Gb 111
Duxberry Av. TW13: Felt 62Y 129
Duxberry Cl. BR2: Broml 71Nc 160
Duxford Cl. RM12: Horn 37Ld 77
Duxford Ho. SE2 47Zc 95
(off Wolvercote Rd.)
Dux Hill TN15: Plax 98Be 205
Dux La. TN15: Plax 98Be 205
Duxons Turn HP2: Hem H 1B 4
Dwelly La. TN8: Eden 9M 211
DW Fitness Ewell 84Ya 174
DW Fitness Waldorf Hotel 3H 223 (44Pb 90)
(off Tavistock St.)
Dwight Rd. WD18: Wat 17T 26
Dyas Rd. TW16: Sun 67W 128
Dye Ho. La. E3 39Cc 72
Dyer Ho. TW12: Hamp 67Da 129
Dyer's Bldgs. EC1 1K 223 (43Qb 90)
Dyers Hall Rd. E11 32Gc 73
Dyers Hall Rd. Sth. E11 33Fc 73
Dyers La. SW15 56Xa 110
Dyers Way RM3: Rom 24Kd 57
Dyke Dr. BR5: Orp 73Yc 161
Dykes Path GU21: Wok 87E 168
Dykes Way BR2: Broml 69Hc 137
Dykewood Cl. DA5: Bexl 62Fd 140
Dylan Cl. WD6: E'tree 17Ma 29
Dylan Rd. DA17: Belv 48Cd 96
Dylan Rd. SE24 56Rb 113
Dylways SE5 56Tb 113
Dymchurch Cl. BR6: Orp 77Uc 160
Dymchurch Cl. IG5: Ilf 26Qc 54
Dymes Path SW19 61Za 132
Dymock St. SW6 55Db 111
Dymoke Rd. RM11: Horn 31Hd 76
Dyneley Rd. SE12 62Lc 137
Dyne Rd. NW6 38Ab 68
Dynes, The TN15: Kems'g 89Md 183
Dynes Rd. TN15: Kems'g 89Md 183
Dynevor Rd. N16 34Ub 71
Dynevor Rd. TW10: Rich 57Na 109
Dynham Rd. NW6 38Cb 69
Dyott St. WC1 2E 222 (44Mb 90)
Dyrham La. EN5: Barn 8Wa 16
Dyrham Pk. Hertfordshire 10Wa 16
Dyrham Pk. Country Club & Golf Course 9Xa 16
Dysart Av. KT2: King T 64La 130
Dysart St. EC2 6G 219 (42Tb 91)
Dyson Cl. SL4: Wind 5F 102
Dyson Ct. HA0: Wemb 35Ja 66
Dyson Ct. NW2 32Ya 68
Dyson Ct. WD17: Wat 14Y 27
Dyson Ho. UB10: Uxb 39N 63
Dyson Ho. SE10 50Hc 93
(off Blackwall La.)
Dyson Rd. E11 30Gc 53
Dyson Rd. E15 37Hc 73
Dysons Cl. EN8: Walt C 5Zb 20
Dysons Rd. N18 22Xb 51
Dytchleys La. CM14: N'side 14Qd 39
Dytchleys Rd. CM14: N'side 14Pd 39

E

Eade Rd. N4 31Sb 71
Eagans Cl. N2 27Fb 49
Eagle Av. RM6: Chad H 30Ad 55
Eagle Cl. EN3: Pond E 14Yb 34
Eagle Cl. EN9: Walt A 6Jc 21
Eagle Cl. RM12: Horn 37Kd 77
Eagle Cl. SE16 50Yb 92
Eagle Cl. SM6: W'gton 79Nb 156
Eagle Ct. E11 28Jc 53
Eagle Ct. EC1 7B 218 (43Rb 91)
Eagle Ct. KT17: Eps 86Wa 174
Eagle Dr. NW9 26Ua 48
Eagle Dwellings EC1 4E 218 (41Sb 91)
(off City Rd.)
Eagle Hgts. SW11 55Gb 111
Eagle Heights Wildlife Pk. 75Kd 163
Eagle Hill SE19 65Tb 135
Eagle Ho. E1 42Xb 91
(off Headlam St.)
Eagle Ho. EC1 3F 219 (41Tb 91)
(off City Rd.)
Eagle Ho. N1 1F 219 (40Tb 71)
(off Eagle Wharf Rd.)
Eagle Ho. RM17: Grays 51Be 121
Eagle Ho. M. SW4 57Lb 112
Eagle La. E11 28Jc 53
Eagle Lodge NW11 31Bb 69
Eagle Mans. N16 36Vb 71
(off Salcombe Rd.)
Eagle M. N1 37Ub 71
Eagle Pl. SW1 5C 222 (45Lb 90)
(off Piccadilly)
Eagle Pl. SW7 7A 226 (50Eb 89)
(off Ellis Way)
Eagle Point EC1 3F 219 (41Tb 91)
(off City Rd.)
Eagle Rd. HA0: Wemb 38Ma 67
Eagle Rd. SL1: Slou 5D 80
Eagle Rd. TW6: H'row A 55V 106
Eagles Dr. TN16: Tats 90Mc 179
Eaglesfield Equestrian Cen. 76Zd 165
Eaglesfield Rd. SE18 53Rc 116
Eagles Rd. DA9: Ghithe 56Xd 120
Eaglestone Cl. TN15: Bor G 91Ce 205
Eagle St. WC1 1H 223 (43Pb 90)
Eagle Ter. IG8: Wfd G 24Kc 53
Eagle Trad. Est. CR4: Mitc 72Hb 155
Eagle Way CM13: Gt War 23Xd 58
Eagle Way DA11: Nflt 57Ce 121
Eagle Wharf Ct. SE1 7K 225 (46Vb 91)
(off Lafone St.)
Eagle Wharf E. E14 45Ac 92
(off Narrow St.)
Eagle Wharf Rd. N1 1E 218 (40Sb 71)
Eagle Wharf W. E14 45Ac 92
(off Narrow St.)
Eagle Works E. E1 6K 219 (42Vb 91)
(off Quaker St.)
Eagle Works W. E1 6K 219 (42Vb 91)
(off Quaker St.)
Eagling Cl. E3 41Cc 92
Ealdham Sq. SE9 56Lc 115
EALING 45Ma 87
Ealing B'way. Cen. 45Ma 87
Ealing Broadway Station (Rail, Underground & Crossrail) 45Ma 87
Ealing Cl. WD6: Bore 11Ta 29
EALING COMMON 45Pa 87
Ealing Common Station (Underground) 46Pa 87
EALING CYGNET HOSPITAL 43Na 87
Ealing Golf Course 41Ka 86
Ealing Grn. W5 46Ma 87
EALING HOSPITAL 46Fa 86
Ealing Lawn Tennis Club & Indoor Tennis Cen. 45Pa 87
Ealing Pk. Gdns. W5 49La 86
Ealing Pk. Mans. W5 48Ma 87
(off Sth. Ealing Rd.)
Ealing Rd. HA0: Wemb 37Na 67
Ealing Rd. TW8: Bford 50Ma 87
Ealing Rd. UB5: N'olt 39Ca 65
Ealing Squash & Fitness Club 44Na 87
Ealing Studios 46Ma 87
Ealing Village W5 44Na 87
Eamont Cl. HA4: Ruis 31R 64
Eamont Ct. NW8 1D 214 (40Gb 69)
(off Eamont St.)
Eamont St. NW8 1D 214 (40Gb 69)
Eardemont Cl. DA1: Cray 56Hd 118
Eardley Cres. SW5 50Cb 89
Eardley Point SE18 49Rc 94
(off Wilmount St.)
Eardley Rd. DA17: Belv 50Cd 96
Eardley Rd. SW16 64Lb 134
Eardley Rd. TN13: S'oaks 96Kd 203
Earhart Ho. NW9 27Wa 48
(off East Dr.)
Earhart Way TW6: Cran 55W 106
Earhart Way TW6: H'row A 55W 106
Earl Cl. N11 22Kb 50
Earldom Rd. SW15 56Ya 110
Earle Gdns. KT2: King T 66Na 131
Earle Ho. SW1 6E 228 (49Mb 90)
(off Montaigne Cl.)
Earleswood KT11: Cobh 84Aa 171
Earleswood Ct. HP3: Hem H 6M 3
Earleydene SL5: Asc 4A 146
Earlham Ct. E11 31Hc 73
Earlham Gro. E7 36Hc 73
Earlham Gro. N22 24Pb 50
Earlham St. WC2 3F 223 (44Nb 90)
Earl Ho. NW1 6E 214 (42Gb 89)
(off Lisson Gro.)
Earlom Ho. WC1 4K 217 (41Qb 90)
(off Margery St.)
Earl Ri. SE18 50Tc 94
Earl Rd. DA11: Nflt 1A 144
Earl Rd. SW14 56Sa 109
Earlsbrook Rd. RH1: Redh 8P 207
(not continuous)
Earlsbury Gdns. HA8: Edg 21Qa 47
Earls Cnr. EN6: S Mim 5Wa 16
EARL'S COURT 49Db 89
Earl's Ct. Gdns. SW5 49Db 89
Earl's Ct. Rd. SW5 49Cb 89
Earl's Ct. Rd. W8 48Cb 89
Earl's Ct. Sq. SW5 50Db 89
Earl's Court Station (Underground) 49Cb 89
Earls Cres. HA1: Harr 28Ga 46
Earlsdown Ho. IG11: Bark 40Tc 74
Earlsferry Way N1 38Nb 70
(not continuous)
EARLSFIELD 60Eb 111
Earlsfield Ho. KT2: King T 67Ma 131
(off Seven Kings Way)
Earlsfield Rd. SW18 60Eb 111
Earlsfield Station (Rail) 60Eb 111
Earlshall Rd. SE9 56Pc 116
Earls Ho. TW9: Kew 52Ra 109
Earls La. EN6: Ridge 4Ua 16
Earls La. EN6: S Mim 4Ua 16
Earls La. SL1: Slou 6D 80
Earlsmead HA2: Harr 35Ba 65
Earlsmead Rd. N15 29Vb 51
Earlsmead Rd. NW10 41Ya 88
Earlsmead Stadium 35Ba 65
Earl's Path IG10: Lough 12Lc 35
Earls Ter. W8 48Bb 89
Earlsthorpe M. SW12 58Jb 112
Earlsthorpe Rd. SE26 63Zb 136
Earlstoke St. EC1 3B 218 (41Rb 91)
Earlston Gro. E9 39Xb 71
Earl St. EC2 7G 219 (43Tb 91)
Earl St. WD17: Wat 13Y 27
Earl's Wlk. RM8: Dag 35Xc 75
Earls Wlk. W8 48Cb 89
Earls Way SE1 7J 225 (46Ub 91)
(off Duchess Wlk.)
Earlswell Wlk. RM3: Rom 22Ld 57
EARLSWOOD 8P 207
Earlswood Av. CR7: Thor H 71Qb 156
Earlswood Cl. SE10 51Gc 115
Earlswood Common 9M 207
Earlswood Ct. RH1: Redh 7P 207
Earlswood Gdns. IG5: Ilf 27Qc 54
Earlswood Rd. RH1: Redh 7P 207
Earlswood Station (Rail) Surrey 8P 207
Earlswood St. SE10 50Gc 93
Early M. NW1 39Kb 70
Early Rivers Ho. E20 36Ec 72
Earnshaw Ho. EC1 5C 218 (42Rb 91)
(off Percival St.)
Earnshaw St. WC2 2E 222 (44Mb 90)
Earsby St. W14 49Ab 88
(not continuous)
Easby Cres. SM4: Mord 72Db 155
Easebourne Rd. RM8: Dag 36Yc 75
Easedale Dr. RM12: Horn 36Jd 76
Easedale Ho. TW7: Isle 57Ha 108
Eashing Point SW15 60Xa 110
(off Wanborough Dr.)
Easington Way RM15: S Ock 43Wd 98
Easleys M. W1 2J 221 (44Jb 90)
East 10 Ent. Pk. E10 32Ac 72
EAST ACTON 45Va 88
East Acton Arc. W3 44Ua 88
East Acton Ct. W3 45Ua 88
East Acton La. W3 46Ua 88
East Acton Station (Underground) 44Va 88
East Arbour St. E1 44Zb 92
East Av. E12 38Nc 74
East Av. E17 28Dc 52
East Av. KT12: W Vill 82V 170
East Av. SM6: W'gton 78Pb 156
East Av. UB1: S'hall 45Ba 85
East Av. UB3: Hayes 46V 84
East Bank N16 31Ub 71
Eastbank Cl. E17 29Dc 52
Eastbank Rd. TW12: Hamp H 64Ea 130
EAST BARNET 16Gb 31
East Barnet Rd. EN4: E Barn 14Fb 31
East Bay La. E20 36Cc 72
East Beckton Bus Station 44Qc 94
East Beckton District Cen. 43Pc 94
EAST BEDFONT 59U 106
East Block SE1 1J 229 (47Pb 90)
(off York Rd.)
Eastbourne Av. W3 44Ta 87
Eastbourne Gdns. SW14 55Sa 109
Eastbourne M. W2 2A 220 (44Eb 89)
Eastbourne Rd. E15 39Gc 73
Eastbourne Rd. E6 41Qc 94
(not continuous)
Eastbourne Rd. N15 30Ub 51
Eastbourne Rd. RH9: G'stone 4A 210
Eastbourne Rd. RH9: S God 4A 210
Eastbourne Rd. SL1: Slou 4E 80
Eastbourne Rd. SW17 65Jb 134
Eastbourne Rd. TW13: Felt 61Z 129
Eastbourne Rd. TW8: Bford 50La 86
Eastbourne Rd. W4 51Sa 109
Eastbourne Ter. W2 2A 220 (44Eb 89)
Eastbournia Av. N9 20Xb 33
Eastbridge SL2: Slou 7M 81
Eastbrook Av. N9 17Yb 34
Eastbrook Av. RM10: Dag 35Ed 76
Eastbrook Cl. GU21: Wok 88C 168
Eastbrook Cl. RM10: Dag 35Ed 76
Eastbrook Dr. RM7: Rush G 34Gd 76
Eastbrookend Country Pk. 34Fd 76
Eastbrook Rd. EN9: Walt A 5Gc 21
Eastbrook Rd. SE3 53Kc 115
Eastbrook Way HP2: Hem H 2N 3
EAST BURNHAM 8E 60
East Burnham La. SL2: Farn R 9E 60
EASTBURY 21V 44
Eastbury Av. EN1: Enf 11Vb 33
Eastbury Av. HA6: Nwood 22U 44
Eastbury Av. IG11: Bark 39Uc 74
Eastbury Ct. AL1: St A 1D 6
Eastbury Ct. EN5: New Bar 15Eb 31
(off Lyonsdown Rd.)
Eastbury Ct. IG11: Bark 39Uc 74
Eastbury Ct. WD19: Wat 17Y 27
Eastbury Farm Cl. HA6: Nwood 21U 44
Eastbury Gro. W4 50Ua 88
Eastbury Manor House 39Vc 75
Eastbury Pl. HA6: Nwood 22V 44
Eastbury Rd. BR5: Pet W 72Tc 160
Eastbury Rd. E6 42Qc 94
Eastbury Rd. HA6: Nwood 23U 44
Eastbury Rd. KT2: King T 66Na 131
Eastbury Rd. RM7: Rom 30Fd 56
Eastbury Rd. WD19: Wat 17X 27
Eastbury Sq. IG11: Bark 39Vc 75
Eastbury Ter. E1 42Zb 92
East Carriage Ho. SE18 48Rc 94
(off Royal Carriage M.)
Eastcastle St. W1 2B 222 (44Lb 90)
Eastcheap EC3 4H 225 (45Ub 91)
East Churchfield Rd. W3 46Ta 87
Eastchurch Rd. TW6: H'row A 54U 106
East Cl. AL2: Chis G 7P 5
East Cl. EN4: Cockf 14Jb 32
East Cl. RM13: Rain 42Kd 97
East Cl. UB6: G'frd 40Ea 66
East Cl. W5 42Qa 87
Eastcombe Av. SE7 51Kc 115
East Comn. SL9: Ger X 30A 42
Eastcote BR6: Orp 74Vc 161
EASTCOTE 31X 65
Eastcote Av. HA2: Harr 33Da 65
Eastcote Av. KT8: W Mole 71Ba 151
Eastcote Av. UB6: G'frd 36Ja 66
Eastcote Hockey & Badminton Club 30V 44
Eastcote Ho. KT17: Eps 84Ua 174
Eastcote Ind. Est. HA4: Ruis 31Y 65
Eastcote La. HA2: Harr 35Aa 65
Eastcote La. UB5: N'olt 36Ba 65
(not continuous)
Eastcote La. Nth. UB5: N'olt 37Ba 65
Eastcote Pl. HA5: Eastc 30X 45
Eastcote Rd. DA16: Well 54Tc 116
Eastcote Rd. HA2: Harr 34Ea 66
Eastcote Rd. HA4: Ruis 31U 64
Eastcote Rd. HA5: Pinn 29Z 45
Eastcote Station (Underground) 31Y 65
Eastcote St. SW9 54Pb 112
Eastcote Vw. HA5: Pinn 28Y 45
EASTCOTE VILLAGE 29X 45
Eastcott Cl. KT2: King T 64Sa 131
East Ct. HA0: Wemb 33La 66
East Cres. EN1: Enf 15Vb 33
East Cres. N11 21Hb 49
East Cres. SL4: Wind 3D 102
East Cres. Rd. DA12: Grav'nd 8E 122
Eastcroft SL2: Slou 2F 80
East Cft. Ho. HA2: Harr 33Ea 66
Eastcroft Rd. KT19: Ewe 80Ua 154
East Cross Route E9: Crowfoot Cl. 36Bc 72
East Cross Route E9: Wansbeck Rd. 38Bc 72
East Croydon Bus Station 75Tb 157
East Croydon Station (Rail & London Tramlink) 75Tb 157
Eastdean Av. KT18: Eps 85Ra 173
East Dene Dr. RM3: Rom 22Md 57
Eastdown Ct. SE13 56Fc 115
Eastdown Ho. E8 35Wb 71
Eastdown Pk. SE13 56Fc 115
East Dr. AL4: St A 1J 7
East Dr. BR5: St M Cry 72Xc 161
East Dr. GU25: Vir W 3L 147
East Dr. HA6: Nwood 19U 26
East Dr. NW9 27Wa 48
East Dr. SL2: Stoke P 1J 81
East Dr. SM5: Cars 81Gb 175
East Dr. WD25: Wat 8X 13
East Duck Lees La. EN3: Pond E 14Ac 34
EAST DULWICH 56Vb 113
East Dulwich Est. SE22 55Ub 113
(off Albrighton Rd.)
East Dulwich Gro. SE22 57Ub 113
East Dulwich Rd. SE15 56Wb 113
East Dulwich Rd. SE22 56Vb 113
(not continuous)

East Dulwich Station (Rail) 56Ub 113
East End Farm HA5: Pinn 27Ba 45
East End Rd. N2 27Db 49
East End Rd. N3 26Cb 49
East End Way HA5: Pinn 27Aa 45
East Entrance RM10: Dag 40Dd 76
Easter Ind. Pk. RM13: Rain 45Gd 96
Eastern App. IG11: Bark 39Wc 75
Eastern Av. E11 30Kc 53
Eastern Av. EN8: Walt C 5Bc 20
Eastern Av. HA5: Pinn 31Z 65
Eastern Av. IG2: Ilf 30Rc 54
Eastern Av. IG4: Ilf 30Nc 54
Eastern Av. KT16: Chert 69J 127
Eastern Av. RM15: Avel 46Sd 98
Eastern Av. RM20: W Thur 50Vd 98
Eastern Av. RM6: Chad H 28Xc 55
Eastern Av. E. RM1: Rom 27Fd 56
Eastern Av. E. RM2: Rom 26Gd 56
Eastern Av. E. RM3: Rom 25Jd 56
Eastern Av. Retail Pk. Romford 28Ed 56
Eastern Av. W. RM1: Rom 27Fd 56
Eastern Av. W. RM6: Chad H 28Ad 55
Eastern Av. W. RM7: Chad H 28Cd 56
Eastern Av. W. RM7: Mawney 28Cd 56
Eastern Av. W. RM7: Rom 28Cd 56
Eastern Bus. Pk. TW6: H'row A 54U 106
Eastern Ct. E15 37Gc 73
(off Gt. Eastern Rd.)
Eastern Gateway E16 45Lc 93
Eastern Ho. E2 41Xb 91
(off Bethnal Grn. Rd.)
Eastern Ind. Est. DA18: Erith 47Cd 96
Eastern Path RM13: Rain 39Ld 77
Eastern Perimeter Rd. TW6: H'row A 54V 106
Eastern Quay Apts. E16 46Kc 93
(off Portsmouth M.)
Eastern Rd. E13 40Kc 73
Eastern Rd. E17 29Ec 52
Eastern Rd. N2 27Hb 49
Eastern Rd. N22 25Nb 50
Eastern Rd. RM1: Rom 29Gd 56
(not continuous)
Eastern Rd. RM17: Grays 49Fe 99
Eastern Rd. SE4 56Cc 114
EASTERN RDBT. 33Sc 74
Eastern Vw. TN16: Big H 89Lc 179
Easternville Gdns. IG2: Ilf 30Sc 54
Eastern Way DA17: Belv 47Dd 96
Eastern Way DA18: Erith 45Ad 95
Eastern Way RM17: Grays 51Ce 121
Eastern Way SE2 45Ad 95
Eastern Way SE28 47Wc 95
Easter Way RH9: S God 9C 210
EAST EWELL 82Ya 174
East Ferry Rd. E14 49Dc 92
Eastfield Av. WD24: Wat 11Z 27
Eastfield Cl. SL1: Slou 8L 81
Eastfield Gdns. RM10: Dag 35Cd 76
Eastfield Rd. CM14: B'wood 19Zd 41
Eastfield Rd. E17 28Cc 52
Eastfield Rd. EN3: Enf W 10Zb 20
Eastfield Rd. EN8: Walt C 3Bc 20
Eastfield Rd. N8 27Nb 50
Eastfield Rd. RH1: Redh 7C 208
Eastfield Rd. RM10: Dag 35Cd 76
Eastfield Rd. RM9: Dag 35Bd 75
Eastfields HA5: Eastc 29Y 45
Eastfields Av. SW18 56Cb 111
Eastfields Rd. CR4: Mitc 68Jb 134
Eastfields Rd. W3 43Sa 87
Eastfield St. E14 43Ac 92
EAST FINCHLEY 28Gb 49
East Finchley Station (Underground) 28Gb 49
East Flint HP1: Hem H 1H 3
East Gdns. GU22: Wok 89E 168
East Gdns. SW17 65Gb 133
Eastgate GU22: Wok 89B 168
Eastgate SM7: Bans 86Bb 175
Eastgate Bus. Pk. E10 32Ac 72
Eastgate Cl. SE28 44Zc 95
Eastglade HA5: Pinn 27Ba 45
Eastglade HA6: Nwood 22V 44
East Grn. HP3: Hem H 7P 3
East Gro. SE17 5E 230 (49Sb 91)
East Hall La. RM13: Wenn 44Md 97
East Hall Rd. BR5: St M Cry 73Ad 161
EAST HAM 40Pc 74
East Ham & Barking By-Pass IG11: Bark 40Tc 74
EAST HAM CARE CENTRE & DAY HOSPITAL 38Mc 73
Eastham Cl. EN5: Barn 15Ab 30
Eastham Cres. CM13: B'wood 21Ce 59
East Ham Leisure Cen. 39Pc 74
East Ham Mnr. Way E6 44Qc 94
East Ham Nature Reserve 42Pc 94
East Ham Nature Reserve Vis. Cen. 42Pc 94
East Ham Station (Underground) 38Nc 74
East Handyside Canopy 39Nb 70
East Harding St. EC4 2A 224 (44Qb 90)
East Heath Rd. NW3 34Eb 69
East Hill CR2: Sande 82Ub 177
East Hill DA1: Dart 59Pd 119
East Hill DA4: S Dar 67Sd 142
East Hill GU22: Wok 88E 168
East Hill HA9: Wemb 33Qa 67
East Hill RH8: Oxt 1J 211
East Hill SW18 57Db 111
East Hill TN16: Big H 90Kc 179
EAST HILL 82Rd 183
East Hill Ct. RH8: Oxt 2J 211
East Hill Dr. DA1: Dart 59Pd 119
East Hill Farm Cvn. Pk. TN15: Knat 82Rd 183
East Hill Pk. TN15: Knat 82Rd 183
East Hill Rd. RH8: Oxt 1J 211
East Hill Rd. TN15: Knat 82Rd 183
Eastholm NW11 28Db 49
East Holme DA8: Erith 53Fd 118
Eastholme UB3: Hayes 46W 84
EAST HORNDON 28Fe 59
EAST HORSLEY 100V 190
East India Bldgs. E14 45Cc 92
(off Saltwell St.)
East India Ct. SE16 47Yb 92
(off St Marychurch St.)
East India Dock Basin Local Nature Reserve 45Gc 93
East India Dock Ho. E14 44Ec 92
East India Dock Rd. E14 44Cc 92
East India Station (DLR) 45Fc 93
East India Way CR0: C'don 74Vb 157
East Kent Av. DA11: Nflt 58Ee 121
Eastlake Ho. NW8 6C 214 (42Fb 89)
(off Frampton St.)
Eastlake Rd. SE5 54Sb 113
Eastlands Cl. RH8: Oxt 99Fc 199
Eastlands Cres. SE21 58Vb 113
Eastlands Way RH8: Oxt 99Fc 199
East La. SE16: Chambers St. 47Wb 91
East La. SE16: Scott Lidgett Cres. 47Wb 91
East La. DA4: S Dar 68Td 142
East La. HA0: Wemb 34Ka 66
East La. HA9: Wemb 34Ma 67
East La. KT1: King T 69Ma 131
East La. KT24: W Hor 98S 190
East La. WD25: Wat 2X 13
East La. WD5: Ab L 10F 4
East La. WD5: Bedm 10F 4
East La. Bus. Pk. HA9: Wemb 33Ma 67
Eastlea Av. WD25: Wat 9Aa 13
Eastlea M. E16 42Gc 93
Eastleigh Av. HA2: Harr 33Da 65
Eastleigh Cl. NW2 34Ua 68
Eastleigh Cl. SM2: Sutt 80Db 155
Eastleigh Rd. DA7: Bex 55Ed 118
Eastleigh Rd. E17 26Bc 52
Eastleigh Wlk. SW15 59Wa 110
Eastleigh Way TW14: Felt 60W 106
East Lodge E16 46Jc 93
(off Wesley Av.)
East Lodge La. EN2: Crew H 8Mb 18
East Lodge La. EN2: Enf 8Mb 18
East London Crematorium 41Hc 93
East London Gymnastic Cen. 44Pc 94
EASTLY END 69G 126
East Mall RM17: Grays 51De 121
(off Grays Shop. Cen.)
East Mall TW18: Staines 63H 127
(within The Elmsleigh Cen.)
EASTMAN DENTAL HOSPITAL New Site 6D 216 (42Mb 90)
EASTMAN DENTAL HOSPITAL & DENTAL INSTITUTE Expected Close 2020 5H 217 (42Pb 90)
Eastman Ho. SW4 58Lb 112
Eastman Rd. W3 47Ta 87
Eastman Way KT19: Eps 82Ra 173
East Mead HA4: Ruis 34Z 65
Eastmead GU21: Wok 9M 167
Eastmead Av. UB6: G'frd 41Da 85
Eastmead Cl. BR1: Broml 68Nc 138
Eastmearn Rd. SE21 61Sb 135
East Mill DA11: Grav'nd 8B 122
East Milton Rd. DA12: Grav'nd 9F 122
East Mimms HP2: Hem H 1N 3
EAST MOLESEY 70Fa 130
East Molesey Memorial Cricket Ground 69Fa 130
Eastmont Rd. KT10: Hin W 75Ga 152
Eastmoor Pl. SE7 48Mc 93
Eastmoor St. SE7 48Mc 93
East Mt. St. E1 43Xb 91
(not continuous)
Eastney Rd. CR0: C'don 74Rb 157
Eastney St. SE10 50Fc 93
Eastnor HP3: Bov 10C 2
Eastnor Cl. RH2: Reig 8H 207
Eastnor Pl. RH2: Reig 8J 207
Eastnor Rd. RH2: Reig 9J 207
Eastnor Rd. SE9 60Sc 116
Eastone Apts. E1 1K 225 (43Vb 91)
(off Lolesworth Cl.)
Easton Gdns. WD6: Bore 14Ua 30
Easton St. WC1 5K 217 (42Qb 90)
East Park Cl. RM6: Chad H 29Ad 55
East Parkside CR6: W'ham 88Cc 178
East Parkside SE10 47Gc 93
East Pk. Wlk. E20 36Ec 72
East Pas. EC1 7D 218 (43Sb 91)
(off Cloth St.)
East Pl. SE27 63Sb 135
East Point SE1 50Wb 91
East Pole Cotts. N14 14Mb 32
East Poultry Av. EC1 1B 224 (43Rb 91)
East Putney Station (Underground) 57Ab 110
East Ramp TW6: H'row A 53R 106
East Ridgeway EN6: Cuff 1Nb 18
East Rd. DA16: Well 54Xc 117
East Rd. E15 39Jc 73
East Rd. EN3: Enf W 10Yb 20
East Rd. EN4: E Barn 18Jb 32
East Rd. HA1: Harr 31Ka 66
East Rd. HA8: Edg 25Ra 47
East Rd. KT13: Weyb 80T 150
East Rd. KT2: King T 67Na 131
East Rd. N1 4F 219 (41Tb 91)
East Rd. RH2: Reig 5H 207
East Rd. RM6: Chad H 29Ad 55
East Rd. RM7: Rush G 31Fd 76
East Rd. SW19 65Eb 133
East Rd. SW3 7H 227 (50Jb 90)
East Rd. TW14: Bedf 59T 106
East Rd. UB7: W Dray 49P 83
East Rochester Way DA15: Sidc 57Uc 116
East Rochester Way DA5: Bexl 58Bd 117
East Row E11 30Jc 53
East Row W10 42Ab 88
Eastry Av. BR2: Hayes 72Hc 159
Eastry Ho. SW8 52Nb 112
(off Hartington Rd.)
Eastry Ho. W11 45Ab 88
(off Walmer Rd.)
Eastry Rd. DA8: Erith 52Cd 118
EAST SHEEN 56Sa 109
East Sheen Av. SW14 57Ta 109
East Shop. Cen. 38Kc 73
East Side W12 47Ya 88
(off Shepherd's Bush Mkt.)
Eastside Halls SW7 3C 226 (48Fb 89)
(off Prince's Gdns.)
Eastside M. E3 40Cc 72
(off Morville St.)
Eastside Rd. NW11 28Bb 49
East Smithfield E1 5K 225 (45Vb 91)
East Stand N5 34Rb 71
East St. BR1: Broml 68Jc 137
East St. DA7: Bex 56Cd 118
East St. HP2: Hem H 2M 3
East St. IG11: Bark 39Sc 74
East St. KT17: Eps 85Ua 174
East St. KT23: Bookh 97Da 191
East St. RM17: Grays 51Ee 121
East St. RM20: Grays 51Ae 121
East St. SE17 7E 230 (50Sb 91)
(not continuous)
East Surrey Gro. SE15 52Vb 113
EAST SURREY HOSPITAL 10B 208
East Surrey Mus. 96Vb 197
East Tenter St. E1 44Vb 91
East Ter. DA12: Grav'nd 8E 122
East Ter. DA15: Sidc 60Uc 116
East Ter. SL4: Wind 3J 103
East Thamesmead Bus. Pk. DA18: Erith 47Bd 95
East Thurrock Rd. RM17: Grays 51Ee 121
EAST TILBURY 9K 101
East Tilbury Rd. SS17: Linf 7J 101
East Tilbury Station (Rail) 9K 101
East Twr. E14 47Dc 92
(off Pan Peninsula Sq.)
East Twr. SW10 52Fb 111
East Towers HA5: Pinn 29Z 45
East Va. W3 46Va 88
East Vw. E4 22Ec 52
East Vw. EN5: Barn 12Bb 31
Eastview Av. SE18 52Uc 116
EAST VILLAGE 36Ec 72
Eastville Av. NW11 30Bb 49
East Wlk. EN4: E Barn 17Jb 32
East Wlk. RH2: Reig 6K 207
East Wlk. UB3: Hayes 46W 84
East Way BR2: Hayes 73Jc 159
East Way CR0: C'don 75Ac 158
East Way E11 29Kc 53
East Way HA4: Ruis 32W 64
East Way UB3: Hayes 46W 84
Eastway E9 37Bc 72
(not continuous)
Eastway KT19: Eps 83Ta 173
Eastway SM4: Mord 71Za 154
Eastway SM6: W'gton 77Lb 156
Eastway Cres. HA2: Harr 33Da 65
Eastwell Cl. BR3: Beck 66Ac 136
Eastwell Ho. SE1 3G 231 (48Tb 91)
(off Weston St.)
East Wick E20 36Cc 72
Eastwick Cres. WD3: Rick 19H 25
Eastwick Dr. KT23: Bookh 95Ca 191
EAST WICKHAM 53Yc 117
Eastwick Pk. Av. KT23: Bookh 96Da 191
Eastwick Rd. KT12: Hers 79X 151
Eastwick Rd. KT23: Bookh 97Da 191
Eastwick Row HP2: Hem H 3A 4
East Wing DA2: Dart 58Sd 120
East Wintergarden 46Dc 92
(off Bank St.)
East Wood WD25: A'ham 8Da 13
Eastwood KT13: Weyb 79T 150
(off Bridgewater Rd.)
Eastwood Cl. E18 26Jc 53
Eastwood Cl. N17 24Xb 51
Eastwood Cl. N7 36Qb 70
Eastwood Ct. HP2: Hem H 1A 4
Eastwood Dr. RM13: Rain 44Kd 97
Eastwood Ho. DA11: Grav'nd 4E 144
(off Hither Flds.)
Eastwood Ho. E3 43Cc 92
(off Bow Comn. La.)
Eastwood Rd. E18 26Jc 53
Eastwood Rd. IG3: Ilf 32Wc 75
Eastwood Rd. N10 26Jb 50
Eastwood Rd. UB7: W Dray 47Q 84
East Woodside DA5: Bexl 59Ad 117
Eastwood St. SW16 65Lb 134
EASTWORTH 74K 149
Eastworth Rd. KT16: Chert 74J 149
easyGym Fulham 51Bb 111
easyGym Slough 7L 81
UB3: Harl
Eatington Rd. E10 29Fc 53
Eaton Av. SL1: Slou 5A 80
Eaton Cl. HA7: Stan 21Ka 46
Eaton Cl. SW1 6H 227 (49Jb 90)
Eaton Ct. E18 26Jc 53
Eaton Ct. HA8: Edg 21Qa 47
Eaton Dr. KT2: King T 66Qa 131
Eaton Dr. RM5: Col R 24Dd 56
Eaton Dr. SW9 56Rb 113
Eaton Gdns. RM9: Dag 38Ad 75
Eaton Ga. HA6: Nwood 23S 44
Eaton Ga. SW1 5H 227 (49Jb 90)
Eaton Ho. E14 45Bc 92
(off Westferry Cir.)
Eaton Ho. SW11 53Fb 111
Eaton La. SW1 4A 228 (48Kb 90)
Eaton Mans. SW1 6H 227 (49Jb 90)
(off Bourne St.)
Eaton M. Nth. SW1 5H 227 (49Jb 90)
Eaton M. Sth. SW1 5J 227 (49Jb 90)
Eaton M. W. SW1 5J 227 (49Jb 90)
Eaton Pk. KT11: Cobh 86Aa 171
Eaton Pk. Rd. KT11: Cobh 86Aa 171
Eaton Pk. Rd. N13 19Qb 32
Eaton Pl. CR3: Cat'm 94Tb 197
Eaton Pl. SW1 5H 227 (49Jb 90)
Eaton Ri. E11 29Lc 53
Eaton Ri. W5 43Ma 87
Eaton Rd. AL1: St A 2F 6
Eaton Rd. DA14: Sidc 61Zc 139
Eaton Rd. EN1: Enf 14Ub 33
Eaton Rd. NW4 29Ya 48
Eaton Rd. RM14: Upm 33Ud 78
Eaton Rd. SM2: Sutt 79Fb 155
Eaton Rd. TW3: Houn 56Fa 108
Eaton Row SW1 4K 227 (48Kb 90)
Eatons Mead E4 19Cc 34
Eaton Sq. DA3: Lfield 69Zd 143
Eaton Sq. SW1 5H 227 (49Jb 90)
Eaton Ter. E3 41Ac 92
Eaton Ter. SW1 5H 227 (49Jb 90)
Eaton Ter. M. SW1 5H 227 (49Jb 90)
(off Eaton Ter.)
Eatonville Rd. SW17 61Hb 133
Eatonville Vs. SW17 61Hb 133
Eaton Way WD6: Bore 11Pa 29
Eaves Cl. KT15: Add 79L 149
Ebbage Ct. GU22: Wok 90A 168
Ebbas Way KT18: Eps 87Ra 173
Ebb Ct. E16 45Sc 94
Ebberns Rd. HP3: Hem H 5M 3
Ebbett Ct. W3 43Ta 87
Ebbisham Dr. SW8 51Pb 112
Ebbisham La. KT20: Tad 93Va 194
Ebbisham La. KT20: Walt H 93Va 194
Ebbisham Rd. KT18: Eps 86Ra 173
Ebbisham Rd. KT4: Wor Pk 75Ya 154
Ebbisham Sports Club 83Sa 173
EBBSFLEET 59Ce 121
Ebbsfleet Bus. Pk. DA11: Nflt 57Ce 121
Ebbsfleet Ct. DA11: Nflt 58De 121
Ebbsfleet Gateway DA10: Ebbs 60Ce 121
Ebbsfleet Grn. DA10: Swans 60Be 121
Ebbsfleet International Station (Rail) 58Ce 121
Ebbsfleet Rd. NW2 36Ab 68
Ebbsfleet United FC 57Ce 121
EBBSFLEET VALLEY 60Zd 121
Ebbsfleet Wlk. DA11: Nflt 58De 121
Ebdon Way SE3 55Kc 115
Ebenezer Ho. SE11 6B 230 (49Rb 91)
Ebenezer Mussel Ho. E2 40Yb 72
(off Patriot Sq.)
Ebenezer St. N1 3F 219 (41Tb 91)
Ebenezer Wlk. SW16 67Lb 134
Ebley Cl. SE15 51Vb 113
Ebner St. SW18 57Db 111
Ebony Cres. EN4: E Barn 15Jb 32
Ebony Ho. E2 41Wb 91
(off Buckfast St.)
Ebor Cotts. SW15 62Ua 132
Ebor St. E1 5K 219 (42Vb 91)
Ebrington Rd. HA3: Kenton 30Ma 47
Ebsworth St. SE23 59Zb 114
Eburne Rd. N7 34Nb 70
Ebury App. WD3: Rick 18M 25
Ebury Bri. SW1 7K 227 (50Kb 90)
Ebury Bri. Est. SW1 7K 227 (50Kb 90)
Ebury Bri. Rd. SW1 7J 227 (50Jb 90)
Ebury Cl. BR2: Kes 76Nc 160
Ebury Cl. HA6: Nwood 22S 44
Ebury M. N5 34Sb 71
Ebury M. SE27 62Rb 135
Ebury M. SW1 5K 227 (49Kb 90)
Ebury M. E. SW1 4K 227 (48Kb 90)
Ebury Rd. WD17: Wat 13Y 27
Ebury Rd. WD3: Rick 18M 25
EBURY RDBT. 18M 25
Ebury Sq. SW1 6J 227 (49Jb 90)
Ebury St. SW1 6J 227 (49Jb 90)
Ecclesbourne Apts. N1 38Tb 71
(off Ecclesbourne Rd.)
Ecclesbourne Cl. N13 22Qb 50
Ecclesbourne Gdns. N13 22Qb 50
Ecclesbourne Rd. CR7: Thor H 71Sb 157
Ecclesbourne Rd. N1 38Sb 71
Eccleshill BR2: Broml 70Hc 137
(off Durham Rd.)
Eccles Rd. SW11 56Hb 111
Eccleston Bri. SW1 5A 228 (49Kb 90)
Eccleston Cl. BR6: Orp 74Tc 160
Eccleston Cl. EN4: Cockf 14Hb 31
Eccleston Cres. RM6: Chad H 31Xc 75
Ecclestone Ct. HA9: Wemb 36Na 67
Ecclestone M. HA9: Wemb 36Na 67
Ecclestone Pl. HA9: Wemb 36Pa 67
Eccleston Ho. SW2 58Qb 112
Eccleston M. SW1 4J 227 (48Jb 90)
Eccleston Pl. SW1 5K 227 (49Kb 90)
Eccleston Rd. W13 45Ja 86
Eccleston Sq. SW1 6A 228 (49Kb 90)
(not continuous)
Eccleston Sq. M. SW1 6A 228 (49Kb 90)
Eccleston St. SW1 4K 227 (48Kb 90)
Eccleston Yd. SW1 5K 227 (49Kb 90)
(off Eccleston Pl.)
Echelforde Dr. TW15: Ashf 63Q 128
Echo Ct. DA12: Grav'nd 1E 144
Echo Ct. E16 47Lc 93
(off Admiralty Av.)
Echo Hgts. E4 18Dc 34
Echo Sq. DA12: Grav'nd 1E 144
Eckford St. N1 1K 217 (40Qb 70)
Eckington Cl. SE14 52Zb 114
Eckington Ho. N15 30Tb 51
(off Fladbury Rd.)
Eckington La. SE14 52Zb 114
Eckstein Rd. SW11 56Gb 111
Eclipse, The KT10: Esh 75Da 151
Eclipse Bldg. N1 37Qb 70
(off Laycock St.)
Eclipse Ho. N22 26Pb 50
(off Station Rd.)
Eclipse Ind. Est. KT19: Eps 85Sa 173
Eclipse Rd. E13 43Kc 93
Ecology Cen. & Arts Pavilion 41Ac 92
Eco Va. SE23 58Xb 113
Ecton Rd. KT15: Add 77K 149
Ector Rd. SE6 61Gc 137
Edam Ct. DA15: Sidc 62Wc 139
Edans Ct. W12 47Va 88
Edbrooke Rd. W9 42Cb 89
Eddington Ct. E16 44Hc 93
(off Silvertown Sq.)
Eddinton Cl. CR0: New Ad 79Ec 158
Eddisbury Ho. SE26 62Wb 135
Eddiscombe Rd. SW6 54Bb 111
Eddy Cl. RM7: Rom 30Dd 56
Eddystone Ho. SE4 57Ac 114
Eddystone Twr. SE8 50Ac 92
Eddystone Wlk. TW19: Stanw 59N 105
Ede Cl. TW3: Houn 55Ba 107
Ede Ct. KT17: Eps 84Va 174
(off East St.)
Edeleny Cl. N2 27Eb 49
Eden Apts. E14 49Ec 92
(off Glengarnock Av.)
Eden Av. IG7: Chig 21Tc 54
Edenbridge Cl. BR5: St M Cry 70Zc 139
Edenbridge Cl. SE16 50Xb 91
(off Masters Dr.)
Edenbridge Rd. E9 38Zb 72
Edenbridge Rd. EN1: Enf 16Ub 33
Edenbridge Rd. SL2: Slou 2D 80
Eden Cl. DA5: Bexl 63Fd 140
Eden Cl. EN3: Enf L 10Cc 20
Eden Cl. HA0: Wemb 39Ma 67
Eden Cl. KT15: New H 82K 169
Eden Cl. NW3 33Cb 69
Eden Cl. SL3: L'ly 50C 82
Eden Cl. W8 48Cb 89
Eden Ct. IG6: Ilf 24Tc 54
Edencourt Rd. SW16 65Kb 134
Edendale W3 45Ra 87
Edendale Rd. DA7: Bex 53Fd 118
Eden Dr. HP2: Hem H 2N 3
Edenfield Gdns. KT4: Wor Pk 76Va 154
Eden Grn. RM15: S Ock 43Xd 98
Eden Gro. E17 29Dc 52
Eden Gro. N7 36Pb 70
Eden Gro. NW10 37Xa 68
Eden Gro. Rd. KT14: Byfl 85N 169
Edenhall Cl. HP2: Hem H 3D 4
Edenhall Cl. RM3: Rom 22Ld 57
Edenhall Glen RM3: Rom 22Ld 57
Edenhall Rd. RM3: Rom 22Ld 57
Edenham Way W10 42Bb 89
Eden Ho. NW8 6D 214 (42Gb 89)
(off Church St.)
Eden Ho. SE16 47Zb 92
(off Water Gdns. Sq.)
Eden Ho. SE8 52Cc 114
(off Deptford High St.)
Edenhurst TN13: S'oaks 97Jd 202
Edenhurst Av. SW6 55Bb 111
Eden Lodge NW6 38Za 68
Eden Lodges IG7: Chig 22Tc 54
Eden M. SW17 62Eb 133
EDEN PARK 71Cc 158
Eden Pk. Av. BR3: Beck 70Ac 136
Eden Park Station (Rail) 71Cc 158
Eden Pl. DA12: Grav'nd 9D 122
Eden Pl. SL5: S'dale 3E 146
Eden Rd. BR3: Beck 70Ac 136
Eden Rd. CR0: C'don 77Tb 157
Eden Rd. DA5: Bexl 63Ed 140
Eden Rd. E17 29Dc 52
Eden Rd. SE27 63Rb 135
Eden Rd. TN14: Dun G 92Gd 202
Edenside Rd. KT23: Bookh 96Ba 191
Edensor Gdns. W4 52Ua 110
Edensor Rd. W4 52Ua 110
Eden St. KT1: King T 68Ma 131
Edenvale EN7: G Oak 1Xb 19
Edenvale Cl. CR4: Mitc 66Jb 134
Edenvale Rd. CR4: Mitc 66Jb 134
Edenvale St. SW6 54Eb 111
Eden Wlk. KT1: King T 68Na 131
Eden Walk Shop. Cen. 68Na 131
Eden Way BR3: Beck 71Bc 158
Eden Way CR6: W'ham 90Ac 178
Eden Way E3 39Bc 72
Ederline Av. SW16 69Pb 134
Edes Flds. RH2: Reig 8G 206
Edgar Cl. BR8: Swan 69Hd 140
Edgar Ct. KT3: N Mald 68Ua 132
Edgar Ho. E11 31Jc 73
Edgar Ho. E9 36Ac 72
(off Homerton Rd.)
Edgar Ho. SW8 52Nb 112
(off Wyvil Rd.)
Edgar Kail Way SE22 56Ub 113
Edgarley Ter. SW6 53Ab 110
Edgar Myles Ho. E16 43Hc 93
(off Ordnance Rd.)
Edgar Rd. CR2: Sande 81Tb 177
Edgar Rd. E3 41Dc 92
Edgar Rd. RM6: Chad H 31Zc 75
Edgar Rd. TN15: Kems'g 89Nd 183
Edgar Rd. TN16: Tats 93Mc 199
Edgar Rd. TW4: Houn 59Ba 107
Edgar Rd. UB7: Yiew 45N 83
Edgar Wallace Cl. SE15 52Ub 113
Edgar Wright Ct. SW6 52Bb 111
(off Dawes Rd.)
Edgbaston Dr. WD7: Shenl 4Na 15
Edgbaston Rd. WD19: Wat 20X 27
Edgcott Ho. W10 43Ya 88
(off Sutton Way)
Edge Finsbury, The 31Sb 71
Edge, The HP3: Hem H 5M 3
Edge, The SE8 52Bc 114
(off Glenville Gro.)
Edge Apts. E15 38Fc 73
Edgeborough Way BR1: Broml 66Mc 137
Edgebury BR7: Chst 63Rc 138
Edgebury Wlk. BR7: Chst 63Sc 138
Edge Bus. Cen., The NW2 33Xa 68
Edge Cl. KT13: Weyb 80Q 150
Edgecombe Ho. SE5 54Ub 113
Edgecombe Ho. SW19 60Ab 110
Edgecoombe CR2: Sels 80Yb 158
Edgecoombe Cl. KT2: King T 66Ta 131
Edgecote Cl. W3 46Sa 87
Edgecot Gro. N15 29Ub 51
Edgecumbe Av. NW9 25Va 48
Edgefield Av. IG11: Bark 38Vc 75
Edgefield Cl. BR3: Beck 65Cc 136
Edgefield Cl. DA1: Dart 60Rd 119
Edgefield Cl. RH1: Redh 10A 208
Edgefield Ct. IG11: Bark 38Vc 75
(off Edgefield Av.)
Edge Hill SE18 51Rc 116
Edge Hill SW19 66Za 132
Edge Hill Av. N3 28Cb 49
Edge Hill Ct. DA14: Sidc 63Vc 139
Edge Hill Ct. SW19 66Za 132
Edgehill Ct. KT12: Walt T 74Y 151
Edgehill Gdns. DA13: Ist R 7B 144
Edgehill Gdns. RM10: Dag 35Cd 76
Edgehill Ho. SW9 54Rb 113
Edgehill Rd. BR7: Chst 62Sc 138
Edgehill Rd. CR4: Mitc 67Kb 134
Edgehill Rd. CR8: Purl 82Qb 176
Edgehill Rd. W13 43La 86
Edgeley KT23: Bookh 96Aa 191
Edgeley La. SW4 55Mb 112
Edgeley Rd. SW4 55Mb 112
Edgell Cl. GU25: Vir W 69B 126
Edgell Rd. TW18: Staines 64H 127
Edgel St. SW18 56Db 111
Edge Point Cl. SE27 64Rb 135
Edge St. W8 46Cb 89
Edgewood Dr. BR6: Chels 78Vc 161
Edgewood Grn. CR0: C'don 74Zb 158
Edgeworth Av. NW4 29Wa 48
Edgeworth Cl. CR3: Whyt 90Wb 177
Edgeworth Cl. NW4 29Wa 48
Edgeworth Cl. SL3: L'ly 7P 81
Edgeworth Ct. EN4: Cockf 14Gb 31
(off Fordham Rd.)
Edgeworth Cres. NW4 29Wa 48
Edgeworth Ho. NW8 39Eb 69
(off Boundary Rd.)
Edgeworth Rd. EN4: Cockf 14Gb 31
Edgeworth Rd. SE9 56Lc 115
Edge Youth Cen., The 77Sa 153
Edgington Rd. SW16 65Mb 134
Edgington Way DA14: Sidc 66Yc 139
Edgson Ho. SW1 7K 227 (50Kb 90)
(off Ebury Bri. Rd.)
EDGWARE 23Qa 47
EDGWAREBURY 18Pa 29
Edgwarebury Gdns. HA8: Edg 22Qa 47
Edgwarebury La. HA8: Edg 18Pa 29
(not continuous)
Edgwarebury La. WD6: E'tree 17Na 29
Edgware Bus Station 23Ra 47
EDGWARE COMMUNITY HOSPITAL 24Ra 47
Edgware Ct. HA8: Edg 23Qa 47
Edgware Rd. NW2 32Xa 68
Edgware Rd. NW9 26Sa 47
Edgware Rd. W2 6B 214 (42Fb 89)
Edgware Road Station (Underground) 1D 220 (43Gb 89)
Edgware Station (Underground) 23Ra 47
Edgware Way HA8: Edg 21Pa 47
Edgware Way NW7 21Ta 47
Edgware Way WD6: E'tree 18Ma 29
Edicule Sq. E3 41Ac 92
Edinburgh Av. SL1: Slou 3E 80
Edinburgh Av. WD3: Rick 16J 25
Edinburgh Cl. E2 40Yb 72
Edinburgh Cl. HA5: Pinn 31Z 65
Edinburgh Cl. UB10: Ick 35R 64
Edinburgh Ct. DA8: Erith 52Fd 118
Edinburgh Ct. KT1: King T 69Na 131
(off Watersplash Cl.)
Edinburgh Ct. SE16 46Zb 92
(off Rotherhithe St.)
Edinburgh Ct. SW20 71Za 154
Edinburgh Cres. EN8: Walt C 5Ac 20
Edinburgh Dr. TW18: Staines 65M 127
Edinburgh Dr. UB10: Ick 35R 64
Edinburgh Dr. WD5: Ab L 4W 12

Edinburgh Gdns. SL4: Wind 5H 103
Edinburgh Ga. SW1 1F 227 (47Hb 89)
Edinburgh Ga. UB9: Den 30H 43
Edinburgh Ho. NW4 27Ya 48
Edinburgh Ho. RM2: Rom 27Ld 57
Edinburgh Ho. W9 41Db 89
(off Maida Vale)
Edinburgh M. RM18: Tilb 4D 122
Edinburgh M. WD19: Wat 16Z 27
Edinburgh Rd. E13 40Kc 73
Edinburgh Rd. E17 29Cc 52
Edinburgh Rd. N18 22Wb 51
Edinburgh Rd. SM1: Sutt 75Eb 155
Edinburgh Rd. W7 47Ha 86
Edington NW5 37Jb 70
Edington Rd. EN3: Enf H 12Yb 34
Edington Rd. SE2 48Xc 95
Edison Av. RM12: Horn 32Hd 76
Edison Bldg. E14 47Cc 92
Edison Cl. AL4: St A 3G 6
Edison Cl. E17 29Cc 52
Edison Cl. RM12: Horn 32Gd 76
Edison Cl. UB7: W Dray 47P 83
Edison Ct. CR0: C'don 73Rb 157
(off Campbell Rd.)
Edison Ct. SE10 48Hc 93
(off Schoolbank Rd.)
Edison Ct. W3 47Ua 88
Edison Ct. WD18: Wat 16W 26
Edison Dr. HA9: Wemb 34Na 67
Edison Dr. UB1: S'hall 44Da 85
Edison Gro. SE18 52Vc 117
Edison Hgts. E1 5K 219 (42Vb 91)
(off Cygnet St.)
Edison Ho. HA9: Wemb 34Sa 67
(off Barnhill Rd.)
Edison Ho. SE1 5F 231 (49Tb 91)
(off New Kent Rd.)
Edison M. SW18 58Db 111
Edison Rd. BR2: Broml 68Jc 137
Edison Rd. DA16: Well 53Vc 117
Edison Rd. EN3: Brim 12Bc 34
Edison Rd. N8 30Mb 50
Edison's Pk. DA2: Dart 55Td 120
Edis St. NW1 39Jb 70
Ediswan Way EN3: Pond E 15Yb 34
Editha Mans. SW10 51Eb 111
(off Edith Gro.)
Edith Bell Ho. SL9: Chal P 23A 42
Edith Brinson Ho. E14 44Fc 93
(off Oban St.)
Edith Cavell Cl. N19 31Nb 70
Edith Cavell Ho. E14 44Dc 92
(off Sturry St.)
Edith Cavell Way SE18 53Nc 116
Edith Gdns. KT5: Surb 73Ra 153
Edith Gro. SW10 51Eb 111
Edith Ho. EN9: Walt A 5Dc 20
Edith Ho. W6 50Ya 88
(off Queen Caroline St.)
Edithna St. SW9 55Nb 112
Edith Nesbit Wlk. SE9 57Pc 116
Edith Neville Cotts. NW1 3D 216 (41Mb 90)
(off Drummond Cres.)
Edith Ramsay Ho. E1 43Ac 92
(off Duckett St.)
Edith Rd. BR6: Chels 78Wc 161
Edith Rd. E15 36Fc 73
Edith Rd. E6 38Mc 73
Edith Rd. N11 24Mb 50
Edith Rd. RM6: Chad H 31Zc 75
Edith Rd. SE25 71Tb 157
Edith Rd. SW19 65Db 133
Edith Rd. W14 49Ab 88
Edith Row SW6 53Db 111
Edith St. E2 40Wb 71
Edith Summerskill Ho. SW6 52Bb 111
(off Clem Attlee Ct.)
Edith Ter. SW10 52Eb 111
Edith Vs. W14 49Bb 89
Edith Yd. SW10 52Eb 111
Edmansons Cl. N17 25Vb 51
Edmeston Cl. E9 37Ac 72
Edmond Beaufort Dr. AL3: St A 1B 6
Edmond Ct. SE14 53Yb 114
Edmonds Ct. KT8: W Mole 71Da 151
Edmonscote W13 43Ja 86
EDMONTON 20Wb 33
Edmonton Bus Station 19Xb 33
Edmonton Ct. SE16 48Yb 92
(off Canada Est.)
Edmonton Grn. Shop. Cen. 19Wb 33
Edmonton Green Station (Overground) 19Wb 33
Edmonton Leisure Cen. 20Wb 33
Edmund Cl. DA13: Meop 10C 144
Edmund Gro. TW13: Hanw 61Ba 129
Edmund Halley Way SE10 47Gc 93
Edmund Ho. SE17 51Rb 113
Edmund Hurst Dr. E6 43Rc 94
Edmund M. WD4: K Lan 1Q 12
Edmund Rd. BR5: St M Cry 72Yc 161
Edmund Rd. CR4: Mitc 69Gb 133
Edmund Rd. DA16: Well 55Wc 117
Edmund Rd. RM13: Rain 40Gd 76
Edmund Rd. RM16: Chaf H 47Zd 99
Edmunds Av. BR5: St P 69Zc 139
Edmundsbury Ct. Est. SW9 56Pb 112
Edmunds Cl. UB4: Yead 43Y 85
Edmund St. SE5 52Tb 113
Edmunds Wlk. N2 28Gb 49
Edmunds Way SL2: Slou 3M 81
Ednam Ho. SE15 51Wb 113
(off Haymerle Rd.)
Edna Rd. SW20 68Za 132
Edna St. SW11 53Gb 111
Edred Ho. E9 35Ac 72
(off Lindisfarne Way)
Edrich Ho. SW4 53Nb 112
Edric Ho. SW1 5E 228 (49Mb 90)
(off Page St.)
Edrick Rd. HA8: Edg 23Sa 47
Edrick Wlk. HA8: Edg 23Sa 47
Edric Rd. SE14 52Zb 114
Edridge Cl. RM12: Horn 36Md 77
Edridge Cl. WD23: Bush 15Ea 28
Edridge Rd. CR0: C'don 76Sb 157
EDRIDGE ROAD COMMUNITY HEALTH CEN. 76Sb 157
Edson Cl. WD25: Wat 5V 12
Education Sq. E1 44Wb 91
(off Alder St.)
Edulf Rd. WD6: Bore 11Ra 29
Edward II Av. KT14: Byfl 86P 169
Edward VII Mans. NW10 41Za 88
(off Chamberlayne Rd.)
Edward Alderton Theatre 55Zc 117
Edward Amey Cl. WD25: Wat 8Y 13
Edward Av. E4 23Dc 52
Edward Av. SM4: Mord 71Fb 155
Edward Bond Ho. WC1 4G 217 (41Nb 90)
(off Cromer St.)
Edward Clifford Ho. SE17 6G 231 (49Tb 91)
(off Elsted St.)
Edward Cl. AL1: St A 3D 6
Edward Cl. N9 17Vb 33
Edward Cl. NW2 35Za 68
Edward Cl. RM16: Chaf H 47Zd 99
Edward Cl. RM2: Rom 27Ld 57
Edward Cl. TW12: Hamp H 64Ea 130
Edward Cl. WD5: Ab L 4V 12
Edward Ct. E16 43Jc 93
Edward Ct. EN9: Walt A 5Hc 21
Edward Ct. HP3: Hem H 6M 3
Edward Ct. TW18: Staines 65L 127
Edward Dodd Ct. N1 3G 219 (41Tb 91)
(off Chart St.)
Edward Edward's Ho. SE1 7B 224 (46Rb 91)
(off Nicholson St.)
Edwardes Pl. W8 48Bb 89
Edwardes Sq. W8 48Bb 89
Edward Gro. EN4: E Barn 15Fb 31
Edward Heylin Ct. E15 40Dc 72
(off High St.)
Edward Heylyn Ho. E3 40Cc 72
(off Thomas Fyre Dr.)
Edward Ho. RH1: Redh 9A 208
Edward Ho. SE11 7J 229 (50Pb 90)
(off Newburn St.)
Edward Ho. W2 6B 214 (42Fb 89)
(off Hall Pl.)
Edward Kennedy Ho. W10 42Ab 88
(off Wornington Rd.)
Edward Mann Cl. E. E1 44Zb 92
(off Pitsea St.)
Edward Mann Cl. W. E1 44Zb 92
(off Pitsea St.)
Edward M. NW1 3A 216 (41Kb 90)
Edward Mills Way E14 44Cc 92
Edward Pl. SE8 51Bc 114
Edward Rd. BR1: Broml 66Kc 137
Edward Rd. BR7: Chst 64Rc 138
Edward Rd. CR0: C'don 73Ub 157
Edward Rd. CR5: Coul 87Mb 176
Edward Rd. E17 28Zb 52
Edward Rd. EN4: E Barn 15Fb 31
Edward Rd. GU20: W'sham 9B 146
Edward Rd. HA2: Harr 27Ea 46
Edward Rd. RM6: Chad H 30Ad 55
Edward Rd. SE20 66Zb 136
Edward Rd. TN16: Big H 90Nc 180
Edward Rd. TW12: Hamp H 64Ea 130
Edward Rd. TW14: Felt 57T 106
Edward Rd. UB5: N'olt 40Y 65
Edward's Av. HA4: Ruis 37X 65
Edwards Cl. CM13: Hut 16Fe 41
Edwards Cl. KT15: New H 81L 169
Edwards Cl. KT4: Wor Pk 75Za 154
Edward's Cotts. N1 37Rb 71
Edwards Ct. CR0: C'don 77Ub 157
(off South Pk. Hill Rd.)
Edwards Ct. DA4: Eyns 76Nd 163
Edwards Ct. EN8: Chesh 2Ac 20
Edwards Ct. SL1: Slou 7J 81
Edwards Dr. N11 24Mb 50
Edwards Gdns. BR8: Swan 70Fd 140
Edward's La. N16 33Ub 71
Edwards Mans. IG11: Bark 38Vc 75
(off Upney La.)
Edwards M. N1 38Rb 71
Edwards M. W1 3H 221 (44Jb 90)
Edwards Pas. E1 43Yb 92
(off Trinity Grn.)
Edwards Pl. KT12: Walt T 74W 150
Edward Sq. N1 1H 217 (39Pb 70)
Edward Sq. SE16 46Ac 92
Edwards Rd. DA17: Belv 49Cd 96
Edward St. E16 42Jc 93
(not continuous)
Edward St. SE14 52Ac 114
Edward St. SE8 51Bc 114
Edwards Way CM13: Hut 16Fe 41
Edwards Way SE4 57Cc 114
Edwards Yd. HA0: Wemb 39Na 67
Edward Temme Av. E15 38Hc 73
Edward Tyler Rd. SE12 61Lc 137
Edward Way TW15: Ashf 61P 127
Edwick Ct. EN8: Chesh 1Zb 20
Edwina Gdns. IG4: Ilf 29Nc 54
Edwin Arnold Ct. DA14: Sidc 63Vc 139
Edwin Av. E6 40Qc 74
(not continuous)
Edwin Cl. DA7: Bex 51Bd 117
Edwin Cl. KT24: W Hor 97T 190
Edwin Cl. RM13: Rain 41Hd 96
Edwin Hall Pl. SE13 58Fc 115
Edwin Ho. SE15 52Wb 113
Edwin Petty Pl. DA2: Dart 59Sd 120
Edwin Pl. CR0: C'don 74Ub 157
(off Cross Rd.)
Edwin Rd. DA2: Wilm 62Kd 141
Edwin Rd. HA8: Edg 23Ta 47
Edwin Rd. KT24: W Hor 97S 190
Edwin Rd. TW1: Twick 60Ha 108
Edwin Rd. TW2: Twick 60Ga 108
Edwin's Mead E9 35Ac 72
Edwin Stray Ho. TW13: Hanw 61Ca 129
Edwin St. DA12: Grav'nd 9D 122
Edwin St. E1 42Yb 92
Edwin St. E16 43Jc 93
Edwin Ware Ct. HA5: Pinn 26Y 45
Edwy Ho. E9 35Bc 72
(off Homerton Rd.)
Edwyn Cl. EN5: Barn 16Ya 30
Edwyn Ho. SW18 58Db 111
(off Neville Gill Cl.)
Eel Brook Cl. SW6 53Db 111
Eel Pie Island TW1: Twick 60Ja 108
Effie Pl. SW6 52Cb 111
Effie Rd. SW6 52Cb 111
EFFINGHAM 99Z 191
Effingham Cl. SM2: Sutt 80Db 155
EFFINGHAM COMMON 96V 190
Effingham Comn. Rd. KT24: Eff 95W 190
Effingham Comn. Rd. KT24: Eff J 95W 190
Effingham Community Sports Cen. 99Aa 191
Effingham Ct. GU22: Wok 91A 188
(off Constitution Hill)
Effingham Golf Course 100Z 191
EFFINGHAM JUNCTION 95W 190
Effingham Junction Station (Rail) 95W 190
Effingham Lodge KT1: King T 70Ma 131
Effingham Pl. KT24: Eff 99Z 191
Effingham Rd. CR0: C'don 73Pb 156
Effingham Rd. KT6: Surb 73Ka 152
Effingham Rd. N8 29Qb 50
Effingham Rd. RH2: Reig 7K 207
Effingham Rd. SE12 57Gc 115
Effort St. SW17 64Gb 133
Effra Cl. SW19 65Db 133
Effra Ct. SW2 57Pb 112
(off Brixton Hill)
Effra Pde. SW2 57Qb 112
Effra Rd. SW19 65Db 133
Effra Rd. SW2 56Qb 112
Effra Rd. Retail Pk. 57Qb 112
Egan Cl. CR8: Kenley 92Tb 197
Egan Way UB3: Hayes 45U 84
Egbert Cl. RM12: Horn 36Hd 76
Egbert Ho. E9 36Ac 72
(off Homerton Rd.)
Egbert St. NW1 39Jb 70
Egbury Ho. SW15 58Va 110
(off Tangley Gro.)
Egdean Wlk. TN13: S'oaks 95Ld 203
Egeremont Rd. SE13 54Dc 114
Egerton Av. BR8: Hext 66Hd 140
Egerton Cl. DA1: Dart 60Kd 119
Egerton Cl. DA17: Belv 50Ed 96
Egerton Cl. HA5: Eastc 28W 44
Egerton Cl. WD23: Bush 16Ba 27
Egerton Ct. E11 31Fc 73
Egerton Cres. SW3 5E 226 (49Gb 89)
Egerton Dr. SE10 53Dc 114
Egerton Dr. TW7: Isle 55Ka 108
Egerton Gdns. IG3: Ilf 34Vc 75
Egerton Gdns. NW10 39Ya 68
Egerton Gdns. NW4 28Xa 48
Egerton Gdns. SW3 5D 226 (49Gb 89)
Egerton Gdns. W13 44Ka 86
Egerton Gdns. M. SW3 4E 226 (48Gb 89)
Egerton Pl. KT13: Weyb 79S 150
Egerton Pl. SW3 4E 226 (48Gb 89)
Egerton Rd. HA0: Wemb 38Pa 67
Egerton Rd. KT13: Weyb 79S 150
Egerton Rd. KT3: N Mald 70Va 132
Egerton Rd. N16 31Vb 71
Egerton Rd. SE25 69Ub 135
Egerton Rd. SL2: Slou 2C 80
Egerton Rd. TW2: Twick 59Ga 108
Egerton Ter. SW3 4E 226 (48Gb 89)
Egerton Way UB3: Harl 52R 106
Eggardon Ct. UB5: N'olt 37Da 65
Egg Farm La. WD4: K Lan 2R 12
Egg Farm La. WD5: Ab L 2T 12
Egg Hall CM16: Epp 1Wc 23
EGHAM 64C 126
Egham Bus. Pk. TW20: Thorpe 67E 126
Egham Bus. Village TW20: Thorpe 68E 126
Egham By-Pass TW20: Egh 64B 126
Egham Cl. SM3: Cheam 75Ab 154
Egham Cl. SW19 61Ab 132
Egham Cres. SM3: Cheam 76Ab 154
Egham Hill TW20: Egh 5P 125
Egham Hill TW20: Eng G 5P 125
EGHAM HYTHE 64G 126
Egham Mus. 64C 126
Egham Orbit 65D 126
Egham Rd. E13 43Kc 93
EGHAM RDBT. 64G 126
Egham Station (Rail) 64C 126
EGHAM WICK 6L 125
Eglantine La. DA4: Farni 73Qd 163
Eglantine La. DA4: Hort K 73Qd 163
Eglantine Rd. SW18 57Eb 111
Egleston Rd. SM4: Mord 72Db 155
Egleton Ho. SW15 59Wa 110
Egley Dr. GU22: Wok 4P 187
Egley Rd. GU22: Wok 4P 187
Eglington Ct. SE17 51Sb 113
Eglington Rd. E4 17Fc 35
Eglinton Hill SE18 51Rc 116
Eglinton Rd. DA10: Swans 58Be 121
Eglinton Rd. SE18 51Qc 116
Eglise Rd. CR6: W'ham 89Ac 178
Egliston M. SW15 55Ya 110
Egliston Rd. SW15 55Ya 110
Eglon M. NW1 38Hb 69
Egmont Av. KT6: Surb 74Pa 153
Egmont Ct. KT12: Walt T 73X 151
(off Egmont Rd.)
Egmont M. KT19: Ewe 77Ta 153
Egmont Pk. Rd. KT20: Walt H 97Wa 194
Egmont Rd. KT12: Walt T 73X 151
Egmont Rd. KT3: N Mald 70Va 132
Egmont Rd. KT6: Surb 74Pa 153
Egmont Rd. SM2: Sutt 80Eb 155
Egmont St. SE14 52Zb 114
Egmont Way KT20: Tad 91Ab 194
Egremont Gdns. SL1: Slou 6E 80
Egremont Ho. E20 36Ec 72
(off Medals Way)
Egremont Ho. SE13 54Dc 114
(off Russett Way)
Egremont Rd. SE27 62Qb 134
Egret Dr. HP3: Hem H 7L 3
Egret Hgts. N17 27Xb 51
(off Waterside Way)
Egret Ho. SE16 49Zb 92
(off Tawny Way)
Egret Ho. UB7: Yiew 45M 83
(off Wraysbury Dr.)
Egret Way UB4: Yead 43Z 85
EGYPT 5F 60
Egypt La. SL2: Farn C 3F 60
Eider Cl. E7 36Hc 73
Eider Cl. UB4: Yead 43Z 85
Eider Ct. SE8 51Bc 114
(off Pilot Cl.)
Eighteenth Rd. CR4: Mitc 70Nb 134
Eighth Av. E12 35Pc 74
Eighth Av. KT20: Lwr K 98Ab 194
Eighth Av. UB3: Hayes 46W 84
Eileen Lenton Ct. N15 28Ub 51
(off Tottenham Grn. E.)
Eileen Rd. SE25 71Tb 157
Eileen Tozer Cen. 77L 149
Eilmer Cl. KT15: Add 76M 149
Eindhoven Cl. SM5: Cars 74Jb 156
Einstein Ho. HA9: Wemb 34Sa 67
Eisenhower Dr. E6 43Nc 94
Ekarro Ho. SW8 53Nb 112
(off Guildford Rd.)
Ekman Cl. DA10: Swans 60Ce 121
Elaine Gro. NW5 36Jb 70
Elam Cl. SE5 54Rb 113
Elam St. SE5 54Rb 113
Eland Ct. E1 43Xb 91
Eland Pl. CR0: Wadd 76Rb 157
Eland Rd. CR0: Wadd 76Rb 157
Eland Rd. SW11 55Hb 111
Eland Rd. RM15: S Ock 43Wd 98
Elba Pl. SE17 5E 230 (49Sb 91)
Elberon Av. CR0: Bedd 72Lb 156
Elbe St. SW6 54Eb 111
Elborough Rd. SE25 71Wb 157
Elborough St. SW18 60Cb 111
Elbourne Ct. SE16 48Zb 92
(off Worgan St.)
Elbourne Trad. Est. DA17: Belv 48Dd 96
Elbourn Ho. SW3 7D 226 (50Gb 89)
(off Cale St.)
Elbow Mdw. SL3: Poyle 53H 105
Elbury Dr. E16 44Jc 93
Elcho Rd. GU24: Brkwd 1A 186
Elcho St. SW11 52Gb 111
Elcot Av. SE15 52Xb 113
Elden Ho. SW3 5D 226 (49Gb 89)
(off Sloane Av.)
Eldenwall Ind. Est. RM8: Dag 32Bd 75
Elder Av. N8 29Nb 50
Elderbek Cl. EN7: Chesh 1Wb 19
Elderberry Cl. AL2: Brick W 3Ba 13
Elderberry Cl. IG6: Ilf 24Rc 54
Elderberry Cl. RM3: Hrld W 26Nd 57
Elderberry Gro. SE27 63Sb 135
Elderberry Rd. W5 47Na 87
Elderberry Way E6 41Pc 94
Elderberry Way WD25: Wat 7X 13
Elder Cl. DA15: Sidc 60Vc 117
Elder Cl. KT17: Eps D 87Ya 174
Elder Cl. N20 19Db 31
Elder Cl. UB7: Yiew 45N 83
Elder Ct. WD23: B Hea 19Ga 28
Elderfield Ho. E14 45Cc 92
(off Pennyfields)
Elderfield Pl. SW17 63Kb 134
Elderfield Rd. E5 35Zb 72
Elderfield Rd. SL2: Stoke P 7K 61
Elderfield Wlk. E11 29Kc 53
Elderflower Way E15 38Gc 73
Elder Gdns. SE27 64Sb 135
Elder Ho. E15 41Gc 93
(off Manor Rd.)
Elder Ho. KT1: King T 67Ma 131
(off Water La.)
Elder Ho. SE16 48Ac 92
Elder Oak Cl. SE20 67Xb 135
Elder Oak Ct. SE20 67Xb 135
(off Anerley Rd.)
Elder Pl. CR2: S Croy 79Rb 157
Elder Rd. GU24: Bisl 7E 166
Elder Rd. SE27 64Sb 135
Eldersley Cl. RH1: Redh 4P 207
Eldersley Gdns. RH1: Redh 4P 207
Elderslie Cl. BR3: Beck 71Cc 158
Elderslie Rd. SE9 57Qc 116
Elder St. E1 6K 219 (42Vb 91)
(not continuous)
Elderton Rd. SE26 63Ac 136
Eldertree Pl. CR4: Mitc 67Lb 134
Eldertree Way CR4: Mitc 67Lb 134
Elder Wlk. N1 39Rb 71
(off Popham St.)
Elder Wlk. SE13 55Ec 114
Elder Way RM13: Rain 41Md 97
Elder Way SL3: L'ly 47B 82
Elderwood Pl. SE27 64Sb 135
Eldon Av. CR0: C'don 75Yb 158
Eldon Av. TW5: Hest 52Ca 107
Eldon Av. WD6: Bore 12Qa 29
Eldon Ct. KT13: Weyb 78S 150
Eldon Ct. NW6 39Cb 69
Eldon Gro. NW3 36Fb 69
Eldon Ho. NW9 27Wa 48
(off East Dr.)
Eldon Pk. SE25 70Xb 135
Eldon Rd. CR3: Cat'm 93Tb 197
Eldon Rd. E17 28Bc 52
Eldon Rd. N22 25Rb 51
Eldon Rd. N9 18Yb 34
Eldon Rd. W8 48Db 89
Eldons Pas. E1 45Wb 91
(off Cable St.)
Eldon St. EC2 1G 225 (43Tb 91)
Eldon Way NW10 41Ra 87
Eldred Dr. BR5: Orp 74Yc 161
Eldred Gdns. RM14: Upm 31Ud 78
Eldred Rd. IG11: Bark 39Uc 74
Eldrick Ct. TW14: Bedf 60T 106
Eldridge Cl. TW14: Felt 60W 106
Eldridge Ct. RM10: Dag 37Dd 76
Eldridge Ct. SE16 48Wb 91
Eleanor Av. KT19: Ewe 82Ta 173
Eleanor Cl. DA1: Dart 56Md 119
Eleanor Cl. N15 27Vb 51
Eleanor Cl. SE16 47Zb 92
Eleanor Ct. E2 39Wb 71
(off Whiston Rd.)
Eleanor Cres. NW7 22Za 48
Eleanor Cross Rd. EN8: Walt C 6Ac 20
Eleanore Pl. AL3: St A 1B 6
Eleanor Gdns. EN5: Barn 15Za 30
Eleanor Gdns. RM8: Dag 33Bd 75
Eleanor Gro. SW13 55Ua 110
Eleanor Gro. UB10: Ick 34R 64
Eleanor Ho. AL1: St A 4D 6
Eleanor Ho. SL9: Chal P 22A 42
(off Micholls Av.)
Eleanor Ho. W6 50Ya 88
(off Queen Caroline St.)
Eleanor Rathbone Ho. N6 31Mb 70
(off Avenue Rd.)
Eleanor Rd. E15 37Hc 73
Eleanor Rd. E8 37Xb 71
Eleanor Rd. EN8: Walt C 5Ac 20
Eleanor Rd. N11 23Nb 50
Eleanor Rd. SW9 53Qb 112
Eleanor St. E3 41Cc 92
Eleanor Wlk. DA9: Ghithe 56Yd 120
Eleanor Wlk. SE18 49Nc 94
Eleanor Way CM14: W'ley 22Zd 59
Eleanor Way EN8: Walt C 6Bc 20
Electra Av. TW6: H'row A 55V 106
Electra Bus. Pk. E16 43Fc 93
Electric Av. EN3: Enf L 8Bc 20
Electric Av. SW9 56Qb 112
Electric Cinema Portobello 44Bb 89
Electric Cinema Shoreditch 5K 219 (42Vb 91)
(off Club Row)
Electric Empire, The SE14 53Zb 114
(off New Cross Rd.)
Electric Ho. E3 41Cc 92
(off Bow Rd.)
Electric La. SW9 56Qb 112
(not continuous)
Electric Pde. E18 26Jc 53
(off George La.)
Electric Pde. IG3: Ilf 33Uc 74
Electric Pde. KT6: Surb 72Ma 153
Electron Trade Cen. BR5: St M Cry 70Xc 139
Elektron Twr. E14 45Fc 93
Eleonora Ter. SM1: Sutt 78Eb 155
(off Lind Rd.)
Elephant & Castle SE1 5C 230 (49Rb 91)
ELEPHANT & CASTLE 4C 230 (48Rb 91)
Elephant & Castle Station (Rail & Underground) 5D 230 (49Sb 91)
Elephant La. SE16 47Yb 92
Elephant Pk. Development SE17 5D 230 (49Sb 91)
Elephant Rd. SE17 5D 230 (49Sb 91)
Elers Rd. UB3: Harl 49T 84
Elers Rd. W13 47La 86
Eleven Acre Ri. IG10: Lough 13Pc 36
Eleventh Av. KT20: Lwr K 98Bb 195
Eley Pl. WD19: Wat 17Z 27
Eley Rd. N18 21Zb 52
Eley Rd. Retail Pk. 22Yb 52
Eleys Est. N18 21Zb 52
(not continuous)
Elfindale Rd. SE24 57Sb 113
Elfin Gro. TW11: Tedd 64Ha 130
Elfin Oak 46Db 89
Elford Cl. SE3 56Lc 115
Elford M. SW4 57Lb 112
Elfort Rd. N5 35Qb 70
Elfrida Cl. IG8: Wfd G 24Jc 53
Elfrida Cres. SE6 63Cc 136
Elfrida Rd. WD18: Wat 15Y 27
Elf Row E1 45Yb 92
Elfwine Rd. W7 43Ga 86
Elgal Cl. BR6: Farnb 78Rc 160
Elgar N8 27Nb 50
(off Boyton Cl.)
Elgar Av. KT5: Surb 74Qa 153
Elgar Av. NW10 37Ta 67
(not continuous)
Elgar Av. SW16 69Nb 134
Elgar Av. W5 47Na 87
Elgar Cl. E13 40Lc 73
Elgar Cl. IG9: Buck H 19Mc 35
Elgar Cl. SE8 52Cc 114
Elgar Cl. UB10: Ick 33Q 64
Elgar Cl. WD6: E'tree 17Ma 29
Elgar Ct. NW6 41Cb 89
Elgar Ct. W14 48Ab 88
(off Blythe Rd.)
Elgar Gdns. RM18: Tilb 3C 122
Elgar Ho. NW6 38Eb 69
(off Fairfax Rd.)
Elgar Ho. SW1 50Kb 90
(off Churchill Gdns.)
Elgar St. SE16 48Ac 92
Elgin Av. HA3: Kenton 26Ka 46
Elgin Av. RM3: Hrld W 24Rd 57
Elgin Av. TW15: Ashf 65S 128
Elgin Av. W12 47Wa 88
Elgin Av. W9 42Bb 89
Elgin Cl. W12 47Xa 88
Elgin Ct. CR2: S Croy 77Sb 157
(off Bramley Hill)
Elgin Ct. W9 42Db 89
Elgin Cres. CR3: Cat'm 94Wb 197
Elgin Cres. TW6: H'row A 54U 106
Elgin Cres. W11 45Ab 88
Elgin Dr. HA6: Nwood 24U 44
Elgin Est. W9 42Cb 89
(off Elgin Av.)
Elgin Ho. CM14: W'ley 21Zd 59
Elgin Ho. E14 44Dc 92
(off Ricardo St.)
Elgin Ho. RM6: Chad H 30Bd 55
(off High Rd.)
Elgin Mans. W9 41Db 89
Elgin M. W11 44Ab 88
Elgin M. Nth. W9 41Db 89
Elgin M. Sth. W9 41Db 89
Elgin Pl. KT13: Weyb 79S 150
Elgin Rd. CR0: C'don 75Vb 157
Elgin Rd. EN8: Chesh 2Yb 20
Elgin Rd. IG3: Ilf 32Uc 74
Elgin Rd. KT13: Weyb 78Q 150
Elgin Rd. N22 26Lb 50
Elgin Rd. SM1: Sutt 76Eb 155
Elgin Rd. SM6: W'gton 79Lb 156
Elgood Av. HA6: Nwood 23V 44
Elgood Cl. W11 45Ab 88
Elgood Ho. NW8 2C 214 (40Fb 69)
(off Wellington Rd.)
Elgood Ho. SE1 2F 231 (47Tb 91)
(off Tabard St.)
Elham Cl. BR1: Broml 66Mc 137
Elham Cres. DA2: Dart 58Rd 119
Elham Ho. E5 36Xb 71
Elia M. N1 2B 218 (40Rb 71)
Elias Ho. CM14: B'wood 19Zd 41
Elias Pl. SW8 51Qb 112
Elia St. N1 2B 218 (40Rb 71)
Elibank Rd. SE9 56Pc 116
Elim Est. SE1 3H 231 (48Ub 91)
Elim St. SE1 3G 231 (48Tb 91)
(not continuous)
Elim Way E13 41Hc 93
Elinor Va. DA10: Swans 59Be 121
Eliot Bank SE23 61Xb 135
Eliot Cotts. SE3 54Gc 115
Eliot Ct. SW18 58Db 111
Eliot Dr. HA2: Harr 33Da 65
Eliot Gdns. SW15 56Wa 110
Eliot Hill SE13 54Ec 114
Eliot M. NW8 2A 214 (40Eb 69)
Eliot Pk. SE13 54Ec 114
Eliot Pl. SE3 54Gc 115
Eliot Rd. DA1: Dart 57Rd 119
Eliot Rd. RM9: Dag 35Zc 75
Eliot Va. SE3 54Fc 115
Elis David Almshouses CR0: C'don 76Rb 157
Elis Way E20 36Ec 72
Elizabethan Cl. TW19: Stanw 59M 105
Elizabethan Way TW19: Stanw 59M 105
Elizabeth Av. EN2: Enf 13Rb 33
Elizabeth Av. IG1: Ilf 33Tc 74
Elizabeth Av. N1 39Sb 71
Elizabeth Av. TW18: Staines 65L 127
Elizabeth Barnes Ct. SW6 54Db 111
(off Marinefield Rd.)
Elizabeth Bates Ct. E1 43Yb 92
(off Fulneck Pl.)
Elizabeth Blackwell Ho. N22 25Qb 50
(off Progress Way)
Elizabeth Blount Ct. E14 44Ac 92
(off Carr St.)
Elizabeth Bri. SW1 6K 227 (49Kb 90)
Elizabeth Cl. E14 44Dc 92
Elizabeth Cl. EN5: Barn 13Za 30
Elizabeth Cl. RM18: Tilb 4D 122
Elizabeth Cl. RM7: Mawney 25Dd 56
Elizabeth Cl. SM1: Sutt 77Bb 155
Elizabeth Cl. W9 6A 214 (42Eb 89)
Elizabeth Clyde Cl. N15 28Ub 51
Elizabeth Cotts. TW9: Kew 53Pa 109
Elizabeth Ct. BR1: Broml 67Hc 137
(off Highland Rd.)
Elizabeth Ct. CR0: C'don 76Ub 157
(off The Avenue)

Elizabeth Ct. CR3: Cat'm 94Sb **197**
Elizabeth Ct. CR3: Whyt. 90Vb **177**
Elizabeth Ct. DA11: Grav'nd 8C **122**
Elizabeth Ct. E10 ....31Dc **72**
Elizabeth Ct. E4 ....22Bc **52**
Elizabeth Ct. IG8: Wfd G 24Lc **53**
Elizabeth Ct. KT13: Weyb 77T **150**
Elizabeth Ct. KT2: King T. 67Na **131**
Elizabeth Ct. NW1 ....5E **214** (42Gb **89**)
.... *(off Palgrave Gdns.)*
Elizabeth Ct. SL1: Slou 7L **81**
Elizabeth Ct. SL4: Wind 4G **102**
.... *(off Beaumont Rd.)*
Elizabeth Ct. SW1 ....4E **228** (48Mb **90**)
.... *(off Milmans Ct.)*
Elizabeth Ct. SW10 ....51Fb **111**
.... *(off Milman's St.)*
Elizabeth Ct. TW11: Tedd 64Ga **130**
Elizabeth Ct. TW16: Sun 69Y **129**
.... *(off Elizabeth Gdns.)*
Elizabeth Ct. WD17: Wat 10V **12**
Elizabeth Croll Ho. WC1 ....3J **217** (41Pb **90**)
.... *(off Penton Ri.)*
Elizabeth Dr. CM16: They B. 8Uc **22**
Elizabeth Dr. SM7: Bans 90Eb **175**
Elizabeth Fry Apts. IG11: Bark 38Sc **74**
.... *(off Kings Rd.)*
Elizabeth Fry Ho. UB3: Harl 49V **84**
Elizabeth Fry M. E8 ....38Xb **71**
Elizabeth Fry Pl. SE18 ....53Nc **116**
Elizabeth Gdns. HA7: Stan 23La **46**
Elizabeth Gdns. TW16: Sun 69Y **129**
Elizabeth Gdns. TW7: Isle 56Ja **108**
Elizabeth Gdns. W3 ....46Va **88**
Elizabeth Garrett Anderson Ho.
DA17: Belv 48Cd **96**
.... *(off Ambrooke Rd.)*
Elizabeth Hart Ct. KT13: Weyb 78P **149**
Elizabeth Ho. CR3: Cat'm 96Wb **197**
Elizabeth Ho. E3 ....41Dc **92**
.... *(off St Leonard's St.)*
Elizabeth Ho. HP2: Hem H 1M **3**
.... *(off Chapel St.)*
Elizabeth Ho. RM16: Grays 46De **99**
Elizabeth Ho. RM2: Rom 28Ld **57**
Elizabeth Ho. SE11 ....6A **230** (49Qb **90**)
.... *(off Reedworth St.)*
Elizabeth Ho. SM3: Cheam 79Ab **154**
.... *(off Park La.)*
Elizabeth Ho. SM7: Bans 90Eb **175**
Elizabeth Ho. W6 ....50Ya **88**
.... *(off Queen Caroline St.)*
Elizabeth Ho. WD24: Wat 12Y **27**
Elizabeth Huggins Cotts. DA11: Grav'nd 1D **144**
Elizabeth Ind. Est. SE14 ....51Zb **114**
Elizabeth M. E10 ....32Ec **72**
Elizabeth M. E2 ....40Wb **71**
.... *(off Kay St.)*
Elizabeth M. HA1: Harr 30Ga **46**
Elizabeth M. NW3 ....37Gb **69**
Elizabeth Newcomen Ho. SE1 ....1F **231** (47Tb **91**)
.... *(off Newcomen St.)*
Elizabeth Pl. DA4: Farni 72Pd **163**
Elizabeth Pl. N15 ....28Tb **51**
Elizabeth Pl. SL4: Eton W 9D **80**
Elizabeth Ride N9 ....17Xb **33**
Elizabeth Rd. CM15: Pil H 16Xd **40**
Elizabeth Rd. E6 ....39Mc **73**
Elizabeth Rd. N15 ....29Ub **51**
Elizabeth Rd. RM13: Rain 43Kd **97**
Elizabeth Rd. RM16: Grays 47Be **99**
Elizabeth Sq. SE16 ....45Ac **92**
.... *(off Sovereign Cres.)*
Elizabeth St. DA9: Ghithe 57Ud **120**
Elizabeth St. SW1 ....5J **227** (49Jb **90**)
Elizabeth Ter. SE9 ....58Pc **116**
Elizabeth Way BR5: St M Cry 71Yc **161**
Elizabeth Way SE19 ....66Tb **135**
Elizabeth Way SL2: Stoke P 9K **61**
Elizabeth Way TW13: Hanw 63Y **129**
Eliza Cook Cl. DA9: Ghithe 56Xd **120**
Eliza Palmer Hub KT12: W Vill 81V **170**
Elkanette M. N20 ....19Eb **31**
Elkington Point SE11 ....6K **229** (49Qb **90**)
.... *(off Lollard St.)*
Elkington Rd. E13 ....42Kc **93**
Elkins, The RM1: Rom 26Gd **56**
Elkins Rd. SL2: Hedg 3J **61**
Elkstone Rd. W10 ....43Bb **89**
Ella Cl. BR3: Beck 68Cc **136**
Ellacott M. SW16 ....61Mb **134**
Ellaline Rd. W6 ....51Za **110**
Ella M. NW3 ....35Hb **69**
Ellanby Cres. N18 ....21Xb **51**
Elland Cl. EN5: New Bar 15Fb **31**
Elland Ho. E14 ....44Bc **92**
.... *(off Copenhagen Pl.)*
Elland Rd. KT12: Walt T 75Z **151**
Elland Rd. SE15 ....56Yb **114**
Ella Rd. N8 ....31Nb **70**
Ellement Cl. HA5: Pinn 29Z **45**
Ellena Ct. N14 ....20Nb **32**
.... *(off Conway Rd.)*
Ellenborough Ho. W12 ....45Xa **88**
.... *(off White City Est.)*
Ellenborough Pl. SW15 ....56Wa **110**
Ellenborough Rd. DA14: Sidc 64Zc **139**
Ellenborough Rd. N22 ....25Sb **51**
Ellenbridge Way CR2: Sande 81Ub **177**
ELLENBROOK 1P **7**
Ellenbrook Cl. WD24: Wat 11Y **27**
Ellenbrook Cres. AL10: Hat 1P **7**
Ellenbrook La. AL10: Hat 1P **7**
Ellen Cl. BR1: Broml 69Mc **137**
Ellen Cl. HP2: Hem H 1P **3**
Ellen Ct. E4 ....18Ec **34**
.... *(off The Ridgeway)*
Ellen Ct. N9 ....19Yb **34**
Ellen Julia Ct. E1 ....44Yb **92**
.... *(off James Voller Way)*
Ellen M. HP2: Hem H 1P **3**
ELLENOR HOSPICE 59Pd **119**
ELLENOR LIONS HOSPICE 3B **144**
Ellen Phillips La. E2 ....40Wb **71**
Ellen St. E1 ....44Wb **91**
Ellen Terry Ct. NW1 ....38Kb **70**
.... *(off Farrier St.)*
Ellen Webb Dr. HA3: W'stone 27Ga **46**
Ellen Wilkinson Ho. E2 ....41Zb **92**
.... *(off Usk St.)*
Ellen Wilkinson Ho. RM10: Dag 34Cd **76**
Ellen Wilkinson Ho. SW6 ....51Bb **111**
.... *(off Clem Attlee Ct.)*
Elleray Rd. TW11: Tedd 65Ha **130**
Ellerby St. SW6 ....53Za **110**
Ellerdale Cl. NW3 ....35Eb **69**
Ellerdale Rd. NW3 ....36Eb **69**
Ellerdale St. SE13 ....56Dc **114**
Ellerdine Rd. TW3: Houn 56Ea **108**
Ellerker Gdns. TW10: Rich 58Na **109**
Ellerman Av. TW2: Whitt 60Ba **107**
Ellerman Rd. RM18: Tilb 4B **122**
ELLERN MEDE CEN. 18Ya **30**
Ellerslie Ct. DA12: Grav'nd 9F **122**
.... *(off Copper Beech Cl.)*
Ellerslie Gdns. NW10 ....39Wa **68**
Ellerslie Rd. W12 ....46Xa **88**
Ellerslie Sq. Ind. Est. SW2 ....57Nb **112**
Ellerton Gdns. RM9: Dag 38Yc **75**
Ellerton Lodge N3 ....26Cb **49**
Ellerton Rd. KT6: Surb 75Pa **153**
Ellerton Rd. RM9: Dag 38Yc **75**
Ellerton Rd. SW13 ....53Wa **110**
Ellerton Rd. SW18 ....60Fb **111**
Ellerton Rd. SW20 ....66Wa **132**
Ellery Ho. SE17 ....6G **231** (49Tb **91**)
Ellery Rd. SE19 ....66Tb **135**
Ellery St. SE15 ....54Xb **113**
Ellesborough Cl. WD19: Wat 22Y **45**
Ellesmere Av. BR3: Beck 68Ec **136**
Ellesmere Av. NW7 ....20Ta **29**
Ellesmere Cl. E11 ....29Hc **53**
Ellesmere Cl. HA4: Ruis 31S **64**
Ellesmere Cl. SL3: Dat 1L **103**
Ellesmere Ct. KT13: Weyb 79U **150**
Ellesmere Ct. SE12 ....60Jc **115**
Ellesmere Ct. W4 ....50Ta **87**
Ellesmere Dr. CR2: Sande 86Xb **177**
Ellesmere Gdns. IG4: Ilf 29Nc **54**
Ellesmere Gro. EN5: Barn 15Bb **31**
Ellesmere Ho. SW10 ....51Eb **111**
.... *(off Fulham Rd.)*
Ellesmere Mans. NW6 ....37Eb **69**
.... *(off Canfield Gdns.)*
Ellesmere Pl. KT12: Hers 78U **150**
Ellesmere Rd. E3 ....40Ac **72**
Ellesmere Rd. HP4: Berk 1A **2**
Ellesmere Rd. KT13: Weyb 80U **150**
Ellesmere Rd. NW10 ....36Wa **68**
Ellesmere Rd. TW1: Twick 58La **108**
Ellesmere Rd. UB6: G'frd 42Ea **86**
Ellesmere Rd. W4 ....51Ta **109**
Ellesmere St. E14 ....44Dc **92**
Ellice Rd. RH8: Oxt 1K **211**
Ellie Cl. SS17: Stan H 1L **101**
Ellie M. E13 ....40Lc **73**
Ellie M. TW15: Ashf 61N **127**
Elliman Av. SL2: Slou 5J **81**
Elliman Sq. SL1: Slou 7K **81**
.... *(within Queensmere Shop. Cen.)*
Ellingfort Rd. E8 ....38Xb **71**
Ellingham GU22: Wok 91A **188**
Ellingham Cl. HP2: Hem H 1A **4**
Ellingham Rd. E15 ....35Fc **73**
Ellingham Rd. HP2: Hem H 1P **3**
Ellingham Rd. KT9: Chess 79Ma **153**
Ellingham Rd. W12 ....47Wa **88**
Ellingham Vw. DA1: Dart 55Qd **119**
Ellington Ct. N14 ....19Mb **32**
Ellington Ho. SE1 ....4E **230** (48Sb **91**)
Ellington Ho. SE18 ....52Qc **116**
Ellington Rd. N10 ....28Kb **50**
Ellington Rd. TW13: Felt 63V **128**
Ellington Rd. TW3: Houn 54Da **107**
Ellington St. N7 ....37Qb **70**
Ellington Way KT18: Tatt C 89Xa **174**
Elliot Cl. E15 ....38Gc **73**
Elliot Cl. IG8: Wfd G 23Mc **53**
Elliot Ho. SW17 ....62Fb **133**
.... *(off Grosvenor Way)*
Elliot Ho. W1 ....1E **220** (43Gb **89**)
.... *(off Cato St.)*
Elliot Rd. NW4 ....30Xa **48**
Elliot Rd. WD17: Wat 10W **12**
Elliott Av. HA4: Ruis 33X **65**
Elliott Cl. HA9: Wemb 34Pa **67**
Elliott Ct. NW10 ....37Ya **68**
Elliott Gdns. RM3: Rom 25Kd **57**
Elliott Gdns. TW17: Shep 70Q **128**
Elliott Rd. BR2: Broml 70Mc **137**
Elliott Rd. CR7: Thor H 70Rb **135**
Elliott Rd. HA7: Stan 23Ja **46**
Elliott Rd. SW9 ....53Rb **113**
Elliott Rd. W4 ....49Ua **88**
Elliotts Cl. UB8: Cowl 43L **83**
Elliotts La. TN16: Bras 96Yc **201**
Elliott's Pl. N1 ....39Rb **71**
Elliott Sq. NW3 ....38Gb **69**
Elliotts Row SE11 ....5C **230** (49Rb **91**)
Elliott St. DA12: Grav'nd 9F **122**
Ellis Av. RM13: Rain 43Jd **96**
Ellis Av. RM8: Dag 32Ad **75**
Ellis Av. SL1: Slou 7J **81**
Ellis Av. SL9: Chal P 25B **42**
Ellis Cl. BR8: Swan 70Fd **140**
Ellis Cl. CR5: Coul 92Pb **196**
Ellis Cl. HA4: Eastc 30W **44**
Ellis Cl. HA8: Edg 23Ua **48**
Ellis Cl. NW10 ....37Xa **68**
Ellis Cl. RM16: Ors 4F **100**
Ellis Cl. SE9 ....61Sc **138**
Elliscombe Mt. SE7 ....51Lc **115**
Elliscombe Rd. SE7 ....51Lc **115**
Ellis Ct. E1 ....44Yb **92**
.... *(off James Voller Way)*
Ellis Ct. W7 ....43Ha **86**
Ellis Farm Cl. GU22: Wok 4P **187**
Ellisfield Dr. SW15 ....59Wa **110**
Ellis Franklin Ct. NW8 ....1A **214** (40Eb **69**)
.... *(off Abbey Rd.)*
Ellis Ho. AL1: St A 3D **6**
Ellis Ho. SE17 ....7F **231** (50Tb **91**)
.... *(off Brandon St.)*
Ellison Apts. E3 ....41Cc **92**
.... *(off Merchant St.)*
Ellison Cl. SL4: Wind 5D **102**
Ellison Gdns. UB2: S'hall 49Ba **85**
Ellison Ho. SE13 ....54Dc **114**
.... *(off Lewisham Rd.)*
Ellison Ho. SL4: Wind 3H **103**
.... *(off Victoria St.)*
Ellison Rd. DA15: Sidc 60Tc **116**
Ellison Rd. SW13 ....54Va **110**
Ellison Rd. SW16 ....66Mb **134**
Ellis Rd. CR4: Mitc 72Hb **155**
Ellis Rd. CR5: Coul 92Pb **196**
Ellis Rd. UB2: S'hall 46Ea **86**
Ellis St. SW1 ....5H **227** (49Jb **90**)
Ellis Ter. SE11 ....52Qb **112**
Elliston Ho. SE18 ....49Qc **94**
.... *(off Wellington St.)*
Elliston Way KT21: Asht 91Na **193**
Ellis Way DA1: Dart 61Pd **141**
Ellmore Cl. RM3: Rom 25Kd **57**
Ellora Rd. SW16 ....64Mb **134**
Ellswood Cl. KT6: Surb 73Ma **153**
Ellsworth St. E2 ....41Xb **91**
Ellwood Ct. W9 ....42Db **89**
.... *(off Clearwell Dr.)*
Ellwood Ct. WD25: Wat 6X **13**
Ellwood Gdns. WD25: Wat 6Y **13**
Ellwood Rd. SL9: Chal P 25A **42**
Elmar Grn. SL2: Slou 1E **80**
Elmar Rd. N15 ....28Tb **51**
Elm Av. HA4: Ruis 32W **64**
Elm Av. RM14: Upm 34Rd **77**
Elm Av. TW19: Stanw 61N **127**
Elm Av. W5 ....46Na **87**
Elm Av. WD19: Wat 17Aa **27**
Elmbank N14 ....17Nb **32**
Elmbank Av. EN5: Barn 14Ya **30**
Elmbank Av. TW20: Eng G 5M **125**
Elm Bank Dr. BR1: Broml 68Mc **137**
Elm Bank Gdns. SW13 ....54Ua **110**
Elmbank Way W7 ....43Fa **86**
Elmbourne Dr. DA17: Belv 49Dd **96**
Elmbourne Rd. SW17 ....62Kb **134**
Elmbridge Av. KT5: Surb 71Ra **153**
Elmbridge Cl. HA4: Ruis 30W **44**
Elmbridge Dr. HA4: Ruis 29V **44**
Elmbridge Est. GU22: Wok 91B **188**
Elmbridge La. GU22: Wok 91B **188**
Elmbridge Rd. IG6: Ilf 23Wc **55**
Elmbridge Wlk. E8 ....38Wb **71**
Elmbridge Xcel Leisure Complex 71X **151**
Elmbridge Xcel Sports Hub 71X **151**
Elmbrook Cl. TW16: Sun 67X **129**
Elmbrook Gdns. SE9 ....56Nc **116**
Elmbrook Rd. SM1: Sutt 77Bb **155**
Elm Cl. CR2: S Croy 79Ub **157**
Elm Cl. CR6: W'ham 89Zb **178**
Elm Cl. DA1: Dart 60Ld **119**
Elm Cl. E11 ....30Kc **53**
Elm Cl. EN9: Walt A 6Fc **20**
Elm Cl. GU21: Wok 7P **167**
Elm Cl. GU23: Rip 96J **189**
Elm Cl. HA2: Harr 30Da **45**
Elm Cl. IG9: Buck H 19Mc **35**
Elm Cl. KT22: Lea 94Ka **192**
Elm Cl. KT5: Surb 73Sa **153**
Elm Cl. N19 ....33Lb **70**
Elm Cl. NW4 ....29Za **48**
Elm Cl. RM7: Mawney 25Dd **56**
Elm Cl. SL2: Farn C 7G **60**
Elm Cl. SM5: Cars 74Hb **155**
Elm Cl. SW20 ....70Ya **132**
Elm Cl. TW19: Stanw 60M **105**
Elm Cl. TW2: Twick 61Da **129**
Elm Cl. UB3: Hayes 44W **84**
ELM CORNER 90Q **170**
Elmcote HA5: Pinn 26Z **45**
Elmcote Way WD3: Crox G 16P **25**
Elm Cotts. CR4: Mitc 68Hb **133**
Elm Ct. CR4: Mitc 68Hb **133**
Elm Ct. EC4 ....3K **223** (44Qb **90**)
.... *(off King's Bench Wlk.)*
Elm Ct. EN4: E Barn 17Gb **31**
Elm Ct. GU21: Knap 9H **167**
Elm Ct. KT8: W Mole 70Da **129**
Elm Ct. SE1 ....2J **231** (47Ub **91**)
.... *(off Royal Oak Yd.)*
Elm Ct. SE13 ....55Fc **115**
Elm Ct. SW9 ....53Qb **112**
.... *(off Cranworth Gdns.)*
Elm Ct. TW16: Sun 66V **128**
.... *(off Grangewood Dr.)*
Elm Ct. W9 ....43Cb **89**
.... *(off Admiral Wlk.)*
Elm Ct. WD17: Wat 13X **27**
Elmcourt Rd. SE27 ....61Rb **135**
Elm Cres. KT2: King T 67Na **131**
Elm Cres. W5 ....46Na **87**
Elm Cft. SL3: Dat 3N **103**
Elmcroft GU22: Wok 90B **168**
.... *(off Fairview Av.)*
Elmcroft KT23: Bookh 96Ca **191**
Elmcroft N6 ....31Lb **70**
Elmcroft N8 ....29Pb **50**
Elmcroft Av. DA15: Sidc 59Vc **117**
Elmcroft Av. E11 ....29Kc **53**
Elmcroft Av. N9 ....16Xb **33**
Elmcroft Av. NW11 ....31Bb **69**
Elmcroft Cl. E11 ....28Kc **53**
Elmcroft Cl. KT9: Chess 76Na **153**
Elmcroft Cl. TW14: Felt 58V **106**
Elmcroft Cl. W5 ....44Ma **87**
Elmcroft Cres. HA2: Harr 27Ca **45**
Elmcroft Cres. NW11 ....31Ab **68**
Elmcroft Dr. KT9: Chess 76Na **153**
Elmcroft Dr. TW15: Ashf 64Q **128**
Elmcroft Gdns. NW9 ....28Qa **47**
Elmcroft Rd. BR6: Orp 73Wc **161**
Elmcroft St. E5 ....35Yb **72**
Elmcroft Ter. UB8: Hil 44Q **84**
Elmdale Rd. N13 ....22Pb **50**
Elmdene KT5: Surb 74Sa **153**
Elmdene Av. RM11: Horn 29Pd **57**
Elmdene Cl. BR3: Beck 72Bc **158**
Elmdene Ct. GU22: Wok 91A **188**
.... *(off Constitution Hill)*
Elmdene Rd. SE18 ....50Rc **94**
Elmdon Rd. RM15: S Ock 43Wd **98**
Elmdon Rd. TW4: Houn 54Z **107**
Elmdon Rd. TW6: H'row A 55V **106**
Elm Dr. AL10: Hat 1C **8**
Elm Dr. AL4: St A 2G **6**
Elm Dr. BR8: Swan 68Fd **140**
Elm Dr. EN8: Chesh 1Ac **20**
Elm Dr. GU24: Chob 2K **167**
Elm Dr. HA2: Harr 30Da **45**
Elm Dr. KT22: Lea 95Ka **192**
Elm Dr. TW16: Sun 68Y **129**
Elmer Av. RM4: Have B. 20Gd **38**
Elmer Cl. EN2: Enf 13Pb **32**
Elmer Cl. RM13: Rain 38Jd **76**
Elmer Cotts. KT22: Fet 95Ja **192**
Elmer Gdns. HA8: Edg 24Ra **47**
Elmer Gdns. RM13: Rain 38Jd **76**
Elmer Gdns. TW7: Isle 55Fa **108**
Elmer Ho. NW8 ....7D **214** (43Gb **89**)
.... *(off Penfold St.)*
Elmer M. KT22: Fet 94Ja **192**
Elmer Rd. SE6 ....59Ec **114**
Elmers Dr. TW11: Tedd 65Ka **130**
ELMERS END 70Ac **136**
Elmers End Rd. BR3: Beck 68Yb **136**
Elmers End Rd. SE20 ....68Yb **136**
Elmers End Station (Rail & London Tramlink) 70Zb **136**
Elmerside Rd. BR3: Beck 70Ac **136**
Elmers Lodge BR3: Beck 70Zb **136**
Elmers Rd. SE25 ....73Wb **157**
Elm Farm Cvn. Pk. KT16: Lyne 73D **148**
Elmfield HA1: Harr 31Ha **66**
Elmfield KT23: Bookh 95Ca **191**
Elmfield Av. CR4: Mitc 67Jb **134**
Elmfield Av. N8 ....29Nb **50**
Elmfield Av. TW11: Tedd 64Ha **130**
Elmfield Cl. DA11: Grav'nd 10D **122**
Elmfield Cl. EN6: Pot B 5Ab **16**
Elmfield Cl. HA1: Harr 33Ga **66**
Elmfield Ct. DA16: Well 53Xc **117**
Elmfield Ho. N2 ....26Fb **49**
.... *(off The Grange)*
Elmfield Ho. NW8 ....40Db **69**
.... *(off Carlton Hill)*
Elmfield Ho. W9 ....42Cb **89**
.... *(off Goldney Rd.)*
Elmfield Pk. BR1: Broml 69Jc **137**
Elmfield Rd. BR1: Broml 68Jc **137**
Elmfield Rd. E17 ....30Zb **52**
Elmfield Rd. E4 ....19Ec **34**
Elmfield Rd. EN6: Pot B 4Ab **16**
Elmfield Rd. N2 ....27Fb **49**
Elmfield Rd. SW17 ....61Jb **134**
Elmfield Rd. UB2: S'hall 48Aa **85**
Elmfield Way CR2: Sande 81Vb **177**
Elmfield Way W9 ....43Cb **89**
Elm Friars Wlk. NW1 ....38Mb **70**
Elm Gdns. CM15: Shenf 13Ee **41**
Elm Gdns. CR4: Mitc 70Mb **134**
Elm Gdns. EN2: Enf 10Tb **19**
Elm Gdns. KT10: Clay 79Ha **152**
Elm Gdns. KT18: Tatt C 91Ya **194**
Elm Gdns. N2 ....27Eb **49**
Elmgate Av. TW13: Felt 62X **129**
Elmgate Gdns. HA8: Edg 22Sa **47**
Elm Grn. W3 ....44Ua **88**
Elmgreen Cl. E15 ....39Gc **73**
Elm Gro. BR6: Orp 74Vc **161**
Elm Gro. CR3: Cat'm 94Ub **197**
Elm Gro. DA8: Erith 52Fd **118**
Elm Gro. GU24: Bisl 8E **166**
Elm Gro. HA2: Harr 31Ca **65**
Elm Gro. IG8: Wfd G 22Hc **53**
Elm Gro. KT12: Walt T 74W **150**
Elm Gro. KT18: Eps 86Sa **173**
Elm Gro. KT2: King T 67Na **131**
Elm Gro. N8 ....30Nb **50**
Elm Gro. NW2 ....35Za **68**
Elm Gro. RM11: Horn 30Nd **57**
Elm Gro. SE15 ....54Vb **113**
Elm Gro. SM1: Sutt 77Db **155**
Elm Gro. SW19 ....66Ab **132**
Elm Gro. UB7: Yiew 45P **83**
Elm Gro. WD24: Wat 9W **12**
Elmgrove Cl. GU21: Wok 1H **187**
Elmgrove Cres. HA1: Harr 29Ha **46**
Elmgrove Gdns. HA1: Harr 29Ja **46**
Elmgrove M. KT13: Weyb 76R **150**
Elm Gro. Pde. SM6: W'gton 76Jb **156**
Elmgrove Point SE18 ....49Tc **94**
Elm Gro. Rd. KT11: Cobh 88Z **171**
Elm Gro. Rd. SW13 ....53Wa **110**
Elm Gro. Rd. W5 ....47Na **87**
Elmgrove Rd. CR0: C'don 73Xb **157**
Elmgrove Rd. HA1: Harr 29Ha **46**
Elmgrove Rd. KT13: Weyb 77Q **150**
Elm Hall Gdns. E11 ....29Kc **53**
.... (not continuous)
Elm Hatch HA5: Hat E 24Ba **45**
Elm Ho. E14 ....47Ec **92**
.... *(off E. Ferry Rd.)*
Elm Ho. E3 ....39Bc **72**
.... *(off Sycamore Av.)*
Elm Ho. KT2: King T 66Pa **131**
.... *(off Elm Rd.)*
Elm Ho. W10 ....42Ab **88**
.... *(off Briar Wlk.)*
Elmhurst DA17: Belv 51Ad **117**
Elmhurst DA9: Ghithe 58Xd **120**
Elmhurst Av. CR4: Mitc 66Kb **134**
Elmhurst Av. N2 ....27Fb **49**
Elmhurst Cl. WD23: Bush 14Aa **27**
Elmhurst Ct. CR0: C'don 77Tb **157**
Elmhurst Dr. E18 ....26Jc **53**
Elmhurst Dr. RM11: Horn 32Ld **77**
Elmhurst Lodge SM2: Sutt 80Eb **155**
Elmhurst Mans. SW4 ....55Mb **112**
Elmhurst Rd. E7 ....38Kc **73**
Elmhurst Rd. EN3: Enf W 9Xb **20**
Elmhurst Rd. N17 ....26Vb **51**
Elmhurst Rd. SE9 ....61Nc **138**
Elmhurst Rd. SL3: L'ly 48C **82**
Elmhurst St. SW4 ....55Mb **112**
Elmhurst Way IG10: Lough 17Pc **36**
Elmington Cl. DA5: Bexl 58Dd **118**
Elmington Est. SE5 ....52Tb **113**
Elmington Rd. SE5 ....52Tb **113**
Elmira St. SE13 ....55Dc **114**
Elm La. GU23: Ock 91Q **190**
Elm La. KT11: Cobh 90R **170**
Elm La. SE6 ....61Bc **136**
Elm Lawn Cl. UB8: Uxb 38N **63**
Elm Lawns Cl. AL1: St A 1C **6**
Elmlea Dr. UB3: Hayes 43U **84**
Elm Lea Trad. Est. N17 ....23Xb **51**
Elmlee Cl. BR7: Chst 65Pc **138**
Elmley Cl. E6 ....43Nc **94**
Elmley St. SE18 ....50Tc **94**
.... (not continuous)
Elm Lodge SW6 ....53Ya **110**
Elmore Cl. HA0: Wemb 40Na **67**
Elmore Cl. HA3: Kenton 26Ka **46**
Elmore Dr. HA3: Kenton 26Ka **46**
Elmore Ho. N1 ....38Tb **71**
.... *(off Elmore St.)*
Elmore Ho. SW9 ....54Rb **113**
Elmore Rd. CR5: Chips 93Hb **195**
Elmore Rd. CR5: Coul 93Hb **195**
Elmore Rd. E11 ....34Ec **72**
Elmore Rd. EN3: Enf W 10Zb **20**
Elmores IG10: Lough 13Qc **36**
Elmore St. N1 ....38Sb **71**
Elm Pde. DA14: Sidc 63Wc **139**
Elm Pde. RM12: Horn 35Kd **77**
Elm Pk. HA7: Stan 22Ka **46**
ELM PARK 35Kd **77**
Elm Pk. SL5: S'dale 4C **146**
Elm Pk. SW2 ....58Pb **112**
Elm Pk. Av. N15 ....29Vb **51**
Elm Pk. Av. RM12: Horn 35Jd **76**
Elm Pk. Chambers SW10 ....50Fb **89**
.... *(off Fulham Rd.)*
Elm Pk. Ct. HA5: Pinn 27Y **45**
Elm Pk. Gdns. CR2: Sels 82Yb **178**
Elm Pk. Gdns. NW4 ....29Za **48**
Elm Pk. Gdns. SW10 ....7B **226** (50Fb **89**)
Elm Pk. Ho. SW10 ....7B **226** (50Fb **89**)
Elm Pk. La. SW3 ....50Fb **89**
Elm Pk. Mans. SW10 ....51Eb **111**
Elm Pk. Rd. E10 ....32Ac **72**
Elm Pk. Rd. HA5: Pinn 26Y **45**
Elm Pk. Rd. N21 ....17Sb **33**
Elm Pk. Rd. N3 ....24Bb **49**
Elm Pk. Rd. SE25 ....69Vb **135**
Elm Pk. Rd. SW3 ....51Fb **111**
Elm Park Station (Underground) 35Kd **77**
Elm Pas. EN5: Barn 14Bb **31**
Elm Pl. SW7 ....7B **226** (50Fb **89**)
Elm Pl. TW15: Ashf 64Q **128**
Elm Quay Ct. SW11 ....51Mb **112**
Elm Rd. GU21: Wok Heath Rd. 87B **168**
Elm Rd. GU21: Wok The Mount 10P **167**
Elm Rd. BR3: Beck 68Bc **136**
Elm Rd. BR6: Chels 80Wc **161**
Elm Rd. CR6: W'ham 89Zb **178**
Elm Rd. CR7: Thor H 70Tb **135**
Elm Rd. CR8: Purl 85Rb **177**
Elm Rd. DA1: Dart 60Md **119**
Elm Rd. DA12: Grav'nd 2E **144**
Elm Rd. DA14: Sidc 63Wc **139**
Elm Rd. DA8: Erith 53Jd **118**
Elm Rd. DA9: Ghithe 58Ud **120**
Elm Rd. E11 ....33Fc **73**
Elm Rd. E17 ....29Ec **52**
Elm Rd. E7 ....37Hc **73**
Elm Rd. EN5: Barn 14Bb **31**
Elm Rd. HA9: Wemb 36Na **67**
Elm Rd. KT10: Clay 79Ha **152**
Elm Rd. KT17: Ewe 79Va **154**
Elm Rd. KT2: King T 67Pa **131**
Elm Rd. KT22: Lea 94Ka **192**
Elm Rd. KT3: N Mald 68Ta **131**
Elm Rd. KT9: Chess 77Na **153**
Elm Rd. N22 ....25Rb **51**
Elm Rd. RH1: Redh 6N **207**
Elm Rd. RM15: Avel 46Td **98**
Elm Rd. RM17: Grays 51Ee **121**
Elm Rd. RM7: Mawney 26Dd **56**
Elm Rd. SL4: Wind 5F **102**
Elm Rd. SM6: W'gton 74Jb **156**
Elm Rd. SW14 ....55Sa **109**
Elm Rd. TN16: Westrm 97Uc **200**
Elm Rd. TW14: Bedf 60T **106**
Elm Rd. W. SM3: Sutt 73Bb **155**
Elm Row NW3 ....34Eb **69**
Elmroyd Av. EN6: Pot B 5Bb **17**
Elmroyd Cl. EN6: Pot B 5Bb **17**
Elms, The CR0: C'don 74Sb **157**
.... *(off Tavistock Rd.)*
Elms, The CR6: W'ham 87Yb **178**
Elms, The E12 ....37Mc **73**
Elms, The EN9: Walt A 7Lc **21**
.... *(within Woodbine Cl. Caravan Pk.)*
Elms, The IG10: Lough 12Hc **35**
Elms, The KT10: Clay 80Ha **152**
Elms, The SW13 ....55Va **110**
Elms, The TW15: Ashf 64Q **128**
Elms Av. N10 ....27Kb **50**
Elms Av. NW4 ....29Za **48**
Elms Cl. RM11: Horn 31Kd **77**
Elmscott Gdns. N21 ....16Sb **33**
Elmscott Rd. BR1: Broml 64Gc **137**
Elms Ct. HA0: Wemb 35Ja **66**
Elms Cres. SW4 ....58Lb **112**
Elmscroft Gdns. EN6: Pot B 4Bb **17**
Elmsdale Rd. E17 ....28Bc **52**
Elmsdene M. HA6: Nwood 23S **44**
Elms Farm Rd. RM12: Horn 36Ld **77**
Elms Gdns. HA0: Wemb 35Ja **66**
Elms Gdns. RM9: Dag 35Bd **75**
Elmshall Pl. AL1: St A 3D **6**
Elmshaw Rd. SW15 ....57Wa **110**
Elmshorn KT17: Eps D 88Ya **174**
Elmshott La. SL1: Slou 5C **80**
Elmshurst Cres. N2 ....28Fb **49**
Elmside CR0: New Ad 79Dc **158**
Elmside Rd. HA9: Wemb 34Qa **67**
Elms Ind. Est. RM3: Hrld W 24Rd **57**
Elms La. HA0: Wemb 34Ja **66**
Elmsleigh Av. HA3: Kenton 28Ka **46**
Elmsleigh Bus Station 64H **127**
Elmsleigh Cen., The 63H **127**
Elmsleigh Ct. SM1: Sutt 76Db **155**
Elmsleigh Ho. TW2: Twick 61Fa **130**
.... *(off Staines Rd.)*
Elmsleigh Rd. TW18: Staines 64H **127**
Elmsleigh Rd. TW2: Twick 61Fa **130**
Elmslie Cl. IG8: Wfd G 23Pc **54**
Elmslie Cl. KT18: Eps 86Sa **173**
Elmslie Point E3 ....43Bc **92**
.... *(off Leopold St.)*
Elms M. W2 ....4B **220** (45Fb **89**)
Elms Pk. Av. HA0: Wemb 35Ja **66**
Elms Rd. HA3: Hrw W 24Ga **46**
Elms Rd. SL9: Chal P 24A **42**
Elms Rd. SW4 ....57Lb **112**
ELMSTEAD 65Pc **138**
Elmstead Av. BR7: Chst 64Pc **138**
Elmstead Av. HA9: Wemb 32Na **67**
Elmstead Cl. KT19: Ewe 78Ua **154**
Elmstead Cl. N20 ....19Cb **31**
Elmstead Cl. TN13: Riv 94Gd **202**
Elmstead Gdns. KT4: Wor Pk 76Wa **154**
Elmstead Glade BR7: Chst 65Pc **138**
Elmstead La. BR7: Chst 66Nc **138**
Elmstead Rd. DA8: Erith 53Gd **118**
Elmstead Rd. IG3: Ilf 33Uc **74**
Elmstead Rd. KT14: W Byf 85J **169**
Elmstead Woods Station (Rail) 65Nc **138**
Elmsted Cres. DA16: Well 51Yc **117**
Elmstone Rd. SW6 ....53Cb **111**
Elmstone Ter. BR5: St M Cry 70Yc **139**
Elm St. WC1 ....6J **217** (42Pb **90**)
Elmsway TW15: Ashf 64Q **128**
Elmswood IG7: Chig 23Tc **54**
Elmswood KT23: Bookh 96Ba **191**
Elmsworth Av. TW3: Houn 54Da **107**
Elm Ter. HA3: Hrw W 25Fa **46**
Elm Ter. NW2 ....34Cb **69**
Elm Ter. NW3 ....35Gb **69**
Elm Ter. RM20: W Thur 51Xd **120**
Elm Ter. SE9 ....58Qc **116**
Elmton Ct. NW8 ....5B **214** (42Fb **89**)
.... *(off Cunningham Pl.)*
Elm Tree Av. KT10: Esh 73Fa **152**
Elm Tree Cl. KT16: Chert 75G **148**
Elm Tree Cl. NW8 ....3B **214** (41Fb **89**)
Elm Tree Cl. TW15: Ashf 64R **128**
Elm Tree Cl. UB5: N'olt 40Ba **65**
Elmtree Cl. KT14: Byfl 85N **169**
Elm Tree Ct. NW8 ....3B **214** (41Fb **89**)
.... *(off Elm Tree Rd.)*
Elm Tree Ct. SE7 ....51Lc **115**
Elm Tree Rd. NW8 ....3B **214** (41Fb **89**)
Elmtree Rd. TW11: Tedd 63Ga **130**
Elm Tree Wlk. WD3: Chor 13H **25**
Elm Vw. Ct. UB2: S'hall 49Ca **85**
Elm Vw. Ho. UB3: Harl 49T **84**
Elm Wlk. BR6: Farnb 76Pc **160**
Elm Wlk. NW3 ....33Cb **69**
Elm Wlk. RM2: Rom 27Jd **56**
Elm Wlk. SW20 ....70Ya **132**
Elm Wlk. WD7: R'lett 8Ha **14**
Elm Way CM14: B'wood 21Wd **58**

Elm Way KT19: Ewe 78Ta 153
Elm Way KT4: Wor Pk 76Ya 154
Elm Way N11 23Jb 50
Elm Way NW10 35Ua 68
Elm Way WD3: Rick 18K 25
Elmway RM16: Grays 45Ee 99
Elmwood Av. HA3: Kenton 29Ja 46
Elmwood Av. N13 22Nb 50
Elmwood Av. TW13: Felt 61W 128
Elmwood Av. TW13: Hanw 61W 128
Elmwood Av. WD6: Bore 14Ra 29
Elmwood Cl. KT17: Ewe 80Wa 154
Elmwood Cl. KT21: Asht 89Ma 173
Elmwood Cl. SM6: W'gton 75Kb 156
Elmwood Ct. E10 32Cc 72
(off Goldsmith Rd.)
Elmwood Ct. HA0: Wemb 34Ja 66
Elmwood Ct. KT21: Asht 89Ma 173
Elmwood Ct. SW11 53Kb 112
Elmwood Cres. NW9 28Sa 47
Elmwood Dr. DA5: Bexl 59Ad 117
Elmwood Dr. KT17: Ewe 79Wa 154
Elmwood Gdns. KT12: Hers 77Y 151
Elmwood Gdns. W7 44Ga 86
Elmwood Gro. HP3: Hem H 5P 3
Elmwood Ho. NW10 40Xa 68
(off All Souls Av.)
Elmwood Pk. SL9: Ger X 32A 62
Elmwood Rd. CR0: C'don 73Rb 157
Elmwood Rd. CR4: Mitc 69Hb 133
Elmwood Rd. GU21: Wok 1H 187
Elmwood Rd. RH1: Redh 2A 208
Elmwood Rd. SE24 57Tb 113
Elmwood Rd. SL2: Slou 5M 81
Elmwood Rd. W4 51Sa 109
Elmworth Gro. SE21 61Tb 135
Elnathan M. W9 42Db 89
Elphinstone Cl. GU24: Brkwd 3D 186
Elphinstone Ct. SW16 65Nb 134
Elphinstone Rd. E17 26Bc 52
Elphinstone St. N5 35Rb 71
Elrick Cl. DA8: Erith 51Gd 118
Elrington Rd. E8 37Wb 71
Elrington Rd. IG8: Wfd G 22Jc 53
Elruge Cl. UB7: W Dray 48M 83
Elsa Cotts. E14 43Ac 92
(off Halley St.)
Elsa Ct. BR3: Beck 67Bc 136
Elsa Rd. DA16: Well 54Xc 117
Elsa St. E1 43Ac 92
Elsdale St. E9 37Yb 72
Elsden M. E2 40Yb 72
Elsden Rd. N17 25Vb 51
Elsdon Rd. GU21: Wok 10L 167
Elsenham Rd. E12 36Qc 74
Elsenham St. SW18 60Bb 111
Elsham Rd. E11 34Gc 73
Elsham Rd. W14 47Ab 88
Elsham Ter. W14 48Ab 88
(off Elsham Rd.)
Elsiedene Rd. N21 17Sb 33
Elsie La. Ct. W2 43Cb 89
(off Westbourne Pk. Vs.)
Elsiemaud Rd. SE4 57Bc 114
Elsie Rd. SE22 56Vb 113
Elsinge Rd. EN1: Enf 8Xb 19
Elsinore Av. TW19: Stanw 59N 105
Elsinore Gdns. NW2 34Ab 68
Elsinore Ho. N1 1K 217 (39Qb 70)
(off Denmark Gro.)
Elsinore Ho. SE5 54Sb 113
(off Denmark Rd.)
Elsinore Ho. SE7 49Nc 94
Elsinore Ho. W6 50Za 88
(off Fulham Pal. Rd.)
Elsinore Rd. SE23 60Ac 114
Elsinore Way TW9: Rich 55Ra 109
Elsley Ct. HA9: Wemb 37Ra 67
Elsley Rd. SW11 55Hb 111
Elspeth Rd. HA0: Wemb 36Na 67
Elspeth Rd. SW11 56Hb 111
Elsrick Av. SM4: Mord 71Cb 155
Elstan Way CR0: C'don 73Ac 158
Elstar Ct. RM13: Rain 40Fd 76
(off Lowen Rd.)
Elstar M. DA9: Ghithe 58Wd 120
Elstead Ct. SM3: Sutt 74Ab 154
Elstead Ho. SW2 59Pb 112
(off Redlands Way)
Elsted St. SE17 6G 231 (49Tb 91)
Elstow Cl. HA4: Ruis 31Z 65
Elstow Cl. SE9 57Pc 116
(not continuous)
Elstow Gdns. RM9: Dag 39Ad 75
Elstow Grange NW6 38Za 68
Elstow Rd. RM9: Dag 39Ad 75
ELSTREE 16Ma 29
Elstree Aerodrome 13Ha 28
Elstree & Borehamwood Mus. 13Qa 29
Elstree & Borehamwood Station (Rail) 14Qa 29
Elstree Cl. RM12: Horn 38Kd 77
Elstree Distribution Pk. WD6: Bore 13Ta 29
Elstree Gdns. DA17: Belv 49Ad 95
Elstree Gdns. IG1: Ilf 36Sc 74
Elstree Gdns. N9 18Xb 33
Elstree Ga. WD6: Bore 12Ta 29
Elstree Hill BR1: Broml 66Gc 137
Elstree Hill Nth. WD6: E'tree 15Ma 29
Elstree Hill Sth. WD6: E'tree 17La 28
Elstree Ho. WD6: Bore 12Ta 29
(off Elstree Way)
Elstree Pk. WD6: Bore 16Ta 29
Elstree Rd. WD23: B Hea 17Fa 28
Elstree Rd. WD6: E'tree 16Ja 28
Elstree Studios WD6: Bore 13Ra 29
Elstree Twr. WD6: Bore 12Ta 29
(off Elstree Way)
Elstree Way WD6: Bore 13Ra 29
Elswick Rd. SE13 54Dc 114
Elswick St. SW6 54Eb 111
Elsworth Cl. TW14: Bedf 60U 106
Elsworthy KT7: T Ditt 72Ga 152
Elsworthy Ct. NW3 38Hb 69
(off Primrose Hill Rd.)
Elsworthy Ri. NW3 38Gb 69
Elsworthy Rd. NW3 39Gb 69
Elsworthy Ter. NW3 38Gb 69
Elsynge Rd. SW18 57Fb 111
ELTHAM 58Pc 116
Eltham Av. SL1: Slou 6C 80
Eltham Cen. 57Qc 116
ELTHAM COMMUNITY HOSPITAL 58Pc 116
Eltham Crematorium 56Tc 116
Eltham Grn. SE9 57Mc 115
Eltham Grn. Rd. SE9 56Lc 115
Eltham High St. SE9 58Pc 116
Eltham Hill SE9 57Mc 115
Eltham Palace & Gdns. 59Nc 116
Eltham Pal. Rd. SE9 58Lc 115
ELTHAM PARK 56Qc 116
Eltham Pk. Gdns. SE9 56Qc 116
Eltham Rd. SE12 57Hc 115
Eltham Rd. SE9 57Lc 115
Eltham Station (Rail) 57Pc 116
Eltham Warren Golf Course 57Rc 116
Elthiron Rd. SW6 53Cb 111
Elthorne Av. W7 47Ha 86
Elthorne Ct. TW13: Felt 60Y 107
ELTHORNE HEIGHTS 43Ga 86
Elthorne Pk. Rd. W7 47Ha 86
Elthorne Rd. N19 33Mb 70
Elthorne Rd. NW9 31Ta 67
Elthorne Rd. UB8: Uxb 40M 63
Elthorne Sports Cen. 48Ha 86
Elthorne Way NW9 30Ta 47
Elthruda Rd. SE13 58Fc 115
Eltisley Rd. IG1: Ilf 35Rc 74
Elton Av. EN5: Barn 15Bb 31
Elton Av. HA0: Wemb 36Ka 66
Elton Av. UB6: G'frd 37Ga 66
Elton Cl. KT1: Hamp W 66La 130
Elton Ho. E3 39Bc 72
(off Candy St.)
Elton Pk. WD17: Wat 12W 26
Elton Pl. N16 36Ub 71
Elton Rd. CR8: Purl 84Lb 176
Elton Rd. KT2: King T 67Pa 131
Elton Way WD25: A'ham 11Ca 27
Eltringham St. SW18 56Eb 111
Eluna Apts. E1 45Xb 91
(off Wapping La.)
Elvaston M. SW7 4A 226 (48Eb 89)
Elvaston Pl. SW7 4A 226 (48Eb 89)
Elveden Cl. GU22: Pyr 89K 169
Elveden Ho. SE24 57Rb 113
Elveden Pl. GU22: Pyr 89K 169
Elveden Pl. NW10 40Qa 67
Elveden Rd. NW10 40Qa 67
Elvedon Rd. KT11: Cobh 83X 171
Elvedon Rd. TW13: Felt 62V 128
Elvendon Rd. N13 23Nb 50
Elven M. SE15 53Yb 114
Elver Gdns. E2 41Wb 91
Elverson Rd. SE8 54Dc 114
Elverson Road Station (DLR) 54Dc 114
Elverton St. SW1 5D 228 (49Mb 90)
Elvet Av. RM2: Rom 28Ld 57
Elvin Ct. NW9 31Sa 67
Elvin Dr. RM16: N Stif 46Zd 99
Elvin Gdns. HA9: Wemb 35Qa 67
Elvington Grn. BR2: Broml 71Hc 159
Elvington La. NW9 25Ua 48
Elvino Rd. SE26 64Ac 136
Elvis Rd. NW2 37Ya 68
Elwell Cl. TW20: Egh 65C 126
Elwick Ct. DA1: Cray 56Kd 119
Elwick Rd. RM15: S Ock 44Yd 98
Elwill Way BR3: Beck 70Ec 136
Elwill Way DA13: Ist R 7B 144
Elwin St. E2 41Wb 91
Elwood Ct. EN5: New Bar 14Eb 31
Elwood St. N5 34Rb 71
Elworth Ho. SW8 52Pb 112
(off Oval Pl.)
Elwyn Gdns. SE12 59Jc 115
Ely Av. SL1: Slou 3G 80
Ely Cl. DA8: Erith 54Hd 118
Ely Cl. KT3: N Mald 68Va 132
Ely Ct. EC1 1A 224 (43Qb 90)
(off Ely Pl.)
Ely Ct. KT1: King T 68Qa 131
Ely Gdns. IG1: Ilf 31Nc 74
Ely Gdns. RM10: Dag 34Ed 76
Ely Gdns. WD6: Bore 15Ta 29
Ely Ho. SE15 52Wb 113
(off Friary Est.)
Elyne Rd. N4 30Qb 50
Ely Pl. EC1 1A 224 (43Qb 90)
Ely Pl. IG8: Wfd G 23Qc 54
Ely Pl. SW8 52Pb 112
Ely Rd. AL1: St A 3F 6
Ely Rd. CR0: C'don 71Tb 157
Ely Rd. E10 30Ec 52
Ely Rd. TW4: Houn 55Y 107
Ely Rd. TW6: H'row A 54T 106
(off Esher Cres.)
Elysian Av. BR5: St M Cry 72Vc 161
ELYSIAN HOUSE 26Ua 48
Elysian M. N7 37Pb 70
Elysian Pl. CR2: S Croy 80Sb 157
Elysium Apts. E1 42Yb 92
(off Theven St.)
Elysium Pl. SW6 54Bb 111
(off Elysium St.)
Elysium St. SW6 54Bb 111
Elystan Bus. Cen. UB4: Yead 45Y 85
Elystan Cl. SM6: W'gton 81Lb 176
Elystan Ho. SW3 6D 226 (49Gb 89)
(off Elystan St.)
Elystan Pl. SW3 7E 226 (50Gb 89)
Elystan St. SW3 6D 226 (49Gb 89)
Elystan Wlk. N1 1K 217 (39Qb 70)
Ely's Yd. E1 7K 219 (43Vb 91)
Emanuel Av. W3 44Sa 87
Emanuel Dr. TW12: Hamp 64Ba 129
Emanuel Ho. SW1 4D 228 (48Mb 90)
Embankment SW15 54Za 110
Embankment, The HP3: Hem H 7P 3
Embankment, The TW1: Twick 60Ja 108
Embankment, The TW19: Wray 9N 103
Embankment Galleries 4J 223 (45Pb 90)
(within Somerset House)
Embankment Gdns. SW3 51Hb 111
Embankment Ho. KT16: Chert 74L 149
Embankment Pier WC2 6G 223 (46Nb 90)
Embankment Pl. WC2 6G 223 (46Nb 90)
Embankment Station (Underground) 6G 223 (46Nb 90)
Embassy Apts. SE5 54Sb 113
(off Coldharbour La.)
Embassy Ct. DA14: Sidc 62Xc 139
Embassy Ct. DA16: Well 55Xc 117
Embassy Ct. E2 42Xb 91
(off Brady St.)
Embassy Ct. N11 23Mb 50
(off Bounds Grn. Rd.)
Embassy Ct. NW8 2C 214 (40Fb 69)
(off Wellington Rd.)
Embassy Ct. SM6: W'gton 79Kb 156
Embassy Ct. W5 45Pa 87
Embassy Gdns. BR3: Beck 67Bc 136
Embassy Ho. NW6 38Db 69
Embassy Lodge N3 26Bb 49
(off Cyprus Rd.)
Embassy of United States of America 51Mb 112
Embassy Theatre Central School of Speech & Drama 38Fb 69
Embassy Way SW11 51Mb 112
Emba St. SE16 47Wb 91
Ember Cen. KT12: Walt T 75Aa 151
Ember Cl. BR5: Pet W 73Sc 160
Ember Cl. KT15: Add 78N 149
Ember Ct. NW9 26Va 48
Embercourt Rd. KT7: T Ditt 72Ga 152
Ember Farm Av. KT8: E Mos 72Fa 152
Ember Farm Way KT8: E Mos 72Fa 152
Ember Gdns. KT7: T Ditt 73Ga 152
Ember La. KT10: Esh 73Fa 152
Ember La. KT8: E Mos 72Fa 152
Ember Rd. SL3: L'ly 48D 82
Emberton SE5 51Ub 113
(off Albany Rd.)
Emberton Ct. EC1 4B 218 (41Rb 91)
(off Tompion St.)
Embleton Rd. SE13 56Dc 114
Embleton Rd. WD19: Wat 20W 26
Embleton Wlk. TW12: Hamp 64Ba 129
Embroidery World Bus. Cen. IG8: Wfd G 26Mc 53
Embry Cl. HA7: Stan 21Ja 46
Embry Dr. HA7: Stan 23Ja 46
Embry Rd. SE9 56Lc 115
Embry Way HA7: Stan 22Ja 46
Emden Cl. UB7: W Dray 47Q 84
Emden St. SW6 53Db 111
Emerald Cl. E16 44Mc 93
Emerald Ct. CR5: Coul 87Mb 176
Emerald Ct. HA4: Ruis 35Y 65
Emerald Ct. SL1: Slou 7J 81
Emerald Ct. WD6: Bore 10Pa 15
(off Aycliffe Rd.)
Emerald Gdns. RM8: Dag 32Cd 76
Emerald Rd. NW10 39Ta 67
Emerald Sq. SW15 57Wa 110
Emerald Sq. UB2: S'hall 48Z 85
Emerald St. WC1 7H 217 (43Pb 90)
Emerson Apts. N8 27Pb 50
Emerson Dr. RM11: Horn 31Md 77
Emerson Gdns. HA3: Kenton 30Pa 47
Emerson Ho. RM11: Horn 30Md 57
Emerson M. KT3: N Mald 70Ua 132
EMERSON PARK 31Md 77
Emerson Pk. Ct. RM11: Horn 31Md 77
Emerson Park Station (Rail) 31Nd 77
Emerson Rd. IG1: Ilf 31Qc 74
Emersons Av. BR8: Hext 66Hd 140
Emerson St. SE1 6D 224 (46Sb 91)
Emerton Cl. DA6: Bex 56Ad 117
Emerton Rd. KT22: Fet 93Ea 192
Emery Hill St. SW1 4C 228 (48Lb 90)
Emery St. SE1 3A 230 (48Qb 90)
Emery Walker Trust 50Wa 88
(off Hammersmith Ter.)
Emes Rd. DA8: Erith 52Ed 118
Emilia Cl. EN3: Pond E 15Xb 33
Emily Bowes Ct. N17 27Xb 51
Emily Ct. DA2: Wilm 63Ld 141
Emily Ct. SE1 2D 230 (47Sb 91)
(off Sudrey St.)
Emily Davison Dr. KT18: Tatt C 90Xa 174
Emily Duncan Pl. E7 35Kc 73
Emily Ho. W10 42Ab 88
(off Kensal Rd.)
Emily Jackson Cl. TN13: S'oaks 96Kd 203
Emily St. E16 44Hc 93
(off Jude St.)
Emirates Air Line 47Hc 93
Emirates Greenwich Peninsula 47Hc 93
Emirates Royal Docks 45Jc 93
Emirates Stadium 35Qb 70
Emley Rd. KT15: Add 76J 149
Emlyn Bldgs. SL4: Eton 2G 102
Emlyn Gdns. W12 47Ua 88
Emlyn La. KT22: Lea 94Ja 192
Emlyn Rd. RH1: Redh 8A 208
Emlyn Rd. W12 47Ua 88
Emma Ho. RM1: Rom 28Gd 56
Emmanuel Ct. E10 31Dc 72
Emmanuel Ho. SE11 6K 229 (49Qb 90)
Emmanuel Lodge EN8: Chesh 2Yb 20
Emmanuel Rd. HA6: Nwood 24V 44
Emmanuel Rd. SW12 60Lb 112
Emma Rd. E13 40Hc 73
Emma St. E2 40Xb 71
Emmaus Way IG7: Chig 22Qc 54
Emmeline Ct. KT12: Walt T 73Y 151
Emmett Cl. WD7: Shenl 5Na 15
Emmetts Cl. GU21: Wok 9P 167
Emminster NW6 39Db 69
(off Abbey Rd.)
Emmott Av. IG6: Ilf 29Sc 54
Emmott Cl. E1 42Ac 92
Emmott Cl. NW11 30Eb 49
Emms Pas. KT1: King T 68Ma 131
Emperor Ho. E20 36Ec 72
(off Napa Cl.)
Emperor Ho. SE4 55Ac 114
(off Dragonfly Pl.)
Emperor's Ga. SW7 48Eb 89
Empingham Ho. SE8 49Zb 92
(off Chilton Gro.)
Empire Av. N18 22Sb 51
Empire Cen. WD24: Wat 11Y 27
Empire Cinema Haymarket 5D 222 (45Mb 90)
(off Haymarket)
Empire Cinema Slough 7K 81
Empire Cinema Sutton 78Db 155
Empire Cinema Walthamstow 28Cc 52
Empire Ct. SE7 51Kc 115
Empire Ct. HA9: Wemb 34Ra 67
Empire Ho. N18 23Tb 51
Empire Ho. SW7 4D 226 (48Gb 89)
(off Thurloe Pl.)
Empire M. SW16 64Nb 134
Empire Pde. HA9: Wemb 34Qa 67
Empire Pde. N18 23Tb 51
Empire Reach SE10 51Dc 114
(off Dowells St.)
Empire Rd. UB6: G'frd 39Ka 66
Empire Sq. N7 34Nb 70
Empire Sq. SE1 2F 231 (47Tb 91)
(off Tabard St.)
Empire Sq. SE20 66Zb 136
(off High St.)
Empire Sq. E. SE1 2F 231 (47Tb 91)
(off Long La.)
Empire Sq. Sth. SE1 2F 231 (47Tb 91)
(off Sterry St.)
Empire Sq. W. SE1 2F 231 (47Tb 91)
(off Tabard St.)
Empire Wlk. DA9: Ghithe 56Yd 120
Empire Way HA9: Wemb 35Pa 67
Empire Wharf E3 39Ac 72
(off Old Ford Rd.)
Empire Wharf Rd. E14 49Fc 93
Empress App. SW6 50Cb 89
Empress Av. E12 33Lc 73
Empress Av. E4 24Dc 52
Empress Av. IG1: Ilf 33Pc 74
Empress Av. IG8: Wfd G 24Hc 53
Empress Dr. BR7: Chst 65Rc 138
Empress M. SE5 54Sb 113
Empress Pde. E4 24Cc 52
Empress Pl. SW6 50Cb 89
Empress Rd. DA12: Grav'nd 9G 122
Empress State Bldg. SW6 50Cb 89
Empress St. SE17 51Sb 113
Empson St. E3 42Dc 92
Emslie Horniman Pleasance 42Ab 88
(off Bosworth Road)
Emsworth Cl. N9 18Yb 34
Emsworth Ct. SW16 62Nb 134
Emsworth Rd. IG6: Ilf 26Rc 54
Emsworth St. SW2 61Pb 134
EMT Ho. E6 43Qc 94
Emu Rd. SW8 54Kb 112
Enard Ho. E3 40Bc 72
(off Cardigan Rd.)
Ena Rd. SW16 69Nb 134
Enborne Grn. RM15: S Ock 43Wd 98
Enbrook St. W10 41Ab 88
Enclave, The SW13 54Va 110
Enclave Ct. EC1 5C 218 (42Rb 91)
(off Dallington St.)
Endale Cl. SM5: Cars 75Hb 155
Endeavour Ho. E14 47Cc 92
(off Cuba St.)
Endeavour Ho. SE16 49Ac 92
(off Ashton Reach)
Endeavour Sq. E20 38Ec 72
Endeavour Way CR0: Bedd 73Nb 156
Endeavour Way IG11: Bark 40Wc 75
Endeavour Way SW19 63Db 133
Endell St. WC2 2F 223 (44Nb 90)
Enderby St. SE10 50Fc 93
Enderley Cl. HA3: Hrw W 26Ga 46
Enderley Rd. HA3: Hrw W 25Ga 46
Endersby Rd. EN5: Barn 15Ya 30
Enders Cl. EN2: Enf 10Qb 18
Endersleigh Gdns. NW4 28Wa 48
Endlebury Rd. E4 19Ec 34
Endlesham Rd. SW12 59Jb 112
Endsleigh Cl. CR2: Sels 82Yb 178
Endsleigh Ct. WC1 5E 216 (42Mb 90)
(off Endsleigh St.)
Endsleigh Gdns. IG1: Ilf 33Pc 74
Endsleigh Gdns. KT12: Hers 78Y 151
Endsleigh Gdns. KT6: Surb 72La 152
Endsleigh Gdns. WC1 5D 216 (42Mb 90)
Endsleigh Ind. Est. UB2: S'hall 49Aa 85
Endsleigh Pl. WC1 5D 216 (42Mb 90)
Endsleigh Rd. RH1: Mers 1C 208
Endsleigh Rd. UB2: S'hall 49Aa 85
Endsleigh Rd. W13 45Ja 86
Endsleigh St. WC1 5D 216 (42Mb 90)
Endway KT5: Surb 73Ra 153
Endwell Rd. SE4 54Ac 114
Endymion Rd. N4 31Qb 70
Endymion Rd. SW2 58Pb 112
Energen Cl. NW10 37Ua 68
Énergie Fitness Old Street 5F 219 (42Tb 91)
Energy Cen., The N1 3H 219 (41Ub 91)
(off Bowling Grn. Wlk.)
ENFIELD 13Tb 33
Enfield Bus. Cen. EN3: Enf H 12Yb 34
Enfield Chase Station (Rail) 13Sb 33
Enfield Cloisters N1 3H 219 (41Ub 91)
(off Fanshaw St.)
Enfield Cl. UB8: Uxb 40M 63
Enfield Crematorium 9Xb 19
Enfield Golf Course 14Rb 33
ENFIELD HIGHWAY 13Yb 34
Enfield Ho. RM3: Rom 24Nd 57
(off Leyburn Cres.)
Enfield Ho. SW9 54Nb 112
(off Stockwell Rd.)
ENFIELD ISLAND VILLAGE 9Cc 20
Enfield Lock EN3: Enf L 10Cc 20
Enfield Lock 10Cc 20
(off Sth. Ordnance Rd.)
ENFIELD LOCK 9Bc 20
Enfield Lock Station (Rail) 9Ac 20
Enfield Mus. 14Tb 33
(off London Rd.)
Enfield Retail Pk. 13Wb 33
Enfield Rd. EN2: Enf 14Mb 32
Enfield Rd. N1 38Ub 71
Enfield Rd. TW6: H'row A 54U 106
Enfield Rd. TW8: Bford 50Ma 87
Enfield Rd. W3 47Ra 87
ENFIELD ROAD RDBT. 54U 106
ENFIELD TOWN 14Tb 33
Enfield Town FC 12Vb 33
Enfield Town Station (Overground) 13Ub 33
Enfield Wlk. TW8: Bford 50Ma 87
ENFIELD WASH 9Zb 20
Enford St. W1 7F 215 (43Hb 89)
Engadine Cl. CR0: C'don 76Vb 157
Engadine St. SW18 60Bb 111
Engate St. SE13 56Ec 114
Engayne Gdns. RM14: Upm 32Rd 77
Engel Pk. NW7 23Ya 48
Engelsine Ct. DA9: Ghithe 58Xd 120
Engine Ct. SW1 7C 222 (46Lb 90)
(off Ambassador's Ct.)
Engineer Cl. SE18 51Qc 116
Engineers Row SE18 49Qc 94
(off Woolwich New Rd.)
Engineers Way HA9: Wemb 35Qa 67
Engineers Wharf UB5: N'olt 42Ba 85
England's La. NW3 37Hb 69
Englands La. IG10: Lough 12Qc 36
England Way KT3: N Mald 70Ra 131
Englefield NW1 4B 216 (41Lb 90)
(off Clarence Gdns.)
Englefield Cl. BR5: St M Cry 71Vc 161
Englefield Cl. CR0: C'don 72Sb 157
Englefield Cl. EN2: Enf 12Qb 32
Englefield Cl. TW20: Eng G 5N 125
Englefield Cres. BR5: St M Cry 70Vc 139
ENGLEFIELD GREEN 4N 125
Englefield Path BR5: St M Cry 70Wc 139
Englefield Rd. GU21: Knap 9G 166
Englefield Rd. N1 38Tb 71
Engleheart Dr. TW14: Felt 58V 106
Engleheart Rd. SE6 59Dc 114
Englehurst TW20: Eng G 5N 125
Englemere Pk. KT22: Oxs 85Da 171
Englewood Rd. SW12 58Kb 112
Engliff La. GU22: Pyr 88J 169
English Gdns. TW19: Wray 6P 103
English Grounds SE1 7H 225 (46Ub 91)
English St. E3 42Bc 92
Enid Cl. AL2: Brick W 3Ba 13
Enid St. SE16 3K 231 (48Vb 91)
Enmore Av. SE25 71Wb 157
Enmore Gdns. SW14 57Ta 109
Enmore Rd. SE25 71Wb 157
Enmore Rd. SW15 56Ya 110
Enmore Rd. UB1: S'hall 42Ca 85
Ennerdale NW1 3B 216 (41Lb 90)
(off Varndell St.)
Ennerdale Av. HA7: Stan 27La 46
Ennerdale Av. RM12: Horn 36Jd 76
Ennerdale Cl. AL1: St A 4F 6
Ennerdale Cl. SM1: Sutt 77Bb 155
Ennerdale Cl. TW14: Felt 60V 106
Ennerdale Ct. E11 31Jc 73
(off Cambridge Rd.)
Ennerdale Cres. SL1: Slou 3A 80
Ennerdale Dr. NW9 29Ua 48
Ennerdale Dr. WD25: Wat 5Y 13
Ennerdale Gdns. HA9: Wemb 32La 66
Ennerdale Ho. E3 42Bc 92
Ennerdale Rd. DA7: Bex 53Cd 118
Ennerdale Rd. TW9: Kew 54Pa 109
Ennerdale Rd. TW9: Rich 54Pa 109
Ennersdale Rd. SE13 57Fc 115
Ennis Ho. E14 44Dc 92
(off Vesey Path)
Ennismore Av. UB6: G'frd 37Ga 66
Ennismore Av. W4 49Va 88
Ennismore Gdns. KT7: T Ditt 72Ga 152
Ennismore Gdns. SW7 2D 226 (47Gb 89)
Ennismore Gdns. M. SW7 3D 226 (48Gb 89)
Ennismore M. SW7 3D 226 (48Gb 89)
Ennismore St. SW7 3D 226 (48Gb 89)
Ennis Rd. N4 32Qb 70
Ennis Rd. SE18 51Sc 116
Ennor Ct. SM3: Cheam 77Ya 154
Ensbury Ho. SW8 52Pb 112
(off Carroun Rd.)
Ensham Ho. SW17 64Hb 133
Ensign Cl. CR8: Purl 82Qb 176
Ensign Cl. TW19: Stanw 60M 105
Ensign Ct. TW6: H'row A 55U 106
Ensign Dr. N13 20Sb 33
Ensign Est. RM19: Purf 49Rd 97
Ensign Ho. E14 47Cc 92
(off Admirals Way)
Ensign Ho. NW9 27Wa 48
(off East Dr.)
Ensign Ho. RM17: Grays 51Be 121
Ensign Ho. SW18 55Eb 111
Ensign St. E1 45Wb 91
Ensign St. SE3 56Kc 115
Ensign Way SM6: W'gton 80Nb 156
Ensign Way TW19: Stanw 60M 105
Enslin Rd. SE9 58Qc 116
Ensor M. SW7 7B 226 (50Fb 89)
Enstone Rd. EN3: Enf H 13Ac 34
Enstone Rd. UB10: Ick 34P 63
Enterdent, The RH9: G'stone 5B 210
Enterdent Cotts. RH9: G'stone 5B 210
Enterdent Rd. RH9: G'stone 6A 210
Enterprise Bus. Pk. E14 47Dc 92
Enterprise Cen., The BR3: Beck 64Ac 136
(off Cricket La.)
Enterprise Cen., The EN6: Pot B 2Ab 16
Enterprise Cl. CR0: C'don 74Qb 156
Enterprise Ct. RH1: Redh 7P 207
(off Mill St.)
Enterprise Ho. E14 50Dc 92
(off St Davids Sq.)
Enterprise Ho. E4 17Ec 34
Enterprise Ho. E9 38Yb 72
(off Tudor Gro.)
Enterprise Ho. IG11: Bark 41Vc 95
Enterprise Ho. KT12: Walt T 73X 151
Enterprise Ind. Est. SE16 50Yb 92
Enterprise Row N15 29Vb 51
Enterprise Trad. Est. UB2: S'hall 47Da 85
Enterprise Way HP2: Hem H 1C 4
Enterprise Way NW10 41Va 88
Enterprise Way SW18 56Cb 111
Enterprise Way TW11: Tedd 65Ha 130
Enterprize Way SE8 49Bc 92
Entertainment Av. SE10 46Gc 93
Enville Ho. WD19: Wat 20Y 27
Envoy Av. TW6: H'row A 55V 106
Envoy Ho. NW9 27Wa 48
(off East Dr.)
ENVOY RDBT. 55V 106
Eothen Cl. CR3: Cat'm 96Wb 197
Epcot M. NW10 41Za 88
Epirus M. SW6 52Cb 111
Epirus Rd. SW6 52Bb 111
EPPING 2Wc 23
Epping Cl. E14 49Cc 92
Epping Cl. RM7: Mawney 27Dd 56
Epping Forest District Mus. 5Ec 20
Epping Forest Shop. Pk. 15Sc 36
Epping Glade E4 16Ec 34
Epping Golf Course 5Xc 23
Epping La. RM4: Abr 12Xc 37
Epping La. RM4: Stap T 12Xc 37
Epping New Rd. IG10: H Beech 12Lc 35
Epping New Rd. IG10: Lough 12Lc 35
Epping New Rd. IG9: Buck H 20Jc 35
Epping Pl. N1 37Qb 70
Epping Rd. CM16: Epp 1Yc 23
Epping Rd. CM16: Epp 8Pc 22
Epping Rd. CM16: N Weald 1Yc 23
Epping Sports Cen. 3Vc 23
Epping Station (Underground) 3Wc 23
Epping Way E4 16Dc 34
Epple Rd. SW6 53Bb 111
EPSOM 85Ta 173
Epsom Bus. Pk. KT17: Eps 83Ua 174
Epsom Cl. DA12: Grav'nd 5E 144
Epsom Cl. DA7: Bex 55Dd 118
Epsom Cl. UB5: N'olt 36Ba 65
Epsom College Fitness Cen. 87Wa 174
Epsom Common Local Nature Reserve 85Pa 173
Epsom Ct. WD3: Rick 18K 25
EPSOM DAY SURGERY CEN. 85Va 174
EPSOM DOWNS 90Ua 174
Epsom Downs Metro Cen. KT20: Tad 92Xa 194
Epsom Downs Racecourse 90Va 174
Epsom Downs Station (Rail) 87Xa 174
Epsom Gap KT9: Lea 87Ka 172
EPSOM GENERAL HOSPITAL 87Sa 173
Epsom Golf Course 87Wa 174
Epsom Ho. RM3: Rom 22Nd 57
(off Dagnam Pk. Dr.)
Epsom Ho. SL4: Wind 3D 102
(off Paddock Cl.)
Epsom Ho. TW18: Staines 63H 127
(off Fairfield Av.)
Epsom La. Nth. KT18: Tad 90Xa 174
Epsom La. Nth. KT18: Tatt C 90Xa 174
Epsom La. Nth. KT20: Tad 92Xa 194
Epsom La. Sth. KT20: Tad 93Ya 194
Epsom Playhouse 85Ta 173
Epsom Polo Club & Equestrian Cen. 83Pa 173

Epsom Rd. CR0: Wadd 77Qb 156
Epsom Rd. E10 30Ec 52
Epsom Rd. IG3: Ilf 30Vc 55
Epsom Rd. KT17: Ewe 83Va 174
Epsom Rd. KT21: Asht 90Pa 173
Epsom Rd. KT22: Lea 93Ka 192
Epsom Rd. SM3: Sutt 73Bb 155
Epsom Rd. SM4: Mord 72Bb 155
Epsom Sports Club 87Ta 173
Epsom Sq. TW6: H'row A 54V 106
Epsom Square 85Ta 173
Epsom Station (Rail) 85Ta 173
Epsom Trade Pk. KT19: Eps 83Ta 173
Epsom Way RM12: Horn 35Pd 77
Epstein Ct. N1 39Rb 71
(off Gaskin St.)
Epstein Rd. SE28 46Wc 95
Epstein Sq. E14 44Cc 92
(off Upper Nth. St.)
Epworth Rd. TW7: Isle 52Ka 108
Epworth St. EC2 6G 219 (42Tb 91)
Equana Apts. SE8 50Ac 92
(off Evelyn St.)
Equestrian Statue George III 1H 125
Equiano Ho. SW9 53Pb 112
(off Lett Rd.)
Equinox Ct. IG2: Ilf 29Rc 54
Equinox Ho. IG11: Bark 37Sc 74
(off Wakering Rd.)
Equinox Sq. E14 44Dc 92
Equity M. W5 46Ma 87
Equity Sq. E2 4K 219 (41Vb 91)
(off Shacklewell St.)
Equus Cl. SL9: Ger X 2N 61
Erasmus St. SW1 6E 228 (49Mb 90)
Erconwald St. W12 44Va 88
Erebus Dr. SE28 48Sc 94
Eresby Dr. BR3: Beck 74Cc 158
Eresby Ho. SW7 2E 226 (47Gb 89)
(off Rutland Ga.)
Eresby Pl. NW6 38Cb 69
Erica Cl. GU24: W End 5C 166
Erica Cl. SL1: Slou 5C 80
Erica Ct. BR8: Swan 70Gd 140
Erica Ct. GU22: Wok 10P 167
Erica Gdns. CR0: C'don 76Dc 158
Erica Ho. N22 25Qb 50
(off Acacia Rd.)
Erica Ho. SE4 55Bc 114
Erica St. W12 45Wa 88
Eric Clarke La. IG11: Bark 42Rc 94
Eric Cl. E7 35Jc 73
Ericcson Cl. SW18 57Cb 111
Eric Fletcher Ct. N1 38Sb 71
(off Essex Rd.)
Erickson Gdns. BR2: Broml 72Nc 160
Eric Liddell Sports Cen. 61Mc 137
Eric Rd. E7 35Jc 73
Eric Rd. NW10 37Va 68
Eric Rd. RM6: Chad H 31Zc 75
Eric Shipman Ter. E13 42Jc 93
(off Balaam St.)
Ericson Ho. SE13 56Fc 115
(off Blessington Rd.)
Eric Steele Ho. AL2: Park 9P 5
Eric St. E3 42Bc 92
(not continuous)
Eric Wilkins Ho. SE1 50Wb 91
(off Old Kent Rd.)
Eridge Ct. WD25: Wat 8Z 13
(off Ley Farm Cl.)
Eridge Grn. Cl. BR5: Orp 74Yc 161
Eridge Rd. W4 48Ta 87
Erin Cl. BR1: Broml 66Gc 137
Erin Cl. IG3: Ilf 30Wc 55
Erin Cl. SW6 52Cb 111
Erin Ct. NW2 37Ya 68
Erindale SE18 51Tc 116
Erindale Ter. SE18 51Tc 116
Erin M. N22 25Rb 51
Erin's Vs. SE14 53Zb 114
(off New Cross Rd.)
Eriswell Cres. KT12: Hers 79U 150
Eriswell Rd. KT12: Hers 77V 150
ERITH 50Hd 96
ERITH & DISTRICT HOSPITAL 51Fd 118
Erith Ct. RM19: Purf 49Qd 97
Erith Cres. RM5: Col R 25Ed 56
Erith High St. DA8: Erith 50Gd 96
(not continuous)
Erith Leisure Cen. 51Gd 118
Erith Playhouse 50Hd 96
(off Erith High St.)
Erith Quarry DA8: Erith 51Ed 118
Erith Rd. DA8: Erith Picardy Rd. 50Ed 96
Erith Rd. DA8: Erith Watling St. 53Ed 118
Erith Rd. DA17: Belv 50Cd 96
Erith Rd. DA17: Erith 50Cd 96
Erith Rd. DA7: Bex 56Dd 118
ERITH RDBT. 50Gd 96
Erith School Community Sports Cen. 52Ed 118
Erith Stadium 51Gd 118
Erith Station (Rail) 50Gd 96
Erith Yacht Club 51Kd 119
Erkenwald Cl. KT16: Chert 73G 148
Erlanger Rd. SE14 53Zb 114
Erlesmere Gdns. W13 48Ja 86
Erlich Cotts. E1 43Yb 92
(off Sidney St.)
Ermine Cl. AL3: St A 3N 5
Ermine Cl. EN7: Chesh 3Xb 19
Ermine Cl. TW4: Houn 54Y 107
Ermine Ho. E3 39Bc 72
(off Parnell Rd.)
Ermine Ho. N17 24Vb 51
(off Moselle St.)
Ermine M. E2 1K 219 (39Vb 71)
Ermine Rd. N15 30Vb 51
Ermine Rd. SE13 56Dc 114
Ermine Side EN1: Enf 15Wb 33
Ermington Rd. SE9 61Sc 138
Ermyn Cl. KT22: Lea 93Ma 193
Ermyn Way KT22: Lea 93Ma 193
Ernald Av. E6 40Nc 74
Ernan Cl. RM15: S Ock 43Wd 98
Ernan Rd. RM15: S Ock 43Wd 98
Erncroft Way TW1: Twick 58Ha 108
Ernest Av. SE27 63Rb 135
Ernest Cl. BR3: Beck 71Cc 158
Ernest Cotts. KT17: Ewe 80Va 154
Ernest Gdns. W4 51Ra 109
Ernest Gro. BR3: Beck 71Bc 158
Ernest Harriss Ho. W9 42Cb 89
(off Elgin Av.)
Ernest Rd. KT1: King T 68Ra 131
Ernest Rd. RM11: Horn 30Nd 57
Ernest Shackleton Lodge SE10 49Gc 93
(off Christchurch Way)
Ernest Sq. KT1: King T 68Ra 131
Ernest St. E1 42Zb 92
Ernle Rd. SW20 66Xa 132
Ernshaw Pl. SW15 57Ab 110
Ernst Bldg. SE1 7D 224 (46Sb 91)
(off Union St.)
Eros 5D 222 (45Mb 90)
Eros Ho. Shops SE6 59Dc 114
(off Brownhill Rd.)
Erpingham Rd. SW15 55Ya 110
Erridge Rd. SW19 68Cb 133
Erriff Dr. RM15: S Ock 43Vd 98
Errington Cl. RM16: Grays 8D 100
Errington Dr. SL4: Wind 3E 102
Errington Rd. W9 42Bb 89
Errol Gdns. KT3: N Mald 70Wa 132
Errol Gdns. UB4: Yead 42X 85
Erroll Rd. RM1: Rom 28Hd 56
Errol St. EC1 6E 218 (42Sb 91)
Erskine Cl. SM1: Sutt 76Gb 155
Erskine Cres. N17 28Xb 51
Erskine Hill NW11 28Cb 49
Erskine Ho. SW1 50Lb 90
(off Churchill Gdns.)
Erskine Ho. TN13: S'oaks 97Jd 202
Erskine Ho. WD19: Wat 20Z 27
Erskine M. NW3 38Hb 69
(off Erskine Rd.)
Erskine Rd. E17 28Bc 52
Erskine Rd. NW3 38Hb 69
Erskine Rd. SM1: Sutt 77Fb 155
Erwin Ho. NW9 27Wa 48
(off Commander Av.)
Erwood Rd. SE7 50Nc 94
Esam Way SW16 64Qb 134
Escombe Ct. CR3: Whyt 91Wb 197
(off Godstone Rd.)
Escot Rd. TW16: Sun 66U 128
Escott Gdns. SE9 63Nc 138
Escott Pl. KT16: Ott 79E 148
Escot Way EN5: Barn 15Ya 30
Escreet Gro. SE18 49Qc 94
Esdaile Gdns. RM14: Upm 31Td 78
ESHER 77Da 151
Esher Av. KT12: Walt T 73W 150
Esher Av. RM7: Rom 30Ed 56
Esher Av. SM3: Cheam 76Za 154
Esher By-Pass KT11: Cobh 85V 170
Esher Cl. DA5: Bexl 60Ad 117
Esher Cl. KT10: Esh 78Da 151
Esher Common 82Ca 171
ESHER COMMON 82Ea 172
Esher Cres. TW6: H'row A 54V 106
Esher Gdns. SW19 61Za 132
Esher Grn. KT10: Esh 77Da 151
Esher Grn. Dr. KT10: Esh 76Ca 151
Esher M. CR4: Mitc 69Jb 134
Esher Pk. Av. KT10: Esh 77Da 151
Esher Pl. Av. KT10: Esh 77Ca 151
Esher Rd. IG3: Ilf 34Uc 74
Esher Rd. KT12: Hers 78Z 151
Esher Rd. KT8: E Mos 72Fa 152
Esher RUFC 75Aa 151
Esher Station (Rail) 75Fa 152
Eskdale AL2: Lon C 9K 7
Eskdale NW1 2B 216 (40Lb 70)
(off Stanhope St.)
Eskdale Av. UB5: N'olt 39Ba 65
Eskdale Cl. DA2: Dart 61Sd 142
Eskdale Cl. HA9: Wemb 33Ma 67
Eskdale Gdns. CR8: Purl 86Tb 177
Eskdale Rd. DA7: Bex 54Cd 118
Eskdale Rd. UB8: Uxb 40K 63
Eskdale Rd. Ind. Est. UB8: Uxb 40K 63
Esker Pl. E2 40Xb 71
Esk Ho. E3 42Bc 92
(off British St.)
Eskley Gdns. RM15: S Ock 43Xd 98
(not continuous)
Eskmont Ridge SE19 66Tb 135
Esk Rd. E13 42Jc 93
Esk Way RM1: Rom 24Fd 56
Esmar Cres. NW9 31Wa 68
Esmat Cl. E11 29Hc 53
Esmeralda Rd. SE1 49Wb 91
Esmond Cl. RM13: Rain 38Kd 77
Esmond Ct. W8 48Db 89
(off Thackeray St.)
Esmond Gdns. W4 49Ta 87
Esmond Rd. NW6 39Bb 69
Esmond Rd. W4 49Ta 87
Esmond St. SW15 56Ab 110
Esparto St. SW18 59Db 111
Esparto Way DA4: S Dar 67Sd 142
Esprit Ct. E1 1K 225 (43Vb 91)
(off Brune St.)
Esquiline La. CR4: Mitc 69Kb 134
Essan Ho. W5 43Ka 86
Essence E3 40Bc 72
(off Cardigan Rd.)
Essence Ct. HA9: Wemb 33Pa 67
Essendene Cl. CR3: Cat'm 95Ub 197
Essendene Rd. CR3: Cat'm 95Ub 197
Essenden Rd. CR2: S Croy 80Ub 157
Essenden Rd. DA17: Belv 50Cd 96
Essendine Mans. W9 41Cb 89
Essendine Rd. W9 41Cb 89
Essendon Pl. AL9: Ess 1N 9
Essex Av. SL2: Slou 3G 80
Essex Av. TW7: Isle 55Ga 108
Essex Cl. E17 28Ac 52
Essex Cl. HA4: Ruis 32Z 65
Essex Cl. KT15: Add 77L 149
Essex Cl. RM7: Mawney 28Dd 56
Essex Cl. SM4: Mord 73Za 154
Essex Ct. EC4 3K 223 (44Qb 90)
(off Brick Ct.)
Essex Ct. SW13 54Va 110
Essex Ct. W6 48Ya 88
(off Hammersmith Gro.)
Essex Fire Mus. 48Ce 99
Essex Gdns. N4 30Rb 51
Essex Gdns. RM11: Horn 29Qd 57
Essex Gdns. SS17: Linf 7J 101
Essex Gro. SE19 65Tb 135
Essex Hall E17 25Zb 52
Essex Ho. E14 44Dc 92
(off Giraud St.)
Essex La. WD4: Hunt C 5T 12
Essex Mans. E11 32Fc 73
Essex M. SE19 65Ub 135
Essex Pk. N3 23Db 49
Essex Pk. M. W3 46Ua 88
Essex Pl. W4 49Sa 87
(not continuous)
Essex Pl. Sq. W4 49Ta 87
Essex Rd. DA1: Dart 58Md 119
(not continuous)
Essex Rd. DA11: Grav'nd 10C 122
Essex Rd. DA3: Lfield 68Zd 143
Essex Rd. E10 30Ec 52
Essex Rd. E12 36Nc 74
Essex Rd. E17 30Ac 52
Essex Rd. E18 26Kc 53
Essex Rd. E4 18Gc 35
Essex Rd. EN2: Enf 14Tb 33
Essex Rd. IG11: Bark 38Tc 74
Essex Rd. N1 1B 218 (39Rb 71)
Essex Rd. NW10 38Ua 68
Essex Rd. RM10: Dag 36Ed 76
Essex Rd. RM20: W Thur 51Wd 120
Essex Rd. RM6: Chad H 31Yc 75
Essex Rd. RM7: Mawney 28Dd 56
Essex Rd. W3 45Sa 87
Essex Rd. W4 49Ta 87
Essex Rd. WD17: Wat 12W 26
Essex Rd. WD6: Bore 13Qa 29
Essex Rd. Sth. E11 31Fc 73
Essex Road Station (Rail) 38Sb 71
Essex St. AL1: St A 1C 6
Essex St. E7 36Jc 73
Essex St. WC2 3K 223 (44Qb 90)
Essex Twr. SE20 67Xb 135
(off Jasmine Gro.)
Essex Vs. W8 47Cb 89
Essex Way CM13: Gt War 23Yd 58
Essex Wharf E5 33Zb 72
Essian St. E1 43Ac 92
Essoldo Ct. WD18: Wat 14Y 27
(off Granville Rd.)
Essoldo Way HA8: Edg 27Pa 47
Estate Cotts. RH5: Mick 99Ma 193
Estate Way E10 32Bc 72
Estcourt Rd. SE25 72Xb 157
Estcourt Rd. SW6 52Bb 111
Estcourt Rd. WD17: Wat 13Y 27
Estella Apts. E15 37Fc 73
(off Grove Cres. Rd.)
Estella Av. KT3: N Mald 70Xa 132
Estella Ho. W11 45Za 88
(off St Ann's Rd.)
Estelle Rd. NW3 35Hb 69
Esterbrooke St. SW1 6D 228 (49Mb 90)
Este Rd. SW11 55Gb 111
Esther Anne Pl. N1 39Rb 71
Esther Cl. N21 17Qb 32
Esther M. BR1: Broml 67Kc 137
(off Freelands Rd.)
Esther Randall Ct. NW1 5A 216 (42Kb 90)
(off Lit. Albany St.)
Esther Rd. E11 31Gc 73
Estoria Cl. SW2 59Qb 112
Estorick Collection of Modern Italian Art 37Rb 71
Estreham Rd. SW16 65Mb 134
Estridge Cl. TW3: Houn 56Ca 107
Estuary Cl. IG11: Bark 41Xc 95
Estuary Ho. E16 46Mc 93
(off Agnes George Wlk.)
Eswarah Ho. KT17: Eps 83Va 174
(off Epsom Rd.)
Eswyn Rd. SW17 63Hb 133
Etal Ho. N1 38Rb 71
(off The Sutton Est.)
Etcetera Theatre 38Kb 70
(off Camden High St.)
Etchingham Ct. N3 24Db 49
Etchingham Pk. Rd. N3 24Db 49
Etchingham Rd. E15 35Ec 72
Eternit Wlk. SW6 53Ya 110
Etfield Gro. DA14: Sidc 64Xc 139
Ethan Dr. N2 27Db 49
Ethel Bailey Cl. KT19: Eps 84Qa 173
Ethelbert Cl. BR1: Broml 68Jc 137
Ethelbert Ct. BR1: Broml 69Jc 137
(off Ethelbert Rd.)
Ethelbert Gdns. IG2: Ilf 29Pc 54
Ethelbert Ho. E9 35Ac 72
(off Homerton Rd.)
Ethelbert Rd. BR1: Broml 69Jc 137
Ethelbert Rd. BR5: St P 69Zc 139
Ethelbert Rd. DA2: Hawl 63Nd 141
Ethelbert Rd. DA8: Erith 52Ed 118
Ethelbert Rd. SW20 67Za 132
Ethelbert St. SW12 60Kb 112
Ethel Brooks Ho. SE18 51Rc 116
Ethelburga Rd. RM3: Hrld W 25Pd 57
Ethelburga St. SW11 53Gb 111
Ethelburga Twr. SW11 53Gb 111
(off Rosenau Rd.)
Etheldene Av. N10 28Lb 50
Ethelden Rd. W12 46Xa 88
Ethelred Ct. CR3: Whyt 91Wb 197
(off Godstone Rd.)
Ethelred Ct. HA3: Kenton 29Qa 47
Ethel Rd. E16 44Kc 93
Ethel Rd. TW15: Ashf 64N 127
Ethel St. SE17 6E 230 (49Sb 91)
Ethel Ter. BR6: Prat B 81Yc 181
Ethelwine Pl. WD5: Ab L 2V 12
Etheridge Grn. IG10: Lough 13Sc 36
Etheridge Rd. IG10: Lough 12Sc 36
Etheridge Rd. NW4 31Ya 68
(not continuous)
Etherley Rd. N15 29Sb 51
Etherow St. SE22 59Wb 113
Etherstone Grn. SW16 63Qb 134
Etherstone Rd. SW16 63Qb 134
Ethnard Rd. SE15 51Xb 113
Ethorpe Cl. SL9: Ger X 29A 42
Ethorpe Cres. SL9: Ger X 29A 42
Ethorpe Ho. SL9: Ger X 29A 42
Ethos Sport Imperial 2C 226 (47Fb 89)
Ethronvi Rd. DA7: Bex 55Ad 117
Etloe Ho. E10 32Cc 72
Etloe Rd. E10 33Cc 72
Etna Rd. AL3: St A 1B 6
ETON 1H 103
Eton Av. EN4: E Barn 16Gb 31
Eton Av. HA0: Wemb 35Ka 66
Eton Av. KT3: N Mald 71Ta 153
Eton Av. N12 24Eb 49
Eton Av. NW3 38Fb 69
Eton Av. TW5: Hest 51Ba 107
Eton Cl. SL3: Dat 1M 103
Eton Cl. SW18 59Db 111
Eton Coll. Rd. NW3 37Hb 69
Eton College Rowing Cen. 2B 102
Eton Ct. HA0: Wemb 35La 66
Eton Ct. NW3 38Fb 69
Eton Ct. SL4: Eton 2H 103
Eton Ct. TW18: Staines 64H 127
Eton Garages NW3 37Gb 69
Eton Gro. NW9 27Qa 47
Eton Gro. SE13 55Gc 115
Eton Hall NW3 37Hb 69
Eton Ho. KT18: Eps 85Sa 173
(off Dalmeny Way)
Eton Ho. N5 35Rb 71
(off Leigh Rd.)
Eton Ho. RH1: Redh 4A 208
(off Old School Cl.)
Eton Ho. UB7: W Dray 47P 83
Eton Ho. WD24: Wat 12Y 27
(off Anglian Cl.)
ETON MANOR 35Dc 72
Eton Mnr. Ct. E10 33Cc 72
(off Leyton Grange Est.)
Eton M. N1 38Pb 70
Eton Pl. NW3 38Jb 70
Eton Ri. NW3 37Hb 69
Eton Riverside SL4: Eton 2H 103
Eton Rd. BR6: Chels 77Xc 161
Eton Rd. IG1: Ilf 35Sc 74
Eton Rd. NW3 38Hb 69
Eton Rd. SL3: Dat 10K 81
Eton Rd. UB3: Harl 52V 106
Eton Sq. SL4: Eton 2H 103
Eton St. TW9: Rich 57Na 109
Eton Vs. NW3 37Hb 69
Eton Wlk. SL1: Slou 8J 81
(off Upton Pk.)
Eton Way DA1: Dart 56Ld 119
ETON WICK 9D 80
Eton Wick Rd. SL4: Eton 9C 80
Eton Wick Rd. SL4: Eton W 9C 80
Etta St. SE8 51Ac 114
Etton Cl. RM12: Horn 33Nd 77
Ettrick St. E14 44Ec 92
(not continuous)
Etwell Pl. KT5: Surb 72Pa 153
Eucalyptus M. SW16 65Mb 134
Euclid Way RM20: W Thur 50Vd 98
Euesden Cl. N9 20Xb 33
Eugene Bann Indoor Tennis Cen. 10F 208
Eugene Cl. RM2: Rom 28Ld 57
Eugene Cotter Ho. SE17 6G 231 (49Tb 91)
(off Tatum St.)
Eugenia Rd. SE16 49Yb 92
Eugenie M. BR7: Chst 67Rc 138
Eureka Rd. KT1: King T 68Qa 131
Eurobet Ho. GU21: Wok 89A 168
(off Church St. W.)
Euro Cl. NW10 37Wa 68
Eurolink Bus. Cen. SW2 56Qb 112
Europa Gym Cen. 57Jd 118
Europa Pk. RM20: Grays 50Zd 99
Europa Pl. EC1 4D 218 (41Sb 91)
Europa Trade Pk. E16 42Gc 93
Europa Trad. Est. DA8: Erith 50Fd 96
European Bus. Cen. NW9 27Sa 47
(not continuous)
European Design Cen. NW9 27Ta 47
Europe Rd. SE18 48Pc 94
Euro Trade Cen. DA17: Belv 47Ed 96
Eustace Bldg. SW11 51Kb 112
Eustace Ho. SE11 5H 229 (49Pb 90)
(off Old Paradise St.)
Eustace Pl. SE18 49Pc 94
Eustace Rd. E6 41Nc 94
Eustace Rd. RM6: Chad H 31Zc 75
Eustace Rd. SW6 52Cb 111
Euston Av. WD18: Wat 15V 26
Euston Cen. NW1 5B 216 (42Lb 90)
Euston Gro. NW1 4D 216 (41Mb 90)
(off Euston Sq.)
Euston Rd. CR0: C'don 74Qb 156
Euston Rd. N1 3F 217 (41Nb 90)
Euston Rd. NW1 6A 216 (42Kb 90)
Euston Sq. NW1 4D 216 (41Mb 90)
(not continuous)
Euston Square Station (Underground) 5C 216 (42Lb 90)
Euston Sta. Colonnade NW1 4D 216 (41Mb 90)
Euston Station (Rail, Underground & Overground) 4C 216 (41Lb 90)
Euston St. NW1 4C 216 (41Lb 90)
Euston Twr. NW1 5B 216 (42Lb 90)
EUSTON UNDERPASS 5B 216 (42Lb 90)
Eva Ct. CR2: S Croy 79Ub 157
Evan Cook Cl. SE15 53Yb 114
Evandale Rd. SW9 54Qb 112
Evangelist Ho. EC4 3B 224 (44Rb 91)
(off Black Friars La.)
Evangelist Rd. NW5 35Kb 70
Evan Ho. E16 43Jc 93
(off Exeter Rd.)
Evan's SL4: Eton 1H 103
(off Keates La.)
Evans Apts. E2 41Zb 92
Evans Av. WD25: Wat 7V 12
Evans Cl. DA9: Ghithe 57Wd 120
Evans Cl. E8 37Vb 71
Evans Cl. WD3: Crox G 15Q 26
Evansdale RM13: Rain 41Hd 96
Evans Gro. TW13: Hanw 61Ca 129
Evans Ho. HP3: Hem H 7P 3
(off The Embankment)
Evans Ho. SW8 52Mb 112
(off Wandsworth Rd.)
Evans Ho. TW13: Hanw 61Ca 129
Evans Ho. W12 45Xa 88
(off White City Est.)
Evans Rd. SE6 61Gc 137
Evanston Av. E4 24Ec 52
Evanston Gdns. IG4: Ilf 30Nc 54
Evans Wharf HP3: Hem H 6N 3
Eva Rd. RM6: Chad H 31Yc 75
Evedon Ho. N1 39Ub 71
(off Halcomb St.)
EVELINA CHILDREN'S HOSPITAL 3H 229 (48Pb 90)
(within St Thomas' Hospital)
Evelina Ct. W6 48Xa 88
(off Vinery Way)
Evelina Mans. SE5 52Tb 113
Evelina Rd. SE15 55Yb 114
Evelina Rd. SE20 66Yb 136
Eveline Lowe Est. SE16 48Wb 91
Eveline Rd. CR4: Mitc 67Hb 133
Evelyn Av. HA4: Ruis 31U 64
Evelyn Av. NW9 28Ta 47
Evelyn Av. RH8: T'sey 96Lc 199
Evelyn Cl. GU22: Wok 2P 187
Evelyn Cl. TW2: Whitt 59Da 107
Evelyn Cotts. RH3: S God 9C 210
Evelyn Ct. E3 43Bc 92
(off Burdett Rd.)
Evelyn Ct. E8 35Wb 71
Evelyn Ct. N1 2F 219 (40Tb 71)
(off Evelyn Wlk.)
Evelyn Cres. TW16: Sun 67V 128
Evelyn Denington Ct. N1 38Rb 71
(off The Sutton Est.)
Evelyn Denington Rd. E6 42Nc 94
Evelyn Dr. HA5: Pinn 24Z 45
Evelyn Fox Ct. W10 43Ya 88
Evelyn Gdns. RH9: G'stone 2A 210
Evelyn Gdns. SW7 7A 226 (50Eb 89)
Evelyn Gdns. TW9: Rich 56Na 109
Evelyn Gro. UB1: S'hall 44Ba 85
Evelyn Gro. W5 46Pa 87
Evelyn Ho. SE14 53Ac 114
(off Loring Rd.)
Evelyn Ho. W12 47Va 88
(off Cobbold Rd.)
Evelyn Ho. W8 47Db 89
(off Hornton Pl.)
Evelyn Mans. SW1 4B 228 (48Lb 90)
(off Carlisle Pl.)
Evelyn Mans. W14 51Ab 110
(off Queen's Club Gdns.)
Evelyn Rd. E16 46Jc 93
Evelyn Rd. E17 28Ec 52
Evelyn Rd. EN4: Cockf 14Hb 31
Evelyn Rd. SW19 64Db 133
Evelyn Rd. TN14: Otf 88Ld 183
Evelyn Rd. TW10: Ham 62La 130
Evelyn Rd. TW9: Rich 55Na 109
Evelyn Rd. W4 48Ta 87
Evelyns Cl. UB8: Hil 44Q 84
Evelyn Sharp Cl. RM2: Rom 27Md 57
Evelyn Sharp Ho. HP2: Hem H 3B 4
Evelyn Sharp Ho. RM2: Rom 27Md 57
Evelyn St. SE8 49Ac 92
Evelyn Ter. TW9: Rich 55Na 109
Evelyn Wlk. CM13: Gt War 23Yd 58
Evelyn Wlk. DA9: Ghithe 56Wd 120
Evelyn Wlk. N1 2F 219 (40Tb 71)
Evelyn Way KT11: Stoke D 88Ba 171
Evelyn Way KT19: Eps 83Qa 173
Evelyn Way SM6: Bedd 77Mb 156
Evelyn Way TW16: Sun 67V 128
Evelyn Yd. W1 2D 222 (44Mb 90)
Eve Martin Ct. EC1 4C 218 (41Rb 91)
(off Goswell Road)
Evening Hill BR3: Beck 66Ec 136
Evenlode Ho. SE2 47Yc 95
Evensyde WD18: Wat 16S 26
Evenwood Cl. SW15 57Ab 110
Everall Cl. HP1: Hem H 2L 3
Everall Ct. E4 23Bc 52
Everard Av. BR2: Hayes 74Jc 159
Everard Av. SL1: Slou 7J 81
Everard Cl. AL1: St A 4B 6
Everard Ct. N13 20Pb 32
Everard Ho. E1 44Wb 91
(off Boyd St.)
Everard La. CR3: Cat'm 94Xb 197
Everard Way HA9: Wemb 34Na 67
Everatt Cl. SW18 58Bb 111
Everdon Rd. SW13 51Wa 110
Everest Cl. DA11: Nflt 2A 144
Everest Ct. GU21: Wok 8J 167
Everest Pl. BR8: Swan 70Fd 140
Everest Pl. E14 43Ec 92
Everest Rd. SE9 57Pc 116
Everest Rd. TW19: Stanw 59M 105
Everest Way HP2: Hem H 1A 4
Everett Cl. HA5: Eastc 27V 44
Everett Cl. WD23: B Hea 18Ga 28
Everett Ct. WD7: R'lett 6Ja 14
Everett Ho. SE17 7G 231 (50Tb 91)
(off East St.)
Everett Wlk. DA17: Belv 50Bd 95
(off Osborne Rd.)
Everglade TN16: Big H 90Mc 179
Everglade Cl. DA3: Hartl 70Ae 143
Everglade Ho. E17 26Bc 52
Everglades, The TW3: Houn 55Ea 108
Everglade Strand NW9 25Va 48
Evergreen Apts. IG8: Wfd G 25Hc 53
(off High Rd. Woodford Grn.)
Evergreen Cl. SE20 66Yb 136
Evergreen Ct. TW19: Stanw 59M 105
Evergreen Dr. UB10: Hil 40R 64
Evergreen Dr. UB7: W Dray 47P 83
Evergreen Oak Av. SL4: Wind 5L 103
Evergreen Sq. E8 38Vb 71
Evergreen Wlk. HP3: Hem H 4N 3
Evergreen Way TW19: Stanw 59M 105
Evergreen Way UB3: Hayes 45V 84
Everilda St. N1 39Pb 70
Evering Rd. E5 34Wb 71
Evering Rd. N16 34Vb 71
Everington Rd. N10 26Hb 49
Everington St. W6 51Za 110
Everitt Rd. NW10 41Ta 87
Everlands Cl. GU22: Wok 90A 168
Everlasting La. AL3: St A 1A 6
(not continuous)
Everleigh St. N4 32Pb 70
Eve Rd. E11 35Gc 73
Eve Rd. E15 40Gc 73
Eve Rd. GU21: Wok 87D 168
Eve Rd. N17 27Ub 51
Eve Rd. TW7: Isle 56Ja 108
Eversfield Gdns. NW7 23Ua 48
Eversfield Rd. RH2: Reig 6K 207
Eversfield Rd. TW9: Kew 54Pa 109
Evershed Ho. E1 2K 225 (44Vb 91)
(off Old Castle St.)
Evershed Wlk. W4 48Sa 87
Eversholt Ct. EN5: New Bar 15Eb 31
Eversholt St. NW1 1B 216 (40Lb 70)
Evershot Rd. N4 32Pb 70
Eversleigh Ct. N3 24Bb 49
Eversleigh Gdns. RM14: Upm 32Td 78
Eversleigh Pl. BR3: Beck 65Cc 136
Eversleigh Rd. E6 39Mc 73
Eversleigh Rd. EN5: New Bar 15Eb 31
Eversleigh Rd. N3 24Bb 49
Eversleigh Rd. SW11 55Hb 111
Eversley Av. DA7: Bex 54Fd 118
Eversley Av. HA9: Wemb 33Qa 67
Eversley Cl. IG10: Lough 13Sc 36
Eversley Cl. N21 16Pb 32
Eversley Cres. HA4: Ruis 33U 64
Eversley Cres. N21 16Qb 32
Eversley Cres. TW7: Isle 53Fa 108
Eversley Cross DA7: Bex 54Gd 118
Eversley Ho. E2 41Wb 91
(off Gosset St.)
Eversley Mt. N21 16Pb 32
Eversley Pk. SW19 65Xa 132
Eversley Pk. Rd. N21 16Pb 32
Eversley Rd. KT5: Surb 70Pa 131
Eversley Rd. SE19 66Tb 135
Eversley Rd. SE7 51Kc 115
Eversley Way CR0: C'don 76Cc 158
Eversley Way TW20: Thorpe 68E 126
Everthorpe Rd. SE15 55Vb 113
Everton Ct. HA7: Stan 27Na 47
(off Honeypot La.)
Everton Dr. HA7: Stan 27Na 47
Everton M. NW1 4B 216 (41Lb 90)

Everton Rd. CR0: C'don 74Wb 157
Everyman Cinema Baker St. 7G 215 (43Hb 89)
(off Baker St.)
Everyman Cinema Barnet 15Cb 31
Everyman Cinema Belsize Pk. 36Gb 69
Everyman Cinema Esher 77Da 151
Everyman Cinema Gerrards Cross 29A 42
Everyman Cinema King's Cross 39Nb 70
Everyman Cinema Muswell Hill 28Kb 50
Everyman Cinema Oxted 1J 211
Everyman Cinema Reigate 6J 207
Everyman Cinema Walton-on-Thames 74W 150
Everyman on the Corner Cinema 39Mb 70
Everyman Screen on the Green Cinema 39Rb 71
(off Upper St.)
Everyone Active Abbs Cross 34Ld 77
Evesham Av. E17 26Cc 52
Evesham Cl. RH2: Reig 5H 207
Evesham Cl. SM2: Sutt 80Cb 155
Evesham Cl. UB6: G'frd 40Da 65
Evesham Ct. TW10: Rich 58Pa 109
Evesham Ct. W13 46Ja 86
(off Tewkesbury Rd.)
Evesham Grn. SM4: Mord 72Db 155
Evesham Ho. E2 40Yb 72
(off Old Ford Rd.)
Evesham Ho. NW8 39Eb 69
(off Abbey Rd.)
Evesham Ho. SW1 7A 228 (50Kb 90)
(part of Abbots Mnr.)
Evesham Rd. DA12: Grav'nd 1F 144
Evesham Rd. E15 38Hc 73
Evesham Rd. N11 22Lb 50
Evesham Rd. RH2: Reig 5H 207
Evesham Rd. SM4: Mord 72Db 155
Evesham Rd. Nth. RH2: Reig 5H 207
Evesham St. W11 45Za 88
Evesham Ter. KT6: Surb 72Ma 153
Evesham Wlk. SE5 54Tb 113
Evesham Wlk. SW9 54Qb 112
Evesham Way IG5: Ilf 27Qc 54
Evesham Way SW11 55Jb 112
Evette M. IG5: Ilf 25Qc 54
Evolution WD25: Wat 6Z 13
Evreham Rd. SL0: Iver 44G 82
Evreham Sports Cen. 43F 82
Evry Rd. DA14: Sidc 65Yc 139
Ewald Rd. SW6 54Bb 111
Ewanrigg Ter. IG8: Wfd G 22Lc 53
Ewan Rd. RM3: Hrld W 26Md 57
Ewart Gro. N22 25Pb 50
Ewart Ho. HA1: Harr 29Ja 46
Ewart Pl. E3 40Bc 72
Ewart Rd. SE23 59Zb 114
Ewe Cl. N7 37Nb 70
EWELL 81Va 174
Ewell By-Pass KT17: Ewe 80Wa 154
Ewell Ct. Av. KT19: Ewe 78Ua 154
Ewell Downs Rd. KT17: Ewe 83Wa 174
Ewell East Station (Rail) 82Xa 174
Ewell Gro. Ct. KT17: Ewe 81Va 174
(off West St.)
Ewell Ho. KT17: Ewe 82Va 174
(off Ewell Ho. Gro.)
Ewell Ho. Gro. KT17: Ewe 82Va 174
Ewell Ho. Pde. KT17: Ewe 82Va 174
(off Epsom Rd.)
Ewellhurst Rd. IG5: Ilf 26Nc 54
Ewell Pk. Gdns. KT17: Ewe 80Wa 154
Ewell Pk. Way KT17: Ewe 79Wa 154
Ewell Rd. KT6: Surb Mount Holme 73Ka 152
Ewell Rd. KT6: Surb South Ter. 72Na 153
Ewell Rd. SM3: Cheam 79Za 154
Ewell West Station (Rail) 81Ua 174
Ewelme Rd. SE23 60Yb 114
Ewen Cres. SW2 59Qb 112
Ewen Henderson Ct. SE14 52Ac 114
(off Goodwood Rd.)
Ewen Ho. N1 39Pb 70
(off Barnsbury Est.)
Ewer St. SE1 7D 224 (46Sb 91)
Ewhurst Av. CR2: Sande 81Vb 177
Ewhurst Cl. E1 43Yb 92
Ewhurst Cl. SM2: Cheam 81Ya 174
Ewhurst Ct. CR4: Mitc 69Fb 133
Ewhurst Rd. SE4 58Bc 114
Exbury Ho. E9 38Yb 72
Exbury Ho. SW1 7D 228 (50Mb 90)
(off Rampayne St.)
Exbury Rd. SE6 61Cc 136
Excalibur Dr. SE6 61Gc 137
Excel Ct. WC2 5E 222 (45Mb 90)
(off Whitcomb St.)
Excel Marina E16 45Jc 93
Excelsior Cl. KT1: King T 68Qa 131
Excelsior Gdns. SE13 54Ec 114
Excelsior Ind. Est. SE15 51Yb 114
Excel Waterfront E16 45Kc 93
Exchange, The CR0: C'don 76Sb 157
(off Surrey St.)
Exchange, The GU22: Wok 89B 168
(off Oriental Rd.)
Exchange, The IG1: Ilf 33Rc 74
Exchange Apts. BR2: Broml 70Kc 137
(off Sparkes Cl.)
Exchange Arc. EC2 7J 219 (43Ub 91)
Exchange Bldg. E1 6K 219 (42Vb 91)
(off Commercial St.)
Exchange Cl. N11 19Jb 32
Exchange Ct. CR0: C'don 74Sb 157
(off Bedford Pk.)
Exchange Ct. WC2 5G 223 (45Nb 90)
Exchange Garages N13 21Qb 50
Exchange Gdns. SW8 51Nb 112
Exchange Ho. E17 30Dc 52
Exchange Ho. EC2 7J 219 (43Ub 91)
(off Exchange Sq.)
Exchange Ho. NW10 38Xa 68
Exchange Ho. SW1 6D 228 (49Mb 90)
(off Vauxhall Bri. Rd.)
Exchange Mans. NW11 31Bb 69
Exchange Pl. EC2 7H 219 (43Ub 91)
Exchange Rd. SL5: S'hill 1A 146
(not continuous)
Exchange Rd. WD18: Wat 13X 27
Exchange Sq. EC2 7H 219 (43Ub 91)
Exchange St. EC1 4D 218 (41Sb 91)
Exchange St. RM1: Rom 29Gd 56
Exchange Wlk. HA5: Pinn 31Aa 65
Executive Pk. AL1: St A 2F 6
Exedown Rd. TN15: Wro 86Xd 184
Exeforde Av. TW15: Ashf 63Q 128
Exeter Cl. E6 44Pc 94
Exeter Cl. WD24: Wat 12Y 27
Exeter Ct. KT6: Surb 71Na 153
(off Maple Rd.)
Exeter Ct. NW6 40Cb 69
(off Cambridge Rd.)
Exeter Gdns. IG1: Ilf 32Nc 74
Exeter Ho. E14 43Ec 92
(off St Ives Pl.)
Exeter Ho. IG11: Bark 38Wc 75
(off Margaret Bondfield Av.)
Exeter Ho. N1 39Ub 71
(off New Era Est.)
Exeter Ho. RM8: Dag 35Wc 75
Exeter Ho. SE15 51Wb 113
(off Friary Est.)
Exeter Ho. SW15 58Ya 110
Exeter Ho. TW13: Hanw 61Ba 129
(off Watermill Way)
Exeter Ho. W2 44Eb 89
(off Hallfield Est.)
Exeter Ho. WD6: Bore 12Qa 29
Exeter Mans. NW2 37Ab 68
Exeter M. NW6 37Db 69
Exeter M. SW6 52Cb 111
Exeter Pl. SE26 62Wb 135
Exeter Rd. CR0: C'don 73Ub 157
Exeter Rd. DA12: Grav'nd 2F 144
Exeter Rd. DA16: Well 54Vc 117
Exeter Rd. E16 43Jc 93
Exeter Rd. E17 29Cc 52
Exeter Rd. EN3: Pond E 13Zb 34
Exeter Rd. HA2: Harr 33Aa 65
Exeter Rd. N14 18Kb 32
Exeter Rd. N9 19Yb 34
Exeter Rd. NW2 36Ab 68
Exeter Rd. RM10: Dag 37Dd 76
Exeter Rd. TW13: Hanw 62Ba 129
Exeter Rd. TW6: H'row A 55T 106
Exeter St. WC2 4G 223 (45Nb 90)
Exeter Way SE14 52Bc 114
Exeter Way TW6: H'row A 54U 106
Exford Ct. SW11 53Fb 111
(off Bolingbroke Wlk.)
Exford Gdns. SE12 60Kc 115
Exford Rd. SE12 61Kc 137
Exhibition Cl. W12 45Ya 88
Exhibition Grounds HA9: Wemb 35Ra 67
Exhibition Rd. SW7 2C 226 (47Fb 89)
Exhibition Way HA9: Wemb 35Qa 67
Exit Rd. N2 26Fb 49
Exmoor Cl. IG6: Ilf 25Sc 54
Exmoor Ho. DA17: Belv 47Dd 96
Exmoor Ho. E3 40Ac 72
(off Gernon Rd.)
Exmoor St. W10 42Za 88
Exmouth Ho. E14 49Dc 92
(off Cahir St.)
Exmouth Ho. EC1 5A 218 (42Qb 90)
(off Pine St.)
Exmouth Mkt. EC1 5K 217 (42Qb 90)
Exmouth M. NW1 4C 216 (41Lb 90)
Exmouth Pl. E8 38Xb 71
Exmouth Rd. DA16: Well 53Yc 117
Exmouth Rd. E17 29Bc 52
Exmouth Rd. HA4: Ruis 34Y 65
Exmouth Rd. RM17: Grays 51De 121
Exmouth Rd. UB4: Hayes 41U 84
Exmouth St. E1 44Yb 92
Exning Rd. E16 42Hc 93
Exon Apts. RM1: Rom 28Hd 56
(off Mercury Gdns.)
Exonbury NW8 39Db 69
(off Abbey Rd.)
Exon St. SE17 7H 231 (50Ub 91)
Explorer Av. TW19: Stanw 60N 105
Explorer Dr. WD18: Wat 16V 26
Explorers Ct. E14 45Fc 93
(off Newport Av.)
Export Ho. SE1 2J 231 (47Ub 91)
(off Tower Bri. Rd.)
Express Dr. IG3: Ilf 32Xc 75
Express Ho. SE8 51Ac 114
(off Rolt St.)
Express Newspapers SE1 6B 224 (46Rb 91)
(off Blackfriars Rd.)
Express Wharf E14 47Cc 92
(off Hutchings St.)
Exton Gdns. RM8: Dag 36Yc 75
Exton Rd. NW10 38Sa 67
Exton St. SE1 7K 223 (46Qb 90)
Eyebright Cl. CR0: C'don 74Zb 158
Eyhurst Av. RM12: Horn 34Jd 76
Eyhurst Cl. KT20: Kgswd 95Bb 195
Eyhurst Cl. NW2 33Wa 68
Eyhurst Pk. KT20: Kgswd 95Eb 195
Eyhurst Pl. CR5: Coul 88Kb 176
Eyhurst Spur KT20: Kgswd 96Bb 195
Eylewood Rd. SE27 64Sb 135
Eynella Rd. SE22 59Vb 113
Eynham Rd. W12 44Ya 88
EYNSFORD 75Nd 163
Eynsford Castle 75Nd 163
Eynsford Cl. BR5: Pet W 73Sc 160
Eynsford Cres. DA5: Bexl 60Yc 117
Eynsford Ho. SE1 2F 231 (47Tb 91)
(off Crosby Row)
Eynsford Ho. SE15 51Yb 114
Eynsford Ho. SE17 6H 231 (49Ub 91)
(off East St.)
Eynsford Ri. DA4: Eyns 77Md 163
Eynsford Rd. BR8: Crock 72Fd 162
Eynsford Rd. DA4: Eyns 74Pd 163
Eynsford Rd. DA4: Farni 74Pd 163
Eynsford Rd. DA9: Ghithe 57Yd 120
Eynsford Rd. IG3: Ilf 33Uc 74
Eynsford Rd. TN14: Eyns 81Kd 183
Eynsford Rd. TN14: S'ham 81Kd 183
Eynsford Station (Rail) 77Md 163
Eynsford Ter. UB7: Yiew 44P 83
Eynsham Dr. SE2 49Wc 95
Eynswood Dr. DA14: Sidc 64Xc 139
Eyot Gdns. W6 50Va 88
Eyot Grn. W4 50Va 88
Eyot Ho. SE16 48Wb 91
(off Frean St.)
Eyre Cl. RM2: Rom 28Kd 57
Eyre Ct. NW8 1B 214 (40Fb 69)
Eyre Grn. SL2: Slou 1E 80
Eyre Rd. NW8 3A 214 (41Eb 89)
Eyre St. Hill EC1 6K 217 (42Qb 90)
Eysham Ct. EN5: New Bar 15Db 31
Eyston Dr. KT13: Weyb 82Q 170
Eythorne Rd. SW9 53Qb 112
Ezra St. E2 41Vb 91

F

Faber Gdns. NW4 29Wa 48
Fabian Bell Twr. E3 40Cc 72
(off Pancras Way)
Fabian Rd. SW6 52Bb 111
Fabian St. E6 42Pc 94
Fable Apts. N1 3D 218 (41Sb 91)
Facade, The RH2: Reig 5J 207
Facade, The SE23 61Yb 136
Fackenden La. TN14: S'ham 85Kd 183
Factory La. CR0: C'don 74Qb 156
Factory La. N17 26Vb 51
Factory Rd. DA11: Nflt 58Ee 121
Factory Rd. E16 46Mc 93
Factory Sq. SW16 65Nb 134
Factory Yd. W7 46Ga 86
Faesten Way DA5: Bexl 62Gd 140
Faggotts Cl. WD7: R'lett 7La 14
Faggs Rd. TW14: Felt 56V 106
Fagus Av. RM13: Rain 41Md 97
Faints Cl. EN7: Chesh 1Vb 19
Fairacre HA5: Eastc 28W 44
Fairacre HP3: Hem H 6P 3
Fairacre KT3: N Mald 69Ua 132
Faircare Ct. HA6: Nwood 24U 44
Fairacre Pl. DA3: Hartl 69Ae 143
Fair Acres BR2: Broml 71Jc 159
Fair Acres CR0: Sels 81Bc 178
Fairacres HA4: Ruis 31V 64
Fairacres KT11: Cobh 84Z 171
Fairacres KT20: Tad 93Ya 194
Fairacres SW15 56Va 110
Fairacres Ct. EN6: Pot B 5Bb 17
Fairacres Ind. Est. SL4: Wind 4B 102
Fairbairn Cl. CR8: Purl 85Qb 176
Fairbairn Grn. SW9 53Rb 113
Fairbank Av. BR6: Farnb 75Rc 160
Fairbank Est. N1 2F 219 (40Tb 71)
Fairbanks Ct. HA0: Wemb 39Na 67
Fairbanks Lodge WD6: Bore 14Qa 29
Fairbanks Rd. N17 27Vb 51
Fairbourne KT11: Cobh 85Z 171
Fairbourne Cl. GU21: Wok 10L 167
Fairbourne Ho. UB3: Harl 48S 84
Fairbourne La. CR3: Cat'm 94Sb 197
Fairbourne Rd. N17 27Ub 51
Fairbourne Rd. SW4 58Mb 112
Fairbriar Ct. KT18: Eps 85Ua 174
(off Hereford Cl.)
Fairbriar Residence SW7 5A 226 (49Eb 89)
(off Stanhope Gdns.)
Fairbridge Rd. N19 33Mb 70
Fairbrook Cl. N13 22Qb 50
Fairbrook Rd. N13 23Qb 50
Fairburn Cl. WD6: Bore 11Qa 29
Fairburn Ct. SW15 57Ab 110
Fairburn Ho. W14 50Bb 89
(off Ivatt Pl.)
Fairby Grange DA3: Hartl 72Ae 165
Fairby Ho. SE1 6K 231 (49Vb 91)
(off Longfield Est.)
Fairby La. DA3: Hartl 72Ae 165
Fairby Rd. SE12 57Kc 115
Fairchild Cl. SW11 54Fb 111
Fairchildes Av. CR0: New Ad 84Fc 179
Fairchildes La. CR6: W'ham 86Fc 179
Fairchild Ho. E2 40Xb 71
(off Cambridge Cres.)
Fairchild Ho. E9 38Yb 72
(off Frampton Pk. Rd.)
Fairchild Ho. N1 3H 219 (41Ub 91)
(off Fanshaw St.)
Fairchild Ho. N3 25Cb 49
Fairchild Pl. EC2 6J 219 (42Ub 91)
(off Gt. Eastern St.)
Fairchild St. EC2 6J 219 (42Ub 91)
Fair Cl. WD23: Bush 17Da 27
Fairclough Cl. UB5: N'olt 42Ba 85
Fairclough St. E1 44Wb 91
Faircroft SL2: Slou 2F 80
Faircroft Ct. TW11: Tedd 65Ja 130
FAIR CROSS 36Uc 74
Faircross Av. IG11: Bark 37Sc 74
Faircross Av. RM5: Col R 24Fd 56
Faircross Ho. WD17: Wat 13X 27
(off The Parade)
Faircross Pde. IG11: Bark 36Uc 74
Faircross Way AL1: St A 1E 6
Fairdale Gdns. SW15 56Xa 110
Fairdale Gdns. UB3: Hayes 47W 84
Fairdene Rd. CR5: Coul 90Mb 176
Fairey Av. UB3: Harl 49V 84
Fairfax Av. KT17: Ewe 81Xa 174
Fairfax Av. RH1: Redh 5N 207
Fairfax Cl. KT12: Walt T 74X 151
Fairfax Cl. RH8: Oxt 2H 211
Fairfax Ct. DA1: Dart 58Qd 119
Fairfax Ct. NW6 38Eb 69
(off Fairfax Rd.)
Fairfax Gdns. SE3 53Lc 115
Fairfax Ho. KT1: King T 69Pa 131
(off Livesey Cl.)
Fairfax Mans. NW6 38Eb 69
(off Finchley Rd.)
Fairfax M. E16 46Kc 93
Fairfax M. N8 28Rb 51
Fairfax M. SW15 56Ya 110
Fairfax Pl. NW6 38Eb 69
Fairfax Pl. W14 48Ab 88
Fairfax Rd. GU22: Wok 92D 188
Fairfax Rd. N8 28Qb 50
Fairfax Rd. NW6 38Eb 69
Fairfax Rd. RM17: Grays 50De 99
Fairfax Rd. RM18: Tilb 3B 122
Fairfax Rd. TW11: Tedd 65Ja 130
Fairfax Rd. W4 48Ua 88
Fairfax Way N10 24Jb 50
Fairfield E1 43Yb 92
(off Redman's Rd.)
Fairfield KT1: King T 68Pa 131
FAIRFIELD 93Ka 192
Fairfield N20 17Fb 31
Fairfield NW1 1B 216 (39Lb 70)
(off Arlington Rd.)
Fairfield App. TW19: Wray 8P 103
Fairfield Av. HA4: Ruis 31S 64
Fairfield Av. HA8: Edg 23Ra 47
Fairfield Av. NW4 30Xa 48
Fairfield Av. RM14: Upm 34Sd 78
Fairfield Av. RM16: Grays 46Ee 99
Fairfield Av. SL3: Dat 2N 103
Fairfield Av. TW18: Staines 63H 127
Fairfield Av. TW2: Whitt 60Da 107
Fairfield Av. WD19: Wat 20Y 27
Fairfield Cl. CR4: Mitc 66Gb 133
Fairfield Cl. DA15: Sidc 58Vc 117
Fairfield Cl. EN3: Pond E 14Zb 34
Fairfield Cl. HA6: Nwood 22S 44
Fairfield Cl. KT19: Ewe 78Ua 154
Fairfield Cl. N12 21Eb 49
Fairfield Cl. RM12: Horn 32Jd 76
Fairfield Cl. SL3: Dat 2P 103
Fairfield Cl. TN15: Kems'g 90Qd 183
Fairfield Cl. WD7: R'lett 9Ja 14
Fairfield Cotts. KT23: Bookh 97Da 191
Fairfield Ct. HA4: Ruis 32T 64
Fairfield Ct. HA6: Nwood 26W 44
Fairfield Ct. KT22: Lea 93Ka 192
(off Leret Way)
Fairfield Ct. NW10 39Wa 68
Fairfield Cres. HA8: Edg 23Ra 47
Fairfield Dr. HA2: Harr 27Ea 46
Fairfield Dr. SW18 57Db 111
Fairfield Dr. UB6: G'frd 39La 66
Fairfield E. KT1: King T 68Na 131
Fairfield Gdns. N8 29Nb 50
Fairfield Gro. SE7 51Mc 115
Fairfield Halls Croydon 76Tb 157
Fairfield La. GU24: W End 4E 166
Fairfield La. SL2: Farn R 10F 60
Fairfield Nth. KT1: King T 68Na 131
Fairfield Pk. KT11: Cobh 86Z 171
Fairfield Path CR0: C'don 76Tb 157
Fairfield Pl. KT1: King T 69Na 131
Fairfield Pl. SL2: Farn R 9F 60
Fairfield Pool & Leisure Cen. 59Nd 119
Fairfield Rd. BR1: Broml 66Jc 137
Fairfield Rd. BR3: Beck 68Cc 136
Fairfield Rd. BR5: Pet W 72Tc 160
Fairfield Rd. CM14: B'wood 20Yd 40
Fairfield Rd. CM16: Epp 1Xc 23
Fairfield Rd. CR0: C'don 76Tb 157
Fairfield Rd. DA7: Bex 54Bd 117
Fairfield Rd. E17 26Ac 52
Fairfield Rd. E3 40Cc 72
Fairfield Rd. IG1: Ilf 37Rc 74
Fairfield Rd. IG8: Wfd G 23Jc 53
Fairfield Rd. KT1: King T 68Na 131
Fairfield Rd. KT22: Lea 93Ka 192
Fairfield Rd. N18 21Wb 51
Fairfield Rd. N8 29Nb 50
Fairfield Rd. SL1: Burn 1A 80
Fairfield Rd. TN15: Bor G 91Be 205
Fairfield Rd. TW19: Wray 8P 103
Fairfield Rd. UB1: S'hall 44Ba 85
Fairfield Rd. UB7: Yiew 45N 83
Fairfield Rd. UB8: Uxb 37M 63
Fairfield Road Bus Station 68Na 131
Fairfields DA12: Grav'nd 4G 144
Fairfields KT16: Chert 74J 149
Fairfields Cl. NW9 29Sa 47
Fairfields Cres. NW9 28Sa 47
Fairfield Sth. KT1: King T 68Na 131
Fairfield Sq. DA11: Grav'nd 8C 122
Fairfields Rd. TW3: Houn 55Ea 108
Fairfield St. SW18 57Db 111
Fairfield Trade Pk. KT1: King T 69Pa 131
Fairfield Wlk. EN8: Chesh 1Ac 20
Fairfield Wlk. KT22: Lea 93Ka 192
(off Fairfield Rd.)
Fairfield Way CR5: Coul 86Mb 176
Fairfield Way EN5: Barn 15Cb 31
Fairfield Way KT19: Ewe 78Ua 154
Fairfield W. KT1: King T 68Na 131
Fairfolds WD25: Wat 8Aa 13
Fairfoot Rd. E3 42Cc 92
Fairford SE6 60Cc 114
Fairford Av. CR0: C'don 71Zb 158
Fairford Av. DA7: Bex 53Fd 118
Fairford Cl. CR0: C'don 71Ac 158
Fairford Cl. KT14: W Byf 86H 169
Fairford Cl. RH2: Reig 4L 207
Fairford Cl. RM3: Rom 23Rd 57
Fairford Ct. SM2: Sutt 80Db 155
Fairford Gdns. KT4: Wor Pk 75Va 154
Fairford Ho. SE11 6A 230 (49Qb 90)
Fairford Way RM3: Rom 23Rd 57
Fairgreen EN4: Cockf 13Hb 31
Fairgreen Ct. EN4: Cockf 13Hb 31
Fairgreen E. EN4: Cockf 13Hb 31
Fairgreen Rd. CR7: Thor H 71Rb 157
Fairhall Ct. KT5: Surb 73Pa 153
Fairham Av. RM15: S Ock 45Wd 98
Fairhaven AL2: Park 9B 6
Fairhaven TW20: Egh 64B 126
Fairhaven Av. CR0: C'don 72Zb 158
Fairhaven Ct. CR2: S Croy 78Sb 157
(off Warham Rd.)
Fairhaven Ct. TW18: Staines 65G 126
(off Bowes Rd.)
Fairhaven Ct. TW20: Egh 64B 126
Fairhaven Cres. WD19: Wat 20W 26
Fairhaven Rd. RH1: Redh 2A 208
Fairhazel Gdns. NW6 37Db 69
Fairhazel Mans. NW6 38Eb 69
(off Fairhazel Gdns.)
Fairhill HP3: Hem H 6P 3
Fairholme TW14: Bedf 59T 106
Fairholme Av. RM2: Rom 29Jd 56
Fairholme Cl. N3 28Ab 48
Fairholme Ct. HA5: Hat E 23Ba 45
Fairholme Cres. KT21: Asht 89La 172
Fairholme Cres. UB4: Hayes 42V 84
Fairholme Gdns. N3 27Ab 48
Fairholme Gdns. RM14: Upm 31Vd 78
Fairholme Rd. CR0: C'don 73Qb 156
Fairholme Rd. HA1: Harr 29Ha 46
Fairholme Rd. IG1: Ilf 31Pc 74
Fairholme Rd. SM1: Sutt 79Bb 155
Fairholme Rd. TW15: Ashf 64N 127
Fairholme Rd. W14 50Ab 88
Fairholt Cl. N16 32Ub 71
Fairholt Rd. N16 32Tb 71
Fairholt St. SW7 3E 226 (48Gb 89)
Fairkytes Av. RM11: Horn 32Md 77
Fairland Ho. BR2: Broml 70Kc 137
Fairland Rd. E15 37Hc 73
Fairlands Av. CR7: Thor H 70Pb 134
Fairlands Av. IG9: Buck H 19Jc 35
Fairlands Av. SM1: Sutt 75Cb 155
Fairlands Ct. SE9 58Qc 116
Fair La. CR5: Coul 98Jb 196
Fairlane Dr. RM15: S Ock 42Xd 98
Fairlawn KT13: Weyb 78U 150
Fairlawn KT2: King T 65Sa 131
Fairlawn KT23: Bookh 96Ba 191
Fairlawn SE7 52Lc 115
Fairlawn Av. DA7: Bex 54Zc 117
Fairlawn Av. N2 28Gb 49
Fairlawn Av. W4 49Sa 87
Fairlawn Cl. KT10: Clay 79Ha 152
Fairlawn Cl. KT2: King T 65Sa 131
Fairlawn Cl. N14 16Lb 32
Fairlawn Cl. TW13: Hanw 63Ba 129
Fairlawn Ct. SE7 52Lc 115
(not continuous)
Fairlawn Ct. W4 49Sa 87
Fairlawn Dr. IG8: Wfd G 24Jc 53
Fairlawn Dr. RH1: Redh 8N 207
FAIRLAWNE 99Yd 204
Fairlawnes SM6: W'gton 78Kb 156
Fairlawn Gdns. UB1: S'hall 45Ba 85
Fairlawn Gro. SM7: Bans 85Fb 175
Fairlawn Gro. W4 49Sa 87
Fairlawn Mans. SE14 53Zb 114
Fairlawn Pk. GU21: Wok 86A 168
Fairlawn Pk. SE26 64Ac 136
Fairlawn Pk. SL4: Wind 6C 102
Fairlawn Rd. SM5: Cars 83Eb 175
Fairlawn Rd. SM7: Bans 84Gb 175
Fairlawn Rd. SW19 66Bb 133
Fairlawns CM14: B'wood 20Wd 40
Fairlawns CM16: Epp 1Xc 23
Fairlawns HA5: Pinn 26Z 45
Fairlawns KT15: Add 78K 149
Fairlawns KT15: Wdhm 83H 169
Fairlawns TW1: Twick 58La 108
Fairlawns TW16: Sun 69W 128
Fairlawns WD17: Wat 10V 12
Fairlawns Cl. RM11: Horn 31Pd 77
Fairlawns Cl. TW18: Staines 65K 127
Fairlead Ho. E14 48Cc 92
(off Alpha Gro.)
Faireads Ho. E17 28Zb 52
(off Wickford Way)
Fairlea Pl. W5 42La 86
Fairley Way EN7: Chesh 1Xb 19
Fairlie Ct. E3 41Dc 92
(off Stroudley Wlk.)
Fairlie Gdns. SE23 59Yb 114
Fairlie Rd. SL1: Slou 4E 80
Fairlight TW12: Hamp H 64Da 129
Fairlight Av. E4 19Fc 35
Fairlight Av. IG8: Wfd G 23Jc 53
Fairlight Av. NW10 40Ua 68
Fairlight Av. SL4: Wind 4H 103
Fairlight Cl. E4 19Fc 35
Fairlight Cl. KT4: Wor Pk 77Ya 154
Fairlight Ct. NW10 40Ua 68
Fairlight Ct. UB6: G'frd 40Ea 66
Fairlight Cross DA3: Lfield 69De 143
Fairlight Dr. UB8: Uxb 37M 63
Fairlight Rd. SW17 63Fb 133
Fairline Ct. BR3: Beck 68Ec 136
FAIRLOP 25Uc 54
Fairlop Cl. RM12: Horn 37Kd 77
Fairlop Ct. E11 32Fc 73
Fairlop Gdns. IG6: Ilf 24Sc 54
Fairlop Outdoor Activity Cen. 25Vc 55
Fairlop Rd. E11 31Fc 73
Fairlop Rd. IG6: Ilf 26Sc 54
Fairlop Station (Underground) 25Tc 54
Fairlop Waters Country Pk. 26Vc 55
Fairlop Waters Golf Course 25Uc 54
Fairmark Dr. UB10: Hil 37Q 64
Fairmead BR1: Broml 70Pc 138
Fairmead GU21: Wok 10N 167
Fairmead KT5: Surb 74Ra 153
Fairmead Cl. BR1: Broml 70Pc 138
Fairmead Cl. KT3: N Mald 69Ta 131
Fairmead Cl. TW5: Hest 52Z 107
Fairmead Ct. TW9: Rich 54Ra 109
Fairmead Cres. HA8: Edg 20Sa 29
Fairmead Gdns. IG4: Ilf 29Nc 54
Fairmead Ho. E9 35Ac 72
Fairmead Rd. CR0: C'don 73Pb 156
Fairmead Rd. IG10: H Beech 15Kc 35
Fairmead Rd. IG10: Lough 15Kc 35
Fairmead Rd. N19 34Mb 70
Fairmeads IG10: Lough 12Rc 36
Fairmeads KT11: Cobh 85Ba 171
Fairmeadside IG10: Lough 15Lc 35
FAIRMILE 84Ba 171
Fairmile Av. KT11: Cobh 86Aa 171
Fairmile Av. SW16 64Mb 134
Fairmile Ct. KT11: Cobh 84Aa 171
Fairmile Ho. TW11: Tedd 63Ja 130
Fairmile La. KT11: Cobh 84Z 171
Fairmile Pk. Copse KT11: Cobh 85Ba 171
Fairmile Pk. Rd. KT11: Cobh 85Ba 171
Fairmont Av. E14 46Fc 93
Fairmont Cl. DA17: Belv 50Bd 95
Fairmont Ho. E3 42Cc 92
(off Wellington Way)
Fairmont Ho. SE16 47Zb 92
(off Needleman St.)
Fairmont M. NW2 33Cb 69
Fairmount Rd. SW2 58Pb 112
Fairoak Cl. BR5: Pet W 73Rc 160
Fairoak Cl. CR8: Kenley 87Rb 177
Fairoak Cl. KT22: Oxs 84Fa 172
Fairoak Dr. SE9 57Tc 116
Fairoak Gdns. RM1: Rom 26Gd 56
Fairoak Gro. EN3: Enf W 9Zb 20
Fairoak La. KT22: Oxs 84Ea 172
Fairoak La. KT9: Chess 83Ka 172
Fair Oak Pl. IG6: Ilf 26Sc 54
Fairoaks Airport 82A 168
Fairoaks Ct. KT15: Add 78K 149
(off Liberty La.)
Fairoaks Gro. EN3: Enf W 9Zb 20
FAIRSEAT 84Ee 185
Fairseat Cl. WD23: B Hea 19Ga 28
Fairseat La. TN15: Stans 81Ce 185
Fairseat La. TN15: Wro 87De 185
Fairs Rd. KT22: Lea 91Ja 192
Fairstead Lodge IG8: Wfd G 23Jc 53
(off Snakes La. W.)
Fairstead Wlk. N1 39Sb 71
(off Popham St.)
Fair St. SE1 1J 231 (47Ub 91)
Fair St. TW3: Houn 55Ea 108
Fairthorne Vs. SE7 50Jc 93
(off Felltram Way)
Fairthorn Rd. SE7 50Jc 93
Fairtrough Rd. BR6: Prat B 84Xc 181
Fairview DA3: Fawk 76Xd 164
Fairview DA8: Erith 52Hd 118
Fairview EN6: Pot B 1Db 17
Fairview HA4: Ruis 35Y 65
Fairview KT17: Ewe 83Ya 174
Fairview Av. CM13: Hut 17Fe 41
Fairview Av. GU22: Wok 90A 168
Fairview Av. HA0: Wemb 37Ma 67
Fairview Av. RM13: Rain 40Md 77
Fairview Av. SS17: Stan H 2L 101
Fairview Chase SS17: Stan H 3L 101
Fairview Cl. E17 25Ac 52
Fairview Cl. GU22: Wok 90B 168
Fairview Cl. IG7: Chig 21Uc 54
Fairview Cl. SE26 64Ac 136
Fairview Ct. NW4 26Za 48
Fairview Ct. TW15: Ashf 64Q 128
Fairview Cres. HA2: Harr 32Ca 65
Fairview Dr. BR6: Orp 77Tc 160
Fairview Dr. IG7: Chig 21Uc 54
Fairview Dr. TW17: Shep 71P 149
Fairview Dr. WD17: Wat 8U 12
Fairview Est. NW10 41Sa 87
Fairview Gdns. IG8: Wfd G 25Kc 53
Fairview Ho. SW2 59Pb 112
Fairview Ind. Pk. RM13: Rain 43Fd 96
Fairview Pl. SW2 59Pb 112
Fairview Rd. DA13: Ist R 66Fe 143
Fairview Rd. DA13: Nflt G 66Fe 143

Fairview Rd. EN2: Enf....11Qb 32
Fairview Rd. IG7: Chig....21Uc 54
Fairview Rd. KT17: Ewe....83Va 174
Fairview Rd. N15....29Vb 51
Fairview Rd. SL2: Slou....2D 80
Fairview Rd. SM1: Sutt....78Fb 155
Fairview Rd. SW16....67Pb 134
Fairviews RH8: Oxt....5L 211
Fairview Vs. E4....24Cc 52
Fairview Way HA8: Edg....21Qa 47
Fairwall Ho. SE5....53Ub 113
Fairwater Av. DA16: Well....56Wc 117
Fairwater Dr. KT15: New H....81M 169
Fairwater Dr. TW17: Shep....71S 150
Fairwater Ho. E16....47Kc 93
(off Bonnet St.)
Fairwater Ho. TW11: Tedd....63Ja 130
Fairway BR5: Pet W....71Tc 160
Fairway DA6: Bex....57Ad 117
Fairway GU25: Vir W....2N 147
Fairway HP3: Hem H....6P 3
Fairway IG8: Wfd G....22Lc 53
Fairway KT16: Chert....74K 149
Fairway RM16: Grays....46De 99
Fairway SM5: Cars....83Eb 175
Fairway SW20....69Ya 132
Fairway, The BR1: Broml....71Pc 160
Fairway, The DA11: Grav'nd....1C 144
Fairway, The EN5: New Bar....16Db 31
Fairway, The GU3: Worp....6G 186
Fairway, The HA0: Wemb....34Ka 66
Fairway, The HA4: Ruis....35Y 65
Fairway, The HA6: Nwood....21U 44
Fairway, The KT13: Weyb....83Q 170
Fairway, The KT22: Lea....90Ja 172
Fairway, The KT3: N Mald....67Ta 131
Fairway, The KT8: W Mole....69Da 129
Fairway, The N13....20Sb 33
Fairway, The N14....16Kb 32
Fairway, The NW7....20Ta 29
Fairway, The RM14: Upm....31Sd 78
Fairway, The SL1: Burn....10A 60
Fairway, The UB10: Hil....41P 83
Fairway, The UB5: N'olt....37Ea 66
Fairway, The W3....44Ua 88
Fairway, The WD5: Ab L....4T 12
Fairway Av. NW9....27Ra 47
Fairway Av. UB7: W Dray....46L 83
Fairway Av. WD6: Bore....12Ra 29
Fairway Cl. TW4: Houn Amberley Way....57Y 107
Fairway Cl. TW4: Houn Islay Gdns....57Z 107
Fairway Cl. AL2: Park....9A 6
Fairway Cl. CR0: C'don....71Ac 158
Fairway Cl. GU22: Wok....1M 187
Fairway Cl. KT10: Surb....76Ka 152
Fairway Cl. KT19: Ewe....77Sa 153
Fairway Cl. NW11....31Eb 69
Fairway Cl. UB7: W Dray....46M 83
Fairway Ct. E3....41Ec 92
(off Culvert Dr.)
Fairway Ct. EN5: New Bar....16Db 31
Fairway Ct. HP3: Hem H....6P 3
Fairway Ct. NW7....20Ta 29
Fairway Ct. SE16....47Zb 92
(off Christopher Cl.)
Fairway Dr. DA2: Dart....59Rd 119
Fairway Dr. SE28....44Zc 95
Fairway Dr. UB6: G'frd....38Da 65
Fairway Gdns. BR3: Beck....72Fc 159
Fairway Gdns. IG1: Ilf....36Sc 74
Fairway Ho. WD6: Bore....13Ra 29
(off Eldon Av.)
Fairways CR8: Kenley....89Sb 177
Fairways E17....28Ec 52
Fairways EN9: Walt A....6Gc 21
Fairways HA7: Stan....26Na 47
Fairways KT24: Eff J....95V 190
Fairways TW11: Tedd....66Ma 131
Fairways TW15: Ashf....65R 128
Fairways TW7: Isle....53Fa 108
Fairways, The RH1: Redh....9M 207
Fairways Bus. Pk. E10....33Ac 72
FAIRWAYS DAY CEN....64J 127
Fairway Trad. Est. TW4: Houn....57Y 107
Fairweather Cl. DA16: Well....57Wc 117
Fairweather Cl. N15....28Ub 51
Fairweather Ct. N13....20Pb 32
Fairweather Ho. N7....35Nb 70
Fairweather Rd. N16....30Wb 51
Fairwell La. KT24: W Hor....100R 190
Fairwyn Rd. SE26....63Ac 136
Faith Cl. CR5: Coul....89Lb 176
Faith Cl. RM2: Rom....29Kd 57
Faith Ct. E3....40Cc 72
(off Lefevre Wlk.)
Faith Ct. SE1....50Vb 91
(off Cooper's Rd.)
Faithfield WD23: Bush....16Aa 27
Faith M. E12....35Mc 73
Fakenham Cl. NW7....24Wa 48
Fakenham Cl. UB5: N'olt....37Ba 65
Fakruddin St. E1....42Wb 91
Falaise TW20: Egh....64A 126
Falcon WC1....7G 217 (43Nb 90)
(off Old Gloucester St.)
Falcon Av. BR1: Broml....70Nc 138
Falcon Av. RM15: S Ock....42Xd 98
Falcon Av. RM17: Grays....52De 121
Falconberg M. W1....2D 222 (44Mb 90)
Falcon Bus. Cen. RM3: Rom....24Nd 57
Falcon Cl. AL10: Hat....2C 8
Falcon Cl. DA1: Dart....57Pd 119
Falcon Cl. EN9: Walt A....6Jc 21
Falcon Cl. HA6: Nwood....24U 44
Falcon Cl. W4....51Sa 109
Falcon Ct. E18....27Kc 53
(off Albert Rd.)
Falcon Ct. EC4....3A 224 (44Qb 90)
Falcon Ct. EN5: New Bar....14Eb 31
Falcon Ct. GU21: Wok....85E 168
Falcon Ct. HA4: Ruis....33U 64
Falcon Ct. N1....2C 218 (40Rb 71)
(off City Gdn. Row)
Falcon Cres. EN3: Pond E....15Zb 34
Falcondal Ct. NW10....41Qa 87
Falcon Dr. TW19: Stanw....58M 105
Falconer Ct. N17....24Sb 51
(off Compton Cres.)
Falconer Rd. IG6: Ilf....22Xc 55
Falconer Rd. WD23: Bush....16Ba 27
Falconer Wlk. N7....33Pb 70
Falconet Ct. E1....46Xb 91
(off Wapping High St.)
Falcon Gro. SW11....55Gb 111
Falcon Highwalk EC2....1D 224 (43Sb 91)
(off Aldersgate St.)
Falcon Ho. BR1: Broml....67Hc 137
Falcon Ho. E14....50Dc 92
(off St Davids Sq.)
Falcon Ho. NW6....39Db 69
(off Springfield Wlk.)
Falcon Ho. SW5....50Db 89
(off Old Brompton Rd.)
Falconhurst KT22: Oxs....87Fa 172
Falcon La. SW11....55Gb 111
Falcon Lodge W9....43Cb 89
(off Admiral Wlk.)
Falcon M. DA11: Nflt....10A 122
Falcon Pk. Community Sports Cen.....54Hb 111
Falcon Pk. Ind. Est. NW10....35Ua 68
Falcon Point SE1....5C 224 (45Rb 91)
Falcon Rd. EN3: Pond E....15Zb 34
Falcon Rd. SW11....54Gb 111
Falcon Rd. TW12: Hamp....66Ba 129
Falconry Ct. CM16: Epp....2Vc 23
Falconry Ct. KT1: King T....69Na 131
(off Fairfield Sth.)
Falcons Cl. TN16: Big H....89Mc 179
Falcon St. E13....42Jc 93
Falcon Ter. SW11....55Gb 111
Falcon Way E11....28Jc 53
Falcon Way E14....49Dc 92
Falcon Way HA3: Kenton....29Na 47
Falcon Way NW9....26Ua 48
Falcon Way RM12: Horn....38Jd 76
Falcon Way TW14: Felt....57X 107
Falcon Way TW16: Sun....68U 128
Falcon Way WD25: Wat....6Aa 13
Falcon Wharf SW11....54Fb 111
Falcon Wood KT22: Lea....92Ha 192
FALCONWOOD....56Vc 117
Falconwood KT24: E Hor....96V 190
FALCONWOOD....56Tc 116
Falconwood TW20: Egh....64A 126
Falconwood Av. DA16: Well....54Tc 116
Falconwood Ct. SE3....54Hc 115
(off Montpelier Row)
Falconwood Pde. DA16: Well....56Uc 116
Falconwood Rd. CR0: Sels....81Bc 178
Falconwood Station (Rail)....56Tc 116
Falcourt Cl. SM1: Sutt....78Db 155
Faldo Ct. CM14: B'wood....20Xd 40
Falkirk Cl. RM11: Horn....32Qd 77
Falkirk Ct. SE16....46Zb 92
(off Rotherhithe St.)
Falkirk Gdns. WD19: Wat....22Z 45
Falkirk Ho. W9....40Db 69
(off Maida Vale)
Falkirk St. N1....2J 219 (40Ub 71)
Falkland Av. N11....21Kb 50
Falkland Av. N3....24Cb 49
Falkland Ho. SE6....63Ec 136
Falkland Ho. W14....50Bb 89
(off Edith Vs.)
Falkland Ho. W8....48Db 89
Falkland Pk. Av. SE25....69Ub 135
Falkland Pl. NW5....36Lb 70
Falkland Rd. EN5: Barn....12Ab 30
Falkland Rd. N8....28Qb 50
Falkland Rd. NW5....36Lb 70
Fallaize Av. IG1: Ilf....35Rc 74
Falling La. UB7: Yiew....45N 83
Fallodon Way NW11....28Cb 49
Fallodon Ho. W11....43Bb 89
(off Tavistock Cres.)
Fallow Cl. IG7: Chig....22Vc 55
FALLOW CORNER....24Eb 49
Fallow Ct. SE16....50Wb 91
(off Argyle Way)
Fallow Ct. Av. N12....24Eb 49
Fallowfield DA2: Bean....62Xd 142
Fallowfield HA7: Stan....21Ja 46
Fallowfield Cl. UB9: Hare....25L 43
Fallowfield Ct. HA7: Stan....20Ja 28
Fallow Flds. IG10: Lough....16Lc 35
Fallowfields Dr. N12....23Gb 49
Fallowhurst Path N3....24Eb 49
Fallow Pl. TW11: Tedd....64Ga 130
Fallows Cl. N2....26Fb 49
Fallsbrook Rd. SW16....65Kb 134
Falman Cl. N9....18Wb 33
Falmer Rd. E17....27Dc 52
Falmer Rd. EN1: Enf....14Ub 33
Falmer Rd. N15....29Sb 51
Falmouth Av. E4....22Fc 53
Falmouth Cl. N22....24Pb 50
Falmouth Cl. SE12....57Hc 115
Falmouth Ct. AL3: St A....1A 6
Falmouth Gdns. IG4: Ilf....28Mc 53
Falmouth Ho. HA5: Hat E....24Ba 45
Falmouth Ho. KT2: King T....67Ma 131
(off Skerne Rd.)
Falmouth Ho. SE11....7A 230 (50Qb 90)
(off Seaton Cl.)
Falmouth Ho. W2....4D 220 (45Gb 89)
(off Clarendon Pl.)
Falmouth Rd. KT12: Hers....77Y 151
Falmouth Rd. SE1....4E 230 (48Sb 91)
Falmouth Rd. SL1: Slou....4E 80
Falmouth St. E15....36Fc 73
Falmouth Wlk. SW15....58Wa 110
Falmouth Way E17....29Bc 52
Falstaff Bldg. E1....45Xb 91
(off Cannon St. Rd.)
Falstaff Cl. DA1: Cray....59Gd 118
Falstaff Ct. SE11....6B 230 (49Rb 91)
(off Opal St.)
Falstaff Gdns. AL1: St A....5P 5
Falstaff Ho. N1....2H 219 (40Ub 71)
(off Regan Way)
Falstaff M. DA9: Ghithe....58Xd 120
Falstaff M. TW12: Hamp H....64Fa 130
(off High St.)
Falstone GU21: Wok....10M 167
Fambridge Cl. SE26....63Bc 136
Fambridge Ct. RM7: Rom....29Fd 56
(off Marks Rd.)
Fambridge Rd. RM8: Dag....32Cd 76
Famet Av. CR8: Purl....85Sb 177
Famet Cl. CR8: Purl....85Sb 177
Famet Gdns. CR8: Kenley....85Sb 177
Famet Wlk. CR8: Purl....85Sb 177
Family Court East London....46Cc 92
Family Court West London....58W 106
Fancourt M. BR1: Broml....69Qc 138
Fane St. W14....51Bb 111
Fangrove Pk. KT16: Lyne....74C 148
Fan Mus., The....52Ec 114
Fanns Ri. RM19: Purf....49Qd 97
Fann St. EC1....6D 218 (42Sb 91)
Fann St. EC2....6D 218 (42Sb 91)
(not continuous)
Fanshawe Av. IG11: Bark....37Sc 74
Fanshawe Cres. RM11: Horn....30Md 57
Fanshawe Cres. RM9: Dag....36Ad 75
Fanshawe Rd. RM16: Grays....8C 100
Fanshawe Rd. TW10: Ham....63La 130
Fanshaw St. N1....3H 219 (41Ub 71)
FANTAIL, THE....76Pc 160
Fantail Cl. SE28....44Yc 95
Fantasia Ct. CM14: W'ley....22Xd 58
Fanthorpe St. SW15....55Ya 110
Faraday Av. DA14: Sidc....61Wc 139
Faraday Cl. N7....37Pb 70
Faraday Cl. SL2: Slou....3F 80
Faraday Cl. WD18: Wat....16T 26
Faraday Ct. WD18: Wat....16W 26
Faraday Ho. E14....45Bc 92
(off Brightlingsea Pl.)
Faraday Ho. EN3: Enf L....9Bc 20
(off Velocity Way)
Faraday Ho. HA9: Wemb....34Sa 67
Faraday Ho. SE1....2F 231 (47Tb 91)
(off Cole St.)
Faraday Ho. SW11....51Kb 112
Faraday Ho. WD18: Wat....16T 26
Faraday Lodge SE10....48Hc 93
Faraday Mans. W14....51Ab 110
(off Queen's Club Gdns.)
Faraday Mus., The....5B 222 (45Lb 90)
Faraday Pl. KT8: W Mole....70Ca 129
Faraday Rd. DA16: Well....55Wc 117
Faraday Rd. E15....37Hc 73
Faraday Rd. KT8: W Mole....70Ca 129
Faraday Rd. SL2: Slou....3F 80
Faraday Rd. SW19....65Cb 133
Faraday Rd. UB1: S'hall....45Da 85
Faraday Rd. W10....43Ab 88
Faraday Rd. W3....45Sa 87
Faraday Way BR5: St M Cry....70Xc 139
Faraday Way CR0: Wadd....74Pb 156
Faraday Way SE18....48Mc 93
Fardell Ct. AL1: St A....2D 6
(off Newsom Pl.)
Fareham Ho. HP1: Hem H....3M 3
Fareham Rd. TW14: Felt....59Y 107
Far End AL10: Hat....3D 8
Farewell Pl. CR4: Mitc....67Gb 133
Fari Ct. E17....28Cc 52
(off Tower M.)
Faringdon Av. BR2: Broml....73Qc 160
Faringdon Av. RM3: Rom....25Ld 57
Faringford Cl. EN6: Pot B....3Fb 17
Faringford Rd. E15....38Gc 73
Farington Acres KT13: Weyb....76T 150
Faris Barn Dr. KT15: Wdhm....84H 169
Faris La. KT15: Wdhm....83H 169
Farjeon Ho. NW6....38Fb 69
(off Hilgrove Rd.)
Farjeon Rd. SE3....53Mc 115
Farland Rd. HP2: Hem H....2B 4
FARLEIGH....86Bc 178
Farleigh Av. BR2: Hayes....73Hc 159
FARLEIGH COMMON....86Ac 178
Farleigh Ct. CR2: S Croy....78Sb 157
Farleigh Court Golf Course....84Cc 178
Farleigh Ct. Rd. CR6: W'ham....86Bc 178
Farleigh Dean Cres. CR0: Sels....83Dc 178
Farleigh Ho. N1....38Rb 71
(off Halton Rd.)
Farleigh Pl. N16....35Vb 71
Farleigh Rd. CR6: W'ham....90Zb 178
Farleigh Rd. KT15: New H....83J 169
Farleigh Rd. N16....35Vb 71
Farleton Cl. KT13: Weyb....79T 150
FARLEY COMMON....98Rc 200
Farley Ct. NW1....6G 215 (42Hb 89)
(off Allsop Pl.)
Farley Ct. W14....48Bb 89
Farleycroft TN16: Westrm....98Sc 200
Farley Dr. IG3: Ilf....32Uc 74
Farley Ho. SE26....62Xb 135
Farley La. TN16: Westrm....98Rc 200
Farley M. SE6....59Ec 114
Farley Nursery TN16: Westrm....99Sc 200
Farley Pk. RH8: Oxt....2H 211
Farley Pl. SE25....70Wb 135
Farley Rd. CR2: Sels....80Xb 157
Farley Rd. DA12: Grav'nd....10H 123
Farley Rd. SE6....59Dc 114
Farleys Cl. KT24: W Hor....98S 190
Farlington Pl. SW15....59Xa 110
Farlow Cl. DA11: Nflt....2B 144
Farlow Rd. SW15....55Za 110
Farlton Rd. SW18....60Db 111
Farman Gro. UB5: N'olt....41Z 85
Farman Ter. HA3: Kenton....28Ma 47
Farm Av. BR8: Swan....69Ed 140
Farm Av. HA0: Wemb....37La 66
Farm Av. HA2: Harr....31Ba 65
Farm Av. NW2....34Ab 68
Farm Av. SW16....63Nb 134
Farmborough Cl. HA1: Harr....31Fa 66
Farm Cl. BR4: W W'ck....76Hc 159
Farm Cl. CM13: Hut....17Ee 41
Farm Cl. CR5: Chips....92Hb 195
Farm Cl. EN5: Barn....15Ya 30
Farm Cl. EN8: Chesh....2Yb 20
Farm Cl. GU3: Worp....10G 186
Farm Cl. IG9: Buck H....20Lc 35
Farm Cl. KT14: Byfl....84P 169
Farm Cl. KT16: Lyne....72C 148
Farm Cl. KT22: Fet....96Fa 192
Farm Cl. KT24: E Hor....100V 190
Farm Cl. RM10: Dag....38Ed 76
Farm Cl. SL5: S'hill....1A 146
Farm Cl. SM2: Sutt....80Fb 155
Farm Cl. SM6: W'gton....82Lb 176
Farm Cl. SW6....52Cb 111
Farm Cl. TW17: Shep....73Q 150
Farm Cl. TW18: Staines....64G 126
Farm Cl. UB1: S'hall....45Da 85
Farm Cl. UB10: Ick....33R 64
Farm Cl. WD6: Bore....10Ma 15
Farm Cl. WD7: Shenl....2Na 15
Farmcote Rd. SE12....60Jc 115
Farm Cotts. BR8: Crock....71Hd 162
Farm Ct. NW4....27Wa 48
Farm Cres. AL2: Lon C....8E 6
Farm Cres. SL2: Slou....3M 81
Farmcroft DA11: Grav'nd....1C 144
Farmdale Rd. SE10....50Jc 93
Farmdale Rd. SM5: Cars....80Gb 155
Farm Dr. CR0: C'don....75Bc 158
Farm Dr. CR8: Purl....84Mb 176
Farm Dr. RM16: Grays....45Ee 99
Farm Dr. SL4: Old Win....8M 103
Farm End E4....15Gc 35
Farm End HA6: Nwood....25R 44
Farm Rd. E10....32Dc 72
Farmer Cl. WD25: Wat....5X 13
Farmers Ct. EN9: Walt A....5Jc 21
Farmer's Rd. SE5....52Rb 113
Farmers Rd. TW18: Staines....64G 126
Farmer St. W8....46Cb 89
Farm Fld. WD17: Wat....10U 12
Farmfield Rd. BR1: Broml....64Gc 137
Farm Flds. CR2: Sande....83Ub 177
Farm Hill Rd. EN9: Walt A....5Fc 21
Farm Holt DA3: New A G....74Be 165
Farmhouse Cl. GU22: Pyr....87F 168
Farm Ho. Ct. NW7....24Wa 48
Farmhouse Rd. SW16....66Lb 134
Farmilo Rd. E17....31Bc 72
Farmington Av. SM1: Sutt....76Fb 155
Farmlands EN2: Enf....11Qb 32
Farmlands HA5: Eastc....28W 44
Farmlands, The UB5: N'olt....37Ba 65
Farmland Wlk. BR7: Chst....64Rc 138
Farm La. CR0: C'don....75Bc 158
Farm La. CR8: Purl....82Lb 176
Farm La. GU23: Send....96E 188
Farm La. KT15: Add....80J 149
Farm La. KT18: Eps D....91Ra 193
Farm La. KT21: Asht....89Qa 173
Farm La. KT24: E Hor....100V 190
Farm La. N14....17Kb 32
Farm La. SL1: Slou....5H 81
Farm La. SW6....51Cb 111
Farm La. WD3: Loud....13L 25
Farmleigh N14....17Lb 32
Farmleigh Gro. KT12: Hers....78V 150
Farmleigh Ho. SW9....57Rb 113
Farm M. CR4: Mitc....68Kb 134
Farm Pl. DA1: Cray....56Jd 118
Farm Pl. W8....46Cb 89
Farm Rd. AL1: St A....1F 6
Farm Rd. CR6: W'ham....91Ac 198
Farm Rd. GU22: Wok....92D 188
Farm Rd. HA6: Nwood....22R 44
Farm Rd. HA8: Edg....23Ra 47
Farm Rd. KT10: Esh....74Da 151
Farm Rd. N21....18Sb 33
Farm Rd. NW10....39Ta 67
Farm Rd. RM13: Rain....41Ld 97
Farm Rd. RM16: Ors....7B 100
Farm Rd. RM18: E Til....9L 101
Farm Rd. SM2: Sutt....80Fb 155
Farm Rd. SM4: Mord....71Db 155
Farm Rd. TN14: S'oaks....92Ld 203
Farm Rd. TW18: Staines....65K 127
Farm Rd. TW4: Houn....60Aa 107
Farm Rd. WD3: Chor....14C 24
Farmside Pl. KT19: Eps....84Pa 173
Farmstead KT19: Eps....81Qa 173
Farmstead Ct. SM6: W'gton....78Kb 156
(off Melbourne Rd.)
Farmstead Rd. HA3: Hrw W....25Fa 46
Farmstead Rd. SE6....63Dc 136
Farm St. W1....5K 221 (45Kb 90)
Farm Va. DA5: Bexl....58Dd 118
Farm Vw. KT11: Cobh....88Z 171
Farm Vw. KT20: Lwr K....99Bb 195
Farm Wlk. NW11....29Bb 49
Farm Way HA6: Nwood....21U 44
Farm Way HP2: Hem H....1P 3
Farm Way IG9: Buck H....21Lc 53
Farm Way KT4: Wor Pk....76Ya 154
Farm Way RM12: Horn....35Ld 77
Farm Way TW19: Stanw M....58H 105
Farm Way WD23: Bush....14Da 27
Farmway RM8: Dag....34Yc 75
Farm Yd. SL4: Wind....2H 103
Farmyard Funworld....13Ba 27
Farnaby Dr. TN13: S'oaks....98Hd 202
Farnaby Ho. W10....41Bb 89
(off Bruckner St.)
Farnaby Rd. BR1: Broml....66Gc 137
Farnaby Rd. BR2: Broml....66Gc 137
Farnaby Rd. SE9....56Lc 115
Farnan Av. E17....26Cc 52
Farnan Lodge SW16....64Nb 134
Farnan Rd. SW16....64Nb 134
FARNBOROUGH....78Sc 160
Farnborough Av. CR2: Sels....81Zb 178
Farnborough Av. E17....27Ac 52
Farnborough Cl. HA9: Wemb....33Ra 67
Farnborough Comn. BR6: Farnb....76Pc 160
Farnborough Cres. BR2: Hayes....74Hc 159
Farnborough Cres. CR2: Sels....81Ac 178
Farnborough Hill BR6: Chels....78Tc 160
Farnborough Hill BR6: Farnb....78Tc 160
Farnborough Ho. SW15....60Wa 110
Farnborough Way BR6: Chels....78Sc 160
Farnborough Way BR6: Farnb....78Sc 160
Farnburn Av. SL1: Slou....3F 80
Farncombe St. SE16....47Wb 91
Farndale Av. N13....19Rb 33
Farndale Ct. SE18....52Nc 116
Farndale Cres. UB6: G'frd....41Ea 86
Farndale Ho. NW6....39Db 69
(off Kilburn Vale)
Farne Ho. WD18: Wat....16V 26
(off Scammell Way)
Farnell M. KT13: Weyb....76R 150
Farnell M. SW5....50Db 89
Farnell Pl. W3....45Ra 87
Farnell Rd. TW18: Staines....62J 127
Farnell Rd. TW7: Isle....55Fa 108
Farnes Dr. RM2: Rom....26Ld 57
Farnfield Ct. CR2: S Croy....78Rb 157
Farnham Cl. HP3: Bov....10C 2
Farnham Cl. N20....17Eb 31
FARNHAM COMMON....7G 60
Farnham Ct. SM3: Cheam....79Ab 154
Farnham Ct. UB1: S'hall....45Ea 86
(off Redcroft Rd.)
Farnham Gdns. SW20....68Xa 132
Farnham Ho. NW1....6E 214 (42Gb 89)
(off Harewood Av.)
Farnham Ho. SE1....1D 230 (47Sb 91)
(off Union St.)
Farnham La. SL2: Slou....1C 80
Farnham Pk. Golf Course
Buckinghamshire....9J 61
Farnham Pk. La. SL2: Farn R....8G 60
Farnham Pl. SE1....7C 224 (46Rb 91)
Farnham Rd. DA16: Well....54Yc 117
Farnham Rd. IG3: Ilf....31Vc 75
Farnham Rd. RM3: Rom....22Md 57
Farnham Rd. SL1: Slou....3G 80
Farnham Rd. SL2: Farn R....1F 80
Farnham Rd. SL2: Slou....1F 80
Farnham Royal SE11....50Pb 90
FARNHAM ROYAL....1G 80
FARNINGHAM....73Pd 163
Farningham Ct. SW16....66Mb 134
Farningham Cres. CR3: Cat'm....95Wb 197
Farningham Hill Rd. DA4: Farni....71Ld 163
Farningham Ho. N4....31Tb 71
Farningham Rd. CR3: Cat'm....95Wb 197
Farningham Rd. N17....24Wb 51
Farningham Road Station (Rail)....68Rd 141
Farnley GU21: Wok....9K 167
Farnley Ho. SW8....54Mb 112
Farnley Rd. E4....17Gc 35
Farnley Rd. SE25....70Tb 135
Farnol Rd. DA1: Dart....57Qd 119
Farnsworth Ct. SE10....48Hc 93
(off West Parkside)
Farnsworth Dr. HA8: Edg....21Na 47
Farnworth Ho. E14....49Fc 93
(off Manchester Rd.)
Faro Cl. BR1: Broml....68Qc 138
Faroe Rd. W14....48Za 88
Farorna Wlk. EN2: Enf....11Qb 32
Farquhar Rd. SE19....64Vb 135
Farquhar Rd. SW19....62Cb 133
Farquharson Rd. CR0: C'don....74Sb 157
Farraline Rd. WD18: Wat....14X 27
Farrance Rd. RM6: Chad H....31Ad 75
Farrance St. E14....44Cc 92
Farrans Ct. HA3: Kenton....31Ka 66
Farrant Av. N22....26Qb 50
Farrant Cl. BR6: Chels....80Wc 161
Farrant Way WD6: Bore....11Na 29
Farr Av. IG11: Bark....40Wc 75
Farrell Ct. DA2: Wilm....62Ld 141
Farrell Ho. E1....44Yb 92
(off Ronald St.)
Farren Rd. SE23....61Ac 136
Farrer Ct. TW1: Twick....59Ma 109
Farrer Ho. SE8....52Cc 114
Farrer Ho. SL4: Eton....10G 80
(off Common La.)
Farrer M. N8....28Lb 50
Farrer Rd. HA3: Kenton....29Na 47
Farrer Rd. N8....28Lb 50
Farrer's Pl. CR0: C'don....77Zb 158
Farrier Cl. BR1: Broml....69Mc 137
Farrier Cl. TW16: Sun....70W 128
Farrier Cl. UB8: Hil....44Q 84
Farrier Ct. E1....43Zb 92
(off White Horse La.)
Farrier Ct. RM13: Rain....40Fd 76
(off Lowen Rd.)
Farrier Pl. SM1: Sutt....76Db 155
Farrier Rd. UB5: N'olt....40Ca 65
Farriers Cl. DA12: Grav'nd....10H 123
Farriers Cl. HP3: Bov....10D 2
Farriers Cl. KT17: Eps....84Ua 174
Farriers Ct. WD25: Wat....4X 13
Farriers Ho. EC1....6E 218 (42Sb 91)
(off Errol St.)
Farriers M. SE15....55Yb 114
Farriers M. SW9....56Qb 112
Farriers Rd. KT17: Eps....83Ua 174
Farrier St. NW1....38Kb 70
Farriers Way WD6: Bore....15Ta 29
Farriers Yd. W6....50Za 88
(off Smiths Sq.)
Farrier Wlk. SW10....51Eb 111
Farringdon Ho. TW9: Kew....52Ra 109
Farringdon Ho. UB7: W Dray....47P 83
Farringdon La. EC1....6A 218 (42Qb 90)
Farringdon Rd. EC1....5K 217 (42Qb 90)
Farringdon Station (Rail, Underground & Crossrail)....7B 218 (43Rb 91)
Farringdon St. EC4....1B 224 (43Rb 91)
Farringford Ct. AL2: Chis G....8N 5
Farrington Av. BR5: St P....69Xc 139
Farrington Av. WD23: Bush....14Da 27
Farrington Ct. BR1: Broml....68Kc 137
(off Widmore Rd.)
Farrington Pl. BR7: Chst....66Tc 138
Farrington Pl. HA6: Nwood....21V 44
Farrins Rents SE16....46Ac 92
Farrow Gdns. RM16: Grays....46De 99
Farrow Ho. NW9....27Wa 48
Farrow La. SE14....52Yb 114
Farrow Pl. SE16....48Ac 92
Farr Rd. EN2: Enf....11Tb 33
Farrs M. WD3: Rick....18M 25
Farsby Ho. Apts. IG11: Bark....42Wc 95
(off Manwell La.)
Farthingale Ct. EN9: Walt A....6Jc 21
Farthingale La. EN9: Walt A....6Jc 21
Farthingale Wlk. E15....38Fc 73
Farthing All. SE1....47Wb 91
Farthing Barn La. BR6: Downe....81Qc 180
Farthing Cl. DA1: Dart....56Pd 119
Farthing Cl. KT22: Lea....92Ha 192
Farthing Cl. WD18: Wat....15Y 27
Farthing Ct. NW7....24Ab 48
Farthing Ct. SL2: S'dale....3F 146
(off Halfpenny La.)
Farthingfield TN15: Wro....88Ce 185
Farthing Flds. E1....46Xb 91
Farthing Grn. La. SL2: Stoke P....10L 61
Farthings GU21: Knap....8J 167
Farthings, The HP1: Hem H....2K 3
Farthings, The KT2: King T....67Qa 131
Farthings Cl. E4....20Gc 35
Farthings Cl. HA5: Eastc....30X 45
Farthing St. BR6: Downe....80Pc 160
FARTHING STREET....81Pc 180
Farthing Way CR5: Coul....89Mb 176
Farwell Rd. DA14: Sidc....63Xc 139
Farwig La. BR1: Broml....67Hc 137
Fashion & Textile Mus....1J 231 (47Ub 91)
Fashion St. E1....1K 225 (43Vb 91)
Fashoda Rd. BR2: Broml....70Mc 137
Fassett Rd. E8....37Wb 71
Fassett Rd. KT1: King T....70Ma 131
Fassett Sq. E8....37Wb 71
Fastrack Mnr. Way DA9: Ghithe....56Zd 121
Fathom Ct. E16....45Kc 93
(off Basin App.)
Fauconberg Ct. W4....51Sa 109
(off Fauconberg Rd.)
Fauconberg Rd. W4....51Sa 109
Faulkner Cl. RM8: Dag....31Zc 75
Faulkner Ct. AL1: St A....1C 6
(off Boundary Rd.)
Faulkner Ho. BR7: Chst....66Tc 138
Faulkner Ho. W6....51Ya 110
Faulkner M. E17....25Ac 52
Faulkners All. EC1....7B 218 (43Rb 91)
Faulkners Rd. KT12: Hers....79Y 151
Faulkner St. SE14....53Yb 114
Fauna Cl. HA7: Stan....21Ma 47
Fauna Cl. RM6: Chad H....30Yc 55
Faunce Ho. SE17....51Rb 113
(off Doddington Gro.)
Faunce St. SE17....50Rb 91
Favart Rd. SW6....53Cb 111
Faversham Av. E4....18Gc 35
Faversham Av. EN1: Enf....16Tb 33
Faversham Cl. IG7: Chig....19Xc 37
Faversham Ct. SE6....61Bc 136
Faversham Ho. CR2: Sels....82Yb 178
Faversham Ho. NW1....1C 216 (39Lb 70)
(off Bayham Pl.)
Faversham Ho. SE17....7H 231 (50Ub 91)
(off Kinglake St.)
Faversham Rd. BR3: Beck....68Bc 136

Faversham Rd. SE6 59Bc 114
Faversham Rd. SM4: Mord 72Db 155
Fawcett Cl. SW11 54Fb 111
Fawcett Cl. SW16 64Qb 134
Fawcett Ct. SW10 51Eb 111
(off Fawcett St.)
Fawcett Est. E5 32Wb 71
Fawcett Rd. CR0: C'don 76Sb 157
Fawcett Rd. NW10 38Va 68
Fawcett Rd. SL4: Wind 3F 102
Fawcett St. SW10 51Eb 111
Fawcus Cl. KT10: Clay 79Ga 152
Fawe Pk. M. SW15 56Bb 111
Fawe Pk. Rd. SW15 56Bb 111
Fawe St. E14 43Dc 92
FAWKE COMMON 99Qd 203
Fawke Comn. TN15: God G 98Qd 203
Fawke Comn. TN15: Under 98Qd 203
Fawkes Av. DA1: Dart 61Pd 141
Fawke Wood Rd. TN15: Under 100Qd 203
FAWKHAM 73Xd 164
Fawkham Av. DA3: Lfield 69Ee 143
FAWKHAM GREEN 76Xd 164
Fawkham Grn. Rd. DA3: Fawk 76Xd 164
Fawkham Ho. SE1 6K 231 (49Vb 91)
(off Longfield Est.)
FAWKHAM MANOR BMI HOSPITAL 74Yd 164
Fawkham Rd. DA3: Fawk 70Zd 143
Fawkham Rd. DA3: Fawk 77Vd 164
Fawkham Rd. DA3: Lfield 70Zd 143
Fawkham Rd. TN15: W King 81Wd 184
Fawley Lodge E14 49Fc 93
(off Millennium Dr.)
Fawley Rd. NW6 36Db 69
Fawnbrake Av. SE24 57Rb 113
Fawn Hgts. IG9: Buck H 19Kc 35
(off Stag La.)
Fawn Rd. E13 40Lc 73
Fawn Rd. IG7: Chig 22Vc 55
Fawns Mnr. Cl. TW14: Bedf 60S 106
Fawns Mnr. Rd. TW14: Bedf 60T 106
Fawood Av. NW10 38Sa 67
Fawsley Cl. SL3: Poyle 52G 104
Fawters Cl. CM13: Hut 16Fe 41
Fayerfield EN6: Pot B 3Fb 17
Faygate Cres. DA6: Bex 57Cd 118
Faygate Rd. SW2 61Pb 134
Fay Grn. WD5: Ab L 5T 12
Fayland Av. SW16 64Lb 134
Faymore Gdns. RM15: S Ock 44Wd 98
Fazeley Ct. W9 43Cb 89
(off Elmfield Way)
Fazeley Ho. UB5: N'olt 41Ba 85
(off Taywood Rd.)
Feacey Down HP1: Hem H 1J 3
Fearn Cl. KT24: E Hor 100U 190
Fearney Mead WD3: Rick 18J 25
Fearnley Cres. TW12: Hamp 64Aa 129
Fearnley Ho. SE5 54Ub 113
Fearnley St. WD18: Wat 14X 27
Fearns Ho. W13 46Ja 86
Fearns Mead CM14: W'ley 22Yd 58
Fearon St. SE10 50Jc 93
Featherbed La. AL2: Pot C 7K 5
Featherbed La. CR0: Sels 80Bc 158
Featherbed La. CR6: W'ham 85Fc 179
Featherbed La. HP3: Hem H 7J 3
(not continuous)
Featherbed La. RM4: Abr 14Zc 37
(not continuous)
Featherbed La. WD5: Bedm 8H 5
Feather M. E1 43Wb 91
Feathers La. TW19: Wray 61C 126
Feathers Pl. SE10 51Fc 115
Featherstone Av. SE23 61Xb 135
Featherstone Ct. UB2: S'hall 48Z 85
Featherstone Gdns. WD6: Bore 14Ta 29
Featherstone Ho. UB4: Yead 43Y 85
Featherstone Ind. Est. UB2: S'hall 48Aa 85
(off Feather Rd.)
Featherstone Rd. NW7 23Xa 48
Featherstone Rd. UB2: S'hall 48Aa 85
Featherstone Sports Cen. 49Z 85
Featherstone St. EC1 5F 219 (42Tb 91)
Featherstone Ter. UB2: S'hall 48Aa 85
Featley Rd. SW9 55Rb 113
Federal Rd. UB6: G'frd 39La 66
Federal Way WD24: Wat 11Y 27
Federation Rd. SE2 49Xc 95
Fee Farm Rd. KT10: Clay 80Ha 152
Feenan Highway RM18: Tilb 2D 122
Feeny Cl. NW10 35Va 68
Felbridge Av. HA7: Stan 25Ja 46
Felbridge Cl. SM2: Sutt 81Db 175
Felbridge Cl. SW16 63Qb 134
Felbridge Ct. TW13: Felt 60X 107
(off High St.)
Felbridge Ct. UB3: Harl 51T 106
Felbridge Ho. SE22 55Ub 113
Felbrigge Rd. IG3: Ilf 33Vc 75
Felcott Cl. KT12: Hers 76Y 151
Felcott Rd. KT12: Hers 76Y 151
Felday Rd. SE13 58Dc 114
FELDEN 6J 3
Felden Cl. HA5: Hat E 24Aa 45
Felden Cl. WD25: Wat 6Z 13
Felden Dr. HP3: Hem H 6J 3
Felden La. HP3: Hem H 5H 3
Felden Lawns HP3: Hem H 6J 3
Felden St. SW6 53Bb 111
Feldman Cl. N16 32Wb 71
Feldspar Ct. EN3: Enf H 13Ac 34
(off Enstone Rd.)
Feldspar M. N13 22Rb 51
Felgate M. W6 49Xa 88
Felhampton Rd. SE9 61Rc 138
Felhurst Cres. RM10: Dag 35Dd 76
Felicia Way RM16: Grays 9D 100
Feline Ct. EN4: E Barn 16Gb 31
Felipe Rd. RM16: Chaf H 48Yd 98
Felix Av. N8 30Nb 50
Felix Ct. E17 29Dc 52
Felix Ct. NW9 26Ua 48
Felix Dr. GU4: W Cla 100J 189
Felix Ho. E16 45Qc 94
(off University Way)
Felix La. TW17: Shep 72U 150
Felix Mnr. BR7: Chst 65Uc 138
Felix Neubergh Ho. EN1: Enf 14Ub 33
Felix Pl. SW2 57Qb 112
(off Talma Rd.)
Felix Point E14 44Cc 92
(off Upper Nth. St.)
Felix Rd. KT12: Walt T 72W 150
Felix Rd. W13 45Ja 86
Felixstowe Ct. E16 46Rc 94
Felixstowe Rd. N17 27Vb 51
Felixstowe Rd. N9 20Wb 33
Felixstowe Rd. NW10 41Xa 88
Felixstowe Rd. SE2 48Xc 95
Felix St. E2 40Xb 71
Felland Way RH2: Reig 10M 207
Fellbrigg Rd. SE22 57Vb 113
Fellbrigg St. E1 42Xb 91
Fellbrook TW10: Ham 62Ka 130
Fellmongers Path SE1 2K 231 (47Vb 91)
(off Tower Bri. Rd.)
Fellmongers Yd. CR0: C'don 76Sb 157
Fellowes Cl. UB4: Yead 42Z 85
Fellowes Cl. WD25: Wat 8Y 13
Fellowes La. AL4: Col H 5P 7
Fellowes Rd. SM5: Cars 76Gb 155
Fellow Grn. GU24: W End 5D 166
Fellow Grn. Rd. GU24: W End 5D 166
Fellows Ct. E2 1K 219 (40Vb 71)
(not continuous)
Fellowship Cl. RM8: Dag 35Wc 75
Fellowship Ho. E6 40Nc 74
(off St Bartholomew's Rd.)
Fellows Rd. NW3 38Fb 69
Fell Path WD6: Bore 15Ta 29
(off Clydesdale Cl.)
Fell Rd. CR0: C'don 76Sb 157
Felltram M. SE7 50Jc 93
Felltram Way SE7 50Jc 93
Fell Wlk. HA8: Edg 25Sa 47
Felmersham Cl. SW4 56Nb 112
Felmingham Rd. SE20 68Yb 136
Felnex Av. SM6: W'gton 75Jb 156
Felnex Trad. Est. NW10 40Ta 67
Felnex Trad. Est. SM6: W'gton 75Jb 156
Felsberg Rd. SW2 58Nb 112
Fels Cl. RM10: Dag 34Dd 76
Fels Farm Av. RM10: Dag 34Ed 76
Felsham M. SW15 55Za 110
(off Felsham Rd.)
Felsham Rd. SW15 55Ya 110
Felspar Cl. SE18 50Vc 95
Felstead Av. IG5: Ilf 25Qc 54
Felstead Cl. CM13: Hut 16Ee 41
Felstead Cl. N13 22Qb 50
Felstead Gdns. E14 50Ec 92
Felstead Rd. BR6: Chels 75Wc 161
Felstead Rd. E11 31Jc 73
Felstead Rd. E9 37Bc 72
Felstead Rd. EN8: Walt C 4Ac 20
Felstead Rd. IG10: Lough 17Nc 36
Felstead Rd. KT19: Eps 83Ta 173
Felstead Rd. RM5: Col R 24Ed 56
Felstead St. E9 37Bc 72
Felstead Way SL2: Slou 2E 80
Felstead Wharf E14 50Ec 92
Felsted Rd. E16 44Mc 93
FELTHAM 60X 107
Feltham Av. KT8: E Mos 70Ga 130
Felthambrook Ind. Est. TW13: Felt 62X 129
Felthambrook Way TW13: Felt 62X 129
Feltham Bus. Complex TW13: Felt 61X 129
Feltham Corporate Cen. TW13: Felt 62X 129
FELTHAMHILL 64V 128
Feltham Hill Rd. TW15: Ashf 64Q 128
Feltham Rd. CR4: Mitc 68Hb 133
Feltham Rd. RH1: Redh 10P 207
Feltham Rd. TW15: Ashf 63Q 128
Feltham Station (Rail) 60X 107
Feltham Wlk. RH1: Redh 10P 207
Felton Cl. BR5: Pet W 72Rc 160
Felton Cl. WD6: Bore 10Na 15
Felton Gdns. IG11: Bark 39Uc 74
Felton Hall Ho. SE16 47Wb 91
(off George Row)
Felton Ho. N1 39Tb 71
(off Colville Est.)
Felton Lea DA14: Sidc 64Vc 139
Felton Rd. IG11: Bark 40Uc 74
Felton Rd. W13 47La 86
Felton St. N1 39Tb 71
Fenbridge Ct. RH1: Redh 4B 208
Fencepiece Rd. IG6: Chig 23Sc 54
Fencepiece Rd. IG6: Ilf 23Sc 54
Fencepiece Rd. IG7: Chig 22Sc 54
Fenchurch Av. EC3 3H 225 (44Ub 91)
Fenchurch Bldgs. EC3 3J 225 (44Ub 91)
Fenchurch Ho. EC3 3K 225 (44Vb 91)
(off Minories)
Fenchurch M. E3 43Bc 92
(off St Paul's Way)
Fenchurch Pl. EC3 3J 225 (44Ub 91)
Fenchurch St. EC3 4H 225 (45Ub 91)
Fenchurch Street Station (Rail) 4K 225 (45Vb 91)
Fen Cl. CM15: Shenf 14Ee 41
Fen Ct. EC3 3H 225 (44Ub 91)
Fendall Rd. KT19: Ewe 78Sa 153
Fendall St. SE1 4J 231 (48Ub 91)
(not continuous)
Fendt Cl. E16 44Hc 93
Fendyke Rd. DA17: Belv 49Zc 95
Fenelon Pl. W14 49Bb 89
Fenemore Rd. CR8: Kenley 92Tb 197
Fengate Cl. KT9: Chess 79Ma 153
Fengates Rd. RH1: Redh 6N 207
Fen Gro. DA15: Sidc 57Vc 117
Fenham Rd. SE15 52Wb 113
Fenland Ho. E5 33Yb 72
Fen La. RM14: Bulp 36Yd 78
Fen La. RM14: N Ock 36Yd 78
Fen La. RM16: Ors 1B 100
Fen La. SW13 53Xa 110
Fenman Ct. N17 25Xb 51
Fenman Gdns. IG3: Ilf 32Xc 75
Fen Mdw. TN15: Igh 90Yd 184
Fenn Cl. BR1: Broml 65Jc 137
Fennel Apts. SE1 7K 225 (46Vb 91)
(off Cayenne Ct.)
Fennel Cl. CR0: C'don 74Zb 158
Fennel Cl. E16 42Gc 93
Fennells Mead KT17: Ewe 81Va 174
Fennell St. SE18 51Qc 116
Fenner Cl. SE16 49Xb 91
Fenner Ho. E1 46Xb 91
(off Watts St.)
Fenner Ho. KT12: Hers 77W 150
Fenner Rd. RM16: Chaf H 49Yd 98
Fenners Marsh DA12: Grav'nd 10H 123
Fenner Sq. SW11 55Fb 111
Fenn Ho. TW7: Isle 53Ka 108
Fennings Rd. SW4 58Mb 112
Fenning St. SE1 1H 231 (47Ub 91)
Fennscombe Ct. GU24: W End 5C 166
Fenns La. GU24: W End 5C 166
Fenn St. E9 36Zb 72
Fenns Way GU21: Wok 87A 168
Fen Pond Cotts. TN15: Igh 90Yd 184
Fen Pond Rd. TN15: Igh 88Yd 184
Fen Pond Rd. TN15: Wro 88Yd 184
Fensomes All. HP2: Hem H 1M 3
Fensomes Cl. HP2: Hem H 1M 3
Fenstanton N4 32Pb 70
(off Marquis Rd.)
Fenstanton Av. N12 23Fb 49
Fen St. E16 45Hc 93
Fens Way BR8: Hext 65Jd 140
Fenswood Cl. DA5: Bexl 58Cd 118
Fentiman Rd. SW8 51Nb 112
Fentiman Way HA2: Harr 33Da 65
Fentiman Way RM11: Horn 32Nd 77
Fenton Av. TW18: Staines 65L 127
Fenton Cl. BR7: Chst 64Pc 138
Fenton Cl. E8 37Vb 71
Fenton Cl. RH1: Redh 6A 208
Fenton Cl. SW9 54Pb 112
Fenton Ho. SE14 52Ac 114
Fenton Ho. TW5: Hest 51Ca 107
Fenton House 34Eb 69
(off Hampstead Gro.)
Fenton Pde. SE10 50Hc 93
(off Woolwich Rd.)
Fenton Rd. HA2: Harr 27Ea 46
Fenton Rd. N17 24Sb 51
Fenton Rd. RH1: Redh 6A 208
Fenton Rd. RM16: Chaf H 48Ae 99
Fentons Av. E13 41Kc 93
Fenton St. E1 44Xb 91
Fenwick Cl. GU21: Wok 10M 167
Fenwick Cl. SE18 51Qc 116
Fenwick Gro. SE15 55Wb 113
Fenwick Ho. EC1 1C 224 (43Rb 91)
(off Little Britain)
Fenwick Path WD6: Bore 10Pa 15
Fenwick Pl. CR2: S Croy 80Rb 157
Fenwick Pl. SW9 55Nb 112
Fenwick Rd. SE15 55Wb 113
Ferby Ct. DA14: Sidc 63Vc 139
(off Main Rd.)
Ferdinand Ct. SE6 59Cc 114
(off Adenmore Rd.)
Ferdinand Dr. SE15 52Ub 113
Ferdinand Ho. NW1 38Jb 70
(off Ferdinand Pl.)
Ferdinand Magellan Ct. E16 47Kc 93
(Ferdinand Magellan Court)
Ferdinand Pl. NW1 38Jb 70
Ferdinand St. NW1 38Jb 70
Ferguson Av. DA12: Grav'nd 3E 144
Ferguson Av. KT5: Surb 71Pa 153
Ferguson Av. RM2: Rom 26Ld 57
FERGUSON CEN., THE 30Ac 52
Ferguson Cl. BR2: Broml 69Fc 137
Ferguson Cl. E14 49Cc 92
Ferguson Ct. RM2: Rom 26Md 57
Ferguson Dr. W3 44Ta 87
Ferguson Gro. EN8: Chesh 1Zb 20
Fergus Rd. N5 36Rb 71
Fergusson M. SW4 54Nb 112
Fergus St. SE10 48Jc 93
Ferial Ct. SE15 52Wb 113
(off Fenham Rd.)
Fermain Ct. E. N1 39Ub 71
(off Hertford Rd.)
Fermain Ct. Nth. N1 39Ub 71
(off De Beauvoir Est.)
Fermain Ct. W. N1 39Ub 71
(off De Beauvoir Est.)
Ferme Pk. Rd. N4 30Pb 50
Ferme Pk. Rd. N8 29Nb 50
Fermor Rd. SE23 60Ac 114
Fermoy Ho. W9 42Bb 89
(off Fermoy Rd.)
Fermoy Rd. UB6: G'frd 42Da 85
Fermoy Rd. W9 42Bb 89
(not continuous)
Fern Av. CR4: Mitc 70Mb 134
Fernbank DA4: Eyns 75Pd 163
Fernbank IG9: Buck H 18Kc 35
Fernbank Av. HA0: Wemb 35Ha 66
Fernbank Av. KT12: Walt T 73Aa 151
Fernbank Av. RM12: Horn 35Ld 77
Fernbank M. SW12 58Lb 112
Fernbank Rd. KT15: Add 78J 149
Fernbrook Av. DA15: Sidc 57Uc 116
Fernbrook Cres. SE13 58Gc 115
(off Leahurst Rd.)
Fernbrook Dr. HA2: Harr 31Da 65
Fernbrook Rd. SE13 57Gc 115
Ferncliff Rd. E8 36Wb 71
Fern Cl. CR6: W'ham 90Ac 178
Fern Cl. DA8: Erith 53Kd 119
Fern Cl. N1 1H 219 (40Ub 71)
Fern Copse KT23: Bookh 97Ba 191
Fern Ct. DA7: Bex 56Cd 118
Fern Ct. RM5: Col R 24Cd 56
Fern Ct. RM7: Rom 29Fd 56
Fern Ct. SE14 54Zb 114
Fern Ct. SS17: Stan H 1M 101
Ferncroft Av. HA4: Ruis 33Y 65
Ferncroft Av. N12 23Hb 49
Ferncroft Av. NW3 34Cb 69
Ferndale BR1: Broml 68Lc 137
Ferndale TN13: S'oaks 94Ld 203
Ferndale Av. E17 29Fc 53
Ferndale Av. KT16: Chert 76G 148
Ferndale Av. TW4: Houn 55Aa 107
Ferndale Cl. DA7: Bex 53Ad 117
Ferndale Community Sports Cen. 55Pb 112
Ferndale Cres. SM5: Cars 74Hb 155
Ferndale Cres. UB8: Cowl 41L 83
Ferndale Rd. DA12: Grav'nd 1D 144
Ferndale Rd. E11 33Gc 73
Ferndale Rd. E7 38Kc 73
Ferndale Rd. EN3: Enf W 9Ac 20
Ferndale Rd. GU21: Wok 88B 168
Ferndale Rd. N15 30Vb 51
Ferndale Rd. RM5: Col R 26Ed 56
Ferndale Rd. SE25 71Xb 157
Ferndale Rd. SM7: Bans 88Bb 175
Ferndale Rd. SW4 56Nb 112
Ferndale Rd. SW9 56Pb 112
Ferndale Rd. TW15: Ashf 64M 127
Ferndale St. E6 45Rc 94
Ferndale Ter. HA1: Harr 28Ha 46
Ferndale Way BR6: Farnb 78Tc 160
Ferndell Av. DA5: Bexl 62Fd 140
Fern Dells AL10: Hat 1B 8
Fern Dene W13 43Ka 86
Ferndene AL2: Brick W 3Ba 13
Ferndene DA3: Lfield 69Fe 143
Ferndene Rd. SE24 56Sb 113
Fernden Way RM7: Rom 30Dd 56
Ferndown HA6: Nwood 26W 44
Ferndown NW1 38Mb 70
(off Camley St.)
Ferndown RM11: Horn 30Pd 57
Ferndown Av. BR6: Orp 74Tc 160
Ferndown Cl. HA5: Pinn 24Aa 45
Ferndown Cl. SM2: Sutt 79Fb 155
Ferndown Cl. SS17: Stan H 3K 101
Ferndown Ct. UB1: S'hall 44Ea 86
(off Haldane Rd.)
Ferndown Gdns. KT11: Cobh 85Y 171
Ferndown Lodge E14 48Ec 92
(off Manchester Rd.)
Ferndown Rd. SE9 59Mc 115
Ferndown Rd. WD19: Wat 21Y 45
Fern Dr. HP3: Hem H 3N 3
Fernecroft AL1: St A 5B 6
Fernery, The TW18: Staines 64G 126
Fernes Cl. UB8: Cowl 44L 83
Ferney Ct. KT14: Byfl 83M 169
Ferney Meade Way TW7: Isle 54Ja 108
Ferney Rd. EN4: E Barn 17Jb 32
Ferney Rd. KT14: Byfl 84M 169
Fern Gro. TW14: Felt 59X 107
Ferngrove Cl. KT22: Fet 95Ga 192
Fern Hall AL10: Hat 2B 8
(off Bishops Ri.)
Fernhall Dr. IG4: Ilf 29Mc 53
Fernhall La. EN9: Walt A 3Mc 21
Fernham Rd. CR7: Thor H 69Sb 135
Fernhead Rd. W9 41Bb 89
Fernheath Way DA2: Wilm 64Fd 140
Fernhill KT22: Oxs 86Fa 172
Fernhill Cl. GU22: Wok 2N 187
Fernhill Ct. E17 26Fc 53
Fernhill Gdns. KT2: King T 64Ma 131
Fernhill La. GU22: Wok 2N 187
Fernhill Pk. GU22: Wok 2N 187
Fern Hill Pl. BR6: Farnb 78Sc 160
Fernhills WD4: Hunt C 6T 12
Fernhill St. E16 46Pc 94
Fernholme Rd. SE15 57Zb 114
Fernhurst Gdns. HA8: Edg 23Qa 47
Fernhurst Rd. CR0: C'don 73Xb 157
Fernhurst Rd. SW6 53Ab 110
Fernhurst Rd. TW15: Ashf 63S 128
Fernie Cl. IG7: Chig 22Wc 55
Fernie Way IG7: Chig 22Wc 55
Fernihough Cl. KT13: Weyb 82Q 170
Fernlands Cl. KT16: Chert 76G 148
Fern La. TW5: Hest 50Ba 85
Fernlea KT23: Bookh 96Da 191
Fernlea Pl. KT11: Cobh 83Z 171
Fernlea Rd. CR4: Mitc 68Jb 134
Fernlea Rd. SW12 60Kb 112
Fernleigh Cl. CR0: Wadd 77Qb 156
Fernleigh Cl. KT12: Walt T 76X 151
Fernleigh Cl. W9 41Bb 89
Fernleigh Ct. HA2: Harr 26Da 45
Fernleigh Ct. HA9: Wemb 33Na 67
Fernleigh Ct. RM7: Rom 29Ed 56
Fernleigh Rd. N21 19Qb 32
Fernly Cl. HA5: Eastc 28W 44
Ferns, The TN15: Plat 92Ee 205
Fernsbury St. WC1 4K 217 (41Qb 90)
Ferns Cl. CR2: Sande 82Xb 177
Ferns Cl. EN3: Enf W 8Ac 20
Fernshaw Cl. SW10 51Eb 111
Fernshaw Mans. SW10 51Eb 111
(off Fernshaw Rd.)
Fernshaw Rd. SW10 51Eb 111
Fernside IG9: Buck H 18Kc 35
Fernside KT7: T Ditt 74Ka 152
Fernside NW11 33Cb 69
Fernside SL2: Slou 5M 81
Fernside Av. NW7 20Ta 29
Fernside Av. TW13: Felt 63X 129
Fernside Ct. NW4 26Za 48
Fernside Rd. SW12 60Hb 111
Fernsleigh Cl. SL9: Chal P 23A 42
Ferns Rd. E15 37Hc 73
Fern St. E3 42Cc 92
Fernthorpe Rd. SW16 65Lb 134
Ferntower Rd. N5 36Tb 71
Fern Towers CR3: Cat'm 97Wb 197
Fernville La. HP2: Hem H 2M 3
Fern Wlk. SE16 50Wb 91
Fern Wlk. TW15: Ashf 64M 127
Fern Way WD25: Wat 7X 13
Fernways IG1: Ilf 35Rc 74
Fernwood CR0: Sels 81Ac 178
Fernwood SW19 60Bb 111
Fernwood Av. HA0: Wemb 37La 66
Fernwood Av. SW16 63Mb 134
Fernwood Cl. BR1: Broml 68Lc 137
Fernwood Ct. N14 17Lb 32
Fernwood Cres. N20 20Hb 31
Fernwood Pl. KT10: Hin W 75Ha 152
Ferny Hill EN4: Had W 9Gb 17
Ferranti Cl. SE18 48Mc 93
Ferraro Cl. TW5: Hest 51Ca 107
Ferrers Av. SM6: Bedd 77Mb 156
Ferrers Av. UB7: W Dray 47M 83
Ferrers Cl. SL1: Slou 6C 80
Ferrers Rd. SW16 64Mb 134
Ferrestone Rd. N8 28Pb 50
Ferrey M. SW9 54Qb 112
Ferriby Cl. N1 38Qb 70
Ferrie Cl. NW9 25Va 48
Ferrier Ind. Est. SW18 56Db 111
(off Ferrier St.)
Ferrier Point E16 43Jc 93
(off Forty Acre La.)
Ferrier St. SW18 56Db 111
Ferriers Way KT18: Tatt C 90Ya 174
Ferring Cl. HA2: Harr 32Ea 66
Ferrings SE21 62Ub 135
Ferris Av. CR0: C'don 76Bc 158
Ferris Rd. SE22 56Wb 113
Ferron Rd. E5 34Xb 71
Ferro Rd. RM13: Rain 42Jd 96
Ferry Av. TW18: Staines 66G 126
Ferrybridge Ho. SE11 4K 229 (48Qb 90)
(off Lambeth Wlk.)
Ferrydale Lodge NW4 28Ya 48
(off Parson St.)
Ferryhills Cl. WD19: Wat 20Y 27
Ferry Ho. E5 32Xb 71
(off Harrington Hill)
Ferry Island Retail Pk. 27Wb 51
Ferry La. KT16: Chert 72J 149
Ferry La. N17 28Wb 51
Ferry La. RM13: Rain 44Gd 96
Ferry La. SW13 51Va 110
Ferry La. TW17: Shep 74Q 150
Ferry La. TW18: Lale 69L 127
Ferry La. TW19: Wray 61D 126
Ferry La. TW8: Bford 51Na 109
Ferry La. TW9: Kew 51Pa 109
Ferry La. Ind. Est. RM13: Rain 43Hd 96
Ferryman's Quay SW6 54Eb 111
Ferrymead Av. UB6: G'frd 41Ca 85
Ferrymead Dr. UB6: G'frd 40Ca 65
Ferrymead Gdns. UB6: G'frd 40Ea 66
Ferrymoor TW10: Ham 62Ka 130
Ferry Pl. SE18 48Qc 94
Ferry Quays TW8: Bford Ferry La. 51Na 109
Ferry Quays TW8: Bford Point Wharf La. 52Ma 109
(off Point Wharf La.)
Ferry Rd. KT7: T Ditt 72Ka 152
Ferry Rd. KT8: W Mole 69Ca 129
Ferry Rd. RM18: Tilb 5C 122
Ferry Rd. SW13 52Wa 110
Ferry Rd. TW1: Twick 60Ka 108
Ferry Rd. TW11: Tedd 64Ka 130
Ferry Sq. TW8: Bford 52Na 109
Ferry St. E14 50Ec 92
Ferry Wharf TW8: Bford 52Na 109
Feryby Rd. RM16: Grays 8D 100
Festing Rd. SW15 55Za 110
Festival Av. DA3: Lfield 69Fe 143
Festival Cl. DA5: Bexl 60Zc 117
Festival Cl. DA8: Erith 52Hd 118
Festival Cl. UB10: Hil 39R 64
Festival Ct. E8 38Vb 71
(off Holly St.)
Festival Ct. SM1: Sutt 73Db 155
Festival Wlk. SM5: Cars 77Hb 155
Festival Way E4 23Ec 52
Festive Mans. E20 36Ec 72
(off Napa Cl.)
Festive Wlk. SW15 54Za 110
Festoon Way E16 45Mc 93
Festubert Pl. E3 40Bc 72
Festuca Ho. E20 36Ec 72
(off Mirabelle Gdns.)
FETCHAM 95Fa 192
Fetcham Comn. La. KT22: Fet 93Da 191
FETCHAM DOWNS 98Fa 192
Fetcham Pk. Dr. KT22: Fet 95Ga 192
Fetherstone Cl. EN6: Pot B 4Fb 17
Fetherstone Ct. RM6: Chad H 30Bd 55
(off High Rd.)
Fetherston Rd. SS17: Stan H 1M 101
Fetherton Ct. IG11: Bark 40Sc 74
(off Spring Pl.)
Fetter La. EC4 3A 224 (44Qb 90)
Fetter La. Apts. EC4 3K 223 (44Qb 90)
(off Fetter La.)
Fettes Ho. NW8 2C 214 (40Fb 69)
(off Wellington Rd.)
Fettle Ct. SE14 51Ac 114
(off Moulding La.)
Fews Lodge RM6: Chad H 28Zc 55
Ffinch St. SE8 52Cc 114
Fiador Apts. SE10 49Gc 93
(off Telegraph Av.)
FICKLESHOLE 86Fc 179
Fiddicroft Av. SM7: Bans 86Db 175
Fiddler's Cl. DA9: Ghithe 56Xd 120
FIDDLERS HAMLET 4Yc 23
Fidelis Ho. E1 1K 225 (43Vb 91)
(off Gun St.)
Fidgeon Cl. BR1: Broml 69Qc 138
Fidler Pl. WD23: Bush 16Da 27
Field Cl. BR1: Broml 68Lc 137
Field Cl. CR2: Sande 86Xb 177
Field Cl. E4 23Dc 52
Field Cl. HA4: Ruis 32S 64
Field Cl. IG9: Buck H 20Lc 35
Field Cl. KT8: W Mole 71Da 151
Field Cl. KT9: Chess 78La 152
Field Cl. NW2 33Wa 68
Field Cl. RM4: Abr 13Xc 37
Field Cl. TW4: Cran 53X 107
Field Cl. UB10: Ick 33R 64
Field Cl. UB3: Harl 52S 106
FIELDCOMMON 73Ba 151
Fieldcommon La. KT12: Walt T 74Aa 151
Field Ct. DA11: Nflt 1B 144
Field Ct. RH8: Oxt 99Gc 199
Field Ct. SW19 62Cb 133
Field Ct. WC1 1J 223 (43Pb 90)
Field End CR5: Coul 86Mb 176
Field End EN5: Ark 14Xa 30
Field End GU24: W End 5D 166
Field End HA4: Ruis 37Y 65
Field End UB5: N'olt 37Z 65
Fieldend TW1: Twick 63Ha 130
Field End Cl. WD19: Wat 17Aa 27
Field End M. WD19: Wat 17Aa 27
Fieldend Rd. SW16 67Lb 134
Field End Rd. HA4: Ruis 32Z 65
Field End Rd. HA5: Eastc 29X 45
Fielden Ter. DA11: Nflt 60Ee 121
Fielder Apts. E3 42Bc 92
(off Heath Pl.)
Fielders Cl. EN1: Enf 14Ub 33
Fielders Cl. HA2: Harr 32Ea 66
Fielders Cres. IG11: Bark 42Yc 95
Fielders Way WD7: Shenl 5Na 15
Fieldfare Cl. HP3: Hem H 7M 3
Fieldfare La. DA9: Ghithe 59Wd 120
Fieldfare Rd. SE28 45Yc 95
Fieldfares AL2: Lon C 9H 7
Fieldgate Ct. KT11: Cobh 86W 170
Fieldgate La. CR4: Mitc 68Gb 133
Fieldgate Mans. E1 43Wb 91
(off Fieldgate St.)
Fieldgate St. E1 43Wb 91
Field Ho. NW6 41Za 88
(off Harvist Rd.)
Field Ho. SM4: Mord 71Db 155
(off School Ga. Dr.)
Fieldhouse Cl. E18 25Jc 53
Fieldhouse Rd. SW12 60Lb 112
Fieldhouse Vs. SM7: Bans 87Gb 175
Fieldhurst SL3: L'ly 50B 82
Fieldhurst Cl. KT15: Add 78K 149
Fielding Av. RM18: Tilb 3C 122
Fielding Av. TW2: Twick 62Ea 130
Fielding Ct. WC2 3F 223 (44Nb 90)
(off Earlham St.)
Fielding Gdns. SL3: L'ly 7N 81
Fielding Ho. NW8 39Eb 69
(off Ainsworth Way)
Fielding Ho. W4 51Ua 110
(off Devonshire Rd.)
Fielding La. BR2: Broml 70Lc 137
Fielding M. SW13 51Xa 110
Fielding Rd. W14 48Za 88
Fielding Rd. W4 48Ta 87
Fieldings, The GU21: Wok 8K 167
Fieldings, The SE23 60Yb 114
Fieldings, The SM7: Bans 89Bb 175
Fieldings Rd. EN8: Chesh 1Bc 20
Fielding St. SE17 51Sb 113
Fielding Ter. W5 45Pa 87
Fielding Wlk. W13 48Ka 86
Fielding Way CM13: Hut 16Ee 41
Field La. TW11: Tedd 64Ja 130
Field La. TW8: Bford 52La 108
Field Maple M. RM5: Col R 24Cd 56
Field Mead NW7 24Ua 48
Field Mead NW9 24Va 48

Fieldoaks Way RH1: Mers 1C 208
Fieldpark Gdns. CR0: C'don 74Ac 158
Field Pl. KT3: N Mald 72Va 154
Field Pl. SW19 67Fb 133
Field Point E7 35Jc 73
Field Rd. E7 35Hc 73
Field Rd. HP2: Hem H 3A 4
Field Rd. N17 27Tb 51
Field Rd. RM15: Avel 46Sd 98
Field Rd. TW14: Felt 58X 107
Field Rd. UB9: Den 35F 62
(not continuous)
Field Rd. W6 50Ab 88
Field Rd. WD19: Wat 16Aa 27
Fields, The SL1: Slou 7H 81
Fields Ct. EN6: Pot B 5Fb 17
FIELDS END 1F 2
Fields End La. HP1: Hem H 1F 2
Fieldsend Rd. SM3: Cheam 78Ab 154
Fields Est. E8 38Wb 71
Fieldside Cl. BR6: Farnb 77Sc 160
Fieldside Rd. BR1: Broml 64Fc 137
Fields Pk. Cres. RM6: Chad H 29Zc 55
Field St. WC1 3H 217 (41Pb 90)
Fields Wlk. SW11 52Kb 112
Fieldsway Ho. N5 36Qb 70
Field Vw. TW13: Felt 63T 128
Field Vw. TW20: Egh 64E 126
Fieldview SW18 60Fb 111
Field Vw. Cl. RM7: Mawney 27Cd 56
Fieldview Cotts. N14 19Mb 32
(off Balaams La.)
Fieldview Ct. TW18: Staines 64J 127
Field Vw. Ri. AL2: Brick W 1Aa 13
Field Vw. Rd. EN6: Pot B 5Cb 17
Field Way GU23: Rip 97H 189
Field Way HA4: Ruis 32S 64
Field Way HP3: Bov 9C 2
Field Way NW10 38Sa 67
Field Way UB6: G'frd 39Da 65
Field Way UB8: Cowl 42M 83
Field Way WD3: Rick 18K 25
Fieldway BR5: Pet W 72Tc 160
Fieldway CR0: New Ad 80Dc 158
Fieldway HP4: Berk 3A 2
Fieldway RM16: Grays 46Ce 99
Fieldway RM8: Dag 34Yc 75
Fieldway SL9: Chal P 24A 42
Fieldway Cres. N5 36Qb 70
Fieldway Stop (London Tramlink) 80Dc 158
Fiennes Cl. RM8: Dag 32Yc 75
Fiennes Way TN13: S'oaks 99Ld 203
Fiesta Dr. RM9: Dag 42Ed 96
Fifehead Cl. TW15: Ashf 65N 127
Fife Rd. E16 43Jc 93
Fife Rd. KT1: King T 68Na 131
Fife Rd. N22 24Rb 51
Fife Rd. SW14 57Sa 109
Fife Ter. N1 1H 217 (40Pb 70)
Fife Way KT23: Bookh 97Ca 191
Fifield Path SE23 62Zb 136
Fifteenth Av. KT20: Lwr K 98Bb 195
Fifth Av. E12 35Pc 74
Fifth Av. KT20: Lwr K 97Ab 194
Fifth Av. RM20: W Thur 51Wd 120
Fifth Av. UB3: Hayes 46V 84
Fifth Av. W10 41Ab 88
Fifth Av. WD25: Wat 7Z 13
Fifth Cross Rd. TW2: Twick 61Fa 130
Fifth Way HA9: Wemb 35Ra 67
Figges Rd. CR4: Mitc 66Jb 134
Figgswood CR5: Coul 94Lb 196
Fight for Peace Academy 46Qc 94
Fig St. TN14: S'oaks 100Hd 202
FIG STREET 100Jd 202
Fig Tree Cl. NW10 39Ua 68
Figtree Hill HP2: Hem H 1M 3
Figure Ct. SW3 50Hb 89
(off West Rd.)
Filament Wlk. SW18 57Cb 111
(off Spectrum Way)
Filanco Ct. W7 46Ha 86
Filbert Cl. AL10: Hat 3B 8
Filborough Way DA12: Grav'nd 1K 145
Filby Cl. DA8: Erith 52Gd 118
Filby Rd. KT9: Chess 79Pa 153
Filey Av. N16 32Wb 71
Filey Cl. SM2: Sutt 80Eb 155
Filey Cl. TN16: Big H 91Kc 199
Filey Spur SL1: Slou 7F 80
Filey Waye HA4: Ruis 33W 64
Filigree Ct. SE16 46Bc 92
Fillebrook Av. EN1: Enf 12Ub 33
Fillebrook Rd. E11 32Fc 73
Filmer Chambers SW6 53Ab 110
(off Filmer Rd.)
Filmer Ho. SW6 53Bb 111
(off Filmer Rd.)
Filmer La. TN14: S'oaks 93Nd 203
Filmer M. SW6 53Bb 111
Filmer Rd. SL4: Wind 4B 102
Filmer Rd. SW6 53Ab 110
Filston La. TN14: Otf 88Fd 182
Filston La. TN14: S'ham 88Fd 182
Filston Rd. DA8: Erith 50Ed 96
Filton Cl. NW9 25Ua 48
Filton Ct. SE14 52Yb 114
(off Farrow La.)
Filton Ho. WD19: Wat 20Z 27
Finborough Ho. SW10 51Eb 111
(off Finborough Rd.)
Finborough Rd. SW10 50Db 89
Finborough Rd. SW17 65Hb 133
Finborough Theatre, The 51Db 111
(off Finborough Rd.)
Finchale Rd. SE2 48Wc 95
Fincham Cl. UB10: Ick 34S 64
Finch Av. SE27 63Tb 135
Finch Cl. AL10: Hat 2C 8
Finch Cl. EN5: Barn 15Cb 31
Finch Cl. GU21: Knap 9G 166
Finch Cl. NW10 37Ta 67
Finch Ct. DA14: Sidc 62Xc 139
Finchdale HP1: Hem H 2J 3
Finchdean Ho. SW15 59Va 110
Finch Dr. TW14: Felt 59Z 107
Finches, The UB9: Den 29H 43
Finches Av. WD3: Crox G 13P 25
Finch Gdns. E4 22Cc 52
Finch Grn. WD3: Chor 14H 25
Finch Ho. E3 39Bc 72
(off Jasmine Sq.)
Finch Ho. SE8 52Dc 114
(off Bronze St.)
Finchingfield Av. IG8: Wfd G 24Lc 53
Finch La. EC3 3G 225 (44Tb 91)
Finch La. WD23: Bush 13Ba 27
FINCHLEY 25Cb 49
Finchley Central Station (Underground) 25Cb 49
Finchley Cl. DA1: Dart 58Qd 119
Finchley Ct. N3 23Db 49
Finchley Golf Course 23Bb 49
Finchley Ind. Est. N12 21Eb 49
Finchley La. NW4 28Ya 48
Finchley Lido Leisure Cen. 24Fb 49
Finchley Manor Club 25Bb 49
FINCHLEY MEMORIAL HOSPITAL 24Eb 49
Finchley Pk. N12 21Eb 49
Finchley Pl. NW8 1B 214 (40Fb 69)
Finchley Rd. NW11 28Bb 49
Finchley Rd. NW2 34Cb 69
Finchley Rd. NW3 35Cb 69
Finchley Rd. NW8 1B 214 (39Fb 69)
Finchley Rd. RM17: Grays 51De 121
Finchley Road & Frognal Station (Overground) 36Eb 69
Finchley Road Station (Underground) 37Eb 69
Finchley Way N3 24Cb 49
Finch Lodge W9 43Cb 89
(off Admiral Wlk.)
Finch M. SE15 53Vb 113
Finch's Ct. E14 45Dc 92
Finch's Ct. M. E14 45Dc 92
(off Finch's Ct.)
Finden Rd. E7 36Lc 73
Findhorn Av. UB4: Yead 43X 85
Findhorn St. E14 44Ec 92
Findlay Ho. E3 41Cc 92
(off Trevithick Way)
Findon Cl. HA2: Harr 34Da 65
Findon Cl. SW18 58Cb 111
Findon Ct. KT15: Add 78H 149
Findon Gdns. RM13: Rain 43Jd 96
Findon Rd. N9 18Xb 33
Findon Rd. W12 47Wa 88
Fine Bush La. UB9: Hare 30R 44
Finefield Wlk. SL1: Slou 7H 81
Fingal St. SE10 50Hc 93
Fingest Ho. NW8 5D 214 (42Gb 89)
(off Lilestone St.)
Finglesham Cl. BR5: Orp 74Zc 161
Finians Cl. UB10: Uxb 38P 63
Finland Rd. SE4 55Ac 114
Finland St. SE16 48Ac 92
Finlay Gdns. KT15: Add 77L 149
Finlays Cl. KT9: Chess 78Qa 153
Finlay St. SW6 53Za 110
Finley Ct. SE5 52Sb 113
(off Redcar St.)
Finmere Ho. N4 31Sb 71
Finnart Cl. KT13: Weyb 77S 150
Finnart Ho. Dr. KT13: Weyb 77S 150
Finnemore Ho. N1 39Sb 71
(off Britannia Row)
Finney Dr. GU20: W'sham 9B 146
Finney La. TW7: Isle 53Ja 108
Finn Ho. N1 3G 219 (41Tb 91)
(off Bevenden St.)
Finnis St. E2 41Xb 91
Finnymore Rd. RM9: Dag 38Ad 75
FINSBURY 3K 217 (41Qb 90)
Finsbury Av. EC2 1G 225 (43Tb 91)
Finsbury Av. Sq. EC2 7H 219 (43Ub 91)
Finsbury Cir. EC2 1G 225 (43Tb 91)
Finsbury Cotts. N22 24Nb 50
Finsbury Ct. EN8: Walt C 6Ac 20
Finsbury Est. EC1 4A 218 (41Qb 90)
Finsbury Ho. N22 25Nb 50
Finsbury Leisure Cen. 4D 218 (41Sb 91)
Finsbury Mkt. EC2 6H 219 (42Ub 91)
(not continuous)
FINSBURY PARK 32Qb 70
Finsbury Pk. Av. N4 30Sb 51
Finsbury Park Interchange (Bus Station) 33Qb 70
Finsbury Pk. Rd. N4 33Rb 71
Finsbury Park Station (Rail & Underground) 33Qb 70
Finsbury Pavement EC2 7G 219 (43Tb 91)
Finsbury Rd. N22 24Pb 50
(not continuous)
Finsbury Sq. EC2 6G 219 (42Tb 91)
Finsbury St. EC2 7F 219 (43Sb 91)
Finsbury Way DA5: Bexl 58Bd 117
Finsen Rd. SE5 56Sb 113
Finstock Rd. W10 44Za 88
Finucane Ct. TW9: Rich 55Pa 109
(off Lwr. Mortlake Rd.)
Finucane Dr. BR5: Orp 73Yc 161
Finucane Gdns. RM13: Rain 37Jd 76
Finucane Ri. WD23: B Hea 19Ea 28
Finway Ct. WD18: Wat 15V 26
Finwhale Ho. E14 48Dc 92
(off Glengall Gro.)
Fiona Cl. KT23: Bookh 96Ca 191
Fiona Ct. EN2: Enf 13Rb 33
Fiona Ct. NW6 40Cb 69
Firbank Cl. E16 43Mc 93
Firbank Cl. EN2: Enf 14Sb 33
Firbank Dr. GU21: Wok 1M 187
Firbank Dr. WD19: Wat 17Aa 27
Firbank La. GU21: Wok 1M 187
Firbank Pl. TW20: Eng G 5M 125
Firbank Rd. RM5: Col R 22Dd 56
Firbank Rd. SE15 54Xb 113
Fir Cl. KT12: Walt T 73W 150
Fircroft Cl. GU22: Wok 90B 168
Fircroft Cl. SL2: Stoke P 7L 61
Fircroft Ct. GU22: Wok 90B 168
Fircroft Gdns. HA1: Harr 34Ga 66
Fircroft Rd. KT9: Chess 77Pa 153
Fircroft Rd. SW17 61Hb 133
Fircroft Rd. TW20: Eng G 6N 125
Fir Dene BR6: Farnb 76Pc 160
Firdene KT5: Surb 74Sa 153
Fireball Hill SL5: S'dale 3B 146
Fire Bell All. KT6: Surb 72Na 153
Firecrest Cl. DA3: Lfield 69De 143
Firecrest Dr. NW3 34Db 69
Firefly Cl. UB3: Hayes 45V 84
Firefly Cl. UB5: N'olt 41Z 85
Firefly Gdns. E6 42Nc 94
Firefly Rd. KT16: Vir W 5L 147
Firehorn Ho. E15 41Gc 93
(off Teasel Way)
Firemans Flats N22 24Nb 50
Fire Station All. EN5: Barn 13Ab 30
Firestation Cen. for Arts & Culture Windsor, The 4G 102
Fire Station M. BR3: Beck 67Cc 136
Fire Station Sq. SE1 7J 225 (46Ub 91)
(off Abbots La.)
Firestone Ho. TW8: Bford 50Na 87
Firethorn Cl. HA8: Edg 21Sa 47
Firewatch Ct. E1 43Ac 92
(off Candle St.)
Firfield Rd. KT15: Add 77J 149
Firfields KT13: Weyb 79R 150
Fir Grange Av. KT13: Weyb 78R 150
Fir Gro. KT3: N Mald 72Va 154
Firgrove GU21: Wok 1M 187
Firgrove Ct. SE6 59Cc 114
Fir Gro. Rd. SW9 54Qb 112
Firham Pk. Av. RM3: Hrld W 24Qd 57
Firhill Rd. SE6 63Cc 136
Fir Ho. W10 42Ab 88
(off Droop St.)
Firlands KT13: Weyb 79U 150
Firle Ct. KT17: Eps 84Va 174
Firle Ho. W10 43Ya 88
(off Sutton Way)
Firman Cl. KT3: N Mald 70Ua 132
Firmans Ct. E17 28Fc 53
Firmingers Rd. BR6: Well H 78Dd 162
Firmin Rd. DA1: Dart 57Ld 119
Fir Rd. SM3: Sutt 74Bb 155
Fir Rd. TW13: Hanw 64Z 129
Firs, The SE26: Border Rd 64Xb 135
Firs, The SE26: Waverley Ct 64Yb 136
Firs, The AL1: St A 6F 6
Firs, The CM15: Pil H 16Wd 40
Firs, The CR3: Cat'm 94Tb 197
Firs, The DA15: Sidc 61Vc 139
Firs, The DA5: Bexl 60Fd 118
Firs, The E6 38Nc 74
Firs, The EN9: Walt A 7Lc 21
(within Woodbine Cl. Caravan Pk.)
Firs, The GU24: Bisl 8E 166
Firs, The HA8: Edg 21Ta 47
(off Stoneyfields La.)
Firs, The IG8: Wfd G 24Lc 53
Firs, The KT14: Byfl 84M 169
Firs, The KT23: Bookh 96Ea 192
Firs, The N20 18Fb 31
Firs, The RM16: Grays 46Ee 99
Firs, The W5 43Ma 87
Firs Av. N10 27Jb 50
Firs Av. N11 23Hb 49
Firs Av. SL4: Wind 5D 102
Firs Av. SW14 56Sa 109
Firsby Av. CR0: C'don 74Zb 158
Firsby Rd. N16 32Wb 71
Firs Cl. AL10: Hat 1D 8
Firs Cl. CR4: Mitc 68Kb 134
Firs Cl. KT10: Clay 79Ga 152
Firs Cl. N10 28Jb 50
Firs Cl. SE23 59Ac 114
Firs Cl. SL0: Iver H 39E 62
Firscroft N13 20Sb 33
Firsdene Cl. KT16: Ott 79F 148
Firs Dr. IG10: Lough 11Qc 36
Firs Dr. SL3: L'ly 46B 82
Firs Dr. TW5: Cran 52X 107
Firs End SL9: Chal P 27A 42
Firsgrove Cres. CM14: W'ley 21Xd 58
Firsgrove Rd. CM14: W'ley 21Xd 58
Firs Ho. N22 25Qb 50
(off Acacia Rd.)
Firside Gro. DA15: Sidc 60Vc 117
Firs La. EN6: Pot B 5Db 17
Firs La. N13 20Sb 33
Firs La. N21 17Sb 33
Firs Pk., The AL9: Hat 5J 9
Firs Pk. Av. N21 18Tb 33
Firs Pk. Gdns. N21 18Sb 33
Firs Rd. CR8: Kenley 87Rb 177
First Av. DA11: Nflt 10A 122
First Av. DA7: Bex 52Yc 117
First Av. E12 35Nc 74
First Av. E13 41Jc 93
First Av. E17 29Cc 52
First Av. EN1: Enf 15Vb 33
First Av. EN9: Walt A 1Kc 21
First Av. HA9: Wemb 33Ma 67
First Av. KT12: Walt T 72X 151
First Av. KT19: Ewe 81Ua 174
First Av. KT20: Lwr K 97Ab 194
First Av. KT8: W Mole 70Ba 129
First Av. N18 21Yb 52
First Av. NW4 28Ya 48
First Av. RM10: Dag 40Dd 76
First Av. RM20: W Thur 51Wd 120
First Av. RM6: Chad H 29Yc 55
First Av. SW14 55Ua 110
First Av. UB3: Hayes 46V 84
First Av. W10 42Bb 89
First Av. W3 46Va 88
First Av. WD25: Wat 7Y 13
First Central Bus. Pk. NW10 41Pa 87
First Cl. KT8: W Mole 69Ea 130
First Cres. SL1: Slou 3F 80
First Cross Rd. TW2: Twick 61Ga 130
First Dr. NW10 38Sa 67
First Quarter KT19: Eps 83Ua 174
First Slip KT22: Lea 90Ja 172
First St. SW3 5E 226 (49Gb 89)
First Way HA9: Wemb 35Ra 67
Firstway SW20 68Ya 132
Firs Wlk. HA6: Nwood 23T 44
Firs Wlk. IG8: Wfd G 22Jc 53
Firswood Av. KT19: Ewe 78Ua 154
Firs Wood Cl. EN6: N'thaw 4Hb 17
Firth Gdns. SW6 53Ab 110
Firth Ho. E2 41Wb 91
(off Turin St.)
Fir Tree Av. SL2: Stoke P 2K 81
Fir Tree Av. UB7: W Dray 48Q 84
Firtree Av. CR4: Mitc 68Jb 134
Fir Tree Cl. BR6: Chels 78Vc 161
Fir Tree Cl. HP3: Hem H 3A 4
Fir Tree Cl. KT10: Esh 78Ea 152
Fir Tree Cl. KT17: Eps D 87Ya 174
Fir Tree Cl. KT19: Ewe 77Va 154
Fir Tree Cl. KT22: Lea 95La 192
Fir Tree Cl. RM1: Rom 27Fd 56
Fir Tree Cl. RM17: Grays 51Fe 121
Fir Tree Cl. SW16 64Lb 134
Fir Tree Cl. W5 44Na 87
Fir Tree Ct. WD6: E'tree 14Pa 29
Fir Tree Gdns. CR0: C'don 77Cc 158
Fir Tree Gro. SM5: Cars 80Hb 155
Fir Tree Hill WD3: Chan C 10P 11
Firtree Ho. SE13 59Fc 115
(off Birdwood Av.)
Fir Tree Pl. TW15: Ashf 64Q 128
Fir Tree Rd. KT17: Eps D 88Xa 174
Fir Tree Rd. KT22: Lea 95La 192
Fir Tree Rd. SM7: Bans 86Ya 174
Fir Tree Rd. TW4: Houn 56Aa 107
Fir Trees CM16: Epp 1Xc 23
Fir Trees RM4: Abr 13Xc 37
Fir Trees Cl. SE16 46Ac 92
Fir Tree Wlk. EN1: Enf 13Tb 33
Fir Tree Wlk. RH2: Reig 6M 207
Fir Tree Wlk. RM10: Dag 34Ed 76
Fir Wlk. SM3: Cheam 79Za 154
Firwood Av. AL4: St A 2J 7
Firwood Cl. GU21: Wok 1J 187
Firwood La. RM3: Hrld W 26Nd 57
Firwood Rd. GU25: Vir W 2J 147
Fisgard Ct. DA12: Grav'nd 8F 122
Fisher Cl. CR0: C'don 74Vb 157
Fisher Cl. E9 36Zb 72
Fisher Cl. EN3: Enf L 9Dc 20
Fisher Cl. KT12: Hers 77X 151
Fisher Cl. SE16 46Zb 92
Fisher Cl. UB6: G'frd 41Ca 85
Fisher Cl. WD4: K Lan 1Q 12
Fisher Ho. E1 45Yb 92
(off Cable St.)
Fisher Ho. N1 1K 217 (39Qb 70)
(off Barnsbury Est.)
Fisherman KT16: Chert 74L 149
Fisherman Cl. TW10: Ham 63Ka 130
Fishermans Dr. SE16 47Zb 92
Fisherman's Pl. W4 51Va 110
Fisherman's Wlk. E14 46Cc 92
Fishermans Wlk. SE28 47Uc 94
Fishermens Hill DA11: Nflt 57De 121
Fisher Rd. HA3: W'stone 26Ha 46
Fisher's Cl. SW16 62Mb 134
FISHERS GREEN 1Dc 20
Fishers Grn. La. EN9: Walt A 1Dc 20
Fisher's Ind. Est. WD18: Wat 15Y 27
Fisher's La. W4 49Ta 87
Fishers La. CM16: Epp 4Uc 22
Fishers Oak TN14: S'oaks 93Ld 203
Fisher St. E16 43Jc 93
Fisher St. WC1 1H 223 (43Pb 90)
Fishers Way DA17: Belv 46Ed 96
Fishers Way HA0: Wemb 36Ka 66
Fishers Wood SL5: S'dale 4G 146
Fishers Wood Gro. BR2: Broml 73Nc 160
Fisherton St. NW8 6B 214 (42Fb 89)
Fishery Cotts. HP1: Hem H 4J 3
Fishery Pas. HP1: Hem H 4J 3
Fishery Rd. HP1: Hem H 4J 3
Fishguard Spur SL1: Slou 7M 81
Fishguard Way E16 46Rc 94
Fishmongers Hall Wharf EC4 5F 225 (45Tb 91)
(off Swan La.)
Fishponds Rd. BR2: Kes 78Mc 159
Fishponds Rd. SW17 63Gb 133
Fishpool St. AL3: St A 2P 5
Fish St. Hill EC3 5G 225 (45Tb 91)
Fish Wharf EC3 5G 225 (45Tb 91)
Fisk Cl. TW16: Sun 65V 128
Fiske Ct. IG11: Bark 40Tc 74
Fiske Ct. N17 25Wb 51
Fiske Ct. SM2: Sutt 80Eb 155
Fitch Ct. SW2 57Qb 112
Fitness4Less Canning Town 43Hc 93
Fitness4Less Sutton 79Cb 155
Fitness First Angel 2B 218 (40Rb 71)
Fitness First Baker Street 1G 221 (43Hb 89)
Fitness First Beckenham 65Cc 136
Fitness First Berkeley Square 5A 222 (45Kb 90)
Fitness First Brentwood 22Yd 58
Fitness First Brixton 55Qb 112
Fitness First Camden 39Kb 70
Fitness First Clapham Junction 56Gb 111
Fitness First Covent Garden 5G 223 (45Nb 90)
(off Bedford St.)
Fitness First Fetter Lane 3A 224 (44Qb 90)
(off Fetter La.)
Fitness First Gracechurch Street 4H 225 (45Ub 91)
(off Gracechurch St.)
Fitness First Great Marlborough Street 3B 222 (44Lb 90)
(off Gt. Marlborough St.)
Fitness First Hammersmith 49Ya 88
Fitness First Harringay 30Rb 51
(off Arena Shop. Pk.)
Fitness First High Holborn 2H 223 (44Pb 90)
(off High Holborn)
Fitness First Highbury 34Rb 71
Fitness First Ilford 33Sc 74
Fitness First Kilburn 39Bb 69
Fitness First Kingly Street 4B 222 (45Lb 90)
(off Kingly St.)
Fitness First Leyton Mills 34Ec 72
Fitness First London Bridge 7G 225 (46Tb 91)
(off London Bri. St.)
Fitness First London Bridge, Cottons 6H 225 (46Ub 91)
(off Tooley St.)
Fitness First Paternoster Square 2C 224 (44Rb 91)
(off Paternoster Sq.)
Fitness First Queen Victoria Street 4D 224 (45Sb 91)
(off Queen Victoria St.)
Fitness First Romford 30Gd 56
Fitness First St Albans 2B 6
(off Verulam Rd.)
Fitness First Streatham 61Nb 134
Fitness First Thomas More Square 45Wb 91
(off Thomas More Sq.)
Fitness First Tooting Bec 62Jb 134
Fittleton Gdns. E3 42Dc 92
Fitzalan Ho. KT17: Ewe 82Va 174
Fitzalan Rd. KT10: Clay 80Ga 152
Fitzalan Rd. N3 27Ab 48
Fitzalan St. SE11 5K 229 (49Qb 90)
Fitzclarence Ho. W11 47Ab 88
(off Holland Park Av.)
Fitzgeorge Av. KT3: N Mald 67Ta 131
Fitzgeorge Av. W14 49Ab 88
Fitzgerald Av. SW14 55Ua 110
Fitzgerald Ct. E10 32Dc 72
(off Leyton Grange Est.)
Fitzgerald Ho. E14 44Dc 92
(off E. India Dock Rd.)
Fitzgerald Ho. SW17 62Fb 133
Fitzgerald Ho. SW9 54Qb 112
Fitzgerald Ho. UB3: Hayes 46X 85
Fitzgerald Rd. E11 29Jc 53
Fitzgerald Rd. KT7: T Ditt 72Ja 152
Fitzgerald Rd. SW14 55Ta 109
Fitzhardinge Ho. W1 2H 221 (44Jb 90)
(off Portman Sq.)
Fitzhardinge St. W1 2H 221 (44Jb 90)
Fitzherbert Ct. IG8: Ilf 25Nc 54
Fitzherbert Cl. IG8: Wfd G 25Nc 54
Fitzherbert Wlk. UB1: S'hall 47Fa 86
Fitzhugh Gro. SW18 58Fb 111
Fitzilian Av. RM3: Hrld W 25Pd 57
Fitzjames Av. CR0: C'don 75Wb 157
Fitzjames Av. W14 49Ab 88
Fitzjohn Av. EN5: Barn 15Ab 30
Fitzjohn's Av. NW3 35Eb 69
Fitzmaurice Ho. SE16 49Xb 91
(off Rennie Est.)
Fitzmaurice Pl. W1 6A 222 (46Kb 90)
Fitzneal St. W12 44Va 88
Fitzpatrick Rd. SW9 53Rb 113
Fitzrobert Pl. TW20: Egh 65C 126
Fitzrovia Apts. W1 6A 216 (42Kb 90)
(off Bolsover St.)
Fitzroy Bri. 39Jb 70
Fitzroy Bus. Pk. BR5: St P 66Zc 139
Fitzroy Cl. N6 32Hb 69
Fitzroy Ct. CR0: C'don 73Tb 157
Fitzroy Ct. DA1: Dart 60Rd 119
(off Churchill Cl.)
Fitzroy Ct. N6 30Lb 50
Fitzroy Ct. W1 6C 216 (42Lb 90)
(off Tottenham Ct. Rd.)
Fitzroy Cres. W4 52Ta 109
Fitzroy Gdns. SE19 66Ub 135
Fitzroy Ho. E14 43Bc 92
(off Wallwood St.)
Fitzroy Ho. SE1 50Vb 91
(off Cooper's Rd.)
Fitzroy House Mus. 6B 216 (42Lb 90)
Fitzroy M. W1 6B 216 (42Lb 90)
(off Cleveland St.)
Fitzroy Pk. N6 32Hb 69
Fitzroy Pl. RH2: Reig 6M 207
Fitzroy Rd. NW1 39Jb 70
Fitzroy Sq. W1 6B 216 (42Lb 90)
FITZROY SQUARE HOSPITAL 6B 216 (42Lb 90)
Fitzroy St. W1 6B 216 (42Lb 90)
(not continuous)
Fitzroy Yd. NW1 39Jb 70
Fitzstephen Rd. RM8: Dag 36Xc 75
Fitzwarren Gdns. N19 32Lb 70
Fitzwilliam Av. TW9: Rich 54Pa 109
Fitzwilliam Cl. N20 18Jb 32
Fitzwilliam Ct. AL1: St A 2C 6
(off St Peter's St.)
Fitzwilliam Ct. WD6: Bore 11Pa 29
(off Lyndhurst Wlk.)
Fitzwilliam Hgts. SE23 61Yb 136
Fitzwilliam Ho. TW9: Rich 56Ma 109
Fitzwilliam M. E16 46Jc 93
Fitzwilliam Rd. SW4 55Lb 112
Fitz Wygram Cl. TW12: Hamp H 64Ea 130
Five Acre NW9 26Va 48
Fiveacre Cl. CR7: Thor H 72Qb 156
Five Acres AL2: Lon C 7H 7
Five Acres WD4: K Lan 1P 11
Five Acres Av. AL2: Brick W 1Ba 13
Five Arches Bus. Pk. DA14: Sidc 64Zc 139
Five Ash Rd. DA11: Grav'nd 9B 122
Five Bell All. E14 44Bc 92
(off Three Colt St.)
Five Elms BR2: Hayes 76Kc 159
Five Elms Rd. BR2: Hayes 76Kc 159
Five Elms Rd. RM9: Dag 34Bd 75
Five Flds. Cl. WD19: Wat 19Ba 27
Five Oaks AL10: Hat 3D 8
(off Sandifield)
Five Oaks Cl. GU21: Wok 1H 187
Five Oaks La. IG7: Chig 23Ad 55
Five Oaks M. BR1: Broml 62Jc 137
Fives Ct. SE11 4B 230 (48Rb 91)
FIVEWAYS New Eltham 61Rc 138
Five Ways Bus. Cen. TW13: Felt 62X 129
FIVEWAYS CORNER Croydon 77Qb 156
FIVEWAYS CORNER Hendon 25Wa 48
Fiveways Rd. SW9 54Qb 112
Five Wents BR8: Swan 68Jd 140
Fixie Bldg. E17 29Bc 52
(off Track St.)
Flack Ct. E10 31Dc 72
Fladbury Rd. N15 30Tb 51
Fladgate Rd. E11 30Gc 53
Flag Cl. CR0: C'don 74Zb 158
Flagon Ct. CR0: C'don 77Sb 157
(off St Andrew's Rd.)
Flags, The HP2: Hem H 2B 4
Flagship Ho. E16 47Kc 93
(off Royal Crest Av.)
Flagstaff Cl. EN9: Walt A 5Dc 20
Flagstaff Rd. EN9: Walt A 5Dc 20
Flag Wlk. HA5: Eastc 30W 44
Flambard Rd. HA1: Harr 30Ja 46
Flamborough Cl. TN16: Big H 91Kc 199
Flamborough Ho. SE15 53Wb 113
(off Clayton Rd.)
Flamborough Rd. HA4: Ruis 34W 64
Flamborough Spur SL1: Slou 7E 80
Flamborough St. E14 44Ac 92
Flamborough Wlk. E14 44Ac 92
(off Flamborough St.)
Flamingo Ct. SE17 7D 230 (50Sb 91)
(off Crampton St.)
Flamingo Ct. SE8 52Cc 114
(off Hamilton St.)
Flamingo Gdns. UB5: N'olt 41Aa 85
Flamingo Wlk. RM12: Horn 37Jd 76
Flamstead Gdns. RM9: Dag 38Yc 75
Flamstead Ho. SW3 7D 226 (50Gb 89)
(off Cale St.)
Flamstead Rd. RM9: Dag 38Yc 75
Flamsted Av. HA9: Wemb 37Qa 67
Flamsteed Rd. SE7 50Nc 94
Flanaghan Apts. E3 42Bc 92
(off Portia Way)
Flanchford Ho. RH2: Reig 5J 207
(off Somers Cl.)
Flanchford Rd. W12 48Va 88
Flanders Ct. DA1: Dart 60Md 119
Flanders Ct. E17 31Ac 72
Flanders Ct. TW20: Egh 64D 126
Flanders Cres. SW17 66Hb 133
Flanders Mans. W4 49Va 88
Flanders Rd. E6 40Pc 74
Flanders Rd. W4 49Ua 88
Flanders Way E9 37Zb 72
Flandrian Cl. EN3: Enf L 10Dc 20
Flank St. E1 45Wb 91
Flannery Ct. SE16 48Xb 91
Flansham Ho. E14 44Bc 92
(off Clemence St.)
Flash La. EN2: Enf 9Rb 19
Flask Wlk. NW3 35Eb 69
Flatfield Rd. HP3: Hem H 4A 4
Flatford Ho. SE6 63Ec 136
Flather Cl. SW16 64Lb 134
Flat Iron Sq. SE1 7E 224 (46Sb 91)
(off Southwark Bri. Rd.)
Flatiron Yd. SE1 7E 224 (46Sb 91)
(off Ayres St.)
Flats, The DA9: Ghithe 57Yd 120
Flats, The HP8: Chal G 15A 24
FLAUNDEN 4D 10
Flaunden Bottom HP5: Lat 9A 10

Flaunden Hill HP3: Flau 5B 10
Flaunden Ho. WD18: Wat 16U 26
Flaunden La. HP3: Bov 4D 10
Flaunden La. HP3: Flau 4D 10
Flaunden La. HP3: Hem H 4D 10
Flaunden La. WD3: Sarr 5G 10
Flaunden Pk. HP3: Flau 3C 10
Flavell M. SE10 50Gc 93
Flavian Cl. AL3: St A 4M 5
Flaxen Cl. E4 20Dc 34
Flaxen Rd. E4 20Dc 34
Flaxley Ho. SW1 7K 227 (50Kb 90)
(part of Abbots Mnr.)
Flaxley Rd. SM4: Mord 73Db 155
Flaxman Ct. DA17: Belv 50Cd 96
(off Hoddesdon Rd.)
Flaxman Ct. W1 3D 222 (44Mb 90)
Flaxman Ct. WC1 4E 216 (41Mb 90)
(off Flaxman Ter.)
Flaxman Ho. SE1 3B 230 (48Rb 91)
(off London Rd.)
Flaxman Ho. W4 50Ua 88
(off Devonshire St.)
Flaxman Rd. SE5 55Rb 113
Flaxman Sports Cen. 54Sb 113
Flaxman Ter. WC1 4E 216 (41Mb 90)
Flaxton Rd. SE18 53Tc 116
Flecker Cl. HA7: Stan 22Ha 46
Flecker Ho. SE5 52Tb 113
(off Lomond Gro.)
Flecknoe Ct. KT16: Chert 74H 149
Fleece Dr. N9 21Wb 51
Fleece Rd. KT6: Surb 74La 152
Fleece Wlk. N7 37Nb 70
Fleeming Cl. E17 26Bc 52
Fleeming Rd. E17 26Bc 52
Fleet Av. DA2: Dart 60Sd 120
Fleet Av. RM14: Upm 30Td 58
Fleetbank Ho. EC4 3A 224 (44Qb 90)
(off Salisbury Sq.)
Fleetbrook Ho. SL3: Dat 3P 103
Fleet Cl. HA4: Ruis 30S 44
Fleet Cl. KT8: W Mole 71Ba 151
Fleet Cl. RM14: Upm 30Td 58
Fleetdale Pde. DA2: Dart 60Sd 120
FLEET DOWNS 60Sd 120
Fleetfield WC1 3G 217 (41Nb 90)
(off Birkenhead St.)
Fleethall Gro. RM16: Grays 46Ce 99
Fleet Ho. E14 45Ac 92
(off Victory Pl.)
Fleet Ho's. DA13: Sflt 65De 143
Fleet La. KT8: W Mole 72Ba 151
Fleet Pl. EC4 2B 224 (44Rb 91)
Fleet Rd. DA11: Nflt 62Ee 143
Fleet Rd. DA2: Dart 60Rd 119
Fleet Rd. IG11: Bark 39Rc 74
Fleet Rd. NW3 36Gb 69
Fleetside KT8: W Mole 71Ba 151
Fleet Sq. WC1 4H 217 (41Pb 90)
Fleet St. EC4 3K 223 (44Qb 90)
Fleet Ter. DA11: Nflt 60Fe 121
FLEETVILLE 2F 6
Fleetway TW20: Thorpe 69E 126
Fleetway WC1 3G 217 (41Nb 90)
(off Birkenhead St.)
Fleetway W. UB6: G'frd 40Ka 66
Fleetwood Cl. CR0: C'don 76Vb 157
Fleetwood Cl. E16 43Mc 93
Fleetwood Cl. KT20: Tad 92Za 194
Fleetwood Cl. KT9: Chess 80Ma 153
Fleetwood Ct. E6 43Pc 94
(off Evelyn Dennington Rd.)
Fleetwood Ct. KT14: W Byf 85J 169
Fleetwood Ct. TW19: Stanw 58M 105
(off Douglas Rd.)
Fleetwood Rd. KT1: King T 69Ra 131
Fleetwood Rd. NW10 36Wa 68
Fleetwood Rd. SL2: Slou 6K 81
Fleetwood Sq. KT1: King T 69Ra 131
Fleetwood St. N16 33Ub 71
Fleetwood Way WD19: Wat 21Y 45
Fleming N8 27Nb 50
(off Boyton Cl.)
Fleming Cl. SW10 51Eb 111
(off Winterton Pl.)
Fleming Cl. W9 42Cb 89
Fleming Ct. CR0: Wadd 78Qb 156
Fleming Ct. DA11: Nflt 60Ee 121
Fleming Ct. W2 7B 214 (43Fb 89)
(off St Mary's Sq.)
Fleming Dr. N21 15Pb 32
Fleming Gdns. RM18: Tilb 3E 122
Fleming Gdns. RM3: Hrld W 26Md 57
Fleming Ho. HA9: Wemb 34Sa 67
(off Barnhill Rd.)
Fleming Ho. N4 32Sb 71
Fleming Ho. SE16 47Wb 91
(off George Row)
Fleming Ho. SW17 62Fb 133
Fleming Lodge W9 43Cb 89
(off Admiral Wlk.)
Fleming Mead CR4: Mitc 66Gb 133
Fleming Rd. EN9: Walt A 7Dc 20
Fleming Rd. RM16: Chaf H 49Yd 98
Fleming Rd. SE17 51Rb 113
Fleming Rd. UB1: S'hall 44Da 85
Flemings CM13: Gt War 23Yd 58
Fleming Wlk. NW9 27Ua 48
Fleming Way SE28 45Zc 95
Fleming Way TW7: Isle 56Ha 108
Flemish Flds. KT16: Chert 73J 149
Flemming Av. HA4: Ruis 32X 65
Flempton Rd. E10 32Ac 72
Fletcher Bldgs. WC2 3G 223 (44Nb 90)
(off Martlett Ct.)
Fletcher Cl. E6 44Rc 94
Fletcher Cl. GU21: Wok 10K 167
(off Robin Hood Rd.)
Fletcher Cl. KT16: Ott 79G 148
Fletcher Ct. NW9 27Ua 48
Fletcher Ho. N1 1J 219 (39Ub 71)
(off Nuttall St.)
Fletcher Ho. SE15 52Yb 114
(off Clifton Way)
Fletcher La. E10 31Ec 72
Fletcher Path SE8 52Cc 114
Fletcher Rd. IG7: Chig 22Vc 55
Fletcher Rd. KT16: Ott 79F 148
Fletcher Rd. W4 48Sa 87
Fletchers Cl. BR2: Broml 70Kc 137
Fletcher St. E1 45Wb 91
Fletcher Way HP2: Hem H 1L 3
Fletching Apts. E3 42Bc 92
(off Siyah Gdn.)
Fletching Rd. E5 34Yb 72
Fletching Rd. SE7 51Lc 115
Flete Ho. WD18: Wat 16U 26
Fletton Rd. N11 24Nb 50
Fleur de Lis St. E1 6K 219 (42Vb 91)
Fleur Gates SW19 59Za 110
Flexlands La. GU24: Chob 2F 166
Flexmere Gdns. N17 25Tb 51
Flexmere Rd. N17 25Tb 51
Flight App. NW9 26Va 48
Flight Ho. N1 1H 219 (39Ub 71)
(off Phillipp St.)
Flimwell Cl. BR1: Broml 64Gc 137
Flinders Cl. AL1: St A 4E 6
Flinders Ho. E1 46Xb 91
(off Green Bank)
Flint Cl. BR6: Chels 79Vc 161
Flint Cl. CR0: C'don 72Pb 156
Flint Cl. E15 38Hc 73
Flint Cl. KT23: Bookh 98Ea 192
Flint Cl. RH1: Redh 5P 207
Flint Cl. SM7: Bans 86Db 175
Flint Cotts. KT22: Lea 93Ka 192
(off Gravel Hill)
Flint Down Cl. BR5: St P 67Wc 139
Flintlock Cl. E1 44Xb 91
Flintlock Cl. TW19: Stanw M 56J 105
Flintmill Cres. SE3 54Nc 116
Flinton St. SE17 7J 231 (50Ub 91)
Flint Ri. DA10: Swans 60Ae 121
Flint St. RM20: W Thur 51Xd 120
Flint St. SE17 6G 231 (49Tb 91)
Flip Out Brent Cross 32Xa 68
Flip Out East Ham 39Nc 74
Flip Out Wandsworth 60Db 111
Flitcroft St. WC2 3E 222 (44Mb 90)
Flitton Ho. N1 38Rb 71
(off The Sutton Est.)
Floathaven Cl. SE28 46Wc 95
Floats, The TN13: Riv 93Gd 202
Flock Mill Pl. SW18 60Db 111
Flockton Ho. KT13: Weyb 75Q 150
Flockton St. SE16 47Wb 91
Flodden Rd. SE5 53Sb 113
Flood La. TW1: Twick 60Ja 108
Flood Pas. SE18 48Nc 94
Flood St. SW3 7E 226 (50Gb 89)
Flood Wlk. SW3 51Gb 111
Flora Cl. E14 44Dc 92
Flora Cl. HA7: Stan 20Na 29
Flora Gdns. CR0: New Ad 83Ec 178
Flora Gdns. RM6: Chad H 30Yc 55
Flora Gdns. W6 49Xa 88
(off Albion Gdns.)
Flora Gro. AL1: St A 3D 6
Flora Ho. E3 39Cc 72
(off Garrison Rd.)
Floral Ct. KT21: Asht 90La 172
Floral Ct. WC2 4F 223 (45Nb 90)
Floral Dr. AL2: Lon C 8H 7
Floral Ho. KT16: Chert 74H 149
(off Fox La. Sth.)
Floral Pl. N1 36Tb 71
Floral St. WC2 4F 223 (45Nb 90)
Flora Rd. WD23: Bush 13Z 27
Flora St. DA17: Belv 50Bd 95
Florence Av. EN2: Enf 13Sb 33
Florence Av. KT15: New H 83J 169
Florence Av. SM4: Mord 71Eb 155
Florence Cantwell Wlk. N19 31Nb 70
(off Jessie Blythe La.)
Florence Cl. CM13: Gt War 23Yd 58
Florence Cl. KT12: Walt T 73X 151
Florence Cl. KT2: King T 64Pa 131
Florence Cl. RM12: Horn 33Nd 77
Florence Cl. RM20: Grays 51Ae 121
Florence Cl. WD25: Wat 7W 12
Florence Ct. AL1: St A 2C 6
(off Alma Rd.)
Florence Ct. E11 28Kc 53
Florence Ct. GU21: Knap 10G 166
Florence Ct. N1 38Rb 71
(off Florence St.)
Florence Ct. SW19 65Ab 132
Florence Ct. W9 4A 214 (41Eb 89)
(off Maida Vale)
Florence Dr. EN2: Enf 13Sb 33
Florence Elson Cl. E12 35Qc 74
Florence Farm Mobile Home Pk.
TN15: W King 79Td 164
Florence Gdns. RM6: Chad H 31Yc 75
Florence Gdns. TW18: Staines 66K 127
Florence Gdns. W4 51Sa 109
Florence Ho. KT2: King T 66Pa 131
(off Florence Rd.)
Florence Ho. SE16 50Xb 91
(off Rotherhithe New Rd.)
Florence Ho. W11 45Za 88
(off St Ann's Rd.)
Florence Ho. WD18: Wat 14U 26
Florence Longman Ho. HP3:
Hem H 6M 3
(off Weymouth St.)
Florence Mans. NW4 29Xa 48
(off Vivian Av.)
Florence Mans. SW6 53Bb 111
(off Rostrevor Rd.)
Florence M. SL3: L'ly 10N 81
Florence Nightingale Mus. 2J 229 (47Pb 90)
Florence Rd. BR1: Broml 67Jc 137
Florence Rd. BR3: Beck 68Ac 136
Florence Rd. CR2: Sande 81Tb 177
Florence Rd. E13 40Jc 73
Florence Rd. E6 39Lc 73
Florence Rd. KT12: Walt T 73X 151
Florence Rd. KT2: King T 66Pa 131
Florence Rd. N4 31Pb 70
(not continuous)
Florence Rd. SE14 53Bc 114
Florence Rd. SE2 49Yc 95
Florence Rd. SW19 65Db 133
Florence Rd. TW13: Felt 60X 107
Florence Rd. UB2: S'hall 49Z 85
Florence Rd. W4 48Ta 87
Florence Rd. W5 45Na 87
Florence Root Ho. IG4: Ilf 29Nc 54
(off Ashgrove Rd.)
Florence Sq. E3 42Dc 92
Florence St. E16 42Hc 93
Florence St. N1 38Rb 71
Florence St. NW4 28Ya 48
Florence Ter. SE14 53Bc 114
Florence Ter. SW15 62Ua 132
Florence Way GU21: Knap 10G 166
Florence Way SW12 60Hb 111
Florence Way UB8: Uxb 38L 63
Florey Lodge W9 43Cb 89
(off Admiral Wlk.)
Florey Sq. N21 15Pb 32
Florfield Pas. E8 37Xb 71
(off Reading La.)
Florfield Rd. E8 37Xb 71
Florian SE5 53Ub 113
Florian Av. SM1: Sutt 77Fb 155
Florian Ct. E16 43Jc 93
(off Hastings Rd.)
Florian Rd. SW15 56Ab 110
Florida Cl. WD23: B Hea 19Fa 28
Florida Ct. BR2: Broml 70Hc 137
(off Westmoreland Rd.)
Florida Ct. TW18: Staines 63J 127
Florida Rd. CR7: Thor H 67Rb 135
Florida St. E2 41Wb 91
Florin Ct. EC1 7D 218 (43Sb 91)
(off Charterhouse Sq.)
Florin Ct. N18 21Ub 51
Florin Ct. SE1 2K 231 (47Vb 91)
(off Tanner St.)
Floris Pl. SW4 55Lb 112
Floriston Av. UB10: Hil 38S 64
Floriston Cl. HA7: Stan 25Ka 46
Floriston Ct. UB5: N'olt 36Da 65
Floriston Gdns. HA7: Stan 25Ka 46
Florys Ct. SW19 60Ab 110
Floss St. SW15 54Ya 110
Flotilla Ho. E16 47Kc 93
(off Cable St.)
Flotilla Ho. SW18 55Eb 111
Flounder Ho. SE8 52Dc 114
(off Creative Rd.)
Flower & Dean Wlk. E1 1K 225 (43Vb 91)
Flower Cres. KT16: Ott 79D 148
Flowerdown Ct. HA4: Eastc 30W 44
(off Lidgould Gro.)
Flowerfield TN14: Otf 89Hd 182
Flowerhill Way DA13: Ist R 6A 144
Flower La. NW7 22Va 48
Flower La. RH9: G'stone 2B 210
Flower M. NW11 30Ab 48
Flower Pot Cl. N15 30Vb 51
Flowers Av. HA4: Eastc 30W 44
Flowers Av. HA4: Ruis 30W 44
Flowers Cl. NW2 34Wa 68
Flowersmead SW17 61Jb 134
Flowers M. N19 33Lb 70
Flower Wlk., The SW7 2A 226 (47Eb 89)
Floyd Rd. SE7 50Lc 93
Floyd's La. GU22: Pyr 88J 169
Floyer Cl. TW10: Rich 57Pa 109
Fludyer St. SE13 56Gc 115
Flutemakers M. SW4 57Mb 112
Flux's La. CM16: Epp 5Wc 23
Flyers Way, The TN16: Westrm 98Tc 200
Flying Angel Ho. E16 45Kc 93
(off Victoria Dock Rd.)
Flynn Ct. E14 45Cc 92
(off Garford St.)
Fogerty Cl. EN3: Enf L 9Dc 20
Fold M. SL2: Farn C 7G 60
Foley Ct. DA1: Dart 60Rd 119
(off Churchill Cl.)
Foley Ho. E1 44Yb 92
(off Tarling St.)
Foley M. KT10: Clay 79Ga 152
Foley Rd. KT10: Clay 80Ga 152
Foley Rd. TN16: Big H 90Mc 179
Foley St. W1 1B 222 (43Lb 90)
Foley Wood KT10: Clay 80Ha 152
Folgate St. E1 7J 219 (43Ub 91)
(not continuous)
Foliot Ho. N1 1H 217 (40Pb 70)
(off Priory Grn. Est.)
Foliot St. W12 44Va 88
Folkes La. RM14: Upm 29Vd 58
Folkestone Ct. SL3: L'ly 50C 82
Folkestone Ct. UB5: N'olt 36Ca 66
(off Newmarket Av.)
Folkestone Ho. SE17 7J 231 (50Ub 91)
(off Upnor Way)
Folkestone Rd. E17 28Dc 52
Folkestone Rd. E6 40Qc 74
Folkestone Rd. N18 21Wb 51
Folkingham La. NW9 25Ta 47
Folkington Cnr. N12 22Bb 49
Folland NW9 26Va 48
(off Hundred Acre)
Follett Cl. SL4: Old Win 8M 103
Follett Dr. WD5: Ab L 3V 12
Follett Ho. SW10 52Fb 111
(off Worlds End Est.)
Follett St. E14 44Ec 92
Follingham Ct. N1 3J 219 (41Ub 91)
(off Drysdale Pl.)
Folly, The GU18: Light 4A 166
Folly Av. AL3: St A 1A 6
Folly Brook & Darland's Lake
Nature Reserve 20Ab 30
Folly Cl. WD7: R'lett 8Ha 14
Folly Ct. AL3: St A 1B 6
(off Folly Av.)
Follyfield Rd. SM7: Bans 86Cb 175
Folly La. AL3: St A 1A 6
Folly La. E17 25Ac 52
Folly La. E4 23Bc 52
Folly M. W11 44Bb 89
Folly Pathway WD7: R'lett 7Ha 14
Folly Wall E14 47Ec 92
Fonda Ct. E14 45Cc 92
(off Premiere Pl.)
Fondant Ct. E3 40Dc 72
(off Taylor Pl.)
Fontaine Ho. E17 28Cc 52
(off Hoe St.)
Fontaine Rd. SW16 66Pb 134
Fontarabia Rd. SW11 56Jb 112
Fontayne Av. IG7: Chig 21Sc 54
Fontayne Av. RM1: Rom 26Gd 56
Fontayne Av. RM13: Rain 38Gd 76
Fontenelle SE5 53Ub 113
Fontenoy Ho. SE11 6B 230 (49Rb 91)
(off Kennington La.)
Fontenoy Rd. SW12 61Kb 134
Fonteyne Gdns. IG8: Wfd G 26Mc 53
Fonthill Cl. SE20 68Wb 135
Fonthill Gdns. DA1: Dart 55Qd 119
(off Halcrow Av.)
Fonthill Ho. SW1 7A 228 (50Kb 90)
(part of Abbots Mnr.)
Fonthill Ho. W14 48Ab 88
(off Russell Rd.)
Fonthill M. N4 33Pb 70
Fonthill Rd. N4 32Pb 70
Font Hills N2 26Eb 49
Fontley Way SW15 59Wa 110
Fontmell Cl. TW15: Ashf 64Q 128
Fontmell Pk. TW15: Ashf 64P 127
Fontwell Cl. HA3: Hrw W 24Ga 46
Fontwell Cl. UB5: N'olt 37Ca 65
Fontwell Dr. BR2: Broml 71Qc 160
Fontwell Pk. Gdns. RM12: Horn 35Nd 77
Foord Cl. DA2: Dart 61Ud 142
Football La. HA1: Harr 32Ha 66
Footbury Hill Rd. BR6: St M Cry 72Wc 161
Footpath, The SW15 58Wa 110
FOOTS CRAY 65Yc 139
Foots Cray High St. DA14: Sidc 65Yc 139
Foots Cray La. DA14: Sidc 60Yc 117
Foots Cray Meadows 63Zc 139
Footscray Rd. SE9 58Qc 116
Forbench Cl. GU23: Rip 94K 189
Forber Ho. E2 41Yb 92
(off Cornwall Av.)
Forbes Av. EN6: Pot B 5Fb 17
Forbes Cl. NW2 34Wa 68
Forbes Cl. RM11: Horn 32Kd 77
Forbes Ho. E7 36Lc 73
(off Romford Rd.)
Forbes Ho. W4 50Qa 87
(off Stonehill Rd.)
Forbes St. E1 44Wb 91
Forbes Way HA4: Ruis 33X 65
Forburg Rd. N16 32Wb 71
Forbury Rd. SE13 55Gc 115
FORCE GREEN 96Tc 200
Force Grn. La. TN16: Westrm 96Tc 200
Fordbridge Cl. KT16: Chert 74K 149
Fordbridge Ct. TW15: Ashf 65N 127
Fordbridge Pk. TW16: Sun 72V 150
Fordbridge Rd. TW15: Ashf 65N 127
Fordbridge Rd. TW16: Sun 72U 150
Fordbridge Rd. TW17: Shep 72U 150
FORDBRIDGE RDBT. 65N 127
Ford Cl. CR7: Thor H 71Rb 157
Ford Cl. E3 40Ac 72
Ford Cl. HA1: Harr 31Fa 66
Ford Cl. RM13: Rain 38Hd 76
Ford Cl. TW15: Ashf 65N 127
Ford Cl. TW17: Shep 70Q 128
Ford Cl. WD23: Bush 14Ea 28
Fordcroft Rd. BR5: St M Cry 71Xc 161
Forde Av. BR1: Broml 69Lc 137
Fordel Rd. SE6 60Ec 114
Ford End IG8: Wfd G 23Kc 53
Ford End UB9: Den 33H 63
Fordgate Bus. Pk. DA17: Belv 47Ed 96
Fordham KT1: King T 68Qa 131
(off Excelsior Cl.)
Fordham Cl. EN4: Cockf 13Gb 31
Fordham Cl. KT4: Wor Pk 74Xa 154
Fordham Cl. RM11: Horn 31Qd 77
Fordham Ho. SE14 52Ac 114
(off Angus St.)
Fordham Rd. EN4: Cockf 13Fb 31
Fordhams Row RM16: Ors 3D 100
Fordham St. E1 44Wb 91
Fordhook Av. W5 46Pa 87
Ford Ho. EN5: New Bar 15Db 31
Fordie Ho. SW1 4G 227 (48Hb 89)
(off Sloane St.)
Ford Ind. Pk. RM9: Dag 42Dd 96
Fordingley Rd. W9 41Bb 89
Fordington Ho. SE26 62Wb 135
Fordington Rd. N6 29Hb 49
Ford La. RM13: Rain 38Hd 76
Ford La. SL0: Iver 44J 83
Ford Lodge RM7: Rom 28Fd 56
Fordmill Rd. SE6 61Cc 136
Ford Pl. RM15: S Ock 45Wd 98
Ford Rd. DA11: Nflt 57De 121
Ford Rd. E3 40Bc 72
Ford Rd. GU22: Wok 92D 188
Ford Rd. GU24: Bisl 6C 166
Ford Rd. GU24: Chob 2G 166
Ford Rd. GU24: W End 6C 166
Ford Rd. KT16: Chert 74K 149
Ford Rd. RM10: Dag 38Cd 76
Ford Rd. RM9: Dag 38Bd 75
Ford Rd. TW15: Ashf 63P 127
Fords Gro. N21 18Sb 33
Fords Pk. Rd. E16 43Jc 93
Fords Pl. HA6: Nwood 26W 44
Ford Sq. E1 43Xb 91
Ford St. E16 44Hc 93
Ford St. E3 39Ac 72
Fordview Ind. Est. RM13: Rain 41Fd 96
Fordwater Rd. KT16: Chert 74K 149
Fordwater Trad. Est. KT16: Chert 74L 149
Fordwich Cl. BR6: Orp 73Vc 161
Fordwych Rd. NW2 35Ab 68
Fordyce Cl. RM11: Horn 31Pd 77
Fordyce Rd. SE13 58Ec 114
Fordyke Rd. RM8: Dag 33Bd 75
Forefield AL2: Chis G 9N 5
Foreign St. SE5 54Rb 113
Foreland Ct. NW4 25Za 48
Foreland Ho. W11 45Ab 88
(off Walmer Rd.)
Foreland St. SE18 49Tc 94
Forelle Way SM5: Cars 81Hb 175
Foreman Ct. TW1: Twick 60Ha 108
Foreman Ho. SE4 56Zb 114
(off Billingford Cl.)
Foremark Cl. IG6: Ilf 22Vc 55
Foreshore SE8 49Bc 92
Forest, The E11 28Gc 53
Forest App. E4 17Gc 35
Forest App. IG8: Wfd G 24Jc 53
Forest Av. E4 17Gc 35
Forest Av. HP3: Hem H 4M 3
Forest Av. IG7: Chig 22Qc 54
Forest Bus. Pk. E10 31Zb 72
Forest Cl. BR7: Chst 67Qc 138
Forest Cl. E11 29Jc 53
Forest Cl. EN9: Walt A 9Kc 21
Forest Cl. GU22: Pyr 87F 168
Forest Cl. IG8: Wfd G 20Kc 35
Forest Cl. KT24: E Hor 97V 190
Forest Cl. N10 25Kb 50
Forest Cl. NW6 38Ab 68
Forest Cl. SL2: Wex 3M 81
Forest Ct. E11 28Gc 53
Forest Ct. E4 18Hc 35
Forest Ct. N12 22Db 49
Forest Cres. KT21: Asht 88Qa 173
Forest Cft. SE23 61Xb 135
FORESTDALE 81Bc 178
Forestdale N14 21Mb 50
Forestdale Cen., The 80Bc 158
Forest Dene Ct. SM2: Sutt 79Eb 155
Forest Dr. BR2: Kes 77Nc 160
Forest Dr. BR3: Beck 73Bc 158
Forest Dr. CM16: They B 8Uc 22
Forest Dr. E12 34Mc 73
Forest Dr. IG8: Wfd G 24Fc 53
Forest Dr. KT20: Kgswd 93Bb 195
Forest Dr. TW16: Sun 66V 128
Forest Dr. E. E11 31Fc 73
Forest Dr. W. E11 31Ec 72
Forest Edge IG9: Buck H 21Lc 53
Forester Ho. E14 45Ac 92
(off Victory Pl.)
Forester Rd. SE15 55Xb 113
Foresters Cl. GU21: Wok 10K 167
Foresters Cl. SM6: W'gton 80Mb 156
Foresters Cl. IG10: Lough 12Qc 36
Foresters Cres. DA7: Bex 56Dd 118
Foresters Dr. E17 28Fc 53
Foresters Dr. SM6: W'gton 80Mb 156
Forest Gdns. N17 26Vb 51
Forest Ga. KT24: E Hor 96V 190
Forest Ga. NW9 28Ua 48
FOREST GATE 36Jc 73
Forest Gate Learning Zone 35Jc 73
(off Woodford Rd.)
Forest Ga. Retreat E7 36Jc 73
(off Odessa Rd.)
Forest Gate Station (Rail & Crossrail) 36Jc 73
Forest Glade E11 30Gc 53
Forest Glade E4 21Gc 53
Forest Gro. E8 37Vb 71
Forest Hgts. IG9: Buck H 19Jc 35
FOREST HILL 61Yb 136
Forest Hill Bus. Cen. SE23 61Yb 136
(off Clyde Va.)
Forest Hill Ind. Est. SE23 61Yb 136
Forest Hill Pools 61Yb 136
Forest Hill Rd. SE22 57Xb 113
Forest Hill Rd. SE23 58Yb 114
Forest Hill School Sports Cen. 62Zb 136
Forest Hill Station (Rail &
Overground) 61Yb 136
Forestholme Cl. SE23 61Yb 136
Forest Ind. Pk. IG6: Ilf 25Uc 54
Forest La. E15 36Gc 73
Forest La. E7 36Hc 73
Forest La. IG7: Chig 22Qc 54
Forest La. KT24: E Hor 96V 190
Forest La. WD7: Shenl 3La 14
Forest Lodge SE23 62Yb 136
(off Dartmouth Rd.)
Forest Mt. Rd. IG8: Wfd G 24Fc 53
Forest Nature Reserve, The 96V 190
Forest Pk. Crematorium 23Yc 55
Forest Point E7 36Kc 73
(off Windsor Rd.)
Fore St. EC2 1E 224 (43Sb 91)
Fore St. HA5: Eastc 28V 44
Fore St. N18 23Vb 51
Fore St. N9 21Wb 51
Fore St. Av. EC2 1F 225 (43Tb 91)
Fore St. Library 22Wb 51
Forest Ridge BR2: Kes 77Nc 160
Forest Ridge BR3: Beck 69Cc 136
Forest Ri. E17 27Fc 53
Forest Rd. SL4: Wind Ash La. 4B 102
Forest Rd. SL4: Wind Plain Ride 10B 102
Forest Rd. IG10: Lough Staple's Rd. 13Mc 35
Forest Rd. IG10: Lough The Ditches Ride 9Qc 22
Forest Rd. CM16: Epp 7Qc 22
Forest Rd. DA8: Erith 53Jd 118
Forest Rd. E11 31Fc 73
Forest Rd. E17 27Dc 52
Forest Rd. E7 35Jc 73
Forest Rd. E8 37Vb 71
Forest Rd. EN3: Enf W 8Ac 20
Forest Rd. EN8: Chesh 1Zb 20
Forest Rd. EN9: Epp 8Nc 22
Forest Rd. EN9: Walt A 8Nc 22
Forest Rd. GU22: Pyr 87F 168
Forest Rd. IG6: Chig 26Tc 54
Forest Rd. IG6: Ilf 26Tc 54
Forest Rd. IG8: Wfd G 20Jc 35
Forest Rd. KT24: E Hor 99V 190
Forest Rd. KT24: Eff J 99V 190
Forest Rd. N17 28Yb 52
Forest Rd. N9 18Xb 33
Forest Rd. RM7: Mawney 27Dd 56
Forest Rd. SM3: Sutt 74Cb 155
Forest Rd. TW13: Felt 61Y 129
Forest Rd. TW9: Kew 52Qa 109
Forest Rd. WD25: Wat 5W 12
Forest Side CM16: Epp 5Tc 22
Forest Side E4 17Hc 35
Forest Side E7 35Kc 73
Forest Side EN9: Walt A 8Lc 21
Forest Side IG9: Buck H 18Lc 35
Forest Side KT4: Wor Pk 74Va 154
Forest St. E7 36Jc 73
Forest Ter. IG7: Chig 22Qc 54
Forest Trad. Est. E17 27Zb 52
Forest Vw. E11 31Hc 73
Forest Vw. E4 17Fc 35
Forest Vw. Av. E10 29Fc 53
Forest Vw. Rd. E12 35Nc 74
Forest Vw. Rd. E17 25Ec 52
Forest Vw. Rd. IG10: Lough 14Mc 35
Forest Wlk. WD23: Bush Aldenham 11Ba 27
Forest Wlk. N10: Bounds Grn. 25Kb 50
Forest Way BR5: St M Cry 71Vc 161
Forest Way DA15: Sidc 59Tc 116
Forest Way IG10: Lough 13Nc 36
Forest Way IG8: Wfd G 21Kc 53
Forest Way KT21: Asht 89Pa 173
Forest Way N19 33Lb 70
Forfar Ho. WD19: Wat 20Y 27
Forfar Rd. N22 25Rb 51
Forfar Rd. SW11 53Jb 112
Forge, The EN6: N'thaw 2Gb 17
Forge Av. CR5: Coul 92Qb 196
Forge Bri. La. CR5: Coul 94Kb 196
Forge Cl. BR2: Hayes 74Jc 159
Forge Cl. CM13: Gt War 25Wd 58
Forge Cl. UB3: Harl 51T 106
Forge Cl. WD4: Chfd 3J 11
Forge Cotts. W5 46Ma 87
Forge Dr. KT10: Clay 80Ja 152
Forge Dr. SL2: Farn C 7G 60
Forge End AL2: Chis G 8N 5
Forge End GU21: Wok 89A 168
Forgefield TN16: Big H 88Mc 179
Forge La. DA12: Grav'nd 1H 145
Forge La. DA12: Shorne 4N 145
Forge La. DA4: Hort K 70Sd 142
Forge La. HA6: Nwood 24U 44
Forge La. SM3: Cheam 80Ab 154
Forge La. TN15: W King 82Wd 184
Forge La. TW10: Ham 60Na 109
Forge La. TW13: Hanw 64Aa 129
Forge La. TW16: Sun 69W 128
Forge M. CR0: Addtn 78Cc 158
Forge M. TW16: Sun 69W 128
(off Forge La.)
Forge Pl. DA12: Grav'nd 10H 123
Forge Pl. NW1 37Jb 70
Forge Pl. WD6: E'tree 16Ma 29
(off New Rd.)
Forge Sq. E14 49Dc 92
Forge Steading SM7: Bans 87Db 175
Forge Way TN14: S'ham 83Hd 182
Forlong Path UB5: N'olt 37Aa 65
(off Cowings Mead)

Forman Pl. N16 35Vb 71
Formation, The E16 47Rc 94
(off Woolwich Mnr. Way)
Formby Av. HA7: Stan 27La 46
Formby Cl. SL3: L'ly 49E 82
Formby Ct. N7 36Qb 70
(off Morgan Rd.)
Formosa Ho. E1 42Ac 92
(off Ernest St.)
Formosa St. W9 42Db 89
Formunt Cl. E16 43Hc 93
Forres Gdns. NW11 30Cb 49
Forres Ho. CM14: W'ley 21Yd 58
Forrester Ho. AL1: St A 2B 6
(off St Peter's St.)
Forrester Path SE26 63Yb 136
Forresters Apts. IG11: Bark 38Sc 74
(off Linton Rd.)
Forrester Way E15 37Fc 73
Forrest Gdns. SW16 69Pb 134
Forrest Shaw DA10: Swans 59Ae 121
Forris Av. UB3: Hayes 46V 84
Forset Ct. W2 2E 220 (44Gb 89)
(off Edgware Rd.)
Forset St. W1 2E 220 (44Gb 89)
(not continuous)
Forstal Cl. BR2: Broml 69Jc 137
Forster Cl. IG8: Wfd G 24Fc 53
Forster Ho. BR1: Broml 63Fc 137
Forster Ho. SW17 62Fb 133
(off Grosvenor Way)
Forster Rd. BR3: Beck 69Ac 136
Forster Rd. E17 30Ac 52
Forster Rd. N17 27Vb 51
Forster Rd. SW2 59Nb 112
Forsters Cl. RM6: Chad H 30Bd 55
Forsters Way UB4: Yead 44X 85
Forston Apts. SW11 56Gb 111
(off Monarch Square)
Forston St. N1 1E 218 (40Sb 71)
Forsyte Cres. SE19 67Ub 135
Forsyte Ho. SW3 7E 226 (50Gb 89)
(off Chelsea Mnr. St.)
Forsythe Shades Ct. BR3: Beck 67Ec 136
Forsyth Gdns. SE17 51Rb 113
Forsyth Ho. E9 38Yb 72
(off Frampton Pk. Rd.)
Forsyth Ho. SW1 7C 228 (50Lb 90)
(off Tachbrook St.)
Forsythia Cl. IG1: Ilf 36Rc 74
Forsythia Gdns. SL3: L'ly 48A 82
Forsyth Path GU21: Wok 85F 168
Forsyth Pl. EN1: Enf 15Ub 33
Forsyth Rd. GU21: Wok 87E 168
Forterie Gdns. IG3: Bark 34Wc 75
Forterie Gdns. IG3: Ilf 34Wc 75
Fortescue Av. E8 38Xb 71
Fortescue Av. TW2: Twick 62Ea 130
Fortescue Rd. HA8: Edg 25Ta 47
Fortescue Rd. KT13: Weyb 77P 149
Fortescue Rd. SW19 66Fb 133
Fortess Gro. NW5 36Lb 70
Fortess Rd. NW5 36Kb 70
Fortess Wlk. NW5 36Kb 70
Fortess Yd. NW5 35Kb 70
Forte St. SE18 53Nc 116
(off Tellson Av.)
Forthbridge Rd. SW11 56Jb 112
Forth Ho. E3 40Bc 72
(off Tredegar Rd.)
Forth Rd. RM14: Upm 30Td 58
Forties Experience 13Ba 27
(off Lincoln Fld.)
Fortin Cl. RM15: S Ock 45Wd 98
Fortin Path RM15: S Ock 45Wd 98
Fortin Way RM15: S Ock 45Wd 98
Fortis Cl. E16 44Lc 93
Fortis Ct. N10 27Jb 50
FORTIS GREEN 28Hb 49
Fortis Grn. N10 27Jb 50
Fortis Grn. N2 28Gb 49
Fortis Grn. Av. N2 27Hb 49
Fortis Grn. Rd. N10 27Jb 50
Fortismere Av. N10 27Jb 50
Fortius Apts. E3 40Cc 72
(off Tredegar La.)
Fortius Wlk. E20 37Ec 72
Fort La. RH2: Reig 2K 207
Fortnam Rd. N19 33Mb 70
Fortnum's Acre HA7: Stan 23Ha 46
Fortress Distribution Pk. RM18: Tilb 6C 122
Fort Rd. RM18: Tilb 6C 122
Fort Rd. RM18: W Til 6C 122
Fort Rd. SE1 49Vb 91
Fort Rd. TN14: Hals 87Ed 182
Fort Rd. UB5: N'olt 38Ca 65
Fortrose Cl. E14 44Fc 93
Fortrose Gdns. SW2 60Nb 112
(not continuous)
Fortrye Cl. DA11: Nflt 1A 144
Fort St. E1 1J 225 (43Ub 91)
Fort St. E16 46Kc 93
Fortuna Cl. DA3: Hartl 70Be 143
Fortuna Cl. N7 37Pb 70
Fortuna Ho. E20 36Dc 72
(off Scarlet Cl.)
Fortune Av. HA8: Edg 25Ra 47
Fortune Ct. E8 38Vb 71
(off Queensbridge Rd.)
Fortune Ct. IG11: Bark 40Yc 75
Fortune Ga. Rd. NW10 39Ua 68
FORTUNE GREEN 36Cb 69
Fortune Grn. Rd. NW6 35Cb 69
Fortune Ho. EC1 6E 218 (42Sb 91)
(off Fortune St.)
Fortune Ho. SE11 6K 229 (49Qb 90)
(off Marylee Way)
Fortune La. WD6: E'tree 16Ma 29
Fortune Pl. SE1 50Vb 91
Fortunes Mead UB5: N'olt 37Aa 65
Fortune St. EC1 6E 218 (42Sb 91)
Fortune St. Pk. 6E 218 (42Sb 91)
Fortunes Wlk. E20 36Ec 72
Fortune Theatre 3G 223 (44Nb 90)
(off Russell St.)
Fortune Wlk. SE28 48Tc 94
(off Broadwater Rd.)
Fortune Way NW10 41Wa 88
Forty Acre La. E16 43Jc 93
Forty Av. HA9: Wemb 34Pa 67
Forty Cl. HA9: Wemb 34Pa 67
Forty Footpath SW14 55Sa 109
Fortyfoot Rd. KT22: Lea 93La 192
Forty Foot Way SE9 59Sc 116
Forty Hall Estate 9Vb 19
FORTY HILL 10Ub 19
Forty Hill EN2: Enf 10Ub 19
Forty Hill Country Pk. 9Vb 19
Forty La. HA9: Wemb 33Ra 67
Forum, The KT16: Chert 74H 149
Forum, The KT8: W Mole 70Da 129
Forum Cl. E3 39Cc 72
Forum Ho. HA9: Wemb 35Qa 67
Forum Magnum Sq. SE1 1J 229 (47Pb 90)
(off York Rd.)
Forumside HA8: Edg 23Qa 47
Forum Way HA8: Edg 23Qa 47
Forval Cl. CR4: Mitc 71Hb 155
Forward Bus. Cen., The E16 42Fc 93
Forward Dr. HA3: W'stone 28Ha 46
Fosbrooke Ho. SW8 52Nb 112
(off Davidson Gdns.)
Fosbury M. W2 45Db 89
Foscote Ct. W9 42Cb 89
(off Amberley Rd.)
Foscote M. W9 42Cb 89
Foscote Rd. NW4 30Xa 48
Foskett Ho. N2 26Fb 49
(off The Grange)
Foskett M. E8 36Vb 71
Foskett Rd. SW6 54Bb 111
Foss Av. CR0: Wadd 78Qb 156
Fossdene Rd. SE7 50Kc 93
Fossdyke Cl. UB4: Yead 43Aa 85
Fosse Ho. AL3: St A 3N 5
Fosset Lodge DA7: Bex 53Ed 118
Fosse Way KT14: W Byf 85H 169
Fosse Way W13 43Ja 86
Foss Ho. NW8 40Db 69
(off Carlton Hill)
Fossil Ct. SE1 3H 231 (48Ub 91)
(off Long La.)
Fossil Rd. SE13 55Cc 114
Fossington Rd. DA17: Belv 49Zc 95
Foss Rd. SW17 63Fb 133
Fossway RM8: Dag 33Yc 75
Foster Av. SL4: Wind 5C 102
Foster Cl. EN8: Chesh 2Ac 20
Foster Ct. E16 45Hc 93
(off Tarling Rd.)
Foster Ct. NW1 38Lb 70
(off Royal College St.)
Foster Ct. NW4 28Ya 48
Fosterdown RH9: G'stone 1P 209
Foster Dr. DA1: Dart 57Pd 119
Foster Ho. SE14 53Bc 114
Foster Ho. WD6: Bore 13Sa 29
Foster La. EC2 2D 224 (44Sb 91)
Foster Rd. E13 42Jc 93
Foster Rd. HP1: Hem H 4K 3
Foster Rd. W3 45Ua 88
Foster Rd. W4 50Ta 87
Fosters Cl. BR7: Chst 64Pc 138
Fosters Cl. E18 25Kc 53
Fosters Gro. GU20: W'sham 7A 146
Fosters La. GU21: Knap 9G 166
Fosters M. DA3: Lfield 68Ae 143
Fosters Path SL2: Slou 2D 80
Foster St. NW4 28Ya 48
Foster's Way SW18 60Db 111
Foster Wlk. NW4 28Ya 48
Fothergill Cl. E13 40Jc 73
Fothergill Dr. N21 15Nb 32
Fotheringay Gdns. SL1: Slou 5E 80
Fotheringham Rd. EN1: Enf 14Vb 33
(not continuous)
Fotherley Rd. WD3: Rick 19H 25
Foubert's Pl. W1 3B 222 (44Lb 90)
Foulden Rd. N16 35Vb 71
Foulden Ter. N16 35Vb 71
(off South Gro.)
Foulis Ter. SW7 7C 226 (50Fb 89)
Foulser Rd. SW17 62Hb 133
Foulsham Rd. CR7: Thor H 69Sb 135
Foundation Pl. SE9 57Qc 116
(off Archery Rd.)
Founder Cl. E6 44Rc 94
Founders Cl. UB5: N'olt 41Ba 85
Founders Ct. EC2 2F 225 (44Tb 91)
(off Lothbury)
Founders Gdns. SE19 66Sb 135
Founders Ho. SW1 7D 228 (50Mb 90)
(off Aylesford St.)
Foundling Ct. WC1 5F 217 (42Nb 90)
(off Brunswick Cen.)
Foundling Mus., The 5G 217 (42Nb 90)
(off Brunswick Sq.)
Foundry, The EC2 4J 219 (41Ub 91)
(off Dereham Pl.)
Foundry Cl. SE16 46Ac 92
Foundry Ct. KT16: Chert 73J 149
Foundry Ct. SL2: Slou 6K 81
Foundry Ga. EN8: Walt C 5Ac 20
Foundry Ho. E14 43Dc 92
(off Morris Rd.)
Foundry Ho. E15 37Fc 73
(off Forrester Way)
Foundry Ho. SW8 53Kb 112
(off Lockington Rd.)
Foundry La. SL3: Hort 55D 104
Foundry M. E17 27Ec 52
Foundry M. KT16: Chert 73J 149
(off Gogmore La.)
Foundry M. NW1 5C 216 (42Lb 90)
Foundry M. SW13 54Va 110
Foundry M. TW3: Houn 56Da 107
Foundry Pl. E1 43Yb 92
(off Jubilee St.)
Foundry Pl. SW18 59Db 111
Founes Dr. RM16: Chaf H 48Ae 99
Fountain Cl. E5 35Xb 71
Fountain Cl. SE18 50Rc 94
Fountain Cl. UB8: Hil 43S 84
Fountain Ct. DA15: Sidc 58Xc 117
Fountain Ct. DA4: Eyns 75Nd 163
Fountain Ct. EC4 4K 223 (45Qb 90)
Fountain Ct. EN8: Chesh 2Zb 20
(off Westbury)
Fountain Ct. SE23 61Zb 136
Fountain Ct. SW1 6K 227 (49Kb 90)
(off Buckingham Pal. Rd.)
Fountain Ct. W11 47Za 88
(off Clearwater Ter.)
Fountain Ct. WD6: Bore 12Qa 29
Fountain Dr. SE19 63Vb 135
Fountain Dr. SM5: Cars 81Hb 175
Fountain Gdns. SL4: Wind 5H 103
Fountain Grn. Sq. SE16 47Wb 91
Fountain Ho. CR4: Mitc 68Hb 133
Fountain Ho. E2 5K 219 (41Vb 91)
(off Redchurch St.)
Fountain Ho. NW6 38Ab 68
Fountain Ho. SE16 47Wb 91
(off Bermondsey Wall E.)
Fountain Ho. SW6 54Eb 111
Fountain Ho. W1 6H 221 (46Jb 90)
(off Park St.)
Fountain M. N5 35Sb 71
(off Highbury Grange)
Fountain M. NW3 37Hb 69
Fountain Park Way W12 45Ya 88
Fountain Pl. EN9: Walt A 6Ec 20
Fountain Pl. SW9 53Qb 112
Fountain Rd. CR7: Thor H 67Sb 135
Fountain Rd. RH1: Redh 8N 207
Fountain Rd. SW17 64Fb 133
FOUNTAIN RDBT. New Malden 70Ua 132
Fountains, The IG10: Lough 17Mc 35
Fountains, The N3 24Db 49
(off Ballards La.)
Fountains Av. TW13: Hanw 62Ba 129
Fountains Cl. TW13: Hanw 61Ba 129
(not continuous)
Fountains Cres. N14 17Nb 32
Fountain Sq. SW1 5A 228 (49Kb 90)
Fountain Wlk. DA11: Nflt 8A 122
Fountayne Bus. Cen. N15 28Wb 51
Fountayne Rd. N15 28Wb 51
Fountayne Rd. N16 33Wb 71
Fount St. SW8 52Mb 112
Fouracre NW1 4B 216 (41Lb 90)
(off Stanhope St.)
Fouracre Path SE25 72Ub 157
Four Acres KT11: Cobh 85Aa 171
Four Acres N12 20Db 31
Fouracres EN3: Enf H 11Ac 34
Fouracres Dr. HP3: Hem H 4P 3
Fouracres Wlk. HP3: Hem H 4P 3
Four Casson Sq. SE1 7J 223 (46Pb 90)
Fourdrinier Way HP3: Hem H 5M 3
Fourfield Cl. KT18: Head 94Sa 193
Fourier Gro. DA1: Dart 54Nd 119
Fourland Wlk. HA8: Edg 23Sa 47
Fournier St. E1 7K 219 (43Vb 91)
Four Oaks CM15: B'wood 20Ae 41
Four Oaks RH8: Oxt 5M 211
Fourscore Mans. E8 38Wb 71
(off Shrubland Rd.)
Four Seasons Cl. E3 40Cc 72
Four Seasons Cres. SM3: Sutt 75Bb 155
Four Seasons Ter. UB7: W Dray 47Q 84
Four Sq. Ct. TW3: Houn 58Ca 107
Fourteenth Av. KT20: Lwr K 98Ab 194
Fourth Av. DA11: Nflt 10A 122
Fourth Av. E12 35Pc 74
Fourth Av. KT20: Lwr K 97Ab 194
Fourth Av. RM20: W Thur 51Wd 120
Fourth Av. RM7: Rush G 32Fd 76
Fourth Av. UB3: Hayes 46V 84
Fourth Av. W10 42Ab 88
Fourth Av. WD25: Wat 7Z 13
Fourth Cross Rd. TW2: Twick 61Fa 130
Fourth Dr. CR5: Coul 88Lb 176
Fourth Way HA9: Wemb 35Ra 67
Four Trees AL2: Chis G 7P 5
Four Tubs, The WD23: Bush 17Fa 28
Fourways Mkt. AL9: Wel G 6E 8
(off Dixons Hill Rd.)
Four Wents KT11: Cobh 86Y 171
Four Wents, The E4 18Fc 35
Four Wents Cl. TN15: Bor G 92Be 205
Fovant Ct. SW8 54Lb 112
Fowey Av. IG4: Ilf 29Mc 53
Fowey Cl. E1 46Xb 91
Fowey Ho. SE11 7A 230 (50Qb 90)
(off Kennings Way)
Fowey Pl. SM2: Sutt 81Cb 175
Fowler Cl. SW11 55Fb 111
Fowler Ho. N15 29Tb 51
(off South Gro.)
Fowler Rd. CR4: Mitc 68Jb 134
Fowler Rd. E7 35Jc 73
Fowler Rd. IG6: Ilf 23Xc 55
Fowler Rd. N1 39Rb 71
Fowlers Cl. DA14: Sidc 64Ad 139
Fowlers Mead GU24: Chob 1J 167
Fowlers M. N19 33Lb 70
(off Holloway Rd.)
Fowler's Wlk. W5 42Ma 87
Fowler Way UB10: Uxb 40N 63
Fowley Cl. EN8: Walt C 6Bc 20
Fowley Mead Pk. EN8: Walt C 6Cc 20
Fownes St. SW11 55Gb 111
Foxacre CR3: Cat'm 94Ub 197
Fox All. WD18: Wat 15Y 27
Fox & Knot St. EC1 7C 218 (43Rb 91)
(off Charterhouse Sq.)
Foxberry Rd. SE4 55Ac 114
Foxberry Wlk. DA11: Nflt 62Fe 143
(off Ashmore Gdns.)
Foxboro Rd. RH1: Redh 4B 208
Foxborough Cl. SL3: L'ly 50C 82
Foxborough Gdns. SE4 57Cc 114
Foxbourne Rd. SW17 61Jb 134
Fox Burrow Rd. IG7: Chig 21Zc 55
Foxbury DA3: Nw A G 76Ae 165
Foxbury TN15: Plat 92De 205
Foxbury Av. BR7: Chst 65Tc 138
Foxbury Cl. BR1: Broml 65Kc 137
Foxbury Cl. BR6: Chels 78Wc 161
Foxbury Dr. BR6: Chels 79Wc 161
Foxbury Rd. BR1: Broml 65Jc 137
Fox Cl. BR6: Chels 78Wc 161
Fox Cl. E1 42Yb 92
Fox Cl. E16 43Jc 93
Fox Cl. GU22: Pyr 87F 168
Fox Cl. KT13: Weyb 78T 150
Fox Cl. RM5: Col R 22Dd 56
Fox Cl. WD23: Bush 14Da 27
Fox Cl. WD6: E'tree 16Ma 29
Foxcombe CR0: New Ad 79Dc 158
(not continuous)
Foxcombe Cl. E6 40Mc 73
Foxcombe Rd. SW15 60Wa 110
FOX CORNER 7G 186
Fox Corner Community Wildlife Area 7G 186
Foxcote SE5 50Ub 91
Fox Covert KT22: Fet 96Fa 192
Fox Covert Cl. SL5: S'hill 1A 146
Foxcroft AL1: St A 4E 6
Foxcroft SL0: Iver 45H 83
Foxcroft WC1 2J 217 (40Pb 70)
(off Penton Ri.)
Foxcroft Rd. SE18 53Rc 116
Foxdell HA6: Nwood 23T 44
Foxdell Way SL9: Chal P 22A 42
Foxdene Cl. E18 27Kc 53
Foxearth Cl. TN16: Big H 90Nc 180
Foxearth Rd. CR2: Sels 82Yb 178
Foxearth Spur CR2: Sels 81Yb 178
Foxes Dale BR2: Broml 69Fc 137
Foxes Dale SE3 55Jc 115
Foxes Dr. EN7: Chesh 1Wb 19
Foxes Grn. RM16: Ors 7C 100
Foxes La. AL9: Wel G 5F 8
Foxes Pde. EN9: Walt A 6Ec 20
(off Stewardstone Rd.)
Foxes Path GU4: Sut G 98B 188
Foxfield NW1 1A 216 (39Kb 70)
(off Arlington Rd.)
Fox Fld. Cl. RM20: W Thur 51Wd 120
Foxfield Cl. HA6: Nwood 23V 44
Foxfield Rd. BR6: Orp 75Tc 160
Foxglove Cl. AL10: Hat 1D 8
Foxglove Cl. DA15: Sidc 58Wc 117
Foxglove Cl. KT16: Chert 74L 149
Foxglove Cl. N9 18Yb 34
Foxglove Cl. SL2: Wex 3M 61
Foxglove Cl. TW19: Stanw 60M 105
Foxglove Cl. UB1: S'hall 45Aa 85
Foxglove Cl. UB7: W Dray 47P 83
Foxglove Ct. E3 40Cc 72
(off Four Seasons Cl.)
Foxglove Ct. HA0: Wemb 40Na 67
Foxglove Gdns. CR8: Purl 83Nb 176
Foxglove Gdns. E11 28Lc 53
Foxglove Gdns. IG7: Chig 22Ad 55
Foxglove La. KT9: Chess 77Qa 153
Foxglove Path SE28 46Uc 94
(off Martins Pl.)
Foxglove Pl. RM3: Rom 23Ld 57
Foxglove Rd. RM15: S Ock 44Yd 98
Foxglove Rd. RM7: Rush G 33Gd 76
Foxgloves, The HP1: Hem H 3G 2
Foxglove St. W12 45Va 88
Foxglove Way SM6: W'gton 74Kb 156
Fox Gro. KT12: Walt T 73X 151
Foxgrove N14 20Nb 32
Foxgrove Av. BR3: Beck 66Dc 136
Foxgrove Dr. GU21: Wok 87C 168
Foxgrove Path WD19: Wat 22Z 45
Foxgrove Rd. BR3: Beck 66Dc 136
Foxhall Rd. RM14: Upm 36Sd 78
Foxham Rd. N19 34Mb 70
Foxhanger Gdns. GU22: Wok 88C 168
Foxherne SL1: Slou 7N 81
Foxherne SL3: L'ly 7N 81
Fox Hill SE19 66Vb 135
Foxhill WD24: Wat 8W 12
Fox Hill Gdns. SE19 66Vb 135
Foxhills GU21: Wok 9N 167
Foxhills Golf Course 77B 148
Foxhills KT16: Ott 79D 148
Foxhills M. KT16: Longc 76D 148
Fox Hills Rd. RM16: Grays 46Fe 99
Foxhills Rd. KT16: Ott 77C 148
Foxhole Rd. SE9 57Nc 116
Foxholes KT13: Weyb 78T 150
Fox Hollow Cl. SE18 50Uc 94
Foxhollow Cl. KT12: Walt T 75Z 151
Fox Hollow Dr. DA7: Bex 55Zc 117
Foxhollow Dr. SL2: Farn C 6G 60
Foxhollows AL2: Lon C 8G 6
Foxholt Gdns. NW10 38Sa 67
Foxhome Cl. BR7: Chst 65Qc 138
Foxhounds La. DA13: Sflt 62Ce 143
Fox Ho. KT16: Chert 74H 149
(off Fox La. Nth.)
Fox Ho. Rd. DA17: Belv 50Dd 96
(not continuous)
Foxlake Rd. KT14: Byfl 84P 169
Foxlands Cl. WD25: Wat 6W 12
Foxlands Cres. RM10: Dag 36Ed 76
Foxlands La. RM10: Dag 36Fd 76
Foxlands Rd. RM10: Dag 36Ed 76
Fox La. BR2: Kes 78Kc 159
Fox La. CR3: Cat'm 93Rb 197
Fox La. KT23: Bookh 96Aa 191
Fox La. N13 19Pb 32
Fox La. RH2: Reig 3K 207
Fox La. W5 42Na 87
Fox La. Nth. KT16: Chert 74H 149
Fox La. Sth. KT16: Chert 74H 149
Fox Lea TN15: Bor G 92Be 205
Foxleas Ct. BR1: Broml 66Gc 137
Foxlees HA0: Wemb 35Ja 66
Foxleigh Grange GU24: Bisl 9F 166
Foxley Cl. E8 36Wb 71
Foxley Cl. IG10: Lough 12Rc 36
Foxley Cl. RH1: Redh 10A 208
Foxley Ct. SM2: Sutt 80Eb 155
Foxley Gdns. CR8: Purl 85Rb 177
Foxley Gro. SL1: Burn 1A 80
Foxley Hall CR8: Purl 84Qb 176
Foxley Hill Rd. CR8: Purl 84Qb 176
Foxley Ho. E3 41Dc 92
(off Bow Rd.)
Foxley La. CR8: Purl 83Lb 176
Foxley M. N20 19Fb 31
Foxley Rd. CR7: Thor H 70Rb 135
Foxley Rd. CR8: Kenley 86Rb 177
Foxley Rd. SL2: Slou 2D 80
Foxley Rd. SW9 52Qb 112
Foxleys WD19: Wat 20Aa 27
Foxley Sq. SW9 52Rb 113
Foxley Wood 85Qb 176
Fox Mnr. Way RM20: W Thur 51Xd 120
Foxmead Cl. EN2: Enf 13Pb 32
Foxmoor Ct. UB9: Den 30J 43
Foxmore St. SW11 53Hb 111
Foxon Cl. CR3: Cat'm 93Ub 197
Foxon La. CR3: Cat'm 93Tb 197
Foxon La. Gdns. CR3: Cat'm 93Ub 197
Fox Rd. E16 43Hc 93
Fox Rd. SL3: L'ly 9P 81
Fox's Path CR4: Mitc 68Gb 133
Fox's Yd. E2 5K 219 (42Vb 91)
(off Rhoda St.)
Foxton Gro. CR4: Mitc 68Fb 133
Foxton Ho. E16 47Qc 94
(off Albert Rd.)
Foxton M. TW10: Rich 58Na 109
Foxton Rd. RM20: Grays 51Zd 121
Foxton Way SM3: Sutt 74Cb 155
Foxtree Ho. WD25: Wat 8Aa 13
Foxtrot Way KT20: Tad 95Ab 194
Foxwarren KT10: Clay 81Ha 172
Foxwell M. SE4 55Ac 114
Foxwell St. SE4 55Ac 114
Fox Wood KT12: W Vill 80V 150
Foxwood Chase EN9: Walt A 7Ec 20
Foxwood Cl. NW7 21Ua 48
Foxwood Cl. TW13: Felt 62X 129
Foxwood Grn. Cl. EN1: Enf 16Ub 33
Foxwood Gro. BR6: Prat B 82Yc 181
Foxwood Gro. DA11: Nflt 10A 122
Fox Wood Nature Reserve 42Na 87
Foxwood Rd. DA2: Bean 62Xd 142
Foxwood Rd. SE3 56Hc 115
Foxwood Way DA3: Lfield 68Fe 143
Foyle Dr. RM15: S Ock 43Wd 98
Foyle Rd. N17 25Wb 51
Foyle Rd. SE3 51Hc 115
Frailey Cl. GU22: Wok 88D 168
Frailey Hill GU22: Wok 88D 168
Framewood Rd. SL2: Stoke P 8N 61
Framewood Rd. SL2: Wex 8N 61
Framewood Rd. SL3: Stoke P 8N 61
Framewood Rd. SL3: Wex 8N 61
Framfield Cl. N12 20Cb 31
Framfield Ct. EN1: Enf 16Ub 33
(off Queen Anne's Gdns.)
Framfield Rd. CR4: Mitc 66Jb 134
Framfield Rd. N5 36Rb 71
Framfield Rd. W7 44Ga 86
Framlingham Cl. E5 33Yb 72
Framlingham Ct. RM6: Chad H 29Xc 55
(off Norwich Cres.)
Framlingham Cres. SE9 63Nc 138
Frampton NW1 38Mb 70
(off Wrotham Rd.)
Frampton Cl. IG6: Ilf 28Tc 54
Frampton Cl. SM2: Sutt 80Cb 155
Frampton Ct. UB9: Den 30H 43
Frampton Ct. W3 47Sa 87
(off Avenue Rd.)
Frampton Ho. NW8 6C 214 (42Fb 89)
(off Frampton St.)
Frampton Pk. Est. E9 38Yb 72
Frampton Pk. Rd. E9 37Yb 72
Frampton Rd. CM16: Epp 1Wc 23
Frampton Rd. EN6: Pot B 2Eb 17
Frampton Rd. TW4: Houn 57Aa 107
Frampton St. NW8 6C 214 (42Fb 89)
Frampton Ter. SE9 62Rc 138
Francemary Rd. SE4 57Cc 114
Frances Av. RM16: Chaf H 49Yd 98
Frances Ct. E17 30Cc 52
Frances Ct. SE25 69Vb 135
Frances Gdns. RM15: S Ock 44Vd 98
Frances Ho. HP3: Hem H 7N 3
Frances M. HP3: Hem H 7A 4
Frances M. RM3: Rom 23Nd 57
Frances Rd. E4 23Cc 52
Frances Rd. SL4: Wind 5G 102
Frances St. SE18 48Pc 94
Frances Wharf E14 44Bc 92
Franche Ct. Rd. SW17 62Eb 133
Francis & Dick James Ct. NW7 24Ab 48
Francis Av. DA7: Bex 54Cd 118
Francis Av. IG1: Ilf 33Tc 74
Francis Av. TW13: Felt 62W 128
Francis Bacon Ct. SE16 49Xb 91
(off Galleywall Rd.)
Francis Barber Cl. SW16 64Pb 134
Francis Bentley M. SW4 55Lb 112
Franciscan Rd. SW17 64Hb 133
Francis Chichester Cl. SL5: Asc 10A 124
Francis Chichester Way SW11 53Jb 112
Francis Cl. E14 49Fc 93
Francis Cl. KT19: Ewe 77Ta 153
Francis Cl. SS17: Horn H 1H 101
Francis Cl. TW17: Shep 70Q 128
Francisco Cl. RM16: Chaf H 48Yd 98
Francis Ct. AL1: St A 3C 6
Francis Ct. DA8: Erith 50Gd 96
Francis Ct. EC1 7B 218 (43Rb 91)
(off Briset St.)
Francis Ct. KT5: Surb 70Na 131
(off Cranes Pk. Av.)
Francis Ct. NW7 22Va 48
(off Watford Way)
Francis Ct. SE14 51Zb 114
(off Myers La.)
Francis Greene Ho. EN9: Walt A 5Dc 20
(off Grove Ct.)
Francis Gro. SW19 65Bb 133
Francis Harvey Way SE8 53Bc 114
Francis Ho. N1 1H 219 (39Ub 71)
(off Colville Est.)
Francis Ho. SW10 52Db 111
(off Coleridge Gdns.)
Francis Ho. SW18 56Eb 111
(off Eltringham St.)
Francis M. SE12 59Jc 115
Francis Pl. N6 31Kb 70
(off Shepherd's Cl.)
Francis Rd. BR5: St P 69Zc 139
Francis Rd. CR0: C'don 73Rb 157
Francis Rd. CR3: Cat'm 94Tb 197
Francis Rd. DA1: Dart 57Md 119
Francis Rd. E10 32Ec 72
Francis Rd. HA1: Harr 29Ja 46
Francis Rd. HA5: Eastc 29Y 45
Francis Rd. IG1: Ilf 33Tc 74
Francis Rd. N2 28Hb 49
Francis Rd. SM6: W'gton 79Lb 156
Francis Rd. TW4: Houn 54Z 107
Francis Rd. UB6: G'frd 40Ka 66
Francis Rd. WD18: Wat 14X 27
Francis St. E15 36Gc 73
Francis St. IG1: Ilf 33Tc 74
Francis St. SW1 5B 228 (49Lb 90)
Francis Ter. N19 34Lb 70
Francis Ter. M. N19 34Lb 70
Francis Wlk. N1 39Pb 70
Francis Way SL1: Slou 5B 80
Franck Ho. EC1 4C 218 (41Rb 91)
(off Goswell Road)
Francklyn Gdns. HA8: Edg 20Qa 29
Franco Av. NW9 25Va 48
Francombe Gdns. RM1: Rom 30Jd 56
Franconia Rd. SW4 57Mb 112
Frank Bailey Wlk. E12 37Qc 74
Frank Beswick Ho. SW6 51Bb 111
(off Clem Attlee Ct.)
Frank Burton Cl. SE7 50Kc 93
Frank Dixon Cl. SE21 59Ub 113
Frank Dixon Way SE21 60Ub 113
Frank Foster Ho. CM16: They B 9Uc 22
Frankfurt Rd. SE24 57Sb 113
Frank Godley Ct. DA14: Sidc 64Xc 139
Frankham Ho. SE8 52Cc 114
(off Frankham St.)
Frankham St. SE8 52Cc 114
Frank Ho. SW8 52Nb 112
(off Wyvil Rd.)
Frankland Cl. IG8: Wfd G 22Lc 53
Frankland Cl. SE16 48Xb 91
Frankland Cl. WD3: Crox G 17Q 26
Frankland Rd. E4 22Cc 52
Frankland Rd. SW7 4B 226 (48Fb 89)
Frankland Rd. WD3: Crox G 16R 26
Franklands Dr. KT15: Add 80H 149
Franklin Av. EN7: Chesh 2Wb 19
Franklin Av. SL2: Slou 3F 80
Franklin Av. WD18: Wat 16W 26
Franklin Bldg. E14 47Cc 92
Franklin Cl. AL4: Col H 4A 8
Franklin Cl. HP3: Hem H 5N 3
Franklin Cl. KT1: King T 69Qa 131
Franklin Cl. N20 17Eb 31
Franklin Cl. SE13 53Cc 114
Franklin Cl. SE27 62Rb 135
Franklin Cotts. HA7: Stan 21Ka 46
Franklin Ct. WD6: Bore 13Ra 29
Franklin Cres. CR4: Mitc 70Lb 134

Franklin Ho. BR2: Broml 69Gc 137
Franklin Ho. E1 46Xb 91
(off Watts St.)
Franklin Ho. E14 44Fc 93
(off E. India Dock Rd.)
Franklin Ho. EN3: Enf L 9Bc 20
Franklin Ho. NW6 41Cb 89
(off Carlton Va.)
Franklin Ind. Est. SE20 67Yb 136
(off Franklin Rd.)
Franklin Pas. SE9 55Nc 116
Franklin Pl. SE13 53Dc 114
Franklin Rd. DA12: Grav'nd 4F 144
Franklin Rd. DA2: Wilm 61Gd 140
Franklin Rd. DA7: Bex 53Ad 117
Franklin Rd. RM12: Horn 37Ld 77
Franklin Rd. SE20 66Yb 136
Franklin Rd. WD17: Wat 12X 27
Franklins WD3: Map C 21G 42
Franklins M. HA2: Harr 33Ea 66
Franklin Sq. W14 50Bb 89
Franklin's Row SW3 7G 227 (50Hb 89)
Franklin St. E3 41Dc 92
Franklin St. N15 30Ub 51
Franklin Way CR0: Wadd 73Nb 156
Franklyn Cres. SL4: Wind 5B 102
Franklyn Gdns. IG6: Ilf 23Tc 54
Franklyn Rd. KT12: Walt T 72W 150
Franklyn Rd. NW10 37Va 68
Frank Martin Ct. EN7: Chesh 2Xb 19
Frank M. SE1 49Xb 91
Franks Av. KT3: N Mald 70Sa 131
Franks Cotts. RM14: Upm 32Wd 78
Franks La. DA4: Hort K 71Rd 163
Frank Soskice Ho. SW6 51Bb 111
(off Clem Attlee Ct.)
Frank St. E13 42Jc 93
Frank Sutton Way SL1: Slou 5H 81
Frankswood Av. BR5: Pet W 71Rc 160
Frankswood Av. UB7: Yiew 44P 83
Frank Towell Ct. TW14: Felt 59W 106
Frank Whipple Pl. E14 44Ac 92
(off Repton St.)
Frank Whymark Ho. SE16 47Yb 92
(off Rupack St.)
Franlaw Cres. N13 21Sb 51
Franmil Rd. RM11: Horn 31Hd 76
Franmil Rd. RM12: Horn 32Jd 76
Fransfield Gro. SE26 62Xb 135
Frans Hals Ct. E14 48Fc 93
Franshams WD23: B Hea 19Ga 28
(off Hartsbourne Rd.)
Frant Cl. SE20 66Yb 136
Franthorne Way SE6 61Dc 136
Frant Rd. CR7: Thor H 71Rb 157
Fraserburgh Ho. E3 40Bc 72
(off Vernon Rd.)
Fraser Cl. DA5: Bexl 60Ed 118
Fraser Cl. E6 44Nc 94
Fraser Ct. E14 50Ec 92
(off Ferry St.)
Fraser Ct. SE1 3E 230 (48Sb 91)
(off Brockham St.)
Fraser Ct. SW11 53Gb 111
(off Surrey La. Est.)
Fraser Cres. WD25: Wat 3X 13
Fraser Ho. TW8: Bford 50Pa 87
Fraser Rd. DA8: Erith 50Fd 96
Fraser Rd. E17 29Dc 52
Fraser Rd. N9 20Xb 33
Fraser Rd. UB6: G'frd 39Ka 66
Fraser St. W4 50Ua 88
Frating Cres. IG8: Wfd G 23Kc 53
Frays Av. UB7: W Dray 47M 83
Frays Cl. UB7: W Dray 48M 83
Frays Island & Mabey's Meadow Nature Reserve 48L 83
Frayslea UB8: Uxb 40L 63
Frays Valley Local Nature Reserve 32M 63
Frays Waye UB8: Uxb 39L 63
Frazer Av. HA4: Ruis 36Y 65
Frazer Cl. RM1: Rom 31Hd 76
Frazer Nash Cl. TW7: Isle 53Ha 108
Frazier St. SE1 2K 229 (47Qb 90)
Frean St. SE16 48Wb 91
Frearson Ho. WC1 3J 217 (41Pb 90)
(off Penton Ri.)
Freda Corbett Cl. SE15 52Wb 113
Freda St. SE16 48Wb 91
Frederica Ct. SW2 60Rb 113
Frederica Rd. E4 17Fc 35
Frederica St. N7 38Pb 70
Frederick Andrews Ct. RM17: Grays 51Fe 121
Frederick Charrington Ho. E1 42Yb 92
(off Wickford St.)
Frederick Cl. SM1: Sutt 77Bb 155
Frederick Cl. W2 4F 221 (45Hb 89)
Frederick Ct. SW3 6G 227 (49Hb 89)
(off Duke of York Sq.)
Frederick Cres. EN3: Enf H 12Yb 34
Frederick Cres. SW9 52Rb 113
Frederick Dobson Ho. W11 45Ab 88
(off Cowling Cl.)
Frederick Gdns. CR0: C'don 72Rb 157
Frederick Gdns. SM1: Sutt 78Bb 155
Frederick Ho. SE18 49Nc 94
(off Pett St.)
Frederick Pl. AL2: F'mre 9C 6
Frederick Pl. N8 30Mb 50
(off Crouch Hall Rd.)
Frederick Pl. SE18 50Rc 94
Frederick Rd. RM13: Rain 40Fd 76
Frederick Rd. SE17 51Rb 113
Frederick Rd. SM1: Sutt 78Bb 155
Frederick's Pl. EC2 3F 225 (44Tb 91)
Fredericks Pl. N12 21Eb 49
Frederick Sq. SE16 45Ac 92
(off Sovereign Cres.)
Frederick's Row EC1 3B 218 (41Rb 91)
Frederick St. WC1 4H 217 (41Pb 90)
Frederick Ter. E8 38Vb 71
Frederick Vs. W7 46Ga 86
(off Lwr. Boston Rd.)
Frederic M. SW1 2G 227 (47Hb 89)
(off Kinnerton St.)
Frederic St. E17 29Ac 52
Fred Mead DA13: Sflt 65Ce 143
Fredora Av. UB4: Hayes 42V 84
Fred Styles Ho. SE7 51Lc 115
Fred Tibble Ct. RM9: Dag 35Ad 75
Fred White Wlk. N7 37Nb 70
Freeborne Gdns. RM13: Rain 37Jd 76
Freedom Cl. E17 28Ac 52
Freedom Rd. N17 26Tb 51
Freedom St. SW11 54Hb 111
Freedown La. SM2: Sutt 85Db 175
Freegrove Rd. N7 36Nb 70
(not continuous)
Freehold Ind. Cen. TW4: Houn 57Y 107
Freeland Ct. DA15: Sidc 62Wc 139
Freeland Pk. NW4 26Ab 48
Freeland Rd. W5 45Pa 87
Freelands Av. CR2: Sels 81Zb 178
Freelands Gro. BR1: Broml 67Kc 137
Freelands Rd. BR1: Broml 67Kc 137
Freelands Rd. KT11: Cobh 86X 171
Freeland Way DA8: Erith 53Jd 118
Freeling Ho. NW8 39Fb 69
(off Dorman Way)
Freeling St. N1: Carnoustie Dr. 38Pb 70
Freeling St. N1: Pembroke St. 38Nb 70
Freeman Cl. TW17: Shep 70U 128
Freeman Cl. UB5: N'olt 38Aa 65
Freeman Ct. N7 34Nb 70
Freeman Ct. SW16 68Nb 134
Freeman Dr. KT8: W Mole 69Ba 129
Freeman Ho. SE11 5B 230 (49Rb 91)
(off George Mathers Rd.)
Freeman Rd. DA12: Grav'nd 2G 144
Freeman Rd. SM4: Mord 71Fb 155
Freemans Cl. SL2: Stoke P 7K 61
Freemans La. UB3: Hayes 45U 84
Freemantle Av. EN3: Pond E 15Zb 34
Freemantle St. SE17 7H 231 (50Ub 91)
Freeman Wlk. SE9 57Lc 115
Freeman Way RM11: Horn 30Pd 57
Freemasons' Hall 3G 223 (44Nb 90)
(off Gt. Queen St.)
Freemasons Pl. CR0: C'don 74Ub 157
(off Freemasons Rd.)
Freemasons Rd. CR0: C'don 74Ub 157
Freemasons Rd. E16 43Kc 93
Free Prae Rd. KT16: Chert 74J 149
Freesia Cl. BR6: Chels 78Vc 161
Freesia Dr. GU24: Bisl 8E 166
Freestone Yd. SL3: Coln 52F 104
Freethorpe Cl. SE19 66Tb 135
Free Trade Wharf E1 45Zb 92
Freezeland Way UB10: Hil 37R 64
FREEZY WATER 8Zb 20
Freight La. N1 38Mb 70
Freightliners City Farm 37Qb 70
Freightmaster Est. RM13: Rain 44Gd 96
Freke Rd. SW11 55Jb 112
Frelford Cl. WD25: Wat 6Y 13
Fremantle Ho. E1 42Xb 91
(off Somerford St.)
Fremantle Ho. RM18: Tilb 3B 122
Fremantle Rd. DA17: Belv 49Cd 96
Fremantle Rd. IG6: Ilf 26Rc 54
Fremantle Way GU24: Brkwd 1B 186
Fremantle Way UB3: Hayes 45V 84
Fremont St. E9 39Xb 71
(not continuous)
French Apts., The CR8: Purl 84Qb 176
Frenchaye KT15: Add 78L 149
Frenches, The RH1: Redh 4A 208
Frenches Ct. RH1: Redh 4A 208
Frenches Rd. RH1: Redh 4A 208
French Gdns. KT11: Cobh 86Y 171
Frenchlands Ga. KT24: E Hor 99U 190
French Ordinary Ct. EC3 4J 225 (45Ub 91)
(off Crutched Friars)
French Pl. E1 4J 219 (41Ub 91)
French Row AL3: St A 2B 6
French St. TN16: Westrm 100Uc 200
French St. TW16: Sun 68Y 129
FRENCH STREET 100Vc 201
French's Wells GU21: Wok 9M 167
Frenchum Gdns. SL1: Slou 5C 80
Frendsbury Rd. SE4 56Ac 114
Frensham Cl. UB1: S'hall 42Ba 85
Frensham Ct. SW19 69Fb 133
Frensham Dr. CR0: New Ad 80Ec 158
Frensham Dr. SW15 62Va 132
Frensham Rd. CR8: Kenley 86Rb 177
Frensham Rd. SE9 61Tc 138
Frensham St. SE15 51Wb 113
Frensham Wlk. SL2: Farn C 6G 60
Frensham Way KT17: Eps D 88Ya 174
Frere St. SW11 54Gb 111
Fresham Ho. BR2: Broml 69Hc 137
(off Durham Rd.)
Freshfield Av. E8 38Vb 71
Freshfield Cl. SE13 56Fc 115
Freshfield Dr. N14 17Kb 32
Freshfield Flats KT20: Lwr K 99Bb 195
Freshfields CR0: C'don 74Bc 158
Freshfields Av. RM14: Upm 36Rd 77
Freshford St. SW18 62Eb 133
Fresh Mill La. UB1: S'hall 42Ca 85
Freshmount Gdns. KT19: Eps 83Ra 173
Freshwater Cl. SW17 65Jb 134
Freshwater Ct. UB1: S'hall 41Ca 85
Freshwater Ct. W1 1E 220 (43Gb 89)
(off Crawford St.)
Freshwater Rd. RM8: Dag 32Zc 75
Freshwater Rd. SW17 65Jb 134
Freshwell Av. RM6: Chad H 28Yc 55
Freshwell Gdns. CM13: W H'dn 30Fe 59
Fresh Wharf Est. IG11: Bark 40Rc 74
Fresh Wharf Rd. IG11: Bark 39Rc 74
Freshwood Cl. BR3: Beck 67Dc 136
Freshwood Way SM6: W'gton 81Kb 176
Freston Gdns. EN4: Cockf 15Jb 32
Freston Pk. N3 26Bb 49
Freston Rd. W10 45Za 88
Freston Rd. W11 45Za 88
Freswick Ho. SE8 49Ac 92
(off Chilton Gro.)
Freta Rd. DA6: Bex 57Bd 117
Freud Mus. 37Eb 69
Frewell Ho. EC1 7K 217 (43Qb 90)
(off Bourne Est.)
Frewin Rd. SW18 60Fb 111
Friar M. SE27 62Rb 135
Friar Rd. BR5: St M Cry 71Wc 161
Friar Rd. EN2: Enf 11Qb 32
Friar Rd. UB4: Yead 42Z 85
Friars, The IG7: Chig 21Uc 54
Friars Av. CM15: Shenf 18Ge 41
Friars Av. N20 20Gb 31
Friars Av. SW15 62Va 132
Friar's Cl. CM15: Shenf 17Ce 41
Friars Cl. E4 20Ec 34
Friars Cl. IG1: Ilf 32Tc 74
Friars Cl. SE1 7C 224 (46Rb 91)
(off Bear La.)
Friars Cl. UB5: N'olt 41Z 85
Friars Ct. E17 25Bc 52
Friars Ct. RM7: Rush G 30Fd 56
Friars Ct. SM6: W'gton 77Kb 156
Friars Gdns. W3 44Ta 87
Friarsgate NW10 40Ra 67
Friars Ga. Cl. IG8: Wfd G 21Jc 53
Friars La. TW9: Rich 57Ma 109
Friars Mead E14 48Ec 92
Friars Mead WD4: K Lan 2Q 12
Friars M. SE9 57Qc 116
Friars Mt. Ho. E2 4K 219 (41Vb 91)
(off Rochelle St.)
Friars Orchard KT22: Fet 93Fa 192
Friars Pl. La. W3 45Ta 87
Friars Ri. GU22: Wok 90C 168
Friars Rd. E6 39Mc 73
Friars Rd. GU25: Vir W 10P 125
Friars Stile Pl. TW10: Rich 58Na 109
Friars Stile Rd. TW10: Rich 58Na 109
Friar St. EC4 3C 224 (44Rb 91)
Friars Wlk. N14 17Kb 32
Friars Wlk. SE2 50Zc 95
Friars Way KT16: Chert 72J 149
Friars Way W3 44Ta 87
Friars Way WD23: Bush 11Ba 27
Friars Way WD4: K Lan 2Q 12
Friars Wood CR0: Sels 81Ac 178
Friary, The EN8: Walt C 5Bc 20
Friary, The SL4: Old Win 8N 103
Friary Cl. N12 22Gb 49
Friary Ct. GU21: Wok 10K 167
Friary Ct. SW1 7C 222 (46Lb 90)
(off Marlborough Rd.)
Friary Est. SE15 51Wb 113
(not continuous)
Friary Island TW19: Wray 8N 103
FRIARY ISLAND 8N 103
Friary La. IG8: Wfd G 21Jc 53
Friary Pk. W3 44Ta 87
Friary Pk. Ct. W3 44Sa 87
Friary Rd. N12 21Fb 49
Friary Rd. N20 21Gb 49
Friary Rd. SE15 52Wb 113
Friary Rd. TW19: Wray 9N 103
Friary Rd. W3 44Sa 87
Friary Way N12 21Gb 49
Friday Hill E4 19Gc 35
FRIDAY HILL 19Gc 35
Friday Hill E. E4 20Gc 35
Friday Hill W. E4 19Gc 35
Friday Rd. CR4: Mitc 66Hb 133
Friday Rd. DA8: Erith 50Fd 96
Friday St. EC4 4D 224 (45Sb 91)
Frideswide Pl. NW5 36Lb 70
Friendly Pl. SE13 53Dc 114
Friendly St. SE8 54Cc 114
Friendly St. M. SE8 54Cc 114
Friends Av. EN8: Chesh 4Zb 20
Friendship Ho. SE1 2C 230 (47Rb 91)
(off Belvedere Pl.)
Friendship Wlk. UB5: N'olt 41Z 85
Friendship Way E15 39Ec 72
Friends Rd. CR0: C'don 76Tb 157
Friends Rd. CR8: Purl 84Rb 177
Friend St. EC1 3B 218 (41Rb 91)
Friends Wlk. TW18: Staines 64H 127
Friends Wlk. UB8: Uxb 38M 63
FRIERN BARNET 22Hb 49
Friern Barnet La. N11 21Gb 49
Friern Barnet La. N20 19Fb 31
Friern Barnet Rd. N11 22Hb 49
Friern Bri. Retail Pk. 23Kb 50
Friern Ct. N20 20Fb 31
Friern Mt. Dr. N20 17Eb 31
Friern Pk. N12 22Eb 49
Friern Rd. SE22 59Wb 113
(not continuous)
Friern Watch Av. N12 21Eb 49
Frigate Ho. E14 49Ec 92
(off Stebondale St.)
Frigate M. SE8 51Cc 114
Frimley Av. RM11: Horn 32Od 77
Frimley Av. SM6: W'gton 78Nb 156
Frimley Cl. CR0: New Ad 80Ec 158
Frimley Cl. SW19 61Ab 132
Frimley Ct. DA14: Sidc 64Yc 139
Frimley Cres. CR0: New Ad 80Ec 158
Frimley Dr. SL1: Slou 7D 80
Frimley Gdns. CR4: Mitc 69Gb 133
Frimley Rd. HP1: Hem H 1G 2
Frimley Rd. IG3: Ilf 34Uc 74
Frimley Rd. KT9: Chess 78Ma 153
Frimley St. E1 42Zb 92
(off Frimley Way)
Frimley Way E1 42Zb 92
Fringewood Cl. HA6: Nwood 25R 44
Frinstead Gro. BR5: St M Cry 70Zc 139
Frinstead Ho. W10 45Za 88
(off Freston Rd.)
Frinsted Rd. DA8: Erith 52Fd 118
Frinton Cl. WD19: Wat 19X 27
Frinton Ct. W13 43Ka 86
(off Hardwick Grn.)
Frinton Dr. IG8: Wfd G 24Fc 53
Frinton M. IG2: Ilf 30Qc 54
Frinton Rd. DA14: Sidc 61Ad 139
Frinton Rd. E6 41Mc 93
Frinton Rd. N15 30Ub 51
Frinton Rd. RM5: Col R 24Bd 55
Frinton Rd. SW17 65Jb 134
Friston Path IG7: Chig 22Uc 54
Friston St. SW6 54Db 111
Friswell Pl. DA6: Bex 56Cd 118
Fritham Cl. KT3: N Mald 72Ua 154
Frith Ct. NW7 24Ab 48
Frithe, The SL2: Slou 4M 81
Frith Ho. NW8 6C 214 (42Fb 89)
(off Frampton St.)
Frith Knowle KT12: Hers 78X 151
Frith La. NW7 24Ab 48
Frith Rd. CR0: C'don 75Sb 157
Frith Rd. E11 35Ec 72
Friths Dr. RH2: Reig 3K 207
Frith St. W1 3D 222 (44Mb 90)
Frithville Ct. W12 46Ya 88
(off Frithville Gdns.)
Frithville Gdns. W12 46Ya 88
Frithwald Rd. KT16: Chert 73H 149
Frithwood Av. HA6: Nwood 23U 44
Frizlands La. RM10: Dag 33Dd 76
Frobisher Cl. CR8: Kenley 89Sb 177
Frobisher Cl. HA5: Pinn 31Z 65
Frobisher Cl. WD23: Bush 16Ca 27
Frobisher Ct. NW9 26Ua 48
Frobisher Ct. SE10 51Fc 115
(off Old Woolwich Rd.)
Frobisher Ct. SE23 61Xb 135
Frobisher Ct. SE8 50Ac 92
(off Evelyn St.)
Frobisher Ct. SM3: Cheam 80Ab 154
Frobisher Ct. W12 47Ya 88
(off Lime Gro.)
Frobisher Cres. EC2 7E 218 (43Sb 91)
(off Silk St.)
Frobisher Cres. TW19: Stanw 59N 105
Frobisher Gdns. E10 31Dc 72
Frobisher Gdns. RM16: Chaf H 48Ae 99
Frobisher Gdns. TW19: Stanw 59N 105
Frobisher Ho. E1 46Xb 91
(off Watts St.)
Frobisher Ho. SW1 51Mb 112
(off Dolphin Sq.)
Frobisher M. EN2: Enf 14Tb 33
Frobisher Pas. E14 46Cc 92
Frobisher Pl. SE15 53Yb 114
Frobisher Rd. AL1: St A 4G 6
Frobisher Rd. DA8: Erith 52Hd 118
Frobisher Rd. E6 44Pc 94
Frobisher Rd. N8 28Qb 50
Frobisher St. SE10 51Gc 115
Frobisher Way DA12: Grav'nd 4G 144
Frobisher Way DA9: Ghithe 56Xd 120
Frobisher Yd. E16 45Sc 94
Froebel Coll. 58Va 110
Froggy La. UB9: Den 34F 62
Froghall La. IG7: Chig 21Tc 54
Frog La. GU4: Sut G 97A 188
Frog La. RM13: Rain 44Fd 96
Frogley Rd. SE22 56Vb 113
Frogmoor Ct. WD3: Rick 19M 25
Frogmoor La. WD3: Rick 19M 25
Frogmore AL2: F'mre 9B 6
FROGMORE 10C 6
FROGMORE 4J 103
Frogmore SW18 57Cb 111
Frogmore Av. UB4: Hayes 42U 84
Frogmore Border SL4: Wind 5J 103
Frogmore Bus. Pk. AL2: F'mre 10C 6
Frogmore Cl. SL1: Slou 7E 80
Frogmore Cl. SM3: Cheam 76Za 154
Frogmore Cotts. WD17: Wat 15Z 27
Frogmore Ct. UB2: S'hall 49Ba 85
Frogmore Dr. SL4: Wind 3J 103
FROGMORE END 5M 3
Frogmore Gdns. SM3: Cheam 77Ab 154
Frogmore Gdns. UB4: Hayes 42U 84
Frogmore Home Pk. AL2: F'mre 9B 6
Frogmore House 4K 103
Frogmore Ind. Est. N5 36Sb 71
Frogmore Ind. Est. NW10 41Sa 87
Frogmore Ind. Est. UB3: Hayes 47U 84
Frogmore Ind. Pk. RM20: W Thur 50Wd 98
Frogmore Paper Mill & Visitor Cen., The 5M 3
Frogmore Rd. HP3: Hem H 5M 3
Frogmore Rd. Ind. Est. HP3: Hem H 6M 3
Frognal NW3 35Eb 69
Frognal Av. DA14: Sidc 65Wc 139
Frognal Av. HA1: Harr 28Ha 46
Frognal Cl. NW3 36Eb 69
FROGNAL CORNER 65Vc 139
Frognal Ct. NW3 37Eb 69
Frognal Gdns. NW3 35Eb 69
Frognal La. NW3 36Db 69
Frognal Pde. NW3 37Eb 69
Frognal Pl. DA14: Sidc 65Wc 139
Frognal Ri. NW3 34Eb 69
Frognal Way NW3 35Eb 69
Frog St. CM15: Dodd 11Ud 40
Frog St. CM15: Kel H 11Ud 40
Frogwell Cl. N15 30Tb 51
Froissart Rd. SE9 57Mc 115
Frome RM18: E Til 9L 101
Frome Ho. SE15 56Xb 113
Frome Rd. N22 27Rb 51
Frome St. N1 1D 218 (40Sb 71)
Fromondes Rd. SM3: Cheam 78Ab 154
Fromow Gdns. GU20: W'sham 9B 146
Fromows Cnr. W4 50Sa 87
Frontenac NW10 38Xa 68
Frontier Works N17 23Ub 51
Front La. RM14: Upm 33Ud 78
Frost Cl. DA10: Swans 60Be 121
Frost Ct. NW9 26Ua 48
(off Salk Cl.)
Frostic Wlk. E1 43Wb 91
Froude St. SW8 54Kb 112
Frowick Cl. AL9: Wel G 5D 8
Frowyke Cres. EN6: S Mim 4Wa 16
Fruen Rd. TW14: Felt 59V 106
Fryatt Rd. N17 24Tb 51
(not continuous)
Fry Cl. RM5: Col R 22Cd 56
Fryday Gro. M. SW12 59Lb 112
(off Weir Rd.)
Frye Ct. E3 41Bc 92
(off Benworth St.)
Frye Ho. E20 37Ec 72
(off Penny Brookes St.)
Fryent Cl. NW9 30Qa 47
Fryent Country Pk. 31Qa 67
Fryent Cres. NW9 30Ua 48
Fryent Flds. NW9 30Ua 48
Fryent Gro. NW9 30Ua 48
Fryent Way NW9 29Qa 47
Fryern Wood CR3: Cat'm 96Sb 197
Fryers Vw. SE4 56Zb 114
(off Frendsbury Rd.)
Fry Ho. E6 38Lc 73
Frying Pan All. E1 1K 225 (43Vb 91)
(off Bell La.)
Frylands Ct. CR0: New Ad 83Ec 178
Fry La. HA8: Edg 21Pa 47
Frymley Vw. SL4: Wind 3B 102
Fry Rd. E6 38Mc 73
Fry Rd. NW10 39Va 68
Frys Ct. SE10 51Ec 114
(off Durnford St.)
Fryston Av. CR0: C'don 75Wb 157
Fryston Av. CR5: Coul 86Kb 176
Fryth Mead AL3: St A 1P 5
Fuchsia Cl. RM7: Rush G 33Gd 76
Fuchsia St. SE2 50Xc 95
Fuchsia Way GU24: W End 5C 166
Fulbeck Dr. NW9 25Ua 48
Fulbeck Ho. N7 37Pb 70
(off Sutterton St.)
Fulbeck Rd. N19 35Lb 70
Fulbeck Wlk. HA8: Edg 19Ra 29
Fulbeck Way HA2: Harr 26Ea 46
Fulbourn KT1: King T 68Na 131
(off Eureka Rd.)
Fulbourne Cl. RH1: Redh 4N 207
Fulbourne Rd. E17 25Ec 52
Fulbourne St. E1 43Xb 91
Fulbrook Av. KT15: New H 83J 169
Fulbrook La. RM15: S Ock 45Vd 98
Fulbrook M. N19 35Lb 70
Fulcher Ho. N1 39Ub 71
(off Colville Est.)
Fulcher Ho. SE8 50Bc 92
Fulford Gro. WD19: Wat 19X 27
Fulford Rd. KT19: Ewe 80Ta 153
Fulford Rd. CR3: Cat'm 93Tb 197
Fulford Rd. KT19: Ewe 80Ta 153
Fulford St. SE16 47Xb 91
FULHAM 53Ab 110
Fulham B'way. SW6 52Cb 111
FULHAM BROADWAY 52Cb 111
Fulham B'way. Shop. Cen. 52Cb 111
Fulham Broadway Station (Underground) 52Cb 111
Fulham Bus. Exchange SW6 53Eb 111
(off The Boulevard)
Fulham Cl. UB10: Hil 42S 84
Fulham Ct. SW6 53Cb 111
Fulham FC 53Za 110
Fulham High St. SW6 54Ab 110
Fulham Island SW6 52Cb 111
(off Farm La.)
Fulham Palace 54Ab 110
Fulham Pal. Rd. SW6 51Za 110
Fulham Pal. Rd. W6 50Ya 88
Fulham Pk. Gdns. SW6 54Bb 111
Fulham Pk. Rd. SW6 54Bb 111
Fulham Pools Virgin Active 51Ab 110
Fulham Rd. SW10 52Db 111
Fulham Rd. SW3 7B 226 (50Fb 89)
Fulham Rd. SW6 54Ab 110
(not continuous)
Fullarton Cres. RM15: S Ock 44Vd 98
Fullbrooks Av. KT4: Wor Pk 74Va 154
Fullbrook School Sports Cen. 84J 169
Fuller Cl. BR6: Chels 78Vc 161
Fuller Cl. E2 42Wb 91
(off Cheshire St.)
Fuller Cl. WD23: Bush 17Fa 28
Fuller Ct. N8 29Mb 50
Fuller Gdns. WD24: Wat 9X 13
Fullerian Cres. WD18: Wat 14V 26
Fuller Rd. RM8: Dag 34Xc 75
Fuller Rd. WD24: Wat 9X 13
Fullers Av. IG8: Wfd G 24Hc 53
Fullers Av. KT6: Surb 75Pa 153
Fullers Cl. EN9: Walt A 5Jc 21
Fullers Cl. RM5: Col R 24Ed 56
Fuller's Griffin Brewery 51Va 110
FULLERS HILL 93Sd 204
Fullers Hill TN16: Westrm 98Tc 200
Fullers La. RM5: Col R 24Ed 56
Fullers Rd. E18 24Hc 53
Fuller St. NW4 28Ya 48
Fuller St. TN15: Seal 92Rd 203
Fullers Way Nth. KT6: Surb 76Pa 153
Fullers Way Sth. KT9: Chess 77Na 153
Fuller's Wood CR0: C'don 78Cc 158
Fullers Wood La. RH1: S Nut 7C 208
Fullerton Av. RM8: Dag 32Ad 75
Fullerton Cl. KT14: Byfl 86P 169
Fullerton Ct. TW11: Tedd 65Ja 130
Fullerton Dr. KT14: Byfl 86N 169
Fullerton Rd. CR0: C'don 73Vb 157
Fullerton Rd. KT14: Byfl 86N 169
Fullerton Rd. SM5: Cars 81Gb 175
Fullerton Rd. SW18 57Db 111
Fullerton Way KT14: Byfl 86N 169
Fuller Way UB3: Harl 50V 84
Fuller Way WD3: Crox G 15Q 26
Fullmer Way KT15: Wdhm 82H 169
Fullwell Av. IG5: Ilf 25Pc 54
Fullwell Av. IG6: Ilf 25Rc 54
FULLWELL CROSS 26Sc 54
FULLWELL CROSS 26Tc 54
Fullwell Cross Leisure Cen. 26Sc 54
Fullwell Pde. IG5: Ilf 25Qc 54
Fullwood's M. N1 3G 219 (41Tb 91)
Fulmar Cl. KT5: Surb 72Pa 153
Fulmar Cres. HP1: Hem H 2J 3
Fulmar Ho. SE16 49Zb 92
(off Tawny Way)
Fulmar Rd. RM12: Horn 38Jd 76
Fulmead St. SW6 53Db 111
FULMER 5P 61
Fulmer Cl. TW12: Hamp 64Aa 129
Fulmer Comn. Rd. SL0: Iver H 37C 62
Fulmer Comn. Rd. SL3: Ful 6P 61
Fulmer Cnr. SL9: Ger X 32D 62
Fulmer Dr. SL9: Ger X 3P 61
Fulmer Ho. NW8 6D 214 (42Gb 89)
(off Mallory St.)
Fulmer Ho. UB8: Uxb 37N 63
Fulmer La. SL3: Ful 34B 62
Fulmer La. SL9: Ger X 34C 62
Fulmer Pl. SL3: Ful 35A 62
Fulmer Ri. SL3: Ful 37B 62
Fulmer Rd. E16 43Mc 93
Fulmer Rd. SL3: Ful 35A 62
Fulmer Rd. SL9: Ger X 31A 62
Fulmer Way SL9: Ger X 30A 42
Fulmer Way W13 48Ka 86
Fulneck Pl. E1 42Yb 92
Fulready Rd. E10 29Fc 53
Fulstone Cl. TW4: Houn 56Ba 107
Fulthorp Rd. SE3 54Hc 115
Fulton Ct. WD6: Bore 10Pa 15
(off Aycliffe Rd.)
Fulton Dr. GU24: Bisl 9E 166
Fulton M. W2 45Eb 89
Fulton Rd. HA9: Wemb 34Qa 67
FULWELL 63Fa 130
Fulwell Ct. IG5: Ilf 25Qc 54
Fulwell Ct. UB1: S'hall 45Ea 86
(off Baird Av.)
Fulwell Golf Course 63Fa 130
Fulwell Pk. Av. TW2: Twick 61Da 129
Fulwell Rd. TW11: Tedd 63Fa 130
Fulwell Station (Rail) 63Fa 130
Fulwich Rd. DA1: Dart 58Pd 119
Fulwood Av. HA0: Wemb 40Pa 67
Fulwood Cl. UB3: Hayes 44V 84
Fulwood Ct. HA3: Kenton 30Ja 46
Fulwood Gdns. TW1: Twick 58Ha 108
Fulwood Pl. WC1 1J 223 (43Pb 90)
Fulwood Wlk. SW19 60Ab 110
Funky Footprints Nature Reserve 72Q 150
Furber St. W6 48Xa 88
Furham Feild HA5: Hat E 24Ca 45
Furley Ho. SE15 52Wb 113
(off Peckham Pk. Rd.)
Furley Rd. SE15 52Wb 113
Furlong Av. CR4: Mitc 69Gb 133
Furlong Cl. CR0: C'don 74Xb 157
Furlong Cl. SM6: W'gton 74Kb 156
Furlong Rd. N7 37Qb 70
Furlongs HP1: Hem H 1J 3
Furlongs, The KT10: Esh 76Da 151
Furlough, The GU22: Wok 88C 168
Furlow Ho. NW9 27Wa 48
Furmage St. SW18 59Db 111
Furneaux Av. SE27 64Rb 135
Furner Cl. DA1: Cray 55Hd 118
Furners Cl. DA8: Erith 52Gd 118
Furness SL4: Wind 4A 102
Furness Cl. RM16: Grays 10D 100
(not continuous)

Furness Ho. SW1 7K 227 (50Kb 90)
(part of Abbots Mnr.)
Furness Pl. SL4: Wind 4A 102
(off Furness)
Furness Rd. HA2: Harr 31Da 65
Furness Rd. NW10 40Wa 68
Furness Rd. SM4: Mord 72Db 155
Furness Rd. SW6 54Db 111
Furness Row SL4: Wind 4A 102
Furness Sq. SL4: Wind 4A 102
Furness Wlk. SL4: Wind 4A 102
Furness Way RM12: Horn 36Jd 76
Furness Way SL4: Wind 4A 102
Furnival Av. SL2: Slou 3F 80
Furnival Cl. GU25: Vir W 2P 147
Furnival Ct. E3 40Cc 72
(off Four Seasons Cl.)
Furnival Mans. W1 1B 222 (43Lb 90)
(off Wells St.)
Furnival St. EC4 2K 223 (44Qb 90)
Furrow Ho. E4 23Ec 52
Furrow La. E9 36Yb 72
Furrows, The KT12: Walt T 75Y 151
Furrows, The UB9: Hare 29L 43
Furrows Pl. CR3: Cat'm 95Vb 197
Fursby Av. N3 23Cb 49
Fursecroft W1 2F 221 (44Hb 89)
(off George St.)
Furtherfield WD5: Ab L 4U 12
Furtherfield Cl. CR0: C'don 72Qb 156
Further Grn. Rd. SE6 59Gc 115
Furtherground HP2: Hem H 3N 3
Furzebank SL5: S'hill 10B 124
Furzebushes La. AL2: Chis G 7L 5
Furze Cl. RH1: Redh 5P 207
Furze Cl. WD19: Wat 22Y 45
FURZEDOWN 64Kb 134
Furzedown Cl. TW20: Egh 65A 126
Furzedown Dr. SW17 64Kb 134
Furzedown Recreation Cen. 64Kb 134
Furzedown Rd. SM2: Sutt 83Eb 175
Furzedown Rd. SW17 64Kb 134
Furze Farm Cl. RM6: Chad H 26Ad 55
Furze Fld. KT22: Oxs 85Fa 172
Furzefield EN8: Chesh 1Xb 19
Furzefield Cen. 3Za 16
Furzefield Cl. BR7: Chst 65Rc 138
Furze Fld. Ct. WD19: Wat 20W 26
Furzefield Ct. EN6: Pot B 3Ab 16
Furzefield Cres. RH2: Reig 8L 207
Furzefield Rd. RH2: Reig 8L 207
Furzefield Rd. SE3 51Kc 115
Furzeground Way UB11: Stock P 46S 84
Furze Gro. KT20: Kgswd 93Bb 195
Furze Hall KT20: Kgswd 93Bb 195
Furzeham Rd. UB7: W Dray 47N 83
Furze Hill CR8: Purl 83Nb 176
Furze Hill KT20: Kgswd 92Bb 195
FURZE HILL 93Bb 195
Furze Hill RH1: Redh 5N 207
Furzehill Cotts. GU24: Pirb 4A 186
Furzehill Pde. WD6: Bore 13Qa 29
Furzehill Rd. WD6: Bore 14Qa 29
(not continuous)
Furzehill Sq. BR5: St M Cry 70Xc 139
Furze La. CR8: Purl 83Nb 176
Furzen Cl. SL2: Slou 1E 80
Furzen Cres. AL10: Hat 3B 8
Furze Pl. RH1: Redh 5P 207
Furze Rd. CR7: Thor H 69Sb 135
Furze Rd. HP1: Hem H 3G 2
Furze Rd. KT15: Add 79H 149
Furze St. E3 43Cc 92
Furze Vw. WD3: Chor 16E 24
Furzewood TW16: Sun 67W 128
Fusedale Way RM15: S Ock 45Vd 98
Fusilier Mus., The 5K 225 (45Vb 91)
Fusiliers Way TW4: Houn 55Y 107
(not continuous)
Fusion RH1: Redh 5P 207
Fusion Apts. SE14 51Ac 114
(off Moulding La.)
Fuzzens Wlk. SL4: Wind 4C 102
Fye Foot La. EC4 4D 224 (45Sb 91)
(off Queen Victoria St.)
Fyfe Apts. N8 27Pb 50
Fyfe Way BR1: Broml 68Jc 137
Fyfield N4 33Qb 70
(off Six Acres Est.)
Fyfield Cl. BR2: Broml 70Fc 137
Fyfield Cl. CM13: W H'dn 30Fe 59
Fyfield Cl. KT17: Eps 86Va 174
Fyfield Ct. E7 37Jc 73
Fyfield Dr. RM15: S Ock 45Vd 98
Fyfield Ho. E6 39Nc 74
(off Ron Leighton Way)
Fyfield Rd. E17 27Fc 53
Fyfield Rd. EN1: Enf 13Ub 33
Fyfield Rd. IG8: Wfd G 24Lc 53
Fyfield Rd. RM13: Rain 39Hd 76
Fyfield Rd. SW9 55Qb 112
Fynes St. SW1 5D 228 (49Mb 90)

G

Gabion Av. RM19: Purf 49Td 98
Gable Cl. DA1: Cray 57Jd 118
Gable Cl. HA5: Hat E 24Ca 45
Gable Cl. WD5: Ab L 4U 12
Gable Ct. RH1: Redh 5A 208
(off St Anne's Mt.)
Gable Ct. SE26 63Xb 135
Gable M. BR2: Broml 75Nc 160
Gable M. SL1: Burn 2A 80
Gables, The BR1: Broml 66Kc 137
Gables, The CM13: Gt War 23Yd 58
Gables, The DA3: Lfield 68Ee 143
Gables, The HA9: Wemb 34Qa 67
Gables, The HP2: Hem H 1M 3
Gables, The IG11: Bark 37Sc 74
Gables, The KT13: Weyb 78S 150
Gables, The KT22: Oxs 84Ea 172
Gables, The N10 27Jb 50
(off Fortis Grn.)
Gables, The RM17: Grays 49Be 99
Gables, The SM7: Bans 89Bb 175
Gables, The WD19: Wat 17Y 27
Gables, The WD25: Wat 5Z 13
Gables Av. TW15: Ashf 64P 127
Gables Av. WD6: Bore 13Pa 29
Gables Cl. GU22: Wok 92B 188
Gables Cl. SE12 60Jc 115
Gables Cl. SE5 53Ub 113
Gables Cl. SL3: Dat 1L 103
Gables Cl. SL9: Chal P 21A 42
Gables Ct. CR8: Purl 84Rb 177
Gables Ct. GU22: Wok 92B 188
Gables Lodge EN4: Had W 10Eb 17
Gables Way SM7: Bans 89Bb 175
Gabriel Cl. RM16: Chaf H 48Yd 98
Gabriel Cl. RM5: Col R 24Ed 56
Gabriel Cl. TW13: Hanw 63Aa 129
Gabriel Ct. E1 43Ac 92
(off Elsa Street)
Gabriel Ct. NW9 26Ua 48
Gabriel Gdns. DA12: Grav'nd 4G 144
Gabriel Ho. N1 1B 218 (39Rb 71)
(off Islington Grn.)
Gabriel Ho. SE11 5H 229 (49Pb 90)
(off Villa St.)
Gabriel Ho. SE16 48Bc 92
(off Odessa St.)
Gabrielle Cl. HA9: Wemb 34Pa 67
Gabrielle Ct. NW3 37Fb 69
Gabriel M. NW2 33Bb 69
Gabriel's M. BR3: Beck 67Zb 136
Gabriel Spring Rd. DA3: Fawk 75Td 164
Gabriel Spring Rd. E. DA3: Fawk 75Ud 164
Gabriel Spring Rd. E. DA3: Hort K 75Ud 164
Gabriel Sq. AL1: St A 3C 6
Gabriel St. SE23 59Zb 114
Gabriels Wharf SE1 6K 223 (46Qb 90)
Gad Cl. E13 41Kc 93
Gaddesden Av. HA9: Wemb 37Pa 67
Gaddesden Cres. WD25: Wat 6Z 13
Gaddesden Ho. EC1 4G 219 (41Tb 91)
(off Cranwood St.)
Gade Av. WD18: Wat 14U 26
Gade Bank WD3: Crox G 14T 26
GADEBRIDGE 1J 3
Gadebridge Ct. HP1: Hem H 1L 3
Gadebridge Ho. SW3 7D 226 (50Gb 89)
(off Cale St.)
Gadebridge La. HP1: Hem H 1J 3
(not continuous)
Gadebridge Point HP1: Hem H 4L 3
(off Cotterells)
Gade Cl. UB3: Hayes 46X 85
Gade Cl. WD18: Wat 14U 26
Gade Pl. HP1: Hem H 4L 3
(off Cotterells)
Gadesden Rd. KT19: Ewe 79Sa 153
Gade Side WD25: Wat 7U 12
(not continuous)
Gade Twr. HP3: Hem H 7A 4
Gade Valley Cl. WD4: K Lan 10A 4
Gadeview HP1: Hem H 2L 3
Gade Vw. Gdns. WD4: Hunt C 4S 12
Gadeview Rd. HP3: Hem H 6L 3
Gadsbury Cl. NW9 30Va 48
Gadsden Cl. RM14: Upm 30Ud 58
Gadsden Ho. W10 42Ab 88
(off Hazlewood Cres.)
Gadswell Cl. WD25: Wat 8Z 13
Gadwall Cl. E16 44Kc 93
Gadwall Ho. NW9 30Wa 48
(off Perryfield Way)
Gadwall Way SE28 47Tc 94
Gage Brown Ho. W10 44Za 88
(off Bridge Cl.)
Gage M. CR2: S Croy 78Rb 157
Gage Rd. E16 43Gc 93
Gage St. WC1 7G 217 (43Nb 90)
Gainford Ho. E2 41Xb 91
(off Ellsworth St.)
Gainford St. N1 39Qb 70
Gainsboro Gdns. UB6: G'frd 36Ga 66
Gainsborough Av. AL1: St A 1D 6
Gainsborough Av. DA1: Dart 57Ld 119
Gainsborough Av. E12 36Qc 74
Gainsborough Av. RM18: Tilb 3C 122
Gainsborough Cl. BR3: Beck 66Cc 136
Gainsborough Cl. KT10: Esh 74Ga 152
Gainsborough Ct. BR2: Broml 70Lc 137
Gainsborough Ct. CM14: W'ley 21Yd 58
(off Gt. Eastern Rd.)
Gainsborough Ct. KT12: Walt T 77W 150
Gainsborough Ct. KT19: Ewe 79Va 154
Gainsborough Ct. N12 22Db 49
Gainsborough Ct. SE16 50Xb 91
(off Stubbs Dr.)
Gainsborough Ct. SE21 61Ub 135
Gainsborough Ct. W12 47Ya 88
Gainsborough Ct. W4 50Ra 87
(off Chaseley Dr.)
Gainsborough Dr. CR2: Sande 85Wb 177
Gainsborough Dr. DA11: Nflt 62Fe 143
Gainsborough Gdns. HA8: Edg 26Pa 47
Gainsborough Gdns. NW11 31Bb 69
Gainsborough Gdns. NW3 34Fb 69
Gainsborough Gdns. TW7: Isle 57Fa 108
Gainsborough Ho. E14: Cassilis Rd. 47Cc 92
(off Cassilis Rd.)
Gainsborough Ho. E14: Victory Pl. 45Ac 92
(off Victory Pl.)
Gainsborough Ho. EN1: Enf 15Wb 33
Gainsborough Ho. RM8: Dag 35Xc 75
(off Longbridge Rd.)
Gainsborough Ho. SW1 6E 228 (49Mb 90)
(off Erasmus St.)
Gainsborough Lodge HA1: Harr 29Ha 46
(off Hindes Rd.)
Gainsborough Mans. W14 51Ab 110
(off Queen's Club Gdns.)
Gainsborough M. SE26 62Xb 135
Gainsborough Pl. CM13: Hut 18Fe 41
Gainsborough Pl. IG7: Chig 20Vc 37
Gainsborough Pl. KT11: Cobh 87Aa 171
Gainsborough Rd. E11 31Gc 73
Gainsborough Rd. E15 41Gc 93
Gainsborough Rd. IG8: Wfd G 23Nc 54
Gainsborough Rd. KT19: Eps 82Sa 173
Gainsborough Rd. KT3: N Mald 72Ta 153
Gainsborough Rd. N12 22Db 49
Gainsborough Rd. RM13: Rain 39Jd 76
Gainsborough Rd. RM8: Dag 35Xc 75
Gainsborough Rd. TW9: Rich 54Pa 109
Gainsborough Rd. UB4: Hayes 40S 64
Gainsborough Rd. W4 49Va 88
Gainsborough Sq. DA6: Bex 55Zc 117
Gainsborough St. E9 37Bc 72
Gainsborough Studios E. N1 1F 219 (39Tb 71)
(off Poole St.)
Gainsborough Studios Nth. N1 1F 219 (39Tb 71)
(off Poole St.)
Gainsborough Studios Sth. N1 1F 219 (39Tb 71)
(off Poole St.)
Gainsborough Studios W. N1 1F 219 (39Tb 71)
(off Poole St.)
Gainsborough Ter. SM2: Sutt 80Bb 155
(off Belmont Ri.)
Gainsborough Twr. UB5: N'olt 40Z 65
(off Academy Gdns.)
Gainsfield Ct. E11 34Gc 73
Gainsford Pl. RH8: C'rst 10G 210
Gainsford Rd. E17 28Bc 52
Gainsford St. SE1 1K 231 (47Vb 91)
Gairloch Ho. NW1 38Mb 70
(off Stratford Vs.)
Gairloch Rd. SE5 54Ub 113
Gaisford St. NW5 37Lb 70
Gaist Av. CR3: Cat'm 94Xb 197
Gaitskell Ct. SW11 54Gb 111
Gaitskell Ho. E17 27Dc 52
Gaitskell Ho. E6 39Mc 73
Gaitskell Ho. RM16: Grays 46De 99
(off Crammavill St.)
Gaitskell Ho. SE17 51Ub 113
(off Villa St.)
Gaitskell Ho. WD6: Bore 14Ta 29
(off Howard Dr.)
Gaitskell Rd. SE9 60Sc 116
Gaitskell Way SE1 1E 230 (47Sb 91)
(off Weller St.)
Gala Av. BR8: Swan 70Hd 140
Gala Bingo Borehamwood 13Qa 29
(within The Point)
Gala Bingo Surrey Quays 48Zb 92
Gala Bingo Tooting 64Gb 133
Gala Bingo Woking 89B 168
(within The Big Apple)
Gala Ct. CR7: Thor H 71Qb 156
Galahad Cl. SL1: Slou 7E 80
Galahad M. E3 40Bc 72
Galahad Rd. BR1: Broml 63Jc 137
Galahad Rd. N9 20Wb 33
Galata Rd. SW13 52Wa 110
Galatea Sq. SE15 55Xb 113
Galaxy Bldg. E14 49Cc 92
(off Crews St.)
Galaxy Ho. EC2 5G 219 (42Tb 91)
(off Leonard St.)
Galba Ct. TW8: Bford 52Ma 109
Galbraith St. E14 48Ec 92
Galdana Av. EN5: New Bar 13Eb 31
Galeborough Av. IG8: Wfd G 24Fc 53
Gale Cl. CR4: Mitc 69Fb 133
Gale Cl. TW12: Hamp 65Aa 129
Gale Cres. SM7: Bans 89Cb 175
Galena Arches W6 49Xa 88
(off Galena Rd.)
Galena Hgts. E20 37Ec 72
(off Mirabelle Gdns.)
Galena Ho. SE18 50Vc 95
(off Grosmont Rd.)
Galena Rd. W6 49Xa 88
Galen Cl. KT19: Eps 83Qa 173
Galen Pl. WC1 1G 223 (43Nb 90)
Galesbury Rd. SW18 58Eb 111
Gales Gdns. E2 41Xb 91
Gale St. E3 43Cc 92
Gale St. RM9: Dag 36Yc 75
Gales Way IG8: Wfd G 24Nc 54
Galey Grn. RM15: S Ock 43Xd 98
Galgate Cl. SW19 60Za 110
Galileo Dr. GU23: Send 95E 188
Gallants Farm Rd. EN4: E Barn 17Gb 31
Galleon Blvd. DA2: Dart 56Td 120
Galleon Cl. DA8: Erith 49Fd 96
Galleon Cl. SE16 47Zb 92
Galleon Ho. E14 49Ec 92
(off Glengarnock Av.)
Galleon M. DA11: Nflt 9A 122
Galleon Rd. RM16: Chaf H 49Yd 98
Galleons Dr. IG11: Bark 41Wc 95
Galleons La. SL3: Wex 2N 81
(not continuous)
Galleons Vw. E14 47Ec 92
Galleria Ct. SE15 51Vb 113
Galleria Shop. Mall, The 26Jc 53
Galleries, The CM14: W'ley 22Xd 58
Galleries, The NW8 2A 214 (40Eb 69)
(off Abbey Rd.)
Gallery, The E20 37Ec 72
(within Westfield Shop. Cen.)
Gallery, The 39Sc 74
(off Clockhouse Av.)
Gallery, The SE14 52Bc 114
(off New Cross Rd.)
Gallery Apts. E1 44Yb 92
(off Commercial Rd.)
Gallery Apts. SE1 2H 231 (47Ub 91)
(off Lamb Wlk.)
Gallery at London Glassblowing, The 1H 231 (47Ub 91)
(off Bermondsey St.)
Gallery By Pool, The 48Yb 92
Gallery Ct. E17 26Ec 52
(off Fulbourne Rd.)
Gallery Ct. SE1 2F 231 (47Tb 91)
(off Pilgrimage St.)
Gallery Ct. SW10 51Eb 111
(off Gunter Gro.)
Gallery Gdns. UB5: N'olt 40Z 65
Gallery Ho. E8 37Xb 71
(off Hackney Gro.)
Gallery Rd. SE21 60Tb 113
Galley, The E16 45Rc 94
GALLEY HILL 3Hc 21
Galley Hill HP1: Hem H 1H 3
Galley Hill Ind. Est. DA10: Swans 57Ae 121
Galley Hill Rd. DA10: Swans 57Be 121
Galley Hill Rd. DA11: Nflt 57Be 121
Galleyhill Rd. EN9: Walt A 5Gc 21
(not continuous)
Galley La. EN5: Barn 10Wa 16
Galleymead Rd. SL3: Poyle 53H 105
Galleywall Rd. SE16 49Xb 91
Galleywall Rd. Trad. Est. SE16 49Xb 91
Galleywood Cres. RM5: Col R 23Fd 56
Galleywood Ho. W10 43Ya 88
(off Sutton Way)
Galliant Cl. SL0: Iver 44G 82
Galliard Cl. N9 16Yb 34
Galliard Ct. N9 16Wb 33
Galliard Rd. N9 18Wb 33
Gallia Rd. N5 36Rb 71
Gallica Ct. SM1: Sutt 74Db 155
Gallions Cl. IG11: Bark 41Wc 95
Gallions Reach Shop. Pk. E6 43Sc 94
Gallions Reach Station (DLR) 45Rc 94
Gallions Rd. E16 45Rc 94
Gallions Rd. SE7 49Kc 93
(not continuous)
GALLIONS RDBT. 45Rc 94
Gallions Vw. Rd. SE28 47Uc 94
Gallipoli Pl. RM9: Dag 39Xc 75
Gallon Cl. SE7 49Lc 93
Gallop, The CR2: Sels 80Xb 157
Gallop, The SL4: Wind 9G 102
Gallop, The SM2: Sutt 81Fb 175
Gallops, The KT10: Esh 76Da 151
Gallosson Rd. SE18 49Uc 94
Galloway Chase SL2: Slou 5L 81
Galloway Dr. DA1: Cray 59Gd 118
Galloway Path CR0: C'don 77Tb 157
Galloway Rd. W12 46Wa 88
GALLOWS CORNER 26Ld 57
GALLOWS CORNER 25Ld 57
Gallows Hill WD4: Hunt C 4S 12
Gallows Hill La. WD5: Ab L 4S 12
Gallows Wood DA3: Fawk 77Wd 164
Gallus Cl. N21 16Pb 32
Gallus Sq. SE3 55Kc 115
Gallys Rd. SL4: Wind 4B 102
Galpins Rd. CR7: Thor H 71Nb 156
Galsworthy Av. E14 44Ac 92
Galsworthy Av. RM6: Chad H 31Xc 75
Galsworthy Cl. NW2 35Ab 68
Galsworthy Cl. SE28 46Xc 95
Galsworthy Cres. SE3 52Lc 115
Galsworthy Ho. W11 44Ab 88
(off Elgin Cres.)
Galsworthy Rd. KT16: Chert 73J 149
Galsworthy Rd. KT2: King T 66Ra 131
Galsworthy Rd. NW2 35Ab 68
Galsworthy Rd. RM18: Tilb 3E 122
Galsworthy Ter. N16 34Tb 71
Galton Ct. NW9 27Ua 48
(off Joslin Av.)
Galton Rd. SL5: S'dale 2D 146
Galton St. W10 41Ab 88
Galva Cl. EN4: Cockf 14Jb 32
Galvani Way CR0: Wadd 74Pb 156
Galveston Ho. E1 42Ac 92
(off Harford St.)
Galveston Rd. SW15 57Bb 111
Galvin Rd. SL1: Slou 6G 80
Galway Cl. SE16 50Xb 91
(off Masters Dr.)
Galway Ho. E1 43Zb 92
(off White Horse La.)
Galway Ho. EC1 4E 218 (41Sb 91)
(off Galway St.)
Galway St. EC1 4E 218 (41Sb 91)
Gambado Beckenham 65Cc 136
Gambado Chelsea 53Eb 111
(off Station Ct.)
Gambado Watford 6X 13
Gambetta St. SW8 54Kb 112
Gambia St. SE1 7C 224 (46Rb 91)
Gambier Ho. EC1 4E 218 (41Sb 91)
(off Mora St.)
Gambles La. GU23: Rip 96L 189
Gambole Rd. SW17 63Gb 133
Games Rd. EN4: Cockf 13Gb 31
Gamlen Rd. SW15 56Za 110
Gamma Ct. CR0: C'don 74Tb 157
(off Sydenham Rd.)
Gammon Cl. HP3: Hem H 3A 4
Gammon Fld. RM16: Grays 5A 100
Gammons Farm Cl. WD24: Wat 8V 12
Gammons La. WD24: Wat 8U 12
(not continuous)
Gamuel Cl. E17 30Cc 52
Gander Grn. Cres. TW12: Hamp 67Ca 129
Gander Grn. La. SM1: Sutt 76Bb 155
Gander Grn. La. SM3: Cheam 75Ab 154
Ganders Ash WD25: Wat 5W 12
Gandhi Cl. E17 30Cc 52
Gandhi Ct. WD24: Wat 12Z 27
Gandolfi St. SE15 51Ub 113
Gangers Hill CR3: Wold 98Cc 198
Gangers Hill RH9: G'stone 100Ac 198
Ganley Ct. SW11 55Fb 111
(off Winstanley Est.)
Gant Ct. EN9: Walt A 6Hc 21
Ganton St. W1 4B 222 (45Lb 90)
Ganton Wlk. WD19: Wat 21Z 45
GANTS HILL 30Qc 54
GANTS HILL 30Qc 54
Gantshill Cres. IG2: Ilf 29Qc 54
Gants Hill Station (Underground) 30Qc 54
GANWICK 8Cb 17
GANWICK CORNER 7Cb 17
Gapp Cl. TN15: W King 80Ud 164
Gap Rd. SW19 64Cb 133
Garage Rd. W3 44Qa 87
Garand Ct. N7 36Pb 70
Garbett Ho. SE17 51Rb 113
(off Doddington Gro.)
Garbrand Wlk. KT17: Ewe 81Va 174
Garbutt Pl. W1 7J 215 (43Jb 90)
Garbutt Rd. RM14: Upm 33Sd 78
Garda Ho. SE10 49Gc 93
(off Cable Wlk.)
Garden Av. AL10: Hat 4C 8
Garden Av. CR4: Mitc 66Kb 134
Garden Av. DA7: Bex 55Bd 117
Garden City HA8: Edg 23Qa 47
Garden Cl. AL1: St A 1F 6
Garden Cl. E4 22Cc 52
Garden Cl. EN5: Ark 14Ya 30
Garden Cl. HA4: Ruis 33U 64
Garden Cl. KT15: Add 77M 149
Garden Cl. KT22: Lea 96La 192
Garden Cl. KT3: N Mald 70Ua 132
Garden Cl. SE12 62Kc 137
Garden Cl. SM6: W'gton 78Nb 156
Garden Cl. SM7: Bans 87Cb 175
Garden Cl. SW15 59Ya 110
Garden Cl. TW12: Hamp 64Ba 129
Garden Cl. TW15: Ashf 65S 128
Garden Cl. UB5: N'olt 39Aa 65
Garden Cl. WD17: Wat 12V 26
Garden Cotts. AL2: F'mre 9B 6
Garden Cotts. BR5: St P 68Yc 139
Garden Ct. CR0: C'don 75Vb 157
Garden Ct. EC4 4K 223 (45Qb 90)
(off Fountain Ct.)
Garden Ct. HA7: Stan 22La 46
Garden Ct. N12 22Db 49
Garden Ct. NW8 3B 214 (41Fb 89)
(off Garden Rd.)
Garden Ct. TN13: S'oaks 94Md 203
(off Garden Rd.)
Garden Ct. TW12: Hamp 64Ba 129
Garden Ct. TW9: Kew 53Pa 109
Garden Ct. W11 45Ab 88
(off Clarendon Rd.)
Garden Ct. W4 48Sa 87
Garden Dr. GU24: Chob 1K 167
Gardener Gro. TW13: Hanw 61Ba 129
Gardeners Cl. N11 19Jb 32
Gardeners Cl. SE9 62Nc 138
Gardeners Cotts. TN14: Hals 83Ad 181
Gardeners Rd. CR0: C'don 74Rb 157
Gardener's Wlk. KT23: Bookh 98Da 191
Garden Farm Cl. KT20: Kgswd 92Ab 194
Garden Fld. La. HP4: Berk 3B 2
Gardenfields KT20: Tad 91Ab 194
Garden Halls, The WC1 4F 217 (41Nb 90)
Garden Ho. N2 26Fb 49
(off The Grange)
Garden Ho. NW6 40Db 69
(off Oxford Rd.)
Garden Ho. SW7 48Db 89
(off Cornwall Gdns.)
Garden Ho's., The W6 51Za 110
(off Bothwell St.)
Gardenia Dr. GU24: W End 5D 166
Gardenia Rd. BR1: Broml 69Qc 138
Gardenia Rd. EN1: Enf 16Ub 33
Gardenia Way IG8: Wfd G 23Jc 53
Garden La. BR1: Broml 65Kc 137
Garden La. SW2 60Pb 112
Garden M. SE10 49Gc 93
Garden M. SL1: Slou 6K 81
Garden M. W2 45Cb 89
Garden Mus., The 4H 229 (48Pb 90)
Garden Pl. DA2: Wilm 62Md 141
Garden Pl. E8 39Vb 71
Garden Reach HP8: Chal G 13A 24
Garden Rd. BR1: Broml 66Kc 137
Garden Rd. KT12: Walt T 72X 151
Garden Rd. NW8 3A 214 (41Eb 89)
Garden Rd. SE20 67Yb 136
Garden Rd. TN13: S'oaks 94Md 203
Garden Rd. TW9: Rich 55Qa 109
Garden Rd. WD5: Ab L 3U 12
Garden Row DA11: Nflt 2B 144
Garden Row SE1 4B 230 (48Rb 91)
Garden Royal SW15 58Za 110
Gardens, The AL9: Brk P 9G 8
Gardens, The BR3: Beck 67Ec 136
Gardens, The E5 31Vb 71
Gardens, The GU24: Pirb 4D 186
Gardens, The HA1: Harr 30Ea 46
Gardens, The HA5: Pinn 30Ba 45
Gardens, The KT10: Esh 77Ca 151
Gardens, The KT11: Cobh 91S 190
Gardens, The N8 28Nb 50
(not continuous)
Gardens, The SE22 56Wb 113
Gardens, The TW14: Felt 57T 106
Gardens, The WD17: Wat 12V 26
Garden Sq. SW1 7J 227 (50Jb 90)
Garden St. E1 43Zb 92
Garden Ter. SW1 7D 228 (50Mb 90)
Garden Ter. SW7 2E 226 (47Gb 89)
(off Trevor Pl.)
Garden Ter. TN15: Seal 93Qd 203
Garden Wlk. BR3: Beck 67Bc 136
Garden Wlk. CR5: Coul 95Kb 196
Garden Wlk. EC2 4H 219 (41Ub 91)
Garden Way IG10: Lough 10Qc 22
Garden Way NW10 37Sa 67
Gardiner Av. NW2 36Ya 68
Gardiner Cl. BR5: St P 68Yc 139
Gardiner Cl. EN3: Pond E 16Zb 34
Gardiner Cl. RM8: Dag 35Zc 75
Gardiner Ct. CR2: S Croy 79Tb 157
Gardiner Ho. SW11 53Gb 111
Gardner Cl. E11 30Kc 53
Gardner Ct. EC1 6B 218 (42Rb 91)
(off Brewery Sq.)
Gardner Ct. N5 35Sb 71
Gardner Ct. WD25: Wat 7Y 13
Gardner Ho. TW13: Hanw 61Ba 129
Gardner Ho. UB1: S'hall 45Z 85
(off The Broadway)
Gardner Ind. Est. BR3: Beck 64Bc 136
Gardner Pl. TW14: Felt 58X 107
Gardner Rd. E13 42Kc 93
Gardners La. EC4 4D 224 (45Sb 91)
Gardner's Way RM20: W Thur 52Wd 120
Gardnor Rd. NW3 35Fb 69
Gard St. EC1 3C 218 (41Rb 91)
Garendon Gdns. SM4: Mord 73Db 155
Garendon Rd. SM4: Mord 73Db 155
Garenne Ct. E4 18Ec 34
Gareth Cl. KT4: Wor Pk 75Za 154
Gareth Ct. SW16 62Mb 134
Gareth Ct. WD6: Bore 10Pa 15
(off Aycliffe Rd.)
Gareth Dr. N9 19Wb 33
Gareth Gro. BR1: Broml 63Jc 137
Garfield EN2: Enf 15Tb 33
(off London Rd.)
Garfield Ct. NW6 38Ab 68
(off Willesden La.)
Garfield M. SW11 55Jb 112
Garfield Pl. KT15: Add 77L 149
Garfield Pl. SL4: Wind 4H 103
Garfield Rd. E13 42Hc 93
Garfield Rd. E4 18Fc 35
Garfield Rd. EN3: Pond E 14Yb 34
Garfield Rd. KT15: Add 78L 149
Garfield Rd. SW11 55Jb 112
Garfield Rd. SW19 64Eb 133
Garfield Rd. TW1: Twick 60Ja 108
Garfield St. WD24: Wat 10X 13
Garford St. E14 45Cc 92
Garganey Ct. NW10 37Ta 67
(off Elgar Av.)
Garganey Wlk. SE28 45Yc 95
Gargery Cl. DA12: Grav'nd 10J 123
Garibaldi Rd. RH1: Redh 7P 207
Garibaldi St. SE18 49Uc 94
Garland Cl. EN8: Chesh 3Ac 20
Garland Cl. HP2: Hem H 1M 3
Garland Cl. SE1 4E 230 (48Sb 91)
Garland Ct. AL1: St A 2C 6
(off Victoria St.)
Garland Ct. E14 45Cc 92
(off Premiere Pl.)
Garland Ct. SE17 6E 230 (49Sb 91)
(off Wansey St.)
Garland Dr. TW3: Houn 54Ea 108
Garland Ho. KT2: King T 67Na 131
(off Skerne Rd.)
Garland Ho. UB7: W Dray 47P 83
Garland Rd. HA7: Stan 25Na 47
Garland Rd. SE18 52Tc 116
Garlands Ct. CR0: C'don 77Tb 157
(off Chatsworth Rd.)
Garlands Ho. NW8 40Eb 69
(off Carlton Hill)
Garlands La. HA1: Harr 32Ha 66
Garlands Rd. KT22: Lea 93Ka 192
Garlands Rd. RH1: Redh 7P 207
Garland Way CR3: Cat'm 94Tb 197
Garland Way RM11: Horn 28Nd 57
Garlichill Rd. KT18: Tatt C 89Xa 174
Garlick Hill EC4 4E 224 (45Sb 91)
Garlies Rd. SE23 62Ac 136
Garlinge Ho. SW9 53Qb 112
(off Gosling Way)
Garlinge Rd. NW2 37Bb 69
Garman Cl. N18 22Tb 51
Garman Rd. N17 24Xb 51
(not continuous)
Garnault M. EC1 4A 218 (41Qb 90)
(off Rosebery Av.)
Garnault Pl. EC1 4A 218 (41Qb 90)

Garnault Rd. EN1: Enf 10Vb 19
Garner Cl. RM8: Dag 32Zc 75
Garner Cl. TW19: Stanw 58M 105
(off Douglas Rd.)
Garner Rd. E17 25Ec 52
Garners Cl. SL9: Chal P 23B 42
Garners End SL9: Chal P 23A 42
Garners Rd. SL9: Chal P 23A 42
Garner St. E2 40Wb 71
Garnet Cl. SL1: Slou 7E 80
Garnet Ho. E1 46Yb 92
(off Garnet St.)
Garnet Pl. UB7: Yiew 46M 83
Garnet Rd. CR7: Thor H 70Sb 135
Garnet Rd. DA8: Erith 52Jd 118
Garnet Rd. NW10 37Ua 68
Garnet St. E1 45Yb 92
Garnett Cl. SE9 55Pc 116
Garnett Cl. WD24: Wat 9Z 13
Garnett Dr. AL2: Brick W 1Ba 13
Garnett Rd. NW3 36Hb 69
Garnett Way E17 25Ac 52
(off McEntee Av.)
Garnet Wlk. E6 43Nc 94
Garnham Cl. N16 33Vb 71
Garnham St. N16 33Vb 71
Garnies Cl. SE15 52Vb 113
Garnon Mead CM16: Coop 1Zc 23
Garrad's Rd. SW16 62Mb 134
Garrard Cl. BR7: Chst 64Rc 138
Garrard Cl. DA7: Bex 55Cd 118
Garrard Rd. SL2: Slou 2C 80
Garrard Rd. SM7: Bans 88Cb 175
Garrard Wlk. NW10 37Ua 68
Garratt Cl. CR0: Bedd 77Nb 156
Garratt Cl. CR7: Thor H 68Sb 135
Garratt Ct. SW18 59Db 111
Garratt La. SW17 62Eb 133
Garratt La. SW18 58Db 111
Garratt Rd. HA8: Edg 24Qa 47
Garratts La. SM7: Bans 88Bb 175
Garratts Rd. WD23: Bush 17Ea 28
Garratt Ter. SW17 63Gb 133
Garraway Ct. SW13 52Ya 110
(off Wyatt Dr.)
Garrett Cl. W3 43Ta 87
Garrett Ho. SE1 1B 230 (47Rb 91)
(off Burrows M.)
Garrett St. EC1 5E 218 (42Sb 91)
Garrick Av. NW11 30Ab 48
Garrick Cl. KT12: Hers 77X 151
Garrick Cl. SW18 56Eb 111
Garrick Cl. TW18: Staines 66J 127
Garrick Cl. TW9: Rich 57Ma 109
Garrick Cl. W5 42Na 87
Garrick Ct. E8 38Vb 71
(off Jacaranda Gro.)
Garrick Cres. CR0: C'don 75Ub 157
Garrick Dr. NW4 26Ya 48
Garrick Dr. SE28 48Tc 94
Garrick Gdns. KT8: W Mole 69Ca 129
Garrick Ho. KT1: King T 70Na 131
(off Surbiton Rd.)
Garrick Ho. W1 7K 221 (46Kb 90)
(off Carrington St.)
Garrick Ho. W4 51Ua 110
Garrick Ind. Cen. NW9 29Va 48
Garrick Pk. NW4 26Ya 48
Garrick Rd. NW9 30Va 48
Garrick Rd. TW9: Rich 54Qa 109
Garrick Rd. UB6: G'frd 42Da 85
Garricks Ho. KT1: King T 68Ma 131
(off Wadbrook St.)
Garrick St. DA11: Grav'nd 8D 122
Garrick St. WC2 4F 223 (45Nb 90)
Garrick Theatre 5E 222 (45Mb 90)
(off Charing Cross Rd.)
Garrick Way NW4 28Za 48
Garrick Yd. WC2 4F 223 (45Nb 90)
(off St Martin's La.)
Garrison Cl. SE18 52Qc 116
Garrison Cl. TW4: Houn 57Ba 107
Garrison La. KT9: Chess 80Ma 153
Garrison Pde. RM19: Purf 49Qd 97
Garrison Rd. E3 39Cc 72
Garrison Sq. SW1 7J 227 (50Jb 90)
Garrolds Cl. BR8: Swan 68Fd 140
Garron La. RM15: S Ock 44Vd 98
Garrow DA3: Lfield 69De 143
Garrowsfield EN5: Barn 16Bb 31
Garry Cl. RM1: Rom 24Gd 56
Garry Way RM1: Rom 24Gd 56
Garsdale Cl. N11 23Jb 50
Garsdale Ter. W14 50Bb 89
(off Aisgill Av.)
Garside Cl. SE28 48Tc 94
Garside Cl. TW12: Hamp 65Da 129
Garside Ct. TW11: Hamp W 67La 130
Garsington M. SE4 55Bc 114
Garsmouth Way WD25: Wat 8Z 13
Garson Cl. KT10: Esh 78Ba 151
Garson Ct. WD6: Bore 12Sa 29
Garson Ho. W2 4B 220 (45Fb 89)
(off Gloucester Ter.)
Garson La. TW19: Wray 9P 103
Garson Rd. KT10: Esh 79Ba 151
GARSTON 7Y 13
Garston Cres. WD25: Wat 6Y 13
Garston Dr. WD25: Wat 6Y 13
Garston Gdns. CR8: Kenley 87Tb 177
Garston Ho. N1 38Rb 71
(off The Sutton Est.)
Garston La. CR8: Kenley 86Tb 177
Garston La. WD25: Wat 6Z 13
Garston Pk. Pde. WD25: Wat 6Z 13
Garstons, The KT23: Bookh 97Ca 191
Garston Station (Rail) 7Z 13
Garter Way SE16 47Zb 92
Garth, The HA3: Kenton 30Pa 47
Garth, The KT11: Cobh 85Aa 171
Garth, The TW12: Hamp H 65Da 129
Garth, The WD5: Ab L 5T 12
Garth Cl. HA4: Ruis 32Z 65
Garth Cl. KT2: King T 64Pa 131
Garth Cl. SM4: Mord 73Za 154
Garth Ct. HA1: Harr 30Ha 46
(off Northwick Pk. Rd.)
Garth Ct. W4 50Ta 87
Garth Ho. NW2 33Bb 69
Garthland Dr. EN5: Barn 15Xa 30
Garth M. W5 42Na 87
Garthorne Rd. SE23 59Zb 114
Garthorne Road Nature Reserve 59Zb 114
Garth Rd. KT2: King T 64Pa 131
Garth Rd. NW2 33Bb 69
Garth Rd. RM15: S Ock 42Yd 98
Garth Rd. SM4: Mord 72Ya 154
Garth Rd. TN13: S'oaks 100Ld 203
Garth Rd. W4 50Ta 87
Garth Rd. Ind. Cen., The SM4: Mord 74Za 154
Garthside TW10: Ham 64Na 131
Garthway N12 23Gb 49
Gartlet Rd. WD17: Wat 13Y 27
Gartmoor Gdns. SW19 60Bb 111
Gartmore Rd. IG3: Ilf 33Vc 75
Garton Bank SM7: Bans 89Cb 175
Garton La. RM15: S Ock 44Vd 98
Garton Pl. SW18 58Eb 111
Gartons Cl. EN3: Pond E 14Yb 34
Gartons Way SW11 55Eb 111
Garvary Rd. E16 44Kc 93
Garvock Dr. TN13: S'oaks 98Jd 202
Garway Ct. E3 40Cc 72
(off Matilda Gdns.)
Garway Rd. W2 44Db 89
Garwood Cl. N17 25Xb 51
Gascoigne Cl. N17 25Vb 51
Gascoigne Gdns. IG8: Wfd G 23Jc 53
Gascoigne Pl. E2 4K 219 (41Vb 91)
(not continuous)
Gascoigne Rd. CR0: New Ad 82Ec 178
Gascoigne Rd. IG11: Bark 39Sc 74
Gascoigne Rd. KT13: Weyb 76R 150
Gascon's Gro. SL2: Slou 2E 80
Gascony Av. NW6 38Cb 69
Gascony Pl. W12 46Za 88
Gascoyne Cl. EN6: S Mim 4Wa 16
Gascoyne Cl. RM3: Rom 24Md 57
Gascoyne Dr. DA1: Cray 55Hd 118
Gascoyne Ho. E9 38Ac 72
Gascoyne Rd. E9 38Zb 72
Gaselee St. E14 46Ec 92
(off Baffin Way)
Gasholder Pk. 1E 216 (39Mb 70)
Gaskarth Rd. HA8: Edg 25Sa 47
Gaskarth Rd. SW12 58Kb 112
Gaskell Ct. SE20 66Zb 136
Gaskell Rd. N6 30Hb 49
Gaskell St. SW4 54Nb 112
Gaskin St. N1 39Rb 71
Gasoline All. TN15: Wro 89Fe 185
Gaspar Cl. SW5 49Db 89
Gaspar M. SW5 49Db 89
Gassiot Rd. SW17 63Hb 133
Gassiot Way SM1: Sutt 76Fb 155
Gasson Ho. SE14 51Zb 114
(off John Williams Cl.)
Gasson Rd. DA10: Swans 58Ae 121
Gastein Rd. W6 51Za 110
Gastigny Ho. EC1 4E 218 (41Sb 91)
(off Pleydell Est.)
Gaston Bell Cl. TW9: Rich 55Pa 109
Gaston Bri. Rd. TW17: Shep 72T 150
Gaston Rd. CR4: Mitc 69Jb 134
Gaston Way TW17: Shep 71T 150
Gataker Ho. SE16 48Xb 91
(off Slippers Pl.)
Gataker St. SE16 48Xb 91
Gatcombe Ct. AL1: St A 3E 6
(off Dexter Cl.)
Gatcombe Ct. BR3: Beck 66Cc 136
Gatcombe Ho. SE22 55Ub 113
Gatcombe M. W5 45Pa 87
Gatcombe Rd. E16 46Jc 93
Gatcombe Rd. N19 34Mb 70
Gatcombe Way EN4: Cockf 13Hb 31
Gateacre Ct. DA14: Sidc 63Xc 139
Gate Cen., The TW8: Bford 52Ja 108
Gate Cinema 45Cb 89
(off Notting Hill Ga.)
Gate Cl. WD6: Bore 11Sa 29
Gate Cotts. WD3. Chor 14F 24
Gatecroft HP3: Hem H 4P 3
(not continuous)
Gate End HA6: Nwood 24W 44
Gatefold Bldg., The UB3: Hayes 48U 84
Gateforth St. NW8 6D 214 (42Gb 69)
Gate Hill Ct. W11 46Bb 89
(off Ladbroke Ter.)
Gatehill Rd. HA6: Nwood 24V 44
Gatehope Dr. RM15: S Ock 44Vd 98
Gate Ho. E3 39Ac 72
(off Gunmakers La.)
Gate Ho. N1 38Tb 71
(off Ufton Rd.)
Gate Ho. NW6 40Db 69
(off Oxford Rd.)
Gatehouse Cl. KT2: King T 66Sa 131
Gatehouse Cl. SL4: Wind 6F 102
Gate Ho. Pl. WD18: Wat 13W 26
Gatehouse Sq. SE1 6E 224 (46Sb 91)
(off Southwark Bri. Rd.)
Gateley Ho. SE4 56Zb 114
(off Coston Wlk.)
Gateley Rd. SW9 55Pb 112
Gate Lodge W9 43Cb 89
(off Admiral Wlk.)
Gately Ct. SE15 52Vb 113
Gate M. SW7 2E 226 (47Gb 89)
(off Rutland Ga.)
Gater Dr. EN2: Enf 11Tb 33
Gatesborough St. EC2 5H 219 (42Ub 91)
Gates Cnr. Cl. E18 25Jc 53
Gates Ct. SE17 7D 230 (50Sb 91)
Gatesden WC1 3G 217 (41Nb 90)
Gatesden Cl. KT22: Fet 95Ea 192
Gatesden Rd. KT22: Fet 94Ea 192
Gates Grn. Rd. BR2: Kes 77Jc 159
Gates Grn. Rd. BR4: W W'ck 76Hc 159
Gateshead Rd. WD6: Bore 11Pa 29
Gateside Rd. SW17 62Hb 133
Gates La. WD25: Wat 3X 13
Gatestone Ct. SE19 65Ub 135
(off Central Hill)
Gatestone Rd. SE19 65Ub 135
Gate St. WC2 2H 223 (44Pb 90)
Gate Theatre, The 46Cb 89
(off Pembridge Rd.)
Gateway KT13: Weyb 76R 150
Gateway SE17 51Sb 113
Gateway, The GU21: Wok 86D 168
Gateway, The WD18: Wat 15U 26
Gateway Apts. E17 29Cc 52
Gateway Arc. N1 1B 218 (40Rb 71)
(off Upper St.)
Gateway Bus. Cen. SE26 65Ac 136
Gateway Bus. Cen. SE28 48Tc 94
Gateway Bus. Pk. CR5: Coul 87Mb 176
Gateway Cl. HA6: Nwood 23S 44
Gateway Ct. AL2: Brick W 2Aa 13
(off The Uplands)
Gateway Ct. IG2: Ilf 30Qc 54
(off Parham Dr.)
Gateway Ho. IG11: Bark 39Sc 74
Gateway Ind. Est. NW10 41Va 88
Gateway M. E8 36Vb 71
Gateway M. N11 23Lb 50
Gateway Pde. DA12: Grav'nd 3H 145
Gateway Retail Pk. 42Rc 94
Gateway Rd. E10 34Dc 72
Gateways KT6: Surb 71Na 153
(off Surbiton Hill Rd.)
Gateways, The EN7: G Oak 1Tb 19
Gateways, The SW3 6E 226 (49Gb 89)
Gateways, The TW9: Rich 56Ma 109
(off Park La.)
Gateways Ct. SM6: W'gton 78Kb 156
Gateway Sq. N18 23Yb 52
GATEWAY SURGICAL CEN. 42Mc 93
Gateway Trad. Est. BR8: Swan 70Jd 140
Gatewick Cl. SL1: Slou 6J 81
Gatfield Gro. TW13: Hanw 61Ca 129
Gatfield Ho. TW13: Hanw 61Ba 129
Gathorne Rd. N22 26Qb 50
Gathorne St. E2 40Zb 72
Gatley Av. KT19: Ewe 78Ra 153
Gatliff Cl. SW1 7K 227 (50Kb 90)
(off Ebury Bri. Rd.)
Gatliff Rd. SW1 7K 227 (50Kb 90)
Gatling Rd. SE2 50Wc 95
Gatonby St. SE15 53Vb 113
Gatting Cl. HA8: Edg 24Sa 47
Gatting Way UB8: Uxb 37N 63
Gattis Wharf N1 1G 217 (40Nb 70)
(off New Wharf Rd.)
GATTON 100Hb 195
Gatton Bottom RH1: Mers 99Hb 195
GATTON BOTTOM 98Kb 196
Gatton Bottom RH2: Reig 1L 207
Gatton Cl. RH2: Reig 3L 207
Gatton Cl. SM2: Sutt 81Db 175
Gatton Pk. Bus. Cen. RH1: Mers 1B 208
Gatton Pk. Ct. RH1: Redh 2P 207
Gatton Pk. Rd. RH1: Redh 3N 207
Gatton Pk. Rd. RH2: Reig 4M 207
Gatton Pl. RH1: Redh 3A 208
Gatton Rd. RH2: Reig 4L 207
Gatton Rd. SW17 63Gb 133
Gattons Way DA14: Sidc 63Bd 139
Gatward Cl. N21 16Rb 33
Gatward Grn. N9 19Vb 33
Gatward Pl. IG11: Bark 41Vc 95
Gatwick Ho. E14 44Bc 92
(off Clemence St.)
Gatwick Rd. DA12: Grav'nd 2D 144
Gatwick Rd. SW18 59Bb 111
Gatwick Way RM12: Horn 34Pd 77
Gauden Cl. SW4 55Mb 112
Gauden Rd. SW4 54Mb 112
Gaudi Apts. N8 27Pb 50
(off Gt. Amwell La.)
Gaugin Ct. SE16 50Xb 91
(off Stubbs Dr.)
Gauging Locks 52La 108
(off Tallow Rd.)
Gaugin Sq. E1 45Wb 91
Gaumont App. WD17: Wat 13X 27
Gaumont Pl. SW2 61Nb 134
Gaumont Ter. W12 47Ya 88
(off Lime Gro.)
Gaumont Twr. E8 37Vb 71
(off Dalston Sq.)
Gauntlet NW9 26Va 48
(off Five Acre)
Gauntlet Cl. UB5: N'olt 38Aa 65
Gauntlett Ct. HA0: Wemb 36Ka 66
Gauntlett Rd. SM1: Sutt 78Fb 155
Gaunt St. SE1 3D 230 (48Sb 91)
Gautrey Rd. SE15 54Yb 114
Gautrey Sq. E6 44Pc 94
Gavell Rd. KT11: Cobh 85W 170
Gavel St. SE17 5G 231 (49Tb 91)
Gavenny Path RM15: S Ock 44Vd 98
Gaverick M. E14 49Cc 92
Gaveston Cl. KT14: Byfl 85P 169
Gavestone Cres. SE12 59Lc 115
Gavestone Rd. SE12 59Kc 115
Gaveston Rd. KT22: Lea 92Ja 192
Gaveston Rd. SL2: Slou 1D 80
Gaviller Pl. E5 35Xb 71
Gavina Cl. SM4: Mord 71Gb 155
Gavin Ho. SE18 49Uc 94
Gaviots Cl. SL9: Ger X 32B 62
Gaviots Grn. SL9: Ger X 31A 62
(not continuous)
Gaviots Way SL9: Ger X 31A 62
Gawain Wlk. N9 20Wb 33
Gawber St. E2 41Yb 92
Gawsworth Cl. E15 36Hc 73
Gawthorne Ct. E3 40Cc 72
Gawton Cres. CR5: Coul 94Lb 196
Gay Cl. NW2 36Xa 68
Gaydon Ho. W2 43Db 89
(off Bourne Ter.)
Gaydon La. NW9 25Ua 48
Gayfere Pl. SE25 68Ub 135
(off Grange Hill)
Gayfere Rd. IG5: Ilf 27Pc 54
Gayfere Rd. KT17: Ewe 78Wa 154
Gayfere St. SW1 4F 229 (48Nb 90)
Gayford Rd. W12 47Va 88
Gay Gdns. RM10: Dag 35Ed 76
Gay Ho. N16 36Ub 71
Gayhurst SE17 51Tb 113
(off Hopwood Rd.)
Gayhurst Ct. UB5: N'olt 41Y 85
Gayhurst Ho. NW8 5D 214 (42Gb 69)
(off Mallory St.)
Gayhurst Rd. E8 38Wb 71
Gaylor Cl. RH1: Blet 5M 209
Gaylor Rd. RM18: Tilb 3B 122
Gaylor Rd. UB5: N'olt 36Ba 65
Gaymead NW8 39Db 69
(off Abbey Rd.)
Gaynes Ct. RM14: Upm 35Rd 77
Gaynesford Rd. SE23 61Zb 136
Gaynesford Rd. SM5: Cars 80Hb 155
Gaynes Hill Rd. IG8: Wfd G 23Nc 54
Gaynes Pk. Est. CM16: Coop 3Ad 23
Gaynes Pk. Rd. RM14: Upm 35Qd 77
Gaynes Rd. RM14: Upm 33Rd 77
Gay Rd. E15 40Fc 73
Gaysham Av. IG2: Ilf 29Qc 54
Gaysham Hall IG5: Ilf 27Rc 54
Gaysley Ho. SE11 6K 229 (49Qb 90)
(off Hotspur St.)
Gay St. SW15 55Za 110
Gayton HA1: Harr 31Ga 66
(off Grove Hill)
Gayton Cl. KT21: Asht 90Na 173
Gayton Ct. HA1: Harr 30Ha 46
Gayton Ct. RH2: Reig 5J 207
Gayton Cres. NW3 35Fb 69
Gayton Ho. E3 42Cc 92
(off Chiltern Rd.)
Gayton Rd. HA1: Harr 30Ha 46
Gayton Rd. NW3 35Fb 69
Gayton Rd. SE2 48Yc 95
Gayville Rd. SW11 58Hb 111
Gaywood Av. EN8: Chesh 2Zb 20
Gaywood Cl. SW2 60Pb 112
Gaywood Rd. E17 27Cc 52
Gaywood Rd. KT21: Asht 90Pa 173
Gaywood St. SE1 4B 230 (48Rb 91)
Gaza St. SE17 7B 230 (50Rb 91)
Gazelle Glade DA12: Grav'nd 4H 145
Gazelle Ho. E15 37Gc 73
Gean Ct. E11 35Fc 73
Gean Ct. N11 23Lb 50
(off Cline Rd.)
Gean Wlk. AL10: Hat 3C 8
Geariesville Gdns. IG6: Ilf 28Rc 54
Gearing Cl. SW17 63Jb 134
Geary Ct. CM14: B'wood 18Yd 40
Geary Dr. CM14: B'wood 18Yd 40
Geary Dr. CM15: B'wood 18Yd 40
Geary Rd. NW10 36Wa 68
Geary St. N7 36Pb 70
Geddes Pl. DA6: Bex 56Cd 118
(off Arnsberg Way)
Geddes Rd. WD23: Bush 14Ea 28
Geddington Ct. EN8: Walt C 6Cc 20
Geddy Ct. RM2: Rom 27Kd 57
Gedeney Rd. N17 25Sb 51
Gedling Ct. SE1 2K 231 (47Vb 91)
(off Sweeney Cres.)
Gedling Pl. SE1 3K 231 (48Vb 91)
Geere Rd. E15 39Hc 73
Geerings, The SS17: Stan H 1P 101
Gees Ct. W1 3J 221 (44Jb 90)
Gee St. EC1 5D 218 (42Sb 91)
Geffery's Ct. SE9 62Nc 138
Geffrye Ct. N1 2J 219 (40Ub 71)
Geffrye Est. N1 2J 219 (40Ub 71)
Geffrye Mus. 2K 219 (40Vb 71)
Geffrye St. E2 1K 219 (40Vb 71)
Geisthorp Ct. EN9: Walt A 5Jc 21
Geldart Rd. SE15 52Xb 113
Geldeston Rd. E5 33Wb 71
Gellatly Rd. SE14 54Yb 114
Gell Cl. UB10: Ick 34P 63
Gelsthorpe Rd. RM5: Col R 24Dd 56
Gem Ct. SE10 52Dc 114
(off Merryweather Pl.)
Gemini Apts. E1 5K 219 (42Vb 91)
(off Sclater St.)
Gemini Bus. Cen. E16 42Fc 93
Gemini Bus. Est. SE14 50Zb 92
Gemini Bus. Pk. E6 43Tc 94
Gemini Ct. E1 45Wb 91
(off Vaughan Way)
Gemini Gro. UB5: N'olt 41Aa 85
Gemini Ho. E3 39Cc 72
(off Garrison Rd.)
Gemini Pl. TW15: Ashf 65U 128
Gemmell Cl. CR8: Purl 86Pb 176
Genas Cl. IG6: Ilf 25Rc 54
General Gordon Pl. SE18 49Rc 94
General Gordon Sq. SE18 49Rc 94
(off Woolwich New Rd.)
General's Wlk., The EN3: Enf W 9Ac 20
General Wolfe Rd. SE10 53Fc 115
Genesis Bus. Pk. GU21: Wok 87E 168
Genesis Bus. Pk. NW10 40Ra 67
Genesis Cl. TW19: Stanw 60P 105
Genesta Glade DA12: Grav'nd 4J 145
Genesta Rd. SE18 51Rc 116
Geneva Cl. TW17: Shep 68U 128
Geneva Ct. NW9 29Va 48
Geneva Dr. SW9 56Qb 112
Geneva Gdns. RM6: Chad H 29Ad 55
Geneva Rd. CR7: Thor H 71Sb 157
Geneva Rd. KT1: King T 70Na 131
Genever Cl. E4 22Cc 52
Genista Rd. N18 22Xb 51
Genoa Av. SW15 57Ya 110
Genoa Ho. E1 42Zb 92
(off Ernest St.)
Genoa Ho. SW18 55Eb 111
Genoa Rd. SE20 67Yb 136
Genotin M. RM12: Horn 36Ld 77
Genotin Rd. EN1: Enf 13Tb 33
Genotin Ter. EN1: Enf 13Tb 33
Gentlemans Row EN2: Enf 13Sb 33
Gentry Cl. SS17: Stan H 1L 101
Gentry Gdns. E13 42Jc 93
Geoff Cade Way E3 43Bc 92
Geoffrey Av. RM3: Hrld W 23Qd 57
Geoffrey Chaucer Way E3 43Bc 92
Geoffrey Cl. SE5 54Sb 113
Geoffrey Ct. SE4 54Bc 114
Geoffrey Gdns. E6 40Nc 74
Geoffrey Ho. SE1 3G 231 (48Tb 91)
(off Pardoner St.)
Geoffrey Jones Ct. NW10 39Wa 68
Geoffrey Rd. SE4 55Bc 114
Geoffrey Whitworth Theatre 56Jd 118
George V Av. HA5: Pinn 26Ba 45
George V Cl. HA5: Pinn 27Ca 45
George V Cl. WD18: Wat 14V 26
George V Way UB6: G'frd 39Ka 66
George V Way WD3: Sarr 8K 11
George Beard Rd. SE8 49Bc 92
George Belt Ho. E2 41Zb 92
(off Smart St.)
George Comberton Wlk. E12 36Qc 74
George Ct. TW15: Ashf 63P 127
(off Church Rd.)
George Ct. UB3: Hayes 43V 84
George Ct. WC2 5G 223 (45Nb 90)
(off John Adam St.)
George Cres. N10 24Jb 50
George Crooks Ho. RM17: Grays 51De 121
(off New Rd.)
George Davies Lodge IG6: Ilf 29Sc 54
(off Veronique Gdns.)
George Downing Est. N16 33Vb 71
George Eliot Ho. SE17 7D 230 (50Sb 91)
(off Thrush St.)
George Eliot Ho. SW1 6C 228 (49Lb 90)
(off Vauxhall Bri. Rd.)
George Elliston Ho. SE1 50Wb 91
(off Old Kent Rd.)
George Eyre Ho. NW8 2C 214 (40Fb 69)
(off Cochrane St.)
George Fld. Ho. WD3: Rick 17M 25
(off Northway)
George Furness Ho. NW10 37Xa 68
(off Grange Rd.)
George Gange Way HA3: W'stone 27Ga 46
George Gillett Ct. EC1 5E 218 (42Sb 91)
(off Banner St.)
GEORGE GREEN 44A 82
George Grn. Dr. SL3: Geor G 44A 82
George Grn. Rd. SL3: Geor G 4P 81
George Groves Rd. SE20 67Wb 135
George Hilsdon Ct. E14 44Ac 92
(off Repton St.)
George Ho. NW6 40Bb 69
(off Albert Rd.)
George Hudson Twr. E15 40Dc 72
(off High St.)
George Inn Yd. SE1 7F 225 (46Tb 91)
Georgelands GU23: Rip 93K 189
George La. BR2: Hayes 74Kc 159
George La. E18 26Jc 53
George La. SE13 58Dc 114
George La. SE6 58Dc 114
George Lansbury Ho. E3 41Bc 92
(off Bow Rd.)
George Lansbury Ho. N22 25Qb 50
(off Progress Way)
George Lansbury Ho. NW10 38Ua 68
George Leybourne Ho. E1 45Wb 91
(off Fletcher St.)
George Lindgren Ho. SW6 52Bb 111
(off Clem Attlee Ct.)
George Livings Cl. IG7: Chig 21Xc 55
George Loveless Ho. E2 3K 219 (41Vb 91)
(off Diss St.)
George Lovell Dr. EN3: Enf L 9Cc 20
George Lowe Ct. W2 43Db 89
(off Bourne Ter.)
George Mathers Rd. SE11 5B 230 (49Rb 91)
George M. EN2: Enf 13Tb 33
George M. NW1 4C 216 (41Lb 90)
(off Drummond St.)
George M. SW9 54Qb 112
George Padmore Ho. E8 39Wb 71
(off Brougham Rd.)
George Peabody Ct. NW1 7D 214 (43Gb 89)
(off Burne St.)
George Peabody St. E13 40Lc 73
George Pl. N17 27Ub 51
George Potter Ho. SW11 54Fb 111
(off George Potter Way)
George Potter Way SW11 54Fb 111
George Rd. E4 23Cc 52
George Rd. KT2: King T 66Ra 131
George Rd. KT3: N Mald 70Va 132
George Row SE16 47Wb 91
Georges Cl. BR5: St P 69Yc 139
George Scott Ho. E1 44Zb 92
(off W. Arbour St.)
Georges Dr. CM15: Pil H 15Vd 40
Georges Mead WD6: E'tree 16Na 29
George Sq. SW19 69Cb 133
George's Rd. N7 36Pb 70
George's Rd. TN16: Tats 92Mc 199
George's Sq. SW6 51Bb 111
(off North End Rd.)
Georges Ter. CR3: Cat'm 94Tb 197
George St. AL3: St A 2B 6
George St. CR0: C'don 75Sb 157
George St. E14 46Dc 92
George St. E16 44Hc 93
George St. HP2: Hem H 1M 3
George St. HP4: Berk 1A 2
George St. IG11: Bark 38Sc 74
George St. RM1: Rom 30Hd 56
George St. RM17: Grays 51Ce 121
George St. TW18: Staines 63H 127
George St. TW3: Houn 54Ba 107
George St. TW9: Rich 57Ma 109
George St. UB2: S'hall 49Aa 85
George St. UB8: Uxb 38M 63
George St. W1 2F 221 (44Hb 89)
George St. W7 46Ga 86
George St. WD18: Wat 14Y 27
George Street Stop (London Tramlink) 75Sb 157
Georges Wood Rd. AL9: Brk P 8J 9
George Tilbury Ho. RM16: Grays 7D 100
Georgetown Cl. SE19 64Ub 135
Georgette Pl. SE10 52Ec 114
George Vale Ho. E2 40Wb 71
(off Mansford St.)
George Vw. Ho. SW18 60Db 111
(off Knaresborough Dr.)
Georgeville Gdns. IG6: Ilf 28Rc 54
George Walter Ct. SE16 49Yb 92
(off Millender Wlk.)
Georgewood Rd. HP3: Hem H 7P 3
George Wyver Cl. SW19 59Ab 110
George Yd. EC3 3G 225 (44Tb 91)
George Yd. W1 4J 221 (45Jb 90)
George Yeomans Cl. EN8: Chesh 2Yb 20
Georgia Ct. SE16 48Wb 91
(off Priter Rd.)
Georgiana St. NW1 39Lb 70
Georgian Cl. BR2: Hayes 74Kc 159
Georgian Cl. HA7: Stan 24Ja 46
Georgian Cl. TW18: Staines 63K 127
Georgian Cl. UB10: Ick 35N 63
Georgian Ct. CR0: C'don 74Tb 157
(off Cross Rd.)
Georgian Ct. E9 39Yb 72
Georgian Ct. EN5: New Bar 14Eb 31
Georgian Ct. HA9: Wemb 37Qa 67
Georgian Ct. N3 25Bb 49
Georgian Ct. NW4 29Xa 48
Georgian Ct. SW16 63Nb 134
Georgian Ho. E16 46Jc 93
(off Capulet M.)
Georgian Ho. N1 39Ub 71
(off Hertford Rd.)
Georgian Way HA1: Harr 33Fa 66
Georgia Rd. CR7: Thor H 67Rb 135
Georgia Rd. KT3: N Mald 70Sa 131
Georgina Gdns. E2 3K 219 (41Vb 91)
Geotgette Ct. SW18 57Db 111
(off Courthouse Way)
Geraint Rd. BR1: Broml 63Jc 137
Geraldine Rd. SW18 57Eb 111
Geraldine Rd. W4 51Qa 109
Geraldine St. SE11 4B 230 (48Rb 91)
Gerald M. SW1 5J 227 (49Jb 90)
(off Gerald Rd.)
Gerald Pl. E8 37Wb 71
(off Dalston Sq.)
Gerald Rd. DA12: Grav'nd 9G 122
Gerald Rd. E16 42Hc 93
Gerald Rd. RM8: Dag 32Bd 75
Gerald Rd. SW1 5J 227 (49Jb 90)
Gerald's Gro. SM7: Bans 86Za 174
Gerard Av. TW4: Houn 59Ca 107
Gerard Gdns. RM13: Rain 40Gd 76
Gerard Pl. E9 38Zb 72
Gerard Rd. HA1: Harr 30Ja 46
Gerard Rd. SW13 53Va 110
Gerards Cl. SE16 50Yb 92
Gerards Pl. SW4 56Mb 112
Gerda Rd. SE9 61Sc 138
Gerdview Dr. DA2: Wilm 63Ld 141
Germander Dr. GU24: Bisl 7E 166

Germander Way E15 41Gc 93
Gernigan Ho. SW18 58Fb 111
Gernon Bushes Nature Reserve 1Ad 23
Gernon Cl. RM13: Rain 40Md 77
Gernon Rd. E3 40Ac 72
Geron Way NW2 32Xa 68
Gerpins La. RM13: Upm 40Pd 77
Gerpins La. RM14: Upm 40Pd 77
Gerrard Cres. CM14: B'wood 20Yd 40
Gerrard Gdns. HA5: Eastc 29W 44
Gerrard Ho. SE14 52Yb 114
(off Briant St.)
Gerrard Pl. W1 4E 222 (45Mb 90)
Gerrard Rd. N1 1B 218 (40Rb 71)
Gerrards Cl. N14 15Lb 32
Gerrards Ct. W5 48Ma 87
GERRARDS CROSS 29A 42
Gerrards Cross Golf Course 27B 42
Gerrards Cross Rd. SL2: Stoke P 7L 61
Gerrards Cross Station (Rail) 29A 42
Gerrards Mead SM7: Bans 88Bb 175
Gerrard St. W1 4D 222 (45Mb 90)
Gerrard Way SE3 56Lc 115
Gerridge Ct. SE1 3A 230 (48Qb 90)
(off Gerridge St.)
Gerridge St. SE1 3A 230 (48Qb 90)
Gerry Raffles Sq. E15 37Fc 73
Gertrude Rd. DA17: Belv 49Cd 96
Gertrude St. SW10 51Eb 111
Gervaise Cl. SL1: Slou 6D 80
Gervase Cl. HA9: Wemb 34Sa 67
Gervase Rd. HA8: Edg 25Sa 47
Gervase St. SE15 52Xb 113
Gervis Ct. TW7: Isle 52Ea 108
Gews Cnr. EN8: Chesh 1Zb 20
Ghent St. SE6 61Cc 136
Ghent Way E8 37Vb 71
Giant Arches Rd. SE24 59Sb 113
Giant Tree Hill WD23: B Hea 18Fa 28
Gibbfield Cl. RM6: Chad H 27Ad 55
Gibbings Ho. SE1 2C 230 (47Rb 91)
(off King James St.)
Gibbins Rd. E15 38Ec 72
Gibbon Ho. NW8 6C 214 (42Fb 89)
(off Fisherton St.)
Gibbon Rd. KT2: King T 67Na 131
Gibbon Rd. SE15 54Yb 114
Gibbon Rd. W3 45Ua 88
Gibbons Cl. WD6: Bore 11Na 29
Gibbons La. DA1: Dart 58Md 119
Gibbons M. NW11 29Bb 49
Gibbon's Rents SE1 7H 225 (46Ub 91)
(off Magdalen St.)
Gibbons Rd. NW10 37Ua 68
Gibbon Wlk. SW15 56Wa 110
Gibb's Acre GU24: Pirb 5D 186
Gibbs Av. SE19 64Tb 135
Gibbs Brook La. RH8: Oxt 8H 211
Gibbs Cl. EN8: Chesh 1Zb 20
Gibbs Cl. SE19 65Tb 135
Gibbs Couch WD19: Wat 20Z 27
Gibbs Grn. HA8: Edg 21Sa 47
Gibbs Grn. W14 50Bb 89
(not continuous)
Gibbs Ho. BR1: Broml 67Hc 137
(off Longfield)
Gibbs La. E2 40Wb 71
Gibb's Rd. N18 21Yb 52
Gibbs Sq. SE19 64Tb 135
Gibney Ter. BR1: Broml 63Hc 137
Gibraltar Cl. CM13: Gt War 23Yd 58
Gibraltar Cres. KT19: Ewe 82Ua 174
Gibraltar Wlk. E2 41Vb 91
(off Shackwell St.)
Gibson Cl. DA11: Nflt 2B 144
Gibson Cl. E1 42Yb 92
Gibson Cl. KT9: Chess 78La 152
Gibson Cl. N21 16Qb 32
Gibson Cl. TW7: Isle 55Ga 108
Gibson Ct. KT10: Hin W 75Ha 152
Gibson Ct. RM1: Rom 30Gd 56
Gibson Ct. SE9 57Lc 115
Gibson Ct. SL3: L'ly 50B 82
Gibson Gdns. N16 33Vb 71
Gibson Ho. SM1: Sutt 77Cb 155
Gibson M. TW1: Twick 58La 108
Gibson Pl. TW19: Stanw 58L 105
Gibson Rd. RM8: Dag 32Yc 75
Gibson Rd. SE11 6J 229 (49Pb 90)
Gibson Rd. SM1: Sutt 78Db 155
Gibson Rd. UB10: Ick 35P 63
Gibsons Hill SW16 66Qb 134
(not continuous)
Gibsons Pl. DA4: Eyns 75Nd 163
Gibson's Pl. TW8: Bford 51Ma 109
(off Sidney Gdns.)
Gibson Sq. N1 39Qb 70
Gibson Sq. Gdns. 39Qb 70
(off Gibson Sq.)
Gibson St. SE10 50Gc 93
Gibson Way CR3: Cat'm 95Tb 197
Gidd Hill CR5: Coul 88Jb 176
Gidea Av. RM2: Rom 27Jd 56
Gidea Cl. RM15: S Ock 41Yd 98
(off Benyon Path)
Gidea Cl. RM2: Rom 27Jd 56
Gidea Lodge RM2: Rom 27Kd 57
GIDEA PARK 27Kd 57
Gidea Park Station (Rail & Crossrail) 28Kd 57
Gideon Cl. DA17: Belv 49Dd 96
Gideon Cl. HA8: Edg 23Qa 47
Gideon M. W5 47Ma 87
Gideon Rd. SW11 55Jb 112
Gidian Ct. AL2: Park 9B 6
Gielgud Theatre 4D 222 (45Mb 90)
(off Shaftesbury Av.)
Giesbach Rd. N19 33Mb 70
Giffard Rd. N18 23Ub 51
Giffin Sq. Mkt. 52Cc 114
(off Giffin St.)
Giffin St. SE8 52Cc 114
Gifford Gdns. W7 43Fa 86
Gifford Ho. SE10 51Ec 93
(off Eastney St.)
Gifford Ho. SW1 50Lb 90
(off Churchill Gdns.)
Gifford Pl. CM14: W'ley 22Zd 59
Gifford Rd. NW10 38Ua 68
Giffords Cross Rd. SS17: Corr 1P 101
Giffordside RM16: Grays 10D 100
Gifford St. N1 38Nb 70
Gift La. E15 39Gc 73
Giggs Hill BR5: St P 68Wc 139
GIGGSHILL 73Ja 152
Giggs Hill Gdns. KT7: T Ditt 74Ja 152
Giggs Hill Rd. KT7: T Ditt 73Ja 152
Gilbert Bri. 7E 218 (43Sb 91)
(off Wood St.)
Gilbert Burnet Ho. HP3: Hem H 4P 3
Gilbert Cl. DA10: Swans 58Zd 121
Gilbert Cl. SE18 53Pc 116
Gilbert Cl. SW19 67Db 133
(off Morden Rd.)
Gilbert Ct. W5 44Pa 87
(off Green Va.)
Gilbert Gro. HA8: Edg 25Ta 47
Gilbert Ho. E17 27Ec 52
Gilbert Ho. E2 41Zb 92
(off Usk St.)
Gilbert Ho. EC2 7E 218 (43Sb 91)
(off Wood St.)
Gilbert Ho. SE8 51Cc 114
Gilbert Ho. SW1 50Kb 90
(off Churchill Gdns.)
Gilbert Ho. SW13 52Xa 110
(off Trinity Chu. Rd.)
Gilbert Ho. SW8 52Nb 112
(off Wyvil Rd.)
Gilbert Pl. WC1 1F 223 (43Nb 90)
Gilbert Rd. BR1: Broml 66Jc 137
Gilbert Rd. DA17: Belv 48Cd 96
Gilbert Rd. HA5: Pinn 28Z 45
Gilbert Rd. RM1: Rom 28Hd 56
Gilbert Rd. RM16: Chaf H 48Yd 98
Gilbert Rd. SE11 6A 230 (49Qb 90)
Gilbert Rd. SW19 66Eb 133
Gilbert Rd. UB9: Hare 26M 43
Gilbert Row DA11: Nflt 1B 144
Gilbert Scott Bldg. SW15 58Ab 110
Gilbert Scott Cl. HA0: Wemb 36Ma 67
Gilbert Scott Ho. W14 49Bb 89
(off Warwick La.)
Gilbert Sheldon Ho. W2 7C 214 (43Fb 89)
(off Edgware Rd.)
Gilberts Lodge KT17: Eps 84Ua 174
Gilbertson Ho. E14 48Cc 92
(off Mellish St.)
Gilbert St. E15 35Gc 73
Gilbert St. EN3: Enf W 9Yb 20
Gilbert St. TW3: Houn 55Ea 108
Gilbert St. W1 3J 221 (44Jb 90)
Gilbert Way CR0: Wadd 75Pb 156
Gilbert Way SL3: L'ly 50B 82
Gilbert White Cl. UB6: G'frd 39Ja 66
Gilbey Cl. UB10: Ick 35R 64
Gilbey Ho. NW1 38Kb 70
Gilbey Rd. SW17 63Gb 133
Gilbeys Yd. NW1 38Jb 70
Gilbourne Rd. SE18 51Vc 117
Gilby Ho. E9 37Zb 72
Gilda Av. EN3: Pond E 15Ac 34
Gilda Ct. NW7 25Wa 48
Gilda Cres. N16 32Wb 71
Gildea Cl. HA5: Hat E 24Ca 45
Gildea St. W1 1A 222 (43Kb 90)
Gilden Cres. NW5 36Jb 70
Gildenhill Rd. BR8: Swan 66Ld 141
Gildersome St. SE18 51Qc 116
Gilders Rd. KT9: Chess 80Pa 153
Giles Cl. RM13: Rain 40Md 77
Giles Coppice SE19 63Vb 135
Giles Cres. UB10: Uxb 39P 63
(off St Andrews Rd.)
Giles Dr. DA10: Swans 59Ae 121
Giles Fld. DA12: Grav'nd 10H 123
Gilesfield Cl. RM13: Rain 39Gd 76
Giles Ho. E15 37Fc 73
(off Forrester Way)
Giles Ho. SL2: Stoke P 8L 61
(off Bells Hill Grn.)
Giles Ho. W11 44Cb 89
(off Westbourne Gro.)
Gilesmead KT18: Eps 86Ua 174
(off Downside)
Gilesmead SE5 53Tb 113
Giles Travers Cl. TW20: Thorpe 69E 126
Gilford Ho. IG1: Ilf 33Rc 74
(off Clements Rd.)
Gilfrid Cl. UB8: Hil 44R 84
Gilhams Av. SM7: Bans 84Za 174
Gilkes Cres. SE21 58Ub 113
Gilkes Pl. SE21 58Ub 113
Gillam Way RM13: Rain 37Jd 76
Gillan Ct. SE12 62Kc 137
Gillan Grn. WD23: B Hea 19Ea 28
Gillards M. E17 28Cc 52
Gillards Way E17 28Cc 52
Gill Av. E16 44Jc 93
Gill Cl. WD18: Wat 16S 26
Gill Cres. DA11: Nflt 2B 144
Gillender St. E14 42Ec 92
Gillender St. E3 42Ec 92
Gillespie Ho. GU25: Vir W 70A 126
(off Holloway Dr.)
Gillespie Pk. Local Nature
Reserve 34Qb 70
Gillespie Rd. N5 34Qb 70
Gillett Av. E6 40Nc 74
GILLETTE CORNER 52Ja 108
Gillett Ho. N8 27Nb 50
(off Campsfield Rd.)
Gillett Pl. N16 36Ub 71
Gillett Rd. CR7: Thor H 70Tb 135
Gillett Sq. N16 36Ub 71
(off Gillett St.)
Gillett St. N16 36Ub 71
Gillfoot NW1 2B 216 (40Lb 70)
(off Hampstead Rd.)
Gillham Ter. N17 23Wb 51
Gilliam Gro. CR8: Purl 82Qb 176
Gillian Av. AL1: St A 6A 6
Gillian Cres. RM2: Rom 26Ld 57
Gillian Ho. HA3: Hrw W 23Ga 46
Gillian Lynne Theatre 2G 223 (44Nb 90)
(off Parker St.)
Gillian Pk. Rd. SM3: Sutt 74Bb 155
Gillian St. SE13 57Dc 114
Gilliat Rd. SL1: Slou 5J 81
Gilliats Grn. WD3: Chor 14F 24
Gillies Ho. NW6 38Fb 69
(off Hilgrove Rd.)
Gillies Rd. TN15: W King 78Ud 164
Gillies St. NW5 36Jb 70
Gilling Ct. NW3 37Gb 69
Gillingham Ho. RM3: Rom 22Nd 57
(off Lindfield Rd.)
Gillingham M. SW1 5B 228 (49Lb 90)
Gillingham Rd. NW2 34Ab 68
Gillingham Row SW1 5B 228 (49Lb 90)
Gillingham St. SW1 5B 228 (49Lb 90)
Gillings Ct. EN5: Barn 14Ab 30
(off Wood St.)
Gillison Wlk. SE16 48Xb 91
Gillis Sq. SW15 58Wa 110
Gillman Dr. E15 39Hc 73
Gillman Ho. E2 40Wb 71
(off Pritchard's Rd.)
Gillmans Rd. BR5: Orp 74Xc 161
Gillray Ho. SW10 51Fb 111
(off Ann La.)
Gills Hill WD7: R'lett 7Ha 14
Gills Hill La. WD7: R'lett 8Ha 14
Gills Hollow WD7: R'lett 8Ha 14
Gills Rd. DA2: G St G 67Wd 142
Gills Rd. DA4: S Dar 67Ud 142
Gill St. E14 44Bc 92
Gillum Cl. EN4: E Barn 18Hb 31
Gilmais KT23: Bookh 97Ea 192
Gilman Cres. SL4: Wind 5B 102
Gilman Ho. N1 38Qb 70
(off Drummond Way)
Gilmore Cl. SL3: L'ly 7N 81
Gilmore Cl. UB10: Ick 34Q 64
Gilmore Ct. N11 22Hb 49
Gilmore Cres. TW15: Ashf 64Q 128
Gilmore Rd. SE13 56Fc 115
Gilmour Cl. EN2: Enf 7Wb 19
Gilmour Cl. EN2: Walt C 7Wb 19
Gilmour Ho. NW9 26Wa 48
Gilpin Av. SW14 56Ta 109
Gilpin Cl. CR4: Mitc 68Gb 133
Gilpin Cl. W2 7A 214 (43Eb 89)
(off Porteus Rd.)
Gilpin Cres. N18 22Vb 51
Gilpin Cres. TW2: Whitt 59Da 107
Gilpin Rd. E5 35Ac 72
Gilpin's Ride HP4: Berk 1A 2
Gilpin Way UB3: Harl 52T 106
Gilray Ho. W2 4B 220 (45Fb 89)
(off Gloucester Ter.)
Gilroy Cl. RM13: Rain 37Hd 76
Gilroy Rd. HP2: Hem H 1M 3
Gilroy Way BR5: Orp 73Xc 161
Gilsland EN9: Walt A 6Hc 21
Gilsland Pl. CR7: Thor H 70Tb 135
Gilsland Rd. CR7: Thor H 70Tb 135
Gilson Pl. N10 24Hb 49
Gilstead Rd. SW6 54Db 111
Gilston Rd. SW10 50Eb 89
Gilton Rd. SE6 62Gc 137
Giltspur St. EC1 2C 224 (44Rb 91)
Gilwell Cl. E4 14Dc 34
Gilwell La. E4 14Ec 34
(not continuous)
Gilwell Pk. E4 13Fc 35
GILWELL PARK 14Fc 35
Ginger Apts. SE1 1K 231 (47Vb 91)
(off Cayenne Ct.)
Ginsburg Yd. NW3 35Eb 69
Gippeswyck Cl. HA5: Pinn 25Z 45
Gipsy Hill SE19 63Ub 135
Gipsy Hill Station (Rail) 64Ub 135
Gipsy La. RM17: Grays 51Ee 121
Gipsy La. SW15 55Xa 110
Gipsy Rd. DA16: Well 52Zc 117
Gipsy Rd. SE27 63Sb 135
Gipsy Rd. Gdns. SE27 63Sb 135
Giralda Cl. E16 43Mc 93
Giraud St. E14 44Dc 92
Girdler's Rd. W14 49Za 88
Girdlestone Wlk. N19 33Lb 70
Girdwood Rd. SW18 59Ab 110
Girling Ho. N1 39Ub 71
(off Colville Est.)
Girling Way TW14: Felt 55W 106
Girona Cl. RM16: Chaf H 48Yd 98
Gironde Rd. SW6 52Bb 111
Girtin Ho. UB5: N'olt 40Z 65
(off Academy Gdns.)
Girton Av. NW9 27Qa 47
Girton Cl. UB5: N'olt 37Ea 66
Girton Ct. EN8: Chesh 2Ac 20
Girton Gdns. CR0: C'don 76Cc 158
Girton Rd. SE26 64Zb 136
Girton Rd. UB5: N'olt 37Ea 66
Girton Vs. W10 44Za 88
Girton Way WD3: Crox G 15S 26
Gisborne Gdns. RM13: Rain 41Hd 96
Gisbourne Cl. SM6: Bedd 76Mb 156
Gisburne Way WD24: Wat 9W 12
Gisburn Ho. SE15 51Wb 113
(off Friary Est.)
Gisburn Rd. N8 28Pb 50
Gissing Wlk. N1 38Qb 70
Gittens Cl. BR1: Broml 63Hc 137
Given Wilson Wlk. E13 40Hc 73
Giverny Ho. SE16 47Zb 92
(off Water Gdns. Sq.)
Givons Gro. KT22: Lea 97Ka 192
GIVONS GROVE 98La 192
GIVONS GROVE RDBT. 96Ka 192
Glacier Ho. SW11 52Mb 112
(off Ponton Rd.)
Glacier Pl. E2 40Xb 71
(off Clare St.)
Glacier Way HA0: Wemb 40Ma 67
Gladbeck Way EN2: Enf 14Rb 33
Gladding Rd. E12 35Mc 73
Glade, The BR1: Broml 68Mc 137
Glade, The BR4: W W'ck 76Dc 158
Glade, The CM13: Hut 18Ce 41
Glade, The CR0: C'don 71Zb 158
Glade, The CR5: Coul 91Qb 196
Glade, The E8 37Wb 71
Glade, The EN2: Enf 13Qb 32
Glade, The IG5: Ilf 25Pc 54
Glade, The IG8: Wfd G 20Kc 35
Glade, The KT14: W Byf 85G 168
Glade, The KT17: Ewe 79Wa 154
Glade, The KT20: Kgswd 93Cb 195
Glade, The KT22: Fet 94Ca 191
Glade, The N12 20Fb 31
Glade, The N21 16Pb 32
Glade, The RM14: Upm 36Sd 78
Glade, The SE7 52Lc 115
Glade, The SL5: S'hill 1A 146
Glade, The SL9: Ger X 2P 61
Glade, The SM2: Cheam 81Ab 174
Glade, The TN13: S'oaks 95Kd 203
Glade, The TW18: Staines 65K 127
Glade, The W12 47Xa 88
(off Coningham Rd.)
Glade Apts. E14 49Ec 92
(off Stebondale St.)
Glade Bus. Cen., The RM20: W Thur 50Vd 98
Glade Cl. KT6: Surb 75Ma 153
Glade Ct. IG5: Ilf 25Pc 54
Glade Ct. UB8: Uxb 37L 63
Glade Gdns. CR0: C'don 73Ac 158
Glade La. UB2: S'hall 48Da 85
Glade Path SE1 2B 230 (47Rb 91)
(off Blackfriars Rd.)
Glades, The 68Jc 137
Glades, The DA12: Grav'nd 5F 144
Glades, The HP1: Hem H 1G 2
Glades, The KT6: Surb 73Na 153
Glades Cl. RM1: Rom 29Kd 57
Gladeside CR0: C'don 72Zb 158
Gladeside N21 16Pb 32
Gladeside Cl. KT9: Chess 80Ma 153
Gladeside Ct. CR6: W'ham 92Xb 197
Gladesmere Ct. WD24: Wat 8X 13
Gladesmore Community School &
Sports Cen. 29Wb 51
Gladesmore Rd. N15 30Vb 51
Glades Pl. BR1: Broml 68Jc 137
Glade Spur KT20: Kgswd 93Db 195
Gladeswood Rd. DA17: Belv 49Dd 96
Glade Wlk. E20 37Dc 72
Gladeway, The EN9: Walt A 5Fc 21
Gladiator St. SE23 59Ac 114
Glading Ter. N16 34Vb 71
Gladioli Cl. TW12: Hamp 65Ca 129
Gladsaxe Ho. SM1: Sutt 77Db 155
Gladsdale Dr. HA5: Eastc 28W 44
Gladsmuir Cl. KT12: Walt T 75Y 151
Gladsmuir Rd. EN5: Barn 12Ab 30
Gladsmuir Rd. N19 32Lb 70
Gladstone Av. E12 38Nc 74
Gladstone Av. N22 26Qb 50
Gladstone Av. TW14: Felt 58W 106
Gladstone Av. TW2: Twick 60Fa 108
Gladstone Ct. NW6 38Bb 69
(off Fairfax Rd.)
Gladstone Ct. SW1 6E 228 (49Mb 90)
(off Regency St.)
Gladstone Ct. Bus. Cen. SW8 53Kb 112
(off Pagden St.)
Gladstone Gdns. TW3: Houn 53Ea 108
Gladstone Ho. CR4: Mitc 68Hb 133
Gladstone Ho. E14 44Cc 92
(off E. India Dock Rd.)
Gladstone M. N22 26Qb 50
Gladstone M. NW6 38Bb 69
(off Cavendish Rd.)
Gladstone M. SE20 66Yb 136
Gladstone Pde. NW2 33Ya 68
Gladstone Pk. Gdns. NW2 35Xa 68
Gladstone Pl. E3 40Bc 72
Gladstone Pl. EN5: Barn 14Za 30
Gladstone Pl. KT8: E Mos 71Ga 152
Gladstone Pl. RM13: Rain 41Ld 97
Gladstone Rd. BR6: Farnb 78Sc 160
Gladstone Rd. CR0: C'don 73Tb 157
Gladstone Rd. DA1: Dart 58Pd 119
Gladstone Rd. IG9: Buck H 18Lc 35
Gladstone Rd. KT1: King T 69Qa 131
Gladstone Rd. KT21: Asht 90Ma 173
Gladstone Rd. KT6: Surb 75Ma 153
Gladstone Rd. SW19 66Cb 133
Gladstone Rd. UB2: S'hall 47Aa 85
Gladstone Rd. W4 48Ta 87
Gladstone Rd. WD17: Wat 13Y 27
Gladstone St. SE1 3B 230 (48Rb 91)
Gladstone Ter. SE27 64Sb 135
(off Bentons La.)
Gladstone Ter. SW8 53Kb 112
Gladstone Way HA3: W'stone 27Ga 46
Gladstone Way SL1: Slou 6E 80
Gladwell Rd. BR1: Broml 65Jc 137
Gladwell Rd. N8 30Pb 50
Gladwin Ho. NW1 2C 216 (40Lb 70)
(off Werrington St.)
Gladwyn Rd. SW15 55Za 110
Gladys Ct. BR1: Broml 61Hc 137
Gladys Dimson Ho. E7 36Hc 73
Gladys Rd. NW6 38Cb 69
Glaisher St. SE8 51Cc 114
Glaisyer Way SL0: Iver H 40E 62
Glamis Cl. EN7: Chesh 1Wb 19
Glamis Ct. W3 47Ra 87
Glamis Cres. UB3: Harl 48S 84
Glamis Dr. RM11: Horn 32Nd 77
Glamis Pl. E1 45Yb 92
Glamis Pl. HP2: Hem H 1N 3
Glamis Rd. E1 45Yb 92
Glamis Way UB5: N'olt 37Ea 66
Glamorgan Cl. CR4: Mitc 69Nb 134
Glamorgan Ct. W7 43Ha 86
(off Copley Cl.)
Glamorgan Rd. KT1: Hamp W 66La 130
Glandford Way RM6: Chad H 29Xc 55
Glanfield Rd. BR3: Beck 70Bc 136
Glanleam Rd. HA7: Stan 21Ma 47
Glanmead CM15: Shenf 18Ae 41
Glanmor Rd. SL2: Slou 5M 81
Glanthams Cl. CM15: Shenf 19Be 41
Glanthams Rd. CM15: Shenf 19Be 41
GLANTY 63E 126
Glanty, The TW20: Egh 63D 126
Glanville Dr. RM11: Horn 32Pd 77
Glanville M. HA7: Stan 22Ja 46
Glanville Rd. BR2: Broml 69Kc 137
Glanville Rd. SW2 57Nb 112
Glanville Way KT19: Eps 84Na 173
Glasbrook Av. TW2: Whitt 60Ba 107
Glasbrook Rd. SE9 59Mc 115
Glaserton Rd. N16 31Ub 71
Glasford St. SW17 65Hb 133
Glasfryn Ct. HA2: Harr 33Fa 66
(off Roxeth Hill)
Glasfryn Ho. HA2: Harr 33Fa 66
(off Roxeth Hill)
Glasgow Ho. W9 40Db 69
(off Maida Vale)
Glasgow Rd. E13 40Kc 73
Glasgow Rd. N18 22Xb 51
Glasgow Ter. SW1 7B 228 (50Lb 90)
Glasier Ct. E15 38Gc 73
Glaskin M. E9 37Ac 72
Glass Blowers Ho. E14 44Fc 93
(off Valencia Cl.)
Glass Bldg., The NW1 39Kb 70
(off Jamestown Rd.)
Glasse Cl. W13 45Ja 86
Glass Foundry Yd. E13 43Kc 93
(off Denmark St.)
Glasshill St. SE1 1C 230 (47Rb 91)
Glass Ho. WC2 3F 223 (44Nb 90)
(off Shaftesbury Av.)
Glass Ho., The SE1 2H 231 (47Ub 91)
(off Royal Oak Yd.)
Glasshouse Cl. UB8: Hil 43R 84
Glasshouse Flds. E1 45Zb 92
(not continuous)
Glasshouse Gdns. E20 38Ec 72
Glasshouse Gdns. Development E20 38Ec 72
Glasshouse St. W1 5C 222 (45Lb 90)
Glasshouse Wlk. SE11 7G 229 (50Nb 90)
Glasshouse Yd. EC1 6D 218 (42Sb 91)
Glasslyn Rd. N8 29Mb 50
Glassmill Ho. HP4: Berk 1A 2
(off Robertson Rd.)
Glassmill La. BR2: Broml 68Hc 137
Glass Mill Leisure Cen. 55Ec 114
Glass St. E2 42Xb 91
Glassworks Studios E2 3J 219 (41Ub 91)
(off Basing Pl.)
Glass Yd. SE18 48Qc 94
Glastonbury Av. IG8: Wfd G 24Mc 53
Glastonbury Cl. BR5: Orp 74Yc 161
Glastonbury Ct. SE14 52Yb 114
(off Farrow La.)
Glastonbury Ho. SE12 57Hc 115
(off Wantage Rd.)
Glastonbury Ho. SW1 7K 227 (50Kb 90)
(part of Abbots Mnr.)
Glastonbury Pl. E1 44Yb 92
Glastonbury Rd. N9 18Wb 33
Glastonbury Rd. SM4: Mord 73Cb 155
Glastonbury St. NW6 36Bb 69
Glaston Ct. W5 46Ma 87
(off Grange Rd.)
Glaucus St. E3 43Dc 92
Glazbury Rd. W14 49Ab 88
Glazebrook Cl. SE21 61Tb 135
Glazebrook Rd. TW11: Tedd 66Ha 130
Gleave Cl. AL1: St A 1F 6
Glebe, The BR7: Chst 67Sc 138
Glebe, The KT4: Wor Pk 74Va 154
Glebe, The SE3 55Gc 115
Glebe, The SW16 63Mb 134
Glebe, The UB7: W Dray 49P 83
Glebe, The WD25: Wat 5Z 13
Glebe, The WD4: K Lan 1Q 12
Glebe Av. CR4: Mitc 68Gb 133
Glebe Av. EN2: Enf 13Rb 33
Glebe Av. HA3: Kenton 28Na 47
Glebe Av. HA4: Ruis 37X 65
Glebe Av. IG8: Wfd G 23Jc 53
Glebe Av. UB10: Ick 34S 64
Glebe Cl. CR2: Sande 83Vb 177
Glebe Cl. GU18: Light 2A 166
Glebe Cl. HP3: Hem H 5N 3
Glebe Cl. KT23: Bookh 98Ca 191
Glebe Cl. UB10: Ick 35S 64
Glebe Cl. W4 50Ua 88
Glebe Cotts. TW13: Hanw 62Ca 129
(off Twickenham Rd.)
Glebe Ct. CR4: Mitc 69Hb 133
Glebe Ct. E3 41Dc 92
(off Rainhill Way)
Glebe Ct. EN8: Chesh 1Zb 20
Glebe Ct. HA7: Stan 22La 46
Glebe Ct. N13 20Qb 32
Glebe Ct. SE3 55Gc 115
Glebe Ct. TN13: S'oaks 98Kd 203
Glebe Ct. W5 46Ma 87
Glebe Ct. W7 45Fa 86
Glebe Ct. WD25: Wat 5Z 13
Glebe Cres. HA3: Kenton 27Na 47
Glebe Cres. NW4 28Ya 48
Glebe Farm Bus. Pk. BR2: Kes 81Mc 179
Glebefield, The TN13: Riv 95Hd 202
Glebe Gdns. CM13: Heron 24Fe 59
Glebe Gdns. KT14: Byfl 86M 169
Glebe Gdns. KT3: N Mald 73Ua 154
Glebe Ho. SE16 48Xb 91
(off Slippers Pl.)
Glebe Ho. Dr. BR2: Hayes 74Kc 159
Glebe Hyrst CR2: Sande 84Vb 177
Glebe Hyrst SE19 63Ub 135
Glebe Knoll BR2: Broml 68Hc 137
Glebeland Gdns. TW17: Shep 72S 150
Glebelands DA1: Cray 56Hd 118
Glebelands E10 33Dc 72
Glebelands IG7: Chig 20Xc 37
Glebelands KT10: Clay 81Ha 172
Glebelands KT8: W Mole 71Da 151
Glebelands Av. E18 26Jc 53
Glebelands Av. IG2: Ilf 31Tc 74
Glebelands Cl. N12 25Fb 49
Glebelands Cl. SE5 55Ub 113
Glebelands Rd. TW14: Felt 60W 106
Glebe La. EN5: Ark 15Wa 30
Glebe La. HA3: Kenton 28Na 47
Glebe La. TN13: S'oaks 99Kd 203
Glebe M. DA15: Sidc 58Vc 117
Glebe Path CR4: Mitc 69Hb 133
Glebe Pl. DA4: Hort K 70Sd 142
Glebe Pl. SW3 51Gb 111
Glebe Rd. BR1: Broml 67Jc 137
Glebe Rd. CR6: W'ham 89Zb 178
Glebe Rd. DA11: Grav'nd 10B 122
Glebe Rd. E8 38Vb 71
Glebe Rd. HA7: Stan 22La 46
Glebe Rd. KT21: Asht 90Ma 173
Glebe Rd. N3 25Eb 49
Glebe Rd. N8 28Pb 50
Glebe Rd. NW10 37Wa 68
Glebe Rd. RH1: Mers 96Kb 196
Glebe Rd. RM10: Dag 37Dd 76
Glebe Rd. RM13: Rain 41Ld 97
Glebe Rd. SL4: Old Win 7M 103
Glebe Rd. SM2: Cheam 81Ab 174
Glebe Rd. SM5: Cars 79Hb 155
Glebe Rd. SW13 54Wa 110
Glebe Rd. TW18: Staines 64K 127
Glebe Rd. TW20: Egh 64E 126
Glebe Rd. UB3: Hayes 46V 84
Glebe Rd. UB8: Uxb 40L 63
Glebe Side TW1: Twick 58Ha 108
Glebe Sq. CR4: Mitc 69Hb 133
Glebe St. W4 50Ua 88
Glebe Ter. W4 50Ua 88
Glebe Way BR4: W W'ck 75Ec 158
Glebe Way CR2: Sande 83Vb 177
Glebe Way DA8: Erith 51Gd 118
Glebe Way IG8: Wfd G 22Lc 53
Glebe Way RM11: Horn 31Nd 77
Glebe Way TW13: Hanw 62Ca 129
Gledhow Gdns. SW5 6A 226 (49Eb 89)
Gledhow Wood KT20: Kgswd 93Db 195
Gledstanes Rd. W14 50Ab 88
Gledwood Av. UB4: Hayes 43V 84
Gledwood Ct. UB4: Hayes 43V 84
Gledwood Cres. UB4: Hayes 43V 84
Gledwood Dr. UB4: Hayes 43V 84
Gledwood Gdns. UB4: Hayes 43V 84
Gleed Av. WD23: B Hea 19Fa 28
Glen Gdns. TW19: Stanw 60M 105
Gleeson Dr. BR6: Chels 78Vc 161
Gleeson M. KT15: Add 77L 149
Glegg Pl. SW15 56Za 110
Glen, The BR2: Broml 68Gc 137
Glen, The BR6: Farnb 76Pc 160
Glen, The CR0: C'don 76Zb 158
Glen, The EN2: Enf 14Rb 33
Glen, The HA5: Eastc 29X 45
Glen, The HA5: Pinn 31Aa 65
Glen, The HA6: Nwood 24T 44
Glen, The HA9: Wemb 35Na 67
Glen, The KT15: Add 78H 149
Glen, The RH1: Redh 8P 207
Glen, The RM13: Rain 42Ld 97

Glen, The SL3: L'ly 9N 81
Glen, The SL5: S'hill 10B 124
Glen, The SS17: Stan H 1P 101
Glen, The UB2: S'hall 50Ba 85
Glenaffric Av. E14 49Ec 92
Glen Albyn Rd. SW19 61Za 132
Glenallan Ho. W14 49Bb 89
(off North End Cres.)
Glenalla Rd. HA4: Ruis 31V 64
Glenalmond Ho. TW15: Ashf 62N 127
Glenalmond Rd. HA3: Kenton 28Na 47
Glenalvon Way SE18 49Nc 94
Glena Mt. SM1: Sutt 77Eb 155
Glenarm Rd. E5 35Yb 72
Glen Av. TW15: Ashf 63Q 128
Glenavon Cl. KT10: Clay 79Ja 152
Glenavon Ct. KT4: Wor Pk 75Xa 154
Glenavon Gdns. SL3: L'ly 9N 81
Glenavon Lodge BR3: Beck 66Cc 136
Glenavon Rd. E15 38Gc 73
Glenbarr Cl. SE9 55Rc 116
Glenbower Ct. AL4: St A 2H 7
Glenbow Rd. BR1: Broml 65Gc 137
Glenbrook Nth. EN2: Enf 14Pb 32
Glenbrook Rd. NW6 36Cb 69
Glenbrook Sth. EN2: Enf 14Pb 32
Glenbuck Ct. KT6: Surb 72Na 153
Glenbuck Rd. KT6: Surb 72Ma 153
Glenburnie Rd. SW17 62Hb 133
Glencairn Dr. W5 42La 86
Glencairne Cl. E16 43Mc 93
Glencairn Rd. SW16 67Nb 134
Glencar Ct. SE19 65Rb 135
Glen Chess WD3: Loud 14L 25
Glen Cl. KT20: Kgswd 95Ab 194
Glen Cl. TW17: Shep 70Q 128
Glencoe Av. IG2: Ilf 31Tc 74
Glencoe Dr. RM10: Dag 35Cd 76
Glencoe Mans. SW9 52Qb 112
(off Mowll St.)
Glencoe Rd. KT13: Weyb 76Q 150
Glencoe Rd. UB4: Yead 43Z 85
Glencoe Rd. WD23: Bush 16Ca 27
Glencorse Grn. WD19: Wat 21Z 45
Glen Ct. BR1: Broml 66Hc 137
(off Bromley Av.)
Glen Ct. DA15: Sidc 63Wc 139
Glen Ct. GU21: Wok 1L 187
Glen Ct. KT14: Byfl 83M 169
Glen Ct. KT15: Add 78H 149
Glen Ct. TW18: Staines 66H 127
(off Riverside Rd.)
Glen Cres. IG8: Wfd G 23Kc 53
Glendale BR8: Swan 70Hd 140
Glendale HP1: Hem H 2K 3
Glendale Av. HA8: Edg 21Pa 47
Glendale Av. N22 24Qb 50
Glendale Av. RM6: Chad H 31Yc 75
Glendale Cl. CM15: Shenf 18Ae 41
Glendale Cl. GU21: Wok 10N 167
Glendale Cl. SE9 55Qc 116
Glendale Dr. SW19 64Bb 133
Glendale Gdns. HA9: Wemb 32Ma 67
Glendale M. BR3: Beck 67Dc 136
Glendale Ri. CR8: Kenley 87Rb 177
Glendale Rd. DA11: Nflt 3A 144
Glendale Rd. DA8: Erith 49Ed 96
Glendale Wlk. EN8: Chesh 2Ac 20
Glendale Way SE28 45Yc 95
Glendall St. SW9 56Pb 112
Glendarvon St. SW15 55Za 110
Glendean Ct. EN3: Enf L 8Ac 20
Glendene Av. KT24: E Hor 98U 190
Glendevon Cl. HA8: Edg 20Ra 29
Glendish Rd. N17 25Xb 51
Glendor Gdns. NW7 21Ta 47
Glendower Cres. BR6: St M Cry 72Wc 161
Glendower Gdns. SW14 55Ta 109
Glendower Pl. SW7 5B 226 (49Fb 89)
Glendower Rd. E4 18Fc 35
Glendower Rd. SW14 55Ta 109
Glendown Ho. E8 36Wb 71
Glendown Rd. SE2 50Wc 95
Glendun Ct. W3 45Ua 88
Glen Dunlop Ho., The TN13: S'oaks 94Kd 203
Glendun Rd. W3 45Ua 88
Gleneagle M. SW16 64Mb 134
Gleneagle Rd. SW16 64Mb 134
Gleneagles HA7: Stan 24Ka 46
Gleneagles W13 43Ka 86
(off Malvern Way)
Gleneagles Cl. BR6: Orp 74Tc 160
Gleneagles Cl. RM3: Hrld W 24Pd 57
Gleneagles Cl. SE16 50Xb 91
Gleneagles Cl. TW19: Stanw 58L 105
Gleneagles Cl. WD19: Wat 21Z 45
Gleneagles Grn. BR6: Orp 74Tc 160
Gleneagles Twr. UB1: S'hall 44Ea 86
(off Fleming Rd.)
Gleneldon M. SW16 63Nb 134
Gleneldon Rd. SW16 63Nb 134
Glenelg Rd. SW2 57Nb 112
Glenesk Rd. SE9 55Qc 116
Glenfarg Rd. SE6 60Ec 114
Glenferrie Rd. AL1: St A 2E 6
Glenfield Cres. HA4: Ruis 31T 64
Glenfield Rd. SM7: Bans 87Db 175
Glenfield Rd. SW12 60Lb 112
Glenfield Rd. TW15: Ashf 65R 128
Glenfield Rd. W13 47Ka 86
Glenfields SL2: Stoke P 9K 61
Glenfield Ter. W13 47Ka 86
Glenfinlas Way SE5 52Rb 113
Glenforth St. SE10 50Hc 93
Glengall Bus. Cen. SE15 51Vb 113
Glengall Gro. E14 48Dc 92
Glengall Pas. NW6 39Cb 69
(off Priory Pk. Rd.)
Glengall Pl. AL1: St A 5C 6
Glengall Rd. DA7: Bex 55Ad 117
Glengall Rd. HA8: Edg 20Ra 29
Glengall Rd. IG8: Wfd G 23Jc 53
Glengall Rd. NW6 39Bb 69
Glengall Rd. SE15 50Vb 91
Glengall Ter. SE15 51Vb 113
Glen Gdns. CR0: Wadd 76Qb 156
Glengariff Mans. SW9 52Qb 112
(off Sth. Island Pl.)
Glengarnock Av. E14 49Ec 92
Glengarry Rd. SE22 57Ub 113
Glenham Dr. IG2: Ilf 29Rc 54
Glenhaven Av. WD6: Bore 13Qa 29
Glenhaven Dr. TW19: Stanw M 57J 105
Glenhead Cl. SE9 55Rc 116
Glenheadon Cl. KT22: Lea 95Ma 193
Glenheadon Ri. KT22: Lea 95Ma 193
Glenhill Cl. N3 26Cb 49
Glen Ho. E16 46Qc 94
(off Storey St.)
Glenhouse Rd. SE9 57Qc 116
Glenhurst BR3: Beck 67Ec 136
Glenhurst Av. DA5: Bexl 60Bd 117
Glenhurst Av. HA4: Ruis 31S 64
Glenhurst Av. NW5 35Jb 70
Glenhurst Ct. SE19 64Vb 135
Glenhurst Ri. SE19 66Sb 135
Glenhurst Rd. N12 22Fb 49
Glenhurst Rd. TW8: Bford 51La 108
Glenilla Rd. NW3 37Gb 69
Glenister Gdns. UB3: Hayes 47X 85
Glenister Ho. UB3: Hayes 46X 85
(off Avondale Dr.)
Glenister Pk. Rd. SW16 66Mb 134
Glenister Rd. SE10 50Hc 93
Glenister St. E16 46Qc 94
Glenkerry Ho. E14 44Ec 92
(off Burcham St.)
Glenlea Rd. SE9 57Pc 116
Glenlee GU22: Wok 1N 187
Glenloch Rd. EN3: Enf H 12Yb 34
Glenloch Rd. NW3 37Gb 69
Glen Luce EN8: Chesh 3Zb 20
Glenluce Rd. SE3 51Jc 115
Glenlyn Av. AL1: St A 3F 6
Glenlyon Rd. SE9 57Qc 116
Glenmead IG9: Buck H 18Lc 35
Glenmere Av. NW7 24Wa 48
Glenmere Row SE12 58Jc 115
Glen M. E17 29Bc 52
Glenmill TW12: Hamp 64Ba 129
Glenmore Cl. KT15: Add 76K 149
Glenmore Gdns. WD5: Ab L 4W 12
Glenmore Lawns W13 44Ja 86
Glenmore Lodge BR3: Beck 67Dc 136
Glenmore Pde. HA0: Wemb 39Na 67
Glenmore Rd. DA16: Well 52Vc 117
Glenmore Rd. NW3 37Gb 69
Glenmore Way IG11: Bark 40Wc 75
Glenmount Path SE18 50Sc 94
Glenn Av. CR8: Purl 83Rb 177
Glennie Ct. SE22 60Wb 113
Glennie Rd. SE27 62Qb 134
Glenny Rd. IG11: Bark 37Sc 74
Glenorchy Cl. UB4: Yead 43Aa 85
Glenpark Ct. W13 45Ja 86
Glenparke Rd. E7 37Kc 73
Glenridding NW1 2C 216 (40Lb 70)
(off Ampthill Est.)
Glen Ri. IG8: Wfd G 23Kc 53
Glen Rd. E13 42Lc 93
Glen Rd. E17 29Bc 52
Glen Rd. KT9: Chess 77Pa 153
Glen Rd. End SM6: W'gton 81Kb 176
Glenrosa Gdns. DA12: Grav'nd 4H 145
Glenrosa St. SW6 54Eb 111
Glenrose Cl. SL2: Slou 4N 81
Glenrose Ct. DA14: Sidc 64Xc 139
Glenrose Ct. SE1 3H 231 (48Ub 91)
(off Long La.)
Glenroy St. W12 44Ya 88
Glensdale Rd. SE4 55Bc 114
Glenshaw Mans. SW9 52Qb 112
(off Brixton Rd.)
Glenshee Cl. HA6: Nwood 23S 44
Glenshiel Rd. SE9 57Qc 116
Glenside IG7: Chig 23Rc 54
Glenside Cl. CR8: Kenley 87Tb 177
Glenside Cotts. SE18 51Rc 116
Glentanner Way SW17 62Fb 133
Glen Ter. E14 47Ec 92
(off Manchester Rd.)
Glentham Gdns. SW13 51Xa 110
Glentham Rd. SW13 51Wa 110
Glenthorne Av. CR0: C'don 74Xb 157
Glenthorne Cl. SM3: Sutt 74Cb 155
Glenthorne Cl. UB10: Hil 41Q 84
Glenthorne Gdns. IG6: Ilf 27Qc 54
Glenthorne Gdns. SM3: Sutt 74Cb 155
Glenthorne M. W6 49Xa 88
Glenthorne Rd. E17 29Ac 52
Glenthorne Rd. KT1: King T 70Pa 131
Glenthorne Rd. N11 22Hb 49
Glenthorne Rd. W6 49Xa 88
Glenthorpe Av. SW15 56Wa 110
Glenthorpe Gdns. HA7: Stan 20Ha 28
Glenthorpe Rd. SM4: Mord 71Za 154
Glenton Cl. RM1: Rom 24Gd 56
Glenton M. SE15 54Yb 114
Glenton Rd. SE13 56Gc 115
Glenton Way RM1: Rom 24Gd 56
Glentrammon Av. BR6: Chels 79Vc 161
Glentrammon Cl. BR6: Chels 78Vc 161
Glentrammon Gdns. BR6: Chels 79Vc 161
Glentrammon Rd. BR6: Chels 79Vc 161
Glentworth Pl. SL1: Slou 6D 80
Glentworth St. NW1 6G 215 (42Hb 89)
Glenure Rd. SE9 57Qc 116
Glenvern Ct. TW7: Isle 53Ja 108
(off White Lodge Cl.)
Glen Vw. DA12: Grav'nd 10E 122
Glenview SE2 51Zc 117
Glenview Gdns. HP1: Hem H 2K 3
Glenview Rd. BR1: Broml 68Mc 137
Glenview Rd. HP1: Hem H 2K 3
Glenville Av. EN2: Enf 10Sb 19
Glenville Gro. SE8 52Bc 114
Glenville M. SW18 59Db 111
Glenville M. Ind. Est. SW18 59Cb 111
Glenville Rd. KT2: King T 67Qa 131
Glen Wlk. TW7: Isle 57Fa 108
(not continuous)
Glen Way WD17: Wat 10U 12
Glenwood Av. NW9 32Ua 68
Glenwood Av. RM13: Rain 42Jd 96
Glenwood Cl. HA1: Harr 29Ha 46
Glenwood Ct. DA14: Sidc 63Wc 139
Glenwood Ct. E18 27Jc 53
Glenwood Dr. RM2: Rom 29Jd 56
Glenwood Gdns. IG2: Ilf 29Qc 54
Glenwood Gro. NW9 32Sa 67
Glenwood Rd. KT17: Ewe 79Wa 154
Glenwood Rd. N15 29Rb 51
Glenwood Rd. NW7 20Ua 30
Glenwood Rd. SE6 60Bc 114
Glenwood Rd. TW3: Houn 55Fa 108
Glenwood Way CR0: C'don 72Zb 158
Glenworth Av. E14 49Fc 93
Glevum Cl. AL3: St A 4M 5
Gliddon Dr. E5 35Xb 71
Gliddon Rd. W14 49Ab 88
Glimpsing Grn. DA18: Erith 48Ad 95
Glisson Rd. UB10: Hil 40Q 64
Gload Cres. BR5: Orp 75Zc 161
Global App. E3 40Ec 72
Globe Apts. SE8 51Bc 114
(off Evelyn St.)
Globe Ho. E14 44Gc 93
Globe Ho. WD3: Chor 14E 24
Globe Ind. Est. RM17: Grays 50Ee 99
Globe Pond Rd. SE16 46Ac 92
Globe Rd. E1 41Yb 92
Globe Rd. E15 36Hc 73
Globe Rd. E2 41Yb 92
Globe Rd. IG8: Wfd G 23Lc 53
Globe Rd. RM11: Horn 30Jd 56
Globe St. SE1 3F 231 (48Tb 91)
Globe Ter. E2 41Yb 92
GLOBE TOWN 41Zb 92
Globe Town Mkt. 41Zb 92
Globe Vw. EC4 4D 224 (45Sb 91)
(off High Timber St.)
Globe Wharf SE16 45Zb 92
Gloria Gdns. RM13: Rain 39Jd 76
Glossop Ho. RM3: Rom 22Nd 57
(off Lindfield Rd.)
Glossop Rd. CR2: Sande 81Tb 177
Gloster Ct. GU21: Wok 88B 168
(off Walton Rd.)
Gloster Ridley Ct. E14 44Bc 92
(off St Anne's Row)
Gloster Rd. GU22: Wok 92C 188
Gloster Rd. KT3: N Mald 70Ua 132
Gloucester W14 49Bb 89
(off Kensington Village)
Gloucester Arc. SW7 5A 226 (49Eb 89)
Gloucester Av. DA15: Sidc 61Uc 138
Gloucester Av. DA16: Well 56Vc 117
Gloucester Av. EN8: Walt C 5Ac 20
Gloucester Av. NW1 38Jb 70
Gloucester Av. RM11: Horn 28Qd 57
Gloucester Av. RM16: Grays 47Ee 99
Gloucester Av. RM18: E Til 10L 101
Gloucester Av. SL1: Slou 3G 80
Gloucester Cir. SE10 52Ec 114
Gloucester Cl. GU21: Brkwd 1E 186
Gloucester Cl. KT7: T Ditt 74Ja 152
Gloucester Cl. NW10 38Ta 67
Gloucester Ct. SE1: Rolls Rd. 7K 231 (50Vb 91)
(off Rolls Rd.)
Gloucester Ct. SE1: Swan St. 3E 230 (48Sb 91)
(off Swan St.)
Gloucester Ct. CR4: Mitc 71Nb 156
Gloucester Ct. EC3 5J 225 (45Ub 91)
Gloucester Ct. HA1: Harr 27Ga 46
Gloucester Ct. NW11 31Bb 69
(off Golders Grn. Rd.)
Gloucester Ct. RH1: Redh 5P 207
(off Gloucester Rd.)
Gloucester Ct. RM18: Tilb 4B 122
Gloucester Ct. SE22 60Wb 113
Gloucester Ct. TW9: Kew 52Qa 109
Gloucester Ct. UB9: Den 31J 63
Gloucester Ct. W7 43Ha 86
(off Copley Cl.)
Gloucester Ct. WD18: Wat 14V 26
Gloucester Ct. WD3: Crox G 14R 26
Gloucester Cres. NW1 39Kb 70
Gloucester Cres. TW18: Staines 65M 127
Gloucester Cres. TW20: Eng G 2P 125
Gloucester Dr. N4 33Rb 71
Gloucester Dr. NW11 28Cb 49
Gloucester Dr. TW18: Staines 62E 126
Gloucester Gdns. EN4: Cockf 14Jb 32
Gloucester Gdns. IG1: Ilf 31Nc 74
Gloucester Gdns. NW11 31Bb 69
Gloucester Gdns. SM1: Sutt 75Db 155
Gloucester Gdns. W2 44Eb 89
Gloucester Ga. NW1 1K 215 (40Kb 70)
(not continuous)
Gloucester Ga. Bri. 1K 215 (39Kb 70)
(off Gloucester Gate)
Gloucester Ga. M. NW1 1K 215 (40Kb 70)
Gloucester Gro. HA8: Edg 25Ta 47
Gloucester Ho. E16 46Jc 93
(off Gatcombe Rd.)
Gloucester Ho. NW6 40Cb 69
(off Cambridge Rd.)
Gloucester Ho. SW9 52Qb 112
Gloucester Ho. TW10: Rich 57Qa 109
Gloucester Ho. WD6: Bore 12Qa 29
Gloucester M. E10 31Cc 72
Gloucester M. W2 3A 220 (44Eb 89)
Gloucester M. W. W2 3A 220 (44Eb 89)
Gloucester Pde. DA15: Sidc 57Wc 117
Gloucester Pde. UB3: Harl 48S 84
Gloucester Pk. Apts. SW7 5A 226 (49Eb 89)
(off Ashburn Pl.)
Gloucester Pl. NW1 5F 215 (42Hb 89)
Gloucester Pl. SL4: Wind 4H 103
Gloucester Pl. W1 7G 215 (43Hb 89)
Gloucester Pl. M. W1 1G 221 (43Hb 89)
Gloucester Rd. CM15: Pil H 15Xd 40
Gloucester Rd. CR0: C'don 74Tb 157
Gloucester Rd. DA1: Dart 59Kd 119
Gloucester Rd. DA12: Grav'nd 3E 144
Gloucester Rd. DA17: Belv 50Bd 95
Gloucester Rd. E10 31Cc 72
Gloucester Rd. E11 29Kc 53
Gloucester Rd. E12 34Pc 74
Gloucester Rd. E17 26Zb 52
Gloucester Rd. EN2: Enf 10Sb 19
Gloucester Rd. EN5: New Bar 15Db 31
Gloucester Rd. HA1: Harr 29Da 45
Gloucester Rd. KT1: King T 68Qa 131
Gloucester Rd. N17 26Tb 51
Gloucester Rd. N18 22Vb 51
Gloucester Rd. RH1: Redh 5P 207
Gloucester Rd. RM1: Rom 30Gd 56
Gloucester Rd. SW7 3A 226 (48Eb 89)
Gloucester Rd. TW11: Tedd 64Ga 130
Gloucester Rd. TW12: Hamp 66Da 129
Gloucester Rd. TW13: Felt 60Y 107
Gloucester Rd. TW2: Twick 60Ea 108
Gloucester Rd. TW4: Houn 56Aa 107
Gloucester Rd. TW9: Kew 52Qa 109
Gloucester Rd. W3 47Sa 87
Gloucester Rd. W5 47La 86
Gloucester Road Station (Underground) 5A 226 (49Eb 89)
Gloucester Sq. E2 39Wb 71
Gloucester Sq. GU21: Wok 89A 168
Gloucester Sq. W2 3C 220 (44Fb 89)
(not continuous)
Gloucester St. SW1 7B 228 (50Lb 90)
Gloucester Ter. KT13: Weyb 78S 150
Gloucester Ter. N14 18Mb 32
(off Crown La.)
Gloucester Ter. W2 44Db 89
Gloucester Wlk. W8 47Cb 89
Gloucester Way EC1 4A 218 (41Qb 90)
Glover Cl. DA10: Swans 60Ce 121
Glover Cl. SE2 49Yc 95
Glover Dr. N18 23Yb 52
Glover Ho. NW6 38Eb 69
(off Harben Rd.)
Glover Ho. SE15 56Xb 113
Glover Rd. HA5: Pinn 30Z 45
Glovers Cl. TN16: Big H 88Kc 179
Glovers Gro. HA4: Ruis 31R 64
Glover's Rd. RH2: Reig 7K 207
Gloxinia Rd. DA13: Sflt 65De 143
Gloxinia Wlk. TW12: Hamp 65Ca 129
Glycena Rd. SW11 55Hb 111
Glyn Av. EN4: E Barn 14Fb 31
Glyn Cl. KT17: Ewe 81Wa 174
Glyn Cl. SE25 68Ub 135
Glyn Ct. HA7: Stan 23Ka 46
Glyn Ct. SW16 62Qb 134
Glyncroft SL1: Slou 7D 80
Glyndale Grange SM2: Sutt 79Db 155
Glyn Davies Cl. TN13: Dun G 92Gd 202
Glyndebourne Ct. UB5: N'olt 41Y 85
(off Canberra Dr.)
Glyndebourne Pk. BR6: Farnb 75Rc 160
Glynde M. SW3 4E 226 (48Gb 89)
(off Walton St.)
Glynde Reach WC1 4G 217 (41Nb 90)
(off Harrison St.)
Glynde Rd. DA7: Bex 55Zc 117
Glynde St. SE4 58Bc 114
Glyndon Rd. SE18 49Sc 94
(not continuous)
Glyn Dr. DA14: Sidc 63Xc 139
Glynfield Rd. NW10 38Ua 68
Glyn Mans. W14 49Ab 88
(off Hammersmith Rd.)
Glynne Rd. N22 26Qb 50
Glyn Rd. E5 35Zb 72
Glyn Rd. EN3: Pond E 14Yb 34
Glyn Rd. KT4: Wor Pk 75Za 154
Glyn St. SE11 50Pb 90
Glynswood SL9: Chal P 24B 42
Glynswood Pl. HA6: Nwood 24R 44
Glynwood Ct. SE23 61Yb 136
Goals Soccer Cen. Bexleyheath 55Bd 117
Goals Soccer Cen. Chingford 23Cc 52
Goals Soccer Cen. Dagenham 39Zc 75
Goals Soccer Cen. Dartford 60Qd 119
Goals Soccer Cen. Eltham 58Lc 115
Goals Soccer Cen. Gillette Corner 51Ha 108
Goals Soccer Cen. Hayes 46Y 85
Goals Soccer Cen. Heathrow 49S 84
Goals Soccer Cen. Ruislip 36Z 65
Goals Soccer Cen. Sutton 76Za 154
Goals Soccer Cen. Tolworth 75Sa 153
Goals Soccer Cen. Wimbledon 69Wa 132
Go Ape! Alexandra Palace 26Mb 50
Go Ape! Battersea Park 53Hb 111
Go Ape! Black Park 40B 62
Go Ape! Trent Park 13Jb 32
Goater's All. SW6 52Bb 111
(off Dawes Rd.)
Goat Ho. Bri. 69Wb 135
Goat La. EN1: Enf 10Vb 19
Goat Rd. CR4: Cars 73Hb 155
Goat Rd. CR4: Mitc 73Hb 155
Goatsfield Rd. TN16: Tats 92Lc 199
Goatswood La. RM4: N'side 17Kd 39
Goatswood La. RM4: Noak H 17Kd 39
Goat Wharf TW8: Bford 51Na 109
Gobions Av. RM5: Col R 24Fd 56
Gobions Way EN6: Pot B 10K 9
Goby Ho. SE8 52Dc 114
(off Creative Rd.)
Godalming Av. SM6: W'gton 78Nb 156
Godalming Rd. E14 43Dc 92
Godbold Rd. E15 42Gc 93
Goddard Cl. TW17: Shep 69P 127
Goddard Ct. HA3: Kenton 26Ja 46
Goddard Dr. WD23: Bush 15Ea 28
Goddard Ho. KT8: E Mos 71Fa 152
Goddard Ho. SE11 6B 230 (49Rb 91)
(off George Mathers Rd.)
Goddard Pl. N19 34Lb 70
Goddard Rd. BR3: Beck 70Zb 136
Goddard Rd. RM16: Grays 46Ce 99
Goddards Way IG1: Ilf 32Tc 74
Goddarts Ho. E17 27Cc 52
GODDEN GREEN 96Qd 203
GODDINGTON 76Yc 161
Goddington Chase BR6: Chels 77Xc 161
Goddington La. BR6: Chels 76Wc 161
Godfree Ct. SE1 1F 231 (47Tb 91)
(off Long La.)
Godfrey Av. TW2: Whitt 59Fa 108
Godfrey Av. UB5: N'olt 39Aa 65
Godfrey Hill SE18 49Nc 94
Godfrey Ho. EC1 4F 219 (41Tb 91)
(off St Luke's Est.)
Godfrey Pl. E2 4K 219 (41Vb 91)
(off Austin St.)
Godfrey Rd. SE18 49Pc 94
Godfrey St. E15 40Ec 72
Godfrey St. SW3 7E 226 (50Gb 89)
Godfrey Way TW4: Houn 59Aa 107
Goding St. SE11 7G 229 (50Nb 90)
Godley Cl. SE14 53Yb 114
Godley Rd. KT14: Byfl 86P 169
Godley Rd. SW18 60Fb 111
Godliman St. EC4 3C 224 (44Rb 91)
Godman Rd. RM16: Grays 8D 100
Godman Rd. SE15 54Xb 113
Godman Wlk. DA1: Dart 54Pd 119
Godolphin Cl. N13 23Rb 51
Godolphin Cl. SM2: Cheam 83Bb 175
Godolphin Ho. NW3 38Gb 69
(off Fellows Rd.)
Godolphin Ho. SL4: Eton 1H 103
(off Common La.)
Godolphin Pl. W3 45Ta 87
Godolphin Rd. KT13: Weyb 79T 150
Godolphin Rd. SL1: Slou 5H 81
Godolphin Rd. W12 46Xa 88
(not continuous)
Godric Cres. CR0: New Ad 82Fc 179
Godson Rd. CR0: Wadd 76Qb 156
Godson St. N1 1K 217 (40Qb 70)
Godson Yd. NW6 41Cb 89
GODSTONE 3A 210
Godstone By-Pass RH9: G'stone 1A 210
Godstone Farm & Playbarn 4A 210
Godstone Golf Course 2C 210
Godstone Grn. RH9: G'stone 3P 209
Godstone Hill RH9: G'stone 99Wb 197
Godstone Ho. SE1 3G 231 (48Tb 91)
(off Pardoner St.)
GODSTONE INTERCHANGE 1A 210
Godstone Mt. CR8: Purl 84Rb 177
Godstone Rd. CR3: Cat'm 96Wb 197
Godstone Rd. CR3: W'ham 90Vb 177
Godstone Rd. CR3: Whyt 90Vb 177
Godstone Rd. CR8: Kenley 84Rb 177
Godstone Rd. CR8: Purl 84Rb 177
Godstone Rd. RH1: Blet 5K 209
Godstone Rd. RH8: Oxt 3E 210
Godstone Rd. SM1: Sutt 77Eb 155
Godstone Rd. TW1: Twick 58Ka 108
Godstone Station (Rail) 10C 210
Godstone Vineyards 100Yb 198
Godstow Rd. SE2 47Xc 95
Godward Sq. E1 42Zb 92
Godwin Cl. E4 10Ec 20
Godwin Cl. KT19: Ewe 79Sa 153
Godwin Cl. N1 1E 218 (40Sb 71)
Godwin Ct. NW1 1C 216 (40Lb 70)
(off Chalton St.)
Godwin Ho. E2 1K 219 (40Vb 71)
(off Thurtle Rd.)
Godwin Ho. SE1 7K 225 (46Vb 91)
(off Duchess Wlk.)
Godwin Rd. BR2: Broml 69Lc 137
Godwin Rd. E7 35Kc 73
Godwin Ter. RM3: Hrld W 25Nd 57
Goffers Rd. SE3 53Gc 115
Goffs Cres. EN7: G Oak 1Sb 19
Goff's La. EN7: G Oak 1Sb 19
GOFF'S OAK 1Sb 19
Goffs Oak Av. EN7: G Oak 1Rb 19
Goffs Rd. TW15: Ashf 65T 128
Goff's Sports & Arts Cen. 1Wb 19
Gogmore Farm Cl. KT16: Chert 73H 149
Gogmore La. KT16: Chert 73J 149
Goidel Cl. SM6: Bedd 77Mb 156
Golborne Gdns. W10 42Ab 88
(not continuous)
Golborne M. W10 43Ab 88
Golborne Rd. W10 43Ab 88
Goldace RM17: Grays 51Be 121
Golda Cl. EN5: Barn 16Za 30
Golda Ct. N3 26Bb 49
Goldbeaters Gro. HA8: Edg 23Ua 48
Goldbeaters Ho. W1 3E 222 (44Mb 90)
(off Manette St.)
Goldcliff Cl. SM4: Mord 73Cb 155
Goldcrest Cl. E16 43Mc 93
Goldcrest Cl. SE28 45Yc 95
Goldcrest M. N16 34Wb 71
Goldcrest M. W5 43Ma 87
Goldcrest Way CR0: New Ad 81Fc 179
Goldcrest Way CR8: Purl 82Mb 176
Goldcrest Way RM3: Hrld W 26Nd 57
Goldcrest Way WD23: Bush 18Ea 28
Goldcroft HP3: Hem H 4A 4
Golden Anchor Ho. SE10 48Jc 93
(off Latimer Square)
Golden Bus. Pk. E10 32Ac 72
Golden Ct. EN4: E Barn 14Gb 31
Golden Ct. TW7: Isle 54Fa 108
Golden Ct. TW9: Rich 57Ma 109
Golden Cres. UB3: Hayes 46V 84
Golden Cross M. W11 44Bb 89
(off Portobello Rd.)
Golden Hinde 6F 225 (46Tb 91)
Golden Hind Pl. SE8 49Bc 92
(off Grove St.)
Golden Jubilee Bridges 7H 223 (46Pb 90)
(off Belvedere Rd.)
Golden La. BR4: W W'ck 76Ec 158
Golden La. EC1 5D 218 (42Sb 91)
Golden La. Campus EC1 6E 218 (42Sb 91)
(off Golden La.)
Golden La. Est. EC1 6D 218 (42Sb 91)
Golden Lane Sport & Fitness Cen. 6D 218 (42Sb 91)
(off Golden La. Est.)
Golden Mnr. W7 45Ga 86
Golden M. SE20 67Yb 136
Golden Mile Ho. TW8: Bford 50Na 87
(off Clayponds La.)
Golden Oak Cl. SL2: Farn C 7G 60
Golden Pde. E17 27Ec 52
(off Wood St.)
Golden Plover Cl. E16 44Jc 93
Golden Sq. W1 4C 222 (45Lb 90)
Golden Yd. NW3 35Eb 69
(off Holly Mt.)
Golders Cl. HA8: Edg 22Ra 47
Golders Ct. NW11 31Bb 69
Golders Gdns. NW11 31Ab 68
GOLDERS GREEN 30Ab 48
Golders Grn. Crematorium 31Cb 69
Golders Grn. Cres. NW11 31Bb 69
Golders Grn. Rd. NW11 30Ab 48
Golders Green Station (Underground) 32Cb 69
Goldersleа NW11 32Cb 69
Golders Mnr. Dr. NW11 30Za 48
Golders Pk. Cl. NW11 32Cb 69
Golders Ri. NW4 29Za 48
Golders Way NW11 31Bb 69
Golderton NW4 28Xa 48
(off Prince of Wales Cl.)
Goldfinch Cl. BR6: Chels 78Wc 161
Goldfinch Ct. E3 40Cc 72
(off Four Seasons Cl.)
Goldfinch Rd. CR2: Sels 82Ac 178
Goldfinch Rd. SE28 48Tc 94
Goldfinch Way WD6: Bore 14Qa 29
Goldfort Wlk. GU21: Knap 8J 167
Goldhawk Ho. NW9 26Wa 48
Goldhawk M. W12 47Xa 88
Goldhawk Rd. W12 48Wa 88
Goldhawk Rd. W6 49Va 88
Goldhawk Road Station (Underground) 47Ya 88
Goldhaze Cl. IG8: Wfd G 24Lc 53
Gold Hill HA8: Edg 23Ta 47
Goldhurst Ho. W6 51Ya 110
Goldhurst Mans. NW6 37Eb 69
(off Goldhurst Ter.)
Goldhurst Ter. NW6 38Db 69
Goldie Ho. N19 31Mb 70
GOLDIE LEIGH HOSPITAL 51Yc 117
Golding Cl. KT9: Chess 79La 152
Golding Ct. N18 23Tb 51
Golding Ct. IG1: Ilf 34Qc 74
Goldingham Av. IG10: Lough 12Sc 36
Golding Ho. NW9 26Wa 48
Golding Rd. TN13: S'oaks 94Ld 203
Goldings, The GU21: Wok 8K 167
Goldings Hill IG10: Lough 8Pc 22
Goldings Ri. IG10: Lough 11Qc 36
Goldings Rd. IG10: Lough 11Qc 36
Golding St. E1 44Wb 91
(not continuous)
Golding Ter. E1 44Wb 91
(off Rope Wlk. Gdns.)
Golding Ter. SW11 53Jb 112
Goldington Bldgs. NW1 1D 216 (39Mb 70)
(off Royal College St.)
Goldington Cres. NW1 1D 216 (40Mb 70)
Goldington St. NW1 1D 216 (40Mb 70)
Gold La. HA8: Edg 23Ta 47
Goldman Cl. E2 42Wb 91
Goldney Rd. W9 42Cb 89
Goldrill Dr. N11 19Jb 32
Goldrings Rd. KT22: Oxs 85Da 171

Goldring Way AL2: Lon C 9F 6
Goldsboro' Rd. SW8 53Mb 112
Goldsborough Cres. E4 19Dc 34
Goldsborough Ho. E14 50Dc 92
(off St Davids Sq.)
Goldsdown Cl. EN3: Enf H 12Ac 34
Goldsdown Rd. EN3: Enf H 12Zb 34
Goldsel Rd. BR8: Crock 71Fd 162
Goldsel Rd. BR8: Swan 71Fd 162
Goldsmere Ct. RM11: Horn 32Nd 77
Goldsmid St. SE18 50Uc 94
Goldsmith Av. E12 37Nc 74
Goldsmith Av. NW9 29Ua 48
Goldsmith Av. RM7: Rush G 31Cd 76
Goldsmith Av. W3 45Ta 87
Goldsmith Cl. HA2: Harr 32Ca 65
Goldsmith Cl. TN16: Big H 89Nc 180
Goldsmith Ct. WC2 2G 223 (44Nb 90)
(off Stukeley St.)
Goldsmith Est. SE15 53Wb 113
Goldsmith La. NW9 28Ra 47
Goldsmith Rd. E10 32Cc 72
Goldsmith Rd. E17 26Zb 52
Goldsmith Rd. N11 22Hb 49
Goldsmith Rd. SE15 53Wb 113
Goldsmith Rd. W3 46Ta 87
Goldsmiths RM17: Grays 51Be 121
Goldsmiths Av. SS17: Corr 1P 101
Goldsmiths Av. SS17: Stan H 1P 101
Goldsmith's Bldgs. W3 46Ta 87
Goldsmiths Cl. GU21: Wok 10N 167
Goldsmiths Cl. W3 46Ta 87
Goldsmiths Coll. 53Ac 114
Goldsmith's Pl. NW6 39Db 69
(off Springfield La.)
Goldsmith's Row E2 40Wb 71
Goldsmith's Sq. E2 40Wb 71
Goldsmith St. EC2 2E 224 (44Sb 91)
Goldsmith Way AL3: St A 1A 6
Goldstone Farm Vw. KT23: Bookh 99Ca 191
Gold St. DA12: Lud'n 10F 144
Gold St. DA12: Sole S 10F 144
GOLD STREET 10G 144
GOLDSWORTH 10P 167
Goldsworth Orchard GU21: Wok 10L 167
GOLDSWORTH PARK 9L 167
Goldsworth Pk. Cen., The 9L 167
Goldsworth Pk. Trad. Est. GU21: Wok 8L 167
Goldsworth Rd. GU21: Wok 10N 167
Goldsworth Rd. Ind. Est. GU21: Wok 9P 167
Goldsworthy Gdns. SE16 50Yb 92
Goldsworthy Way SL1: Slou 4A 80
Goldthorpe NW1 39Lb 70
(off Camden St.)
Goldvale Ho. GU21: Wok 89A 168
(off Church St. W.)
Goldwell Ho. SE22 55Ub 113
(off Quorn Rd.)
Goldwell Rd. CR7: Thor H 70Pb 134
Goldwin Cl. SE14 53Yb 114
Goldwing Cl. E16 44Jc 93
Gole Rd. GU24: Pirb 3A 186
Golf Cl. CR7: Thor H 67Qb 134
Golf Cl. GU22: Pyr 86G 168
Golf Cl. HA7: Stan 24La 46
Golf Cl. WD23: Bush 13Z 27
Golf Club Cotts. SL5: S'dale 4G 146
Golf Club Dr. KT2: King T 66Ta 131
Golf Club Rd. AL9: Brk P 8J 9
Golf Club Rd. GU22: Wok 2L 187
Golf Club Rd. KT13: Weyb 81R 170
Golfe Rd. IG1: Ilf 34Tc 74
Golf Ho. Rd. RH8: Limp 1N 211
Golf Kingdom Barking 29Bd 55
Golf Links Av. DA11: Grav'nd 4D 144
Golf Ride EN2: Crew H 7Qb 18
Golf Rd. BR1: Broml 69Qc 138
Golf Rd. CR8: Kenley 90Tb 177
Golf Rd. W5 44Pa 87
Golf Side SM2: Cheam 83Ab 174
Golf Side TW2: Twick 62Fa 130
Golfside Cl. KT3: N Mald 68Ua 132
Golfside Cl. N20 20Gb 31
Gollogly Ter. SE7 50Lc 93
Gombards AL3: St A 1B 6
Gombard's All. AL3: St A 2B 6
Gomer Gdns. TW11: Tedd 65Ja 130
Gomer Pl. TW11: Tedd 65Ja 130
Gomm Rd. SE16 48Yb 92
Gomshall Av. SM6: W'gton 78Nb 156
Gomshall Gdns. CR8: Kenley 87Ub 177
Gomshall Rd. SM2: Cheam 82Ya 174
Gondar Gdns. NW6 36Bb 69
Gonnerston AL3: St A 1P 5
Gonson St. SE8 51Dc 114
Gonston Cl. SW19 61Ab 132
Gonville Av. WD3: Crox G 16R 26
Gonville Cres. UB5: N'olt 37Da 65
Gonville Rd. CR7: Thor H 71Pb 156
Gonville St. SW6 55Ab 110
Gooch Ho. E5 34Xb 71
Gooch Ho. EC1 7K 217 (43Qb 90)
(off Portpool La.)
Gooch Ho. SW11 52Mb 112
(off Malthouse Rd.)
Goodacre Cl. EN6: Pot B 4Db 17
Goodacre Cl. KT13: Weyb 78S 150
Goodacre Ct. EN6: Pot B 4Db 17
Goodall Ho. SE4 56Zb 114
Goodall Rd. E11 34Ec 72
Goodbury Rd. TN15: Knat 85Rd 183
Goodchild Rd. N4 32Sb 71
Gooden Ct. HA1: Harr 34Ga 66
Goodenough Cl. CR5: Coul 92Qb 196
Goodenough Rd. SW19 66Bb 133
Goodenough Way CR5: Coul 92Pb 196
Gooderham Ho. RM16: Grays 7D 100
Goodey Rd. IG11: Bark 38Vc 75
Goodfaith Ho. E14 45Dc 92
(off Simpson's Rd.)
Goodge Pl. W1 1C 222 (43Lb 90)
Goodge St. W1 1C 222 (43Lb 90)
Goodge Street Station (Underground) 7D 216 (43Mb 90)
Goodhall Cl. HA7: Stan 23Ja 46
Goodhall St. NW10 41Va 88
(not continuous)
Goodhart Ho. SM7: Bans 90Eb 175
Goodhart Pl. E14 45Ac 92
Goodhart Way BR4: W W'ck 73Gc 159
Goodhew Rd. CR0: C'don 72Wb 157
Goodhope Ho. E14 45Dc 92
(off Poplar High St.)
Gooding Cl. KT3: N Mald 70Sa 131
Goodinge Cl. N7 37Nb 70
Goodinge Rd. N7 37Nb 70
Gooding Ho. SE7 50Lc 93
Goodison Cl. WD23: Bush 15Ea 28
Goodlake Ct. UB9: Den 31H 63
Goodman Cres. CR0: C'don 72Rb 157
Goodman Cres. SW2 61Nb 134
Goodman Pk. SL2: Slou 5L 81
Goodman Pl. TW18: Staines 63H 127
Goodman Rd. E10 31Ec 72
Goodman's Ct. E1 4K 225 (45Vb 91)
Goodmans Ct. HA0: Wemb 35Ma 67
Goodman's Stile E1 44Wb 91
Goodmans Yd. E1 4K 225 (45Vb 91)
GOODMAYES 33Wc 75
Goodmayes Av. IG3: Ilf 32Wc 75
GOODMAYES HOSPITAL 29Wc 55
Goodmayes La. IG3: Ilf 35Wc 75
Goodmayes Lodge RM8: Dag 35Wc 75
Goodmayes Retail Pk. 32Xc 75
Goodmayes Rd. IG3: Ilf 32Wc 75
Goodmayes Station (Rail & Crossrail) 32Wc 75
Goodmead Rd. BR6: Orp 73Wc 161
Goodmead Rd. BR6: St M Cry 73Wc 161
Goodrich Cl. WD25: Wat 7W 12
Goodrich Ct. W10 44Za 88
Goodrich Ho. E2 40Yb 72
(off Sewardstone Rd.)
Goodrich Rd. SE22 58Vb 113
Goodridge Ho. E4 24Ec 52
Goodson Ho. SM4: Mord 73Eb 155
(off Green La.)
Goodson Rd. NW10 38Ua 68
Goodspeed Ho. E14 45Dc 92
(off Simpson's Rd.)
Goods Way N1 1F 217 (40Nb 70)
Goodway Gdns. E14 44Fc 93
Goodwill Dr. HA2: Harr 32Ca 65
Goodwill Ho. E14 45Dc 92
(off Simpson's Rd.)
Goodwin Cl. CR4: Mitc 69Fb 133
Goodwin Cl. SE16 48Vb 91
Goodwin Ct. EN4: E Barn 16Gb 31
Goodwin Ct. EN8: Chesh 1Ac 20
Goodwin Ct. N8 27Nb 50
(off Campsbourne Rd.)
Goodwin Ct. SW19 66Gb 133
Goodwin Dr. DA14: Sidc 62Zc 139
Goodwin Gdns. CR0: Wadd 79Rb 157
Goodwin Ho. N9 18Yb 34
Goodwin Ho. WD18: Wat 16U 26
Goodwin Rd. CR0: Wadd 78Rb 157
Goodwin Rd. N9 18Zb 34
Goodwin Rd. SL2: Slou 1D 80
Goodwin Rd. W12 47Wa 88
Goodwins Ct. WC2 4F 223 (45Nb 90)
Goodwin St. N4 33Qb 70
Goodwin Way RM3: Rom 22Md 57
Goodwood Apts. E4 24Dc 52
Goodwood Av. EN3: Enf W 9Yb 20
Goodwood Av. RM12: Horn 35Nd 77
Goodwood Av. WD24: Wat 7U 12
Goodwood Cl. HA7: Stan 22La 46
Goodwood Cl. SM4: Mord 70Cb 133
Goodwood Ct. W1 7A 216 (43Kb 90)
(off Devonshire St.)
Goodwood Cres. DA12: Grav'nd 4E 144
Goodwood Dr. UB5: N'olt 37Ca 65
Goodwood Ho. SE14 52Ac 114
(off Goodwood Rd.)
Goodwood Ho. SL4: Wind 3D 102
(off Paddock Cl.)
Goodwood Pde. BR3: Beck 70Ac 136
Goodwood Pde. WD24: Wat 8U 12
Goodwood Path WD6: Bore 12Qa 29
Goodwood Rd. RH1: Redh 4P 207
Goodwood Rd. SE14 52Ac 114
Goodworth Rd. RH1: Redh 4B 208
Goodworth Rd. TN15: Wro 88Be 185
Goodwyn Av. NW7 22Ua 48
Goodwyns Va. N10 25Jb 50
Goodyear Ho. N2 26Fb 49
(off The Grange)
Goodyear Pl. SE5 51Sb 113
Goodyer Ho. SW1 7D 228 (50Mb 90)
(off Tachbrook St.)
Goodyers Av. WD7: R'lett 5Ha 14
Goodyers Gdns. NW4 29Za 48
Goosander Way SE28 48Tc 94
Gooseacre La. HA3: Kenton 29Ma 47
Goosecroft HP1: Hem H 1H 3
Goosefields WD3: Rick 16L 25
Goose Grn. KT11: D'side 91W 190
Goose Grn. SL2: Farn R 10F 60
Goose Grn. Cl. BR5: St P 68Wc 139
Goose Grn. Trad. Est. SE22 56Vb 113
Goose La. GU22: Wok 4M 187
Gooseley La. E6: Claps Ga. La. 42Rc 94
Gooseley La. E6: Folkestone Rd. 41Qc 94
Goosens Cl. SM1: Sutt 78Eb 155
Goosepool KT16: Chert 73H 149
Goose Rye Rd. GU3: Worp 8H 187
Goose Sq. E6 44Pc 94
Gooshays Dr. RM3: Rom 22Nd 57
Gooshays Gdns. RM3: Rom 23Nd 57
Gophir La. EC4 4F 225 (45Tb 91)
Gopsall St. N1 1G 219 (39Tb 71)
Goral Mead WD3: Rick 18M 25
Gordian Apts. SE10 49Gc 93
(off Cable Wlk.)
Gordon Av. CR2: Sande 82Sb 177
Gordon Av. E4 23Gc 53
Gordon Av. HA7: Stan 24Ha 46
Gordon Av. RM12: Horn 33Hd 76
Gordon Av. SW14 56Ua 110
Gordon Av. TW1: Twick 57Ja 108
Gordonbrock Rd. SE4 57Cc 114
Gordon Cl. AL1: St A 3F 6
Gordon Cl. E17 30Cc 52
Gordon Cl. KT16: Chert 76G 148
Gordon Cl. N19 32Lb 70
Gordon Cl. RM18: E Til 2M 123
Gordon Cl. TW18: Staines 64K 127
Gordon Cotts. W8 47Db 89
(off Dukes La.)
Gordon Ct. HA8: Edg 22Pa 47
Gordon Ct. RH1: Redh 8P 207
(off St John's Ter. Rd.)
Gordon Cres. CR0: C'don 74Ub 157
Gordon Cres. UB3: Hayes 49W 84
Gordondale Rd. SW19 61Cb 133
Gordon Dr. KT16: Chert 76G 148
Gordon Dr. TW17: Shep 73T 150
Gordon Gdns. HA8: Edg 26Ra 47
Gordon Gro. SE5 54Rb 113
Gordon Hill EN2: Enf 11Sb 33
Gordon Hill Station (Rail) 11Rb 33
GORDON HOSPITAL 6D 228 (49Mb 90)
Gordon Ho. AL1: St A 3F 6
Gordon Ho. E1 45Yb 92
(off Glamis Rd.)
Gordon Ho. SW1 4C 228 (48Lb 90)
(off Greencoat Pl.)
Gordon Ho. W5 41Na 87
Gordon Ho. Rd. NW5 35Jb 70
Gordon Mans. W14 48Za 88
(off Addison Gdns.)
Gordon Mans. WC1 6D 216 (42Mb 90)
(off Torrington Pl.)
Gordon Pl. DA12: Grav'nd 8E 122
Gordon Pl. W8 47Cb 89
Gordon Prom. DA12: Grav'nd 8E 122
Gordon Prom. E. DA12: Grav'nd 8E 122
Gordon Rd. BR3: Beck 69Bc 136
Gordon Rd. CM15: Shenf 18Ce 41
Gordon Rd. CR3: Cat'm 93Tb 197
Gordon Rd. DA1: Dart 59Md 119
Gordon Rd. DA11: Nflt 9A 122
Gordon Rd. DA15: Sidc 57Uc 116
Gordon Rd. DA17: Belv 49Ed 96
Gordon Rd. E11 30Jc 53
Gordon Rd. E15 35Ec 72
Gordon Rd. E18 25Kc 53
Gordon Rd. E4 17Gc 35
Gordon Rd. EN2: Enf 11Sb 33
Gordon Rd. EN9: Walt A 6Cc 20
Gordon Rd. HA3: W'stone 27Ga 46
Gordon Rd. IG1: Ilf 34Tc 74
Gordon Rd. IG11: Bark 39Uc 74
Gordon Rd. KT10: Clay 80Ga 152
Gordon Rd. KT2: King T 67Pa 131
Gordon Rd. KT5: Surb 73Pa 153
Gordon Rd. N11 24Mb 50
Gordon Rd. N3 24Bb 49
Gordon Rd. N9 19Xb 33
Gordon Rd. RH1: Redh 3A 208
Gordon Rd. RM16: Grays 7A 100
Gordon Rd. RM6: Chad H 30Bd 55
Gordon Rd. SE15 54Xb 113
Gordon Rd. SL4: Wind 4D 102
Gordon Rd. SM5: Cars 79Hb 155
Gordon Rd. TN13: S'oaks 97Kd 203
Gordon Rd. TW15: Ashf 62N 127
Gordon Rd. TW17: Shep 72T 150
Gordon Rd. TW18: Staines 63E 126
Gordon Rd. TW3: Houn 56Ea 108
Gordon Rd. TW9: Rich 54Pa 109
Gordon Rd. UB2: S'hall 49Aa 85
Gordon Rd. UB7: Yiew 45N 83
Gordon Rd. W13 45Ka 86
Gordon Rd. W4 51Ra 109
Gordon Rd. W5 45La 86
Gordon Sq. WC1 5D 216 (42Mb 90)
Gordon Sq. Gdn. 5D 216 (42Mb 90)
(off Gordon Square)
Gordon St. E13 41Jc 93
Gordon St. WC1 5D 216 (42Mb 90)
Gordons Way RH8: Oxt 100Fc 199
Gordon Way BR1: Broml 67Jc 137
Gordon Way EN5: Barn 14Bb 31
Gore Cl. UB9: Hare 28K 43
Gore Cotts. DA2: Daren 62Sd 142
Gore Ct. NW9 29Qa 47
Gorefield Ho. NW6 40Cb 69
(off Gorefield Pl.)
Gorefield Pl. NW6 40Cb 69
Gore Ho. N1 38Qb 70
(off Drummond Way)
Gorelands La. HP8: Chal G 18A 24
Gore Rd. DA2: Dart 61Sd 142
Gore Rd. E9 39Yb 72
Gore Rd. SW20 68Ya 132
GORESBROOK INTERCHANGE 40Bd 75
Goresbrook Rd. RM9: Dag 39Xc 75
Gore St. SW7 3A 226 (48Eb 89)
Gorhambury House 1J 5
Gorham Dr. AL1: St A 5C 6
Gorham Ho. SE16 47Zb 92
(off Wolfe Cres.)
Gorham Pl. W11 45Ab 88
Goring Cl. RM5: Col R 25Ed 56
Goring Gdns. RM8: Dag 35Yc 75
Goring Mdw. TN15: Wro 88Be 185
Goring Pl. TN15: Wro 88Ce 185
Goring Rd. N11 23Nb 50
Goring Rd. RM10: Dag 37Fd 76
Goring Rd. TW18: Staines 64F 126
Gorings Sq. TW18: Staines 63G 126
Goring St. EC3 2J 225 (44Ub 91)
(off Houndsditch)
Goring Way UB6: G'frd 40Ea 66
Gorle Cl. WD25: Wat 7W 12
Gorleston Rd. N15 29Tb 51
Gorleston St. W14 49Ab 88
(not continuous)
Gorman Rd. SE18 49Pc 94
Gorringe Av. DA4: S Dar 68Td 142
Gorringe Pk. Av. CR4: Mitc 66Hb 133
Gorse Cl. AL10: Hat 3B 8
Gorse Cl. E16 44Jc 93
Gorse Cl. KT20: Tad 92Xa 194
Gorse Cnr. AL3: St A 1B 6
Gorsefield Ho. E14 45Cc 92
(off E. India Dock Rd.)
Gorse Hill DA4: Farni 73Qd 163
Gorse Hill GU22: Wok 3K 187
Gorse Hill La. GU25: Vir W 10P 125
Gorse Hill Rd. GU25: Vir W 10P 125
Gorselands Cl. KT14: W Byf 83L 169
Gorse La. GU24: Chob 10J 147
Gorse Meade SL1: Slou 6F 80
Gorse Ri. SW17 64Jb 134
Gorse Rd. BR5: St M Cry 75Cd 162
Gorse Rd. CR0: C'don 77Cc 158
Gorse Wlk. UB7: Yiew 44N 83
Gorse Way DA3: Hartl 71Be 165
Gorseway RM7: Rush G 32Gd 76
Gorsewood Rd. DA3: Hartl 71Be 165
(not continuous)
Gorsewood Rd. DA3: Lfield 71Be 165
(not continuous)
Gorsewood Rd. GU21: Wok 1H 187
Gorst Rd. NW10 42Sa 67
Gorst Rd. SW11 58Hb 111
Gorsuch Pl. E2 3K 219 (41Vb 91)
Gorsuch St. E2 2K 219 (40Vb 71)
Gosberton Rd. SW12 60Hb 111
Gosbury Hill KT9: Chess 77Na 153
Gosden Rd. GU24: W End 5D 166
Gosfield Rd. KT19: Eps 84Ta 173
Gosfield Rd. RM8: Dag 33Cd 76
Gosfield St. W1 7B 216 (43Lb 90)
Gosford Gdns. IG4: Ilf 29Pc 54
Gosford Ho. E3 40Bc 72
(off Tredegar Rd.)
Gosford Ho. WD18: Wat 16U 26
Gosforth La. WD19: Wat 20W 26
Gosforth Path WD19: Wat 20W 26
Gosforth Pl. SL1: Slou 5D 80
Goshawk Ct. NW9 31Va 68
Goshawk Gdns. UB4: Hayes 41U 84
Goslar Way SL4: Wind 4F 102
Goslett Ct. WD23: Bush 15Ca 27
Goslett Yd. WC2 3E 222 (44Mb 90)
Gosling Cl. UB6: G'frd 41Ca 85
Gosling Grn. SL3: L'ly 48A 82
Gosling Ho. E1 45Yb 92
(off Sutton St.)
Gosling Rd. SL3: L'ly 48A 82
Gosling Way SW9 53Qb 112
Gospatrick Rd. N17 24Sb 51
GOSPEL OAK 35Jb 70
Gospel Oak Station (Overground) 35Jb 70
Gosport Dr. RM12: Horn 37Ld 77
Gosport Rd. E17 29Bc 52
Gosport Wlk. N17 28Xb 51
Gossage Rd. SE18 50Tc 94
Gossage Rd. UB10: Uxb 38P 63
Gossamer Gdns. E2 40Xb 71
Gossamers, The WD25: Wat 6Aa 13
Gosse Ct. N1 39Ub 71
(off Downham Rd.)
Gosset St. E2 3K 219 (41Vb 91)
Goss Hill BR8: Swan 65Ld 141
Gosshill Rd. BR7: Chst 68Qc 138
Gossington Cl. BR7: Chst 63Rc 138
Gosterwood St. SE8 51Ac 114
Gostling Rd. TW2: Whitt 60Ca 107
Goston Gdns. CR7: Thor H 69Qb 134
Goston Ga. SW8 53Pb 112
(off Hampson Way)
Goswell Hill SL4: Wind 3H 103
Goswell Pl. EC1 4C 218 (41Rb 91)
(off Goswell Rd.)
Goswell Rd. EC1 2B 218 (40Rb 71)
Goswell Rd. SL4: Wind 3H 103
Gothenburg Ct. SE8 49Ac 92
(off Bailey St.)
Gothic Cl. DA1: Dart 62Md 141
Gothic Cotts. EN2: Enf 12Sb 33
(off Chase Grn. Av.)
Gothic Ct. SE5 52Sb 113
(off Wyndham Rd.)
Gothic Ct. UB3: Harl 51T 106
Gothic Rd. TW2: Twick 61Fa 130
Gottfried M. NW5 35Lb 70
Goudhurst Rd. BR1: Broml 64Gc 137
Gouge Av. DA11: Nflt 10A 122
Gough Ho. KT1: King T 68Na 131
(off Eden St.)
Gough Ho. N1 39Rb 71
(off Windsor St.)
Gough Rd. E15 35Hc 73
Gough Rd. EN1: Enf 12Xb 33
Gough Sq. EC4 2A 224 (44Qb 90)
Gough St. WC1 6J 217 (42Pb 90)
Gough Wlk. E14 44Cc 92
Gould Ct. AL9: Wel G 6D 8
Goulden Ho. SW11 54Gb 111
Goulden Ho. App. SW11 54Gb 111
Goulding Gdns. CR7: Thor H 68Sb 135
Gouldman Ho. E1 42Yb 92
(off Wyllen Cl.)
Gould Rd. TW14: Felt 59U 106
Gould Rd. TW2: Twick 60Ga 108
Goulds Cotts. RM4: Abr 13Xc 37
Gould's Grn. UB8: Hil 45R 84
GOULDS GREEN 43R 84
Gould Ter. E8 36Xb 71
Gould Way HA8: Edg 24Ra 47
Goulston St. E1 2K 225 (44Vb 91)
Goulton Rd. E5 35Xb 71
Gourley Pl. N15 29Ub 51
Gourley St. N15 29Ub 51
Gourney Gro. RM16: Grays 45De 99
Gourock Rd. SE9 57Qc 116
Govan St. E2 39Wb 71
Gover Ct. SW4 54Nb 112
Gover Hill TN11: Roug 100Fe 205
GOVER HILL 100Fe 205
Government Row EN3: Enf L 10Cc 20
Govett Av. TW17: Shep 71S 150
Govett Gro. GU20: W'sham 8B 146
Govier Cl. E15 38Gc 73
Gowan Av. SW6 53Ab 110
Gowan Ho. E2 4K 219 (41Vb 91)
(off Chambord St.)
Gowan Rd. NW10 37Xa 68
Gowar Fld. EN6: S Mim 4Wa 16
Gower, The TW20: Thorpe 69D 126
Gower Cl. SW4 58Lb 112
Gower Ct. WC1 5D 216 (42Mb 90)
Gower Ho. E17 27Dc 52
Gower Ho. KT13: Weyb 79T 150
Gower Ho. SE17 7E 230 (50Sb 91)
(off Morecambe St.)
Gower Lodge KT13: Weyb 79T 150
(off St George's Rd.)
Gower M. WC1 1D 222 (43Mb 90)
Gower M. Mans. WC1 7E 216 (43Mb 90)
(off Gower M.)
Gower Pl. RM16: Chaf H 48Xd 98
Gower Pl. WC1 5D 216 (42Mb 90)
Gower Rd. E7 37Jc 73
Gower Rd. KT13: Weyb 79T 150
Gower Rd. TW7: Isle 51Ha 108
Gowers La. RM16: Ors 7B 100
Gower St. WC1 5C 216 (42Lb 90)
Gower's Wlk. E1 44Wb 91
Gowings Grn. SL1: Slou 7C 80
Gowland Pl. BR3: Beck 68Bc 136
Gowlett Rd. SE15 55Wb 113
Gowlland Cl. CR0: C'don 73Wb 157
Gowrie Pl. CR3: Cat'm 94Sb 197
Gowrie Rd. SW11 55Jb 112
Grabex Bus. Cen. BR5: St P 69Xc 139
Graburn Way KT8: E Mos 69Fa 130
Grace Av. DA7: Bex 54Bd 117
Grace Av. WD7: Shenl 5Ma 15
Grace Bus. Cen. CR4: Mitc 72Hb 155
Gracechurch St. EC3 4G 225 (45Tb 91)
Grace Cl. HA8: Edg 24Sa 47
Grace Cl. IG6: Ilf 23Vc 55
Grace Cl. SE9 62Mc 137
Grace Cl. WD6: Bore 11Ta 29
Grace Ct. CR0: C'don 76Rb 157
(off Waddon Rd.)
Grace Ct. SL1: Burn 2A 80
Grace Ct. SL1: Slou 6G 80
Grace Ct. SM2: Sutt 81Db 175
Gracedale Rd. SW16 64Kb 134
Gracefield Gdns. SW16 62Nb 134
Gracehill E1 43Yb 92
(off Hannibal Rd.)
Grace Ho. SE11 51Pb 112
(off Vauxhall St.)
Grace Jones Cl. E8 37Wb 71
Grace M. BR3: Beck 65Cc 136
Grace M. SE20 68Yb 136
(off Marlow Rd.)
Grace Path SE26 63Yb 136
Grace Pl. E3 41Dc 92
Grace Rd. CR0: C'don 72Sb 157
Graces All. E1 45Wb 91
Grace's M. SE5 54Tb 113
Graces M. NW8 2A 214 (40Eb 69)
Grace's Rd. SE5 54Ub 113
Grace St. E3 41Dc 92
Gracious Pond Rd. GU24: Chob 10L 147
Grade Cl. WD6: E'tree 14Qa 29
Gradient, The SE26 63Wb 135
Graduate Pl. SE1 3H 231 (48Ub 91)
(off Long La.)
Graeme Rd. EN1: Enf 12Tb 33
Graemesdyke Av. SW14 55Ra 109
Graftonbury M. EN2: Crew H 6Rb 19
Grafton Chambers NW1 4E 216 (41Mb 90)
(off Grafton Pl.)
Grafton Cl. AL4: St A 3H 7
Grafton Cl. KT14: W Byf 85H 169
Grafton Cl. KT4: Wor Pk 76Ua 154
Grafton Cl. SL3: Geor G 44A 82
Grafton Cl. TW4: Houn 60Aa 107
Grafton Cl. W13 44Ja 86
Grafton Ct. TW14: Bedf 60T 106
Grafton Cres. NW1 37Kb 70
Grafton Gdns. N4 30Sb 51
Grafton Gdns. RM8: Dag 33Ad 75
Grafton Ho. E3 41Cc 92
(off Wellington Way)
Grafton Ho. SE8 50Bc 92
Grafton M. W1 6B 216 (42Lb 90)
Grafton Pk. Rd. KT4: Wor Pk 75Ua 154
Grafton Pl. NW1 4E 216 (41Mb 90)
Grafton Rd. CR0: C'don 74Qb 156
Grafton Rd. EN2: Enf 13Pb 32
Grafton Rd. HA1: Harr 29Ea 46
Grafton Rd. KT3: N Mald 69Ua 132
Grafton Rd. KT4: Wor Pk 76Ta 153
Grafton Rd. NW5 36Jb 70
Grafton Rd. RM8: Dag 33Ad 75
Grafton Rd. W3 45Sa 87
Graftons, The NW2 34Cb 69
Grafton Sq. SW4 55Lb 112
Grafton St. W1 5A 222 (45Kb 90)
Grafton Ter. NW5 36Hb 69
Grafton Way KT8: W Mole 70Ba 129
Grafton Way W1 6B 216 (42Lb 90)
(not continuous)
Grafton Way WC1 6C 216 (42Lb 90)
Grafton Yd. NW5 37Kb 70
Graham Av. CR4: Mitc 67Jb 134
Graham Av. W13 47Ka 86
Graham Cl. AL1: St A 4B 6
Graham Cl. CM13: Hut 15Ee 41
Graham Cl. CR0: C'don 75Cc 158
Graham Ct. AL3: St A 1B 6
(off Grange St.)
Graham Ct. SE14 51Zb 114
(off Myers La.)
Graham Ct. UB5: N'olt 36Aa 65
Grahame Ho. RH1: Redh 4N 207
GRAHAME PARK 25Ua 48
Grahame Pk. Way NW7 24Va 48
Grahame Pk. Way NW9 26Va 48
Grahame White Ho. HA3: Kenton 27Ma 47
Graham Gdns. KT6: Surb 74Na 153
Graham Ho. KT23: Bookh 96Ba 191
Graham Ho. N9 18Yb 34
(off Cumberland Rd.)
Graham Lodge NW4 30Xa 48
Graham Mans. IG11: Bark 38Wc 75
(off Lansbury Av.)
Graham Rd. CR4: Mitc 67Jb 134
Graham Rd. CR8: Purl 85Qb 176
Graham Rd. DA6: Bex 56Bd 117
Graham Rd. E13 42Jc 93
Graham Rd. E8 37Wb 71
Graham Rd. GU20: W'sham 9A 146
Graham Rd. HA3: W'stone 27Ga 46
Graham Rd. N15 27Rb 51
Graham Rd. NW4 30Xa 48
Graham Rd. SW19 66Bb 133
Graham Rd. TW12: Hamp H 63Ca 129
Graham Rd. W4 48Ta 87
Graham St. N1 2C 218 (40Rb 71)
Graham Ter. DA15: Sidc 58Xc 117
(off Westerham Dr.)
Graham Ter. SW1 6H 227 (49Jb 90)
Grail Rd. SE6 61Gc 137
Grainger Cl. UB5: N'olt 36Da 65
Grainger Ct. SE5 52Sb 113
Grainger Rd. N22 25Sb 51
Grainger Rd. TW7: Isle 54Ha 108
Grainges Yd. UB8: Uxb 38L 63
Grainstore, The E16 45Jc 93
Gramer Cl. E11 33Fc 73
Gramophone La. UB3: Hayes 47U 84
Grampian Cl. BR6: St M Cry 72Vc 161
Grampian Cl. SM2: Sutt 80Eb 155
Grampian Cl. UB3: Harl 52T 106
Grampian Gdns. NW2 32Ab 68
Grampians, The W6 47Za 88
(off Shepherd's Bush Rd.)
Grampian Way SL3: L'ly 50C 82
Gramsci Way SE6 62Dc 136
Granada St. SW17 64Hb 133
Granard Av. SW15 57Xa 110
Granard Bus. Cen. NW7 23Ua 48
Granard Ho. E9 37Zb 72
Granard Rd. SW12 59Hb 111
Granaries, The EN9: Walt A 6Gc 21
Granary Cl. N9 17Yb 34
Granary Ct. E15 37Fc 73
(off Millstone Cl.)
Granary Mans. SE28 47Sc 94
Granary Rd. E1 42Xb 91
Granary Sq. N1 1F 217 (39Nb 70)
Granary St. NW1 39Mb 70
Granby Pk. Rd. EN7: Chesh 1Vb 19
Granby Pl. SE1 2K 229 (47Qb 90)
(off Lwr. Marsh)
Granby Rd. DA11: Nflt 58Ee 121
Granby Rd. SE9 54Pc 116
Granby St. E2 42Wb 91
(not continuous)
Granby Ter. NW1 2B 216 (40Lb 70)
Grand Arc. N12 22Eb 49
Grand Av. EC1 7C 218 (43Rb 91)
(not continuous)
Grand Av. HA9: Wemb 36Qa 67
Grand Av. KT5: Surb 71Ra 153
Grand Av. N10 28Jb 50
Grand Av. E. HA9: Wemb 36Ra 67
Grand Canal Apts. N1 39Ub 71
(off De Beauvoir Cres.)
Grand Canal Av. SE16 49Ac 92
Grand Connaught Rooms 2H 223 (44Pb 90)
(off Gt. Queen St.)
Grand Courts RM8: Dag 34Ad 75
Grand Depot Rd. SE18 50Qc 94

Grand Dr. SW20 68Ya 132
Grand Dr. UB2: S'hall 47Ea 86
Granden Rd. SW16 68Nb 134
Grandfield Av. WD17: Wat 11V 26
Grandfield Ct. W4 51Ta 109
Grandis Cotts. GU23: Rip 94K 189
Grandison Rd. KT4: Wor Pk 75Ya 154
Grandison Rd. SW11 57Hb 111
Grand Junc. Pl. UB8: Uxb 40K 63
Grand Junc. Wharf E2 41Zb 92
Grand Junc. Wharf N1 2D 218 (40Sb 71)
Grand Pde. HA9: Wemb 33Qa 67
Grand Pde. KT6: Surb 74Qa 153
Grand Pde. N4 29Rb 51
Grand Pde. SW14 56Sa 109
(off Up. Richmond Rd. W.)
Grand Pde. M. SW15 57Ab 110
Grand Regent Twr. E2 41Zb 92
(off Palmer's Rd.)
Grandstand Rd. KT17: Eps D 89Va 174
Grandstand Way UB5: N'olt 36Ba 65
Grand Twr. SW15 57Ab 110
(off Plaza Gdns.)
Grand Union Cen. W10 42Za 88
(off West Row)
Grand Union Cl. W9 43Bb 89
Grand Union Cres. E8 39Wb 71
Grand Union Ent. Pk. UB2: S'hall 48Ca 85
Grand Union Hgts. HA0: Wemb 39Ma 67
Grand Union Ho. N1 39Ub 71
(off Hertford Rd.)
Grand Union Ho. SL0: Iver 45H 83
Grand Union Ind. Est. NW10 40Ra 67
Grand Union Village UB5: N'olt 41Ba 85
Grand Union Wlk. NW1 38Kb 70
(off Kentish Town Rd.)
Grand Union Way UB2: S'hall 47Ca 85
Grand Union Way WD4: K Lan 1R 12
Grand Vw. Av. TN16: Big H 89Lc 179
Grand Vitesse Ind. Cen. SE1 7C 224 (46Rb 91)
(off Gt. Suffolk St.)
Grand Wlk. E1 42Ac 92
Granfield St. SW11 53Fb 111
Grange, The N20: Grangeview Rd 18Eb 31
Grange, The N20: Oxford Gdns. 18Fb 31
Grange, The AL4: Col H 5P 7
Grange, The CR0: C'don 75Bc 158
Grange, The DA4: S Dar 67Td 142
Grange, The E17 29Ac 52
(off Lynmouth Rd.)
Grange, The EN9: Walt A 9Fc 21
Grange, The GU24: Chob 2J 167
Grange, The GU25: Vir W 70A 126
(off Holloway Dr.)
Grange, The HA0: Wemb 38Qa 67
Grange, The KT12: Walt T 75X 151
Grange, The KT3: N Mald 71Va 154
Grange, The KT4: Wor Pk 77Ta 153
Grange, The N2 26Fb 49
Grange, The SE1 3K 231 (48Vb 91)
Grange, The SL1: Burn 1A 80
(off Green La.)
Grange, The SL4: Old Win 7M 103
Grange, The SW19 65Za 132
Grange, The TN15: W King 81Vd 184
Grange, The W13 43La 86
Grange, The W14 49Bb 89
Grange, The W3 47Ra 87
Grange, The W4 50Ra 87
Grange, The WD3: Rick 17M 25
Grange, The WD5: Ab L 3U 12
Grange Av. EN4: E Barn 18Gb 31
Grange Av. HA7: Stan 26Ka 46
Grange Av. IG8: Wfd G 23Jc 53
Grange Av. N12 22Eb 49
Grange Av. N20 17Ab 30
Grange Av. SE25 68Ub 135
Grange Av. TW2: Twick 61Ga 130
Grangecliffe Gdns. SE25 68Ub 135
Grange Cl. CM13: Ingve 22Ee 59
Grange Cl. CR5: Chips 92Hb 195
Grange Cl. DA15: Sidc 62Wc 139
Grange Cl. HA8: Edg 22Sa 47
Grange Cl. HP2: Hem H 3A 4
Grange Cl. IG8: Wfd G 24Jc 53
Grange Cl. KT22: Lea 92Ma 193
Grange Cl. KT8: W Mole 70Da 129
Grange Cl. RH1: Blet 5K 209
Grange Cl. RH1: Mers 100Kb 196
Grange Cl. SL9: Chal P 25A 42
Grange Cl. TN16: Westrm 98Sc 200
Grange Cl. TW19: Wray 58A 104
Grange Cl. TW5: Hest 51Ba 107
Grange Cl. UB3: Hayes 43U 84
Grange Cl. WD17: Wat 11W 26
Grange Ct. AL3: St A 1B 6
(not continuous)
Grange Ct. EN9: Walt A 6Ec 20
Grange Ct. HA1: Harr 35Ha 66
Grange Ct. HA5: Pinn 27Aa 45
Grange Ct. IG10: Lough 15Mc 35
Grange Ct. KT12: Walt T 75W 150
Grange Ct. NW10 34Ua 68
(off Neasden La.)
Grange Ct. RH1: Mers 100Kb 196
Grange Ct. RH9: S God 10C 210
Grange Ct. SM2: Sutt 80Db 155
Grange Ct. SM6: W'gton 76Kb 156
Grange Ct. TW17: Shep 70Q 128
Grange Ct. TW18: Staines 64J 127
Grange Ct. TW20: Egh 64B 126
Grange Ct. UB5: N'olt 40Y 65
Grange Ct. WC2 3J 223 (44Pb 90)
Grangecourt Rd. N16 32Ub 71
Grange Cres. DA2: Dart 58Rd 119
Grange Cres. IG7: Chig 22Tc 54
Grange Cres. SE28 44Yc 95
Grangedale Cl. HA6: Nwood 25U 44
Grange Dr. BR6: Prat B 81Yc 181
Grange Dr. BR7: Chst 65Nc 138
Grange Dr. GU21: Wok 87A 168
Grange Dr. RH1: Mers 100Kb 196
Grange Farm Cl. HA2: Harr 33Ea 66
Grangefield NW1 38Mb 70
(off Marquis Rd.)
Grange Flds. SL9: Chal P 25A 42
Grangefields Rd. GU4: Jac W 10P 187
Grange Gdns. HA5: Pinn 27Aa 45
Grange Gdns. N14 18Mb 32
Grange Gdns. NW3 34Db 69
Grange Gdns. SE25 68Ub 135
Grange Gdns. SL2: Farn C 6H 61
Grange Gdns. SM7: Bans 85Db 175
Grange Gro. N1 37Sb 71
Grange Hill HA8: Edg 22Sa 47
GRANGE HILL 23Tc 54
Grange Hill SE25 68Ub 135
Grange Hill TN15: Plax 99Ae 205
Grangehill Pl. SE9 55Pc 116
Grangehill Rd. SE9 56Pc 116
Grange Hill Station (Underground) 21Tc 54
Grange Ho. DA11: Grav'nd 9C 122
Grange Ho. DA8: Erith 54Jd 118
Grange Ho. NW10 38Xa 68
Grange Ho. SE1 4K 231 (48Vb 91)
Grange La. DA3: Hartl 73Ce 165
Grange La. SE21 61Vb 135
Grange La. WD25: Let H 11Fa 28
Grange Lodge SW19 65Za 132
Grange Mans. KT17: Ewe 80Va 154
Grange Mdw. SM7: Bans 85Db 175
Grange M. N21 16Rb 33
Grange M. TW13: Felt 63W 128
Grangemill Rd. SE6 62Cc 136
Grangemill Way SE6 61Cc 136
Grangemount KT22: Lea 92Ma 193
Grange Pk. GU21: Wok 87A 168
GRANGE PARK 16Rb 33
Grange Pk. W5 46Na 87
Grange Pk. Av. N21 16Sb 33
Grange Pk. Pl. SW20 66Xa 132
Grange Pk. Rd. CR7: Thor H 70Tb 135
Grange Pk. Rd. E10 32Dc 72
Grange Park Station (Rail) 15Rb 33
Grange Pl. KT12: Walt T 75W 150
Grange Pl. NW6 38Cb 69
Grange Pl. TW18: Lale 68L 127
Granger Ct. WD6: Bore 14Qa 29
(off Whitehall Cl.)
Grange Rd. BR6: Orp 75Tc 160
Grange Rd. CR2: S Croy 82Sb 177
Grange Rd. CR3: Cat'm 97Wb 197
Grange Rd. CR7: Thor H 70Tb 135
Grange Rd. DA11: Grav'nd 9C 122
Grange Rd. E10 32Cc 72
Grange Rd. E13 41Hc 93
Grange Rd. E17 29Ac 52
(not continuous)
Grange Rd. GU2: Guild 10M 187
Grange Rd. GU21: Wok 86A 168
Grange Rd. GU24: Pirb 4A 186
Grange Rd. HA1: Harr 29Ja 46
Grange Rd. HA2: Harr 33Fa 66
Grange Rd. HA8: Edg 23Ta 47
Grange Rd. IG1: Ilf 35Rc 74
Grange Rd. KT1: King T 69Na 131
Grange Rd. KT12: Hers 77Aa 151
Grange Rd. KT15: New H 82J 169
Grange Rd. KT22: Lea 92Ma 193
Grange Rd. KT8: W Mole 70Da 129
Grange Rd. KT9: Chess 77Na 153
Grange Rd. N17 23Wb 51
Grange Rd. N18 23Wb 51
Grange Rd. N6 30Jb 50
Grange Rd. NW10 37Xa 68
Grange Rd. RM15: Avel 46Sd 98
Grange Rd. RM17: Grays 51De 121
Grange Rd. RM3: Rom 23Kd 57
Grange Rd. SE1 4J 231 (48Ub 91)
Grange Rd. SE19 68Tb 135
Grange Rd. SE25 69Tb 135
Grange Rd. SL9: Chal P 25A 42
Grange Rd. SM2: Sutt 80Cb 155
Grange Rd. SW13 53Wa 110
Grange Rd. TN13: S'oaks 99Jd 202
Grange Rd. TN15: Plat 92Ee 205
Grange Rd. TW20: Egh 64B 126
Grange Rd. UB1: S'hall 47Aa 85
Grange Rd. UB3: Hayes 44U 84
Grange Rd. W4 50Ra 87
Grange Rd. W5 46Ma 87
Grange Rd. WD23: Bush 15Aa 27
Grange Rd. WD6: E'tree 15Pa 29
Granger Way RM1: Rom 30Jd 56
Grange St. AL3: St A 1B 6
Grange St. N1 1G 219 (39Tb 71)
Grange St. M. AL3: St A 1B 6
Grange Va. SM2: Sutt 80Db 155
Grange Vw. Rd. N20 18Eb 31
Grange Wlk. SE1 3J 231 (48Ub 91)
Grange Wlk. M. SE1 4J 231 (48Ub 91)
(off Grange Wlk.)
Grange Way DA3: Hartl 72Be 165
Grange Way DA8: Erith 52Kd 119
Grange Way SL0: Iver 44H 83
Grangeway IG8: Wfd G 21Lc 53
Grangeway N12 21Db 49
Grangeway NW6 38Cb 69
Grangeway, The N21 16Rb 33
Grangeway Gdns. IG4: Ilf 29Nc 54
Grangeways Cl. DA11: Nflt 3B 144
Grangewick Rd. RM16: Grays 8A 100
Grangewood DA5: Bexl 60Bd 117
Grangewood EN6: Pot B 2Db 17
Grangewood SL3: Wex 3N 81
Grangewood Av. RM13: Rain 42Ld 97
Grangewood Av. RM16: Grays 8A 100
Grangewood Cl. CM13: B'wood 20Be 41
Grangewood Cl. HA5: Eastc 29W 44
Grangewood Dr. TW16: Sun 66V 128
Grangewood La. BR3: Beck 65Bc 136
Grangewood St. E6 39Mc 73
Grangewood Ter. SE25 68Tb 135
Grange Yd. SE1 4K 231 (48Vb 91)
Granham Gdns. N9 19Vb 33
Granite Apts. E15 37Gc 73
Granite Apts. SE10 50Gc 93
Granite St. SE18 50Vc 95
Granleigh Rd. E11 33Gc 73
Gransden Av. E8 38Xb 71
Gransden Ho. SE8 50Bc 92
Gransden Rd. W12 47Va 88
Grant Av. KT12: Walt T 71Y 151
Grant Av. SL1: Slou 4J 81
Grantbridge St. N1 1C 218 (40Rb 71)
Grantchester KT1: King T 68Qa 131
(off St Peters Rd.)
Grantchester Cl. HA1: Harr 34Ha 66
Grant Cl. DA17: Belv 50Bd 95
Grant Cl. N14 17Lb 32
Grant Cl. N17 26Ub 51
Grant Cl. TW17: Shep 72R 150
Grant Ct. E4 18Ec 34
(off The Ridgeway)
Grant Ct. NW9 26Va 48
(off Hazel Cl.)
Grantham Cl. HA8: Edg 20Na 29
Grantham Ct. KT2: King T 64Ma 131
Grantham Ct. RM6: Chad H 31Bd 75
Grantham Ct. SE16 47Zb 92
(off Eleanor Cl.)
Grantham Gdns. RM6: Chad H 30Bd 55
Grantham Grn. WD6: Bore 15Sa 29
Grantham Ho. E14 44Gc 93
Grantham Ho. SE15 51Wb 113
(off Friary Est.)
Grantham Ho. TW16: Sun 66U 128
Grantham Ho. UB5: N'olt 41Ba 85
(off Taywood Rd.)
Grantham M. HP4: Berk 1A 2
Grantham Pl. W1 7K 221 (46Kb 90)
Grantham Rd. E12 35Qc 74
Grantham Rd. SW9 54Nb 112
Grantham Rd. W4 52Ua 110
Grantham Way RM16: Grays 46Ce 99
Grant Ho. E17 28Cc 52
(off High St.)
Grant Ho. SW9 53Pb 112
(off Liberty St.)
Grantley Ho. SE14 51Zb 114
(off Myers La.)
Grantley Pl. KT10: Esh 78Ea 152
Grantley Rd. TW4: Cran 54Y 107
Grantley St. E1 41Zb 92
Grant Mus. of Zoology 6D 216 (42Mb 90)
Grantock Rd. E17 25Fc 53
Granton Av. RM14: Upm 34Pd 77
Granton Rd. DA14: Sidc 65Yc 139
Granton Rd. IG3: Ilf 32Wc 75
Granton Rd. SW16 67Lb 134
Grant Pl. CR0: C'don 74Vb 157
Grant Rd. CR0: C'don 74Vb 157
Grant Rd. HA3: W'stone 27Ha 46
Grant Rd. SW11 56Fb 111
Grants Cl. NW7 24Ya 48
Grants La. RH8: Limp 4N 211
Grants La. TN8: Eden 9N 211
Grants La. TN8: Limp 9N 211
Grants Quay Wharf EC3 5G 225 (45Tb 91)
Grant St. E13 41Jc 93
Grant St. N1 1K 217 (40Qb 70)
Grant Ter. N16 31Wb 71
(off Castlewood Rd.)
Grantully Rd. W9 41Db 89
Grant Wlk. SL5: S'dale 4C 146
Grant Way TW7: Isle 51Ja 108
Grantwood Cl. RH1: Redh 10A 208
Granville Arc. SW9 56Qb 112
Granville Av. N9 20Yb 34
Granville Av. SL2: Slou 3H 81
Granville Av. TW13: Felt 61W 128
Granville Av. TW3: Houn 57Ca 107
Granville Cl. CR0: C'don 75Ub 157
Granville Cl. KT13: Weyb 79S 150
Granville Cl. KT14: Byfl 85P 169
Granville Ct. AL1: St A 2D 6
(off Granville Rd.)
Granville Ct. N1 39Ub 71
Granville Ct. N4 30Pb 50
Granville Ct. SE14 52Ac 114
(off Nynehead St.)
Granville Dene HP3: Bov 9C 2
Granville Gdns. SW16 67Pb 134
Granville Gdns. W5 46Pa 87
Granville Gro. SE13 55Ec 114
Granville Ho. E14 44Cc 92
(off E. India Dock Rd.)
Granville Mans. W12 47Ya 88
(off Shepherd's Bush Grn.)
Granville M. DA14: Sidc 63Wc 139
Granville Pk. SE13 55Ec 114
Granville Pl. HA5: Pinn 27Z 45
Granville Pl. N12 24Eb 49
Granville Pl. SW6 52Db 111
Granville Pl. W1 3H 221 (44Jb 90)
Granville Point NW2 33Bb 69
Granville Rd. AL1: St A 2D 6
Granville Rd. CM16: Epp 1Xc 23
Granville Rd. DA11: Grav'nd 9B 122
Granville Rd. DA14: Sidc 63Wc 139
Granville Rd. DA16: Well 55Yc 117
Granville Rd. E17 30Dc 52
Granville Rd. E18 26Kc 53
Granville Rd. EN5: Barn 14Ya 30
Granville Rd. GU22: Wok 92B 188
Granville Rd. IG1: Ilf 32Rc 74
Granville Rd. KT13: Weyb 80S 150
Granville Rd. N12 24Eb 49
Granville Rd. N13 23Pb 50
Granville Rd. N22 25Rb 51
Granville Rd. N4 30Pb 50
Granville Rd. NW2 33Bb 69
Granville Rd. NW6 40Cb 69
(not continuous)
Granville Rd. RH8: Oxt 1K 211
Granville Rd. SW18 59Bb 111
Granville Rd. SW19 66Cb 133
Granville Rd. TN13: S'oaks 96Jd 202
Granville Rd. TN16: Westrm 98Sc 200
Granville Rd. UB10: Hil 37R 64
Granville Rd. UB3: Harl 49V 84
Granville Rd. WD18: Wat 14Y 27
Granville Sq. SE15 52Ub 113
Granville Sq. WC1 4J 217 (41Pb 90)
Granville St. WC1 4J 217 (41Pb 90)
Granwood Ct. TW7: Isle 53Ga 108
Grape St. WC2 2F 223 (44Nb 90)
Graphic Ho. WD24: Wat 10W 12
Graphite Apts., The N1 2F 219 (40Tb 71)
(off Provost St.)
Graphite Point E2 41Zb 92
(off Palmer's Rd.)
Graphite Sq. SE11 7H 229 (50Pb 90)
Grapsome Cl. KT9: Chess 80La 152
Grasdene Rd. SE18 52Wc 117
Grasholm Way SL3: L'ly 49E 82
Grasmere NW1 4A 216 (41Kb 90)
(off Osnaburgh St.)
Grasmere SL4: Wind 2C 102
Grasmere Av. BR6: Farnb 76Rc 160
Grasmere Av. HA4: Ruis 31S 64
Grasmere Av. HA9: Wemb 31La 66
Grasmere Av. SL2: Slou 5L 81
Grasmere Av. SW15 63Ta 131
Grasmere Av. SW19 69Cb 133
Grasmere Av. TW3: Houn 58Da 107
Grasmere Av. W3 45Ta 87
Grasmere Cl. HP3: Hem H 4B 4
Grasmere Cl. IG10: Lough 12Pc 36
Grasmere Cl. TW14: Felt 60V 106
Grasmere Cl. TW20: Egh 66D 126
Grasmere Cl. WD25: Wat 4X 13
Grasmere Ct. N22 23Pb 50
Grasmere Ct. SE26 64Wb 135
Grasmere Ct. SM2: Sutt 79Eb 155
Grasmere Ct. SW13 51Wa 110
(off Verdun Rd.)
Grasmere Gdns. BR6: Farnb 76Rc 160
Grasmere Gdns. HA3: W'stone 26Ja 46
Grasmere Gdns. IG4: Ilf 29Pc 54
Grasmere Pde. SL2: Slou 5M 81
Grasmere Point SE15 52Yb 114
(off Old Kent Rd.)
Grasmere Rd. AL1: St A 4F 6
Grasmere Rd. BR1: Broml 67Hc 137
Grasmere Rd. BR6: Farnb 76Rc 160
Grasmere Rd. CR8: Purl 83Rb 177
Grasmere Rd. DA7: Bex 54Ed 118
Grasmere Rd. E13 40Jc 73
Grasmere Rd. N10 25Kb 50
Grasmere Rd. N17 23Wb 51
Grasmere Rd. SE25 72Xb 157
Grasmere Rd. SW16 64Nb 134
Grasmere Way KT14: Byfl 84P 169
Grassbanks DA1: Dart 60Pd 119
Grassfield Cl. CR5: Coul 91Kb 196
Grasshaven Way SE28 46Vc 95
(not continuous)
Grassingham End SL9: Chal P 24A 42
Grassingham Rd. SL9: Chal P 24A 42
Grassington Cl. AL2: Brick W 2Ca 13
Grassington Cl. N11 23Jb 50
Grassington Rd. DA14: Sidc 63Wc 139
Grassmere Rd. RM11: Horn 28Pd 57
Grassmount CR8: Purl 82Lb 176
Grassmount SE23 61Xb 135
Grass Pk. N3 25Bb 49
Grass Rd. RM18: E Til 1K 123
Grassway SM6: W'gton 77Lb 156
Grassy Cl. HP1: Hem H 1J 3
Grassy La. TN13: S'oaks 98Kd 203
Grasvenor Av. EN5: Barn 15Cb 31
Gratton Dr. SL4: Wind 6C 102
Gratton Rd. W14 48Ab 88
Gratton Ter. NW2 34Za 68
Gravel Cl. IG7: Chig 19Wc 37
Graveley Av. WD6: Bore 14Sa 29
Gravel Hill CR0: Addtn 79Zb 158
Gravel Hill DA6: Bex 56Dd 118
Gravel Hill HP1: Hem H 2J 3
Gravel Hill IG10: H Beech 10Jc 21
Gravel Hill KT22: Lea 93Ka 192
Gravel Hill N3 26Bb 49
Gravel Hill SL9: Chal P 23A 42
GRAVEL HILL 23B 42
Gravel Hill UB8: Uxb 36M 63
Gravel Hill Cl. DA6: Bex 57Dd 118
Gravel Hill Stop (London Tramlink) 79Ac 158
Gravelhill Ter. HP1: Hem H 3J 3
Gravel La. E1 2K 225 (44Vb 91)
Gravel La. HP1: Hem H 2J 3
Gravel La. IG7: Chig 15Vc 37
Gravelly Hill CR3: Cat'm 100Ub 197
Gravel Path HP1: Hem H 2J 3
Gravel Path HP4: Berk 1A 2
Gravel Pit La. SE9 57Rc 116
Gravel Pit Way BR6: Orp 75Wc 161
Gravel Rd. BR2: Broml 76Nc 160
Gravel Rd. DA4: Sut H 66Rd 141
Gravel Rd. TW2: Twick 60Ga 108
Gravelwood Cl. BR7: Chst 62Sc 138
Gravely Ct. HP2: Hem H 3C 4
Gravely Ho. SE8 49Ac 92
(off Chilton Gro.)
Gravenel Gdns. SW17 64Gb 133
(off Nutwell St.)
Graveney Gro. SE20 66Yb 136
Graveney Rd. SW17 63Gb 133
GRAVESEND 8D 122
Gravesend Golf Cen. 3J 145
Gravesend Rd. DA12: Shorne 2M 145
Gravesend Rd. TN15: Stans 87De 185
Gravesend Rd. TN15: Wro 87De 185
Gravesend Rd. W12 45Wa 88
Gravesend Sailing Club 8F 122
Gravesend Station (Rail) 8D 122
Gravesend Visitor Information 8D 122
GRAVESHAM COMMUNITY HOSPITAL 8C 122
Gravesham Ct. DA12: Grav'nd 9D 122
Gravesham Way BR3: Beck 73Bc 158
Gray Av. RM8: Dag 32Bd 75
Gray Cl. KT15: Add 78K 149
Gray Ct. E1 43Ac 92
Gray Ct. HA5: Pinn 28Aa 45
Gray Ct. SL4: Wind 4E 102
Gray Gdns. RM13: Rain 37Jd 76
Grayham Cres. KT3: N Mald 70Ta 131
Grayham Rd. KT3: N Mald 70Ta 131
Gray Ho. SE17 7E 230 (50Sb 91)
(off King & Queen St.)
Gray Ho. SL2: Stoke P 8L 61
(off Bells Hill Grn.)
Grayland Cl. BR1: Broml 67Mc 137
Graylands CM16: They B 9Tc 22
Graylands GU21: Wok 88A 168
Graylands RM17: Grays 51Ae 121
Graylands Cl. GU21: Wok 88A 168
Graylands Cl. SL1: Slou 6D 80
Grayling KT16: Chert 74L 149
Grayling Cl. E16 42Gc 93
Grayling Ct. W5 46Ma 87
(off Grange Rd.)
Grayling Rd. N16 33Tb 71
Graylings, The WD5: Ab L 5T 12
Grayling Sq. E2 41Wb 91
(off Nelson Gdns.)
Gray Pl. KT16: Ott 79F 148
GRAYS 49De 99
Grays Athletic FC 45Ud 98
GRAYS COURT COMMUNITY HOSPITAL 38Dd 76
Grayscroft Rd. SW16 66Mb 134
Gray's End Cl. RM17: Grays 48Ce 99
Grays Farm Rd. BR5: St P 67Xc 139
Grayshott Rd. SW11 54Jb 112
Gray's Inn Bldgs. EC1 6K 217 (42Qb 90)
(off Rosebery Av.)
Gray's Inn Pl. WC1 1J 223 (43Pb 90)
Gray's Inn Rd. WC1 3G 217 (41Nb 90)
Gray's Inn Sq. WC1 7J 217 (43Pb 90)
Gray's La. KT18: Eps D 92Qa 193
Gray's La. KT21: Asht 91Pa 193
Grays La. TW15: Ashf 63R 128
Grayson Ho. EC1 4E 218 (41Sb 91)
(off Radnor St.)
Grays Pk. Rd. SL2: Stoke P 10L 61
Grays Pl. SL2: Slou 6K 81
Grays Rd. SL1: Slou 6K 81
Grays Rd. TN16: Westrm 93Rc 200
Grays Rd. UB10: Uxb 38N 63
Grays Shop. Cen. 51Ce 121
Grays Station (Rail) 51Ce 121
Grays Ter. E7 37Lc 73
Grayston Ho. SE3 56Lc 115
Gray St. SE1 2A 230 (47Qb 90)
Grays Wlk. CM13: Hut 17Fe 41
Grayswood Gdns. SW20 68Xa 132
Grayswood Point SW15 60Wa 110
Gray's Yd. W1 3J 221 (44Jb 90)
(off James St.)
Graywood Ct. N12 24Eb 49
Grazebrook Rd. N16 33Tb 71
Grazeley Cl. DA6: Bex 57Ed 118
Grazeley Ct. SE19 64Ub 135
Grazings, The HP2: Hem H 1P 3
Great Acre Ct. SW4 56Mb 112
Great Amwell La. N8 27Pb 50
Great Arthur Ho. EC1 6D 218 (42Sb 91)
(off Golden La. Est.)
Great Barn Cres. GU24: W End 4E 166
Great Bell All. EC2 2F 225 (44Tb 91)
Great Benty UB7: W Dray 49N 83
GREAT BOOKHAM 98Da 191
Great Bookham Common 93Ba 191
Great Brownings SE21 63Vb 135
GREAT BURGH 89Ya 174
Great Bushey Dr. N20 18Db 31
Great Cambridge Ind. Est. EN1: Enf 15Xb 33
GREAT CAMBRIDGE JUNC. 21Tb 51
Great Cambridge Rd. EN1: Enf 16Wb 33
Great Cambridge Rd. EN8: Chesh 6Yb 20
Great Cambridge Rd. EN8: Walt C 6Yb 20
Great Cambridge Rd. N17 23Tb 51
Great Cambridge Rd. N18 21Tb 51
Great Cambridge Rd. N9 20Tb 33
Great Castle St. W1 2A 222 (44Kb 90)
Great Central Av. HA4: Ruis 36Y 65
Great Central St. NW1 7F 215 (43Hb 89)
Great Central Way HA9: Wemb 35Sa 67
Great Central Way NW10 36Ua 68
Great Chapel St. W1 2D 222 (44Mb 90)
Great Charta Cl. TW20: Eng G 2N 125
Great Charter Cl. RM12: Horn 36Jd 76
Great Chart St. SW11 56Fb 111
Great Chertsey Rd. TW13: Hanw 62Ba 129
Great Chertsey Rd. TW13: Twick 62Ba 129
Great Chertsey Rd. TW2: Twick 61Da 129
Great Chertsey Rd. W4 53Sa 109
(not continuous)
Great Church La. W6 49Za 88
Great Clayne Rd. DA12: Grav'nd 10J 123
Great Cockcrow Railway 74F 148
Great College St. SW1 3F 229 (48Nb 90)
Great Comp Gdn. 89Pe 205
Great Cft. WC1 4G 217 (41Nb 90)
(off Cromer St.)
Great Cross Av. SE10 52Fc 115
Great Cullings RM7: Rush G 33Gd 76
Great Cumberland M. W1 3F 221 (44Hb 89)
Great Cumberland Pl. W1 2F 221 (44Hb 89)
Great Dover St. SE1 2E 230 (47Sb 91)
Greatdown Rd. W7 42Ha 86
Great Eastern Ent. Cen. E14 47Dc 92
Great Eastern Mkt. 37Ec 72
(within Westfield Shop. Cen.)
Great Eastern Rd. CM14: W'ley 21Yd 58
Great Eastern Rd. E15 38Fc 73
Great Eastern Rd. EN8: Walt C 5Yb 20
Great Eastern St. EC2 4H 219 (41Ub 91)
Great Eastern Wharf SW11 52Gb 111
Great Ellshams SM7: Bans 88Cb 175
Great Elms Rd. BR2: Broml 70Lc 137
Great Elms Rd. HP3: Hem H 6P 3
Greater London Ho. NW1 1B 216 (40Lb 70)
(off Hampstead Rd.)
Great Fld. NW9 25Ua 48
Greatfield NW5 36Lb 70
Greatfield Av. E6 42Pc 94
Greatfield Cl. N19 35Lb 70
Greatfield Cl. SE4 56Cc 114
Greatfields Dr. UB8: Hil 43Q 84
Greatfields Rd. IG11: Bark 39Tc 74
Great Fleete Way IG11: Bark 40Yc 75
Great Galley Cl. IG11: Bark 41Xc 95
Great Gdns. Rd. RM11: Horn 30Kd 57
Great Gatton Cl. CR0: C'don 73Ac 158
Great George St. SW1 2E 228 (47Mb 90)
Great Gregories La. CM16: Epp 5Uc 22
Great Gro. WD23: Bush 14Da 27
Great Guildford Bus. Sq. SE1 7D 224 (46Sb 91)
Great Guildford St. SE1 6D 224 (46Sb 91)
Great Hall SW11 53Jb 112
(off Battersea Pk. Rd.)
Greatham Rd. WD23: Bush 13Z 27
Greatham Rd. Ind. Est. WD23: Bush 13Z 27
Greatham Wlk. SW15 60Wa 110
Great Harry Dr. SE9 62Qc 138
Great Heart HP2: Hem H 1N 3
Greathurst End KT23: Bookh 96Ba 191
Great James St. WC1 7H 217 (43Pb 90)
Great Leys Ct. AL2: Lon C 9E 6
Great Marlborough St. W1 3B 222 (44Lb 90)
Great Maze Pond SE1 1G 231 (47Tb 91)
Great Mill Apts. E2 1K 219 (39Vb 71)
(off Whiston Rd.)
Great Minster Ho. SW1 5E 228 (49Mb 90)
(off Marsham St.)
Great Nelmes Chase RM11: Horn 29Pd 57
GREATNESS 92Ld 203
Greatness La. TN14: S'oaks 93Ld 203
Greatness Mill Ct. TN14: S'oaks 93Ld 203
Greatness Rd. TN14: S'oaks 93Ld 203
Great Newport St. WC2 4F 223 (45Nb 90)
Great New St. EC4 2A 224 (44Qb 90)
(off New Fetter La.)
Great Norman St. TN14: Ide H 100Bd 201
Great Nth. Leisure Pk. N12 24Fb 49
Great Nth. Rd. AL9: Brk P 2D 8
Great Nth. Rd. AL9: Hat 2D 8
Great Nth. Rd. AL9: Wel G 2D 8
Great Nth. Rd. EN5: Barn 12Bb 31
Great Nth. Rd. EN5: New Bar 15Cb 31
Great Nth. Rd. EN6: Pot B 10L 9
Great Nth. Rd. N2 29Gb 49
Great Nth. Rd. N6 29Hb 49
Great Nth. Way NW4 26Xa 48
Great Oaks CM13: Hut 16De 41
Great Oaks IG7: Chig 21Sc 54
Greatorex Ho. E1 43Wb 91
(off Greatorex St.)
Greatorex St. E1 43Wb 91
Great Ormond St. WC1 7G 217 (43Nb 90)
GREAT ORMOND STREET HOSPITAL FOR CHILDREN 6G 217 (42Nb 90)
Great Owl Rd. IG7: Chig 20Qc 36
Great Pk. WD4: K Lan 2P 11
Great Pk. Cl. UB10: Hil 38Q 64
Great Percy St. WC1 3J 217 (41Pb 90)
Great Peter St. SW1 4D 228 (48Mb 90)
Great Portland St. W1 6A 216 (42Kb 90)
Great Portland Street Station (Underground) 6A 216 (42Kb 90)
Great Pulteney St. W1 4C 222 (45Lb 90)
Great Queen St. DA1: Dart 59Pd 119
Great Queen St. WC2 3G 223 (44Nb 90)
Great Rd. HP2: Hem H 1P 3
Great Ropers La. CM13: Gt War 23Wd 58
Great Russell Mans. WC1 1F 223 (43Nb 90)
(off Gt. Russell St.)
Great Russell St. WC1 2E 222 (44Mb 90)
Great St Helen's EC3 2H 225 (44Ub 91)
Great St Thomas Apostle EC4 4E 224 (45Sb 91)
Great Scotland Yd. SW1 7F 223 (46Nb 90)

Great Slades EN6: Pot B 5Bb 17
Great Smith St. SW1 3E 228 (48Mb 90)
Great Sth. W. Rd. TW14: Bedf 59S 106
Great Sth. W. Rd. TW14: Felt 59S 106
Great Sth. W. Rd. TW4: Houn 55X 107
Great Spilmans SE22 57Ub 113
Great Strand NW9 25Va 48
Great Sturgess Rd. HP1: Hem H 2H 3
Great Suffolk St. SE1 7C 224 (46Rb 91)
Great Sutton St. EC1 6C 218 (42Rb 91)
Great Swan All. EC2 2F 225 (44Tb 91)
Great Tattenhams KT18: Tatt C 90Xa 174
Great Thrift BR5: Pet W 70Sc 138
Great Till Cl. TN14: Otf 88Gd 182
Great Titchfield St. W1 6A 216 (42Kb 90)
Great Tower St. EC3 4H 225 (45Ub 91)
Great Trinity La. EC4 4E 224 (45Sb 91)
Great Turnstile WC1 1J 223 (43Pb 90)
Great Turnstile Ho. WC1 1J 223 (43Pb 90)
(off Great Turnstile)
GREAT WARLEY 25Wd 58
Great Warley St. CM13: Gt War 25Wd 58
Great Western Ind. Pk. UB2: S'hall 47Da 85
Great Western Rd. W11 43Bb 89
Great Western Rd. W2 43Bb 89
Great Western Rd. W9 43Bb 89
Great West Rd. W4: Cedars Rd 50Ra 87
Great West Rd. W4: Dorchester Gro. 51Va 110
Great West Rd. TW5: Hest 54Z 107
Great West Rd. TW7: Bford 52Ga 108
Great West Rd. TW7: Isle 52Ga 108
Great West Rd. TW8: Bford 52Ja 108
Great West Rd. W6 50Va 88
Great W. Trad. Est. TW8: Bford 51Ka 108
Great Whites Rd. HP3: Hem H 4P 3
Great Winchester St. EC2 2G 225 (44Tb 91)
Great Windmill St. W1 4D 222 (45Mb 90)
Greatwood BR7: Chst 66Qc 138
Greatwood Cl. KT16: Ott 81E 168
Great Woodcote Dr. CR8: Purl 82Mb 176
Great Woodcote Pk. CR8: Purl 82Mb 176
Great Wood Country Pk. 8P 9
Great Wood Vis. Cen. 8P 9
Great Yd. SE1 1J 231 (47Ub 91)
(off Crucifix La.)
Greaves Cl. IG11: Bark 38Tc 74
Greaves Cotts. E14 43Ac 92
(off Maroon St.)
Greaves Pl. SW17 63Gb 133
Greaves Twr. SW10 52Eb 111
(off Worlds End Est.)
Grebe Av. UB4: Yead 44Z 85
Grebe Cl. E17 24Ac 52
Grebe Cl. E7 36Hc 73
Grebe Cl. IG11: Bark 42Wc 95
Grebe Ct. E14 47Ec 92
(off River Barge Cl.)
Grebe Ct. SE8 51Bc 114
(off Dorking Cl.)
Grebe Ct. SM1: Sutt 78Bb 155
Grebe Crest RM20: W Thur 49Wd 98
Grebe Ter. KT1: King T 69Na 131
Grecian Cres. SE19 65Rb 135
Greding Wlk. CM13: Hut 19De 41
Greek Ct. W1 3E 222 (44Mb 90)
Greek Orthodox Cathedral Kimisis Panayias, The 25Pb 50
Greek Orthodox Cathedral of St Sophia 45Db 89
Greek St. W1 3E 222 (44Mb 90)
Green, The KT20: Tad Oatlands Rd. 91Ab 194
Green, The KT20: Tad Stokes Riding 95Za 194
Green, The AL2: Lon C 9H 7
Green, The AL3: St A 4P 5
Green, The BR1: Broml 62Jc 137
(not continuous)
Green, The BR2: Hayes 73Jc 159
Green, The BR5: St P 66Xc 139
Green, The BR6: Prat B 82Yc 181
Green, The CM16: They B 8Uc 22
(not continuous)
Green, The CR0: Sels 81Bc 178
Green, The CR3: Wold 95Cc 198
Green, The CR6: W'ham 89Zb 178
Green, The DA14: Sidc 63Wc 139
Green, The DA16: Well 56Uc 116
Green, The DA2: Dart 61Td 142
Green, The DA7: Bex 53Cd 118
Green, The E11 30Kc 53
Green, The E15 37Gc 73
Green, The E4 18Ec 34
Green, The EN8: Chesh 1Yb 20
Green, The EN9: Walt A 6Ec 20
Green, The GU23: Rip 93L 189
Green, The HA0: Wemb 33Ja 66
Green, The IG8: Wfd G 22Jc 53
Green, The IG9: Buck H 18Kc 35
Green, The KT10: Clay 79Ha 152
Green, The KT12: Hers 78Z 151
Green, The KT12: W Vill 82U 170
Green, The KT17: Ewe 83Wa 174
Green, The KT22: Fet 96Fa 192
Green, The KT3: N Mald 69Ta 131
Green, The N14 19Mb 32
Green, The N17 23Sb 51
Green, The N21 17Qb 32
Green, The N9 19Wb 33
Green, The RH3: Bkld 5C 206
Green, The RH9: G'stone 4P 209
Green, The RM13: Wenn 45Nd 97
Green, The RM15: S Ock 41Zd 99
Green, The RM16: Ors 3C 100
Green, The RM18: W Til 1G 122
Green, The RM3: Rom 19Ld 39
Green, The RM4: Have B 20Gd 38
Green, The SL1: Burn 3A 80
Green, The SL1: Slou 7H 81
Green, The SL3: Dat 2M 103
Green, The SM1: Sutt 76Db 155
Green, The SM4: Mord 70Ab 132
Green, The SM5: Cars 77Jb 156
Green, The SM6: W'gton 75Jb 156
Green, The SS17: Stan H 2M 101
Green, The SW14 55Sa 109
Green, The SW19 64Za 132
Green, The TN13: S'oaks 94Md 203
Green, The TN14: Otf 88Kd 183
Green, The TN15: Seal 93Pd 203
(off Church Rd.)
Green, The TN16: Westrm 98Tc 200
Green, The TW13: Felt 61X 129
Green, The TW15: Ashf 64M 127
Green, The TW17: Shep 70U 128
Green, The TW19: Wray 58A 104
Green, The TW2: Twick 60Ga 108
Green, The TW20: Eng G 3N 125
Green, The TW5: Hest 51Ca 107
Green, The TW9: Rich 57Ma 109
Green, The UB10: Ick 33S 64
Green, The UB2: S'hall 48Aa 85
Green, The UB7: W Dray 48M 83
Green, The UB9: Hare 25L 43
Green, The W3 44Ua 88
Green, The W5 46Ma 87
Green, The WD25: Let H 11Ga 28
Green, The WD3: Crox G 16P 25
Green, The WD3: Sarr 7J 11
Greenacre DA1: Dart 61Md 141
Greenacre GU21: Knap 8J 167
Greenacre SL4: Wind 4C 102
Greenacre Cl. BR8: Swan 70Gd 140
Greenacre Cl. EN5: Barn 10Bb 17
Greenacre Cl. UB5: N'olt 36Ba 65
Greenacre Ct. TW20: Eng G 5N 125
Greenacre Gdns. E17 28Ec 52
Greenacre Pl. SM6: W'gton 75Kb 156
Greenacres CM16: Epp 1Vc 23
GREENACRES 58Rd 119
Greenacres HP2: Hem H 3D 4
Greenacres KT20: Lwr K 100Bb 195
Greenacres KT23: Bookh 96Da 191
Greenacres N3 26Bb 49
Greenacres RH8: Oxt 99Gc 199
Greenacres SE9 58Qc 116
Greenacres WD23: B Hea 19Fa 28
Green Acres CR0: C'don 76Vb 157
Green Acres DA14: Sidc 63Wc 139
Greenacres Av. UB10: Ick 34P 63
Greenacres Cl. BR6: Farnb 77Sc 160
Greenacres Cl. RM13: Rain 41Nd 97
Greenacres Dr. HA7: Stan 23Ka 46
Greenacres Ho. SW18 60Db 111
(off Knaresborough Dr.)
Greenacre Sq. SE16 47Zb 92
Greenacre Wlk. N14 20Mb 32
Greenall Cl. EN8: Chesh 2Ac 20
Green All. SL2: Farn C 6G 60
Greenan Ct. E2 41Zb 92
(off Meath Cres.)
Green Arbour Ct. EC1 2C 224 (44Rb 91)
(off Old Bailey)
Green Av. NW7 21Ta 47
Green Av. W13 48Ka 86
Greenaway Gdns. NW3 35Db 69
Greenaway Ho. NW8 39Eb 69
(off Boundary Rd.)
Greenaway Ho. WC1 4K 217 (41Qb 90)
(off Fernsbury St.)
Greenaway Ter. TW19: Stanw 60N 105
(off Victory Cl.)
Green Bank E1 46Xb 91
Green Bank N12 21Db 49
Greenbank Av. HA0: Wemb 36Ja 66
Greenbank Cl. E4 19Ec 34
Greenbank Cl. RM3: Rom 20Md 39
Greenbank Ct. TW7: Isle 54Ha 108
(off Lanadron Cl.)
Greenbank Cres. NW4 28Ab 48
Greenbank Lodge BR7: Chst 68Qc 138
(off Forest Cl.)
Greenbank Rd. WD17: Wat 8T 12
Greenbanks AL1: St A 4D 6
Greenbanks DA1: Dart 61Nd 141
Greenbanks HA1: Harr 35Ga 66
Greenbanks RM14: Upm 33Ud 78
Greenbanks Cl. SE13 55Dc 114
Greenbay Rd. SE7 52Mc 115
Greenberry St. NW8 2D 214 (40Gb 69)
Greenbrook Av. EN4: Had W 11Eb 31
Greenbury Cl. WD3: Chor 14E 24
Green Bus. Cen., The TW18: Staines 63E 126
Green Cl. AL9: Brk P 8G 8
Green Cl. BR2: Broml 69Gc 137
Green Cl. EN8: Chesh 3Ac 20
Green Cl. NW11 31Eb 69
Green Cl. NW9 30Sa 47
Green Cl. SM5: Cars 75Hb 155
Green Cl. TW13: Hanw 63Ba 129
Greencoat Mans. SW1 4C 228 (48Lb 90)
(off Greencoat Row)
Greencoat Pl. SW1 5C 228 (49Lb 90)
Greencoat Row SW1 4C 228 (48Lb 90)
Green Ct. TW16: Sun 65V 128
Greencourt Av. CR0: C'don 75Xb 157
Greencourt Av. HA8: Edg 25Ra 47
Greencourt Gdns. CR0: C'don 74Xb 157
Greencourt Ho. E1 42Zb 92
(off Mile End Rd.)
Green Ct. Rd. BR8: Crock 71Fd 162
Greencourt Rd. BR5: Pet W 71Tc 160
Greencrest Pl. NW2 34Wa 68
Greencroft HA8: Edg 22Sa 47
Greencroft Av. HA4: Ruis 33Y 65
Greencroft Cl. E6 43Mc 93
Greencroft Gdns. EN1: Enf 13Ub 33
Greencroft Gdns. NW6 38Db 69
Greencroft Rd. TW5: Hest 53Ba 107
Green Curve SM7: Bans 86Bb 175
Greendale NW7 21Ua 48
Green Dale SE22 57Ub 113
Green Dale SE5 56Tb 113
Green Dale Cl. SE22 57Ub 113
Greendale M. SL2: Slou 5L 81
Greendale Wlk. DA11: Nflt 2A 144
Green Dell Way HP3: Hem H 3C 4
Green Dragon Ct. SE1 7F 225 (46Tb 91)
(off Bedale St.)
Green Dragon Ho. CR0: C'don 76Sb 157
(off High St.)
Green Dragon Ho. WC2 2G 223 (44Nb 90)
(off Stukeley St.)
Green Dragon La. N21 16Qb 32
Green Dragon La. TW8: Bford 50Na 87
Green Dragons Airsports 92Dc 198
Green Dragon Yd. E1 43Wb 91
Green Dr. GU23: Rip 95H 189
Green Dr. SL3: L'ly 49A 82
(not continuous)
Green Dr. UB1: S'hall 46Ca 85
Greene Ct. SE14 51Zb 114
(off Samuel Cl.)
Green Edge WD25: Wat 7X 13
Greene Fielde End TW18: Staines 66M 127
Greene Ho. SE1 4F 231 (48Tb 91)
(off Burbage Cl.)
Greene Ho. SL9: Chal P 21A 42
GREEN END 3J 3
Green End KT9: Chess 77Na 153
Green End N21 19Rb 33
Green End Bus. Cen. WD3: Sarr 9J 11
Green End Gdns. HP1: Hem H 3J 3
Green End La. HP1: Hem H 2H 3
Greenend Rd. W4 47Ua 88
Green End Rd. HP1: Hem H 2J 3
(not continuous)
Greener Ct. CR0: C'don 72Sb 157
(off Goodman Cres.)
Greener Ho. SW4 55Mb 112
Greene Wlk. HP4: Berk 2A 2
Green Farm Cl. BR6: Chels 78Vc 161
Green Farm La. DA12: Shorne 5N 145
Greenfell Mans. SE8 51Dc 114
Greenfern Av. SL1: Slou 4A 80
Green Ferry Way E17 28Zb 52
Greenfield Av. KT5: Surb 73Ra 153
Greenfield Av. WD19: Wat 19Z 27
Greenfield Ct. SE9 62Nc 138
Greenfield Dr. BR1: Broml 68Lc 137
Greenfield Dr. N2 28Hb 49
Greenfield End SL9: Chal P 24B 42
Greenfield Gdns. BR5: Pet W 73Tc 160
Greenfield Gdns. NW2 33Ab 68
Greenfield Gdns. RM9: Dag 39Zc 75
Greenfield Ho. SW19 60Za 110
Greenfield Ho. TW20: Eng G 5M 125
(off Kings La.)
Greenfield Link CR5: Coul 87Nb 176
Greenfield Pl. UB3: Hayes 45V 84
Greenfield Rd. DA2: Wilm 64Fd 140
Greenfield Rd. E1 43Wb 91
Greenfield Rd. N15 29Ub 51
Greenfield Rd. RM9: Dag 39Yc 75
Greenfields EN6: Cuff 2Nb 18
Greenfields IG10: Lough 14Qc 36
Greenfields UB1: S'hall 44Ca 85
Greenfields Cl. CM13: Gt War 23Yd 58
Greenfields Cl. IG10: Lough 14Qc 36
Greenfield St. EN9: Walt A 6Ec 20
Greenfield Way HA2: Harr 27Da 45
Greenfinches DA3: Lfield 69De 143
GREENFORD 41Ca 85
Greenford Av. UB1: S'hall 45Ba 85
Greenford Av. W7 42Ga 86
Greenford Bus. Cen. UB6: G'frd 38Fa 66
Greenford Gdns. UB6: G'frd 41Ea 86
GREENFORD GREEN 37Ga 66
Greenford Ind. Est. UB6: G'frd 38Da 65
Greenford Pk. UB6: G'frd 38Fa 66
Greenford Rd. HA1: Harr 35Ga 66
Greenford Rd. SM1: Sutt 77Db 155
(not continuous)
Greenford Rd. UB1: S'hall 46Ea 86
Greenford Rd. UB6: G'frd 39Fa 66
GREENFORD RDBT. 40Fa 66
Greenford Sports Cen. 41Ca 85
Greenford Station (Rail & Underground) 39Fa 66
Green Gdns. BR6: Farnb 78Sc 160
Greengate UB6: G'frd 37Ka 66
Greengate Lodge E13 40Kc 73
(off Hollybush St.)
Greengate Pde. IG2: Ilf 30Tc 54
Greengate St. E13 40Kc 73
Green Glade CM16: They B 9Uc 22
Green Glades RM11: Horn 30Pd 57
Greenhalgh Wlk. N2 28Eb 49
Greenham Cl. SE1 2K 229 (47Qb 90)
Greenham Cres. E4 23Bc 52
Greenham Ho. E9 39Yb 72
(off Templecombe Rd.)
Greenham Ho. TW7: Isle 55Fa 108
Greenham Rd. N10 26Jb 50
Greenham Wlk. GU21: Wok 10N 167
Greenhaven Dr. SE28 44Xc 95
Greenhayes Av. SM7: Bans 86Cb 175
Greenhayes Cl. RH2: Reig 6L 207
Greenhayes Gdns. SM7: Bans 87Cb 175
Green Hedges TW1: Twick 57La 108
Greenheys Cl. HA6: Nwood 25U 44
Greenheys Dr. E18 27Hc 53
Greenheys Pl. GU22: Wok 90B 168
GREENHILL 29Ga 46
Greenhill HA9: Wemb 33Ra 67
Greenhill IG9: Buck H 18Lc 35
Greenhill NW3 35Fb 69
Greenhill SE18 50Pc 94
Greenhill SM1: Sutt 75Eb 155
Green Hill BR6: Downe 84Pc 180
Greenhill Av. CR3: Cat'm 93Xb 197
Greenhill Ct. EN5: New Bar 15Db 31
Greenhill Ct. HP1: Hem H 3K 3
Greenhill Ct. SE18 50Pc 94
Greenhill Cres. WD18: Wat 16U 26
Greenhill Gdns. UB5: N'olt 40Ba 65
Greenhill Gro. E12 35Nc 74
Green Hill La. CR6: W'ham 89Ac 178
Greenhill Pde. EN5: New Bar 15Db 31
Greenhill Pk. EN5: New Bar 15Db 31
Greenhill Pk. NW10 39Ua 68
Greenhill Rd. DA11: Nflt 1B 144
Greenhill Rd. HA1: Harr 30Ga 46
Greenhill Rd. NW10 39Ua 68
Greenhill Rd. TN14: Otf 87Ld 183
Greenhills Cl. WD3: Rick 15K 25
Greenhill's Rents EC1 7B 218 (43Rb 91)
Greenhills Ter. N1 37Tb 71
Greenhill Ter. SE18 50Pc 94
Greenhill Ter. UB5: N'olt 40Ba 65
Greenhill Way HA1: Harr 30Ga 46
Greenhill Way HA9: Wemb 33Ra 67
GREENHITHE 57Xd 120
Greenhithe Cl. DA15: Sidc 59Uc 116
Greenhithe for Bluewater Station (Rail) 57Wd 120
Greenholm Rd. SE9 57Rc 116
Green Hundred Rd. SE15 51Wb 113
Greenhurst La. RH8: Oxt 4K 211
Greenhurst Rd. SE27 64Qb 134
Greening St. SE2 49Yc 95
Green Lake Gro. RM3: Rom 21Md 57
Greenlake Ter. TW18: Staines 66J 127
Greenland Cres. UB2: S'hall 48Y 85
Greenland Ho. E1 42Ac 92
(off Ernest St.)
Greenland M. SE8 50Zb 92
Greenland Pl. NW1 39Kb 70
Greenland Quay SE16 49Zb 92
Greenland Rd. EN5: Barn 16Ya 30
Greenland Rd. NW1 39Lb 70
Greenlands KT16: Chert 76E 148
Greenlands KT19: Ewe 78Ra 153
Greenlands TN15: Plat 92De 205
Greenlands La. NW4 25Xa 48
Greenlands Rd. KT13: Weyb 76R 150
Greenlands Rd. TN15: Kems'g 91Rd 203
Greenlands Rd. TW18: Staines 63J 127
Greenland St. NW1 39Kb 70
Greenland Way CR0: Bedd 73Mb 156
Green La. SM4: Mord Central Rd. 72Cb 155
Green La. HP3: Bov Chesham Rd. 10B 2
Green La. SM4: Mord Lwr. Morden La. 73Ya 154
Green La. TW20: Egh The Avenue 63D 126
Green La. RH1: Redh Timperley Gdns. 4N 207
Green La. TW20: Egh Vicarage Cres. 64D 126
Green La. AL1: St A 6E 6
Green La. BR7: Chst 63Rc 138
Green La. CM14: B'wood 18Wd 40
Green La. CM14: Gt War 24Wd 58
Green La. CM14: Kel H 12Sd 40
Green La. CM15: Pil H 15Yd 40
Green La. CR3: Cat'm 94Sb 197
Green La. CR5: Coul 98Cb 195
Green La. CR6: W'ham 88Ac 178
Green La. CR7: Thor H 67Rb 135
Green La. CR8: Purl 83Lb 176
Green La. DA12: Shorne 5M 145
Green La. E4 11Gc 35
Green La. EN9: Walt A 6Lc 21
Green La. GU22: Wok 3M 187
Green La. GU23: Ock 96R 190
Green La. GU24: Chob 2K 167
Green La. GU4: W Cla 99J 189
Green La. HA1: Harr 34Ga 66
Green La. HA6: Nwood 23T 44
Green La. HA7: Stan 21Ka 46
Green La. HA8: Edg 21Pa 47
(not continuous)
Green La. HP2: Hem H 3C 4
Green La. IG1: Ilf 33Tc 74
Green La. IG3: Ilf 32Xc 75
Green La. IG7: Chig 18Sc 36
Green La. KT11: Cobh 84Aa 171
Green La. KT12: Hers 79X 151
Green La. KT14: Byfl 84P 169
Green La. KT15: Add 76H 149
Green La. KT16: Chert 75G 148
Green La. KT20: Lwr K 98Bb 195
Green La. KT21: Asht 89La 172
Green La. KT22: Lea 93Ma 193
(not continuous)
Green La. KT3: N Mald 71Sa 153
Green La. KT4: Wor Pk 74Wa 154
Green La. KT8: W Mole 71Da 151
Green La. KT9: Chess 81Ma 173
Green La. NW4 28Za 48
Green La. RH1: Blet 3L 209
Green La. RH2: Reig 6H 207
Green La. RM14: Avel 39Td 78
Green La. RM14: Upm 37Td 78
Green La. RM16: N Stif 43Ee 99
Green La. RM16: Ors 43Ee 99
Green La. RM8: Dag 32Xc 75
Green La. SE20 66Zb 136
Green La. SE9 60Rc 116
Green La. SL1: Burn 1A 80
Green La. SL2: Farn C 7F 60
Green La. SL3: Dat 3M 103
Green La. SL4: Wind 4E 102
Green La. SL5: Asc 7C 124
Green La. SW16 66Pb 134
Green La. TW13: Hanw 64Aa 129
Green La. TW16: Sun 66V 128
Green La. TW17: Shep 72S 150
Green La. TW18: Staines 67G 126
Green La. TW20: Thorpe 68E 126
Green La. TW4: Houn 55X 107
Green La. UB8: Hil 43S 84
Green La. W7 47Ga 86
Green La. WD19: Wat 17Y 27
Green La. WD3: Crox G 15P 25
Green La. Av. KT12: Hers 78Y 151
Green La. Bus. Pk. SE9 61Qc 138
Green La. Cl. KT14: Byfl 84P 169
Green La. Cl. KT16: Chert 75G 148
Green La. Cotts. HA7: Stan 21Ka 46
Green La. Ct. SL1: Burn 1A 80
Green La. Gdns. CR7: Thor H 68Sb 135
Green Lanes KT19: Ewe 81Ua 174
Green Lanes N13 23Pb 50
Green Lanes N15 27Rb 51
Green Lanes N16 34Sb 71
Green Lanes N21 19Rb 33
Green Lanes N4 32Sb 71
Green Lanes N8 27Rb 51
Green Lanes Wlk. N4 32Sb 71
Green La. W. KT24: W Hor 97Q 190
Greenlaw Ct. W5 44Ma 87
(off Mount Pk. Rd.)
Greenlaw Gdns. KT3: N Mald 73Va 154
Greenlawn La. TW8: Bford 49Ma 87
Greenlawns N12 23Db 49
Green Lawns HA4: Ruis 32Y 65
Greenlaw St. SE18 48Qc 94
Green Leaf Av. SM6: Bedd 77Mb 156
Greenleaf Cl. SW2 59Qb 112
Greenleafe Dr. IG6: Ilf 27Rc 54
Greenleaf Ho. Bus. Cen. EN6: Pot B 4Cb 17
Greenleaf Rd. E17 27Bc 52
Greenleaf Rd. E6 39Lc 73
Greenleaf Way HA3: W'stone 27Ha 46
Greenlea Pk. SW19 66Fb 133
Greenleas EN9: Walt A 6Gc 21
Green Leas KT1: King T 69Na 131
(off Mill St.)
Green Leas TW16: Sun 65V 128
Green Leas Cl. TW16: Sun 65V 128
Greenleaves Ct. TW15: Ashf 65R 128
Greenleigh Av. BR5: St P 70Xc 139
Green Line Coach Station 6A 228 (49Kb 90)
(off Bulleid Way)
Greenlink Wlk. TW9: Kew 53Ra 109
Green Man Gdns. W13 45Ja 86
Green Man La. TW14: Felt 56W 106
(not continuous)
Green Man La. W13 45Ja 86
Green Mnr. Way DA11: Nflt 56Be 121
Green Man Pas. W13 45Ka 86
(not continuous)
GREEN MAN RDBT. 31Hc 73
Greenman St. N1 38Sb 71
Greenmead DA18: Erith 48Ad 95
Green Mead KT10: Esh 79Ba 151
Greenmead Cl. SE25 71Wb 157
Green Mdw. EN6: Pot B 2Cb 17
Greenmeads GU22: Wok 94A 188
Green M. N1 3G 219 (41Tb 91)
Green Moor Link N21 17Rb 33
Greenmoor Rd. EN3: Enf H 12Yb 34
Greenoak Pl. EN4: Cockf 12Hb 31
Greenoak Ri. TN16: Big H 90Lc 179
Green Oaks UB2: S'hall 49Z 85
Greenoak Way SW19 63Za 132
Greenock Rd. SL1: Slou 4E 80
Greenock Rd. SW16 67Mb 134
Greenock Rd. W3 48Ra 87
Greenock Way RM1: Rom 24Gd 56
Greeno Cres. TW17: Shep 71Q 150
Green Pde. TW3: Houn 57Da 107
Green Pk. London 1A 228 (47Kb 90)
Green Pk. TW18: Staines 62G 126
Greenpark Ct. HA0: Wemb 38La 66
Green Park Station (Underground) 7A 222 (46Kb 90)
Green Pk. Way UB6: G'frd 38Ga 66
Green Pl. DA1: Cray 57Gd 118
Green Pl. SE10 47Gc 93
Green Point E15 37Gc 73
Green Pond Cl. E17 27Bc 52
Green Pond Rd. E17 27Ac 52
Green Ride CM16: Epp 7Rc 22
Green Ride IG10: Lough 14Lc 35
(not continuous)
Green Rd. GU23: Ock 96S 190
Green Rd. N14 16Kb 32
Green Rd. N20 20Eb 31
Green Rd. TW20: Thorpe 70C 126
Green Rd. Nth. EN3: Pond E 14Ac 34
Greenrod Pl. TW8: Bford 50Na 87
(off Claypoonds La.)
Greenroof Way SE10 48Hc 93
Greensand Cl. RH1: Mers 100Mb 196
Green Sand Rd. RH1: Redh 5A 208
Greens Cl., The IG10: Lough 12Qc 36
Green's Ct. W1 4D 222 (45Mb 90)
(off Brewer St.)
Green's Ct. W11 46Bb 89
(off Lansdowne M.)
Green's End SE18 49Rc 94
Greenshank Cl. E17 24Ac 52
Greenshank Ho. NW9 30Va 48
Greenshaw CM14: B'wood 18Xd 40
Greenshaw School Sports Cen. 75Eb 155
Greenshields Ind. Est. E16 47Jc 93
Greenside BR8: Swan 68Fd 140
Greenside DA5: Bexl 60Ad 117
Greenside RM8: Dag 32Yc 75
Greenside SL2: Slou 3E 80
Greenside WD6: Bore 10Qa 15
Greenside Cl. IG6: Ilf 23Sc 54
Greenside Cl. N20 19Fb 31
Greenside Cl. SE6 61Fc 137
Greenside Cotts. GU23: Rip 93L 189
Greenside Dr. KT21: Asht 90Ka 172
Greenside Rd. CR0: C'don 73Qb 156
Greenside Rd. W12 48Wa 88
Greenside Wlk. TN16: Big H 90Kc 179
Greenslade Av. KT21: Asht 91Ra 193
Greenslade Rd. IG11: Bark 38Tc 74
Greensleeves Cl. AL4: St A 3G 6
Greensleeves Dr. CM14: W'ley 22Xd 58
Greenstead Av. IG8: Wfd G 24Lc 53
Greenstead Cl. IG8: Wfd G 23Lc 53
Greenstead Gdns. IG8: Wfd G 23Lc 53
Greenstead Gdns. SW15 57Xa 110
Greensted Ct. CR3: Whyt 91Wb 197
(off Godstone Rd.)
Greensted Rd. IG10: Lough 17Nc 36
Greenstone M. E11 30Jc 53
Green St. AL9: Hat 2J 9
Green St. E13 38Lc 73
Green St. E7 37Kc 73
Green St. EN3: Brim 12Yb 34
Green St. EN3: Enf H 12Yb 34
Green St. TW16: Sun 67W 128
Green St. W1 4H 221 (45Jb 90)
Green St. WD3: Chen 11E 24
Green St. WD3: Chor 11E 24
Green St. WD6: Bore 9Qa 15
Green St. WD7: Shenl 7Qa 15
GREEN STREET GREEN 79Vc 161
GREEN STREET GREEN 65Wd 142
Green St. Grn. Rd. DA1: Dart 60Rd 119
Green St. Grn. Rd. DA2: Daren 62Sd 142
Green St. Grn. Rd. DA2: Dart 62Sd 142
Green St. Grn. Rd. DA2: G St G 62Sd 142
Greenstreet Hill SE14 54Zb 114
Greensward WD23: Bush 16Da 27
Green Ter. EC1 4A 218 (41Qb 90)
Green Tiles UB9: Den 31H 63
Green Tiles La. UB9: Den 30H 43
Green Trees CM16: Epp 3Wc 23
Green Va. DA6: Bex 57Zc 117
Green Va. W5 44Pa 87
Greenvale Rd. GU21: Knap 10H 167
Greenvale Rd. SE9 56Pc 116
Green Verges HA7: Stan 24Ma 47
Green Vw. GU22: Wok 91B 188
Green Vw. KT9: Chess 80Pa 153
Green Vw. RH9: G'stone 3P 209
Greenview Av. BR3: Beck 72Ac 158
Greenview Av. CR0: C'don 72Ac 158
Green Vw. Cl. HP3: Bov 1C 10
Greenview Cl. W3 46Ua 88
Green Vw. Ct. WD5: Ab L 4T 12
Greenview Ct. TW15: Ashf 63P 127
Greenview Dr. SW20 69Ya 132
Green Vw. Ho. DA11: Grav'nd 4E 144
(off Southfields Grn.)
Green Wlk. DA1: Cray 57Hd 118
Green Wlk. HA4: Ruis 32V 64
Green Wlk. IG10: Lough 17Nc 36
Green Wlk. IG8: Wfd G 23Nc 54
Green Wlk. NW4 29Za 48
Green Wlk. SE1 4H 231 (48Ub 91)
Green Wlk. TW12: Hamp 65Ba 129
Green Wlk. UB2: S'hall 50Ca 85
Green Wlk., The E4 18Fc 35
Greenwatt Way SL1: Slou 8H 81
Greenway BR7: Chst 64Qc 138
Greenway CM13: Hut 17Ce 41
Greenway E3 39Cc 72
Greenway E6 42Qc 94
Greenway HA3: Kenton 29Na 47
Greenway HA5: Pinn 26X 45
Greenway HP2: Hem H 2B 4
Greenway IG8: Wfd G 22Lc 53
Greenway KT23: Bookh 95Da 191
Greenway N14 19Nb 32
Greenway N20 19Cb 31
Greenway RM3: Hrld W 23Rd 57
Greenway RM8: Dag 33Yc 75
Greenway SL1: Burn 10A 60
Greenway SM6: W'gton 77Lb 156
Greenway SW20 70Ya 132
Greenway TN16: Tats 92Lc 199
Greenway UB4: Yead 41W 84
Green Way BR2: Broml 72Nc 160
Green Way DA3: Hartl 71Ae 165
Green Way RH1: Redh 4N 207
Green Way SE9 57Mc 115
Green Way TW16: Sun 70W 128
Greenway, The BR5: St M Cry 72Xc 161
Greenway, The EN3: Enf W 7Zb 20
Greenway, The EN6: Pot B 5Cb 17
Greenway, The HA3: W'stone 25Ga 46
Greenway, The HA5: Pinn 30Ba 45
Greenway, The KT18: Eps 86Qa 173
Greenway, The NW9 26Ta 47
Greenway, The RH8: Oxt 5M 211
Greenway, The SL1: Slou 6B 80
Greenway, The SL9: Chal P 27A 42
Greenway, The TW4: Houn 56Ba 107
Greenway, The UB10: Ick 33R 64
Greenway, The UB8: Uxb 40L 63
Greenway, The WD3: Rick 17J 25

Greenway Av. E17 28Fc 53
Greenway Cl. KT14: W Byf 85J 169
Greenway Cl. N11 23Jb 50
Greenway Cl. N15 28Vb 51
Greenway Cl. N20 19Cb 31
Greenway Cl. N4 33Sb 71
Greenway Cl. NW9 26Ta 47
Greenway Ct. IG1: Ilf 32Qc 74
Greenway Dr. TW18: Staines 67M 127
Greenway Gdns. CR0: C'don 76Bc 158
Greenway Gdns. HA3: W'stone 26Ga 46
Greenway Gdns. NW9 26Ta 47
Greenway Gdns. UB6: G'frd 41Ca 85
Greenways BR3: Beck 69Cc 136
Greenways DA3: Lfield 69Fe 143
Greenways EN7: G Oak 1Rb 19
Greenways KT10: Hin W 77Ga 152
Greenways KT20: Walt H 97Xa 194
Greenways TW20: Egh 64A 126
Greenways WD5: Ab L 4U 12
Greenways, The TW1: Twick 58Ja 108
Greenways Ct. RM11: Horn 30Md 57
Greenways Dr. SL5: S'dale 4C 146
Greenways Dr. TW4: Houn 56Aa 107
Greenwell Cl. RH9: G'stone 2P 209
Greenwell St. W1 6A 216 (42Kb 90)
GREENWICH 52Ec 114
GREENWICH & BEXLEY COMMUNITY HOSPICE 50Yc 95
Greenwich Av. CM14: B'wood 18Xd 40
Greenwich Bus. Pk. SE10 52Dc 114
Greenwich Cen., The 50Hc 93
(off Lambarde Sq.)
Greenwich Chu. St. SE10 51Ec 114
Greenwich Ct. AL1: St A 3E 6
Greenwich Ct. E1 44Xb 91
(off Cavell St.)
Greenwich Ct. EN8: Walt C 6Ac 20
Greenwich Cres. E6 43Nc 94
Greenwich Foot Tunnel 50Ec 92
Greenwich Hgts. SE18 52Nc 116
Greenwich High Rd. SE10 53Dc 114
Greenwich Ho. SE13 58Fc 115
Greenwich Mkt. SE10 51Ec 114
GREENWICH MILLENNIUM VILLAGE 48Hc 93
Greenwich Pk. 52Fc 115
Greenwich Pk. St. SE10 51Fc 115
Greenwich Peninsula Ecology Pk. 48Jc 93
Greenwich Peninsula Golf Driving Range 47Fc 93
Greenwich Picturehouse 52Ec 114
Greenwich Quay SE8 51Dc 114
Greenwich Shop. Pk. 49Kc 93
Greenwich Sth. St. SE10 53Dc 114
Greenwich Station (Rail & DLR) 52Dc 114
Greenwich Theatre 52Ec 114
Greenwich Tourist Info. Cen. 51Ec 114
Greenwich Vw. Pl. E14 48Dc 92
Greenwich Way EN9: Walt A 8Ec 20
Greenwich Yacht Club 48Jc 93
Greenwood NW5 36Lb 70
(off Osney Cres.)
Greenwood Av. EN3: Enf H 12Ac 34
Greenwood Av. EN7: Chesh 3Xb 19
Greenwood Av. RM10: Dag 35Dd 76
Greenwood Bus. Cen. CR0: C'don 73Vb 157
Greenwood Cl. BR5: Pet W 72Uc 160
Greenwood Cl. DA15: Sidc 61Wc 139
Greenwood Cl. EN7: Chesh 3Xb 19
Greenwood Cl. KT15: Wdhm 83H 169
Greenwood Cl. KT7: T Ditt 74Ja 152
Greenwood Cl. SM4: Mord 70Ab 132
Greenwood Cl. UB3: Hayes 46W 84
Greenwood Cl. WD23: B Hea 17Ga 28
Greenwood Cotts. SL5: S'dale 2G 146
Greenwood Dr. E4 22Fc 53
Greenwood Dr. WD25: Wat 6X 13
Greenwood Gdns. CR3: Cat'm 97Wb 197
Greenwood Gdns. IG6: Ilf 24Sc 54
Greenwood Gdns. N13 20Rb 33
Greenwood Gdns. RH8: Oxt 6L 211
Greenwood Gdns. WD7: Shenl 5Na 15
Greenwood Ho. EC1 4K 217 (41Qb 90)
(off Rosebery Av.)
Greenwood Ho. N22 25Pb 50
Greenwood Ho. RM17: Grays 51De 121
(off Argent St.)
Greenwood Ho. SE4 56Zb 114
Greenwood La. TW12: Hamp H 64Da 129
Greenwood Mans. IG11: Bark 38Wc 75
(off Lansbury Av.)
Greenwood Pk. KT2: King T 66Ua 132
Greenwood Pk. Leisure Cen. 7P 5
Greenwood Pl. KT12: Hers 76Aa 151
Greenwood Pl. NW5 36Kb 70
Greenwood Pl. TN15: Wro 89Ce 185
Greenwood Rd. CR0: C'don 73Rb 157
Greenwood Rd. CR4: Mitc 69Mb 134
Greenwood Rd. DA5: Bexl 63Fd 140
Greenwood Rd. E13 40Hc 73
Greenwood Rd. E8 37Wb 71
Greenwood Rd. GU21: Wok 2J 187
Greenwood Rd. GU24: Brkwd 3A 186
Greenwood Rd. IG7: Chig 21Xc 55
Greenwood Rd. KT7: T Ditt 74Ja 152
Greenwood Rd. TW7: Isle 55Ha 108
Greenwoods, The HA2: Harr 34Ea 66
Greenwood Ter. NW10 39Ta 67
Greenwood Theatre 1G 231 (47Tb 91)
Greenwood Way TN13: S'oaks 97Hd 202
Green Wrythe Cres. SM5: Cars 74Gb 155
Green Wrythe La. SM5: Cars 72Fb 155
Green Yd. WC1 5J 217 (42Pb 90)
Greenyard EN9: Walt A 5Ec 20
Green Yd., The EC3 3H 225 (44Ub 91)
(off Leadenhall Pl.)
Greer Garson Rd. UB9: Den 29J 43
Greer Rd. HA3: Hrw W 25Ea 46
Greet Ho. SE1 2A 230 (47Qb 90)
(off Frazier St.)
Greet St. SE1 7A 224 (46Qb 90)
Greg Cl. E10 30Ec 52
Gregor M. SE3 52Jc 115
Gregory Av. EN6: Pot B 5Eb 17
Gregory Cl. BR2: Broml 70Gc 137
Gregory Cl. GU21: Wok 9N 167
Gregory Cl. TN14: S'ham 83Hd 182
Gregory Cres. SE9 59Mc 115
Gregory Dr. SL4: Old Win 8M 103
Gregory M. EN9: Walt A 4Dc 20
Gregory Pl. KT15: Add 80L 149
Gregory Pl. W8 47Db 89
Gregory Rd. RM6: Chad H 28Zc 55
Gregory Rd. SL2: Hedg 3H 61
Gregory Rd. UB2: S'hall 48Ca 85
Gregson Cl. WD6: Bore 11Sa 29
Gregson's Ride IG10: Lough 10Qc 22
Grehan M. SE14 52Bc 114
Greig Cl. N8 29Nb 50
Greig Ter. SE17 51Rb 113
Grenaby Av. CR0: C'don 73Tb 157
Grenaby Rd. CR0: C'don 73Tb 157
Grenada Ho. E14 45Bc 92
(off Limehouse C'way.)
Grenada Rd. SE7 52Lc 115
Grenade St. E14 45Bc 92
Grenadier Cl. AL4: St A 3G 6
Grenadier Pl. CR3: Cat'm 94Sb 197
Grenadier St. E16 46Qc 94
Grenard Cl. SE15 52Wb 113
Grena Gdns. TW9: Rich 56Pa 109
Grena Rd. TW9: Rich 56Pa 109
Grendon Gdns. HA9: Wemb 33Qa 67
Grendon Ho. E9 38Yb 72
(off Shore Pl.)
Grendon Ho. N1 2H 217 (40Pb 70)
(off Calshot St.)
Grendon Lodge HA8: Edg 19Sa 29
Grendon St. NW8 5D 214 (42Gb 89)
Grenfell Av. RM12: Horn 32Hd 76
Grenfell Cl. WD6: Bore 11Sa 29
Grenfell Ct. E3 42Dc 92
(off Barry Blandford Way)
Grenfell Ct. NW7 23Xa 48
Grenfell Gdns. HA3: Kenton 31Na 67
Grenfell Gdns. IG3: Ilf 29Vc 55
Grenfell Ho. SE5 52Sb 113
Grenfell Rd. CR4: Mitc 65Hb 133
Grenfell Rd. W11 45Za 88
Grenfell Wlk. W11 45Za 88
Grenier Apts. SE15 52Xb 113
Grennan Cl. CM13: Ingve 23Fe 59
Grennell Cl. SM1: Sutt 75Fb 155
Grennell Rd. SM1: Sutt 75Eb 155
Grenoble Gdns. N13 23Qb 50
Grenside Rd. KT13: Weyb 76R 150
Grenville Cl. EN8: Walt C 4Zb 20
Grenville Cl. KT11: Cobh 85Z 171
Grenville Cl. KT5: Surb 74Sa 153
Grenville Cl. N3 25Ab 48
Grenville Cl. SL1: Burn 10A 60
Grenville Cl. TW2: Twick 62Fa 130
Grenville Cl. W13 43Ka 86
Grenville Ct. WD3: Chor 14E 24
Grenville Gdns. IG8: Wfd G 25Lc 53
Grenville Ho. E3 40Ac 72
(off Arbery Rd.)
Grenville Ho. SE8 51Cc 114
(off New King St.)
Grenville Ho. SW1 51Mb 112
(off Dolphin Sq.)
Grenville M. N19 32Nb 70
Grenville M. SW7 6A 226 (49Eb 89)
Grenville M. TW12: Hamp H 64Da 129
Grenville Pl. NW7 22Ta 47
Grenville Pl. SW7 4A 226 (48Eb 89)
Grenville Rd. CR0: New Ad 81Ec 178
Grenville Rd. N19 32Nb 70
Grenville Rd. RM16: Chaf H 49Xd 98
Grenville St. WC1 6G 217 (42Nb 90)
Gresford Cl. AL4: St A 2H 7
Gresham Av. CR6: W'ham 90Ac 178
Gresham Av. DA3: Hartl 70Be 143
Gresham Av. N20 21Hb 49
Gresham Cl. CM14: B'wood 20Yd 40
Gresham Cl. DA5: Bexl 58Ad 117
Gresham Cl. EN2: Enf 13Sb 33
Gresham Cl. RH8: Oxt 1K 211
Gresham Cl. TN16: Tats 93Lc 199
Gresham Ct. CM14: B'wood 20Yd 40
Gresham Ct. CR8: Purl 83Qb 176
Gresham Ct. TW18: Staines 64J 127
Gresham Dr. RM6: Chad H 29Xc 55
Gresham Gdns. NW11 32Ab 68
Gresham Ho. GU22: Wok 94C 188
Gresham Lodge E17 29Dc 52
Gresham Pk. Rd. GU22: Wok 93D 188
Gresham Pl. E3 43Cc 92
Gresham Pl. N19 33Mb 70
Gresham Pl. RH8: Oxt 1K 211
Gresham Rd. BR3: Beck 68Ac 136
Gresham Rd. CM14: B'wood 20Yd 40
Gresham Rd. E16 44Kc 93
Gresham Rd. E6 40Pc 74
Gresham Rd. HA8: Edg 23Pa 47
Gresham Rd. NW10 36Ta 67
Gresham Rd. RH8: Oxt 100Hc 199
Gresham Rd. SE25 70Wb 135
Gresham Rd. SL1: Slou 4E 80
Gresham Rd. SW9 55Qb 112
Gresham Rd. TW12: Hamp 65Ca 129
Gresham Rd. TW18: Staines 64H 127
Gresham Rd. TW3: Houn 53Ea 108
Gresham Rd. UB10: Hil 40Q 64
Gresham St. EC2 2D 224 (44Sb 91)
Gresham Way SW19 62Db 133
Gresham Way Ind. Est. SW19 62Db 133
(off Gresham Way)
Gresley Cl. E17 30Ac 52
Gresley Cl. N15 28Tb 51
Gresley Ct. EN1: Enf 7Yb 20
Gresley Ct. EN6: Pot B 2Eb 17
Gresley Ho. SW8 52Mb 112
Gresley Rd. N19 32Lb 70
Gressenhall Rd. SW18 58Bb 111
Gresse St. W1 2D 222 (44Mb 90)
Gresswell Cl. DA14: Sidc 62Wc 139
Greswell St. SW6 53Za 110
Greta Bank KT24: W Hor 98S 190
Gretton Ho. E2 41Yb 92
(off Globe Rd.)
Gretton Rd. N17 24Vb 51
Greville Av. CR2: Sels 82Zb 178
Greville Cl. AL9: Wel G 6E 8
Greville Cl. KT21: Asht 91Na 193
Greville Cl. TW1: Twick 59Ka 108
Greville Ct. E5 34Xb 71
(off Napoleon Rd.)
Greville Ct. HA1: Harr 35Ga 66
Greville Ct. KT21: Asht 90Na 173
(off Greville Pk. Rd.)
Greville Ct. KT23: Bookh 97Da 191
Greville Hall NW6 40Db 69
Greville Ho. SW1 3G 227 (48Hb 89)
(off Kinnerton St.)
Greville Lodge E13 39Kc 73
Greville Lodge HA8: Edg 21Ra 47
(off Broadhurst Av.)
Greville Lodge N12 22Db 49
Greville M. NW6 39Db 69
(off Greville Rd.)
Greville Pk. Av. KT21: Asht 90Na 173
Greville Pk. Rd. KT21: Asht 90Na 173
Greville Pl. NW6 40Db 69
Greville Rd. E17 28Ec 52
Greville Rd. NW6 40Db 69
Greville Rd. TW10: Rich 58Pa 109
Greville St. EC1 1K 223 (43Qb 90)
(not continuous)
Grey Alders SM7: Bans 86Ya 174
Greycaine Rd. WD24: Wat 9Z 13
Greycaine Trad. Est. WD24: Wat 9Z 13
Grey Cl. NW11 30Eb 49
Greycoat Gdns. SW1 4D 228 (48Mb 90)
(off Greycoat St.)
Greycoat Pl. SW1 4D 228 (48Mb 90)
Greycoat St. SW1 4D 228 (48Mb 90)
Greycot Rd. BR3: Beck 64Cc 136
Grey Eagle St. E1 6K 219 (42Vb 91)
Greyfell Cl. HA7: Stan 22Ka 46
Greyfields Cl. CR8: Purl 85Rb 177
Greyford Cl. KT22: Lea 95La 192
Greyfriars CM13: Hut 17De 41
Greyfriars SE26 62Wb 135
(off Wells Pk. Rd.)
Greyfriars Dr. GU24: Bisl 7E 166
Greyfriars Dr. SL5: Asc 1A 146
Greyfriars Ho. RM11: Horn 30Md 57
Greyfriars Pas. EC1 2C 224 (44Rb 91)
Greyfriars Rd. GU23: Rip 96J 189
Greyhound Commercial Cen., The DA1: Cray 57Gd 118
Greyhound Ct. WC2 4J 223 (45Pb 90)
Greyhound Hill NW4 27Wa 48
Greyhound La. EN6: S Mim 5Wa 16
Greyhound La. RM16: Ors 7C 100
Greyhound La. SW16 65Mb 134
Greyhound Mans. W6 51Ab 110
(off Greyhound Rd.)
Greyhound Rd. N17 27Ub 51
Greyhound Rd. NW10 41Xa 88
Greyhound Rd. SM1: Sutt 78Eb 155
Greyhound Rd. W14 51Ab 110
Greyhound Rd. W6 51Za 110
Greyhound Ter. SW16 67Lb 134
Greyhound Way DA1: Cray 57Gd 118
Grey Ho. W12 45Xa 88
(off White City Est.)
Grey Ho., The WD17: Wat 12W 26
Greyladies Gdns. SE10 54Ec 114
Greys Pk. Cl. BR2: Kes 78Mc 159
Greystead Rd. SE23 59Yb 114
Greystoke Av. HA5: Pinn 27Ca 45
Greystoke Ct. W5 42Na 87
Greystoke Dr. HA4: Ruis 30R 44
Greystoke Gdns. EN2: Enf 14Mb 32
Greystoke Gdns. W5 42Na 87
Greystoke Ho. SE15 51Wb 113
(off Peckham Pk. Rd.)
Greystoke Ho. W5 42Na 87
Greystoke Lodge W5 42Pa 87
(off Hanger La.)
Greystoke Pk. Ter. W5 41Ma 87
Greystoke Pl. EC4 2K 223 (44Qb 90)
Greystoke Rd. SL2: Slou 3D 80
Greystone Cl. CR2: Sels 83Yb 178
Greystone Gdns. HA3: Kenton 30La 46
Greystone Gdns. IG6: Ilf 26Sc 54
Greystone Pk. TN14: Sund 97Ad 201
Greystones Cl. RH1: Redh 8M 207
Greystones Cl. TN15: Kems'g 89Nd 183
Greystones Dr. RH2: Reig 4L 207
Greyswood St. SW16 65Kb 134
Greythorne Rd. GU21: Wok 10L 167
Grey Towers Av. RM11: Horn 31Md 77
Grey Towers Gdns. RM11: Horn 31Ld 77
Grey Turner Ho. W12 44Wa 88
Grice Av. TN16: Big H 85Kc 179
Gridiron Pl. RM14: Upm 34Rd 77
Grierson Ho. SW16 63Lb 134
Grierson Rd. SE23 59Zb 114
Grieves Rd. DA11: Nflt 2B 144
Griffen Ct. BR3: Beck 67Dc 136
Griffin Av. RM14: Upm 30Ud 58
Griffin Cen. TW14: Felt 57X 107
Griffin Cen., The KT1: King T 68Ma 131
(off Market Pl.)
Griffin Cl. NW10 36Xa 68
Griffin Cl. SL1: Slou 7G 80
Griffin Ct. DA11: Nflt 57Ce 121
Griffin Ct. KT21: Asht 91Pa 193
Griffin Ct. KT23: Bookh 98Da 191
Griffin Ct. TW8: Bford 51Na 109
Griffin Ct. W4 50Va 88
Griffin Ho. CR0: C'don 73Rb 157
Griffin Ho. E14 44Dc 92
(off Ricardo St.)
Griffin Ho. N1 39Ub 71
(off Halcomb St.)
Griffin Ho. W6 49Za 88
(off Hammersmith Rd.)
Griffin Mnr. Way SE28 48Tc 94
Griffin M. SW12 60Lb 112
Griffin Pk. 51Ma 109
Griffin Rd. N17 26Ub 51
Griffin Rd. SE18 50Tc 94
Griffins, The RM16: Grays 47De 99
Griffins Cl. N21 17Tb 33
Griffin's Wood Cotts. CM16: Epp 4Tc 22
Griffin Wlk. DA9: Ghithe 57Vd 120
Griffin Way KT23: Bookh 98Ca 191
Griffin Way TW16: Sun 68W 128
Griffith Cl. E17 25Fc 53
Griffith Cl. RM8: Dag 31Yc 75
Griffiths Cl. KT4: Wor Pk 75Xa 154
Griffiths Cl. WD23: Bush 15Fa 28
Griffiths Rd. RM19: Purf 49Sd 98
Griffiths Rd. SW19 66Cb 133
Griffiths Way AL1: St A 4A 6
Griffon Way WD25: Wat 6V 12
Grifon Cl. RM16: Chaf H 48Yd 98
Grifon Rd. RM16: Chaf H 49Yd 98
Griggs App. IG1: Ilf 33Sc 74
Griggs Cl. IG3: Ilf 35Uc 74
Griggs Ct. SE1 4J 231 (48Ub 91)
(off Grigg's Pl.)
Griggs Gdns. RM12: Horn 36Ld 77
Grigg's Pl. SE1 3J 231 (48Ub 91)
Griggs Rd. E10 30Ec 52
Griggs Way TN15: Bor G 92Ce 205
Grilse Cl. N9 21Xb 51
Grimaldi Ho. N1 1H 217 (40Pb 70)
(off Calshot St.)
Grimsby Gro. E16 47Rc 94
Grimsby Rd. SL1: Slou 7D 80
Grimsby St. E2 42Vb 91
Grim's Ditch 22Da 45
Grimsdyke Cres. EN5: Barn 13Ya 30
Grim's Dyke Golf Course 22Ca 45
Grimsdyke Lodge AL1: St A 2E 6
Grimsdyke Rd. HA5: Hat E 24Aa 45
Grimsel Path SE5 52Rb 113
Grimshaw Cl. N6 31Jb 70
Grimshaw Way RM1: Rom 29Hd 56
Grimstone Cl. RM5: Col R 23Dd 56
Grimston Rd. AL1: St A 3D 6
Grimston Rd. SW6 54Bb 111
Grimthorpe Ho. EC1 5B 218 (42Rb 91)
(off Agdon St.)
Grimwade Av. CR0: C'don 76Wb 157
Grimwade Cl. SE15 55Yb 114
Grimwood Rd. TW1: Twick 59Ha 108
Grindall Cl. CR0: Wadd 77Rb 157
Grindall Ho. E1 42Xb 91
(off Darling Row)
Grindal St. SE1 2K 229 (47Qb 90)
Grindcobbe AL1: St A 5B 6
Grindleford Av. N11 19Jb 32
Grindley Gdns. CR0: C'don 72Vb 157
Grindley Ho. E3 43Bc 92
(off Leopold St.)
Grindstone Cres. GU21: Knap 10F 166
GRINDSTONE HANDLE CORNER 10F 166
Grinling Pl. SE8 51Cc 114
Grinstead Rd. SE8 50Ac 92
Grisedale NW1 3B 216 (41Lb 90)
(off Cumberland Mkt.)
Grisedale Cl. CR8: Purl 86Ub 177
Grisedale Gdns. CR8: Purl 86Ub 177
Grittleton Av. HA9: Wemb 37Ra 67
Grittleton Rd. W9 42Cb 89
Grizedale Ter. SE23 61Xb 135
Grobars Av. GU21: Wok 7N 167
Grocer's Hall Ct. EC2 3F 225 (44Tb 91)
Grocer's Hall Gdns. EC2 3F 225 (44Tb 91)
(off Prince's St.)
Grogan Cl. TW12: Hamp 65Ba 129
Groombridge Cl. DA16: Well 57Wc 117
Groombridge Cl. KT12: Hers 78X 151
Groombridge Ho. SE17 7J 231 (50Ub 91)
(off Upnor Way)
Groombridge Rd. E9 38Zb 72
Groom Cl. BR2: Broml 70Kc 137
Groom Cres. SW18 59Fb 111
Groome Ho. SE11 6J 229 (49Pb 90)
Groomfield Cl. SW17 63Jb 134
Groom Pl. SW1 3J 227 (48Jb 90)
Grooms Dr. HA5: Eastc 29W 44
Grosmont Rd. SE18 50Vc 95
Grosse Way SW15 58Xa 110
Grosvenor Av. HA2: Harr 30Da 45
Grosvenor Av. N5 36Sb 71
Grosvenor Av. SM5: Cars 79Hb 155
Grosvenor Av. SW14 55Ua 110
Grosvenor Av. TW10: Rich 57Na 109
Grosvenor Av. UB4: Hayes 40V 64
Grosvenor Av. WD4: K Lan 10C 4
Grosvenor Cl. IG10: Lough 11Rc 36
Grosvenor Cl. SL0: Iver H 41F 82
Grosvenor Cotts. SW1 5H 227 (49Jb 90)
Grosvenor Ct. E10 32Dc 72
Grosvenor Ct. E14 44Bc 92
(off Wharf La.)
Grosvenor Ct. N14 17Lb 32
Grosvenor Ct. NW6 39Za 68
Grosvenor Ct. NW7 22Ta 47
(off Hale La.)
Grosvenor Ct. SE5 51Sb 113
Grosvenor Ct. SL1: Slou 4J 81
Grosvenor Ct. SL9: Ger X 28A 42
Grosvenor Ct. SM2: Sutt 79Db 155
Grosvenor Ct. SM4: Mord 70Cb 133
Grosvenor Ct. TW11: Tedd 65Ja 130
Grosvenor Ct. W14 48Za 88
(off Irving Rd.)
Grosvenor Ct. W3 46Qa 87
Grosvenor Ct. W5 45Na 87
(off The Grove)
Grosvenor Ct. WD3: Crox G 15T 26
Grosvenor Ct. Mans. W2 3F 221 (44Hb 89)
(off Edgware Rd.)
Grosvenor Cres. DA1: Dart 57Md 119
Grosvenor Cres. NW9 28Qa 47
Grosvenor Cres. SW1 2J 227 (47Jb 90)
Grosvenor Cres. UB10: Hil 38R 64
Grosvenor Cres. M. SW1 2H 227 (47Jb 90)
Grosvenor Dr. IG10: Lough 12Rc 36
Grosvenor Dr. RM11: Horn 32Ld 77
Grosvenor Est. SW1 5E 228 (49Mb 90)
Grosvenor Gdns. E6 41Mc 93
Grosvenor Gdns. IG8: Wfd G 23Jc 53
Grosvenor Gdns. KT2: King T 65Ma 131
Grosvenor Gdns. N10 27Lb 50
Grosvenor Gdns. N14 14Mb 32
Grosvenor Gdns. NW11 30Bb 49
Grosvenor Gdns. NW2 37Ya 68
Grosvenor Gdns. RM14: Upm 32Td 78
Grosvenor Gdns. SM6: W'gton 80Lb 156
Grosvenor Gdns. SW1 3K 227 (48Kb 90)
Grosvenor Gdns. SW14 55Ua 110
Grosvenor Gdns. M. E. SW1 3A 228 (48Kb 90)
(off Beeston Pl.)
Grosvenor Gdns. M. Nth. SW1 4K 227 (48Kb 90)
(off Grosvenor Gdns.)
Grosvenor Gdns. M. Sth. SW1 4A 228 (48Kb 90)
(off Ebury St.)
Grosvenor Ga. W1 5H 221 (45Jb 90)
Grosvenor Hgts. E4 17Gc 35
Grosvenor Hill SW19 65Ab 132
Grosvenor Hill W1 4K 221 (45Kb 90)
Grosvenor Hill Ct. W1 4K 221 (45Kb 90)
(off Bourdon St.)
Grosvenor Ho. SM1: Sutt 78Db 155
(off West St.)
Grosvenor M. KT18: Eps D 91Ta 193
Grosvenor M. RH2: Reig 9K 207
Grosvenor Pde. W5 46Qa 87
(off Uxbridge Rd.)
Grosvenor Pk. SE5 52Sb 113
Grosvenor Pk. Rd. E17 29Cc 52
Grosvenor Path IG10: Lough 11Rc 36
Grosvenor Pl. GU21: Wok 89B 168
(off Stanley Rd.)
Grosvenor Pl. KT13: Weyb 76T 150
Grosvenor Pl. SW1 2J 227 (47Jb 90)
Grosvenor Ri. E. E17 29Dc 52
Grosvenor Rd. AL1: St A 3C 6
Grosvenor Rd. BR4: W W'ck 74Dc 158
Grosvenor Rd. BR5: St M Cry 72Uc 160
Grosvenor Rd. DA17: Belv 51Cd 118
Grosvenor Rd. DA6: Bex 57Zc 117
Grosvenor Rd. E10 32Ec 72
Grosvenor Rd. E11 29Kc 53
Grosvenor Rd. E6 39Mc 73
Grosvenor Rd. E7 37Kc 73
Grosvenor Rd. GU24: Chob 5H 167
Grosvenor Rd. HA6: Nwood 22V 44
Grosvenor Rd. IG1: Ilf 34Sc 74
Grosvenor Rd. KT18: Eps D 91Ta 193
Grosvenor Rd. N10 25Kb 50
Grosvenor Rd. N3 24Bb 49
Grosvenor Rd. N9 18Xb 33
Grosvenor Rd. RM16: Ors 4F 100
Grosvenor Rd. RM7: Rush G 31Fd 76
Grosvenor Rd. RM8: Dag 32Bd 75
Grosvenor Rd. SE25 70Vb 135
Grosvenor Rd. SM6: W'gton 79Kb 156
Grosvenor Rd. SW1 51Kb 112
Grosvenor Rd. TW1: Twick 60Ja 108
Grosvenor Rd. TW10: Rich 57Na 109
Grosvenor Rd. TW18: Staines 66J 127
Grosvenor Rd. TW3: Houn 55Ba 107
Grosvenor Rd. TW8: Bford 51Ma 109
Grosvenor Rd. UB2: S'hall 48Ba 85
Grosvenor Rd. W4 50Ra 87
Grosvenor Rd. W7 46Ja 86
Grosvenor Rd. WD17: Wat 14Y 27
Grosvenor Rd. WD6: Bore 13Qa 29
Grosvenor Sq. DA3: Lfield 69Ae 143
(off Park Dr.)
Grosvenor Sq. W1 4J 221 (45Jb 90)
Grosvenor Sq. WD4: K Lan 10C 4
Grosvenor St. W1 4K 221 (45Kb 90)
Grosvenor Studios SW1 5H 227 (49Jb 90)
(off Eaton Ter.)
Grosvenor Ter. HP1: Hem H 3J 3
Grosvenor Ter. SE5 52Sb 113
Grosvenor Va. HA4: Ruis 33V 64
Grosvenor Vale Stadium 33V 64
Grosvenor Way E5 33Yb 72
Grosvenor Way SW17 62Fb 133
Grosvenor Wharf Rd. E14 49Fc 93
Grotes Bldgs. SE3 54Gc 115
Grote's Pl. SE3 54Gc 115
Groton Rd. SW18 61Db 133
Grotto Ct. SE1 1C 230 (47Rb 91)
Grotto Pas. W1 7H 215 (43Jb 90)
Grotto Rd. KT13: Weyb 76R 150
Grotto Rd. TW1: Twick 61Ha 130
Groundsel Wlk. HP2: Hem H 2D 4
Grove, The AL9: Brk P 8J 9
Grove, The BR4: W W'ck 76Dc 158
Grove, The BR8: Swan 69Hd 140
Grove, The CM14: B'wood 21Vd 58
Grove, The CR3: Cat'm 93Rb 197
Grove, The CR5: Coul 87Mb 176
Grove, The DA10: Swans 57Be 121
Grove, The DA12: Grav'nd 9D 122
Grove, The DA14: Sidc 64Ad 139
Grove, The DA6: Bex 56Zc 117
Grove, The E15 37Gc 73
Grove, The EN2: Enf 12Qb 32
Grove, The EN6: Pot B 4Eb 17
Grove, The GU21: Wok 88B 168
Grove, The HA1: Harr 31Ga 66
Grove, The HA7: Stan 19Ja 28
Grove, The HA8: Edg 21Ra 47
Grove, The HP5: Lat 8A 10
Grove, The KT12: Walt T 73X 151
Grove, The KT15: Add 78K 149
Grove, The KT17: Eps 85Ua 174
Grove, The KT17: Ewe 82Va 174
Grove, The KT24: Eff 100Z 191
Grove, The N13 21Qb 50
(not continuous)
Grove, The N3 25Cb 49
Grove, The N4 31Pb 70
Grove, The N6 32Jb 70
Grove, The N8 29Mb 50
Grove, The NW11 31Ab 68
Grove, The NW9 29Ta 47
Grove, The RM14: Upm 35Rd 77
GROVE, THE 60Wb 113
Grove, The SL1: Slou 7L 81
Grove, The SS17: Stan H 3M 101
Grove, The TN15: W King 82Vd 184
Grove, The TN16: Big H 90Mc 179
Grove, The TW1: Twick 58Ka 108
Grove, The TW11: Tedd 63Ja 130
Grove, The TW20: Egh 64C 126
Grove, The TW7: Isle 53Ga 108
Grove, The UB10: Ick 36Q 64
Grove, The UB6: G'frd 44Ea 86
Grove, The W5 46Ma 87
Grove, The WD3: Crox G 14Q 26
(off Dugdales)
Grove, The WD4: K Lan 2L 11
Grove, The WD7: R'lett 6Ja 14
Grove Av. HA5: Pinn 28Aa 45
Grove Av. KT17: Eps 85Ua 174
Grove Av. N10 26Lb 50
Grove Av. N3 24Cb 49
Grove Av. SM1: Sutt 79Cb 155
Grove Av. TW1: Twick 60Ha 108
Grove Av. W7 45Ga 86
Grove Bank WD19: Wat 18Z 27
Grovebarns TW18: Staines 65J 127
Grovebury Cl. DA8: Erith 51Fd 118
Grovebury Ct. DA6: Bex 57Dd 118
Grovebury Ct. N14 17Mb 32
Grovebury Gdns. AL2: Park 9A 6
Grovebury Rd. SE2 47Xc 95
Grove Cl. BR2: Hayes 75Jc 159
Grove Cl. KT1: King T 70Pa 131
Grove Cl. KT19: Eps 82Qa 173
Grove Cl. N14 17Lb 32
Grove Cl. SE23 60Ac 114
Grove Cl. SL1: Slou 8L 81
Grove Cl. SL4: Old Win 9M 103
Grove Cl. TW13: Hanw 63Aa 129
Grove Cl. UB10: Ick 36Q 64
Grove Cnr. KT23: Bookh 98Da 191
Grove Cotts. SW3 51Gb 111
(off Chelsea Mnr. St.)
Grove Cotts. W4 51Ua 110
Grove Cotts. WD23: Bush 16Ca 27
(off Falconer Rd.)
Grove Ct. EN5: Barn 13Bb 31
(off Hadley Ridge)
Grove Ct. EN9: Walt A 5Dc 20
Grove Ct. KT1: King T 69Na 131
(off Grove Cres.)
Grove Ct. KT8: E Mos 71Fa 152
Grove Ct. NW8 3B 214 (41Fb 89)
(off Grove End Rd.)
Grove Ct. RH1: Redh 4B 208
(off Gumbrell M.)
Grove Ct. RM14: Upm 35Qd 77
Grove Ct. SE15 52Ub 113
(off Peckham Rd.)
Grove Ct. SW10 7A 226 (50Eb 89)
(off Drayton Gdns.)
Grove Ct. TW20: Egh 64C 126
Grove Ct. TW3: Houn 56Ca 107
Grove Ct. W5 46Na 87
Grove Craft Workshops, The DA3: Fawk 73Wd 164
Grove Cres. E18 26Hc 53
Grove Cres. KT1: King T 69Na 131
Grove Cres. KT12: Walt T 73X 151
Grove Cres. NW9 28Sa 47
Grove Cres. TW13: Hanw 63Aa 129
Grove Cres. WD3: Crox G 14Q 26
Grove Cres. Rd. E15 37Fc 73
Grovedale Cl. EN7: Chesh 2Vb 19
Grovedale Rd. N19 33Mb 70

Grove Dwellings E1 43Yb 92
Grove End E18 26Hc 53
Grove End NW5 35Kb 70
Grove End Gdns. NW8 2B 214 (40Fb 69)
Grove End Ho. NW8 4B 214 (41Fb 89)
(off Grove End Rd.)
Grove End La. KT10: Esh 74Fa 152
Grove End Rd. NW8 2B 214 (40Fb 69)
Grove Farm Pk. HA6: Nwood 22T 44
Grove Farm Retail Pk. 31Yc 75
Grovefield N11 21Kb 50
(off Coppies Gro.)
Grove Footpath KT5: Surb 70Na 131
Grove Gdns. EN3: Enf W 10Zb 20
Grove Gdns. NW4 29Wa 48
Grove Gdns. NW8 4E 214 (41Gb 89)
Grove Gdns. RM10: Dag 34Ed 76
Grove Gdns. TW10: Rich 58Pa 109
Grove Gdns. TW11: Tedd 63Ja 130
Grove Golf Course Watford, The 9S 12
Grove Grn. HA6: Nwood 22T 44
Grove Grn. Rd. E11 34Ec 72
Grove Hall Ct. E3 40Cc 72
(off Jebb St.)
Grove Hall Ct. NW8 3A 214 (41Eb 89)
Grove Hall Rd. WD23: Bush 14Aa 27
GROVE HEATH 95K 189
Grove Heath Ct. GU23: Rip 96L 189
Grove Heath Nth. GU23: Rip 94K 189
Grove Heath Rd. GU23: Rip 95K 189
Groveherst Rd. DA1: Dart 55Pd 119
Grove Hill E18 26Hc 53
Grove Hill HA1: Harr 31Ga 66
Grovehill Ct. BR1: Broml 65Hc 137
Grove Hill Rd. HA1: Harr 31Ha 66
Grove Hill Rd. SE5 55Ub 113
Grovehill Rd. RH1: Redh 6P 207
Grove Ho. CM14: W'ley 21Xd 58
Grove Ho. EN8: Chesh 2Xb 19
Grove Ho. KT17: Eps 85Ua 174
(off The Grove)
Grove Ho. N3 27Za 48
Grove Ho. RH1: Redh 6P 207
(off Huntingdon Rd.)
Grove Ho. SW3 51Gb 111
(off Chelsea Mnr. St.)
Grove Ho. WD23: Bush 16Ba 27
Grove Ho. Rd. N8 28Nb 50
Groveland Av. SW16 66Pb 134
Groveland Ct. EC4 3E 224 (44Sb 91)
(off Bow La.)
Groveland Rd. BR3: Beck 69Bc 136
Grovelands AL2: Park 9P 5
Grovelands KT1: King T 70Ma 131
(off Palace Rd.)
Grovelands KT8: W Mole 70Ca 129
Grovelands Cl. HA2: Harr 34Da 65
Grovelands Cl. SE5 54Ub 113
Grovelands Ct. N14 17Mb 32
Grovelands Rd. BR5: St P 66Wc 139
Grovelands Rd. CR8: Purl 84Nb 176
Grovelands Rd. N13 21Pb 50
Grovelands Rd. N15 30Wb 51
Grovelands Way RM17: Grays 50Be 99
Groveland Way KT3: N Mald 71Sa 153
Grove La. CM16: Epp 2Wc 23
Grove La. CR5: Bans 86Kb 176
Grove La. CR5: Coul 86Kb 176
Grove La. IG7: Chig 20Vc 37
Grove La. KT1: King T 70Na 131
Grove La. SE5 53Tb 113
Grove La. UB8: Hil 42P 83
Grove La. Ter. SE5 54Tb 113
Grove Lea AL10: Hat 3C 8
Groveley Rd. TW13: Felt 63W 128
Groveley Rd. TW16: Sun 64U 128
Grove Mans. W6 47Ya 88
(off Hammersmith Gro.)
Grove Mead AL10: Hat 1B 8
Grove M. W6 48Ya 88
Grove Mill La. WD17: Wat 9R 12
Grove Mill Pl. SM5: Cars 76Jb 156
Grove Nature Reserve, The 42P 83
Grove Pde. SL1: Slou 7L 81
Grove Pk. E11 30Kc 53
Grove Pk. NW9 28Sa 47
GROVE PARK 62Kc 137
Grove Pk. SE5 54Ub 113
GROVE PARK 53Sa 109
Grove Pk. Av. E4 24Dc 52
Grove Pk. Bri. 52Sa 109
Grove Pk. Gdns. W4 52Ra 109
Grove Pk. M. W4 52Sa 109
Grove Pk. Nature Reserve 60Hc 115
Grove Pk. Rd. N15 28Ub 51
Grove Pk. Rd. RM13: Rain 39Jd 76
Grove Pk. Rd. SE9 62Lc 137
Grove Pk. Rd. W4 52Ra 109
Grove Park Station (Rail) 62Kc 137
Grove Pk. Ter. W4 52Ra 109
Grove Pas. E2 40Xb 71
Grove Path EN7: Chesh 3Wb 19
Grove Pl. AL9: Wel G 6E 8
Grove Pl. IG11: Bark 38Sc 74
Grove Pl. KT13: Weyb 78S 150
Grove Pl. NW3 34Fb 69
Grove Pl. SE9 58Pc 116
Grove Pl. SW12 59Kb 112
Grove Pl. W3 46Sa 87
Grove Pl. WD25: A'ham 11Da 27
Grover Cl. HP2: Hem H 1M 3
Grover Ct. SE13 54Dc 114
Grover Ho. HP3: Hem H 8P 3
Grover Ho. SE11 7J 229 (50Pb 90)
Grove Rd. AL1: St A 3B 6
Grove Rd. CR4: Mitc 69Jb 134
(not continuous)
Grove Rd. CR7: Thor H 70Qb 134
Grove Rd. DA11: Nflt 57De 121
Grove Rd. DA17: Belv 51Bd 117
Grove Rd. DA7: Bex 56Ed 118
Grove Rd. E11 31Hc 73
Grove Rd. E17 30Dc 52
Grove Rd. E18 26Hc 53
Grove Rd. E3 39Zb 72
Grove Rd. E4 21Ec 52
Grove Rd. EN4: Cockf 13Gb 31
Grove Rd. GU21: Wok 88B 168
Grove Rd. HA5: Pinn 29Ba 45
Grove Rd. HA6: Nwood 22T 44
Grove Rd. HA8: Edg 23Qa 47
Grove Rd. HP1: Hem H 4J 3
Grove Rd. KT16: Chert 72H 149
Grove Rd. KT17: Eps 85Ua 174
Grove Rd. KT21: Asht 90Pa 173
Grove Rd. KT6: Surb 71Ma 153
Grove Rd. KT8: E Mos 70Fa 130
Grove Rd. N11 22Kb 50
Grove Rd. N12 22Fb 49
Grove Rd. N15 29Ub 51
Grove Rd. NW2 37Ya 68
Grove Rd. RH1: Redh 6P 207
Grove Rd. RH8: Tand 5G 210
Grove Rd. RM17: Grays 51De 121
Grove Rd. RM6: Chad H 31Xc 75
Grove Rd. SL1: Burn 1B 80
Grove Rd. SL4: Wind 4G 102
Grove Rd. SM1: Sutt 79Cb 155
Grove Rd. SS17: Stan H 3M 101
Grove Rd. SW13 54Va 110
Grove Rd. SW19 66Eb 133
Grove Rd. TN14: S'oaks 93Ld 203
Grove Rd. TN15: Seal 93Qd 203
Grove Rd. TN16: Tats 92Lc 199
Grove Rd. TW10: Rich 58Pa 109
Grove Rd. TW17: Shep 72S 150
Grove Rd. TW2: Twick 62Fa 130
Grove Rd. TW3: Houn 56Ca 107
Grove Rd. TW7: Isle 53Ga 108
Grove Rd. TW8: Bford 50La 86
Grove Rd. UB8: Uxb 38M 63
Grove Rd. W3 46Sa 87
Grove Rd. W5 45Ma 87
Grove Rd. WD3: Rick 19J 25
Grove Rd. WD6: Bore 11Qa 29
Grove Rd. W. EN3: Enf W 9Yb 20
Grover Rd. WD19: Wat 17Z 27
Grovers Farm Cotts. KT15: Wdhm 83G 168
Groves Cl. RM15: S Ock 45Vd 98
Grove Shaw KT20: Kgswd 96Ab 194
Groveside KT23: Bookh 98Ca 191
Groveside Cl. KT23: Bookh 99Ca 191
Groveside Cl. SM5: Cars 75Gb 155
Groveside Cl. W3 43Qa 87
Groveside Ct. SW11 54Fb 111
Groveside Rd. E4 19Gc 35
Grovestile Waye TW14: Bedf 59T 106
Grove St. N18 22Vb 51
Grove St. SE8 49Bc 92
Grove Ter. NW5 34Kb 70
Grove Ter. TW11: Tedd 63Ja 130
Grove Ter. UB1: S'hall 45Ca 85
Grove Ter. M. NW5 34Kb 70
Grove Va. BR7: Chst 65Qc 138
Grove Va. SE22 56Vb 113
Grove Vs. E14 45Dc 92
Grove Way HA9: Wemb 36Ra 67
Grove Way KT10: Esh 73Ea 152
Grove Way UB8: Uxb 38M 63
Grove Way WD3: Chor 15D 24
Groveway RM8: Dag 34Zc 75
Groveway SW9 53Pb 112
Grovewood TW9: Kew 53Qa 109
Grove Wood Cl. BR1: Broml 69Qc 138
Grovewood Cl. WD3: Chor 15D 24
Grove Wood Hill CR5: Coul 86Lb 176
Grovewood Pl. IG8: Wfd G 23Pc 54
Grubbs La. AL9: Hat 4J 9
GRUBB STREET 67Xd 142
Grub St. RH8: Limp 100Lc 199
Grummant Rd. SE15 53Vb 113
Grundy St. E14 44Dc 92
Gruneisen Rd. N3 24Db 49
Grunwick Cl. NW2 36Wa 68
Gtec Ho. E15 40Fc 73
(off Canning Rd.)
Guardhouse Way NW7 22Za 48
Guardian Apts. E3 42Bc 92
(off Kevtar Gdn.)
Guardian Av. NW9 27Ua 48
Guardian Av. RM16: N Stif 47Zd 99
Guardian Bus. Cen. RM3: Rom 24Md 57
Guardian Cl. RM11: Horn 33Kd 77
Guardian Ct. SE12 57Gc 115
Guards Av. CR3: Cat'm 94Sb 197
Guards Ct. SL5: S'dale 3F 146
Guardsman Cl. CM14: W'ley 22Zd 59
Guards Memorial 7E 222 (46Mb 90)
Guards' Mus., The 2C 228 (47Lb 90)
Guards Polo Club 7J 125
Guards Rd. SL4: Wind 4A 102
Guards Wlk. SL4: Wind 4A 102
Guards Way CR3: Cat'm 94Sb 197
Gubbins La. RM3: Hrld W 24Pd 57
Gubyon Av. SE24 57Rb 113
Guerin Sq. E3 41Bc 92
Guernsey Cl. TW5: Hest 52Ca 107
Guernsey Farm Dr. GU21: Wok 7P 167
Guernsey Gro. SE24 59Sb 113
Guernsey Ho. EN3: Enf W 10Zb 20
(off Eastfield Rd.)
Guernsey Ho. N1 37Sb 71
(off Channel Island Est.)
Guernsey Ho. WD18: Wat 16W 26
Guernsey Rd. E11 32Fc 73
Guernsey Way GU21: Knap 10E 166
Guglielmo Marconi M. E3 40Bc 72
Guibal Rd. SE12 59Kc 115
Guildersfield Rd. SW16 66Nb 134
Guildford Av. TW13: Felt 61V 128
Guildford Ct. SW8 52Nb 112
(off Guildford Rd.)
Guildford Gdns. RM3: Rom 23Nd 57
Guildford Gro. SE10 53Dc 114
Guildford La. GU22: Wok 1P 187
Guildford Rd. GU22: Wok Bourne Way 4P 187
Guildford Rd. GU22: Wok Wych Hill La. 91A 188
Guildford Rd. AL1: St A 3F 6
Guildford Rd. CR0: C'don 72Tb 157
Guildford Rd. E17 25Ec 52
Guildford Rd. E6 44Pc 94
Guildford Rd. GU21: Knap 9F 166
Guildford Rd. GU21: Wok 83D 168
Guildford Rd. GU24: Bisl 4C 166
Guildford Rd. GU24: Chob 6H 167
Guildford Rd. GU24: Pirb 5D 186
Guildford Rd. GU24: W End 4C 166
Guildford Rd. GU3: Worp 7G 186
Guildford Rd. GU4: Sut G 9P 187
Guildford Rd. IG3: Ilf 33Uc 74
Guildford Rd. KT16: Chert 78E 148
Guildford Rd. KT16: Ott 78E 148
Guildford Rd. KT22: Fet 97Fa 192
Guildford Rd. KT23: Bookh 99Aa 191
Guildford Rd. RM3: Rom 23Nd 57
Guildford Rd. SW8 53Nb 112
Guildford St. KT16: Chert 73H 149
Guildford St. TW18: Staines 65J 127
Guildford Way SM6: W'gton 78Nb 156
Guildhall London 2E 224 (44Sb 91)
(off Aldermanbury)
Guildhall Art Gallery 2F 225 (44Tb 91)
(off Aldermanbury)
Guildhall Bldgs. EC2 2F 225 (44Tb 91)
(off Gresham St.)
Guildhall Coll. 44Xb 91
Guildhall Library 2E 224 (44Sb 91)
(off Aldermanbury)
Guildhall Offices EC2 2E 224 (44Sb 91)
(off Basinghall St.)
Guildhall Yd. EC2 2E 224 (44Sb 91)
Guildhouse, The WD3: Crox G 15Q 26
Guildhouse St. SW1 5B 228 (49Lb 90)
Guildown Av. N12 21Db 49
Guild Rd. DA8: Erith 52Hd 118
Guild Rd. SE7 51Mc 115
Guildsway E17 25Bc 52
Guileshill La. GU23: Ock 95N 189
Guilford Av. KT5: Surb 71Pa 153
Guilford Pl. WC1 6H 217 (42Pb 90)
Guilford St. WC1 6F 217 (42Nb 90)
Guilfoyle NW9 26Va 48
Guillemot Ct. SE8 51Bc 114
(off Alexandra Cl.)
Guillemot Pl. N22 26Pb 50
Guinea Ct. E1 45Wb 91
(off Royal Mint St.)
Guinea Point E14 44Ac 92
(off Repton St.)
Guinery Gro. HP3: Hem H 6P 3
Guinevere Gdns. EN8: Chesh 3Ac 20
Guinness Cl. E9 38Ac 72
Guinness Cl. UB3: Harl 48T 84
Guinness Ct. CR0: C'don 75Vb 157
Guinness Ct. E1 3K 225 (44Vb 91)
(off Mansell St.)
Guinness Ct. EC1 4E 218 (41Sb 91)
(off Lever St.)
Guinness Ct. GU21: Wok 10K 167
Guinness Ct. SE1 1H 231 (47Ub 91)
(off Snowsfields)
Guinness Ct. SW3 6F 227 (49Hb 89)
Guinness Sq. SE1 5H 231 (49Ub 91)
Guinness Trust SW3 6F 227 (49Hb 89)
(off Cadogan St.)
Guinness Trust Bldgs. SE11 7B 230 (50Rb 91)
Guinness Trust Bldgs. W6 50Ya 88
(off Fulham Pal. Rd.)
Guinness Trust Est., The N16 32Ub 71
Guion Rd. SW6 54Bb 111
Gulderose Rd. RM3: Hrld W 26Nd 57
Gulland Wlk. N1 37Sb 71
(off Church Rd.)
Gullane Ho. E3 40Bc 72
(off Shetland Rd.)
Gullbrook HP1: Hem H 2J 3
Gullet Wood Rd. WD25: Wat 7W 12
Gulliver Cl. UB5: N'olt 39Ba 65
Gulliver Rd. DA15: Sidc 61Tc 138
Gulliver's Ho. EC1 6D 218 (42Sb 91)
(off Goswell Rd.)
Gulliver St. SE16 48Ac 92
Gullivers Wlk. SE8 49Bc 92
Gull Wlk. RM12: Horn 38Kd 77
Gulston Wlk. SW3 6G 227 (49Hb 89)
(off Blackland Ter.)
Gumbrell M. RH1: Redh 4B 208
Gumleigh Rd. W5 49La 86
Gumley Cl. RM20: Grays 51Zd 121
Gumley Gdns. TW7: Isle 55Ja 108
Gumley Rd. RM20: Grays 51Zd 121
Gumping Rd. BR5: Farnb 75Sc 160
Gundulf St. SE11 6K 229 (49Qb 90)
Gundulph Rd. BR2: Broml 69Lc 137
Gunfleet Cl. DA12: Grav'nd 9G 122
Gun Hill RM18: W Til 1F 122
Gun Ho. E1 46Xb 91
(off Wapping High St.)
Gunmakers La. E3 39Ac 72
Gun M. IG9: Buck H 21Mc 53
Gunnel Ct. E3 41Ec 92
(off Bolinder Way)
Gunnell Cl. CR0: C'don 72Wb 157
Gunnell Cl. SE25 72Wb 157
(off Backley Gdns.)
Gunnell Cl. SE26 63Wb 135
Gunner Dr. EN3: Enf L 9Cc 20
Gunner La. SE18 50Qc 94
GUNNERSBURY 50Ra 87
Gunnersbury Av. W3 48Qa 87
Gunnersbury Av. W4 49Qa 87
Gunnersbury Av. W5 46Pa 87
Gunnersbury Cl. W4 50Ra 87
Gunnersbury Ct. W3 47Ra 87
Gunnersbury Cres. W3 47Qa 87
Gunnersbury Dr. W5 47Pa 87
Gunnersbury Gdns. W3 47Qa 87
Gunnersbury La. W3 48Qa 87
Gunnersbury Mnr. W5 46Pa 87
Gunnersbury M. W4 50Ra 87
GUNNERSBURY PARK 48Qa 87
Gunnersbury Pk. Mus. 48Qa 87
Gunnersbury Station (Underground & Overground) 50Ra 87
Gunnersbury Triangle Nature Reserve 49Sa 87
Gunners Gro. E4 20Ec 34
Gunners Rd. SW18 61Fb 133
Gunnery Ter. SE18 48Sc 94
Gunning Pl. DA8: Erith 52Hd 118
Gunning Rd. RM17: Grays 50Fe 99
Gunning St. SE18 49Uc 94
Gunn Rd. DA10: Swans 58Ae 121
Gunpowder Sq. EC4 2A 224 (44Qb 90)
(off E. Harding St.)
Gunstor Rd. N16 35Ub 71
Gun St. E1 1K 225 (43Vb 91)
Gunter Gro. HA8: Edg 25Ta 47
Gunter Gro. SW10 51Eb 111
Gunter Hall Studios SW10 51Eb 111
(off Gunter Gro.)
Gunters Mead KT10: Esh 83Ea 172
(not continuous)
Gunters Mead KT22: Oxs 83Ea 172
Gunterstone Rd. W14 49Ab 88
Gunthorpe St. E1 1K 225 (43Vb 91)
Gunton M. SE13 57Fc 115
Gunton Rd. E5 34Xb 71
Gunton Rd. SW17 65Jb 134
Gunwhale Cl. SE16 46Zb 92
Gun Wharf E1 46Yb 92
(off Wapping High St.)
Gunyard M. SE18 52Nc 116
Gurdon Ho. E14 44Cc 92
(off Dod St.)
Gurdon Rd. SE7 50Jc 93
Gurnard Cl. UB7: Yiew 45M 83
Gurnell Gro. W13 42Ha 86
Gurnell Leisure Cen. 41Ha 86
Gurney Cl. E15 36Gc 73
Gurney Cl. E17 25Zb 52
Gurney Cl. IG11: Bark 37Rc 74
Gurney Ct. Rd. AL1: St A 1D 6
Gurney Cres. CR0: C'don 74Pb 156
Gurney Dr. N2 28Eb 49
Gurney Ho. E2 40Wb 71
(off Goldsmiths Row)
Gurney Ho. UB3: Harl 50U 84
Gurney Rd. E15 36Gc 73
Gurney Rd. SM5: Cars 77Jb 156
Gurney Rd. SW6 55Eb 111
Gurney Rd. UB5: N'olt 41X 85
Gurney's Cl. RH1: Redh 7P 207
Guru Nanak Marg DA12: Grav'nd 9E 122
Gutenberg Ct. SE1 4K 231 (48Vb 91)
(off Grange Rd.)
Guthridge Cl. E14 43Cc 92
Guthrie Ct. SE1 3A 230 (48Qb 90)
(off Morley St.)
Guthrie St. SW3 7C 226 (50Fb 89)
Gutteridge La. RM4: Stap A 16Ed 38
Gutter La. EC2 3D 224 (44Sb 91)
Guyatt Gdns. CR4: Mitc 68Jb 134
Guy Barnett Gro. SE3 55Jc 115
Guy Rd. SM6: Bedd 76Mb 156
Guyscliff Rd. SE13 57Ec 114
Guysfield Cl. RM13: Rain 39Jd 76
Guysfield Dr. RM13: Rain 39Jd 76
GUY'S HOSPITAL 1F 231 (47Tb 91)
GUY'S NUFFIELD HOUSE 1F 231 (47Tb 91)
(within Guy's Hospital)
Guys Retreat IG9: Buck H 17Lc 35
Guy St. SE1 1G 231 (47Tb 91)
Guy St. Pk. 1G 231 (47Tb 91)
Guy Townsley Sq. E3 43Dc 92
Gwalior Ho. N14 16Lb 32
Gwalior Rd. SW15 56Za 110
Gwendolen Av. SW15 56Za 110
Gwendolen Cl. SW15 57Za 110
Gwendolen Ho. TW19: Stanw 60N 105
(off Yeoman Dr.)
Gwendoline Av. E13 39Kc 73
Gwendoline Ct. EN8: Walt C 6Bc 20
Gwendwr Rd. W14 50Ab 88
Gweneth Cotts. HA8: Edg 23Qa 47
Gwen Morris Ho. SE5 52Sb 113
Gwennap Pl. KT21: Asht 91Na 193
Gwent Cl. WD25: Wat 6Z 13
Gwent Ct. SE16 46Zb 92
(off Rotherhithe St.)
Gwillim Cl. DA15: Sidc 57Wc 117
Gwilym Maries Ho. E2 41Xb 91
(off Blythe St.)
Gwydor Rd. BR3: Beck 69Zb 136
Gwydyr Rd. BR2: Broml 69Hc 137
Gwyn Cl. SW6 52Eb 111
Gwynedd Cl. TN16: Tats 94Mc 199
Gwynne Av. CR0: C'don 73Zb 158
Gwynne Cl. SL4: Wind 3C 102
Gwynne Cl. W4 51Va 110
Gwynne Ho. E1 43Xb 91
(off Turner St.)
Gwynne Ho. SW1 7H 227 (50Jb 90)
(off Lwr. Sloane St.)
Gwynne Ho. WC1 4K 217 (41Qb 90)
(off Lloyd Baker St.)
Gwynne Pk. Av. IG8: Wfd G 23Pc 54
Gwynne Pl. WC1 4J 217 (41Pb 90)
Gwynne Rd. CR3: Cat'm 95Tb 197
Gwynne Rd. SW11 54Fb 111
Gwynn Rd. DA11: Nflt 61Ee 143
Gyfford Wlk. EN7: Chesh 3Xb 19
Gylcote Cl. SE5 56Tb 113
Gyles Pk. HA7: Stan 25La 46
Gyllyngdune Gdns. IG3: Ilf 33Vc 75
Gym Bloomsbury, The 6F 217 (42Nb 90)
Gym Hemel Hempstead, The 3P 3
Gym Holborn Circus, The 2A 224 (44Qb 90)
(off Thavie's Inn)
Gym Hounslow, The 55Da 107
Gym Kingsbury, The 29Qa 47
Gym London Monument, The 5G 225 (45Tb 91)
Gym Walworth Road, The 50Sb 91
Gymnasium, The N1 2F 217 (40Nb 70)
GYPSY CORNER 43Ta 87
Gypsy La. SL2: Stoke P 5J 61
Gypsy La. WD4: Hunt C 7T 12

H

Haarlem Rd. W14 48Za 88
Haberdasher Est. N1 3G 219 (41Tb 91)
Haberdasher Pl. N1 3G 219 (41Tb 91)
Haberdashers Ct. SE14 55Zb 114
Haberdasher St. N1 3G 219 (41Tb 91)
Habgood Rd. IG10: Lough 13Nc 36
Habitat Cl. SE15 54Xb 113
Habitat Sq. SE10 48Hc 93
(off Teal St.)
Haccombe Rd. SW19 65Eb 133
HACKBRIDGE 74Jb 156
Hackbridge Pk. Gdns. SM5: Cars 75Hb 155
Hackbridge Rd. SM6: W'gton 75Jb 156
Hackbridge Station (Rail) 75Kb 156
Hacketts La. GU22: Pyr 86H 169
Hackford Rd. SW9 53Pb 112
Hackford Wlk. SW9 53Qb 112
Hackforth Cl. EN5: Barn 15Xa 30
Hackington Cres. BR3: Beck 65Cc 136
HACKNEY 37Xb 71
Hackney Central Station (Overground) 37Xb 71
Hackney City Farm 40Wb 71
Hackney Cl. WD6: Bore 15Ta 29
Hackney Downs Station (Overground) 36Xb 71
Hackney Empire Theatre 37Xb 71
Hackney Fashion Hub E9 37Yb 72
Hackney Gro. E8 37Xb 71
Hackney Marshes Cen. 35Bc 72
Hackney Mus. 37Xb 71
Hackney Picturehouse 37Xb 71
(off Mare St.)
Hackney Rd. E2 3K 219 (41Vb 91)
Hackney University Technical Coll. 3J 219 (41Ub 91)
HACKNEY WICK 37Ac 72
HACKNEY WICK 37Bc 72
Hackney Wick Station (Overground) 37Cc 72
Hackworth Point E3 41Dc 92
(off Rainhill Way)
Hacon Sq. E8 38Xb 71
(off Mare St.)
HACTON 36Pd 77
Hacton Dr. RM12: Horn 35Md 77
Hacton La. RM12: Horn 33Pd 77
Hacton La. RM12: Upm 33Pd 77
Hacton La. RM14: Upm 36Pd 77
Hacton Pde. RM12: Horn 34Pd 77
Hadar Cl. N20 18Cb 31
Haddenham Ct. WD19: Wat 20Z 27
Hadden Rd. SE28 48Uc 94
Hadden Way UB6: G'frd 37Fa 66
Haddington Rd. BR1: Broml 62Fc 137
Haddo Ho. SE10 51Dc 114
(off Haddo St.)
Haddon Cl. EN1: Enf 16Wb 33
Haddon Cl. HP3: Hem H 3A 4
Haddon Cl. KT13: Weyb 76U 150
Haddon Cl. KT3: N Mald 71Va 154
Haddon Cl. WD6: Bore 12Qa 29
Haddon Ct. NW4 27Ya 48
Haddon Ct. W3 45Va 88
Haddonfield SE8 49Zb 92
Haddon Gro. DA15: Sidc 59Vc 117
Haddon Rd. BR5: St M Cry 71Yc 161
Haddon Rd. SM1: Sutt 77Db 155
(not continuous)
Haddon Rd. WD3: Chor 15E 24
Haddo St. SE10 51Dc 114
Haden Ct. N4 33Qb 70
Haden La. N11 21Lb 50
Hadfield Cl. UB1: S'hall 41Ba 85
Hadfield Ho. E1 44Wb 91
(off Ellen St.)
Hadfield Rd. SS17: Stan H 2M 101
Hadfield Rd. TW19: Stanw 58M 105
Hadland Cl. HP3: Bov 8C 2
Hadleigh Cl. E1 42Yb 92
Hadleigh Cl. HA2: Harr 35Ca 65
Hadleigh Cl. SW20 68Bb 133
Hadleigh Cl. WD7: Shenl 2Ma 15
Hadleigh Ct. CM14: B'wood 20Wd 40
Hadleigh Ct. E4 17Gc 35
Hadleigh Ct. NW2 37Ya 68
Hadleigh Dr. SM2: Sutt 81Cb 175
Hadleigh Gro. CR5: Coul 88Mb 176
Hadleigh Ho. E1 42Yb 92
(off Hadleigh Cl.)
Hadleigh Lodge IG8: Wfd G 23Jc 53
(off Snakes La. W.)
Hadleigh Rd. N9 17Xb 33
Hadleigh St. E2 41Yb 92
Hadleigh Wlk. E6 44Nc 94
HADLEY 13Bb 31
Hadley Cl. N21 16Qb 32
Hadley Cl. WD6: E'tree 16Pa 29
Hadley Comn. EN5: Barn 12Cb 31
Hadley Comn. EN5: New Bar 12Cb 31
Hadley Ct. EN5: New Bar 13Db 31
Hadley Ct. N16 32Wb 71
Hadley Ct. SL3: Poyle 53G 104
(off Coleridge Cres.)
Hadley Gdns. UB2: S'hall 50Ba 85
Hadley Gdns. W4 50Ta 87
Hadley Grn. EN5: Barn 12Bb 31
Hadley Grn. Rd. EN5: Barn 12Bb 31
Hadley Grn. W. EN5: Barn 12Bb 31
Hadley Gro. EN5: Barn 12Ab 30
Hadley Highstone EN5: Barn 11Bb 31
Hadley M. EN5: Barn 13Bb 31
Hadley Pde. EN5: Barn 13Ab 30
(off High St.)
Hadley Pl. KT13: Weyb 80Q 150
Hadley Ridge EN5: Barn 13Bb 31
Hadley Rd. CR4: Mitc 70Mb 134
Hadley Rd. DA17: Belv 49Bd 95
Hadley Rd. EN2: Enf 10Nb 18
Hadley Rd. EN4: Had W 10Jb 18
Hadley Rd. EN5: New Bar 12Db 31
Hadley St. NW1 37Kb 70
(not continuous)
Hadley Way N21 16Qb 32
HADLEY WOOD 10Eb 17
Hadley Wood Golf Course 11Gb 31
Hadley Wood Lawn Tennis Club 10Eb 17
Hadley Wood Nature Reserve 12Eb 31
Hadley Wood Ri. CR8: Kenley 86Rb 177
Hadley Wood Rd. EN4: Cockf 12Eb 31
Hadley Wood Rd. EN5: Cockf 12Eb 31
Hadley Wood Rd. EN5: New Bar 12Eb 31
Hadley Wood Station (Rail) 10Eb 17
Hadlow Ct. SL1: Slou 6G 80
Hadlow Ho. SE17 7J 231 (50Ub 91)
(off Kinglake Est.)
Hadlow Pl. SE19 66Wb 135
Hadlow Rd. DA14: Sidc 63Wc 139
Hadlow Rd. DA16: Well 52Yc 117
Hadlow Way DA13: Isl R 6A 144
Hadrian Cl. AL3: St A 4M 5
Hadrian Cl. E3 39Cc 72
(off Garrison Rd.)
Hadrian Cl. TW19: Stanw 59N 105
Hadrian Ct. SM2: Sutt 80Db 155
Hadrian Est. E2 40Wb 71
Hadrian M. CR4: Mitc 69Kb 134
Hadrian M. N7 38Pb 70
Hadrians Ride EN1: Enf 15Vb 33
Hadrian St. SE10 50Gc 93
Hadrian Way TW19: Stanw 59M 105
(not continuous)
Hadstock Ho. NW1 3E 216 (41Mb 90)
(off Ossulston St.)
Hadyn Pk. Ct. W12 47Wa 88
(off Curwen Rd.)
Hadyn Pk. Rd. W12 47Wa 88
Hafer Rd. SW11 56Hb 111
Hafton Rd. SE6 60Gc 115
Hagden La. WD18: Wat 14V 26
Haggard Rd. TW1: Twick 59Ka 108
Hagger Ct. E17 27Fc 53
HAGGERSTON 38Vb 71
Haggerston Rd. E8 38Vb 71
Haggerston Rd. WD6: Bore 10Na 15
Haggerston Station (Overground) 39Vb 71
Haggerston Studios E8 39Vb 71
(off Kingsland Rd.)
Hague St. E2 41Wb 91
Ha Ha Rd. SE18 51Pc 116
Haider Cl. NW2 33Za 68
Haig Cl. AL1: St A 3F 6
Haig Dr. SL1: Slou 7F 80
Haig Gdns. DA12: Grav'nd 9E 122
Haigh Cres. RH1: Redh 8B 208
Haig Ho. AL1: St A 3F 6
Haig Ho. E2 40Wb 71
(off Shipton St.)
Haig Pl. SM4: Mord 72Cb 155
Haig Rd. HA7: Stan 22La 46
Haig Rd. RM16: Grays 8C 100
Haig Rd. TN16: Big H 89Nc 180
Haig Rd. UB8: Hil 43R 84
Haig Rd. E. E13 41Lc 93
Haig Rd. W. E13 41Lc 93
Haigville Gdns. IG6: Ilf 28Rc 54
Hailes Cl. CR0: Bedd 75Mb 156
Hailes Cl. SW19 65Eb 133
Haileybury Av. EN1: Enf 16Vb 33
Haileybury Rd. BR6: Chels 77Wc 161
Hailey Rd. DA18: Erith 47Cd 96
Hailey Rd. Bus. Pk. DA18: Erith 47Cd 96
Hailing M. BR2: Broml 69Kc 137
(off Wendover Rd.)
Hailsham Av. SW2 61Pb 134
Hailsham Cl. KT6: Surb 73Ma 153
Hailsham Cl. RM3: Rom 22Ld 57
Hailsham Dr. HA1: Harr 27Fa 46

Hailsham Gdns. RM3: Rom 22Ld 57
Hailsham Ho. NW8 6D 214 (42Gb 89)
(off Salisbury St.)
Hailsham Rd. RM3: Rom 22Ld 57
Hailsham Rd. SW17 65Jb 134
Hailsham Ter. N18 22Tb 51
Haimo Rd. SE9 57Mc 115
HAINAULT 22Wc 55
Hainault Bri. Pde. IG1: Ilf 33Rc 74
(off Hainault St.)
Hainault Bus. Pk. IG6: Ilf 22Yc 55
(not continuous)
Hainault Ct. E17 28Fc 53
(off Forest Ri.)
Hainault Forest Children's Zoo 21Zc 55
Hainault Forest Country Pk. 20Zc 37
Hainault Forest Country Pk. Vis. Cen. 21Ad 55
Hainault Forest Golf Course 23Zc 55
Hainault Gore RM6: Chad H 29Ad 55
Hainault Gro. IG7: Chig 21Sc 54
Hainault Lodge Local Nature Reserve 23Yc 55
Hainault Rd. RM6: Chad H Forest Rd. 24Xc 55
Hainault Rd. RM6: Chad H Sylvan Av. 30Bd 55
Hainault Rd. E11 32Ec 72
Hainault Rd. IG7: Chig 20Rc 36
Hainault Rd. RM5: Col R 26Ed 56
Hainault Rd. RM5: Rom 26Ed 56
Hainault Station (Underground) 24Uc 54
Hainault St. IG1: Ilf 33Sc 74
Hainault St. SE9 60Rc 116
Haines Cl. N1 38Ub 71
Haines Ct. KT13: Weyb 78T 150
Haines Ho. SW11 52Mb 112
(off Ponton Rd.)
Haines St. SW11 52Lb 112
Haines Wlk. SM4: Mord 73Db 155
Haines Way WD25: Wat 6W 12
Hainford Cl. SE4 56Zb 114
Haining Cl. W4 50Qa 87
Hainthorpe Rd. SE27 62Rb 135
Hainton Cl. E1 44Xb 91
Halberd M. E5 33Xb 71
Halbutt Gdns. RM9: Dag 34Bd 75
Halbutt St. RM9: Dag 35Bd 75
Halcomb St. N1 1H 219 (39Ub 71)
Halcot Av. DA6: Bex 57Dd 118
Halcrow Av. DA1: Dart 55Qd 119
Halcrow St. E1 43Xb 91
Halcyon EN1: Enf 15Ub 33
(off Private Rd.)
Halcyon Cl. KT22: Oxs 87Fa 172
Halcyon Cl. SW13 55Wa 110
Halcyon Way RM11: Horn 32Pd 77
Halcyon Wharf E1 46Wb 91
(off Hermitage Wall)
Haldane Cl. EN3: Enf L 10Dc 20
Haldane Cl. N10 24Kb 50
Haldane Gdns. DA11: Nflt 60Ee 121
Haldane Pl. SW18 60Db 111
Haldane Rd. E6 41Mc 93
Haldane Rd. SE28 45Zc 95
Haldane Rd. SW6 52Bb 111
Haldane Rd. UB1: S'hall 45Ea 86
Haldan Rd. E4 23Ec 52
Haldon Cl. IG7: Chig 22Uc 54
Haldon Rd. SW18 58Bb 111
Hale SL4: Wind 3C 102
Hale, The E4 24Fc 53
HALE, THE 22Ta 47
Hale, The N17 27Wb 51
Halebourne La. GU24: Chob 10E 146
Halebourne La. GU24: W End 10E 146
Hale Cl. BR6: Farnb 77Sc 160
Hale Cl. E4 20Ec 34
Hale Cl. HA8: Edg 22Sa 47
Hale Ct. HA8: Edg 22Sa 47
Hale Dr. NW7 23Sa 47
HALE END 23Gc 53
Hale End RM3: Rom 23Kd 57
Hale End Cl. HA4: Ruis 30W 44
Hale End Rd. E17 25Ec 52
Hale End Rd. E4 23Fc 53
Hale End Rd. IG8: Wfd G 24Fc 53
Hale Ends GU22: Wok 3M 187
Halefield Rd. N17 25Xb 51
Hale Gdns. N17 28Wb 51
Hale Gdns. W3 46Qa 87
Hale Gro. Gdns. NW7 22Ua 48
Hale Ho. RM11: Horn 30Jd 56
(off Benjamin Cl.)
Hale Ho. SW1 7E 228 (50Mb 90)
(off Lindsay Sq.)
Hale La. HA8: Edg 22Ra 47
Hale La. NW7 22Ta 47
Hale La. TN14: Otf 89Hd 182
Hale Path SE27 63Rb 135
Hale Pit Rd. KT23: Bookh 98Ea 192
Hale Rd. E6 42Nc 94
Hale Rd. N17 27Wb 51
Hales Ct. WD25: Wat 8Y 13
Hales Oak KT23: Bookh 98Ea 192
Halesowen Rd. SM4: Mord 73Db 155
Hales Pk. HP2: Hem H 1C 4
Hales Pk. Cl. HP2: Hem H 1C 4
Hales Prior N1 2H 217 (40Pb 70)
(off Calshot St.)
Hales St. SE8 52Cc 114
Hale St. E14 45Dc 92
Hale St. TW18: Staines 63G 126
Haleswood KT11: Cobh 86X 171
Haleswood Rd. HP2: Hem H 1B 4
Halesworth Cl. E5 33Yb 72
Halesworth Cl. RM3: Rom 24Nd 57
Halesworth Rd. RM3: Rom 23Nd 57
Halesworth Rd. SE13 55Dc 114
Hale Wlk. W7 43Ga 86
Haley Rd. NW4 30Ya 48
Half Acre HA7: Stan 22La 46
Half Acre TW8: Bford 51Ma 109
Halfacre Hill SL9: Chal P 25B 42
Half Acre Rd. W7 46Ga 86
Halfhides EN9: Walt A 5Fc 21
Half Moon Cotts. GU23: Rip 93L 189
Half Moon Ct. CR0: C'don 73Rb 157
Half Moon Ct. EC1 1D 224 (43Sb 91)
(off Bartholomew Cl.)
Half Moon Cres. N1 1J 217 (40Pb 70)
(not continuous)
Half Moon La. CM16: Epp 3Vc 23
Half Moon La. SE24 58Sb 113
Half Moon M. AL1: St A 2B 6
Half Moon Pas. E1 3K 225 (44Vb 91)
(not continuous)
Half Moon St. W1 6A 222 (46Kb 90)
Halford Cl. HA8: Edg 26Ra 47
Halford Pl. W7 46Ha 86
Halford Rd. E10 29Fc 53
Halford Rd. SW6 51Cb 111
Halford Rd. TW10: Rich 57Na 109
Halford Rd. UB10: Ick 35Q 64
Halfpence La. DA12: Cobh 10J 145
Halfpenny La. SL5: S'dale 3E 146
Halfway Ct. RM19: Purf 49Qd 97
Halfway Grn. KT12: Walt T 76X 151
Halfway St. DA15: Sidc 59Tc 116
Haliburton Rd. TW1: Twick 57Ja 108
Haliday Ho. N1 37Tb 71
(off Mildmay St.)
Haliday Wlk. N1 37Tb 71
Halidon Cl. E9 36Yb 72
Halidon Ri. RM3: Hrld W 23Rd 57
Halifax NW9 26Va 48
Halifax Cl. AL2: Brick W 3Ba 13
Halifax Cl. TW11: Tedd 65Ga 130
Halifax Cl. WD25: Wat 6V 12
Halifax Ho. RM3: Rom 22Nd 57
(off Lindfield Rd.)
Halifax Rd. EN2: Enf 12Sb 33
Halifax Rd. UB6: G'frd 39Da 65
Halifax Rd. WD3: Herons 17E 24
Halifax St. SE26 62Xb 135
Halifield Dr. DA17: Belv 48Ad 95
Haling Down Pas. CR2: S Croy 81Sb 177
Haling Down Pas. CR8: Purl 82Rb 177
(not continuous)
Haling Gro. CR2: S Croy 80Sb 157
Haling Pk. Gdns. CR2: S Croy 79Rb 157
Haling Pk. Rd. CR2: S Croy 78Rb 157
Haling Rd. CR2: S Croy 79Tb 157
Halings La. UB9: Den 28F 42
Haliwell Ho. NW6 39Db 69
(off Mortimer Cres.)
Halkett Ho. E2 39Yb 72
(off Waterloo Gdns.)
Halkin Arc. SW1 3G 227 (48Hb 89)
Halkingcroft SL3: L'ly 7N 81
Halkin M. SW1 3H 227 (48Jb 90)
Halkin Pl. SW1 3H 227 (48Jb 90)
Halkin St. SW1 2J 227 (47Jb 90)
Hall, The SE3 55Jc 115
Hallam Cl. BR7: Chst 64Pc 138
Hallam Cl. WD24: Wat 12Y 27
Hallam Ct. W1 7A 216 (43Kb 90)
(off Hallam St.)
Hallam Gdns. HA5: Hat E 24Aa 45
Hallam Ho. SW1 50Lb 90
(off Churchill Gdns.)
Hallam M. W1 7A 216 (43Kb 90)
Hallam Rd. N15 28Rb 51
Hallam Rd. SW13 55Xa 110
Hallam St. W1 7A 216 (43Kb 90)
Halland Way HA6: Nwood 23T 44
Hallane Ho. SE27 64Sb 135
Hall Apts. E3 43Bc 92
(off Geoff Cade Way)
Hall Av. RM15: Avel 46Sd 98
Hall Barns, The CM16: Epp 3Pc 22
Hall Cl. W5 43Na 87
Hall Cl. WD3: Rick 18J 25
Hall Ct. SL3: Dat 2M 103
Hall Ct. TW11: Tedd 64Ha 130
Hall Cres. RM15: Avel 47Sd 98
Hall Dr. SE26 64Yb 136
Hall Dr. UB9: Hare 25L 43
Hall Dr. W7 44Ga 86
Halley Gdns. SE13 56Fc 115
Halley Ho. E2 40Wb 71
(off Pritchards Rd.)
Halley Ho. SE10 50Hc 93
(off Armitage Rd.)
Halley Rd. E12 37Mc 73
Halley Rd. E7 37Lc 73
Halley Rd. EN9: Walt A 8Dc 20
Halley's App. GU21: Wok 9L 167
Halley's Ct. GU21: Wok 10L 167
Halley St. E14 43Ac 92
Halley's Wlk. KT15: Add 80L 149
Hall Farm Cl. HA7: Stan 21Ka 46
Hall Farm Dr. TW2: Whitt 59Fa 108
Hallfield Est. W2 44Eb 89
(not continuous)
Hallford Way DA1: Dart 58Ld 119
Hall Gdns. AL4: Col H 5P 7
Hall Gdns. E4 21Bc 52
Hall Ga. NW8 3B 214 (41Fb 89)
Hall Grn. La. CM13: Hut 17Ee 41
Hall Heath Cl. AL1: St A 1F 6
Hall Hill RH8: Oxt 3H 211
Hall Hill TN15: Seal 95Rd 203
Halliards, The KT12: Walt T 72W 150
Halliday Cl. KT17: Ewe 82Wa 174
Halliday Cl. WD7: Shenl 4Na 15
Halliday Ho. E1 44Wb 91
(off Christian St.)
Halliday Sq. UB2: S'hall 46Fa 86
Halliford Cl. TW17: Shep 70T 128
Halliford Rd. TW16: Sun 71V 150
Halliford Rd. TW17: Shep 71U 150
Halliford St. N1 38Sb 71
Halliloo Valley Rd. CR3: Wold 92Zb 198
Hallingbury Ct. E17 27Dc 52
Halling Ho. SE1 2G 231 (47Tb 91)
(off Long La.)
Hallings Wharf Studios E15 39Fc 73
Hallington Cl. GU21: Wok 9M 167
Hallington Ct. HA8: Edg 21Pa 47
(off Brannigan Way)
Halliwell Ct. SE22 57Wb 113
Halliwell Rd. SW2 58Pb 112
Halliwick Ct. Pde. N12 23Hb 49
(off Woodhouse Rd.)
Halliwick Rd. N10 25Jb 50
Hall La. CM15: Shenf 13Ae 41
Hall La. E4 22Ac 52
Hall La. NW4 25Wa 48
Hall La. RM14: Upm 26Sd 58
Hall La. RM15: S Ock 40Zd 79
Hall La. UB3: Harl 52T 106
HALL LANE 22Zb 52
Hallmark Ho. E14 43Cc 92
(off Ursula Gould Way)
Hallmark Trad. Est. HA9: Wemb 35Sa 67
Hall Mdw. SL1: Burn 10A 60
Hallmead Rd. SM1: Sutt 76Db 155
Hall Oak Wlk. NW6 37Bb 69
Hallowell Av. CR0: Bedd 77Nb 156
Hallowell Cl. CR4: Mitc 69Jb 134
Hallowell Gdns. CR7: Thor H 68Sb 135
Hallowell Rd. HA6: Nwood 24U 44
Hallowes Cres. WD19: Wat 20W 26
Hallowfield Way CR4: Mitc 69Fb 133
Hallows Gro. TW16: Sun 64V 128
Hall Pk. HP4: Berk 2A 2
Hall Pk. Ga. HP4: Berk 3A 2
Hall Pk. Hill HP4: Berk 3A 2
Hall Pk. Rd. RM14: Upm 36Sd 78
Hall Pl. AL1: St A 1C 6
Hall Pl. GU21: Wok 88C 168
Hall Pl. W2 6B 214 (42Fb 89)
(not continuous)
Hall Place & Gdns. 58Ed 118
Hall Pl. Cl. AL1: St A 1C 6
Hall Pl. Cres. DA5: Bexl 57Ed 118
Hall Pl. Dr. KT13: Weyb 78U 150
Hall Pl. Gdns. AL1: St A 1C 6
Hall Place Sports Pavilion 58Ed 118
Hall Rd. DA1: Dart 56Pd 119
Hall Rd. DA11: Nflt 62Ee 143
Hall Rd. E15 35Fc 73
Hall Rd. E6 39Pc 74
Hall Rd. NW8 4A 214 (41Eb 89)
Hall Rd. RM15: Avel 47Sd 98
Hall Rd. RM2: Rom 27Kd 57
Hall Rd. RM6: Chad H 30Yc 55
Hall Rd. SM6: W'gton 81Kb 176
Hall Rd. TW7: Isle 57Fa 108
Halls Farm Cl. GU21: Knap 9H 167
Hallside Rd. EN1: Enf 10Vb 19
Hallsland Way RH8: Oxt 5K 211
Halls Ter. UB10: Hil 42R 84
Hall St. EC1 3C 218 (41Rb 91)
Hall St. N12 22Eb 49
Hallsville Rd. E16 44Hc 93
Hallswelle Pde. NW11 29Bb 49
Hallswelle Rd. NW11 29Bb 49
Hall Ter. RM15: Avel 47Td 98
Hall Ter. RM3: Hrld W 24Qd 57
Hall Twr. W2 7C 214 (43Fb 89)
(off Hall Pl.)
Hall Vw. SE9 61Mc 137
Hall Wlk., The HP4: Berk 1A 2
(off Little Bri. Rd.)
Hall Way CR8: Purl 85Rb 177
Hallwood Cres. CM15: Shenf 17Ae 41
Hallywell Cres. E6 43Pc 94
Halo E15 39Ec 72
Halons Rd. SE9 59Qc 116
Halpin Bldg. SE10 48Jc 93
(off Rennie Street)
Halpin Pl. SE17 6G 231 (49Tb 91)
Halsbrook Rd. SE3 55Lc 115
Halsbury Cl. HA7: Stan 21Ka 46
Halsbury Ct. HA7: Stan 22Ka 46
Halsbury Ho. N7 35Pb 70
(off Biddestone Rd.)
Halsbury Rd. W12 46Xa 88
Halsbury Rd. E. UB5: N'olt 35Ea 66
Halsbury Rd. W. UB5: N'olt 36Da 65
Halse Dr. SL2: Farn C 5C 60
Halsend UB3: Hayes 46X 85
Halsey Ho. WC1 1H 223 (43Pb 90)
(off Red Lion Sq.)
Halsey M. SW3 5F 227 (49Hb 89)
Halsey Pk. AL2: Lon C 10K 7
Halsey Pl. WD24: Wat 10X 13
Halsey Rd. WD18: Wat 13X 27
Halsey St. SW3 5F 227 (49Hb 89)
Halsham Cres. IG11: Bark 36Vc 75
Halsmere Rd. SE5 53Rb 113
HALSTEAD 84Bd 181
Halstead Cl. CR0: C'don 76Sb 157
Halstead Ct. E17 31Bc 72
Halstead Ct. N1 2G 219 (40Tb 71)
(off Murray Gro.)
Halstead Gdns. N21 18Tb 33
Halstead Hill EN7: G Oak 1Tb 19
Halstead Ho. RM3: Rom 23Md 57
(off Dartfields)
Halstead La. TN14: Hals 87Ad 181
Halstead La. TN14: Knock 87Ad 181
Halstead Pl. TN14: Hals 84Ad 181
Halstead Rd. DA8: Erith 53Gd 118
Halstead Rd. E11 29Jc 53
Halstead Rd. EN1: Enf 14Ub 33
Halstead Rd. N21 18Sb 33
Halstead Way CM13: Hut 16Ee 41
Halston Cl. SW11 58Hb 111
Halstow Rd. NW10 41Za 88
Halstow Rd. SE10 50Jc 93
Halsway UB3: Hayes 46W 84
Halt Dr. SS17: Linf 9J 101
Halter Cl. WD6: Bore 15Ta 29
Halton Cl. AL2: Park 10A 6
Halton Cl. N11 23Hb 49
Halton Ct. SE3 55Kc 115
Halton Cross St. N1 39Rb 71
Halton Ho. N1 38Rb 71
(off Halton Rd.)
Halton Mans. N1 38Rb 71
Halton Pl. N1 39Sb 71
Halton Rd. CR8: Kenley 92Ub 197
Halton Rd. N1 38Rb 71
Halton Rd. RM16: Grays 8E 100
Halt Robin La. DA17: Belv 49Dd 96
Halt Robin Rd. DA17: Belv 49Cd 96
(not continuous)
Haltside AL10: Hat 1A 8
Halwick Cl. HP1: Hem H 4K 3
Halyard Ho. E14 48Ec 92
(off Manchester Rd.)
Halyard Pl. E16 47Kc 93
Halyard St. RM9: Dag 42Ad 95
HAM 62La 130
Ham, The TW8: Bford 52La 108
Hamara Ghar E13 39Lc 73
Hambalt Rd. SW4 57Lb 112
Hamble Cl. GU21: Wok 9L 167
Hamble Cl. HA4: Ruis 33U 64
Hamble Ct. WD18: Wat 14W 26
Hambleden Pl. WD7: R'lett 7Ha 14
Hambledon SE17 51Tb 113
(off Villa St.)
Hambledon Cl. UB8: Hil 42R 84
Hambledon Ct. SE22 56Ub 113
Hambledon Ct. W5 45Na 87
Hambledon Gdns. SE25 69Vb 135
Hambledon Hill KT18: Eps 88Sa 173
Hambledon Pl. KT23: Bookh 95Ca 191
Hambledon Pl. SE21 60Ub 113
Hambledon Rd. CR3: Cat'm 95Tb 197
Hambledon Rd. SW18 59Bb 111
Hambledon Va. KT18: Eps 88Sa 173
Hambledown Rd. DA15: Sidc 59Tc 116
Hamble Dr. UB3: Hayes 45V 84
Hamblehyrst BR3: Beck 68Dc 136
Hamble La. RM15: S Ock 43Vd 98
Hamble St. SW6 55Db 111
Hambleton SL4: Old Win 10N 103
(off Burfield Rd.)
Hambleton Cl. KT4: Wor Pk 75Ya 154
Hambleton Cl. UB9: Hare 26L 43
Hamble Wlk. GU21: Wok 10L 167
Hamble Wlk. UB5: N'olt 40Ca 65
(off Brabazon Rd.)
Hambley Ho. SE16 49Xb 91
(off Camilla Rd.)
Hamblings Cl. WD7: Shenl 5Ma 15
Hamblin Ho. UB1: S'hall 45Aa 85
(off The Broadway)
Hambridge Way SW2 59Qb 112
Hambro Av. BR2: Hayes 74Jc 159
Hambrook Rd. SE25 69Xb 135
Hambro Rd. SW16 65Mb 134
Hambrough Ho. UB4: Yead 43Y 85
Hambrough Rd. UB1: S'hall 46Aa 85
Ham Cl. TW10: Ham 62La 130
(not continuous)
Ham Common 63Na 131
Ham Comn. TW10: Ham 62Ma 131
Ham Ct. NW9 26Ua 48
Ham Ct. RM2: Rom 28Ld 57
Ham Cft. Cl. TW13: Felt 62W 128
Hamden Cres. RM10: Dag 34Dd 76
Hamel Cl. HA3: Kenton 28Ma 47
Hamella Ho. E9 36Ac 72
(off Sadler Pl.)
Hamer Cl. HP3: Bov 10C 2
Hamerton Rd. DA11: Nflt 57De 121
Hameway E6 41Qc 94
Ham Farm Rd. TW10: Ham 63Ma 131
Hamfield Cl. RH8: Oxt 99Ec 198
Ham Flds. TW10: Ham 62Ka 130
Hamfrith Rd. E15 37Hc 73
Ham Ga. Av. TW10: Ham 62Ma 131
Hamhaugh Island TW17: Shep 75Q 150
Ham House & Gdn. 60La 108
Hamilton Av. GU22: Pyr 87G 168
Hamilton Av. IG6: Ilf 28Rc 54
Hamilton Av. KT11: Cobh 85W 170
Hamilton Av. KT6: Surb 75Qa 153
Hamilton Av. N9 17Wb 33
Hamilton Av. RM1: Rom 26Fd 56
Hamilton Av. SM3: Cheam 75Ab 154
Hamilton Cl. AL2: Brick W 3Ca 13
Hamilton Cl. CR8: Purl 84Rb 177
Hamilton Cl. EN4: Cockf 14Gb 31
Hamilton Cl. EN6: S Mim 5Wa 16
Hamilton Cl. HA7: Stan 19Ga 28
Hamilton Cl. KT16: Chert 74H 149
Hamilton Cl. KT19: Eps 84Sa 173
Hamilton Cl. N17 27Vb 51
Hamilton Cl. NW8 4B 214 (41Fb 89)
Hamilton Cl. SE16 47Ac 92
Hamilton Cl. TW11: Tedd 65Ka 130
Hamilton Cl. TW13: Felt 64V 128
Hamilton Ct. AL10: Hat 2D 8
Hamilton Ct. CR0: C'don 74Wb 157
Hamilton Ct. DA8: Erith 52Hd 118
(off Frobisher Rd.)
Hamilton Ct. KT11: Cobh 85W 170
Hamilton Ct. KT16: Chert 73J 149
Hamilton Ct. KT23: Bookh 97Da 191
Hamilton Ct. SE6 60Hc 115
Hamilton Ct. SW15 55Ab 110
Hamilton Ct. TW3: Houn 56Da 107
(off Hanworth Rd.)
Hamilton Ct. W5 45Na 87
Hamilton Ct. W9 41Eb 89
(off Maida Vale)
Hamilton Cres. CM14: W'ley 21Yd 58
Hamilton Cres. HA2: Harr 34Ba 65
Hamilton Cres. N13 21Qb 50
Hamilton Cres. TW3: Houn 57Da 107
Hamilton Dr. GU2: Guild 10L 187
Hamilton Dr. RM3: Hrld W 26Nd 57
Hamilton Dr. SL5: S'dale 3C 146
Hamilton Gdns. NW8 3A 214 (41Eb 89)
Hamilton Gdns. WD23: Bush 14Ba 27
Hamilton Hall NW8 40Eb 69
(off Hamilton Ter.)
Hamilton Ho. E14: St Davids Sq. 50Dc 92
(off St Davids Sq.)
Hamilton Ho. E14: Victory Pl. 45Bc 92
(off Victory Pl.)
Hamilton Ho. E3 41Bc 92
(off British St.)
Hamilton Ho. NW8 3B 214 (41Fb 89)
(off Hall Rd.)
Hamilton Ho. W4 51Ua 110
Hamilton Ho. W8 47Db 89
(off Vicarage Ga.)
Hamilton La. N5 35Rb 71
Hamilton Lodge E1 42Yb 92
(off Cleveland Gro.)
Hamilton Mead HP3: Bov 9C 2
Hamilton M. KT13: Weyb 77Q 150
(off Holstein Av.)
Hamilton M. SW18 60Cb 111
Hamilton M. SW19 66Db 133
Hamilton M. W1 1K 227 (47Kb 90)
Hamilton Pde. TW13: Felt 63V 128
Hamilton Pk. N5 35Rb 71
Hamilton Pk. W. N5 35Rb 71
Hamilton Pl. KT14: W Byf 85H 169
Hamilton Pl. KT20: Kgswd 94Bb 195
Hamilton Pl. N19 34Mb 70
Hamilton Pl. SL9: Ger X 29A 42
Hamilton Pl. TW16: Sun 66X 129
Hamilton Pl. W1 7J 221 (46Jb 90)
Hamilton Rd. AL1: St A 1E 6
Hamilton Rd. CR7: Thor H 69Tb 135
Hamilton Rd. DA15: Sidc 63Wc 139
Hamilton Rd. DA7: Bex 54Ad 117
Hamilton Rd. E15 41Gc 93
Hamilton Rd. E17 26Ac 52
Hamilton Rd. EN4: Cockf 14Gb 31
Hamilton Rd. HA1: Harr 29Ga 46
Hamilton Rd. IG1: Ilf 35Rc 74
Hamilton Rd. N2 27Eb 49
Hamilton Rd. N9 17Wb 33
Hamilton Rd. NW10 36Wa 68
Hamilton Rd. NW11 31Za 68
Hamilton Rd. RM2: Rom 29Kd 57
Hamilton Rd. SE27 63Tb 135
Hamilton Rd. SL1: Slou 4E 80
Hamilton Rd. SW19 66Db 133
Hamilton Rd. TW13: Felt 63V 128
Hamilton Rd. TW2: Twick 60Ga 108
Hamilton Rd. TW8: Bford 51Ma 109
Hamilton Rd. UB1: S'hall 46Ba 85
Hamilton Rd. UB3: Hayes 45X 85
Hamilton Rd. UB8: Cowl 42M 83
Hamilton Rd. W4 47Ua 88
Hamilton Rd. W5 45Na 87
Hamilton Rd. WD19: Wat 20X 27
Hamilton Rd. WD4: Hunt C 5S 12
Hamilton Rd. Ind. Est. SE27 63Tb 135
Hamilton Rd. M. SW19 66Db 133
Hamilton Sq. N12 23Fb 49
Hamilton Sq. SE1 1G 231 (47Tb 91)
Hamilton St. SE8 51Cc 114
Hamilton St. WD18: Wat 15Y 27
Hamilton Ter. NW8 40Db 69
Hamilton Wlk. DA8: Erith 52Hd 118
Hamilton Way N13 21Rb 51
Hamilton Way N3 23Cb 49
Hamilton Way SL2: Farn C 6G 60
Hamilton Way SM6: W'gton 81Mb 176
HAM ISLAND 6N 103
Ham Lands Nature Reserve 61Ja 130
Ham La. SL4: Old Win 7N 103
Ham La. TW20: Eng G 3M 125
Hamlea Cl. SE12 57Jc 115
Hamlet, The SE5 55Tb 113
Hamlet Cl. AL2: Brick W 2Ba 13
Hamlet Cl. RM5: Col R 24Cd 56
Hamlet Cl. SE13 56Gc 115
Hamlet Cl. SE6 59Dc 114
Hamlet Ct. E3 41Cc 92
(off Tomlin's Gro.)
Hamlet Ct. EN1: Enf 15Ub 33
Hamlet Ct. SE11 7B 230 (50Rb 91)
(off Opal St.)
Hamlet Ct. W6 49Wa 88
Hamlet Gdns. W6 49Wa 88
Hamlet Ind. Est. E9 38Cc 72
Hamlet Intl. Ind. Est. DA8: Erith 50Fd 96
Hamlet Lodge UB10: Hil 37R 64
Hamlet M. SE21 60Tb 113
Hamleton Ter. RM9: Dag 38Yc 75
(off Flamstead Rd.)
Hamlet Rd. RM5: Col R 24Cd 56
Hamlet Rd. SE19 66Vb 135
Hamlet Sq. NW2 34Ab 68
Hamlets Way E3 42Bc 92
Hamlet Way SE1 1G 231 (47Tb 91)
Hamlin Cres. HA5: Eastc 29Y 45
Hamlin Rd. TN13: Riv 94Gd 202
Hamlyn Cl. HA8: Edg 20Na 29
Hamlyn Ct. TN13: Dun G 93Gd 202
Hamlyn Gdns. SE19 66Ub 135
Hamlyn Ho. TW13: Felt 60X 107
Hamm Ct. KT13: Weyb 75N 149
Hammelton Ct. BR1: Broml 67Hc 137
(off London Rd.)
Hammelton Rd. BR1: Broml 67Hc 137
HAMMERFIELD 2K 3
Hammerfield Ho. SW3 7E 226 (50Gb 89)
(off Cale St.)
Hammer La. HP2: Hem H 1P 3
Hammer Pde. WD25: Wat 5W 12
Hammers Ga. AL2: Chis G 8N 5
Hammers La. NW7 22Wa 48
Hammersley Ho. SE14 52Yb 114
(off Pomeroy St.)
Hammersley Rd. E16 43Jc 93
HAMMERSMITH 49Ya 88
Hammersmith Apollo Eventim Apollo, The 50Ya 88
Hammersmith Bri. 51Xa 110
Hammersmith Bri. Rd. W6 50Ya 88
Hammersmith B'way. W6 49Ya 88
HAMMERSMITH BROADWAY 49Ya 88
Hammersmith Bus Station 49Ya 88
Hammersmith Emb. W6 51Ya 110
Hammersmith Fitness & Squash Cen. 49Za 88
(off Chalk Hill Rd.)
Hammersmith Flyover W6 50Ya 88
HAMMERSMITH FLYOVER 50Ya 88
Hammersmith Gro. W6 47Ya 88
HAMMERSMITH HOSPITAL 44Wa 88
Hammersmith Info. Cen. 49Ya 88
(within The Broadway Cen.)
Hammersmith Rd. W14 49Za 88
Hammersmith Rd. W6 49Za 88
Hammersmith Station (Underground) 49Ya 88
Hammersmith Ter. W6 50Wa 88
Hammerton Cl. DA5: Bexl 62Gd 140
Hammet Cl. UB4: Yead 43Z 85
Hammett St. EC3 4K 225 (45Vb 91)
Hamm Moor La. KT15: Add 78N 149
Hammond Av. CR4: Mitc 68Kb 134
Hammond Cl. EN5: Barn 15Ab 30
Hammond Cl. GU21: Wok 7N 167
Hammond Cl. TW12: Hamp 67Ca 129
Hammond Cl. UB6: G'frd 36Fa 66
Hammond Ct. E10 33Dc 72
(off Leyton Grange Est.)
Hammond Ct. RM12: Horn 32Hd 76
Hammond Ct. SE11 7K 229 (50Qb 90)
(off Hotspur St.)
Hammond End SL2: Farn C 5F 60
Hammond Ho. E14 48Cc 92
(off Tiller Rd.)
Hammond Ho. SE14 52Yb 114
(off Lubbock St.)
Hammond Lodge W9 43Cb 89
(off Admiral Wlk.)
Hammond Rd. EN1: Enf 12Xb 33
Hammond Rd. GU21: Wok 7N 167
Hammond Rd. UB2: S'hall 48Aa 85
Hammonds Cl. RM8: Dag 34Yc 75
Hammonds La. CM13: Gt War 23Xd 58
Hammond St. NW5 37Lb 70
Hammond Way SE28 45Xc 95
HAM MOOR 77N 149
Hamond Cl. CR2: S Croy 81Rb 177
Hamonde Cl. HA8: Edg 19Ra 29
Hamond Sq. N1 1H 219 (40Ub 71)
Ham Pk. Rd. E15 38Hc 73
Ham Pk. Rd. E7 38Jc 73
Hampden Av. BR3: Beck 68Ac 136
Hampden Cl. NW1 2E 216 (40Mb 70)
Hampden Cl. SL2: Stoke P 1L 81
Hampden Ct. N10 24Jb 50
Hampden Cres. CM14: W'ley 21Yd 58
Hampden Cres. EN7: Chesh 3Xb 19
Hampden Gurney St. W1 3F 221 (44Hb 89)
Hampden Ho. SW9 54Qb 112
(off Overton Rd.)
Hampden La. N17 25Vb 51
Hampden Pl. AL2: F'mre 1Ga 14
Hampden Rd. BR3: Beck 68Ac 136
Hampden Rd. HA3: Hrw W 25Ea 46
Hampden Rd. KT1: King T 69Qa 131
Hampden Rd. N10 24Jb 50
Hampden Rd. N17 25Wb 51
Hampden Rd. N19 33Mb 70
Hampden Rd. N8 28Qb 50
Hampden Rd. RM17: Grays 50De 99
Hampden Rd. RM5: Col R 24Dd 56
Hampden Rd. SL3: L'ly 4B 82
Hampden Rd. SL9: Chal P 25A 42
Hampden Sq. N14 18Kb 32
Hampden Way N14 18Kb 32
Hampden Way WD17: Wat 8U 12
Hampermill La. WD19: Wat 19V 26
Hampshire Av. SL1: Slou 3G 80
Hampshire Cl. N18 22Xb 51
Hampshire Ct. KT15: Add 78L 149
Hampshire Gdns. SS17: Linf 7J 101
Hampshire Hog La. W6 50Xa 88
Hampshire Ho. SL9: Chal P 22A 42
Hampshire Rd. N22 24Pb 50

Hampshire Rd. RM11: Horn 28Qd 57
Hampshire St. NW5 37Mb 70
Hampson Way SW8 53Pb 112
HAMPSTEAD 35Fb 69
Hampstead Av. IG8: Wfd G 24Qc 54
Hampstead Cl. AL2: Brick W 3Ba 13
Hampstead Cl. SE28 46Xc 95
Hampstead Cricket Club 36Db 69
(off Lymington Rd.)
Hampstead Gdns. NW11 30Cb 49
Hampstead Gdns. RM6: Chad H 29Xc 55
HAMPSTEAD GARDEN SUBURB 29Eb 49
Hampstead Ga. NW3 36Eb 69
Hampstead Golf Course 31Fb 69
Hampstead Grn. NW3 36Gb 69
Hampstead Gro. NW3 34Eb 69
Hampstead Heath 33Fb 69
Hampstead Heath Station (Overground) 35Gb 69
Hampstead Hgts. N2 27Eb 49
Hampstead High St. NW3 35Fb 69
Hampstead Hill Gdns. NW3 35Fb 69
Hampstead Ho. NW1 4B 216 (41Lb 90)
(off William Rd.)
Hampstead La. N6 31Fb 69
Hampstead La. NW3 32Fb 69
Hampstead Lodge NW1 7D 214 (43Gb 89)
(off Bell St.)
Hampstead M. BR3: Beck 70Dc 136
Hampstead Mus. 35Fb 69
(off New End Sq.)
Hampstead Rd. NW1 2B 216 (40Lb 70)
Hampstead Sq. NW3 34Eb 69
Hampstead Station (Underground) 35Eb 69
Hampstead Theatre 38Fb 69
Hampstead Wlk. E3 39Bc 72
Hampstead Way NW11 29Bb 49
Hampstead W. NW6 37Cb 69
HAMPTON 67Da 129
Hampton & Richmond Borough FC 67Da 129
Hampton Bus. Pk. TW13: Hanw 62Aa 129
Hampton Cl. GU21: Knap 1G 186
Hampton Cl. N11 22Kb 50
Hampton Cl. NW6 41Cb 89
Hampton Cl. RM16: Chaf H 48Yd 98
Hampton Cl. SW20 66Ya 132
Hampton Cl. WD6: Bore 15Sa 29
HAMPTON COURT 69Ga 130
HAMPTON COURT 70Ga 130
Hampton Ct. N1 37Rb 71
Hampton Ct. N22 25Lb 50
Hampton Ct. SE14 52Ac 114
(off Batavia Rd.)
Hampton Ct. SE16 45Zb 92
(off King & Queen Wharf)
Hampton Ct. Av. KT8: E Mos 72Fa 152
Hampton Ct. Bri. 70Ga 130
Hampton Ct. Cres. KT8: E Mos 69Fa 130
Hampton Ct. Est. KT7: T Ditt 70Ga 130
Hampton Ct. M. KT8: E Mos 70Ga 130
(off Feltham Av.)
Hampton Court Palace 70Ha 130
Hampton Court Palace Golf Course 71La 152
Hampton Ct. Pde. KT8: E Mos 70Ga 130
Hampton Ct. Rd. KT1: Hamp W 69Ka 130
Hampton Ct. Rd. KT8: E Mos 69Ha 130
Hampton Ct. Rd. TW12: E Mos 68Ea 130
Hampton Ct. Rd. TW12: Hamp 68Ea 130
Hampton Court Station (Rail) 70Ga 130
Hampton Ct. Way KT10: T Ditt 75Ga 152
Hampton Ct. Way KT7: T Ditt 75Ga 152
Hampton Ct. Way KT8: E Mos 72Ga 152
Hampton Cres. DA12: Grav'nd 1G 144
Hampton Golf Course 62Ca 129
Hampton Grange BR1: Broml 66Lc 137
Hampton Gro. KT17: Ewe 83Va 174
HAMPTON HILL 64Ea 130
Hampton Hill Bus. Pk. TW12: Hamp H 64Ea 130
(off High St.)
Hampton Hill Theatre 64Ea 130
Hampton Ho. DA7: Bex 54Dd 118
(off Erith Rd.)
Hampton Ho. SW8 52Lb 112
(off Ascalon St.)
Hampton La. TW13: Hanw 63Aa 129
Hampton Mead IG10: Lough 13Rc 36
Hampton M. EN3: Enf H 13Yb 34
Hampton M. NW10 41Ta 87
Hampton M. WD23: B Hea 18Ea 28
Hampton Open Air Pool 66Ea 130
Hampton Ri. HA3: Kenton 30Na 47
Hampton Rd. CR0: C'don 72Sb 157
Hampton Rd. E11 32Fc 73
Hampton Rd. E4 22Bc 52
Hampton Rd. E7 36Kc 73
Hampton Rd. HA7: Stan 20Ga 28
Hampton Rd. IG1: Ilf 35Sc 74
Hampton Rd. KT4: Wor Pk 75Wa 154
Hampton Rd. RH1: Redh 10P 207
Hampton Rd. TW11: Tedd 64Fa 130
Hampton Rd. TW12: Hamp H 64Fa 130
Hampton Rd. TW2: Twick 62Fa 130
Hampton Rd. E. TW13: Hanw 63Ba 129
Hampton Rd. Ind. Pk. CR0: C'don 72Sb 157
Hampton Rd. W. TW13: Hanw 62Aa 129
Hampton Sports & Fitness Cen. 64Ca 129
Hampton Station (Rail) 67Ca 129
Hampton St. SE1 6D 230 (49Sb 91)
Hampton St. SE17 6C 230 (49Rb 91)
HAMPTON WICK 67La 130
Hampton Wick Library 67La 130
Hampton Wick Station (Rail) 67La 130
Hampton Youth Project (Sports Hall) 65Ba 129
Ham Ridings TW10: Ham 64Pa 131
HAMSEY GREEN 88Yb 178
Hamsey Grn. Gdns. CR6: W'ham 88Xb 177
Hamsey Way CR2: Sande 87Xb 177
Hamshades Cl. DA15: Sidc 62Vc 139
Hamston Ho. W8 48Db 89
(off Kensington Ct. Pl.)
Ham St. TW10: Ham 60Ka 108
Ham Vw. CR0: C'don 72Ac 158
Ham Yd. W1 4D 222 (45Mb 90)
Hanah Ct. SW19 66Za 132
Hanameel St. E16 46Kc 93
Hana M. E5 35Xb 71
Hanbury Cl. EN8: Chesh 1Ac 20
Hanbury Cl. NW4 27Ya 48
Hanbury Ct. HA1: Harr 30Ha 46
Hanbury Dr. E11 31Hc 73
Hanbury Dr. N21 15Pb 32
Hanbury Dr. TN16: Big H 85Kc 179
Hanbury Ho. E1 43Wb 91
(off Hanbury St.)
Hanbury Ho. SW8 52Nb 112
(off Regent's Bri. Gdns.)
Hanbury M. CR0: C'don 74Ac 158
Hanbury M. N1 1E 218 (39Sb 71)
Hanbury Path GU21: Wok 86F 168
Hanbury Rd. N17 26Xb 51
Hanbury Rd. W3 47Ra 87
Hanbury St. E1 7K 219 (43Vb 91)
Hanbury Wlk. DA5: Bexl 62Gd 140
Hancock Ct. WD6: Bore 11Sa 29
Hancock Nunn Ho. NW3 37Hb 69
(off Fellows Rd.)
Hancock Rd. E3 41Ec 92
Hancock Rd. SE19 65Tb 135
Hancocks Mt. SL5: S'hill 2B 146
Hancroft Rd. HP3: Hem H 4P 3
Hancross Cl. AL2: Brick W 2Aa 13
Handa Cl. HP3: Hem H 5B 4
Handa Wlk. N1 37Tb 71
Hand Axe Yd. WC1 3G 217 (41Nb 90)
HAND CLINIC, THE 3A 102
Hand Ct. WC1 1J 223 (43Pb 90)
Handcroft Rd. CR0: C'don 73Rb 157
Handel & Hendrix in London 4A 222 (45Kb 90)
(off Brook St.)
Handel Bus. Cen. SW8 51Nb 112
Handel Cl. HA8: Edg 23Pa 47
Handel Cres. RM18: Tilb 2C 122
Handel Mans. SW13 52Ya 110
Handel Mans. WC1 5G 217 (42Nb 90)
(off Handel St.)
Handel Pde. HA8: Edg 24Qa 47
(off Whitchurch La.)
Handel Pl. NW10 37Ta 67
Handel St. WC1 5F 217 (42Nb 90)
Handel Way HA8: Edg 24Qa 47
Handen Rd. SE12 57Gc 115
Handford Ct. WD25: Wat 6Z 13
Handforth Rd. IG1: Ilf 34Rc 74
Handforth Rd. SW9 52Qb 112
Handley Dr. SE3 55Kc 115
Handley Ga. AL2: Brick W 1Ba 13
Handley Gro. NW2 34Za 68
Handley Page Rd. IG11: Bark 42Wc 95
Handley Page Rd. SM6: W'gton 80Pb 156
Handley Page Way AL2: Col S 2Ha 14
Handley Rd. E9 38Yb 72
Handleys Ct. HP2: Hem H 3M 3
(off Selden Hill)
Handowe Cl. NW4 28Wa 48
Handpost Lodge Gdns. HP2: Hem H 3D 4
Handside Cl. KT4: Wor Pk 74Za 154
Hands Wlk. E16 44Jc 93
Handsworth Av. E4 23Fc 53
Handsworth Rd. N17 27Tb 51
Handsworth Way WD19: Wat 20W 26
Handtrough Way IG11: Bark 40Rc 74
Handyside St. N1 39Nb 70
Hanford Cl. SW18 60Cb 111
Hanford Rd. RM15: Avel 46Sd 98
Hanford Row SW19 65Ya 132
Hangar Ruding WD19: Wat 20Ba 27
Hangboy Slade IG10: Lough 9Pc 22
Hanger Cl. HP1: Hem H 3K 3
Hanger Ct. GU21: Knap 9J 167
Hanger Ct. W5 42Pa 87
Hanger Grn. W5 42Qa 87
Hanger Hill KT13: Weyb 79R 150
HANGER HILL 42Pa 87
Hanger La. W5 40Na 67
HANGER LANE 41Na 87
Hanger Lane Station (Underground) 41Na 87
Hanger Va. La. W5 44Pa 87
(not continuous)
Hanger Vw. Way W3 44Qa 87
Hanging Hill La. CM13: B'wood 20De 41
Hanging Hill La. CM13: Hut 20De 41
Hanging Sword All. EC4 3A 224 (44Qb 90)
(off Whitefriars St.)
Hangrove Hill BR6: Downe 85Rc 180
Hankey Ho. SE1 2F 231 (47Tb 91)
(off Hankey Pl.)
Hankey Pl. SE1 2G 231 (47Tb 91)
Hankins Ho. SE10 48Jc 93
(off Peartree Way)
Hankins La. NW7 19Ua 30
Hanley Cl. SL4: Wind 3B 102
Hanley Gdns. N4 32Pb 70
Hanley Pl. BR3: Beck 66Cc 136
Hanley Rd. N4 32Nb 70
Hanmer Wlk. N7 33Pb 70
Hannaford Wlk. E3 42Dc 92
Hannah Barlow Ho. SW8 53Pb 112
Hannah Bldg. E1 44Xb 91
(off Watney St.)
Hannah Cl. BR3: Beck 69Ec 136
Hannah Cl. NW10 35Sa 67
Hannah Ct. E15 40Hc 73
Hannah Mary Way SE1 49Wb 91
Hannah M. SM6: W'gton 80Lb 156
Hannards Way IG6: Ilf 22Xc 55
Hannay Ho. SW15 58Ab 110
Hannay La. N8 31Mb 70
Hannay Wlk. SW16 61Mb 134
Hannell Rd. SW6 52Ab 110
Hannen Rd. SE27 62Rb 135
Hannibal Rd. E1 43Yb 92
Hannibal Rd. TW19: Stanw 59M 105
Hannibal Way CR0: Wadd 78Pb 156
Hannington Rd. SW4 55Kb 112
Hanno Cl. SM6: W'gton 80Mb 156
Hannover Ho. GU25: Vir W 71A 148
Hanover Pl. RM1: Rom 29Hd 56
Hanover Av. E16 46Jc 93
Hanover Av. TW13: Felt 60W 106
Hanover Circ. UB3: Hayes 44S 84
Hanover Cl. RH1: Mers 100Lb 196
Hanover Cl. SL1: Slou 8L 81
Hanover Cl. SL4: Wind 3D 102
Hanover Cl. SM3: Cheam 77Ab 154
Hanover Cl. TW15: Ashf 63N 127
Hanover Cl. TW20: Eng G 5M 125
Hanover Cl. TW9: Kew 52Qa 109
Hanover Ct. E8 39Vb 71
(off Stean St.)
Hanover Ct. EN9: Walt A 5Ec 20
(off Quakers La.)
Hanover Ct. GU22: Wok 91A 188
Hanover Ct. HA4: Ruis 34W 64
Hanover Ct. NW9 27Ua 48
Hanover Ct. SE19 66Wb 135
(off Anerley Rd.)
Hanover Ct. SW15 56Va 110
Hanover Ct. W12 46Wa 88
(off Uxbridge Rd.)
Hanover Ct. WD3: Crox G 15Q 26
Hanover Dr. BR7: Chst 63Sc 138
Hanover Flats W1 4J 221 (45Jb 90)
(off Binney St.)
Hanover Gdns. IG6: Ilf 24Sc 54
Hanover Gdns. SE11 51Qb 112
Hanover Gdns. WD5: Ab L 2V 12
Hanover Ga. NW1 4E 214 (41Gb 89)
Hanover Ga. SL1: Slou 6E 80
Hanover Ga. Mans. NW1 5E 214 (42Gb 89)
Hanover Grn. HP1: Hem H 4J 3
Hanover Ho. E14 46Bc 92
(off Westferry Cir.)
Hanover Ho. NW8 2D 214 (40Gb 69)
(off St John's Wood High St.)
Hanover Ho. SE16 47Zb 92
(off Dominion Dr.)
Hanover Ho. SW9 55Qb 112
Hanover Mans. SW2 57Qb 112
(off Barnwell Rd.)
Hanover Mead NW11 29Ab 48
Hanover Pk. SE15 53Wb 113
Hanover Pl. CM14: W'ley 22Xd 58
Hanover Pl. DA3: Nw A G 75Be 165
Hanover Pl. E3 41Bc 92
Hanover Pl. WC2 3G 223 (44Nb 90)
Hanover Rd. N15 28Vb 51
Hanover Rd. NW10 38Ya 68
Hanover Rd. SW19 66Eb 133
Hanover Sq. W1 3A 222 (44Kb 90)
Hanover Steps W2 3E 220 (44Gb 89)
(off St George's Flds.)
Hanover St. CR0: C'don 76Rb 157
Hanover St. W1 3A 222 (44Kb 90)
Hanover Ter. NW1 4F 215 (41Hb 89)
Hanover Ter. TW7: Isle 53Ja 108
Hanover Ter. M. NW1 4E 214 (41Gb 89)
Hanover Trad. Est. N7 36Nb 70
Hanover Wlk. AL10: Hat 3B 8
Hanover Wlk. KT13: Weyb 76T 150
Hanover Way DA6: Bex 55Zc 117
Hanover Way SL4: Wind 4D 102
Hanover W. Ind. Est. NW10 41Ta 87
Hanover Yd. N1 1D 218 (40Sb 71)
(off Noel Rd.)
Hansa Cl. UB2: S'hall 48Y 85
Hansard M. W14 47Za 88
Hansart Way EN2: Enf 11Qb 32
Hanscomb M. SW4 56Lb 112
Hans Ct. SW3 3F 227 (48Hb 89)
(off Hans Rd.)
Hans Cres. SW1 3F 227 (48Hb 89)
Hanselin Cl. HA7: Stan 22Ha 46
Hansel Rd. NW6 41Cb 89
Hansen Dr. N21 15Pb 32
Hanshaw Dr. HA8: Edg 25Ta 47
Hansler Ct. SW19 60Ab 110
(off Princes Way)
Hansler Gro. KT8: E Mos 70Fa 130
Hansler Rd. SE22 57Vb 113
Hansol Rd. DA6: Bex 57Ad 117
Hansom M. SE11 7J 229 (50Pb 90)
Hansom Ter. BR1: Broml 67Kc 137
(off Freelands Gro.)
Hanson Cl. BR3: Beck 65Dc 136
Hanson Cl. IG10: Lough 12Sc 36
Hanson Cl. SW12 59Kb 112
Hanson Cl. SW14 55Sa 109
(not continuous)
Hanson Cl. UB7: W Dray 48P 83
Hanson Ct. E17 30Dc 52
Hanson Dr. IG10: Lough 12Sc 36
Hanson Gdns. UB1: S'hall 47Aa 85
Hanson Grn. IG10: Lough 12Sc 36
Hanson Ho. E1 45Wb 91
(off Pinchin St.)
Hanson St. W1 7B 216 (43Lb 90)
Hans Pl. SW1 3G 227 (48Hb 89)
Hans Rd. SW3 3F 227 (48Hb 89)
Hans St. SW1 4G 227 (48Hb 89)
Hanway Pl. W1 2D 222 (44Mb 90)
Hanway Rd. W7 44Fa 86
Hanway St. W1 2D 222 (44Mb 90)
HANWELL 46Ha 86
Hanwell Ho. W2 43Cb 89
(off Gt. Western Rd.)
Hanwell Station (Rail & Crossrail) 45Ga 86
HANWORTH 63Z 129
Hanworth Air Pk. Leisure Cen. 61Z 129
Hanworth Ho. SE5 52Rb 113
Hanworth La. KT16: Chert 74H 149
Hanworth Rd. RH1: Redh 10P 207
Hanworth Rd. TW12: Hamp 63Ba 129
Hanworth Rd. TW13: Felt 60X 107
Hanworth Rd. TW16: Sun 66W 128
(not continuous)
Hanworth Rd. TW3: Houn 57Ca 107
Hanworth Rd. TW4: Houn 60Aa 107
Hanworth Ter. TW3: Houn 56Da 107
Hanworth Trad. Est. KT16: Chert 74H 149
Hanworth Trad. Est. TW13: Hanw 62Aa 129
Hapgood Cl. UB6: G'frd 36Fa 66
Happy Valley Ind. Est. WD4: K Lan 10B 4
Harad's Pl. E1 45Wb 91
Harbans Ct. SL3: Poyle 53G 104
Harben Pde. NW3 38Eb 69
(off Finchley Rd.)
Harben Rd. NW6 38Eb 69
Harberson Rd. E15 39Hc 73
Harberson Rd. SW12 60Kb 112
Harbert Gdns. AL2: Park 1Da 13
Harberton Rd. N19 32Lb 70
Harbet Rd. E4 23Ac 52
Harbet Rd. N18 22Ac 52
Harbet Rd. W2 1C 220 (43Fb 89)
Harbex Cl. DA5: Bexl 59Dd 118
Harbinger Rd. E14 49Dc 92
Harbledown Ho. SE1 2F 231 (47Tb 91)
(off Manciple St.)
Harbledown Pl. BR5: St M Cry 70Yc 139
Harbledown Rd. CR2: Sande 83Wb 177
Harbledown Rd. SW6 53Cb 111
Harbord Cl. SE5 54Tb 113
Harbord Ho. SE16 49Zb 92
(off Cope St.)
Harbord St. SW6 53Za 110
Harborne Cl. WD19: Wat 22Y 45
Harborough Av. DA15: Sidc 59Uc 116
Harborough Cl. SL1: Slou 6B 80
Harborough Ho. UB5: N'olt 41Ba 85
(off Taywood Rd.)
Harborough Rd. SW16 63Pb 134
Harbour Av. SW10 53Eb 111
Harbour Cl. CR4: Mitc 67Jb 134
Harbour Club Chelsea 54Eb 111
Harbour Club Kensington 49Db 89
(off Point West)
Harbour Club Notting Hill 43Cb 89
Harbour Ct. WD23: Bush 15Da 27
Harbourer Cl. IG6: Ilf 22Xc 55
Harbourer Rd. IG6: Ilf 22Xc 55
Harbour Exchange Sq. E14 47Dc 92
Harbourfield Rd. SM7: Bans 87Db 175
Harbour Reach SW6 53Eb 111
Harbour Rd. SE5 55Sb 113
Harbourside Ct. SE8 49Ac 92
(off Plough Way)
Harbour Vw. DA11: Nflt 57De 121
Harbour Way E14 47Dc 92
Harbour Yd. SW10 53Eb 111
Harbridge Av. SW15 59Va 110
Harbury Rd. SM5: Cars 81Gb 175
Harbut Rd. SW11 56Fb 111
Harbutt Rd. HA9: Wemb 35Qa 67
Harcombe Rd. N16 34Ub 71
Harcourt TW19: Wray 58A 104
Harcourt Av. DA15: Sidc 58Yc 117
Harcourt Av. E12 35Pc 74
Harcourt Av. HA8: Edg 20Sa 29
Harcourt Av. SM6: W'gton 77Kb 156
HAREFIELD 25L 43
Harcourt Bldgs. EC4 4K 223 (45Qb 90)
(off Middle Temple La.)
Harcourt Cl. TW20: Egh 65E 126
Harcourt Cl. TW7: Isle 55Ja 108
Harcourt Fld. SM6: W'gton 77Kb 156
Harcourt Ho. W1 2K 221 (44Kb 90)
(off Cavendish Sq.)
Harcourt Lodge SM6: W'gton 77Kb 156
Harcourt M. RM2: Rom 29Hd 56
Harcourt Rd. CR7: Thor H 72Pb 156
Harcourt Rd. DA6: Bex 56Ad 117
Harcourt Rd. E15 40Hc 73
Harcourt Rd. N22 25Mb 50
Harcourt Rd. SE4 55Bc 114
Harcourt Rd. SL4: Wind 3C 102
Harcourt Rd. SM6: W'gton 77Kb 156
Harcourt Rd. SW19 66Db 133
Harcourt Rd. WD23: Bush 15Da 27
Harcourt St. W1 1E 220 (43Gb 89)
Harcourt Ter. SW10 50Db 89
Harcourt Way RH9: S God 9C 210
Hardcastle Cl. CR0: C'don 72Wb 157
Hardcastle Ho. SE14 53Ac 114
(off Loring Rd.)
Hardcourts Cl. BR4: W W'ck 76Dc 158
Hardegray Cl. SM2: Sutt 81Cb 175
Hardell Cl. TW20: Egh 64C 126
Hardel Ri. SW2 60Rb 113
Hardel Wlk. SW2 59Qb 112
Harden Ct. SE7 49Nc 94
Harden Farm Cl. CR5: Coul 93Lb 196
Harden Ho. SE5 54Ub 113
Harden Rd. DA11: Nflt 2B 144
Harden's Manorway SE7 48Mc 93
(not continuous)
Harders Rd. SE15 54Xb 113
Hardess St. SE24 55Sb 113
Hardie Cl. NW10 36Ta 67
Hardie Rd. RM10: Dag 34Ed 76
Hardie Rd. SS17: Stan H 1M 101
Harding Cl. CR0: C'don 76Vb 157
Harding Cl. SE17 51Sb 113
Harding Cl. WD25: Wat 5Y 13
Harding Dr. RM8: Dag 32Ad 75
Hardinge Cl. UB8: Hil 43R 84
Hardinge Cres. SE18 48Sc 94
Hardinge La. E1 44Yb 92
(not continuous)
Hardinge Rd. N18 23Ub 51
Hardinge Rd. NW10 39Xa 68
Hardinge St. E1: Johnson St. 45Yb 92
Hardinge St. E1: Steel's La. 44Yb 92
Harding Ho. SW13 51Xa 110
(off Wyatt Dr.)
Harding Ho. UB3: Hayes 44X 85
Harding Rd. DA7: Bex 54Bd 117
Harding Rd. KT18: Eps D 91Ua 194
Harding Rd. RM16: Grays 8C 100
Harding's Cl. KT2: King T 67Pa 131
Hardings Cl. HP3: Hem H 5K 3
Hardings Cl. SL0: Iver H 41E 82
Hardings La. SE20 65Zb 136
Harding Spur SL3: L'ly 51B 104
Hardings Row SL0: Iver H 41E 82
Hardingstone Ct. EN8: Walt C 6Bc 20
Hardington NW1 38Jb 70
(off Belmont St.)
Hardley Cres. RM11: Horn 28Md 57
Hardman Rd. KT2: King T 68Na 131
Hardman Rd. SE7 50Kc 93
Hardres Ter. BR5: Orp 74Zc 161
Hardwick Cl KT19: Eps 78Ra 153
Hardwick Cl. HA7: Stan 22La 46
Hardwick Cl. KT22: Oxs 87Ea 172
Hardwick Ct. DA8: Erith 51Fd 118
Hardwick Cres. DA2: Dart 58Rd 119
Hardwicke Av. TW5: Hest 53Ca 107
Hardwicke M. WC1 4J 217 (41Pb 90)
(off Lloyd Baker M.)
Hardwicke Pl. AL2: Lon C 9H 7
Hardwicke Rd. N13 23Nb 50
Hardwicke Rd. RH2: Reig 5J 207
Hardwicke Rd. TW10: Ham 63La 130
Hardwicke Rd. W4 49Ta 87
Hardwicke St. IG11: Bark 39Sc 74
Hardwick Grn. W13 43Ka 86
Hardwick Ho. DA2: Dart 58Sd 120
(off Stone Ho. La.)
Hardwick Ho. NW8 5E 214 (42Gb 89)
(off Lilestone St.)
Hardwick La. KT16: Lyne 73E 148
Hardwick Pl. SW16 66Lb 134
Hardwick Rd. RH1: Redh 8M 207
Hardwicks Sq. SW18 57Cb 111
Hardwick St. EC1 4A 218 (41Qb 90)
Hardwidge St. SE1 1H 231 (47Ub 91)
Hardy Av. DA1: Dart 59Ld 119
Hardy Av. DA11: Nflt 1A 144
Hardy Av. E16 46Jc 93
Hardy Av. HA4: Ruis 36X 65
Hardy Cl. EN5: Barn 16Ab 30
Hardy Cl. HA5: Pinn 31Z 65
Hardy Cl. SE16 47Zb 92
Hardy Cl. SL1: Slou 6E 80
Hardy Cotts. SE10 51Fc 115
Hardy Ct. DA8: Erith 52Hd 118
Hardy Gro. DA1: Dart 56Qd 119
Hardy Ho. SW17 62Fb 133
(off Grosvenor Way)
Hardy Ho. SW18 59Db 111
Hardy Ho. SW4 59Lb 112
Hardying Ho. E17 28Ac 52
Hardy M. UB8: Uxb 39L 63
Hardy Pas. N22 25Pb 50
Hardy Rd. E4 23Bc 52
Hardy Rd. HP2: Hem H 1P 3
Hardy Rd. SE3 52Hc 115
Hardy Rd. SW19 66Db 133
Hardy's M. KT8: E Mos 70Ga 130
Hardy's Yd. TN13: Dun G 93Gd 202
Hardy Way EN2: Enf 11Qb 32
Hare & Billet Rd. SE3 53Fc 115
Harebell Dr. E6 43Qc 94
Harebell Hill KT11: Cobh 86Z 171
Harebell Way RM3: Rom 24Md 57
Harebreaks, The WD24: Wat 8W 12
Harecastle Cl. UB4: Yead 42Aa 85
Hare Ct. EC4 3K 223 (44Qb 90)
(off Church Ct.)
Harecourt Rd. N1 37Sb 71
Hare Cres. WD25: Wat 4W 12
Harecroft KT22: Fet 96Da 191
Harecroft La. UB10: Ick 34S 64
Haredale Ho. SE16 47Wb 91
(off East La.)
Haredale Rd. SE24 56Sb 113
Haredon Cl. SE23 59Zb 114
Harefield KT10: Hin W 76Ga 152
HAREFIELD 25L 43
Harefield Av. SM2: Cheam 81Ab 174
Harefield Cl. EN2: Enf 11Qb 32
Harefield Grn. NW7 23Ya 48
HAREFIELD HOSPITAL 25L 43
Harefield M. SE4 55Bc 114
Harefield Rd. DA14: Sidc 62Zc 139
Harefield Rd. N8 29Mb 50
Harefield Rd. SE4 55Bc 114
Harefield Rd. SW16 66Pb 134
Harefield Rd. UB8: Uxb 38L 63
Harefield Rd. WD3: Rick 19M 25
Hare Hall La. RM2: Rom 28Kd 57
Harehatch La. SL1: Burn 2C 60
Harehatch La. SL2: Burn 2D 60
Harehatch La. SL2: Farn C 2D 60
Hare Hill KT15: Add 79G 148
Hare Hill Cl. GU22: Pyr 87J 169
Harelands Cl. GU21: Wok 9N 167
Harelands La. GU21: Wok 10N 167
(not continuous)
Hare La. AL10: Hat 2D 8
Hare La. KT10: Clay 78Fa 152
Hare Marsh E2 42Wb 91
Harendon KT20: Tad 93Ya 194
Harepark Cl. HP1: Hem H 1H 3
Harepit Cl. CR2: S Croy 80Rb 157
Hare Pl. EC4 3A 224 (44Qb 90)
(off Fleet St.)
Hare Row E2 40Xb 71
Hares Bank CR0: New Ad 82Fc 179
Haresfield Rd. RM10: Dag 37Cd 76
Harestone Dr. CR3: Cat'm 96Vb 197
Harestone Hill CR3: Cat'm 98Vb 197
Harestone La. CR3: Cat'm 97Ub 197
Harestone Valley Rd. CR3: Cat'm 98Ub 197
Hare St. SE18 48Qc 94
Hare Ter. RM20: Grays 50Zd 99
Hare Wlk. N1 2J 219 (40Ub 71)
(not continuous)
Harewood WD3: Rick 14K 25
(not continuous)
Harewood Av. NW1 6E 214 (42Gb 89)
Harewood Av. NW7 23Ab 48
Harewood Av. UB5: N'olt 38Ba 65
Harewood Cl. RH2: Reig 3L 207
Harewood Cl. UB5: N'olt 38Ba 65
Harewood Ct. CR6: W'ham 90Ac 178
Harewood Dr. IG5: Ilf 26Pc 54
Harewood Gdns. CR2: Sande 87Xb 177
Harewood Hill CM16: They B 7Uc 22
Harewood Pl. SL1: Slou 8L 81
Harewood Pl. W1 3A 222 (44Kb 90)
Harewood Rd. CM15: Pil H 16Xd 40
Harewood Rd. CR2: S Croy 79Ub 157
Harewood Rd. SW19 65Gb 133
Harewood Rd. TW7: Isle 52Ha 108
Harewood Rd. WD19: Wat 20X 27
Harewood Row NW1 7E 214 (43Gb 89)
Harewood Ter. UB2: S'hall 49Ba 85
Harfield Gdns. SE5 55Ub 113
Harfield Rd. TW16: Sun 68Z 129
Harfleur Ct. SE11 6B 230 (49Rb 91)
(off Opal St.)
Harford Cl. E4 17Dc 34
Harford Dr. WD17: Wat 10U 12
Harford Ho. SE5 51Sb 113
(off Bethwin Rd.)
Harford Ho. W11 43Bb 89
Harford M. N19 34Mb 70
Harford Rd. E4 17Dc 34
Harford St. E1 42Ac 92
Harford Wlk. N2 28Fb 49
Harfst Way BR8: Swan 67Ed 140
Hargood Cl. HA3: Kenton 30Na 47
Hargood Rd. SE3 53Lc 115
Hargrave Mans. N19 33Mb 70
Hargrave Pk. N19 33Lb 70
Hargrave Pl. N7 36Mb 70
Hargrave Rd. N19 33Lb 70
Hargraves Ho. W12 45Xa 88
(off White City Est.)
Hargreaves Av. EN7: Chesh 3Xb 19
Hargreaves Cl. EN7: Chesh 3Xb 19
Hargreaves Ct. E3 41Ec 92
(off Bolinder Way)
Hargwyne St. SW9 55Pb 112
Hari Cl. UB5: N'olt 36Da 65
Haringey Independent Cinema 28Sb 51
Haringey Pk. N8 30Nb 50
Haringey Pas. N8 28Qb 50
Haringey Rd. N8 28Nb 50
Harington Ter. N18 20Tb 33
Harington Ter. N9 20Tb 33
Harkett Cl. HA3: W'stone 26Ha 46
Harkett Ct. HA3: W'stone 26Ha 46
Harkness EN7: Chesh 1Xb 19
Harkness Cl. KT17: Eps D 88Ya 174
Harkness Cl. RM3: Rom 22Pd 57
Harkness Ct. SM1: Sutt 74Db 155
(off Cleeve Way)
Harkness Ho. E1 44Wb 91
(off Christian St.)
Harkness Rd. HP2: Hem H 1M 3
Harland Av. CR0: C'don 76Vb 157
Harland Av. DA15: Sidc 62Tc 138
Harland Cl. SW19 69Db 133
Harland Rd. SE12 60Jc 115
Harlands Gro. BR6: Farnb 77Rc 160
Harlands Gro. CR0: New Ad 82Gc 179
Harlech Gdns. HA5: Pinn 31Z 65
Harlech Gdns. TW5: Hest 51Y 107
Harlech Rd. N14 20Nb 32
Harlech Rd. WD5: Ab L 3V 12
Harlech Twr. W3 47Sa 87
Harlequin, The 14Y 27
Harlequin Av. TW8: Bford 51Ja 108
Harlequin Cl. IG11: Bark 42Wc 95
Harlequin Cl. TW7: Isle 57Ga 108
Harlequin Cl. UB4: Yead 43Z 85
Harlequin Ct. E1 45Wb 91
(off Thomas More St.)
Harlequin Ct. NW10 37Ta 67
(off Mitchellbrook Way)
Harlequin Ct. W5 45La 86
Harlequin FC 59Ga 108
Harlequin Ho. DA18: Erith 48Ad 95
(off Kale Rd.)
Harlequin Rd. TW11: Tedd 66Ka 130
Harlequin Theatre & Cinema 5P 207

Harlescott Rd. SE15 56Zb 114
HARLESDEN 40Va 68
Harlesden Cl. RM3: Rom 23Pd 57
Harlesden Gdns. NW10 39Va 68
Harlesden La. NW10 39Wa 68
Harlesden Plaza NW10 40Va 68
Harlesden Rd. AL1: St A 2E 6
Harlesden Rd. NW10 39Wa 68
Harlesden Rd. RM3: Rom 23Pd 57
Harlesden Station (Underground & Overground) 40Ta 67
Harlesden Wlk. RM3: Rom 24Pd 57
Harleston Cl. E5 33Yb 72
Harle Way RM13: Rain 42Ld 97
Harley Cl. HA0: Wemb 37Ma 67
Harley Ct. E11 31Jc 73
Harley Ct. HA1: Harr 28Fa 46
Harley Ct. N20 20Eb 31
Harley Cres. HA1: Harr 28Fa 46
Harleyford BR1: Broml 67Kc 137
Harleyford Ct. SE11 51Pb 112 (off Harleyford Rd.)
Harleyford Mnr. W3 46Sa 87 (off Edgecote Cl.)
Harleyford Rd. SE11 51Pb 112
Harleyford St. SE11 51Qb 112
Harley Gdns. BR6: Orp 77Uc 160
Harley Gdns. SW10 7A 226 (50Eb 89)
Harley Gro. E3 41Bc 92
Harley Ho. E11 31Fc 73
Harley Ho. E14 44Bc 92 (off Frances Wharf)
Harley Ho. NW1 6J 215 (42Jb 90) (off Marylebone Rd.)
Harley Ho. WD6: Bore 12Ra 29 (off Brook Cl.)
Harley Pl. W1 1K 221 (43Kb 90)
Harley Rd. HA1: Harr 28Fa 46
Harley Rd. NW10 40Ua 68
Harley Rd. NW3 38Fb 69
Harley St. W1 6K 215 (42Kb 90)
HARLEY STREET CLINIC, THE 7K 215 (43Kb 90)
Harley Vs. NW10 40Ua 68
Harlie Ct. SE6 58Cc 114
Harling Ct. SW11 54Hb 111
Harlinger St. SE18 48Nc 94
HARLINGTON 51T 106
Harlington Cl. UB3: Harl 52S 106
HARLINGTON CORNER 53T 106
HARLINGTON HOSPICE 50T 84
Harlington Rd. DA7: Bex 55Ad 117
Harlington Rd. UB8: Hil 41Q 84
Harlington Rd. E. TW13: Felt 60Y 107
Harlington Rd. E. TW14: Felt 59X 107
Harlington Rd. W. TW14: Felt 58X 107
Harlington Sports Cen., The 49T 84 (off Pinkwell La.)
Harlington Young People's Cen. 49T 84
Harlow Ct. RH2: Reig 6M 207 (off Wray Comn. Rd.)
Harlow Gdns. RM5: Col R 23Ed 56
Harlow Mans. IG11: Bark 38Rc 74 (off Whiting Av.)
Harlow Rd. N13 20Tb 33
Harlow Rd. RM13: Rain 39Hd 76
Harlton Ct. EN9: Walt A 6Hc 21
Harlyn Dr. HA5: Eastc 27X 45
Harlynwood SE5 52Sb 113 (off Wyndham Rd.)
Harman Av. DA11: Grav'nd 4D 144
Harman Av. IG8: Wfd G 23Hc 53
Harman Cl. E4 21Fc 53
Harman Cl. NW2 34Ab 68
Harman Cl. SE1 50Wb 91
Harman Dr. DA15: Sidc 58Vc 117
Harman Dr. NW2 34Ab 68
Harman Pl. CR8: Purl 83Rb 177
Harman Ri. IG3: Ilf 35Uc 74
Harman Rd. EN1: Enf 15Vb 33
Harmer Rd. DA10: Swans 58Be 121
Harmer St. DA12: Grav'nd 8E 122
HARMONDSWORTH 51M 105
Harmondsworth La. UB7: Harm 51N 105
Harmondsworth La. UB7: Sip 51N 105
Harmondsworth Moor Waterside 51K 105
Harmondsworth Moor Waterside Vis. Cen. 51K 105
Harmondsworth Rd. UB7: W Dray 50N 83
Harmon Ho. SE8 49Bc 92
Harmonia Ct. WD17: Wat 10W 12
Harmont Ho. W1 1K 221 (43Kb 90) (off Harley St.)
Harmony Apts. BR1: Broml 68Jc 137 (off High St.)
Harmony Cl. NW11 29Ab 48 (not continuous)
Harmony Cl. SM6: W'gton 81Nb 176
Harmony Pl. SE1 50Vb 91
Harmony Pl. SE8 51Dc 114 (off Dancers Way)
Harmony Ter. HA2: Harr 32Da 65
Harmony Way BR1: Broml 68Jc 137
Harmony Way NW4 28Ya 48
Harmood Gro. NW1 38Kb 70
Harmood Ho. NW1 38Kb 70 (off Harmood St.)
Harmood Pl. NW1 38Kb 70
Harmood St. NW1 37Kb 70
Harmsworth M. SE11 4B 230 (48Rb 91)
Harmsworth St. SE17 50Rb 91
Harmsworth Way N20 18Bb 31
Harnetts Cl. BR8: Crock 73Fd 162
Harold Av. DA17: Belv 50Bd 95
Harold Av. UB3: Hayes 48V 84
Harold Campbell Ct. DA1: Dart 59Md 119 (off North St.)
Harold Ct. EN8: Walt C 6Bc 20 (off Holdbrook Sth.)
Harold Ct. RM3: Hrld W 24Rd 57
Harold Ct. SE16 47Zb 92 (off Christopher Cl.)
Harold Ct. Rd. RM3: Hrld W 23Rd 57
Harold Cres. EN9: Walt A 4Ec 20
Harold Est. SE1 4J 231 (48Ub 91)
Harold Gibbons Ct. SE7 51Lc 115
HAROLD HILL 23Pd 57
Harold Hill Ind. Est. RM3: Rom 24Md 57
Harold Ho. E2 40Zb 72 (off Mace St.)
Harold Laski Ho. EC1 4C 218 (41Rb 91) (off Percival St.)
Harold Maddison Ho. SE17 7C 230 (50Rb 91) (off Penton Pl.)
Harold Mugford Ter. E6 44Qc 94 (off Pearl Cl.)
HAROLD PARK 23Rd 57
Harold Pinter Theatre 5D 222 (45Mb 90) (off Panton St.)
Harold Pl. SE11 50Qb 90
Harold Rd. DA2: Hawl 63Pd 141
Harold Rd. E11 32Gc 73
Harold Rd. E13 39Kc 73
Harold Rd. E4 21Ec 52
Harold Rd. IG8: Wfd G 25Jc 53
Harold Rd. N15 29Vb 51
Harold Rd. N8 29Pb 50
Harold Rd. NW10 41Ta 87
Harold Rd. SE19 66Tb 135
Harold Rd. SM1: Sutt 77Fb 155
Harold's Bridge 5Ec 20
Haroldstone Rd. E17 29Zb 52
Harold Vw. RM3: Hrld W 26Pd 57
Harold Wilson Ho. SE28 46Xc 95
Harold Wilson Ho. SW6 51Bb 111 (off Clem Attlee Ct.)
HAROLD WOOD 25Pd 57
Harold Wood Hall RM3: Rom 25Md 57 (off Neave Cres.)
Harold Wood Station (Rail & Crossrail) 25Pd 57
Harp All. EC4 2B 224 (44Rb 91)
Harp Bus. Cen., The NW2 33Wa 68
Harpenden Rd. E12 33Lc 73
Harpenden Rd. SE27 62Rb 135
Harpenmead Point NW2 33Bb 69
Harper Cl. N14 15Lb 32
Harper Cl. RM16: Chaf H 50Yd 98
Harper Ho. SW9 55Rb 113
Harper La. WD7: R'lett 4Ha 14
Harper La. WD7: Shenl 4Ha 14
Harper M. SW17 62Eb 133
Harper Rd. E6 44Pc 94
Harper Rd. SE1 3E 230 (48Sb 91)
Harper's Yd. N17 25Vb 51
Harpers Yd. TW7: Isle 54Ga 108 (off Rennels Way)
Harpesford Av. GU25: Vir W 1M 147
Harp Island Cl. NW10 33Ta 67
Harp La. EC3 5H 225 (45Ub 91)
Harpley Sq. E1 42Zb 92
Harpour Rd. IG11: Bark 37Sc 74
Harp Rd. W7 42Ha 86
Harpsden St. SW11 53Jb 112
Harps Oak La. RH1: Mers 97Hb 195
Harpswood Cl. CR5: Coul 94Lb 196
Harpur M. WC1 7H 217 (43Pb 90)
Harpurs KT20: Tad 94Za 194
Harpur St. WC1 7H 217 (43Pb 90)
Harraden Rd. SE3 53Lc 115
Harrap Chase RM17: Grays 50Be 99
Harrier Av. E11 30Kc 53
Harrier Cen., The 79Ta 153
Harrier Cl. HP3: Hem H 7M 3
Harrier Cl. RM12: Horn 37Kd 77
Harrier Ct. TW4: Houn 55Aa 107
Harrier M. SE28 47Tc 94
Harrier Rd. NW9 26Ua 48
Harriers Cl. W5 45Na 87
Harrier Way E6 43Pc 94
Harrier Way EN9: Walt A 6Jc 21
Harriescourt EN9: Walt A 4Jc 21
Harries Rd. UB4: Yead 42Y 85
Harriet Cl. E8 39Wb 71
Harriet Ct. SE14 52Yb 114 (off Pomeroy St.)
Harriet Gdns. CR0: C'don 75Wb 157
Harriet Ho. HP3: Hem H 7N 3
Harriet Ho. SW6 52Db 111 (off Wandon Rd.)
Harriet M. DA16: Well 54Xc 117
Harriet St. SW1 2G 227 (47Hb 89)
Harriet Tubman Cl. SW2 59Pb 112
Harriet Wlk. SW1 2G 227 (47Hb 89)
Harriet Walker Way WD3: Rick 17H 25
Harriet Way WD23: Bush 17Fa 28
HARRINGAY 29Rb 51
Harringay Gdns. N8 28Rb 51
Harringay Green Lanes Station (Overground) 30Rb 51
Harringay Rd. N15 29Rb 51 (not continuous)
Harringay Station (Rail) 30Qb 50
Harrington Cl. CR0: Bedd 75Nb 156
Harrington Cl. NW10 34Ta 67
Harrington Cl. SL4: Wind 6D 102
Harrington Ct. CR0: C'don 75Tb 157
Harrington Ct. SW7 5C 226 (49Fb 89) (off Harrington Rd.)
Harrington Ct. W10 41Bb 89
Harrington Cres. RM16: N Stif 46Zd 99
Harrington Gdns. SW7 49Db 89
Harrington Hill E5 32Xb 71
Harrington Ho. NW1 3B 216 (41Lb 90) (off Harrington St.)
Harrington Ho. UB10: Ick 35R 64
Harrington Rd. E11 32Gc 73
Harrington Rd. SE25 70Wb 135
Harrington Rd. SW7 5B 226 (49Fb 89)
Harrington Road Stop (London Tramlink) 69Yb 136
Harrington Sq. NW1 1B 216 (40Lb 70)
Harrington St. NW1 2B 216 (41Lb 70) (not continuous)
Harrington Way SE18 48Mc 93
Harriott Cl. SE10 49Hc 93
Harriott Ho. E1 43Yb 92 (off Jamaica St.)
Harriott's Cl. KT21: Asht 92La 192
Harriott's La. KT21: Asht 91La 192
Harris Academy Sports Cen. 74Yc 161
Harris Bldgs. E1 44Wb 91 (off Burslem St.)
Harris Cl. DA11: Nflt 2B 144
Harris Cl. EN2: Enf 11Rb 33
Harris Cl. N11 22Hb 49
Harris Cl. RM3: Rom 24Nd 57
Harris Cl. TW3: Houn 53Ca 107
Harris Ct. HA9: Wemb 34Pa 67
Harris Gdns. SL1: Slou 7G 80
Harris Ho. E11 32Gc 73
Harris Ho. E3 41Cc 92 (off Alfred St.)
Harris Ho. SW9 55Qb 112 (off St James's Cres.)
Harris La. WD7: Shenl 6Qa 15
Harris Lodge SE6 60Ec 114
Harrison Av. DA3: Lfield 69Be 143
Harrison Cl. CM13: Hut 15Fe 41
Harrison Cl. HA6: Nwood 23S 44
Harrison Cl. N20 18Gb 31
Harrison Cl. RH2: Reig 7K 207
Harrison Cl. RM7: Mawney 27Cd 56
Harrison Ct. E18 25Jc 53 (off Queen Mary Av.)
Harrison Dr. BR1: Broml 70Qc 138
Harrison Ho. E1 44Xb 91
Harrison Ho. HP3: Hem H 7P 3 (off The Embankment)
Harrison Ho. SE17 7F 231 (50Tb 91) (off Brandon St.)
Harrison Rd. EN9: Walt A 7Ec 20
Harrison Rd. NW10 39Ta 67
Harrison Rd. RM10: Dag 37Dd 76
Harrison Rd. TN15: Bor G 92Be 205
Harrisons Ct. SE14 51Zb 114 (off Myers La.)
Harrison's Ri. CR0: Wadd 76Rb 157
Harrison St. WC1 4G 217 (41Nb 90)
Harrisons Wharf RM19: Purf 50Qd 97
Harrison Wlk. EN8: Chesh 2Zb 20
Harrison Way SL1: Slou 6B 80
Harrison Way TN13: S'oaks 94Jd 202
Harrison Way TW17: Shep 71R 150
Harris Rd. DA7: Bex 53Ad 117
Harris Rd. RM9: Dag 36Bd 75
Harris Rd. WD25: Wat 7W 12
Harris Sports Cen. 57Yb 114
Harris St. E17 31Bc 72
Harris St. SE5 52Tb 113
Harris Way TW16: Sun 67U 128
Harrod Ct. NW9 28Sa 47
Harrods 3F 227 (48Hb 89)
Harrogate Ct. N11 23Jb 50
Harrogate Ct. SE12 59Jc 115
Harrogate Ct. SE26 62Wb 135 (off Droitwich Cl.)
Harrogate Ct. SL3: L'ly 50C 82
Harrogate Rd. WD19: Wat 20Y 27
Harrold Ho. NW3 38Fb 69
Harrold Rd. RM8: Dag 36Xc 75
Harrovian Bus. Village HA1: Harr 31Ga 66
HARROW 30Ga 46
Harrow & Wealdstone Station (Rail, Underground & Overground) 28Ga 46
Harrow Arts Cen. 24Ca 45
Harrow Av. EN1: Enf 16Vb 33
Harroway Mnr. KT22: Fet 94Ha 192
Harroway Rd. SW11 54Fb 111
Harrow Borough FC 35Ba 65
Harrow Bottom Rd. GU25: Vir W 72B 148
Harrow Bus Station 30Ga 46
Harrowby Gdns. DA11: Nflt 1A 144
Harrowby Ho. W1 2F 221 (44Hb 89) (off Harrowby St.)
Harrowby St. W1 2E 220 (44Gb 89)
Harrow Cl. KT15: Add 75K 149
Harrow Cl. KT9: Chess 80Ma 153
Harrow Cl. RM11: Horn 32Kd 77
Harrow Club W10 45Za 88
Harrow Cres. RM3: Rom 24Kd 57
HARROW CYGNET HOSPITAL 33Ga 66
Harrowdene Cl. HA0: Wemb 35Ma 67
Harrowdene Gdns. TW11: Tedd 65Ja 130
Harrowdene Rd. HA0: Wemb 34Ma 67
Harrow Dr. N9 18Vb 33
Harrow Dr. RM11: Horn 30Kd 57
Harrowes Meade HA8: Edg 20Qa 29
Harrow Flds. Gdns. HA1: Harr 34Ga 66
Harrow Gdns. BR6: Chels 77Xc 161
Harrow Gdns. CR6: W'ham 88Bc 178
Harrow Gdns. KT8: E Mos 69Fa 130
Harrowgate Ho. E9 37Zb 72
Harrowgate Rd. E9 37Ac 72
Harrow Grn. E11 34Gc 73
Harrow High School Sports Cen. 30Ja 46
Harrow Hill Golf Course 31Ha 66
Harrow Ho. RH1: Redh 4A 208 (off Old School Ct.)
Harrow La. E14 45Dc 92
Harrow La. RM14: Bulp 36Ee 79
Harrow Leisure Cen. 27Ha 46
Harrow Lodge NW8 5B 214 (42Fb 89) (off Northwick Ter.)
Harrow Manorway SE2 46Yc 95
Harrow Mnr. Way SE28 45Yc 95
Harrow Mkt. SL3: L'ly 48C 82
Harrow Mus. 27Ea 46
HARROW ON THE HILL 32Ga 66
Harrow-on-the-Hill Station (Rail & Underground) 30Ga 46
Harrow Pk. HA1: Harr 33Ga 66
Harrow Pl. E1 2J 225 (44Ub 91)
Harrow Rd. CR6: W'ham 87Bc 178
Harrow Rd. E11 34Gc 73
Harrow Rd. E6 39Nc 74
Harrow Rd. HA0: Wemb 35Ha 66
Harrow Rd. HA9: Wemb 36Qa 67
Harrow Rd. IG1: Ilf 35Sc 74
Harrow Rd. IG11: Bark 39Uc 74
Harrow Rd. NW10 41Xa 88
Harrow Rd. SL3: L'ly 48B 82
Harrow Rd. SM5: Cars 79Gb 155
Harrow Rd. TN14: Knock 87Ad 181
Harrow Rd. TW14: Bedf 61Q 128
Harrow Rd. W10 42Ab 88
Harrow Rd. W2 7A 214 (43Eb 89) (not continuous)
Harrow Rd. W9 42Bb 89
HARROW ROAD 38Ra 67
Harrow Rd. Bri. 7A 214 (43Eb 89)
Harrow School Golf Course 33Ha 66
Harrow Sports Hall 31Ja 66
Harrow St. NW1 7E 214 (43Gb 89) (off Daventry St.)
Harrow Vw. HA1: Harr 28Fa 46
Harrow Vw. HA2: Harr 26Ea 46
Harrow Vw. UB10: Hil 41S 84
Harrow Vw. UB3: Hayes 44W 84
Harrow Vw. Rd. W5 42Ka 86
Harrow Vw. W. HA2: Harr 27Ea 46
Harrow Way TW17: Shep 68S 128
Harrow Way WD19: Wat 20Aa 27
HARROW WEALD 25Ga 46
Harrow Weald Lawn Tennis Club 25Ga 46
Harrow Weald Pk. HA3: Hrw W 23Fa 46
Harry Cl. CR0: C'don 72Sb 157
Harry Cole Ct. SE17 7G 231 (50Tb 91) (off Thurlow St.)
Harry Day M. SE27 62Sb 135
Harry Hinkins Ho. SE17 7E 230 (50Sb 91) (off Bronti Cl.)
Harry Lambourn Ho. SE15 52Xb 113 (off Gervase St.)
Harrys Pl. RM15: S Ock 43Zd 99
Harry Zeital Way E5 33Yb 72
Harston Dr. EN3: Enf L 10Cc 20
Harston Wlk. E3 42Dc 92
Hartcliff Ct. W7 47Ha 86
Hart Cl. CR0: C'don 76Rb 157
Hart Cl. RH1: Blet 5L 209
Hart Cnr. RM20: Grays 50Zd 99
Hart Ct. E6 38Qc 74
Hart Cres. IG7: Chig 22Vc 55
Hartcroft Cl. HP3: Hem H 3B 4
Hart Dyke Cres. BR8: Swan 69Fd 140
Hart Dyke Rd. BR5: Orp 75Zc 161
Hart Dyke Rd. BR8: Swan 69Fd 140
Harte Rd. TW3: Houn 54Ba 107
Hartfield Av. UB5: N'olt 40X 65
Hartfield Av. WD6: E'tree 14Qa 29
Hartfield Cl. WD6: E'tree 15Qa 29
Hartfield Cres. BR4: W W'ck 76Jc 159
Hartfield Cres. SW19 66Bb 133
Hartfield Gro. SE20 67Yb 136
Hartfield Ho. UB5: N'olt 40X 65 (off Hartfield Av.)
Hartfield Pl. DA11: Nflt 59Fe 121
Hartfield Rd. BR4: W W'ck 77Jc 159
Hartfield Rd. KT22: Lea 92Ha 192
Hartfield Rd. KT9: Chess 78Ma 153
Hartfield Rd. SW19 66Bb 133
Hartfield Ter. E3 40Cc 72
Hartford Av. HA3: Kenton 27Ja 46
Hartforde Rd. WD6: Bore 12Qa 29
Hartford Rd. DA5: Bexl 58Cd 118
Hartford Rd. KT19: Ewe 79Ra 153
Hart Gro. UB1: S'hall 43Ca 85
Hart Gro. W5 46Qa 87
Hart Gro. Ct. W5 46Qa 87
Harthall La. HP3: Hem H 9D 4
Harthall La. WD4: K Lan 10B 4
Hartham Cl. N7 36Nb 70
Hartham Cl. TW7: Isle 53Ja 108
Hartham Rd. N17 26Vb 51
Hartham Rd. N7 36Nb 70
Hartham Rd. TW7: Isle 53Ha 108
Harting Rd. SE9 62Nc 138
Hartington Cl. BR6: Farnb 78Sc 160
Hartington Cl. HA1: Harr 35Ga 66
Hartington Cl. RH2: Reig 4J 207
Hartington Ct. SW8 53Nb 112
Hartington Ct. W4 52Ra 109
Hartington Ho. SW1 7E 228 (50Mb 90) (off Drummond Ga.)
Hartington Rd. E16 44Kc 93
Hartington Rd. E17 30Ac 52
Hartington Rd. SW8 53Nb 112
Hartington Rd. TW1: Twick 59Ka 108
Hartington Rd. UB2: S'hall 48Aa 85
Hartington Rd. W13 45Ka 86
Hartington Rd. W4 52Ra 109
Hartismere Rd. SW6 52Bb 111
Hartlake Rd. E9 37Zb 72
Hartland NW1 1C 216 (39Lb 70) (off Royal College St.)
Hartland Cl. HA8: Edg 19Qa 29
Hartland Cl. KT15: New H 82L 169
Hartland Cl. N21 16Sb 33
Hartland Cl. SL1: Slou 6H 81
Hartland Ct. N11 22Hb 49 (off Hartland Rd.)
Hartland Dr. HA4: Ruis 34X 65
Hartland Dr. HA8: Edg 19Qa 29
Hartland Rd. CM16: Epp 3Wc 23
Hartland Rd. E15 38Hc 73
Hartland Rd. EN8: Chesh 2Zb 20
Hartland Rd. KT15: Add 80J 149
Hartland Rd. N11 22Hb 49
Hartland Rd. NW1 38Kb 70
Hartland Rd. NW6 40Bb 69
Hartland Rd. RM12: Horn 33Jd 76
Hartland Rd. SM4: Mord 73Cb 155
Hartland Rd. TW12: Hamp H 63Da 129
Hartland Rd. TW7: Isle 55Ja 108
Hartlands, The TW5: Cran 51X 107
Hartlands Cl. DA5: Bexl 58Bd 117
Hartland Way CR0: C'don 76Ac 158
Hartland Way SM4: Mord 73Bb 155
Hartlepool Ct. E16 46Rc 94
HARTLEY 71Ae 165
Hartley Av. E6 39Nc 74
Hartley Av. NW7 22Va 48
Hartley Bottom Rd. DA3: Hartl 76De 165
Hartley Bottom Rd. DA3: Lfield 76De 165
Hartley Bottom Rd. DA3: Nw A G 77Ce 165
Hartley Bottom Rd. TN15: Ash 79Ce 165
Hartley Cl. BR1: Broml 68Pc 138
Hartley Cl. NW7 22Va 48
Hartley Cl. SL3: Stoke P 9N 61
Hartley Copse SL4: Old Win 8L 103
Hartley Down CR8: Purl 87Pb 176
Hartley Farm CR8: Purl 87Pb 176
HARTLEY GREEN 71Ae 165
Hartley Hill CR8: Purl 87Pb 176
Hartley Hill DA3: Hartl 74Ce 165
HARTLEY HILL 74Ce 165
Hartley Ho. SE1 5K 231 (49Vb 91) (off Longfield Est.)
Hartley Old Rd. CR8: Purl 87Pb 176
Hartley Rd. CR0: C'don 73Sb 157
Hartley Rd. DA16: Well 52Yc 117
Hartley Rd. DA3: Lfield 68Ae 143
Hartley Rd. E11 32Hc 73
Hartley Rd. TN16: Westrm 97Tc 200
Hartley St. E2 41Yb 92 (not continuous)
Hartley Way CR8: Purl 87Pb 176
Hart Lodge EN5: Barn 13Ab 30
Hartmann Rd. E16 46Mc 93
Hartmoor M. EN3: Enf W 9Zb 20
Hartnoll St. N7 36Pb 70
Harton Cl. BR1: Broml 67Mc 137
Harton Lodge SE8 53Cc 114 (off Harton St.)
Harton Rd. N9 19Xb 33
Harton St. SE8 53Cc 114
Hartop Point SW6 52Ab 110 (off Pellant Rd.)
Hart Rd. AL1: St A 3B 6
Hart Rd. KT14: Byfl 85N 169
Hartsbourne Av. WD23: B Hea 19Fa 28
Hartsbourne Cl. WD23: B Hea 19Fa 28
Hartsbourne Country Club & Golf Course 19Ea 28
Hartsbourne Ct. UB1: S'hall 44Ea 86 (off Fleming Rd.)
Hartsbourne Pk. WD23: B Hea 19Ga 28
Hartsbourne Rd. WD23: B Hea 19Fa 28
Hartsbourne Way HP2: Hem H 3C 4
Harts Cl. WD23: Bush 12Ca 27
Hartscroft CR0: Sels 81Ac 178
Harts Gro. IG8: Wfd G 22Jc 53
Hartshaw DA3: Lfield 68De 143
Hartshill Cl. UB10: Hil 38R 64
Hartshill Rd. DA11: Nflt 1B 144
Hartshill Wlk. GU21: Wok 8M 167
Hartshorn All. EC3 3J 225 (44Ub 91) (off Leadenhall St.)
Hartshorn Gdns. E6 42Qc 94
Hartslands Rd. TN13: S'oaks 95Ld 203
Hart's La. RH9: S God 8B 210
Hart's La. SE14 53Ac 114
Harts La. IG11: Bark 37Rc 74
Hartslock Dr. SE2 47Zc 95
Hartsmead Rd. SE9 61Pc 138
Hartspiece Rd. RH1: Redh 8A 208
Hartspring La. WD23: Bush 12Ca 27
Hartspring La. WD25: A'ham 12Ca 27
Hart Sq. SM4: Mord 72Cb 155
Hart St. CM14: B'wood 19Yd 40
Hart St. EC3 4J 225 (45Ub 91)
Hartsway EN3: Pond E 14Yb 34
Hartswood Av. RH2: Reig 10J 207
Hartswood Cl. CM14: W'ley 21Ae 59
Hartswood Cl. WD23: Bush 12Ca 27
Hartswood Gdns. W12 48Va 88
Hartswood Golf Course 21Ae 59
Hartswood Grn. WD23: B Hea 19Fa 28
Hartswood Rd. CM13: Gt War 23Ae 59
Hartswood Rd. CM13: W'ley 23Ae 59
Hartswood Rd. CM14: W'ley 21Ae 59
Hartswood Rd. W12 47Va 88
HARTSWOOD SPIRE HOSPITAL 23Xd 58
Hartsworth Cl. E13 40Hc 73
Hartville Rd. SE18 49Uc 94
Hartwell Cl. SW2 60Pb 112
Hartwell Dr. E4 23Ec 52
Hartwell Ho. SE7 50Kc 93 (off Troughton Rd.)
Hartwell St. E8 37Vb 71
Harty Cl. RM16: Grays 46De 99
Harvard Ct. NW6 36Db 69
Harvard Hill W4 51Ra 109
Harvard Ho. SE17 51Rb 113 (off Doddington Gro.)
Harvard La. W4 50Sa 87
Harvard Rd. SE13 57Ec 114
Harvard Rd. TW7: Isle 53Ga 108
Harvard Rd. W4 50Ra 87
Harvard Wlk. RM12: Horn 35Jd 76
Harvel Cl. BR5: St P 69Wc 139
Harvel Cres. SE2 50Zc 95
Harvel Rd. TN15: Fair 84Fe 185
Harvest Bank Rd. BR4: W W'ck 76Hc 159
Harvest Ct. KT10: Esh 75Ca 151
Harvest Ct. RM13: Rain 40Fd 76 (off Broadis Way)
Harvest Ct. TW17: Shep 70Q 128
Harvest End WD25: Wat 8Z 13
Harvester Rd. KT19: Eps 82Ta 173
Harvesters Cl. TW7: Isle 57Fa 108
Harvest La. IG10: Lough 17Mc 35
Harvest La. KT7: T Ditt 72Ja 152
Harvest Rd. TW13: Felt 63W 128
Harvest Rd. TW20: Eng G 4P 125
Harvest Rd. WD23: Bush 14Da 27
Harvest Way BR8: Crock 73Fd 162
Harvey RM16: Grays 47De 99
Harvey Cl. NW9 26Ua 48
Harvey Ct. E17 29Cc 52
Harvey Ct. KT19: Eps 81Ra 173
Harvey Dr. TW12: Hamp 67Da 129
Harveyfields EN9: Walt A 6Ec 20
Harvey Gdns. E11 32Hc 73
Harvey Gdns. IG10: Lough 13Rc 36
Harvey Gdns. SE7 50Lc 93
Harvey Ho. E1 42Xb 91 (off Brady St.)
Harvey Ho. EN3: Enf L 9Bc 20
Harvey Ho. N1 39Tb 71 (off Colville Est.)
Harvey Ho. RM6: Chad H 28Zc 55
Harvey Ho. SW1 50Mb 90 (off Aylesford St.)
Harvey Ho. TW8: Bford 50Na 87
Harvey Lodge W9 43Cb 89 (off Admiral Wlk.)
Harvey M. N8 29Pb 50 (off Harvey Rd.)
Harvey Rd. AL2: Lon C 8G 6
Harvey Rd. E11 32Gc 73
Harvey Rd. IG1: Ilf 36Rc 74
Harvey Rd. KT12: Walt T 73V 150
Harvey Rd. N8 29Pb 50
Harvey Rd. SE5 53Tb 113 (not continuous)
Harvey Rd. SL3: L'ly 48D 82
Harvey Rd. TW4: Houn 59Ba 107
Harvey Rd. UB10: Hil 40Q 64
Harvey Rd. UB5: N'olt 38Y 65
Harvey Rd. WD3: Crox G 16Q 26
Harvey's Bldgs. WC2 5G 223 (45Nb 90)
Harveys La. RM7: Rush G 33Fd 76
Harvey St. N1 39Tb 71
Harvil Ct. NW9 27Ua 48 (off Mornington Cl.)
Harvill Rd. DA14: Sidc 64Ad 139
Harvil Rd. UB10: Ick 32N 63
Harvil Rd. UB9: Hare 28L 43
Harvington Wlk. E8 38Wb 71
Harvingwell Pl. HP2: Hem H 1B 4
Harvist Est. N7 35Qb 70
Harvist Rd. NW6 40Za 68
Harwater Dr. IG10: Lough 12Pc 36
Harwell Cl. HA4: Ruis 32T 64
Harwell Pas. N2 28Hb 49
Harwich Rd. SL1: Slou 4E 80
Harwicke Ho. E3 41Dc 92 (off Bow Rd.)
Harwood Av. BR1: Broml 68Kc 137
Harwood Av. CR4: Mitc 69Gb 133
Harwood Av. RM11: Horn 27Nd 57
Harwood Cl. HA0: Wemb 35Ma 67
Harwood Cl. N12 23Gb 49
Harwood Ct. N1 39Tb 71 (off Colville Est.)
Harwood Ct. SW15 56Ya 110
Harwood Dr. UB10: Hil 39P 63
Harwood Gdns. SL4: Old Win 9M 103
Harwood Hall La. RM14: Upm 37Rd 77
Harwood M. SW6 52Cb 111
Harwood Point SE16 47Bc 92
Harwood Rd. SW6 52Cb 111
Harwoods Rd. WD18: Wat 14W 26
Harwoods Yd. N21 17Qb 32
Harwood Ter. SW6 53Db 111
Hascombe Ter. SE5 54Tb 113 (off Love Wlk.)
Hasedines Rd. HP1: Hem H 1J 3
Haselbury Rd. N18 21Ub 51
Haselbury Rd. N9 20Ub 33
Haseldine Mdws. AL10: Hat 1B 8
Haseldine Rd. AL2: Lon C 8H 7
Haseley End SE23 59Yb 114
Hasell Pk. Pl. BR2: Broml 75Pc 160
Haselrigge Rd. SW4 56Mb 112
Haseltine Rd. SE26 63Bc 136
Haselwood Dr. EN2: Enf 14Rb 33
Haskard Rd. RM9: Dag 35Zc 75
Hasker St. SW3 5E 226 (49Gb 89)
Haskins SS17: Stan H 1P 101
Haslam Av. SM3: Sutt 74Ab 154
Haslam Cl. N1 38Qb 70
Haslam Cl. UB10: Ick 33S 64
Haslam Ct. N11 21Kb 50

Haslam Ho. N1 .....38Sb 71
..... (off Canonbury Rd.)
Haslam St. SE15 .....52Vb 113
Haslemere Av. CR4: Mitc 68Fb 133
Haslemere Av. EN4: E Barn 18Hb 31
Haslemere Av. NW4 .....30Za 48
Haslemere Av. SW18 .....61Db 133
Haslemere Av. TW5: Cran 54Y 107
Haslemere Av. W13 .....48Ja 86
Haslemere Av. W7 .....48Ja 86
Haslemere Bus. Cen. EN1: Enf 14Xb 33
Haslemere Cl. SM6: W'gton 78Nb 156
Haslemere Cl. TW12: Hamp 64Ba 129
Haslemere Gdns. N3 .....27Bb 49
Haslemere Heathrow Est., The TW4: Cran 54X 107
Haslemere Ind. Est. SW18 .....61Db 133
Haslemere Rd. CR7: Thor H 71Rb 157
Haslemere Rd. DA7: Bex 54Bd 117
Haslemere Rd. IG3: Ilf 33Vc 75
Haslemere Rd. N21 .....19Rb 33
Haslemere Rd. N8 .....31Mb 70
Haslemere Rd. SL4: Wind 3E 102
Hasler Cl. SE28 .....45Xc 95
Haslers Wharf E3 .....39Ac 72
..... (off Old Ford Rd.)
Haslett Rd. TW17: Shep 68U 128
Haslingden Ho. RM3: Rom 22Nd 57
..... (off Dagnam Pk. Dr.)
Hasluck Gdns. EN5: New Bar 16Db 31
Hassall Ct. GU22: Wok 93C 188
Hassard St. E2 .....2K 219 (40Vb 71)
Hassenbrook Rd. SS17: Stan H 1N 101
Hassendean Rd. SE3 .....52Kc 115
Hassett Rd. E9 .....37Zb 72
Hassocks Cl. SE26 .....62Xb 135
Hassocks Rd. SW16 .....67Mb 134
Hassock Wood BR2: Kes 77Mc 159
Hassop Rd. NW2 .....35Za 68
Hassop Wlk. SE9 .....63Nc 138
Hasted Cl. DA9: Ghithe 58Yd 120
Hasted Rd. ME2: High'm 7P 145
Hasted Rd. ME2: Strood 7P 145
Hasted Rd. SE7 .....50Mc 93
Haste Hill Golf Course 26U 44
Hastings Av. EN7: Chesh 1Vb 19
Hastings Av. IG6: Ilf 28Sc 54
Hastings Cl. EN5: New Bar 14Eb 31
Hastings Cl. HA0: Wemb 35La 66
Hastings Cl. RM17: Grays 51Ae 121
Hastings Cl. SE15 .....52Wb 113
Hastings Ct. TW11: Tedd 64Fa 130
Hastings Dr. KT6: Surb 72La 152
Hastings Ho. EN3: Enf H 12Yb 34
Hastings Ho. SE18 .....49Pc 94
..... (off Mulgrave Rd.)
Hastings Ho. W12 .....45Xa 88
..... (off White City Est.)
Hastings Ho. W13 .....45Ka 86
Hastings Ho. WC1 .....4F 217 (41Nb 90)
..... (off Hastings St.)
Hastings Mdw. SL2: Stoke P 9K 61
Hastings Pl. CR0: C'don 74Vb 157
..... (off Hastings Rd.)
Hastings Rd. BR2: Broml 74Nc 160
Hastings Rd. CR0: C'don 74Vb 157
Hastings Rd. E16 .....43Jc 93
Hastings Rd. N11 .....22Lb 50
Hastings Rd. N17 .....27Tb 51
Hastings Rd. RM2: Rom 29Kd 57
Hastings Rd. W13 .....45Ka 86
Hastings St. SE18 .....48Sc 94
Hastings St. WC1 .....4F 217 (41Nb 90)
Hastings Way WD23: Bush 14Aa 27
Hastings Way WD3: Crox G 14S 26
Hastingwood Ct. E17 .....29Dc 52
Hastoe Cl. UB4: Yead 42Aa 85
Hasty Cl. CR4: Mitc 67Kb 134
Haswell Cres. SL1: Slou 7D 80
Hat & Mitre Ct. EC1 .....6C 218 (42Rb 91)
..... (off St John St.)
Hatch, The EN3: Enf H 11Zb 34
Hatch, The SL4: Wind 2A 102
Hatcham Mews Bus. Cen. SE14 .....53Zb 114
..... (off Hatcham Pk. Rd.)
Hatcham Pk. M. SE14 .....53Zb 114
Hatcham Pk. Rd. SE14 .....53Zb 114
Hatcham Rd. SE15 .....51Yb 114
Hatcham St. SE26 .....64Bc 136
Hatchard Rd. N19 .....33Mb 70
Hatch Cl. KT15: Add 76K 149
Hatchcroft NW4 .....27Xa 48
Hatch End GU20: W'sham 9A 146
HATCH END 24Ba 45
Hatch End Lawn Tennis Club 23Ca 45
Hatch End Station (Overground) 24Ca 45
Hatch End Swimming Pool 24Ca 45
Hatchers M. SE1 .....2J 231 (47Ub 91)
..... (off Bermondsey St.)
Hatchett Rd. TW14: Bedf 60S 106
Hatch Farm M. KT15: Add 76L 149
Hatchfield Ho. N15 .....30Ub 51
..... (off Albert Rd.)
HATCHFORD 91U 190
HATCHFORD END 91S 190
Hatchford Mnr. KT11: Cobh 90U 170
Hatch Gdns. KT20: Tad 92Za 194
Hatchgate Gdns. SL1: Burn 1B 80
Hatch Gro. RM6: Chad H 28Ad 55
Hatchingtan, The GU3: Worp 8N 187
Hatchlands Rd. RH1: Redh 6N 207
Hatch La. CR5: Bans 87Hb 175
Hatch La. CR5: Coul 87Hb 175
Hatch La. E4 .....21Fc 53
Hatch La. GU23: Ock 92R 190
Hatch La. KT11: Cobh 90R 170
Hatch La. SL4: Wind 5E 102
Hatch La. UB7: Harm 52M 105
Hatch Pl. KT2: King T 64Pa 131
Hatch Rd. CM15: Pil H 15Wd 40
Hatch Rd. SW16 .....68Nb 134
Hatch Side IG7: Chig 22Qc 54
Hatchwood Cl. IG8: Wfd G 21Hc 53
Hatcliffe Almshouses SE10 .....50Gc 93
..... (off Tuskar St.)
Hatcliffe Cl. SE3 .....55Hc 115
Hatcliffe St. SE10 .....50Hc 93
Hatfeild Cl. CR4: Mitc 70Fb 133
Hatfeild Mead SM4: Mord 71Cb 155
Hatfield Cl. CM13: Hut 17Fe 41
Hatfield Cl. IG6: Ilf 27Rc 54
Hatfield Cl. KT14: W Byf 84K 169
Hatfield Cl. RM12: Horn 36Md 77
Hatfield Cl. SE14 .....52Zb 114
Hatfield Cl. SM2: Sutt 81Db 175
Hatfield Ct. SE3 .....52Jc 115
Hatfield Ct. UB5: N'olt 41Y 85
..... (off Canberra Dr.)
Hatfield Ho. EC1 .....6D 218 (42Sb 91)
..... (off Golden La. Est.)

Hatfield Ho. SE10 .....52Dc 114
Hatfield Leisure Cen. 2D 8
Hatfield M. RM9: Dag 38Ad 75
Hatfield Pk. 1G 8
Hatfield Rd. AL1: St A 2C 6
Hatfield Rd. AL4: S'ford 2H 7
Hatfield Rd. AL4: St A 2H 7
Hatfield Rd. E15 .....36Gc 73
Hatfield Rd. EN6: Pot B 2Eb 17
Hatfield Rd. KT21: Asht 91Pa 193
Hatfield Rd. RM16: Chaf H 50Zd 99
Hatfield Rd. RM9: Dag 37Ad 75
Hatfield Rd. SL1: Slou 7L 81
Hatfield Rd. W13 .....46Ja 86
Hatfield Rd. W4 .....47Ta 87
Hatfield Rd. WD24: Wat 11X 27
Hatfields IG10: Lough 13Rc 36
Hatfields SE1 .....6A 224 (46Qb 90)
Hatham Grn. La. TN15: Stans 82Zd 185
Hathaway Cl. BR2: Broml 74Pc 160
Hathaway Cl. HA4: Ruis 35V 64
Hathaway Cl. HA7: Stan 22Ja 46
Hathaway Cl. IG6: Ilf 23Rc 54
Hathaway Ct. AL4: St A 2J 7
Hathaway Ct. RH1: Redh 5A 208
..... (off St Anne's Ri.)
Hathaway Cres. E12 .....37Pc 74
Hathaway Gdns. RM17: Grays 48Ce 99
Hathaway Gdns. RM6: Chad H 29Zc 55
Hathaway Gdns. W13 .....43Ja 86
Hathaway Ho. N1 .....2H 219 (40Ub 71)
Hathaway Rd. CR0: C'don 73Rb 157
Hathaway Rd. RM17: Grays 48De 99
Hatherleigh Cl. KT9: Chess 78Ma 153
Hatherleigh Cl. NW7 .....23Za 48
Hatherleigh Cl. SM4: Mord 70Cb 133
Hatherleigh Gdns. EN6: Pot B 4Fb 17
Hatherleigh Rd. HA4: Ruis 33W 64
Hatherleigh Way RM3: Rom 25Md 57
Hatherley Ct. W2 .....44Db 89
..... (off Hatherley Gro.)
Hatherley Cres. DA14: Sidc 61Wc 139
Hatherley Gdns. E6 .....41Mc 93
Hatherley Gdns. N8 .....30Nb 50
Hatherley Gro. W2 .....44Db 89
Hatherley Ho. E17 .....28Cc 52
Hatherley M. E17 .....28Cc 52
Hatherley Rd. DA14: Sidc 63Wc 139
Hatherley Rd. E17 .....28Bc 52
Hatherley Rd. TW9: Kew 53Pa 109
Hatherley St. SW1 .....6C 228 (49Lb 90)
Hathern Gdns. SE9 .....63Qc 138
Hatherop Rd. TW12: Hamp 66Ba 129
Hathersage Ct. N1 .....36Tb 71
Hatherwood KT22: Lea 93Ma 193
Hathorne Cl. SE15 .....54Xb 113
Hathway St. SE14 .....54Zb 114
Hathway Ter. SE14 .....54Zb 114
..... (off Hathway St.)
Hatley Av. IG6: Ilf 28Sc 54
Hatley Cl. N11 .....22Hb 49
Hatley Rd. N4 .....33Pb 70
Hatteraick St. SE16 .....47Yb 92
Hattersfield Cl. DA17: Belv 49Bd 95
Hatters La. WD18: Wat 16T 26
HATTON 56V 106
Hatton Av. SL2: Slou 2H 81
Hatton Cl. DA11: Nflt 2A 144
Hatton Cl. RM16: Chaf H 48Zd 99
Hatton Cl. SE18 .....52Tc 116
Hatton Ct. SL4: Wind 4G 102
HATTON CROSS 56V 106
Hatton Cross Cen. TW6: H'row A 55V 106
Hatton Cross Station (Underground) 56V 106
Hatton Gdn. EC1 .....7A 218 (43Qb 90)
Hatton Gdns. CR4: Mitc 71Hb 155
Hatton Grn. TW14: Felt 56W 106
Hatton Gro. UB7: W Dray 47M 83
Hatton Hill GU20: W'sham 7A 146
HATTON HILL 8A 146
Hatton Ho. E1 .....45Wb 91
..... (off Hindmarsh Cl.)
Hatton Ho. EN8: Chesh 1Zb 20
..... (off Church La.)
Hatton Ho. KT1: King T 68Pa 131
..... (off Victoria Rd.)
Hatton M. DA9: Ghithe 56Yd 120
Hatton Pl. EC1 .....7A 218 (43Qb 90)
Hatton Rd. CR0: C'don 74Qb 156
Hatton Rd. EN8: Chesh 1Zb 20
Hatton Rd. HA0: Wemb 39Na 67
Hatton Rd. TW14: Bedf 59S 106
Hatton Rd. TW14: Felt 59S 106
Hatton Rd. Sth. TW14: Felt 56V 106
Hatton Row NW8 .....6C 214 (42Fb 89)
..... (off Hatton St.)
Hatton St. NW8 .....6C 214 (42Fb 89)
Hatton Wlk. EN2: Enf 13Tb 33
Hatton Wall EC1 .....7A 218 (43Qb 90)
Haughmond N12 .....21Db 49
Haunch of Venison Yd. W1 .....3K 221 (44Kb 90)
Hauteville Ct. Gdns. W6 .....48Va 88
..... (off South Side)
Havana Cl. RM1: Rom 29Gd 56
Havana Rd. SW19 .....61Cb 133
Havanna Dr. NW11 .....29Ab 48
Havannah St. E14 .....47Cc 92
Havant Ho. RM3: Rom 24Nd 57
..... (off Kingsbridge Cir.)
Havant Rd. E17 .....27Ec 52
Havelock Cl. W12 .....45Xa 88
Havelock Ct. UB2: S'hall 48Ba 85
..... (off Havelock Rd.)
Havelock Dr. DA9: Ghithe 58Xd 120
Havelock Ho. SE1 .....6K 231 (49Vb 91)
..... (off Fort Rd.)
Havelock Ho. SE23 .....60Yb 114
Havelock Pl. HA1: Harr 30Ga 46
Havelock Rd. BR2: Broml 70Lc 137
Havelock Rd. CR0: C'don 75Vb 157
Havelock Rd. DA1: Dart 59Kd 119
Havelock Rd. DA11: Grav'nd 10B 122
Havelock Rd. DA17: Belv 49Bd 95
Havelock Rd. HA3: W'stone 27Ga 46
Havelock Rd. N17 .....26Wb 51
Havelock Rd. SW19 .....64Eb 133
Havelock Rd. UB2: S'hall 48Aa 85
Havelock Rd. WD4: K Lan 10P 3
Havelock St. IG1: Ilf 33Rc 74
Havelock St. N1 .....39Nb 70
Havelock Ter. SW8 .....53Kb 112
Havelock Ter. Arches SW8 .....53Kb 112
..... (off Havelock Ter.)
Havelock Wlk. SE23 .....60Yb 114
Haven, The RM16: Grays 10C 100
Haven, The TW16: Sun 66W 128
Haven, The TW9: Rich 55Qa 109
Haven Cl. BR8: Swan 68Hd 140
Haven Cl. DA13: Ist R 7B 144

Haven Cl. DA14: Sidc 65Yc 139
Haven Cl. KT10: Esh 75Ga 152
Haven Cl. SE9 .....62Pc 138
Haven Cl. SW19 .....62Za 132
Haven Cl. UB4: Hayes 42U 84
Haven Cl. App. IG7: Chig 20Xc 37
Haven Ct. BR3: Beck 68Ec 136
Haven Ct. KT10: Esh 75Ga 152
Haven Ct. KT5: Surb 72Pa 153
Haven Ct. RH1: Redh 4B 208
Haven Dr. KT19: Eps 82Ra 173
Havengore Av. DA12: Grav'nd 9G 122
Haven Grn. W5 .....44Ma 87
Haven Grn. Ct. W5 .....44Ma 87
Haven Hill TN15: Ash 79Ce 165
HAVEN HOUSE CHILDREN'S HOSPICE 23Hc 53
Havenhurst Ri. EN2: Enf 12Qb 32
Haven La. W5 .....44Na 87
Haven Lodge EN1: Enf 16Ub 33
..... (off Village Rd.)
Haven Lodge SE18 .....49Rc 94
..... (off Vincent Rd.)
Haven M. E3 .....43Bc 92
Haven M. N1 .....38Qb 70
Haven Pl. KT10: Esh 75Ga 152
Haven Pl. RM16: Grays 47Ee 99
Haven Pl. W5 .....45Ma 87
Havenpool NW8 .....39Db 69
..... (off Abbey Rd.)
Haven Rd. TW15: Ashf 63R 128
Havensfield WD4: Chfd 3K 11
Haven St. NW1 .....38Kb 70
Haven Way KT19: Eps 83Ra 173
Haven Way SE1 .....3K 231 (48Vb 91)
Havenwood HA9: Wemb 34Ra 67
Havenwood Cl. CM13: Gt War 23Yd 58
Havercroft Cl. AL3: St A 4P 5
Haverfield Gdns. TW9: Kew 52Qa 109
Haverfield Rd. E3 .....41Ac 92
Haverford Way HA8: Edg 25Pa 47
Haverhill Rd. E4 .....18Ec 34
Haverhill Rd. SW12 .....60Lb 112
Havering NW1 .....38Kb 70
..... (off Castlehaven Rd.)
HAVERING-ATTE-BOWER 20Gd 38
Havering Country Pk. 21Ed 56
Havering Dr. RM1: Rom 28Gd 56
Havering Gdns. RM6: Chad H 29Yc 55
Havering Mus. 29Gd 56
HAVERING PARK 22Dd 56
Havering Rd. RM1: Have B 22Fd 56
Havering Rd. RM1: Rom 22Fd 56
Havering St. E1 .....44Zb 92
Havering Way IG11: Bark 41Xc 95
Haverley St. SE26 .....64Bc 136
Havers Av. KT12: Hers 78Z 151
Haversham Cl. TW1: Twick 58Ma 109
Haversham Ct. UB6: G'frd 37Ha 66
Haversham Pl. N6 .....33Hb 69
Haverstock Cl. HA1: Harr 31Ea 66
Haverstock Ct. BR5: St P 68Xc 139
Haverstock Hill NW3 .....36Gb 69
Haverstock Pl. N1 .....3C 218 (41Rb 91)
..... (off Haverstock St.)
Haverstock Rd. NW5 .....36Jb 70
Haverstock St. N1 .....2C 218 (40Rb 71)
Haverthwaite Rd. BR6: Orp 75Tc 160
Havilland M. W12 .....47Xa 88
Havil St. SE5 .....52Ub 113
Havisham Apts. E15 .....37Fc 73
..... (off Grove Cres. Rd.)
Havisham Ho. SE16 .....47Wb 91
Havisham Pl. SE19 .....66Rb 135
Havisham Rd. DA12: Grav'nd 10J 123
Hawarden Gro. SE24 .....59Sb 113
Hawarden Hill NW2 .....34Wa 68
Hawarden Rd. CR3: Cat'm 93Sb 197
Hawarden Rd. E17 .....28Zb 52
Hawbridge Rd. E11 .....32Fc 73
Hawbush Ct. RM6: Ilf 28Xc 55
Hawes Cl. HA6: Nwood 24V 44
Hawes Ho. E17 .....28Zb 52
Hawes La. BR4: W W'ck 74Ec 158
Hawes La. E4 .....10Ec 20
Hawes Rd. BR1: Broml 67Kc 137
..... (not continuous)
Hawes Rd. KT20: Tad 92Za 194
Hawes Rd. N18 .....23Xb 51
Hawes St. N1 .....38Rb 71
Haweswater Dr. WD25: Wat 6Y 13
Haweswater Ho. TW7: Isle 57Ha 108
Hawfield Bank BR6: Chels 76Zc 161
Hawfield Gdns. AL2: Park 8B 6
Hawfinch Gdns. RM3: Hrld W 26Nd 57
Hawfinch Ho. NW9 .....31Va 68
Hawgood St. E3 .....43Cc 92
Hawk Cl. EN9: Walt A 6Jc 21
Hawk Cnr. RM20: Grays 50Zd 99
Hawkdene E4 .....16Dc 34
Hawke Ct. UB4: Yead 42Y 85
..... (off Perth Av.)
Hawke Ho. E1 .....42Zb 92
..... (off Ernest St.)
Hawke Pk. Rd. N22 .....27Rb 51
Hawke Pl. SE16 .....47Zb 92
Hawker NW9 .....25Va 48
..... (off Everglade Strand)
Hawker Cl. TN16: Big H 88Nc 180
Hawker Ct. E3 .....41Ec 92
..... (off Bolinder Way)
Hawker Ct. KT1: King T 68Pa 131
..... (off Church Rd.)
Hawker Ct. SL3: L'ly 48C 82
Hawker Dr. KT15: Add 76M 149
Hawke Rd. SE19 .....65Tb 135
Hawker Pl. E17 .....26Ec 52
Hawker Rd. CR0: Wadd 79Qb 156
Hawkesbury Cl. IG6: Ilf 21Xc 55
Hawkesbury Rd. SW15 .....57Xa 110
Hawkes Cl. SL3: L'ly 48D 82
Hawkesfield Rd. SE23 .....61Ac 136
Hawkes Leap GU20: W'sham 7A 146
Hawkesley Cl. TW1: Twick 63Ja 130
Hawkesley Ct. WD7: R'lett 7Ha 14
Hawkesley Ho. E17 .....30Dc 52
Hawkes Pl. TN13: S'oaks 99Jd 202
Hawkes Rd. CR4: Mitc 67Hb 133
Hawkes Rd. TW14: Felt 59W 106
Hawkesworth Cl. HA6: Nwood 24U 44
Hawkes Yd. KT7: T Ditt 72Ha 152
Hawke Twr. SE14 .....51Ac 114
Hawkewood Rd. TW16: Sun 69W 128
Hawkfield Ct. TW7: Isle 54Ga 108
Hawkhirst Rd. CR3: Kenley 89Ub 177
Hawkhirst Rd. CR8: Kenley 87Tb 177
Hawkhurst KT11: Cobh 86Ca 171
Hawkhurst Gdns. KT9: Chess 77Na 153
Hawkhurst Gdns. RM5: Col R 23Fd 56
Hawkhurst Rd. SW16 .....67Mb 134

Hawkhurst Way BR4: W W'ck 75Dc 158
Hawkhurst Way KT3: N Mald 71Ta 153
Hawkinge N17 .....26Tb 51
..... (off Gloucester Rd.)
Hawkinge Wlk. BR5: St P 69Xc 139
Hawkinge Way RM12: Horn 37Ld 77
Hawkins Av. DA12: Grav'nd 3E 144
Hawkins Cl. HA1: Harr 31Fa 66
Hawkins Cl. NW7 .....22Ta 47
Hawkins Cl. WD6: Bore 12Sa 29
Hawkins Ct. SE18 .....49Nc 94
Hawkins Dr. RM16: Chaf H 47Zd 99
Hawkins Ho. SE8 .....51Cc 114
..... (off New King St.)
Hawkins Ho. SW1 .....51Lb 112
..... (off Dolphin Sq.)
Hawkins Rd. NW10 .....38Ua 68
Hawkins Rd. TW11: Tedd 65Ka 130
Hawkins Ter. SE7 .....50Nc 94
Hawkins Way HP3: Bov 8C 2
Hawkins Way SE6 .....64Cc 136
Hawkley Gdns. SE27 .....61Rb 135
Hawkridge Cl. RM6: Chad H 30Yc 55
Hawkridge Dr. RM17: Grays 50Fe 99
Hawksbrook La. BR3: Beck 72Dc 158
Hawkshaw Cl. SW2 .....59Nb 112
Hawkshead NW1 .....3B 216 (41Lb 90)
..... (off Stanhope St.)
Hawkshead Cl. BR1: Broml 66Gc 137
Hawkshead Ct. EN8: Walt C 6Bc 20
..... (off Eleanor Way)
Hawkshead La. AL9: Brk P 10E 8
Hawkshead La. AL9: N Mym 10E 8
Hawkshead Rd. EN6: Pot B 10H 9
Hawkshead Rd. NW10 .....38Va 68
Hawkshead Rd. W4 .....47Ua 88
Hawk's Hill KT22: Fet 95Ha 192
Hawkshill AL1: St A 3E 6
Hawks Hill Cl. KT22: Fet 95Ha 192
Hawkshill Cl. KT10: Esh 79Ca 151
Hawk's Hill Ct. KT22: Fet 95Ha 192
Hawkshill Dr. HP3: Hem H 5H 3
Hawk's Hill Ho. KT22: Fet 96Ha 192
Hawkshill Pl. KT10: Esh 79Ca 151
Hawkshill Rd. SL2: Slou 1E 80
Hawkshill Way KT10: Esh 79Ba 151
Hawkslade Rd. SE15 .....57Zb 114
Hawksley Rd. N16 .....34Ub 71
Hawksmead Cl. EN3: Enf W 8Zb 20
Hawks M. SE10 .....52Ec 114
Hawksmoor WD7: Shenl 5Qa 15
Hawksmoor Cl. E6 .....44Nc 94
Hawksmoor Cl. SE18 .....50Uc 94
Hawksmoor Grn. CM13: Hut 15Fe 41
..... (not continuous)
Hawksmoor Grn. BR2: Broml 72Mc 159
Hawksmoor M. E1 .....45Xb 91
Hawksmoor Pl. E2 .....42Wb 91
..... (off Cheshire St.)
Hawksmoor St. W6 .....51Za 110
Hawksmouth E4 .....17Ec 34
Hawks Pas. KT1: King T 68Pa 131
..... (off London Rd.)
Hawks Rd. KT1: King T 68Pa 131
Hawkstone Rd. SE16 .....49Yb 92
Hawksview KT11: Cobh 85Ba 171
Hawksway TW18: Staines 62H 127
Hawkswell Cl. GU21: Wok 9K 167
Hawkswell Wlk. GU21: Wok 9K 167
Hawkswood Gro. SL3: Ful 37B 62
Hawkswood La. SL3: Ful 36B 62
Hawkswood La. SL9: Ger X 35B 62
Hawksworth Ho. BR1: Broml 68Jc 137
Hawkwell Ct. E4 .....20Ec 34
Hawkwell Ho. RM8: Dag 32Cd 76
Hawkwell Wlk. N1 .....39Sb 71
..... (off Maldon Cl.)
Hawkwood Cres. E4 .....16Dc 34
Hawkwood Dell KT23: Bookh 98Ca 191
Hawkwood La. BR7: Chst 67Sc 138
Hawkwood Mt. E5 .....32Xb 71
Hawkwood Ri. KT23: Bookh 98Ca 191
Hawlands Dr. HA5: Pinn 31Aa 65
HAWLEY 64Qd 141
Hawley Cl. TW12: Hamp 65Ba 129
Hawley Cres. NW1 .....38Kb 70
Hawley M. NW1 .....38Kb 70
Hawley Rd. DA1: Dart 61Nd 141
Hawley Rd. DA2: Hawl 63Qd 141
Hawley Rd. N18 .....22Zb 52
Hawley Rd. NW1 .....38Kb 70
..... (not continuous)
HAWLEY'S CORNER 93Rc 200
Hawley St. NW1 .....38Kb 70
Hawley Ter. DA2: Hawl 64Qd 141
Hawley Va. DA2: Hawl 64Qd 141
Hawley Way TW15: Ashf 64Q 128
Haws La. TW19: Stanw M 58J 105
Hawstead La. BR6: Chels 78Bd 161
Hawstead La. BR6: Well H 78Bd 161
Hawstead Rd. SE6 .....58Dc 114
Hawsted IG9: Buck H 17Kc 35
Hawthorn Av. CM13: B'wood 20Be 41
Hawthorn Av. CR7: Thor H 67Rb 135
Hawthorn Av. E3 .....39Bc 72
Hawthorn Av. N13 .....22Nb 50
Hawthorn Av. RM13: Rain 42Kd 97
Hawthorn Cen., The HA1: Harr 29Ha 46
Hawthorn Cl. BR5: Pet W 72Tc 160
Hawthorn Cl. CR6: W'ham 90Ac 178
Hawthorn Cl. GU22: Wok 92A 188
Hawthorn Cl. RH1: Redh 10A 208
Hawthorn Cl. SL0: Iver H 40E 62
Hawthorn Cl. SM7: Bans 86Ab 174
Hawthorn Cl. TW12: Hamp 64Ca 129
Hawthorn Cl. TW5: Cran 52X 107
Hawthorn Cl. WD17: Wat 10V 12
Hawthorn Cl. WD5: Ab L 4W 12
Hawthorn Cotts. DA16: Well 55Wc 117
..... (off Hook La.)
Hawthorn Ct. HA5: Pinn 26Y 45
..... (off Rickmansworth Rd.)
Hawthorn Ct. TW15: Ashf 66S 128
Hawthorn Ct. TW9: Kew 53Ra 109
Hawthorn Cres. CR2: Sels 83Yb 178
Hawthorn Cres. IG5: Ilf 25Pc 54
Hawthorn Cres. SW17 .....64Jb 134
Hawthornden Cl. N12 .....23Gb 49
Hawthorndene Cl. BR2: Hayes 75Hc 159
Hawthorndene Rd. BR2: Hayes 75Hc 159
Hawthorn Dr. BR4: W W'ck 77Gc 159
Hawthorn Dr. HA2: Harr 30Ca 45
Hawthorn Dr. UB9: Den 37L 63
Hawthorne Av. CR4: Mitc 68Fb 133
Hawthorne Av. EN7: Chesh 3Xb 19
Hawthorne Av. HA3: Kenton 30Ja 46
Hawthorne Av. HA4: Ruis 30X 45
Hawthorne Av. SM5: Cars 80Jb 156
Hawthorne Av. TN16: Big H 87Mc 179

Hawthorne Cl. BR1: Broml 69Pc 138
Hawthorne Cl. DA12: Grav'nd 3D 144
Hawthorne Cl. EN7: Chesh 3Xb 19
Hawthorne Cl. N1 .....37Ub 71
Hawthorne Cl. SM1: Sutt 75Eb 155
Hawthorne Ct. HA6: Nwood 26W 44
Hawthorne Ct. KT12: Walt T 74Z 151
Hawthorne Ct. TW19: Stanw 59M 105
..... (off Hawthorne Way)
Hawthorne Ct. W5 .....46Na 87
Hawthorne Cres. SE10 .....50Hc 93
Hawthorne Cres. SL1: Slou 4J 81
Hawthorne Cres. UB7: W Dray 47P 83
Hawthorne Dr. SL4: Wink 1A 124
Hawthorne Gdns. CR3: Cat'm 93Ub 197
Hawthorne Gro. NW9 .....31Sa 67
Hawthorne Ho. N15 .....29Wb 51
Hawthorne Ho. SW1 .....50Lb 90
..... (off Churchill Gdns.)
Hawthorne M. UB6: G'frd 44Ea 86
Hawthorne Pl. KT17: Eps 84Ua 174
Hawthorne Pl. UB3: Hayes 45V 84
Hawthorne Rd. BR1: Broml 69Nc 138
Hawthorne Rd. E17 .....27Cc 52
Hawthorne Rd. N18 .....23Vb 51
Hawthorne Rd. TW18: Staines 64E 126
Hawthorne Rd. WD7: R'lett 6Ja 14
Hawthornes AL10: Hat 2B 8
Hawthorne Way N9 .....19Vb 33
Hawthorne Way TW19: Stanw 59M 105
Hawthorn Farm Av. UB5: N'olt 39Aa 65
Hawthorn Gdns. W5 .....48Ma 87
Hawthorn Gro. EN2: Enf 10Tb 19
Hawthorn Gro. EN5: Ark 16Va 30
Hawthorn Gro. SE20 .....66Xb 135
Hawthorn Hatch TW8: Bford 52Ka 108
Hawthorn Ho. E15 .....37Fc 73
..... (off Forrester Way)
Hawthorn Ho. SE16 .....47Ac 92
..... (off Blondin Way)
Hawthorn Ho. WD23: Bush 15Z 27
..... (off Plantation Cl.)
Hawthorn La. HP1: Hem H 1H 3
Hawthorn La. SL2: Farn C 8D 60
Hawthorn La. TN13: S'oaks 94Hd 202
Hawthorn M. NW7 .....25Ab 48
Hawthorn Pk. BR8: Swan 68Jd 140
Hawthorn Pl. DA8: Erith 50Ed 96
Hawthorn Rd. DA1: Dart 61Md 141
Hawthorn Rd. DA6: Bex 56Bd 117
Hawthorn Rd. GU22: Wok 2P 187
Hawthorn Rd. GU23: Rip 96J 189
Hawthorn Rd. IG9: Buck H 21Mc 53
Hawthorn Rd. N8 .....27Mb 50
Hawthorn Rd. NW10 .....38Wa 68
Hawthorn Rd. SM1: Sutt 79Gb 155
Hawthorn Rd. SM6: W'gton 80Kb 156
Hawthorn Rd. TW13: Felt 60W 106
Hawthorn Rd. TW8: Bford 52Ka 108
Hawthorn Row KT22: Lea 89Ja 172
Hawthorns CR2: S Croy 77Sb 157
..... (off Bramley Hill)
Hawthorns DA3: Hartl 70Be 143
Hawthorns IG8: Wfd G 20Jc 35
Hawthorns, The EN6: Ridge 5Ua 16
Hawthorns, The HP3: Hem H 6H 3
Hawthorns, The IG10: Lough 14Qc 36
Hawthorns, The KT17: Ewe 80Va 154
Hawthorns, The RH8: Oxt 5L 211
Hawthorns, The SL3: Poyle 53H 105
Hawthorns, The WD3: Map C 22F 42
Hawthorns School Sports Cen., The 3H 209
Hawthorn Ter. DA15: Sidc 57Vc 117
Hawthorn Wlk. W10 .....42Ab 88
Hawthorn Way AL2: Chis G 6N 5
Hawthorn Way GU24: Bisl 8E 166
Hawthorn Way IG7: Chig 21Tc 54
Hawthorn Way KT15: New H 82L 169
Hawthorn Way RH1: Redh 8B 208
Hawthorn Way TW17: Shep 70T 128
Hawtrees WD7: R'lett 7Ha 14
Hawtrey Av. UB5: N'olt 40Z 65
Hawtrey Cl. SL1: Slou 7M 81
Hawtrey Dr. HA4: Ruis 31W 64
Hawtrey Ho. SL4: Eton 1H 103
..... (off Slough Rd.)
Hawtrey Rd. NW3 .....38Gb 69
Haxell Ct. SL4: Wind 4G 102
Haxted Rd. BR1: Broml 67Kc 137
Hayburn Mead HP1: Hem H 3K 3
Hayburn Way RM12: Horn 32Hd 76
Hay Cl. E15 .....38Gc 73
Hay Cl. WD6: Bore 12Sa 29
Haycroft Cl. CR5: Coul 90Rb 177
Haycroft Gdns. NW10 .....39Wa 68
Haycroft Rd. KT6: Surb 75Ma 153
Haycroft Rd. SW2 .....57Nb 112
Hay Currie St. E14 .....44Dc 92
Hayday Rd. E16 .....43Jc 93
..... (not continuous)
Hayden Cl. EN5: Ark 15Wa 30
Hayden Ct. KT15: New H 83K 169
Hayden Ct. TW13: Felt 63U 128
Hayden Piper Ho. SW3 .....51Hb 111
..... (off Caversham St.)
Hayden Rd. EN9: Walt A 7Ec 20
Haydens Cl. BR5: Orp 72Yc 161
Haydens M. W3 .....44Sa 87
Hayden's Pl. W11 .....44Bb 89
Hayden Twr. SW8 .....52Mb 112
Hayden Way RM5: Col R 26Ed 56
Haydn Av. CR8: Purl 86Qb 176
Haydock Av. UB5: N'olt 37Ca 65
Haydock Cl. RM12: Horn 35Pd 77
Haydock Grn. UB5: N'olt 37Ca 65
Haydock Grn. Flats UB5: N'olt 37Ca 65
..... (off Haydock Grn.)
Haydon Cl. EN1: Enf 16Ub 33
Haydon Cl. NW9 .....28Sa 47
Haydon Cl. RM3: Rom 24Kd 57
Haydon Ct. NW9 .....28Sa 47
Haydon Dell WD23: Bush 16Ba 27
Haydon Dr. HA5: Eastc 28W 44
Haydon Hill Ho. WD23: Bush 17Ba 27
Haydon Pk. Rd. SW19 .....64Cb 133
Haydon Rd. RM8: Dag 33Yc 75
Haydon Rd. WD19: Wat 16Aa 27
Haydons Rd. SW19 .....64Db 133
Haydons Road Station (Rail) 64Eb 133
Haydon St. EC3 .....4K 225 (45Vb 91)
Haydon Wlk. E1 .....3K 225 (44Vb 91)
Haydon Way SW11 .....56Fb 111
Hay Dr. CR4: Mitc 68Gb 133
HAYES 74Jc 159
HAYES 44U 84
Hayes, The KT18: Eps D 91Ua 194
Hayes & Harlington Station (Rail & Crossrail) 48V 84
Hayes & Yeading United FC 46Y 85

Hayes Barton GU22: Pyr 88F 168
Hayes Bri. Retail Pk. 45Y 85
Hayes Chase BR4: W W'ck 72Fc 159
Hayes Cl. BR2: Hayes 75Jc 159
Hayes Cl. RM20: Grays 51Yd 120
HAYES COTTAGE NURSING HOME 44U 84
Hayes Ct. BR2: Hayes 76Jc 159
Hayes Ct. HA0: Wemb 39Na 67
Hayes Ct. SE5 52Sb 113
(off Camberwell New Rd.)
Hayes Ct. SW2 60Nb 112
Hayes Cres. NW11 29Bb 49
Hayes Cres. SM3: Cheam 77Za 154
Hayes Dr. RM13: Rain 38Kd 77
HAYES END 42T 84
Hayes End Cl. UB4: Hayes 42T 84
Hayes End Dr. UB4: Hayes 42T 84
Hayes End Rd. UB4: Hayes 42T 84
Hayesens Ho. SW17 63Eb 133
Hayesford Pk. Dr. BR2: Broml 71Hc 159
Hayes Gdn. BR2: Hayes 74Jc 159
Hayes Gro. SE22 55Vb 113
HAYES GROVE PRIORY HOSPITAL 75Jc 159
Hayes Hill BR2: Hayes 74Gc 159
Hayes Hill Rd. BR2: Hayes 74Hc 159
Hayes La. BR2: Broml 71Kc 159
Hayes La. BR2: Hayes 71Kc 159
Hayes La. BR3: Beck 69Ec 136
Hayes La. CR8: Kenley 88Rb 177
Hayes Mead Rd. BR2: Hayes 74Gc 159
Hayes Metro Cen. UB4: Yead 45Y 85
Hayes M. SE8 53Bc 114
Hayes Pk. Lodge UB4: Hayes 42T 84
Hayes Pl. NW1 6E 214 (42Gb 89)
Hayes Rd. BR2: Broml 70Jc 137
Hayes Rd. DA9: Ghithe 59Ud 120
Hayes Rd. UB2: S'hall 49X 85
Hayes Station (Rail) 74Jc 159
Hayes St. BR2: Hayes 74Kc 159
Hayes Ter. DA12: Shorne 4N 145
HAYES TOWN 47V 84
Hayes Wlk. EN6: Pot B 5Db 17
Hayes Way BR3: Beck 70Ec 136
Hayes Wood Av. BR2: Hayes 74Kc 159
Hayfield Cl. WD23: Bush 14Da 27
Hayfield Pas. E1 42Yb 92
Hayfield Rd. BR5: St M Cry 71Wc 161
Hayfield Yd. E1 42Yb 92
Haygarth Pl. SW19 64Za 132
Hay Grn. RM11: Horn 30Qd 57
Haygreen Cl. KT2: King T 65Ra 131
Hay Hill W1 5A 222 (45Kb 90)
Hayhurst Ct. N1 39Rb 71
(off Dibden St.)
Hayhurst Gro. GU24: Bisl 9E 166
Hayland Cl. NW9 28Ta 47
Haylands Ct. TW8: Bford 51La 108
Hay La. NW9 28Sa 47
Hay La. SL3: Ful 5P 61
Hayle RM18: E Til 8L 101
Hayles Bldgs. SE11 5C 230 (49Rb 91)
(off Elliotts Row)
Hayles St. SE11 5B 230 (49Rb 91)
Haylett Gdns. KT1: King T 70Ma 131
Hayley Cl. RM16: Chaf H 48Yd 98
Hayling Av. TW13: Felt 62W 128
Hayling Cl. N16 36Ub 71
Hayling Cl. SL1: Slou 6F 80
Hayling Ct. SM3: Cheam 77Ya 154
Hayling Rd. WD19: Wat 20W 26
Hayling Way HA8: Edg 21Pa 47
Haymaker Cl. UB10: Uxb 38P 63
Hayman Cres. UB4: Hayes 40T 64
Haymans Point SE11 6H 229 (49Pb 90)
Hayman St. N1 38Rb 71
Haymarket SW1 5D 222 (45Mb 90)
Haymarket Arc. SW1 5D 222 (45Mb 90)
(off Haymarket)
Haymarket Ct. E8 38Vb 71
(off Jacaranda Gro.)
Haymarket Theatre Royal 5E 222 (45Mb 90)
(off Haymarket)
Haymeads Dr. KT10: Esh 79Ea 152
Haymer Gdns. KT4: Wor Pk 76Wa 154
Haymerle Ho. SE15 51Wb 113
(off Haymerle Rd.)
Haymerle Rd. SE15 51Wb 113
Hay M. NW3 37Hb 69
Haymill Cl. UB6: G'frd 41Ha 86
Haymill Rd. SL1: Slou 2B 80
Haymill Rd. SL2: Slou 2B 80
Hayne Ho. W11 46Ab 88
(off Penzance Pl.)
Hayne Rd. BR3: Beck 68Bc 136
Haynes Cl. GU23: Rip 94K 189
Haynes Cl. N11 20Jb 32
Haynes Cl. N17 24Xb 51
Haynes Cl. SE3 55Gc 115
Haynes Cl. SL3: L'ly 50B 82
Haynes Dr. N9 20Xb 33
Haynes La. SE19 65Ub 135
Haynes Pk. Ct. RM11: Horn 29Ld 57
Haynes Rd. DA11: Nflt 2B 144
Haynes Rd. HA0: Wemb 38Na 67
Haynes Rd. RM11: Horn 28Md 57
Hayne St. EC1 7C 218 (43Rb 91)
Haynt Wlk. SW20 69Ab 132
Hayre Dr. UB2: S'hall 50Aa 85
Hay's Ct. SE16 47Yb 92
(off Rotherhithe St.)
Hayse Hill SL4: Wind 3B 102
Hay's Galleria SE1 6H 225 (46Ub 91)
Hays La. SE1 6H 225 (46Ub 91)
Haysleigh Gdns. SE20 68Wb 135
Hay's M. W1 6K 221 (46Kb 90)
Haysoms Cl. RM1: Rom 28Gd 56
Hays Rd. DA12: Grav'nd 10J 123
Haystall Cl. UB4: Hayes 40U 64
Hay St. E2 39Wb 71
Hays Wlk. SM2: Cheam 82Za 174
Hayter Ct. E11 33Kc 73
Hayter Rd. SW2 57Nb 112
Hayton Cl. E8 37Vb 71
Hayton Cres. KT20: Tad 92Ya 194
Haywain RH8: Oxt 2H 211
Hayward Cl. DA1: Cray 57Fd 118
Hayward Cl. SW19 66Db 133
Hayward Copse WD3: Loud 14M 25
Hayward Ct. SW9 54Nb 112
(off Studley Rd.)
Hayward Dr. DA1: Dart 62Pd 141
Hayward Gallery 6J 223 (46Pb 90)
Hayward Gdns. SW15 58Ya 110
Hayward Ho. N1 1K 217 (40Qb 70)
(off Penton St.)
Hayward M. SE4 57Bc 114
Hayward Rd. KT7: T Ditt 74Ha 152
Hayward Rd. N20 19Eb 31
Haywards Cl. RM6: Chad H 29Xc 55
Haywards Mead SL4: Eton W 10D 80
Hayward's Pl. EC1 5B 218 (42Rb 91)
Haywood Cl. HA5: Pinn 26Z 45
Haywood Ct. EN9: Walt A 6Hc 21
Haywood Cres. WD17: Wat 10W 12
Haywood Dr. HP3: Hem H 5H 3
Haywood Dr. WD3: Chor 15H 25
Haywood Lodge N11 23Nb 50
(off York Rd.)
Haywood M. RM3: Rom 23Nd 57
Haywood Pk. WD3: Chor 15H 25
Haywood Pl. RM16: Grays 7E 100
Haywood Ri. BR6: Orp 78Uc 160
Haywood Rd. BR2: Broml 70Mc 137
Hazel Av. UB7: W Dray 480 84
Hazel Bank SE25 68Ub 135
Hazelbank KT5: Surb 74Sa 153
Hazelbank WD3: Crox G 16S 26
Hazelbank Ct. KT16: Chert 74L 149
Hazelbank Rd. KT16: Chert 74L 149
Hazelbank Rd. SE6 61Fc 137
Hazelbourne Av. TN15: Bor G 93Ae 205
Hazelbourne Rd. SW12 58Kb 112
Hazelbrouck Gdns. IG6: Ilf 24Tc 54
Hazelbury Av. WD5: Ab L 4S 12
Hazelbury Cl. SW19 68Cb 133
Hazelbury Grn. N9 20Ub 33
Hazelbury La. N9 20Ub 33
Hazel Cl. CR0: C'don 73Zb 158
Hazel Cl. CR4: Mitc 70Mb 134
Hazel Cl. KT19: Eps 82Ta 173
Hazel Cl. N13 20Tb 33
Hazel Cl. N19 33Lb 70
Hazel Cl. NW9 26Ua 48
Hazel Cl. RH2: Reig 8L 207
Hazel Cl. RM12: Horn 34Kd 77
Hazel Cl. SE15 54Wb 113
(off Bournemouth Cl.)
Hazel Cl. TW2: Whitt 59Ea 108
Hazel Cl. TW20: Eng G 5M 125
Hazel Cl. TW8: Bford 52Ka 108
Hazel Cotts. TN14: Hals 85Bd 181
Hazel Ct. CR6: W'ham 89Ac 178
Hazel Ct. IG10: Lough 15Qc 36
Hazel Ct. KT11: Cobh 85W 170
Hazel Ct. W5 45Na 87
Hazel Ct. WD7: Shenl 5Pa 15
Hazelcroft HA5: Hat E 23Da 45
Hazelcroft Cl. UB10: Hil 38P 63
Hazeldean Rd. NW10 38Ta 67
Hazeldell Link HP1: Hem H 3G 2
Hazeldell Rd. HP1: Hem H 3G 2
Hazelden Cl. TN15: W King 81Wd 184
Hazeldene EN8: Walt C 4Ac 20
Hazeldene KT15: Add 78L 149
Hazeldene Ct. CR8: Kenley 87Tb 177
Hazeldene Dr. HA5: Pinn 27Y 45
Hazeldene Gdns. UB10: Hil 39S 64
Hazeldene Rd. DA16: Well 54Yc 117
Hazeldene Rd. IG3: Ilf 33Xc 75
Hazeldon Rd. SE4 57Ac 114
Hazel Dr. DA8: Erith 53Jd 118
Hazel Dr. GU23: Rip 97H 189
Hazel Dr. RM15: S Ock 41Zd 99
Hazeleigh CM13: B'wood 20De 41
Hazeleigh Gdns. IG8: Wfd G 22Nc 54
Hazeleigh Ho. RM1: Rom 28Gd 56
(off Market Link)
Hazel End BR8: Swan 71Gd 162
Hazel Gdns. HA8: Edg 21Ra 47
Hazel Gdns. RM16: Grays 8A 100
Hazelgreen Cl. N21 18Rb 33
Hazel Gro. AL10: Hat 3B 8
Hazel Gro. BR6: Farnb 75Rc 160
Hazel Gro. EN1: Enf 16Wb 33
Hazel Gro. HA0: Wemb 39Na 67
Hazel Gro. RM6: Chad H 27Ad 55
Hazel Gro. SE26 63Zb 136
Hazel Gro. TW13: Felt 60W 106
Hazel Gro. TW18: Staines 65K 127
Hazel Gro. WD25: Wat 7X 13
Hazel Gro. Ho. AL10: Hat 2B 8
Hazel Ho. E3 39Bc 72
(off Barge La.)
Hazelhurst BR3: Beck 67Fc 137
Hazelhurst Ct. SE6 64Ec 136
(off Beckenham Hill Rd.)
Hazelhurst Rd. SL1: Burn 10A 60
Hazelhurst Rd. SW17 63Eb 133
Hazel La. IG6: Ilf 23Rc 54
Hazel La. SE10 50Hc 93
Hazel La. TW10: Ham 61Na 131
Hazell Cres. RM5: Col R 25Dd 56
Hazellville Rd. N19 31Mb 70
Hazell Way SL2: Stoke P 8K 61
Hazel Mead EN5: Ark 15Xa 30
Hazel Mead KT17: Ewe 82Wa 174
Hazelmere Cl. KT22: Lea 91Ka 192
Hazelmere Cl. TW14: Felt 58U 106
Hazelmere Cl. UB5: N'olt 40Ba 65
Hazelmere Ct. SW2 60Pb 112
Hazelmere Dr. UB5: N'olt 40Ba 65
Hazelmere Gdns. RM11: Horn 29Ld 57
Hazelmere Rd. BR5: Pet W 70Sc 138
Hazelmere Rd. NW6 39Bb 69
Hazelmere Rd. UB5: N'olt 40Ba 65
Hazelmere Wlk. UB5: N'olt 40Ba 65
(not continuous)
Hazelmere Way BR2: Hayes 72Jc 159
Hazel M. DA10: Swans 59Be 121
Hazel M. N22 27Qb 50
(off High Rd.)
Hazelnut Ct. RM3: Hrld W 26Nd 57
(off Firwood La.)
Hazelnut Ho. BR8: Swan 69Hd 140
(off Squirrels Cl.)
Hazel Pde. KT22: Fet 94Ea 192
Hazel Ri. RM11: Horn 30Ld 57
Hazel Rd. AL2: Park 10P 5
Hazel Rd. DA1: Dart 61Md 141
Hazel Rd. DA8: Erith 53Jd 118
Hazel Rd. E15 36Gc 73
Hazel Rd. KT14: W Byf 86J 169
Hazel Rd. NW10 41Xa 88
(not continuous)
Hazel Rd. RH2: Reig 8L 207
Hazeltree La. UB5: N'olt 41Aa 85
Hazel Tree Rd. WD24: Wat 9X 13
Hazel Wlk. BR2: Broml 72Qc 160
Hazel Way CR5: Chips 91Hb 195
Hazel Way E4 23Bc 52
Hazel Way KT22: Fet 94Ea 192
Hazel Way SE1 5K 231 (49Vb 91)
Hazelway Cl. KT22: Fet 95Ea 192
Hazelwood IG10: Lough 15Mc 35
Hazelwood SS17: Linf 9J 101
HAZELWOOD 83Tc 180
Hazelwood Av. SM4: Mord 70Db 133
Hazelwood Cl. W5 47Na 87
Hazelwood Ct. KT6: Surb 72Na 153
Hazelwood Ct. N13 21Qb 50
(off Hazelwood La.)
Hazelwood Ct. NW10 34Ua 68
Hazelwood Cres. N13 21Qb 50
Hazelwood Dr. AL4: St A 1G 6
Hazelwood Dr. HA5: Pinn 26X 45
Hazelwood Dr. TW16: Sun 69W 128
Hazelwood Gdns. CM15: Pil H 16Wd 40
Hazelwood Gro. CR2: Sande 85Xb 177
Hazelwood Hgts. RH8: Oxt 3L 211
Hazelwood Ho. SE8 49Ac 92
Hazelwood Ho. TW16: Sun 67W 128
Hazelwood Ho's. BR2: Broml 69Gc 137
Hazelwood La. CR5: Chips 90Gb 175
Hazelwood La. N13 21Qb 50
Hazelwood La. WD5: Ab L 4S 12
Hazelwood Pk. Cl. IG7: Chig 22Uc 54
Hazelwood Rd. E17 29Ac 52
Hazelwood Rd. EN1: Enf 16Vb 33
Hazelwood Rd. GU21: Knap 10J 167
Hazelwood Rd. RH8: Oxt 4M 211
Hazelwood Rd. TN14: Cud 84Tc 180
Hazelwood Rd. WD3: Crox G 16S 26
Hazelwood Sports Club 17Sb 33
Hazlebury Rd. SW6 54Db 111
Hazledean Rd. CR0: C'don 75Tb 157
Hazledene Rd. W4 51Sa 109
Hazlemere Gdns. KT4: Wor Pk 74Wa 154
Hazlemere Marina EN9: Walt A 6Dc 20
Hazlemere Rd. SL2: Slou 6M 81
Hazle's Pottery Barn 29Xd 58
Hazlewell Rd. SW15 57Ya 110
Hazlewood Cl. E5 34Ac 72
Hazlewood Cl. HA2: Harr 28Da 45
Hazlewood Cres. W10 42Ab 88
Hazlewood M. SW9 55Nb 112
Hazlewood Twr. W10 42Ab 88
(off Golborne Gdns.)
Hazlitt Cl. TW13: Hanw 63Aa 129
Hazlitt M. W14 48Ab 88
Hazlitt Rd. W14 48Ab 88
Hazon Way KT19: Eps 84Sa 173
Heacham Av. UB10: Ick 34S 64
Headbourne Ho. E17 27Ac 52
(off Sutherland Rd.)
Headbourne Ho. SE1 3G 231 (48Tb 91)
Headcorn Pl. CR7: Thor H 70Pb 134
Headcorn Rd. BR1: Broml 64Hc 137
Headcorn Rd. CR7: Thor H 70Pb 134
Headcorn Rd. N17 24Vb 51
Headfort Pl. SW1 2J 227 (47Jb 90)
Headingley Cl. IG6: Ilf 23Vc 55
Headingley Cl. WD7: Shenl 4Na 15
Headingley Dr. BR3: Beck 65Cc 136
Headington Ct. CR0: C'don 77Sb 157
(off Tanfield Rd.)
Headington Pl. SL2: Slou 6K 81
(off Mill St.)
Headington Rd. SW18 61Eb 133
Headlam Rd. SW4 58Mb 112
(not continuous)
Headlam St. E1 42Xb 91
HEADLEY 98Ta 193
Headley App. IG2: Ilf 29Rc 54
Headley Av. SM6: W'gton 78Pb 156
Headley Chase CM14: W'ley 21Yd 58
Headley Cl. KT19: Ewe 79Qa 153
Headley Comn. CM13: Gt War 24Xd 58
Headley Comn. Rd. KT18: Head 99Ta 193
Headley Comn. Rd. KT18: Walt H 99Ta 193
Headley Ct. KT18: Head 95Ra 193
Headley Ct. SE26 64Yb 136
HEADLEY COURT DEFENCE MEDICAL
REHABILITATION CEN. 95Ra 193
Headley Dr. CR0: New Ad 80Dc 158
Headley Dr. IG2: Ilf 30Rc 54
Headley Dr. KT18: Tatt C 91Xa 194
Headley Gro. KT20: Tad 93Xa 194
Headley Heath 100Ra 193
Headley M. SW18 57Db 111
Headley Rd. KT18: Eps 90Ra 173
Headley Rd. KT18: Eps D 90Ra 173
Headley Rd. KT18: Eps D 92Sa 193
Headley Rd. KT18: Eps D 94Qa 193
Headley Rd. KT18: Head 94Qa 193
Headley Rd. KT22: Lea 94La 192
Head's M. W11 44Cb 89
HEADSTONE 28Ea 46
Headstone Dr. HA1: Harr 27Fa 46
Headstone Dr. HA3: W'stone 27Ga 46
Headstone Gdns. HA2: Harr 28Ea 46
Headstone La. HA2: Harr 28Ca 45
Headstone La. HA3: Hrw W 24Da 45
Headstone Lane Station (Overground) 25Da 45
Headstone Manor 27Ea 46
Headstone Pde. HA1: Harr 28Fa 46
Headstone Rd. HA1: Harr 29Ga 46
Head St. E1 44Zb 92
(not continuous)
Headway, The KT17: Ewe 81Va 174
Headway Cl. TW10: Ham 63La 130
Headway Gdns. E17 25Cc 52
Heald St. SE14 53Cc 114
Healey Ho. E3 42Cc 92
(off Wellington Way)
Healey Ho. SW9 52Qb 112
Healey Rd. WD18: Wat 16V 26
Healey St. NW1 37Kb 70
Healy Ct. EN5: Barn 16Za 30
Healy Dr. BR6: Orp 77Vc 161
Heards La. CM15: Shenf 13Be 41
Hearne Rd. W4 51Qa 109
Hearn Pl. SW16 63Qb 134
Hearn Ri. UB5: N'olt 39Z 65
Hearn Rd. RM1: Rom 30Gd 56
Hearn's Bldgs. SE17 6G 231 (49Tb 91)
Hearnshaw St. E14 44Ac 92
Hearn's Rd. BR5: St P 70Yc 139
Hearn St. EC2 6J 219 (42Ub 91)
Hearnville Rd. SW12 60Jb 112
Hearsall Av. SS17: Stan H 1N 101
Heart, The KT12: Walt T 74W 150
Heart, The CR3: Cat'm 96Sb 197
HEATH, THE 79R 150
Heath, The W7 46Ga 86
Heath, The WD7: R'lett 5Ja 14
Heathacre SL3: Coln 53G 104
Heatham Pk. TW2: Twick 59Ha 108
Heath Av. AL3: St A 1B 6
Heath Av. DA7: Bex 51Zc 117
Heathbourne Rd. HA7: Stan 19Ga 28
Heathbourne Rd. WD23: B Hea 18Ga 28
Heathbridge KT13: Weyb 80Q 150
Heathbridge App. KT13: Weyb 79Q 150
Heath Brow HP1: Hem H 4L 3
Heath Brow NW3 34Eb 69
Heath Bus. Cen. TW3: Houn 56Ea 108
Heath Cl. BR5: Orp 73Yc 161
Heath Cl. BR8: Swan 68Gd 140
Heath Cl. CR2: S Croy 79Rb 157
Heath Cl. EN6: Pot B 2Db 17
Heath Cl. GU25: Vir W 10P 125
Heath Cl. HP1: Hem H 3L 3
Heath Cl. NW11 31Db 69
Heath Cl. RH2: Reig 6G 206
Heath Cl. RM2: Rom 27Jd 56
Heath Cl. SM7: Bans 86Db 175
Heath Cl. TW19: Stanw 58L 105
Heath Cl. UB3: Harl 52T 106
Heath Cl. W5 42Pa 87
Heathclose Av. DA1: Dart 59Kd 119
Heathclose Rd. DA1: Dart 60Jd 118
Heathcock Ct. WC2 5G 223 (45Nb 90)
(off Exchange Ct.)
Heathcote KT20: Tad 93Za 194
Heathcote Av. IG5: Ilf 26Pc 54
Heathcote Ct. IG5: Ilf 25Pc 54
(not continuous)
Heathcote Ct. SL4: Wind 5H 103
(off Osbourne Rd.)
Heathcote Ga. SW6 55Db 111
Heathcote Gro. E4 20Ec 34
Heathcote Rd. KT18: Eps 86Ta 173
Heathcote Rd. TW1: Twick 58Ka 108
Heathcote St. WC1 5H 217 (42Pb 90)
Heathcote Way UB7: Yiew 46M 83
Heath Ct. CR0: C'don 77Tb 157
(off Heathfield Rd.)
Heath Ct. SE9 60Sc 116
Heath Ct. TW4: Houn 56Ba 107
Heath Ct. UB8: Uxb 38M 63
Heath Cft. NW11 32Db 69
Heathcroft W5 42Pa 87
Heathcroft Av. TW16: Sun 66V 128
Heathcroft Gdns. E17 25Fc 53
Heathdale Av. TW4: Houn 55Aa 107
Heathdene KT20: Tad 90Ab 174
Heathdene Dr. DA17: Belv 49Dd 96
Heathdene Mnr. WD17: Wat 11V 26
Heathdene Rd. SM6: W'gton 80Kb 156
Heathdene Rd. SW16 66Pb 134
Heathdown Rd. GU22: Pyr 87F 168
Heath Dr. CM16: They B 8Uc 22
Heath Dr. EN6: Pot B 2Cb 17
Heath Dr. GU23: Send 94D 188
Heath Dr. GU24: Brkwd 2E 186
Heath Dr. KT20: Walt H 97Wa 194
Heath Dr. NW3 35Db 69
Heath Dr. RM2: Rom 25Jd 56
Heath Dr. SM2: Sutt 81Eb 175
Heath Dr. SW20 70Ya 132
Heathedge SE26 61Xb 135
Heath End Rd. DA5: Bexl 60Gd 118
Heather Av. RM1: Rom 26Fd 56
Heatherbank BR7: Chst 68Qc 138
Heatherbank SE9 54Pc 116
Heatherbank Cl. DA1: Cray 58Gd 118
Heatherbank Cl. KT11: Cobh 83Z 171
Heather Cl. CM15: Pil H 15Xd 40
Heather Cl. E6 44Qc 94
Heather Cl. GU21: Wok 7N 167
Heather Cl. KT15: New H 82K 169
Heather Cl. KT20: Kgswd 94Ab 194
Heather Cl. N7 34Pb 70
Heather Cl. RH1: Redh 3B 208
Heather Cl. RM1: Rom 25Fd 56
Heather Cl. SE13 59Fc 115
Heather Cl. SW8 55Kb 112
Heather Cl. TW12: Hamp 67Ba 129
Heather Cl. TW7: Isle 57Fa 108
Heather Cl. UB8: Hil 43P 83
Heather Cl. WD5: Ab L 4W 12
HEATHER CLOSE REHABILITATION
CEN. 59Fc 115
Heather Ct. AL2: Lon C 9H 7
Heather Ct. DA14: Sidc 65Zc 139
Heatherdale Cl. KT2: King T 65Qa 131
Heatherdene KT24: W Hor 97T 190
Heatherdene Cl. CR4: Mitc 70Fb 133
Heatherdene Cl. N12 24Eb 49
Heatherden Grn. SL0: Iver H 39E 62
Heatherden Ho. SL0: Iver H 38D 62
Heather Dr. DA1: Dart 59Jd 118
Heather Dr. EN2: Enf 12Rb 33
Heather Dr. RM1: Rom 26Fd 56
Heather Dr. SL5: S'dale 3F 146
Heather End BR8: Swan 70Fd 140
Heatherfield La. KT13: Weyb 78U 150
Heatherfields KT15: New H 82K 169
Heatherfold Way HA5: Eastc 27V 44
Heather Gdns. EN9: Walt A 8Ec 20
Heather Gdns. NW11 30Ab 48
Heather Gdns. RM1: Rom 26Fd 56
Heather Gdns. SM2: Sutt 79Cb 155
Heather Glen RM1: Rom 26Fd 56
Heather Ho. E14 44Ec 92
(off Dee St.)
Heatherlands TW16: Sun 65W 128
Heather La. UB7: Yiew 44N 83
Heather La. WD24: Wat 7V 12
Heatherlea Gro. KT4: Wor Pk 74Xa 154
Heatherley Ct. E5 34Wb 71
Heatherley Dr. IG5: Ilf 27Nc 54
Heather Pk. Dr. HA0: Wemb 38Qa 67
Heather Pk. Pde. HA0: Wemb 38Pa 67
(off Heather Pk. Dr.)
Heather Pl. KT10: Esh 77Da 151
Heather Ri. WD23: Bush 12Ba 27
Heather Rd. E4 23Bc 52
Heather Rd. NW2 33Va 68
Heather Rd. SE12 61Jc 137
Heathers, The TW19: Stanw 59P 105
Heatherset Cl. KT10: Esh 78Ea 152
Heatherset Gdns. SW16 66Pb 134
Heatherside Cl. KT23: Bookh 97Ba 191
Heatherside Dr. GU25: Vir W 2L 147
Heatherside Gdns. SL2: Farn C 4H 61
Heatherside Rd. DA14: Sidc 62Yc 139
Heatherside Rd. KT19: Ewe 80Ta 153
Heatherton Ter. N3 26Db 49
Heathervale Cvn. Pk. KT15: New H 82L 169
Heathervale Rd. KT15: New H 82K 169
Heathervale Way KT15: New H 82L 169
Heather Wlk. GU24: Brkwd 3B 186
Heather Wlk. HA8: Edg 22Ra 47
Heather Wlk. KT12: W Vill 82U 170
Heather Wlk. TW2: Whitt 59Ca 107
(off Stephenson Rd.)
Heather Wlk. W10 42Ab 88
Heather Way CR2: Sels 81Zb 178
Heather Way EN6: Pot B 4Bb 17
Heather Way GU24: Chob 10J 147
Heather Way HA7: Stan 23Ha 46
Heather Way HP2: Hem H 1M 3
Heather Way RM1: Rom 26Fd 56
Heatherwood Cl. E12 33Lc 73
Heatherwood Dr. UB4: Hayes 40T 64
Heath Farm Ct. WD17: Wat 9T 12
Heath Farm La. AL3: St A 1C 6
Heathfield BR7: Chst 65Sc 138
Heathfield E4 20Ec 34
Heathfield HA1: Harr 31Ha 66
Heathfield KT11: Cobh 86Ca 171
Heathfield Av. SL5: S'dale 1C 146
Heathfield Av. SW18 59Fb 111
Heathfield Cl. BR2: Kes 78Lc 159
Heathfield Cl. E16 43Mc 93
Heathfield Cl. EN6: Pot B 2Db 17
Heathfield Cl. GU22: Wok 90C 168
Heathfield Cl. WD19: Wat 17Y 27
Heathfield Ct. AL1: St A 1C 6
(off Avenue Rd.)
Heathfield Ct. E3 40Cc 72
(off Tredegar Rd.)
Heathfield Ct. SE14 52Yb 114
Heathfield Ct. SE20 66Yb 136
Heathfield Ct. TW15: Ashf 62N 127
Heathfield Ct. W4 50Ta 87
Heathfield Dr. CR4: Mitc 67Gb 133
Heathfield Dr. RH1: Redh 10N 207
Heathfield Gdns. CR0: C'don 77Tb 157
Heathfield Gdns. NW11 30Za 48
Heathfield Gdns. SE3 54Gc 115
(off Baizdon Rd.)
Heathfield Gdns. SW18 58Fb 111
Heathfield Gdns. W4 50Sa 87
Heathfield Ho. SE3 54Gc 115
Heathfield La. BR7: Chst 65Sc 138
Heathfield Nth. TW2: Twick 59Ga 108
Heathfield Pk. NW2 37Ya 68
Heathfield Pk. Dr. RM6: Chad H 29Xc 55
Heathfield Ri. HA4: Ruis 31S 64
Heathfield Rd. BR1: Broml 66Hc 137
Heathfield Rd. BR2: Kes 78Lc 159
Heathfield Rd. CR0: C'don 77Tb 157
Heathfield Rd. DA6: Bex 56Bd 117
Heathfield Rd. GU22: Wok 90C 168
Heathfield Rd. KT12: Hers 77Aa 151
Heathfield Rd. SW18 58Eb 111
Heathfield Rd. TN13: S'oaks 94Hd 202
Heathfield Rd. W3 47Ra 87
Heathfield Rd. WD23: Bush 14Aa 27
Heathfields Cl. KT21: Asht 90La 172
Heathfields Ct. TW4: Houn 57Aa 107
Heathfield Sth. TW2: Twick 59Ha 108
Heathfield Sq. SW18 59Fb 111
Heathfield Ter. BR8: Swan 68Fd 140
Heathfield Ter. SE18 51Uc 116
Heathfield Ter. W4 50Sa 87
Heathfield Va. CR2: Sels 81Zb 178
Heath Gdns. DA1: Dart 60Ld 119
Heath Gdns. TW1: Twick 60Ha 108
Heathgate NW11 30Db 49
Heathgate Pl. NW3 36Hb 69
Heath Gro. SE20 66Yb 136
Heath Gro. TW16: Sun 66V 128
Heath Ho. DA15: Sidc 63Vc 139
Heath Ho. Rd. GU22: Wok 4G 186
Heath Hurst Rd. NW3 35Gb 69
Heathhurst Rd. CR2: Sande 81Tb 177
Heathland M. KT18: Tatt C 91Za 194
Heathland Rd. N16 32Ub 71
Heathlands KT20: Tad 94Za 194
Heathlands Cl. GU21: Wok 86A 168
Heathlands Cl. TW1: Twick 61Ha 130
Heathlands Cl. TW16: Sun 68W 128
Heathlands Ri. DA1: Dart 58Kd 119
Heathlands Way TW4: Houn 57Aa 107
Heathland Way RM16: Grays 8A 100
Heath La. DA1: Dart 60Ld 119
Heath La. HP1: Hem H 4L 3
Heath La. SE3 54Fc 115
(not continuous)
Heath La. (Up.) DA1: Dart 61Jd 140
Heathlee Rd. DA1: Cray 58Gd 118
Heathlee Rd. SE3 56Hc 115
Heathley End BR7: Chst 65Sc 138
Heath Lodge WD23: B Hea 18Ga 28
Heathmans Rd. SW6 53Bb 111
Heath Mead SW19 62Za 132
Heath M. GU23: Rip 95K 189
Heath Mill La. GU3: Worp 7F 186
HEATH PARK 30Jd 56
Heath Pk. Ct. RM2: Rom 29Jd 56
Heath Pk. Dr. BR1: Broml 69Nc 138
Heathpark Dr. GU20: W'sham 9C 146
Heath Pk. Ho. HP1: Hem H 4L 3
Heath Pk. Rd. RM1: Rom 29Jd 56
Heath Pk. Rd. RM2: Rom 29Jd 56
Heath Pas. NW3 33Db 69
Heath Pl. E3 42Bc 92
Heathpool Ct. E1 42Xb 91
Heath Ridge Grn. KT11: Cobh 85Ca 171
Heath Ri. BR2: Hayes 72Hc 159
Heath Ri. GU23: Rip 95K 189
Heath Ri. GU25: Vir W 10P 125
Heath Ri. SW15 58Za 110
Heath Rd. AL1: St A 1C 6
Heath Rd. CR3: Cat'm 95Tb 197
Heath Rd. CR7: Thor H 69Sb 135
Heath Rd. DA1: Cray 58Hd 118
Heath Rd. DA5: Bexl 60Ed 118
Heath Rd. EN6: Pot B 2Cb 17
Heath Rd. GU21: Wok 87B 168
Heath Rd. HA1: Harr 31Ea 66
Heath Rd. KT13: Weyb 77Q 150
Heath Rd. KT22: Oxs 84Ea 172
Heath Rd. RM16: Grays 6B 100
Heath Rd. RM16: Ors 6B 100
Heath Rd. RM6: Chad H 31Zc 75
Heath Rd. SW8 54Kb 112
Heath Rd. TW1: Twick 60Ha 108
Heath Rd. TW2: Twick 60Ha 108
Heath Rd. TW3: Houn 56Da 107
Heath Rd. TW3: Isle 56Da 107
Heath Rd. TW7: Isle 56Fa 108
Heath Rd. UB10: Hil 42S 84
Heath Rd. WD19: Wat 17Z 27
Heath Robinson Mus. 28Z 45
HEATHROW AIRPORT 57R 106
HEATHROW AIRPORT 55L 105
HEATHROW AIRPORT 55Q 106
Heathrow Blvd. UB7: Sip 52P 105
(not continuous)
Heathrow Causeway Cen. TW4: Houn 55X 107
Heathrow Central Bus Station 55R 106
Heathrow Central Station (Rail) 55Q 106
Heathrow Central Tourist Info. Cen. 55R 106
Heathrow Cl. UB7: Lford 53K 105
Heathrow Gateway TW4: Houn 59Aa 107
UB7: Harm
Heathrow Interchange UB4: Yead 46Y 85
Heathrow Intl. Trad. Est. TW4: Houn 55X 107
UB7: Lford
Heathrow Prologis Pk. UB3: Harl 48R 84
Heathrow Terminal 4 Station (Rail) 58S 106

Heathrow Terminal 4 Station (Underground & Crossrail) 57S 106
Heathrow Terminal 5 Station (Rail, Underground & Crossrail) 55L 105
Heathrow Terminals 2 & 3 Station (Rail, Underground & Crossrail) 55Q 106
Heath Royal SW15 58Za 110
Heaths Cl. EN1: Enf 12Ub 33
Heath Side BR5: Pet W 74Sc 160
HEATH SIDE 62Hd 140
Heath Side NW3 35Fb 69
Heathside AL4: Col H 5M 7
Heathside KT10: Hin W 76Ga 152
Heathside KT13: Weyb 78R 150
Heathside NW11 32Cb 69
Heathside SE13 54Ec 114
Heathside TW4: Houn 59Ba 107
Heathside Av. DA7: Bex 53Ad 117
Heathside Cl. HA6: Nwood 22T 44
Heathside Cl. IG2: Ilf 29Tc 54
Heathside Cl. KT10: Hin W 76Ga 152
Heathside Ct. KT20: Tad 95Xa 194
Heathside Cres. GU22: Wok 89B 168
Heathside Gdns. GU22: Wok 89C 168
Heathside Pk. Rd. GU22: Wok 90B 168
Heathside Pl. KT18: Tatt C 90Za 174
Heathside Rd. GU22: Wok 90B 168
Heathside Rd. HA6: Nwood 21T 44
Heathstan Rd. W12 44Wa 88
Heath St. DA1: Dart 59Md 119
Heath St. NW3 35Eb 69
Heath Ter. RM6: Chad H 31Zc 75
Heath Vw. KT24: E Hor 97V 190
Heath Vw. N2 28Eb 49
Heathview NW5 35Jb 70
Heathview Av. DA1: Cray 58Gd 118
Heath Vw. Cl. N2 28Eb 49
Heathview Cres. DA1: Dart 60Jd 118
HEATHVIEW DAY CEN. 51Yc 117
Heathview Dr. SE2 51Zc 117
Heathview Gdns. RM16: Grays 48Ee 99
Heathview Gdns. SW15 59Ya 110
Heathview Rd. CR7: Thor H 70Qb 134
Heathview Rd. RM16: Grays 47Ee 99
Heath Vs. NW3 34Fb 69
Heath Vs. SE18 50Vc 95
Heathville Rd. N19 31Nb 70
Heathwall St. SW11 55Hb 111
Heath Way DA8: Erith 53Ed 118
Heath Way WD7: Shenl 2La 14
Heathway CR0: C'don 76Bc 158
Heathway CR3: Cat'm 97Sb 197
Heathway IG8: Wfd G 22Lc 53
Heathway KT24: E Hor 96V 190
Heathway RM10: Dag 34Bd 75
Heathway RM9: Dag 34Bd 75
HEATHWAY 39Cd 76
Heathway SE3 52Jc 115
Heathway SL0: Iver H 40F 62
Heathway UB2: S'hall 49Z 85
Heathway Ct. NW3 33Cb 69
Heathway Ind. Est. RM10: Dag 35Dd 76
Heathwood Gdns. BR8: Swan 68Ed 140
Heathwood Gdns. SE7 49Nc 94
Heathwood Pde. BR8: Swan 68Ed 140
Heathwood Point SE23 62Zb 136
Heathwood Wlk. DA5: Bexl 60Gd 118
Heaton Av. RM3: Rom 24Kd 57
Heaton Cl. E4 20Ec 34
Heaton Cl. RM3: Rom 24Ld 57
Heaton Ct. EN8: Chesh 1Zb 20
Heaton Ct. WD17: Wat 10W 12
Heaton Grange Rd. RM2: Rom 26Hd 56
Heaton Ho. SW10 51Eb 111
(off Fulham Rd.)
Heaton Rd. CR4: Mitc 66Jb 134
Heaton Rd. SE15 54Xb 113
Heaton Way RM3: Rom 24Ld 57
Heaven Tree Cl. N1 37Sb 71
Heaver Cl. DA3: Lfield 69Ce 143
HEAVERHAM 89Ud 184
Heaverham Rd. TN15: Kems'g 89Rd 183
Heaver Rd. SW11 55Fb 111
Heaver Trad. Est. TN15: Ash 77Zd 165
Heavitree Cl. SE18 50Tc 94
Heavitree Rd. SE18 50Tc 94
(not continuous)
Hebden St. E2 1K **219** (39Vb 71)
Hebden St. SW8 52Mb 112
Hebden Ter. N17 23Ub 51
Hebdon Rd. SW17 62Gb 133
Heberden Ct. RM19: Purf 50Rd 97
(off Wingrove Dr.)
Heber Mans. W14 51Ab 110
(off Queen's Club Gdns.)
Heber Rd. NW2 36Za 68
Heber Rd. SE22 58Vb 113
Hebrides Ct. E1 43Ac 92
(off Ocean Est.)
Hebron Rd. W6 48Ya 88
Hecham Cl. E17 26Ac 52
Heckets Ct. KT10: Esh 82Ea 172
Heckfield Pl. SW6 52Cb 111
Heckford Cl. SE18 50Uc 94
Heckford Cl. WD18: Wat 16S 26
Heckford Ho. E14 44Dc 92
(off Grundy St.)
Heckford St. E1 45Zb 92
Heckford St. Bus. Cen. E1 45Zb 92
(off Heckford St.)
Hector NW9 25Va 48
(off Five Acre)
Hector Cl. N9 19Wb 33
Hector Ct. SW9 52Qb 112
(off Caldwell St.)
Hector Ho. E2 40Xb 71
(off Old Bethnal Grn. Rd.)
Hector St. SE18 49Uc 94
Heddington Gro. N7 36Pb 70
Heddon Cl. TW7: Isle 56Ja 108
Heddon Ct. Av. EN4: Cockf 15Hb 31
Heddon Ct. Pde. EN4: Cockf 15Jb 32
Heddon Rd. EN4: Cockf 15Hb 31
Heddon St. W1 4B **222** (45Lb 90)
Hedera Pl. TW4: Houn 56Ba 107
Hedgecroft Cotts. GU23: Rip 93K 189
Hedgegate Ct. W11 44Bb 89
(off Powis Ter.)
Hedge Hill EN2: Enf 11Rb 33
Hedge La. N13 20Rb 33
Hedgeley IG4: Ilf 28Pc 54
Hedgemans Rd. RM9: Dag 38Zc 75
Hedgemans Way RM9: Dag 37Ad 75
Hedge Pl. Rd. DA9: Ghithe 58Vd 120
HEDGERLEY 2H 61
Hedgerley Ct. GU21: Wok 9N 167
Hedgerley Gdns. UB6: G'frd 40Ea 66
HEDGERLEY GREEN 1J 61
Hedgerley Hill SL2: Hedg 4H 61
HEDGERLEY HILL 3H 61
Hedgerley La. SL2: Ger X 1J 61
Hedgerley La. SL2: Hedg 1J 61
Hedgerley La. SL9: Ger X 2N 61
Hedge Row HP1: Hem H 1J 3
Hedgerow SL9: Chal P 23A 42
Hedgerow Ct. E6 39Pc 74
(off Nelson St.)
Hedgerow La. EN5: Ark 15Xa 30
Hedgerows, The DA11: Nflt 1A 144
Hedgerow Wlk. EN8: Chesh 2Zb 20
Hedgers Cl. IG10: Lough 14Qc 36
Hedgers Gro. E9 37Ac 72
Hedger St. SE11 5B **230** (49Rb 91)
Hedges Cl. TW14: Felt 58X 107
Hedgeside Rd. HA6: Nwood 22S 44
Hedges Way WD3: Crox G 17P 25
Hedge Wlk. SE6 64Dc 136
Hedgewood Gdns. IG5: Ilf 29Qc 54
Hedgley M. SE12 57Hc 115
Hedgley St. SE12 57Hc 115
Hedingham Cl. N1 38Sb 71
Hedingham Dr. TW20: Eng G 2N 125
Hedingham Ho. KT2: King T 67Na 131
(off Royal Quarter)
Hedingham Rd. RM11: Horn 32Qd 77
Hedingham Rd. RM16: Chaf H 50Yd 98
Hedingham Rd. RM8: Dag 36Xc 75
Hedley Av. RM20: Grays 52Yd 120
Hedley Cl. RM1: Rom 29Gd 56
Hedley Ho. E14 48Ec 92
(off Stewart St.)
Hedley Rd. AL1: St A 2F 6
Hedley Rd. TW2: Whitt 59Ca 107
Hedley Row N5 36Tb 71
Hedley Vs. AL1: St A 2F 6
Hedsor Ho. E2 5K **219** (42Vb 91)
(off Ligonier St.)
Hedworth Av. EN8: Walt C 5Zb 20
Heenan Cl. IG11: Bark 37Sc 74
Heene Rd. EN2: Enf 11Tb 33
Heer M. E2 40Wb 71
(off Hackney Rd.)
Hega Ho. E14 43Ec 92
(off Ullin St.)
Heideck Gdns. CM13: Hut 19De 41
Heidegger Cres. SW13 52Xa 110
Heigham Rd. E6 38Nc 74
Heighton Gdns. CR0: Wadd 78Rb 157
Heights, The BR3: Beck 66Ec 136
(not continuous)
Heights, The IG10: Lough 12Pc 36
Heights, The KT13: Weyb 82Q 170
Heights, The SE7 50Lc 93
Heights, The UB5: N'olt 36Ba 65
Heights Cl. SM7: Bans 88Ab 174
Heights Cl. SW20 66Xa 132
Heiron St. SE17 51Rb 113
Helby Rd. SW4 58Mb 112
Heldar Ct. SE1 2G **231** (47Tb 91)
Heldefirth Cl. DA10: Swans 60Be 121
Helder Gro. SE12 59Hc 115
Helder St. CR2: S Croy 79Tb 157
Heldmann Cl. TW3: Houn 56Fa 108
Helegan Cl. BR6: Chels 77Vc 161
Helena Cl. EN4: Had W 10Fb 17
Helena Cl. SW19 59Za 110
Helena Ct. NW6 38Db 69
(off Compayne Gdns.)
Helena Ct. W5 43Ma 87
Helena Pl. E9 39Xb 71
Helena Pl. HP2: Hem H 1M 3
Helena Rd. E13 40Hc 73
Helena Rd. E17 29Cc 52
Helena Rd. NW10 36Xa 68
Helena Rd. SL4: Wind 4H 103
Helena Rd. W5 43Ma 87
Helena Sq. SE16 45Ac 92
(off Sovereign Cres.)
Helen Av. TW14: Felt 59X 107
Helen Cl. DA1: Dart 59Kd 119
Helen Cl. KT8: W Mole 70Da 129
Helen Cl. N2 27Eb 49
Helen Gladstone Ho. SE1 1B **230** (47Rb 91)
(off Surrey Row)
Helen Ho. E2 40Xb 71
(off Old Bethnal Grn. Rd.)
Helen Peele Cotts. SE16 48Yb 92
(off Lower Rd.)
Helen Rd. RM11: Horn 27Md 57
Helenslea Av. NW11 32Cb 69
Helen's Pl. E2 41Yb 92
Helen St. SE18 49Rc 94
Helen Taylor Ho. SE16 48Wb 91
(off Evelyn Lowe Est.)
Helford Cl. HA4: Ruis 33U 64
Helford Ct. RM15: S Ock 45Xd 98
Helford Wlk. GU21: Wok 10L 167
Helford Way RM14: Upm 30Td 58
Helgiford Gdns. TW16: Sun 66U 128
Heligan Ho. SE16 47Zb 92
(off Water Gdns. Sq.)
Helios, The 45Ya 88
Helios Rd. SM6: W'gton 74Jb 156
Helios Way EN5: Barn 15Bb 31
Heliport Ind. Est. SW11 54Fb 111
Helix Ct. W11 46Za 88
(off Swanscombe Rd.)
Helix Gdns. SW2 58Pb 112
Helix Rd. SW2 58Pb 112
Helix Ter. SW19 61Za 132
Helleborine RM17: Grays 50Be 99
Hellen Way WD19: Wat 21Z 45
Hellings St. E1 46Wb 91
Helm, The E16 45Rc 94
Helm Cl. KT19: Eps 84Qa 173
Helme Cl. SW19 64Bb 133
Helmer Row EC1 5E **218** (42Sb 91)
Helmore Rd. IG11: Bark 38Vc 75
Helmsdale GU21: Wok 10M 167
Helmsdale Apts. SW11 56Gb 111
(off Monarch Square)
Helmsdale Cl. RM1: Rom 24Gd 56
Helmsdale Cl. UB4: Yead 42Aa 85
Helmsdale Ho. NW6 40Db 69
(off Carlton Vale)
Helmsdale Rd. RM1: Rom 24Gd 56
Helmsdale Rd. SW16 67Mb 134
Helmsley Ho. RM3: Rom 24Nd 57
(off Leyburn Cres.)
Helmsley Pl. E8 38Xb 71
Helmsley St. E8 38Xb 71
Helperby Rd. NW10 38Ua 68
Helsby Ct. NW8 5B **214** (42Fb 89)
(off Pollitt Dr.)
Helsinki Sq. SE16 48Ac 92
Helston NW1 1C **216** (39Lb 70)
(off Camden St.)
Helston Cl. HA5: Hat E 24Ba 45
Helston Ct. N15 29Ub 51
(off Culvert Rd.)
Helston Ho. SE11 7A **230** (50Qb 90)
(off Kennings Way)
Helston La. SL4: Wind 3F 102
Helston Pl. WD5: Ab L 4V 12
Helvellyn Cl. TW20: Egh 66D 126
Helvetia St. SE6 61Bc 136
Helwys Ct. E4 23Dc 52
Hemans St. SW8 52Mb 112
Hemans St. Est. SW8 52Nb 112
Hemberton Rd. SW9 55Nb 112
HEMEL HEMPSTEAD 1M 3
HEMEL HEMPSTEAD GENERAL HOSPITAL 3M 3
Hemel Hempstead Ind. Est. HP2: Hem H Maylands Ct. 1B 4
Hemel Hempstead Rd. AL3: St A 3L 5
Hemel Hempstead Rd. HP3: Hem H 4D 4
Hemel Hempstead Station (Rail) 5J 3
Hemel Hempstead Town FC 2A 4
Hemery Rd. UB6: G'frd 36Fa 66
Hemingford Cl. N12 22Fb 49
Hemingford Ct. WD17: Wat 13Y 27
Hemingford Rd. N1 39Pb 70
Hemingford Rd. SM3: Cheam 77Ya 154
Hemingford Rd. WD17: Wat 8U 12
Heming Rd. HA8: Edg 24Ra 47
Hemington Av. N11 22Hb 49
Hemingway Cl. NW5 35Jb 70
Hemley Rd. RM16: Ors 4F 100
Hemlock Cl. KT20: Kgswd 95Ab 194
Hemlock Cl. SW16 68Lb 134
Hemlock Ho. SE16 48Ac 92
Hemlock Rd. W12 45Va 88
(not continuous)
Hemmen La. UB3: Hayes 44V 84
Hemming Cl. TW12: Hamp 67Ca 129
Hemmings Cl. DA14: Sidc 61Xc 139
Hemmings Mead KT19: Ewe 79Sa 153
Hemming St. E1 42Wb 91
Hemming Way SL2: Slou 1F 80
Hemming Way WD25: Wat 7W 12
Hemnall M. CM16: Epp 2Wc 23
(off Hemnall St.)
Hemnall St. CM16: Epp 3Vc 23
Hempshaw Av. SM7: Bans 88Hb 175
Hempson Av. SL3: L'ly 8N 81
Hempstead Cl. IG9: Buck H 19Jc 35
Hempstead Rd. E17 27Fc 53
Hempstead Rd. HP3: Bov 9C 2
Hempstead Rd. WD17: Wat 8T 12
(not continuous)
Hempstead Rd. WD4: K Lan 8P 3
Hemp Wlk. SE17 5G **231** (49Tb 91)
Hemsby Rd. KT9: Chess 79Pa 153
Hemsley Rd. WD4: K Lan 1R 12
Hemstal Rd. NW6 38Cb 69
Hemsted Rd. DA8: Erith 52Gd 118
Hemswell Dr. NW9 25Ua 48
Hemsworth Ct. N1 1H **219** (40Ub 71)
Hemsworth St. N1 1H **219** (40Ub 71)
Hemus Pl. SW3 7E **226** (50Gb 89)
Hemwood Rd. SL4: Wind 5B 102
Henage La. GU22: Wok 92E 188
Hen & Chicken Ct. EC4 3K **223** (44Qb 90)
(off Fleet St.)
Hen & Chickens Theatre 37Rb 71
(off St Paul's Rd.)
Henbane Path RM3: Rom 24Md 57
Henbit Cl. KT20: Tad 91Xa 194
Henbury Way WD19: Wat 20Z 27
Henchman St. W12 44Va 88
Hencroft St. Nth. SL1: Slou 7K 81
Hencroft St. Sth. SL1: Slou 8K 81
Hendale Av. NW4 27Xa 48
Henderson Cl. NW10 37Sa 67
Henderson Cl. RM11: Horn 33Kd 77
Henderson Ct. N12 21Db 49
Henderson Ct. NW3 36Fb 69
(off Fitzjohn's Av.)
Henderson Ct. SE14 51Zb 114
(off Myers La.)
Henderson Cres. KT22: Lea 92Ha 192
Henderson Dr. DA1: Dart 56Pd 119
Henderson Dr. NW8 5B **214** (42Fb 89)
Henderson Gro. TN16: Big H 84Lc 179
Henderson Ho. RM10: Dag 34Cd 76
(off Kershaw Rd.)
Henderson Pl. WD5: Bedm 9F 4
Henderson Rd. CR0: C'don 72Tb 157
Henderson Rd. E7 37Lc 73
Henderson Rd. N9 18Xb 33
Henderson Rd. SW18 59Gb 111
Henderson Rd. UB4: Yead 41W 84
Hendfield Ct. SM6: W'gton 79Kb 156
Hendham Rd. SW17 61Gb 133
HENDON 28Ya 48
Hendon Av. N3 25Ab 48
HENDON BMI HOSPITAL 27Ya 48
Hendon Central Station (Underground) 29Xa 48
Hendon Crematorium 25Za 48
Hendon FC 35Ba 65
Hendon Gdns. RM5: Col R 23Ed 56
Hendon Golf Course 24Ya 48
Hendon Gro. KT19: Eps 81Qa 173
Hendon Hall Ct. NW4 27Za 48
Hendon Ho. NW4 29Za 48
Hendon La. N3 27Ab 48
Hendon Leisure Cen. 31Za 68
Hendon Lodge NW4 27Xa 48
Hendon Pk. Mans. NW4 29Ya 48
Hendon Pk. Row NW11 30Bb 49
Hendon Rd. N9 19Wb 33
Hendon Station (Rail) 30Wa 48
Hendon St. SE10 48Hc 93
Hendon Ter. TW15: Ashf 65T 128
Hendon Way NW2 31Za 68
Hendon Way NW4 30Xa 48
Hendon Way TW19: Stanw 58M 105
Hendon Wood La. NW7 16Va 30
Hendre Ho. SE1 6J **231** (49Ub 91)
(off Hendre Rd.)
Hendren Cl. UB6: G'frd 36Fa 66
Hendre Rd. SE1 6J **231** (49Ub 91)
Hendrick Av. SW12 59Hb 111
Heneage Cres. CR0: New Ad 82Ec 178
Heneage La. EC3 2J **225** (44Ub 91)
Heneage Pl. EC3 3J **225** (44Ub 91)
Heneage St. E1 43Vb 91
Henfield Cl. DA5: Bexl 58Cd 118
Henfield Cl. N19 32Lb 70
Henfield Rd. SW19 67Bb 133
Hengelo Gdns. CR4: Mitc 70Fb 133
Hengest Av. KT10: Surb 76Ja 152
Hengist Rd. DA8: Erith 52Dd 118
Hengist Rd. SE12 59Kc 115
Hengist Way BR2: Broml 70Gc 137
Hengist Way SM6: W'gton 80Mb 156
Hengrave Rd. SE23 58Yb 114
Hengrove Ct. DA5: Bexl 60Ad 117
Hengrove Cres. TW15: Ashf 62M 127
Henham Ct. RM5: Col R 25Ed 56
HENHURST 7G 144
Henhurst Rd. DA12: Cobh 6F 144
Henley Av. SM3: Cheam 76Ab 154
Henley Bus. Pk. GU3: Norm 10A 186
Henley Cl. SE16 47Yb 92
(off St Marychurch St.)
Henley Cl. TW7: Isle 53Ha 108
Henley Cl. UB6: G'frd 40Ea 66
Henley Ct. GU22: Wok 92D 188
Henley Ct. N14 17Lb 32
Henley Ct. NW2 37Za 68
Henley Ct. TW20: Egh 63C 126
Henley Deane DA11: Nflt 3A 144
Henley Dr. KT2: King T 66Va 132
Henley Dr. SE1 5K **231** (49Vb 91)
Henley Gdns. HA5: Eastc 27X 45
Henley Gdns. RM6: Chad H 29Ad 55
Henley Ga. GU3: Norm 9A 186
Henley Hgts. N1 38Pb 70
(off Caledonian Rd.)
Henley Ho. AL1: St A 3C 6
(off Lattimore Rd.)
Henley Ho. E2 5K **219** (42Vb 91)
(off Swanfield St.)
Henley Prior N1 2H **217** (40Pb 70)
(off Affleck St.)
Henley Rd. E16 47Pc 94
Henley Rd. IG1: Ilf 35Sc 74
Henley Rd. N18 21Ub 51
Henley Rd. NW10 39Ya 68
Henley Rd. SL1: Slou 4C 80
Henley St. SW11 54Jb 112
Henley Way TW13: Hanw 64Z 129
Henlow Pl. TW10: Ham 61Ma 131
HENLYS CORNER 28Bb 49
HENLYS RDBT. 54Y 107
Henman Way CM14: B'wood 18Xd 40
Henneker Cl. RM5: Col R 23Ed 56
Hennel Cl. SE23 62Yb 136
Hennessey M. RM8: Dag 32Ad 75
Hennessy Ct. E10 30Ec 52
Hennessy Ct. GU21: Wok 85E 168
Hennessy Rd. N9 19Yb 34
Henniker Gdns. E6 41Mc 93
Henniker M. SW3 51Fb 111
Henniker Point E15 36Gc 73
(off Leytonstone Rd.)
Henniker Rd. E15 36Fc 73
Henningham Rd. N17 25Tb 51
Henning St. SW11 53Gb 111
Henrietta Barnet Wlk. NW11 30Cb 49
Henrietta Cl. KT15: Add 80J 149
Henrietta Cl. KT18: Eps 87Qa 173
Henrietta Cl. SE8 51Cc 114
Henrietta Ct. TW1: Twick 59La 108
(off Richmond Rd.)
Henrietta Gdns. N21 18Qb 32
Henrietta Ho. N15 30Ub 51
(off St Ann's Rd.)
Henrietta Ho. W6 50Ya 88
(off Queen Caroline St.)
Henrietta M. WC1 5G **217** (42Nb 90)
Henrietta Pl. W1 3K **221** (44Kb 90)
Henrietta St. WC2 4G **223** (45Nb 90)
Henriques St. E1 44Wb 91
Henry Addlington Cl. E6 43Rc 94
Henry Chester Bldg. SW15 54Ya 110
Henry Cl. EN2: Enf 10Ub 19
Henry Cooper Way SE9 62Mc 137
Henry Ct. HA7: Stan 24Ma 47
Henry Darlot Dr. NW7 23Za 48
Henry De Grey Cl. RM17: Grays 49Be 99
Henry Dent Cl. SE5 55Tb 113
Henry Dickens Ct. W11 45Za 88
Henry Doulton Dr. SW17 63Jb 134
Henry Gdns. DA11: Nflt 1B 144
Henry Hatch Ct. SM2: Sutt 80Eb 155
Henry Ho. SE1 7A **224** (46Qb 90)
Henry Ho. SW8 52Nb 112
(off Wyvil Rd.)
Henry Hudson Apts. SE10 50Gc 93
(off Banning St.)
Henry Jackson Rd. SW15 55Za 110
Henry Lodge KT12: Hers 79Y 151
Henry Macaulay Av. KT2: King T 67Ma 131
Henry Moore Ct. SW3 7D **226** (50Gb 89)
Henry Peters Dr. TW11: Tedd 64Ga 130
(off Somerset Gdns.)
Henry Purcell Ho. E16 46Kc 93
(off Evelyn Rd.)
Henry Rd. E6 40Nc 74
Henry Rd. EN4: E Barn 15Fb 31
Henry Rd. N4 32Sb 71
Henry Rd. SL1: Slou 7H 81
Henry Rd. SW9 53Qb 112
Henrys Av. IG8: Wfd G 22Hc 53
Henrys Grant AL1: St A 3C 6
Henryson Rd. SE4 57Cc 114
Henry St. BR1: Broml 67Kc 137
Henry St. HP3: Hem H 6M 3
Henry St. RM17: Grays 51Ee 121
Henry's Wlk. IG6: Ilf 24Tc 54
Henry Tate M. SW16 64Pb 134
Henry Tudor Ct. SE9 59Sc 116
Henry Wlk. DA1: Dart 55Od 119
Henry Wise Ho. SW1 6C **228** (49Lb 90)
(off Vauxhall Bri. Rd.)
Hensby M. WD19: Wat 16Aa 27
Hensford Gdns. SE26 63Xb 135
Henshall Point E3 41Dc 92
(off Bromley High St.)
Henshall St. N1 37Tb 71
Henshawe Rd. RM8: Dag 34Zc 75
Henshaw St. SE17 5F **231** (49Tb 91)
Henslow Cres. DA1: Dart 54Pd 119
Henslowe Rd. SE22 57Wb 113
Henslow Ho. SE15 52Wb 113
(off Peckham Pk. Rd.)
Henslow Way GU21: Wok 86F 168
Henson Av. NW2 36Ya 68
Henson Cl. BR6: Farnb 75Rc 160
Henson Path HA3: Kenton 27Ma 47
Henson Pl. UB5: N'olt 39Y 65
Henson Rd. RM8: Dag 32Ad 75
Henstridge Pl. NW8 1D **214** (39Gb 69)
Hensworth Rd. TW15: Ashf 64M 127
Henty Cl. SW11 52Gb 111
Henty Wlk. SW15 57Xa 110
Henville Rd. BR1: Broml 67Kc 137
Henwick Rd. SE9 55Mc 115
Henwood Side IG8: Wfd G 23Pc 54
Hepburn Cl. RM16: Chaf H 49Zd 99
Hepburn Ct. EN6: S Mim 4Wa 16
Hepburn Ct. WD6: Bore 14Qa 29
(off Whitehall Cl.)
Hepburn Gdns. BR2: Hayes 74Gc 159
Hepburn M. SW11 57Hb 111
Hepburn Pl. W3 45Ra 87
Hepdon M. SW17 64Fb 133
Hepple Cl. TW7: Isle 54Ka 108
Hepplestone Cl. SW15 58Xa 110
Hepscott Rd. E9 37Cc 72
Hepworth Ct. N1 39Rb 71
(off Gaskin St.)
Hepworth Ct. NW3 36Gb 69
Hepworth Ct. SM3: Sutt 74Cb 155
Hepworth Ct. SW1 7K **227** (50Kb 90)
Hepworth Gdns. IG11: Bark 36Wc 75
Hepworth Rd. SW16 66Nb 134
Hepworth Way KT12: Walt T 74V 150
Hera Av. EN5: Barn 15Bb 31
Heracles NW9 25Va 48
(off Five Acre)
Heracles Cl. AL2: Park 10A 6
Hera Ct. E14 49Cc 92
(off Homer Dr.)
Herald Gdns. SM6: W'gton 75Kb 156
Herald's Pl. SE11 5B **230** (49Rb 91)
Herald St. E2 42Xb 91
Herald Wlk. DA1: Dart 57Pd 119
Herbal Hill EC1 6A **218** (42Qb 90)
Herbal Hill Gdns. EC1 6A **218** (42Qb 90)
(off Herbal Hill)
Herbal Pl. EC1 6A **218** (42Qb 90)
(off Herbal Hill)
Herbert Ct. KT5: Surb 72Pa 153
(off Fulmar Cl.)
Herbert Cres. GU21: Knap 10J 167
Herbert Cres. SW1 3G **227** (48Hb 89)
Herbert Gdns. NW10 40Xa 68
Herbert Gdns. RM6: Chad H 31Zc 75
Herbert Gdns. W4 51Ra 109
Herbert Ho. E1 2K **225** (44Vb 91)
(off Old Castle St.)
Herbert M. SW2 58Qb 112
Herbert Morrison Ho. SW6 51Bb 111
(off Clem Attlee Ct.)
Herbert Pl. SE18 51Rc 116
Herbert Pl. TW7: Isle 54Fa 108
Herbert Rd. BR2: Broml 71Mc 159
Herbert Rd. BR8: Hext 65Kd 141
Herbert Rd. DA10: Swans 58Be 121
Herbert Rd. DA7: Bex 54Ad 117
Herbert Rd. E12 35Nc 74
Herbert Rd. E17 31Bc 72
Herbert Rd. IG3: Ilf 33Uc 74
Herbert Rd. KT1: King T 69Pa 131
Herbert Rd. N11 24Nb 50
Herbert Rd. N15 29Vb 51
Herbert Rd. NW9 30Wa 48
Herbert Rd. RM11: Horn 31Nd 77
Herbert Rd. SE18 52Qc 116
(not continuous)
Herbert Rd. SW19 66Bb 133
(not continuous)
Herbert Rd. UB1: S'hall 46Ba 85
Herbert St. E13 40Jc 73
Herbert St. HP2: Hem H 1M 3
Herbert St. NW5 37Jb 70
Herbrand Est. WC1 5F **217** (42Nb 90)
Herbrand St. WC1 5F **217** (42Nb 90)
Hercies Rd. UB10: Hil 38P 63
Hercules Ct. SE14 51Ac 114
Hercules Ho. E14 44Gc 93
Hercules Pl. N7 34Nb 70
(not continuous)
Hercules Rd. SE1 4J **229** (48Pb 90)
Hercules St. N7 34Nb 70
Hercules Way WD25: Wat 6V 12
Hercules Wharf E14 45Gc 93
(off Orchard Pl.)
Hercules Yd. N7 34Nb 70
Here E. E20 36Cc 72
Hereford Av. EN4: E Barn 18Hb 31
Hereford Bldgs. SW3 51Fb 111
(off Old Church St.)
Hereford Cl. GU21: Knap 1E 186
Hereford Cl. KT18: Eps 85Ta 173
Hereford Cl. TW18: Staines 67K 127
Hereford Copse GU22: Wok 1M 187
Hereford Ct. HA1: Harr 28Ga 46
Hereford Ct. SM2: Sutt 80Cb 155
Hereford Ct. W7 43Ha 86
(off Copley Cl.)
Hereford Gdns. HA5: Pinn 29Aa 45
Hereford Gdns. IG1: Ilf 31Nc 74
Hereford Gdns. SE13 57Gc 115
Hereford Gdns. TW2: Twick 60Ea 108
Hereford Ho. N18 22Xb 51
(off Cameron Cl.)
Hereford Ho. NW6 40Cb 69
(off Carlton Vale)
Hereford Ho. SW10 52Db 111
(off Fulham Rd.)
Hereford Ho. SW3 3E **226** (48Gb 89)
(off Ovington Gdns.)
Hereford Mans. W2 44Cb 89
(off Hereford Rd.)
Hereford M. W2 44Cb 89
Hereford Pl. SE14 52Bc 114
Hereford Retreat SE15 52Wb 113
Hereford Rd. E11 29Kc 53
Hereford Rd. E3 40Bc 72
Hereford Rd. TW13: Felt 60Y 107
Hereford Rd. W2 44Cb 89
Hereford Rd. W3 45Ra 87
Hereford Rd. W5 48La 86
Hereford Sq. SW7 6A **226** (49Eb 89)
Hereford St. E2 42Wb 91
Hereford Way KT9: Chess 78La 152
Herent Dr. IG5: Ilf 28Nc 54
Herent Gdns. IG5: Ilf 28Pc 54
Hereward Av. CR8: Purl 83Qb 176
Hereward Cl. EN9: Walt A 4Fc 21
Hereward Gdns. N13 22Qb 50
Hereward Grn. IG10: Lough 11Sc 36
Hereward Lincoln Ho. DA11: Nflt 58Fe 121
(off London Rd.)
Hereward Rd. SW17 63Hb 133
Herga Ct. HA1: Harr 34Ga 66
Herga Ct. WD17: Wat 12W 26
Herga Hyll RM16: Ors 3C 100
Herga Rd. HA3: W'stone 28Ha 46
Herington Gro. CM13: Hut 17Ce 41
Heriot Av. E4 19Cc 34
Heriot Cl. KT16: Chert 73H 149
Heriot Rd. KT16: Chert 73J 149
Heriot Rd. NW4 29Ya 48
Heriots Cl. HA7: Stan 21Ja 46
Heritage Av. NW9 27Va 48
Heritage Cl. AL3: St A 2B 6
Heritage Cl. SW9 55Rb 113
Heritage Cl. TW16: Sun 67W 128
Heritage Cl. UB8: Cowl 42L 83

Heritage Ct. SE8 ....50Zb 92
Heritage Ct. TW20: Egh 64C 126
(off Station Rd.)
Heritage Ga. SL9: Ger X 27A 42
Heritage Hill BR2: Kes 78Lc 159
Heritage La. NW6 ....37Cb 69
Heritage Pl. SW18 ....60Eb 111
Heritage Pl. TW8: Bford 50Pa 87
(off Heritage Wlk.)
Heritage Quay DA12: Grav'nd 8E 122
Heritage Vw. HA1: Harr 34Ha 66
Heritage Wlk. TW8: Bford 50Pa 87
(off Kew Bri. Rd.)
Heritage Wlk. WD3: Chor 13G 24
Herkomer Cl. WD23: Bush 16Da 27
Herkomer Rd. WD23: Bush 15Ca 27
Herlwyn Av. HA4: Ruis 33U 64
Herlwyn Gdns. SW17 ....63Hb 133
Her Majesty's Theatre 6D 222 (46Mb 90)
(off Haymarket)
Herm Cl. TW7: Isle 52Ea 108
Hermes Cl. EN5: Barn 15Bb 31
Hermes Cl. W9 ....42Cb 89
Hermes Ct. SW2 ....58Pb 112
Hermes Ct. SW9 ....53Qb 112
(off Southey Rd.)
Hermes St. N1 ....2K 217 (40Qb 70)
Hermes Wlk. UB5: N'olt 40Ca 65
Herm Ho. EN3: Enf W 10Zb 20
Herm Ho. N1 ....37Sb 71
(off Clifton Rd.)
Hermiston Av. N8 ....29Nb 50
Hermitage, The KT1: King T 70Ma 131
Hermitage, The SE13 ....54Ec 114
Hermitage, The SE23 ....60Yb 114
Hermitage, The SW13 ....53Va 110
Hermitage, The TW10: Rich 57Na 109
Hermitage, The TW13: Felt 62V 128
Hermitage, The UB8: Uxb 37M 63
Hermitage Basin 46Wb 91
(off Cromwell Cl.)
Hermitage Bri. GU21: Wok 2H 187
Hermitage Bri. Cotts. GU21: Wok 1G 186
Hermitage Cl. E18 ....28Hc 53
Hermitage Cl. EN2: Enf 12Rb 33
Hermitage Cl. KT10: Clay 79Ja 152
Hermitage Cl. SE2 ....48Yc 95
Hermitage Cl. SL3: L'ly 8N 81
Hermitage Cl. TW17: Shep 70Q 128
Hermitage Ct. E1 ....46Wb 91
(off Knighten St.)
Hermitage Ct. E18 ....28Jc 53
Hermitage Ct. EN6: Pot B 5Eb 17
Hermitage Ct. NW2 ....34Cb 69
Hermitage Ct. TW18: Staines 64H 127
Hermitage Gdns. NW2 ....34Cb 69
Hermitage Gdns. SE19 ....66Sb 135
Hermitage Grn. SW16 ....67Nb 134
Hermitage Ho. N1 ....1B 218 (40Rb 71)
(off Gerrard Rd.)
Hermitage La. CR0: C'don 73Wb 157
Hermitage La. N18 ....22Tb 51
Hermitage La. NW2 ....34Cb 69
Hermitage La. SE25 ....72Wb 157
Hermitage La. SL4: Wind 6E 102
Hermitage La. SW16 ....66Pb 134
Hermitage Moorings E1 ....46Wb 91
Hermitage Path SW16 ....67Nb 134
Hermitage Rd. CR8: Kenley 87Sb 177
Hermitage Rd. GU21: Wok 1H 187
Hermitage Rd. N15 ....30Tb 51
Hermitage Rd. N4 ....31Rb 71
Hermitage Rd. SE19 ....66Sb 135
Hermitage Row E8 ....36Wb 71
Hermitage St. W2 ....1B 220 (43Fb 89)
Hermitage Vs. SW6 ....51Cb 111
(off Lillie Rd.)
Hermitage Wlk. E18 ....28Hc 53
Hermitage Wall E1 ....46Wb 91
Hermitage Waterside E1 ....46Wb 91
(off Thomas More St.)
Hermitage Way HA7: Stan 25Ja 46
Hermitage Woods Cres. GU21: Wok 1J 187
Hermitage Woods Est. GU21: Wok 1J 187
Hermit Pl. NW6 ....39Db 69
Hermit Rd. E16 ....43Hc 93
Hermit St. EC1 ....3B 218 (41Rb 91)
Hermon Gro. UB3: Hayes 46W 84
Hermon Hill E11 ....29Jc 53
Hermon Hill E18 ....28Kc 53
Hern, The TN15: Crou 94Ee 205
Herndon Cl. TW20: Egh 63C 126
Herndon Rd. SW18 ....57Eb 111
Herne Cl. NW10 ....36Ta 67
Herne Cl. UB3: Hayes 44V 84
Herne Cl. WD23: Bush 17Fa 28
Herne Ct. WD23: Bush 17Ea 28
Herne Hill SE24 ....58Sb 113
HERNE HILL 57Sb 113
Herne Hill Ho. SE24 ....58Rb 113
(off Railton Rd.)
Herne Hill Rd. SE24 ....55Sb 113
Herne Hill Station (Rail) 58Rb 113
Herne Hill Velodrome 58Tb 113
Herne M. N18 ....21Wb 51
Herne Pl. SE24 ....57Rb 113
Herne Rd. KT6: Surb 75Ma 153
Herne Rd. WD23: Bush 16Da 27
Hernes Cl. TW18: Staines 67K 127
Herneshaw AL10: Hat 2B 8
Hernshaw CM13: Heron 24Fe 59
Herold Cl. RM13: Rain 39Jd 76
Heron Chase CM13: Heron 24Fe 59
Heron Cl. E17 ....26Bc 52
Heron Cl. HP3: Hem H 7P 3
Heron Cl. IG9: Buck H 18Jc 35
Heron Cl. NW10 ....37Ua 68
Heron Cl. SM1: Sutt 78Bb 155
Heron Cl. UB8: Uxb 37M 63
Heron Cl. WD3: Rick 19M 25
Heron Ct. BR2: Broml 70Lc 137
Heron Ct. CM13: Heron 25Fe 59
Heron Ct. E14 ....48Fc 93
(off New Union Cl.)
Heron Ct. HA4: Ruis 33T 64
Heron Ct. KT1: King T 69Na 131
Heron Ct. KT17: Eps 86Wa 174
Heron Ct. NW9 ....26Ua 48
Heron Ct. TW19: Stanw 60N 105
Heron Cres. DA14: Sidc 62Uc 138
Heron Dale KT15: Add 78M 149
Herondale CR2: Sels 81Zb 178
Herondale Av. SW18 ....60Fb 111
Heron Dr. N4 ....33Sb 71
Heron Dr. SL3: L'ly 49D 82
Heronfield EN6: Pot B 2Eb 17
Heronfield TW20: Eng G 5N 125
Heron Flight Av. RM12: Horn 38Jd 76
HERONGATE 24Fe 59
Herongate N1 ....39Sb 71
(off Ridgewell Cl.)
Herongate Rd. BR8: Hext 65Gd 140
Herongate Rd. E12 ....33Lc 73
Heron Hill DA17: Belv 49Cc 96
Heron Ho. DA14: Sidc 62Xc 139
Heron Ho. E3 ....39Bc 72
(off Sycamore Av.)
Heron Ho. E6 ....38Nc 74
Heron Ho. NW8 ....2D 214 (40Gb 69)
(off Newcourt St.)
Heron Ho. SW11 ....52Gb 111
(off Searles Cl.)
Heron Ho. UB7: Yiew 45M 83
(off Wraysbury Dr.)
Heron Ho. W13 ....42Ja 86
Heron Ind. Est. E15 ....40Dc 72
Heron Lake Rd. TW19: Staines 61E 126
Heron Mead EN3: Enf L 10Cc 20
Heron M. IG1: Ilf 33Rc 74
Heron Pl. E16 ....46Lc 93
(off Bramwell Way)
Heron Pl. SE16 ....46Ac 92
Heron Pl. UB9: Hare 23J 43
Heron Pl. W1 ....2J 221 (44Jb 90)
(off Thayer St.)
Heron Quay E14 ....46Cc 92
Heron Quays Station (DLR) 46Cc 92
Heron Rd. CR0: C'don 75Ub 157
Heron Rd. SE24 ....56Sb 113
Heron Rd. TW1: Twick 56Ja 108
Heronry, The KT12: Hers 79W 150
Herons, The E11 ....30Hc 53
Herons, The RM12: Horn 32Md 77
Heronsbrook SL5: Asc 8C 124
Heronscourt GU18: Light 3A 166
Herons Cft. KT13: Weyb 79S 150
Heronsforde W13 ....44La 86
Heronsgate HA8: Edg 22Qa 47
HERONSGATE 17E 24
Heronsgate Rd. WD3: Chor 16D 24
Heron's Lea N6 ....30Hb 49
Heronslea WD25: Wat 8Y 13
Heronslea Dr. HA7: Stan 22Na 47
Heron's Pl. TW7: Isle 55Ka 108
Heron Sq. TW9: Rich 57Ma 109
Herons Ri. EN4: E Barn 14Gb 31
Herons Way AL1: St A 6E 6
(not continuous)
Herons Way GU24: Brkwd 3B 186
Heronswood EN9: Walt A 6Gc 21
Heron Trad. Est. W3 ....43Ra 87
Heron Vw. TW8: Bford 52La 108
(off Commerce Rd.)
Heron Wlk. GU21: Wok 86E 168
Heron Wlk. HA6: Nwood 21U 44
Heron Way AL10: Hat 2C 8
Heron Way RM14: Upm 32Ud 78
Heron Way RM20: W Thur 50Xd 98
Heron Way SM6: W'gton 80Mb 156
Heron Way TW14: Felt 56W 106
Heronway CM13: Hut 18De 41
Heronway IG8: Wfd G 21Lc 53
Herrick Ct. W3 ....48Sa 87
(off Bollo Bri. Rd.)
Herrick Ho. N16 ....35Tb 71
(off Howard Rd.)
Herrick Ho. SE5 ....52Tb 113
(off Elmington Est.)
Herrick Rd. N5 ....34Sb 71
Herrick St. SW1 ....6E 228 (49Mb 90)
Herries St. W10 ....40Ab 68
Herringham Rd. SE7 ....48Lc 93
Herrings La. GU20: W'sham 8B 146
Herrings La. KT16: Chert 72J 149
Herron Ct. BR2: Broml 70Hc 137
Herrongate Cl. EN1: Enf 12Vb 33
Hersant Cl. NW10 ....39Wa 68
Herschell M. SE5 ....55Sb 113
Herschell Rd. SE23 ....59Ac 114
Herschel Pk. Dr. SL1: Slou 7K 81
Herschel Sports 5H 81
Herschel St. SL1: Slou 7K 81
HERSHAM 78Z 151
Hersham By-Pass KT12: Hers 78X 151
Hersham Cl. SW15 ....59Wa 110
Hersham Gdns. KT12: Hers 77X 151
Hersham Golf Course 76Aa 151
HERSHAM GREEN 78Z 151
Hersham Grn. Shop. Cen. 78Z 151
Hersham Pl. KT12: Hers 78Z 151
Hersham Rd. KT12: Hers 74W 150
Hersham Rd. KT12: Walt T 74W 150
Hersham Station (Rail) 76Aa 151
Hersham Trad. Est. KT12: Walt T 75Aa 151
Hershell Ct. SW14 ....56Ra 109
Hertford Av. SW14 ....57Ta 109
Hertford Cl. EN4: Cockf 13Fb 31
Hertford Cl. WD3: Crox G 14R 26
Hertford Ct. E6 ....41Pc 94
Hertford Ct. N13 ....20Qb 32
Hertford End Ct. HA6: Nwood 22U 44
Hertford Ho. UB5: N'olt 42Ba 85
Hertford Lock Ho. E3 ....39Bc 72
(off Parnell Rd.)
Hertford M. EN6: Pot B 3Eb 17
Hertford Pl. W1 ....6B 216 (42Lb 90)
Hertford Pl. WD3: Map C 21H 43
Hertford Rd. EN3: Enf H 13Yb 34
Hertford Rd. EN3: Enf W 13Yb 34
Hertford Rd. EN4: Cockf 13Eb 31
Hertford Rd. EN8: Walt C 7Ac 20
Hertford Rd. IG11: Bark 38Qc 74
Hertford Rd. IG2: Ilf 30Uc 54
Hertford Rd. N1 ....39Ub 71
(not continuous)
Hertford Rd. N2 ....27Gb 49
Hertford Rd. N9 ....19Xb 33
Hertfordshire Fire Mus. 15Z 27
Hertford St. W1 ....7K 221 (46Kb 90)
Hertford Wlk. DA17: Belv 50Cd 96
Hertford Way CR4: Mitc 70Nb 134
Hertford Wharf N1 ....39Ub 71
(off Hertford Rd.)
Herts Bus. Cen. AL2: Lon C 8H 7
Hertshill Gdns. WD6: Bore 12Sa 29
Hertslet Rd. N7 ....34Pb 70
Hertsmere Ind. Pk. WD6: Bore 13Ta 29
Hertsmere Rd. E14 ....46Cc 92
Hertswood Cen. 11Sa 29
Hertswood Ct. EN5: Barn 14Ab 30
Herts Young Mariners Base 2Bc 20
Hervey Cl. N3 ....25Cb 49
Hervey Pk. Rd. E17 ....28Ac 52
Hervey Rd. SE3 ....53Kc 115
Hervey Way N3 ....25Cb 49
Hesa Rd. UB3: Hayes 44W 84
Hesewall Cl. SW4 ....54Lb 112
Hesiers Hill CR6: W'ham 89Gc 179
Hesiers Rd. CR6: W'ham 88Gc 179
Hesketh Av. DA2: Dart 60Rd 119
Hesketh Pl. W11 ....45Ab 88
Hesketh Rd. E7 ....34Jc 73
Heslop Rd. SW12 ....60Hb 111
Hesper M. SW5 ....49Db 89
Hesperus Cres. E14 ....49Dc 92
Hessel Rd. W13 ....47Ja 86
Hessel St. E1 ....44Xb 91
Hesselyn Dr. RM13: Rain 38Kd 77
Hessle Gro. KT17: Ewe 83Va 174
Hestercombe Av. SW6 ....54Ab 110
Hesterman Way CR0: Wadd 74Pb 156
Hester M. RM5: Col R 24Cd 56
Hester Rd. N18 ....22Wb 51
Hester Rd. SW11 ....52Gb 111
Hester Ter. TW9: Rich 55Qa 109
Hestia Ho. SE1 ....2H 231 (47Ub 91)
(off City Wlk.)
HESTON 52Ca 107
Heston Av. TW5: Hest 51Aa 107
Heston Cen., The TW5: Cran 50Y 85
Heston Community Sports Hall 52Ca 107
Heston Grange TW5: Hest 51Ba 107
Heston Grange La. TW5: Hest 51Ba 107
Heston Ho. SE8 ....53Cc 114
Heston Ind. Mall TW5: Hest 52Ba 107
Heston Phoenix Distribution Pk. TW5: Hest 51Y 107
Heston Pool 51Ba 107
Heston Rd. RH1: Redh 10P 207
Heston Rd. TW5: Hest 51Ca 107
HESTON SERVICE AREA 51Z 107
Heston St. SE14 ....53Cc 114
Heswell Grn. WD19: Wat 20W 26
Hetherington Cl. SL2: Slou 1D 80
Hetherington Rd. SW4 ....56Nb 112
Hetherington Rd. TW17: Shep 68S 128
Hetherington Way UB10: Ick 35N 63
Hethersett Cl. RH2: Reig 3L 207
Hethpool Ho. W2 ....6B 214 (42Fb 89)
(off Hall Pl.)
Hetley Gdns. SE19 ....66Vb 135
Hetley Rd. W12 ....46Xa 88
Heton Gdns. NW4 ....28Xa 48
Heusden Way SL9: Ger X 32B 62
Hevelius Cl. SE10 ....50Hc 93
Hever Av. TN15: W King 78Ud 164
Hever Cotts. DA12: Sole S 10G 144
Hever Ct. Rd. DA12: Grav'nd 5E 144
Hever Cft. SE9 ....63Qc 138
Hever Gdns. BR1: Broml 68Qc 138
Heverham Rd. SE18 ....49Uc 94
Hever Ho. SE15 ....51Zb 114
(off Lovelinch Cl.)
Hever Pl. KT8: E Mos 69Ea 130
Hever Rd. TN15: W King 79Ud 164
Heversham Ho. SE15 ....51Yb 114
Heversham Rd. DA7: Bex 54Cd 118
Hever Wood Rd. TN15: W King 80Ud 164
Hevingham Dr. RM6: Chad H 29Yc 55
Hevingham Vw. SE26 ....64Xb 135
Hewens Rd. UB10: Hil 42S 84
Hewens Rd. UB4: Hil 42S 84
Hewer St. W10 ....43Za 88
Hewers Way KT20: Tad 92Xa 194
Hewett Cl. HA7: Stan 21Ka 46
Hewett Pl. BR8: Swan 70Fd 140
Hewett Rd. RM8: Dag 36Zc 75
Hewetts Quay IG11: Bark 39Rc 74
Hewett St. EC2 ....6J 219 (42Ub 91)
Hewins Cl. EN9: Walt A 4Gc 21
Hewish Rd. N18 ....21Ub 51
Hewison St. E3 ....40Bc 72
Hewitt Av. N22 ....26Rb 51
Hewitt Cl. CR0: C'don 76Cc 158
Hewitt Rd. N8 ....29Qb 50
Hewitts Rd. BR6: Well H 80Bd 161
HEWITTS RDBT. 80Bd 161
Hewlett Cl. KT15: Add 76M 149
Hewlett Ho. SW8 ....52Kb 112
(off Havelock Ter.)
Hewlett Rd. E3 ....40Ac 72
Hewson Way SE17 ....5D 230 (49Sb 91)
Hew Watt Cl. RM16: Ors 3C 100
Hexagon, The N6 ....32Hb 69
Hexagon Bus. Cen. UB4: Yead 45Y 85
Hexagon Ho. RM1: Rom 29Hd 56
(off Mercury Gdns.)
Hexal Rd. SE6 ....62Gc 137
Hexham Gdns. TW7: Isle 52Ja 108
Hexham Gdns. UB5: N'olt 36Ba 65
Hexham Rd. EN5: New Bar 14Db 31
Hexham Rd. SE27 ....61Sb 135
Hexham Rd. SM4: Mord 74Db 155
HEXTABLE 66Hd 140
Hextable Heritage Cen. & Gdns. 66Gd 140
Hextalls La. CR3: Blet 100Ub 197
Hextalls La. CR3: Cat'm 100Ub 197
Hextalls La. RH1: Blet 100Tb 197
Heybourne Cres. NW9 ....25Ua 48
Heybourne Rd. N17 ....24Xb 51
Heybridge NW1 ....37Kb 70
(off Lewis St.)
Heybridge Av. SW16 ....66Nb 134
Heybridge Dr. IG6: Ilf 26Tc 54
Heybridge Way E10 ....31Ac 72
Heydon Ct. BR4: W W'ck 75Gc 159
(off Deer Pk. Way)
Heydon Ho. SE14 ....53Yb 114
(off Kender St.)
Heyford Av. SW20 ....69Bb 133
Heyford Av. SW8 ....52Nb 112
Heyford End AL2: Park 10A 6
Heyford Rd. CR4: Mitc 68Gb 133
Heyford Rd. WD7: R'lett 9Ha 14
Heyford Ter. SW8 ....52Nb 112
(off Old Sth. Lambeth Rd.)
Heygate St. SE17 ....6D 230 (49Sb 91)
Heylyn Sq. E3 ....41Bc 92
Heymede KT22: Lea 95La 192
Heynes Rd. RM8: Dag 35Yc 75
Heysham Dr. WD19: Wat 22Y 45
Heysham La. NW3 ....34Db 69
Heysham Rd. N15 ....30Tb 51
Heythorp Cl. GU21: Wok 9K 167
Heythorp St. SW18 ....60Bb 111
Heythrop Dr. UB10: Ick 35P 63
Heywood Av. NW9 ....25Ua 48
Heywood Ct. HA7: Stan 22La 46
Heywood Ho. SE14 ....51Zb 114
(off Myers La.)
Heyworth Rd. E15 ....36Hc 73
Heyworth Rd. E5 ....35Xb 71
Hibbert Av. WD24: Wat 10Z 13
Hibbert Ho. E14 ....48Cc 92
(off Tiller Rd.)
Hibbert Lodge SL9: Chal P 26A 42
Hibbert Rd. E17 ....31Bc 72
Hibbert Rd. HA3: W'stone 26Ha 46
Hibbert's All. SL4: Wind 3H 103
Hibbert St. SW11 ....55Fb 111
Hibberts Way SL9: Ger X 27A 42
Hibbs Cl. BR8: Swan 68Fd 140
Hibernia Ct. DA9: Ghithe 56Wd 120
Hibernia Dr. DA12: Grav'nd 2H 145
Hibernia Gdns. TW3: Houn 56Ca 107
Hibernia Point SE2 ....47Zc 95
(off Wolvercote Rd.)
Hibernia Rd. TW3: Houn 56Ca 107
Hibiscus Cl. HA8: Edg 21Sa 47
Hibiscus Ho. E17 ....27Dc 52
Hibiscus Ho. TW13: Felt 60W 106
Hibiscus Lodge E15 ....38Gc 73
(off Glenavon Rd.)
Hichisson Rd. SE15 ....57Yb 114
Hicken Rd. SW2 ....57Pb 112
Hickes Ho. NW6 ....38Fb 69
Hickey's Almshouses TW9: Rich 56Pa 109
Hickin Cl. SE7 ....49Mc 93
Hickin St. E14 ....48Ec 92
Hickleton NW1 ....39Lb 70
(off Camden St.)
Hickling Ho. SE16 ....48Xb 91
(off Slippers Pl.)
Hickling Rd. IG1: Ilf 36Rc 74
Hickman Av. E4 ....23Ec 52
Hickman Cl. E16 ....43Mc 93
Hickman Rd. RM6: Chad H 31Yc 75
Hickmans Cl. RH9: G'stone 4A 210
Hickman Wlk. SE18 ....50Qc 94
(off Brumwell Av.)
Hickmore Wlk. SW4 ....55Mb 112
Hickory Cl. N9 ....17Wb 33
Hicks Av. UB6: G'frd 41Fa 86
Hicks Bolton Ho. NW6 ....40Bb 69
(off Denmark Ho.)
Hicks Cl. SW11 ....55Gb 111
Hicks Ct. RM10: Dag 34Dd 76
Hicks Gallery 64Cb 133
Hicks Ho. SE16 ....48Wb 91
(off Spa Rd.)
Hicks St. SE8 ....50Ac 92
Hidcote Apts. SW11 ....56Gb 111
(off Danvers Av.)
Hidcote Cl. GU22: Wok 88D 168
Hidcote Gdns. SW20 ....69Xa 132
Hidden Cl. KT8: W Mole 70Ea 130
Hide E6 ....44Qc 94
Hideaway, The WD5: Ab L 3V 12
Hide Pl. SW1 ....6D 228 (49Mb 90)
Hider Ct. SE3 ....52Lc 115
Hide Rd. HA1: Harr 28Ea 46
Hides St. N7 ....37Pb 70
Hide Twr. SW1 ....6D 228 (49Mb 90)
(off Regency St.)
Hierro Ct. E1 ....43Ac 92
(off Ocean Est.)
Higgins Ho. N1 ....39Ub 71
(off Colville Est.)
Higginson Ho. NW3 ....38Hb 69
(off Fellows Rd.)
Higgins Wlk. TW12: Hamp 65Aa 129
(off Abbott Cl.)
Higgs Ind. Est. SE24 ....55Rb 113
High Acre Cl. KT22: Fet 96Fa 192
High Acres EN2: Enf 13Rb 33
High Acres WD5: Ab L 4T 12
HIGHAM HILL 26Ac 52
Higham Hill Rd. E17 ....25Ac 52
Higham M. UB5: N'olt 42Ba 85
Higham Path E17 ....27Ac 52
Higham Pl. E17 ....27Ac 52
Higham Rd. IG8: Wfd G 23Jc 53
Higham Rd. N17 ....27Tb 51
Highams, The E17 ....25Ec 52
Highams Ct. E4 ....20Ec 34
Highams Hill CR6: W'ham 84Jc 179
Highams La. GU24: Chob 9E 146
Highams Lodge Bus. Cen. E17 ....27Zb 52
HIGHAMS PARK 23Fc 53
Highams Park Station (Overground) 23Fc 53
Higham Sta. Av. E4 ....23Cc 52
Higham St. E17 ....27Ac 52
High Ash Cl. SS17: Linf 8J 101
High Ashton KT2: King T 66Ra 131
Highbanks Cl. DA16: Well 52Xc 117
Highbanks Rd. HA5: Hat E 23Da 45
Highbank Way N8 ....30Qb 50
HIGH BARNET 12Za 30
High Barnet Station (Underground) 14Cb 31
High Barn Rd. KT24: Eff 100Z 191
High Barn Rd. KT24: Ran C 100Z 191
Highbarns HP3: Hem H 7A 4
Highbarrow Cl. CR8: Purl 82Pb 176
Highbarrow Rd. CR0: C'don 74Wb 157
High Beech CR2: S Croy 80Ub 157
HIGH BEECH 10Kc 21
High Beech N21 ....16Pb 32
High Beeches BR6: Chels 79Wc 161
High Beeches DA14: Sidc 64Ad 139
High Beeches KT13: Weyb 79U 150
High Beeches SL9: Ger X 2P 61
High Beeches SM7: Bans 86Ya 174
High Beeches Cl. CR8: Purl 82Mb 176
High Beech Golf Course 10Kc 21
High Beech Rd. IG10: Lough 14Nc 36
High Birch Ct. EN4: E Barn 14Gb 31
(off Park Rd.)
High Bri. SE10 ....50Fc 93
Highbridge Cl. WD7: R'lett 5Ja 14
Highbridge Ct. SE14 ....52Yb 114
(off Farrow La.)
Highbridge Ind. Est. UB8: Uxb 38L 63
Highbridge Retail Pk. 6Dc 20
Highbridge Rd. IG11: Bark 39Rc 74
Highbridge St. EN9: Walt A 5Dc 20
(not continuous)
High Bri. Wharf SE10 ....50Fc 93
(off High Bri.)
Highbrook Rd. SE3 ....55Mc 115
High Broom Cres. BR4: W W'ck 73Dc 158
HIGHBURY 35Rb 71
Highbury & Islington Station (Rail, Underground & Overground) 37Rb 71
Highbury Av. CR7: Thor H 68Qb 134
Highbury Cl. BR4: W W'ck 75Dc 158
Highbury Cl. KT3: N Mald 70Sa 131
HIGHBURY CORNER 37Rb 71
Highbury Cres. N5 ....36Rb 71
Highbury Dr. KT22: Lea 93Ja 192
Highbury Est. N5 ....36Sb 71
Highbury Gdns. IG3: Ilf 33Uc 74
Highbury Gdns. N7 ....36Qb 70
Highbury Grange N5 ....35Sb 71
Highbury Gro. N5 ....36Rb 71
Highbury Gro. Ct. N5 ....37Sb 71
Highbury Hill N5 ....34Qb 70
Highbury Leisure Cen. 37Rb 71
Highbury Mans. N1 ....38Rb 71
(off Upper St.)
Highbury New Pk. N5 ....36Sb 71
Highbury Pk. N5 ....34Rb 71
Highbury Pl. N5 ....37Rb 71
Highbury Quad. N5 ....34Rb 71
Highbury Rd. SW19 ....64Ab 132
Highbury Sq. N14 ....18Lb 32
Highbury Stadium Sq. N5 ....34Rb 71
Highbury Sta. Rd. N1 ....37Qb 70
Highbury Ter. N5 ....36Rb 71
Highbury Ter. M. N5 ....36Rb 71
High Canons WD6: Bore 9Sa 15
High Cedar Dr. SW20 ....66Ya 132
Highclere SL5: S'hill 1B 146
Highclere Cl. CR8: Kenley 87Sb 177
Highclere Ct. AL1: St A 1C 6
(off Avenue Rd.)
Highclere Ct. GU21: Knap 9G 166
Highclere Dr. HP3: Hem H 6A 4
Highclere Gdns. GU21: Knap 9G 166
Highclere Rd. GU21: Knap 9G 166
Highclere Rd. KT3: N Mald 69Ta 131
Highclere St. SE26 ....63Ac 136
Highcliffe W13 ....43Ka 86
(off Clivedon Ct.)
Highcliffe Dr. SW15 ....58Va 110
Highcliffe Gdns. IG4: Ilf 29Nc 54
High Cl. WD3: Rick 15L 25
Highcombe SE7 ....51Kc 115
Highcombe Cl. SE9 ....60Mc 115
High Coombe Pl. KT2: King T 65Ta 131
Highcotts La. GU4: W Cla 98H 189
Highcroft NW9 ....29Ua 48
Highcroft Av. HA0: Wemb 38Qa 67
High Cft. Cotts. BR8: Swan 70Jd 140
Highcroft Ct. KT23: Bookh 95Ca 191
Highcroft Est. N19 ....31Nb 70
Highcroft Gdns. NW11 ....30Bb 49
Highcroft Rd. HP3: Hem H 6J 3
Highcroft Rd. N19 ....31Nb 70
Highcroft Trailer Gdns. HP3: Bov 8D 2
High Cross WD25: A'ham 9Fa 14
HIGH CROSS 9Fa 14
High Cross Cen., The N15 ....28Wb 51
Highcross Pl. KT16: Chert 74H 149
High Cross Rd. N17 ....27Wb 51
High Cross Rd. TN15: Ivy H 98Xd 204
Highcross Rd. DA13: Sflt 64Zd 143
Highcross Way SW15 ....60Wa 110
Highdaun Dr. SW16 ....70Pb 134
High Dells AL10: Hat 1B 8
Highdene GU22: Wok 90B 168
(off Fairview Av.)
Highdown KT4: Wor Pk 75Ua 154
Highdown Cl. SM7: Bans 88Bb 175
Highdown La. SM2: Sutt 83Db 175
Highdown Rd. SW15 ....58Xa 110
High Dr. CR3: Wold 94Bc 198
High Dr. KT22: Oxs 86Fa 172
High Dr. KT3: N Mald 67Sa 131
High Elms IG7: Chig 21Uc 54
High Elms IG8: Wfd G 22Jc 53
High Elms RM14: Upm 32Ud 78
High Elms Cl. HA6: Nwood 23S 44
High Elms Country Pk. 80Tc 160
High Elms Golf Course 80Sc 160
High Elms La. WD25: Wat 5Y 13
High Elms Rd. BR6: Downe 83Qc 180
HIGHER DENHAM 31E 62
Higher Dr. CR8: Purl 85Qb 176
Higher Dr. KT24: E Hor 99U 190
Higher Dr. SM7: Bans 84Za 174
Higher Grn. KT17: Eps 85Wa 174
Highfield HP8: Chal G 19A 24
Highfield SM7: Bans 89Gb 175
Highfield WD19: Wat 20Ba 27
Highfield WD23: B Hea 19Ga 28
Highfield WD25: Wat 6V 12
Highfield WD4: K Lan 10N 3
Highfield Av. BR6: Chels 78Vc 161
Highfield Av. DA8: Erith 51Dd 118
Highfield Av. HA5: Pinn 29Ba 45
Highfield Av. HA9: Wemb 34Pa 67
Highfield Av. NW11 ....31Za 68
Highfield Av. NW9 ....29Sa 47
Highfield Av. UB6: G'frd 36Ga 66
Highfield Cl. EN9: Walt A 4Jc 21
Highfield Cl. HA6: Nwood 25U 44
Highfield Cl. KT14: W Byf 85J 169
Highfield Cl. KT22: Oxs 83Fa 172
Highfield Cl. KT6: Surb 74La 152
Highfield Cl. N22 ....25Qb 50
Highfield Cl. NW9 ....29Sa 47
Highfield Cl. RM5: Col R 23Ed 56
Highfield Cl. SE13 ....58Fc 115
Highfield Cl. TW20: Eng G 5N 125
Highfield Cotts. DA2: Wilm 65Kd 141
Highfield Ct. N14 ....16Lb 32
Highfield Ct. NW11 ....30Ab 48
Highfield Ct. SL2: Farn R 9F 60
Highfield Ct. SL9: Chal P 22B 42
Highfield Ct. TW20: Eng G 5P 125
(off Highfield Rd.)
Highfield Cres. HA6: Nwood 25U 44
Highfield Cres. RM12: Horn 33Pd 77
Highfield Dr. BR2: Broml 70Gc 137
Highfield Dr. BR4: W W'ck 75Dc 158
Highfield Dr. CR3: Cat'm 94Wb 197
Highfield Dr. KT19: Ewe 79Va 154
Highfield Dr. UB10: Ick 35N 63
Highfield Gdns. NW11 ....30Ab 48
Highfield Gdns. RM16: Grays 47Fe 99
Highfield Grn. CM16: Epp 3Uc 22
Highfield Hall AL4: St A 6J 7
Highfield Hill SE19 ....66Tb 135
Highfield La. AL4: St A 4G 6
Highfield La. HP2: Hem H 1P 3
Highfield Link RM5: Col R 23Ed 56
Highfield Mnr. AL4: St A 6J 7
Highfield M. NW6 ....38Db 69
(off Compayne Gdns.)
Highfield Pk. KT15: Add 79J 149
Highfield Pk. Cen. 3G 6
Highfield Pk. Dr. AL4: St A 5F 6
Highfield Pl. CM16: Epp 3Uc 22
Highfield Rd. BR1: Broml 70Pc 138
Highfield Rd. BR7: Chst 69Vc 139
Highfield Rd. CR3: Cat'm 94Wb 197
Highfield Rd. CR8: Purl 82Pb 176
Highfield Rd. DA1: Dart 59Md 119
Highfield Rd. DA6: Bex 57Bd 117
Highfield Rd. HA6: Nwood 25U 44
Highfield Rd. IG8: Wfd G 24Nc 54
Highfield Rd. KT12: Walt T 74W 150
Highfield Rd. KT14: W Byf 85J 169
Highfield Rd. KT16: Chert 74J 149

Highfield Rd. KT5: Surb 73Sa 153
Highfield Rd. N21 19Rb 33
Highfield Rd. NW11 30Ab 48
Highfield Rd. RM12: Horn 33Pd 77
Highfield Rd. RM5: Col R 24Ed 56
Highfield Rd. SL4: Wind 5D 102
Highfield Rd. SM1: Sutt 78Gb 155
Highfield Rd. TN15: Kems'g 88Nd 183
Highfield Rd. TN16: Big H 89Lc 179
Highfield Rd. TW13: Felt 61W 128
Highfield Rd. TW16: Sun 71V 150
Highfield Rd. TW20: Eng G 5N 125
Highfield Rd. TW7: Isle 53Ha 108
Highfield Rd. W3 43Ra 87
Highfield Rd. WD23: Bush 15Aa 27
Highfield Rd. Nth. DA1: Dart 58Md 119
Highfield Rd. Sth. DA1: Dart 59Md 119
High Flds. SL5: S'dale 1D 146
Highfields KT21: Asht 91Ma 193
Highfields KT22: Fet 96Fa 192
Highfields KT24: E Hor 100V 190
Highfields SM1: Sutt 75Cb 155
Highfields WD7: R'lett 7Ha 14
Highfields Gro. N6 32Hb 69
Highfield Towers RM5: Col R 22Fd 56
Highfield Way EN6: Pot B 4Db 17
Highfield Way RM12: Horn 33Pd 77
Highfield Way WD3: Rick 16J 25
High Firs BR8: Swan 70Gd 140
High Firs WD7: R'lett 7Ha 14
High Foleys KT10: Clay 80Ka 152
High Gables BR2: Broml 68Gc 137
High Gables IG10: Lough 15Mc 35
High Gdns. GU22: Wok 1M 187
High Gth. KT10: Esh 79Ea 152
HIGHGATE 30Jb 50
Highgate Av. N6 31Kb 70
Highgate Cemetery 32Jb 70
Highgate Cl. N6 31Jb 70
Highgate Edge N2 29Gb 49
Highgate Golf Course 30Gb 49
Highgate Hgts. N6 30Lb 50
Highgate High St. N6 32Jb 70
Highgate Hill N19 32Kb 70
Highgate Hill N6 32Kb 70
HIGHGATE HOSPITAL 30Hb 49
Highgate Ho. SE26 62Wb 135
HIGHGATE MENTAL HEALTH CEN. 33Kb 70
Highgate Rd. NW5 34Jb 70
Highgate Spinney N8 30Mb 50
Highgate Station (Underground) 30Kb 50
Highgate Wlk. SE23 61Yb 136
Highgate W. Hill N6 32Jb 70
High Gro. BR1: Broml 67Mc 137
High Gro. SE18 52Tc 116
Highgrove CM15: Pil H 16Xd 40
Highgrove Cl. BR7: Chst 67Nc 138
Highgrove Cl. N11 22Jb 50
Highgrove Ct. BR3: Beck 66Cc 136
Highgrove Ct. EN8: Walt C 6Yb 20
Highgrove Ct. SM1: Sutt 79Cb 155
Highgrove Ho. CM14: B'wood 19Yd 40
(off Regency Ct.)
Highgrove Ho. HA4: Ruis 30W 44
Highgrove M. RM17: Grays 50Ee 99
Highgrove M. SM5: Cars 76Hb 155
Highgrove Pool & Fitness Cen. 30W 44
Highgrove Rd. RM8: Dag 36Yc 75
Highgrove Ter. E4 19Fc 35
Highgrove Way HA4: Ruis 30W 44
High Hill Est. E5 32Xb 71
High Hill Ferry E5 32Xb 71
High Hill Rd. CR6: W'ham 87Ec 178
High Holborn WC1 2F 223 (44Nb 90)
High Ho. La. RM18: W Til 8F 100
High Ho. M. N16 33Ub 71
High Ho. Production Pk. RM19: Purf 50Td 98
Highland Av. CM15: B'wood 18Yd 40
Highland Av. IG10: Lough 16Nc 36
Highland Av. RM10: Dag 34Ed 76
Highland Av. W7 44Ga 86
Highland Cotts. SM6: W'gton 77Lb 156
Highland Ct. BR1: Broml 67Hc 137
Highland Ct. E18 25Kc 53
Highland Cft. BR3: Beck 64Dc 136
Highland Dr. HP3: Hem H 2B 4
Highland Dr. WD23: Bush 17Da 27
Highland Pk. TW13: Felt 63Y 128
Highland Rd. BR1: Broml 67Hc 137
Highland Rd. BR2: Broml 67Hc 137
Highland Rd. CR8: Purl 86Qb 176
Highland Rd. DA6: Bex 57Cd 118
Highland Rd. HA6: Nwood 26V 44
Highland Rd. SE19 65Ub 135
Highland Rd. TN14: Bad M 82Dd 182
Highlands KT21: Asht 91La 192
Highlands N20 19Fb 31
Highlands SL2: Farn C 6G 60
Highlands WD19: Wat 18Y 27
Highlands, The EN5: New Bar 14Cb 31
Highlands, The EN6: Pot B 2Eb 17
Highlands, The HA8: Edg 26Ra 47
Highlands, The KT24: E Hor 97U 190
Highlands, The WD3: Rick 17K 25
Highlands Av. KT22: Lea 94La 192
Highlands Av. N21 15Pb 32
Highlands Av. W3 45Sa 87
Highlands Cl. KT22: Lea 94Ka 192
Highlands Cl. N4 31Nb 70
Highlands Cl. SL9: Chal P 24B 42
Highlands Cl. TW3: Houn 53Da 107
Highlands Ct. SE19 65Ub 135
Highlands End SL9: Chal P 24B 42
Highlands Farm Bus. Pk. BR8: Swan 67Jd 140
Highlands Gdns. IG1: Ilf 32Pc 74
Highlands Heath SW15 59Ya 110
Highlands Hill BR8: Swan 67Jd 140
Highlands La. GU22: Wok 93A 188
Highlands La. SL9: Chal P 24B 42
Highlands Pk. KT22: Lea 95Ma 193
Highlands Pk. TN15: Seal 93Nd 203
Highlands Rd. BR5: Orp 73Xc 161
Highlands Rd. EN5: New Bar 15Cb 31
Highlands Rd. KT22: Lea 94Ka 192
Highlands Rd. RH2: Reig 5M 207
Highland St. E15 40Dc 72
HIGHLANDS VILLAGE 15Pb 32
Highland Ter. SE13 55Dc 114
(off Algernon Rd.)
Highland Vw. Pk. Homes UB7: W Dray 49L 83
High La. CR3: Wold 91Bc 198
High La. CR6: W'ham 90Bc 178
High La. CR6: Wold 90Bc 178
High La. W7 43Fa 86
Highlawn Hall HA1: Harr 34Ga 66
Highlea Cl. NW9 24Ua 48
High Level Dr. SE26 63Wb 135
Highlever Rd. W10 43Ya 88
High Mead BR4: W W'ck 75Fc 159
High Mead HA1: Harr 29Ga 46
High Mead IG7: Chig 19Sc 36
Highmead SE18 52Vc 117
Highmead Ct. CM15: B'wood 18Zd 41
Highmead Cres. HA0: Wemb 38Pa 67
High Mdw. Cl. HA5: Eastc 28Y 45
Highmeadow Cres. NW9 29Ta 47
High Mdw. Pl. KT16: Chert 72H 149
High Mdws. IG7: Chig 22Tc 54
High Meads Rd. E16 44Mc 93
Highmore Rd. SE3 52Gc 115
High Mt. NW4 30Wa 48
High Oaks EN2: Enf 10Pb 18
High Oaks HA6: Nwood 22V 44
High Oaks Cl. CR5: Coul 91Kb 196
High Pde., The SW16 62Nb 134
High Pk. Av. KT24: E Hor 98V 190
High Pk. Av. TW9: Kew 53Qa 109
High Pk. Rd. TW9: Kew 53Qa 109
High Path SW19 67Db 133
High Pine Cl. KT13: Weyb 78S 150
High Pines CR6: W'ham 91Yb 198
High Point N6 31Jb 70
High Point SE9 62Rc 138
Highpoint KT13: Weyb 78Q 150
High Ridge N10 25Kb 50
High Ridge Cl. HP3: Hem H 7M 3
Highridge Cl. KT18: Eps 86Ua 174
Highridge Pl. EN2: Enf 10Pb 18
(off Oak Av.)
High Ridge Rd. HP3: Hem H 7M 3
High Rd. AL9: Ess 3N 9
High Rd. CM16: Epp 5Rc 22
High Rd. CR5: Chips 97Gb 195
High Rd. CR5: Coul 97Gb 195
High Rd. DA2: Wilm 62Ld 141
High Rd. E18 25Jc 53
High Rd. HA0: Wemb 36Ma 67
High Rd. HA3: Hrw W 24Ga 46
High Rd. HA5: Eastc 30W 44
High Rd. HA9: Wemb 36Na 67
High Rd. IG1: Ilf 34Rc 74
(not continuous)
High Rd. IG10: Lough 16Lc 35
High Rd. IG3: Ilf 32Vc 75
High Rd. IG7: Chig 22Qc 54
High Rd. IG9: Buck H 19Kc 35
High Rd. KT14: Byfl 84M 169
High Rd. N11 22Kb 50
High Rd. N15 28Vb 51
High Rd. N17 26Vb 51
High Rd. N2 25Pb 50
High Rd. N22 25Pb 50
High Rd. NW10 37Ua 68
High Rd. RH2: Reig 99Eb 195
High Rd. RM16: N Stif 46Zd 99
High Rd. RM16: Ors 4A 100
High Rd. RM6: Chad H 31Zc 75
High Rd. SS17: Corr 2P 101
(not continuous)
High Rd. SS17: Stan H 2P 101
(not continuous)
High Rd. UB10: Ick 34R 64
High Rd. UB4: Hayes 43U 84
High Rd. UB8: Cowl 43L 83
High Rd. WD23: B Hea 18Fa 28
High Rd. WD25: Wat 7V 12
High Rd. E. Finchley N2 25Fb 49
High Rd. Leyton E10 30Dc 52
High Rd. Leyton E15 35Ec 72
High Rd. Leytonstone E11 34Gc 73
High Rd. Leytonstone E15 35Gc 73
High Rd. Nth. Finchley N12 20Eb 31
(not continuous)
High Rd. Whetstone N20 17Eb 31
High Rd. Woodford Grn. E18 25Jc 53
High Rd. Woodford Grn. IG8: Wfd G 23Hc 53
High Sheldon N6 30Hb 49
Highshore Rd. SE15 54Vb 113
(not continuous)
High Silver IG10: Lough 14Mc 35
High Standing CR3: Cat'm 97Sb 197
Highstead Cres. DA8: Frith 53Gd 118
Highstone Av. E11 30Jc 53
Highstone Ct. E11 30Hc 53
(off New Wanstead)
Highstone Mans. NW1 38Lb 70
(off Camden Rd.)
High St. SL1: Slou Brammas Cl. 8G 80
High St. GU21: Wok Commercial Way 89A 168
High St. GU21: Wok Horsell Birch 7M 167
High St. SL1: Slou Wellington St. 6J 81
High St. SL1: Slou William St. 7K 81
(not continuous)
High St. AL2: Lon C 7G 6
High St. AL3: St A 2B 6
High St. AL4: Col H 4M 7
High St. BR1: Broml 68Jc 137
(not continuous)
High St. BR3: Beck 68Cc 136
High St. BR4: W W'ck 74Dc 158
High St. BR5: St M Cry 72Yc 161
(not continuous)
High St. BR6: Chels 80Vc 161
High St. BR6: Downe 83Qc 180
High St. BR6: Farnb 78Rc 160
High St. BR6: Orp 75Wc 161
High St. BR7: Chst 65Rc 138
High St. BR8: Swan 69Hd 140
High St. CM14: B'wood 19Yd 40
High St. CM16: Epp 3Vc 23
High St. CR0: C'don 76Sb 157
High St. CR3: Cat'm 95Ub 197
High St. CR7: Thor H 70Sb 135
High St. CR8: Purl 83Qb 176
High St. DA1: Dart 58Nd 119
High St. DA10: Swans 57Be 121
High St. DA11: Grav'nd 8D 122
High St. DA11: Nflt 58De 121
High St. DA2: Bean 62Xd 142
High St. DA4: Eyns 75Nd 163
High St. DA4: Farni 72Pd 163
High St. DA9: Ghithe 56Xd 120
High St. E11 29Jc 53
High St. E13 40Jc 73
High St. E15 40Ec 72
High St. E17 29Ac 52
High St. EN3: Pond E 16Yb 34
High St. EN5: Barn 13Ab 30
High St. EN6: Pot B 5Db 17
High St. EN8: Chesh 1Zb 20
High St. EN8: Walt C 5Ac 20
High St. GU21: Knap 9G 166
High St. GU22: Wok 93C 188
High St. GU23: Rip 93L 189
High St. GU24: Chob 3J 167
High St. GU24: W End 4D 166
High St. HA1: Harr 32Ga 66
High St. HA3: Hrw W 26Ga 46
High St. HA3: W'stone 26Ga 46
High St. HA4: Ruis 31U 64
High St. HA5: Pinn 27Aa 45
High St. HA6: Nwood 25V 44
High St. HA8: Edg 23Qa 47
High St. HA9: Wemb 35Pa 67
High St. HP1: Hem H 1L 3
High St. HP3: Bov 9C 2
High St. IG6: Ilf 27Sc 54
High St. KT1: Hamp W 67La 130
High St. KT1: King T 69Ma 131
High St. KT10: Clay 79Ha 152
High St. KT10: Esh 77Da 151
High St. KT11: Cobh 86X 171
High St. KT12: Walt T 74W 150
High St. KT13: Weyb 77Q 150
High St. KT15: Add 77K 149
High St. KT17: Eps 85Ta 173
High St. KT17: Ewe 81Va 174
High St. KT19: Eps 85Ta 173
High St. KT20: Tad 95Ya 194
High St. KT22: Lea 94Ka 192
High St. KT22: Oxs 85Fa 172
High St. KT23: Bookh 97Da 191
High St. KT3: N Mald 70Ua 132
High St. KT7: T Ditt 72Ja 152
High St. KT8: W Mole 70Ca 129
High St. N14 18Mb 32
High St. N8 28Nb 50
High St. NW7 22Xa 48
High St. RH1: Blet 5J 209
High St. RH1: Mers 100Kb 196
High St. RH1: Nutf 5F 208
High St. RH1: Redh 6P 207
High St. RH2: Reig 6J 207
High St. RH8: Limp 100Jc 199
High St. RH8: Oxt 2H 211
High St. RH9: G'stone 2A 210
High St. RM1: Rom 29Gd 56
High St. RM11: Horn 32Md 77
High St. RM12: Horn 32Md 77
High St. RM15: Avel 46Sd 98
High St. RM17: Grays 51Ce 121
(not continuous)
High St. RM19: Purf 50Qd 97
High St. SE20 65Yb 136
High St. SE25 70Vb 135
High St. SL0: Iver 44G 82
High St. SL1: Burn 1A 80
High St. SL3: Coln 52E 104
High St. SL3: Dat 3M 103
High St. SL3: L'ly 50B 82
High St. SL4: Eton 1H 103
High St. SL4: Wind 3H 103
High St. SL5: S'dale 1E 146
High St. SL5: S'hill 1B 146
High St. SL9: Chal P 25A 42
High St. SM1: Sutt 77Db 155
High St. SM3: Cheam 79Ab 154
High St. SM5: Cars 78Jb 156
High St. SM7: Bans 87Cb 175
High St. SS17: Stan H 2L 101
High St. SW19 64Za 132
High St. TN13: Chips 94Ed 202
High St. TN13: S'oaks 97Ld 203
High St. TN14: Otf 88Jd 182
High St. TN14: S'ham 82Hd 182
High St. TN15: Bor G 92Be 205
High St. TN15: Kems'g 89Rd 183
High St. TN15: Seal 93Pd 203
High St. TN15: Wro 88Ce 185
High St. TN16: Bras 96Xc 201
High St. TN16: Westrm 99Sc 200
High St. TW11: Tedd 64Ha 130
High St. TW12: Hamp 67Ea 130
High St. TW12: Hamp H 67Ea 130
High St. TW13: Felt 62V 128
High St. TW17: Shep 72R 150
High St. TW18: Staines 63G 126
(not continuous)
High St. TW19: Stanw 58M 105
High St. TW19: Wray 58A 104
High St. TW2: Whitt 59Ea 108
High St. TW20: Egh 64B 126
High St. TW3: Houn 55Da 107
High St. TW5: Cran 53W 106
High St. TW8: Bford 52La 108
High St. UB1: S'hall 46Ba 85
High St. UB3: Harl 51T 106
High St. UB7: Harm 51M 105
High St. UB7: Yiew 45M 83
High St. UB8: Cowl 42L 83
High St. UB8: Uxb 38L 63
High St. UB9: Hare 26L 43
High St. W3 46Ra 87
High St. W5 46Ma 87
High St. WD17: Wat 13X 27
High St. WD23: Bush 16Ca 27
High St. WD3: Rick 18M 25
High St. WD4: K Lan 1Q 12
High St. WD5: Ab L 3U 12
High St. WD5: Bedm 9F 4
High St. WD6: E'tree 16Ma 29
High St. Colliers Wood SW19 66Fb 133
High St. Harlesden NW10 40Va 68
High Street Kensington Station (Underground) 47Db 89
High St. M. SW19 64Ab 132
High St. Nth. E12 36Nc 74
High St. Nth. E6 39Nc 74
High St. Sth. E6 40Pc 74
High St. W. SL1: Slou 7J 81
High Timber St. EC4 4D 224 (45Sb 91)
High Tor Cl. BR1: Broml 66Kc 137
High Tor Vw. SE28 46Uc 94
High Tree Cl. CR8: Purl 82Pb 176
High Tree Cl. KT15: Add 78J 149
High Trees CR0: C'don 74Ac 158
High Trees DA2: Dart 58Rd 119
High Trees EN4: E Barn 15Gb 31
High Trees HP2: Hem H 2M 3
High Trees N20 20Eb 31
High Trees SW2 60Qb 112
High Trees Cl. CR3: Cat'm 94Vb 197
High Trees Ct. CR3: Cat'm 95Vb 197
Hightrees Ct. CM14: W'ley 21Yd 58
Hightrees Ct. W7 45Ga 86
Hightrees Ho. SW12 58Jb 112
High Trees Rd. RH2: Reig 7L 207
High Vw. AL10: Hat 2B 8
High Vw. HA5: Pinn 28Y 45
High Vw. HP8: Chal G 19A 24
High Vw. SM2: Cheam 83Bb 175
High Vw. WD18: Wat 16V 26
High Vw. WD3: Chor 14J 25
Highview CR3: Cat'm 96Ub 197
Highview GU21: Knap 9J 167
Highview N6 30Lb 50
Highview NW7 20Ta 29
Highview UB5: N'olt 41Aa 85
High Vw. Av. RM17: Grays 50Ee 99
Highview Av. HA8: Edg 21Sa 47
Highview Av. SM6: W'gton 78Pb 156
High Vw. Cl. IG10: Lough 15Lc 35
High Vw. Cl. SE19 68Vb 135
Highview Cl. EN6: Pot B 5Eb 17
Highview Cl. KT20: Tad 93Wa 194
High Vw. Ct. HA3: Hrw W 24Ga 46
Highview Ct. IG10: Lough 15Mc 35
(off High Rd.)
Highview Ct. RH2: Reig 6M 207
(off Wray Comn. Rd.)
Highview Cres. CM13: Hut 16Ee 41
High Vw. Gdns. RM17: Grays 50Ee 99
Highview Gdns. EN6: Pot B 5Eb 17
Highview Gdns. HA8: Edg 21Sa 47
Highview Gdns. N11 22Lb 50
Highview Gdns. N3 27Ab 48
Highview Gdns. RM14: Upm 33Rd 77
Highview Ho. RM6: Chad H 28Ad 55
Highview Lodge EN2: Enf 13Rb 33
(off The Ridgeway)
High Vw. Pde. IG4: Ilf 29Pc 54
High Vw. Pk. WD4: K Lan 10D 4
Highview Path SM7: Bans 87Cb 175
High Vw. Rd. BR6: Downe 82Qc 180
High Vw. Rd. E18 26Hc 53
High Vw. Rd. SE19 65Tb 135
Highview Rd. DA14: Sidc 63Xc 139
Highview Rd. W13 43Ja 86
Highway, The BR6: Chels 78Xc 161
Highway, The E1 45Wb 91
Highway, The HA7: Stan 25Ha 46
Highway, The SM2: Sutt 81Eb 175
Highwayman's Ridge GU20: W'sham 7A 146
Highway Trad. Cen., The E1 45Zb 92
(off Heckford St.)
Highwold CR5: Chips 90Jb 176
Highwood BR2: Broml 69Fc 137
Highwood Av. N12 21Eb 49
Highwood Av. WD23: Bush 11Ba 27
Highwood Cl. BR6: Farnb 75Sc 160
Highwood Cl. CM14: B'wood 17Xd 40
Highwood Cl. CR8: Kenley 89Sb 177
Highwood Cl. SE22 60Wb 113
Highwood Cl. EN5: New Bar 15Cb 31
Highwood Ct. N12 20Eb 31
Highwood Dr. BR6: Farnb 75Sc 160
Highwood Gdns. IG5: Ilf 29Pc 54
Highwood Gro. NW7 22Ta 47
Highwoodhall La. HP3: Hem H 7A 4
Highwood Hill NW7 19Va 30
HIGHWOOD HILL 20Va 30
Highwood La. IG10: Lough 15Qc 36
Highwood Rd. N19 34Nb 70
Highwoods CR3: Cat'm 97Ub 197
Highwoods KT22: Lea 93La 192
High Worple HA2: Harr 31Ba 65
Highworth Rd. N11 23Mb 50
Highworth St. NW1 7E 214 (43Gb 89)
(off Daventry St.)
Hi-Gloss Cen. SE8 50Ac 92
Hilary Av. CR4: Mitc 69Jb 134
Hilary Cl. DA8: Erith 53Dd 118
Hilary Cl. RM12: Horn 36Md 77
Hilary Cl. SW6 52Db 111
Hilary Dennis Ct. E11 28Jc 53
Hilary M. SE1 1E 230 (47Sb 91)
Hilary Rd. W12 44Va 88
(not continuous)
Hilberry Ct. WD23: Bush 17Da 27
Hilbert Rd. SM3: Cheam 76Za 154
Hilborough Cl. SW19 66Eb 133
Hilborough Ct. E8 38Vb 71
Hilborough Way BR6: Farnb 78Tc 160
Hilbury AL10: Hat 1B 8
Hilda Ct. KT6: Surb 73Ma 153
Hilda Lockert Wlk. SW9 54Rb 113
(off Loughborough Rd.)
Hilda May Av. BR8: Swan 68Gd 140
Hilda Rd. E16 42Gc 93
Hilda Rd. E6 38Mc 73
Hilda Rd. UB2: S'hall 47Ea 86
Hilda Ter. SW9 54Qb 112
Hilda Va. Cl. BR6: Farnb 77Rc 160
Hilda Va. Rd. BR6: Farnb 77Qc 160
Hildenborough Gdns. BR1: Broml 65Gc 137
Hildenborough Ho. BR3: Beck 66Bc 136
(off Bethersden Cl.)
Hilden Dr. DA8: Erith 52Kd 119
Hildenlea Pl. BR2: Broml 68Fc 137
Hildenley Cl. RH1: Mers 100Mb 196
Hilderley Ho. KT1: King T 69Pa 131
(off Winery La.)
Hilders, The KT21: Asht 89Ra 173
Hildreth St. SW12 60Kb 112
Hildreth St. M. SW12 60Kb 112
Hildyard Rd. SW6 51Cb 111
Hiley Rd. NW10 41Ya 88
Hilfield La. WD25: A'ham 11Da 27
Hilfield La. Sth. WD23: Bush 16Ha 28
Hilgrove Rd. NW6 38Eb 69
Hiliary Gdns. HA7: Stan 26La 46
Hiljon Cres. SL9: Chal P 25A 42
Hill, The CR3: Cat'm 96Vb 197
Hill, The DA11: Nflt 58Ee 121
Hillacre CR3: Cat'm 97Ub 197
Hillars Heath Rd. CR5: Coul 87Nb 176
Hillary N8 27Nb 50
(off Boyton Cl.)
Hillary Av. DA11: Nflt 2A 144
Hillary Ct. W12 47Ya 88
(off Titmuss St.)
Hillary Cres. KT12: Walt T 74Y 151
Hillary Dr. TW7: Isle 57Ha 108
Hillary Ho. WD6: Bore 13Ra 29
(off Eldon Av.)
Hillary M. E14 43Dc 92
Hillary Ri. EN5: New Bar 14Cb 31
Hillary Rd. HP2: Hem H 1A 4
Hillary Rd. SL3: L'ly 47A 82
Hillary Rd. UB2: S'hall 48Ca 85
Hill Barn CR2: Sande 83Ub 177
Hillbeck Cl. SE15 52Yb 114
(not continuous)
Hillbeck Way UB6: G'frd 39Fa 66
Hillborne Cl. UB3: Harl 50W 84
Hillboro Ct. E11 31Fc 73
Hillborough Av. TN13: S'oaks 94Md 203
Hillbrook Gdns. KT13: Weyb 80Q 150
Hillbrook Rd. SW17 62Hb 133
Hill Brow BR1: Broml 67Mc 137
Hill Brow DA1: Cray 58Hd 118
Hillbrow KT3: N Mald 69Va 132
Hillbrow RH2: Reig 6L 207
Hillbrow Cl. DA5: Bexl 63Fd 140
Hillbrow Cotts. RH9: G'stone 4A 210
Hillbrow Ct. RH9: G'stone 4A 210
Hillbrow Rd. BR1: Broml 66Gc 137
Hillbrow Rd. KT10: Esh 77Ea 152
Hillbury Av. HA3: Kenton 29Ka 46
Hillbury Cl. CR6: W'ham 90Yb 178
Hillbury Cres. CR6: W'ham 90Yb 178
Hillbury Gdns. CR6: W'ham 90Yb 178
Hillbury Rd. CR3: Whyt 89Wb 177
Hillbury Rd. CR6: W'ham 89Wb 177
Hillbury Rd. SW17 62Kb 134
Hill Cl. BR7: Chst 64Rc 138
Hill Cl. CR8: Purl 85Sb 177
Hill Cl. DA13: Ist R 6A 144
Hill Cl. EN5: Barn 15Ya 30
Hill Cl. GU21: Wok 8P 167
Hill Cl. HA1: Harr 34Ga 66
Hill Cl. HA7: Stan 21Ka 46
Hill Cl. KT11: Cobh 84Ca 171
Hill Cl. NW11 30Cb 49
Hill Cl. NW2 34Xa 68
Hill Comn. HP3: Hem H 6A 4
Hillcote Av. SW16 66Qb 134
Hill Ct. EN4: E Barn 14Gb 31
Hill Ct. EN6: Pot B 6Eb 17
Hill Ct. RM1: Rom 28Hd 56
Hill Ct. UB5: N'olt 36Ca 65
Hill Ct. W5 42Pa 87
Hillcourt Av. N12 23Db 49
Hillcourt Est. N16 32Tb 71
Hillcourt Rd. SE22 58Xb 113
Hill Cres. DA5: Bexl 60Ed 118
Hill Cres. HA1: Harr 29Ja 46
Hill Cres. KT4: Wor Pk 75Ya 154
Hill Cres. KT5: Surb 71Pa 153
Hill Cres. N20 19Db 31
Hill Cres. RM11: Horn 30Ld 57
Hill Crest DA15: Sidc 59Wc 117
Hill Crest EN6: Pot B 6Eb 17
Hill Crest KT6: Surb 73Na 153
Hill Crest TN13: S'oaks 94Jd 202
Hillcrest AL10: Hat 1C 8
Hillcrest AL3: St A 4P 5
Hillcrest KT13: Weyb 77R 150
Hillcrest N21 17Qb 32
Hillcrest N6 31Jb 70
Hillcrest SE24 56Tb 113
Hillcrest W11 45Bb 89
(off St John's Gdns.)
Hillcrest Av. HA5: Pinn 28Z 45
Hillcrest Av. HA8: Edg 21Ra 47
Hillcrest Av. KT16: Chert 76G 148
Hillcrest Av. NW11 29Bb 49
Hillcrest Av. RM20: W Thur 51Wd 120
Hillcrest Cl. BR3: Beck 72Bc 158
Hillcrest Cl. EN7: G Oak 1Sb 19
Hillcrest Cl. KT18: Eps 87Va 174
Hillcrest Cl. SE26 63Wb 135
Hillcrest Ct. KT13: Weyb 77R 150
Hillcrest Ct. RM5: Col R 25Fd 56
Hillcrest Ct. SM2: Sutt 79Fb 155
(off Eaton Rd.)
Hillcrest Dr. DA9: Ghithe 57Wd 120
Hillcrest Gdns. KT10: Hin W 76Ha 152
Hillcrest Gdns. N3 28Ab 48
Hillcrest Gdns. NW2 34Wa 68
Hillcrest Pde. CR5: Coul 86Kb 176
Hillcrest Rd. BR1: Broml 64Jc 137
Hillcrest Rd. BR6: Chels 75Wc 161
Hillcrest Rd. CR3: Whyt 89Vb 177
Hillcrest Rd. CR8: Purl 82Pb 176
Hillcrest Rd. DA1: Dart 59Gd 118
Hillcrest Rd. E17 26Fc 53
Hillcrest Rd. E18 26Hc 53
Hillcrest Rd. IG10: Lough 16Mc 35
Hillcrest Rd. RM11: Horn 31Jd 76
Hillcrest Rd. TN16: Big H 88Mc 179
Hillcrest Rd. W3 46Ra 87
Hillcrest Rd. W5 43Na 87
Hillcrest Rd. WD7: Shenl 5Qa 15
Hillcrest Vw. BR3: Beck 72Bc 158
Hillcrest Way CM16: Epp 3Wc 23
Hillcrest Waye SL9: Ger X 31B 62
Hill Cft. WD7: R'lett 5Ja 14
Hillcroft IG10: Lough 12Qc 36
Hillcroft Av. CR8: Purl 85Lb 176
Hillcroft Av. HA5: Pinn 30Ba 45
Hillcroft Ct. CR3: Cat'm 95Ub 197
Hillcroft Cres. HA4: Ruis 34Z 65
Hillcroft Cres. HA9: Wemb 35Pa 67
Hillcroft Cres. W5 44Ma 87
Hillcroft Cres. WD19: Wat 18X 27
Hillcroft Rd. E6 43Rc 94
Hillcroome Rd. SM2: Sutt 79Fb 155
Hillcross Av. SM4: Mord 72Za 154
Hilldale Rd. SM1: Sutt 77Bb 155
Hilldeane Rd. CR8: Purl 81Qb 176
Hilldene Av. RM3: Rom 23Ld 57
Hilldene Cl. RM3: Rom 22Md 57
Hilldown Ct. SW16 66Nb 134
Hilldown Rd. BR2: Hayes 74Gc 159
Hilldown Rd. SW16 66Nb 134
Hill Dr. NW9 32Sa 67
Hill Dr. SW16 69Pb 134
Hilldrop Cres. N7 36Mb 70
Hilldrop Est. N7 36Mb 70
(not continuous)
Hilldrop La. N7 36Mb 70
Hilldrop Rd. BR1: Broml 65Kc 137
Hilldrop Rd. N7 36Mb 70
Hille Bus. Cen. WD24: Wat 11X 27
Hill End BR6: Orp 75Vc 161
HILL END 23K 43
Hillend SE18 53Qc 116
Hillend Grn. NW7 21Ya 48
Hill End La. AL4: St A 5F 6
Hill End Rd. UB9: Hare 24L 43
Hillersden Ho. SW1 7K 227 (50Kb 90)
(off Ebury Bri. Rd.)
Hillersdon SL2: Slou 3M 81
Hillersdon Av. HA8: Edg 22Pa 47
Hillersdon Av. SW13 54Wa 110
Hillery Cl. SE17 6G 231 (49Tb 91)
Hilley Fld. La. KT22: Fet 94Ea 192
Hill Farm Av. WD25: Wat 5W 12
(not continuous)
Hill Farm Cl. WD25: Wat 5W 12
Hill Farm Cotts. HA4: Ruis 31S 64
Hill Farm Ind. Est. WD25: Wat 5W 12
Hill Farm Rd. SL9: Chal P 24A 42
Hill Farm Rd. UB10: Ick 35T 64
Hill Farm Rd. W10 43Ya 88
Hillfield Av. HA0: Wemb 38Na 67
Hillfield Av. N8 29Nb 50
Hillfield Av. NW9 29Ua 48
Hillfield Av. SM4: Mord 72Gb 155
Hillfield Cl. HA2: Harr 28Ea 46
Hillfield Cl. RH1: Redh 6A 208
Hillfield Ct. HP2: Hem H 2N 3
Hillfield Ct. KT10: Esh 78Da 151
Hillfield Ct. NW3 36Gb 69
Hillfield Ho. N5 36Sb 71

Hillfield M. N8 28Pb 50
Hillfield Pde. SM4: Mord 72Fb 155
Hillfield Pk. N10 28Kb 50
Hillfield Pk. N21 19Qb 32
Hillfield Pk. M. N10 28Kb 50
Hillfield Pl. TN13: Dun G 92Gd 202
Hill Fld. Rd. TW12: Hamp 66Ba 129
Hillfield Rd. HP2: Hem H 2M 3
Hillfield Rd. NW6 36Bb 69
Hillfield Rd. RH1: Redh 6A 208
Hillfield Rd. SL9: Chal P 24A 42
Hillfield Rd. TN13: Dun G 92Gd 202
Hillfield Sq. SL9: Chal P 24A 42
Hillfoot Av. RM5: Col R 25Ed 56
Hillfoot Rd. RM5: Col R 25Ed 56
Hillgate Pl. SW12 59Kb 112
Hillgate Pl. W8 46Cb 89
Hillgate St. W8 46Cb 89
Hill Ga. Wlk. N6 30Lb 50
Hillground Gdns. CR2: S Croy 82Rb 177
Hill Gro. RM1: Rom 27Gd 56
Hill Gro. TW13: Hanw 61Ba 129
Hillgrove SL9: Chal P 25A 42
Hill Hall 8Bd 23
Hillhampton Pl. SL5: S'dale 3D 146
Hill Ho. BR2: Broml 68Hc 137
Hill Ho. E5 32Xb 71
(off Harrington Hill)
Hill Ho. N19 33Lb 70
(off Highgate Hill)
Hill Ho. SE28 46Tc 94
Hillhouse EN9: Walt A 5Hc 21
Hill Ho. Apts. N1 2J 217 (40Pb 70)
(off Pentonville Rd.)
Hill Ho. Av. HA7: Stan 24Ha 46
Hill Ho. Cl. N21 17Qb 32
Hill Ho. Cl. SL9: Chal P 24A 42
Hill Ho. Dr. KT13: Weyb 83Q 170
Hill Ho. Dr. RH2: Reig 8K 207
Hill Ho. Dr. RM16: Grays 10E 100
Hill Ho. Dr. TW12: Hamp 67Ca 129
Hill Ho. M. BR2: Broml 68Hc 137
Hill Ho. Rd. DA2: Dart 59Sd 120
Hill Ho. Rd. SW16 64Pb 134
Hillhurst Gdns. CR3: Cat'm 92Ub 197
Hilliard Ho. E1 46Xb 91
(off Prusom St.)
Hilliard Rd. HA6: Nwood 25V 44
Hilliards Ct. E1 46Yb 92
Hilliards Rd. UB8: Cowl 44M 83
Hillier Cl. EN5: New Bar 16Db 31
Hillier Gdns. CR0: Wadd 78Qb 156
Hillier Ho. NW1 38Mb 70
(off Camden Sq.)
Hillier Lodge TW11: Tedd 64Fa 130
Hillier Pl. KT9: Chess 79Ma 153
Hillier Rd. SW11 58Hb 111
Hilliers Av. UB8: Hil 41Q 84
Hilliers La. CR0: Bedd 76Nb 156
Hillier Way SL3: L'ly 49C 82
Hillingdale TN16: Big H 90Kc 179
HILLINGDON 41Q 84
Hillingdon Athletic Club 29T 44
Hillingdon Athletics Stadium 37N 63
Hillingdon Av. TN13: S'oaks 93Ld 203
Hillingdon Av. TW19: Stanw 60N 105
HILLINGDON CIRCUS 37R 64
Hillingdon Ct. HA3: Kenton 28Ma 47
Hillingdon Cycle Circuit 46Y 85
Hillingdon Golf Course 40P 63
HILLINGDON HEATH 42R 84
Hillingdon Hill UB10: Hil 40N 63
HILLINGDON HOSPITAL 43P 83
Hillingdon Outdoor Activities Cen. 31L 63
Hillingdon Pde. UB10: Hil 42R 84
(off Uxbridge Rd.)
Hillingdon Ri. TN13: S'oaks 94Md 203
Hillingdon Rd. DA11: Grav'nd 1D 144
Hillingdon Rd. DA7: Bex 54Ed 118
Hillingdon Rd. UB10: Uxb 39N 63
Hillingdon Rd. UB8: Uxb 39M 63
Hillingdon Rd. WD25: Wat 6W 12
Hillingdon Sports & Leisure Complex 37N 63
Hillingdon Station (Underground) 36R 64
Hillingdon St. SE17 51Rb 113
Hillington Gdns. IG8: Wfd G 26Mc 53
Hill La. HA4: Ruis 32S 64
Hill La. KT20: Kgswd 93Ab 194
Hill Ley AL10: Hat 1B 8
Hillman Cl. RM11: Horn 27Md 57
Hillman Cl. UB8: Uxb 36N 63
Hillman Dr. W10 42Ya 88
Hillman St. E8 37Xb 71
Hillmarton Rd. N7 36Nb 70
Hillmarton Ter. N7 36Nb 70
(off Hillmarton Rd.)
Hillmead Dr. SW9 56Rb 113
Hillmont Rd. KT10: Hin W 76Ga 152
Hillmore Ct. SE13 55Fc 115
(off Belmont Hill)
Hillmore Gro. SE26 64Ac 136
Hillmount GU22: Wok 90A 168
(off Constitution Hill)
Hill Pk. 94Nc 200
HILL PARK 95Rc 200
Hill Pk. Ct. KT22: Lea 91Ha 192
Hill Pk. Dr. KT22: Lea 91Ha 192
Hill Pk. Sth. KT22: Lea 92Ha 192
Hill Path SW16 64Pb 134
Hill Pl. SL2: Farn C 8F 60
Hillpoint WD3: Loud 15L 25
Hillreach SE18 50Pc 94
Hill Ri. DA2: Daren 64Td 142
Hill Ri. EN6: Pot B 6Eb 17
Hill Ri. HA4: Ruis 32S 64
Hill Ri. KT10: Hin W 75Ka 152
Hill Ri. N9 16Xb 33
Hill Ri. NW11 28Db 49
Hill Ri. RM14: Upm 33Qd 77
Hill Ri. SE23 60Xb 113
Hill Ri. SL9: Chal P 26A 42
Hill Ri. TW10: Rich 57Ma 109
Hill Ri. UB6: G'frd 38Ea 66
Hill Ri. WD3: Rick 16K 25
Hillrise KT12: Walt T 73V 150
Hillrise SL3: L'ly 51C 104
Hillrise Av. WD24: Wat 10Z 13
Hill Rise Ct. KT22: Lea 93Ka 192
(off Park Rise)
Hill Rise Cres. SL9: Chal P 26A 42
Hillrise Mans. N19 31Nb 70
(off Warltersville Rd.)
Hillrise Rd. N19 31Nb 70
Hillrise Rd. RM5: Col R 23Ed 56
Hill Rd. CM14: B'wood 20Wd 40
Hill Rd. CM16: They B 10Uc 22
Hill Rd. CR4: Mitc 67Kb 134
Hill Rd. CR8: Purl 84Pb 176
Hill Rd. DA2: Wilm 61Nd 141
Hill Rd. HA0: Wemb 34Ka 66
Hill Rd. HA1: Harr 29Ja 46
Hill Rd. HA5: Pinn 29Aa 45
Hill Rd. HA6: Nwood 23T 44
Hill Rd. KT22: Fet 94Da 191
Hill Rd. N10 25Hb 49
Hill Rd. NW8 2A 214 (40Eb 69)
Hill Rd. SM1: Sutt 78Db 155
Hill Rd. SM5: Cars 79Gb 155
Hillsboro' Rd. SE22 57Ub 113
Hillsborough Ct. NW6 39Db 69
(off Mortimer Cres.)
Hillsborough Grn. WD19: Wat 20W 26
Hill's Chace CM14: W'ley 21Yd 58
Hillsgrove Cl. DA16: Well 52Yc 117
Hillside AL10: Hat 1C 8
HILLSIDE 49Ed 96
Hillside DA2: Daren 64Ud 142
Hillside DA4: Farni 73Pd 163
Hillside DA8: Erith 49Ed 96
Hillside EN5: New Bar 15Eb 31
Hillside GU22: Wok 2P 187
Hillside GU25: Vir W 2N 147
Hillside KT10: Esh 78Da 151
Hillside N8 30Mb 50
Hillside NW10 38Sa 67
Hillside NW5 34Jb 70
Hillside NW9 28Ta 47
Hillside RM17: Grays 49Fe 99
Hillside RM3: Rom 21Md 57
Hillside SE10 52Fc 115
(off Croom's Hill)
Hillside SL1: Slou 7J 81
Hillside SL5: S'hill 1A 146
Hillside SM7: Bans 87Ab 174
Hillside SW19 65Za 132
Hillside UB9: Hare 29L 43
Hillside, The BR6: Prat B 81Xc 181
Hillside Av. CR8: Purl 85Rb 177
Hillside Av. DA12: Grav'nd 1F 144
Hillside Av. EN8: Chesh 3Zb 20
Hillside Av. HA9: Wemb 35Pa 67
Hillside Av. IG8: Wfd G 23Lc 53
Hillside Av. N11 23Hb 49
Hillside Av. SE10 53Ec 114
Hillside Av. WD6: Bore 14Ra 29
Hillside Cl. GU21: Knap 9H 167
Hillside Cl. IG8: Wfd G 22Lc 53
Hillside Cl. NW8 40Db 69
Hillside Cl. SL9: Chal P 23A 42
Hillside Cl. SM4: Mord 70Ab 132
Hillside Cl. SM7: Bans 88Ab 174
Hillside Cl. WD5: Ab L 4U 12
Hillside Cotts. HP3: Hem H 3C 4
Hillside Ct. AL1: St A 1C 6
(off Hillside Rd.)
Hillside Ct. BR8: Swan 70Jd 140
Hillside Ct. EN8: Chesh 3Zb 20
Hillside Cres. EN2: Enf 10Tb 19
Hillside Cres. EN8: Chesh 3Zb 20
Hillside Cres. HA2: Harr 32Ea 66
Hillside Cres. HA6: Nwood 25W 44
Hillside Cres. WD19: Wat 16Aa 27
Hillside Dr. DA12: Grav'nd 1F 144
Hillside Dr. HA8: Edg 23Qa 47
Hillside Gdns. E17 27Fc 53
Hillside Gdns. EN5: Barn 14Ab 30
Hillside Gdns. HA3: Kenton 31Na 67
Hillside Gdns. HA6: Nwood 24W 44
Hillside Gdns. HA8: Edg 21Pa 47
Hillside Gdns. HP4: Berk 2A 2
Hillside Gdns. KT15: Add 78H 149
Hillside Gdns. N11 23Lb 50
Hillside Gdns. N6 30Kb 50
Hillside Gdns. SM6: W'gton 80Lb 156
Hillside Gdns. SW2 61Qb 134
Hillside Ga. AL1: St A 1C 6
(off Hillside Rd.)
Hillside Gro. N14 17Mb 32
Hillside Gro. NW7 24Wa 48
Hillside Ho. CR0: Wadd 77Rb 157
(off Duppas Av.)
Hillside La. BR2: Hayes 75Hc 159
(not continuous)
Hillside Mans. EN5: Barn 14Bb 31
Hillside Mans. WD23: Bush 18Ca 27
Hillside Pk. SL5: S'dale 4D 146
Hillside Pas. SW2 61Pb 134
Hillside Path CR5: Coul 90Nb 176
Hillside Ri. HA6: Nwood 24W 44
Hillside Rd. AL1: St A 1C 6
Hillside Rd. BR2: Broml 69Hc 137
Hillside Rd. CR0: Wadd 78Rb 157
Hillside Rd. CR3: Whyt 90Wb 177
Hillside Rd. CR5: Coul 90Nb 176
Hillside Rd. DA1: Cray 58Jd 118
Hillside Rd. HA5: Pinn 24X 45
Hillside Rd. HA6: Nwood 24W 44
Hillside Rd. KT17: Ewe 82Ya 174
Hillside Rd. KT21: Asht 89Pa 173
Hillside Rd. KT5: Surb 70Pa 131
Hillside Rd. N15 31Ub 71
Hillside Rd. SM2: Sutt 80Bb 155
Hillside Rd. SW2 61Pb 134
Hillside Rd. TN13: S'oaks 95Md 203
Hillside Rd. TN15: Kems'g 89Pd 183
Hillside Rd. TN16: Tats 91Nc 200
Hillside Rd. UB1: S'hall 42Ca 85
Hillside Rd. W5 43Na 87
Hillside Rd. WD23: Bush 15Aa 27
Hillside Rd. WD3: Chor 15E 24
Hillside Rd. WD7: R'lett 7Ka 14
Hillside Wlk. CM14: B'wood 20Vd 40
Hills La. HA6: Nwood 25U 44
Hills La. TN15: Knat 85Rd 183
Hillsleigh Rd. W8 46Bb 89
Hillsmead Way CR2: Sande 85Wb 177
Hills M. W5 45Na 87
Hills Pl. W1 3B 222 (44Lb 90)
Hills Rd. IG9: Buck H 18Kc 35
Hillstone Ct. E3 42Dc 92
(off Empson St.)
Hillstowe St. E5 34Yb 72
Hill St. AL3: St A 2A 6
Hill St. TW9: Rich 57Ma 109
Hill St. W1 6J 221 (46Jb 90)
Hillswood Dr. KT16: Chert 77D 148
Hillthorpe Cl. CR8: Purl 82Pb 176
Hill Top IG10: Lough 12Qc 36
Hill Top NW11 28Db 49
Hill Top SM3: Sutt 73Bb 155
Hill Top SM4: Mord 72Cb 155
Hilltop E17 27Dc 52
Hilltop Av. NW10 38Sa 67
(not continuous)
Hill Top Cl. IG10: Lough 13Qc 36
Hilltop Cl. KT22: Lea 95La 192
Hilltop Cl. SL5: Asc 8C 124
Hill Top Ct. IG8: Wfd G 23Pc 54
Hilltop Ct. NW8 38Eb 69
(off Alexandra Rd.)
Hilltop Farm WD4: K Lan 9D 4
Hilltop Gdns. BR6: Orp 75Uc 160
Hilltop Gdns. DA1: Dart 57Pd 119
Hilltop Gdns. NW4 26Xa 48
Hilltop Ho. N6 31Mb 70
Hilltop La. CR3: Cat'm 98Qb 196
Hilltop La. RH1: Mers 98Qb 196
Hill Top Pl. IG10: Lough 13Qc 36
Hilltop Ri. KT23: Bookh 98Ea 192
Hilltop Rd. CR3: Whyt 89Ub 177
Hilltop Rd. NW6 38Cb 69
Hilltop Rd. RH2: Reig 8K 207
Hilltop Rd. RM20: W Thur 51Xd 120
Hilltop Rd. WD4: K Lan 9D 4
Hill Top Vw. IG8: Wfd G 23Pc 54
Hilltop Wlk. CR3: Wold 92Ac 198
Hilltop Way HA7: Stan 20Ja 28
Hill Vw. CR3: Whyt 89Vb 177
Hill Vw. DA1: Dart 57Pd 119
Hill Vw. GU24: Chob 7G 146
Hill Vw. NW3 39Hb 69
(off Ainger Rd.)
Hill Vw. SL2: Hedg 2H 61
Hill Vw. TN15: Bor G 92Ce 205
Hillview SW20 66Xa 132
Hillview TN15: Bor G 95Be 205
Hillview Av. HA3: Kenton 29Na 47
Hillview Av. RM11: Horn 30Ld 57
Hill Vw. Cl. KT20: Tad 93Ya 194
Hill Vw. Cl. TN15: Bor G 92Ce 205
Hillview Cl. CR8: Purl 83Rb 177
Hillview Cl. HA5: Hat E 23Ba 45
Hillview Cl. HA9: Wemb 33Pa 67
Hill Vw. Ct. GU22: Wok 90B 168
Hillview Cres. BR6: Orp 74Uc 160
Hillview Cres. IG1: Ilf 30Pc 54
Hill Vw. Dr. DA16: Well 54Uc 116
Hill Vw. Dr. SE28 46Uc 94
Hillview Dr. RH1: Redh 7A 208
Hill Vw. Gdns. NW9 29Ta 47
Hillview Gdns. HA2: Harr 27Ca 45
Hillview Gdns. NW4 28Za 48
Hillview Ho. DA12: Grav'nd 10E 122
Hill Vw. Pl. KT11: Cobh 85Ba 171
Hill Vw. Rd. DA3: Lfield 69De 143
Hill Vw. Rd. GU22: Wok 90B 168
Hill Vw. Rd. KT10: Clay 80Ja 152
Hill Vw. Rd. TW1: Twick 58Ja 108
Hill Vw. Rd. TW19: Wray 8P 103
Hillview Rd. BR6: Orp 74Vc 161
Hillview Rd. BR7: Chst 64Qc 138
Hillview Rd. HA5: Hat E 24Ba 45
Hillview Rd. NW7 21Za 48
Hillview Rd. SM1: Sutt 76Eb 155
Hillway N6 33Jb 70
Hillway NW9 32Ua 68
Hillway, The CM15: Mount 11Fe 41
Hill Waye SL9: Ger X 30B 42
Hillwood Cl. CM13: Hut 18De 41
Hillwood Gro. CM13: Hut 18De 41
Hill-Wood Ho. NW1 2C 216 (40Lb 70)
(off Polygon Rd.)
Hillworth BR3: Beck 68Dc 136
Hillworth Rd. SW2 59Qb 112
Hillyard Ho. SW9 53Qb 112
Hillyard Pl. SW20 66Xa 132
Hillyard Rd. W7 43Ga 86
Hillyard St. SW9 53Qb 112
Hillydeal Rd. TN14: Otf 87Ld 183
Hillyfield E17 26Ac 52
Hillyfield Cl. E9 36Ac 72
Hillyfields IG10: Lough 12Qc 36
Hilly Flds. Cres. SE13 55Cc 114
Hilly Flds. Cres. SE4 55Cc 114
Hilmay Dr. HP1: Hem H 3K 3
Hilperton Rd. SL1: Slou 7J 81
Hilsea Point SW15 60Xa 110
Hilsea St. E5 35Yb 72
Hilton Av. N12 22Fb 49
Hilton Cl. UB8: Uxb 40K 63
Hilton Ho. SE4 56Zb 114
Hilton Ho. W13 44La 86
Hilton Way CR2: Sande 87Xb 177
Hilversum Cres. SE22 57Ub 113
Himalayan Way WD18: Wat 16V 26
Himalaya Palace Cinema 46Ba 85
Himley Rd. SW17 64Gb 133
Hinchinbrook Ho. NW6 39Db 69
(off Mortimer Cres.)
Hinchley Cl. KT10: Hin W 77Ha 152
Hinchley Dr. KT10: Hin W 76Ha 152
Hinchley Mnr. KT10: Hin W 76Ha 152
Hinchley Way KT10: Hin W 76Ja 152
HINCHLEY WOOD 76Ha 152
Hinchley Wood Memorial Gdns. 76Ha 152
(off Station App.)
Hinchley Wood Station (Rail) 76Ha 152
Hinckley Rd. SE15 56Wb 113
Hind Cl. IG7: Chig 22Vc 55
Hind Ct. EC4 3A 224 (44Qb 90)
Hind Cres. DA8: Erith 51Fd 118
Hinde Ho. W1 2J 221 (44Jb 90)
(off Hinde St.)
Hinde M. W1 2J 221 (44Jb 90)
(off Marlebone La.)
Hindes Rd. HA1: Harr 29Fa 46
Hinde St. W1 2J 221 (44Jb 90)
Hind Gro. E14 44Cc 92
Hindhead Cl. N16 32Ub 71
Hindhead Cl. UB8: Hil 43R 84
Hindhead Gdns. UB5: N'olt 39Aa 65
Hindhead Grn. WD19: Wat 22Y 45
Hindhead Point SW15 60Xa 110
Hindhead Way SM6: W'gton 78Nb 156
Hind Ho. N7 35Qb 70
Hind Ho. SE14 51Zb 114
(off Myers La.)
Hindle Ho. E8 36Vb 71
Hindlip Ho. SW8 53Mb 112
Hindmans Rd. SE22 57Wb 113
Hindmans Way RM9: Dag 42Bd 95
Hindmarsh Cl. E1 45Wb 91
Hindmarsh Cres. DA11: Nflt 61Ee 143
Hindon Ct. SW1 5B 228 (49Lb 90)
(off Guildhouse St.)
Hindrey Rd. E5 36Xb 71
Hindsley's Pl. SE23 61Yb 136
Hind Ter. RM20: Grays 50Zd 99
Hine Cl. CR5: Coul 94Lb 196
Hine Cl. KT19: Eps 83Ra 173
Hine Ho. AL4: St A 3H 7
Hinkler Rd. HA3: Kenton 27Ma 47
Hinkley Cl. UB9: Hare 28L 43
Hinksey Cl. SL3: L'ly 48D 82
Hinksey Path SE2 47Zc 95
Hinstock NW6 39Db 69
(off Belsize Rd.)
Hinstock Rd. SE18 51Sc 116
Hinton Av. TW4: Houn 56Z 107
Hinton Cl. SE9 60Nc 116
Hinton Ct. E10 33Dc 72
(off Leyton Grange Est.)
Hinton Rd. N18 21Ub 51
Hinton Rd. SE24 55Sb 113
Hinton Rd. SL1: Slou 5C 80
Hinton Rd. SM6: W'gton 79Lb 156
Hinton Rd. SW9 55Rb 113
Hinton Rd. UB8: Uxb 39L 63
Hipley St. GU22: Wok 92D 188
Hippisley Ct. TW7: Isle 55Ha 108
Hippodrome M. W11 45Ab 88
Hippodrome Pl. W11 45Ab 88
Hiroshima Prom. SE7 48Lc 93
Hirst Ct. SW1 51Kb 112
Hirst Cres. HA9: Wemb 34Na 67
Hispano M. EN3: Enf L 9Cc 20
Hitcham Rd. E17 31Bc 72
Hitchcock Cl. TW17: Shep 69P 127
Hitchcock Cl. WD6: Bore 12Sa 29
Hitchcock La. E20 37Ec 72
Hitchen Hatch La. TN13: S'oaks 96Jd 202
Hitchen Hatch Pl. TN13: S'oaks 95Kd 203
Hitchens Cl. HP1: Hem H 1H 3
Hitchin Cl. RM3: Rom 21Ld 57
Hitchings Way RH2: Reig 10J 207
Hitchin La. HA7: Stan 24Ma 47
Hitchin Rd. HA7: Stan 24Na 47
Hitchin Sq. E3 40Ac 72
Hitch St. RM9: Dag 41Ad 95
Hithe Gro. SE16 48Yb 92
Hitherbroom Rd. UB3: Hayes 46W 84
Hither Farm Rd. SE3 55Lc 115
Hitherfield Rd. RM8: Dag 33Ad 75
Hitherfield Rd. SW16 61Pb 134
Hither Flds. DA11: Grav'nd 4E 144
HITHER GREEN 58Fc 115
Hither Green Crematorium 61Hc 137
Hither Grn. La. SE13 57Ec 114
Hither Green Station (Rail) 58Gc 115
Hither Mdw. SL9: Chal P 25A 42
Hithermoor Rd. TW19: Stanw M 58H 105
Hitherwell Dr. HA3: Hrw W 25Fa 46
Hitherwood Cl. RH2: Reig 4M 207
Hitherwood Cl. RM12: Horn 35Md 77
Hitherwood Ct. NW9 26Ua 48
(off Charcot Rd.)
Hitherwood Dr. SE19 63Vb 135
Hittard Ct. SE17 51Tb 113
(off Red Lion Row)
Hive, The DA11: Nflt 58De 121
Hive, The 25Na 47
Hive Cl. CM14: B'wood 19Wd 40
Hive Cl. WD23: B Hea 19Fa 28
Hive Football & Fitness Cen., The 25Na 47
Hive La. DA11: Nflt 58De 121
Hive Rd. WD23: B Hea 19Fa 28
Hixberry La. AL4: St A 3H 7
HMP Belmarsh 47Uc 94
HMP Brixton 58Nb 112
HMP Bronzefield 63L 127
HMP Coldingley 9D 166
HMP Downview 84Eb 175
HMP High Down 84Eb 175
HMP Isis 47Vc 95
HMP Pentonville 37Pb 70
HMP Send 99L 189
HMP Thameside 48Uc 94
HMP Mount, The 8C 2
HMP Wandsworth 59Fb 111
HMP Wormwood Scrubs 44Wa 88
HMS Belfast 6J 225 (46Ub 91)
HMYOI Feltham 62T 128
Hoad Cres. GU22: Wok 94B 188
Hoadly Ho. SE1 7D 224 (46Sb 91)
(off Union St.)
Hoadly Rd. SW16 62Mb 134
Hobart Cl. N20 19Gb 31
Hobart Cl. UB4: Yead 42Z 85
Hobart Ct. CR2: S Croy 78Tb 157
(off South Pk. Hill Rd.)
Hobart Ct. IG8: Wfd G 21Hc 53
Hobart Dr. UB4: Yead 42Z 85
Hobart Gdns. CR7: Thor H 69Tb 135
Hobart La. UB4: Yead 42Z 85
Hobart Pl. SW1 3K 227 (48Kb 90)
Hobart Pl. TW10: Rich 59Pa 109
Hobart Rd. IG6: Ilf 26Sc 54
Hobart Rd. KT4: Wor Pk 76Xa 154
Hobart Rd. RM18: Tilb 3C 122
Hobart Rd. RM9: Dag 35Zc 75
Hobart Rd. UB4: Yead 42Z 85
Hobbayne Rd. W7 44Fa 86
Hobbes Wlk. SW15 57Xa 110
Hobbledown Adventure Farm Pk. & Zoo 82Qa 173
Hobbs Cl. AL4: St A 3J 7
Hobbs Cl. EN8: Chesh 1Zb 20
Hobbs Cl. KT14: W Byf 85K 169
Hobbs Ct. SE1 47Vb 91
(off Mill St.)
HOBBS CROSS 8Zc 23
Hobbs Cross Golf Course 6Zc 23
Hobbs Cross Rd. CM16: Epp 6Yc 23
Hobbs Cross Rd. CM16: Fidd H 6Yc 23
Hobbs Cross Rd. CM16: They G 6Yc 23
Hobbs Grn. N2 27Eb 49
Hobbs Hill Rd. HP3: Hem H 6N 3
Hobbs La. EN8: Chesh 1Zb 20
Hobbs M. IG3: Ilf 33Vc 75
Hobbs Pl. N1 1H 219 (39Ub 71)
Hobbs Pl. Est. N1 1H 219 (40Ub 71)
(off Hobbs Pl.)
Hobbs Rd. SE27 63Sb 135
Hobby Ho. SE1 49Xb 91
Hobby St. EN3: Pond E 15Zb 34
Hobday St. E14 44Dc 92
Hobhouse Ct. SW1 5E 222 (45Mb 90)
(off Suffolk St.)
Hobill Wlk. KT5: Surb 72Pa 153
Hoblands End BR7: Chst 65Uc 138
Hobletts Rd. HP2: Hem H 1P 3
Hobson's Pl. E1 43Wb 91
Hobury St. SW10 51Eb 111
HOCKENDEN 69Cd 140
Hockenden La. BR8: Swan 69Cd 140
Hockering Est. GU22: Wok 90D 168
Hockering Gdns. GU22: Wok 90C 168
Hockering Rd. GU22: Wok 90C 168
Hocker St. E2 4K 219 (41Vb 91)
Hockett Cl. SE8 49Bc 92
(off Grove St.)
Hockford Cl. GU24: Pirb 8F 186
Hockington Ct. EN5: New Bar 14Db 31
Hockley Av. E6 40Nc 74
Hockley Bottom HP5: Ley H 5A 10
Hockley Ct. E18 25Jc 53
Hockley Dr. RM2: Rom 26Kd 57
HOCKLEY HOLE 9M 61
Hockley La. SL2: Stoke P 8M 61
Hockley M. IG11: Bark 41Uc 94
Hockliffe Ho. W10 43Ya 88
(off Sutton Way)
Hockney Ct. SE16 50Xb 91
(off Rossetti Rd.)
Hocroft Av. NW2 34Bb 69
Hocroft Ct. NW2 34Bb 69
Hocroft Rd. NW2 35Bb 69
Hocroft Wlk. NW2 34Bb 69
Hodder Dr. UB6: G'frd 40Ha 66
Hoddesdon Rd. DA17: Belv 50Cd 96
Hodes Row NW3 35Jb 70
Hodford Rd. NW11 32Bb 69
Hodges Cl. RM16: Chaf H 50Zd 99
Hodges Way WD18: Wat 16W 26
Hodgkin Cl. SE28 45Zc 95
Hodgkin Ct. SE5 52Tb 113
(off Dobson Wlk.)
Hodgkins M. HA7: Stan 22Ka 46
Hodister Cl. SE5 52Sb 113
Hodnet Gro. SE16 49Zb 92
Hodsoll Ct. BR5: St M Cry 71Zc 161
Hodsoll St. TN15: Hod S 81Fe 185
HODSOLL STREET 81Fe 185
Hodson Cl. HA2: Harr 34Ba 65
Hodson Cres. BR5: St M Cry 71Zc 161
Hodson Pl. EN3: Enf L 10Cc 20
Hoe, The WD19: Wat 19Z 27
Hoebridge Golf Course 91E 188
Hoebrook Cl. GU22: Wok 3P 187
Hoe Ct. GU22: Wok 91A 188
Hoecroft Ct. EN3: Enf W 10Yb 20
(off Hoe La.)
Hoe La. EN1: Enf 10Wb 19
Hoe La. EN3: Enf W 10Yb 20
Hoe La. RM4: Abr 13Xc 37
Hoe St. E17 28Cc 52
Hoever Ho. SE6 63Ec 136
Hoey Ct. E3 42Dc 92
(off Barry Blandford Way)
Hoffmann Gdns. CR2: Sels 80Xb 157
Hoffman Sq. N1 3G 219 (41Tb 91)
(off Chart St.)
Hoffmans Rd. E17 27Zb 52
Hofland Rd. W14 48Ab 88
Hoford Rd. RM18: W Til 9F 100
Hoford Rd. SS17: Linf 6H 101
Hogan Bus. Cen. GU21: Wok 89A 168
Hogan M. W2 7A 214 (43Eb 89)
Hogan Way E5 33Wb 71
Hogarth Av. CM15: B'wood 20Ae 41
Hogarth Av. TW15: Ashf 65S 128
Hogarth Bus. Pk. W4 51Ua 110
Hogarth Cl. E16 43Mc 93
Hogarth Cl. SL1: Slou 5C 80
Hogarth Cl. UB8: Uxb 41L 83
Hogarth Cl. W5 43Na 87
Hogarth Ct. E1 44Wb 91
(off Batty St.)
Hogarth Ct. NW1 38Lb 70
(off St Pancras Way)
Hogarth Ct. SE19 63Vb 135
Hogarth Ct. TW5: Hest 52Aa 107
Hogarth Ct. WD23: Bush 17Da 27
Hogarth Cres. CR0: C'don 73Sb 157
Hogarth Cres. SW19 67Fb 133
Hogarth Gdns. TW5: Hest 52Ca 107
Hogarth Health Club, The 49Va 88
Hogarth Hill NW11 28Bb 49
Hogarth Ho. EC1 1C 224 (43Rb 91)
(off Bartholomew Cl.)
Hogarth Ho. SW1 6E 228 (49Mb 90)
(off Erasmus St.)
Hogarth Ho. UB5: N'olt 40Z 65
(off Gallery Gdns.)
Hogarth Ind. Est. NW10 42Wa 88
Hogarth La. W4 51Ua 110
Hogarth Pl. SW5 49Db 89
(off Hogarth Rd.)
Hogarth Reach IG10: Lough 15Pc 36
Hogarth Rd. HA8: Edg 26Qa 47
Hogarth Rd. RM16: Grays 46Ce 99
Hogarth Rd. RM8: Dag 36Xc 75
Hogarth Rd. SW5 49Db 89
HOGARTH RDBT. 51Ua 110
Hogarth's House 51Ua 110
(off Hogarth La.)
Hogarth Way TW12: Hamp 67Ea 130
Hogden Cl. KT20: Kgswd 97Bb 195
Hogfair La. SL1: Burn 1A 80
Hogg La. RM16: Grays 48Ce 99
Hogg La. RM17: Grays 47Ce 99
(not continuous)
Hogg La. WD6: E'tree 14Ja 28
Hog Hill Rd. RM5: Col R 24Bd 55
Hognore La. TN15: Wro 86Fe 185
HOGPITS BOTTOM 4D 10
Hogscross La. CR5: Coul 95Hb 195
Hogshill La. KT11: Cobh 86X 171
(not continuous)
Hogs La. DA11: Nflt 62Fe 143
Hogsmill Ho. KT1: King T 69Pa 131
(off Vineyard Cl.)
Hogsmill La. KT1: King T 69Pa 131
Hogsmill Local Nature Reserve 78Ra 153
Hogsmill Wlk. KT1: King T 69Na 131
(off Penrhyn Rd.)
Hogsmill Way KT19: Ewe 78Sa 153
Hogs Orchard BR8: Swan 67Kd 141
Hogtrough Hill TN16: Bras 92Vc 201
Hogtrough Hill TN16: S'oaks 92Vc 201
Hogtrough La. RH1: S Nut 7C 208
Hogtrough La. RH8: Oxt 100Dc 198
Hogtrough La. RH9: G'stone 99Cc 198
Holbeach Cl. NW9 25Ua 48
Holbeach Gdns. DA15: Sidc 58Uc 116
Holbeach M. SW12 60Kb 112
Holbeach Rd. SE6 59Cc 114
Holbeck Rd. W13 46Ka 86
Holbeck Row SE15 52Wb 113
Holbein Ga. HA6: Nwood 22U 44
Holbein Ho. SW1 7H 227 (50Jb 90)
(off Holbein M.)
Holbein M. SW1 7H 227 (50Jb 90)
Holbein Pl. SW1 6H 227 (49Jb 90)
Holbein Ter. RM8: Dag 35Yc 75
(off Marlborough Rd.)
Holberton Gdns. NW10 41Xa 88
Holborn EC1 1K 223 (43Qb 90)
HOLBORN 2G 223 (44Nb 90)
Holborn Bars EC1 1K 223 (43Qb 90)
(off Holborn)
Holborn Cir. EC1 1A 224 (43Qb 90)
Holborn Cl. NW7 21Va 48
Holborn Ho. W12 44Xa 88
Holborn Pl. WC1 1H 223 (43Pb 90)

Holborn Rd. E13 42Kc 93
Holborn Station (Underground) 2H 223 (44Pb 90)
Holborn Viaduct EC1 1B 224 (43Rb 91)
Holborn Way CR4: Mitc 68Hb 133
Holbreck Pl. GU22: Wok 90B 168
Holbrook Cl. EN1: Enf 11Vb 33
Holbrook Cl. N19 32Kb 70
Holbrook Ct. TW20: Egh 64E 126
Holbrooke Ct. N7 35Nb 70
Holbrooke Pl. TW10: Rich 57Ma 109
Holbrook Gdns. WD25: A'ham 8Da 13
Holbrook Ho. BR7: Chst 67Tc 138
Holbrook La. BR7: Chst 66Tc 138
Holbrook Mdw. TW20: Egh 65E 126
Holbrook Rd. E15 40Hc 73
Holbrook Way BR2: Broml 72Pc 160
Holburne Cl. SE3 53Lc 115
Holburne Gdns. SE3 53Mc 115
Holburne Rd. SE3 53Lc 115
Holcombe Cl. TN16: Westrm 98Tc 200
Holcombe Hill NW7 20Wa 30
Holcombe Ho. SW9 55Nb 112
(off Landor Rd.)
Holcombe Ho. Gdns. SL5: S'dale 4E 146
Holcombe Pl. SE4 55Ac 114
(off St Asaph Rd.)
Holcombe Rd. IG1: Ilf 31Qc 74
Holcombe Rd. N17 27Vb 51
Holcombe St. W6 49Xa 88
Holcon Ct. RH1: Redh 3A 208
Holcote Cl. DA17: Belv 48Ad 95
Holcroft Ct. W1 7B 216 (43Lb 90)
(off Clipstone St.)
Holcroft Ho. SW11 55Fb 111
Holcroft Rd. E9 38Yb 72
HOLDBROOK 6Bc 20
Holdbrook Nth. EN8: Walt C 5Bc 20
Holdbrook Sth. EN8: Walt C 6Bc 20
Holdbrook Way RM3: Hrld W 26Pd 57
Holden Av. N12 22Db 49
Holden Av. NW9 32Sa 67
Holdenby Rd. SE4 57Ac 114
Holden Cl. RM8: Dag 34Xc 75
Holden Cl. KT13: Weyb 80Q 150
Holden Gdns. CM14: W'ley 22Zd 59
Holden Ho. N1 39Sb 71
(off Prebend St.)
Holden Ho. SE8 52Cc 114
Holdenhurst Av. N12 24Eb 49
Holden Pl. DA11: Nflt 61Ee 143
Holden Pl. KT11: Cobh 86X 171
Holden Point E15 37Fc 73
(off Waddington St.)
Holden Rd. N12 22Db 49
Holden St. SW11 54Jb 112
Holden Way RM14: Upm 32Td 78
Holder Cl. N3 24Db 49
Holdernesse Cl. TW7: Isle 53Ja 108
Holdernesse Rd. SW17 62Hb 133
Holderness Ho. SE5 55Ub 113
Holderness Way SE27 64Rb 135
HOLDERS HILL 26Za 48
Holder's Hill Av. NW4 26Za 48
HOLDERS HILL CIR. 24Ab 48
Holders Hill Cres. NW4 26Za 48
Holders Hill Dr. NW4 27Za 48
Holder's Hill Gdns. NW4 26Ab 48
Holders Hill Pde. NW7 25Ab 48
Holders Hill Rd. NW4 26Za 48
Holders Hill Rd. NW7 25Ab 48
Holecroft EN9: Walt A 6Gc 21
Hole Farm La. CM13: Gt War 27Xd 58
Hole in the Wall All. DA11: Grav'nd 8D 122
(off High St.)
Holford Ho. SE16 49Xb 91
(off Camilla Rd.)
Holford Ho. WC1 3J 217 (41Pb 90)
(off Gt. Percy St.)
Holford M. WC1 3K 217 (41Qb 90)
(off Cruikshank St.)
Holford Pl. WC1 3J 217 (41Pb 90)
Holford Rd. NW3 34Eb 69
Holford St. WC1 3J 217 (41Pb 90)
Holford Way SW15 58Wa 110
Holford Yd. WC1 2K 217 (40Qb 70)
(off Cruikshank St.)
Holgate Av. SW11 55Fb 111
Holgate Ct. RM1: Rom 29Gd 56
(off Western Rd.)
Holgate Gdns. RM10: Dag 37Cd 76
Holgate Rd. RM10: Dag 36Cd 76
Holgate St. SE7 48Mc 93
Holinser Ter. W5 46Ma 87
Hollam Ho. N8 28Pb 50
HOLLAND 5L 211
Holland Av. SM2: Sutt 81Cb 175
Holland Av. SW20 67Va 132
Holland Cl. BR2: Hayes 75Hc 159
Holland Cl. EN5: New Bar 17Fb 31
Holland Cl. HA7: Stan 22Ka 46
Holland Cl. KT19: Eps 83Sa 173
Holland Cl. RH1: Redh 6P 207
Holland Cl. RM7: Rom 29Ed 56
Holland Ct. E17 28Ec 52
(off Evelyn Rd.)
Holland Ct. KT6: Surb 73Ma 153
Holland Ct. NW7 23Wa 48
Holland Cres. RH8: Oxt 5L 211
Holland Dr. SE23 62Ac 136
Holland Dwellings WC2 2G 223 (44Nb 90)
(off Newton St.)
Holland Gdns. SE9: Sidc 60Tc 116
Holland Gdns. TW20: Thorpe 68H 127
Holland Gdns. TW8: Bford 51Na 109
Holland Gdns. W14 48Ab 88
Holland Gdns. WD25: Wat 7Y 13
Hollandgreen Pl. W8 48Cb 89
Holland Gro. SW9 52Qb 112
Holland Ho. E4 21Ec 52
Holland Ho. NW10 40Xa 68
(off Holland Rd.)
Holland Ho. SL4: Eton 10H 81
(off Common La.)
Holland La. RH8: Oxt 5L 211
Holland Pk. CR3: Cat'm 95Tb 197
Holland Pk. W11 46Ab 88
HOLLAND PARK 46Bb 89
Holland Pk. 47Bb 89
Holland Pk. Av. IG3: Ilf 30Uc 54
Holland Pk. Av. W11 47Ab 88
Holland Pk. Ct. W14 47Ab 88
(off Holland Pk. Gdns.)
Holland Pk. Gdns. W14 47Ab 88
Holland Pk. Mans. W14 46Ab 88
(off Holland Pk. Gdns.)
Holland Pk. M. W11 46Ab 88
Holland Pk. Rd. W14 48Bb 89
HOLLAND PARK RDBT. 47Za 88
Holland Park Station (Underground) 46Bb 89
Holland Pk. Tennis Ground 46Ab 88
Holland Pk. Ter. W11 46Ab 88
(off Portland Rd.)
Holland Pk. Theatre (Open Air) 47Bb 89
Holland Pas. N1 39Sb 71
(off Basire St.)
Holland Pl. W8 47Db 89
(off Kensington Chu. St.)
Holland Pl. Chambers W8 47Db 89
(off Holland Pl.)
Holland Ri. Ho. SW9 52Pb 112
(off Clapham Rd.)
Holland Rd. E15 41Gc 93
Holland Rd. E6 39Pc 74
Holland Rd. HA0: Wemb 37Ma 67
Holland Rd. NW10 39Wa 68
Holland Rd. RH8: Oxt 5L 211
Holland Rd. SE25 71Wb 157
Holland Rd. W14 47Za 88
Hollands, The GU22: Wok 90A 168
Hollands, The KT4: Wor Pk 74Va 154
Hollands, The TW13: Hanw 63Z 129
Hollands Cl. DA12: Shorne 4N 145
Holland St. SE1 6C 224 (46Rb 91)
Holland St. W8 47Cb 89
Holland Vs. Rd. W14 47Ab 88
Holland Wlk. W8: Kensington 46Bb 89
Holland Wlk. N19: Up. Holloway 32Mb 70
Holland Wlk. HA7: Stan 22Ja 46
Holland Wlk. N19 32Mb 70
(off Calverley Gro.)
Hollar Way BR2: Hayes 75Hc 159
Hollar Rd. N16 34Vb 71
Hollen St. W1 2D 222 (44Mb 90)
Holles Cl. TW12: Hamp 65Ca 129
Holles Ho. SW9 54Qb 112
Holles St. W1 2A 222 (44Kb 90)
Holley Rd. W3 47Ua 88
Hollickwood Av. N12 23Hb 49
Holliday Sq. SW11 55Fb 111
(off Fowler Cl.)
Hollidge Way RM10: Dag 38Dd 76
Holliers Way AL10: Hat 1C 8
Hollies, The AL3: St A 1C 6
(off Carlisle Av.)
Hollies, The DA12: Grav'nd 5F 144
Hollies, The DA3: Lfield 69Ee 143
Hollies, The E11 29Jc 53
(off New Wanstead)
Hollies, The EN9: Walt A 7Lc 21
(within Woodbine Cl. Caravan Pk.)
Hollies, The HA3: W'stone 28Ja 46
Hollies, The HP3: Bov 1C 10
Hollies, The KT15: Add 78L 149
(off Bourne Way)
Hollies, The KT23: Bookh 96Ea 192
Hollies, The N20 18Fb 31
Hollies, The RH8: Oxt 5M 211
Hollies, The SS17: Stan H 2L 101
Hollies, The WD18: Wat 14V 26
Hollies Av. DA15: Sidc 61Vc 139
Hollies Av. KT14: W Byf 85H 169
Hollies Cl. SW16 65Qb 134
Hollies Cl. TW1: Twick 61Ha 130
Hollies Ct. KT15: Add 78L 149
Hollies End NW7 22Xa 48
Hollies Rd. W5 49La 86
Hollies Way EN6: Pot B 3Eb 17
Hollies Way SW12 59Jb 112
Holligrave Rd. BR1: Broml 67Jc 137
Hollingbourne Av. DA7: Bex 53Bd 117
Hollingbourne Gdns. W13 43Ka 86
Hollingbourne Rd. SE24 57Sb 113
Hollingsworth Ct. KT6: Surb 73Ma 153
Hollingsworth M. WD25: Wat 6W 12
Hollingsworth Rd. CR0: C'don 79Xb 157
Hollington Ct. BR7: Chst 65Rc 138
Hollington Cres. KT3: N Mald 72Va 154
Hollington Rd. E6 41Pc 94
Hollington Rd. N17 26Wb 51
Hollingworth Cl. KT8: W Mole 70Ba 129
Hollingworth Rd. BR5: Pet W 72Rc 160
Hollingworth Way TN16: Westrm 98Tc 200
Hollins Ho. N7 35Nb 70
Hollisfield WC1 4G 217 (41Nb 90)
(off Cromer St.)
Hollis Pl. RM17: Grays 49Ce 99
Hollis Row RH1: Redh 8P 207
Hollister Ho. NW6 41Cb 89
(off Kilburn Pk. Rd.)
Hollman Gdns. SW16 65Rb 135
Hollow, The IG8: Wfd G 21Hc 53
HOLLOWAY 34Nb 70
Holloway Cl. UB7: Harm 50N 83
Holloway Dr. GU25: Vir W 70A 126
Holloway Hill KT16: Chert 76E 148
Holloway Hill KT16: Lyne 76E 148
Holloway Ho. NW2 34Ya 68
(off Stoll Cl.)
Holloway Ho. TW20: Egh 64B 126
(off Stoneylands Rd.)
Holloway La. UB7: Harm 51M 105
Holloway La. UB7: W Dray 51M 105
Holloway La. WD3: Chen 10D 10
Holloway La. WD3: Sarr 10D 10
Holloway M. TW20: Eng G 5N 125
Holloway Rd. E11 34Fc 73
Holloway Rd. E6 41Pc 94
Holloway Rd. N19 33Mb 70
Holloway Rd. N7 35Pb 70
Holloway Road Station (Underground) 36Pb 70
Holloways La. AL9: Wel G 5F 8
Holloway St. TW3: Houn 55Da 107
Hollow Cotts. RM19: Purf 50Qd 97
Hollowfield Av. RM17: Grays 49Fe 99
Hollowfield Wlk. UB5: N'olt 37Aa 65
Hollow Hill La. SL0: Iver 45D 82
Hollow La. GU25: Vir W 9N 125
Hollows, The TW8: Bford 51Pa 109
Hollowtree M. WD3: Crox G 15Q 26
Holly Av. HA7: Stan 26Na 47
Holly Av. KT12: Walt T 74Z 151
Holly Av. KT15: New H 82J 169
Hollybank GU24: W End 5D 166
Hollybank HP3: Hem H 7N 3
Hollybank Cl. TW12: Hamp 64Ca 129
Holly Bank Rd. GU22: Wok 3M 187
Hollybank Rd. KT14: W Byf 86J 169
Hollyberry La. NW3 35Eb 69
Hollybrake Cl. BR7: Chst 66Tc 138
Hollybush Av. AL2: Chis G 6N 5
Hollybush Cl. E11 29Jc 53
Hollybush Cl. HA3: Hrw W 25Ga 46
Hollybush Cl. RM5: Col R 22Ed 56
Hollybush Cl. TN13: S'oaks 96Ld 203
Hollybush Cl. WD19: Wat 17Y 27
Hollybush Ct. TN13: S'oaks 96Ld 203
Hollybush Gdns. E2 41Xb 91
Holly Bush Hill NW3 35Eb 69
Hollybush Hill E11 30Hc 53
Hollybush Hill SL2: Stoke P 8L 61
Hollybush Ho. E2 41Xb 91
Holly Bush La. TN13: S'oaks 96Ld 203
Holly Bush La. TW12: Hamp 66Ba 129
Hollybush La. BR6: Well H 79Cd 162
Hollybush La. GU23: Rip 91M 189
Hollybush La. HP1: Hem H 1H 3
Hollybush La. SL0: Iver 44D 82
Hollybush La. UB9: Den 33E 62
Hollybush Pl. E2 41Xb 91
Hollybush Rd. DA12: Grav'nd 1E 144
Hollybush Rd. KT2: King T 64Na 131
Hollybush Sports Complex 95Md 203
Holly Bush Steps NW3 35Eb 69
(off Holly Mt.)
Hollybush St. E13 41Kc 93
Holly Bush Va. NW3 35Eb 69
Hollybush Wlk. SW9 56Rb 113
Hollybush Way EN7: Chesh 1Wb 19
Holly Cl. AL10: Hat 1B 8
Holly Cl. BR3: Beck 70Ec 136
Holly Cl. GU21: Wok 1M 187
Holly Cl. IG9: Buck H 20Mc 35
Holly Cl. KT16: Longc 6L 147
Holly Cl. KT19: Eps 81Ta 173
Holly Cl. SL2: Farn C 5G 60
Holly Cl. SM6: W'gton 80Kb 156
Holly Cl. TW13: Hanw 64Aa 129
Holly Cl. TW16: Sun 69X 129
Holly Cl. TW20: Eng G 5M 125
Hollycombe TW20: Eng G 3N 125
Holly Cott. M. UB8: Hil 43Q 84
Holly Ct. DA11: Nflt 57De 121
Holly Ct. DA14: Sidc 63Xc 139
(off Sidcup Hill)
Holly Ct. KT16: Chert 74H 149
(off King St.)
Holly Ct. KT22: Lea 94Ja 192
(off Belmont Rd.)
Holly Ct. N15 28Ub 51
Holly Ct. RM1: Rom 28Hd 56
(off Dolphin App.)
Holly Ct. SE10 48Hc 93
Holly Ct. SM2: Sutt 80Cb 155
Holly Cres. BR3: Beck 71Bc 158
Holly Cres. IG8: Wfd G 24Fc 53
Holly Cres. SL4: Wind 4B 102
Hollycroft Av. HA9: Wemb 33Pa 67
Hollycroft Av. NW3 34Cb 69
Hollycroft Cl. CR2: S Croy 78Ub 157
Hollycroft Cl. UB7: Sip 51Q 106
Hollycroft Gdns. UB7: Sip 51Q 106
Hollydale Cl. UB5: N'olt 35Da 65
Hollydale Dr. BR2: Broml 76Pc 160
Hollydale Rd. SE15 53Yb 114
Hollydene BR2: Broml 67Hc 137
(off Beckenham La.)
Hollydene SE13 58Fc 115
Hollydene SE15 53Xb 113
Hollydown Way E11 34Fc 73
Holly Dr. E4 17Dc 34
Holly Dr. EN6: Pot B 5Db 17
Holly Dr. RM15: S Ock 41Zd 99
Holly Dr. SL4: Old Win 8J 103
Holly Farm Rd. UB2: S'hall 50Aa 85
Hollyfield AL10: Hat 3C 8
Hollyfield Av. N11 22Hb 49
Hollyfield Rd. KT5: Surb 73Pa 153
Holly Gdns. DA7: Bex 56Ed 118
Holly Gdns. UB7: W Dray 47P 83
Holly Ga. KT15: Add 77K 149
Holly Grn. KT13: Weyb 77T 150
Holly Gro. HA5: Pinn 25Aa 45
Holly Gro. NW9 31Sa 67
Holly Gro. SE15 54Vb 113
Hollygrove WD23: Bush 17Fa 28
Hollygrove Cl. TW3: Houn 56Ba 107
Hollyhedge Rd. KT11: Cobh 86X 171
Holly Hedges La. HP3: Bov 2E 10
Holly Hedge Ter. SE13 57Fc 115
Holly Hill N21 16Pb 32
Holly Hill NW3 35Eb 69
Holly Hill Dr. SM7: Bans 88Cb 175
Holly Hill Pk. SM7: Bans 89Cb 175
Holly Hill Rd. DA17: Belv 50Dd 96
Holly Hill Rd. DA17: Erith 50Dd 96
Holly Hill Rd. DA8: Erith 50Ed 96
Hollyhock Cl. HP1: Hem H 1G 2
Hollyhock Dr. GU24: Bisl 7E 166
Holly Ho. CM15: B'wood 18Zd 41
Holly Ho. TW8: Bford 51La 108
Holly Ho. W10 42Ab 88
(off Hawthorn Wlk.)
Holly Ind. Pk. WD24: Wat 11Y 27
Holly La. GU3: Worp 10G 186
Holly La. IG3: Ilf 33Wc 75
Holly La. SM7: Bans 88Cb 175
Holly La. E. SM7: Bans 88Cb 175
Holly La. W. SM7: Bans 89Cb 175
Holly Lea GU4: Jac W 10P 187
Holly Lodge GU22: Wok 89B 168
(off Heathside Cres.)
Holly Lodge HA1: Harr 29Fa 46
Holly Lodge KT20: Lwr K 98Ab 194
Holly Lodge SL1: Burn 2A 80
Holly Lodge W8 47Cb 89
(off Thornwood Gdns.)
Holly Lodge Gdns. N6 33Jb 70
Holly Lodge Mans. N6 33Jb 70
Hollymead SM5: Cars 76Hb 155
Hollymead Rd. CR5: Chips 90Jb 176
Hollymeoak Rd. CR5: Coul 91Kb 196
Holly M. SW10 7A 226 (50Eb 89)
Hollymoor La. KT19: Ewe 82Ta 173
Holly Mt. NW3 35Eb 69
Hollymount Cl. SE10 53Ec 114
Holly Pde. KT11: Cobh 86X 171
(off High St.)
Holly Pde. TW13: Felt 62V 128
(off High St.)
Holly Pk. N3 27Bb 49
Holly Pk. N4 31Nb 70
(not continuous)
Holly Pk. Est. N4 31Pb 70
Holly Pk. Gdns. N3 27Cb 49
Holly Pk. Rd. N11 22Jb 50
Holly Pk. Rd. W7 46Ha 86
Holly Pl. NW3 35Eb 69
(off Holly Berry La.)
HOLLY PRIVATE HOSPITAL, THE 19Kc 35
Holly Rd. BR6: Chels 80Wc 161
Holly Rd. DA1: Dart 60Md 119
Holly Rd. E11 31Hc 73
Holly Rd. EN3: Enf W 8Zb 20
Holly Rd. RH2: Reig 8K 207
Holly Rd. TW1: Twick 60Ha 108
Holly Rd. TW12: Hamp H 65Ea 130
Holly Rd. TW3: Houn 56Da 107
Holly Rd. W4 49Ta 87
Holly St. E8 38Vb 71
Holly Ter. N6 32Jb 70
Hollytree Av. BR8: Swan 68Gd 140
Holly Tree Cl. SW19 60Za 110
Hollytree Cl. SL9: Chal P 22A 42
Holly Tree Ct. HP2: Hem H 2B 4
Holly Tree Cres. SM5: Cars 74Hb 155
Holly Tree Ho. SE4 55Bc 114
(off Brockley Rd.)
Hollytree Ho. WD24: Wat 8U 12
Hollytree Pde. DA14: Sidc 65Yc 139
(off Sidcup Hill)
Holly Tree Rd. CR3: Cat'm 94Ub 197
Hollyview Cl. NW4 30Wa 48
Holly Village N6 33Kb 70
Holly Vs. W6 48Xa 88
(off Wellesley Av.)
Holly Wlk. EN2: Enf 13Sb 33
Holly Wlk. NW3 35Eb 69
Holly Wlk. SL4: Wind 3C 124
Holly Way CR4: Mitc 70Mb 134
Hollywell Gro. RM12: Horn 32Hd 76
Hollywood Bowl Dagenham 39Ad 75
Hollywood Bowl Finchley 24Fb 49
Hollywood Bowl Lakeside 48Wd 98
Hollywood Bowl Surrey Quays 48Zb 92
Hollywood Bowl O2, The 46Gc 93
Hollywood Bowl Tolworth 75Ra 153
Hollywood Bowl Watford 5Y 13
Hollywood Bowl intu Watford 14Y 27
Hollywood Ct. SW10 51Eb 111
(off Hollywood Rd.)
Hollywood Ct. W5 45Pa 87
Hollywood Ct. WD6: E'tree 14Qa 29
Hollywood Gdns. UB4: Yead 44X 85
Hollywood La. TN15: W King 83Vd 184
Hollywood M. SW10 51Eb 111
Hollywood Rd. E4 22Ac 52
Hollywood Rd. SW10 51Eb 111
Hollywoods CR0: Sels 81Bc 178
Hollywood Way DA8: Erith 52Kd 119
Hollywood Way IG8: Wfd G 24Fc 53
Holman Ct. KT17: Ewe 81Wa 174
Holman Dr. UB2: S'hall 46Fa 86
Holman Ho. E2 41Zb 92
(off Roman Rd.)
Holman Hunt Ho. W6 50Ab 88
(off Field Rd.)
Holman Rd. KT19: Ewe 78Sa 153
Holman Rd. SW11 54Fb 111
Holmbank Dr. TW17: Shep 70U 128
Holmbridge Gdns. EN3: Pond E 14Zb 34
Holmbrook NW1 2C 216 (40Lb 70)
(off Eversholt St.)
Holmbrook Dr. NW4 29Za 48
Holmbury Ct. CR2: S Croy 78Ub 157
Holmbury Ct. SW17 62Hb 133
Holmbury Ct. SW19 66Gb 133
Holmbury Gdns. UB3: Hayes 46V 84
Holmbury Gro. CR0: Sels 80Bc 158
Holmbury Ho. SE24 57Rb 113
Holmbury Mnr. DA14: Sidc 63Wc 139
Holmbury Pk. BR1: Broml 66Nc 138
Holmbury Vw. E5 32Xb 71
Holmbush Rd. SW15 58Ab 110
Holm Cl. KT15: Wdhm 84G 168
Holmcote Gdns. N5 36Sb 71
Holm Ct. SE12 62Kc 137
Holmcroft KT20: Walt H 97Xa 194
Holmcroft Ho. E17 28Dc 52
Holmcroft Way BR2: Broml 71Pc 160
Holmdale Cl. WD6: Bore 12Pa 29
Holmdale Gdns. NW4 29Za 48
Holmdale Rd. BR7: Chst 64Sc 138
Holmdale Rd. NW6 36Cb 69
Holmdale Ter. N15 30Ub 51
Holmdene N12 22Db 49
Holmdene Av. HA2: Harr 27Da 45
Holmdene Av. NW7 23Wa 48
Holmdene Av. SE24 57Sb 113
Holmdene Cl. BR3: Beck 68Ec 136
Holmdene Ct. BR1: Broml 69Nc 138
Holmead Rd. SW6 52Db 111
Holmebury Cl. WD23: B Hea 19Ga 28
Holme Chase KT13: Weyb 79S 150
Holme Cl. EN8: Chesh 3Ac 20
Holme Ct. TW7: Isle 55Ja 108
Holmedale SL2: Slou 5N 81
Holmefield Ho. W10 42Ab 88
(off Hazlewood Cres.)
Holmefield Pl. KT15: New H 82K 169
Holme Ho. SE15 52Xb 113
(off Studholme St.)
Holme Lacey Rd. SE12 58Hc 115
Holme Lea WD25: Wat 6Y 13
Holmeoak Av. RM13: Rain 40Fd 76
Holme Pk. WD6: Bore 12Pa 29
Holme Pl. HP2: Hem H 1C 4
Holme Rd. E6 39Nc 74
Holme Rd. RM11: Horn 32Qd 77
Holmes Av. E17 27Bc 52
Holmes Cl. CR8: Purl 85Pb 176
Holmes Cl. GU22: Wok 93B 188
Holmes Cl. SE22 56Wb 113
Holmes Cl. SL5: S'hill 2A 146
Holmes Ct. AL3: St A 1C 6
(off Carlisle Av.)
Holmes Ct. DA12: Grav'nd 10H 123
Holmesdale EN8: Walt C 7Yb 20
Holmesdale KT13: Weyb 79T 150
(off Bridgewater Rd.)
Holmesdale Av. RH1: Mers 3C 208
Holmesdale Av. SW14 55Sa 109
Holmesdale Cl. SE25 69Vb 135
Holmesdale Hill DA4: S Dar 67Sd 142
Holmesdale Ho. NW6 39Cb 69
(off Kilburn Vale)
Holmesdale Mnr. RH1: Redh 4A 208
Holmesdale Natural History Mus. 6K 207
Holmesdale Pk. RH1: Nutf 6F 208
Holmesdale Rd. CR0: C'don 71Tb 157
Holmesdale Rd. DA4: S Dar 67Sd 142
Holmesdale Rd. DA7: Bex 54Zc 117
Holmesdale Rd. N6 31Kb 70
Holmesdale Rd. RH1: S Nut 8F 208
Holmesdale Rd. RH2: Reig 5J 207
Holmesdale Rd. SE25 71Tb 157
Holmesdale Rd. TN13: S'oaks 95Ld 203
Holmesdale Rd. TW11: Tedd 66La 130
Holmesdale Rd. TW9: Kew 53Pa 109
Holmesdale Tunnel 6Zb 20
Holmesley Rd. SE23 58Ac 114
Holmesley Rd. WD6: Bore 12Ra 29
Holmes Mead GU22: Pyr 87H 169
Holmes Pl. SW10 51Eb 111
Holmes Rd. NW5 36Kb 70
Holmes Rd. SW19 66Eb 133
Holmes Rd. TW1: Twick 61Ha 130
Holmes Ter. SE1 1K 229 (47Qb 90)
(off Waterloo Rd.)
Holmeswood SM2: Sutt 79Db 155
Holmeswood Ct. N22 26Qb 50
HOLMETHORPE 3B 208
Holmethorpe Av. RH1: Redh 3B 208
Holmethorpe Ind. Est. RH1: Redh 3B 208
Holmethorpe Lagoons Nature Reserve 3C 208
Holme Way HA7: Stan 23Ha 46
Holmewood Gdns. SW2 59Pb 112
Holmewood Rd. SE25 69Ub 135
Holmewood Rd. SW2 59Pb 112
Holmfield Av. NW4 29Za 48
Holmfield Ct. NW3 36Gb 69
Holm Gro. UB10: Hil 38Q 64
Holmgrove Ho. CR8: Purl 84Qb 176
Holmhurst SE13 58Fc 115
Holmhurst Rd. DA17: Belv 50Dd 96
Holmlea Ct. CR0: C'don 77Tb 157
(off Chatsworth Rd.)
Holmlea Rd. SL3: Dat 3P 103
Holmlea Wlk. SL3: Dat 3N 103
Holmleigh Av. DA1: Dart 57Md 119
Holmleigh Ct. EN3: Pond E 14Yb 34
Holmleigh Rd. N16 32Ub 71
Holmleigh Rd. Est. N16 32Ub 71
Holm Oak Cl. SW15 58Bb 111
Holmoak Cl. CR8: Purl 82Pb 176
Holm Oak M. SW4 57Nb 112
Holm Oak Pk. WD18: Wat 15V 26
Holmoaks Ho. BR3: Beck 68Ec 136
Holmsdale Cl. SL0: Iver 44H 83
Holmsdale Gro. DA7: Bex 54Gd 118
Holmsdale Ho. E14 45Dc 92
(off Poplar High St.)
Holmsdale Ho. N11 21Kb 50
(off Coppies Gro.)
Holmshaw Cl. SE26 63Ac 136
Holmshill La. WD6: Bore 8Ua 16
Holmside Ri. WD19: Wat 20X 27
Holmside Rd. SW12 58Jb 112
Holmsley Cl. KT3: N Mald 72Va 154
Holmsley Ho. SW15 59Va 110
(off Tangley Gro.)
Holmstall Av. HA8: Edg 27Sa 47
Holmstall Pde. HA8: Edg 26Sa 47
Holmstead Ct. CR2: S Croy 78Tb 157
Holm Wlk. SE3 54Jc 115
Holmwood Av. CM15: Shenf 16Ce 41
Holmwood Av. CR2: Sande 85Vb 177
Holmwood Cl. HA2: Harr 27Ea 46
Holmwood Cl. KT15: Add 78J 149
Holmwood Cl. KT24: E Hor 100U 190
Holmwood Cl. SM2: Cheam 81Za 174
Holmwood Cl. UB5: N'olt 37Da 65
Holmwood Gdns. N3 26Cb 49
Holmwood Gdns. SM6: W'gton 79Kb 156
Holmwood Gro. NW7 22Ta 47
Holmwood Rd. EN3: Enf W 8Zb 20
Holmwood Rd. IG3: Ilf 33Uc 74
Holmwood Rd. KT9: Chess 78Ma 153
Holmwood Rd. SM2: Cheam 81Ya 174
Holmwood Vs. SE7 50Jc 93
Holne Chase N2 30Eb 49
Holne Chase SM4: Mord 72Bb 155
Holness Rd. E15 37Hc 73
Holocaust Memorial Gdn., The 1G 227 (47Hb 89)
Holroyd Cl. KT10: Clay 81Ha 172
Holroyd Rd. KT10: Clay 81Ha 172
Holroyd Rd. SW15 56Ya 110
Holsart Cl. KT20: Tad 94Xa 194
Holsgrove Ho. W3 46Ua 88
Holst Ct. SE1 3K 229 (48Qb 90)
(off Kennington Rd.)
Holstein Av. KT13: Weyb 77Q 150
Holstein Way DA18: Erith 48Zc 95
Holst Ho. W12 44Xa 88
(off Du Cane Rd.)
Holst Mans. SW13 51Xa 110
Holstock Rd. IG1: Ilf 33Sc 74
Holsworth Cl. HA2: Harr 29Ea 46
Holsworthy Ho. E3 41Dc 92
(off Talwin St.)
Holsworthy Ho. RM3: Rom 25Ld 57
Holsworthy Sq. WC1 6J 217 (42Pb 90)
(off Elm St.)
Holsworthy Way KT9: Chess 78La 152
Holt, The HP2: Hem H 3N 3
Holt, The IG6: Ilf 23Sc 54
Holt, The SM4: Mord 70Cb 133
Holt, The SM6: W'gton 77Lb 156
Holt Cl. BR7: Chst 64Pc 138
Holt Cl. DA14: Sidc 63Ad 139
Holt Cl. IG7: Chig 22Vc 55
Holt Cl. KT16: Ott 79F 148
Holt Cl. N10 28Jb 50
Holt Cl. SE28 45Xc 95
Holt Cl. WD6: E'tree 14Pa 29
Holt Ct. SE10 51Ec 114
(off Horseferry Pl.)
Holt Ho. EN8: Chesh 1Xb 19
Holt Ho. SW2 58Qb 112
Holton St. E1 42Zb 92
Holt Pl. CR8: Purl 83Rb 177
Holt Rd. E16 46Nc 94
Holt Rd. HA0: Wemb 34Ka 66
Holt Rd. RM3: Rom 24Nd 57
Holtsmere Cl. WD25: Wat 7Y 13
Holt Way IG7: Chig 22Vc 55
Holtwhite Av. EN2: Enf 12Sb 33
Holtwhite's Hill EN2: Enf 11Rb 33
Holt Wood CR6: W'ham 88Cc 178
Holtwood Rd. KT22: Oxs 85Ea 172
Holwell Pl. HA5: Pinn 28Aa 45
Holwood Cl. KT12: Walt T 75Y 151
Holwood Est. BR2: Kes 79Nc 160
Holwood Pk. Av. BR6: Farnb 77Pc 160
Holwood Pl. SW4 56Mb 112
Holybourne Av. SW15 59Wa 110
Holycross Cl. SM4: Mord 73Cb 155
HOLYFIELD 1Fc 21
Holyfield Rd. EN9: Walt A 1Ec 20
Holyhead Cl. E6 43Pc 94
Holyhead Ct. KT1: King T 70Ma 131
(off Anglesea Rd.)
Holyhead M. SL1: Slou 4B 80
(off Kelpatrick Rd.)
Holyoake Av. GU21: Wok 9N 167
Holyoake Ct. SE16 47Bc 92
Holyoake Cres. GU21: Wok 9N 167
Holyoake Ho. W5 42La 86
Holyoake Mt. DA12: Grav'nd 10F 122
Holyoake Ter. TN13: S'oaks 96Jd 202
Holyoake Wlk. N2 27Eb 49
Holyoake Wlk. W5 42La 86

Holyoak Rd. SE11 6B 230 (49Rb 91)
Holyport Rd. SW6 52Za 110
Holyrood Av. HA2: Harr 35Aa 65
Holyrood Ct. NW1 39Kb 70
(off Gloucester Av.)
Holyrood Ct. WD18: Wat 14X 27
(off Marlborough Rd.)
Holyrood Cres. AL1: St A 5C 6
Holyrood Gdns. HA8: Edg 27Ra 47
Holyrood Gdns. RM16: Grays 9E 100
Holyrood M. E16 46Jc 93
Holyrood Rd. EN5: New Bar 16Eb 31
Holyrood St. SE1 7H 225 (46Ub 91)
HOLYWELL 16V 26
Holywell Cen. EC2 5H 219 (42Ub 91)
(off Phipp St.)
Holywell Cl. BR6: Chels 77Wc 161
Holywell Cl. SE16 50Xb 91
Holywell Cl. SE3 51Jc 115
Holywell Cl. TW19: Stanw 60N 105
Holywell Hill AL1: St A 3B 6
Holywell La. EC2 5J 219 (42Ub 91)
Holywell M. AL1: St A 2B 6
(off Holywell Hill)
Holywell Rd. WD18: Wat 15W 26
Holywell Row EC2 6H 219 (42Ub 91)
Holywell Way TW19: Stanw 60N 105
Homan Ct. N12 21Fb 49
Homebeech Ho. GU22: Wok 90A 168
(off Mt. Hermon Rd.)
Homebush Ho. E4 17Dc 34
Homecedars Ho. WD23: B Hea 18Fa 28
Home Cl. GU25: Vir W 2P 147
Home Cl. KT22: Fet 93Fa 192
Home Cl. SM5: Cars 75Hb 155
Home Cl. UB5: N'olt 41Ba 85
Home Ct. KT6: Surb 71Ma 153
Homecroft Gdns. IG10: Lough 14Rc 36
Homecroft Rd. N22 25Sb 51
Homecroft Rd. SE26 64Yb 136
Homedean Rd. TN13: Chips 94Ed 202
Home Farm BR6: Chels 78Cd 162
Home Farm Cl. KT10: Esh 79Da 151
Home Farm Cl. KT16: Ott 80C 148
Home Farm Cl. KT20: Tad 89Za 174
Home Farm Cl. KT7: T Ditt 73Ha 152
Home Farm Cl. RH3: B'wth 7A 206
Home Farm Cl. TW17: Shep 70U 128
Home Farm Cl. HP3: Bov 2A 10
Home Farm Gdns. KT12: Walt T 75Y 151
Home Farm Pl. RH1: Mers 100Kb 196
Home Farm Rd. AL9: N Mym 8D 8
Home Farm Rd. CM13: B'wood 25Ae 59
Home Farm Rd. CM13: L War 25Ae 59
Home Farm Rd. WD3: Rick 21Q 44
Homefarm Rd. W7 44Ga 86
Home Farm Way SL3: Stoke P 9N 61
Homefield EN9: Walt A 4Jc 21
Homefield HP3: Bov 10D 2
Homefield SM4: Mord 70Cb 133
Homefield Av. IG2: Ilf 29Uc 54
Homefield Av. KT12: Hers 77Z 151
Homefield Cl. BR5: St P 70Xc 139
Homefield Cl. BR8: Swan 69Hd 140
Homefield Cl. CM16: Epp 2Wc 23
Homefield Cl. KT15: Wdhm 84G 168
Homefield Cl. KT22: Lea 93La 192
Homefield Cl. NW10 37Sa 67
Homefield Cl. UB4: Yead 42Z 85
Homefield Farm Rd. DA4: S Dar 67Qd 141
Homefield Farm Rd. DA4: Sut H 67Qd 141
Homefield Gdns. CR4: Mitc 68Eb 133
Homefield Gdns. KT20: Tad 92Ya 194
Homefield Gdns. N2 27Fb 49
Homefield Ho. SE23 62Yb 136
Homefield M. BR3: Beck 67Cc 136
Homefield Pk. SM1: Sutt 79Db 155
Homefield Pl. CR0: C'don 75Vb 157
Homefield Ri. BR6: Orp 74Wc 161
Homefield Rd. BR1: Broml 67Lc 137
Homefield Rd. CR3: Coul 92Rb 197
Homefield Rd. CR5: Coul 91Rb 197
Homefield Rd. CR6: W'ham 91Yb 198
Homefield Rd. HA0: Wemb 35Ja 66
Homefield Rd. HA8: Edg 23Ta 47
Homefield Rd. HP2: Hem H 2A 4
Homefield Rd. KT12: Walt T 73Aa 151
Homefield Rd. SW19 65Za 132
Homefield Rd. TN13: Riv 94Gd 202
Homefield Rd. W4 50Wa 88
Homefield Rd. WD23: Bush 14Ca 27
Homefield Rd. WD3: Chor 14E 24
Homefield Rd. WD7: R'lett 9Ha 14
Homefield St. N1 2H 219 (40Ub 71)
Homefirs Ho. HA9: Wemb 34Pa 67
Home Gdns. DA1: Dart 58Nd 119
Home Gdns. RM10: Dag 34Ed 76
Homeheather Ho. IG4: Ilf 29Pc 54
Home Hill BR8: Hext 66Hd 140
Homehurst Ho. CM15: B'wood 18Zd 41
Homeland Dr. SM2: Sutt 81Db 175
Homelands KT22: Lea 93La 192
Homelands SL4: Wink 1A 124
Homelands Dr. SE19 66Ub 135
Homelands Pl. TN16: Big H 89Nc 180
Home Lea BR6: Chels 78Vc 161
Homeleigh Ct. EN8: Chesh 1Xb 19
Homeleigh Ct. SW16 62Nb 134
Homeleigh Rd. SE15 57Zb 114
Homeleigh St. EN8: Chesh 2Xb 19
Homeleigh Ter. SL2: Slou 4L 81
Homemanor Ho. WD18: Wat 13X 27
(off Cassio Rd.)
Home Mead HA7: Stan 25La 46
Homemead DA12: Grav'nd 9D 122
Homemead DA9: Ghithe 58Xd 120
Homemead Cl. DA12: Grav'nd 9D 122
Home Mead Local Nature Reserve 11Rc 36
Home Mdw. SL2: Farn R 10G 60
Home Mdw. SM7: Bans 88Cb 175
Home Mdw. M. SE22 57Wb 113
Homemead Rd. BR2: Broml 71Pc 160
Homemead Rd. CR0: C'don 72Lb 156
Home Orchard DA1: Dart 58Nd 119
Home Pk. KT1: E Mos 71Ka 152
Home Pk. RH8: Oxt 3L 211
Home Pk. Cotts. WD4: K Lan 2R 12
Home Pk. Ct. KT1: King T 70Ma 131
(off Palace Rd.)
Home Pk. Ind. Est. WD4: K Lan 2R 12
Home Pk. Lock 2R 12
(off Mill Link Rd.)
Home Pk. Mill Link WD4: K Lan 2R 12
Home Pk. Pde. KT1: Hamp W 68Ma 131
(off High St.)
Home Pk. Rd. SW19 63Bb 133
Home Pk. Ter. KT1: Hamp W 68Ma 131
(off Hampton Ct. Rd.)
Home Pk. Wlk. KT1: King T 70Ma 131
Homer Cl. DA7: Bex 53Ed 118
Homer Dr. E14 49Cc 92
Homer Rd. SW11 54Gb 111
Homer Rd. CR0: C'don 72Zb 158
Homer Rd. E9 37Ac 72
Homer Row W1 1E 220 (43Gb 89)
Homersham Rd. KT1: King T 68Qa 131
Homers Rd. SL4: Wind 3B 102
Homer St. W1 1E 220 (43Gb 89)
HOMERTON 36Zb 72
Homerton Gro. E9 36Zb 72
Homerton High St. E9 36Zb 72
Homerton Rd. E9 36Ac 72
Homerton Row E9 36Yb 72
Homerton Station (Overground) 37Zb 72
Homerton Ter. E9 37Yb 72
(not continuous)
HOMERTON UNIVERSITY HOSPITAL 36Zb 72
Homesdale Cl. E11 29Jc 53
Homesdale Rd. BR1: Broml 69Lc 137
Homesdale Rd. BR2: Broml 70Lc 137
Homesdale Rd. BR5: Pet W 73Uc 160
Homesdale Rd. CR3: Cat'm 95Tb 197
Homesfield NW11 29Cb 49
Homestall Rd. SE22 57Yb 114
Homestead, The DA1: Cray 57Gd 118
(off Crayford High St.)
Homestead, The DA1: Dart 58Ld 119
Homestead Cl. AL2: Park 9A 6
Homestead Ct. EN5: New Bar 15Cb 31
Homestead Gdns. KT10: Clay 78Ga 152
Homestead Paddock N14 15Kb 32
Homestead Pk. NW2 34Va 68
Homestead Rd. BR6: Chels 80Xc 161
Homestead Rd. CR3: Cat'm 95Tb 197
Homestead Rd. RM8: Dag 33Bd 75
Homestead Rd. SW6 52Bb 111
Homestead Rd. TW18: Staines 65K 127
Homestead Rd. WD3: Rick 17M 25
Homesteads, The N11 21Kb 50
Homestead Way CR0: New Ad 83Ec 178
Homevale Cl. BR2: Hayes 73Hc 159
Homewalk Ho. SE26 63Xb 135
Homewater Ho. KT17: Eps 85Ua 174
Homewaters Av. TW16: Sun 67V 128
Home Way WD3: Rick 18H 25
Homeway RM3: Hrld W 23Rd 57
Homewillow Cl. N21 16Rb 33
Homewood SL3: Geor G 4P 81
Homewood Cl. TW12: Hamp 65Ba 129
Homewood Ct. WD3: Chor 14H 25
Homewood Cres. BR7: Chst 65Uc 138
Homewoods SW12 59Lb 112
Homeworth Ho. GU22: Wok 90A 168
(off Ashdown Cl.)
Homildon Ho. SE26 62Wb 135
Homlesdale Bus. Cen. TN15: Plat 91Ee 205
Honduras St. EC1 5D 218 (42Sb 91)
Honeybourne Rd. NW6 36Db 69
Honeybourne Way BR5: Pet W 74Tc 160
Honey Brook EN9: Walt A 5Gc 21
Honeybrook Rd. SW12 59Lb 112
Honey Cl. RM10: Dag 37Dd 76
Honeycroft IG10: Lough 14Rc 36
Honeycroft Dr. AL4: St A 5G 6
Honeycroft Hill UB10: Uxb 38N 63
Honeycross Rd. HP1: Hem H 3G 2
Honeyden Rd. DA14: Sidc 65Ad 139
Honeyfield Rd. SE23 62Zb 136
Honeyghan Ct. SE17 7H 231 (50Ub 91)
(off Sedan Way)
Honey Hill UB10: Uxb 38P 63
Honey La. EC2 3E 224 (44Sb 91)
(off Trump St.)
Honey La. EN9: Walt A 5Gc 21
Honey La. Ho. EN9: Walt A 6Jc 21
Honey La. Ho. SW10 51Db 111
(off Finborough Rd.)
Honeyman Cl. NW6 38Za 68
Honeymead N8 27Nb 50
(off Campsfield Rd.)
Honey M. RM7: Rom 29Ed 56
Honey M. SE27 63Sb 135
(off Norwood High St.)
Honeypot Bus. Cen. HA7: Stan 25Na 47
Honeypot Cl. HP2: Hem H 1M 3
Honeypot Cl. NW9 28Pa 47
Honey Pot La. TN15: Kems'g 91Rd 203
Honeypot La. CM14: B'wood 20Wd 40
Honeypot La. EN9: Walt A 6Lc 21
Honeypot La. HA7: Stan 24Ma 47
Honeypot La. NW9 28Pa 47
Honeypot La. TN15: Hod S 81Ee 185
Honeypot La. TN8: Eden 10P 211
Honeypots Rd. GU22: Wok 4P 187
Honeysett Rd. N17 26Vb 51
Honeysuckle Bottom KT24: E Hor 100W 190
Honeysuckle Cl. CM15: Pil H 15Xd 40
Honeysuckle Cl. RM3: Rom 23Ld 57
Honeysuckle Cl. SL0: Iver 44E 82
Honeysuckle Cl. UB1: S'hall 45Aa 85
Honeysuckle Ct. IG1: Ilf 37Rc 74
Honeysuckle Ct. IG9: Buck H 20Mc 35
Honeysuckle Gdns. AL10: Hat 1D 8
Honeysuckle Gdns. CR0: C'don 73Zb 158
Honeysuckle Gdns. E17 27Cc 52
Honeysuckle La. N22 26Sb 51
Honeysuckle Pl. KT17: Eps D 88Xa 174
Honeytree Ct. IG10: Lough 12Rc 36
Honeywell Rd. SW11 58Hb 111
Honeywood Cl. EN6: Pot B 5Fb 17
Honeywood Ho. SE15 53Wb 113
(off Goldsmith Rd.)
Honeywood Mus. 78Hb 155
Honeywood Rd. NW10 40Va 68
Honeywood Rd. TW7: Isle 56Ja 108
Honeywood Wlk. SM5: Cars 77Hb 155
Honister Cl. HA7: Stan 25Ka 46
Honister Gdns. HA7: Stan 24Ka 46
Honister Hgts. CR8: Purl 86Tb 177
Honister Pl. HA7: Stan 25Ka 46
Honiton Gdns. NW7 24Za 48
Honiton Gdns. SE15 54Yb 114
(off Gibbon Rd.)
Honiton Ho. EN3: Pond E 13Zb 34
Honiton Rd. DA16: Well 54Vc 117
Honiton Rd. NW6 40Bb 69
Honiton Rd. RM7: Rom 30Fd 56
Honley Rd. SE6 59Dc 114
Honnor Gdns. TW7: Isle 54Fa 108
Honnor Rd. TW18: Staines 66M 127
HONOR OAK 58Yb 114
Honor Oak Crematorium 57Zb 114
Honor Oak Pk. SE23 58Yb 114
HONOR OAK PARK 59Ac 114
Honor Oak Park Station (Rail & Overground) 58Zb 114
Honor Oak Ri. SE23 58Yb 114
Honor Oak Rd. SE23 60Yb 114
Honour Gdns. RM8: Dag 35Wc 75
Honour Lea Av. E20 36Dc 72
Honours Mead HP3: Bov 9C 2
Hood Av. BR5: St M Cry 71Xc 161
Hood Av. N14 16Kb 32
Hood Av. SW14 57Sa 109
Hood Cl. CR0: C'don 74Rb 157
Hoodcote Gdns. N21 17Rb 33
Hood Ct. EC4 3A 224 (44Qb 90)
(off Fleet St.)
Hood Ho. SE5 52Tb 113
(off Elmington Est.)
Hood Ho. SW1 50Mb 90
(off Dolphin Sq.)
Hood Point SE16 47Bc 92
(off Rotherhithe St.)
Hood Rd. RM13: Rain 40Gd 76
Hood Rd. SW20 66Va 132
Hood Wlk. RM7: Mawney 25Dd 56
HOOK 77Ma 153
Hook, The EN5: New Bar 16Fb 31
Hooke Ct. SE10 53Ec 114
(off Winforton St.)
Hooke Ho. E3 40Ac 72
(off Gernon Rd.)
Hooke Rd. KT24: E Hor 97V 190
Hookers Rd. E17 27Zb 52
Hook Farm Rd. BR2: Broml 71Mc 159
Hookfield KT19: Eps 85Sa 173
Hookfield M. KT19: Eps 85Sa 173
Hookfields DA11: Nflt 2A 144
Hook Ga. EN1: Enf 8Xb 19
HOOK GREEN 65Ce 143
HOOK GREEN 63Jd 140
Hook Grn. La. DA2: Wilm 62Hd 140
Hook Grn. Rd. DA13: Sflt 66Ae 143
Hookham Ct. SW8 53Mb 112
HOOK HEATH 2M 187
Hook Heath Av. GU22: Wok 1M 187
Hook Heath Gdns. GU22: Wok 3K 187
Hook Heath Rd. GU22: Wok 3J 187
Hook Hill CR2: Sande 82Ub 177
Hook Hill La. GU22: Wok 3M 187
Hook Hill Pk. GU22: Wok 3M 187
Hooking Grn. HA2: Harr 29Da 45
HOOK JUNC. 76Na 153
Hook La. DA16: Well 56Vc 117
Hook La. EN6: N'thaw 4Hb 17
Hook La. GU24: W End 5A 166
Hook La. RM4: Abr 16Bd 37
Hook La. RM4: Stap A 16Bd 37
Hook Mill La. GU18: Light 1B 166
Hook Ri. Nth. KT6: Surb 76Na 153
Hook Ri. Sth. KT6: Surb 76Na 153
Hook Ri. Sth. Ind. Pk. KT6: Surb 76Pa 153
Hook Rd. KT19: Eps 80Sa 153
Hook Rd. KT19: Ewe 80Sa 153
Hook Rd. KT6: Surb 75Na 153
Hook Rd. KT9: Chess 78Ma 153
Hooks Cl. SE15 53Xb 113
Hooks Hall Dr. RM10: Dag 34Ed 76
Hookstone La. GU24: W End 3D 166
Hookstone Way IG8: Wfd G 24Mc 53
Hook Wlk. HA8: Edg 23Sa 47
Hookwood Cnr. RH8: Limp 100Kc 199
Hookwood Cotts. BR6: Prat B 83Yc 181
Hookwood Cotts. KT18: Head 96Sa 193
Hookwood Pk. RH8: Limp 1M 211
HOOKWOOD PARK 100Kc 199
Hookwood Rd. BR6: Prat B 83Yc 181
Hool Cl. NW9 29Sa 47
HOOLEY 93Kb 196
Hooley La. RH1: Redh 7P 207
Hooper Dr. UB8: Hil 43R 84
Hooper Ho. TW15: Ashf 62N 127
Hooper Rd. E16 44Jc 93
Hooper's Ct. SW3 2F 227 (47Hb 89)
Hooper's M. W3 46Sa 87
Hoopers M. WD23: Bush 18Da 27
Hooper Sq. E1 44Wb 91
(off Hooper St.)
Hooper St. E1 44Wb 91
Hoopers Yd. NW6 39Bb 69
(off Kimberley Rd.)
Hoopers Yd. TN13: S'oaks 98Ld 203
Hoop La. NW11 31Bb 69
Hop Ct. HA0: Wemb 36Ja 66
(off Brewery Cl.)
Hope Cl. CM15: Mount 11Fe 41
Hope Cl. IG8: Wfd G 23Lc 53
Hope Cl. N1 37Sb 71
Hope Cl. NW4 26Ya 48
Hope Cl. RM6: Chad H 28Zc 55
Hope Cl. SE12 62Kc 137
Hope Cl. SM1: Sutt 78Eb 155
Hope Cl. TW8: Bford 50Na 87
Hope Ct. NW10 41Za 88
(off Chamberlayne Rd.)
Hope Ct. SE1 50Wb 91
(off Avocet Cl.)
Hopedale Rd. SE7 51Kc 115
Hopefield Animal Sanctuary 16Yd 40
Hopefield Av. NW6 40Ab 68
Hope Gdns. W3 47Ra 87
Hope Grn. WD25: Wat 5W 12
Hope Ho. CR0: C'don 77Ub 157
(off Steep Hill)
Hope La. SE9 61Rc 138
Hope Pk. BR1: Broml 66Hc 137
Hope Rd. DA10: Swans 58Be 121
Hope Rd. SS17: Stan H 3M 101
Hopes Cl. TW5: Hest 51Ca 107
Hope Sq. EC2 1H 225 (43Ub 91)
(off Sun St. Pas.)
Hope St. E14 44Gc 93
Hope St. SW11 55Fb 111
Hope Taylor Wlk. KT20: Tad 92Ya 194
Hope Ter. RM20: Grays 50Zd 99
Hopetown St. E1 43Vb 91
Hopewell Cl. RM16: Chaf H 50Zd 99
Hopewell Dr. DA12: Grav'nd 4H 145
Hopewell St. SE5 52Tb 113
Hopewell Yd. SE5 52Tb 113
(off Hopewell St.)
Hope Wharf SE16 47Yb 92
Hopfield GU21: Wok 88A 168
Hopfield Av. KT14: Byfl 84N 169
Hopfield Cl. TN14: Otf 88Ld 183
Hopgarden, The SL4: Eton 10H 81
(off Common La.)
Hopgarden La. TN13: S'oaks 100Jd 202
Hop Gdns. WC2 5F 223 (45Nb 90)
Hop Gdn. Way WD25: Wat 3Y 13
Hopgood St. W12 46Ya 88
Hopground Cl. AL1: St A 4E 6
Hopground Ho. E20 37Ec 72
(off De Coubertin St.)
Hopkins Cl. DA1: Dart 54Qd 119
Hopkins Cl. N10 24Jb 50
Hopkins Cl. RM2: Rom 27Ld 57
Hopkins Ho. E14 44Cc 92
(off Canton St.)
Hopkins M. E15 39Hc 73
Hopkinsons Pl. NW1 39Jb 70
Hopkins Rd. E10 31Dc 72
Hopkins St. W1 3C 222 (44Lb 90)
Hopkins Yd. AL1: St A 3C 6
Hoppers Rd. N13 19Qb 32
Hoppers Rd. N21 19Qb 32
Hoppett Rd. E4 19Gc 35
Hoppety, The KT20: Tad 94Za 194
Hopping La. N1 37Rb 71
Hoppingwood Av. KT3: N Mald 69Ua 132
Hoppner Rd. UB4: Hayes 40T 64
Hopps Ct. NW9 26Ua 48
(off Salk Cl.)
Hops Ho. E17 29Bc 52
(off Old Brewery Way)
Hop St. SE10 49Hc 93
Hopton Ct. BR2: Hayes 74Kc 159
Hopton Gdns. KT3: N Mald 72Wa 154
Hopton Rd. SE18 48Rc 94
Hopton Rd. SW16 64Nb 134
Hopton's Gdns. SE1 6C 224 (46Rb 91)
(off Hopton St.)
Hopton St. SE1 5B 224 (45Rb 91)
Hoptree Cl. N12 22Db 49
Hopwood Cl. SW17 62Eb 133
Hopwood Cl. WD17: Wat 8U 12
Hopwood Rd. SE17 51Tb 113
Hopwood Wlk. E8 38Wb 71
Horace Av. RM7: Rush G 32Ed 76
Horace Bldg. SW11 52Kb 112
Horace Jones Ho. SE1 7K 225 (46Vb 91)
(off Duchess Wlk.)
Horace Rd. E7 35Kc 73
Horace Rd. IG6: Ilf 27Sc 54
Horace Rd. KT1: King T 69Pa 131
Horatio Ct. SE16 46Yb 92
(off Rotherhithe St.)
Horatio Ho. E2 40Vb 71
(off Horatio St.)
Horatio Ho. W6 50Za 88
(off Fulham Pal. Rd.)
Horatio Pl. E14 46Ec 92
(off Managers St.)
Horatio Pl. SW19 67Cb 133
Horatio St. E2 40Vb 71
Horatius Way CR0: Wadd 78Pb 156
Horbury Cres. W11 45Cb 89
Horbury M. W11 45Bb 89
Horder Rd. SW6 53Ab 110
Hordle Gdns. AL1: St A 4D 6
Hordle Prom. Sth. SE15 52Vb 113
(off Quarley Way)
Horizon Bldg. E14 45Cc 92
(off Hertsmere Rd.)
Horizon Bus. Cen. N9 19Zb 34
(off Goodwin Rd.)
Horizon Bus. Village KT13: Weyb 84Q 170
Horizon Cl. TN16: Westrm 97Wc 201
Horizon Ct. SM2: Cheam 80Ab 154
(off Up. Mulgrave Rd.)
Horizon Ho. BR8: Swan 70Gd 140
Horizon Ho. SW18 55Eb 111
(off Juniper Dr.)
Horizon Ind. Est. SE15 51Wb 113
Horksley Gdns. CM13: Hut 16Ee 41
Horle Wlk. SE5 54Rb 113
Horley Cl. DA6: Bex 57Cd 118
Horley Rd. RH1: Redh 8P 207
Horley Rd. SE9 63Nc 138
Hormead Rd. W9 42Bb 89
Hornbeam Av. RM14: Upm 35Qd 77
Hornbeam Chase RM15: S Ock 41Zd 99
Hornbeam Cl. CM13: B'wood 20De 41
Hornbeam Cl. CM16: They B 9Tc 22
Hornbeam Cl. IG1: Ilf 36Tc 74
Hornbeam Cl. IG11: Bark 41Wc 95
Hornbeam Cl. IG9: Buck H 20Mc 35
Hornbeam Cl. KT17: Bans 87Ya 174
Hornbeam Cl. NW7 20Va 30
Hornbeam Cl. SE11 5K 229 (49Qb 90)
Hornbeam Cl. UB5: N'olt 36Ba 65
Hornbeam Cl. WD6: Bore 11Qa 29
Hornbeam Cres. TW8: Bford 52Ka 108
Hornbeam Gdns. KT3: N Mald 72Wa 154
Hornbeam Gdns. SL1: Slou 8L 81
Hornbeam Gro. E4 20Gc 35
Hornbeam Ho. IG9: Buck H 20Nc 36
Hornbeam Ho. 47Ac 92
Hornbeam La. AL9: Ess 3N 9
Hornbeam La. DA7: Bex 54Ed 118
Hornbeam La. E4 15Gc 35
Hornbeam Rd. CM16: They B 9Tc 22
Hornbeam Rd. IG9: Buck H 20Mc 35
Hornbeam Rd. RH2: Reig 9K 207
Hornbeam Rd. UB4: Yead 43Y 85
Hornbeams AL2: Brick W 2Ba 13
Hornbeams Av. EN1: Enf 7Yb 20
Hornbeam Sq. E3 39Bc 72
Hornbeams Ri. N11 23Jb 50
Hornbeam Ter. SM5: Cars 74Gb 155
Hornbeam Wlk. KT12: W Vill 81V 170
Hornbeam Wlk. TW10: Rich 61Pa 131
Hornbeam Way BR2: Broml 72Qc 160
Hornbeam Way EN7: Chesh 1Vb 19
Hornbean Ho. E15 41Gc 93
(off Manor Rd.)
Hornbill Cl. UB8: Cowl 44M 83
Hornblower Cl. SE16 48Ac 92
Hornbuckle Cl. HA2: Harr 33Fa 66
Hornby Cl. NW3 38Fb 69
Hornby Ct. NW10 37Va 68
Hornby Ho. SE11 51Qb 112
(off Clayton St.)
Horncastle Cl. SE12 59Jc 115
Horncastle Rd. SE12 59Jc 115
Hornchurch N17 26Tb 51
(off Gloucester Rd.)
HORNCHURCH 32Md 77
Hornchurch Cl. KT2: King T 63Ma 131
Hornchurch Country Pk. 38Ld 77
Hornchurch Hill CR3: Whyt 90Vb 177
Hornchurch Rd. RM11: Horn 32Jd 76
Hornchurch Rd. RM12: Horn 32Jd 76
Hornchurch Sports Cen. 33Ld 77
Hornchurch Stadium 33Qd 77
Hornchurch Station (Underground) 34Md 77
Horndean Cl. SW15 60Wa 110
Horndon Cl. RM5: Col R 25Ed 56
Horndon Grn. RM5: Col R 25Ed 56
Horndon Ind. Pk. CM13: W H'dn 30Ee 59
Horndon Rd. RM5: Col R 25Ed 56
Horndon Rd. SS17: Horn H 2J 101
Horner Ho. N1 1J 219 (39Ub 71)
(off Nuttall St.)
Horner La. CR4: Mitc 68Fb 133
Horne Rd. TW17: Shep 70Q 128
Horner Sq. E1 7K 219 (43Vb 91)
(within Old Spitalfields Mkt.)
Hornet Bus. Est. TN15: Bor G 93Ae 205
Hornets, The WD18: Wat 14X 27
Horne Way E6 43Tc 94
Horne Way SW15 54Ya 110
Hornfair Rd. SE7 51Lc 115
Hornford Way RM7: Rush G 31Gd 76
HORN HILL 21D 42
Horn Hill La. SL9: Chal P 22B 42
Hornhill Rd. WD3: Map C 22D 42
Horniman Dr. SE23 60Xb 113
Horniman Gdns. 60Xb 113
Horniman Mus. 60Xb 113
Horning Cl. SE9 63Nc 138
Horn La. IG8: Wfd G 23Jc 53
Horn La. SE10 50Jc 93
(not continuous)
Horn La. W3 45Sa 87
(not continuous)
Horn Link Way SE10 49Jc 93
Hornminster Glen RM11: Horn 33Qd 77
HORN PARK 57Kc 115
Horn Pk. Cl. SE12 57Kc 115
Horn Pk. La. SE12 57Kc 115
Hornsby La. RM16: Ors 5C 100
Hornscroft Cl. IG11: Bark 38Uc 74
HORNS CROSS 58Ud 120
Horns End Pl. HA5: Eastc 28Y 45
HORNSEY 28Nb 50
Hornsey Cricket Club 29Mb 50
Hornsey La. N6 32Kb 70
Hornsey La. Est. N19 31Mb 70
Hornsey La. Gdns. N6 31Lb 70
Hornsey Pk. Rd. N8 27Pb 50
Hornsey Ri. N19 31Mb 70
Hornsey Ri. Gdns. N19 31Mb 70
Hornsey Rd. N19 32Nb 70
Hornsey Rd. N7 33Pb 70
Hornsey Station (Rail) 28Pb 50
Hornsey St. N7 36Pb 70
HORNSEY VALE 29Pb 50
HORNS GREEN 89Uc 180
Hornshay St. SE15 51Yb 114
Horns Lodge Rd. TN15: Stans 82Ce 185
Horns Rd. IG2: Ilf 29Sc 54
Horns Rd. IG6: Ilf 28Tc 54
Hornton Ct. W8 47Cb 89
(off Kensington High St.)
Hornton Pl. W8 47Db 89
Hornton St. W8 47Cb 89
Horn Yd. DA12: Grav'nd 8D 122
Horsa Rd. DA8: Erith 52Dd 118
Horsa Rd. SE12 59Lc 115
Horse & Dolphin Yd. W1 4E 222 (45Mb 90)
(off Macclesfield St.)
Horsebridge Cl. RM9: Dag 39Ad 75
Horsecroft SM7: Bans 89Bb 175
Horsecroft Cl. BR6: Orp 74Xc 161
Horsecroft Mdws. SM7: Bans 88Bb 175
Horsecroft Rd. HA8: Edg 24Ta 47
Horsecroft Rd. HP1: Hem H 4J 3
Horse Fair KT1: King T 68Ma 131
Horseferry Pl. SE10 51Ec 114
Horseferry Rd. E14 45Ac 92
Horseferry Rd. SW1 4D 228 (48Mb 90)
Horseferry Rd. Est. SW1 4D 228 (48Mb 90)
(off Horseferry Rd.)
Horse Guards Av. SW1 7F 223 (46Nb 90)
Horse Guards Parade 7E 222 (46Mb 90)
Horse Guards Rd. SW1 7E 222 (46Mb 90)
Horse Leaze E6 44Qc 94
Horselers HP3: Hem H 5A 4
Horseley Ct. E1 43Ac 92
HORSELL 8N 167
Horsell Birch GU21: Wok 7L 167
HORSELL COMMON 5P 167
Horsell Comn. Rd. GU21: Wok 6N 167
Horsell Ct. KT16: Chert 73K 149
Horsell Moor GU21: Wok 9P 167
Horsell Pk. GU21: Wok 8P 167
Horsell Pk. Cl. GU21: Wok 8P 167
Horsell Ri. GU21: Wok 7P 167
Horsell Ri. Cl. GU21: Wok 7P 167
Horsell Rd. BR5: St P 67Xc 139
Horsell Rd. N5 36Qb 70
(not continuous)
Horsell Va. GU21: Wok 88A 168
Horsell Way GU21: Wok 8N 167
Horselydown La. SE1 1K 231 (47Vb 91)
Horselydown Mans. SE1 1K 231 (47Vb 91)
(off Lafone St.)
Horseman Side CM14: N'side 15Md 39
HORSEMAN SIDE 14Nd 39
Horseman Side RM4: N'side 17Kd 39
Horseman Side RM4: Stap A 17Kd 39
Horsemans Ride AL2: Chis G 8N 5
Horsemongers M. SE1 2E 230 (47Sb 91)
(off Cole St.)
Horsemoor Cl. SL3: L'ly 49C 82
Horsenden Av. UB6: G'frd 36Ha 66
Horsenden Cres. UB6: G'frd 36Ha 66
Horsenden Hill Footgolf Cen. 37Ja 66
Horsenden Hill Golf Course 38Ja 66
Horsenden La. Nth. UB6: G'frd 37Ga 66
Horsenden La. Sth. UB6: G'frd 39Ja 66
Horse Ride CM16: Epp 7Rc 22
Horse Ride CM16: They B 7Rc 22
Horse Ride SM5: Cars 82Gb 175
Horse Ride SW1 7C 222 (46Lb 90)
Horseshoe, The CR5: Coul 85Mb 176
Horseshoe, The HP3: Hem H 4C 4
Horseshoe, The SM7: Bans 87Bb 175
Horseshoe Bus. Pk. AL2: Brick W 2Da 13
Horseshoe Cloister, The SL4: Wind 3H 103
(within Windor Castle)
Horseshoe Cl. E14 50Ec 92
Horseshoe Cl. EN9: Walt A 6Jc 21
Horseshoe Cl. NW2 33Xa 68
Horseshoe Ct. EC1 5C 218 (42Rb 91)
(off Brewhouse Yd.)
Horse Shoe Cres. UB5: N'olt 40Ca 65
Horseshoe Dr. UB8: Hil 44Q 84
Horse Shoe Grn. SM1: Sutt 75Db 155
Horseshoe Hill EN9: Walt A 5Lc 21
Horseshoe Hill SL1: Burn 5A 60
Horseshoe La. EN2: Enf 13Sb 33
Horseshoe La. N20 18Za 30
Horseshoe La. WD25: Wat 4X 13
Horseshoe M. SW2 56Nb 112
Horseshoe Wharf SE1 6F 225 (46Tb 91)
(off Clink St.)
Horses Ride IG10: Lough 9Pc 22
Horse Yd. N1 39Rb 71
(off Essex Rd.)
Horsfeld Gdns. SE9 57Nc 116
Horsfeld Rd. SE9 57Mc 115
Horsfield Cl. DA2: Dart 59Sd 120

Horsfield Ho. N1 38Sb 71
(off Northampton St.)
Horsford Rd. SW2 57Pb 112
Horsham Av. N12 22Gb 49
Horsham Ct. N17 25Wb 51
(off Lansdowne Rd.)
Horsham Rd. DA6: Bex 57Cd 118
Horsham Rd. TW14: Bedf 58S 106
Horsley Camping & Caravanning Club Site KT24: W Hor 96S 190
Horsley Cl. KT19: Eps 85Ta 173
Horsley Ct. KT24: E Hor 98U 190
Horsley Ct. SW1 6E 228 (49Mb 90)
(off Vincent St.)
Horsley Dr. CR0: New Ad 80Ec 158
Horsley Dr. KT2: King T 64Ma 131
Horsley Rd. BR1: Broml 67Kc 137
Horsley Rd. E4 19Ec 34
Horsley Rd. KT11: D'side 94W 190
Horsleys WD3: Map C 22F 42
Horsley Station (Rail) 97U 190
Horsley St. SE17 51Tb 113
Horsman Ho. SE5 51Sb 113
(off Bethwin Rd.)
Horsmans Pl. DA1: Dart 59Md 119
(off Instone Rd.)
Horsman St. SE5 51Sb 113
Horsmonden Cl. BR6: Orp 73Vc 161
Horsmonden Rd. SE4 57Bc 114
Horsnell Cl. SE5 52Tb 113
Hortensia Ho. SW10 52Eb 111
(off Gunter Gro.)
Hortensia Rd. SW10 52Eb 111
Horticultural Pl. W4 50Ta 87
HORTON 83Sa 173
HORTON 55C 104
Horton Av. NW2 35Ab 68
Horton Bri. Rd. UB7: Yiew 46P 83
Horton Cl. UB7: Yiew 46Q 84
Horton Country Pk. 82Pa 173
Horton Country Pk. Local Nature Reserve 81Pa 173
Horton Cres. KT19: Eps 83Qa 173
Horton Footpath KT19: Eps 83Sa 173
Horton Gdns. KT19: Eps 83Sa 173
Horton Gdns. SL3: Hort 55B 104
Horton Halls SW17 62Fb 133
Horton Hill KT19: Eps 83Sa 173
Horton Ho. SE15 51Yb 114
Horton Ho. SW8 52Pb 112
Horton Ho. W6 50Ab 88
(off Field Rd.)
Horton Ind. Pk. UB7: Yiew 46P 83
HORTON KIRBY 70Sd 142
Horton Kirby Trad. Est. DA4: S Dar 68Sd 142
Horton La. KT19: Eps 84Qa 173
Horton Pde. UB7: Yiew 46N 83
Horton Pk. Golf Course 80Sa 153
Horton Pl. TN16: Westrm 98Tc 200
Horton Rd. DA4: Hort K 70Sd 142
Horton Rd. DA4: S Dar 70Sd 142
Horton Rd. E8 37Xb 71
Horton Rd. SL3: Coln 54C 104
Horton Rd. SL3: Dat 2M 103
Horton Rd. SL3: Hort 2M 103
Horton Rd. SL3: Hort 54C 104
Horton Rd. SL3: Poyle 55G 104
Horton Rd. TW19: Stanw M 56H 105
Horton Rd. UB11: Stock P 46Q 84
Horton Rd. UB7: Yiew 46N 83
Horton Rd. Ind. Est. UB7: Yiew 46P 83
Hortons Way TN16: Westrm 98Tc 200
Horton Trad. Est. SL3: Hort 55E 104
Horton Way CR0: C'don 71Zb 158
Horton Way DA4: Farni 73Pd 163
Hortus Rd. E4 19Ec 34
Hortus Rd. UB2: S'hall 47Ba 85
Horvath Cl. KT13: Weyb 77T 150
Horwood Cl. WD3: Rick 17J 25
Horwood Ct. WD24: Wat 9Z 13
Horwood Ho. E2 41Xb 91
(off Pott St.)
Horwood Ho. NW8 5E 214 (42Gb 89)
(off Paveley St.)
Hosack Rd. SW17 61Jb 134
Hoselands Vw. DA3: Hartl 70Ae 143
Hoser Av. SE12 61Jc 137
Hosey Hill TN16: Westrm 99Uc 200
HOSEY HILL 100Uc 200
Hosier La. EC1 1B 224 (43Rb 91)
Hoskins, The RH8: Oxt 1J 211
(off Station Rd. W.)
Hoskins Cl. E16 44Lc 93
Hoskins Cl. UB3: Harl 50V 84
Hoskins Rd. RH8: Oxt 1J 211
(not continuous)
Hoskins St. SE10 50Fc 93
Hoskins Wlk. RH8: Oxt 1J 211
(off Station Rd. W.)
Hospital Bri. Rd. TW2: Twick 59Da 107
Hospital Bri. Rd. TW2: Whitt 59Da 107
HOSPITAL BRIDGE RDBT. 61Da 129
HOSPITAL FOR TROPICAL DISEASES 6C 216 (42Lb 90)
(off Mortimer Markey)
HOSPITAL OF ST JOHN & ST ELIZABETH 2B 214 (40Fb 69)
Hospital Rd. E11 29Fc 53
Hospital Rd. E9 36Zb 72
Hospital Rd. TN13: S'oaks 93Ld 203
Hospital Rd. TW3: Houn 55Ca 107
Hospital Way SE13 59Fc 115
Hotham Cl. BR8: Swan 67Kd 141
Hotham Cl. DA4: Sut H 66Rd 141
Hotham Cl. KT8: W Mole 69Ca 129
Hotham Rd. SW15 55Ya 110
Hotham Rd. SW19 66Eb 133
Hotham Rd. M. SW19 66Eb 133
Hotham St. E15 39Gc 73
Hothfield Pl. SE16 48Yb 92
Hotspur Ind. Est. N17 23Xb 51
Hotspur Rd. UB5: N'olt 40Ca 65
Hotspur St. SE11 6K 229 (49Qb 90)
Hotspur Way EN2: Enf 7Vb 19
Hottsfield DA3: Hartl 69Ae 143
Houblon Rd. TW10: Rich 57Na 109
Houblons Hill CM16: Coop 3Yc 23
Houghton Cl. E8 37Vb 71
Houghton Cl. TW12: Hamp 65Aa 129
Houghton Ct. EC1 6D 218 (42Sb 91)
(off Glasshouse Yd.)
Houghton Rd. N15 28Vb 51
Houghton Sq. SW9 54Nb 112
(off Clapham Rd.)
Houghton St. WC2 3J 223 (44Pb 90)
(not continuous)
Houlder Cres. CR0: Wadd 79Rb 157
Houlton Ho. SW3 7F 227 (50Hb 89)
(off Walpole St.)
Houlton Pl. E3 42Bc 92
(off Hamlets Way)
Houndsden Rd. N21 16Pb 32
Houndsditch EC3 2J 225 (44Ub 91)
Houndsfield Rd. N9 17Xb 33
HOUNSLOW 55Da 107
Hounslow & District Indoor Bowls Club 54Ba 107
Hounslow Av. TW3: Houn 57Da 107
Hounslow Bus. Pk. TW3: Houn 56Ca 107
Hounslow Central Station (Underground) 55Da 107
Hounslow Cen. TW3: Houn 55Da 107
Hounslow East Station (Underground) 54Ea 108
Hounslow Gdns. TW3: Houn 57Da 107
Hounslow Heath Local Nature Reserve 58Aa 107
Hounslow Rd. TW13: Hanw 63Z 129
Hounslow Rd. TW14: Felt 60X 107
Hounslow Rd. TW2: Whitt 58Da 107
Hounslow Station (Rail) 57Da 107
Hounslow Urban Farm 57W 106
HOUNSLOW WEST 55Aa 107
Hounslow West Station (Underground) 54Aa 107
Housefield Way AL4: St A 5G 6
Household Cavalry Mus., The 7E 222 (46Mb 90)
House Mill, The 41Ec 92
House of Illustration 1F 217 (39Nb 70)
Houses of Parliament 3G 229 (48Nb 90)
Houston Bus. Pk. UB4: Yead 46Y 85
Houston Pl. KT10: Esh 74Ga 152
Houston Rd. KT6: Surb 72Ka 152
Houston Rd. SE23 61Ac 136
Houstoun Ct. TW5: Hest 52Ba 107
Hove Av. E17 29Bc 52
Hove Cl. CM13: Hut 19Ee 41
Hove Cl. RM17: Grays 51Ce 121
Hoveden Rd. NW2 36Ab 68
Hove Gdns. SM1: Sutt 74Db 155
Hove St. SE15 52Yb 114
(off Culmore Rd.)
Hoveton Rd. SE28 44Yc 95
Hoveton Way IG6: Ilf 24Rc 54
Howard Agne Cl. HP3: Bov 9C 2
Howard Av. DA5: Bexl 60Yc 117
Howard Av. KT17: Ewe 82Wa 174
Howard Av. SL2: Slou 3H 81
Howard Bldg. SW11 51Kb 112
Howard Bus. Pk. EN9: Walt A 6Fc 21
Howard Cl. AL1: St A 4G 6
Howard Cl. EN9: Walt A 5Fc 21
Howard Cl. IG10: Lough 16Nc 36
Howard Cl. KT20: Walt H 97Va 194
Howard Cl. KT21: Asht 90Pa 173
Howard Cl. KT22: Lea 95La 192
Howard Cl. KT24: W Hor 97T 190
Howard Cl. N11 19Jb 32
Howard Cl. NW2 35Ab 68
Howard Cl. TW12: Hamp 66Ea 130
Howard Cl. TW16: Sun 65V 128
Howard Cl. W3 44Ra 87
Howard Cl. WD23: B Hea 17Ga 28
Howard Cl. WD24: Wat 9W 12
Howard Ct. AL1: St A 4B 6
(off Cottonmill La.)
Howard Ct. GU21: Knap 1G 186
(off Tudor Way)
Howard Ct. IG11: Bark 39Tc 74
Howard Ct. RH2: Reig 5L 207
Howard Dr. WD6: Bore 14Ta 29
Howard Ho. E16 46Kc 93
(off Wesley Av.)
Howard Ho. SE8 51Bc 114
(off Evelyn St.)
Howard Ho. SW1 50Lb 90
(off Dolphin Sq.)
Howard Ho. SW9 55Rb 113
(off Barrington Rd.)
Howard Ho. W1 6A 216 (42Kb 90)
(off Cleveland St.)
Howard M. N5 35Rb 71
Howard Pl. KT13: Weyb 78S 150
Howard Pl. RH2: Reig 4J 207
Howard Rd. BR1: Broml 66Jc 137
Howard Rd. CR5: Coul 87Lb 176
Howard Rd. DA1: Dart 58Qd 119
Howard Rd. E11 34Gc 73
Howard Rd. E17 27Cc 52
Howard Rd. E6 40Pc 74
Howard Rd. HA7: Stan 25Ma 47
Howard Rd. IG1: Ilf 35Rc 74
Howard Rd. IG11: Bark 39Tc 74
Howard Rd. KT23: Bookh 99Da 191
Howard Rd. KT24: Eff J 95W 190
Howard Rd. KT3: N Mald 69Ua 132
Howard Rd. KT5: Surb 72Pa 153
Howard Rd. N15 30Ub 51
Howard Rd. N16 35Tb 71
Howard Rd. NW2 35Za 68
Howard Rd. RH2: Reig 7K 207
Howard Rd. RM14: Upm 33Sd 78
Howard Rd. RM16: Chaf H 48Yd 98
Howard Rd. SE20 67Yb 136
Howard Rd. SE25 71Wb 157
Howard Rd. TW7: Isle 55Ha 108
Howard Rd. UB1: S'hall 44Da 85
Howards Cl. GU22: Wok 92C 188
Howards Cl. HA5: Pinn 26X 45
Howards Crest Cl. BR3: Beck 68Ec 136
Howards Ga. SL2: Farn R 1F 80
Howards Ho. RH2: Reig 5K 207
Howard's La. SW15 56Xa 110
Howards La. KT15: Add 79H 149
Howards Rd. E13 41Jc 93
Howards Rd. GU22: Wok 92B 188
Howards Thicket SL9: Ger X 3N 61
Howard St. KT7: T Ditt 73Ka 152
Howards Wood Dr. SL9: Ger X 3P 61
Howard Venue, The 66Hd 140
Howard Wlk. N2 28Eb 49
Howard Way EN5: Barn 15Za 30
Howarth Rd. SE2 50Wc 95
Howberry Cl. HA8: Edg 23Ma 47
Howberry Rd. CR7: Thor H 67Tb 135
Howberry Rd. HA7: Stan 23Ma 47
Howberry Rd. HA8: Edg 23Ma 47
Howburgh Ct. RM19: Purf 50Rd 97
(off Wingrove Dr.)
Howbury La. DA8: Erith 54Jd 118
Howbury Rd. SE15 55Yb 114
Howcroft Cres. N3 24Cb 49
Howcroft Ho. E3 41Bc 92
(off Benworth St.)
Howcroft La. UB6: G'frd 41Fa 86
Howden Cl. SE28 45Zc 95
Howden Dr. KT15: Add 80J 149
Howden Rd. SE25 68Vb 135
Howden St. SE15 55Wb 113
Howe Cl. RM7: Mawney 25Cd 56
Howe Cl. WD7: Shenl 4Na 15
Howe Dr. CR3: Cat'm 94Tb 197
Howell Cl. RM6: Chad H 29Zc 55
Howell Ct. E10 31Dc 72
Howell Hill SM2: Cheam 82Ya 174
Howell Hill Cl. KT17: Ewe 83Ya 174
Howell Hill Gro. KT17: Ewe 83Ya 174
Howell Hill Nature Reserve 83Za 174
Howells Cl. TN15: W King 79Ud 164
Howell Wlk. SE1 6C 230 (49Rb 91)
Howerd Way SE18 53Nc 116
(not continuous)
Howe Rd. HP3: Hem H 4A 4
Howes Cl. N3 27Cb 49
Howeth Ct. N11 23Hb 49
(off Ribblesdale Av.)
Howfield Pl. N17 27Vb 51
Howgate Rd. SW14 55Ta 109
Howick Pl. SW1 4C 228 (48Lb 90)
Howie St. SW11 52Gb 111
Howitt Cl. N16 35Ub 71
Howitt Cl. NW3 37Gb 69
Howitt Rd. NW3 37Gb 69
Howitts Cl. KT10: Esh 79Ca 151
Howland Cl. HA5: Hat E 23Ca 45
Howland Est. SE16 48Yb 92
Howland Gth. AL1: St A 6A 6
Howland Ho. SW16 62Nb 134
Howland M. E. W1 7C 216 (43Lb 90)
Howlands TN15: Wro 87Be 185
Howlands Ct. TN15: Wro 87Be 185
(off Howlands)
Howland St. W1 7C 216 (43Lb 90)
Howland Way SE16 47Ac 92
How La. CR5: Chips 89Jb 176
Howlett Apts. N1 38Nb 70
(off Caledonian Rd.)
Howletts La. HA4: Ruis 29S 44
Howlett's Rd. SE24 58Sb 113
Howley Pl. W2 7A 214 (43Eb 89)
Howley Rd. CR0: C'don 76Rb 157
Hows Cl. UB8: Uxb 39L 63
Howse Rd. EN9: Walt A 7Dc 20
Howsman Rd. SW13 51Wa 110
Howson Rd. SE4 56Ac 114
Howson Ter. TW10: Rich 58Na 109
Hows Rd. UB8: Uxb 39L 63
How's St. E2 1K 219 (40Vb 71)
Howton Pl. WD23: B Hea 18Fa 28
How Wood AL2: Park 10P 5
HOW WOOD 9A 6
How Wood Station (Rail) 10A 6
Hox Pk. Dr. TW20: Eng G 3P 125
HOXTON 2H 219 (40Ub 71)
Hoxton Hall Theatre 2J 219 (40Ub 71)
(off Hoxton St.)
Hoxton Mkt. N1 4H 219 (41Ub 91)
(off Coronet St.)
Hoxton Sq. N1 4H 219 (41Ub 91)
Hoxton Station (Overground) 2K 219 (40Vb 71)
Hoxton St. N1 1H 219 (39Ub 71)
Hoylake Cl. SL1: Slou 7C 80
Hoylake Cres. UB10: Ick 33Q 64
Hoylake Gdns. CR4: Mitc 69Lb 134
Hoylake Gdns. HA4: Ruis 32X 65
Hoylake Gdns. RM3: Hrld W 24Qd 57
Hoylake Gdns. WD19: Wat 21Z 45
Hoylake Rd. W3 44Ua 88
Hoyland Cl. SE15 52Xb 113
Hoyle Rd. SW17 64Gb 133
Hoy St. E16 44Hc 93
Hoy Ter. RM20: Grays 50Zd 99
HQS Wellington 5K 223 (45Qb 90)
(off Victoria Embankment)
Hub Westminster, The 2G 215 (40Hb 69)
Hub, The TW20: Egh 64B 126
Hubbard Ct. IG10: Lough 17Nc 36
Hubbard Dr. KT9: Chess 79Ma 153
Hubbard Ho. SW10 52Fb 111
(off World's End Pas.)
Hubbard Rd. SE27 63Sb 135
Hubbards Chase RM11: Horn 29Qd 57
Hubbards Cl. RM11: Horn 29Qd 57
Hubbards Cl. UB8: Hil 44R 84
Hubbards Rd. WD3: Chor 15F 24
Hubbard St. E15 39Gc 73
Hubbinet Ind. Est. RM7: Mawney 27Ed 56
Huberd Ho. SE1 3G 231 (48Tb 91)
(off Manciple St.)
Hubert Cres. WD23: Bush 15Fa 28
Hubert Gro. SW9 55Nb 112
Hubert Ho. NW8 6D 214 (42Gb 89)
(off Ashbridge St.)
Hubert Rd. CM14: B'wood 20Xd 40
Hubert Rd. E6 41Mc 93
Hubert Rd. RM13: Rain 41Hd 96
Hubert Rd. SL3: L'ly 8P 81
Hucknall Cl. RM3: Rom 23Pd 57
Hucknall Ct. NW8 5B 214 (42Fb 89)
(off Cunningham Pl.)
Huddart St. E3 43Bc 92
(not continuous)
Huddleston Cl. E2 40Yb 72
Huddleston Cres. RH1: Mers 100Mb 196
Huddlestone Rd. E7 35Hc 73
Huddlestone Rd. NW2 37Xa 68
Huddleston Rd. N7 34Lb 70
Hudson NW9 25Va 48
(off Five Acre)
Hudson Apts. N8 27Pb 50
Hudson Bldg. E1 43Wb 91
(off Chicksand St.)
Hudson Cl. AL1: St A 5B 6
Hudson Cl. DA12: Grav'nd 3E 144
Hudson Cl. E15 39Jc 73
Hudson Cl. W12 45Xa 88
Hudson Cl. WD24: Wat 3V 12
Hudson Ct. E14 50Cc 92
(off Maritime Quay)
Hudson Gdns. BR6: Chels 79Vc 161
Hudson Ho. KT19: Eps 85Ta 173
Hudson Ho. SW10 52Eb 111
(off Hortensia Rd.)
Hudson Ho. W11 44Ab 88
(off Ladbroke Gro.)
Hudson Pl. SE18 50Sc 94
Hudson Pl. SL3: L'ly 50B 82
Hudson Rd. DA7: Bex 54Bd 117
Hudson Rd. UB3: Harl 51T 106
Hudsons KT20: Tad 93Za 194
Hudsons Cl. SS17: Stan H 1M 101
Hudsons Ct. EN6: Pot B 3Cb 17
Hudson's Pl. SW1 5A 228 (49Kb 90)
Hudson Way E16 45Sc 94
Hudson Way N9 20Yb 34
Hudson Way NW2 34Za 68
Hugero Point SE10 48Jc 93
Huggens Coll. DA11: Nflt 57De 121
Huggin Ct. EC4 4E 224 (45Sb 91)
(off Huggin Hill)
Huggin Hill EC4 4E 224 (45Sb 91)
Huggins Ho. E3 41Cc 92
(off Alfred St.)
Huggins La. AL9: Wel G 5E 8
Huggins Pl. SW2 60Pb 112
Hughan Rd. E15 36Fc 73
Hugh Astor Ct. SE1 3C 230 (48Rb 91)
(off Keyworth St.)
Hugh Clark Ho. W13 46Ja 86
(off Singapore Rd.)
Hugh Cubitt Ho. N1 2J 217 (40Pb 70)
(off Collier St.)
Hugh Dalton Av. SW6 51Bb 111
Hughenden Av. HA3: Kenton 29Ka 46
Hughenden Gdns. UB5: N'olt 41Y 85
(not continuous)
Hughenden Ho. NW8 5D 214 (42Gb 89)
(off Jerome Cres.)
Hughenden Rd. KT4: Wor Pk 73Wa 154
Hughenden Rd. SL1: Slou 4H 81
Hughendon EN5: New Bar 14Db 31
Hughendon Ct. UB3: Hayes 45V 84
(off Chamberlain Cl.)
Hughendon Ter. E15 35Ec 72
Hughes Cl. N12 22Eb 49
Hughes Ct. N7 36Mb 70
Hughes Ho. E2 41Yb 92
(off Sceptre Ho.)
Hughes Ho. E3 41Cc 92
(off Trevithick Way)
Hughes Ho. SE17 6C 230 (49Rb 91)
(off Peacock St.)
Hughes Ho. SE5 53Sb 113
(off Flodden Rd.)
Hughes Ho. SE8 51Cc 114
(off Benbow St.)
Hughes Mans. E1 42Wb 91
Hughes Rd. IG6: Ilf 23Tc 54
Hughes Rd. RM16: Grays 8C 100
Hughes Rd. TW15: Ashf 66S 128
Hughes Rd. UB3: Hayes 45X 85
Hughes Ter. SW9 55Rb 113
(off Styles Gdns.)
Hughes Wlk. CR0: C'don 73Sb 157
Hugh Gaitskell Cl. SW6 51Bb 111
Hugh Gaitskell Ho. N16 33Vb 71
Hugh Herland Ho. KT1: King T 69Na 131
Hugh M. SW1 6A 228 (49Kb 90)
Hugh Platt Ho. E2 40Xb 71
(off Patriot Sq.)
Hugh St. SW1 6K 227 (49Kb 90)
Hugo Cl. WD18: Wat 14U 26
Hugo Gdns. RM13: Rain 37Jd 76
Hugo Gryn Way WD7: Shenl 3Na 15
Hugo Ho. SW1 3G 227 (48Hb 89)
(off Sloane St.)
Hugon Rd. SW6 55Db 111
Hugo Rd. N19 35Lb 70
Huguenot Dr. N13 22Qb 50
Huguenot Pl. E1 43Vb 91
Huguenot Pl. SW18 57Eb 111
Huguenot Sq. SE15 55Xb 113
HULBERRY 75Jd 162
Hullbridge M. N1 39Tb 71
Hull Cl. SE16 47Zb 92
Hull Cl. SL1: Slou 7G 80
Hullett's La. CM15: Pil H 13Ud 40
Hull Pl. E16 46Sc 94
Hull St. EC1 4D 218 (41Sb 91)
Hulme Pl. SE1 2E 230 (47Sb 91)
Hulse Av. IG11: Bark 37Tc 74
Hulse Av. RM7: Mawney 25Dd 56
Hulse Ter. IG1: Ilf 37Sc 74
Hulsewood Cl. DA2: Wilm 62Kd 141
Hult Intl. Studios E1 44Wb 91
(off Alder St.)
Hulton Cl. KT22: Lea 95La 192
Hult Twr. E1 44Wb 91
(off Alder St.)
Hulverston Cl. SM2: Sutt 82Db 175
Humber Av. RM15: S Ock 44Vd 98
Humber Cl. UB7: W Dray 46M 83
Humber Ct. W7 44Fa 86
(off Hobbayne Rd.)
Humber Dr. RM14: Upm 30Td 58
Humber Dr. W10 42Za 88
Humber Rd. DA1: Dart 57Md 119
Humber Rd. NW2 33Xa 68
Humber Rd. SE3 51Hc 115
Humberstone Cl. WD25: Wat 6W 12
Humberstone Rd. E13 41Lc 93
Humberton Cl. E9 36Ac 72
Humber Trad. Est. NW2 33Xa 68
Humber Way SL3: L'ly 49C 82
Humbleward Pl. RM3: Rom 22Nd 57
Humbolt Rd. W6 51Ab 110
Hume Av. RM18: Tilb 5D 122
Hume Cl. RM18: Tilb 5C 122
Hume Ct. N1 38Rb 71
(off Hawes St.)
Hume Ho. W11 46Za 88
(off Queensdale Cres.)
Hume M. RM18: Tilb 4D 122
Humes Av. W7 48Ga 86
Hume Ter. E16 43Kc 93
Hume Way HA4: Ruis 30W 44
Hummer Rd. TW20: Egh 63C 126
Humphrey Cl. IG5: Ilf 25Pc 54
Humphrey Cl. KT22: Fet 94Ea 192
Humphrey St. SE1 7K 231 (50Vb 91)
Humphries Cl. RM9: Dag 35Bd 75
Humphry Repton Way HA9: Wemb 35Qa 67
Hundred Acre NW9 26Va 48
Hundred Acres EN8: Walt C 6Ac 20
Hungerdown E4 18Ec 34
Hungerford Av. SL2: Slou 3J 81
Hungerford Bridge 6H 223 (46Pb 90)
Hungerford Ho. SW1 51Lb 112
(off Churchill Gdns.)
Hungerford La. WC2 6F 223 (46Nb 90)
(off Craven St.)
Hungerford Rd. N7 37Mb 70
Hungerford Sq. KT13: Weyb 77T 150
Hungerford St. E1 44Xb 91
Hungry Hill La. GU23: Rip 98L 189
Hungry Hill La. GU23: Send 98L 189
Hunsdon Cl. RM9: Dag 37Ad 75
Hunsdon Dr. TN13: S'oaks 95Kd 203
Hunsdon Rd. SE14 52Zb 114
Hunslett St. E2 41Yb 92
Hunstanton Cl. SL3: Coln 52E 104
Hunstanton Ho. NW1 7E 214 (43Gb 89)
(off Cosway St.)
Hunston Rd. SM4: Mord 74Db 155
Hunt Cl. W11 46Za 88
Hunt Ct. N14 17Kb 32
Hunt Ct. RM7: Rush G 30Gd 56
(off Union Rd.)
Hunt Ct. UB5: N'olt 40Z 65
(off Gallery Gdns.)
Hunter Av. CM15: Shenf 16Ce 41
Hunter Cl. EN6: Pot B 5Db 17
Hunter Cl. SE1 4G 231 (48Tb 91)
Hunter Cl. SM6: W'gton 80Nb 156
Hunter Cl. SW12 60Jb 112
Hunter Cl. WD6: Bore 15Sa 29
Huntercombe Cl. SL6: Tap 4A 80
HUNTERCOMBE HOSPITAL MAIDENHEAD 5A 80
HUNTERCOMBE HOSPITAL ROEHAMPTON 59Wa 110
Huntercombe La. Nth. SL1: Burn 3A 80
Huntercombe La. Nth. SL1: Slou 3A 80
Huntercombe La. Nth. SL6: Tap 4A 80
Huntercombe La. Sth. SL6: Tap 6A 80
Hunter Ct. KT19: Eps 82Qa 173
Hunter Ct. SL1: Slou 3A 80
Huntercrombe Gdns. WD19: Wat 21Y 45
Hunter Dr. RM12: Horn 35Ld 77
Hunter Ho. SE1 2C 230 (47Rb 91)
(off King James St.)
Hunter Ho. SW5 50Cb 89
(off Old Brompton Rd.)
Hunter Ho. SW8 52Mb 112
(off Fount St.)
Hunter Ho. TW13: Felt 60W 106
(off Hazel Gro.)
Hunter Ho. WC1 5F 217 (42Nb 90)
(off Hunter St.)
Hunterian Mus. 2J 223 (44Pb 90)
Hunter Lodge W9 43Cb 89
(off Admiral Wlk.)
Hunter Rd. CR7: Thor H 69Tb 135
Hunter Rd. IG1: Ilf 36Rc 74
Hunter Rd. SW20 67Ya 132
Hunters Chase RH9: S God 9D 210
Hunters Cl. DA5: Bexl 62Gd 140
Hunters Cl. HP3: Bov 1C 10
Hunters Cl. KT19: Eps 85Sa 173
Hunters Ct. TW9: Rich 57Ma 109
Huntersfield Cl. RH2: Reig 3K 207
Hunters Ga. RH1: Nutf 5F 208
Hunters Ga. WD25: Wat 5W 12
Hunters Gro. BR6: Farnb 77Sc 160
Hunters Gro. HA3: Kenton 28La 46
Hunters Gro. RM5: Col R 22Dd 56
Hunters Gro. UB3: Hayes 46W 84
Hunters Hall Rd. RM10: Dag 35Cd 76
Hunters Hill HA4: Ruis 34Y 65
Hunter's La. WD25: Wat 5V 12
Hunters Mdw. SE19 63Ub 135
Hunters M. SL4: Wind 3G 102
Hunters Reach EN7: Chesh 1Vb 19
Hunters Ride AL2: Brick W 3Ca 13
Hunter's Rd. KT9: Chess 76Na 153
Hunters Sq. RM10: Dag 35Cd 76
Hunter St. WC1 5G 217 (42Nb 90)
Hunters Wlk. TN14: Knock 86Ad 181
Hunter's Way CR0: C'don 77Ub 157
Hunters Way EN2: Enf 11Qb 32
Hunters Way SL1: Slou 6C 80
Hunter Wlk. E13 40Jc 73
Hunter Wlk. WD6: Bore 15Ta 29
Hunting Cl. KT10: Esh 77Ca 151
Huntingdon Cl. CR4: Mitc 69Nb 134
Huntingdon Cl. UB5: N'olt 37Ca 65
Huntingdon Dr. RM3: Hrld W 25Nd 57
Huntingdon Gdns. KT4: Wor Pk 76Ya 154
Huntingdon Gdns. W4 52Sa 109
Huntingdon Rd. GU21: Wok 9K 167
Huntingdon Rd. N2 27Gb 49
Huntingdon Rd. N9 19Yb 34
Huntingdon Rd. RH1: Redh 6P 207
Huntingdon St. E16 44Hc 93
Huntingdon St. N1 38Pb 70
Huntingfield CR0: Sels 80Bc 158
Huntingfield Rd. SW15 56Wa 110
Huntingfield Way TW20: Egh 66F 126
Hunting Ga. Cl. EN2: Enf 13Qb 32
Hunting Ga. Dr. KT9: Chess 80Na 153
Hunting Ga. M. SM1: Sutt 76Db 155
Hunting Ga. M. TW2: Twick 60Ga 108
Hunting Pl. TW5: Hest 51Ba 107
Huntings Farm IG1: Ilf 33Uc 74
Huntings Rd. RM10: Dag 37Cd 76
Huntington Cl. DA5: Bexl 60Dd 118
Huntington Ho. SW11 52Kb 112
(off Palmer Rd.)
Huntington Pl. SL3: L'ly 48D 82
Huntland Cl. RM13: Rain 43Kd 97
Huntley Av. DA11: Nflt 58De 121
Huntley Cl. SE10 50Gc 93
Huntley Cl. TW19: Stanw 59N 105
Huntley Ho. KT12: W Vill 81V 170
Huntley St. WC1 6C 216 (42Lb 90)
Huntley Way SW20 68Wa 132
Huntloe Ho. SE14 53Yb 114
(off Kender St.)
Huntly Dr. N3 23Cb 49
Huntly Rd. SE25 70Ub 135
HUNTON BRIDGE 5S 12
HUNTON BRIDGE 7T 12
Hunton Bri. Hill WD4: Hunt C 5S 12
Hunton Bridge Lock 6S 12
(off Old Mill Rd.)
Hunton Cl. WD4: Hunt C 5S 12
Hunton St. E1 43Wb 91
Hunt Rd. DA11: Nflt 2A 144
Hunt Rd. UB2: S'hall 48Ca 86
Hunt's Cl. SE3 54Jc 115
Hunt's Ct. WC2 5E 222 (45Mb 90)
Hunts Farm Cl. TN15: Bor G 92Ce 205
Huntshaw Ho. E3 41Dc 92
(off Devons Rd.)
Hunts La. E15 40Ec 72
Huntsman La. TN15: Wro H 90Fe 185
Huntsman Rd. IG6: Ilf 22Xc 55
Huntsmans Cl. CR6: W'ham 91Yb 198
Huntsmans Cl. KT22: Fet 96Fa 192
Huntsmans Cl. TW13: Felt 63X 129
Huntsmans Ct. CR3: Cat'm 93Sb 197
(off Coulsdon Rd.)
Huntsmans Dr. RM14: Upm 36Sd 78
Huntsman St. SE17 6H 231 (49Ub 91)
Hunts Mead EN3: Enf H 13Zb 34
Hunts Mead Cl. BR7: Chst 66Pc 138
Huntsmill Rd. HP1: Hem H 3G 2
Huntsmoor Rd. KT19: Ewe 78Ta 153
Huntspill St. SW17 62Eb 133
Hunts Slip Rd. SE21 62Ub 135
Huntsworth M. NW1 5F 215 (42Hb 89)
Hurdwick Ho. NW1 1B 216 (40Lb 70)
(off Harrington Sq.)

Hurdwick Pl. NW1 1B 216 (40Lb 70)
(off Hampstead Rd.)
Hurleston Ho. SE8 50Bc 92
Hurley Cl. KT12: Walt T 75X 151
Hurley Cl. SM7: Bans 88Bb 175
Hurley Cl. SW17 65Jb 134
(off Mitcham Rd.)
Hurley Ct. W5 44La 86
Hurley Cres. SE16 47Zb 92
Hurley Ho. SE11 6B 230 (49Rb 91)
Hurley Ho. UB7: W Dray 47P 83
(off Park Lodge Av.)
Hurley Rd. UB6: G'frd 44Da 86
Hurlfield DA2: Wilm 62Ld 141
Hurlford GU21: Wok 9L 167
HURLINGHAM 55Db 111
Hurlingham Bus. Pk. SW6 55Cb 111
Hurlingham Club, The 55Cb 111
Hurlingham Ct. SW6 55Bb 111
Hurlingham Gdns. SW6 55Bb 111
Hurlingham Pk. 54Bb 111
Hurlingham Retail Pk. 55Db 111
Hurlingham Rd. DA7: Bex 52Bd 117
Hurlingham Rd. SW6 54Bb 111
Hurlingham Sq. SW6 55Cb 111
Hurlingham Yacht Club 55Ab 110
Hurlock St. N5 34Rb 71
Hurlstone Rd. SE25 71Ub 157
Hurn Ct. TW4: Houn 54Z 107
Hurn Ct. Rd. TW4: Houn 54Z 107
Hurnford Cl. CR2: Sande 82Ub 177
Huron Cl. BR6: Chels 79Uc 160
Huron Rd. SW17 61Jb 134
Hurrell Dr. HA2: Harr 27Ea 46
Hurren Cl. SE3 55Gc 115
Hurricane Rd. SM6: W'gton 80Nb 156
Hurricane Trad. Cen. NW9 25Wa 48
Hurricane Way SL3: L'ly 50D 82
Hurricane Way WD5: Ab L 4W 12
Hurry Cl. E15 38Gc 73
Hursley Rd. IG7: Chig 22Vc 55
Hurst, The TN11: Roug 100Fe 205
Hurst, The TN15: Crou 97Ee 205
Hurst Av. E4 21Cc 52
Hurst Av. N6 30Lb 50
Hurstbourne KT10: Clay 79Ha 152
Hurstbourne Gdns. IG11: Bark 37Uc 74
Hurstbourne Ho. SW15 58Va 110
(off Tangley Gro.)
Hurstbourne Rd. SE23 60Ac 114
Hurst Cl. BR2: Hayes 74Hc 159
Hurst Cl. E4 20Cc 34
Hurst Cl. GU22: Wok 2N 187
Hurst Cl. KT18: Head 96Sa 193
Hurst Cl. KT9: Chess 78Qa 153
Hurst Cl. NW11 30Db 49
Hurst Cl. UB5: N'olt 37Ba 65
Hurstcombe IG9: Buck H 19Jc 35
Hurst Ct. DA15: Sidc 61Wc 139
Hurst Ct. E6 43Mc 93
(off Tollgate Rd.)
Hurst Ct. IG8: Wfd G 23Kc 53
(off Snakes La. W.)
Hurst Ct. WD17: Wat 10W 12
Hurstcourt Rd. SM1: Sutt 75Db 155
Hurstdene Av. BR2: Hayes 74Hc 159
Hurstdene Av. TW18: Staines 65K 127
Hurstdene Gdns. N15 31Ub 71
Hurst Dr. EN8: Walt C 6Zb 20
Hurst Dr. KT20: Walt H 98Wa 194
Hurstfield BR2: Broml 71Jc 159
Hurstfield Cres. UB4: Hayes 42U 84
Hurstfield Dr. SL6: Tap 4A 80
Hurstfield Rd. KT8: W Mole 69Ca 129
HURST GREEN 4L 211
Hurst Grn. Cl. RH8: Oxt 4L 211
Hurst Grn. Rd. RH8: Oxt 4K 211
Hurst Green Station (Rail) 4K 211
Hurst Gro. KT12: Walt T 74V 150
Hurst Ho. WC1 2J 217 (40Pb 70)
(off Penton Ri.)
Hurstlands RH8: Oxt 4L 211
Hurstlands Cl. RM11: Horn 31Ld 77
Hurstlands Dr. BR6: Chels 76Yc 161
Hurst La. KT18: Head 96Sa 193
Hurst La. KT8: E Mos 70Ea 130
Hurst La. SE2 50Zc 95
Hurst La. TW20: Egh 68C 126
Hurst La. Est. SE2 50Zc 95
Hurstleigh WD3: Chor 15E 24
Hurstleigh Cl. RH1: Redh 4P 207
Hurstleigh Dr. RH1: Redh 4P 207
Hurstleigh Gdns. IG5: Ilf 25Pc 54
Hurst Lodge KT13: Weyb 79T 150
(off Gower Rd.)
Hurstmead Ct. HA8: Edg 21Ra 47
HURST PARK 68Ea 130
Hurst Pk. Av. RM12: Horn 35Nd 77
Hurst Pl. DA1: Dart 58Ld 119
Hurst Pl. HA6: Nwood 25R 44
Hurst Pool 69Da 129
Hurst Ri. EN5: New Bar 13Cb 31
Hurst Rd. CR0: C'don 78Tb 157
Hurst Rd. DA15: Bexl 61Wc 139
Hurst Rd. DA15: Sidc 61Wc 139
Hurst Rd. DA5: Bexl 60Zc 117
Hurst Rd. DA8: Erith 53Ed 118
Hurst Rd. E17 27Dc 52
Hurst Rd. IG9: Buck H 18Mc 35
Hurst Rd. KT12: Walt T 71Y 151
Hurst Rd. KT18: Head 95Ta 193
Hurst Rd. KT19: Eps 83Ta 173
Hurst Rd. KT20: Walt H 95Va 194
Hurst Rd. KT8: E Mos 69Da 129
Hurst Rd. KT8: W Mole 69Da 129
Hurst Rd. N21 18Qb 32
Hurst Rd. SL1: Slou 3B 80
Hurst Springs DA5: Bexl 60Ad 117
Hurst St. SE24 58Rb 113
Hurstview Grange CR2: S Croy 80Rb 157
Hurst Vw. Rd. CR2: S Croy 80Ub 157
Hurst Way CR2: S Croy 79Ub 157
Hurst Way GU22: Pyr 86G 168
Hurst Way TN13: S'oaks 99Ld 203
Hurstway Rd. W11 45Za 88
(off Hurstway Wlk.)
Hurstway Wlk. W11 45Za 88
Hurstwood Av. CM15: Pil H 17Xd 40
Hurstwood Av. DA5: Bexl 60Ad 117
Hurstwood Av. DA7: Bex 53Gd 118
Hurstwood Av. DA8: Erith 53Gd 118
Hurstwood Av. E18 28Kc 53
Hurstwood Ct. N12 23Gb 49
Hurstwood Ct. NW11 28Bb 49
(off Finchley Rd.)
Hurstwood Ct. RM14: Upm 32Sd 78
Hurstwood Dr. BR1: Broml 69Pc 138
Hurstwood Rd. NW11 28Ab 48
Hurtwood Rd. KT12: Walt T 73Ba 151
Hurworth Av. SL3: L'ly 8N 81
Husborne Ho. SE8 49Ac 92
(off Chilton Gro.)
Huskards RM14: Upm 33Rd 77
Huson Cl. NW3 38Gb 69
Hussain Cl. HA1: Harr 35Ha 66
Hussars Cl. TW4: Houn 55Aa 107
Husseywell Cres. BR2: Hayes 74Jc 159
Hutchings Lodge WD3: Rick 18N 25
Hutchingsons Rd. CR0: New Ad 83Ec 178
Hutchings St. E14 47Cc 92
Hutchings Wlk. NW11 28Db 49
Hutchings Wharf E14 47Cc 92
(off Hutchings St.)
Hutchins Cl. E15 38Ec 72
Hutchins Cl. RM12: Horn 34Nd 77
Hutchinson Ct. RM6: Chad H 28Zc 55
Hutchinson Ho. NW3 38Hb 69
Hutchinson Ho. SE14 52Yb 114
Hutchinson's Bank Nature Reserve 84Ec 178
Hutchinson Ter. HA9: Wemb 34Ma 67
Hutchins Rd. SE28 45Wc 95
Hutson Ter. RM19: Purf 51Td 120
HUTTON 15Ee 41
Hutton Cl. GU20: W'sham 10B 146
Hutton Cl. IG8: Wfd G 23Kc 53
Hutton Cl. KT12: Hers 78X 151
Hutton Cl. UB6: G'frd 36Fa 66
Hutton Ct. N4 32Pb 70
(off Victoria Rd.)
Hutton Ct. N9 17Yb 34
(off Tramway Av.)
Hutton Ct. W5 43Ka 86
Hutton Dr. CM13: Hut 17Ee 41
Hutton Gdns. HA3: Hrw W 24Ea 46
Hutton Ga. CM13: Hut 17De 41
Hutton Gro. N12 22Db 49
Hutton La. HA3: Hrw W 24Ea 46
Hutton M. SW15 57Xa 110
HUTTON MOUNT 18De 41
Hutton Pl. CM13: Hut Cedar Rd 16Fe 41
Hutton Pl. CM13: Hut Yew Tree Cl. 16De 41
Hutton Rd. CM15: Shenf 17Be 41
Hutton Row HA8: Edg 23Sa 47
Hutton St. EC4 3B 224 (44Rb 91)
Hutton Wlk. HA3: Hrw W 24Ea 46
Huxbear St. SE4 57Bc 114
Huxley Cl. EN7: Chesh 1Vb 19
Huxley Cl. SL3: Wex 2M 81
Huxley Cl. UB5: N'olt 40Aa 65
Huxley Cl. UB8: Cowl 42M 83
Huxley Dr. RH8: Oxt 5L 211
Huxley Dr. RM6: Chad H 31Xc 75
Huxley Gdns. NW10 41Pa 87
Huxley Ho. NW8 6C 214 (42Fb 89)
(off Fisherton St.)
Huxley Pde. N18 22Tb 51
Huxley Pl. N13 20Rb 33
Huxley Rd. DA16: Well 55Vc 117
Huxley Rd. E10 33Ec 72
Huxley Rd. N18 21Tb 51
Huxley Sayze N18 22Tb 51
Huxley Sth. N18 22Tb 51
Huxley St. W10 41Ab 88
Hyacinth Cl. IG1: Ilf 37Rc 74
Hyacinth Cl. TW12: Hamp 65Ca 129
Hyacinth Dr. UB10: Uxb 38N 63
Hyacinth Ho. E17 27Dc 52
(off Vine St.)
Hyacinth Rd. SW15 60Wa 110
Hybrid Ho. W3 46Ua 88
Hyburn Cl. AL2: Brick W 2Ba 13
Hyburn Cl. HP3: Hem H 3B 4
Hycliffe Gdns. IG7: Chig 21Sc 54
Hyde, The NW9 28Ua 48
(not continuous)
HYDE, THE 29Va 48
Hyde Av. EN6: Pot B 5Db 17
Hyde Cl. E13 40Jc 73
Hyde Cl. EN5: Barn 13Bb 31
Hyde Cl. RM1: Rom 23Fd 56
Hyde Cl. RM16: Chaf H 48Zd 99
Hyde Cl. TW15: Ashf 65U 128
Hyde Ct. AL2: Lon C 9F 6
Hyde Ct. EN8: Walt C 6Ac 20
Hyde Ct. N20 20Fb 31
Hyde Ct. NW9 29Ua 48
Hyde Dr. BR5: St P 70Xc 139
Hyde Est. Rd. NW9 29Va 48
Hyde Farm M. SW12 60Mb 112
Hydefield Cl. N21 18Tb 33
Hydefield Ct. N9 19Ub 33
Hyde Gro. DA1: Dart 54Pd 119
Hyde Ho. E3 43Cc 92
(off Furze St.)
Hyde Ho. TW3: Houn 55Ea 108
Hyde Ho. UB8: Uxb 37N 63
Hyde Ho. W13 46Ja 86
(off Singapore Rd.)
Hyde Ind. Est., The NW9 29Va 48
Hyde La. AL2: F'mre 10A 6
(not continuous)
Hyde La. AL2: Park 10A 6
(not continuous)
Hyde La. GU23: Ock 92R 190
Hyde La. HP3: Bov 9B 2
Hyde La. HP3: Hem H 9A 4
Hyde La. SW11 53Gb 111
Hyde Mdws. HP3: Bov 10C 2
Hyde M. RM1: Rom 23Fd 56
Hyde Pk. 6F 221 (46Hb 89)
Hyde Pk. Av. N21 19Sb 33
HYDE PARK CORNER 1K 227 (47Kb 90)
Hyde Pk. Cnr. W1 1J 227 (47Jb 90)
Hyde Park Corner Station (Underground) 1J 227 (47Jb 90)
Hyde Pk. Cres. W2 3D 220 (44Gb 89)
Hyde Pk. Gdns. N21 18Sb 33
Hyde Pk. Gdns. W2 4C 220 (45Fb 89)
(not continuous)
Hyde Pk. Gdns. M. W2 4C 220 (45Fb 89)
Hyde Pk. Ga. SW7 2A 226 (47Eb 89)
(not continuous)
Hyde Pk. Ga. M. SW7 2A 226 (47Eb 89)
Hyde Pk. Mans. NW1 1E 220 (43Gb 89)
(off Cabbell St.)
Hyde Pk. Pl. W2 4E 220 (45Gb 89)
Hyde Pk. Sq. W2 3D 220 (44Gb 89)
Hyde Pk. Sq. M. W2 3D 220 (44Gb 89)
(off Southwick Pl.)
Hyde Pk. St. W2 3D 220 (44Gb 89)
Hyde Pk. Towers W2 5A 220 (45Eb 89)
Hyderabad Way E15 38Gc 73
Hyde Rd. CR2: Sande 85Ub 177
Hyde Rd. DA7: Bex 54Bd 117
Hyde Rd. N1 1H 219 (39Ub 71)
Hyde Rd. TW10: Rich 57Pa 109
Hyde Rd. WD17: Wat 12W 26
Hyder Rd. RM16: Grays 8E 100
Hyders Forge TN15: Plax 99Ce 205
(not continuous)
Hydeside Gdns. N9 19Vb 33
Hyde's Pl. N1 38Rb 71
Hyde St. SE8 51Cc 114
Hyde Ter. HP3: Hem H 6E 4
Hyde Ter. TW15: Ashf 65U 128
Hydethorpe Av. N9 19Vb 33
Hydethorpe Hgts. CR2: S Croy 85Tb 177
Hydethorpe Rd. SW12 60Lb 112
Hyde Va. SE10 52Ec 114
Hyde Wlk. SM4: Mord 73Cb 155
Hyde Way N9 19Vb 33
Hyde Way UB3: Harl 49V 84
Hydon Ct. N11 22Hb 49
Hydra Bldg., The EC1 4A 218 (41Qb 90)
(off Hardwick St.)
Hydro Ho. KT16: Chert 74L 149
Hyland Cl. RM11: Horn 31Kd 77
Hylands Cl. KT18: Eps 87Sa 173
Hylands M. KT18: Eps 87Sa 173
Hylands Rd. E17 26Fc 53
Hylands Rd. KT18: Eps 87Sa 173
Hyland Way RM11: Horn 31Kd 77
Hylle Cl. SL4: Wind 3C 102
Hylton Pl. RH1: Mers 3C 208
Hylton St. SE18 49Vc 95
Hyndewood SE23 62Zb 136
Hyndford Cres. DA9: Ghithe 57Yd 120
(off Calcroft Av.)
Hyndman Ho. RM10: Dag 34Cd 76
(off Kershaw Rd.)
Hyndman St. SE15 51Xb 113
Hynton Rd. RM8: Dag 33Yc 75
Hyperion Ct. E16 43Jc 93
(off Robertson Rd.)
Hyperion Ho. E3 40Ac 72
(off Arbery Rd.)
Hyperion Ho. SW2 58Pb 112
Hyperion Pl. KT19: Ewe 81Ta 173
Hyrstdene CR2: S Croy 77Rb 157
Hyson Rd. SE16 50Xb 91
Hythe, The TW18: Staines 64G 126
Hythe Av. DA7: Bex 52Ad 117
Hythe Cl. BR5: St M Cry 70Yc 139
Hythe Cl. N18 21Wb 51
HYTHE END 61D 126
Hythe End Rd. TW19: Wray 61B 126
Hythe Fld. Av. TW20: Egh 65F 126
Hythe Ho. SE16 47Yb 92
(off Swan Rd.)
Hythe Ho. W6 49Ya 88
(off Shepherd's Bush Rd.)
Hythe Pk. Rd. TW20: Egh 64E 126
Hythe Rd. CR7: Thor H 68Tb 135
Hythe Rd. KT6: Surb 72Na 153
Hythe Rd. NW10 41Va 88
Hythe Rd. TW18: Staines 64F 126
Hythe Rd. Ind. Est. NW10 41Wa 88
Hythe St. DA1: Dart 58Nd 119
Hythe St. (Lower) DA1: Dart 57Nd 119
Hyver Hill NW7 16Ta 29

I

Ian Bowater Ct. N1 3G 219 (41Tb 91)
(off East Rd.)
Ian Ct. SE23 61Yb 136
Ian Sq. EN3: Enf H 11Zb 34
Ibberton Ho. SW8 52Pb 112
(off Meadow Rd.)
Ibberton Ho. W14 48Ab 88
(off Russell Rd.)
Ibbetson Path IG10: Lough 13Rc 36
Ibbotson Av. E16 44Hc 93
Ibbotson Ct. SL3: Poyle 53G 104
Ibbott St. E1 42Yb 92
Iberia Ho. N19 31Mb 70
Iberian Av. SM6: Bedd 77Mb 156
Ibex Ho. E15 36Gc 73
(off Forest La.)
Ibis Ct. BR3: Beck 67Cc 137
(off Albemarle Rd.)
Ibis Ct. SE8 51Bc 114
(off Edward Pl.)
Ibis La. W4 53Sa 109
Ibis Way UB4: Yead 44Z 85
Ibrox Ct. IG9: Buck H 19Lc 35
Ibscott Cl. RM10: Dag 37Ed 76
Ibsley Gdns. SW15 60Wa 110
Ibsley Way EN4: Cockf 15Gb 31
ICA Cinema 7E 222 (46Mb 90)
(within ICA)
Icarus Ho. E3 41Bc 92
(off British St.)
ICA Theatre 6E 222 (46Mb 90)
(within ICA)
Icehouse Wood RH8: Oxt 3J 211
Iceland Rd. E3 39Cc 72
Iceland Wharf SE16 49Ac 92
Iceni Ct. E3 39Bc 72
(off Parnell Rd.)
Iceni Ct. IG9: Buck H 18Kc 35
Ice Wharf N1 1G 217 (40Nb 70)
Ice Wharf Marina N1 1G 217 (40Nb 70)
(off New Wharf Rd.)
Ice Works, The NW1 38Kb 70
(off Jamestown Rd.)
Ickburgh Est. E5 33Xb 71
Ickburgh Rd. E5 34Xb 71
ICKENHAM 34R 64
Ickenham Cl. HA4: Ruis 33T 64
Ickenham Grn. UB10: Ick 32R 64
Ickenham Rd. HA4: Ruis 33S 64
Ickenham Station (Underground) 35S 64
Ickleton Rd. SE9 63Nc 138
Icklingham Ga. KT11: Cobh 84Y 171
Icklingham Rd. KT11: Cobh 84Y 171
Icknield Cl. AL3: St A 4M 5
Icknield Dr. IG2: Ilf 29Rc 54
Icknield Ho. SW3 7E 226 (50Gb 89)
(off Cale St.)
Ickworth Pk. Rd. E17 28Ac 52
Icon Apts. SE1 3H 231 (48Ub 91)
(off Cluny Pl.)
Icona Point E15 39Ec 72
(off Warton Rd.)
Icon College of Technology & Management 44Wb 91
(off Adler St.)
Iconia Ho. BR2: Broml 70Lc 137
Idaho Bldg. SE13 53Dc 114
(off Deal's Gateway)
Ida Rd. N15 28Tb 51
Ida St. E14 44Ec 92
(not continuous)
Idea Store (Whitechapel) Library 43Xb 91
Ide Mans. E1 45Zb 92
(off Cable St.)
Iden Cl. BR2: Broml 69Gc 137
Idlecombe Rd. SW17 65Jb 134
Idleigh Ct. Rd. DA13: Meop 75De 165
Idmiston Rd. E15 35Hc 73
Idmiston Rd. KT4: Wor Pk 73Va 154
Idmiston Rd. SE27 62Sb 135
Idmiston Sq. KT4: Wor Pk 73Va 154
Idol La. EC3 5H 225 (45Ub 91)
Idonia St. SE8 52Cc 114
Iffley Cl. UB8: Uxb 38M 63
Iffley Ct. TW18: Staines 64H 127
Iffley Rd. W6 48Xa 88
IFIELD 6E 144
Ifield Cl. RH1: Redh 8N 207
Ifield Ho. SE17 7J 231 (50Ub 91)
(off Madron St.)
Ifield Rd. DA13: Meop 78Fe 165
Ifield Rd. SW10 51Db 111
Ifield Way DA12: Grav'nd 5F 144
Ifold Rd. RH1: Redh 8A 208
Ifor Evans Pl. E1 42Zb 92
IGHTHAM 93Yd 204
Ightham By-Pass TN15: Igh 93Yd 204
Ightham Cl. DA3: Lfield 69Ae 143
IGHTHAM COMMON 95Xd 204
Ightham Ho. BR3: Beck 66Bc 136
(off Bethersden Cl.)
Ightham Ho. SE17 6H 231 (49Ub 91)
(off Beckway St.)
Ightham Mote 99Wd 204
Ightham Rd. DA8: Erith 52Cd 118
Ightham Rd. TN11: S'brne 100Yd 204
Ightham Rd. TN15: S'brne 100Yd 204
Ikona Ct. KT13: Weyb 78S 150
Ikon Ho. E1 45Yb 92
(off Devonport St.)
Ilbert St. W10 41Za 88
Ilchester Gdns. W2 45Db 89
Ilchester Mans. W8 48Cb 89
(off Abingdon Rd.)
Ilchester Pl. W14 48Bb 89
Ilchester Rd. RM8: Dag 36Xc 75
Ildersly Gro. SE21 61Tb 135
Ilderton Rd. SE15 51Yb 114
Ilderton Rd. SE16 50Yb 92
Ilderton Wharf SE15 51Yb 114
(off Rollins St.)
Ilex Cl. TW16: Sun 68Y 129
Ilex Cl. TW20: Eng G 6M 125
Ilex Ho. KT15: New H 82J 169
Ilex Rd. NW10 37Va 68
Ilex Way SW16 64Qb 134
ILFORD 34Rc 74
Ilford Bldg. IG1: Ilf 34Qc 74
Ilford Golf Course 32Pc 74
Ilford Hill IG1: Ilf 34Qc 74
Ilford Ho. N1 37Tb 71
(off Dove Rd.)
Ilford La. IG1: Ilf 34Rc 74
Ilford Sports Club 33Uc 74
Ilford Station (Overground & Crossrail) 34Qc 74
Ilfracombe Cres. RM12: Horn 35Ld 77
Ilfracombe Flats SE1 1E 230 (47Sb 91)
(off Marshalsea Rd.)
Ilfracombe Gdns. RM6: Chad H 31Xc 75
Ilfracombe Rd. BR1: Broml 62Hc 137
Iliffe St. SE17 7C 230 (50Rb 91)
Iliffe Yd. SE17 7C 230 (50Rb 91)
(off Crampton St.)
Ilkeston Ct. E5 35Zb 72
(off Overbury St.)
Ilkley Cl. SE19 65Tb 135
Ilkley Rd. E16 43Lc 93
Ilkley Rd. WD19: Wat 22Z 45
Illingworth SL4: Wind 5C 102
Illingworth Cl. CR4: Mitc 69Fb 133
Illingworth Way EN1: Enf 15Ub 33
Illumina Ho. SW18 57Db 111
(off Broomhill Rd.)
Ilmington Rd. HA3: Kenton 30Ma 47
Ilminster Gdns. SW11 56Gb 111
Ilsley Ct. SW8 54Lb 112
Image Ct. RM7: Rush G 30Gd 56
Images IG2: Ilf 30Qc 54
Imani Mans. SW11 54Fb 111
IMAX (BFI) 7K 223 (46Qb 90)
Imber Cl. KT10: Esh 74Fa 152
Imber Cl. N14 17Lb 32
Imber Court 73Fa 152
Imber Cross KT7: T Ditt 72Ha 152
Imber Gro. KT10: Esh 73Fa 152
Imber Pk. Rd. KT10: Esh 74Fa 152
Imber St. N1 39Tb 71
Impact Bus. Pk. UB6: G'frd 40Ka 66
Impact Ct. SE20 68Xb 135
Impact Ho. CR0: C'don 76Sb 157
Imperial Av. N16 35Ub 71
Imperial Bus. Est. DA11: Nflt 8B 122
Imperial Cl. HA2: Harr 30Ca 45
Imperial Cl. NW2 36Xa 68
Imperial College London Charing Cross Campus 51Za 110
Imperial College London Chelsea & Westminster Campus 51Eb 111
(within Chelsea & Westminster Hospital)
Imperial College London Hamersmith Campus 44Wa 88
Imperial College London Royal Brompton Campus, Emmanuel Kaye 50Gb 89
(off Manresa Rd.)
Imperial College London Royal Brompton Campus, Guy Scadding Bldg. 7C 226 (50Fb 89)
(off Dovehouse St.)
Imperial College London Silwood Pk. Campus 9D 124
Imperial College London St Mary's Campus 2C 220 (44Fb 89)
(off Norfolk Pl.)
Imperial College London Sth. Kensington Campus 3B 226 (48Fb 89)
Imperial College London Sth. Kensington Campus, Ennismore Gdns. M. 3C 226 (48Fb 89)
(off Ennismore Gdns. M.)
Imperial College London Sth. Kensington Campus, Kensington Gore 3B 226 (48Fb 89)
Imperial Coll. Rd. SW7 4B 226 (48Fb 89)
Imperial Ct. HA2: Harr 31Ca 65
Imperial Ct. N20 20Eb 31
Imperial Ct. N6 30Lb 50
Imperial Ct. NW8 1E 214 (40Gb 69)
(off Prince Albert Rd.)
Imperial Ct. SE11 7K 229 (50Qb 90)
Imperial Ct. SL4: Wind 5E 102
Imperial Cres. SW6 54Eb 111
Imperial Dr. DA12: Grav'nd 4H 145
Imperial Dr. HA2: Harr 31Ca 65
Imperial Gdns. CR4: Mitc 69Kb 134
Imperial Gro. EN4: Had W 11Db 31
Imperial Hgts. E18 25Jc 53
(off Queen Mary Av.)
Imperial Ho. E14 45Bc 92
(off Victory Pl.)
Imperial Ho. E3 41Ac 92
(off Grove Rd.)
Imperial M. E6 40Mc 73
Imperial M. SW9 56Pb 112
(off Brighton Ter.)
Imperial Pk. KT22: Lea 92Ja 192
Imperial Pk. WD24: Wat 11Y 27
Imperial Pl. BR7: Chst 67Qc 138
Imperial Pl. WD6: Bore 13Ra 29
Imperial Retail Pk. 8C 122
Imperial Rd. N22 24Nb 50
Imperial Rd. SL4: Wind 5E 102
Imperial Rd. SW6 53Db 111
Imperial Rd. TW14: Felt 59U 106
Imperial Sq. SW6 53Db 111
Imperial St. E3 41Ec 92
Imperial Trad. Est. RM13: Rain 42Ld 97
Imperial War Mus. London All Saints Annexe 4A 230 (48Qb 90)
(off Austral St.)
Imperial War Mus. London Main Museum 4A 230 (48Qb 90)
Imperial Way BR7: Chst 62Sc 138
Imperial Way CR0: Wadd 79Pb 156
Imperial Way HA3: Kenton 30Na 47
Imperial Way HP3: Hem H 6N 3
Imperial Way WD24: Wat 11Y 27
Imperial Way WD3: Crox G 17R 26
Imperial Wharf E2 40Xb 71
(off Darwen Pl.)
Imperial Wharf SW6 54Eb 111
Imperial Wharf Station (Overground) 53Eb 111
Imperium Ho. E1 44Xb 91
(off Cannon St. Rd.)
Imprimo Pk. IG10: Lough 14Sc 36
Impulse Leisure Belhus Pk. 44Ud 98
Impulse Leisure Blackshots 47Fe 99
Imre Cl. W12 46Xa 88
Inca Dr. SE9 59Rc 116
Inca Ter. N15 27Rb 51
Ince Rd. KT12: Hers 79U 150
Inchmery Rd. SE6 61Dc 136
Inchwood BR4: Addtn 77Dc 158
Indells AL10: Hat 1B 8
Independence Ho. SW19 67Fb 133
(off Chapter Way)
Independent Ind. Est. UB7: Yiew 46N 83
Independent Pl. E8 36Vb 71
Independents Rd. SE3 55Hc 115
Inderwick Rd. N8 29Pb 50
Indescon Ct. E14 47Dc 92
Indescon Sq. E14 47Cc 92
Index Apts. RM1: Rom 28Hd 56
(off Mercury Gdns.)
India Gdns. UB5: N'olt 38Y 65
Indiana Bldg. SE13 53Cc 114
(off Deal's Gateway)
India Pl. WC2 4H 223 (45Pb 90)
(off Montreal Pl.)
India Rd. SL1: Slou 7M 81
India St. EC3 3K 225 (44Vb 91)
India Way SW15 58Wa 110
India Way W12 45Xa 88
indigo at the O2 46Gc 93
Indigo M. E14 45Ec 92
Indigo M. N16 34Tb 71
Indigo Wlk. N2 28Hb 49
Indigo Wlk. N6 28Hb 49
Indus Ct. SE15 52Vb 113
(off Amstel Ct.)
Indus Rd. SE7 52Lc 115
Infirmary Ct. SW3 51Hb 111
(off West Rd.)
Inforum M. SE15 52Wb 113
Infrastructure Way IG11: Bark 41Zc 95
Ingal Rd. E13 42Jc 93
Ingate Pl. SW8 53Kb 112
Ingatestone Rd. E12 32Lc 73
Ingatestone Rd. IG8: Wfd G 24Kc 53
Ingatestone Rd. SE25 70Xb 135
Ingelow Ho. W8 47Db 89
(off Holland St.)
Ingelow Rd. SW8 54Kb 112
Ingels Mead CM16: Epp 1Vc 23
Ingersoll Rd. EN3: Enf W 10Yb 20
Ingersoll Rd. W12 46Xa 88
Ingestre Pl. W1 3C 222 (44Lb 90)
Ingestre Rd. E7 35Jc 73
Ingestre Rd. NW5 35Kb 70
Ingham Cl. CR2: Sels 81Zb 178
Ingham Rd. CR2: Sels 81Yb 178
Ingham Rd. NW6 35Cb 69
Inglebert St. EC1 3K 217 (41Qb 90)
Ingleboro Dr. CR8: Purl 85Tb 177
Ingleborough La. TN15: Plat 91Fe 205
Ingleborough St. SW9 54Qb 112
Ingleby Dr. HA1: Harr 34Fa 66
Ingleby Gdns. IG7: Chig 20Xc 37
Ingleby Rd. IG1: Ilf 32Rc 74
Ingleby Rd. N7 34Nb 70
Ingleby Rd. RM10: Dag 37Dd 76
Ingleby Rd. RM16: Grays 8D 100
Ingleby Way BR7: Chst 64Qc 138
Ingleby Way SM6: W'gton 81Mb 176
Ingle Cl. HA5: Pinn 27Aa 45
Ingledene Cl. NW4 30Wa 48
Ingledew Rd. SE18 50Tc 94
Inglefield EN6: Pot B 2Cb 17
Inglefield Sq. E1 46Xb 91
(off Prusom St.)
Ingleglen RM11: Horn 31Qd 77
Ingleglen SL2: Farn C 6F 60
Inglehurst KT15: New H 82K 169
Inglehurst Gdns. IG4: Ilf 29Pc 54
Inglemere Rd. CR4: Mitc 66Hb 133
Inglemere Rd. SE23 62Zb 136
Ingle M. EC1 3K 217 (41Qb 90)
Inglenorth Ct. BR8: Crock 72Ed 162
Inglesham Wlk. E9 37Bc 72
Ingleside SL3: Poyle 53G 104
Ingleside Cl. BR3: Beck 66Cc 136
Ingleside Gro. SE3 51Hc 115
Inglethorpe St. SW6 53Za 110
Ingleton Av. DA16: Well 57Wc 117
Ingleton Rd. N18 23Wb 51
Ingleton Rd. SM5: Cars 81Gb 175
Ingleton St. SW9 54Qb 112
Ingleway N12 23Fb 49
Inglewood BR7: Chst 65Tc 138
Inglewood BR8: Swan 68Gd 140
Inglewood CR0: Sels 81Ac 178
Inglewood GU21: Wok 10M 167
Inglewood KT16: Chert 76H 149
Inglewood Cl. E14 49Cc 92

Inglewood Cl. IG6: Ilf 23Vc 55
Inglewood Cl. RM12: Horn 35Md 77
Inglewood Copse BR1: Broml 68Nc 138
Inglewood Ct. BR1: Broml 66Gc 137
Inglewood Gdns. AL2: St A 7C 6
Inglewood M. KT6: Surb 74Qa 153
Inglewood M. SE27 64Sb 135
(off Elder Rd.)
Inglewood Rd. DA7: Bex 56Fd 118
Inglewood Rd. NW6 36Cb 69
Inglis Barracks NW7 23Za 48
Inglis Rd. CR0: C'don 74Vb 157
Inglis Rd. W5 45Pa 87
Inglis St. SE5 53Rb 113
Inglis Way NW7 23Za 48
Ingoldisthorpe Gro. SE15 51Vb 113
Ingoldsby Rd. DA12: Grav'nd 10G 122
Ingot Twr. E14 43Cc 92
(off Ursula Gould Way)
Ingram Av. NW11 31Eb 69
Ingram Cl. HA7: Stan 22La 46
Ingram Cl. SE11 5J 229 (49Pb 90)
Ingram Ct. CR0: C'don 74Rb 157
Ingram Ho. E3 39Ac 72
Ingram Ho. CR7: Thor H 67Sb 135
Ingram Rd. DA1: Dart 60Nd 119
Ingram Rd. N2 28Gb 49
Ingram Rd. RM17: Grays 49Ee 99
Ingrams Cl. KT12: Hers 78Y 151
Ingram Way UB6: G'frd 39Fa 66
INGRAVE 22Ee 59
INGRAVE COMMON 21Ce 59
Ingrave Rd. CM13: B'wood 20Ae 41
Ingrave Rd. CM14: B'wood 19Zd 41
Ingrave Rd. CM15: B'wood 19Zd 41
Ingrave Rd. RM1: Rom 28Gd 56
Ingrave St. SW11 55Fb 111
Ingrebourne Apts. SW6 55Db 111
Ingrebourne Av. RM3: Rom 21Md 57
Ingrebourne Ct. E4 20Dc 34
Ingrebourne Gdns. RM14: Upm 32Sd 78
Ingrebourne Hill 40Jd 76
Ingrebourne Ho. BR1: Broml 64Fc 137
(off Brangbourne Rd.)
Ingrebourne Ho. NW8 7C 214 (43Fb 89)
(off Broadley St.)
Ingrebourne Rd. RM13: Rain 42Kd 97
Ingress Abbey 56Yd 120
Ingress Gdns. DA9: Ghithe 57Zd 121
INGRESS PARK 56Yd 120
Ingress Pk. Av. DA9: Ghithe 56Yd 120
Ingress St. W4 50Ua 88
Ingress Ter. DA13: Sflt 63Ae 143
Ingreway RM3: Hrld W 23Rd 57
Inigo Jones Rd. SE7 52Nc 116
Inigo Pl. WC2 4F 223 (45Nb 90)
(off Bedford St.)
Ink Bldg. W10 43Za 88
Inkerman Rd. AL1: St A 3C 6
Inkerman Rd. GU21: Knap 10J 167
Inkerman Rd. NW5 37Kb 70
Inkerman Rd. SL4: Eton W 9D 80
Inkerman Ter. W8 48Cb 89
(off Allen St.)
Inkerman Way GU21: Wok 10J 167
Inks Grn. E4 22Ec 52
Inkster Ho. SW11 55Gb 111
Inkwell Cl. N12 20Eb 31
Ink Works Ct. SE1 2J 231 (47Ub 91)
(off Bell Yd. M.)
Inman Rd. NW10 39Ua 68
Inman Rd. SW18 59Eb 111
Inmans Row IG8: Wfd G 21Jc 53
Inner Circ. NW1 4H 215 (41Jb 90)
Inner Ct. SW3 51Gb 111
Innerd Ct. CR0: C'don 72Sb 157
(off Harry Cl.)
Inner Pk. Rd. SW19 60Za 110
Inner Ring E. TW6: H'row A 55R 106
Inner Ring W. TW6: H'row A 55Q 106
Inner Temple La. EC4 3K 223 (44Qb 90)
(off Fleet St.)
Innes Cl. SW20 68Ab 132
Innes Ct. HP3: Hem H 5M 3
Innes Gdns. SW15 58Xa 110
Innes St. SE15 52Ub 113
Innes Yd. CR0: C'don 76Sb 157
Innis Ho. SE17 7H 231 (50Ub 91)
(off East St.)
Inniskilling Rd. E13 40Lc 73
Innova Bus. Pk. EN3: Enf L 8Bc 20
Innova Ct. CR0: C'don 74Ub 157
Innova Pas. E1 5K 219 (42Vb 91)
(off Sclater St.)
Innovation Cen., The E14 47Ec 92
(off Marsh Wall)
Innovation Cl. HA0: Wemb 39Na 67
Innova Way EN3: Enf L 8Bc 20
Inns of Court & City Yeomanry Mus. 1J 223 (43Pb 90)
Insignia Point E20 36Ec 72
Inskip Cl. E10 33Dc 72
Inskip Dr. RM11: Horn 32Nd 77
Inskip Rd. RM8: Dag 32Zc 75
Insley Ho. E3 41Dc 92
(off Bow Rd.)
Inspirations Way BR6: Orp 73Wc 161
Institute for Arts in Therapy & Education, The 39Rb 71
(off Britannia Row)
Institute of Archaeology Collections 5D 216 (42Mb 90)
Institute of Commonwealth Studies 7E 216 (43Mb 90)
(off Russell Sq.)
Institute of Contemporary Arts (ICA) 7E 222 (46Mb 90)
(off Carlton Ho. Ter.)
Institute of Germanic & Romance Studies 7E 216 (43Mb 90)
(off Russell Sq.)
Institute of Ophthalmology 4F 219 (41Tb 91)
(off Peerless St.)
Institute of Psychoanalysis, The 42Cb 89
(off Elgin Av.)
Institute Pl. E8 36Xb 71
Institute Rd. CM16: Coop 1Zc 23
Instone Rd. DA1: Dart 59Md 119
Integer Gdns. E11 31Fc 73
Interchange, The BR8: Swan 72Jd 162
Interchange, The NW1 38Kb 70
(off Camden Lock Pl.)
Interface Ho. TW3: Houn 55Ca 107
(off Staines Rd.)
International Av. TW5: Cran 50Y 85
International Bus. Pk. E15 39Fc 73
International Hall WC1 6G 217 (42Nb 90)
(off Lansdowne Ter.)
International Ho. E1 5K 225 (45Vb 91)
(off St Katharine's Way)
International Ho. TW8: Bford 50Na 87
(off Gt. West Rd.)
INTERNATIONAL QUARTER, THE 37Ec 72
International Sq. E20 37Ec 72
(within Westfield Shop. Cen.)
International Trad. Est. UB2: S'hall 48X 85
International Way DA10: Ebbs 59Ce 121
International Way E20 37Ec 72
International Way TW16: Sun 67U 128
intu Lakeside 48Xd 98
intu Watford 14Y 27
Inverary Pl. SE18 51Tc 116
Inver Cl. E5 33Yb 72
Inverclyde Gdns. RM6: Chad H 28Yc 55
(not continuous)
Inver Ct. W2 44Db 89
Inver Ct. W6 48Wa 88
Inveresk Gdns. KT4: Wor Pk 76Wa 154
Inverforth Cl. NW3 33Eb 69
Inverforth Rd. N11 22Kb 50
Invergarry Ho. NW6 40Db 69
(off Carlton Vale)
Inverine Rd. SE7 50Kc 93
Invermead Cl. W6 48Wa 88
Invermore Pl. SE18 49Sc 94
Inverness Av. EN1: Enf 11Ub 33
Inverness Ct. SE6 60Hc 115
Inverness Dr. IG6: Ilf 23Uc 54
Inverness Gdns. W8 46Db 89
Inverness M. E16 46Sc 94
Inverness M. W2 45Db 89
Inverness Pl. W2 45Db 89
Inverness Rd. KT4: Wor Pk 74Za 154
Inverness Rd. N18 22Xb 51
Inverness Rd. TW3: Houn 56Ba 107
Inverness Rd. UB2: S'hall 49Aa 85
Inverness St. NW1 39Kb 70
Inverness Ter. W2 44Db 89
Inverton Rd. SE15 56Zb 114
Invicta Bus. Pk. TN15: Wro 89Ee 185
Invicta Cen., The IG11: Bark 39Wc 75
Invicta Cl. BR7: Chst 64Qc 138
Invicta Cl. E3 43Cc 92
Invicta Cl. TW14: Felt 60V 106
Invicta Gro. UB5: N'olt 41Ba 85
Invicta Pde. DA14: Sidc 63Xc 139
Invicta Pk. DA1: Dart 56Ld 119
Invicta Plaza SE1 6B 224 (46Rb 91)
Invicta Rd. DA2: Dart 58Rd 119
Invicta Rd. SE3 52Jc 115
In Vw. Ct. KT12: Hers 77W 150
Inville Rd. SE17 7G 231 (50Tb 91)
(not continuous)
Inville Wlk. SE17 7G 231 (50Tb 91)
Invito IG2: Ilf 30Qc 54
Inwen Ct. SE8 50Ac 92
(not continuous)
Inwood Av. CR5: Coul 92Qb 196
Inwood Av. TW3: Houn 55Ea 108
Inwood Bus. Pk. TW3: Houn 56Da 107
Inwood Cl. CR0: C'don 75Ac 158
Inwood Ct. GU22: Wok 94B 188
Inwood Ct. KT12: Walt T 75Y 151
Inwood Ct. NW1 38Lb 70
(off Rochester Sq.)
Inwood Ho. N1 39Rb 71
(off Elliott's Pl.)
Inwood Rd. TW3: Houn 56Da 107
Inworth St. SW11 54Gb 111
Inworth Wlk. N1 39Sb 71
(off Popham St.)
IO Cen. EN9: Walt A 7Cc 20
IO Cen. SE18 48Sc 94
(not continuous)
Iona Cl. SE6 59Cc 114
Iona Cl. SM4: Mord 73Db 155
Iona Cres. SL1: Slou 4C 80
Ion Ct. E2 40Wb 71
Ionian Bldg. E14 45Ac 92
(off Narrow St.)
Ionian Ho. E1 42Zb 92
(off Duckett St.)
Ionia Wlk. DA12: Grav'nd 2H 145
Ion Sq. E2 40Wb 71
IO Trade Cen. CR0: Bedd 77Pb 156
Ipsden Bldgs. SE1 1A 230 (47Qb 90)
(off Windmill Wlk.)
Ipswich Rd. SL1: Slou 4D 80
Ipswich Rd. SW17 65Jb 134
Ira Ct. SE27 61Rb 135
Ireland Cl. E6 43Pc 94
Ireland Pl. N22 24Nb 50
Ireland Yd. EC4 3C 224 (44Rb 91)
Irene M. W7 46Ha 86
(off Uxbridge Rd.)
Irene Rd. BR6: Orp 73Vc 161
Irene Rd. KT11: Stoke D 86Da 171
Irene Rd. SW6 53Cb 111
Irene Stebbings Ho. AL4: St A 1G 6
Ireton Av. KT12: Walt T 75U 150
Ireton Cl. N10 24Jb 50
Ireton Ho. SW15 57Ab 110
(off Stamford Sq.)
Ireton Pl. RM17: Grays 49Ce 99
Ireton St. E3 42Cc 92
Iris Av. DA5: Bexl 57Ad 117
Iris Cl. BR5: St M Cry 71Yc 161
Iris Cl. CM15: Pil H 15Xd 40
Iris Cl. CR0: C'don 74Zb 158
Iris Cl. E6 43Nc 94
Iris Cl. KT6: Surb 73Pa 153
Iris Cl. N14 17Mb 32
Iris Ct. SE14 53Yb 114
(off Briant St.)
Iris Cres. DA7: Bex 51Bd 117
Iris Dr. GU24: Bisl 7E 166
Iris Gdns. KT7: T Ditt 73Ga 152
Iris M. TW4: Houn 58Ca 107
Iris Path RM3: Rom 24Ld 57
Iris Rd. GU24: Bisl 7E 166
Iris Rd. KT19: Ewe 78Ra 153
Iris Wlk. HA8: Edg 21Sa 47
Iris Way E4 23Bc 52
Irkdale Av. EN1: Enf 11Vb 33
Iron Bri. Cl. NW10 36Ua 68
Ironbridge Cl. UB2: S'hall 46Ea 86
Iron Bri. Ho. NW1 38Hb 69
Iron Bri. Rd. Nth. UB11: Stock P 47Q 84
Iron Bri. Rd. Sth. UB7: W Dray 47Q 84
Iron Mill La. DA1: Cray 56Gd 118
Iron Mill Pl. DA1: Cray 56Hd 118
Iron Mill Pl. SW18 58Db 111
Iron Mill Rd. SW18 58Db 111
Ironmonger La. EC2 3E 224 (44Sb 91)
Ironmonger Pas. EC1 4E 218 (41Sb 91)
(off Ironmonger Row)
Ironmonger Row EC1 4D 218 (41Sb 91)
Ironmonger Row Baths 4E 218 (41Sb 91)
(off Ironmonger Row)
Ironmongers Pl. E14 49Cc 92
Iron Railway Cl. CR5: Coul 88Mb 176
Ironside Cl. SE16 47Zb 92
Ironside Ct. CR2: S Croy 79Tb 157
Ironside Ct. TW11: Hamp W 67La 130
Ironside Ho. E9 35Ac 72
Irons Way RM5: Col R 24Ed 56
Iron Works E3 39Cc 72
Ironworks, The N1 2G 217 (40Nb 70)
(off Albion Wlk.)
Ironworks Way E13 40Mc 73
Irvine Av. HA3: Kenton 27Ja 46
Irvine Cl. E14 43Dc 92
Irvine Cl. N20 19Gb 31
Irvine Ct. W1 6C 216 (42Lb 90)
(off Whitfield St.)
Irvine Gdns. RM15: S Ock 44Vd 98
Irvine Ho. N7 37Pb 70
(off Caledonian Rd.)
Irvine Pl. GU25: Vir W 71A 148
Irvine Way BR6: Orp 73Vc 161
Irving Av. UB5: N'olt 39Z 65
Irving Gro. SW9 54Pb 112
Irving Ho. SE17 50Rb 91
(off Doddington Gro.)
Irving Mans. W14 51Ab 110
(off Queen's Club Gdns.)
Irving M. N1 37Sb 71
Irving Rd. W14 48Za 88
Irving St. WC2 5E 222 (45Mb 90)
Irving Wlk. DA10: Swans 59Ae 121
Irving Way BR8: Swan 68Fd 140
Irving Way NW9 29Wa 48
Irwell Ct. W7 44Fa 86
(off Hobbayne Rd.)
Irwell Est. SE16 48Yb 92
Irwin Av. SE18 52Uc 116
Irwin Cl. NW7 22Ab 48
Irwin Cl. UB10: Ick 34Q 64
Irwin Gdns. NW10 39Xa 68
Isaac Way SE1 1E 230 (47Sb 91)
(off Sanctuary St.)
Isabel Hill Cl. TW12: Hamp 67Da 129
Isabella Cl. N14 17Lb 32
Isabella Ct. TW10: Rich 58Pa 109
(off Kingsmead)
Isabella Dr. BR6: Farnb 77Sc 160
Isabella Ho. SE11 7B 230 (50Rb 91)
(off Othello Cl.)
Isabella Ho. W6 50Ya 88
(off Queen Caroline St.)
Isabella M. N1 37Ub 71
Isabella Pl. KT2: King T 64Pa 131
Isabella Plantation Gdn. 62Ra 131
Isabella Rd. E9 36Yb 72
Isabella St. SE1 7B 224 (46Rb 91)
Isabelle Cl. EN7: G Oak 1Sb 19
Isabel St. SW9 53Pb 112
Isambard Cl. UB8: Cowl 42M 83
Isambard M. E14 48Ec 92
Isambard Pl. SE16 46Yb 92
Isbell Gdns. RM1: Rom 24Gd 56
Isbells Dr. RH2: Reig 7K 207
Isel Way SE22 57Ub 113
Isham Rd. SW16 68Nb 134
Isis Cl. HA4: Ruis 30S 44
Isis Cl. SW15 56Ya 110
Isis Ct. W4 52Ra 109
Isis Dr. RM14: Upm 30Ud 58
Isis Ho. KT16: Chert 73L 149
Isis Ho. N18 23Vb 51
Isis Ho. NW8 6C 214 (42Fb 89)
(off Church St. Est.)
Isis Ho. SE20 66Wb 135
Isis St. SW18 61Eb 133
Island, The KT11: D'side 90X 171
Island, The KT13: Weyb 79N 149
Island, The KT7: T Ditt 72Ja 152
Island, The TW19: Wray 62C 126
Island, The UB7: Lford 52L 105
Island Apts. N1 39Sb 71
Island Barn Reservoir Sailing Club 72Da 151
Island Cen. Way EN3: Enf L 9Cc 20
Island Cl. TW18: Staines 63G 126
Island Farm Av. KT8: W Mole 71Ba 151
Island Farm Rd. KT8: W Mole 71Ba 151
Island Gardens Station (DLR) 50Ec 92
Island Ho. E3 41Ec 92
Island Rd. CR4: Mitc 66Hb 133
Island Rd. SE16 49Zb 92
Island Row E14 44Bc 92
Isla Rd. SE18 51Sc 116
Islay Gdns. TW4: Houn 57Z 107
Islay Ho. WD18: Wat 16V 26
(off Old Hill)
Islay Wlk. N1 37Sb 71
(off Douglas Rd. Sth.)
Isleden Ho. N1 39Sb 71
(off Prebend St.)
Isledon Rd. N7 34Qb 70
ISLEDON VILLAGE 34Qb 70
Islehurst Cl. BR7: Chst 67Qc 138
ISLE OF DOGS 47Dc 92
Isles Quarry Rd. TN15: Bor G 93Ae 205
ISLEWORTH 55Ja 108
Isleworth Ait Nature Reserve 55Ka 108
Isleworth Bus. Complex TW7: Isle 54Ha 108
Isleworth Prom. TW1: Twick 56Ka 108
Isleworth Recreation Cen. 56Ha 108
Isleworth Station (Rail) 54Ha 108
Isley Ct. E14 42Ec 92
(off Teviot St.)
Isley Ct. SW8 54Lb 112
ISLINGTON 38Rb 71
Islington Bus. Cen. N1 39Sb 71
(off Coleman Flds.)
Islington Crematorium 25Hb 49
Islington Ecology Cen., The 34Qb 70
Islington Grn. N1 1B 218 (39Rb 71)
(not continuous)
Islington High St. N1 2A 218 (40Qb 70)
(not continuous)
Islington Mus. 4B 218 (41Rb 91)
Islington Pk. M. N1 38Rb 71
Islington Pk. St. N1 38Qb 70
Islington Pl. N1 1K 217 (39Qb 71)
Islington Sq. N1 39Rb 71
Islington Tennis Cen. 37Nb 70
Islip Gdns. HA8: Edg 24Ta 47
Islip Gdns. UB5: N'olt 38Aa 65
Islip Mnr. Rd. UB5: N'olt 38Aa 65
Islip St. NW5 36Lb 70
Ismailia Rd. E7 38Kc 73
Ismay Ct. SL2: Slou 4J 81
Ismays Rd. TN15: Igh 97Xd 204
Ismays Rd. TN15: Ivy H 97Xd 204
Isobel Ho. HA1: Harr 29Ha 46
Isobel Pl. N15 28Vb 51
Isola Ct. N1 39Sb 71
(off Popham Rd.)
Isom Cl. E13 41Kc 93
Isopad Ho. WD6: Bore 13Ra 29
(off Shenley Rd.)
Issa Rd. TW3: Houn 56Ba 107
Issigonis Ho. W3 46Va 88
(off Cowley Rd.)
Istead Ri. DA13: Ist R 6B 144
ISTEAD RISE 6B 144
Istra Ho. E20 36Ec 72
(off Logan Cl.)
Itaska Cotts. WD23: B Hea 18Ga 28
ITCHINGWOOD COMMON 5N 211
Itchingwood Comn. Rd. RH8: Limp 5N 211
Ithell Ct. HA0: Wemb 36Ma 67
Ivanhoe Cl. UB8: Cowl 43M 83
Ivanhoe Dr. HA3: Kenton 27Ja 46
Ivanhoe Ho. E3 40Ac 72
(off Grove Rd.)
Ivanhoe Rd. SE5 55Vb 113
Ivanhoe Rd. TW4: Houn 55Z 107
Ivaro Cl. RM5: Col R 25Ed 56
Ivatt Pl. W14 50Bb 89
Ivatt Way N17 27Rb 51
Iveagh Av. NW10 40Qa 67
Iveagh Cl. E9 39Zb 72
Iveagh Cl. HA6: Nwood 25R 44
Iveagh Cl. NW10 40Qa 67
Iveagh Ct. BR3: Beck 69Ec 136
Iveagh Ct. E1 3K 225 (44Vb 91)
(off Haydon St.)
Iveagh Ct. HP2: Hem H 1M 3
Iveagh Ho. SW10 52Eb 111
(off King's Rd.)
Iveagh Ho. SW9 54Rb 113
Iveagh Ter. NW10 40Qa 67
(off Iveagh Av.)
Ivedon Rd. DA16: Well 54Yc 117
Ive Farm Cl. E10 33Cc 72
Ive Farm La. E10 33Cc 72
Iveley Rd. SW4 54Lb 112
IVER 44G 82
Iver Ct. SL0: Iver 44H 83
Iverdale Cl. SL0: Iver 45E 82
Ivere Dr. EN5: New Bar 16Db 31
Iver Golf Course 46D 82
IVER HEATH 40F 62
Iver Ho. N1 39Ub 71
(off Halcomb St.)
Iverhurst Cl. DA6: Bex 57Zc 117
Iver La. SL0: Iver 44J 83
Iver La. UB8: Cowl 42K 83
Iver Lodge SL0: Iver 43H 83
Iverna Ct. W8 48Cb 89
Iverna Gdns. TW14: Felt 57T 106
Iverna Gdns. W8 48Cb 89
Iver Rd. CM15: Pil H 16Xd 40
Iverson Rd. NW6 37Bb 69
Iver Station (Rail & Crossrail) 47H 83
Ivers Way CR0: New Ad 80Dc 158
Ives Gdns. RM1: Rom 28Hd 56
Ives Rd. E16 43Gc 93
Ives Rd. SL3: L'ly 48B 82
Ives St. SW3 5E 226 (49Gb 89)
Ivestor Ter. SE23 59Yb 114
Ivimey St. E2 41Wb 91
Ivinghoe Cl. EN1: Enf 11Ub 33
Ivinghoe Cl. WD25: Wat 7Z 13
Ivinghoe Ho. N7 36Mb 70
Ivinghoe Rd. RM8: Dag 36Xc 75
Ivinghoe Rd. WD23: Bush 17Fa 28
Ivinghoe Rd. WD3: Rick 17J 25
Ivo Pl. N19 34Mb 70
Ivor Ct. N8 30Nb 50
Ivor Ct. NW1 5F 215 (42Hb 89)
(off Gloucester Pl.)
Ivor Gro. SE9 60Rc 116
Ivories, The N1 38Sb 71
(off Northampton St.)
Ivor Pl. NW1 6F 215 (42Hb 89)
Ivor St. NW1 38Lb 70
Ivory Cl. AL4: St A 4G 6
Ivory Ct. E18 25Jc 53
Ivory Ct. HP3: Hem H 5N 3
Ivory Ct. TW13: Felt 60W 106
Ivorydown BR1: Broml 63Jc 137
Ivory Ho. E1 6K 225 (46Vb 91)
Ivory Pl. W11 45Ab 88
(off Treadgold St.)
Ivory Sq. SW11 55Eb 111
Ivy Bower Cl. DA9: Ghithe 57Xd 120
Ivybridge Cl. TW1: Twick 59Ja 108
Ivybridge Cl. UB8: Uxb 41N 83
Ivybridge Ct. BR7: Chst 67Qc 138
(off Old Hill)
Ivybridge Ct. NW1 38Kb 70
(off Lewis St.)
Ivybridge La. WC2 5G 223 (45Nb 90)
Ivy Bri. Retail Pk. 57Ha 108
IVY CHIMNEYS 5Wc 23
Ivy Chimneys Rd. CM16: Epp 4Uc 22
Ivychurch Cl. SE20 66Yb 136
Ivychurch La. SE17 7K 231 (50Vb 91)
Ivy Cl. DA1: Dart 58Qd 119
Ivy Cl. DA12: Grav'nd 3E 144
Ivy Cl. HA2: Harr 35Ba 65
Ivy Cl. HA5: Eastc 31Y 65
Ivy Cl. TW16: Sun 68Y 129
Ivy Cl. WD25: Wat 5V 12
Ivy Cotts. E14 45Ec 92
Ivy Cotts. UB10: Hil 41Q 84
Ivy Ct. SE16 50Wb 91
(off Argyle Way)
Ivy Cres. SL1: Slou 5D 80
Ivy Cres. W4 49Sa 87
Ivydale Rd. SE15 55Zb 114
Ivydale Rd. SM5: Cars 75Hb 155
Ivyday Gro. SW16 62Pb 134
Ivydene GU21: Knap 10F 166
Ivydene KT8: W Mole 71Ba 151
Ivydene Cl. SM1: Sutt 77Eb 155
Ivydene Ct. IG9: Buck H 19Lc 35
(off Queen's Rd.)
Ivy Gdns. CR4: Mitc 69Mb 134
Ivy Gdns. N8 30Nb 50
IVY HATCH 98Xd 204
Ivy Ho. WD19: Wat 16Z 27
Ivy Ho. La. HP4: Berk 1A 2
Ivy Ho. La. TN13: Dun G 90Fd 182
Ivy Ho. La. TN13: Otf 90Fd 182
Ivy Ho. La. TN14: Otf 89Gd 182
Ivy Ho. Rd. UB10: Ick 34R 64
Ivyhouse Rd. RM9: Dag 37Zc 75
Ivy La. GU22: Wok 89D 168
Ivy La. TN14: Knock 88Ad 181
Ivy La. TW4: Houn 56Ba 107
Ivy Lea WD3: Rick 18J 25
Ivy Lodge W11 46Cb 89
(off Notting Hill Ga.)
Ivy Lodge La. RM3: Hrld W 25Rd 57
Ivy Mill Cl. RH9: G'stone 4P 209
Ivy Mill La. RH9: G'stone 4N 209
Ivymount Rd. SE27 62Qb 134
Ivy Rd. E16 44Jc 93
Ivy Rd. E17 30Cc 52
Ivy Rd. KT6: Surb 74Qa 153
Ivy Rd. N14 17Lb 32
Ivy Rd. NW2 35Ya 68
Ivy Rd. SE4 56Bc 114
Ivy Rd. SW17 64Gb 133
Ivy Rd. TW3: Houn 56Da 107
Ivy St. N1 1H 219 (40Ub 71)
Ivy Vs. DA9: Ghithe 57Wd 120
Ivy Wlk. HA6: Nwood 25U 44
Ivy Wlk. RM9: Dag 37Ad 75
Ixworth Pl. SW3 7D 226 (50Gb 89)
Izane Rd. DA6: Bex 56Bd 117

J

Jacana Ct. E1 45Vb 91
(off Star Pl.)
Jacaranda Cl. KT3: N Mald 69Ua 132
Jacaranda Gro. E8 38Vb 71
Jacaranda Ho. KT15: Add 75M 149
Jackass La. BR2: Kes 78Kc 159
Jackass La. RH8: G'stone 3D 210
Jackass La. RH8: Tand 3D 210
Jack Barnett Way N22 26Pb 50
Jack Clow Rd. E15 40Gc 73
Jack Cook Ho. IG11: Bark 38Rc 74
Jack Cornwell St. E12 35Qc 74
Jack Dash Way E6 42Nc 94
Jackdaw Cl. RM3: Hrld W 25Pd 57
Jack Dimmer Cl. SW16 68Lb 134
Jackets La. HA6: Nwood 25R 44
Jackets La. UB9: Hare 24Q 44
Jacketts Fld. WD5: Ab L 3V 12
Jack Evans Ct. RM15: S Ock 44Wd 98
Jack Goodchild Way KT1: King T 69Ra 131
Jack Jones Way RM9: Dag 39Bd 75
Jacklin Grn. IG8: Wfd G 21Jc 53
Jackman Ho. E1 46Xb 91
(off Watts St.)
Jackman M. NW2 34Ua 68
Jackmans La. GU21: Wok 1L 187
Jackman St. E8 39Xb 71
Jacks Farm Way E4 23Ec 52
Jacks La. UB9: Hare 25J 43
Jackson & Joseph Bldg. E1 7K 219 (43Vb 91)
(off Princelet St.)
Jackson Cl. DA9: Ghithe 57Vd 120
Jackson Cl. E9 38Yb 72
Jackson Cl. KT18: Eps 86Ta 173
Jackson Cl. RM11: Horn 28Pd 57
Jackson Cl. SL3: L'ly 48A 82
Jackson Cl. UB10: Uxb 38N 63
Jackson Ct. E7 37Kc 73
Jackson Ho. N11 22Lb 50
Jackson Rd. BR2: Broml 75Pc 160
Jackson Rd. EN4: E Barn 16Gb 31
Jackson Rd. IG11: Bark 39Tc 74
Jackson Rd. N7 35Pb 70
Jackson Rd. UB10: Uxb 38N 63
Jacksons La. N6 31Jb 70
Jacksons Lane Theatre 30Kb 50
(off Archway Rd.)
Jacksons Pl. CR0: C'don 74Tb 157
Jackson St. SE18 51Qc 116
Jackson's Way CR0: C'don 76Cc 158
Jackson Way KT19: Eps 81Qa 173
Jackson Way UB2: S'hall 47Da 85
Jacks Pl. E1 7K 219 (43Vb 91)
(off Corbet Pl.)
Jack the Ripper Mus. 45Wb 91
Jack Walker Ct. N5 35Rb 71
Jacob Cl. SL4: Wind 3C 102
Jacob Ct. AL4: St A 4G 6
Jacob Ho. DA18: Erith 47Zc 95
(off Kale Rd.)
Jacobin Lodge N7 36Nb 70
Jacob Mans. E1 44Xb 91
(off Commercial Rd.)
Jacob M. HA7: Stan 19Ja 28
Jacobs Av. RM3: Hrld W 26Nd 57
Jacobs Cl. RM10: Dag 35Dd 76
Jacobs Ct. E1 44Wb 91
(off Plumber's Row)
Jacobs Ho. E13 41Lc 93
(off New City Rd.)
Jacobs Island Ho. SE16 48Vb 91
(off Spa Rd.)
Jacobs Island Pier 47Wb 91
(off Bermondsey Wall W.)
Jacob's Ladder CR6: W'ham 91Wb 197
Jacobs La. DA4: Hort K 69Sd 142
Jacobs M. SW15 56Ab 110
Jacob St. SE1 47Wb 91
JACOBS WELL 10P 187
Jacob's Well M. W1 1J 221 (43Jb 90)
Jacotts Ho. W10 42Ya 88
(off Sutton Way)
Jacquard Ct. SW18 57Db 111
(off Courthouse Way)
Jacqueline Cl. UB5: N'olt 39Aa 65
Jacqueline Creft Ter. N6 30Jb 50
(off Grange Rd.)
Jacqueline Ho. NW1 39Hb 69
(off Regent's Pk. Rd.)
Jacqueline Vs. E17 29Ec 52
(off Shernhall St.)
Jade Cl. E16 44Mc 93
Jade Cl. NW2 31Za 68
Jade Cl. RM8: Dag 32Yc 75
Jade Ter. NW6 38Eb 69
Jaffe Rd. IG1: Ilf 32Tc 74
Jaffray Pl. SE27 63Rb 135
Jaffray Rd. BR2: Broml 70Mc 137
Jaggard Way SW12 59Hb 111
Jagger Cl. DA2: Dart 59Sd 120
Jagger Ho. SW11 53Hb 111
(off Rosenau Rd.)
Jago Cl. SE18 51Sc 116
Jago Wlk. SE5 52Tb 113
Jail La. TN16: Big H 87Mc 179
Jake Russell Wlk. E16 45Lc 93
Jakes Vw. AL2: Park 9A 6
Jamaica Rd. CR7: Thor H 72Rb 157
Jamaica Rd. SE1 2K 231 (47Vb 91)
Jamaica Rd. SE16 48Wb 91
Jamaica St. E1 44Yb 92
James Allens School Swimming Pool 57Ub 113
James Anderson Ct. E2 1J 219 (40Ub 71)
(off Kingsland Rd.)
James Av. NW2 36Ya 68

James Av. RM8: Dag 32Bd 75
James Av. TN15: W King 80Ud 164
(off London Rd.)
James Bedford Cl. HA5: Pinn 26Y 45
James Boswell Cl. SW16 63Pb 134
James Brine Ho. E2 3K 219 (41Vb 91)
(off Ravenscroft St.)
James Campbell Ho. E2 40Yb 72
(off Old Ford Rd.)
James Clavell Sq. SE18 48Rc 94
James Cl. E13 40Jc 73
James Cl. NW11 30Ab 48
James Cl. RM2: Rom 29Jd 56
James Cl. WD23: Bush 15Aa 27
James Clubb Way DA1: Dart 61Pd 141
James Collins Cl. W9 42Bb 89
James Ct. AL2: Lon C 9F 6
James Ct. HA6: Nwood 25V 44
James Ct. N1 39Sb 71
(off Raynor Pl.)
James Ct. NW9 26Ua 48
James Ct. UB5: N'olt 40Aa 65
(off Church Rd.)
James Docherty Ho. E2 40Xb 71
(off Patriot Sq.)
James Dudson Ct. NW10 38Sa 67
James Est. CR4: Mitc 68Hb 133
James Gdns. N22 24Rb 51
James Hammett Ho. E2 3K 219 (41Vb 91)
(off Ravenscroft St.)
James Hill Ho. W10 42Ab 88
(off Kensal Rd.)
James Ho. E1 42Ac 92
(off Solebay St.)
James Ho. SE16 47Zb 92
(off Wolfe Cres.)
James Ho. SW8 52Nb 112
(off Wyvil Rd.)
James Ho. W10 42Ab 88
James Joyce Wlk. SE24 56Rb 113
James La. E10 31Ec 72
James La. E11 30Fc 53
James Leal Cen. Woodford 22Mc 53
James Lee Sq. EN3: Enf L 10Cc 20
James Lighthill Ho. WC1 2J 217 (40Pb 70)
(off Penton Ri.)
James Lind Ho. SE8 49Bc 92
(off Grove St.)
James Martin Cl. UB9: Den 30J 43
James Mdw. SL3: L'ly 51B 104
James Middleton Ho. E2 41Xb 91
(off Middleton St.)
James Morgan M. N1 1E 218 (39Sb 71)
James Newman Ct. SE9 62Qc 138
Jameson Cl. SL2: Slou 5K 81
Jameson Cl. W3 47Sa 87
Jameson Cl. AL1: St A 1D 6
(off Avenue Rd.)
Jameson Ct. E2 40Yb 72
(off Russia La.)
Jameson Ho. SE11 7H 229 (50Pb 90)
(off Glasshouse Wlk.)
Jameson Lodge N6 30Lb 50
Jameson St. W8 46Cb 89
James Pl. N17 25Vb 51
James Riley Point E15 39Ec 72
(off Carpenters Rd.)
James Rd. DA1: Dart 59Jd 118
James's Cotts. TW9: Kew 52Qa 109
James Stewart Ho. NW6 38Bb 69
James St. CM16: Epp 1Wc 23
James St. EN1: Enf 15Vb 33
James St. IG11: Bark 38Sc 74
James St. SL4: Wind 3H 103
James St. TW3: Houn 55Fa 108
James St. W1 2J 221 (44Jb 90)
James St. WC2 4G 223 (45Nb 90)
James Stroud Ho. SE17 7E 230 (50Sb 91)
(off Walworth Pl.)
James Ter. SW14 55Ta 109
(off Church Path)
Jameston Lodge HA4: Ruis 32V 64
Jamestown Rd. NW1 39Kb 70
Jamestown Way E14 45Fc 93
James Voller Way E1 44Yb 92
James Watt Way DA8: Erith 51Hd 118
James Way WD19: Wat 21Z 45
James Yd. E4 23Fc 53
Jam Factory, The SE1 3H 231 (48Ub 91)
(off Green Wlk.)
Jamieson Ho. TW4: Houn 58Ba 107
Jamilah Ho. E16 45Qc 94
(off University Way)
Jamnagar Cl. TW18: Staines 65H 127
Jamuna Cl. E14 43Ac 92
Jane Austen Hall E16 46Kc 93
(off Wesley Av.)
Jane Austen Ho. SW1 50Lb 90
(off Churchill Gdns.)
Jane Seymour Ct. SE9 59Tc 116
Jane St. E1 44Xb 91
Janet Adegoke Swimming Pool 45Wa 88
Janet St. E14 48Cc 92
Janeway Pl. SE16 47Xb 91
Janeway St. SE16 47Wb 91
Janice M. IG1: Ilf 33Rc 74
Janmead CM13: Hut 17De 41
Janoway Hill La. GU21: Wok 1N 187
Jansen Wlk. SW11 55Fb 111
Janson Cl. E15 36Gc 73
Janson Cl. NW10 34Ua 68
Janson Rd. E15 36Gc 73
Jansons Rd. N15 27Ub 51
Japan Cres. N4 31Pb 70
Japan Rd. RM6: Chad H 30Zc 55
Japonica Cl. GU21: Wok 10N 167
Japonica Ho. KT15: Add 75M 149
Jaquard Ct. E2 40Yb 72
(off Bishop's Way)
Jardine Rd. E1 45Zb 92
Jarman Cl. HP3: Hem H 4N 3
Jarman Ho. E1 43Yb 92
(off Jubilee St.)
Jarman Ho. SE16 49Zb 92
(off Hawkstone Rd.)
Jarman Pk. HP2: Hem H 3P 3
Jarman Square 3P 3
Jarman Way HP2: Hem H 3P 3
Jarrah Cotts. RM19: Purf 51Td 120
Jarratt Ho. SL4: Wind 5F 102
(off St Leonard's Rd.)
Jarret Ho. E3 41Cc 92
(off Bow Rd.)
Jarrett Cl. SW2 60Rb 113
Jarrett Cl. SM4: Mord 71Db 155
Jarrow Rd. N17 28Xb 51
Jarrow Rd. RM6: Chad H 30Yc 55
Jarrow Rd. SE16 49Yb 92
Jarrow Way E9 35Bc 72
Jarvis Cl. EN5: Barn 15Za 30
Jarvis Cl. IG11: Bark 39Tc 74
Jarvis Ho. SE15 53Wb 113
(off Goldsmith Rd.)
Jarvis Rd. CR2: S Croy 79Tb 157
Jarvis Rd. SE22 56Ub 113
Jarvis Way RM3: Hrld W 26Nd 57
Jashoda Ho. SE18 50Qc 94
(off Connaught M.)
Jasmin Cl. HA6: Nwood 25V 44
Jasmin Ct. SE12 58Jc 115
Jasmine Cl. BR6: Farnb 75Rc 160
Jasmine Cl. CM13: Gt War 23Xd 58
Jasmine Cl. GU21: Wok 8K 167
Jasmine Cl. IG1: Ilf 36Rc 74
Jasmine Cl. UB1: S'hall 45Aa 85
Jasmine Ct. SW19 64Cb 133
Jasmine Ct. TW15: W King 80Ud 164
Jasmine Gdns. CR0: C'don 76Dc 158
Jasmine Gdns. HA2: Harr 33Ca 65
Jasmine Gro. SE20 67Xb 135
Jasmine Ho. KT15: Add 75M 149
Jasmine Ho. SW18 55Eb 111
Jasmine Rd. RM7: Rush G 33Gd 76
Jasmine Sq. E3 39Bc 72
(off Hawthorn Av.)
Jasmine Ter. CM15: Pil H 15Vd 40
Jasmine Ter. UB7: W Dray 47Q 84
Jasmine Way KT8: E Mos 70Ga 130
Jasmin Lodge SE16 50Xb 91
(off Sherwood Gdns.)
Jasmin Rd. KT19: Ewe 78Ra 153
Jasmin Way HP1: Hem H 1G 2
Jason Cl. CM14: B'wood 21Vd 58
Jason Cl. KT13: Weyb 78S 150
Jason Cl. RH1: Redh 10N 207
Jason Cl. RM16: Ors 4G 100
Jason Ct. SW9 53Qb 112
(off Southey Rd.)
Jason Ct. W1 2J 221 (44Jb 90)
(off Wigmore St.)
Jason Wlk. SE9 63Qc 138
Jasper Av. W7 47Ha 86
Jasper Cl. EN3: Enf W 10Zb 20
Jasper Pas. SE19 65Vb 135
Jasper Rd. E16 44Mc 93
Jasper Rd. SE19 64Vb 135
Jasper Wlk. N1 3F 219 (41Tb 91)
Java Ho. E14 44Gc 93
Java Wharf SE1 1K 231 (47Vb 91)
(off Shad Thames)
Javelin Ct. HA7: Stan 22Ka 46
(off William Dr.)
Javelin Way UB5: N'olt 41Z 85
Jay Av. KT15: Add 76N 149
Jaycroft EN2: Enf 11Qb 32
Jay Gdns. BR7: Chst 63Pc 138
Jay Ho. E3 39Bc 72
(off Hawthorn Av.)
Jay M. SM5: Cars 76Hb 155
Jay M. SW7 2A 226 (47Eb 89)
Jays Cl. AL2: Brick W 3Ca 13
Jays St. N1 1J 217 (39Pb 70)
Jazzfern Ter. HA0: Wemb 36Ja 66
Jeal Oakwood Ct. KT18: Eps 86Ua 174
Jean Batten Cl. SM6: W'gton 80Pb 156
Jean Brown Indoor Arena,
The 25Tc 54
Jean Darling Ho. SW10 51Fb 111
(off Milman's St.)
Jean Ho. SW17 64Gb 133
Jeanne Ct. E14 43Ac 92
(off Pechora Way)
Jean Pardies Ho. E1 43Yb 92
(off Jubilee St.)
Jebb Av. SW2 58Nb 112
(not continuous)
Jebb Cl. SL4: Wind 4B 102
Jebb St. E3 40Cc 72
Jedburgh Rd. E13 41Lc 93
Jedburgh St. SW11 56Jb 112
Jeddo M. W12 47Va 88
Jeddo Rd. W12 47Va 88
Jeeyas Apts. E16 43Hc 93
Jefferies Way SS17: Stan H 1P 101
Jefferson Bldg. E14 47Cc 92
Jefferson Cl. IG2: Ilf 29Rc 54
Jefferson Cl. SL3: L'ly 49C 82
Jefferson Cl. W13 48Ka 86
Jefferson Ho. TW8: Bford 51Na 109
Jefferson Ho. UB7: W Dray 47P 83
(off Park Lodge Av.)
Jefferson Pl. BR2: Broml 72Mc 159
Jefferson Plaza E3 42Ec 92
(off Hannaford Wlk.)
Jefferson Rd. GU24: Brkwd 2A 186
Jefferson Wlk. SE18 51Qc 116
Jeffery Harrison Vis. Cen. 93Hd 202
Jeffery Row SE12 57Kc 115
Jeffrey's Pl. NW1 38Lb 70
Jeffreys Rd. EN3: Brim 14Ac 34
Jeffreys Rd. SW4 54Nb 112
Jeffrey's St. NW1 38Lb 70
Jeffreys Wlk. SW4 54Nb 112
Jeffs Cl. TW12: Hamp 65Da 129
Jeffs Rd. SM1: Sutt 77Bb 155
Jeger Av. E2 39Vb 71
Jeken Rd. SE9 56Lc 115
Jelf Rd. SW2 57Qb 112
Jelico Point SE16 47Bc 92
(off Rotherhithe St.)
Jelley Way GU22: Wok 94B 188
Jellicoe Av. DA12: Grav'nd 2E 144
Jellicoe Av. W. DA12: Grav'nd 2E 144
Jellicoe Cl. SL1: Slou 7F 80
Jellicoe Gdns. HA7: Stan 23Ha 46
Jellicoe Ho. E2 40Wb 71
(off Ropley St.)
Jellicoe Rd. E13 42Jc 93
Jellicoe Rd. N17 24Tb 51
Jellicoe Rd. WD18: Wat 16W 26
Jemma Knowles Cl. SW2 60Qb 112
(off Tulse Hill)
Jemmett Cl. KT2: King T 67Ra 131
Jemotts Ct. SE14 51Zb 114
(off Myers La.)
Jem Paterson Ct. HA1: Harr 35Ga 66
Jengar Cl. SM1: Sutt 77Db 155
Jenkins Av. AL2: Brick W 2Aa 13
Jenkins Cl. DA11: Nflt 2B 144
Jenkins La. IG11: Bark 40Sc 74
Jenkinson Ho. E2 41Zb 92
(off Usk St.)
Jenkins Rd. E13 42Kc 93
Jenner Av. W3 43Ta 87
Jenner Cl. CR5: Coul 89Lb 176
Jenner Cl. DA14: Sidc 63Wc 139
Jenner Ct. N21 15Nb 32
(off Pennington Dr.)
Jenner Dr. GU24: W End 5E 166
Jenner Ho. WC1 5G 217 (42Nb 90)
(off Hunter St.)
Jenner Pl. SW13 51Xa 110
Jenner Rd. N16 34Vb 71
Jenner Way KT19: Eps 81Qa 173
Jennery La. SL1: Burn 1A 80
Jennett Rd. CR0: Wadd 76Qb 156
Jennifer Ho. E14 44Ec 92
Jennifer Ho. SE11 6A 230 (49Qb 90)
(off Reedworth St.)
Jennifer Rd. BR1: Broml 62Hc 137
Jenningham Dr. RM16: Grays 46Ce 99
Jenningsbury Ho. SW3 7E 226 (50Gb 89)
(off Cale St.)
Jennings Cl. DA1: Dart 54Qd 119
Jennings Cl. KT15: New H 81L 169
Jennings Cl. KT6: Surb 73La 152
Jennings Cl. RM8: Dag 32Ad 75
Jennings Ho. SE10 50Fc 93
(off Old Woolwich Rd.)
Jennings Rd. AL1: St A 1D 6
Jennings Rd. SE22 58Vb 113
Jennings Way EN5: Barn 13Ya 30
Jennings Way HP3: Hem H 4N 3
Jenningtree Rd. DA8: Erith 52Kd 119
Jenningtree Way DA17: Belv 47Ed 96
Jenny Hammond Cl. E11 34Hc 73
Jenny Path RM3: Rom 24Md 57
Jennys Way CR5: Coul 94Lb 196
Jensen Ho. E3 42Cc 92
(off Wellington Way)
Jenson Hgts. N1 38Pb 70
(off Caledonian Rd.)
Jenson Way SE19 66Vb 135
Jenton Av. DA7: Bex 53Ad 117
Jephson Ct. SW4 54Nb 112
Jephson Ho. SE17 51Rb 113
(off Doddington Gro.)
Jephson Rd. E7 38Lc 73
Jephson St. SE5 53Tb 113
Jephtha Rd. SW18 58Cb 111
Jeppos La. CR4: Mitc 70Hb 133
Jepson Dr. DA2: Dart 58Sd 120
Jepson Ho. SW6 53Db 111
(off Pearscroft Rd.)
Jerdan Ho. SW6 52Cb 111
(off North End Rd.)
Jerdan Pl. SW6 52Cb 111
Jeremiah Ct. RH1: Mers 3C 208
Jeremiah St. E14 44Dc 92
Jeremy Bentham Ho. E2 41Wb 91
(off Mansford St.)
Jeremy's Grn. N18 21Xb 51
Jermyn St. SW1 6B 222 (46Lb 90)
Jermyn Street Theatre 5D 222 (45Mb 90)
(off Jermyn St.)
Jerningham Av. IG5: Ilf 26Rc 54
Jerningham Ct. SE14 53Ac 114
Jerningham Rd. SE14 54Ac 114
Jerome Cres. NW8 5D 214 (42Gb 89)
Jerome Dr. AL3: St A 4N 5
Jerome Ho. KT1: Hamp W 68Ma 131
(off Old Bridge St.)
Jerome Ho. NW1 7E 214 (43Gb 89)
(off Lisson Gro.)
Jerome Ho. SW7 5B 226 (49Fb 89)
(off Glendower Pl.)
Jerome Pl. KT1: King T 68Ma 131
(off Wadbrook St.)
Jerome St. E1 6K 219 (42Vb 91)
Jerome Twr. W3 47Ra 87
(off Howley Pl.)
Jerrard St. SE13 55Dc 114
Jerrold St. N1 2J 219 (40Vb 71)
Jersey Av. HA7: Stan 26Ka 46
Jersey Cl. GU21: Brkwd 10E 166
Jersey Cl. KT16: Chert 76H 149
Jersey Dr. BR5: Pet W 72Tc 160
Jersey Ho. EN3: Enf W 10Zb 20
(off Eastfield Rd.)
Jersey Ho. N1 37Sb 71
(off Jersey Rd.)
Jersey Ho. WD18: Wat 16V 26
Jersey Pl. SL5: S'hill 2B 146
Jersey Rd. E11 32Fc 73
Jersey Rd. E16 44Lc 93
Jersey Rd. IG1: Ilf 35Rc 74
Jersey Rd. N1 37Sb 71
Jersey Rd. RM13: Rain 38Jd 76
Jersey Rd. SW17 65Kb 134
Jersey Rd. TW3: Houn 53Da 107
Jersey Rd. TW5: Hest 53Da 107
Jersey Rd. TW5: Isle 53Da 107
Jersey Rd. TW7: Isle 52Fa 108
Jersey Rd. W7 47Ja 86
Jersey St. E2 41Xb 91
Jerusalem Pas. EC1 6B 218 (42Rb 91)
Jervis Av. EN3: Enf W 7Ac 20
Jervis Ct. RM10: Dag 37Dd 76
Jervis Ct. SE10 53Ec 114
(off Blissett St.)
Jervis Ct. W1 3A 222 (44Kb 90)
(off Princes St.)
Jervis Rd. SW6 51Bb 111
Jerviston Gdns. SW16 65Qb 134
Jerwood Space Art Gallery 1C 230 (47Rb 91)
(off Union St.)
Jeskyns 8G 144
Jeskyns Rd. DA12: Cobh 8F 144
Jeskyns Rd. DA13: Cobh 9E 144
Jeskyns Rd. DA13: Sole S 9E 144
Jesmond Av. HA9: Wemb 37Pa 67
Jesmond Cl. CR4: Mitc 69Kb 134
Jesmond Dene CM16: They B 8Tc 22
Jesmond Dene NW3 37Eb 69
Jesmond Rd. CR0: C'don 73Vb 157
Jesmond Rd. RM16: Grays 46Fe 99
Jesmond Way HA7: Stan 22Na 47
Jessam Av. E5 32Xb 71
Jessamine Pl. DA2: Dart 59Sd 120
Jessamine Rd. W7 46Ha 86
Jessamine Ter. BR8: Swan 67Ed 140
Jessamy Rd. KT13: Weyb 75Q 150
Jessel Dr. IG10: Lough 11Sc 36
Jessel Ho. SW1 5E 228 (49Mb 90)
(off Page St.)
Jessel Ho. WC1 4F 217 (41Nb 90)
(off Judd St.)
Jessel Mans. W14 51Ab 110
(off Queen's Club Gdns.)
Jesse Rd. E10 32Ec 72
Jessett Cl. DA8: Erith 49Fd 96
Jessica Rd. SW18 58Eb 111
Jessie Blythe La. N19 31Nb 70
Jessie Duffett Ho. SE5 52Sb 113
(off Pitman St.)
Jessie Wood Ct. SW9 52Qb 112
(off Caldwell St.)
Jessiman Ter. TW17: Shep 71Q 150
Jesson Ho. SE17 6F 231 (49Tb 91)
(off Orb St.)
Jessop Av. UB2: S'hall 49Ba 85
Jessop Ct. N1 2C 218 (40Rb 71)
Jessop Ho. W4 49Ta 87
(off Kirton Cl.)
Jessop Lodge CR0: C'don 75Sb 157
(off Tamworth Rd.)
Jessopp Ct. EN9: Walt A 6Hc 21
Jessop Pl. W7 48Ga 86
Jessop Rd. SE24 56Rb 113
Jessop Sq. E14 46Cc 92
Jessops Way CR0: Bedd 72Lb 156
Jessops Way CR0: Mitc 72Lb 156
Jessup Cl. SE18 49Sc 94
Jetstar Way UB5: N'olt 41Aa 85
Jetty Ho. KT16: Chert 73L 149
Jetty Wlk. RM17: Grays 51Ce 121
Jevington Way SE12 60Kc 115
Jevons Ho. NW8 38Fb 69
(off Hilgrove Rd.)
Jewell Ct. WD25: Wat 6Z 13
Jewel Rd. E17 27Cc 52
Jewels Hill TN16: Big H 84Jc 179
Jewel Sq. E1 45Xb 91
Jewel Tower 3F 229 (48Nb 90)
(off College M.)
Jewish Mus. 1A 216 (39Kb 70)
Jewry St. EC3 3K 225 (44Vb 91)
Jew's Row SW18 56Db 111
Jews' Wlk. SE26 63Xb 135
Jeymer Av. NW2 36Xa 68
Jeymer Dr. UB6: G'frd 39Da 65
(not continuous)
Jeypore Rd. SW18 59Eb 111
Jeypore Rd. Pas. SW18 58Eb 111
JFK Ho. WD23: Bush 14Ba 27
Jhumat Pl. IG1: Ilf 34Qc 74
(off Roden St.)
Jigger Mast Ho. SE18 48Qc 94
Jillian Cl. TW12: Hamp 66Ca 129
Jim Bradley Cl. SE18 49Qc 94
Jim Griffiths Ho. SW6 51Bb 111
(off Clem Attlee Ct.)
Jim Veal Dr. N7 37Nb 70
JJ's Clay Shooting Club 62Vd 142
Joan Cres. SE9 59Mc 115
Joan Gdns. RM8: Dag 33Ad 75
Joanna Ho. W6 50Ya 88
(off Queen Caroline St.)
Joan Rd. RM8: Dag 33Ad 75
Joan St. SE1 7B 224 (46Rb 91)
Job Drain Pl. IG11: Bark 40Wc 75
Jocelin Ho. N1 1J 217 (39Pb 70)
(off Barnsbury Est.)
Jocelyn Rd. TW9: Rich 55Na 109
Jocelyn St. SE15 53Wb 113
Jocketts Hill HP1: Hem H 2H 3
Jocketts Rd. HP1: Hem H 3H 3
Jockey's Flds. WC1 7J 217 (43Pb 90)
Jodane St. SE8 49Bc 92
Jodrell Cl. TW7: Isle 53Ja 108
Jodrell Rd. E3 39Bc 72
Jodrell Way RM20: W Thur 50Vd 98
Joe Hunte Ct. SE27 64Rb 135
Joel St. HA5: Eastc 27W 44
Joel St. HA6: Nwood 26W 44
Johanna St. SE1 2K 229 (47Qb 90)
John Adams Ct. N9 19Vb 33
John Adam St. WC2 5G 223 (45Nb 90)
John Aird Ct. W2 7A 214 (43Eb 89)
(off Howley Pl.)
John Archer Way SW18 58Fb 111
John Ashby Cl. SW2 58Nb 112
John Austin Cl. KT2: King T 67Pa 131
John Baird Ct. SE26 63Yb 136
John Barker Ct. NW6 38Ab 68
John Barnes Wlk. E15 37Hc 73
John Bell Twr. E. E3 40Cc 72
(off Pancras Way)
John Bell Twr. W. E3 40Cc 72
(off Pancras Way)
John Betts' Ho. W12 48Va 88
John Bond Ho. E3 40Bc 72
(off Wright's Rd.)
John Bowles Ct. E1 45Zb 92
(off Schoolhouse La.)
John Bradshaw Rd. N14 18Mb 32
John Brent Ho. SE8 49Zb 92
(off Haddonfield)
John Buck Ho. NW10 39Va 68
John Bull Pl. W4 49Ra 87
John Bunn Mill KT15: Add 78N 149
(off Bourneside Rd.)
John Burns Dr. IG11: Bark 38Uc 74
Johnby Cl. EN3: Enf W 9Ac 20
John Campbell Rd. N16 36Ub 71
John Carpenter St. EC4 4B 224 (45Rb 91)
John Cartwright Ho. E2 41Xb 91
(off Old Bethnal Grn. Rd.)
John Clay Gdns. RM16:
Grays 46De 99
John Cobb Rd. KT13: Weyb 80Q 150
John Crane St. SE17 51Tb 113
John Donne Way SE10 52Dc 114
(off Norman Rd.)
John Drinkwater Cl. E11 31Hc 73
John Fearon Wlk. W10 41Ab 88
(off Dart St.)
John Fielden Ho. E2 41Xb 91
(off Canrobert St.)
John Fisher St. E1 45Wb 91
John F Kennedy Memorial 1P 125
John Gale Ct. KT17: Ewe 81Va 174
(off West St.)
John Goddard Way TW13: Felt 61X 129
John Gooch Dr. EN2: Enf 11Rb 33
John Harrison Way SE10 48Hc 93
John Horner M. N1 1D 218 (40Sb 71)
JOHN HOWARD CEN. 36Ac 72
John Hunter Av. SW17 62Gb 133
John Islip St. SW1 6E 228 (49Mb 90)
John Kaye Ct. TW17: Shep 71Q 150
John Keats Ho. N22 24Pb 50
John Keats Lodge EN2: Enf 11Tb 33
John Kennedy Ct. N1 37Tb 71
(off Newington Grn. Rd.)
John Kennedy Ho. SE16 49Zb 92
(off Rotherhithe Old Rd.)
John Knight Lodge SW6 52Cb 111
John Lamb Ct. HA3: W'stone 25Ga 46
John McDonald Ho. E14 48Ec 92
(off Glengall Gro.)
John McKenna Wlk. SE16 48Wb 91
John Masefield Ho. N15 30Tb 51
(off Fladbury Rd.)
John Maurice Cl. SE17 5F 231 (49Tb 91)
John Mills Ct. UB9: Den 30H 43
(off Stewarts Ct.)
John Nash M. E14 44Ac 92
(off Commercial Rd.)
John Newton Ct. DA16: Well 55Xc 117
John Norman Gro. GU18: Light 2A 166
Johnny Andrews Ho. E1 44Zb 92
(off Boulcott St.)
John Orwell Sports Cen. 46Xb 91
John Parker Cl. RM10: Dag 38Dd 76
John Parker Sq. SW11 55Fb 111
John Parry Ct. N1 1J 219 (40Ub 71)
(off Hare Wlk.)
John Penn Ho. SE14 52Bc 114
(off Amersham Va.)
John Penn St. SE13 53Dc 114
John Penry Ho. SE1 50Wb 91
(off Marlborough Gro.)
John Perrin Pl. HA3: Kenton 31Na 67
John Prince's St. W1 2A 222 (44Kb 90)
John Pritchard Ho. E1 42Wb 91
(off Buxton St.)
John Ratcliffe Ho. NW6 41Cb 89
(off Chippenham Gdns.)
John Rennie Wlk. E1 45Xb 91
John Riley Ho. E3 43Bc 92
(off Geoffrey Chaucer Way)
John Roll Way SE16 48Wb 91
John Ruskin St. SE5 52Rb 113
John's Av. NW4 28Ya 48
John Sayer Cl. IG11: Bark 41Vc 95
John's Cl. TW15: Ashf 63S 128
Johns Cl. DA3: Hartl 71Be 165
John Scurr Ho. E14 44Ac 92
(off Ratcliffe La.)
Johnsdale RH8: Oxt 1K 211
John Sessions Sq. E1 44Wb 91
(off Alie St.)
JOHN'S HOLE 59Sd 120
John Silkin La. SE8 50Zb 92
John's La. SM4: Mord 71Eb 155
John's M. WC1 6J 217 (42Pb 90)
John Smith Av. SW6 52Bb 111
John Smith M. E14 45Fc 93
Johnson Cl. DA11: Nflt 62Fe 143
Johnson Cl. E8 39Wb 71
Johnson Cl. HP3: Hem H 4N 3
Johnson Ct. SE9 56Lc 115
Johnson Dr. CR5: Coul 89Lb 176
Johnson Ho. E2 41Wb 91
(off Roberta St.)
Johnson Ho. NW1 2C 216 (40Lb 70)
(off Cranleigh St.)
Johnson Ho. NW3 38Hb 69
(off Adelaide Rd.)
Johnson Ho. SW1 6J 227 (49Jb 90)
(off Cundy St.)
Johnson Ho. SW8 52Mb 112
(off Wandsworth Rd.)
Johnson Ho. W8 47Cb 89
(off Campden Hill)
Johnson Lock Ct. E1 43Ac 92
Johnson Lodge W9 43Cb 89
(off Admiral Wlk.)
Johnson Mans. W14 51Ab 110
(off Queen's Club Gdns.)
Johnson Rd. BR2: Broml 71Mc 159
Johnson Rd. CR0: C'don 73Tb 157
Johnson Rd. NW10 39Ta 67
Johnson Rd. TW5: Hest 52Y 107
Johnson's Av. TN14: Bad M 82Dd 182
Johnsons Cl. SM5: Cars 75Hb 155
Johnson's Ct. EC4 3A 224 (44Qb 90)
Johnsons Ct. TN15: Seal 93Pd 203
Johnsons Dr. TW12: Hamp 67Ea 130
Johnsons Ind. Est. UB3: Hayes 47V 84
Johnson's Pl. SW1 7B 228 (50Lb 90)
Johnson St. E1 45Yb 92
Johnson St. UB2: S'hall 48Y 85
Johnson's Way DA9: Ghithe 58Yd 120
Johnsons Way NW10 42Ra 87
Johnson's Yd. UB8: Uxb 38L 63
John Spencer Sq. N1 37Rb 71
John's Pl. E1 44Xb 91
John's Rd. DA13: Meop 10B 144
John's Rd. TN16: Tats 92Mc 199
John's Ter. CR0: C'don 74Ub 157
John's Ter. RM3: Hrld W 23Rd 57
Johnston Cl. SW9 53Pb 112
Johnston Ct. E10 34Dc 72
Johnstone Cl. WD19: Wat 20Z 27
Johnstone Ho. SE13 55Fc 115
(off Belmont Hill)
Johnstone Rd. E6 41Pc 94
Johnston Rd. IG8: Wfd G 23Jc 53
Johnston Ter. NW2 34Za 68
John Strachey Ho. SW6 51Bb 111
(off Clem Attlee Ct.)
John St. E15 39Hc 73
John St. EN1: Enf 15Vb 33
John St. RM17: Grays 51Ee 121
John St. SE25 70Wb 135
John St. TW3: Houn 54Aa 107
John St. WC1 6J 217 (42Pb 90)
John Strype Ct. E10 32Dc 72
Johns Wlk. CR3: W'ham 91Wb 197
John's Way RM15: S Ock 44Zd 99
John Trundle Ct. EC2 7D 218 (43Sb 91)
(off Aldersgate St.)
John Trundle Highwalk EC2 7D 218 (43Sb 91)
(off Aldersgate St.)
John Tucker Ho. E14 48Cc 92
(off Mellish St.)
John Watkin Cl. KT19: Eps 81Ra 173
John Wesley Cl. E6 41Pc 94
John Wesley Ct. TW1: Twick 60Ja 108
John Wesley Highwalk EC2 1D 224 (43Sb 91)
(off Aldersgate St.)
John Wetherby Ct. E15 40Dc 72
(off High St.)
John Wetherby Ct. E. E15 40Dc 72
(off High St.)
John Wetherby Ct. W. E15 40Dc 72
(off High St.)
John Wheatley Ho. SW6 51Bb 111
(off Clem Attlee Ct.)
John William Cl. RM16: Chaf H 50Zd 99
John Williams Cl. KT2: King T 67Ma 131
John Williams Cl. SE14 51Zb 114
John Wilson St. SE18 48Qc 94
John Woolley Cl. SE13 56Gc 115
Joiners Arms Yd. SE5 53Tb 113
Joiners Cl. SL9: Chal P 24B 42
Joiner's La. SL9: Chal P 25A 42
Joiners M. E11 34Gc 73
Joiners Pl. N5 35Tb 71
Joiner St. SE1 7G 225 (46Tb 91)
Joiners Way SL9: Chal P 24A 42
Joiners Yd. N1 2G 217 (40Nb 70)
(off Caledonia St.)
Joinville Pl. KT15: Add 77M 149

Jolles Ho. E3 41Dc 92
(off Bromley High St.)
Jolliffe Rd. RH1: Mers 98Lb 196
Jolly M. SW16 68Lb 134
Jollys La. HA2: Harr 32Fa 66
Jollys La. UB4: Yead 43Z 85
Jonathans RM11: Horn 32Nd 77
(off High St.)
Jonathan St. SE11 7H 229 (50Pb 90)
Jones Cotts. EN5: Ark 15Wa 30
Jones Ho. E14 44Fc 93
(off Blair St.)
Jones M. SW15 56Ab 110
Jones Rd. E13 42Kc 93
Jones Rd. EN7: G Oak 2Rb 19
Jones St. W1 5K 221 (45Kb 90)
Jones Wlk. TW10: Rich 58Pa 109
Jones Way SL2: Hedg 3H 61
Jonquil Gdns. TW12: Hamp 65Ca 129
Jonson Cl. CR4: Mitc 70Kb 134
Jonson Cl. UB4: Hayes 43W 84
Jonson Ho. SE1 4G 231 (48Tb 91)
(off Burbage Cl.)
Jonzen Wlk. E14 44Cc 92
Jopling Rd. GU24: Bisl 9E 166
Jordan Cl. CR2: Sande 83Vb 177
Jordan Cl. HA2: Harr 34Ba 65
Jordan Cl. WD25: Wat 7V 12
Jordan Ct. SW15 56Za 110
Jordan Ho. N1 39Tb 71
(off Colville Est.)
Jordan Ho. SE4 56Zb 114
(off St Norbert Rd.)
Jordan Rd. UB6: G'frd 39Ka 66
Jordans Cl. RM10: Dag 35Dd 76
Jordans Cl. TW19: Stanw 59L 105
Jordans Cl. TW7: Isle 53Ga 108
Jordans Ho. NW8 5C 214 (42Fb 89)
(off Capland St.)
Jordans M. TW2: Twick 61Ga 130
Jordans Rd. WD3: Rick 17J 25
Jordan's Way AL2: Brick W 2Aa 13
Jordans Way RM13: Rain 40Md 77
Joscoyne Ho. E1 44Xb 91
(off Philpot St.)
Joseph Av. W3 44Ta 87
Joseph Cl. N4 33Rb 71
Joseph Conrad Ho. SW1 6C 228 (49Lb 90)
(off Tachbrook St.)
Joseph Ct. AL1: St A 3F 6
(off Cambridge Rd.)
Joseph Ct. CM14: W'ley 22Xd 58
Joseph Ct. N16 30Ub 51
(off Amhurst Pk.)
Joseph Grimaldi Pk. 2J 217 (40Pb 70)
Joseph Hardcastle Cl. SE14 52Zb 114
Josephine Av. KT20: Lwr K 98Bb 195
Josephine Av. SW2 57Pb 112
Josephine Cl. KT20: Lwr K 99Bb 195
Joseph Irwin Ho. E14 45Bc 92
(off Gill St.)
Joseph Lister Ct. E7 38Jc 73
Joseph Locke Way KT10: Esh 75Ca 151
Joseph M. N7 37Qb 70
(off Westbourne Rd.)
Joseph M. SE15 57Zb 114
Joseph Powell Cl. SW12 58Lb 112
Joseph Priestley Ho. E2 41Xb 91
(off Canrobert St.)
Joseph Ray Rd. E11 33Gc 73
Joseph St. E3 42Bc 92
Joseph Trotter Cl. EC1 4A 218 (41Qb 90)
(off Finsbury Est.)
Joshua Cl. CR2: S Croy 80Rb 157
Joshua Cl. N10 24Kb 50
Joshua Pedley M. E3 40Cc 72
Joshua St. E14 44Ec 92
Joshua Wlk. EN8: Walt C 6Cc 20
Josiah Dr. UB10: Ick 33S 64
Joslin Av. NW9 27Ua 48
Josling Cl. RM17: Grays 51Be 121
Joslings Cl. W12 45Wa 88
Joslin Rd. RM19: Purf 50Sd 98
Joslyn Cl. EN3: Enf L 10Cc 20
Jossiline Ct. E3 40Ac 72
(off Ford St.)
Joubert Mans. SW3 7E 226 (50Gb 89)
(off Jubilee Pl.)
Joubert St. SW11 54Hb 111
Jourdelay's SL4: Eton 1H 103
(off Jourdelay's Pas.)
Jourdelay's Pas. SL4: Eton 1H 103
Journeys End SL2: Stoke P 3J 81
Jove Gdns. AL4: S'ford 1L 7
Jowett St. SE15 52Vb 113
Jowitt Ho. E2 41Zb 92
(off Morpeth St.)
Joyce Av. N18 22Vb 51
Joyce Butler Ho. N22 25Pb 50
Joyce Ct. EN9: Walt A 6Fc 21
Joyce Dawson Way SE28 45Wc 95
JOYCE GREEN 56Pd 119
Joyce Grn. DA1: Dart 55Nd 119
Joyce Grn. La. DA1: Dart 53Md 119
(not continuous)
Joyce Grn. Wlk. DA1: Dart 56Pd 119
Joyce Latimore Ct. N9 20Xb 33
(off Colthurst Dr.)
Joyce Page Cl. SE7 51Mc 115
Joyce Wlk. SW2 58Qb 112
Joydens Wood 63Dd 140
JOYDENS WOOD 63Fd 140
Joydens Wood Rd. DA5: Bexl 63Fd 140
Joydon Dr. RM6: Chad H 30Xc 55
Joyes Cl. RM3: Rom 21Md 57
Joyners Cl. RM9: Dag 35Bd 75
Joy of Life Fountain 6H 221 (46Jb 90)
Joy Rd. DA12: Grav'nd 10E 122
Joystone Ct. EN4: E Barn 14Gb 31
(off Park Rd.)
Jubb Powell Ho. N15 30Ub 51
Jubilee, The SE10 52Dc 114
Jubilee Arch SL4: Wind 3H 103
Jubilee Av. AL2: Lon C 8H 7
Jubilee Av. E4 23Ec 52
Jubilee Av. RM7: Rom 29Dd 56
Jubilee Av. TW2: Whitt 60Ea 108
Jubilee Bldgs. NW8 1B 214 (39Fb 69)
Jubilee Cl. DA9: Ghithe 58Yd 120
Jubilee Cl. HA5: Pinn 26Y 45
Jubilee Cl. HP2: Hem H 1P 3
Jubilee Cl. KT1: Hamp W 67La 130
Jubilee Cl. NW10 40Ua 68
Jubilee Cl. NW9 30Ta 47
Jubilee Cl. RM7: Rom 29Dd 56
Jubilee Cl. TW19: Stanw 59L 105
Jubilee Cotts. HP5: Whel H 8A 2
Jubilee Cotts. SL3: L'ly 50E 82
Jubilee Cotts. TN14: S'oaks 92Kd 203
Jubilee Country Pk. 70Rc 138
Jubilee Country Pk. Local Nature
Reserve 71Qc 160
Jubilee Ct. BR4: W W'ck 74Ec 158
Jubilee Ct. DA1: Dart 59Md 119
(off Spring Vale Sth.)
Jubilee Ct. E16 44Hc 93
(off Silvertown Sq.)
Jubilee Ct. E18 25Jc 53
Jubilee Ct. EN9: Walt A 5Hc 21
Jubilee Ct. HA3: Kenton 31Na 67
Jubilee Ct. N10 27Jb 50
Jubilee Ct. SE10 51Dc 114
(off Dowells St.)
Jubilee Ct. TW18: Staines 64J 127
Jubilee Ct. TW3: Houn 55Da 107
(off Bristow Rd.)
Jubilee Cres. DA12: Grav'nd 1G 144
Jubilee Cres. E14 48Ec 92
Jubilee Cres. KT15: Add 78M 149
Jubilee Cres. N9 18Wb 33
Jubilee Cres. TN15: Igh 93Yd 204
Jubilee Dr. HA4: Ruis 35Z 65
Jubilee Gdns. UB1: S'hall 43Ca 85
Jubilee Hall Gym 4G 223 (45Nb 90)
(within Jubilee Hall)
Jubilee Hgts. SE10 53Ec 114
(off Parkside Av.)
Jubilee Ho. HA7: Stan 22Ma 47
Jubilee Ho. SE11 6A 230 (49Qb 90)
(off Reedworth St.)
Jubilee Ho. WC1 5H 217 (42Pb 90)
(off Gray's Inn Rd.)
Jubilee La. W5 44Na 87
Jubilee Lodge IG7: Chig 17Rc 36
Jubilee Mans. E1 44Yb 92
(off Jubilee St.)
Jubilee Mkt. IG8: Wfd G 23Lc 53
Jubilee Mkt. WC2 4G 223 (45Nb 90)
(off Covent Garden)
Jubilee Mkt. 4G 223 (45Nb 90)
(off Covent Gdn.)
Jubilee Pde. IG8: Wfd G 23Lc 53
Jubilee Pl. SW3 7E 226 (50Gb 89)
Jubilee Pl. Shop. Mall 46Dc 92
(off Bank St.)
Jubilee Ri. TN15: Seal 93Pd 203
Jubilee Rd. BR6: Well H 79Cd 162
Jubilee Rd. RM20: W Thur 51Xd 120
Jubilee Rd. SM3: Cheam 80Za 154
Jubilee Rd. UB6: G'frd 39Ka 66
Jubilee Rd. WD24: Wat 10W 12
Jubilee Sq. GU21: Wok 89A 168
Jubilee Statue 3E 124
Jubilee St. E1 44Yb 92
Jubilee Vs. KT10: Esh 74Fa 152
Jubilee Wlk. WD19: Wat 21X 45
Jubilee Wlk. WD4: K Lan 2Q 12
Jubilee Wlk. WD6: E'tree 16Ma 29
(off High St.)
Jubilee Walkway SE1 5B 224 (45Rb 91)
Jubilee Way CR5: Coul 90Pb 176
Jubilee Way DA14: Sidc 61Wc 139
Jubilee Way KT9: Chess 77Qa 153
Jubilee Way SL3: Dat 2N 103
Jubilee Way SW19 67Db 133
Jubilee Way TW14: Felt 60W 106
Jubilee Way Training Track 76Ra 153
Jubilee Yd. SE1 1K 231 (47Vb 91)
(off Lafone St.)
Judd Apts. N8 27Pb 50
(off Gt. Amwell La.)
Judd St. WC1 3F 217 (41Nb 90)
Jude St. E16 44Hc 93
Judeth Gdns. DA12: Grav'nd 4G 144
Judge Heath La. UB3: Hayes 44S 84
Judge Heath La. UB8: Hil 44S 84
Judge's Hill EN6: N'thaw 1Gb 17
Judge St. WD24: Wat 10X 13
Judges Wlk. NW3 34Eb 69
Judge Wlk. KT10: Clay 79Ga 152
Judith Anne Ct. RM14: Upm 33Ud 78
Judith Av. RM5: Col R 23Dd 56
Judy's Pas. SL4: Eton 10G 80
Juer St. SW11 52Gb 111
Juett Lodge SE10 49Jc 93
(off Peartree Way)
Jug Hill TN16: Big H 88Mc 179
Juglans Rd. BR6: Orp 74Wc 161
Jules Thorn Av. EN1: Enf 14Wb 33
Julia Ct. E17 29Dc 52
Julia Gdns. IG11: Bark 40Zc 75
Julia Garfield M. E16 46Kc 93
(not continuous)
Juliana Cl. N2 26Eb 49
Julian Av. W3 45Ra 87
Julian Cl. EN5: New Bar 13Db 31
Julian Cl. GU21: Wok 10N 167
Julian Ct. NW1 38Lb 70
(off Rochester Sq.)
Julian Hill HA1: Harr 33Ga 66
Julian Hill KT13: Weyb 80Q 150
Julian Ho. SE21 63Ub 135
Julian Pl. E14 50Dc 92
Julian Rd. BR6: Chels 79Wc 161
Julians Cl. TN13: S'oaks 99Jd 202
Julians Way TN13: S'oaks 99Jd 202
Julian Taylor Path SE23 61Xb 135
Julia Scurr St. E3 43Cc 92
Julia St. NW5 35Jb 70
Julien Rd. CR5: Coul 87Mb 176
Julien Rd. W5 48La 86
Juliet Ho. N1 1H 219 (40Ub 71)
(off Arden Est.)
Juliette Cl. RM15: Avel 47Pd 97
Juliette M. RM1: Rom 29Hd 56
Juliette Rd. E13 40Jc 73
Juliet Way RM15: Avel 47Pd 97
Julius Caesar Way HA7:
Stan 20Ma 29
Julius Ho. E14 44Fc 93
(off E. India Dock Rd.)
Julius Nyerere Cl. N1 39Pb 70
(off Copenhagen St.)
Jump Giants 50Wd 98
Junction App. SE13 55Ec 114
Junction App. SW11 55Gb 111
Junction M. W2 2D 220 (44Gb 89)
Junction Pl. W2 2C 220 (44Fb 89)
(off Praed St.)
Junction Rd. CM14: W'ley 21Yd 58
Junction Rd. CR2: S Croy 78Tb 157
Junction Rd. DA1: Dart 58Md 119
Junction Rd. E13 40Kc 73
Junction Rd. GU18: Light 2A 166
Junction Rd. HA1: Harr 30Fa 46
(not continuous)
Junction Rd. N17 27Wb 51
Junction Rd. N19 35Lb 70
Junction Rd. N9 18Wb 33
Junction Rd. RM1: Rom 28Hd 56
Junction Rd. TW15: Ashf 64S 128
Junction Rd. TW8: Bford 49Ma 87
Junction Rd. W5 49La 86
Junction Rd. E. RM6: Chad H 31Ad 75
Junction Rd. W. RM6: Chad H 31Ad 75
Junction Shop. Cen., The 56Gb 111
June Cl. CR5: Coul 86Kb 176
Junewood Cl. KT15: Wdhm 83H 169
Jungle Falls Adventure
Golf 14Lb 32
Juniper Av. AL2: Brick W 3Ca 13
Juniper Cl. EN5: Barn 15Za 30
Juniper Cl. HA2: Harr 33Da 65
Juniper Cl. HA9: Wemb 36Qa 67
Juniper Cl. KT19: Eps 81Ta 173
Juniper Cl. KT9: Chess 78Pa 153
Juniper Cl. RH2: Reig 8L 207
Juniper Cl. RH8: Oxt 5M 211
Juniper Cl. TN16: Big H 89Nc 180
Juniper Cl. TW13: Felt 62X 129
Juniper Cl. WD3: Rick 20M 25
Juniper Ct. CM13: B'wood 20Be 41
(off The Limes)
Juniper Ct. HA3: Hrw W 25Ha 46
Juniper Ct. HA6: Nwood 25W 44
Juniper Ct. KT8: W Mole 70Da 129
Juniper Ct. RM6: Chad H 30Xc 55
Juniper Ct. SL1: Slou 7L 81
Juniper Ct. TW3: Houn 56Da 107
(off Grove Rd.)
Juniper Ct. W8 48Db 89
(off St Mary's Pl.)
Juniper Ct. WD3: Chor 16E 24
Juniper Cres. NW1 38Jb 70
Juniper Dr. GU24: Bisl 7E 166
Juniper Dr. RM15: S Ock 41Ae 99
Juniper Dr. SW18 56Eb 111
Juniper Gdns. SW16 67Lb 134
Juniper Gdns. TW16: Sun 65V 128
Juniper Gdns. WD7: Shenl 5Na 15
Juniper Ga. WD3: Rick 20M 25
Juniper Grn. HP1: Hem H 2G 2
Juniper Gro. WD17: Wat 10W 12
Juniper Ho. SE14 52Yb 114
Juniper Ho. TW9: Kew 53Ra 109
Juniper Ho. W10 42Ab 88
(off Fourth Av.)
Juniper La. E6 43Nc 94
Juniper Pl. KT17: Bans 87Ya 174
Juniper Rd. IG1: Ilf 34Qc 74
Juniper Rd. RH2: Reig 8L 207
Juniper St. E1 45Yb 92
Juniper Wlk. BR8: Swan 68Fd 140
Juniper Way RM3: Hrld W 25Nd 57
Juniper Way UB3: Hayes 45T 84
Juno Ct. AL3: St A 4N 5
Juno Ct. SW9 52Qb 112
(off Caldwell St.)
Juno Ent. Cen. SE14 51Zb 114
Juno Ho. E3 39Cc 72
(off Garrison Rd.)
Juno Way SE14 51Zb 114
Juno Way Ind. Est. SE14 51Zb 114
Jupiter Ct. AL3: St A 4N 5
Jupiter Ct. E3 40Cc 72
(off Four Seasons Cl.)
Jupiter Ct. SL1: Slou 5C 80
Jupiter Ct. SW9 52Qb 112
(off Caldwell St.)
Jupiter Ct. UB5: N'olt 41Z 85
(off Seasprite Cl.)
Jupiter Dr. HP2: Hem H 1P 3
Jupiter Hgts. UB10: Uxb 39P 63
Jupiter Ho. E14 50Dc 92
(off St Davids Sq.)
Jupiter Ho. E16 44Hc 93
(off Turner St.)
Jupiter Ho. HA2: Harr 32Ca 65
(off Cotterells)
Jupiter Way N7 37Pb 70
Jupp Rd. E15 38Fc 73
Jupp Rd. W. E15 39Fc 73
Jura Ho. SE16 49Zb 92
(off Plough Way)
Jurassic Encounter 69Wa 132
Jurgens Rd. RM19: Purf 51Td 120
Jurston Ct. SE1 2A 230 (47Qb 90)
(off Gerridge St.)
Jury St. DA11: Grav'nd 8D 122
Justice Apts. E1 44Zb 92
(off Aylward St.)
Justice Wlk. SW3 51Gb 111
Justin Cl. TW8: Bford 52Ma 109
Justines Pl. E2 41Zb 92
Justin Pl. N22 24Pb 50
Justin Plaza CR4: Mitc 70Gb 133
Justin Rd. E4 23Bc 52
Jute La. EN3: Brim 12Ac 34
Jutland Cl. N19 32Nb 70
Jutland Gdns. CR5: Coul 92Pb 196
Jutland Ho. SE18 49Nc 94
(off Prospect Va.)
Jutland Ho. SE5 54Sb 113
Jutland Ho. SL4: Wind 4D 102
Jutland Pl. TW20: Egh 64E 126
Jutland Rd. E13 42Jc 93
Jutland Rd. SE6 59Ec 114
Jutsums Av. RM7: Rom 30Dd 56
Jutsums Ct. RM7: Rom 30Dd 56
Jutsums La. RM7: Rom 30Dd 56
Jutsums La. RM7: Rush G 30Dd 56
Juxon Cl. HA3: Hrw W 25Da 45
Juxon Ho. EC4 3C 224 (44Rb 91)
(off St Paul's Chyd.)
Juxon St. SE11 5J 229 (49Pb 90)
JVC Bus. Pk. NW2 32Wa 68

K

Kaduna Cl. HA5: Eastc 29W 44
Kaine Pl. CR0: C'don 73Ac 158
KALEIDOSCOPE 58Dc 114
Kaleidoscope 97Ld 203
Kaleidoscope Ho. E20 36Ec 72
(off Mirabelle Gdns.)
Kale Rd. DA18: Erith 47Zc 95
Kalima Cvn. Site GU24: Chob 2M 167
Kambala Rd. SW11 55Fb 111
Kamen Ho. SE1 7H 225 (46Ub 91)
(off Magdalen St.)
Kamrans Pl. HA8: Edg 26Pa 47
Kandlewood CM13: Hut 17De 41
Kane Ct. SE10 49Jc 93
(off Peartree Way)
Kangley Bri. Rd. SE26 65Bc 136
Kangley Bus. Cen. SE26 64Bc 136
Kanli M. SE6 62Dc 136
Kaplan Dr. N21 15Pb 32
Kapuvar Cl. SE15 54Wb 113
Karachi Ho. E15 37Gc 73
(off Well St.)
Kara Way NW2 35Za 68
Kareena Cl. RM12: Horn 32Nd 77
Karen Cl. CM15: B'wood 17Yd 40
Karen Cl. RM13: Rain 40Gd 76
Karen Cl. SS17: Stan H 1L 101
Karen Ct. BR1: Broml 67Hc 137
Karen Ter. E11 33Hc 73
Karenza Ct. HA9: Wemb 31La 66
Kariba Cl. N9 20Yb 34
Karim M. E17 28Ac 52
Karina Cl. IG7: Chig 22Uc 54
Karma Way HA2: Harr 32Ca 65
Karner Ho. E20 36Dc 72
(off Logan Cl.)
Karoline Gdns. UB6: G'frd 40Fa 66
Kashgar Rd. SE18 49Vc 95
Kashmir Cl. KT15: New H 81M 169
Kashmir Rd. SE7 52Mc 115
Kassala Rd. SW11 53Hb 111
Katana GU22: Wok 91A 188
(off Brooklyn Rd.)
Katella Trad. Est. IG11: Bark 41Uc 94
Kates Cl. EN5: Ark 15Wa 30
Katharine Ho. CR0: C'don 76Sb 157
(off Katharine St.)
Katharine St. CR0: C'don 76Sb 157
Katherine Bell Twr. E3 40Cc 72
(off Pancras Way)
Katherine Cl. HP3: Hem H 5N 3
Katherine Cl. KT15: Add 79J 149
Katherine Cl. N4 31Sb 71
Katherine Cl. SE16 46Zb 92
Katherine Ct. GU21: Knap 1G 186
(off Tudor Way)
Katherine Ct. SE23 60Xb 113
Katherine Gdns. IG6: Ilf 24Sc 54
Katherine Gdns. SE9 56Mc 115
Katherine Ho. W10 42Ab 88
(off Portobello Rd.)
Katherine M. CR3: Whyt 89Vb 177
Katherine Pl. WD5: Ab L 4W 12
Katherine Rd. E6 38Mc 73
Katherine Rd. E7 36Lc 73
Katherine Rd. TW1: Twick 60Ja 108
Katherine Sq. W11 46Ab 88
Kathleen Av. HA0: Wemb 38Na 67
Kathleen Av. W3 43Sa 87
Kathleen Godfree Ct. SW19 65Cb 133
Kathleen Rd. SW11 55Hb 111
Katial Ho. EC1 3C 218 (41Rb 91)
(off Goswell Road)
Kavanagh Ct. CM14: W'ley 22Xd 58
Kavanagh Ct. KT18: Eps 86Ua 174
(off Martin's Av.)
Kavanaghs Rd. CM14: B'wood 20Wd 40
Kavanaghs Ter. CM14: B'wood 20Xd 40
Kavan Gdns. TW5: Cran 52W 106
Kavsan Pl. TW5: Cran 52W 106
Kayani Av. N4 32Sb 71
Kayani Ho. E16 44Kc 93
(off Burrard Rd.)
Kay Av. KT15: Add 76N 149
Kaye Don Way KT13: Weyb 82Q 170
Kayemoor Rd. SM2: Sutt 79Fb 155
Kay Rd. SW9 54Nb 112
Kaysland Pk. TN15: W King 80Ud 164
Kays Ter. E18 25Hc 53
Kay St. DA16: Well 53Xc 117
Kay St. E15 38Fc 73
(off New Mount St.)
Kay St. E2 40Wb 71
Kay Wlk. AL4: St A 2H 7
Kaywood Cl. SL3: L'ly 8N 81
K D Plaza HP1: Hem H 4L 3
KD Twr. HP1: Hem H 4L 3
(off Cotterells)
Kean Cres. RM8: Dag 32Ad 75
Kean Ho. SE17 51Rb 113
Kean Ho. TW1: Twick 58Ma 109
(off Arosa Rd.)
Kean St. WC2 3H 223 (44Pb 90)
Kearton Cl. CR8: Kenley 89Sb 177
Kearton Pl. CR3: Cat'm 94Wb 197
Keary Rd. DA10: Swans 59Ae 121
Keate Ho. SL4: Eton 1H 103
(off Keates La.)
Keates La. SL4: Eton 1G 102
Keatley Grn. E4 23Bc 52
Keats CR0: C'don 74Sb 157
(off Saffron Central Sq.)
Keats Apts. E3 42Bc 92
(off Wraxall Rd.)
Keats Av. E16 46Kc 93
Keats Av. RH1: Redh 4A 208
Keats Av. RM3: Rom 24Kd 57
Keats Cl. E11 29Kc 53
Keats Cl. EN3: Pond E 15Zb 34
Keats Cl. IG7: Chig 23Sc 54
Keats Cl. NW3 35Gb 69
Keats Cl. SE1 6K 231 (49Vb 91)
Keats Cl. SE7 52Lc 115
Keats Cl. SW19 65Fb 133
Keats Cl. UB4: Hayes 43W 84
Keats Cl. WD6: Bore 14Qa 29
Keats Est. N16 33Vb 71
(off Kyverdale Rd.)
Keats Gdns. RM18: Tilb 4D 122
Keats Gro. NW3 35Gb 69
Keats Ho. DA1: Cray 57Gd 118
Keats Ho. E2 41Yb 92
(off Roman Rd.)
Keats Ho. HA2: Harr 33Ga 66
Keats Ho. SE5 52Sb 113
(off Elmington Est.)
Keats Ho. SW1 51Lb 112
(off Churchill Gdns.)
Keats House 35Gb 69
Keats Pde. N9 19Wb 33
(off Church St.)
Keats Pl. EC2 1F 225 (43Tb 91)
(off Moorfields)
Keats Rd. DA16: Well 53Uc 116
Keats Rd. DA17: Belv 48Ed 96
Keats Rd. E10 31Dc 72
Keats Wlk. CM13: Hut 17Fe 41
Keats Way CR0: C'don 72Yb 158
Keats Way CR5: Coul 89Lb 176
Keats Way UB6: G'frd 43Da 85
Keats Way UB7: W Dray 49P 83
Kebbell Ter. E7 36Kc 73
(off Claremont Rd.)
Keble Cl. KT4: Wor Pk 74Va 154
Keble Cl. UB5: N'olt 36Ea 66
Keble Ct. AL1: St A 2C 6
(off Newsom Pl.)
Keble Ct. WD6: Bore 11Pa 29
(off Gateshead Rd.)
Keble Pl. SW13 51Xa 110
Keble St. SW17 63Eb 133
Keble Ter. WD5: Ab L 4V 12
Kebony Cl. UB7: W Dray 47Q 84
Kechill Gdns. BR2: Hayes 73Jc 159
Kedeston Ct. SM1: Sutt 74Db 155
Kedge Ho. E14 48Cc 92
(off Tiller Rd.)
Kedleston Dr. BR5: St M Cry 71Vc 161
Kedleston Wlk. E2 41Xb 91
Kedyngton Ho. HA8: Edg 26Sa 47
(off Burnt Oak B'way.)
Keeble Cl. SE18 51Rc 116
Keedonwood Rd. BR1: Broml 64Gc 137
Keel KT16: Chert 74L 149
Keel Cl. IG11: Bark 40Yc 75
Keel Cl. N18 23Ub 51
Keel Cl. SE16 46Zb 92
Keel Ct. E14 45Fc 93
(off Newport Av.)
Keel Dr. SL1: Slou 7F 80
Keele Cl. WD24: Wat 12Y 27
Keele Ho. RM8: Dag 35Wc 75
Keeler Cl. SL4: Wind 5C 102
Keeley Rd. CR0: C'don 75Sb 157
Keeley St. WC2 3H 223 (44Pb 90)
Keeling Ho. E2 40Xb 71
(off Claredale St.)
Keeling Rd. SE9 57Mc 115
Keelson Ho. E14 48Cc 92
(off Mellish St.)
Keely Cl. EN4: E Barn 15Gb 31
Keemor Cl. SE18 52Qc 116
Keen's Acre SL2: Stoke P 9L 61
Keensacre SL0: Iver H 40F 62
Keens Cl. SW16 64Mb 134
Keens Rd. CR0: C'don 77Sb 157
Keen's Yd. N1 37Rb 71
Keep, The KT2: King T 65Pa 131
Keep, The SE3 54Jc 115
Keepers Ct. CR2: S Croy 78Sb 157
(off Warham Rd.)
Keepers Farm Cl. SL4: Wind 4C 102
(not continuous)
Keepers M. TW11: Tedd 65La 130
Keepers Wlk. GU25: Vir W 1P 147
Keepier Wharf E14 45Zb 92
(off Narrow St.)
Keeping Ct. BR2: Broml 69Jc 137
(off St Mark's Sq.)
Keeton's Rd. SE16 48Xb 91
(not continuous)
Keevil Dr. SW19 59Za 110
Keibs Way SE17 7H 231 (50Ub 91)
Keighley Cl. N7 36Nb 70
Keighley Rd. RM3: Rom 24Nd 57
Keightley Dr. SE9 60Sc 116
Keilder Cl. UB10: Hil 40Q 64
Keildon Rd. SW11 56Hb 111
Keiller Ho. E16 46Pc 94
(off Kennard St.)
Keir, The SW19 64Ya 132
Keir Hardie Est. E5 32Xb 71
Keir Hardie Ho. N19 31Mb 70
Keir Hardie Ho. NW10 38Va 68
Keir Hardie Ho. W6 51Za 110
(off Fulham Pal. Rd.)
Keir Hardie Way IG11: Bark 38Wc 75
Keir Hardie Way UB4: Yead 41W 84
Keirin Rd. E20 36Dc 72
Keith Av. DA4: Sut H 65Rd 141
Keith Axon Cen. 30Yc 55
(off Grove Rd.)
Keith Connor Cl. SW8 55Kb 112
Keith Gro. W12 47Wa 88
Keith Ho. NW6 40Db 69
(off Carlton Vale)
Keith Ho. SW8 52Nb 112
(off Wheatsheaf La.)
Keith Pk. Cres. TN16: Big H 85Kc 179
Keith Pk. Rd. UB10: Uxb 38P 63
Keith Rd. E17 25Bc 52
Keith Rd. IG11: Bark 40Tc 74
Keith Rd. UB3: Hayes 48U 84
Keiths Rd. HP3: Hem H 3A 4
Keith Way RM11: Horn 31Nd 77
Kelbrook Rd. SE3 54Nc 116
Kelburn Way RM13: Rain 41Jd 96
Kelby Ho. N7 37Pb 70
(off Sutterton St.)
Kelby Path SE9 62Rc 138
Kelceda Cl. NW2 33Wa 68
Kelday Hgts. E1 44Xb 91
(off Spencer Way)
Kelf Gro. UB3: Hayes 44V 84
Kelfield Ct. W10 44Za 88
Kelfield Gdns. W10 44Ya 88
Kelfield M. W10 44Za 88
Kelland Cl. N8 29Mb 50
Kelland Rd. E13 42Jc 93
Kellaway Rd. SE3 54Mc 115
Keller Cres. E12 35Mc 73
Keller Gdns. CR5: Coul 89Lb 176
Kellerton Rd. SE13 57Gc 115
Kellet Ho's. WC1 4G 217 (41Nb 90)
(off Tankerton St.)
Kellett Ho. N1 39Ub 71
(off Colville Est.)
Kellett Rd. SW2 56Qb 112
Kelling Gdns. CR0: C'don 73Rb 157
Kellino St. SW17 63Hb 133
Kellner Rd. SE28 48Vc 95
Kellogg Twr. UB6: G'frd 36Ga 66
Kellow Ho. SE1 1F 231 (47Tb 91)
(off Tennis St.)
Kell St. SE1 3C 230 (48Rb 91)
Kelly Av. SE15 52Vb 113
Kelly Cl. NW10 34Ta 67
Kelly Cl. TW17: Shep 68U 128
Kelly Ct. E14 45Cc 92
(off Garford St.)
Kelly Ct. WD6: Bore 12Ta 29
Kelly M. W9 42Bb 89
Kelly St. NW1 37Kb 70
Kelly Ter. E17 28Cc 52
Kelly Way RM6: Chad H 29Ad 55
Kelman Cl. EN8: Chesh 3Zb 20
Kelman Cl. SW4 54Mb 112
Kelmore Gro. SE22 56Wb 113
Kelmscott Cl. E17 25Bc 52
Kelmscott Cl. WD18: Wat 15W 26
Kelmscott Cres. WD18: Wat 15W 26
Kelmscott Gdns. W12 48Wa 88
Kelmscott House 50Xa 88
Kelmscott Pl. KT21: Asht 89La 172
Kelmscott Rd. SW11 57Gb 111
Kelpatrick Rd. SL1: Slou 4B 80

Kelross Pas. N5 35Sb 71
Kelross Rd. N5 35Sb 71
Kelsall Cl. SE3 54Kc 115
Kelsall M. TW9: Kew 53Ra 109
Kelsall Pl. SL5: Asc 3H 146
Kelsey Cl. KT9: Chess 80Ma 153
Kelsey Ga. BR3: Beck 68Dc 136
Kelsey La. BR3: Beck 68Cc 136
Kelsey Pk. Av. BR3: Beck 68Dc 136
Kelsey Pk. Rd. BR3: Beck 68Cc 136
Kelsey Rd. BR5: St P 68Xc 139
Kelsey Sq. BR3: Beck 68Cc 136
Kelsey St. E2 42Wb 91
Kelsey Way BR3: Beck 69Cc 136
Kelshall WD25: Wat 8Aa 13
Kelsie Way IG6: Ilf 23Uc 54
Kelso Dr. DA12: Grav'nd 3H 145
Kelson Ho. E14 48Ec 92
Kelson Ho. E16 47Kc 93
Kelso Pl. W8 48Db 89
Kelso Rd. SM5: Cars 73Eb 155
Kelston Rd. IG6: Ilf 26Rc 54
Kelvedon Av. KT12: Hers 80U 150
Kelvedon Cl. CM13: Hut 16Fe 41
Kelvedon Cl. KT2: King T 65Qa 131
KELVEDON COMMON 11Td 40
Kelvedon Ho. SW8 53Nb 112
Kelvedon Rd. SW6 52Bb 111
Kelvedon Wlk. RM13: Rain 39Hd 76
Kelvedon Way IG8: Wfd G 23Pc 54
Kelvin Av. KT22: Lea 91Ha 192
Kelvin Av. N13 23Pb 50
Kelvin Av. TW11: Tedd 65Ga 130
Kelvinbrook KT8: W Mole 69Da 129
Kelvin Cl. KT19: Ewe 79Qa 153
Kelvin Ct. SE20 67Xb 135
Kelvin Ct. TW7: Isle 54Ga 108
Kelvin Ct. W11 45Cb 89
(off Kensington Pk. Rd.)
Kelvin Cres. HA3: Hrw W 24Ga 46
Kelvin Dr. TW1: Twick 58Ka 108
Kelvin Gdns. CR0: Wadd 73Nb 156
Kelvin Gdns. UB1: S'hall 44Ca 85
Kelvin Gro. KT9: Chess 76Ma 153
Kelvin Gro. SE26 62Xb 135
Kelvington Cl. CR0: C'don 73Ac 158
Kelvington Rd. SE15 57Zb 114
Kelvin Ho. SE26 63Bc 136
(off Worsley Bri. Rd.)
Kelvin Ind. Est. UB6: G'frd 38Da 65
Kelvin Pde. BR6: Orp 74Uc 160
Kelvin Rd. DA16: Well 55Wc 117
Kelvin Rd. N5 35Sb 71
Kelvin Rd. RM18: Tilb 4C 122
Kelway Ho. W14 50Bb 89
Kember St. N1 38Pb 70
Kemble Av. KT15: Add 80H 149
Kemble Cl. EN6: Pot B 5Fb 17
Kemble Cl. KT13: Weyb 77T 150
Kemble Cotts. KT15: Add 77J 149
Kemble Dr. BR2: Broml 76Nc 160
Kemble Ho. SW9 55Rb 113
(off Barrington Rd.)
Kemble Pde. EN6: Pot B 4Eb 17
Kemble Rd. CR0: Wadd 76Rb 157
Kemble Rd. N17 25Wb 51
Kemble Rd. SE23 60Zb 114
Kembleside Rd. TN16: Big H 90Lc 179
Kemble St. WC2 3H 223 (44Pb 90)
Kemerton Rd. BR3: Beck 68Dc 136
Kemerton Rd. CR0: C'don 73Vb 157
Kemerton Rd. SE5 55Sb 113
Kemey's St. E9 36Ac 72
Kemishford GU22: Wok 5L 187
Kemmel Rd. RM9: Dag 39Xc 75
Kemnal Rd. BR7: Chst 63Tc 138
Kemp NW9 25Va 48
(off Quakers Course)
Kemp Ct. SW8 52Nb 112
(off Hartington Rd.)
Kempe Cl. AL1: St A 6A 6
Kempe Cl. SL3: L'ly 49E 82
Kempe Ho. SE1 4G 231 (48Tb 91)
(off Burbage Cl.)
Kempe Rd. EN1: Enf 8Xb 19
Kempe Rd. NW6 40Za 68
Kemp Gdns. CR0: C'don 72Sb 157
Kemp Ho. E2 40Zb 72
(off Sewardstone Rd.)
Kemp Ho. E6 37Qc 74
Kemp Ho. W1 4D 222 (45Mb 90)
(off Berwick St.)
Kempis Way SE22 57Ub 113
Kemplay Rd. NW3 35Fb 69
Kempley Ct. RM17: Grays 51Fe 121
Kemp Pl. WD23: Bush 16Ca 27
Kemp Rd. RM8: Dag 32Zc 75
Kemprow WD25: A'ham 8Fa 14
KEMPROW 8Fa 14
Kemps Ct. W1 3D 222 (44Mb 90)
(off Hopkins St.)
Kemps Dr. E14 45Cc 92
Kemps Dr. HA6: Nwood 24V 44
Kempsford Gdns. SW5 50Cb 89
Kempsford Rd. SE11 6A 230 (49Qb 90)
(not continuous)
Kemps Gdns. SE13 57Ec 114
Kempshott Rd. SW16 66Mb 134
Kempson Rd. SW6 53Cb 111
Kempster Way RM3: Rom 20Ld 39
Kempthorne Rd. SE8 49Bc 92
Kempthorne St. DA11: Grav'nd 8D 122
Kempton Av. RM12: Horn 35Pd 77
Kempton Av. TW16: Sun 67X 129
Kempton Av. UB5: N'olt 37Ca 65
Kempton Cl. DA8: Erith 51Ed 118
Kempton Cl. UB10: Ick 35S 64
Kempton Ct. E1 43Xb 91
Kempton Ct. TW16: Sun 67X 129
Kempton Ga. Bus. Cen. TW12: Hamp 67Ba 129
Kempton Ho. N1 1H 219 (39Ub 71)
(off Hoxton St.)
Kempton Ho. SL4: Wind 3D 102
(off Paddock Cl.)
Kempton Ho. TW18: Staines 63H 127
Kempton Nature Reserve 65Z 129
Kempton Pk. Racecourse 66Y 129
Kempton Park Station (Rail) 66X 129
Kempton Rd. E6 39Pc 74
Kemptons, The TW15: Ashf 61Q 128
Kempton Wlk. CR0: C'don 72Ac 158
Kempt St. SE18 51Qc 116
KEMSING 89Rd 183
Kemsing Cl. BR2: Hayes 75Hc 159
Kemsing Cl. CR7: Thor H 70Sb 135
Kemsing Cl. DA5: Bexl 59Ad 117
Kemsing Down Nature Reserve 88Pd 183
Kemsing Heritage Cen. 89Rd 183
Kemsing Ho. SE1 2G 231 (47Tb 91)
(off Long La.)
Kemsing Rd. SE10 50Jc 93
Kemsing Rd. TN15: Kems'g 88Vd 184
Kemsing Rd. TN15: Wro 88Vd 184
Kemsing Station (Rail) 91Td 204
Kemsley SE13 57Dc 114
Kemsley Chase SL2: Farn R 9G 60
Kemsley Cl. DA11: Nflt 3B 144
Kemsley Cl. DA9: Ghithe 58Xd 120
Kemsley Ct. W13 46La 86
Kemsley Rd. TN16: Tats 91Mc 199
Kenbrook Ho. NW5 36Lb 70
Kenbrook Ho. W14 48Bb 89
Kenbury Cl. UB10: Ick 34Q 64
Kenbury Dr. SL1: Slou 7D 80
Kenbury Gdns. SE5 54Sb 113
Kenbury Mans. SE5 54Sb 113
(off Kenbury St.)
Kenbury St. SE5 54Sb 113
Kenchester Cl. SW8 52Nb 112
Kencot Way DA18: Erith 47Bd 95
Kendal NW1 3A 216 (41Kb 90)
(off Augustus St.)
Kendal RM19: Purf 50Sd 98
Kendal Av. CM16: Epp 2Wc 23
Kendal Av. IG11: Bark 39Uc 74
Kendal Av. N18 21Tb 51
Kendal Av. W3 42Qa 87
Kendal Cl. IG8: Wfd G 19Hc 35
Kendal Cl. N20 19Gb 31
Kendal Cl. RH2: Reig 5M 207
Kendal Cl. SL2: Slou 5L 81
Kendal Cl. SW9 52Rb 113
Kendal Cl. TW14: Felt 60V 106
Kendal Cl. UB4: Hayes 40U 64
Kendal Ct. W3 43Qa 87
Kendal Cft. RM12: Horn 36Jd 76
Kendal Dr. SL2: Slou 5L 81
Kendale HP3: Hem H 3B 4
Kendale RM16: Grays 8D 100
Kendale Rd. BR1: Broml 64Gc 137
Kendal Gdns. N18 21Tb 51
Kendal Gdns. SM1: Sutt 75Eb 155
Kendal Ho. E9 39Yb 72
Kendal Ho. N1 2J 217 (40Pb 70)
(off Priory Grn. Est.)
Kendal Ho. SE20 68Wb 135
(off Derwent Rd.)
Kendall Av. BR3: Beck 68Ac 136
Kendall Av. CR2: Sande 81Tb 177
Kendall Av. Sth. CR2: Sande 82Sb 177
Kendall Ct. CM14: W'ley 22Xd 58
Kendall Ct. DA15: Sidc 62Wc 139
Kendall Ct. SW19 65Fb 133
Kendall Ct. WD6: Bore 12Sa 29
Kendall Gdns. DA11: Grav'nd 9B 122
Kendall Ho. W14 49Bb 89
(off Warwick La.)
Kendall Lodge BR1: Broml 67Kc 137
(off Willow Tree Wlk.)
Kendall Lodge CM16: Epp 2Wc 23
Kendall Mnr. HA6: Nwood 24R 44
Kendall Pl. W1 1H 221 (43Jb 90)
Kendall Rd. BR3: Beck 68Ac 136
Kendall Rd. SE18 53Nc 116
Kendall Rd. TW7: Isle 54Ja 108
Kendalmere Cl. N10 25Kb 50
Kendal Pde. N18 21Tb 51
Kendal Pl. SW15 57Bb 111
Kendal Rd. EN9: Walt A 7Ec 20
Kendal Rd. NW10 35Wa 68
Kendals Cl. WD7: R'lett 8Ga 14
Kendal Steps W2 3E 220 (44Gb 89)
(off St George's Flds.)
Kendal St. W2 3E 220 (44Gb 89)
Kender Est. SE14 53Yb 114
(off Queen's Rd.)
Kender St. SE14 52Yb 114
Kendoa Rd. SW4 56Mb 112
Kendon Cl. E11 29Kc 53
Kendon Ho. E15 38Fc 73
(off Bryant St.)
Kendor Av. KT19: Eps 83Sa 173
Kendra Hall Rd. CR2: S Croy 80Rb 157
Kendrey Gdns. TW2: Whitt 59Ga 108
Kendrick Ct. SE15 53Xb 113
(off Colmore M.)
Kendrick M. SW7 6B 226 (49Fb 89)
Kendrick Pl. SW7 6B 226 (49Fb 89)
Kendrick Rd. SL3: Slou 8M 81
Kenelm Cl. HA1: Harr 34Ja 66
Kenerne Dr. EN5: Barn 15Ab 30
Kenford Cl. WD25: Wat 4X 13
Ken Friar Bri. 35Qb 70
Kenia Wlk. DA12: Grav'nd 2H 145
Kenilford Rd. SW12 59Kb 112
Kenilworth Av. E17 26Cc 52
Kenilworth Av. HA2: Harr 35Ba 65
Kenilworth Av. KT11: Stoke D 86Da 171
Kenilworth Av. RM3: Rom 23Rd 57
Kenilworth Av. SW19 64Cb 133
Kenilworth Cl. HP2: Hem H 3N 3
Kenilworth Cl. SL1: Slou 8K 81
Kenilworth Cl. SM7: Bans 88Db 175
Kenilworth Cl. WD6: Bore 13Sa 29
Kenilworth Ct. DA2: Dart 58Rd 119
(off Osbourne Rd.)
Kenilworth Ct. SW15 55Ab 110
(off Lwr. Richmond Rd.)
Kenilworth Ct. WD17: Wat 11W 26
Kenilworth Cres. EN1: Enf 11Ub 33
Kenilworth Dr. KT12: Walt T 76Z 151
Kenilworth Dr. WD3: Crox G 14R 26
Kenilworth Dr. WD6: Bore 13Sa 29
Kenilworth Gdns. IG10: Lough 16Pc 36
Kenilworth Gdns. IG3: Ilf 33Vc 75
Kenilworth Gdns. RM12: Horn 34Ld 77
Kenilworth Gdns. SE18 54Rc 116
Kenilworth Gdns. TW18: Staines 64L 127
Kenilworth Gdns. UB1: S'hall 41Ba 85
Kenilworth Gdns. UB4: Hayes 43V 84
Kenilworth Gdns. WD19: Wat 22Y 45
Kenilworth Rd. BR5: Pet W 72Sc 160
Kenilworth Rd. E3 40Ac 72
Kenilworth Rd. HA8: Edg 20Sa 29
Kenilworth Rd. KT17: Ewe 78Wa 154
Kenilworth Rd. NW6 39Bb 69
Kenilworth Rd. SE20 67Zb 136
Kenilworth Rd. TW15: Ashf 62M 127
Kenilworth Rd. W5 46Na 87
KENLEY 86Sb 177
Kenley N17 26Tb 51
(off Gloucester Rd.)
Kenley Airfield 91Tb 197
Kenley Av. NW9 25Ua 48
Kenley Cl. BR7: Chst 69Uc 138
Kenley Cl. DA5: Bexl 59Cd 118
Kenley Cl. EN4: E Barn 14Gb 31
Kenley Gdns. CR7: Thor H 70Rb 135
Kenley Gdns. RM12: Horn 33Pd 77
Kenley La. CR8: Kenley 86Sb 177
Kenley Pl. UB10: Uxb 39N 63
Kenley Rd. KT1: King T 68Ra 131
Kenley Rd. SW19 68Cb 133
Kenley Rd. TW1: Twick 58Ka 108
Kenley Station (Rail) 86Sb 177
Kenley Wlk. SM3: Cheam 77Za 154
Kenley Wlk. W11 45Ab 88
Kenlor Rd. SW17 64Fb 133
Kenmare Cl. UB10: Ick 33T 64
Kenmare Dr. CR4: Mitc 66Hb 133
Kenmare Dr. N17 26Vb 51
Kenmare Gdns. N13 21Sb 51
Kenmare Rd. CR7: Thor H 72Qb 156
Kenmere Gdns. HA0: Wemb 39Qa 67
Kenmere Rd. DA16: Well 54Yc 117
Kenmont Gdns. NW10 41Xa 88
(not continuous)
Kenmore Av. HA3: Kenton 28Ja 46
Kenmore Av. HA3: W'stone 28Ja 46
Kenmore Cl. KT17: Eps D 89Xa 174
Kenmore Cl. TW9: Kew 52Qa 109
Kenmore Ct. NW6 38Db 69
(off Acol Rd.)
Kenmore Cres. UB4: Hayes 41V 84
Kenmore Gdns. HA8: Edg 26Ra 47
Kenmore Rd. CR8: Kenley 86Rb 177
Kenmore Rd. HA3: Kenton 27Ma 47
Kenmure Rd. E8 36Xb 71
Kenmure Yd. E8 36Xb 71
Kennacraig Cl. E16 46Jc 93
Kennard Ho. SW11 54Jb 112
Kennard Rd. E15 38Fc 73
Kennard Rd. N11 22Hb 49
Kennard St. E16 46Pc 94
Kennard St. SW11 53Jb 112
Kenneally SL4: Wind 4A 102
Kenneally Cl. SL4: Wind 4A 102
(off Kenneally)
Kenneally Pl. SL4: Wind 4A 102
(off Kenneally)
Kenneally Row SL4: Wind 4A 102
(off Kenneally)
Kenneally Wlk. SL4: Wind 4A 102
Kennedy Av. EN3: Pond E 16Yb 34
Kennedy Cl. AL2: Lon C 8H 7
Kennedy Cl. BR5: Pet W 74Tc 160
Kennedy Cl. CR4: Mitc 67Jb 134
Kennedy Cl. E13 40Jc 73
Kennedy Cl. HA5: Hat E 23Ba 45
Kennedy Cl. SL2: Farn C 7G 60
Kennedy Ct. TW15: Ashf 64S 128
Kennedy Ct. WD23: B Hea 19Fa 28
Kennedy Cox Ho. E16 43Hc 93
(off Burke St.)
Kennedy Gdns. TN13: S'oaks 95Ld 203
Kennedy Ho. DA11: Nflt 2A 144
Kennedy Ho. SE11 7H 229 (50Pb 90)
(off Vauxhall Wlk.)
Kennedy Ho. SL1: Slou 6B 80
(off Harrison Way)
Kennedy Path W7 42Ha 86
Kennedy Rd. IG11: Bark 39Uc 74
Kennedy Rd. W7 43Ga 86
Kennedy Wlk. SE17 6G 231 (49Tb 91)
(off Elsted St.)
Kennel Cl. KT22: Fet 96Ea 192
Kennel Cotts. HP3: Hem H 7M 3
Kennel La. CM15: Dodd 11Ud 40
Kennel La. CM15: Kel H 11Ud 40
Kennel La. GU20: W'sham 8A 146
Kennel La. KT22: Fet 94Da 191
Kennelwood Cres. CR0: New Ad 83Fc 179
Kennet Cl. E17 27Fc 53
Kennet Cl. RM14: Upm 30Ud 58
Kennet Cl. SW11 56Fb 111
Kennet Ct. W9 43Cb 89
(off Elmfield Way)
Kennet Grn. RM15: S Ock 45Xd 98
Kenneth Av. IG1: Ilf 35Rc 74
Kenneth Campbell Ho. NW8 5C 214 (42Fb 89)
(off Orchardson St.)
Kenneth Chambers Ct. IG8: Wfd G 23Nc 54
Kenneth Ct. SE11 5A 230 (49Qb 90)
Kenneth Cres. NW2 36Xa 68
Kenneth Gdns. HA7: Stan 23Ja 46
Kenneth More Rd. IG1: Ilf 34Rc 74
Kenneth More Theatre 34Rc 74
Kennet Ho. NW8 6C 214 (42Fb 89)
(off Church St. Est.)
Kennet Ho. SW18 56Cb 111
(off Enterprise Way)
Kenneth Rd. RM6: Chad H 31Zc 75
Kenneth Rd. SM7: Bans 87Fb 175
Kenneth Robbins Ho. N17 24Xb 51
Kenneth Way W5 45Pa 87
Kenneth Younger Ho. SW6 51Bb 111
(off Clem Attlee Ct.)
Kennet Rd. DA1: Cray 55Jd 118
Kennet Rd. TW7: Isle 55Ha 108
Kennet Rd. W9 42Bb 89
Kennet Sq. CR4: Mitc 67Gb 133
Kennet St. E1 46Wb 91
Kennett Ct. BR8: Swan 69Gd 140
(off Oakleigh Ct.)
Kennett Ct. W4 52Ra 109
Kennett Ct. WD18: Wat 14X 27
(off Whippendell Rd.)
Kennett Dr. UB4: Yead 43Aa 85
Kennett La. KT16: Chert 74J 149
Kennett Rd. SL3: L'ly 48D 82
Kennett Wharf La. EC4 4E 224 (45Sb 91)
KENNINGHALL 22Yb 52
Kenninghall Rd. E5 34Wb 71
Kenninghall Rd. N18 22Yb 52
Kenning Ho. N1 39Ub 71
(off Colville Est.)
Kenning St. SE16 47Yb 92
Kennings Way SE11 7A 230 (50Qb 90)
Kenning Ter. N1 39Ub 71
KENNINGTON 51Qb 112
Kennington Grn. SE11 50Qb 90
Kennington La. SE11 50Pb 90
Kennington Oval SE11 51Pb 112
KENNINGTON OVAL 51Qb 112
Kennington Pal. Ct. SE11 7K 229 (50Qb 90)
(off Sancroft St.)
Kennington Pk. Gdns. SE11 51Rb 113
Kennington Pk. Ho. SE11 50Qb 90
(off Kennington Pk. Pl.)
Kennington Pk. Pl. SE11 51Qb 112
Kennington Pk. Rd. SE11 51Qb 112
Kennington Rd. SE1 3K 229 (48Qb 90)
Kennington Rd. SE11 4K 229 (48Qb 90)
Kennington Station (Underground) 7B 230 (50Rb 91)
Kennistoun Ho. NW5 36Lb 70
Kennoldes SE21 61Tb 135
(off Croxted Rd.)
Kenny Dr. SM5: Cars 81Jb 176
Kennyland Ct. NW4 30Xa 48
(off Hendon Way)
Kennylands Rd. IG6: Ilf 24Wc 55
Kenrick Pl. W1 1H 221 (43Jb 90)
Kenrick Sq. RH1: Blet 5L 209
KENSAL GREEN 41Ya 88
Kensal Green Station (Underground & Overground) 41Ya 88
Kensal Ho. W10 42Za 88
(off Ladbroke Gro.)
KENSAL RISE 40Za 68
Kensal Rise Station (Overground) 40Za 68
Kensal Rd. W10 42Ab 88
KENSAL TOWN 42Ab 88
Kensal Wharf W10 42Za 88
KENSINGTON 48Cb 89
Kensington Arc. W8 47Db 89
(off Kensington High St.)
Kensington Av. CR7: Thor H 67Qb 134
Kensington Av. E12 37Nc 74
Kensington Av. WD18: Wat 14V 26
Kensington Bus. Cen. SW3 3E 226 (48Gb 89)
(off Brompton Rd.)
Kensington Cen. W14 49Ab 88
(not continuous)
Kensington Chu. Ct. W8 47Db 89
Kensington Chu. St. W8 46Cb 89
Kensington Chu. Wlk. W8 47Db 89
(not continuous)
Kensington Cl. AL1: St A 4E 6
Kensington Cl. N11 23Jb 50
Kensington Ct. RM17: Grays 51Ee 121
Kensington Ct. SE16 46Zb 92
(off King & Queen Wharf)
Kensington Ct. W8 47Db 89
Kensington Ct. Gdns. W8 48Db 89
(off Kensington Ct. Pl.)
Kensington Ct. Mans. W8 47Db 89
(off Kensington Ct.)
Kensington Ct. M. W8 48Db 89
(off Kensington Ct. Pl.)
Kensington Ct. Pl. W8 48Db 89
Kensington Dr. IG8: Wfd G 26Mc 53
Kensington Gdns. IG1: Ilf 32Pc 74
Kensington Gdns. KT1: King T 69Ma 131
Kensington Gdns. RM18: E Til 9L 101
(off Queen Mary Av.)
Kensington Gdns. 6A 220 (46Eb 89)
Kensington Gdns. Sq. W2 44Db 89
Kensington Ga. W8 3A 226 (48Eb 89)
Kensington Gore SW7 2A 226 (47Eb 89)
Kensington Hall Gdns. W14 50Bb 89
Kensington Hgts. HA1: Harr 30Ha 46
(off Sheepcote Rd.)
Kensington Hgts. W8 46Cb 89
Kensington High St. W14 48Bb 89
Kensington High St. W8 48Bb 89
Kensington Ho. IG8: Wfd G 24Qc 54
Kensington Ho. SW11 52Kb 112
(off Palmer Rd.)
Kensington Ho. UB7: W Dray 47P 83
(off Park Lodge Av.)
Kensington Ho. W14 47Za 88
Kensington Ho. W8 47Db 89
(off Kensington Ct.)
Kensington Leisure Cen. 45Ab 88
Kensington Mall W8 46Cb 89
Kensington Mans. SW5 50Cb 89
(off Trebovir Rd.)
Kensington Memorial Pk. 43Za 88
(off St Mark's Road)
Kensington M. SL4: Wind 5G 102
Kensington Olympia Station (Rail, Underground & Overground) 48Ab 88
Kensington Palace 46Db 89
Kensington Pal. Gdns. W8 46Db 89
Kensington Pk. RM4: Stap A 17Gd 38
Kensington Pk. Gdns. W11 45Bb 89
Kensington Pk. M. W11 44Bb 89
Kensington Pk. Rd. W11 44Bb 89
Kensington Path E10 33Dc 72
(off Osborne Rd.)
Kensington Pl. W8 46Cb 89
Kensington Rd. CM15: Pil H 16Wd 40
Kensington Rd. RM7: Rom 30Ed 56
Kensington Rd. SW7 2A 226 (47Eb 89)
Kensington Rd. UB5: N'olt 41Ca 85
Kensington Rd. W8 47Db 89
Kensington Sq. W8 48Db 89
Kensington Ter. CR2: S Croy 80Tb 157
Kensington Village W14 49Bb 89
Kensington Way CM14: B'wood 18Yd 40
Kensington Way WD6: Bore 13Ta 29
Kensington W. W14 49Ab 88
Kensworth Ho. EC1 4G 219 (41Tb 91)
(off Cranwood St.)
Kent Av. DA16: Well 57Vc 117
Kent Av. RM9: Dag 42Cd 96
Kent Av. SL1: Slou 3G 80
Kent Av. W13 43Ka 86
Kent Bldg. E14 44Gc 93
Kent Cl. BR6: Chels 79Uc 160
Kent Cl. CR4: Mitc 70Nb 134
Kent Cl. TN15: W King 80Ud 164
Kent Cl. TW18: Staines 65M 127
Kent Cl. UB8: Uxb 37L 63
Kent Cl. WD6: Bore 10Ta 15
Kent Ct. E2 40Vb 71
Kent Ct. NW9 26Ua 48
Kent Dr. EN4: Cockf 14Jb 32
Kent Dr. RM12: Horn 35Md 77
Kent Dr. TW11: Tedd 64Ga 130
Kent Firefighting Mus. 83Yd 184
Kentford Way UB5: N'olt 39Aa 65
Kent Gdns. HA4: Ruis 30W 44
Kent Gdns. W13 43Ka 86
Kent Ga. Way CR0: Addtn 79Bc 158
Kent Hatch Rd. RH8: Crock H 1N 211
Kent Hatch Rd. RH8: Limp 1N 211
Kent Ho. SE1 7K 231 (50Vb 91)
Kent Ho. SW1 7D 228 (50Mb 90)
(off Aylesford St.)
Kent Ho. W11 45Ab 88
(off Boyne Ter. M.)
Kent Ho. W4 50Ua 88
(off Devonshire St.)
Kent Ho. W8 47Db 89
(off Kensington Ct.)
Kent Ho. WD23: Bush 14Ba 27
Kent Ho. La. BR3: Beck 65Ac 136
Kent Ho. Rd. BR3: Beck 67Zb 136
Kent Ho. Rd. SE26 64Ac 136
Kent Ho. Station App. BR3: Beck 67Ac 136
Kent House Station (Rail) 67Ac 136
Kentish Bldgs. SE1 1F 231 (47Tb 91)
Kentish La. AL9: Brk P 8L 9
Kentish La. AL9: Hat 8L 9
Kentish Pl. SE2 48Wc 95
Kentish Rd. DA17: Belv 49Cd 96
KENTISH TOWN 36Kb 70
Kentish Town Ind. Est. NW5 36Kb 70
Kentish Town Rd. NW1 38Kb 70
Kentish Town Rd. NW5 37Kb 70
Kentish Town Sports Cen. 37Kb 70
Kentish Town Station (Rail & Underground) 36Lb 70
Kentish Town West Station (Overground) 37Kb 70
Kentish Way BR1: Broml 68Kc 137
Kentish Way BR2: Broml 69Kc 137
Kent Kraft Ind. Est. DA11: Nflt 57Ee 121
Kentlea Rd. SE28 47Uc 94
Kentmere Ho. SE15 51Yb 114
Kentmere Mans. W5 42Ka 86
Kentmere Rd. SE18 49Uc 94
KENTON 29La 46
Kenton Av. HA1: Harr 31Ha 66
Kenton Av. TW16: Sun 68Z 129
Kenton Av. UB1: S'hall 45Ca 85
Kenton Ct. HA3: Kenton 30Ka 46
Kenton Ct. SE26 63Ac 136
(off Adamsrill Rd.)
Kenton Ct. TW1: Twick 58Ma 109
Kenton Ct. W14 48Bb 89
Kentone Ct. SE25 70Xb 135
Kenton Gdns. AL1: St A 3D 6
Kenton Gdns. HA3: Kenton 29La 46
Kenton Ho. E1 42Yb 92
(off Mantus Cl.)
Kentonian Cl. HA3: Kenton 29Ka 46
Kenton La. HA3: Hrw W 23Ha 46
Kenton La. HA3: Kenton 23Ha 46
Kenton La. HA3: W'stone 23Ha 46
Kenton Pk. Av. HA3: Kenton 29Ma 47
Kenton Pk. Cl. HA3: Kenton 28La 46
Kenton Pk. Cres. HA3: Kenton 28Ma 47
Kenton Pk. Mans. HA3: Kenton 29La 46
(off Kenton Rd.)
Kenton Pk. Pde. HA3: Kenton 29La 46
Kenton Pk. Rd. HA3: Kenton 28La 46
Kenton Rd. E9 37Zb 72
Kenton Rd. HA1: Harr 31Ha 66
Kenton Rd. HA3: Kenton 30Ja 46
Kentons La. SL4: Wind 4C 102
Kenton Station (Underground & Overground) 30Ka 46
Kenton St. WC1 5F 217 (42Nb 90)
Kenton Way GU21: Wok 9K 167
Kenton Way UB4: Hayes 41U 84
Kent Pk. Ind. Est. SE15 51Xb 113
Kent Pas. NW1 5F 215 (42Hb 89)
Kent Rd. BR4: W W'ck 74Dc 158
Kent Rd. BR5: St M Cry 72Xc 161
Kent Rd. DA1: Dart 58Md 119
Kent Rd. DA11: Grav'nd 10C 122
Kent Rd. DA3: Lfield 68Zd 143
Kent Rd. GU20: W'sham 8B 146
Kent Rd. GU22: Wok 88D 168
Kent Rd. KT1: King T 69Ma 131
Kent Rd. KT8: E Mos 70Ea 130
Kent Rd. N21 18Tb 33
Kent Rd. RM10: Dag 36Dd 76
Kent Rd. RM17: Grays 51Ee 121
Kent Rd. TW9: Kew 52Qa 109
Kent Rd. W4 48Sa 87
Kents Av. HP3: Hem H 6M 3
Kent's Pas. TW12: Hamp 67Ba 129
Kent St. E13 41Lc 93
Kent St. E2 1K 219 (40Vb 71)
Kent Ter. NW1 4E 214 (41Gb 89)
Kent Vw. RM13: Wenn 45Md 97
Kent Vw. RM15: Avel 47Sd 98
Kent Vw. Gdns. IG3: Ilf 33Uc 74
Kent Wlk. SW9 56Rb 113
Kent Way KT6: Surb 76Na 153
Kentwell Cl. SE4 56Ac 114
Kent Wharf SE8 52Dc 114
(off Creekside)
Kentwode Grn. SW13 52Wa 110
Kentwyns Ri. RH1: S Nut 7F 208
Kent Yd. SW7 2E 226 (47Gb 89)
Kenver Av. N12 23Fb 49
Kenward Rd. SE9 57Lc 115
Kenward Way SW11 54Jb 112
Ken Way HA9: Wemb 34Sa 67
Kenway RM13: Rain 41Md 97
Kenway RM5: Col R 26Ed 56
Kenway Cl. RM13: Rain 41Ld 97
Kenway Rd. SW5 49Db 89
Kenway Wlk. RM13: Rain 41Md 97
Ken Wilson Ho. E2 40Wb 71
(off Pritchards Rd.)
Kenwood Av. DA3: Lfield 69Ee 143
Kenwood Av. N14 15Mb 32
Kenwood Cl. NW3 32Fb 69
Kenwood Cl. UB7: Sip 51Q 106
Kenwood Ct. NW9 28Sa 47
(off Elmwood Cres.)
Kenwood Dr. BR3: Beck 69Ec 136
Kenwood Dr. KT12: Hers 79X 151
Kenwood Dr. WD3: Rick 19H 25
Kenwood Gdns. E18 27Kc 53
Kenwood Gdns. IG2: Ilf 29Qc 54
Kenwood Gdns. IG5: Ilf 28Qc 54
Kenwood Ho. SW9 56Rb 113
Kenwood Ho. WD18: Wat 17T 26
Kenwood House 32Gb 69
Kenwood Pk. KT13: Weyb 79T 150
Kenwood Pl. N6 32Hb 69
Kenwood Ridge CR8: Kenley 89Rb 177
Kenwood Rd. N6 30Hb 49
Kenwood Rd. N9 18Wb 33
Kenworth Cl. EN8: Walt C 5Zb 20
Kenworthy Rd. E9 36Ac 72
Kenwrick Ho. N1 1J 217 (39Pb 70)
(off Barnsbury Est.)
Kenwyn Dr. NW2 33Ua 68
Kenwyn Lodge N2 28Hb 49
Kenwyn Rd. DA1: Dart 57Md 119
Kenwyn Rd. SW20 67Ya 132
Kenwyn Rd. SW4 56Mb 112
Kenya Rd. SE7 52Mc 115
Kenyngton Ct. TW16: Sun 64W 128
Kenyngton Dr. TW16: Sun 64W 128
Kenyngton Pl. HA3: Kenton 29La 46
Kenyon Ho. SE5 52Sb 113
(off Camberwell Rd.)
Kenyon Mans. W14 51Ab 110
(off Queen's Club Gdns.)
Kenyons KT24: W Hor 100R 190
Kenyon St. SW6 53Za 110
Kenyon Way SL3: L'ly 48B 82
Keogh Rd. E15 37Gc 73

Kepler Ho. SE10 50Hc 93
(off Armitage Rd.)
Kepler Rd. SW4 56Nb 112
Keppel Cl. DA9: Ghithe 56Xd 120
Keppel Ho. SE8 50Bc 92
Keppel Ho. SW3 6D 226 (49Gb 89)
(off Elystan St.)
Keppel Rd. E6 38Pc 74
Keppel Rd. RM9: Dag 35Ad 75
Keppel Row SE1 7D 224 (46Sb 91)
Keppel Spur SL4: Old Win 9M 103
Keppel St. SL4: Wind 4H 103
Keppel St. WC1 7E 216 (43Mb 90)
Kepplestone M. BR3: Beck 68Ec 136
Kerbela St. E2 42Wb 91
Kerbey St. E14 44Dc 92
Kerdistone Cl. EN6: Pot B 2Db 17
Kerfield Cres. SE5 53Tb 113
Kerfield Pl. SE5 53Tb 113
Kerlin Vw. SW16 68Lb 134
Kernow Cl. RM12: Horn 33Nd 77
Kerr Cl. CR2: Sels 80Ac 158
Kerria Way GU24: W End 5C 166
Kerri Cl. EN5: Ark 14Ya 30
Kerridge Ct. N1 37Ub 71
(off Balls Pond Rd.)
Kerrier Ho. SW10 52Eb 111
(off Stadium St.)
Kerrill Av. CR5: Coul 91Qb 196
Kerrington Ct. W10 42Ab 88
(off Wornington Rd.)
Kerrington Ct. W12 47Ya 88
(off Uxbridge Rd.)
Kerris Ho. SE11 7A 230 (50Qb 90)
(off Tavy Cl.)
Kerrison Pl. W5 46Ma 87
Kerrison Rd. E15 39Fc 73
Kerrison Rd. SW11 55Gb 111
Kerrison Rd. W5 46Ma 87
Kerrison Vs. W5 46Ma 87
Kerry Av. HA7: Stan 21La 46
Kerry Av. RM15: Avel 47Pd 97
Kerry Cl. E16 44Kc 93
Kerry Cl. N13 19Pb 32
Kerry Cl. RM14: Upm 31Vd 78
Kerry Ct. HA7: Stan 21Ma 47
Kerry Dr. RM14: Upm 31Vd 78
Kerry Ho. E1 44Yb 92
(off Sidney St.)
Kerry Path SE14 51Bc 114
Kerry Rd. RM16: Grays 46Fe 99
Kerry Rd. SE14 51Bc 114
Kerry Ter. GU21: Wok 88D 168
Kerscott Ho. E3 41Dc 92
(off Rainhill Way)
Kersey Dr. CR2: Sels 84Yb 178
Kersey Gdns. RM3: Hrld W 24Nd 57
Kersey Gdns. SE9 63Nc 138
Kersfield Ho. SW15 58Za 110
Kersfield Rd. SW15 58Za 110
Kershaw Cl. RM11: Horn 31Nd 77
Kershaw Cl. RM16: Chaf H 49Yd 98
Kershaw Cl. SW18 58Fb 111
Kershaw Rd. RM10: Dag 34Cd 76
Kerslake Ho. SL9: Chal P 22Az 42
Kersley M. SW11 53Hb 111
Kersley Rd. N16 33Ub 71
Kersley St. SW11 54Hb 111
Kerstin Cl. UB3: Hayes 45V 84
Kerswell Cl. N15 29Ub 51
Kerswell Cl. SL2: Slou 2D 80
Kerwick Cl. N7 38Nb 70
Keslake Mans. NW10 40Za 68
(off Station Ter.)
Keslake Rd. NW6 40Za 68
Kessock Cl. N17 29Xb 51
Kesteven Cl. IG6: Ilf 23Vc 55
Kestlake Rd. DA5: Bexl 58Yc 117
KESTON 78Lc 159
Keston Av. BR2: Kes 78Lc 159
Keston Av. CR5: Coul 91Rb 197
Keston Av. KT15: New H 83J 169
Keston Cl. DA16: Well 52Yc 117
Keston Cl. N18 20Tb 33
Keston Ct. DA5: Bexl 59Bd 117
Keston Ct. KT5: Surb 71Pa 153
(off Cranes Pk.)
Keston Gdns. BR2: Kes 77Lc 159
Keston Ho. SE17 7J 231 (50Ub 91)
(off Kinglake Est.)
KESTON MARK 76Nc 160
KESTON MARK 77Nc 160
Keston M. WD17: Wat 12X 27
Keston Pk. Cl. BR2: Kes 76Pc 160
Keston Rd. CR7: Thor H 72Qb 156
Keston Rd. N17 27Tb 51
Keston Rd. SE15 55Wb 113
Keston Windmill 78Mc 159
Kestrel Av. E6 43Nc 94
Kestrel Av. SE24 57Rb 113
Kestrel Av. TW18: Staines 62H 127
Kestrel Cl. IG6: Ilf 21Xc 55
Kestrel Cl. KT19: Eps 84Qa 173
Kestrel Cl. KT2: King T 63Ma 131
Kestrel Cl. NW10 36Ta 67
Kestrel Cl. NW9 26Ua 48
Kestrel Cl. RM12: Horn 38Kd 77
Kestrel Cl. WD25: Wat 6Aa 13
Kestrel Ct. CR2: S Croy 79Sb 157
Kestrel Ct. E17 26Zb 52
Kestrel Ct. E3 40Cc 72
(off Four Seasons Cl.)
Kestrel Ct. HA4: Ruis 33U 64
Kestrel Ct. SM6: W'gton 78Lb 156
Kestrel Grn. AL10: Hat 1C 8
Kestrel Ho. EC1 3D 218 (41Sb 91)
(off Pickard St.)
Kestrel Ho. EN3: Pond E 15Ac 34
Kestrel Ho. SE10 53Ec 114
(off Parkside Av.)
Kestrel Pk. CM13: Hut 15Fe 41
Kestrel Path SL2: Slou 2C 80
Kestrel Pl. DA9: Ghithe 58Wd 120
(off Woodpecker Dr.)
Kestrel Pl. SE14 51Ac 114
Kestrel Rd. EN9: Walt A 6Jc 21
Kestrels, The AL2: Brick W 3Ba 13
Kestrels, The UB9: Den 29H 43
(off Patrons Way E.)
Kestrel Way CR0: New Ad 81Fc 179
Kestrel Way GU21: Wok 7M 167
Kestrel Way UB3: Hayes 47T 84
Keswick Av. RM11: Horn 32Md 77
Keswick Av. SW15 64Ua 132
Keswick Av. SW19 68Cb 133
Keswick Av. TW17: Shep 69U 128
Keswick B'way. SW15 57Bb 111
(off Up. Richmond Rd.)
Keswick Cl. AL1: St A 3F 6
Keswick Cl. SM1: Sutt 77Eb 155
Keswick Ct. BR2: Broml 70Hc 137
Keswick Ct. SE13 57Dc 114
Keswick Ct. SE6 60Hc 115
Keswick Ct. SL2: Slou 5K 81
Keswick Dr. EN3: Enf W 8Yb 20
Keswick Dr. GU18: Light 3A 166
Keswick Gdns. HA4: Ruis 30T 44
Keswick Gdns. HA9: Wemb 35Na 67
Keswick Gdns. IG4: Ilf 28Nc 54
Keswick Gdns. RM19: Purf 51Sd 120
Keswick Ho. RM3: Rom 23Md 57
(off Dartfields)
Keswick Ho. SE5 54Sb 113
Keswick M. W5 46Na 87
Keswick Rd. BR4: W W'ck 75Gc 159
Keswick Rd. BR6: Orp 74Vc 161
Keswick Rd. DA7: Bex 53Cd 118
Keswick Rd. KT22: Fet 96Ea 192
Keswick Rd. KT23: Bookh 97Da 191
Keswick Rd. SW15 57Ab 110
Keswick Rd. TW2: Whitt 58Ea 108
Keswick Rd. TW20: Egh 66D 126
Ketch St. IG11: Bark 39Tc 74
Kettering Ct. CR7: Thor H 70Sb 135
Kettering Rd. EN3: Enf W 9Zb 20
Kettering Rd. RM3: Rom 24Nd 57
Kettering St. SW16 65Lb 134
Kett Gdns. SW2 57Pb 112
Kettlebaston Rd. E10 32Bc 72
Kettleby Ho. SW9 55Rb 113
(off Barrington Rd.)
Kettlewell Cl. GU21: Wok 6P 167
Kettlewell Cl. N11 23Jb 50
Kettlewell Ct. BR8: Swan 68Hd 140
Kettlewell Dr. GU21: Wok 86A 168
Kettlewell Hill GU21: Wok 86A 168
Ketton Grn. RH1: Mers 100Mb 196
Ketton Ho. W10 42Ya 88
(off Sutton Way)
Ketts Pl. IG7: Chig 23Ad 55
Kevan Ct. E17 28Cc 52
Kevan Dr. GU23: Send 96G 188
Kevan Ho. SE5 52Sb 113
Kevelioc Rd. N17 25Sb 51
Kevere Ct. HA6: Nwood 22R 44
Kevin Cl. TW4: Houn 54Z 107
KEVINGTON 72Ad 161
Kevington Cl. BR5: St P 70Vc 139
Kevington Dr. BR5: St P 70Vc 139
Kevington Dr. BR7: Chst 70Vc 139
Kevtar Gdn. E3 41Bc 92
KEW 53Qa 109
Kew Bridge Bridge 51Qa 109
Kew Bri. Arches TW9: Kew 51Qa 109
Kew Bri. Ct. W4 50Qa 87
Kew Bri. Distribution Cen. TW8: Bford 50Pa 87
KEW BRIDGE JUNCTION Junction 50Pa 87
Kew Bri. Rd. TW8: Bford 51Pa 109
Kew Bridge Station (Rail) 50Pa 87
Kew Cl. RM1: Rom 23Gd 56
Kew Cl. UB8: Uxb 40M 63
Kew Ct. KT2: King T 67Na 131
Kew Cres. SM3: Cheam 76Ab 154
Kewferry Dr. HA6: Nwood 22R 44
Kewferry Rd. HA6: Nwood 23S 44
Kew Foot Rd. TW9: Rich 56Na 109
Kew Gdns. 52Na 109
Kew Gdns. Rd. TW9: Kew 52Pa 109
Kew Gardens Station (Underground & Overground) 53Qa 109
KEW GREEN 52Qa 109
Kew Grn. TW9: Kew 51Pa 109
Kew Mdw. Path TW9: Kew Clifford Av. 54Sa 109
Kew Mdw. Path TW9: Kew Magnolia Ct. 53Ra 109
Kew Palace 52Na 109
Kew Retail Pk. Kew 53Ra 109
Kew Riverside Pk. TW9: Kew 52Ra 109
Kew Rd. TW9: Kew 51Qa 109
Kew Rd. TW9: Rich 51Qa 109
Key Cl. E1 42Xb 91
Keyes Ho. SW1 50Mb 90
(off Dolphin Sq.)
Keyes Rd. DA1: Dart 56Pd 119
Keyes Rd. NW2 36Za 68
Keyfield Ter. AL1: St A 3B 6
(not continuous)
Keyham Ho. W2 43Cb 89
(off Westbourne Pk. Rd.)
Key Ho. SE11 51Qb 112
Keymer Cl. TN16: Big H 88Lc 179
Keymer Pl. E14 44Bc 92
Keymer Rd. SW2 61Pb 134
Keynes Cl. N2 28Hb 49
Keynes Ct. SE28 45Xc 95
(off Attlee Rd.)
Keynsham Av. IG8: Wfd G 21Gc 53
Keynsham Gdns. SE9 57Nc 116
Keynsham Rd. SE9 57Mc 115
Keynsham Rd. SM4: Mord 74Db 155
Keynsham Wlk. SM4: Mord 74Db 155
Keys Ct. CR0: C'don 76Tb 157
(off Beech Ho. Rd.)
Keyse Rd. SE1 4K 231 (48Vb 91)
Keyser Pl. WD23: Bush 15Aa 27
Keysham Av. TW5: Cran 53W 106
Keystone Cres. N1 2G 217 (40Nb 70)
Keystone Pas. WD6: Bore 13Qa 29
Key W. Ct. IG7: Chig 21Rc 54
Keywood Dr. KT12: Hers 77Y 151
Keywood Dr. TW16: Sun 65W 128
Keyworth Cl. E5 35Ac 72
Keyworth Pl. SE1 3C 230 (48Rb 91)
(off Keyworth St.)
Keyworth St. SE1 3C 230 (48Rb 91)
Kezia M. SE8 50Ac 92
Kezia St. SE8 50Ac 92
Khalsa Av. DA12: Grav'nd 9E 122
Khalsa Ct. N22 25Rb 51
Khama Rd. SW17 63Gb 133
Khartoum Pl. DA12: Grav'nd 8E 122
Khartoum Rd. E13 41Kc 93
Khartoum Rd. IG1: Ilf 36Rc 74
Khartoum Rd. SW17 63Fb 133
Khyber Rd. SW11 54Gb 111
Kia Oval, The 51Pb 112
Kibble Cl. RM6: Chad H 31Yc 75
Kibworth St. SW8 52Pb 112
Kidabulous 67U 128
Kidborough Down KT23: Bookh 99Ca 191
KIDBROOKE 54Kc 115
Kidbrooke Est. SE3 55Lc 115
Kidbrooke Gdns. SE3 54Jc 115
Kidbrooke Green Nature Reserve 55Lc 115
Kidbrooke Gro. SE3 53Jc 115
Kidbrooke La. SE9 56Nc 116
Kidbrooke Pk. Cl. SE3 53Kc 115
Kidbrooke Pk. Rd. SE3 53Kc 115
Kidbrooke Station (Rail) 55Kc 115
Kidbrooke Way SE3 54Kc 115
Kidderminster Pl. CR0: C'don 74Rb 157
Kidderminster Rd. CR0: C'don 74Rb 157
Kidderminster Rd. SL2: Slou 1E 80
Kidderpore Av. NW3 35Cb 69
Kidderpore Gdns. NW3 35Cb 69
Kidd Pl. SE7 50Nc 94
Kidman Cl. RM2: Rom 27Ld 57
Kidspace Croydon 79Qb 156
Kiebs Way SE17 7G 231 (50Tb 91)
Kielder Cl. IG6: Ilf 23Vc 55
Kier Hardie Ho. RM16: Grays 6A 100
Kier Pk. SL5: Asc 9A 124
Kiffen St. EC2 5G 219 (42Tb 91)
Kilberry Cl. TW7: Isle 53Fa 108
Kilbrennan Ho. E14 44Ec 92
(off Findhorn St.)
KILBURN 40Bb 69
Kilburn Bri. NW6 39Cb 69
Kilburn Ga. NW6 40Db 69
Kilburn High Rd. NW6 38Bb 69
Kilburn High Road Station (Overground) 39Db 69
Kilburn Ho. NW6 40Bb 69
(off Malvern Pl.)
Kilburn La. W10 41Za 88
Kilburn La. W9 40Ab 68
Kilburn Library & Youth Cen. 39Db 69
Kilburn Pk. Rd. NW6 41Cb 89
Kilburn Park Station (Underground) 40Cb 69
Kilburn Pl. NW6 39Cb 69
Kilburn Priory NW6 39Db 69
Kilburn Sq. NW6 39Cb 69
Kilburn Station (Underground) 37Bb 69
Kilburn Va. NW6 39Db 69
Kilburn Va. Est. NW6 39Db 69
(off Kilburn Vale)
Kilby Cl. WD25: Wat 7Z 13
Kilby Ct. SE10 48Hc 93
(off Greenroof Way)
Kilcorral Cl. KT17: Eps 86Wa 174
Kildare Cl. HA4: Ruis 32Y 65
Kildare Ct. W2 44Cb 89
(off Kildare Ter.)
Kildare Gdns. W2 44Cb 89
Kildare Rd. E16 43Jc 93
Kildare Ter. W2 44Cb 89
Kildare Wlk. E14 44Cc 92
Kildonan Cl. WD17: Wat 11V 26
Kildoran Rd. SW2 57Nb 112
Kildowan Rd. IG3: Ilf 32Wc 75
Kilgour Rd. SE23 58Ac 114
Kilkie St. SW6 54Eb 111
Killarney Rd. SW18 58Eb 111
Killasser Ct. KT20: Tad 95Ya 194
Killburns Mill Cl. SM6: W'gton 75Kb 156
Killearn Rd. SE6 60Fc 115
Killester Gdns. KT4: Wor Pk 77Xa 154
Killewarren Way BR5: Orp 72Yc 161
Killick Cl. TN13: Dun G 93Gd 202
Killick Ho. SM1: Sutt 77Cb 155
Killick M. SM3: Cheam 79Ab 154
Killick St. N1 1H 217 (40Pb 70)
Killick Way E1 43Zb 92
Killieser Av. SW2 61Nb 134
Killigarth Ct. DA14: Sidc 63Wc 139
Killigrew Ho. TW16: Sun 66U 128
Killip Cl. E16 44Hc 93
Killoran Ho. E14 48Ec 92
(off Galbraith St.)
Killowen Av. UB5: N'olt 36Ea 66
Killowen Cl. KT20: Tad 94Za 194
Killowen Rd. E9 37Zb 72
Killy Hill GU24: Chob 10J 147
Killyon Rd. SW8 54Lb 112
Killyon Ter. SW8 54Lb 112
Kilmaine Rd. SW6 52Ab 110
Kilmarnock Gdns. RM8: Dag 34Yc 75
Kilmarnock Pk. RH2: Reig 5K 207
Kilmarnock Rd. WD19: Wat 21Z 45
Kilmarsh Rd. W6 49Ya 88
Kilmartin Av. SW16 69Qb 134
Kilmartin Rd. IG3: Ilf 33Wc 75
Kilmartin Way RM12: Horn 36Kd 77
Kilmington Cl. CM13: Hut 19De 41
Kilmington Rd. SW13 51Wa 110
Kilmiston Av. TW17: Shep 72S 150
Kilmiston Ho. TW17: Shep 72S 150
Kilmore Ho. E14 44Dc 92
(off Vesey Path)
Kilmorey Gdns. TW1: Twick 57Ka 108
Kilmorey Rd. TW1: Twick 56Ka 108
Kilmorie Rd. SE23 60Ac 114
Kilmuir Ho. KT17: Eps 85Ua 174
(off Depot Rd.)
Kilmuir Ho. SW1 6J 227 (49Jb 90)
(off Bury St.)
Kiln Cinema 38Bb 69
Kiln Cl. UB3: Harl 51T 106
Kiln Cotts. HP2: Hem H 1B 4
Kiln Ct. E14 45Bc 92
(off Newell St.)
Kilncroft HP3: Hem H 4B 4
Kilndown DA12: Grav'nd 5F 144
Kilner Ho. E16 43Kc 93
(off Freemasons Rd.)
Kilner Ho. SE11 51Qb 112
(off Clayton St.)
Kilner St. E14 43Cc 92
Kilnfields BR6: Well H 79Cd 162
Kiln Ground HP3: Hem H 4A 4
Kiln Ho. E1 43Zb 92
(off Duckett St.)
Kiln Ho. UB2: S'hall 48Ca 85
(off Lockwood Rd.)
Kiln Ho. UB7: View 45M 83
Kiln La. GU23: Rip 96J 189
Kiln La. GU24: Bisl 9F 166
Kiln La. KT17: Eps 83Ua 174
Kiln La. SL2: Hedg 2G 60
Kiln La. SL5: S'dale 1E 146
Kiln M. SW17 64Fb 133
Kiln Pl. NW5 36Jb 70
Kilns, The RH1: Mers 3B 208
Kilns, The RH1: Redh 3B 208
Kilnside KT10: Clay 80Ja 152
Kiln Theatre 38Bb 69
Kiln Wlk. RH1: Redh 10A 208
Kiln Way HA6: Nwood 23U 44
Kiln Way RM17: Grays 50Be 99
Kilnwood TN14: Hals 85Bd 181
Kiln Wood La. RM4: Have B 22Fd 56
Kilpatrick Way UB4: Yead 43Aa 85
Kilravock St. W10 41Ab 88
Kilronan W3 44Ta 87
Kilross Rd. TW14: Bedf 60T 106
Kilrue La. KT12: Hers 77V 150
Kilrush Ter. GU21: Wok 88C 168
Kilsby Wlk. RM9: Dag 37Xc 75
Kilsha Rd. KT12: Walt T 72Y 151
Kilsmore La. EN8: Chesh 1Zb 20
Kilvinton Dr. EN2: Enf 10Tb 19
Kilworth Av. CM15: Shenf 16Ce 41
Kimbell Gdns. SW6 53Ab 110
Kimbell Pl. SE3 56Lc 115
Kimber Cl. SL4: Wind 6D 102
Kimber Ct. SE1 3H 231 (48Ub 91)
(off Long La.)
Kimberley Av. E6 40Nc 74
Kimberley Av. IG2: Ilf 31Tc 74
Kimberley Av. RM7: Rom 29Ed 56
Kimberley Av. SE15 54Xb 113
Kimberley Bus. Pk. BR2: Kes 81Lc 179
Kimberley Cl. SL3: L'ly 49B 82
Kimberley Ct. NW6 39Ab 68
(off Kimberley Rd.)
Kimberley Dr. DA14: Sidc 61Zc 139
Kimberley Gdns. EN1: Enf 13Vb 33
Kimberley Gdns. N4 29Rb 51
Kimberley Ga. BR1: Broml 66Gc 137
Kimberley Ho. E14 48Ec 92
(off Galbraith St.)
Kimberley Pl. CR8: Purl 83Qb 176
Kimberley Ride KT11: Cobh 85Da 171
Kimberley Rd. AL3: St A 1A 6
Kimberley Rd. BR3: Beck 68Zb 136
Kimberley Rd. CR0: C'don 72Rb 157
Kimberley Rd. E11 33Fc 73
Kimberley Rd. E16 42Hc 93
Kimberley Rd. E17 25Ac 52
Kimberley Rd. E4 18Gc 35
Kimberley Rd. N17 26Wb 51
Kimberley Rd. N18 23Xb 51
Kimberley Rd. NW6 39Ab 68
Kimberley Rd. SW9 54Nb 112
Kimberley Wlk. KT12: Walt T 73X 151
Kimberley Way E4 18Gc 35
Kimber Pl. TW4: Houn Conway Rd. 59Ba 107
Kimber Pl. TW4: Houn Marryat Cl. 56Ba 107
Kimber Rd. SW18 59Cb 111
Kimbers Dr. SL1: Burn 1B 80
Kimble Cl. WD18: Wat 17U 26
Kimble Cres. WD23: Bush 17Ea 28
Kimble Ho. NW8 5E 214 (42Gb 89)
(off Lilestone St.)
Kimble Rd. SW19 65Fb 133
Kimblewick WD19: Wat 18Aa 27
Kimbolton Cl. SE12 58Hc 115
Kimbolton Ct. SW3 6D 226 (49Gb 89)
(off Fulham Rd.)
Kimbolton Grn. WD6: Bore 14Sa 29
Kimbolton Row SW3 6D 226 (49Gb 89)
(off Fulham Rd.)
Kimmeridge Gdns. SE9 63Nc 138
Kimmeridge Rd. SE9 63Nc 138
Kimmins Ct. SE16 48Wb 91
(off Old Jamaica Rd.)
Kimps Way HP3: Hem H 5A 4
Kimpton Av. CM15: B'wood 17Xd 40
Kimpton Ho. SW15 59Wa 110
Kimpton Ind. Est. SM3: Sutt 75Bb 155
Kimpton Link Bus. Cen. SM3: Sutt 75Bb 155
Kimpton Pk. Way SM1: Sutt 75Bb 155
Kimpton Pk. Way SM3: Sutt 75Ab 154
Kimpton Pl. WD25: Wat 6Z 13
Kimpton Rd. SE5 53Tb 113
Kimpton Rd. SM3: Sutt 75Bb 155
Kimptons Cl. EN6: Pot B 4Za 16
Kimptons Mead EN6: Pot B 5Za 16
Kimpton Trade & Bus. Cen. SM3: Sutt 75Bb 155
Kinburn Dr. TW20: Egh 64A 126
Kinburn St. SE16 47Zb 92
Kincaid Rd. SE15 52Xb 113
Kincardine Gdns. W9 42Cb 89
(off Harrow Rd.)
Kincha Lodge KT2: King T 67Pa 131
(off Elm Rd.)
Kinch Gro. HA9: Wemb 31Pa 67
Kincraig Dr. TN13: S'oaks 96Jd 202
Kinder Cl. SE28 45Zc 95
Kinder Ho. N1 1G 219 (40Tb 71)
(off Cranston Est.)
Kinderscout HP3: Hem H 4A 4
Kindersley Ho. E1 44Wb 91
(off Pinchin St.)
Kindersley Way WD5: Ab L 3S 12
Kinder St. E1 44Xb 91
Kinderton Cl. N14 18Lb 32
Kindred Ho. CR0: C'don 76Sb 157
Kinefold Ho. N7 37Nb 70
(off York Way Est.)
Kinesis Gym & Fitness Cen. 56Nc 116
(off Well Hall Rd.)
Kinetic Bus. Cen. WD6: Bore 13Qa 29
Kinetic Cres. EN3: Enf L 8Bc 20
Kinfauns Av. RM11: Horn 30Ld 57
Kinfauns Rd. IG3: Ilf 32Wc 75
Kinfauns Rd. SW2 61Qb 134
Kingaby Gdns. RM13: Rain 38Jd 76
King Acre Ct. TW18: Staines 62G 126
King Alfred Av. SE6 63Cc 136
(not continuous)
King Alfred Rd. RM3: Hrld W 26Pd 57
King & Queen Cl. SE9 63Nc 138
King & Queen St. SE17 7E 230 (50Sb 91)
King & Queen Wharf SE16 45Zb 92
King Arthur Cl. SE15 52Yb 114
King Arthur Ct. EN8: Chesh 3Ac 20
KING CHARLES I ISLAND 6F 223 (46Nb 90)
(end of Whitehall)
King Charles Ct. SE17 51Rb 113
(off Royal Rd.)
King Charles Cres. KT5: Surb 73Pa 153
King Charles Ho. SW6 52Db 111
(off Wandon Rd.)
King Charles Rd. KT5: Surb 71Pa 153
King Charles Rd. WD7: Shenl 4Na 15
King Charles's Ct. SE10 51Ec 114
(off Park Row)
King Charles St. SW1 1E 228 (47Mb 90)
King Charles Ter. E1 45Xb 91
(off Sovereign Cl.)
King Charles Wlk. SW19 60Ab 110
King Ct. E10 31Dc 72
Kingcup Av. HP2: Hem H 2D 4
Kingcup Cl. CR0: C'don 73Zb 158
Kingcup Dr. GU24: Bisl 7E 166
King David La. E1 45Yb 92
Kingdom St. W2 1A 220 (43Eb 89)
Kingdom Way UB8: Cowl 43M 83
Kingdon Ho. E14 48Ec 92
(off Galbraith St.)
Kingdon Rd. NW6 37Cb 69
King Edward VII Av. SL4: Wind 2J 103
KING EDWARD VII HOSPITAL 5G 102
King Edward Av. DA1: Dart 58Md 119
King Edward Av. RM13: Rain 40Md 77
King Edward Bldg. EC1 2C 224 (44Rb 91)
(off King Edward St.)
King Edward Ct. HA9: Wemb 36Na 67
(off Elm Rd.)
King Edward Ct. Shop. Cen. 3H 103
King Edward Dr. KT9: Chess 76Na 153
King Edward Dr. RM16: Grays 7A 100
King Edward Ho. WD23: Bush 14Ba 27
King Edward Mans. E8 39Xb 71
(off Mare St.)
King Edward M. SW13 53Wa 110
King Edward Pl. WD23: Bush 14Ba 27
King Edward Rd. CM14: B'wood 20Yd 40
King Edward Rd. DA9: Ghithe 57Wd 120
(not continuous)
King Edward Rd. E10 32Ec 72
King Edward Rd. E17 27Ac 52
King Edward Rd. EN5: New Bar 14Cb 31
King Edward Rd. EN8: Walt C 5Ac 20
King Edward Rd. RM1: Rom 30Hd 56
King Edward Rd. SS17: Stan H 3M 101
King Edward Rd. WD19: Wat 16Aa 27
King Edward Rd. WD7: Shenl 5Pa 15
King Edward's Gdns. W3 46Qa 87
King Edwards Gro. TW11: Tedd 65Ka 130
King Edwards Mans. SW6 52Cb 111
(off Fulham Rd.)
King Edward's Pl. W3 46Qa 87
King Edward's Rd. E9 39Xb 71
King Edward's Rd. EN3: Pond E 14Zb 34
King Edward's Rd. HA4: Ruis 32T 64
King Edward's Rd. N9 17Xb 33
King Edwards Rd. IG11: Bark 39Tc 74
King Edward St. EC1 2D 224 (44Sb 91)
King Edward St. HP3: Hem H 6L 3
King Edward St. SL1: Slou 7H 81
King Edward the Third M. SE16 47Xb 91
KING EDWARD VII'S HOSPITAL SISTER AGNES 7J 215 (43Jb 90)
King Edward Wlk. SE1 3A 230 (48Qb 90)
KINGFIELD 92C 188
Kingfield Cl. GU22: Wok 92B 188
Kingfield Dr. GU22: Wok 92B 188
Kingfield Gdns. GU22: Wok 92B 188
KINGFIELD GREEN 92B 188
Kingfield Grn. GU22: Wok 92B 188
Kingfield Rd. GU22: Wok 92A 188
Kingfield Rd. W5 42Ma 87
Kingfield Stadium 92B 188
Kingfield St. E14 49Ec 92
Kingfisher Av. E11 30Kc 53
Kingfisher Cl. BR5: St P 70Zc 139
Kingfisher Cl. CM13: Hut 17Ce 41
Kingfisher Cl. HA3: Hrw W 24Ha 46
Kingfisher Cl. HA6: Nwood 25R 44
Kingfisher Cl. KT12: Hers 78Aa 151
Kingfisher Cl. KT16: Chert 74L 149
Kingfisher Cl. KT22: Lea 92La 192
Kingfisher Cl. SE28 45Yc 95
Kingfisher Ct. GU21: Wok Woodlands Pk. 86E 168
Kingfisher Ct. CR0: C'don 76Sb 157
(off Wandle Rd.)
Kingfisher Ct. E14 47Ec 92
(off River Barge Cl.)
Kingfisher Ct. EN2: Enf 10Pb 18
Kingfisher Ct. GU21: Wok 89A 168
(off Vale Farm Rd.)
Kingfisher Ct. KT8: E Mos 70Ga 130
Kingfisher Ct. SE1 2E 230 (47Sb 91)
(off Swan St.)
Kingfisher Ct. SL2: Slou 2F 80
Kingfisher Ct. SM1: Sutt 78Bb 155
Kingfisher Ct. SW19 61Za 132
Kingfisher Ct. TN15: W King 80Ud 164
Kingfisher Ct. TW3: Houn 57Da 107
Kingfisher Ct. TW7: Isle 54Fa 108
Kingfisher Dr. DA9: Ghithe 57Wd 120
Kingfisher Dr. HP3: Hem H 7P 3
Kingfisher Dr. RH1: Redh 3A 208
Kingfisher Dr. TW10: Ham 63Ka 130
Kingfisher Dr. TW18: Staines 63H 127
Kingfisher Gdns. CR2: Sels 83Zb 178
Kingfisher Hgts. E16 46Lc 93
(off Bramwell Way)
Kingfisher Hgts. N17 27Xb 51
(off Waterside Way)
Kingfisher Hgts. RM17: Grays 50Ce 99
Kingfisher Ho. SW18 55Eb 111
Kingfisher Ho. W14 48Bb 89
(off Melbury Rd.)
Kingfisher Leisure Cen. Kingston upon Thames 68Na 131
Kingfisher Lure WD3: Loud 14K 25
Kingfisher Lure WD4: K Lan 1R 12
Kingfisher M. SE13 56Dc 114
Kingfisher Pl. DA4: S Dar 68Sd 142
Kingfisher Pl. N22 26Pb 50
Kingfisher Rd. RM14: Upm 32Vd 78
Kingfishers, The UB9: Den 30J 43
(off Patrons Way E.)
Kingfisher Sq. SE8 51Bc 114
(off Clyde St.)
Kingfisher St. E6 43Nc 94
Kingfisher Wlk. NW9 26Ua 48
Kingfisher Way BR3: Beck 71Zb 158
Kingfisher Way NW10 37Ta 67
King Frederick IX Twr. SE16 48Bc 92
King Gdns. CR0: Wadd 78Rb 157
King George IV Ct. SE17 7F 231 (50Tb 91)
(off Dawes St.)
King George VI Av. CR4: Mitc 70Hb 133
King George VI Av. RM18: E Til 9K 101
King George VI Av. TN16: Big H 88Mc 179
King George VI Memorial 7D 222 (46Mb 90)
King George V Station (DLR) 46Qc 94
King George Av. E16 44Lc 93
King George Av. IG2: Ilf 29Tc 54
King George Av. KT12: Walt T 74Z 151
King George Av. WD23: Bush 16Da 27
King George Cl. RM7: Mawney 27Ed 56
King George Cl. TW16: Sun 64U 128
King George Cres. HA0: Wemb 36Ma 67
KING GEORGE HOSPITAL 28Wc 55
King George M. SW17 64Hb 133
King George Rd. EN9: Walt A 6Ec 20
King George Sailing Club 16Dc 34
King George's Av. WD18: Wat 15U 26
King George's Dr. KT15: New H 82J 169
King George's Dr. UB1: S'hall 43Ba 85
King George's Field 76Qa 153
King George's Fld. 48Yb 92
(off Lower Road)
King George's Fld. 54Pc 116
King George Sq. TW10: Rich 58Pa 109
King Georges Rd. CM15: Pil H 16Xd 40
King George's Trad. Est. KT9: Chess 77Qa 153
King George St. SE10 52Ec 114
King Georges Wlk. KT10: Esh 77Ea 152
King George Way E4 14Dc 39

Kingham Cl. SW18 ....59Eb 111
Kingham Cl. W11 ....47Ab 88
King Harold Ct. EN9: Walt A 5Ec 20
(off Sun St.)
King Harold Lodge EN9: Walt A 5Fc 21
King Harolds Way DA17: Belv 51Ad 117
King Harolds Way DA7: Belv 52Zc 117
King Harolds Way DA7: Bex 52Zc 117
King Harry La. AL3: St A 3N 5
King Harry St. HP1: Hem H 3M 3
King Harry St. HP2: Hem H 3M 3
King Henry Ct. EN9: Walt A 8Ec 20
King Henry Lodge E4 ....21Cc 52
King Henry M. BR6: Chels 78Vc 161
King Henry M. HA2: Harr 32Ga 66
King Henry's Dr. CR0: New Ad 81Dc 178
King Henry's Drive Stop (London Tramlink) 81Dc 178
King Henry's M. EN3: Enf L 9Cc 20
(off Shepley M.)
King Henry's Reach W6 ....51Ya 110
King Henry's Rd. KT1: King T 69Ra 131
King Henry's Rd. NW3 ....38Gb 69
King Henry's Stairs E1 ....46Xb 91
King Henry St. N16 ....36Ub 71
King Henry's Wlk. N1 ....37Ub 71
King Henrys Wlk. CM16: Epp 1Xc 23
(off Boleyn Row)
King Henry Ter. E1 ....45Xb 91
(off Sovereign Cl.)
Kinghorn St. EC1 ....1D 224 (43Sb 91)
King Ho. W12 ....44Xa 88
Kingisholt Ct. NW10 ....41Za 88
(off Wellington Rd.)
King James' Av. EN6: Cuff 1Nb 18
King James Ct. SE1 ....2C 230 (47Rb 91)
(off King James St.)
King James St. SE1 ....2C 230 (47Rb 91)
King John Ct. EC2 ....5J 219 (42Ub 91)
King John La. TW19: Wray 7P 103
King John's Cl. TW19: Wray 7P 103
King Johns Pl. TW20: Egh 64A 126
King John Sq. TW20: Eng G 2N 125
King John St. E1 ....43Zb 92
King John's Wlk. SE9 ....59Nc 116
Kinglake Ct. GU21: Wok 10J 167
Kinglake Est. SE17 ....7J 231 (50Ub 91)
Kinglake St. SE17 ....7H 231 (50Ub 91)
(not continuous)
Kinglet Cl. E7 ....37Jc 73
Kingley Pk. WD4: K Lan 1R 12
Kingly Ct. W1 ....4C 222 (45Lb 90)
(off Beak St.)
Kingly St. W1 ....3B 222 (44Lb 90)
Kings Acre RH1: S Nut 9F 208
Kingsand Rd. SE12 ....61Jc 137
Kings Arbour UB2: S'hall 50Aa 85
King's Arms All. TW8: Bford 51Ma 109
Kings Arms Ct. E1 ....43Wb 91
King's Arms Yd. SW18 ....57Db 111
Kings Arms Yd. EC2 ....2F 225 (44Tb 91)
Kings Arms Yd. RM1: Rom 29Gd 56
Kingsash Dr. UB4: Yead 42Aa 85
King's Av. IG9: Buck H Langfords 19Mc 35
King's Av. IG9: Buck H The Broadway 21Lc 53
King's Av. GU24: Brkwd 1B 186
King's Av. IG8: Wfd G 23Kc 53
King's Av. N10 ....27Jb 50
King's Av. SM5: Cars 80Gb 155
King's Av. TW16: Sun 64V 128
King's Av. UB6: G'frd 43Da 85
King's Av. WD18: Wat 14V 26
Kings Av. BR1: Broml 65Hc 137
Kings Av. HP3: Hem H 6P 3
Kings Av. KT14: Byfl 84M 169
Kings Av. KT3: N Mald 70Ua 132
Kings Av. N21 ....18Rb 33
Kings Av. RH1: Redh 8N 207
Kings Av. RM6: Chad H 30Bd 55
Kings Av. SW12 ....60Mb 112
Kings Av. SW4 ....59Mb 112
Kings Av. TW3: Houn 53Da 107
Kings Av. W5 ....44Ma 87
King's Bench St. SE1 ....1C 230 (47Rb 91)
King's Bench Wlk. EC4 ....3A 224 (44Qb 90)
King's Blvd. N1 ....2F 217 (40Nb 70)
Kingsbridge Av. W3 ....47Pa 87
Kingsbridge Cir. RM3: Rom 23Nd 57
Kingsbridge Cl. RM3: Rom 23Nd 57
Kingsbridge Ct. E14 ....48Cc 92
(off Dockers Tanner Rd.)
Kingsbridge Ct. NW1 ....38Kb 70
(off Castlehaven Rd.)
Kingsbridge Cres. UB1: S'hall 43Ba 85
Kingsbridge Dr. NW7 ....24Za 48
Kingsbridge Rd. IG11: Bark 40Tc 74
Kingsbridge Rd. KT12: Walt T 73X 151
Kingsbridge Rd. RM3: Rom 23Nd 57
Kingsbridge Rd. SM4: Mord 72Za 154
Kingsbridge Rd. UB2: S'hall 49Ba 85
Kingsbridge Rd. W10 ....44Ya 88
Kingsbridge Way UB4: Hayes 41U 84
Kingsbridge Wharf IG11: Bark 41Uc 94
Kingsbrook KT22: Lea 90Ja 172
KINGSBURY 29Ra 47
Kingsbury Av. AL3: St A 1A 6
Kingsbury Circ. NW9 ....29Qa 47
Kingsbury Cres. TW18: Staines 63F 126
Kingsbury Dr. SL4: Old Win 9L 103
KINGSBURY GREEN 30Ta 47
Kingsbury M. AL3: St A 1P 5
Kingsbury Rd. N1 ....37Ub 71
Kingsbury Rd. NW9 ....29Qa 47
Kingsbury Station (Underground) 29Qa 47
Kingsbury Ter. N1 ....37Ub 71
Kingsbury Trad. Est. NW9 ....30Ta 47
Kingsbury Watermill Mus. 1P 5
Kings Chase CM14: B'wood 20Yd 40
Kings Chase KT8: E Mos 69Ea 130
Kings Chase Vw. EN2: Enf 12Qb 32
Kingsclere Cl. SW15 ....59Wa 110
Kingsclere Ct. N12 ....22Gb 49
Kingsclere Pl. EN2: Enf 12Sb 33
Kingscliffe Gdns. SW19 ....60Bb 111
King's Cl. DA1: Cray 56Gd 118
King's Cl. NW4 ....28Za 48
King's Cl. WD18: Wat 14X 27
King's Cl. WD4: Chfd 3K 11
Kings Cl. E10 ....31Dc 72
Kings Cl. GU24: W End 5E 166
Kings Cl. HA6: Nwood 23V 44
Kings Cl. KT12: Walt T 74X 151
Kings Cl. KT7: T Ditt 72Ja 152
Kings Cl. TW18: Staines 66M 127
King's Club, The 65Ya 132
Kings Coll. Ct. NW3 ....38Gb 69
KING'S COLLEGE DENTAL INSTITUTE 54Tb 113
KING'S COLLEGE HOSPITAL 54Tb 113
King's College London Denmark Hill Campus 54Tb 113
King's College London Guy's Campus 7F 225 (46Tb 91)
(within Guy's Hospital)
King's College London Institute of Psychiatry, De Crespigny Park 54Tb 113
King's College London Maughan Library 2K 223 (44Qb 90)
King's College London St Thomas' Campus - Lambeth Pal. Rd. 4H 229 (48Pb 90)
(off Lambeth Pal. Rd.)
King's College London St Thomas' Campus - St Thomas' House 3H 229 (48Pb 90)
(off Lambeth Pal. Rd.)
King's College London Strand Campus 4J 223 (45Pb 90)
King's College London Waterloo Campus 7K 223 (46Qb 90)
Kings Coll. Rd. NW3 ....38Gb 69
Kings Coll. Rd. HA4: Ruis 30V 44
King's College School of Medicine & Dentistry 54Sb 113
Kingscote Rd. CR0: C'don 73Xb 157
Kingscote Rd. KT3: N Mald 69Ta 131
Kingscote Rd. W4 ....48Ta 87
Kingscote St. EC4 ....4B 224 (45Rb 91)
King's Ct. E13 ....39Kc 73
King's Ct. KT20: Tad 94Xa 194
King's Ct. SE1 ....1C 230 (47Rb 91)
Kings Ct. GU24: W End 6E 166
Kings Ct. HA9: Wemb 33Ra 67
Kings Ct. IG9: Buck H 19Mc 35
Kings Ct. KT12: Walt T 76X 151
Kings Ct. KT14: Byfl 83M 169
Kings Ct. N7 ....38Pb 70
(off Caledonian Rd.)
Kings Ct. NW8 ....1F 215 (39Hb 69)
(off Prince Albert Rd.)
Kings Ct. W6 ....49Wa 88
Kings Ct. WD6: Bore 11Pa 29
(off Bennington Dr.)
Kings Ct. Mans. SW6 ....53Bb 111
(off Fulham Rd.)
Kings Ct. M. KT8: E Mos 71Fa 152
Kings Ct. Nth. SW3 ....50Gb 89
Kingscourt Rd. SW16 ....62Mb 134
Kings Ct. Sth. SW3 ....50Gb 89
(off Chelsea Mnr. Gdns.)
King's Cres. N4 ....34Sb 71
Kings Cres. Est. N4 ....33Sb 71
Kingscroft SW4 ....58Nb 112
Kingscroft Rd. KT22: Lea 92Ka 192
Kingscroft Rd. NW2 ....37Bb 69
Kingscroft Rd. SM7: Bans 87Fb 175
KING'S CROSS 1F 217 (40Nb 70)
King's Cross Bri. N1 ....3G 217 (41Nb 90)
(off Gray's Inn Rd.)
Kings Cross La. RH1: S Nut 8D 208
Kings Cross Rd. WC1 ....3H 217 (41Pb 90)
King's Cross St Pancras Station (Underground) 3F 217 (41Nb 90)
King's Cross Sq. N1 ....3G 217 (41Nb 90)
(off Euston Rd.)
King's Cross Station (Rail & Underground) 2F 217 (40Nb 70)
Kingsdale Ct. DA10: Swans 58Ae 121
Kingsdale Ct. EN9: Walt A 6Jc 21
(off Lamplighters Cl.)
Kingsdale Gdns. W11 ....46Za 88
Kingsdale Rd. SE18 ....52Vc 117
Kingsdale Rd. SE20 ....66Zb 136
Kingsdene KT20: Tad 93Xa 194
Kingsdown Av. CR2: S Croy 82Rb 177
Kingsdown Av. W13 ....47Ka 86
Kingsdown Av. W3 ....45Ua 88
Kingsdown Cl. DA12: Grav'nd 10H 123
Kingsdown Cl. SE16 ....50Xb 91
(off Masters Dr.)
Kingsdown Cl. W10 ....44Za 88
Kingsdowne Rd. KT6: Surb 73Na 153
Kingsdown Ho. E8 ....36Wb 71
Kingsdown Point SW2 ....61Qb 134
Kingsdown Rd. E11 ....34Gc 73
Kingsdown Rd. KT17: Eps 85Wa 174
Kingsdown Rd. N19 ....33Nb 70
Kingsdown Rd. SM3: Cheam 78Ab 154
Kingsdown Way BR2: Hayes 73Jc 159
King's Dr. HA8: Edg 21Pa 47
Kings Dr. DA12: Grav'nd 2D 144
Kings Dr. HA9: Wemb 33Ra 67
Kings Dr. KT12: W Vill 81V 170
Kings Dr. KT5: Surb 73Qa 153
Kings Dr. KT7: T Ditt 73Ka 152
Kings Dr. TW11: Tedd 64Fa 130
Kingsend HA4: Ruis 32T 64
Kingsend Ct. HA4: Ruis 32U 64
KINGS FARM 2E 144
Kings Farm E17 ....25Dc 52
Kings Farm Av. TW10: Rich 56Qa 109
Kings Farm Rd. WD3: Chor 16F 24
Kingsfield SL4: Wind 3B 102
Kingsfield Av. HA2: Harr 28Da 45
Kingsfield Bus. Cen. RH1: Redh 7A 208
Kingsfield Ct. WD19: Wat 17Z 27
Kingsfield Dr. EN3: Enf W 7Zb 20
Kingsfield Ho. SE9 ....62Mc 137
King's Fld. Recreation Ground, The 68La 130
Kingsfield Rd. HA1: Harr 31Fa 66
Kingsfield Rd. WD19: Wat 17Z 27
Kingsfield Ter. DA1: Dart 57Md 119
Kingsfield Ter. HA1: Harr 31Fa 66
Kingsfield Way EN3: Enf W 7Zb 20
Kingsfield Way RH1: Redh 7A 208
Kingsford St. NW5 ....36Hb 69
Kingsford Way E6 ....43Pc 94
King's Gdns. NW6 ....38Cb 69
Kings Gdns. IG1: Ilf 32Tc 74
Kings Gdns. KT12: Walt T 74X 151
Kings Gdns. RM14: Upm 31Ud 78
Kings Gth. M. SE23 ....61Yb 136
Kings Ga. KT15: Add 77K 149
Kingsgate AL3: St A 4P 5
Kingsgate HA9: Wemb 34Sa 67
Kingsgate Av. N3 ....27Cb 49
Kingsgate Bus. Cen. KT2: King T 67Na 131
(off Kingsgate Rd.)
Kingsgate Cl. BR5: St P 68Yc 139
Kingsgate Cl. DA7: Bex 53Ad 117
Kingsgate Est. N1 ....37Ub 71
Kingsgate Ho. SW9 ....53Qb 112
Kingsgate Mans. WC1 ....1H 223 (43Pb 90)
(off Red Lion Sq.)
Kings Ga. M. N8 ....29Pb 50
(off Spencer Rd.)
Kingsgate Pde. SW1 ....3C 228 (48Lb 90)
(off Victoria St.)
Kingsgate Pl. NW6 ....38Cb 69
Kingsgate Rd. KT1: King T 67Na 131
Kingsgate Rd. KT2: King T 67Na 131
Kingsgate Rd. NW6 ....38Cb 69
Kings Ga. Wlk. SW1 ....3C 228 (48Lb 90)
(off Victoria St.)
Kings Grange HA4: Ruis 32V 64
Kings Grn. IG10: Lough 13Nc 36
Kingsground SE9 ....59Mc 115
King's Gro. SE15 ....52Xb 113
(not continuous)
Kings Gro. RM1: Rom 29Jd 56
Kingsgrove Cl. DA14: Sidc 63Vc 139
Kings Hall Leisure Cen. 36Yb 72
Kings Hall M. SE13 ....55Ec 114
Kings Hall Rd. BR3: Beck 66Ac 136
Kings Head All. KT22: Lea 94Ka 192
(off High St.)
Kings Head Hill E4 ....17Dc 34
Kingshead Ho. NW7 ....21Xa 48
Kings Head La. KT14: Byfl 83M 169
Kings Head Pas. SW4 ....56Mb 112
(off Clapham Pk. Rd.)
Kings Head Theatre 39Rb 71
(off Upper St.)
King's Head Yd. SE1 ....7F 225 (46Tb 91)
King's Highway SE18 ....51Uc 116
King's Hill IG10: Lough 12Nc 36
Kingshill Av. HA3: Kenton 28Ka 46
Kingshill Av. KT4: Wor Pk 73Wa 154
Kingshill Av. RM5: Col R 23Ed 56
Kingshill Av. UB4: Hayes 41U 84
Kingshill Av. UB4: Yead 41U 84
Kingshill Av. UB5: N'olt 41W 84
Kingshill Cl. UB4: Hayes 41W 84
Kingshill Ct. WD23: Bush 16Ea 28
Kingshill Ct. EN5: Barn 14Ab 30
Kingshill Dr. HA3: Kenton 26Ka 46
Kingshold Rd. E9 ....38Yb 72
Kingsholm Gdns. SE9 ....56Mc 115
Kingshott Ho. KT17: Eps 84Ua 174
(off East St.)
King's Ho. SW10 ....51Fb 111
(off King's Rd.)
Kings Ho. SW8 ....52Nb 112
(off Sth. Lambeth Rd.)
King's Ho. Studios SW10 ....51Fb 111
(off Lamont Rd. Pas.)
Kingshurst Rd. SE12 ....59Jc 115
Kingside SE18 ....48Nc 94
Kingsingfield Cl. TN15: W King 80Ud 164
Kingsingfield Rd. TN15: W King 81Ud 184
Kings Keep BR2: Broml 68Gc 137
Kings Keep KT1: King T 70Na 131
Kings Keep SW15 ....57Za 110
Kingsland EN6: Pot B 5Bb 17
KINGSLAND 37Ub 71
Kingsland NW8 ....39Gb 69
Kingsland Basin 39Ub 71
Kingsland Grn. E8 ....37Ub 71
Kingsland High St. E8 ....37Vb 71
Kingsland Pas. E8 ....37Ub 71
Kingsland Rd. E13 ....41Lc 93
Kingsland Rd. E2 ....3J 219 (41Ub 91)
Kingsland Rd. E8 ....39Ub 71
Kingsland Rd. HP1: Hem H 4J 3
Kingsland Shop. Cen. 37Vb 71
King's La. WD4: Chfd 3J 11
Kings La. GU20: W'sham 8C 146
Kings La. SM1: Sutt 79Fb 155
Kings La. TW20: Eng G 4L 125
KINGS LANGLEY 1Q 12
Kings Langley By-Pass HP1: Hem H 4F 2
Kings Langley Lock 1R 12
(off Waterside Cl.)
Kings Langley Station (Rail) 2S 12
Kingslawn Cl. SW15 ....57Xa 110
Kingslea KT22: Lea 92Ja 192
Kingslee Ct. SM2: Sutt 80Db 155
Kingsleigh Cl. TW8: Bford 51Ma 109
Kingsleigh Pl. CR4: Mitc 69Hb 133
Kingsleigh Wlk. BR2: Broml 70Hc 137
(off Vicarage Rd.)
Kingsley Av. DA1: Dart 57Qd 119
Kingsley Av. EN8: Chesh 1Xb 19
Kingsley Av. SM1: Sutt 77Fb 155
Kingsley Av. SM7: Bans 87Cb 175
Kingsley Av. TW20: Eng G 5M 125
Kingsley Av. TW3: Houn 54Ea 108
Kingsley Av. UB1: S'hall 45Ca 85
Kingsley Av. W13 ....43Ja 86
Kingsley Av. WD6: Bore 12Pa 29
Kingsley Cl. N2 ....29Eb 49
Kingsley Cl. RM10: Dag 35Dd 76
Kingsley Ct. DA6: Bex 56Cd 118
Kingsley Ct. HA8: Edg 20Ra 29
Kingsley Ct. KT12: Walt T 76W 150
(off Ashley Pk. Rd.)
Kingsley Ct. KT4: Wor Pk 75Va 154
(off The Avenue)
Kingsley Ct. NW2 ....37Xa 68
Kingsley Ct. RM2: Rom 30Kd 57
Kingsley Dr. KT4: Wor Pk 75Va 154
Kingsley Flats SE1 ....5J 231 (49Ub 91)
(off Old Kent Rd.)
Kingsley Gdns. E4 ....22Cc 52
Kingsley Gdns. KT16: Ott 79F 148
Kingsley Gdns. RM11: Horn 28Md 57
KINGSLEY GREEN 3La 14
Kingsley Gro. RH2: Reig 9J 207
Kingsley Ho. SW3 ....51Fb 111
(off Beaufort St.)
Kingsley Ho. W14 ....49Ab 88
(off Avonmore Pl.)
Kingsley Mans. W14 ....51Ab 110
(off Greyhound Rd.)
Kingsley M. BR7: Chst 65Rc 138
Kingsley M. E1 ....45Xb 91
Kingsley M. W8 ....48Db 89
Kingsley Path SL2: Slou 2B 80
Kingsley Pl. N6 ....31Jb 70
Kingsley Rd. BR6: Chels 80Vc 161
Kingsley Rd. CM13: Hut 17Fe 41
Kingsley Rd. CR0: C'don 74Qb 156
Kingsley Rd. E17 ....26Ec 52
Kingsley Rd. E7 ....38Jc 73
Kingsley Rd. HA2: Harr 35Ea 66
Kingsley Rd. HA5: Pinn 28Ba 45
Kingsley Rd. IG10: Lough 13Tc 36
Kingsley Rd. IG6: Ilf 25Sc 54
Kingsley Rd. N13 ....21Qb 50
Kingsley Rd. NW6 ....39Bb 69
Kingsley Rd. SW19 ....64Db 133
Kingsley Rd. TW3: Houn 53Da 107
Kingsley St. SW11 ....55Hb 111
Kingsley Wlk. RM16: Grays 9C 100
Kingsley Way N2 ....29Eb 49
Kingsley Wood Dr. SE9 ....62Pc 138
Kings Lodge HA4: Ruis 32U 64
(off Pembroke Rd.)
Kings Lodge N12 ....23Eb 49
Kingslyn Cres. SE19 ....67Ub 135
Kings Lynn Cl. RM3: Rom 23Md 57
Kings Lynn Dr. RM3: Rom 23Md 57
Kings Lynn Path RM3: Rom 23Md 57
Kings Mall W6 ....49Ya 88
Kingsman Dr. RM16: Grays 45De 99
Kingsman Pde. SE18 ....48Pc 94
Kingsman Rd. SS17: Stan H 2K 101
Kings Mans. SW3 ....51Gb 111
(off Lawrence St.)
Kingsman St. SE18 ....48Pc 94
Kings Mead RH1: S Nut 8E 208
Kingsmead EN5: New Bar 14Cb 31
Kingsmead EN6: Cuff 1Nb 18
Kingsmead EN8: Chesh 1Zb 20
Kingsmead GU21: Wok 88C 168
Kingsmead KT13: Weyb 79T 150
Kingsmead TN16: Big H 88Mc 179
Kingsmead TW10: Rich 58Pa 109
Kingsmead Av. CR4: Mitc 69Lb 134
Kingsmead Av. KT4: Wor Pk 75Xa 154
Kingsmead Av. KT6: Surb 75Qa 153
Kingsmead Av. N9 ....18Xb 33
Kingsmead Av. NW9 ....31Ta 67
Kingsmead Av. RM1: Rom 30Gd 56
Kingsmead Av. TW16: Sun 68Y 129
Kingsmead Cl. DA15: Sidc 61Wc 139
Kingsmead Cl. KT19: Ewe 80Ta 153
Kingsmead Cl. TW11: Tedd 65Ka 130
Kingsmead Cotts. BR2: Broml 74Nc 160
Kingsmead Ct. N6 ....31Mb 70
Kingsmead Dr. UB5: N'olt 38Ba 65
Kingsmead Ho. E9 ....35Ac 72
Kingsmead Ho. SL1: Slou 6G 80
Kingsmead Lodge SM2: Sutt 79Fb 155
Kingsmead Mans. RM1: Rom 30Hd 56
(off Kingsmead Av.)
Kings Mdw. WD4: K Lan 10A 4
Kingsmeadow 69Qa 131
Kingsmeadow Athletics Cen. 69Qa 131
Kings Mdw. Ct. EN9: Walt A 6Jc 21
(off Horseshoe Cl.)
Kings Mead Pk. KT10: Clay 80Ga 152
Kingsmead Rd. SW2 ....61Qb 134
Kingsmead Way E9 ....35Ac 72
Kingsmere Cl. SW15 ....55Za 110
Kingsmere Pk. NW9 ....32Ra 67
Kingsmere Pl. N16 ....32Tb 71
Kingsmere Rd. SW19 ....61Za 132
King's M. SW4 ....57Nb 112
King's M. WC1 ....6J 217 (42Pb 90)
Kings M. HP2: Hem H 1M 3
(off George St.)
Kings M. IG7: Chig 19Sc 36
Kingsmill NW8 ....1C 214 (40Fb 69)
(off Kingsmill Ter.)
Kingsmill Bus. Pk. KT1: King T 69Pa 131
Kingsmill Ct. AL10: Hat 2D 8
Kingsmill Gdns. RM9: Dag 36Bd 75
Kingsmill Ho. SW3 ....7E 226 (50Gb 89)
(off Cale St.)
Kingsmill Rd. RM9: Dag 36Bd 75
Kingsmill Ter. NW8 ....1C 214 (40Fb 69)
Kings Mill Way UB9: Den 37L 63
Kingsnorth Ho. W10 ....44Za 88
Kingsnympton Pk. KT2: King T 66Ra 131
King's Oak WD3: Crox G 14Q 26
Kings Oak RM7: Mawney 27Cd 56
KING'S OAK BMI HOSPITAL 10Qb 18
Kingsoak Ho. GU21: Wok 88C 168
King's Orchard SE9 ....58Nc 116
King's Paddock TW12: Hamp 67Ea 130
King's Pde. SM5: Cars 76Hb 155
(off Wrythe La.)
Kings Pde. N17 ....27Vb 51
Kings Pde. NW10 ....39Ya 68
Kings Pde. SS17: Stan H 2L 101
(off King St.)
Kings Pde. W12 ....48Wa 88
Kings Pde. WD18: Wat 15X 27
(off Vicarage Rd.)
Kings Pk. SL3: Coln 52F 104
Kingspark Bus. Cen. KT3: N Mald 70Sa 131
Kingspark Ct. E18 ....27Jc 53
Kings Pk. Ind. Est. WD4: K Lan 1R 12
King's Pas. KT2: King T 67Ma 131
Kings Pas. E11 ....31Gc 73
Kings Pas. KT1: King T 68Ma 131
King's Pl. SE1 ....2E 230 (47Sb 91)
King's Pl. W4 ....50Sa 87
Kings Pl. GU24: W End 6E 166
Kings Pl. IG10: Lough 17Mc 35
Kings Pl. IG9: Buck H 19Lc 35
Kings Place 1G 217 (40Nb 70)
King Sq. EC1 ....4D 218 (41Sb 91)
Kings Quarter Apts. N1 ....39Pb 70
(off Copenhagen St.)
King's Quay SW10 ....53Eb 111
(off Chelsea Harbour Dr.)
Kings Reach SL3: L'ly 9N 81
Kings Reach Twr. SE1 ....6A 224 (46Qb 90)
(off Stamford St.)
Kings Ride Ga. TW10: Rich 56Qa 109
Kingsridge SW19 ....61Ab 132
Kingsridge Gdns. DA1: Dart 58Md 119
King's Rd. AL2: Lon C 8G 6
King's Rd. AL3: St A 2P 5
King's Rd. BR6: Orp 77Vc 161
King's Rd. CM14: B'wood 19Yd 40
King's Rd. CM14: W'ley 19Yd 40
King's Rd. E6 ....39Lc 73
King's Rd. KT2: King T 66Na 131
King's Rd. KT6: Surb 74La 152
King's Rd. N17 ....25Vb 51
King's Rd. RM1: Rom 29Jd 56
King's Rd. SL1: Slou 8J 81
King's Rd. SL4: Wind 4H 103
King's Rd. SL5: S'dale 1B 146
King's Rd. SL5: S'hill 1B 146
King's Rd. SM2: Sutt 82Cb 175
King's Rd. SW10 ....52Eb 111
King's Rd. SW19 ....65Cb 133
King's Rd. SW3 ....51Fb 111
King's Rd. SW6 ....52Db 111
King's Rd. TW11: Tedd 64Fa 130
King's Rd. UB7: W Dray 47P 83
King's Rd. UB8: Uxb 40M 63
King's Rd. CR4: Mitc 69Jb 134
Kings Rd. E11 ....31Gc 73
Kings Rd. E4 ....18Fc 35
Kings Rd. EN5: Barn 13Ya 30
Kings Rd. EN8: Walt C 5Ac 20
Kings Rd. GU21: Wok 88C 168
Kings Rd. GU24: W End 6E 166
Kings Rd. HA2: Harr 33Ba 65
Kings Rd. IG11: Bark 38Sc 74
Kings Rd. KT12: Walt T 75X 151
Kings Rd. KT15: New H 82K 169
Kings Rd. N18 ....22Wb 51
Kings Rd. N22 ....25Pb 50
Kings Rd. NW10 ....38Xa 68
Kings Rd. SE25 ....69Wb 135
Kings Rd. SW14 ....55Ta 109
Kings Rd. TN16: Big H 88Lc 179
Kings Rd. TW1: Twick 58Ka 108
Kings Rd. TW10: Rich 58Pa 109
Kings Rd. TW13: Felt 60Y 107
Kings Rd. TW20: Egh 63C 126
Kings Rd. W5 ....43Ma 87
King's Scholars' Pas. SW1 ....4B 228 (48Lb 90)
(off Carlisle Pl.)
King's Shade Wlk. KT19: Eps 85Ta 173
King Stable St. SL4: Eton 2H 103
King Stairs Cl. SE16 ....47Xb 91
King's Ter. NW1 ....1B 216 (39Lb 70)
King's Ter. TW7: Isle 56Ja 108
Kings Ter. SL3: L'ly 51D 104
Kingsthorpe Rd. SE26 ....63Zb 136
Kingston Av. UB7: View Ash Gro 45P 83
Kingston Av. UB7: View Whitethorn Av. 46P 83
Kingston Av. KT22: Lea 93Ka 192
Kingston Av. KT24: E Hor 98U 190
Kingston Av. SM3: Cheam 76Ab 154
Kingston Av. TW14: Felt 58U 106
Kingston Bri. 68Ma 131
Kingston Bus. Cen. KT9: Chess 76Na 153
Kingston By-Pass KT6: Surb 76Ma 153
Kingston By-Pass Rd. KT10: Hin W 75Ja 152
Kingston By-Pass Rd. KT10: Surb 75Ga 152
Kingston Cl. RM6: Chad H 27Ad 55
(not continuous)
Kingston Cl. TW11: Tedd 65Ka 130
Kingston Cl. UB5: N'olt 39Ba 65
Kingston Ct. DA11: Nflt 57De 121
Kingston Crematorium 69Qa 131
Kingston Cres. BR3: Beck 67Bc 136
Kingston Cres. TW15: Ashf 64L 127
Kingston Gdns. CR0: Bedd 76Nb 156
Kingston Hall Rd. KT1: King T 69Ma 131
Kingston Hill KT2: King T 67Qa 131
Kingston Hill Av. RM6: Chad H 27Ad 55
Kingston Hill Pl. KT2: King T 63Sa 131
KINGSTON HOSPITAL 67Ra 131
Kingston Ho. KT1: King T 70Ma 131
(off Surbiton Rd.)
Kingston Ho. NW1 ....1C 216 (39Lb 70)
(off Camden St.)
Kingston Ho. NW6 ....38Ab 68
Kingston Ho. E. SW7 ....2D 226 (47Gb 89)
(off Prince's Ga.)
Kingston Ho. Est. KT6: Surb 72Ka 152
Kingston Ho. Gdns. KT22: Lea 93Ka 192
Kingston Ho. Nth. SW7 ....2D 226 (47Gb 89)
(off Prince's Ga.)
Kingston Ho. Sth. SW7 ....2D 226 (47Gb 89)
(off Ennismore Gdns.)
Kingstonian FC 76Qa 153
Kingston La. KT24: W Hor 99Q 190
Kingston La. TW11: Tedd 64Ja 130
Kingston La. UB7: W Dray 47P 83
Kingston La. UB8: Hil 41N 83
Kingston Lodge KT3: N Mald 70Ua 132
Kingston Mans. SW9 ....53Pb 112
(off Clapham Rd.)
Kingston Mus. 68Na 131
Kingston Pl. HA3: Hrw W 24Ha 46
Kingston Ri. KT15: New H 82J 169
Kingston Rd. SW19: Norstead Pl. 60Xa 110
Kingston Rd. SW19: Rothesay Av. 67Bb 133
Kingston Rd. EN4: E Barn 15Fb 31
Kingston Rd. IG1: Ilf 35Rc 74
Kingston Rd. KT1: King T 69Ra 131
Kingston Rd. KT17: Ewe 81Va 174
Kingston Rd. KT19: Ewe 76Sa 153
Kingston Rd. KT22: Lea 90Ja 172
(not continuous)
Kingston Rd. KT3: N Mald 69Ra 131
Kingston Rd. KT4: Wor Pk 76Sa 153
Kingston Rd. KT5: Surb 75Ra 153
Kingston Rd. N9 ....19Wb 33
Kingston Rd. RM1: Rom 28Hd 56
Kingston Rd. SW15 ....61Wa 132
Kingston Rd. SW20 ....68Za 132
Kingston Rd. TW11: Tedd 64Ka 130
Kingston Rd. TW15: Ashf 65N 127
Kingston Rd. TW18: Staines 63H 127
Kingston Rd. UB2: S'hall 47Ba 85
Kingston Sq. KT22: Lea 91Ja 192
(off Buffers La.)
Kingston Sq. SE19 ....64Tb 135
Kingston Station (Rail) 67Na 131
Kingston University Kingston Hill Campus 64Ta 131
Kingston University Knights Pk. Campus 69Na 131
Kingston University Penrhyn Rd. Campus, Reg Bailey Bldg. 69Ma 131
Kingston University Penrhyn Road Campus 70Na 131
Kingston University Roehampton Vale Cen. 62Va 132
KINGSTON UPON THAMES 68Ma 131
Kingston upon Thames Tourist Info. Cen. 68Ma 131
Kingston Va. SW15 ....63Ta 131
KINGSTON VALE 63Ua 132
Kingstown St. NW1 ....39Jb 70
(not continuous)
King St. DA12: Grav'nd 8D 122
King St. E13 ....42Jc 93
King St. EC2 ....2E 224 (44Sb 91)
King St. KT16: Chert 74J 149
King St. N17 ....25Vb 51
King St. N2 ....27Fb 49
King St. SS17: Stan H 2L 101
King St. SW1 ....7C 222 (46Lb 90)
King St. TW1: Twick 60Ja 108
King St. TW9: Rich 57Ma 109
King St. UB2: S'hall 48Aa 85
King St. W3 ....46Sa 87
King St. W6 ....49Wa 88
King St. WC2 ....4F 223 (45Nb 90)
King St. WD18: Wat 14Y 27
King St. Cloisters W6 ....49Xa 88
(off King St.)
King St. M. N2 ....27Fb 49
King St. Pde. TW1: Twick 60Ja 108
(off King St.)
Kingsville Ct. UB7: Yiew 45M 83
Kings Wlk. CR2: Sande 86Xb 177
Kings Wlk. RM17: Grays 51Ce 121
Kings Wlk. Shop. Cen. Chelsea 7F 227 (50Hb 89)
Kings Warren KT22: Oxs 83Ea 172
Kingswater Pl. SW11 ....52Gb 111
King's Way CR0: Wadd 78Pb 156
Kings Way HA1: Harr 28Ga 46
Kingsway BR4: W W'ck 76Gc 159
Kingsway BR5: Pet W 71Tc 160

Kingsway EN3: Pond E 15Xb 33
Kingsway EN6: Cuff 2Nb 18
Kingsway GU21: Wok 10P 167
Kingsway HA9: Wemb 35Na 67
Kingsway IG8: Wfd G 22Lc 53
Kingsway KT3: N Mald 70Ya 132
Kingsway N12 .....23Eb 49
Kingsway SL0: Iver 44G 82
Kingsway SL2: Farn C 7F 60
Kingsway SL9: Chal P 27A 42
Kingsway SW14 .....55Ra 109
Kingsway TW19: Stanw 60M 105
Kingsway UB3: Hayes 43S 84
Kingsway WC2 .....2H 223 (44Pb 90)
Kingsway, The KT17: Ewe 83Ua 174
Kingsway Av. CR2: Sels 81Yb 178
Kingsway Av. GU21: Wok 10P 167
Kingsway Bus. Pk. TW12: Hamp 67Ba 129
Kingsway Cres. HA2: Harr 28Ea 46
Kingsway Est. N18 .....23Zb 52
Kingsway Mans. WC1 .....7H 217 (43Pb 90)
.....(off Red Lion Sq.)
Kingsway M. SL2: Farn C 7F 60
Kingsway Nth. Orbital Rd. WD25: Wat 7V 12
Kingsway Pde. N16 .....34Tb 71
.....(off Albion Rd.)
Kingsway Pl. EC1 .....5A 218 (42Qb 90)
.....(off Sans Wlk.)
Kingsway Rd. SM3: Cheam 80Ab 154
Kingsway Ter. KT13: Weyb 81Q 170
Kingswear Rd. HA4: Ruis 33W 64
Kingswear Rd. NW5 .....34Kb 70
Kingswell Ride EN6: Cuff 2Nb 18
Kingswey Bus. Pk. GU21: Wok 86E 168
King's Wharf SE10 .....51Dc 114
.....(off Wood Wharf)
Kings Wharf E8 .....39Ub 71
.....(off Kingsland Rd.)
Kingswick Cl. SL5: S'hill 10C 124
Kingswick Dr. SL5: S'hill 10B 124
Kingswood E2 .....40Yb 72
.....(off Cyprus St.)
KINGSWOOD 96Ab 194
KINGSWOOD 6X 13
Kingswood Av. BR2: Broml 69Gc 137
Kingswood Av. BR8: Swan 70Hd 140
Kingswood Av. CR2: Sande 87Xb 177
Kingswood Av. CR7: Thor H 71Qb 156
Kingswood Av. DA17: Belv 49Bd 95
Kingswood Av. NW6 .....39Ab 68
Kingswood Av. TW12: Hamp 65Da 129
Kingswood Av. TW3: Houn 53Ba 107
KINGSWOOD CEN. 28Qa 47
Kingswood Cl. BR6: Orp 73Uc 160
Kingswood Cl. DA1: Dart 58Ld 119
Kingswood Cl. EN1: Enf 15Ub 33
Kingswood Cl. KT13: Weyb 80R 150
Kingswood Cl. KT3: N Mald 72Va 154
Kingswood Cl. KT6: Surb 73Na 153
Kingswood Cl. N20 .....17Eb 31
Kingswood Cl. SW8 .....52Nb 112
Kingswood Cl. TW15: Ashf 64T 128
Kingswood Cl. TW20: Eng G 3P 125
Kingswood Ct. E4 .....22Cc 52
Kingswood Ct. GU21: Wok 88A 168
Kingswood Ct. KT20: Kgswd 96Ab 194
Kingswood Ct. NW6 .....38Cb 69
.....(off West End La.)
Kingswood Ct. SE13 .....58Fc 115
Kingswood Ct. TN15: W King 80Ud 164
Kingswood Ct. TW10: Rich 57Pa 109
Kingswood Creek TW19: Wray 7P 103
Kingswood Dr. SE19 .....63Ub 135
Kingswood Dr. SM2: Sutt 81Db 175
Kingswood Dr. SM5: Cars 74Hb 155
Kingswood Est. SE21 .....63Ub 135
Kingswood Flds. Bus. Pk. KT20: Kgswd 97Cb 195
Kingswood Golf Course 96Cb 195
Kingswood Grange KT20: Lwr K 100Cb 195
Kingswood Hgts. E18 .....25Jc 53
.....(off Queen Mary Av.)
Kingswood Ho. KT20: Kgswd 92Bb 195
Kingswood Ho. SL2: Slou 3G 80
Kingswood La. CR2: Sande 86Zb 178
Kingswood La. CR6: W'ham 87Yb 178
Kingswood Library & Community Cen. 63Ub 135
Kingswood M. N15 .....28Rb 51
Kings Wood Pk. CM16: Epp 1Xc 23
Kingswood Pk. KT20: Kgswd 93Ab 194
Kingswood Pk. N3 .....26Bb 49
Kingswood Pl. CR3: Cat'm 95Vb 197
Kingswood Pl. SE13 .....56Gc 115
Kingswood Pl. UB4: Hayes 43U 84
Kingswood Ri. TW20: Eng G 4P 125
Kingswood Rd. BR2: Broml 70Fc 137
Kingswood Rd. E11 .....31Gc 73
Kingswood Rd. HA9: Wemb 34Qa 67
Kingswood Rd. IG3: Ilf 32Wc 75
Kingswood Rd. KT20: Tad 93Xa 194
Kingswood Rd. SE20 .....65Yb 136
Kingswood Rd. SW19 .....66Bb 133
Kingswood Rd. SW2 .....58Nb 112
Kingswood Rd. TN13: Dun G 92Gd 202
Kingswood Rd. W4 .....48Sa 87
Kingswood Rd. WD25: Wat 6X 13
Kingswood Station (Rail) 93Bb 195
Kingswood Ter. W4 .....48Sa 87
Kingswood Way CR2: Sande 85Yb 178
.....(not continuous)
Kingswood Way CR2: Sels 85Yb 178
.....(not continuous)
Kingswood Way SM6: W'gton 78Nb 156
Kingsworth Cl. BR3: Beck 71Ac 158
Kingsworthy Cl. KT1: King T 69Pa 131
Kings Yd. SW15 .....55Ya 110
.....(off Lwr. Richmond Rd.)
Kingthorpe Gdns. 78K 149
.....(off Burleigh Rd.)
Kingthorpe Rd. NW10 .....38Ta 67
Kingthorpe Ter. NW10 .....37Ta 67
Kington Ho. NW6 .....39Db 69
.....(off Mortimer Cres.)
Kingward Ho. E1 .....43Wb 91
.....(off Hanbury St.)
King Wardrobe Apts. EC4 .....3C 224 (44Rb 91)
.....(off Carter La.)
Kingwell Rd. EN4: Had W 10Fb 17
Kingweston Cl. NW2 .....34Ab 68
King William IV Gdns. SE20 .....65Yb 136
King William Ct. EN9: Walt A 7Ec 20
.....(off Kendal Rd.)
King William La. SE10 .....50Gc 93
King William's Ct. SE10 .....51Fc 115
.....(off Park Row)
King William St. EC4 .....3G 225 (44Tb 91)
King William Wlk. SE10 .....51Ec 114
.....(not continuous)
Kingwood Gdns. E1 .....44Wb 91
.....(off Piazza Wlk.)
Kingwood Rd. SW6 .....53Ab 110
Kinlet Rd. SE18 .....53Sc 116
Kinloch Dr. NW9 .....31Ta 67
Kinloch Rd. KT20: Tad 91Ya 194
Kinloch St. N7 .....34Pb 70
Kinloss Ct. N3 .....28Bb 49
Kinloss Gdns. N3 .....27Bb 49
Kinloss Rd. SM5: Cars 73Eb 155
Kinnaird Av. BR1: Broml 65Hc 137
Kinnaird Av. W4 .....52Sa 109
Kinnaird Cl. BR1: Broml 65Hc 137
Kinnaird Cl. SL1: Slou 4A 80
Kinnaird Ho. SE17 .....5G 231 (49Tb 91)
Kinnaird Way IG8: Wfd G 23Pc 54
Kinnear Apts. N8 .....27Pb 50
Kinnear Rd. W12 .....47Va 88
Kinnersley Wlk. RH2: Reig 10J 207
Kinnerton Pl. Nth. SW1 .....2G 227 (47Hb 89)
.....(off Kinnerton St.)
Kinnerton Pl. Sth. SW1 .....2G 227 (47Hb 89)
.....(off Kinnerton St.)
Kinnerton St. SW1 .....2H 227 (47Jb 90)
Kinnerton Yd. SW1 .....2H 227 (47Jb 90)
.....(off Kinnerton St.)
Kinnoul Rd. W6 .....51Ab 110
Kino Bermondsey 3J 231 (48Ub 91)
.....(off Bermondsey Sq.)
Kinross Av. KT4: Wor Pk 75Wa 154
Kinross Cl. HA3: Kenton 29Pa 47
Kinross Cl. HA8: Edg 19Ra 29
Kinross Cl. TW16: Sun 64V 128
Kinross Ct. BR1: Broml 67Hc 137
.....(off Highland Rd.)
Kinross Ct. SE6 .....60Hc 115
Kinross Dr. TW16: Sun 64V 128
Kinross Ho. N1 .....39Pb 70
.....(off Bemerton Est.)
Kinross Ter. E17 .....26Bc 52
Kinsale Cl. NW7 .....23Za 48
Kinsale Rd. SE15 .....55Wb 113
Kinsella Gdns. SW19 .....64Xa 132
Kinsham Ho. E2 .....42Wb 91
.....(off Ramsey St.)
Kinsheron Pl. KT8: E Mos 70Ea 130
Kintore Way SE1 .....5K 231 (49Vb 91)
Kintyre Cl. SW16 .....68Pb 134
Kintyre Ct. SW2 .....59Nb 112
Kintyre Ho. E14 .....46Ec 92
.....(off Coldharbour)
Kintyre Ho. WD18: Wat 16V 26
.....(off Explorer Dr.)
Kinveachy Gdns. SE7 .....50Nc 94
Kinver Ho. N19 .....33Mb 70
Kinver Rd. SE26 .....63Yb 136
Kipings KT20: Tad 93Za 194
Kipling Av. RM18: Tilb 3D 122
Kipling Cl. CM14: W'ley 22Xd 58
Kipling Ct. SL4: Wind 4F 102
Kipling Ct. W7 .....45Ha 86
Kipling Dr. SW19 .....65Fb 133
Kipling Est. SE1 .....2G 231 (47Tb 91)
Kipling Ho. N19 .....32Nb 70
.....(off Charles St.)
Kipling Ho. SE5 .....52Tb 113
.....(off Elmington Est.)
Kipling Pl. HA7: Stan 23Ha 46
Kipling Rd. DA1: Dart 57Rd 119
Kipling Rd. DA7: Bex 53Ad 117
Kipling St. SE1 .....2G 231 (47Tb 91)
Kipling Ter. N9 .....20Tb 33
Kipling Twr. W3 .....48Sa 87
.....(off Palmerston Rd.)
Kipling Towers RM3: Rom 24Kd 57
KIPPINGTON 98Jd 202
Kippington Cl. TN13: S'oaks 96Hd 202
Kippington Dr. SE9 .....60Mc 115
Kippington Rd. TN13: S'oaks 96Jd 202
Kira Bldg. E3 .....41Bc 92
Kiran Apts. E1 .....43Vb 91
.....(off Chicksand St.)
Kirby Cl. HA6: Nwood 23V 44
Kirby Cl. IG10: Lough 17Nc 36
Kirby Cl. IG6: Ilf 23Uc 54
Kirby Cl. KT19: Ewe 78Va 154
Kirby Cl. RM3: Rom 22Qd 57
Kirby Est. SE16 .....48Xb 91
Kirby Est. UB7: Yiew 45M 83
Kirby Gro. SE1 .....1H 231 (47Ub 91)
Kirby Rd. DA2: Dart 59Td 120
Kirby Rd. GU21: Wok 9N 167
Kirby St. EC1 .....7A 218 (43Qb 90)
Kirby Way KT12: Walt T 72Y 151
Kirby Way UB8: Hil 42P 83
Kirchen Rd. W13 .....45Ka 86
Kirkby Apts. E3 .....43Bc 92
.....(off St Paul's Way)
Kirkby Cl. N11 .....23Jb 50
Kirkcaldy Grn. WD19: Wat 20Y 27
Kirkcourt TN13: S'oaks 95Jd 202
Kirkdale SE26 .....61Xb 135
Kirkdale Cnr. SE26 .....63Yb 136
Kirkdale Rd. E11 .....32Gc 73
Kirkeby Ho. EC1 .....7K 217 (43Qb 90)
.....(off Leather La.)
Kirkfield Cl. W13 .....46Ka 86
Kirkgate, The KT17: Eps 85Ua 174
Kirkham Apts. IG11: Bark 38Sc 74
.....(off Linton Rd.)
Kirkham Ho. RM3: Rom 22Md 57
.....(off Montgomery Cres.)
Kirkham Rd. E6 .....44Nc 94
Kirkham St. SE18 .....51Uc 116
Kirk Ho. HA9: Wemb 34Na 67
Kirkland Av. GU21: Wok 8J 167
Kirkland Av. IG5: Ilf 26Qc 54
Kirkland Cl. DA15: Sidc 58Uc 116
Kirkland Dr. EN2: Enf 11Sb 33
Kirkland Ho. E14: St Davids Sq. 50Dc 92
.....(off St Davids Sq.)
Kirkland Ho. E14: Westferry Rd. 50Dc 92
.....(off Westferry Rd.)
Kirkland Ter. BR3: Beck 65Cc 136
Kirkland Wlk. E8 .....37Vb 71
Kirk La. SE18 .....51Sc 116
Kirkleas Rd. KT6: Surb 74Na 153
Kirklees Rd. CR7: Thor H 71Qb 156
Kirklees Rd. RM8: Dag 36Yc 75
Kirkley Rd. SW19 .....67Cb 133
Kirkly Cl. CR2: Sande 81Ub 177
Kirkman Pl. W1 .....1D 222 (43Mb 90)
.....(off Tottenham Ct. Rd.)
Kirkmichael Rd. E14 .....44Ec 92
Kirk Ri. SM1: Sutt 76Db 155
Kirk Rd. E17 .....30Bc 52
Kirkside Rd. SE3 .....51Jc 115
Kirk's Place 43Bc 92
Kirkstall Av. N17 .....28Tb 51
Kirkstall Gdns. SW2 .....60Nb 112
Kirkstall Ho. SW1 .....7K 227 (50Kb 90)
.....(part of Abbots Mnr.)
Kirkstall Rd. SW2 .....60Mb 112
Kirkstead Ct. E5 .....35Zb 72
Kirksted Rd. SM4: Mord 74Db 155
Kirkstone NW1 .....3B 216 (41Lb 90)
.....(off Harrington St.)
Kirkstone Way BR1: Broml 66Gc 137
Kirk St. WC1 .....6J 217 (42Pb 90)
.....(off Northington St.)
Kirkton Rd. N15 .....28Ub 51
Kirkwall Pl. E2 .....41Yb 92
Kirkwall Spur SL1: Slou 3J 81
Kirkwood Pl. NW1 .....38Jb 70
Kirkwood Rd. SE15 .....54Xb 113
Kirn Rd. W13 .....45Ka 86
Kirrane Cl. KT3: N Mald 71Va 154
Kirtley Ho. SW8 .....53Lb 112
Kirtley Rd. SE26 .....63Ac 136
Kirtling St. SW11 .....52Lb 112
Kirton Cl. RM12: Horn 37Ld 77
Kirton Cl. W4 .....49Ta 87
Kirton Gdns. E2 .....4K 219 (41Vb 91)
.....(not continuous)
Kirton Lodge SW18 .....58Db 111
Kirton Rd. E13 .....40Lc 73
Kirton Wlk. HA8: Edg 24Sa 47
Kirwyn Way SE5 .....52Rb 113
Kitcat Ter. E3 .....41Cc 92
.....(not continuous)
Kitchen Ct. E10 .....33Dc 72
Kitchener Av. DA12: Grav'nd 3E 144
Kitchener Cl. AL1: St A 3F 6
Kitchener Ho. SE18 .....52Qc 116
Kitchener Ho. SL9: Chal P 21A 42
Kitchener Rd. CR7: Thor H 69Tb 135
Kitchener Rd. E17 .....25Dc 52
Kitchener Rd. E7 .....37Kc 73
Kitchener Rd. N17 .....27Ub 51
Kitchener Rd. N2 .....27Gb 49
Kitchener Rd. RM10: Dag 37Dd 76
Kite Ho. SE1 .....49Xb 91
Kite Ho. SE3 .....56Kc 115
Kite Pl. E2 .....41Wb 91
.....(off Warner Pl.)
Kite Yd. SW11 .....53Hb 111
.....(off Cambridge Rd.)
Kitley Gdns. SE19 .....67Vb 135
Kitsmead La. KT16: Longc 4N 147
Kitsmead La. KT16: Vir W 4N 147
Kitson Rd. SE5 .....52Tb 113
Kitson Rd. SW13 .....53Wa 110
Kitswell Way WD7: R'lett 5Ha 14
Kitters Grn. WD5: Ab L 3U 12
Kittiwake Cl. CR2: Sels 82Ac 178
Kittiwake Ct. SE1 .....2E 230 (47Sb 91)
.....(off Gt. Dover St.)
Kittiwake Ct. SE8 .....51Bc 114
.....(off Abinger Gro.)
Kittiwake Ho. SL1: Slou 6J 81
Kittiwake Pl. SM1: Sutt 78Bb 155
Kittiwake Rd. UB5: N'olt 41Z 85
Kittiwake Way UB4: Yead 43Z 85
Kitto Rd. SE14 .....54Zb 114
KITT'S END 9Ab 16
Kitts End Rd. EN5: Barn 8Za 16
Kiver Rd. N19 .....33Mb 70
Klea Av. SW4 .....58Lb 112
Kleine Wharf N1 .....39Ub 71
Klein's Wharf E14 .....48Cc 92
.....(off Westferry Rd.)
Knapdale Cl. SE23 .....61Xb 135
KNAPHILL 9H 167
Knapmill Rd. SE6 .....61Cc 136
Knapmill Way SE6 .....61Dc 136
Knapp Cl. NW10 .....37Ua 68
Knapp Rd. E3 .....42Cc 92
Knapp Rd. TW15: Ashf 63P 127
Knapton M. SW17 .....65Jb 134
Knaresborough Dr. SW18 .....60Db 111
Knaresborough Pl. SW5 .....49Db 89
Knatchbull Rd. NW10 .....39Ta 67
Knatchbull Rd. SE5 .....54Rb 113
Knatts La. TN15: Knat 83Td 184
Knatts La. TN15: W King 83Td 184
KNATTS VALLEY 83Td 184
Knatts Valley Cvn. Pk. TN15: Knat 82Ud 184
Knatts Valley Rd. TN15: Knat 79Sd 164
Knave Wood Rd. TN15: Kems'g 89Nd 183
Knebworth Av. E17 .....25Cc 52
Knebworth Cl. EN5: New Bar 14Db 31
Knebworth Ho. SW8 .....54Mb 112
Knebworth Path WD6: Bore 14Ta 29
Knebworth Rd. N16 .....35Ub 71
Knee Hill SE2 .....49Yc 95
Knee Hill Cres. SE2 .....49Yc 95
Kneller Gdns. TW7: Isle 58Fa 108
Kneller Ho. UB5: N'olt 40Z 65
.....(off Academy Gdns.)
Kneller Rd. KT3: N Mald 73Ua 154
Kneller Rd. SE4 .....56Ac 114
Kneller Rd. TW2: Whitt 58Ea 108
Knevett Ter. TW3: Houn 56Ca 107
Knifton Ct. RM6: Pot B 3Zc 15
Knight Cl. RM8: Dag 33Yc 75
Knight Ct. E4 .....18Ec 34
.....(off The Ridgeway)
Knight Ct. N15 .....29Ub 51
Knighten St. E1 .....46Xb 91
Knighthead Point E14 .....47Cc 92
Knight Ho. SE17 .....6H 231 (49Ub 91)
.....(off Tatum St.)
Knightland Rd. E5 .....33Xb 71
Knightleas Ct. NW2 .....37Ya 68
Knightleys Ct. E10 .....32Ac 72
.....(off Wellington Rd.)
Knightley Wlk. SW18 .....56Cb 111
Knighton Cl. CR2: S Croy 81Rb 177
Knighton Cl. IG8: Wfd G 21Kc 53
Knighton Cl. RM7: Rom 30Fd 56
Knighton Dr. IG8: Wfd G 21Kc 53
Knighton Grn. IG9: Buck H 19Kc 35
Knighton La. IG9: Buck H 19Kc 35
Knighton Pk. Rd. SE26 .....64Zb 136
Knighton Pl. IG9: Buck H 19Kc 35
.....(off Knighton La.)
Knighton Pl. KT11: Cobh 88Aa 171
Knighton Rd. E7 .....34Jc 73
Knighton Rd. RH1: Redh 8A 208
Knighton Rd. RM7: Rom 30Ed 56
Knighton Rd. TN14: Otf 88Hd 182
Knighton Way La. UB9: Den 37K 63
Knightrider Ct. EC4 .....4D 224 (45Sb 91)
.....(off Knightrider St.)
Knightrider St. EC4 .....3C 224 (44Rb 91)
Knights Arc. SW1 .....2F 227 (47Hb 89)
.....(off Knightsbridge)
Knights Av. W5 .....47Na 87
Knightsbridge SW1 .....2F 227 (47Hb 89)
Knightsbridge SW7 .....2F 227 (47Hb 89)
KNIGHTSBRIDGE 2D 226 (47Gb 89)
Knightsbridge Apts., The SW7 .....2F 227 (47Hb 89)
.....(off Knightsbridge)
Knightsbridge Ct. BR2: Broml 72Nc 160
.....(off Wells Vw. Dr.)
Knightsbridge Ct. SL3: L'ly 49C 82
.....(off High St.)
Knightsbridge Ct. SW1 .....2G 227 (47Hb 89)
.....(off Sloane St.)
Knightsbridge Ct. WD18: Wat 14U 26
Knightsbridge Cres. TW18: Staines 65K 127
Knightsbridge Gdns. RM7: Rom 29Fd 56
Knightsbridge Grn. SW1 .....2F 227 (47Hb 89)
.....(not continuous)
Knightsbridge Station (Underground) 2G 227 (47Hb 89)
Knightsbridge Way HP2: Hem H 1N 3
Knights Cl. E9 .....36Yb 72
Knights Cl. KT8: W Mole 71Ba 151
Knights Cl. SL4: Wind 3B 102
Knights Cl. TW20: Egh 65F 126
Knights Community Stadium, The 77Cb 155
Knightscote Cl. UB9: Hare 26M 43
Knights Ct. BR1: Broml 62Hc 137
Knights Ct. KT1: King T 69Na 131
Knights Ct. WD23: B Hea 18Fa 28
Knights Cft. DA3: Nw A G 76Be 165
.....(not continuous)
Knights Fld. DA4: Eyns 76Nd 163
Knights Grn. WD3: Chor 13G 24
Knights Hill SE27 .....64Rb 135
Knight's Hill Sq. SE27 .....63Rb 135
Knight's Ho. SW10 .....52Eb 111
.....(off Hortensia Rd.)
Knight's Ho. W14 .....50Bb 89
.....(off Baron's Ct. Rd.)
Knights Ho. SW8 .....52Nb 112
.....(off Sth. Lambeth Rd.)
Knights Mnr. Way DA1: Dart 57Pd 119
Knights Mead KT16: Chert 73K 149
Knight's Orchard AL3: St A 2A 6
Knight's Pk. KT1: King T 69Na 131
Knight's Pl. TW2: Twick 60Ga 108
Knights Pl. RH1: Redh 5A 208
Knights Pl. SL4: Wind 4G 102
Knights Place Farm Equestrian Cen. 9P 145
Knights Ridge BR6: Chels 78Xc 161
Knight's Rd. E16 .....47Jc 93
Knights Rd. HA7: Stan 21La 46
Knights Twr. SE8 .....50Cc 92
Knight's Wlk. SE11 .....6B 230 (49Rb 91)
.....(not continuous)
Knights Wlk. RM4: Abr 13Xc 37
Knights Way CM13: B'wood 20Ce 41
Knights Way IG6: Ilf 23Sc 54
Knights Way RH9: G'stone 3N 209
Knightswood GU21: Wok 10K 167
Knightswood Cl. HA8: Edg 19Sa 29
Knightswood Ct. N6 .....31Mb 70
Knightswood Ho. N12 .....23Eb 49
Knightswood Rd. RM13: Rain 40Jd 76
Knightwood Cl. RH2: Reig 8J 207
Knightwood Cres. KT3: N Mald 72Ua 154
Knipp Hill KT11: Cobh 85Ba 171
Knivet Rd. SW6 .....51Cb 111
KNOCKHALL 57Yd 120
Knockhall Chase DA9: Ghithe 57Xd 120
Knockhall Rd. DA9: Ghithe 58Yd 120
KNOCKHOLT 89Xc 181
Knockholt Cl. SM2: Sutt 82Db 175
Knockholt Main Rd. TN14: Knock 91Vc 201
KNOCKHOLT POUND 87Ad 181
Knockholt Rd. SE9 .....57Mc 115
Knockholt Rd. TN14: Hals 86Bd 181
Knockholt Station (Rail) 81Ad 181
KNOCKMILL 84Vd 184
Knock Mill La. TN15: W King 85Wd 184
Knole 98Nd 203
Knole, The DA13: Ist R 6A 144
Knole, The SE9 .....63Qc 138
Knole Academy Sports & Leisure Cen. 93Jd 202
Knole Cl. CR0: C'don 72Yb 158
Knole Ct. UB5: N'olt 41Y 85
.....(off Broomcroft Av.)
Knole Ga. DA15: Sidc 62Uc 138
Knole La. TN13: S'oaks 98Ld 203
Knole Pk. 97Md 203
Knole Pk. Golf Course 96Md 203
Knole Rd. DA1: Dart 59Jd 118
Knole Rd. TN13: S'oaks 95Md 203
Knole Way TN13: S'oaks 97Ld 203
Knole Wood SL5: S'dale 4C 146
Knoll, The BR2: Hayes 75Jc 159
Knoll, The BR3: Beck 67Dc 136
Knoll, The HA1: Harr 32Ha 66
Knoll, The HA5: Pinn 26Z 45
Knoll, The KT11: Cobh 85Ca 171
Knoll, The KT16: Chert 74H 149
Knoll, The KT22: Lea 93La 192
Knoll, The W13 .....43La 86
Knoll Ct. SE19 .....64Vb 135
.....(off Farquhar Rd.)
Knoll Cres. HA6: Nwood 26U 44
.....(not continuous)
Knoll Dr. N14 .....17Jb 32
Knolles Cres. AL9: Wel G 5D 8
Knoll Ho. NW8 .....40Eb 69
.....(off Carlton Hill)
Knollmead KT5: Surb 74Sa 153
Knoll Pk. Rd. KT16: Chert 74H 149
Knoll Ri. BR6: Orp 74Vc 161
Knoll Rd. DA14: Sidc 64Xc 139
Knoll Rd. DA5: Bexl 59Cd 118
Knoll Rd. SW18 .....57Eb 111
KNOLL RDBT. 93La 192
Knolls, The KT17: Eps D 88Ya 174
Knolls Cl. KT4: Wor Pk 76Xa 154
Knollys Cl. SW16 .....62Qb 134
Knolly's Ho. WC1 .....5F 217 (42Nb 90)
.....(off Tavistock Pl.)
Knollys Rd. SW16 .....62Pb 134
Knolton Way SL2: Slou 4M 81
Knot Ho. SE1 .....7K 225 (46Vb 91)
.....(off Brewery Sq.)
Knotley Way BR4: W W'ck 75Dc 158
Knottisford St. E2 .....41Yb 92
Knotts Grn. M. E10 .....30Dc 52
Knotts Grn. Rd. E10 .....30Dc 52
Knotts Pl. TN13: S'oaks 96Jd 202
Knowland M. CR7: Thor H 70Tb 135
Knowlden Ho. E1 .....45Yb 92
.....(off Cable St.)
Knowle, The KT20: Tad 93Ya 194
Knowle Av. DA7: Bex 52Ad 117
Knowle Cl. SW9 .....55Qb 112
Knowledge Ct. SW16 .....69Pb 134
Knowle Gdns. KT14: W Byf 85H 169
KNOWLE GREEN 64K 127
Knowle Grn. TW18: Staines 64J 127
Knowle Gro. GU25: Vir W 3N 147
Knowle Gro. Cl. GU25: Vir W 3N 147
Knowle Hill GU25: Vir W 3M 147
KNOWLE HILL 3M 147
Knowle Lodge CR3: Cat'm 95Wb 197
Knowle Pk. KT11: Cobh 88Aa 171
Knowle Pk. Av. TW18: Staines 65K 127
Knowle Rd. BR2: Broml 75Pc 160
Knowle Rd. TW2: Twick 60Ga 108
Knowles Cl. UB7: Yiew 46N 83
Knowles Ct. HA1: Harr 30Ha 46
Knowles Hill Cres. SE13 .....57Fc 115
Knowles Ho. SW18 .....58Db 111
.....(off Neville Gill Cl.)
Knowles Wlk. SW4 .....55Lb 112
Knowles Wharf NW1 .....39Lb 70
.....(off St Pancras Way)
Knowl Hill GU22: Wok 91D 188
Knowl Pk. WD6: E'tree 15Na 29
Knowlton Cotts. RM15: S Ock 43Yd 98
Knowlton Grn. BR2: Broml 71Hc 159
Knowlton Ho. SW9 .....53Qb 112
.....(off Cowley Rd.)
Knowl Way WD6: E'tree 15Pa 29
Knowl Wood La. BR6: Farnb 75Qc 160
Knowsley Av. UB1: S'hall 46Da 85
Knowsley Rd. SW11 .....54Hb 111
Knox Ct. SW4 .....54Nb 112
Knox Rd. E7 .....37Hc 73
Knox St. W1 .....7F 215 (43Hb 89)
Knoyle Ho. W14 .....48Ab 88
.....(off Russell Rd.)
Knoyle St. SE14 .....51Ac 114
Knutsford Av. WD24: Wat 10Z 13
Koblenz Ho. N8 .....27Nb 50
.....(off Newland Rd.)
Kohat Rd. SW19 .....64Db 133
Koh-I-Noor Av. WD23: Bush 16Ca 27
Kola Ct. SL2: Slou 4M 81
Koonowla Cl. TN16: Big H 87Mc 179
Koops Mill M. SE1 .....48Vb 91
Kooringa CR6: W'ham 91Xb 197
Korda Cl. TW17: Shep 69P 127
Korda Cl. WD6: Bore 12Sa 29
Korea Cotts. KT11: Cobh 88Z 171
Kossuth St. SE10 .....50Gc 93
Kotan Dr. TW18: Staines 63E 126
Kotata Ho. E20 .....37Ec 72
.....(off Ravens Wlk.)
Kotree Way SE1 .....49Wb 91
Kramer M. SW5 .....50Cb 89
Kreedman Wlk. E8 .....36Wb 71
Kreisel Wlk. TW9: Kew 51Pa 109
Kristina Ct. SM2: Sutt 79Cb 155
.....(off Overton Rd.)
Krithia Rd. RM9: Dag 39Xc 75
Krupnik Pl. EC2 .....5J 219 (42Ub 91)
.....(shown as Curtain Pl.)
Kuala Gdns. SW16 .....67Pb 134
Kubrick Bus. Est. E7 .....35Kc 73
.....(off Station App.)
Kuflink Stadium 57Ce 121
Kuhn Way E7 .....36Jc 73
Kurdish Mus. 49Wa 88
Kwame Ho. E16 .....45Rc 94
.....(off University Way)
Kwesi M. SE27 .....64Qb 134
Kydbrook Cl. BR5: Pet W 73Sc 160
Kyle Ho. NW6 .....39Cb 69
Kylemore Cl. E6 .....40Mc 73
Kylemore Rd. NW6 .....38Cb 69
Kylestrome Ho. SW1 .....6J 227 (49Jb 90)
.....(off Cundy St.)
Kymberley Rd. HA1: Harr 30Ga 46
Kyme Rd. RM11: Horn 30Hd 56
Kymes Ct. HA2: Harr 33Fa 66
Kynance Cl. RM3: Rom 21Ld 57
Kynance Gdns. HA7: Stan 25La 46
Kynance M. SW7 .....48Db 89
Kynance Pl. SW7 .....4A 226 (48Eb 89)
Kynaston Av. CR7: Thor H 71Sb 157
Kynaston Av. N16 .....34Vb 71
Kynaston Cl. HA3: Hrw W 24Fa 46
Kynaston Ct. CR3: Cat'm 97Ub 197
Kynaston Cres. CR7: Thor H 71Sb 157
Kynaston Rd. BR1: Broml 64Jc 137
Kynaston Rd. BR5: Orp 73Xc 161
Kynaston Rd. CR7: Thor H 71Sb 157
Kynaston Rd. EN2: Enf 11Tb 33
Kynaston Rd. N16 .....34Ub 71
Kynaston Wood HA3: Hrw W 24Fa 46
Kynersley Cl. SM5: Cars 76Hb 155
Kynoch Ct. SS17: Stan H 2N 101
Kynoch Rd. N18 .....21Yb 52
Kyrkly Ct. RM19: Purf 50Rd 97
.....(off Linnet Way)
Kyrle Rd. SW11 .....58Jb 112
Kytes Dr. WD25: Wat 5Z 13
Kytes Est. WD25: Wat 5Z 13
Kyverdale Rd. N16 .....31Vb 71

L

Laban Cen. 51Dc 114
Laban Wlk. SE8 .....51Dc 114
.....(off Copperas St.)
Laboratory Spa & Health Club, The 27Lb 50
Laboratory Sq. SE18 .....48Rc 94
Labour in Vain Rd. TN15: Stans 85Zd 185
Labour in Vain Rd. TN15: Wro 85Zd 185
Laburnham Cl. EN5: Barn 13Bb 31
Laburnham Cl. RM14: Upm 31Wd 78
Laburnham Gdns. RM14: Upm 31Wd 78
Laburnum Av. BR8: Swan 69Ed 140
Laburnum Av. DA1: Dart 60Ld 119
Laburnum Av. N17 .....24Tb 51
Laburnum Av. N9 .....19Vb 33
Laburnum Av. RM12: Horn 33Jd 76
Laburnum Av. SM1: Sutt 76Gb 155
Laburnum Av. UB7: Yiew 45P 83
Laburnum Cl. E4 .....23Bc 52
Laburnum Cl. EN8: Chesh 3Zb 20
Laburnum Cl. HA0: Wemb 39Qa 67
Laburnum Cl. N11 .....23Jb 50
Laburnum Cl. SE15 .....52Yb 114
Laburnum Ct. E2 .....1K 219 (39Vb 71)
Laburnum Ct. HA1: Harr 29Da 45
Laburnum Ct. HA7: Stan 21La 46
Laburnum Ct. SE16 .....47Yb 92
.....(off Albion St.)
Laburnum Ct. UB8: Uxb 37L 63
.....(off Harefield Rd.)
Laburnum Cres. TW16: Sun 67X 129
Laburnum Gdns. CR0: C'don 73Zb 158
Laburnum Gdns. N21 .....19Sb 33
Laburnum Gro. AL2: Chis G 7P 5

Laburnum Gro. DA11: Nflt 59Fe 121
Laburnum Gro. HA4: Ruis 30T 44
Laburnum Gro. KT3: N Mald 68Ta 131
Laburnum Gro. N21 19Sb 33
Laburnum Gro. NW9 31Sa 67
Laburnum Gro. RM15: S Ock 41Yd 98
Laburnum Gro. SL3: L'ly 51D 104
Laburnum Gro. TW3: Houn 56Ba 107
Laburnum Gro. UB1: S'hall 42Ba 85
LABURNUM HEALTH CEN. 33Cd 76
(off Bradwell Av.)
Laburnum Ho. BR2: Broml 67Fc 137
Laburnum Ho. RM10: Dag 33Cd 76
Laburnum La. E2 1K 219 (40Vb 71)
Laburnum Lodge N3 26Bb 49
Laburnum Pl. SE9 57Qc 116
Laburnum Pl. TW20: Eng G 5M 125
Laburnum Rd. CM16: Coop 1Yc 23
Laburnum Rd. CR4: Mitc 68Jb 134
Laburnum Rd. GU22: Wok 2P 187
Laburnum Rd. KT16: Chert 74J 149
Laburnum Rd. KT18: Eps 85Ua 174
Laburnum Rd. SW19 66Eb 133
Laburnum Rd. UB3: Harl 49V 84
Laburnums, The E6 42Nc 94
Laburnum St. E2 1K 219 (39Vb 71)
Laburnum Wlk. RM12: Horn 36Ld 77
Laburnum Way BR2: Broml 73Qc 160
Laburnum Way TW19: Stanw 60P 105
Labyrinth Twr. E8 37Vb 71
(off Dalston Sq.)
Lacebark Cl. DA15: Sidc 59Vc 117
Lace Cl. SM6: W'gton 75Kb 156
Lace Ct. E1 43Zb 92
(off Master's St.)
Laceman M. RM5: Col R 23Ed 56
Lacewing Cl. E13 41Jc 93
Lacey Av. CR5: Coul 92Qb 196
Lacey Cl. N9 19Wb 33
Lacey Cl. TW20: Egh 66F 126
Lacey Dr. CR5: Coul 92Rb 197
Lacey Dr. HA8: Edg 21Na 47
Lacey Dr. RM8: Dag 34Yc 75
Lacey Dr. TW12: Hamp 67Ba 129
Lacey Grn. CR5: Coul 92Qb 196
Lacey Gro. UB10: Uxb 40N 63
Lacey M. E3 40Cc 72
Lacine Ct. SE16 47Zb 92
(off Christopher Cl.)
Lackford Rd. CR5: Chips 90Hb 175
Lackington St. EC2 7G 219 (43Tb 91)
Lackland Ho. SE1 7K 231 (50Vb 91)
(off Rowcross St.)
Lackmore Rd. EN1: Enf 7Yb 20
Lacland Ho. SW10 52Fb 111
(off Worlds End Est.)
Lacock Cl. SW19 65Eb 133
Lacock Ct. W13 46Ja 86
Lacon Ho. WC1 7H 217 (43Pb 90)
(off Theobald's Rd.)
Lacon Rd. SE22 56Wb 113
Lacrosse Way SW16 67Mb 134
Lacy Rd. SW15 56Za 110
Ladas Rd. SE27 63Sb 135
Ladbroke Cotts. RH1: Redh 5A 208
(off Ladbroke Rd.)
Ladbroke Ct. E1 2K 225 (44Vb 91)
Ladbroke Ct. RH1: Redh 4A 208
Ladbroke Cres. W11 44Ab 88
Ladbroke Gdns. W11 45Bb 89
Ladbroke Gro. RH1: Redh 5A 208
Ladbroke Gro. W10 42Za 88
Ladbroke Gro. W11 44Ab 88
Ladbroke Gro. Ho. W11 45Bb 89
(off Ladbroke Gro.)
Ladbroke Grove Memorial 42Za 88
Ladbroke Grove Station (Underground) 44Ab 88
Ladbroke M. W11 46Ab 88
Ladbroke Rd. EN1: Enf 16Vb 33
Ladbroke Rd. KT18: Eps 86Ta 173
Ladbroke Rd. RH1: Redh 5A 208
Ladbroke Rd. W11 46Bb 89
Ladbroke Sq. W11 45Bb 89
Ladbroke Ter. W11 45Bb 89
Ladbroke Wlk. W11 46Bb 89
Ladbrook Cl. BR1: Broml 65Gc 137
Ladbrook Cl. HA5: Pinn 29Ba 45
Ladbrooke Cl. EN6: Pot B 4Cb 17
Ladbrooke Cres. DA14: Sidc 62Zc 139
Ladbrooke Dr. EN6: Pot B 4Cb 17
Ladbrooke Rd. SL1: Slou 8G 80
Ladbrook Rd. SE25 70Tb 135
Ladderstile Ride KT2: King T 64Ra 131
Ladderswood Way N11 22Lb 50
Ladds Way BR8: Swan 70Fd 140
Ladies Gro. AL3: St A 1P 5
(not continuous)
Ladlands SE22 59Wb 113
Lady Anne Ct. E18 25Jc 53
(off Queen Mary Av.)
Lady Astor Ct. SL1: Slou 7J 81
Lady Aylesford Av. HA7: Stan 22Ja 46
Lady Booth Rd. KT1: King T 68Na 131
Lady Craig Ct. UB8: Hil 43R 84
Ladycroft Gdns. BR6: Farnb 78Sc 160
Ladycroft Rd. SE13 55Dc 114
Ladycroft Wlk. HA7: Stan 25Ma 47
Ladycroft Way BR6: Farnb 78Sc 160
Ladyday Pl. SL1: Slou 6G 80
Lady Dock Path SE16 47Ac 92
Lady Elizabeth Ho. SW14 55Sa 109
Ladyfern Ho. E3 43Cc 92
(off Gail St.)
Ladyfields DA11: Nflt 3B 144
Ladyfields IG10: Lough 14Sc 36
Ladyfields Cl. IG10: Lough 14Sc 36
Lady Florence Ctyd. SE8 52Cc 114
(off Reginald Sq.)
Lady Forsdyke Way KT19: Eps 81Qa 173
Ladygate Bowls Club 30S 44
(off Ladygate La.)
Ladygate La. HA4: Ruis 30R 44
Ladygrove CR0: Sels 81Ac 178
Lady Harewood Way KT19: Eps 81Qa 173
Lady Hay KT4: Wor Pk 75Va 154
Lady Jane Ct. KT2: King T 68Pa 131
(off Cambridge Rd.)
Lady Jane Pl. DA1: Dart 54Pd 119
Lady Margaret Ho. SE17 51Tb 113
(off Queen's Row)
Lady Margaret Rd. N19 35Lb 70
Lady Margaret Rd. NW5 36Lb 70
Lady Margaret Rd. SL5: S'dale 4D 146
Lady Margaret Rd. UB1: S'hall 45Ba 85
Lady May Ho. SE5 52Sb 113
(off Pitman St.)
Lady Mdw. WD4: K Lan 9M 3
Lady Micos Almshouses E1 44Yb 92
(off Aylward St.)
Lady Sarah Cohen Ho. N11 23Hb 49
(off Asher Loftus Way)
Lady's Chapel's Lock 6S 12
(off Old Mill Rd.)
Lady's Cl. WD18: Wat 14Y 27
Lady Shaw Ct. N13 19Pb 32
Ladyship Ter. SE22 59Wb 113
Ladysmith Av. E6 40Nc 74
Ladysmith Av. IG2: Ilf 31Uc 74
Ladysmith Cl. NW7 24Wa 48
Ladysmith Rd. AL3: St A 1B 6
Ladysmith Rd. E16 41Hc 93
Ladysmith Rd. EN1: Enf 13Ub 33
(not continuous)
Ladysmith Rd. HA3: W'stone 26Ga 46
Ladysmith Rd. N17 26Wb 51
Ladysmith Rd. N18 22Xb 51
Ladysmith Rd. SE9 58Qc 116
Lady Somerset Rd. NW5 35Kb 70
Lady Spencer's Gro. AL1: St A 3A 6
Lady Spencer's Gro. AL3: St A 3A 6
Lady's Wlk. TN15: Igh 96Wd 204
Ladythorpe Cl. KT15: Add 77K 149
Ladywalk WD3: Map C 22G 42
LADYWELL 57Dc 114
Ladywell Arena (Running Track) 58Cc 114
Ladywell Cl. SE4 57Cc 114
Ladywell Hgts. SE4 58Bc 114
Ladywell Rd. SE13 57Cc 114
Ladywell Station (Rail) 57Dc 114
Ladywell St. E15 39Hc 73
Ladywell Water Twr. SE4 57Cc 114
Ladywood Av. BR5: Pet W 71Uc 160
Ladywood Cl. WD3: Loud 13K 25
Ladywood Rd. DA2: Daren 64Ud 142
Ladywood Rd. KT6: Surb 75Qa 153
Lady Yorke Pk. SL0: Iver H 37F 62
Laelia Ho's. AL1: St A 3E 6
Lafone Av. TW13: Felt 61Y 129
Lafone St. SE1 1K 231 (47Vb 91)
Lagado M. SE16 46Zb 92
Lagare Apts. SE1 1C 230 (47Rb 91)
(off Surrey Row)
Lagham Pk. RH9: S God 9C 210
Lagham Rd. RH9: S God 10C 210
Laglands Cl. RH2: Reig 4L 207
Lagonda Av. IG6: Ilf 23Vc 55
Lagonda Ho. E3 42Cc 92
(off Tidworth Rd.)
Lagonda Way DA1: Dart 56Ld 119
Lagonier Ho. EC1 4E 218 (41Sb 91)
(off Ironmonger Row)
Lagoon Rd. BR5: St M Cry 71Yc 161
Laguna Ct. AL1: St A 2C 6
(off Beaconsfield Rd.)
Laharna Trad. Est. WD24: Wat 10Y 13
Laidlaw Dr. N21 15Pb 32
Laing Cl. IG6: Ilf 23Tc 54
Laing Dean UB5: N'olt 39Y 65
Laing Ho. SE5 52Sb 113
Laings Av. CR4: Mitc 68Hb 133
Lainlock Pl. TW3: Houn 53Da 107
Lainson St. SW18 59Cb 111
Lairdale Cl. SE21 60Sb 113
Laird Av. RM16: Grays 47Fe 99
Laird Ho. SE5 52Sb 113
(off Redcar St.)
Lairs Cl. N7 36Nb 70
Lait Ho. BR3: Beck 67Dc 136
Laitwood Rd. SW12 60Kb 112
Lakanal SE5 53Ub 113
(off Sceaux Gdns.)
Lake, The WD23: B Hea 18Fa 28
Lake Av. BR1: Broml 65Jc 137
Lake Av. RM13: Rain 40Md 77
Lake Av. SL1: Slou 5H 81
Lake Bus. Cen. N17 24Wb 51
Lake Cl. KT14: Byfl 84M 169
Lake Cl. RM8: Dag 34Zc 75
Lake Cl. SW19 64Bb 133
Lakedale Ct. IG11: Bark 42Xc 95
Lakedale Rd. SE18 51Uc 116
Lake Dr. WD23: B Hea 19Fa 28
LAKE END 7A 80
Lake End Rd. SL4: Dor 7A 80
Lake Farm Country Pk. 46U 84
Lakefield Cl. SE20 66Xb 135
Lakefield Rd. N22 26Rb 51
Lakefields Cl. RM13: Rain 40Md 77
Lake Gdns. RM10: Dag 36Cd 76
Lake Gdns. SM6: W'gton 76Kb 156
Lake Gdns. TW10: Ham 61Ka 130
Lakehall Gdns. CR7: Thor H 71Rb 157
Lakehall Rd. CR7: Thor H 71Rb 157
Lake Ho. SE1 2D 230 (47Sb 91)
(off Southwark Bri. Rd.)
Lake Ho. Rd. E11 34Jc 73
Lakehurst Rd. KT19: Ewe 78Ua 154
Lakeland Cl. HA3: Hrw W 23Fa 46
Lakeland Cl. IG7: Chig 21Xc 55
Lakeman Ho. SL9: Chal P 22B 42
Lakenham Pl. E3 42Bc 92
Lakenheath N14 15Lb 32
Laker Ct. SW4 53Nb 112
Laker Ho. E16 47Kc 93
Laker Ind. Est. BR3: Beck 64Ac 136
Lake Ri. RM1: Rom 26Hd 56
Lake Ri. RM20: W Thur 49Wd 98
Lake Rd. CR0: C'don 75Bc 158
Lake Rd. E10 31Dc 72
Lake Rd. GU25: Vir W 1M 147
Lake Rd. RM6: Chad H 28Zc 55
Lake Rd. RM9: Dag 41Dd 96
Lake Rd. SW19 64Bb 133
Laker Pl. SW15 58Ab 110
Lakers Ri. SM7: Bans 88Gb 175
Lakeside BR3: Beck 69Dc 136
Lakeside EN2: Enf 14Mb 32
Lakeside GU21: Wok 1J 187
Lakeside KT13: Weyb 75U 150
Lakeside KT19: Ewe 79Ua 154
Lakeside KT2: King T 66Ra 131
Lakeside N3 26Db 49
Lakeside RH1: Redh 4A 208
Lakeside RM13: Rain 40Nd 77
Lakeside SM6: W'gton 77Kb 156
Lakeside W13 44La 86
Lakeside Av. IG4: Ilf 28Mc 53
Lakeside Av. SE28 46Wc 95
Lakeside Bus. Village RM16: Chaf H 49Xd 98
(off Fleming Rd.)
Lakeside Bus Station 48Xd 98
Lakeside Cl. DA15: Sidc 57Yc 117
Lakeside Cl. GU21: Wok 1J 187
Lakeside Cl. HA4: Ruis 28T 44
Lakeside Cl. IG7: Chig 21Vc 55
Lakeside Cl. SE25 68Wb 135
Lakeside Ct. N4 33Sb 71
Lakeside Ct. WD6: E'tree 15Qa 29
Lakeside Cres. CM14: B'wood 20Zd 41
Lakeside Cres. EN4: E Barn 15Hb 31
Lakeside Dr. BR2: Broml 76Nc 160
Lakeside Dr. GU24: Chob 5J 167
Lakeside Dr. KT10: Esh 79Ea 152
Lakeside Dr. NW10 41Pa 87
Lakeside Dr. SL2: Stoke P 9J 61
Lakeside Grange KT13: Weyb 76S 150
Lakeside Ind. Est. SL3: Coln 51J 105
Lakeside Karting 47Xd 98
Lakeside Leisure Pk. 49Wd 98
Lakeside Pk. KT16: Chert 74K 149
Lakeside Pl. AL2: Lon C 9H 7
Lakeside Retail Pk. Thurrock 49Wd 98
Lakeside Rd. N13 21Pb 50
Lakeside Rd. SL0: Rich P 51J 105
Lakeside Rd. SL3: Coln 52H 105
Lakeside Rd. SL3: Rich P 52H 105
Lakeside Rd. W14 48Za 88
Lakeside Ter. EC2 7E 218 (43Sb 91)
(off Silk St.)
Lakeside Way HA9: Wemb 35Qa 67
Lakes Rd. BR2: Kes 78Lc 159
Lakestreet Grn. RH8: Limp 1P 211
Lakeswood Rd. BR5: Pet W 72Rc 160
Lake Vw. EN6: Pot B 5Eb 17
(not continuous)
Lake Vw. HA8: Edg 22Pa 47
Lake Vw. WD4: K Lan 10B 4
Lakeview Ct. SE28 45Xc 95
Lake Vw. Est. E3 40Ac 72
Lakeview Pk. RM3: Rom 20Ld 39
Lake Vw. Rd. TN13: S'oaks 95Jd 202
Lakeview Rd. DA16: Well 56Xc 117
Lakeview Rd. SE27 64Qb 134
Lake Vw. Ter. N18 21Vb 51
(off Sweet Briar Wlk.)
Lakewood KT10: Esh 83Ba 171
Lakin Cl. SM5: Cars 77Jb 156
Lakis Cl. NW3 35Eb 69
LALEHAM 69L 127
Laleham Abbey TW18: Lale 70L 127
Laleham Av. NW7 20Ta 29
Laleham Camping Club TW18: Lale 71L 149
Laleham Cl. TW18: Staines 67K 127
Laleham Ct. GU21: Wok 88A 168
Laleham Ct. SM1: Sutt 78Eb 155
Laleham Ho. E2 5K 219 (42Vb 91)
(off Camlet St.)
Laleham Pk. 70L 127
Laleham Reach KT16: Chert 69J 127
LALEHAM REACH 69J 127
Laleham Rd. SE6 59Ec 114
Laleham Rd. TW17: Shep 70P 127
Laleham Rd. TW18: Staines 64H 127
Lalor St. SW6 54Ab 110
Lalsham Ho. WD19: Wat 20Y 27
Lamb All. AL1: St A 2B 6
(off Chequer St.)
Lambarde Av. SE9 63Qc 138
Lambarde Dr. TN13: S'oaks 95Jd 202
Lambarde Rd. TN13: S'oaks 94Jd 202
Lambardes DA3: Nw A G 76Be 165
Lambardes Cl. BR6: Prat B 83Yc 181
Lambarde Sq. SE10 50Hc 93
Lambard Ho. SE10 52Ec 114
(off Langdale Rd.)
Lamb Cl. AL10: Hat 1D 8
Lamb Cl. RM18: Tilb 4E 122
Lamb Cl. UB5: N'olt 41Aa 85
Lamb Cl. WD25: Wat 6Y 13
Lamb Ct. E14 45Ac 92
(off Narrow St.)
Lamberhurst Cl. BR5: Orp 74Zc 161
Lamberhurst Ho. SE15 51Yb 114
Lamberhurst Rd. RM8: Dag 32Bd 75
Lamberhurst Rd. SE27 63Qb 134
Lambert Av. SL3: L'ly 48A 82
Lambert Av. TW9: Rich 55Qa 109
Lambert Cl. TN16: Big H 88Mc 179
Lambert Cotts. RH1: Blet 5L 209
Lambert Ct. DA8: Erith 51Ed 118
(off Park Cres.)
Lambert Ct. KT13: Weyb 80P 149
Lambert Ct. WD23: Bush 14Z 27
Lambert Ho. SW11 55Fb 111
(off Gartons Way)
Lambert Jones M. EC2 7D 218 (43Sb 91)
(off Beech St.)
Lambert Lodge TW8: Bford 50Ma 87
(off Layton Rd.)
Lambert M. DA13: Sflt 65Ce 143
Lambert M. N12 22Eb 49
(off Lambert Way)
Lamberton Ct. WD6: Bore 11Qa 29
(off Gateshead Rd.)
Lambert Rd. E16 44Kc 93
Lambert Rd. N12 22Fb 49
Lambert Rd. SM7: Bans 86Cb 175
Lambert Rd. SW2 57Nb 112
Lambert's Pl. CR0: C'don 74Tb 157
Lamberts Rd. KT5: Surb 71Na 153
Lambert St. N1 38Qb 70
Lambert Wlk. HA9: Wemb 34Na 67
Lambert Way N12 22Eb 49
LAMBETH 4J 229 (48Pb 90)
Lambeth Bri. SW1 5G 229 (49Nb 90)
LAMBETH COMMUNITY CARE CEN. 5A 230 (49Qb 90)
(off Monkton St.)
Lambeth Crematorium 63Eb 133
Lambeth High St. SE1 6H 229 (49Pb 90)
Lambeth Hill EC4 4D 224 (45Sb 91)
LAMBETH HOSPITAL 55Pb 112
Lambeth North Station (Underground) 3K 229 (48Qb 90)
Lambeth Palace 4H 229 (48Pb 90)
Lambeth Pal. Rd. SE1 4H 229 (48Pb 90)
Lambeth Pier 4G 229 (48Nb 90)
Lambeth Rd. CR0: C'don 73Qb 156
Lambeth Rd. SE1 5H 229 (49Pb 90)
Lambeth Rd. SE11 4J 229 (48Pb 90)
Lambeth Towers SE11 4K 229 (48Pb 90)
(off Kennington Rd.)
Lambeth Wlk. SE11 6J 229 (49Pb 90)
(not continuous)
Lambfold Ho. N7 37Nb 70
(off North Rd.)
Lamb Ho. SE10 51Ec 114
(off Haddo St.)
Lamb Ho. SE5 52Sb 113
(off Elmington Est.)
Lambkins M. E17 28Ec 52
Lamb La. E8 38Xb 71
Lamble St. NW5 36Jb 70
Lambley Rd. RM9: Dag 37Xc 75
Lambly Hill GU25: Vir W 69A 126
Lambolle Pl. NW3 37Gb 69
Lambolle Rd. NW3 37Gb 69
Lamborne Pl. UB10: Ick 34R 64
Lambourn Chase WD7: R'lett 8Ha 14
Lambourn Cl. CR2: S Croy 81Rb 177
Lambourn Cl. NW5 35Lb 70
Lambourn Cl. W7 47Ha 86
Lambourne RM18: E Til 9L 101
LAMBOURNE 14Zc 37
Lambourne Av. SW19 63Bb 133
Lambourne Cl. DA4: Farni 73Qd 163
Lambourne Cl. IG7: Chig 20Xc 37
Lambourne Cl. SL1: Burn 10A 60
Lambourne Cl. SM7: Bans 87Za 174
Lambourne Ct. IG8: Wfd G 24Lc 53
Lambourne Ct. UB8: Uxb 39K 63
Lambourne Cres. GU21: Wok 85F 168
Lambourne Cres. IG7: Chig 19Xc 37
Lambourne Dr. CM13: Hut 17Fe 41
Lambourne Dr. KT11: Cobh 87Z 171
LAMBOURNE END 17Ad 37
Lambourne End Outdoor Cen. 17Ad 37
Lambourne Gdns. E4 19Cc 34
Lambourne Gdns. EN1: Enf 12Vb 33
Lambourne Gdns. IG11: Bark 38Vc 75
Lambourne Gdns. RM12: Horn 33Md 77
Lambourne Golf Course, The 7A 60
Lambourne Gro. SE16 50Zb 92
Lambourne Ho. NW8 7C 214 (43Fb 89)
(off Broadley St.)
Lambourne Ho. SE20 67Wb 135
Lambourne Pl. SE3 53Kc 115
Lambourne Rd. E11 31Ec 72
Lambourne Rd. IG11: Bark 38Uc 74
Lambourne Rd. IG3: Ilf 33Uc 74
Lambourne Rd. IG7: Chig 21Vc 55
Lambourne Sq. RM4: Abr 18Yc 37
Lambourn Gro. KT1: King T 68Ra 131
LAMBOURN GROVE 3H 7
Lambourn Rd. SW4 55Kb 112
Lambrook Ho. SE15 53Wb 113
Lambrook Ter. SW6 53Ab 110
Lamb's Bldgs. EC1 6F 219 (42Tb 91)
Lambs Bus. Pk. RH9: S God 10P 209
Lamb's Cl. N9 19Wb 33
Lambs Cl. EN6: Cuff 1Pb 18
Lamb's Conduit Pas. WC1 7H 217 (43Pb 90)
Lamb's Conduit St. WC1 6H 217 (42Pb 90)
(not continuous)
Lambscroft Av. SE9 62Lc 137
Lambscroft Way SL9: Chal P 26A 42
Lamb's La. Nth. RM13: Rain 42Md 97
Lamb's La. Sth. RM13: Rain 43Kd 97
Lambs Mdw. IG8: Wfd G 26Mc 53
Lamb's M. N1 1B 218 (39Rb 71)
Lambs Pas. EC1 6F 219 (42Tb 91)
Lambs Ter. N9 19Tb 33
Lamb St. E1 7K 219 (43Vb 91)
Lamb's Wlk. EN2: Enf 12Sb 33
Lambton Av. EN8: Walt C 5Zb 20
Lambton Ho. SL4: Wind 5E 102
Lambton M. N19 32Nb 70
(off Lambton Rd.)
Lambton Pl. W11 45Bb 89
Lambton Rd. N19 32Nb 70
Lambton Rd. SW20 67Ya 132
Lamb Wlk. SE1 2H 231 (47Ub 91)
Lamerock Rd. BR1: Broml 63Hc 137
Lamerton Rd. IG6: Ilf 26Rc 54
Lamerton St. SE8 51Cc 114
Lamford Cl. N17 24Tb 51
Lamington St. W6 49Xa 88
Lamlash St. SE11 5B 230 (49Rb 91)
Lamley Ho. SE10 52Dc 114
(off Ashburnham Pl.)
Lammas Av. CR4: Mitc 68Jb 134
Lammas Av. SL4: Wind 4G 102
Lammas Cl. TW18: Staines 62G 126
Lammas Ct. SL4: Wind 4G 102
Lammas Ct. TW19: Staines 61F 126
Lammas Dr. TW18: Staines 63F 126
Lammas Grn. SE26 62Xb 135
Lammas Hill KT10: Esh 77Da 151
Lammas La. KT10: Esh 77Ba 151
Lammas La. KT10: Hers 77Ba 151
Lammas Pk. Gdns. W5 46La 86
Lammas Pk. Rd. W5 47Ma 87
Lammas Rd. E10 33Ac 72
Lammas Rd. E9 38Zb 72
Lammas Rd. SL1: Slou 3B 80
Lammas Rd. TW10: Ham 63La 130
Lammas Rd. WD18: Wat 15Y 27
Lammermoor Rd. SW12 59Kb 112
Lamont Rd. SW10 51Fb 111
Lamont Rd. Pas. SW10 51Fb 111
(off Lamont Rd.)
LAMORBEY 60Vc 117
Lamorbey Cl. DA15: Sidc 60Vc 117
Lamorbey Pk. 60Xc 117
Lamorna Av. DA12: Grav'nd 1F 144
Lamorna Cl. BR6: Orp 73Wc 161
Lamorna Cl. E17 26Ec 52
Lamorna Cl. WD7: R'lett 6Ka 14
Lamorna Gro. HA7: Stan 25Ma 47
Lampard Gro. N16 32Vb 71
Lampern Sq. E2 41Wb 91
Lampeter Cl. GU22: Wok 90A 168
Lampeter Cl. NW9 30Ua 48
Lampeter Ho. RM3: Rom 24Nd 57
(off Kingsbridge Cir.)
Lampeter Sq. W6 51Ab 110
Lamplighter Cl. E1 42Yb 92
Lamplighters Cl. DA1: Dart 58Pd 119
Lamplighters Cl. EN9: Walt A 6Jc 21
Lampmead Rd. SE12 57Hc 115
Lamp Office Ct. WC1 6H 217 (42Pb 90)
(off Conduit St.)
Lamport Cl. SE18 49Pc 94
LAMPTON 53Da 107
Lampton Av. TW3: Houn 53Da 107
Lampton Ct. TW3: Houn 53Da 107
Lampton Ho. Cl. SW19 63Za 132
Lampton Pk. Rd. TW3: Houn 54Da 107
Lampton Rd. TW3: Houn 54Ca 107
Lamsey Rd. HP3: Hem H 4M 3
Lamson Rd. RM13: Rain 42Hd 96
(not continuous)
Lanacre Av. NW9 25Ta 47
Lanadron Cl. TW7: Isle 54Ha 108
Lanain Ct. SE12 59Hc 115
Lanark Cl. W5 43La 86
Lanark Ct. UB5: N'olt 36Ca 65
(off Newmarket Av.)
Lanark Ho. SE1 50Wb 91
(off Old Kent Rd.)
Lanark Mans. W12 47Ya 88
(off Pennard Rd.)
Lanark Mans. W9 5A 214 (42Eb 89)
(off Lanark Rd.)
Lanark M. W9 4A 214 (41Eb 89)
Lanark Pl. W9 5A 214 (42Eb 89)
Lanark Rd. W9 40Db 69
Lanark Sq. E14 48Dc 92
Lanata Wlk. UB4: Yead 42Z 85
(off Alba Cl.)
Lanbury Rd. SE15 56Zb 114
Lancashire Ct. W1 4K 221 (45Kb 90)
Lancaster Av. CR4: Mitc 71Nb 156
Lancaster Av. E18 28Kc 53
Lancaster Av. EN4: Had W 10Eb 17
Lancaster Av. IG11: Bark 38Uc 74
Lancaster Av. SE27 61Rb 135
Lancaster Av. SL2: Slou 2G 80
Lancaster Av. SW19 64Za 132
Lancaster Cl. BR2: Broml 70Hc 137
Lancaster Cl. CM15: Pil H 15Wd 40
Lancaster Cl. GU21: Wok 88C 168
Lancaster Cl. KT2: King T 64Ma 131
Lancaster Cl. N1 38Ub 71
Lancaster Cl. N17 24Wb 51
Lancaster Cl. NW9 24Va 48
Lancaster Cl. TW15: Ashf 63N 127
Lancaster Cl. TW19: Stanw 58N 105
Lancaster Cl. TW20: Eng G 4P 125
Lancaster Cl. W2 45Db 89
(off St Petersburgh Pl.)
Lancaster Cotts. TW10: Rich 58Na 109
Lancaster Ct. DA12: Grav'nd 2E 144
Lancaster Ct. KT12: Walt T 73W 150
Lancaster Ct. KT19: Ewe 82Ta 173
Lancaster Ct. SE27 61Rb 135
Lancaster Ct. SM2: Sutt 80Cb 155
(off Mulgrave Rd.)
Lancaster Ct. SM7: Bans 86Bb 175
Lancaster Ct. SW6 52Bb 111
Lancaster Ct. TW19: Stanw 60N 105
Lancaster Ct. W2 4A 220 (45Eb 89)
(off Lancaster Ga.)
Lancaster Dr. E14 46Ec 92
Lancaster Dr. HP3: Bov 9B 2
Lancaster Dr. IG10: Lough 16Nc 36
Lancaster Dr. NW3 37Gb 69
Lancaster Dr. RM12: Horn 36Kd 77
Lancaster Gdns. BR1: Broml 71Nc 160
Lancaster Gdns. KT2: King T 64Ma 131
Lancaster Gdns. SW19 64Ab 132
Lancaster Gdns. W13 47Ka 86
Lancaster Ga. W2 5A 220 (45Eb 89)
Lancaster Gate Station (Underground) 4B 220 (45Fb 89)
Lancaster Gro. NW3 37Fb 69
Lancaster Hall E16 46Jc 93
(off Wesley Av.)
Lancaster Ho. E11 33Hc 73
Lancaster Ho. EN2: Enf 11Tb 33
Lancaster Ho. RH1: Redh 9N 207
Lancaster Ho. RM8: Dag 35Xc 75
Lancaster Ho. TW7: Isle 52Ha 108
Lancaster House 1C 228 (47Lb 90)
(off Stable Yd. Rd.)
Lancaster Lodge W11 44Ab 88
(off Lancaster Rd.)
Lancaster M. SW18 57Db 111
Lancaster M. TW10: Rich 58Na 109
Lancaster M. W2 4A 220 (45Eb 89)
Lancaster Pk. TW10: Rich 57Na 109
Lancaster Pl. IG1: Ilf 36Sc 74
Lancaster Pl. SW19 64Za 132
Lancaster Pl. TW1: Twick 58Ja 108
Lancaster Pl. TW4: Houn 54Y 107
Lancaster Pl. WC2 4H 223 (45Pb 90)
Lancaster Rd. AL1: St A 1D 6
Lancaster Rd. E11 33Gc 73
Lancaster Rd. E17 26Zb 52
Lancaster Rd. E7 38Jc 73
Lancaster Rd. EN2: Enf 11Tb 33
Lancaster Rd. EN4: E Barn 15Fb 31
Lancaster Rd. HA2: Harr 29Ca 45
Lancaster Rd. N11 23Mb 50
Lancaster Rd. N18 22Vb 51
Lancaster Rd. N4 31Pb 70
Lancaster Rd. NW10 36Wa 68
Lancaster Rd. RM16: Chaf H 50Zd 99
Lancaster Rd. SE25 68Vb 135
Lancaster Rd. SW19 64Za 132
Lancaster Rd. UB1: S'hall 45Aa 85
Lancaster Rd. UB5: N'olt 37Ea 66
Lancaster Rd. UB8: Uxb 37M 63
Lancaster Rd. W11 44Ab 88
Lancaster Rd. Ind. Est. EN4: E Barn 15Fb 31
Lancaster Stables NW3 37Gb 69
Lancaster St. SE1 2B 230 (47Rb 91)
Lancaster Ter. W2 4B 220 (45Fb 89)
Lancaster Wlk. UB3: Hayes 44S 84
Lancaster Wlk. W2 6A 220 (46Eb 89)
Lancaster Way KT4: Wor Pk 73Xa 154
Lancaster Way WD5: Ab L 3V 12
Lancastrian Rd. SM6: W'gton 80Nb 156
Lance Cft. DA3: Nw A G 75Be 165
Lancefield Ho. SE15 56Xb 113
Lancefield St. W10 41Bb 89
Lancell St. N16 33Ub 71
Lancelot Av. HA0: Wemb 35Ma 67
Lancelot Cl. SL1: Slou 7E 80
Lancelot Ct. BR6: Orp 75Xc 161
Lancelot Cres. HA0: Wemb 35Ma 67
Lancelot Gdns. EN4: E Barn 17Jb 32
Lancelot Pl. SW7 2F 227 (47Hb 89)
Lancelot Rd. DA16: Well 56Wc 117
Lancelot Rd. HA0: Wemb 35Ma 67
Lancelot Rd. IG6: Ilf 23Uc 54
Lance Rd. HA1: Harr 31Ea 66
Lancer Sq. W8 47Db 89
(off Kensington Chu. St.)
Lancey Cl. SE7 49Nc 94
Lanchester Ct. W2 3F 221 (44Hb 89)
(off Seymour St.)
Lanchester Rd. N6 29Hb 49
Lanchester Way SE14 53Yb 114
Lancing Gdns. N9 18Vb 33
Lancing Ho. CR0: C'don 77Tb 157
(off Coombe Rd.)
Lancing Ho. WD24: Wat 12Y 27
(off Hallam Cl.)
Lancing Rd. BR6: Orp 75Wc 161
Lancing Rd. CR0: C'don 73Pb 156
Lancing Rd. IG2: Ilf 30Tc 54
Lancing Rd. RM3: Rom 24Nd 57
Lancing Rd. TW13: Felt 61V 128
Lancing Rd. W13 45Ka 86
Lancing St. NW1 4D 216 (41Mb 90)
Lancing Way WD3: Crox G 15R 26
Lancresse Cl. UB8: Uxb 37M 63
Lancresse Ct. N1 39Ub 71
(off De Beauvoir Est.)
Landale Gdns. DA1: Dart 59Ld 119
Landale Ho. SE16 48Yb 92
(off Lower Rd.)
Landau Apts. SW6 51Cb 111

Landau Ct. CR2: S Croy 78Sb 157
(off Warham Rd.)
Landau Way DA8: Erith 50Md 97
Landcroft Rd. SE22 57Vb 113
Landells Rd. SE22 58Vb 113
Lander Cl. DA12: Grav'nd 3F 144
Lander Rd. RM17: Grays 50Fe 99
Landford Cl. WD3: Rick 19N 25
Landford Rd. SW15 55Ya 110
Landgrove Rd. SW19 64Cb 133
Landing Waiters Ho. E14 44Fc 93
(off New Village Av.)
Landin Ho. E14 44Cc 92
(off Thomas Rd.)
Landleys Fld. N7 36Mb 70
(off Long Mdw.)
Landmann Ho. SE16 49Xb 91
(off Rennie Est.)
Landmann Point SE10 48Jc 93
Landmann Way SE14 50Zb 92
Landmark Arts Cen. 64Ka 130
Landmark Commercial Cen. N18 23Ub 51
Landmark East Twr. E14 47Cc 92
(off Marsh Wall)
Landmark Hgts. E5 35Ac 72
Landmark Ho. IG10: Lough 14Sc 36
Landmark Ho. W6 50Ya 88
(off Hammersmith Bri. Rd.)
Landmark Pl. UB10: Hil 40R 64
Landmark Pl. UB9: Den 31J 63
Landmark Row SL3: L'ly 51D 104
Landmark Sq. E14 47Cc 92
Landmark West Twr. E14 47Cc 92
(off Marsh Wall)
Landmead Rd. EN8: Chesh 1Ac 20
Landon Pl. SW1 3F 227 (48Hb 89)
Landon's Cl. E14 46Ec 92
Landon Wlk. E14 45Dc 92
Landon Way TW15: Ashf 65R 128
Landor Ho. SE5 52Tb 113
(off Elmington Est.)
Landor Ho. W2 43Cb 89
(off Westbourne Pk. Rd.)
Landor Rd. SW9 55Nb 112
Landor Space 55Nb 112
Landor Wlk. W12 47Wa 88
Landra Gdns. N21 16Rb 33
Landrake NW1 1C 216 (39Lb 70)
(off Plender St.)
Landridge Dr. EN1: Enf 10Xb 19
Landridge Rd. SW6 54Bb 111
Landrock Rd. N8 30Nb 50
Landscape Rd. CR6: W'ham 91Xb 197
Landscape Rd. IG8: Wfd G 24Kc 53
Landsdown Cl. EN5: New Bar 14Eb 31
Landsdowne Ct. N19 32Nb 70
(off Fairbridge Rd.)
Landseer Av. DA11: Nflt 62Fe 143
Landseer Av. E12 36Qc 74
Landseer Cl. HA8: Edg 26Qa 47
Landseer Cl. RM11: Horn 32Kd 77
Landseer Cl. SW19 67Eb 133
Landseer Ct. UB4: Hayes 40T 64
Landseer Ho. NW8 5C 214 (42Fb 89)
(off Frampton St.)
Landseer Ho. SW1 6E 228 (49Mb 90)
(off Herrick St.)
Landseer Ho. SW11 53Jb 112
Landseer Ho. UB5: N'olt 40Z 65
(off Parkfield Dr.)
Landseer Rd. EN1: Enf 15Wb 33
Landseer Rd. KT3: N Mald 73Ta 153
Landseer Rd. N19 34Nb 70
(not continuous)
Landseer Rd. SM1: Sutt 79Cb 155
Lands End WD6: E'tree 16Ma 29
Landstead Rd. SE18 52Tc 116
Landulph Ho. SE11 7A 230 (50Qb 90)
(off Kennings Way)
Landward Ct. W1 2E 220 (44Gb 89)
(off Harrowby St.)
Landway TN15: Seal 92Pd 203
Landway, The BR5: St P 69Yc 139
Landway, The TN15: Bor G 92Be 205
Landway, The TN15: Kems'g 89Qd 183
Lane, The GU25: Vir W 69A 126
Lane, The KT15: Add 77L 149
Lane, The KT16: Chert 69J 127
Lane, The NW8 40Eb 69
Lane, The SE3 55Jc 115
Lane Av. DA9: Ghithe 58Yd 120
Lane Cl. KT15: Add 78K 149
Lane Cl. NW2 34Xa 68
Lane End AL10: Hat 3B 8
Lane End DA7: Bex 55Dd 118
Lane End KT18: Eps 86Ra 173
Lane End SW15 58Za 110
LANE END 63Td 142
Lane End Dr. GU21: Knap 9G 166
Lane Gdns. KT10: Clay 80Ha 152
Lane Gdns. WD23: B Hea 17Ga 28
Lane M. E12 34Pc 74
Lanercost Cl. SW2 61Qb 134
Lanercost Gdns. N14 17Nb 32
Lanercost Rd. SW2 61Qb 134
Lanes Av. DA11: Nflt 2B 144
Lanesborough Ct. N1 3H 219 (41Ub 91)
(off Fanshaw St.)
Lanesborough Pl. SW1 1J 227 (47Jb 90)
(off Grosvenor Pl.)
Lanesborough Way SW17 62Fb 133
Laneside BR7: Chst 64Rc 138
Laneside HA8: Edg 22Sa 47
Laneside Av. RM8: Dag 31Bd 75
Laneway SW15 57Xa 110
Laney Ho. EC1 7K 217 (43Qb 90)
(off Leather La.)
Lanfranc Ct. HA1: Harr 33Ha 66
Lanfranc Rd. E3 40Ac 72
Lanfrey Pl. W14 50Bb 89
Langafel Cl. DA3: Lfield 68Ae 143
Langaller La. KT22: Fet 94Da 191
Langan Ho. E14 44Bc 92
(off Keymer Pl.)
Langbourne Av. N6 33Jb 70
Langbourne Ct. E17 30Ac 52
Langbourne Mans. N6 33Jb 70
Langbourne Pl. E14 50Dc 92
Langbourne Way KT10: Clay 79Ja 152
Langbrook Rd. SE3 55Mc 115
Lang Cl. KT22: Fet 95Da 191
Langcroft Cl. SM5: Cars 76Hb 155
Langdale NW1 3B 216 (41Lb 90)
(off Stanhope St.)
Langdale Av. CR4: Mitc 69Hb 133
Langdale Cl. BR6: Farnb 76Rc 160
Langdale Cl. GU21: Wok 8N 167
Langdale Cl. RM8: Dag 32Yc 75
Langdale Cl. SE17 51Sb 113
Langdale Cl. SW14 56Ra 109
Langdale Cres. DA7: Bex 52Cd 118
Langdale Dr. UB4: Hayes 40U 64
Langdale Gdns. EN8: Walt C 7Zb 20
Langdale Gdns. RM12: Horn 36Jd 76
Langdale Gdns. UB6: G'frd 41Ka 86
Langdale Ho. SW1 7B 228 (50Lb 90)
(off Churchill Gdns.)
Langdale Lodge WD3: Rick 17M 25
(off Parsonage Rd.)
Langdale Pde. CR4: Mitc 69Hb 133
Langdale Rd. CR7: Thor H 70Qb 134
Langdale Rd. SE10 52Ec 114
Langdale St. E1 44Xb 91
Langdale Ter. WD6: Bore 13Sa 29
Langdale Wlk. DA11: Nflt 2A 144
(off Landseer Av.)
Langdon Ct. EC1 2C 218 (40Rb 71)
(off City Rd.)
Langdon Ct. NW10 39Ua 68
Langdon Cres. E6 40Qc 74
Langdon Dr. NW9 32Sa 67
Langdon Ho. E14 44Ec 92
(off Ida St.)
Langdon Pk. TW11: Tedd 66La 130
Langdon Pk. Rd. N6 31Lb 70
Langdon Park Station (DLR) 44Dc 92
Langdon Pl. SW14 55Sa 109
Langdon Rd. BR2: Broml 69Kc 137
Langdon Rd. E6 39Qc 74
Langdon Rd. SM4: Mord 71Eb 155
Langdons Ct. UB2: S'hall 48Ca 85
Langdon Shaw DA14: Sidc 64Vc 139
Langdon Wlk. SM4: Mord 71Eb 155
Langdon Way SE1 49Wb 91
Langford Cl. AL4: St A 1G 6
Langford Cl. E8 36Wb 71
Langford Cl. N15 30Ub 51
Langford Cl. NW8 1A 214 (40Eb 69)
Langford Cl. W3 47Ra 87
Langford Ct. NW8 2A 214 (40Eb 69)
(off Abbey Rd.)
Langford Cres. EN4: Cockf 14Hb 31
Langford Gdn. 53Db 111
(off Pearscroft Road)
Langford Grn. CM13: Hut 15Ee 41
Langford Grn. SE5 55Ub 113
Langford Ho. SE8 51Cc 114
Langford M. N1 38Qb 70
Langford M. SW11 56Fb 111
(off St John's Hill)
Langford Pl. DA14: Sidc 62Wc 139
Langford Pl. NW8 1A 214 (40Eb 69)
Langford Rd. EN4: Cockf 14Hb 31
Langford Rd. IG8: Wfd G 23Lc 53
Langford Rd. SW6 54Db 111
Langfords IG9: Buck H 19Mc 35
Langham Cl. BR2: Broml 75Nc 160
Langham Cl. N15 27Rb 51
(off Langham Rd.)
Langham Ct. HA4: Ruis 36X 65
Langham Ct. NW4 29Za 48
Langham Ct. RH9: S God 10C 210
Langham Ct. RM11: Horn 31Md 77
Langham Ct. SW20 68Ya 132
Langham Dene CR8: Kenley 87Rb 177
Langham Dr. RM6: Chad H 30Xc 55
Langham Gdns. HA0: Wemb 33La 66
Langham Gdns. HA8: Edg 24Sa 47
Langham Gdns. N21 15Qb 32
Langham Gdns. TW10: Ham 63La 130
Langham Gdns. W13 45Ka 86
Langham Ho. E15 37Fc 73
(off Forrester Way)
Langham Ho. Cl. TW10: Ham 63Ma 131
Langham Mans. SW5 50Db 89
(off Earl's Ct. Sq.)
Langham Pk. Pl. BR2: Broml 70Hc 137
Langham Pl. N15 27Rb 51
Langham Pl. TW20: Egh 64B 126
Langham Pl. W1 1A 222 (43Kb 90)
Langham Pl. W4 51Ua 110
Langham Rd. HA8: Edg 23Sa 47
Langham Rd. N15 27Rb 51
Langham Rd. SW20 67Ya 132
Langham Rd. TW11: Tedd 64Ka 130
Langham St. W1 1A 222 (43Kb 90)
Langhedge Cl. N18 23Vb 51
Langhedge La. N18 23Vb 51
Langhedge La. Ind. Est. N18 23Vb 51
Langholm Cl. SW12 59Mb 112
Langholme WD23: Bush 18Ea 28
Langhorn Dr. TW2: Twick 59Ga 108
Langhorne Ct. NW8 38Fb 69
(off Dorman Way)
Langhorne Rd. RM10: Dag 38Cd 76
Langhorne St. SE18 52Pc 116
Lang Ho. SW8 52Nb 112
(off Hartington Rd.)
Lang Ho. TW19: Stanw 60N 105
LANGHURST 8P 211
Langland Ct. HA6: Nwood 24S 44
Langland Cres. HA7: Stan 26Ma 47
Langland Dr. HA5: Pinn 24Aa 45
Langland Gdns. CR0: C'don 75Bc 158
Langland Gdns. NW3 36Db 69
Langland Ho. SE5 52Tb 113
(off Edmund St.)
Langlands Dr. DA2: Daren 64Ud 142
Langlands Ri. KT19: Eps 85Sa 173
Langler Rd. NW10 40Ya 68
LANGLEY 48C 82
Langley Av. HA4: Ruis 33X 65
Langley Av. HP3: Hem H 5N 3
Langley Av. KT4: Wor Pk 74Za 154
Langley Av. KT6: Surb 74Ma 153
LANGLEY BOTTOM 91Ta 193
Langley Broom SL3: L'ly 50B 82
LANGLEYBURY 6R 12
Langleybury Flds. WD4: Lang 6P 11
Langleybury La. WD17: Wat 9R 12
Langleybury La. WD4: Lang 8R 12
Langley Bus. Cen. SL3: L'ly 47C 82
Langley Bus. Pk. SL3: L'ly 47C 82
Langley Cl. KT18: Eps D 91Ta 193
Langley Cl. RM3: Rom 24Md 57
Langley Ct. EN7: G Oak 1Sb 19
Langley Ct. RH2: Reig 5K 207
Langley Ct. WC2 4F 223 (45Nb 90)
Langley Cres. E11 31Lc 73
Langley Cres. HA8: Edg 20Sa 29
Langley Cres. RM9: Dag 39Yc 75
Langley Cres. UB3: Harl 52V 106
Langley Cres. WD4: K Lan 2Q 12
Langley Dr. CM14: B'wood 20Wd 40
Langley Dr. E11 31Kc 73
Langley Dr. W3 47Ra 87
Langley Gdns. BR2: Broml 70Lc 137
Langley Gdns. BR5: Pet W 72Rc 160
Langley Gdns. RM9: Dag 38Zc 75
Langley Gro. KT3: N Mald 68Ua 132
Langley Hill WD4: K Lan 1P 11
Langley Hill Cl. WD4: K Lan 1Q 12
Langley Ho. W2 43Cb 89
(off Alfred Rd.)
Langley La. KT18: Head 97Ra 193
Langley La. SW8 51Pb 112
Langley La. WD25: Wat 5W 12
Langley La. WD5: Ab L 3V 12
Langley Leisure Cen. 49D 82
Langley Lodge La. WD4: K Lan 3Q 12
Langley Mans. SW8 51Pb 112
(off Langley La.)
Langley Mdw. IG10: Lough 12Tc 36
Langley M. RM9: Dag 38Zc 75
Langley Oaks Av. CR2: Sande 82Wb 177
Langley Pk. NW7 23Ua 48
Langley Pk. SL3: L'ly 47B 82
Langley Pk. Country Pk. 42B 82
Langley Pk. Golf Course 72Fc 159
Langley Pk. La. SL0: Iver 45D 82
Langley Pk. Rd. SL0: Iver 45D 82
Langley Pk. Rd. SL3: L'ly 47C 82
Langley Pk. Rd. SM1: Sutt 78Eb 155
Langley Pk. Rd. SM2: Sutt 79Eb 155
Langley Pk. Sports Cen. 72Ec 158
Langley Quay SL3: L'ly 47C 82
Langley Rd. BR3: Beck 70Ac 136
Langley Rd. CR2: Sels 81Zb 178
Langley Rd. DA16: Well 51Yc 117
Langley Rd. KT6: Surb 73Na 153
Langley Rd. SL3: L'ly 7N 81
Langley Rd. SW19 67Bb 133
Langley Rd. TW18: Staines 65H 127
Langley Rd. TW7: Isle 54Ha 108
Langley Rd. WD17: Wat 11V 26
Langley Rd. WD4: Chfd 2K 11
Langley Rd. WD5: Ab L 3U 12
LANGLEY RDBT. 50C 82
Langley Row EN5: Barn 11Bb 31
Langley Station (Rail & Crossrail) 47C 82
Langley St. WC2 3F 223 (44Nb 90)
Langley Tennis Club 7B 4
LANGLEY VALE 91Ua 194
Langley Va. Rd. KT18: Eps D 92Sa 193
Langley Wlk. GU22: Wok 91A 188
Langley Way BR4: W W'ck 74Fc 159
Langley Way WD17: Wat 12U 26
Langley Wharf WD4: K Lan 9A 4
Langmans La. GU21: Wok 10M 167
Langmans Way GU21: Wok 8J 167
Langmead Dr. WD23: B Hea 18Ga 28
Langmead Ho. E3 41Dc 92
(off Bruce Rd.)
Langmead St. SE27 63Rb 135
Langmore Ct. DA6: Bex 55Zc 117
Langmore Ho. E1 44Wb 91
(off Stutfield St.)
Langport Ct. KT12: Walt T 74Y 151
Langport Ho. RM3: Rom 24Nd 57
(off Leyburn Rd.)
Langport Ho. SW9 54Rb 113
Langridge M. TW12: Hamp 65Ba 129
Langroyd Rd. SW17 61Hb 133
Langshott Cl. KT15: Wdhm 83G 168
Langside Av. SW15 56Wa 110
Langside Cres. N14 20Mb 32
Langstone M. UB1: S'hall 46Ca 85
Langstone Way NW7 24Za 48
Langston Hughes Cl. SE24 56Rb 113
Langston Rd. IG10: Lough 15Sc 36
Lang St. E1 42Yb 92
Langthorn Ct. EC2 2F 225 (44Tb 91)
Langthorne Ct. BR1: Broml 63Ec 136
Langthorne Cres. RM17: Grays 49Ee 99
Langthorne Ho. E3 41Bc 92
(off Merchant St.)
Langthorne Ho. UB3: Harl 49U 84
Langthorne Rd. E11 34Ec 72
(not continuous)
Langthorne St. SW6 52Za 110
Langton Av. E6 41Qc 94
Langton Av. KT17: Ewe 83Va 174
Langton Av. N20 17Eb 31
Langton Cl. GU21: Wok 9K 167
Langton Cl. KT15: Add 76K 149
Langton Cl. SL1: Slou 6B 80
Langton Cl. WC1 5J 217 (42Pb 90)
Langton Gro. HA6: Nwood 22S 44
Langton Ho. SE11 5J 229 (49Pb 90)
(off Lambeth Wlk.)
Langton Lodge TW18: Staines 64F 126
Langton Pl. SW18 60Cb 111
Langton Ri. SE23 59Xb 113
Langton Rd. HA3: Hrw W 24Ea 46
Langton Rd. KT8: W Mole 70Ea 130
Langton Rd. NW2 34Ya 68
Langton Rd. SW9 52Rb 113
Langton's Mdw. SL2: Farn C 7G 60
Langton St. SW10 51Eb 111
Langton Way CR0: C'don 76Ub 157
Langton Way RM16: Grays 9E 100
Langton Way SE3 53Hc 115
Langton Way TW20: Egh 65E 126
Langtree Av. SL1: Slou 7D 80
Langtry Ct. TW7: Isle 54Ha 108
Langtry Ho. KT2: King T 67Qa 131
(off London Rd.)
Langtry Pl. SW6 51Cb 111
Langtry Rd. NW8 39Db 69
Langtry Rd. UB5: N'olt 40Z 65
Langtry Wlk. NW8 39Db 69
Langwood WD17: Wat 11W 26
Langwood Chase TW11: Tedd 65La 130
Langwood Cl. KT21: Asht 89Qa 173
Langwood Gdns. WD17: Wat 11W 26
Langworth Cl. DA2: Wilm 62Md 141
Langworth Dr. UB4: Yead 44X 85
Langworthy HA5: Hat E 23Ca 45
Lanhill Rd. W9 42Cb 89
Lanier Rd. SE13 58Fc 115
Lanigan Dr. TW3: Houn 57Da 107
Lankaster Gdns. N2 25Fb 49
Lankers Dr. HA2: Harr 30Ba 45
Lankester Sq. RH8: Oxt 100Fc 199
Lankton Cl. BR3: Beck 67Ec 136
Lannock Rd. UB3: Hayes 46V 84
Lannoy Point SW6 52Ab 110
(off Pellant Rd.)
Lannoy Rd. SE9 60Sc 116
Lanrick Rd. E14 44Fc 93
Lanridge Rd. SE2 48Zc 95
Lansbury Av. IG11: Bark 38Wc 75
Lansbury Av. N18 22Tb 51
Lansbury Av. RM6: Chad H 29Ad 55
Lansbury Av. TW14: Felt 58X 107
Lansbury Cl. NW10 36Sa 67
Lansbury Ct. SE28 45Xc 95
(off Saunders Way)
Lansbury Cres. DA1: Dart 57Qd 119
Lansbury Dr. UB4: Hayes 40U 64
Lansbury Est. E14 44Dc 92
Lansbury Est. GU21: Knap 10H 167
Lansbury Gdns. E14 44Fc 93
Lansbury Gdns. RM18: Tilb 3C 122
Lansbury Rd. EN3: Enf H 11Zb 34
Lansbury Way N18 22Ub 51
Lanscombe Wlk. SW8 53Nb 112
Lansdell Ho. SW2 58Qb 112
Lansdell Rd. CR4: Mitc 68Jb 134
Lansdown Cl. GU21: Wok 1K 187
Lansdown Cl. KT12: Walt T 74Y 151
Lansdowne Av. BR6: Farnb 74Rc 160
Lansdowne Av. DA7: Bex 52Zc 117
Lansdowne Av. SL1: Slou 6J 81
Lansdowne Cl. KT6: Surb 75Ra 153
Lansdowne Cl. SW20 66Za 132
Lansdowne Cl. TW1: Twick 60Ha 108
Lansdowne Cl. WD25: Wat 7Z 13
Lansdowne Ct. CR8: Purl 82Rb 177
Lansdowne Ct. IG5: Ilf 27Nc 54
Lansdowne Ct. KT4: Wor Pk 75Wa 154
Lansdowne Ct. SL1: Slou 6J 81
Lansdowne Ct. W11 45Ab 88
(off Lansdowne Ri.)
Lansdowne Cres. W11 45Ab 88
Lansdowne Dr. E8 37Wb 71
Lansdowne Gdns. SW8 53Nb 112
Lansdowne Grn. SW8 53Nb 112
Lansdowne Gro. NW10 35Ua 68
Lansdowne Hill SE27 62Rb 135
Lansdowne Ho. KT18: Eps 86Sa 173
(off Dalmeny Way)
Lansdowne Ho. W11 46Bb 89
(off Ladbroke Rd.)
Lansdowne La. SE7 51Mc 115
Lansdowne M. SE7 50Mc 93
Lansdowne M. W11 46Bb 89
Lansdowne Pl. SE1 3G 231 (48Tb 91)
Lansdowne Pl. SE19 66Vb 135
Lansdowne Ri. W11 45Ab 88
Lansdowne Rd. BR1: Broml 66Jc 137
Lansdowne Rd. CR0: C'don 75Tb 157
Lansdowne Rd. CR8: Purl 84Qb 176
Lansdowne Rd. E11 33Hc 73
Lansdowne Rd. E17 30Cc 52
Lansdowne Rd. E18 27Jc 53
Lansdowne Rd. E4 19Cc 34
Lansdowne Rd. HA1: Harr 31Ga 66
Lansdowne Rd. HA7: Stan 23La 46
Lansdowne Rd. IG3: Ilf 32Vc 75
Lansdowne Rd. KT19: Ewe 80Sa 153
Lansdowne Rd. N10 26Lb 50
Lansdowne Rd. N17 25Vb 51
Lansdowne Rd. N3 24Cb 49
Lansdowne Rd. RM18: Tilb 4B 122
Lansdowne Rd. SW20 66Ya 132
Lansdowne Rd. TN13: S'oaks 94Md 203
Lansdowne Rd. TW18: Staines 66K 127
Lansdowne Rd. TW3: Houn 55Da 107
Lansdowne Rd. UB8: Hil 44S 84
Lansdowne Rd. W11 45Ab 88
Lansdowne Row W1 6A 222 (46Kb 90)
Lansdowne Sq. DA11: Nflt 8B 122
Lansdowne Ter. WC1 6G 217 (42Nb 90)
Lansdowne Wlk. W11 46Bb 89
Lansdowne Way SW8 53Mb 112
Lansdowne Wood Cl. SE27 62Rb 135
Lansdowne Workshops SE7 50Lc 93
Lansdown Pl. DA11: Nflt 10B 122
Lansdown Rd. DA11: Nflt 10B 122
Lansdown Rd. DA14: Sidc 62Xc 139
Lansdown Rd. E7 38Lc 73
Lansfield Av. N18 21Wb 51
Lanson Apts. SW11 52Kb 112
Lantan Hgts. E20 38Ec 72
Lantern SE1 1D 230 (47Sb 91)
(off Lant St.)
Lantern Cl. BR6: Farnb 77Rc 160
Lantern Cl. HA0: Wemb 36Ma 67
Lantern Cl. SW15 56Wa 110
Lantern Ho. UB3: Harl 48S 84
(off Nine Acres Cl.)
Lanterns Way E14 47Cc 92
Lantern Way UB7: W Dray 47N 83
Lant Ho. SE1 2D 230 (47Sb 91)
(off Toulmin St.)
Lantry Ct. W3 46Ra 87
Lant St. SE1 1D 230 (47Sb 91)
Lanvanor Rd. SE15 54Yb 114
Lanward Apts. N1 38Pb 70
(off Caledonian Rd.)
Lanyard Ho. SE8 49Bc 92
Lapford Cl. W9 42Bb 89
Lapis Cl. DA12: Grav'nd 10K 123
Lapis Cl. NW10 41Qa 87
Lapis M. E15 39Ec 72
La Plata Gro. CM14: B'wood 20Xd 40
Lapponum Wlk. UB4: Yead 42Z 85
Lapse Wood Wlk. SE23 60Xb 113
Lapstone Gdns. HA3: Kenton 30La 46
Lapwin Cl. RM18: E Til 9K 101
Lapwing Cl. CR2: Sels 82Ac 178
Lapwing Cl. DA8: Erith 52Kd 119
Lapwing Ct. KT6: Surb 76Qa 153
Lapwing Ct. SE1 2E 230 (47Sb 91)
(off Swan St.)
Lapwing Pl. WD25: Wat 4Y 13
Lapwings DA3: Lfield 69De 143
Lapwings, The DA12: Grav'nd 1F 144
Lapwing Ter. E7 36Mc 73
Lapwing Twr. SE8 51Bc 114
(off Taylor Cl.)
Lapwing Way UB4: Yead 44Z 85
Lapwing Way WD5: Ab L 3W 12
Lapworth N11 21Kb 50
(off Coppies Gro.)
Lapworth Cl. BR6: Chels 75Vc 161
Lapworth Ct. W2 43Db 89
(off Delamere Ter.)
Lara Cl. KT9: Chess 80Na 153
Lara Cl. SE13 58Ec 114
Larbert Rd. SW16 66Lb 134
Larby Pl. KT17: Ewe 82Ua 174
Larch Av. AL2: Brick W 2Aa 13
Larch Av. SL5: S'dale 1C 146
Larch Av. W3 46Ua 88
Larch Cl. CR6: W'ham 91Ac 198
Larch Cl. E13 42Kc 93
Larch Cl. KT20: Kgswd 93Eb 195
Larch Cl. N11 24Jb 50
Larch Cl. N19 33Lb 70
Larch Cl. RH1: Redh 8L 207
Larch Cl. SE8 51Bc 114
Larch Cl. SL2: Slou 3F 80
Larch Cl. SW12 61Kb 134
Larch Ct. SE1 2H 231 (47Ub 91)
(off Royal Oak Yd.)
Larch Ct. W9 43Cb 89
(off Admiral Wlk.)
Larch Cres. KT19: Ewe 79Ra 153
Larch Cres. UB4: Yead 42Y 85
Larch Dene BR6: Farnb 75Qc 160
Larch Dr. W4 50Qa 87
Larches, The GU21: Wok 88A 168
Larches, The HA6: Nwood 23S 44
Larches, The N13 20Sb 33
Larches, The UB10: Hil 41R 84
Larches, The WD23: Bush 15Aa 27
Larches Av. EN1: Enf 7Yb 20
Larches Av. SW14 56Ta 109
Larchfield Cl. KT13: Weyb 76V 150
Larch Grn. NW9 25Ua 48
Larch Gro. DA15: Sidc 60Vc 117
Larch Ho. BR2: Broml 67Gc 137
Larch Ho. SE16 47Yb 92
(off Ainsty Est.)
Larch Ho. UB4: Yead 43Y 85
Larch Ho. W10 42Ab 88
(off Rowan Wlk.)
Larchmoor Pk. SL2: Stoke P 6L 61
Larch Pl. RM3: Hrld W 26Nd 57
Larch Rd. DA1: Dart 59Md 119
Larch Rd. E10 33Cc 72
Larch Rd. NW2 35Ya 68
Larch Tree Way CR0: C'don 76Cc 158
Larchvale Ct. SM2: Sutt 80Db 155
Larch Vw. HP1: Hem H 3K 3
Larch Wlk. BR8: Swan 68Fd 140
Larch Way BR2: Broml 73Qc 160
Larchwood Av. RM5: Col R 23Dd 56
Larchwood Cl. RM5: Col R 23Ed 56
Larchwood Cl. SM7: Bans 87Ab 174
Larchwood Dr. TW20: Eng G 5M 125
Larchwood Gdns. CM15: Pil H 16Wd 40
Larchwood Ho. UB7: W Dray 47P 83
(off Park Lodge Av.)
Larchwood Rd. GU21: Wok 2H 187
Larchwood Rd. HP2: Hem H 1P 3
Larchwood Rd. SE9 61Rc 138
Larcombe Cl. CR0: C'don 77Vb 157
Larcombe Ct. SM2: Sutt 80Db 155
(off Worcester Rd.)
Larcom St. SE17 6E 230 (49Sb 91)
Larden Rd. W3 46Ua 88
Largewood Av. KT6: Surb 75Qa 153
Largo Wlk. DA8: Erith 53Gd 118
Lariat Apts. SE10 49Gc 93
(off Cable Wlk.)
Larissa St. SE17 7G 231 (50Tb 91)
Lark Av. TW18: Staines 62H 127
Larkbere Rd. SE26 63Ac 136
Lark Cl. CM14: W'ley 21Xd 58
Lark Ct. NW9 25Ua 48
(off Lanacre Av.)
Larken Cl. WD23: Bush 18Ea 28
Larken Dr. WD23: Bush 18Ea 28
Larkfield KT11: Cobh 85W 170
Larkfield Av. HA3: Kenton 27Ka 46
Larkfield Cl. BR2: Hayes 75Hc 159
Larkfield Rd. DA14: Sidc 62Vc 139
Larkfield Rd. TN13: Bes G 95Ed 202
Larkfield Rd. TW9: Rich 56Na 109
Larkfields DA11: Nflt 2A 144
Larkhall Cl. KT12: Hers 79Y 151
Larkhall La. SW4 54Mb 112
Larkhall Ri. SW4 55Lb 112
(not continuous)
Larkham Cl. TW13: Felt 62U 128
Larkin Cl. CM13: Hut 17Ee 41
Larkin Cl. CR5: Coul 89Pb 176
Larkings La. SL2: Stoke P 9M 61
Lark Ri. AL10: Hat 2C 8
Lark Row E2 39Yb 72
Larks Fld. DA3: Hartl 70Be 143
Larksfield TW20: Eng G 6M 125
Larksfield Gro. EN1: Enf 11Xb 33
Larks Gro. IG11: Bark 38Uc 74
Larkshall Ct. RM7: Mawney 26Ed 56
Larkshall Cres. E4 21Ec 52
Larkshall Rd. E4 22Ec 52
Larkspur Cl. BR6: Chels 75Yc 161
Larkspur Cl. E6 43Nc 94
Larkspur Cl. HA4: Ruis 31S 64
Larkspur Cl. HP1: Hem H 1G 2
Larkspur Cl. N17 24Tb 51
Larkspur Cl. NW9 29Ra 47
Larkspur Cl. RM15: S Ock 41Yd 98
Larkspur Gro. HA8: Edg 21Sa 47
Larkspur Lodge DA14: Sidc 62Xc 139
Larkspur Way KT19: Ewe 78Sa 153
Larks Ridge AL2: Chis G 9N 5
Larks Way GU21: Knap 8G 166
Larkswood Cl. DA8: Erith 53Jd 118
Larkswood Ct. E4 22Fc 53
Larkswood Ri. HA5: Eastc 28Y 45
Larkswood Rd. E4 21Cc 52
Lark Way SM5: Cars 73Gb 155
Larkway Cl. NW9 28Ta 47
Larkwell La. DA3: Hartl 70Be 143
Larkwood Av. SE10 53Ec 114
Larmans Rd. EN3: Enf W 8Yb 20
Larnach Rd. W6 51Za 110
Larne Rd. HA4: Ruis 31V 64
Larner Rd. DA8: Erith 52Gd 118
La Roche Cl. SL3: L'ly 8N 81
Larpent Av. SW15 57Ya 110
Larsen Dr. EN9: Walt A 6Fc 21
Larson Wlk. E14 48Cc 92
Larwood Cl. UB6: G'frd 36Fa 66
Lascar Cl. TW3: Houn 55Ba 107
Lascar Wharf Bldg. E14 44Ac 92
(off Parnham St.)
Lascelles Av. HA1: Harr 31Fa 66
Lascelles Cl. CM15: Pil H 15Wd 40
Lascelles Cl. E11 33Fc 73
Lascelles Ho. NW1 6E 214 (42Gb 89)
(off Harewood Av.)
Lascelles Rd. SL3: Slou 8M 81
Lascotts Rd. N22 23Pb 50
Laseron Ho. N15 28Vb 51
(off Tottenham Grn. E.)
Laserquest Romford 26Bd 55
Laserquest Woking 89B 168
(within The Big Apple)
Las Palmas Est. TW17: Shep 73Q 150
Lassa Rd. SE9 57Nc 116
Lassell St. SE10 50Fc 93
Lasseter Pl. SE3 51Hc 115
Lasswade Ct. KT16: Chert 73G 148
Lasswade Rd. KT16: Chert 73G 148
Lastingham Ct. TW18: Staines 65J 127
Latchett Rd. E18 25Kc 53
Latchford Pl. HP1: Hem H 3J 3
Latchford Pl. IG7: Chig 21Xc 55

Latching Cl. RM3: Rom 21Md 57
Latchingdon Ct. E17 28Zb 52
Latchingdon Gdns. IG8: Wfd G 23Nc 54
Latchmere Cl. TW10: Ham 64Na 131
Latchmere La. KT2: King T 65Pa 131
Latchmere La. TW10: Ham 64Pa 131
Latchmere Leisure Cen. 54Hb 111
Latchmere Pas. SW11 54Gb 111
Latchmere Pl. TW15: Ashf 61N 127
Latchmere Rd. KT2: King T 66Na 131
Latchmere Rd. SW11 54Hb 111
Latchmere St. SW11 54Hb 111
Latchmoor Av. SL9: Chal P 28A 42
Latchmoor Gro. SL9: Chal P 28A 42
Latchmoor Way SL9: Chal P 28A 42
Lateward Rd. TW8: Bford 51Ma 109
Latham Cl. DA2: Dart 61Ud 142
Latham Cl. E6 43Nc 94
Latham Cl. TN16: Big H 88Lc 179
Latham Cl. TW1: Twick 59Ja 108
Latham Ct. N11 23Nb 50
(off Brownlow Rd.)
Latham Ct. SW5 49Cb 89
(off W. Cromwell Rd.)
Latham Ct. UB5: N'olt 41Z 85
(off Delta Gro.)
Latham Ho. E1 44Zb 92
(off Chudleigh St.)
Latham Pl. RM14: Upm 32Sd 78
Latham Rd. DA6: Bex 57Cd 118
Latham Rd. TW1: Twick 59Ha 108
Latham's Way CR0: Wadd 74Pb 156
Lathkill Cl. EN1: Enf 17Wb 33
Lathkill Ct. BR3: Beck 67Bc 136
Lathom Rd. E6 38Nc 74
LATIMER 9A 10
Latimer Av. E6 39Pc 74
Latimer Chase WD3: Chen 12E 24
Latimer Cl. GU22: Wok 88D 168
Latimer Cl. HA5: Pinn 25Y 45
Latimer Cl. KT4: Wor Pk 77Xa 154
Latimer Cl. WD18: Wat 17U 26
Latimer Ct. BR2: Broml 70Hc 137
(off Durham Rd.)
Latimer Ct. EN8: Walt C 6Bc 20
Latimer Ct. RH1: Redh 8P 207
Latimer Dr. RM12: Horn 34Md 77
Latimer Gdns. HA5: Pinn 25Y 45
Latimer Ho. E9 37Zb 72
Latimer Ho. W11 45Bb 89
(off Kensington Pk. Rd.)
Latimer Ind. Est. W10 44Ya 88
Latimer Pl. W10 44Ya 88
Latimer Rd. CR0: C'don 76Rb 157
Latimer Rd. E7 35Kc 73
Latimer Rd. EN5: New Bar 13Db 31
Latimer Rd. N15 30Ub 51
Latimer Rd. SW19 65Db 133
Latimer Rd. TW11: Tedd 64Ha 130
Latimer Rd. W10 43Ya 88
(not continuous)
Latimer Rd. WD3: Chen 9C 10
Latimer Road Station (Underground) 45Za 88
Latimer Sq. SE10 48Jc 93
Latitude KT16: Chert 74L 149
(off Bridge Wharf)
Latitude Apts. CR0: C'don 76Tb 157
(off Fairfield Rd.)
Latitude Ct. E16 45Sc 94
Latitude Ho. NW1 39Kb 70
(off Oval Rd.)
Latium Cl. AL1: St A 3B 6
Latona Ct. SW9 52Qb 112
(off Caldwell St.)
Latona Dr. DA12: Grav'nd 4H 145
Latona Rd. SE15 51Wb 113
La Tourne Gdns. BR6: Farnb 76Sc 160
Lattimer Pl. W4 52Ua 110
Lattimore Ho. AL1: St A 2C 6
(off Lattimore Rd.)
Lattimore Rd. AL1: St A 3C 6
Latton Cl. KT10: Esh 77Da 151
Latton Cl. KT12: Walt T 73Aa 151
Latvia Ct. SE17 50Sb 91
(off Macleod St.)
Latymer Cl. KT13: Weyb 77S 150
Latymer Ct. W6 49Za 88
Latymer Gdns. N3 26Ab 48
Latymer Rd. N9 18Ub 33
Latymer Way N9 19Ub 33
Laubin Cl. TW1: Twick 56Ka 108
Lauder Cl. UB5: N'olt 40Z 65
Lauder Ct. N14 17Nb 32
Lauderdale Dr. TW10: Ham 62Ma 131
Lauderdale Ho. SW9 53Qb 112
(off Gosling Way)
Lauderdale Ho. TW18: Staines 64H 127
(off Gresham Rd.)
Lauderdale House Community
Arts Cen. 32Kb 70
(within Lauderdale House)
Lauderdale Mans. W9 41Db 89
(off Lauderdale Rd.)
Lauderdale Pde. W9 42Db 89
Lauderdale Pl. EC2 7D 218 (43Sb 91)
(off Beech St.)
Lauderdale Rd. W9 41Db 89
Lauderdale Rd. WD4: Hunt C 5S 12
Lauderdale Twr. EC2 7D 218 (43Sb 91)
(off Beech St.)
Laud St. CR0: C'don 76Sb 157
Laud St. SE11 7H 229 (50Pb 90)
Laugan Wlk. SE17 7E 230 (50Sb 91)
Laughton Ct. WD6: Bore 12Sa 29
Laughton Rd. UB5: N'olt 39Z 65
Launcelot Rd. BR1: Broml 63Jc 137
Launcelot St. SE1 2K 229 (47Qb 90)
Launceston WD3: Chor 16D 24
Launceston Cl. RM3: Rom 25Ld 57
Launceston Gdns. UB6: G'frd 39La 66
Launceston Pl. W8 48Eb 89
Launceston Rd. UB6: G'frd 39La 66
Launch St. E14 48Ec 92
Launders Ga. W3 47Ra 87
Launder's La. RM13: Rain 41Pd 97
Launder's La. RM13: Wenn 41Pd 97
Laundress La. N16 34Wb 71
Laundry Cl. CR0: C'don 73Tb 157
Laundry La. CM15: Mount 11Fe 41
Laundry La. N1 39Sb 71
Laundry M. SE23 59Ac 114
Laundry Rd. W6 51Ab 110
Launton Dr. DA6: Bex 56Zc 117
Laura Cl. E11 29Lc 53
Laura Cl. EN1: Enf 15Ub 33
Lauradale Rd. N2 28Hb 49
Laura Dr. BR8: Hext 66Jd 140
Laura Pl. E5 35Yb 72
Laura Ter. N4 33Rb 71
Laura Trott Leisure Cen. 2Ac 20
Laureate Way HP1: Hem H 1K 3
Laurel Apts. SE17 5H 231 (49Ub 91)
(off Townsend St.)
Laurel Av. DA12: Grav'nd 1E 144
Laurel Av. EN6: Pot B 4Bb 17
Laurel Av. SL3: L'ly 47A 82
Laurel Av. TW1: Twick 60Ha 108
Laurel Av. TW20: Eng G 4M 125
Laurel Bank GU24: Chob 3J 167
(off Bagshot Rd.)
Laurel Bank HP3: Hem H 5H 3
Laurel Bank N12 21Eb 49
Laurel Bank Cl. TN16: Big H 88Lc 179
Laurel Bank Gdns. SW6 54Bb 111
Laurel Bank Rd. EN2: Enf 11Sb 33
Laurel Bank Vs. W7 47Ga 86
(off Lwr. Boston Rd.)
Laurel Cl. CM13: Hut 15De 41
Laurel Cl. DA1: Dart 60Ld 119
Laurel Cl. DA14: Sidc 62Wc 139
Laurel Cl. HP2: Hem H 1P 3
Laurel Cl. IG6: Ilf 23Sc 54
Laurel Cl. N19 33Lb 70
Laurel Cl. SL3: Poyle 52G 104
Laurel Cl. SW17 64Gb 133
Laurel Cl. WD19: Wat 17Z 27
Laurel Ct. CM13: Hut 16Ee 41
(off The Spinney)
Laurel Ct. CM16: Epp 3Wc 23
Laurel Ct. CR2: S Croy 77Ub 157
(off South Pk. Hill Rd.)
Laurel Ct. EN6: Cuff 1Pb 18
Laurel Ct. HA0: Wemb 40Na 67
Laurel Ct. RM13: Rain 42Ld 97
Laurel Ct. SE1 4F 231 (48Tb 91)
(off Garland Cl.)
Laurel Ct. SL0: Iver H 38F 62
Laurel Cres. CR0: C'don 76Cc 158
Laurel Cres. GU21: Wok 85E 168
Laurel Cres. RM7: Rush G 32Gd 76
Laurel Dr. N21 17Qb 32
Laurel Dr. RH8: Oxt 3K 211
Laurel Dr. RM15: S Ock 42Zd 99
Laurel Edge AL1: St A 1D 6
(off Avenue Rd.)
Laurel Flds. EN6: Pot B 3Bb 17
Laurel Gdns. BR1: Broml 70Nc 138
Laurel Gdns. E4 17Dc 34
Laurel Gdns. KT15: New H 82K 169
Laurel Gdns. NW7 20Ta 29
Laurel Gdns. TW15: Ashf 64S 128
Laurel Gdns. TW4: Houn 56Aa 107
Laurel Gdns. W7 46Ga 86
Laurel Gro. SE20 66Yb 136
Laurel Gro. SE26 63Zb 136
Laurel Ho. BR2: Broml 67Gc 137
Laurel Ho. E3 39Bc 72
(off Hornbeam Sq.)
Laurel Ho. SE8 51Bc 114
Laurel La. RM12: Horn 33Nd 77
Laurel La. UB7: W Dray 49N 83
Laurel Lodge La. EN5: Barn 8Ya 16
Laurells, The KT22: Fet 96Fa 192
Laurel Mnr. SM2: Sutt 80Eb 155
Laurel M. SE5 55Sb 113
(off Harbour Rd.)
Laurel Pk. HA3: Hrw W 24Ha 46
Laurel Rd. AL1: St A 2D 6
Laurel Rd. SW13 54Wa 110
Laurel Rd. SW20 67Xa 132
Laurel Rd. TW12: Hamp H 64Fa 130
Laurels, The AL2: Brick W 10N 5
Laurels, The BR1: Broml 67Kc 137
Laurels, The BR2: Broml 70Jc 137
Laurels, The DA2: Wilm 62Ld 141
Laurels, The DA3: Lfield 69Ee 143
Laurels, The IG9: Buck H 18Lc 35
Laurels, The KT11: Cobh 87Aa 171
Laurels, The KT13: Weyb 76T 150
Laurels, The NW10 39Xa 68
Laurels, The SM7: Bans 89Bb 175
Laurels, The SW9 52Rb 113
(off Langton Rd.)
Laurels, The WD23: B Hea 19Ga 28
Laurels, The WD6: Bore 11Qa 29
Laurelsfield AL3: St A 5P 5
Laurels Rd. SL0: Iver H 40F 62
Laurel St. E8 37Vb 71
Laurel Vw. N12 20Db 31
Laurel Way E18 28Hc 53
Laurel Way N20 20Cb 31
Laurence Calvert Cl. IG11: Bark 40Uc 74
Laurence Ct. E10 31Dc 72
Laurence Ct. W11 45Ab 88
(off Lansdowne Rd.)
Laurence M. W12 47Wa 88
Laurence Pountney Hill EC4 4F 225 (45Tb 91)
Laurence Pountney La. EC4 4F 225 (45Tb 91)
Laurence Ri. DA2: Dart 58Sd 120
Laurence Rd. TW3: Houn 55Ea 108
Laurie Cl. TN14: S'oaks 92Kd 203
Laurie Gro. SE14 53Ac 114
Laurie Ho. SE1 4C 230 (48Rb 91)
(off St George's Rd.)
Laurie Ho. W8 46Cb 89
(off Airlie Gdns.)
Laurie Rd. W7 43Ga 86
Laurier Rd. CR0: C'don 73Vb 157
Laurier Rd. NW5 34Kb 70
Lauries Cl. HP1: Hem H 4E 2
Laurie Wlk. RM1: Rom 29Gd 56
Laurimel Cl. HA7: Stan 23Ka 46
Laurino Pl. WD23: B Hea 19Ea 28
Lauriston Cl. GU21: Knap 9H 167
Lauriston Ho. E9 38Yb 72
(off Lauriston Rd.)
Lauriston Rd. E9 38Yb 72
Lauriston Rd. SW19 65Za 132
Lausanne Rd. N8 28Qb 50
Lausanne Rd. SE15 53Yb 114
Lauser Rd. TW19: Stanw 59L 105
Laval Ho. TW8: Bford 50Na 87
(off Ealing Rd.)
Lavell St. N16 35Tb 71
Lavender Av. CM15: Pil H 15Xd 40
Lavender Av. CR4: Mitc 67Gb 133
Lavender Av. KT4: Wor Pk 76Ya 154
Lavender Av. NW9 32Sa 67
Lavender Cl. BR2: Broml 72Nc 160
Lavender Cl. CR3: Cat'm 97Sb 197
Lavender Cl. CR5: Coul 91Lb 196
Lavender Cl. E4 21Cc 52
Lavender Cl. KT22: Lea 94La 192
Lavender Cl. RM15: S Ock 42Ae 99
Lavender Cl. RM3: Rom 24Md 57
Lavender Cl. SM5: Cars 77Kb 156
Lavender Cl. SW3 51Fb 111
Lavender Ct. KT22: Lea 94La 192
Lavender Ct. KT8: W Mole 69Da 129
Lavender Ct. SM2: Sutt 80Eb 155
Lavender Ct. TW14: Felt 58X 107
Lavender Gdns. EN2: Enf 11Rb 33
Lavender Gdns. HA3: Hrw W 23Ga 46
Lavender Gdns. SW11 56Hb 111
Lavender Ga. KT22: Oxs 86Da 171
Lavender Gro. CR4: Mitc 67Gb 133
Lavender Gro. E8 38Wb 71
Lavender Hill BR8: Swan 69Fd 140
Lavender Hill EN2: Enf 11Qb 32
Lavender Hill SW11 56Gb 111
Lavender Ho. SE16 46Zb 92
(off Rotherhithe St.)
Lavender Ho. TW9: Kew 53Ra 109
Lavender M. SS17: Stan H 1N 101
Lavender M. TW12: Hamp H 65Ea 130
Lavender Pk. Rd. KT14: W Byf 84J 169
Lavender Pl. IG1: Ilf 36Rc 74
Lavender Pond Nature Pk. 46Ac 92
Lavender Ri. UB7: W Dray 47Q 84
Lavender Rd. CR0: C'don 72Pb 156
Lavender Rd. EN2: Enf 11Tb 33
Lavender Rd. GU22: Wok 88D 168
Lavender Rd. KT19: Ewe 78Ra 153
Lavender Rd. SE16 46Ac 92
Lavender Rd. SM1: Sutt 77Fb 155
Lavender Rd. SM5: Cars 77Jb 156
Lavender Rd. SW11 55Fb 111
Lavender Rd. UB8: Hil 43P 83
Lavender Sq. SW9 53Pb 112
(off Printers Rd.)
Lavender St. E15 37Gc 73
Lavender Sweep SW11 56Hb 111
Lavender Ter. SW11 55Gb 111
Lavender Va. SM6: W'gton 79Mb 156
Lavender Wlk. CR4: Mitc 69Jb 134
Lavender Wlk. HP2: Hem H 1M 3
Lavender Wlk. SW11 56Hb 111
Lavender Way CR0: C'don 72Zb 158
Lavendon Ho. NW8 5E 214 (42Gb 89)
(off Paveley St.)
Lavengro Rd. SE27 61Sb 135
Lavenha Ct. CM15: B'wood 18Zd 41
Lavenham Rd. SW18 61Bb 133
Lavernock Rd. DA7: Bex 54Cd 118
Lavers Rd. N16 34Ub 71
Laverstoke Gdns. SW15 59Va 110
Laverton M. SW5 49Db 89
Laverton Pl. SW5 49Db 89
Lavette Ho. E3 41Cc 92
(off Rainhill Way)
Lavidge Rd. SE9 61Nc 138
Lavina Gro. N1 1H 217 (40Pb 70)
Lavington Cl. E9 37Bc 72
Lavington Rd. CR0: Bedd 76Pb 156
Lavington Rd. W13 46Ka 86
Lavington St. SE1 7C 224 (46Rb 91)
Lavinia Av. WD25: Wat 6Z 13
Lavinia Rd. DA1: Dart 58Pd 119
Lavisham Ho. BR1: Broml 64Kc 137
Lavrock La. WD3: Crox G 17P 25
Lawdons Gdns. CR0: Wadd 77Rb 157
Lawes Ho. W10 41Bb 89
(off Lancefield St.)
Lawes Way IG11: Bark 41Vc 95
Lawford Av. WD3: Chor 16E 24
Lawford Cl. RM12: Horn 35Ld 77
Lawford Cl. WD3: Chor 16E 24
Lawford Gdns. CR8: Kenley 88Sb 177
Lawford Gdns. DA1: Dart 57Ld 119
Lawford Rd. N1 38Ub 71
Lawford Rd. NW5 37Lb 70
Lawford Rd. W4 52Sa 109
Lawford's Hill Cl. GU3: Worp 6G 186
Lawford's Hill Rd. GU3: Worp 6G 186
Lawfords Wharf NW1 38Lb 70
(off Lyme St.)
Law Ho. IG11: Bark 40Wc 75
Lawkland SL2: Farn R 1G 80
Lawless Ho. E14 45Ec 92
(off Bazely St.)
Lawless St. E14 45Dc 92
Lawley Ho. TW1: Twick 58Ma 109
Lawley Rd. N14 17Kb 32
Lawley St. E5 35Yb 72
Lawlor Cl. TW16: Sun 67X 129
Lawn, The SL3: Dat 3N 103
Lawn, The UB2: S'hall 50Ca 85
Lawn Av. UB7: W Dray 47L 83
Lawn Cl. BR1: Broml 65Kc 137
Lawn Cl. BR8: Swan 68Ed 140
Lawn Cl. HA4: Ruis 34V 64
Lawn Cl. KT3: N Mald 68Ua 132
Lawn Cl. N9 17Vb 33
Lawn Cl. SL3: Dat 2N 103
Lawn Cres. TW9: Kew 54Qa 109
Lawn Farm Gro. RM6: Chad H 28Ad 55
Lawnfield Ct. NW6 38Za 68
(off Coverdale Rd.)
Lawn Gdns. W7 46Ga 86
Lawn Ho. Cl. E14 47Ec 92
Lawn La. HP3: Hem H 4M 3
Lawn La. SW8 51Pb 112
Lawn Pk. TN13: S'oaks 99Kd 203
Lawn Rd. BR3: Beck 66Bc 136
Lawn Rd. DA11: Nflt 58Ee 121
Lawn Rd. NW3 36Hb 69
Lawn Rd. UB8: Uxb 38L 63
Lawns, The AL3: St A 1A 6
Lawns, The CM14: W'ley 22Ae 59
(off Uplands Rd.)
Lawns, The DA14: Sidc 63Xc 139
Lawns, The E4 22Cc 52
Lawns, The HA5: Hat E 24Da 45
Lawns, The HP1: Hem H 1G 2
Lawns, The SE19 67Tb 135
Lawns, The SE3 55Hc 115
Lawns, The SL3: Poyle 53G 104
Lawns, The SM2: Cheam 80Ab 154
Lawns, The SW19 64Bb 133
Lawns, The WD7: Shenl 5Na 15
Lawns Ct. HA9: Wemb 33Pa 67
Lawns Cres. RM17: Grays 51Fe 121
Lawnside SE3 56Hc 115
Lawns Pl. RM17: Grays 51Fe 121
Lawns Way RM5: Col R 24Ed 56
Lawnswood EN5: Barn 15Ab 30
Lawn Ter. SE3 55Gc 115
Lawn Va. HA5: Pinn 26Aa 45
Lawrance Gdns. EN8: Chesh 1Zb 20
Lawrance Sq. DA11: Nflt 2B 144
Lawrence Av. E12 35Qc 74
Lawrence Av. E17 25Zb 52
Lawrence Av. KT3: N Mald 72Ta 153
Lawrence Av. N13 21Rb 51
Lawrence Av. NW10 39Ta 67
Lawrence Av. NW7 21Ua 48
Lawrence Bldgs. N16 34Vb 71
Lawrence Campe Cl. N20 20Fb 31
Lawrence Cl. E3 41Cc 92
Lawrence Cl. N15 28Ub 51
Lawrence Cl. W12 45Xa 88
Lawrence Ct. N10 27Lb 50
Lawrence Ct. N16 34Vb 71
(off Smalley Rd. Est.)
Lawrence Ct. NW7 22Ua 48
Lawrence Ct. SE6 58Cc 114
Lawrence Ct. SL4: Wind 4G 102
Lawrence Ct. W3 48Sa 87
(off Stanley Rd.)
Lawrence Ct. WD19: Wat 20Z 27
Lawrence Cres. GU20: W'sham 9B 146
Lawrence Cres. HA8: Edg 26Qa 47
Lawrence Cres. RM10: Dag 34Dd 76
Lawrence Dr. DA12: Cobh 10J 145
Lawrence Dr. UB10: Ick 35S 64
Lawrence Est. TW4: Houn 56Y 107
Lawrence Gdns. NW7 20Va 30
Lawrence Gdns. RM18: Tilb 2D 122
Lawrence Gro. UB10: Uxb 40N 63
Lawrence Hill E4 19Cc 34
Lawrence Hill Gdns. DA1: Dart 58Ld 119
Lawrence Hill Rd. DA1: Dart 58Ld 119
Lawrence Ho. NW1 38Kb 70
(off Hawley Cres.)
Lawrence Ho. SW1 6E 228 (49Mb 90)
(off Cureton St.)
Lawrence La. EC2 3E 224 (44Sb 91)
Lawrence La. RH3: Bkld 4D 206
Lawrence Mans. SW3 51Gb 111
(off Lordship Pl.)
Lawrence M. SW15 56Ya 110
Lawrence M. SW8 52Nb 112
Lawrence Pde. TW7: Isle 55Ka 108
(off Lower Sq.)
Lawrence Pl. N1 39Nb 70
(off Brydon Wlk.)
Lawrence Rd. BR4: W W'ck 77Jc 159
Lawrence Rd. DA8: Erith 52Dd 118
Lawrence Rd. E13 39Kc 73
Lawrence Rd. E6 39Nc 74
Lawrence Rd. HA5: Pinn 30Z 45
Lawrence Rd. N15 28Ub 51
Lawrence Rd. N18 21Xb 51
(not continuous)
Lawrence Rd. RM2: Rom 29Kd 57
Lawrence Rd. SE25 70Vb 135
Lawrence Rd. TW10: Ham 63La 130
Lawrence Rd. TW12: Hamp 66Ba 129
Lawrence Rd. TW4: Houn 56Y 107
Lawrence Rd. UB4: Hayes 40S 64
Lawrence Rd. W5 49Ma 87
Lawrences Cl. CR5: Coul 91Rb 197
Lawrence St. E16 43Hc 93
Lawrence St. NW7 22Va 48
Lawrence St. SW3 51Gb 111
Lawrence Trad. Est. RM17: Grays 50Ae 99
Lawrence Trad. Est. SE10 49Gc 93
Lawrence Way NW10 34Sa 67
Lawrence Way SL1: Slou 3A 80
Lawrence Weaver Cl. SM4: Mord 72Cb 155
Lawrence Yd. N15 28Ub 51
Lawrie Ho. SW19 64Db 133
(off Durnsford Rd.)
Lawrie Pk. Av. SE26 64Xb 135
Lawrie Pk. Cres. SE26 64Xb 135
Lawrie Pk. Gdns. SE26 63Xb 135
Lawrie Pk. Rd. SE26 65Xb 135
Laws Cl. SE25 70Tb 135
Lawson Cl. E16 43Lc 93
Lawson Cl. IG1: Ilf 36Tc 74
Lawson Cl. SW19 62Za 132
Lawson Ct. KT6: Surb 73Ma 153
Lawson Ct. N11 23Lb 50
(off Ring Way)
Lawson Ct. N4 32Pb 70
(off Lorne Rd.)
Lawson Gdns. DA1: Dart 57Md 119
Lawson Gdns. HA5: Eastc 27X 45
Lawson Ho. SE18 51Qc 116
(off Nightingale Pl.)
Lawson Ho. W12 45Xa 88
(off White City Est.)
Lawson Rd. DA1: Dart 56Md 119
Lawson Rd. EN3: Enf H 11Yb 34
Lawson Rd. UB1: S'hall 42Ca 85
Lawson Ter. SE15 56Yb 114
Lawson Wlk. SM5: Cars 81Jb 176
Lawson Way SL5: S'dale 2F 146
Law St. SE1 3G 231 (48Tb 91)
Lawton Grn. IG10: Lough 12Rc 36
Lawton Rd. E10 32Ec 72
Lawton Rd. E3 41Ac 92
(not continuous)
Lawton Rd. EN4: Cockf 13Fb 31
Lawton Rd. IG10: Lough 12Rc 36
Laxcon Cl. NW10 36Ta 67
Laxey Rd. BR6: Chels 79Vc 161
Laxfield Ct. E8 39Wb 71
(off Pownall Rd.)
Laxford Ho. SW1 6J 227 (49Jb 90)
(off Cundy St.)
Laxley Cl. SE5 52Rb 113
Laxton Ct. CR7: Thor H 70Sb 135
Laxton Gdns. RH1: Mers 100Mb 196
Laxton Gdns. WD7: Shenl 4Na 15
Laxton Pl. NW1 5A 216 (42Kb 90)
Layard Rd. CR7: Thor H 68Tb 135
Layard Rd. EN1: Enf 11Vb 33
Layard Rd. SE16 49Xb 91
Layard Sq. SE16 49Xb 91
Layborne Av. RM3: Rom 19Ld 39
Laybourne Ho. E14 47Cc 92
(off Admirals Way)
Laybrook Lodge E18 28Hc 53
Layburn Cres. SL3: L'ly 51D 104
Laycock St. N1 37Qb 70
Layer Gdns. W3 45Qa 87
Layfield Cl. NW4 31Xa 68
Layfield Cres. NW4 31Xa 68
Layfield Ho. SE10 50Jc 93
(off Kemsing Rd.)
Layfield Rd. NW4 31Xa 68
Layhams Rd. BR2: Kes 78Hc 159
Layhams Rd. BR4: W W'ck 77Gc 159
Layhams Rd. CR6: W'ham 84Hc 179
Layhill HP2: Hem H 1M 3
Laymarsh Cl. DA17: Belv 48Bd 95
Laymead Cl. UB5: N'olt 37Aa 65
Laystall Ct. WC1 6K 217 (42Qb 90)
(off Mt. Pleasant)
Laystall St. EC1 6K 217 (42Qb 90)
Layton Ct. KT13: Weyb 77R 150
Layton Ct. TW8: Bford 50Ma 87
Layton Cres. CR0: Wadd 78Qb 156
Layton Cres. SL3: L'ly 47B 82
Layton Pl. TW9: Kew 53Qa 109
Layton Rd. TW3: Houn 56Da 107
Layton Rd. TW8: Bford 50Ma 87
Layton's La. TW16: Sun 68V 128
Layzell Wlk. SE9 60Mc 115
Lazare Ct. TW18: Staines 64H 127
(off Gresham Rd.)
Lazar Wlk. N7 33Pb 70
Lazenby Ct. WC2 4F 223 (45Nb 90)
(off Floral St.)
Lea, The TW20: Egh 66E 126
Leabank Cl. HA1: Harr 34Ga 66
Leabank Sq. E9 37Cc 72
Leabank Vw. N15 30Wb 51
Lea Bon Ct. E15 39Hc 73
(off Plaistow Gro.)
Leabourne Rd. N16 31Wb 51
LEA BRIDGE 34Zb 72
Lea Bri. Ind. Cen. E10 32Ac 72
Lea Bri. Rd. E10 31Cc 72
Lea Bri. Rd. E17 29Fc 53
Lea Bri. Rd. E5 34Yb 72
Lea Bridge Station (Rail) 32Ac 72
Lea Bushes WD25: Wat 7Aa 13
Leach Gro. KT22: Lea 94La 192
Lea Cl. TW2: Whitt 59Ba 107
Lea Cl. WD23: Bush 15Da 27
Lea Ct. E13 41Jc 93
Lea Ct. E4 19Ec 34
Lea Ct. N15 28Wb 51
Lea Cres. HA4: Ruis 35V 64
Leacroft SL1: Slou 7D 80
Leacroft SL5: S'dale 1E 146
Leacroft TW18: Staines 64J 127
Leacroft Av. SW12 59Hb 111
Leacroft Cl. CR8: Kenley 88Sb 177
Leacroft Cl. N21 19Rb 33
Leacroft Cl. TW18: Staines 63K 127
Leacroft Cl. UB7: Yiew 44N 83
Leacroft Rd. SL0: Iver 44G 82
Leadale Av. E4 19Cc 34
Leadale Rd. N15 30Wb 51
Leadale Rd. N16 30Wb 51
Leadbeaters Cl. N11 22Hb 49
Leadbetter Ct. NW10 38Ta 67
(off Melville Rd.)
Leaden Cl. IG10: Lough 13Rc 36
Leadenhall Mkt. 3H 225 (44Ub 91)
Leadenhall Pl. EC3 3H 225 (44Ub 91)
Leadenhall St. EC3 3H 225 (44Ub 91)
Leadenham Ct. E3 42Cc 92
Leaden Hill CR5: Coul 87Mb 176
Leader Av. E12 36Qc 74
Leadings, The HA9: Wemb 35Sa 67
Leadmill La. E20 35Dc 72
Leaf Cl. HA6: Nwood 24T 44
Leaf Cl. KT7: T Ditt 71Ga 152
Leaf Gro. SE27 64Qb 134
Leaf Hill Dr. RM3: Rom 21Ld 57
Leaf Ho. HA1: Harr 29Ha 46
(off Catherine Pl.)
Leafield Cl. GU21: Wok 10N 167
Leafield Cl. SW16 65Rb 135
Leafield La. DA14: Sidc 62Bd 139
Leafield Rd. SM1: Sutt 75Cb 155
Leafield Rd. SW20 69Bb 133
Leaford Ct. WD24: Wat 8V 12
Leaford Cres. WD24: Wat 9V 12
Leaf Wlk. N7 35Mb 70
Leafy Way AL1: St A 5B 6
Leafy Gro. BR2: Kes 78Lc 159
Leafy Oak Rd. SE12 63Lc 137
Leafy Way CM13: Hut 18Fe 41
Leafy Way CR0: C'don 75Vb 157
Lea Gdns. HA9: Wemb 35Pa 67
Leagrave St. E5 34Yb 72
Lea Hall Gdns. E10 32Cc 72
Lea Hall Rd. E10 32Cc 72
Leaholme Gdns. SL1: Slou 3A 80
Leaholme Way HA4: Ruis 30S 44
Lea Ho. NW8 6D 214 (42Gb 89)
(off Salisbury St.)
Leahurst Rd. SE13 57Fc 115
LEA INTERCHANGE 36Cc 72
Leake Ct. SE1 2J 229 (47Pb 90)
Leake St. SE1 1J 229 (47Pb 90)
(not continuous)
Lealand Rd. N15 30Vb 51
Leamington Av. BR1: Broml 64Lc 137
Leamington Av. BR6: Orp 77Uc 160
Leamington Av. E17 29Cc 52
Leamington Av. SM4: Mord 70Ab 132
Leamington Cl. BR1: Broml 63Lc 137
Leamington Cl. E12 36Nc 74
Leamington Cl. RM3: Rom 23Qd 57
Leamington Cl. TW3: Houn 57Ea 108
Leamington Ct. SE3 51Gc 115
Leamington Cres. HA2: Harr 34Aa 65
Leamington Gdns. IG3: Ilf 33Vc 75
Leamington Ho. HA8: Edg 22Pa 47
Leamington Ho. W11 43Bb 89
(off Tavistock Rd.)
Leamington Pk. W3 43Ta 87
Leamington Pl. UB4: Hayes 42V 84
Leamington Rd. RM3: Rom 22Qd 57
Leamington Rd. UB2: S'hall 49Z 85
Leamington Rd. Vs. W11 43Bb 89
Leamore Ct. E2 41Zb 92
Leamore St. W6 49Ya 88
Lea Mt. EN7: G Oak 1Ub 19
LEAMOUTH 45Gc 93
Leamouth Rd. E14 44Fc 93
Leamouth Rd. E6 43Nc 94
Leander Ct. E9 38Zb 72
(off Lauriston Rd.)
Leander Ct. KT6: Surb 73Ma 153
Leander Ct. NW9 25Ua 48
Leander Ct. SE8 53Cc 114
Leander Dr. DA12: Grav'nd 3H 145
Leander Gdns. WD25: Wat 9Aa 13
Leander Rd. CR7: Thor H 70Pb 134
Leander Rd. SW2 58Pb 112
Leander Rd. UB5: N'olt 40Ca 65
Lea Pk. Trad. Est. E10 31Bc 72
Learner Dr. HA2: Harr 33Ca 65
Lea Rd. BR3: Beck 68Cc 136
Lea Rd. EN2: Enf 11Tb 33
Lea Rd. EN9: Walt A 6Cc 20
Lea Rd. RM16: Grays 10C 100
Lea Rd. TN13: S'oaks 99Ld 203
Lea Rd. UB2: S'hall 49Aa 85
Lea Rd. WD24: Wat 10X 13
Lea Rd. Ind. Pk. EN9: Walt A 6Cc 20
Lea Rd. Trad. Est. EN9: Walt A 6Cc 20
Learoyd Gdns. E6 45Pc 94
Leary Ho. SE11 7J 229 (50Pb 90)
Leas, The HP3: Hem H 6A 4
Leas, The RM14: Upm 31Td 78
Leas, The TW18: Staines 63J 127
Leas, The WD23: Bush 11Ba 27

Leas Cl. KT9: Chess 80Pa 153
Leas Dale SE9 62Qc 138
Leas Dr. SL0: Iver 44G 82
Leas Grn. BR7: Chst 65Vc 139
Leaside HP2: Hem H 3C 4
Leaside KT23: Bookh 95Ca 191
Leaside Av. N10 27Jb 50
Leaside Bus. Cen. EN3: Brim 12Bc 34
Leaside Ct. UB10: Hil 41R 84
Leaside Mans. N10 27Jb 50
(off Fortis Grn.)
Leaside Rd. E5 32Yb 72
Leas La. CR6: W'ham 90Zb 178
Leasowes Rd. E10 32Cc 72
Lea Sq. E3 39Bc 72
Leas Rd. CR6: W'ham 90Zb 178
Leasway CM14: B'wood 20Zd 41
Leasway RM14: Upm 35Sd 78
Leasway RM16: Grays 46Ee 99
Leathart Cl. RM12: Horn 38Kd 77
Leatherbottle Grn. DA18: Erith 48Bd 95
Leather Bottle La. DA17: Belv 49Ad 95
Leather Cl. CR4: Mitc 68Jb 134
Leatherdale St. E1: Portelet Rd. 41Zb 92
Leather Gdns. E15 39Gc 73
LEATHERHEAD 94Ka 192
Leatherhead Bus. Pk. KT22: Lea 91Ha 192
Leatherhead Cl. N16 32Vb 71
LEATHERHEAD COMMON 91Ja 192
LEATHERHEAD COMMUNITY HOSPITAL 94La 192
Leatherhead Golf Course 88Ja 172
Leatherhead Leisure Cen. 95Ja 192
Leatherhead Mus. of Local History 94Ka 192
Leatherhead Rd. KT21: Asht 92Ma 193
Leatherhead Rd. KT22: Lea 93Ma 193
Leatherhead Rd. KT22: Oxs 86Fa 172
Leatherhead Rd. KT23: Bookh 98Da 191
Leatherhead Rd. KT9: Chess 86Ka 172
Leatherhead Station (Rail) 93Ja 192
Leatherhead Theatre 94Ka 192
Leatherhead Trade Pk. KT22: Lea 93Ja 192
Leather La. EC1 7K 217 (43Qb 90)
(not continuous)
Leather La. RM11: Horn 32Md 77
Leather Mkt., The SE1 2H 231 (47Ub 91)
(off Weston St.)
Leathermarket Ct. SE1 2H 231 (47Ub 91)
Leathermarket St. SE1 2H 231 (47Ub 91)
Leather Pl. SE1 4J 231 (48Ub 91)
(off Crimscott St.)
Leather Rd. SE16 49Zb 92
Leathersellers Cl. EN5: Barn 13Ab 30
(off The Avenue)
Leather St. E1 45Zb 92
Leathsail Rd. HA2: Harr 34Da 65
Leathwaite Rd. SW11 56Hb 111
Leathwell Rd. SE8 54Dc 114
Lea Va. DA1: Cray 56Fd 118
Lea Valley Bus. Pk. E10 33Ac 72
Lea Valley Rd. E4 16Bc 34
Lea Valley Rd. EN3: Pond E 15Ac 34
Lea Valley Trad. Est. N18 23Zb 52
Lea Valley Viaduct N18 22Zb 52
Leaveland Cl. BR3: Beck 70Cc 136
Leaver Gdns. UB6: G'frd 40Fa 66
LEAVESDEN 4V 12
Leavesden Country Pk. 4W 12
Leavesden Ct. WD5: Ab L 3W 12
Leavesden Film Studios 5U 12
LEAVESDEN GREEN 5W 12
LEAVESDEN GREEN 7U 12
Leavesden Pk. WD25: Wat 5V 12
Leavesden Rd. HA7: Stan 23Ja 46
Leavesden Rd. KT13: Weyb 78R 150
Leavesden Rd. WD24: Wat 10X 13
LEAVES GREEN 83Mc 179
Leaves Grn. Cres. BR2: Kes 83Lc 179
Leaves Grn. Rd. BR2: Kes 83Mc 179
Lea Vw. EN9: Walt A 5Dc 20
Lea Vw. Ho. E5 32Xb 71
Leaway E10 32Zb 72
Leazes Av. CR3: Cat'm 95Qb 196
Lebanon Av. TW13: Hanw 64Z 129
Lebanon Cl. WD17: Wat 8T 12
Lebanon Ct. TW1: Twick 59Ka 108
Lebanon Dr. KT11: Cobh 85Ca 171
Lebanon Gdns. SW18 58Cb 111
Lebanon Gdns. TN16: Big H 89Mc 179
Lebanon Pk. TW1: Twick 59Ka 108
Lebanon Rd. CR0: C'don 74Ub 157
Lebanon Rd. SW18 57Cb 111
Lebanon Road Stop (London Tramlink) 75Ub 157
Leben Ct. SM1: Sutt 79Eb 155
Lebus Ho. NW8 2D 214 (40Gb 69)
(off Cochrane St.)
Lebus St. N17 27Xb 51
Le Chateau CR0: C'don 76Tb 157
(off Chatsworth Rd.)
Lechmere App. IG8: Wfd G 26Lc 53
Lechmere Av. IG7: Chig 21Sc 54
Lechmere Av. IG8: Wfd G 26Mc 53
Lechmere Rd. NW2 37Xa 68
Leckford Rd. SW18 61Eb 133
Leckhampton Pl. SW2 59Qb 112
Leckwith Av. DA7: Bex 51Ad 117
Lecky St. SW7 7B 226 (50Fb 89)
Leconfield Av. SW13 55Va 110
Leconfield Ho. SE5 56Ub 113
Leconfield Rd. N5 35Tb 71
Leconfield Wlk. RM12: Horn 37Ld 77
Le Corte Cl. WD4: K Lan 1P 11
Lectern La. AL1: St A 6C 6
Leda Av. EN3: Enf W 10Zb 20
Leda Ct. SW9 52Qb 112
(off Caldwell St.)
Leda M. HP2: Hem H 1P 3
Ledam Ho. EC1 7K 217 (43Qb 90)
(off Bourne Est.)
Leda Rd. SE18 48Pc 94
Ledbury Ho. SE22 55Ub 113
Ledbury Ho. W11 44Bb 89
(off Colville Rd.)
Ledbury M. Nth. W11 45Cb 89
Ledbury M. W. W11 45Cb 89
Ledbury Pl. CR0: C'don 77Sb 157
Ledbury Rd. CR0: C'don 77Tb 157
Ledbury Rd. RH2: Reig 6J 207
Ledbury Rd. W11 44Bb 89
Ledbury St. SE15 52Wb 113
Ledger Dr. KT15: Add 78H 149
Ledger M. E17 30Cc 52
Ledgers La. CR6: W'ham 89Dc 178
Ledgers Rd. CR6: W'ham 88Cc 178
Ledgers Rd. SL1: Slou 7H 81
Ledrington Rd. SE19 65Wb 135
Ledway Dr. HA9: Wemb 31Pa 67
LEE 58Jc 115
Lee, The HA6: Nwood 22V 44
Lee Av. RM6: Chad H 30Ad 55
Lee Bri. SE13 55Ec 114
Leechcroft Av. BR8: Swan 69Hd 140
Leechcroft Av. DA15: Sidc 57Vc 117
Leechcroft Rd. SM6: W'gton 76Jb 156
Leech La. KT18: Head 98Ra 193
Lee Chu. St. SE13 56Gc 115
Lee Cl. E17 25Zb 52
Lee Cl. EN5: New Bar 14Eb 31
Lee Conservancy Rd. E9 36Bc 72
Lee Ct. E3 42Ec 92
(off Navigation Rd.)
Lee Ct. SE13 56Fc 115
(off Lee High Rd.)
Leecroft Rd. EN5: Barn 15Ab 30
Leeds Cl. BR6: Chels 75Zc 161
Leeds Ct. EC1 5B 218 (42Rb 91)
(off St John St.)
Leeds Pl. N4 32Pb 70
Leeds Rd. IG1: Ilf 32Tc 74
Leeds Rd. SL1: Slou 5J 81
Leeds St. N18 22Wb 51
Leefern Rd. W12 47Wa 88
Leefe Way EN6: Cuff 1Mb 18
Lee Gdns. Av. RM11: Horn 32Qd 77
Leegate SE12 57Hc 115
Leegate Cl. GU21: Wok 8M 167
LEE GREEN 57Hc 115
Lee Grn. BR5: St M Cry 71Wc 161
Lee Grn. La. KT18: Head 96Ra 193
Lee Gro. IG7: Chig 19Rc 36
Lee High Rd. SE12 56Hc 115
Lee High Rd. SE13 55Ec 114
Leeke St. WC1 3H 217 (41Pb 90)
Leeland Rd. W13 46Ja 86
Leeland Ter. W13 46Ja 86
Leeland Way NW10 35Va 68
Lee M. BR3: Beck 69Ac 136
Leeming Rd. WD6: Bore 11Pa 29
Leemount Ho. NW4 28Za 48
Lee Pk. SE3 56Hc 115
Lee Pk. Way N18 21Zb 52
Lee Pk. Way N9 21Zb 52
Leerdam Dr. E14 48Ec 92
Lee Rd. EN1: Enf 16Wb 33
Lee Rd. NW7 24Za 48
Lee Rd. SE3 55Hc 115
Lee Rd. SW19 67Db 133
Lee Rd. UB6: G'frd 39La 66
Lees, The CR0: C'don 75Bc 158
Lees Av. HA6: Nwood 25V 44
Lees Ct. W1 4H 221 (45Jb 90)
(off Lees Pl.)
Lees Ho. SE17 7G 231 (50Tb 91)
(off Inville Rd.)
Leeside EN5: Barn 15Ab 30
Leeside EN6: Pot B 4Fb 17
Leeside Ct. SE16 46Zb 92
(off Rotherhithe St.)
Leeside Cres. NW11 30Ab 48
Leeside Ind. Est. N17 24Yb 52
Leeside Rd. N17 23Xb 51
Leeside Works N17 24Yb 52
Leeson Gdns. SL4: Eton W 9C 80
Leeson Ho. TW1: Twick 59Ka 108
Leeson Rd. SE24 56Qb 112
Leesons Hill BR5: St P 69Wc 139
Leesons Hill BR7: Chst 69Uc 138
Leeson's Way BR5: St P 68Vc 139
Lees Pde. UB10: Hil 42R 84
Lees Pl. W1 4H 221 (45Jb 90)
Lees Rd. UB8: Hil 42R 84
Lee Station (Rail) 58Jc 115
Lee St. E8 39Vb 71
Lee Ter. SE13 55Gc 115
Lee Ter. SE3 55Gc 115
Lee Valley Athletics Cen. 18Ac 34
Lee Valley Golf Course 17Ac 34
Lee Valley Hockey & Tennis Cen. 35Dc 72
Lee Valley Ice Cen. 33Zb 72
Lee Valley Pk. 24Zb 52
Lee Valley Pk. Info. Cen. 8Wb 19
Lee Valley Regional Pk. 3Cc 20
Lee Valley Technopark N17 27Wb 51
Lee Valley VeloPark 36Dc 72
Lee Valley White Water Cen. 5Cc 20
Leeve Ho. W10 41Bb 89
(off Lancefield St.)
Lee Vw. EN2: Enf 11Rb 33
Leeward Ct. E1 46Wb 91
Leeward Ct. SE8 49Ac 92
(off Yeoman St.)
Leeward Gdns. SW19 64Ab 132
Leeward Ho. N1 39Ub 71
(off Halcomb St.)
Leeway SE8 50Bc 92
Leeway Cl. HA5: Hat E 24Ba 45
Leeways, The SM3: Cheam 79Ab 154
Leewood Cl. SE12 58Jc 115
Leewood Pl. BR8: Swan 70Fd 140
Leewood Way KT24: Eff 99Y 191
Lefa Bus. & Ind. Pk. DA14: Sidc 65Zc 139
Lefevre Wlk. E3 40Cc 72
Leff Ho. NW6 39Ab 68
Lefroy Ho. SE1 2D 230 (47Sb 91)
(off Southwark Bri. Rd.)
Lefroy Rd. W12 47Wa 88
Left Side N14 18Mb 32
(off Statione Pde.)
Legacy Bldg. SW11 51Mb 112
Legacy Wharf E15 40Dc 72
Legard Rd. N5 34Rb 71
Legatt Rd. SE9 57Mc 115
Leggatt Rd. E15 40Ec 72
Leggatts Cl. WD24: Wat 8V 12
Leggatts Ri. WD25: Wat 7W 12
Leggatts Way WD24: Wat 8V 12
Leggatts Wood Av. WD24: Wat 8X 13
Legge St. SE13 57Ec 114
Leggfield Ter. HP1: Hem H 2H 3
Leghorn Rd. NW10 40Va 68
Leghorn Rd. SE18 50Tc 94
Legion Cl. N1 38Qb 70
Legion Ct. SM4: Mord 72Cb 155
Legion Rd. UB6: G'frd 39Ea 66
Legion Ter. E3 39Bc 72
Legion Way N12 24Gb 49
Legoland 7B 102
Leg O'Mutton Reservoir Local Nature Reserve 52Va 110
Legon Av. RM7: Rush G 32Ed 76
Legrace Av. TW4: Houn 54Z 107
Leicester Av. CR4: Mitc 70Nb 134
Leicester Cl. KT4: Wor Pk 77Ya 154
Leicester Ct. TW1: Twick 58Ma 109
(off Clevedon Rd.)
Leicester Ct. W9 43Cb 89
(off Elmfield Way)
Leicester Ct. WC2 4E 222 (45Mb 90)
(off Lisle St.)
Leicester Flds. 5E 222 (45Mb 90)
(off Leicester Sq.)
Leicester Gdns. IG3: Ilf 31Uc 74
Leicester Ho. N18 22Xb 51
(off Cavendish Cl.)
Leicester Ho. SW9 55Rb 113
(off Loughborough Rd.)
Leicester M. N2 27Gb 49
Leicester Pl. WC2 4E 222 (45Mb 90)
Leicester Rd. CR0: C'don 73Ub 157
Leicester Rd. E11 29Kc 53
Leicester Rd. EN5: New Bar 15Db 31
Leicester Rd. N2 27Gb 49
Leicester Rd. NW10 38Ta 67
Leicester Rd. RM18: Tilb 3B 122
Leicester Sq. WC2 5E 222 (45Mb 90)
Leicester Square Station (Underground) 4F 223 (45Nb 90)
Leicester Square Theatre 4E 222 (45Mb 90)
(off Leicester Pl.)
Leicester St. WC2 4E 222 (45Mb 90)
Leigh, The KT2: King T 66Ua 132
Leigham Av. SW16 62Nb 134
Leigham Cl. SW16 62Pb 134
Leigham Ct. SM6: W'gton 79Lb 156
Leigham Ct. Rd. SW16 61Nb 134
Leigham Dr. TW7: Isle 52Ga 108
Leigham Hall Pde. SW16 62Nb 134
(off Streatham High Rd.)
Leigham Va. SW16 62Pb 134
Leigham Va. SW2 61Qb 134
Leigh Av. IG4: Ilf 28Mc 53
Leigh Cl. KT15: Add 80H 149
Leigh Cl. KT3: N Mald 70Sa 131
Leigh Cl. Ind. Est. KT3: N Mald 70Ta 131
Leigh Cnr. KT11: Cobh 87Y 171
Leigh Ct. HA2: Harr 32Ga 66
Leigh Ct. KT11: Cobh 86Z 171
Leigh Ct. W14 49Bb 89
(off Avonmore Pl.)
Leigh Ct. WD6: Bore 12Ta 29
Leigh Ct. Cl. KT11: Cobh 86Y 171
Leigh Cres. CR0: New Ad 80Dc 158
Leigh Dr. RM3: Rom 21Md 57
Leigh Gdns. NW10 40Ya 68
Leigh Hill Rd. KT11: Cobh 87Y 171
Leigh Hunt Dr. N14 18Mb 32
Leigh M. KT16: Chert 74J 149
Leigh Orchard Cl. SW16 62Pb 134
Leigh Pk. SL3: Dat 2M 103
Leigh Pl. DA16: Well 54Wc 117
Leigh Pl. DA2: Hawl 63Qd 141
Leigh Pl. EC1 7K 217 (43Qb 90)
Leigh Pl. KT11: Cobh 87Y 171
Leigh Pl. TW13: Felt 60Y 107
Leigh Pl. La. RH9: G'stone 4B 210
Leigh Rd. DA11: Grav'nd 1D 144
Leigh Rd. E10 31Ec 72
Leigh Rd. E6 37Qc 74
Leigh Rd. KT11: Cobh 86X 171
Leigh Rd. N5 35Rb 71
Leigh Rd. SL1: Slou 5F 80
Leigh Rd. TW3: Houn 56Fa 108
Leigh Rodd WD19: Wat 20Ba 27
Leigh Sq. SL4: Wind 4B 102
Leigh St. WC1 4F 217 (41Nb 90)
Leigh Ter. BR5: St P 69Xc 139
Leighton Av. E12 36Qc 74
Leighton Av. HA5: Pinn 27Aa 45
Leighton Cl. HA8: Edg 26Qa 47
Leighton Ct. EN8: Chesh 1Zb 20
Leighton Cres. NW5 36Lb 70
Leighton Gdns. CR0: C'don 74Rb 157
Leighton Gdns. CR2: Sande 85Xb 177
Leighton Gdns. NW10 40Xa 68
Leighton Gdns. RM18: Tilb 2C 122
Leighton Gro. NW5 36Lb 70
Leighton Ho. EN6: Pot B 4Cb 17
Leighton Ho. SW1 6E 228 (49Mb 90)
(off Herrick St.)
Leighton House Mus. 48Bb 89
Leighton Mans. W14 51Ab 110
(off Greyhound Rd.)
Leighton Pl. NW5 36Lb 70
Leighton Rd. EN1: Enf 15Vb 33
Leighton Rd. HA3: Hrw W 26Fa 46
Leighton Rd. NW5 36Lb 70
Leighton Rd. W13 47Ja 86
Leighton St. CR0: C'don 74Rb 157
Leighton Way KT18: Eps 86Ta 173
Leila Parnell Pl. SE7 51Lc 115
Leinster Av. SW14 55Sa 109
Leinster Gdns. W2 3A 220 (44Eb 89)
Leinster M. EN5: Barn 13Ab 30
Leinster M. W2 4A 220 (45Eb 89)
Leinster Pl. W2 44Eb 89
Leinster Rd. N10 28Kb 50
Leinster Sq. W2 44Cb 89
(not continuous)
Leinster Ter. W2 4A 220 (45Eb 89)
Leirum St. N1 1J 217 (39Pb 70)
Leiston Spur SL1: Slou 4J 81
Leisure La. KT14: W Byf 84K 169
Leisure Way N12 24Fb 49
Leisure W. 61X 129
Leitch Ho. NW8 38Fb 69
(off Hilgrove Rd.)
Leith Cl. KT15: Add 79L 149
Leith Cl. NW9 32Ta 67
Leith Cl. SL1: Slou 6L 81
Leithcote Gdns. SW16 63Pb 134
Leithcote Path SW16 62Pb 134
Leith Hill BR5: St P 67Wc 139
Leith Hill Grn. BR5: St P 67Wc 139
Leith Mans. W9 41Db 89
(off Grantully Rd.)
Leith Pk. Rd. DA12: Grav'nd 10D 122
Leith Rd. KT17: Eps 84Ua 174
Leith Rd. N22 25Rb 51
Leith Towers SM2: Sutt 80Db 155
Leith Yd. NW6 39Cb 69
(off Quex Rd.)
Lela Av. TW4: Houn 54Y 107
Lelitia Cl. E8 39Wb 71
Lely Ho. UB5: N'olt 40Z 65
(off Academy Gdns.)
Leman St. E1 2K 225 (44Vb 91)
Le Mare Ter. E3 43Cc 92
Le May Av. SE12 62Kc 137
Lemmon Rd. SE10 51Gc 115
Lemna Rd. E11 31Hc 73
Le Moal Ho. E1 43Yb 92
(off Stepney Way)
Lemonade Bldg. IG11: Bark 38Sc 74
(off Ripple Rd.)
Lemonfield Dr. WD25: Wat 4Aa 13
Lemon Gro. TW13: Felt 60W 106
Lemon Tree Ho. E3 41Bc 92
(off Bow Rd.)
Lemonwell Dr. SE9 57Sc 116
Lemsford Cl. N15 30Wb 51
Lemsford Ct. N4 33Sb 71
Lemsford Ct. WD6: Bore 14Sa 29
Lemsford Rd. AL1: St A 2D 6
Lemuel St. SW18 58Eb 111
Lena Cres. N9 19Yb 34
Lena Gdns. W6 48Ya 88
Lena Kennedy Cl. E4 23Ec 52
Lenanton Steps E14 47Cc 92
(off Manilla St.)
Len Bishop Ct. E1 45Zb 92
(off Schoolhouse La.)
Len Clifton Ho. SE18 49Pc 94
(off Cambridge Barracks Rd.)
Lendal Ter. SW4 55Mb 112
Lenderyou Ct. DA1: Dart 59Md 119
(off Phoenix Pl.)
Lendon Rd. TN15: Bor G 93Be 205
Lendy Pl. TW16: Sun 70W 128
Lenelby Rd. KT6: Surb 74Qa 153
Len Freeman Pl. SW6 51Bb 111
Lenham Ho. SE1 3G 231 (48Tb 91)
(off Staple St.)
Lenham Rd. CR7: Thor H 68Tb 135
Lenham Rd. DA7: Bex 51Bd 117
Lenham Rd. SE12 56Hc 115
Lenham Rd. SM1: Sutt 77Db 155
Lenmore Av. RM17: Grays 48Ee 99
Lennard Av. BR4: W W'ck 75Gc 159
Lennard Cl. BR4: W W'ck 75Gc 159
Lennard Rd. BR2: Broml 74Pc 160
Lennard Rd. BR3: Beck 65Ac 136
Lennard Rd. CR0: C'don 74Sb 157
Lennard Rd. SE20 65Zb 136
Lennard Rd. TN13: Dun G 92Gd 202
Lennard Row RM15: Avel 46Td 98
Lennon Rd. NW2 36Ya 68
Lennox Av. DA11: Grav'nd 8B 122
Lennox Cl. RM1: Rom 30Hd 56
Lennox Cl. RM16: Chaf H 49Yd 98
Lennox Ct. RH1: Redh 5A 208
(off St Anne's Ri.)
Lennox Gdns. CR0: Wadd 77Rb 157
Lennox Gdns. IG1: Ilf 32Pc 74
Lennox Gdns. NW10 35Va 68
Lennox Gdns. SW1 4F 227 (48Hb 89)
Lennox Gdns. M. SW1 4F 227 (48Hb 89)
Lennox Ho. DA17: Belv 48Cd 96
(off Ambrooke Rd.)
Lennox Ho. TW1: Twick 58Ma 109
(off Clevedon Rd.)
Lennox Rd. DA11: Grav'nd 8B 122
Lennox Rd. E17 30Bc 52
Lennox Rd. N4 33Pb 70
Lennox Rd. SW9 53Rb 113
Lennox Rd. E. DA11: Grav'nd 9C 122
Le Noke Av. RM3: Rom 22Ld 57
Lenor Cl. DA6: Bex 56Ad 117
Lensbury Av. SW6 54Eb 111
Lensbury Cl. EN8: Chesh 1Ac 20
Lensbury Way SE2 48Yc 95
Lens Rd. E7 38Lc 73
LENT 2A 80
Len Taylor Cl. UB4: Hayes 42U 84
Lenthall Av. RM17: Grays 47Ce 99
Lenthall Ho. SW1 50Lb 90
(off Churchill Gdns.)
Lenthall Rd. E8 38Wb 71
Lenthall Rd. IG10: Lough 14Tc 36
Lenthorp Rd. SE10 49Hc 93
Lentmead Rd. BR1: Broml 62Hc 137
Lenton Path SE18 51Tc 116
Lenton Ri. TW9: Rich 55Na 109
Lenton St. SE18 49Tc 94
Lenton Ter. N4 33Qb 70
Len Williams Ct. NW6 40Cb 69
Leo Ct. TW8: Bford 52Ma 109
Leof Cres. SE6 64Dc 136
Leominster Rd. SM4: Mord 72Eb 155
Leominster Wlk. SM4: Mord 72Eb 155
Leonard Av. DA10: Swans 59Ae 121
Leonard Av. RM7: Rush G 32Fd 76
Leonard Av. SM4: Mord 71Eb 155
Leonard Av. TN14: Otf 88Kd 183
Leonard Cir. EC2 5G 219 (42Tb 91)
Leonard Ct. HA3: Hrw W 25Ga 46
Leonard Ct. W8 48Cb 89
Leonard Ct. WC1 5E 216 (42Mb 90)
UB7: Sip
Leonard Pl. N16 35Ub 71
Leonard Rd. E4 23Cc 52
Leonard Rd. E7 35Jc 73
Leonard Rd. N9 20Vb 33
Leonard Rd. SW16 67Lb 134
Leonard Rd. UB2: S'hall 48Z 85
Leonard Robbins Path SE28 45Xc 95
(off Tawney Rd.)
Leonard St. E16 46Nc 94
Leonard St. EC2 5G 219 (42Tb 91)
Leonard Way CM14: B'wood 21Ud 58
Leon Ho. CR0: C'don 76Sb 157
Leonora Ho. W9 5A 214 (42Eb 89)
(off Lanark Rd.)
Leonora Tyson M. SE21 61Tb 135
Leontine Cl. SE15 52Wb 113
Leopards Ct. EC1 7K 217 (43Qb 90)
(off Baldwins Gdns.)
Leopold Av. SW19 64Bb 133
Leopold Bldgs. E2 3K 219 (41Vb 91)
(off Columbia Rd.)
Leopold Ct. KT10: Esh 79Ea 152
(off Princess Sq.)
Leopold M. E9 39Yb 72
Leopold Rd. E17 29Cc 52
Leopold Rd. N18 22Xb 51
Leopold Rd. N2 27Fb 49
Leopold Rd. NW10 38Ua 68
Leopold Rd. SW19 63Bb 133
Leopold Rd. W5 46Pa 87
Leopold St. E3 43Bc 92
Leopold Ter. SW19 64Bb 133
Leo St. SE15 52Xb 113
Leo Yd. EC1 6C 218 (42Rb 91)
(off St John St.)
Le Personne Homes CR3: Cat'm 94Tb 197
(off Banstead Rd.)
Le Personne Rd. CR3: Cat'm 94Tb 197
Leppoc Rd. SW4 57Mb 112
Leret Way KT22: Lea 93Ka 192
Leroy St. SE1 5H 231 (49Ub 91)
Lerry Cl. W14 51Bb 111
Lerwick Ct. EN1: Enf 15Ub 33
Lerwick Dr. SL1: Slou 3J 81
Lesbourne Rd. RH2: Reig 7K 207
Lescombe Cl. SE23 62Ac 136
Lescombe Rd. SE23 62Ac 136
Lescot Pl. BR2: Broml 72Nc 160
Lesley Cl. BR8: Swan 69Fd 140
Lesley Cl. DA13: Ist R 7B 144
Lesley Cl. DA5: Bexl 59Dd 118
Lesley Ct. SW1 4D 228 (48Mb 90)
(off Strutton Ground)
Leslie Dunne Ho. SL4: Wind 4C 102
Leslie Foster Cl. IG11: Bark 40Vc 75
Leslie Gdns. SM2: Sutt 79Cb 155
Leslie Gro. CR0: C'don 74Ub 157
Leslie Gro. Pl. CR0: C'don 74Ub 157
Leslie Ho. SW8 52Nb 112
(off Wheatsheaf La.)
Leslie Pk. Rd. CR0: C'don 74Ub 157
Leslie Prince Ct. SE5 52Tb 113
Leslie Rd. E11 35Ec 72
Leslie Rd. E16 44Kc 93
Leslie Rd. GU24: Chob 2J 167
Leslie Rd. N2 27Fb 49
Leslie Smith Sq. SE18 51Qc 116
Lesnes Abbey Woods 49Zc 95
Lesney Av. E20 36Cc 72
Lesney Farm Est. DA8: Erith 52Fd 118
Lesney Pk. DA8: Erith 51Fd 118
Lesney Pk. Rd. DA8: Erith 51Fd 118
Lessar Av. SW4 58Lb 112
Lessingham Av. IG5: Ilf 27Qc 54
Lessingham Av. SW17 63Hb 133
Lessing St. SE23 59Ac 114
Lessington Av. RM7: Rom 30Ed 56
Lessness Av. DA7: Bex 52Zc 117
LESSNESS HEATH 50Cd 96
Lessness Pk. DA17: Belv 50Bd 95
Lessness Rd. DA17: Belv 51Cd 118
Lessness Rd. SM4: Mord 72Eb 155
Lester Av. E15 42Gc 93
Lester Ct. E3 41Dc 92
(off Bruce Rd.)
Lester Ct. WD24: Wat 9Y 13
Lestock Cl. SE25 69Wb 135
(off Manor Rd.)
Leston Cl. RM13: Rain 41Kd 97
Leswin Pl. N16 34Vb 71
Leswin Rd. N16 34Vb 71
Letchford Gdns. NW10 41Wa 88
Letchford Ho. E3 40Cc 72
(off Thomas Fyre Dr.)
Letchford M. NW10 41Wa 88
Letchford Ter. HA3: Hrw W 25Da 45
LETCHMORE HEATH 11Ga 28
Letchmore Ho. W10 42Ya 88
(off Sutton Way)
Letchmore Rd. WD7: R'lett 8Ja 14
Letchworth Av. TW14: Felt 59V 106
Letchworth Cl. BR2: Broml 71Jc 159
Letchworth Cl. WD19: Wat 22Z 45
Letchworth Dr. BR2: Broml 71Jc 159
Letchworth Rd. HA7: Stan 24Na 47
Letchworth St. SW17 63Hb 133
Lethbridge Cl. SE10 53Ec 114
Letterstone Rd. SW6 52Bb 111
Lettice St. SW6 53Bb 111
Lett La. DA10: Swans 60Be 121
Lett Rd. E15 38Fc 73
Lett Rd. SW9 53Pb 112
LETT'S GREEN 88Vc 181
Lettsom St. SE5 54Ub 113
Lettsom Wlk. E13 40Jc 73
Leucha Rd. E17 29Ac 52
Levana Cl. SW19 60Ab 110
Levant Ho. E1 42Zb 92
(off Ernest St.)
Levehurst Ho. SE27 64Sb 135
Leven Cl. EN8: Walt C 5Zb 20
Leven Cl. WD19: Wat 22Z 45
Levendale Rd. SE23 61Ac 136
Leven Dr. EN8: Walt C 5Zb 20
Levenhurst Way SW4 54Nb 112
Leven Rd. E14 43Ec 92
Leven Way UB3: Hayes 44U 84
Leveret Cl. CR0: New Ad 83Fc 179
Leveret Cl. WD25: Wat 6W 12
Leverett St. SW3 5E 226 (49Gb 89)
Leverholme Gdns. SE9 63Qc 138
Leverington Pl. N1 4G 219 (41Tb 91)
Leverson St. SW16 65Lb 134
Lever Sq. RM16: Grays 9B 100
LEVERSTOCK GREEN 3C 4
Leverstock Grn. Rd. HP2: Hem H 1A 4
Leverstock Grn. Rd. HP3: Hem H 2B 4
(not continuous)
Leverstock Grn. Way HP3: Hem H 2C 4
Leverstock Ho. SW3 7E 226 (50Gb 89)
(off Cale St.)
Lever St. EC1 4C 218 (41Rb 91)
Leverton Cl. N22 25Pb 50
Leverton Pl. NW5 36Lb 70
Leverton St. NW5 36Lb 70
Leverton Way EN9: Walt A 5Ec 20
Leveson Rd. RM16: Grays 8D 100
Levett Building, The EC1 1C 224 (43Rb 91)
(off Little Britain)
Levett Gdns. IG3: Ilf 35Vc 75
Levett Rd. IG11: Bark 37Uc 74
Levett Rd. KT22: Lea 92Ka 192
Levett Rd. SS17: Stan H 1N 101
Levett Sq. TW9: Kew 52Ra 109
Levine Gdns. IG11: Bark 40Zc 75
Levison Way N19 32Mb 70
Levita Ho. NW1 3E 216 (41Mb 90)
(off Ossulston St.)
Levyne Ct. EC1 5K 217 (42Qb 90)
(off Pine St.)
Lewen Cl. CR0: C'don 74Tb 157
Lewes Cl. RM17: Grays 51Ce 121
Lewes Cl. UB5: N'olt 37Ca 65
Lewes Ct. CR4: Mitc 69Hb 133
(off Chatsworth Pl.)
Lewes Ct. SL1: Slou 8G 80
Lewesdon Cl. SW19 60Za 110
Lewes Ho. SE1 1J 231 (47Ub 91)
(off Druid St.)
Lewes Ho. SE15 51Wb 113
(off Friary Est.)
Lewes Rd. BR1: Broml 68Mc 137
Lewes Rd. N12 22Gb 49
Lewes Rd. RM3: Rom 21Md 57
Leweston Pl. N16 31Vb 71
Lewes Way WD3: Crox G 14S 26
Lew Evans Ho. SE22 57Wb 113
Lewey Ho. E3 42Bc 92
(off Joseph St.)
Lewgars Av. NW9 30Sa 47
Lewing Cl. BR6: Orp 74Uc 160
Lewington Apts. SE16 49Yb 92
(off Alpine Rd.)
Lewington Cen. SE16 49Yb 92
(off Alpine Rd.)

Lewington Ct. EN3: Enf W 9Zb 20
Lewin Rd. DA6: Bex 56Ad 117
Lewin Rd. SW14 55Ta 109
Lewin Rd. SW16 65Mb 134
Lewins Farm Ct. SL1: Slou 5D 80
Lewins Rd. KT18: Eps 86Ra 173
Lewins Way SL1: Slou 5D 80
Lewin Ter. TW14: Bedf 59T 106
Lewis Av. E17 25Cc 52
Lewis Cl. CM15: Shenf 17Be 41
Lewis Cl. KT15: Add 77L 149
Lewis Cl. N14 17Lb 32
Lewis Cl. UB9: Hare 26L 43
Lewis Ct. DA11: Nflt 1B 144
Lewis Ct. KT22: Lea 93Ja 192
(off Highbury Dr.)
Lewis Ct. SE16 50Xb 91
(off Stubbs Dr.)
Lewis Cres. NW10 36Ta 67
Lewis Cubitt Pk. 39Nb 70
Lewis Cubitt Sq. N1 39Nb 70
Lewis Cubitt Wlk. N1 39Nb 70
Lewis Gdns. N16 30Vb 51
Lewis Gdns. N2 26Fb 49
Lewis Gro. SE13 55Ec 114
LEWISHAM 55Ec 114
Lewisham Cen. 56Ec 114
Lewisham Hgts. SE23 60Yb 114
Lewisham High St. SE13 55Ec 114
(not continuous)
Lewisham Hill SE13 54Ec 114
Lewisham Indoor Bowls Cen. 64Bc 136
Lewisham Lions Cen. 50Yb 92
Lewisham Model Mkt. 56Ec 114
(off Lewisham High St.)
Lewisham Pk. SE13 57Ec 114
Lewisham Rd. SE13 53Dc 114
Lewisham Station (Rail & DLR) 55Ec 114
Lewisham St. SW1 2E 228 (47Mb 90)
Lewisham Way SE14 53Bc 114
Lewisham Way SE4 53Bc 114
Lewis Ho. E14 46Ec 92
(off Coldharbour)
Lewis Ho. N1 37Rb 71
(off Canonbury Rd.)
Lewis Ho. WD18: Wat 16V 26
Lewis La. SL9: Chal P 25A 42
Lewis M. BR7: Chst 64Pc 138
Lewis Pl. E8 36Wb 71
Lewis Rd. CR4: Mitc 68Fb 133
Lewis Rd. DA10: Swans 58Ae 121
Lewis Rd. DA13: Ist R 7B 144
Lewis Rd. DA14: Sidc 62Yc 139
Lewis Rd. DA16: Well 55Yc 117
Lewis Rd. RM11: Horn 30Ld 57
Lewis Rd. SM1: Sutt 77Db 155
Lewis Rd. TW10: Rich 57Ma 109
Lewis Rd. UB1: S'hall 47Aa 85
Lewis Silkin Ho. SE15 51Yb 114
(off Lovelinch Cl.)
Lewis Sports & Leisure Cen. 67Vb 135
Lewis St. NW1 37Kb 70
(not continuous)
Lewiston Cl. KT4: Wor Pk 73Xa 154
Lewis Way RM10: Dag 37Dd 76
Leworth Pl. SL4: Wind 3H 103
Lexden Dr. RM6: Chad H 30Xc 55
Lexden Rd. CR4: Mitc 70Mb 134
Lexden Rd. W3 45Ra 87
Lexden Ter. EN9: Walt A 6Ec 20
(off Sewardstone Rd.)
Lexham Gdns. W8 49Cb 89
Lexham Gdns. M. W8 48Db 89
Lexham Ho. W8 49Db 89
(off Lexham Gdns.)
Lexham M. W8 49Cb 89
Lexham Wlk. W8 48Db 89
Lexicon Apts. RM1: Rom 28Hd 56
(off Mercury Gdns.)
Lexington Apts. EC1 5F 219 (42Tb 91)
Lexington Bldg. E3 40Cc 72
Lexington Ct. WD6: Bore 13Pa 29
Lexington Ct. CR8: Purl 82Sb 177
Lexington Ct. EN6: Pot B 3Za 16
(off Mimms Hall Rd.)
Lexington Ho. UB7: W Dray 47P 83
(off Park Lodge Av.)
Lexington Pl. KT1: Hamp W 66Ma 131
Lexington St. W1 4C 222 (45Lb 90)
Lexington Way EN5: Barn 14Za 30
Lexington Way RM14: Upm 30Vd 58
Lexton Gdns. SW12 60Mb 112
Leyborne Av. W13 47Ka 86
Leyborne Pk. TW9: Kew 53Qa 109
Leybourne Av. KT14: Byfl 85P 169
Leybourne Cl. BR2: Broml 72Jc 159
Leybourne Cl. KT14: Byfl 85P 169
Leybourne Ho. E14 44Bc 92
(off Dod St.)
Leybourne Ho. SE15 51Yb 114
Leybourne Rd. E11 32Hc 73
Leybourne Rd. NW1 38Kb 70
Leybourne Rd. NW9 29Qa 47
Leybourne Rd. UB10: Hil 39S 64
Leybourne St. NW1 38Kb 70
Leybridge Ct. SE12 57Jc 115
Leyburn Cl. E17 28Dc 52
Leyburn Cres. RM3: Rom 24Nd 57
Leyburn Gdns. CR0: C'don 75Ub 157
Leyburn Gro. N18 23Wb 51
Leyburn Rd. N18 23Wb 51
Leyburn Rd. RM3: Rom 24Nd 57
Leycester Cl. GU20: W'sham 7A 146
Leycroft Cl. IG10: Lough 15Qc 36
Leycroft Gdns. DA8: Erith 53Kd 119
Leydenhatch La. BR8: Swan 67Ed 140
Leyden Mans. N19 31Nb 70
Leyden St. E1 1K 225 (43Vb 91)
Leydon Cl. SE16 46Zb 92
Leyes Rd. E16 44Mc 93
Ley Farm Cl. WD25: Wat 8Y 13
Leyfield KT4: Wor Pk 74Ua 154
Leyhill Cl. BR8: Swan 70Gd 140
Ley Hill Rd. HP3: Bov 1A 10
Ley Ho. SE1 2D 230 (47Sb 91)
(off Scovell Rd.)
Leyland Av. AL1: St A 4B 6
Leyland Av. EN3: Enf H 12Ac 34
Leyland Cl. EN8: Chesh 1Yb 20
Leyland Ct. SE15 52Vb 113
(off Shield Street)
Leyland Gdns. IG8: Wfd G 22Lc 53
Leyland Ho. E14 45Dc 92
(off Hale St.)
Leyland Rd. SE12 57Jc 115
Leylands SW18 58Bb 111
Leylands La. TW19: Stanw M 56H 105
(not continuous)
Leylang Rd. SE14 52Zb 114
Leys, The HA3: Kenton 30Pa 47
Leys, The KT12: Hers 77Ba 151
Leys, The N2 28Eb 49
Leys, The WD7: R'lett 9Ja 14
Leys Av. RM10: Dag 39Ed 76
Leys Cl. HA1: Harr 29Fa 46
Leys Cl. RM10: Dag 38Ed 76
Leys Cl. UB9: Hare 25M 43
Leys Ct. SW9 54Qb 112
Leysdown Av. DA7: Bex 56Ed 118
Leysdown Ho. SE17 7J 231 (50Ub 91)
(off Madron St.)
Leysdown Rd. SE9 61Nc 138
Leysfield Rd. W12 48Wa 88
Leys Gdns. EN4: Cockf 15Jb 32
Leyspring Rd. E11 32Hc 73
Leys Rd. HP3: Hem H 4N 3
Leys Rd. KT22: Oxs 84Fa 172
Leys Rd. E. EN3: Enf H 11Ac 34
Leys Rd. W. EN3: Enf H 11Ac 34
Leys Sq. N3 25Db 49
Ley St. IG1: Ilf 33Rc 74
Ley St. IG2: Ilf 32Tc 74
Leyswood Dr. IG2: Ilf 29Uc 54
Leythe Rd. W3 47Sa 87
LEYTON 33Ec 72
Leyton Bus. Cen. E10 33Cc 72
Leyton Ct. SE23 60Yb 114
LEYTON CROSS 61Jd 140
Leyton Cross Rd. DA2: Wilm 62Hd 140
Leyton Grange E10 32Cc 72
Leyton Grn. Rd. E10 30Ec 52
Leyton Grn. Twr. E10 30Ec 52
(off Leyton Grn. Rd.)
Leyton Ho. E2 4K 219 (41Vb 91)
(off Calvert Av.)
Leyton Ind. Village E10 31Zb 72
Leyton Leisure Cen. 31Ec 72
Leyton Link Est. E10 31Ac 72
Leyton Mnr. Pk. 31Dc 72
Leyton Midland Road Station (Overground) 31Ec 72
Leyton Mills E10 34Ec 72
Leyton Orient FC 34Dc 72
Leyton Pk. Rd. E10 34Ec 72
Leyton Rd. E15 36Ec 72
Leyton Rd. SW19 66Eb 133
Leyton Station (Underground) 34Ec 72
LEYTONSTONE 32Gc 73
Leytonstone High Road Station (Overground) 33Gc 73
Leytonstone Ho. E11 31Hc 73
(off Hanbury Dr.)
Leytonstone Leisure Cen. 33Gc 73
Leytonstone Rd. E15 37Gc 73
Leytonstone Station (Underground) 32Gc 73
Leyton Way E11 31Gc 73
Leywick St. E15 40Gc 73
Lezayre Rd. BR6: Chels 79Vc 161
Lianne Gro. SE9 62Lc 137
Liardet St. SE14 51Ac 114
Libari Mans. E20 36Ec 72
(off Victory Pde.)
Liberia Rd. N5 37Rb 71
Liberties Pl. TW20: Eng G 4N 125
Liberty, The RM1: Rom 29Gd 56
Liberty Av. SW19 67Fb 133
Liberty Bri. Rd. E15 36Ec 72
Liberty Bri. Rd. E20 36Ec 72
Liberty Cen. HA0: Wemb 39Pa 67
Liberty Cl. KT4: Wor Pk 74Ya 154
Liberty Cl. N18 21Vb 51
Liberty Ct. BR1: Broml 69Mc 137
Liberty Ct. CR8: Purl 84Qb 176
Liberty Ct. IG11: Bark 40Xc 75
Liberty Hall Rd. KT15: Add 78J 149
Liberty Ho. CR7: Thor H 71Qb 156
(off Thornton Rd.)
Liberty Ho. E1 45Wb 91
(off Ensign St.)
Liberty Ho. E13 40Jc 73
Liberty Ho. KT16: Chert 74H 149
(off Guildford St.)
Liberty La. KT15: Add 78J 149
Liberty M. N22 25Rb 51
Liberty M. SW12 58Kb 112
Liberty Point CR0: C'don 73Wb 157
(off Blackhorse La.)
Liberty Ri. KT15: Add 79J 149
Liberty St. SW9 53Pb 112
Liberty Wlk. AL1: St A 3G 6
Libra Mans. E3 40Bc 72
(off Libra Rd.)
Libra Rd. E13 40Jc 73
Libra Rd. E3 39Bc 72
Library & Mus. of Freemasonry, The 3G 223 (44Nb 90)
(off Gt. Queen St.)
Library Ct. N17 27Vb 51
Library Hill CM14: B'wood 19Zd 41
Library Mans. W12 47Ya 88
(off Pennard Rd.)
Library M. TW12: Hamp H 65Ea 130
Library Pde. NW10 39Ua 68
(off Craven Pk. Rd.)
Library Pl. E1 45Xb 91
Library Sq. E1 42Ac 92
Library St. SE1 2B 230 (47Rb 91)
Library Way TW2: Whitt 59Ea 108
Libro Ct. E4 21Cc 52
Lichfield Cl. EN4: Cockf 13Hb 31
Lichfield Ct. KT6: Surb 71Na 153
(off Claremont Rd.)
Lichfield Ct. TW9: Rich 57Na 109
Lichfield Gdns. TW9: Rich 56Na 109
Lichfield Gro. N3 25Cb 49
Lichfield Ho. WD6: Bore 12Qa 29
(off Stratfield Rd.)
Lichfield La. TW2: Whitt 60Ea 108
Lichfield Pl. AL1: St A 1D 6
Lichfield Rd. E3 41Ac 92
Lichfield Rd. E6 41Mc 93
Lichfield Rd. HA6: Nwood 27W 44
Lichfield Rd. IG8: Wfd G 21Gc 53
Lichfield Rd. N9 19Wb 33
Lichfield Rd. NW2 35Ab 68
Lichfield Rd. RM8: Dag 35Xc 75
Lichfield Rd. TW4: Houn 55Y 107
Lichfield Rd. TW9: Kew 53Pa 109
Lichfield Ter. RM14: Upm 33Ud 78
Lichfield Ter. TW9: Rich 57Na 109
(off Sheen Rd.)
Lichfield Way CR2: Sels 82Zb 178
Lichlade Cl. BR6: Orp 77Vc 161
Lickey Ho. W14 51Bb 111
(off North End Rd.)
Lidcote Gdns. SW9 54Pb 112
Liddall Way UB7: Yiew 46P 83
Liddell SL4: Wind 5A 102
Liddell Cl. HA3: Kenton 27Ma 47
Liddell Gdns. NW10 40Ya 68
Liddell Pl. NW6 37Cb 69
Liddell Pl. SL4: Wind 4A 102
Liddell Rd. NW6 37Cb 69
Liddell Sq. SL4: Wind 4A 102
Liddell Way SL4: Wind 5A 102
Liddiard Ho. W11 45Ab 88
(off Lansdowne Rd.)
Lidding Rd. HA3: Kenton 29Ma 47
Liddington Rd. E15 39Hc 73
Liddon Rd. BR1: Broml 69Lc 137
Liddon Rd. E13 41Kc 93
Liden Cl. E17 31Bc 72
Lidfield Rd. N16 35Tb 71
Lidgate Rd. SE15 52Vb 113
Lidget Gro. CR5: Coul 88Lb 176
Lidgould Gro. HA4: Ruis 30W 44
Lidiard Rd. SW18 61Eb 133
Lidlington Pl. NW1 2B 216 (40Lb 70)
Lido Sq. N17 26Tb 51
Lidstone Cl. GU21: Wok 9M 167
Lidstone Ct. SL3: Geor G 4P 81
Lidyard Rd. N19 32Lb 70
Lieutenant Ellis Way EN7: Chesh 2Vb 19
Lieutenant Ellis Way EN7: G Oak 2Vb 19
Lieutenant Ellis Way EN8: Chesh 4Xb 19
Lieutenant Ellis Way EN8: Walt C 4Xb 19
Lifestyle Club @ Charlton, The 52Mc 115
Liffler Rd. SE18 50Uc 94
Liffords Pl. SW13 54Va 110
Lifford St. SW15 56Za 110
Lightbox, The 88A 168
Light Cinema Addlestone, The 77L 149
Lightcliffe Rd. N13 21Qb 50
Lighterage Ct. TW8: Bford 51Na 109
Lighter Cl. SE16 49Ac 92
Lighterman Ho. E14 45Ec 92
Lighterman M. E1 44Zb 92
Lighterman Point E14 44Fc 93
(off New Village Av.)
Lighterman Rd. RM13: Rain 38Gd 76
(off Ongar Way)
Lighterman's M. DA11: Nflt 9A 122
Lightermans Rd. E14 47Cc 92
Lightermans Wlk. SW18 56Cb 111
Lightermans Way DA9: Ghithe 56Yd 120
Lightfoot Rd. N8 29Nb 50
Lightfoot Vs. N1 38Qb 70
(off Augustas La.)
Light Horse Ct. SW3 7H 227 (50Jb 90)
(off Royal Hospital Rd.)
Lighthouse, The 52Sb 113
Lighthouse Apts. E1 44Yb 92
(off Commercial Rd.)
Lighthouse Vw. SE10 46Gc 93
Lightley Cl. HA0: Wemb 39Na 67
LIGHTWATER 2A 166
Lightwater Mdw. GU18: Light 3A 166
Lightwater Rd. GU18: Light 3A 166
Ligonier St. E2 5K 219 (42Vb 91)
Lilac Av. EN1: Enf 8Yb 20
Lilac Av. GU22: Wok 2P 187
Lilac Cl. AL10: Hat 4B 8
Lilac Cl. CM15: Pil H 15Xd 40
Lilac Cl. E4 23Bc 52
Lilac Cl. EN7: Chesh 3Xb 19
Lilac Ct. E13 39Lc 73
Lilac Ct. SL2: Slou 1D 80
Lilac Ct. TW11: Tedd 63Ha 130
Lilac Gdns. BR8: Swan 69Fd 140
Lilac Gdns. CR0: C'don 76Cc 158
Lilac Gdns. RM7: Rush G 32Gd 76
Lilac Gdns. UB3: Hayes 44U 84
Lilac Gdns. W5 48Ma 87
Lilac Ho. SE4 55Cc 114
Lilac La. E2 41Zb 92
Lilac M. N22 27Qb 50
(off Perth Rd.)
Lilac Pl. SE11 6H 229 (49Pb 90)
Lilac Pl. UB7: Yiew 45P 83
Lilac St. W12 45Wa 88
Lilah M. BR2: Broml 68Gc 137
Lila Pl. BR8: Swan 70Gd 140
Lilburne Gdns. SE9 57Nc 116
Lilburne Rd. SE9 57Nc 116
Lilburne Wlk. NW10 37Sa 67
Lile Cres. W7 43Ga 86
Lilestone Ho. NW8 5C 214 (42Fb 89)
(off Frampton St.)
Lilestone St. NW8 5D 214 (42Gb 89)
Lilford Ho. SE5 54Sb 113
Lilford Rd. SE5 54Rb 113
Lilian Barker Cl. SE12 57Jc 115
Lilian Board Way UB6: G'frd 36Fa 66
Lilian Cl. N16 34Ub 71
Lilian Cres. CM13: Hut 19Ee 41
Lilian Gdns. IG8: Wfd G 25Kc 53
Lilian Knowles Ho. E1 1K 225 (43Vb 91)
(off Crispin St.)
Lilian Rd. SW16 67Lb 134
Lilium M. E10 31Cc 72
Lillechurch Rd. RM8: Dag 37Xc 75
Lilleshall Rd. SM4: Mord 72Fb 155
Lilley Cl. CM14: B'wood 21Vd 58
Lilley Cl. E1 46Wb 91
Lilley Dr. KT20: Kgswd 94Db 195
Lilley La. NW7 22Ta 47
Lilley Mead RH1: Mers 3C 208
Lilley Way SL1: Slou 6C 80
Lillian Av. W3 47Qa 87
Lillian Rd. SW13 51Wa 110
Lillie Bri. Dpt. W14 50Bb 89
(off Aisgill Av.)
Lillie Mans. SW6 51Ab 110
(off Lillie Rd.)
Lillie Rd. SW6 51Ab 110
Lillie Rd. TN16: Big H 90Mc 179
Lillie Road Fitness Cen. 52Za 110
Lillieshall Rd. SW4 55Kb 112
Lillie Sq. SW6 51Cb 111
Lillie Yd. SW6 51Cb 111
Lillingston Ho. N7 35Qb 70
Lillington Gdns. Est. SW1 6C 228 (49Lb 90)
(off Vauxhall Bri. Rd.)
Lilliot's La. KT22: Lea 91Ja 192
Lilliput Av. UB5: N'olt 39Aa 65
Lilliput Ct. SE12 57Kc 115
Lilliput Rd. E15 39Gc 73
Lilliput Rd. RM7: Rush G 31Fd 76
Lily Cl. HA5: Pinn 26Y 45
Lily Cl. RM13: Rain 42Ld 97
Lily Cl. W14 49Za 88
Lily Dr. UB7: W Dray 49M 83
Lily Gdns. HA0: Wemb 40La 66
Lily M. SE11 5B 230 (49Rb 91)
Lily Nichols Ho. E16 46Mc 93
(off Connaught Rd.)
Lily Pl. EC1 7A 218 (43Qb 90)
Lily Rd. E17 30Cc 52
Lilyville Rd. SW6 53Bb 111
Lily Way N13 22Nb 50
Limasol St. SE16 3K 231 (48Vb 91)
Limborough Ho. E14 43Cc 92
(off Thomas Rd.)
Limbourne Av. RM8: Dag 31Bd 75
Limburg Rd. SW11 56Gb 111
Lime Av. SL4: Wind Adelaide Rd. 3K 103
Lime Av. SL4: Wind Holly Wlk. 2D 124
Lime Av. CM13: B'wood 20Be 41
Lime Av. DA11: Nflt 59Fe 121
Lime Av. RM14: Upm 35Qd 77
Lime Av. UB7: Yiew 45P 83
Limeburner La. EC4 3B 224 (44Rb 91)
Limebush Cl. KT15: New H 81L 169
Lime Cl. BR1: Broml 70Nc 138
Lime Cl. E1 46Wb 91
Lime Cl. GU4: W Cla 100K 189
Lime Cl. HA3: W'stone 26Ja 46
Lime Cl. HA5: Eastc 27V 44
Lime Cl. IG9: Buck H 19Mc 35
Lime Cl. RH2: Reig 9K 207
Lime Cl. RM15: S Ock 41Yd 98
Lime Cl. RM7: Rom 28Ed 56
Lime Cl. SM5: Cars 75Hb 155
Lime Cl. WD19: Wat 17Z 27
Lime Ct. CR4: Mitc 68Fb 133
Lime Ct. E11 33Gc 73
(off Trinity Cl.)
Lime Ct. E17 29Ec 52
Lime Ct. HA1: Harr 30Ha 46
(off Gayton Rd.)
Lime Ct. HA4: Ruis 31X 65
Lime Ct. SE9 61Rc 138
Lime Cres. TW16: Sun 68Y 129
Limecroft Cl. KT19: Ewe 80Ta 153
Limecroft Rd. GU21: Knap 9F 166
Limedene Cl. HA5: Pinn 25Z 45
Lime Gro. BR6: Farnb 75Rc 160
Lime Gro. CR6: W'ham 90Ac 178
Lime Gro. DA15: Sidc 58Vc 117
Lime Gro. E4 23Bc 52
Lime Gro. GU22: Wok 93A 188
Lime Gro. GU4: W Cla 100J 189
Lime Gro. HA4: Ruis 30X 45
Lime Gro. IG6: Ilf 23Vc 55
Lime Gro. KT15: Add 77J 149
Lime Gro. KT3: N Mald 69Ta 131
Lime Gro. N20 18Bb 31
Lime Gro. TW1: Twick 58Ha 108
Lime Gro. UB3: Hayes 45T 84
Lime Gro. W12 47Ya 88
Limeharbour E14 48Dc 92
Lime Ho. TW9: Kew 53Ra 109
LIMEHOUSE 44Bc 92
Limehouse C'way. E14 45Bc 92
Lime Ho. Ct. E14 44Bc 92
(off Wharf La.)
Limehouse Ct. E14 44Cc 92
Limehouse Cut E14 43Dc 92
(off Morris Rd.)
Limehouse Flds. Est. E14 43Ac 92
Limehouse Link E14 44Ac 92
Limehouse Lodge E5 33Yb 72
(off Harry Zeital Way)
Limehouse Station (Rail & DLR) 44Ac 92
Lime Kiln Dr. SE7 51Kc 115
Limekiln Pl. SE19 66Vb 135
Limekiln Wharf E14 45Bc 92
Limelight Ho. SE11 6B 230 (49Rb 91)
(off Dugard Way)
Lime Lodge TW16: Sun 66V 128
(off Forest Dr.)
Lime Mdw. Av. CR2: Sande 85Wb 177
Lime Pit La. TN13: Dun G 89Ed 182
Lime Quay E14 42Ec 92
Limerick Cl. SW12 59Lb 112
Limerick Gdns. RM14: Upm 31Vd 78
Limerick M. N2 27Gb 49
Lime Rd. BR8: Swan 69Fd 140
Lime Rd. TW9: Rich 56Pa 109
Lime Row DA18: Erith 48Bd 95
Limerston St. SW10 51Eb 111
Limes, The GU21: Wok Ridgeway 7P 167
Limes, The AL1: St A 1C 6
Limes, The BR2: Broml 75Nc 160
Limes, The CM13: B'wood 20Be 41
Limes, The DA1: Dart 59Pd 119
Limes, The GU21: Wok 88C 168
(off Maybury Rd.)
Limes, The HP3: Hem H 5M 3
Limes, The KT19: Eps 82Ra 173
Limes, The KT22: Lea 95Ka 192
Limes, The KT8: W Mole 70Da 129
Limes, The RM11: Horn 27Md 57
Limes, The RM19: Purf 50Qd 97
Limes, The SL4: Wind 4A 102
Limes, The SW18 58Cb 111
Limes, The W2 45Cb 89
Limes, The WD3: Rick 18L 25
Limes, The WD4: Hunt C 5S 12
(off Bridge Rd.)
Limes Av. CR0: Wadd 76Qb 156
Limes Av. E11 28Kc 53
Limes Av. EN8: Walt C 5Bc 20
Limes Av. IG7: Chig 22Sc 54
Limes Av. N12 21Eb 49
Limes Av. NW11 31Ab 68
Limes Av. NW7 23Ua 48
Limes Av. SE20 66Xb 135
Limes Av. SM5: Cars 74Hb 155
Limes Av. SW13 54Va 110
Limes Av., The N11 22Kb 50
Limes Cl. KT22: Lea 93La 192
Limes Cl. N11 22Lb 50
Limes Cl. RH1: Redh 10P 207
Limes Cl. TW15: Ashf 64Q 128
Limes Ct. BR3: Beck 68Dc 136
Limes Ct. CM15: B'wood 18Zd 41
Limes Ct. NW6 38Ab 68
(off Brondesbury Pk.)
Limesdale Gdns. HA8: Edg 26Sa 47
Limes Fld. Rd. SW13 55Ua 110
Limesford Rd. SE15 56Zb 114
Limes Gdns. SW18 58Cb 111
Limes Gro. SE13 56Ec 114
Limes M. TW20: Egh 64B 126
Limes Pl. CR0: C'don 73Tb 157
Limes Rd. BR3: Beck 68Dc 136
Limes Rd. CR0: C'don 73Tb 157
Limes Rd. EN8: Chesh 4Ac 20
Limes Rd. KT13: Weyb 77Q 150
Limes Rd. TW20: Egh 64B 126
Limes Row BR6: Farnb 78Rc 160
Limestone Wlk. DA18: Erith 47Zc 95
Lime St. E17 28Ac 52
Lime St. EC3 4H 225 (45Ub 91)
Lime St. Pas. EC3 3H 225 (44Ub 91)
Lime St. Sq. EC3 3H 225 (44Ub 91)
Limes Wlk. SE15 56Yb 114
Limes Wlk. W5 47Ma 87
Lime Ter. W7 45Ga 86
Lime Tree Av. CR5: Coul 89Lb 176
Lime Tree Av. DA9: Bluew 59Wd 120
Lime Tree Av. KT10: Esh 74Fa 152
Lime Tree Av. KT7: T Ditt 74Ga 152
Lime Tree Cl. E18 28Lc 53
Lime Tree Cl. KT23: Bookh 96Ca 191
Lime Tree Cl. WD23: Bush 16Ea 28
Limetree Cl. SW2 60Pb 112
Lime Tree Ct. AL10: Hat 4C 8
Lime Tree Ct. AL2: Lon C 8F 6
Lime Tree Ct. CR2: S Croy 79Sb 157
Lime Tree Ct. E3 43Cc 92
(off Whitehorn St.)
Lime Tree Ct. HA5: Hat E 24Ca 45
(off The Avenue)
Lime Tree Ct. KT21: Asht 90Na 173
Lime Tree Ct. SE17 7E 230 (50Sb 91)
(off Walworth Rd.)
Lime Tree Gro. CR0: C'don 76Bc 158
Lime Tree Ho. DA4: Farni 72Nd 163
Lime Tree Pl. AL1: St A 3D 6
Lime Tree Pl. BR5: St M Cry 73Ad 161
Lime Tree Pl. CR4: Mitc 67Kb 134
Lime Tree Rd. TW5: Hest 53Da 107
Limetree Ter. DA16: Well 55Wc 117
Limetree Ter. SE6 60Bc 114
Lime Tree Wlk. BR4: W W'ck 77Hc 159
Lime Tree Wlk. EN2: Enf 10Sb 19
Lime Tree Wlk. GU21: Wok 8P 167
Lime Tree Wlk. GU25: Vir W 70A 126
Lime Tree Wlk. TN13: S'oaks 97Kd 203
Lime Tree Wlk. WD23: B Hea 18Ga 28
Lime Tree Wlk. WD3: Rick 15K 25
Limetree Wlk. SW17 64Jb 134
Lime View Apts. E14 44Ac 92
(off Commerical Rd.)
Lime Wlk. E15 39Gc 73
Lime Wlk. HP3: Hem H 4P 3
Lime Wlk. KT8: E Mos 70Ha 130
Lime Wlk. UB9: Den 36L 63
Lime Way WD7: Shenl 3La 14
Limewood Cl. BR3: Beck 71Ec 158
Limewood Cl. E17 28Bc 52
Limewood Cl. GU21: Wok 2H 187
Limewood Cl. W13 44Ka 86
Limewood Ct. IG4: Ilf 29Pc 54
Limewood Ho. KT19: Eps 81Ta 173
Limewood M. SE20 66Wb 135
(off Lullington Rd.)
Limewood Rd. DA8: Erith 52Ed 118
Lime Works Rd. RH1: Mers 98Lb 196
LIMPSFIELD 1M 211
Limpsfield Av. CR7: Thor H 71Pb 156
Limpsfield Av. SW19 61Za 132
LIMPSFIELD CHART 2P 211
Limpsfield Chart Golf Course 1N 211
Limpsfield Rd. CR2: Sande 84Wb 177
Limpsfield Rd. CR6: W'ham 88Yb 178
Limscott Ho. E3 41Dc 92
(off Bruce Rd.)
Linacre Cl. SE15 55Xb 113
Linacre Ct. W6 50Za 88
Linacre Rd. NW2 37Xa 68
Linale Ho. N1 2F 219 (40Tb 71)
(off Murray Gro.)
Linberry Wlk. SE8 49Bc 92
Linchfield Rd. SL3: Dat 3N 103
Linchmere Rd. SE12 59Hc 115
Lincoln Apts. W12 45Ya 88
(off Fountain Park Way)
Lincoln Av. N14 20Lb 32
Lincoln Av. RM7: Rush G 33Fd 76
Lincoln Av. SW19 62Za 132
Lincoln Av. TW2: Twick 61Ea 130
Lincoln Cl. DA8: Erith 54Hd 118
Lincoln Cl. HA2: Harr 29Ba 45
Lincoln Cl. RM11: Horn 29Qd 57
Lincoln Cl. SE25 72Wb 157
Lincoln Cl. UB6: G'frd 39Ea 66
Lincoln Ct. CR2: S Croy 78Sb 157
(off Warham Rd.)
Lincoln Ct. IG2: Ilf 30Sc 54
Lincoln Ct. KT13: Weyb 79T 150
(off Old Av.)
Lincoln Ct. N16 31Tb 71
Lincoln Ct. SE12 62Kc 137
Lincoln Ct. SL1: Slou 8J 81
Lincoln Ct. UB9: Den 30H 43
Lincoln Ct. WD25: Wat 5X 13
Lincoln Ct. WD6: Bore 15Ta 29
Lincoln Cres. EN1: Enf 15Ub 33
Lincoln Dr. GU22: Pyr 87G 168
Lincoln Dr. WD19: Wat 20Y 27
Lincoln Dr. WD3: Crox G 14R 26
Lincoln Fld. WD23: Bush 13Ba 27
Lincoln Gdns. IG1: Ilf 31Nc 74
Lincoln Grn. Rd. BR5: St M Cry 71Vc 161
Lincoln Gro. KT13: Weyb 76R 150
Lincoln Hatch La. SL1: Burn 2A 80
Lincoln Ho. SE5 52Qb 112
Lincoln Ho. SW3 2F 227 (47Hb 89)
Lincoln Ho. TW8: Bford 50Ma 87
(off Ealing Rd.)
Lincoln M. AL3: St A 3A 6
Lincoln M. N15 28Sb 51
Lincoln M. NW6 39Bb 69
Lincoln M. SE21 61Tb 135
Lincoln Pde. HA9: Wemb 32Na 67
Lincoln Pde. N2 27Gb 49
(off Lincoln Rd.)
Lincoln Plaza E14 47Dc 92
(off Lightermans Rd.)
Lincoln Rd. CR4: Mitc 71Nb 156
Lincoln Rd. DA14: Sidc 64Xc 139
Lincoln Rd. DA8: Erith 54Hd 118
Lincoln Rd. E13 42Kc 93
Lincoln Rd. E18 25Jc 53
Lincoln Rd. E7 37Mc 73
Lincoln Rd. EN1: Enf 14Ub 33
Lincoln Rd. EN3: Pond E 15Xb 33
Lincoln Rd. HA0: Wemb 37Ma 67
Lincoln Rd. HA2: Harr 29Ba 45
Lincoln Rd. HA6: Nwood 27V 44
Lincoln Rd. KT3: N Mald 69Sa 131
Lincoln Rd. KT4: Wor Pk 74Xa 154
Lincoln Rd. N2 27Gb 49
Lincoln Rd. SE25 69Xb 135
Lincoln Rd. SL9: Chal P 25A 42
Lincoln Rd. TW13: Hanw 62Ba 129
Lincolns, The NW7 20Va 30
Lincolns Fld. CM16: Epp 2Vc 23
Lincolnshire Ter. DA2: Daren 63Td 142
Lincoln's Inn Flds. WC2 2H 223 (44Pb 90)
Lincoln's Inn Hall 2J 223 (44Pb 90)
Lincolns La. CM14: Pil H 15Sd 40

Lincolns La. CM14: S Weald 15Sd 40
Lincoln St. E11 33Gc 73
Lincoln St. SW3 6F 227 (49Hb 89)
Lincoln Ter. SM2: Sutt 80Cb 155
Lincoln Wlk. KT19: Ewe 82Ta 173
Lincoln Way EN1: Enf 15Xb 33
Lincoln Way SL1: Slou 5B 80
Lincoln Way TW16: Sun 67U 128
Lincoln Way WD3: Crox G 14R 26
Lincombe Ct. KT15: Add 78K 149
Lincombe Rd. BR1: Broml 62Hc 137
Lind Ct. E18 25Hc 53
Lindal Cres. EN2: Enf 14Nb 32
Lindale Cl. GU25: Vir W 10K 125
Lindal Cl. IG4: Ilf 29Pc 54
Lindales, The N17 23Wb 51
(off Grasmere Rd.)
Lindal Rd. SE4 57Bc 114
Lindbergh Rd. SM6: W'gton 80Nb 156
Linden SL3: L'ly 50D 82
Linden Av. CR5: Coul 88Kb 176
Linden Av. CR7: Thor H 70Rb 135
Linden Av. DA1: Dart 60Ld 119
Linden Av. EN1: Enf 11Wb 33
Linden Av. HA4: Ruis 32W 64
Linden Av. HA9: Wemb 36Pa 67
Linden Av. NW10 40Za 68
Linden Av. TW3: Houn 57Da 107
Linden Av. WD18: Wat 14U 26
Linden Chase TN13: S'oaks 94Kd 203
Linden Cl. BR6: Chels 78Wc 161
Linden Cl. EN7: Chesh 2Xb 19
Linden Cl. HA4: Ruis 32W 64
Linden Cl. HA7: Stan 22Ka 46
Linden Cl. KT15: New H 83J 169
Linden Cl. KT20: Tad 92Za 194
Linden Cl. KT7: T Ditt 73Ha 152
Linden Cl. N14 16Lb 32
Linden Cl. RM19: Purf 51Sd 120
Linden Cl. SL0: Iver H 40F 62
Linden Ct. DA14: Sidc 63Uc 138
Linden Ct. KT22: Lea 93Ka 192
Linden Ct. SE20 66Xb 135
(off Anerley Pk.)
Linden Ct. TW20: Eng G 5M 125
Linden Ct. W12 46Ya 88
Linden Cres. AL1: St A 2G 6
Linden Cres. IG8: Wfd G 23Kc 53
Linden Cres. KT1: King T 68Pa 131
Linden Cres. UB6: G'frd 37Ha 66
Linden Dr. CR3: Cat'm 96Sb 197
Linden Dr. SL2: Farn R 9G 60
Linden Dr. SL9: Chal P 25A 42
Lindenfield BR7: Chst 68Rc 138
Linden Gdns. EN1: Enf 11Wb 33
Linden Gdns. KT22: Lea 93La 192
Linden Gdns. W2 45Cb 89
Linden Gdns. W4 50Ua 88
Linden Glade HP1: Hem H 3K 3
Linden Gro. CR6: W'ham 90Ac 178
Linden Gro. KT12: Walt T 75V 150
Linden Gro. KT3: N Mald 69Ua 132
Linden Gro. SE15 55Xb 113
Linden Gro. SE26 65Yb 136
Linden Gro. TW11: Tedd 64Ha 130
Linden Ho. CM16: Epp 3Wc 23
Linden Ho. SE8 51Bc 114
(off Abinger Gro.)
Linden Ho. TW12: Hamp 65Ca 129
Linden Lawns HA9: Wemb 35Pa 67
Linden Lea HA5: Hat E 24Ba 45
Linden Lea N2 29Eb 49
Linden Lea WD25: Wat 5W 12
Linden Leas BR4: W W'ck 75Fc 159
Linden Mans. N6 32Kb 70
(off Hornsey La.)
Linden M. N1 36Tb 71
Linden M. W2 45Cb 89
Linden Pit Path KT22: Lea Kingfisher Cl. 92La 192
Linden Pit Path KT22: Lea Linden Rd. 93Ka 192
Linden Pl. CR4: Mitc 70Gb 133
Linden Pl. KT17: Eps 84Ua 174
Linden Pl. KT24: E Hor 98U 190
Linden Pl. TW18: Staines 63J 127
Linden Ri. CM14: W'ley 22Zd 59
Linden Rd. KT13: Weyb 81S 170
Linden Rd. KT22: Lea 93Ka 192
Linden Rd. N10 28Kb 50
Linden Rd. N11 19Hb 31
Linden Rd. N15 28Sb 51
Linden Rd. TW12: Hamp 66Ca 129
Lindens, The CR0: New Ad 79Ec 158
Lindens, The E17 28Dc 52
(off Prospect Hill)
Lindens, The EN9: Walt A 7Lc 21
(within Woodbine Cl. Caravan Pk.)
Lindens, The HP3: Hem H 5H 3
Lindens, The IG10: Lough 15Pc 36
Lindens, The N12 22Fb 49
Lindens, The W4 53Sa 109
Lindens Cl. KT24: Eff 100Aa 191
Linden Sq. TN13: Riv 94Gd 202
Linden Sq. UB9: Hare 23J 43
Linden St. RM7: Rom 28Fd 56
Linden Wlk. N19 33Lb 70
Linden Way CR8: Purl 82Lb 176
Linden Way GU22: Wok 93B 188
Linden Way GU23: Rip 97H 189
Linden Way N14 16Lb 32
Linden Way TW17: Shep 71S 150
Linder's Field Local Nature Reserve 17Mc 35
Lindeth Cl. HA7: Stan 23Ka 46
Lindfield Gdns. NW3 36Db 69
Lindfield Rd. CR0: C'don 72Vb 157
Lindfield Rd. RM3: Rom 22Nd 57
Lindfield Rd. W5 42La 86
Lindfield St. E14 44Cc 92
Lindhill Cl. EN3: Enf H 12Zb 34
Lindholme Ct. NW9 25Ua 48
(off Pageant Av.)
Lindie Gdns. UB8: Uxb 38N 63
Lindisfarne Cl. DA12: Grav'nd 1G 144
Lindisfarne Ct. HA8: Edg 21Pa 47
Lindisfarne Rd. RM8: Dag 34Yc 75
Lindisfarne Rd. SW20 66Wa 132
Lindisfarne Way E9 35Ac 72
Lindiswara Ct. WD3: Crox G 16Q 26
Lindley Ct. KT1: Hamp W 67La 130
Lindley Est. SE15 52Wb 113
Lindley Ho. E1 43Yb 92
(off Lindley St.)
Lindley Ho. SE15 52Wb 113
(off Peckham Pk. Rd.)
Lindley Pl. TW9: Kew 53Qa 109
Lindley Rd. E10 33Ec 72
Lindley Rd. KT12: Walt T 76Z 151
Lindley Rd. RH9: G'stone 2A 210
Lindley St. E1 43Yb 92
Lindlings HP1: Hem H 3G 2
Lindop Ho. E1 42Ac 92
(off Mile End Rd.)
Lindore Rd. SW11 56Hb 111
Lindores Rd. SM5: Cars 73Eb 155
Lindo St. SE15 54Yb 114
Lindrick Ho. WD19: Wat 20Y 27
Lind Rd. SM1: Sutt 78Eb 155
Lindrop St. SW6 54Eb 111
Lindsay Cl. KT19: Eps 85Sa 173
Lindsay Cl. KT9: Chess 80Na 153
Lindsay Cl. TW19: Stanw 57M 105
Lindsay Ct. AL3: St A 2A 6
(off Verulam Rd.)
Lindsay Ct. CR0: C'don 77Tb 157
(off Eden Rd.)
Lindsay Ct. SE13 55Dc 114
(off Loampit Va.)
Lindsay Ct. SW11 53Fb 111
(off Battersea High St.)
Lindsay Dr. HA3: Kenton 30Na 47
Lindsay Dr. TW17: Shep 72T 150
Lindsay Ho. SW7 3A 226 (48Eb 89)
(off Gloucester Rd.)
Lindsay Pl. EN7: Chesh 2Xb 19
Lindsay Rd. KT15: New H 82J 169
Lindsay Rd. KT4: Wor Pk 75Xa 154
Lindsay Rd. TW12: Hamp H 63Da 129
Lindsay Sq. SW1 7E 228 (50Mb 90)
Lindsell St. SE10 53Ec 114
Lindsey Cl. BR1: Broml 69Mc 137
Lindsey Cl. CM14: B'wood 21Wd 58
Lindsey Cl. CR4: Mitc 70Nb 134
Lindsey Ct. N13 20Qb 32
(off Green Lanes)
Lindsey Gdns. TW14: Bedf 59T 106
Lindsey Ho. W5 49Ma 87
Lindsey M. N1 38Sb 71
Lindsey Rd. RM8: Dag 35Yc 75
Lindsey Rd. UB9: Den 34J 63
Lindsey St. EC1 7C 218 (43Rb 91)
Lindsey Way RM11: Horn 29Ld 57
Lind St. SE8 54Cc 114
Lindum Pl. AL3: St A 4M 5
Lindum Rd. TW11: Tedd 66La 130
Lindvale GU21: Wok 87A 168
Lindway SE27 64Rb 135
Lindwood Cl. E6 44Pc 94
Linear Pk. 51Mb 112
Linen Ho., The W10 40Ab 68
Linen M. W12 47Wa 88
Liner Ho. E16 47Lc 93
Linfield WC1 4H 217 (41Pb 90)
(off Sidmouth St.)
Linfield Cl. KT12: Hers 78X 151
Linfield Cl. NW4 27Ya 48
LINFORD 7J 101
Linford Christie Stadium 43Wa 88
Linford Ho. E16 43Jc 93
(off Hammersley Rd.)
Linford Ho. E2 39Wb 71
(off Whiston Rd.)
Linford Rd. E17 27Ec 52
Linford Rd. RM16: Grays 9D 100
Linford Rd. RM16: W Til 9D 100
Linford St. SW8 53Lb 112
Linford St. Bus. Est. SW8 53Lb 112
(off Linford St.)
Linford Wood Local Nature Reserve 7J 101
Lingard Av. NW9 26Ua 48
Lingard Ho. E14 48Ec 92
(off Marshfield St.)
Lingards Rd. SE13 56Ec 114
Lingey Cl. DA15: Sidc 61Vc 139
Lingfield Apts. E4 24Dc 52
Lingfield Av. DA2: Dart 59Rd 119
Lingfield Av. KT1: King T 70Na 131
Lingfield Av. RM14: Upm 34Pd 77
Lingfield Cl. EN1: Enf 16Ub 33
Lingfield Cl. HA6: Nwood 24U 44
Lingfield Ct. UB5: N'olt 40Ca 65
Lingfield Cres. SE9 56Tc 116
Lingfield Gdns. CR5: Coul 91Rb 197
Lingfield Gdns. N9 17Xb 33
Lingfield Ho. SE1 2C 230 (47Rb 91)
(off Lancaster St.)
Lingfield Rd. DA12: Grav'nd 1D 144
Lingfield Rd. KT4: Wor Pk 76Ya 154
Lingfield Rd. SW19 64Za 132
Lingfield Rd. TN15: Bor G 92De 205
Lingfield Way WD17: Wat 10V 12
Lingham Ct. SW9 54Nb 112
Lingham St. SW9 54Nb 112
Lingholm Way EN5: Barn 15Za 30
Lingmere Cl. IG7: Chig 19Sc 36
Lingmoor Dr. WD25: Wat 5Y 13
Ling Rd. DA8: Erith 51Ed 118
Ling Rd. E16 43Jc 93
Lingrove Gdns. IG9: Buck H 19Kc 35
Lings Coppice SE21 61Tb 135
Lingwell Rd. SW17 62Gb 133
Lingwood DA7: Bex 54Dd 118
Lingwood Ct. N2 27Gb 49
(off Norfolk Cl.)
Lingwood Gdns. TW7: Isle 52Ga 108
Lingwood Rd. E5 31Wb 71
Linhope St. NW1 5F 215 (42Hb 89)
Link, The DA3: Nw A G 75Be 165
Link, The EN3: Enf H 11Ac 34
Link, The HA0: Wemb 32La 66
Link, The HA5: Eastc 31Y 65
Link, The NW2 32Wa 68
Link, The SE9 62Qc 138
(off William Barefoot Dr.)
Link, The SL2: Slou 4M 81
Link, The UB5: N'olt 36Ba 65
Link, The W3 44Ra 87
Link Av. GU22: Pyr 87F 168
Linkenholt Mans. W6 49Va 88
(off Stamford Brook Av.)
Linkfield BR2: Hayes 72Jc 159
Linkfield KT8: W Mole 69Da 129
Linkfield Cnr. RH1: Redh 5N 207
Linkfield Gdns. RH1: Redh 6N 207
Linkfield La. RH1: Redh 5N 207
Linkfield Lodge RH1: Redh 5N 207
Linkfield Rd. TW7: Isle 54Ha 108
Linkfield St. RH1: Redh 6N 207
Link Ho. E3 41Dc 92
Link Ho. W10 44Za 88
(off Kingsdown Cl.)
Link La. SM6: W'gton 79Mb 156
Linklea Cl. NW9 24Ua 48
Link Pk. Heathrow SL0: Thorn 48L 83
Link Pl. IG6: Ilf 23Vc 55
Link Rd. E1 45Wb 91
Link Rd. KT15: Add 77N 149
Link Rd. N11 21Jb 50
Link Rd. RM9: Dag 40Dd 76
Link Rd. SL3: Dat 3N 103
Link Rd. SM6: W'gton 74Jb 156
Link Rd. TW14: Felt 59V 106
Link Rd. WD23: Bush 12Z 27
Link Rd. WD24: Bush 12Z 27
Link Rd. WD24: Wat 12Z 27
Links, The E17 28Ac 52
Links, The KT12: Walt T 75W 150
Links Av. RM2: Rom 26Kd 57
Links Av. SM4: Mord 70Cb 133
Links Brow KT22: Fet 96Ga 192
Links Bus. Cen. GU22: Wok 91E 188
Links Cl. KT21: Asht 89La 172
Linkscroft Av. TW15: Ashf 65R 128
Links Dr. N20 18Cb 31
Links Dr. WD6: E'tree 13Pa 29
Links Dr. WD7: R'lett 5Ha 14
Links Gdns. SW16 66Qb 134
Links Grn. Way KT11: Cobh 86Ca 171
Linkside IG7: Chig 22Sc 54
Linkside KT3: N Mald 68Ua 132
Linkside N12 23Cb 49
Linkside Cl. EN2: Enf 13Pb 32
Linkside Gdns. EN2: Enf 13Pb 32
Links Pl. KT21: Asht 89Ma 173
Links Rd. BR4: W W'ck 74Ec 158
Links Rd. IG8: Wfd G 22Jc 53
Links Rd. KT17: Eps 85Wa 174
Links Rd. KT21: Asht 90La 172
Links Rd. NW2 33Va 68
Links Rd. SW17 65Jb 134
Links Rd. TW15: Ashf 64N 127
Links Rd. W3 44Qa 87
Links Side EN2: Enf 13Pb 32
Link St. E9 37Yb 72
Links Vw. AL3: St A 1P 5
Links Vw. DA1: Dart 60Kd 119
Links Vw. N3 24Bb 49
Linksview N2 29Hb 49
(off Great Nth. Rd.)
Links Vw. Cl. HA7: Stan 24Ja 46
Linksview Ct. TW12: Hamp H 63Fa 130
Links Vw. Rd. CR0: C'don 76Cc 158
Links Vw. Rd. TW12: Hamp H 64Ea 130
Links Way BR3: Beck 72Cc 158
Links Way KT23: Bookh 100Aa 191
Links Way WD3: Crox G 13S 26
Linksway HA6: Nwood 24S 44
Linksway NW4 26Za 48
Linkswood Rd. SL1: Burn 10A 60
Links Yd. E1 43Wb 91
(off Spelman St.)
Link Way BR2: Broml 73Nc 160
Link Way HA5: Pinn 25Z 45
Link Way RM11: Horn 32Nd 77
Link Way TW10: Ham 61Ka 130
Link Way TW18: Staines 65K 127
Link Way UB9: Den 30J 43
Linkway GU22: Wok 89E 168
Linkway N4 31Sb 71
Linkway RM8: Dag 35Yc 75
Linkway SW20 69Xa 132
Linkway, The EN5: Barn 16Db 31
Linkway, The SM2: Sutt 81Eb 175
Linkway Rd. CM14: B'wood 20Vd 40
Linkwood Wlk. NW1 38Mb 70
Linley Cl. RM18: E Til 2M 123
Linley Ct. SM1: Sutt 77Eb 155
Linley Cres. RM7: Mawney 27Dd 56
Linley Ho. N1 39Sb 71
Linley Rd. N17 26Ub 51
Linnell Cl. NW11 30Db 49
Linnell Cl. RH1: Redh 10P 207
Linnell Dr. NW11 30Db 49
Linnell Ho. E1 7K 219 (43Vb 91)
(off Folgate St.)
Linnell Ho. NW8 39Eb 69
(off Ainsworth Way)
Linnell Rd. N18 22Wb 51
Linnell Rd. RH1: Redh 7A 208
Linnell Rd. SE5 54Ub 113
Linnet Cl. CR2: Sels 82Zb 178
Linnet Cl. N9 18Zb 34
Linnet Cl. SE28 45Yc 95
Linnet Cl. WD23: Bush 17Ea 28
Linnet Ct. SW15 58Ya 110
Linnet Ho. DA9: Ghithe 58Wd 120
(off Waterstone Way)
Linnet M. SW12 59Jb 112
Linnet Rd. HP3: Hem H 6M 3
Linnet Rd. WD5: Ab L 3W 12
Linnett Cl. E4 21Ec 52
Linnet Wlk. AL10: Hat 2C 8
Linnet Way RM19: Purf 50Rd 97
Linom Rd. SW4 56Nb 112
Linscott Rd. E5 35Yb 72
Linsdell Rd. IG11: Bark 39Sc 74
Linsey Cl. HP3: Hem H 6A 4
Linsey Ct. E10 32Cc 72
(off Grange Rd.)
Linsey St. SE16 49Wb 91
(not continuous)
Linslade Cl. HA5: Eastc 28X 45
Linslade Cl. TW4: Houn 57Aa 107
Linslade Ho. E2 39Wb 71
Linslade Ho. NW8 5E 214 (42Gb 89)
(off Paveley St.)
Linstead Rd. BR6: Chels 79Wc 161
Linstead St. NW6 38Cb 69
Linstead Way SW18 59Ab 110
Linsted Ct. SE9 58Uc 116
Linster Gro. WD6: Bore 15Sa 29
Lintaine Cl. W6 51Ab 110
Linters Ct. RH1: Redh 4P 207
Linthorpe Av. HA0: Wemb 37La 66
Linthorpe Rd. EN4: Cockf 13Gb 31
Linthorpe Rd. N16 31Ub 71
Linton Av. WD6: Bore 11Pa 29
Linton Cl. CR4: Mitc 73Hb 155
Linton Cl. DA16: Well 53Xc 117
Linton Cl. SE7 50Lc 93
Linton Ct. NW1 37Lb 70
(off Agar Gro.)
Linton Ct. RM1: Rom 26Gd 56
Linton Gdns. E6 44Nc 94
Linton Glade CR0: Sels 81Ac 178
Linton Gro. SE27 64Rb 135
Linton Ho. E3 43Cc 92
(off St Paul's Way)
Linton Rd. IG11: Bark 38Sc 74
Lintons La. KT17: Eps 84Ua 174
Linton St. N1 1E 218 (39Sb 71)
(not continuous)
Lintott Ct. TW19: Stanw 58M 105
Linver Rd. SW6 54Cb 111
Linwood Cl. SE5 54Vb 113
Linwood Cres. EN1: Enf 11Wb 33
Linzee Rd. N8 28Nb 50
Lion & Lamb Ct. CM14: B'wood 19Yd 40
(off High St.)
Lion Apts. SE16 50Xb 91
(off Rotherhithe New Rd.)
Lion Av. TW1: Twick 60Ha 108
Lion Bus. Pk. DA12: Grav'nd 9H 123
Lion Cl. SE4 58Cc 114
Lion Cl. TW17: Shep 69N 127
Lion Ct. CM16: Epp 2Wc 23
Lion Ct. E1 45Zb 92
(off The Highway)
Lion Ct. HP3: Hem H 7P 3
Lion Ct. N1 39Pb 70
(off Copenhagen St.)
Lion Ct. SE1 7J 225 (46Ub 91)
(off Magdalen St.)
Lion Ct. WD6: Bore 11Sa 29
Lionel Gdns. SE9 57Mc 115
Lionel Mans. W14 48Za 88
(off Haarlem Rd.)
Lionel M. W10 43Ab 88
Lionel Oxley Ho. RM17: Grays 51De 121
(off New Rd.)
Lionel Rd. SE9 57Mc 115
Lionel Rd. Nth. TW8: Bford 48Na 87
Lionel Rd. Sth. TW8: Bford 50Pa 87
Liongate Ent. Pk. CR4: Mitc 70Fb 133
Lion Ga. Gdns. TW9: Rich 55Pa 109
Lion Ga. M. SW18 59Cb 111
Liongate M. KT8: E Mos 69Ha 130
Lion Grn. Rd. CR5: Coul 88Mb 176
Lion Head Ct. CR0: C'don 77Sb 157
(off St Andrew's Rd.)
Lion La. RH1: Redh 5P 207
Lion Mills E2 40Wb 71
Lion Pk. Av. KT9: Chess 77Qa 153
Lion Retail Pk. 88D 168
Lion Rd. CR0: C'don 71Sb 157
Lion Rd. DA6: Bex 56Ad 117
Lion Rd. E6 43Pc 94
Lion Rd. N9 19Wb 33
Lion Rd. TW1: Twick 60Ha 108
Lions Cl. SE9 62Lc 137
Lions Row CM14: W'ley 21Yd 58
Lion Way TW8: Bford 52Ma 109
Lion Wharf Rd. TW7: Isle 55Ka 108
Lion Yd. SW4 56Mb 112
Liphook Cl. RM12: Horn 35Hd 76
Liphook Cres. SE23 59Yb 114
Liphook Rd. WD19: Wat 21Z 45
Lippitts Hill IG10: Lough 11Gc 35
Lipsham Cl. SM7: Bans 85Fb 175
Lipton Cl. SE28 45Yc 95
Lipton Rd. E1 44Zb 92
Liquorice La. GU22: Wok 94B 188
Lisbon Av. TW2: Twick 61Ea 130
Lisbon Cl. E17 26Bc 52
Lisburne Rd. NW3 35Hb 69
Lisford St. SE15 53Vb 113
Lisgar Ter. W14 49Bb 89
Liskeard Cl. BR7: Chst 65Sc 138
Liskeard Gdns. SE3 53Jc 115
Liskeard Ho. SE11 7A 230 (50Qb 90)
(off Kennings Way)
Liskeard Lodge CR3: Cat'm 98Wb 197
Lisle Cl. DA12: Grav'nd 1K 145
Lisle Cl. SW17 63Kb 134
Lisle Ct. NW2 34Ab 68
Lisle Pl. RM17: Grays 48Ce 99
Lisle St. WC2 4E 222 (45Mb 90)
Lismirrane Ind. Pk. WD6: E'tree 16Ka 28
Lismore HP3: Hem H 4C 4
Lismore SW19 64Bb 133
(off Woodside)
Lismore Blvd. NW9 27Va 48
Lismore Cir. NW5 36Jb 70
Lismore Cl. TW7: Isle 54Ja 108
Lismore Pk. SL2: Slou 4K 81
Lismore Rd. CR2: S Croy 79Ub 157
Lismore Rd. N17 27Tb 51
Lismore Wlk. N1 37Sb 71
(off Clephane Rd. Nth.)
Lissant Cl. KT6: Surb 73Ma 153
Lisselton Ho. NW4 28Za 48
(off Belle Vue Est.)
Lissenden Gdns. NW5 35Jb 70
(not continuous)
Lissenden Mans. NW5 35Jb 70
Lissoms Rd. CR5: Chips 90Jb 176
Lisson Grn. Est. NW8 5D 214 (42Gb 89)
(off Tresham Cres.)
Lisson Gro. NW1 6D 214 (42Gb 89)
Lisson Gro. NW8 5C 214 (42Fb 89)
LISSON GROVE 7E 214 (43Gb 89)
Lisson Ho. NW1 7D 214 (43Gb 89)
(off Lisson St.)
Lisson St. NW1 7D 214 (43Gb 89)
Lister Av. RM3: Hrld W 26Md 57
Lister Cl. CR4: Mitc 67Gb 133
Lister Cl. W3 43Ta 87
Lister Cotts. WD6: E'tree 15Ja 28
Lister Ct. HA1: Harr 31Ka 66
Lister Ct. N16 33Ub 71
Lister Ct. NW9 26Ua 48
Lister Dr. DA11: Nflt 60Ee 121
Lister Gdns. N18 22Sb 51
Listergate Ct. SW15 56Ya 110
LISTER HOSPITAL, THE 50Kb 90
Lister Ho. E1 43Wb 91
(off Lomas St.)
Lister Ho. HA9: Wemb 34Sa 67
(off Barnhill Rd.)
Lister Ho. UB3: Harl 49U 84
Lister Lodge W9 43Cb 89
(off Admiral Wlk.)
Lister Rd. E11 32Gc 73
Lister Rd. RM18: Tilb 4C 122
Lister Wlk. SE28 45Zc 95
Liston Rd. N17 25Wb 51
Liston Rd. SW4 55Lb 112
Liston Way IG8: Wfd G 24Lc 53
Listowel Cl. SW9 52Qb 112
Listowel Rd. RM10: Dag 34Cd 76
Listria Pk. N16 33Ub 71
Litcham Ho. E1 41Zb 92
(off Longnor Rd.)
Litcham Spur SL1: Slou 4H 81
Litchfield Av. E15 37Gc 73
Litchfield Av. SM4: Mord 73Bb 155
Litchfield Ct. E17 30Cc 52
Litchfield Gdns. KT11: Cobh 86X 171
Litchfield Gdns. NW10 37Wa 68
Litchfield Rd. SM1: Sutt 77Eb 155
Litchfield St. WC2 4E 222 (45Mb 90)
Litchfield Way NW11 29Db 49
Lithgow's Rd. TW6: H'row A 56U 106
Lithos Rd. NW3 37Db 69
Littel Dormers SL9: Ger X 29B 42
Litten Cl. RM5: Col R 23Cd 56
Litten Cl. SL3: L'ly 9N 81
Litten Nature Reserve 41Ea 86
Little Acre BR3: Beck 69Cc 136
Little Albany St. NW1 5A 216 (42Kb 90)
(off Longford St.)
Little Angel Theatre 39Rb 71
(off Dagmar Ter.)
Little Argyll St. W1 3B 222 (44Lb 90)
Little Aston Rd. RM3: Hrld W 24Qd 57
Little Belhus Cl. RM15: S Ock 42Xd 98
Little Benty UB7: W Dray 50M 83
Little Birch Cl. KT15: New H 81M 169
Little Birches DA15: Sidc 61Uc 138
Little Boltons, The SW10 50Db 89
Little Boltons, The SW5 50Db 89
LITTLE BOOKHAM 96Ba 191
LITTLE BOOKHAM COMMON 94Aa 191
Little Bookham St. KT23: Bookh 95Ba 191
Little Bornes SE21 63Ub 135
Littlebourne SE13 59Gc 115
Littlebourne Ho. SE17 7J 231 (50Ub 91)
(off Upnor Way)
Little Bow La. RM13: Rain 39Hd 76
Little Bridge Rd. HP4: Berk 1A 2
Little Brights Rd. DA17: Belv 47Dd 96
Little Britain EC1 1C 224 (43Rb 91)
LITTLE BRITAIN 44L 83
Littlebrook Av. SL2: Slou 2C 80
Littlebrook Bus. Cen. DA1: Dart 54Rd 119
Littlebrook Cl. CR0: C'don 72Zb 158
Littlebrook Gdns. EN8: Chesh 2Zb 20
LITTLE BROOK HOSPITAL 58Sd 120
LITTLEBROOK INTERCHANGE 56Rd 119
Littlebrook Mnr. Way DA1: Dart Rennie Dr. 55Rd 119
Littlebrook Mnr. Way DA1: Dart St Vincents Av. 57Qd 119
Little Brownings SE23 61Xb 135
Little Buntings SL4: Wind 5D 102
Littlebury Ct. WD18: Wat 14W 26
Littlebury Rd. SW4 55Mb 112
Little Bury St. N9 18Tb 33
Little Bushey La. WD23: Bush 12Ca 27
Little Cedars N12 21Eb 49
LITTLE CHALFONT 11A 24
Little Chapels Way SL1: Slou 7E 80
Little Chelsea Ho. SW10 51Eb 111
(off Edith Gro.)
Little Chesters KT20: Walt H 97Wa 194
Little Chester St. SW1 3K 227 (48Kb 90)
Little Cloisters SW1 3F 229 (48Nb 90)
Little College La. EC4 4F 225 (45Tb 91)
(off College St.)
Little College St. SW1 3F 229 (48Nb 90)
Littlecombe SE7 51Kc 115
Littlecombe Cl. SW15 58Za 110
Little Comn. HA7: Stan 20Ja 28
Little Common La. RH1: Blet 4H 209
Littlecote Cl. SW19 59Ab 110
Littlecote Pl. HA5: Hat E 25Aa 45
Little Cottage Pl. SE10 52Dc 114
Little Ct. BR4: W W'ck 75Gc 159
Littlecourt Rd. TN13: S'oaks 96Jd 202
Little Cranmore La. KT24: W Hor 100R 190
Little Cft. DA13: Ist R 6A 144
Littlecroft SE9 55Qc 116
Little Croft Cl. AL2: Lon C 9E 6
Littlecroft Rd. TW20: Egh 64B 126
Littledale DA2: Daren 62Sd 142
Littledale SE2 51Xc 117
Little Dean's Yd. SW1 3F 229 (48Nb 90)
(off Dean's Yd.)
Little Dimocks SW12 61Kb 134
Little Dorrit Ct. SE1 1E 230 (47Sb 91)
Littledown Rd. SL1: Slou 6K 81
Little Dragons IG10: Lough 14Mc 35
LITTLE EALING 48Ma 87
Little Ealing La. W5 49La 86
Little East Fld. CR5: Coul 93Mb 196
Little Edward St. NW1 3A 216 (41Kb 90)
Little Elms UB3: Harl 52T 106
Little Essex St. WC2 4K 223 (45Qb 90)
(off Essex St.)
Little Ferry Rd. TW1: Twick 60Ka 108
Littlefield Cl. KT1: King T 68Na 131
Littlefield Cl. N19 35Lb 70
Littlefield Ho. KT1: King T 68Na 131
(off Littlefield Cl.)
Littlefield Rd. HA8: Edg 24Sa 47
Little Friday Rd. E4 19Gc 35
Little Gdn. Cl. RM3: Rom 23Nd 57
Little Gaynes Gdns. RM14: Upm 35Rd 77
Little Gaynes La. RM14: Upm 35Pd 77
Little Gearies IG6: Ilf 28Rc 54
Little George St. SW1 2F 229 (47Nb 90)
Little Gerpins La. RM14: Upm 39Pd 77
Little Goldings Est. IG10: Lough 11Qc 36
Little Grange UB6: G'frd 41Ja 86
Little Graylings WD5: Ab L 5U 12
Little Grebe Ho. UB7: Yiew 45M 83
(off Wraysbury Dr.)
LITTLE GREEN 13Q 26
Little Grn. TW9: Rich 56Ma 109
Little Green La. KT16: Chert 76G 148
Little Green La. WD3: Crox G 13P 25
Little Green St. NW5 35Kb 70
Little Gregories La. CM16: They B 7Cc 22
Little Gro. WD23: Bush 14Da 27
Littlegrove EN4: E Barn 16Gb 31
Little Halliards KT12: Walt T 72W 150
Littlehayes WD4: K Lan 1Q 12
Little Hay Golf Course 6E 2
Little Heath RM6: Chad H 28Xc 55
Little Heath SE7 51Nc 116
LITTLE HEATH 2Eb 17
LITTLE HEATH 29Xc 55
Little Heath La. HP4: Berk 3D 2
Little Heath La. HP4: Pott E 3D 2
Littleheath La. KT11: Cobh 86Ca 171
Littleheath La. KT11: Stoke D 86Ca 171
Little Heath Rd. DA7: Bex 53Bd 117
Little Heath Rd. GU24: Chob 1J 167
Littleheath Rd. CR2: Sels 80Xb 157
Little Highwood Way CM14: B'wood 17Xd 40
Little Hill WD3: Herons 17E 24
Little Holland House 80Hb 155
Little Holt E11 29Jc 53
Little How Cft. WD5: Ab L 3S 12
LITTLE ILFORD 36Pc 74
Little Ilford La. E12 35Pc 74
Littlejohn Rd. BR5: St M Cry 72Wc 161
Littlejohn Rd. W7 44Ha 86
Little Julians Hill TN13: S'oaks 99Jd 202
Little Larkins EN5: Barn 16Ab 30
Little London Cl. UB8: Hil 43R 84
Little London Ct. SE1 47Vb 91
(off Wolseley St.)
Little Marlborough St. W1 3B 222 (44Lb 90)
(off Kingly St.)
Little Martins WD23: Bush 15Da 27
Littlemead GU21: Wok 8K 167

Littlemead KT10: Esh 77Fa 152
Littlemede SE9 62Pc 138
Little Mimms HP2: Hem H 1M 3
Littlemoor Gdns. GU22: Wok 89E 168
Littlemoor Rd. IG1: Ilf 34Tc 74
Littlemore Rd. SE2 47Wc 95
Little Moreton Cl. KT14: W Byf 84K 169
Little Moss La. HA5: Pinn 26Aa 45
Little Newport St. WC2 4E 222 (45Mb 90)
Little New St. EC4 2A 224 (44Qb 90)
Little Norman St. TN14: Ide H 100Ad 201
Little Oaks Cl. TW17: Shep 70P 127
Little Orchard GU21: Wok 86C 168
Little Orchard HP2: Hem H 1A 4
Little Orchard KT15: Wdhm 83J 169
Little Orchard Cl. HA5: Pinn 26Aa 45
Little Orchard Cl. WD5: Ab L 4T 12
Little Orchard Pl. KT10: Esh 76Ea 152
Little Orchards KT18: Eps 86Ua 174
(off Worple Rd.)
Little Oxhey La. WD19: Wat 22Z 45
Little Pk. HP3: Bov 10C 2
Little Park Dr. TW13: Felt 61Aa 129
Little Park Gdns. EN2: Enf 13Sb 33
Little Park La. KT17: Ewe 83Ya 174
Little Pastures CM14: B'wood 21Vd 58
Little Pipers Cl. EN7: G Oak 1Rb 19
Little Pluckett's Way IG9: Buck H 18Mc 35
Little Portland St. W1 2B 222 (44Lb 90)
Littleport Spur SL1: Slou 4J 81
Little Potters WD23: Bush 17Fa 28
Little Queen's Rd. TW11: Tedd 65Ha 130
Little Queen St. DA1: Dart 59Pd 119
Little Redlands BR1: Broml 68Nc 138
Little Red Wlk. DA1: Dart 54Pd 119
Little Riding GU22: Wok 88D 168
Little Rd. HP2: Hem H 1N 3
Little Rd. UB3: Hayes 47V 84
Little Roke Av. CR8: Kenley 86Rb 177
Little Roke Rd. CR8: Kenley 86Sb 177
Littlers Cl. SW19 67Fb 133
Little Russell St. WC1 1F 223 (43Nb 90)
Little St James's St. SW1 7B 222 (46Lb 90)
Little St Leonard's SW14 55Sa 109
Little Sanctuary SW1 2E 228 (47Mb 90)
Little Smith St. SW1 3E 228 (48Mb 90)
Little Somerset St. E1 3K 225 (44Vb 91)
Little South St. SE5 53Ub 113
LITTLE STANMORE 24Pa 47
Littlestone Cl. BR3: Beck 65Cc 136
Little Strand NW9 26Va 48
Little Stream Cl. HA6: Nwood 22U 44
Little St. EN9: Walt A 8Ec 20
Little Sutton La. SL3: L'ly Hurricane Way 50E 82
Little Sutton La. SL3: L'ly Kings Ter. 51D 104
Little Thames Wlk. SE8 51Dc 114
(off Dancers Way)
Little Thrift BR5: Pet W 70Sc 138
LITTLE THURROCK 48Fe 99
Little Titchfield St. W1 1B 222 (43Lb 90)
LITTLETON 69Q 128
Littleton Av. E4 18Hc 35
LITTLETON COMMON 66S 128
Littleton Cres. HA1: Harr 33Ha 66
Littleton Ho. RH2: Reig 5J 207
(off Somers Cl.)
Littleton Ho. SW1 7B 228 (50Lb 90)
(off Lupus St.)
Littleton La. RH2: Reig 8F 206
Littleton La. TW17: Shep 73M 149
Littleton Rd. HA1: Harr 33Ha 66
Littleton Rd. TW15: Ashf 66S 128
Littleton St. SW18 61Eb 133
Little Trinity La. EC4 4E 224 (45Sb 91)
Little Turnstile WC1 1H 223 (43Pb 90)
LITTLE VENICE 7A 214 (43Eb 89)
Little Venice Sports Cen. 7B 214 (43Fb 89)
LITTLE WARLEY 26Ae 59
Little Warley Hall La. CM13: L War 26Ae 59
LITTLEWICK 8J 167
LITTLEWICK COMMON 7J 167
Littlewick Rd. GU21: Knap 8J 167
Littlewick Rd. GU21: Wok 8J 167
Little Windmill Hill WD4: Chfd 4G 10
Littlewold Dr. CR6: W'ham 91Zb 198
Little Wood TN13: S'oaks 94Ld 203
Littlewood SE13 58Ec 114
Little Wood Cl. BR5: St P 67Wc 139
Littlewood Cl. W13 48Ka 86
LITTLE WOODCOTE 83Kb 176
Little Woodcote Est. SM5: Cars 83Kb 176
Little Woodcote Est. SM6: W'gton 83Kb 176
Little Woodcote La. CR8: Purl 83Kb 176
Little Woodcote La. SM5: Cars 84Kb 176
Little Woodlands SL4: Wind 5D 102
Little Wood St. KT1: King T 68Ma 131
Littleworth Av. KT10: Esh 78Fa 152
Littleworth Bus. Cen. DA5: Bexl 60Dd 118
LITTLEWORTH COMMON 4B 60
Littleworth Comn. Rd. KT10: Esh 76Fa 152
Littleworth La. KT10: Esh 77Fa 152
Littleworth Pl. KT10: Esh 77Fa 152
Littleworth Rd. KT10: Esh 78Fa 152
Littleworth Rd. SL1: Burn 4A 60
Livermere Ct. E8 39Vb 71
(off Queensbridge Rd.)
Livermere Rd. E8 39Vb 71
Liverpool Gro. SE17 50Sb 91
Liverpool Rd. AL1: St A 2C 6
Liverpool Rd. CR7: Thor H 69Sb 135
Liverpool Rd. E10 30Ec 52
Liverpool Rd. E16 43Gc 93
Liverpool Rd. KT2: King T 66Qa 131
Liverpool Rd. N1 38Qb 70
Liverpool Rd. N7 36Qb 70
Liverpool Rd. SL1: Slou 4F 80
Liverpool Rd. W5 47Ma 87
Liverpool Rd. WD18: Wat 15X 27
Liverpool St. EC2 1H 225 (43Ub 91)
Liverpool Street Station (Rail, Underground & Crossrail) 1H 225 (43Ub 91)
Liverymen Wlk. DA9: Ghithe 56Yd 120
Livery Stables Cl. BR2: Broml 76Mc 159
Livesey Cl. KT1: King T 69Pa 131
Livesey Cl. SE28 48Sc 94
Livesey Pl. SE15 51Wb 113
Livingstone Ct. E10 30Ec 52
Livingstone Ct. EN5: Barn 12Ab 30
(off Church La.)
Livingstone Ct. HA3: W'stone 27Ha 46
Livingstone Gdns. DA12: Grav'nd 4F 144
LIVINGSTONE HOSPITAL 59Pd 119
Livingstone Ho. KT17: Eps 84Ua 174
(off Winter Cl.)
Livingstone Ho. NW10 38Ta 67
Livingstone Ho. SE5 52Sb 113
(off Wyndham Rd.)
Livingstone Lodge W9 43Cb 89
(off Admiral Wlk.)
Livingstone Mans. W14 51Ab 110
(off Queen's Club Gdns.)
Livingstone Pl. E14 50Ec 92
Livingstone Rd. CR3: Cat'm 94Tb 197
Livingstone Rd. CR7: Thor H 68Sb 135
Livingstone Rd. DA12: Grav'nd 4F 144
Livingstone Rd. E17 30Dc 52
Livingstone Rd. N13 23Nb 50
Livingstone Rd. SW11 55Fb 111
Livingstone Rd. TW3: Houn 56Ea 108
Livingstone Rd. UB1: S'hall 45Z 85
Livingstone Ter. RM13: Rain 39Gd 76
LivingWell Health Club Dartford 56Sd 120
LivingWell Health Club London Heathrow Hilton 58S 106
LivingWell Health Club Wembley 35Qa 67
Livonia St. W1 3C 222 (44Lb 90)
Lizard St. EC1 4E 218 (41Sb 91)
Lizban St. SE3 52Kc 115
Lizmans Ter. W8 48Cb 89
(off Earl's Ct. Rd.)
Llanbury Cl. SL9: Chal P 24A 42
Llandovery Ho. E14 47Ec 92
(off Chipka St.)
Llanelly Rd. NW2 33Bb 69
Llanover Rd. HA9: Wemb 34Ma 67
Llanover Rd. SE18 51Qc 116
Llanthony Rd. SM4: Mord 71Fb 155
Llanvanor Rd. NW2 33Bb 69
Llewellyn Ct. SE20 67Yb 136
Llewellyn Mans. W14 49Ab 88
(off Hammersmith Rd.)
Llewellyn St. SE16 47Wb 91
Lloyd Av. CR5: Coul 86Jb 176
Lloyd Av. SW16 67Nb 134
Lloyd Baker St. WC1 4J 217 (41Pb 90)
(not continuous)
Lloyd Cl. EN7: G Oak 2Vb 19
Lloyd Ct. HA5: Pinn 29Z 45
Lloyd Ct. SW11 54Hb 111
(off Roydon Cl.)
Lloyd Ho. BR3: Beck 65Dc 136
Lloyd Ho. CR0: C'don 74Tb 157
(off Tavistock Rd.)
Lloyd M. EN3: Enf L 10Cc 20
Lloyd Pk. 26Cc 52
Lloyd Pk. Av. CR0: C'don 77Vb 157
Lloyd Pk. Ho. E17 27Cc 52
Lloyd Park Stop (London Tramlink) 77Vb 157
Lloyd Rd. E17 28Zb 52
Lloyd Rd. E6 39Pc 74
Lloyd Rd. KT4: Wor Pk 76Ya 154
Lloyd Rd. RM9: Dag 38Bd 75
Lloyd's Av. EC3 3J 225 (44Ub 91)
Lloyd's Pl. SE3 54Gc 115
Lloyd Sq. WC1 3K 217 (41Qb 90)
Lloyd's Row EC1 4A 218 (41Qb 90)
Lloyd St. WC1 3K 217 (41Qb 90)
Lloyds Way BR3: Beck 71Ac 158
Lloyds Wharf SE1 2K 231 (47Vb 91)
(off Mill St.)
Lloyd Thomas Ct. N22 24Pb 50
Lloyd Vs. E6 42Nc 94
Lloyd Vs. SE4 54Cc 114
Loam Ct. DA1: Dart 60Nd 119
Loampit Hill SE13 54Cc 114
Loampit Rd. DA8: Erith 51Ed 118
Loampit Va. SE13 55Dc 114
LOAMPIT VALE 55Ec 114
Loanda Cl. E8 39Vb 71
Loates La. WD17: Wat 13Y 27
Loats Rd. SW2 58Nb 112
Lobelia Cl. E6 43Nc 94
Lobelia Rd. GU24: Bisl 7E 166
Local Board Rd. WD17: Wat 15Z 27
Locarno Cl. SW16 64Lb 134
Locarno Rd. UB6: G'frd 42Fa 86
Locarno Rd. W3 46Sa 87
Lochaber Rd. SE13 56Gc 115
Lochaline St. W6 51Ya 110
Lochan Cl. UB4: Yead 42Aa 85
Loch Cres. HA8: Edg 21Pa 47
Lochinvar Cl. SL1: Slou 7F 80
Lochinvar St. SW12 59Kb 112
Lochleven Ho. N2 26Fb 49
(off The Grange)
Lochmere Cl. DA8: Erith 51Dd 118
Lochmore Ho. SW1 6J 227 (49Jb 90)
(off Cundy St.)
Lochnagar St. E14 43Ec 92
Lockbridge Ct. W9 43Cb 89
(off Woodfield Rd.)
Lock Bldg., The E15 40Ec 72
Lock Chase SE3 55Gc 115
Lock Cl. KT15: Wdhm 84G 168
Lock Cl. UB2: S'hall 47Fa 86
Lock Cotts. UB1: S'hall 47Fa 86
Lock Ct. E5 33Zb 72
Locke Apts. CR0: C'don 75Tb 157
Locke Cl. RM13: Rain 37Hd 76
Locke Gdns. SL3: L'ly 7N 81
Locke Hgts. N1 38Pb 70
(off Caledonian Rd.)
Locke Ho. SW8 53Lb 112
(off Wadhurst Rd.)
Locke King Cl. KT13: Weyb 80Q 150
Locke King Rd. KT13: Weyb 80Q 150
Lockers, The HP1: Hem H 1L 3
Lockers Pk. La. HP1: Hem H 2K 3
Lockes End E17 28Ec 52
Lockesfield Pl. E14 50Dc 92
Lockesley Dr. BR5: St M Cry 72Vc 161
Lockesley Sq. KT6: Surb 72Ma 153
Lockestone KT13: Weyb 79P 149
Lockestone Cl. KT13: Weyb 79P 149
Locket Rd. HA3: W'stone 27Ga 46
Locket Rd. M. HA3: W'stone 26Ga 46
Lockets Cl. SL4: Wind 3B 102
Locke Way GU21: Wok 89B 168
Lockfield Av. EN3: Brim 12Ac 34
Lockfield Dr. GU21: Wok 8J 167
Lockgate Cl. E9 36Bc 72
Lockhart Cl. EN3: Pond E 15Xb 33
Lockhart Cl. N7 37Pb 70
Lockhart Rd. KT11: Cobh 85Y 171
Lockhart Rd. WD17: Wat 11W 26
Lockhart St. E3 42Bc 92
Lock Ho. HP3: Hem H 6N 3
Lockhouse, The NW1 39Jb 70
Lockhurst St. E5 35Zb 72
Lockie Pl. SE25 69Wb 135
Lockier Wlk. HA9: Wemb 34Ma 67
Lockington Rd. SW8 53Kb 112
Lock Island TW17: Shep 75Q 150
Lock Keepers Hgts. SE16 48Zb 92
(off Brunswick Quay)
Lock La. GU22: Pyr 88K 169
Lockmead Rd. N15 30Wb 51
Lockmead Rd. SE13 55Ec 114
Lock M. NW1 37Mb 70
(off Northpoint Sq.)
Lock Mill Apts. E2 1K 219 (39Vb 71)
(off Whiston Rd.)
Lock Path SL4: Dor 1B 102
Lock Path SL4: Eton W 1B 102
Lock Rd. TW10: Ham 63La 130
LOCKSBOTTOM 76Qc 160
Locksfields SE17 6G 231 (49Tb 91)
(off Catesby St.)
Lockside E14 45Ac 92
(off Narrow St.)
Lock Side Way E16 45Sc 94
Locks La. CR4: Mitc 67Jb 134
Locksley Dr. GU21: Wok 9K 167
Locksley Est. E14 44Bc 92
Locksley Ho. E15 37Fc 73
(off Forrester Way)
Locksley St. E14 43Bc 92
Locksmeade Rd. TW10: Ham 63La 130
Locksons Cl. E14 43Dc 92
Lockswood GU24: Brkwd 2F 186
Lockswood Cl. EN4: Cockf 14Hb 31
Locks Yd. TN13: S'oaks 97Ld 203
(off High St.)
Lockton St. W10 45Za 88
(off Bramley Rd.)
Lock Vw. Ct. E14 45Ac 92
(off Narrow St.)
Lockview Ct. HP3: Hem H 6N 3
Lockwell Rd. RM10: Dag 34Cd 76
Lockwood Cl. SE26 63Zb 136
Lockwood Ct. WD6: Bore 12Sa 29
Lockwood Ho. E5 33Yb 72
Lockwood Ho. SE11 51Qb 112
Lockwood Ind. Pk. N17 27Xb 51
Lockwood Path GU21: Wok 85F 168
Lockwood Pl. DA1: Dart 55Qd 119
Lockwood Pl. E4 23Cc 52
Lockwood Rd. UB2: S'hall 48Ca 85
Lockwood Sq. SE16 48Xb 91
Lockwood Wlk. RM1: Rom 29Gd 56
Lockwood Way E17 26Zb 52
Lockwood Way KT9: Chess 78Qa 153
Lockworks Ho. E17 28Zb 52
(off Wickford Way)
Lockyer Est. SE1 2G 231 (47Tb 91)
(off Kipling St.)
Lockyer Ho. SE10 50Hc 93
(off Armitage Rd.)
Lockyer Ho. SW15 55Za 110
Lockyer Ho. SW8 52Mb 112
(off Wandsworth Rd.)
Lockyer M. EN3: Enf L 10Dc 20
Lockyer Rd. RM19: Purf 51Sd 120
Lockyer St. SE1 2G 231 (47Tb 91)
Locomotive Dr. TW14: Felt 60W 106
Locton Grn. E3 39Bc 72
Loddiges Ho. E9 38Yb 72
Loddiges Rd. E9 38Yb 72
Lodding Salts Rd. DA12: Grav'nd 10J 123
Loddon Ho. NW8 6C 214 (42Fb 89)
(off Church St. Est.)
Loddon Spur SL1: Slou 5J 81
Loder Cl. GU21: Wok 85F 168
Loder St. SE15 52Yb 114
Lodge, The CM16: Epp 1Yc 23
Lodge, The SM7: Bans 89Eb 175
Lodge, The W12 47Za 88
(off Richmond Way)
Lodge, The WD24: Wat 12Y 27
Lodge Av. CR0: Wadd 76Qb 156
Lodge Av. DA1: Dart 58Ld 119
Lodge Av. HA3: Kenton 28Na 47
Lodge Av. RM2: Rom 28Jd 56
Lodge Av. RM8: Dag 37Xc 75
Lodge Av. RM9: Dag 39Wc 75
Lodge Av. SW14 55Ua 110
Lodge Av. WD6: E'tree 15Pa 29
LODGE AVENUE FLYOVER JUNC. 39Wc 75
Lodgebottom Rd. KT18: Head 99Qa 193
Lodgebottom Rd. RH5: Mick 99Qa 193
Lodge Cl. BR6: Orp 74Xc 161
Lodge Cl. HA8: Edg 23Pa 47
Lodge Cl. IG7: Chig 20Wc 37
Lodge Cl. KT11: Stoke D 88Ba 171
Lodge Cl. KT17: Ewe 82Ya 174
Lodge Cl. KT22: Fet 94Fa 192
Lodge Cl. N18 22Sb 51
Lodge Cl. SL1: Slou 7G 80
Lodge Cl. SL5: Asc 8A 124
Lodge Cl. SM6: W'gton 74Jb 156
Lodge Cl. TW20: Eng G 4P 125
Lodge Cl. TW7: Isle 53Ka 108
Lodge Cl. UB8: Cowl 42L 83
Lodge Cl. WD25: Wat 4Aa 13
Lodge Ct. HA0: Wemb 36Na 67
(off Station Gro.)
Lodge Ct. RM12: Horn 33Nd 77
Lodge Cres. BR6: Orp 74Xc 161
Lodge Cres. EN8: Walt C 6Zb 20
Lodge Dr. N13 21Qb 50
Lodge Dr. WD3: Loud 14L 25
Lodge End WD3: Crox G 14T 26
Lodge End WD7: R'lett 6Ka 14
Lodge Gdns. BR3: Beck 71Bc 158
Lodge Hill CR8: Purl 87Qb 176
Lodge Hill DA16: Well 52Xc 117
Lodge Hill IG4: Ilf 28Nc 54
Lodge Hill SE2 52Xc 117
Lodgehill Pk. Cl. HA2: Harr 33Da 65
Lodge La. CR0: New Ad 79Cc 158
Lodge La. DA12: Cobh 10J 145
Lodge La. DA5: Bexl 58Zc 117
Lodge La. EN9: Walt A 7Fc 21
Lodge La. HP8: Chal G 11A 24
Lodge La. N12 22Eb 49
Lodge La. RM16: Grays 47Ce 99
Lodge La. RM17: Grays 47Ce 99
Lodge La. RM5: Col R 24Cd 56
Lodge La. TN16: Westrm 99Sc 200
Lodge M. N5 35Sb 71
Lodge Pl. SM1: Sutt 78Db 155
Lodge Rd. BR1: Broml 66Lc 137
Lodge Rd. CM16: Epp 6Qc 22
Lodge Rd. CM16: Walt A 6Qc 22
Lodge Rd. CR0: C'don 72Rb 157
Lodge Rd. KT22: Fet 94Ea 192
Lodge Rd. NW4 28Ya 48
Lodge Rd. NW8 4C 214 (41Fb 89)
Lodge Rd. SM6: W'gton 78Kb 156
Lodge Vs. IG8: Wfd G 23Hc 53
Lodge Wlk. CR6: W'ham 88Cc 178
Lodge Way SL4: Wind 5C 102
Lodge Way TW15: Ashf 61N 127
Lodge Way TW17: Shep 68S 128
Lodore Gdns. NW9 29Ua 48
Lodore Grn. UB10: Ick 34N 63
Lodore St. E14 44Ec 92
Lodysons Cl. RM16: Ors 3C 100
Loewen Rd. RM16: Grays 8C 100
Loft, The W12 46Za 88
Loft Ho. E1 2K 225 (44Vb 91)
(off Middlesex St.)
Lofthouse Pl. KT9: Chess 79La 152
Loftie St. SE16 47Wb 91
Lofting Ho. N1 38Qb 70
(off Liverpool Rd.)
Lofting Rd. N1 38Pb 70
Lofts on the Pk. E9 37Zb 72
(off Cassland Rd.)
Loftus Rd. IG11: Bark 37Sc 74
Loftus Rd. W12 46Xa 88
Loftus Vs. W12 46Xa 88
(off Loftus Rd.)
Logan Cl. E20 36Ec 72
Logan Cl. EN3: Enf H 11Zb 34
Logan Cl. TW4: Houn 55Ba 107
Logan Ct. AL2: Lon C 9F 6
Logan Ct. RM1: Rom 29Gd 56
LOGANDENE ELDERLY MENTAL ILLNESS (EMI) UNIT 4P 3
Logan M. RM1: Rom 29Gd 56
Logan M. W8 49Cb 89
Logan Pl. W8 49Cb 89
Logan Rd. HA9: Wemb 33Ma 67
Logan Rd. N9 19Xb 33
Log Cabin, The 48La 86
(off Northfield Av.)
Loggetts SE21 61Ub 135
Logs Hill BR1: Broml 67Nc 138
Logs Hill BR7: Chst 66Nc 138
Logs Hill Cl. BR7: Chst 67Nc 138
Lohmann Ho. SE11 51Qb 112
(off Kennington Oval)
Lois Dr. TW17: Shep 71R 150
Lolesworth Cl. E1 1K 225 (43Vb 91)
Lolland Ho. SE7 49Nc 94
Lollard St. SE11 5J 229 (49Pb 90)
(not continuous)
Lollesworth La. KT24: W Hor 98S 190
Loman Path RM15: S Ock 44Vd 98
Loman St. SE1 1C 230 (47Rb 91)
Lomas Cl. CR0: New Ad 80Ec 158
Lomas Dr. E8 38Vb 71
Lomas St. E1 43Wb 91
Lombard Av. EN3: Enf H 11Yb 34
Lombard Av. IG3: Ilf 32Uc 74
Lombard Bus. Pk. CR0: C'don 73Pb 156
Lombard Bus. Pk. SW19 68Db 133
Lombard Ct. EC3 4G 225 (45Tb 91)
Lombard Ct. RM7: Rom 28Ed 56
Lombard Ct. W3 46Ra 87
Lombard La. EC4 3A 224 (44Qb 90)
Lombard Pl. E3 40Ac 72
Lombard Rd. N11 22Kb 50
Lombard Rd. SW11 54Fb 111
Lombard Rd. SW19 68Db 133
LOMBARD RDBT. 73Pb 156
Lombards, The RM11: Horn 31Pd 77
Lombards Chase CM13: W H'dn 30Fe 59
Lombard St. DA4: Hort K 71Sd 164
Lombard St. EC3 3G 225 (44Tb 91)
Lombard Trad. Est. SE7 49Kc 93
Lombard Wall SE7 48Kc 93
Lombard Wharf SW11 54Fb 111
Lombardy Cl. GU21: Wok 9K 167
Lombardy Cl. HP2: Hem H 3D 4
Lombardy Cl. IG6: Ilf 24Rc 54
Lombardy Dr. HP4: Berk 2A 2
Lombardy Pl. W2 45Db 89
Lombardy Retail Pk. 45X 85
Lombardy Way WD6: Bore 11Na 29
Lomond Cl. HA0: Wemb 38Pa 67
Lomond Cl. N15 29Ub 51
Lomond Gdns. CR2: Sels 80Ac 158
Lomond Gro. SE5 52Tb 113
Lomond Ho. SE5 52Tb 113
Loncin Mead Av. KT15: New H 81L 169
Loncroft Rd. SE5 51Ub 113
Londesborough Rd. N16 35Ub 71
Londinium Twr. E1 4K 225 (45Vb 91)
(off W. Tenter St.)
LONDON 6F 223 (46Nb 90)
London Aquatics Cen. Queen Elizabeth Olympic Pk. 38Ec 72
London Biggin Hill Airport 84Mc 179
London Bri. EC4 5G 225 (45Tb 91)
London Bri. SE1 6G 225 (46Tb 91)
London Bridge Experience 6G 225 (46Tb 91)
(off Tooley St.)
LONDON BRIDGE HOSPITAL 6G 225 (46Tb 91)
London Bridge Station (Rail & Underground) 7G 225 (46Tb 91)
London Bri. St. SE1 7G 225 (46Tb 91)
London Bri. Wlk. SE1 6G 225 (46Tb 91)
(off Duke St. Hill)
London Broncos RLFC 25Na 47
London Business School 5F 215 (42Hb 89)
London Bus Mus., The 80P 149
London Canal Mus. 1G 217 (40Nb 70)
London Central Markets 1B 224 (43Rb 91)
(off West Smithfield)
LONDON CITY AIRPORT 46Nc 94
London City Airport Station (DLR) 46Nc 94
London City Coll. 7K 223 (46Qb 90)
(off Waterloo Rd.)
LONDON CLINIC 6J 215 (42Jb 90)
London Coliseum 5F 223 (45Nb 90)
(off St Martin's La.)
London College of Fashion, The John Prince's St. 2A 222 (44Kb 90)
London College of Fashion, The Mare St. 38Yb 72
LONDON COLNEY 8H 7
London Colney By-Pass AL2: Lon C 7H 7
London Ct. CM14: W'ley 22Xd 58
Londonderry Pde. DA8: Erith 52Fd 118
London Designer Outlet HA9: Wemb 35Qa 67
London Distribution Pk. RM18: Tilb 3A 122
London Dock E1 45Wb 91
London Dungeon 1H 229 (47Pb 90)
London E. Leisure Pk. RM9: Dag 39Ad 75
London Eye 1H 229 (47Pb 90)
LONDON EYE HOSPITAL 2K 221 (44Kb 90)
(off Harley St.)
London Flds. East Side E8 38Xb 71
London Fields Lido 38Xb 71
London Fields Station (Overground) 38Xb 71
London Flds. West Side E8 38Wb 71
London Film Mus. 4G 223 (45Nb 90)
(off Wellington St.)
London Fruit Exchange E1 1K 225 (43Vb 91)
(off Brushfield St.)
London Gateway Dr. SS17: Stan H 1P 101
LONDON GATEWAY SERVICE AREA 18Sa 29
London Golf Course, The 80Yd 164
London Group Bus. Pk. NW2 32Va 68
LONDON HEATHROW AIRPORT 57R 106
LONDON HEATHROW AIRPORT 55L 105
LONDON HEATHROW AIRPORT 55Q 106
London Heliport, The 54Eb 111
London Ho. EC1 1D 224 (43Sb 91)
(off Aldersgate St.)
London Ho. NW8 1E 214 (40Gb 69)
(off Avenue Rd.)
London Ho. WC1 5H 217 (42Pb 90)
LONDON INDEPENDENT BMI HOSPITAL 43Zb 92
London Ind. Pk., The E6 43Rc 94
(not continuous)
London Info. Cen. 5E 222 (45Mb 90)
(off Leicester Sq.)
London Intl. Cruise Terminal 6C 122
London International Gallery of Children's Art 32Kb 70
(within Waterlow Pk. Cen.)
London Irish RFC 69V 128
London La. BR1: Broml 66Hc 137
London La. E8 38Xb 71
London Marathon Community Track 39Dc 72
London Master Bakers Almshouses E10 30Dc 52
London Mercantile Court 2K 223 (44Qb 90)
London Metropolitan Archives 5A 218 (42Qb 90)
(off Northampton Rd.)
London Metropolitan University London City Campus, Calcutta House & Goulston St. 2K 225 (44Vb 91)
London Metropolitan University London City Campus, Central House 44Wb 91
(off Whitechapel High St.)
London Metropolitan University London City Campus, Commercial Rd. 44Wb 91
London Metropolitan University London City Campus, Moorgate 1G 225 (43Tb 91)
(off Moorgate)
London Metropolitan University North London Campus, Eden Grove 36Pb 70
London Metropolitan University North London Campus, Stapleton House 36Pb 70
London Metropolitan University North London Campus, Tower Bldg. & Graduate Cen. 36Qb 70
London M. W2 3C 220 (44Fb 89)
London Mill Apts. E2 1K 219 (39Vb 71)
(off Whiston Rd.)
London Mithraeum 3F 225 (44Tb 91)
(off Queen Victoria Street)
London Motorcycle Mus. Ravenor Farm 41Ea 86
London Oxford St. Youth Hostel W1 3C 222 (44Lb 90)
(off Noel St.)
London Palace, The 5D 230 (49Sb 91)
(within Elephant & Castle Shopping Centre)
London Palladium 3B 222 (44Lb 90)
(off Argyll St.)
London Pavilion 5D 222 (45Mb 90)
(off Piccadilly Circus)
London Plane Ho. E15 41Gc 93
(off Teasel Way)
London Regatta Cen. 45Mc 93
London Rd. BR8: Swan Birchwood Rd. 67Ed 140
(not continuous)
London Rd. CR4: Mitc Brookfields Av. 71Gb 155
London Rd. BR8: Swan High St. 70Hd 140
London Rd. RM19: Purf Juliette Way 48Qd 97
London Rd. CR4: Mitc Mill Grn. Rd. 73Jb 156
London Rd. TN14: Hals Old London Rd. 81Bd 181
London Rd. TN14: Hals Shacklands Rd. 84Dd 182
London Rd. RM19: Purf Tank Hill Rd. 50Qd 97
London Rd. AL1: St A 3C 6
London Rd. BR1: Broml 66Hc 137
London Rd. CM14: B'wood 21Vd 58
London Rd. CR0: C'don 73Rb 157
London Rd. CR3: Cat'm 95Tb 197
London Rd. CR7: Thor H 71Qb 156
London Rd. DA1: Bexl 57Fd 118
London Rd. DA1: Cray 57Fd 118
London Rd. DA10: Swans 57Zd 121
London Rd. DA11: Nflt 58Fe 121
London Rd. DA2: Dart 59Rd 119
London Rd. DA4: Farni 72Nd 163
London Rd. DA9: Ghithe 57Xd 120
London Rd. E13 40Jc 73
London Rd. EN2: Enf 13Tb 33
London Rd. GU23: Send 98G 188
London Rd. GU25: Vir W 8L 125
London Rd. HA1: Harr 33Ga 66
London Rd. HA7: Stan 22La 46
London Rd. HA9: Wemb 36Na 67
London Rd. HP1: Hem H 4G 2
London Rd. HP3: Hem H 5K 3
London Rd. HP4: Berk 2A 2
London Rd. IG11: Bark 38Rc 74
London Rd. KT17: Ewe 81Va 174
London Rd. KT2: King T 68Pa 131
London Rd. RH1: Redh 5P 207
London Rd. RH2: Reig 6J 207
London Rd. RM15: Avel 46Pd 97
London Rd. RM15: Purf 46Pd 97
London Rd. RM17: Grays 50Ce 99
London Rd. RM18: Tilb 4D 122
London Rd. RM20: Grays 51Ud 120
London Rd. RM20: W Thur 51Ud 120
London Rd. RM4: Abr 14Vc 37
London Rd. RM4: Stap T 11Ed 38
London Rd. RM6: Chad H 30Cd 56
London Rd. RM7: Chad H 30Cd 56
London Rd. RM7: Rom 30Cd 56
London Rd. SE1 3B 230 (48Rb 91)
London Rd. SE23 60Xb 113
London Rd. SL3: Dat 2M 103
(not continuous)
London Rd. SL3: L'ly 8N 81
London Rd. SL5: Asc 9A 124
London Rd. SL5: S'dale 4C 146
London Rd. SL5: S'hill 9A 124
London Rd. SM3: Cheam 76Za 154
London Rd. SM4: Mord 71Cb 155
London Rd. SM6: W'gton 77Kb 156
London Rd. SS17: Stan H 2K 101
London Rd. SW16 67Pb 134
London Rd. SW17 66Hb 133
London Rd. TN13: Dun G 89Fd 182
(not continuous)
London Rd. TN13: Riv 89Fd 182
(not continuous)
London Rd. TN13: S'oaks 95Hd 202
London Rd. TN15: Bor G 78Td 164
London Rd. TN15: W King 78Td 164
London Rd. TN15: Wro 78Td 164
London Rd. TN15: Wro H 78Td 164
London Rd. TN16: Westrm 95Sc 200
London Rd. TW1: Twick 59Ja 108
London Rd. TW14: Bedf 61Q 128

London Rd. TW15: Ashf 62L 127
London Rd. TW18: Staines 63J 127
London Rd. TW20: Eng G 8M 125
London Rd. TW3: Houn 55Ea 108
London Rd. TW7: Bford 54Ha 108
London Rd. TW7: Isle 54Ha 108
London Rd. TW7: Isle 57Ja 108
London Rd. TW7: Twick 57Ja 108
London Rd. TW8: Bford 52Ka 108
London Rd. WD23: Bush 16Aa 27
London Rd. WD3: Rick 19N 25
London Rd. WD6: Bore 8Ra 15
London Rd. WD7: Shenl 5Pa 15
LONDON ROAD MOOR LANE RDBT. 19N 25
London Rd. Nth. RH1: Mers 98Kb 196
London Rd. Retail Pk. Hemel Hempstead 5L 3
LONDON ROAD RDBT. 58Ja 108
London Rd. Sth. RH1: Mers 2A 208
London Rd. Sth. RH1: Redh 2A 208
London School of Economics & Political Science 3H 223 (44Pb 90)
Londons Cl. RM14: Upm 36Sd 78
London Scottish FC 55Ma 109
London Scottish Golf Course 62Xa 132
London Shootfighters Gym 42Qa 87
London South Bank University Havering 26Nd 57
London South Bank University Keyworth St. 3C 230 (48Rb 91)
London South Bank University Southwark Campus 3C 230 (48Rb 91)
London South Bank University Technopark 3C 230 (48Rb 91)
(off London Rd.)
London's Roman Amphitheatre 2F 225 (44Tb 91)
(off Aldermanbury)
London Stadium Queen Elizabeth Olympic Pk. 38Dc 72
London Stile W4 50Qa 87
London St. EC3 4J 225 (45Ub 91)
London St. KT16: Chert 73J 149
London St. W2 2B 220 (44Fb 89)
London Ter. E2 40Wb 71
London Transport Mus. 4G 223 (45Nb 90)
London Transport Mus. Depot 47Qa 87
London Trocadero 5D 222 (45Mb 90)
London Wall EC2 1D 224 (43Sb 91)
London Wall Bldgs. EC2 1G 225 (43Tb 91)
(off London Wall)
London Wall Pl. EC2 1E 224 (43Sb 91)
(off St Alphage Gdn.)
LONDON WELBECK HOSPITAL 1J 221 (43Jb 90)
(off Welbeck St.)
London Wetland Cen. 53Xa 110
London Wetland Cen. Vis. Cen. 53Xa 110
London Wharf E2 39Xb 71
(off Wharf Pl.)
London Zoo 1H 215 (40Jb 70)
Londrina Ct. HP4: Berk 1A 2
(off Londrina Ter.)
Londrina Ter. HP4: Berk 1A 2
Loneacre GU20: W'sham 9C 146
LONESOME 67Lb 134
Lonesome Cvn. Site SW16 67Kb 134
Lonesome La. RH2: Reig 10K 207
Lonesome La. RH2: Sid 10K 207
Lonesome Way SW16 67Kb 134
Long Acre BR6: Orp 75Zc 161
Long Acre WC2 4F 223 (45Nb 90)
Long Acre Ct. W13 43Ja 86
Longacre Pl. SM5: Cars 79Jb 156
Longacre Rd. E17 25Fc 53
Longacres AL4: St A 2H 7
(not continuous)
Longaford Way CM13: Hut 18Ee 41
Long Barn Cl. WD25: Wat 4W 12
Longbeach Rd. SW11 55Hb 111
Longberrys NW2 34Bb 69
Longboat Row UB1: S'hall 44Ba 85
Longbourn SL4: Wind 5E 102
Longbourne Way KT16: Chert 72H 149
Longbow Apts. E3 42Bc 92
(off St Clements Av.)
Longbow Ho. EC1 7F 219 (43Tb 91)
(off Chiswell St.)
Longboyds KT11: Cobh 87X 171
Longbridge Ho. E16 45Rc 94
(off University Way)
Longbridge Ho. RM8: Dag 35Xc 75
(off Longbridge Rd.)
Longbridge Rd. IG11: Bark 37Tc 74
Longbridge Rd. RM8: Dag 35Wc 75
Longbridge Way SE13 57Ec 114
Longbridge Way UB8: Uxb 40K 63
Longbury Cl. BR5: St P 69Xc 139
Longbury Dr. BR5: St P 69Xc 139
Long Chaulden HP1: Hem H 2G 2
Longcliffe Path WD19: Wat 20W 26
Long Cl. SL2: Farn C 8F 60
Long Copse Cl. KT23: Bookh 95Da 191
Long Ct. RM19: Purf 49Qd 97
Longcourt M. E11 28Lc 53
Longcroft SE9 62Pc 138
Longcroft WD19: Wat 17X 27
Longcroft Av. SM7: Bans 86Eb 175
Longcroft Dr. EN8: Walt C 6Bc 20
Longcrofte Rd. HA8: Edg 24Ma 47
Longcroft La. HP3: Bov 10E 2
Longcroft La. HP3: Hem H 10E 2
Longcroft Ri. IG10: Lough 15Qc 36
Longcroft Rd. WD3: Map C 22F 42
Longcrofts EN9: Walt A 6Gc 21
LONGCROSS 6L 147
Longcross Grange KT16: Vir W 5L 147
Longcross Rd. KT16: Longc 6K 147
Longcross Station (Rail) 4K 147
Long Deacon Rd. E4 18Gc 35
Longdean Pk. HP3: Hem H 6A 4
Long Deans Nature Reserve 7B 4
Longden Av. KT15: Add 80H 149
LONG DITTON 74La 152
Longdon Ct. RM1: Rom 29Hd 56
Longdon Wood BR2: Kes 76Nc 160
Longdown La. Nth. KT17: Eps 86Wa 174
Longdown La. Sth. KT17: Eps 86Wa 174
Longdown La. Sth. KT17: Eps D 86Wa 174
Longdown Rd. KT17: Eps 86Wa 174
Longdown Rd. SE6 63Cc 136
Long Dr. HA4: Ruis 36Y 65
Long Dr. SL1: Burn 1A 80
Long Dr. UB6: G'frd 39Da 65
Long Dr. UB7: W Dray 47N 83
Long Dr. W3 44Ua 88
Long Elmes HA3: Hrw W 25Da 45
Long Elms WD5: Ab L 5T 12
Long Elms Cl. WD5: Ab L 5T 12
Long Fallow AL2: Chis G 9N 5
Longfellow Dr. CM13: Hut 17Ee 41
Longfellow Rd. E17 30Bc 52
Longfellow Rd. KT4: Wor Pk 75Wa 154
Longfellow Way SE1 6K 231 (49Vb 91)
Long Fld. NW9 24Ua 48
Longfield BR1: Broml 67Hc 137
LONGFIELD 68Ae 143
Longfield HP3: Hem H 4B 4
Longfield IG10: Lough 15Mc 35
Longfield SL2: Hedg 4H 61
Longfield Av. DA13: Meop 9A 144
Longfield Av. DA3: Lfield 68Ee 143
Longfield Av. E17 28Ac 52
Longfield Av. EN3: Enf W 9Yb 20
Longfield Av. HA9: Wemb 32Na 67
Longfield Av. NW7 24Wa 48
Longfield Av. RM11: Horn 31Hd 76
Longfield Av. SM6: W'gton 74Jb 156
Longfield Av. W5 45La 86
Longfield Chalk Bank Local Nature Reserve 68Yd 142
Longfield Cres. KT20: Tad 92Ya 194
Longfield Cres. SE26 62Yb 136
Longfield Dr. CR4: Mitc 67Gb 133
Longfield Dr. SW14 57Ra 109
Longfield Est. SE1 6K 231 (49Vb 91)
LONGFIELD HILL 70Fe 143
Longfield Ho. E17 29Bc 52
Longfield Ho. W5 45La 86
Longfield Rd. DA3: Long H 71Fe 165
Longfield Rd. W5 44La 86
Longfield Station (Rail) 69Ae 143
Longfield St. SW18 59Cb 111
Longfield Wlk. W5 44La 86
LONGFORD 92Gd 202
LONGFORD 53K 105
Longford Av. TW14: Felt 58U 106
Longford Av. TW19: Stanw 60N 105
Longford Av. UB1: S'hall 45Da 85
LONGFORD CIR. 53K 105
Longford Cl. TW12: Hamp H 63Ca 129
Longford Cl. TW13: Hanw 62Aa 129
Longford Cl. UB4: Yead 45Z 85
Longford Ct. E5 35Zb 72
(off Pedro St.)
Longford Ct. KT19: Ewe 77Sa 153
Longford Ct. NW4 28Za 48
Longford Ct. TN13: Dun G 92Gd 202
Longford Ct. TW12: Hamp H 65Da 129
Longford Ct. UB1: S'hall 46Ca 85
(off Uxbridge Rd.)
Longford Gdns. SM1: Sutt 76Eb 155
Longford Gdns. UB4: Yead 45Z 85
Longford Ho. BR1: Broml 64Fc 137
(off Brangbourne Rd.)
Longford Ho. E1 44Yb 92
(off Jubilee St.)
Longford Ho. TW12: Hamp H 63Ca 129
Longford Ind. Est. TW12: Hamp 65Da 129
LONGFORDMOOR 53J 105
Longford Rd. TW2: Whitt 60Ca 107
Longford St. NW1 5A 216 (42Kb 90)
Longford Wlk. SW2 59Qb 112
Longford Way TW19: Stanw 60N 105
Long Furlong Dr. SL2: Slou 2B 80
Long Gables SL9: Ger X 29B 42
Long Grn. IG7: Chig 21Uc 54
Long Gro. RM3: Hrld W 26Nd 57
Long Gro. Rd. KT19: Eps 83Sa 173
Longhayes Av. RM6: Chad H 28Zc 55
Longhayes Ct. RM6: Chad H 28Zc 55
Long Heath Dr. KT23: Bookh 96Aa 191
Longheath Gdns. CR0: C'don 71Yb 158
Longhedge Ho. SE26 63Wb 135
(off High Level Dr.)
Long Hedges TW3: Houn 54Ca 107
Longhedge St. SW11 54Jb 112
Long Hill CR3: Wold 93Zb 198
Longhill Rd. SE6 61Fc 137
Longhook Gdns. UB5: N'olt 40W 64
Longhope Cl. SE15 51Ub 113
Longhouse Rd. RM16: Grays 8D 100
Long Ho's. GU24: Pirb 6B 186
Longhurst Ho. W10 41Bb 89
(off Lancefield St.)
Longhurst Rd. CR0: C'don 72Xb 157
Longhurst Rd. SE13 57Fc 115
Longitude Apts. CR0: C'don 75Tb 157
(off Addiscombe Gro.)
Long John HP3: Hem H 4P 3
Longland Ct. E9 36Ac 72
(off Mabley St.)
Longland Ct. SE1 50Wb 91
Longland Dr. N20 20Db 31
Longland Pl. KT19: Eps 84Na 173
LONGLANDS 62Sc 138
Longlands HP2: Hem H 2P 3
Longlands Av. CR5: Coul 86Jb 176
Longlands Cl. EN8: Chesh 4Zb 20
Longlands Ct. CR4: Mitc 67Jb 134
Longlands Ct. DA15: Sidc 61Vc 139
Longlands Ct. W11 45Bb 89
Longlands Pk. Cres. DA15: Sidc 62Uc 138
Longlands Rd. DA15: Sidc 62Uc 138
Long La. CR0: C'don 72Yb 158
Long La. DA7: Bex 52Zc 117
Long La. EC1 1C 224 (43Rb 91)
Long La. HP3: Bov 3B 10
Long La. HP3: Flau 3B 10
Long La. N2 26Eb 49
Long La. N3 25Db 49
Long La. RM16: Grays 47Ce 99
Long La. SE1 2F 231 (47Tb 91)
Long La. TW19: Stanw 61P 127
Long La. UB10: Hil 41Q 84
Long La. UB10: Ick 41Q 84
Long La. WD3: Herons 16E 24
Long La. WD3: Rick 18G 24
Longleat Ho. SW1 7D 228 (50Mb 90)
(off Rampayne St.)
Longleat M. BR5: St M Cry 70Yc 139
Longleat Rd. EN1: Enf 15Ub 33
Longleat Way TW14: Bedf 59T 106
Longlees WD3: Map C 22F 42
Longleigh Ho. SE5 53Ub 113
(off Peckham Rd.)
Longleigh La. DA7: Bex 51Yc 117
Longleigh La. SE2 51Yc 117
Long Lents Ho. NW10 39Ta 67
(off Shrewsbury Rd.)
Longley Av. HA0: Wemb 39Pa 67
Longley Ct. SW8 53Nb 112
Longley M. RM16: Ors 7B 100
Longley Rd. CR0: C'don 73Rb 157
Longley Rd. HA1: Harr 29Ea 46
Longley Rd. SW17 65Gb 133
Long Leys E4 23Dc 52
Longley St. SE1 49Wb 91
Longley Way NW2 34Ya 68
Long Lodge Dr. KT12: Walt T 76Y 151
Longman Ct. HP3: Hem H 7N 3
Longman Ho. E2 40Zb 72
(off Mace St.)
Longman Ho. E8 39Vb 71
(off Haggerston Rd.)
Longman Ho. HP3: Hem H 8P 3
Longmans Cl. WD18: Wat 16S 26
Longmans M. SW20 68Ab 132
Long Mark Rd. E16 43Mc 93
Longmarsh La. SE28 46Uc 94
Longmarsh Vw. DA4: Sut H 67Rd 141
Long Mead NW9 25Va 48
Longmead BR7: Chst 68Qc 138
Longmead SL4: Wind 3C 102
Longmead Bus. Cen. KT19: Eps 83Ta 173
Longmead Cl. CM15: Shenf 18Ae 41
Longmead Cl. CR3: Cat'm 94Ub 197
Longmead Dr. DA14: Sidc 61Zc 139
Longmeade DA12: Grav'nd 10H 123
Longmead Ho. SE27 64Sb 135
Longmead La. SL1: Burn 8B 60
Long Mdw. CM13: Hut 19Ee 41
Long Mdw. NW5 36Mb 70
Long Mdw. RM3: Rom 19Ld 39
Long Mdw. TN13: Riv 93Fd 202
Longmeadow KT23: Bookh 97Ba 191
Long Mdw. Cl. BR4: W W'ck 73Ec 158
Longmeadow Rd. DA15: Sidc 60Uc 116
Longmead Rd. KT19: Eps 83Ta 173
Longmead Rd. KT19: Ewe 83Ta 173
Longmead Rd. KT7: T Ditt 73Ga 152
Longmead Rd. SW17 64Hb 133
Longmead Rd. UB3: Hayes 45V 84
Longmere Gdns. KT20: Tad 91Za 194
Long Mill SE10 53Dc 114
(off Greenwich High Rd.)
Long Mill La. TN11: Dun G 100Ce 205
Long Mill La. TN15: Crou 92De 205
Long Mill La. TN15: Plat 92De 205
Long Mill La. TN15: Plax 92De 205
Long Mill La. Crouch TN15: Crou 94Ee 205
Long Mimms HP2: Hem H 1N 3
Long Moor EN8: Chesh 1Ac 20
Longmoore St. SW1 6B 228 (49Lb 90)
Longmoor Point SW15 62Xa 132
(off Norley Va.)
Longmore Av. EN4: E Barn 16Fb 31
Longmore Av. EN5: New Bar 16Eb 31
Longmore Cl. WD3: Map C 22G 42
Longmore Gdns. Est. SW1 6C 228 (49Lb 90)
(off Vauxhall Bri. Rd.)
Longmore Rd. KT12: Hers 77Aa 151
Longnor Est. E1 41Zb 92
Longnor Rd. E1 41Zb 92
Long Orchards KT20: Kgswd 92Ab 194
Long Pightle Mobile Home Pk. WD3: Chan C 9P 11
Long Pond Rd. SE3 53Gc 115
Longport Cl. IG6: Ilf 23Wc 55
Long Reach GU23: Ock 95Q 190
Long Reach KT24: W Hor 96R 190
Longreach DA1: Dart 55Qd 119
(off Vickers La.)
Longreach Ct. IG11: Bark 40Tc 74
Long Reach Rd. IG11: Bark 42Vc 95
Longreach Rd. DA8: Erith 52Kd 119
Long Readings La. SL2: Slou 1F 80
Longridge WD7: R'lett 6Ka 14
Longridge Gro. GU22: Pyr 86H 169
Longridge Ho. SE1 4E 230 (48Sb 91)
Longridge La. UB1: S'hall 44Da 85
Longridge Rd. IG11: Bark 38Sc 74
Longridge Rd. SW5 49Cb 89
Long Ridges N2 27Jb 50
(off Fortis Grn.)
Longridge Vw. CR5: Chips 92Hb 195
Long Ridings Av. CM13: Hut 15De 41
Long Rd. SW4 56Kb 112
Long Room, The UB9: Hare 24J 43
Longroyd KT24: E Hor 98U 190
(off Cobham Way)
Longs Cl. GU22: Pyr 88J 169
Long's Ct. WC2 5E 222 (45Mb 90)
(off Orange St.)
Longs Ct. TW9: Rich 56Pa 109
Longsdon Way CR3: Cat'm 96Wb 197
Longshaw KT22: Lea 92Ja 192
Longshaw Rd. E4 20Fc 35
Longshore SE8 49Bc 92
Longshott Ct. SW5 49Cb 89
(off W. Cromwell Rd.)
Longside Cl. TW20: Egh 67E 126
Longspring WD24: Wat 9X 13
Longspring Wood Nature Reserve 10E 4
Longstaff Cres. SW18 58Cb 111
Longstaff Rd. SW18 58Cb 111
Longstone Av. NW10 38Va 68
Longstone Ct. SE1 2E 230 (47Sb 91)
(off Gt. Dover St.)
Longstone Rd. SL0: Iver H 40E 62
Longstone Rd. SW17 64Kb 134
Long St. E2 3K 219 (41Vb 91)
Long St. EN9: Walt A 2Mc 21
Longthornton Rd. SW16 68Lb 134
Longthorpe Ct. W6 49Wa 88
Longton Av. SE26 63Wb 135
Longton Gro. SE26 63Xb 135
Longtown Cl. RM3: Rom 22Ld 57
Longtown Ct. DA2: Dart 58Rd 119
(off Osbourne Rd.)
Longtown Rd. RM3: Rom 22Ld 57
Longview Vs. RM5: Col R 25Bd 55
Longview Way RM5: Col R 25Fd 56
Longville Rd. SE11 5B 230 (49Rb 91)
Long Wlk. SW13: Barnes 54Ua 110
Long Wlk. SE18: Woolwich 51Rc 116
Long Wlk. DA13: Ist R 7A 144
Long Wlk. EN9: Walt A 2Cc 20
Long Wlk. HP8: Chal G 13A 24
Long Wlk. KT14: W Byf 86L 169
Long Wlk. KT18: Tatt C 91Ya 194
Long Wlk. KT3: N Mald 69Sa 131
Long Wlk. SE1 3J 231 (48Ub 91)
Long Wlk. SL4: Wind 7H 103
Long Wlk., The SL4: Wind 10H 103
Longwalk Rd. UB11: Stock P 46R 84
Long Wall E15 41Fc 93
Longwater Ho. KT1: King T 69Ma 131
(off Portsmouth Rd.)
Longwood Av. SL3: L'ly 50D 82
Longwood Bus. Pk. TW16: Sun 71V 150
Longwood Cl. RM14: Upm 36Sd 78
Longwood Ct. RM14: Upm 36Sd 78
(off Corbets Tey Rd.)
Longwood Dr. SW15 58Wa 110
Longwood Gdns. IG5: Ilf 28Pc 54
Longwood Gdns. IG6: Ilf 27Rc 54
Longwood Rd. CR8: Kenley 88Tb 177
Longworth Cl. SE28 44Zc 95
Long Yd. WC1 6H 217 (42Pb 90)
Loning, The EN3: Enf W 10Yb 20
Loning, The NW9 28Ua 48
Lonsdale Av. CM13: Hut 16Fe 41
Lonsdale Av. E6 42Mc 93
Lonsdale Av. HA9: Wemb 36Na 67
Lonsdale Av. RM7: Rom 30Ed 56
Lonsdale Cl. E6 42Nc 94
Lonsdale Cl. HA5: Hat E 24Aa 45
Lonsdale Cl. HA8: Edg 22Pa 47
Lonsdale Cl. SE9 62Mc 137
Lonsdale Cl. UB8: Hil 43S 84
Lonsdale Ct. KT6: Surb 73Ma 153
Lonsdale Cres. DA2: Dart 60Sd 120
Lonsdale Cres. IG2: Ilf 30Rc 54
Lonsdale Dr. EN2: Enf 14Mb 32
Lonsdale Dr. Nth. EN2: Enf 15Nb 32
Lonsdale Gdns. CR7: Thor H 70Pb 134
Lonsdale Ho. W11 44Bb 89
(off Lonsdale Rd.)
Lonsdale M. TW9: Kew 53Qa 109
Lonsdale M. W11 44Bb 89
(off Colville M.)
Lonsdale Pl. N1 38Qb 70
Lonsdale Rd. DA7: Bex 54Bd 117
Lonsdale Rd. E11 30Hc 53
Lonsdale Rd. KT13: Weyb 80Q 150
Lonsdale Rd. NW6 40Bb 69
Lonsdale Rd. SE25 70Xb 135
Lonsdale Rd. SW13 51Wa 110
Lonsdale Rd. UB2: S'hall 48Z 85
Lonsdale Rd. W11 44Bb 89
Lonsdale Rd. W4 49Va 88
Lonsdale Road Reservoir Local Nature Reserve 52Va 110
Lonsdale Sq. N1 38Qb 70
Lonsdale Yd. W11 45Cb 89
Loobert Rd. N15 27Ub 51
Looe Gdns. IG6: Ilf 27Rc 54
Look Ahead SL1: Slou 7J 81
Lookout La. E14 44Gc 93
Loom Gro. RM1: Rom 29Jd 56
Loom La. WD7: R'lett 9Ha 14
Loom Pl. WD7: R'lett 8Ja 14
Loop Ct. SE10 49Gc 93
(off Telegraph Av.)
Loop Rd. BR7: Chst 65Sc 138
Loop Rd. GU22: Wok 92B 188
Loop Rd. KT18: Eps 88Sa 173
Lopen Rd. N18 21Ub 51
Lopez Ho. SW9 55Nb 112
Lorac Ct. SM2: Sutt 80Cb 155
Loraine Cl. EN3: Pond E 15Yb 34
Loraine Cotts. N7 35Pb 70
Loraine Ct. BR7: Chst 64Rc 138
Loraine Gdns. KT21: Asht 89Na 173
Loraine Ho. SM6: W'gton 77Kb 156
Loraine Rd. N7 35Pb 70
Loraine Rd. W4 51Ra 109
Lorane Ct. WD17: Wat 12W 26
Lord Admiral's Vw. SE18 49Pc 94
(off Frances St.)
Lord Alexander Ho. HP1: Hem H 3L 3
Lord Amory Way E14 47Ec 92
Lord Av. IG5: Ilf 28Pc 54
Lord Chancellor Wlk. KT2: King T 67Sa 131
Lord Ct. IG5: Ilf 28Pc 54
Lord Darby M. TN14: Cud 87Tc 180
Lordell Pl. SW19 65Ya 132
Lorden Wlk. E2 41Wb 91
Lord Gdns. IG5: Ilf 28Pc 54
Lord Graham M. N18 22Wb 51
Lord Hills Bri. W2 43Db 89
Lord Hills Rd. W2 43Db 89
Lord Holland La. SW9 54Qb 112
Lord Kensington Ho. W14 49Bb 89
(off Radnor Ter.)
Lord Knyvett Cl. TW19: Stanw 58M 105
Lord Knyvetts Ct. TW19: Stanw 58M 105
Lord Mayor's Dr. SL2: Farn C 7D 60
Lord Napier Pl. W6 50Wa 88
Lord Nth. St. SW1 4F 229 (48Nb 90)
Lord Raglan Ho. SL4: Wind 5G 102
Lord Roberts M. SW6 52Db 111
Lord Robert's Ter. SE18 50Qc 94
Lord Rosebery Lodge KT18: Eps 86Sa 173
Lord's 3C 214 (41Fb 89)
Lordsbury Fld. SM6: W'gton 82Lb 176
Lords Cl. SE21 61Sb 135
Lords Cl. TW13: Hanw 61Aa 129
Lords Cl. WD7: Shenl 4Na 15
Lordsgrove Cl. KT20: Tad 92Xa 194
Lordship Cl. CM13: Hut 17Fe 41
Lordship Gro. N16 33Tb 71
Lordship La. N17 25Sb 51
Lordship La. N22 26Qb 50
Lordship La. SE22 56Vb 113
Lordship La. Est. SE21 59Wb 113
Lordship Pk. N16 33Sb 71
Lordship Pk. M. N16 33Sb 71
Lordship Pl. SW3 51Gb 111
Lordship Rd. EN7: Chesh 2Xb 19
Lordship Rd. N16 32Tb 71
Lordship Rd. UB5: N'olt 38Aa 65
Lordship Ter. N16 33Tb 71
Lordsmead Rd. N17 25Ub 51
Lord St. DA12: Grav'nd 9D 122
Lord St. E16 46Nc 94
Lord St. WD17: Wat 13Y 27
Lords Vw. NW8 4C 214 (41Fb 89)
(not continuous)
Lordswood Cl. DA2: Daren 63Ud 142
Lordswood Cl. DA6: Bex 57Ad 117
Lords Wood Ho. CR5: Coul 94Mb 196
Lord Warwick St. SE18 48Pc 94
Loreburn Ho. N7 35Pb 70
Lorenzo Ho. IG3: Ilf 30Wc 55
Lorenzo St. WC1 3H 217 (41Pb 90)
Loretto Gdns. HA3: Kenton 28Na 47
Lorian Cl. N12 21Db 49
Lorian Dr. RH2: Reig 5L 207
Lorimer Row BR2: Broml 72Mc 159
Loriners Cl. KT11: Cobh 86W 170
Loriners Link HP2: Hem H 1N 3
Loring Rd. N20 19Gb 31
Loring Rd. SE14 53Ac 114
Loring Rd. SL4: Wind 3D 102
Loring Rd. TW7: Isle 54Ha 108
Loris Rd. W6 48Ya 88
Lorn Ct. SW9 54Qb 112
Lorne, The KT23: Bookh 98Ca 191
Lorne Av. CR0: C'don 73Zb 158
Lorne Cl. NW8 4E 214 (41Gb 89)
Lorne Cl. SL1: Slou 8F 80
Lorne Ct. SL1: Slou 8G 80
Lorne Gdns. CR0: C'don 73Zb 158
Lorne Gdns. E11 28Lc 53
Lorne Gdns. GU21: Knap 1H 187
Lorne Gdns. W11 47Za 88
Lorne Ho. E1 43Ac 92
(off Ben Jonson Rd.)
Lorne Rd. CM14: W'ley 21Yd 58
Lorne Rd. E17 29Cc 52
Lorne Rd. E7 35Kc 73
Lorne Rd. HA3: W'stone 26Ha 46
Lorne Rd. N4 32Pb 70
Lorne Rd. TW10: Rich 57Pa 109
Lorne Ter. N3 26Bb 49
Lorn Rd. SW9 54Pb 112
Lorraine Ct. NW1 38Kb 70
Lorraine Pk. HA3: Hrw W 24Ga 46
Lorrimore Rd. SE17 51Rb 113
Lorrimore Sq. SE17 51Rb 113
Lorton Cl. DA12: Grav'nd 1G 144
Lorton Ho. NW6 39Cb 69
(off Kilburn Vale)
Loseberry Rd. KT10: Clay 78Fa 152
Losfield Rd. SL4: Wind 3C 102
Lossie Dr. SL0: Iver 45D 82
Lost Theatre 53Mb 112
Lothair Rd. W5 47Ma 87
Lothair Rd. Nth. N4 30Rb 51
Lothair Rd. Sth. N4 31Qb 70
Lothair St. SW11 55Gb 111
Lothbury EC2 2F 225 (44Tb 91)
Lothian Av. UB4: Yead 43X 85
Lothian Cl. HA0: Wemb 34Ja 66
Lothian Rd. SW9 53Rb 113
Lothian Wood KT20: Tad 94Xa 194
Lothrop St. W10 41Ab 88
Lots Rd. SW10 52Eb 111
Lotus Cl. IG7: Chig 23Ad 55
Lotus Cl. SE21 62Tb 135
Lotus M. N19 33Nb 70
Lotus Pk. TW18: Staines 63F 126
Lotus Rd. TN16: Big H 90Pc 180
Loubet St. SW17 65Hb 133
Loudoun Av. IG6: Ilf 29Rc 54
Loudoun Rd. NW8 39Eb 69
LOUDWATER 13L 25
Loudwater Cl. TW16: Sun 70W 128
Loudwater Dr. WD3: Loud 14L 25
Loudwater Hgts. WD3: Loud 13K 25
Loudwater Ho. WD3: Loud 14L 25
Loudwater La. WD3: Crox G 15L 25
Loudwater La. WD3: Loud 15L 25
Loudwater Ridge WD3: Loud 14L 25
Loudwater Rd. TW16: Sun 70W 128
Loughborough Est. SW9 55Rb 113
Loughborough Ho. RM8: Dag 35Wc 75
(off Academy Way)
Loughborough Junction Station (Rail) 55Rb 113
Loughborough Pk. SW9 56Rb 113
Loughborough Rd. SW9 54Qb 112
Loughborough St. SE11 7J 229 (50Pb 90)
Lough Rd. N7 36Pb 70
LOUGHTON 14Nc 36
Loughton Bus. Cen. IG10: Lough 14Sc 36
Loughton Ct. EN9: Walt A 5Kc 21
Loughton Golf Course 11Rc 36
Loughton La. CM16: They B 10Tc 22
Loughton Leisure Cen. 14Nc 36
Loughton Seedbed Cen. IG10: Lough 14Tc 36
Loughton Station (Underground) 15Nc 36
Loughton Way IG9: Buck H 18Mc 35
Louisa Cl. E9 39Zb 72
Louisa Ct. TW2: Twick 61Ga 130
Louisa Gdns. E1 42Zb 92
Louisa Ho. IG3: Ilf 30Wc 55
Louisa Oakes Cl. E4 21Bc 52
Louisa St. E1 42Zb 92
Louise Aumonier Wlk. N19 31Nb 70
(off Jessie Blythe La.)
Louise Bennett Cl. SE24 56Rb 113
Louise Ct. E11 29Kc 53
Louise Ct. N22 25Qb 50
Louise De Marillac Ho. E1 43Yb 92
(off Smithy St.)
Louise Gdns. RM13: Rain 41Gd 96
Louise Rd. E15 37Gc 73
Louise Wlk. HP3: Bov 10C 2
Louise White Ho. N19 32Mb 70
Louis Gdns. BR7: Chst 63Pc 138
Louis M. N10 25Kb 50
Louisville Rd. SW17 62Jb 134
Lourdes Cl. SE13 55Fc 115
Lousada Lodge N14 16Lb 32
(off Avenue Rd.)
Louvaine Rd. SW11 56Fb 111
Louvain Rd. DA9: Ghithe 59Ud 120
Louvain Way WD25: Wat 4X 13
Lovage App. E6 43Nc 94
Lovat Cl. E14 44Bc 92
Lovat Cl. NW2 34Va 68
Lovat La. EC3 5H 225 (45Ub 91)
Lovatt Cl. HA8: Edg 23Ra 47
Lovatt Ct. SW12 60Kb 112
Lovatt Dr. HA4: Ruis 29W 44
Lovatts WD3: Crox G 14Q 26
Lovatts Cotts. AL2: Park 9B 6
Lovat Wlk. TW5: Hest 52Aa 107
Loveday Rd. W13 47Ka 86
Love Grn. La. SL0: Iver 43F 82
Lovegrove Cl. CR2: Sande 82Tb 177
Lovegrove Dr. SL2: Slou 2D 80
Lovegrove St. SE1 50Wb 91
Lovegrove Wlk. E14 46Ec 92
Lovegrove Way N20 19Fb 31
Love Hill La. SL3: L'ly 45C 82
Lovejoy Ct. SE11 51Qb 112
(off Pownell Ter.)
Lovejoy La. SL4: Wind 4B 102
Lovekyn Cl. KT2: King T 68Na 131
Lovelace Av. BR2: Broml 72Qc 160
Lovelace Cl. KT24: Eff J 95W 190
Lovelace Cl. TN15: W King 79Ud 164
Lovelace Dr. GU22: Pyr 88G 168
Lovelace Gdns. IG11: Bark 35Wc 75
Lovelace Gdns. KT12: Hers 78Y 151
Lovelace Gdns. KT6: Surb 73Ma 153
Lovelace Grn. SE9 55Pc 116
Lovelace Ho. W13 45Ka 86
Lovelace Rd. EN4: E Barn 17Gb 31
Lovelace Rd. KT6: Surb 73La 152
Lovelace Rd. SE21 61Sb 135
Lovelace St. E8 39Vb 71
Lovelace Vs. KT7: T Ditt 73Ka 152
(off Portsmouth Rd.)
Loveland Ct. SE1 47Vb 91
(off Jamaica Rd.)
Loveland Mans. IG11: Bark 38Vc 75
(off Upney La.)
Lovelands La. GU24: Chob 5G 166
Lovelands La. KT20: Lwr K 99Db 195
Love La. AL9: N Mym 10D 8
Love La. BR1: Broml 69Kc 137
Love La. CR4: Mitc 69Gb 133
(not continuous)

Love La. DA12: Grav'nd 9E 122
Love La. DA5: Bexl 58Bd 117
Love La. EC2 2E 224 (44Sb 91)
Love La. HA5: Pinn 26Z 45
Love La. IG8: Wfd G 23Pc 54
Love La. KT20: Walt H 99Va 194
Love La. KT6: Surb 75La 152
Love La. N17 24Vb 51
Love La. RH9: G'stone 4A 210
Love La. RM15: Avel 47Sd 98
Love La. RM18: E Til 1K 123
Love La. SE18 49Rc 94
Love La. SE25 69Xb 135
Love La. SL0: Iver 44F 82
Love La. SM1: Sutt 79Bb 155
Love La. SM3: Cheam 79Ab 154
Love La. SM3: Sutt 79Ab 154
Love La. SM4: Mord 73Cb 155
Love La. WD4: K Lan 1N 11
Love La. WD5: Ab L 2V 12
Lovel Av. DA16: Well 54Wc 117
Lovel Cl. HP1: Hem H 2J 3
Lovelinch Cl. SE15 51Yb 114
Lovell Ho. E8 39Wb 71
(off Shrubland Rd.)
Lovell Pl. SE16 48Ac 92
Lovell Rd. EN1: Enf 7Xb 19
Lovell Rd. TW10: Ham 62La 130
Lovell Rd. UB1: S'hall 44Da 85
Lovells Cl. GU18: Light 2A 166
Lovell Wlk. RM13: Rain 37Jd 76
Lovelock Cl. CR8: Kenley 89Sb 177
Loveridge M. NW6 37Bb 69
Loveridge Rd. NW6 37Bb 69
Lovers La. DA9: Ghithe 56Zd 121
Lovers Wlk. N3 24Cb 49
Lovers Wlk. NW7 23Bb 49
Lovers Wlk. RM5: Col R 22Fd 56
Lovers Wlk. SE10 51Fc 115
Lovers' Wlk. W1 6H 221 (46Jb 90)
Lovett Dr. SM5: Cars 73Eb 155
Lovett Rd. AL2: Lon C 8E 6
Lovett Rd. TW18: Staines 63D 126
Lovett Rd. UB9: Hare 27L 43
Lovett's Pl. SW18 56Db 111
Lovett Way NW10 36Sa 67
Love Wlk. SE5 54Tb 113
Lovibond La. SE10 52Dc 114
(off Norman Rd.)
Lovibonds Av. BR6: Farnb 77Rc 160
Lovibonds Av. UB7: Yiew 44P 83
Lowbell La. AL2: Lon C 9J 7
Lowbrook Rd. IG1: Ilf 35Rc 74
Low Cl. DA9: Ghithe 57Wd 120
Low Cross Wood La. SE21 62Vb 135
Lowdell Cl. UB7: Yiew 44N 83
Lowden Rd. N9 18Xb 33
Lowden Rd. SE24 56Rb 113
Lowden Rd. UB1: S'hall 45Aa 85
Lowder Ho. E1 46Xb 91
(off Wapping La.)
Lowe, The IG7: Chig 21Wc 55
Lowe Av. E16 43Jc 93
Lowe Cl. IG7: Chig 22Wc 55
Lowell Ho. SE5 52Sb 113
(off Wyndham Est.)
Lowell St. E14 44Ac 92
Lowen Rd. RM13: Rain 40Fd 76
Lower Addiscombe Rd. CR0: C'don 74Ub 157
Lower Addison Gdns. W14 47Ab 88
Lower Adeyfield Rd. HP2: Hem H 1M 3
Lower Alderton Hall La. IG10: Lough 15Qc 36
Lower Ash Est. TW17: Shep 72V 150
LOWER ASHTEAD 91Ma 193
Lower Barn HP3: Hem H 5P 3
Lower Barn Rd. CR8: Purl 84Sb 177
Lower Bedfords Rd. RM1: Have B 23Gd 56
Lower Bedfords Rd. RM1: Rom 23Gd 56
Lower Belgrave St. SW1 4K 227 (48Kb 90)
LOWER BITCHET 98Sd 204
Lower Boston Rd. W7 46Ga 86
Lower Bridge Rd. RH1: Redh 6P 207
Lower Britwell Rd. SL2: Slou 2B 80
(not continuous)
Lower Broad St. RM10: Dag 39Cd 76
Lower Bury La. CM16: Epp 3Uc 22
Lower Camden BR7: Chst 66Pc 138
Lower Church Hill DA9: Ghithe 57Ud 120
Lower Church St. CR0: C'don 75Rb 157
Lower Cippenham La. SL1: Slou 6C 80
LOWER CLAPTON 35Xb 71
Lower Clapton Rd. E5 34Xb 71
Lower Clarendon Wlk. W11 44Ab 88
(off Clarendon Rd.)
Lower Common Sth. SW15 55Xa 110
Lower Coombe St. CR0: C'don 77Sb 157
Lower Ct. Rd. KT19: Eps 83Sa 173
Lower Cres. SS17: Linf 8J 101
Lower Cft. BR8: Swan 70Hd 140
Lower Dagnall St. AL3: St A 2A 6
Lower Derby Rd. WD17: Wat 14Y 27
Lower Dock Wlk. E16 45Sc 94
(off Lock Side Way)
Lower Downs Rd. SW20 67Za 132
Lower Drayton Pl. CR0: C'don 75Rb 157
Lower Dunnymans SM7: Bans 86Bb 175
LOWER EDMONTON 19Wb 33
Lower Farm Rd. KT24: Eff 96X 191
LOWER FELTHAM 62W 128
Lower Fosters NW4 29Ya 48
(off New Brent St.)
Lower George St. TW9: Rich 57Ma 109
Lower Gravel Rd. BR2: Broml 74Nc 160
LOWER GREEN 75Da 151
Lower Grn. Gdns. KT4: Wor Pk 74Wa 154
Lower Grn. Rd. KT10: Esh 75Da 151
Lower Grn. W. CR4: Mitc 69Gb 133
Lower Grosvenor Pl. SW1 3K 227 (48Kb 90)
Lower Gro. Rd. TW10: Rich 58Pa 109
Lower Guildford Rd. GU21: Knap 9H 167
Lower Guild Hall DA9: Bluew 59Vd 120
LOWER HALLIFORD 73T 150
Lower Hall La. E4 22Ac 52
(not continuous)
Lower Hampton Rd. TW16: Sun 69Y 129
Lower Ham Rd. KT2: King T 64Ma 131
Lower Higham Rd. DA12: Grav'nd 10H 123
Lower Higham Rd. DA12: Shorne 10H 123
Lower High St. WD17: Wat 14Y 27
Lower Hill Rd. KT19: Eps 84Ra 173
LOWER HOLLOWAY 36Pb 70
Lower Hook Bus. Pk. BR6: Downe 80Pc 160
Lower James St. W1 4C 222 (45Lb 90)
Lower John St. W1 4C 222 (45Lb 90)
Lower Kenwood Av. EN2: Enf 15Nb 32
Lower King's Rd. KT2: King T 67Na 131
LOWER KINGSWOOD 99Bb 195
Lower Lea Crossing E14 45Gc 93
Lower Lea Crossing E16 45Gc 93
Lower Lees Rd. SL2: Slou 1E 80
Lower Maidstone Rd. N11 23Lb 50
Lower Mall W6 50Xa 88
Lower Mardyke Av. RM13: Rain 40Ed 76
Lower Marsh SE1 2K 229 (47Qb 90)
Lower Marsh La. KT1: King T 70Pa 131
Lower Mead RH1: Redh 4P 207
Lower Mead SL0: Iver H 41F 82
Lower Merton Ri. NW3 38Gb 69
Lower Mill KT17: Ewe 80Va 154
Lower Morden La. SM4: Mord 72Ya 154
Lower Mortlake Rd. TW9: Rich 56Na 109
Lower New Change Pas. EC4 3E 224 (44Sb 91)
(off One New Change)
Lower Noke Cl. CM14: Rom 19Nd 39
Lower Noke Cl. CM14: S Weald 19Nd 39
Lower Northfield SM7: Bans 86Bb 175
Lower Nursery SL5: S'dale 1E 146
Lower Paddock Rd. WD19: Wat 16Aa 27
Lower Pk. Rd. CR5: Chips 90Gb 175
Lower Pk. Rd. DA17: Belv 49Cd 96
Lower Pk. Rd. IG10: Lough 15Mc 35
Lower Pk. Rd. N11 22Lb 50
Lower Park Trad. Est. NW10 42Sa 87
Lower Paxton Rd. AL1: St A 3C 6
Lower Peryers KT24: E Hor 100U 190
Lower Pillory Down CR5: Coul 85Kb 176
LOWER PLACE 40Sa 67
Lower Pl. Bus. Cen. NW10 40Ta 67
(off Steele Rd.)
Lower Plantation WD3: Loud 13L 25
Lower Pyrford Rd. GU22: Pyr 88K 169
Lower Queen's Rd. IG9: Buck H 19Mc 35
Lower Range Rd. DA12: Grav'nd 9G 122
Lower Richmond Rd. SW14 55Ra 109
Lower Richmond Rd. SW15 55Xa 110
Lower Richmond Rd. TW9: Rich 55Qa 109
Lower Rd. BR5: St M Cry 72Xc 161
Lower Rd. BR8: Hext 66Jd 140
Lower Rd. BR8: Swan 66Jd 140
Lower Rd. CM15: Mount 12Fe 41
Lower Rd. CR8: Kenley 85Rb 177
Lower Rd. DA11: Nflt 56Ce 121
Lower Rd. DA12: High'm 1N 145
Lower Rd. DA12: Shorne 1N 145
Lower Rd. DA17: Belv 48Dd 96
Lower Rd. DA8: Erith 49Ed 96
Lower Rd. HA2: Harr 32Fa 66
Lower Rd. HP3: Hem H 8A 4
Lower Rd. IG10: Lough 11Qc 36
Lower Rd. KT22: Fet 96Fa 192
Lower Rd. KT23: Bookh 97Ca 191
Lower Rd. KT24: Eff 99Z 191
Lower Rd. RH1: Redh 8M 207
Lower Rd. SE1 1K 229 (47Qb 90)
Lower Rd. SE16 47Yb 92
(not continuous)
Lower Rd. SE8 49Zb 92
Lower Rd. SL9: Chal P 25A 42
(not continuous)
Lower Rd. SL9: Ger X 25A 42
(not continuous)
Lower Rd. SM1: Sutt 77Eb 155
Lower Rd. UB9: Den 31E 62
Lower Rd. WD3: Chor 14E 24
Lower Robert St. WC2 5G 223 (45Nb 90)
(off Robert St.)
Lower Rose Gallery DA9: Bluew 59Vd 120
Lower Sales HP1: Hem H 3H 3
Lower Sandfields GU23: Send 96F 188
Lower Sand Hills KT6: Surb 73La 152
Lower Sawley Wood SM7: Bans 86Bb 175
LOWER SHORNE 3N 145
Lower Shott KT23: Bookh 98Ca 191
Lower Sloane St. SW1 6H 227 (49Jb 90)
Lower Sq. TW7: Isle 55Ka 108
Lower Sq., The SM1: Sutt 78Db 155
(off St Nicholas Way)
Lower Stable St. N1 1F 217 (39Nb 70)
(off Stable Street)
Lower Station Rd. DA1: Cray 58Gd 118
Lower Strand NW9 26Va 48
Lower Sunbury Rd. TW12: Hamp 68Ba 129
Lower Swaines CM16: Epp 2Uc 22
LOWER SYDENHAM 63Zb 136
Lower Sydenham Ind. Est. SE26 64Bc 136
Lower Sydenham Station (Rail) 64Bc 136
Lower Tail WD19: Wat 20Aa 27
Lower Teddington Rd. KT1: Hamp W 67Ma 131
Lower Ter. NW3 34Eb 69
Lower Ter. SE27 64Rb 135
(off Woodcote Pl.)
Lower Thames St. EC3 5G 225 (45Tb 91)
Lower Thames Wlk. DA9: Bluew 60Vd 120
Lower Tub WD23: Bush 17Fa 28
Lower Village Rd. SL5: S'hill 1A 146
Lowerwood Ct. W11 44Ab 88
(off Westbourne Pk. Rd.)
Lower Wood Rd. KT10: Clay 79Ka 152
LOWER WOODSIDE 4H 9
Lower Yott HP2: Hem H 3P 3
Lowestoft Cl. E5 33Yb 72
(off Theydon Rd.)
Lowestoft Dr. SL1: Slou 4B 80
Lowestoft M. E16 47Rc 94
Lowestoft Rd. WD24: Wat 11X 27
Loweswater Cl. HA9: Wemb 33Ma 67
Loweswater Cl. WD25: Wat 5Y 13
Loweswater Ho. E3 42Bc 92
Lowewood Rd. RM3: Rom 22Ld 57
Lowfield Rd. NW6 38Cb 69
Lowfield Rd. W3 44Ra 87
Lowfield St. DA1: Dart 59Nd 119
Low Hall Cl. E4 17Dc 34
Low Hall La. E17 30Ac 52
Low Hall Mnr. Bus. Cen. E17 30Ac 52
Lowick Rd. HA1: Harr 28Ga 46
Lowlands Dr. TW19: Stanw 57M 105
Lowlands Gdns. RM7: Rom 30Dd 56
Lowlands Rd. HA1: Harr 30Ga 46
Lowlands Rd. HA5: Eastc 31Y 65
Lowlands Rd. RM15: Avel 46Sd 98
Lowman Rd. N7 35Pb 70
Lownde M. SW16 61Nb 134
Lowndes Cl. SW1 4J 227 (48Jb 90)
Lowndes Ct. SW1 3G 227 (48Hb 89)
Lowndes Ct. W1 3B 222 (44Lb 90)
(off Carnaby St.)
Lowndes Lodge SW1 3G 227 (48Hb 89)
(off Cadogan Pl.)
Lowndes M. SW16 61Nb 134
Lowndes Pl. SW1 4H 227 (48Jb 90)
Lowndes Sq. SW1 2G 227 (47Hb 89)
Lowndes St. SW1 3H 227 (48Jb 90)
Lownds Ct. BR1: Broml 68Jc 137
Lowood Ct. SE19 64Vb 135
(off Farquhar Rd.)
Lowood Ho. E1 45Yb 92
(off Bewley St.)
Lowood St. E1 45Xb 91
Lowry Cl. DA8: Erith 49Fd 96
Lowry Ct. SE16 50Xb 91
(off Stubbs Dr.)
Lowry Cres. CR4: Mitc 68Gb 133
Lowry Ho. E14 47Cc 92
(off Cassilis Rd.)
Lowry Ho. N17 25Vb 51
(off Pembury Rd.)
Lowry Ho. W3 48Sa 87
(off Palmerston Rd.)
Lowry Rd. RM8: Dag 36Xc 75
Lowshoe La. RM5: Col R 25Cd 56
Lowson Gro. WD19: Wat 17Aa 27
LOW STREET 1H 123
Low St. La. RM18: W Til 9H 101
Lowswood Cl. HA6: Nwood 25S 44
Lowther Cl. KT16: Chert 76G 148
Lowther Cl. WD6: E'tree 15Pa 29
Lowther Dr. EN2: Enf 14Nb 32
Lowther Hill SE23 59Ac 114
Lowther Ho. SW1 50Lb 90
(off Churchill Gdns.)
Lowther Rd. E17 26Ac 52
Lowther Rd. HA7: Stan 27Pa 47
Lowther Rd. KT2: King T 67Pa 131
Lowther Rd. N7 36Qb 70
Lowther Rd. SW13 53Va 110
Lowthorpe GU21: Wok 10L 167
Lowth Rd. SE5 53Sb 113
LOXFORD 36Sc 74
Loxford Av. E6 40Mc 73
Loxford Cl. CR3: Cat'm 97Vb 197
Loxford Gdns. N5 35Rb 71
Loxford Ho. KT17: Eps 84Ua 174
Loxford La. IG1: Ilf 36Sc 74
Loxford La. IG3: Ilf 35Uc 74
Loxford Rd. CR3: Cat'm 97Vb 197
Loxford Rd. IG11: Bark 37Rc 74
Loxford Ter. IG11: Bark 37Sc 74
Loxford Way CR3: Cat'm 97Vb 197
Loxham Rd. E4 24Dc 52
Loxham St. WC1 4G 217 (41Nb 90)
Loxley Cl. KT14: Byfl 86N 169
Loxley Cl. SE26 64Zb 136
Loxley Ho. HA9: Wemb 34Na 67
Loxley Rd. SW18 60Fb 111
Loxley Rd. TW12: Hamp 63Ba 129
Loxton Rd. SE23 60Zb 114
Loxwood KT13: Weyb 76U 150
Loxwood Cl. BR5: Orp 75Zc 161
Loxwood Cl. HP3: Hem H 5H 3
Loxwood Cl. TW14: Bedf 60T 106
Loxwood Ho. CR8: Purl 83Rb 177
Loxwood Rd. N17 27Ub 51
Loyd Ct. AL4: St A 4G 6
LSO St Lukes 5E 218 (42Sb 91)
(off Old St.)
Lubbock Ho. E14 45Dc 92
(off Poplar High St.)
Lubbock Rd. BR7: Chst 66Pc 138
Lubbock St. SE14 52Yb 114
Lucan Dr. TW18: Staines 66M 127
Lucan Ho. N1 39Tb 71
(off Colville Est.)
Lucan Pl. SW3 6D 226 (49Gb 89)
Lucan Rd. EN5: Barn 13Ab 30
Lucas Av. E13 39Kc 73
Lucas Av. HA2: Harr 33Ca 65
Lucas Cl. NW10 38Wa 68
Lucas Ct. EN9: Walt A 5Hc 21
Lucas Ct. SE26 64Ac 136
Lucas Ct. SW11 53Jb 112
Lucas Cres. DA9: Ghithe 56Yd 120
(off Ingress Pk. Av.)
Lucas Gdns. N2 26Eb 49
LUCAS GREEN 7B 166
Lucas Grn. Rd. GU24: W End 7B 166
Lucas Ho. SW10 52Db 111
(off Coleridge Gdns.)
Lucas Ho. WC1 4G 217 (41Nb 90)
(off Argyle Wlk.)
Lucas Pk. Dr. KT20: Walt H 95Wa 194
Lucas Rd. RM17: Grays 48Ce 99
Lucas Rd. SE20 65Yb 136
Lucas Sq. NW11 30Cb 49
Lucas St. SE8 53Cc 114
Lucent Ho. SW18 57Cb 111
(off Hardwicks Sq.)
Lucerna Ct. DA3: Lfield 69Be 143
(off Harrison Av.)
Lucerne Cl. GU22: Wok 91A 188
Lucerne Cl. N13 20Nb 32
Lucerne Ct. DA18: Erith 48Ad 95
Lucerne Gro. E17 28Fc 53
Lucerne M. W8 46Cb 89
Lucerne Rd. BR6: Orp 74Vc 161
Lucerne Rd. CR7: Thor H 71Rb 157
Lucerne Rd. N5 35Rb 71
Lucerne Way RM3: Rom 23Md 57
Lucey Rd. SE16 48Wb 91
Lucey Way SE16 48Wb 91
Lucia Hgts. E20 36Ec 72
(off Logan Cl.)
Lucida Ct. WD18: Wat 15U 26
(off Whippendell Rd.)
Lucie Av. TW15: Ashf 65R 128
Lucien Rd. SW17 63Jb 134
Lucien Rd. SW19 61Db 133
Lucinda Ct. E17 26Zb 52
Lucinda Ct. EN1: Enf 14Ub 33
Lucknow St. SE18 52Uc 116
Lucks Hill HP1: Hem H 2G 2
Lucorn Cl. SE12 58Hc 115
Lucton M. IG10: Lough 14Rc 36
Luctons Av. IG9: Buck H 18Lc 35
Lucy Brown Ho. SE1 7E 224 (46Sb 91)
(off Park St.)
Lucy Cres. W3 43Sa 87
Lucy Gdns. RM8: Dag 34Bd 75
Luddesdon Rd. DA8: Erith 52Cd 118
Luddington Av. GU25: Vir W 68B 126
Ludford Cl. CR0: Wadd 76Rb 157
Ludgate B'way. EC4 3B 224 (44Rb 91)
Ludgate Cir. EC4 3B 224 (44Rb 91)
Ludgate Hill EC4 3B 224 (44Rb 91)
Ludgate Sq. EC4 3C 224 (44Rb 91)
Ludham NW5 36Hb 69
Ludham Cl. IG6: Ilf 25Sc 54
Ludham Cl. SE28 44Yc 95
Ludlow Cl. BR2: Broml 69Jc 137
Ludlow Cl. HA2: Harr 35Ba 65
Ludlow Ct. W3 47Sa 87
Ludlow Mead WD19: Wat 20X 27
Ludlow Pl. RM17: Grays 48De 99
Ludlow Rd. TW13: Felt 63W 128
Ludlow Rd. W5 42La 86
Ludlow St. EC1 5D 218 (42Sb 91)
Ludlow Way N2 28Eb 49
Ludlow Way WD3: Crox G 14S 26
Ludovick Wlk. SW15 56Ua 110
Ludwell Ho. W14 48Ab 88
(off Russell Rd.)
Ludwick M. SE14 52Ac 114
Luff Cl. SL4: Wind 5C 102
Luff Ct. E3 43Cc 92
(off Shelmerdine Cl.)
Luffenham Ho. WD19: Wat 20Z 27
Luffield Rd. SE2 48Xc 95
Luffman Rd. SE12 62Kc 137
Lugard Ho. W12 46Xa 88
(off Bloemfontein Rd.)
Lugard Rd. SE15 54Xb 113
Lugg App. E12 34Qc 74
Luke Allsopp Sq. RM10: Dag 34Dd 76
Luke Ho. E1 44Xb 91
(off Tillman St.)
Lukes Cl. NW2 33Wa 68
Luke St. EC2 5H 219 (42Ub 91)
Lukin Cres. E4 20Fc 35
Lukin St. E1 44Yb 92
Lukintone Cl. IG10: Lough 16Nc 36
Luli Ct. SE14 51Bc 114
Lullarook Cl. TN16: Big H 88Lc 179
LULLINGSTONE 78Ld 163
Lullingstone Av. BR8: Swan 69Hd 140
Lullingstone Castle 78Ld 163
Lullingstone Cl. BR5: St P 66Xc 139
Lullingstone Country Pk. 78Gd 162
Lullingstone Country Pk. Visitor Cen. 79Kd 163
Lullingstone Cres. BR5: St P 66Wc 139
Lullingstone Ho. SE15 51Yb 114
(off Lovelinch Cl.)
Lullingstone La. DA4: Eyns 76Ld 163
Lullingstone La. SE13 58Fc 115
Lullingstone Pk. Golf Course 77Fd 162
Lullingstone Rd. DA17: Belv 51Bd 117
Lullingstone Roman Villa 76Kd 163
Lullington Gth. BR1: Broml 66Gc 137
Lullington Gth. N12 22Bb 49
Lullington Gth. WD6: Bore 15Ra 29
Lullington Rd. RM9: Dag 38Ad 75
Lullington Rd. SE20 66Wb 135
Lulot Gdns. N19 33Kb 70
Lulsgate M. E3 42Bc 92
Lulworth NW1 38Mb 70
(off Wrotham Rd.)
Lulworth SE17 7F 231 (50Tb 91)
(off Portland St.)
Lulworth Av. EN7: G Oak 1Rb 19
Lulworth Av. HA9: Wemb 31La 66
Lulworth Av. TW5: Hest 53Da 107
Lulworth Cl. HA2: Harr 34Ba 65
Lulworth Cl. SS17: Stan H 3K 101
Lulworth Ct. N1 38Ub 71
(off St Peter's Way)
Lulworth Cres. CR4: Mitc 68Gb 133
Lulworth Dr. HA5: Pinn 30Z 45
Lulworth Dr. RM5: Col R 22Dd 56
Lulworth Gdns. HA2: Harr 33Aa 65
Lulworth Ho. SW8 52Pb 112
Lulworth Pl. KT19: Eps 84Na 173
Lulworth Rd. DA16: Well 54Vc 117
Lulworth Rd. SE15 54Xb 113
Lulworth Rd. SE9 61Nc 138
Lulworth Waye UB4: Yead 44X 85
Lumen Rd. HA9: Wemb 33Ma 67
Lumiere Apts. SW11 56Fb 111
Lumiere Bldg., The E7 36Mc 73
(off Romford Rd.)
Lumiere Ct. SW17 61Jb 134
Lumina Bldgs. E14 46Ec 92
(off Prestons Rd.)
Lumina Bus. Pk. EN1: Enf 15Wb 33
Lumina Loft Apts. SE1 2J 231 (47Ub 91)
(off Tower Bri. Rd.)
Lumina Way EN1: Enf 15Wb 33
Luminosity Ct. W13 45Ka 86
Lumley Cl. DA17: Belv 50Cd 96
Lumley Ct. WC2 5G 223 (45Nb 90)
Lumley Flats SW1 7H 227 (50Jb 90)
(off Holbein Pl.)
Lumley Gdns. SM3: Cheam 78Ab 154
Lumley Rd. SM3: Cheam 78Ab 154
Lumley St. W1 3J 221 (44Jb 90)
Lumsdon NW8 39Db 69
(off Abbey Rd.)
Luna Ho. SE16 47Wb 91
Lunan Ho. E3 40Bc 72
(off Shetland Rd.)
Luna Pl. AL1: St A 2E 6
Lunar Cl. TN16: Big H 88Mc 179
Lunaria Ho. E20 36Ec 72
(off Elis Way)
Luna Rd. CR7: Thor H 69Sb 135
Lundin Wlk. WD19: Wat 21Z 45
Lund Point E15 39Ec 72
Lundy Ct. SL1: Slou 5C 80
Lundy Dr. UB3: Harl 49U 84
Lundy Ho. WD18: Wat 15V 26
Lundy Wlk. N1 37Sb 71
Lunedale Rd. DA2: Dart 60Sd 120
Lunghurst Rd. CR3: Wold 92Bc 198
Lunham Rd. SE19 65Ub 135
Luntley Pl. E1 43Wb 91
(off Chicksand St.)
Lupin Cl. CR0: C'don 74Zb 158
Lupin Cl. RM7: Rush G 33Fd 76
Lupin Cl. SW2 61Rb 135
Lupin Cl. UB7: W Dray 50M 83
Lupin Cres. IG1: Ilf 37Rc 74
Lupin M. E17 27Ec 52
Lupino Ct. SE11 5J 229 (49Pb 90)
Lupin Point SE1 47Vb 91
(off Abbey St.)
Luppit Cl. CM13: Hut 18Ce 41
Lupton Cl. SE12 62Kc 137
Lupton St. NW5 35Lb 70
(not continuous)
Lupus St. SW1 50Kb 90
Luralda Wharf E14 50Fc 93
Lurgan Av. W6 51Za 110
Lurline Gdns. SW11 53Jb 112
Luscombe Ct. BR2: Broml 68Gc 137
Luscombe Way SW8 52Nb 112
Lushes Ct. IG10: Lough 15Rc 36
Lushes Rd. IG10: Lough 15Rc 36
Lushington Dr. KT11: Cobh 86X 171
Lushington Ho. KT12: Walt T 72Y 151
Lushington Rd. NW10 40Xa 68
Lushington Rd. SE6 63Dc 136
Lushington Ter. E8 36Wb 71
Lusted Hall La. TN16: Big H 92Kc 199
Lusted Hall La. TN16: Tats 92Kc 199
Lusted Rd. TN13: Dun G 92Gd 202
Lutea Ho. SM2: Sutt 80Eb 155
(off Walnut M.)
Luther Cl. HA8: Edg 19Sa 29
Luther King Cl. E17 30Bc 52
Luther M. TW11: Tedd 64Ha 130
Luther Rd. TW11: Tedd 64Ha 130
Luton Ho. E13 42Jc 93
(off Luton Rd.)
Luton Ho. RM3: Rom 22Nd 57
(off Lindfield Rd.)
Luton Pl. SE10 52Ec 114
Luton Rd. DA14: Sidc 62Yc 139
Luton Rd. E13 42Jc 93
Luton Rd. E17 27Bc 52
Luton St. NW8 6C 214 (42Fb 89)
Lutton Ter. NW3 35Eb 69
(off Lakis Cl.)
Luttrell Av. SW15 57Xa 110
Lutwyche M. SE6 61Bc 136
Lutwyche Rd. SE6 61Bc 136
Lutyens Cl. KT24: Eff 99Z 191
Lutyens Ho. SW1 50Lb 90
(off Churchill Gdns.)
Lux Apts. SW18 57Cb 111
(off Broomhill Rd.)
Luxborough Ho. W1 7H 215 (43Jb 90)
(off Luxborough St.)
Luxborough La. IG7: Chig 20Nc 36
Luxborough St. W1 7H 215 (43Jb 90)
Luxborough Twr. W1 7H 215 (43Jb 90)
(off Luxborough St.)
Lux Bldg., The RM7: Rush G 30Gd 56
Luxembourg M. E15 36Gc 73
Luxemburg Gdns. W6 49Za 88
Luxfield Rd. SE9 60Nc 116
Luxford St. SE16 49Zb 92
Luxmore St. SE4 53Bc 114
Luxor St. SE5 55Sb 113
LUXTED 86Qc 180
Luxted Rd. BR6: Downe 84Qc 180
Lyall Av. SE21 63Ub 135
Lyall M. SW1 4H 227 (48Jb 90)
Lyall M. W. SW1 4H 227 (48Jb 90)
Lyall St. SW1 4H 227 (48Jb 90)
Lyal Rd. E3 40Ac 72
Lycaste Cl. AL1: St A 3D 6
Lycett Pl. W12 47Wa 88
Lyceum Theatre London 4H 223 (45Pb 90)
(off Wellington St.)
Lych Ga. WD25: Wat 5Z 13
Lychgate Ct. N12 22Fb 49
Lychgate Mnr. HA1: Harr 31Ga 66
Lych Ga. Rd. BR6: Orp 74Wc 161
Lych Ga. Wlk. UB3: Hayes 45V 84
(not continuous)
Lych Way GU21: Wok 8P 167
Lyconby Gdns. CR0: C'don 73Ac 158
Lydd Cl. DA14: Sidc 62Uc 138
Lydden Ct. SE9 58Uc 116
Lydden Gro. SW18 59Db 111
Lydden Rd. SW18 59Db 111
Lydd Rd. DA7: Bex 52Bd 117
Lydeard Rd. E6 38Pc 74
Lydele Cl. GU21: Wok 87B 168
Lydford NW1 1C 216 (39Lb 70)
(off Royal College St.)
Lydford Av. SL2: Slou 3H 81
Lydford Cl. N16 36Ub 71
(off Pellerin Rd.)
Lydford Ct. DA2: Dart 58Rd 119
(off Osbourne Rd.)
Lydford Rd. N15 29Tb 51
Lydford Rd. NW2 37Ya 68
Lydford Rd. W9 42Bb 89
Lydger Cl. GU22: Wok 92D 188
Lydhurst Av. SW2 61Pb 134
Lydia Cotts. DA11: Grav'nd 9D 122
Lydia Ct. AL9: Wel G 6E 8
Lydia Ct. KT1: King T 69Na 131
(off Grove Cres.)
Lydia Ct. N12 23Eb 49
Lydia M. AL9: Wel G 6E 8
Lydia Rd. DA8: Erith 51Hd 118
Lydney Cl. SW19 61Ab 132
Lydon Rd. SW4 55Lb 112
Lydsey Cl. SL2: Slou 1E 80
Lydstep Rd. BR7: Chst 63Qc 138
Lye, The KT20: Tad 94Ya 194
Lye La. AL2: Brick W 10N 5
Lyell Pl. E. SL4: Wind 5A 102
Lyell Pl. W. SL4: Wind 5A 102
Lyell Rd. SL4: Wind 5A 102
Lyell St. E14 44Gc 93
Lyell Wlk. E. SL4: Wind 5A 102
Lyell Wlk. W. SL4: Wind 5A 102
Lyfield KT22: Oxs 86Da 171
Lyfield Ct. KT23: Bookh 97Ca 191
Lyford Rd. SW18 59Fb 111
Lyford St. SE7 49Nc 94
Lygon Ho. E2 41Vb 91
(off Gosset St.)
Lygon Ho. SW6 53Ab 110
(off Fulham Pal. Rd.)
Lygon Pl. SW1 4K 227 (48Kb 90)
Lyham Cl. SW2 58Nb 112
Lyham Rd. SW2 57Nb 112
Lyle Cl. CR4: Mitc 73Jb 156
Lyle Ct. SM4: Mord 72Fb 155
Lyle Pk. TN13: S'oaks 95Kd 203
Lyly Ho. SE1 4G 231 (48Tb 91)
(off Burbage Cl.)
Lymbourne Cl. SM2: Sutt 82Cb 175
Lymden Gdns. RH2: Reig 7K 207
Lyme Farm Rd. SE12 56Jc 115
Lyme Gro. E9 38Yb 72
Lyme Gro. Ho. E9 38Yb 72
(off Lyme Gro.)
Lymer Av. SE19 64Vb 135
Lyme Regis Rd. SM7: Bans 89Bb 175
Lyme Rd. DA16: Well 53Xc 117
Lymescote Gdns. SM1: Sutt 75Cb 155
Lyme St. NW1 38Lb 70
Lyme Ter. NW1 38Lb 70
Lyminge Cl. DA14: Sidc 63Vc 139
Lyminge Gdns. SW18 60Gb 111
Lymington Av. N22 26Qb 50
Lymington Cl. E6 43Pc 94
Lymington Cl. SW16 68Mb 134
Lymington Ct. SM1: Sutt 76Db 155
Lymington Ct. WD25: Wat 6W 12
Lymington Dr. HA4: Ruis 33T 64
Lymington Gdns. KT19: Ewe 78Va 154
Lymington Lodge E14 48Fc 93
(off Schooner Cl.)
Lymington Rd. NW6 37Db 69
Lymington Rd. RM8: Dag 32Zc 75
Lyminster Cl. UB4: Yead 43Aa 85
Lympne N17 26Tb 51
(off Gloucester Rd.)
Lympstone Gdns. SE15 52Wb 113
Lynbridge Gdns. N13 21Rb 51

Lynbrook Cl. RM13: Rain 40Fd 76
Lynbrook Gro. SE15 52Ub 113
Lynbury Ct. WD18: Wat 13W 26
Lynceley Grange CM16: Epp 1Wc 23
Lynch, The UB8: Uxb 38L 63
Lynch Cl. SE3 54Hc 115
Lynch Cl. UB8: Uxb 38L 63
Lynch Ct. AL4: St A 3G 6
Lynchen Cl. TW5: Cran 53X 107
LYNCH HILL 1C 80
Lynch Hill La. SL2: Slou 2C 80
Lynch Wlk. SE8 51Bc 114
(off Prince St.)
Lyncott Cres. SW4 56Kb 112
Lyncourt SE3 54Fc 115
Lyncroft Av. HA5: Pinn 29Aa 45
Lyncroft Gdns. KT17: Ewe 81Va 174
Lyncroft Gdns. NW6 36Cb 69
Lyncroft Gdns. TW3: Houn 57Ea 108
Lyncroft Gdns. W13 47La 86
Lyncroft Mans. NW6 36Cb 69
Lyndale KT7: T Ditt 73Ga 152
Lyndale NW2 35Bb 69
Lyndale Av. NW2 34Bb 69
Lyndale Cl. SE3 51Hc 115
Lyndale Ct. KT14: W Byf 85J 169
Lyndale Ct. RH1: Redh 3A 208
Lyndale Est. RM20: W Thur 51Xd 120
Lyndale Rd. RH1: Redh 3P 207
Lyndean Ind. Est. SE2 48Yc 95
Lynde Ho. KT12: Walt T 72Y 151
Lynde Ho. SW4 55Mb 112
Lynden Ho. E1 41Zb 92
(off Westfield Way)
Lynden Hyrst CR0: C'don 75Vb 157
Lynden Way BR8: Swan 69Ed 140
Lyndhurst Av. HA5: Pinn 25X 45
Lyndhurst Av. KT5: Surb 74Ra 153
Lyndhurst Av. N12 23Hb 49
Lyndhurst Av. NW7 23Ua 48
Lyndhurst Av. SW16 68Mb 134
Lyndhurst Av. TW16: Sun 69W 128
Lyndhurst Av. TW2: Whitt 60Ba 107
Lyndhurst Av. UB1: S'hall 46Da 85
Lyndhurst Cl. BR6: Farnb 77Rc 160
Lyndhurst Cl. CR0: C'don 76Vb 157
Lyndhurst Cl. DA7: Bex 55Dd 118
Lyndhurst Cl. GU21: Wok 7P 167
Lyndhurst Cl. NW10 34Ta 67
Lyndhurst Ct. E18 25Jc 53
Lyndhurst Ct. NW8 39Fb 69
(off Finchley Rd.)
Lyndhurst Ct. SM2: Sutt 80Cb 155
(off Grange Rd.)
Lyndhurst Dr. E10 31Ec 72
Lyndhurst Dr. KT3: N Mald 73Ua 154
Lyndhurst Dr. RM11: Horn 32Ld 77
Lyndhurst Dr. TN13: S'oaks 96Gd 202
Lyndhurst Gdns. EN1: Enf 14Ub 33
Lyndhurst Gdns. HA5: Pinn 25X 45
Lyndhurst Gdns. IG11: Bark 37Uc 74
Lyndhurst Gdns. IG2: Ilf 30Tc 54
Lyndhurst Gdns. N3 25Ab 48
Lyndhurst Gdns. NW3 36Fb 69
Lyndhurst Gro. SE15 54Ub 113
Lyndhurst Lodge E14 49Fc 93
(off Millennium Dr.)
Lyndhurst Ri. IG7: Chig 21Qc 54
Lyndhurst Rd. CR5: Coul 88Jb 176
Lyndhurst Rd. CR7: Thor H 70Qb 134
Lyndhurst Rd. DA7: Bex 55Dd 118
Lyndhurst Rd. E4 24Ec 52
Lyndhurst Rd. N18 21Wb 51
Lyndhurst Rd. N22 23Qb 50
Lyndhurst Rd. NW3 36Fb 69
Lyndhurst Rd. RH2: Reig 9J 207
Lyndhurst Rd. UB6: G'frd 42Da 85
Lyndhurst Sq. SE15 53Vb 113
Lyndhurst Ter. NW3 36Fb 69
Lyndhurst Vs. RH1: Redh 3P 207
Lyndhurst Wlk. WD6: Bore 11Pa 29
Lyndhurst Way CM13: Hut 17Ee 41
Lyndhurst Way DA13: Ist R 7A 144
Lyndhurst Way KT16: Chert 76G 148
Lyndhurst Way SE15 53Vb 113
Lyndhurst Way SM2: Sutt 81Cb 175
Lyndon Av. DA15: Sidc 57Vc 117
Lyndon Av. HA5: Hat E 23Aa 45
Lyndon Av. SM6: W'gton 76Jb 156
Lyndon Ho. E18 25Jc 53
(off Queen Mary Av.)
Lyndon Rd. DA17: Belv 49Cd 96
Lyndon Yd. SW17 63Eb 133
Lyndwood Dr. SL4: Old Win 8L 103
Lyndwood Pde. SL4: Old Win 8L 103
(off St Luke's Rd.)
LYNE 74C 148
Lyne Cl. GU25: Vir W 72B 148
Lyne Cres. E17 25Bc 52
Lyne Crossing Rd. KT16: Lyne 72C 148
Lyne Gdns. TN16: Big H 90Nc 180
Lynegrove Av. TW15: Ashf 64S 128
Lyneham Dr. NW9 25Ua 48
Lyneham Wlk. E5 36Ac 72
Lyneham Wlk. HA5: Eastc 27V 44
Lyne La. GU25: Vir W 72C 148
Lyne La. KT16: Lyne 72C 148
Lyne La. TW20: Thorpe 70C 126
Lyne Rd. GU25: Vir W 2P 147
Lynette Av. SW4 58Kb 112
Lyne Way HP1: Hem H 1H 3
Lynford Cl. EN5: Ark 15Va 30
Lynford Cl. HA8: Edg 25Sa 47
Lynford Ct. CR0: C'don 77Ub 157
(off Coombe Rd.)
Lynford French Ho. SE17 7D 230 (50Sb 91)
(off Thrush St.)
Lynford Gdns. HA8: Edg 20Ra 29
Lynford Gdns. IG3: Ilf 33Vc 75
Lynford Ter. N9 18Vb 33
Lyngarth Cl. KT23: Fet 97Fa 192
Lynhurst KT13: Weyb 79R 150
Lynhurst Cres. UB10: Hil 38S 64
Lynhurst Rd. UB10: Hil 38S 64
Lynmere Rd. DA16: Well 54Xc 117
Lyn M. E3 41Bc 92
Lyn M. N16 35Ub 71
Lynmouth Av. EN1: Enf 16Vb 33
Lynmouth Av. SM4: Mord 72Za 154
Lynmouth Av. Gdns. 74Za 154
Lynmouth Dr. HA4: Ruis 33X 65
Lynmouth Gdns. TW5: Hest 52Z 107
Lynmouth Gdns. UB6: G'frd 39Ka 66
Lynmouth Ho. RM3: Rom 22Nd 57
(off Dagnam Pk. Dr.)
Lynmouth Ri. BR5: St M Cry 70Xc 139
Lynmouth Rd. E17 30Ac 52
Lynmouth Rd. N16 32Vb 71
Lynmouth Rd. N2 27Hb 49
Lynmouth Rd. UB6: G'frd 39Ka 66
Lynn Cl. HA3: Hrw W 26Fa 46
Lynn Cl. TW15: Ashf 64T 128
Lynn Ct. CR3: Whyt 90Vb 177
Lynne Cl. BR6: Chels 79Vc 161
Lynne Cl. CR2: Sels 83Yb 178
Lynne Cl. SE23 59Ac 114
Lynne Ct. CR2: S Croy 77Ub 157
(off Birdhurst Rd.)
Lynne Ct. NW6 38Db 69
(off Priory Rd.)
Lynnett Ct. E9 37Ac 72
(off Annis Rd.)
Lynnett Rd. RM8: Dag 33Zc 75
Lynne Wlk. KT10: Esh 78Ea 152
Lynne Way UB5: N'olt 40Z 65
Lynn Ho. SE15 51Xb 113
(off Friary Est.)
Lynn M. E11 33Gc 73
Lynn Rd. E11 33Gc 73
Lynn Rd. IG2: Ilf 31Tc 74
Lynn Rd. SW12 59Kb 112
Lynn St. EN2: Enf 11Tb 33
Lynn Wlk. RH2: Reig 9K 207
Lynross Cl. RM3: Hrld W 26Pd 57
Lynscott Way CR2: S Croy 81Rb 177
Lynstead Ct. BR3: Beck 68Ac 136
Lynsted Cl. BR1: Broml 68Lc 137
Lynsted Cl. DA6: Bex 57Dd 118
Lynsted Gdns. SE9 56Mc 115
Lynton Av. AL1: St A 3G 6
Lynton Av. BR5: St M Cry 70Xc 139
Lynton Av. N12 21Fb 49
Lynton Av. NW9 28Va 48
Lynton Av. RM7: Mawney 25Cd 56
Lynton Av. W13 44Ja 86
Lynton Cl. KT9: Chess 77Na 153
Lynton Cl. NW10 36Ua 68
Lynton Cl. TW7: Isle 56Ha 108
Lynton Ct. KT17: Ewe 83Va 174
Lynton Cres. IG2: Ilf 30Rc 54
Lynton Crest EN6: Pot B 4Cb 17
Lynton Est. SE1 49Wb 91
Lynton Gdns. EN1: Enf 17Ub 33
Lynton Gdns. N11 23Mb 50
Lynton Grange N2 27Hb 49
Lynton Ho. GU22: Wok 90A 168
(off Station App.)
Lynton Ho. IG1: Ilf 33Sc 74
(off High Rd.)
Lynton Ho. W2 44Eb 89
(off Hallfield Est.)
Lynton Mans. SE1 3K 229 (48Qb 90)
(off Westminster Bri. Rd.)
Lynton Mead N20 20Cb 31
Lynton Pde. EN8: Chesh 2Ac 20
Lynton Rd. CR0: C'don 72Qb 156
Lynton Rd. DA11: Grav'nd 10C 122
Lynton Rd. E4 22Dc 52
Lynton Rd. HA2: Harr 33Aa 65
Lynton Rd. KT3: N Mald 71Ta 153
Lynton Rd. N8 29Mb 50
(not continuous)
Lynton Rd. NW6 40Bb 69
Lynton Rd. SE1 6K 231 (49Vb 91)
Lynton Rd. W3 45Qa 87
Lynton Rd. Sth. DA11: Grav'nd 10C 122
Lynton Ter. W3 44Sa 87
Lynton Wlk. UB4: Hayes 41U 84
Lynwood Av. CR5: Coul 87Kb 176
Lynwood Av. KT17: Eps 86Va 174
Lynwood Av. SL3: L'ly 8P 81
Lynwood Av. TW20: Egh 65A 126
Lynwood Cl. E18 25Lc 53
Lynwood Cl. GU21: Wok 85F 168
Lynwood Cl. HA2: Harr 34Aa 65
Lynwood Cl. RM5: Col R 23Dd 56
Lynwood Ct. KT1: King T 68Ra 131
Lynwood Ct. KT17: Eps 85Va 174
Lynwood Cres. SL5: S'dale 2C 146
Lynwood Dr. HA6: Nwood 25V 44
Lynwood Dr. KT4: Wor Pk 75Wa 154
Lynwood Dr. RM5: Col R 23Dd 56
Lynwood Gdns. CR0: Wadd 77Pb 156
Lynwood Gdns. UB1: S'hall 44Ba 85
Lynwood Gro. BR6: Orp 73Uc 160
Lynwood Gro. N21 18Qb 32
Lynwood Hgts. WD3: Rick 15K 25
Lynwood Rd. KT17: Eps 86Va 174
Lynwood Rd. KT7: T Ditt 75Ha 152
Lynwood Rd. RH1: Redh 4A 208
Lynwood Rd. SW17 62Hb 133
Lynwood Rd. W5 41Ma 87
Lynx Hill KT24: E Hor 100V 190
Lynx Way E16 45Mc 93
Lyon Bus. Pk. IG11: Bark 40Uc 74
Lyon Ct. HA4: Ruis 32V 64
Lyon Ct. KT12: Walt T 75Aa 151
Lyon Ho. NW8 6D 214 (42Gb 89)
(off Broadley St.)
Lyon Ind. Est. NW2 33Xa 68
Lyon Meade HA7: Stan 25La 46
Lyon Pk. Av. HA0: Wemb 37Na 67
(not continuous)
Lyon Rd. HA1: Harr 30Ha 46
Lyon Rd. KT12: Walt T 75Aa 151
Lyon Rd. RM1: Rom 31Hd 76
Lyon Rd. SW19 67Eb 133
LYONSDOWN 15Eb 31
Lyonsdown Av. EN5: New Bar 16Eb 31
Lyonsdown Rd. EN5: New Bar 16Eb 31
Lyons Dr. GU2: Guild 10L 187
Lyons Ind. Est. UB8: Cowl 45M 83
Lyons Pl. HA7: Stan 20Ga 28
Lyons Pl. NW8 6B 214 (42Fb 89)
(off Central St.)
Lyon St. N1 38Pb 70
Lyons Wlk. W14 49Ab 88
Lyon Way AL4: St A 2L 7
Lyon Way UB6: G'frd 39Ga 66
Lyoth Rd. BR5: Farnb 75Sc 160
Lyrical Way HP1: Hem H 1K 3
Lyric Ct. E8 38Vb 71
(off Holly St.)
Lyric Dr. UB6: G'frd 42Da 85
Lyric M. SE26 63Yb 136
Lyric Rd. SW13 53Va 110
Lyric Sq. W6 49Ya 88
(off King St.)
Lyric Theatre Hammersmith 49Ya 88
Lyric Theatre Westminster 4D 222 (45Mb 90)
(off Shaftesbury Av.)
Lysander NW9 25Va 48
Lysander Cl. HP3: Bov 9B 2
Lysander Gdns. KT6: Surb 72Pa 153
Lysander Gro. N19 32Mb 70
Lysander Ho. E2 40Xb 71
(off Temple St.)
Lysander M. N19 32Lb 70
Lysander Rd. CR0: Wadd 79Pb 156
Lysander Rd. HA4: Ruis 33T 64
Lysander Way BR6: Farnb 76Sc 160
Lysander Way WD5: Ab L 4W 12
Lysia Ct. SW6 52Za 110
(off Lysia St.)
Lysias Rd. SW12 58Kb 112
Lysia St. SW6 52Za 110
Lysley Pl. AL9: Brk P 9L 9
Lysons Wlk. SW15 56Wa 110
Lyster M. KT11: Cobh 85Y 171
Lytchet Rd. BR1: Broml 66Jc 137
Lytchet Way EN3: Enf H 11Yb 34
Lytchgate Cl. CR2: S Croy 80Ub 157
Lytcott Dr. KT8: W Mole 69Ba 129
Lytcott Gro. SE22 57Ub 113
Lytham Av. WD19: Wat 22Z 45
Lytham Cl. SE28 44Ad 95
Lytham Ct. SL5: S'hill 1A 146
Lytham Gro. W5 41Pa 87
Lytham St. SE17 50Tb 91
Lyttelton Cl. NW3 38Gb 69
Lyttelton Ct. N2 29Eb 49
Lyttelton Ho. E9 38Yb 72
(off Well St.)
Lyttelton Rd. E10 34Dc 72
Lyttelton Rd. N2 29Eb 49
Lyttelton Theatre 6K 223 (46Qb 90)
(within National Theatre)
Lyttleton Ct. UB4: Yead 42Y 85
(off Dunedin Way)
Lyttleton Rd. N8 27Qb 50
Lytton Av. EN3: Enf L 10Ac 20
Lytton Av. N13 19Qb 32
Lytton Cl. IG10: Lough 13Tc 36
Lytton Cl. N2 30Fb 49
Lytton Cl. UB5: N'olt 38Ba 65
Lytton Ct. WC1 1G 223 (43Nb 90)
(off Barter St.)
Lytton Gdns. SM6: Bedd 77Mb 156
Lytton Gro. SW15 57Za 110
Lytton Pk. KT11: Cobh 84Ba 171
Lytton Rd. E11 31Gc 73
Lytton Rd. EN5: New Bar 14Eb 31
Lytton Rd. GU22: Wok 88D 168
Lytton Rd. HA5: Pinn 24Aa 45
Lytton Rd. RM16: Grays 9C 100
Lytton Rd. RM2: Rom 29Kd 57
Lytton Strachey Path SE28 45Xc 95
Lytton Ter. E12 37Pc 74
Lyveden Rd. SE3 52Kc 115
Lyveden Rd. SW17 65Hb 133
Lywood Cl. KT20: Tad 94Ya 194

M

Mabbett Ho. SE18 51Qc 116
(off Nightingale Pl.)
Mabbotts KT20: Tad 93Za 194
Mabbutt Cl. AL2: Brick W 2Aa 13
Mabel Evetts Ct. UB3: Hayes 45Y 85
Mabel Rd. BR8: Hext 65Jd 140
Mabel St. GU21: Wok 10P 167
Maberley Cres. SE19 66Wb 135
Maberley Rd. BR3: Beck 69Zb 136
Maberley Rd. SE19 67Vb 135
Mabledon Pl. WC1 4E 216 (41Mb 90)
Mablethorpe Rd. SW6 52Ab 110
Mabley St. E9 36Ac 72
Mablin Lodge IG9: Buck H 18Lc 35
McAdam Dr. EN2: Enf 12Rb 33
McAllister Gro. IG11: Bark 41Wc 95
McArdle Way SL3: Coln 52F 104
Macaret Cl. N20 17Db 31
Macarthur Cl. DA8: Erith 50Gd 96
Macarthur Cl. E7 37Jc 73
Macarthur Cl. HA9: Wemb 37Ra 67
Macarthur Ter. SE7 51Mc 115
Macartney Ho. SE10 52Fc 115
(off Chesterfield Wlk.)
Macartney Ho. SW9 53Qb 112
(off Gosling Way)
Macaulay Av. KT10: Hin W 75Ha 152
Macaulay Ct. SW4 55Kb 112
Macaulay Rd. CR3: Cat'm 94Ub 197
Macaulay Rd. E6 40Mc 73
Macaulay Rd. SW4 55Kb 112
Macaulay Sq. SW4 56Kb 112
Macaulay Wlk. SW9 53Pb 112
(off Lett Rd.)
Macaulay Way SE28 46Xc 95
McAuley Cl. SE1 3K 229 (48Qb 90)
McAuley Cl. SE9 57Rc 116
Macauley Ho. W10 43Ab 88
(off Portobello Rd.)
Macauley M. SE13 54Ec 114
McAuliffe Dr. SL2: Farn C 5D 60
McAusland Ho. E3 40Bc 72
(off Wright's Rd.)
Macbean St. SE18 48Rc 94
Macbeth Ho. N1 1H 219 (40Ub 71)
Macbeth St. W6 50Xa 88
McBride Ho. E3 40Bc 72
(off Libra Rd.)
McCabe Ct. E16 43Hc 93
(off Barking Rd.)
McCall Cl. SW4 54Nb 112
McCall Cres. SE7 50Nc 94
McCall Ho. N7 35Nb 70
McCarthy Rd. TW13: Hanw 64Z 129
McClaren Technology Cen.
GU21: Wok 83C 168
Macclesfield Apts. N1 39Tb 71
(off Branch Pl.)
Macclesfield Ho. EC1 4D 218 (41Sb 91)
(off Central St.)
Macclesfield Ho. RM3:
Rom 22Nd 57
(off Dagnam Pk. Dr.)
Macclesfield Rd. EC1 3D 218 (41Sb 91)
Macclesfield Rd. SE25 71Yb 158
Macclesfield St. W1 4E 222 (45Mb 90)
McClintock Pl. EN3: Enf L 10Dc 20
McCoid Way SE1 2D 230 (47Sb 91)
McCall Ct. CR2: S Croy 81Sb 177
McCrone M. NW3 37Fb 69
McCudden Rd. DA1: Dart 55Pd 119
McCullum Rd. E3 39Bc 72
McDermott Cl. SW11 55Gb 111
McDermott Rd. SE15 55Wb 113
McDermott Rd. TN15: Bor G 92Be 205
Macdonald Av. RM10: Dag 34Dd 76
Macdonald Av. RM11: Horn 27Nd 57
McDonald Cl. AL10: Hat 2C 8
(not continuous)
Macdonald Ho. SW11 54Jb 112
(off Dagnall St.)
McDonald Ho. NW6 41Cb 89
(off Malvern Rd.)
Macdonald Rd. E17 26Ec 52
(not continuous)
Macdonald Rd. E7 35Jc 73
Macdonald Rd. N11 22Hb 49
Macdonald Rd. N19 33Lb 70
Macdonald Way RM11: Horn 28Nd 57
Macdonnell Gdns. WD25: Wat 7V 12
McDonough Cl. KT9: Chess 77Na 153
McDougall Ct. TW9: Rich 54Qa 109
McDougall Rd. HP4: Berk 1A 2
McDowall Cl. E16 43Hc 93
McDowall Rd. SE5 53Sb 113
Macduff Rd. SW11 53Jb 112
Mace Cl. E1 46Xb 91
Mace Ct. RM17: Grays 1A 122
Mace Ho. TW7: Isle 53Ka 108
Mace La. TN14: Cud 85Tc 180
McEntee Av. E17 25Ac 52
Mace St. E2 40Zb 72
McEwan Ho. E3 40Bc 72
(off Roman Rd.)
McEwen Way E15 39Fc 73
(off Rokeby St.)
Macey Ho. SW11 53Gb 111
Macey St. SE10 51Ec 114
(off Thames St.)
McFadden Ct. E10 34Dc 72
(off Buckingham Rd.)
Macfarland Gro. SE15 52Ub 113
Macfarlane La. TW7: Isle 51Ha 108
Macfarlane Rd. W12 46Ya 88
Macfarren Pl. NW1 6J 215 (42Jb 90)
Macfarron Ho. W10 41Ab 88
(off Parry Rd.)
McGlashon Ho. E1 42Wb 91
(off Hunton St.)
McGrath Rd. E15 36Hc 73
McGredy EN7: Chesh 1Xb 19
McGregor Ct. N1 3J 219 (41Ub 91)
(off Hoxton St.)
Macgregor Rd. E16 43Lc 93
McGregor Rd. W11 44Bb 89
Machell Rd. SE15 55Yb 114
McIndoe Ct. N1 39Tb 71
(off Sherborne St.)
McIntosh Cl. RM1: Rom 27Gd 56
McIntosh Cl. SM6: W'gton 80Nb 156
Macintosh Ho. W1 7J 215 (43Jb 90)
(off Beaumont St.)
McIntosh Ho. SE16 49Yb 92
(off Millender Wlk.)
McIntosh Ho. SE20 67Wb 135
McIntosh Rd. RM1: Rom 27Gd 56
McIntyre Ct. SE18 49Nc 94
(off Prospect Va.)
Mackay Ho. W12 45Xa 88
(off White City Est.)
Mackay Rd. SW4 55Kb 112
McKay Rd. SW20 66Xa 132
McKay Trad. Est. SL3: Poyle 54G 104
McKeever Cl. EN9: Walt A 5Dc 20
McKeever Ho. E16 43Jc 93
(off Hammersley Rd.)
McKellar Cl. WD23: B Hea 19Ea 28
McKenna Ho. E3 40Bc 72
(off Wright's Rd.)
Mackennal St. NW8 2E 214 (40Gb 69)
Mackenzie Cl. W12 45Xa 88
McKenzie Cl. CR5: Coul 88Mb 176
Mackenzie Ho. N8 28Nb 50
(off Pembroke Rd.)
Mackenzie Ho. NW2 34Wa 68
Mackenzie Mall SL1: Slou 7K 81
(within Queensmere Shop. Cen.)
Mackenzie Rd. BR3: Beck 68Yb 136
Mackenzie Rd. N7 37Pb 70
Mackenzie St. SL1: Slou 7K 81
Mackenzie Wlk. E14 46Cc 92
Mackenzie Way DA12: Grav'nd 5F 144
McKenzie Way KT19: Eps 81Qa 173
McKerrell Rd. SE15 53Wb 113
Mackeson Rd. NW3 35Hb 69
Mackie Rd. SW2 59Qb 112
McKillop Way DA14: Sidc 66Yc 139
Mackintosh Cl. SL9: Ger X 28A 42
Mackintosh La. E9 36Zb 72
Mackintosh St. BR2: Broml 72Mc 159
Macklin St. WC2 2G 223 (44Nb 90)
Mackonochie Ho. EC1 7K 217 (43Qb 90)
(off Baldwins Gdns.)
Mackrells RH1: Redh 9L 207
Mackrow Wlk. E14 45Ec 92
Mack's Rd. SE16 49Wb 91
Mackworth Ho. NW1 3B 216 (41Lb 90)
(off Augustus St.)
Mackworth St. NW1 3B 216 (41Lb 90)
McLaren Ho. SE1 3B 230 (48Rb 91)
(off St Georges Cir.)
Maclaren M. SW15 56Ya 110
Maclean Rd. SE23 58Ac 114
Maclean Ter. DA12: Grav'nd 1H 145
Maclennan Av. RM13: Rain 41Md 97
Macleod Cl. RM17: Grays 49Fe 99
McLeod Ct. SE22 60Wb 113
Macleod Rd. N21 15Nb 32
McLeod Rd. SE2 49Xc 95
McLeod's M. SW7 49Db 89
Macleod St. SE17 50Sb 91
Maclise Ho. SW1 6F 229 (49Nb 90)
(off Marsham St.)
Maclise Rd. W14 48Ab 88
Macmahon Cl. GU24: Chob 3J 167
McMillan Cl. DA12: Grav'nd 3E 144
Macmillan Ct. HA2: Harr 32Ca 65
Macmillan Ct. UB6: G'frd 42Fa 86
Macmillan Gdns. DA1: Dart 55Pd 119
McMillan Ho. NW8 4E 214 (41Gb 89)
(off Lorne Cl.)
Macmillan Ho. SM7: Bans 86Bb 175
(off Basing Rd.)
McMillan Ho. SE14 53Ac 114
McMillan Ho. SE4 55Ac 114
(off Arica Rd.)
Macmillan Rd. TN14: Dun G 92Hd 202
McMillan St. SE8 51Cc 114
McMillan Student Village SE8 51Cc 114
Macmillan Way SW17 63Kb 134
McNair Rd. UB2: S'hall 48Da 85
Macnamara Ho. SW10 52Fb 111
(off Worlds End Est.)
McNeil Rd. SE5 54Ub 113
McNicol Dr. NW10 40Sa 67
Macoma Rd. SE18 51Tc 116
Macoma Ter. SE18 51Tc 116
Maconochies Rd. E14 50Dc 92
Macon Way RM14: Upm 31Ud 78
MacOwan Theatre 49Cb 89
Macquarie Way E14 49Dc 92
McRae La. CR4: Mitc 73Hb 155
Macready Ho. W1 1E 220 (43Gb 89)
(off Crawford St.)
Macready Pl. N7 35Nb 70
(not continuous)
Mcready Rd. N20 19Fb 31
Macrea Ho. E3 41Bc 92
(off Bow Rd.)
Macroom Ho. W9 41Bb 89
(off Macroom Rd.)
Macroom Rd. W9 41Bb 89
Macs Ho. E17 27Dc 52
Mac's Pl. EC4 2A 224 (44Qb 90)
(off Greystoke Pl.)
Madame Tussaud's 6H 215 (42Jb 90)
Madan Cl. TN16: Westrm 97Uc 200
Madan Rd. TN16: Westrm 97Tc 200
Madans Wlk. KT18: Eps 87Ta 173
(not continuous)
Mada Rd. BR6: Farnb 76Rc 160
Maddams St. E3 42Dc 92
Madden Cl. DA10: Swans 58Zd 121
Madderfields Ct. N11 25Mb 50
Maddison Cl. N2 26Eb 49
Maddison Cl. TW11: Tedd 65Ha 130
Maddison Ct. E16 43Jc 93
(off Hastings Rd.)
Maddison Hgts. WD18: Wat 14V 26
(off Chiltern Cl.)
Maddocks Cl. DA14: Sidc 64Ad 139
Maddocks Ho. E1 45Xb 91
(off Cornwall St.)
Maddock Way SE17 51Rb 113
Maddox La. KT23: Bookh 94Aa 191
Maddox Pk. KT23: Bookh 95Aa 191
Maddox Rd. HP2: Hem H 2B 4
Maddox St. W1 4A 222 (45Kb 90)
Madeira Av. BR1: Broml 66Gc 137
Madeira Cl. KT14: W Byf 85J 169
Madeira Cres. KT14: W Byf 85H 169
Madeira Gro. IG8: Wfd G 23Lc 53
Madeira Rd. CR4: Mitc 70Hb 133
Madeira Rd. E11 32Fc 73
Madeira Rd. KT14: W Byf 85H 169
Madeira Rd. N13 21Rb 51
Madeira Rd. SW16 64Nb 134
Madeira St. E14 43Dc 92
Madeira Twr. SW11 52Mb 112
Madeira Wlk. CM15: B'wood 20Ae 41
Madeira Wlk. RH2: Reig 5M 207
Madeira Wlk. SL4: Wind 3H 103
Madeleine Cl. RM6: Chad H 30Yc 55
Madeleine Ct. HA7: Stan 24Na 47
(off Letchworth Rd.)
Madeleine Ct. IG7: Chig 22Uc 54
Madeleine Ho. CR2: S Croy 82Vb 177
Madeley Rd. W5 44Ma 87
Madeline Gro. IG1: Ilf 36Tc 74
Madeline Rd. SE20 66Wb 135
Madells CM16: Epp 3Vc 23
Madge Gill Way E6 39Nc 74
(off High St. Nth.)
Madge Hill W7 45Ga 86
Madinah Rd. E8 37Wb 71
Madison, The SE1 1F 231 (47Tb 91)
(off Long La.)
Madison Bldg. SE10 53Dc 114
(off Blackheath Rd.)
Madison Cl. SM2: Sutt 80Fb 155
Madison Ct. RM10: Dag 37Dd 76
Madison Cres. DA7: Bex 52Yc 117
Madison Gdns. BR2: Broml 69Hc 137
Madison Gdns. DA7: Bex 52Yc 117
Madison Ho. E14 45Bc 92
(off Victory Pl.)
Madison Wlk. RM16: Chaf H 47Be 99
Madison Way E20 36Dc 72
Madison Way TN13: S'oaks 95Hd 202
Madoc Cl. NW2 33Cb 69
Madras Pl. N7 37Qb 70
Madras Rd. IG1: Ilf 35Rc 74
Madresfield Ct. WD7: Shenl 4Ma 15
Madrid Rd. SW13 53Wa 110
Madrigal La. SE5 52Rb 113
Madron St. SE17 7J 231 (50Ub 91)
Maesmaur Rd. TN16: Tats 93Mc 199
Mafeking Av. E6 40Nc 74
Mafeking Av. IG2: Ilf 31Tc 74
Mafeking Av. TW8: Bford 51Na 109
Mafeking Rd. E16 42Hc 93
Mafeking Rd. EN1: Enf 13Vb 33
Mafeking Rd. N17 26Wb 51
Mafeking Rd. TW19: Wray 61D 126
Magazine Ga. W2 6E 220 (46Gb 89)
Magazine Pl. KT22: Lea 94Ka 192
Magazine Rd. CR3: Cat'm 94Rb 197
Magdala Av. N19 33Lb 70
Magdala Rd. CR2: S Croy 80Tb 157
Magdala Rd. TW7: Isle 55Ja 108
Magdalen Cl. KT14: Byfl 86N 169
Magdalen Ct. AL1: St A 1C 6
(off Newsom Pl.)
Magdalen Cres. KT14: Byfl 86N 169
Magdalene Cl. SE15 54Xb 113
Magdalene Gdns. E6 42Qc 94
Magdalene Gdns. N20 18Hb 31
Magdalene Rd. TW17: Shep 70P 127
Magdalen Gro. BR6: Chels 77Xc 161
Magdalen Ho. E16 46Kc 93
(off Keats Av.)
Magdalen M. NW3 37Eb 69
(off Frognal)
Magdalen Pas. E1 45Vb 91
Magdalen Rd. SW18 60Eb 111
Magdalen St. SE1 7H 225 (46Ub 91)
Magee St. SE11 51Qb 112
Magellan Blvd. E16 45Sc 94
Magellan Ct. NW10 38Ta 67
(off Brentfield Rd.)
Magellan Ho. E1 42Zb 92
(off Ernest St.)
Magellan Pl. E14 49Cc 92
Magic Circle 5C 216 (42Lb 90)
(off Stephenson Way)
Magisters Lodge WD3: Crox G 16Q 26
Magistrates' Court Barkingside 27Sc 54
Magistrates' Court Belmarsh 47Uc 94
Magistrates' Court Bexley 56Cd 118
Magistrates' Court Bromley 67Hc 137
Magistrates' Court City of
London 3F 225 (44Tb 91)
(off Queen Victoria St.)
Magistrates' Court Croydon 76Tb 157
Magistrates' Court Ealing 46Ja 86
(off Green Man La.)
Magistrates' Court East Berkshire, Slough 7J 81
Magistrates' Court Hendon 30Va 48
Magistrates' Court Highbury Corner 37Qb 70
Magistrates' Court Lavender Hill 55Hb 111
Magistrates' Court Romford 28Hd 56

Magistrates' Court Sevenoaks 95Hd 202
Magistrates' Court St Albans 2B 6
Magistrates' Court Staines Upon Thames 64J 127
Magistrates' Court Stratford 38Fc 73
Magistrates' Court Thames 41Cc 92
Magistrates' Court Uxbridge 38L 63
Magistrates' Court Westminster 7E 214 (43Gb 89)
Magistrates' Court Willesden 37Va 68
Magistrates' Court Wimbledon 65Cb 133
Magna Carta La. TW19: Wray 10P 103
Magna Carta Memorial 1P 125
Magna Ct. TW18: Staines 64G 126
Magna Rd. TW20: Eng G 5M 125
Magna Sq. SW14 55Sa 109
(off Moore Cl.)
Magnaville Rd. WD23: B Hea 17Ga 28
Magnetic Cres. EN3: Enf L 9Bc 20
Magnet Point Est. RM20: Grays 51Yd 120
Magnet Rd. HA9: Wemb 33Ma 67
Magnet Rd. RM20: Grays 51Yd 120
Magnin Cl. E8 39Wb 71
Magnolia Av. WD5: Ab L 4W 12
Magnolia Cl. AL2: Park 8B 6
Magnolia Cl. E10 33Cc 72
Magnolia Cl. KT2: King T 65Ra 131
Magnolia Cl. RM15: S Ock 42Ae 99
Magnolia Cl. HA3: Kenton 31Pa 67
Magnolia Ct. SM2: Sutt 80Cb 155
(off Grange Rd.)
Magnolia Ct. SM6: W'gton 78Kb 156
Magnolia Ct. TW13: Felt 60W 106
(off Plum Cl.)
Magnolia Ct. TW9: Kew 53Ra 109
Magnolia Ct. UB10: Hil 37R 64
Magnolia Ct. UB5: N'olt 42Aa 85
Magnolia Cres. CM13: Gt War 23Xd 58
Magnolia Dr. SM7: Bans 88Bb 175
Magnolia Dr. TN16: Big H 88Mc 179
Magnolia Gdns. AL1: St A 4E 6
Magnolia Gdns. E10 33Cc 72
Magnolia Gdns. HA8: Edg 21Sa 47
Magnolia Gdns. SL3: L'ly 8N 81
Magnolia Ho. SE8 51Bc 114
(off Evelyn St.)
Magnolia Ho. TW16: Sun 66V 128
Magnolia Lodge E4 20Dc 34
Magnolia Lodge W8 48Db 89
(off St Mary's Ga.)
Magnolia Pl. HA2: Harr 26Fa 46
Magnolia Pl. SW4 57Nb 112
Magnolia Pl. W5 43Ma 87
Magnolia Rd. W4 51Ra 109
Magnolia St. UB7: W Dray 50M 83
Magnolia Vw. TW18: Staines 66K 127
Magnolia Way CM15: Pil H 15Xd 40
Magnolia Way EN8: Chesh 1Yb 20
Magnolia Way KT19: Ewe 78Sa 153
Magnolia Wharf W4 51Qa 109
Magnum Cl. RM13: Rain 42Kd 97
Magpie All. EC4 3A 224 (44Qb 90)
Magpie Bottom TN15: Knat 85Md 183
Magpie Cl. CR5: Coul 90Lb 176
Magpie Cl. E7 36Hc 73
Magpie Cl. EN1: Enf 10Wb 19
Magpie Cl. NW9 26Ua 48
Magpie Hall Cl. BR2: Broml 72Nc 160
Magpie Hall La. BR2: Broml 71Pc 160
Magpie Hall Rd. WD23: B Hea 19Ga 28
Magpie Ho. E3 39Bc 72
(off Sycamore Av.)
Magpie La. CM13: L War 26Zd 59
Magpie Pl. SE14 51Ac 114
Magpie Pl. WD25: Wat 4Y 13
Magpie Wlk. AL10: Hat 2C 8
Magpie Way SL2: Slou 2D 80
Magri Wlk. E1 43Yb 92
Maguire Apts. E3 43Bc 92
(off Geoff Cade Way)
Maguire Dr. TW10: Ham 63La 130
Maguire St. SE1 1K 231 (47Vb 91)
Maha Bldg. E3 41Cc 92
(off Merchant St.)
Mahatma Gandhi Ind. Est. SE24 56Rb 113
Mahlon Av. HA4: Ruis 36X 65
(not continuous)
Mahogany Cl. SE16 46Ac 92
Mahon Cl. EN1: Enf 11Vb 33
Mahoney Ho. SE14 53Bc 114
(off Heald St.)
Mahonia Cl. GU24: W End 5D 166
Maibeth Gdns. BR3: Beck 70Ac 136
Maida Av. E4 17Dc 34
Maida Av. W2 7A 214 (43Eb 89)
MAIDA HILL 42Bb 89
Maida Rd. DA17: Belv 48Cd 96
Maida Va. W9 40Db 69
MAIDA VALE 42Db 89
Maida Va. Rd. DA1: Dart 57Jd 118
Maida Vale Station (Underground) 41Db 89
Maida Way E4 17Dc 34
Maiden Erlegh Av. DA5: Bexl 60Ad 117
Maidenhead Rd. SL4: Wind 2A 102
Maiden La. DA1: Cray 55Jd 118
Maiden La. NW1 38Mb 70
Maiden La. SE1 7E 224 (46Sb 91)
Maiden La. WC2 5G 223 (45Nb 90)
Maiden Pl. NW5 34Lb 70
Maiden Rd. E15 38Gc 73
Maidens Bri. EN2: Enf 9Wb 19
Maidenshaw Rd. KT19: Eps 84Ta 173
Maidenstone Hill SE10 53Ec 114
Maids of Honour Row TW9: Rich 57Ma 109
Maidstone Av. RM5: Col R 26Ed 56
Maidstone Bldgs. M. SE1 7E 224 (46Sb 91)
Maidstone Ho. E14 44Dc 92
(off Carmen St.)
Maidstone Rd. DA14: Sidc 65Zc 139
Maidstone Rd. DA14: Swan 65Zc 139
Maidstone Rd. N11 23Lb 50
Maidstone Rd. RM17: Grays 51Ce 121
Maidstone Rd. TN13: Riv 94Gd 202
Maidstone Rd. TN15: Bor G 92Ce 205
Maidstone Rd. TN15: Plat 92Ce 205
Maidstone Rd. TN15: Seal 93Qd 203
Maidstone Rd. TN15: Wro H 92Ce 205
Mailcoach Yd. E2 3J 219 (41Ub 91)
Main Av. EN1: Enf 15Vb 33
Main Av. HA6: Nwood 20S 26
Main Dr. HA9: Wemb 34Ma 67
Main Dr. SL0: Rich P 49G 82
Maine Twr. E14 47Dc 92
Main Mill SE10 52Dc 114
(off Greenwich High St.)
Main Pde. WD3: Chor 14E 24
Main Pde. Flats WD3: Chor 14E 24
Mainridge Rd. BR7: Chst 63Qc 138
Main Rd. DA4: Farni Donkey La. 75Sd 164
Main Rd. DA4: Farni London Rd. 72Nd 163
Main Rd. BR2: Kes 84Lc 179
Main Rd. BR5: St P 67Yc 139
Main Rd. BR8: Crock 72Fd 162
Main Rd. BR8: Hext 66Hd 140
Main Rd. DA14: Sidc 62Tc 138
Main Rd. DA3: Lfield 68Zd 143
Main Rd. DA3: Long H 68Zd 143
Main Rd. DA4: Sut H 65Rd 141
Main Rd. RM1: Rom 28Hd 56
Main Rd. RM2: Rom 28Hd 56
Main Rd. SL4: Wind 2A 102
Main Rd. TN14: Sund 96Zc 201
Main Rd. TN16: Big H 85Lc 179
Main Rd. TN16: Westrm 85Lc 179
Main Rd. Cotts. BR6: Prat B 81Yc 181
Mainstone Cres. GU24: Brkwd 3B 186
Mainstone Rd. GU24: Bisl 8D 166
Main St. KT15: Add 76N 149
Main St. TW13: Hanw 64Z 129
Mainwaring Ct. CR4: Mitc 68Jb 134
Mais Ho. SE26 61Xb 135
Maisie Webster Cl. TW19: Stanw 59M 105
Maismore St. SE15 51Wb 113
Maisonettes, The SM1: Sutt 78Bb 155
Maison Ho. N20 18Eb 31
Maitland Cl. KT12: Walt T 75Aa 151
Maitland Cl. KT14: W Byf 85J 169
Maitland Cl. SE10 52Dc 114
Maitland Cl. TW4: Houn 55Ba 107
Maitland Ct. W2 4B 220 (45Fb 89)
(off Lancaster Ter.)
Maitland Ho. E2 40Yb 72
(off Waterloo Gdns.)
Maitland Ho. SW1 51Lb 112
(off Churchill Gdns.)
Maitland Pk. Est. NW3 37Hb 69
Maitland Pk. Rd. NW3 37Hb 69
Maitland Pk. Vs. NW3 37Hb 69
Maitland Pl. E5 35Xb 71
Maitland Rd. E15 37Hc 73
Maitland Rd. SE26 65Zb 136
Maitlands IG10: Lough 13Pc 36
Maitland Yd. W13 46Ja 86
Maize Row E14 45Bc 92
Maizey Ct. CM15: Pil H 15Wd 40
Majendie Rd. SE18 50Tc 94
Majestic Way CR4: Mitc 68Hb 133
Major Cl. SW9 55Rb 113
Major Draper St. SE18 48Rc 94
Major Rd. E15 36Fc 73
Major Rd. SE16 48Wb 91
Majors Farm Rd. SL3: Dat 2P 103
Makepeace Av. N6 33Jb 70
Makepeace Mans. N6 33Jb 70
Makepeace Rd. E11 28Jc 53
Makepeace Rd. UB5: N'olt 40Aa 65
Makers' Yd. E20 37Cc 72
Makinen Ho. IG9: Buck H 18Lc 35
Makins St. SW3 6E 226 (49Gb 89)
Malabar Ct. W12 45Xa 88
(off India Way)
Malabar St. E14 47Cc 92
Malacca Farm GU4: W Cla 99K 189
Malam Ct. SE11 6K 229 (49Qb 90)
Malam Gdns. E14 45Dc 92
Malan Apts. CR8: Purl 87Qb 176
Malan Cl. TN16: Big H 89Nc 180
Malan Sq. RM13: Rain 37Kd 77
Malbrook Rd. SW15 56Xa 110
Malcolm Cl. SE20 66Yb 136
Malcolm Ct. E7 37Hc 73
Malcolm Ct. HA7: Stan 22La 46
Malcolm Ct. NW4 30Wa 48
Malcolm Cres. NW4 30Wa 48
Malcolm Dr. KT6: Surb 74Na 153
Malcolm Ho. N1 2H 219 (40Ub 71)
(off Arden Est.)
Malcolm Pl. E2 42Yb 92
Malcolm Rd. CR5: Coul 87Mb 176
Malcolm Rd. E1 42Yb 92
Malcolm Rd. SE20 66Yb 136
Malcolm Rd. SE25 72Wb 157
Malcolm Rd. SW19 65Ab 132
Malcolm Rd. UB10: Ick 35P 63
Malcolm Sargent Ho. E16 46Kc 93
(off Evelyn Rd.)
Malcolmson Ho. SW1 7D 228 (50Mb 90)
(off Aylesford St.)
Malcolms Way N14 15Lb 32
Malcolm Way E11 29Jc 53
Malden Av. SE25 70Xb 135
Malden Av. UB6: G'frd 36Ga 66
Malden Ct. KT3: N Mald 69Xa 132
Malden Ct. N4 30Sb 51
Malden Cres. NW1 37Jb 70
Malden Flds. WD23: Bush 15Z 27
Malden Golf Course 68Ua 132
MALDEN GREEN 74Wa 154
Malden Grn. Av. KT4: Wor Pk 74Va 154
Malden Grn. M. KT4: Wor Pk 74Wa 154
Malden Hill KT3: N Mald 69Ua 132
Malden Hill Gdns. KT3: N Mald 69Va 132
Malden Ho. WD19: Wat 20Y 27
MALDEN JUNC. 71Va 154
Malden Lodge WD17: Wat 12X 27
Malden Manor Station (Rail) 73Ua 154
Malden Pk. KT3: N Mald 72Va 154
Malden Pl. NW5 36Jb 70
Malden Rd. KT3: N Mald 71Ua 154
Malden Rd. KT4: Wor Pk 73Va 154
Malden Rd. NW5 36Hb 69
Malden Rd. SM3: Cheam 77Za 154
Malden Rd. WD17: Wat 12X 27
Malden Rd. WD6: Bore 13Qa 29
MALDEN RUSHETT 83La 172
Malden Way KT3: N Mald 73Ta 153
Maldon & District Society of Model Engineers 74Ja 152
Maldon Cl. E15 36Gc 73
Maldon Cl. N1 39Sb 71
Maldon Cl. SE5 55Ub 113
Maldon Ct. E6 39Qc 74
Maldon Ct. SM6: W'gton 78Lb 156
Maldon Rd. N9 20Vb 33
Maldon Rd. RM7: Rush G 31Ed 76
Maldon Rd. SM6: W'gton 78Kb 156
Maldon Rd. W3 45Sa 87
Maldon Wlk. IG8: Wfd G 23Lc 53
Malet Cl. TW20: Egh 65F 126
Malet Pl. WC1 6D 216 (42Mb 90)
Malet St. WC1 6D 216 (42Mb 90)
Maley Av. SE27 61Rb 135
Malford Ct. E18 26Jc 53
Malford Gro. E18 28Hc 53
Malfort Rd. SE5 55Ub 113
Malham Cl. N11 23Jb 50
Malham Rd. SE23 60Zb 114
Malham Rd. Ind. Est. SE23 60Zb 114
Malham Ter. N18 22Xb 51
Malibu Ct. SE26 62Xb 135
Malin Ct. HP3: Hem H 5L 3
Malins Cl. EN5: Barn 15Xa 30
Malkin Way WD18: Wat 14U 26
Mall, The AL2: Park 9A 6
Mall, The BR1: Broml 69Jc 137
(off Old Brewery Way)
Mall, The BR8: Swan 69Gd 140
Mall, The CR0: C'don 75Sb 157
Mall, The DA6: Bex 56Cd 118
Mall, The E15 38Fc 73
Mall, The HA3: Kenton 30Pa 47
Mall, The KT12: Hers 78Z 151
(off Hersham Grn. Shop. Cen.)
Mall, The KT6: Surb 71Ma 153
Mall, The N14 20Nb 32
Mall, The RM10: Dag 37Cd 76
Mall, The RM11: Horn 32Kd 77
(not continuous)
Mall, The RM17: Grays 51Ce 121
(off Grays Shop. Cen.)
Mall, The SW1 1C 228 (47Lb 90)
Mall, The SW14 57Sa 109
Mall, The TW8: Bford 51Ma 109
Mall, The W5 45Na 87
Mallams M. SW9 55Rb 113
Mallard Cl. DA1: Dart 57Pd 119
Mallard Cl. E9 37Bc 72
Mallard Cl. EN5: New Bar 16Fb 31
Mallard Cl. NW6 40Cb 69
Mallard Cl. RH1: Redh 3A 208
Mallard Cl. RM14: Upm 31Vd 78
Mallard Cl. SL1: Burn 10A 60
Mallard Cl. TW2: Whitt 59Ca 107
Mallard Cl. W7 47Ga 86
Mallard Ct. E17 27Fc 53
Mallard Ct. WD3: Rick 17M 25
(off Swan Cl.)
Mallard Dr. SL1: Slou 5D 80
Mallard Ho. NW8 2D 214 (40Gb 69)
(off Bridgeman St.)
Mallard Ho. SW6 53Eb 111
(off Station Ct.)
Mallard Path SE28 48Tc 94
Mallard Pl. N22 26Pb 50
Mallard Pl. TW1: Twick 62Ja 130
Mallard Point E3 41Dc 92
(off Rainhill Way)
Mallard Rd. CR2: Sels 82Zb 178
Mallard Rd. WD5: Ab L 3W 12
Mallards E11 31Jc 73
(off Blake Hall Rd.)
Mallards, The HP3: Hem H 7P 3
Mallards, The TW18: Lale 68K 127
Mallards Ct. WD19: Wat 20Ba 27
(off Hangar Ruding)
Mallards Reach KT13: Weyb 75T 150
Mallards Rd. IG11: Bark 42Wc 95
Mallards Rd. IG8: Wfd G 24Kc 53
Mallard Wlk. BR3: Beck 71Zb 158
Mallard Wlk. DA14: Sidc 65Yc 139
Mallard Way CM13: Hut 17De 41
Mallard Way HA6: Nwood 24S 44
Mallard Way NW9 31Sa 67
Mallard Way SM6: W'gton 81Lb 176
Mallard Way WD25: Wat 9Aa 13
Mall Chambers W8 46Cb 89
(off Kensington Mall)
Mallet Dr. UB5: N'olt 36Ba 65
Mallet Rd. SE13 58Fc 115
Mall Galleries 6E 222 (46Mb 90)
(off The Mall)
Malling SE13 57Dc 114
Malling Cl. CR0: C'don 72Yb 158
Malling Gdns. SM4: Mord 72Eb 155
Malling Way BR2: Hayes 73Hc 159
Mallinson Cl. RM12: Horn 36Ld 77
Mallinson Rd. CR0: Bedd 76Mb 156
Mallinson Rd. SW11 57Gb 111
Mallinson Sports Cen. 31Hb 69
Mallins Way RM16: Grays 9B 100
Mallion Ct. EN9: Walt A 5Hc 21
Mallon Gdns. E1 2K 225 (44Vb 91)
(off Commercial St.)
Mallord St. SW3 51Fb 111
Mallory Bldgs. EC1 6B 218 (42Rb 91)
(off St John St.)
Mallory Cl. DA12: Grav'nd 3E 144
Mallory Cl. E14 43Dc 92
Mallory Cl. SE4 56Ac 114
Mallory Ct. N17 23Vb 51
(off Cannon Rd.)
Mallory Ct. SE12 59Kc 115
Mallory Gdns. EN4: E Barn 17Jb 32
Mallory St. NW8 5D 214 (42Gb 89)
Mallory Way SE3 56Kc 115
Mallow Cl. CR0: C'don 74Zb 158
Mallow Cl. DA11: Nflt 3A 144
Mallow Cl. KT20: Tad 92Xa 194
Mallow Ct. RM17: Grays 51Fe 121
Mallow Mead NW7 24Ab 48
Mallows, The UB10: Ick 34R 64
Mallow St. EC1 5F 219 (42Tb 91)
Mall Rd. W6 50Xa 88
Mall Vs. W6 50Xa 88
(off Mall Rd.)
Mallys Pl. DA4: S Dar 67Sd 142
Malmains Cl. BR3: Beck 70Fc 137
Malmains Way BR3: Beck 70Ec 136
Malm Cl. WD3: Rick 19M 25
Malmesbury E2 40Yb 72
(off Cyprus St.)
Malmesbury Cl. HA5: Eastc 28V 44
Malmesbury Rd. E16 43Gc 93
Malmesbury Rd. E18 25Hc 53
Malmesbury Rd. E3 41Bc 92
Malmesbury Rd. SM4: Mord 73Eb 155
Malmesbury Ter. E16 43Hc 93
Malmes Cft. HP3: Hem H 4C 4
Malmo Twr. SE8 49Ac 92
Malmsey Ho. SE11 7J 229 (50Pb 90)
Malmsmead Ho. E9 36Bc 72
(off Homerton Rd.)
Malmstone Av. RH1: Mers 100Lb 196
Malory Cl. BR3: Beck 68Ac 136
Malpas Dr. HA5: Pinn 29Z 45
Malpas Rd. E8 36Xb 71
Malpas Rd. RM16: Grays 8E 100
Malpas Rd. RM9: Dag 37Zc 75
Malpas Rd. SE4 54Bc 114
Malpas Rd. SL2: Slou 5M 81
Malswick Ct. SE15 52Ub 113
(off Tower Mill Rd.)
Malta Rd. E10 32Cc 72
Malta Rd. RM18: Tilb 4B 122
Malta St. EC1 5C 218 (42Rb 91)
Maltby Cl. BR6: Orp 74Wc 161
Maltby Dr. EN1: Enf 10Xb 19
Maltby Ho. SE1 3K 231 (48Vb 91)
(off Maltby St.)
Maltby Ho. SE3 56Lc 115
Maltby Rd. KT9: Chess 79Qa 153
Maltby St. SE1 2K 231 (47Vb 91)
Malt Hill TW20: Egh 64A 126
Malt Ho. E17 29Bc 52
(off Old Brewery Way)
Malt Ho. Cl. SL4: Old Win 9M 103
Malthouse Ct. AL1: St A 3B 6
(off Sopwell La.)
Malthouse Ct. GU24: W End 4D 166
Malthouse Ct. TW8: Bford 51Na 109
(off High St.)
Malthouse Dr. RM17: Grays 50Be 99
Malthouse Dr. TW13: Hanw 64Z 129
Malthouse Dr. W4 51Va 110
Malthouse Fld. DA12: Grav'nd 10J 123
Malthouse La. DA12: Shorne 4N 145
Malthouse La. GU24: Pirb 5F 186
Malthouse La. GU24: W End 5D 166
Malthouse La. GU3: Worp 6G 186
Malthouse La. TW20: Egh 64C 126
Malthouse M. UB9: Hare 26L 43
Malthouse Pas. SW13 54Va 110
(off Clevelands Gdns.)
Malt Ho. Pl. RM1: Rom 29Gd 56
Malthouse Pl. WD7: R'lett 6Ja 14
Malthouse Rd. SW11 52Mb 112
Malthouse Rd. TN15: Ash 82Be 185
Malthouse Rd. TN15: Stans 82Be 185
Malthus Path SE28 46Yc 95
Malting Ho. E14 45Bc 92
(off Oak La.)
Malting La. RM16: Ors 2C 100
Maltings, The AL1: St A 2B 6
Maltings, The BR6: Orp 74Vc 161
Maltings, The DA11: Grav'nd 8C 122
(off West St.)
Maltings, The HP2: Hem H 1M 3
Maltings, The KT14: Byfl 85P 169
Maltings, The RH8: Oxt 3K 211
Maltings, The RM1: Rom 31Hd 76
Maltings, The TW18: Staines 63G 126
Maltings, The W4 50Qa 87
(off Spring Gro.)
Maltings, The WD4: Hunt C 6S 12
Maltings Arts Theatre 2B 6
Maltings Cl. E3 41Ec 92
Maltings Cl. SW13 54Va 110
Maltings Dr. CM16: Epp 1Wc 23
Maltings Ent. Cen., The DA12: Grav'nd 10H 123
Maltings La. CM16: Epp 1Wc 23
Maltings Lodge W4 51Ua 110
(off Corney Reach Way)
Maltings M. DA15: Sidc 62Wc 139
Maltings Pl. SE1 1J 231 (47Ub 91)
(off Roper La.)
Maltings Pl. SW6 53Db 111
Malting Way TW7: Isle 55Ha 108
Malt Kiln Pl. DA2: Dart 58Rd 119
Malt La. WD7: R'lett 7Ja 14
Malton Av. SL1: Slou 4F 80
Malton M. SE18 51Uc 116
Malton M. W10 44Ab 88
Malton Rd. W10 44Ab 88
Malton St. SE18 51Uc 116
Maltravers St. WC2 4J 223 (45Pb 90)
Malt Shovel Cotts. DA4: Eyns 76Md 163
Malt St. SE1 51Wb 113
Malus Cl. HP2: Hem H 1A 4
Malus Cl. KT15: Add 80H 149
Malus Dr. KT15: Add 80H 149
Malva Cl. SW18 57Db 111
Malvern Av. DA7: Bex 52Ad 117
Malvern Av. E4 24Fc 53
Malvern Av. HA2: Harr 34Aa 65
Malvern Cl. CR4: Mitc 69Lb 134
Malvern Cl. KT16: Ott 79E 148
Malvern Cl. KT6: Surb 74Na 153
Malvern Cl. SE20 68Wb 135
Malvern Cl. UB10: Ick 33Q 64
Malvern Cl. W10 43Bb 89
Malvern Cl. WD23: Bush 16Ea 28
Malvern Ct. KT18: Eps 86Ta 173
Malvern Ct. SL3: L'ly 51C 104
Malvern Ct. SM2: Sutt 80Cb 155
Malvern Ct. SW7 5C 226 (49Fb 89)
(off Onslow Sq.)
Malvern Ct. W12 47Wa 88
(off Hadyn Pk. Rd.)
Malvern Dr. IG3: Bark 35Vc 75
Malvern Dr. IG3: Ilf 35Vc 75
Malvern Dr. IG8: Wfd G 22Lc 53
Malvern Dr. TW13: Hanw 64Z 129
Malvern Gdns. HA3: Kenton 28Na 47
Malvern Gdns. IG10: Lough 16Pc 36
Malvern Gdns. NW2 33Ab 68
Malvern Ho. CR8: Kenley 86Rb 177
(off Foxley Rd.)
Malvern Ho. DA11: Nflt 58Fe 121
(off Laburnum Gro.)
Malvern Ho. N16 32Vb 71
Malvern Ho. SE17 7E 230 (50Sb 91)
(off Liverpool Gro.)
Malvern Ho. WD18: Wat 16T 26
Malvern M. NW6 41Cb 89
Malvern Pl. NW6 41Bb 89
Malvern Rd. BR6: Chels 77Xc 161
Malvern Rd. CR7: Thor H 70Qb 134
Malvern Rd. E11 33Gc 73
Malvern Rd. E6 39Nc 74
Malvern Rd. E8 38Wb 71
Malvern Rd. EN3: Enf W 9Ac 20
Malvern Rd. KT6: Surb 75Na 153
Malvern Rd. N17 27Wb 51
Malvern Rd. N8 27Qb 50
Malvern Rd. NW6 40Bb 69
(not continuous)
Malvern Rd. RM11: Horn 30Jd 56
Malvern Rd. RM17: Grays 9A 100
Malvern Rd. TW12: Hamp 66Ca 129
Malvern Rd. UB3: Harl 52U 106
Malvern Ter. N1 39Qb 70
Malvern Ter. N9 18Vb 33
Malvern Way W13 43Ka 86
Malvern Way WD3: Crox G 15R 26
Malvina Av. DA12: Grav'nd 1D 144
Malwood Rd. SW12 58Kb 112
Malyons, The TW17: Shep 72T 150
Malyons Rd. BR8: Hext 66Hd 140
Malyons Rd. SE13 58Dc 114
Malyons Ter. SE13 57Dc 114
Managers St. E14 46Ec 92
Manan Cl. HP3: Hem H 4C 4
Manatee Pl. SM6: Bedd 76Mb 156
Manaton Cl. SE15 55Xb 113
Manaton Cres. UB1: S'hall 44Ca 85
Manbey Gro. E15 37Gc 73
Manbey M. E15 37Gc 73
Manbey Pk. Rd. E15 37Gc 73
Manbey Rd. E15 37Gc 73
Manbey St. E15 37Gc 73
Manbre Rd. W6 51Ya 110
Manbrough Av. E6 41Pc 94
Manby Wlk. E17 29Ac 52
Manchester Ct. E16 44Kc 93
(off Garvary Rd.)
Manchester Dr. W10 42Ab 88
Manchester Gro. E14 50Ec 92
Manchester Ho. SE17 7E 230 (50Sb 91)
(off East St.)
Manchester M. W1 1H 221 (43Jb 90)
(off Manchester St.)
Manchester Rd. CR7: Thor H 69Sb 135
Manchester Rd. E14 50Ec 92
Manchester Rd. N15 30Tb 51
Manchester Sq. W1 2H 221 (44Jb 90)
Manchester St. W1 1H 221 (43Jb 90)
Manchester Way RM10: Dag 35Dd 76
Manchuria Rd. SW11 58Jb 112
Manciple St. SE1 2F 231 (47Tb 91)
Mancroft Ct. NW8 39Fb 69
(off St John's Wood Pk.)
Mandalay Rd. SW4 57Lb 112
Mandara Pl. SE8 49Ac 92
(off Yeoman St.)
Mandarin Ct. NW10 37Ta 67
(off Mitchellbrook Way)
Mandarin Ct. SE8 51Bc 114
Mandarin St. E14 45Cc 92
Mandarin Way UB4: Yead 44Z 85
Mandarin Wharf N1 39Ub 71
(off De Beauvoir Cres.)
Mandela Cl. NW10 38Sa 67
Mandela Cl. W12 45Xa 88
Mandela Ct. UB8: Cowl 43L 83
Mandela Ho. E2 4K 219 (41Vb 91)
(off Virginia Rd.)
Mandela Ho. SE5 54Rb 113
Mandela Pl. WD24: Wat 12Z 27
Mandela Rd. E16 44Jc 93
Mandela St. NW1 39Lb 70
Mandela St. SW9 52Qb 112
(not continuous)
Mandela Way SE1 5H 231 (49Ub 91)
Mandeley M. E11 32Hc 73
Mandel Ho. SW18 56Db 111
Manderley W14 48Bb 89
(off Oakwood La.)
Mandeville Cl. SE3 52Hc 115
Mandeville Cl. SW20 66Ab 132
Mandeville Cl. WD17: Wat 10V 12
Mandeville Ct. E4 22Ac 52
Mandeville Ct. TW20: Egh 63C 126
Mandeville Dr. AL1: St A 5B 6
Mandeville Dr. KT6: Surb 74Ma 153
Mandeville Ho. SE1 7K 231 (50Vb 91)
(off Rolls Rd.)
Mandeville Ho. SW4 57Lb 112
Mandeville M. SW4 56Nb 112
Mandeville Pl. W1 2J 221 (44Jb 90)
Mandeville Rd. EN3: Enf W 8Ac 20
Mandeville Rd. EN6: Pot B 4Eb 17
Mandeville Rd. N14 19Kb 32
Mandeville Rd. TW17: Shep 71Q 150
Mandeville Rd. TW7: Isle 54Ja 108
Mandeville Rd. UB5: N'olt 38Ca 65
Mandeville St. E5 34Ac 72
Mandeville Wlk. CM13: Hut 17Fe 41
Mandir La. IG7: Chig 18Sc 36
Mandrake Rd. SW17 62Hb 133
Mandrake Way E15 38Gc 73
Mandrell Rd. SW2 57Nb 112
Manesty Ct. N14 17Mb 32
(off Ivy Rd.)
Manet Gdns. W3 45Ua 88
Manette St. W1 3E 222 (44Mb 90)
Manfield Cl. SL2: Slou 1E 80
Manford Cl. IG7: Chig 21Wc 55
Manford Ct. IG7: Chig 22Vc 55
(off Manford Way)
Manford Cross IG7: Chig 22Wc 55
Manford Ind. Est. DA8: Erith 51Jd 118
Manford Way IG7: Chig 22Uc 54
Manfred Rd. SW15 57Bb 111
Manger Rd. N7 37Nb 70
Mangold Way DA18: Erith 48Zc 95
Manhattan Av. WD18: Wat 14V 26
Manhattan Bldg. E3 40Cc 72
Manhattan Bus. Pk. W5 41Na 87
Manhattan Loft Gdns. E20 37Ec 72
Manilla Ct. RM6: Chad H 30Xc 55
(off Quarles Pk. Rd.)
Manilla St. E14 47Cc 92
Manilla Wlk. SE10 49Gc 93
Manister Rd. SE2 48Wc 95
Manitoba Ct. SE16 47Yb 92
(off Canada Est.)
Manitoba Gdns. BR6: Chels 79Vc 161
Manley Ct. N16 34Vb 71
Manley Ho. SE11 6K 229 (49Qb 90)
Manley Rd. HP2: Hem H 1N 3
Manley St. NW1 39Jb 70
Manly Dixon Dr. EN3: Enf W 9Ac 20
Manna Ho. E20 37Ec 72
(off Glade Wlk.)
Mannamead KT18: Eps D 91Ua 194
Mannamead Cl. KT18: Eps D 91Ua 194
Mannan Ho. E3 40Bc 72
(off Roman Rd.)
Mann Cl. CR0: C'don 76Sb 157
Manneby Prior N1 2J 217 (40Pb 70)
(off Cumming St.)
Mannequin Ho. E17 27Zb 52
Manning Ct. SE28 46Xc 95
(off Titmuss Av.)
Manning Ct. WD19: Wat 16Z 27
Manningford Cl. EC1 3B 218 (41Rb 91)
Manning Gdns. CR0: C'don 73Xb 157
Manning Gdns. HA3: Kenton 31Ma 67
Manning Ho. W11 44Ab 88
(off Westbourne Pk. Rd.)
Manning Pl. TW10: Rich 58Pa 109
Manning Rd. BR5: St M Cry 71Zc 161
Manning Rd. E17 29Ac 52
Manning Rd. RM10: Dag 37Cd 76
Manning St. RM15: Avel 46Sd 98
Manningtree Cl. SW19 60Ab 110
Manningtree Rd. HA4: Ruis 35X 65
Manningtree St. E1 44Wb 91
Mannin Rd. RM6: Chad H 31Xc 75
Mannock Cl. NW9 27Ta 47
Mannock Dr. IG10: Lough 12Sc 36
Mannock M. E18 25Lc 53
Mannock Rd. DA1: Dart 55Pd 119
Mannock Rd. N22 27Rb 51
Mann's Cl. TW7: Isle 57Ha 108
Manns Rd. HA8: Edg 23Qa 47
Manns Ter. SE27 62Sb 135

Manny Shinwell Ho. SW6 51Bb 111
(off Clem Attlee Ct.)
Manoel Rd. TW2: Twick 61Ea 130
Manor, The IG8: Wfd G 24Qc 54
Manor Av. CR3: Cat'm 96Ub 197
Manor Av. HP3: Hem H 5M 3
Manor Av. RM11: Horn 29Ld 57
Manor Av. SE4 54Bc 114
Manor Av. TW4: Houn 55Z 107
Manor Av. UB5: N'olt 38Ba 65
Manorbrook SE3 56Jc 115
Manor Chase KT13: Weyb 78R 150
MANOR CIRCUS 55Qa 109
Manor Cl. CR6: W'ham 89Ac 178
Manor Cl. DA1: Cray 56Fd 118
Manor Cl. DA12: Grav'nd 1K 145
Manor Cl. DA2: Wilm 62Jd 140
Manor Cl. E17 25Ac 52
Manor Cl. EN5: Barn 14Ab 30
Manor Cl. GU22: Pyr 89H 169
Manor Cl. HA4: Ruis 32V 64
Manor Cl. KT24: E Hor 100U 190
Manor Cl. KT4: Wor Pk 74Ua 154
Manor Cl. NW7 22Ta 47
Manor Cl. NW9 29Ra 47
Manor Cl. RH9: S God 10D 210
Manor Cl. RM1: Rom 29Jd 56
Manor Cl. RM10: Dag 37Fd 76
Manor Cl. RM15: Avel 46Sd 98
Manor Cl. SE28 45Yc 95
Manor Cl. Sth. RM15: Avel 46Sd 98
Manor Cotts. HA6: Nwood 25V 44
Manor Cotts. N2 26Eb 49
(off Manor Cotts. App.)
Manor Cotts. WD3: Chor 16D 24
Manor Cotts. App. N2 26Eb 49
Manor Ct. BR4: W W'ck 74Dc 158
Manor Ct. DA13: Sole S 10F 144
Manor Ct. DA7: Bex 56Dd 118
Manor Ct. E10 32Dc 72
Manor Ct. E4 18Gc 35
Manor Ct. EN1: Enf 8Xb 19
Manor Ct. EN6: Pot B 4Bb 17
Manor Ct. EN8: Chesh 3Zb 20
Manor Ct. HA1: Harr 30Ha 46
Manor Ct. HA9: Wemb 36Na 67
Manor Ct. IG11: Bark 38Vc 75
Manor Ct. KT13: Weyb 77R 150
Manor Ct. KT2: King T 67Qa 131
Manor Ct. KT8: W Mole 70Ca 129
Manor Ct. N14 19Mb 32
Manor Ct. N2 29Hb 49
Manor Ct. N20 20Hb 31
(off York Way)
Manor Ct. SL1: Slou 6D 80
Manor Ct. SM5: Cars 76Jb 156
Manor Ct. SW16 62Nb 134
Manor Ct. SW2 57Pb 112
Manor Ct. SW3 7E 226 (50Gb 89)
(off Hemus Pl.)
Manor Ct. SW6 53Db 111
Manor Ct. TW18: Staines 64F 126
Manor Ct. TW2: Twick 61Ea 130
Manor Ct. UB9: Hare 26L 43
Manor Ct. W3 49Qa 87
Manor Ct. WD7: R'lett 10Ha 14
Manor Ct. Rd. W7 45Ga 86
Manor Cres. GU24: Brkwd 2B 186
Manor Cres. KT14: Byfl 85P 169
Manor Cres. KT19: Eps 84Qa 173
Manor Cres. KT5: Surb 72Qa 153
Manor Cres. RM11: Horn 29Ld 57
Manorcroft Pde. EN8: Chesh 2Zb 20
Manorcrofts Rd. TW20: Egh 65C 126
Manor Dene SE28 44Yc 95
Manordene Cl. KT7: T Ditt 74Ja 152
Manordene Rd. SE28 44Yc 95
Manor Dr. AL2: Chis G 9N 5
Manor Dr. DA3: Hartl 72Ce 165
Manor Dr. HA9: Wemb 35Pa 67
Manor Dr. KT10: Hin W 75Ha 152
Manor Dr. KT15: New H 82J 169
Manor Dr. KT19: Ewe 79Ua 154
Manor Dr. KT5: Surb 72Pa 153
Manor Dr. N14 18Kb 32
Manor Dr. N20 21Hb 49
Manor Dr. NW7 22Ta 47
Manor Dr. TW13: Hanw 64Z 129
Manor Dr. TW16: Sun 68W 128
Manor Dr., The KT4: Wor Pk 74Ua 154
Manor Dr. Nth. KT3: N Mald 73Ta 153
Manor Dr. Nth. KT4: Wor Pk 74Ua 154
Manor Est. SE16 49Xb 91
Manor Farm Ruislip 31U 64
Manor Farm TW20: Egh 64C 126
Manor Farm Av. TW17: Shep 72R 150
Manor Farm Cl. KT4: Wor Pk 74Ua 154
Manor Farm Cl. SL4: Wind 5D 102
Manor Farm Cotts. SL4: Old Win 7L 103
Manor Farm Cotts. TN15: Igh 93Wd 204
Manor Farm Ct. E6 41Pc 94
Manor Farm Ct. TW20: Egh 64C 126
Manor Farm Dr. E4 20Gc 35
MANOR FARM ESTATE 9N 103
Manor Farm Ho. SL4: Wind 5D 102
Manor Farm La. TW20: Egh 64C 126
Manor Farm Rd. EN1: Enf 7Xb 19
Manor Farm Rd. HA0: Wemb 40Ma 67
Manor Farm Rd. SW16 68Qb 134
Manor Fld. DA12: Shorne 4N 145
Manorfield Cl. N19 35Lb 70
(off Fulbrook M.)
Manor Flds. SW15 58Za 110
Manorfields Cl. BR7: Chst 69Vc 139
Manor Forstal DA3: Nw A G 76Be 165
Manor Gdns. CR2: S Croy 79Vb 157
Manor Gdns. HA4: Ruis 36Y 65
Manor Gdns. KT24: Eff 100Z 191
Manor Gdns. N7 34Nb 70
Manor Gdns. SW20 68Bb 133
Manor Gdns. SW4 54Lb 112
(off Larkhall Ri.)
Manor Gdns. TW12: Hamp 66Da 129
Manor Gdns. TW16: Sun 67W 128
Manor Gdns. TW9: Rich 56Pa 109
Manor Gdns. W3 49Qa 87
Manor Gdns. W4 50Ua 88
Manor Ga. UB5: N'olt 38Aa 65
Manor Ga. La. DA2: Wilm 62Jd 140
Manorgate Rd. KT2: King T 67Qa 131
Manor Grn. Rd. KT19: Eps 85Ra 173
Manor Gro. BR3: Beck 68Dc 136
Manor Gro. SE15 51Yb 114
Manor Gro. TW9: Rich 56Qa 109
Manor Gro. Av. EN8: Chesh 2Xb 19
Manor Hall IG7: Chig 22Sc 54
Manor Hall Av. NW4 26Za 48
Manor Hall Dr. NW4 26Za 48
Manorhall Gdns. E10 32Cc 72
Manor Hill SM7: Bans 87Hb 175
Manor Ho. DA3: Nw A G 76Be 165
Manor Ho. NW1 7E 214 (43Gb 89)
(off Lisson Gro.)
Manor Ho. SL4: Eton 1H 103
(off Common La.)
Manor Ho. UB2: S'hall 48Aa 85
MANOR HOUSE 31Sb 71
MANOR HOUSE 31Tb 71
Manor Ho., The KT20: Kgswd 95Eb 195
Manor Ho. Ct. KT18: Eps 85Sa 173
Manor Ho. Ct. TW17: Shep 73R 150
Manor Ho. Ct. W9 42Eb 89
(off Warrington Gdns.)
Manor Ho. Dr. HA6: Nwood 24R 44
Manor Ho. Dr. KT12: Hers 78V 150
Manor Ho. Dr. NW6 38Za 68
Manor Ho. Est. HA7: Stan 23Ka 46
Manor Ho. Gdn. E11 30Kc 53
Manor Ho. Gdns. WD5: Ab L 3T 12
Manor Ho. La. KT23: Bookh 98Aa 191
Manor Ho. La. SL3: Dat 3M 103
Manor House Station (Underground) 31Sb 71
Manor Ho. Way TW7: Isle 55Ka 108
Manor La. DA3: Fawk 73Yd 164
Manor La. DA3: Hartl 72Ce 165
Manor La. KT20: Lwr K 1J 207
Manor La. SE12 57Gc 115
Manor La. SE13 57Gc 115
Manor La. SL9: Ger X 1P 61
Manor La. SM1: Sutt 78Eb 155
Manor La. TN15: Ash 74Yd 164
Manor La. TW13: Felt 61W 128
Manor La. TW16: Sun 68W 128
Manor La. UB3: Harl 51T 106
Manor La. Ter. SE13 56Gc 115
Manor Leaze TW20: Egh 64D 126
Manor Lodge NW6 38Za 68
(off Willesden La.)
Manor M. NW6 40Cb 69
(off Cambridge Av.)
Manor M. SE4 54Bc 114
Manor Mt. SE23 60Yb 114
Manor Pde. HA1: Harr 30Ha 46
Manor Pde. N16 33Vb 71
Manor Pde. NW10 40Va 68
(off High St. Harlesden)
Manor Pk. BR7: Chst 68Tc 138
Manor Pk. DA8: Erith 51Jd 118
MANOR PARK 35Mc 73
Manor Pk. SE13 56Fc 115
MANOR PARK 3H 81
Manor Pk. TW13: Felt 61W 128
Manor Pk. TW18: Staines 62F 126
Manor Pk. TW9: Rich 56Pa 109
Manor Pk. Cl. BR4: W W'ck 74Dc 158
Manor Pk. Crematorium 35Lc 73
Manor Pk. Cres. HA8: Edg 23Qa 47
Manor Pk. Dr. HA2: Harr 27Da 45
Manor Pk. Gdns. HA8: Edg 22Qa 47
Manor Pk. Pde. SE13 56Fc 115
(off Lee High Rd.)
Manor Pk. Rd. BR4: W W'ck 74Dc 158
Manor Pk. Rd. BR7: Chst 67Sc 138
Manor Pk. Rd. E12 35Mc 73
(not continuous)
Manor Pk. Rd. N2 27Eb 49
Manor Pk. Rd. NW10 39Va 68
Manor Pk. Rd. SM1: Sutt 78Eb 155
Manor Park Station (Rail & Crossrail) 35Mc 73
Manor Pl. BR1: Broml 67Nc 138
Manor Pl. BR7: Chst 68Tc 138
Manor Pl. CR4: Mitc 69Lb 134
Manor Pl. DA1: Dart 60Nd 119
Manor Pl. KT12: Walt T 73V 150
(not continuous)
Manor Pl. KT20: Kgswd 93Bb 195
Manor Pl. KT23: Bookh 98Ca 191
Manor Pl. SE17 7C 230 (50Rb 91)
Manor Pl. SM1: Sutt 77Db 155
Manor Pl. TW14: Felt 60W 106
Manor Pl. TW18: Staines 64K 127
Manor Pl. Ind. Est. WD6: Bore 13Sa 29
Manor Rd. AL1: St A 1C 6
Manor Rd. AL2: Lon C 8G 6
Manor Rd. BR3: Beck 68Dc 136
Manor Rd. BR4: W W'ck 75Dc 158
Manor Rd. CR4: Mitc 70Lb 134
Manor Rd. DA1: Cray 56Gd 118
Manor Rd. DA10: Swans 58Zd 121
Manor Rd. DA12: Grav'nd 8D 122
Manor Rd. DA13: Sole S 10E 144
Manor Rd. DA15: Sidc 62Vc 139
Manor Rd. DA3: Long H 71Ee 165
Manor Rd. DA5: Bexl 60Dd 118
Manor Rd. DA8: Erith 51Hd 118
Manor Rd. E10 31Cc 72
Manor Rd. E15 40Gc 73
Manor Rd. E16 42Gc 93
Manor Rd. E17 26Ac 52
Manor Rd. EN2: Enf 12Sb 33
Manor Rd. EN5: Barn 14Ab 30
Manor Rd. EN6: Pot B 3Bb 17
Manor Rd. EN9: Walt A 5Fc 21
Manor Rd. GU21: Wok 8N 167
Manor Rd. GU23: Rip 95H 189
Manor Rd. HA1: Harr 30Ja 46
Manor Rd. HA4: Ruis 32T 64
Manor Rd. IG10: H Beech 11Kc 35
Manor Rd. IG10: Lough 16Kc 35
Manor Rd. IG11: Bark 37Vc 75
Manor Rd. IG7: Chig 22Sc 54
Manor Rd. IG8: Wfd G 23Pc 54
Manor Rd. KT12: Walt T 73V 150
Manor Rd. KT8: E Mos 70Fa 130
Manor Rd. N16 33Tb 71
Manor Rd. N17 25Wb 51
Manor Rd. N22 23Nb 50
Manor Rd. RH1: Mers 1C 208
Manor Rd. RH2: Reig 4H 207
Manor Rd. RM1: Rom 29Jd 56
Manor Rd. RM10: Dag 37Ed 76
Manor Rd. RM17: Grays 51Ee 121
Manor Rd. RM18: Tilb 4C 122
Manor Rd. RM20: W Thur 51Yd 120
Manor Rd. RM4: Abr 19Yc 37
Manor Rd. RM4: Stap A 19Yc 37
Manor Rd. RM6: Chad H 30Zc 55
Manor Rd. SE25 70Wb 135
Manor Rd. SL4: Wind 4C 102
Manor Rd. SM2: Cheam 80Bb 155
Manor Rd. SM6: W'gton 77Kb 156
Manor Rd. SS17: Stan H 2M 101
Manor Rd. SW20 68Bb 133
Manor Rd. TN14: Sund 96Zc 201
Manor Rd. TN15: W King 83Vd 184
Manor Rd. TN16: Tats 92Nc 200
Manor Rd. TW11: Tedd 64Ja 130
(not continuous)
Manor Rd. TW15: Ashf 64P 127
Manor Rd. TW2: Twick 61Ea 130
Manor Rd. TW9: Rich 56Qa 109
Manor Rd. UB3: Hayes 44W 84
Manor Rd. W13 45Ja 86
Manor Rd. WD17: Wat 11X 27
Manor Rd. Ho. HA1: Harr 30Ja 46
Manor Rd. Nth. KT10: Hin W 76Ha 152
Manor Rd. Nth. KT10: T Ditt 76Ha 152
Manor Rd. Nth. KT7: T Ditt 75Ja 152
Manor Rd. Nth. SM6: W'gton 77Kb 156
Manor Rd. Sth. KT10: Hin W 77Ga 152
Manorside EN5: Barn 14Ab 30
Manorside Cl. SE2 49Yc 95
Manor Sq. RM8: Dag 33Yc 75
Manor Va. TW8: Bford 50La 86
Manor Vw. DA3: Hartl 72Ce 165
Manor Vw. N3 26Db 49
Manorville Rd. HP3: Hem H 6L 3
Manor Wlk. KT13: Weyb 78R 150
Manorway EN1: Enf 17Ub 33
Manorway IG8: Wfd G 22Lc 53
Manor Way DA11: Nflt Botany Rd. 56Ce 121
Manor Way DA11: Nflt Pilgrims Rd. 56Ae 121
Manor Way BR2: Broml 72Nc 160
Manor Way BR3: Beck 68Cc 136
Manor Way BR5: Pet W 70Sc 138
Manor Way CM14: B'wood 20Wd 40
Manor Way CR2: S Croy 79Ub 157
Manor Way CR4: Mitc 69Lb 134
Manor Way CR8: Purl 84Nb 176
Manor Way DA10: Nflt 56Zd 121
Manor Way DA10: Swans 56Zd 121
Manor Way DA5: Bexl 60Cd 118
Manor Way DA7: Bex 55Fd 118
Manor Way E4 21Fc 53
Manor Way EN6: Pot B 2Cb 17
Manor Way EN8: Chesh 3Ac 20
Manor Way GU22: Wok 93D 188
Manor Way HA2: Harr 28Da 45
Manor Way HA4: Ruis 31U 64
Manor Way KT22: Oxs 87Ea 172
Manor Way KT4: Wor Pk 74Ua 154
Manor Way NW9 28Ua 48
Manor Way RM13: Rain 42Gd 96
Manor Way RM17: Grays 52De 121
Manor Way SE23 59Yb 114
Manor Way SE3 56Hc 115
Manor Way SM7: Bans 88Hb 175
Manor Way SS17: Stan H 1P 101
Manor Way TW20: Egh 65B 126
Manor Way UB2: S'hall 49Z 85
Manor Way WD3: Crox G 14Q 26
Manor Way WD6: Bore 13Sa 29
Manor Way, The SM6: W'gton 77Kb 156
Manorway, The SS17: Corr 1L 101
Manorway, The SS17: Cory 1L 101
Manorway, The SS17: Stan H 1L 101
Manor Way Bus. Cen. RM13: Rain 43Fd 96
Manor Way Bus. Pk. DA10: Nflt 57Be 121
Manor Waye UB8: Uxb 39M 63
Manor Wood Rd. CR8: Purl 85Nb 176
Manor Youth Cen., The 3D 102
Manpreet Ct. E12 36Pc 74
Manresa Rd. SW3 7D 226 (50Gb 89)
Mansard Beeches SW17 64Jb 134
Mansard Cl. HA5: Pinn 27Z 45
Mansard Cl. RM12: Horn 33Jd 76
Mansards, The AL1: St A 1C 6
Mansbridge Ho. SW8 53Kb 112
(off Patcham Ter.)
Manse Cl. UB3: Harl 51T 106
Mansel Cl. SL2: Slou 3M 81
Mansel Gro. E17 25Cc 52
Mansell Cl. SL4: Wind 3C 102
Mansell Rd. UB6: G'frd 43Da 85
Mansell Rd. W3 47Ua 88
Mansell St. E1 3K 225 (44Vb 91)
Mansell Way CR3: Cat'm 94Tb 197
Mansel Rd. SW19 65Ab 132
Manse Pde. BR8: Swan 70Jd 140
Manser Ct. RM13: Rain 41Gd 96
Mansergh Cl. SE18 52Nc 116
Manse Rd. N16 34Vb 71
Manser Rd. RM13: Rain 41Gd 96
Manse Way BR8: Swan 70Jd 140
Mansfield Av. EN4: E Barn 16Hb 31
Mansfield Av. HA4: Ruis 32X 65
Mansfield Av. N15 28Tb 51
Mansfield Cl. BR5: St M Cry 73Zc 161
Mansfield Cl. N9 16Wb 33
Mansfield Ct. E2 1K 219 (39Vb 71)
(off Whiston Rd.)
Mansfield Ct. SE15 52Vb 113
(off Sumner Rd.)
Mansfield Dr. RH1: Mers 100Mb 196
Mansfield Dr. UB4: Hayes 42U 84
Mansfield Gdns. RM12: Horn 33Md 77
Mansfield Hgts. N2 29Gb 49
Mansfield Hill E4 17Dc 34
Mansfield Ho. N1 39Ub 71
(off Halcomb St.)
Mansfield M. W1 1K 221 (43Kb 90)
Mansfield Pl. CR2: S Croy 79Tb 157
Mansfield Pl. EN6: Cuff 1Pb 18
Mansfield Pl. NW3 35Eb 69
Mansfield Rd. BR8: Hext 65Gd 140
Mansfield Rd. CR2: S Croy 79Tb 157
Mansfield Rd. E11 30Kc 53
Mansfield Rd. E17 28Bc 52
Mansfield Rd. IG1: Ilf 33Qc 74
Mansfield Rd. KT9: Chess 78La 152
Mansfield Rd. NW3 36Hb 69
Mansfield Rd. W3 42Ra 87
Mansfield St. W1 1K 221 (43Kb 90)
Mansford St. E2 40Wb 71
Manship Rd. CR4: Mitc 66Jb 134
Mansion Cvn. Site SL0: Iver 46E 82
Mansion Cl. SW9 53Qb 112
(not continuous)
Mansion Gdns. NW3 34Db 69
Mansion Ho. Dr. HA7: Stan 19Ga 28
Mansion Ho. Pl. EC4 3F 225 (44Tb 91)
Mansion House Station (Underground) 4E 224 (45Sb 91)
Mansion Ho. St. EC4 3F 225 (44Tb 91)
(off Poultry)
Mansion La. SL0: Iver 46E 82
Mansion Lock Ho. NW1 38Kb 70
(off Hawley Cres.)
Mansion Ri. DA10: Swans 59Be 121
Mansions, The SW5: Earl's Ct. Rd. 50Db 89
Mansions, The SW5: Old Brompton Rd. 50Db 89
(off Old Brompton Rd.)
Mansion Vw. E15 39Fc 73
(off High St.)
Manson Ho. N1 38Qb 70
(off Drummond Way)
Manson M. SW7 6B 226 (49Fb 89)
Manson Pl. SW7 6B 226 (49Fb 89)
Manstead Gdns. RM13: Rain 44Kd 97
Mansted Gdns. RM6: Chad H 31Yc 75
Manston N17 26Tb 51
(off Adams Rd.)
Manston NW1 38Lb 70
(off Agar Gro.)
Manston Av. UB2: S'hall 49Ca 85
Manston Cl. EN8: Chesh 2Yb 20
Manston Cl. SE20 67Yb 136
Manston Ct. E17 25Bc 52
Manstone Rd. NW2 36Ab 68
Manston Gro. KT2: King T 64Ma 131
Manston Ho. W14 48Ab 88
(off Russell Rd.)
Manston Way AL4: St A 3H 7
Manston Way RM12: Horn 37Kd 77
Manthorpe Rd. SE18 50Sc 94
Mantilla Rd. SW17 63Jb 134
Mantle Cl. SL4: Wind 5B 102
Mantle Ct. SW18 58Db 111
(off Mapleton Rd.)
Mantle Rd. SE4 55Ac 114
Mantlet Cl. SW16 66Lb 134
Mantle Way E15 38Gc 73
Manton Av. W7 47Ha 86
Manton Cl. UB3: Hayes 45U 84
Manton Rd. EN3: Enf L 9Cc 20
Manton Rd. SE2 49Wc 95
Manton Way EN3: Enf L 10Dc 20
Mantua St. SW11 55Fb 111
Mantus Cl. E1 42Yb 92
Mantus Rd. E1 42Yb 92
Manuka Cl. W7 46Ja 86
Manuka Hgts. E20 36Ec 72
(off Napa Cl.)
Manus Way N20 19Eb 31
Manville Gdns. SW17 62Kb 134
Manville Rd. SW17 61Jb 134
Manwell La. IG11: Bark 42Wc 95
Manwood Rd. SE4 57Bc 114
Manwood St. E16 46Pc 94
Manygate La. TW17: Shep 73S 150
Manygate Pk. TW17: Shep 72T 150
(off Mitre Cl.)
Manygates SW12 61Kb 134
Mapesbury Ct. NW2 36Ab 68
Mapesbury M. NW4 30Wa 48
Mapesbury Rd. NW2 38Ab 68
Mapeshill Pl. NW2 37Ya 68
Mapes Ho. NW6 38Ab 68
Mape St. E2 42Xb 91
(not continuous)
Maple Av. E4 22Bc 52
Maple Av. HA2: Harr 33Da 65
Maple Av. RM14: Upm 34Rd 77
Maple Av. UB7: Yiew 45N 83
Maple Av. W3 46Ua 88
Maple Cl. AL10: Hat 1C 8
Maple Cl. BR5: Pet W 71Tc 160
Maple Cl. BR8: Swan 68Gd 140
Maple Cl. CM13: B'wood 20Be 41
Maple Cl. CM16: They B 9Tc 22
Maple Cl. CR3: Whyt 89Vb 177
Maple Cl. CR4: Mitc 67Kb 134
Maple Cl. HA4: Ruis 30X 45
Maple Cl. IG6: Ilf 22Uc 54
Maple Cl. IG9: Buck H 20Mc 35
Maple Cl. KT19: Eps 81Ta 173
Maple Cl. N16 30Wb 51
Maple Cl. N3 23Cb 49
Maple Cl. RM12: Horn 34Kd 77
Maple Cl. SW4 58Mb 112
Maple Cl. TW12: Hamp 65Ba 129
Maple Cl. UB4: Yead 41Z 85
Maple Cl. WD23: Bush 12Aa 27
Maple Ct. CR0: C'don Lwr. Coombe St. 77Sb 157
(off Lwr. Coombe St.)
Maple Ct. CR0: C'don The Waldrons 77Sb 157
(off The Waldrons)
Maple Ct. DA8: Erith 52Hd 118
Maple Ct. DA9: Ghithe 59Ud 120
Maple Ct. E3 40Cc 72
(off Four Seasons Cl.)
Maple Ct. E6 43Qc 94
Maple Ct. GU21: Wok 8N 167
Maple Ct. KT22: Lea 92Ha 192
Maple Ct. KT3: N Mald 69Ta 131
Maple Ct. SE6 60Dc 114
Maple Ct. SL4: Wind 5G 102
Maple Ct. SL9: Ger X 29B 42
Maple Ct. TW15: Ashf 66T 128
Maple Ct. TW20: Eng G 5M 125
Maple Ct. WD25: Wat 8Z 13
Maple Ct. WD6: Bore 14Qa 29
(off Drayton Rd.)
Maple Cres. DA15: Sidc 58Wc 117
Maple Cres. SL2: Slou 5M 81
Maplecroft Cl. E6 44Mc 93
MAPLE CROSS 22F 42
Maple Cross Ind. Est. WD3: Map C 21H 43
Mapledale Av. CR0: C'don 75Wb 157
Mapledene BR7: Chst 65Sc 138
Mapledene Est. E8 38Wb 71
Mapledene Rd. E8 38Vb 71
Maple Dr. IG7: Chig 21Tc 54
Maple Dr. KT23: Bookh 97Da 191
Maple Dr. RM15: S Ock 42Qd 99
Maplefield AL2: Park 1Da 13
Maple Gdns. HA8: Edg 24Ua 48
Maple Gdns. KT17: Eps 85Ua 174
Maple Gdns. TW19: Stanw 61N 127
Maple Ga. IG10: Lough 12Qc 36
Maple Grn. HP1: Hem H 1G 2
Maple Gro. GU22: Wok 93A 188
Maple Gro. KT23: Bookh 99Ca 191
Maple Gro. NW9 31Sa 67
Maple Gro. TW8: Bford 52Ka 108
Maple Gro. UB1: S'hall 43Ba 85
Maple Gro. W5 48Ma 87
Maple Gro. WD17: Wat 11W 26
Maple Gro. Bus. Cen. TW4: Houn 56Y 107
MAPLE HILL 2A 10
Maple Ho. E17 27Dc 52
Maple Ho. EN6: Pot B 5Db 17
Maple Ho. HA9: Wemb 35Qa 67
(off Empire Way)
Maple Ho. KT1: King T 71Na 153
(off Maple Rd.)
Maple Ho. N19 34Lb 70
Maple Ho. RH1: Redh 6P 207
(off Chapel Rd.)
Maple Ho. SE8 52Bc 114
(off Idonia St.)
Maple Ho. TW9: Kew 53Ra 109
Maplehurst BR2: Broml 68Gc 137
Maplehurst KT22: Fet 95Fa 192
Maplehurst Cl. DA2: Wilm 61Gd 140
Maplehurst Cl. KT1: King T 70Na 131
Maple Ind. Est. TW13: Felt 62W 128
Maple Leaf Cl. TN16: Big H 88Mc 179
Maple Leaf Cl. WD5: Ab L 4W 12
Mapleleaf Cl. CR2: Sels 83Zb 178
Maple Leaf Dr. DA15: Sidc 60Vc 117
Mapleleafe Gdns. IG6: Ilf 27Rc 54
Maple Leaf Sq. SE16 47Zb 92
Maple Lodge W8 48Db 89
(off Abbots Wlk.)
Maple Lodge Cl. WD3: Map C 21G 42
Maple Lodge Nature Reserve 22H 43
Maple M. NW6 40Db 69
Maple M. SE16 47Zb 92
Maple M. SW16 64Pb 134
Maple Pl. KT20: Kgswd 96Bb 195
Maple Pl. N17 24Wb 51
Maple Pl. SM7: Bans 86Za 174
Maple Pl. UB7: Yiew 46N 83
Maple Pl. W1 6C 216 (42Lb 90)
Maple Rd. CR3: Whyt 89Vb 177
Maple Rd. DA1: Dart 60Ld 119
Maple Rd. DA12: Grav'nd 3E 144
Maple Rd. E11 30Gc 53
Maple Rd. GU23: Rip 96J 189
Maple Rd. KT21: Asht 91Ma 193
Maple Rd. KT6: Surb 72Ma 153
Maple Rd. RH1: Redh 10P 207
Maple Rd. RM17: Grays 51Ee 121
Maple Rd. SE20 67Xb 135
Maple Rd. UB4: Yead 41Y 85
Maples SS17: Stan H 1N 101
Maples, The DA3: Lfield 68De 143
Maples, The EN7: G Oak 1Ub 19
Maples, The EN9: Walt A 7Lc 21
Maples, The KT1: Hamp W 66La 130
Maples, The KT10: Clay 80Ja 152
Maples, The KT16: Ott 79D 148
Maples, The SM7: Bans 86Db 175
Maples, The WD6: Bore 11Qa 29
MAPLESCOMBE 78Sd 164
Maplescombe La. DA4: Farni 76Qd 163
Maples Pl. E1 43Xb 91
Maple Springs EN9: Walt A 5Jc 21
Maplestead Rd. RM9: Dag 39Xc 75
Maplestead Rd. SW2 59Pb 112
Maple St. E2 40Xb 71
Maple St. RM7: Rom 28Ed 56
Maple St. W1 7B 216 (43Lb 90)
Maplethorpe Rd. CR7: Thor H 70Qb 134
Mapleton Cl. BR2: Broml 72Jc 159
Mapleton Cres. EN3: Enf W 10Yb 20
Mapleton Cres. SW18 58Db 111
Mapleton Rd. E4 20Ec 34
Mapleton Rd. EN1: Enf 12Xb 33
Mapleton Rd. SW18 58Cb 111
(not continuous)
Maple Tree Pl. SE3 53Nc 116
Maple Wlk. SM2: Sutt 82Db 175
Maple Wlk. W10 41Za 88
Maple Way CR5: Coul 93Kb 196
Maple Way EN9: Walt A 1Kc 21
Maple Way SE16 47Ac 92
Maple Way TW13: Felt 62W 128
Maplewood Apts. N4 31Sb 71
(off Katherine Cl.)
Maplewood Ct. HA6: Nwood 22V 44
(off Eastbury Av.)
Maplewood Ct. TW15: Ashf 63N 127
Maplin Cl. N21 16Pb 32
Maplin Ho. SE2 47Xc 95
(off Wolvercote Rd.)
Maplin Pk. SL3: L'ly 47D 82
Maplin Rd. E16 44Jc 93
Maplin St. E3 41Bc 92
Mapperley Cl. E11 30Hc 53
Mapperley Dr. IG8: Wfd G 24Gc 53
Marabou Cl. E12 36Nc 74
Mara Ho. E20 36Dc 72
(off Victory Pde.)
Maran Way DA18: Erith 47Zc 95
Maraschino Apartment CR0: C'don 74Tb 157
(off Cherry Orchard Rd.)
Marathon Ho. NW1 7F 215 (43Hb 89)
(off Marylebone Rd.)
Marathon Way SE28 47Vc 95
Marbaix Gdns. TW7: Isle 53Fa 108
Marban Rd. W9 41Bb 89
Marbeck Cl. SL4: Wind 3B 102
Marble Arch W1 4F 221 (45Hb 89)
Marble Arch 4G 221 (45Hb 89)
(off Cumberland Ga.)
MARBLE ARCH 4F 221 (45Hb 89)
Marble Arch Apts. W1 2F 221 (44Hb 89)
(off Harrowby St.)
Marble Arch Station (Underground) 3G 221 (44Hb 89)
Marble Cl. W3 46Ra 87
Marble Dr. NW2 32Za 68
Marble Hill Cl. TW1: Twick 59Ka 108
Marble Hill Gdns. TW1: Twick 59Ka 108
Marble Hill House 59La 108
Marble Ho. SE18 50Vc 95
Marble Ho. W9 42Bb 89
Marble Quay E1 46Wb 91
Marbles Ho. SE5 51Sb 113
(off Grosvenor Ter.)
Marbles Way KT20: Tad 91Za 194
Marbrook Ct. SE12 62Lc 137
Marcella Rd. SW9 54Qb 112
Marcellina Way BR6: Orp 76Uc 160
Marcet Rd. DA1: Dart 57Ld 119
March NW9 25Va 48
(off Long Mead)
Marchant Cl. NW7 23Ua 48
Marchant Ho. N1 39Ub 71
(off Halcomb St.)
Marchant Rd. E11 33Fc 73
Marchant St. SE14 51Ac 114
Marchbank Rd. W14 51Bb 111
March Ct. SW15 56Xa 110
Marchmant Cl. RM12: Horn 34Ld 77
Marchmont Gdns. TW10: Rich 57Pa 109
Marchmont Rd. SM6: W'gton 80Lb 156
Marchmont Rd. TW10: Rich 57Pa 109
Marchmont St. WC1 5F 217 (42Nb 90)
March Rd. KT13: Weyb 78Q 150
March Rd. TW1: Twick 59Ja 108
Marchside Cl. TW5: Hest 53Z 107
Marchwood Cl. SE5 52Ub 113
Marchwood Cres. W5 44La 86
Marcia Ct. SE1 6J 231 (49Ub 91)
(off Marcia Rd.)
Marcia Ct. SL1: Slou 6D 80
Marcia Rd. SE1 6J 231 (49Ub 91)
Marcilly Rd. SW18 57Fb 111
Marco Dr. HA5: Hat E 24Ba 45
Marcon Ct. E8 36Xb 71
(off Amhurst Rd.)
Marconi Gdns. CM15: Pil H 15Yd 40

Marconi Pl. N11 21Kb 50
Marconi Rd. DA11: Nflt 62Fe 143
Marconi Rd. E10 32Cc 72
Marconi Way AL4: St A 2H 7
Marconi Way UB1: S'hall 44Da 85
Marcon Pl. E8 37Xb 71
Marco Rd. W6 48Ya 88
Marcourt Lawns W5 42Na 87
Marcus Ct. E15 39Gc 73
Marcus Ct. GU22: Wok 90B 168
Marcuse Rd. CR3: Cat'm 95Tb 197
Marcus Garvey M. SE22 58Xb 113
Marcus Garvey Way SE24 56Qb 112
Marcus Rd. DA1: Dart 59Jd 118
Marcus St. E15 39Hc 73
Marcus St. SW18 58Db 111
Marcus Ter. SW18 58Db 111
Mardale Ct. NW7 24Wa 48
Mardale Dr. NW9 29Ta 47
Mardell Rd. CR0: C'don 71Zb 158
Marden Av. BR2: Hayes 72Jc 159
Marden Cl. IG7: Chig 19Xc 37
Marden Cres. CR0: C'don 72Pb 156
Marden Cres. DA5: Bexl 57Ed 118
Marden Ho. E8 36Xb 71
MARDEN PARK 95Ac 198
Marden Rd. CR0: C'don 72Pb 156
Marden Rd. N17 26Ub 51
Marden Rd. RM1: Rom 30Gd 56
Marden Sq. SE16 48Xb 91
Marder Rd. W13 47Ja 86
Mardon HA5: Hat E 24Ba 45
Mardyke Cl. RM13: Rain 40Ed 76
Mardyke Ho. SE17 5G 231 (49Tb 91)
(off Mason St.)
Mardyke Valley Golf Course 45Zd 99
Mardyke Vw. RM19: Purf 48Ud 98
Mardyke Wlk. RM16: Grays 46Ce 99
Marechal Niel Av. DA15: Sidc 62Tc 138
Marechal Niel Pde. DA14: Sidc 62Tc 138
(off Main Rd.)
Maresby Ho. E4 19Dc 34
Marescroft Rd. SL2: Slou 2C 80
Maresfield CR0: C'don 76Ub 157
Maresfield Gdns. NW3 36Eb 69
Mare St. E8 36Xb 71
Marfleet Cl. SM5: Cars 75Gb 155
Margaret Av. CM15: Shenf 17Be 41
Margaret Av. E4 16Dc 34
Margaret Barr Row DA10: Swans 59Ae 121
Margaret Bondfield Av. IG11: Bark 38Wc 75
Margaret Bondfield Ho. E3 40Ac 72
(off Driffield Rd.)
MARGARET CENTRE (HOSPICE) 30Gc 53
Margaret Cl. CM16: Epp 1Vc 23
Margaret Cl. EN6: Pot B 5Eb 17
Margaret Cl. EN9: Walt A 5Fc 21
Margaret Cl. RM2: Rom 29Kd 57
Margaret Cl. TW18: Staines 65M 127
Margaret Cl. WD5: Ab L 4V 12
Margaret Ct. EN4: E Barn 14Fb 31
Margaret Ct. W1 2B 222 (44Lb 90)
(off Margaret St.)
Margaret Dr. RM11: Horn 32Pd 77
Margaret Gardner Dr. SE9 61Pc 138
Margaret Herbison Ho. SW6 51Bb 111
(off Clem Attlee Ct.)
Margaret Ho. W6 50Ya 88
(off Queen Caroline St.)
Margaret Ho. WD5: Ab L 2U 12
Margaret Ingram Cl. SW6 51Bb 111
Margaret Lockwood Cl. KT1: King T 70Pa 131
Margaret McMillan Ho. E16 44Lc 93
Margaret McMillan Pk. 52Bc 114
Margaret Rd. CM16: Epp 1Wc 23
Margaret Rd. DA5: Bexl 58Zc 117
Margaret Rd. E11 30Gc 53
Margaret Rd. EN4: E Barn 14Fb 31
Margaret Rd. N16 32Vb 71
Margaret Rd. RM2: Rom 29Kd 57
Margaret Rutherford Pl. SW12 60Lb 112
Margarets Ct. HA8: Edg 22Ra 47
Margaret St. W1 2A 222 (44Kb 90)
Margaretta Ter. SW3 51Gb 111
Margaretting Rd. E12 32Lc 73
Margaret Way CR5: Coul 91Rb 197
Margaret Way IG4: Ilf 30Nc 54
Margaret White Ho. NW1 3D 216 (41Mb 90)
(off Chalton St.)
Margate Rd. SW2 57Nb 112
Margeholes WD19: Wat 19Aa 27
Margerie Ct. E2 40Xb 71
(off Esker Pl.)
MARGERY 1J 207
Margery Fry Ct. N7 34Nb 70
Margery Gro. KT20: Lwr K 1G 206
Margery La. KT20: Lwr K 1H 207
Margery Pk. Rd. E7 37Jc 73
Margery Rd. RM8: Dag 34Zc 75
Margery St. WC1 4K 217 (41Qb 90)
Margery Ter. E7 37Jc 73
(off Margery Pk. Rd.)
Margery Wood La. KT20: Lwr K 1H 207
Margery Wood La. KT20: Reig 1H 207
Margherita Pl. EN9: Walt A 6Hc 21
Margherita Rd. EN9: Walt A 6Jc 21
Margin Dr. SW19 64Za 132
Margravine Gdns. W6 50Za 88
Margravine Rd. W6 50Za 88
Marham Dr. NW9 25Ua 48
Marham Gdns. SM4: Mord 72Eb 155
Marham Gdns. SW18 60Gb 111
Mar Ho. NW9 27Ta 47
Maria Cl. SE1 49Xb 91
Maria Ct. SE25 68Ub 135
Mariam Gdns. RM12: Horn 33Pd 77
Marian Cl. RM16: N Stif 46Ae 99
Marian Cl. UB4: Yead 42Z 85
Marian Ct. E9 36Yb 72
Marian Ct. SM1: Sutt 78Db 155
Marian Gdns. BR1: Broml 66Lc 137
Marian Gdns. WD25: Wat 5X 13
Marian Lawson Ct. IG7: Chig 22Wc 55
Marianne Cl. SE5 53Ub 113
Marianne North Gallery 54Pa 109
Marian Pl. E2 40Xb 71
Marian Rd. SW16 67Lb 134
Marian St. E2 40Xb 71
Marian Way NW10 38Va 68
Maria Ter. E1 43Zb 92
Maria Theresa Cl. KT3: N Mald 71Ta 153
Maribor SE10 52Ec 114
(off Burney St.)
Maricas Av. HA3: Hrw W 25Fa 46
Marie Curie SE5 53Vb 113
MARIE CURIE HOSPICE Hampstead 36Fb 69
Marie Lloyd Gdns. N19 31Nb 70
Marie Lloyd Ho. N1 2F 219 (40Tb 71)
(off Murray Gro.)
Marie Lloyd Wlk. E8 37Vb 71
Marie Mnr. Way DA2: Dart 56Ud 120
Mariener Ct. E4 22Fc 53
Mariette Way SM6: W'gton 81Nb 176
Marigold All. SE1 5B 224 (45Rb 91)
Marigold Cl. UB1: S'hall 45Aa 85
Marigold Dr. GU24: Bisl 7E 166
Marigold Rd. N17 24Yb 52
Marigold St. SE16 47Xb 91
Marigold Way CR0: C'don 74Zb 158
MARILLAC NURSING HOME, THE 23Zd 59
Marina App. UB4: Yead 43Aa 85
Marina Av. KT3: N Mald 71Xa 154
Marina Cl. BR2: Broml 69Jc 137
Marina Cl. KT16: Chert 74L 149
Marina Ct. E3 41Bc 92
(off Alfred St.)
Marina Ct. EN9: Walt A 6Cc 20
Marina Dr. DA1: Dart 60Qd 119
Marina Dr. DA11: Nflt 9B 122
Marina Dr. DA16: Well 54Uc 116
Marina Gdns. EN8: Chesh 2Yb 20
Marina Gdns. RM7: Rom 29Ed 56
Marina One N1 1G 217 (40Nb 70)
(off New Wharf Rd.)
Marina Pde. TW18: Staines 66K 127
Marina Pl. KT1: Hamp W 67Ma 131
Marina Point E14 48Dc 92
(off Lanark Sq.)
Marina Point SW6 54Eb 111
Marina Vw. Ter. HP3: Hem H 6P 3
Marina Way SL0: Iver 45J 83
Marina Way SL1: Slou 5B 80
Marina Way TW11: Tedd 66Ma 131
Marine Cl. DA8: Erith 52Hd 118
Marine Ct. E11 33Gc 73
Marine Ct. RM19: Purf 49Pd 97
Marine Cres. IG6: Ilf 23Tc 54
Marine Dr. IG11: Bark 42Wc 95
Marine Dr. SE18 49Pc 94
Marinefield Rd. SW6 54Db 111
Marinel Ho. SE5 52Sb 113
Mariner Bus. Cen. CR0: Wadd 78Qb 156
Mariner Gdns. TW10: Ham 62La 130
Mariner Rd. E12 35Qc 74
Mariners Cl. EN4: E Barn 15Fb 31
Mariners Ct. DA9: Ghithe 56Xd 120
(off High St.)
Mariners M. E14 49Fc 93
Mariners Pl. SE16 49Ac 92
(off Plough Way)
Mariners Wlk. DA8: Erith 51Hd 118
Mariners Way DA11: Nflt 9A 122
Mariner Way HP2: Hem H 3A 4
Marine St. SE16 48Wb 91
Marine Twr. SE8 51Bc 114
(off Abinger Gro.)
Marion Av. TW17: Shep 71R 150
Marion Cl. IG6: Ilf 24Tc 54
Marion Cl. WD23: Bush 11Ba 27
Marion Cres. BR5: St M Cry 71Wc 161
Marion Gro. IG8: Wfd G 22Gc 53
Marion Ho. NW1 39Hb 69
(off Regent's Pk. Rd.)
Marion M. SE21 62Tb 135
Marion Rd. CR7: Thor H 71Sb 157
Marion Rd. NW7 22Wa 48
Marischal Rd. SE13 55Fc 115
Marisco Cl. RM16: Grays 9D 100
Marish Ct. SL3: L'ly 48C 82
Marish La. UB9: Den 29E 42
Marish Wharf SL3: L'ly 47A 82
Maritime Cl. DA9: Ghithe 57Xd 120
Maritime Ga. DA11: Nflt 9A 122
Maritime Ho. SE18 49Rc 94
Maritime Quay E14 50Cc 92
Maritime St. E3 42Bc 92
Maritime St. SE16 48Zb 92
Marius Mans. SW17 61Jb 134
Marius Rd. SW17 61Jb 134
Marjoram Cl. SS17: Stan H 1M 101
Marjorams Av. IG10: Lough 12Qc 36
Marjorie Fosters Way GU24: Brkwd 1B 186
Marjorie Gro. SW11 56Hb 111
Marjorie M. E1 44Zb 92
Markab Rd. HA6: Nwood 22V 44
Mark Av. E4 16Dc 34
Mark Cl. DA7: Bex 53Ad 117
Mark Cl. UB1: S'hall 45Da 85
Mark Dr. SL9: Chal P 21A 42
Marke Cl. BR2: Kes 77Nc 160
Markedge La. CR5: Coul 96Hb 195
Markedge La. RH1: Mers 97Hb 195
Markeston Grn. WD19: Wat 21Z 45
Market, The SM1: Sutt 74Eb 155
Market, The SM5: Cars 74Eb 155
Market All. DA12: Grav'nd 8D 122
Market App. W12 47Ya 88
Market Chambers EN2: Enf 13Tb 33
(off Church St.)
Market Cl. KT23: Bookh 97Ea 192
Market Ct. W1 2B 222 (44Lb 90)
(off Market Pl.)
Market Dr. W4 52Ua 110
Market Entrance SW8 52Lb 112
Market Est. N7 37Nb 70
Marketfield Rd. RH1: Redh 6P 207
Marketfield Way RH1: Redh 6P 207
Market Hall N22 26Qb 50
Market Hill SE18 48Qc 94
Market Ho. SL9: Chal P 25A 42
Market La. HA8: Edg 25Sa 47
Market La. SL0: Iver 47E 82
Market La. SL3: L'ly 48E 82
Market La. W12 47Ya 88
Market Link RM1: Rom 28Gd 56
Market Mdw. BR5: St M Cry 70Yc 139
Market M. W1 7K 221 (46Kb 90)
Market Oak La. HP3: Hem H 6A 4
Market Pde. BR1: Broml 67Jc 137
(off East St.)
Market Pde. DA14: Sidc 63Xc 139
Market Pde. E10 30Ec 52
(off High Rd. Leyton)
Market Pde. E17 27Bc 52
(off Higham Hill Rd.)
Market Pde. KT17: Ewe 81Va 174
(off High St.)
Market Pde. N16 32Wb 71
(off Oldhill St.)
Market Pde. N9 19Wb 33
(off Winchester Rd.)
Market Pde. SE25 70Wb 135
Market Pde. TW13: Hanw 62Aa 129
Market Pav. E10 34Cc 72
Market Pl. AL3: St A 2B 6
Market Pl. DA1: Dart 59Nd 119
Market Pl. DA6: Bex 56Cd 118
Market Pl. EN2: Enf 13Tb 33
Market Pl. KT1: King T 68Ma 131
Market Pl. N2 27Fb 49
Market Pl. RM1: Rom 29Gd 56
Market Pl. RM18: Tilb 4C 122
Market Pl. RM4: Abr 13Xc 37
Market Pl. SE16 49Wb 91
(not continuous)
Market Pl. SL3: Coln 52E 104
Market Pl. SL9: Chal P 25A 42
Market Pl. TW8: Bford 52La 108
Market Pl. UB1: S'hall 46Ba 85
Market Pl. W1 2B 222 (44Lb 90)
Market Pl. W3 46Sa 87
Market Pl., The NW11 28Eb 49
Market Rd. N7 37Nb 70
Market Rd. TW9: Rich 55Qa 109
Market Row SW9 56Qb 112
Market Sq. BR1: Broml 68Jc 137
(not continuous)
Market Sq. E14 44Dc 92
Market Sq. EN9: Walt A 5Ec 20
Market Sq. GU21: Wok 89A 168
Market Sq. KT1: King T 68Ma 131
(off Market Pl.)
Market Sq. TN16: Westrm 98Tc 200
Market Sq. TW18: Staines 64G 126
Market Sq. UB8: Uxb 38L 63
Market Sq., The N9 18Xb 33
(within Edmonton Grn. Shop. Cen.)
Market St. DA1: Dart 59Nd 119
Market St. E1 7K 219 (43Vb 91)
Market St. E6 40Pc 74
Market St. KT15: Add 77L 149
Market St. SE18 49Qc 94
Market St. SL4: Wind 3H 103
Market St. WD18: Wat 14X 27
Market Ter. TW8: Bford 51Na 109
(off Albany Rd.)
Market Trad. Est. UB2: S'hall 49X 85
Market Wlk. GU21: Wok 89A 168
Market Way E14 44Dc 92
Market Way HA0: Wemb 36Na 67
Market Way TN16: Westrm 98Tc 200
Market Yd. SE8 52Cc 114
Market Yd. M. SE1 2H 231 (47Ub 91)
Markfield CR0: Sels 82Bc 178
(not continuous)
Markfield Beam Engine & Mus. 29Wb 51
Markfield Gdns. E4 17Dc 34
Markfield Rd. CR3: Cat'm 98Xb 197
Markfield Rd. N15 28Wb 51
Markham Cl. WD6: Bore 12Pa 29
Markham Ho. RM10: Dag 34Cd 76
(off Uvedale Rd.)
Markham Pl. SW3 7F 227 (50Hb 89)
Markhams SS17: Stan H 1P 101
Markham Sq. SW3 7F 227 (50Hb 89)
Markham St. SE17 7H 231 (50Ub 91)
Markham St. SW3 7E 226 (50Gb 89)
Markhole Cl. TW12: Hamp 66Ba 129
Mark Ho. E2 40Zb 72
(off Sewardstone Rd.)
Markhouse Av. E17 30Ac 52
Markhouse Pas. E17 30Bc 52
(off Downsfield Rd.)
Markhouse Rd. E17 30Bc 52
Markland Ho. W10 45Za 88
(off Darfield Way)
Mark La. DA12: Grav'nd 9G 122
(not continuous)
Mark La. EC3 4J 225 (45Ub 91)
Mark Lodge EN4: Cockf 14Gb 31
(off Edgeworth Rd.)
Markmanor Av. E17 31Ac 72
Mark Oak La. KT22: Fet 94Ca 191
Mark Rd. N22 26Rb 51
Marksbury Av. TW9: Rich 55Qa 109
MARKS GATE 25Ad 55
MARKS GATE HEALTH CEN. 27Ad 55
(off Lawn Farm Gro.)
Mark Sq. EC2 5H 219 (42Ub 91)
Marks Rd. CR6: W'ham 90Ac 178
Marks Rd. RM7: Rom 29Ed 56
(not continuous)
Marks Sq. DA11: Nflt 3B 144
Markstone Ho. SE1 2B 230 (47Rb 91)
(off Lancaster St.)
Markstone Ter. BR6: Orp 73Wc 161
Mark St. E15 38Gc 73
Mark St. EC2 5H 219 (42Ub 91)
Mark St. RH2: Reig 5K 207
Mark Twain Dr. NW2 35Xa 68
Markville Gdns. CR3: Cat'm 97Wb 197
Mark Wade Cl. E12 32Mc 73
Mark Way BR8: Swan 71Jd 162
Markway TW16: Sun 68Y 129
Markwell Cl. SE26 63Xb 135
Markwick Av. EN8: Chesh 2Yb 20
Markyate Ho. W10 42Ya 88
(off Sutton Way)
Markyate Rd. RM8: Dag 36Xc 75
Marland Ho. SW1 3G 227 (48Hb 89)
(off Sloane St.)
Marlands Rd. IG5: Ilf 27Nc 54
Marlborough SW19 60Za 110
(off Inner Pk. Rd.)
Marlborough W9 3A 214 (41Eb 69)
(off Maida Vale)
Marlborough Av. E8 39Wb 71
(not continuous)
Marlborough Av. HA4: Ruis 30S 44
Marlborough Av. HA8: Edg 20Ra 29
Marlborough Av. N14 20Lb 32
Marlborough Bldgs. AL1: St A 2C 6
Marlborough Bus. Cen. KT16: Chert 75H 149
Marlborough Cl. BR6: Orp 72Vc 161
Marlborough Cl. KT12: Hers 76Z 151
Marlborough Cl. N20 20Hb 31
Marlborough Cl. RM14: Upm 32Ud 78
Marlborough Cl. RM16: Grays 47Ee 99
Marlborough Cl. SE17 6D 230 (49Sb 91)
Marlborough Cl. SW19 65Gb 133
Marlborough Ct. CR2: S Croy 79Ub 157
(off Birdhurst Rd.)
Marlborough Ct. EN1: Enf 15Ub 33
Marlborough Ct. HA1: Harr 28Fa 46
Marlborough Ct. HA6: Nwood 24V 44
Marlborough Ct. IG9: Buck H 19Lc 35
Marlborough Ct. N17 25Wb 51
(off Kemble Rd.)
Marlborough Ct. SM6: W'gton 80Lb 156
Marlborough Ct. TN16: Westrm 98Sc 200
(off Croydon Rd.)
Marlborough Ct. W1 4B 222 (45Lb 90)
(off Carnaby St.)
Marlborough Ct. W8 49Cb 89
(off Pembroke Rd.)
Marlborough Cres. TN13: S'oaks 96Gd 202
Marlborough Cres. UB3: Harl 52T 106
Marlborough Cres. W4 48Ta 87
Marlborough Dr. IG5: Ilf 27Nc 54
Marlborough Dr. KT13: Weyb 76S 150
Marlborough Dr. WD23: Bush 14Ba 27
Marlborough Flats SW3 5E 226 (49Gb 89)
(off Walton St.)
Marlborough Gdns. KT6: Surb 73Ma 153
Marlborough Gdns. N20 20Hb 31
Marlborough Gdns. RM14: Upm 32Td 78
Marlborough Ga. AL1: St A 2C 6
Marlborough Ga. Ho. W2 4B 220 (45Fb 89)
(off Elms M.)
Marlborough Gro. SE1 50Wb 91
Marlborough Hill HA1: Harr 28Fa 46
Marlborough Hill NW8 1A 214 (40Eb 69)
Marlborough Ho. E16 46Jc 93
(off Hardy Av.)
Marlborough Ho. UB7: W Dray 47P 83
(off Park Lodge Av.)
Marlborough Ho. WD23: Bush 14Ca 27
Marlborough House 7C 222 (46Lb 90)
Marlborough La. SE7 51Lc 115
Marlborough Lodge NW8 2A 214 (40Eb 69)
(off Hamilton Ter.)
Marlborough Mans. NW6 36Db 69
(off Canon Hill)
Marlborough M. SM7: Bans 87Cb 175
Marlborough M. SW2 56Pb 112
Marlborough Pde. UB10: Hil 42R 84
Marlborough Pk. Av. DA15: Sidc 59Wc 117
Marlborough Pl. NW8 40Eb 69
Marlborough Rd. AL1: St A 2C 6
Marlborough Rd. BR2: Broml 70Lc 137
Marlborough Rd. CM15: Pil H 16Wd 40
Marlborough Rd. CR2: S Croy 80Sb 157
Marlborough Rd. DA1: Dart 58Ld 119
Marlborough Rd. DA7: Bex 55Zc 117
Marlborough Rd. E15 35Gc 73
Marlborough Rd. E18 26Jc 53
Marlborough Rd. E4 23Dc 52
Marlborough Rd. E7 38Lc 73
Marlborough Rd. GU21: Wok 88C 168
Marlborough Rd. N19 33Mb 70
Marlborough Rd. N22 24Nb 50
Marlborough Rd. N9 18Wb 33
Marlborough Rd. RM7: Mawney 28Cd 56
Marlborough Rd. RM8: Dag 35Xc 75
Marlborough Rd. SE18 48Rc 94
Marlborough Rd. SE28 48Sc 94
Marlborough Rd. SL3: L'ly 9P 81
Marlborough Rd. SM1: Sutt 76Cb 155
Marlborough Rd. SW1 7C 222 (46Lb 90)
Marlborough Rd. SW19 65Gb 133
Marlborough Rd. TW10: Rich 58Pa 109
Marlborough Rd. TW12: Hamp 65Ca 129
Marlborough Rd. TW13: Felt 61Z 129
Marlborough Rd. TW15: Ashf 64M 127
Marlborough Rd. TW7: Isle 53Ka 108
Marlborough Rd. UB10: Hil 42R 84
Marlborough Rd. UB2: S'hall 48Y 85
Marlborough Rd. W4 50Sa 87
Marlborough Rd. W5 47Ma 87
Marlborough Rd. WD18: Wat 14X 27
Marlborough Sq. EN8: Chesh 2Xb 19
Marlborough St. SW3 6D 226 (49Gb 89)
Marlborough Yd. N19 33Mb 70
Marlbury NW8 39Db 69
(off Abbey Rd.)
Marld, The KT21: Asht 90Pa 173
Marle Gdns. EN9: Walt A 4Ec 20
Marler Ho. DA8: Erith 54Hd 118
Marler Rd. SE23 60Ac 114
Marlescroft Way IG10: Lough 15Rc 36
Marley Av. DA7: Bex 51Zc 117
Marley Cl. KT15: Add 79H 149
Marley Cl. N15 28Rb 51
Marley Cl. UB6: G'frd 41Ca 85
Marley Ho. E16 45Rc 94
(off University Way)
Marley Ho. W11 45Za 88
(off St Ann's Rd.)
Marley St. SE16 49Zb 92
Marley Wlk. NW2 36Ya 68
Marl Fld. Cl. KT4: Wor Pk 74Wa 154
Marlin Cl. TW16: Sun 65U 128
Marling Ct. TW12: Hamp 65Ba 129
Marlingdene Cl. TW12: Hamp 65Ca 129
MARLING PARK 66Ba 129
Marlings Cl. BR7: Chst 70Uc 138
Marlings Cl. CR3: Whyt 89Ub 177
Marlings Pk. Av. BR7: Chst 70Uc 138
Marling Way DA12: Grav'nd 5G 144
Marlin Ho. WD18: Wat 16T 26
Marlin Pk. TW14: Felt 57X 107
Marlins, The HA6: Nwood 22V 44
Marlins Cl. SM1: Sutt 78Eb 155
Marlins Cl. WD3: Chor 12G 24
Marlins Mdw. WD18: Wat 16T 26
Marlin Sq. WD5: Ab L 3V 12
Marloes Cl. HA0: Wemb 35Ma 67
Marloes Rd. W8 48Db 89
Marlow Av. RM19: Purf 49Qd 97
Marlow Cl. SE20 69Xb 135
Marlow Ct. N14 17Lb 32
Marlow Ct. NW6 38Za 68
Marlow Ct. NW9 27Va 48
Marlow Cres. TW1: Twick 58Ha 108
Marlow Dr. SM3: Cheam 75Za 154
Marlowe Cl. BR7: Chst 65Tc 138
Marlowe Cl. DA11: Nflt 61De 143
Marlowe Cl. IG6: Ilf 25Sc 54
Marlowe Ct. SE19 64Vb 135
Marlowe Ct. SW3 6E 226 (49Gb 89)
(off Petyward)
Marlowe Gdns. RM3: Rom 25Ld 57
Marlowe Gdns. SE9 58Qc 116
Marlowe Ho. IG8: Wfd G 24Qc 54
Marlowe Ho. KT1: King T 70Ma 131
(off Portsmouth Rd.)
Marlowe Path SE8 51Cc 114
Marlowe Rd. E17 28Ec 52
Marlowes HP1: Hem H 2M 3
Marlowes, The DA1: Cray 56Fd 118
Marlowes, The NW8 39Fb 69
Marlowes Cen., The 3M 3
Marlowe Sq. CR4: Mitc 70Lb 134
Marlowe Way CR0: Bedd 75Nb 156
Marlow Gdns. UB3: Harl 48T 84
Marlow Ho. E2 4K 219 (41Vb 91)
(off Calvert Av.)
Marlow Ho. KT5: Surb 71Na 153
(off Cranes Pk.)
Marlow Ho. SE1 3K 231 (48Vb 91)
(off Abbey St.)
Marlow Ho. TW11: Tedd 63Ja 130
Marlow Ho. W2 44Db 89
(off Hallfield Est.)
Marlow Rd. E6 41Pc 94
Marlow Rd. RM8: Dag 32Ad 75
Marlow Rd. SE20 69Xb 135
Marlow Rd. UB2: S'hall 48Ba 85
Marlow Way SE16 47Zb 92
Marlow Workshops E2 4K 219 (41Vb 91)
(off Virginia Rd.)
Marlpit Av. CR5: Coul 89Nb 176
Marlpit La. CR5: Coul 88Mb 176
Marl Rd. SW18 56Eb 111
Marlston NW1 4A 216 (41Kb 90)
(off Munster Sq.)
Marlton St. SE10 50Hc 93
Marlu Ct. SE14 53Zb 114
(off Hatcham Pk. M.)
Marlu Ho. SE14 53Zb 114
(off Hatcham Pk. M.)
Marlwood Cl. DA15: Sidc 61Uc 138
Marlyon Rd. IG6: Ilf 22Xc 55
Marmadon Rd. SE18 49Vc 95
Marmara Apts. E16 45Jc 93
(off Western Gateway)
Marmion App. E4 21Cc 52
Marmion Av. E4 21Bc 52
Marmion Cl. E4 21Bc 52
Marmion M. SW11 55Jb 112
Marmion Rd. SW11 56Jb 112
Marmont Rd. SE15 53Wb 113
Marmora Rd. SE22 58Yb 114
Marmot Rd. TW4: Houn 55Z 107
Marncrest Cl. KT12: Hers 78X 151
Marne Av. DA16: Well 55Wc 117
Marne Av. N11 21Kb 50
Marnell Way TW4: Houn 55Z 107
Marnel Rd. RM9: Dag 39Xc 75
Marner Point E3 42Ec 92
Marne St. W10 41Ab 88
Marney Rd. SW11 56Jb 112
Marneys Cl. KT18: Eps 87Qa 173
Marnfield Cres. SW2 60Qb 112
Marnham Av. NW2 35Ab 68
Marnham Ct. HA0: Wemb 36La 66
Marnham Cres. UB6: G'frd 41Da 85
Marnham Ri. HP1: Hem H 1J 3
Marnock Ho. SE17 7F 231 (50Tb 91)
(off Brandon St.)
Marnock Rd. SE4 57Bc 114
Maroon St. E14 43Ac 92
Maroons Way SE6 64Cc 136
Marqueen Ct. W8 47Db 89
(off Kensington Chu. St.)
Marqueen Towers SW16 66Pb 134
Marquess Hgts. E18 25Kc 53
Marquess Rd. N1 37Tb 71
Marquis Cl. HA0: Wemb 38Pa 67
Marquis Ct. IG11: Bark 36Uc 74
Marquis Ct. KT1: King T 70Ma 131
(off Anglesea Rd.)
Marquis Ct. KT19: Eps 85Ta 173
Marquis Ct. N4 32Pb 70
(off Marquis Rd.)
Marquis Ct. TW19: Stanw 60N 105
Marquis Rd. N22 23Pb 50
Marquis Rd. N4 32Pb 70
Marquis Rd. NW1 37Mb 70
Marrabon Cl. DA15: Sidc 60Wc 117
Marram Ct. RM17: Grays 1A 122
Marrick Cl. SW15 56Wa 110
Marrick Ho. NW6 39Db 69
(off Mortimer Cres.)
Marriett Ho. SE6 63Ec 136
Marrilyne Av. EN3: Enf L 10Bc 20
Marriot Ct. UB3: Hayes 45U 84
(off Barra Hall Rd.)
Marriott Cl. TW14: Felt 58T 106
Marriot Ter. WD3: Chor 14H 25
Marriott Lodge Cl. KT15: Add 77L 149
Marriott Rd. DA1: Dart 59Pd 119
Marriott Rd. E15 39Gc 73
Marriott Rd. EN5: Barn 13Za 30
Marriott Rd. N10 25Hb 49
Marriott Rd. N4 32Pb 70
Marriotts Cl. NW9 30Va 48
Marriotts Way HP3: Hem H 4M 3
Marriotts Wharf DA11: Grav'nd 7D 122
Mar Rd. RM15: S Ock 42Yd 98
Marrowells KT13: Weyb 76V 150
Marryat Cl. TW4: Houn 56Ba 107
Marryat Ho. SW1 7B 228 (50Lb 90)
(off Churchill Gdns.)
Marryat Pl. SW19 63Ab 132
Marryat Rd. EN1: Enf 7Xb 19
Marryat Rd. SW19 64Za 132
Marryat Sq. SW6 53Ab 110
Marsala Rd. SE13 56Dc 114
Marsalis Ho. E3 41Cc 92
(off Rainhill Way)
Marsault Ct. TW9: Rich 56Na 109
(off Kew Foot Rd.)
Marsden Gdns. DA1: Dart 54Pd 119
Marsden Rd. N9 19Xb 33
Marsden Rd. SE15 55Vb 113
Marsden St. NW5 37Jb 70
Marsden Way BR6: Orp 77Vc 161
Marshall Bldg. W2 1B 220 (43Fb 89)
(off Hermitage St.)
Marshall Cl. CR2: Sande 85Wb 177
Marshall Cl. CR3: Cat'm 96Sb 197
Marshall Cl. HA1: Harr 31Fa 66
Marshall Cl. SW18 58Eb 111
Marshall Cl. TW4: Houn 57Ba 107
Marshall Ct. NW6 38Za 68
(off Coverdale Rd.)
Marshall Ct. SE20 66Xb 135
(off Anerley Pk.)
Marshall Dr. UB4: Hayes 43V 84
Marshall Est. NW7 21Wa 48
Marshall Ho. N1 1G 219 (40Tb 71)
(off Cranston Est.)
Marshall Ho. SE1 4J 231 (48Ub 91)
(off Page's Wlk.)
Marshall Ho. SE17 7F 231 (50Tb 91)
(off East St.)
Marshall Pde. GU22: Pyr 87H 169
Marshall Path SE28 45Xc 95
Marshall Pl. KT15: New H 81L 169
Marshall Rd. E10 34Dc 72
Marshall Rd. N17 25Tb 51
Marshalls Cl. KT19: Eps 85Sa 173
Marshalls Cl. N11 21Kb 50
Marshalls Ct. AL1: St A 1F 6
Marshalls Dr. RM1: Rom 27Gd 56
Marshalls Gro. SE18 49Nc 94
Marshall's Pl. SE16 4K 231 (48Vb 91)
Marshall's Rd. SM1: Sutt 77Db 155
Marshalls Rd. RM7: Rom 28Fd 56
Marshall St. NW10 38Ta 67
Marshall St. W1 3C 222 (44Lb 90)
Marshall Street Leisure Cen. 3C 222 (44Lb 90)
Marshall Wlk. TW20: Eng G 2N 125
Marshalsea Rd. SE1 1E 230 (47Sb 91)

Marsham Cl. BR7: Chst 64Rc 138
Marsham Ct. SW1 5E 228 (49Mb 90)
Marsham Ho. SL9: Ger X 29B 42
Marsham La. SL9: Ger X 30A 42
Marsham Lodge SL9: Ger X 30A 42
Marsham St. SW1 4E 228 (48Mb 90)
Marsham Way SL9: Ger X 29A 42
Marsh Av. CR4: Mitc 68Hb 133
Marsh Av. KT19: Ewe 82Ua 174
Marshbrook Cl. SE3 55Mc 115
Marsh Cen., The E1 2K 225 (44Vb 91)
(off Whitechapel High St.)
Marsh Cl. EN8: Walt C 5Ac 20
Marsh Cl. KT15: Add 77K 149
Marsh Cl. NW7 20Va 30
Marsh Ct. E8 37Wb 71
Marsh Ct. SE17 7G 231 (50Tb 91)
(off Thurlow St.)
Marsh Ct. SW19 67Eb 133
Marshcroft Dr. EN8: Chesh 2Ac 20
Marsh Dr. NW9 30Va 48
Marshe Cl. EN6: Pot B 4Fb 17
Marsh Farm Rd. TW2: Twick 60Ha 108
Marshfield SL3: Dat 3N 103
Marshfield St. E14 48Ec 92
Marshfoot Cl. RM15: Avel 45Td 98
Marshfoot Rd. RM16: Grays 10C 100
Marshfoot Rd. RM17: Grays 10A 100
Marshgate Bus. Cen. E15 39Ec 72
Marshgate La. E15 39Dc 72
Marshgate La. E20 38Cc 72
Marshgate Path SE28 48Sc 94
Marsh Grn. Rd. RM10: Dag 39Cd 76
Marsh Hall HA9: Wemb 34Pa 67
Marsh Hill E9 36Ac 72
Marsh Ho. SW1 50Mb 90
(off Aylesford St.)
Marsh Ho. SW8 53Lb 112
Marsh La. E10 33Bc 72
Marsh La. HA7: Stan 22La 46
Marsh La. KT15: Add 77K 149
Marsh La. N17 24Xb 51
Marsh La. NW7 20Ua 30
MARSHMOOR 4E 8
Marshmoor Cres. AL9: Wel G 4F 8
Marshmoor La. AL9: Wel G 4E 8
Marsh Rd. HA0: Wemb 41Ma 87
Marsh Rd. HA5: Pinn 28Aa 45
Marshside Cl. N9 18Yb 34
Marsh St. DA1: Dart Bob Dunn Way 55Qd 119
(not continuous)
Marsh St. DA1: Dart Hilltop Gdns. 57Pd 119
Marsh St. E14 49Dc 92
Marsh St. Nth. DA1: Dart 55Qd 119
Marsh Vw. DA12: Grav'nd 10H 123
Marsh Wall E14 46Cc 92
Marsh Way RM13: Rain 41Fd 96
Marshwood Apts. SW11 56Gb 111
(off Eckstein Road)
Marshwood Ho. NW6 39Cb 69
(off Kilburn Vale)
Marshwood Rd. GU18: Light 3B 166
Marsland Cl. SE17 7C 230 (50Rb 91)
Marsom Ho. N1 2F 219 (40Tb 71)
(off Provost St.)
Marston KT19: Eps 83Sa 173
Marston Av. KT9: Chess 79Na 153
Marston Av. RM10: Dag 33Cd 76
Marston Cl. HP3: Hem H 3A 4
Marston Cl. NW6 38Eb 69
Marston Cl. RM10: Dag 34Cd 76
Marston Ct. DA9: Ghithe 56Wd 120
Marston Ct. KT12: Walt T 74X 151
Marston Dr. CR6: W'ham 90Ac 178
Marston Ho. SW9 54Qb 112
Marston Rd. GU21: Wok 9M 167
Marston Rd. IG5: Ilf 25Nc 54
Marston Rd. TW11: Tedd 64Ka 130
Marston Way SE19 66Rb 135
Marsworth Av. HA5: Pinn 25Z 45
Marsworth Cl. UB4: Yead 43Aa 85
Marsworth Cl. WD18: Wat 16U 26
Marsworth Ho. E2 39Wb 71
(off Whiston Rd.)
Marsworth Ho. HA0: Wemb 39Na 67
Martaban Rd. N16 33Vb 71
Martara M. SE17 7D 230 (50Sb 91)
Marta Rose Ct. SE20 68Xb 135
(off Wadhurst Cl.)
Martello Cl. RM17: Grays 51Fe 121
Martello St. E8 38Xb 71
Martello Ter. E8 38Xb 71
Martell Rd. SE21 62Tb 135
Martel Pl. E8 37Vb 71
Marten Rd. E17 26Cc 52
Martens Av. DA7: Bex 56Dd 118
Martens Cl. DA7: Bex 56Ed 118
Martham Cl. IG6: Ilf 25Rc 54
Martham Cl. SE28 45Zc 95
Martha Rd. E15 37Gc 73
Martha's Bldgs. EC1 5F 219 (42Tb 91)
Martha St. E1 44Yb 92
Marthorne Cres. HA3: Hrw W 26Fa 46
Martina Ter. IG7: Chig 22Uc 54
Martin Bowes Rd. SE9 55Pc 116
Martinbridge Trad. Est. EN1: Enf 15Wb 33
Martin Cl. AL10: Hat 2C 8
Martin Cl. CR2: Sels 83Zb 178
Martin Cl. CR6: W'ham 88Xb 177
Martin Cl. N9 18Zb 34
Martin Cl. SL4: Wind 3A 102
Martin Cl. UB10: Uxb 40N 63
Martin Ct. AL1: St A 2C 6
(off St Peter's St.)
Martin Ct. CR2: S Croy 78Ub 157
(off Birdhurst Rd.)
Martin Ct. E14 47Ec 92
(off River Barge Cl.)
Martin Cres. CR0: C'don 74Qb 156
Martindale SL0: Iver 42F 82
Martindale SW14 57Sa 109
Martindale Av. BR6: Chels 78Wc 161
Martindale Av. E16 45Jc 93
Martindale Ho. E14 45Dc 92
(off Poplar High St.)
Martin Dale Ind. Est. EN1: Enf 13Xb 33
Martindale Rd. GU21: Wok 10L 167
Martindale Rd. HP1: Hem H 1H 3
Martindale Rd. SW12 59Kb 112
Martindale Rd. TW4: Houn 55Aa 107
Martin Dene DA6: Bex 57Bd 118
Martin Dr. DA2: Dart 58Sd 120
Martin Dr. RM13: Rain 42Kd 97
Martin Dr. UB5: N'olt 36Ba 65
Martineau Cl. KT10: Esh 77Fa 152
Martineau Dr. TW1: Twick 56Ka 108
Martineau Est. E1 45Yb 92
Martineau Ho. SL9: Chal P 22A 42
Martineau Ho. SW1 7B 228 (50Lb 90)
(off Churchill Gdns.)
Martineau M. N5 35Rb 71
Martineau Rd. N5 35Rb 71
Martineau Sq. E1 45Wb 91
Martingale Cl. TW16: Sun 70W 128
Martingale Ho. E1 46Xb 91
(off Raine St.)
Martingales Cl. TW10: Ham 62Ma 131
Martin Gdns. RM8: Dag 35Yc 75
Martin Gro. SM4: Mord 69Cb 133
Martin Ho. DA11: Nflt 2C 144
Martin Ho. DA2: Dart 59Sd 120
Martin Ho. E3 39Bc 72
(off Old Ford Rd.)
Martin Ho. SE1 4E 230 (48Sb 91)
Martin Ho. SW8 52Nb 112
(off Wyvil Rd.)
Martini Dr. EN3: Enf L 9Cc 20
Martin Kinggett Gdns. RM9: Dag 39Ad 75
Martin La. EC4 4G 225 (45Tb 91)
(not continuous)
Martin Ri. DA6: Bex 57Bd 117
Martin Rd. DA2: Wilm 62Ld 141
Martin Rd. RM15: Avel 46Td 98
Martin Rd. RM8: Dag 35Yc 75
Martin Rd. SL1: Slou 8J 81
Martins, The HA9: Wemb 34Pa 67
Martins, The SE26 64Xb 135
Martins Cl. BR4: W W'ck 74Fc 159
Martins Cl. BR5: St P 69Zc 139
Martins Cl. SS17: Stan H 1M 101
Martins Cl. WD7: R'lett 8Ga 14
Martins Ct. AL1: St A 5F 6
Martins Dr. EN8: Chesh 1Ac 20
Martinsfield Cl. IG7: Chig 21Uc 54
Martin's Mt. EN5: New Bar 14Cb 31
Martins Pl. SE28 46Uc 94
Martin's Plain SL2: Stoke P 1K 81
Martin's Rd. BR2: Broml 68Gc 137
Martins Shaw TN13: Chips 94Ed 202
Martinstown Cl. RM11: Horn 30Qd 57
Martin St. SE28 46Uc 94
Martins Wlk. WD6: Bore Borehamwood 14Qa 29
Martins Wlk. N22: Noel Pk. 27Qb 50
Martins Wlk. N10 25Jb 50
Martins Wlk. SE28 46Uc 94
Martinsyde GU22: Wok 89E 168
Martin Way GU21: Wok 10L 167
Martin Way SM4: Mord 69Ab 132
Martin Way SW20 68Za 132
Martlands Ind. Est. GU22: Wok 5L 187
Martlesham N17 26Ub 51
(off Adams Rd.)
Martlesham Cl. RM12: Horn 36Ld 77
Martlesham Wlk. NW9 26Ua 48
Martlet Gro. UB5: N'olt 41Z 85
Martlett Ct. WC2 3G 223 (44Nb 90)
Martley Dr. IG2: Ilf 29Rc 54
Martock Cl. HA3: W'stone 28Ja 46
Martock Gdns. N11 22Hb 49
Marton Cl. SE6 62Cc 136
Marton Rd. N16 33Ub 71
Martynside NW9 25Va 48
Martyr Cl. AL1: St A 6B 6
MARTYR'S GREEN 92T 190
Martyr's La. GU21: Wok 92D 168
Martys Yd. NW3 35Fb 69
Marula Ho. E1 44Wb 91
(off Boulevard Walkway)
Marunden Grn. SL2: Slou 1D 80
Marvell Av. UB4: Hayes 43W 84
Marvell Ct. RM6: Chad H 30Xc 55
(off Quarles Pk. Rd.)
Marvell Ho. SE5 52Tb 113
(off Camberwell Rd.)
Marvels Cl. SE12 61Kc 137
Marvels La. SE12 61Kc 137
Marville Rd. SW6 52Bb 111
Marvin St. E8 37Xb 71
Marwell TN16: Westrm 98Rc 200
Marwell Cl. BR4: W W'ck 75Hc 159
Marwell Cl. RM1: Rom 29Jd 56
Marwood Cl. DA16: Well 55Xc 117
Marwood Cl. WD4: K Lan 1P 11
Marwood Dr. NW7 24Za 48
Marwood Sq. N10 28Jb 50
Mary Adelaide Cl. SW15 63Ua 132
Mary Ann Gdns. SE8 51Cc 114
Maryatt Av. HA2: Harr 33Da 65
Marybank SE18 49Pc 94
Mary Bayly Ho. W11 46Ab 88
(off Wilsham St.)
Mary Boast Wlk. SE5 54Tb 113
Mary Burrows Gdns. TN15: Kems'g 89Rd 183
Mary Cl. HA7: Stan 28Pa 47
Mary Datchelor Cl. SE5 53Tb 113
Mary Datchelor Ho. SE5 53Tb 113
(off Grove La.)
Mary Drew Almshouses TW20: Eng G 5P 125
Maryfield Cl. DA5: Bexl 62Gd 140
Mary Flux Ct. SW5 50Db 89
(off Bramham Gdns.)
Mary Grn. NW8 39Db 69
Maryhill Cl. CR8: Kenley 89Sb 177
Mary Holben Ho. SW16 64Lb 134
Mary Ho. W6 50Ya 88
(off Queen Caroline St.)
Mary Jones Ct. E14 45Cc 92
(off Garford St.)
Maryland AL10: Hat 1B 8
Maryland Ind. Est. E15 36Gc 73
Maryland Pk. E15 36Gc 73
(not continuous)
Maryland Point E15 37Gc 73
(off The Grove)
Maryland Rd. CR7: Thor H 67Rb 135
Maryland Rd. E15 36Fc 73
Maryland Rd. N22 23Pb 50
MARYLANDS INTERCHANGE 13Ee 41
Maryland Sq. E15 36Gc 73
Marylands Rd. W9 42Cb 89
(not continuous)
Maryland Station (Rail & Crossrail) 37Gc 73
Maryland St. E15 36Fc 73
Maryland Wlk. N1 39Sb 71
(off Popham St.)
Maryland Way TW16: Sun 68W 128
Mary Lawrenson Pl. SE3 52Jc 115
MARYLEBONE 1J 221 (43Jb 90)
Marylebone Cricket Club 3C 214 (41Fb 89)
(shown as Lord's)
Marylebone Fly-Over W2 1C 220 (43Fb 89)
MARYLEBONE FLYOVER 1D 220 (43Gb 89)
Marylebone Gdns. TW9: Rich 56Qa 109
Marylebone High St. W1 7J 215 (43Jb 90)
Marylebone La. W1 1J 221 (43Jb 90)
Marylebone M. W1 1K 221 (43Kb 90)
Marylebone Pas. W1 2C 222 (44Lb 90)
Marylebone Rd. NW1 7E 214 (43Gb 89)
Marylebone Station (Rail & Underground) 6F 215 (42Hb 89)
Marylebone St. W1 1J 221 (43Jb 90)
Mary Le Bow Way E3 43Dc 92
Marylee Way SE11 6J 229 (49Pb 90)
Mary Macarthur Ho. E2 41Zb 92
(off Warley St.)
Mary Macarthur Ho. RM10: Dag 34Cd 76
(off Wythenshawe Rd.)
Mary Macarthur Ho. W6 51Ab 110
Mary Morgan Ct. SL2: Slou 2H 81
Mary Neuner Rd. N22 27Pb 50
Mary Neuner Rd. N8 27Pb 50
Maryon Gro. SE7 49Nc 94
Maryon Ho. NW6 38Eb 69
(off Goldhurst Ter.)
Maryon M. NW3 35Gb 69
Maryon Rd. SE18 49Nc 94
Maryon Rd. SE7 49Nc 94
Mary Peters Dr. UB6: G'frd 36Fa 66
Mary Pl. W11 45Ab 88
Mary Rose Cl. RM16: Chaf H 49Yd 98
Mary Rose Cl. TW12: Hamp 67Ca 129
Mary Rose Mall E6 43Pc 94
Mary Rose Sq. SE16 49Ac 92
(off Cary Av.)
Maryrose Way N20 18Fb 31
Marys Ct. NW1 5E 214 (42Gb 89)
(off Palgrave Gdns.)
Mary Seacole Cl. E8 39Vb 71
Mary Seacole Ho. W6 48Wa 88
(off Invermead Cl.)
Maryside SL3: L'ly 47A 82
Mary Smith Ct. SW5 49Cb 89
(off Trebovir Rd.)
Marysmith Ho. SW1 7E 228 (50Mb 90)
(off Cureton St.)
Mary's Ter. TW1: Twick 59Ja 108
Mary St. E16 43Hc 93
Mary St. N1 39Sb 71
Mary Ter. NW1 1A 216 (39Kb 70)
Maryville DA16: Well 54Vc 117
Mary Wallace Theatre Twickenham 60Ja 108
Mary Way WD19: Wat 21Y 45
Mary Wharrie Ho. NW3 38Hb 69
(off Fellows Rd.)
Marzell Ho. W14 50Bb 89
(off North End Rd.)
Marzena Ct. TW3: Houn 58Ea 108
Masbro' Rd. W14 48Za 88
Mascalls Ct. SE7 51Lc 115
Mascalls Gdns. CM14: B'wood 21Vd 58
Mascalls La. CM13: Gt War 23Xd 58
Mascalls La. CM14: B'wood 21Vd 58
Mascalls La. CM14: Gt War 21Vd 58
Mascalls Rd. SE7 51Lc 115
Mascoll Path SL2: Slou 1D 80
Mascotte Rd. SW15 56Za 110
Mascotts Cl. NW2 34Xa 68
Masefield Av. HA7: Stan 22Ha 46
Masefield Av. UB1: S'hall 45Ca 85
Masefield Av. WD6: Bore 15Ra 29
Masefield Cl. DA8: Erith 53Hd 118
Masefield Cl. RM3: Rom 25Kd 57
Masefield Ct. CM14: W'ley 21Yd 58
Masefield Ct. EN5: New Bar 14Eb 31
Masefield Ct. KT6: Surb 73Ma 153
Masefield Cres. N14 15Lb 32
Masefield Cres. RM3: Rom 25Ld 57
Masefield Dr. RM14: Upm 31Sd 78
Masefield Gdns. E6 42Qc 94
Masefield Ho. NW6 41Cb 89
(off Stafford Rd.)
Masefield La. UB4: Yead 42X 85
Masefield Rd. DA1: Dart 57Rd 119
Masefield Rd. DA11: Nflt 62Fe 143
Masefield Rd. RM16: Grays 7A 100
Masefield Rd. TW12: Hamp 63Ba 129
Masefield Vw. BR6: Farnb 76Sc 160
Masefield Way TW19: Stanw 60P 105
Masey M. SW2 57Qb 112
Mashie Rd. W3 44Ua 88
Mashiters Hill RM1: Rom 25Fd 56
Mashiters Wlk. RM1: Rom 27Gd 56
Masjid La. E14 43Bc 92
Maskall Cl. SW2 60Qb 112
Maskani Wlk. SW16 66Lb 134
Maskell Rd. SW17 62Eb 133
Maskelyne Cl. SW11 53Gb 111
Maslen Rd. AL4: St A 5G 6
Mason Av. DA10: Swans 61Be 143
Mason Cl. DA7: Bex 55Dd 118
Mason Cl. E16 45Jc 93
Mason Cl. EN9: Walt A 6Hc 21
Mason Cl. SE16 50Wb 91
Mason Cl. SW20 67Za 132
Mason Cl. TW12: Hamp 67Ba 129
Mason Cl. WD6: Bore 12Ta 29
Mason Dr. RM3: Hrld W 26Nd 57
Mason Ho. E9 38Yb 72
(off Frampton Pk. Rd.)
Mason Ho. SE1 49Wb 91
(off Simms Rd.)
Masonic Hall Rd. KT16: Chert 72H 149
Mason Rd. IG8: Wfd G 21Gc 53
Mason Rd. SM1: Sutt 78Db 155
Masonry Ho. SE14 53Zb 114
(off Fishers Ct.)
Mason's Arms M. W1 3A 222 (44Kb 90)
Mason's Av. CR0: C'don 76Sb 157
Mason's Av. EC2 2F 225 (44Tb 91)
Masons Av. HA3: W'stone 28Ha 46
Mason's Bri. Rd. RH1: Redh 10B 208
Mason's Bri. Rd. RH1: Salf 10B 208
Masons Ct. SL1: Slou 5C 80
Masons Grn. La. W3 42Qa 87
Masons Grn. La. W5 42Qa 87
Masons Hill BR1: Broml 69Jc 137
Masons Hill BR2: Broml 69Kc 137
Masons Hill SE18 49Rc 94
Mason's Pde. EN7: G Oak 1Sb 19
Masons Pl. CR4: Mitc 67Hb 133
Masons Pl. EC1 3D 218 (41Sb 91)
Masons Rd. EN1: Enf 8Xb 19
Masons Rd. HP2: Hem H 1B 4
Masons Rd. SL1: Slou 5C 80
Mason St. SE17 5G 231 (49Tb 91)
Mason's Yd. SW1 6C 222 (46Lb 90)
Mason's Yd. SW19 64Za 132
Masons Yd. EC1 3C 218 (41Rb 91)
Mason Way EN9: Walt A 5Gc 21
Massey Cl. N11 22Kb 50
Massey Ct. E6 39Lc 73
(off Florence Rd.)
Massie Rd. E8 37Wb 71
Massingberd Way SW17 63Kb 134
Massinger St. SE17 6H 231 (49Ub 91)
Massingham St. E1 42Zb 92
Masson Av. HA4: Ruis 37Y 65
Masson Ho. TW8: Bford 51Pa 109
Mast, The E16 45Sc 94
Mast Ct. SE16 49Ac 92
(off Boat Lifter Way)
Master Cl. RH8: Oxt 1J 211
Master Gunner Pl. SE18 52Nc 116
Masterhead Ho. E16 47Kc 93
(off Royal Crest Av.)
Masterman Ho. SE5 52Tb 113
(off Elmington Est.)
Masterman Rd. E6 41Nc 94
TW5: Hest
Masters Cl. SW16 65Lb 134
Masters Ct. RM2: Rom 29Kd 57
(off Academy Flds. Rd.)
Masters Dr. SE16 50Xb 91
Masters Lodge E1 44Yb 92
(off Johnson St.)
Masters St. E1 43Zb 92
Masthead Cl. DA2: Dart 56Sd 120
Mast Ho. Ter. E14 49Cc 92
(not continuous)
Mastin M. DA12: Grav'nd 8F 122
Mastmaker Ct. E14 47Cc 92
Mastmaker Rd. E14 47Cc 92
Mast Quay SE18 48Pc 94
Mast St. IG11: Bark 39Tc 74
MASWELL PARK 57Ea 108
Maswell Pk. Cres. TW3: Houn 57Ea 108
Maswell Pk. Rd. TW3: Houn 57Da 107
Matcham Ct. TW1: Twick 58Ma 109
(off Clevedon Rd.)
Matcham Rd. E11 34Gc 73
Match Ct. E3 40Cc 72
(off Blondin St.)
Matching Ct. E3 41Bc 92
(off Merchant St.)
Matchless Dr. SE18 52Qc 116
Material Store, The UB3: Hayes 48U 84
(off Material Wlk.)
Material Wlk. UB3: Hayes 47U 84
Matfield Cl. BR2: Broml 71Jc 159
Matfield Rd. DA17: Belv 51Cd 118
Matha Ct. BR1: Broml 67Lc 137
Matham Gro. SE22 56Vb 113
Matham Rd. KT8: E Mos 71Fa 152
Mathecombe Rd. SL1: Slou 7D 80
Matheson Lang Ho. SE1 2K 229 (47Qb 90)
(off Baylis Rd.)
Matheson Rd. W14 49Bb 89
Mathews Av. E6 40Qc 74
Mathews Pk. Av. E15 37Hc 73
Mathews Yd. WC2 3F 223 (44Nb 90)
Mathias Cl. KT18: Eps 85Sa 173
Mathieson Ct. SE1 2C 230 (47Rb 91)
(off King James St.)
Mathisen Way SL3: Poyle 53G 104
Mathison Ho. SW10 52Eb 111
(off Coleridge Gdns.)
Matilda Cl. SE19 66Tb 135
Matilda Gdns. E3 40Cc 72
Matilda Ho. E1 46Wb 91
(off St Katherine's Way)
Matilda St. N1 39Pb 70
Matisse Ct. EC1 5F 219 (42Tb 91)
(off Featherstone St.)
Matisse Rd. TW3: Houn 55Da 107
Matlock Cl. EN5: Barn 15Za 30
Matlock Cl. SE24 56Sb 113
Matlock Ct. NW8 1A 214 (40Eb 69)
(off Abbey Rd.)
Matlock Ct. SE5 56Tb 113
Matlock Ct. W11 45Cb 89
(off Kensington Pk. Rd.)
Matlock Cres. SM3: Cheam 77Ab 154
Matlock Cres. WD19: Wat 20Y 27
Matlock Gdns. RM12: Horn 34Nd 77
Matlock Gdns. SM3: Cheam 77Ab 154
Matlock Ho. E15 37Fc 73
(off Forrester Way)
Matlock Pl. SM3: Cheam 77Ab 154
Matlock Rd. CR3: Cat'm 93Ub 197
Matlock Rd. E10 30Ec 52
Matlock St. E14 44Ac 92
Matlock Way KT3: N Mald 67Ta 131
Maton Ho. SW6 52Bb 111
(off Estcourt Rd.)
Matrimony Pl. SW8 54Lb 112
Matson Ct. IG8: Wfd G 24Gc 53
Matson Ho. SE16 48Xb 91
Matthew Arnold Cl. KT11: Cobh 86W 170
Matthew Arnold Cl. TW18: Staines 65L 127
Matthew Arnold Sports Cen. 65L 127
Matthew Cl. W10 42Za 88
Matthew Ct. CR4: Mitc 71Mb 156
Matthew Ct. E17 27Ec 52
Matthew Parker St. SW1 2E 228 (47Mb 90)
Matthews Cl. HA9: Wemb 34Qa 67
Matthews Cl. RM3: Hrld W 25Pd 57
Matthews Ct. SL5: S'hill 10B 124
Matthews Gdns. CR0: New Ad 83Fc 179
Matthews Ho. E14 43Cc 92
(off Burgess St.)
Matthews La. TW18: Staines 63H 127
Matthews Lodge KT15: Add 77M 149
Matthews Rd. UB6: G'frd 36Fa 66
Matthews St. RH2: Reig 10J 207
Matthews St. SW11 54Hb 111
Matthews Wlk. E17 25Cc 52
(off Chingford Rd.)
Matthews Yd. CR0: C'don 76Sb 157
(off Surrey St.)
Matthias Apts. N1 38Tb 71
(off Northchurch Rd.)
Matthias Ct. TW10: Rich 57Na 109
Matthias Rd. N16 36Ub 71
Mattison Rd. N4 30Qb 50
Mattock La. W13 46Ka 86
Mattock La. W5 46La 86
Maud Cashmore Way SE18 48Pc 94
Maud Chadburn Pl. SW4 58Kb 112
Maude Cres. WD24: Wat 9X 13
Maude Ho. E2 40Wb 71
(off Ropley St.)
Maude Rd. BR8: Hext 65Jd 140
Maude Rd. E17 29Ac 52
Maude Rd. SE5 53Ub 113
Maudesville Cotts. W7 46Ga 86
Maude Ter. E17 29Ac 52
Maud Gdns. E13 39Hc 73
Maud Gdns. IG11: Bark 40Vc 75
Maudlins Grn. E1 46Wb 91
Maud Rd. E10 34Ec 72
Maud Rd. E13 40Hc 73
Maudslay Rd. SE9 55Pc 116
MAUDSLEY HOSPITAL, THE 54Tb 113
Maudsley Ho. TW8: Bford 50Na 87
Maud St. E16 43Hc 93
Maud Wilkes Cl. NW5 36Lb 70
Maugham Way W3 48Sa 87
Mauleverer Rd. SW2 57Nb 112
Maundeby Wlk. NW10 37Ua 68
Maunder Cl. RM16: Chaf H 49Zd 99
Maunder Rd. W7 46Ha 86
Maunsel St. SW1 5D 228 (49Mb 90)
Maureen Campbell Ct. TW17: Shep 71R 150
(off Harrison Way)
Maureen Ct. BR3: Beck 68Yb 136
Maurer Ct. SE10 48Hc 93
Mauretania Bldg. E1 45Zb 92
(off Jardine Rd.)
Maurice Av. CR3: Cat'm 94Tb 197
Maurice Av. N22 26Rb 51
Maurice Browne Av. NW7 23Za 48
Maurice Ct. E1 41Zb 92
Maurice Ct. N22 25Pb 50
Maurice Ct. TW8: Bford 52Ma 109
Maurice Drummond Ho. SE10 53Dc 114
(off Catherine Gro.)
Maurice St. W12 44Xa 88
Maurice Wlk. NW11 28Eb 49
Maurier Cl. UB5: N'olt 39Y 65
Mauritius Rd. SE10 49Gc 93
Maury Rd. N16 33Wb 71
Mausoleum Windsor 5K 103
Mauveine Gdns. TW3: Houn 56Ca 107
Mavelstone Cl. BR1: Broml 67Nc 138
Mavelstone Rd. BR1: Broml 67Mc 137
Maverton Rd. E3 39Cc 72
Mavery Ct. BR1: Broml 66Hc 137
(off Bromley Av.)
Mavis Av. KT19: Ewe 78Ua 154
Mavis Cl. KT19: Ewe 78Ua 154
Mavis Gro. RM12: Horn 33Nd 77
Mavis Wlk. E6 43Nc 94
(off Greenwich Cres.)
Mavor Ho. N1 1J 217 (39Pb 70)
(off Barnsbury Est.)
Mawbery Gro. RM3: Rom 22Qd 57
Mawbey Ho. SE1 50Wb 91
Mawbey Pl. SE1 50Vb 91
Mawbey Rd. KT16: Ott 79F 148
Mawbey Rd. SE1 50Vb 91
Mawbey St. SW8 52Nb 112
Mawdley Ho. SE1 2B 230 (47Rb 91)
(off Webber Row)
MAWNEY 27Dd 56
Mawney Cl. RM7: Mawney 26Dd 56
Mawney Rd. RM7: Mawney 26Dd 56
Mawney Rd. RM7: Rom 26Dd 56
Maws Mdw. TN15: W King 81Wd 184
Mawson Cl. SW20 68Ab 132
Mawson Ct. N1 1G 219 (39Tb 71)
(off Gopsall St.)
Mawson Ho. EC1 7K 217 (43Qb 90)
(off Baldwins Gdns.)
Mawson La. W4 51Va 110
Maxden Ct. SE15 55Vb 113
Maxey Gdns. RM9: Dag 35Ad 75
Maxey Rd. RM9: Dag 35Ad 75
Maxey Rd. SE18 49Sc 94
Maxfield Cl. N20 17Eb 31
Maxilla Wlk. W10 44Za 88
Maxim Apts. BR2: Broml 70Kc 137
(off Tiger La.)
Maximfeldt Rd. DA8: Erith 50Gd 96
Maxim Rd. DA1: Cray 57Gd 118
Maxim Rd. DA8: Erith 49Gd 96
Maxim Rd. N21 16Qb 32
Maxim Twr. RM1: Rom 28Hd 56
(off Mercury Gdns.)
Maxted Pk. HA1: Harr 31Ga 66
Maxted Rd. SE15 55Vb 113
Maxwell Cl. CR0: Wadd 74Nb 156
Maxwell Cl. UB3: Hayes 45W 84
Maxwell Cl. WD3: Rick 19J 25
Maxwell Ct. IG7: Chig 20Xc 37
Maxwell Ct. SE22 60Wb 113
Maxwell Ct. SW4 57Mb 112
Maxwell Dr. KT14: W Byf 83L 169
Maxwell Gdns. BR6: Orp 76Vc 161
Maxwell Ri. WD19: Wat 17Aa 27
Maxwell Rd. AL1: St A 3F 6
Maxwell Rd. DA16: Well 55Vc 117
Maxwell Rd. HA6: Nwood 24T 44
Maxwell Rd. RM7: Rush G 30Gd 56
Maxwell Rd. SW6 52Db 111
Maxwell Rd. TW15: Ashf 65S 128
Maxwell Rd. UB7: W Dray 49P 83
Maxwell Rd. WD6: Bore 13Ra 29
Maxwells W. EN8: Chesh 3Yb 20
Maxwelton Av. NW7 22Ta 47
Maxwelton Cl. NW7 22Ta 47
Maya Angelou Ct. E4 21Ec 52
Maya Apts. E20 36Ec 72
(off Victory Pde.)
Maya Cl. SE15 54Xb 113
Mayall Cl. EN3: Enf L 10Cc 20
Mayall Rd. SE24 57Rb 113
Maya Pl. N11 24Mb 50
Maya Rd. N2 28Eb 49
May Av. BR5: St M Cry 71Xc 161
May Av. DA11: Nflt 10B 122
May Av. Ind. Est. DA11: Nflt 10B 122
(off May Av.)
Maybank Av. E18 26Kc 53
Maybank Av. HA0: Wemb 36Ha 66
Maybank Av. RM12: Horn 36Kd 77
Maybank Gdns. HA5: Eastc 29W 44
Maybank Lodge RM12: Horn 36Ld 77
Maybank Rd. E18 25Kc 53
May Bate Av. KT2: King T 67Ma 131
Maybells Commercial Est. IG11: Bark 40Zc 75
Mayberry Ct. BR3: Beck 66Bc 136
(off Copers Cope Rd.)
Mayberry Pl. KT5: Surb 73Pa 153
Maybourne Cl. SE26 65Xb 135
Maybourne Ri. GU22: Wok 6P 187
Maybrick Rd. RM11: Horn 30Ld 57
MAYBURY 88E 168
Maybury Av. DA2: Dart 60Sd 120
Maybury Av. EN8: Chesh 1Xb 19
Maybury Cl. BR5: Pet W 71Rc 160
Maybury Cl. EN1: Enf 10Xb 19
Maybury Cl. IG10: Lough 14Rc 36
Maybury Cl. KT20: Tad 91Ab 194
Maybury Cl. SL1: Slou 4B 80
Maybury Ct. CR2: S Croy 78Rb 157
(off Haling Pk. Rd.)
Maybury Ct. HA1: Harr 30Fa 46
Maybury Ct. W1 1J 221 (43Jb 90)
(off Marylebone St.)
Maybury Est. GU22: Wok 88E 168
Maybury Gdns. NW10 37Xa 68
Maybury Hill GU22: Wok 88D 168

Maybury M. N6 31Lb 70
Maybury Rd. E13 42Lc 93
Maybury Rd. GU21: Wok 89B 168
Maybury Rd. IG11: Bark 40Vc 75
Maybury Rough GU22: Wok 89D 168
Maybury St. SW17 64Gb 133
Maybush Rd. RM11: Horn 31Nd 77
Maychurch Cl. HA7: Stan 24Ma 47
May Cl. AL3: St A 1B 6
May Cl. KT9: Chess 79Pa 153
Maycock Gro. HA6: Nwood 23V 44
May Cotts. WD18: Wat 15Y 27
(off Lammas Rd.)
May Ct. RM17: Grays 1A 122
May Ct. SW19 67Eb 133
(off Pincott Rd.)
Maycroft HA5: Pinn 26X 45
Maycroft Av. RM17: Grays 50Fe 99
Maycroft Gdns. RM17: Grays 50Fe 99
Maycross Av. SM4: Mord 70Bb 133
Mayday Gdns. SE3 54Nc 116
Mayday Rd. CR7: Thor H 72Rb 157
Maydeb Ct. RM6: Chad H 30Bd 55
Maydew Ho. SE16 49Yb 92
(off Abbeyfield Est.)
Maydwell Ho. E14 43Cc 92
(off Thomas Rd.)
Maydwell Lodge WD6: Bore 12Pa 29
Mayell Cl. KT22: Lea 95La 192
Mayerne Rd. SE9 57Mc 115
Mayer Rd. EN9: Walt A 8Dc 20
Mayesbrook Pk. Arena 36Wc 75
Mayesbrook Rd. IG11: Bark 39Vc 75
Mayesbrook Rd. IG3: Ilf 34Wc 75
Mayesbrook Rd. RM8: Dag 34Xc 75
Mayes Cl. BR8: Swan 70Jd 140
Mayes Cl. CR0: New Ad 80Fc 159
Mayes Cl. CR6: W'ham 90Zb 178
Mayesford Rd. RM6: Chad H 31Yc 75
Mayes Rd. N22 26Pb 50
Mayeswood Rd. SE12 63Lc 137
MAYFAIR 5K 221 (45Kb 90)
Mayfair Av. DA7: Bex 53Zc 117
Mayfair Av. IG1: Ilf 33Pc 74
Mayfair Av. KT4: Wor Pk 74Wa 154
Mayfair Av. RM6: Chad H 30Zc 55
Mayfair Av. TW2: Whitt 59Ea 108
Mayfair Cl. BR3: Beck 67Dc 136
Mayfair Cl. KT6: Surb 74Na 153
Mayfair Ct. HA8: Edg 22Pa 47
Mayfair Ct. WD18: Wat 14U 26
Mayfair Gdns. IG8: Wfd G 24Jc 53
Mayfair Gdns. N17 23Sb 51
Mayfair M. NW1 38Hb 69
(off Regents Pk. Rd.)
Mayfair M. SW12 60Kb 112
Mayfair Pl. KT13: Weyb 78S 150
Mayfair Pl. W1 6A 222 (46Kb 90)
Mayfair Rd. DA1: Dart 57Md 119
Mayfair Row W1 7K 221 (46Kb 90)
Mayfair Ter. N14 17Mb 32
Mayfare WD3: Crox G 15T 26
Mayfield DA7: Bex 55Bd 117
Mayfield EN9: Walt A 6Fc 21
Mayfield KT22: Lea 93La 192
(not continuous)
Mayfield Av. BR6: Orp 74Vc 161
Mayfield Av. HA3: Kenton 29Ka 46
Mayfield Av. IG8: Wfd G 23Jc 53
Mayfield Av. KT15: New H 82K 169
Mayfield Av. N12 21Eb 49
Mayfield Av. N14 19Mb 32
Mayfield Av. W13 48Ka 86
Mayfield Av. W4 49Ua 88
Mayfield Cvn. Pk. UB7: W Dray 48L 83
Mayfield Cl. E8 37Vb 71
Mayfield Cl. KT12: Hers 77W 150
Mayfield Cl. KT15: New H 82L 169
Mayfield Cl. KT7: T Ditt 74Ka 152
Mayfield Cl. SE20 67Xb 135
Mayfield Cl. SW4 57Mb 112
Mayfield Cl. TW15: Ashf 65R 128
Mayfield Cl. UB10: Hil 41R 84
Mayfield Ct. EN9: Walt A 6Jc 21
(off Lamplighters Cl.)
Mayfield Ct. RH1: Redh 10P 207
Mayfield Cres. CR7: Thor H 70Pb 134
Mayfield Cres. N9 16Xb 33
Mayfield Dr. HA5: Pinn 28Ba 45
Mayfield Dr. SL4: Wind 5E 102
Mayfield Gdns. CM14: B'wood 18Xd 40
Mayfield Gdns. KT12: Hers 77W 150
Mayfield Gdns. KT15: New H 82K 169
Mayfield Gdns. NW4 30Za 48
Mayfield Gdns. TW18: Staines 65H 127
Mayfield Gdns. W7 44Fa 86
Mayfield Grn. KT23: Bookh 99Ca 191
Mayfield Gro. RM13: Rain 41Ld 97
Mayfield Ho. E2 40Xb 71
(off Cambridge Heath Rd.)
Mayfield Mans. SW15 57Bb 111
Mayfield Pl. SL4: Wink 2A 124
Mayfield Rd. BR1: Broml 71Nc 160
Mayfield Rd. CR2: Sande 81Tb 177
Mayfield Rd. CR7: Thor H 70Pb 134
Mayfield Rd. DA11: Grav'nd 9B 122
Mayfield Rd. DA17: Belv 49Ed 96
Mayfield Rd. E13 42Hc 93
Mayfield Rd. E17 26Ac 52
Mayfield Rd. E4 19Ec 34
Mayfield Rd. E8 38Vb 71
Mayfield Rd. EN3: Enf H 12Zb 34
Mayfield Rd. KT12: Hers 77W 150
Mayfield Rd. KT13: Weyb 78P 149
Mayfield Rd. N8 29Pb 50
Mayfield Rd. RM8: Dag 32Yc 75
Mayfield Rd. SM2: Sutt 79Fb 155
Mayfield Rd. SW19 67Bb 133
Mayfield Rd. W12 47Ua 88
Mayfield Rd. W3 45Ra 87
Mayfield Rd. Flats N8 30Pb 50
Mayfields DA10: Swans 58Ae 121
Mayfields HA9: Wemb 33Qa 67
Mayfields RM16: Grays 47Ee 99
Mayfields Cl. HA9: Wemb 33Qa 67
Mayfield Vs. DA14: Sidc 65Yc 139
Mayflower Av. HP2: Hem H 2M 3
Mayflower Cl. HA4: Ruis 30S 44
Mayflower Cl. RM15: S Ock 42Yd 98
Mayflower Cl. SE16 49Zb 92
Mayflower Ho. CM13: Gt War 23Yd 58
Mayflower Ho. E14 47Cc 92
(off Westferry Rd.)
Mayflower Ho. IG11: Bark 39Tc 74
(off Westbury Rd.)
Mayflower Path CM13: Gt War 23Yd 58
Mayflower Rd. AL2: Park 9P 5
Mayflower Rd. RM16: Chaf H 50Yd 98
Mayflower Rd. SW9 55Nb 112
Mayflower St. SE16 47Yb 92
Mayflower Way SL2: Farn C 6G 60
Mayfly Cl. BR5: St P 70Zc 139
Mayfly Cl. HA5: Eastc 31Y 65
Mayfly Gdns. UB5: N'olt 41Z 85
MAYFORD 4N 187
Mayford NW1 1C 216 (40Lb 70)
(not continuous)
Mayford Cl. BR3: Beck 69Zb 136
Mayford Cl. GU22: Wok 4P 187
Mayford Cl. SW12 59Hb 111
Mayford Grn. GU22: Wok 4N 187
Mayford Meadows Local Nature Reserve 4P 187
Mayford Rd. SW12 59Hb 111
May Gdns. HA0: Wemb 41La 86
May Gdns. WD6: E'tree 16Ma 29
Maygood Ho. N1 1J 217 (40Pb 70)
(off Maygood St.)
Maygoods Cl. UB8: Cowl 43M 83
Maygoods Grn. UB8: Cowl 43M 83
Maygoods La. UB8: Cowl 43M 83
Maygood St. N1 1K 217 (40Qb 70)
Maygoods Vw. UB8: Cowl 43L 83
Maygreen Cres. RM11: Horn 31Jd 76
Maygrove Rd. NW6 37Bb 69
Mayhew Cl. E4 20Cc 34
Mayhew Ct. SE5 56Tb 113
Mayhill Ct. SE15 52Ub 113
(off Tower Mill Rd.)
Mayhill Rd. EN5: Barn 16Ab 30
Mayhill Rd. SE7 51Kc 115
May Ho. E3 40Cc 72
(off Thomas Fyre Dr.)
Mayhurst Av. GU22: Wok 88E 168
Mayhurst Cl. GU22: Wok 88E 168
Mayhurst Cres. GU22: Wok 88E 168
Mayhurst M. GU22: Wok 88E 168
Mayland Mans. IG11: Bark 38Rc 74
(off Whiting Av.)
Maylands Av. RM12: Horn 35Kd 77
Maylands Ct. HP2: Hem H 1B 4
Maylands Dr. DA14: Sidc 62Zc 139
Maylands Dr. UB8: Uxb 37M 63
Maylands Golf Course 22Rd 57
Maylands Ho. SW3 6E 226 (49Gb 89)
(off Cale St.)
Maylands Rd. WD19: Wat 21Y 45
Maylands Way RM3: Hrld W 23Sd 58
May La. HA3: Kenton 31Pa 67
Maylie Ho. SE16 47Xb 91
(off Marigold St.)
Maynard Cl. DA8: Erith 52Hd 118
Maynard Cl. N15 29Ub 51
Maynard Cl. SW6 52Db 111
Maynard Ct. EN3: Enf L 10Cc 20
Maynard Ct. EN9: Walt A 6Hc 21
Maynard Ct. SL4: Wind 3E 102
Maynard Ct. TW18: Staines 63J 127
Maynard Dr. AL1: St A 5B 6
Maynard Path E17 29Ec 52
Maynard Pl. EN6: Cuff 1Pb 18
Maynard Rd. E17 29Ec 52
Maynard Rd. HP2: Hem H 3M 3
Maynards RM11: Horn 31Nd 77
Maynards Quay E1 45Yb 92
Mayne Av. AL3: St A 4M 5
Mayne Ct. SE26 64Xb 135
Maynooth Gdns. SM5: Cars 73Hb 155
Mayo Cl. EN8: Chesh 1Yb 20
Mayo Ct. W13 48Ka 86
Mayo Gdns. HP1: Hem H 3K 3
Mayo Ho. E1 43Yb 92
(off Lindley St.)
Mayola Rd. E5 35Yb 72
Mayo Rd. CR0: C'don 71Tb 157
Mayo Rd. KT12: Walt T 73W 150
Mayo Rd. NW10 37Ua 68
Mayor's & City of London Court, The 2F 225 (44Tb 91)
(off Basinghall St.)
Mayor's La. DA2: Wilm 63Ld 141
Mayow Rd. SE23 62Zb 136
Mayow Rd. SE26 63Zb 136
Mayplace Av. DA1: Cray 56Jd 118
Mayplace Cl. DA7: Bex 55Dd 118
Mayplace La. SE18 51Rc 116
(not continuous)
Mayplace Rd. E. DA1: Cray 56Fd 118
Mayplace Rd. E. DA7: Bex 55Dd 118
Mayplace Rd. W. DA7: Bex 56Cd 118
MAYPOLE 79Cd 162
MAYPOLE 60Gd 118
Maypole Ct. UB2: S'hall 47Ba 85
(off Merrick Rd.)
Maypole Cres. DA8: Erith 51Md 119
Maypole Cres. IG6: Ilf 24Tc 54
Maypole Dr. IG7: Chig 20Wc 37
Maypole Rd. BR6: Chels 78Bd 161
Maypole Rd. BR6: Well H 78Bd 161
Maypole Rd. DA12: Grav'nd 10H 123
May Rd. DA2: Hawl 63Pd 141
May Rd. E13 40Jc 73
May Rd. E4 23Cc 52
May Rd. TW2: Twick 60Ga 108
Mayroyd Av. KT6: Surb 75Qa 153
May's Bldgs. M. SE10 52Ec 114
Mays Cl. KT13: Weyb 82P 169
May's Ct. SE10 52Fc 115
Mays Ct. WC2 5F 223 (45Nb 90)
Maysfield Rd. GU23: Send 95F 188
MAY'S GREEN 92U 190
Mays Gro. GU23: Send 95F 188
Mays Hill Rd. BR2: Broml 68Gc 137
Mays La. EN5: Ark 17Xa 30
Mays La. EN5: Barn 17Xa 30
Maysoule Rd. SW11 56Fb 111
Mays Rd. TW11: Tedd 64Fa 130
Mayston M. SE10 50Jc 93
(off Ormiston Rd.)
May St. W14: Kelway Ho 50Bb 89
May St. W14: Orchard Sq 50Bb 89
Mayswood Gdns. RM10: Dag 37Ed 76
Maythorne Cl. WD18: Wat 14U 26
Maythorne Cotts. SE13 57Fc 115
Mayton St. N7 34Pb 70
Maytree Cl. HA8: Edg 20Sa 29
Maytree Cl. RM13: Rain 40Gd 76
Maytree Ct. CR4: Mitc 69Jb 134
Maytree Ct. UB5: N'olt 41Aa 85
Maytree Cres. WD24: Wat 7V 12
Maytree Gdns. W5 47Ma 87
May Tree Ho. SE4 55Bc 114
(off Wickham Rd.)
Maytree La. HA7: Stan 24Ja 46
Maytrees GU21: Knap 9G 166
Maytrees WD7: R'lett 9Ja 14
Maytree Wlk. SW2 61Qb 134
Mayville Est. N16 36Ub 71
Mayville Rd. E11 33Gc 73
(not continuous)
Mayville Rd. IG1: Ilf 36Rc 74
May Wlk. E13 40Kc 73
Mayward Ho. SE5 53Ub 113
(off Peckham Rd.)
Maywater Cl. CR2: Sande 83Tb 177
Maywin Dr. RM11: Horn 32Pd 77
Maywood Cl. BR3: Beck 66Dc 136
May Wynne Ho. E16 45Kc 93
(off Murray Sq.)
Mazal Ct. KT8: W Mole 70Da 129
Maze Hill SE10 51Gc 115
Maze Hill SE3 52Hc 115
Maze Hill Lodge SE10 51Fc 115
(off Park Vista)
Maze Hill Station (Rail) 51Gc 115
Mazenod Av. NW6 38Cb 69
Maze Rd. TW9: Kew 52Qa 109
MCC Cricket Mus. & Tours 4B 214 (41Fb 89)
Mead, The BR3: Beck 67Ec 136
Mead, The BR4: W W'ck 74Fc 159
Mead, The DA3: Nw A G 75Ae 165
Mead, The EN8: Chesh 1Yb 20
Mead, The KT21: Asht 91Na 193
Mead, The N2 26Eb 49
Mead, The SM6: W'gton 79Mb 156
Mead, The UB10: Ick 33Q 64
Mead, The W13 43Ka 86
Mead, The WD19: Wat 20Aa 27
Mead Av. SL3: L'ly 47D 82
Meadbank Studios SW11 52Gb 111
(off Parkgate Rd.)
Mead Cl. BR8: Swan 71Jd 162
Mead Cl. HA3: Hrw W 25Fa 46
Mead Cl. IG10: Lough 12Rc 36
Mead Cl. NW1 37Jb 70
Mead Cl. RH1: Redh 3A 208
Mead Cl. RM16: Grays 47De 99
Mead Cl. RM2: Rom 26Jd 56
Mead Cl. SL3: L'ly 47D 82
Mead Cl. TW20: Egh 65D 126
Mead Cl. UB9: Den 33J 63
Mead Ct. EN9: Walt A 6Dc 20
Mead Ct. GU21: Knap 8J 167
Mead Ct. KT16: Chert 73J 149
Mead Ct. NW9 29Sa 47
Mead Ct. TW20: Egh 65E 126
Mead Cres. DA1: Dart 60Md 119
Mead Cres. E4 21Ec 52
Mead Cres. KT23: Bookh 97Ca 191
Mead Cres. SM1: Sutt 76Gb 155
Meadcroft Rd. SE11 51Rb 113
(not continuous)
Meadcroft Rd. SE17 51Rb 113
Meade Cl. W4 51Qa 109
Meade Ct. KT20: Walt H 96Wa 194
Meade Ho. E14 44Gc 93
(off Lyell St.)
Meade M. SW1 7E 228 (50Mb 90)
(off Causton St.)
Mead End KT21: Asht 89Pa 173
Meader Ct. SE14 52Zb 114
Meades, The KT13: Weyb 79S 150
Meadfarm Cl. RM3: Rom 22Nd 57
Mead Fld. HA2: Harr 34Ba 65
Meadfield HA8: Edg 19Ra 29
(not continuous)
Meadfield Av. SL3: L'ly 47C 82
Meadfield Grn. HA8: Edg 19Ra 29
Meadfield Rd. SL3: L'ly 48C 82
Meadfoot Rd. SW16 66Lb 134
Meadgate Av. IG8: Wfd G 22Nc 54
Mead Gro. RM6: Chad H 27Zc 55
Mead Ho. SL0: Iver H 41F 82
Mead Ho. W11 46Bb 89
(off Ladbroke Rd.)
Mead Ho. La. UB4: Hayes 42T 84
Meadhurst Club 64V 128
Meadhurst Pk. TW16: Sun 65U 128
Meadhurst Rd. KT16: Chert 74K 149
Meadlands Dr. TW10: Ham 61Ma 131
Mead La. KT16: Chert 73K 149
Mead Lodge W4 47Ta 87
Meadow, The BR7: Chst 65Sc 138
Meadow, The N10 27Jb 50
Meadow Av. CR0: C'don 72Zb 158
Meadow Av. WD7: Shenl 3La 14
Meadow Bank KT24: E Hor 99V 190
Meadow Bank N21 16Pb 32
Meadowbank KT5: Surb 72Pa 153
Meadowbank NW3 38Hb 69
Meadowbank SE3 55Hc 115
Meadowbank WD19: Wat 17Y 27
Meadowbank WD4: K Lan 2Q 12
Meadow Bank Cl. TN15: W King 81Vd 184
Meadowbank Cl. HP3: Bov 10D 2
Meadowbank Cl. SW6 52Ya 110
Meadowbank Cl. TW7: Isle 53Ga 108
Meadowbank Gdns. TW5: Cran 53W 106
Meadowbank Rd. GU18: Light 2A 166
Meadowbank Rd. NW9 31Ta 67
Meadowbanks EN5: Ark 15Wa 30
Meadowbridge Ct. CR0: C'don 71Tb 157
(off Princess Rd.)
Meadowbrook RH8: Oxt 2G 210
Meadowbrook Cl. SL3: Poyle 53H 105
Meadowbrook Ct. TW7: Isle 55Ga 108
Meadow Cl. AL2: Brick W 1Ca 13
Meadow Cl. AL2: Lon C 9H 7
Meadow Cl. AL9: Wel G 6F 8
Meadow Cl. BR7: Chst 64Rc 138
Meadow Cl. CR8: Purl 85Mb 176
Meadow Cl. DA6: Bex 57Bd 117
Meadow Cl. E4 18Dc 34
Meadow Cl. E9 37Bc 72
Meadow Cl. EN3: Enf W 10Ac 20
Meadow Cl. EN5: Barn 16Bb 31
Meadow Cl. HA4: Ruis 30V 44
Meadow Cl. IG11: Bark 38Wc 75
Meadow Cl. KT10: Hin W 76Ha 152
Meadow Cl. KT12: Hers 77Ba 151
Meadow Cl. RM3: Rom 20Ld 39
Meadow Cl. SE6 64Cc 136
Meadow Cl. SL4: Old Win 7M 103
Meadow Cl. SM1: Sutt 75Eb 155
Meadow Cl. SS17: Linf 8J 101
Meadow Cl. SW20 70Ya 132
Meadow Cl. TN13: S'oaks 95Jd 202
Meadow Cl. TW10: Ham 60Na 109
Meadow Cl. TW4: Houn 58Ca 107
Meadow Cl. UB5: N'olt 40Ca 65
Meadow Cl. WD7: Shenl 2La 14
Meadow Cotts. GU24: W End 4D 166
Meadow Ct. E16 43Kc 93
(off Booth Rd.)
Meadow Ct. KT18: Eps 85Sa 173
Meadow Ct. N1 1H 219 (40Ub 71)
Meadow Ct. RH1: Mers 2C 208
Meadow Ct. TW18: Staines 62G 126
Meadow Ct. TW3: Houn 58Da 107
Meadowcourt Rd. SE3 56Hc 115
Meadow Cft. TW18: Staines 65G 126
(off Bowes Rd.)
Meadowcroft AL1: St A 5E 6
Meadowcroft BR1: Broml 69Pc 138
Meadowcroft SL9: Chal P 26A 42
Meadowcroft W4 50Qa 87
(off Brooks Rd.)
Meadowcroft WD23: Bush 16Da 27
Meadowcroft Cl. E10 31Dc 72
Meadowcroft Cl. N13 19Qb 32
Meadowcroft Farm DA2: G St G 66Xd 142
Meadowcroft M. E1 45Wb 91
(off Cable St.)
Meadowcroft M. SE6 58Dc 114
Meadowcroft Rd. N13 19Qb 32
Meadow Cross EN9: Walt A 6Gc 21
Meadow Dr. GU23: Rip 95H 189
Meadow Dr. HP3: Bov 10B 2
Meadow Dr. N10 27Kb 50
Meadow Dr. NW4 26Ya 48
Meadow Dr. RM15: Avel 45Td 98
Meadowford Cl. SE28 45Wc 95
Meadow Gdns. HA8: Edg 23Ra 47
Meadow Gdns. TW18: Staines 64F 126
Meadow Gth. NW10 37Sa 67
Meadow Ga. HA2: Harr 34Da 65
Meadow Ga. KT21: Asht 89Na 173
Meadowgate Cl. NW7 22Va 48
Meadow Hill CR5: Coul 86Lb 176
Meadow Hill CR8: Purl 85Lb 176
Meadow Hill KT3: N Mald 72Ua 154
MEADOW HOUSE HOSPICE 47Fa 86
Meadowlands KT11: Cobh 85W 170
Meadowlands RH8: Oxt 6L 211
Meadowlands RM11: Horn 31Nd 77
Meadowlands TN15: Seal 92Pd 203
Meadowlands Pk. KT15: Add 76N 149
Meadow La. DA3: Nw A G 75Be 165
Meadow La. KT22: Fet 94Ea 192
Meadow La. SE12 62Kc 137
Meadow La. SL4: Eton 1F 102
Meadow La. SL4: Eton W 1F 102
Meadowlark Ho. NW9 31Va 68
Meadowlea Cl. UB7: Harm 51M 105
Meadow M. SW8 51Pb 112
Meadow Pk. CM13: Gt War 21Ra 29
Meadow Pl. SW8 52Nb 112
Meadow Pl. W4 52Ua 110
Meadow Ri. CR5: Coul 85Mb 176
Meadow Ri. GU21: Knap 9G 166
Meadow Rd. BR2: Broml 68Gc 137
Meadow Rd. CM16: Epp 1Vc 23
Meadow Rd. DA11: Grav'nd 1C 144
Meadow Rd. GU25: Vir W 1J 147
Meadow Rd. HA5: Pinn 28Z 45
Meadow Rd. HP3: Hem H 6A 4
Meadow Rd. IG10: Lough 15Nc 36
Meadow Rd. IG11: Bark 38Vc 75
Meadow Rd. KT10: Clay 79Ga 152
Meadow Rd. KT21: Asht 89Na 173
Meadow Rd. RM16: Grays 46Ee 99
Meadow Rd. RM7: Rush G 32Ed 76
Meadow Rd. RM9: Dag 37Bd 75
Meadow Rd. SL3: L'ly 48A 82
Meadow Rd. SM1: Sutt 77Gb 155
Meadow Rd. SW19 66Eb 133
Meadow Rd. SW8 52Pb 112
Meadow Rd. TW13: Felt 61Aa 129
Meadow Rd. TW15: Ashf 64T 128
Meadow Rd. UB1: S'hall 45Ba 85
Meadow Rd. WD23: Bush 15Da 27
Meadow Rd. WD25: Wat 6W 12
Meadow Rd. WD6: Bore 12Ra 29
Meadow Row SE1 4D 230 (48Sb 91)
Meadows, The BR6: Chels 79Yc 161
Meadows, The CM13: Ingve 23Ee 59
Meadows, The CR6: W'ham 89Zb 178
Meadows, The E4 21Fc 53
Meadows, The EN6: S Mim 5Wa 16
Meadows, The HP1: Hem H 1G 2
Meadows, The TN14: Hals 85Bd 181
Meadows, The WD25: Wat 8Z 13
MEADOWS, THE 10Pa 15
Meadows Cl. CM13: Ingve 23Ee 59
Meadows Cl. E10 33Cc 72
Meadows Ct. DA14: Sidc 65Xc 139
Meadows End TW16: Sun 67W 128
Meadows Ho. KT12: Walt T 74W 150
Meadowside DA1: Dart 60Md 119
Meadowside KT12: Walt T 75Y 151
Meadowside KT23: Bookh 95Ca 191
Meadowside SE3 56Lc 115
Meadowside SE9 56Lc 115
Meadowside TW1: Twick 59Ma 109
Meadowside TW18: Staines 64J 127
Meadowside WD25: Wat 3X 13
Meadowside Rd. RM14: Upm 36Sd 78
Meadowside Rd. SM2: Cheam 81Ab 174
Meadows Leigh Cl. KT13: Weyb 76R 150
Meadow Stile CR0: C'don 76Sb 157
Meadows Way SE4 55Ac 114
Meadowsweet Cl. E16 43Mc 93
Meadowsweet Cl. SW20 70Ya 132
MEADOWS WEST PARK HOSPITAL, THE 83Na 173
Meadow Vw. BR5: St P 69Yc 139
Meadow Vw. CM16: Epp 1Wc 23
Meadow Vw. DA15: Sidc 59Xc 117
Meadow Vw. GU22: Wok 93B 188
(not continuous)
Meadow Vw. HA1: Harr 32Ga 66
Meadow Vw. KT16: Chert 74L 149
Meadow Vw. SL3: L'ly 46B 82
Meadow Vw. TN13: Dun G 90Fd 182
Meadow Vw. TW19: Stanw M 57H 105
Meadow Vw. UB8: Cowl 43L 83
Meadow Vw. Rd. CR7: Thor H 71Rb 157
Meadow Vw. Rd. SW20 70Ya 132
Meadow Vw. Rd. UB4: Hayes 42T 84
Meadowview Rd. DA5: Bexl 58Ad 117
Meadowview Rd. KT19: Ewe 81Ua 174
Meadowview Rd. SE6 64Bc 136
Meadow Wlk. DA2: Wilm 63Ld 141
(not continuous)
Meadow Wlk. E18 28Jc 53
Meadow Wlk. KT17: Ewe 80Va 154
Meadow Wlk. KT19: Ewe 79Ua 154
Meadow Wlk. KT20: Walt H 96Xa 194
Meadow Wlk. RM9: Dag 37Bd 75
Meadow Wlk. SM6: W'gton 76Kb 156
Meadow Way BR6: Farnb 76Qc 160
Meadow Way DA2: Dart 59Sd 120
Meadow Way EN6: Pot B 6Cb 17
Meadow Way GU24: W End 4D 166
Meadow Way HA4: Ruis 30X 45
Meadow Way HA9: Wemb 35Ma 67
Meadow Way HP3: Hem H 5H 3
Meadow Way IG7: Chig 20Sc 36
Meadow Way KT15: Add 77K 149
Meadow Way KT20: Tad 89Ab 174
Meadow Way KT23: Bookh 95Da 191
Meadow Way KT24: W Hor 97T 190
Meadow Way KT9: Chess 78Na 153
Meadow Way NW9 29Ta 47
Meadow Way RH2: Reig 10K 207
Meadow Way RM14: Upm 34Sd 78
Meadow Way SL4: Old Win 8M 103
Meadow Way WD3: Rick 17L 25
Meadow Way WD4: K Lan 2Q 12
Meadow Way WD5: Bedm 9F 4
Meadow Way, The HA3: Hrw W 25Ga 46
Meadow Waye TW5: Hest 51Aa 107
Meadow Works EN5: New Bar 16Db 31
Mead Path SW17 63Eb 133
Mead Pl. CR0: C'don 74Sb 157
Mead Pl. E9 37Yb 72
Mead Pl. WD3: Rick 18K 25
Mead Plat NW10 37Sa 67
Mead Rd. BR7: Chst 65Sc 138
Mead Rd. CR3: Cat'm 95Vb 197
Mead Rd. DA1: Dart 60Md 119
Mead Rd. DA11: Grav'nd 1D 144
Mead Rd. HA8: Edg 23Qa 47
Mead Rd. KT12: Hers 77Aa 151
Mead Rd. TW10: Ham 62La 130
Mead Rd. UB8: Uxb 38M 63
Mead Rd. WD7: Shenl 5Qa 15
Mead Row SE1 3K 229 (48Qb 90)
Meads, The AL2: Brick W 1Ba 13
Meads, The HA8: Edg 23Ta 47
Meads, The RM14: Upm 33Ud 78
Meads, The SL4: Wind 4E 102
Meads, The SM3: Cheam 76Ab 154
Meads, The SM4: Mord 71Gb 155
Meads, The UB8: Cowl 42N 83
Meads Ct. E15 37Hc 73
Meadside GU22: Wok 90B 168
(off Park Dr.)
Meadside KT18: Eps 86Ta 173
(off South St.)
Meadside Cl. BR3: Beck 67Ac 136
Meads La. IG3: Ilf 31Uc 74
Meads Rd. EN3: Enf H 11Ac 34
Meads Rd. N22 26Rb 51
Meadsway CM13: Gt War 23Xd 58
Mead Ter. HA9: Wemb 35Ma 67
MEAD VALE 8M 207
Meadvale Rd. CR0: C'don 73Vb 157
Meadvale Rd. W5 42Ka 86
Mead Wlk. SL3: L'ly 47D 82
Mead Way BR2: Hayes 72Hc 159
Mead Way CR0: C'don 75Ac 158
Mead Way CR5: Coul 90Nb 176
Mead Way HA4: Ruis 30T 44
Mead Way SL1: Slou 3B 80
Mead Way WD23: Bush 12Aa 27
Meadway AL4: Col H 5P 7
Meadway BR3: Beck 67Ec 136
Meadway CR6: W'ham 88Yb 178
Meadway EN3: Enf W 8Yb 20
Meadway EN5: Barn 14Cb 31
Meadway EN5: New Bar 14Cb 31
Meadway HP4: Berk 1A 2
Meadway IG3: Bark 35Uc 74
Meadway IG3: Ilf 35Uc 74
Meadway IG8: Wfd G 22Lc 53
Meadway KT10: Esh 81Da 171
Meadway KT19: Eps 84Sa 173
Meadway KT22: Oxs 86Ga 172
Meadway KT24: Eff 100Aa 191
Meadway KT5: Surb 74Sa 153
Meadway N14 19Mb 32
Meadway NW11 30Cb 49
Meadway RM17: Grays 49Fe 99
Meadway RM2: Rom 26Jd 56
Meadway SW20 70Ya 132
Meadway TN14: Hals 85Bd 181
Meadway TW15: Ashf 63Q 128
Meadway TW18: Staines 66J 127
Meadway TW2: Twick 60Fa 108
Meadway, The BR6: Chels 78Xc 161
Meadway, The EN6: Cuff 1Pb 18
Meadway, The IG10: Lough 16Pc 36
Meadway, The IG9: Buck H 18Mc 35
Meadway, The SE3 54Fc 115
Meadway, The TN13: S'oaks 94Hd 202
Meadway Cl. EN5: Barn 14Cb 31
Meadway Cl. HA5: Hat E 23Da 45
Meadway Cl. NW11 30Db 49
Meadway Cl. TW18: Staines 66H 127
Meadway Ct. RM8: Dag 33Bd 75
Meadway Ct. TW11: Tedd 64La 130
Meadway Ct. W5 42Pa 87
Meadway Dr. GU21: Wok 8N 167
Meadway Dr. KT15: Add 80L 149
Meadway Gdns. HA4: Ruis 30T 44
Meadway Ga. NW11 30Cb 49
Meadway Pk. SL9: Ger X 2P 61
Mead Way Path CR5: Coul 90Pb 176
Meaford Way SE20 66Xb 135
Meakin Est. SE1 3H 231 (48Ub 91)
Meander Ho. E20 36Dc 72
(off Logan Cl.)
Meanley Rd. E12 35Nc 74
Meard St. W1 3D 222 (44Mb 90)
(not continuous)
Meare Cl. KT20: Tad 95Ya 194
Mears Cl. E1 43Wb 91
(off Settles St.)
Meath Cl. BR5: St M Cry 71Xc 161
Meath Cres. E2 41Zb 92
Meath Ho. SE24 58Rb 113
(off Dulwich Rd.)
Meath Rd. E15 40Hc 73
Meath Rd. IG1: Ilf 34Sc 74
Meath St. SW11 53Kb 112
Meautys AL3: St A 4M 5
Mecca Bingo Camden 39Kb 70
(off Arlington Rd.)
Mecca Bingo Catford 59Dc 114
Mecca Bingo Croydon 75Sb 157
Mecca Bingo Dagenham 39Ad 75
Mecca Bingo Hayes 44W 84
Mecca Bingo Morden 73Eb 155
Mecca Bingo Romford 28Hd 56
Mecca Bingo Wood Green 26Qb 50
(off Lordship La.)
Mecklenburgh Pl. WC1 5H 217 (42Pb 90)
Mecklenburgh Sq. WC1 5H 217 (42Pb 90)
Mecklenburgh St. WC1 5H 217 (42Pb 90)
Medals Way E20 36Ec 72
Medbree Ct. RM16: Ors 3C 100
Medburn St. NW1 1D 216 (40Mb 70)
Medbury Rd. DA12: Grav'nd 10H 123
Medcalf Rd. EN3: Enf L 9Bc 20

Medcroft Gdns. SW14 56Sa 109
Medd Ct. AL1: St A 2C 6
(off St Peter's St.)
Medebourne Cl. SE3 55Jc 115
Medebridge Rd. RM16: N Stif 45Be 99
Medebridge Rd. RM16: Ors 45Be 99
Mede Cl. TW19: Wray 10P 103
Mede Ct. TW18: Staines 62G 126
Medefield KT22: Fet 96Fa 192
Mede Ho. BR1: Broml 64Kc 137
(off Pike Cl.)
Medesenge Way N13 23Rb 51
Medfield St. SW15 59Wa 110
Medhurst Cl. E3 40Ac 72
(not continuous)
Medhurst Cl. GU24: Chob 1K 167
Medhurst Cres. DA12: Grav'nd 2G 144
Medhurst Dr. BR1: Broml 64Gc 137
Medhurst Gdns. DA12: Grav'nd 2H 145
Median Rd. E5 36Yb 72
Medici Cl. IG3: Ilf 30Wc 55
(not continuous)
Medick Ct. RM17: Grays 1A 122
Medina Av. KT10: Hin W 76Ga 152
Medina Gro. N7 34Qb 70
Medina Pl. E14 43Cc 92
Medina Rd. N7 34Qb 70
Medina Rd. RM17: Grays 50Fe 99
Medina Sq. KT19: Eps 81Qa 173
Medlake Pl. TW20: Egh 66E 126
Medlake Rd. TW20: Egh 65E 126
Medland Cl. SM6: W'gton 74Jb 156
Medland Ho. E14 45Ac 92
Medland M. KT16: Chert 74J 149
Medlar Cl. UB5: N'olt 40Z 65
Medlar Ct. SL2: Slou 6N 81
Medlar Dr. RM15: S Ock 42Ae 99
Medlar Ho. DA15: Sidc 62Wc 139
Medlar Rd. RM17: Grays 51Fe 121
Medlar St. SE5 53Sb 113
Medley Rd. NW6 37Cb 69
Medman Cl. UB8: Uxb 40L 63
Medora Rd. RM7: Rom 28Fd 56
Medora Rd. SW2 59Pb 112
Medow Mead WD7: R'lett 5Ha 14
Medusa Ct. DA12: Grav'nd 8F 122
(off Admirals Way)
Medusa Rd. SE6 58Dc 114
Medway Bldgs. E3 40Ac 72
(off Medway Rd.)
Medway Cl. CR0: C'don 72Yb 158
Medway Cl. IG1: Ilf 36Sc 74
Medway Cl. RM7: Mawney 25Dd 56
Medway Cl. WD25: Wat 6Y 13
Medway Ct. NW11 30Db 49
Medway Ct. WC1 4F 217 (41Nb 90)
(off Judd St.)
Medway Dr. UB6: G'frd 40Ha 66
Medway Gdns. HA0: Wemb 35Ja 66
Medway Ho. KT2: King T 67Ma 131
Medway Ho. NW8 6D 214 (42Gb 89)
(off Penfold St.)
Medway Ho. SE1 2G 231 (47Tb 91)
(off Hankey Pl.)
Medway M. E3 40Ac 72
Medway Pde. UB6: G'frd 40Ha 66
Medway Rd. DA1: Cray 55Jd 118
Medway Rd. E3 40Ac 72
Medway St. SW1 4D 228 (48Mb 90)
Medwin St. SW4 56Pb 112
Meerbrook Rd. SE3 55Lc 115
Meeson Rd. E15 38Hc 73
Meesons La. RM17: Grays 49Be 99
Meeson St. E5 35Ac 72
Meeson's Wharf E15 40Ec 72
Meeting Fld. Path E9 37Yb 72
Meeting Ho. All. E1 46Xb 91
Meeting Ho. La. SE15 53Xb 113
Megan Ct. SE14 52Yb 114
(off Pomeroy St.)
Megelish M. UB1: S'hall 41Ca 85
Megg La. WD4: Chfd 1K 11
Mehetabel Rd. E9 36Yb 72
Meister Cl. IG1: Ilf 32Tc 74
Melancholy Wlk. TW10: Ham 61La 130
Melanda Cl. BR7: Chst 64Pc 138
Melanie Cl. DA7: Bex 53Ad 117
Melba Gdns. RM18: Tilb 2C 122
Melba Way SE13 53Dc 114
Melbourne Av. HA5: Pinn 27Da 45
Melbourne Av. N13 23Pb 50
Melbourne Av. SL1: Slou 4G 80
Melbourne Av. W13 46Ja 86
Melbourne Cl. BR6: Orp 73Uc 160
Melbourne Cl. SE20 66Wb 135
Melbourne Cl. SM6: W'gton 78Lb 156
Melbourne Cl. UB10: Ick 35Q 64
Melbourne Ct. E5 35Ac 72
(off Daubeney Rd.)
Melbourne Ct. EN8: Walt C 6Bc 20
(off Holdbrook Sth.)
Melbourne Ct. N10 24Kb 50
Melbourne Ct. W9 5A 214 (42Eb 89)
(off Randolph Av.)
Melbourne Gdns. RM6: Chad H 29Ad 55
Melbourne Gro. SE22 56Ub 113
Melbourne Ho. UB4: Yead 42Y 85
Melbourne Ho. W8 46Cb 89
(off Kensington Pl.)
Melbourne Mans. W14 51Ab 110
(off Musard Rd.)
Melbourne M. SE6 59Ec 114
Melbourne M. SW9 53Qb 112
Melbourne Pl. WC2 3J 223 (44Pb 90)
Melbourne Quay DA11: Grav'nd 8D 122
Melbourne Rd. E10 31Dc 72
Melbourne Rd. E17 28Ac 52
Melbourne Rd. E6 40Pc 74
Melbourne Rd. IG1: Ilf 32Rc 74
Melbourne Rd. RM18: Tilb 3A 122
Melbourne Rd. SM6: W'gton 78Kb 156
Melbourne Rd. SW19 67Cb 133
Melbourne Rd. TW11: Tedd 65La 130
Melbourne Rd. WD23: Bush 15Da 27
Melbourne Sq. SW9 53Qb 112
Melbourne Ter. SW6 52Db 111
(off Moore Pk. Rd.)
Melbourne Way EN1: Enf 16Vb 33
Melbourne Yd. SE19 65Ub 135
Melbray M. SW6 54Bb 111
Melbreak Ho. SE22 55Ub 113
Melbury Av. UB2: S'hall 48Da 85
Melbury Cl. BR7: Chst 65Nc 138
Melbury Cl. KT10: Clay 79Ka 152
Melbury Cl. KT14: W Byf 86J 169
Melbury Cl. KT16: Chert 73J 149
Melbury Ct. W8 48Bb 89
Melbury Dr. SE5 52Ub 113
Melbury Gdns. CR2: Sande 83Ub 177
Melbury Gdns. SW20 67Xa 132
Melbury Ho. SW8 52Pb 112
(off Richborne Ter.)
Melbury Rd. HA3: Kenton 29Pa 47
Melbury Rd. W14 48Bb 89
Melchester W11 44Bb 89
(off Ledbury Rd.)
Melchester Ho. N19 34Mb 70
(off Wedmore St.)
Melcombe Ct. NW1 7F 215 (43Hb 89)
(off Melcombe Pl.)
Melcombe Gdns. HA3: Kenton 30Pa 47
Melcombe Ho. SW8 52Pb 112
(off Dorset Rd.)
Melcombe Pl. NW1 7F 215 (43Hb 89)
Melcombe Regis Ct. W1 1J 221 (43Jb 90)
(off Weymouth St.)
Melcombe St. NW1 6G 215 (42Hb 89)
Meldex Cl. NW7 23Ya 48
Meldon Cl. SW6 53Db 111
Meldone Cl. KT5: Surb 73Ra 153
Meldon Vw. DA1: Dart 55Qd 119
Meldrum Cl. BR5: Orp 72Yc 161
Meldrum Cl. RH8: Oxt 4K 211
Meldrum Rd. IG3: Ilf 33Wc 75
Melfield Gdns. SE6 63Ec 136
Melford Av. IG11: Bark 37Uc 74
Melford Cl. KT9: Chess 78Pa 153
Melford Ct. SE1 3J 231 (48Ub 91)
(off Fendall St.)
Melford Ct. SE22 60Wb 113
(not continuous)
Melford Pas. SE22 59Wb 113
Melford Pl. CM14: B'wood 18Yd 40
Melford Pl. CM15: B'wood 18Yd 40
Melford Rd. E11 33Gc 73
Melford Rd. E17 28Ac 52
Melford Rd. E6 42Pc 94
Melford Rd. IG1: Ilf 33Tc 74
Melford Rd. SE22 59Wb 113
Melford Rd. SL2: Slou 2D 80
Melfort Av. CR7: Thor H 69Rb 135
Melfort Rd. CR7: Thor H 69Rb 135
Melgund Rd. N5 36Qb 70
Melia Cl. WD25: Wat 7Y 13
Melina Cl. UB3: Hayes 43T 84
Melina Ct. NW8 4B 214 (41Fb 89)
(off Grove End Rd.)
Melina Ct. SW15 55Wa 110
Melina Pl. NW8 4B 214 (41Fb 89)
Melina Rd. W12 47Xa 88
Melior Ct. N6 30Lb 50
Melior Pl. SE1 1H 231 (47Ub 91)
Melior St. SE1 1H 231 (47Ub 91)
Meliot Rd. SE6 61Fc 137
Melksham Cl. RM3: Rom 24Pd 57
Melksham Dr. RM3: Rom 24Pd 57
Melksham Gdns. RM3: Rom 24Nd 57
Melksham Grn. RM3: Rom 24Pd 57
Meller Cl. CR0: Bedd 76Nb 156
Meller Ho. E20 37Ec 72
(off Champions Wlk.)
Mellifont Cl. SM5: Cars 73Fb 155
Melling Dr. EN1: Enf 11Wb 33
Melling St. SE18 51Uc 116
Mellish Cl. IG11: Bark 39Vc 75
Mellish Cl. N20 19Fb 31
Mellish Flats E10 31Cc 72
Mellish Gdns. IG8: Wfd G 22Jc 53
Mellish Ho. E1 44Xb 91
(off Varden St.)
Mellish Ind. Est. SE18 48Mc 93
Mellish St. E14 48Cc 92
Mellish Way RM11: Horn 29Ld 57
Mellison Rd. SW17 64Gb 133
Melliss Av. TW9: Kew 53Ra 109
Mellitus St. W12 43Va 88
Mellonde Cl. RM3: Rom 24Kd 57
Mellor Cl. KT12: Walt T 73Ba 151
Mellor Wlk. SL4: Wind 3H 103
(off Bachelors Acre)
Mellow Cl. SM7: Bans 86Eb 175
Mellowes Rd. RM11: Horn 30Jd 56
Mellow La. E. UB4: Hayes 41S 84
Mellow La. W. UB10: Hil 41S 84
Mellows Rd. IG5: Ilf 27Pc 54
Mellows Rd. SM6: W'gton 78Mb 156
Mells Cres. SE9 63Pc 138
Mell St. SE10 50Gc 93
Melody La. N5 36Sb 71
Melody Rd. SW18 57Eb 111
Melody Rd. TN16: Big H 90Lc 179
Melon Pl. W8 47Cb 89
Melon Rd. E11 34Gc 73
Melon Rd. SE15 53Wb 113
Melrose Av. CR4: Mitc 66Kb 134
Melrose Av. DA1: Cray 59Gd 118
Melrose Av. EN6: Pot B 4Cb 17
Melrose Av. N22 25Rb 51
Melrose Av. NW2 36Xa 68
Melrose Av. SW16 69Pb 134
Melrose Av. SW19 61Bb 133
Melrose Av. TW2: Whitt 59Da 107
Melrose Av. UB6: G'frd 40Da 65
Melrose Av. WD6: Bore 15Ra 29
Melrose Cl. SE12 60Jc 115
Melrose Cl. UB4: Hayes 43W 84
Melrose Cl. UB6: G'frd 40Da 65
Melrose Ct. EN8: Chesh 1Zb 20
Melrose Ct. W13 46Ja 86
(off Williams Rd.)
Melrose Cres. BR6: Orp 77Tc 160
Melrose Dr. UB1: S'hall 46Ca 85
Melrose Gdns. HA8: Edg 27Ra 47
Melrose Gdns. KT12: Hers 78Y 151
Melrose Gdns. KT3: N Mald 69Ta 131
Melrose Gdns. W6 48Ya 88
Melrose Ho. NW6 41Cb 89
(off Carlton Vale)
Melrose Ho. SW1 7K 227 (50Kb 90)
(part of Abbots Mnr.)
Melrose Pl. WD17: Wat 10V 12
Melrose Rd. CR5: Coul 87Kb 176
Melrose Rd. HA5: Pinn 28Ba 45
Melrose Rd. KT13: Weyb 78Q 150
Melrose Rd. SW13 54Ua 110
Melrose Rd. SW18 58Bb 111
Melrose Rd. SW19 68Cb 133
Melrose Rd. TN16: Big H 88Lc 179
Melrose Rd. W3 48Sa 87
Melrose Ter. W6 48Ya 88
Melrose Tudor SM6: W'gton 78Nb 156
(off Plough La.)
Melsa Rd. SM4: Mord 72Eb 155
Melsted Rd. HP1: Hem H 2K 3
Melstock Av. RM14: Upm 35Sd 78
Melthorne Dr. HA4: Ruis 34Y 65
Melthorpe Gdns. SE3 53Nc 116
Melton Cl. HA4: Ruis 32Y 65
Melton Ct. SM2: Sutt 80Eb 155
Melton Ct. SW7 5C 226 (49Fb 89)
Melton Flds. KT19: Ewe 81Ta 173
Melton Gdns. RM1: Rom 31Hd 76
Melton Pl. KT19: Ewe 81Ta 173
Melton Rd. RH1: Mers 2C 208
Melton St. NW1 4C 216 (41Lb 90)
Melville Av. CR2: S Croy 78Vb 157
Melville Av. SW20 66Wa 132
Melville Av. UB6: G'frd 36Ha 66
Melville Cl. UB10: Ick 33T 64
Melville Ct. RM3: Rom 24Nd 57
Melville Ct. SE8 49Ac 92
Melville Ct. W12 48Xa 88
(off Goldhawk Rd.)
Melville Ct. W4 50Qa 87
(off Haining Cl.)
Melville Gdns. N13 22Rb 51
Melville Ho. EN5: New Bar 15Fb 31
Melville Pl. N1 38Sb 71
Melville Rd. DA14: Sidc 61Yc 139
Melville Rd. E17 27Bc 52
Melville Rd. NW10 38Ta 67
Melville Rd. RM13: Rain 42Jd 96
Melville Rd. RM5: Col R 24Dd 56
Melville Rd. SW13 53Wa 110
Melville Vs. Rd. W3 46Sa 87
Melvin Rd. SE20 67Yb 136
Melvinshaw KT22: Lea 93La 192
Melwood Ho. E1 44Xb 91
(off Watney Mkt.)
Melyn Cl. N7 35Lb 70
Memel Ct. EC1 6D 218 (42Sb 91)
(off Memel St.)
Memel St. EC1 6D 218 (42Sb 91)
Memess Path SE18 51Qc 116
Memorial Av. E15 41Gc 93
Memorial Cl. RH8: Oxt 99Fc 199
Memorial Cl. TW5: Hest 51Ba 107
Memorial Hgts. IG2: Ilf 30Tc 54
MEMORIAL HOSPITAL 54Qc 116
Menai Pl. E3 40Cc 72
(off Blondin St.)
Menard Ct. EC1 4E 218 (41Sb 91)
(off Galway St.)
Mendez Way SW15 58Wa 110
Mendham Ho. SE1 3H 231 (48Ub 91)
(off Cluny Pl.)
Mendip Cl. KT4: Wor Pk 74Ya 154
Mendip Cl. SE26 63Yb 136
Mendip Cl. SL3: L'ly 50C 82
Mendip Cl. UB3: Harl 52T 106
Mendip Ct. SE14 51Yb 114
(off Avonley Rd.)
Mendip Ct. SW11 55Eb 111
Mendip Dr. NW2 33Ab 68
Mendip Ho. N9 19Wb 33
(within Edmonton Grn. Shop. Cen.)
Mendip Ho's. E2 41Yb 92
(off Welwyn St.)
Mendip Rd. DA7: Bex 53Gd 118
Mendip Rd. IG2: Ilf 29Uc 54
Mendip Rd. RM11: Horn 31Jd 76
Mendip Rd. SW11 55Eb 111
Mendip Rd. WD23: Bush 16Ea 28
Mendora Rd. SW6 52Ab 110
Mendoza Cl. RM11: Horn 29Nd 57
Menelik Rd. NW2 35Ab 68
Menier Chocolate Factory 7E 224 (46Sb 91)
(off Southwark St.)
Menlo Gdns. SE19 66Tb 135
Menlo Lodge N13 20Pb 32
(off Crothall Cl.)
Menon Dr. N9 20Xb 33
Menotti St. E2 42Wb 91
Menteath Ho. E14 44Cc 92
(off Dod St.)
Menthone Pl. RM11: Horn 31Md 77
Mentmore Cl. HA3: Kenton 30La 46
Mentmore Ho. KT18: Eps 86Sa 173
(off Dalmeny Way)
Mentmore Rd. AL1: St A 4B 6
Mentmore Ter. E8 38Xb 71
Mentone Mans. SW10 52Db 111
(off Fulham Rd.)
Meon Cl. KT20: Tad 94Xa 194
Meon Ct. TW7: Isle 54Ga 108
Meon Rd. W3 47Sa 87
Meopham Rd. CR4: Mitc 67Lb 134
Meopham Station (Rail) 10C 144
Mepham Cres. HA3: Hrw W 24Ea 46
Mepham Gdns. HA3: Hrw W 24Ea 46
Mepham St. SE1 7K 223 (46Qb 90)
Mera Dr. DA7: Bex 56Cd 118
Meranti Ho. E1 44Wb 91
(off Goodman's Stile)
Merantun Way SW19 67Db 133
Merbury Cl. SE13 57Ec 114
Merbury Cl. SE28 46Tc 94
Merbury Rd. SE28 46Tc 94
Mercator Pl. E14 50Cc 92
Mercator Rd. SE13 56Fc 115
Mercedes-Benz World 81P 169
Mercer Av. DA10: Swans 60Ae 121
Mercer Bldg. EC2 5J 219 (42Ub 91)
(off New Inn Yd.)
Mercer Cl. KT7: T Ditt 73Ha 152
Mercer Ct. E1 43Ac 92
Mercer Ho. SW1 7K 227 (50Kb 90)
(off Ebury Bri. Rd.)
Mercer Ho. SW8 53Kb 112
(off Gladstone Ter.)
Merceron Ho's. E2 41Yb 92
(off Globe Rd.)
Merceron St. E1 42Xb 91
Mercer Pl. HA5: Pinn 26Y 45
Mercers HP2: Hem H 1N 3
Mercers Cl. SE10 49Hc 93
Mercer's Cotts. E1 44Ac 92
(off White Horse Rd.)
Mercers Country Pk. 3D 208
Mercers M. N19 34Mb 70
Mercers Pl. W6 49Za 88
Mercers Rd. N19 34Mb 70
(not continuous)
Mercers Row AL1: St A 4A 6
Mercer St. WC2 3F 223 (44Nb 90)
Mercer Wlk. UB8: Uxb 38L 63
Mercer Wlk. WC2 3F 223 (44Nb 90)
(off Mercer St.)
Merchant Cl. KT19: Ewe 78Ta 153
Merchant Ct. E1 46Yb 92
(off Wapping Wall)
Merchant Ho. E14 48Dc 92
(off Selsdon Way)
Merchant Ind. Ter. NW10 42Sa 87
Merchant Navy Memorial 5K 225 (45Vb 91)
Merchants Cl. GU21: Knap 9G 166
Merchants Cl. SE25 70Wb 135
Merchants Ho. E14 44Fc 93
(off New Village Av.)
Merchants Ho. E15 37Fc 73
(off Forrester Way)
Merchants Ho. SE10 50Fc 93
(off Collington St.)
Merchants Lodge E17 28Cc 52
(off Westbury Rd.)
Merchants Pl. SE13 56Gc 115
Merchant Sq. W2 1C 220 (43Fb 89)
(off Nth. Wharf Rd.)
Merchant Sq. E. W2 1C 220 (43Fb 89)
(off Harbet Road)
Merchant Sq. W. W2 1C 220 (43Fb 89)
(off North Wharf Road)
Merchants Row SE10 50Fc 93
(off Hoskins St.)
Merchant St. E3 41Bc 92
Merchiston Rd. SE6 61Fc 137
Merchland Rd. SE9 60Sc 116
Mercia Gro. SE13 56Ec 114
Mercia Ho. SE5 54Sb 113
(off Denmark Rd.)
Mercia Ho. TW15: Ashf 67S 128
Mercian Way SL1: Slou 6B 80
Mercia Wlk. GU21: Wok 89B 168
Mercier Ct. E16 46Lc 93
(off Starboard Way)
Mercier Rd. SW15 57Ab 110
Mercury NW9 25Va 48
(off Quakers Course)
Mercury Cen. TW14: Felt 57W 106
Mercury Ct. E14 49Cc 92
(off Homer Dr.)
Mercury Ct. RM1: Rom 29Hd 56
Mercury Ct. SW9 53Qb 112
(off Southey Rd.)
Mercury Gdns. RM1: Rom 29Gd 56
Mercury Ho. E16 44Hc 93
(off Jude St.)
Mercury Ho. E3 39Cc 72
(off Garrison Rd.)
Mercury Ho. KT17: Ewe 82Wa 174
(off Cheam Rd.)
Mercury Ho. TW8: Bford 51La 108
(off Glenhurst Rd.)
Mercury Ho. W5 42Pa 87
Mercury Mall, The RM1: Rom 29Hd 56
Mercury Rd. TW8: Bford 51La 108
Mercury Way SE14 51Zb 114
Mercy Ter. SE13 57Dc 114
Merebank Cl. RH9: G'stone 2A 210
Merebank La. CR0: Wadd 78Pb 156
Mere Cl. BR6: Farnb 75Qc 160
Mere Cl. SW15 59Za 110
Mereden Ct. AL1: St A 5A 6
(off Tavistock Av.)
Meredith Av. NW2 36Ya 68
Meredith Cl. HA5: Pinn 24Z 45
Meredith Ct. EN8: Chesh 3Zb 20
Meredith Ho. N16 36Ub 71
Meredith M. SE4 56Bc 114
Meredith Rd. RM16: Grays 9C 100
Meredith St. E13 41Jc 93
Meredith St. EC1 4B 218 (41Rb 91)
Meredith Wlk. RM8: Dag 32Ad 75
(off Ellis Av.)
Meredyth Rd. SW13 54Wa 110
Mere End CR0: C'don 73Zb 158
Merefield Gdns. KT20: Tad 91Za 194
Mere Rd. KT13: Weyb 76T 150
Mere Rd. KT20: Tad 96Xa 194
Mere Rd. SE2 47Zc 95
Mere Rd. SL1: Slou 8K 81
Mere Rd. TW10: Dun G 93Gd 202
Mere Rd. TW17: Shep 72R 150
Mereside BR6: Farnb 75Qc 160
Mereside Pk. TW15: Ashf 63S 128
Mereside Pl. GU25: Vir W 4L 147
Meretone Cl. SE4 56Ac 114
Mereton Mans. SE8 53Cc 114
(off Brookmill Rd.)
Merevale Cres. SM4: Mord 72Eb 155
Mereway Rd. TW2: Twick 60Fa 108
Merewood Cl. BR1: Broml 68Qc 138
Merewood Gdns. CR0: C'don 73Zb 158
Merewood Rd. DA7: Bex 54Ed 118
Mereworth Cl. BR2: Broml 71Hc 159
Mereworth Dr. SE18 52Rc 116
Mereworth Ho. SE15 51Yb 114
Merganser Ct. E1 45Wb 91
(off Star Pl.)
Merganser Ct. SE8 51Bc 114
(off Edward St.)
Merganser Gdns. SE28 48Tc 94
MERIDEN 8Aa 13
Meriden Cl. BR1: Broml 66Mc 137
Meriden Cl. IG6: Ilf 25Sc 54
Meriden Ct. SW3 7D 226 (50Gb 89)
(off Chelsea Mnr. St.)
Meriden Ho. N1 1J 219 (39Ub 71)
(off Wilmer Gdns.)
Meriden Way WD25: Wat 8Aa 13
Meredith Ct. KT1: King T 68Pa 131
Meridia Ct. E15 39Ec 72
(off Biggerstaff Rd.)
Meridian Bus. Pk. EN3: Pond E 16Ac 34
Meridian Bus. Pk. EN9: Walt A 7Dc 20
Meridian Cen. CR0: New Ad 82Gc 179
Meridian Cl. NW7 21Ta 47
Meridian Ct. RM17: Grays 52De 121
Meridian Ct. SE15 52Xb 113
(off Gervase St.)
Meridian Ct. SE16 47Wb 91
(off East La.)
Meridian Ct. UB4: Yead 42Y 85
Meridian Ga. E14 47Dc 92
Meridian Ho. SE10: Azof St. 49Gc 93
(off Azof St.)
Meridian Ho. SE10: Royal Hill 52Ec 114
(off Royal Hill)
Meridian Ho. NW1 38Lb 70
(off Baynes St.)
Meridian Ho. SW18 56Eb 111
(off Juniper Dr.)
MERIDIAN PARK 8Ec 20
Meridian Pl. E14 47Dc 92
Meridian Point SE8 51Dc 114
Meridian Rd. SE7 52Mc 115
Meridian Sq. E15 38Fc 73
Meridian Trad. Est. SE7 49Kc 93
Meridian Wlk. N17 23Ub 51
Meridian Water Development N18 23Zb 52
Meridian Water Station (Rail) 23Yb 52
Meridian Way EN3: Pond E 16Zb 34
Meridian Way EN9: Walt A 6Dc 20
Meridian Way N18 22Yb 52
Meridian Way N9 20Yb 34
Meriel Wlk. DA9: Ghithe 56Xd 120
Merifield Rd. SE9 56Lc 115
Merileys Cl. DA3: Lfield 69Ee 143
Merino Cl. E11 28Lc 53
Merino Cl. SM6: W'gton 75Jb 156
Merino Ct. EC1 4E 218 (41Sb 91)
(off Lever St.)
Merino Pl. DA15: Sidc 58Wc 117
Merioneth Ct. W7 43Ha 86
(off Copley Cl.)
Merita Ho. E1 46Wb 91
(off Nesham St.)
Merivale Rd. HA1: Harr 31Ea 66
Merivale Rd. SW15 56Ab 110
Merland Cl. KT20: Tad 92Ya 194
Merland Grn. KT20: Tad 92Ya 194
Merland Ri. KT18: Tatt C 91Ya 194
Merland Ri. KT20: Tad 91Ya 194
Merle Av. UB9: Hare 26K 43
MERLE COMMON 8M 211
Merle Comn. Rd. RH8: Oxt 7L 211
Merle Mans. E20 37Ec 72
(off Glade Wlk.)
Merlewood TN13: S'oaks 95Kd 203
Merlewood Cl. CR3: Cat'm 92Tb 197
Merlewood Dr. BR7: Chst 67Pc 138
Merley Ct. NW9 32Sa 67
Merlin NW9 25Va 48
(off Near Acre)
Merlin Cen., The AL4: St A 2K 7
Merlin Cl. CR0: C'don 77Ub 157
Merlin Cl. CR4: Mitc 69Gb 133
Merlin Cl. EN9: Walt A 6Jc 21
Merlin Cl. IG6: Ilf 22Yc 55
Merlin Cl. RM16: Chaf H 48Ae 99
Merlin Cl. RM5: Col R 23Fd 56
Merlin Cl. SL3: L'ly 51D 104
Merlin Cl. SM6: W'gton 79Pb 156
Merlin Cl. UB5: N'olt 41Y 85
Merlin Ct. BR2: Broml 69Hc 137
Merlin Ct. DA9: Ghithe 58Wd 120
(off Waterstone Way)
Merlin Ct. GU21: Wok 86E 168
Merlin Ct. HA4: Ruis 33T 64
Merlin Ct. HA7: Stan 22Ka 46
(off William Dr.)
Merlin Ct. SE3 56Kc 115
Merlin Cres. HA8: Edg 25Pa 47
Merlin Gdns. BR1: Broml 62Jc 137
Merlin Gdns. RM5: Col R 23Fd 56
Merling Cl. KT9: Chess 78La 152
Merlin Gro. BR3: Beck 70Bc 136
Merlin Gro. IG6: Ilf 24Rc 54
Merlin Hgts. N17 27Xb 51
(off Daneland Wlk.)
Merlin Ho. EN3: Pond E 15Zb 34
Merlin Rd. DA16: Well 56Wc 117
Merlin Rd. E12 33Mc 73
Merlin Rd. RM5: Col R 23Fd 56
Merlin Rd. Nth. DA16: Well 56Wc 117
Merlins Av. HA2: Harr 34Ba 65
Merlins Ct. WC1 4K 217 (41Qb 90)
(off Margery St.)
Merlin St. WC1 4K 217 (41Qb 90)
Merlin Way WD25: Wat 6V 12
Mermagen Dr. RM13: Rain 38Kd 77
Mermaid Cl. DA11: Nflt 59Fe 121
Mermaid Ct. E8 38Vb 71
(off Celandine Dr.)
Mermaid Ct. SE1 1F 231 (47Tb 91)
Mermaid Ct. SE16 46Bc 92
Mermaid Ho. E14 45Ec 92
(off Bazely St.)
Mermaid Twr. SE8 51Bc 114
(off Abinger Gro.)
Mermerus Gdns. DA12: Grav'nd 3H 145
Meroe Ct. N16 33Ub 71
Mero Way NW7 24Ab 48
Merredene St. SW2 58Pb 112
Merriall Cl. DA10: Swans 59Be 121
Merriam Av. E9 37Bc 72
Merriam Cl. E4 22Ec 52
Merrick Rd. UB2: S'hall 47Ba 85
Merrick Sq. SE1 3F 231 (48Tb 91)
Merridene N21 16Rb 33
Merrielands Cres. RM9: Dag 40Bd 75
Merrielands Retail Pk. 39Bd 75
Merrilands Rd. KT4: Wor Pk 74Ya 154
Merrilees Rd. DA15: Sidc 59Uc 116
Merrilyn Cl. KT10: Clay 79Ja 152
Merriman Rd. SE3 53Lc 115
Merrington Rd. SW6 51Cb 111
Merrin Hill CR2: Sande 83Ub 177
Merrion Av. HA7: Stan 22Ma 47
Merrion Ct. HA4: Ruis 32V 64
(off Pembroke Rd.)
Merrist Wood Golf Course 10F 186
Merritt Gdns. KT9: Chess 79La 152
Merritt Ho. RM1: Rom 31Hd 76
(off South St.)
Merritt Rd. SE4 57Bc 114
Merritt Wlk. AL9: Wel G 5D 8
Merrivale N14 16Mb 32
Merrivale NW1 1C 216 (39Lb 70)
(off Camden St.)
Merrivale Av. IG4: Ilf 28Mc 53
Merrivale Gdns. GU21: Wok 9N 167
Merrivale M. HA8: Edg 25Sa 47
Merrivale M. UB7: Yiew 46M 83
Merrow Bldgs. SE1 1C 230 (47Rb 91)
(off Rushworth St.)
Merrow Ct. CR4: Mitc 68Fb 133
Merrow Dr. HP1: Hem H 1G 2
Merrow La. GU4: Burp 100E 188
Merrow La. GU4: Guild 101E 188
Merrow Rd. SM2: Cheam 82Za 174
Merrows Cl. HA6: Nwood 25S 44
Merrow St. SE17 50Tb 91
Merrow Wlk. SE17 7G 231 (50Tb 91)
Merrow Way CR0: New Ad 79Ec 158
Merrydown Way BR7: Chst 67Nc 138
Merryfield SE3 54Hc 115
Merryfield Ct. SW11 54Hb 111
Merryfield Gdns. HA7: Stan 22La 46
Merryfield Ho. SE9 62Lc 137
(off Grove Pk. Rd.)
Merryfields AL4: St A 2J 7
Merryfields UB8: Uxb 40M 63
Merryfields WD24: Wat 10W 12
Merryfields Cl. DA3: Hartl 70Be 143
Merryfields Way SE6 59Dc 114
MERRY HILL 18Da 27
Merryhill Cl. E4 17Dc 34
Merry Hill Mt. WD23: Bush 18Da 27
Merry Hill Rd. WD23: Bush 16Ba 27
Merryhills Cl. TN16: Big H 88Mc 179
Merryhills Ct. N14 15Lb 32
Merryhills Dr. EN2: Enf 14Mb 32
Merrylands KT16: Chert 76G 148
Merrylands Rd. KT23: Bookh 95Ba 191

Merrymeade Chase CM15: B'wood 18Zd 41
Merrymeade Country Pk. 16Zd 41
Merrymeet SM7: Bans 86Hb 175
Merryweather Cl. DA1: Dart 58Pd 119
Merryweather Ct. KT3: N Mald 71Ua 154
Merryweather Ct. N19 34Lb 70
Merryweather Pl. SE10 52Dc 114
Merrywood Gro. KT20: Lwr K 2G 206
Merrywood Pk. RH2: Reig 4K 207
Mersea Ho. IG11: Bark 37Rc 74
Mersey Av. RM14: Upm 30Td 58
Mersey Ct. KT2: King T 67Ma 131
(off Samuel Gray Gdns.)
Mersey Rd. E17 27Bc 52
Mersey Wlk. UB5: N'olt 40Ca 65
Mersham Dr. NW9 29Qa 47
Mersham Pl. CR7: Thor H 68Tb 135
(off Livingstone Rd.)
Mersham Pl. SE20 67Xb 135
Mersham Rd. CR7: Thor H 69Tb 135
MERSTHAM 100Lb 196
Merstham Rd. RH1: Blet 1G 208
Merstham Station (Rail) 100Lb 196
Merten Rd. RM6: Chad H 31Ad 75
Merthyr Ter. SW13 51Xa 110
MERTON 66Eb 133
Merton Abbey Mills SW19 67Eb 133
Merton Av. DA3: Hartl 70Ae 143
(not continuous)
Merton Av. UB10: Hil 38R 64
Merton Av. UB5: N'olt 36Ea 66
Merton Av. W4 49Va 88
Merton Ct. DA16: Well 54Xc 117
Merton Ct. IG1: Ilf 30Nc 54
Merton Ct. WD6: Bore 11Pa 29
Merton Gdns. BR5: Pet W 71Rc 160
Merton Gdns. KT20: Tad 91Za 194
Merton Hall Gdns. SW20 67Ab 132
Merton Hall Rd. SW19 66Ab 132
Merton High St. SW19 66Db 133
Merton Ind. Pk. SW19 67Db 133
Merton La. N6 33Hb 69
Merton Lodge EN5: New Bar 15Eb 31
Merton Mans. SW20 68Za 132
MERTON PARK 68Cb 133
Merton Pk. Pde. SW19 67Bb 133
Merton Park Stop
(London Tramlink) 67Cb 133
Merton Pl. RM16: Grays 9C 100
Merton Pl. SW19 67Eb 133
(off Nelson Gro. Rd.)
Merton Ri. NW3 38Gb 69
Merton Rd. E17 29Ec 52
Merton Rd. EN2: Enf 10Tb 19
Merton Rd. HA2: Harr 32Ea 66
Merton Rd. IG11: Bark 38Vc 75
Merton Rd. IG3: Ilf 31Vc 75
Merton Rd. SE25 71Vb 157
Merton Rd. SL1: Slou 8L 81
Merton Rd. SW18 58Cb 111
Merton Rd. SW19 66Db 133
Merton Rd. WD18: Wat 14X 27
Merton Wlk. KT22: Lea 90Ja 172
Merton Way KT22: Lea 91Ja 192
Merton Way KT8: W Mole 70Da 129
Merton Way UB10: Hil 38R 64
Mertoun Ter. W1 1F 221 (43Hb 89)
(off Seymour Pl.)
Merttins Rd. SE15 57Zb 114
Meru Cl. NW5 35Jb 70
Mervan Rd. SW2 56Qb 112
Mervyn Av. SE9 62Sc 138
Mervyn Rd. TW17: Shep 73S 150
Mervyn Rd. W13 48Ja 86
Merwin Way SL4: Wind 4B 102
Meryfield Cl. WD6: Bore 12Pa 29
Meryton Ho. SL4: Wind 5E 102
Mesne Way TN14: S'ham 84Hd 182
Messaline Av. W3 44Sa 87
Messant Cl. RM3: Hrld W 26Md 57
Messenger Ct. SE16 48Wb 91
(off Spa Rd.)
Messent Rd. SE9 57Lc 115
Messeter Pl. SE9 58Qc 116
Messina Av. NW6 38Cb 69
Messina Way RM9: Dag 40Bd 76
Messiter Ho. N1 1J 217 (39Pb 70)
(off Barnsbury Est.)
Messom M. TW1: Twick 60Ja 108
Metcalfe Av. SM5: Cars 82Gb 175
Metcalfe Ct. SE10 48Hc 93
Metcalf Rd. TW15: Ashf 64R 128
Metcalf Wlk. TW13: Hanw 63Aa 129
Meteor St. SW11 56Jb 112
Meteor Way SM6: W'gton 80Nb 156
Metford Cres. EN3: Enf L 10Cc 20
Methley St. SE11 7A 230 (50Qb 90)
Methuen Cl. HA8: Edg 24Qa 47
Methuen Pk. N10 27Kb 50
Methuen Rd. DA17: Belv 49Dd 96
Methuen Rd. DA6: Bex 56Bd 117
Methuen Rd. HA8: Edg 24Qa 47
Methven Ct. N9 20Wb 33
(off The Broadway)
Methwold Rd. W10 43Za 88
Metro Bus. Cen. SE26 65Bc 136
Metro Central Hgts. SE1 4D 230 (48Sb 91)
(off Newington C'way.)
Metro Cen. BR5: St M Cry 72Xc 161
Metro Cen. WD18: Wat 17S 26
Metro Golf Cen. 24Ya 48
Metro Ind. Cen. TW7: Isle 54Ga 108
Metropolis SE11 4C 230 (48Rb 91)
(off Oswin St.)
Metropolis Plaza E3 42Ac 92
(off Copperfield Rd.)
Metropolitan Bldg. SW10 52Eb 111
Metropolitan Bus. Cen. N1 38Ub 71
(off Enfield Rd.)
Metropolitan Cen., The UB6: G'frd 39Da 65
Metropolitan Cl. E14 43Cc 92
Metropolitan Cres. SW4 57Mb 112
Metropolitan Ho. TW8: Bford 51Pa 109
Metropolitan M. WD18: Wat 14U 26
Metropolitan Pl. WD18: Wat 14V 26
Metropolitan Police FC 72Fa 152
Metropolitan Police Sports Club 13Ca 27
Metropolitan Sta. App. WD18: Wat 13V 26
Metropolitan Sta. Bldgs. W6 49Ya 88
(off Beadon Rd.)
Metropolitan Wharf E1 46Yb 92
Metro Trad. Est. HA9: Wemb 35Ra 67
Meux Cl. EN7: Chesh 3Wb 19
Mews, The TN13: S'oaks
Hartslands Rd. 95Ld 203
Mews, The TN13: S'oaks
Hitchen Hatch La. 96Jd 202
Mews, The AL1: St A 1C 6
Mews, The DA14: Sidc 63Wc 139
Mews, The DA3: Lfield 69Ae 143
(off Bramblefield Cl.)
Mews, The IG4: Ilf 29Mc 53
Mews, The KT10: Clay 79Ga 152
Mews, The N1 39Sb 71
Mews, The N8 27Qb 50
Mews, The RH2: Reig 5K 207
Mews, The RM1: Rom 28Gd 56
Mews, The RM17: Grays 49Ee 99
Mews, The SE16 67Pb 134
Mews, The TW1: Twick 58Ka 108
Mews, The TW12: Hamp H 65Ea 130
Mews, The TW20: Eng G 3P 125
Mews, The WD18: Wat 14Y 27
(off Smith St.)
Mews, The WD25: A'ham 10Ea 14
Mews, The WD7: R'lett 8Ka 14
Mews Cotts. CM13: Ingve 23De 59
Mews End TN16: Big H 90Mc 179
Mews Pl. IG8: Wfd G 21Jc 53
Mews St. E1 46Wb 91
Mexborough NW1 39Lb 70
Mexfield Rd. SW15 57Bb 111
Meyer Grn. EN1: Enf 10Wb 19
Meyer Rd. DA8: Erith 51Fd 118
Meyers Cl. SL3: Slou 7N 81
Meymott St. SE1 7B 224 (46Rb 91)
Meynell Cres. E9 38Zb 72
Meynell Gdns. E9 38Zb 72
Meynell Rd. E9 38Zb 72
Meynell Rd. RM3: Rom 24Kd 57
Meyrick Cl. GU21: Knap 8J 167
Meyrick Ct. AL1: St A 2F 6
Meyrick Ho. E14 43Cc 92
(off Burgess St.)
Meyrick Rd. NW10 37Wa 68
Meyrick Rd. SW11 55Fb 111
Mezen Cl. HA6: Nwood 22T 44
MFA Bowl Lewisham 55Ec 114
Miah Ter. E1 46Wb 91
Miall Wlk. SE26 63Ac 136
Mia M. N13 22Qb 50
Mica Ho. N1 38Qb 70
Micawber Av. UB8: Hil 42Q 84
Micawber Ct. N1 3E 218 (41Sb 91)
(off Windsor Ter.)
Micawber Ho. SE16 47Wb 91
(off Llewellyn St.)
Micawber St. N1 3E 218 (41Sb 91)
Michael Cliffe Ho. EC1 4B 218 (41Rb 91)
(off Finsbury Est.)
Michael Cl. E3 42Bc 92
Michael Edwards Studio Theatre,
The 51Ec 114
(within Cutty Sark)
Michael Faraday Ho. SE17 50Ub 91
(off Beaconsfield Rd.)
Michael Gdns. DA12: Grav'nd 4G 144
Michael Gdns. RM11: Horn 28Md 57
Michael Gaynor Cl. W7 46Ha 86
Michael Haines Ho. SW9 52Qb 112
(off Sth. Island Pl.)
Michael Manley Ind. Est. SW8 53Lb 112
Michaelmas Cl. SW20 69Ya 132
Michael Rd. E11 32Gc 73
Michael Rd. SE25 69Ub 135
Michael Rd. SW6 53Db 111
Michael Robbins Way NW7 22Ab 48
Michaels Cl. SE13 56Gc 115
Michaels La. DA3: Fawk 75Xd 164
Michaels La. TN15: Ash 75Yd 164
MICHAEL SOBELL HOUSE (HOSPICE) 23R 44
Michael Stewart Ho. SW6 51Bb 111
(off Clem Attlee Ct.)
Michelangelo Ct. SE16 50Xb 91
(off Stubbs Dr.)
Micheldever Rd. SE12 58Gc 115
Michelet Cl. GU18: Light 2A 166
Michelham Gdns. KT20: Tad 92Ya 194
Michelham Gdns. TW1: Twick 62Ha 130
Michelle Ct. BR1: Broml 67Hc 137
(off Blyth Rd.)
Michelle Ct. N12 22Eb 49
Michelle Ct. W3 45Ta 87
Michelsdale Dr. TW9: Rich 56Na 109
Michelson Ho. SE11 6J 229 (49Pb 90)
(off Black Prince Rd.)
Michel's Row TW9: Rich 56Na 109
(off Michelsdale Dr.)
Michel Wlk. SE18 50Rc 94
Michie Ct. GU24: Bisl 9E 166
Michigan Av. E12 35Pc 74
Michigan Bldg. E14 46Fc 93
(off Biscayne Av.)
Michigan Ho. E14 48Cc 92
Mickleham Down N12 21Bb 49
Micholls Av. SL9: Chal P 21A 42
Micholls Cotts. SL9: Chal P 22B 42
(off Micholls Av.)
Mick Jagger Cen., The 58Ld 119
(off Shepherd's La.)
Mickledore NW1 2C 216 (40Lb 70)
(off Ampthill Est.)
MICKLEFIELD GREEN 10K 11
Micklefield Rd. HP2: Hem H 2C 4
Micklefield Way WD6: Bore 10Na 15
MICKLEHAM 100La 192
Mickleham By-Pass RH5: Mick 99Ka 192
Mickleham Cl. BR5: St P 68Vc 139
MICKLEHAM DOWNS 98Ma 193
Mickleham Dr. RH5: Mick 98La 192
Mickleham Gdns. SM3: Cheam 79Ab 154
Mickleham Rd. BR5: St P 67Vc 139
Mickleham Way CR0: New Ad 80Fc 159
Micklem Dr. HP1: Hem H 1H 3
Micklethwaite Rd. SW6 51Cb 111
Mickleton Ho. W2 43Cb 89
(off Westbourne Pk. Rd.)
Midas Bus. Cen. RM10: Dag 35Dd 76
Midas Ind. Est. UB8: Uxb 40K 63
Midas Metropolitan Ind. Est.
SM4: Mord 73Ya 154
MID BECKTON 43Pc 94
Midcroft HA4: Ruis 32U 64
Midcroft SL2: Slou 2F 80
Mid Cross La. SL9: Chal P 22B 42
Middle Boy RM4: Abr 13Yc 37
Middle Cl. CR5: Coul 92Qb 196
Middle Cl. KT17: Eps 84Ua 174
Middle Cres. UB9: Den 31F 62
Middle Dartrey Wlk. SW10 52Eb 111
(off Worlds End Est.)
Middle Dene NW7 20Ta 29
Middle Down WD25: A'ham 8Da 13
Middle Farm Cl. KT24: Eff 99Z 191
Middle Farm Pl. KT24: Eff 99Y 191
Middlefield NW8 39Fb 69
Middlefield Cl. CR5: Chips 91Gb 195
Middlefielde W13 43Ka 86
Middlefield Gdns. IG2: Ilf 30Rc 54
Middlefields CR0: Sels 81Ac 178
Middle Furlong WD23: Bush 14Da 27
MIDDLE GREEN 46A 82
Middle Grn. SL3: L'ly 46A 82
Middle Grn. TW18: Staines 66M 127
Middle Grn. Cl. KT5: Surb 72Pa 153
Middlegreen Rd. SL3: L'ly 7P 81
Middleham Ct. DA2: Dart 58Rd 119
(off Osbourne Rd.)
Middleham Gdns. N18 23Wb 51
Middleham Rd. N18 23Wb 51
Middle Hill HP1: Hem H 2G 2
Middle Hill TW20: Egh 3N 125
Middle Hill TW20: Eng G 3N 125
Middle La. HP3: Bov 1C 10
Middle La. KT17: Eps 84Ua 174
Middle La. N8 29Nb 50
Middle La. TN15: Seal 93Pd 203
Middle La. TW11: Tedd 65Ha 130
Middle La. M. N8 29Nb 50
Middlemarch Lodge WD3: Rick 17N 25
Middlemead Cl. KT23: Bookh 97Ca 191
Middlemead Rd. KT23: Bookh 97Ba 191
Middle Mill Halls of Residence
KT1: King T 69Pa 131
Middle New St. EC4 2A 224 (44Qb 90)
(off Pemberton Row)
Middle Ope WD24: Wat 9X 13
Middle Pk. Av. SE9 58Mc 115
Middle Path HA2: Harr 32Fa 66
Middle Rd. CM13: Ingve 22Ee 59
Middle Rd. E13 40Jc 73
Middle Rd. EN4: E Barn 16Gb 31
Middle Rd. HA2: Harr 33Fa 66
Middle Rd. KT22: Lea 93Ka 192
Middle Rd. SW16 68Mb 134
Middle Rd. UB9: Den 31E 62
Middle Row W10 42Ab 88
Middlesborough Ho. RM3: Rom 24Nd 57
(off Kingsbridge Cir.)
Middlesbrough Rd. N18 23Wb 51
Middlesex Bldg., The E1 1J 225 (43Ub 91)
(off Artillery La.)
Middlesex Bus. Cen. UB2: S'hall 47Ca 85
Middlesex CCC 3C 214 (41Fb 89)
Middlesex Cl. UB1: S'hall 42Da 85
Middlesex Ct. HA1: Harr 29Ha 46
Middlesex Ct. KT15: Add 78L 149
(off Bush Cl.)
Middlesex Ct. TW8: Bford 50La 86
(off Glenhurst Rd.)
Middlesex Ct. W4 49Va 88
Middlesex Filter Beds Nature
Reserve 34Zb 72
Middlesex Guildhall 2F 229 (47Nb 90)
(off Lit. George St.)
Middlesex Ho. HA0: Wemb 39Ma 67
Middlesex Ho. UB8: Uxb 38L 63
(off Mercer Wlk.)
Middlesex Pl. E9 37Yb 72
(off Elsdale St.)
Middlesex Rd. CR4: Mitc 71Nb 156
Middlesex St. E1 1J 225 (43Ub 91)
Middlesex University Archway Campus 32Lb 70
Middlesex University Hendon Campus 28Xa 48
Middlesex Wharf E5 33Yb 72
Middle St. CR0: C'don 75Sb 157
(not continuous)
Middle St. EC1 7D 218 (43Sb 91)
Middle Temple Hall 4K 223 (45Qb 90)
Middle Temple La. EC4 3K 223 (44Qb 90)
Middleton Av. DA14: Sidc 65Xc 139
Middleton Av. E4 21Bc 52
Middleton Av. UB6: G'frd 40Fa 66
Middleton Cl. E4 20Bc 34
Middleton Dr. HA5: Eastc 27W 44
Middleton Dr. SE16 47Zb 92
Middleton Gdns. IG2: Ilf 30Rc 54
Middleton Green 41Xb 91
(off Old Bethnal Green Road)
Middleton Gro. IG11: Bark 41Wc 95
Middleton Gro. N7 36Nb 70
Middleton Hall La. CM15: B'wood 19Ae 41
Middleton Ho. E8 38Wb 71
Middleton Ho. SE1 4F 231 (48Tb 91)
(off Burbage Cl.)
Middleton Ho. SW1 6E 228 (49Mb 90)
(off Causton St.)
Middleton M. N7 36Nb 70
Middleton Pl. W1 1B 222 (43Lb 90)
(off Langham St.)
Middleton Rd. CM15: Shenf 18Ae 41
Middleton Rd. E8 38Vb 71
Middleton Rd. KT11: D'side 91X 191
Middleton Rd. KT19: Ewe 82Ta 173
Middleton Rd. NW11 31Cb 69
Middleton Rd. SM4: Cars 72Db 155
Middleton Rd. SM4: Mord 72Db 155
Middleton Rd. SM5: Cars 73Fb 155
Middleton Rd. UB3: Hayes 43T 84
Middleton Rd. WD3: Rick 18J 25
Middleton St. E2 41Xb 91
Middleton Way SE13 56Fc 115
Middle Wlk. GU21: Wok 89A 168
Middle Wlk. SL1: Burn 1A 80
Middle Way DA18: Erith 48Ad 95
Middle Way SW16 68Mb 134
Middle Way UB4: Yead 42Y 85
Middle Way WD24: Wat 9W 12
Middleway NW11 29Db 49
Middle Way, The HA3: W'stone 26Ha 46
Middle Yd. SE1 6H 225 (46Ub 91)
Middlings, The TN13: S'oaks 98Hd 202
Middlings Ri. TN13: S'oaks 97Hd 202
Middlings Wood TN13: S'oaks 97Jd 202
Midfield Av. BR8: Hext 65Jd 140
Midfield Av. DA7: Bex 55Ed 118
Midfield Pde. DA7: Bex 55Ed 118
Midfield Way BR5: St P 67Wc 139
Midford Ho. NW4 28Ya 48
(off Belle Vue Est.)
Midford Pl. W1 6C 216 (42Lb 90)
Midgarth Cl. KT22: Oxs 86Ea 172
Midholm HA9: Wemb 32Qa 67
Midholm NW11 28Db 49
Midholm Cl. NW11 28Db 49
Midholm Rd. CR0: C'don 75Ac 158
Midhope Cl. GU22: Wok 91A 188
Midhope Gdns. GU22: Wok 91A 188
Midhope Ho. WC1 4G 217 (41Nb 90)
(off Midhope St.)
Midhope Rd. GU22: Wok 91A 188
Midhope St. WC1 4G 217 (41Nb 90)
Midhurst SE26 65Yb 136
Midhurst Av. CR0: C'don 73Qb 156
Midhurst Av. N10 27Jb 50
Midhurst Cl. RM12: Horn 35Jd 76
Midhurst Gdns. UB10: Hil 39S 64
Midhurst Hill DA6: Bex 58Cd 118
Midhurst Ho. E14 44Bc 92
(off Salmon La.)
Midhurst Pde. N10 27Jb 50
(off Fortis Grn.)
Midhurst Rd. W13 47Ja 86
Midhurst Way E5 35Wb 71
Mid Kent Golf Course 2D 144
Midland Goods Shed N1 1F 217 (39Nb 70)
(off Handyside St.)
Midland Ho. AL1: St A 2C 6
(off Alma Rd.)
Midland Pde. NW6 37Db 69
Midland Pl. E14 50Ec 92
Midland Rd. E10 31Ec 72
Midland Rd. HP2: Hem H 2M 3
Midland Rd. NW1 2E 216 (40Mb 70)
Midland Ter. NW10 42Ua 88
Midland Ter. NW2 34Za 68
Midleton Rd. KT3: N Mald 69Sa 131
Midlothian Rd. E3 42Bc 92
(off Burdett Rd.)
Midmoor Rd. SW12 60Lb 112
Midmoor Rd. SW19 67Za 132
Midnight Av. SE5 52Rb 113
Midship Cl. SE16 46Zb 92
Midship Point E14 47Cc 92
(off The Quarterdeck)
Midstrath Rd. NW10 35Ua 68
Mid St. RH1: S Nut 9F 208
Midsummer Av. TW4: Houn 56Ba 107
Midsummer Ct. TW18: Staines 63J 127
(off Kingston Rd.)
Midsummer Wlk. GU21: Wok 8P 167
Midway AL3: St A 5P 5
Midway KT12: Walt T 75X 151
Midway SM3: Sutt 73Bb 155
Midway Av. KT16: Chert 69J 127
Midway Av. TW20: Thorpe 69D 126
Midway Cl. TW18: Staines 62K 127
(not continuous)
Midway Ho. EC1 3C 218 (41Rb 91)
(off Manningford Cl.)
Midwinter Cl. DA16: Well 55Wc 117
Midwood Cl. NW2 34Xa 68
Miena Way KT21: Asht 89Ma 173
Miers Cl. E6 39Qc 74
Mighell Av. IG4: Ilf 29Mc 53
Mikado Cl. UB9: Hare 26M 43
Mikardo Ct. E14 45Ec 92
(off Poplar High St.)
Mike Spring Ct. DA12: Grav'nd 3F 144
Milan Ct. N11 25Mb 50
Milan Ho. AL1: St A 3D 6
Milan Rd. UB1: S'hall 47Ba 85
Milan Wlk. CM14: B'wood 18Xd 40
Milborne Gro. SW10 50Eb 89
Milborne St. E9 37Yb 72
Milborough Cres. SE12 58Gc 115
Milbourne Ct. WD17: Wat 12W 26
Milbourne Ho. KT1: King T 67Qa 131
(off Coombe Rd.)
Milbourne La. KT10: Esh 79Ea 152
Milbourne Pl. KT19: Ewe 77Ta 153
Milbrook KT10: Esh 79Ea 152
Milburn Dr. UB7: Yiew 45N 83
Milburn Wlk. KT18: Eps 87Ua 174
Milby Ct. WD6: Bore 11Pa 29
Milcombe Cl. GU21: Wok 10N 167
Milcote St. SE1 2B 230 (47Rb 91)
Mildenhall Ho. RM3: Rom 22Qd 57
(off Redcar Rd.)
Mildenhall Rd. E5 35Yb 72
Mildenhall Rd. SL1: Slou 4J 81
Mildmay Av. N1 37Tb 71
Mildmay Gro. Nth. N1 36Tb 71
Mildmay Gro. Sth. N1 36Tb 71
MILDMAY HOSPITAL 4K 219 (41Vb 91)
Mildmay Pk. N1 36Tb 71
Mildmay Pl. N16 36Ub 71
Mildmay Pl. TN14: S'ham 83Hd 182
Mildmay Rd. IG1: Ilf 34Rc 74
Mildmay Rd. N1 36Tb 71
Mildmay Rd. RM7: Rom 29Ed 56
Mildmay St. N1 37Tb 71
Mildred Av. UB3: Harl 49T 84
Mildred Av. UB5: N'olt 36Da 65
Mildred Av. WD18: Wat 14V 26
Mildred Av. WD6: Bore 14Qa 29
(not continuous)
Mildred Cl. DA1: Dart 58Qd 119
Mildred Ct. CR0: C'don 74Wb 157
Mildred Rd. DA8: Erith 50Gd 96
Mildrose Ct. NW6 41Cb 89
(off Malvern M.)
Mildura Ct. N8 28Pb 50
Mile Cl. EN9: Walt A 5Ec 20
MILE END 42Bc 92
Mile End, The E17 25Zb 52
Mile End Climbing Wall 41Ac 92
MILE END GREEN 68Zd 143
MILE END HOSPITAL 42Zb 92
Mile End Pk. 40Ac 72
Mile End Pk. Leisure Cen. 43Ac 92
Mile End Pl. E1 42Zb 92
Mile End Rd. E1 43Yb 92
Mile End Rd. E3 42Ac 92
Mile End Stadium 43Bc 92
Mile End Station (Underground) 42Bc 92
Milehams Ind. Est. RM19: Purf 48Qd 97
Mile Ho. Cl. AL1: St A 5E 6
Mile Ho. La. AL1: St A 5E 6
Mile Path GU22: Wok 3K 187
Mile Rd. SM6: Bedd 74Kb 156
(not continuous)
Mile Rd. SM6: W'gton 74Kb 156
(not continuous)
Miles Bldgs. NW1 7D 214 (43Gb 89)
(off Penfold Pl.)
Miles Cl. SE28 46Tc 94
Miles Ct. CR0: C'don 75Rb 157
(off Cuthbert Rd.)
Miles Ct. E1 44Xb 91
(off Tillman St.)
Miles Dr. SE28 46Uc 94
MILES GREEN 9D 166
Miles Ho. SE10 50Gc 93
(off Tuskar St.)
Miles Ho. UB9: Den 29J 43
Miles La. KT11: Cobh 85Aa 171
Miles La. RH8: Tand 8D 210
Miles La. RH9: S God 8D 210
Miles Lodge E15 36Fc 73
(off Colegrave Rd.)
Miles Lodge HA1: Harr 29Fa 46
Milespit Hill NW7 22Xa 48
Miles Pl. KT5: Surb 70Pa 131
Miles Pl. NW8 7C 214 (43Fb 89)
(off Broadley St.)
Miles Rd. CR4: Mitc 69Gb 133
Miles Rd. KT19: Eps 84Ta 173
Miles Rd. N8 27Nb 50
Miles St. SW8 51Nb 112
(not continuous)
Milestone Cl. GU23: Rip 94J 189
Milestone Cl. N9 19Wb 33
Milestone Cl. SM2: Sutt 79Fb 155
Milestone Ct. E10 31Dc 72
Milestone Dr. CR8: Purl 86Pb 176
MILESTONE GREEN 56Sa 109
Milestone Ho. KT1: King T 69Ma 131
(off Surbiton Rd.)
Milestone Rd. DA2: Dart 58Rd 119
Milestone Rd. SE19 65Vb 135
Miles Way N20 19Gb 31
Milfoil St. W12 45Wa 88
Milford Cl. SE2 51Ad 117
Milford Ct. SL1: Slou 7L 81
Milford Ct. UB1: S'hall 46Ca 85
Milford Gdns. CR0: C'don 71Yb 158
Milford Gdns. HA0: Wemb 35Ma 67
Milford Gdns. HA8: Edg 24Qa 47
Milford Gro. SM1: Sutt 77Eb 155
Milford La. WC2 4K 223 (45Qb 90)
Milford M. SW16 62Pb 134
Milford Rd. RM16: Grays 46Fe 99
Milford Rd. UB1: S'hall 45Ca 85
Milford Rd. W13 46Ka 86
Milford Towers SE6 59Dc 114
Milking La. BR2: Kes 83Mc 179
Milking La. BR6: Downe 84Nc 180
Milk St. BR1: Broml 65Kc 137
Milk St. E16 46Rc 94
Milk St. EC2 2E 224 (44Sb 91)
Milkwell Gdns. IG8: Wfd G 24Kc 53
Milkwell Yd. SE5 53Sb 113
Milkwood Rd. SE24 57Rb 113
Milk Yd. E1 45Yb 92
Mill, The KT13: Weyb 75Q 150
Millais Av. E12 36Qc 74
Millais Ct. UB5: N'olt 40Z 65
(off Academy Gdns.)
Millais Cres. KT19: Ewe 78Ua 154
Millais Gdns. HA8: Edg 26Qa 47
Millais Ho. SW1 6F 229 (49Nb 90)
(off Marsham St.)
Millais Pl. RM18: Tilb 2C 122
Millais Rd. E11 35Ec 72
Millais Rd. EN1: Enf 15Vb 33
Millais Rd. KT3: N Mald 73Ua 154
Millais Way KT19: Ewe 77Sa 153
Millan Cl. KT15: New H 82K 169
Millard Cl. N16 36Ub 71
Millard Rd. SE8 50Bc 92
Millard Ter. RM10: Dag 37Cd 76
Mill Av. UB8: Uxb 40L 63
Millbank HP3: Hem H 6M 3
Millbank SM6: W'gton 78Mb 156
Millbank SW1 4F 229 (48Nb 90)
Millbank Ct. SW1 5F 229 (49Nb 90)
(off John Islip St.)
Millbank Twr. SW1 6F 229 (49Nb 90)
Millbank Way SE12 57Jc 115
Millbourne Ho. KT10: Esh 79Ea 152
(off Princess Sq.)
Millbourne Rd. TW13: Hanw 63Aa 129
Mill Bri. EN5: Barn 16Bb 31
Mill Bri. Pl. UB8: Uxb 40L 63
Millbro BR8: Hext 67Jd 140
Millbroke M. IG8: Wfd G 23Gc 53
Millbrook KT13: Weyb 77U 150
Millbrook Av. DA16: Well 56Tc 116
Millbrooke Ct. SW15 57Ab 110
(off Keswick Rd.)
Millbrook Gdns. RM2: Rom 26Gd 56
Millbrook Gdns. RM6: Chad H 30Bd 55
Millbrook Ho. SE15 51Wb 113
(off Peckham Pk. Rd.)
Millbrook Pk. NW7 23Ab 48
Millbrook Pas. SW9 55Rb 113
Millbrook Pl. NW1 1B 216 (40Lb 70)
(off Hampstead Rd.)
Mill Brook Rd. BR5: St M Cry 70Yc 139
Millbrook Rd. N9 18Xb 33
Millbrook Rd. SW9 55Rb 113
Millbrook Rd. WD23: Bush 11Ba 27
Millbrook Way SL3: Poyle 54G 104
Mill Cleave KT14: W Byf 84J 169
(off Claremont Rd.)
Mill Cl. HP3: Hem H 7A 4
Mill Cl. KT23: Bookh 96Ca 191
Mill Cl. NW7 25Ab 48
Mill Cl. SM5: Cars 75Jb 156
Mill Cl. UB7: W Dray 48M 83
Mill Cnr. EN5: Barn 11Bb 31
Mill Ct. DA4: S Dar 68Sd 142
Mill Ct. E10 34Ec 72
Mill Ct. E5 33Yb 72
Mill Ct. SE28 45Xc 95
(off Titmuss Av.)
Mill Ct. SL2: Slou 6K 81
Millcroft Ho. SE6 63Ec 136
(off Melfield Gdns.)
Mill Dr. HA4: Ruis 31T 64
Millen Ct. DA4: Hort K 70Sd 142
MILL END 18H 25
Millender Wlk. SE16 49Yb 92
(off New Rotherhithe Rd.)
Millennium Bridge Southwark 5D 224 (45Sb 91)
Millennium Bri. Ho. EC4 4D 224 (45Sb 91)
(off Up. Thames St.)
Millennium Bus. Cen. NW2 33Xa 68
Millennium Cen., The 35Fd 76
Millennium Cl. E16 44Kc 93
Millennium Cl. UB8: Uxb 40L 63
Millennium Ct. EN8: Chesh 1Xb 19
Millennium Dr. E14 49Fc 93
Millennium Ho. E17 29Zb 52
Millennium Ho. SW15 57Ab 110
(off Plaza Gdns.)
Millennium Pl. E2 40Xb 71
Millennium Sq. SE1 1K 231 (47Vb 91)
Millennium Wlk. CM14: B'wood 19Zd 41
(off High St.)
Millennium Way SE10 47Gc 93
Millennium Wharf WD3: Rick 17N 25
Miller Av. EN3: Enf L 10Cc 20
Miller Cen., The 96Wb 197
Miller Cl. BR1: Broml 64Kc 137
Miller Cl. CR4: Mitc 73Hb 155
Miller Cl. HA5: Pinn 26Y 45
Miller Cl. RM5: Col R 24Cd 56
Miller Ct. DA7: Bex 55Ed 118
Miller Ho. E1 43Zb 92
(off Shandy St.)
Miller Ho. W10 42Bb 89
(off Harrow Rd.)

Miller Pl. KT19: Eps 84Na 173
Miller Pl. SL9: Ger X 29A 42
Miller Rd. CR0: C'don 74Pb 156
Miller Rd. DA12: Grav'nd 1J 145
Miller Rd. SW19 65Fb 133
Miller's Av. E8 36Vb 71
Millers Cl. DA1: Dart 59Md 119
Millers Cl. IG7: Chig 19Xc 37
Millers Cl. KT12: Hers 77Y 151
Millers Cl. NW7 21Wa 48
Millers Cl. TW18: Staines 64K 127
Millers Cl. WD3: Chor 13H 25
Millers Copse KT18: Eps D 91Ta 193
Millers Ct. HA0: Wemb 40Na 67
(off Vicars Bri. Cl.)
Millers Ct. TW20: Egh 65F 126
Millers Grn. Cl. EN2: Enf 13Rb 33
Miller's House Visitor Cen. 41Ec 92
(off Three Mill La.)
Miller's La. IG7: Chig 18Xc 37
Miller's La. SL4: Old Win 8K 103
Millers Mdw. Cl. SE3 56Hc 115
Miller Smith Cl. KT20: Tad 95Ya 194
Millers Ri. AL1: St A 3C 6
Millers Row E20 36Dc 72
Miller's Ter. E8 36Vb 71
Miller St. NW1 1B 216 (40Lb 70)
(not continuous)
Millers Way W6 47Ya 88
Millers Wharf Ho. E1 46Wb 91
(off St Katherine's Way)
Millers Yd. N3 25Db 49
Miller Wlk. SE1 7A 224 (46Qb 90)
Mills Sq. SW9 56Qb 112
Millet Rd. UB6: G'frd 40Da 65
Mill Farm Av. TW16: Sun 66U 128
Mill Farm Bus. Pk. TW4: Houn 59Aa 107
Mill Farm Cl. HA5: Pinn 26Y 45
Mill Farm Cres. TW4: Houn 60Aa 107
Millfield DA3: Nw A G 75Ae 165
Millfield HP4: Berk 1A 2
Millfield KT1: King T 69Pa 131
Millfield N4 33Qb 70
Millfield TW16: Sun 67T 128
Millfield Av. E17 25Ac 52
Millfield Cl. AL2: Lon C 8H 7
Millfield Cl. RM11: Horn 31Jd 76
Millfield Dr. DA11: Nflt 1A 144
Millfield Ho. WD18: Wat 16T 26
Millfield La. DA3: Nw A G 75Ae 165
Millfield La. KT20: Kgswd 97Bb 195
Millfield La. N6 32Gb 69
Millfield Pl. N6 33Jb 70
Millfield Rd. HA8: Edg 26Sa 47
Millfield Rd. TN15: W King 79Td 164
Millfield Rd. TW4: Houn 60Aa 107
Millfields Cl. BR5: St P 70Xc 139
Millfields Cotts. BR5: St M Cry 70Yc 139
Millfields Rd. E5 35Yb 72
Millfield Theatre 21Tb 51
Millfield Wlk. HP3: Hem H 5A 4
Mill Fld. Way RM15: Avel 45Sd 98
Mill Footpath RM1: Rom 30Hd 56
(off Thurloe Gdns.)
Millford GU21: Wok 9M 167
Mill Gdns. SE26 62Xb 135
Mill Grn. CR4: Mitc 73Jb 156
Mill Grn. Bus. Pk. CR4: Mitc 73Jb 156
Mill Grn. Rd. CR4: Mitc 73Hb 155
Millgrove St. SW11 53Jb 112
Millharbour E14 47Dc 92
Millhaven Cl. RM6: Chad H 30Xc 55
Millhedge Cl. KT11: Cobh 88Aa 171
Mill Hill CM15: Shenf 17Ae 41
MILL HILL 22Ua 48
Mill Hill SW13 54Wa 110
Mill Hill Broadway Bus Station 23Ua 48
Mill Hill Broadway Station (Rail) 23Ua 48
Mill Hill Cir. NW7 22Va 48
22Va 48
Mill Hill East Station (Underground) 24Ab 48
Mill Hill Golf Course 19Ua 30
Mill Hill Gro. W3 46Sa 87
Mill Hill Ind. Est. NW7 23Va 48
Mill Hill La. DA12: Shorne 4M 145
Mill Hill Old Railway Nature Reserve 23Sa 47
Mill Hill Rd. SW13 54Wa 110
Mill Hill Rd. W3 47Ra 87
Mill Hill School Sports Cen. 21Wa 48
Mill Hill Ter. W3 46Ra 87
Millhoo Ct. EN9: Walt A 6Hc 21
Mill Ho. IG8: Wfd G 22Hc 53
Mill Ho. Cl. DA4: Eyns 74Nd 163
Mill Ho. La. TW20: Thorpe 70D 126
Millhouse La. WD5: Bedm 9F 4
Millhouse Pl. SE27 63Rb 135
Millicent Fawcett Ct. N17 25Vb 51
Millicent Gro. N13 22Rb 51
Millicent Preston Ho. IG11: Bark 39Tc 74
(off Ripple Rd.)
Millicent Rd. E10 32Bc 72
Milligan St. E14 45Bc 92
Milliner Ho. SW10 52Eb 111
(off Hortensia Rd.)
Milliner's Ct. AL1: St A 2C 6
Milliners Ct. IG10: Lough 12Qc 36
Milliners Ho. SE1 2J 231 (47Ub 91)
(off Bermondsey St.)
Milliners Ho. SW18 56Cb 111
Milling Rd. HA8: Edg 24Ta 47
Millington Cl. SL1: Slou 6D 80
Millington Ho. N16 34Tb 71
Millington Rd. UB3: Harl 48U 84
Mill La. BR6: Downe 82Qc 180
Mill La. CR0: Wadd 76Pb 156
Mill La. DA4: Eyns 74Nd 163
Mill La. E4 13Dc 34
Mill La. GU23: Rip 91M 189
Mill La. GU24: Pirb 6B 186
Mill La. IG8: Wfd G 22Hc 53
Mill La. KT14: Byfl 85P 169
Mill La. KT17: Ewe 81Va 174
Mill La. KT20: Walt H 96Za 194
Mill La. KT22: Fet 94Ja 192
Mill La. NW6 36Bb 69
Mill La. RH1: Mers 3C 208
Mill La. RH8: Oxt 4K 211
Mill La. RM16: Chaf H 49Zd 99
Mill La. RM16: Ors 3C 100
(not continuous)
Mill La. RM20: Chaf H 50Zd 99
Mill La. RM20: Grays 50Zd 99
Mill La. RM4: Nave 11Kd 39
Mill La. RM6: Chad H 30Ad 55
Mill La. SE18 50Qc 94
Mill La. SE8 53Cc 114
(off Deptford Bri.)
Mill La. SL3: Hort 55D 104
Mill La. SL4: Wind 2E 102
Mill La. SL5: S'hill 8D 124
Mill La. SL9: Ger X 30B 42
Mill La. SM5: Cars 77Hb 155
Mill La. TN14: S'ham 82Hd 182
Mill La. TN14: S'oaks 93Ld 203
Mill La. TN15: Bor G 94Zd 205
Mill La. TN15: Igh 94Zd 205
Mill La. TN16: Westrm 99Sc 200
Mill La. TW20: Thorpe 70E 126
Mill La. WD3: Crox G 16S 26
Mill La. WD4: K Lan 1Q 12
Mill La. Trad. Est. CR0: Wadd 76Pb 156
Millman Ct. WC1 6H 217 (42Pb 90)
(off Millman St.)
Millman M. WC1 6H 217 (42Pb 90)
Millman Pl. WC1 6J 217 (42Pb 90)
(off Millman St.)
Millman Rd. E16 45Pc 94
Millman St. WC1 6H 217 (42Pb 90)
Millmark Gro. SE14 54Ac 114
Millmarsh La. EN3: Brim 12Ac 34
Mill Mead TW18: Staines 63H 127
Millmead KT10: Esh 75Ca 151
Millmead KT14: Byfl 84P 169
Millmead Ind. Cen. N17 26Xb 51
Mill Mead Rd. N17 27Xb 51
MILL MEADS 40Fc 73
Mill Pk. Av. RM12: Horn 33Nd 77
Mill Pl. BR7: Chst 67Qc 138
Mill Pl. DA1: Cray 56Jd 118
Mill Pl. E14 44Ac 92
Mill Pl. KT1: King T 69Pa 131
Mill Pl. SL3: Dat 4P 103
Mill Pl. Cvn. Pk. SL3: Dat 4N 103
Mill Plat TW7: Isle 54Ja 108
(not continuous)
Mill Plat Av. TW7: Isle 54Ja 108
Mill Pond Cl. SW8 52Mb 112
Mill Pond Cl. TN14: S'oaks 93Md 203
Millpond Ct. KT15: Add 78N 149
Millpond Est. SE16 47Xb 91
Millpond Pl. SM5: Cars 76Jb 156
Mill Pond Rd. DA1: Dart 58Nd 119
Mill Ridge HA8: Edg 22Pa 47
Mill River Trad. Est. EN3: Pond E 13Ac 34
Mill Rd. DA11: Nflt 9A 122
Mill Rd. DA2: Hawl 63Pd 141
Mill Rd. DA8: Erith 52Ed 118
Mill Rd. E16 46Kc 93
Mill Rd. IG1: Ilf 34Qc 74
Mill Rd. KT10: Esh 75Ca 151
Mill Rd. KT11: Cobh 87Y 171
Mill Rd. KT17: Eps 84Va 174
Mill Rd. KT20: Tad 95Za 194
Mill Rd. RM15: Avel 45Sd 98
Mill Rd. RM19: Purf 51Rd 119
Mill Rd. SW19 66Eb 133
Mill Rd. TN13: Dun G 92Gd 202
Mill Rd. TW2: Twick 61Ea 130
Mill Rd. UB7: W Dray 48L 83
Mill Row DA5: Bexl 60Dd 118
Mill Row N1 1J 219 (39Ub 71)
Mills Cl. UB10: Hil 40Q 64
Mills Ct. EC2 4H 219 (41Ub 91)
(off Curtain Rd.)
Mills Ct. WD6: Bore 12Sa 29
Mills Cres. TN15: Seal 91Pd 203
Mills Gro. E14 44Ec 92
Mills Gro. NW4 27Za 48
Mill Shaw RH8: Oxt 4K 211
Millshott Cl. SW6 53Ya 110
Mills Ho. SW8 53Lb 112
(off Thessaly Rd.)
Millside SM5: Cars 75Hb 155
Millside Ct. KT23: Bookh 97Ca 191
Millside Ct. SL0: Thorn 47K 83
Millside Ind. Est. DA1: Dart 56Md 119
Millside Pl. TW7: Isle 54Ka 108
Millsmead Way IG10: Lough 12Pc 36
Mills Rd. KT12: Hers 78Y 151
Mills Row W4 49Ta 87
Mills Spur SL4: Old Win 9M 103
Millstead Cl. KT20: Tad 94Xa 194
Millstone Cl. DA4: S Dar 68Sd 142
Millstone Cl. E15 37Fc 73
Millstone M. DA4: S Dar 67Sd 142
Millstream Cl. N13 22Qb 50
Millstream Ho. SE16 47Xb 91
(off Jamaica Rd.)
Millstream La. SL1: Slou 6C 80
Mill Stream Lodge WD3: Rick 19J 25
Millstream Rd. SE1 2K 231 (47Vb 91)
Mill St. HP3: Hem H 5M 3
(off Durrants Hill Rd.)
Mill St. KT1: King T 69Na 131
Mill St. RH1: Redh 7N 207
Mill St. SE1 2K 231 (47Vb 91)
Mill St. SL2: Slou 6K 81
Mill St. SL3: Coln 52H 105
Mill St. TN16: Westrm 99Tc 200
Mill St. W1 4A 222 (45Kb 90)
Mills Way CM13: Hut 18Ee 41
Mills Yd. SW6 55Db 111
Mill Trad. Est., The NW10 41Sa 87
Mill Va. BR2: Broml 68Hc 137
Mill Vw. AL2: Park 9B 6
Mill Vw. Cl. KT17: Ewe 80Va 154
Millview Cl. RH2: Reig 2L 207
Mill Vw. Gdns. CR0: C'don 76Zb 158
Millview Pk. TN15: W King 82Wd 184
MILLWALL 49Dc 92
Millwall Dock Rd. E14 48Cc 92
Millwall FC 50Yb 92
Millwall Pk. 49Ec 92
Millwards AL10: Hat 3D 8
Mill Way KT18: Head 97Qa 193
Mill Way KT18: Lea 97Qa 193
Mill Way KT22: Lea 96Pa 193
Mill Way TW14: Felt 57X 107
Mill Way WD23: Bush 12Aa 27
Mill Way WD3: Rick 18H 25
Millway NW7 21Ua 48
Millway RH2: Reig 6M 207
Millway Gdns. UB5: N'olt 37Ba 65
Millwell Cres. IG7: Chig 22Tc 54
Mill W. SL2: Slou 6K 81
Millwood Rd. BR5: St P 69Yc 139
Millwood Rd. TW3: Houn 57Ea 108
Millwood St. W10 43Ab 88
Millwrights Wlk. HP3: Hem H 7P 3
(off Belswains La.)
Mill Yd. E1 45Wb 91
Mill Yd. Ind. Est. HA8: Edg 25Ra 47
Milman Cl. HA5: Pinn 27Z 45
Milman Rd. NW6 40Za 68
Milman's Ho. SW10 51Fb 111
(off Milman's St.)
Milman's St. SW10 51Fb 111
Milne Ct. E18 25Jc 53
Milne Feild HA5: Hat E 24Ca 45
Milne Gdns. SE9 57Nc 116
Milne Ho. SE18 49Pc 94
(off Ogilby St.)
Milne Pk. E. CR0: New Ad 83Fc 179
Milne Pk. W. CR0: New Ad 83Fc 179
Milner App. CR3: Cat'm 93Wb 197
Milner Cl. CR3: Cat'm 94Vb 197
Milner Cl. WD25: Wat 6X 13
Milner Ct. SE15 52Vb 113
(off Colegrove Rd.)
Milner Ct. WD23: Bush 16Da 27
Milner Dr. KT11: Cobh 84Ba 171
Milner Dr. TW2: Whitt 59Fa 108
Milner Pl. N1 39Qb 70
Milner Pl. SE1 1J 229 (47Pb 90)
Milner Pl. SM5: Cars 77Jb 156
Milner Rd. CR3: Cat'm 94Wb 197
Milner Rd. CR7: Thor H 69Tb 135
Milner Rd. E15 41Gc 93
Milner Rd. KT1: King T 69Ma 131
Milner Rd. RM8: Dag 33Yc 75
Milner Rd. SM4: Mord 71Fb 155
Milner Rd. SW19 67Db 133
Milner Sq. N1 38Rb 71
Milner St. SW3 5F 227 (49Hb 89)
Milner Wlk. SE9 61Tc 138
Milne Way UB9: Hare 25K 43
Milnthorpe Rd. W4 51Ta 109
Milo Gdns. SE22 58Vb 113
Milo Rd. SE22 58Vb 113
Milrood Ho. E1 43Zb 92
(off Stepney Grn.)
Milroy Av. DA11: Nflt 1A 144
Milroy Wlk. SE1 6B 224 (46Rb 91)
Milson Rd. W14 48Za 88
Milstead Ho. E5 36Xb 71
Milthorne Cl. WD3: Crox G 15P 25
MILTON 8E 122
Milton Av. CR0: C'don 73Tb 157
Milton Av. DA12: Grav'nd 10E 122
Milton Av. E6 38Mc 73
Milton Av. EN5: Barn 15Bb 31
Milton Av. N6 31Lb 70
Milton Av. NW10 39Sa 67
Milton Av. NW9 27Sa 47
Milton Av. RM12: Horn 33Hd 76
Milton Av. SM1: Sutt 76Fb 155
Milton Av. TN14: Bad M 82Dd 182
Milton Chantry Heritage Cen. 8E 122
Milton Cl. N2 29Eb 49
Milton Cl. SE1 6K 231 (49Vb 91)
Milton Cl. SL3: Hort 55C 104
Milton Cl. SM1: Sutt 76Fb 155
Milton Cl. UB4: Hayes 44W 84
Milton Ct. DA12: Grav'nd 10E 122
Milton Ct. E17 28Cc 52
Milton Ct. EC2 7F 219 (43Tb 91)
Milton Ct. EN9: Walt A 6Ec 20
Milton Ct. RM6: Chad H 31Yc 75
Milton Ct. SE14 51Bc 114
(not continuous)
Milton Ct. SW18 57Cb 111
Milton Ct. TW2: Twick 62Ga 130
Milton Ct. UB10: Ick 34R 64
Milton Court Concert Hall 7F 219 (43Tb 91)
(off Milton Ct.)
Milton Ct. Rd. SE14 51Ac 114
Milton Cres. IG2: Ilf 31Rc 74
Milton Dr. TW17: Shep 70N 127
Milton Dr. WD6: Bore 15Ra 29
Milton Gdn. Est. N16 35Tb 71
Milton Gdns. KT18: Eps 86Ua 174
Milton Gdns. RM18: Tilb 3D 122
Milton Gdns. TW19: Stanw 60P 105
Milton Gro. N11 22Lb 50
Milton Gro. N16 35Tb 71
Milton Hall Rd. DA12: Grav'nd 10F 122
Milton Ho. AL3: St A 4P 5
Milton Ho. E17 28Cc 52
Milton Ho. E2 41Yb 92
(off Roman Rd.)
Milton Ho. SE5 52Tb 113
(off Elmington Est.)
Milton Ho. SL9: Chal P 21A 42
Milton Ho. SM1: Sutt 76Cb 155
Milton Ho. TW20: Egh 64C 126
(off Station Rd.)
Milton Lodge DA14: Sidc 63Wc 139
Milton Lodge TW1: Twick 59Ha 108
Milton Mans. W14 51Ab 110
(off Queen's Club Gdns.)
Milton Pk. N6 31Lb 70
Milton Pk. TW20: Egh 66C 126
Milton Pl. DA12: Grav'nd 8E 122
Milton Pl. N7 36Qb 70
(off Eastwood Cl.)
Milton Rd. CM14: W'ley 21Yd 58
Milton Rd. CR0: C'don 73Tb 157
Milton Rd. CR3: Cat'm 93Tb 197
Milton Rd. CR4: Mitc 66Jb 134
Milton Rd. DA10: Swans 58Ae 121
Milton Rd. DA12: Grav'nd 8D 122
Milton Rd. DA16: Well 53Vc 117
Milton Rd. DA17: Belv 49Cd 96
Milton Rd. E17 28Cc 52
Milton Rd. HA1: Harr 28Ga 46
Milton Rd. KT12: Walt T 76Z 151
Milton Rd. KT15: Add 79J 149
Milton Rd. N15 28Rb 51
Milton Rd. N6 31Lb 70
Milton Rd. NW7 22Wa 48
Milton Rd. NW9 31Wa 68
Milton Rd. RM1: Rom 30Jd 56
Milton Rd. RM17: Grays 50De 99
Milton Rd. SE24 57Rb 113
Milton Rd. SL2: Slou 2H 81
Milton Rd. SM1: Sutt 76Cb 155
Milton Rd. SM6: W'gton 79Lb 156
Milton Rd. SW14 55Ta 109
Milton Rd. SW19 65Eb 133
Milton Rd. TN13: Dun G 93Gd 202
Milton Rd. TW12: Hamp 66Ca 129
Milton Rd. TW20: Egh 64B 126
Milton Rd. UB10: Ick 35R 64
Milton Rd. W3 46Ta 87
Milton Rd. W7 45Ha 86
Milton Rd. Bus. Pk. DA12: Grav'nd 9E 122
Milton St. DA10: Swans 58Zd 121
Milton St. EC2 7F 219 (43Tb 91)
Milton St. EN9: Walt A 6Ec 20
Milton St. WD24: Wat 10X 13
Milton Way KT22: Fet 97Ea 192
Milton Way UB7: W Dray 49P 83
Milverton Dr. UB10: Ick 35S 64
Milverton Gdns. IG3: Ilf 33Vc 75
Milverton Ho. SE6 62Ac 136
Milverton Pl. BR1: Broml 64Lc 137
Milverton Rd. NW6 38Ya 68
Milverton St. SE11 50Qb 90
Milverton Way SE9 63Qc 138
Milward St. E1 43Xb 91
Milward Wlk. SE18 51Qc 116
MIMBRIDGE 5L 167
Mimms Hall Rd. EN6: Pot B 3Za 16
Mimms La. EN6: Ridge 5Sa 15
Mimms La. WD7: Shenl 5Qa 15
Mimosa Cl. BR6: Chels 75Yc 161
Mimosa Cl. CM15: Pil H 15Xd 40
Mimosa Cl. KT17: Eps D 88Ya 174
Mimosa Cl. RM3: Rom 24Ld 57
Mimosa Dr. CR3: Wold 95Bc 198
Mimosa Ho. E20 36Ec 72
(off Liberty Bri. Rd.)
Mimosa Ho. UB4: Yead 43Y 85
Mimosa Lodge NW10 36Va 68
Mimosa Rd. UB4: Yead 43Y 85
Mimosa St. SW6 53Bb 111
Mina Av. SL3: L'ly 7P 81
Minard Rd. SE6 59Gc 115
Mina Rd. SE17 7J 231 (50Ub 91)
Mina Rd. SW19 67Cb 133
Mina Ter. N9 17Wb 33
Minchenden Ct. N14 19Mb 32
Minchenden Cres. N14 20Lb 32
Minchin Cl. KT22: Lea 94Ja 192
Minchin Ho. E14 44Cc 92
(off Dod St.)
Mincing La. EC3 4H 225 (45Ub 91)
Mincing La. GU24: Chob 10K 147
Minden Gdns. IG11: Bark 41Xc 95
Minden Rd. SE20 67Xb 135
Minden Rd. SM3: Sutt 75Ab 154
Minehead Ho. RM3: Rom 22Nd 57
(off Dagnam Pk. Dr.)
Minehead Rd. HA2: Harr 34Ca 65
Minehead Rd. SW16 64Pb 134
Mineral Cl. EN5: Barn 16Ya 30
Mineral St. SE18 49Uc 94
Minera M. SW1 5H 227 (49Jb 90)
Minerva Cl. DA14: Sidc 62Uc 138
Minerva Cl. SW9 52Qb 112
Minerva Cl. TW19: Stanw M 57J 105
Minerva Ct. EC1 6A 218 (42Qb 90)
(off Bowling Grn. La.)
Minerva Dr. WD24: Wat 8U 12
Minerva Lodge N7 37Pb 70
Minerva Rd. E4 24Dc 52
Minerva Rd. KT1: King T 68Pa 131
Minerva Rd. NW10 42Sa 87
Minerva St. E2 40Xb 71
Minerva Wlk. EC1 2C 224 (44Rb 91)
Minerva Way EN5: Barn 15Bb 31
Minet Av. NW10 40Ua 68
Minet Country Pk. 47Y 85
Minet Dr. UB3: Hayes 46W 84
Minet Gdns. NW10 40Ua 68
Minet Gdns. UB3: Hayes 46X 85
Minet Rd. SW9 54Rb 113
Minford Gdns. W14 47Za 88
Minford Ho. W14 47Za 88
(off Minford Gdns.)
Mingard Wlk. N7 33Pb 70
Ming St. E14 45Cc 92
Minimax Cl. TW14: Felt 58W 106
Minima Yacht Club 69Ma 131
(off High St.)
Minister Ct. AL2: F'mre 10C 6
Ministry Way SE9 61Pc 138
Miniver Pl. EC4 4E 224 (45Sb 91)
(off Garlick Hill)
Mink Ct. TW4: Houn 54Y 107
Minley Ct. RH2: Reig 5J 207
Minnie Baldock St. E16 44Hc 93
Minniecroft Rd. SL1: Burn 1A 80
Minniedale KT5: Surb 71Pa 153
Minnow St. SE17 6J 231 (49Ub 91)
Minnow Wlk. SE17 6H 231 (49Ub 91)
Minoan Dr. HP3: Hem H 6N 3
Minorca Rd. KT13: Weyb 77Q 150
Minories EC3 3K 225 (44Vb 91)
MINOR INJURIES UNIT (CHESHUNT) 3Ac 20
MINOR INJURIES UNIT (GRAVESHAM) 8C 122
MINOR INJURIES UNIT
(GUY'S HOSPITAL) 1G 231 (47Tb 91)
MINOR INJURIES UNIT (NORTHWOOD) 23R 44
MINOR INJURIES UNIT
(ORSETT HOSPITAL) 3C 100
MINOR INJURIES UNIT
(PARKWAY HEALTH CENTRE) 82Ec 178
MINOR INJURIES UNIT (PURLEY) 83Qb 176
MINOR INJURIES UNIT
(ROEHAMPTON) 58Wa 110
MINOR INJURIES UNIT (ST ALBANS) 1A 6
MINOR INJURIES UNIT (ST BARTHOLOMEW'S
HOSPITAL) 1C 224 (43Rb 91)
MINOR INJURIES UNIT
(SEVENOAKS HOSPITAL) 93Ld 203
Minotaur Dr. EN5: Barn 15Bb 31
Minshaw Ct. DA14: Sidc 63Vc 139
Minshill St. SW8 53Mb 112
Minshull Pl. BR3: Beck 66Cc 136
Minson Rd. E9 39Zb 72
Minstead Gdns. SW15 59Va 110
Minstead Way KT3: N Mald 72Ua 154
Minster Av. SM1: Sutt 75Cb 155
Minster Cl. AL10: Hat 2C 8
Minster Ct. EC3 4J 225 (45Ub 91)
(off Mincing La.)
Minster Ct. RM11: Horn 33Qd 77
Minster Ct. W5 42Na 87
Minster Dr. CR0: C'don 77Ub 157
Minster Gdns. KT8: W Mole 70Ba 129
Minster Ho. AL10: Hat 2C 8
Minsterley Av. TW17: Shep 70U 128
Minster Pavement EC3 4J 225 (45Ub 91)
(off Mincing La.)
Minster Rd. BR1: Broml 66Kc 137
Minster Rd. NW2 36Ab 68
Minster Wlk. N8 28Nb 50
Minster Way RM11: Horn 32Pd 77
Minster Way SL3: L'ly 47B 82
Minstrel Cl. HP1: Hem H 1K 3
Minstrel Gdns. KT5: Surb 70Pa 131
Mint Bus. Pk. E16 43Kc 93
Mint Cl. UB10: Hil 41R 84
Mintern Cl. N13 20Rb 33
Minterne Av. UB2: S'hall 49Ca 85
Minterne Rd. HA3: Kenton 29Pa 47
Minterne Way UB4: Yead 44Y 85
Mintern St. N1 1G 219 (40Tb 71)
Minter Rd. IG11: Bark 42Wc 95
Minters Orchard TN15: Plat 92De 205
Mint La. KT20: Lwr K 1J 207
Minton Apts. SW8 52Nb 112
Minton Ho. SE11 5K 229 (49Qb 90)
(off Walnut Tree Wlk.)
Minton M. NW6 37Db 69
Minton Ri. SL6: Tap 4A 80
Mint Rd. SM6: W'gton 77Kb 156
Mint Rd. SM7: Bans 88Eb 175
Mint St. E2 42Xb 91
(off Three Colts La.)
Mint St. SE1 1D 230 (47Sb 91)
Mint Wlk. CR0: C'don 76Sb 157
Mint Wlk. CR6: W'ham 89Zb 178
Mint Wlk. GU21: Knap 9J 167
Mintwater Cl. KT17: Ewe 82Wa 174
Mirabelle Gdns. E20 36Ec 72
Mirabel Rd. SW6 52Bb 111
Mirador Cres. SL2: Slou 5M 81
Mira Ho. E20 36Ec 72
(off Prize Wlk.)
Miramar Way RM12: Horn 36Md 77
Miranda Cl. E1 43Yb 92
Miranda Ct. W3 44Pa 87
Miranda Ho. N1 2H 219 (40Ub 71)
(off Crondall St.)
Miranda Rd. N19 32Lb 70
Mirfield Cl. RM3: Rom 23Nd 57
Mirfield St. SE7 49Mc 93
Miriam La. AL2: Chis G 8M 5
Miriam Rd. SE18 50Uc 94
Mirravale Trad. Est. RM8: Dag 31Ad 75
Mirren Cl. HA2: Harr 35Ba 65
Mirrie La. UB9: Den 29E 42
Mirror Path SE9 62Lc 137
Misbourne Av. SL9: Chal P 22A 42
Misbourne Cl. SL9: Chal P 22A 42
Misbourne Ct. SL3: L'ly 49C 82
Misbourne Mdws. UB9: Den 32E 62
Misbourne Rd. UB10: Hil 39Q 64
Misbourne Va. SL9: Chal P 22A 42
Miskin Rd. DA1: Dart 59Ld 119
(not continuous)
Miskin Theatre 61Ld 141
Miskin Way DA12: Grav'nd 5F 144
Missden Dr. HP3: Hem H 4C 4
Missenden SE17 7G 231 (50Tb 91)
(off Roland Way)
Missenden Cl. TW14: Felt 60V 106
Missenden Gdns. SL1: Burn 4A 80
Missenden Gdns. SM4: Mord 72Eb 155
Missenden Ho. NW8 5D 214 (42Gb 89)
(off Jerome Cres.)
Missenden Ho. WD18: Wat 17U 26
(off Chenies Way)
Mission, The E14 44Bc 92
(off Commercial Rd.)
Mission Gro. E17 29Ac 52
Mission Pl. SE15 53Wb 113
Mission Sq. TW8: Bford 51Na 109
Missouri Ct. HA5: Eastc 30Y 45
Mistletoe Cl. CR0: C'don 74Zb 158
Mistley Ct. KT18: Eps 85Ta 173
(off Ashley Rd.)
Mistral SE5 53Ub 113
Mistral Ct. AL1: St A 3E 6
(off Bakers Ct.)
Misty's Fld. KT12: Walt T 74Y 151
Mitali Pas. E1 44Wb 91
MITCHAM 69Hb 133
Mitcham Eastfields Station (Rail) 68Jb 134
Mitcham Gdn. Village CR4: Mitc 71Jb 156
Mitcham Golf Course 71Jb 156
Mitcham Ho. SE5 53Sb 113
Mitcham Ind. Est. CR4: Mitc 67Jb 134
Mitcham Junction Station
(Rail & London Tramlink) 71Jb 156
Mitcham La. SW16 65Lb 134
Mitcham Pk. CR4: Mitc 70Gb 133
Mitcham Rd. CR0: C'don 72Nb 156
Mitcham Rd. E6 41Nc 94
Mitcham Rd. IG3: Ilf 31Vc 75
Mitcham Rd. SW17 64Hb 133
Mitcham Stop (London Tramlink) 70Gb 133
Mitchell NW9 25Va 48
(off Quakers Course)
Mitchell Av. DA11: Nflt 61Fe 143
Mitchellbrook Way NW10 37Ta 67
Mitchell Cl. AL1: St A 6B 6
Mitchell Cl. DA1: Dart 61Nd 141
Mitchell Cl. DA17: Belv 48Ed 96
Mitchell Cl. HP3: Bov 9B 2
Mitchell Cl. RM13: Rain 40Ld 77
Mitchell Cl. RM8: Dag 34Yc 75
Mitchell Cl. SE2 49Yc 95
Mitchell Cl. SL1: Slou 8E 80
Mitchell Cl. WD5: Ab L 4W 12
Mitchell Ho. N1 38Rb 71
(off College Cross)
Mitchell Ho. W12 45Xa 88
(off White City Est.)
Mitchell Rd. BR6: Orp 77Vc 161
Mitchell Rd. N13 22Sb 51
Mitchell's Pl. SE21 58Ub 113
(off Aysgarth Rd.)
Mitchell St. EC1 5D 218 (42Sb 91)
(not continuous)
Mitchell Wlk. E6: Allhallows Rd. 43Nc 94
(off Allhallows Rd.)
Mitchell Wlk. E6: Elmley Cl. 43Pc 94
Mitchell Wlk. DA10: Swans 59Ae 121
Mitchell Way BR1: Broml 67Jc 137
Mitchell Way NW10 37Sa 67
Mitchem Cl. TN15: W King 80Ud 164
Mitchener's La. RH1: Blet 6K 209
Mitchison Ct. TW16: Sun 67W 128
(off Downside)
Mitchison Rd. N1 37Tb 71
Mitchley Av. CR2: Sande 85Ub 177
Mitchley Av. CR8: Purl 85Sb 177
Mitchley Gro. CR2: Sande 85Wb 177
Mitchley Hill CR2: Sande 85Vb 177
Mitchley Rd. N17 27Wb 51
Mitchley Vw. CR2: Sande 85Wb 177
Mitford Bldgs. SW6 52Cb 111
(off Dawes Rd.)
Mitford Cl. KT9: Chess 79La 152
Mitford Rd. N19 33Nb 70
Mitman Wlk. DA1: Dart 54Qd 119
Mitre, The E14 45Bc 92
Mitre Av. E17 27Cc 52
Mitre Bri. Ind. Pk. W10 42Xa 88
(not continuous)
Mitre Cl. BR2: Broml 68Hc 137
Mitre Cl. SM2: Sutt 80Eb 155
Mitre Cl. TW17: Shep 72T 150
Mitre Ho. SW3 7F 227 (50Hb 89)
(off King's Rd.)
Mitre Pas. EC3 3J 225 (44Ub 91)
(off Mitre Sq.)
Mitre Pas. SE10 47Gc 93
Mitre Rd. E15 40Gc 73
Mitre Rd. SE1 1A 230 (47Qb 90)
Mitre Sq. EC3 3J 225 (44Ub 91)

Mitre St. EC3 3J 225 (44Ub 91)
Mitre Way W10 42Xa 88
Mitre Yd. SW3 5E 226 (49Gb 89)
Mitten Ho. SE8 52Dc 114
(off Creative Rd.)
Mixbury Gro. KT13: Weyb 79T 150
Mixnams La. KT16: Chert 69J 127
Mizen Cl. KT11: Cobh 86Z 171
Mizen Ct. E14 47Cc 92
(off Alpha Gro.)
Mizens Railway 8H 167
Mizen Way KT11: Cobh 87Y 171
Mizzen Mast Ho. SE18 48Pc 94
Mizzen St. IG11: Bark 39Tc 74
Moat, The KT3: N Mald 67Ua 132
Moat Cl. BR6: Chels 79Vc 161
Moat Cl. TN13: Chips 94Ed 202
Moat Cl. WD23: Bush 15Da 27
Moat Ct. DA15: Sidc 62Vc 139
Moat Ct. KT16: Ott 79E 148
Moat Ct. KT21: Asht 89Na 173
Moat Ct. SE9 58Pc 116
Moat Cres. N3 27Db 49
Moat Cft. DA16: Well 55Yc 117
Moat Dr. E13 40Lc 73
Moat Dr. HA1: Harr 28Ea 46
Moat Dr. HA4: Ruis 31U 64
Moat Dr. SL2: Slou 3N 81
Moated Farm Dr. KT15: Add 80L 149
Moated Farm Dr. KT15: New H 80L 149
Moat Farm Rd. UB5: N'olt 37Ba 65
Moatfield NW6 38Ab 68
Moatfield Rd. WD23: Bush 15Da 27
Moatlands Ho. WC1 4G 217 (41Nb 90)
(off Cromer St.)
Moat La. DA8: Erith 53Jd 118
Moat La. KT8: E Mos 69Ha 130
Moat Lodge, The HA2: Harr 33Ga 66
Moat Pl. SW9 55Pb 112
Moat Pl. UB9: Den 35K 63
Moat Pl. W3 44Ra 87
Moat Side EN3: Pond E 14Zb 34
Moat Side TW13: Hanw 63Y 129
Moat Vw. Ct. WD23: Bush 15Da 27
Mobbs Cl. SL2: Stoke P 1L 81
Moberly Rd. SW4 59Mb 112
Moberly Sports Cen. 41Za 88
Moberly Way CR8: Kenley 92Tb 197
Mobil Ct. WC2 3J 223 (44Pb 90)
(off Clement's Inn)
MOBY DICK 28Ad 55
Mocatta Ho. E1 42Xb 91
(off Brady St.)
Mocatta M. RH1: Mers 3C 208
Mocha Ct. E3 40Dc 72
(off Taylor Pl.)
Mockford M. RH1: Mers 3C 208
MoDA 26Va 48
Modbury Gdns. NW5 37Jb 70
Modder Pl. SW15 56Za 110
Model Cotts. GU24: Pirb 3B 186
Model Cotts. SW14 56Sa 109
Model Cotts. W13 47Ka 86
Model Farm Cl. SE9 62Nc 138
Modena Ho. E14 44Gc 93
(off Lyell St.)
Modena M. WD18: Wat 14U 26
Modern Ct. EC4 2B 224 (44Rb 91)
(off Farringdon St.)
Modling Ho. E2 40Zb 72
(off Mace St.)
Moelwyn N7 36Mb 70
Moelyn M. HA1: Harr 29Ja 46
Moffat Ct. SW19 64Cb 133
Moffat Ho. SE5 52Sb 113
Moffat Rd. CR7: Thor H 68Sb 135
Moffat Rd. N13 23Nb 50
Moffat Rd. SW17 63Hb 133
Moffats Cl. AL9: Brk P 8J 9
Moffats La. AL9: Brk P 9G 8
MOGADOR 100Ab 194
Mogador Rd. KT20: Lwr K 100Ab 194
Mogden La. TW7: Isle 57Ha 108
Mogul Bldg. E15 36Ec 72
(off Property Row)
Mohammedi Pk. UB5: N'olt 39Ca 65
Mohawk Ho. E3 40Ac 72
(off Gernon Rd.)
Mohmmad Khan Rd. E11 32Hc 73
Moineau NW9 25Va 48
(off Long Mead)
Moira Cl. N17 26Ub 51
Moira Ho. SW9 53Qb 112
(off Gosling Way)
Moira Rd. SE9 56Pc 116
Moir Cl. CR2: Sande 81Wb 177
Mokswell Ct. N10 25Jb 50
Molash Rd. BR5: St M Cry 70Zc 139
Molasses Ho. SW11 55Eb 111
(off Clove Hitch Quay)
Molasses Row SW11 55Eb 111
Mole Abbey Gdns. KT8: W Mole 69Da 129
Mole Bus. Pk. KT22: Lea 93Ja 192
Mole Ct. KT19: Ewe 77Sa 153
Mole Ho. NW8 6C 214 (42Fb 89)
(off Church St. Est.)
Molember Ct. KT8: E Mos 70Ga 130
Molember Rd. KT8: E Mos 71Ga 152
Mole Pl. KT8: W Mole 70Da 129
Mole Rd. KT12: Hers 78Z 151
Mole Rd. KT22: Fet 93Fa 192
Molescroft SE9 62Sc 138
Molesey Av. KT8: W Mole 71Ba 151
Molesey Cl. KT12: Hers 77Aa 151
Molesey Dr. SM3: Cheam 75Ab 154
Molesey Heath Local Nature Reserve 72Ca 151
MOLESEY HOSPITAL 71Ca 151
Molesey Pk. Av. KT8: W Mole 71Da 151
Molesey Pk. Cl. KT8: E Mos 71Ea 152
Molesey Pk. Rd. KT8: E Mos 71Da 151
Molesey Pk. Rd. KT8: W Mole 71Da 151
Molesey Rd. KT12: Hers 78Z 151
Molesey Rd. KT12: Walt T 78Z 151
Molesey Rd. KT8: W Mole 72Ba 151
Molesey Road 75Aa 151
Molesford Rd. SW6 53Cb 111
Molesham Cl. KT8: W Mole 69Da 129
Molesham Way KT8: W Mole 69Da 129
Moles Hill KT22: Oxs 83Fa 172
Molesworth Ho. SE17 51Rb 113
(off Brandon Est.)
Molesworth Rd. KT11: Cobh 85W 170
Molesworth St. SE13 56Ec 114
Mole Valley Pl. KT21: Asht 91Ma 193
Moliner Ct. BR3: Beck 66Cc 136
Mollands Ct. RM15: S Ock 42Ae 99
Mollands La. RM15: S Ock 42Yd 98
Mollis Ho. E3 43Cc 92
(off Gale St.)
Mollison Av. EN3: Brim 7Ac 20
Mollison Av. EN3: Enf L 7Ac 20
Mollison Av. EN3: Enf W 7Ac 20
Mollison Av. EN3: Pond E 7Ac 20
Mollison Dr. SM6: W'gton 80Mb 156
Mollison Ri. DA12: Grav'nd 4G 144
Mollison Sq. SM6: W'gton 80Mb 156
(off Mollison Dr.)
Mollison Way HA8: Edg 26Pa 47
Molloy Ct. GU21: Wok 88C 168
Molly Huggins Cl. SW12 59Lb 112
Molten Ct. SE14 51Bc 114
(off Moulding La.)
Molteno Rd. WD17: Wat 10W 12
Molton Ho. N1 1J 217 (39Pb 70)
(off Barnsbury Est.)
Molyneaux Av. HP3: Bov 9B 2
Molyneux Dr. SW17 63Kb 134
Molyneux Rd. GU20: W'sham 9B 146
Molyneux Rd. KT13: Weyb 78Q 150
Molyneux St. W1 1E 220 (43Gb 89)
Molyns M. SL1: Slou 6C 80
Monaco Works WD4: K Lan 2R 12
Monahan Av. CR8: Purl 84Pb 176
Monarch Cl. BR4: W W'ck 77Hc 159
Monarch Cl. RM13: Rain 40Jd 76
Monarch Cl. RM18: Tilb 4D 122
Monarch Cl. TW14: Felt 59U 106
Monarch Ct. HA7: Stan 24Ma 47
(off Howard Rd.)
Monarch Ct. N2 29Fb 49
Monarch Dr. E16 43Mc 93
Monarch Dr. UB3: Hayes 45V 84
Monarch Ho. W8 48Cb 89
(off Kensington High St.)
Monarch M. E17 30Dc 52
Monarch M. SW16 64Qb 134
Monarch Pde. CR4: Mitc 68Hb 133
Monarch Pl. IG9: Buck H 19Lc 35
Monarch Point SW6 54Eb 111
Monarch Rd. DA17: Belv 48Cd 96
Monarch Sq. SW11 56Gb 111
Monarchs Way EN8: Walt C 5Ac 20
Monarchs Way HA4: Ruis 32T 64
Monarch Wlk. WD7: Shenl 5Pa 15
Monarch Way IG2: Ilf 30Tc 54
Mona Rd. SE15 54Yb 114
Monastery Cl. AL3: St A 2A 6
Monastery Gdns. EN2: Enf 12Tb 33
Monastery Gdns. SL3: Dat 5M 103
Mona St. E16 43Hc 93
Monaveen Gdns. KT8: W Mole 69Da 129
Monck Ho. SE1 2E 230 (47Sb 91)
(off Cole St.)
Moncks Row SW18 58Bb 111
Monck St. SW1 4E 228 (48Mb 90)
Monckton Ct. W14 48Bb 89
(off Strangways Ter.)
Monckton Rd. TN15: Bor G 92Be 205
Monclar Rd. SE5 56Tb 113
Moncorvo Cl. SW7 2D 226 (47Gb 89)
Moncrieff Cl. E6 44Nc 94
Moncrieff Pl. SE15 54Wb 113
Moncrieff St. SE15 54Wb 113
Monday All. N16 33Vb 71
(off High St.)
Mondial Way UB3: Harl 52S 106
Mondragon Ho. SW8 53Nb 112
(off Guildford Rd.)
Monds Cotts. TN14: Sund 96Ad 201
Monega Rd. E12 37Mc 73
Monega Rd. E7 37Lc 73
Monet Ct. SE16 50Xb 91
(off Stubbs Dr.)
Money Av. CR3: Cat'm 94Ub 197
Moneyer Ho. N1 3F 219 (41Tb 71)
(off Provost St.)
MONEYHILL 18L 25
Moneyhill Ct. WD3: Rick 18K 25
Moneyhill Pde. WD3: Rick 18K 25
Money Hill Rd. WD3: Rick 18L 25
Money La. UB7: W Dray 48M 83
Money Rd. CR3: Cat'm 94Tb 197
Mongers Almshouses E9 38Zb 72
(off Church Cres.)
Mongers La. KT17: Ewe 82Va 174
(not continuous)
Monica Cl. WD24: Wat 12Y 27
Monica Ct. EN1: Enf 15Ub 33
Monica James Ho. DA14: Sidc 62Wc 139
Monica Shaw Ct. NW1 2E 216 (40Mb 70)
(off Purchese St.)
Monier Rd. E3 38Cc 72
Monivea Rd. BR3: Beck 66Bc 136
Monkchester Cl. IG10: Lough 11Pc 36
Monk Ct. W12 46Wa 88
Monk Dr. E16 45Jc 93
MONKEN HADLEY 12Bb 31
Monkey Puzzle Way SM5: Cars 81Hb 175
Monkfrith Av. N14 16Kb 32
Monkfrith Cl. N14 17Kb 32
Monkfrith Way N14 17Jb 32
Monkhams EN9: Walt A 2Ec 20
Monkham's Av. IG8: Wfd G 22Kc 53
Monkham's Dr. IG8: Wfd G 22Kc 53
Monkham's La. IG8: Wfd G 22Jc 53
Monkham's La. IG9: Buck H 20Kc 35
Monkleigh Rd. SM4: Mord 69Ab 132
Monks Av. EN5: New Bar 16Eb 31
Monks Av. KT8: W Mole 71Ba 151
Monks Chase CM13: Ingve 22Ee 59
Monks Cl. AL1: St A 4C 6
Monks Cl. EN2: Enf 12Sb 33
Monks Cl. HA2: Harr 33Da 65
Monks Cl. HA4: Ruis 35Z 65
Monks Cl. SE2 49Zc 95
Monks Cl. SL5: Asc 2A 146
Monks Ct. RH2: Reig 6K 207
Monks Cres. KT12: Walt T 74X 151
Monks Cres. KT15: Add 78K 149
Monksdene Gdns. SM1: Sutt 76Db 155
Monks Dr. SL5: Asc 2A 146
Monks Dr. W3 43Qa 87
Monksfarm Pl. SE2 47Xc 95
Monksfield Way SL2: Slou 2E 80
Monks Ga. AL1: St A 4C 6
Monks Grn. KT22: Fet 93Ea 192
Monksgrove IG10: Lough 15Qc 36
Monks Haven SS17: Stan H 1N 101
Monks Horton Way AL1: St A 1E 6
Monks La. TN8: Eden 9P 211
Monksmead WD6: Bore 14Sa 29
MONKS ORCHARD 73Ac 158
Monks Orchard DA1: Dart 61Md 141
Monks Orchard Rd. BR3: Beck 74Cc 158
Monks Pk. HA9: Wemb 37Ra 67
Monks Pk. Gdns. HA9: Wemb 38Ra 67
Monks Pl. CR3: Cat'm 94Xb 197
Monks Rd. EN2: Enf 12Rb 33
Monks Rd. GU25: Vir W 10P 125
Monks Rd. SL4: Wind 4B 102
Monks Rd. SM7: Bans 89Cb 175
Monk St. SE18 49Qc 94
Monk's Wlk. KT16: Chert 70G 126
Monk's Wlk. RH2: Reig 6K 207
Monk's Wlk. TW20: Thorpe 69F 126
Monks Wlk. DA13: Sflt 65Ce 143
Monks Way BR3: Beck 72Cc 158
Monks Way BR5: Farnb 74Sc 160
Monks Way NW11 28Bb 49
Monks Way TW18: Staines 66M 127
Monks Way UB7: Harm 51N 105
Monks Well DA9: Ghithe 56Yd 120
(off Watermans Way)
Monkswell La. CR5: Coul 96Db 195
Monkswood Av. EN9: Walt A 5Fc 21
Monkswood Gdns. IG5: Ilf 27Qc 54
Monkswood Gdns. WD6: Bore 15Ta 29
Monkton Ho. E5 36Xb 71
Monkton Ho. SE16 47Zb 92
(off Wolfe Cres.)
Monkton Rd. DA16: Well 54Vc 117
Monkton St. SE11 5A 230 (49Qb 90)
Monkville Av. NW11 28Bb 49
Monkville Pde. NW11 28Bb 49
Monkwell Sq. EC2 1E 224 (43Sb 91)
Monkwood Cl. RM1: Rom 29Jd 56
Monmouth Av. E18 27Kc 53
Monmouth Av. KT1: Hamp W 66La 130
Monmouth Cl. CR4: Mitc 70Nb 134
Monmouth Cl. DA16: Well 56Wc 117
Monmouth Cl. W4 48Sa 87
Monmouth Ct. DA12: Grav'nd 8F 122
(off Romulus Rd.)
Monmouth Ct. W7 43Ha 86
(off Copley Cl.)
Monmouth Gro. TW8: Bford 49Na 87
Monmouth Pl. W2 44Db 89
(off Monmouth Rd.)
Monmouth Rd. E6 41Pc 94
Monmouth Rd. N9 19Xb 33
Monmouth Rd. RM9: Dag 36Bd 75
Monmouth Rd. UB3: Harl 49U 84
Monmouth Rd. W2 44Cb 89
Monmouth Rd. WD17: Wat 13X 27
Monmouth St. WC2 3F 223 (44Nb 90)
Monnery Rd. N19 34Lb 70
Monnow Grn. RM15: Avel 45Sd 98
Monnow Rd. RM15: Avel 45Sd 98
Monnow Rd. SE1 50Wb 91
Mono La. TW13: Felt 61X 129
Monolulu Ct. SE17 7G 231 (50Tb 91)
(off East St.)
Monoux Almshouses E17 28Dc 52
Monoux Gro. E17 25Cc 52
Monro Ct. E16 42Hc 93
Monroe Cres. EN1: Enf 11Xb 33
Monroe Dr. SW14 57Ra 109
Monroe Ho. NW8 4E 214 (41Gb 89)
(off Lorne Cl.)
Monro Gdns. HA3: Hrw W 24Ga 46
Monro Ind. Est. EN8: Walt C 6Ac 20
Monro Pl. KT19: Eps 81Qa 173
Monro Way E5 35Wb 71
Monsell Ct. N4 34Rb 71
Monsell Gdns. TW18: Staines 64G 126
Monsell Rd. N4 34Qb 70
Monsey Pl. E1 42Ac 92
Monson Rd. NW10 40Wa 68
Monson Rd. RH1: Redh 3P 207
Monson Rd. SE14 52Zb 114
Mons Wlk. TW20: Egh 64E 126
Mons Way BR2: Broml 72Nc 160
Montacute Rd. CR0: New Ad 81Ec 178
Montacute Rd. SE6 59Bc 114
Montacute Rd. SM4: Mord 72Fb 155
Montacute Rd. WD23: B Hea 17Ga 28
Montagu Ct. W1 1G 221 (43Hb 89)
(off Montagu Pl.)
Montagu Cres. N18 21Xb 51
Montague Av. CR2: Sande 84Ub 177
Montague Av. SE4 56Bc 114
Montague Av. W7 46Ha 86
Montague Cl. EN5: Barn 14Bb 31
Montague Cl. KT12: Walt T 73X 151
Montague Cl. SE1 6F 225 (46Tb 91)
Montague Cl. SL2: Farn R 10F 60
Montague Ct. DA15: Sidc 62Wc 139
Montague Ct. N7 38Qb 70
(off St Clements St.)
Montague Dr. CR3: Cat'm 94Sb 197
Montague Gdns. W3 45Qa 87
Montague Hall Pl. WD23: Bush 16Ca 27
Montague Ho. E16 46Kc 93
(off Wesley Av.)
Montague Ho. IG3: Ilf 32Wc 75
Montague Ho. N1 39Ub 71
(off Halcomb St.)
Montague M. E3 41Bc 92
(off Tredegar Ter.)
Montague M. SE20 65Yb 136
Montague Pas. UB8: Uxb 38M 63
Montague Pl. BR8: Swan 70Hd 140
Montague Pl. WC1 7E 216 (43Mb 90)
Montague Rd. CR0: C'don 74Rb 157
Montague Rd. E11 33Hc 73
Montague Rd. E8 36Wb 71
Montague Rd. N15 28Wb 51
Montague Rd. N8 29Pb 50
Montague Rd. SL1: Slou 5H 81
Montague Rd. SW19 66Db 133
Montague Rd. TW10: Rich 58Na 109
Montague Rd. TW3: Houn 55Da 107
Montague Rd. UB2: S'hall 49Aa 85
Montague Rd. UB8: Uxb 38M 63
Montague Rd. W13 44Ka 86
Montague Rd. W7 46Ha 86
Montague Sq. SE15 52Yb 114
Montague St. EC1 1D 224 (43Sb 91)
Montague St. WC1 7F 217 (43Nb 90)
Montague Ter. BR2: Broml 70Hc 137
Montague Walks HA0: Wemb 39Pa 67
Montague Waye UB2: S'hall 48Aa 85
Montagu Gdns. DA1: Dart 54Pd 119
Montagu Gdns. N18 21Xb 51
Montagu Gdns. SM6: W'gton 77Lb 156
Montagu Ind. Est. N18 21Yb 52
Montagu Mans. W1 7G 215 (43Hb 89)
Montagu M. Nth. W1 1G 221 (43Hb 89)
Montagu M. Sth. W1 2G 221 (44Hb 89)
Montagu M. W. W1 2G 221 (44Hb 89)
Montagu Pl. W1 1F 221 (43Hb 89)
Montagu Rd. N18 22Xb 51
Montagu Rd. N9 20Yb 34
Montagu Rd. NW4 30Wa 48
Montagu Rd. SL3: Dat 3M 103
(not continuous)
Montagu Row W1 1G 221 (43Hb 89)
Montagu Sq. W1 1G 221 (43Hb 89)
Montagu St. W1 2G 221 (44Hb 89)
Montaigne Cl. SW1 6E 228 (49Mb 90)
Montalt Rd. IG8: Wfd G 21Hc 53
Montana HA9: Wemb 35Qa 67
(off Exhibition Way)
Montana Bldg. SE13 53Dc 114
(off Deal's Gateway)
Montana Cl. CR2: Sande 82Tb 177
Montana Gdns. SE26 64Bc 136
Montana Gdns. SM1: Sutt 78Eb 155
Montana Rd. SW17 62Jb 134
Montana Rd. SW20 67Ya 132
Montanaro Ct. N1 39Sb 71
(off Coleman Flds.)
Montayne Rd. EN8: Chesh 4Zb 20
Montbazon Ct. CM14: B'wood 18Xd 40
Montbelle Rd. SE9 62Rc 138
Montbretia Cl. BR5: St M Cry 70Yc 139
Montcalm Cl. BR2: Hayes 72Jc 159
Montcalm Cl. UB4: Yead 41X 85
Montcalm Ho. E14 49Bc 92
Montcalm Rd. SE7 52Mc 115
Montclare St. E2 5K 219 (42Vb 91)
Monteagle Av. IG11: Bark 37Sc 74
Monteagle Ct. N1 1J 219 (40Ub 71)
Monteagle Way E5 34Wb 71
Monteagle Way SE15 55Xb 113
Montefiore Ct. N16 32Vb 71
Montefiore St. SW8 54Kb 112
Montego Cl. SE24 56Qb 112
Montem La. SL1: Slou 6H 81
Montem Leisure Cen. 7H 81
Montem Rd. KT3: N Mald 70Ua 132
Montem Rd. SE23 59Bc 114
Montem St. N4 32Pb 70
Montenotte Rd. N8 29Lb 50
Monterey Apts. N15 29Tb 51
Monterey Cl. DA5: Bexl 61Ed 140
Monterey Cl. NW7 22Ua 48
(off The Broadway)
Monterey Cl. UB10: Hil 38Q 64
Monterey Pl. KT22: Oxs 86Fa 172
Monterey Studios W10 40Ab 68
Montesole Ct. HA5: Pinn 26Y 45
Montevetro SW11 53Fb 111
Montfichet Rd. E20 38Ec 72
Montford Pl. E15 40Ec 72
Montford Pl. SE11 50Qb 90
Montford Rd. TW16: Sun 70W 128
Montfort Gdns. IG6: Ilf 23Sc 54
Montfort Ho. E14 48Ec 93
(off Galbraith St.)
Montfort Ho. E2 41Yb 92
(off Victoria Pk. Sq.)
Montfort Pl. SW19 60Za 110
Montfort Rd. TN15: Kems'g 89Nd 183
Montgolfier Wlk. UB5: N'olt 41Aa 85
Montgomerie M. SE23 59Yb 114
Montgomery Av. HP2: Hem H 1A 4
Montgomery Av. KT10: Hin W 75Ga 152
Montgomery Cl. CR4: Mitc 70Nb 134
Montgomery Cl. DA12: Grav'nd 3F 144
Montgomery Cl. DA15: Sidc 58Vc 117
Montgomery Cl. RM16: Grays 47Ee 99
Montgomery Ct. CR2: S Croy 78Ub 157
(off Birdhurst Rd.)
Montgomery Ct. KT22: Lea 92Ka 192
(off Levett Rd.)
Montgomery Ct. W4 52Sa 109
Montgomery Cres. RM3: Rom 22Ld 57
Montgomery Gdns. SM2: Sutt 80Fb 155
Montgomery Ho. UB5: N'olt 41Ba 85
(off Taywood Rd.)
Montgomery Ho. W2 1B 220 (43Fb 89)
(off Harrow Rd.)
Montgomery Lodge E1 42Yb 92
(off Cleveland Gro.)
Montgomery Pl. SL2: Slou 4N 81
Montgomery Rd. DA4: S Dar 67Td 142
Montgomery Rd. GU22: Wok 90A 168
Montgomery Rd. HA8: Edg 23Pa 47
Montgomery Rd. W4 49Sa 87
Montgomery Sq. E14 46Dc 92
Montgomery St. E14 46Dc 92
Montgomery Way CR8: Kenley 92Tb 197
Montholme Rd. SW11 58Hb 111
Monthope Rd. E1 43Wb 91
Montolieu Gdns. SW15 57Xa 110
Montpelier Av. DA5: Bexl 59Zc 117
Montpelier Av. W5 43La 86
Montpelier Cl. UB10: Hil 39Q 64
Montpelier Ct. BR2: Broml 70Hc 137
(off Westmoreland Rd.)
Montpelier Ct. SL4: Wind 4G 102
Montpelier Ct. W5 43Ma 87
Montpelier Gdns. E6 41Mc 93
Montpelier Gdns. RM6: Chad H 31Yc 75
Montpelier Gro. NW5 36Lb 70
Montpelier M. SW7 3E 226 (48Gb 89)
Montpelier Pl. E1 44Yb 92
Montpelier Pl. SW7 3E 226 (48Gb 89)
Montpelier Ri. HA9: Wemb 32Ma 67
Montpelier Ri. NW11 31Ab 68
(not continuous)
Montpelier Rd. CR8: Purl 82Rb 177
Montpelier Rd. N3 25Eb 49
Montpelier Rd. SE15 53Xb 113
Montpelier Rd. SM1: Sutt 77Eb 155
Montpelier Rd. W5 43Ma 87
Montpelier Row SE3 54Hc 115
Montpelier Row TW1: Twick 59La 108
Montpelier Sq. SW7 2E 226 (47Gb 89)
Montpelier St. SW7 3E 226 (48Gb 89)
Montpelier Ter. SW7 2E 226 (47Gb 89)
Montpelier Va. SE3 54Hc 115
Montpelier Wlk. SW7 3E 226 (48Gb 89)
Montpelier Way NW11 31Ab 68
Montpellier Ct. KT12: Walt T 72W 150
Montpellier Ho. IG7: Chig 22Sc 54
Montrave Rd. SE20 65Yb 136
Montreal Ho. SE16 47Zb 92
(off Maple M.)
Montreal Ho. UB4: Yead 41X 85
(off Ayles Rd.)
Montreal Pl. WC2 4H 223 (45Pb 90)
Montreal Rd. IG1: Ilf 31Sc 74
Montreal Rd. RM18: Tilb 5C 122
Montreal Rd. TN13: Riv 95Gd 202
Montrell Rd. SW2 60Nb 112
Montrose SL1: Slou 4F 80
Montrose Av. DA15: Sidc 59Wc 117
Montrose Av. DA16: Well 55Tc 116
Montrose Av. HA8: Edg 26Sa 47
Montrose Av. NW6 40Ab 68
Montrose Av. RM2: Rom 26Ld 57
Montrose Av. SL1: Slou 4F 80
(not continuous)
Montrose Av. SL3: Dat 2N 103
Montrose Av. TW2: Whitt 59Da 107
Montrose Cl. DA16: Well 55Vc 117
Montrose Cl. IG8: Wfd G 21Jc 53
Montrose Cl. TW15: Ashf 65S 128
Montrose Ct. HA1: Harr 29Da 45
Montrose Ct. NW11 28Bb 49
Montrose Ct. NW9 26Sa 47
Montrose Ct. SE6 61Hc 137
Montrose Ct. SW7 2C 226 (47Fb 89)
Montrose Cres. HA0: Wemb 37Na 67
Montrose Cres. N12 23Eb 49
Montrose Gdns. CR4: Mitc 68Hb 133
Montrose Gdns. KT22: Oxs 84Fa 172
Montrose Gdns. SM1: Sutt 75Db 155
Montrose Ho. E14 48Cc 92
Montrose Ho. SW1 2J 227 (47Jb 90)
(off Montrose Pl.)
Montrose Pl. SW1 2J 227 (47Jb 90)
Montrose Rd. HA3: W'stone 26Ga 46
Montrose Rd. TW14: Bedf 58T 106
Montrose Wlk. HA7: Stan 23Ka 46
Montrose Wlk. KT13: Weyb 76R 150
Montrose Way SE23 60Zb 114
Montrose Way SL3: Dat 3P 103
Montrouge Cres. KT17: Eps D 88Ya 174
Montserrat Av. IG8: Wfd G 24Fc 53
Montserrat Cl. SE19 64Tb 135
Montserrat Rd. SW15 56Ab 110
Monument, The 4G 225 (45Tb 91)
Monument Bri. Ind. Est. E. GU21: Wok 87D 168
Monument Bri. Ind. Est. W. GU21: Wok 87C 168
Monument Bus. Cen. GU21: Wok 87D 168
Monument Gdns. SE13 57Ec 114
Monument Grn. KT13: Weyb 76R 150
Monument Hill KT13: Weyb 77R 150
Monument La. SL9: Chal P 23A 42
Monument Pl. AL3: St A 1B 6
(off Ashwell St.)
Monument Rd. GU21: Wok 86C 168
Monument Rd. KT13: Weyb 77R 150
Monument Station (Underground) 4G 225 (45Tb 91)
Monument St. EC3 4G 225 (45Tb 91)
Monument Way N15 27Vb 51
Monument Way N17 27Vb 51
Monument Way E. GU21: Wok 87D 168
Monument Way W. GU21: Wok 87C 168
Monza St. E1 45Yb 92
Moodkee St. SE16 48Yb 92
Moody Rd. SE15 53Vb 113
Moody St. E1 41Zb 92
Moon Ct. SE12 56Jc 115
Moon Ho. HA1: Harr 28Ga 46
Moon La. EN5: Barn 13Bb 31
Moonlight Dr. SE23 59Xb 113
Moonraker Point SE1 1D 230 (47Sb 91)
(off Pocock St.)
Moon St. N1 39Rb 71
MOOR, THE 61F 126
Moorcroft HA8: Edg 25Ra 47
Moorcroft Gdns. BR2: Broml 71Nc 160
Moorcroft La. UB8: Hil 43Q 84
Moorcroft Rd. SW16 62Nb 134
Moorcroft Way HA5: Pinn 29Aa 45
Moordown SE18 52Rc 116
Moore Av. RM18: Tilb 4D 122
Moore Av. RM20: Grays 50Ae 99
Moore Cl. CR4: Mitc 68Kb 134
Moore Cl. DA2: Dart 61Td 142
Moore Cl. KT15: Add 78K 149
Moore Cl. SL1: Slou 7F 80
Moore Cl. SW14 55Sa 109
Moore Ct. HA0: Wemb 37Na 67
Moore Ct. HA7: Stan 25Ma 47
Moore Ct. N1 39Rb 71
(off Gaskin St.)
Moore Cres. RM9: Dag 39Xc 75
Moorefield Rd. N17 26Vb 51
Moore Gro. Cres. TW20: Egh 65B 126
Moorehead Way SE3 55Jc 115
Moore Ho. AL3: St A 4N 5
Moore Ho. E1 45Yb 92
(off Cable St.)
Moore Ho. E14 47Cc 92
Moore Ho. E2 41Yb 92
(off Roman Rd.)
Moore Ho. N8 28Nb 50
(off Pembroke Rd.)
Moore Ho. RM11: Horn 30Jd 56
(off Benjamin Cl.)
Moore Ho. SE10 50Hc 93
(off Armitage Rd.)
Moore Ho. SW1 7K 227 (50Kb 90)
(off Gatliff Rd.)
Mooreland Rd. BR1: Broml 66Hc 137
Moor End Rd. HP1: Hem H 3L 3
Moore Pk. Rd. SW6 52Cb 111
Moore Place Golf Course 78Ca 151
Moore Rd. DA10: Swans 58Ae 121
Moore Rd. GU24: Brkwd 3A 186
Moore Rd. SE19 65Sb 135
Moore Rd. WD25: Wat 3X 13
Moores La. SL4: Eton W 9D 80
Moores Pl. CM14: B'wood 19Zd 41
Moore St. SW3 5F 227 (49Hb 89)
Moore Wlk. E7 35Jc 73
Moore Way SM2: Sutt 81Cb 175
Moorey Cl. E15 39Hc 73
Moorfield Av. W5 42Ma 87
Moorfield Rd. BR6: Orp 73Wc 161
Moorfield Rd. EN3: Enf H 11Yb 34
Moorfield Rd. KT9: Chess 78Na 153
Moorfield Rd. UB8: Cowl 44M 83
Moorfield Rd. UB9: Den 31J 63
Moorfields EC2 1F 225 (43Tb 91)
Moorfields Cl. TW18: Staines 67G 126
MOORFIELDS EYE HOSPITAL 4F 219 (41Tb 91)
Moorfields Highwalk EC2 1F 225 (43Tb 91)
(off New Union St.)
Moor Furlong SL1: Slou 6B 80
Moorgate EC2 2F 225 (44Tb 91)
Moorgate Pl. EC2 2F 225 (44Tb 91)
(off Moorgate)
Moorgate Station (Rail & Underground) 1F 225 (43Tb 91)
Moorgreen Ho. EC1 3B 218 (41Rb 91)
(off Wynyatt St.)
Moorhall Rd. UB9: Hare 30K 43
Moorhayes Dr. TW18: Lale 69L 127
Moorhen Cl. DA8: Erith 51Kd 119
Moorhen Dr. NW9 30Va 48
Moorhen Ho. E3 39Bc 72
(off Old Ford Rd.)
Moorhen Wlk. DA9: Ghithe 58Wd 120
(off Sanderling Way)
Moorholme GU22: Wok 91A 188
Moorhouse NW9 25Va 48
MOORHOUSE 99Pc 200
MOORHOUSE BANK 100Qc 200
Moorhouse Rd. HA3: Kenton 27Ma 47

Moorhouse Rd. TN16: Westrm 100Qc 200
Moorhouse Rd. W2 44Db 89
Moorhurst Av. EN7: G Oak 1Qb 18
Moorings, The AL1: St A 1D 6
(off Althorp Rd.)
Moorings, The E16 43Lc 93
(off Prince Regent La.)
Moorings, The HP3: Hem H 7N 3
(off Dickinson Quay)
Moorings, The KT14: W Byf 84L 169
Moorings, The KT23: Bookh 97Ca 191
Moorings, The SL4: Wind 2A 102
Moorings, The WD23: Bush 14Aa 27
Moorings Ho. TW8: Bford 52La 108
MOOR JUNC. 52K 105
Moorland Cl. RM5: Col R 24Dd 56
Moorland Cl. TW2: Whitt 59Ca 107
Moorland Rd. HP1: Hem H 4J 3
Moorland Rd. SW9 56Rb 113
Moorland Rd. UB7: Harm 51L 105
Moorlands AL2: F'mre 10C 6
Moorlands KT12: Walt T 76W 150
(off Ashley Pk. Rd.)
Moorlands UB5: N'olt 39Aa 65
Moorlands, The GU22: Wok 93B 188
Moorlands Av. NW7 23Xa 48
Moor La. EC2 1F 225 (43Tb 91)
(not continuous)
Moor La. GU22: Wok 94A 188
Moor La. KT9: Chess 77Na 153
Moor La. RM14: Upm 32Ud 78
Moor La. TW18: Staines 62G 126
Moor La. TW19: Staines 60F 104
Moor La. UB7: Harm 51L 105
Moor La. WD3: Rick 18P 25
Moor La. WD3: Sarr 8G 10
Moor La. Crossing WD18: Wat 17S 26
Moormead Dr. KT19: Ewe 78Ua 154
Moor Mead Rd. TW1: Twick 58Ja 108
Moormede Cres. TW18: Staines 63H 127
Moor Mill La. AL2: Col S 1Ga 14
(not continuous)
Moor Pk. Rickmansworth 20Q 26
MOOR PARK 20S 26
Moor Pk. Gdns. KT2: King T 66Ua 132
Moor Pk. Golf Course 20Q 26
Moor Pk. Ind. Cen. WD18: Wat 17S 26
Moor Pk. Rd. HA6: Nwood 22T 44
Moor Park Station (Underground) 20T 26
Moor Pl. EC2 1F 225 (43Tb 91)
Moor Pl. GU20: W'sham 8A 146
Moor Rd. TN14: S'oaks 92Kd 203
Moors, The RH1: Mers 3C 208
Moorside HP3: Hem H 5K 3
Moorside Rd. BR1: Broml 62Gc 137
Moors Nature Reserve Redhill, The 4B 208
Moorsom Way CR5: Coul 89Mb 176
Moorstown Ct. SL1: Slou 7J 81
Moor St. W1 3E 222 (44Mb 90)
Moortown Rd. WD19: Wat 21Y 45
Moor Vw. WD18: Wat 17W 26
Moorview Ho. HP2: Hem H 3M 3
(off The Spires)
Moot Ct. NW9 29Qa 47
Moran Cl. AL2: Brick W 3Ba 13
Moran Ho. E1 46Xb 91
(off Wapping La.)
Morant Gdns. RM5: Col R 22Dd 56
Morant Pl. N22 25Pb 50
Morant Rd. RM16: Grays 8D 100
Morants Ct. Rd. TN14: Dun G 90Ed 182
Morant St. E14 45Cc 92
Mora Rd. NW2 35Ya 68
Mora St. EC1 4E 218 (41Sb 91)
Morat St. SW9 53Pb 112
Moravian Cl. SW10 51Fb 111
Moravian Pl. SW10 51Fb 111
Moravian St. E2 40Yb 72
Moray Av. UB3: Hayes 46V 84
Moray Cl. HA8: Edg 19Ra 29
Moray Cl. RM1: Rom 24Gd 56
Moray Ct. CR2: S Croy 78Sb 157
(off Warham Rd.)
Moray Dr. SL2: Slou 4L 81
Moray Ho. E1 42Ac 92
(off Harford St.)
Moray M. N7 33Pb 70
Moray Rd. N4 33Pb 70
Moray Way RM1: Rom 24Fd 56
Mordaunt Gdns. RM9: Dag 38Ad 75
Mordaunt Ho. NW10 39Ta 67
(off Stracey Rd.)
Mordaunt Rd. NW10 39Ta 67
Mordaunt St. SW9 55Pb 112
MORDEN 69Db 133
Morden Cl. KT20: Tad 92Za 194
Morden Ct. SM4: Mord 70Db 133
Morden Ct. Pde. SM4: Mord 70Db 133
Morden Gdns. CR4: Mitc 70Fb 133
Morden Gdns. UB6: G'frd 36Ha 66
Morden Hall Pk. 69Eb 133
Morden Hall Rd. SM4: Mord 69Db 133
Morden Hill SE13 54Ec 114
Morden Ho. SM4: Mord 70Cb 133
Morden La. SE13 54Ec 114
Morden Leisure Cen. 71Bb 155
MORDEN PARK 72Ab 154
Morden Rd. CR4: Mitc 70Eb 133
Morden Rd. RM6: Chad H 31Ad 75
Morden Rd. SE3 54Jc 115
Morden Rd. SM4: Mord 70Eb 133
Morden Rd. SW19 67Db 133
Morden Rd. M. SE3 54Jc 115
Morden Road Stop (London Tramlink) 68Db 133
Morden South Station (Rail) 71Cb 155
Morden Station (Underground) 69Db 133
Morden St. SE13 53Dc 114
Morden Way SM3: Sutt 73Cb 155
Morden Wharf SE10 48Gc 93
(off Morden Wharf Rd.)
Morden Wharf Rd. SE10 48Gc 93
Mordern Ho. NW1 6E 214 (42Gb 89)
(off Harewood Av.)
Mordon Rd. IG3: Ilf 31Vc 75
Moreau M. N5 36Sb 71
Moreau Wlk. SL3: Geor G 44Ba 64
Morecambe Cl. E1 43Zb 92
Morecambe Cl. RM12: Horn 36Kd 77
Morecambe Gdns. HA7: Stan 21Ma 47
Morecambe Ho. RM3: Rom 22Nd 57
(off Chudleigh Rd.)
Morecambe St. SE17 6E 230 (49Sb 91)
Morecambe Ter. N18 21Tb 51
(off Gt. Cambridge Rd.)
More Cl. CR8: Purl 83Qb 176
More Cl. E16 44Hc 93
More Cl. W14 49Za 88
Morecoombe Cl. KT2: King T 66Ra 131
More Copper Ho. SE1 7H 225 (46Ub 91)
(off Magdalen St.)
Moree Way N18 21Wb 51
Moreing Dr. DA9: Ghithe 56Yd 120
Moreland Av. RM16: Grays 47Ee 99
Moreland Av. SL3: Coln 52E 104
Moreland Cl. SL3: Coln 52E 104
Moreland Cotts. E3 40Cc 72
(off Fairfield Rd.)
Moreland Ct. NW2 34Cb 69
Moreland Dr. SL9: Ger X 31B 62
Moreland St. EC1 3C 218 (41Rb 91)
Moreland Way E4 20Dc 34
More La. KT10: Esh 75Da 151
Morel Ct. TN13: S'oaks 94Kd 203
Morella Cl. GU25: Vir W 10P 125
Morella Rd. SW12 59Hb 111
Morell Cl. EN5: New Bar 13Eb 31
Morello Av. UB8: Hil 43R 84
Morello Cl. BR8: Swan 70Fd 140
Morello Dr. SL3: L'ly 46B 82
Morel M. RM8: Dag 32Zc 75
More London Pl. SE1 7H 225 (46Ub 91)
(not continuous)
More London Riverside SE1 7J 225 (46Ub 91)
(not continuous)
Moremead EN9: Walt A 5Fc 21
Moremead Rd. SE6 63Bc 136
Morena St. SE6 59Dc 114
Moreno Ho. E2 40Xb 71
(off Esker Pl.)
Moresby Av. KT5: Surb 73Ra 153
Moresby Rd. E5 32Xb 71
Moresby Wlk. SW8 54Lb 112
More's Gdn. SW3 51Fb 111
(off Cheyne Wlk.)
Mores La. CM14: Pil H 14Sd 40
Moretaine Rd. TW15: Ashf 62M 127
Moreton Almshouses TN16: Westrm 98Tc 200
Moreton Av. TW7: Isle 53Ga 108
Moreton Bay Ind. Est. RM2: Rom 26Md 57
Moreton Cl. BR8: Swan 68Gd 140
Moreton Cl. E5 33Xb 71
Moreton Cl. N15 30Tb 51
Moreton Cl. NW7 23Ya 48
Moreton Cl. SW1 7C 228 (50Lb 90)
(off Moreton Ter.)
Moreton Ct. DA1: Cray 55Hd 118
Moreton Gdns. IG8: Wfd G 22Nc 54
Moreton Ho. SE16 48Xb 91
Moreton Ind. Est. BR8: Swan 70Jd 140
Moreton Pl. SW1 7C 228 (50Lb 90)
Moreton Rd. CR2: S Croy 78Tb 157
Moreton Rd. KT4: Wor Pk 75Wa 154
Moreton Rd. N15 30Tb 51
Moretons HA1: Harr 32Ga 66
Moreton St. SW1 7C 228 (50Lb 90)
Moreton Ter. SW1 7C 228 (50Lb 90)
Moreton Ter. M. Nth. SW1 7C 228 (50Lb 90)
Moreton Ter. M. Sth. SW1 7C 228 (50Lb 90)
Moreton Twr. W3 46Ra 87
Moreton Way SL1: Slou 6B 80
Morewood Cl. TN13: S'oaks 95Hd 202
Morewood Cl. Ind. Est. TN13: S'oaks 95Hd 202
Morford Cl. HA4: Ruis 31X 65
Morford Way HA4: Ruis 31X 65
Morgan Av. E17 28Fc 53
Morgan Cl. HA6: Nwood 23V 44
Morgan Cl. RM10: Dag 38Cd 76
Morgan Ct. SM5: Cars 77Hb 155
Morgan Ct. TW15: Ashf 64R 128
Morgan Cres. CM16: They B 8Tc 22
Morgan Cres. RM8: Dag 32Ad 75
Morgan Dr. DA9: Ghithe 59Ud 120
Morgan Gdns. WD25: A'ham 10Da 13
Morgan Ho. SW1 6C 228 (49Lb 90)
(off Vauxhall Bri. Rd.)
Morgan Ho. SW8 53Lb 112
(off Wadhurst Rd.)
Morgan Mans. N7 36Qb 70
(off Morgan Rd.)
Morgan Rd. BR1: Broml 66Jc 137
Morgan Rd. N7 36Qb 70
Morgan Rd. W10 43Bb 89
Morgan's La. UB3: Hayes 43T 84
Morgans La. SE1 7H 225 (46Ub 91)
(off Tooley St.)
Morgan St. E16 43Hc 93
Morgan St. E3 41Ac 92
(not continuous)
Morgan Wlk. BR3: Beck 70Dc 136
Morgan Way IG8: Wfd G 23Nc 54
Morgan Way RM13: Rain 41Ld 97
Moriarty Cl. BR1: Broml 70Qc 138
Moriatry Cl. N7 35Nb 70
Morie St. SW18 57Db 111
Morieux Rd. E10 32Bc 72
Moring Rd. SW17 63Jb 134
Morkyns Wlk. SE21 62Ub 135
Morland Av. CR0: C'don 74Ub 157
Morland Av. DA1: Dart 57Kd 119
Morland Cl. CR4: Mitc 69Gb 133
Morland Cl. NW11 32Db 69
Morland Cl. TW12: Hamp 64Ba 129
Morland Ct. W12 47Xa 88
(off Coningham Rd.)
Morland Est. E8 38Wb 71
Morland Gdns. NW10 38Ta 67
Morland Gdns. UB1: S'hall 46Da 85
Morland Ho. NW1 2C 216 (40Lb 70)
(off Werrington St.)
Morland Ho. NW6 39Cb 69
Morland Ho. SW1 5F 229 (49Nb 90)
(off Marsham St.)
Morland Ho. W11 44Ab 88
(off Lancaster Rd.)
Morland M. N1 38Qb 70
Morland Pl. N15 28Ub 51
Morland Rd. CR0: C'don 74Ub 157
Morland Rd. E17 29Zb 52
Morland Rd. HA3: Kenton 29Na 47
Morland Rd. IG1: Ilf 33Rc 74
Morland Rd. RM10: Dag 38Cd 76
Morland Rd. SE20 65Zb 136
Morland Rd. SM1: Sutt 78Eb 155
Morland Way EN8: Chesh 1Ac 20
Morley Av. E4 24Fc 53
Morley Av. N18 21Wb 51
Morley Av. N22 26Qb 50
Morley Cl. BR6: Farnb 75Rc 160
Morley Cl. SL3: L'ly 47B 82
Morley Ct. BR2: Broml 70Hc 137
Morley Ct. E4 22Bc 52
Morley Ct. KT22: Fet 93Fa 192
Morley Cres. HA4: Ruis 33Y 65
Morley Cres. HA8: Edg 19Sa 29
Morley Cres. E. HA7: Stan 26La 46
Morley Cres. W. HA7: Stan 27La 46
Morley Hill EN2: Enf 10Tb 19
Morley Ho. SE15 52Vb 113
(off Commercial Way)
Morley Rd. BR7: Chst 67Sc 138
Morley Rd. CR2: Sande 82Vb 177
Morley Rd. E10 32Ec 72
Morley Rd. E15 40Hc 73
Morley Rd. IG11: Bark 39Tc 74
Morley Rd. RM6: Chad H 29Ad 55
Morley Rd. SE13 56Ec 114
Morley Rd. SM3: Sutt 74Bb 155
Morley Rd. TW1: Twick 58Ma 109
Morley Sq. RM16: Grays 9C 100
Morley St. SE1 3A 230 (48Qb 90)
Morna Rd. SE5 54Sb 113
Morning La. E9 37Yb 72
Morningside Cl. GU25: Vir W 3P 147
Morningside Rd. KT4: Wor Pk 75Ya 154
Mornington Av. BR1: Broml 69Lc 137
Mornington Av. IG1: Ilf 31Qc 74
Mornington Av. W14 49Bb 89
Mornington Av. Mans. W14 49Bb 89
(off Mornington Av.)
Mornington Cl. IG8: Wfd G 21Jc 53
Mornington Cl. NW9 27Ua 48
Mornington Cl. TN16: Big H 89Mc 179
Mornington Ct. DA5: Bexl 60Fd 118
Mornington Ct. NW1 1B 216 (40Lb 70)
(off Mornington Cres.)
Mornington Cres. NW1 1B 216 (40Lb 70)
Mornington Cres. TW5: Cran 53X 107
Mornington Crescent Station (Underground) 1B 216 (40Lb 70)
Mornington Gro. E3 41Cc 92
Mornington M. SE5 53Sb 113
Mornington Pl. NW1 1B 216 (40Lb 70)
Mornington Pl. SE8 52Bc 114
(off Mornington Rd.)
Mornington Rd. E11 31Hc 73
Mornington Rd. E4 17Fc 35
Mornington Rd. IG10: Lough 13Sc 36
Mornington Rd. IG8: Wfd G 21Hc 53
Mornington Rd. SE8 52Bc 114
Mornington Rd. TW15: Ashf 64S 128
Mornington Rd. UB6: G'frd 43Da 85
Mornington Rd. WD7: R'lett 6Ja 14
Mornington St. NW1 1A 216 (40Kb 70)
Mornington Ter. NW1 1A 216 (39Kb 70)
Mornington Wlk. TW10: Ham 63La 130
Moro Apts. E14 44Ec 92
(off New Festival Av.)
Morocco St. SE1 2H 231 (47Ub 91)
Morocco Wharf E1 46Xb 91
(off Wapping High St.)
Morpeth Av. WD6: Bore 10Pa 15
Morpeth Cl. HP2: Hem H 3N 3
Morpeth Gro. E9 39Zb 72
Morpeth Mans. SW1 5B 228 (49Lb 90)
(off Morpeth Ter.)
Morpeth Rd. E9 39Zb 72
Morpeth St. E2 41Yb 92
Morpeth Ter. SW1 4B 228 (48Lb 90)
Morpeth Wlk. N17 24Xb 51
Morphou Rd. NW7 23Ab 48
Morrab Gdns. IG3: Ilf 34Vc 75
Morrel Ct. E2 40Wb 71
(off Goldsmiths Row)
Morrells Yd. SE11 7A 230 (50Qb 90)
(off Cleaver St.)
Morrice Cl. SL3: L'ly 49B 82
Morris Av. E12 36Pc 74
Morris Av. UB8: Uxb 37N 63
Morris Blitz Ct. N16 35Vb 71
Morris Cl. BR6: Orp 76Uc 160
Morris Cl. CR0: C'don 71Ac 158
Morris Cl. SL9: Chal P 25B 42
Morris Ct. E4 20Dc 34
Morris Ct. EN9: Walt A 6Hc 21
Morris Ct. SL4: Wind 3C 102
Morris Dr. DA17: Erith 49Ed 96
Morris Gdns. DA1: Dart 57Qd 119
Morris Gdns. SW18 59Cb 111
Morris Ho. E2 41Yb 92
(off Roman Rd.)
Morris Ho. NW8 6D 214 (42Gb 89)
(off Salisbury St.)
Morris Ho. W3 47Va 88
Morrish Rd. SW2 59Nb 112
Morris M. SW19 64Eb 133
Morrison Av. E4 23Cc 52
Morrison Av. N17 27Ub 51
Morrison Bldgs. Nth. E1 44Wb 91
(off Commercial Rd.)
Morrison Ct. EN5: Barn 14Ab 30
(off Manor Way)
Morrison Ct. N12 24Gb 49
Morrison Ct. SW1 4E 228 (48Mb 90)
(off Gt. Smith St.)
Morrison Ho. RM16: Grays 47Fe 99
Morrison Ho. SW2 60Qb 112
(off Tulse Hill)
Morrison Rd. IG11: Bark 40Ad 75
Morrison Rd. RM9: Bark 40Ad 75
Morrison Rd. RM9: Dag 40Ad 75
Morrison Rd. SW9 54Qb 112
Morrison Rd. UB4: Yead 41X 85
Morrison St. SW11 55Jb 112
Morris Pl. N4 33Qb 70
Morris Rd. E14 43Dc 92
Morris Rd. E15 35Gc 73
Morris Rd. RH1: S Nut 8E 208
Morris Rd. RM3: Rom 24Kd 57
Morris Rd. RM8: Dag 33Bd 75
Morris Rd. TW7: Isle 55Ha 108
Morriss Ho. SE16 47Xb 91
(off Cherry Gdn. St.)
Morris St. E1 44Xb 91
Morriston Cl. WD19: Wat 22Y 45
Morris Wlk. DA1: Dart 54Qd 119
Morris Way AL2: Lon C 8H 7
Morritt Ho. HA0: Wemb 36Ma 67
(off Talbot Rd.)
Morse Cl. E13 41Jc 93
Morse Cl. UB9: Hare 26L 43
Morshead Mans. W9 41Cb 89
(off Morshead Rd.)
Morshead Rd. W9 41Cb 89
Morson Rd. EN3: Pond E 16Ac 34
Morston Cl. KT20: Tad 92Xa 194
Morston Gdns. SE9 63Pc 138
Mortain Ho. SE16 49Xb 91
(off Roseberry St.)
Morten Cl. SW4 58Mb 112
Morten Gdns. UB9: Den 31J 63
Morteyne Rd. N17 25Tb 51
Mortgramit Sq. SE18 48Qc 94
Mortham St. E15 39Gc 73
Mortimer Cl. NW2 33Bb 69
Mortimer Cl. SW16 61Mb 134
Mortimer Cl. WD23: Bush 16Da 27
Mortimer Ct. NW8 2A 214 (40Eb 69)
(off Abbey Rd.)
Mortimer Cres. AL3: St A 4N 5
Mortimer Cres. KT4: Wor Pk 76Ta 153
Mortimer Cres. NW6 39Db 69
Mortimer Dr. EN1: Enf 15Ub 33
Mortimer Dr. RM3: Rom 22Ld 57
Mortimer Dr. TN16: Big H 84Lc 179
Mortimer Est. NW6 39Db 69
(off Mortimer Pl.)
Mortimer Ho. W11 46Za 88
Mortimer Ho. W14 49Ab 88
(off North End Rd.)
Mortimer Mkt. WC1 6C 216 (42Lb 90)
Mortimer Pl. NW6 39Db 69
Mortimer Rd. BR6: Orp 74Wc 161
Mortimer Rd. CR4: Mitc 67Hb 133
Mortimer Rd. DA8: Erith 51Fd 118
Mortimer Rd. E6 41Pc 94
Mortimer Rd. N1 38Ub 71
(not continuous)
Mortimer Rd. NW10 41Ya 88
Mortimer Rd. SL3: L'ly 8P 81
Mortimer Rd. W13 44La 86
Mortimer Sq. DA10: Swans 61Ce 143
Mortimer Sq. W11 45Za 88
Mortimer St. W1 2B 222 (44Lb 90)
Mortimer Ter. NW5 35Kb 70
MORTLAKE 55Ta 109
Mortlake Cl. CR0: Bedd 76Nb 156
Mortlake Crematorium 54Ra 109
Mortlake Dr. CR4: Mitc 67Gb 133
Mortlake High St. SW14 55Ta 109
Mortlake Rd. E16 44Kc 93
Mortlake Rd. IG1: Ilf 35Sc 74
Mortlake Rd. TW9: Kew 52Qa 109
Mortlake Rd. TW9: Rich 52Qa 109
Mortlake Station (Rail) 55Sa 109
Mortlake Ter. TW9: Kew 52Qa 109
(off Mortlake Rd.)
Mortlock Cl. SE15 53Xb 113
Mortlock Ct. E7 35Mc 73
Morton KT20: Tad 93Za 194
Morton Apts. E16 45Kc 94
(off Lock Side Way)
Morton Cl. E1 44Yb 92
Morton Cl. GU21: Wok 7N 167
Morton Cl. SM6: W'gton 80Pb 156
Morton Cl. UB8: Hil 42P 83
Morton Ct. UB5: N'olt 36Ea 66
Morton Cres. N14 21Mb 50
Morton Dr. SL2: Farn C 5C 60
Morton Gdns. SM6: W'gton 78Lb 156
Morton Ho. SE17 51Rb 113
Morton M. SW5 49Db 89
Morton Pl. SE1 4K 229 (48Qb 90)
Morton Rd. E15 38Hc 73
Morton Rd. GU21: Wok 7P 167
Morton Rd. N1 38Sb 71
Morton Rd. SM4: Mord 71Fb 155
Morton Way N14 20Lb 32
Morvale Cl. DA17: Belv 49Bd 95
Morval Rd. SW2 57Qb 112
Morven Cl. EN6: Pot B 3Eb 17
Morven Rd. SW17 62Hb 133
Morville Ho. SW18 58Fb 111
(off Fitzhugh Gro.)
Morville St. E3 40Cc 72
Morwell St. WC1 1D 222 (43Mb 90)
Mosaic Ho. HP2: Hem H 2M 3
Mosbach Gdns. CM13: Hut 19De 41
Moscow Mans. SW5 49Cb 89
(off Cromwell Rd.)
Moscow Pl. W2 45Db 89
Moscow Rd. W2 45Cb 89
Mosedale NW1 4B 216 (41Lb 90)
(off Cumberland Mkt.)
Moseley Row SE10 49Hc 93
Moselle Av. N22 26Qb 50
Moselle Cl. N8 27Pb 50
Moselle Ho. N17 24Vb 51
(off William St.)
Moselle Pl. N17 24Vb 51
Moselle Rd. TN16: Big H 90Nc 180
Moselle St. N17 24Vb 51
Mospey Cres. KT17: Eps 87Va 174
Mosque Ter. E1 43Wb 91
(off Whitechapel Rd.)
Mosque Twr. E1 43Wb 91
(off Fieldgate St.)
Mosque Twr. E3 40Ac 72
(off Ford St.)
Mosquito Cl. SM6: W'gton 80Nb 156
Moss Bank RM17: Grays 49Be 99
(not continuous)
Mossborough Cl. N12 23Db 49
Mossbury Rd. SW11 55Gb 111
Moss Cl. E1 43Wb 91
Moss Cl. HA5: Pinn 26Ba 45
Moss Cl. N9 18Wb 33
Moss Cl. WD3: Rick 19M 25
Mossdown Cl. DA17: Belv 49Cd 96
Mossendew Cl. UB9: Hare 25M 43
Mossfield KT11: Cobh 85W 170
Mossford Ct. IG6: Ilf 27Rc 54
Mossford Grn. IG6: Ilf 27Rc 54
Mossford La. IG6: Ilf 26Rc 54
Mossford St. E3 42Bc 92
Moss Gdns. CR2: Sels 80Zb 158
Moss Gdns. TW13: Felt 61W 128
Moss Hall Ct. N12 23Db 49
Moss Hall Cres. N12 23Db 49
Moss Hall Gro. N12 23Db 49
Mossington Gdns. SE16 49Yb 92
Moss La. HA5: Pinn 25Aa 45
Moss La. RM1: Rom 30Hd 56
Mosslea Rd. BR2: Broml 71Mc 159
Mosslea Rd. BR6: Farnb 76Sc 160
Mosslea Rd. CR3: Whyt 88Vb 177
Mosslea Rd. SE20 65Yb 136
(not continuous)
Mossop St. SW3 5E 226 (49Gb 89)
Moss Rd. RM10: Dag 38Cd 76
Moss Rd. RM15: S Ock 43Yd 98
Moss Rd. WD25: Wat 6X 13
Moss Side AL2: Brick W 2Ba 13
Mossville Gdns. SM4: Mord 69Bb 133
Moss Way DA2: Daren 63Td 142
Mosswell Ho. N10 25Jb 50
Moston Cl. UB3: Harl 50V 84
Mostyn Av. HA9: Wemb 36Pa 67
Mostyn Gdns. NW10 41Za 88
Mostyn Gro. E3 40Cc 72
Mostyn Rd. HA8: Edg 24Ua 48
Mostyn Rd. SW19 67Bb 133
Mostyn Rd. SW9 53Qb 112
Mostyn Rd. WD23: Bush 15Ea 28
Mostyn Ter. RH1: Redh 7A 208
Mosul Way BR2: Broml 72Nc 160
Mosyer Dr. BR5: Orp 75Zc 161
Motcomb St. SW1 3H 227 (48Jb 90)
Mote, The DA3: Nw A G 75Be 165
Mote Rd. TN15: Ivy H 100Wd 204
Mote Rd. TN15: S'brne 100Wd 204
Moth Cl. SM6: W'gton 80Nb 156
Mothers Sq. E5 35Xb 71
Motherwell Way RM20: W Thur 50Wd 98
Motley Av. EC2 5H 219 (42Ub 91)
(off Phipp St.)
Motspur Pk. KT3: N Mald 72Va 154
MOTSPUR PARK 72Wa 154
Motspur Park Station (Rail) 71Xa 154
Motte La. DA10: Swans 60Ae 121
MOTTINGHAM 61Nc 138
Mottingham Gdns. SE9 60Mc 115
Mottingham La. SE12 60Lc 115
Mottingham La. SE9 60Lc 115
Mottingham Rd. N9 16Zb 34
Mottingham Rd. SE9 61Nc 138
Mottingham Station (Rail) 60Pc 116
Mottisfont Rd. SE2 48Wc 95
Mottistone Gro. SM2: Sutt 81Db 175
Motts Hill La. KT20: Tad 95Wa 194
Motts Hill La. KT20: Walt H 95Wa 194
Motts La. RM8: Dag 33Bd 75
Mott St. E4 10Fc 21
Mott St. IG10: H Beech 11Gc 35
Mott St. IG10: Lough 11Gc 35
Mouchotte Cl. TN16: Big H 84Kc 179
Moulding La. SE14 51Ac 114
Moules Ct. SE5 52Sb 113
Moulins Rd. E9 38Yb 72
Moulsford Ho. N7 36Mb 70
Moulsford Ho. W2 43Cb 89
(off Westbourne Pk. Rd.)
Moultain Hill BR8: Swan 70Jd 140
Moulton Av. TW3: Houn 54Aa 107
Moultrie Way RM14: Upm 30Ud 58
Mound, The SE9 62Qc 138
Moundfield Rd. N16 30Wb 51
Moundsfield Way SL1: Slou 7C 80
Mounsey Ho. W10 41Ab 88
(off Third Av.)
Mount, The GU21: Wok Elm Rd. 10P 167
Mount, The GU21: Wok St John's Hill 1L 187
Mount, The BR1: Broml 67Nc 138
Mount, The CM14: B'wood 20Yd 40
Mount, The CR2: S Croy 78Sb 157
(off Warham Rd.)
Mount, The CR5: Coul 87Jb 176
Mount, The CR6: W'ham 91Wb 197
Mount, The DA6: Bex 57Dd 118
Mount, The E5 33Xb 71
(not continuous)
Mount, The EN6: Pot B 2Db 17
Mount, The GU21: Knap 1G 186
Mount, The GU25: Vir W 2P 147
Mount, The HA9: Wemb 33Ra 67
Mount, The IG10: Lough 14Nc 36
Mount, The KT10: Esh 79Ca 151
Mount, The KT13: Weyb 75U 150
Mount, The KT17: Ewe 82Va 174
Mount, The KT20: Lwr K 98Bb 195
Mount, The KT22: Fet 95Ga 192
Mount, The KT3: N Mald 69Va 132
Mount, The KT4: Wor Pk 77Xa 154
Mount, The N20 19Eb 31
Mount, The NW3 34Eb 69
Mount, The RM3: Rom 20Ld 39
Mount, The SS17: Stan H 1P 101
Mount, The UB5: N'olt 36Da 65
Mount, The W3 46Sa 87
Mount, The W8 46Cb 89
(off Bedford Gdns.)
Mount, The WD3: Rick 16L 25
Mountacre Cl. SE26 63Vb 135
Mount Adon Pk. SE22 59Wb 113
Mountague Pl. E14 45Ec 92
Mountain Cl. TN15: Wro 88Be 185
Mountain Ho. SE11 6J 229 (49Pb 90)
Mount Angelus Rd. SW15 59Va 110
Mount Ararat Rd. TW10: Rich 57Na 109
Mount Arlington BR2: Broml 68Gc 137
(off Park Hill Rd.)
Mount Ash Rd. SE26 62Xb 135
Mount Av. CM13: Hut 17De 41
Mount Av. CR3: Cat'm 96Sb 197
Mount Av. E4 20Cc 34
Mount Av. RM3: Hrld W 23Sd 58
Mount Av. UB1: S'hall 44Ca 85
Mount Av. W5 43La 86
Mountbatten Cl. AL1: St A 5F 6
Mountbatten Cl. SE18 51Uc 116
Mountbatten Cl. SE19 64Ub 135
Mountbatten Cl. SL1: Slou 8L 81
Mountbatten Ct. IG9: Buck H 19Mc 35
Mountbatten Ct. SE16 46Yb 92
(off Rotherhithe St.)
Mountbatten Ct. SL3: L'ly 48A 82
Mountbatten Gdns. BR3: Beck 70Ac 136
Mountbatten Ho. N6 31Jb 70
(off Hillcrest)
Mountbatten Ho. RM2: Rom 28Ld 57
Mountbatten M. SW18 59Eb 111
Mountbatten Sq. SL4: Wind 3G 102
Mountbel Rd. HA7: Stan 25Ja 46
Mount Carmel Chambers W8 47Cb 89
(off Dukes La.)
Mount Cl. BR1: Broml 67Nc 138
Mount Cl. CR8: Kenley 88Tb 177
Mount Cl. EN4: Cockf 14Jb 32
Mount Cl. GU22: Wok 3N 187
Mount Cl. HP1: Hem H 2H 3
Mount Cl. KT22: Fet 95Ga 192
Mount Cl. SL2: Farn C 5G 60
Mount Cl. SM5: Cars 81Jb 176
Mount Cl. TN13: S'oaks 95Hd 202
Mount Cl. W5 43La 86
Mount Cl., The GU25: Vir W 2P 147
Mountcombe Cl. KT6: Surb 73Na 153
Mount Ct. BR4: W W'ck 75Gc 159
Mount Ct. SW15 55Ab 110
Mount Cres. CM14: W'ley 21Zd 59
Mount Culver Av. DA14: Sidc 65Zc 139
Mount Dr. AL2: Park 7B 6
Mount Dr. DA6: Bex 57Ad 117
Mount Dr. HA2: Harr 29Ba 45
Mount Dr. HA9: Wemb 33Sa 67
Mount Dr., The RH2: Reig 4M 207
Mountearl Gdns. SW16 62Pb 134
Mount Eaton Ct. W5 43Ka 86
(off Mount Av.)
Mount Echo Av. E4 19Dc 34
Mount Echo Dr. E4 18Dc 34
MOUNT END 4Bd 23
Mount Ephraim La. SW16 62Mb 134
Mount Ephraim Rd. SW16 62Mb 134
Mount Felix KT12: Walt T 74V 150

Mountfield TN15: Bor G 92Ce 205
Mountfield Cl. SE6 59Fc 115
Mountfield Cl. SS17: Stan H 1N 101
Mountfield Rd. E6 40Qc 74
Mountfield Rd. HP2: Hem H 2N 3
Mountfield Rd. N3 27Bb 49
Mountfield Rd. W5 44Ma 87
Mountfield Ter. SE6 59Fc 115
Mountfield Way BR5: St M Cry 70Yc 139
Mountford Mans. SW11 53Jb 112
(off Battersea Pk. Rd.)
Mountfort Cres. N1 38Qb 70
Mountfort Ho. N1 38Qb 70
(off Barnsbury Sq.)
Mountfort Ter. N1 38Qb 70
Mount Gdns. SE26 62Xb 135
Mount Grace Rd. EN6: Pot B 3Cb 17
Mount Gro. HA8: Edg 20Sa 29
Mountgrove Rd. N5 34Rb 71
Mount Harry Rd. TN13: S'oaks 95Jd 202
MOUNT HERMON 1N 187
Mount Hermon Cl. GU22: Wok 1P 187
Mount Hermon Rd. GU22: Wok 1P 187
Mount Hill TN14: Knock 89Wc 181
Mount Hill La. SL9: Ger X 2M 61
Mount Holme KT7: T Ditt 73Ka 152
Mounthurst Rd. BR2: Hayes 73Hc 159
Mountington Pk. Cl. HA3: Kenton 30Ma 47
Mountjoy Cl. EC2 1E 224 (43Sb 91)
(off Monkwell Sq.)
Mountjoy Cl. SE2 47Xc 95
Mountjoy Ho. EC2 1D 224 (43Sb 91)
(off Monkwell Sq.)
Mount La. UB9: Den 33F 62
Mount Lee TW20: Egh 64B 126
Mount Lodge N6 30Lb 50
Mount M. TW12: Hamp 67Da 129
Mount Mills EC1 4C 218 (41Rb 91)
MOUNTNESSING 11Fe 41
Mountnessing By-Pass CM15: Mount 13Fe 41
Mountnessing By-Pass CM15: Shenf 13Fe 41
Mountnessing La. CM15: Dodd 11Zd 41
Mountnessing Windmill 11Fe 41
Mount Nod DA9: Ghithe 57Yd 120
Mount Nod Rd. SW16 62Pb 134
Mount Olive Ct. W7 47Ga 86
Mount Pde. EN4: Cockf 14Gb 31
Mount Pk. SM5: Cars 80Jb 156
Mount Pk. Av. CR2: S Croy 81Rb 177
Mount Pk. Av. HA1: Harr 33Fa 66
Mount Pk. Cres. W5 44Ma 87
Mount Pk. Rd. HA1: Harr 34Fa 66
Mount Pk. Rd. HA5: Eastc 29W 44
Mount Pk. Rd. W5 43Ma 87
Mount Pl. W3 46Ra 87
MOUNT PLEASANT 25J 43
Mount Pleasant AL3: St A 1P 5
Mount Pleasant EN4: Cockf 14Gb 31
Mount Pleasant HA0: Wemb 39Na 67
Mount Pleasant HA4: Ruis 33Y 65
Mount Pleasant IG1: Ilf 36Sc 74
Mount Pleasant KT13: Weyb 76Q 150
Mount Pleasant KT17: Ewe 82Va 174
Mount Pleasant KT24: Eff 100Aa 191
Mount Pleasant KT24: W Hor 100R 190
Mount Pleasant SE27 63Sb 135
Mount Pleasant SL9: Chal P 22A 42
Mount Pleasant TN16: Big H 89Mc 179
Mount Pleasant UB9: Hare 25J 43
Mount Pleasant WC1 6K 217 (42Qb 90)
Mount Pleasant Av. CM13: Hut 16Fe 41
Mount Pleasant Cotts. N14 17Mb 32
(off The Wells)
Mount Pleasant Cres. N4 32Pb 70
Mount Pleasant Hill E5 33Xb 71
Mount Pleasant La. AL2: Brick W 2Aa 13
Mount Pleasant La. E5 32Xb 71
Mount Pleasant M. N4 32Pb 70
(off Mt. Pleasant Cres.)
Mount Pleasant Pl. SE18 49Tc 94
Mount Pleasant Rd. CR3: Cat'm 95Wb 197
Mount Pleasant Rd. DA1: Dart 58Pd 119
Mount Pleasant Rd. E17 26Ac 52
Mount Pleasant Rd. IG7: Chig 21Tc 54
Mount Pleasant Rd. KT3: N Mald 69Sa 131
Mount Pleasant Rd. N17 26Ub 51
Mount Pleasant Rd. NW10 38Ya 68
Mount Pleasant Rd. RM5: Col R 23Fd 56
Mount Pleasant Rd. SE13 58Dc 114
Mount Pleasant Rd. W5 42La 86
Mount Pleasant Vs. N4 31Pb 70
Mount Pleasant Wlk. DA5: Bexl 57Ed 118
Mount Ri. RH1: Redh 8M 207
Mount Rd. CM16: Fidd H 4Zc 23
Mount Rd. CM16: They M 4Zc 23
Mount Rd. CR4: Mitc 68Fb 133
Mount Rd. DA1: Cray 58Hd 118
Mount Rd. DA6: Bex 57Zc 117
Mount Rd. EN4: E Barn 15Gb 31
Mount Rd. GU22: Wok 3M 187
Mount Rd. GU24: Chob 4M 167
Mount Rd. KT3: N Mald 69Ta 131
Mount Rd. KT9: Chess 78Pa 153
Mount Rd. NW2 34Xa 68
Mount Rd. NW4 30Wa 48
Mount Rd. RM4: Stap T 9Bd 23
Mount Rd. RM8: Dag 32Bd 75
Mount Rd. SE19 65Tb 135
Mount Rd. SW19 61Cb 133
Mount Rd. TW13: Hanw 62Aa 129
Mount Rd. UB3: Hayes 47W 84
Mount Row W1 5K 221 (45Kb 90)
Mountsfield Cl. TW19: Stanw M 58J 105
Mountsfield Ct. SE13 58Fc 115
Mountsfield Ho. SE10 53Ec 114
(off Primrose Way)
Mounts Hill SL4: Wink 1A 124
Mountside CR3: Cat'm 96Vb 197
Mountside HA7: Stan 25Ja 46
Mountside Cl. DA11: Nflt 59Ee 121
Mount Side Pl. GU22: Wok 90B 168
Mounts Pond Rd. SE3 54Fc 115
(not continuous)
Mount Sq., The NW3 34Eb 69
Mounts Rd. DA9: Ghithe 57Xd 120
Mount Stewart Av. HA3: Kenton 31Ma 67
Mount St. W1 5H 221 (45Jb 90)
Mount St. M. W1 5K 221 (45Kb 90)
Mountstuart Ct. TW11: Hamp W 67Ka 130
Mount Ter. E1 43Xb 91
Mount Vernon NW3 35Eb 69
MOUNT VERNON HOSPITAL Northwood 23R 44
Mount Vw. AL2: Lon C 9J 7
Mount Vw. EN2: Enf 10Pb 18
Mount Vw. HA6: Nwood 23V 44
Mount Vw. NW7 20Ta 29
Mount Vw. UB2: S'hall 49Z 85
Mount Vw. W5 42Ma 87
Mount Vw. WD3: Rick 18K 25
Mountview Cl. NW11 32Db 69
Mountview Ct. N8 28Rb 51
Mountview Dr. RH1: Redh 8N 207
Mount Vw. Rd. E4 17Fc 35
Mount Vw. Rd. KT10: Clay 80Ka 152
Mount Vw. Rd. N4 31Nb 70
Mount Vw. Rd. NW9 29Ta 47
Mountview Rd. BR6: Orp 73Wc 161
(not continuous)
Mountview Rd. BR6:
St M Cry 73Wc 161
(not continuous)
Mount Vs. SE27 62Rb 135
Mountway EN6: Pot B 2Cb 17
Mount Way SM5: Cars 81Jb 176
Mountwood KT8: W Mole 69Da 129
Mountwood Cl. CR2: Sande 82Xb 177
Movers La. IG11: Bark 39Tc 74
MOVERS LANE 40Uc 74
Mowat Ct. KT4: Wor Pk 75Va 154
(off The Avenue)
Mowat Ind. Est. WD24: Wat 10Y 13
Mowatt Cl. N19 32Mb 70
Mowbray Av. KT14: Byfl 85N 169
Mowbray Ct. CM16: Epp 2Uc 22
Mowbray Ct. N22 25Qb 50
Mowbray Ct. SE19 66Vb 135
Mowbray Cres. TW20: Egh 64C 126
Mowbray Gdns. UB5: N'olt 39Ca 65
Mowbray Hall TW20: Eng G 2P 125
Mowbray Ho. N2 26Fb 49
(off The Grange)
Mowbray Pde. HA8: Edg 21Qa 47
Mowbray Rd. EN5: New Bar 15Eb 31
Mowbray Rd. HA8: Edg 21Qa 47
Mowbray Rd. NW6 38Ab 68
Mowbray Rd. SE19 67Vb 135
Mowbray Rd. TW10: Ham 62La 130
Mowbrays Cl. RM5: Col R 25Ed 56
Mowbrays Rd. RM5: Col R 26Ed 56
Mowbrey Gdns. IG10: Lough 11Sc 36
(not continuous)
Mowlem St. E2 40Xb 71
Mowlem Trad. Est. N17 24Yb 52
Mowll St. SW9 52Qb 112
Moxey Cl. TN16: Big H 85Lc 179
Moxom Av. EN8: Chesh 2Ac 20
Moxon Cl. E13 40Hc 73
Moxon Pl. UB10: Uxb 39P 63
Moxon St. EN5: Barn 13Bb 31
Moxon St. W1 1H 221 (43Jb 90)
E16: Lon
Moye Cl. E2 40Wb 71
Moyers Rd. E10 31Ec 72
Moy La. SE18 50Rc 94
Moylan Rd. W6 51Ab 110
Moyle Ho. DA17: Belv 47Dd 96
Moyle Ho. SW1 50Lb 90
(off Churchill Gdns.)
Moyne Ct. GU21: Wok 10K 167
Moyne Ho. SW9 57Rb 113
Moyne Pl. NW10 40Qa 67
Moynihan Dr. N21 15Nb 32
Moys Cl. CR0: C'don 72Nb 156
Moyser Rd. SW16 64Kb 134
Mozart St. W10 41Bb 89
Mozart Ter. SW1 6J 227 (49Jb 90)
MTV Europe 38Kb 70
Muchelney Rd. SM4: Mord 72Eb 155
Muckhatch La. TW20: Thorpe 69D 126
MUCKING 4M 101
MUCKINGFORD 8J 101
Muckingford Rd. RM18: W Til 9F 100
Muckingford Rd. SS17: Linf 9H 101
Mucking Wharf Rd. SS17: Stan H 4L 101
Mudchute Pk. & Farm 49Ec 92
Mudchute Station (DLR) 49Dc 92
Muddy La. SL2: Slou 3J 81
Mudlands Ind. Est. RM13: Rain 41Gd 96
Mud La. AL1: St A 3A 6
Mudlarks Blvd. SE10 48Hc 93
Mudlarks Way SE10 48Jc 93
Muggeridge Cl. CR2: S Croy 78Tb 157
Muggeridge Rd. RM10: Dag 35Dd 76
Muggins La. DA12: Shorne 3L 145
MUGSWELL 97Db 195
Muirdown Av. SW14 56Ta 109
Muir Dr. DA1: Dart 55Pd 119
Muir Dr. SW18 58Gb 111
Muirfield W3 44Ua 88
Muirfield Cl. SE16 50Xb 91
Muirfield Cl. WD19: Wat 22Y 45
Muirfield Cres. E14 48Dc 92
Muirfield Grn. WD19: Wat 21X 45
Muirfield Rd. GU21: Wok 10L 167
Muirfield Rd. WD19: Wat 21X 45
Muirhead Quay IG11: Bark 40Sc 74
Muirkirk Rd. SE6 60Ec 114
Muir Rd. E5 34Wb 71
Muir St. E16 46Nc 94
(not continuous)
Mulberry Apts. N4 32Sb 71
(off Coster Av.)
Mulberry Av. SL4: Wind 5K 103
Mulberry Av. TW19: Stanw 60N 105
Mulberry Cl. AL2: Park 10P 5
Mulberry Cl. E4 19Cc 34
Mulberry Cl. EN4: E Barn 14Fb 31
Mulberry Cl. GU21: Wok 86A 168
Mulberry Cl. KT13: Weyb 76R 150
Mulberry Cl. KT19: Eps 81Sa 173
Mulberry Cl. N8 29Nb 50
Mulberry Cl. NW3 35Fb 69
Mulberry Cl. NW4 27Ya 48
Mulberry Cl. RH2: Reig 7K 207
Mulberry Cl. RM2: Rom 28Ld 57
Mulberry Cl. SE22 57Wb 113
Mulberry Cl. SE7 51Mc 115
Mulberry Cl. SW16 63Lb 134
Mulberry Cl. SW3 51Fb 111
(off Beaufort St.)
Mulberry Cl. TW13: Felt 62X 129
Mulberry Cl. UB5: N'olt 40Aa 65
Mulberry Cl. WD17: Wat 8U 12
Mulberry Ct. DA1: Cray 57Gd 118
Mulberry Ct. E11 35Fc 73
(off Langthorne Rd.)
Mulberry Ct. EC1 4C 218 (41Rb 91)
(off Tompion St.)
Mulberry Ct. HP1: Hem H 1M 3
Mulberry Ct. IG11: Bark 37Vc 75
Mulberry Ct. KT6: Surb 73Ma 153
Mulberry Ct. N2 27Gb 49
(off Bedford Rd.)
Mulberry Ct. SW3 51Fb 111
(not continuous)
Mulberry Ct. TW1: Twick 62Ha 130
Mulberry Ct. W9 41Bb 89
(off Ashmore Rd.)
Mulberry Cres. TW8: Bford 52Ka 108
Mulberry Cres. UB7: W Dray 47Q 84
Mulberry Dr. RM19: Purf 49Pd 97
Mulberry Dr. SL3: L'ly 50A 82
Mulberry Gdns. WD7: Shenl 5Na 15
Mulberry Ga. SM7: Bans 88Bb 175
Mulberry Hill CM15: Shenf 17Be 41
Mulberry Ho. BR2: Broml 67Gc 137
Mulberry Ho. E2 41Yb 92
(off Victoria Pk. Sq.)
Mulberry Ho. SE10 53Ec 114
(off Parkside Av.)
Mulberry Ho. SE8 51Bc 114
Mulberry Housing Co-operative
SE1 6A 224 (46Qb 90)
(off Upper Rd.)
Mulberry La. CR0: C'don 74Vb 157
Mulberry Lodge WD19: Wat 16Z 27
(off Eastbury Rd.)
Mulberry M. SE14 53Bc 114
Mulberry M. SM6: W'gton 79Lb 156
Mulberry Pde. UB7: W Dray 48Q 84
Mulberry Pl. E14 45Ec 92
(off Clove Cres.)
Mulberry Pl. HA2: Harr 26Fa 46
Mulberry Pl. SE9 56Mc 115
Mulberry Pl. TN16: Bras 96Yc 201
(not continuous)
Mulberry Pl. W6 50Wa 88
Mulberry Rd. DA11: Nflt 2A 144
Mulberry Rd. E8 38Vb 71
Mulberry St. E1 44Wb 91
Mulberry Tree M. W4 47Sa 87
Mulberry Trees TW17: Shep 73T 150
Mulberry Wlk. SW3 51Fb 111
Mulberry Way DA17: Belv 47Ed 96
Mulberry Way E18 26Kc 53
Mulberry Way IG6: Ilf 28Sc 54
Mulberry Way KT21: Asht 91Na 193
Mulberry Way RM18: Tilb 3C 122
Mulgrave Ct. SM2: Sutt 79Db 155
(off Mulgrave Rd.)
Mulgrave Rd. CR0: C'don 76Tb 157
Mulgrave Rd. HA1: Harr 33Ja 66
Mulgrave Rd. NW10 35Va 68
Mulgrave Rd. SE18 49Pc 94
Mulgrave Rd. SM2: Sutt 80Bb 155
Mulgrave Rd. SW6 51Bb 111
Mulgrave Rd. W5 41Ma 87
Mulgrave Way GU21: Knap 10J 167
Mulholland Cl. CR4: Mitc 68Kb 134
Mulkern Rd. N19 32Mb 70
(not continuous)
Mullards Cl. CR4: Mitc 74Hb 155
Mullein Ct. RM17: Grays 51Fe 121
Mullender Ct. DA12: Grav'nd 10J 123
Mullens Rd. TW20: Egh 64D 126
Mullen Twr. WC1 6K 217 (42Qb 90)
(off Mt. Pleasant)
Mullet Gdns. E2 41Wb 91
Mulletsfield WC1 4G 217 (41Nb 90)
(off Cromer St.)
Mull Ho. E3 40Bc 72
(off Stafford Rd.)
Mull Ho. WD18: Wat 15V 26
Mulligans Apts. NW6 38Cb 69
(off Kilburn High Rd.)
Mullins Path SW14 55Ta 109
Mullins Pl. SW4 59Nb 112
Mullion Cl. HA3: Hrw W 25Da 45
Mullion Wlk. WD19: Wat 21Z 45
Mull Wlk. N1 37Sb 71
(off Clephane Rd.)
Mulready Ho. SW1 6F 229 (49Nb 90)
(off Marsham St.)
Mulready St. NW8 6D 214 (42Gb 89)
Mulready Wlk. HP3: Hem H 6N 3
Multi Way W3 47Ua 88
Multon Ho. E9 38Yb 72
Multon Rd. SW18 59Fb 111
Multon Rd. TN15: W King 79Ud 164
Mulvaney Way SE1 2G 231 (47Tb 91)
(not continuous)
Mumford Mills SE10 53Dc 114
(off Greenwich High Rd.)
Mumford Rd. SE24 57Rb 113
Muncaster Cl. TW15: Ashf 63Q 128
Muncaster Rd. SW11 57Hb 111
Muncaster Rd. TW15: Ashf 64R 128
Muncies M. SE6 61Ec 136
Mundania Ct. SE22 58Xb 113
Mundania Rd. SE22 58Xb 113
Munday Ho. SE1 4F 231 (48Tb 91)
(off Burbage Cl.)
Munday Rd. E16 45Jc 93
MUNDEN 6Da 13
Munden Dr. WD25: Wat 9Aa 13
Munden Gro. WD24: Wat 10Y 13
Munden Ho. E3 41Dc 92
(off Bromley High St.)
Munden St. W14 49Ab 88
Munden Vw. WD25: Wat 8Z 13
Mundesley Cl. WD19: Wat 21Y 45
Mundesley Spur SL1: Slou 4J 81
Mundford Rd. E5 33Yb 72
Mundon Gdns. IG1: Ilf 32Tc 74
Mund St. W14 50Bb 89
Mundy Ct. SL4: Eton 1H 103
Mundy Ho. W10 41Ab 88
(off Dart St.)
Mundy St. N1 3H 219 (41Ub 91)
Munford Dr. DA10: Swans 59Ae 121
Mungo Pk. Cl. WD23: B Hea 19Ea 28
Mungo Pk. Rd. DA12: Grav'nd 4F 144
Mungo Pk. Rd. RM13: Rain 37Jd 76
Mungo Pk. Way BR5: Orp 73Yc 161
Munkenbeck Bldg. W2 1B 220 (43Fb 89)
(off Hermitage St.)
Munnery Way BR6: Farnb 76Qc 160
Munnings Gdns. TW7: Isle 57Fa 108
Munnings Ho. E16 46Kc 93
(off Portsmouth M.)
Munro Dr. N11 23Lb 50
Munro Ho. KT11: Cobh 84Z 171
Munro Ho. SE1 2K 229 (47Qb 90)
Munro M. W10 43Ab 88
(not continuous)
Munro Ter. SW10 52Fb 111
Munslow Gdns. SM1: Sutt 77Fb 155
Munster Av. TW4: Houn 57Aa 107
Munster Ct. SW6 54Bb 111
Munster Ct. TW11: Tedd 65La 130
Munster Gdns. N13 21Rb 51
Munster M. SW6 52Ab 110
Munster Rd. SW6 52Ab 110
Munster Rd. TW11: Tedd 65Ka 130
Munster Sq. NW1 4A 216 (41Kb 90)
Munster Ter. E13 41Lc 93
Munton Rd. SE17 5E 230 (49Sb 91)
Muratori Ho. WC1 4K 217 (41Qb 90)
(off Margery St.)
Murchison Av. DA5: Bexl 60Zc 117
Murchison Ho. W10 43Ab 88
(off Ladbroke Gro.)
Murchison Rd. E10 33Ec 72
Murdoch Cl. TW18: Staines 64J 127
Murdoch Ho. SE16 48Yb 92
(off Moodkee St.)
Murdock Cl. E16 44Hc 93
Murdock St. SE15 51Xb 113
Murfett Cl. SW19 61Ab 132
Murfitt Way RM14: Upm 35Qd 77
Muriel Av. WD18: Wat 15Y 27
Muriel St. N1 1J 217 (40Pb 70)
(not continuous)
Murillo Rd. SE13 56Fc 115
Muro Ct. SE1 2C 230 (47Rb 91)
(off Milcote St.)
Murphy Ho. SE1 3C 230 (48Rb 91)
(off Borough Rd.)
Murphy St. SE1 2K 229 (47Qb 90)
Murrain Rd. N4 33Sb 71
Murray Av. BR1: Broml 69Kc 137
Murray Av. TW3: Houn 57Da 107
Murray Bus. Cen. BR5: St P 69Xc 139
Murray Cl. SE28 46Uc 94
Murray Ct. E3 43Bc 92
(off Geoff Cade Way)
Murray Ct. HA1: Harr 30Ha 46
Murray Ct. SL5: S'hill 2A 146
Murray Ct. TW2: Twick 61Fa 130
Murray Ct. W7 47Ga 86
Murray Cres. HA5: Pinn 25Z 45
Murray Dr. DA2: Dart 58Sd 120
Murray Grn. GU21: Wok 86E 168
Murray Gro. N1 2E 218 (40Sb 71)
Murray Ho. KT16: Ott 79E 148
Murray Ho. SE18 49Pc 94
(off Rideout St.)
Murray M. NW1 38Mb 70
Murray Rd. BR5: St P 69Xc 139
Murray Rd. HA6: Nwood 25U 44
Murray Rd. KT16: Ott 79E 148
Murray Rd. SW19 65Za 132
Murray Rd. TW10: Ham 61Ka 130
Murray Rd. W5 49La 86
Murray's La. KT14: Byfl 86M 169
Murray's La. KT14: W Byf 86M 169
Murray Sq. E16 44Jc 93
Murray St. NW1 38Lb 70
Murrays Yd. SE18 49Rc 94
Murray Ter. NW3 35Eb 69
Murray Ter. W5 49Ma 87
Murrell's Wlk. KT23: Bookh 95Ca 191
Murreys, The KT21: Asht 90La 172
Murreys Ct. KT21: Asht 90Ma 173
Mursell Est. SW8 53Pb 112
Murthering La. RM4: Nave 17Hd 38
Murthering La. RM4: Stap A 17Hd 38
Murton Ct. AL1: St A 1C 6
Murtwell Dr. IG7: Chig 23Sc 54
Musard Rd. W14 51Ab 110
Musard Rd. W6 51Ab 110
Musbury St. E1 44Yb 92
Muscal W6 51Ab 110
(off Field Rd.)
Muscatel Pl. SE5 53Ub 113
Muschamp Rd. SE15 55Vb 113
Muschamp Rd. SM5: Cars 75Gb 155
Muscott Ho. E2 39Wb 71
(off Whiston Rd.)
Muscovy Ho. DA18: Erith 47Ad 95
(off Kale Rd.)
Muscovy Pl. HA6: Nwood 24S 44
Muscovy St. EC3 4J 225 (45Ub 91)
Museum Chambers WC1 1F 223 (43Nb 90)
(off Bury Pl.)
Museum Ho. E2 41Yb 92
(off Burnham St.)
Museum La. SW7 4C 226 (48Fb 89)
Museum Mans. WC1 1F 223 (43Nb 90)
(off Gt. Russell St.)
Museum of Brands, Packaging &
Advertising 44Ab 88
(off Lancaster Rd.)
Museum of Comedy 1F 223 (43Nb 90)
(off Bloomsbury Way)
Museum of Croydon 76Sb 157
(off High St.)
Museum of Domestic Design &
Architecture 26Va 48
Museum of Eton Life 1H 103
Museum of London 1D 224 (43Sb 91)
Museum of London Docklands 45Cc 92
Museum of Richmond 57Ma 109
Museum of the Nat. Rifle Assocation,
The 1B 186
Museum of Order of St John,
The 6B 218 (42Rb 91)
Museum of Water & Steam 50Pa 87
Museum of Wimbledon 65Ab 132
Museum Pas. E2 41Yb 92
Museum St. WC1 1F 223 (43Nb 90)
Museum Way W3 47Qa 87
Musgrave Cl. EN4: Had W 11Eb 31
Musgrave Ct. SW11 53Gb 111
Musgrave Cres. SW6 52Cb 111
Musgrave Rd. TW7: Isle 53Ha 108
Musgrove Cl. CR8: Purl 85Pb 176
Musgrove Rd. SE14 53Zb 114
Musical Mus., The 51Na 109
Musjid Rd. SW11 54Fb 111
Musket Cl. EN4: E Barn 16Fb 31
Musk Hill HP1: Hem H 3G 2
Musquash Way TW4: Houn 54Y 107
Mussenden La. DA3: Fawk 73Ud 164
Mussenden La. DA4: Hort K 71Sd 164
Mustang Ho. N1 37Rb 71
(off Canonbury Rd.)
Mustard Mill Rd. TW18: Staines 63H 127
Mustians SL4: Eton 1G 102
(off Eton Wick Rd.)
Muston Rd. E5 33Xb 71
Mustow Pl. SW6 54Bb 111
Muswell Av. N10 25Kb 50
Muswell Hill N10 27Kb 50
MUSWELL HILL 27Kb 50
Muswell Hill B'way. N10 27Kb 50
Muswell Hill Golf Course 25Lb 50
Muswell Hill Pl. N10 28Kb 50
Muswell Hill Rd. N10 29Kb 50
Muswell Hill Rd. N6 30Jb 50
Muswell M. N10 27Kb 50
Muswell Rd. N10 27Kb 50
Mutchetts Cl. WD25: Wat 5Aa 13
Mutrix Rd. NW6 39Cb 69
Mutton La. EN6: Pot B 3Ya 16
Mutton La. EN6: S Mim 3Ya 16
Mutton Pl. NW1 37Jb 70
Muybridge Rd. KT3: N Mald 68Sa 131
Muybridge Yd. KT5: Surb 73Pa 153
Myatt Rd. SW9 53Rb 113
Myatts Fld. Sth. SW9 54Qb 112
Mycenae Rd. SE3 52Jc 115
Myddelton Av. EN1: Enf 10Ub 19
Myddelton Cl. EN1: Enf 11Vb 33
Myddelton Gdns. N21 17Sb 33
Myddelton House Gdns. 8Vb 19
Myddelton Pk. N20 20Fb 31
Myddelton Pas. EC1 3A 218 (41Qb 90)
Myddelton Rd. N8 28Nb 50
(not continuous)
Myddelton Sq. EC1 3A 218 (41Qb 90)
Myddelton St. EC1 4A 218 (41Qb 90)
Myddleton Av. N4 33Sb 71
Myddleton Cl. HA7: Stan 19Ja 28
Myddleton Ct. RM11: Horn 31Hd 76
Myddleton Ho. N1 2K 217 (40Qb 70)
(off Pentonville Rd.)
Myddleton M. N22 24Nb 50
Myddleton Path EN7: Chesh 3Xb 19
Myddleton Rd. N22 24Nb 50
Myddleton Rd. UB8: Uxb 39L 63
Myers Cl. WD7: Shenl 4Na 15
Myers Ct. CM14: W'ley 22Yd 58
Myers Dr. SL2: Farn C 6F 60
Myers Ho. SE5 52Sb 113
(off Bethwin Rd.)
Myers La. SE14 51Zb 114
Myers Wlk. E14 46Ec 92
Mygrove Cl. RM13: Rain 40Md 77
Mygrove Gdns. RM13: Rain 40Md 77
Mygrove Rd. RM13: Rain 40Md 77
Myla Ho. CR8: Kenley 88Rb 177
Myles Ct. EN7: G Oak 1Sb 19
Myles Ct. SE16 48Yb 92
(off Neptune St.)
Mylis Cl. SE26 63Xb 135
Mylius Cl. SE14 52Yb 114
Mylne Cl. W6 50Wa 88
Mylne St. EC1 3K 217 (41Qb 90)
Mylor Cl. GU21: Wok 86A 168
Mymms Dr. AL9: Brk P 8J 9
Mymms Ho. AL9: Wel G 5E 8
Mynn's Cl. KT18: Eps 86Ra 173
Myra St. SE2 49Wc 95
Myrdle Ct. E1 44Wb 91
(off Myrdle St.)
Myrdle St. E1 43Wb 91
MYRKE 10K 81
Myrke, The SL3: Dat 9K 81
Myrna Cl. SW19 66Gb 133
Myron Pl. SE13 55Ec 114
Myrtle Av. HA4: Ruis 31W 64
Myrtle Av. TW14: Felt 57U 106
Myrtleberry Cl. E8 37Vb 71
(off Beechwood Rd.)
Myrtle Cl. DA8: Erith 52Gd 118
Myrtle Cl. EN4: E Barn 18Hb 31
Myrtle Cl. GU18: Light 3A 166
Myrtle Cl. SL3: Poyle 53G 104
Myrtle Cl. UB7: W Dray 48P 83
Myrtle Cl. UB8: Hil 43P 83
Myrtle Cres. SL2: Slou 5K 81
Myrtledene Rd. SE2 50Wc 95
Myrtle Gdns. W7 46Ga 86
Myrtle Grn. HP1: Hem H 1G 2
Myrtle Gro. EN2: Enf 10Tb 19
Myrtle Gro. KT3: N Mald 68Sa 131
Myrtle Gro. RM15: Avel 47Sd 98
Myrtle Pl. DA2: Dart 59Td 120
Myrtle Rd. CM14: W'ley 21Yd 58
Myrtle Rd. CR0: C'don 76Cc 158
Myrtle Rd. DA1: Dart 60Md 119
Myrtle Rd. E17 30Ac 52
Myrtle Rd. E6 39Pc 74
Myrtle Rd. IG1: Ilf 33Rc 74
Myrtle Rd. N13 20Sb 33
Myrtle Rd. RM3: Rom 23Ld 57
Myrtle Rd. SM1: Sutt 78Eb 155
Myrtle Rd. TW12: Hamp H 65Ea 130
Myrtle Rd. TW3: Houn 54Ea 108
Myrtle Rd. W3 46Sa 87
Myrtleside Cl. HA6: Nwood 24T 44
Myrtle Wlk. N1 2H 219 (40Ub 71)
Mysore Rd. SW11 55Hb 111
Myton Rd. SE21 62Tb 135
Mytton Ho. SW8 52Pb 112
(off St Stephens Ter.)

N

N1 Shop. Cen. 1A 218 (40Qb 70)
Nacton Ct. RM6: Chad H 29Yc 55
(off Hevingham Dr.)
Nadine Ct. SM6: W'gton 81Lb 176
Nadine St. SE7 50Lc 93
Nafferton Ri. IG10: Lough 15Mc 35
Nagasaki Wlk. SE7 48Kc 93
Nagle Cl. E17 26Fc 53
NAG'S HEAD 34Nb 70
Nags Head Ct. EC1 6E 218 (42Sb 91)
(off Golden La.)
Nags Head La. CM14: B'wood 24Sd 58
Nags Head La. CM14: Upm 24Sd 58
Nags Head La. DA16: Well 55Xc 117
Nags Head La. RM14: Upm 25Sd 58
Nags Head Rd. EN3: Pond E 14Yb 34
Nags Head Shop. Cen. 35Pb 70
Nailsworth Cres. RH1: Mers 1D 208
Nalzee Cl. SL9: Ger X 31A 62
Nainby Ho. SE11 6K 229 (49Qb 90)
(off Hotspur St.)
Nairn Ct. RM18: Tilb 4B 122
Nairne Gro. SE24 57Tb 113
Nairn Grn. WD19: Wat 20W 26
Nairn Ho. CM14: W'ley 21Yd 58
(off Cameron Cl.)
Nairn Rd. HA4: Ruis 37Y 65
Nairn St. E14 43Ec 92
Naldera Gdns. SE3 51Jc 115
Nallhead Rd. TW13: Hanw 64Y 129
Nalton Ho. NW6 38Eb 69
(off Belsize Rd.)
Namba Roy Cl. SW16 63Pb 134
Namco Funscape Romford 30Gd 56
Namton Dr. CR7: Thor H 70Pb 134
Nan Clark's La. NW7 19Ua 30
Nancy Downs WD19: Wat 17Y 27
Nankin St. E14 44Cc 92
Nanscott Ho. WD19: Wat 20Z 27
Nansen Ho. NW10 38Ta 67
(off Stonebridge Pk.)
Nansen Rd. DA12: Grav'nd 3F 144
Nansen Rd. SW11 55Jb 112
Nansen Village N12 21Db 49
Nant Ct. NW2 33Bb 69

Nanterre Ct. WD17: Wat 12W 26
Nantes Cl. SW18 56Eb 111
Nantes Pas. E1 7K 219 (43Vb 91)
Nant Rd. NW2 33Bb 69
Nant St. E2 41Xb 91
Nantwich Ho. RM3: Rom 22Nd 57
(off Lindfield Rd.)
Naomi St. SE8 49Ac 92
Naoroji St. WC1 4K 217 (41Qb 90)
Nap, The WD4: K Lan 1Q 12
Napa Cl. E20 36Ec 72
Napier NW9 25Va 48
Napier Av. E14 50Cc 92
Napier Av. SW6 55Bb 111
Napier Cl. AL2: Lon C 7H 7
Napier Cl. RM11: Horn 32Kd 77
Napier Cl. SE8 52Bc 114
Napier Cl. UB7: W Dray 48P 83
Napier Cl. W14 48Ab 88
Napier Ct. BR2: Broml 70Kc 137
(off Napier Rd.)
Napier Ct. CR3: Cat'm 94Ub 197
Napier Ct. GU21: Wok 88A 168
Napier Ct. N1 1F 219 (40Tb 71)
(off Cropley St.)
Napier Ct. SE12 62Kc 137
Napier Ct. SW6 55Bb 111
(off Ranelagh Gdns.)
Napier Ct. UB4: Yead 42Y 85
(off Dunedin Way)
Napier Dr. WD23: Bush 14Aa 27
Napier Gro. N1 2E 218 (40Sb 71)
Napier Ho. E3 41Cc 92
(off Campbell Rd.)
Napier Ho. RM13: Rain 41Hd 96
(off Dunedin Rd.)
Napier Ho. SE17 51Rb 113
(off Cooks Rd.)
Napier Ho. W3 46Ua 88
Napier Lodge TW15: Ashf 65T 128
Napier Pl. W14 48Bb 89
Napier Rd. BR2: Broml 70Kc 137
Napier Rd. CR2: S Croy 80Tb 157
Napier Rd. DA11: Nflt 10B 122
Napier Rd. DA17: Belv 49Bd 95
Napier Rd. E11 35Gc 73
Napier Rd. E15 40Gc 73
(not continuous)
Napier Rd. E6 39Qc 74
Napier Rd. EN3: Pond E 15Zb 34
Napier Rd. HA0: Wemb 37Ma 67
Napier Rd. N17 27Ub 51
Napier Rd. NW10 41Xa 88
Napier Rd. SE25 70Xb 135
Napier Rd. TW15: Ashf 66T 128
Napier Rd. TW7: Isle 56Ja 108
Napier Rd. W14 48Bb 89
Napier St. SE8 52Bc 114
(off Napier Cl.)
Napier Ter. N1 38Rb 71
Napier Wlk. TW15: Ashf 66T 128
Napolean Ho. BR1: Broml 69Lc 137
Napoleon La. SE18 52Pc 116
Napoleon Rd. E5 34Xb 71
Napoleon Rd. TW1: Twick 59Ka 108
NAPSBURY 9G 6
Napsbury Av. AL2: Lon C 8G 6
Napsbury La. AL1: St A 5E 6
Napton Cl. UB4: Yead 42Aa 85
Nara SE13 54Dc 114
Narbonne Av. SW4 57Lb 112
Narboro Ct. RM1: Rom 29Jd 56
Narborough Cl. UB10: Ick 33S 64
Narborough St. SW6 54Db 111
Narcissus Rd. NW6 36Cb 69
Nardini NW9 25Va 48
(off Long Mead)
Nare Rd. RM15: Avel 45Sd 98
Naresby Fold HA7: Stan 23La 46
Nares Cl. TW19: Stanw 60N 105
Narev Dr. EN4: E Barn 15Gb 31
Narford Rd. E5 34Wb 71
Narrowboat Av. TW8: Bford 52La 108
Narrow Boat Cl. SE28 47Tc 94
Narrow La. CR6: W'ham 91Xb 197
Narrow St. E14 45Ac 92
Narrow St. W3 46Ra 87
Narrow Way BR2: Broml 72Nc 160
Narvic Ho. SE5 54Sb 113
Narwhal Inuit Art Gallery 50Ta 87
Nascot Pl. WD17: Wat 12X 27
Nascot Rd. WD17: Wat 12X 27
Nascot St. W12 44Ya 88
Nascot St. WD17: Wat 12X 27
Nascot Wood Rd. WD17: Wat 9V 12
Naseby Cl. NW6 38Eb 69
Naseby Cl. TW7: Isle 53Ga 108
Naseby Ct. DA14: Sidc 63Vc 139
Naseby Ct. KT12: Walt T 75Y 151
Naseby Rd. IG5: Ilf 25Pc 54
Naseby Rd. RM10: Dag 34Cd 76
Naseby Rd. SE19 65Tb 135
NASH 79Jc 159
Nash Bank DA13: Ist R 8B 144
Nash Bank DA13: Meop 8B 144
Nash Cl. AL9: Wel G 5F 8
Nash Cl. SM1: Sutt 76Fb 155
Nash Cl. WD6: E'tree 14Pa 29
Nash Ct. HA3: Kenton 30Ka 46
Nash Cft. DA11: Nflt 3A 144
Nash Dr. RH1: Redh 4P 207
Nashe Ho. SE1 4F 231 (48Tb 91)
(off Burbage Cl.)
Nash Gdns. RH1: Redh 4P 207
Nash Grn. BR1: Broml 65Jc 137
Nash Grn. HP3: Hem H 7P 3
Nash Ho. E14 47Cc 92
(off Alpha Gro.)
Nash Ho. E17 27Dc 52
Nash Ho. NW1 1A 216 (40Kb 70)
(off Park Village E.)
Nash Ho. SW1 7A 228 (50Kb 90)
(off Lupus St.)
Nash La. BR2: Kes 80Jc 159
NASH MILLS 8P 3
Nash Mills La. HP3: Hem H 8P 3
Nash Mills Recreation Cen. 8P 3
Nash Pl. E14 46Dc 92
Nash Rd. N9 19Yb 34
Nash Rd. RM6: Chad H 28Zc 55
Nash Rd. SE4 56Ac 114
Nash Rd. SL3: L'ly 49B 82
Nash St. DA13: Meop 8B 144
Nash St. NW1 3A 216 (41Kb 90)
NASH STREET 8B 144
Nash's Yd. UB8: Uxb 38M 63
Nash Way HA3: Kenton 30Ka 46
Nasmyth St. W6 48Xa 88
Nassau Path SE28 46Yc 95
Nassau Rd. SW13 53Va 110
Nassau St. W1 1B 222 (43Lb 90)
Nassington Rd. NW3 35Hb 69
NAST HYDE 1A 8
Nasturtium Dr. GU24: Bisl 7E 166
Natalie Cl. TW14: Bedf 59T 106
Natalie M. N22 23Nb 50
Natalie M. TW2: Twick 62Fa 130
Natal Rd. CR7: Thor H 69Tb 135
Natal Rd. IG1: Ilf 35Rc 74
Natal Rd. N11 23Nb 50
Natal Rd. SW16 65Mb 134
Natasha Ct. RM3: Rom 24Ld 57
Natasha M. SE15 56Yb 114
Nathan Cl. RM14: Upm 32Ud 78
Nathan Ct. N9 17Yb 34
(off Causeyware Rd.)
Nathan Ho. SE11 6A 230 (49Qb 90)
(off Reedworth St.)
Nathaniel Cl. E1 1K 225 (43Vb 91)
Nathaniel Ct. E17 31Ac 72
Nathans Rd. HA0: Wemb 32La 66
Nathan Way SE28 49Uc 94
National Archives, The 52Ra 109
National Army Mus. 51Hb 111
National Gallery 5E 222 (45Mb 90)
NATIONAL HOSPITAL FOR NEUROLOGY & NEUROSURGERY 6G 217 (42Nb 90)
National Maritime Mus. 51Fc 115
National Portrait Gallery 5E 222 (45Mb 90)
(off St Martin's Pl.)
National Rifle Association & Nat. Shooting Cen. Headquarters 1B 186
(off Queen's Way)
NATIONAL SOCIETY FOR EPILEPSY, THE 21B 42
National Tennis Cen. 57Ua 110
National Ter. SE16 47Xb 91
(off Bermondsey Wall E.)
National Theatre 6K 223 (46Qb 90)
National Works TW4: Houn 55Ba 107
Nation Way E4 18Ec 34
Natural History Mus. Eton 1H 103
Natural History Mus. London 4B 226 (48Fb 89)
Nature Gdn. 50Yb 92
(off Bramcote Gro.)
Naunton Way RM12: Horn 34Md 77
Nautical Dr. E16 47Kc 93
Nautical Ho. SW18 56Eb 111
(off Juniper Dr.)
Nauticus Wlk. E14 49Dc 92
Nautilus Bldg., The EC1 3A 218 (41Qb 90)
(off Myddelton Pas.)
Naval Ho. E14 45Fc 93
(off Quixley St.)
Naval Ho. SE18 48Rc 94
Naval Row E14 45Ec 92
Navarino Gro. E8 37Wb 71
Navarino Mans. E8 37Wb 71
Navarino Rd. E8 37Wb 71
Navarre Ct. WD4: K Lan 10B 4
Navarre Gdns. RM5: Col R 22Dd 56
Navarre Rd. E6 40Nc 74
Navarre Rd. SW9 53Rb 113
Navarre St. E2 5K 219 (42Vb 91)
Navenby Wlk. E3 42Cc 92
NAVESTOCK 12Md 39
Navestock Cl. E4 20Ec 34
Navestock Cres. IG8: Wfd G 24Lc 53
NAVESTOCK HEATH 12Md 39
NAVESTOCK SIDE 12Sd 40
Navestockside CM14: Kel C 12Sd 40
Navestockside CM14: N'side 12Sd 40
Navigation Ct. E16 45Sc 94
Navigation Dr. EN3: Enf L 10Cc 20
Navigation Ho. KT15: Add 77N 149
Navigation Ho. SE16 49Ac 92
(off Grand Canal Av.)
Navigation M. IG6: Ilf 23Tc 54
Navigation Rd. E3 42Ec 92
Navigator Dr. UB2: S'hall 47Ea 86
Navigator Pk. UB2: S'hall 49Y 85
Navigator Sq. N19 33Lb 70
Navy St. SW4 55Mb 112
Naxos Bldg. E14 47Bc 92
Nayim Pl. E8 36Xb 71
Nayland Ho. SE6 63Ec 136
Naylor Bldg. E. E1 44Wb 91
(off Assam St.)
Naylor Bldg. W. E1 44Wb 91
(off Adler St.)
Naylor Gro. EN3: Pond E 15Zb 34
Naylor Ho. SE17 6G 231 (49Tb 91)
(off Flint St.)
Naylor Ho. W10 41Ab 88
(off Dart St.)
Naylor Rd. N20 19Eb 31
Naylor Rd. SE15 52Xb 113
Nazareth Gdns. SE15 54Xb 113
Nazeing Wlk. RM13: Rain 38Hd 76
Nazrul St. E2 3K 219 (41Vb 91)
NCR Bus. Cen. NW10 36Ua 68
Neagle Cl. WD6: Bore 11Sa 29
Neagle Ho. NW2 34Ya 68
(off Stoll Cl.)
Neal Av. UB1: S'hall 42Ba 85
Neal Cl. HA6: Nwood 25W 44
Neal Cl. SL9: Ger X 32D 62
Neal Ct. EN9: Walt A 5Hc 21
Nealden St. SW9 55Pb 112
Neale Cl. N2 27Eb 49
Neale Ct. RM9: Dag 37Xc 75
Neal Rd. TN15: W King 79Ud 164
Neal St. SE10 48Hc 93
Neal St. WC2 3F 223 (44Nb 90)
Neal St. WD18: Wat 15Y 27
Neal's Yd. WC2 3F 223 (44Nb 90)
Neap Ct. E3 42Ec 92
(off Navigation Rd.)
Near Acre NW9 25Va 48
NEASDEN 34Ua 68
Neasden Cl. NW10 36Ua 68
NEASDEN JUNC. 35Ua 68
Neasden La. NW10 34Ua 68
Neasden La. Nth. NW10 34Ta 67
Neasden Station (Underground) 36Ua 68
Neasham Rd. RM8: Dag 36Xc 75
Neate Ho. SW1 7C 228 (50Lb 90)
(off Lupus St.)
Neate St. SE5 51Ub 113
Neath Gdns. SM4: Mord 72Eb 155
Neathouse Pl. SW1 5B 228 (49Lb 90)
Neats Acre HA4: Ruis 31T 64
Neatscourt Rd. E6 43Mc 93
Neave Cres. RM3: Rom 25Ld 57
Neb La. RH8: Oxt 3G 210
Nebraska Bldg. SE13 53Dc 114
(off Deal's Gateway)
Nebraska St. SE1 2F 231 (47Tb 91)
Nebula Ct. E13 40Jc 73
(off Umbriel Pl.)
Neckinger SE16 3K 231 (48Vb 91)
Neckinger Est. SE16 3K 231 (48Vb 91)
Neckinger St. SE1 47Vb 91
Nectarine Way SE13 54Dc 114
Needham Cl. SL4: Wind 3C 102
Needham Ho. SE11 7K 229 (50Qb 90)
(off Marylee Way)
Needham Rd. W11 44Cb 89
Needham Ter. NW2 34Za 68
Needleman Cl. NW9 26Ua 48
Needleman St. SE16 47Zb 92
Needles Bank RH9: G'stone 3P 209
(not continuous)
Needwood Ho. N4 32Sb 71
Neela Cl. UB10: Ick 35R 64
Neeld Cres. HA9: Wemb 36Qa 67
Neeld Cres. NW4 29Xa 48
Neeld Pde. HA9: Wemb 36Pa 67
Neeld Pl. W9 43Bb 89
Neil Cl. TW15: Ashf 64S 128
Neild Way WD3: Rick 17H 25
Neilson Cl. WD18: Wat 15V 26
Neil Wates Cres. SW2 60Qb 112
Nelgarde Rd. SE6 59Cc 114
Nella Rd. W6 51Za 110
Nelldale Rd. SE16 49Yb 92
Nellgrove Rd. UB10: Hil 42R 84
Nell Gwynn Av. TW17: Shep 72T 150
Nell Gwynn Cl. WD7: Shenl 4Na 15
Nell Gwynne Av. SL5: S'hill 10B 124
Nell Gwynne Cl. KT19: Eps 83Qa 173
Nell Gwynne Cl. SL5: S'hill 10B 124
Nell Gwynn Ho. SW3 6E 226 (49Gb 89)
(off Sloane Av.)
Nello Cressall Way E3 43Bc 92
Nello James Gdns. SE27 63Tb 135
Nelmes Cl. RM11: Horn 29Pd 57
Nelmes Cres. RM11: Horn 29Nd 57
Nelmes Rd. RM11: Horn 31Nd 77
Nelmes Way RM11: Horn 28Md 57
Nelson Arc. SE10 51Ec 114
(off Nelson Rd.)
Nelson Av. AL1: St A 5F 6
Nelson Cl. CM14: W'ley 22Zd 59
Nelson Cl. CR0: C'don 74Rb 157
Nelson Cl. KT12: Walt T 74X 151
Nelson Cl. NW6 41Cb 89
Nelson Cl. RM7: Mawney 25Dd 56
Nelson Cl. SL3: L'ly 9P 81
Nelson Cl. TN16: Big H 89Nc 180
Nelson Cl. TW14: Felt 60V 106
Nelson Cl. UB10: Hil 41R 84
Nelson Ct. DA8: Erith 52Hd 118
(off Frobisher Rd.)
Nelson Ct. KT16: Chert 74J 149
Nelson Ct. SE16 46Yb 92
(off Brunel Rd.)
Nelson Gdns. E2 41Wb 91
Nelson Gdns. TW3: Houn 58Ca 107
Nelson Gro. Rd. SW19 67Db 133
Nelson Ho. DA9: Ghithe 57Zd 121
Nelson Ho. RM3: Rom 22Nd 57
(off Lindfield Rd.)
Nelson Ho. SW1 51Lb 112
(off Dolphin Sq.)
Nelson La. UB10: Hil 41R 84
Nelson Mandela Cl. N10 26Jb 50
Nelson Mandela Ho. N16 33Wb 71
Nelson Mandela Rd. SE3 55Lc 115
Nelson Pas. EC1 4E 218 (41Sb 91)
Nelson Pl. N1 2C 218 (40Rb 71)
Nelson Rd. BR2: Broml 70Lc 137
Nelson Rd. CR3: Cat'm 95Tb 197
Nelson Rd. DA1: Dart 58Ld 119
Nelson Rd. DA11: Nflt 1B 144
Nelson Rd. DA14: Sidc 63Wc 139
Nelson Rd. DA17: Belv 50Bd 95
Nelson Rd. E11 28Jc 53
Nelson Rd. E4 23Dc 52
Nelson Rd. EN3: Pond E 16Zb 34
Nelson Rd. HA1: Harr 33Fa 66
Nelson Rd. HA7: Stan 23La 46
Nelson Rd. KT3: N Mald 71Ta 153
Nelson Rd. N15 28Ub 51
Nelson Rd. N8 29Pb 50
Nelson Rd. N9 19Xb 33
Nelson Rd. RM13: Rain 40Hd 76
Nelson Rd. RM15: S Ock 40Yd 78
Nelson Rd. RM16: Ors 4F 100
Nelson Rd. SE10 51Ec 114
Nelson Rd. SL4: Wind 5D 102
Nelson Rd. SW19 66Db 133
Nelson Rd. TW15: Ashf 64N 127
Nelson Rd. TW2: Whitt 59Da 107
Nelson Rd. TW3: Houn 58Ca 107
Nelson Rd. TW6: H'row A 53P 105
Nelson Rd. UB10: Hil 41R 84
Nelson Rd. M. SW19 66Db 133
(off Nelson Rd.)
Nelson's Column 6F 223 (46Nb 90)
Nelson Sq. SE1 1B 230 (47Rb 91)
Nelson's Row SW4 56Mb 112
Nelson St. E1 44Xb 91
Nelson St. E16 44Hc 93
Nelson St. E6 40Pc 74
(not continuous)
Nelsons Yd. NW1 1B 216 (40Lb 70)
(off Mornington Cres.)
Nelson Ter. N1 2C 218 (40Rb 71)
Nelson Trad. Est. SW19 67Db 133
Nelson Wlk. E3: Bromley-by-Bow 42Dc 92
Nelson Wlk. SE16: Canada Water 46Ac 92
Nelson Wlk. KT19: Eps 81Qa 173
Nelwyn Av. RM11: Horn 29Pd 57
Nemoure Rd. W3 45Sa 87
Nemus Apts. SE8 49Zb 92
Nene Gdns. TW13: Hanw 62Ba 129
Nene Rd. TW6: H'row A 53R 106
NENE ROAD RDBT. 53R 106
Neo Apts. SL1: Slou 7L 81
Nepaul Rd. SW11 54Gb 111
Nepean St. SW15 58Wa 110
Nepicar La. TN15: Wro 87Ee 185
Neptune Bus. Pk. RM19: Purf 50Ud 98
Neptune Cl. RM13: Rain 40Hd 76
Neptune Cl. DA8: Erith 52Hd 118
(off Frobisher Rd.)
Neptune Ct. E14 49Cc 92
(off Homer Dr.)
Neptune Ct. E16 43Jc 93
(off Hammersley Rd.)
Neptune Ct. WD6: Bore 13Qa 29
Neptune Development SE8 50Ac 92
Neptune Dr. HP2: Hem H 1N 3
Neptune Ho. E3 39Cc 72
(off Garrison Rd.)
Neptune Ho. SE16 48Yb 92
(off Moodkee St.)
Neptune Rd. HA1: Harr 30Fa 46
Neptune Rd. TW6: H'row A 53T 106
Neptune St. SE16 48Yb 92
Neptune Wlk. DA8: Erith 49Fd 96
Neptune Way SL1: Slou 7C 80
Nero Ct. TW8: Bford 52Ma 109
Nero Ho. AL1: St A 3D 6
Nero Ho. E20 37Ec 72
(off Anthems Way)
Neroli Ho. E1 44Wb 91
(off Boulevard Walkway)
Nesbit Ct. SE17 51Rb 113
(off Cook's Rd.)
Nesbit Rd. SE9 56Mc 115
Nesbitt Cl. SE3 55Gc 115
Nesbitts All. EN5: Barn 13Bb 31
Nesbitt Sq. SE19 66Ub 135
Nescot Sports Cen. 83Wa 174
Nesham Ho. N1 1H 219 (39Ub 71)
(off Hoxton St.)
Nesham St. E1 45Wb 91
Ness Rd. DA8: Erith 51Md 119
Ness St. SE16 48Wb 91
Nesta Rd. IG8: Wfd G 23Gc 53
Nestles Av. UB3: Hayes 48V 84
Neston Rd. WD24: Wat 9Y 13
Nestor Av. N21 16Rb 33
Nestor Ho. E2 40Xb 71
(off Old Bethnal Grn. Rd.)
Nethan Dr. RM15: Avel 45Sd 98
Netheravon Rd. W4 49Va 88
Netheravon Rd. W7 46Ha 86
Netheravon Rd. Sth. W4 50Va 88
Netherbury Rd. W5 48Ma 87
Netherby Gdns. EN2: Enf 14Nb 32
Netherby Pk. KT13: Weyb 78U 150
Netherby Rd. SE23 59Yb 114
Nether Cl. N3 24Cb 49
Nethercote Av. GU21: Wok 9K 167
Nethercott Ho. E3 41Dc 92
(off Bruce Rd.)
Nethercourt Av. N3 23Cb 49
Netherene La. CR5: Coul 95Lb 196
Netherfield Gdns. IG11: Bark 37Tc 74
Netherfield Rd. N12 22Db 49
Netherfield Rd. SW17 62Jb 134
Netherford Rd. SW4 54Lb 112
Netherhall Gdns. NW3 37Eb 69
Netherhall Way NW3 36Eb 69
Netherheys Dr. CR2: S Croy 80Rb 157
Netherlands, The CR5: Coul 91Lb 196
Netherlands Rd. EN5: New Bar 16Fb 31
Netherleigh Cl. N6 32Kb 70
Netherleigh Pk. RH1: S Nut 9E 208
Nethern Ct. Rd. CR3: Wold 95Cc 198
Netherne Dr. CR5: Coul 93Kb 196
Netherne La. CR5: Coul 93Lb 196
Netherne La. RH1: Coul 95Lb 196
Netherne Leisure Cen. 94Mb 196
NETHERNE-ON-THE-HILL 94Mb 196
Netherpark Dr. RM2: Rom 26Hd 56
Nether St. N12 22Eb 49
(not continuous)
Nether St. N3 24Cb 49
Netherton Gro. SW10 51Eb 111
Netherton Rd. N15 30Tb 51
Netherton Rd. TW1: Twick 57Ja 108
Netherway AL3: St A 5N 5
Netherwood N2 26Fb 49
Netherwood Pl. W14 48Za 88
(off Netherwood Rd.)
Netherwood Rd. W14 48Za 88
Netherwood St. NW6 38Bb 69
Nethewode Ct. DA17: Belv 48Dd 96
(off Lower Pk. Rd.)
Netley SE5 53Ub 113
(off Redbridge Gdns.)
Netley Cl. CR0: New Ad 80Ec 158
Netley Cl. SM3: Cheam 78Za 154
Netley Dr. KT12: Walt T 73Ba 151
Netley Gdns. SM4: Mord 73Eb 155
Netley Rd. E17 29Bc 52
Netley Rd. IG2: Ilf 29Tc 54
Netley Rd. SM4: Mord 73Eb 155
Netley Rd. TW8: Bford 51Na 109
Netley St. NW1 4B 216 (41Lb 90)
Nettlecombe NW1 38Mb 70
(off Agar Gro.)
Nettlecombe Cl. SM2: Sutt 81Db 175
Nettlecroft HP1: Hem H 3K 3
Nettleden Av. HA9: Wemb 37Qa 67
Nettleden Ho. SW3 6E 226 (49Gb 89)
(off Cale St.)
Nettlefold Pl. SE27 62Rb 135
Nettlefold Pl. TW16: Sun 71W 150
Nettlefold Wlk. KT12: Walt T 74V 150
Nettlestead Cl. BR3: Beck 66Bc 136
Nettleton Ct. EC2 1D 224 (43Sb 91)
(off London Wall)
Nettleton Rd. SE14 53Zb 114
Nettleton Rd. TW6: H'row A 53R 106
Nettleton Rd. UB10: Ick 35P 63
Nettlewood Rd. SW16 66Mb 134
Neuchatel Rd. SE6 61Bc 136
Neutron Twr. E14 45Fc 93
Nevada Bldg. SE10 53Dc 114
(off Blackheath Rd.)
Nevada Cl. KT3: N Mald 70Sa 131
Nevada St. SE10 51Ec 114
Nevell Rd. RM16: Grays 8D 100
Nevern Mans. SW5 50Cb 89
(off Warwick Rd.)
Nevern Pl. SW5 49Cb 89
Nevern Rd. SW5 49Cb 89
Nevern Sq. SW5 49Cb 89
Nevil Cl. HA6: Nwood 22T 44
Nevil Ho. SW9 54Rb 113
(off Loughborough Est.)
Nevill Ct. SW10 52Eb 111
(off Edith Ter.)
Neville Av. KT3: N Mald 67Ta 131
Neville Cl. DA15: Sidc 63Vc 139
Neville Cl. E11 34Hc 73
Neville Cl. EN6: Pot B 3Bb 17
Neville Cl. KT10: Esh 79Ba 151
Neville Cl. NW1 2E 216 (40Mb 70)
Neville Cl. NW6 40Bb 69
Neville Cl. SE15 53Wb 113
Neville Cl. SL2: Stoke P 7K 61
Neville Cl. SM7: Bans 86Db 175
Neville Cl. TW3: Houn 54Da 107
Neville Cl. W3 47Sa 87
Neville Ct. NW8 2B 214 (40Fb 69)
(off Abbey Rd.)
Neville Ct. SL1: Burn 1A 80
Neville Dr. N2 30Eb 49
Neville Gdns. RM8: Dag 34Zc 75
Neville Gill Cl. SW18 58Cb 111
Neville Ho. N11 21Jb 50
Neville Ho. N22 25Pb 50
(off Neville Pl.)
Neville Ho. NW6 40Bb 69
(off Denmark Rd.)
Neville Ho. Yd. KT1: King T 68Na 131
Neville Pl. N22 25Pb 50
Neville Rd. CR0: C'don 73Tb 157
Neville Rd. E7 38Jc 73
Neville Rd. IG6: Ilf 25Sc 54
Neville Rd. KT1: King T 68Qa 131
Neville Rd. NW6 40Bb 69
Neville Rd. RM8: Dag 34Zc 75
Neville Rd. TW10: Ham 62La 130
Neville Rd. W5 42Ma 87
Nevilles Ct. NW2 34Wa 68
Neville St. SW7 7B 226 (50Fb 89)
Neville Ter. SW7 7B 226 (50Fb 89)
Neville Wlk. SM5: Cars 73Gb 155
Nevill Gro. WD24: Wat 11X 27
Nevill La. EC4 2A 224 (44Qb 90)
(off New Fetter La.)
Nevill Rd. N16 35Ub 71
Nevill Way IG10: Lough 16Nc 36
Nevin Dr. E4 18Dc 34
Nevin Ho. UB3: Harl 48S 84
Nevinson Cl. SW18 58Fb 111
Nevis Cl. E13 40Kc 73
Nevis Cl. RM1: Rom 23Gd 56
Nevis Rd. SW17 61Jb 134
Nevitt Ho. N1 2G 219 (40Tb 71)
(off Cranston Est.)
New Acres Rd. SE28 47Uc 94
NEW ADDINGTON 82Ec 178
New Addington Leisure Cen. 82Ec 178
New Addington Stop (London Tramlink) 82Ec 178
Newall Cl. UB10: Uxb 39P 63
Newall Ho. SE1 3E 230 (48Sb 91)
(off Bath Ter.)
Newall Rd. TW6: H'row A 53S 106
New Arc. UB8: Uxb 39M 63
Newark Cl. GU23: Rip 93J 189
Newark Cotts. GU23: Rip 93J 189
Newark Ct. KT12: Walt T 74Y 151
Newark Cres. NW10 41Ta 87
Newark Grn. WD6: Bore 13Ta 29
Newark Ho. SW9 54Rb 113
Newark Knok E6 44Qc 94
Newark La. GU22: Pyr 91H 189
Newark La. GU23: Rip 91H 189
Newark Rd. CR2: S Croy 79Tb 157
Newark St. E1 43Xb 91
(not continuous)
Newark Way NW4 28Wa 48
New Ash Cl. N2 27Fb 49
NEW ASH GREEN 75Be 165
New Atlas Wharf E14 48Cc 92
(off Arnhem Pl.)
New Baltic Wharf SE8 50Ac 92
(off Evelyn St.)
NEW BARN 69Ee 143
New Barn Cl. SM6: W'gton 79Pb 156
New Barnes Av. AL1: St A 5E 6
NEW BARNET 14Fb 31
New Barnet Station (Rail) 15Fb 31
New Barn La. CR3: Whyt 88Ub 177
New Barn La. TN14: Cud 88Sc 180
New Barn La. TN16: Cud 90Sc 180
New Barn La. TN16: Westrm 90Sc 180
New Barn Rd. BR8: Swan 67Gd 140
New Barn Rd. DA13: Ist R 62Ee 143
New Barn Rd. DA13: Lfield 62Ee 143
New Barn Rd. DA13: Nflt G 62Ee 143
New Barn Rd. DA13: Sflt 62Ee 143
New Barn Rd. DA3: Ist R 69De 143
New Barn Rd. DA3: Lfield 69De 143
New Barns Av. CR4: Mitc 70Mb 134
New Barn St. E13 42Jc 93
New Barns Way IG7: Chig 20Rc 36
New Battlebridge La. RH1: Mers 2B 208
Newbeach Ho. SL2: Slou 1F 80
Newbeck Ct. BR3: Beck 66Bc 136
NEW BECKENHAM 65Bc 136
New Beckenham Station (Rail) 66Bc 136
New Bentham Ct. N1 38Sb 71
(off Ecclesbourne Rd.)
Newberries Av. WD7: R'lett 7Ka 14
Newberry Cres. SL4: Wind 4B 102
New Berry La. KT12: Hers 78Z 151
Newberry M. SW4 56Nb 112
Newbery Cl. CR3: Cat'm 97Ub 197
Newbery Ho. N1 38Sb 71
(off Northampton St.)
Newbery Rd. DA8: Erith 53Hd 118
Newbery Way SL1: Slou 7H 81
Newbiggin Path WD19: Wat 21Y 45
Newbold Cotts. E1 44Yb 92
Newbolt Av. SM3: Cheam 78Ya 154
Newbolt Ho. SE17 7F 231 (50Tb 91)
(off Brandon St.)
Newbolt Rd. HA7: Stan 22Ha 46
New Bond St. W1 3K 221 (44Kb 90)
Newborough Grn. KT3: N Mald 70Ta 131
New Brent St. NW4 29Ya 48
Newbridge Point SE23 62Zb 136
(off Windrush La.)
New Bri. St. EC4 3B 224 (44Rb 91)
New Broad St. EC2 1H 225 (43Ub 91)
New B'way. TW12: Hamp H 64Fa 130
New B'way. UB10: Hil 41R 84
New B'way. W5 45Ma 87
New Broadway Bldgs. W5 45Ma 87
New Bldgs. SL4: Eton 1H 103
(off Westons Yd.)
Newburgh Rd. RM17: Grays 50Fe 99
Newburgh Rd. W3 46Sa 87
Newburgh St. W1 3C 222 (44Lb 90)
New Burlington M. W1 4B 222 (45Lb 90)
New Burlington Pl. W1 4B 222 (45Lb 90)
New Burlington St. W1 4B 222 (45Lb 90)
Newburn Ho. SE11 7J 229 (50Pb 90)
(off Newburn St.)
Newburn St. SE11 7J 229 (50Pb 90)
Newbury Av. EN3: Enf L 10Bc 20
Newbury Cl. DA2: Dart 59Rd 119
Newbury Cl. RM10: Dag 33Cd 76
Newbury Cl. RM3: Rom 23Ld 57
Newbury Cl. UB5: N'olt 37Ba 65
Newbury Ct. DA14: Sidc 63Vc 139
Newbury Ct. E5 36Ac 72
(off Daubeney Rd.)
Newbury Gdns. KT19: Ewe 77Va 154
Newbury Gdns. RM14: Upm 34Pd 77
Newbury Gdns. RM3: Rom 23Md 57
Newbury Ho. N22 25Nb 50
Newbury Ho. SW9 54Rb 113

Newbury Ho. W2 44Db 89
(off Hallfield Est.)
Newbury M. NW5 37Jb 70
NEWBURY PARK 29Tc 54
Newbury Park Station (Underground) 30Tc 54
Newbury Rd. BR2: Broml 69Jc 137
Newbury Rd. E4 23Ec 52
Newbury Rd. IG2: Ilf 30Uc 54
Newbury Rd. RM3: Rom 22Md 57
Newbury Rd. TW6: H'row A 53P 105
Newbury St. EC1 7D 218 (43Sb 91)
Newbury Wlk. RM3: Rom 22Md 57
Newbury Way UB5: N'olt 37Aa 65
New Bus. Cen., The NW10 41Va 88
New Butt La. SE8 52Cc 114
New Butt La. Nth. SE8 52Cc 114
(off Hales St.)
Newby NW1 4B 216 (41Lb 90)
(off Robert St.)
Newby Cl. EN1: Enf 12Ub 33
Newby Ho. E14 45Ec 92
(off Newby Pl.)
Newby Pl. E14 45Ec 92
Newby St. SW8 55Kb 112
New Caledonian Mkt. 3J 231 (48Ub 91)
(off Bermondsey Sq.)
New Caledonian Wharf SE16 48Bc 92
Newcastle Av. IG6: Ilf 23Wc 55
Newcastle Cl. EC4 2B 224 (44Rb 91)
Newcastle Ct. EC4 4E 224 (45Sb 91)
(off College Hill)
Newcastle Ho. W1 7H 215 (43Jb 90)
(off Luxborough St.)
Newcastle Pl. W2 7C 214 (43Fb 89)
Newcastle Row EC1 5A 218 (42Qb 90)
New C'way. RH2: Reig 9K 207
New Cavendish St. W1 1J 221 (43Jb 90)
New Century Ho. E16 44Hc 93
(off Jude St.)
New Change EC4 3D 224 (44Sb 91)
New Change Pas. EC4 3D 224 (44Sb 91)
(off New Change)
Newchapel Ho. KT15: Add 78L 149
New Chapel Sq. TW13: Felt 60X 107
New Charles St. EC1 3C 218 (41Rb 91)
NEW CHARLTON 49Lc 93
New Chiswick Pool 52Ua 110
New Church Rd. SE5 52Sb 113
(not continuous)
Newchurch Rd. SL2: Slou 3D 80
New City Rd. E13 41Lc 93
New Claremont Apts. SE1 5K 231 (49Vb 91)
(off Setchell Rd.)
New Clocktower Pl. N7 37Nb 70
New Cl. SW19 69Eb 133
New Cl. TW13: Hanw 64Aa 129
New Colebrooke Ct. SM5: Cars 80Hb 155
New College Ct. NW3 37Eb 69
(off College Cres.)
New College M. N1 38Qb 70
New College Pde. NW3 37Fb 69
(off Finchley Rd.)
Newcombe Gdns. SW16 63Nb 134
Newcombe Gdns. TW4: Houn 56Ba 107
Newcombe Ho. E5 34Xb 71
Newcombe Pk. HA0: Wemb 39Pa 67
Newcombe Pk. NW7 22Ua 48
Newcombe Ri. UB7: Yiew 44N 83
Newcombe St. W8 46Cb 89
Newcomen Rd. E11 34Hc 73
Newcomen Rd. SW11 55Fb 111
Newcomen St. SE1 1F 231 (47Tb 91)
Newcome Path WD7: Shenl 6Qa 15
Newcome Rd. WD7: Shenl 6Qa 15
New Compton St. WC2 3E 222 (44Mb 90)
New Concordia Wharf SE1 47Wb 91
New Coppice GU21: Wok 1J 187
New Cotts. GU24: Pirb 4B 186
New Cotts. KT23: Bookh 98Da 191
(off Dorking Rd.)
New Cotts. RM13: Wenn 44Ld 97
New Ct. EC4 4K 223 (45Qb 90)
(off Fountain Ct.)
New Ct. KT15: Add 76L 149
New Ct. UB5: N'olt 36Da 65
Newcourt UB8: Cowl 43L 83
Newcourt Ho. E2 41Xb 91
(off Pott St.)
Newcourt St. NW8 2D 214 (40Gb 69)
New Covent Garden Market 52Mb 112
New Crane Pl. E1 46Yb 92
New Crane Wharf E1 46Yb 92
(off New Crane Pl.)
New Cres. Yd. NW10 40Va 68
Newcroft Cl. UB8: Hil 43P 83
Newcroft Ho. CR0: C'don 75Vb 157
(off Homefield Pl.)
NEW CROSS 52Bc 114
NEW CROSS 53Bc 114
NEW CROSS GATE 53Zb 114
NEW CROSS GATE 53Zb 114
New Cross Gate Station (Rail & Overground) 52Ac 114
New Cross Rd. SE14 52Yb 114
New Cross Station (Rail & Overground) 52Bc 114
Newdales Cl. N9 19Wb 33
Newdene Av. UB5: N'olt 40Z 65
NEW DENHAM 37K 63
Newdigate Grn. UB9: Hare 25M 43
Newdigate Ho. E14 44Bc 92
(off Norbiton Rd.)
Newdigate Rd. UB9: Hare 25L 43
Newdigate Rd. E. UB9: Hare 25M 43
New Diorama Theatre 5A 216 (42Kb 90)
New Drum St. E1 44Vb 91
(off Aldgate Pl.)
Newell Ri. HP3: Hem H 5N 3
Newell Rd. HP3: Hem H 5N 3
Newell St. E14 44Bc 92
NEW ELTHAM 61Sc 138
New Eltham Station (Rail) 60Rc 116
New End NW3 34Eb 69
New End Sq. NW3 35Fb 69
New England Hill GU24: W End 4B 166
New England Ind. Est. IG11: Bark 40Sc 74
New England St. AL3: St A 2A 6
Newenham Rd. KT23: Bookh 98Ca 191
Newent Cl. SE15 52Ub 113
Newent Cl. SM5: Cars 74Hb 155
NEW EPSOM & EWELL COTTAGE HOSPITAL 83Na 173
New Era Est. N1 39Ub 71
(off Halcomb St.)
New Era Ho. N1 39Ub 71
(off Halcomb St.)
New Farm Av. BR2: Broml 70Jc 137
New Farm Cl. TW18: Staines 67L 127
New Farm Dr. RM4: Abr 13Yc 37
New Farm La. HA6: Nwood 25U 44
New Ferry App. SE18 48Qc 94
New Festival Av. E14 44Cc 92
New Fetter La. EC4 2A 224 (44Qb 90)
Newfield Cl. TW12: Hamp 67Ca 129
Newfield La. HP2: Hem H 2N 3
Newfield Ri. NW2 34Wa 68
Newfield Way AL4: St A 4G 6
Newford Cl. HP2: Hem H 1B 4
New Ford Rd. EN8: Walt C 6Bc 20
New Forest La. IG7: Chig 23Qc 54
Newgale Gdns. HA8: Edg 25Pa 47
New Gdn. Dr. UB7: W Dray 47N 83
Newgate CR0: C'don 74Sb 157
Newgate Cl. TW13: Hanw 61Aa 129
Newgate St. E4 20Gc 35
(not continuous)
Newgate St. EC1 2C 224 (44Rb 91)
New Globe Wlk. SE1 6D 224 (46Sb 91)
New Goulston St. E1 2K 225 (44Vb 91)
New Grn. Pl. SE19 65Ub 135
New Gun Wharf E3 39Ac 72
(off Gunmaker's La.)
New Hall Cl. HP3: Bov 9C 2
Newhall Ct. EN9: Walt A 5Hc 21
Newhall Ct. N1 39Sb 71
(off Popham Rd.)
New Hall Dr. RM3: Hrld W 25Nd 57
Newhall Gdns. KT12: Walt T 75Y 151
Newham Academy of Music 39Nc 74
NEWHAM CENTRE FOR MENTAL HEALTH 42Mc 93
Newham City Farm 44Mc 93
Newham Dockside E16 45Nc 94
Newham Leisure Cen. 42Lc 93
Newham's Row SE1 3J 231 (48Ub 91)
NEWHAM UNIVERSITY HOSPITAL 42Lc 93
Newham Way DA8: Erith 52Hd 118
Newham Way E16 43Hc 93
Newham Way E6 42Mc 93
Newhaven Cl. UB3: Harl 49V 84
Newhaven Cres. TW15: Ashf 64T 128
Newhaven Gdns. SE9 56Mc 115
Newhaven La. E16 42Hc 93
Newhaven Rd. SE25 71Tb 157
Newhaven Spur SL2: Slou 2F 80
NEW HAW 81K 169
New Haw Lock 80L 149
(off Byfleet Rd.)
New Haw Rd. KT15: Add 78L 149
New Heston Rd. TW5: Hest 52Ba 107
Newholme Ct. KT13: Weyb 76U 150
New Hope Ct. NW10 41Xa 88
New Horizons Ct. TW8: Bford 51Ja 108
NEW HOUSE 1A 144
Newhouse Av. RM6: Chad H 27Zc 55
Newhouse Cl. KT3: N Mald 73Ua 154
Newhouse Cres. WD25: Wat 4X 13
New Ho. La. DA11: Grav'nd 2B 144
New Ho. La. DA11: Nflt 2B 144
New Ho. La. TN15: Wro 88Ae 185
New Ho. Pk. AL1: St A 5E 6
Newhouse Rd. HP3: Bov 8C 2
Newhouse Wlk. SM4: Mord 73Eb 155
Newick Cl. DA5: Bexl 58Dd 118
Newick Rd. E5 35Xb 71
Newing Grn. BR1: Broml 66Mc 137
NEWINGTON 3D 230 (48Sb 91)
Newington Barrow Way N7 34Pb 70
Newington Butts SE1 5C 230 (49Rb 91)
Newington Butts SE11 6C 230 (49Rb 91)
Newington C'way. SE1 4C 230 (48Rb 91)
Newington Ct. N16 35Sb 71
(off Green Lanes)
Newington Ct. Bus. Cen. SE1 3D 230 (48Sb 91)
(off Newington C'way.)
Newington Grn. N1 36Tb 71
Newington Grn. N16 36Tb 71
Newington Grn. Community Gdns. 36Tb 71
(off Newington Grn.)
Newington Grn. Mans. N16 36Tb 71
Newington Grn. Rd. N1 37Tb 71
Newington Ind. Est. SE17 6C 230 (49Rb 91)
New Inn B'way. EC2 5J 219 (42Ub 91)
New Inn Pas. WC2 3J 223 (44Pb 90)
(off Houghton St.)
New Inn Sq. EC2 5J 219 (42Ub 91)
(off Bateman's Row)
New Inn St. EC2 5J 219 (42Ub 91)
New Inn Yd. EC2 5J 219 (42Ub 91)
New Jubilee Ct. IG8: Wfd G 24Jc 53
New Jubilee Wharf E1 46Yb 92
(off Wapping Wall)
New Kelvin Av. TW11: Tedd 65Ga 130
New Kent Rd. AL1: St A 2B 6
New Kent Rd. SE1 4D 230 (48Sb 91)
New Kings Rd. SW6 54Bb 111
New King St. SE8 51Cc 114
Newland Cl. AL1: St A 5E 6
Newland Cl. HA5: Hat E 23Aa 45
Newland Ct. EC1 5F 219 (42Tb 91)
(off St Luke's Est.)
Newland Ct. HA9: Wemb 33Qa 67
Newland Dr. EN1: Enf 11Xb 33
Newland Gdns. W13 47Ja 86
Newland Ho. N8 27Nb 50
(off Newland Rd.)
Newland Ho. SE14 51Zb 114
(off John Williams Cl.)
Newland Rd. N8 27Nb 50
Newlands HA1: Harr 32Ga 66
NEWLANDS 20Na 29
Newlands NW1 3B 216 (41Lb 90)
(off Harrington St.)
NEWLANDS 57Zb 114
Newlands, The KT7: T Ditt 74Ga 152
Newlands, The SM6: W'gton 80Lb 156
Newlands Av. GU22: Wok 93B 188
Newlands Av. KT7: T Ditt 74Ga 152
Newlands Av. WD7: R'lett 6Ha 14
Newlands Cl. CM13: Hut 17Ee 41
Newlands Cl. HA0: Wemb 37La 66
Newlands Cl. HA8: Edg 20Na 29
Newlands Cl. KT12: Hers 77Aa 151
Newlands Cl. UB2: S'hall 50Aa 85
Newlands Ct. CR3: Cat'm 93Sb 197
(off Coulsdon Rd.)
Newlands Ct. KT15: Add 78K 149
(off Church Rd.)
Newlands Ct. SE9 58Qc 116
Newlands Dr. SL3: Poyle 55G 104
Newlands Pk. SE26 65Yb 136
Newlands Pk. WD5: Bedm 8F 4
Newlands Pl. EN5: Barn 15Za 30
Newlands Quay E1 45Yb 92
Newlands Rd. HP1: Hem H 1G 2
Newlands Rd. IG8: Wfd G 19Hc 35
Newlands Rd. SW16 68Nb 134
Newland St. E16 46Nc 94
Newlands Wlk. WD25: Wat 5Z 13
Newlands Way EN6: Pot B 2Db 17
Newlands Way KT9: Chess 78La 152
Newlands Woods CR0: Sels 81Bc 178
New La. GU4: Sut G 94A 188
Newling Cl. E6 44Pc 94
New Lodge Dr. RH8: Oxt 100Hc 199
New London Performing Arts Cen. 28Kb 50
New London St. EC3 4J 225 (45Ub 91)
(off Hart St.)
New Lydenburg Commercial Est. SE7 48Lc 93
New Lydenburg St. SE7 48Lc 93
Newlyn KT13: Weyb 77V 150
Newlyn NW1 1C 216 (39Lb 70)
(off Plender St.)
Newlyn Cl. AL2: Brick W 2Aa 13
Newlyn Cl. BR6: Chels 77Wc 161
Newlyn Cl. UB8: Hil 43Q 84
Newlyn Gdns. HA2: Harr 31Ba 65
Newlyn Ho. HA5: Hat E 24Ba 45
Newlyn Rd. DA16: Well 54Vc 117
Newlyn Rd. EN5: Barn 14Bb 31
Newlyn Rd. N17 25Vb 51
NEW MALDEN 70Ua 132
New Malden Station (Rail) 69Ua 132
New Maltings RM15: Avel 46Td 98
Newman Cl. NW10 37Xa 68
Newman Cl. RM11: Horn 29Nd 57
Newman Cl. SE26 63Yb 136
Newman Ct. BR1: Broml 67Jc 137
(off North St.)
Newman Ct. TW15: Ashf 65R 128
Newman Ho. SE1 4B 230 (48Rb 91)
Newman Pas. W1 1C 222 (43Lb 90)
Newman Rd. BR1: Broml 67Jc 137
Newman Rd. CR0: C'don 74Pb 156
Newman Rd. E13 41Kc 93
Newman Rd. E17 29Zb 52
Newman Rd. UB3: Hayes 45X 85
Newman Rd. Ind. Est. CR0: C'don 73Pb 156
Newmans Cl. IG10: Lough 13Qc 36
Newman's Ct. EC3 3G 225 (44Tb 91)
(off Cornhill)
Newmans Dr. CM13: Hut 17Ee 41
Newmans Ga. CM13: Hut 17Ee 41
Newmans La. IG10: Lough 13Qc 36
Newmans La. KT6: Surb 72Ma 153
Newmans Pl. SL5: S'dale 3F 146
Newmans Rd. DA11: Nflt 1B 144
Newman's Row WC2 1J 223 (43Pb 90)
Newman St. W1 1C 222 (43Lb 90)
Newmans Way EN4: Had W 11Eb 31
Newman Ter. SE1 3K 229 (48Qb 90)
Newman Way HA3: Kenton 29Na 47
New Nth. Pl. EC2 5H 219 (42Ub 91)
New Nth. Rd. IG6: Ilf 24Tc 54
New Nth. Rd. N1 38Sb 71
New Nth. Rd. RH2: Reig 9H 207
New Nth. St. WC1 7H 217 (43Pb 90)
Newnton Cl. N4 31Tb 71
New Oak Rd. N2 26Eb 49
New Orleans Wlk. N19 31Mb 70
New Oxford St. WC1 2E 222 (44Mb 90)
New Pde. KT23: Bookh 97Ea 192
New Pde. TW15: Ashf 63P 127
New Pde. UB7: Yiew 46N 83
New Pde. WD3: Chor 14E 24
New Pde. Flats WD3: Chor 14E 24
New Paragon Wlk. SE17 5F 231 (49Tb 91)
New Pk. Av. N13 20Sb 33
New Pk. Cl. UB5: N'olt 37Aa 65
New Pk. Dr. HP2: Hem H 1B 4
New Pk. Est. N18 22Yb 52
New Pk. Ho. N13 21Pb 50
New Pk. Pde. SW2 59Nb 112
(off New Pk. Rd.)
New Pk. Rd. SW2 60Mb 112
New Pk. Rd. TW15: Ashf 64S 128
New Pk. Rd. UB9: Hare 25L 43
New Peachey La. UB8: Cowl 44M 83
Newpiece IG10: Lough 13Rc 36
New Pl. CR0: Addtn 79Cc 158
New Pl. SL5: S'dale 5C 146
New Pl. Gdns. RM14: Upm 33Td 78
New Pl. Sq. SE16 48Xb 91
New Plaistow Rd. E15 39Gc 73
New Plymouth Ho. RM13: Rain 41Hd 96
(off Dunedin Rd.)
New Pond Pde. HA4: Ruis 34W 64
Newport Av. E13 42Kc 93
Newport Av. E14 45Fc 93
Newport Cl. EN3: Enf W 9Ac 20
Newport Ct. WC2 4E 222 (45Mb 90)
Newport Ho. E3 41Ac 92
(off Strahan Rd.)
Newport Lodge EN1: Enf 15Ub 33
(off Village Rd.)
Newport Mead WD19: Wat 21Z 45
Newport Pl. WC2 4E 222 (45Mb 90)
Newport Rd. E10 33Ec 72
Newport Rd. E17 28Ac 52
Newport Rd. SL2: Slou 2C 80
Newport Rd. SW13 53Wa 110
Newport Rd. TW6: H'row A 53Q 106
Newport Rd. UB4: Hayes 43T 84
Newport Rd. W3 47Sa 87
Newports BR8: Crock 73Fd 162
Newport St. SE11 6H 229 (49Pb 90)
Newport Street Gallery 5J 229 (49Pb 90)
New Priory Ct. NW6 38Cb 69
(off Mazenod Av.)
New Providence Wharf E14 46Fc 93
New Provident Pl. HP4: Berk 1A 2
Newquay Cres. HA2: Harr 33Aa 65
Newquay Gdns. WD19: Wat 19X 27
Newquay Ho. SE11 7K 229 (50Qb 90)
Newquay Rd. SE6 61Dc 136
New Quebec St. W1 3G 221 (44Hb 89)
New Rathmore Rd. DA11: Grav'nd 9D 122
New Ride SW7 1E 226 (47Gb 89)
New River Av. N8 27Pb 50
New River Cl. EN7: Chesh 3Xb 19
New River Ct. N5 35Sb 71
New River Cres. N13 21Rb 51
New River Head EC1 4A 218 (41Qb 90)
New River Sports & Fitness Cen. 24Rb 51
New River Wlk. N1 37Sb 71
(not continuous)
New River Way N4 31Tb 71
New Rd. NW7: Bittacy Ct. 24Ab 48
New Rd. NW7: Hendon Wood La. 17Va 30
New Rd. HP4: Berk White Hill 1A 2
New Rd. BR6: Orp 73Wc 161
New Rd. BR8: Hext 66Hd 140
New Rd. BR8: Swan 69Hd 140
New Rd. CM14: B'wood 19Zd 41
New Rd. CR4: Mitc 74Hb 155
New Rd. DA11: Grav'nd 8C 122
New Rd. DA16: Well 54Xc 117
New Rd. DA4: S Dar 68Sd 142
New Rd. E1 43Xb 91
New Rd. E4 21Dc 52
New Rd. EN6: S Mim 5Wa 16
New Rd. HA1: Harr 35Ha 66
New Rd. HP8: Chal G 13A 24
New Rd. IG3: Ilf 33Uc 74
New Rd. KT10: Esh 76Ea 152
New Rd. KT13: Weyb 78S 150
New Rd. KT2: King T 66Qa 131
New Rd. KT20: Tad 95Ya 194
New Rd. KT22: Oxs 83Ha 172
New Rd. KT8: W Mole 70Ca 129
New Rd. N17 25Vb 51
New Rd. N22 25Sb 51
New Rd. N8 29Nb 50
New Rd. N9 19Xb 33
New Rd. RH8: Limp 2M 211
New Rd. RH8: Tand 8E 210
New Rd. RM10: Dag 40Cd 76
New Rd. RM13: Rain 41Jd 96
New Rd. RM13: Wenn 41Jd 96
New Rd. RM17: Grays 51Ce 121
(not continuous)
New Rd. RM4: Abr 16Zc 37
New Rd. RM9: Dag 40Cd 76
New Rd. SE2 49Zc 95
New Rd. SL3: Dat 3P 103
New Rd. SL3: L'ly 48C 82
New Rd. TN14: Sund 96Zc 201
New Rd. TW10: Ham 63La 130
New Rd. TW13: Hanw 64Aa 129
New Rd. TW14: Bedf 58T 106
New Rd. TW14: Felt 60X 107
New Rd. TW17: Shep 69Q 128
New Rd. TW18: Staines 64E 126
New Rd. TW3: Houn 56Da 107
New Rd. TW8: Bford 51Ma 109
New Rd. UB3: Harl 52S 106
New Rd. UB8: Hil 42S 84
New Rd. WD17: Wat 14Y 27
New Rd. WD25: Let H 11Ga 28
New Rd. WD3: Crox G 15Q 26
New Rd. WD3: Sarr 11H 25
New Rd. WD4: Chfd 2H 11
New Rd. WD6: E'tree 16Ma 29
New Rd. WD7: R'lett 8Ga 14
New Rd. WD7: Shenl 6Qa 15
New Rd. Hill BR2: Kes 81Nc 180
New Rd. Hill BR6: Downe 81Nc 180
New Rochford St. NW5 36Hb 69
New Row NW2 34Ya 68
New Row WC2 4F 223 (45Nb 90)
Newry Rd. TW1: Twick 57Ja 108
Newsam Av. N15 29Tb 51
Newsham Rd. GU21: Wok 9K 167
Newsholme Dr. N21 15Pb 32
New Site KT15: Add 76N 149
Newsom Pl. AL1: St A 1C 6
NEW SOUTHGATE 22Kb 50
New Southgate Crematorium 20Kb 32
New Southgate Ind. Est. N11 22Lb 50
New Southgate Station (Rail) 22Kb 50
New Spitalfields Mkt. E10 34Cc 72
New Spring Gdns. Wlk. SE11 7G 229 (50Nb 90)
New Sq. SL1: Slou 7J 81
New Sq. TW14: Bedf 60S 106
New Sq. WC2 2K 223 (44Qb 90)
New Sq. Pk. TW14: Bedf 60S 106
New Sq. Pas. WC2 2K 223 (44Qb 90)
(off Star Yd.)
Newstead AL10: Hat 3B 8
Newstead Av. BR6: Orp 76Tc 160
Newstead Cl. N12 23Gb 49
Newstead Ct. UB5: N'olt 41Aa 85
Newstead Ho. CR3: Cat'm 98Xb 197
Newstead Ho. N1 1A 218 (40Qb 70)
(off Tolpuddle St.)
Newstead Ho. RM3: Rom 21Md 57
(off Troopers Dr.)
Newstead Ri. CR3: Cat'm 98Xb 197
Newstead Rd. SE12 59Hc 115
Newstead Wlk. SM5: Cars 73Eb 155
Newstead Way SW19 63Za 132
New St. EC2 1J 225 (43Ub 91)
New St. HP4: Berk 1A 2
New St. TN16: Westrm 99Sc 200
New St. TW18: Staines 63J 127
New St. WD18: Wat 14Y 27
NEW STREET 78Fe 165
New St. Hill BR1: Broml 64Kc 137
New St. Rd. DA13: Meop 76Ee 165
New St. Rd. TN15: Ash 77Ee 165
New St. Rd. TN15: Nw A G 77Ee 165
New St. Sq. EC4 2A 224 (44Qb 90)
(not continuous)
New Swan Yd. DA12: Grav'nd 8D 122
New Tank Hill Rd. RM15: Avel 48Qd 97
New Tank Hill Rd. RM19: Avel 48Qd 97
New Tank Hill Rd. RM19: Purf 48Qd 97
New Tavern Fort 8E 122
(off Commercial Pl.)
Newteswell Dr. EN9: Walt A 4Fc 21
Newton Abbot Rd. DA11: Nflt 1B 144
Newton Av. N10 25Jb 50
Newton Av. W3 47Sa 87
Newton Cl. E17 30Ac 52
Newton Cl. HA2: Harr 33Ca 65
Newton Cl. SL3: L'ly 47B 82
Newton Ct. E3 43Cc 92
Newton Ct. NW6 38Eb 69
(off Fairfax Rd.)
Newton Ct. SL4: Old Win 8L 103
Newton Ct. SW17 62Fb 133
(off Grosvenor Way)
Newton Ct. W8 47Cb 89
(off Kensington Chu. St.)
Newton Cres. WD6: Bore 14Sa 29
Newton Gro. W4 49Ua 88
Newton Ho. E1 45Xb 91
(off Cornwall St.)
Newton Ho. E17 27Dc 52
(off Prospect Hill)
Newton Ho. EN3: Enf H 13Zb 34
Newton Ho. NW8 39Db 69
(off Abbey Rd.)
Newton Ho. SE20 66Zb 136
Newton Ind. Est. RM6: Chad H 28Zc 55
Newton La. SL4: Old Win 8M 103
Newton Lodge SE10 48Hc 93
(off Teal St.)
Newton Mans. W14 51Ab 110
(off Queen's Club Gdns.)
Newton Pk. Pl. BR7: Chst 66Pc 138
Newton Pl. E14 49Cc 92
Newton Rd. CR8: Purl 84Lb 176
Newton Rd. DA16: Well 55Wc 117
Newton Rd. E15 36Fc 73
Newton Rd. HA0: Wemb 38Pa 67
Newton Rd. HA3: Hrw W 26Ga 46
Newton Rd. IG7: Chig 22Xc 55
Newton Rd. IG7: Ilf 22Xc 55
Newton Rd. N15 29Wb 51
Newton Rd. NW2 35Ya 68
Newton Rd. RM18: Tilb 5C 122
Newton Rd. SW19 66Ab 132
Newton Rd. TW7: Isle 54Ha 108
Newton Rd. W2 44Db 89
Newtons Cl. RM13: Rain 38Hd 76
Newtons Ct. DA2: Dart 56Td 120
Newtonside Orchard SL4: Old Win 8L 103
Newton St. WC2 2G 223 (44Nb 90)
Newton's Yd. SW18 57Cb 111
Newton Ter. BR2: Broml 72Mc 159
Newton Wlk. HA8: Edg 25Ra 47
Newton Way N18 22Sb 51
Newton Wood Rd. KT21: Asht 88Pa 173
New Tower Bldgs. E1 46Xb 91
NEW TOWN 58Pd 119
Newtown Rd. UB9: Den 37K 63
Newtown St. SW11 53Kb 112
New Trinity Rd. N2 27Fb 49
New Turnstile WC1 1H 223 (43Pb 90)
(off High Holborn)
New Union Cl. E14 48Ec 92
New Union Sq. SW11 51Mb 112
New Union St. EC2 1F 225 (43Tb 91)
NEW VICTORIA HOSPITAL Kingston upon Thames 67Ua 132
New Victoria Theatre Woking 89A 168
New Village Av. E11 44Fc 93
New Wlk. TN15: Wro 88Be 185
New Wanstead E11 30Hc 53
New Warren La. SE18 48Rc 94
New Way Rd. NW9 28Ua 48
New Wharf Rd. N1 1G 217 (40Nb 70)
New Wickham La. TW20: Egh 66C 126
New Willow Ho. E13 40Jc 73
(off Plaistow Rd.)
NEW WINDSOR 5H 103
New Windsor St. UB8: Uxb 39L 63
NEWYEARS GREEN 31P 63
New Years Grn. La. UB9: Hare 30N 43
New Years La. TN14: Knock 88Vc 181
New Zealand Av. KT12: Walt T 74V 150
New Zealand Golf Course 84F 168
New Zealand Way RM13: Rain 41Hd 96
New Zealand Way W12 45Xa 88
Nexus KT16: Chert 73H 149
Nexus Apts. BR1: Broml 69Kc 137
(off Elmfield Rd.)
Nexus Cl. TW14: Felt 57W 106
Nexus Ct. AL1: St A 3B 6
Nexus Ct. E11 32Gc 73
Nexus Ct. NW6 41Cb 89
NHS WALK-IN CENTRE (ASHFORD) 61N 127
(within Ashford Hospital)
NHS WALK-IN CENTRE (BARKING HOSPITAL) 38Vc 75
NHS WALK-IN CENTRE (BELMONT HEALTH CENTRE) 26Ja 46
NHS WALK-IN CENTRE (CLAPHAM JUNCTION) 55Gb 111
NHS WALK-IN CENTRE (CRICKLEWOOD HEALTH CENTRE) 35Za 68
NHS WALK-IN CENTRE (EARL'S COURT) 49Db 89
NHS WALK-IN CENTRE (EDGWARE) 24Ra 47
NHS WALK-IN CENTRE (FINCHLEY) 24Eb 49
NHS WALK-IN CENTRE (HAROLD WOOD) 26Pd 57
NHS WALK-IN CENTRE (ISLE OF DOGS) 47Cc 92
NHS WALK-IN CENTRE (PARSONS GREEN) 53Cb 111
NHS WALK-IN CENTRE (PINNER) 27Aa 45
NHS WALK-IN CENTRE (SOHO) 3D 222 (44Mb 90)
(off Frith St.)
NHS WALK-IN CENTRE (SOUTH HORNCHURCH) 39Jd 76
NHS WALK-IN CENTRE (TEDDINGTON) 65Ga 130
NHS WALK-IN CENTRE (THAMESMEAD) 47Uc 94
NHS WALK-IN CENTRE (UPTON HOSPITAL) 8K 81
NHS WALK-IN CENTRE (WEMBLEY) 37Ma 67
NHS WALK-IN CENTRE (WHITE HORSE SURGERY) 60Fe 121
NHS WALK-IN CENTRE (WOKING) 90B 168
(within Woking Community Hospital)
Niagara Av. W5 49La 86
Niagara Cl. EN8: Chesh 1Zb 20
Niagara Cl. N1 1E 218 (40Sb 71)
Niagra Ct. SE16 48Yb 92
(off Canada Est.)
Nibthwaite Rd. HA1: Harr 29Ga 46
Nice Bus. Pk. SE15 51Xb 113
Nicholas Cl. RM15: S Ock 41Yd 98
Nicholas Cl. UB6: G'frd 40Da 65
Nicholas Cl. WD24: Wat 9X 13
Nicholas Ct. E13 41Kc 93
Nicholas Ct. N7 36Pb 70
Nicholas Ct. SE12 60Jc 115
Nicholas Ct. W4 51Ua 110
(off Corney Reach Way)

Nicholas Gdns. GU22: Pyr 88H 169
Nicholas Gdns. SL1: Slou 6C 80
Nicholas Gdns. W5 47Ma 87
Nicholas Hawksmoor Dr. WD6: Bore 12Ra 29
Nicholas Ho. AL4: St A 3H 7
Nicholas La. EC4 4G 225 (45Tb 91)
(not continuous)
Nicholas Lodge KT10: Esh 75Ca 151
Nicholas M. W4 51Ua 110
Nicholas Pas. EC4 4G 225 (45Tb 91)
(off Nicholas La.)
Nicholas Rd. CR0: Bedd 77Nb 156
Nicholas Rd. E1 42Yb 92
Nicholas Rd. RM8: Dag 33Bd 75
Nicholas Rd. W11 45Za 88
Nicholas Rd. WD6: E'tree 16Pa 29
Nicholas Stacey Ho. SE7 50Kc 93
(off Frank Burton Cl.)
Nicholas Way HA6: Nwood 25S 44
Nicholas Way HP2: Hem H 1P 3
Nicholay Rd. N19 32Mb 70
(not continuous)
Nichol Cl. N14 18Mb 32
Nicholes Rd. TW3: Houn 56Ca 107
Nichol La. BR1: Broml 66Jc 137
Nicholl Ho. N4 32Sb 71
Nicholl Rd. CM16: Epp 3Vc 23
Nicholls SL4: Wind 5A 102
Nicholls Av. UB8: Hil 42Q 84
Nicholls Cl. CR3: Cat'm 94Sb 197
Nichollsfield Wlk. N7 36Pb 70
Nicholls M. SW16 63Nb 134
Nicholls Point E15 39Jc 73
(off Park Gro.)
Nicholl St. E2 39Wb 71
Nicholls Wlk. SL4: Wind 5A 102
Nichols Cl. KT9: Chess 79La 152
Nichols Cl. N4 32Qb 70
(off Osborne Rd.)
Nichols Ct. E2 2K 219 (40Vb 71)
Nichols Grn. W5 43Na 87
Nicholson Ct. E17 28Zb 52
Nicholson Ct. N17 27Vb 51
Nicholson Ho. SE17 7F 231 (50Tb 91)
Nicholson M. KT1: King T 70Na 131
Nicholson M. TW20: Egh 64C 126
(off Station Rd.)
Nicholson Rd. CR0: C'don 74Vb 157
Nicholson Rd. TN14: Dun G 92Hd 202
Nicholson Sq. E3 41Ec 92
(off Bolinder Way)
Nicholson St. SE1 7B 224 (46Rb 91)
Nicholson Wlk. TW20: Egh 64C 126
Nicholson Wlk. UB10: Uxb 39N 63
Nickelby Apts. E15 37Fc 73
(off Grove Cres. Rd.)
Nickelby Cl. SE28 44Yc 95
Nickleby Cl. UB8: Hil 44R 84
Nickleby Ho. SE16 47Wb 91
(off Parkers Row)
Nickleby Ho. W11 46Za 88
(off St Ann's Rd.)
Nickleby Rd. DA12: Grav'nd 10J 123
Nickols Wlk. SW18 56Db 111
Nicky La. HP2: Hem H 1N 3
Nicola Cl. CR2: S Croy 79Sb 157
Nicola Cl. HA3: Hrw W 26Fa 46
Nicola M. IG6: Ilf 23Rc 54
Nicolas Wlk. RM16: Grays 7D 100
Nicola Ter. DA7: Bex 53Ad 117
Nicol Cl. TW1: Twick 58Ka 108
Nicoll Cir. NW7 23Ab 48
Nicoll Ct. N10 24Kb 50
Nicoll Ct. NW10 39Ua 68
Nicoll Pl. NW4 30Xa 48
Nicoll Rd. NW10 39Ua 68
Nicoll Way WD6: Bore 15Ta 29
Nicolson NW9 25Ua 48
Nicolson Dr. WD23: B Hea 18Ea 28
Nicolson Rd. BR5: Orp 73Zc 161
Nicolson Way TN13: S'oaks 94Md 203
Nicosia Rd. SW18 59Gb 111
Niederwald Rd. SE26 63Ac 136
Nield Rd. UB3: Hayes 47V 84
Nigel Cl. UB5: N'olt 39Aa 65
Nigel Ct. N3 24Db 49
Nigel Fisher Way KT9: Chess 80La 152
Nigel Ho. EC1 7K 217 (43Qb 90)
(off Portpool La.)
Nigel M. IG1: Ilf 35Rc 74
Nigel Playfair Av. W6 49Xa 88
Nigel Rd. E7 36Lc 73
Nigel Rd. SE15 55Wb 113
Nigeria Rd. SE7 52Lc 115
Nighthawk NW9 25Ua 48
Nightingale Av. E4 22Gc 53
Nightingale Av. HA1: Harr 31Ka 66
Nightingale Av. KT24: W Hor 96T 190
Nightingale Av. RM14: Upm 32Vd 78
Nightingale Cl. DA11: Nflt 2A 144
Nightingale Cl. E4 21Fc 53
Nightingale Cl. HA5: Eastc 29Y 45
Nightingale Cl. KT11: Cobh 83Z 171
Nightingale Cl. KT19: Eps 80Qa 173
Nightingale Cl. SM5: Cars 75Jb 156
Nightingale Cl. TN16: Big H 87Lc 179
Nightingale Cl. W4 51Sa 109
Nightingale Cl. WD5: Ab L 3W 12
Nightingale Cl. WD7: R'lett 8Ha 14
Nightingale Cnr. BR5: St M Cry 70Zc 139
Nightingale Ct. BR2: Broml 68Gc 137
Nightingale Ct. E14 47Ec 92
(off Ovex Cl.)
Nightingale Ct. GU21: Wok 10J 167
Nightingale Ct. HA1: Harr 30Ha 46
Nightingale Ct. N4 33Pb 70
(off Tollington Pk.)
Nightingale Ct. RH1: Redh 5A 208
(off St Anne's Mt.)
Nightingale Ct. SL1: Slou 8L 81
Nightingale Ct. SM1: Sutt 78Eb 155
Nightingale Ct. WD3: Rick 17L 25
Nightingale Ct. WD7: R'lett 7Ja 14
Nightingale Cres. KT24: W Hor 97S 190
Nightingale Cres. RM3: Hrld W 26Nd 57
Nightingale Dr. KT19: Ewe 79Ra 153
Nightingale Gro. DA1: Dart 56Pd 119
Nightingale Gro. SE13 57Fc 115
Nightingale Hgts. SE18 51Rc 116
NIGHTINGALE HOSPITAL 7E 214 (43Gb 89)
Nightingale Ho. KT17: Eps East St. 84Ua 174
Nightingale Ho. BR8: Swan 69Gd 140
(off London Rd.)
Nightingale Ho. E1 46Wb 91
(off Thomas More St.)
Nightingale Ho. E2 1J 219 (39Ub 71)
(off Kingsland Rd.)
Nightingale Ho. KT17: Eps 84Ua 174
(off Winter Cl.)
Nightingale Ho. NW8 6D 214 (42Gb 89)
(off Samford St.)
Nightingale Ho. SE18 50Qc 94
(off Connaught M.)
Nightingale Ho. UB7: W Dray 47P 83
Nightingale Ho. W12 44Ya 88
(off Du Cane Rd.)
Nightingale La. AL1: St A 5G 6
Nightingale La. AL4: St A 6G 6
Nightingale La. BR1: Broml 68Lc 137
Nightingale La. E11 28Kc 53
Nightingale La. N8 28Nb 50
Nightingale La. SW12 59Hb 111
Nightingale La. SW4 58Kb 112
Nightingale La. TW10: Rich 59Na 109
Nightingale Lodge W9 43Cb 89
(off Admiral Wlk.)
Nightingale M. E11 29Jc 53
Nightingale M. E3 40Zb 72
Nightingale M. KT1: King T 69Ma 131
(off South La.)
Nightingale M. SE11 5B 230 (49Rb 91)
Nightingale Pk. SL2: Farn C 8D 60
Nightingale Pl. SE18 51Qc 116
Nightingale Pl. SW10 51Eb 111
Nightingale Pl. WD3: Rick 17M 25
Nightingale Ri. KT15: Add 76N 149
Nightingale Rd. BR5: Pet W 72Sc 160
Nightingale Rd. CR2: Sels 83Zb 178
Nightingale Rd. E5 34Xb 71
Nightingale Rd. KT10: Esh 78Ba 151
Nightingale Rd. KT12: Walt T 73Y 151
Nightingale Rd. KT24: E Hor 97V 190
Nightingale Rd. KT8: W Mole 71Da 151
Nightingale Rd. N1 37Sb 71
Nightingale Rd. N22 25Nb 50
Nightingale Rd. N9 16Yb 34
Nightingale Rd. NW10 40Va 68
Nightingale Rd. SM5: Cars 76Hb 155
Nightingale Rd. TN15: Kems'g 89Md 183
Nightingale Rd. TW12: Hamp 64Ca 129
Nightingale Rd. W7 46Ha 86
Nightingale Rd. WD23: Bush 15Ca 27
Nightingale Rd. WD3: Rick 17L 25
Nightingales EN9: Walt A 6Gc 21
Nightingales, The TW19: Stanw 60P 105
Nightingale Shott TW20: Egh 65B 126
Nightingale Sq. SW12 59Jb 112
Nightingale Va. SE18 51Qc 116
Nightingale Wlk. N1 37Sb 71
Nightingale Wlk. SL4: Wind 5G 102
Nightingale Wlk. SW4 58Kb 112
Nightingale Way BR8: Swan 69Gd 140
Nightingale Way E6 43Nc 94
Nightingale Way RH1: Blet 6L 209
Nightingale Way UB9: Den 31H 63
Nihill Pl. CR0: C'don 74Vb 157
Nile Cl. N16 34Vb 71
Nile Dr. N9 19Yb 34
Nile Ho. N1 3F 219 (41Tb 91)
(off Nile St.)
Nile Path SE18 51Qc 116
Nile Rd. E13 40Lc 73
Nile St. N1 3E 218 (41Sb 91)
Nile Ter. SE15 7K 231 (50Vb 91)
Nimbus Rd. KT19: Eps 82Ta 173
Nimegen Way SE22 57Ub 113
Nimmo Dr. WD23: B Hea 17Fa 28
Nimrod NW9 25Ua 48
Nimrod Cl. UB5: N'olt 41Z 85
Nimrod Ho. E16 43Kc 93
(off Vanguard Cl.)
Nimrod Pas. N1 37Ub 71
Nimrod Rd. SW16 65Kb 134
Nina Mackay Cl. E15 39Gc 73
Nine Acre La. AL10: Hat 1B 8
Nine Acres SL1: Slou 6D 80
Nine Acres Cl. E12 36Nc 74
Nine Acres Cl. UB3: Harl 48S 84
Nineacres Way CR5: Coul 88Nb 176
Ninedells Pl. AL1: St A 2D 6
NINE ELMS 52Lb 112
Nine Elms Av. UB8: Cowl 43M 83
Nine Elms Cl. TW14: Felt 60V 106
Nine Elms Cl. UB8: Cowl 44M 83
Nine Elms Gro. DA11: Grav'nd 9C 122
Nine Elms La. SW11 52Lb 112
Nine Elms La. SW8 51Mb 112
Ninefields EN9: Walt A 5Hc 21
Ninehams Cl. CR3: Cat'm 92Tb 197
Ninehams Gdns. CR3: Cat'm 92Tb 197
Ninehams Rd. CR3: Cat'm 93Tb 197
Ninehams Rd. TN16: Tats 93Lc 199
Nine Stiles Cl. UB9: Den 37K 63
Nineteenth Rd. CR4: Mitc 70Nb 134
Ninhams Wood BR6: Farnb 77Qc 160
Ninnings Rd. SL9: Chal P 24B 42
Ninnings Way SL9: Chal P 24B 42
Ninth Av. KT20: Lwr K 97Bb 195
Ninth Av. UB3: Hayes 46W 84
Nipper All. KT1: King T 68Na 131
(off Clarence St.)
Nipponzan Myohoji Peace Pagoda 51Hb 111
Nisbet Ho. E9 36Zb 72
Nisbett Dr. TN15: Bor G 93Ae 205
Nisbett Wlk. DA14: Sidc 63Wc 139
Nita Ct. SE12 60Jc 115
Nita Rd. CM14: W'ley 22Yd 58
Nithdale Rd. SE18 52Rc 116
Nithsdale Gro. UB10: Ick 34S 64
Niton Cl. EN5: Barn 16Za 30
Niton Ct. SS17: Stan H 3L 101
Niton Rd. TW9: Rich 55Qa 109
Niton St. SW6 52Za 110
Niveda Cl. W12 47Wa 88
Niven Cl. WD6: Bore 11Sa 29
Nixey Cl. SL1: Slou 7L 81
Nizam Ct. EN3: Enf W 10Yb 20
Noah's Ark TN15: Kems'g 90Rd 183
NOAH'S ARK 91Rd 203
NOAH'S ARK CHILDREN'S HOSPICE 14Bb 31
Noah's Yd. N1 2G 217 (40Nb 70)
Noakes Ind. Site RM13: Wenn 45Pd 97
NOAK HILL 19Nd 39
Noak Hill Rd. RM3: Rom 22Kd 57
Nobel Cl. NW9 26Ta 47
Nobel Dr. UB3: Harl 53T 106
Nobel Ho. RH1: Redh 5P 207
Nobel Ho. SE5 54Sb 113
Nobel Rd. N18 21Yb 52
Nobel Vs. EN9: Walt A 6Fc 21
Noble Cnr. TW5: Hest 53Ca 107
Noble Ct. CR4: Mitc 68Fb 133
Noble Ct. E1 45Xb 91
Noble Ct. SL2: Slou 6K 81
(off Mill St.)
Noblefield Hgts. N2 29Gb 49
Noble M. N16 34Tb 71
(off Albion Rd.)
Noble St. EC2 2D 224 (44Sb 91)
Noble St. KT12: Walt T 76Y 151
Nobles Way TW20: Egh 66A 126
Noble Yd. N1 1B 218 (39Rb 71)
(off Camden Pas.)
Nocavia Ho. SW6 54Eb 111
(off Townmead Rd.)
Noel NW9 25Ua 48
Noel Ct. CR0: C'don 73Tb 157
Noel Ct. TW4: Houn 55Ba 107
Noel Coward Av. UB9: Den 29J 43
Noel Coward Ho. SW1 6C 228 (49Lb 90)
(off Vauxhall Bri. Rd.)
Noel Coward Theatre 4F 223 (45Nb 90)
(off St Martin's La.)
Noel Ho. NW6 38Fb 69
(off Harben Rd.)
NOEL PARK 26Rb 51
Noel Pk. Rd. N22 26Qb 50
Noel Rd. E6 42Nc 94
Noel Rd. N1 1B 218 (40Rb 71)
Noel Rd. W3 45Qa 87
Noel Sq. RM8: Dag 35Yc 75
Noel Sq. TW11: Tedd 64Ha 130
Noel St. W1 3C 222 (44Lb 90)
Noel Ter. DA14: Sidc 63Xc 139
Noel Ter. SE23 61Yb 136
Noke Dr. RH1: Redh 5A 208
Noke La. AL2: Chis G 8L 5
Noke La. Bus. Cen. AL2: Chis G 9M 5
Noke Side AL2: Chis G 9N 5
Noko W10 41Za 88
Nolands Cl. RM5: Col R 26Ed 56
Nolan Mans. E20 36Dc 72
(off Honour Lea Av.)
Nolan Path WD6: Bore 11Pa 29
(off Bennington Dr.)
Nolan Way E5 35Wb 71
Noll Ho. N7 33Pb 70
(off Tomlins Wlk.)
Nolton Pl. HA8: Edg 25Pa 47
Nomad Theatre 100V 190
Nonsuch Abbeyfield KT17: Ewe 81Va 174
Nonsuch Cl. IG6: Ilf 23Rc 54
Nonsuch Ct. Av. KT17: Ewe 82Xa 174
Nonsuch Ho. SW19 67Fb 133
(off Chapter Way)
Nonsuch Ind. Est. KT17: Eps 83Ua 174
Nonsuch Pl. SM3: Cheam 80Za 154
(off Ewell Rd.)
Nonsuch Wlk. SM2: Cheam 82Ya 174
(not continuous)
Nook Apts. E1 44Vb 91
(off Scarborough St.)
Nora Gdns. NW4 28Za 48
Nora Leverton Ct. NW1 38Lb 70
(off Randolph St.)
NORBITON 68Qa 131
Norbiton Av. KT1: King T 67Qa 131
Norbiton Comn. Rd. KT1: King T 69Ra 131
Norbiton Hall KT2: King T 68Pa 131
Norbiton Ho. NW1 1C 216 (39Lb 70)
(off Camden St.)
Norbiton Rd. E14 44Bc 92
Norbiton Station (Rail) 67Qa 131
Norbreck Gdns. NW10 41Pa 87
Norbreck Pde. NW10 41Na 87
Norbroke St. W12 45Va 88
Norburn St. W10 43Ab 88
NORBURY 67Pb 134
Norbury Av. CR7: Thor H 67Qb 134
Norbury Av. SW16 67Pb 134
Norbury Av. TW3: Houn 56Fa 108
Norbury Av. WD24: Wat 11Y 27
Norbury Cl. SW16 67Qb 134
Norbury Ct. Rd. SW16 69Nb 134
Norbury Cres. SW16 67Pb 134
Norbury Cross SW16 69Nb 134
Norbury Gdns. RM6: Chad H 29Zc 55
Norbury Gro. NW7 20Ua 30
Norbury Hill SW16 66Qb 134
NORBURY PARK 98Ka 192
Norbury Pl. KT22: Fet 94Ga 192
Norbury Ri. SW16 69Nb 134
Norbury Rd. CR7: Thor H 68Sb 135
Norbury Rd. E4 22Cc 52
Norbury Rd. RH2: Reig 6H 207
Norbury Rd. TW13: Felt 62Y 128
Norbury Station (Rail) 67Pb 134
Norbury Trad. Est. SW16 68Pb 134
Norbury Way KT23: Bookh 97Ea 192
Norcombe Gdns. HA3: Kenton 30La 46
Norcombe Ho. N19 34Mb 70
(off Wedmore St.)
Norcott Cl. UB4: Yead 42Y 85
Norcott Rd. N16 33Wb 71
Norcroft Gdns. SE22 59Wb 113
Norcutt Rd. TW2: Twick 60Ga 108
Nordenfeldt Rd. DA8: Erith 50Fd 96
Norden Ho. E2 41Xb 91
(off Pott St.)
Nordmann Pl. RM15: S Ock 42Zd 99
Nore Hill Pinnacle 91Dc 198
Norelands Dr. SL1: Burn 10A 60
Norfield Rd. DA2: Wilm 63Ed 140
Norfolk Apts. E4 24Dc 52
Norfolk Av. CR2: Sande 82Vb 177
Norfolk Av. N13 23Rb 51
Norfolk Av. N15 30Vb 51
Norfolk Av. SL1: Slou 3G 80
Norfolk Av. WD24: Wat 10Y 13
Norfolk Cl. DA1: Dart 58Qd 119
Norfolk Cl. EN4: Cockf 14Jb 32
Norfolk Cl. N13 23Rb 51
Norfolk Cl. N2 27Gb 49
Norfolk Cl. SL0: Iver 44F 82
Norfolk Cl. TW1: Twick 58Ka 108
Norfolk Cotts. RH1: S Nut 8E 208
Norfolk Ct. EN5: Barn 14Ab 30
Norfolk Ct. RM6: Chad H 29Xc 55
(off Norwich Cres.)
Norfolk Cres. DA15: Sidc 59Uc 116
Norfolk Cres. W2 2D 220 (44Gb 89)
Norfolk Farm Cl. GU22: Pyr 88F 168
Norfolk Farm Rd. GU22: Pyr 87F 168
Norfolk Gdns. DA7: Bex 53Bd 117
Norfolk Gdns. WD6: Bore 14Ta 29
Norfolk Ho. BR2: Broml 70Hc 137
(off Westmoreland Rd.)
Norfolk Ho. EC4 4D 224 (45Sb 91)
(off Trig La.)
Norfolk Ho. SE20 67Yb 136
Norfolk Ho. SE8 53Cc 114
(off Brookmill Rd.)
Norfolk Ho. SL2: Farn C 4G 60
Norfolk Ho. SW1 5E 228 (49Mb 90)
(off Page St.)
Norfolk Ho. Rd. SW16 62Mb 134
Norfolk Mans. SW11 53Hb 111
(off Prince of Wales Dr.)
Norfolk M. W10 43Ab 88
(off Blagrove Rd.)
Norfolk Pl. DA16: Well 54Wc 117
Norfolk Pl. RM16: Chaf H 50Yd 98
Norfolk Pl. W2 2C 220 (44Fb 89)
(not continuous)
Norfolk Rd. CR7: Thor H 69Sb 135
Norfolk Rd. DA12: Grav'nd 8F 122
(not continuous)
Norfolk Rd. E17 26Zb 52
Norfolk Rd. E6 39Pc 74
Norfolk Rd. EN3: Pond E 16Xb 33
Norfolk Rd. EN5: New Bar 13Cb 31
Norfolk Rd. HA1: Harr 29Da 45
Norfolk Rd. IG11: Bark 38Uc 74
Norfolk Rd. IG3: Ilf 32Uc 74
Norfolk Rd. KT10: Clay 78Ga 152
Norfolk Rd. NW10 38Ua 68
Norfolk Rd. NW8 1C 214 (39Fb 69)
Norfolk Rd. RM10: Dag 36Dd 76
Norfolk Rd. RM14: Upm 34Qd 77
Norfolk Rd. RM7: Rom 30Ed 56
Norfolk Rd. SW19 66Gb 133
Norfolk Rd. TW13: Felt 60Y 107
Norfolk Rd. UB8: Uxb 37M 63
Norfolk Rd. WD3: Rick 18N 25
Norfolk Row SE1 5H 229 (49Pb 90)
Norfolk Sq. W2 3C 220 (44Fb 89)
Norfolk Sq. M. W2 3C 220 (44Fb 89)
(off London St.)
Norfolk St. E7 36Jc 73
Norfolk Ter. W6 50Ab 88
Norgrove Pk. SL9: Ger X 28A 42
Norgrove St. SW12 59Jb 112
Norham Ct. DA2: Dart 58Rd 119
(off Osbourne Rd.)
Norheads La. CR6: W'ham 91Jc 199
Norheads La. TN16: Big H 90Jc 179
Norhyrst Av. SE25 69Vb 135
NORK 87Za 174
Nork Gdns. SM7: Bans 86Ab 174
Nork Ri. SM7: Bans 88Za 174
Nork Way SM7: Bans 88Ya 174
Norland Ho. W11 46Za 88
(off Queensdale Cres.)
Norland Pl. W11 46Ab 88
Norland Rd. W11 46Za 88
(not continuous)
Norlands Cres. BR7: Chst 67Rc 138
Norlands Ga. BR7: Chst 67Rc 138
Norlands La. TW20: Thorpe 69G 126
Norland Sq. W11 46Ab 88
Norland Sq. Mans. W11 46Ab 88
(off Norland Sq.)
Norlem Ct. SE8 49Ac 92
(off Seafarer Way)
Norley Va. SW15 60Wa 110
Norlington Rd. E10 32Ec 72
Norlington Rd. E11 32Fc 73
Norman Av. CR2: Sande 82Sb 177
Norman Av. KT17: Eps 84Va 174
Norman Av. N22 25Rb 51
Norman Av. TW1: Twick 59La 108
Norman Av. TW13: Hanw 61Aa 129
Norman Av. UB1: S'hall 45Aa 85
Norman Butler Ho. W10 42Ab 88
(off Ladbroke Gro.)
Normanby Cl. SW15 57Bb 111
Normanby Rd. NW10 35Va 68
Norman Cl. AL1: St A 5C 6
Norman Cl. BR6: Farnb 76Sc 160
Norman Cl. EN9: Walt A 5Fc 21
Norman Cl. KT18: Tatt C 91Xa 194
Norman Cl. N22 25Sb 51
Norman Cl. RM5: Col R 25Dd 56
Norman Cl. TN15: Kems'g 89Md 183
Norman Colyer Ct. KT19: Eps 82Ta 173
Norman Ct. EN6: Pot B 3Eb 17
Norman Ct. IG2: Ilf 31Tc 74
Norman Ct. N3 25Cb 49
(off Nether St.)
Norman Ct. N4 31Qb 70
Norman Ct. NW10 38Wa 68
Norman Ct. W13 46Ka 86
(off Kirkfield Cl.)
Norman Cres. CM13: B'wood 20Ce 41
Norman Cres. HA5: Pinn 25Y 45
Norman Cres. TW5: Hest 52Z 107
Normand Gdns. W14 51Ab 110
(off Greyhound Rd.)
Normand Mans. W14 51Ab 110
(off Normand M.)
Normand M. W14 51Ab 110
Normand Rd. W14 51Bb 111
Normandy Av. EN5: Barn 15Bb 31
Normandy Bus. Pk. GU3: Norm 10C 186
Normandy Cl. SE26 62Ac 136
Normandy Ct. HP2: Hem H 1M 3
Normandy Dr. UB3: Hayes 44S 84
Normandy Ho. E14 47Ec 92
(off Plevna St.)
Normandy Ho. EN2: Enf 10Sb 19
Normandy Pl. W12 46Za 88
Normandy Rd. AL3: St A 1B 6
Normandy Rd. SW9 53Qb 112
Normandy Ter. E16 44Kc 93
Normandy Wlk. TW20: Egh 64E 126
Normandy Way DA8: Erith 53Gd 118
Norman Gro. E3 40Ac 72
Norman Hay Trad. Est., The UB7: Sip 52P 105
Norman Ho. SE1 3J 231 (48Ub 91)
(off Riley Rd.)
Norman Ho. SW8 52Nb 112
(off Wyvil Rd.)
Norman Ho. TW13: Hanw 61Ba 129
(off Watermill Way)
Normanhurst CM13: Hut 16Ee 41
Normanhurst TW15: Ashf 64Q 128
Normanhurst Av. DA7: Bex 53Zc 117
Normanhurst Dr. TW1: Twick 57Ja 108
Normanhurst Rd. BR5: St P 68Xc 139
Normanhurst Rd. KT12: Walt T 75Z 151
Normanhurst Rd. SW2 61Pb 134
Normanhurst Rd. TN15: Bor G 92Ce 205
Norman Leddy Memorial Gdns. 44Qb 90
Norman Pk. Athletics Track 72Kc 159
Norman Rd. CR7: Thor H 71Rb 157
Norman Rd. DA1: Dart 60Nd 119
Norman Rd. DA17: Belv 48Dd 96
(not continuous)
Norman Rd. E11 33Fc 73
Norman Rd. E6 42Pc 94
Norman Rd. IG1: Ilf 36Rc 74
Norman Rd. N15 29Vb 51
Norman Rd. RM11: Horn 31Jd 76
Norman Rd. SE10 52Dc 114
Norman Rd. SM1: Sutt 78Cb 155
Norman Rd. SW19 66Eb 133
Norman Rd. TW15: Ashf 65T 128
Normans, The SL2: Slou 4M 81
Norman's Cl. DA11: Grav'nd 9C 122
Norman's Cl. NW10 37Ta 67
Normans Cl. UB8: Hil 43P 83
Normansfield Av. TW11: Tedd 66La 130
Normansfield Cl. WD23: Bush 17Da 27
Normanshire Dr. E4 21Cc 52
Norman's Mead NW10 37Ta 67
Norman St. EC1 4D 218 (41Sb 91)
Norman Ter. NW6 36Bb 69
Normanton Av. SW19 61Cb 133
Normanton Ct. CR2: S Croy 78Ub 157
(off Croham Rd.)
Normanton Pk. E4 19Gc 35
Normanton Rd. CR2: S Croy 78Ub 157
Normanton St. SE23 61Zb 136
Norman Way N14 19Nb 32
Norman Way W3 43Ra 87
Normington Cl. SW16 64Qb 134
Norrels Dr. KT24: E Hor 98V 190
Norrels Ride KT24: E Hor 97V 190
Norrice Lea N2 29Fb 49
Norris NW9 25Va 48
(off Withers Mead)
Norris Cl. AL2: Lon C 8F 6
Norris Cl. KT19: Eps 83Ra 173
Norris Ho. E9 39Yb 72
(off Handley Rd.)
Norris Ho. N1 39Ub 71
(off Colville Est.)
Norris Ho. SE8 50Bc 92
(off Grove St.)
Norris Ho. TW7: Isle 54Ja 108
Norris Rd. TW18: Staines 63H 127
Norris St. SW1 5D 222 (45Mb 90)
Norris Way DA1: Cray 55Hd 118
Norroy Rd. SW15 56Za 110
Norry's Cl. EN4: Cockf 14Hb 31
Norry's Rd. EN4: Cockf 14Hb 31
Norseman Cl. IG3: Ilf 32Xc 75
Norseman Way UB6: G'frd 39Da 65
Norstead Pl. SW15 61Wa 132
Norsted La. BR6: Prat B 84Wc 181
North Access Rd. E17 30Zb 52
North Acre NW9 25Ua 48
North Acre SM7: Bans 88Bb 175
NORTH ACTON 42Ta 87
North Acton Bus. Pk. W3 43Ta 87
North Acton Rd. NW10 40Ta 67
North Acton Station (Underground) 43Ta 87
Northallerton Way RM3: Rom 22Md 57
Northall Rd. DA7: Bex 54Ed 118
Northampton Av. SL1: Slou 4G 80
Northampton Gro. N1 36Tb 71
Northampton Ho. RM3: Rom 21Nd 57
(off Broseley Rd.)
Northampton Pk. N1 37Sb 71
Northampton Pl. SL1: Slou 4G 80
(off Northampton Av.)
Northampton Rd. CR0: C'don 75Wb 157
Northampton Rd. EC1 5A 218 (42Qb 90)
Northampton Rd. EN3: Pond E 14Ac 34
Northampton Row EC1 5A 218 (42Qb 90)
(off Rosoman Pl.)
Northampton Sq. EC1 4B 218 (41Rb 91)
Northampton St. N1 38Sb 71
Northanger Rd. SW16 65Nb 134
North App. HA6: Nwood 19S 26
North App. WD25: Wat 7V 12
North Ash Rd. DA3: Nw A G 76Ae 165
North Audley St. W1 4H 221 (45Jb 90)
North Av. HA2: Harr 30Da 45
North Av. KT12: W Vill 81U 170
North Av. N18 21Wb 51
North Av. SM5: Cars 80Jb 156
North Av. TW9: Kew 53Qa 109
North Av. UB1: S'hall 45Ba 85
North Av. UB3: Hayes 45W 84
North Av. W13 43Ka 86
North Av. WD7: Shenl 4Na 15
NORTHAW 2Hb 17
Northaw Ho. W10 42Ya 88
(off Sutton Way)
NORTHAW PARK 4Hb 17
Northaw Pl. EN6: N'thaw 2Fb 17
Northaw Rd. E. EN6: Cuff 3Mb 18
Northaw Rd. W. EN6: N'thaw 2Hb 17
North Bank NW8 4D 214 (41Gb 89)
Northbank Rd. E17 26Ec 52
NORTH BECKTON 43Pc 94
North Birkbeck Rd. E11 34Fc 73
North Block RM2: Rom 27Md 57
North Block SE1 1J 229 (47Pb 90)
(off Chicheley St.)
Northborough Rd. SL2: Slou 2E 80
Northborough Rd. SW16 69Mb 134
Northbourne BR2: Hayes 73Jc 159
Northbourne Rd. SW4 57Mb 112
Northbrook Dr. HA6: Nwood 25U 44
Northbrook Rd. CR0: C'don 71Tb 157
Northbrook Rd. EN5: Barn 16Ab 30
Northbrook Rd. IG1: Ilf 33Qc 74
Northbrook Rd. N22 24Nb 50
Northbrook Rd. SE13 57Gc 115
Northburgh St. EC1 6C 218 (42Rb 91)
North Burnham Cl. SL1: Burn 10A 60
Northbury Cl. IG11: Bark 38Sc 74
North Carriage Dr. W2 4E 220 (45Gb 89)
(off Bayswater Rd.)
NORTH CHEAM 76Ya 154
Northchurch SE17 7G 231 (50Tb 91)
(not continuous)
Northchurch Ho. E2 39Wb 71
(off Whiston Rd.)
Northchurch Rd. HA9: Wemb 37Qa 67
Northchurch Rd. N1 38Tb 71
(not continuous)
Northchurch Ter. N1 38Ub 71
North Circular Rd. E12 34Qc 74
North Circular Rd. E18 26Lc 53
North Circular Rd. E4 23Bc 52
North Circular Rd. IG1: Ilf 33Pc 74
North Circular Rd. IG11: Bark 36Qc 74
North Circular Rd. N12 25Fb 49
North Circular Rd. N13 22Qb 50
North Circular Rd. N3 27Cb 49
North Circular Rd. NW10 40Pa 67
North Circular Rd. NW11 31Za 68
North Circular Rd. NW2 34Ua 68
North Circular Rd. NW4 31Ya 68
Northcliffe Cl. KT4: Wor Pk 76Ua 154
Northcliffe Dr. N20 18Bb 31
North Cl. AL2: Chis G 7P 5
North Cl. DA6: Bex 56Zc 117
North Cl. EN5: Barn 15Ya 30
North Cl. IG7: Chig 22Wc 55
North Cl. RM10: Dag 39Cd 76

North Cl. SL4: Wind ... 3D 102
North Cl. SM4: Mord ... 70Ab 132
North Cl. TW14: Bedf ... 58T 106
North Colonnade, The E14 ... 46Cc 92
... (not continuous)
North Comn. KT13: Weyb ... 77S 150
North Comn. Rd. UB8: Uxb ... 36M 63
North Comn. Rd. W5 ... 45Na 87
Northcote HA5: Pinn ... 26Y 45
Northcote KT15: Add ... 77M 149
Northcote Av. KT5: Surb ... 73Ra 153
Northcote Av. TW7: Isle ... 57Ja 108
Northcote Av. UB1: S'hall ... 45Aa 85
Northcote Av. W5 ... 45Na 87
Northcote Cl. KT24: W Hor ... 97S 190
Northcote Cres. KT24: W Hor ... 97S 190
Northcote Pk. KT22: Oxs ... 86Ea 172
Northcote Rd. CR0: C'don ... 72Tb 157
Northcote Rd. DA11: Grav'nd ... 10B 122
Northcote Rd. DA14: Sidc ... 63Uc 138
Northcote Rd. E17 ... 28Ac 52
Northcote Rd. KT24: W Hor ... 97S 190
Northcote Rd. KT3: N Mald ... 69Sa 131
Northcote Rd. NW10 ... 38Ua 68
Northcote Rd. SW11 ... 57Gb 111
Northcote Rd. TW1: Twick ... 57Ja 108
North Cotts. AL2: Lon C ... 7E 6
Northcott Av. N22 ... 25Nb 50
Northcotts Long Elms Cl. WD5: Ab L ... 5T 12
... (off Long Elms Cl.)
North Countess Rd. E17 ... 26Bc 52
North Ct. BR1: Broml ... 67Kc 137
... (off Palace Gro.)
North Ct. SE24 ... 55Rb 113
North Ct. SW1 ... 4F 229 (48Nb 90)
... (off Gt. Peter St.)
North Ct. W1 ... 7C 216 (43Lb 90)
Northcourt WD3: Rick ... 18J 25
NORTH CRAY ... 64Ad 139
North Cray Rd. DA14: Sidc ... 65Ad 139
North Cray Rd. DA5: Bexl ... 60Dd 118
North Cray Woods ... 63Zc 139
North Cres. E16 ... 42Fc 93
North Cres. N3 ... 26Bb 49
North Cres. WC1 ... 7D 216 (43Mb 90)
Northcroft SL2: Slou ... 2F 80
Northcroft Cl. TW20: Eng G ... 4M 125
Northcroft Ct. W12 ... 47Wa 88
Northcroft Gdns. TW20: Eng G ... 4M 125
Northcroft Rd. KT19: Ewe ... 80Ua 154
Northcroft Rd. TW20: Eng G ... 4M 125
Northcroft Rd. W13 ... 47Ka 86
North Crofts SE23 ... 60Xb 113
Northcroft Ter. W13 ... 47Ka 86
Northcroft Vs. TW20: Eng G ... 4M 125
North Cross Rd. IG6: Ilf ... 28Sc 54
North Cross Rd. SE22 ... 57Vb 113
Northdale Ct. SE25 ... 69Vb 135
North Dene IG7: Chig ... 22Tc 54
North Dene NW7 ... 20Ta 29
North Dene TW3: Houn ... 53Da 107
Northdene Gdns. N15 ... 30Vb 51
North Down CR2: Sande ... 83Ub 177
Northdown Cl. HA4: Ruis ... 34V 64
Northdown Cl. RH9: G'stone ... 2A 210
Northdown Gdns. IG2: Ilf ... 29Uc 54
Northdown Rd. AL10: Hat ... 3C 8
Northdown Rd. CR3: Wold ... 96Cc 198
Northdown Rd. DA16: Well ... 54Xc 117
Northdown Rd. DA3: Lfield ... 68Zd 143
Northdown Rd. RM11: Horn ... 31Kd 77
Northdown Rd. SL9: Chal P ... 23A 42
Northdown Rd. SM2: Sutt ... 82Cb 175
Northdown Rd. TN15: Kems'g ... 89Nd 183
North Downs Bus. Pk. TN13: Dun G ... 88Ed 182
North Downs Cres. CR0: New Ad ... 81Dc 178
North Downs Golf Course ... 97Cc 198
NORTH DOWNS PRIVATE HOSPITAL ... 97Vb 197
North Downs Rd. CR0: New Ad ... 82Dc 178
Northdown St. N1 ... 1G 217 (40Nb 70)
North Dr. BR3: Beck ... 70Dc 136
North Dr. BR6: Orp ... 77Uc 160
North Dr. GU24: Brkwd ... 3A 186
North Dr. GU25: Vir W ... 2J 147
North Dr. HA4: Ruis ... 31U 64
North Dr. RM2: Rom ... 27Ld 57
North Dr. SL2: Stoke P ... 1J 81
North Dr. SW16 ... 63Lb 134
North Dr. TW3: Houn ... 54Ea 108
North Dulwich Station (Rail) ... 57Tb 113
North Ealing Station (Underground) ... 44Pa 87
NORTH EAST LONDON NHS TREATMENT CEN. ... 28Wc 55
... (within King George Hospital)
North E. Surrey Crematorium ... 72Ya 154
NORTH END ... 53Hd 118
NORTH END ... 33Eb 69
Northend CM14: W'ley ... 22Yd 58
Northend HP3: Hem H ... 4B 4
North End CR0: C'don ... 75Sb 157
North End IG9: Buck H ... 17Lc 35
North End NW3 ... 33Eb 69
North End RM3: Rom ... 19Ld 39
North End Av. NW3 ... 33Eb 69
North End Cres. W14 ... 49Bb 89
North End Ho. W14 ... 49Ab 88
North End La. BR6: Downe ... 83Qc 180
North End La. SL5: S'dale ... 3F 146
North End Pde. W14 ... 49Ab 88
... (off North End Rd.)
Northend Pl. KT12: Hers ... 77Aa 151
Northend Rd. DA1: Erith ... 54Hd 118
Northend Rd. DA8: Erith ... 52Hd 118
North End Rd. HA9: Wemb ... 34Qa 67
North End Rd. NW11 ... 32Cb 69
North End Rd. SW6 ... 51Bb 111
North End Rd. W14 ... 49Ab 88
Northend Trad. Est. DA8: Erith ... 52Gd 118
North End Way NW3 ... 33Eb 69
Northern Av. N9 ... 19Ub 33
Northernhay Wlk. SM4: Mord ... 70Ab 132
Northern Hgts. N8 ... 31Mb 70
... (off Crescent Rd.)
Northern Perimeter Rd. TW6: H'row A ... 53R 106
Northern Perimeter Rd. (W.) TW6: H'row A ... 53M 105
Northern Pct. RM20: W Thur ... 49Vd 98
Northern Rd. E13 ... 40Kc 73
Northern Rd. SL2: Slou ... 2H 81
Northesk Ho. E1 ... 42Xb 91
... (off Tent St.)
Northey Av. SM2: Cheam ... 82Za 174
North Eyot Gdns. W6 ... 50Va 88
Northey St. E14 ... 45Ac 92
NORTH FELTHAM ... 58X 107
North Feltham Trad. Est. TW14: Felt ... 57X 107
Northfield DA3: Hartl ... 69Be 143
Northfield GU18: Light ... 3A 166
Northfield IG10: Lough ... 14Mc 35
Northfield Av. BR5: Orp ... 72Yc 161
Northfield Av. HA5: Pinn ... 28Z 45
Northfield Av. W13 ... 46Ka 86
Northfield Av. W5 ... 48La 86
Northfield Cl. BR1: Broml ... 67Nc 138
Northfield Cl. UB3: Harl ... 48V 84
Northfield Ct. TW18: Staines ... 67K 127
Northfield Cres. SM3: Cheam ... 77Ab 154
Northfield Farm M. KT11: Cobh ... 85W 170
Northfield Gdns. RM9: Dag ... 35Bd 75
Northfield Gdns. WD24: Wat ... 9Y 13
Northfield Ho. SE15 ... 51Wb 113
Northfield Pde. UB3: Harl ... 48U 84
Northfield Pk. UB3: Harl ... 48V 84
Northfield Path RM9: Dag ... 35Bd 75
Northfield Pl. KT13: Weyb ... 80R 150
Northfield Recreation Ground W5 ... 49Ka 86
Northfield Rd. E6 ... 38Pc 74
Northfield Rd. EN3: Pond E ... 15Xb 33
Northfield Rd. EN4: Cockf ... 13Gb 31
Northfield Rd. EN8: Walt C ... 4Ac 20
Northfield Rd. KT11: Cobh ... 85W 170
Northfield Rd. N16 ... 31Ub 71
Northfield Rd. RM9: Dag ... 35Bd 75
Northfield Rd. SL4: Eton W ... 9D 80
Northfield Rd. TW18: Staines ... 67K 127
Northfield Rd. TW5: Hest ... 51Z 107
Northfield Rd. W13 ... 47Ka 86
Northfield Rd. WD6: Bore ... 11Ra 29
Northfields KT17: Eps ... 83Ua 174
Northfields KT21: Asht ... 90Na 173
... (not continuous)
Northfields RM17: Grays ... 49Ee 99
Northfields SW18 ... 56Cb 111
NORTHFIELDS ... 48Ka 86
Northfields Ind. Est. HA0: Wemb ... 39Qa 67
Northfields Prospect Bus. Cen. SW18 ... 56Cb 111
Northfields Rd. W3 ... 43Ra 87
Northfields Station (Underground) ... 48La 86
NORTH FINCHLEY ... 22Eb 49
North Finchley Bus Station ... 22Eb 49
NORTHFLEET ... 58De 121
NORTHFLEET GREEN ... 64Ee 143
Northfleet Grn. Rd. DA13: Nflt G ... 65Ee 143
Northfleet Ho. SE1 ... 1F 231 (47Tb 91)
... (off Tennis St.)
Northfleet Ind. Est. DA11: Nflt ... 56Be 121
... (not continuous)
Northfleet Lodge GU22: Wok ... 91A 188
Northfleet Station (Rail) ... 58De 121
Northfleet Urban Country Pk. ... 60Fe 121
Northflock St. SE16 ... 47Wb 91
North Gdn. E14 ... 46Bc 92
North Gdns. SW19 ... 66Fb 133
North Ga. NW8 ... 2D 214 (40Gb 69)
... (off Prince Albert Rd.)
Northgate HA6: Nwood ... 24S 44
Northgate Bus. Cen. EN1: Enf ... 13Xb 33
Northgate Ct. SW9 ... 55Qb 112
Northgate Dr. NW9 ... 30Ua 48
Northgate Ho. E14 ... 45Cc 92
... (off E. India Dock Rd.)
Northgate Ho. EN8: Chesh ... 1Ac 20
... (off Turner's Hill)
Northgate Ind. Pk. RM5: Col R ... 25Bd 55
Northgate Path WD6: Bore ... 10Pa 15
Northgate Rd. IG11: Bark ... 41Yc 95
North Gates N12 ... 25Eb 49
... (off Bow La.)
North Glade, The DA5: Bexl ... 59Bd 117
North Gower St. NW1 ... 4C 216 (41Lb 90)
North Grn. NW9 ... 24Ua 48
North Grn. SL1: Slou ... 5J 81
North Greenwich Station (Underground) ... 47Gc 93
North Gro. KT16: Chert ... 72H 149
North Gro. N15 ... 29Tb 51
North Gro. N6 ... 31Jb 70
NORTH HARROW ... 29Da 45
North Harrow Station (Underground) ... 29Da 45
North Hatton Rd. TW6: H'row A ... 53T 106
North Hill N6 ... 30Hb 49
North Hill WD3: Chor ... 12G 24
North Hill Av. N6 ... 30Jb 50
North Hill Dr. RM3: Rom ... 20Md 39
North Hill Grn. RM3: Rom ... 21Md 57
NORTH HILLINGDON ... 38S 64
North Ho. SE8 ... 50Bc 92
North Hyde Gdns. UB3: Harl ... 49W 84
North Hyde Gdns. UB3: Hayes ... 49W 84
North Hyde La. TW5: Hest ... 50Aa 85
North Hyde La. UB2: S'hall ... 50Z 85
North Hyde Rd. UB3: Harl ... 48U 84
North Hyde Rd. UB3: Hayes ... 48U 84
North Hyde Wharf UB2: S'hall ... 49Y 85
Northiam N12 ... 21Cb 49
Northiam WC1 ... 4G 217 (41Nb 90)
... (off Cromer St.)
Northiam St. E9 ... 39Xb 71
Northington St. WC1 ... 6J 217 (42Pb 90)
NORTH KENSINGTON ... 43Ya 88
North Kent Av. DA11: Nflt ... 58Ee 121
Northlands EN6: Pot B ... 3Fb 17
Northlands Av. BR6: Orp ... 77Uc 160
Northlands St. SE5 ... 54Sb 113
North La. DA11: Grav'nd ... 4E 144
North La. TW11: Tedd ... 65Ha 130
North Lawns DA11: Nflt ... 58Ee 121
... (off Lawn Rd.)
Northleigh Ho. E3 ... 41Dc 92
... (off Powis Rd.)
North Lodge E16 ... 46Kc 93
... (off Wesley Av.)
North Lodge EN5: New Bar ... 15Eb 31
North Lodge Cl. SW15 ... 57Za 110
North London Bus. Pk. N11 ... 19Jb 32
NORTH LONDON CLINIC ... 19Wb 33
NORTH LONDON HOSPICE (FINCHLEY) ... 20Eb 31
NORTH LONDON HOSPICE (WINCHMORE HILL) ... 20Sb 33
NORTH LONDON PRIORY HOSPITAL ... 18Nb 32
NORTH LOOE ... 85Ya 174
North Mall N9 ... 19Xb 33
... (within Edmonton Grn. Shop. Cen.)
North Mall RM17: Grays ... 51De 121
... (off Grays Shop. Cen.)
North Mall SW18 ... 57Db 111
... (off Southside Shop. Cen.)
North Mall TW18: Staines ... 63H 127
... (within The Elmsleigh Cen.)
Northmead RH1: Redh ... 3P 207
Northmead Rd. SL2: Slou ... 2D 80
North M. WC1 ... 6J 217 (42Pb 90)
North Middlesex Golf Course ... 20Fb 31
NORTH MIDDLESEX UNIVERSITY HOSPITAL ... 22Ub 51
North Mill Apts. E8 ... 39Vb 71
... (off Lovelace St.)
Northmoor Hill Wood Nature Reserve ... 28H 43
North Mt. N20 ... 19Eb 31
... (off High Rd.)
NORTH MYMMS ... 7C 8
NORTH OCKENDON ... 37Xd 78
Northolm HA8: Edg ... 21Ta 47
Northolme Cl. RM16: Grays ... 48Ee 99
Northolme Gdns. HA8: Edg ... 25Qa 47
Northolme Ri. BR6: Orp ... 75Uc 160
Northolme Rd. N5 ... 35Sb 71
Northolt N17 ... 26Ub 51
... (off Griffin Rd.)
NORTHOLT ... 38Ca 65
Northolt Av. HA4: Ruis ... 36X 65
Northolt Gdns. UB6: G'frd ... 36Ha 66
Northolt Golf Course ... 40Aa 65
Northolt Leisure Cen. ... 37Ca 65
Northolt Park Station (Rail) ... 35Da 65
Northolt Rd. HA2: Harr ... 35Da 65
Northolt Rd. TW6: H'row A ... 53N 105
... (not continuous)
Northolt Station (Underground) ... 37Ca 65
Northolt Trad. Est. UB5: N'olt ... 38Da 65
Northolt Way RM12: Horn ... 37Ld 77
North Orbital Commercial Pk. AL1: St A ... 6E 6
North Orbital Rd. AL1: St A ... 6B 6
North Orbital Rd. AL2: Brick W ... 2Z 13
North Orbital Rd. AL2: Chis G ... 2Z 13
North Orbital Rd. AL2: St A ... 7C 6
North Orbital Rd. AL4: S'ford ... 6H 7
North Orbital Rd. AL4: St A ... 6H 7
North Orbital Rd. UB9: Den ... 29J 43
North Orbital Rd. WD3: Map C ... 22G 42
North Orbital Rd. WD3: W Hyd ... 25G 42
Northover BR1: Broml ... 62Hc 137
North Pde. HA8: Edg ... 26Qa 47
North Pde. KT9: Chess ... 78Pa 153
North Pde. UB1: S'hall ... 44Ca 85
... (off North Rd.)
North Pk. SE9 ... 58Pc 116
North Pk. SL0: Rich P ... 48E 82
North Pk. SL9: Chal P ... 27A 42
North Pk. SL9: Ger X ... 27A 42
North Pk. La. RH9: G'stone ... 1N 209
North Pas. SW18 ... 57Cb 111
North Pl. CR4: Mitc ... 66Hb 133
North Pl. EN9: Walt A ... 5Dc 20
North Pl. TW11: Tedd ... 65Ha 130
North Point N8 ... 29Pb 50
Northpoint Cl. SM1: Sutt ... 76Eb 155
Northpoint Ho. N1 ... 37Tb 71
... (off Essex Rd.)
Northpoint Sq. NW1 ... 37Mb 70
North Pole La. BR2: Kes ... 79Hc 159
North Pole Rd. W10 ... 43Ya 88
Northport St. N1 ... 1G 219 (39Tb 71)
North Quay Pl. E14 ... 45Dc 92
North Ride W2 ... 5D 220 (45Gb 89)
Northridge Rd. DA12: Grav'nd ... 2E 144
Northridge Way HP1: Hem H ... 3H 3
North Riding AL2: Brick W ... 2Ca 13
North Riding DA3: Lfield ... 69Fe 143
North Ri. W2 ... 3E 220 (44Gb 89)
North Rd. BR1: Broml ... 67Kc 137
North Rd. BR4: W W'ck ... 74Dc 158
North Rd. CM14: B'wood ... 18Yd 40
North Rd. DA1: Dart ... 58Hd 118
North Rd. DA17: Belv ... 48Dd 96
North Rd. EN8: Walt C ... 5Ac 20
North Rd. GU21: Wok ... 88C 168
North Rd. HA1: Harr ... 31Ja 66
North Rd. HA8: Edg ... 25Ra 47
North Rd. IG3: Ilf ... 33Uc 74
North Rd. KT12: Hers ... 78Y 151
North Rd. KT6: Surb ... 72Ma 153
North Rd. N6 ... 31Jb 70
North Rd. N7 ... 37Nb 70
North Rd. N9 ... 18Xb 33
North Rd. RH2: Reig ... 9H 207
North Rd. RM15: S Ock ... 38Yd 78
North Rd. RM15: Upm ... 38Yd 78
North Rd. RM19: Purf ... 49Sd 98
North Rd. RM4: Have B ... 20Gd 38
North Rd. RM6: Chad H ... 29Ad 55
North Rd. SE18 ... 49Uc 94
North Rd. SW19 ... 65Eb 133
North Rd. TW14: Bedf ... 58T 106
North Rd. TW8: Bford ... 51Na 109
North Rd. TW9: Kew ... 55Qa 109
North Rd. TW9: Rich ... 55Qa 109
North Rd. UB1: S'hall ... 44Ca 85
North Rd. UB3: Hayes ... 43T 84
North Rd. UB7: W Dray ... 48P 83
North Rd. W5 ... 48Ma 87
North Rd. WD3: Chor ... 15F 24
North Rd. Av. CM14: B'wood ... 18Yd 40
Northrop Rd. TW6: H'row A ... 53U 106
North Row SL3: Ful ... 35A 62
North Row W1 ... 4G 221 (45Hb 89)
North Row Bldgs. W1 ... 4H 221 (45Jb 90)
... (off North Row)
North Service Rd. CM14: B'wood ... 19Yd 40
North Several SE3 ... 54Fc 115
NORTH SHEEN ... 55Qa 109
North Sheen Station (Rail) ... 56Qa 109
North Side EN9: Walt A ... 2Kc 21
Northside Rd. BR1: Broml ... 67Jc 137
Northside Studios E8 ... 39Xb 71
... (off Andrew's Rd.)
North Side Wandsworth Comn. SW18 ... 57Fb 111
Northspur Rd. SM1: Sutt ... 76Cb 155
North Sq. DA3: Nw A G ... 75Be 165
North Sq. N9 ... 19Xb 33
... (off New Rd.)
North Sq. NW11 ... 29Cb 49
North Stand N5 ... 34Rb 71
North Star Blvd. DA9: Ghithe ... 56Wd 120
... (off Evelyn Wlk.)
North Station App. RH1: S Nut ... 8F 208
Northstead Rd. SW2 ... 61Qb 134
NORTH STIFFORD ... 46Ae 99
North St. BR1: Broml ... 67Jc 136
North St. DA1: Dart ... 59Md 119
North St. DA12: Grav'nd ... 9D 122
North St. DA7: Bex ... 56Cd 118
North St. E13 ... 40Kc 73
North St. IG11: Bark ... 37Rc 74
North St. KT22: Lea ... 93Ja 192
North St. NW4 ... 29Ya 48
North St. RH1: Redh ... 5P 207
North St. RM1: Rom ... 27Fd 56
North St. RM11: Horn ... 31Md 77
North St. RM5: Rom ... 27Fd 56
North St. SM5: Cars ... 76Hb 155
North St. SW4 ... 55Lb 112
North St. TW20: Egh ... 64B 126
North St. TW7: Isle ... 55Ja 108
North St. Pas. E13 ... 40Kc 73
North Tenter St. E1 ... 3K 225 (44Vb 91)
North Ter. SL4: Wind ... 2H 103
North Ter. SW3 ... 4D 226 (48Gb 89)
North Ter. WC2 ... 6E 222 (46Mb 90)
... (off Trafalgar Sq.)
Northumberland All. EC3 ... 3J 225 (44Ub 91)
... (not continuous)
Northumberland Av. DA16: Well ... 56Tc 116
Northumberland Av. E12 ... 32Lc 73
Northumberland Av. EN1: Enf ... 11Xb 33
Northumberland Av. RM11: Horn ... 29Ld 57
Northumberland Av. TW7: Isle ... 53Ha 108
Northumberland Av. WC2 ... 6F 223 (46Nb 90)
Northumberland Cl. DA8: Erith ... 52Ed 118
Northumberland Cl. TW19: Stanw ... 58N 105
Northumberland Cres. TW14: Felt ... 58U 106
Northumberland Gdns. BR1: Broml ... 70Qc 138
Northumberland Gdns. CR4: Mitc ... 71Mb 156
Northumberland Gdns. N9 ... 20Vb 33
Northumberland Gdns. TW7: Isle ... 52Ja 108
Northumberland Gro. N17 ... 24Xb 51
Northumberland Hall AL9: N Mym ... 10F 8
NORTHUMBERLAND HEATH ... 52Ed 118
Northumberland Ho. IG8: Wfd G ... 24Qc 54
Northumberland Ho. SW1 ... 6F 223 (46Nb 90)
... (off Northumberland Av.)
Northumberland Pk. DA8: Erith ... 52Ed 118
Northumberland Pk. N17 ... 24Vb 51
Northumberland Pk. Ind. Est. N17 ... 24Xb 51
Northumberland Pk. School Sports Cen. ... 24Wb 51
Northumberland Park Station (Rail) ... 24Xb 51
Northumberland Pl. TW10: Rich ... 57Ma 109
Northumberland Pl. W2 ... 44Cb 89
Northumberland Rd. DA13: Ist R ... 6B 144
Northumberland Rd. E17 ... 31Cc 72
Northumberland Rd. E6 ... 44Nc 94
Northumberland Rd. EN5: New Bar ... 16Eb 31
Northumberland Rd. HA2: Harr ... 29Ba 45
Northumberland Rd. SS17: Linf ... 7J 101
Northumberland St. WC2 ... 6F 223 (46Nb 90)
Northumberland Wlk. SL0: Rich P ... 47G 82
Northumberland Way DA8: Erith ... 53Ed 118
Northumbria St. E14 ... 44Cc 92
North Verbena Gdns. W6 ... 50Wa 88
North Vw. HA5: Eastc ... 31Y 65
North Vw. SW19 ... 64Ya 132
North Vw. W5 ... 42La 86
Northview BR8: Swan ... 68Gd 140
Northview HP1: Hem H ... 4F 2
Northview Av. RM18: Tilb ... 3C 122
North Vw. Cvn. Site IG6: Ilf ... 24Wc 55
North Vw. Cres. KT18: Tatt C ... 89Ya 174
Northview Cres. NW10 ... 35Va 68
North Vw. Dr. IG8: Wfd G ... 26Mc 53
Northview Pde. N7 ... 34Nb 70
North Vw. Rd. N8 ... 28Mb 50
North Vw. Rd. TN14: S'oaks ... 93Ld 203
North Vs. NW1 ... 37Mb 70
North Wlk. CR0: New Ad ... 79Dc 158
... (not continuous)
Northwall Rd. E20 ... 36Cc 72
NORTH WATFORD ... 9X 13
Northway NW11 ... 29Db 49
Northway SM4: Mord ... 69Ab 132
Northway SM6: W'gton ... 77Lb 156
Northway WD3: Rick ... 17M 25
North Way HA5: Pinn ... 28Z 45
North Way N11 ... 23Lb 50
North Way N9 ... 19Zb 34
North Way NW9 ... 27Ra 47
North Way UB10: Uxb ... 38N 63
Northway Cir. NW7 ... 21Ta 47
Northway Ct. NW7 ... 21Ua 48
Northway Cres. NW7 ... 21Ta 47
Northway Gdns. ... 29Db 49
Northway Ho. N20 ... 18Eb 31
Northway Rd. CR0: C'don ... 72Vb 157
Northway Rd. SE5 ... 55Sb 113
Northways NW3 ... 38Fb 69
... (off College Cres.)
Northways Pde. NW3 ... 38Fb 69
... (off College Cres.)
North Weald Cl. RM12: Horn ... 38Kd 77
Northweald La. KT2: King T ... 64Ma 131
NORTH WEMBLEY ... 34Ma 67
North Wembley Station (Underground & Overground) ... 34Ma 67
North Western Av. WD24: Wat ... 7V 12
North Western Av. WD25: A'ham ... 7T 12
... (not continuous)
North Western Av. WD25: Wat ... 7T 12
... (not continuous)
Northwest Pl. N1 ... 1A 218 (40Qb 70)
North Weylands Ind. Est. KT12: Walt T ... 75Aa 151
North Wharf E14 ... 46Ec 92
... (off Coldharbour)
North Wharf Rd. W2 ... 1B 220 (43Fb 89)
Northwick Av. HA3: Kenton ... 30Ja 46
Northwick Circ. HA3: Kenton ... 30La 46
Northwick Cl. HA1: Harr ... 32Ka 66
Northwick Cl. NW8 ... 5B 214 (42Fb 89)
NORTHWICK DAY CEN. ... 21Y 45
Northwick Ho. NW8 ... 5A 214 (42Eb 89)
... (off St John's Wood Rd.)
NORTHWICK PARK HOSPITAL ... 31Ja 66
Northwick Pk. Playgolf ... 32Ja 66
Northwick Pk. Rd. HA1: Harr ... 30Ha 46
NORTHWICK PARK RDBT. ... 31Ja 66
Northwick Park Station (Underground) ... 31Ka 66
Northwick Rd. HA0: Wemb ... 39Ma 67
Northwick Rd. WD19: Wat ... 21Y 45
Northwick Ter. NW8 ... 5B 214 (42Fb 89)
Northwick Wlk. HA1: Harr ... 31Ha 66
Northwold Dr. HA5: Pinn ... 26Y 45
Northwold Rd. E5 ... 33Wb 71
Northwold Rd. N16 ... 33Vb 71
NORTHWOOD ... 23U 44
Northwood RM16: Grays ... 7D 100
Northwood Av. CR8: Purl ... 84Qb 176
Northwood Av. GU21: Knap ... 10H 167
Northwood Av. RM12: Horn ... 35Jd 76
Northwood Club, The ... 26V 44
North Wood Ct. SE25 ... 69Wb 135
Northwood Dr. DA9: Ghithe ... 58Wd 120
Northwood Est. E5 ... 33Wb 71
Northwood Gdns. IG5: Ilf ... 28Qc 54
Northwood Gdns. N12 ... 22Fb 49
Northwood Gdns. UB6: G'frd ... 36Ha 66
Northwood Golf Course ... 24T 44
Northwood Hall N6 ... 31Lb 70
NORTHWOOD HILLS ... 26W 44
NORTHWOOD HILLS CIR. ... 25W 44
Northwood Hills Station (Underground) ... 26W 44
Northwood Ho. KT2: King T ... 68Qa 131
... (off Coombe Rd.)
Northwood Ho. SE27 ... 63Tb 135
Northwood Pl. DA18: Erith ... 48Bd 95
Northwood Rd. CR7: Thor H ... 68Rb 135
Northwood Rd. N6 ... 31Kb 70
Northwood Rd. SE23 ... 60Bc 114
Northwood Rd. SM5: Cars ... 79Jb 156
Northwood Rd. TW6: H'row A ... 53M 105
Northwood Rd. UB9: Hare ... 25L 43
Northwood Station (Underground) ... 24U 44
Northwood Way HA6: Nwood ... 24V 44
Northwood Way SE19 ... 65Tb 135
Northwood Way UB9: Hare ... 25M 43
NORTH WOOLWICH ... 47Qc 94
North Woolwich Rd. E16 ... 46Hc 93
North Worple Way SW14 ... 55Ta 109
Nortoft Rd. SL9: Chal P ... 23B 42
Norton Almshouses EN8: Chesh ... 2Zb 20
... (off Turner's Hill)
Norton Av. KT5: Surb ... 73Ra 153
Norton Cl. E4 ... 22Cc 52
Norton Cl. EN1: Enf ... 12Xb 33
Norton Cl. GU3: Worp ... 9H 187
Norton Cl. WD6: Bore ... 11Qa 29
Norton Ct. BR3: Beck ... 67Bc 136
Norton Folgate E1 ... 7J 219 (43Ub 91)
Norton Folgate Ho. E1 ... 7K 219 (43Vb 91)
... (off Puma Ct.)
Norton Gdns. SW16 ... 68Nb 134
Norton Ho. E1 ... 44Xb 91
... (off Bigland St.)
Norton Ho. E2 ... 40Zb 72
... (off Mace St.)
Norton Ho. SW1 ... 4E 228 (48Mb 90)
... (off Arneway St.)
Norton Ho. SW9 ... 54Pb 112
... (off Aytoun Rd.)
Norton La. DA9: Ghithe ... 57Yd 120
Norton La. KT11: Cobh ... 91V 190
Norton Pk. SL5: S'hill ... 1A 146
Norton Rd. E10 ... 32Bc 72
Norton Rd. HA0: Wemb ... 37Ma 67
Norton Rd. RM10: Dag ... 37Fd 76
Norton Rd. UB8: Uxb ... 41M 83
Norval Grn. SW9 ... 54Qb 112
Norval Rd. HA0: Wemb ... 33Ka 66
Norway Dr. SL2: Slou ... 3M 81
Norway Ga. SE16 ... 48Ac 92
Norway Ho. N1 ... 39Ub 71
... (off Hertford Rd.)
Norway Pl. E14 ... 44Bc 92
Norway St. SE10 ... 51Dc 114
Norway Wlk. RM13: Rain ... 42Ld 97
Norway Wharf E14 ... 44Bc 92
Norwegian War Memorial ... 7E 220 (46Gb 89)
Norwich Cres. RM6: Chad H ... 29Xc 55
Norwich Ho. E14 ... 44Dc 92
... (off Cordelia St.)
Norwich Ho. WD6: Bore ... 12Qa 29
... (off Stratfield Rd.)
Norwich M. IG3: Ilf ... 32Wc 75
Norwich Pl. DA6: Bex ... 56Cd 118
Norwich Rd. CR7: Thor H ... 69Sb 135
Norwich Rd. E7 ... 36Jc 73
Norwich Rd. HA6: Nwood ... 27V 44
Norwich Rd. UB6: G'frd ... 39Da 65
Norwich St. EC4 ... 2K 223 (44Qb 90)
Norwich Wlk. HA8: Edg ... 24Sa 47
Norwich Way WD3: Crox G ... 13R 26
NORWOOD ... 65Ub 135
Norwood Av. HA0: Wemb ... 39Pa 67
Norwood Av. RM7: Rush G ... 31Gd 76
Norwood Bottom Lock ... 47Da 85
... (off Poplar Av.)
Norwood Cl. KT24: Eff ... 100Aa 191
Norwood Cl. NW2 ... 34Ab 68
Norwood Cl. TW2: Twick ... 61Fa 130
Norwood Cl. UB2: S'hall ... 49Ca 85
Norwood Ct. DA1: Dart ... 57Qd 119
... (off Farnol Rd.)
Norwood Dr. HA2: Harr ... 30Ba 45
Norwood Farm La. KT11: Cobh ... 83W 170
Norwood Farm La. KT12: Cobh ... 83X 171
Norwood Gdns. UB2: S'hall ... 49Ba 85
Norwood Gdns. UB4: Yead ... 42Y 85
NORWOOD GREEN ... 49Ba 85
Norwood Grn. Rd. UB2: S'hall ... 49Ca 85
Norwood High St. SE27 ... 62Rb 135
Norwood Ho. E14 ... 45Dc 92
... (off Poplar High St.)
Norwood Junction Station (Rail & Overground) ... 70Wb 135
Norwood La. SL0: Iver ... 42F 82
NORWOOD NEW TOWN ... 65Sb 135
Norwood Pk. Rd. SE27 ... 64Sb 135
Norwood Rd. EN8: Chesh ... 2Ac 20
Norwood Rd. KT24: Eff ... 100Aa 191
Norwood Rd. SE24 ... 60Rb 113
Norwood Rd. SE27 ... 61Rb 135
Norwood Rd. TW5: Hest ... 51Y 107
Norwood Rd. UB2: S'hall ... 48Aa 85
Norwood Ter. UB2: S'hall ... 49Da 85
Norwood Top Lock ... 48Da 85
... (off Barge Dr.)
Notley End TW20: Eng G ... 6N 125
Notley Pl. SW4 ... 59Nb 112
Notley St. SE5 ... 52Tb 113
Notson Rd. SE25 ... 70Xb 135
Notting Barn Rd. W10 ... 42Za 88
Nottingdale Sq. W11 ... 46Ab 88
Nottingham Av. E16 ... 43Lc 93
Nottingham Cl. GU21: Wok ... 10K 167
Nottingham Cl. WD25: Wat ... 5W 12
Nottingham Ct. GU21: Wok ... 10K 167
... (off Nottingham Cl.)
Nottingham Ct. WC2 ... 3F 223 (44Nb 90)
Nottingham Ho. WC2 ... 3F 223 (44Nb 90)
... (off Shorts Gdns.)
Nottingham Pl. W1 ... 7H 215 (43Jb 90)
Nottingham Rd. CR2: S Croy ... 77Sb 157
Nottingham Rd. E10 ... 30Ec 52
Nottingham Rd. SW17 ... 60Hb 111
Nottingham Rd. TW7: Isle ... 54Ha 108
Nottingham Rd. WD3: Herons ... 17E 24
Nottingham St. W1 ... 7H 215 (43Jb 90)
Nottingham Ter. NW1 ... 6H 215 (42Jb 90)
... (off York Ter. W.)
NOTTING HILL ... 45Bb 89
Notting Hill Ga. W11 ... 46Cb 89
Notting Hill Gate Station (Underground) ... 46Cb 89
Nottingwood Ho. W11 ... 45Ab 88
... (off Clarendon Rd.)
Nova Bldg. E14 ... 49Cc 92
Nova Ct. E. E14 ... 46Ec 92
... (off Yabsley St.)
Nova Ct. W. E14 ... 46Ec 92
... (off Yabsley St.)
Nova M. SM3: Sutt ... 74Ab 154
Novar Cl. BR6: Orp ... 73Vc 161
Nova Rd. CR0: C'don ... 74Rb 157

Novar Rd. SE9 ....60Sc 116
Novello Cl. TN15: Bor G 91Be 205
Novello Ct. N1 ....39Sb 71
(off Dibden St.)
Novello St. SW6 ....53Cb 111
Novello Theatre Sunninghill 1B 146
Novello Theatre West End 4H 223 (45Pb 90)
(off Aldwych)
Novello Way WD6: Bore 11Ta 29
Novellus Ct. KT18: Eps 86Ta 173
(off South St.)
Novem Ho. E1 ....43Wb 91
(off Chicksand St.)
Nowell Rd. SW13 ....51Wa 110
Nower, The TN14: Knock 91Vc 201
Nower Cl. KT18: Tatt C 89Za 174
Nower Ct. HA5: Pinn 28Ba 45
Nower Hill HA5: Pinn 28Ba 45
Noyna Rd. SW17 ....62Hb 133
NRG Gym Gravesend 8D 122
(off Garrick St.)
Nubia Way BR1: Broml 62Gc 137
Nucleus Apts. SW15 ....59Za 110
(off W. Hill)
Nucleus Bus. & Innovation Cen.,
The DA1: Dart 55Qd 119
Nuding Cl. SE13 ....55Cc 114
Nuffield Ct. TW5: Hest 52Ba 107
Nuffield Health Baker Street 12Tb 33
Nuffield Health Battersea 54Hb 111
(within Latchmere Leisure Cen.)
Nuffield Health Bloomsbury 5H 217 (42Pb 90)
(off Mecklenburgh Pl.)
Nuffield Health Bromley 71Kc 159
Nuffield Health Cannon
Street 5F 225 (45Tb 91)
(off Cousin La.)
Nuffield Health Cheam 80Ab 154
Nuffield Health Chingford 21Ec 52
Nuffield Health Chislehurst 64Uc 138
Nuffield Health Covent
Garden 3G 223 (44Nb 90)
(off Endell St.)
Nuffield Health Ealing 45Ma 87
(within Ealing Broadway Cen.)
Nuffield Health Friern Barnet 22Jb 50
Nuffield Health Fulham 53Za 110
Nuffield Health Hemel Hempstead 1C 4
Nuffield Health Hendon 29Ya 48
Nuffield Health Ilford 34Rc 74
(off Clements Rd.)
Nuffield Health Islington 39Rb 71
Nuffield Health Kingston upon Thames,
Richmond Rd. 67Na 131
Nuffield Health Leatherhead 94Ka 192
(off The Crescent)
Nuffield Health Merton Abbey 67Fb 133
(off Watermill Way)
Nuffield Health Norbury 68Pb 134
Nuffield Health Paddington 1A 220 (43Eb 89)
Nuffield Health Purley Way 79Qb 156
Nuffield Health Romford 29Gd 56
Nuffield Health St Albans 5F 6
Nuffield Health Stockley Park 45S 84
Nuffield Health Stoke Poges 10M 61
Nuffield Health Sunbury 67W 128
Nuffield Health Surbiton 72La 152
Nuffield Health Surrey Street 76Sb 157
(off Surrey St.)
Nuffield Health Sydney Rd. 14Tb 33
Nuffield Health Twickenham 59Ga 108
Nuffield Health Wandsworth 59Db 111
Nuffield Health Wandsworth,
Southside Shop. Cen. 57Db 111
Nuffield Health West Byfleet 86K 169
Nuffield Health Willesden Green 38Ya 68
Nuffield Health Wimbledon 65Bb 133
Nuffield Lodge N6 ....30Lb 50
Nuffield Lodge W9 ....43Cb 89
(off Admiral Wlk.)
Nuffield Rd. BR8: Hext 65Jd 140
Nugent Rd. N19 ....32Nb 70
Nugent Rd. SE25 ....69Vb 135
Nugents Ct. HA5: Pinn 25Aa 45
Nugent Shop. Pk. 70Yc 139
Nugent's Pk. HA5: Hat E 25Aa 45
Nugent Ter. NW8 ....2A 214 (40Eb 69)
Numa Ct. TW8: Bford 52Ma 109
No. 1 St. SE18 ....48Rc 94
No. One EC1 ....6E 218 (42Sb 91)
Numbers Farm WD4: K Lan 1S 12
Nunappleton Way RH8: Oxt 4L 211
Nun Ct. EC2 ....2F 225 (44Tb 91)
(off Coleman St.)
Nuneaton Rd. RM9: Dag 38Ad 75
Nunfield WD4: Chfd 3K 11
NUNHEAD 55Xb 113
Nunhead Cemetery Local Nature
Reserve 56Yb 114
Nunhead Cres. SE15 ....55Xb 113
Nunhead Est. SE15 ....56Xb 113
Nunhead Grn. SE15 ....55Xb 113
Nunhead Gro. SE15 ....55Xb 113
Nunhead La. SE15 ....55Xb 113
Nunhead Pas. SE15 ....55Wb 113
Nunhead Station (Rail) 54Yb 114
Nunnery Cl. AL1: St A 4C 6
Nunnery Stables AL1: St A 4B 6
Nunnington Cl. SE9 ....62Nc 138
Nunns Rd. EN2: Enf 12Sb 33
Nunns Way RM17: Grays 49Fe 99
Nuns La. AL1: St A 6C 6
Nuns Wlk. GU25: Vir W 1P 147
NUPER'S HATCH 17Gd 38
Nupton Dr. EN5: Barn 16Ya 30
Nuralite Ind. Cen. ME3: High'm 9P 123
Nurse Cl. HA8: Edg 25Sa 47
Nursery, The DA8: Erith 52Hd 118
Nursery App. N12 ....23Gb 49
Nursery Av. CR0: C'don 75Zb 158
Nursery Av. DA7: Bex 55Bd 117
Nursery Av. N3 ....26Eb 49
Nursery Cl. BR6: Orp 73Wc 161
Nursery Cl. BR8: Swan 68Ed 140
Nursery Cl. CR0: C'don 75Zb 158
Nursery Cl. DA2: Dart 59Sd 120
Nursery Cl. EN3: Enf H 11Zb 34
Nursery Cl. GU21: Wok 8N 167
Nursery Cl. IG8: Wfd G 22Kc 53
Nursery Cl. KT15: Wdhm 82H 169
Nursery Cl. KT17: Ewe 82Ua 174
Nursery Cl. KT20: Walt H 97Xa 194
Nursery Cl. RM15: S Ock 42Yd 98
Nursery Cl. RM6: Chad H 30Zc 55
Nursery Cl. SE4 ....54Bc 114
Nursery Cl. SW15 ....56Za 110
Nursery Cl. TN13: S'oaks 94Jd 203
Nursery Cl. TW14: Felt 59X 107
(not continuous)
Nursery Cl. WD19: Wat 18X 27
Nursery Cotts. AL1: St A 5P 5
Nursery Ct. N17 ....24Vb 51
Nursery Ct. W13 ....43Ja 86
Nursery Gdns. BR7: Chst 65Rc 138
Nursery Gdns. EN3: Enf H 11Zb 34
Nursery Gdns. TW12: Hamp 63Ba 129
Nursery Gdns. TW16: Sun 68V 128
Nursery Gdns. TW18: Staines 65K 127
Nursery Gdns. TW4: Houn 57Ba 107
Nursery Gro. DA11: Grav'nd 4D 144
Nursery La. E2 ....1K 219 (39Vb 71)
Nursery La. E7 ....37Jc 73
Nursery La. SL3: L'ly 6P 81
Nursery La. UB8: Cowl 42M 83
Nursery La. W10 ....43Ya 88
Nurserymans Rd. N11 ....19Jb 32
Nursery M. DA11: Grav'nd 4E 144
Nursery Pl. SL4: Old Win 8M 103
Nursery Pl. TN13: Chips 94Fd 202
Nursery Ri. EN9: Walt A 4Jc 21
Nursery Rd. SW19: Elm Gro. 66Ab 132
Nursery Rd. SW19: Parkleigh Rd. 68Db 133
Nursery Rd. CR4: Mitc 69Gb 133
Nursery Rd. CR7: Thor H 70Tb 135
Nursery Rd. E9 ....37Yb 72
Nursery Rd. EN9: Walt A 3Fc 21
Nursery Rd. GU21: Knap 9H 167
Nursery Rd. HA5: Pinn 27Y 45
Nursery Rd. IG10: Lough 15Lc 35
Nursery Rd. KT20: Walt H 97Wa 194
Nursery Rd. N14 ....17Lb 32
Nursery Rd. N2 ....25Fb 49
Nursery Rd. SM1: Sutt 77Eb 155
Nursery Rd. SS17: Stan H 1N 101
Nursery Rd. SW9 ....56Pb 112
Nursery Rd. TW16: Sun 68U 128
Nursery Row EN5: Barn 13Ab 30
Nursery Row SE17 ....6F 231 (49Tb 91)
Nursery St. N17 ....24Vb 51
Nursery Wlk. NW4 ....27Ya 48
Nursery Wlk. RM7: Rush G 31Fd 76
Nursery Way RH8: Oxt 1J 211
Nursery Way TW19: Wray 8P 103
Nursery Waye UB8: Uxb 39M 63
NURSTEAD 9B 144
Nurstead Av. DA3: Lfield 70Fe 143
Nurstead Chu. La. DA13: Meop 10B 144
Nurstead Chu. La. DA13: Sole S 10B 144
Nurstead Hill Farm DA3: Long H 70Fe 143
Nurstead La. DA13: Meop 10A 144
Nurstead La. DA3: Long H 70Fe 143
Nurstead Rd. DA8: Erith 52Cd 118
Nut Ash La. KT18: Head 98Ta 193
Nutberry Av. RM16: Grays 47Ce 99
Nutberry Cl. RM16: Grays 47Ce 99
Nutbourne Ct. TW18: Staines 66H 127
(off Riverside Rd.)
Nutbourne St. W10 ....41Ab 88
Nutbrook St. SE15 ....55Wb 113
Nutbrowne Rd. RM9: Dag 39Bd 75
Nutcroft Gro. KT22: Fet 93Ga 192
Nutcroft Rd. SE15 ....52Xb 113
NUTFIELD 5F 208
Nutfield Cl. N18 ....23Wb 51
Nutfield Cl. SM5: Cars 76Gb 155
Nutfield Ct. BR1: Broml 69Jc 137
Nutfield Ct. RH1: Nutf 4F 208
Nutfield Ct. RH1: Redh 4B 208
(off Goodworth Rd.)
Nutfield Gdns. IG3: Ilf 33Vc 75
Nutfield Gdns. UB5: N'olt 40Y 65
Nutfield Marsh Rd. RH1: Nutf 3D 208
NUTFIELD PARK 9G 208
Nutfield Pas. CR7: Thor H 70Rb 135
(off Nutfield Rd.)
Nutfield Rd. CR5: Coul 88Jb 176
Nutfield Rd. CR7: Thor H 70Rb 135
Nutfield Rd. E15 ....35Ec 72
Nutfield Rd. NW2 ....34Wa 68
Nutfield Rd. RH1: Mers 1C 208
Nutfield Rd. RH1: Nutf 6B 208
Nutfield Rd. RH1: Redh 6B 208
Nutfield Rd. SE22 ....56Vb 113
Nutfields TN15: Igh 95Xd 204
Nutfield Station (Rail) 8E 208
Nutfield Way BR6: Farnb 75Rc 160
Nutford Pl. W1 ....2F 221 (44Hb 89)
Nuthatch DA3: Lfield 69De 143
Nuthatch Cl. RM3: Rom 21Md 57
Nuthatch Cl. TW19: Stanw 60P 105
Nuthatch Gdns. RH2: Reig 10L 207
Nuthatch Gdns. SE28 ....47Tc 94
(not continuous)
Nuthatch M. IG5: Ilf 25Pc 54
Nuthatch Pl. KT17: Ewe 82Wa 174
Nuthatch Row KT10: Clay 79Ha 152
Nuthurst Av. SW2 ....61Pb 134
Nutkin Wlk. UB8: Uxb 38N 63
Nutley Cl. BR8: Hext 67Hd 140
Nutley Ct. RH2: Reig 6H 207
(off Nutley La.)
Nutley Gro. RH2: Reig 6J 207
Nutley La. RH2: Reig 5H 207
Nutley Ter. NW3 ....37Eb 69
Nutmead Cl. DA5: Bexl 60Ed 118
Nutmeg Cl. E16 ....42Gc 93
Nutmeg La. E14 ....44Fc 93
Nuttall St. N1 ....1J 219 (40Ub 71)
Nutter La. E11 ....30Lc 53
Nuttfield Cl. WD3: Crox G 16S 26
Nutt Gro. HA8: Edg 19Ma 29
Nut Tree Cl. BR6: Chels 76Zc 161
Nutt St. SE15 ....52Vb 113
Nutty La. TW17: Shep 69S 128
Nutwell St. SW17 ....64Gb 133
Nuxley Rd. DA17: Belv 51Bd 117
Nyall Ct. RM2: Rom 27Ld 57
(off Nairn St.)
Nyanza St. SE18 ....51Tc 116
Nye Bevan Est. E5 ....34Zb 72
Nye Bevan Ho. SW6 ....52Bb 111
(off St Thomas's Way)
Nyefield Pk. KT20: Walt H 98Wa 194
Nye Way HP3: Bov 10C 2
Nyland Ct. SE8 ....49Ac 92
(off Naomi St.)
Nylands Av. TW9: Kew 53Qa 109
Nymans Gdns. SW20 ....69Xa 132
Nynehead St. SE14 ....52Ac 114
Nyon Gro. SE6 ....61Bc 136
Nyssa Cl. IG8: Wfd G 23Pc 54
Nyth Cl. RM14: Upm 30Td 58
Nyton Cl. N19 ....32Nb 70

O

O2, The 46Gc 93
O2 Brixton Academy 55Qb 112
O2 Cen. 37Eb 69
O2 Forum Kentish Town 36Kb 70
O2 Shepherd's Bush Empire Theatre 47Ya 88
Oakapple Cl. CR2: Sande 86Xb 177
Oak Apple Ct. SE12 ....60Jc 115
Oak Av. AL2: Brick W 2Ca 13
Oak Av. CR0: C'don 74Cc 158
Oak Av. EN2: Enf 10Pb 18
Oak Av. N10 ....24Kb 50
Oak Av. N17 ....24Tb 51
Oak Av. N8 ....28Nb 50
Oak Av. RM14: Upm 34Rd 77
Oak Av. TN13: S'oaks 100Kd 203
Oak Av. TW12: Hamp 64Aa 129
Oak Av. TW20: Egh 66E 126
Oak Av. TW5: Hest 52Z 107
Oak Av. UB10: Ick 33R 64
Oak Av. UB7: W Dray 48Q 84
Oak Avenue Local Nature Reserve 64Aa 129
Oak Bank CR0: New Ad 79Ec 158
Oakbank CM13: Hut 15Fe 41
Oakbank GU22: Wok 91A 188
Oakbank KT22: Fet 95Ea 192
Oakbank WD7: R'lett 8Ka 14
Oakbank Av. KT12: Walt T 73Ba 151
Oakbank Gro. SE24 ....56Sb 113
Oakbark Ho. TW8: Bford 52La 108
(off High St.)
Oakbourne Av. GU24: W End 6E 166
Oakbrook Cl. BR1: Broml 63Kc 137
Oakbury Rd. SW6 ....54Db 111
Oak Cl. DA1: Cray 56Hd 118
Oak Cl. EN9: Walt A 6Fc 21
Oak Cl. HP3: Hem H 6P 3
Oak Cl. N14 ....17Kb 32
Oak Cl. RH8: Oxt 4L 211
Oak Cl. SM1: Sutt 75Eb 155
Oakcombe Cl. KT3: N Mald 67Ua 132
Oak Cott. Cl. SE6 ....60Hc 115
Oak Cotts. W7 ....47Ga 86
Oak Ct. HA6: Nwood 23T 44
Oak Ct. RM15: S Ock 40Yd 78
Oak Ct. SE15 ....52Vb 113
(off Sumner Rd.)
Oak Cres. E16 ....43Gc 93
Oakcroft SL2: Slou 1E 80
Oakcroft Bus. Cen. KT9: Chess 77Pa 153
Oakcroft Cl. HA5: Pinn 26X 45
Oakcroft Cl. KT14: W Byf 86H 169
Oakcroft Rd. KT14: W Byf 86H 169
Oakcroft Rd. KT9: Chess 77Pa 153
Oakcroft Rd. SE13 ....54Fc 115
Oakcroft Vs. KT9: Chess 77Pa 153
Oakdale N14 ....18Kb 32
Oakdale Av. HA3: Kenton 29Na 47
Oakdale Av. HA6: Nwood 26W 44
Oakdale Cl. WD19: Wat 21Y 45
Oakdale Ct. E4 ....22Ec 52
Oakdale Gdns. E4 ....22Ec 52
Oakdale La. TN8: C Hil 97Lc 199
Oakdale Rd. E11 ....33Fc 73
Oakdale Rd. E18 ....26Kc 53
Oakdale Rd. E7 ....38Kc 73
Oakdale Rd. KT13: Weyb 76Q 150
Oakdale Rd. KT19: Ewe 81Ta 173
Oakdale Rd. N4 ....30Sb 51
Oakdale Rd. SE15 ....55Yb 114
Oakdale Rd. SW16 ....64Nb 134
Oakdale Rd. WD19: Wat 20Y 27
Oakdale Way CR4: Mitc 73Jb 156
Oak Dene W13 ....43Ka 86
Oakdene EN8: Chesh 2Ac 20
Oakdene GU24: Chob 2K 167
Oakdene KT20: Tad 92Ab 194
Oakdene RM3: Hrld W 26Pd 57
Oakdene SE15 ....53Xb 113
Oakdene SL5: S'dale 2D 146
Oakdene Av. BR7: Chst 64Qc 138
Oakdene Av. DA8: Erith 51Ed 118
Oakdene Av. KT7: T Ditt 74Ja 152
Oakdene Cl. HA5: Hat E 24Ba 45
Oakdene Cl. KT23: Bookh 99Ea 192
Oakdene Cl. RM11: Horn 30Kd 57
Oakdene Ct. KT11: Cobh 86X 171
(off Between Streets)
Oakdene Ct. KT12: Walt T 76X 151
Oakdene Ct. KT13: Weyb 77Q 150
Oakdene Dr. KT5: Surb 73Sa 153
Oakdene M. SM3: Sutt 74Bb 155
Oakdene Pde. KT11: Cobh 86X 171
Oakdene Pk. N3 ....24Bb 49
Oakdene Rd. BR5: St M Cry 71Vc 161
Oakdene Rd. HP3: Hem H 6P 3
Oakdene Rd. KT11: Cobh 86X 171
Oakdene Rd. KT23: Bookh 96Ba 191
Oakdene Rd. RH1: Redh 6P 207
Oakdene Rd. TN13: S'oaks 94Jd 202
Oakdene Rd. UB10: Hil 40R 64
Oakdene Rd. WD24: Wat 8X 13
Oakdene Way AL1: St A 2G 6
Oakden St. SE11 ....5A 230 (49Qb 90)
Oake Ct. SW15 ....57Ab 110
Oakeford Ho. W14 ....48Ab 88
(off Russell Rd.)
Oaken Coppice KT21: Asht 91Qa 193
Oak End Ct. SL9: Ger X 29A 42
Oak End Dr. SL0: Iver H 40E 62
Oaken Dr. KT10: Clay 79Ha 152
Oak End Way KT15: Wdhm 84G 168
Oak End Way SL9: Ger X 29B 42
Oakenholt Ho. SE2 ....46Zc 95
Oaken La. KT10: Clay 77Ga 152
Oakenshaw Cl. CR6: W'ham 87Yb 178
Oakenshaw Cl. KT6: Surb 73Na 153
Oakenwood KT23: Bookh 97Ba 191
Oakes Cl. E6 ....44Pc 94
Oakes Cres. DA1: Dart 57Pd 119
Oakeshott Av. N6 ....33Jb 70
Oakes M. E14 ....43Ec 92
(off St Anne's Pas.)
Oakey La. SE1 ....3K 229 (48Qb 90)
Oak Farm WD6: Bore 15Sa 29
Oak Farm La. TN15: Fair 82Fe 185
Oakfield E4 ....22Dc 52
Oakfield GU21: Wok 9J 167
Oakfield WD3: Rick 17H 25
Oakfield Av. HA3: Kenton 27Ka 46
Oakfield Av. SL1: Slou 6F 80
Oakfield Cl. EN6: Pot B 3Bb 17
Oakfield Cl. HA4: Ruis 30V 44
Oakfield Cl. KT13: Weyb 77S 150
Oakfield Cl. KT3: N Mald 71Va 154
Oakfield Ct. KT13: Weyb 77S 150
Oakfield Ct. N8 ....31Nb 70
Oakfield Ct. NW2 ....31Za 68
Oakfield Ct. WD6: Bore 13Ra 29
Oakfield Dr. RH2: Reig 4J 207
Oakfield Gdns. BR3: Beck 71Dc 158
Oakfield Gdns. N18 ....21Ub 51
Oakfield Gdns. SE19 ....64Ub 135
(not continuous)
Oakfield Gdns. SM5: Cars 74Gb 155
Oakfield Gdns. UB6: G'frd 42Fa 86
Oakfield Glade KT13: Weyb 77S 150
Oakfield Ho. E3 ....43Cc 92
(off Gale St.)
Oakfield La. DA1: Dart Old Bexley La 61Gd 140
Oakfield La. BR2: Kes 77Lc 159
Oakfield La. DA2: Dart 61Kd 141
Oakfield La. DA2: Wilm 61Kd 141
Oakfield Lodge IG1: Ilf 34Rc 74
(off Albert Rd.)
Oakfield Pk. Rd. DA1: Dart 61Md 141
Oakfield Pl. DA1: Dart 61Md 141
Oakfield Rd. BR6: Orp 73Wc 161
Oakfield Rd. CR0: C'don 74Sb 157
Oakfield Rd. E17 ....26Ac 52
Oakfield Rd. E6 ....39Nc 74
Oakfield Rd. IG1: Ilf 34Rc 74
Oakfield Rd. KT11: Cobh 86X 171
Oakfield Rd. KT21: Asht 89Ma 173
Oakfield Rd. N14 ....19Nb 32
Oakfield Rd. N3 ....25Db 49
Oakfield Rd. N4 ....30Qb 50
Oakfield Rd. SE20 ....66Xb 135
Oakfield Rd. SW19 ....62Za 132
Oakfield Rd. TW15: Ashf 64R 128
Oakfield Rd. Ind. Est. SE20 ....66Xb 135
Oakfields IG10: Lough 15Qc 36
Oakfields KT12: Walt T 74W 150
Oakfields KT14: W Byf 86K 169
Oakfields TN13: S'oaks 98Kd 203
Oakfields Rd. NW11 ....30Ab 48
Oakfield St. SW10 ....51Eb 111
Oakford Rd. NW5 ....35Lb 70
Oak Gdns. CR0: C'don 75Cc 158
Oak Gdns. HA8: Edg 26Sa 47
Oak Glade CM16: Coop 1Zc 23
Oak Glade HA6: Nwood 25R 44
Oak Glade KT19: Eps 84Qa 173
Oak Glen RM11: Horn 27Nd 57
Oak Grange Rd. GU4: W Cla 100K 189
Oak Grn. WD5: Ab L 4U 12
Oak Grn. Way WD5: Ab L 4U 12
Oak Gro. AL10: Hat 1B 8
Oak Gro. BR4: W W'ck 74Ec 158
Oak Gro. HA4: Ruis 31X 65
Oak Gro. NW2 ....35Ab 68
Oak Gro. RM16: Grays 10D 100
Oak Gro. TW16: Sun 66X 129
Oakgrove CR3: Cat'm 95Tb 197
Oak Gro. Rd. SE20 ....67Yb 136
Oakhall Ct. E11 ....30Kc 53
Oakhall Ct. TW16: Sun 64V 128
Oakhall Dr. TW16: Sun 64V 128
Oak Hall Rd. E11 ....30Kc 53
Oakham Cl. EN4: Cockf 13Hb 31
Oakham Cl. SE6 ....61Bc 136
Oakham Dr. BR2: Broml 70Hc 137
Oakham Ho. W10 ....42Ya 88
(off Sutton Way)
Oakhampton Rd. NW7 ....24Za 48
Oak Hill IG8: Wfd G 24Fc 53
Oak Hill KT18: Eps 88Ta 173
Oak Hill KT6: Surb 73Na 153
Oak Hill TN13: S'oaks 96Jd 202
Oakhill KT10: Clay 79Ja 152
Oakhill Av. HA5: Pinn 26Aa 45
Oakhill Av. NW3 ....35Db 69
Oak Hill Cl. IG8: Wfd G 24Fc 53
Oakhill Cl. HP2: Hem H 3D 4
Oakhill Cl. KT21: Asht 90La 172
Oakhill Cl. WD3: Map C 21G 42
Oak Hill Ct. IG8: Wfd G 24Fc 53
Oakhill Ct. SE23 ....58Yb 114
Oakhill Ct. SW19 ....66Za 132
Oak Hill Cres. IG8: Wfd G 24Fc 53
Oak Hill Cres. KT6: Surb 73Na 153
Oakhill Dr. KT6: Surb 73Na 153
Oak Hill Gdns. IG8: Wfd G 25Gc 53
Oakhill Gdns. KT13: Weyb 75U 150
Oak Hill Gro. KT6: Surb 72Na 153
Oak Hill Pk. NW3 ....35Db 69
Oak Hill Pk. M. NW3 ....35Eb 69
Oak Hill Path KT6: Surb 72Na 153
Oakhill Pl. SW15 ....57Cb 111
Oak Hill Rd. KT6: Surb 72Na 153
Oak Hill Rd. RM4: Stap A 17Fd 38
Oakhill Rd. BR3: Beck 68Ec 136
Oakhill Rd. BR6: Orp 74Vc 161
Oakhill Rd. KT15: Add 79H 149
Oakhill Rd. KT21: Asht 90La 172
Oakhill Rd. RH2: Reig 7K 207
Oakhill Rd. RM19: Purf 50Rd 97
Oakhill Rd. SM1: Sutt 76Db 155
Oakhill Rd. SW15 ....57Bb 111
Oakhill Rd. SW16 ....67Pb 134
Oakhill Rd. TN13: S'oaks 96Jd 202
Oakhill Rd. WD3: Map C 21F 42
Oak Hill Way NW3 ....35Db 69
(not continuous)
Oak Hill Woods Nature Reserve 16Hb 31
Oak Ho. E14 ....47Ec 92
(off Stewart St.)
Oak Ho. KT15: Add 77M 149
(off Victory Pk. Rd.)
Oak Ho. KT19: Eps 82Sa 173
Oak Ho. KT22: Lea 92Ha 192
Oak Ho. N2 ....26Fb 49
Oak Ho. RM7: Rom 29Fd 56
Oak Ho. TN13: S'oaks 96Kd 203
Oak Ho. TW9: Kew 53Ra 109
Oak Ho. W10 ....42Ab 88
(off Sycamore Wlk.)
Oakhouse Rd. DA6: Bex 57Cd 118
Oakhurst GU21: Wok 8P 167
Oakhurst GU24: Chob 1J 167
Oakhurst Av. DA7: Bex 52Ad 117
Oakhurst Av. EN4: E Barn 18Gb 31
Oakhurst Cl. BR7: Chst 67Pc 138
Oakhurst Cl. E17 ....28Gc 53
Oakhurst Cl. IG6: Ilf 25Sc 54
Oakhurst Cl. KT2: King T 65Pa 131
Oakhurst Cl. TW11: Tedd 64Ga 130
Oakhurst Ct. E17 ....28Gc 53
(off Woodford New Rd.)
Oakhurst Gdns. DA7: Bex 52Ad 117
Oakhurst Gdns. E17 ....28Gc 53
Oakhurst Gdns. E4 ....18Hc 35
Oakhurst Gro. SE22 ....56Wb 113
Oakhurst Pl. WD18: Wat 14V 26
Oakhurst Ri. SM5: Cars 82Gb 175
Oakhurst Rd. EN3: Enf W 8Zb 20
Oakhurst Rd. KT19: Ewe 79Sa 153
Oakington Av. HA2: Harr 31Ca 65
Oakington Av. HA9: Wemb 34Pa 67
Oakington Av. HP6: L Chal 11A 24
Oakington Av. UB3: Harl 49T 84
Oakington Cl. TW16: Sun 68Y 129
Oakington Ct. EN2: Enf 12Rb 33
Oakington Dr. TW16: Sun 68Y 129
Oakington Mnr. Dr. HA9: Wemb 36Qa 67
Oakington Rd. W9 ....42Cb 89
Oakington Way N8 ....31Nb 70
Oakland Ct. KT15: Add 76K 149
Oakland Gdns. CM13: Hut 15Ee 41
Oakland Pl. IG9: Buck H 19Jc 35
Oakland Quay E14 ....48Dc 92
Oakland Rd. E15 ....35Fc 73
Oaklands BR3: Beck 67Dc 136
Oaklands CR8: Kenley 86Sb 177
Oaklands KT22: Fet 96Fa 192
Oaklands N21 ....19Pb 32
Oaklands RH9: S God 9C 210
Oaklands W13 ....43Ja 86
Oaklands, The WD5: Bedm 1V 12
Oaklands Av. AL9: Brk P 9G 8
Oaklands Av. BR4: W W'ck 76Dc 158
Oaklands Av. CR7: Thor H 70Qb 134
Oaklands Av. DA15: Sidc 59Vc 117
Oaklands Av. KT10: Esh 74Fa 152
Oaklands Av. N9 ....16Xb 33
Oaklands Av. RM1: Rom 27Gd 56
Oaklands Av. TW7: Isle 51Ha 108
Oaklands Av. WD19: Wat 18X 27
Oaklands Cl. BR5: Pet W 72Uc 160
Oaklands Cl. DA6: Bex 57Bd 117
Oaklands Cl. GU22: Wok 94B 188
Oaklands Cl. HA0: Wemb 36Ma 67
Oaklands Cl. KT9: Chess 77La 152
Oaklands Cl. TN15: W King 79Ud 164
Oaklands Ct. HA0: Wemb 36Ma 67
Oaklands Ct. NW10 ....39Ua 68
(off Nicoll Rd.)
Oaklands Ct. SE20 ....66Yb 136
(off Chestnut Gro.)
Oaklands Ct. WD17: Wat 11W 26
Oaklands Dr. RH1: Redh 8B 208
Oaklands Dr. RM15: S Ock 43Yd 98
Oaklands Dr. TW2: Whitt 59Ea 108
Oaklands Est. SW4 ....58Lb 112
Oaklands Gdns. CR8: Kenley 86Sb 177
Oaklands Ga. HA6: Nwood 23U 44
Oaklands Gro. W12 ....46Wa 88
Oaklands Hamlet IG7: Chig 22Ad 55
Oaklands La. EN5: Ark 14Xa 30
Oaklands La. TN16: Big H 85Kc 179
Oaklands M. NW2 ....35Za 68
(off Oaklands Rd.)
Oaklands Pk. CM13: Hut 18De 41
Oaklands Pk. Av. IG1: Ilf 33Sc 74
Oaklands Pas. NW2 ....35Za 68
(off Oaklands Rd.)
Oaklands Pl. SW4 ....56Lb 112
Oaklands Rd. BR1: Broml 66Gc 137
Oaklands Rd. DA11: Nflt 3B 144
Oaklands Rd. DA2: Dart 60Rd 119
Oaklands Rd. DA6: Bex 56Bd 117
Oaklands Rd. N20 ....17Bb 31
Oaklands Rd. NW2 ....35Za 68
Oaklands Rd. SW14 ....55Ta 109
Oaklands Rd. W13 ....46Ja 86
Oaklands Rd. W7 ....47Ha 86
(not continuous)
Oaklands Way KT20: Tad 94Ya 194
Oaklands Way SM6: W'gton 80Mb 156
Oakland Way KT19: Ewe 79Ua 154
Oak La. E14 ....45Bc 92
Oak La. EN6: Cuff 1Pb 18
Oak La. GU22: Wok 88D 168
Oak La. IG8: Wfd G 21Hc 53
Oak La. N11 ....23Mb 50
Oak La. N2 ....26Fb 49
Oak La. SL4: Wind 3E 102
Oak La. TN13: S'oaks 100Hd 202
Oak La. TW1: Twick 59Ja 108
Oak La. TW20: Eng G 2N 125
Oak La. TW7: Isle 56Ga 108
Oaklawn Rd. KT22: Lea 90Ga 172
Oak Leaf Cl. KT19: Eps 84Sa 173
Oakleafe Gdns. IG6: Ilf 27Rc 54
Oaklea Lodge IG3: Ilf 34Wc 75
Oaklea Pas. KT1: King T 69Ma 131
Oakleigh GU18: Light 3A 166
Oakleigh KT18: Eps 86Ua 174
Oakleigh RH9: G'stone 2A 210
Oakleigh Av. HA8: Edg 26Ra 47
Oakleigh Av. KT6: Surb 74Qa 153
Oakleigh Av. N20 ....19Fb 31
Oakleigh Cl. BR8: Swan 69Gd 140
Oakleigh Cl. N20 ....20Hb 31
Oakleigh Ct. EN4: E Barn 16Gb 31
Oakleigh Ct. HA8: Edg 26Sa 47
Oakleigh Ct. KT23: Bookh 97Ca 191
Oakleigh Ct. N1 ....2F 219 (40Tb 71)
(off Murray Gro.)
Oakleigh Ct. RH8: Oxt 1J 211
Oakleigh Ct. UB1: S'hall 46Ba 85
Oakleigh Cres. N20 ....19Gb 31
Oakleigh Dr. WD3: Crox G 16S 26
Oakleigh Gdns. BR6: Orp 77Uc 160
Oakleigh Gdns. HA8: Edg 22Pa 47
Oakleigh Gdns. N20 ....18Eb 31
Oakleigh M. N20 ....18Eb 31
OAKLEIGH PARK 18Eb 31
Oakleigh Pk. Av. BR7: Chst 67Qc 138
Oakleigh Pk. Lawn Tennis &
Squash Club 19Fb 31
Oakleigh Pk. Nth. N20 ....18Fb 31
Oakleigh Pk. Sth. N20 ....17Gb 31
Oakleigh Park Station (Rail) 17Fb 31
Oakleigh Ri. CM16: Epp 4Wc 23
Oakleigh Rd. HA5: Hat E 23Ba 45
Oakleigh Rd. UB10: Hil 38S 64
Oakleigh Rd. Nth. N20 ....19Fb 31
Oakleigh Rd. Sth. N11 ....20Jb 32
Oakleigh Way CR4: Mitc 67Kb 134
Oakleigh Way KT6: Surb 74Qa 153
Oakley Av. CR0: Bedd 77Pb 156
Oakley Av. IG11: Bark 38Vc 75
Oakley Av. W5 ....45Qa 87
Oakley Cl. E4 ....20Ec 34
Oakley Cl. E6 ....44Nc 94
Oakley Cl. KT15: Add 77M 149
Oakley Cl. RM20: Grays 51Yd 120
Oakley Cl. TW7: Isle 53Fa 108
Oakley Cl. W7 ....45Ga 86
Oakley Ct. CR4: Mitc 73Jb 156
Oakley Ct. IG10: Lough 12Qc 36
Oakley Ct. RH1: Redh 5A 208
(off St Anne's Ri.)
Oakley Cres. EC1 ....2C 218 (40Rb 71)
Oakley Cres. SL1: Slou 5J 81
Oakley Dr. BR2: Broml 76Nc 160
Oakley Dr. RM3: Rom 22Qd 57
Oakley Dr. SE13 ....58Fc 115
Oakley Dr. SE9 ....60Tc 116
Oakley Gdns. N8 ....29Pb 50
Oakley Gdns. SM7: Bans 87Db 175

Oakley Gdns. SW3 51Gb 111
Oakley Grange HA1: Harr 33Fa 66
OAKLEY GREEN 4A 102
Oakley Ho. SE11 6K 229 (49Qb 90)
(off Hotspur St.)
Oakley Ho. SW1 5G 227 (49Hb 89)
Oakley Ho. W5 45Qa 87
Oakley M. EN2: Enf 12Qb 32
Oakley Pk. DA5: Bexl 59Yc 117
Oakley Pl. SE1 7K 231 (50Vb 91)
Oakley Rd. BR2: Broml 76Nc 160
Oakley Rd. CR6: W'ham 90Wb 177
Oakley Rd. HA1: Harr 30Ga 46
Oakley Rd. N1 38Tb 71
Oakley Rd. SE25 71Xb 157
Oakley Sq. NW1 2C 216 (40Lb 70)
Oakley St. SW3 51Gb 111
Oakley Studios SW3 51Gb 111
(off Up. Cheyne Row)
Oakley Wlk. W6 51Za 110
Oakley Yd. E2 42Vb 91
Oak Lock M. W4 50Ua 88
Oak Lodge E11 30Jc 53
Oak Lodge KT11: Cobh 87Y 171
(off Leigh Cnr.)
Oak Lodge SM1: Sutt 77Eb 155
Oak Lodge TN13: S'oaks 96Jd 202
Oak Lodge TW16: Sun 66V 128
(off Forest Dr.)
Oak Lodge W8 48Db 89
(off Chantry Sq.)
Oak Lodge Av. IG7: Chig 22Tc 54
Oak Lodge Cl. HA7: Stan 22La 46
Oak Lodge Cl. KT12: Hers 78Y 151
Oak Lodge Dr. BR4: W W'ck 73Dc 158
Oak Lodge La. TN16: Westrm 97Tc 200
Oaklodge Way NW7 22Va 48
Oakman Ho. SW19 60Za 110
Oakmead Av. BR2: Hayes 72Jc 159
Oakmead Ct. HA7: Stan 21La 46
Oakmeade HA5: Hat E 23Ca 45
Oakmead Gdns. HA8: Edg 21Ta 47
Oakmead Grn. KT18: Eps 87Sa 173
Oakmead Pl. CR4: Mitc 67Gb 133
Oakmead Rd. CR0: C'don 72Mb 156
Oakmead Rd. SW12 60Jb 112
Oakmede EN5: Barn 14Za 30
OAKMERE 4Eb 17
Oakmere Av. EN6: Pot B 5Eb 17
Oakmere Cl. EN6: Pot B 3Fb 17
Oakmere La. EN6: Pot B 4Eb 17
Oakmere M. EN6: Pot B 4Eb 17
Oakmere Rd. SE2 51Wc 117
Oakmont Pl. BR6: Orp 74Tc 160
Oakmoor Way IG7: Chig 22Uc 54
Oak Pk. KT14: W Byf 85G 168
Oak Pk. Gdns. SW19 60Za 110
Oak Pk. M. N16 34Vb 71
Oak Path WD23: Bush 16Da 27
(off Mortimer Cl.)
Oak Pl. SW18 57Db 111
Oakridge AL2: Brick W 1Ba 13
Oakridge GU24: W End 5D 166
Oakridge Av. WD7: R'lett 6Ha 14
Oakridge Dr. N2 27Fb 49
Oakridge Ho. SL9: Ger X 29B 42
Oakridge La. BR1: Broml 64Fc 137
Oakridge La. WD25: A'ham 6Fa 14
Oakridge La. WD25: R'lett 6Fa 14
Oakridge La. WD7: R'lett 5Ha 14
Oakridge Pl. SL2: Farn C 5G 60
Oakridge Pl. SL9: Ger X 29B 42
Oakridge Rd. BR1: Broml 63Fc 137
Oak Ri. IG9: Buck H 20Mc 35
Oak Rd. DA8: Erith Mill Rd. 52Ed 118
Oak Rd. DA8: Erith Moat La. 54Jd 118
Oak Rd. BR6: Chels 80Wc 161
Oak Rd. CM16: Epp 2Vc 23
Oak Rd. CR3: Cat'm 94Ub 197
Oak Rd. DA12: Grav'nd 2E 144
Oak Rd. DA9: Ghithe 58Ud 120
Oak Rd. KT11: Cobh 87Z 171
Oak Rd. KT22: Lea 90Ja 172
Oak Rd. KT3: N Mald 68Ta 131
Oak Rd. RH2: Reig 5K 207
Oak Rd. RM17: Grays 51Ee 121
Oak Rd. RM3: Hrld W 25Pd 57
Oak Rd. TN16: Westrm 97Tc 200
Oak Rd. W5 45Ma 87
Oak Row SW16 68Lb 134
Oakroyd Av. EN6: Pot B 5Bb 17
Oakroyd Cl. EN6: Pot B 5Bb 17
Oaks, The BR2: Broml 72Qc 160
Oaks, The BR8: Swan 68Gd 140
Oaks, The DA2: Dart 58Rd 119
Oaks, The EN2: Enf 13Rb 33
Oaks, The EN9: Walt A 7Lc 21
(within Woodbine Cl. Cvn. Pk.)
Oaks, The HA4: Ruis 31T 64
Oaks, The IG8: Wfd G 23Gc 53
Oaks, The KT14: W Byf 86J 169
Oaks, The KT18: Eps 86Va 174
Oaks, The KT20: Tad 95Ya 194
Oaks, The KT22: Fet 95Ea 192
Oaks, The N12 21Db 49
Oaks, The NW10 38Xa 68
Oaks, The NW6 38Za 68
(off Brondesbury Pk.)
Oaks, The SE18 50Sc 94
Oaks, The SM4: Mord 70Ab 132
Oaks, The TW13: Felt 61Z 129
Oaks, The TW18: Staines 63H 127
Oaks, The UB4: Hayes 40S 64
Oaks, The WD19: Wat 18Y 27
Oaks, The WD6: Bore 11Qa 29
Oaks Av. KT4: Wor Pk 76Xa 154
Oaks Av. RM5: Col R 26Ed 56
Oaks Av. SE19 64Ub 135
Oaks Av. TW13: Felt 61Aa 129
Oaks Cvn. Pk., The KT9: Chess 76La 152
Oaks Cl. KT22: Lea 93Ja 192
Oaks Cl. WD7: R'lett 7Ha 14
Oaksend Cl. KT22: Oxs 83Ea 172
Oaksford Av. SE26 62Xb 135
Oaks Golf Cen., The 83Hb 175
Oaks Gro. E4 19Gc 35
Oakshade Rd. BR1: Broml 63Fc 137
Oakshade Rd. KT22: Oxs 86Ea 172
Oakshaw RH8: Oxt 99Fc 199
Oakshaw Rd. SW18 59Db 111
Oakshott Ct. NW1 2D 216 (40Mb 70)
(not continuous)
Oakside UB9: Den 36K 63
Oakside Ct. IG6: Ilf 26Tc 54
Oakside Ter. NW10 34Ta 67
Oaks La. CR0: C'don 76Xb 157
Oaks La. IG2: Ilf 29Uc 54
Oaks La. KT23: Bookh 95Ba 191
Oaks Pk. 83Hb 175
Oaks Path WD25: Wat 4Ba 13
Oaks Pavilion M. SE19 64Ub 135
Oak Sq. SW9 54Pb 112
Oak Sq. TN13: S'oaks 98Ld 203
Oaks Rd. CR0: C'don 78Xb 157
Oaks Rd. CR8: Kenley 86Rb 177
Oaks Rd. GU21: Wok 89A 168
Oaks Rd. RH2: Reig 5M 207
Oaks Rd. TW19: Stanw 58M 105
Oaks Sq. KT19: Eps 85Ta 173
Oaks Track SM5: Cars 83Hb 175
Oaks Track SM6: W'gton 82Jb 176
Oak St. HP3: Hem H 6P 3
Oak St. RM7: Rom 29Ed 56
Oaks Vw. KT23: Bookh 97Ba 191
Oaks Way CR8: Kenley 86Sb 177
Oaks Way GU23: Rip 96J 189
Oaks Way KT18: Tatt C 91Xa 194
Oaks Way KT6: Surb 74Ma 153
Oaks Way SM5: Cars 80Hb 155
Oakthorpe Ct. N13 22Sb 51
Oakthorpe Est. N13 22Sb 51
Oakthorpe Rd. N13 22Qb 50
Oak Tree Av. DA9: Bluew 59Vd 120
Oaktree Av. N13 20Rb 33
Oak Tree Cl. GU21: Knap 10F 166
Oak Tree Cl. GU25: Vir W 2P 147
Oak Tree Cl. GU4: Jac W 10P 187
Oak Tree Cl. HA7: Stan 24La 46
Oak Tree Cl. IG10: Lough 11Sc 36
Oak Tree Cl. KT19: Ewe 79Ra 153
Oak Tree Cl. KT23: Bookh 97Ba 191
Oak Tree Cl. TN13: S'oaks 98Ld 203
Oak Tree Cl. W5 44La 86
Oak Tree Cl. WD5: Ab L 4T 12
Oaktree Cl. CM13: B'wood 20Be 41
Oak Tree Ct. UB5: N'olt 40Y 65
Oak Tree Ct. W3 45Ra 87
Oaktree Ct. KT15: Add 78K 149
Oaktree Ct. WD6: E'tree 16Ma 29
Oak Tree Dell NW9 29Sa 47
Oak Tree Dr. N20 18Db 31
Oak Tree Dr. SL3: L'ly 50D 82
Oak Tree Dr. TW20: Eng G 4N 125
Oak Tree Gdns. AL1: St A 1C 6
Oak Tree Gdns. BR1: Broml 64Kc 137
Oaktree Gdns. SE9 62Rc 138
Oaktree Gro. IG1: Ilf 36Tc 74
Oak Tree Ho. W9 42Cb 89
(off Shirland Rd.)
Oak Tree M. NW2 37Wa 68
Oak Tree Pl. KT10: Esh 75Da 151
Oak Tree Rd. GU21: Knap 10F 166
Oak Tree Rd. NW8 4D 214 (41Gb 89)
Oaktree Wlk. CR3: Cat'm 94Ub 197
Oak Vw. HP3: Bov 10D 2
Oak Vw. TW20: Egh 65E 126
Oak Vw. WD18: Wat 13U 26
Oakview Apts. SM1: Sutt 77Fb 155
Oakview Cl. EN7: Chesh 1Xb 19
Oakview Cl. WD19: Wat 16Y 27
Oak Vw. Gdns. SL3: L'ly 49B 82
Oakview Gdns. N2 28Fb 49
Oakview Gro. CR0: C'don 74Ac 158
Oakview Lodge NW11 31Bb 69
(off Beechcroft Av.)
Oakview Rd. SE6 64Dc 136
Oak Village NW5 35Jb 70
Oak Vs. NW11 30Bb 49
(off Hendon Pk. Row)
Oakville Ho. SE16 47Zb 92
(off Dominion Dr.)
Oak Wlk. SM6: W'gton 81Kb 176
(off Helios Rd.)
Oak Way CR0: C'don 72Zb 158
Oak Way KT21: Asht 88Qa 173
Oak Way N14 17Kb 32
Oak Way RH2: Reig 7M 207
Oak Way TW14: Felt 60U 106
Oak Way W3 46Ua 88
Oakway BR2: Broml 68Fc 137
Oakway GU21: Wok 1J 187
Oakway RM16: Grays 46De 99
Oakway SW20 70Ya 132
Oakway Cl. DA5: Bexl 58Ad 117
Oakway Pl. WD7: R'lett 6Ja 14
Oakways SE9 58Rc 116
Oakwell Dr. EN6: N'thaw 4Kb 18
Oakwood EN9: Walt A 7Gc 21
Oakwood KT18: Eps 86Ua 174
(off Worple Rd.)
OAKWOOD 14Mb 32
Oakwood SM6: W'gton 81Kb 176
Oakwood Av. BR2: Broml 69Kc 137
Oakwood Av. BR3: Beck 68Ec 136
Oakwood Av. CR4: Mitc 68Fb 133
Oakwood Av. CR8: Purl 84Rb 177
Oakwood Av. KT19: Eps 81Qa 173
Oakwood Av. N14 17Mb 32
Oakwood Av. UB1: S'hall 45Ca 85
Oakwood Av. WD6: Bore 14Ra 29
Oakwood Bus. Pk. NW10 42Ta 87
Oakwood Chase RM11: Horn 30Pd 57
Oakwood Cl. BR7: Chst 65Pc 138
Oakwood Cl. DA1: Dart 60Rd 119
Oakwood Cl. IG8: Wfd G 23Nc 54
Oakwood Cl. KT24: E Hor 99U 190
Oakwood Cl. N14 16Lb 32
Oakwood Cl. RH1: Redh 6A 208
Oakwood Cl. RH1: S Nut 8F 208
Oakwood Cl. RM3: Rom 23Nd 57
Oakwood Cl. SE13 58Fc 115
Oakwood Ct. BR8: Swan 68Ed 140
(off Lawn Cl.)
Oakwood Ct. E6 39Nc 74
Oakwood Ct. GU24: Bisl 8E 166
Oakwood Ct. HA1: Harr 30Fa 46
Oakwood Ct. SL1: Slou 4J 81
(off Mildenhall Rd.)
Oakwood Ct. W14 48Bb 89
Oakwood Cres. N21 16Nb 32
Oakwood Cres. UB6: G'frd 37Ja 66
Oakwood Dr. AL4: St A 1G 6
Oakwood Dr. DA7: Bex 56Ed 118
Oakwood Dr. HA8: Edg 23Sa 47
Oakwood Dr. KT24: E Hor 99U 190
Oakwood Dr. SE19 65Tb 135
Oakwood Dr. TN13: S'oaks 95Kd 203
Oakwood Farm TN15: Fair 84Ee 185
Oakwood Gdns. BR6: Farnb 75Sc 160
Oakwood Gdns. GU21: Knap 10E 166
Oakwood Gdns. IG3: Ilf 33Vc 75
Oakwood Gdns. SM1: Sutt 75Cb 155
Oakwood Grange KT13: Weyb 76U 150
Oakwood Hall KT20: Kgswd 95Eb 195
Oakwood Hill IG10: Lough 16Pc 36
Oakwood Hill Ind. Est. IG10: Lough 15Sc 36
Oakwood Ho. E9 37Yb 72
(off Frampton Pk. Rd.)
Oakwood Ind. Est. DA11: Nflt 57Ce 121
Oakwood Ind. Est. IG10: Lough 15Rc 36
Oakwood La. W14 48Bb 89
Oakwood Lodge N14 16Lb 32
(off Avenue Rd.)
Oakwood Mans. W14 48Bb 89
(off Oakwood Ct.)
Oakwood Pde. IG10: Lough 16Pc 36
Oakwood Pde. N14 15Lb 32
Oakwood Pk. Rd. N14 17Mb 32
Oak Wood Pl. SL9: Ger X 33A 62
Oakwood Pl. CR0: C'don 72Qb 156
Oakwood Ri. CR3: Cat'm 97Ub 197
Oakwood Ri. DA3: Lfield 69Ae 143
Oakwood Rd. AL2: Brick W 1Ba 13
Oakwood Rd. BR6: Farnb 75Sc 160
Oakwood Rd. CR0: C'don 72Qb 156
Oakwood Rd. GU20: W'sham 9C 146
Oakwood Rd. GU21: Wok 1J 187
Oakwood Rd. GU25: Vir W 1N 147
Oakwood Rd. HA5: Pinn 26X 45
Oakwood Rd. NW11 28Cb 49
Oakwood Rd. RH1: Mers 1G 208
Oakwood Rd. SW20 67Wa 132
Oakwood Station (Underground) 15Lb 32
Oakwood Vw. N14 16Mb 32
Oakworth Rd. W10 43Ya 88
Oak Yd. WD17: Wat 13Y 27
Oarsman Pl. KT8: E Mos 70Ga 130
Oasis, The BR1: Broml 68Lc 137
Oasis Academy Sports Hall 10Zb 20
Oasis Sports Cen. 2F 223 (44Nb 90)
Oast Cotts. TN13: S'oaks 94Jd 202
Oast Ct. E14 45Bc 92
(off Newell St.)
Oast Ho. Cl. TW19: Wray 59A 104
Oasthouse Way BR5: St M Cry 70Xc 139
Oast Lodge W4 52Ua 110
(off Corney Reach Way)
Oast Rd. RH8: Oxt 3K 211
Oast Way DA3: Hartl 72Ae 165
Oates Cl. BR2: Broml 69Fc 137
Oates Rd. RM5: Col R 22Dd 56
Oatfield Ho. N15 30Ub 51
(off Perry Ct.)
Oatfield Rd. BR6: Orp 74Vc 161
Oatfield Rd. KT20: Tad 92Xa 194
Oatland Ri. E17 26Ac 52
Oatlands Av. KT13: Weyb 78T 150
Oatlands Chase KT13: Weyb 76U 150
Oatlands Cl. KT13: Weyb 77S 150
Oatlands Ct. KT13: Weyb 77T 150
Oatlands Dr. KT13: Weyb 77S 150
Oatlands Dr. SL1: Slou 4H 81
Oatlands Grn. KT13: Weyb 76T 150
Oatlands Mere KT13: Weyb 76T 150
OATLANDS PARK 76T 150
Oatlands Rd. EN3: Enf H 11Yb 34
Oatlands Rd. KT20: Tad 91Ab 194
Oat La. EC2 2D 224 (44Sb 91)
Oatridge Gdns. HP2: Hem H 1B 4
Oatwell Ho. SW3 7E 226 (50Gb 89)
(off Cale St.)
Oban Cl. E13 42Lc 93
Oban Ct. SL1: Slou 7H 81
Oban Ho. E14 44Fc 93
(off Oban St.)
Oban Ho. IG11: Bark 40Tc 74
Oban Rd. E13 41Lc 93
Oban Rd. SE25 70Tb 135
Oban St. E14 44Fc 93
Obelisk Ride TW20: Eng G 5J 125
Oberon Cl. WD6: Bore 11Sa 29
Oberon Ct. E6 38Mc 73
Oberon Ct. UB9: Den 30H 43
Oberon Ho. N1 1H 219 (40Ub 71)
(off Arden Est.)
Oberon Way TW17: Shep 69N 127
Oberstein Rd. SW11 56Fb 111
Oborne Cl. SE24 57Rb 113
O'Brien Ho. E2 41Zb 92
(off Roman Rd.)
Observatory Gdns. W8 47Cb 89
Observatory M. E14 49Fc 93
Observatory Rd. SW14 56Sa 109
Observatory Rd. SW7 4B 226 (48Fb 89)
Observatory Shop. Cen., The 7L 81
Observatory Wlk. RH1: Redh 6P 207
Observer Cl. NW9 27Ua 48
Observer Dr. WD18: Wat 14U 26
Occupation La. SE18 53Rc 116
Occupation La. W5 49Ma 87
Occupation Rd. KT19: Ewe 80Ta 153
Occupation Rd. SE17 7D 230 (50Sb 91)
Occupation Rd. W13 47Ka 86
Occupation Rd. WD18: Wat 15X 27
Ocean Est. E1: Ben Jonson Rd. 43Ac 92
Ocean Est. E1: Ernest St. 42Zb 92
Oceanis Apts. E16 45Jc 93
(off Seagull La.)
Ocean St. E1 43Zb 92
Ocean Wharf E14 47Bc 92
Ockbrook E1 43Yb 92
(off Hannibal Rd.)
Ockenden GU22: Wok 91A 188
(off Constitution Hill)
Ockenden Cl. GU22: Wok 90B 168
Ockenden Gdns. GU22: Wok 90B 168
Ockenden Rd. GU22: Wok 90B 168
Ockendon Leisure Cen. 43Wd 98
Ockendon M. N1 37Tb 71
Ockendon Rd. N1 37Tb 71
Ockendon Rd. RM14: N Ock 36Sd 78
Ockendon Rd. RM14: Upm 36Sd 78
Ockendon Station (Rail) 41Xd 98
OCKHAM 93R 190
Ockham Bldg. SE16 4K 231 (48Vb 91)
(off Deal's Gateway)
Ockham Dr. BR5: St P 66Wc 139
Ockham Dr. KT24: W Hor 96T 190
Ockham Dr. UB6: G'frd 38Ea 66
Ockham La. GU23: Ock 93Q 190
Ockham La. KT11: Cobh 92U 190
Ockham Rd. Nth. GU23: Ock 92N 189
Ockham Rd. Nth. KT24: E Hor 96S 190
Ockham Rd. Nth. KT24: W Hor 96S 190
Ockham Rd. Sth. KT24: E Hor 98U 190
Ockley Ct. DA14: Sidc 62Uc 138
Ockley Ct. SM1: Sutt 77Eb 155
Ockley Rd. CR0: C'don 73Pb 156
Ockley Rd. SW16 63Nb 134
Ockleys Mead RH9: G'stone 1A 210
Octagon, The SW10 52Db 111
(off Coleridge Gdns.)
Octagon Arc. EC2 1H 225 (43Ub 91)
Octagon Ct. SE16 46Zb 92
(off Rotherhithe St.)
Octagon Rd. KT12: W Vill 81U 170
Octavia Cl. CR4: Mitc 71Gb 155
Octavia Ct. WD24: Wat 12Y 27
Octavia Ho. SW1 4D 228 (48Mb 90)
(off Medway St.)
Octavia Ho. W10 42Ab 88
Octavia M. W9 42Bb 89
Octavia Rd. TW7: Isle 55Ga 108
Octavia St. SW11 53Gb 111
Octavia Way SE28 45Xc 95
Octavia Way TW18: Staines 65J 127
Octavius St. SE8 52Cc 114
October Pl. NW4 27Za 48
Odard Rd. KT8: W Mole 70Ca 129
Oddesey Rd. WD6: Bore 11Ra 29
Oddmark Ho. IG11: Bark 40Tc 74
Odelia Ct. E15 39Ec 72
(off Biggerstaff Rd.)
Odell Cl. IG11: Bark 38Vc 75
Odell Ho. E14 44Cc 92
(off New Festival Av.)
Odell Wlk. SE13 55Ec 114
Odencroft Rd. SL2: Slou 1E 80
Odeon, The IG11: Bark 38Tc 74
Odeon Cinema Beckenham 68Cc 136
Odeon Cinema Camden Town 39Kb 70
(off Parkway)
Odeon Cinema Covent Gdn. 3E 222 (44Mb 90)
(off Shaftesbury Av.)
Odeon Cinema Edmonton 17Zb 34
Odeon Cinema Epsom 85Ua 174
Odeon Cinema Greenwich 49Hc 93
Odeon Cinema Haymarket, Panton St. 5E 222 (45Mb 90)
(off Panton St.)
Odeon Cinema Holloway 34Nb 70
Odeon Cinema IMAX (BFI) 7K 223 (46Qb 90)
Odeon Cinema Kingston upon Thames 68Na 131
(within The Rotunda Cen.)
Odeon Cinema Leicester Sq. 5E 222 (45Mb 90)
(off Leicester Sq.)
Odeon Cinema Putney 55Ab 110
Odeon Cinema Richmond upon Thames, Hill Street 57Ma 109
Odeon Cinema Richmond upon Thames, Red Lion Street 57Ma 109
Odeon Cinema South Woodford 26Jc 53
Odeon Cinema Streatham 62Nb 134
Odeon Cinema Surrey Quays 48Zb 92
Odeon Cinema Swiss Cottage 38Fb 69
Odeon Cinema Tottenham Ct. Rd. 1D 222 (43Mb 90)
(off Tottenham Ct. Rd.)
Odeon Cinema Uxbridge 39M 63
Odeon Cinema Whiteleys 44Db 89
Odeon Cinema Wimbledon 65Bb 133
Odeon Ct. E16 43Jc 93
Odeon Ct. NW10 39Ua 68
(off St Albans Rd.)
Odeon Pde. N7 34Nb 70
(off Holloway Rd.)
Odeon Pde. SE9 56Nc 116
(off Well Hall Rd.)
Odeon Pde. UB6: G'frd 37Ka 66
(off Allendale Rd.)
Odessa Rd. E7 34Hc 73
Odessa Rd. NW10 40Wa 68
Odessa St. SE16 47Bc 92
Odessa Wharf SE16 48Bc 92
(off Odessa St.)
Odessey Ho. E15 37Fc 73
(off Leyton Rd.)
Odette Ct. WD6: Bore 14Qa 29
(off Whitehall Cl.)
Odette Duval Ho. E1 43Xb 92
(off Stepney Way)
Odger St. SW11 54Hb 111
Odhams Trad. Est. WD24: Wat 9Y 13
Odhams Wlk. WC2 3G 223 (44Nb 90)
Odiam Hall AL9: N Mym 10F 8
Odin Ho. SE5 54Sb 113
O'Donaghue Ho's. SS17: Stan H 1N 101
O'Donnell Ct. WC1 5G 217 (42Nb 90)
O'Driscoll Ho. W12 44Xa 88
Odyssey Bus. Pk. HA4: Ruis 36X 65
Offa Cl. RM12: Horn 36Hd 76
Offa Rd. AL3: St A 2A 6
Offa's Mead E9 35Bc 72
Offenbach Ho. E2 40Zb 72
(off Mace St.)
Offenham Rd. SE9 63Pc 138
Offenham Rd. SW9 53Qb 112
Offers Ct. KT1: King T 69Pa 131
Offerton Ho. WD19: Wat 20Y 27
Offerton Rd. SW4 55Lb 112
Offham Ho. SE17 6H 231 (49Ub 91)
(off Beckway St.)
Offham Slope N12 22Bb 49
Office Pk., The KT22: Lea 91Ha 192
Offley Pl. TW7: Isle 54Fa 108
Offley Rd. SW9 52Qb 112
Offord Cl. N17 23Wb 51
Offord Gro. WD25: Wat 5V 12
Offord Rd. N1 38Pb 70
Offord St. N1 38Pb 70
Ogden Ho. TW13: Hanw 62Aa 129
Ogilby St. SE18 49Pc 94
Ogilvie Ho. E1 44Zb 92
(off Stepney C'way.)
Oglander Rd. SE15 56Vb 113
Ogle St. W1 7B 216 (43Lb 90)
Oglethorpe Rd. RM10: Dag 34Bd 75
O'Gorman Ho. SW10 52Eb 111
(off King's Rd.)
O'Grady Ho. E17 27Dc 52
Ohio Bldg. SE13 53Dc 114
(off Deal's Gateway)
Ohio Rd. E13 42Hc 93
Oil Mill La. W6 50Wa 88
Okeburn Rd. SW17 64Jb 134
Okehampton Cl. N12 22Fb 49
Okehampton Cres. DA16: Well 53Xc 117
Okehampton Rd. NW10 39Ya 68
Okehampton Rd. RM3: Rom 23Ld 57
Okehampton Sq. RM3: Rom 23Ld 57
Okemore Gdns. BR5: St M Cry 70Yc 139
Olaf Ct. W8 47Cb 89
(off Kensington Chu. St.)
Olaf St. W11 45Za 88
Olave Ct. CR5: Coul 88Lb 176
Old Abbey La. SE16 4K 231 (48Vb 91)
(off Vauban St.)
Old Acre GU22: Pyr 86J 169
Oldacre GU24: W End 4D 166
Oldacre M. SW12 59Jb 112
Old Aeroworks, The NW8 6C 214 (42Fb 89)
(off Hatton St.)
Old Amersham Rd. SL9: Ger X 32D 62
Old Av. KT13: Weyb 80S 150
Old Av. KT14: W Byf 85G 168
Old Av. Cl. KT14: W Byf 85G 168
Old Bailey EC4 3C 224 (44Rb 91)
Old Bailey 2C 224 (44Rb 91)
Old Bakery Ct. SL0: Iver 44H 83
Old Barge Ho. All. SE1 5A 224 (45Qb 90)
(off Barge Ho. St.)
Old Barn Cl. SM2: Cheam 80Ab 154
Old Barn Cl. TN15: Kems'g 89Qd 183
Old Barn La. CR8: Kenley 88Vb 177
Old Barn La. WD3: Crox G 15P 25
Old Barn M. WD3: Crox G 15P 25
Old Barn Rd. KT18: Eps 89Sa 173
Old Barn Way DA7: Bex 55Fd 118
Old Barracks W8 47Db 89
Old Barrack Yd. SW1 2H 227 (47Jb 90)
(not continuous)
Old Barrowfield E15 39Gc 73
Old Beaconsfield Rd. SL2: Farn C 7G 60
Old Bell Ct. HP2: Hem H 1M 3
Old Bellgate Pl. E14 48Cc 92
Oldberry Rd. HA8: Edg 23Ta 47
Old Bethnal Grn. Rd. E2 41Wb 91
OLD BEXLEY 59Dd 118
Old Bexley Bus. Pk. DA5: Bexl 59Dd 118
Old Bexley La. DA1: Dart 60Hd 118
Old Bexley La. DA5: Bexl 61Fd 140
Old Bexley La. DA5: Dart 61Fd 140
Old Billingsgate Mkt. EC3 5H 225 (45Ub 91)
(off Lwr. Thames St.)
Old Billingsgate Wlk. EC3 5H 225 (45Ub 91)
Old Bond St. W1 5B 222 (45Lb 90)
Oldborough Rd. HA0: Wemb 34La 66
OLD BRENTFORD 52Ma 109
Old Brewer's Yd. WC2 3F 223 (44Nb 90)
Old Brewery M. NW3 35Fb 69
Old Brewery Way E17 29Bc 52
Old Brickworks Ind. Est., The RM3: Hrld W 25Qd 57
Old Bri. Cl. UB5: N'olt 40Ca 65
Old Bridge La. KT17: Eps 84Va 174
Old Bridge St. KT1: Hamp W 68Ma 131
Old Broad St. EC2 2G 225 (44Tb 91)
OLD BROAD STREET PRIVATE MEDICAL CEN. 2H 225 (44Ub 91)
(off Old Broad St.)
Old Bromley Rd. BR1: Broml 64Fc 137
Old Brompton Rd. SW5 50Cb 89
Old Brompton Rd. SW7 7A 226 (50Eb 89)
Old Bldgs. WC2 2K 223 (44Qb 90)
(off Chancery La.)
Old Burlington St. W1 5B 222 (45Lb 90)
OLDBURY 94Xd 204
Oldbury Cl. BR5: St M Cry 70Zc 139
Oldbury Cl. TN15: Igh 94Xd 204
Oldbury Cotts. TN15: Igh 93Xd 204
Oldbury Ct. E9 36Ac 72
(off Mabley St.)
Oldbury Hill 93Wd 204
Oldbury Hill Camping & Caravanning Club Site TN15: Seal 94Vd 204
Oldbury Hillfort 94Wd 204
Oldbury Ho. W2 43Db 89
(off Harrow Rd.)
Oldbury La. TN15: Igh 93Xd 204
Oldbury Pl. W1 7J 215 (43Jb 90)
Oldbury Rd. EN1: Enf 12Wb 33
Oldbury Rd. KT16: Chert 73G 148
Oldbury Vs. TN15: Igh 94Xd 204
Old Canal M. SE15 7K 231 (50Vb 91)
(off Trafalgar Av.)
Old Carriageway, The TN13: Chips 94Ed 202
Old Castle St. E1 2K 225 (44Vb 91)
Old Cavendish St. W1 2K 221 (44Kb 90)
Old Change Ct. EC4 3D 224 (44Sb 91)
(off Distaff La.)
Old Chapel Pl. SW9 54Qb 112
Old Chapel Rd. BR8: Crock 73Ed 162
Old Charlton Rd. TW17: Shep 71S 150
Old Chelsea M. SW3 51Gb 111
Old Chertsey Rd. GU24: Chob 2M 167
Old Chestnut Av. KT10: Esh 79Ca 151
Old Chiswick Yd. W4 51Ua 110
(off Pumping Sta. Rd.)
Old Chorleywood Rd. WD3: Rick 16M 25
Old Church Cl. BR6: Orp 74Xc 161
Old Church Ct. N11 22Kb 50
Oldchurch Gdns. RM7: Rush G 31Fd 76
Old Church La. HA7: Stan 22Ka 46
Old Church La. NW9 33Ta 67
Old Church La. UB6: G'frd 41Ja 86
Old Church Path KT10: Esh 77Da 151
Oldchurch Ri. RM7: Rush G 31Gd 76
Old Church Rd. E1 44Zb 92
Old Church Rd. E4 21Cc 52
Oldchurch Rd. RM7: Rush G 31Fd 76
Old Church St. SW3 7C 226 (50Fb 89)
Old Claygate La. KT10: Clay 79Ja 152
Old Clem Sq. SE18 51Qc 116
(off Woolwich Comn.)
Old Coach Rd. KT16: Chert 71F 148
Old Coach Rd. TN15: Wro 86Ae 185
Old Coal Yd. SE28 49Tc 94
Old Coalyard, The TW20: Egh 65B 126
Old College Ct. DA17: Belv 50Dd 96
Old Common 84X 171
OLD COMMON 85X 171
Old Common Rd. KT11: Cobh 84X 171
Old Common Rd. WD3: Chor 14F 24
Old Compton St. W1 4D 222 (45Mb 90)
Old Cople La. SE19 64Vb 135
Old Cote Dr. TW5: Hest 51Ca 107
Old Cotts. AL2: Lon C 1Na 15
Old Cotts. TN15: Igh 94Xd 204
OLD COULSDON 91Qb 196
Old Ct. KT21: Asht 91Na 193
Old Ct. Ho. W8 47Db 89
(off Old Court Pl.)
Old Courthouse, The KT18: Eps 85Ta 173
(off The Parade)
Old Ct. Pl. W8 47Db 89
Old Ctyd., The BR1: Broml 67Kc 137
Old Crabtree La. HP2: Hem H 3N 3
Old Crown La. CM14: Kel H 12Td 40
Old Curiosity Shop 2J 223 (44Pb 90)
(off Portsmouth St.)
Old Dairy N15 30Ub 51
Old Dairy Gdns. UB2: S'hall 50Ca 85
Old Dairy La. HA4: Ruis 35X 65
Old Dairy M. NW5 37Kb 70
Old Dairy M. SW12 60Jb 112
Old Dairy M. SW4 56Nb 112
(off Tintern St.)
Old Dairy Sq. N21 17Qb 32
(off Wade Hill)
Old Dartford Rd. DA4: Farni 72Pd 163
Old Dean HP3: Bov 10C 2
Old Deer Pk. 54La 108
Old Deer Pk. Gdns. TW9: Rich 55Na 109
Old Devonshire Rd. SW12 59Kb 112

Old Dock Cl. TW9: Kew 51Qa 109
Old Dover Rd. SE3 52Jc 115
Old Downs DA3: Hartl 71Ae 165
Oldegate Ho. E6 38Mc 73
Olden La. CR8: Purl 84Qb 176
Old Esher Cl. KT12: Hers 78Z 151
Old Esher Rd. KT12: Hers 78Z 151
Old Farleigh Rd. CR2: Sels 82Yb 178
Old Farleigh Rd. CR6: W'ham 85Ac 178
Old Farm Av. DA15: Sidc 60Tc 116
Old Farm Av. N14 17Lb 32
Old Farm Cl. SW17 61Gb 133
Old Farm Cl. TW20: Thorpe 69D 126
Old Farm Cl. TW4: Houn 56Ba 107
Old Farm Gdns. BR8: Swan 69Hd 140
Old Farmhouse Dr. KT22: Oxs 87Fa 172
Old Farmhouse M. AL9: Wel G 5E 8
Old Farm La. SW19 64Db 133
Old Farm Pas. TW12: Hamp 67Ea 130
Old Farm Rd. N2 25Fb 49
Old Farm Rd. TW12: Hamp 65Ba 129
Old Farm Rd. UB7: W Dray 47M 83
Old Farm Rd. E. DA15: Sidc 61Wc 139
Old Farm Rd. W. DA15: Sidc 61Vc 139
Old Ferry Dr. TW19: Wray 8N 103
Old Field Cl. HP6: L Chal 11A 24
Oldfield Cl. BR1: Broml 70Pc 138
Oldfield Cl. HA7: Stan 22Ja 46
Oldfield Cl. UB6: G'frd 36Ga 66
Oldfield Ct. AL1: St A 3C 6
Oldfield Ct. KT5: Surb 70Pa 131
(off Cranes Pk. Cres.)
Oldfield Farm Gdns. UB6: G'frd 39Fa 66
Oldfield Gdns. KT21: Asht 91Ma 193
Oldfield Gro. SE16 49Zb 92
Oldfield Ho. W4 50Ua 88
(off Devonshire Rd.)
Oldfield La. Nth. UB6: G'frd 40Fa 66
Oldfield La. Sth. UB6: G'frd 42Ea 86
Oldfield M. N6 31Lb 70
Oldfield Pl. DA1: Dart 58Nd 119
Oldfield Rd. AL2: Lon C 7H 7
Oldfield Rd. BR1: Broml 70Pc 138
Oldfield Rd. DA7: Bex 54Ad 117
Oldfield Rd. HP1: Hem H 3G 2
Oldfield Rd. N16 34Ub 71
Oldfield Rd. NW10 38Va 68
Oldfield Rd. SW19 65Ab 132
Oldfield Rd. TW12: Hamp 67Ba 129
Oldfield Rd. W3 47Va 88
Oldfields CM14: W'ley 21Yd 58
Oldfields Cir. UB5: N'olt 37Ea 66
Oldfields Rd. SM1: Sutt 76Bb 155
Oldfield Wood GU22: Wok 89D 168
Old Fire Sta., The SE18 52Rc 116
Old Fishery La. HP1: Hem H 5H 3
(not continuous)
Old Fish St. Hill EC4 4D 224 (45Sb 91)
(off Queen Victoria St.)
Old Fives Ct. SL1: Burn 1A 80
Old Fleet La. EC4 2B 224 (44Rb 91)
Old Fold Cl. EN5: Barn 11Bb 31
Old Fold La. EN5: Barn 11Bb 31
Old Fold Manor Golf Course 11Ab 30
Old Fold Vw. EN5: Barn 13Ya 30
OLD FORD 39Bc 72
Old Ford 39Cc 72
Old Ford Rd. E2 41Yb 92
Old Ford Rd. E3 40Ac 72
Old Ford Trading Cen. E3 39Cc 72
(off Maverton Rd.)
Old Forge Cl. HA7: Stan 21Ja 46
Old Forge Cl. WD25: Wat 5W 12
Old Forge Ct. EN9: Walt A 6Jc 21
(off Lamplighters Cl.)
Old Forge Cres. TW17: Shep 72R 150
Old Forge M. W12 47Xa 88
Old Forge Rd. EN1: Enf 10Vb 19
Old Forge Rd. N19 33Mb 70
Old Forge Way DA14: Sidc 63Xc 139
Old Fox Cl. CR3: Cat'm 93Rb 197
Old Gannon Cl. HA6: Nwood 21S 44
Old Gdn., The TN13: Chips 95Ed 202
Old Garden Ct. AL3: St A 2A 6
Old Gloucester St. WC1 7G 217 (43Nb 90)
Old Goods Yd., The W2 1A 220 (43Eb 89)
Old Gorhambury House 1H 5
Old Hall Cl. HA5: Pinn 25Aa 45
Old Hall Dr. HA5: Pinn 25Aa 45
Oldham Ter. W3 46Sa 87
(not continuous)
Old Harrow La. TN16: Westrm 91Sc 200
Old Hatch Mnr. HA4: Ruis 31V 64
Old Hat Factory, The AL1: St A 3C 6
(off Inkerman Rd.)
Old Haven Cl. DA8: Erith 54Hd 118
Old Heath Rd. KT13: Weyb 79Q 150
Old Highwayman Pl. SW15 60Xa 110
Old Hill BR6: Downe 79Tc 160
Old Hill BR7: Chst 67Qc 138
Old Hill GU22: Wok 2P 187
Old Hill Est. GU22: Wok 2P 187
Oldhill St. N16 32Wb 71
Old Homesdale Rd. BR2: Broml 70Lc 137
Old Hospital Cl. SW12 60Hb 111
Old Ho. Cl. KT17: Ewe 82Va 174
Old Ho. Cl. SW19 64Ab 132
Old Ho. Ct. HP2: Hem H 2P 3
Old Ho. Ct. SL3: Wex 3P 81
Old Ho. Gdns. TW1: Twick 58La 108
Old Ho. La. WD4: Bucks 7N 11
Old Ho. La. WD4: Lang 7N 11
Oldhouse La. GU20: W'sham 10A 146
(not continuous)
Oldhouse La. GU24: Bisl 6E 166
Old Ho. Rd. HP2: Hem H 2P 3
Old Howlett's La. HA4: Ruis 30T 44
OLD ISLEWORTH 55Ka 108
Old Jamaica Bus. Est. SE16 48Vb 91
Old Jamaica Rd. SE16 48Wb 91
Old James St. SE15 55Xb 113
Old Jenkins Cl. SS17: Stan H 2K 101
Old Jewry EC2 3F 225 (44Tb 91)
Old Kenton La. NW9 29Ra 47
Old Kent Rd. SE1 5H 231 (49Ub 91)
Old Kent Rd. SE15 51Wb 113
Old Kingston Rd. KT4: Wor Pk 75Sa 153
Old La. KT11: Cobh 89R 170
Old La. RH8: Oxt 1K 211
(not continuous)
Old La. TN15: Igh 95Xd 204
Old La. TN16: Tats 92Mc 199
Old La. Gdns. KT11: Cobh 94W 190
Old Laundry, The BR7: Chst 67Sc 138
Old Leys AL10: Hat 3C 8
Old Library Ct. HA4: Ruis 33W 64
Old Library Ho. E3 40Ac 72
(off Roman Rd.)
Old Lodge La. CR8: Kenley 85Pb 176
Old Lodge La. CR8: Purl 85Pb 176
Old Lodge Pl. TW1: Twick 58Ka 108
Old Lodge Way HA7: Stan 22Ja 46
Old London Rd. AL1: St A 3B 6
Old London Rd. DA14: Sidc 66Cd 140
Old London Rd. DA14: Swan 66Cd 140
Old London Rd. KT18: Eps D 91Wa 194
(not continuous)
Old London Rd. KT2: King T 68Na 131
Old London Rd. KT24: E Hor 98W 190
Old London Rd. RH5: Mick 99La 192
Old London Rd. TN14: Bad M 81Bd 181
Old London Rd. TN14: Hals 81Bd 181
Old London Rd. TN14: Knock 87Ad 181
Old London Rd. TN15: Wro 87Be 185
Old Lyonian Sports Club 29Ea 46
Old MacDonald's Farm 18Pd 39
Old Maidstone Rd. DA14: Sidc 66Bd 139
OLD MALDEN 74Ua 154
Old Malden La. KT4: Wor Pk 75Ta 153
Old Malt Way GU21: Wok 9P 167
Old Mnr. Ct. NW8 1A 214 (40Eb 69)
Old Mnr. Dr. DA12: Grav'nd 10E 122
Old Mnr. Dr. TW7: Isle 58Ea 108
Old Mnr. Ho. M. TW17: Shep 69Q 128
Old Mnr. Rd. UB2: S'hall 49Z 85
Old Mnr. Way BR7: Chst 64Pc 138
Old Mnr. Way DA7: Bex 54Fd 118
Old Manor Yd. SW5 49Db 89
Old Mkt. Ct. SM1: Sutt 77Db 155
Old Mkt. Sq. E2 3K 219 (41Vb 91)
Old Marylebone Rd. NW1 1E 220 (43Gb 89)
Old Mead SL9: Chal P 23A 42
Oldmead Cl. RM3: Rom 23Nd 57
Oldmead Ho. RM10: Dag 37Dd 76
Old M. HA1: Harr 29Ga 46
Old Mile Ho. Ct. AL1: St A 5E 6
Old Mill Cl. DA4: Eyns 74Nd 163
Old Mill Cl. UB8: Cowl 43K 83
Old Mill Ct. E18 27Lc 53
Old Mill Gdns. HP4: Berk 1A 2
Old Mill La. RH1: Mers 100Kb 196
Old Mill La. UB8: Cowl 44K 83
Old Mill Pde. RM1: Rom 29Hd 56
Old Mill Pl. RM7: Rom 30Fd 56
Old Mill Pl. TW19: Wray 58D 104
Old Mill Rd. SE18 51Tc 116
Old Mill Rd. UB9: Den 34J 63
Old Mill Rd. WD4: Hunt C 5S 12
Old Mitre Ct. EC4 3A 224 (44Qb 90)
Old Moat M. RM3: Rom 22Md 57
Old Montague St. E1 43Wb 91
Old Nichol St. E2 5K 219 (42Vb 91)
Old North St. WC1 7H 217 (43Pb 90)
(off Theobald's Rd.)
Old Nurseries La. KT11: Cobh 85X 171
Old Nursery DA13: Ist R 6B 144
Old Nursery Cl. WD7: Shenl 6Qa 15
Old Nursery Ct. E2 2K 219 (40Vb 71)
(off Dawson St.)
Old Nursery Ct. SL2: Hedg 3G 60
Old Nursery Pl. TW15: Ashf 64R 128
Old Oak AL1: St A 5C 6
Old Oak Av. CR5: Chips 91Gb 195
Old Oak Cl. KT11: Cobh 85X 171
Old Oak Cl. KT9: Chess 77Pa 153
OLD OAK COMMON 43Ua 88
Old Oak Comn. La. NW10 43Ua 88
Old Oak Comn. La. W3 43Ua 88
Old Oak Gdns. GU21: Wok 7P 167
Old Oak La. NW10 41Ua 88
Old Oak Rd. W3 45Va 88
Old Oaks EN9: Walt A 4Gc 21
Old Operating Theatre Mus. & Herb Garret, The 7G 225 (46Tb 91)
(off St Thomas St.)
Old Orchard AL2: Park 8A 6
Old Orchard KT14: Byfl 84P 169
Old Orchard TW16: Sun 68Y 129
Old Orchard, The NW3 35Hb 69
Old Orchard, The SL0: Iver 44H 83
Old Orchard Cl. EN4: Had W 10Hb 17
Old Orchard Cl. UB8: Hil 44Q 84
Old Otford Rd. TN14: Otf 89Kd 183
(not continuous)
Old Otford Rd. TN14: S'oaks 89Kd 183
(not continuous)
OLD OXTED 2H 211
Old Palace La. TW9: Rich 57La 108
Old Palace Rd. CR0: C'don 76Rb 157
Old Palace Rd. KT13: Weyb 76R 150
Old Palace Ter. TW9: Rich 57Ma 109
Old Palace Yd. SW1 3F 229 (48Nb 90)
Old Palace Yd. TW9: Rich 57La 108
Old Paradise St. SE11 5H 229 (49Pb 90)
Old Pk. Av. EN2: Enf 14Sb 33
Old Pk. Av. SW12 58Jb 112
Old Parkbury La. AL2: Col S 1Ha 14
Old Pk. Gro. EN2: Enf 14Sb 33
Old Pk. La. W1 7K 221 (46Kb 90)
Old Pk. M. TW5: Hest 52Ba 107
Old Park Ride EN7: Walt C 3Rb 19
Old Pk. Ridings N21 16Rb 33
Old Pk. Rd. EN2: Enf 13Rb 33
Old Pk. Rd. N13 21Pb 50
Old Pk. Rd. SE2 50Wc 95
Old Pk. Rd. Sth. EN2: Enf 14Rb 33
Old Pk. Vw. EN2: Enf 13Qb 32
Old Pk. Wood Nature Reserve 24K 43
Old Parsonage Yd. DA4: Hort K 69Sd 142
Old Parvis Rd. KT14: W Byf 84L 169
Old Pearson St. SE10 52Dc 114
Old Perry St. BR7: Chst 65Uc 138
Old Perry St. DA11: Nflt 1A 144
Old Polhill TN14: Hals 86Ed 182
Old Polhill TN14: Otf 86Ed 182
Old Police Sta., The SW17 61Hb 133
Old Police Station M. SE20 66Zb 136
Old Post Office La. SE3 55Kc 115
Old Post Office M. GU21: Wok 10L 167
Old Post Office Wlk. KT6: Surb 72Ma 153
(off Victoria Rd.)
Old Pottery Cl. RH2: Reig 8K 207
Old Pound Cl. TW7: Isle 53La 108
Old Priory Av. BR6: Orp 73Xc 161
Old Priory Pk. AL1: St A 3C 6
(off Old London Rd.)
Old Pye St. SW1 3D 228 (48Mb 90)
Old Pye St. Est. SW1 4D 228 (48Mb 90)
(off Old Pye St.)
Old Quebec St. W1 3G 221 (44Hb 89)
Old Queen St. SW1 2E 228 (47Mb 90)
Old Rectory, The KT23: Bookh 99Ba 191
Old Rectory Cl. KT20: Wat H 96Wa 194
Old Rectory Dr. AL10: Hat 1D 8
Old Rectory Gdns. HA8: Edg 23Qa 47
Old Rectory Gdns. KT11: Cobh 87X 171
Old Rectory La. KT24: E Hor 98U 190
Old Rectory La. UB9: Den 31G 62
Old Redding HA3: Hrw W 22Da 45
Old Red Lion Theatre 2A 218 (40Qb 70)
(off St John St.)
Old Redstone Dr. RH1: Redh 7A 208
Oldridge Rd. SW12 59Jb 112
Old Rd. CM14: N'side 14Nd 39
Old Rd. CM14: Nave 14Nd 39
Old Rd. DA1: Cray 57Fd 118
Old Rd. EN3: Enf H 11Yb 34
Old Rd. KT15: Add 80H 149
Old Rd. RH3: Bkld 6A 206
Old Rd. RM4: Nave 12Md 39
Old Rd. SE13 56Gc 115
Old Rd. E. DA12: Grav'nd 10D 122
Old Rd. W. DA11: Grav'nd 10B 122
Old Rope Wlk. TW16: Sun 69X 129
Old Royal Free Pl. N1 39Qb 70
Old Royal Free Sq. N1 39Qb 70
Old Royal Naval College 50Fc 93
Old Ruislip Rd. UB5: N'olt 40Y 65
Old St Mary's KT24: W Hor 100R 190
Old's App. WD18: Wat 18R 26
Old Saw Mill, The TN15: Plat 94Fe 205
Old School, The WC1 7J 217 (43Pb 90)
(off Princeton St.)
Old School Cl. BR3: Beck 68Zb 136
Old School Cl. RH1: Redh 4A 208
Old School Cl. SE10 48Hc 93
Old School Cl. SW19 68Cb 133
Old School Cotts. HP5: Whel H 8A 2
Old School Ct. BR8: Swan 68Gd 140
(off Bonney Way)
Old School Ct. KT22: Lea 94Ka 192
Old School Ct. N17 27Vb 51
Old School Ct. TN13: S'oaks 94Ld 203
Old School Ct. TW19: Wray 59A 104
Old School Cres. E7 37Jc 73
Old School Gdns. CR3: Cat'm 95Wb 197
Old School Ho's. TN15: God G 96Qd 203
Old School M. KT13: Weyb 77T 150
Old School M. TW18: Staines 64F 126
Old School Pl. CR0: Wadd 77Qb 156
Old School Pl. GU22: Wok 93A 188
Old School Rd. UB8: Hil 42P 83
Old Schools La. KT17: Ewe 81Va 174
Old School Sq. E14 44Cc 92
(off Pelling St.)
Old School Sq. KT7: T Ditt 72Ha 152
Old School Ter. SM3: Cheam 80Za 154
Old School Wlk. TN13: S'oaks 97Kd 203
(off London Rd.)
Old School Yd. DA12: Grav'nd 9G 122
Old School Yd. RH1: Nutf 5F 208
Old's Cl. WD18: Wat 18R 26
Old Seacoal La. EC4 2B 224 (44Rb 91)
Old Sessions Ho., The EC1 6B 218 (42Rb 91)
Old Shire La. EN9: Walt A 7Jc 21
Old Shire La. WD3: Chor 16C 24
Old Slade La. SL0: Rich P 48G 82
Old Slade La. SL3: Coln 50H 83
Old Soar Manor 98De 205
Old Soar Rd. TN15: Plax 99De 205
Old Solesbridge La. WD3: Chor 13J 25
Old Sopwell Gdns. AL1: St A 4C 6
Old Sth. Cl. HA5: Pinn 25Z 45
Old Sth. Lambeth Rd. SW8 52Nb 112
Old Speech Room Gallery 32Ga 66
Old Spitalfields Market 7K 219 (43Vb 91)
Old Sq. WC2 2J 223 (44Pb 90)
Old Stable M. N5 34Sb 71
Old Stable Row SE18 49Qc 94
(off Woolwich New Rd.)
Old Stables Ct. SE5 53Sb 113
(off Camberwell New Rd.)
Old Stable Yd. DA2: Bean 62Zd 143
Old Stable Yd., The BR8: Swan 67Ld 141
Old Station App. KT22: Lea 93Ja 192
Old Station Bus. Cen., The AL1: St A 4D 6
Old Station Gdns. TW11: Tedd 65Ja 130
(off Victoria Rd.)
Old Station Ho. SE17 7E 230 (50Sb 91)
Old Station La. RM13: Rain 42Jd 96
Old Station Pas. TW9: Rich 56Ma 109
(off Little Green)
Old Station Rd. IG10: Lough 15Nc 36
Old Station Rd. UB3: Harl 48V 84
Old Sta. Way SW4 55Mb 112
(off Voltaire Rd.)
Old Station Way SW4 55Mb 112
(Off Voltaire Road)
Old Station Yd., The E17 28Ec 52
Oldstead Rd. BR1: Broml 63Ec 136
Old Stede Cl. KT21: Asht 89Pa 173
Old Stockley Rd. UB7: W Dray 47R 84
Old St. E13 40Kc 73
Old St. EC1 5D 218 (42Sb 91)
OLD STREET 5G 219 (42Tb 91)
Old St., The KT22: Fet 95Fa 192
Old Street Station (Rail & Underground) 5G 219 (42Tb 91)
Old St. Yd. EC1 5F 219 (42Tb 91)
(off City Rd.)
Old Studio Cl. CR0: C'don 73Tb 157
Old Sungate Cotts. RM5: Col R 25Bd 55
Old Sun Wharf E14 45Ac 92
(off Narrow St.)
Old Surrey M., The RH9: G'stone 2A 210
Old Swan Wharf SW11 53Fb 111
Old Swan Yd. SM5: Cars 77Hb 155
Old Terry's Lodge Rd. TN15: Kems'g 88Vd 184
Old Theatre Ct. SE1 6E 224 (46Sb 91)
(off Porter St.)
Old Tilburstow Rd. RH9: S God 6A 210
Old Town CR0: C'don 76Rb 157
Old Town SW4 55Lb 112
Old Town Hall Apts. SE16 4K 231 (48Vb 91)
Old Town Hall Arts Cen., The 1L 3
Old Tramyard SE18 49Uc 94
Old Trowley WD5: Ab L 3V 12
Old Twelve Cl. W7 42Ga 86
Old Tye Av. TN16: Big H 88Nc 180
Old Uxbridge Rd. WD3: Map C 22G 42
Old Uxbridge Rd. WD3: W Hyd 22G 42
Old Vic Theatre, The 1A 230 (47Qb 90)
(off The Cut)
Old Vinyl Factory, The UB3: Hayes 47U 84
Old Wlk., The TN14: Otf 89Ld 183
Old Watercress Wlk. SM5: Cars 77Jb 156
Old Watford Rd. AL2: Brick W 2Aa 13
Old Watling St. DA11: Grav'nd 4C 144
Oldway La. SL1: Slou 5B 80
(not continuous)
Old Westhall Cl. CR6: W'ham 91Yb 198
Old Wharf Way KT13: Weyb 77P 149
Old Willow Cl. E3 41Cc 92
OLD WINDSOR 8L 103
Old Windsor Lock SL4: Old Win 7N 103
OLD WOKING 93D 188
Old Woking Rd. GU22: Pyr 91D 188
Old Woking Rd. GU22: Wok 91D 188
Old Woking Rd. KT14: W Byf 86J 169
Old Woolwich Rd. SE10 51Fc 115
Old Works, The AL1: St A 3C 6
(off Black Cut)
Old Yd., The RH1: Blet 5J 209
Old Yd., The TN16: Bras 96Yc 201
Old Yews, The DA3: Lfield 69De 143
Old York Rd. SW18 57Db 111
Oleander Cl. BR6: Farnb 78Tc 160
Oleander Ho. SE15 50Vb 91
O'Leary Sq. E1 43Yb 92
Olga St. E3 40Ac 72
Olinda Rd. N16 30Vb 51
Oliphant St. E14 45Ec 92
(off Bullivant St.)
Oliphant St. W10 41Za 88
Olive Blythe Ho. W10 42Ab 88
(off Ladbroke Gro.)
Olive Cl. AL1: St A 3F 6
Olive Ct. E5 33Yb 72
(off Woodmill Rd.)
Olive Ct. N1 1A 218 (39Qb 70)
(off Liverpool Rd.)
Olive Gro. IG7: Chig 22Ad 55
Olive Gro. N15 28Sb 51
Olive Haines Lodge SW15 57Bb 111
Olive Ho. EC1 5A 218 (42Qb 90)
(off Bowling Grn. La.)
Oliver Av. SE25 69Vb 135
Oliver Bus. Pk. NW10 40Sa 67
Oliver Cl. AL2: Park 9B 6
Oliver Cl. HP3: Hem H 6N 3
Oliver Cl. KT15: Add 77J 149
Oliver Cl. RM20: W Thur 52Vd 120
Oliver Cl. W4 51Ra 109
Oliver Cl. Ind. Est. RM20: W Thur 52Vd 120
Oliver Ct. SE18 49Sc 94
Oliver Ct. WD25: Wat 8Z 13
Oliver Cres. DA4: Farni 73Pd 163
Oliver Gdns. E6 43Nc 94
Oliver Goldsmith Est. SE15 53Wb 113
Oliver Gro. SE25 70Vb 135
Oliver Ho. SE14 52Ac 114
(off New Cross Rd.)
Oliver Ho. SE16 47Wb 91
(off George Row)
Oliver Ho. SW8 52Nb 112
(off Wyvil Rd.)
Oliver M. SE15 54Wb 113
Olive Rd. DA1: Dart 60Md 119
Olive Rd. E13 41Lc 93
Olive Rd. NW2 35Xa 68
Olive Rd. SW19 66Eb 133
Olive Rd. W5 48Ma 87
Oliver Ri. HP3: Hem H 6N 3
Oliver Rd. BR8: Swan 69Fd 140
Oliver Rd. CM15: Shenf 15Ce 41
Oliver Rd. E10 33Dc 72
Oliver Rd. E17 29Ec 52
Oliver Rd. HP3: Hem H 6N 3
Oliver Rd. KT3: N Mald 68Sa 131
Oliver Rd. NW10 40Sa 67
Oliver Rd. RM13: Rain 39Hd 76
Oliver Rd. RM20: W Thur 52Vd 120
Oliver Rd. SM1: Sutt 77Fb 155
Olivers Mill DA3: Nw A G 75Ae 165
Olivers Row N8 30Nb 50
Olivers Wharf E1 46Xb 91
(off Wapping High St.)
Olivers Yd. EC1 5G 219 (42Tb 91)
Olive St. RM7: Rom 29Fd 56
Olive Tree Ho. SE15 51Yb 114
(off Sharratt St.)
Olivette St. SW15 55Za 110
Olive Waite Ho. NW6 38Cb 69
Olivia Ct. CR0: C'don 73Sb 157
(off Whitehorse Rd.)
Olivia Ct. EN2: Enf 11Sb 33
(off Chase Side)
Olivia Dr. SL3: L'ly 50B 82
Olivia Gdns. UB9: Hare 25L 43
Olivia M. HA3: Hrw W 24Ga 46
Olivier Ct. UB9: Den 30H 43
(off Patrons Way E.)
Olivier Theatre 6K 223 (46Qb 90)
(within National Theatre)
Ollard's Ct. IG10: Lough 15Mc 35
Ollard's Gro. IG10: Lough 14Mc 35
Olleberrie La. WD3: Sarr 3F 10
Ollerton Grn. E3 39Bc 72
Ollerton Rd. N11 22Mb 50
Olley Cl. SM6: W'gton 80Nb 156
Ollgar Cl. W12 46Va 88
Olliffe St. E14 48Ec 92
Olmar St. SE1 51Wb 113
Olmstead Cl. N10 28Jb 50
Olney Ho. NW8 5E 214 (42Gb 89)
(off Tresham Cres.)
Olney Rd. SE17 51Rb 113
(not continuous)
Olron Cres. DA6: Bex 57Zc 117
Olven Rd. SE18 51Sc 116
Olveston Wlk. SM5: Cars 72Fb 155
Olwen M. HA5: Pinn 26Z 45
Olyffe Av. DA16: Well 53Wc 117
Olyffe Dr. BR3: Beck 67Ec 136
Olympia 48Ab 88
Olympia Ind. Est. N22 27Pb 50
Olympia M. W2 45Db 89
Olympian Cl. TW19: Stanw 58M 105
Olympian Ct. E14 49Cc 92
(off Homer Dr.)
Olympian Ct. E3 40Cc 72
(off Wick La.)
Olympian Way SE10 46Gc 93
(not continuous)
Olympia Way W14 48Ab 88
Olympic Golf Driving Range, The 70Kd 141
Olympic Ho. N16 35Vb 71
Olympic M. SW18 57Eb 111
Olympic Pk. Av. E20 36Dc 72
Olympic Way HA9: Wemb 34Qa 67
Olympic Way UB6: G'frd 39Ea 66
Olympus Gro. N22 25Qb 50
Olympus Sq. E5 34Wb 71
O'Mahoney Ct. SW17 62Eb 133
Oman Av. NW2 35Ya 68
Oman Way E14 44Ac 92
O'Meara St. SE1 7E 224 (46Sb 91)
Omega Cl. E14 48Dc 92
Omega Ct. RM7: Rom 29Fd 56
Omega Ct. WD18: Wat 15U 26
Omega Ho. SW10 52Eb 111
(off King's Rd.)
Omega Pl. N1 2G 217 (40Nb 70)
(off Caledonian Rd.)
Omega Rd. GU21: Wok 88C 168
Omega St. SE14 53Cc 114
Omega Way TW20: Thorpe 67E 126
Omega Works E3 38Cc 72
Ommaney Rd. SE14 53Zb 114
Omnibus Bldg. RH2: Reig 7K 207
Omnibus Ho. N22 26Qb 50
(off Lordship La.)
Omnibus Way E17 26Cc 52
Omnium Ct. WC1 7H 217 (43Pb 90)
(off Princeton St.)
Ondine Rd. SE15 56Vb 113
One Blackfriars SE1 6B 224 (46Rb 91)
(off Rennie St.)
One Casson Sq. SE1 7J 223 (46Pb 90)
(off Casson Sq.)
Onedin Ct. E1 45Wb 91
(off Ensign St.)
Onega Ga. SE16 48Ac 92
One Hyde Pk. SW1 1F 227 (47Hb 89)
O'Neill Ho. NW8 2D 214 (40Gb 69)
(off Cochrane St.)
O'Neill Path SE18 51Qc 116
One Lillie Sq. SW6 51Cb 111
One New Change 3D 224 (44Sb 91)
One Owen St. EC1 2B 218 (40Rb 71)
(off Goswell St.)
One Pin La. SL2: Farn C 5G 60
One Pin Pl. SL2: Farn C 4H 61
One Southbank Pl. SE1 1J 229 (47Pb 90)
(off York Rd.)
One Elephant, The SE1 5C 230 (49Rb 91)
(off Brook Dr.)
One Tree Cl. SE23 58Yb 114
One Tree Hill Local Nature Reserve 58Yb 114
Ongar Cl. KT15: Add 79H 149
Ongar Cl. RM6: Chad H 29Yc 55
Ongar Hill KT15: Add 79J 149
Ongar Pde. KT15: Add 79J 149
Ongar Pl. CM14: B'wood 19Zd 41
Ongar Pl. KT15: Add 79J 149
Ongar Rd. KT15: Add 78J 149
Ongar Rd. RM4: Abr 13Xc 37
Ongar Rd. RM4: Stap T 13Xc 37
Ongar Rd. SW6 51Cb 111
Ongar Way RM13: Rain 39Gd 76
Onra Rd. E17 31Cc 72
Onslow Av. SM2: Cheam 82Bb 175
Onslow Av. TW10: Rich 57Na 109
Onslow Cl. E4 19Ec 34
Onslow Cl. GU22: Wok 89C 168
Onslow Cl. KT7: T Ditt 74Ga 152
Onslow Cl. W10 41Bb 89
Onslow Ct. SW10 50Eb 89
(off Drayton Gdns.)
Onslow Cres. BR7: Chst 67Rc 138
Onslow Cres. GU22: Wok 89C 168
Onslow Cres. SW7 6C 226 (49Fb 89)
Onslow Dr. DA14: Sidc 61Zc 139
Onslow Gdns. CR2: Sande 84Wb 177
Onslow Gdns. E18 27Kc 53
Onslow Gdns. KT7: T Ditt 74Ga 152
Onslow Gdns. N10 29Kb 50
Onslow Gdns. N21 15Qb 32
Onslow Gdns. SM6: W'gton 79Lb 156
Onslow Gdns. SW7 6B 226 (49Fb 89)
Onslow Ho. KT2: King T 67Pa 131
(off Acre Rd.)
Onslow M. KT16: Chert 72H 149
Onslow M. E. SW7 6B 226 (49Fb 89)
Onslow M. W. SW7 6B 226 (49Fb 89)
Onslow Pde. N14 18Kb 32
Onslow Pl. GU24: Bisl 9F 166
Onslow Rd. CR0: C'don 73Pb 156
Onslow Rd. KT12: Hers 77V 150
Onslow Rd. KT3: N Mald 70Wa 132
Onslow Rd. SL5: S'dale 3F 146
Onslow Rd. TW10: Rich 57Na 109
Onslow Sq. SW7 5C 226 (49Fb 89)
Onslow St. EC1 6A 218 (42Qb 90)
Onslow Way GU22: Pyr 87H 169
Onslow Way KT7: T Ditt 74Ga 152
Ontario Point SE16 47Yb 92
(off Surrey Quays Rd.)
Ontario St. SE1 4C 230 (48Rb 91)
Ontario Twr. E14 45Fc 93
Ontario Way E14 45Cc 92
(not continuous)
On the Hill WD19: Wat 19Aa 27
Onyx M. E15 37Hc 73
Opal Apts. W2 44Cb 89
(off Hereford Rd.)
Opal Cl. E16 44Mc 93
Opal Ct. E15 39Ec 72
Opal Ct. SL3: Wex 2N 81
Opal M. IG1: Ilf 33Rc 74
Opal M. NW6 39Bb 69
Opal St. SE11 7B 230 (50Rb 91)
Opeck's Cl. SL2: Stoke P 2M 81
Opeck's Cl. SL2: Wex 2M 81
Open Air Stage 32Gb 69
Opendale Rd. SL1: Burn 3A 80
Openshaw Rd. SE2 49Xc 95
Openview SW18 60Eb 111
Opera Ct. N19 34Mb 70
(off Wedmore St.)
Ophelia Gdns. NW2 34Ab 68
Ophelia Ho. W6 50Za 88
(off Fulham Pal. Rd.)
Ophir Ter. SE15 53Wb 113
Opie Ho. NW8 1D 214 (40Gb 69)
(off Townshend Est.)
Opossum Way TW4: Houn 55Y 107
Oppenheim Rd. SE13 54Ec 114
Oppidan Apts. NW6 38Cb 69
(off Netherwood St.)
Oppidans Rd. NW3 38Hb 69
Optima Pk. DA1: Cray 54Jd 118
Opulens Pl. HA6: Nwood 24S 44
Opus Ct. WD6: Bore 13Qa 29
Oram Pl. HP3: Hem H 5M 3
Orange Ct. La. BR6: Downe 81Qc 180
Orange Gro. E11 34Gc 73
Orange Gro. IG7: Chig 23Sc 54
Orange Hill Rd. HA8: Edg 24Sa 47
Orange Pl. SE16 48Yb 92
Orangery, The TW10: Ham 61La 130
Orangery Gallery, The 47Bb 89
Orangery La. SE9 57Pc 116
Orange St. WC2 5E 222 (45Mb 90)
Orange Tree Ct. SE5 52Ub 113
(off Havil St.)
Orange Tree Hill RM4: Have B 22Fd 56
Orange Tree Theatre 56Na 109
Orange Yd. W1 3E 222 (44Mb 90)
(off Manette St.)
Oransay Rd. N1 37Sb 71
Oratory La. SW3 7C 226 (50Fb 89)
Orbain Rd. SW6 52Ab 110
Orbel St. SW11 53Gb 111

Orbis Wharf SW11 55Fb 111
Orbital 25 Bus. Pk. WD18: Wat 17T 26
Orbital Bus. Cen. EN9: Walt A 6Ec 20
Orbital Cen. IG8: Wfd G 26Mc 53
Orbital Cres. WD25: Wat 7V 12
Orbital One DA1: Dart 61Rd 141
Orbital One Ind. Est. DA1: Dart 61Qd 141
Orb St. SE17 6F 231 (49Tb 91)
Orchard, The KT17: Ewe Meadow Wlk. 80Va 154
Orchard, The KT17: Ewe Tayles Hill Dr. 82Va 174
Orchard, The BR8: Swan 68Fd 140
Orchard, The GU21: Wok 8M 167
Orchard, The GU22: Wok 94A 188
Orchard, The GU25: Vir W 71A 148
Orchard, The KT13: Weyb 77R 150
Orchard, The KT16: Ott 79F 148
Orchard, The N14 15Kb 32
Orchard, The N20 18Db 31
Orchard, The N21 16Tb 33
Orchard, The NW11 29Cb 49
Orchard, The SE3 54Fc 115
Orchard, The SM7: Bans 87Cb 175
Orchard, The TN13: Dun G 93Gd 202
Orchard, The TW3: Houn 54Ea 108
Orchard, The W4 49Ta 87
Orchard, The W5 43Ma 87
(off Montpelier Rd.)
Orchard, The WD17: Wat 9U 12
Orchard, The WD4: K Lan 1Q 12
Orchard Av. CM13: B'wood 20Be 41
Orchard Av. CR0: C'don 75Ac 158
Orchard Av. CR4: Mitc 74Jb 156
Orchard Av. DA1: Dart 59Kd 119
Orchard Av. DA11: Grav'nd 4D 144
Orchard Av. DA17: Belv 51Ad 117
Orchard Av. KT15: Wdhm 83H 169
Orchard Av. KT3: N Mald 68Ua 132
Orchard Av. KT7: T Ditt 74Ja 152
Orchard Av. N14 16Lb 32
Orchard Av. N20 19Fb 31
Orchard Av. N3 27Cb 49
Orchard Av. RM13: Rain 42Ld 97
Orchard Av. SL1: Slou 3B 80
Orchard Av. SL4: Wind 3E 102
Orchard Av. TW14: Felt 57T 106
Orchard Av. TW15: Ashf 65S 128
Orchard Av. TW5: Hest 52Aa 107
Orchard Av. UB1: S'hall 46Ba 85
Orchard Av. WD25: Wat 4X 13
Orchard Bungs. SL2: Farn C 8D 60
Orchard Bus. Cen. SE26 64Bc 136
Orchard Cl. AL1: St A 3D 6
Orchard Cl. DA3: Lfield 68De 143
Orchard Cl. DA7: Bex 53Ad 117
Orchard Cl. E11 28Kc 53
Orchard Cl. E4 21Cc 52
Orchard Cl. GU22: Wok 88D 168
Orchard Cl. GU24: W End 5B 166
Orchard Cl. HA0: Wemb 39Na 67
Orchard Cl. HA4: Ruis 31S 64
Orchard Cl. HA8: Edg 23Na 47
Orchard Cl. HP2: Hem H 1P 3
Orchard Cl. KT12: Walt T 73X 151
Orchard Cl. KT19: Ewe 79Ra 153
Orchard Cl. KT22: Fet 94Fa 192
Orchard Cl. KT22: Lea 91Ha 192
Orchard Cl. KT24: E Hor 96V 190
Orchard Cl. KT6: Surb 74Ka 152
Orchard Cl. N1 38Sb 71
Orchard Cl. NW2 34Wa 68
Orchard Cl. RM15: S Ock 42Yd 98
Orchard Cl. SE23 58Yb 114
Orchard Cl. SL1: Burn 2A 80
Orchard Cl. SM7: Bans 86Db 175
Orchard Cl. SW20 70Ya 132
Orchard Cl. TN14: S'oaks 92Ld 203
Orchard Cl. TW15: Ashf 65S 128
Orchard Cl. TW20: Egh 64D 126
Orchard Cl. UB5: N'olt 37Ea 66
Orchard Cl. UB9: Den 37K 63
Orchard Cl. W10 43Bb 89
Orchard Cl. WD17: Wat 12V 26
Orchard Cl. WD23: B Hea 18Fa 28
Orchard Cl. WD3: Chor 14F 24
Orchard Cl. WD6: E'tree 14Pa 29
Orchard Cl. WD7: R'lett 9Ga 14
Orchard Cotts. KT2: King T 67Pa 131
Orchard Cotts. UB3: Hayes 47U 84
Orchard Ct. CR3: Cat'm 96Vb 197
Orchard Ct. E10 32Dc 72
Orchard Ct. EN5: New Bar 13Db 31
Orchard Ct. EN6: Pot B 3Cb 17
Orchard Ct. GU24: Chob 1K 167
Orchard Ct. HA8: Edg 22Pa 47
Orchard Ct. HP3: Bov 9C 2
Orchard Ct. KT12: Walt T 74V 150
(off Bridge St.)
Orchard Ct. KT4: Wor Pk. 74Wa 154
Orchard Ct. N14 16Lb 32
Orchard Ct. SE26 63Bc 136
Orchard Ct. SM6: W'gton 78Kb 156
Orchard Ct. TW2: Twick 61Fa 130
Orchard Ct. TW7: Isle 53Fa 108
Orchard Ct. UB7: Lford 52L 105
Orchard Ct. UB8: Uxb 40N 63
(off The Greenway)
Orchard Ct. W1 2H 221 (44Jb 90)
(off Fitzhardinge St.)
Orchard Cres. EN1: Enf 11Vb 33
Orchard Cres. HA8: Edg 22Sa 47
Orchard Dene KT14: W Byf 85J 169
(off Madeira Rd.)
Orchard Dr. AL2: Park 9P 5
Orchard Dr. CM16: They B 8Uc 22
Orchard Dr. DA13: Meop 10B 144
Orchard Dr. GU21: Wok 87A 168
Orchard Dr. HA8: Edg 22Pa 47
Orchard Dr. KT21: Asht 92Ma 193
Orchard Dr. RM17: Grays 47Ce 99
Orchard Dr. SE3 54Fc 115
Orchard Dr. TW17: Shep 69U 128
Orchard Dr. UB8: Cowl 42M 83
Orchard Dr. WD17: Wat 11V 26
Orchard Dr. WD3: Chor 13E 24
Orchard End CR3: Cat'm 94Ub 197
Orchard End KT13: Weyb 75U 150
Orchard End KT22: Fet 96Ea 192
Orchard Farm Av. KT8: E Mos 72Fa 152
Orchard Gdns. EN9: Walt A 6Ec 20
Orchard Gdns. KT18: Eps 86Sa 173
Orchard Gdns. KT24: Eff 100Aa 191
Orchard Gdns. KT9: Chess 77Na 153
Orchard Gdns. SM1: Sutt 78Cb 155
Orchard Ga. KT10: Esh 74Fa 152
Orchard Ga. NW9 28Ua 48
Orchard Ga. SL2: Farn C 6G 60
Orchard Ga. UB6: G'frd 37Ka 66
Orchard Grn. BR6: Orp 75Uc 160
Orchard Gro. BR6: Orp 75Vc 161
Orchard Gro. CR0: C'don 73Ac 158
Orchard Gro. HA3: Kenton 29Pa 47
Orchard Gro. HA8: Edg 25Qa 47
Orchard Gro. SE20 66Wb 135
Orchard Hgts. CM16: Epp 5Vc 23
Orchard Hill DA1: Cray 57Gd 118
Orchard Hill GU20: W'sham 10B 146
Orchard Hill SE13 54Dc 114
Orchard Hill SM5: Cars 78Hb 155
Orchard Ho. DA8: Erith 53Hd 118
Orchard Ho. SE16 48Yb 92
Orchard Ho. SE5 53Sb 113
(off County Gro.)
Orchard Ho. SW6 52Bb 111
(off Varna Rd.)
Orchard Ho. W12 46Wa 88
Orchard Ho. La. AL1: St A 3B 6
Orchard La. CM15: Pil H 15Vd 40
Orchard La. IG8: Wfd G 21Lc 53
Orchard La. KT8: E Mos 72Fa 152
Orchard La. RH9: G'stone 2P 209
Orchard La. SW20 67Xa 132
Orchard Lea DA13: Sflt 64Be 143
Orchard Lea Cl. GU22: Pyr 87G 168
Orchardleigh KT22: Lea 94Ka 192
Orchardleigh Av. EN3: Enf H 12Yb 34
Orchard Lodge SL1: Slou 6C 80
(off Streamside)
Orchard Mains GU22: Wok 1N 187
Orchard Mead Ho. NW2 33Cb 69
Orchardmede N21 16Tb 33
Orchard M. GU21: Knap 10F 166
Orchard M. N1 38Tb 71
Orchard M. N6 31Kb 70
Orchard M. SW17 62Eb 133
Orchard on the Grn., The WD3: Crox G 15P 25
Orchard Pde. EN6: Pot B 3Za 16
Orchard Pl. BR2: Kes 81Lc 179
Orchard Pl. BR5: St P 69Yc 139
Orchard Pl. E14 45Gc 93
(not continuous)
Orchard Pl. E5 36Xb 71
Orchard Pl. EN8: Chesh 2Zb 20
Orchard Pl. N17 24Vb 51
Orchard Pl. TN14: Sund 96Ad 201
Orchard Pl. UB8: Uxb 38M 63
Orchard Pl. W4 49Ua 88
Orchard Ri. CR0: C'don 74Ac 158
Orchard Ri. HA5: Eastc 27V 44
Orchard Ri. KT2: King T 67Sa 131
Orchard Ri. TW10: Rich 56Ra 109
Orchard Ri. E. DA15: Sidc 57Vc 117
Orchard Ri. W. DA15: Sidc 57Uc 116
Orchard Rd. BR1: Broml 67Lc 137
Orchard Rd. BR6: Farnb 78Rc 160
Orchard Rd. BR6: Prat B 82Yc 181
Orchard Rd. CR2: Sande 86Xb 177
Orchard Rd. CR4: Mitc 74Jb 156
Orchard Rd. DA10: Swans 57Ae 121
Orchard Rd. DA11: Nflt 61Ee 143
Orchard Rd. DA14: Sidc 63Uc 138
Orchard Rd. DA16: Well 55Xc 117
Orchard Rd. DA17: Belv 49Cd 96
Orchard Rd. EN3: Pond E 15Yb 34
Orchard Rd. EN5: Barn 14Bb 31
Orchard Rd. KT1: King T 68Na 131
Orchard Rd. KT9: Chess 77Na 153
Orchard Rd. N6 31Kb 70
Orchard Rd. RH2: Reig 6K 207
Orchard Rd. RM10: Dag 39Cd 76
Orchard Rd. RM15: S Ock 42Yd 98
Orchard Rd. RM7: Mawney 25Dd 56
Orchard Rd. SE18 49Tc 94
Orchard Rd. SE3 54Gc 115
Orchard Rd. SL4: Old Win 8M 103
Orchard Rd. SM1: Sutt 78Cb 155
Orchard Rd. TN13: Riv 94Gd 202
Orchard Rd. TN14: Otf 88Hd 182
Orchard Rd. TW1: Twick 57Ja 108
Orchard Rd. TW12: Hamp 66Ba 129
Orchard Rd. TW13: Felt 60W 106
Orchard Rd. TW16: Sun 66X 129
Orchard Rd. TW4: Houn 57Ba 107
Orchard Rd. TW8: Bford 51La 108
Orchard Rd. TW9: Rich 55Qa 109
Orchard Rd. UB3: Hayes 45W 84
Orchards, The CM16: Epp 4Wc 23
Orchards Cvn. Site, The WD3: Chal P 24D 42
Orchards Cl. KT14: W Byf 86J 169
Orchardson Ho. NW8 6B 214 (42Fb 89)
(off Orchardson St.)
Orchardson St. NW8 6B 214 (42Fb 89)
Orchard Sq. SW1 7J 227 (50Jb 90)
Orchard Sq. W14 50Bb 89
Orchards Res. Pk. SL3: L'ly 46B 82
Orchards Shop. Cen. 58Nd 119
Orchard St. AL3: St A 3A 6
Orchard St. DA1: Dart 58Nd 119
Orchard St. E17 28Ac 52
Orchard St. HP3: Hem H 5M 3
Orchard St. W1 3H 221 (44Jb 90)
Orchard Studios W6 49Za 88
(off Brook Grn.)
Orchard Ter. DA9: Ghithe 57Ud 120
Orchard Ter. EN1: Enf 16Wb 33
Orchard Ter. NW10 35Va 68
Orchard Theatre, The 58Nd 119
Orchard Vw. KT16: Chert 72J 149
Orchard Vw. UB8: Cowl 42M 83
Orchard Vs. DA14: Sidc 65Yc 139
Orchard Wlk. KT2: King T 67Qa 131
(off Gordon Rd.)
Orchard Way BR3: Beck Monks Orchard Rd. 73Cc 158
Orchard Way BR3: Beck Orchard Av. 72Bc 158
Orchard Way CR0: C'don 74Ac 158
Orchard Way DA2: Wilm 62Md 141
Orchard Way EN1: Enf 13Ub 33
Orchard Way EN6: Pot B 10K 9
Orchard Way GU23: Send 97E 188
Orchard Way GU3: Worp 6H 187
Orchard Way HP3: Bov 10C 2
Orchard Way IG7: Chig 20Wc 37
Orchard Way KT10: Esh 79Ea 152
Orchard Way KT15: Add 78K 149
Orchard Way KT20: Lwr K 98Bb 195
Orchard Way RH2: Reig 9K 207
Orchard Way RH8: Oxt 5L 211
Orchard Way SL3: L'ly 46A 82
Orchard Way SM1: Sutt 77Fb 155
Orchard Way TN15: Kems'g 89Qd 183
Orchard Way TW15: Ashf 61P 127
Orchard Way WD3: Rick 17J 25
Orchard Waye UB8: Uxb 40M 63
Orchard Wharf E14 45Gc 93
(off Orchard Pl.)
Orchehill Ct. SL9: Ger X 29A 42
Orchehill Ri. SL9: Ger X 29A 42
Orchestra Ct. HA8: Edg 24Ra 47
Orchid Cl. E6 43Nc 94
Orchid Cl. EN7: G Oak 2Sb 19
Orchid Cl. KT9: Chess 80La 152
Orchid Cl. RM4: Abr 13Xc 37
Orchid Cl. SE13 57Fc 115
Orchid Cl. UB1: S'hall 45Aa 85
Orchid Ct. HA9: Wemb 33Na 67
Orchid Ct. TW20: Egh 63D 126
Orchid Dr. GU24: Bisl 7E 166
Orchid Dr. HP2: Hem H 3M 3
Orchid Gdns. TW3: Houn 56Ba 107
Orchid Grange N14 17Lb 32
Orchid Mead SM7: Bans 86Db 175
Orchid M. NW10 39Ta 67
Orchid Rd. N14 17Lb 32
Orchid St. W12 45Wa 88
Orchis Gro. RM17: Grays 50Be 99
Orchis Way RM3: Rom 23Pd 57
Orde NW9 25Va 48
Orde Hall St. WC1 6H 217 (42Pb 90)
Ordell Ct. E3 40Bc 72
(off Ordell Rd.)
Ordell Rd. E3 40Bc 72
Ordnance Cl. TW13: Felt 61W 128
Ordnance Cres. SE10 47Gc 93
Ordnance Dock Pl. UB2: S'hall 49Z 85
Ordnance Hill NW8 1C 214 (39Fb 69)
Ordnance M. NW8 1C 214 (40Fb 69)
Ordnance Rd. DA12: Grav'nd 8E 122
Ordnance Rd. E16 43Hc 93
Ordnance Rd. EN3: Enf L 9Zb 20
Ordnance Rd. EN3: Enf W 9Zb 20
Ordnance Rd. SE18 51Qc 116
Oregano Cl. UB7: Yiew 44N 83
Oregano Dr. E14 44Fc 93
Oregon Av. E12 35Pc 74
Oregon Bldg. SE13 53Dc 114
(off Deal's Gateway)
Oregon Cl. KT3: N Mald 70Sa 131
Oregon M. W5 43La 86
Oregon Sq. BR6: Orp 74Tc 160
Oreilly St. SE1 5K 231 (49Vb 91)
(off Willow Wlk.)
Orestan La. KT24: Eff 99X 191
Orestes M. NW6 36Cb 69
Oreston Rd. RM13: Rain 41Md 97
Orewell Gdns. RH2: Reig 8K 207
Orford Cl. DA2: Dart 58Rd 119
(off Osbourne Rd.)
Orford Ct. HA7: Stan 23La 46
Orford Ct. SE27 61Rb 135
Orford Gdns. TW1: Twick 61Ha 130
Orford Rd. E17 29Cc 52
Orford Rd. E18 27Kc 53
Orford Rd. SE6 62Dc 136
ORGAN CROSSROADS 80Wa 154
Organ Hall Rd. WD6: Bore 11Na 29
Organ La. E4 19Ec 34
Oriana Ho. E10 33Dc 72
(off Grange Pk. Rd.)
Oriana Ho. E14 45Bc 92
(off Victory Pl.)
Oriel Cl. CR4: Mitc 70Mb 134
Oriel Ct. AL1: St A 1C 6
Oriel Ct. CR0: C'don 74Tb 157
Oriel Ct. NW3 35Eb 69
Oriel Dr. SW13 51Ya 110
Oriel Gdns. IG5: Ilf 27Pc 54
Oriel Ho. NW6 39Cb 69
(off Priory Pk. Rd.)
Oriel Ho. RM7: Rom 30Fd 56
Oriel M. E18 26Jc 53
Oriel Pl. NW3 35Eb 69
(off Heath St.)
Oriel Rd. E9 37Zb 72
Oriel Way UB5: N'olt 38Da 65
Oriens M. E20 36Ec 72
Oriental Cl. GU22: Wok 89B 168
Oriental Rd. E16 46Mc 93
Oriental Rd. GU22: Wok 89B 168
Oriental Rd. SL5: S'hill 10B 124
Oriental St. E14 45Cc 92
(off Pennyfields)
Orient Cl. AL1: St A 4C 6
Orient Ho. SW6 53Eb 111
(off Station Ct.)
Orient Ind. Pk. E10 33Cc 72
Orient St. SE11 5B 230 (49Rb 91)
Orient Way E10 32Ac 72
Orient Way E5 34Zb 72
Orient Wharf E1 46Xb 91
(off Wapping High St.)
Origin Bus. Pk. NW10 41Qa 87
(off Leyburn Rd.)
Oriole Cl. WD5: Ab L 3W 12
Oriole Way SE28 45Xc 95
Orion E14 49Cc 92
(off Crews St.)
Orion Bus. Cen. SE14 50Zb 92
Orion Cen., The CR0: Bedd 75Nb 156
Orion Ho. E1 42Xb 91
(off Coventry Rd.)
Orion M. SM4: Mord 70Cb 133
Orion Pk. RM9: Dag 40Bd 75
Orion Rd. N11 24Jb 50
Orion Way HA6: Nwood 21V 44
Orissa Rd. SE18 50Uc 94
Orkney Ct. E1 43Ac 92
(off Ocean Est.)
Orkney Ho. N1 39Pb 70
(off Bemerton Est.)
Orkney Ho. WD18: Wat 15V 26
(off Himalayan Way)
Orkney St. SW11 54Jb 112
Orlando Gdns. KT19: Ewe 82Ta 173
Orlando Rd. SW4 55Lb 112
Orleans Cl. KT10: Esh 75Fa 152
Orleans Ct. KT12: Walt T 75Y 151
Orleans Ct. TW1: Twick 59Ka 108
Orleans House Gallery 60Ka 108
Orleans Pk. School Sports Cen. 59Ka 108
Orleans Rd. SE19 65Tb 135
Orleans Rd. TW1: Twick 59Ka 108
Orlestone Gdns. BR6: Chels 78Ad 161
Orleston M. N7 37Qb 70
Orleston Rd. N7 37Qb 70
Orley Ct. HA1: Harr 35Ha 66
Orley Farm Rd. HA1: Harr 34Ga 66
Orlick Rd. DA12: Grav'nd 10K 123
Orlop St. SE10 50Gc 93
Ormanton Rd. SE26 63Wb 135
Orme Ct. W2 45Db 89
Orme Ct. M. W2 45Db 89
(off Orme La.)
Orme Ho. E8 39Vb 71
Orme La. W2 45Db 89
Ormeley Rd. SW12 60Kb 112
Orme Rd. KT1: King T 68Ra 131
Orme Rd. SM1: Sutt 79Db 155
Ormerod Gdns. CR4: Mitc 68Jb 134
Ormesby Cl. SE28 45Zc 95
Ormesby Dr. EN6: Pot B 4Za 16
Ormesby Way HA3: Kenton 30Pa 47
Orme Sq. W2 45Db 89
Ormiston Gro. W12 46Xa 88
Ormiston Rd. SE10 50Jc 93
Ormond Av. TW10: Rich 57Ma 109
Ormond Av. TW12: Hamp 67Da 129
Ormond Cl. RM3: Hrld W 26Md 57
Ormond Cl. WC1 7G 217 (43Nb 90)
Ormond Cres. TW12: Hamp 67Da 129
Ormond Dr. TW12: Hamp 66Da 129
Ormonde Av. BR6: Farnb 75Sc 160
Ormonde Av. KT19: Ewe 82Ta 173
Ormonde Ct. NW8 1F 215 (39Hb 69)
(off St Edmund's Cl.)
Ormonde Ct. RM11: Horn 31Md 77
(off Clydesdale Rd.)
Ormonde Ct. SW15 56Ya 110
Ormonde Ga. SW3 7G 227 (50Hb 89)
(not continuous)
Ormonde Mans. WC1 7G 217 (43Nb 90)
(off Southampton Row)
Ormonde Pl. KT13: Weyb 79T 150
Ormonde Pl. SW1 6H 227 (49Jb 90)
Ormonde Ri. IG9: Buck H 18Lc 35
Ormonde Rd. GU21: Wok 8N 167
Ormonde Rd. HA6: Nwood 21T 44
Ormonde Rd. SW14 55Sa 109
Ormonde Ter. NW8 1F 215 (39Hb 69)
Ormond Ho. N16 33Tb 71
Ormond M. WC1 6G 217 (42Nb 90)
Ormond Rd. N19 32Nb 70
Ormond Rd. TW10: Rich 57Ma 109
Ormond Yd. SW1 6C 222 (46Lb 90)
Ormrod Ct. W11 44Ab 88
(off Westbourne Pk. Rd.)
Ormsby SM2: Sutt 80Db 155
Ormsby Gdns. UB6: G'frd 40Ea 66
Ormsby Lodge W4 48Ua 88
Ormsby Pl. N16 34Vb 71
Ormsby Point SE18 49Rc 94
(off Vincent Rd.)
Ormsby St. E2 1K 219 (40Vb 71)
Ormside St. SE15 51Yb 114
Ormside Way RH1: Redh 2B 208
Ormskirk Rd. WD19: Wat 21Z 45
Ornan Rd. NW3 36Gb 69
Oronsay HP3: Hem H 4B 4
Orpen Ho. SW5 49Cb 89
(off Trebovir Rd.)
Orpen Wlk. N16 34Ub 71
Orphanage Rd. WD17: Wat 12Y 27
Orphanage Rd. WD24: Wat 12Y 27
Orpheus Ho. W10 42Bb 89
(off Harrow Rd.)
Orpheus St. SE5 53Tb 113
ORPINGTON 74Wc 161
Orpington Bus. Pk. BR5: St M Cry 70Xc 139
Orpington By-Pass BR6: Orp 75Xc 161
Orpington By-Pass Rd. TN14: Bad M 81Bd 181
Orpington Gdns. N18 20Ub 33
Orpington Golf Cen. 68Ad 139
ORPINGTON HOSPITAL 77Vc 161
Orpington Mans. N21 18Rb 33
Orpington Retail Pk. 70Yc 139
Orpington Rd. BR7: Chst 69Uc 138
Orpington Rd. N21 18Rb 33
Orpington Station (Rail) 75Uc 160
Orpington Superbowl 74Wc 161
Orpington Trade Cen. BR5: St P 69Xc 139
Orpin Rd. RH1: Mers 2B 208
Orpwood Cl. TW12: Hamp 65Ba 129
ORSETT 3C 100
Orsett Golf Course 5G 100
ORSETT HEATH 7B 100
Orsett Heath Cres. RM16: Grays 8C 100
ORSETT HOSPITAL 3C 100
Orsett Ind. Pk. RM16: Ors 3H 101
Orsett M. W2 44Db 89
(not continuous)
Orsett Rd. RM16: Ors 2E 100
Orsett Rd. RM17: Grays 50De 99
Orsett Rd. SS17: Horn H 1G 100
Orsett Smock Mill 4A 100
Orsett St. SE11 7J 229 (50Pb 90)
Orsett Ter. IG8: Wfd G 24Lc 53
Orsett Ter. W2 44Db 89
Orsman Rd. N1 39Ub 71
Ortman Cl. SL9: Ger X 31A 62
Orton Gro. EN1: Enf 11Wb 33
Orton Ho. RM3: Rom 24Nd 57
(off Leyburn Rd.)
Orton Pl. SW19 66Db 133
Orton St. E1 46Wb 91
Orville Rd. SW11 54Fb 111
Orwell RM18: E Til 9L 101
Orwell Cl. RM13: Rain 43Fd 96
Orwell Cl. SL4: Wind 5H 103
Orwell Cl. UB3: Hayes 45U 84
Orwell Ct. E8 39Wb 71
(off Pownall Rd.)
Orwell Ct. N5 35Sb 71
Orwell Ct. SE13 55Dc 114
Orwell Ct. SW17 62Fb 133
(off Grosvenor Way)
Orwell Ct. WD24: Wat 13Z 27
Orwell Rd. E13 40Lc 73
Osbaldeston Rd. N16 33Wb 71
Osberton Rd. SE12 57Jc 115
Osbert St. SW1 6D 228 (49Mb 90)
Osborn Cl. E8 39Wb 71
Osborne Av. TW19: Stanw 60N 105
Osborne Cl. BR3: Beck 70Ac 136
Osborne Cl. EN4: Cockf 13Hb 31
Osborne Cl. KT17: Eps 84Va 174
Osborne Cl. RM11: Horn 30Kd 57
Osborne Cl. TW13: Hanw 64Z 129
Osborne Ct. E10 31Dc 72
Osborne Ct. EN6: Pot B 1Db 17
Osborne Ct. HA2: Harr 28Da 45
Osborne Ct. SL4: Wind 4G 102
Osborne Ct. W5 43Na 87
Osborne Ct. WD6: Bore 15Sa 29
(off Sandringham Cl.)
Osborne Gdns. CR7: Thor H 68Sb 135
Osborne Gdns. EN6: Pot B 2Db 17
Osborne Gro. E17 28Bc 52
Osborne Hgts. CM14: W'ley 21Xd 58
Osborne Ho. E16 46Jc 93
(off Wesley Av.)
Osborne M. E17 28Bc 52
Osborne M. SL4: Wind 4G 102
Osborne Pl. SM1: Sutt 78Fb 155
Osborne Rd. CM15: Pil H 16Wd 40
Osborne Rd. CR7: Thor H 68Sb 135
Osborne Rd. DA17: Belv 50Bd 95
Osborne Rd. E10 33Dc 72
Osborne Rd. E7 36Kc 73
Osborne Rd. E9 37Bc 72
Osborne Rd. EN3: Enf H 12Ac 34
Osborne Rd. EN6: Pot B 2Db 17
Osborne Rd. IG9: Buck H 18Kc 35
Osborne Rd. KT12: Walt T 74W 150
Osborne Rd. KT2: King T 66Na 131
Osborne Rd. N13 20Qb 32
Osborne Rd. N4 32Qb 70
Osborne Rd. NW2 37Xa 68
Osborne Rd. RH1: Redh 3A 208
Osborne Rd. RM11: Horn 30Kd 57
Osborne Rd. RM9: Dag 36Bd 75
Osborne Rd. SL4: Wind 4G 102
Osborne Rd. TW20: Egh 65B 126
Osborne Rd. TW3: Houn 55Ba 107
Osborne Rd. UB1: S'hall 44Ea 86
Osborne Rd. UB8: Uxb 38L 63
Osborne Rd. W3 48Ra 87
Osborne Rd. WD24: Wat 10Y 13
Osborne Sq. RM9: Dag 35Bd 75
Osborne St. SL1: Slou 7K 81
Osborne Ter. SW17 64Hb 133
(off Church La.)
Osborne Way KT19: Eps 84Na 173
Osborne Way KT9: Chess 78Pa 153
Osborn Gdns. NW7 24Za 48
Osborn La. SE23 59Ac 114
Osborn St. E1 43Vb 91
Osborn Ter. SE3 56Hc 115
Osbourne Av. WD4: K Lan 10P 3
Osbourne Ho. IG8: Wfd G 24Qc 54
Osbourne Ho. TW2: Twick 61Ea 130
Osbourne Rd. DA2: Dart 58Rd 119
Oscar Cl. CR8: Purl 82Qb 176
Oscar Ct. SE16 47Ac 92
Oscar Faber Pl. N1 38Ub 71
Oscar St. SE8 54Cc 114
(not continuous)
Oseney Cres. NW5 36Lb 70
Osgood Av. BR6: Chels 78Vc 161
Osgood Gdns. BR6: Chels 78Vc 161
OSIDGE 18Kb 32
Osidge La. N14 18Jb 32
Osier Cl. BR6: Farnb 77Rc 160
Osier Ct. E1 42Zb 92
(off Osier St.)
Osier Ct. RM7: Rom 30Fd 56
Osier Ct. TW8: Bford 51Na 109
(off Ealing Rd.)
Osier Cres. N10 25Hb 49
Osier Ho. SE16 48Ac 92
Osier La. SE10 48Hc 93
Osier M. W4 51Ua 110
Osier Pl. TW20: Egh 65E 126
Osiers, The WD3: Crox G 16S 26
Osiers Ct. KT1: King T 67Ma 131
(off Steadfast Rd.)
Osiers Rd. SW18 56Cb 111
Osiers Twr. SW18 56Cb 111
(off Enterprise Way)
Osier St. E1 42Yb 92
Osier Way CR4: Mitc 71Hb 155
Osier Way E10 34Dc 72
Osier Way SM7: Bans 86Ab 174
Oslac Rd. SE6 64Dc 136
Oslo Ct. NW8 2D 214 (40Gb 69)
(off Prince Albert Rd.)
Oslo Ho. E9 37Bc 72
(off Felstead St.)
Oslo Ho. SE5 54Sb 113
(off Carew St.)
Oslo Sq. SE16 48Ac 92
Oslo Twr. SE8 49Ac 92
(off Naomi St.)
Osman Cl. N15 30Tb 51
Osmani Youth Cen. 43Wb 91
Osman Rd. N9 20Wb 33
Osman Rd. W6 48Ya 88
Osmington Ho. SW8 52Pb 112
(off Dorset Rd.)
Osmond Cl. HA2: Harr 33Ea 66
Osmond Gdns. SM6: W'gton 78Lb 156
Osmunda Ct. E1 44Wb 91
(off Myrdle St.)
Osmund St. W12 43Va 88
Osnaburgh St. NW1: Euston Rd. 5A 216 (42Kb 90)
Osnaburgh St. NW1: Robert St. 4A 216 (41Kb 90)
(off Robert St.)
Osnaburgh Ter. NW1 5A 216 (42Kb 90)
Osney Ho. SE2 47Zc 95
Osney Wlk. SM5: Cars 72Fb 155
Osney Way DA12: Grav'nd 1H 145
Osprey NW9 25Va 48
Osprey Cl. BR2: Broml 74Nc 160
Osprey Cl. E11 28Jc 53
Osprey Cl. E17 24Ac 52
Osprey Cl. E6 43Nc 94
Osprey Cl. HP3: Hem H 7P 3
Osprey Cl. KT22: Fet 94Ea 192
Osprey Cl. SM1: Sutt 78Bb 155
Osprey Cl. UB7: W Dray 47N 83
Osprey Cl. WD25: Wat 6Aa 13
Osprey Ct. BR3: Beck 66Cc 136
Osprey Ct. CM14: B'wood 20Xd 40
Osprey Ct. CR0: C'don 76Sb 157
(off Innes Yd.)
Osprey Ct. E1 45Wb 91
(off Star Pl.)
Osprey Ct. EN9: Walt A 6Jc 21
Osprey Dr. KT18: Tatt C 89Xa 174
Osprey Est. SE16 49Zb 92
Osprey Gdns. CR2: Sels 82Ac 178
Osprey Ho. E14 45Ac 92
(off Victory Pl.)
Osprey Ho. SE1 50Xb 91
(off Lynton Rd.)
Osprey La. HA2: Harr 33Da 65
Osprey M. EN3: Pond E 15Yb 34
Osprey Rd. EN9: Walt A 6Jc 21
Ospringe Cl. SE20 66Yb 136
Ospringe Ct. SE9 58Tc 116
Ospringe Ho. SE1 1A 230 (47Qb 90)
(off Wootton St.)
Ospringe Rd. NW5 35Lb 70
Osram Ct. W6 48Ya 88
Osram Rd. HA9: Wemb 34Ma 67
Osric Path N1 2H 219 (40Ub 71)
Ossel Ct. SE10 49Gc 93
Ossian M. N4 31Pb 70
Ossian Rd. N4 31Pb 70
OSSIE GARVIN RDBT. 45X 85
Ossington Bldgs. W1 7H 215 (43Jb 90)
Ossington Cl. W2 45Cb 89
Ossington St. W2 45Cb 89
Ossory Rd. SE1 51Wb 113
Ossulston St. NW1 2D 216 (40Mb 70)
Ossulton Pl. N2 27Eb 49

Ossulton Way N2 28Eb 49
Ostade Rd. SW2 59Pb 112
Ostell Cres. EN3: Enf L 10Cc 20
Ostend Pl. SE1 5D 230 (49Sb 91)
Osten M. SW7 48Db 89
Osterberg Rd. DA1: Dart 56Pd 119
OSTERLEY 52Fa 108
Osterley Av. TW7: Isle 52Fa 108
Osterley Cl. BR5: St P 67Wc 139
Osterley Ct. TW7: Isle 53Fa 108
Osterley Ct. UB5: N'olt 41Y 85
(off Canberra Dr.)
Osterley Cres. TW7: Isle 53Ga 108
Osterley Gdns. CR7: Thor H 68Sb 135
Osterley Gdns. UB2: S'hall 47Ea 86
Osterley Ho. E14 44Dc 92
(off Giraud St.)
Osterley La. TW7: Isle 49Ea 86
Osterley La. UB2: S'hall 50Ca 85
Osterley Lodge TW7: Isle 52Ga 108
(off Church Rd.)
Osterley Pk. 51Ea 108
Osterley Pk. & House 51Ea 108
Osterley Pk. Rd. UB2: S'hall 48Ba 85
Osterley Pk. Vw. Rd. W7 47Ga 86
Osterley Rd. N16 35Ub 71
Osterley Rd. TW7: Isle 52Ga 108
Osterley Sports & Athletics Cen. 52Ga 108
Osterley Sports Club 48Ea 86
Osterley Station (Underground) 52Fa 108
Osterley Views UB2: S'hall 47Fa 86
Oster St. AL3: St A 1A 6
Oster Ter. E17 29Zb 52
Ostlers Cl. HP3: Hem H 3D 4
Ostlers Dr. TW15: Ashf 64S 128
Ostliffe Rd. N13 22Sb 51
Ostro Twr. E14 47Dc 92
Oswald Bldg. SW11 51Kb 112
Oswald Cl. KT22: Fet 94Ea 192
Oswald Cl. RM12: Horn 36Hd 76
Oswald Rd. AL1: St A 3C 6
Oswald Rd. KT22: Fet 94Ea 192
Oswald Rd. UB1: S'hall 46Aa 85
Oswald's Mead E9 35Ac 72
Oswald St. E5 34Zb 72
Oswald Ter. NW2 34Ya 68
Osward CR0: Sels 81Bc 178
(not continuous)
Osward Pl. N9 19Xb 33
Osward Rd. SW17 61Hb 133
Oswell Ho. E1 46Xb 91
(off Farthing Flds.)
Oswin Cl. BR5: St M Cry 71Xc 161
Oswin St. SE11 5C 230 (49Rb 91)
Oswyth Rd. SE5 54Ub 113
OTFORD 88Kd 183
Otford Cl. BR1: Broml 69Qc 138
Otford Cl. DA5: Bexl 58Dd 118
Otford Cl. SE20 67Yb 136
Otford Cres. SE4 58Bc 114
Otford Heritage Cen. 88Kd 183
Otford Ho. SE1 2G 231 (47Tb 91)
(off Staple St.)
Otford Ho. SE15 51Yb 114
(off Lovelinch Cl.)
Otford La. TN14: Hals 84Bd 181
Otford Rd. TN14: S'oaks 91Kd 203
Otford Station (Rail) 88Ld 183
Othello Cl. SE11 7B 230 (50Rb 91)
Other Place Theatre, The 3B 228 (48Lb 90)
(off Palace St.)
Otho Ct. TW8: Bford 52Ma 109
Otium Leisure Club St Albans 9M 5
Otley App. IG2: Ilf 30Rc 54
Otley Dr. IG2: Ilf 29Rc 54
Otley Ho. N5 34Qb 70
Otley Rd. E16 44Lc 93
Otley Ter. E5 34Zb 72
Otley Way WD19: Wat 20Y 27
Otlinge Rd. BR5: St M Cry 70Zc 139
Ottawa Gdns. RM10: Dag 38Fd 76
Ottawa Ho. SE16 47Yb 92
(off Province Dr.)
Ottawa Ho. UB4: Yead 41X 85
(off Ayles Rd.)
Ottawa Rd. RM18: Tilb 4C 122
Ottaway St. E5 34Wb 71
Otterbourne Rd. CR0: C'don 75Sb 157
Otterbourne Rd. E4 20Fc 35
Otterburn Gdns. TW7: Isle 52Ja 108
Otterburn Ho. SE5 52Sb 113
(off Sultan St.)
Otterburn St. SW17 65Hb 133
Otter Cl. E15 39Ec 72
Otter Cl. KT16: Ott 79D 148
Otterden Cl. BR6: Orp 76Uc 160
Otterden St. SE6 63Cc 136
Otterden Ter. SE1 6K 231 (49Vb 91)
(off Lynton Rd.)
Otter Dr. SM5: Cars 74Hb 155
Otterfield Rd. UB7: Yiew 45N 83
Otter Gdns. AL10: Hat 1D 8
Ottermead La. KT16: Ott 79E 148
Otter Mdw. KT22: Lea 91Ha 192
Otter Rd. UB6: G'frd 42Ea 86
Otters Cl. BR5: St P 70Zc 139
OTTERSHAW 79E 148
Ottershaw Pk. KT16: Ott 80C 148
(not continuous)
Otterspool La. WD25: A'ham 10Aa 13
(not continuous)
Otterspool Way WD25: A'ham 10Ba 13
(not continuous)
Otter Way UB7: Yiew 46N 83
Ottley Dr. SE3 56Lc 115
Otto Cl. SE26 62Xb 135
Otto Dr. UB2: S'hall 48Ca 85
Ottoman Ter. WD17: Wat 13Y 27
Otto St. SE17 51Rb 113
Ott's Yd. N19 35Lb 70
(off Southcote Rd.)
Ottway's Av. KT21: Asht 91Ma 193
Ottways La. KT21: Asht 92Ma 193
Otway Gdns. WD23: Bush 17Ga 28
Otways Cl. EN6: Pot B 4Db 17
Oulton Cl. E5 33Yb 72
Oulton Cl. SE28 44Yc 95
Oulton Cres. EN6: Pot B 3Za 16
Oulton Cres. IG11: Bark 36Vc 75
Oulton Rd. N15 29Tb 51
Oulton Way WD19: Wat 21Ba 45
Oundle Av. WD23: Bush 16Ea 28
Oundle Ho. RM3: Rom 22Md 57
(off Montgomery Cres.)
Our Lady's Cl. SE19 65Tb 135
Ousden Cl. EN8: Chesh 2Ac 20
Ousden Dr. EN8: Chesh 2Ac 20
Ouseley Lodge SL4: Old Win 9N 103
Ouseley Rd. SL4: Old Win 9N 103
Ouseley Rd. SW12 60Hb 111
Ouseley Rd. TW19: Wray 9N 103
Outer Circ. NW1 2E 214 (40Gb 69)
Outgate Rd. NW10 38Va 68
Outlook Pl. WD17: Wat 11W 26
Outram Pl. KT13: Weyb 78S 150
Outram Pl. N1 39Nb 70
Outram Rd. CR0: C'don 75Vb 157
Outram Rd. E6 39Nc 74
Outram Rd. N22 25Mb 50
Outwich St. EC3 2J 225 (44Ub 91)
(off Camomile St.)
Outwood Ho. SW2 59Pb 112
(off Deepdene Gdns.)
Outwood La. CR5: Chips 92Gb 195
Outwood La. CR5: Kgswd 92Gb 195
Outwood La. KT20: Kgswd 94Db 195
Outwood La. RH1: Blet 5K 209
Outwood La. RH1: S Nut 5K 209
Oval Surrey CCC, The 51Pb 112
Oval, The DA15: Sidc 59Wc 117
Oval, The DA3: Lfield 69Ee 143
Oval, The E2 40Xb 71
Oval, The SM7: Bans 86Cb 175
Oval Ct. HA8: Edg 24Sa 47
Oval Gdns. RM17: Grays 48Ee 99
Oval Ho. CR0: C'don 74Ub 157
(off Oval Rd.)
Ovalhouse 51Qb 112
Oval Mans. SE11 51Pb 112
Oval Pl. SW8 52Pb 112
Oval Rd. CR0: C'don 75Tb 157
Oval Rd. NW1 39Kb 70
Oval Rd. Nth. RM10: Dag 39Dd 76
Oval Rd. Sth. RM10: Dag 40Dd 76
Oval Station (Underground) 51Qb 112
Ovaltine Ct. WD4: K Lan 1R 12
Ovaltine Dr. WD4: K Lan 1R 12
Oval Way SE11 50Pb 90
Oval Way SL9: Ger X 28A 42
Ovanna M. N1 37Ub 71
Ovenden Rd. TN14: Chev 92Zc 201
Ovenden Rd. TN14: Sund 92Zc 201
Overbrae BR3: Beck 65Cc 136
Overbrook KT24: W Hor 100R 190
Overbrook Wlk. HA8: Edg 24Qa 47
(not continuous)
Overbury Av. BR3: Beck 69Dc 136
Overbury Cres. CR0: New Ad 82Ec 178
Overbury Rd. N15 30Tb 51
Overbury St. E5 35Zb 72
Overchess Ridge WD3: Chor 13H 25
Overcliffe DA11: Grav'nd 8C 122
Overcliff Rd. RM17: Grays 49Fe 99
(not continuous)
Overcliff Rd. SE13 55Cc 114
Overcourt Cl. DA15: Sidc 58Xc 117
Overdale KT21: Asht 87Na 173
Overdale RH1: Blet 5J 209
Overdale Av. KT3: N Mald 68Sa 131
Overdale Rd. W5 48La 86
Overdown Rd. SE6 63Cc 136
Overhill CR6: W'ham 91Yb 198
Overhill Rd. CR8: Purl 81Qb 176
Overhill Rd. SE22 59Wb 113
Overhill Way BR3: Beck 71Fc 159
Overlea Rd. E5 31Wb 71
Overlord Ct. KT22: Fet 93Fa 192
Overmead BR8: Swan 71Gd 162
Overmead DA15: Sidc 59Tc 116
Over Minnis DA3: Nw A G 76Be 165
Oversley Ho. W2 43Cb 89
(off Alfred Rd.)
Overstand Cl. BR3: Beck 71Cc 158
Overstone Gdns. CR0: C'don 73Bc 158
Overstone Ho. E14 44Cc 92
(off E. India Dock Rd.)
Overstone Rd. W6 48Ya 88
Overstrand Ho. RM12: Horn 34Kd 77
Overstrand Mans. SW11 53Hb 111
Overstream WD3: Loud 14K 25
Over Misbourne Rd., The SL9: Ger X 30C 42
Over Misbourne Rd., The UB9: Den 30D 42
Overthorpe Cl. GU21: Knap 9J 167
Overton Cl. NW10 37Sa 67
Overton Cl. TW7: Isle 53Ha 108
Overton Ct. E11 31Jc 73
Overton Ct. SM2: Sutt 80Cb 155
Overton Dr. E11 31Jc 73
Overton Dr. RM6: Chad H 31Yc 75
Overton Ho. SW15 59Va 110
(off Tangley Gro.)
Overton Rd. E10 32Ac 72
Overton Rd. N14 15Nb 32
Overton Rd. SE2 48Yc 95
Overton Rd. SM2: Sutt 79Cb 155
Overton Rd. SW9 54Qb 112
Overton Rd. E. SE2 48Zc 95
Overton's Yd. CR0: C'don 76Sb 157
Overy Ho. SE1 2B 230 (47Rb 91)
Overy Liberty DA1: Dart 59Nd 119
Overy St. DA1: Dart 58Nd 119
Ovesdon Av. HA2: Harr 32Ba 65
Oveton Way KT23: Bookh 98Ca 191
Ovett Cl. SE19 65Ub 135
Ovex Cl. E14 47Ec 92
Ovington Ct. GU21: Wok 8K 167
Ovington Ct. SW3 4E 226 (48Gb 89)
(off Brompton Rd.)
Ovington Gdns. SW3 4E 226 (48Gb 89)
Ovington M. SW3 4E 226 (48Gb 89)
(off Ovington Gdns.)
Ovington Sq. SW3 4E 226 (48Gb 89)
Ovington St. SW3 5E 226 (49Gb 89)
Owen Cl. UB5: N'olt Arnold Rd. 37Aa 65
Owen Cl. UB5: N'olt Attlee Rd. 41X 85
Owen Cl. CR0: C'don 72Tb 157
Owen Cl. DA10: Swans 60Ce 121
Owen Cl. RM5: Col R 23Dd 56
Owen Cl. SE28 46Yc 95
Owen Cl. SL3: L'ly 50B 82
Owen Cl. TW16: Sun 67U 128
Owen Gdns. IG8: Wfd G 23Nc 54
Owen Ho. TW1: Twick 59Ka 108
Owen Ho. TW14: Felt 59W 106
Owenite St. SE2 49Xc 95
Owen Mans. W14 51Ab 110
(off Queen's Club Gdns.)
Owen Pl. KT22: Lea 94Ka 192
Owen Rd. GU20: W'sham 8B 146
Owen Rd. N13 22Sb 51
Owen Rd. UB4: Yead 41X 85
Owens M. E11 33Gc 73
Owen Sq. WD19: Wat 16Z 27
(off Pinner Rd.)
Owen's Row EC1 3B 218 (41Rb 91)
Owen St. EC1 2B 218 (40Rb 71)
Owens Way SE23 59Ac 114
Owens Way WD3: Crox G 15Q 26
Owen Wlk. SE20 67Wb 135
Owen Way NW10 37Sa 67
Owgan Cl. SE5 52Tb 113
Owl Cl. CR2: Sels 82Zb 178
Owlets Hall Cl. RM11: Horn 27Pd 57
Owlets Hall Cl. RM11: Hrld W 27Pd 57
Owletts 9H 145
Owl Pk. IG10: Lough 13Hc 35
Ownstead Gdns. CR2: Sande 83Vb 177
Ownsted Hill CR0: New Ad 82Ec 178
Oxberry Av. SW6 54Ab 110
Oxborough Ho. SW18 56Eb 111
(off Eltringham St.)
Oxdowne Cl. KT11: Stoke D 86Da 171
Oxenden Wood Rd. BR6: Chels 79Xc 161
Oxendon St. SW1 5D 222 (45Mb 90)
Oxenford St. SE15 55Vb 113
Oxenham Ho. SE8 51Cc 114
(off Benbow St.)
Oxenhill Rd. TN15: Kems'g 89Nd 183
Oxenholme NW1 2C 216 (40Lb 70)
(off Hampstead Rd.)
Oxenpark Av. HA9: Wemb 31Na 67
Oxestall's Rd. SE8 50Ac 92
Oxford & Cambridge Mans.
NW1 1E 220 (43Gb 89)
(off Old Marylebone Rd.)
Oxford Av. AL1: St A 3G 6
Oxford Av. N14 18Lb 32
Oxford Av. RM11: Horn 28Qd 57
Oxford Av. RM16: Grays 9C 100
Oxford Av. SL1: Slou 3D 80
Oxford Av. SW20 68Ab 132
Oxford Av. TW5: Hest 50Ca 85
Oxford Av. UB3: Harl 52V 106
Oxford Cir. W1 3B 222 (44Lb 90)
(off Oxford St.)
Oxford Cir. Av. W1 3B 222 (44Lb 90)
Oxford Circus Station
(Underground) 3B 222 (44Lb 90)
Oxford Cl. CR4: Mitc 69Lb 134
Oxford Cl. DA12: Grav'nd 1H 145
Oxford Cl. EN8: Chesh 1Zb 20
Oxford Cl. HA6: Nwood 21T 44
Oxford Cl. N9 19Xb 33
Oxford Cl. RM2: Rom 29Jd 56
Oxford Cl. TW15: Ashf 66S 128
Oxford Ct. CM14: W'ley 21Zd 59
Oxford Ct. EC4 4F 225 (45Tb 91)
(off Cannon St.)
Oxford Ct. KT18: Eps 86Ua 174
Oxford Ct. KT6: Surb 71Na 153
(off Avenue Elmers)
Oxford Ct. TW13: Hanw 63Z 129
Oxford Ct. W3 44Qa 87
Oxford Ct. W4 50Ra 87
Oxford Ct. W7 43Ha 86
(off Copley Cl.)
Oxford Ct. W9 43Cb 89
(off Elmfield Way)
Oxford Cres. KT3: N Mald 72Ta 153
Oxford Dr. HA4: Ruis 33Y 65
Oxford Dr. SE1 7H 225 (46Ub 91)
Oxford Gdns. N20 18Fb 31
Oxford Gdns. N21 17Sb 33
Oxford Gdns. UB9: Den 34H 63
Oxford Gdns. W10 44Ya 88
Oxford Gdns. W4 50Qa 87
Oxford Ga. W6 49Za 88
Oxford Ho. BR2: Broml 72Mc 159
(off Wells Vw. Dr.)
Oxford Ho. E3 43Bc 92
(off William Whiffin Sq.)
Oxford Ho. NW6 40Cb 69
(off Oxford Rd.)
Oxford Ho. RM8: Dag 35Wc 75
Oxford Ho. WD6: Bore 12Qa 29
(off Stratfield Rd.)
Oxford M. DA5: Bexl 59Cd 118
Oxford Pl. NW10 34Ta 67
(off Press Rd.)
Oxford Rd. DA14: Sidc 64Xc 139
Oxford Rd. E15 37Fc 73
(not continuous)
Oxford Rd. EN3: Pond E 15Xb 33
Oxford Rd. HA1: Harr 30Ea 46
Oxford Rd. HA3: W'stone 27Ha 46
Oxford Rd. IG1: Ilf 36Sc 74
Oxford Rd. IG8: Wfd G 22Mc 53
Oxford Rd. N4 32Qb 70
Oxford Rd. N9 19Xb 33
Oxford Rd. NW6 40Cb 69
Oxford Rd. RH1: Redh 5N 207
Oxford Rd. RM3: Rom 23Pd 57
Oxford Rd. SE19 65Tb 135
Oxford Rd. SL4: Wind 3G 102
Oxford Rd. SM5: Cars 79Gb 155
Oxford Rd. SM6: W'gton 78Lb 156
Oxford Rd. SS17: Stan H 2K 101
Oxford Rd. SW15 56Ab 110
Oxford Rd. TW11: Tedd 64Fa 130
Oxford Rd. UB8: Uxb 37L 63
Oxford Rd. UB9: Den 33F 62
Oxford Rd. W5 45Ma 87
Oxford Rd. E. SL4: Wind 3G 102
Oxford Rd. Nth. W4 50Ra 87
Oxford Rd. Sth. W4 50Qa 87
Oxford Row TW16: Sun 69Y 129
Oxford Sq. W2 3D 220 (44Gb 89)
Oxford St. W1 3H 221 (44Jb 90)
Oxford St. WD18: Wat 15X 27
Oxford Ter. NW6 40Db 69
(off Oxford Rd.)
Oxford Wlk. UB1: S'hall 46Ba 85
Oxford Way TW13: Hanw 63Z 129
Oxgate Cen. NW2 33Xa 68
Oxgate Ct. NW2 33Wa 68
Oxgate Ct. Pde. NW2 33Wa 68
Oxgate Gdns. NW2 34Xa 68
Oxgate La. NW2 33Xa 68
Oxgate Pde. NW2 33Wa 68
Oxhawth Cres. BR2: Broml 71Qc 160
OXHEY 16Y 27
Oxhey Av. WD19: Wat 17Z 27
Oxhey Dr. HA6: Nwood 21X 45
Oxhey Dr. WD19: Wat 21X 45
Oxhey La. HA3: Hrw W 22Ca 45
Oxhey La. HA5: Hat E 21Ca 45
Oxhey La. WD19: Wat 18Aa 27
Oxhey Pk. 16Y 27
Oxhey Ridge Cl. HA6: Nwood 21X 45
Oxhey Rd. WD19: Wat 17Y 27
Oxhey Woods Local Nature Reserve 21W 44
Oxlade Dr. SL3: L'ly 9N 81
Ox La. KT17: Ewe 81Wa 174
Oxleas E6 44Rc 94
Oxleas Cl. DA16: Well 54Tc 116
OXLEASE 1C 8
Oxlease Dr. AL10: Hat 1D 8
Oxleay Rd. HA2: Harr 32Ca 65
Oxleigh Cl. KT3: N Mald 71Ua 154
Oxley Cl. RM2: Rom 26Ld 57
Oxley Cl. SE1 7K 231 (50Vb 91)
Oxley Sq. E3 42Dc 92
(off Truman Wlk.)
Oxleys Rd. EN9: Walt A 4Jc 21
Oxleys Rd. NW2 34Xa 68
Oxlip Cl. CR0: C'don 74Zb 158
Oxlow La. RM10: Dag 35Cd 76
Oxlow La. RM9: Dag 35Bd 75
Oxonian St. SE22 56Vb 113
Oxo Tower Wharf 5A 224 (45Qb 90)
OXSHOTT 85Fa 172
Oxshott Lodge KT22: Oxs 86Fa 172
Oxshott Ri. KT11: Cobh 85Z 171
Oxshott Rd. KT22: Lea 88Ga 172
Oxshott Station (Rail) 86Ea 172
Oxshott Village Sports Club 86Ea 172
Oxshott Way KT11: Cobh 87Aa 171
OXTED 1J 211
Oxted Cl. CR4: Mitc 69Fb 133
Oxted Ct. RH1: Redh 4B 208
(off Reynolds Av.)
Oxted Ho. RM3: Rom 22Pd 57
(off Redcar Rd.)
Oxted Pl. RH8: Oxt 99Fc 199
Oxted Rd. RH9: G'stone 2A 210
Oxted Station (Rail) 1J 211
Oxtoby Way SW16 67Mb 134
Oxygen, The E16 45Jc 93
OYO Bus. Units DA17: Belv 47Ed 96
OYO Bus. Units RM9: Dag 42Bd 95
Oystercatcher Cl. E16 44Kc 93
Oyster Cl. EN5: Barn 16Ab 30
Oyster Ct. SE17 6D 230 (49Sb 91)
(off Crampton St.)
Oysterfields AL3: St A 1P 5
Oyster La. KT14: Byfl 82M 169
Oyster M. E7 37Mc 73
Oyster Row E1 44Yb 92
Oyster Wharf SW11 54Fb 111
Ozolins Way E16 44Jc 93

P

Pablo Neruda Cl. SE24 56Rb 113
Paceheath Cl. RM5: Col R 23Fd 56
Pace Pl. E1 44Xb 91
Pachesham Dr. KT22: Lea 88Ha 172
Pachesham Golf Course 90Ga 172
Pachesham Pk. KT22: Lea 88Ja 172
PACHESHAM PARK 88Ja 172
Pacific Bldg. E15 36Fc 73
(off Property Row)
Pacific Cl. DA10: Swans 57Ae 121
Pacific Cl. TW14: Felt 60V 106
Pacific Ct. E1 43Yb 92
Pacific Ho. E1 42Zb 92
(off Ernest St.)
Pacific M. SW9 56Qb 112
Pacific Rd. E16 44Jc 93
Pacific Wharf SE16 46Zb 92
Packenham Ho. E2 41Vb 91
(off Wellington Row)
Packenham Ho. W10 45Za 88
(off Shalfleet Dr.)
Packet Boat Ho. UB8: Cowl 44L 83
Packet Boat La. UB8: Cowl 44K 83
Packet Boat Marina UB8: Yiew 44L 83
Packham Cl. BR6: Chels 75Yc 161
Packham Ct. KT4: Wor Pk 76Ya 154
Packham Rd. DA11: Nflt 2B 144
Packhorse Cl. AL4: St A 2Ta 15
Packhorse La. EN6: Ridge 2Ta 15
Packhorse La. WD6: Bore 9Ua 16
Packhorse Rd. SL9: Ger X 28A 42
Packhorse Rd. TN13: Bes G 95Ed 202
Packington Rd. W3 48Sa 87
Packington Sq. N1 1D 218 (39Sb 71)
(not continuous)
Packington St. N1 39Rb 71
Packmores Rd. SE9 57Tc 116
Packwell Pl. TW5: Hest 51Y 107
Padbrook RH8: Limp 1L 211
(not continuous)
Padbrook Cl. RH8: Limp 1M 211
Padbury SE17 7J 231 (50Ub 91)
(off Bagshot St.)
Padbury Cl. TW14: Bedf 60T 106
Padbury Ct. E2 4K 219 (41Vb 91)
Padbury Ho. NW8 5E 214 (42Gb 89)
(off Tresham Cres.)
Padbury Oaks UB7: Lford 53K 105
Padcroft Rd. UB7: Yiew 46M 83
Paddenswick Rd. W6 48Wa 88
PADDINGTON 2B 220 (44Fb 89)
Paddington Cl. UB4: Yead 43Aa 85
Paddington Ct. W7 43Ha 86
(off Copley Cl.)
Paddington Gdns. W2 1B 220 (43Fb 89)
Paddington Grn. W2 7B 214 (43Fb 89)
Paddington Sports Club 42Db 89
Paddington Station (Rail, Underground
& Crossrail) 2B 220 (44Fb 89)
Paddington St. W1 7H 215 (43Jb 90)
Paddock, The SL4: Wink Squirrel La. 10A 102
Paddock, The DA2: Dart 61Ud 142
Paddock, The GU18: Light 3A 166
Paddock, The N10 27Jb 50
Paddock, The NW9 29Qa 47
Paddock, The SL3: Dat 3M 103
Paddock, The SL9: Chal P 22A 42
Paddock, The TN16: Westrm 98Sc 200
Paddock, The UB10: Ick 35R 64
Paddock Cl. BR6: Farnb 77Rc 160
Paddock Cl. DA4: S Dar 67Sd 142
Paddock Cl. KT4: Wor Pk 74Ua 154
Paddock Cl. RH8: Oxt 3K 211
Paddock Cl. RM16: Ors 3D 100
Paddock Cl. SE26 63Zb 136
Paddock Cl. SE3 54Jc 115
Paddock Cl. SL4: Wind 4A 102
Paddock Cl. TN15: Plat 93Ee 205
Paddock Cl. UB5: N'olt 40Ca 65
Paddock Cl. WD19: Wat 16Aa 27
Paddock Gdns. SE19 65Ub 135
Paddock Gdns. TN15: Crou 95De 205
Paddock Ho. DA11: Grav'nd 4E 144
Paddock Ho. EN5: Ark 13Ua 30
Paddock Lodge EN1: Enf 15Ub 33
(off Village Rd.)
Paddock Mobile Home Pk. BR2: Kes 81Nc 180
Paddock Pas. SE19 65Ub 135
(off Paddock Gdns.)
Paddock Rd. DA6: Bex 56Ad 117
Paddock Rd. HA4: Ruis 34Z 65
Paddock Rd. NW2 34Wa 68
Paddocks, The AL4: St A 4L 7
Paddocks, The BR8: Swan 67Ld 141
Paddocks, The CR0: Addtn 79Cc 158
Paddocks, The EN4: Cockf 13Hb 31
Paddocks, The GU25: Vir W 72A 148
Paddocks, The HA9: Wemb 33Ra 67
Paddocks, The KT13: Weyb 76U 150
Paddocks, The KT15: New H 82K 169
Paddocks, The KT23: Bookh 98Da 191
Paddocks, The RM16: Ors 3D 100
Paddocks, The RM4: Stap A 16Hd 38
Paddocks, The TN13: S'oaks 96Md 203
Paddocks, The W5 48Ma 87
(off Popes La.)
Paddocks, The WD3: Chor 14H 25
Paddocks Cl. BR5: Orp 75Zc 161
Paddocks Cl. HA2: Harr 35Da 65
Paddocks Cl. KT11: Cobh 86Y 171
Paddocks Cl. KT21: Asht 90Na 173
Paddocks Grn. NW9 32Ra 67
Paddocks Mead GU21: Wok 8J 167
Paddocks Way KT16: Chert 74K 149
Paddocks Way KT21: Asht 90Na 173
Paddock Vw. WD7: R'lett 5Ja 14
Paddock Wlk. CR6: W'ham 91Xb 197
Paddock Way BR7: Chst 66Tc 138
Paddock Way GU21: Wok 86D 168
Paddock Way HP1: Hem H 2G 2
Paddock Way RH8: Oxt 3K 211
Paddock Way SW15 59Ya 110
Padelford La. HA7: Stan 19Ja 28
Padfield Ct. HA9: Wemb 34Pa 67
Padfield Rd. SE5 55Sb 113
Padgate Ho. WD19: Wat 20Y 27
Padgets, The EN9: Walt A 6Gc 21
Padley Cl. KT9: Chess 78Pa 153
Padnall Ct. RM6: Chad H 27Zc 55
Padnall Rd. RM6: Chad H 27Zc 55
Padstone Ho. E3 41Dc 92
(off Talwin St.)
Padstow Cl. BR6: Chels 77Vc 161
Padstow Cl. SL3: L'ly 47A 82
Padstow Ho. E14 45Bc 92
(off Three Colt St.)
Padstow Rd. EN2: Enf 11Rb 33
Padstow Wlk. TW14: Felt 60V 106
Padua Rd. SE20 67Yb 136
Pagden St. SW8 53Kb 112
Pageant Av. NW9 25Ta 47
Pageant Cl. RM18: Tilb 3E 122
Pageant Cres. SE16 46Ac 92
Pageantmaster Ct. EC4 3B 224 (44Rb 91)
(off Ludgate Hill)
Pageant Rd. AL1: St A 3B 6
Pageant Wlk. CR0: C'don 76Ub 157
Page Av. HA9: Wemb 34Sa 67
Page Cl. DA2: Bean 62Yd 142
Page Cl. HA3: Kenton 30Pa 47
Page Cl. RM9: Dag 36Ad 75
Page Cl. TW12: Hamp 65Aa 129
Page Ct. NW7 24Xa 48
Page Cres. CR0: Wadd 78Rb 157
Page Cres. DA8: Erith 52Hd 118
Page Cft. KT15: Add 75K 149
Page Grn. Rd. N15 29Wb 51
Page Grn. Ter. N15 29Vb 51
Page Heath La. BR1: Broml 69Mc 137
Page Heath Vs. BR1: Broml 69Mc 137
Page High N22 26Qb 50
(off Lymington Av.)
Page Ho. SE10 51Ec 114
(off Welland St.)
Pagehurst Rd. CR0: C'don 73Xb 157
Page Mdw. NW7 24Xa 48
Page M. SW11 54Jb 112
Page Pl. AL2: F'mre 10C 6
Page Rd. TW14: Bedf 58T 106
Pages Hill N10 26Jb 50
Pages La. N10 26Jb 50
Pages La. RM3: Hrld W 26Rd 57
Pages La. UB8: Uxb 37L 63
Page St. NW7 25Wa 48
Page St. SW1 5E 228 (49Mb 90)
Page's Wlk. SE1 5H 231 (49Ub 91)
Pages Yd. W4 51Va 110
Paget Av. SM1: Sutt 76Fb 155
Paget Cl. TW12: Hamp H 63Fa 130
Paget Gdns. BR7: Chst 67Rc 138
Paget Ho. E2 40Yb 72
(off Bishop's Way)
Paget Ho. SL9: Chal P 21A 42
(off Micholls Av.)
Paget La. TW7: Isle 55Fa 108
Paget Pl. KT2: King T 65Sa 131
Paget Pl. KT7: T Ditt 74Ha 152
Paget Ri. SE18 51Qc 116
Paget Rd. IG1: Ilf 35Rc 74
Paget Rd. N16 32Tb 71
Paget Rd. SL3: L'ly 49B 82
Paget Rd. UB10: Hil 42S 84
Paget St. EC1 3B 218 (41Rb 91)
Paget Ter. SE18 51Rc 116
Pagette Way RM17: Grays 50Ce 99
Pagham Ho. W10 42Ya 88
(off Sutton Way)
Pagin Ho. N15 29Ub 51
(off Braemar Rd.)
Pagitts Gro. EN4: Had W 11Db 31
Paglesfield CM13: Hut 16Ee 41
Pagnell St. SE14 52Bc 114
Pagoda Av. TW9: Rich 55Pa 109
Pagoda Gdns. SE3 54Fc 115
Pagoda Gro. SE27 61Sb 135
Paignton Cl. RM3: Rom 25Md 57
Paignton Rd. HA4: Ruis 34W 64
Paignton Rd. N15 30Ub 51
Paines Brook Ct. RM3: Rom 23Pd 57
Paines Brook Rd. RM3: Rom 23Pd 57
Paines Brook Way RM3: Rom 23Pd 57
Paines Cl. HA5: Pinn 27Aa 45
Painesfield Dr. KT16: Chert 74J 149
Paines La. HA5: Pinn 25Aa 45
Pain's Cl. CR4: Mitc 68Kb 134
PAINS HILL 86V 170
Pains Hill RH8: Limp 4N 211
PAINS HILL 3N 211
PAINSHILL 85V 170
Pains Hill Ho. KT11: Cobh 86V 170
Painshill Pk. 86V 170
Painsthorpe Rd. N16 34Ub 71
Painted Hall Greenwich 51Ec 114
Painters Ash La. DA11: Nflt 62Fe 143
Painters La. EN3: Enf W 7Ac 20
Painters M. SE16 49Wb 91
Painters Rd. IG2: Ilf 27Vc 55
Paisley Rd. N22 25Rb 51
Paisley Rd. SM5: Cars 74Fb 155
Paisley Ter. SM5: Cars 73Fb 155
Pakefield M. SW2 60Pb 112
Pakeman Ho. SE1 1C 230 (47Rb 91)
(off Surrey Row)

Pakeman St. N7 ....34Pb 70
Pakenham Cl. SW12 ....60Jb 112
Pakenham St. WC1 ....4J 217 (41Pb 90)
Pakes Way CM16: They B 9Uc 22
Pakington Ho. SW9 ....54Nb 112
(off Stockwell Gdns. Est.)
Palace Arts Way HA9: Wemb 35Qa 67
Palace Av. W8 ....47Db 89
Palace Cl. E9 ....37Bc 72
Palace Cl. SL1: Slou 6D 80
Palace Cl. WD4: K Lan 2P 11
Palace Ct. BR1: Broml 67Kc 137
(off Palace Gro.)
Palace Ct. GU21: Wok 88C 168
(off Maybury Rd.)
Palace Ct. HA3: Kenton 30Na 47
Palace Ct. NW3 ....36Db 69
Palace Ct. W2 ....45Db 89
(not continuous)
Palace Ct. Gdns. N10 ....27Lb 50
Palace Dr. KT13: Weyb 76R 150
Palace Exchange 14Tb 33
Palace Gdns. IG9: Buck H 18Mc 35
Palace Gdns. M. W8 ....46Db 89
Palace Gdns. Shop. Cen. 14Tb 33
Palace Gdns. Ter. W8 ....46Cb 89
Palace Ga. W8 ....2A 226 (47Eb 89)
Palace Gates M. N8 ....28Nb 50
(off The Campsbourne)
Palace Gates Rd. N22 ....25Mb 50
Palace Grn. CR0: Sels 80Bc 158
Palace Grn. W8 ....46Db 89
Palace Gro. BR1: Broml 67Kc 137
Palace Gro. SE19 ....66Vb 135
Palace Mans. KT1: King T 70Ma 131
(off Palace Rd.)
Palace Mans. W14 ....49Ab 88
(off Hammersmith Rd.)
Palace M. E17 ....28Bc 52
Palace M. SW1 ....6J 227 (49Jb 90)
(off Eaton Ter.)
Palace M. SW6 ....52Bb 111
Palace Pde. E17 ....28Bc 52
Palace Pl. SW1 ....3B 228 (48Lb 90)
Palace Pl. Mans. W8 ....47Db 89
(off Kensington Ct.)
Palace Rd. BR1: Broml 67Kc 137
Palace Rd. HA4: Ruis 35Aa 65
Palace Rd. KT1: King T 70Ma 131
Palace Rd. KT8: E Mos 69Ea 130
Palace Rd. N11 ....24Nb 50
Palace Rd. N8 ....29Mb 50
(not continuous)
Palace Rd. SE19 ....66Vb 135
Palace Rd. SW2 ....60Pb 112
Palace Rd. TN16: Westrm 93Qc 200
Palace Sq. SE19 ....66Vb 135
Palace St. SW1 ....3B 228 (48Lb 90)
Palace Superbowl 5D 230 (49Sb 91)
(within Elephant & Castle Shop. Cen.)
Palace Theatre London 3E 222 (44Mb 90)
(off Shaftesbury Av.)
Palace Vw. BR1: Broml 69Kc 137
(not continuous)
Palace Vw. CR0: C'don 77Bc 158
Palace Vw. SE12 ....61Jc 137
Palace Vw. Rd. E4 ....22Dc 52
Palace Way GU22: Wok 92D 188
Palace Way KT13: Weyb 76R 150
Palace Wharf W6 ....52Ya 110
(off Rainville Rd.)
Palamos Rd. E10 ....32Cc 72
Palatine Av. N16 ....35Ub 71
Palatine Rd. N16 ....35Ub 71
Palazzo Apts. N1 ....38Ub 71
(off Ardleigh Rd.)
Palemead Cl. SW6 ....53Za 110
Palermo Rd. NW10 ....40Wa 68
Palestine Gro. SW19 ....67Fb 133
Palestra Ho. SE1 ....7B 224 (46Rb 91)
(off Blackfriars Rd.)
Palewell Cl. BR5: St P 68Xc 139
Palewell Comn. Dr. SW14 ....57Ta 109
Palewell Pk. SW14 ....57Ta 109
Paley Gdns. IG10: Lough 13Rc 36
Palfrey Cl. AL3: St A 1B 6
Palfrey Pl. SW8 ....52Pb 112
Palgrave Av. UB1: S'hall 45Ca 85
Palgrave Ct. TW11: Hamp W 66La 130
Palgrave Gdns. NW1 ....5E 214 (42Gb 89)
Palgrave Ho. SE5 ....52Sb 113
(off Wyndham Est.)
Palgrave Ho. TW2: Whitt 59Ea 108
Palgrave Rd. W12 ....48Va 88
Palins Way RM16: Grays 46Ce 99
Palissy St. E2 ....4K 219 (41Vb 91)
(not continuous)
Palladian Cir. DA9: Ghithe 56Yd 120
Palladian Gdns. W4 ....51Ua 110
Palladino Ho. SW17 ....64Gb 133
(off Laurel Cl.)
Palladio Ct. SW18 ....58Db 111
(off Mapleton Cres.)
Palladium Ct. E8 ....38Vb 71
(off Queensbridge Rd.)
Pallant Ho. SE1 ....4G 231 (48Tb 91)
(off Tabard St.)
Pallant Way BR6: Farnb 76Qc 160
Pallet Way SE18 ....53Nc 116
Palliser Ct. W14 ....50Ab 88
(off Palliser Rd.)
Palliser Dr. RM13: Rain 43Jd 96
Palliser Ho. E1 ....42Zb 92
(off Ernest St.)
Palliser Ho. SE10 ....51Fc 115
(off Trafalgar Rd.)
Palliser Rd. W14 ....50Ab 88
Pallister Ter. SW15 ....62Va 132
Pall Mall SW1 ....7C 222 (46Lb 90)
Pall Mall E. SW1 ....6E 222 (46Mb 90)
Pall Mall Pl. SW1 ....7C 222 (46Lb 90)
(off Pall Mall)
Palmadium Cl. N13 ....20Qb 32
Palmar Cres. DA7: Bex 55Cd 118
Palmar Rd. DA7: Bex 54Cd 118
Palmarsh Rd. BR5: St M Cry 70Zc 139
Palm Av. DA14: Sidc 65Zc 139
Palm Cl. E10 ....34Dc 72
Palm Ct. SE15 ....52Vb 113
(off Garnies Cl.)
Palmeira Rd. DA7: Bex 55Zc 117
Palmer Av. DA12: Grav'nd 3F 144
Palmer Av. SM3: Cheam 77Ya 154
Palmer Av. WD23: Bush 15Da 27
Palmer Cl. BR4: W W'ck 76Fc 159
Palmer Cl. RH1: Redh 7A 208
Palmer Cl. TW5: Hest 53Ca 107
Palmer Cl. UB5: N'olt 37Aa 65
Palmer Cres. KT1: King T 69Na 131
Palmer Cres. KT16: Ott 79F 148
Palmer Dr. BR1: Broml 70Rc 138
Palmer Gdns. EN5: Barn 15Za 30
Palmer Ho. SE14 ....52Zb 114
(off Lubbock St.)
Palmer Pl. N7 ....36Qb 70
Palmer Rd. E13 ....42Kc 93
Palmer Rd. RM8: Dag 32Zc 75
Palmer Rd. SW11 ....52Kb 112
Palmers SS17: Stan H 1P 101
Palmers Av. RM17: Grays 50Ee 99
Palmer's Ct. N11 ....22Lb 50
(off Palmer's Rd.)
Palmers Dr. RM17: Grays 49Ee 99
Palmersfield Rd. SM7: Bans 86Cb 175
PALMERS GREEN 20Qb 32
Palmers Green Station (Rail) 21Pb 50
Palmers Gro. KT8: W Mole 70Ca 129
Palmers Hill CM16: Epp 1Wc 23
Palmers La. EN1: Enf 11Xb 33
Palmers La. EN3: Enf H 11Yb 34
Palmer's Moor La. SL0: Iver 42J 83
Palmers Orchard TN14: S'ham 83Hd 182
Palmers Pas. SW14 ....55Sa 109
(off Little St Leonard's)
Palmer's Rd. E2 ....40Zb 72
Palmer's Rd. N11 ....22Lb 50
Palmers Rd. SW14 ....55Sa 109
Palmers Rd. SW16 ....68Pb 134
Palmers Rd. WD6: Bore 11Ra 29
Palmer's Sports & Fitness Cen. 9A 100
Palmerston Av. SL3: Slou 8M 81
Palmerston Cen. HA3: W'stone 27Ha 46
Palmerston Cl. GU21: Wok 86C 168
Palmerston Cl. RH1: Redh 9A 208
Palmerston Ct. E3 ....40Zb 72
(off Old Ford Rd.)
Palmerston Ct. IG9: Buck H 18Lc 35
Palmerston Ct. KT6: Surb 73Ma 153
Palmerston Cres. N13 ....22Pb 50
Palmerston Cres. SE18 ....51Sc 116
Palmerstone Ct. GU25: Vir W 71A 148
(off Ridge Way)
Palmerston Gdns. RM20: Grays 50Zd 99
Palmerston Gro. SW19 ....66Cb 133
Palmerston Ho. SE1 ....2K 229 (47Qb 90)
(off Westminster Bri. Rd.)
Palmerston Ho. SM7: Bans 87Bb 175
(off Basing Rd.)
Palmerston Ho. W8 ....46Cb 89
(off Kensington Pl.)
Palmerston Mans. W14 ....51Ab 110
(off Queen's Club Gdns.)
Palmerston Rd. BR6: Farnb 77Sc 160
Palmerston Rd. CR0: C'don 71Tb 157
Palmerston Rd. E17 ....27Bc 52
Palmerston Rd. E7 ....37Kc 73
Palmerston Rd. HA3: W'stone 27Ga 46
Palmerston Rd. IG9: Buck H 19Kc 35
Palmerston Rd. N22 ....24Pb 50
Palmerston Rd. NW6 ....38Bb 69
(not continuous)
Palmerston Rd. RM13: Rain 40Ld 77
Palmerston Rd. RM20: Grays 51Zd 121
Palmerston Rd. SM1: Sutt 78Eb 155
Palmerston Rd. SM5: Cars 77Hb 155
Palmerston Rd. SW14 ....56Sa 109
Palmerston Rd. SW19 ....66Cb 133
Palmerston Rd. TW2: Twick 58Ga 108
Palmerston Rd. TW3: Houn 53Ea 108
Palmerston Rd. W3 ....48Sa 87
Palmerston Way SW8 ....52Kb 112
Palmer St. SW1 ....3D 228 (48Mb 90)
(not continuous)
Palmers Way EN8: Chesh 1Ac 20
Palmers Wharf KT1: King T 68Ma 131
(off Emms Pas.)
Palm Gro. W5 ....48Na 87
Palm Rd. RM7: Rom 29Ed 56
Palomino Cl. UB4: Hayes 40T 64
Palyn Ho. EC1 ....4E 218 (41Sb 91)
(off Radnor St.)
Pamela Av. HP3: Hem H 5P 3
Pamela Ct. N12 ....23Db 49
Pamela Gdns. HA5: Eastc 29X 45
Pamela St. E8 ....39Vb 71
Pampisford Rd. CR2: S Croy 81Rb 177
Pampisford Rd. CR8: Purl 83Qb 176
Pams Way KT19: Ewe 78Ta 153
Panama Ho. E1 ....43Zb 92
(off Beaumont Sq.)
Panavia Ct. NW9 ....26Va 48
Pancake La. HP2: Hem H 3D 4
Pancras La. EC4 ....3E 224 (44Sb 91)
Pancras Rd. N1 ....1F 217 (40Nb 70)
Pancras Rd. NW1 ....1D 216 (40Mb 70)
Pancras Sq. N1 ....1F 217 (40Nb 70)
Pancras Square Leisure Cen. 1F 217 (40Nb 70)
Pancras Way E3 ....40Cc 72
Pancroft RM4: Abr 13Xc 37
Pandangle Ho. E8 ....39Vb 71
(off Kingsland Rd.)
Pandian Way NW1 ....37Mb 70
Pandora Ct. E16 ....43Jc 93
(off Robertson Rd.)
Pandora Rd. NW6 ....37Cb 69
Panfield M. IG2: Ilf 30Qc 54
Panfield Rd. SE2 ....48Wc 95
Pangbourne NW1 ....4B 216 (41Lb 90)
(off Stanhope St.)
Pangbourne Av. W10 ....43Ya 88
Pangbourne Dr. HA7: Stan 22Ma 47
Panhard Pl. UB1: S'hall 45Da 85
Pank Av. EN5: New Bar 15Eb 31
Pankhurst Av. E16 ....46Kc 93
Pankhurst Av. E3 ....43Dc 92
Pankhurst Cl. SE14 ....52Zb 114
Pankhurst Cl. TW7: Isle 55Ha 108
Pankhurst Green 23Kc 53
(off Snakes Lane West)
Pankhurst Ho. W12 ....44Xa 88
Pankhurst Pl. WD24: Wat 13Y 27
Pankhurst Rd. KT12: Walt T 73Y 151
Pankridge WD19: Wat 18Z 27
Panmuir Rd. SW20 ....67Xa 132
Panmure Cl. N5 ....35Rb 71
Panmure Ct. SE26 ....62Xb 135
Panmure Ct. UB1: S'hall 44Ea 86
(off Osborne Rd.)
Panmure Rd. SE26 ....62Xb 135
Pannells Cl. KT16: Chert 74H 149
Pannells Ct. TW5: Hest 51Ca 107
Panorama Ct. N6 ....30Lb 50
Panoramic Twr. E14 ....44Dc 92
(off Burcham St.)
Pan Peninsula Sq. E14 ....47Dc 92
Pansy Gdns. W12 ....45Wa 88
Panter's BR8: Hext 66Hd 140
Panther Dr. NW10 ....36Ta 67
Pantile Cotts. RM14: Upm 29Td 58
Pantile Rd. KT13: Weyb 77T 150
Pantile Row SL3: L'ly 49C 82
Pantiles, The BR1: Broml 69Nc 138
Pantiles, The DA7: Bex 52Bd 117
Pantiles, The NW11 ....28Bb 49
Pantiles, The WD23: B Hea 18Fa 28
Pantiles Cl. GU21: Wok 10M 167
Pantiles Cl. N13 ....22Rb 51
Pantile Wlk. UB8: Uxb 38L 63
(off The Pavilions)
Panton Cl. CR0: C'don 74Rb 157
Panton St. SW1 ....5D 222 (45Mb 90)
Panxworth Rd. HP3: Hem H 4M 3
Panyer All. EC1 ....2D 224 (44Sb 91)
(off Newgate St.)
Panyers Gdns. RM10: Dag 34Dd 76
Paper Bldgs. EC4 ....4A 224 (45Qb 90)
(off King's Bench Wlk.)
Papercourt La. GU23: Rip 94H 189
Papercourt Sailing Club 94H 189
Papermill Cl. SM5: Cars 77Jb 156
Paper Mill La. DA1: Dart 56Md 119
Paper Mill M. DA9: Ghithe 56Yd 120
Papermill Pl. E17 ....26Ac 52
Paper Mill Wharf E14 ....45Ac 92
Papillons Wlk. SE3 ....54Jc 115
Papworth Gdns. N7 ....36Pb 70
Papworth Way SW2 ....59Qb 112
Papyrus Ho. N1 ....3D 218 (41Sb 91)
Parabola Ct. KT8: E Mos 70Ea 130
Parade, The KT18: Eps Ashley Rd. 85Ta 173
Parade, The WD19: Wat The Mead 20Aa 27
Parade, The CM14: B'wood 20Yd 40
Parade, The CR0: C'don 72Nb 156
Parade, The DA1: Cray 57Hd 118
Parade, The DA10: Swans 57Be 121
Parade, The DA12: Grav'nd 1F 144
Parade, The DA3: Hartl 71Ae 165
Parade, The GU25: Vir W 2P 147
Parade, The KT10: Clay 79Ga 152
Parade, The KT17: Eps 85Ua 174
Parade, The KT18: Eps 86Qa 173
(off Spa Dr.)
Parade, The KT2: King T 68Na 131
(off London Rd.)
Parade, The KT20: Tad 91Ab 194
Parade, The KT22: Lea 92Ja 192
(off Kingston Rd.)
Parade, The KT4: Wor Pk 77Va 154
Parade, The RH1: Redh 7A 208
Parade, The RM15: Avel 47Sd 98
Parade, The RM3: Hrld W 23Rd 57
Parade, The SE26 ....62Xb 135
(off Wells Pk. Rd.)
Parade, The SE4 ....54Bc 114
(off Up. Brockley Rd.)
Parade, The SM1: Sutt 76Bb 155
Parade, The SM5: Cars 78Hb 155
(off Beynon Rd.)
Parade, The SW11 ....52Hb 111
Parade, The TN15: Kems'g 89Nd 183
Parade, The TN16: Tats 93Lc 199
(off Ship Hill)
Parade, The TW12: Hamp H 64Fa 130
Parade, The TW16: Sun 66V 128
Parade, The TW18: Staines 64F 126
(off Meadow Gdns.)
Parade, The UB6: G'frd 36Ka 66
Parade, The WD17: Wat 13X 27
Parade Ct. KT24: E Hor 98U 190
Parade Gdns. E4 ....24Dc 52
Parade Ground Path SE18 ....52Pc 116
Parade Mans. NW4 ....29Xa 48
Parade M. SE27 ....61Rb 135
PARADISE 3M 3
Paradise HP2: Hem H 3M 3
Paradise Gdns. NW9 ....49Xa 88
Paradise Ind. Est. HP2: Hem H 3M 3
Paradise Pas. N7 ....36Qb 70
Paradise Path SE28 ....46Wc 95
Paradise Pl. SE18 ....49Nc 94
Paradise Rd. EN9: Walt A 6Ec 20
Paradise Rd. SW4 ....54Nb 112
Paradise Rd. TW9: Rich 57Ma 109
Paradise Row E2 ....41Xb 91
Paradise St. SE16 ....47Xb 91
Paradise Wlk. SW3 ....51Hb 111
Paragon TW8: Bford 50La 86
(off Boston Pk. Rd.)
Paragon, The SE3 ....54Hc 115
Paragon Cl. E16 ....44Jc 93
Paragon Gro. KT5: Surb 72Pa 153
Paragon M. SE1 ....5G 231 (49Tb 91)
Paragon M. SW8 ....51Pb 112
Paragon Pl. KT5: Surb 72Pa 153
Paragon Pl. SE3 ....54Hc 115
Paragon Rd. E9 ....37Yb 72
Paramount Bldg. EC1 ....5B 218 (42Rb 91)
(off St John St.)
Paramount Ct. WC1 ....6C 216 (42Lb 90)
(off University St.)
Paramount Ind. Est. WD24: Wat 10Y 13
(off Sandown Rd.)
Parbury Ri. KT9: Chess 79Na 153
Parbury Rd. SE23 ....58Ac 114
Parchment Cl. CR4: Mitc 73Jb 156
Parchmore Rd. CR7: Thor H 68Rb 135
Parchmore Way CR7: Thor H 68Rb 135
Pardoe Rd. E10 ....31Dc 72
Pardoner Ho. SE1 ....3G 231 (48Tb 91)
(off Pardoner St.)
Pardoner St. SE1 ....3G 231 (48Tb 91)
(not continuous)
Pardon St. EC1 ....5C 218 (42Rb 91)
Parent Shop. Mall 26Jc 53
(off Marlborough Rd.)
Parents Paradise Watford 13Z 27
Pares Cl. GU21: Wok 8P 167
Parfett St. E1 ....43Wb 91
(not continuous)
Parfitt Cl. NW3 ....32Eb 69
Parfour Dr. CR8: Kenley 88Sb 177
Parfrey St. W6 ....51Ya 110
Pargraves Ct. HA9: Wemb 33Qa 67
Parham Dr. IG2: Ilf 30Rc 54
Parham Way N10 ....26Lb 50
Paris Corte SE13 ....55Dc 114
(off Loampit Va.)
Paris Gdn. SE1 ....6B 224 (46Rb 91)
Parish Cl. RM11: Horn 33Kd 77
Parish Cl. WD25: Wat 6Y 13
Parish Ct. KT6: Surb 71Na 153
Parish Ga. Dr. DA15: Sidc 58Uc 116
Parish La. SE20 ....65Zb 136
Parish La. SL2: Farn C 4F 60
Parish M. SE20 ....66Zb 136
Paris Ho. E2 ....40Xb 71
(off Old Bethnal Grn. Rd.)
Parish Wharf Pl. SE18 ....49Nc 94
Parison Cl. TW9: Rich 55Qa 109
Park, The DA14: Sidc 64Wc 139
Park, The HA1: Harr 32Ga 66
Park, The KT23: Bookh 95Ca 191
Park, The N6 ....30Jb 50
Park, The NW11 ....32Db 69
Park, The SE23 ....60Yb 114
Park, The SL9: Ger X 28B 42
Park, The SM5: Cars 78Hb 155
Park, The W5 ....46Ma 87
Park & Ride Windsor Home Park 1J 103
Park & Ride Windsor Legoland 7C 102
Park App. DA16: Well 56Xc 117
Park App. SE16 ....48Xb 91
Park Av. NW10: Brent Cres. 40Pa 67
(not continuous)
Park Av. NW10: Park Av. Nth. 36Xa 68
Park Av. AL1: St A 1E 6
Park Av. BR1: Broml 65Hc 137
Park Av. BR4: W W'ck 75Ec 158
Park Av. BR6: Chels 75Wc 161
Park Av. BR6: Farnb 76Pc 160
Park Av. CM13: Hut 18Ee 41
Park Av. CR3: Cat'm 96Ub 197
Park Av. CR4: Mitc 66Kb 134
Park Av. DA11: Nflt 10A 122
Park Av. DA12: Grav'nd 10E 122
Park Av. E15 ....37Gc 73
Park Av. E6 ....39Qc 74
Park Av. EN1: Enf 15Tb 33
Park Av. EN6: Pot B 6Eb 17
Park Av. HA4: Ruis 30T 44
Park Av. IG1: Ilf 32Qc 74
Park Av. IG11: Bark 37Sc 74
Park Av. IG8: Wfd G 22Kc 53
Park Av. N13 ....20Qb 32
Park Av. N18 ....21Wb 51
Park Av. N22 ....26Nb 50
Park Av. N3 ....25Db 49
Park Av. NW11 ....32Db 69
Park Av. NW2 ....37Xa 68
Park Av. RM14: Upm 31Ud 78
Park Av. RM20: W Thur 51Wd 120
Park Av. SM5: Cars 79Jb 156
Park Av. SW14 ....56Ta 109
Park Av. TW17: Shep 69U 128
Park Av. TW18: Staines 65H 127
Park Av. TW19: Wray 7P 103
Park Av. TW20: Egh 65E 126
Park Av. TW3: Houn 58Da 107
Park Av. UB1: S'hall 47Ba 85
Park Av. WD18: Wat 14W 26
Park Av. WD23: Bush 12Aa 27
Park Av. WD3: Chor 15J 25
Park Av. WD7: R'lett 5Ka 14
Park Av. E. KT17: Ewe 79Wa 154
Park Av. Maisonettes WD23: Bush 12Ba 27
Park Av. M. CR4: Mitc 66Kb 134
Park Av. Nth. N8 ....27Mb 50
Park Av. Nth. NW10 ....36Xa 68
Park Av. Rd. N17 ....24Xb 51
Park Av. Sth. N8 ....28Mb 50
Park Av. W. KT17: Ewe 79Wa 154
Park Blvd. RM2: Rom 25Hd 56
Park Central SE17 ....5D 230 (49Sb 91)
Park Central Bldg. E3 ....40Cc 72
Park Chase HA9: Wemb 35Pa 67
Park Cliff Rd. DA9: Ghithe 56Yd 120
Park Cl. AL9: Brk P 8J 9
Park Cl. E9 ....39Yb 72
Park Cl. HA3: Hrw W 25Ga 46
Park Cl. KT10: Esh 79Ca 151
Park Cl. KT12: Walt T 75V 150
Park Cl. KT14: Byfl 85M 169
Park Cl. KT15: New H 82K 169
Park Cl. KT2: King T 67Qa 131
Park Cl. KT22: Fet 96Fa 192
Park Cl. N12 ....20Fb 31
Park Cl. NW10 ....41Pa 87
Park Cl. NW2 ....34Xa 68
Park Cl. RH8: Oxt 100Hc 199
Park Cl. SL4: Wind 4H 103
Park Cl. SM5: Cars 79Hb 155
Park Cl. SW1 ....1F 227 (47Hb 89)
Park Cl. TW12: Hamp 67Ea 130
Park Cl. TW3: Houn 57Ea 108
Park Cl. W14 ....48Bb 89
Park Cl. W4 ....51Ta 109
Park Cl. WD23: Bush 13Z 27
Park Cl. WD3: Rick 21R 44
Park Club Acton, The 46Ua 88
Park Cnr. SL4: Wind 5C 102
PARK CORNER 4M 7
Park Cnr. Dr. KT24: E Hor 100U 190
Park Cnr. Rd. DA13: Sflt 63Be 143
Park Ct. CR2: S Croy 78Sb 157
(off Warham Rd.)
Park Ct. E17 ....29Dc 52
Park Ct. E4 ....19Ec 34
Park Ct. GU22: Wok 90B 168
Park Ct. HA3: Kenton 31Na 67
Park Ct. HA9: Wemb 36Na 67
Park Ct. KT1: Hamp W 67La 130
Park Ct. KT14: W Byf 85J 169
Park Ct. KT3: N Mald 70Ta 131
Park Ct. N11 ....24Mb 50
Park Ct. N17 ....24Wb 51
Park Ct. SE21 ....62Sb 135
Park Ct. SE26 ....65Xb 135
Park Ct. SM6: W'gton 78Nb 156
Park Ct. SW11 ....53Kb 112
Park Ct. UB8: Uxb 39M 63
Park Ct. W6 ....49Wa 88
Park Cres. DA8: Erith 51Ed 118
Park Cres. EN2: Enf 14Tb 33
Park Cres. HA3: Hrw W 25Ga 46
Park Cres. N3 ....24Eb 49
Park Cres. RM11: Horn 31Jd 76
Park Cres. SL5: S'dale 2D 146
Park Cres. TW2: Twick 60Fa 108
Park Cres. W1 ....6K 215 (42Kb 90)
Park Cres. WD6: E'tree 13Pa 29
Park Cres. M. E. W1 ....6A 216 (42Kb 90)
Park Cres. M. W. W1 ....6K 215 (42Kb 90)
Park Cres. Rd. DA8: Erith 51Fd 118
Park Cft. HA8: Edg 25Sa 47
Parkcroft Rd. SE12 ....59Hc 115
Parkdale N11 ....23Mb 50
Parkdale Cres. KT4: Wor Pk 76Ta 153
Parkdale Rd. SE18 ....50Uc 94
Park Dr. DA3: Lfield 69Ae 143
Park Dr. E14 ....46Dc 92
Park Dr. EN6: Pot B 3Db 17
Park Dr. GU22: Wok 90B 168
Park Dr. HA2: Harr 31Ca 65
Park Dr. HA3: Hrw W 23Fa 46
Park Dr. KT13: Weyb 78R 150
Park Dr. KT21: Asht 90Qa 173
Park Dr. N21 ....16Sb 33
Park Dr. NW11 ....32Db 69
Park Dr. RM1: Rom 28Fd 56
Park Dr. RM10: Dag 34Ed 76
Park Dr. RM14: Upm 35Rd 77
Park Dr. SE7 ....51Nc 116
Park Dr. SL5: S'dale 2D 146
Park Dr. SW14 ....57Ta 109
Park Dr. TN15: Wro 89De 185
Park Dr. W3 ....48Qa 87
Park Driving Range, The 3L 81
Park Dwellings NW3 ....36Hb 69
Park E. Bldg. E3 ....40Cc 72
(off Fairfield Rd.)
Park End BR1: Broml 67Hc 137
Park End NW3 ....35Gb 69
Park End Rd. RM1: Rom 28Gd 56
Parker Av. RM18: Tilb 3E 122
Parker Bldg. SE16 ....48Wb 91
(off Old Jamaica Rd.)
Parker Cl. E16 ....46Nc 94
Parker Cl. SM5: Cars 79Hb 155
Parker Ct. N1 ....39Sb 71
(off Basire St.)
Parker Gdns. SL4: Old Win 8M 103
Parker Ho. E14 ....47Cc 92
(off Admirals Way)
Parker Ind. Cen. DA2: Dart 59Sd 120
Parker M. WC2 ....2G 223 (44Nb 90)
Parke Rd. SW13 ....53Wa 110
Parke Rd. TW16: Sun 70W 128
Parker Rd. CR0: C'don 77Sb 157
Parker Rd. RM17: Grays 50Be 99
Parker's Cl. KT21: Asht 91Na 193
Parker's Hill KT21: Asht 91Na 193
Parker's La. KT21: Asht 91Na 193
Parkers Row SE1 ....47Wb 91
Parker St. E16 ....46Nc 94
Parker St. WC2 ....2G 223 (44Nb 90)
Parker St. WD24: Wat 11X 27
Parkes Rd. IG7: Chig 22Uc 54
Parkes St. E20 ....37Cc 72
Park Farm Cl. HA5: Eastc 29X 45
Park Farm Cl. N2 ....27Eb 49
Park Farm Ct. UB3: Hayes 45U 84
Park Farm Rd. BR1: Broml 67Mc 137
Park Farm Rd. KT2: King T 66Na 131
Park Farm Rd. RM14: Upm 36Pd 77
Parkfield DA3: Hartl 70Ae 143
PARKFIELD 3Db 17
Parkfield TN15: S'oaks 95Nd 203
Parkfield TN15: Seal 95Nd 203
Parkfield TW7: Isle 53Ga 108
Parkfield WD3: Chor 14H 25
Parkfield Av. HA2: Harr 26Ea 46
Parkfield Av. SW14 ....56Ua 110
Parkfield Av. TW13: Felt 62W 128
Parkfield Av. UB10: Hil 41R 84
Parkfield Av. UB5: N'olt 40Z 65
Parkfield Cl. HA8: Edg 23Ra 47
Parkfield Cl. UB5: N'olt 40Aa 65
Parkfield Ct. SE14 ....53Bc 114
(off Parkfield Rd.)
Parkfield Cres. HA2: Harr 26Ea 46
Parkfield Cres. HA4: Ruis 33Aa 65
Parkfield Cres. TW13: Felt 62W 128
Parkfield Dr. UB5: N'olt 40Z 65
Parkfield Gdns. HA2: Harr 27Da 45
Parkfield Ho. HA2: Harr 25Da 45
Parkfield Ind. Est. SW11 ....54Jb 112
Parkfield Pde. TW13: Felt 62W 128
Parkfield Rd. HA2: Harr 34Ea 66
Parkfield Rd. NW10 ....38Xa 68
Parkfield Rd. SE14 ....53Bc 114
Parkfield Rd. SW4 ....58Mb 112
(not continuous)
Parkfield Rd. TW13: Felt 62W 128
Parkfield Rd. UB10: Ick 33R 64
Parkfield Rd. UB5: N'olt 40Aa 65
Parkfields CR0: C'don 74Bc 158
Parkfields KT22: Oxs 83Fa 172
Parkfields SW15 ....56Ya 110
Parkfields Av. NW9 ....32Ta 67
Parkfields Av. SW20 ....67Xa 132
Parkfields Cl. SM5: Cars 77Jb 156
Parkfields Rd. KT2: King T 64Pa 131
Parkfield St. N1 ....1A 218 (40Qb 70)
Parkfield Vw. EN6: Pot B 4Db 17
Parkfield Way BR2: Broml 72Pc 160
Park Gdns. DA8: Erith 49Fd 96
Park Gdns. E10 ....32Cc 72
Park Gdns. KT2: King T 64Pa 131
Park Gdns. NW9 ....27Ra 47
Park Ga. N2 ....27Fb 49
Park Ga. N21 ....17Pb 32
Park Ga. W5 ....43Ma 87
Parkgate N1 ....39Tb 71
(off Southgate Rd.)
Parkgate SE3 ....55Hc 115
Parkgate SL1: Burn 2A 80
Parkgate Av. EN4: Had W 11Eb 31
Park Ga. Cl. KT2: King T 65Ra 131
Park Ga. Ct. GU22: Wok 90A 168
Park Ga. Ct. TW12: Hamp H 65Ea 130
Parkgate Cres. EN4: Had W 11Eb 31
Parkgate Gdns. SW14 ....57Ta 109
Parkgate M. N6 ....31Lb 70
Parkgate Rd. BR6: Well H 77Dd 162
Parkgate Rd. RH2: Reig 7K 207
Parkgate Rd. SM6: W'gton 78Jb 156
Parkgate Rd. SW11 ....52Gb 111
Parkgate Rd. WD24: Wat 9Y 13
Park Gates HA2: Harr 35Ca 65
Park Grange IG7: Chig 22Tc 54
Park Grange Gdns. TN13: S'oaks 99Ld 203
Park Grn. KT23: Bookh 96Ca 191
Park Gro. BR1: Broml 67Kc 137
Park Gro. DA7: Bex 56Ed 118
Park Gro. E15 ....39Jc 73
Park Gro. HA8: Edg 22Pa 47
Park Gro. HP8: Chal G 13A 24
Park Gro. N11 ....24Mb 50
Park Gro. Rd. E11 ....33Gc 73
Park Hall SE10 ....52Fc 115
(off Croom's Hill)
Park Hall Rd. N2 ....28Gb 49
Park Hall Rd. RH2: Reig 4J 207
Park Hall Rd. SE21 ....62Sb 135
Park Hall Trad. Est. SE21 ....62Sb 135
Parkham Ct. BR2: Broml 68Gc 137
Parkham Ho. RH1: Redh 4B 208
(off Reynolds Av.)
Parkham St. SW11 ....53Gb 111
Park Hgts. GU22: Wok 90A 168
(off Constitution Hill)
Park Hgts. KT18: Eps 86Ta 173
Park Hgts. Ct. E14 ....44Bc 92
(off Wharf La.)
Parkhill KT10: Esh 77Ea 152

Park Hill BR1: Broml 70Nc 138
Park Hill DA13: Meop 10A 144
Park Hill IG10: Lough 15Mc 35
Park Hill SE23 61Xb 135
Park Hill SM5: Cars 79Gb 155
Park Hill SW4 57Mb 112
Park Hill TW10: Rich 58Pa 109
Park Hill W5 43Ma 87
Park Hill Cl. SM5: Cars 78Gb 155
Parkhill Cl. RM12: Horn 33Ld 77
Park Hill Ct. SW17 62Hb 133
Park Hill Dr. KT11: Cobh 87Aa 171
Park Hill M. CR2: S Croy 78Tb 157
Park Hill Ri. CR0: C'don 75Ub 157
Park Hill Rd. BR2: Broml 68Gc 137
Park Hill Rd. CR0: C'don 75Ub 157
Park Hill Rd. HP1: Hem H 2K 3
Park Hill Rd. KT17: Ewe 83Va 174
Park Hill Rd. SM6: W'gton 80Kb 156
Park Hill Rd. TN14: Otf 89Nd 183
Parkhill Rd. DA14: Sidc 62Tc 138
Parkhill Rd. DA15: Sidc 62Tc 138
Parkhill Rd. DA5: Bexl 59Bd 117
Parkhill Rd. E4 18Ec 34
Parkhill Rd. NW3 36Hb 69
Parkhill Wlk. NW3 36Hb 69
Parkholme Rd. E8 37Wb 71
Park Homes AL2: Lon C 8G 6
Park Ho. E9 38Yb 72
(off Shore Rd.)
Park Ho. N21 17Pb 32
Park Ho. SE5 53Tb 113
(off Camberwell Grn.)
Park Ho. TN13: S'oaks 94Ld 203
Park Ho. W1 3H 221 (44Jb 90)
(off Oxford St.)
Park Ho. Dr. RH2: Reig 8H 207
Park Ho. Gdns. TW1: Twick 57La 108
Park Ho. Pas. N6 31Jb 70
Parkhouse St. SE5 52Tb 113
Parkhurst KT19: Eps 82Sa 173
Parkhurst Ct. N7 35Nb 70
Parkhurst Gdns. DA5: Bexl 59Cd 118
Parkhurst Rd. DA5: Bexl 59Cd 118
Parkhurst Rd. E12 35Qc 74
Parkhurst Rd. E17 28Ac 52
Parkhurst Rd. N11 22Jb 50
Parkhurst Rd. N17 26Wb 51
Parkhurst Rd. N22 23Pb 50
Parkhurst Rd. N7 35Nb 70
Parkhurst Rd. SM1: Sutt 77Fb 155
Park Ind. Est. AL2: F'mre 9C 6
Parkinson Ct. N1 4G 219 (41Tb 91)
(off Charles Sq. Est.)
Parkinson Ho. E9 38Yb 72
(off Frampton Pk. Rd.)
Parkinson Ho. SW1 6C 228 (49Lb 90)
(off Tachbrook St.)
Park Lake Dr. RH3: Bkld 5D 206
Parkland Av. RM1: Rom 27Gd 56
Parkland Av. RM14: Upm 36Rd 77
Parkland Av. SL3: L'ly 9P 81
Parkland Cl. IG7: Chig 20Sc 36
Parkland Cl. TN13: S'oaks 100Ld 203
Parkland Ct. E15 36Gc 73
(off Maryland Pk.)
Parkland Ct. W14 47Ab 88
(off Holland Pk. Av.)
Parkland Dr. AL3: St A 3N 5
Parkland Gdns. SW19 60Za 110
Parkland Gro. TW15: Ashf 62Q 128
Parkland Mead BR1: Broml 69Rc 138
(not continuous)
Parkland M. BR7: Chst 66Tc 138
Parkland Rd. IG8: Wfd G 24Kc 53
Parkland Rd. N22 26Pb 50
Parkland Rd. TW15: Ashf 63Q 128
Parklands CM16: Coop 1Zc 23
(not continuous)
Parklands EN9: Walt A 4Fc 21
Parklands IG7: Chig 20Sc 36
Parklands KT15: Add 78L 149
Parklands KT23: Bookh 95Ca 191
Parklands KT5: Surb 71Pa 153
Parklands N6 31Kb 70
Parklands RH1: Redh 4A 208
Parklands RH8: Oxt 3J 211
Parklands WD23: Bush 16Ea 28
Parklands Cl. EN4: Had W 10Fb 17
Parklands Cl. IG2: Ilf 31Sc 74
Parklands Cl. SW14 57Sa 109
Parklands Ct. KT19: Eps 84Na 173
Parklands Ct. TW5: Hest 54Z 107
Parklands Dr. KT16: Chert 77F 148
Parklands Dr. N3 27Ab 48
Parklands Gro. TW7: Isle 53Ha 108
Parklands Pde. TW5: Hest 54Z 107
(off Parklands Ct.)
Parklands Rd. SW16 64Kb 134
Parklands Way KT4: Wor Pk 75Ua 154
Parkland Wlk. Local Nature Reserve 31Nb 70
Park La. AL4: Col H 5M 7
Park La. BR8: Swan 68Ld 141
Park La. CM13: Heron 24Fe 59
Park La. CR0: C'don 76Tb 157
Park La. CR5: Coul 93Mb 196
Park La. CR6: W'ham 89Ac 178
Park La. DA9: Ghithe 58Wd 120
Park La. E15 39Fc 73
Park La. EN8: Walt C 5Yb 20
Park La. HA2: Harr 34Da 65
Park La. HA7: Stan 20Ja 28
Park La. HA9: Wemb 36Na 67
Park La. HP1: Hem H 3M 3
Park La. HP2: Hem H 3M 3
Park La. KT21: Asht 90Pa 173
Park La. N17 24Vb 51
(not continuous)
Park La. N9 20Ub 33
(not continuous)
Park La. RH2: Reig 7G 206
Park La. RM11: Horn 30Hd 56
Park La. RM12: Horn 37Kd 77
Park La. RM15: Avel 46Td 98
(not continuous)
Park La. RM6: Chad H 30Zc 55
Park La. SL1: Burn 3C 60
Park La. SL3: Hort 55C 104
Park La. SL3: Slou 8M 81
Park La. SM3: Cheam 79Ab 154
Park La. SM5: Cars 77Jb 156
Park La. SM6: W'gton 78Jb 156
Park La. TN13: S'oaks 96Ld 203
Park La. TN15: God G 93Qd 203
Park La. TN15: Kems'g 90Qd 183
Park La. TN15: Seal 93Qd 203
Park La. TW11: Tedd 65Ha 130
Park La. TW5: Cran 52W 106
Park La. TW9: Rich 56Ma 109
Park La. UB4: Hayes 43U 84
Park La. UB9: Hare 25J 43
Park La. W1 4G 221 (45Jb 89)
Park La. Cl. N17 24Wb 51
Park La. E. RH2: Reig 9H 207
Park La. Mans. CR0: C'don 76Tb 157
(off Park La.)
PARK LANGLEY 70Ec 136
Parklangley Club, The 70Ec 136
Parklangley Tennis Club 70Ec 136
Park Lawn SL2: Farn R 1G 80
Parklawn Av. KT18: Eps 85Ra 173
Park Lawn Rd. KT13: Weyb 77S 150
Parklea Cl. NW9 25Ua 48
Park Lee Ct. N16 31Ub 71
Parkleigh Rd. SW19 68Db 133
Park Ley Rd. CR3: Wold 92Zb 198
Parkleys TW10: Ham 63Ma 131
Parkleys Pde. TW10: Ham 63Ma 131
Park Lodge NW8 38Fb 69
Park Lodge W14 48Bb 89
(off Melbury Rd.)
Park Lodge Av. UB7: W Dray 47P 83
Park Lofts SW2 57Nb 112
(off Lyham Rd.)
Park Lorne NW8 4E 214 (41Gb 89)
(off Park Rd.)
Park Mnr. SM2: Sutt 80Eb 155
(off Christchurch Pk.)
Park Mans. HA6: Nwood 19S 26
Park Mans. NW4 29Xa 48
Park Mans. NW8 2D 214 (40Gb 69)
(off Allitsen Rd.)
Park Mans. SW1 2F 227 (47Hb 89)
(off Knightsbridge)
Park Mans. SW11 53Hb 111
(off Prince of Wales Dr.)
Park Mans. SW8 51Nb 112
Parkmead IG10: Lough 15Qc 36
Parkmead SW15 58Xa 110
Park Mead DA15: Sidc 57Xc 117
Park Mead HA2: Harr 34Da 65
Parkmead Cl. CR0: C'don 72Zb 158
Parkmead Gdns. NW7 23Va 48
Park M. BR7: Chst 65Rc 138
Park M. N8 30Nb 50
Park M. RH8: Oxt 100Hc 199
Park M. RM13: Rain 37Jd 76
Park M. RM15: Avel 46Td 98
Park M. SE10 50Hc 93
Park M. SE24 59Sb 113
Park M. TW19: Stanw 59P 105
Park M. W10 40Ab 68
Parkmore Cl. IG8: Wfd G 21Jc 53
Park Nook Gdns. EN2: Enf 9Tb 19
Park Pale DA12: Shorne 7M 145
Park Pale ME2: Strood 7P 145
Park Pale Bri. 7N 145
Park Pde. NW10 40Va 68
Park Pde. UB3: Hayes 44U 84
Park Pde. W3 48Qa 87
Park Piazza SE13 58Fc 115
Park Pl. AL2: Park 9B 6
Park Pl. BR1: Broml 67Kc 137
(off Park Rd.)
Park Pl. DA12: Grav'nd 8E 122
Park Pl. E14 46Cc 92
Park Pl. GU22: Wok 90B 168
(off Hill Vw. Rd.)
Park Pl. HA9: Wemb 35Pa 67
Park Pl. N1 38Tb 71
(off Downham Rd.)
Park Pl. SW1 7B 222 (46Lb 90)
Park Pl. TN13: Bes G 95Fd 202
Park Pl. TW12: Hamp H 65Ea 130
Park Pl. UB9: Hare 25L 43
Park Pl. W3 49Qa 87
Park Pl. W5 46Ma 87
Park Pl. Dr. W3 48Qa 87
Park Pl. Vs. W2 7A 214 (43Eb 89)
Park Plaza EN8: Walt C 6Yb 20
Park Ride SL4: Wind 9B 102
Park Ridings N8 27Qb 50
Park Ri. HA3: Hrw W 25Ga 46
Park Ri. KT22: Lea 93Ka 192
Park Ri. SE23 60Ac 114
Park Ri. Cl. KT22: Lea 93Ka 192
Park Ri. Rd. SE23 60Ac 114
Park Rd. SM6: W'gton Clifton Rd. 78Kb 156
Park Rd. SM6: W'gton Elmwood Cl. 75Kb 156
Park Rd. BR1: Broml 67Kc 137
Park Rd. BR3: Beck 66Bc 136
Park Rd. BR5: St M Cry 71Yc 161
Park Rd. BR7: Chst 65Rc 138
Park Rd. BR8: Swan 70Hd 140
Park Rd. CM14: B'wood 18Xd 40
Park Rd. CR3: Cat'm 95Ub 197
Park Rd. CR6: W'ham 86Gc 179
Park Rd. CR8: Kenley 87Rb 177
Park Rd. DA1: Dart 59Qd 119
Park Rd. DA10: Swans 58Ae 121
Park Rd. DA11: Grav'nd 10D 122
Park Rd. E10 32Cc 72
Park Rd. E12 32Kc 73
Park Rd. E15 39Jc 73
Park Rd. E17 29Bc 52
Park Rd. E6 39Lc 73
Park Rd. EN3: Enf W 8Ac 20
Park Rd. EN4: E Barn 14Fb 31
Park Rd. EN5: Barn 14Bb 31
Park Rd. EN6: N'thaw 2Jb 18
Park Rd. EN8: Walt C 5Zb 20
Park Rd. GU22: Wok 89B 168
(not continuous)
Park Rd. HA0: Wemb 37Na 67
Park Rd. HP1: Hem H 4L 3
Park Rd. IG1: Ilf 34Tc 74
Park Rd. KT1: Hamp W 67La 130
Park Rd. KT10: Esh 77Da 151
Park Rd. KT2: King T 64Pa 131
Park Rd. KT21: Asht 90Na 173
Park Rd. KT3: N Mald 70Ta 131
Park Rd. KT5: Surb 72Pa 153
Park Rd. KT8: E Mos 70Ea 130
Park Rd. N11 24Mb 50
Park Rd. N14 18Mb 32
Park Rd. N15 28Rb 51
Park Rd. N18 21Vb 51
Park Rd. N2 27Fb 49
Park Rd. N8 28Lb 50
Park Rd. NW1 5F 215 (42Hb 89)
Park Rd. NW10 39Ua 68
Park Rd. NW4 31Wa 68
Park Rd. NW8 3D 214 (41Gb 89)
Park Rd. NW9 31Ta 67
Park Rd. RH1: Redh 4P 207
(not continuous)
Park Rd. RH8: Oxt 100Hc 199
Park Rd. RM17: Grays 50De 99
Park Rd. SE25 70Ub 135
Park Rd. SL2: Farn R 10G 60
Park Rd. SL2: Stoke P 10G 60
Park Rd. SM3: Cheam 79Ab 154
Park Rd. SM7: Bans 87Db 175
Park Rd. SS17: Stan H 2K 101
Park Rd. SW19 65Fb 133
Park Rd. TW1: Twick 58La 108
Park Rd. TW10: Rich 58Pa 109
Park Rd. TW11: Tedd 65Ha 130
Park Rd. TW12: Hamp H 63Da 129
Park Rd. TW13: Hanw 63Z 129
Park Rd. TW15: Ashf 64R 128
Park Rd. TW16: Sun 66X 129
Park Rd. TW17: Shep 74Q 150
Park Rd. TW19: Stanw 58K 105
Park Rd. TW19: Stanw M 58K 105
Park Rd. TW20: Egh 63C 126
Park Rd. TW3: Houn 57Da 107
Park Rd. TW7: Isle 53Ka 108
Park Rd. UB4: Hayes 43U 84
Park Rd. UB8: Uxb 38N 63
Park Rd. W4 52Sa 109
Park Rd. W7 45Ha 86
Park Rd. WD17: Wat 11W 26
Park Rd. WD23: Bush 16Ca 27
Park Rd. WD3: Rick 17M 25
Park Rd. WD7: R'lett 7Ja 14
Park Rd. E. UB10: Uxb 40M 63
Park Rd. E. W3 47Ra 87
Park Rd. Ho. KT2: King T 66Qa 131
Park Rd. Ind. Est. BR8: Swan 69Hd 140
Park Rd. Nth. W3 47Ra 87
Park Rd. Nth. W4 50Ta 87
Park Rd. Pools & Fitness 29Mb 50
Park Row SE10 51Fc 115
PARK ROYAL 41Ra 87
Park Royal NW10 41Qa 87
PARK ROYAL CENTRE (FOR MENTAL HEALTH) 40Sa 67
PARK ROYAL JUNC. 39Qa 67
Park Royal Metro Cen. NW10 42Ra 87
Park Royal Rd. NW10 41Sa 87
Park Royal Rd. W3 42Sa 87
Park Royal Station (Underground) 42Qa 87
Park St James NW8 1F 215 (39Hb 69)
(off St James's Ter. M.)
Parkshot TW9: Rich 56Ma 109
Parkside DA14: Sidc 61Xc 139
Parkside EN6: Pot B 4Eb 17
Parkside EN8: Walt C 6Ac 20
Parkside HP3: Hem H 9A 4
Parkside IG9: Buck H 19Kc 35
Parkside KT15: New H 83K 169
Parkside N3 25Db 49
Parkside NW7 23Wa 48
Parkside RM16: Grays 48Fe 99
Parkside SE3 52Hc 115
Parkside SL9: Ger X 29B 42
Parkside SM3: Cheam 79Ab 154
Parkside SW1 1G 227 (47Hb 89)
(off Knightsbridge)
Parkside SW19 62Za 132
Parkside TN14: Hals 85Bd 181
Parkside TW12: Hamp H 64Fa 130
Parkside UB3: Hayes 45U 84
Parkside W3 46Ua 88
Parkside W5 45Na 87
Parkside WD19: Wat 16Y 27
Park Side CM16: Epp 1Xc 23
Park Side NW2 34Wa 68
Parkside Av. BR1: Broml 70Nc 138
Parkside Av. DA7: Bex 54Fd 118
Parkside Av. RM1: Rom 27Fd 56
Parkside Av. RM18: Tilb 4D 122
(not continuous)
Parkside Av. SE10 53Ec 114
Parkside Av. SW19 64Za 132
Parkside Bus. Est. SE8: Blackhorse Rd. 51Ac 114
(not continuous)
Parkside Cl. KT24: E Hor 97V 190
Parkside Cl. SE20 66Yb 136
Parkside Ct. E11 30Jc 53
(off Wanstead Pl.)
Parkside Ct. E16 47Lc 93
(off Booth Rd.)
Parkside Ct. KT13: Weyb 77Q 150
Parkside Ct. N22 23Pb 50
Parkside Ct. RH1: Redh 2P 207
Parkside Cres. KT5: Surb 72Sa 153
Parkside Cres. N7 34Qb 70
Parkside Cross DA7: Bex 54Gd 118
Parkside Dr. HA8: Edg 20Qa 29
Parkside Dr. WD17: Wat 12U 26
Parkside Est. E9 39Yb 72
(not continuous)
Parkside Gdns. CR5: Coul 89Kb 176
Parkside Gdns. EN4: E Barn 18Hb 31
Parkside Gdns. SW19 63Za 132
PARKSIDE HOSPITAL 62Za 132
Parkside Ho. RM10: Dag 34Ed 76
Parkside Lodge DA17: Belv 50Ed 96
Parkside Lodge SL3: Slou 8L 81
(off Upton Ct. Rd.)
Parkside M. CR6: W'ham 88Cc 178
Parkside Pde. DA1: Erith 54Hd 118
(off Northend Rd.)
Parkside Pl. KT24: E Hor 97V 190
Parkside Pl. TW18: Staines 65J 127
Parkside Rd. DA17: Belv 49Dd 96
Parkside Rd. HA6: Nwood 22V 44
Parkside Rd. SL5: S'dale 1E 146
Parkside Rd. SW11 53Jb 112
Parkside Rd. TW3: Houn 57Da 107
Parkside Sq. E14 49Ec 92
Parkside Sq. SE10 53Ec 114
Parkside Stadium 45Ud 98
Parkside Ter. BR6: Farnb 76Rc 160
(off Willow Wlk.)
Parkside Ter. N18 21Tb 51
Parkside Wlk. DA11: Nflt 61Ee 143
Parkside Wlk. SL1: Slou 8L 81
Parkside Way HA2: Harr 28Da 45
Parks Info. Cen. 6E 220 (46Gb 89)
Park Sth. SW11 53Jb 112
(off Austin Rd.)
Park Sq. KT10: Esh 77Da 151
Park Sq. RM4: Abr 16Zc 37
Park Sq. E. NW1 5K 215 (42Kb 90)
Park Sq. M. NW1 6K 215 (42Kb 90)
Park Sq. W. NW1 6K 215 (42Kb 90)
Parkstead Rd. SW15 57Wa 110
Park Steps W2 4E 220 (45Gb 89)
(off St George's Flds.)
Parkstone Av. N18 23Vb 51
Parkstone Av. RM11: Horn 30Nd 57
Parkstone Rd. E17 27Ec 52
Parkstone Rd. SE15 54Wb 113
Park St. AL2: Park 8B 6
Park St. CR0: C'don 75Sb 157
Park St. SE1 6D 224 (46Sb 91)
Park St. SL1: Slou 8K 81
Park St. SL3: Coln 53F 104
Park St. SL4: Wind 3H 103
Park St. SW6 53Eb 111
Park St. TW11: Tedd 65Ga 130
Park St. W1 4H 221 (45Jb 90)
PARK STREET 8B 6
Park St. La. AL2: Park 2Da 13
Park Street Station (Rail) 8B 6
Park Ter. DA9: Ghithe 57Yd 120
Park Ter. EN3: Enf H 10Ac 20
Park Ter. KT4: Wor Pk 74Wa 154
Park Ter. SE3 55Lc 115
Park Ter. SM5: Cars 76Gb 155
Park Ter. TN14: Sund 96Zc 201
Park Ter. WD25: Wat 5Z 13
Park Theatre 33Qb 70
Parkthorne Cl. HA2: Harr 30Da 45
Parkthorne Dr. HA2: Harr 30Ca 45
Parkthorne Rd. SW12 59Mb 112
Park Towers W1 7K 221 (46Kb 90)
(off Brick St.)
Park Va. Ct. CM14: B'wood 18Yd 40
Park Vw. BR6: Orp 73Xc 161
Park Vw. CR3: Cat'm 97Wb 197
Park Vw. DA10: Swans 59Be 121
Park Vw. EN6: Pot B 5Eb 17
(not continuous)
Park Vw. HA5: Hat E 25Ba 45
Park Vw. HA9: Wemb 36Ra 67
Park Vw. IG7: Chig 20Nc 36
Park Vw. KT12: Hers 78Y 151
Park Vw. KT15: Add 78L 149
(off Bourne Way)
Park Vw. KT23: Bookh 97Ca 191
Park Vw. KT3: N Mald 69Va 132
Park Vw. N21 17Pb 32
Park Vw. N5 35Sb 71
Park Vw. RM15: Avel 46Td 98
Park Vw. RM6: Chad H 30Zc 55
Park Vw. SE8 50Zb 92
(off Trundleys Rd.)
Park Vw. TN13: S'oaks 96Ld 203
Park Vw. TN15: Hod S 81Ee 185
Park Vw. UB7: Yiew 45N 83
Park Vw. W3 43Sa 87
Parkview UB6: G'frd 41Ja 86
(off Perivale La.)
Parkview WD25: Wat 4Aa 13
Park View Academy Sports Cen. 28Sb 51
Park Vw. Apts. SE16 48Xb 91
(off Banyard Rd.)
Parkview Apts. E14 44Dc 92
(off Chrisp St.)
Parkview Chase SL1: Slou 4C 80
Park Vw. Cl. AL1: St A 3E 6
Parkview Cl. SM5: Cars 80Hb 155
Park Vw. Ct. E3 43Cc 92
Park Vw. Ct. GU22: Wok 91B 188
Park Vw. Ct. N12 21Gb 49
Park Vw. Ct. SE12 62Lc 137
Park Vw. Ct. SE20 67Xb 135
Parkview Ct. HA3: Hrw W 24Ga 46
Parkview Ct. IG2: Ilf 30Uc 54
Parkview Ct. SW18 58Cb 111
Parkview Ct. SW6 54Ab 110
Park Vw. Cres. N11 21Kb 50
Parkview Cres. KT4: Wor Pk 73Ya 154
Park Vw. Dr. CR4: Mitc 68Fb 133
Park Vw. Est. E2 40Zb 72
Park Vw. Gdns. IG4: Ilf 28Pc 54
Park Vw. Gdns. N22 25Qb 50
Park Vw. Gdns. NW4 29Ya 48
Park Vw. Gdns. RM17: Grays 50De 99
Park Vw. Ho. E4 22Cc 52
Park Vw. Ho. SE24 58Rb 113
(off Hurst St.)
Parkview Ho. N9 17Xb 33
Parkview Ho. RM12: Horn 33Kd 77
Parkview Ho. WD19: Wat 16Z 27
Park Vw. Mans. E20 36Dc 72
(off Olympic Pk. Av.)
Park Vw. Mans. N4 30Rb 51
Park Vw. M. SW9 54Pb 112
Parkview M. RM13: Rain 43Kd 97
Park Vw. Ri. HP2: Hem H 1N 3
Park View Road 55Yc 117
Park Vw. Rd. CR3: Wold 94Ac 198
Park Vw. Rd. DA16: Well 55Yc 117
Park Vw. Rd. HA5: Pinn 24X 45
Park Vw. Rd. KT22: Lea 92Ha 192
Park Vw. Rd. N17 27Wb 51
Park Vw. Rd. N3 25Db 49
Park Vw. Rd. NW10 35Va 68
Park Vw. Rd. UB1: S'hall 46Ca 85
Park Vw. Rd. UB8: Hil 44P 83
Park Vw. Rd. W5 43Na 87
Parkview Rd. CR0: C'don 74Wb 157
Parkview Rd. SE9 60Rc 116
Parkview Way KT19: Eps 82Ta 173
Park Village E. NW1 1K 215 (40Kb 70)
Park Village W. NW1 1K 215 (40Kb 70)
Park Vs. RM6: Chad H 30Zc 55
Parkville Rd. SW6 52Bb 111
Park Vista SE10 51Fc 115
Park Wlk. IG1: Ilf 33Rc 74
(within The Exchange)
Park Wlk. KT21: Asht 91Pa 193
Park Wlk. N6 31Jb 70
Park Wlk. SE10 52Fc 115
Park Wlk. SW10 51Eb 111
Parkway CR0: New Ad 81Dc 178
Parkway DA18: Erith 48Ad 95
Parkway IG3: Ilf 34Vc 75
Parkway IG8: Wfd G 22Lc 53
Parkway KT13: Weyb 77T 150
Parkway N14 19Nb 32
Parkway NW1 1K 215 (39Kb 70)
Parkway RM13: Rain 42Jd 96
Parkway RM6: Chad H 28Ad 55
Parkway RM2: Rom 26Hd 56
Parkway SW20 70Za 132
Parkway UB10: Hil 38Q 64
Park Way CM15: Shenf 18Be 41
Park Way DA5: Bexl 62Gd 140
Park Way EN2: Enf 12Qb 32
Park Way HA4: Ruis 32W 64
Park Way HA8: Edg 25Ra 47
Park Way KT23: Bookh 95Ca 191
Park Way KT8: W Mole 69Da 129
Park Way N20 21Hb 49
Park Way NW11 29Ab 48
Park Way TW14: Felt 59X 107
Park Way W3 48Qa 87
Park Way WD3: Rick 18L 25
Parkway, The SL0: Iver H 40E 62
Parkway, The TW5: Cran 50W 84
Park Way Ct. HA4: Ruis 32V 64
Parkway Ct. AL1: St A 5F 6
Parkway Cres. E15 36Ec 72
Parkway Trad. Est. TW5: Hest 51Y 107
Park W. W2 2E 220 (44Gb 89)
(off Park West Pl.)
Park W. Bldg. E3 40Cc 72
Park W. Pl. W2 2E 220 (44Gb 89)
Park Wharf SE8 50Ac 92
(off Evelyn St.)
Parkwood BR3: Beck 66Cc 136
Parkwood CM15: Dodd 11Zd 41
Parkwood N20 20Jb 32
Parkwood NW8 1F 215 (39Hb 69)
(off St Edmund's Ter.)
Parkwood Av. KT10: Esh 74Ea 152
Parkwood Cl. KT22: Fet 95Ea 192
Parkwood Cl. SM7: Bans 87Za 174
Park Wood Ct. HA4: Ruis 28S 44
Parkwood Dr. HP1: Hem H 2H 3
Parkwood Flats N20 20Hb 31
Park Wood Golf Course 94Nc 200
Parkwood Gro. TW16: Sun 69W 128
Parkwood Health & Fitness Cen. 3Cb 17
Parkwood M. N6 30Kb 50
Parkwood Rd. DA5: Bexl 59Bd 117
Parkwood Rd. RH1: Nutf 5E 208
Parkwood Rd. SM7: Bans 87Za 174
Parkwood Rd. SW19 64Bb 133
Parkwood Rd. TN16: Tats 93Nc 200
Parkwood Rd. TW7: Isle 53Ha 108
Parkwood Vw. SM7: Bans 88Ya 174
Park Works Rd. RH1: Nutf 5F 208
Parlaunt Rd. SL3: L'ly 49C 82
Parley Dr. GU21: Wok 9N 167
Parliament Ct. E1 1J 225 (43Ub 91)
(off Artillery La.)
Parliament Hill 34Hb 69
Parliament Hill NW3 35Gb 69
Parliament Hill Fields 34Jb 70
Parliament Hill Lido 35Jb 70
Parliament Hill Mans. NW5 35Jb 70
Parliament M. SW14 54Sa 109
Parliament Sq. SW1 2F 229 (47Nb 90)
Parliament St. SW1 2F 229 (47Nb 90)
Parliament Vw. SE1 5H 229 (49Pb 90)
Parma Cres. SW11 56Hb 111
Parmiter St. E2 40Xb 71
Parmoor Ct. EC1 5D 218 (42Sb 91)
(off Gee St.)
Parndon Ho. IG10: Lough 17Nc 36
Parnell Cl. HA8: Edg 21Ra 47
Parnell Cl. RM16: Chaf H 50Yd 98
Parnell Cl. W12 48Xa 88
Parnell Cl. WD5: Ab L 2V 12
Parnell Gdns. KT13: Weyb 83Q 170
Parnell Ho. WC1 1E 222 (43Mb 90)
Parnelli Ho. N1 37Rb 71
(off Canonbury Rd.)
Parnell Rd. E3 39Bc 72
Parnell Way HA3: Stan 25Ka 46
Parnham Av. GU18: Light 3B 166
Parnham Cl. BR1: Broml 69Rc 138
Parnham St. E14 44Ac 92
(not continuous)
Parolles Rd. N19 32Lb 70
Paroma Rd. DA17: Belv 49Cd 96
Parpins WD4: K Lan 10N 3
Parr Av. KT17: Ewe 81Xa 174
Parr Cl. KT22: Lea 92Ha 192
Parr Cl. N18 21Xb 51
Parr Cl. N9 21Xb 51
Parr Cl. RM16: Chaf H 49Yd 98
Parr Ct. GU21: Knap 1G 186
(off Tudor Way)
Parr Ct. N1 1F 219 (40Tb 71)
(off New North Rd.)
Parr Ct. TW13: Hanw 63Y 129
Parr Ho. E16 46Kc 93
(off Beaulieu Av.)
Parrington Ho. SW4 58Mb 112
Parritt Rd. RH1: Redh 4B 208
Parrock, The DA12: Grav'nd 10E 122
Parrock Av. DA12: Grav'nd 10E 122
PARROCK FARM 2F 144
Parrock Rd. DA12: Grav'nd 10E 122
Parrock St. DA12: Grav'nd 8D 122
Parrotts Cl. WD3: Crox G 14D 26
Parr Rd. E6 39Mc 73
Parr Rd. HA7: Stan 25Ma 47
Parrs Cl. CR2: Sande 81Tb 177
Parrs Pl. TW12: Hamp 66Ca 129
Parr St. N1 1F 219 (40Tb 71)
Parrs Way W6 51Ya 110
Parry Av. E6 44Pc 94
Parry Cl. KT17: Ewe 80Xa 154
Parry Cl. SS17: Stan H 1M 101
Parry Cotts. SL9: Chal P 21A 42
(off Chesham La.)
Parry Dr. KT13: Weyb 82Q 170
Parry Grn. Nth. SL3: L'ly 49B 82
Parry Grn. Sth. SL3: L'ly 49B 82
Parry Ho. E1 46Xb 91
(off Green Bank)
Parry Pl. SE18 49Rc 94
Parry Rd. SE25 69Ub 135
Parry Rd. W10 41Ab 88
Parry St. SW8 51Nb 112
Parsifal Rd. NW6 36Cb 69
Parsley Gdns. CR0: C'don 74Zb 158
Parsloes Av. RM9: Dag 35Zc 75
Parsonage Bank DA4: Eyns 75Nd 163
(off Edwards Ct.)
Parsonage Cl. CR6: W'ham 88Bc 178
Parsonage Cl. UB3: Hayes 44V 84
Parsonage Cl. WD5: Ab L 2U 12
Parsonage Ct. IG10: Lough 13Rc 36
Parsonage Farm WD3: Rick 17L 25
Parsonage Gdns. EN2: Enf 12Sb 33
Parsonage Gro. GU24: W End 4E 166
Parsonage La. AL9: Wel G 5D 8
Parsonage La. DA14: Sidc 63Bd 139
Parsonage La. DA2: Sut H 64Rd 141
Parsonage La. DA4: Sut H 65Rd 141
Parsonage La. EN1: Enf 12Tb 33
Parsonage La. EN2: Enf 12Sb 33
Parsonage La. SL2: Farn C 7G 60
Parsonage La. SL2: Farn R 7G 60
Parsonage La. SL4: Wind 3E 102
Parsonage Manorway DA17: Belv 51Cd 118
Parsonage Rd. AL9: Wel G 5D 8
Parsonage Rd. RM13: Rain 41Ld 77
Parsonage Rd. RM20: Grays 51Yd 120
Parsonage Rd. TW20: Eng G 4P 125
Parsonage Rd. WD3: Rick 17M 25
Parsonage St. E14 49Ec 92
Parsons Cl. DA3: Lfield 69Be 143
Parsons Cl. SM1: Sutt 76Db 155
Parsons Ct. WD18: Wat 16W 28

Parson's Cres. HA8: Edg 20Qa 29
Parsonsfield Cl. SM7: Bans 87Za 174
Parsonsfield Rd. SM7: Bans 88Za 174
Parsons Ga. M. SW6 54Cb 111
Parson's Grn. SW6 53Cb 111
PARSONS GREEN 54Bb 111
Parson's Grn. La. SW6 53Cb 111
Parsons Green Station (Underground) 53Cb 111
Parson's Gro. HA8: Edg 20Qa 29
Parsons Hill SE18 48Qc 94
(off Powis St.)
Parsons Ho. W2 6B 214 (42Fb 89)
(off Hall Pl.)
Parsons La. DA2: Wilm 62Kd 141
Parsons La. TN15: Stans 82Ae 185
Parsons Lodge NW6 38Db 69
(off Priory Rd.)
Parson's Mead CR0: C'don 74Rb 157
Parsons Mead KT8: E Mos 69Ea 130
Parsons M. SW18 57Eb 111
Parson's Rd. E13 40Lc 73
Parsons Rd. SL3: L'ly 50B 82
Parson St. NW4 28Ya 48
Parson's Wood La. SL2: Farn C 8H 61
Parthenia Dr. TW7: Isle 55Ja 108
Parthenia Rd. SW6 53Cb 111
Parthia Cl. KT20: Tad 91Xa 194
Partingdale La. NW7 22Za 48
Partington Cl. N19 32Mb 70
Partridge Cl. E16 43Mc 93
Partridge Cl. EN5: Barn 16Ya 30
Partridge Cl. HA7: Stan 21Na 47
Partridge Cl. UB10: Uxb 39P 63
Partridge Cl. WD23: Bush 18Ea 28
Partridge Ct. EC1 5B 218 (42Rb 91)
(off Cyrus St.)
Partridge Dr. BR6: Farnb 76Sc 160
Partridge Grn. SE9 62Qc 138
Partridge Ho. E3 40Bc 72
(off Stafford Rd.)
Partridge Knoll CR8: Purl 84Rb 177
Partridge La. RM3: Hrld W 26Nd 57
Partridge Mead SM7: Bans 87Ya 174
Partridge M. KT16: Chert 75J 149
Partridge Rd. DA14: Sidc 62Uc 138
Partridge Rd. TW12: Hamp 65Ba 129
Partridges, The HP3: Hem H 4A 4
Partridge Sq. E6 43Nc 94
Partridge Way N22 25Nb 50
Parvills EN9: Walt A 4Fc 21
Parvis Rd. KT14: Byfl 85K 169
Parvis Rd. KT14: W Byf 85K 169
Pasadena Cl. UB3: Hayes 47W 84
Pasadena Cl. Trad. Est. UB3: Hayes 47X 85
Pasadena Pk. TN15: Knat 82Rd 183
Pascall Ho. SE17 51Sb 113
(off Draco St.)
Pascal M. SE19 66Wb 135
Pascal Rd. UB1: S'hall 44Da 85
Pascal St. SW8 52Mb 112
Pascoe Rd. SE13 57Fc 115
Pasfield Ct. EN9: Walt A 5Fc 21
Pasley Cl. SE17 7C 230 (50Rb 91)
Pasquier Rd. E17 27Ac 52
Passage, The TW9: Rich 57Na 109
Passey Pl. SE9 58Pc 116
Passfield Dr. E14 43Dc 92
Passfield Hall WC1 5E 216 (42Mb 90)
(off Endsleigh Pl.)
Passfield Path SE28 45Xc 95
Passfields SE6 62Dc 136
Passfields W14 50Bb 89
(off Star Rd.)
Passing All. EC1 7C 218 (43Rb 91)
(off St John St.)
PASSINGFORD BRIDGE 12Ed 38
Passingham Ho. TW5: Hest 51Ca 107
Passive Cl. RM13: Rain 42Hd 96
Passmore Ct. E14 44Cc 92
(off New Festival Av.)
Passmore Edwards Ho. N11 23Mb 50
Passmore Edwards Ho. SL9: Chal P 22A 42
Passmore Gdns. N11 23Mb 50
Passmore Ho. E2 1K 219 (39Vb 71)
(off Kingsland Rd.)
Passmore St. SW1 7H 227 (50Jb 90)
Pastel Ct. E1 43Zb 92
(off Shandy St.)
Pastens Rd. RH8: Limp 3N 211
Pasteur Cl. NW9 26Ua 48
Pasteur Ct. HA1: Harr 32Ka 66
Pasteur Dr. RM3: Hrld W 26Md 57
Pasteur Gdns. N18 22Rb 51
Paston Cl. E5 34Zb 72
Paston Cl. SM6: W'gton 76Lb 156
Paston Cres. SE12 59Kc 115
Pastoral Way CM14: W'ley 22Xd 58
Pastor Ct. N6 30Lb 50
Pastor St. SE11 5C 230 (49Rb 91)
Pasture Cl. HA0: Wemb 34Ka 66
Pasture Cl. WD23: Bush 17Ea 28
Pasture Rd. HA0: Wemb 33Ka 66
Pasture Rd. RM9: Dag 36Bd 75
Pasture Rd. SE6 60Hc 115
Pastures, The AL10: Hat 1D 8
Pastures, The AL2: Chis G 6N 5
Pastures, The HP1: Hem H 1G 2
Pastures, The N20 18Bb 31
Pastures, The WD19: Wat 17Y 27
Pastures Mead UB10: Hil 37Q 64
Pastures Path E11 32Hc 73
Pasture Vw. AL4: S'ford 1L 7
Patch, The TN13: Riv 94Gd 202
Patcham Ter. SW8 53Kb 112
Patch Cl. UB10: Uxb 39P 63
PATCHETTS GREEN 11Ea 28
Patching Way UB4: Yead 43Aa 85
Patent Ho. E14 43Dc 92
(off Morris Rd.)
Paternoster Cl. EN9: Walt A 5Hc 21
Paternoster Hill EN9: Walt A 4Hc 21
Paternoster La. EC4 3C 224 (44Rb 91)
Paternoster Row EC4 3D 224 (44Sb 91)
Paternoster Row RM4: Noak H 18Ld 39
Paternoster Sq. EC4 3C 224 (44Rb 91)
Paterson Ct. EC1 4F 219 (41Tb 91)
(off St Luke's Est.)
Paterson Rd. TW15: Ashf 64M 127
Pater St. W8 48Cb 89
Pates Mnr. Dr. TW14: Bedf 59T 106
Path, The SW19 67Db 133
Pathfield Rd. SW16 65Mb 134
Pathway, The GU23: Send 97H 189
Pathway, The WD19: Wat 18Z 27
Pathway, The WD7: R'lett 8Ha 14
Patience Rd. SW11 54Gb 111
Patina Mans. E20 36Ec 72
(off Mirabelle Gdns.)
Patio Cl. SW4 58Mb 112
Pat Larner Ho. AL1: St A 3B 6
(off Belmont Hill)
Patmore Est. SW8 53Lb 112
Patmore Ho. N16 36Ub 71
Patmore La. KT12: Hers 79V 150
Patmore Link Rd. HP2: Hem H 2C 4
Patmore Rd. EN9: Walt A 6Gc 21
Patmore St. SW8 53Lb 112
Patmore Way RM5: Col R 22Dd 56
Patmos Lodge SW9 53Rb 113
(off Elliott Rd.)
Patmos Rd. SW9 52Rb 113
Paton Cl. E3 41Cc 92
Paton Ho. SW9 54Pb 112
(off Stockwell Rd.)
Paton St. EC1 4D 218 (41Sb 91)
Patricia Cl. SL1: Slou 5C 80
Patricia Ct. BR7: Chst 67Tc 138
Patricia Ct. DA16: Well 52Xc 117
Patricia Dr. RM11: Horn 32Nd 77
Patricia Gdns. SM2: Sutt 83Cb 175
Patrick Coman Ho. EC1 4B 218 (41Rb 91)
(off St John St.)
Patrick Connolly Gdns. E3 41Dc 92
Patrick Ct. SE1 2C 230 (47Rb 91)
(off Webber St.)
Patrick Cres. RM8: Dag 32Ad 75
Patrick Gro. EN9: Walt A 5Dc 20
Patrick Pas. SW11 54Gb 111
Patrick Rd. E13 41Lc 93
Patrington Cl. UB8: Cowl 41L 83
Patriot Sq. E2 40Xb 71
Patrol Pl. SE6 58Dc 114
Patroni Ct. E15 41Gc 93
(off Durban Rd.)
Patrons Way E. UB9: Den 30H 43
Patrons Way W. UB9: Den 29H 43
Pat Shaw Ho. E1 42Zb 92
(off Globe Rd.)
Patshull Pl. NW5 37Lb 70
Patshull Rd. NW5 37Lb 70
Patten All. TW10: Rich 57Ma 109
Pattenden Rd. SE6 60Bc 114
Patten Ho. N4 32Sb 71
Patten Rd. SW18 59Gb 111
Patterdale NW1 4A 216 (41Kb 90)
(off Osnaburgh St.)
Patterdale Cl. BR1: Broml 65Hc 137
Patterdale Rd. DA2: Dart 60Td 120
Patterdale Rd. SE15 52Yb 114
Pattern Ho. EC1 5B 218 (42Rb 91)
Patterson Ct. DA1: Dart 57Qd 119
Patterson Ct. SE19 66Vb 135
Patterson Rd. SE19 65Vb 135
Pattina Wlk. SE16 46Ac 92
(off Silver Wlk.)
Pattison Ho. E1 44Zb 92
(off Wellesley St.)
Pattison Ho. SE1 1E 230 (47Sb 91)
(off Redcross Way)
Pattison Rd. NW2 34Cb 69
Pattison Wlk. SE18 50Sc 94
Paul Byrne Ho. N2 27Eb 49
Paul Cl. E15 39Gc 73
Paul Ct. N18 21Wb 51
(off Fairfield Rd.)
Paul Ct. RM7: Rom 29Ed 56
Paul Ct. TW20: Egh 65F 126
Paul Daisley Ct. NW6 38Ab 68
(off Christchurch Av.)
Paulet Rd. SE5 54Rb 113
Paulet Way NW10 38Ua 68
Paul Gdns. CR0: C'don 75Vb 157
Paul Greengrass Cinema 9D 122
(off Woodville Pl.)
Paulhan Rd. HA3: Kenton 28Ma 47
Paul Ho. W10 42Ab 88
(off Ladbroke Gro.)
Paulin Dr. N21 17Qb 32
Pauline Cres. TW2: Whitt 60Ea 108
Pauline Ho. E1 43Wb 91
(off Old Montague St.)
Paulinus Cl. BR5: St P 68Yc 139
Paul Julius Cl. E14 45Fc 93
Paul Robeson Cl. E6 41Qc 94
Paul Robeson Ho. WC1 2J 217 (40Pb 70)
(off Penton Ri.)
Paul Robeson Theatre, The 55Da 107
Pauls Grn. EN8: Walt C 5Ac 20
Paul's Nursery Rd. IG10: H Beech 11Kc 35
Paul's Pl. KT21: Asht 91Ra 193
Paul St. E15 39Gc 73
Paul St. EC2 6G 219 (42Tb 91)
Paul's Wlk. EC4 4D 224 (45Sb 91)
Paultons Ho. SW3 51Fb 111
(off Paultons Sq.)
Paultons Sq. SW3 51Fb 111
Paultons St. SW3 51Fb 111
Paul Vanson Ct. KT12: Hers 79Z 151
Pauntley St. N19 32Lb 70
Pavan Ct. E2 41Yb 92
(off Sceptre Rd.)
Paved Ct. TW9: Rich 57Ma 109
Paveley Ct. NW7 24Ab 48
(off Langstone Way)
Paveley Dr. SW11 52Gb 111
Paveley Ho. N1 2H 217 (40Pb 70)
(off Priory Grn. Est.)
Paveley St. NW8 4D 214 (41Gb 89)
Pavement, The E11 32Ec 72
(off Hainault Rd.)
Pavement, The SW19 65Bb 133
(off Worple Rd.)
Pavement, The SW4 56Lb 112
Pavement, The TW11: Tedd 66Ka 130
Pavement, The TW7: Isle 55Ja 108
(off South St.)
Pavement, The W5 48Na 87
Pavement M. RM6: Chad H 31Zc 75
Pavement Sq. CR0: C'don 74Wb 157
Pavers Way E3 40Zb 72
Pavet Cl. RM10: Dag 37Dd 76
Pavilion, The KT20: Kgswd 95Eb 195
Pavilion, The RH2: Reig 4N 207
Pavilion, The SW8 52Mb 112
Pavilion Apts. NW8 4C 214 (41Fb 89)
(off Stafford Rd.)
Pavilion Ct. NW6 41Cb 89
Pavilion Gdns. TN13: S'oaks 96Kd 203
Pavilion Gdns. TW18: Staines 66K 127
Pavilion La. BR3: Beck 65Bc 136
Pavilion Leisure Cen. 68Jc 137
Pavilion Lodge HA2: Harr 32Fa 66
Pavilion M. N3 27Cb 49
Pavilion Pde. W12 44Ya 88
(off Wood La.)
Pavilion Pl. KT8: W Mole 69Ea 130
Pavilion Rd. IG1: Ilf 31Pc 74
Pavilion Rd. SW1 3G 227 (48Hb 89)
Pavilion Rd. SW3 5G 227 (49Hb 89)
Pavilion Rd. TW11: Tedd 66Ha 130
Pavilions, The EN3: Enf L 8Bc 20
Pavilions, The KT14: Byfl 83M 169
Pavilions, The SL4: Wind 3F 102
Pavilions, The 38L 63
Pavilion Sports & Fitness Club, The 69Ea 130
Pavilion Sq. SW17 62Hb 133
Pavilions Shop. Cen. Waltham Cross 5Ac 20
Pavilion St. E13 40Lc 73
Pavilion St. SW1 4G 227 (48Hb 89)
Pavilion Ter. IG2: Ilf 29Uc 54
Pavilion Ter. W12 44Ya 88
(off Wood La.)
Pavilion Wlk. E10 33Cc 72
Pavilion Way HA4: Ruis 33Y 65
Pavilion Way HA8: Edg 24Ra 47
Pavilion Way SE10 47Gc 93
Pavillion Ho. SE16 47Zb 92
(off Water Gdns. Sq.)
Pavillion M. N4 33Pb 70
(off Tollington Pl.)
Pawleyne Cl. SE20 66Yb 136
Pawsey Cl. E13 39Kc 73
Pawsons Rd. CR0: C'don 72Sb 157
Paxfold HA7: Stan 22Ma 47
Paxford Rd. HA0: Wemb 33Ka 66
Paxton Av. SL1: Slou 8G 80
Paxton Cl. KT12: Walt T 73Y 151
Paxton Cl. TW9: Kew 54Pa 109
Paxton Ct. CR4: Mitc 68Hb 133
(off Armfield Cres.)
Paxton Ct. N7 37Qb 70
(off Westbourne Rd.)
Paxton Ct. SE12 62Lc 137
Paxton Ct. SE26 63Ac 136
(off Adamsrill Rd.)
Paxton Ct. WD6: Bore 14Sa 29
Paxton Gdns. GU21: Wok 84F 168
Paxton Gro. CR5: Coul 89Lb 176
Paxton Ho. SE17 7F 231 (50Tb 91)
(off Morecambe St.)
Paxton Ho. SE25 70Vb 135
Paxton M. SE19 66Ub 135
(off Westow St.)
Paxton Pl. SE27 63Ub 135
Paxton Point SE10 52Dc 114
Paxton Rd. AL1: St A 3C 6
Paxton Rd. BR1: Broml 66Jc 137
Paxton Rd. HP4: Berk 1A 2
Paxton Rd. N17 25Ub 51
Paxton Rd. SE23 62Ac 136
Paxton Rd. W4 51Ua 110
Paxton Ter. SW1 51Kb 112
Paymal Ho. E1 43Yb 92
(off Stepney Way)
Payne Cl. IG11: Bark 38Uc 74
Payne Ho. N1 1J 217 (39Pb 70)
(off Barnsbury Est.)
Paynell Ct. SE3 55Gc 115
Payne Rd. E3 40Dc 72
Paynes Cotts. TN13: Dun G 90Fd 182
Paynesfield Av. SW14 55Ta 109
Paynesfield Rd. TN16: Tats 93Lc 199
(not continuous)
Paynesfield Rd. WD23: B Hea 17Ha 28
Payne St. SE8 52Bc 114
Paynes Wlk. W6 51Ab 110
Paynetts Ct. KT13: Weyb 78T 150
Payzes Gdns. IG8: Wfd G 23Hc 53
Peaberry Ct. NW4 27Wa 48
Peabody Av. SW1 7K 227 (50Kb 90)
Peabody Bldgs. E1 45Wb 91
(off John Fisher St.)
Peabody Bldgs. EC1 6E 218 (42Sb 91)
(off Banner St.)
Peabody Bldgs. SW3 51Gb 111
(off Cheyne Row)
Peabody Cl. CR0: C'don 74Yb 158
Peabody Cl. SE10 53Dc 114
Peabody Cl. SW1 50Kb 90
Peabody Cotts. N17 25Ub 51
Peabody Ct. EC1 6E 218 (42Sb 91)
(off Roscoe St.)
Peabody Ct. SE5 53Tb 113
(off Kimpton Rd.)
Peabody Est. SE1: Duchy St. 7A 224 (46Qb 90)
Peabody Est. EC1: Dufferin St. 6E 218 (42Sb 91)
(off Dufferin St.)
Peabody Est. EC1: Farringdon La. 6A 218 (42Qb 90)
(off Farringdon La.)
Peabody Est. SE1: Marshalsea Rd. 1E 230 (47Sb 91)
(off Marshalsea Rd.)
Peabody Est. SE1: Southwark St. 6D 224 (46Sb 91)
Peabody Est. E1 45Zb 92
(off Brodlove La.)
Peabody Est. E2 40Xb 71
(off Minerva St.)
Peabody Est. N1 39Sb 71
Peabody Est. SE24 59Rb 113
Peabody Est. SE5 53Tb 113
(off Camberwell Grn.)
Peabody Est. SW1 5C 228 (49Lb 90)
(off Vauxhall Bri. Rd.)
Peabody Est. SW11 56Gb 111
Peabody Est. SW3 51Gb 111
Peabody Est. SW6 51Cb 111
(off Lillie Rd.)
Peabody Est. W10 43Ya 88
Peabody Est. W6 50Ya 88
Peabody Hill SE21 60Rb 113
Peabody Ho. N1 39Sb 71
(off Greenman St.)
Peabody Sq. N1 39Sb 71
(off Peabody Est.)
Peabody Sq. SE1 2B 230 (47Rb 91)
(not continuous)
Peabody Ter. EC1 6A 218 (42Qb 90)
(off Farringdon La.)
Peabody Twr. EC1 6E 218 (42Sb 91)
(off Golden La.)
Peabody Trust SE17 6F 231 (49Tb 91)
Peabody Yd. N1 39Sb 71
Peace Cl. EN7: Chesh 1Xb 19
Peace Cl. N14 15Kb 32
Peace Cl. SE25 70Ub 135
Peace Cl. UB6: G'frd 39Fa 66
Peace Ct. SE1 50Wb 91
(off Harmony Pl.)
Peace Dr. WD17: Wat 13W 26
Peace Gro. HA9: Wemb 34Ra 67
PEACE HOSPICE 13W 26
Peace Prospect WD17: Wat 13W 26
Peace Rd. SL0: Iver H 39D 62
Peace Rd. SL3: Ful 41B 82
Peace Rd. SL3: Wex 41B 82
Peace St. SE18 51Qc 116
Peach Cft. DA11: Nflt 2A 144
Peaches Cl. SM2: Cheam 80Ab 154
Peachey Cl. UB8: Cowl 44M 83
Peachey Ho. SW18 56Eb 111
(off Eltringham St.)
Peachey La. UB8: Cowl 43M 83
Peach Gro. E11 34Fc 73
Peach Rd. TW13: Felt 60W 106
Peach Rd. W10 41Za 88
Peach Tree Av. UB7: Yiew 44P 83
Peachtree Cl. EN1: Enf 12Wb 33
Peachtree Ct. IG6: Ilf 24Rc 54
Peachum Rd. SE3 51Hc 115
Peachwalk M. E3 40Zb 72
Peachy Cl. HA8: Edg 23Qa 47
Peacock Av. TW14: Bedf 60T 106
Peacock Cl. E4 24Bc 52
Peacock Cl. KT19: Eps 84Pa 173
Peacock Cl. NW7 22Ab 48
Peacock Cl. RM11: Horn 28Nd 57
Peacock Cl. RM8: Dag 32Yc 75
Peacock Gdns. CR2: Sels 82Ac 178
Peacock Ho. SE5 53Ub 113
(off St Giles Rd.)
Peacock Ind. Est. N17 24Vb 51
Peacock Pl. N1 37Qb 70
Peacock St. DA12: Grav'nd 9E 122
Peacock St. SE17 6C 230 (49Rb 91)
Peacock Theatre 3H 223 (44Pb 90)
(off Portugal St.)
Peacock Wlk. N6: Highgate 31Kb 70
Peacock Wlk. WD5: Ab L 3W 12
Peacock Yd. SE17 7C 230 (50Rb 91)
(off Iliffe St.)
Peak, The SE26 62Yb 136
Peake Pl. KT15: Wdhm 83J 169
Peakes Pl. AL1: St A 2D 6
(off Granville Rd.)
Peaketon Av. IG4: Ilf 28Mc 53
Peak Hill SE26 63Yb 136
Peak Hill Av. SE26 63Yb 136
Peak Hill Gdns. SE26 63Yb 136
Peaks Hill CR8: Purl 82Mb 176
Peaks Hill Ri. CR8: Purl 82Nb 176
Pea La. RM14: Upm 37Wd 78
Peal Gdns. W13 42Ja 86
Peall Rd. CR0: C'don 72Pb 156
Peall Rd. Ind. Est. CR0: C'don 72Pb 156
Pearce Cl. CR4: Mitc 68Jb 134
Pearcefield Av. SE23 60Yb 114
Pearce Ho. SW1 6E 228 (49Mb 90)
(off Causton St.)
Pearces Wlk. AL1: St A 3B 6
(off Albert St.)
Pear Cl. NW9 28Ta 47
Pear Cl. SE14 52Ac 114
Pear Ct. SE15 52Vb 113
(off Thruxton Way)
Pearcroft Rd. E11 33Fc 73
Pearcy Cl. RM3: Rom 24Nd 57
Peardon St. SW8 54Kb 112
Peareswood Gdns. HA7: Stan 25Ma 47
Peareswood Rd. DA8: Erith 53Hd 118
Pearfield Rd. SE23 62Ac 136
Pearing Cl. KT4: Wor Pk 75Za 154
Pearl Cl. CR7: Thor H 68Tb 135
Pearl Cl. E6 44Qc 94
Pearl Cl. NW2 31Za 68
Pearl Ct. GU21: Wok 8J 167
Pearl Gdns. SL1: Slou 6F 80
Pearl Rd. E17 27Cc 52
Pearl St. E1 46Xb 91
Pearmain Cl. TW17: Shep 71R 150
Pearmain Ct. W6 48Xa 88
(off Vinery Way)
Pearman Ho. SL9: Chal P 21A 42
(off Micholls Av.)
Pearman St. SE1 3A 230 (48Qb 90)
Pear Pl. SE1 1K 229 (47Qb 90)
Pear Rd. E11 34Fc 73
Pearsall Ho. TN15: Bor G 91Be 205
Pears Av. TW17: Shep 69U 128
Pearscroft Ct. SW6 53Db 111
Pearscroft Rd. SW6 53Db 111
Pearse St. SE15 51Ub 113
Pearson Cl. CR8: Purl 83Rb 177
Pearson Cl. EN5: New Bar 13Db 31
Pearson Cl. SE5 53Sb 113
(off Camberwell New Rd.)
Pearson M. SW4 55Mb 112
(off Edgeley Rd.)
Pearson's Av. SE14 53Cc 114
Pearson Sq. W1 1B 222 (43Lb 90)
Pearson St. E2 1K 219 (40Vb 71)
Pearson Way CR4: Mitc 67Jb 134
Pearson Way DA1: Dart 61Pd 141
Pears Rd. TW3: Houn 55Ea 108
Peartree SE26 64Ac 136
Pear Tree Av. UB7: Yiew 44P 83
Peartree Av. SW17 62Eb 133
Pear Tree Cl. BR2: Broml 71Mc 159
Pear Tree Cl. BR8: Swan 68Fd 140
Pear Tree Cl. CR4: Mitc 68Gb 133
Pear Tree Cl. E2 1K 219 (39Vb 71)
Pear Tree Cl. KT15: Add 78J 149
Pear Tree Cl. KT19: Eps 81Ta 173
Pear Tree Cl. KT9: Chess 78Qa 153
Peartree Cl. CR2: Sande 86Xb 177
Peartree Cl. DA8: Erith 53Fd 118
Peartree Cl. HP1: Hem H 1J 3
Peartree Cl. RM15: S Ock 40Yd 78
Peartree Cl. SL1: Slou 6D 80
Pear Tree Ct. E18 25Kc 53
Pear Tree Ct. EC1 6A 218 (42Qb 90)
Pear Tree Ct. SE26 62Bc 136
Peartree Ct. WD25: Wat 8Z 13
Peartree Gdns. RM7: Mawney 26Dd 56
Peartree Gdns. RM8: Dag 35Xc 75
Pear Tree Ho. SE4 55Bc 114
Pear Tree La. DA12: High'm 6N 145
Pear Tree La. DA12: Shorne 6N 145
Pear Tree La. RM13: Rain 40Fd 76
Peartree La. E1 45Yb 92
Pear Tree Rd. KT15: Add 78J 149
Pear Tree Rd. TW15: Ashf 64S 128
Peartree Rd. EN1: Enf 13Ub 33
Peartree Rd. HP1: Hem H 1J 3
Pear Trees CM13: Ingve 23Ee 59
Peartrees UB7: Yiew 45M 83
Pear Tree St. EC1 5D 218 (42Sb 91)
Peartree Way SE10 49Jc 93
Peary Ho. NW10 38Ta 67
Peary Mead DA1: Dart 57Pd 119
Peary Pl. E2 41Yb 92
Peascod Pl. SL4: Wind 3H 103
(off Peascod St.)
Peascod St. SL4: Wind 3G 102
Peascroft Rd. HP3: Hem H 5A 4
Pease Cl. RM12: Horn 38Kd 77
Pease Hill TN15: Ash 79Ae 165
Peasmead Ter. E4 21Ec 52
Peatfield Cl. DA15: Sidc 62Uc 138
Peatmore Av. GU22: Pyr 88J 169
Peatmore Cl. GU22: Pyr 88J 169
Peatmore Dr. GU24: Brkwd 3A 186
Pebble Cl. KT20: Walt H 1A 206
Pebblehill Rd. RH3: B'wth 1A 206
Pebble La. KT18: Eps D 94Qa 193
Pebble La. KT22: Lea 96Pa 193
Pebble Way W3 46Ra 87
(off Steyne Rd.)
Pebworth Ct. RH1: Redh 4A 208
Pebworth Rd. HA1: Harr 33Ja 66
Peche Way BR5: Orp 74Zc 161
Pechora Way E14 43Ac 92
Peckarmans Wood SE26 62Wb 135
Peckett Sq. N5 35Sb 71
Peckford Pl. SW9 54Qb 112
PECKHAM 53Wb 113
Peckham Bus Station 53Wb 113
Peckham Gro. SE15 52Ub 113
Peckham High St. SE15 53Wb 113
Peckham Hill St. SE15 52Wb 113
Peckham Hurst Rd. TN11: Roug 100Fe 205
Peckham Pk. Rd. SE15 52Wb 113
Peckhamplex 54Wb 113
Peckham Pulse Leisure Cen. 53Wb 113
Peckham Rd. SE15 53Ub 113
Peckham Rd. SE5 53Ub 113
Peckham Rye SE15 55Wb 113
Peckham Rye SE22 56Wb 113
Peckham Rye Station (Rail & Overground) 54Wb 113
Peckham Sq. SE15 53Wb 113
Peckham Wlk. Av. TN15: Plax 100Zd 205
Pecks Yd. E1 7K 219 (43Vb 91)
(off Hanbury St.)
Peckwater St. NW5 36Lb 70
Pedham Place Golf Course 72Ld 163
Pedham Pl. Ind. Est. BR8: Swan 71Jd 162
Pedlar's Wlk. N7 36Pb 70
Pedley Rd. RM8: Dag 32Yc 75
Pedley St. E1 42Wb 91
Pedro St. E5 34Zb 72
Pedworth Gdns. SE16 49Yb 92
Peebles Ct. UB1: S'hall 44Ea 86
(off Haldane Rd.)
Peebles Ho. NW6 40Db 69
(off Carlton Vale)
Peek Cres. SW19 64Za 132
Peel Cl. E4 19Dc 34
Peel Cl. N9 20Wb 33
Peel Cl. SL4: Wind 5F 102
Peel Ct. SL1: Slou 3F 80
Peel Dr. IG5: Ilf 27Nc 54
Peel Dr. NW9 27Va 48
Peelers Cl. EN2: Crew H 6Rb 19
Peel Gro. E2 40Yb 72
Peel Pas. W8 46Cb 89
Peel Pl. IG5: Ilf 26Nc 54
Peel Pl. SE18 53Pc 116
Peel Pl. SW6 51Cb 111
Peel Pct. NW6 40Cb 69
Peel Rd. BR6: Farnb 78Sc 160
Peel Rd. E18 25Hc 53
Peel Rd. HA3: W'stone 27Ha 46
Peel Rd. HA9: Wemb 34Ma 67
Peel Sq. NW9 27Wa 48
Peel St. W8 46Cb 89
Peel Way RM3: Hrld W 26Pd 57
Peel Way UB8: Hil 43N 83
Peerage Way RM11: Horn 31Nd 77
Peerglow Est. EN3: Pond E 15Yb 34
Peerglow Ind. Est. WD18: Wat 18R 26
Peerless Dr. UB9: Hare 29L 43
Peerless St. EC1 4F 219 (41Tb 91)
Pegamoid Rd. N18 20Yb 34
Pegasus Cl. N16 35Tb 71
Pegasus Ct. CR3: Cat'm 95Vb 197
Pegasus Ct. DA12: Grav'nd 2E 144
Pegasus Ct. KT1: King T 69Ma 131
Pegasus Ct. KT22: Lea 93La 192
(off Epsom Rd.)
Pegasus Ct. N21 17Sb 33
Pegasus Ct. NW10 41Xa 88
(off Trenmar Gdns.)
Pegasus Ct. SM7: Bans 87Cb 175
Pegasus Ct. TW20: Egh 64D 126
Pegasus Ct. TW8: Bford 50Pa 87
Pegasus Ct. W3 44Sa 87
(off Horn La.)
Pegasus Ct. WD5: Ab L 4V 12
Pegasus Ho. E1 42Zb 92
(off Beaumont Sq.)
Pegasus Ho. E13 41Kc 93
Pegasus Pl. AL3: St A 1B 6
Pegasus Pl. SE11 51Qb 112
Pegasus Pl. SW6 53Cb 111
Pegasus Rd. CR0: Wadd 79Qb 156
Pegasus Way N11 23Kb 50
Pegelm Gdns. RM11: Horn 31Pd 77
Peggotty Way UB8: Hil 44R 84
Pegg Rd. TW5: Hest 52Z 107
Peggy Bond Cl. KT16: Chert 74K 149
Pegler Sq. SE3 55Kc 115
Pegley Gdns. SE12 61Jc 137
Pegmire La. WD25: A'ham 11Ea 28
Pegrum Dr. AL2: Lon C 9F 6
Pegswood Ct. E1 45Wb 91
(off Cable St.)
Pegwell St. SE18 52Uc 116
Peket Cl. TW18: Staines 67G 126
Pekin Cl. E14 44Cc 92
(off Pekin St.)
Pekin St. E14 44Cc 92
Pelabon Ho. TW1: Twick 58Ma 109
(off Clevedon Rd.)
Peldon Cl. IG8: Wfd G 24Lc 53
Peldon Ct. TW9: Rich 56Pa 109
Peldon Pas. TW10: Rich 56Pa 109
Peldon Wlk. N1 39Rb 71
(off Popham St.)
Pelham Av. IG11: Bark 39Vc 75
Pelham Cl. SE5 55Ub 113
Pelham Cotts. DA5: Bexl 60Dd 118
Pelham Ct. DA14: Sidc 62Wc 139
Pelham Ct. HP2: Hem H 2C 4
Pelham Ct. SW3 6D 226 (49Gb 89)
(off Fulham Rd.)
Pelham Ct. TW18: Staines 64K 127
(off Kingston Rd.)
Pelham Cres. SW7 6D 226 (49Gb 89)
Pelham Ho. CR3: Cat'm 96Vb 197
Pelham Ho. SW1 4E 228 (48Mb 90)
(off Gt. Peter St.)

Pelham Ho. W14 49Bb 89
(off Mornington Av.)
Pelham La. WD25: A'ham 8Da 13
Pelham Pl. SW7 5D 226 (49Gb 89)
Pelham Pl. W13 42Ha 86
Pelham Rd. BR3: Beck 68Yb 136
Pelham Rd. DA11: Grav'nd 10B 122
Pelham Rd. DA7: Bex 55Cd 118
Pelham Rd. E18 27Kc 53
Pelham Rd. IG1: Ilf 33Tc 74
Pelham Rd. N15 28Vb 51
Pelham Rd. N22 26Qb 50
Pelham Rd. SW19 66Cb 133
Pelham Rd. Sth. DA11: Grav'nd 10B 122
Pelham Rd. Sth. DA11: Nflt 10B 122
Pelhams, The WD25: Wat 7Z 13
Pelham's Cl. KT10: Esh 77Ca 151
Pelham St. SW7 5C 226 (49Fb 89)
Pelham's Wlk. KT10: Esh 77Ca 151
Pelham Ter. DA11: Grav'nd 9B 122
Pelham Towers DA11: Grav'nd 9B 122
Pelham Way KT23: Bookh 98Da 191
Pelican Dr. HA2: Harr 33Da 65
Pelican Est. SE15 53Vb 113
Pelican Ho. SE5 53Vb 113
Pelican Ho. SE8 49Bc 92
Pelican Pas. E1 42Yb 92
Pelican Wlk. SW9 56Rb 113
Pelican Wharf E1 46Yb 92
(off Wapping Wall)
Pelier St. SE17 51Sb 113
Pelinore Rd. SE6 61Gc 137
Pella Ho. SE11 7J 229 (50Pb 90)
Pellant Rd. SW6 52Ab 110
Pellatt Gro. N22 25Qb 50
Pellatt Rd. HA9: Wemb 33Ma 67
Pellatt Rd. SE22 57Vb 113
Pellerin Rd. N16 36Ub 71
Pellew Ho. E1 42Xb 91
(off Somerford St.)
Pelling Hill SL4: Old Win 9M 103
Pellings Cl. BR2: Broml 69Gc 137
Pelling St. E14 44Cc 92
Pellipar Cl. N13 20Qb 32
Pellipar Gdns. SE18 50Pc 94
Pellipar Rd. SE18 50Pc 94
Pellow Cl. EN5: Barn 16Bb 31
Pells La. TN15: W King 84Wd 184
Pell St. SE8 49Ac 92
Pelly Ct. CM16: Epp 3Vc 23
Pelly Rd. E13 39Jc 73
(not continuous)
Pelman Ho. KT19: Eps 82Ra 173
Pelman Way KT19: Eps 82Ra 173
Peloton Av. E20 36Dc 72
Pelter St. E2 3K 219 (41Vb 91)
Pelton Av. SM2: Sutt 82Db 175
Pelton Rd. SE10 50Gc 93
Pembar Av. E17 27Ac 52
Pemberley Apts. RM2: Rom 26Ld 57
Pemberley Chase KT19: Ewe 78Ra 153
Pemberley Cl. KT19: Ewe 78Ra 153
Pemberley Ho. KT19: Ewe 78Ra 153
(off Pemberley Chase)
Pemberley Lodge SL4: Wind 5E 102
Pember Rd. NW10 41Za 88
Pemberton Almshouses AL1: St A 1B 6
(off St Peter's St.)
Pemberton Av. RM2: Rom 27Kd 57
Pemberton Cl. AL1: St A 5B 6
Pemberton Cl. TW19: Stanw 60N 105
Pemberton Ct. E1 41Zb 92
(off Portelet Rd.)
Pemberton Ct. EN1: Enf 13Ub 33
Pemberton Gdns. BR8: Swan 69Gd 140
Pemberton Gdns. N19 34Lb 70
Pemberton Gdns. RM6: Chad H 29Ad 55
Pemberton Ho. SE26 63Wb 135
(off High Level Dr.)
Pemberton Pl. E8 38Xb 71
Pemberton Pl. KT10: Esh 76Ea 152
Pemberton Rd. KT8: E Mos 70Ea 130
Pemberton Rd. N4 29Qb 50
Pemberton Rd. SL2: Slou 2C 80
Pemberton Row EC4 2A 224 (44Qb 90)
Pemberton Ter. N19 34Lb 70
Pembrey Way RM12: Horn 37Ld 77
Pembridge Av. TW2: Whitt 60Ba 107
Pembridge Chase HP3: Bov 10B 2
Pembridge Cl. HP3: Bov 10B 2
Pembridge Cres. W11 45Cb 89
Pembridge Gdns. W2 45Cb 89
Pembridge M. W11 45Cb 89
PEMBRIDGE PALLIATIVE CARE CEN. 43Za 88
(within St Charles Hospital)
Pembridge Pl. SW15 57Cb 111
Pembridge Pl. W2 45Cb 89
Pembridge Rd. HP3: Bov 10C 2
Pembridge Rd. W11 45Cb 89
Pembridge Sq. W2 45Cb 89
Pembridge Studios W11 45Cb 89
(off Pembridge Vs.)
Pembridge Vs. W11 45Cb 89
Pembridge Vs. W2 44Cb 89
Pembroke W14 49Bb 89
(off Kensington Village)
Pembroke Av. EN1: Enf 10Xb 19
Pembroke Av. HA3: Kenton 27Ja 46
Pembroke Av. HA5: Pinn 32Z 65
Pembroke Av. KT12: Hers 77Z 151
Pembroke Av. KT5: Surb 71Ra 153
Pembroke Av. N1 39Nb 70
Pembroke Bldgs. NW10 41Wa 88
Pembroke Bus. Cen. BR8: Swan 67Fd 140
Pembroke Cen., The HA4: Ruis 32V 64
Pembroke Cl. RM11: Horn 28Pd 57
Pembroke Cl. SL5: S'hill 1B 146
Pembroke Cl. SM7: Bans 89Db 175
Pembroke Cl. SW1 2J 227 (47Jb 90)
Pembroke Cotts. W8 48Cb 89
(off Pembroke Sq.)
Pembroke Ct. TW20: Eng G 2P 125
Pembroke Ct. W7 44Ha 86
(off Copley Cl.)
Pembroke Ct. W8 48Cb 89
(off Sth. Edwardes Sq.)
Pembroke Dr. EN7: G Oak 1Rb 19
Pembroke Dr. RM15: Avel 47Sd 98
Pembroke Gdns. GU22: Wok 90C 168
Pembroke Gdns. HA4: Ruis 32V 64
Pembroke Gdns. RM10: Dag 34Dd 76
Pembroke Gdns. SW14 56Ra 109
Pembroke Gdns. W8 49Bb 89
Pembroke Gdns. Cl. W8 48Cb 89
Pembroke Hall NW4 27Ya 48
(off Mulberry Cl.)
Pembroke Ho. RM8: Dag 36Wc 75
Pembroke Ho. SW1 4H 227 (48Jb 90)
(off Chesham St.)
Pembroke Ho. W2 44Db 89
(off Hallfield Est.)
Pembroke Ho. W3 47Sa 87
(off Park Rd. E.)
Pembroke Ho. WD6: Bore 14Qa 29
(off Academy Ct.)
Pembroke Lodge HA7: Stan 23Ma 47
Pembroke Mans. NW6 37Eb 69
(off Canfield Gdns.)
Pembroke M. E3 41Ac 92
Pembroke M. N10 25Jb 50
Pembroke M. SL5: S'hill 1B 146
Pembroke M. TN13: S'oaks 97Kd 203
Pembroke M. W8 48Cb 89
Pembroke Pde. DA8: Erith 50Ed 96
Pembroke Pl. DA4: Sut H 67Rd 141
Pembroke Pl. HA8: Edg 24Qa 47
Pembroke Pl. TW7: Isle 54Ga 108
Pembroke Pl. W8 48Cb 89
Pembroke Rd. BR1: Broml 68Lc 137
Pembroke Rd. CR4: Mitc 68Jb 134
Pembroke Rd. DA8: Erith 50Ed 96
Pembroke Rd. E17 29Dc 52
Pembroke Rd. E6 43Pc 94
Pembroke Rd. GU22: Wok 90C 168
Pembroke Rd. HA4: Ruis 32U 64
Pembroke Rd. HA6: Nwood 20S 26
Pembroke Rd. HA9: Wemb 34Ma 67
Pembroke Rd. IG3: Ilf 32Vc 75
Pembroke Rd. N10 25Jb 50
Pembroke Rd. N13 20Sb 33
Pembroke Rd. N15 29Vb 51
Pembroke Rd. N8 28Nb 50
Pembroke Rd. SE25 70Ub 135
Pembroke Rd. TN13: S'oaks 97Kd 203
Pembroke Rd. UB6: G'frd 42Da 85
Pembroke Rd. W8 49Bb 89
Pembroke Sq. W8 48Cb 89
Pembroke St. N1 38Nb 70
(not continuous)
Pembroke Studios W8 48Bb 89
Pembroke Ter. NW8 1B 214 (39Fb 69)
(off Queen's Ter.)
Pembroke Vs. TW9: Rich 56Ma 109
Pembroke Vs. W8 49Cb 89
Pembroke Wlk. W8 49Cb 89
Pembroke Way UB3: Harl 48S 84
Pembrook M. SW11 56Fb 111
Pembry Cl. SW9 53Qb 112
Pembury Av. KT4: Wor Pk 74Wa 154
Pembury Cl. BR2: Hayes 73Hc 159
Pembury Cl. CR5: Coul 86Jb 176
Pembury Cl. E5 36Xb 71
Pembury Ct. UB3: Harl 51T 106
Pembury Cres. DA14: Sidc 61Ad 139
Pembury Pl. E5 36Xb 71
Pembury Rd. DA7: Bex 52Ad 117
Pembury Rd. E5 36Xb 71
Pembury Rd. N17 25Vb 51
Pembury Rd. SE25 70Wb 135
Pemdevon Rd. CR0: C'don 73Qb 156
Pemell Cl. E1 42Yb 92
Pemell Ho. E1 42Yb 92
(off Pemell Cl.)
Pemerich Cl. UB3: Harl 50V 84
Pempath Pl. HA9: Wemb 33Ma 67
Pemsel Ct. HP3: Hem H 4M 3
Penally Pl. N1 39Tb 71
Penang Ho. E1 46Xb 91
(off Prusom St.)
Penang St. E1 46Xb 91
Penard Rd. UB2: S'hall 48Da 85
Penarth Cen. SE15 51Yb 114
Penarth Ct. SM2: Sutt 80Eb 155
Penarth St. SE15 51Yb 114
Penates KT10: Esh 77Fa 152
Penberth Rd. SE6 61Ec 136
Penbury Rd. UB2: S'hall 49Ba 85
Pencombe M. W11 45Bb 89
Pencraig Way SE15 51Xb 113
Pencroft Dr. DA1: Dart 59Ld 119
Pendall Cl. EN4: E Barn 14Gb 31
Penda Rd. DA8: Erith 52Dd 118
Pendarves Rd. SW20 67Ya 132
Penda's Mead E9 35Ac 72
Pendeen Ct. SL1: Slou 6E 80
Pendell Av. UB3: Harl 52W 106
Pendell Rd. RH1: Blet 3H 209
Pendennis Cl. KT14: W Byf 86J 169
Pendennis Ho. SE8 49Ac 92
Pendennis Rd. BR6: Chels 75Yc 161
Pendennis Rd. N17 27Tb 51
Pendennis Rd. SW16 63Nb 134
Pendennis Rd. TN13: S'oaks 95Kd 203
Pendenza KT11: Cobh 88Aa 171
Penderel Rd. TW3: Houn 57Ca 107
Penderry Ri. SE6 61Fc 137
Penderyn Way N7 35Mb 70
Pendlebury Ct. KT5: Surb 70Na 131
(off Cranes Pk.)
Pendle Ct. UB10: Hil 39R 64
Pendle Ho. SE26 62Wb 135
Pendle Rd. SW16 65Kb 134
Pendlestone Rd. E17 29Dc 52
Pendleton Cl. RH1: Redh 7P 207
Pendleton Ho. RH1: Redh 9L 207
Pendleton Rd. RH1: Redh 9L 207
Pendleton Rd. RH2: Reig 9L 207
Pendlewood Cl. W5 43La 86
Pendley Ho. E2 39Wb 71
(off Whiston Rd.)
Pendolino Way NW10 38Qa 67
Pendragon Rd. BR1: Broml 62Hc 137
Pendragon Wlk. NW9 30Ua 48
Pendrell Ho. WC2 3E 222 (44Mb 90)
(off New Compton St.)
Pendrell Rd. SE4 54Ac 114
Pendrell St. SE18 51Tc 116
Pendula Dr. UB4: Yead 42Z 85
Pendulum M. E8 36Vb 71
Penenden DA3: Nw A G 75Be 165
Penerley Rd. RM13: Rain 43Kd 97
Penerley Rd. SE6 60Dc 114
Penfield Ct. NW9 27Ua 48
(off Tanner Cl.)
Penfield Lodge W9 43Cb 89
(off Admiral Wlk.)
Penfields Ho. N7 37Nb 70
(off York Way Est.)
Penfold Cl. CR0: Wadd 76Qb 156
Penfold La. DA5: Bexl 61Zc 139
(not continuous)
Penfold Pl. NW1 7D 214 (43Gb 89)
Penfold Rd. N9 18Zb 34
Penfold St. NW1 7D 214 (43Gb 89)
Penfold St. NW8 6C 214 (42Fb 89)
Penfold Trad. Est. WD24: Wat 11Y 27
Penford Gdns. SE9 55Mc 115
Penford St. SE5 54Rb 113
Pengarth Rd. DA5: Bexl 57Zc 117
PENGE 66Yb 136
Penge East Station (Rail) 65Yb 136
Penge Ho. SW11 55Fb 111
Penge La. SE20 66Yb 136
Pengelly Apts. E14 50Dc 92
(off Bartlett M.)
Pengelly Cl. EN7: Chesh 2Xb 19
Penge Rd. E13 39Lc 73
Penge Rd. SE20 69Wb 135
Penge Rd. SE25 69Wb 135
Penge West Station (Rail & Overground) 65Xb 135
Penhale Cl. BR6: Chels 77Wc 161
Penhale Pl. SL1: Burn 3A 80
Penhall Rd. SE7 49Mc 93
Pen Haven SL9: Ger X 29B 42
Penhill Rd. DA5: Bexl 58Yc 117
Penhurst GU21: Wok 86B 168
Penhurst Mans. SW6 53Bb 111
(off Rostrevor Rd.)
Penhurst Pl. SE1 4J 229 (48Pb 90)
(off Carlisle La.)
Penhurst Rd. IG6: Ilf 24Rc 54
Penifather La. UB6: G'frd 41Fa 86
Peninsula Apts. N1 39Sb 71
(off Basire St.)
Peninsula Apts. W2 1C 220 (43Fb 89)
(off Praed St.)
Peninsula Ct. E14 48Dc 92
(off E. Ferry Rd.)
Peninsula Ct. N1 39Sb 71
(off Basire St.)
Peninsula Hgts. SE1 7G 229 (50Nb 90)
Peninsular Cl. NW7 22Ab 48
Peninsular Cl. TW14: Felt 58T 106
Peninsular Pk. 49Jc 93
Peninsular Pk. Rd. SE7 49Jc 93
Peninsula Sq. SE10 47Gc 93
Penistone Rd. SW16 66Nb 134
Penistone Wlk. RM3: Rom 23Ld 57
Penketh Dr. HA1: Harr 34Fa 66
Penley Ct. WC2 4J 223 (45Pb 90)
Penman Cl. AL2: Chis G 9N 5
Penman's Grn. WD4: Bucks 5H 11
Penman's Hill WD4: Bucks 5H 11
Penmayne Ho. SE11 7A 230 (50Qb 90)
(off Kennings Way)
Penmon Rd. SE2 48Wc 95
Pennack Rd. SE15 51Vb 113
Penn Almshouses SE10 53Ec 114
(off Greenwich Sth. St.)
Pennant M. W8 49Db 89
Pennant Ter. E17 26Bc 52
Pennard Ho. WD19: Wat 20Z 27
Pennard Mans. W12 47Ya 88
(off Goldhawk Rd.)
Pennard Rd. W12 47Ya 88
Pennards, The TW16: Sun 69Y 129
Penn Cl. HA3: Kenton 28La 46
Penn Cl. RM16: Ors 2D 100
Penn Cl. UB6: G'frd 40Da 65
Penn Cl. UB8: Cowl 42M 83
Penn Cl. WD3: Chor 16F 24
Penn Ct. NW9 27Ta 47
Penn Dr. UB9: Den 30H 43
Penne Cl. WD7: R'lett 6Ha 14
Penner Cl. SW19 61Ab 132
Penners Gdns. KT6: Surb 73Na 153
Pennethorne Cl. E9 39Yb 72
Pennethorne Ho. SW11 55Fb 111
Pennethorne Rd. SE15 52Xb 113
Penney Cl. DA1: Dart 59Md 119
Penney Gro. DA11: Nflt 61De 143
Penn Gdns. BR7: Chst 68Rc 138
Penn Gdns. RM5: Col R 24Cd 56
Penn Ho. HA6: Nwood 20T 26
Penn Ho. NW8 6D 214 (42Gb 89)
(off Mallory St.)
Penn Ho. SL1: Burn 1A 80
Penn Ho. SL4: Eton 10H 81
(off Common La.)
Penn Ho. SL9: Chal P 22A 42
Penn Ho. WD19: Wat 20Z 27
Pennine Dr. NW2 33Za 68
Pennine Ho. N9 20Wb 33
(off Plevna Rd.)
Pennine La. NW2 33Ab 68
Pennine Pde. NW2 33Ab 68
Pennine Rd. SL2: Slou 3E 80
Pennine Way DA11: Nflt 2A 144
Pennine Way DA7: Bex 53Gd 118
Pennine Way UB3: Harl 52T 106
Pennington Cl. RM5: Col R 22Cd 56
Pennington Cl. SE27 63Tb 135
Pennington Ct. SE16 46Ac 92
Pennington Dr. KT13: Weyb 76U 150
Pennington Dr. N21 15Nb 32
Pennington Lodge KT5: Surb 71Na 153
(off Cranes Dr.)
Pennington St. E1 45Xb 91
Pennington Way SE12 61Kc 137
Pennis La. DA3: Fawk 72Zd 165
Penniston Cl. N17 26Sb 51
Penniwell Cl. HA8: Edg 21Pa 47
Penn La. DA5: Bexl 57Zc 117
Penn La. TN14: Ide H 99Ad 201
Penn La. TN14: Sund 99Ad 201
Penn Mdw. SL2: Stoke P 9K 61
Penn Pl. WD3: Rick 17M 25
Penn Rd. AL2: Park 9A 6
Penn Rd. N7 36Nb 70
Penn Rd. SL2: Slou 2H 81
Penn Rd. SL3: Dat 3P 103
Penn Rd. SL9: Chal P 25A 42
Penn Rd. WD24: Wat 11X 27
Penn Rd. WD3: Rick 18H 25
Penn St. N1 39Tb 71
Penn Way WD3: Chor 16F 24
Penny Brookes St. E20 37Ec 72
Penny Cl. E4 20Gc 35
Penny Cl. KT22: Lea 92Ha 192
Penny Cl. RM13: Rain 41Kd 97
Penny Ct. AL3: St A 1B 6
(off Worley Rd.)
Penny Ct. WD17: Wat 12X 27
(off Westland Rd.)
Pennycroft CR0: Sels 81Ac 178
Penny Farthing M. TW12: Hamp H 65Ea 130
Pennyfather La. EN2: Enf 13Sb 33
Pennyfield KT11: Cobh 85W 170
Pennyfields CM14: W'ley 21Yd 58
Pennyfields E14 45Cc 92
(not continuous)
Penny Flds. Ho. SE8 53Bc 114
(off Francis Harvey Way)
Pennyford Ct. NW8 5B 214 (42Fb 89)
(off St John's Wood Rd.)
Penny La. TW17: Shep 73U 150
Pennylets Grn. SL2: Stoke P 8K 61
Pennymead Dr. KT24: E Hor 99V 190
Pennymead Pl. KT10: Esh 79Ba 151
Pennymead Ri. KT24: E Hor 99V 190
Penny M. SW12 59Kb 112
Pennymoor Wlk. W9 42Bb 89
(off Fernhead Rd.)
PENNY POT 4G 166
Pennypot La. GU24: Chob 5F 166
Penny Rd. NW10 41Ra 87
Penny Royal SM6: W'gton 79Mb 156
Pennyroyal Av. E6 44Qc 94
Pennyroyal Dr. UB7: W Dray 47P 83
Penpoll Rd. E8 37Xb 71
Penpool La. DA16: Well 55Xc 117
Penrhyn Av. E17 25Bc 52
Penrhyn Cl. CR3: Cat'm 92Tb 197
Penrhyn Cres. E17 25Cc 52
Penrhyn Cres. SW14 56Sa 109
Penrhyn Gdns. KT1: King T 70Ma 131
Penrhyn Gro. E17 25Cc 52
Penrhyn Rd. KT1: King T 70Na 131
Penrith Cl. BR3: Beck 67Dc 136
Penrith Cl. RH2: Reig 5N 207
Penrith Cl. SW15 57Ab 110
Penrith Cl. UB8: Uxb 38M 63
Penrith Cres. RM13: Rain 36Jd 76
Penrith Pl. SE27 61Rb 135
Penrith Rd. CR7: Thor H 68Sb 135
Penrith Rd. IG6: Ilf 23Vc 55
Penrith Rd. KT3: N Mald 70Ta 131
Penrith Rd. N15 29Tb 51
Penrith Rd. RM3: Rom 23Qd 57
Penrith St. SW16 65Lb 134
Penrose Av. WD19: Wat 19Z 27
Penrose Ct. SW12 60Jb 112
Penrose Ct. TW20: Eng G 5N 125
(not continuous)
Penrose Dr. KT19: Eps 83Qa 173
Penrose Gdns. NW3 35Cb 69
Penrose Gro. SE17 7D 230 (50Sb 91)
Penrose Ho. N21 15Pb 32
Penrose Ho. SE17 50Sb 91
Penrose Rd. KT22: Fet 94Ea 192
Penrose St. SE17 7D 230 (50Sb 91)
Penrose Way SE10 46Gc 93
Penryn Ho. RH1: Redh 4A 208
(off London Rd.)
Penryn Ho. SE11 7B 230 (50Rb 91)
(off Seaton Cl.)
Penryn St. NW1 1D 216 (40Mb 70)
Penry St. SE1 6J 231 (49Ub 91)
Pensbury Pl. SW8 54Lb 112
Pensbury St. SW8 54Lb 112
Penscroft Gdns. WD6: Bore 14Ta 29
Pensford Av. TW9: Kew 54Qa 109
Penshurst NW5 37Jb 70
Penshurst Av. DA15: Sidc 58Wc 117
Penshurst Cl. DA3: Lfield 68Fe 143
Penshurst Cl. TN15: W King 79Ud 164
Penshurst Gdns. HA8: Edg 22Ra 47
Penshurst Grn. BR2: Broml 71Hc 159
Penshurst Ho. SE15 51Yb 114
(off Lovelinch Cl.)
Penshurst Rd. CR7: Thor H 71Rb 157
Penshurst Rd. DA7: Bex 53Bd 117
Penshurst Rd. E9 38Zb 72
Penshurst Rd. EN6: Pot B 3Fb 17
Penshurst Rd. N17 24Vb 51
Penshurst Wlk. BR2: Broml 71Hc 159
Penshurst Way BR5: St M Cry 70Yc 139
Penshurst Way SM2: Sutt 80Cb 155
Pensilver Cl. EN4: E Barn 14Gb 31
Penstemon Cl. N3 23Cb 49
Penstock Footpath N8 27Pb 50
Penta Ct. WD6: Bore 14Qa 29
(off Station Rd.)
Pentagram Yd. W11 44Cb 89
(off Needham Rd.)
Pentavia Retail Pk. 24Va 48
Pentelow Gdns. TW14: Felt 58W 106
Pentire Cl. GU21: Wok 86A 168
Pentire Cl. RM14: Upm 30Ud 58
Pentire Rd. E17 25Fc 53
Pentland Av. HA8: Edg 19Ra 29
Pentland Av. TW17: Shep 71Q 150
Pentland Cl. N9 19Yb 34
Pentland Cl. NW11 33Ab 68
Pentland Gdns. SW18 58Eb 111
Pentland Pl. UB5: N'olt 39Aa 65
Pentland Rd. NW6 41Cb 89
Pentland Rd. SL2: Slou 3E 80
Pentland Rd. WD23: Bush 16Ea 28
Pentlands Cl. CR4: Mitc 69Kb 134
Pentland St. SW18 58Eb 111
Pentland Way UB10: Ick 34S 64
Pentlow St. SW15 55Ya 110
Pentlow Way IG9: Buck H 17Nc 36
Pentney Rd. E4 18Fc 35
Pentney Rd. SW12 60Lb 112
Pentney Rd. SW19 67Ab 132
Penton Av. TW18: Staines 66H 127
Penton Ct. TW18: Staines 65H 127
Penton Dr. EN8: Chesh 1Zb 20
Penton Gro. N1 2K 217 (40Qb 70)
Penton Hall TW18: Staines 67J 127
Penton Hall Dr. TW18: Staines 67J 127
Penton Hook Marina 68H 127
Penton Hook Rd. TW18: Staines 66J 127
Penton Hook Yacht Club 69J 127
Penton Ho. N1 2K 217 (40Qb 70)
(off Pentonville Rd.)
Penton Ho. SE2 46Zc 95
Penton Pk. KT16: Chert 69K 127
Penton Pl. SE17 7C 230 (50Rb 91)
Penton Ri. WC1 3J 217 (41Pb 90)
Penton Rd. TW18: Staines 66H 127
Penton St. N1 1K 217 (40Qb 70)
PENTONVILLE 1J 217 (40Pb 70)
Pentonville Rd. N1 2G 217 (40Nb 70)
Pentrich Av. EN1: Enf 10Wb 19
Pentridge St. SE15 52Vb 113
Pentstemon Dr. DA10: Swans 57Ae 121
Pentyre Av. N18 22Tb 51
Penventon Ct. RM18: Tilb 4C 122
(off Dock Rd.)
Penwerris Av. TW7: Isle 52Ea 108
Penwerris Ct. TW5: Hest 52Ea 108
Penwith Rd. SW18 61Cb 133
Penwith Wlk. GU22: Wok 1P 187
Penwood Ct. HA5: Pinn 28Ba 45
Penwood End GU22: Wok 3M 187
Penwood Ho. SW15 58Va 110
Penwortham Ct. N22 26Qb 50
(off Mayes Rd.)
Penwortham Rd. CR2: Sande 82Sb 177
Penwortham Rd. SW16 65Kb 134
Penylan Pl. HA8: Edg 24Qa 47
Penywern Rd. SW5 50Cb 89
Penzance Cl. UB9: Hare 25M 43
Penzance Gdns. RM3: Rom 23Qd 57
(not continuous)
Penzance Pl. W11 46Ab 88
Penzance Rd. RM3: Rom 23Qd 57
Penzance Spur SL2: Slou 2F 80
Penzance St. W11 46Ab 88
Peony Cl. CM15: Pil H 16Xd 40
Peony Ct. IG8: Wfd G 24Gc 53
Peony Ct. SW10 51Eb 111
Peony Gdns. W12 45Wa 88
Peony Rd. RM3: Rom 22Md 57
Peoplebuilding HP2: Hem H 1B 4
Peperfield WC1 4H 217 (41Pb 90)
(off Cromer St.)
Pepler Ho. W10 42Ab 88
(off Wornington Rd.)
Pepler M. SE5 50Vb 91
Pepler Way SL1: Burn 1A 80
Peplins Cl. AL9: Brk P 8G 8
Peplins Way AL9: Brk P 7G 8
Peploe Rd. NW6 40Za 68
Peplow Cl. UB7: Yiew 46M 83
Pepper All. IG10: Lough 12Hc 35
Pepper Cl. CR3: Cat'm 97Ub 197
Pepper Cl. E6 43Pc 94
Peppercorn Cl. CR7: Thor H 68Tb 135
Pepper Hill DA11: Nflt 62Ee 143
PEPPER HILL 62Fe 143
Pepperhill La. DA11: Nflt 62Ee 143
Peppermead Sq. SE13 57Cc 114
Peppermint Cl. CR0: C'don 73Nb 156
Peppermint Pl. E11 34Gc 73
Pepper St. E14 48Dc 92
Pepper St. SE1 1D 230 (47Sb 91)
Peppie Cl. N16 33Ub 71
Pepys Cl. DA1: Dart 56Qd 119
Pepys Cl. DA11: Nflt 62Fe 143
Pepys Cl. KT21: Asht 89Qa 173
Pepys Cl. RM18: Tilb 3E 122
Pepys Cl. SL3: L'ly 51D 104
Pepys Cl. UB10: Ick 35R 64
Pepys Cres. E16 46Jc 93
Pepys Cres. EN5: Barn 15Ya 30
Pepys Ho. E2 41Yb 92
(off Kirkwall Pl.)
Pepys Ri. BR6: Orp 74Vc 161
Pepys Rd. SE14 53Zb 114
Pepys Rd. SW20 67Ya 132
Pepys St. EC3 4J 225 (45Ub 91)
Perceval Av. NW3 36Gb 69
Perceval Ct. UB5: N'olt 36Ca 65
Perceval Ho. W5 45La 86
Percheron Cl. TW7: Isle 55Ha 108
Percheron Dr. GU21: Knap 1G 186
Percheron Rd. WD6: Bore 16Ta 29
Perch St. E8 35Vb 71
Percival Av. NW9 26Va 48
Percival Cl. KT22: Oxs 83Da 171
Percival Ct. EN8: Chesh 2Ac 20
Percival Ct. N17 24Vb 51
Percival Gdns. KT15: Add 80J 149
Percival Gdns. RM6: Chad H 30Yc 55
Percivall Ho. EC1 1C 224 (43Rb 91)
(off Bartholomew Cl.)
Percival M. SE11 51Pb 112
Percival Rd. BR6: Farnb 75Rc 160
Percival Rd. EN1: Enf 14Vb 33
Percival Rd. RM11: Horn 30Ld 57
Percival Rd. SW14 56Sa 109
Percival Rd. TW13: Felt 61V 128
Percival St. EC1 5B 218 (42Rb 91)
Percival Way KT19: Ewe 77Ta 153
Percy Av. TW15: Ashf 64Q 128
Percy Bilton Ct. TW5: Hest 53Da 107
(off Skinners La.)
Percy Bryant Rd. TW16: Sun 66U 128
Percy Bush Rd. UB7: W Dray 48P 83
Percy Cir. WC1 3J 217 (41Pb 90)
Percy Ct. WD18: Wat 14X 27
Percy Gdns. EN3: Pond E 15Zb 34
Percy Gdns. KT4: Wor Pk 74Ta 153
Percy Gdns. TW7: Isle 55Ka 108
Percy Gdns. UB4: Hayes 41U 84
Percy Laurie Ho. SW15 56Za 110
(off Nursery Cl.)
Percy M. W1 1D 222 (43Mb 90)
(off Rathbone Pl.)
Percy Pl. SL3: Dat 3M 103
Percy Rd. CR4: Mitc 73Jb 156
Percy Rd. DA7: Bex 54Ad 117
Percy Rd. E11 31Gc 73
Percy Rd. E16 43Gc 93
Percy Rd. IG3: Ilf 31Wc 75
Percy Rd. N12 22Eb 49
Percy Rd. N21 17Sb 33
Percy Rd. RM7: Mawney 27Dd 56
Percy Rd. SE20 67Zb 136
Percy Rd. SE25 71Wb 157
Percy Rd. TW12: Hamp 66Ca 129
Percy Rd. TW2: Whitt 60Da 107
Percy Rd. TW7: Isle 56Ja 108
Percy Rd. W12 47Wa 88
Percy Rd. WD18: Wat 14X 27
Percy St. RM17: Grays 51Ee 121
Percy St. W1 1D 222 (43Mb 90)
Percy Ter. BR1: Broml 69Rc 138
Percy Way TW2: Whitt 60Ea 108
Percy Yd. WC1 3J 217 (41Pb 90)
Peregrine Cl. NW10 36Ta 67
Peregrine Cl. WD25: Wat 6Aa 13
Peregrine Ct. DA16: Well 53Vc 117
Peregrine Ct. SE8 51Cc 114
(off Edward St.)
Peregrine Ct. SW16 63Pb 134
Peregrine Gdns. CR0: C'don 75Ac 158
Peregrine Ho. EC1 3C 218 (41Rb 91)
(off Hall St.)
Peregrine Rd. EN9: Walt A 6Jc 21
Peregrine Rd. IG6: Ilf 22Xc 55
Peregrine Rd. N17 24Sb 51
Peregrine Rd. TW16: Sun 68V 128
Peregrine Wlk. RM12: Horn 37Kd 77
Peregrine Way AL10: Hat 1C 8
Peregrine Way SW19 66Ya 132
Perendale Dr. TW17: Shep 67S 128
Perham Rd. W14 50Ab 88
Perham Way AL2: Lon C 8H 7
Peridot Ct. E2 4K 219 (41Vb 91)
(off Virginia Rd.)
Peridot St. E6 43Nc 94
Perifield SE21 60Sb 113
Perilla Ho. E1 44Wb 91
(off Boulevard Walkway)
Perimeade Rd. UB6: G'frd 40La 66
Periton Rd. SE9 56Mc 115
PERIVALE 39La 66
Perivale Gdns. W13 42Ka 86
Perivale Gdns. WD25: Wat 6X 13
Perivale Grange UB6: G'frd 41Ja 86

Perivale La. UB6: G'frd 41Ja 86
Perivale Lodge UB6: G'frd 41Ja 86
(off Perivale La.)
Perivale New Bus. Cen. UB6: G'frd 40La 66
Perivale Pk. UB6: G'frd 40Ka 66
Perivale Pk. Athletics Track 41Ga 86
Perivale Pk. Golf Course 41Ga 86
Perivale Station (Underground) 40Ja 66
Perivale Wood Local Nature Reserve 39Ha 66
Periwood Cres. UB6: G'frd 39Ja 66
Perkin Cl. HA0: Wemb 36Ka 66
Perkin Cl. TW3: Houn 56Ca 107
Perkins Cl. DA9: Ghithe 57Wd 120
Perkins Ct. TW15: Ashf 64P 127
Perkins Gdns. UB10: Ick 33S 64
Perkins Ho. E14 43Bc 92
(off Wallwood St.)
Perkin's Rents SW1 4D 228 (48Mb 90)
Perkins Rd. IG2: Ilf 29Tc 54
Perkins Sq. SE1 6E 224 (46Sb 91)
Perks Cl. SE3 55Gc 114
Perkyn Sq. N17 27Xb 51
Perleybrooke La. GU21: Wok 9L 167
Perley Ho. E3 43Bc 92
(off Weatherley Cl.)
Permain Cl. DA9: Ghithe 58Wd 120
Permain Cl. WD7: Shenl 5Na 15
Perpins Rd. SE9 58Uc 116
Perran Cl. DA3: Hartl 70Be 143
Perran Rd. SW2 60Rb 113
Perran Wlk. TW8: Bford 50Na 87
Perren St. NW5 37Kb 70
Perrers Rd. W6 49Xa 88
Perrin Apts. N1 38Nb 70
(off Caledonian Rd.)
Perrin Cl. TW15: Ashf 64P 127
Perrin Cl. WD23: B Hea 18Ga 28
Perrin Ct. GU21: Wok 87D 168
Perrin Ct. TW15: Ashf 63Q 128
Perring Est. E3 43Cc 92
(off Gale St.)
Perrin Ho. NW6 41Cb 89
Perrin Rd. DA1: Dart 57Pd 119
Perrin Rd. HA0: Wemb 35Ka 66
Perrin's Ct. NW3 35Eb 69
Perrin's La. NW3 35Eb 69
Perrin's Wlk. NW3 35Eb 69
Perronet Ho. SE1 4C 230 (48Rb 91)
(off Princess St.)
Perrott St. SE18 49Sc 94
Perry Av. W3 44Ta 87
Perry Cl. IG2: Ilf 30Tc 54
Perry Cl. RM13: Rain 40Fd 76
Perry Cl. UB8: Hil 44R 84
Perry Ct. E14 50Cc 92
(off Maritime Quay)
Perry Ct. KT2: King T 68Na 131
(off Old London Rd.)
Perry Ct. N15 30Ub 51
Perrycroft SL4: Wind 5C 102
Perryfields Way SL1: Burn 2A 80
Perryfield Way NW9 30Va 48
Perryfield Way TW10: Ham 62Ka 130
Perry Gdns. N9 20Tb 33
Perry Gth. UB5: N'olt 39Y 65
Perry Gro. DA1: Dart 56Qd 119
Perry Hall Cl. BR6: St M Cry 73Wc 161
Perry Hall Rd. BR6: St M Cry 72Vc 161
Perry Hill GU3: Worp 9J 187
(not continuous)
Perry Hill SE6 62Bc 136
Perry Ho. SL1: Burn 2A 80
Perry How KT4: Wor Pk 74Va 154
Perry Lodge E12 32Mc 73
Perryman Ho. IG11: Bark 39Sc 74
(off The Shaftesburys)
Perrymans Farm Rd. IG2: Ilf 30Tc 54
Perryman Way SL2: Slou 1D 80
(not continuous)
Perry Mead EN2: Enf 12Rb 33
Perry Mead WD23: Bush 16Ea 28
Perrymead St. SW6 53Cb 111
Perryn Ct. TW1: Twick 58Ja 108
Perryn Ho. W3 45Ua 88
Perryn Rd. SE16 48Xb 91
Perryn Rd. W3 46Ta 87
Perry Ri. SE23 62Ac 136
Perry Rd. RM9: Dag 43Bd 95
Perrys La. BR6: Prat B 85Xc 181
Perry's Pl. W1 2D 222 (44Mb 90)
Perry St. BR7: Chst 65Tc 138
Perry St. DA1: Cray 56Gd 118
Perry St. DA11: Nflt 10A 122
PERRY STREET 10B 122
Perry St. Gdns. BR7: Chst 65Uc 138
Perry St. Shaw BR7: Chst 66Uc 138
Perry's Way RM15: S Ock 43Yd 98
Perry Va. SE23 61Yb 136
Perry Way RM15: Avel 45Sd 98
Persant Rd. SE6 61Gc 137
Perseverance Cotts. GU23: Rip 93L 189
Perseverance Pl. SW9 52Qb 112
Perseverance Pl. TW9: Rich 56Na 109
Perseverance Works E2 3J 219 (41Ub 91)
(off Kingsland Rd.)
Persfield Cl. KT17: Ewe 82Va 174
Persfield M. KT17: Ewe 82Va 174
Pershore Cl. IG2: Ilf 29Rc 54
Pershore Gro. SM5: Cars 72Fb 155
Pershore Ho. W13 46Ja 86
(off Singapore Rd.)
Pert Cl. N10 24Kb 50
Perth Av. NW9 31Ta 67
Perth Av. SL1: Slou 4G 80
Perth Av. UB4: Yead 42Y 85
Perth Cl. SE5 56Tb 113
Perth Cl. SW20 68Va 132
Perth Cl. UB5: N'olt 36Ca 65
Perth Ho. N1 38Pb 70
(off Bemerton Est.)
Perth Ho. RM18: Tilb 4C 122
Perth Rd. BR3: Beck 68Ec 136
Perth Rd. E10 32Ac 72
Perth Rd. E13 40Kc 73
Perth Rd. IG11: Bark 40Tc 74
Perth Rd. IG2: Ilf 30Qc 54
Perth Rd. N22 25Rb 51
Perth Rd. N4 32Qb 70
Perth Ter. IG2: Ilf 31Sc 74
Perth Trad. Est. SL1: Slou 3F 80
Peruvian Wharf E16 46Jc 93
Perwell Av. HA2: Harr 32Ba 65
Perystreete SE23 61Yb 136
Pescot Av. DA3: Lfield 69Ce 143
Petands Ct. RM12: Horn 34Md 77
(off Randall Dr.)
Petavel Rd. TW11: Tedd 65Ga 130
Peter Av. NW10 38Xa 68
Peter Av. RH8: Oxt 1H 211
Peter Best Ho. E1 44Xb 91
(off Nelson St.)
Peterboat Cl. SE10 49Gc 93
Peterborough Av. RM14: Upm 32Ud 78
Peterborough Ct. EC4 2A 224 (44Qb 90)
Peterborough Gdns. IG1: Ilf 31Nc 74
Peterborough Ho. WD6: Bore 12Qa 29
(off Stratfield Rd.)
Peterborough M. SW6 54Cb 111
Peterborough Rd. E10 29Ec 52
Peterborough Rd. HA1: Harr 32Ga 66
Peterborough Rd. SM5: Cars 72Gb 155
Peterborough Rd. SW6 54Cb 111
Peterborough Vs. SW6 53Db 111
Peter Butler Ho. SE1 47Wb 91
(off Wolseley St.)
Peterchurch Ho. SE15 51Xb 113
(off Commercial Way)
Petergate SW11 56Eb 111
PETER GIDNEY NEURODISABILITY CEN. 63Sd 142
Peter Harrison Planetarium 52Fc 115
Peterhead Ct. UB1: S'hall 44Ea 86
(off Osborne Rd.)
Peterhead M. SL3: L'ly 50C 82
Peter Heathfield Ho. E15 39Fc 73
(off Wise Rd.)
Peterhill Cl. SL9: Chal P 22A 42
Peter Hills Ho. SE16 49Wb 91
(off Alexis St.)
Peter Ho. SW8 52Nb 112
(off Luscombe Way)
Peter James Bus. Cen. UB3: Hayes 47W 84
Peter James Ent. Cen. NW10 41Sa 87
Peter Kennedy Ct. CR0: C'don 72Bc 158
Peterley Bus. Cen. E2 40Xb 71
Peter Lyell Ct. HA4: Ruis 33X 65
Peter May Sports Cen. 24Ec 52
Peter Pan Statue 6B 220 (46Fb 89)
Peters Av. AL2: Lon C 8G 6
Peters Cl. DA16: Well 54Uc 116
Peters Cl. HA7: Stan 23Ma 47
Peters Cl. RM8: Dag 32Zc 75
Peters Ct. W2 44Db 89
(off Porchester Rd.)
Petersfield Av. RM3: Rom 23Nd 57
Petersfield Av. SL2: Slou 6L 81
Petersfield Av. TW18: Staines 64L 127
Petersfield Cl. N18 22Sb 51
Petersfield Cl. RM3: Rom 23Qd 57
Petersfield Cres. CR5: Coul 87Nb 176
Petersfield Ri. SW15 60Xa 110
Petersfield Rd. TW18: Staines 64L 127
Petersfield Rd. W3 47Sa 87
Petersgate KT2: King T 65Sa 131
(off Warren Rd.)
PETERSHAM 60Na 109
Petersham Av. KT14: Byfl 84N 169
Petersham Cl. KT14: Byfl 84N 169
Petersham Cl. SM1: Sutt 78Cb 155
Petersham Cl. TW10: Ham 61Ma 131
Petersham Dr. BR5: St P 68Vc 139
Petersham Gdns. BR5: St P 68Vc 139
Petersham Ho. SW7 5B 226 (49Fb 89)
(off Kendrick M.)
Petersham La. SW7 3A 226 (48Eb 89)
Petersham M. SW7 4A 226 (48Eb 89)
Petersham Pl. SW7 3A 226 (48Eb 89)
Petersham Rd. TW10: Ham 58Ma 109
Petersham Rd. TW10: Rich 58Ma 109
Petersham Ter. CR0: Bedd 76Nb 156
(off Richmond Grn.)
Peter's Hill EC4 4D 224 (45Sb 91)
Peter Shore Ct. E1 43Zb 92
(off Beaumont Sq.)
Peter's La. EC1 7B 218 (43Rb 91)
(not continuous)
Peterslea WD4: K Lan 1R 12
Petersmead Cl. KT20: Tad 95Ya 194
Peterson Ct. IG10: Lough 12Qc 36
Peter's Path SE26 63Xb 135
Peterstone Rd. SE2 47Xc 95
Peterstow Cl. SW19 61Ab 132
Peter St. DA12: Grav'nd 9D 122
Peter St. W1 4D 222 (45Mb 90)
Peterwood Pk. CR0: Wadd 75Pb 156
Peterwood Way CR0: Wadd 75Pb 156
Petham Ct. BR8: Crock 72Hd 162
Petherton Ct. HA1: Harr 30Ha 46
(off Gayton Rd.)
Petherton Ct. NW10 39Za 68
(off Tiverton Rd.)
Petherton Rd. N5 36Sb 71
Petiver Cl. E9 38Yb 72
Petley Rd. W6 51Za 110
Peto Pl. NW1 5A 216 (42Kb 90)
Peto St. Nth. E16 44Hc 93
Petra Way E16 43Hc 93
Petre Cl. CM13: W H'dn 30Ee 59
Petresfield Way CM13: W H'dn 30Ee 59
Petridge Rd. RH1: Redh 10P 207
Petrie Cl. NW2 37Ab 68
Petrie Ho. SE18 51Qc 116
(off Woolwich Comn.)
Petrie Mus. of Egyptian Archaeology 6D 216 (42Mb 90)
Petros Gdns. NW3 37Db 69
Pettacre Cl. SE28 48Sc 94
Pett Cl. RM11: Horn 33Kd 77
Petten Cl. BR5: Orp 74Zc 161
Petten Gro. BR5: Orp 74Yc 161
Petters Rd. KT21: Asht 88Pa 173
Petticoat Lane Market 1J 225 (43Ub 91)
Petticoat Sq. E1 2K 225 (44Vb 91)
Petticoat Twr. E1 2K 225 (44Vb 91)
(off Petticoat Sq.)
PETTINGS 80De 165
Pettits Blvd. RM1: Rom 25Gd 56
Pettits Cl. RM1: Rom 26Gd 56
Pettits La. RM1: Rom 26Gd 56
Pettits La. Nth. RM1: Rom 25Fd 56
Pettits Pl. RM10: Dag 36Cd 76
Pettits Rd. RM10: Dag 36Cd 76
Pettiward Cl. SW15 56Ya 110
Pettley Gdns. RM7: Rom 29Fd 56
Pettman Cres. SE28 48Tc 94
Pettsgrove Av. HA0: Wemb 36La 66
Pett's Hill UB5: N'olt 36Da 65
Petts La. TW17: Shep 70Q 128
Pett St. SE18 49Nc 94
PETTS WOOD 71Tc 160
Petts Wood Rd. BR5: Pet W 71Sc 160
Petts Wood Station (Rail) 71Sc 160
Petty Cross SL1: Slou 4C 80
Petty France SW1 3C 228 (48Lb 90)
Pettys Cl. EN8: Chesh 1Zb 20
Petty Wales EC3 5J 225 (45Ub 91)
Petworth Cl. CR5: Coul 91Lb 196
Petworth Cl. UB5: N'olt 38Ba 65
Petworth Ct. SL4: Wind 3E 102
Petworth Gdns. SW20 69Xa 132
Petworth Gdns. UB10: Hil 39S 64
Petworth Rd. DA6: Bex 57Cd 118
Petworth Rd. N12 22Gb 49
Petworth St. SW11 53Gb 111
Petworth Way RM12: Horn 35Hd 76
Petyt Pl. SW3 51Gb 111
Petyward SW3 6E 226 (49Gb 89)
Pevensey Av. EN1: Enf 12Ub 33
Pevensey Av. N11 22Mb 50
Pevensey Cl. TW7: Isle 52Ea 108
Pevensey Ct. SW16 62Qb 134
Pevensey Ct. W3 47Ra 87
Pevensey Ho. E1 43Zb 92
(off Ben Jonson Rd.)
Pevensey Rd. E7 35Hc 73
Pevensey Rd. SL2: Slou 3E 80
Pevensey Rd. SW17 63Fb 133
Pevensey Rd. TW13: Felt 60Aa 107
Pevensey Way WD3: Crox G 14R 26
Peverel E6 44Qc 94
Peverel Ho. RM10: Dag 33Cd 76
Peverett Cl. N11 22Kb 50
Peveril Ct. DA2: Dart 58Rd 119
(off Osbourne Rd.)
Peveril Dr. TW11: Tedd 64Fa 130
Peveril Ho. SE1 4G 231 (48Tb 91)
(off Rephidim St.)
Pewsey Cl. E4 22Cc 52
Peyton Pl. SE10 52Ec 114
Peyton's Cotts. RH1: Nutf 4F 208
Pharamond NW2 37Za 68
Pharaoh Cl. CR4: Mitc 73Hb 155
Pharaoh's Island TW17: Shep 75P 149
Pheasant Cl. CR8: Purl 85Rb 177
Pheasant Cl. E16 44Kc 93
Pheasant Ct. WD25: Wat 7Y 13
Pheasantry Ho. SW3 7E 226 (50Gb 89)
(off Jubilee Pl.)
Pheasantry Welcome Cen., The 67Ha 130
Pheasants Way WD3: Rick 17K 25
Pheasant Wlk. SL9: Chal P 21A 42
Phelps Cl. TN15: W King 79Ud 164
Phelps La. NW7 22Ab 48
Phelps Lodge N1 1J 217 (39Pb 70)
Phelp St. SE17 51Tb 113
Phelps Way UB3: Harl 49V 84
Phene St. SW3 51Gb 111
Philadelphia Ct. SW10 52Eb 111
(off Uverdale Rd.)
Philand La. CR5: Coul 89Kb 176
Philanthropic Rd. RH1: Redh 7A 208
Philbeach Gdns. SW5 50Cb 89
Phil Brown Pl. SW8 55Kb 112
(off Daley Thompson Way)
Philbye M. SL1: Slou 7D 80
Philchurch Pl. E1 44Wb 91
Philia Ho. NW1 38Lb 70
(off Farrier St.)
Philimore Cl. SE18 50Uc 94
Philip Av. BR8: Swan 70Fd 140
Philip Av. RM7: Rush G 32Fd 76
Philip Cl. CM15: Pil H 16Xd 40
Philip Cl. RM7: Rush G 32Fd 76
Philip Ct. W2 7B 214 (43Fb 89)
(off Hall Pl.)
Philip Gdns. CR0: C'don 75Bc 158
Philip Ho. NW6 39Db 69
(off Mortimer Pl.)
Philip Jones Ct. N4 32Pb 70
Philip La. N15 28Tb 51
Philip Mole Ho. W9 42Cb 89
(off Chippenham Rd.)
Philipot Path SE9 58Pc 116
Philippa Gdns. SE9 57Mc 115
Philippa Way RM16: Grays 9D 100
Philip Rd. N15 28Tb 51
Philip Rd. RM13: Rain 41Gd 96
Philip Rd. SE15 55Wb 113
Philip Rd. TW18: Staines 65M 127
Philips Cl. SM5: Cars 74Jb 156
Philip Sidney Ct. RM16: Chaf H 50Zd 99
(off Philip Sidney Rd.)
Philip Sq. SW8 54Kb 112
Philip St. E13 42Jc 93
Philip Sydney Rd. RM16: Chaf H 50Zd 99
Philip Wlk. SE15 55Wb 113
(not continuous)
Phillida Rd. RM3: Hrld W 26Qd 57
Phillimore Ct. W8 47Cb 89
(off Kensington High St.)
Phillimore Ct. WD7: R'lett 8Ga 14
Phillimore Gdns. NW10 39Ya 68
Phillimore Gdns. W3 46Qa 87
Phillimore Gdns. W8 47Cb 89
Phillimore Gdns. Cl. W8 48Cb 89
Phillimore Pl. W8 47Cb 89
Phillimore Pl. WD7: R'lett 8Ga 14
Phillimore Ter. W8 48Cb 89
(off Allen St.)
Phillimore Wlk. W8 48Cb 89
Phillipers WD25: Wat 8Z 13
Phillip Ho. E1 43Vb 91
(off Heneage St.)
Phillippines Shaw TN14: Ide H 100Yc 201
Phillipp St. N1 1H 219 (39Ub 71)
Phillips Cl. DA1: Dart 58Kd 119
Phillips Ct. HA8: Edg 23Qa 47
Phillip's Quad. GU22: Wok 90A 168
Philpot La. EC3 4H 225 (45Ub 91)
Philpot La. GU24: Chob 5M 167
Philpot Path IG1: Ilf 34Sc 74
Philpots Cl. UB7: Yiew 45M 83
Philpot Sq. SW6 55Db 111
Philpot St. E1 44Xb 91
Phineas Pett Rd. SE9 55Nc 116
Phipps Bri. Rd. CR4: Mitc 69Fb 133
Phipps Bri. Rd. SW19 68Eb 133
Phipps Bridge Stop (London Tramlink) 69Fb 133
Phipps Hatch La. EN2: Enf 10Sb 19
Phipps Ho. SE7 50Kc 93
(off Woolwich Rd.)
Phipps Ho. W12 45Xa 88
(off White City Est.)
Phipp's M. SW1 4A 228 (48Kb 90)
(off Buckingham Palace Rd.)
Phipps Rd. SL1: Slou 3B 80
(not continuous)
Phipp St. EC2 5H 219 (42Ub 91)
Phoebeth Rd. SE4 57Cc 114
Phoebe Wlk. E16 44Kc 93
Phoenix Apts. WD17: Wat 15Z 27
(off Lwr. High St.)
Phoenix Av. SE10 47Gc 93
Phoenix Cen. 80Nb 156
Phoenix Cinema East Finchley 28Gb 49
Phoenix Cl. BR4: W W'ck 75Fc 159
Phoenix Cl. CR4: Mitc 69Fb 133
Phoenix Cl. E17 26Bc 52
Phoenix Cl. E8 39Vb 71
Phoenix Cl. HA6: Nwood 21V 44
Phoenix Cl. KT19: Eps 84Qa 173
Phoenix Cl. W12 45Xa 88
Phoenix Ct. CR2: S Croy 78Vb 157
Phoenix Ct. DA11: Nflt 57Ce 121
Phoenix Ct. E1 42Xb 91
(off Buckhurst St.)
Phoenix Ct. E14 49Cc 92
Phoenix Ct. E4 20Dc 34
Phoenix Ct. KT17: Eps 85Ua 174
(off Depot Rd.)
Phoenix Ct. KT3: N Mald 69Va 132
Phoenix Ct. NW1 2E 216 (40Mb 70)
(off Purchese St.)
Phoenix Ct. SE14 51Ac 114
(off Chipley St.)
Phoenix Ct. TW13: Felt 63U 128
Phoenix Ct. TW3: Houn 55Ca 107
Phoenix Ct. TW4: Houn 57Z 107
Phoenix Ct. TW8: Bford 50Na 87
Phoenix Dr. BR2: Kes 77Mc 159
Phoenix Fitness Cen. 45Wa 88
Phoenix Hgts. E. E14 47Cc 92
(off Byng St.)
Phoenix Hgts. W. E14 47Cc 92
(off Mastmaker Ct.)
Phoenix Ho. AL1: St A 3E 6
(off Campfield Rd.)
Phoenix Ind. Est. HA1: Harr 28Ha 46
Phoenix Lodge Mans. W6 49Za 88
(off Brook Grn.)
Phoenix Pk. NW2 33Wa 68
Phoenix Pl. DA1: Dart 59Md 119
Phoenix Pl. WC1 5J 217 (42Pb 90)
Phoenix Point SE28 46Yc 95
Phoenix Rd. NW1 3D 216 (41Mb 90)
Phoenix Rd. SE20 65Yb 136
Phoenix St. WC2 3E 222 (44Mb 90)
Phoenix Theatre 3E 222 (44Mb 90)
(off Charing Cross Rd.)
Phoenix Trad. Est. UB6: G'frd 39La 66
Phoenix Trad. Pk. TW8: Bford 50Ma 87
Phoenix Way E16 43Hc 93
(off Barking Rd.)
Phoenix Way SW18 57Eb 111
Phoenix Way TW5: Hest 51Z 107
Phoenix Wharf E1 46Xb 91
(off Wapping High St.)
Phoenix Wharf Rd. SE1 2K 231 (47Vb 91)
(off Tanner St.)
Phoenix Works HA5: Hat E 24Ba 45
Phoenix Yd. WC1 4J 217 (41Pb 90)
(off King's Cross Rd.)
Photographers' Gallery 3B 222 (44Lb 90)
(off Ramillies St.)
Phygtle, The SL9: Chal P 23A 42
Phyllis Av. KT3: N Mald 71Xa 154
Phyllis Hodges Ho. NW1 2D 216 (40Mb 70)
(off Aldenham St.)
Phyllis Ho. CR0: Wadd 77Rb 157
(off Ashley La.)
Phyppe Way E17 29Fc 53
Physical Energy Statue 7A 220 (46Eb 89)
Physic Pl. SW3 51Hb 111
Piano La. N16 34Tb 71
Piano Works IG11: Bark 38Sc 74
(off Ripple Rd.)
Piano Yd. NW5 36Kb 70
Piazza, The UB8: Uxb 38M 63
Piazza, The WC2 4G 223 (45Nb 90)
(off Covent Gdn.)
Piazza Wlk. E1 44Wb 91
Picardy Ho. EN2: Enf 10Sb 19
Picardy Manorway DA17: Belv 48Dd 96
Picardy Rd. DA17: Belv 50Cd 96
Picardy St. DA17: Belv 48Cd 96
Picasso Ct. WD24: Wat 8W 12
Piccadilly W1 7K 221 (46Kb 90)
Piccadilly Arc. SW1 6B 222 (46Lb 90)
(off Piccadilly)
Piccadilly Circus 5D 222 (45Mb 90)
Piccadilly Circus Station (Underground) 5D 222 (45Mb 90)
Piccadilly Ct. N7 37Pb 70
(off Caledonian Rd.)
Piccadilly Pl. W1 5C 222 (45Lb 90)
(off Piccadilly)
Piccadilly Theatre 4C 222 (45Lb 90)
(off Denman St.)
Pickard Cl. N14 18Mb 32
Pickard Gdns. E3 43Bc 92
Pickard St. EC1 3C 218 (41Rb 91)
Pickering Av. E6 40Qc 74
Pickering Cl. E9 38Zb 72
Pickering Ct. DA2: Dart 58Rd 119
(off Osbourne Rd.)
Pickering Gdns. CR0: C'don 72Vb 157
Pickering Gdns. N11 23Jb 50
Pickering Ho. W2 44Eb 89
(off Hallfield Est.)
Pickering Ho. W5 49La 86
(off Windmill Rd.)
Pickering La. BR5: Farnb 74Tc 160
Pickering M. W2 44Db 89
Pickering Pl. SW1 7C 222 (46Lb 90)
(off St James's St.)
Pickering Rd. IG11: Bark 37Sc 74
Pickering St. N1 39Rb 71
Pickets Cl. WD23: B Hea 18Fa 28
Pickets St. SW12 59Kb 112
Pickett Cft. HA7: Stan 25Ma 47
Picketts Lock La. N9 19Yb 34
Picketts Lock La. Ind. Est. N9 19Ac 34
Picketts Ter. SE22 57Wb 113
Pickford Cl. DA7: Bex 54Ad 117
Pickford Dr. SL3: L'ly 46A 82
Pickford La. DA7: Bex 54Ad 117
Pickford Rd. AL1: St A 2F 6
Pickford Rd. DA7: Bex 55Ad 117
Pickfords Gdns. SL1: Slou 6J 81
Pickfords Wharf N1 2D 218 (40Sb 71)
Pickfords Wharf SE1 6F 225 (46Tb 91)
Pick Hill EN9: Walt A 4Hc 21
Pickhurst Grn. BR2: Hayes 73Hc 159
Pickhurst La. BR2: Hayes 73Hc 159
Pickhurst La. BR4: W W'ck 71Gc 159
Pickhurst Mead BR2: Hayes 73Hc 159
Pickhurst Pk. BR2: Broml 71Gc 159
Pickhurst Ri. BR4: W W'ck 73Ec 158
Pickins Piece SL3: Hort 54C 104
Pickle M. SW9 52Qb 112
Pickle Sq. SE1 4J 231 (48Ub 91)
(off The Tannery)
Pickmoss La. TN14: Otf 88Jd 183
Pickwick Cl. TW4: Houn 57Aa 107
Pickwick Ct. SE9 60Nc 116
Pickwick Gdns. DA11: Nflt 62Fe 143
Pickwick Ho. DA11: Nflt 62Fe 143
Pickwick Ho. SE16 47Wb 91
(off George Row)
Pickwick Ho. W11 46Za 88
(off St Ann's Rd.)
Pickwick M. N18 21Ub 51
Pickwick Pl. HA1: Harr 31Ga 66
Pickwick Rd. SE21 59Tb 113
Pickwick St. SE1 2D 230 (47Sb 91)
Pickwick Way BR7: Chst 65Sc 138
Pickworth Cl. SW8 52Nb 112
Picquets Way SM7: Bans 88Ab 174
Picton Mt. CR6: W'ham 91Wb 197
Picton Pl. KT6: Surb 74Qa 153
Picton Pl. W1 3J 221 (44Jb 90)
Picton St. SE5 52Tb 113
Picture Ho. SW16 61Nb 134
Picturehouse Central 4D 222 (45Mb 90)
(off Shaftesbury Av.)
Picture Ho. M. E17 28Fc 53
Pied Bull Ct. WC1 1G 223 (43Nb 90)
(off Bury Pl.)
Pied Bull Yd. N1 39Rb 71
(off Theberton St.)
Pied Bull Yd. WC1 1F 223 (43Nb 90)
(off Bury Pl.)
Piedmont Rd. SE18 50Tc 94
(not continuous)
PIELD HEATH 42P 83
Pield Heath Av. UB8: Hil 42Q 84
Pield Heath Rd. UB8: Cowl 42N 83
Pield Heath Rd. UB8: Hil 42N 83
Pierce Campion Ct. E17 27Bc 52
Piercing Hill CM16: They B 7Tc 22
Pier Head E1 46Xb 91
(not continuous)
Pierhead Wharf E1 46Xb 91
(off Wapping High St.)
Pier Ho. SW3 51Gb 111
Pierian Spring HP1: Hem H 1K 3
Pieris Ho. TW13: Felt 61W 128
(off High St.)
Piermont Grn. SE22 57Xb 113
Piermont Pl. BR1: Broml 68Nc 138
Piermont Rd. SE22 57Xb 113
Pier Pde. E16 46Qc 94
(off Pier Rd.)
Pierpoint Bldg. E14 47Bc 92
Pierrepoint Rd. W3 45Ra 87
Pierrepont Arc. N1 1B 218 (40Rb 71)
(off Islington High St.)
Pierrepont Row N1 1B 218 (40Rb 71)
(off Camden Pas.)
Pier Rd. DA11: Nflt 8B 122
Pier Rd. DA8: Erith 51Gd 118
(not continuous)
Pier Rd. DA9: Ghithe 56Xd 120
Pier Rd. E16 46Qc 94
Pier Rd. TW14: Felt 57X 107
Pierson Rd. SL4: Wind 3B 102
Pier St. E14 49Ec 92
(not continuous)
Pier Ter. SW18 56Db 111
Pier Wlk. SE10 47Gc 93
Pier Way SE28 47Sc 94
Pier Wharf RM17: Grays 52Ce 121
Pietra Lara Bldg. EC1 5D 218 (42Sb 91)
(off Pear Tree St.)
Pigeon Ho. La. CR5: Coul 97Eb 195
Pigeon La. TW12: Hamp 63Ca 129
Piggott Ho. E2 40Zb 72
(off Sewardstone Rd.)
Piggs Cnr. RM17: Grays 48Ee 99
Piggy La. WD3: Chor 16D 24
Pigott St. E14 44Cc 92
Pike Cl. BR1: Broml 64Kc 137
Pike Cl. UB10: Uxb 39P 63
Pike Cres. TW15: Ashf 63P 127
Pike La. RM14: Upm 36Vd 78
Pikemans Ct. SW5 49Cb 89
(off W. Cromwell Rd.)
Pike Rd. NW7 21Ta 47
Pikes Cotts. EN5: Ark 14Ya 30
Pike's End HA5: Eastc 28X 45
Pikes Hill KT17: Eps 85Ua 174
Pikestone Cl. UB4: Yead 42Aa 85
Pikethorne SE23 61Zc 136
Pilgrimage St. SE1 2F 231 (47Tb 91)
Pilgrim Cl. AL2: Park 9A 6
Pilgrim Cl. SM4: Mord 73Db 155
Pilgrim Hill SE27 63Sb 135
Pilgrim Ho. SE1 4G 231 (48Tb 91)
(off Tabard St.)
Pilgrim Ho. SE16 47Yb 92
(off Brunel Rd.)
Pilgrim M. RH2: Reig 6J 207
Pilgrims Cloisters SE5 52Ub 113
(off Sedgmoor Pl.)
Pilgrim's Cl. CM15: Pil H 15Vd 40
Pilgrims Cl. N13 21Pb 50
Pilgrims Cl. UB5: N'olt 36Ea 66
Pilgrims Cl. WD25: Wat 5Z 13
Pilgrims Cnr. NW6 40Cb 69
(off Chichester Rd.)
Pilgrim's Ct. DA1: Dart 57Qd 119
Pilgrims Ct. EN1: Enf 12Tb 33
PILGRIMS HATCH 16Xd 40
Pilgrim's La. NW3 35Fb 69
Pilgrims La. CM14: Pil H 14Td 40
Pilgrims La. RH8: T'sey 97Kc 199
Pilgrims La. RM16: Chaf H 48Zd 99
Pilgrims La. RM16: N Stif 46Yd 98
Pilgrims La. TN16: Westrm 95Nc 200
Pilgrims' La. CR3: Cat'm 98Pb 196
Pilgrims M. E14 45Gc 93
Pilgrim's Pl. NW3 35Fb 69
Pilgrims Pl. RH2: Reig 4J 207
Pilgrims Ri. EN4: E Barn 15Gb 31
Pilgrims Rd. DA10: Nflt 56Ae 121
Pilgrims Rd. DA10: Swans 56Ae 121
PILGRIMS RDBT. 47Yd 98
Pilgrim St. EC4 3B 224 (44Rb 91)
Pilgrims Vw. DA9: Ghithe 58Yd 120
Pilgrim's Way GU24: Bisl 8E 166
Pilgrim's Way HA9: Wemb 32Ra 67
Pilgrim's Way TN13: Dun G 89Ed 182
(not continuous)
Pilgrims Way CR2: S Croy 79Vb 157
Pilgrims Way DA1: Dart 60Qd 119
Pilgrims Way E6 39Nc 74
Pilgrims Way N19 32Mb 70
Pilgrims Way RH2: Reig 4H 207
Pilgrims Way TN14: Otf 88Md 183
Pilgrims Way TN14: Sund 93Xc 201
Pilgrims Way TN15: Kems'g 88Sd 184
Pilgrims Way TN15: Wro 89Be 185
(not continuous)
Pilgrims Way TN16: Bras 95Pc 200

Pilgrims Way TN16: Westrm 95Pc 200
Pilgrims Way Cotts. TN15: Kems'g 89Qd 183
Pilgrims Way W. TN14: Otf 87Ld 183
Pilgrims Way W. TN14: Dun G 89Fd 182
Pilgrims Way W. TN14: Otf 89Fd 182
Pilkington Rd. BR6: Farnb 76Sc 160
Pilkington Rd. SE15 54Xb 113
Pillar Box La. TN15: Seal 94Ud 204
Pill Box Studios E2 42Xb 91
(off Coventry Rd.)
Pillfold Ho. SE11 5H 229 (49Pb 90)
(off Old Paradise St.)
Pillions La. UB4: Hayes 42T 84
Pilot Cl. SE8 51Bc 114
Pilot Ind. Cen. NW10 42Ta 87
Pilots Pl. DA12: Grav'nd 8E 122
Pilot Wlk. SE10 47Hc 93
Pilsdon Cl. SW19 60Za 110
Piltdown Rd. WD19: Wat 21Z 45
Pilton Est., The CR0: C'don 75Rb 157
Pilton Gdns. SM4: Mord 72Db 155
Pilton Pl. SE17 7E 230 (50Sb 91)
Pimento Ct. W5 48Ma 87
PIMLICO 6E 4
PIMLICO 7B 228 (50Lb 90)
Pimlico Ho. SW1 7K 227 (50Kb 90)
(off Ebury Bri. Rd.)
Pimlico Rd. SW1 7H 227 (50Jb 90)
Pimlico Sq. SW1 7H 227 (50Jb 90)
Pimlico Station (Underground) 7D 228 (50Mb 90)
Pimpernel Way RM3: Rom 23Md 57
Pimp Hall Nature Reserve 19Fc 35
Pinchbeck Rd. BR6: Chels 79Vc 161
Pinchfield WD3: Map C 22F 42
Pinchin & Johnsons Yd. E1 45Wb 91
(off Pinchin St.)
Pinchin St. E1 45Wb 91
Pincombe Ho. SE17 7F 231 (50Tb 91)
(off Orb St.)
Pincott La. KT24: W Hor 100R 190
Pincott Pl. SE4 55Zb 114
Pincott Rd. DA6: Bex 57Cd 118
Pincott Rd. SW19 66Eb 133
Pincroft Wood DA3: Lfield 69Ee 143
Pindar St. EC2 7H 219 (43Ub 91)
PINDEN 68Yd 142
Pindock M. W9 42Db 89
Pindoria M. E1 6K 219 (42Vb 91)
(off Quaker St.)
Pineapple Ct. SW1 3B 228 (48Lb 90)
(off Castle La.)
Pine Av. BR4: W W'ck 74Dc 158
Pine Av. DA12: Grav'nd 10F 122
Pine Av. E15 36Fc 73
Pine Cl. BR8: Swan 70Hd 140
Pine Cl. CR8: Kenley 89Tb 177
Pine Cl. E10 33Dc 72
Pine Cl. EN8: Chesh 1Zb 20
Pine Cl. GU21: Wok 8N 167
Pine Cl. HA7: Stan 21Ka 46
Pine Cl. KT15: New H 83K 169
Pine Cl. KT19: Eps 81Sa 173
Pine Cl. N14 17Lb 32
Pine Cl. N19 33Lb 70
Pine Cl. SE20 67Yb 136
Pine Coombe CR0: C'don 77Zb 158
Pinecote Dr. SL5: S'dale 3D 146
Pine Ct. CM13: Gt War 23Xd 58
Pine Ct. KT13: Weyb 78S 150
Pine Ct. KT15: Add 77K 149
(off Church Rd.)
Pine Ct. N21 15Pb 32
Pine Ct. RM14: Upm 35Rd 77
Pine Ct. UB5: N'olt 42Aa 85
Pine Cres. CM13: Hut 14Fe 41
Pine Cres. SM5: Cars 83Fb 175
Pinecrest Gdns. BR6: Farnb 77Rc 160
Pine Cft. KT13: Weyb 79T 150
(off St George's Rd.)
Pinecroft CM13: Hut 17De 41
Pinecroft HP3: Hem H 6P 3
Pinecroft RM2: Rom 28Ld 57
Pinecroft Ct. DA16: Well 52Wc 117
Pinecroft Ct. HP3: Hem H 6P 3
Pinecroft Cres. EN5: Barn 14Ab 30
Pine Dean KT23: Bookh 97Da 191
Pinedene SE15 53Xb 113
Pinefield Cl. E14 45Cc 92
Pine Gdns. HA4: Ruis 32X 65
Pine Gdns. KT5: Surb 72Qa 153
Pinegate KT23: Bookh 97Da 191
Pine Glade BR6: Farnb 77Pc 160
Pine Gro. AL2: Brick W 2Ba 13
Pine Gro. AL9: Brk P 7K 9
Pine Gro. GU20: W'sham 9B 146
Pine Gro. KT13: Weyb 78R 150
Pine Gro. N20 18Bb 31
Pine Gro. N4 33Nb 70
Pine Gro. SW19 64Bb 133
Pine Gro. WD23: Bush 12Ba 27
Pine Gro. M. KT13: Weyb 78S 150
Pine Hill KT18: Eps 87Ta 173
Pine Ho. E3 39Ac 72
(off Barge La.)
Pine Ho. SE16 47Yb 92
(off Ainsty Est.)
Pine Ho. W10 42Ab 88
(off Droop St.)
Pinehurst GU22: Wok 90B 168
(off Park Dr.)
Pinehurst SL5: S'hill 1B 146
Pinehurst TN14: S'oaks 93Nd 203
Pinehurst TW20: Eng G 6N 125
Pinehurst Cl. KT20: Kgswd 94Cb 195
Pinehurst Cl. SS17: Stan H 3K 101
Pinehurst Ct. WD5: Ab L 4U 12
Pinehurst Ct. W11 44Bb 89
(off Colville Gdns.)
Pinehurst Gdns. KT14: W Byf 84L 169
Pinehurst Wlk. BR6: Orp 74Tc 160
Pinelands Cl. SE3 52Hc 115
Pinel Cl. GU25: Vir W 70A 126
Pine Lodge KT11: Cobh 87Y 171
(off Leigh Cnr.)
Pine Lodge Way KT19: Eps 83Qa 173
Pinemartin Cl. NW2 34Ya 68
Pine M. NW10 40Za 68
(off Clifford Gdns.)
Pine Needle La. HA6: Nwood 24V 44
Pineneedle La. TN13: S'oaks 95Kd 203
Pine Pl. SM7: Bans 86Za 174
Pine Pl. UB4: Hayes 42V 84
Piner Cotts. SL4: Wind 5C 102
Pine Ridge AL1: St A 5E 6
Pine Ridge SM5: Cars 80Jb 156
Pineridge Cl. KT13: Weyb 77U 150
Pineridge Ct. EN5: Barn 14Za 30
Pine Rd. GU22: Wok 2N 187
Pine Rd. N11 19Jb 32
Pine Rd. NW2 35Ya 68
Pines, The CR5: Coul 90Kb 176
Pines, The CR8: Purl 85Sb 177
Pines, The GU21: Wok 86B 168
Pines, The HP3: Hem H 6H 3
Pines, The IG8: Wfd G 20Hc 35
Pines, The KT9: Chess 76Na 153
Pines, The N14 15Lb 32
Pines, The RM16: Grays 46De 99
Pines, The SE19 65Rb 135
Pines, The SL2: Wex 3M 81
Pines, The SL3: L'ly 46B 82
Pines, The TN14: S'oaks 93Md 203
Pines, The TW16: Sun 69W 128
Pines, The WD6: Bore 12Pa 29
Pines Av. EN1: Enf 8Yb 20
Pines Cl. HA6: Nwood 23U 44
Pines Rd. BR1: Broml 68Nc 138
Pine St. EC1 5A 218 (42Qb 90)
Pine Tree Cl. HP2: Hem H 1M 3
Pine Tree Cl. TW2: Whitt 59Da 107
Pine Tree Cl. TW5: Cran 53X 107
Pinetree Gdns. HP3: Hem H 4N 3
Pine Tree Hill GU22: Pyr 88F 168
Pinetree Ho. WD25: Wat 8Aa 13
Pine Tree La. TN15: Ivy H 97Wd 204
Pine Tree Lodge BR2: Broml 70Hc 137
Pine Trees Bus. Pk. TW18: Staines 64G 126
Pine Trees Dr. UB10: Ick 35N 63
Pine Tree Way SE13 55Dc 114
Pine Vw. TN15: Plat 92Ee 205
Pineview Ct. E4 18Ec 34
Pine Vw. Mnr. CM16: Epp 2Wc 23
Pine Wlk. CR3: Cat'm 94Ub 197
Pine Wlk. KT11: Cobh 86Z 171
Pine Wlk. KT23: Bookh 97Da 191
Pine Wlk. KT24: E Hor 100V 190
Pine Wlk. KT5: Surb 72Qa 153
Pine Wlk. SM5: Cars 82Fb 175
Pine Wlk. SM7: Bans 89Hb 175
Pine Wlk. E. SM5: Cars 83Fb 175
Pine Wlk. W. SM5: Cars 82Fb 175
Pine Way TW20: Eng G 5M 125
Pine Wood TW16: Sun 67W 128
Pinewood Av. DA15: Sidc 60Uc 116
Pinewood Av. HA5: Hat E 23Da 45
Pinewood Av. KT15: New H 81L 169
Pinewood Av. RM13: Rain 42Kd 97
Pinewood Av. TN14: S'oaks 93Md 203
Pinewood Av. UB8: Hil 44P 83
Pinewood Cl. AL4: St A 2G 6
Pinewood Cl. BR6: Orp 74Tc 160
Pinewood Cl. CR0: C'don 76Ac 158
Pinewood Cl. GU21: Wok 87C 168
Pinewood Cl. HA5: Hat E 23Da 45
Pinewood Cl. HA6: Nwood 22W 44
Pinewood Cl. SL0: Iver H 38E 62
Pinewood Cl. SL9: Ger X 31A 62
Pinewood Cl. SS17: Linf 8K 101
Pinewood Cl. WD17: Wat 11W 26
Pinewood Cl. WD6: Bore 11Ta 29
Pinewood Ct. EN2: Enf 13Rb 33
Pinewood Ct. KT15: Add 77M 149
Pinewood Ct. SW4 58Mb 112
Pinewood Dr. BR6: Orp 78Uc 160
Pinewood Dr. EN6: Pot B 3Bb 17
Pinewood Dr. KT15: New H 82L 169
Pinewood Dr. TW18: Staines 64J 127
Pinewood Film Studios 38D 62
Pinewood Gdns. HP1: Hem H 2K 3
Pinewood Gdns. TW11: Tedd 64Ka 130
Pinewood Grn. SL0: Iver H 38E 62
Pinewood Gro. KT15: New H 82K 169
Pinewood Gro. W5 44La 86
Pinewood Lodge WD23: B Hea 18Fa 28
Pinewood M. TW19: Stanw 58M 105
Pinewood Pk. KT15: New H 83K 169
Pinewood Pl. DA2: Wilm 61Gd 140
(not continuous)
Pinewood Pl. KT19: Ewe 77Ta 153
Pinewood Rd. BR2: Broml 70Jc 137
Pinewood Rd. GU25: Vir W 10L 125
Pinewood Rd. RM4: Have B 21Ed 56
Pinewood Rd. SE2 51Zc 117
Pinewood Rd. SL0: Iver H 37D 62
Pinewood Rd. TW13: Felt 62X 129
Pinewood Way CM13: Hut 15Fe 41
Pinfold Rd. SW16 63Nb 134
Pinfold Rd. WD23: Bush 12Ba 27
Pinglestone Cl. UB7: Harm 52N 105
Pinkcoat Cl. TW13: Felt 62X 129
Pinkerton Pl. SW16 63Mb 134
Pinkham Mans. W4 50Qa 87
Pinkham Way N11 24Jb 50
Pinks Farm WD7: Shenl 3Sa 15
Pink's Hill BR8: Swan 71Gd 162
Pinkwell Av. UB3: Harl 49T 84
Pinkwell La. UB3: Harl 49S 84
Pinley Gdns. RM9: Dag 39Xc 75
Pinnace Ho. E14 48Ec 92
(off Manchester Rd.)
Pinnacle, The RM6: Chad H 30Ad 55
(off High Rd.)
Pinnacle, The TN13: S'oaks 96Jd 202
Pinnacle Apts. CR0: C'don 74Sb 157
(off Saffron Central Sq.)
Pinnacle Cl. N10 27Kb 50
Pinnacle Hill DA7: Bex 56Dd 118
Pinnacle Hill Nth. DA7: Bex 56Dd 118
Pinnacle Ho. BR5: St M Cry 70Yc 139
(off Ridge Pl.)
Pinnacle Ho. EN1: Enf 13Ub 33
(off Colman Pde.)
Pinnacle Ho. NW9 26Va 48
(off Heritage Av.)
Pinnacle Ho. SW18 56Eb 111
Pinnacle Pl. HA7: Stan 21Ka 46
Pinnacles EN9: Walt A 6Gc 21
Pinnacle Way E14 44Ac 92
(off Commercial Rd.)
Pinnata Cl. EN2: Enf 11Sb 33
Pinn Cl. UB8: Cowl 44M 83
Pinnell Rd. SE9 56Mc 115
PINNER 28Aa 45
Pinner Ct. HA5: Pinn 28Ca 45
Pinner Ct. NW8 5B 214 (42Fb 89)
(off St John's Wood Rd.)
PINNER GREEN 26Y 45
Pinner Grn. HA5: Pinn 26Y 45
Pinner Gro. HA5: Pinn 28Aa 45
Pinner Hill HA5: Pinn 24X 45
Pinner Hill Farm HA5: Pinn 25X 45
Pinner Hill Golf Course 23X 45
Pinner Hill Rd. HA5: Pinn 24X 45
Pinner Pk. HA5: Pinn 26Ca 45
Pinner Pk. 25Ca 45
Pinner Pk. Av. HA2: Harr 27Da 45
Pinner Pk. Gdns. HA2: Harr 26Ea 46
Pinner Rd. HA5: Pinn Nower Hill 28Ba 45
Pinner Rd. HA1: Harr 29Da 45
Pinner Rd. HA2: Harr 28Ca 45
Pinner Rd. HA6: Nwood 25V 44
Pinner Rd. HA6: Pinn 25V 44
Pinner Rd. WD19: Wat 16Z 27
Pinners Cl. SM5: Cars 75Gb 155
Pinner Station (Underground) 28Aa 45
Pinner Vw. HA1: Harr 28Ea 46
Pinner Vw. HA2: Harr 27Ea 46
PINNERWOOD PARK 25Y 45
Pinnocks Av. DA11: Grav'nd 10D 122
Pinn Way HA4: Ruis 31T 64
Pinson Way BR5: Orp 74Zc 161
Pinstone Way SL9: Ger X 33D 62
Pintail Cl. E6 43Nc 94
Pintail Cl. RM18: E Til 9K 101
Pintail Ct. SE8 51Bc 114
(off Pilot Cl.)
Pintail Rd. IG8: Wfd G 24Kc 53
Pintail Way UB4: Yead 43Z 85
Pinter Ho. SW9 54Nb 112
(off Grantham Rd.)
Pintle Pl. E3 41Ec 92
Pinto Cl. WD6: Bore 16Ta 29
Pinto Twr. SW8 52Nb 112
Pinto Way SE3 56Kc 115
Pioneer Cen., The SE15 53Yb 114
Pioneer Cl. E14 43Dc 92
Pioneer Ct. E16 43Jc 93
(off Hammersley Rd.)
Pioneer Ho. WC1 3H 217 (41Pb 90)
(off Britannia St.)
Pioneer Pl. CR0: Sels 81Cc 178
Pioneer Point IG1: Ilf 34Rc 74
Pioneers Ind. Pk. CR0: Bedd 74Nb 156
Pioneer St. SE15 53Wb 113
Pioneer Way BR8: Swan 69Gd 140
Pioneer Way W12 44Xa 88
Pioneer Way WD18: Wat 16V 26
Piper Bldg., The SW6 55Db 111
Piper Cl. N7 37Pb 70
Piper Rd. KT1: King T 69Qa 131
Pipers Cl. KT11: Cobh 87Z 171
Pipers Cl. SL1: Burn 1A 80
Piper's End GU25: Vir W 9P 125
Piper's Gdns. CR0: C'don 73Ac 158
Pipers Grn. NW9 29Sa 47
Pipers Grn. La. HA8: Edg 20Na 29
(not continuous)
Piper's Grn. Rd. TN16: B Char 100Wc 201
Pipers Ho. SE10 50Fc 93
(off Collington St.)
Pipers La. TN16: B Char 99Wc 201
Pipers Rd. CR5: Coul 88Mb 176
Piper Way IG1: Ilf 32Tc 74
Pipewell Rd. SM5: Cars 72Gb 155
Pipit Cl. RM18: E Til 10K 101
Pipit Dr. SW15 58Ya 110
Pipit Wlk. HP3: Hem H 6M 3
Pippenhall SE9 58Rc 116
Pippin Cl. CR0: C'don 74Bc 158
Pippin Cl. NW2 34Wa 68
Pippin Cl. SL2: Slou 3H 81
Pippin Cl. TW13: Hanw 62Ba 129
Pippin Cl. WD7: Shenl 5Na 15
Pippin Ct. SW8 51Pb 112
(off Vauxhall Gro.)
Pippin Ho. W10 45Za 88
(off Freston Rd.)
Pippin Mans. E20 36Ec 72
(off Mirabelle Gdns.)
Pippins, The SL3: L'ly 46B 82
Pippins, The WD25: Wat 6Y 13
Pippins Cl. UB7: W Dray 48M 83
Pippins Ct. TW15: Ashf 65R 128
Pique M. E1 45Zb 92
(off Glasshouse Flds.)
Piquet Rd. SE20 68Yb 136
Pirate Cove Adventure Pk. 59Wd 120
PIRBRIGHT 5D 186
PIRBRIGHT CAMP 3A 186
Pirbright Cres. CR0: New Ad 79Ec 158
Pirbright Grn. GU24: Pirb 5D 186
Pirbright Rd. SW18 60Bb 111
Pirbright Ter. GU24: Pirb 5D 186
Pirie Cl. SE5 55Tb 113
Pirie St. E16 46Kc 93
Pirin Ct. E4 21Cc 52
Pirrip Cl. DA12: Grav'nd 10H 123
Pirton Ct. HP2: Hem H 1P 3
Pisa Pl. WD18: Wat 13U 26
Pisces Ct. HA8: Edg 24Qa 47
Pissarro Ho. N1 38Qb 70
(off Augustas La.)
Pitcairn Cl. RM7: Mawney 28Cd 56
Pitcairn Ho. E9 38Yb 72
Pitcairn Rd. CR4: Mitc 66Hb 133
Pitcairn's Path HA2: Harr 34Ea 66
Pitcher Ct. DA11: Nflt 61Ee 143
Pitcher La. TW15: Ashf 63P 127
Pitchers Yd SL1: Burn 1A 80
Pitchfont La. RH8: Limp 97Hc 199
Pitchfont La. RH8: T'sey 97Hc 199
Pitchford St. E15 38Fc 73
PITCH PLACE 10L 187
Pitfield DA3: Hartl 70Be 143
Pitfield Cres. SE28 46Wc 95
Pitfield Est. N1 3H 219 (41Ub 91)
Pitfield St. N1 4H 219 (41Ub 91)
Pitfield Way EN3: Enf H 11Yb 34
Pitfield Way NW10 37Sa 67
Pitfold Cl. SE12 58Kc 115
Pitfold Rd. SE12 58Jc 115
Pitlake CR0: C'don 75Rb 157
Pitlochry Ho. SE27 61Rb 135
(off Elmcourt Rd.)
Pitman Bldg. SE16 48Wb 91
(off Old Jamaica Rd.)
Pitman Ho. SE8 53Cc 114
Pitman St. SE5 52Sb 113
Pitmaston Ho. SE13 54Ec 114
(off Lewisham Rd.)
Pitmaston Rd. SE13 54Ec 114
Pitsea Pl. E1 44Zb 92
Pitsea St. E1 44Zb 92
Pitshanger La. W5 42Ka 86
Pitshanger Manor House & Gallery 46Ma 87
Pitson Cl. KT15: Add 77M 149
Pitt Cres. SW19 63Db 133
Pitt Dr. AL4: St A 5G 6
Pittman Cl. CM13: Ingve 22Ee 59
Pittman Gdns. IG1: Ilf 36Sc 74
Pitt Pl. KT17: Eps 86Ua 174
Pitt Rd. BR6: Farnb 77Sc 160
Pitt Rd. CR0: C'don 71Sb 157
Pitt Rd. CR7: Thor H 71Sb 157
Pitt Rd. KT17: Eps 86Ua 174
(not continuous)
Pitt's Head M. W1 7J 221 (46Jb 90)
Pittsmead Av. BR2: Hayes 73Jc 159
Pitts Rd. SL1: Slou 6G 80
Pitt St. W8 47Cb 89
Pittville Gdns. SE25 69Wb 135
Pittwood CM15: Shenf 18Ce 41
Pitwell M. E8 37Wb 71
Pitwood Grn. KT20: Tad 92Ya 194
Pitwood Pk. Ind. Est. KT20: Tad 92Xa 194
Pix Farm La. HP1: Hem H 3D 2
Pixfield Ct. BR2: Broml 68Hc 137
(off Beckenham La.)
Pixies Hill Cres. HP1: Hem H 4H 3
(not continuous)
Pixies Hill Rd. HP1: Hem H 3H 3
Pixley M. GU24: Bisl 9E 166
Pixley St. E14 44Bc 92
Pixton Way CR0: Sels 81Ac 178
Place London, The 4E 216 (41Mb 90)
Place, The SE1 7G 225 (46Tb 91)
Place Farm Av. BR6: Orp 74Tc 160
Place Farm Rd. RH1: Blet 2K 209
Placehouse La. CR5: Coul 91Pb 196
Plackett Way SL1: Slou 6B 80
Plain, The CM16: Epp 1Xc 23
Plaines Cl. SL1: Slou 6D 80
Plain Ride SL4: Wind 10A 102
Plaisterers Highwalk EC2 1D 224 (43Sb 91)
(off London Wall)
PLAISTOW 66Jc 137
PLAISTOW 40Jc 73
Plaistow Cl. SS17: Stan H 1M 101
Plaistow Gro. BR1: Broml 66Kc 137
Plaistow Gro. E15 39Hc 73
Plaistow La. BR1: Broml 66Jc 137
Plaistow Pk. Rd. E13 40Kc 73
(not continuous)
Plaistow Rd. E13 40Jc 73
Plaistow Rd. E15 39Hc 73
Plaistow Station (Underground) 40Hc 73
Plaistow Wharf E16 47Jc 93
Plaitford Cl. WD3: Rick 19N 25
Plamer Ct. NW9 26Ua 48
Plane Av. DA11: Nflt 59Fe 121
Plane Ho. BR2: Broml 68Gc 137
Planes, The KT16: Chert 73L 149
Plane St. SE26 62Xb 135
Planet Ice Hemel Hempstead 3P 3
Planetree Ct. W6 49Za 88
(off Brook Grn.)
Plane Tree Cres. TW13: Felt 62X 129
Plane Tree Ho. SE8 51Ac 114
(off Etta St.)
Plane Tree Ho. W8 47Bb 89
(off Duchess of Bedford's Wlk.)
Planetree Path E17 28Cc 52
(off Selborne Rd.)
Plane Tree Wlk. N2 27Gb 49
Plane Tree Wlk. SE19 65Ub 135
Plantaganet Pl. EN9: Walt A 5Dc 20
Plantagenet Cl. KT4: Wor Pk 77Ta 153
Plantagenet Gdns. RM6: Chad H 31Zc 75
Plantagenet Ho. SE18 48Pc 94
(off Leda Rd.)
Plantagenet Pl. RM6: Chad H 31Zc 75
Plantagenet Rd. EN5: New Bar 14Eb 31
Plantain Gdns. E11 34Fc 73
(off Hollydown Way)
Plantain Pl. SE1 1F 231 (47Tb 91)
Plantation, The SE3 54Jc 115
Plantation Cl. DA9: Ghithe 58Vd 120
Plantation Cl. SW4 57Nb 112
Plantation Cl. WD23: Bush 15Z 27
Plantation Dr. BR5: Orp 74Zc 161
Plantation La. CR3: W'ham 91Ac 198
Plantation La. CR3: Wold 91Ac 198
Plantation La. CR6: W'ham 91Ac 198
Plantation La. EC3 4H 225 (45Ub 91)
(off Rood La.)
Plantation Pl. EC3 4H 225 (45Ub 91)
(off Mincing La.)
Plantation Rd. BR8: Hext 66Jd 140
Plantation Rd. DA8: Erith 53Jd 118
Plantation Wharf SW11 55Eb 111
Plants & People Exhibition 52Pa 109
Plasel Ct. E13 39Kc 73
(off Pawsey Cl.)
PLASHET 37Nc 74
Plashet Cl. SS17: Stan H 1M 101
Plashet Gdns. CM13: B'wood 21Ce 59
Plashet Gro. E6 39Lc 73
Plashet Rd. E13 39Jc 73
Plassy Rd. SE6 59Dc 114
Plate Ho. E14 50Dc 92
(off Burrells Wharf Sq.)
Platford Grn. RM11: Horn 28Nd 57
Platform Theatre 39Nb 70
Platina St. EC2 5G 219 (42Tb 91)
(off Tabernacle St.)
Platinum Ct. E1 42Yb 92
(off Cephas Av.)
Platinum Ct. RM7: Mawney 27Dd 56
Platinum M. N15 29Vb 51
Platinum Way TN15: Plat 92Ee 205
Plato Rd. SW2 56Nb 112
Platt, The SW15 55Za 110
Platt Comn. TN15: Plat 92Ee 205
Platt Ct. AL1: St A 2C 6
(off Hatfield Rd.)
Platt Halls NW9 26Va 48
Platt Ho. La. TN15: Fair 85Ee 185
Platt Ho. La. TN15: Wro 85Ee 185
Platt Ind. Est. TN15: Plat 91De 205
Platt Mill Cl. TN15: Plat 92De 205
Platt Mill Ter. TN15: Plat 92De 205
Platt's Eyot TW12: Hamp 68Ca 129
Platt's La. NW3 35Cb 69
Platts Rd. EN3: Enf H 11Yb 34
Platt St. NW1 1D 216 (40Mb 70)
Plawsfield Rd. BR3: Beck 67Zb 136
Plaxdale Grn. Rd. TN15: Stans 85Zd 185
Plaxdale Ho. SE17 6H 231 (49Ub 91)
(off Congreve St.)
PLAXTOL 99Ae 205
Plaxtol Cl. BR1: Broml 67Lc 137
Plaxtol La. TN15: Plax 99Yd 204
Plaxtol Rd. DA8: Erith 52Cd 118
PLAXTOL SPOUTE 99Ce 205
Plaxton Ct. E11 34Hc 73
Playfair Ho. E14 44Cc 92
(off Saracen St.)
Playfair Mans. W14 51Ab 110
(off Queen's Club Gdns.)
Playfair St. W6 50Ya 88
Playfield Av. RM5: Col R 25Ed 56
Playfield Cres. SE22 57Vb 113
Playfield Rd. HA8: Edg 26Sa 47
Playford Rd. N4 33Pb 70
(not continuous)
Playgreen Way SE6 62Cc 136
Playground Cl. BR3: Beck 68Zb 136
Playground Gdns. E2 4K 219 (41Vb 91)
(off Rochelle St.)
Playhouse Ct. SE1 7D 224 (46Sb 91)
(off Southwark Bri. Rd.)
Playhouse Theatre London 6G 223 (46Nb 90)
(off Northumberland Av.)
Playhouse Yd. EC4 3B 224 (44Rb 91)
Plaza DA9: Bluew 60Vd 120
Plaza Bus. Cen. EN3: Brim 12Bc 34
Plaza Gdns. SW15 57Ab 110
Plaza Hgts. E10 34Ec 72
Plaza Pde. HA0: Wemb 37Na 67
(off Ealing Rd.)
Plaza Pde. NW6 40Db 69
Plaza Wlk. NW9 27Sa 47
Pleasance, The SW15 56Xa 110
Pleasance Rd. BR5: St P 68Xc 139
Pleasance Rd. SW15 57Xa 110
Pleasance Theatre London 37Nb 70
(off Carpenters M.)
Pleasant Gro. CR0: C'don 76Bc 158
Pleasant Pl. KT12: Hers 79Y 151
Pleasant Pl. N1 38Rb 71
Pleasant Pl. WD3: W Hyd 24G 42
Pleasant Row NW1 39Kb 70
Pleasant Vw. DA8: Erith 50Gd 96
Pleasant Vw. Pl. BR6: Farnb 78Rc 160
Pleasant Way HA0: Wemb 40La 66
Pleasaunce Mans. SE10 50Jc 93
(off Halstow Rd.)
Pleasure Pit Rd. KT21: Asht 90Ra 173
Plender Ct. NW1 39Lb 70
(off College Pl.)
Plender St. NW1 1B 216 (39Lb 70)
Pleshey Rd. N7 35Mb 70
Plesman Way SM6: W'gton 81Nb 176
Plessey Bldg. E14 44Cc 92
(off Dod St.)
Plevna Cres. N15 30Ub 51
Plevna Rd. N9 20Wb 33
Plevna Rd. TW12: Hamp 67Da 129
Plevna St. E14 48Ec 92
Pleydell Av. SE19 66Vb 135
Pleydell Av. W6 49Va 88
Pleydell Ct. EC4 3A 224 (44Qb 90)
(off Pleydell St.)
Pleydell Est. EC1 4E 218 (41Sb 91)
(off Radnor St.)
Pleydell Gdns. SE19 65Vb 135
(off Anerley Hill)
Pleydell Ho. EC4 3A 224 (44Qb 90)
(off Pleydell St.)
Pleydell St. EC4 3A 224 (44Qb 90)
(off Bouverie St.)
Plimley Pl. W12 47Za 88
(off Shepherd's Bush Pl.)
Plimsoll Cl. E14 44Dc 92
Plimsoll Rd. N4 34Qb 70
Plough Cl. NW10 41Xa 88
Plough Ct. EC3 4G 225 (45Tb 91)
Plough Ct. RM13: Rain 40Fd 76
(off Broadis Way)
Plough Farm Cl. HA4: Ruis 30T 44
Plough Hill EN6: Cuff 1Nb 18
Plough Hill TN15: Bor G 95Be 205
Plough Ind. Est. KT22: Lea 92Ja 192
Plough La. CR8: Purl 81Nb 176
Plough La. KT11: Cobh 89W 170
Plough La. SE22 58Vb 113
Plough La. SL2: Stoke P 9M 61
Plough La. SM6: Bedd 77Nb 156
Plough La. SW17 63Eb 133
Plough La. SW19 64Db 133
Plough La. TW11: Tedd 64Ja 130
Plough La. UB9: Hare 23L 43
Plough La. WD3: Sarr 5H 11
Plough La. Cl. SM6: Bedd 78Nb 156
Ploughlees La. SL1: Slou 5J 81
Ploughmans Cl. NW1 39Mb 70
Ploughmans End TW7: Isle 57Fa 108
Ploughmans Wlk. N2 26Eb 49
(off Long La.)
Plough M. SW11 56Fb 111
Plough Pl. EC4 2A 224 (44Qb 90)
Plough Ri. RM14: Upm 31Ud 78
Plough Rd. KT19: Ewe 81Ta 173
Plough Rd. SW11 55Fb 111
Plough St. E1 44Vb 91
(off Buckle St.)
Plough Ter. SW11 56Fb 111
Plough Way SE16 49Zb 92
Plough Yd. EC2 6J 219 (42Ub 91)
Plover Cl. TW18: Staines 62H 127
Plover Gdns. RM14: Upm 32Vd 78
Plover Ho. SW9 52Qb 112
(off Brixton Rd.)
Plovers Ri. GU24: Brkwd 2C 186
Plover Way SE16 48Ac 92
Plover Way UB4: Yead 44Z 85
Plowden Bldgs. EC4 4K 223 (45Qb 90)
(off Middle Temple La.)
Plowman Cl. N18 22Tb 51
Plowman Way RM8: Dag 32Xc 75
Plumber's Row E1 43Wb 91
Plumbridge St. SE10 53Ec 114
Plum Cl. TW13: Felt 60W 106
Plume Ho. SE10 51Dc 114
(off Creek Rd.)
Plum Gth. TW8: Bford 49Ma 87
Plum La. RM13: Rain 40Fd 76
Plum La. SE18 52Rc 116
Plummer La. CR4: Mitc 68Hb 133
Plummer Rd. SW4 59Mb 112
Plummers Cft. TN13: Dun G 93Gd 202
Plumpton Av. RM12: Horn 35Nd 77
Plumpton Cl. UB5: N'olt 37Ca 65
Plumpton Way SM5: Cars 76Gb 155
PLUMSTEAD 49Uc 94
PLUMSTEAD COMMON 51Rc 116
Plumstead Comn. Rd. SE18 51Rc 116
Plumstead High St. SE18 49Tc 94
Plumstead Rd. SE18 49Rc 94
(not continuous)
Plumstead Station (Rail) 49Tc 94
Plumtree Cl. RM10: Dag 37Dd 76
Plumtree Cl. SM6: W'gton 80Mb 156
Plum Tree Ct. RM3: Rom 21Nd 57
Plumtree Ct. EC4 2B 224 (44Rb 91)
Plumtree Mead IG10: Lough 13Qc 36
Plum Tree M. SW16 65Nb 134
Pluto Cl. SL1: Slou 7C 80
Pluto Ri. HP2: Hem H 1N 3
Plymen Ho. KT8: W Mole 71Ca 151
Plymouth Ct. KT5: Surb 70Na 131
(off Cranes Pk. Av.)
Plymouth Dr. TN13: S'oaks 96Ld 203
Plymouth Ho. IG11: Bark 38Wc 75
(off Margaret Bondfield Av.)

Plymouth Ho. SE10 53Dc 114
(off Devonshire Dr.)
Plymouth Pk. TN13: S'oaks 96Ld 203
Plymouth Rd. BR1: Broml 67Kc 137
Plymouth Rd. E16 43Jc 93
Plymouth Rd. RM16: Chaf H 49Yd 98
Plymouth Rd. SL1: Slou 3C 80
Plymouth Ter. NW2 37Ya 68
(off Sidmouth Rd.)
Plymouth Wharf E14 49Fc 93
Plympton Av. NW6 38Bb 69
Plympton Cl. DA17: Belv 48Ad 95
Plympton Pl. NW8 6D 214 (42Gb 89)
Plympton Rd. NW6 38Bb 69
Plympton St. NW8 6D 214 (42Gb 89)
Plymstock Rd. DA16: Well 52Yc 117
Pocahontas Memorial 8D 122
(off Church St.)
Pocket Hill TN13: S'oaks 100Jd 202
Pocklington Cl. NW9 26Ua 48
Pocklington Cl. W12 48Wa 88
(off Ashchurch Pk. Vs.)
Pocklington Ct. SW15 60Wa 110
Pocklington Lodge W12 48Wa 88
Pocock Av. UB7: W Dray 48P 83
Pococks La. SL4: Eton 10J 81
Pocock St. SE1 1B 230 (47Rb 91)
Podmore Rd. SW18 56Eb 111
Poet Ct. E1 43Zb 92
(off Shandy St.)
Poets Chase HP1: Hem H 1K 3
Poets Ct. SE25 70Wb 135
Poets Ct. W3 46Sa 87
Poets Pl. RM3: Rom 24Kd 57
Poet's Rd. N5 36Tb 71
Poets Way HA1: Harr 28Ga 46
Point Borehamwood, The 13Qa 29
Point, The E17 28Cc 52
(off Tower M.)
Point, The HA4: Ruis 35W 64
Point, The W2 1B 220 (43Fb 89)
(off Nth. Wharf Rd.)
Pointalls Cl. N3 26Eb 49
Point Cl. SE10 53Ec 114
Pointer Cl. SE28 44Zc 95
Pointers, The KT21: Asht 92Na 193
Pointers Cl. E14 50Dc 92
Pointers Cotts. TW10: Ham 61La 130
POINTERS GREEN 90W 170
Pointers Rd. KT11: Cobh 88S 170
Point Hill SE10 52Ec 114
Point Pl. HA9: Wemb 38Ra 67
Point Pleasant SW18 56Cb 111
Point Ter. E7 36Kc 73
(off Claremont Rd.)
Point W. SW7 49Db 89
Point Wharf TW8: Bford 52Na 109
Point Wharf La. TW8: Bford 52Ma 109
Poirier Ho. CR8: Purl 84Pb 176
Poland St. W1 2C 222 (44Lb 90)
Polar Pk. UB7: Harm 52P 105
Poldo Ho. SE10 49Gc 93
(off Cable Wlk.)
Polebrook Rd. SE3 55Lc 115
Pole Cat All. BR2: Hayes 75Hc 159
Polecroft La. SE6 61Bc 136
Polehamptons, The TW12: Hamp 66Ea 130
Polehanger La. HP1: Hem H 1G 2
Pole Hill Rd. E4 17Ec 34
Pole Hill Rd. UB10: Hil 43R 84
Pole Hill Rd. UB4: Hayes 40S 64
Polesden Gdns. SW20 68Xa 132
Polesden La. GU23: Rip 95H 189
Polesden Rd. KT23: Bookh 100Da 191
Polesden Vw. KT23: Bookh 99Da 191
Poles Hill WD3: Sarr 5G 10
Polesteeple Hill TN16: Big H 89Mc 179
Polesworth Ho. W2 43Cb 89
(off Alfred Rd.)
Polesworth Rd. RM9: Dag 38Zc 75
Poley Rd. SS17: Stan H 2L 101
Polhill TN14: Hals 87Ed 182
Polhill Pk. TN14: Hals 88Fd 182
Police Sta. La. WD23: Bush 17Da 27
Police Sta. Rd. KT12: Hers 79Y 151
POLISH WAR MEMORIAL 38Y 65
Polka Theatre for Children 65Db 133
Pollard Av. UB9: Den 30H 43
Pollard Cl. E16 45Jc 93
Pollard Cl. IG7: Chig 22Wc 55
Pollard Cl. N7 35Pb 70
Pollard Cl. SL4: Old Win 7M 103
Pollard Ho. KT4: Wor Pk 77Ya 154
Pollard Ho. N1 2H 217 (40Pb 70)
(off Northdown St.)
Pollard Ho. SE16 48Vb 91
(off Spa Rd.)
Pollard Rd. GU22: Wok 88D 168
Pollard Rd. N20 19Gb 31
Pollard Rd. SM4: Mord 71Fb 155
Pollard Row E2 41Wb 91
Pollards WD3: Map C 22F 42
Pollards Cl. EN7: G Oak 1Sb 19
Pollards Cl. IG10: Lough 15Lc 35
Pollards Cres. SW16 69Nb 134
Pollards Hill E. SW16 69Pb 134
Pollards Hill Nth. SW16 69Nb 134
Pollards Hill Sth. SW16 69Nb 134
Pollards Hill W. SW16 69Pb 134
Pollards Oak Cres. RH8: Oxt 4L 211
Pollards Oak Rd. RH8: Oxt 4L 211
Pollard St. E2 41Wb 91
Pollards Wood Hill RH8: Oxt 2M 211
Pollards Wood Rd. RH8: Oxt 3M 211
Pollards Wood Rd. SW16 69Nb 134
Pollard Wlk. DA14: Sidc 65Yc 139
Pollen St. W1 3B 222 (44Lb 90)
Pollitt Dr. NW8 5B 214 (42Fb 89)
Pollock Ho. W10 42Ab 88
(off Kensal Rd.)
Pollock's Toy Mus. Pollocks 7C 216 (43Lb 90)
Pollyhaugh DA4: Eyns 76Nd 163
Polo Cen., The SL5: S'hill 9G 124
Polo M. BR7: Chst 64Tc 138
Polperro Cl. BR6: St M Cry 72Vc 161
Polperro Ho. W2 43Cb 89
(off Westbourne Pk. Rd.)
Polperro M. SE11 5B 230 (49Rb 91)
Polsted Rd. SE6 59Bc 114
Polsten M. EN3: Enf L 9Cc 20
Polthorne Est. SE18 49Tc 94
(off Polthorne Gro.)
Polthorne Gro. SE18 49Sc 94
Polworth Rd. SW16 64Nb 134
Polychrome Ct. SE1 2B 230 (47Rb 91)
(off Waterloo Rd.)
Polydamas Cl. E3 40Cc 72
Polygon, The NW8 39Fb 69
(off Avenue Rd.)
Polygon, The SW4 56Lb 112
Polygon Bus. Cen. SL3: Poyle 54H 105
Polygon Rd. NW1 2D 216 (40Mb 70)
(not continuous)
Polytechnic St. SE18 49Qc 94
Pomarium Cl. DA11: Grav'nd 4D 144
Pomell Way E1 2K 225 (44Vb 91)
Pomeroy Cl. TW1: Twick 56Ka 108
Pomeroy Cres. WD24: Wat 8X 13
Pomeroy Ho. E2 40Zb 72
(off St James's Av.)
Pomeroy Ho. W11 44Ab 88
(off Lancaster Rd.)
Pomeroy St. SE14 52Yb 114
Pomfret Pl. E14 45Ec 92
(off Bullivant St.)
Pomfret Rd. SE5 55Rb 113
Pomoja La. N19 33Nb 70
Pomona Ho. SE8 49Ac 92
(off Evelyn St.)
Pompadour Cl. CM14: W'ley 22Yd 58
Pompadour Way IG11: Bark 40Xc 75
Pond Cl. CR8: Kenley 88Rb 177
Pond Cl. KT12: Hers 79V 150
(not continuous)
Pond Cl. N12 23Gb 49
Pond Cl. SE3 54Jc 115
Pond Cl. UB9: Hare 26L 43
Pond Cott. La. BR4: W W'ck 74Cc 158
Pond Cotts. SE21 60Ub 113
PONDERS END 15Yb 34
Ponders End Ind. Est. EN3: Pond E 15Ac 34
Ponders End Station (Rail) 15Ac 34
Ponder St. N7 38Pb 70
Pond Farm Cl. KT20: Walt H 96Xa 194
Pond Farm Est. E5 34Yb 72
Pond Fld. End IG10: Lough 16Lc 35
Pondfield Ho. SE27 64Sb 135
Pondfield La. CM13: B'wood 21Ce 59
Pondfield La. DA12: Shorne 6N 145
Pondfield Rd. BR2: Hayes 74Gc 159
Pondfield Rd. BR6: Farnb 76Rc 160
Pondfield Rd. CR8: Kenley 88Rb 177
Pondfield Rd. RM10: Dag 36Dd 76
Pond Grn. HA4: Ruis 33U 64
Pond Hill Gdns. SM3: Cheam 79Ab 154
Pond Ho. HA7: Stan 23Ka 46
Pond Ho. KT16: Chert 73K 149
Pond Ho. SW3 6D 226 (49Gb 89)
Pond La. TN15: Ivy H 97Ud 204
Pond Lees Cl. RM10: Dag 38Fd 76
Pond Mead SE21 58Tb 113
Pond Path BR7: Chst 65Rc 138
Pond Piece KT22: Oxs 85Da 171
Pond Pl. KT21: Asht 89Na 173
Pond Pl. SW3 6D 226 (49Gb 89)
Pond Rd. E15 40Gc 73
Pond Rd. GU22: Wok 2L 187
Pond Rd. HP3: Hem H 7A 4
Pond Rd. SE3 54Hc 115
Pond Rd. TW20: Egh 65E 126
Ponds, The KT13: Weyb 79T 150
Pondside Av. KT4: Wor Pk 74Ya 154
Pondside Cl. UB3: Harl 51T 106
Pond Sq. N6 32Jb 70
Pond St. NW3 36Gb 69
Pond Wlk. RM14: Upm 33Ud 78
Pond Way TW11: Tedd 65La 130
Pondwicks Cl. AL1: St A 3A 6
Pondwood Ri. BR6: Orp 73Uc 160
Ponler St. E1 44Xb 91
Ponsard Rd. NW10 41Xa 88
Ponsford St. E9 37Yb 72
Ponsonby Ho. E2 40Yb 72
(off Bishop's Way)
Ponsonby Pl. SW1 7E 228 (50Mb 90)
Ponsonby Rd. SW15 59Xa 110
Ponsonby Ter. SW1 7E 228 (50Mb 90)
Ponsonby Vs. E2 40Yb 72
(off Lark Row)
Pontefract Ct. UB5: N'olt 36Da 65
(off Newmarket Av.)
Pontefract Rd. BR1: Broml 64Hc 137
Pontes Av. TW3: Houn 56Ba 107
Pontifex Apts. SE1 6F 225 (46Tb 91)
(off Stoney St.)
Pontoise Cl. TN13: S'oaks 94Hd 202
Ponton Rd. SW11 51Mb 112
Pontoon Dock Station (DLR) 46Lc 93
Pontoon Reach E16 46Mc 93
Pont St. SW1 4F 227 (48Hb 89)
Pont St. M. SW1 4F 227 (48Hb 89)
Pontypool Pl. SE1 1B 230 (47Rb 91)
Pontypool Wlk. RM3: Rom 23Ld 57
Pony Chase KT11: Cobh 85Ba 171
Pooja Ct. NW1 38Lb 70
(off Agar Gro.)
Pool Cl. BR3: Beck 64Cc 136
Pool Cl. KT8: W Mole 71Ba 151
Pool Ct. SE6 61Cc 136
Poole Cl. HA4: Ruis 33U 64
Poole Ct. N1 38Ub 71
(off St Peter's Way)
Poole Ct. TW4: Houn 54Aa 107
Poole Ct. Rd. TW4: Houn 54Aa 107
Poole Ho. RM16: Grays 7E 100
Poole Ho. SE11 4K 229 (48Qb 90)
(off Lambeth Wlk.)
Poole La. TW19: Stanw 59P 105
POOL END 71Q 150
Pool End Cl. TW17: Shep 71Q 150
Poole Rd. E9 37Zb 72
Poole Rd. GU21: Wok 90A 168
Poole Rd. KT19: Ewe 79Ta 153
Poole Rd. RM11: Horn 31Pd 77
Pooles Bldgs. WC1 6K 217 (42Qb 90)
(off Mt. Pleasant)
Pooles Cotts. TW10: Ham 61Ma 131
Pooles La. RM9: Dag 40Ad 75
Pooles La. SW10 52Eb 111
Pooles Pk. N4 33Qb 70
Poole St. N1 1F 219 (39Tb 71)
Poole Way UB4: Hayes 41U 84
Pooley Av. TW20: Egh 64D 126
Pooley Dr. SW14 55Sa 109
POOLEY GREEN 64E 126
Pooley Grn. Cl. TW20: Egh 64E 126
Pooley Grn. Rd. TW20: Egh 64D 126
Pooley Ho. E1 41Zb 92
Pooley Ho. E18 25Jc 53
Pooleys La. AL9: Wel G 5D 8
Pool Ho. NW8 7C 214 (43Fb 89)
(off Penfold St.)
Pool in the Pk. 91B 188
Pool La. SL1: Slou 5J 81
Poolmans Rd. SL4: Wind 5B 102
Poolmans St. SE16 47Zb 92
Pool Rd. HA1: Harr 31Fa 66
Pool Rd. KT8: W Mole 71Ba 151
Poolsford Rd. NW9 28Ua 48
Poolside Manor 25Bb 49
Pools on the Pk. 56Ma 109
Pool St. E20 38Ec 72
Poonah St. E1 44Yb 92
Pope Cl. SW19 65Fb 133
Pope Cl. TW14: Felt 60V 106
Pope Ct. CM14: W'ley 22Xd 58
Pope Ho. SE16 49Xb 91
(off Manor Est.)
Pope Ho. SE5 52Tb 113
(off Elmington Est.)
Pope Rd. BR2: Broml 71Mc 159
Popes Av. TW2: Twick 61Ga 130
Popes Cl. SL3: Coln 52D 104
Pope's Ct. WD5: Ab L 3U 12
Popes Ct. TW2: Twick 61Ga 130
Popes Dr. N3 25Cb 49
Popes Gro. CR0: C'don 76Bc 158
Popes Gro. TW1: Twick 61Ha 130
Popes Gro. TW2: Twick 61Ha 130
Pope's Head All. EC3 3G 225 (44Tb 91)
Popes La. RH8: Oxt 6J 211
Popes La. W5 48Ma 87
Popes La. WD24: Wat 9X 13
Pope's Rd. SW9 55Qb 112
Pope's Rd. WD5: Ab L 3U 12
Pope St. SE1 2J 231 (47Ub 91)
Popham Cl. TW13: Hanw 62Ba 129
Popham Gdns. TW9: Rich 55Qa 109
Popham Rd. N1 39Sb 71
Popham St. N1 39Rb 71
(not continuous)
Pop in Commercial Cen. HA9: Wemb 36Ra 67
Popinjays Row SM3: Cheam 78Za 154
(off Netley Cl.)
POPLAR 45Dc 92
Poplar Av. AL10: Hat 1P 7
Poplar Av. BR6: Farnb 75Rc 160
Poplar Av. CR4: Mitc 67Hb 133
Poplar Av. DA12: Grav'nd 3E 144
Poplar Av. KT22: Lea 94Ka 192
Poplar Av. UB2: S'hall 48Da 85
Poplar Av. UB7: Yiew 45P 83
Poplar Baths Leisure Cen. 45Dc 92
Poplar Bath St. E14 44Dc 92
Poplar Bus. Pk. E14 45Ec 92
Poplar Cl. E9 36Bc 72
Poplar Cl. HA5: Pinn 25Z 45
Poplar Cl. KT17: Eps D 87Xa 174
Poplar Cl. RM15: S Ock 42Zd 99
Poplar Cl. SL3: Poyle 53G 104
Poplar Ct. SW19 64Cb 133
Poplar Ct. TW1: Twick 58La 108
Poplar Ct. UB5: N'olt 40Y 65
Poplar Cres. KT19: Ewe 79Sa 153
Poplar Dr. CM13: Hut 16Ee 41
Poplar Dr. SM7: Bans 86Za 174
Poplar Farm Cl. KT19: Ewe 79Sa 153
Poplar Gdns. KT3: N Mald 68Ta 131
Poplar Gdns. GU22: Wok 91A 188
Poplar Gro. HA9: Wemb 34Sa 67
Poplar Gro. KT3: N Mald 68Ta 131
Poplar Gro. N11 23Jb 50
Poplar Gro. W6 47Ya 88
Poplar High St. E14 45Dc 92
Poplar Ho. KT19: Eps 81Ta 173
Poplar Ho. SE16 47Zb 92
(off Woodland Cres.)
Poplar Ho. SE4 56Bc 114
(off Wickham Rd.)
Poplar Ho. SL3: L'ly 50B 82
Poplar M. W12 46Ya 88
Poplar Mt. DA17: Belv 49Dd 96
Poplar Pl. SE28 45Yc 95
Poplar Pl. UB3: Hayes 45W 84
Poplar Pl. W2 45Db 89
Poplar Rd. KT10: Surb 76Ka 152
Poplar Rd. KT22: Lea 94Ka 192
Poplar Rd. SE24 56Sb 113
Poplar Rd. SM3: Sutt 74Bb 155
Poplar Rd. SW19 68Cb 133
Poplar Rd. TW15: Ashf 64S 128
Poplar Rd. UB9: Den 36L 63
Poplar Rd. Sth. SW19 69Cb 133
Poplar Row CM16: They B 9Uc 22
Poplars, The AL1: St A 6F 6
Poplars, The DA12: Grav'nd 9G 122
Poplars, The EN9: Walt A 7Lc 21
Poplars, The HP1: Hem H 3K 3
Poplars, The N14 15Kb 32
Poplars, The RM4: Abr 13Xc 37
Poplars, The UB9: Hare 25L 43
Poplars, The WD6: Bore 11Qa 29
Poplars Av. NW2 37Ya 68
Poplars Cl. AL10: Hat 1P 7
Poplars Cl. DA3: Lfield 69Ee 143
Poplars Cl. HA4: Ruis 32U 64
Poplars Cl. WD25: Wat 4X 13
Poplar Shaw EN9: Walt A 5Hc 21
Poplars Rd. E17 30Dc 52
Poplar Station (DLR) 45Dc 92
Poplar St. RM7: Rom 28Ed 56
Poplar Vw. HA9: Wemb 33Ma 67
Poplar Wlk. CR0: C'don 75Sb 157
Poplar Wlk. CR3: Cat'm 95Ub 197
Poplar Wlk. SE24 55Sb 113
(not continuous)
Poplar Way IG6: Ilf 28Sc 54
Poplar Way TW13: Felt 62W 128
Poppins Cl. KT12: Walt T 71Y 151
Poppins Ct. EC4 3B 224 (44Rb 91)
Poppleton Rd. E11 30Gc 53
Poppy Cl. CM15: Pil H 15Xd 40
Poppy Cl. DA17: Belv 48Dd 96
Poppy Cl. EN5: New Bar 16Eb 31
Poppy Cl. HP1: Hem H 1G 2
Poppy Cl. SL3: L'ly 9N 81
Poppy Cl. SM6: W'gton 74Jb 156
Poppy Cl. UB5: N'olt 37Ba 65
Poppy Cl. UB7: W Dray 48P 83
Poppy Dr. EN3: Pond E 14Xb 33
Poppy La. CR0: C'don 73Yb 158
Poppy M. SE22 57Wb 113
Poppy Pl. SE3 58Fc 115
Porchester Cl. DA3: Hartl 70Be 143
Porchester Cl. RM11: Horn 30Nd 57
Porchester Cl. SE5 56Sb 113
Porchester Ct. W2 45Db 89
(off Porchester Gdns.)
Porchester Gdns. W2 45Db 89
Porchester Gdns. M. W2 44Db 89
Porchester Ga. W2 45Db 89
(off Bayswater Rd.)
Porchester Ho. E1 44Xb 91
(off Philpot St.)
Porchester Leisure Cen. 44Db 89
Porchester Mead BR3: Beck 65Dc 136
Porchester Pl. W2 3E 220 (44Gb 89)
Porchester Rd. KT1: King T 68Ra 131
Porchester Rd. W2 44Db 89
Porchester Sq. W2 44Db 89
Porchester Sq. M. W2 44Db 89
Porchester Ter. W2 45Eb 89
Porchester Ter. Nth. W2 44Db 89
Porchfield Cl. DA12: Grav'nd 1E 144
Porchfield Cl. SM2: Sutt 82Db 175
Porch Way N20 20Hb 31
Porcupine Cl. SE9 61Nc 138
Porden Rd. SW2 56Pb 112
Porlock Av. HA2: Harr 32Ea 66
Porlock Ho. SE26 62Wb 135
Porlock Rd. EN1: Enf 17Vb 33
Porlock St. SE1 1G 231 (47Tb 91)
Porrington Cl. BR7: Chst 67Pc 138
Portal Cl. HA4: Ruis 35W 64
Portal Cl. SE27 62Qb 134
Portal Cl. UB10: Uxb 38N 63
Portal Way W3 43Ta 87
Port Av. DA9: Ghithe 58Xd 120
Portbury Cl. SE15 53Wb 113
Port Cres. E13 42Kc 93
Portcullis Ho. SW1 2F 229 (47Nb 90)
(off Bridge St.)
Portcullis Lodge Rd. EN1: Enf 13Tb 33
Port East Apts. E14 45Cc 92
(off Hertsmere Rd.)
Portelet Ct. N1 39Ub 71
(off De Beauvoir Est.)
Portelet Rd. E1 41Zb 92
Porten Ho's. W14 48Ab 88
(off Porten Rd.)
Porten Rd. W14 48Ab 88
Porter Cl. RM20: Grays 51Yd 120
Porter Rd. E6 44Pc 94
Porters & Walters Almshouses N22 24Pb 50
(off Nightingale Rd.)
Porters Av. RM8: Dag 37Xc 75
Porters Av. RM9: Dag 37Xc 75
Porters Cl. CM14: B'wood 18Wd 40
Porters Lodge, The SW10 52Eb 111
(off Coleridge Gdns.)
Porters M. RM9: Dag 37Xc 75
Porters Pk. Dr. WD7: Shenl 5Ma 15
Porters Pk. Golf Course 6La 14
Porter Sq. N19 32Nb 70
Porter St. SE1 6E 224 (46Sb 91)
Porter St. W1 7G 215 (43Hb 89)
Porters Wlk. E1 45Xb 91
Porters Way N12 23Gb 49
Porters Way UB7: W Dray 48P 83
Porteus Pl. SW4 55Lb 112
Porteus Rd. W2 7A 214 (43Eb 89)
Portfleet Pl. N1 39Ub 71
(off De Beauvoir Rd.)
Portgate Cl. W9 42Bb 89
Porthallow Cl. BR6: Chels 77Vc 161
Porthcawe Rd. SE26 63Ac 136
Port Hill BR6: Prat B 84Xc 181
Porthkerry Av. DA16: Well 56Wc 117
Port Ho. E14 50Dc 92
(off Burrells Wharf Sq.)
Portia Ct. IG11: Bark 38Wc 75
Portia Ct. SE11 7B 230 (50Rb 91)
(off Opal St.)
Portia Way E3 42Bc 92
Porticos, The SW3 51Fb 111
(off King's Rd.)
Portinscale Rd. SW15 57Ab 110
Portishead Ho. W2 43Cb 89
(off Westbourne Pk. Rd.)
Portland Av. DA12: Grav'nd 1D 144
Portland Av. DA15: Sidc 58Wc 117
Portland Av. KT3: N Mald 73Va 154
Portland Av. N16 31Vb 71
Portland Bus. Cen. SL3: Dat 3M 103
Portland Cl. KT4: Wor Pk 73Xa 154
Portland Cl. RM6: Chad H 29Ad 55
Portland Cl. SL2: Slou 2B 80
Portland Commercial Est. IG11: Bark 40Yc 75
Portland Cotts. CR0: Bedd 73Mb 156
Portland Ct. N1 38Ub 71
(off St Peter's Way)
Portland Ct. SE1 3F 231 (48Tb 91)
(off Gt. Dover St.)
Portland Ct. SE14 51Ac 114
(off Whitcher Cl.)
Portland Cres. HA7: Stan 26Ma 47
Portland Cres. SE9 61Nc 138
Portland Cres. TW13: Felt 63T 128
Portland Cres. UB6: G'frd 42Da 85
Portland Dr. EN2: Enf 10Ub 19
Portland Dr. EN7: Chesh 3Wb 19
Portland Dr. RH1: Mers 1D 208
Portland Gdns. N4 30Rb 51
Portland Gdns. RM6: Chad H 29Zc 55
Portland Gro. SW8 53Pb 112
Portland Hgts. HA6: Nwood 21V 44
PORTLAND HOSPITAL FOR WOMEN & CHILDREN 6A 216 (42Kb 90)
Portland Ho. RH1: Mers 1D 208
Portland Ho. SL9: Ger X 29B 42
Portland Ho. SW1 4B 228 (48Lb 90)
(off Bressenden Pl.)
Portland Ho. SW15 57Za 110
Portland Ho. M. KT18: Eps 86Ta 173
(off Caithness Dr.)
Portland Mans. W14 49Bb 89
(off Addison Bri. Pl.)
Portland M. W1 3C 222 (44Lb 90)
Portland Pk. SL9: Ger X 30A 42
Portland Pl. DA3: Lfield 69Ae 143
(off Park Dr.)
Portland Pl. DA9: Ghithe 56Yd 120
Portland Pl. KT17: Eps 84Ua 174
Portland Pl. SE25 70Wb 135
(off Sth. Norwood Hill)
Portland Pl. W1 7K 215 (43Kb 90)
Portland Ri. N4 32Rb 71
(not continuous)
Portland Rd. BR1: Broml 63Lc 137
Portland Rd. CR4: Mitc 68Gb 133
Portland Rd. DA11: Nflt 58Fe 121
Portland Rd. DA12: Grav'nd 10D 122
Portland Rd. KT1: King T 69Na 131
Portland Rd. N15 28Vb 51
Portland Rd. SE25 70Wb 135
Portland Rd. SE9 61Nc 138
Portland Rd. TW15: Ashf 62N 127
Portland Rd. UB2: S'hall 48Ba 85
Portland Rd. UB4: Hayes 41U 84
Portland Rd. W11 45Ab 88
Portlands SL9: Ger X 1P 61
Portland Sq. E1 46Xb 91
Portland St. AL3: St A 2A 6
Portland St. SE17 7F 231 (50Tb 91)
Portland Ter. HA8: Edg 24Qa 47
Portland Ter. TW9: Rich 56Ma 109
Portland Wlk. SE17 51Tb 113
Portley La. CR3: Cat'm 93Ub 197
Portley Wood Rd. CR3: Whyt 92Vb 197
Portmadoc Ho. RM3: Rom 21Nd 57
(off Broseley Rd.)
Portman Av. SW14 55Ta 109
Portman Cl. DA5: Bexl 60Gd 118
Portman Cl. DA7: Bex 55Ad 117
Portman Cl. W1 2G 221 (44Hb 89)
Portman Dr. IG8: Wfd G 26Mc 53
Portman Gdns. NW9 26Ta 47
Portman Gdns. UB10: Hil 38Q 64
Portman Ga. NW1 6E 214 (42Gb 89)
Portman Mans. W1 7G 215 (43Hb 89)
(off Chiltern St.)
Portman M. Sth. W1 3H 221 (44Jb 90)
Portman Pl. E2 41Yb 92
Portman Rd. KT1: King T 68Pa 131
Portman Sq. W1 2H 221 (44Jb 90)
Portman St. W1 3H 221 (44Jb 90)
Portman Towers W1 2G 221 (44Hb 89)
Portmeadow Wlk. SE2 47Zc 95
Portmeers Cl. E17 30Bc 52
Portmore Gdns. RM5: Col R 22Cd 56
Portmore Pk. Rd. KT13: Weyb 77Q 150
Portmore Pl. KT13: Weyb 76T 150
(off Oatlands Dr.)
Portmore Quays KT13: Weyb 77P 149
Portmore Way KT13: Weyb 76Q 150
Portnall Dr. GU25: Vir W 10K 125
Portnall Ho. W9 41Bb 89
(off Portnall Rd.)
Portnall Ri. GU25: Vir W 1K 147
Portnall Rd. GU25: Vir W 1K 147
Portnall Rd. W9 40Bb 69
Portnalls Cl. CR5: Coul 88Kb 176
Portnalls Ri. CR5: Coul 88Lb 176
Portnalls Rd. CR5: Coul 90Kb 176
Portnoi Cl. RM1: Rom 26Fd 56
Portobello Ct. W11 44Bb 89
Portobello Grn. W10 43Ab 88
(off Portobello Rd.)
Portobello Ho. BR2: Broml 72Mc 159
Portobello Lofts W10 42Ab 88
(off Kensal Rd.)
Portobello Mkt. 44Bb 89
Portobello M. W11 45Cb 89
Portobello Pde. TN15: W King 81Wd 184
Portobello Rd. W10 43Ab 88
Portobello Rd. W11 44Bb 89
Portobello Road Market 45Bb 89
(off Portobello Rd.)
Porton Ct. KT6: Surb 72La 152
Portpool La. EC1 7K 217 (43Qb 90)
Portree Cl. N22 24Pb 50
Portree St. E14 44Fc 93
Portrush Ct. UB1: S'hall 44Ea 86
(off Whitecote Rd.)
Portsdown HA8: Edg 22Qa 47
Portsdown Av. NW11 30Bb 49
Portsdown M. NW11 30Bb 49
Portsea Hall W2 3F 221 (44Hb 89)
(off Portsea Pl.)
Portsea M. W2 3E 220 (44Gb 89)
(off Portsea Pl.)
Portsea Pl. W2 3E 220 (44Gb 89)
Portsea Rd. RM18: Tilb 3E 122
Portslade Rd. SW8 54Lb 112
Portsmouth Av. KT7: T Ditt 73Ja 152
Portsmouth Ct. SL1: Slou 5J 81
Portsmouth M. E16 46Kc 93
Portsmouth Rd. KT10: Esh Hawkshill Cl. 79Ca 151
Portsmouth Rd. KT10: Esh Sandown Rd. 77Ea 152
Portsmouth Rd. KT10: Esh Seven Hills Rd. Sth. 83Aa 171
Portsmouth Rd. GU23: Rip 93M 189
Portsmouth Rd. GU23: Rip 97H 189
Portsmouth Rd. GU23: Send 97H 189
Portsmouth Rd. GU23: Wis 93M 189
Portsmouth Rd. KT1: King T 70Ma 131
Portsmouth Rd. KT11: Cobh 85V 170
Portsmouth Rd. KT6: Surb 72Ka 152
Portsmouth Rd. KT7: T Ditt 73Ja 152
Portsmouth Rd. SW15 59Xa 110
Portsmouth St. WC2 3H 223 (44Pb 90)
Portsoken Pav. EC3 3K 225 (44Vb 91)
(off Aldgate Sq.)
Portsoken St. E1 4K 225 (45Vb 91)
Portswood Pl. SW15 58Va 110
Portugal Gdns. TW2: Twick 61Ea 130
Portugal Rd. GU21: Wok 88B 168
Portugal St. WC2 3H 223 (44Pb 90)
Port Way GU24: Bisl 8E 166
Portway E15 39Hc 73
Portway KT17: Ewe 81Wa 174
Portway RM13: Rain 39Jd 76
Portway Cres. KT17: Ewe 81Wa 174
Portway Gdns. SE18 52Mc 115
Pory Ho. SE11 6J 229 (49Pb 90)
Poseidon Ct. E14 49Cc 92
(off Homer Dr.)
Postal Cl. DA5: Bexl 59Dd 118
Postal Mus., The 5J 217 (42Pb 90)
Post Boys Row KT11: Cobh 86W 170
Postern, The EC2 1E 224 (43Sb 91)
(off Wood St.)
Postern Grn. EN2: Enf 12Qb 32
Postgate Way DA11: Nflt 61Ee 143
Post Ho. La. KT23: Bookh 97Ca 191
Post La. TW2: Twick 60Fa 108
Postmark Development EC1 5J 217 (42Pb 90)
Postmasters Lodge HA5: Pinn 31Aa 65
Post Mdw. SL0: Iver H 41F 82
Postmill Cl. CR0: C'don 76Yb 158
Postmill Cl. RM5: Col R 23Cd 56
Post Office All. W4 51Ra 109
(off Thames Rd.)
Post Office App. E7 36Kc 73
Post Office Ct. EC3 3G 225 (44Tb 91)
(off King William St.)
Post Office La. SL3: Geor G 4P 81
Post Office M. GU20: W'sham 9B 146
Post Office Way SW11 52Mb 112
Post Rd. UB2: S'hall 48Da 85
Postway M. IG1: Ilf 34Rc 74
(not continuous)
Potager Pl. CR0: Bedd 76Mb 156
Potash La. TN15: Plat 95Ee 205
Potier St. SE1 4G 231 (48Tb 91)
Potiphar Pl. CM14: W'ley 21Xd 58
Potter Cl. CR4: Mitc 68Kb 134
Potter Cl. SE15 52Ub 113
Potter Cl. SE2 50Wc 95
Potterells AL9: N Mym 7F 8
Potter Ho. E1 43Zb 92
(off Beaufort Gdns.)
Potteries, The EN5: Barn 15Cb 31

Potteries, The KT16: Ott 79G 148
Potterne Cl. SW19 59Za 110
POTTERS BAR 4Bb 17
POTTERS BAR COMMUNITY HOSPITAL 6Eb 17
Potters Bar Mus. 4Bb 17
Potters Bar Station (Rail) 4Bb 17
Potters Bar Station Yd. EN6: Pot B 4Cb 17
(off Darkes La.)
Potters Cl. CR0: C'don 74Ac 158
Potters Cl. IG10: Lough 12Nc 36
Potters Ct. EN6: Pot B 4Cb 17
Potters Ct. SM1: Sutt 79Bb 155
(off Rosebery Rd.)
Potters Cross SL0: Iver H 41G 82
POTTERS CROUCH 6K 5
Potterscrouch La. AL2: Pot C 6K 5
Potterscrouch La. AL3: St A 5M 5
Potters End HA5: Pinn 23X 45
Pottersfield EN1: Enf 14Ub 33
(off Lincoln Rd.)
Potters Flds. SE1 7J 225 (46Ub 91)
Potters Gro. KT3: N Mald 70Sa 131
Potters Hgts. Cl. HA5: Pinn 24X 45
Potter's La. SW16 65Mb 134
Potters La. EN5: New Bar 14Cb 31
Potters La. GU23: Send 95D 188
Potters La. WD6: Bore 11Sa 29
Potters Lodge E14 50Ec 92
(off Ferry St.)
Potters M. WD6: E'tree 16Ma 29
Potter's Rd. EN5: New Bar 14Db 31
Potters Rd. SW6 54Eb 111
Potters Rd. UB2: S'hall 48Ca 85
Potters Row E20 36Dc 72
(off Keirin Rd.)
Potter St. HA5: Pinn 25X 45
Potter St. HA6: Nwood 25W 44
Potter St. Hill HA5: Pinn 23X 45
Potters Way RH2: Reig 10L 207
Pottery Café 53Bb 111
(off Fulham Rd.)
Pottery Cl. SE25 69Wb 135
Pottery Ga. N11 23Mb 50
Pottery La. W11 46Ab 88
Pottery M. SW6 54Bb 111
Pottery Rd. DA5: Bexl 61Ed 140
Pottery Rd. TW8: Bford 51Na 109
Pottery St. SE16 47Xb 91
Pott St. E2 41Xb 91
POUCHEN END 3F 2
Poulcott TW19: Wray 58A 104
Poulett Gdns. TW1: Twick 60Ja 108
Poulett Rd. E6 40Pc 74
Poulter Ct. DA10: Swans 59Be 121
Poulter Pk. 72Gb 155
Poulters Wood BR2: Kes 78Mc 159
Poultney Cl. WD7: Shenl 4Pa 15
Poulton Av. SM1: Sutt 76Fb 155
Poulton Cl. E8 37Xb 71
Poulton Ho. W3 43Ta 87
(off Victoria Rd.)
Poultry EC2 3F 225 (44Tb 91)
Pound, The SL1: Burn 2B 80
Pound Bank Cl. TN15: W King 81Vd 184
Pound Cl. BR6: Orp 75Tc 160
Pound Cl. KT19: Eps 83Ta 173
Pound Cl. KT6: Surb 74La 152
Pound Ct. KT21: Asht 90Pa 173
Pound Ct. Dr. BR6: Orp 75Tc 160
Pound Cres. KT22: Fet 93Fa 192
Pound Farm Cl. KT10: Esh 74Fa 152
Poundfield WD25: Wat 7V 12
Poundfield Ct. GU22: Wok 93E 188
Poundfield Gdns. GU22: Wok 92E 188
(not continuous)
Poundfield Rd. IG10: Lough 15Qc 36
Pound Grn. DA5: Bexl 59Cd 118
Pound La. GU20: W'sham 9A 146
Pound La. KT19: Eps 84Sa 173
Pound La. NW10 37Wa 68
Pound La. RM16: Ors 2C 100
Pound La. TN13: S'oaks 96Ld 203
Pound La. TN14: Knock 87Zc 181
Pound La. WD7: Shenl 5Pa 15
Pound Pk. Rd. SE7 49Mc 93
Pound Path E3 40Zb 72
(off Stoneway Wlk.)
Pound Pl. SE9 58Qc 116
Pound Rd. KT16: Chert 73K 149
Pound Rd. SM7: Bans 89Bb 175
Pound St. SM5: Cars 78Hb 155
Pound Way BR7: Chst 66Sc 138
Pounsley Rd. TN13: Dun G 93Gd 202
Pountney Rd. SW11 55Jb 112
POVEREST 70Wc 139
Poverest Rd. BR5: St M Cry 71Vc 161
Povey Ho. SE17 6H 231 (49Ub 91)
(off Beckway St.)
Powderham Ct. GU21: Knap 10H 167
Powder Mill La. DA1: Dart 61Nd 141
Powder Mill La. TW2: Whitt 59Ba 107
Powdermill La. EN9: Walt A 5Dc 20
Powdermill M. EN9: Walt A 5Dc 20
(off Powdermill La.)
Powell Av. DA2: Dart 61Ud 142
Powell Cl. HA8: Edg 23Pa 47
Powell Cl. KT9: Chess 78Ma 153
Powell Ct. CR2: S Croy 77Rb 157
(off Bramley Hill)
Powell Ct. E17 27Dc 52
Powell Dr. E4 14Dc 34
Powell Gdns. RH1: Redh 4B 208
Powell Gdns. RM10: Dag 35Cd 76
Powell Ho. EN1: Enf 14Ub 33
(off Dunstan M.)
Powell Ho. W2 4B 220 (45Fb 89)
(off Gloucester Ter.)
Powell Rd. E5 34Xb 71
Powell Rd. IG9: Buck H 17Lc 35
Powell's Wlk. W4 51Ua 110
Power Dr. EN3: Enf L 8Bc 20
Powergate Bus. Pk. NW10 41Ta 87
Powerhouse, The 47U 84
Powerhouse La. UB3: Hayes 47U 84
Powerleague Battersea 52Lb 112
Powerleague Colney Hatch 24Jb 50
Powerleague Croydon 79Pb 156
Powerleague Ilford 24Vc 55
Powerleague Mill Hill 24Xa 48
Powerleague Newham 41Sc 94
Powerleague Slough 8H 81
Powerleague Tottenham 23Xb 51
Powerleague Watford 13Ca 27
Powerleague Wembley 35Qa 67
Power Rd. W4 49Qa 87
Powers Ct. TW1: Twick 59Ma 109
Powerscroft Rd. DA14: Sidc 65Yc 139
(not continuous)
Powerscroft Rd. E5 35Yb 72
Power Works Est. DA8: Erith 53Jd 118
Powis Ct. EN6: Pot B 6Eb 17
Powis Ct. W11 44Bb 89
(off Powis Gdns.)
Powis Ct. WD23: B Hea 18Fa 28
(off Rutherford Way)
Powis Gdns. NW11 31Bb 69
Powis Gdns. W11 44Bb 89
Powis Ho. WC2 2G 223 (44Nb 90)
(off Macklin St.)
Powis M. W11 44Bb 89
Powis Pl. WC1 6G 217 (42Nb 90)
Powis Rd. E3 41Dc 92
Powis Sq. W11 44Bb 89
(not continuous)
Powis St. SE18 48Qc 94
Powis Ter. W11 44Bb 89
Powlesland Ct. E1 44Ac 92
(off White Horse Rd.)
Powlett Ho. NW1 37Kb 70
(off Powlett Pl.)
Powlett Pl. NW1 38Jb 70
Pownall Gdns. TW3: Houn 56Da 107
Pownall Rd. E8 39Vb 71
Pownall Rd. TW3: Houn 56Da 107
Pownell Ter. SE11 51Qb 112
Pownsett Ter. IG1: Ilf 36Sc 74
Powster Rd. BR1: Broml 64Jc 137
Powys Cl. DA7: Bex 51Zc 117
Powys Ct. N11 22Nb 50
Powys Ct. WD6: Bore 13Ta 29
Powys La. N13 21Nb 50
Powys La. N14 21Nb 50
POYLE 53G 104
Poyle Ind. Est. SL3: Poyle 55H 105
Poyle New Cotts. SL3: Poyle 54H 105
Poyle Pk. SL3: Poyle 55G 104
Poyle Rd. SL3: Poyle 55G 104
Poyle Technical Cen. SL3: Poyle 54G 104
Poyle Trad. Est. SL3: Poyle 55G 104
Poynder Rd. RM18: Tilb 3D 122
Poynders Ct. SW4 58Lb 112
Poynders Gdns. SW4 59Lb 112
Poynders Hill HP2: Hem H 3C 4
Poynders Pde. SW4 58Mb 112
Poynders Rd. SW4 58Lb 112
Poynings, The SL0: Rich P 49H 83
Poynings Cl. BR6: Chels 75Yc 161
Poynings Rd. N19 34Lb 70
Poynings Way N12 22Db 49
Poynings Way RM3: Hrld W 25Nd 57
Poyntell Cres. BR7: Chst 67Tc 138
Poynter Ct. UB5: N'olt 40Z 65
(off Gallery Gdns.)
Poynter Ho. NW8 5B 214 (42Fb 89)
(off Fisherton St.)
Poynter Ho. W11 46Za 88
(off Queensdale Cres.)
Poynter Rd. EN1: Enf 15Wb 33
Poynton Rd. N17 26Wb 51
Poyntz Rd. SW11 54Hb 111
Poyser St. E2 40Xb 71
Prado Path TW1: Twick 60Ha 108
(off Laurel Av.)
Prae Cl. AL3: St A 1P 5
Praed M. W2 2C 220 (44Fb 89)
Praed St. W2 3B 220 (44Fb 89)
Praetorian Ct. AL1: St A 5A 6
Praetorian Pl. RH1: Redh 3B 208
Pragel St. E13 40Lc 73
Pragnell Rd. SE12 61Kc 137
Prague Pl. SW2 57Nb 112
Prah Rd. N4 33Qb 70
Prairie Bldg. E15 36Fc 73
(off Property Row)
Prairie Cl. KT15: Add 76K 149
Prairie Rd. KT15: Add 76K 149
Prairie St. SW8 54Jb 112
Praline Ct. E3 40Dc 72
(off Taylor Pl.)
Pratchett Ct. WD18: Wat 14U 26
(off Raven Cl.)
Pratt M. NW1 39Lb 70
PRATT'S BOTTOM 81Yc 181
PRATT'S BOTTOM 82Yc 181
Pratts La. KT12: Hers 77Z 151
Pratts Pas. KT1: King T 68Na 131
Pratt St. NW1 39Lb 70
Pratt Wlk. SE11 5J 229 (49Pb 90)
Prayle Gro. NW2 32Za 68
Preachers Ct. EC1 6C 218 (42Rb 91)
(off Charterhouse Sq.)
Prebend Gdns. W4 49Va 88
Prebend Gdns. W6 49Va 88
(not continuous)
Prebend Mans. W4 49Va 88
(off Chiswick High Rd.)
Prebend St. N1 39Sb 71
Precinct, The N1 1D 218 (39Sb 71)
Precinct, The SS17: Stan H 2M 101
Precinct, The TW20: Egh 64C 126
Precinct Rd. UB3: Hayes 45W 84
Precincts, The SL1: Burn 2A 80
Precincts, The SM4: Mord 72Cb 155
Precista Ct. BR6: Orp 73Xc 161
Premier Av. RM16: Grays 47Ee 99
Premier Cnr. W9 40Bb 69
Premier Ct. EN3: Enf W 10Zb 20
Premiere Pl. E14 45Cc 92
Premier Ho. N1 38Rb 71
(off Waterloo Ter.)
Premier Pk. NW10 39Ra 67
(not continuous)
Premier Pk. Rd. NW10 40Ra 67
Premier Pl. SW15 56Ab 110
Premier Pl. WD18: Wat 15V 26
Prendergast Rd. SE3 55Gc 115
Prentice Ct. SW19 64Bb 133
Prentis Rd. SW16 63Mb 134
Prentiss Ct. SE7 49Mc 93
Presburg Rd. KT3: N Mald 71Ua 154
Presburg St. E5 34Zb 72
Presbury Ct. GU21: Wok 10L 167
Presbury Rd. HA8: Edg 25Pa 47
Prescot St. E1 4K 225 (45Vb 91)
Prescott Av. BR5: Pet W 72Rc 160
Prescott Cl. SW16 66Nb 134
Prescott Grn. IG10: Lough 13Sc 36
Prescott Ho. SE17 51Rb 113
(off Hillingdon St.)
Prescott Pl. SW4 55Mb 112
Prescott Rd. SL3: Poyle 54G 104
Presentation M. SW2 61Pb 134
Preshaw Cres. CR4: Mitc 69Gb 133
President Dr. E1 46Xb 91
President Ho. EC1 4C 218 (41Rb 91)
President Quay E1 46Vb 91
(off St Katherine's Way)
President St. EC1 3D 218 (41Sb 91)
(off Central St.)
Prespa Cl. N9 19Yb 34
Press Ct. SE1 50Wb 91
Press Ho. BR5: Pet W 71Sc 160
Press Ho. E1 43Zb 92
(off Trafalgar Gdns.)
Press Ho. NW10 34Ta 67
Pressing La. UB3: Hayes 48U 84
Press Rd. NW10 34Ta 67
Press Rd. UB8: Uxb 37M 63
Prestage Way E14 45Ec 92
Prestbury Cres. SM7: Bans 88Hb 175
Prestbury Rd. E7 38Lc 73
Prestbury Sq. SE9 63Pc 138
Prested Rd. SW11 56Gb 111
Prestige Ho. N20 18Eb 31
(off Acton Wlk.)
Prestige Way NW4 29Ya 48
PRESTON 32Na 67
Preston Av. E4 23Fc 53
Preston Cl. SE1 5H 231 (49Ub 91)
Preston Cl. TW2: Twick 62Ga 130
Preston Ct. DA14: Sidc 63Vc 139
(off The Crescent)
Preston Ct. EN5: New Bar 14Eb 31
Preston Ct. KT12: Walt T 74Y 151
Preston Dr. DA7: Bex 53Zc 117
Preston Dr. E11 29Lc 53
Preston Dr. KT19: Ewe 79Ua 154
Preston Gdns. EN3: Enf L 9Ac 20
Preston Gdns. IG1: Ilf 30Nc 54
Preston Gdns. NW10 37Va 68
Preston Gro. KT21: Asht 89La 172
Preston Hill HA3: Kenton 31Na 67
Preston Ho. SE1: Preston Cl. 5H 231 (49Ub 91)
(off Preston Cl.)
Preston Ho. SE1: St Saviour's Est. 3K 231 (48Vb 91)
(off St Saviour's Est.)
Preston Ho. RM10: Dag 34Cd 76
(off Uvedale Rd.)
Preston La. KT20: Tad 93Xa 194
Preston Mnr. Rd. KT20: Tad 92Ya 194
Preston Pl. NW2 37Wa 68
Preston Pl. TW10: Rich 57Na 109
Preston Rd. DA11: Nflt 10A 122
Preston Rd. E11 30Gc 53
Preston Rd. HA3: Kenton 32Na 67
Preston Rd. HA9: Wemb 32Na 67
Preston Rd. RM3: Rom 21Md 57
Preston Rd. SE19 65Rb 135
Preston Rd. SL2: Slou 5N 81
Preston Rd. SW20 66Va 132
Preston Rd. TW17: Shep 71Q 150
Preston Road Station (Underground) 32Na 67
Preston's Rd. E14 45Ec 92
Prestons Rd. BR2: Hayes 76Jc 159
Preston St. E2 40Zb 72
Preston Waye HA3: Kenton 32Na 67
Prestwich Ter. SW4 57Lb 112
Prestwick Cl. UB2: S'hall 50Aa 85
Prestwick Ct. UB1: S'hall 45Ea 86
(off Baird Av.)
Prestwick Rd. WD19: Wat 18Y 27
Prestwick Road Meadows Local Nature Reserve 22Z 45
Prestwood SL2: Slou 4M 81
Prestwood WD19: Wat 19Aa 27
Prestwood Av. HA3: Kenton 28Ka 46
Prestwood Cl. HA3: Kenton 28Ka 46
Prestwood Cl. SE18 51Wc 117
Prestwood Dr. RM5: Col R 22Ed 56
Prestwood Gdns. CR0: C'don 73Sb 157
Prestwood Ga. AL1: St A 1C 6
Prestwood Ho. SE16 48Xb 91
(off Drummond Rd.)
Prestwood St. N1 2E 218 (40Sb 71)
Pretoria Av. E17 28Ac 52
Pretoria Cres. E4 18Ec 34
Pretoria Rd. E11 32Fc 73
Pretoria Rd. E16 42Hc 93
Pretoria Rd. E4 18Ec 34
Pretoria Rd. IG1: Ilf 36Rc 74
Pretoria Rd. KT16: Chert 74H 149
Pretoria Rd. N17 24Vb 51
Pretoria Rd. RM7: Rom 28Ed 56
Pretoria Rd. SW16 65Kb 134
Pretoria Rd. WD18: Wat 14W 26
Pretoria Rd. Nth. N18 23Vb 51
Pretty La. CR5: Coul 93Lb 196
Prevost Rd. N11 19Jb 32
PREY HEATH 5N 187
Prey Heath Cl. GU22: Wok 6M 187
Prey Heath Rd. GU22: Wok 6M 187
Priam Ho. E2 40Xb 71
(off Old Bethnal Grn. Rd.)
Price Cl. SW17 62Hb 133
Price Ho. N1 39Sb 71
(off Britannia Row)
Price Rd. CR0: Wadd 78Rb 157
Prices Cl. NW7 23Ab 48
Price's Ct. SW11 55Fb 111
Prices La. RH2: Reig 9J 207
Price's M. N1 39Pb 70
Price's St. SE1 7C 224 (46Rb 91)
Price Way TW12: Hamp 65Aa 129
Prichard Ct. N7 36Pb 70
Prichard Ho. SE11 6K 229 (49Qb 90)
(off Kennington Rd.)
Pricklers Hill EN5: New Bar 16Db 31
Prickley Wood BR2: Hayes 74Hc 159
Priddy Pl. RH1: Mers 3C 208
Priddy's Yd. CR0: C'don 75Sb 157
Prideaux Ho. WC1 3J 217 (41Pb 90)
(off Prideaux Pl.)
Prideaux Pl. W3 45Ta 87
Prideaux Pl. WC1 3J 217 (41Pb 90)
Prideaux Rd. SW9 55Nb 112
Pridham Rd. CR7: Thor H 70Tb 135
Priest Cl. TW12: Hamp 63Ca 129
Priestfield Rd. SE23 62Ac 136
Priest Hill RH8: Limp 1M 211
Priest Hill SL4: Old Win 1N 125
Priest Hill TW20: Eng G 2N 125
Priest Hill TW20: Old Win 2N 125
Priest Pk. Cl. KT17: Eps 85Xa 174
Priestlands Pk. Rd. DA15: Sidc 62Vc 139
Priest La. GU24: W End 5A 166
Priestley Cl. N16 31Vb 71
Priestley Ct. RM17: Grays 49Ee 99
Priestley Gdns. RM6: Chad H 30Xc 55
Priestley Ho. EC1 5D 218 (42Sb 91)
(off Old St.)
Priestley Ho. HA9: Wemb 34Sa 67
(off Barnhill Rd.)
Priestley Rd. CR4: Mitc 68Jb 134
Priestley Way E17 27Zb 52
Priestley Way NW2 32Wa 68
Priestly Gdns. GU22: Wok 92C 188
Priestman Point E3 41Dc 92
(off Rainhill Way)
Priest Pk. Av. HA2: Harr 33Ca 65
Priests Av. RM1: Rom 26Fd 56
Priests Bri. SW14 55Ua 110
Priests Bri. SW15 55Ua 110
Priest's Ct. EC2 2D 224 (44Sb 91)
(off Foster La.)
Priest's Fld. CM13: Ingve 22Ee 59
Priests La. CM15: B'wood 17Be 41
Priest's La. CM15: Shenf 17Be 41
Priest's Wlk. DA12: Grav'nd 1J 145
Prima Rd. SW9 52Qb 112
Primary Rd. SL1: Slou 8H 81
Prime Ho. KT22: Lea 91Ka 192
Prime Meridian Line 52Fc 115
Prime Meridian Wlk. E14 45Fc 93
Primeplace M. CR7: Thor H 68Sb 135
Primezone M. N8 30Nb 50
Primmett Cl. TN15: W King 79Ud 164
Primrose Av. EN2: Enf 11Tb 33
Primrose Av. RM6: Chad H 31Xc 75
Primrose Cl. AL10: Hat 1D 8
Primrose Cl. E3 40Cc 72
Primrose Cl. HA2: Harr 34Ba 65
Primrose Cl. HP1: Hem H 3G 2
Primrose Cl. N3 26Db 49
Primrose Cl. SE6 64Ec 136
Primrose Cl. SM6: W'gton 73Kb 156
Primrose Cl. CM14: B'wood 20Yd 40
Primrose Ct. NW8 39Gb 69
(off Prince Albert Rd.)
Primrose Ct. SW12 59Mb 112
Primrose Dr. GU24: Bisl 7E 166
Primrose Dr. UB7: W Dray 49M 83
Primrose Gdns. HA4: Ruis 36Y 65
Primrose Gdns. NW3 37Gb 69
Primrose Gdns. WD23: Bush 17Da 27
Primrose Gdns. WD7: R'lett 7Ja 14
Primrose Glen RM11: Horn 28Nd 57
Primrose Hill CM14: B'wood 20Yd 40
Primrose Hill EC4 3A 224 (44Qb 90)
PRIMROSE HILL 39Jb 70
Primrose Hill WD4: K Lan 10B 4
Primrose Hill Ct. NW3 38Hb 69
Primrose Hill Rd. NW3 38Hb 69
Primrose Hill Studios NW1 39Jb 70
Primrose Ho. KT18: Eps 85Sa 173
(off Dalmeny Way)
Primrose Ho. SE15 53Wb 113
(off Peckham Hill St.)
Primrose Ho. SE16 47Ac 92
(off Blondin Way)
Primrose La. CR0: C'don 74Yb 158
Primrose La. WD25: A'ham 10Fa 14
Primrose Mans. SW11 53Jb 112
Primrose M. NW1 38Hb 69
(off Sharpleshall St.)
Primrose M. SE3 52Jc 115
Primrose Path EN7: Chesh 3Wb 19
Primrose Pl. TW7: Isle 54Ha 108
Primrose Rd. E10 32Dc 72
Primrose Rd. E18 26Kc 53
Primrose Rd. KT12: Hers 78Y 151
Primrose Sq. E9 38Yb 72
Primrose St. EC2 7H 219 (43Ub 91)
Primrose Ter. DA12: Grav'nd 10E 122
Primrose Wlk. KT17: Ewe 80Va 154
Primrose Wlk. SE14 52Ac 114
Primrose Way HA0: Wemb 40Ma 67
Primrose Way SE10 54Ec 114
Primula St. W12 44Wa 88
Primus Cl. RM13: Rain 42Ld 97
Prince Albert Ct. NW8 1F 215 (39Hb 69)
(off Prince Albert Rd.)
Prince Albert Ct. TW16: Sun 66V 128
Prince Albert M. SW11 53Gb 111
Prince Albert Rd. NW1 3D 214 (41Gb 89)
Prince Albert Rd. NW8 3D 214 (41Gb 89)
Prince Albert's Wlk. SL4: Wind 3L 103
Prince Arthur M. NW3 35Eb 69
Prince Arthur Rd. NW3 36Eb 69
Prince Charles Av. DA4: S Dar 68Td 142
Prince Charles Av. RM16: Ors 2D 100
Prince Charles Cinema 4E 222 (45Mb 90)
(off Leicester Pl.)
Prince Charles Dr. NW4 31Ya 68
Prince Charles Rd. SE3 54Hc 115
Prince Charles Way SM6: W'gton 76Kb 156
Prince Consort Cotts. SL4: Wind 4H 103
Prince Consort Dr. BR7: Chst 67Tc 138
Prince Consort Rd. SW7 3A 226 (48Eb 89)
Prince Consort's Dr. SL4: Wind 8D 102
Princedale Rd. W11 46Ab 88
Prince Edward Mans. W2 45Cb 89
(off Moscow Rd.)
Prince Edward Rd. E9 37Bc 72
Prince Edward Theatre 3E 222 (44Mb 90)
(off Old Compton St.)
Prince Eugene Pl. AL1: St A 5A 6
Prince George Av. N14 15Mb 32
Prince George Rd. N16 35Ub 71
Prince George's Av. SW20 68Ya 132
Prince George's Rd. SW19 67Fb 133
Prince Henry Rd. SE7 52Mc 115
Prince Imperial Rd. BR7: Chst 67Rc 138
Prince Imperial Rd. SE18 53Pc 116
Prince John Rd. SE9 57Nc 116
Princelet St. E1 7K 219 (43Vb 91)
Prince Michael of Kent Ct. DA1: Cray 54Jd 118
Prince of Orange Ct. SE16 48Yb 92
(off Lower Rd.)
Prince of Orange La. SE10 52Ec 114
Prince of Wales Cl. DA12: Grav'nd 9F 122
Prince of Wales Cl. NW4 28Xa 48
Prince of Wales Dr. SW11 53Gb 111
Prince of Wales Dr. SW8 52Kb 112
Prince of Wales Footpath EN3: Enf W 10Zb 20
Prince of Wales Mans. SW11 53Jb 112
Prince of Wales Pas. NW1 4B 216 (41Lb 90)
(off Hampstead Rd.)
Prince of Wales Rd. E16 44Lc 93
Prince of Wales Rd. NW5 37Jb 70
Prince of Wales Rd. SE3 54Hc 115
Prince of Wales Rd. SM1: Sutt 75Fb 155
Prince of Wales Ter. W4 50Ua 88
Prince of Wales Ter. W8 47Db 89
Prince of Wales Theatre 5D 222 (45Mb 90)
(off Coventry St.)
Prince Pk. HP1: Hem H 3J 3
Prince Philip Av. RM16: Grays 46Ce 99
Prince Regent Ct. NW8 1E 214 (40Gb 69)
(off Avenue Rd.)
Prince Regent Ct. SE16 45Ac 92
(off Edward Sq.)
Prince Regent La. E13 41Kc 93
Prince Regent La. E16 43Lc 93
Prince Regent M. NW1 4B 216 (41Lb 90)
(off Hampstead Rd.)
Prince Regent Rd. TW3: Houn 55Da 107
Prince Regent Station (DLR) 45Lc 93
Prince Rd. SE25 71Ub 157
Prince Rupert Rd. SE9 56Pc 116
Princes Arc. SW1 6C 222 (46Lb 90)
(off Piccadilly)
Prince's Av. UB6: G'frd 44Da 85
Prince's Av. WD18: Wat 15V 26
Princes Av. BR5: Pet W 71Uc 160
Princes Av. CR2: Sande 87Xb 177
Princes Av. DA2: Dart 60Rd 119
Princes Av. EN3: Enf W 8Ac 20
Princes Av. IG8: Wfd G 21Kc 53
Princes Av. KT6: Surb 74Qa 153
Princes Av. N10 27Kb 50
Princes Av. N13 22Qb 50
Princes Av. N22 25Mb 50
Princes Av. N3 25Cb 49
Princes Av. NW9 28Qa 47
Princes Av. SM5: Cars 80Hb 155
Princes Av. W3 48Qa 87
Princes Cir. WC2 2F 223 (44Nb 90)
Prince's Cl. TW11: Tedd 63Fa 130
Princes Cl. CR2: Sande 87Xb 177
Princes Cl. DA14: Sidc 62Zc 139
Princes Cl. HA8: Edg 22Qa 47
Princes Cl. N4 32Rb 71
Princes Cl. NW9 28Qa 47
Princes Cl. SL4: Eton W 10D 80
Princes Cl. SW4 55Lb 112
Prince's Ct. SE16 48Bc 92
Prince's Ct. SW3 3F 227 (48Hb 89)
(off Brompton Rd.)
Princes Ct. HA9: Wemb 36Na 67
Princes Ct. HP3: Hem H 5K 3
Princes Ct. KT13: Weyb 78R 150
(off Princes Rd.)
Princes Ct. Bus. Cen. E1 45Xb 91
Prince's Dr. KT22: Oxs 84Ga 172
Princes Dr. HA1: Harr 27Ga 46
Princesfield Rd. EN9: Walt A 5Kc 21
Prince's Gdns. SW7 3C 226 (48Fb 89)
Princes Gdns. W3 43Qa 87
Princes Gdns. W5 42La 86
Prince's Ga. SW7 2C 226 (47Fb 89)
(not continuous)
Prince's Ga. Ct. SW7 2C 226 (47Fb 89)
Prince's Ga. M. SW7 3C 226 (48Fb 89)
Princes Ho. W11 45Bb 89
Princes La. N10 27Kb 50
Prince's M. W2 45Db 89
Princes M. TW3: Houn 56Ca 107
Princes M. W6 50Xa 88
(off Down Pl.)
Princes Pde. EN6: Pot B 4Db 17
Princes Pde. NW11 30Ab 48
(off Golders Grn. Rd.)
Princes Pk. RM13: Rain 38Jd 76
Princes Pk. Av. NW11 30Ab 48
Princes Pk. Av. UB3: Hayes 45T 84
Princes Pk. Circ. UB3: Hayes 45T 84
Princes Pk. Cl. UB3: Hayes 45T 84
Princes Pk. Golf Course 60Qd 119
Princes Pk. La. UB3: Hayes 45T 84
Princes Pk. Pde. UB3: Hayes 45T 84
Princes Pk. Stadium 60Qd 119
Princes Pl. SW1 6C 222 (46Lb 90)
(off Duke St.)
Princes Pl. W11 46Ab 88
Prince's Plain BR2: Broml 73Nc 160
Prince's Ri. SE13 54Ec 114
Princes Riverside Rd. SE16 46Zb 92
Prince's Rd. CM14: Kel C 11Nd 39
Prince's Rd. CM14: N'side 11Nd 39
Prince's Rd. RM1: Rom 29Jd 56
Prince's Rd. SW19 65Cb 133
Prince's Rd. TW11: Tedd 63Fa 130
Princes Rd. BR8: Hext 65Jd 140
Princes Rd. DA1: Dart 58Jd 118
Princes Rd. DA12: Grav'nd 3E 144
Princes Rd. DA2: Dart 60Rd 119
(not continuous)
Princes Rd. IG6: Ilf 28Tc 54
Princes Rd. IG9: Buck H 19Lc 35
Princes Rd. KT13: Weyb 78R 150
Princes Rd. KT2: King T 66Qa 131
Princes Rd. N18 21Yb 52
Princes Rd. RH1: Redh 8P 207
Princes Rd. SE20 65Zb 136
Princes Rd. SW14 55Ta 109
Princes Rd. TW10: Rich 57Pa 109
Princes Rd. TW13: Felt 61V 128
Princes Rd. TW15: Ashf 64P 127
Princes Rd. TW20: Egh 65B 126
Princes Rd. TW9: Kew 53Pa 109
Princes Rd. W13 46Ka 86
PRINCES ROAD INTERCHANGE 60Rd 119
Princessa Ct. EN2: Enf 15Tb 33
PRINCESS ALICE HOSPICE, THE 78Ca 151
Princess Alice Ho. W10 42Ya 88
Princess Alice Way SE28 47Tc 94
Princess Av. HA9: Wemb 33Na 67
Princess Av. RM18: E Til 9L 101
Princess Av. SL4: Wind 5F 102
Princess Cl. SE28 44Zc 95
Princess Ct. KT1: King T 69Pa 131
(off Horace Rd.)
Princess Ct. N6 31Lb 70
Princess Ct. NW6 37Db 69
(off Compayne Gdns.)
Princess Ct. W1 1F 221 (43Hb 89)
(off Bryanston Pl.)
Princess Ct. W2 45Db 89
(off Queensway)
Princess Cres. N4 33Rb 71
Princess Diana Dr. AL4: St A 3H 7
Princesses Pde. DA1: Cray 57Gd 118
(off Waterside)
Princess Gdns. GU22: Wok 88D 168
PRINCESS GRACE HOSPITAL 6J 215 (42Jb 90)
Princess Ho. RH1: Redh 5A 208
Princess La. HA4: Ruis 32U 64
Princess Louise Bldg. SE8 52Cc 114
(off Hales St.)
Princess Louise Cl. W2 7C 214 (43Fb 89)
Princess Louise Wlk. W10 43Za 88
PRINCESS MARGARET BMI HOSPITAL 4H 103
Princess Margaret Rd. RM18: E Til 10L 101
Princess Margaret Rd. SS17: E Til 8K 101
Princess Margaret Rd. SS17: Linf 8K 101
Princess Mary Ho. SW1 6E 228 (49Mb 90)
(off Vincent St.)
Princess Marys Rd. KT15: Add 77L 149
Princess May Rd. N16 35Ub 71
Princess M. KT1: King T 69Pa 131
Princess M. NW3 36Fb 69
Princess Pde. BR6: Farnb 76Qc 160

Princess Pde. RM10: Dag ... 40Cd 76
Princess Pk. KT15: Add ... 77J 149
Princess Pk. Mnr. N11 ... 22Jb 50
Princess Sq. W2 ... 45Db 89
(not continuous)
Princess Rd. CR0: C'don ... 72Sb 157
Princess Rd. GU22: Wok ... 88D 168
Princess Rd. NW1 ... 39Jb 70
Princess Rd. NW6 ... 40Cb 69
PRINCESS ROYAL UNIVERSITY HOSPITAL ... 76Qc 160
Princess Sq. KT10: Esh ... 79Ea 152
Princess St. SE1 ... 4C 230 (48Rb 91)
Prince's St. EC2 ... 3F 225 (44Tb 91)
Princes St. DA11: Grav'nd ... 8D 122
Princes St. DA7: Bex ... 55Bd 117
Princes St. N17 ... 23Ub 51
Princes St. SL1: Slou ... 7M 81
Princes St. SM1: Sutt ... 77Fb 155
Princes St. TW9: Rich ... 56Na 109
Princes St. W1 ... 3A 222 (44Kb 90)
Princess Way RH1: Redh ... 5A 208
Princes Ter. E13 ... 39Kc 73
Prince's Twr. SE16 ... 47Yb 92
(off Elephant La.)
Prince St. SE8 ... 51Bc 114
Prince St. WD17: Wat ... 13Y 27
Princes Vw. DA1: Dart ... 60Qd 119
Princes Way BR4: W W'ck ... 77Hc 159
Princes Way CM13: Hut ... 19Ce 41
Princes Way CR0: Wadd ... 78Pb 156
Princes Way HA4: Ruis ... 35Aa 65
Princes Way IG9: Buck H ... 19Lc 35
Princes Way SW19 ... 59Za 110
Princes Way W3 ... 48Qa 87
Prince's Yd. W11 ... 46Ab 88
Princethorpe Ho. W2 ... 43Db 89
(off Woodchester Sq.)
Princethorpe Rd. SE26 ... 63Zb 136
Princeton Ct. SW15 ... 55Za 110
Princeton M. KT2: King T ... 67Qa 131
Princeton St. WC1 ... 1H 223 (43Pb 90)
Prince William Ct. TW15: Ashf ... 64P 127
(off Clarendon Rd.)
Principal Cl. N14 ... 18Lb 32
Principal Pl. EC2 ... 6J 219 (42Ub 91)
(off Shoreditch High St.)
Principal Sq. E9 ... 36Zb 72
Pringle Gdns. CR8: Purl ... 82Pb 176
Pringle Gdns. SW16 ... 63Lb 134
(not continuous)
Printers Av. WD18: Wat ... 15U 26
Printers Ct. AL1: St A ... 3B 6
(off Thorpe Rd.)
Printers Inn Ct. EC4 ... 2K 223 (44Qb 90)
Printers M. E3 ... 39Ac 72
Printers Rd. SW9 ... 53Pb 112
Printer St. EC4 ... 2A 224 (44Qb 90)
Printing Ho. La. UB3: Hayes ... 47U 84
Printing Ho. Yd. E2 ... 3J 219 (41Ub 91)
Printon Ho. E14 ... 43Bc 92
(off Wallwood St.)
Print Room, The ... 44Cb 89
(off Hereford Rd.)
Print Room at the Coronet ... 46Cb 89
(off Notting Hill Ga.)
Print Village SE15 ... 54Vb 113
Printwork Apts. SE1 ... 3H 231 (48Ub 91)
(off Long La.)
Printwork Apts. SE5 ... 53Sb 113
(off Coldharbour La.)
Priolo Rd. SE7 ... 50Lc 93
Prior Av. SM2: Sutt ... 80Gb 155
Prior Bolton St. N1 ... 37Rb 71
Prior Chase RM17: Grays ... 49Ce 99
Prior Ct. KT8: W Mole ... 71Ba 151
Prioress Cres. DA9: Ghithe ... 56Yd 120
Prioress Ho. E3 ... 41Dc 92
(off Bromley High St.)
Prioress Rd. SE27 ... 62Rb 135
Prioress St. SE1 ... 4H 231 (48Ub 91)
Prior Rd. IG1: Ilf ... 34Qc 74
Priors, The KT21: Asht ... 91Ma 193
Priors, The NW3 ... 34Fb 69
Priors Cl. SL1: Slou ... 8L 81
Priors Ct. GU21: Wok ... 10L 167
Prior's Cft. GU22: Wok ... 92C 188
Priors Cft. E17 ... 26Ac 52
Priors Farm La. UB5: N'olt ... 37Aa 65
Priors Fld. UB5: N'olt ... 37Aa 65
Priorsford Av. BR5: St M Cry ... 70Wc 139
Priors Gdns. HA4: Ruis ... 36Y 65
Priors Golf Course, The ... 16Ld 39
Priors Mead EN1: Enf ... 11Ub 33
Priors Mead KT23: Bookh ... 97Ea 192
Priors Pk. RM12: Horn ... 34Ld 77
Priors Rd. SL4: Wind ... 5B 102
Prior St. SE10 ... 52Ec 114
Priors Wood KT10: Hin W ... 75Ha 152
Priory, The CR0: Wadd ... 77Qb 156
Priory, The KT22: Lea ... 94Ka 192
Priory, The N8 ... 28Mb 50
Priory, The RH9: G'stone ... 3P 209
Priory, The SE3 ... 56Hc 115
Priory Apts., The SE6 ... 60Dc 114
Priory Av. BR5: Pet W ... 72Tc 160
Priory Av. E17 ... 29Cc 52
Priory Av. E4 ... 20Bc 34
Priory Av. HA0: Wemb ... 35Ha 66
Priory Av. N8 ... 28Mb 50
Priory Av. SM3: Cheam ... 77Za 154
Priory Av. UB9: Hare ... 28L 43
Priory Av. W4 ... 49Ua 88
Priory Cl. BR3: Beck ... 69Ac 136
Priory Cl. BR7: Chst ... 67Pc 138
Priory Cl. CM15: Pil H ... 15Wd 40
Priory Cl. DA1: Dart ... 57Ld 119
Priory Cl. E18 ... 25Jc 53
Priory Cl. E4 ... 20Bc 34
Priory Cl. GU21: Wok ... 85F 168
Priory Cl. HA0: Wemb ... 35Ha 66
Priory Cl. HA4: Ruis ... 32V 64
Priory Cl. HA7: Stan ... 20Ha 28
Priory Cl. KT12: Walt T ... 76W 150
Priory Cl. N14 ... 15Kb 32
Priory Cl. N20 ... 18Bb 31
Priory Cl. N3 ... 25Bb 49
Priory Cl. SL5: S'dale ... 3E 146
Priory Cl. SW19 ... 67Db 133
Priory Cl. TW12: Hamp ... 67Ba 129
Priory Cl. TW16: Sun ... 66W 128
Priory Cl. UB3: Hayes ... 45X 85
Priory Cl. UB9: Den ... 34J 63
Priory Cl. UB9: Hare ... 28K 43
Priory Ct. AL1: St A ... 3C 6
Priory Ct. DA1: Dart ... 58Md 119
Priory Ct. E17 ... 26Bc 52
Priory Ct. E6 ... 39Lc 73
Priory Ct. E9 ... 36Zb 72
Priory Ct. EC4 ... 3C 224 (44Rb 91)
(off Pilgrim St.)
Priory Ct. HA0: Wemb ... 40Na 67
Priory Ct. KT1: King T ... 69Na 131
(off Denmark Rd.)
Priory Ct. KT17: Ewe ... 81Va 174
Priory Ct. SM3: Cheam ... 77Ab 154
Priory Ct. SW8 ... 53Mb 112
Priory Ct. TW20: Egh ... 65E 126
Priory Ct. TW3: Houn ... 55Da 107
Priory Ct. WD23: Bush ... 18Ea 28
(off Sparrows Herne)
Priory Ct. Est. E17 ... 26Bc 52
Priory Cres. HA0: Wemb ... 34Ja 66
Priory Cres. SE19 ... 66Sb 135
Priory Cres. SM3: Cheam ... 77Za 154
Priory Dr. HA7: Stan ... 20Ha 28
Priory Dr. RH2: Reig ... 8J 207
Priory Dr. SE2 ... 50Zc 95
Priory Fld. Dr. HA8: Edg ... 21Ra 47
Priory Flds. DA4: Eyns ... 75Pd 163
Priory Flds. WD17: Wat ... 11W 26
Priory Gdns. Ramsden ... 73Xc 161
Priory Gdns. DA1: Dart ... 57Md 119
Priory Gdns. HA0: Wemb ... 35Ja 66
Priory Gdns. N6 ... 30Kb 50
Priory Gdns. NW10 ... 41Na 87
Priory Gdns. SE25 ... 70Vb 135
Priory Gdns. SW13 ... 55Va 110
Priory Gdns. TW12: Hamp ... 66Ba 129
Priory Gdns. TW15: Ashf ... 64T 128
Priory Gdns. UB9: Hare ... 28L 43
Priory Gdns. W4 ... 49Ua 88
Priory Gdns. W5 ... 41Na 87
Priory Grange N2 ... 27Hb 49
(off Fortis Grn.)
Priory Grn. N1 ... 1H 217 (40Pb 70)
Priory Grn. TW18: Staines ... 64K 127
Priory Grn. Est. N1 ... 1H 217 (40Pb 70)
Priory Gro. EN5: Barn ... 15Cb 31
Priory Gro. RM3: Rom ... 20Nd 39
Priory Gro. SW8 ... 53Nb 112
Priory Hgts. N1 ... 1J 217 (40Pb 70)
(off Wynford Rd.)
Priory Hill DA1: Dart ... 57Md 119
Priory Hill HA0: Wemb ... 35Ja 66
PRIORY HOSPITAL, HEMEL HEAMPSTEAD ... 7H 3
PRIORY HOSPITAL ROEHAMPTON ... 56Va 110
PRIORY HOSPITAL STURT HOUSE ... 99Ma 194
Priory Ho. E1 ... 7K 219 (43Vb 91)
(off Folgate St.)
Priory Ho. EC1 ... 5B 218 (42Rb 91)
(off Sans Wlk.)
Priory Ho. SW1 ... 7D 228 (50Mb 90)
(off Rampayne St.)
Priory La. DA4: Eyns ... 74Pd 163
Priory La. KT8: W Mole ... 70Da 129
Priory La. SW15 ... 58Ua 110
Priory Leas SE9 ... 60Nc 116
Priory Lodge W4 ... 50Qa 87
(off Kew Bri. Ct.)
Priory Lodge WD3: Rick ... 17M 25
(off Nightingale Pl.)
Priory Mans. SW10 ... 7A 226 (50Eb 89)
(off Drayton Gdns.)
Priory Mkt. Pl. DA1: Dart ... 59Nd 119
Priory M. CR3: Cat'm ... 97Vb 197
Priory M. RM11: Horn ... 32Kd 77
Priory M. SW8 ... 53Nb 112
Priory M. TW18: Staines ... 64K 127
Priory Pk. HA8: Edg ... 21Ra 47
Priory Pk. SE3 ... 55Hc 115
Priory Pk. Rd. HA0: Wemb ... 35Ja 66
Priory Pk. Rd. NW6 ... 39Bb 69
Priory Path RM3: Rom ... 20Nd 39
Priory Pl. DA1: Dart ... 58Md 119
Priory Pl. KT12: Walt T ... 76W 150
Priory Retail Pk. ... 66Fb 133
Priory Rd. CR0: C'don ... 73Qb 156
Priory Rd. E6 ... 39Mc 73
Priory Rd. IG10: Lough ... 14Nc 36
Priory Rd. IG11: Bark ... 38Tc 74
Priory Rd. KT9: Chess ... 76Na 153
Priory Rd. N8 ... 28Lb 50
Priory Rd. NW6 ... 39Db 69
Priory Rd. RH2: Reig ... 8J 207
Priory Rd. RM3: Rom ... 20Nd 39
Priory Rd. SL1: Slou ... 3A 80
Priory Rd. SL5: S'dale ... 3E 146
Priory Rd. SM3: Cheam ... 77Za 154
Priory Rd. SS17: Stan H ... 1N 101
Priory Rd. SW19 ... 66Fb 133
Priory Rd. TW12: Hamp ... 66Ba 129
Priory Rd. TW3: Houn ... 57Ea 108
Priory Rd. TW9: Kew ... 51Qa 109
Priory Rd. W4 ... 48Ta 87
Priory Rd. Nth. DA1: Dart ... 56Md 119
Priory Rd. Sth. DA1: Dart ... 58Md 119
Priory Shop. Cen. ... 58Nd 119
Priory St. E3 ... 41Dc 92
Priory Ter. NW6 ... 39Db 69
Priory Ter. TW16: Sun ... 66W 128
Priory Vw. WD23: B Hea ... 17Ga 28
Priory Vs. N11 ... 23Hb 49
(off Colney Hatch La.)
Priory Wlk. AL1: St A ... 5C 6
Priory Wlk. SW10 ... 7A 226 (50Eb 89)
Priory Wlk. TW16: Sun ... 66W 128
Priory Way HA2: Harr ... 28Da 45
Priory Way SL3: Dat ... 2M 103
Priory Way UB2: S'hall ... 48Z 85
Priory Way UB7: Harm ... 51N 105
Priscilla Cl. N15 ... 29Sb 51
Pritchard Ho. E2 ... 40Xb 71
(off Ada Pl.)
Pritchard's Rd. E2 ... 39Wb 71
Pritchett Cl. EN3: Enf L ... 9Cc 20
Priter Rd. SE16 ... 48Wb 91
Priter Rd. Hostel SE16 ... 48Wb 91
(off Dockley Rd.)
Priter Way SE16 ... 48Wb 91
Private Rd. EN1: Enf ... 15Tb 33
Privet Dr. WD25: Wat ... 6V 12
Privet M. CR8: Purl ... 84Lb 176
Prize Wlk. E20 ... 36Ec 72
Probert Rd. SW2 ... 57Qb 112
Probyn Ho. SW1 ... 5E 228 (49Mb 90)
(off Page St.)
Probyn Rd. SW2 ... 61Rb 135
Procter Ho. SE1 ... 50Wb 91
(off Avondale Sq.)
Procter Ho. SE5 ... 52Tb 113
(off Picton St.)
Procter St. WC1 ... 1H 223 (43Pb 90)
Proctor Cl. CR4: Mitc ... 67Jb 134
Proctor Gdns. KT23: Bookh ... 97Da 191
Proctor Pl. RM13: Rain ... 39Jd 76
Proctors Cl. TW14: Felt ... 60W 106
Proffits Cotts. KT20: Tad ... 94Za 194
Profumo Rd. KT12: Hers ... 78Z 151
Progress Bus. Cen. SL1: Slou ... 4B 80
Progress Bus. Pk. CR0: Wadd ... 75Pb 156
Progress Cen., The EN3: Pond E ... 13Zb 34
Progress Way CR0: Wadd ... 75Pb 156
Progress Way EN1: Enf ... 15Wb 33
Progress Way N22 ... 25Qb 50
Project Pk. E16 ... 42Fc 93
Project Rd. IG11: Bark ... 42Xc 95
Prologis Pk. CR0: Bedd ... 73Mb 156
Prologis Pk. E3 ... 42Ec 92
Prologis Pk. TW4: Houn ... 56Y 107
Promenade, The HA8: Edg ... 22Qa 47
Promenade, The W4 ... 54Ua 110
Promenade App. Rd. W4 ... 52Ua 110
Promenade de Verdun CR8: Purl ... 83Mb 176
Propeller Cres. CR0: Wadd ... 78Qb 156
Propeller Way NW4 ... 27Wa 48
Prospect Av. SS17: Stan H ... 2K 101
Prospect Bus. Pk. IG10: Lough ... 14Tc 36
Prospect Cl. DA17: Belv ... 49Cd 96
Prospect Cl. HA4: Ruis ... 31Z 65
Prospect Cl. SE26 ... 63Xb 135
Prospect Cl. TW3: Houn ... 53Ba 107
Prospect Cl. WD23: Bush ... 17Fa 28
Prospect Cotts. SW18 ... 56Cb 111
Prospect Cres. TW2: Whitt ... 58Ea 108
Prospect Gro. DA12: Grav'nd ... 9F 122
Prospect Hill E17 ... 28Dc 52
Prospect Ho. E17 ... 27Ec 52
(off Prospect Hill)
Prospect Ho. E3 ... 41Cc 92
(off Campbell Rd.)
Prospect Ho. KT19: Eps ... 81Ra 173
Prospect Ho. N1 ... 2K 217 (40Qb 70)
(off Donegal St.)
Prospect Ho. SE1 ... 4C 230 (48Rb 91)
(off Gaywood St.)
Prospect Ho. SE16 ... 48Wb 91
(off Frean St.)
Prospect Ho. SW19 ... 67Fb 133
(off Chapter Way)
Prospect Ho. W10 ... 44Za 88
(off Bridge Cl.)
PROSPECT HOUSE ... 13W 26
Prospect La. TW20: Eng G ... 4L 125
Prospect Pk. ... 42Dc 92
(off Reeves Rd.)
Prospect Pl. BR2: Broml ... 69Kc 137
Prospect Pl. DA1: Dart ... 58Nd 119
Prospect Pl. DA12: Grav'nd ... 9F 122
Prospect Pl. E1 ... 46Yb 92
(not continuous)
Prospect Pl. KT17: Eps ... 84Ua 174
Prospect Pl. N17 ... 24Ub 51
Prospect Pl. N2 ... 28Fb 49
Prospect Pl. N7 ... 35Nb 70
Prospect Pl. NW2 ... 34Bb 69
Prospect Pl. NW3 ... 35Eb 69
Prospect Pl. RM17: Grays ... 51De 121
Prospect Pl. RM5: Col R ... 26Ed 56
Prospect Pl. SE8 ... 51Bc 114
(off Evelyn St.)
Prospect Pl. SL4: Wind ... 5H 103
(off Osborne Rd.)
Prospect Pl. SW11 ... 52Kb 112
Prospect Pl. SW20 ... 66Xa 132
Prospect Pl. TW18: Staines ... 64H 127
Prospect Pl. W4 ... 50Ta 87
Prospect Quay SW18 ... 56Cb 111
(off Lightermans Wlk.)
Prospect Ring N2 ... 27Fb 49
Prospect Rd. AL1: St A ... 4B 6
Prospect Rd. EN5: New Bar ... 14Cb 31
Prospect Rd. EN8: Chesh ... 1Yb 20
Prospect Rd. IG8: Wfd G ... 23Lc 53
Prospect Rd. KT6: Surb ... 72La 152
Prospect Rd. NW2 ... 34Bb 69
Prospect Rd. RM11: Horn ... 27Pd 57
Prospect Rd. TN13: S'oaks ... 95Ld 203
Prospect Row E15 ... 36Ec 72
Prospect Row Ho's. E15 ... 36Ec 72
(off Property Row)
Prospect St. SE16 ... 48Xb 91
Prospectus Pl. CR2: S Croy ... 79Sb 157
Prospect Va. SE18 ... 49Nc 94
Prospect Way CM13: Hut ... 14Fe 41
Prospect Wharf E1 ... 45Yb 92
Prospero Ho. E1 ... 4K 225 (45Vb 91)
(off Portsoken St.)
Prospero Rd. N19 ... 32Mb 70
Prossers KT20: Tad ... 93Za 194
Protea Cl. E16 ... 42Hc 93
Protea Pl. E9 ... 38Yb 72
(off Lyme Gro.)
Protheroe Ho. N17 ... 27Vb 51
Prothero Gdns. NW4 ... 29Xa 48
Prothero Ho. NW10 ... 38Ta 67
(off Fawood Av.)
Prothero Rd. SW6 ... 52Ab 110
Proton Twr. E14 ... 45Fc 93
Proud Ho. E1 ... 44Xb 91
(off Amazon St.)
Prout Gro. NW10 ... 35Ua 68
Prout Rd. E5 ... 34Xb 71
Provence St. N1 ... 1D 218 (40Sb 71)
Providence Av. HA2: Harr ... 32Ca 65
Providence Cl. E9 ... 39Zb 72
Providence Ct. W1 ... 4J 221 (45Jb 90)
Providence Ho. E14 ... 44Bc 92
(off Three Colt St.)
Providence La. UB3: Harl ... 52T 106
Providence Pl. GU22: Pyr ... 86J 169
Providence Pl. KT17: Eps ... 84Ua 174
Providence Pl. N1 ... 39Rb 71
Providence Pl. RM5: Col R ... 25Ed 55
Providence Pl. SE10 ... 50Gc 93
Providence Rd. UB7: Yiew ... 46N 83
Providence Row N1 ... 2H 217 (40Pb 70)
(off Pentonville Rd.)
Providence Row Cl. E2 ... 41Xb 91
Providence Sq. SE1 ... 47Wb 91
Providence St. DA9: Ghithe ... 57Wd 120
Providence Twr. E14 ... 45Fc 93
(off Fairmont Av.)
Providence Twr. SE16 ... 47Wb 91
(off Bermondsey Wall W.)
Providence Yd. E2 ... 41Wb 91
(off Ezra St.)
Provident Ind. Est. UB3: Hayes ... 47W 84
Province Dr. SE16 ... 47Yb 92
Province Sq. E14 ... 46Ec 92
(off Blackwall Way)
Provincial Ter. SE20 ... 66Zb 136
Provost Ct. NW3 ... 37Hb 69
(off Eton Rd.)
Provost Est. N1 ... 2F 219 (40Tb 71)
(off Provost St.)
Provost Rd. NW3 ... 38Hb 69
Provost St. N1 ... 2F 219 (40Tb 71)
Provost Way RM8: Dag ... 35Wc 75
Prowse Av. WD23: B Hea ... 18Ea 28
Prowse Ct. N18 ... 22Wb 51
(off Lord Graham M.)
Prowse Pl. NW1 ... 38Lb 70
Proyers Path HA1: Harr ... 31Ka 66
Prudence La. BR6: Farnb ... 77Qc 160
Pruden Cl. N14 ... 19Lb 32
Prudent Pas. EC2 ... 3E 224 (44Sb 91)
(off King St.)
Prudhoe Ct. DA2: Dart ... 58Rd 119
(off Osbourne Rd.)
Prune Hill TW20: Egh ... 6P 125
Prune Hill TW20: Eng G ... 6P 125
Prunus Cl. GU24: W End ... 5C 166
Prusom's Island E1 ... 46Yb 92
(off Wapping High St.)
Prusom St. E1 ... 46Xb 91
Pryce Ho. E3 ... 41Cc 92
(off Campbell Rd.)
Pryor Cl. WD5: Ab L ... 4V 12
PSYCHIATRIC UNIT (BARNET GENERAL HOSPITAL) ... 14Ya 30
Puccinia Ct. TW19: Stanw ... 60N 105
(off Yeoman Dr.)
Puck La. EN9: Walt A ... 1Fc 21
Pucknells Cl. BR8: Swan ... 67Ed 140
Pucks Hill GU21: Knap ... 9H 167
Pudding La. AL3: St A ... 2B 6
(off Chequer St.)
Pudding La. EC3 ... 5G 225 (45Tb 91)
Pudding La. IG7: Chig ... 16Uc 36
Pudding La. TN15: Seal ... 93Pd 203
Pudding Mill La. E15 ... 39Dc 72
Pudding Mill Lane Station (DLR) ... 39Dc 72
Puddingstone Dr. AL4: St A ... 4G 6
Puddle Dock EC4 ... 4B 224 (45Rb 91)
(not continuous)
PUDDLEDOCK ... 65Gd 140
Puddledock Farm Fishery ... 32Yd 78
Puddledock La. DA2: Wilm ... 64Gd 140
PUDDS CROSS ... 1A 10
Puffin Cl. BR3: Beck ... 71Zb 158
Puffin Cl. IG11: Bark ... 41Xc 95
Pugin Cl. SW19 ... 63Db 133
Pugin Ct. N1 ... 38Qb 70
(off Liverpool Rd.)
Pulborough Ho. RM3: Rom ... 24Nd 57
(off Kingsbridge Cir.)
Pulborough Rd. SW18 ... 59Bb 111
Pulborough Way TW4: Houn ... 56Y 107
Pulford Rd. N15 ... 30Tb 51
Pulham Av. N2 ... 28Eb 49
Pulham Ho. SW8 ... 52Pb 112
(off Dorset Rd.)
Pullen's Bldgs. SE17 ... 7C 230 (50Rb 91)
(off Iliffe St.)
Puller Rd. EN5: Barn ... 12Ab 30
Puller Rd. HP1: Hem H ... 3J 3
Pulleyns Av. E6 ... 41Nc 94
Pulleys Cl. HP1: Hem H ... 1H 3
Pulleys La. HP1: Hem H ... 1G 2
(not continuous)
Pullman Cl. AL1: St A ... 4C 6
Pullman Ct. RM17: Grays ... 51Ce 121
Pullman Ct. SW2 ... 60Nb 112
Pullman Gdns. SW15 ... 58Ya 110
Pullman M. SE12 ... 62Kc 137
Pullman Pl. RH1: Mers ... 100Lb 196
(off Station Rd.)
Pullman Pl. SE9 ... 57Nc 116
Pullmans Pl. TW18: Staines ... 64J 127
Pullman Sq. GU24: Bisl ... 9E 166
Pullman Sq. RM17: Grays ... 51Ce 121
Pulross Rd. SW9 ... 55Pb 112
Pulse Apts. NW6 ... 36Db 69
(off Lymington Rd.)
Pulse Ct. RM7: Rush G ... 30Gd 56
Pulsford Cl. TW1: Twick ... 59Ha 108
Pulteney Cl. E3 ... 39Bc 72
Pulteney Cl. TW7: Isle ... 55Ja 108
Pulteney Gdns. E18 ... 27Jc 53
Pulteney Rd. E18 ... 27Kc 53
Pulteney Ter. N1 ... 39Pb 70
(not continuous)
Pultney St. N1 ... 1J 217 (39Pb 70)
Pulton Ho. SE4 ... 56Ac 114
(off Turnham Rd.)
Pulton Pl. SW6 ... 52Cb 111
Puma Ct. E1 ... 7K 219 (43Vb 91)
Pump All. TW8: Bford ... 52Ma 109
Pump Cl. UB5: N'olt ... 40Ca 65
Pump Ct. EC4 ... 3K 223 (44Qb 90)
Pumphandle Path N2 ... 26Fb 49
(off Oak La.)
Pump Hill IG10: Lough ... 12Pc 36
Pump Ho. SE25 ... 70Vb 135
Pumphouse, The N8 ... 28Pb 50
Pump Ho. Cl. BR2: Broml ... 68Gc 137
Pump Ho. Cl. SE16 ... 47Yb 92
Pump Ho. Cres. TW8: Bford ... 50Na 87
Pumphouse Cres. WD17: Wat ... 15Y 27
Pump House Gallery ... 52Jb 112
Pump Ho. La. SW11 ... 52Lb 112
Pump Ho. M. E1 ... 45Wb 91
(off Hooper St.)
Pump House Theatre & Arts Cen. ... 15Z 27
Pumping Ho. E14 ... 45Fc 93
(off Naval Row)
Pumping Station Rd. W4 ... 52Ua 110
Pumpkin Hill SL1: Burn ... 7C 60
Pump La. BR6: Well H ... 78Dd 162
Pump La. SE14 ... 52Yb 114
Pump La. SL5: Asc ... 7C 124
Pump La. UB3: Hayes ... 47W 84
Pump Pail Nth. CR0: C'don ... 76Sb 157
Pump Pail Sth. CR0: C'don ... 76Sb 157
Pump St. SS17: Horn H ... 1J 101
Punchard Cres. EN3: Enf L ... 10Dc 20
Punch Cft. DA3: Nw A G ... 76Ae 165
Punderson's Gdns. E2 ... 41Xb 91
Punjab La. UB1: S'hall ... 46Ba 85
Purbeck Av. KT3: N Mald ... 72Va 154
Purbeck Cl. RH1: Mers ... 100Mb 196
Purbeck Dr. GU21: Wok ... 86B 168
Purbeck Dr. NW2 ... 33Za 68
Purbeck Gdns. SE26 ... 64Bc 136
Purbeck Ho. SW8 ... 52Pb 112
(off Bolney St.)
Purbeck Ho. WD18: Wat ... 16V 26
(off Scammell Way)
Purbeck Rd. RM11: Horn ... 31Jd 76
Purberry Gro. KT17: Ewe ... 82Va 174
Purberry Shot KT17: Ewe ... 82Va 174
Purbrock Av. WD25: Wat ... 8Y 13
Purbrook Est. SE1 ... 2J 231 (47Ub 91)
Purbrook St. SE1 ... 3J 231 (48Ub 91)
Purcell Cl. CR8: Kenley ... 86Sb 177
Purcell Cl. SS17: Stan H ... 1L 101
Purcell Cl. WD6: Bore ... 11Ma 29
Purcell Cres. SW6 ... 52Za 110
Purcell Ho. EN1: Enf ... 10Wb 19
Purcell Ho. SW10 ... 51Fb 111
(off Milman's St.)
Purcell Mans. W14 ... 51Ab 110
(off Queen's Club Gdns.)
Purcell M. NW10 ... 38Ua 68
Purcell Rd. UB6: G'frd ... 43Da 85
Purcell Room ... 6J 223 (46Pb 90)
(in Southbank Cen.)
Purcells Av. HA8: Edg ... 22Qa 47
Purcell's Cl. KT21: Asht ... 90Pa 173
Purcell St. N1 ... 1H 219 (40Ub 71)
Purcell Way SS17: Stan H ... 1L 101
Purchese St. NW1 ... 1E 216 (40Mb 70)
Purday Ho. W10 ... 41Ab 88
(off Bruckner St.)
Purdon Ho. SE15 ... 53Wb 113
(off Oliver Goldsmith Est.)
Purdy Ct. KT4: Wor Pk ... 75Wa 154
Purdy St. E3 ... 42Dc 92
PureGym Aldgate ... 2K 225 (44Vb 91)
(off Houndsditch)
PureGym Bayswater ... 45Db 89
(off Moscow Pl.)
PureGym Borehamwood ... 13Qa 29
PureGym Burnham ... 4C 80
PureGym Canary Wharf ... 45Cc 92
PureGym Croydon ... 75Sb 157
(off Crown Hill)
PureGym East India Dock ... 45Ec 92
(off Clove Cres.)
PureGym Edgware ... 23Qa 47
PureGym Finchley ... 26Db 49
PureGym Goldsworth ... 9P 167
PureGym Hallam St. ... 7A 216 (43Kb 90)
(off Hallam St.)
PureGym Hammersmith Palais ... 49Ya 88
PureGym Holborn ... 7H 217 (43Pb 90)
(off Theobald's Rd.)
PureGym Limehouse ... 44Ac 92
PureGym London Wall ... 1F 225 (43Tb 91)
(off London Wall)
PureGym Marylebone, Balcombe St. ... 7F 215 (43Hb 89)
PureGym Muswell Hill ... 28Kb 50
PureGym New Barnet ... 14Fb 31
PureGym Northolt ... 39Ca 65
PureGym Piccadilly ... 6D 222 (46Mb 90)
(off Regent St.)
PureGym Purley ... 82Sb 177
PureGym Putney ... 56Ab 110
PureGym South Kensington ... 5D 226 (49Gb 89)
PureGym Southgate ... 18Mb 32
PureGym St Pauls ... 1D 224 (43Sb 91)
(off Little Britain)
PureGym Staines ... 63H 127
PureGym Sydenham ... 63Yb 136
PureGym Tower Hill ... 4K 225 (45Vb 91)
(off America Sq.)
PureGym Wembley ... 36Na 67
Purelake M. SE13 ... 55Fc 115
(off Marischal Rd.)
PURFLEET ... 49Qd 97
Purfleet By-Pass RM19: Purf ... 49Rd 97
Purfleet Deep Wharf RM19: Purf ... 51Sd 120
Purfleet Heritage & Military Cen. ... 49Pd 97
Purfleet Ind. Access Rd. RM15: Avel ... 48Qd 97
Purfleet Ind. Pk. RM15: Avel ... 46Pd 97
(not continuous)
Purfleet Rd. RM15: Avel ... 47Qd 97
Purfleet Station (Rail) ... 50Qd 97
Purfleet Thames Terminal RM19: Purf ... 52Sd 120
Purkis Cl. UB8: Hil ... 45S 84
Purland Cl. RM8: Dag ... 32Bd 75
Purland Rd. SE28 ... 47Vc 95
(not continuous)
Purleigh Av. IG8: Wfd G ... 23Nc 54
PURLEY ... 83Qb 176
Purley Av. NW2 ... 33Ab 68
Purley Bury Av. CR8: Purl ... 83Sb 177
Purley Bury Cl. CR8: Purl ... 83Sb 177
Purley Cl. IG5: Ilf ... 26Qc 54
PURLEY CROSS ... 83Qb 176
Purley Downs Golf Course ... 83Tb 177
Purley Downs Rd. CR2: Sande ... 82Sb 177
Purley Downs Rd. CR8: Purl ... 82Sb 177
Purley Hill CR8: Purl ... 84Rb 177
Purley Knoll CR8: Purl ... 83Pb 176
Purley Leisure Cen. ... 83Qb 176
Purley Oaks Rd. CR2: Sande ... 81Tb 177
Purley Oaks Station (Rail) ... 81Tb 177
Purley Pde. CR8: Purl ... 83Qb 176
Purley Pk. Rd. CR8: Purl ... 82Rb 177
Purley Pl. N1 ... 38Rb 71
Purley Ri. CR8: Purl ... 84Pb 176
Purley Rd. CR2: S Croy ... 80Tb 157
Purley Rd. CR8: Purl ... 83Qb 176
Purley Rd. N9 ... 20Ub 33
Purley Station (Rail) ... 83Qb 176
Purley Va. CR8: Purl ... 85Rb 177
Purley Vw. Ter. CR2: S Croy ... 80Tb 157
(off Sanderstead Rd.)
PURLEY WAR MEMORIAL HOSPITAL ... 83Qb 176
Purley Way CR0: C'don ... 73Pb 156
Purley Way CR0: Wadd ... 73Pb 156
Purley Way CR8: Purl ... 82Qb 176
Purley Way Cen., The ... 75Pb 156
Purley Way Cres. CR0: C'don ... 73Pb 156
Purlieu Way CM16: They B ... 7Uc 22
Purlings Rd. WD23: Bush ... 15Da 27
Purneys Rd. SE9 ... 56Mc 115
Purrett Rd. SE18 ... 50Vc 95
Purser Ho. SW2 ... 58Qb 112
Pursers Ct. SL2: Slou ... 4J 81
Pursers Cross Rd. SW6 ... 53Bb 111
Pursewardens Cl. W13 ... 46La 86
Pursey Cl. TN15: W King ... 79Ud 164
Pursley Gdns. WD6: Bore ... 10Qa 15
Pursley Rd. NW7 ... 24Xa 48
Purton Ct. SL2: Farn R ... 8G 60
Purton La. SL2: Farn C ... 8G 60
Purton La. SL2: Farn R ... 8G 60
Purves Rd. NW10 ... 41Xa 88
Purvis Ho. CR0: C'don ... 73Tb 157
Pusey Ho. E14 ... 44Cc 92
(off Saracen St.)
Puteaux Ho. E2 ... 40Zb 72
(off Mace St.)
PUTNEY ... 56Za 110
Putney Arts Theatre ... 56Za 110
Putney Bri. ... 55Ab 110
Putney Bri. App. SW6 ... 55Ab 110
Putney Bri. Rd. SW15 ... 56Ab 110
Putney Bri. Rd. SW18 ... 56Cb 111
Putney Bridge Station (Underground) ... 55Bb 111

Putney Comn. SW15 55Ya 110
Putney Exchange (Shop. Cen.) 56Za 110
Putney Gdns. RM6: Chad H 29Xc 55
Putney Heath SW15 59Xa 110
PUTNEY HEATH 58Ya 110
Putney Heath La. SW15 58Za 110
Putney High St. SW15 56Za 110
Putney Hill SW15 59Za 110
(not continuous)
Putney Leisure Cen. 56Ya 110
Putney Pk. Av. SW15 56Wa 110
Putney Pk. La. SW15 56Xa 110
(not continuous)
Putney Rd. EN3: Enf W 8Zb 20
Putney Station (Rail) 56Ab 110
PUTNEY VALE 62Wa 132
Putney Va. Crematorium 61Xa 132
Putney Wharf SW15 55Ab 110
Puttenham Cl. WD19: Wat 19Y 27
Putt in the Pk. 56Bb 111
Puttocks Cl. AL9: Wel G 5E 8
Puttocks Dr. AL9: Wel G 5E 8
Pycroft Way N9 21Vb 51
Pycombe Cnr. N12 21Bb 49
Pyecombe Cnr. N12 21Bb 49
Pyghtle, The UB9: Den 31J 63
Pyghtle Footpath UB9: Den 32J 63
Pylbrook Rd. SM1: Sutt 76Cb 155
Pyle Cl. KT15: Add 77L 149
Pyle Hill GU22: Wok 6P 187
PYLE HILL 6P 187
Pylon Way CR0: Bedd 74Nb 156
Pym Cl. EN4: E Barn 15Fb 31
Pymers Mead SE21 60Sb 113
Pymmes Brook Dr. EN4: E Barn 14Gb 31
Pymmes Brook Ho. N10 24Jb 50
Pymmes Cl. N13 22Pb 50
Pymmes Cl. N17 25Xb 51
Pymmes Gdns. Nth. N9 20Wb 33
Pymmes Gdns. Sth. N9 20Wb 33
Pymmes Grn. Rd. N11 21Kb 50
Pymmes Rd. N13 23Nb 50
Pym Orchard TN16: Bras 96Yc 201
Pym Pl. RM17: Grays 49Ce 99
Pynchester Cl. UB10: Ick 33Q 64
Pynefield Ho. WD3: Rick 17H 25
Pyne Rd. KT6: Surb 74Qa 153
Pynest Grn. La. EN9: Lough 10Jc 21
Pynest Grn. La. EN9: Walt A 10Jc 21
Pyne Ter. SW19 60Ab 110
(off Windlesham Gro.)
Pynfolds SE16 47Xb 91
Pynham Cl. SE2 48Xc 95
Pynnacles Cl. HA7: Stan 22Ka 46
Pynnersmead SE24 57Sb 113
Pyramid Ct. KT1: King T 68Pa 131
(off Cambridge Rd.)
Pyramid Ho. TW4: Houn 54Aa 107
Pyrcroft La. KT13: Weyb 78R 150
Pyrcroft Rd. KT16: Chert 73G 148
PYRFORD 88J 169
Pyrford Comn. Rd. GU22: Pyr 88F 168
Pyrford Ct. GU22: Pyr 89G 168
Pyrford Golf Course 89K 169
PYRFORD GREEN 89K 169
Pyrford Heath GU22: Pyr 88H 169
Pyrford Ho. SW9 56Rb 113
Pyrford Lock GU23: Wis 88L 169
Pyrford Rd. GU22: Pyr 86J 169
Pyrford Rd. KT14: W Byf 85J 169
PYRFORD VILLAGE 90J 169
Pyrford Wood Est. GU22: Pyr 88H 169
Pyrford Woods GU22: Pyr 87G 168
Pyrford Woods Cl. GU22: Pyr 87H 169
Pyrian Cl. GU22: Wok 88F 168
Pyrland Rd. N5 36Tb 71
Pyrland Rd. TW10: Rich 58Pa 109
Pyrles Grn. IG10: Lough 11Rc 36
Pyrles La. IG10: Lough 12Rc 36
Pyrmont Gro. SE27 62Rb 135
Pyrmont Rd. W4 51Qa 109
Pytchley Cres. SE19 65Sb 135
Pytchley Rd. SE22 55Ub 113

Q

Q Bldg., The E15 37Gc 73
(off The Grove)
QED - Queen Elizabeth Distribution Pk. RM19: Purf 50Td 98
QPR Training Academy & Sports Complex 48Fa 86
Quad Ct. SE1 3J 231 (48Ub 91)
(off Grigg's Pl.)
Quadrangle, The E15 37Gc 73
Quadrangle, The SE24 57Sb 113
Quadrangle, The SW10 53Eb 111
Quadrangle, The SW6 52Ab 110
Quadrangle, The W2 2D 220 (44Gb 89)
Quadrangle Cl. SE1 5H 231 (49Ub 91)
Quadrangle M. HA7: Stan 24La 46
Quadrant, The DA7: Bex 52Zc 117
Quadrant, The HA2: Harr 27Fa 46
Quadrant, The HA8: Edg 23Qa 47
(off Manor Pk. Cres.)
Quadrant, The KT17: Eps 85Ua 174
Quadrant, The RM19: Purf 49Sd 98
Quadrant, The SM2: Sutt 79Eb 155
Quadrant, The SW20 67Ab 132
Quadrant, The TW9: Rich 56Ma 109
Quadrant, The W10 41Za 88
Quadrant, The WD3: Rick 17N 25
Quadrant Arc. RM1: Rom 29Gd 56
Quadrant Arc. W1 5C 222 (45Lb 90)
(off Regent St.)
Quadrant Bus. Cen. NW6 39Ab 68
Quadrant Cl. NW4 29Xa 48
Quadrant Ct. DA9: Ghithe 56Vd 120
Quadrant Ct. HA9: Wemb 35Pa 67
Quadrant Ctyd., The KT13: Weyb 77Q 150
(off Quadrant Way)
Quadrant Gro. NW5 36Hb 69
Quadrant Ho. E1 45Wb 91
(off Nesham St.)
Quadrant Ho. E15 41Gc 93
(off Durban Rd.)
Quadrant Ho. SE1 6B 224 (46Rb 91)
(off Burrell St.)
Quadrant Rd. CR7: Thor H 70Rb 135
Quadrant Rd. TW9: Rich 56Ma 109
Quadrant Wlk. E14 48Dc 92
(off Lanterns Way)
Quadrant Way KT13: Weyb 77Q 150
Quadrivium Point SL1: Slou 6G 80
Quad Rd. HA9: Wemb 34Ma 67
Quaggy Wlk. SE3 56Jc 115
Quail Gdns. CR2: Sels 82Ac 178
Quain Mans. W14 51Ab 110
(off Queen's Club Gdns.)
Quainton St. NW10 34Ta 67
Quaker Cl. TN13: S'oaks 95Md 203
Quaker Ct. E1 6K 219 (42Vb 91)
(off Quaker St.)
Quaker Ct. EC1 5E 218 (42Sb 91)
(off Banner St.)
Quaker Dr. RM3: Rom 22Ld 57
Quaker La. EN9: Walt A 6Ec 20
Quaker La. UB2: S'hall 48Ca 85
Quakers Cl. DA3: Hartl 69Ae 143
Quakers Course NW9 25Va 48
Quakers Hall La. TN13: S'oaks 94Ld 203
Quakers La. EN6: Pot B 2Db 17
Quakers La. TW7: Isle 52Ja 108
Quakers Pl. E7 36Mc 73
Quaker St. E1 6K 219 (42Vb 91)
Quakers Wlk. N21 15Tb 33
Quality Ct. WC2 2K 223 (44Qb 90)
(off Chancery La.)
Quality St. RH1: Mers 100Kb 196
Quant Bldg. E17 28Cc 52
Quantock Cl. SL3: L'ly 50C 82
Quantock Cl. UB3: Harl 52T 106
Quantock Dr. KT4: Wor Pk 75Ya 154
Quantock Gdns. NW2 33Za 68
Quantock Ho. N16 32Vb 71
Quantock M. SE15 54Wb 113
Quantock Rd. DA7: Bex 54Gd 118
Quantum Ct. E1 45Yb 92
(off King David La.)
Quarles Cl. RM5: Col R 24Cd 56
Quarles Pk. Rd. RM6: Chad H 30Xc 55
Quarrendon St. SW6 54Cb 111
Quarr Rd. SM5: Cars 72Fb 155
Quarry, The RH3: B'wth 4A 206
Quarry Cl. DA11: Grav'nd 9B 122
Quarry Cl. KT22: Lea 93Ma 193
Quarry Cl. RH8: Oxt 2J 211
Quarry Cotts. RH2: Reig 3K 207
Quarry Cotts. TN13: S'oaks 95Jd 202
Quarry Gdns. KT22: Lea 93Ma 193
Quarry Hill RM17: Grays 50Ce 99
Quarry Hill TN15: S'oaks 95Md 203
Quarry Hill Pk. RH2: Reig 3L 207
Quarry Hill Rd. TN15: Bor G 92Be 205
Quarry M. RM19: Purf 49Qd 97
Quarry Pk. Rd. SM1: Sutt 79Bb 155
Quarry Path RH8: Oxt 3J 211
Quarry Ri. SM1: Sutt 79Bb 155
Quarry Rd. RH8: Oxt 2J 211
Quarry Rd. RH9: G'stone 100Yb 198
Quarry Rd. SW18 58Eb 111
Quarryside Bus. Pk. RH1: Redh 3B 208
Quarry Vw. DA9: Ghithe 58Wd 120
(off Woodpecker Dr.)
Quarterdeck, The E14 47Cc 92
Quarter Ho. SW18 56Eb 111
Quartermaine Av. GU22: Wok 94B 188
Quartermass Cl. HP1: Hem H 1J 3
Quartermass Rd. HP1: Hem H 1J 3
Quartermaster La. NW7 22Ab 48
Quarters Apts. CR0: C'don 75Tb 157
(off Wellesley Rd.)
Quartz Apts. SE14 51Ac 114
(off Moulding La.)
Quartz Cl. DA8: Erith 52Jd 118
Quartz Ho. HA2: Harr 32Ca 65
Quasar Hemel Hempstead 3M 3
Quastel Ho. SE1 2F 231 (47Tb 91)
(off Long La.)
Quatre Ports E4 22Fc 53
Quaves Rd. SL3: Slou 8M 81
Quay Ho. E14 47Cc 92
(off Admirals Way)
Quay La. DA9: Ghithe 56Xd 120
Quayle Cres. N20 19Eb 31
Quay Rd. IG11: Bark 39Rc 74
Quayside Cotts. E1 46Wb 91
(off Mews St.)
Quayside Ct. SE16 46Zb 92
(off Abbotshade Rd.)
Quayside Ho. E14 46Bc 92
Quayside Ho. E16 44Hc 93
(off Tarling Rd.)
Quayside Ho. TW8: Bford 51Pa 109
Quayside Ho. W10 42Ab 88
Quayside Wlk. KT1: King T 68Ma 131
(off Wadbrook St.)
Quay Sth. Ct. UB9: Hare 25J 43
Quay Vw. Apts. E14 48Cc 92
(off Arden Cres.)
Quay W. Ct. WD3: W Hyd 24H 43
Quebec Av. TN16: Westrm 98Tc 200
Quebec Cotts. TN16: Westrm 99Tc 200
Quebec House 98Tc 200
Quebec M. W1 3G 221 (44Hb 89)
Quebec Rd. IG1: Ilf 31Rc 74
Quebec Rd. IG2: Ilf 31Sc 74
Quebec Rd. RM18: Tilb 4C 122
Quebec Rd. UB4: Yead 44Y 85
Quebec Sq. TN16: Westrm 98Tc 200
Quebec Way SE16 47Zb 92
Quebec Wharf E14 44Bc 92
Quebec Wharf E8 39Ub 71
(off Kingsland Rd.)
Quedgeley Ct. SE15 51Vb 113
(off Ebley Cl.)
Queen Adelaide Ct. SE20 65Yb 136
Queen Adelaide Rd. SE20 65Yb 136
Queen Adelaide's Ride SL4: Wink 8B 102
Queen Alexandra Mans. WC1 4F 217 (41Nb 90)
(off Bidborough St.)
Queen Alexandra's Ct. SW19 64Bb 133
Queen Alexandra's Way KT19: Eps 83Qa 173
Queen Anne Av. BR2: Broml 69Hc 137
Queen Anne Dr. KT10: Clay 80Ga 152
Queen Anne Ga. DA7: Bex 55Zc 117
Queen Anne Ho. E16 46Jc 93
(off Hardy Av.)
Queen Anne M. W1 1A 222 (43Kb 90)
Queen Anne Rd. E9 37Zb 72
Queen Anne's Cl. SL4: Wind 2F 124
Queen Anne's Cl. TW2: Twick 62Fa 130
Queen Anne's Ct. SE10 50Fc 93
(off Park Row)
Queen Anne's Gdns. CR4: Mitc 69Hb 133
Queen Anne's Gdns. EN1: Enf 16Ub 33
Queen Anne's Gdns. KT22: Lea 93Ka 192
Queen Anne's Gdns. W4 48Ua 88
Queen Anne's Gdns. W5 47Na 87
Queen Anne's Ga. SW1 2D 228 (47Mb 90)
Queen Anne's Gro. EN1: Enf 17Tb 33
Queen Anne's Gro. W4 48Ua 88
Queen Anne's Gro. W5 47Na 87
Queen Anne's Pl. EN1: Enf 16Ub 33
Queen Anne's Ride SL4: Wind 4E 124
Queen Anne's Rd. SL4: Wind 6G 102
Queen Annes Sq. SE1 49Wb 91
(off Monnow Rd.)
Queen Anne's Ter. KT22: Lea 93Ka 192
Queen Anne St. W1 2K 221 (44Kb 90)
Queen Anne Ter. E1 45Xb 91
(off Sovereign Cl.)
Queen Ann's Ct. SL4: Wind 3H 103
(off Peascod St.)
Queenborough Gdns. BR7: Chst 65Tc 138
Queenborough Gdns. IG2: Ilf 28Qc 54
Queenbridge Ind. Pk. RM20: W Thur 51Wd 120
Queen Caroline's Temple 7B 220 (46Fb 89)
Queen Caroline St. W6 50Ya 88
Queen Catherine Ho. SW6 52Db 111
(off Wandon Rd.)
QUEEN CHARLOTTE'S & CHELSEA HOSPITAL 44Wa 88
Queen Charlotte's Cottage 54Ma 109
Queen Charlotte St. SL4: Wind 3H 103
(off Market St.)
Queen Ct. WC1 6G 217 (42Nb 90)
(off Queen Sq.)
Queendale Ct. GU21: Wok 8K 167
Queen Elizabeth II Bri. 55Td 120
Queen Elizabeth II Bri. 54Td 120
Queen Elizabeth II Stadium 12Vb 33
Queen Elizabeth Av. RM18: E Til 9K 101
Queen Elizabeth Bldgs. EC4 4K 223 (45Qb 90)
(off Middle Temple La.)
Queen Elizabeth Ct. EN5: Barn 13Cb 31
Queen Elizabeth Ct. EN9: Walt A 8Ec 20
(off Greenwich Way)
Queen Elizabeth Gdns. SM4: Mord 70Cb 133
Queen Elizabeth Hall 6J 223 (46Pb 90)
QUEEN ELIZABETH HOSPITAL Charlton 52Nc 116
Queen Elizabeth Ho. SW12 59Jb 112
Queen Elizabeth Leisure Cen. 14Bb 31
Queen Elizabeth Olympic Pk. 38Dc 72
Queen Elizabeth Pl. RM18: Tilb 3C 122
Queen Elizabeth Rd. E17 27Ac 52
Queen Elizabeth Rd. KT2: King T 68Pa 131
Queen Elizabeth's Cl. N16 33Tb 71
Queen Elizabeth's Coll. SE10 52Ec 114
Queen Elizabeth's Dr. CR0: New Ad 81Fc 179
Queen Elizabeth's Dr. N14 18Nb 32
Queen Elizabeth's Gdns. CR0: New Ad 82Fc 179
Queen Elizabeth's Hunting Lodge 17Hc 35
Queen Elizabeth St. SE1 1J 231 (47Ub 91)
Queen Elizabeth's Wlk. N16 32Tb 71
Queen Elizabeth's Wlk. SL4: Wind 4J 103
Queen Elizabeth's Wlk. SM6: Bedd 77Mb 156
Queen Elizabeth Wlk. SW13 53Wa 110
Queen Elizabeth Way GU22: Wok 91B 188
Queen Elizabeth Way SL4: Eton 3F 102
Queen Elizabeth Way SL4: Wind 3F 102
Queenhill Rd. CR2: Sels 82Xb 177
Queenhithe EC4 4E 224 (45Sb 91)
Queenhythe Cres. GU4: Jac W 10P 187
Queenhythe Rd. GU4: Jac W 10P 187
Queen Isabella Way EC1 2C 224 (44Rb 91)
(off King Edward St.)
Queen Margaret Flats E2 41Xb 91
(off St Jude's Rd.)
Queen Margaret's Gro. N1 36Ub 71
Queen Mary Av. E18 25Jc 53
Queen Mary Av. RM18: E Til 9L 101
Queen Mary Av. SM4: Mord 71Za 154
Queen Mary Cl. GU22: Wok 88E 168
Queen Mary Cl. KT6: Surb 76Qa 153
Queen Mary Cl. RM1: Rom 30Hd 56
Queen Mary Ct. RM18: E Til 9L 101
Queen Mary Ct. TW19: Stanw 60N 105
Queen Mary Ho. E16 46Kc 93
(off Wesley Av.)
Queen Mary Ho. E18 25Kc 53
Queen Mary Rd. SE19 65Rb 135
Queen Mary Rd. TW17: Shep 68S 128
Queen Mary's Av. SM5: Cars 80Hb 155
Queen Mary's Av. WD18: Wat 14U 26
Queen Marys Bldgs. SW1 5C 228 (49Lb 90)
(off Stillington St.)
Queen Mary's Ct. SE10 51Fc 115
(off Park Row)
Queen Marys Ct. EN9: Walt A 7Ec 20
(off Harrison Rd.)
Queen Mary's Dr. KT15: New H 82H 169
QUEEN MARY'S HOSPITAL FOR CHILDREN 74Eb 155
QUEEN MARY'S HOSPITAL, ROEHAMPTON 58Wa 110
QUEEN MARY'S HOSPITAL, SIDCUP 65Wc 139
Queen Mary's Ho. SW15 58Wa 110
Queen Mary University of London Charterhouse Sq. 6C 218 (42Rb 91)
Queen Mary University of London Lincoln's Inn Flds. Campus 2H 223 (44Pb 90)
(off Remnant St.)
Queen Mary University of London Mile End Campus 42Ac 92
Queen Mary University of London W. Smithfield Campus 1C 224 (43Rb 91)
(off Giltspur St.)
Queen Mothers Dr. UB9: Den 30H 43
Queen Mother Sports Cen., The 5B 228 (49Lb 90)
(off Vauxhall Bri. Rd.)
Queen of Denmark Ct. SE16 48Bc 92
Queens Acre SL4: Wind 6H 103
Queens Acre SM3: Cheam 80Za 154
Queens Acre Ho. SL4: Wind 5H 103
Queens All. CM16: Epp 3Vc 23
Queen's Av. KT14: Byfl 84M 169
Queen's Av. UB6: G'frd 44Da 85
Queen's Av. WD18: Wat 14V 26
Queens Av. HA7: Stan 27La 46
Queens Av. IG8: Wfd G 22Kc 53
Queens Av. N10 27Jb 50
Queens Av. N20 19Fb 31
Queens Av. N21 18Rb 33
Queens Av. N3 24Eb 49
Queens Av. TW13: Hanw 63Y 129
Queensberry Ho. TW9: Rich 57La 108
Queensberry M. W. SW7 5B 226 (49Fb 89)
Queensberry Pl. E12 36Mc 73
Queensberry Pl. SW7 5B 226 (49Fb 89)
Queensberry Pl. TW9: Rich 57Ma 109
(off Friars La.)
Queensberry Way SW7 5B 226 (49Fb 89)
Queensborough Ct. N3 28Bb 49
(off Tillingbourne Gdns.)
Queensborough M. W2 45Eb 89
Queensborough Pas. W2 45Eb 89
(off Queensborough M.)
Queensborough Studios W2 45Eb 89
(off Queensborough M.)
Queensborough Ter. W2 45Db 89
Queensbridge Ct. E2 1K 219 (39Vb 71)
(off Queensbridge Rd.)
Queensbridge Pk. TW7: Isle 57Ga 108
Queensbridge Rd. E2 39Vb 71
Queensbridge Rd. E8 37Vb 71
Queensbridge Sports & Community Cen. 38Vb 71
QUEENSBURY 27Na 47
Queensbury Circ. Pde. HA3: Kenton 27Na 47
Queensbury Circ. Pde. HA7: Kenton 27Na 47
Queensbury Rd. HA0: Wemb 40Pa 67
Queensbury Rd. NW9 31Ta 67
Queensbury Sta. Pde. HA8: Edg 27Pa 47
Queensbury Station (Underground) 27Pa 47
Queensbury St. N1 38Sb 71
Queen's Cir. SW11 52Kb 112
Queen's Cl. KT10: Esh 77Da 151
Queen's Cl. SL4: Old Win 7L 103
Queens Cl. GU24: Bisl 8E 166
Queens Cl. HA8: Edg 22Qa 47
Queens Cl. KT20: Walt H 96Wa 194
Queens Cl. SM6: W'gton 78Kb 156
Queen's Club Gdns. W14 51Ab 110
Queen's Club (Tennis Courts), The 50Ab 88
Queens Club Ter. W14 51Bb 111
(off Normand Rd.)
Queen's Ct. NW8 1B 214 (40Fb 69)
(off Queen's Ter.)
Queen's Ct. SM2: Sutt 83Cb 175
Queen's Ct. WD17: Wat 13Y 27
(off Queen's Rd.)
Queens Ct. AL1: St A 2F 6
Queens Ct. CR2: S Croy 78Sb 157
(off Warham Rd.)
Queens Ct. CR7: Thor H 71Qb 156
Queens Ct. E11 31Gc 73
Queens Ct. GU22: Wok 90B 168
Queens Ct. HA3: Kenton 26Ka 46
Queens Ct. IG9: Buck H 19Mc 35
Queens Ct. KT13: Weyb 78T 150
Queens Ct. KT19: Ewe 82Ua 174
Queens Ct. NW11 29Bb 49
Queens Ct. NW6 36Db 69
Queens Ct. RH1: Redh 5A 208
(off St Anne's Mt.)
Queens Ct. RH1: S Nut 10E 208
Queens Ct. RM11: Horn 32Md 77
Queens Ct. SE16 48Vb 91
(off Old Jamaica Rd.)
Queens Ct. SE23 61Yb 136
Queens Ct. SL1: Slou 5K 81
Queens Ct. TW10: Rich 58Pa 109
Queens Ct. TW18: Staines 65M 127
Queens Ct. W2 45Db 89
(off Queensway)
Queen's Ct. WD6: Bore 11Pa 29
(off Bennington Dr.)
Queenscourt HA9: Wemb 35Na 67
Queen's Ct. Ride KT11: Cobh 85W 170
Queen's Cres. NW5 37Jb 70
Queen's Cres. TW10: Rich 57Pa 109
Queenscroft Rd. SE9 57Mc 115
Queensdale Cres. W11 46Za 88
(not continuous)
Queensdale Pl. W11 46Ab 88
Queensdale Rd. W11 46Za 88
Queensdale Wlk. W11 46Ab 88
Queen's Diamond Jubilee Galleries, The 3F 229 (48Nb 90)
Queensdown Rd. E5 35Xb 71
Queen's Dr. EN8: Walt C 6Cc 20
Queen's Dr. KT22: Oxs 83Ea 172
Queen's Dr. KT5: Surb 73Qa 153
Queen's Dr. KT7: T Ditt 72Ja 152
Queen's Dr. N4 33Rb 71
Queen's Dr. SL3: Ful 38B 62
Queen's Dr. SL3: Wex 38B 62
Queens Dr. E10 31Cc 72
Queens Dr. TN14: S'oaks 92Ld 203
Queens Dr. W3 44Pa 87
Queens Dr. W5 44Pa 87
Queens Dr. WD5: Ab L 4V 12
Queens Dr., The WD3: Rick 17H 25
Queen's Elm Pde. SW3 7C 226 (50Fb 89)
(off Old Church St.)
Queen's Elm Sq. SW3 7C 226 (50Fb 89)
Queen's Farm DA12: Shorne 10N 123
Queen's Farm Rd. DA12: Grav'nd 1N 145
Queen's Farm Rd. DA12: Shorne 1N 145
Queensferry Wlk. N17 28Xb 51
Queensfield Ct. SM3: Cheam 77Ya 154
Queen's Gallery 2A 228 (47Kb 90)
Queen's Gdns. NW4 29Ya 48
Queen's Gdns. RM13: Rain 40Fd 76
Queen's Gdns. TW5: Hest 53Aa 107
Queen's Gdns. W2 4A 220 (45Eb 89)
Queens Gdns. W5 42La 86
Queens Gdns. DA2: Dart 60Rd 119
Queens Gdns. RM14: Upm 30Vd 58
Queen's Ga. SW7 2A 226 (47Eb 89)
Queens Ga. WD17: Wat 13Y 27
(off Lord St.)
Queensgate EN8: Walt C 6Bc 20
Queensgate KT11: Cobh 84Z 171
Queensgate Cen. RM17: Grays 50Ce 99
Queens Ga. Cotts. SL4: Wind 6H 103
Queensgate Ct. N12 22Db 49
Queen's Ga. Gdns. SW7 4A 226 (48Eb 89)
Queens Ga. Gdns. BR7: Chst 67Tc 138
Queens Ga. Gdns. SW15 56Xa 110
Queensgate Ho. E3 40Bc 72
(off Hereford Rd.)
Queen's Ga. M. SW7 3A 226 (48Eb 89)
Queensgate M. BR3: Beck 67Ac 136
Queen's Ga. Pl. SW7 4A 226 (48Eb 89)
Queensgate Pl. NW6 38Cb 69
Queen's Ga. Pl. M. SW7 4A 226 (48Eb 89)
Queen's Ga. Ter. SW7 3A 226 (48Eb 89)
Queens Ga. Vs. E9 38Ac 72
Queen's Gro. NW8 1B 214 (39Fb 69)
Queens Gro. Rd. E4 18Fc 35
Queen's Gro. Studios NW8 1B 214 (39Fb 69)
Queen's Head Pas. EC4 2D 224 (44Sb 91)
Queen's Head St. N1 39Rb 71
Queen's Head Yd. SE1 7F 225 (46Tb 91)
(off Borough High St.)
Queens Hill Ri. SL5: Asc 9A 124
QUEEN'S HOSPITAL 31Gd 76
Queens Ho. SE17 51Tb 113
(off Merrow St.)
Queens Ho. SW8 52Nb 112
(off Sth. Lambeth Rd.)
Queens Ho. TW11: Tedd 65Ha 130
Queens Ho. W2 45Db 89
(off Queensway)
Queen's House, The 51Fc 115
(within National Maritime Mus.)
Queenshurst Sq. KT2: King T 67Na 131
Queen's Ice & Bowl 45Db 89
Queenside M. RM12: Horn 33Nd 77
Queen's Keep TW1: Twick 58La 108
Queensland Av. N18 23Sb 51
Queensland Av. SW19 67Db 133
Queensland Cl. E17 26Bc 52
Queensland Ho. E16 46Qc 94
(off Rymill St.)
Queensland Rd. N7 35Qb 70
Queens La. N10 27Kb 50
Queen's Mans. W6 49Za 88
(off Brook Grn.)
Queen's Mkt. E13 39Lc 73
Queens Mead HA8: Edg 23Pa 47
Queensmead KT22: Oxs 83Ea 172
Queensmead NW8 39Fb 69
Queensmead SL3: Dat 2M 103
Queensmead Av. KT17: Ewe 82Xa 174
Queens Mead Rd. BR2: Broml 68Hc 137
Queensmead Sports Cen. 36Z 65
Queensmere Cl. SW19 61Za 132
Queensmere Ct. SW13 52Va 110
Queensmere Rd. SL1: Slou 7K 81
Queensmere Rd. SW19 61Za 132
Queensmere Shop. Cen. 7K 81
Queen's M. W2 45Db 89
Queensmill Rd. SW6 52Za 110
Queen's Pde. N11 22Hb 49
(off Friern Barnet Rd.)
Queen's Pde. NW2 37Ya 68
(off Willesden La.)
Queens Pde. N8 28Rb 51
Queens Pde. NW4 29Ya 48
(off Queens Rd.)
Queens Pde. W5 44Pa 87
Queen's Pde. Cl. N11 22Hb 49
QUEENS PARK 40Ab 68
Queens Pk. Ct. W10 41Za 88
Queen's Pk. Gdns. TW13: Felt 62V 128
Queen's Pk. Rangers FC 46Xa 88
Queen's Pk. Rd. CR3: Cat'm 95Ub 197
Queen's Pk. Rd. RM3: Hrld W 25Qd 57
Queen's Park Station (Underground & Overground) 40Bb 69
Queens Pas. BR7: Chst 65Rc 138
Queens Pl. GU24: Chob 5E 166
Queens Pl. SM4: Mord 70Cb 133
Queens Pl. WD17: Wat 13Y 27
Queen's Prom. KT1: King T 70Ma 131
Queen's Prom. KT1: Surb 70Ma 131
Queen Sq. WC1 6G 217 (42Nb 90)
Queen Sq. Pl. WC1 6G 217 (42Nb 90)
(off Queen Sq.)
Queen's Quay EC4 4E 224 (45Sb 91)
(off Up. Thames St.)
Queens Reach KT1: King T 68Ma 131
Queens Reach KT8: E Mos 70Ga 130
Queens Ride SW13 55Wa 110
Queens Ri. TW10: Rich 58Pa 109
Queen's Rd. WD17: Wat Carey Pl. 14Y 27
(not continuous)
Queen's Rd. WD17: Wat Orphanage Rd. 12Y 27
Queen's Rd. CR4: Mitc 69Fb 133
Queen's Rd. DA16: Well 54Xc 117
Queen's Rd. DA8: Erith 51Gd 118
Queen's Rd. E17 30Bc 52
Queen's Rd. EN1: Enf 14Ub 33
Queen's Rd. EN8: Walt C 5Ac 20
Queen's Rd. GU21: Knap 10G 166
Queen's Rd. IG10: Lough 13Nc 36
Queen's Rd. IG9: Buck H 19Kc 35
Queen's Rd. KT7: T Ditt 71Ha 152
Queen's Rd. SE14 53Yb 114
Queen's Rd. SE15 53Xb 113
Queen's Rd. SL1: Slou 5K 81
Queen's Rd. SL3: Dat 2M 103
Queen's Rd. SL4: Wind 4G 102
Queen's Rd. SL5: S'hill 1B 146
Queen's Rd. SM2: Sutt 82Cb 175
Queen's Rd. SW14 55Ta 109
Queen's Rd. TW10: Rich 59Pa 109
Queen's Rd. TW11: Tedd 65Ha 130
Queen's Rd. TW12: Hamp H 63Da 129
Queen's Rd. TW13: Felt 60X 107
Queen's Rd. TW20: Egh 64B 126
Queen's Rd. TW3: Houn 55Da 107
Queen's Rd. UB8: Uxb 41L 83
Queen's Rd. W5 44Na 87
Queens Rd. BR1: Broml 68Jc 137
Queens Rd. BR3: Beck 68Ac 136
Queens Rd. BR7: Chst 65Rc 138
Queens Rd. CM14: B'wood 20Yd 40
Queens Rd. CR0: C'don 72Rb 157
Queens Rd. DA12: Grav'nd 2E 144
Queens Rd. E11 31Fc 73
Queens Rd. E13 39Kc 73
Queens Rd. EN5: Barn 13Za 30
Queens Rd. GU24: Bisl 2C 186
Queens Rd. GU24: Brkwd 2C 186
Queens Rd. IG11: Bark 37Sc 74
Queens Rd. KT12: Hers 77V 150
Queens Rd. KT13: Weyb 77S 150
Queens Rd. KT2: King T 66Qa 131
Queens Rd. KT3: N Mald 70Va 132
Queens Rd. N11 24Nb 50
Queens Rd. N3 25Eb 49
Queens Rd. N9 20Xb 33
Queens Rd. NW4 29Ya 48
Queens Rd. SL4: Eton W 10D 80
Queens Rd. SM4: Mord 70Cb 133
Queens Rd. SM6: W'gton 78Kb 156
Queens Rd. SW19 65Bb 133
Queens Rd. TW1: Twick 60Ja 108
Queens Rd. UB2: S'hall 47Z 85
Queens Rd. UB3: Hayes 44U 84
Queens Rd. UB7: W Dray 47P 83
Queens Rd. Est. EN5: Barn 13Za 30
Queens Road Station, Peckham (Rail & Overground) 53Yb 114
Queens Rd. W. E13 40Jc 73
Queen's Row SE17 51Tb 113
Queen's Sq., The HP2: Hem H 2P 3
Queens St. TW15: Ashf 63P 127
Queen's Ter. E13 39Kc 73
Queen's Ter. NW8 1B 214 (39Fb 69)
Queens Ter. E1 42Yb 92
(off Cephas St.)
Queens Ter. KT7: T Ditt 72Ja 152
(off Queens Dr.)
Queens Ter. SL4: Wind 5H 103
Queens Ter. TW7: Isle 56Ja 108
Queen's Ter. Cotts. W7 47Ga 86
Queen's Theatre London 4D 222 (45Mb 90)
(off Shaftesbury Av.)
Queens Theatre Hornchurch 31Md 77
Queensthorpe M. SE26 63Zb 136
Queensthorpe Rd. SE26 63Zb 136
Queen's Tower 3B 226 (48Fb 89)
(within Imperial College London)
Queenstown Gdns. RM13: Rain 41Hd 96
Queenstown M. SW8 54Kb 112
Queenstown Rd. SW11 51Kb 112
Queenstown Rd. SW8 54Kb 112
Queenstown Road Station, Battersea (Rail) 53Kb 112

Queen St. AL3: St A 2A 6
Queen St. CM14: W'ley 22Yd 58
Queen St. CR0: C'don 77Sb 157
Queen St. DA12: Grav'nd 8D 122
Queen St. DA7: Bex 55Bd 117
Queen St. DA8: Erith 51Gd 118
Queen St. EC4 4E 224 (45Sb 91) (not continuous)
Queen St. KT16: Chert 74J 149
Queen St. N17 23Ub 51
Queen St. RM7: Rom 30Fd 56
Queen St. W1 6K 221 (46Kb 90)
Queen St. WD4: Chfd 4J 11
Queen St. Pl. EC4 5E 224 (45Sb 91)
Queensville Rd. SW12 59Mb 112
Queen's Wlk. SW1: Green Pk 7B 222 (46Lb 90)
Queen's Wlk. N5: Highbury 36Rb 71
Queen's Wlk. TW15: Ashf 63M 127
Queen's Wlk. W5 42La 86
Queens Wlk. E4 18Fc 35
Queens Wlk. HA1: Harr 28Ga 46
Queens Wlk. HA4: Ruis 33Y 65
Queens Wlk. NW9 33Sa 67
Queen's Wlk., The SE1 6J 223 (46Pb 90)
Queens Wlk. Ter. HA4: Ruis 34Y 65
Queen's Way GU24: Brkwd 1B 186
Queen's Way NW4 29Ya 48
Queens Way AL3: St A 2B 6 (off Chequer St.)
Queens Way EN8: Walt C 6Bc 20
Queens Way TW13: Hanw 63Y 129
Queens Way WD7: Shenl 4Na 15
Queensway BR4: W W'ck 76Gc 159
Queensway BR5: Pet W 71Sc 160
Queensway CR0: Wadd 79Pb 156
Queensway EN3: Pond E 14Xb 33
Queensway HP1: Hem H 1L 3
Queensway HP2: Hem H 1N 3
Queensway RH1: Redh 5P 207
Queensway TW16: Sun 68X 129
Queensway W2 44Db 89
Queensway, The SL9: Chal P 28A 42
Queensway Bus. Cen. EN3: Pond E 14Xb 33
Queensway Ind. Est. EN3: Pond E 14Yb 34
Queensway M. SE6 63Ec 136 (off Whitefoot La.)
Queensway Nth. KT12: Hers 77Y 151
Queensway Sth. KT12: Hers 78Y 151
Queensway Station (Underground) 45Db 89
Queenswell Av. N20 20Gb 31
Queenswood Av. CM13: Hut 14Fe 41
Queenswood Av. CR7: Thor H 71Qb 156
Queenswood Av. E17 25Ec 52
Queenswood Av. SM6: Bedd 77Mb 156
Queenswood Av. TW12: Hamp 65Da 129
Queenswood Av. TW3: Houn 54Ba 107
Queenswood Ct. KT2: King T 67Qa 131
Queenswood Ct. SE27 63Tb 135
Queenswood Ct. SW4 57Nb 112
Queenswood Cres. TW20: Eng G 6N 125
Queenswood Cres. WD25: Wat 5W 12
Queenswood Gdns. E11 32Kc 73
Queenswood Ho. CM14: B'wood 19Zd 41 (off Eastfield Rd.)
Queenswood Lodge RM2: Rom 27Jd 56
Queenswood Pk. N3 26Ab 48
Queens Wood Rd. N10 30Kb 50
Queenswood Rd. DA15: Sidc 57Vc 117
Queenswood Rd. GU21: Wok 1H 187
Queenswood Rd. SE23 62Zb 136
Queen's Yd. WC1 7C 216 (43Lb 90)
Queens Yd. E9 37Cc 72
QUEEN VICTORIA 77Za 154
Queen Victoria Av. HA0: Wemb 38Ma 67
Queen Victoria Memorial 2B 228 (47Lb 90)
Queen Victoria Rd. GU24: Brkwd 1B 186
Queen Victoria Seaman's Rest E14 44Dc 92 (off E. India Dock Rd.)
Queen Victoria Statue 46Db 89
Queen Victoria St. EC4 4B 224 (45Rb 91)
Queen Victoria Ter. E1 45Xb 91 (off Sovereign Cl.)
Queen Victoria Wlk. SL4: Wind 3J 103
Queenwood Golf Course 79B 148
Queenwood Rd. SS17: Stan H 3K 101
Quelmans Head Ride SL4: Wind 1B 124
Quemerford Rd. N7 36Pb 70
Quendell Wlk. HP2: Hem H 2N 3
Quendon Dr. EN9: Walt A 5Fc 21
Quendon Ho. W10 42Ya 88 (off Sutton Way)
Quenington Ct. SE15 51Vb 113
Quennell Cl. KT21: Asht 91Pa 193
Quennell Way CM13: Hut 17Ee 41
Quentin Ho. SE1 1B 230 (47Rb 91) (off Chaplin Cl.)
Quentin Pl. SE13 55Gc 115
Quentin Rd. SE13 55Gc 115
Quentins Dr. TN16: Big H 88Rc 180
Quentins Wlk. TN16: Big H 88Rc 180 (off St Anns Way)
Quentin Way GU25: Vir W 10M 125
Quernmore Cl. BR1: Broml 65Jc 137
Quernmore Rd. BR1: Broml 65Jc 137
Quernmore Rd. N4 30Qb 50
Querrin St. SW6 54Eb 111
Quest, The W11 45Ab 88 (off Clarendon Rd.)
Quested Ct. E8 36Xb 71 (off Brett Rd.)
Questor DA1: Dart 61Nd 141
Questors Theatre, The 45La 86
Quex Ct. NW6 39Db 69 (off West End La.)
Quex M. NW6 39Cb 69
Quex Rd. NW6 39Cb 69
Quiberon Ct. E13 39Jc 73 (off Pelly Rd.)
Quiberon Ct. TW16: Sun 69W 128
Quickley Brow WD3: Chor 16D 24
Quickley La. WD3: Chor 16D 24
Quickley Ri. WD3: Chor 16E 24
Quickmoor La. WD4: Bucks 5K 11
Quick Rd. W4 50Ua 88
Quicks Rd. SW19 66Db 133
Quick St. N1 2B 218 (40Rb 71)
Quick St. M. N1 2B 218 (40Rb 71)
Quickswood NW3 38Gb 69
Quickwood Cl. WD3: Rick 16J 25
Quiet Cl. KT15: Add 77J 149
Quiet Nook BR2: Hayes 76Mc 159
Quill Ho. E2 42Wb 91 (off Cheshire St.)
Quill La. SW15 56Za 110
Quillot, The KT12: Hers 78V 150
Quill St. N4 34Qb 70
Quill St. W5 41Na 87
Quilp St. SE1 1D 230 (47Sb 91) (not continuous)
Quilter Gdns. BR5: Orp 74Yc 161
Quilter Rd. BR5: Orp 74Yc 161
Quilters Pl. SE9 60Sc 116
Quilter St. E2 41Wb 91
Quilter St. SE18 50Vc 95
Quilter Way RM3: Rom 22Md 57
Quilting Ct. SE16 47Zb 92 (off Garter Way)
Quinbrookes SL2: Slou 4N 81
Quince Cl. SL5: S'hill 10B 124
Quince Dr. GU24: Bisl 7F 166
Quince Ho. SE13 54Dc 114 (off Quince Rd.)
Quince Rd. SE13 54Dc 114
Quinces Cft. HP1: Hem H 1J 3
Quince Tree Cl. RM15: S Ock 41Yd 98
Quincy Rd. TW20: Egh 64C 126
Quinn Cl. E2 40Yb 72
Quinnell Cl. SE18 50Vc 95
Quinta Dr. EN5: Barn 15Xa 30
Quintain Ho. KT1: King T 68Ma 131 (off Wood St.)
Quintet, The KT12: Walt T 74W 150
Quintin Av. SW20 67Bb 133
Quintin Cl. HA5: Eastc 28X 45
Quinton Cl. BR3: Beck 69Ec 136
Quinton Cl. SM6: W'gton 77Kb 156
Quinton Cl. TW5: Cran 52X 107
Quinton Ct. SE16 49Ac 92 (off Plough Way)
Quinton Ho. SW8 52Nb 112 (off Wyvil Rd.)
Quinton Rd. KT7: T Ditt 74Ja 152
Quinton St. SW18 61Eb 133
Quintrell Cl. GU21: Wok 9M 167
Quixley St. E14 45Fc 93
Quoin Ho. UB7: View 45M 83
Quorn Rd. SE22 56Ub 113

R

Rabbit La. KT12: Hers 80W 150
Rabbit Row W8 46Cb 89
Rabbits Rd. DA4: S Dar 68Td 142
Rabbits Rd. E12 35Nc 74
Rabbs Mill Ho. UB8: Uxb 40M 63
Rabies Heath Rd. RH1: Blet 5L 209
RABLEY 4Ta 15
Rablus Pl. DA4: Farni 72Pd 163
Rabournmead Dr. UB5: N'olt 36Aa 65
Raby Rd. KT3: N Mald 70Ta 131
Raby St. E14 44Ac 92
Raccoon Way TW4: Houn 54Y 107
Racefield Cl. DA12: Shorne 6N 145
RAC Golf Course 89Sa 173
Rachel Cl. IG6: Ilf 27Tc 54
Racine SE5 53Ub 113 (off Sceaux Gdns.)
Rackham Cl. DA16: Well 54Xc 117
Rackham M. SW16 65Lb 134
Rackstraw Ho. NW3 38Hb 69
Racton Rd. SW6 51Cb 111
RADA Chenies St. 7D 216 (43Mb 90) (off Chenies St.)
RADA Gower St. 7E 216 (43Mb 90) (off Gower St.)
RADA Studios 7D 216 (43Mb 90) (off Chenies St.)
Radbourne Av. W5 49La 86
Radbourne Cl. E5 35Zb 72
Radbourne Ct. HA3: Kenton 30Ka 46
Radbourne Cres. E17 26Fc 53
Radbourne Rd. SW12 59Lb 112
Radburn Pl. DA10: Swans 57Ae 121
Radcliff Ct. E3 42Bc 92 (off Joseph St.)
Radcliffe Av. EN2: Enf 11Sb 33
Radcliffe Av. NW10 40Wa 68
Radcliffe Gdns. SM5: Cars 81Gb 175
Radcliffe Ho. CM14: B'wood 20Xd 40
Radcliffe Ho. SE16 49Xb 91 (off Anchor St.)
Radcliffe Ho. SE20 67Wb 135
Radcliffe M. TW12: Hamp H 64Ea 130
Radcliffe Path SW8 54Kb 112
Radcliffe Rd. CR0: C'don 75Vb 157
Radcliffe Rd. HA3: W'stone 26Ja 46
Radcliffe Rd. N21 18Rb 33
Radcliffe Rd. SE1 3J 231 (48Ub 91)
Radcliffe Sq. SW15 58Za 110
Radcliffe Way UB5: N'olt 41Z 85
Radcot Av. SL3: L'ly 48D 82
Radcot Point SE23 62Zb 136
Radcot St. SE11 7A 230 (50Qb 90)
Raddington Rd. W10 43Ab 88
Raddon Twr. E8 37Vb 71 (off Dalston Sq.)
Radfield Dr. DA2: Dart 60Sd 120
Radfield Way DA15: Sidc 59Tc 116
Radford Ct. SE15 52Xb 113 (off Old Kent Rd.)
Radford Est. NW10 41Ua 88
Radford Ho. E14 43Dc 92 (off St Leonard's Rd.)
Radford Ho. N7 36Pb 70
Radford Rd. SE13 58Ec 114
Radford Way IG11: Bark 41Vc 95
Radio La. RM8: Dag 34Xc 75
Radipole Rd. SW6 53Bb 111
UB3: Harl
Radisson Ct. SE1 3H 231 (48Ub 91) (off Long La.)
Radius Apts. N1 2H 217 (40Pb 70) (off Omega Pl.)
Radius Pk. TW14: Felt 56V 106
Radland Rd. E16 44Hc 93
Radleigh Pl. BR3: Beck 65Cc 136
Radlet Av. SE26 62Xb 135
RADLETT 7Ja 14
Radlett Cen., The 8Ja 14
Radlett Cl. E7 37Hc 73
Radlett Golf Academy 4Ja 14
Radlett La. WD7: R'lett 6Ma 15
Radlett La. WD7: Shenl 6Ma 15
Radlett Pk. Golf Course 13Ma 29
Radlett Pk. Rd. WD7: R'lett 6Ja 14
Radlett Pl. NW8 39Gb 69
Radlett Rd. AL2: Col S 1Ga 14
Radlett Rd. AL2: F'mre 1Ga 14
Radlett Rd. WD17: Wat 13Y 27
Radlett Rd. WD24: Wat 13Y 27
Radlett Rd. WD25: A'ham 10Da 13
Radlett Station (Rail) 7Ja 14
Radley Av. IG3: Bark 35Wc 75
Radley Av. IG3: Ilf 35Wc 75
Radley Cl. TW14: Felt 60V 106
Radley Ct. AL1: St A 2C 6 (off Newsome Pl.)
Radley Ct. SE16 47Zb 92
Radley Gdns. HA3: Kenton 28Na 47
Radley Ho. NW1 5F 215 (42Hb 89) (off Gloucester Pl.)
Radley Ho. SE2 47Zc 95 (off Wolvercote Rd.)
Radley M. W8 48Cb 89
Radley Rd. N17 26Ub 51
Radley's La. E18 26Jc 53
Radleys Mead RM10: Dag 37Dd 76
Radley Sq. E5 33Yb 72
Radley Ter. E16 43Hc 93 (off Hermit Rd.)
Radlix Rd. E10 32Cc 72
Radnor Av. DA16: Well 57Xc 117
Radnor Av. HA1: Harr 29Ga 46
Radnor Cl. BR7: Chst 65Uc 138
Radnor Cl. CR4: Mitc 70Nb 134
Radnor Cl. HA3: Hrw W 25Ha 46
Radnor Ct. RH1: Redh 6N 207
Radnor Ct. W7 44Ha 86 (off Copley Cl.)
Radnor Cres. IG4: Ilf 29Pc 54
Radnor Cres. SE18 52Wc 117
Radnor Gdns. EN1: Enf 11Ub 33
Radnor Gdns. TW1: Twick 61Ha 130
Radnor Gro. UB10: Hil 40Q 64
Radnor Ho. EC1 4E 218 (41Sb 91) (off Radnor St.)
Radnor Ho. SW16 68Pb 134
Radnor Lodge W2 3C 220 (44Fb 89) (off Sussex Pl.)
Radnor M. W2 3C 220 (44Fb 89)
Radnor Pl. W2 3D 220 (44Gb 89)
Radnor Rd. HA1: Harr 29Fa 46
Radnor Rd. KT13: Weyb 76Q 150
Radnor Rd. NW6 39Ab 68
Radnor Rd. SE15 52Wb 113
Radnor Rd. TW1: Twick 60Ha 108
Radnor St. EC1 4E 218 (41Sb 91)
Radnor Ter. SM2: Sutt 80Cb 155
Radnor Ter. W14 49Bb 89
Radnor Wlk. CR0: C'don 72Ac 158
Radnor Wlk. E14 49Cc 92 (off Barnsdale Av.)
Radnor Wlk. SW3 7E 226 (50Gb 89)
Radnor Way NW10 42Ra 87
Radnor Way SL3: L'ly 49A 82
Radolphs KT20: Tad 94Za 194
Radstock Av. HA3: Kenton 27Ja 46
Radstock Cl. N11 23Jb 50
Radstock Ho. RM3: Rom 22Md 57 (off Darlington Gdns.)
Radstock St. SW11 52Gb 111 (not continuous)
Radstock Way RH1: Mers 100Mb 196
Radstone Ct. GU22: Wok 90B 168
Radway Ho. W2 43Cb 89 (off Alfred Rd.)
Radwin Cl. RM5: Col R 23Fd 56
Radzan Cl. DA2: Wilm 61Gd 140
Raebarn Gdns. EN5: Barn 15Xa 30
Raeburn Av. DA1: Dart 57Kd 119
Raeburn Av. KT5: Surb 74Ra 153
Raeburn Cl. KT1: Hamp W 66Ma 131
Raeburn Cl. NW11 30Eb 49
Raeburn Ct. GU21: Wok 1L 187
Raeburn Gro. GU21: Wok 10L 167
Raeburn Ho. UB5: N'olt 40Z 65 (off Academy Gdns.)
Raeburn Rd. DA15: Sidc 58Uc 116
Raeburn Rd. HA8: Edg 25Qa 47
Raeburn Rd. UB4: Hayes 40T 64
Raeburn St. SW2 56Nb 112
RAF Bomber Command Memorial 1K 227 (47Kb 90)
Rafdene Copse GU22: Wok 1N 187
Raffles Ho. NW4 28Xa 48
Rafford Way BR1: Broml 68Kc 137
RAF Mus. London 26Wa 48
RAF Northolt Aerodrome 37V 64
RAF Uxbridge, Battle of Britain Bunker 39P 63
Ragged Hall La. AL2: Chis G 6K 5
Ragged Hall La. AL2: Pot C 6K 5
Ragged School Mus. 43Ac 92
Ragge Way TN15: Seal 92Pd 203
Raggleswood BR7: Chst 67Qc 138
Rag Hill Cl. TN16: Tats 93Nc 200
Rag Hill Rd. TN16: Tats 93Mc 199
Raglan Av. EN8: Walt C 6Zb 20
Raglan Cl. RH2: Reig 4M 207
Raglan Cl. TW4: Houn 57Ba 107
Raglan Ct. CR2: S Croy 78Rb 157
Raglan Ct. E17 29Ec 52
Raglan Ct. HA9: Wemb 35Pa 67
Raglan Ct. SE12 57Jc 115
Raglan Gdns. WD19: Wat 18X 27
Raglan Pct. CR3: Cat'm 94Ub 197
Raglan Rd. BR2: Broml 70Lc 137
Raglan Rd. DA17: Belv 49Bd 95
Raglan Rd. E17 29Ec 52
Raglan Rd. EN1: Enf 17Vb 33
Raglan Rd. GU21: Knap 10J 167
Raglan Rd. GU21: Wok 10J 167
Raglan Rd. RH2: Reig 3K 207
Raglan Rd. SE18 50Sc 94
Raglan St. NW5 37Kb 70
Raglan Ter. HA2: Harr 35Da 65
Raglan Way UB5: N'olt 37Ea 66
Ragley Cl. W3 47Sa 87
Ragstone Rd. SL1: Slou 8J 81
Ragstones TN15: Seal 92Pd 203
Ragwort Ct. SE26 64Xb 135
Rahere Ct. E1 42Ac 92 (off Toby La.)
Rahn Rd. CM16: Epp 3Wc 23
Raider Cl. RM7: Mawney 25Cd 56
Railey M. NW5 36Lb 70
Railpit La. CR6: W'ham 87Gc 179
Railshead Rd. TW1: Isle 56Ka 108
Railshead Rd. TW7: Isle 56Ka 108
Rails La. GU24: Pirb 7A 186
Railton Cl. KT13: Weyb 80Q 150
Railton Rd. SE24 56Qb 112
Railway & Bicycle Apts., The TN13: S'oaks 96Jd 202
Railway App. HA1: Harr 28Ha 46
Railway App. HA3: Harr 28Ha 46
Railway App. N4 30Qb 50
Railway App. RM7: Rush G 30Fd 56
Railway App. SE1 6G 225 (46Tb 91)
Railway App. SM6: W'gton 78Kb 156
Railway App. TW1: Twick 59Ja 108
Railway Arches E1: Barnardo St. 44Zb 92 (off Barnardo St.)
Railway Arches E1: Chapman St. 45Xb 91 (off Chapman St.)
Railway Arches E2: Cremer St. 2K 219 (40Vb 71) (off Cremer St.)
Railway Arches E2: Geffrye St. 2K 219 (40Vb 71) (off Geffrye St.)
Railway Arches E2: Laburnum St. 1K 219 (39Vb 71) (off Laburnum St.)
Railway Arches E8: Martello Ter. 38Xb 71 (off Martello Ter.)
Railway Arches E8: Mentmore Ter. 38Xb 71 (off Mentmore Ter.)
Railway Arches E10 32Ec 72
Railway Arches E11 32Fc 73 (off Grove Grn. Rd.)
Railway Arches E16 46Jc 93
Railway Arches E3 42Bc 92 (off Cantrell Rd.)
Railway Arches E7 35Jc 73 (off Winchelsea Rd.)
Railway Arches W12 47Ya 88 (off Shepherd's Bush Mkt.)
Railway Arches W6 48Ya 88
Railway Av. SE16 47Yb 92 (not continuous)
Railway Children Wlk. BR1: Broml 61Jc 137
Railway Cotts. E15 40Gc 73 (off Baker's Row)
Railway Cotts. SW19 63Db 133
Railway Cotts. W6 47Ya 88 (off Sulgrave Rd.)
Railway Cotts. WD24: Wat 11X 27
Railway Cotts. WD4: K Lan 2R 12
Railway Cotts. WD7: R'lett 7Ka 14
Railway Fields Local Nature Reserve 30Rb 51
Railway Gro. SE14 52Bc 114
Railway M. W10 44Ab 88
Railway Pde. CM15: Shenf 17Ce 41 (off Hutton Rd.)
Railway Pas. TW11: Tedd 65Ja 130
Railway Pl. DA12: Grav'nd 8D 122
Railway Pl. DA17: Belv 48Cd 96
Railway Ri. SE22 56Ub 113
Railway Rd. EN8: Walt C 5Bc 20
Railway Rd. TW11: Tedd 63Ga 130
Railway Side SW13 55Ua 110
Railway Sidings DA13: Meop 10C 144
Railway Sidings Ind. Est. DA13: Meop 10C 144
Railway Sidings Rd. SE16 48Wb 91
Railway Sq. CM14: B'wood 20Yd 40
Railway Sta. Bri. E7 36Kc 73
Railway St. DA11: Nflt 57Ce 121
Railway St. N1 2G 217 (40Nb 70)
Railway St. RM6: Chad H 31Yc 75
Railway Ter. CR5: Coul 87Mb 176 (off Station App.)
Railway Ter. E17 25Ec 52
Railway Ter. SE13 57Dc 114
Railway Ter. SL2: Slou 6K 81
Railway Ter. TN16: Westrm 97Tc 200
Railway Ter. TW13: Felt 60W 106
Railway Ter. TW18: Staines 64F 126
Railway Ter. WD24: Wat 11X 27
Railway Ter. WD4: K Lan 9A 4
Railway Vw. SL3: Hort 55B 104
Railway Wharf KT1: King T 67Ma 131 (off Thames Side)
Rainbird Cl. HA0: Wemb 40La 66
Rainbird Pl. CM14: Pil H 15Sd 40
Rainborough Cl. NW10 37Sa 67
Rainbow Av. E14 50Dc 92
Rainbow Ct. GU21: Wok 8J 167
Rainbow Ct. SE14 51Ac 114 (off Chipley St.)
Rainbow Ct. WD19: Wat 16Y 27
Rainbow Gdns. DA1: Dart 54Pd 119
Rainbow Ind. Est. SW20 68Xa 132
Rainbow Ind. Est. UB7: Yiew 45M 83
Rainbow La. RM18: W Til 9F 100
Rainbow La. SS17: Stan H 1P 101
Rainbow Leisure Cen. Epsom 84Ua 174
Rainbow Quay SE16 48Ac 92 (not continuous)
Rainbow Rd. DA8: Erith 52Jd 118
Rainbow Rd. RM16: Chaf H 49Yd 98
Rainbow St. SE5 52Ub 113
Raine Gdns. IG8: Wfd G 21Jc 53
Rainer Apartment CR0: C'don 74Tb 157 (off Cherry Orchard Rd.)
Rainer Cl. EN8: Chesh 1Zb 20
Raines Ct. N16 33Vb 71
Raine St. E1 46Xb 91
RAINHAM 42Jd 96
Rainham Cl. SE9 58Uc 116
Rainham Cl. SW11 58Gb 111
Rainham Hall 42Jd 96
Rainham Ho. NW1 1C 216 (39Lb 70) (off Bayham Pl.)
Rainham Marshes 45Jd 96
Rainham Marshes Nature Reserve 48Nd 97
Rainham Marshes Nature Reserve Vis. Cen. 49Pd 97
Rainham Rd. NW10 41Ya 88
Rainham Rd. RM12: Horn 36Hd 76
Rainham Rd. RM12: Rain 36Hd 76
Rainham Rd. RM13: Rain 38Gd 76
Rainham Rd. Nth. RM10: Dag 33Cd 76
Rainham Rd. Sth. RM10: Dag 35Dd 76
Rainham Station (Rail) Essex 42Jd 96
Rainham Trad. Est. RM13: Rain 42Hd 96
Rainhill Way E3 41Cc 92 (not continuous)
Rainsborough Av. SE8 49Ac 92
Rainsborough Ho. SW15 57Ab 110 (off Stamford Sq.)
Rainsborough Sq. SW6 51Cb 111
Rainsford Cl. HA7: Stan 22La 46
Rainsford Rd. NW10 41Qa 87
Rainsford St. W2 2D 220 (44Gb 89)
Rainsford Way RM12: Horn 32Jd 76
Rainton Rd. SE7 50Jc 93
Rainville Rd. W6 51Ya 110
Raisins Hill HA5: Eastc 27Y 45
Raith Av. N14 20Mb 32
Rajsee Apts. E2 41Wb 91 (off Bethnal Grn. Rd.)
Raleana Rd. E14 46Ec 92
Raleigh Av. SM6: Bedd 77Mb 156
Raleigh Av. UB4: Yead 43X 85
Raleigh Cl. DA12: Grav'nd 3E 144
Raleigh Cl. DA8: Erith 51Hd 118
Raleigh Cl. HA4: Ruis 33V 64
Raleigh Cl. HA5: Pinn 31Z 65
Raleigh Cl. NW4 29Ya 48
Raleigh Cl. SL1: Slou 6E 80
Raleigh Ct. BR3: Beck 67Dc 136
Raleigh Ct. DA8: Erith 52Hd 118
Raleigh Ct. SE16 46Zb 92 (off Clarence M.)
Raleigh Ct. SE8 50Ac 92 (off Evelyn St.)
Raleigh Ct. SM6: W'gton 79Kb 156
Raleigh Ct. TW18: Staines 63J 127
Raleigh Ct. W12 47Ya 88 (off Scott's Rd.)
Raleigh Ct. W13 43Ka 86
Raleigh Dr. KT10: Clay 78Fa 152
Raleigh Dr. KT5: Surb 74Sa 153
Raleigh Dr. N20 20Gb 31
Raleigh Gdns. CR4: Mitc 69Hb 133 (not continuous)
Raleigh Gdns. SW2 58Pb 112
Raleigh Ho. BR1: Broml 67Jc 137 (off Hammelton Rd.)
Raleigh Ho. CR0: C'don 73Pb 156 (off Mitcham Rd.)
Raleigh Ho. E14 47Dc 92 (off Admirals Way)
Raleigh Ho. SW1 51Mb 112 (off Dolphin Sq.)
Raleigh M. BR6: Chels 78Vc 161
Raleigh M. N1 39Rb 71 (off Packington St.)
Raleigh Rd. EN2: Enf 14Tb 33
Raleigh Rd. N8 28Qb 50
Raleigh Rd. SE20 66Zb 136
Raleigh Rd. TW13: Felt 62V 128
Raleigh Rd. TW9: Rich 55Pa 109
Raleigh Rd. UB2: S'hall 50Aa 85
Raleigh St. N1 1C 218 (39Rb 71)
Raleigh Way N14 18Mb 32
Raleigh Way TW13: Hanw 64Y 129
Rale La. E4 17Fc 35
Ralliwood Rd. KT21: Asht 91Qa 193
Rally Bldg. E17 29Bc 52
Ralph Bayer Ct. E3 43Cc 92 (off Geoff Cade Way)
Ralph Brook Ct. N1 3G 219 (41Tb 91) (off Chart St.)
Ralph Ct. W2 44Db 89 (off Queensway)
Ralph Perring Ct. BR3: Beck 70Cc 136
Ralston Ct. SL4: Wind 3H 103 (off Russell St.)
Ralston St. SW3 50Hb 89
Ralston Way WD19: Wat 19Z 27
Ramac Ind. Est. SE7 49Kc 93
Rama Cl. SW16 66Nb 134
Rama Ct. HA1: Harr 33Ga 66
Ramac Way SE7 49Kc 93
Rama La. SE19 66Vb 135
Ramar Ho. E1 43Wb 91 (off Hanbury St.)
Rambert 6K 223 (46Qb 90) (off Upper Ground)
Rambler Cl. SL6: Tap 4A 80
Rambler Cl. SW16 63Lb 134
Rambler La. DA1: Dart 54Pd 119
Rambler La. SL3: L'ly 8N 81
Rame Cl. SW17 64Jb 134
Ramilles Cl. SW2 58Nb 112
Ramillies Pl. W1 3B 222 (44Lb 90)
Ramillies Rd. DA15: Sidc 58Xc 117
Ramillies Rd. NW7 19Ua 30
Ramillies Rd. W4 49Ta 87
Ramillies St. W1 3B 222 (44Lb 90)
Ramney Dr. EN3: Enf L 8Ac 20
Ramones Ter. CR4: Mitc 70Nb 134 (off Yorkshire Rd.)
Ramornie Cl. KT12: Hers 77Ba 151
Ramparts, The AL3: St A 3P 5
Rampart St. E1 44Xb 91
Ram Pas. KT1: King T 68Ma 131
Rampayne St. SW1 7D 228 (50Mb 90)
Ram Pl. E9 37Yb 72
Rampton Cl. E4 20Cc 34
Ram Quarter SW18 57Db 111
Ramryge Ct. AL1: St A 4B 6
Ramsay Gdns. RM3: Rom 25Ld 57
Ramsay Ho. NW8 1D 214 (40Gb 69) (off Townshend Est.)
Ramsay M. SW3 51Gb 111
Ramsay Rd. E7 35Gc 73
Ramsay Rd. GU20: W'sham 8C 146
Ramsay Rd. W3 48Sa 87
Ramsbury Rd. AL1: St A 3C 6
Ramscroft Cl. N9 17Ub 33
Ramsdale Rd. SW17 64Jb 134
RAMSDEN 74Yc 161
Ramsden Cl. BR5: Orp 74Yc 161
Ramsden Dr. RM5: Col R 24Cd 56
Ramsden Rd. BR5: Orp 74Xc 161
Ramsden Rd. DA8: Erith 52Fd 118
Ramsden Rd. N11 22Hb 49
Ramsden Rd. SW12 58Jb 112
Ramsey Cl. AL1: St A 4E 6
Ramsey Cl. AL9: Brk P 9M 9
Ramsey Cl. NW9 30Va 48
Ramsey Cl. UB6: G'frd 36Fa 66
Ramsey Ct. CR0: C'don 75Rb 157 (off Church St.)
Ramsey Ct. SL2: Slou 2B 80
Ramsey Ho. SW9 52Qb 112
Ramsey Lodge Ct. AL1: St A 1C 6
Ramsey Pl. CR3: Cat'm 94Sb 197
Ramsey Rd. CR7: Thor H 72Pb 156
Ramsey St. E2 42Wb 91
Ramsey Wlk. N1 37Tb 71
Ramsey Way N14 17Lb 32
Ramsfort Ho. SE16 49Xb 91 (off Camilla Rd.)
Ramsgate Cl. E16 46Kc 93
Ramsgate St. E8 37Vb 71
Ramsgill App. IG2: Ilf 28Vc 55
Ramsgill Dr. IG2: Ilf 29Vc 55
Rams Gro. RM6: Chad H 28Ad 55
Ramshorn Ri. HP1: Hem H 3G 2
Ram St. SW18 57Db 111
Ram Twr. SW18 57Db 111
Ramulis Dr. UB4: Yead 42Z 85
Ramuswood Av. BR6: Chels 78Uc 160
Rancliffe Gdns. SE9 56Nc 116
Rancliffe Rd. E6 40Nc 74
Randal Cres. RH2: Reig 8J 207
Randall Av. NW2 33Ua 68
Randall Cl. DA8: Erith 51Ed 118
Randall Cl. SL3: L'ly 50B 82
Randall Cl. SW11 53Gb 111
Randall Ct. NW7 24Wa 48
Randall Ct. RM13: Rain 42Kd 97
Randall Ct. SL4: Old Win 8L 103 (off Lyndwood Dr.)
Randall Dr. RM12: Horn 35Ld 77
Randall Dr. RM16: Ors 3C 100
Randall Hill Rd. TN15: Wro 88Be 185
Randall Pl. SE10 52Ec 114
Randall Rd. SE11 7H 229 (50Pb 90)
Randall Row SE11 6H 229 (49Pb 90)
Randalls Cres. KT22: Lea 92Ja 192
Randalls Dr. CM13: Hut 16Fe 41

Randalls Pk. Av. KT22: Lea 92Ja 192
Randalls Pk. Crematorium
(Leatherhead) 92Ga 192
Randalls Pk. Dr. KT22: Lea 93Ja 192
Randalls Rents SE16 48Bc 92
(off Gulliver St.)
Randalls Ride HP2: Hem H 1M 3
Randalls Rd. KT22: Lea 91Ga 192
Randalls Wlk. AL2: Brick W 2Ba 13
Randalls Way KT22: Lea 93Ja 192
Randell's Rd. N1 39Nb 70
Randisbourne Gdns. SE6 62Dc 136
Randle Rd. TW10: Ham 63La 130
Randlesdown Rd. SE6 63Cc 136
(not continuous)
Randles La. TN14: Knock 87Zc 181
Randolph App. E16 44Lc 93
Randolph Av. W9 40Db 69
Randolph Cl. DA7: Bex 55Ed 118
Randolph Cl. GU21: Knap 9J 167
Randolph Cl. KT11: Stoke D 87Ca 171
Randolph Cl. KT2: King T 64Sa 131
Randolph Ct. HA5: Hat E 24Ca 45
(off The Avenue)
Randolph Ct. NW8 39Eb 69
Randolph Cres. W9 6A 214 (42Eb 89)
Randolph Gdns. NW6 40Db 69
Randolph Gro. RM6: Chad H 29Yc 55
Randolph Ho. KT18: Eps 86Sa 173
(off Dalmeny Way)
Randolph M. W9 6A 214 (42Eb 89)
Randolph Rd. BR2: Broml 74Pc 160
Randolph Rd. E17 29Dc 52
Randolph Rd. KT17: Eps 86Va 174
Randolph Rd. SL3: L'ly 48A 82
Randolph Rd. UB1: S'hall 47Ba 85
Randolph Rd. W9 6A 214 (42Eb 89)
Randolph's La. TN16: Westrm 98Rc 200
Randolph St. NW1 38Lb 70
Randon Cl. HA2: Harr 26Da 45
Ranelagh Av. SW13 54Wa 110
Ranelagh Av. SW6 55Bb 111
Ranelagh Bri. W2 43Db 89
Ranelagh Cl. HA8: Edg 21Qa 47
Ranelagh Cotts. SW1 7K 227 (50Kb 90)
(off Ebury Bri. Rd.)
Ranelagh Dr. HA8: Edg 21Qa 47
Ranelagh Dr. TW1: Twick 56Ka 108
Ranelagh Gdns. DA11: Nflt 9B 122
Ranelagh Gdns. E11 29Lc 53
Ranelagh Gdns. IG1: Ilf 32Pc 74
Ranelagh Gdns. SW6 55Ab 110
(not continuous)
Ranelagh Gdns. W4 52Sa 109
Ranelagh Gdns. W6 49Va 88
Ranelagh Gdns. Mans. SW6 55Ab 110
(off Ranelagh Gdns.)
Ranelagh Gro. SW1 7J 227 (50Jb 90)
Ranelagh Ho. SW3 7F 227 (50Hb 89)
(off Elystan Pl.)
Ranelagh M. W5 47Ma 87
Ranelagh Pl. KT3: N Mald 71Ua 154
Ranelagh Rd. E11 35Gc 73
Ranelagh Rd. E15 40Gc 73
Ranelagh Rd. E6 39Qc 74
Ranelagh Rd. HA0: Wemb 37Ma 67
Ranelagh Rd. HP2: Hem H 2B 4
Ranelagh Rd. N17 27Ub 51
Ranelagh Rd. N22 25Pb 50
Ranelagh Rd. NW10 40Va 68
Ranelagh Rd. RH1: Redh 6N 207
Ranelagh Rd. SW1 7C 228 (50Lb 90)
Ranelagh Rd. UB1: S'hall 46Z 85
Ranelagh Rd. W5 47Ma 87
Ranfurly Rd. SM1: Sutt 75Cb 155
Rangbourne Ho. N7 36Nb 70
Rangefield Rd. BR1: Broml 64Gc 137
Rangemoor Rd. N15 29Vb 51
Range Rd. DA12: Grav'nd 9G 122
Ranger's House 53Fc 115
Ranger's Rd. E4 17Gc 35
Ranger's Rd. IG10: Lough 17Jc 35
Rangers Sq. SE10 53Fc 115
Ranger Wlk. KT15: Add 78K 149
Range Way TW17: Shep 73Q 150
Rangeworth Pl. DA15: Sidc 62Vc 139
Rangoon St. EC3 3K 225 (44Vb 91)
(off Crutched Friars)
Rankin Cl. NW9 27Ua 48
Rankine Ho. SE1 4D 230 (48Sb 91)
(off Bath Ter.)
Ranleigh Gdns. DA7: Bex 52Bd 117
Ranmere St. SW12 60Kb 112
Ranmoor Cl. HA1: Harr 28Fa 46
Ranmoor Gdns. HA1: Harr 28Fa 46
Ranmore Av. CR0: C'don 76Vb 157
Ranmore Cl. RH1: Redh 3A 208
Ranmore Ct. KT6: Surb 71Ma 153
Ranmore Path BR5: St M Cry 70Wc 139
Ranmore Pl. KT13: Weyb 78S 150
Ranmore Rd. SM2: Cheam 81Za 174
Rannoch Cl. HA8: Edg 19Ra 29
Rannoch Rd. W6 51Ya 110
Rannock Av. NW9 31Ta 67
Ranskill Ct. WD6: Bore 11Qa 29
Ranskill Rd. WD6: Bore 11Qa 29
Ransom Cl. WD19: Wat 17Y 27
Ransome's Dock Bus. Cen. SW11 52Gb 111
Ranson Rd. SE7 49Lc 93
Ranston Cl. UB9: Den 30H 43
Ranston St. NW1 7D 214 (43Gb 89)
Rant Mdw. HP3: Hem H 4A 4
(not continuous)
Ranulf Rd. NW2 35Bb 69
Ranwell Cl. E3 39Bc 72
Ranwell Ho. E3 39Bc 72
(off Ranwell Cl.)
Ranworth Cl. DA8: Erith 54Gd 118
Ranworth Cl. HP3: Hem H 4M 3
Ranworth Gdns. EN6: Pot B 3Za 16
Ranworth Rd. N9 19Yb 34
Ranyard Cl. KT9: Chess 76Pa 153
Raphael Av. RM1: Rom 27Hd 56
Raphael Av. RM18: Tilb 2C 122
Raphael Cl. KT1: King T 70Ma 131
Raphael Cl. WD7: Shenl 4Na 15
Raphael Ct. SE16 50Xb 91
(off Stubbs Dr.)
Raphael Dr. IG10: Lough 12Rc 36
Raphael Dr. KT7: T Ditt 73Ha 152
Raphael Dr. WD24: Wat 12Z 27
Raphael Ho. IG1: Ilf 33Sc 74
Raphael Rd. DA12: Grav'nd 9F 122
Raphael St. SW7 2F 227 (47Hb 89)
Raphen Apts. E3 40Ac 72
(off Medway Rd.)
Rapier Cl. RM19: Purf 49Pd 97
Rapley Ho. E2 41Wb 91
(off Turin St.)
Rapley's Fld. GU24: Pirb 5C 186
Rapsley La. GU21: Knap 10F 166
Raquel Ct. SE1 1H 231 (47Ub 91)
(off Snowfields)
Rasehill Cl. WD3: Rick 15L 25
Rashleigh Ct. SW8 54Kb 112
Rashleigh Ho. WC1 4F 217 (41Nb 90)
(off Thanet St.)
Rashleigh St. SW8 54Kb 112
(off Peardon St.)
Rashleigh Way DA4: Hort K 70Sd 142
Rasper Rd. N20 19Eb 31
Rastell Av. SW2 61Mb 134
RATCLIFF 43Ac 92
Ratcliffe Cl. SE12 59Jc 115
Ratcliffe Cl. UB8: Cowl 41M 83
Ratcliffe Ct. SE1 2E 230 (47Sb 91)
(off Gt. Dover St.)
Ratcliffe Cross St. E1 44Zb 92
Ratcliffe Ho. E14 44Ac 92
(off Barnes St.)
Ratcliffe La. E14 44Ac 92
Ratcliffe Orchard E1 45Zb 92
Ratcliff Rd. E7 36Lc 73
Rathbone Ho. E16 44Hc 93
(off Rathbone St.)
Rathbone Ho. NW6 39Cb 69
Rathbone Mkt. E16 43Hc 93
Rathbone Pl. W1 1D 222 (43Mb 90)
Rathbone Sq. CR0: C'don 77Sb 157
Rathbone Sq. W1 2C 222 (44Lb 90)
Rathbone St. E16 44Hc 93
Rathbone St. W1 1C 222 (43Lb 90)
Rathcoole Av. N8 29Pb 50
Rathcoole Gdns. N8 29Pb 50
Rathfern Rd. SE6 60Bc 114
Rathgar Av. W13 46Ka 86
Rathgar Cl. N3 26Bb 49
Rathgar Cl. RH1: Redh 10A 208
Rathgar Rd. SW9 55Rb 113
Rathlin HP3: Hem H 5B 4
Rathmell Dr. SW4 58Mb 112
Rathmore Rd. DA11: Grav'nd 8D 122
(off New Rathmore Rd.)
Rathmore Rd. SE7 50Kc 93
Rathnew Ct. E2 41Zb 92
(off Meath Cres.)
Rathore Cl. RM6: Chad H 29Zc 55
Rats La. IG10: H Beech 10Kc 21
Rats La. IG10: Lough 10Kc 21
Rattray Ct. SE6 61Hc 137
Rattray Rd. SW2 56Qb 112
Raul Rd. SE15 54Wb 113
Raveley St. NW5 35Lb 70
(not continuous)
Ravel Gdns. RM15: Avel 44Sd 98
Ravel Ho. SW11 55Fb 111
(off York Place)
Ravel Rd. RM15: Avel 44Sd 98
Raven Cl. NW9 26Ua 48
Raven Cl. RM7: Rush G 31Cd 76
Raven Cl. WD18: Wat 15U 26
Raven Cl. WD3: Rick 17L 25
Raven Ct. AL10: Hat 1C 8
Raven Ct. DA12: Grav'nd 8F 122
(off Admirals Way)
Ravencroft RM16: Grays 7D 100
Ravendale Rd. TW16: Sun 68V 128
Ravenet St. SW11 53Kb 112
(not continuous)
Ravenet St. SW8 53Kb 112
Ravenfield TW20: Eng G 5N 125
Ravenfield Rd. SW17 62Hb 133
Ravenhill Rd. E13 40Lc 73
(off Bassington Rd.)
Raven Ho. SE16 49Zb 92
(off Tawny Way)
Ravenings Pde. IG3: Ilf 32Wc 75
Ravenna Rd. SW15 57Za 110
Ravenoak Way IG7: Chig 22Uc 54
Ravenor Ct. UB6: G'frd 42Da 85
Ravenor Pk. Rd. UB6: G'frd 41Da 85
Raven Rd. E18 26Lc 53
Raven Row E1 43Xb 91
Raven Row Contemporary Art
Cen. 1K 225 (43Vb 91)
Raven's Ait 71Ma 153
Ravensbourne Apts. SW6 55Eb 111
(off Central Av.)
Ravensbourne Av. BR2: Broml 66Fc 137
Ravensbourne Av. BR3: Beck 66Fc 137
Ravensbourne Av. TW19: Stanw 60N 105
Ravensbourne Ct. SE6 59Cc 114
Ravensbourne Cres. RM3: Hrld W 27Pd 57
Ravensbourne Gdns. IG5: Ilf 25Qc 54
Ravensbourne Gdns. W13 43Ka 86
Ravensbourne Ho. BR1: Broml 64Fc 137
Ravensbourne Ho. E15 37Fc 73
(off Forrester Way)
Ravensbourne Ho. NW8 7D 214 (43Gb 89)
(off Broadley St.)
Ravensbourne Mans. SE8 51Cc 114
(off Berthon St.)
Ravensbourne Pk. SE6 59Cc 114
Ravensbourne Pk. Cres. SE6 59Bc 114
Ravensbourne Pl. SE13 54Dc 114
Ravensbourne Pl. SE8 53Dc 114
Ravensbourne Rd. BR1: Broml 69Jc 137
Ravensbourne Rd. DA1: Cray 55Jd 118
Ravensbourne Rd. SE6 59Bc 114
Ravensbourne Rd. TW1: Twick 58La 108
Ravensbourne Station (Rail) 66Fc 137
Ravensbourne Ter. TW19: Stanw 60N 105
Ravensbury Av. SM4: Mord 71Eb 155
Ravensbury Ct. CR4: Mitc 70Fb 133
(off Ravensbury Gro.)
Ravensbury Gro. CR4: Mitc 70Fb 133
Ravensbury La. CR4: Mitc 70Fb 133
Ravensbury Path CR4: Mitc 70Fb 133
Ravensbury Rd. BR5: St P 69Vc 139
Ravensbury Rd. SW18 61Db 133
Ravensbury Ter. SW18 61Db 133
Ravenscar NW1 1B 216 (39Lb 70)
(off Bayham St.)
Ravenscar Rd. BR1: Broml 63Gc 137
Ravenscar Rd. KT6: Surb 75Pa 153
Ravens Cl. BR2: Broml 68Hc 137
Ravens Cl. EN1: Enf 12Ub 33
Ravens Cl. GU21: Knap 8G 166
Ravens Cl. KT6: Surb 72Ma 153
Ravens Cl. RH1: Redh 5P 207
Ravens Ct. KT1: King T 71Ma 153
(off Uxbridge Rd.)
Ravenscourt CM15: B'wood 17Yd 40
Ravenscourt TW16: Sun 67V 128
Ravenscourt Av. W6 49Wa 88
Ravenscourt Cl. HA4: Ruis 31S 64
Ravenscourt Cl. RM12: Horn 34Nd 77
Ravenscourt Dr. RM12: Horn 34Nd 77
Ravenscourt Gdns. W6 49Wa 88
Ravenscourt Gro. RM12: Horn 33Nd 77
Ravenscourt Pk. EN5: Barn 14Za 30
Ravenscourt Pk. W6 48Wa 88
Ravenscourt Pk. Mans. W6 48Xa 88
(off Paddenswick Rd.)
Ravenscourt Park Station
(Underground) 49Xa 88
Ravenscourt Pl. W6 49Xa 88
Ravenscourt Rd. BR5: St P 69Wc 139
Ravenscourt Rd. W6 49Xa 88
Ravenscourt Sq. W6 48Wa 88
Ravenscraig Rd. N11 21Lb 50
Ravenscroft WD25: Wat 7Aa 13
Ravenscroft Av. HA9: Wemb 32Na 67
Ravenscroft Av. NW11 31Bb 69
(not continuous)
Ravenscroft Cl. E16 43Jc 93
Ravenscroft Cotts. EN5: New Bar 14Cb 31
Ravenscroft Cres. SE9 62Pc 138
Ravenscroft Pk. EN5: Barn 13Za 30
Ravenscroft Rd. BR3: Beck 68Yb 136
Ravenscroft Rd. E16 43Jc 93
Ravenscroft Rd. KT13: Weyb 83S 170
Ravenscroft Rd. W4 49Sa 87
Ravenscroft St. E2 2K 219 (40Vb 71)
Ravensdale Av. N12 21Eb 49
Ravensdale Gdns. SE19 66Tb 135
Ravensdale Gdns. TW4: Houn 55Aa 107
Ravensdale Ind. Est. N16 30Wb 51
Ravensdale Mans. N8 30Nb 50
(off Haringey Pk.)
Ravensdale M. TW18: Staines 65K 127
Ravensdale Rd. N16 31Vb 71
Ravensdale Rd. TW4: Houn 55Aa 107
Ravensdell HP1: Hem H 1H 3
Ravens Dene BR7: Chst 64Pc 138
Ravensdon St. SE11 7A 230 (50Qb 90)
Ravens Fld. SL3: L'ly 7P 81
Ravensfield Cl. RM9: Dag 35Zc 75
Ravensfield Gdns. KT19: Ewe 78Ua 154
Ravens Ga. M. BR2: Broml 68Gc 137
Ravenshaw St. NW6 36Bb 69
Ravenshead Cl. CR2: Sels 83Yb 178
Ravenshill BR7: Chst 67Rc 138
Ravenshurst Av. NW4 28Ya 48
Ravenside KT1: King T 71Ma 153
(off Portsmouth Rd.)
Ravenside Cl. N18 22Zb 52
Ravenside Retail Pk. London 22Zb 52
Ravenslea Rd. SW12 59Hb 111
Ravensleigh Gdns. BR1: Broml 64Kc 137
Ravensmead SL9: Chal P 22B 42
Ravensmead Rd. BR2: Broml 66Fc 137
Ravensmede Way W4 49Va 88
Ravensmere CM16: Epp 3Wc 23
Ravens M. SE12 57Jc 115
Ravensquay Bus. Cen. BR5: St M Cry 71Xc 161
Ravenstone SE17 7J 231 (50Ub 91)
Ravenstone Rd. N8 27Qb 50
Ravenstone Rd. NW9 30Va 48
Ravenstone St. SW12 60Jb 112
Ravens Wlk. E20 37Dc 72
Ravens Way SE12 57Jc 115
Ravens Wharf HP4: Berk 1A 2
Ravenswold CR8: Kenley 87Sb 177
Ravenswood DA5: Bexl 60Ad 117
Ravenswood Av. BR4: W W'ck 74Ec 158
Ravenswood Av. KT6: Surb 75Pa 153
Ravenswood Cl. KT11: Cobh 87Z 171
Ravenswood Cl. RM5: Col R 22Dd 56
Ravenswood Ct. GU22: Wok 90B 168
Ravenswood Ct. KT2: King T 65Ra 131
Ravenswood Ct. W3 48Sa 87
(off Bassington Rd.)
Ravenswood Cres. BR4: W W'ck 74Ec 158
Ravenswood Cres. HA2: Harr 33Ba 65
Ravenswood Gdns. TW7: Isle 53Ga 108
Ravenswood Ind. Est. E17 28Ec 52
Ravenswood Pk. HA6: Nwood 23W 44
Ravenswood Rd. CR0: Wadd 76Rb 157
Ravenswood Rd. E17 28Ec 52
Ravenswood Rd. SW12 59Kb 112
Ravensworth Ct. SW6 52Db 111
(off Fulham Rd.)
Ravensworth Rd. NW10 41Xa 88
Ravensworth Rd. SE9 62Pc 138
Ravensworth Rd. SL2: Slou 1E 80
Raven Wharf SE1 1K 231 (47Vb 91)
(off Lafone St.)
Ravey St. EC2 5H 219 (42Ub 91)
Ravine Gro. SE18 51Uc 116
Ravine Way SW11 51Mb 112
Rav Pinter Cl. N16 31Ub 71
Rawchester Cl. SW18 60Bb 111
Rawlings Cl. BR3: Beck 71Ec 158
Rawlings Cl. BR6: Chels 78Vc 161
Rawlings Cres. HA9: Wemb 34Ra 67
Rawlings St. SW3 5F 227 (49Hb 89)
Rawlins Cl. CR2: Sels 80Bc 158
Rawlins Cl. N3 27Ab 48
Rawlinson Ct. NW2 31Ya 68
Rawlinson Ho. SE13 56Fc 115
(off Mercator Rd.)
Rawlinson Ter. N17 27Vb 51
Rawlyn Cl. RM16: Chaf H 50Yd 98
Rawnsley Av. CR4: Mitc 71Fb 155
Rawreth Wlk. N1 39Sb 71
(off Basire St.)
Rawson St. SW11 53Jb 112
(not continuous)
Rawsthorne Cl. E16 46Pc 94
Rawsthorne Ct. TW4: Houn 56Ba 107
Rawstone Wlk. E13 40Jc 73
Rawstorne Pl. EC1 3B 218 (41Rb 91)
Rawstorne St. EC1 3B 218 (41Rb 91)
(not continuous)
Raybell Ct. TW7: Isle 54Ha 108
Rayburne Ct. IG9: Buck H 18Lc 35
Rayburne Ct. W14 48Ab 88
Rayburn Rd. RM11: Horn 31Qd 77
Ray Cl. KT9: Chess 79La 152
Raydean Rd. EN5: New Bar 15Db 31
Raydon Rd. EN8: Chesh 4Zb 20
Raydons Gdns. RM9: Dag 36Ad 75
Raydons Rd. RM9: Dag 36Ad 75
Raydon St. N19 33Kb 70
Rayfield CM16: Epp 2Wc 23
Rayfield Cl. BR2: Broml 72Nc 160
Rayford Av. SE12 59Hc 115
Rayford Cl. DA1: Dart 57Ld 119
Ray Gdns. HA7: Stan 22Ka 46
Ray Gdns. IG11: Bark 40Wc 75
Ray Gunter Ho. SE17 7C 230 (50Rb 91)
(off Marsland Cl.)
Ray Ho. N1 39Ub 71
(off Colville Est.)
Ray Ho. W10 44Za 88
(off Cambridge Gdns.)
Ray Lamb Way DA8: Erith 51Kd 119
Rayleas Cl. SE18 53Rc 116
Rayleigh Av. TW11: Tedd 65Ga 130
Rayleigh Cl. CM13: Hut 16Ee 41
Rayleigh Cl. N13 20Tb 33
Rayleigh Ct. KT1: King T 68Qa 131
Rayleigh Ct. N22 25Sb 51
Rayleigh Ho. WD5: Ab L 4V 12
Rayleigh Pde. CM13: Hut 16Ee 41
Rayleigh Ri. CR2: S Croy 79Ub 157
Rayleigh Rd. CM13: Hut 16De 41
Rayleigh Rd. E16 46Kc 93
Rayleigh Rd. IG8: Wfd G 23Lc 53
Rayleigh Rd. N13 20Sb 33
Rayleigh Rd. SS17: Stan H 2K 101
Rayleigh Rd. SW19 67Bb 133
Ray Lodge Rd. IG8: Wfd G 23Lc 53
Ray Massey Way E6 39Nc 74
(off High St. Nth.)
Raymead Av. CR7: Thor H 71Qb 156
Raymead Cl. KT22: Fet 94Ga 192
Raymead Pas. CR7: Thor H 71Qb 156
(off Raymead Av.)
Raymead Way KT22: Fet 94Ga 192
Raymede Towers W10 43Za 88
(off Treverton St.)
Raymer Cl. AL1: St A 1C 6
Raymere Gdns. SE18 52Tc 116
Raymond Av. E18 27Hc 53
Raymond Av. W13 48Ja 86
Raymond Bldgs. WC1 7J 217 (43Pb 90)
Raymond Chadburn Ho. E7 35Kc 73
Raymond Cl. SE26 64Yb 136
Raymond Cl. SL3: Poyle 53G 104
Raymond Cl. WD5: Ab L 4T 12
Raymond Ct. EN6: Pot B 6Eb 17
Raymond Ct. N10 24Kb 50
Raymond Gdns. IG7: Chig 20Xc 37
Raymond Postgate Ct. SE28 45Xc 95
Raymond Rd. BR3: Beck 70Ac 136
Raymond Rd. E13 39Lc 73
Raymond Rd. IG2: Ilf 31Tc 74
Raymond Rd. SL3: L'ly 48C 82
Raymond Rd. SW19 65Ab 132
Raymond Way KT10: Clay 79Ja 152
Raymouth Rd. SE16 49Xb 91
Raynald Ho. SW16 62Nb 134
Rayne Ct. E18 28Hc 53
Rayne Ho. SW12 58Jb 112
Rayne Ho. W9 42Db 89
(off Delaware Rd.)
Rayner Cl. SM5: Cars 78Hb 155
Rayner Ct. W12 47Ya 88
(off Bamborough Gdns.)
Rayners Cl. HA0: Wemb 36Ma 67
Rayners Cl. SL3: Coln 52E 104
Rayner's Ct. DA11: Nflt 57De 121
Rayners Cres. UB5: N'olt 41X 85
Rayners Gdns. UB5: N'olt 40X 65
Rayners La. HA2: Harr 32Ca 65
Rayners La. HA5: Pinn 29Ba 45
RAYNERS LANE 32Ba 65
Rayners Lane Station
(Underground) 31Ba 65
Rayners Rd. SW15 57Ab 110
Rayner Ter. E14 44Ac 92
(off Carr St.)
Rayner Towers E10 31Cc 72
(off Albany Rd.)
Raynes Av. E11 31Lc 73
Raynes Cl. GU21: Knap 1F 186
RAYNES PARK 70Ya 132
Raynes Pk. Bri. 68Ya 132
Raynes Pk. School Sports Cen. 69Xa 132
Raynes Park Station (Rail) 68Ya 132
Raynham W2 2D 220 (44Gb 89)
(off Norfolk Cres.)
Raynham Av. N18 23Wb 51
Raynham Ho. E1 42Zb 92
(off Harpley Sq.)
Raynham Rd. N18 22Wb 51
Raynham Rd. W6 49Xa 88
Raynham Ter. N18 22Wb 51
Raynor Cl. UB1: S'hall 46Ba 85
Raynor Pl. N1 38Sb 71
Raynton Cl. HA2: Harr 32Aa 65
Raynton Cl. UB4: Hayes 42V 84
Raynton Dr. UB4: Hayes 42V 84
Raynton Rd. EN3: Enf W 3Zb 20
Rayon Cl. SM6: W'gton 75Jb 156
Ray Rd. KT8: W Mole 71Da 151
Ray Rd. RM5: Col R 22Dd 56
Ray's Av. SL4: Wind 2D 102
Rays Av. N18 21Yb 52
Rays Hill DA4: Hort K 70Sd 142
Rays Rd. BR4: W W'ck 73Ec 158
Rays Rd. N18 21Yb 52
Ray St. EC1 6A 218 (42Qb 90)
Ray St. Bri. 6A 218 (42Qb 90)
(off Farringdon Rd.)
Ray Wlk. N7 33Pb 70
Raywood Cl. UB3: Harl 52S 106
Raywood Mans. E20 37Ec 72
(off West Pk. Wlk.)
Razia M. E12 36Pc 74
Reachview Cl. NW1 38Lb 70
Read Cl. KT7: T Ditt 73Ja 152
Read Ct. E17 30Cc 52
Read Ct. EN9: Walt A 5Jc 21
Reade Ct. W3 48Sa 87
(off Stanley Rd.)
Readens, The SM7: Bans 88Gb 175
Reader Ho. SE5 53Sb 113
(off Badsworth Rd.)
Read Ho. SE11 51Qb 112
(off Clayton St.)
Read Ho. SE20 66Xb 135
(off Anerley Pk.)
Reading Arch Rd. RH1: Redh 6P 207
Reading Cl. SE22 58Wb 113
Reading Ho. SE15 51Wb 113
(off Friary Est.)
Reading Ho. W2 44Eb 89
(off Hallfield Est.)
Reading La. E8 37Xb 71
Reading Rd. SM1: Sutt 78Eb 155
Reading Rd. UB5: N'olt 36Da 65
Readings, The WD3: Chor 13H 25
Readman Ct. SE20 67Xb 135
Read Rd. KT21: Asht 89Ma 173
Reads Cl. IG1: Ilf 34Rc 74
Reads Rest La. KT20: Tad 91Bb 195
Read Way DA12: Grav'nd 4F 144
Ream Apts. SE23 61Yb 136
(off Clyde Ter.)
Reapers Cl. NW1 39Mb 70
Reapers Way TW7: Isle 57Fa 108
Reardon Ct. N21 19Rb 33
Reardon Ho. E1 46Xb 91
(off Reardon St.)
Reardon Path E1 46Xb 91
(not continuous)
Reardon St. E1 46Xb 91
Reaston St. SE14 52Zb 114
Rebecca Ct. DA14: Sidc 63Xc 139
Rebecca Ho. E3 42Bc 92
(off Brokesley St.)
Rebecca Ho. N12 21Db 49
(off Woodside Pk. Rd.)
Reckitt Rd. W4 50Ua 88
Recognition Ho. SL4: Wind 4E 102
Record St. SE15 51Yb 114
Record Wlk. UB3: Hayes 47U 84
Recovery St. SW17 64Gb 133
Recreation Av. RM3: Hrld W 26Pd 57
Recreation Av. RM7: Rom 29Ed 56
Recreation Rd. BR2: Broml 68Hc 137
Recreation Rd. DA15: Sidc 62Uc 138
Recreation Rd. SE26 63Zb 136
Recreation Rd. UB2: S'hall 49Aa 85
Recreation Way CR4: Mitc 69Mb 134
Rector St. N1 39Sb 71
Rectory Bus. Cen. DA14: Sidc 63Xc 139
Rectory Chambers SW3 51Gb 111
(off Old Church St.)
Rectory Chase CM13: L War 28Zd 59
Rectory Cl. DA1: Cray 56Gd 118
Rectory Cl. DA14: Sidc 63Xc 139
Rectory Cl. E4 20Cc 34
Rectory Cl. HA7: Stan 22Ka 46
Rectory Cl. KT14: Byfl 85M 169
Rectory Cl. KT21: Asht 91Pa 193
Rectory Cl. KT6: Surb 74La 152
Rectory Cl. N3 25Bb 49
Rectory Cl. SL2: Farn R 1G 80
Rectory Cl. SL4: Wind 3E 102
Rectory Cl. SW20 69Ya 132
Rectory Cl. TW17: Shep 69Q 128
Rectory Ct. E18 25Hc 53
Rectory Ct. SM6: W'gton 77Lb 156
Rectory Ct. TW13: Felt 63Y 129
Rectory Cres. E11 30Lc 53
(not continuous)
Rectory Farm 53Y 107
Rectory Farm Rd. EN2: Enf 10Pb 18
(not continuous)
Rectory Field 52Kc 115
Rectory Fld. Cres. SE7 52Lc 115
Rectory Flds. RM16: Ors 3D 100
Rectory Gdns. BR3: Beck 67Cc 136
(off Rectory Rd.)
Rectory Gdns. N8 28Nb 50
Rectory Gdns. RM14: Upm 33Td 78
Rectory Gdns. SW4 55Lb 112
Rectory Gdns. UB5: N'olt 39Ba 65
Rectory Grn. BR3: Beck 67Bc 136
Rectory Gro. CR0: C'don 75Rb 157
Rectory Gro. SW4 55Lb 112
Rectory Gro. TW12: Hamp 63Ba 129
Rectory La. CM13: Heron 24Fe 59
Rectory La. DA14: Sidc 63Xc 139
Rectory La. GU20: W'sham 9A 146
Rectory La. HA7: Stan 22Ka 46
Rectory La. HA8: Edg 23Qa 47
Rectory La. IG10: Lough 12Qc 36
Rectory La. KT14: Byfl 85N 169
Rectory La. KT21: Asht 91Pa 193
Rectory La. KT23: Bookh 98Ba 191
Rectory La. KT6: Surb 74Ka 152
Rectory La. RH3: Bkld 3B 206
Rectory La. SM6: W'gton 77Lb 156
Rectory La. SM7: Bans 86Hb 175
Rectory La. SW17 65Jb 134
Rectory La. TN13: S'oaks 98Ld 203
Rectory La. TN15: Igh 94Yd 204
Rectory La. TN16: Bras 96Yc 201
Rectory La. TN16: Westrm 95Nc 200
Rectory La. WD23: Bush 16Ca 27
Rectory La. WD3: Rick 18M 25
Rectory La. WD4: K Lan 10A 4
Rectory La. WD7: Shenl 5Pa 15
Rectory Mdw. DA13: Sflt 65Ce 143
Rectory Orchard SW19 63Ab 132
Rectory Pk. CR2: Sande 85Ub 177
Rectory Pk. Av. UB5: N'olt 41Ba 85
Rectory Pl. SE18 49Qc 94
Rectory Rd. BR2: Kes 80Mc 159
Rectory Rd. BR3: Beck 67Cc 136
Rectory Rd. CR5: Coul 97Eb 195
Rectory Rd. DA10: Swans 59Ae 121
Rectory Rd. E12 36Pc 74
Rectory Rd. E17 28Dc 52
Rectory Rd. N16 33Vb 71
Rectory Rd. RM10: Dag 37Dd 76
Rectory Rd. RM16: Ors 2C 100
Rectory Rd. RM17: Grays 48Fe 99
Rectory Rd. RM18: W Til 1F 122
Rectory Rd. SM1: Sutt 76Cb 155
Rectory Rd. SS17: Stan H 2L 101
Rectory Rd. SW13 54Wa 110
Rectory Rd. TN15: Ash 78De 165
Rectory Rd. TW4: Cran 54Y 107
Rectory Rd. UB2: S'hall 48Ba 85
Rectory Rd. UB3: Hayes 44W 84
Rectory Rd. W3 46Ra 87
Rectory Rd. WD3: Rick 18M 25
Rectory Road Station (Overground) 34Vb 71
Rectory Sq. E1 43Zb 92
Rectory Ter. SS17: Stan H 2L 101
Rectory Way UB10: Ick 33R 64
Reculver Ho. SE15 51Yb 114
(off Lovelinch Cl.)
Reculver M. N18 21Wb 51
Reculver Rd. SE16 50Zb 114
Red Anchor Cl. SW3 51Fb 111
Redan Pl. W2 44Db 89
Redan St. W14 48Za 88
Redan Ter. SE5 54Rb 113
Redbarn Cl. CR8: Purl 83Rb 177
Red Barracks Rd. SE18 49Pc 94
Redberry Gro. SE26 62Yb 136
Redbourne Av. N3 25Cb 49
Redbourne Dr. SE28 44Zc 95
(not continuous)
Redbourne Ho. E14 44Bc 92
(off Norbiton Rd.)
Redbourn Ho. W10 42Ya 88
(off Sutton Way)
REDBRIDGE 30Nc 54
Redbridge Cycling Cen. 23Yc 55
Redbridge Ent. Cen. IG1: Ilf 33Sc 74
Redbridge Foyer IG1: Ilf 33Sc 74
(off Sylvan Rd.)
Redbridge Gdns. SE5 52Ub 113
Redbridge Ho. E16 45Rc 94
(off University Way)
Redbridge La. E. IG4: Ilf 30Mc 53
Redbridge La. W. E11 30Kc 53
REDBRIDGE RDBT. 30Mc 53

Redbridge Sports & Leisure Cen. 25Tc 54
Redbridge Station (Underground) 30Mc 53
Redbrooke Ct. SS17: Linf 8J 101
Redburn Ind. Est. EN3: Pond E 16Zb 34
Redburn St. SW3 51Hb 111
Redbury Cl. RM13: Rain 42Ld 97
Redcar Cl. UB5: N'olt 36Da 65
Redcar Rd. RM3: Rom 22Pd 57
Redcar St. SE5 52Sb 113
Redcastle Cl. E1 45Yb 92
Red Cedars Rd. BR6: Orp 73Uc 160
Redchurch St. E2 5K 219 (42Vb 91)
Redcliffe Cl. SW5 50Db 89
(off Old Brompton Rd.)
Redcliffe Ct. E5 34Xb 71
(off Napoleon Rd.)
Redcliffe Gdns. IG1: Ilf 32Qc 74
Redcliffe Gdns. SW10 50Db 89
Redcliffe Gdns. SW5 50Db 89
Redcliffe Gdns. W4 52Ra 109
Redcliffe M. SW10 50Db 89
Redcliffe Pl. SW10 51Eb 111
Redcliffe Rd. SW10 50Eb 89
Redcliffe Sq. SW10 50Db 89
Redcliffe St. SW10 51Db 111
Redclose Av. SM4: Mord 71Cb 155
Redclyffe Rd. E6 39Lc 73
Redclyf Ho. E1 42Yb 92
(off Cephas St.)
Red Cottage M. SL3: L'ly 8N 81
Red Ct. SL1: Slou 6J 81
Redcourt CR0: C'don 76Ub 157
Redcourt GU22: Pyr 87F 168
Red Cow La. EC1 5D 218 (42Sb 91)
Redcroft Rd. UB1: S'hall 45Ea 86
Red Cross Cotts. SE1 1E 230 (47Sb 91)
(off Ayres St.)
Redcross Way SE1 1E 230 (47Sb 91)
Redden Ct. Rd. RM3: Hrld W 27Nd 57
Redding Cl. DA2: Dart 61Ud 142
Redding Ho. SE18 48Nc 94
Redding Ho. WD18: Wat 16U 26
Reddings HP3: Hem H 4A 4
Reddings, The NW7 20Va 30
Reddings, The WD6: Bore 13Pa 29
Reddings Av. WD23: Bush 15Da 27
Reddings Cl. NW7 21Va 48
Reddington Cl. CR2: Sande 81Tb 177
Reddington Dr. SL3: L'ly 48A 82
Redding Way GU21: Knap 1F 186
Reddins Rd. SE15 51Wb 113
Reddons Rd. BR3: Beck 66Ac 136
Reddown Rd. CR5: Coul 90Mb 176
Reddy Rd. DA8: Erith 51Hd 118
Rede Ct. KT13: Weyb 76R 150
(off Old Palace Rd.)
Redenham Ho. SW15 59Wa 110
(off Ellisfield Dr.)
Rede Pl. W2 44Cb 89
Redesdale Gdns. TW7: Isle 52Ja 108
Redesdale St. SW3 51Gb 111
Redfern Av. TW4: Houn 59Ca 107
Redfern Cl. UB8: Uxb 39L 63
Redfern Ct. WD18: Wat 15U 26
(off Whippendell Rd.)
Redfern Gdns. RM2: Rom 26Md 57
Redfern Ho. E13 39Hc 73
(off Redriffe Rd.)
Redfern Ho. NW8 39Fb 69
(off Dorman Way)
Redfern Rd. NW10 38Ua 68
Redfern Rd. SE6 59Ec 114
Redfield La. SW5 49Cb 89
Redfield M. SW5 49Db 89
Redford Av. CR5: Coul 87Kb 176
Redford Av. CR7: Thor H 70Pb 134
Redford Av. SM6: W'gton 79Nb 156
Redford Cl. TW13: Felt 61V 128
Redford Rd. SL4: Wind 3B 102
Redford Wlk. N1 39Sb 71
(off Popham St.)
Redford Way UB8: Uxb 38M 63
Redgate Dr. BR2: Hayes 75Kc 159
Redgate Ter. SW15 58Za 110
Redgrave Cl. CR0: C'don 72Vb 157
Redgrave Ct. UB9: Den 29H 43
Redgrave Rd. SW15 55Za 110
Redgrave Ter. E2 41Wb 91
(off Derbyshire St.)
Redgrove Ho. CM16: Epp 1Wc 23
(off Stonards Hill)
Redhall Cl. AL10: Hat 3B 8
Redhall Ct. CR3: Cat'm 95Tb 197
Redhall Dr. AL10: Hat 4B 8
Redhall End AL4: Col H 4A 8
Redhall La. AL4: Col H 4A 8
(not continuous)
Redhall La. WD3: Chan C 11N 25
Redheath Cl. WD25: Wat 7V 12
Red Hill BR7: Chst 64Rc 138
Red Hill UB9: Den 33F 62
REDHILL 5P 207
Redhill & Reigate Golf Course 9M 207
Redhill Comn. RH1: Redh 7N 207
Redhill Ct. SW2 61Qb 134
Redhill Dr. HA8: Edg 26Ra 47
Redhill Ho. RH1: Redh 4P 207
Redhill Rd. DA3: Nw A G 76Ae 165
Redhill Rd. KT11: Cobh 84R 170
Redhill Station (Rail) 5A 208
Redhill St. NW1 2A 216 (40Kb 70)
Redhill Wood DA3: Nw A G 76Ce 165
(not continuous)
Red House 56Ad 117
Red Ho. Cotts. TN13: S'oaks 97Ld 203
Red Ho. La. DA6: Bex 56Zc 117
Red Ho. La. KT12: Walt T 75W 150
Redhouse Rd. CR0: C'don 72Mb 156
Redhouse Rd. TN16: Tats 92Lc 199
Red Ho. Sq. N1 38Sb 71
Redif Ho. RM10: Dag 35Dd 76
Redington Gdns. NW3 35Db 69
Redington Ho. N1 1J 217 (40Pb 70)
(off Priory Grn. Est.)
Redington Rd. NW3 34Db 69
Redknap Ho. TW10: Ham 62La 130
Redland Gdns. KT8: W Mole 70Ba 129
Redlands CR5: Coul 88Nb 176
Redlands N15 28Tb 51
Redlands TW11: Tedd 65Ja 130
Redlands, The BR3: Beck 68Dc 136
Redlands Ct. BR1: Broml 66Hc 137
Redlands Rd. EN3: Enf H 11Ac 34
Redlands Rd. TN13: S'oaks 96Hd 202
Redlands Way SW2 59Pb 112
Red La. KT10: Clay 79Ja 152
Red La. RH8: Oxt 6M 211
Red Leaf Cl. SL3: L'ly 46B 82
Redleaf Cl. DA17: Belv 51Cd 118
Redleaf Cl. KT22: Fet 96Fa 192
Redleaves Av. TW15: Ashf 65R 128
Redlees Cl. TW7: Isle 56Ja 108
Red Leys UB8: Uxb 38N 63
Redlibbets Golf Course 74Zd 165
Redlin Ct. RH1: Redh 4P 207
Red Lion Bus. Pk. KT6: Surb 76Pa 153
Red Lion Cl. BR5: St M Cry 72Yc 161
Red Lion Cl. SE17 51Tb 113
(off Red Lion Row)
Red Lion Ct. WD25: A'ham 9Ea 14
Red Lion Ct. EC4 3A 224 (44Qb 90)
Red Lion Ct. SE1 6E 224 (46Sb 91)
Red Lion Ct. TW3: Houn 55Da 107
(off Alexandra Rd.)
Red Lion Hill N2 26Fb 49
(not continuous)
Red Lion La. GU24: Chob 1J 167
Red Lion La. HP3: Hem H 8A 4
Red Lion La. SE18 52Qc 116
Red Lion La. WD3: Sarr 7J 11
Red Lion Pde. HA5: Pinn 27Aa 45
Red Lion Pl. SE18 53Qc 116
Red Lion Rd. GU24: Chob 1J 167
Red Lion Rd. KT6: Surb 75Pa 153
Red Lion Row SE17 51Sb 113
Red Lion Sq. SW18 57Cb 111
Red Lion Sq. TN15: Bor G 92Be 205
Red Lion Sq. WC1 1H 223 (43Pb 90)
Red Lion St. TW9: Rich 57Ma 109
Red Lion St. WC1 7H 217 (43Pb 90)
Red Lion Wlk. TW3: Houn 55Da 107
(off High St.)
Red Lion Yd. W1 6K 221 (46Kb 90)
(off Waverton St.)
Red Lodge BR4: W W'ck 74Ec 158
Red Lodge WD6: Bore 13Pa 29
Red Lodge Cres. DA5: Bexl 62Fd 140
Red Lodge Rd. BR4: W W'ck 74Ec 158
Red Lodge Rd. DA5: Bexl 62Fd 140
Redlynch Ct. W14 47Ab 88
(off Addison Cres.)
Redlynch Ho. SW9 53Qb 112
(off Gosling Way)
Redman Cl. UB5: N'olt 40Y 65
Redman Ho. EC1 7K 217 (43Qb 90)
(off Bourne Est.)
Redman Ho. SE1 2E 230 (47Sb 91)
(off Borough High St.)
Redmans La. TN14: S'ham 79Ed 162
Redmans Pl. TN13: S'oaks 97Ld 203
(off Akehurst La.)
Redman's Rd. E1 43Yb 92
Redmead La. E1 46Wb 91
Redmead Rd. UB3: Harl 49U 84
Redmill Ho. E1 42Xb 91
(off Headlam St.)
Redmond Ho. N1 1J 217 (39Pb 70)
(off Barnsbury Est.)
Redmore Rd. W6 49Xa 88
Red Oak Cl. BR6: Farnb 76Rc 160
Red Oak Cl. CR0: C'don 75Cc 158
Red Oaks Mead CM16: They B 9Tc 22
Redo Ho. E12 36Qc 74
(off Dore Av.)
Red Path E9 37Ac 72
Redpath Way SE10 47Gc 93
Red Pl. W1 4H 221 (45Jb 90)
Redpoll Way DA18: Erith 48Zc 95
Red Post Hill SE21 57Tb 113
Red Post Hill SE24 56Tb 113
Red Post Ho. E6 38Mc 73
Redriffe Rd. E13 39Hc 73
Redriff Est. SE16 48Bc 92
Redriff Rd. RM7: Mawney 26Dd 56
Redriff Rd. SE16 49Zb 92
Red Rd. CM14: W'ley 21Xd 58
Red Rd. GU18: Light 4A 166
Red Rd. WD6: Bore 13Pa 29
Redroofs Cl. BR3: Beck 67Dc 136
Redrose Trad. Cen. EN4: E Barn 15Fb 31
RED ROVER 56Wa 110
Redrup Ho. SE14 51Zb 114
(off John Williams Cl.)
Redruth Cl. N22 24Pb 50
Redruth Gdns. KT10: Clay 80Ha 152
Redruth Gdns. RM3: Rom 22Pd 57
Redruth Ho. SM2: Sutt 80Db 155
Redruth Rd. E9 39Yb 72
Redruth Rd. RM3: Rom 22Pd 57
Redruth Wlk. RM3: Rom 22Pd 57
Redsan Cl. CR2: S Croy 80Tb 157
Redshank Ho. SE1 50Vb 91
(off Avocet Cl.)
Red Sq. N16 34Tb 71
Redstart Cl. CR0: New Ad 82Fc 179
Redstart Cl. E6 43Nc 94
Redstart Cl. SE14 52Ac 114
Redstart Mans. IG1: Ilf 34Qc 74
(off Mill Rd.)
Redstone Apts. CR2: S Croy 84Vb 177
Redstone Hill RH1: Redh 6A 208
Redstone Hollow RH1: Redh 7A 208
Redstone Mnr. RH1: Redh 6A 208
Redstone Pk. RH1: Redh 6A 208
Redstone Rd. RH1: Redh 7A 208
Redston Rd. N8 28Mb 50
Red St. DA13: Sflt 64Ce 143
REDSTREET 65De 143
Redtiles Gdns. CR8: Kenley 87Rb 177
Redvers Ct. CR6: W'ham 90Zb 178
(off Redvers Rd.)
Redvers Rd. CR6: W'ham 90Zb 178
Redvers Rd. N22 26Qb 50
Redvers St. N1 3J 219 (41Ub 91)
Redwald Rd. E5 35Zb 72
Redway Dr. TW2: Whitt 59Ea 108
REDWELL 94Yd 204
Redwell Cotts. TN15: Igh 95Yd 204
Redwell Ct. EN8: Walt C 6Zb 20
Redwell La. TN15: Igh 95Xd 204
Redwing Cl. CR2: Sels 83Zb 178
Redwing Ct. BR6: Orp 73Wc 161
(off High St.)
Redwing Ct. RM3: Rom 25Md 57
Redwing Ct. SE1 2E 230 (47Sb 91)
(off Swan St.)
Redwing Cres. DA9: Ghithe 58Wd 120
Redwing Dr. HP3: Hem H 6L 3
Redwing Gdns. KT14: W Byf 84K 169
Redwing Gro. WD5: Ab L 3W 12
Redwing M. SE5 54Sb 113
Redwing Path SE28 47Tc 94
Redwing Rd. SM6: W'gton 80Nb 156
Redwood TW20: Thorpe 68G 126
Redwood Av. RM3: Hrld W 26Nd 57
Redwood Chase RM15: S Ock 42Yd 98
Redwood Cl. AL1: St A 2G 6
Redwood Cl. CR8: Kenley 86Sb 177
Redwood Cl. DA15: Sidc 60Wc 117
Redwood Cl. E3 40Cc 72
Redwood Cl. N14 17Mb 32
Redwood Cl. SE16 46Ac 92
Redwood Cl. UB10: Hil 40R 64
Redwood Cl. WD19: Wat 21Y 45
Redwood Ct. DA1: Dart 58Qd 119
Redwood Ct. KT17: Ewe 83Va 174
Redwood Ct. KT22: Lea 92Ha 192
(off Park Vw. Rd.)
Redwood Ct. KT6: Surb 73Ma 153
Redwood Ct. N19 31Mb 70
Redwood Ct. NW6 38Ab 68
Redwood Ct. UB5: N'olt 41Aa 85
Redwood Dr. HP3: Hem H 4N 3
Redwood Dr. KT19: Eps 82Sa 173
Redwood Dr. SL5: S'dale 2F 146
Redwood Est. TW5: Cran 51X 107
Redwood Gdns. E4 16Dc 34
Redwood Gdns. IG7: Chig 22Wc 55
Redwood Gdns. SL1: Slou 5H 81
Redwood Gro. W5 48Ka 86
Redwood Ho. EC1 3D 218 (41Sb 91)
(off Bollinder Pl.)
Redwood Ho. HA9: Wemb 35Qa 67
(off Empire Way)
Redwood Ho. TN16: Westrm 95Rc 200
Redwood Mans. W8 48Db 89
(off Chantry Sq.)
Redwood M. SW4 55Kb 112
Redwood M. TW15: Ashf 66T 128
(off Staines Rd. W.)
Redwood Mt. RH2: Reig 3J 207
Redwood Pl. TN13: S'oaks 95Hd 202
Redwood Ri. WD6: Bore 9Ra 15
Redwoods KT15: Add 79J 149
Redwoods SW15 60Wa 110
Redwoods, The SL4: Wind 5H 103
Redwoods Cl. IG9: Buck H 19Kc 35
Redwood Wlk. KT6: Surb 74Ma 153
Redwood Way EN5: Barn 15Za 30
Reece M. SW7 5B 226 (49Fb 89)
Reed Av. BR6: Orp 76Uc 160
Reed Cl. AL2: Lon C 9J 7
Reed Cl. E16 43Jc 93
Reed Cl. SE12 57Jc 115
Reed Cl. SL0: Iver 44G 82
Reed Ct. DA9: Ghithe 56Yd 120
Reede Dr. RH1: Redh 9A 208
Reede Gdns. RM10: Dag 36Dd 76
Reede Rd. RM10: Dag 37Cd 76
Reede Way RM10: Dag 37Dd 76
Reed Gdns. CR5: Coul 91Rb 197
Reedham Cl. AL2: Brick W 1Ca 13
Reedham Cl. N17 28Xb 51
Reedham Dr. CR8: Purl 85Pb 176
Reedham Pk. Av. CR8: Purl 88Qb 176
Reedham Rd. SL1: Burn 1A 80
Reedham Station (Rail)
Surrey 85Pb 176
Reedham St. SE15 54Wb 113
Reedholm Vs. N16 35Tb 71
Reed Ho. SW19 63Db 133
Reed Pl. KT14: W Byf 85G 168
Reed Pl. SW4 56Mb 112
Reed Pl. TW17: Shep 74P 149
Reed Pond Wlk. RM2: Rom 26Hd 56
Reed Rd. N17 26Vb 51
Reeds Chapel WD24: Wat 12Y 27
(off Keele Cl.)
Reeds Cres. WD24: Wat 12Y 27
Reedsfield Cl. TW15: Ashf 62R 128
Reedsfield Rd. TW15: Ashf 63R 128
Reeds Mdw. RH1: Mers 2C 208
Reed's Pl. NW1 38Lb 70
Reed St. GU22: Wok 94B 188
Reeds Wlk. WD24: Wat 12Y 27
(not continuous)
Reed Way SL1: Slou 5D 80
Reed Way SL4: Wind 3B 102
Reedworth St. SE11 6A 230 (49Qb 90)
Reef Ho. E14 48Ec 92
(off Manchester St.)
Reef St. RM9: Dag 41Ad 95
Reel Cinema Borehamwood 13Qa 29
(within The Point)
Reenglass Rd. HA7: Stan 21Ma 47
Rees Dr. HA7: Stan 21Na 47
Rees Gdns. CR0: C'don 72Vb 157
Reesland Cl. E12 37Qc 74
Rees St. N1 39Sb 71
Reet Gdns. SL1: Slou 6K 81
Reeves Farm Cl. NW9 30Ua 48
Reeve Rd. RH2: Reig 10L 207
Reeves Av. NW9 31Ta 67
Reeves Cnr. CR0: C'don 75Rb 157
Reeves Corner Stop
(London Tramlink) 75Rb 157
Reeves Cres. BR8: Swan 69Fd 140
Reeves Ho. KT22: Lea 92Ha 192
Reeves Ho. SE1 2A 230 (47Qb 90)
(off Baylis Rd.)
Reeves M. W1 5H 221 (45Jb 90)
Reeves Rd. E3 42Dc 92
Reeves Rd. SE18 51Rc 116
Reference Ho. NW9 27Va 48
Reflection, The E16 47Rc 94
(off Woolwich Mnr. Way)
Reflection Ho. E2 42Wb 91
(off Cheshire St.)
Reflex Apts. BR2: Broml 70Kc 137
(off Commercial Rd.)
Reform Row N17 26Vb 51
Reform St. SW11 54Hb 111
Regal Bldg. W10 41Za 88
Regal Cl. E1 43Wb 91
Regal Cl. W5 43Ma 87
Regal Ct. CR4: Mitc 69Hb 133
Regal Ct. N18 22Vb 51
Regal Ct. NW6 40Bb 69
(off Malvern Rd.)
Regal Ct. SW6 52Cb 111
(off Dawes Rd.)
Regal Cres. SM6: W'gton 76Kb 156
Regal Dr. N11 22Kb 50
Regal Ho. IG2: Ilf 30Tc 54
Regal Ho., The SW6 54Eb 111
Regalia Cl. E16 47Kc 93
Regalia Point E2 41Zb 92
(off Palmer's Rd.)
Regal La. NW1 39Jb 70
Regal Pl. E3 41Bc 92
Regal Pl. SW6 52Db 111
Regal Row SE15 53Yb 114
Regal Way HA3: Kenton 30Na 47
Regal Way WD24: Wat 10Y 13
Regal Wharf Apts. N1 39Ub 71
(off De Beauvoir Cres.)
Regan Ho. N18 23Vb 51
Regan Way N1 1H 219 (40Ub 71)
Regarder Rd. IG7: Chig 22Wc 55
Regarder Rd. IG7: Ilf 22Wc 55
Regarth Av. RM1: Rom 30Gd 56
Regatta Ho. TW11: Tedd 63Ja 130
Regatta La. W6 51Ya 110
Regatta Point E14 47Cc 92
(off Westferry Rd.)
Regatta Point TW8: Bford 51Pa 109
Regency Cl. IG7: Chig 22Sc 54
Regency Cl. TN15: W King 79Ud 164
Regency Cl. TW12: Hamp 64Ba 129
Regency Cl. W5 44Na 87
Regency Ct. AL1: St A 2F 6
Regency Ct. BR2: Broml 72Mc 159
Regency Ct. CM14: B'wood 19Xd 40
Regency Ct. E18 26Jc 53
Regency Ct. E3 40Bc 72
(off Norman Gro.)
Regency Ct. E9 39Yb 72
(off Park Cl.)
Regency Ct. EN1: Enf 15Tb 33
Regency Ct. HA7: Stan 24Ma 47
(off Unwin Way)
Regency Ct. HP2: Hem H 2M 3
Regency Ct. KT15: Add 76M 149
(off Albert Rd.)
Regency Ct. SE8 52Bc 114
(off Glenville Gro.)
Regency Ct. SM1: Sutt 77Eb 155
Regency Ct. TW11: Tedd 65Ka 130
Regency Ct. WD17: Wat 11W 26
(off Langley Rd.)
Regency Cres. NW4 26Za 48
Regency Dr. HA4: Ruis 32U 64
Regency Dr. KT14: W Byf 85H 169
Regency Gdns. KT12: Walt T 74Y 151
Regency Gdns. KT13: Weyb 79S 150
Regency Gdns. RM11: Horn 31Ld 77
Regency Ho. E16 46Jc 93
(off Pepys Cres.)
Regency Ho. N3 26Bb 49
Regency Ho. SW1 5E 228 (49Mb 90)
(off Regency St.)
Regency Ho. SW6 53Eb 111
(off The Boulevard)
Regency Lawn NW5 34Kb 70
Regency Lodge IG9: Buck H 19Mc 35
Regency Lodge KT13: Weyb 76U 150
(off Oatlands Chase)
Regency Lodge NW3 38Fb 69
(off Adelaide Rd.)
Regency M. BR3: Beck 67Ec 136
Regency M. NW10 37Wa 68
Regency M. SW9 52Rb 113
Regency M. TW7: Isle 57Ga 108
Regency Pde. NW3 38Fb 69
(off Finchley Rd.)
Regency Pl. SW1 5E 228 (49Mb 90)
Regency St. NW10 42Ua 88
Regency St. SW1 5E 228 (49Mb 90)
Regency Ter. SW7 7B 226 (50Fb 89)
(off Fulham Rd.)
Regency Wlk. CR0: C'don 72Bc 158
Regency Wlk. TW10: Rich 57Na 109
(off The Vineyard)
Regency Way DA6: Bex 55Zc 117
Regeneration House 1F 217 (39Nb 70)
Regeneration Rd. SE16 49Zb 92
Regent Av. UB10: Hil 38R 64
Regent Bus. Cen. UB3: Hayes 47X 85
Regent Cl. HA3: Kenton 30Na 47
Regent Cl. KT15: New H 81M 169
Regent Cl. N12 22Eb 49
Regent Cl. RH1: Mers 1C 208
Regent Cl. RM16: Grays 47Ee 99
Regent Cl. TW4: Cran 53X 107
Regent Cl. WD4: K Lan 1Q 12
Regent Ct. N20 19Fb 31
Regent Ct. N3 24Db 49
Regent Ct. NW6 38Bb 69
(off Cavendish Rd.)
Regent Ct. NW8 4D 214 (41Gb 89)
(off North Bank)
Regent Ct. RM15: Avel 46Td 98
Regent Ct. SL1: Slou 4J 81
Regent Ct. SL4: Wind 3H 103
Regent Ct. W6 48Xa 88
(off Vinery Way)
Regent Ct. W8 48Db 89
(off Wright's La.)
Regent Cres. RH1: Redh 4P 207
Regent Gdns. IG3: Ilf 31Wc 75
Regent Ga. EN8: Walt C 6Zb 20
Regent Ho. CM14: B'wood 20Xd 40
Regent Ho. KT17: Eps 83Ua 174
Regent Ho. RH1: Redh 5P 207
Regent Ho. W14 49Ab 88
(off Windsor Way)
Regent Pde. SM2: Sutt 79Eb 155
Regent Pk. KT22: Lea 90Ja 172
Regent Pl. CR0: C'don 74Vb 157
Regent Pl. SW19 64Eb 133
Regent Pl. W1 4C 222 (45Lb 90)
Regent Rd. CM16: Epp 2Vc 23
Regent Rd. KT5: Surb 71Pa 153
Regent Rd. SE24 58Rb 113
Regents Av. N13 22Qb 50
Regents Bri. Gdns. SW8 52Nb 112
Regents Canal Ho. E14 44Ac 92
(off Commercial Rd.)
Regents Cl. CR2: S Croy 79Ub 157
Regents Cl. CR3: Whyt 90Ub 177
Regents Cl. HA8: Edg 21Na 47
Regents Cl. UB4: Hayes 43V 84
Regents Cl. WD7: R'lett 6Ja 14
Regents Ct. BR1: Broml 66Hc 137
Regents Ct. DA11: Grav'nd 7D 122
Regents Ct. E8 39Wb 71
Regents Ct. HA5: Pinn 26Z 45
Regents Ct. KT13: Weyb 79R 150
Regents Ct. KT2: King T 67Na 131
(off Sopwith Way)
Regents Dr. BR2: Kes 78Mc 159
Regents Dr. IG8: Wfd G 23Qc 54
Regents Ga. Ho. E14 45Ac 92
(off Horseferry Rd.)
Regents Ho. E3 43Cc 92
(off Bow Common La.)
Regents M. NW8 1A 214 (40Eb 69)
Regent's Pk. 2F 215 (40Hb 69)
REGENT'S PARK 4A 216 (41Kb 70)
Regent's Pk. Barracks 2A 216 (40Kb 70)
(off Albany St.)
Regents Pk. Est. NW1 3B 216 (41Lb 90)
(off Robert St.)
Regent's Pk. Gdns. M. NW1 39Hb 69
Regent's Pk. Ho. NW8 4E 214 (41Gb 89)
(off Park Rd.)
Regent's Pk. Open Air
Theatre 4H 215 (41Jb 90)
Regent's Pk. Rd. NW1 38Hb 69
(not continuous)
Regents Pk. Rd. N3 27Bb 49
Regent's Park Station
(Underground) 6K 215 (42Kb 90)
Regent's Pk. Ter. NW1 39Kb 70
Regent's Pl. SE3 54Jc 115
Regents Pl. IG10: Lough 17Mc 35
Regents Pl. KT12: Walt T 76Y 151
Regents Pl. NW1 5B 216 (42Lb 90)
Regent's Pl. Plaza NW1 5B 216 (42Lb 90)
Regents Plaza NW6 40Db 69
(off Kilburn High Rd.)
Regent Sq. DA17: Belv 49Dd 96
Regent Sq. E3 41Dc 92
Regent Sq. WC1 4G 217 (41Nb 90)
Regent's Row E8 39Wb 71
Regents Studios E8 39Xb 71
Regent St. NW10 41Za 88
Regent St. SW1 5D 222 (45Mb 90)
Regent St. W1 2A 222 (44Kb 90)
Regent St. W4 50Qa 87
Regent St. WD24: Wat 10X 13
Regent Street Cinema 2A 222 (44Kb 90)
Regent's University London 5G 215 (42Hb 89)
Regents Wlk. SL5: Asc 2A 146
Regents Wharf E2 39Xb 71
(off Wharf Pl.)
Regents Wharf N1 1H 217 (40Pb 70)
Regent Ter. SW8 51Nb 112
Regent Way CM14: B'wood 17Xd 40
Regiment Hill NW7 23Ab 48
Regina Cl. EN5: Barn 13Za 30
Regina Ho. SE20 67Zb 136
Reginald Ellingworth St.
RM9: Dag 39Xc 75
Reginald Pl. SE8 52Cc 114
(off Deptford High St.)
Reginald Rd. E7 38Jc 73
Reginald Rd. HA6: Nwood 25V 44
Reginald Rd. RM3: Hrld W 25Qd 57
Reginald Rd. SE8 52Cc 114
Reginald Sorenson Ho. E11 31Fc 73
Reginald Sq. SE8 52Cc 114
Regina Point SE16 48Yb 92
(off Canada Est.)
Regina Rd. N4 32Pb 70
Regina Rd. SE25 69Wb 135
Regina Rd. UB2: S'hall 49Aa 85
Regina Rd. W13 46Ja 86
Regina Ter. W13 46Ka 86
Regis Ct. CR4: Mitc 67Gb 133
Regis Ct. N8 28Pb 50
Regis Ct. NW1 7F 215 (43Hb 89)
(off Melcombe Pl.)
Regis Ho. W1 7J 215 (43Jb 90)
(off Beaumont St.)
Regis Pl. SW2 56Pb 112
Regis Rd. NW5 36Kb 70
Regna Cl. RM13: Rain 43Kd 97
Regnart Bldgs. NW1 4C 216 (41Lb 90)
(off Euston St.)
Regnas Ho. E15 37Hc 73
(off Carnarvon Rd.)
Regnolruf Ct. KT12: Walt T 73W 150
Regnum Apts. E1 6K 219 (42Vb 91)
(off Wheler St.)
Reid Av. CR3: Cat'm 93Tb 197
Reid Bldg. E3 42Bc 92
(off Eric St.)
Reid Cl. CR5: Coul 88Kb 176
Reid Cl. HA5: Eastc 28W 44
Reid Cl. UB3: Hayes 44U 84
Reidhaven Rd. SE18 49Uc 94
REIDON HILL 10F 166
Reidonhill Cotts. GU21: Knap 10F 166
REIGATE 6J 207
Reigate Av. SM1: Sutt 74Cb 155
Reigate Bus. M. RH2: Reig 5H 207
Reigate Heath 6F 206
REIGATE HEATH 6F 206
Reigate Heath Golf Course 6E 206
Reigate Heath Windmill 6E 206
Reigate Hill RH2: Reig 2L 207
Reigate Hill Cl. RH2: Reig 3J 207
Reigate Hill Golf Course 100Jb 196
REIGATE HILL INTERCHANGE 1K 207
Reigate Priory Mus. 7J 207
Reigate Rd. RH2: Reig Chart La. 6K 207
Reigate Rd. RH2: Reig Station Rd. 5E 206
Reigate Rd. BR1: Broml 62Hc 137
Reigate Rd. IG3: Ilf 33Vc 75
Reigate Rd. KT17: Eps 82Va 174
Reigate Rd. KT17: Eps D 82Va 174
Reigate Rd. KT17: Ewe 82Va 174
Reigate Rd. KT18: Tatt C 89Za 174
Reigate Rd. KT20: Tad 89Za 174
Reigate Rd. KT22: Lea 95La 192
Reigate Rd. RH1: Redh 6M 207
Reigate Station (Rail) 5J 207
Reigate Way SM6: W'gton 78Nb 156
Reighton Rd. E5 34Wb 71
Reindeer Cl. E13 39Jc 73
Reinickendorf Av. SE9 58Sc 116
Reis Pl. N15 28Vb 51
(off Blenheim Ri.)
Reizel Cl. N16 32Vb 71
Relay Rd. W12 45Za 88
Relf Rd. SE15 55Wb 113
Reliance Arc. SW9 56Qb 112
Reliance Wharf N1 39Ub 71
Relko Ct. KT19: Eps 83Ta 173
Relko Gdns. SM1: Sutt 78Fb 155
Relton M. SW7 3E 226 (48Gb 89)
Rembrandt Cl. E14 48Fc 93
Rembrandt Cl. SW1 7H 227 (50Jb 90)
(off Graham Ter.)
Rembrandt Ct. KT19: Ewe 79Va 154
Rembrandt Ct. SE16 50Xb 91
(off Stubbs Dr.)
Rembrandt Dr. DA11: Nflt 62Fe 143
Rembrandt Rd. HA8: Edg 26Qa 47
Rembrandt Rd. SE13 56Gc 115
Rembrandt Way KT12: Walt T 75X 151
Rembrandt Way WD18: Wat 15V 26
Reminder La. SE10 48Hc 93
(not continuous)
Remington Rd. E6 44Nc 94
Remington Rd. N15 30Tb 51
Remington St. N1 2C 218 (40Rb 71)
Remnant St. WC2 2H 223 (44Pb 90)
Remsted Ho. NW6 39Db 69
(off Mortimer Cres.)
Remus Bldg., The EC1 4A 218 (41Qb 90)
(off Hardwick St.)
Remus Cl. AL1: St A 6B 6
Remus Rd. E3 38Cc 72

Renaissance KT15: Add 77L 149
(off High St.)
Renaissance Ct. SM1: Sutt 74Eb 155
Renaissance Ct. TW3: Houn 55Ea 108
(off Prince Regent Rd.)
Renaissance Ho. KT17: Eps 85Ua 174
(off Up. High St.)
Renaissance Sq. W4 51Ua 110
Renaissance Wlk. SE10 48Hc 93
(off Teal St.)
Renbold Ho. SE10 53Ec 114
(off Blissett St.)
Rendalls HA1: Harr 32Ga 66
(off Grove Hill)
Rendel Apts. E16 45Sc 94
(off Lock Side Way)
Rendel Ho. SM7: Bans 90Eb 175
Rendle Cl. CR0: C'don 71Vb 157
Rendle Ho. W10 42Ab 88
(off Wornington Rd.)
Rendlesham Av. WD7: R'lett 9Ha 14
Rendlesham Rd. E5 35Wb 71
Rendlesham Rd. EN2: Enf 11Rb 33
Rendlesham Way WD3: Chor 16E 24
Renforth St. SE16 48Yb 92
Renfree Way TW17: Shep 73Q 150
Renfrew Cl. E6 45Qc 94
Renfrew Ct. TW4: Houn 54Aa 107
Renfrew Ho. E17 26Bc 52
Renfrew Ho. NW6 40Db 69
(off Carlton Vale)
Renfrew Rd. KT2: King T 66Ra 131
Renfrew Rd. SE11 5B 230 (49Rb 91)
Renfrew Rd. TW4: Houn 54Z 107
Renmans, The KT21: Asht 88Pa 173
Renmuir St. SW17 65Hb 133
Rennell St. SE13 55Ec 114
Rennels Way TW7: Isle 54Ga 108
Renness Rd. E17 27Ac 52
Rennets Cl. SE9 57Uc 116
Rennets Wood Rd. SE9 57Tc 116
Rennie Cl. TW15: Ashf 62M 127
Rennie Cotts. E1 42Yb 92
(off Pemell Cl.)
Rennie Ct. EN3: Enf L 10Cc 20
Rennie Ct. SE1 6B 224 (46Rb 91)
(off Upper Ground)
Rennie Dr. DA1: Dart 54Qd 119
Rennie Est. SE16 49Xb 91
RENNIE GROVE HOSPICE 1A 6
Rennie Ho. SE1 4D 230 (48Sb 91)
(off Bath Ter.)
Rennie St. SE1 6B 224 (46Rb 91)
(not continuous)
Rennie St. SE10 48Jc 93
Rennie Ter. RH1: Redh 7A 208
Renoir Ct. SE16 50Xb 91
(off Stubbs Dr.)
Renovation, The E16 47Rc 94
(off Woolwich Mnr. Way)
Renown Cl. CR0: C'don 74Rb 157
Renown Cl. RM7: Mawney 25Cd 56
Rensburg Rd. E17 29Zb 52
Renshaw Cl. DA17: Belv 51Bd 117
Renshaw Cl. SE6 59Cc 114
Renshaw Ind. Est. TW18: Staines 63H 127
Renters Av. NW4 30Ya 48
Renton Cl. SW2 58Pb 112
Renton Dr. BR5: Orp 73Zc 161
Renton Dr. BR5: St M Cry 73Zc 161
Renwick Dr. BR2: Broml 72Mc 159
Renwick Ind. Est. IG11: Bark 40Xc 75
Renwick Rd. IG11: Bark 42Xc 95
Repens Way UB4: Yead 42Z 85
Rephidim St. SE1 4H 231 (48Ub 91)
Replingham Rd. SW18 60Bb 111
Reporton Rd. SW6 52Ab 110
Repository Rd. SE18 51Pc 116
Repton Av. HA0: Wemb 35La 66
Repton Av. RM2: Rom 27Jd 56
Repton Av. UB3: Harl 49T 84
Repton Cl. SM5: Cars 78Gb 155
Repton Ct. BR1: Broml 65Mc 137
Repton Ct. BR3: Beck 67Dc 136
Repton Dr. RM2: Rom 28Jd 56
Repton Gdns. RM2: Rom 27Jd 56
Repton Gro. IG5: Ilf 25Pc 54
Repton Ho. E16 47Kc 93
(off Royal Crest Av.)
Repton Ho. E4 23Ec 52
Repton Ho. SW1 6C 228 (49Lb 90)
(off Charlwood St.)
REPTON PARK 24Qc 54
Repton Rd. BR6: Chels 76Wc 161
Repton Rd. HA3: Kenton 28Pa 47
Repton St. E14 44Ac 92
Repton Way WD3: Crox G 15Q 26
Repulse Cl. RM5: Col R 25Cd 56
Reservoir Cl. CR7: Thor H 69Tb 135
Reservoir Cl. DA9: Ghithe 58Yd 120
Reservoir Ct. AL10: Hat 1B 8
Reservoir Rd. HA4: Ruis 28S 44
Reservoir Rd. IG10: H Beech 11Kc 35
Reservoir Rd. N14 15Lb 32
Reservoir Rd. SE4 54Ac 114
Reservoir Studios E1 44Zb 92
(off Cable St.)
Reservoir Way IG6: Chig 21Xc 55
Reservoir Way NW10 38Wa 68
Resham Cl. UB2: S'hall 48Y 85
Residence Twr. N4 31Sb 71
(off Goodchild Rd.)
Resolution Plaza E1 2K 225 (44Vb 91)
(off Old Castle St.)
Resolution Wlk. SE18 48Pc 94
Resolution Way SE8 52Cc 114
(off Deptford High St.)
Reson Way HP1: Hem H 3K 3
Restavon Cvn. Site TN16: Big H 88Rc 180
Restell Cl. SE3 51Gc 115
Restmor Way SM6: W'gton 75Jb 156
Reston Cl. WD6: Bore 10Qa 15
Reston Path WD6: Bore 10Qa 15
Reston Pl. SW7 2A 226 (47Eb 89)
Restons Cres. SE9 58Tc 116
Restoration Sq. SW11 53Fb 111
Restormel Cl. TW3: Houn 57Ca 107
Restormel Ho. SE11 6A 230 (49Qb 90)
(off Chester Way)
Retcar Pl. N19 33Kb 70
Retford Cl. RM3: Rom 23Qd 57
Retford Cl. WD6: Bore 10Qa 15
Retford Path RM3: Rom 23Qd 57
Retford Rd. RM3: Rom 23Pd 57
Retford St. N1 2J 219 (40Ub 71)
Retingham Way E4 19Dc 34
Retles Ct. HA1: Harr 31Ga 66
Retreat, The BR6: Chels 79Xc 161
Retreat, The CM13: Hut 16De 41
Retreat, The CM14: B'wood 18Xd 40
Retreat, The CR7: Thor H 70Tb 135
Retreat, The HA2: Harr 31Ca 65
Retreat, The HP6: L Chal 11A 24
Retreat, The KT4: Wor Pk 75Xa 154
Retreat, The KT5: Surb 72Pa 153
Retreat, The NW9 29Ta 47
Retreat, The RM17: Grays 51De 121
Retreat, The SW14 55Ua 110
Retreat, The TN13: S'oaks 97Kd 203
Retreat, The TW20: Eng G 4P 125
Retreat, The WD4: K Lan 3S 12
Retreat Cl. HA3: Kenton 29La 46
Retreat Ho. E9 37Yb 72
Retreat Mobile Home Pk., The IG9: Buck H 19Jc 35
Retreat Pl. E9 37Yb 72
Retreat Rd. TW9: Rich 57Ma 109
Retreat Way IG7: Chig 20Xc 37
Reubens Ct. W4 50Ra 87
(off Chaseley Dr.)
Reubens Rd. CM13: Hut 16De 41
Reuters Plaza E14 46Dc 92
(off The South Colonnade)
Reveley Cotts. WD23: Bush 16Ca 27
Reveley Sq. SE16 47Ac 92
Revell Cl. KT22: Fet 94Da 191
Revell Dr. KT22: Fet 94Da 191
Revell Ri. SE18 51Vc 117
Revell Rd. KT1: King T 68Ra 131
Revell Rd. SM1: Sutt 79Bb 155
Revelon Rd. SE4 56Ac 114
Revelstoke Rd. SW18 61Bb 133
Reventlow Rd. SE9 60Sc 116
Reverdy Rd. SE1 49Wb 91
Reverend Cl. HA2: Harr 34Da 65
Revere Way KT19: Ewe 81Ua 174
Revesby Cl. GU24: W End 5B 166
Revesby Rd. SM5: Cars 72Fb 155
Review Lodge RM10: Dag 39Dd 76
Review Rd. NW2 33Va 68
Review Rd. RM10: Dag 39Dd 76
Revolution Karting 43Bc 92
Rewell St. SW6 52Eb 111
Rewley Rd. SM5: Cars 71Fb 155
Rex Av. TW15: Ashf 65Q 128
Rex Cl. RM5: Col R 24Dd 56
Rex Pl. W1 5J 221 (45Jb 90)
Reydon Av. E11 29Lc 53
Reydon Pl. KT12: Walt T 75X 151
Reynard Cl. BR1: Broml 69Qc 138
Reynard Cl. SE4 55Ac 114
Reynard Dr. SE19 66Vb 135
Reynard Pl. SE14 51Ac 114
Reynardson Rd. N17 24Sb 51
Reynards Way AL2: Brick W 1Ba 13
Reynard Way TW8: Bford 50La 86
Reynolah Gdns. SE7 50Kc 93
Reynolds Av. E12 36Qc 74
Reynolds Av. KT9: Chess 80Na 153
Reynolds Av. RH1: Redh 4B 208
Reynolds Av. RM6: Chad H 31Yc 75
Reynolds Cl. HP1: Hem H 1J 3
Reynolds Cl. NW11 31Db 69
Reynolds Cl. SM5: Cars 74Hb 155
Reynolds Cl. SW19 67Fb 133
Reynolds Ct. RM6: Chad H 27Zc 55
Reynolds Dr. HA8: Edg 27Pa 47
Reynolds Fitness Spa 61Gd 140
Reynolds Ho. E2 40Yb 72
(off Approach Rd.)
Reynolds Ho. NW8 2C 214 (40Fb 69)
(off Wellington Rd.)
Reynolds Ho. SW1 6E 228 (49Mb 90)
(off Erasmus St.)
Reynolds Pl. SE3 52Kc 115
Reynolds Pl. TW10: Rich 58Pa 109
Reynolds Rd. KT3: N Mald 73Ta 153
Reynolds Rd. SE15 56Yb 114
Reynolds Rd. UB4: Yead 42Y 85
Reynolds Rd. W4 48Sa 87
Reynolds Sports Cen. 47Qa 87
Reynolds Way CR0: C'don 77Ub 157
Rhapsody Cres. CM14: W'ley 21Xd 58
Rheidol M. N1 1D 218 (40Sb 71)
Rheidol Ter. N1 1D 218 (39Sb 71)
Rheingold Way SM6: W'gton 81Nb 176
Rhein Ho. N8 27Nb 50
(off Campsfield Rd.)
Rheola Cl. N17 25Vb 51
Rhoda McGaw Theatre 89A 168
Rhoda St. E2 5K 219 (42Vb 91)
Rhodes Av. N22 25Lb 50
Rhodes Cl. TW20: Egh 64D 126
Rhodes Ct. TW20: Egh 64E 126
(off Rhodes Cl.)
Rhodes Ho. N1 3F 219 (41Tb 91)
(off Provost St.)
Rhodesia Rd. E11 33Fc 73
Rhodesia Rd. SW9 54Nb 112
Rhodes Moorhouse Ct. SM4: Mord 72Cb 155
Rhodes St. N7 36Pb 70
Rhodes Way WD24: Wat 12Z 27
Rhodeswell Rd. E14 43Ac 92
(not continuous)
Rhodium Ct. E14 43Cc 92
(off Thomas Rd.)
Rhododendron Ride TW20: Eng G 5K 125
Rhodrons Av. KT9: Chess 78Na 153
Rhondda Gro. E3 41Ac 92
RHS Gdn. Wisley 90N 169
RHS Lawrence Hall 4D 228 (48Mb 90)
RHS Lindley Hall 5D 228 (49Mb 90)
(off Vincent Sq.)
Rhyl Rd. UB6: G'frd 40Ha 66
Rhyl St. NW5 37Jb 70
Rhymes, The HP1: Hem H 1K 3
Rhys Av. N11 24Mb 50
Rialto Rd. CR4: Mitc 68Jb 134
Ribble Cl. IG8: Wfd G 23Lc 53
Ribbledale AL2: Lon C 9K 7
Ribblesdale Av. N11 23Jb 50
Ribblesdale Av. UB5: N'olt 37Da 65
Ribblesdale Ho. NW6 39Cb 69
(off Kilburn Vale)
Ribblesdale Rd. DA2: Dart 60Sd 120
Ribblesdale Rd. N8 28Pb 50
Ribblesdale Rd. SW16 65Kb 134
Ribbon Dance M. SE5 53Tb 113
Ribbons Wlk. E20 36Ec 72
Ribchester Av. UB6: G'frd 41Ha 86
Ribston Cl. BR2: Broml 74Pc 160
Ribston Cl. WD7: Shenl 5Ma 15
Ricardo Path SE28 46Yc 95
Ricardo Rd. SL4: Old Win 8M 103
Ricardo St. E14 44Dc 92
Ricards Rd. SW19 64Bb 133
Riccall Ct. NW9 25Ua 48
(off Pageant Av.)
Ricebridge La. RH2: Reig 9D 206
Rice Cl. HP2: Hem H 1P 3
Rice Pde. BR5: Pet W 71Tc 160
Riceyman Ho. WC1 4K 217 (41Qb 90)
(off Lloyd Baker St.)
Richard Anderson Ct. SE14 52Zb 114
(off Monson Rd.)
Richard Blackburn Ho. RM7: Rush G 33Gd 76
Richard Burbidge Mans. SW13 51Ya 110
(off Brasenose Dr.)
Richard Burton Ct. IG9: Buck H 19Lc 35
(off Palmerston Rd.)
Richard Challoner Sports Cen. 73Ta 153
Richard Cl. SE18 49Nc 94
RICHARD DESMOND CHILDREN'S EYE CEN. 4F 219 (41Tb 91)
(within Moorfields Eye Hospital)
Richard Fell Ho. E12 35Qc 74
(off Walton Rd.)
Richard Fielden Ho. E1 41Ac 92
Richard Ho. SE16 49Yb 92
(off Silwood St.)
RICHARD HOUSE CHILDREN'S HOSPICE 45Mc 93
Richard Ho. Dr. E16 44Mc 93
Richard Neale Ho. E1 45Xb 91
(off Cornwall St.)
Richard Neve Ho. SE18 49Uc 94
(off Plumstead High St.)
Richard Robert Residence, The E15 37Fc 73
(off Salway Rd.)
Richard Ryan Pl. RM9: Dag 39Ad 75
Richards Av. RM7: Rom 30Ed 56
Richards Cl. HA1: Harr 29Ja 46
Richards Cl. UB10: Hil 39Q 64
Richards Cl. UB3: Harl 51T 106
Richards Cl. WD23: Bush 17Fa 28
Richards Fld. KT19: Ewe 81Ta 173
Richard Sharples Ct. SM2: Sutt 80Eb 155
Richardson Cl. AL2: Lon C 9J 7
(not continuous)
Richardson Cl. DA9: Ghithe 57Vd 120
Richardson Cl. E8 39Vb 71
Richardson Ct. SW4 54Nb 112
(off Studley Rd.)
Richardson Gdns. RM10: Dag 37Dd 76
Richardson Ho. HP3: Hem H 8P 3
Richardson Pl. AL4: Col H 4M 7
Richardson Rd. E15 40Gc 73
Richardson's M. W1 6B 216 (42Lb 90)
(off Warren St.)
Richard's Pl. SW3 5E 226 (49Gb 89)
Richards Pl. E17 27Cc 52
Richards Rd. KT11: Stoke D 86Da 171
Richard Stagg Cl. AL1: St A 4G 6
Richard St. E1 44Xb 91
Richards Way SL1: Slou 6C 80
Richard Tress Way E3 42Bc 92
Richbell WC1 7H 217 (43Pb 90)
(off Boswell St.)
Richbell Cl. KT21: Asht 90Ma 173
Richbell Pl. WC1 7H 217 (43Pb 90)
Richborne Ter. SW8 52Pb 112
Richborough Cl. BR5: St M Cry 70Zc 139
Richborough Ho. SE15 51Yb 114
(off Sharratt St.)
Richborough Rd. NW2 35Ab 68
Richbourne Ct. W1 2E 220 (44Gb 89)
(off Harrowby St.)
Riches Cl. TW3: Houn 54Fa 108
Riches Rd. IG1: Ilf 33Sc 74
Richfield Rd. WD23: Bush 17Ea 28
Richford Ga. W6 48Ya 88
Richford Rd. E15 39Hc 73
Richford St. W6 47Ya 88
Rich Ind. Est. SE1 5J 231 (49Ub 91)
Rich Ind. Est. SE15 51Xb 113
RICHINGS PARK 48G 82
Richings Pk. Golf Course 49F 82
Richings Pl. SL0: Rich P 48G 82
Richings Way SL0: Rich P 48G 82
(not continuous)
Richland Av. CR5: Coul 86Jb 176
Richland Ho. SE15 53Wb 113
(off Goldsmith Rd.)
Richlands Av. KT17: Ewe 77Wa 154
Rich La. SW5 50Db 89
Richman Ho. SE8 50Bc 92
(off Grove St.)
Richmer Rd. DA8: Erith 52Jd 118
Richmix Sq. E1 5K 219 (42Vb 91)
(off Bethnal Grn. Rd.)
RICHMOND 57Ma 109
Richmond, American International University in London, The Kensington Campus, Ansdell Street 48Db 89
(off Ansdell St.)
Richmond, American International University in London, The Kensington Campus, St Albans Grove 48Db 89
Richmond, American International University in London, The Kensington Campus, Young Street 47Db 89
Richmond, American International University in London, The Richmond Hill Campus 59Na 109
Richmond & London Scottish RUFC 55Ma 109
Richmond Athletic Ground 55Ma 109
Richmond Av. CM14: B'wood 18Yd 40
Richmond Av. E4 22Fc 53
Richmond Av. N1 39Pb 70
Richmond Av. NW10 37Ya 68
Richmond Av. SW20 67Ab 132
Richmond Av. TW14: Felt 58U 106
Richmond Av. UB10: Hil 37R 64
Richmond Bri. 58Ma 109
Richmond Bldgs. W1 3D 222 (44Mb 90)
RICHMOND CIRCUS 56Na 109
Richmond Cl. E17 30Bc 52
Richmond Cl. EN8: Chesh 1Yb 20
Richmond Cl. KT18: Eps 86Ua 174
Richmond Cl. KT22: Fet 96Ea 192
Richmond Cl. TN16: Big H 91Kc 199
Richmond Cl. WD6: Bore 15Ta 29
Richmond Cotts. W14 49Ab 88
(off Hammersmith Rd.)
Richmond Ct. AL1: St A 2E 6
Richmond Ct. AL10: Hat 2D 8
Richmond Ct. CR4: Mitc 69Fb 133
Richmond Ct. E8 38Xb 71
(off Mare St.)
Richmond Ct. EN6: Pot B 3Eb 17
Richmond Ct. HA9: Wemb 34Pa 67
Richmond Ct. IG10: Lough 15Mc 35
(off High Rd.)
Richmond Ct. N11 23Jb 50
(off Pickering Gdns.)
Richmond Ct. NW6 38Za 68
(off Willesden La.)
Richmond Ct. SW1 2G 227 (47Hb 89)
(off Sloane St.)
Richmond Ct. W14 49Ab 88
(off Hammersmith Rd.)
Richmond Cres. E4 22Fc 53
Richmond Cres. KT19: Eps 84Na 173
Richmond Cres. N1 39Pb 70
Richmond Cres. N9 18Wb 33
Richmond Cres. SL1: Slou 6L 81
Richmond Cres. TW18: Staines 64H 127
Richmond Cricket Ground 55Na 109
Richmond Dr. DA12: Grav'nd 1G 144
Richmond Dr. IG8: Wfd G 24Qc 54
Richmond Dr. TW17: Shep 72T 150
Richmond Dr. WD17: Wat 12U 26
Richmond FC 55Ma 109
Richmond Gdns. HA3: Hrw W 24Ha 46
Richmond Gdns. NW4 29Wa 48
Richmond Golf Course Surrey 61Na 131
Richmond Grn. CR0: Bedd 76Nb 156
Richmond Gro. KT5: Surb 72Pa 153
Richmond Gro. N1 38Rb 71
(not continuous)
Richmond Hill TW10: Rich 58Na 109
Richmond Hill Ct. TW10: Rich 58Na 109
Richmond Ho. CR3: Cat'm 96Vb 197
Richmond Ho. E3 43Cc 92
(off Bow Common La.)
Richmond Ho. NW1 2A 216 (40Kb 70)
(off Park Village E.)
Richmond Ho. SE17 7F 231 (50Tb 91)
(off Portland St.)
Richmond La. DA1: Dart 57Pd 119
Richmond Mans. SW5 50Db 89
(off Old Brompton Rd.)
Richmond Mans. TW1: Twick 58Ma 109
Richmond M. SE6 60Dc 114
Richmond M. TW11: Tedd 64Ha 130
Richmond M. W1 3D 222 (44Mb 90)
Richmond Olympus Gym & Squash Club Richmond 56Ma 109
Richmond Pde. TW1: Twick 58La 108
(off Richmond Rd.)
Richmond Pk. IG10: Lough 17Mc 35
Richmond Pk. 60Qa 109
Richmond Pk. Golf Course 59Ua 110
Richmond Pk. Rd. KT2: King T 67Na 131
Richmond Pk. Rd. SW14 57Sa 109
Richmond Pl. SE18 49Sc 94
Richmond Pl. SL9: Ger X 3N 61
Richmond Rd. CR0: Bedd 76Nb 156
Richmond Rd. CR5: Coul 87Kb 176
Richmond Rd. CR7: Thor H 69Rb 135
Richmond Rd. E11 33Fc 73
Richmond Rd. E4 18Fc 35
Richmond Rd. E7 36Kc 73
Richmond Rd. E8 38Vb 71
Richmond Rd. EN5: New Bar 15Db 31
Richmond Rd. EN6: Pot B 3Eb 17
Richmond Rd. IG1: Ilf 34Sc 74
Richmond Rd. KT2: King T 64Ma 131
Richmond Rd. N11 23Nb 50
Richmond Rd. N15 30Ub 51
Richmond Rd. N2 26Eb 49
Richmond Rd. RM1: Rom 30Hd 56
Richmond Rd. RM17: Grays 51Ee 121
Richmond Rd. SW20 67Xa 132
Richmond Rd. TW1: Twick 59Ka 108
Richmond Rd. TW18: Staines 64H 127
Richmond Rd. TW7: Isle 55Ja 108
Richmond Rd. W5 47Na 87
RICHMOND ROYAL HOSPITAL 55Na 109
Richmond Station (Rail, Underground & Overground) 56Na 109
Richmond St. E13 40Jc 73
Richmond Ter. SW1 1F 229 (47Nb 90)
Richmond Theatre 56Ma 109
Richmond Way E11 33Jc 73
Richmond Way KT22: Fet 95Da 191
(not continuous)
Richmond Way W12 47Za 88
Richmond Way W14 48Za 88
Richmond Way WD3: Crox G 14S 26
Richmondwood SL5: S'dale 4F 146
Richmount Gdns. SE3 55Jc 115
Rich St. E14 45Bc 92
Rickard Cl. NW4 28Xa 48
Rickard Cl. SW2 60Qb 112
Rickard Cl. UB7: W Dray 48M 83
Rickards Cl. KT6: Surb 75Na 153
Ricketts Hill Rd. TN16: Tats 90Mc 179
Rickett St. SW6 51Cb 111
Rickfield Cl. AL10: Hat 2C 8
Rickford GU3: Worp 8H 187
Rickford Hill GU3: Worp 8H 187
Rickman Ct. KT15: Add 76K 149
Rickman Cres. KT15: Add 76K 149
Rickman Hill CR5: Coul 90Kb 176
Rickman Hill Rd. CR5: Chips 90Kb 176
Rickman Hill Rd. CR5: Coul 90Kb 176
Rickman Ho. E1 41Yb 92
(off Rickman St.)
Rickman's La. SL2: Stoke P 7J 61
Rickman St. E1 42Yb 92
RICKMANSWORTH 18M 25
Rickmansworth Aquadrome 19L 25
Rickmansworth Golf Course 19P 25
Rickmansworth La. SL9: Chal P 24A 42
Rickmansworth Pk. 17M 25
Rickmansworth Rd. HA5: Pinn 26X 45
Rickmansworth Rd. HA6: Nwood 22R 44
Rickmansworth Rd. UB9: Hare 25L 43
Rickmansworth Rd. WD17: Wat 13X 27
Rickmansworth Rd. WD18: Wat 14U 26
Rickmansworth Rd. WD3: Chor 13G 24
Rickmansworth Sailing Club 25H 43
Rickmansworth Station (Rail & Underground) 17M 25
Rick Roberts Way E15 39Ec 72
Ricksons La. KT24: W Hor 99R 190
Rickthorne Rd. N19 33Nb 70
Rickyard Path SE9 56Nc 116
Riddell Ct. SE5 7K 231 (50Vb 91)
(off Albany Rd.)
Ridding La. UB6: G'frd 36Ha 66
Riddings, The CR3: Cat'm 97Vb 197
RIDDLESDOWN 85Sb 177
Riddlesdown Av. CR8: Purl 83Sb 177
Riddlesdown Rd. CR8: Kenley 82Sb 177
Riddlesdown Rd. CR8: Purl 82Sb 177
Riddlesdown Station (Rail) 85Sb 177
Riddons Rd. SE12 62Lc 137
Ride, The EN3: Pond E 13Yb 34
Ride, The EN8: Chesh 3Zb 20
Ride, The TW8: Bford 50Ka 86
Rideout St. SE18 49Pc 94
Rider Cl. DA15: Sidc 58Uc 116
Riders Twr. E17 29Bc 52
(off Track St.)
Riders Way RH9: G'stone 3A 210
Ride Hill, Mountainboarding Cen., The 7D 208
Ridgeway Dr. W3 48Qa 87
Ridgdale St. E3 40Dc 72
RIDGE 6Ua 16
Ridge, The BR6: Orp 75Tc 160
Ridge, The CR3: Wold 98Cc 198
Ridge, The CR5: Coul 86Nb 176
Ridge, The CR6: W'ham 96Hc 199
Ridge, The CR8: Purl 82Lb 176
Ridge, The DA5: Bexl 59Bd 117
Ridge, The EN5: Barn 15Bb 31
Ridge, The GU22: Wok 89D 168
Ridge, The HA6: Nwood 24V 44
Ridge, The KT18: Eps 90Sa 173
Ridge, The KT22: Fet 96Fa 192
Ridge, The KT5: Surb 71Qa 153
Ridge, The SL5: S'dale 3E 146
Ridge, The TW2: Whitt 59Fa 108
Ridge Av. DA1: Cray 58Hd 118
Ridge Av. N21 17Sb 33
Ridgebank SL1: Slou 5D 80
Ridgebrook Rd. SE3 55Mc 115
Ridge Cl. GU22: Wok 3M 187
Ridge Cl. NW4 26Za 48
Ridge Cl. NW9 28Ta 47
Ridge Cl. SE28 47Tc 94
Ridge Ct. CR6: W'ham 90Wb 177
Ridge Ct. SE22 59Wb 113
Ridge Ct. SL4: Wind 5G 102
Ridge Crest EN2: Enf 11Pb 32
Ridgecroft Cl. DA5: Bexl 60Ed 118
Ridgefield WD17: Wat 9U 12
Ridgegate Cl. RH2: Reig 4M 207
RIDGE GREEN 9E 208
Ridge Grn. RH1: S Nut 9E 208
Ridge Grn. Cl. RH1: S Nut 9E 208
Ridge Hill AL2: Lon C 10L 7
Ridge Hill NW11 32Ab 68
Ridge Hill WD7: Shenl 1Sa 15
RIDGE HILL 2Sa 15
Ridgehurst Av. WD25: Wat 6V 12
Ridgelands KT22: Fet 96Fa 192
Ridge La. WD17: Wat 9U 12
Ridge Langley CR2: Sande 82Wb 177
Ridge Lea HP1: Hem H 2H 3
Ridgemead Cl. N14 19Nb 32
Ridgemead Rd. TW20: Eng G 2L 125
Ridgemont Gdns. HA8: Edg 21Sa 47
Ridgemont Pl. RM11: Horn 30Md 57
Ridgemount KT13: Weyb 75U 150
Ridgemount Av. CR0: C'don 74Zb 158
Ridgemount Av. CR5: Coul 89Kb 176
Ridgemount Cl. SE20 66Xb 135
Ridgemount End SL9: Chal P 22A 42
Ridgemount Gdns. EN2: Enf 13Rb 33
Ridgemount Rd. SL5: S'dale 5E 146
Ridgemount Way RH1: Redh 8M 207
Ridge Pk. CR8: Purl 82Mb 176
Ridge Pl. BR5: St M Cry 70Yc 139
Ridge Rd. CR4: Mitc 66Kb 134
Ridge Rd. N21 18Sb 33
Ridge Rd. N8 30Pb 50
Ridge Rd. NW2 34Bb 69
Ridge Rd. SM3: Sutt 74Ab 154
(not continuous)
Ridge St. WD24: Wat 10X 13
Ridges Yd. CR0: C'don 76Rb 157
Ridgeview AL2: Lon C 10K 7
Ridgeview Cl. EN5: Barn 16Za 30
Ridgeview Rd. N20 20Db 31
Ridge Way DA1: Cray 58Hd 118
Ridge Way GU25: Vir W 71A 148
Ridge Way SE19 65Ub 135
Ridge Way TW13: Hanw 62Aa 129
Ridgeway BR2: Hayes 75Jc 159
Ridgeway CM13: Hut 18De 41
Ridgeway DA2: Daren 64Ud 142
Ridgeway GU21: Wok 7P 167
Ridgeway IG8: Wfd G 21Lc 53
Ridgeway KT19: Eps 84Sa 173
Ridgeway RM17: Grays 9A 100
Ridgeway WD3: Rick 17K 25
Ridgeway WD7: Shenl 3La 14
Ridgeway, The CR2: Sande 81Ub 177
Ridgeway, The CR0: Wadd 76Pb 156
Ridgeway, The DA12: Shorne 6N 145
Ridgeway, The E4 19Dc 34
Ridgeway, The EN2: Enf 8Mb 18
Ridgeway, The EN6: Cuff 10N 9
Ridgeway, The EN6: N'thaw 10N 9
Ridgeway, The EN6: Pot B 6Fb 17
Ridgeway, The GU24: Brkwd 2D 186
Ridgeway, The HA2: Harr 29Ba 45
(not continuous)
Ridgeway, The HA3: Kenton 30La 46
Ridgeway, The HA4: Ruis 31W 64
Ridgeway, The HA7: Stan 23La 46
Ridgeway, The KT12: Walt T 74V 150
Ridgeway, The KT22: Fet 95Fa 192
Ridgeway, The KT22: Oxs 86Ea 172
Ridgeway, The N11 21Hb 49
Ridgeway, The N14 19Nb 32
Ridgeway, The N3 24Db 49
Ridgeway, The NW11 31Ab 68
Ridgeway, The NW7 20Wa 30
Ridgeway, The NW9 28Ta 47
Ridgeway, The RM2: Rom 28Jd 56
Ridgeway, The RM3: Hrld W 25Pd 57
Ridgeway, The SL0: Iver 46G 82
Ridgeway, The SL9: Chal P 27A 42
Ridgeway, The W3 48Qa 87
Ridgeway, The WD17: Wat 9U 12
Ridgeway, The WD7: R'lett 9Ha 14
Ridgeway Av. DA12: Grav'nd 2D 144
Ridgeway Av. EN4: E Barn 16Hb 31
Ridgeway Bungs. DA12: Shorne 6P 145
Ridgeway Cl. GU21: Wok 7P 167
Ridgeway Cl. HP3: Hem H 8P 3
Ridgeway Cl. KT22: Fet 96Ga 192
Ridgeway Cl. KT22: Oxs 86Ea 172
Ridgeway Ct. HA5: Hat E 24Ca 45
Ridgeway Ct. RH1: Redh 7P 207
Ridgeway Cres. BR6: Orp 76Uc 160
Ridgeway Cres. Gdns. BR6: Orp 75Uc 160
Ridgeway Dr. BR1: Broml 63Kc 137
Ridgeway E. DA15: Sidc 57Vc 117
Ridgeway Gdns. GU21: Wok 7P 167
Ridgeway Gdns. IG4: Ilf 29Nc 54
Ridgeway Gdns. N6 31Lb 70
Ridgeway Rd. RH1: Redh 6P 207
Ridgeway Rd. TW7: Isle 52Ga 108
Ridgeway Rd. Nth. TW7: Isle 51Ga 108
Ridgeway Trad. Est. SL0: Iver 45G 82
Ridgeway Wlk. UB5: N'olt 37Aa 65
(off Cowings Mead)
Ridgeway W. DA15: Sidc 57Uc 116
Ridgewell Av. RM16: Ors 2C 100

Ridgewell Cl. N1 39Sb 71
Ridgewell Cl. RM10: Dag 39Dd 76
Ridgewell Cl. SE26 63Bc 136
Ridgewell Gro. RM12: Horn 38Kd 77
Ridgewood DA3: Lfield 68Ee 143
Ridgewood KT13: Weyb 80Q 150
Ridgmont Plaza AL1: St A 3D 6
(off Ridgmont Rd.)
Ridgmont Rd. AL1: St A 2C 6
(not continuous)
Ridgmount Gdns. WC1 7D 216 (43Mb 90)
Ridgmount Pl. WC1 7D 216 (43Mb 90)
Ridgmount Rd. SW18 57Db 111
Ridgmount St. WC1 7D 216 (43Mb 90)
Ridgway GU22: Pyr 87J 169
RIDGWAY 87J 169
Ridgway SW19 66Ya 132
Ridgway TW10: Rich 58Na 109
Ridgway, The SM2: Sutt 80Fb 155
Ridgway Ct. SW19 65Za 132
Ridgway Gdns. SW19 66Za 132
Ridgway Pl. SW19 65Ab 132
Ridgway Rd. GU22: Pyr 87H 169
Ridgway Rd. SW9 55Rb 113
Ridgwell Rd. E16 43Lc 93
Ridgy Fld. Cl. TN15: Wro 89Ce 185
Riding, The GU21: Wok 86D 168
Riding, The NW11 31Bb 69
Riding Ct. Farm SL3: Dat 1M 103
Riding Ct. Rd. SL3: Dat 2N 103
Riding Ct. Rd. SL3: L'ly 2N 103
Riding Hill CR2: Sande 85Wb 177
Riding Ho. St. W1 1A 222 (43Kb 90)
Ridings, The E11 29Jc 53
Ridings, The EN4: E Barn 17Fb 31
Ridings, The GU23: Rip 95J 189
Ridings, The HP5: Lat 8A 10
Ridings, The IG7: Chig 21Xc 55
Ridings, The KT11: Cobh 84Ca 171
Ridings, The KT15: Add 79G 148
Ridings, The KT17: Ewe 81Va 174
Ridings, The KT18: Eps 87Ua 174
Ridings, The KT20: Tad 92Bb 195
Ridings, The KT21: Asht 89Ma 173
Ridings, The KT24: E Hor 97U 190
Ridings, The KT5: Surb 71Qa 153
Ridings, The RH2: Reig 4M 207
Ridings, The SL0: Rich P 49H 83
Ridings, The SL4: Wind 2A 102
Ridings, The TN16: Big H 89Nc 180
Ridings, The TW16: Sun 67W 128
Ridings, The W5 42Pa 87
Ridings Av. N21 14Rb 33
Ridings Cl. N6 31Lb 70
Ridings La. GU23: Ock 95R 190
Ridings La. UB4: Hayes 40T 64
Ridlands Gro. RH8: Limp 2P 211
Ridlands La. RH8: Limp 2P 211
Ridlands Ri. RH8: Limp 2P 211
Ridler Rd. EN1: Enf 10Ub 19
RIDLEY 78De 165
Ridley Av. W13 48Ka 86
Ridley Cl. IG11: Bark 38Vc 75
Ridley Cl. RM3: Rom 25Kd 57
Ridley Ct. SW16 65Nb 134
Ridley Ho. SW1 4E 228 (48Mb 90)
(off Monck St.)
Ridley Rd. BR2: Broml 69Hc 137
Ridley Rd. CR6: W'ham 90Yb 178
Ridley Rd. DA16: Well 53Xc 117
Ridley Rd. E7 35Lc 73
Ridley Rd. E8 36Vb 71
Ridley Rd. NW10 40Wa 68
Ridley Rd. SW19 66Db 133
Ridsdale Rd. GU21: Wok 9M 167
Ridsdale Rd. SE20 66Xb 135
Riefield Rd. SE9 56Sc 116
Riesco Dr. CR0: C'don 79Yb 158
Riffel Rd. NW2 36Ya 68
Riffhams CM13: B'wood 20De 41
Rifle Butts All. KT18: Eps 86Va 174
Rifle St. E14 43Dc 92
Riga M. E1 44Wb 91
(off Commercial Rd.)
Rigault Rd. SW6 54Ab 110
Rigby Cl. CR0: Wadd 76Qb 156
Rigby Gdns. RM16: Grays 9D 100
Rigby La. UB3: Hayes 47S 84
Rigby Lodge SL1: Slou 4J 81
Rigby M. IG1: Ilf 33Qc 74
Rigby Pl. EN3: Enf L 9Cc 20
Rigden St. E14 44Dc 92
Rigeley Rd. NW10 41Wa 88
Rigg App. E10 32Zb 72
Rigge Pl. SW4 56Mb 112
Riggindale Rd. SW16 64Mb 134
Riggs Way TN15: Wro 88Be 185
Riley Cl. KT19: Eps 83Ra 173
Riley Ho. E3 42Cc 92
(off Ireton St.)
Riley Ho. SW10 52Fb 111
(off Riley St.)
Riley Rd. EN3: Enf W 10Yb 20
Riley Rd. SE1 3K 231 (48Vb 91)
Riley St. SW10 51Fb 111
Rill Ct. IG11: Bark 40Sc 74
(off Spring Pl.)
Rill Ho. SE5 52Tb 113
(off Harris St.)
Rill La. HA8: Edg 23Ta 47
Rima Ho. SW3 51Fb 111
(off Callow St.)
Rinaldo Rd. SW12 59Kb 112
Ring, The SW7 1C 226 (47Fb 89)
Ring, The W2 4C 220 (45Fb 89)
Ring Cl. BR1: Broml 66Kc 137
Ring Ct. SE1 1B 230 (47Rb 91)
(off The Cut)
Ringcroft St. N7 36Qb 70
Ringcross Youth Cen. 36Pb 70
(off Lough Rd.)
Ringers Ct. BR1: Broml 69Jc 137
(off Ringers Rd.)
Ringers Rd. BR1: Broml 69Jc 137
Ringford Rd. SW18 57Bb 111
Ring Ho. E1 45Yb 92
(off Sage St.)
Ringles Ct. E6 39Pc 74
Ringlet Cl. E16 43Kc 93
Ringlewell Cl. EN1: Enf 12Xb 33
Ringley Pk. Av. RH2: Reig 7M 207
Ringley Pk. Rd. RH2: Reig 6L 207
Ringmer Av. SW6 53Ab 110
Ringmer Gdns. N19 33Nb 70
Ringmer Pl. N21 15Tb 33
Ringmer Way BR1: Broml 71Nc 160
Ringmore Ri. SE23 59Xb 113
Ringmore Rd. KT12: Walt T 76Y 151
Ringmore Vw. SE23 59Xb 113
Ring Rd. W12 46Ya 88
(not continuous)
Ringsfield Ho. SE17 7E 230 (50Sb 91)
(off Bronti Cl.)
Ringshall Rd. BR5: St P 69Wc 139
Ringside Ct. SE28 46Yc 95
Ringslade Rd. N22 26Pb 50
Ringstead Rd. SE6 59Dc 114
Ringstead Rd. SM1: Sutt 77Fb 155
Ring Way N11 23Lb 50
Ringway UB2: S'hall 50Z 85
Ringway Rd. AL2: Park 9P 5
Ringwold Cl. BR3: Beck 66Ac 136
Ringwood Av. BR6: Prat B 82Yc 181
Ringwood Av. CR0: C'don 73Nb 156
Ringwood Av. N2 26Hb 49
Ringwood Av. RH1: Redh 3P 207
Ringwood Av. RM12: Horn 33Md 77
Ringwood Cl. HA5: Pinn 27Y 45
Ringwood Gdns. E14 49Cc 92
Ringwood Gdns. SW15 60Wa 110
Ringwood Lodge RH1: Redh 3A 208
Ringwood Rd. E17 30Bc 52
Ringwood Way N21 18Rb 33
Ringwood Way TW12: Hamp H 63Ca 129
Rio Cinema Dalston 36Ub 71
(off Kingsland High St.)
RIPLEY 93L 189
Ripley Av. TW20: Egh 65A 126
Ripley Bldgs. SE1 1C 230 (47Rb 91)
(off Rushworth St.)
Ripley By-Pass GU23: Rip 95L 189
Ripley Cl. BR1: Broml 71Pc 160
Ripley Cl. CR0: New Ad 79Ec 158
Ripley Cl. SL3: L'ly 49A 82
Ripley Ct. CR4: Mitc 68Fb 133
Ripley Ct. IG7: Chig 19Sc 36
Ripley Gdns. SM1: Sutt 77Eb 155
(not continuous)
Ripley Gdns. SW14 55Ta 109
Ripley Ho. SW1 51Lb 112
(off Churchill Gdns.)
Ripley La. GU23: Rip 95N 189
Ripley La. KT24: W Hor 97Q 190
Ripley M. E11 30Gc 53
Ripley Rd. DA17: Belv 49Cd 96
Ripley Rd. E16 44Lc 93
Ripley Rd. EN2: Enf 11Sb 33
Ripley Rd. GU23: Send 98L 189
Ripley Rd. GU4: E Clan 99L 189
Ripley Rd. IG3: Ilf 33Vc 75
Ripley Rd. TW12: Hamp 66Ca 129
Ripleys Mkt. 59Nd 119
RIPLEY SPRINGS 65A 126
Ripley Vw. IG10: Lough 10Rc 22
Ripley Vs. W5 44La 86
Ripley Way EN7: Chesh 2Xb 19
Ripley Way HP1: Hem H 1G 2
Ripley Way KT19: Eps 83Qa 173
Ripon Cl. UB5: N'olt 36Ca 65
Ripon Ct. N11 23Jb 50
(off Ribblesdale Av.)
Ripon Gdns. IG1: Ilf 30Nc 54
Ripon Gdns. KT9: Chess 78Ma 153
Ripon Ho. RM3: Rom 23Md 57
(off Dartfields)
Ripon Rd. N17 27Tb 51
Ripon Rd. N9 17Xb 33
Ripon Rd. SE18 51Rc 116
Ripon Way WD6: Bore 15Ta 29
Rippersley Rd. DA16: Well 53Wc 117
Ripple Nature Reserve, The 41Xc 95
Ripple Rd. IG11: Bark 38Sc 74
Ripple Rd. IG11: Dag 38Sc 74
Ripple Rd. RM9: Dag 39Xc 75
RIPPLESIDE 39Xc 75
Rippleside Commercial Est. IG11: Bark 40Yc 75
Ripplesmere Cl. SL4: Old Win 9N 103
Ripplevale Gro. N1 38Pb 70
Rippolson Rd. SE18 50Vc 95
Ripston Rd. TW15: Ashf 64T 128
Risborough Cl. N10 27Kb 50
Risborough Dr. KT4: Wor Pk 73Wa 154
Risborough Ho. NW8 5E 214 (42Gb 89)
(off Mallory St.)
Risborough St. SE1 1C 230 (47Rb 91)
Risdon Ho. SE16 47Yb 92
(off Risdon St.)
Risdon St. SE16 48Yb 92
Rise, The AL2: Park 7B 6
Rise, The CR2: Sels 81Yb 178
Rise, The DA1: Cray 56Hd 118
Rise, The DA12: Grav'nd 3G 144
Rise, The DA5: Bexl 59Yc 117
Rise, The DA9: Ghithe 58Wd 120
Rise, The E11 29Jc 53
Rise, The EN9: Walt A 2Kc 21
Rise, The HA8: Edg 22Ra 47
Rise, The IG9: Buck H 17Mc 35
Rise, The KT17: Ewe 82Va 174
Rise, The KT20: Tad 92Ya 194
Rise, The KT24: E Hor 98U 190
Rise, The N13 21Qb 50
Rise, The NW10 35Ta 67
Rise, The NW7 23Va 48
Rise, The TN13: S'oaks 100Ld 203
Rise, The UB10: Hil 40P 63
Rise, The UB6: G'frd 36Ja 66
Rise, The WD6: E'tree 15Pa 29
Risebridge Chase RM1: Rom 24Hd 56
Risebridge Golf Course 24Hd 56
Risebridge Rd. RM2: Rom 26Hd 56
Risedale Cl. HP3: Hem H 5N 3
Risedale Hill HP3: Hem H 5N 3
Risedale Rd. DA7: Bex 55Ed 118
Risedale Rd. HP3: Hem H 5N 3
Riseholme Ct. E9 37Bc 72
Riseldine Rd. SE23 58Ac 114
RISE PARK 26Gd 56
Rise Pk. Blvd. RM1: Rom 25Hd 56
Rise Pk. Pde. RM1: Rom 26Gd 56
Rise Rd. SL5: S'dale 1C 146
Riseway CM15: B'wood 20Ae 41
Rising Hill Cl. HA6: Nwood 23S 44
Risinghill St. N1 1J 217 (40Pb 70)
Risingholme Cl. HA3: Hrw W 25Ga 46
Risingholme Cl. WD23: Bush 17Da 27
Risingholme Rd. HA3: Hrw W 26Ga 46
Risings, The E17 28Fc 53
Risings Ter. RM11: Horn 27Pd 57
(off Prospect Rd.)
Rising Sun Ct. EC1 1C 224 (43Rb 91)
(off Cloth Fair)
Risley Av. N17 25Sb 51
Risley Cl. SM4: Mord 71Db 155
Rita Rd. SW8 51Nb 112
Ritches Rd. N15 29Sb 51
Ritchie Ho. E14 44Fc 93
(off Blair St.)
Ritchie Ho. N19 32Mb 70
Ritchie Ho. SE16 48Yb 92
(off Howland Est.)
Ritchie Rd. CR0: C'don 72Xb 157
Ritchie St. N1 1A 218 (40Qb 70)
Ritchings Av. E17 28Ac 52
Ritcroft Cl. HP3: Hem H 3B 4
Ritcroft Dr. HP3: Hem H 3B 4
Ritcroft St. HP3: Hem H 3B 4
Ritherdon Rd. SW17 61Jb 134
Ritson Ho. N1 1H 217 (39Pb 70)
(off Barnsbury Est.)
Ritson Rd. E8 37Wb 71
Ritter St. SE18 51Qc 116
Ritz Ct. EN6: Pot B 3Cb 17
Ritz Pde. W5 42Pa 87
Ritzy Picturehouse 56Qb 112
(off Coldharbour La.)
Riva Bldg., The SE13 56Fc 115
Rivaz Pl. E9 37Yb 72
Riven Ct. W2 44Db 89
(off Inverness Ter.)
Rivenhall Gdns. E18 28Hc 53
River App. HA8: Edg 25Sa 47
RIVER ASH ESTATE 73V 150
River Av. KT7: T Ditt 73Ja 152
River Av. N13 20Rb 33
River Av. Ind. Est. N13 22Qb 50
River Bank KT7: T Ditt 71Ha 152
River Bank KT8: E Mos 69Ga 130
River Bank N21 17Sb 33
River Bank TW12: Hamp 69Ca 129
Riverbank TW18: Staines 65H 127
Riverbank, The SL4: Wind 2F 102
Riverbank Point UB8: Uxb 37L 63
Riverbank Rd. BR1: Broml 62Jc 137
Riverbank Way SM6: W'gton 74Jb 156
Riverbank Way TW8: Bford 51La 108
River Barge Cl. E14 47Ec 92
River Bourne Health Club 73J 149
River Brent Bus. Pk. W7 48Ga 86
River Cl. E11 30Lc 53
River Cl. EN8: Walt C 6Cc 20
River Cl. HA4: Ruis 30V 44
River Cl. RM13: Rain 43Kd 97
River Cl. UB2: S'hall 47Ea 86
River Ct. CM15: Shenf 13Ee 41
River Ct. GU21: Wok 86E 168
River Ct. KT6: Surb 71Ma 153
(off Portsmouth Rd.)
River Ct. RM19: Purf 49Pd 97
River Ct. SE1 5B 224 (45Rb 91)
River Ct. TN13: Riv 94Gd 202
River Ct. TW17: Shep 73S 150
Rivercourt Rd. W6 49Xa 88
River Crane Way TW13: Hanw 61Ba 129
(off Watermill Way)
Riverdale SE13 56Ec 114
Riverdale Cl. IG11: Bark 42Xc 95
Riverdale Ct. N21 15Tb 33
Riverdale Dr. GU22: Wok 93B 188
Riverdale Dr. SW18 60Db 111
Riverdale Gdns. TW1: Twick 58La 108
Riverdale Ho. SE13 56Ec 114
Riverdale Rd. DA5: Bexl 59Bd 117
Riverdale Rd. DA8: Erith 50Dd 96
Riverdale Rd. SE18 50Vc 95
Riverdale Rd. TW1: Twick 58La 108
Riverdale Rd. TW13: Hanw 63Aa 129
Riverdale Shop. Cen. 55Ec 114
Riverdene HA8: Edg 20Sa 29
Riverdene Ind. Est. KT12: Hers 78Z 151
Riverdene Rd. IG1: Ilf 34Qc 74
River Dr. RM14: Upm 30Sd 58
Riverfield Rd. TW18: Staines 65H 127
Riverfleet WC1 3G 217 (41Nb 90)
(off Birkenhead St.)
Riverford Ho. W2 43Cb 89
(off Westbourne Pk. Rd.)
River Front EN1: Enf 13Ub 33
River Gdns. SM5: Cars 75Jb 156
River Gdns. TW14: Felt 57X 107
River Gdns. Bus. Cen. TW14: Felt 57X 107
River Gdns. Wlk. SE10 50Gc 93
River Gro. Pk. BR3: Beck 67Bc 136
RIVERHEAD 94Gd 202
Riverhead Cl. E17 26Zb 52
Riverhead Dr. SM2: Sutt 82Db 175
Riverhead M. TN13: Riv 95Gd 202
River Hgts. E15 39Ec 72
River Hgts. N15 25Vb 51
River Hill KT11: Cobh 87X 171
Riverhill KT4: Wor Pk 75Ta 153
Riverhill M. KT4: Wor Pk 76Ta 153
Riverhill Mobile Home Pk. KT4: Wor Pk 75Ta 153
Riverholme Dr. KT19: Ewe 81Ta 173
Riverhope Mans. SE18 48Nc 94
River Ho. SE26 62Xb 135
Riverhouse Barn 73V 150
River Island Cl. KT22: Fet 92Fa 192
River La. KT11: Cobh 88Aa 171
River La. KT22: Fet 93Fa 192
River La. KT22: Lea 93Fa 192
River La. TW10: Ham 59Ma 109
Riverleigh Ct. E4 22Bc 52
Riverlight Quay SW11 51Lb 112
River Lodge SW1 51Lb 112
(off Grosvenor Rd.)
Rivermead KT1: King T 71Ma 153
Rivermead KT14: Byfl 85P 169
Rivermead KT8: E Mos 69Ea 130
Rivermead Cl. KT15: Add 80L 149
Rivermead Cl. TW11: Tedd 64Ka 130
Rivermead Ct. SW6 55Bb 111
Rivermead Ho. E9 36Ac 72
Rivermead Ho. TW16: Sun 69Y 129
(off Thames St.)
Rivermead Rd. N18 23Zb 52
Rivermeads Av. TW2: Twick 62Ca 129
Rivermede Ct. TW20: Egh 63D 126
River Mill One SE13 55Ec 114
(off Station Rd.)
River Mill Two SE13 55Ec 114
(off Station Rd.)
River Mole Bus. Pk. KT10: Esh 75Ca 151
River Mole Local Nature Reserve 91Fa 192
River Mt. KT12: Walt T 73V 150
Rivernook Cl. KT12: Walt T 71Y 151
River Pde. TN13: Riv 94Gd 202
River Pk. HP1: Hem H 4J 3
River Pk. Av. TW18: Staines 63F 126
River Pk. Gdns. BR2: Broml 66Fc 137
River Pk. Rd. N22 26Pb 50
River Pk. Vw. BR6: Orp 73Xc 161
River Pl. N1 38Sb 71
River Reach TW11: Tedd 64La 130
River Rd. CM14: B'wood 21Vd 58
River Rd. IG11: Bark 40Uc 74
River Rd. IG9: Buck H 18Nc 36
River Rd. SL4: Wind 2A 102
River Rd. TW18: Staines 67H 127
River Rd. Bus. Pk. IG11: Bark 41Vc 95
Rivers Apts. N17 23Vb 51
(off Cannon Rd.)
Riversdale DA11: Nflt 1A 144
Riversdale Gdns. N22 25Qb 50
Riversdale Rd. KT7: T Ditt 71Ja 152
Riversdale Rd. N5 34Rb 71
Riversdale Rd. RM5: Col R 24Dd 56
Riversdell Cl. KT16: Chert 73H 149
Riversend Rd. HP3: Hem H 5L 3
Riversfield Rd. EN1: Enf 13Ub 33
Rivers Ho. TW7: Isle 56Ka 108
(off Richmond Rd.)
Rivers Ho. TW8: Bford 50Qa 87
(off Aitman Dr.)
Riverside AL2: Lon C 9J 7
Riverside DA4: Eyns 75Md 163
Riverside E3 39Cc 72
Riverside HP1: Hem H 3M 3
Riverside KT16: Chert 68J 127
Riverside NW4 31Xa 68
Riverside SE7 48Kc 93
Riverside SW11 52Gb 111
Riverside TW1: Twick 60Ka 108
Riverside TW10: Rich 57Ma 109
Riverside TW16: Sun 68Z 129
Riverside TW17: Shep 73Q 150
Riverside TW18: Staines 67H 127
Riverside TW19: Wray 9N 103
Riverside TW20: Egh 62C 126
Riverside TW9: Rich 57Ma 109
Riverside W6 51Ya 110
Riverside WC1 3G 217 (41Nb 90)
(off Birkenhead St.)
Riverside, The KT8: E Mos 69Fa 130
Riverside Apts. N13 22Pb 50
Riverside Apts. N4 31Tb 71
(off Goodchild Rd.)
Riverside App. RM18: Tilb 6B 122
Riverside Arts Cen. 69Y 129
Riverside Av. GU18: Light 3A 166
Riverside Av. KT8: E Mos 71Fa 152
Riverside Bus. Cen. SW18 60Db 111
Riverside Bus. Pk. SW19 67Eb 133
Riverside Cl. AL1: St A 3C 6
Riverside Cl. BR5: St P 68Yc 139
Riverside Cl. E5 32Yb 72
Riverside Cl. GU24: Brkwd 2D 186
Riverside Cl. KT1: King T 70Ma 131
Riverside Cl. RM1: Rom 28Fd 56
Riverside Cl. SM6: W'gton 76Kb 156
Riverside Cl. TW18: Staines 67H 127
Riverside Cl. W7 42Ga 86
Riverside Cl. WD4: K Lan 1R 12
Riverside Cotts. IG11: Bark 40Tc 74
Riverside Ct. AL1: St A 4C 6
Riverside Ct. E4 16Cc 34
Riverside Ct. KT22: Fet 94Ja 192
Riverside Ct. SE3 56Hc 115
Riverside Ct. SW8 51Mb 112
Riverside Ct. TW14: Felt 59U 106
Riverside Ct. TW7: Isle 54Ha 108
(off Woodlands Rd.)
Riverside Dr. CR4: Mitc 71Gb 155
Riverside Dr. KT10: Esh 77Ca 151
Riverside Dr. NW11 30Ab 48
Riverside Dr. TW10: Ham 62Ka 130
Riverside Dr. TW18: Staines 64G 126
Riverside Dr. W4 52Ta 109
Riverside Dr. WD3: Rick 18M 25
Riverside Gdns. EN2: Enf 12Sb 33
Riverside Gdns. GU22: Wok 93D 188
Riverside Gdns. HA0: Wemb 40Na 67
Riverside Gdns. N3 27Ab 48
Riverside Gdns. W6 50Xa 88
Riverside Hgts. RM18: Tilb 4C 122
Riverside Ho. N1 38Sb 71
(off Canonbury St.)
Riverside Ind. Est. AL2: Lon C 9J 7
Riverside Ind. Est. DA1: Dart 57Nd 119
Riverside Ind. Est. EN3: Pond E 16Ac 34
Riverside Ind. Est. IG11: Bark 41Wc 95
Riverside Leisure Area 8E 122
Riverside Mans. E1 46Yb 92
(off Milk Yd.)
Riverside M. CR0: Bedd 76Nb 156
Riverside Pk. 15Ub 33
(off Park Avenue)
Riverside Pk. KT13: Weyb 78N 149
Riverside Pk. SL3: Poyle 54G 104
Riverside Path EN8: Chesh 1Yb 20
Riverside Pl. N11 20Lb 32
Riverside Pl. TW19: Stanw 58M 105
Riverside Rail Freight Terminal
RM18: Tilb 6C 122
Riverside Retail Pk. Sevenoaks 91Ld 203
Riverside Rd. AL1: St A 3C 6
Riverside Rd. DA14: Sidc 62Ad 139
Riverside Rd. E15 40Ec 72
Riverside Rd. KT12: Hers 77Z 151
Riverside Rd. N15 30Wb 51
Riverside Rd. SW17 63Db 133
Riverside Rd. TW18: Staines 66H 127
Riverside Rd. TW19: Stanw 57M 105
Riverside Rd. WD19: Wat 16X 27
Riverside Shop. Cen. Erith 51Hd 118
Riverside Sq. N18 23Zb 52
Riverside Studios 50Ya 88
Riverside Twr. SW6 54Eb 111
(off The Boulevard)
Riverside Vs. KT6: Surb 72La 152
Riverside Wlk. W4: Chiswick 51Va 110
(off Chiswick Wharf)
Riverside Wlk. EN5: Barn Ducks Island 16Za 30
(not continuous)
Riverside Wlk. KT1: King T Kingston 69Ma 131
Riverside Wlk. N12: W. Finchley 23Cb 49
(not continuous)
Riverside Wlk. BR4: W W'ck 74Dc 158
Riverside Wlk. SL4: Wind 2H 103
Riverside Wlk. TW7: Isle 55Ga 108
Riverside Wlk. Local Nature Reserve 1N 147
Riverside Way DA1: Dart 57Nd 119
Riverside Way UB8: Uxb 39K 63
Riverside Way Ind. Est. UB8: Uxb 38K 63
Riverside Wharf DA1: Dart 56Md 119
Riverside Wharf E3 39Cc 72
Riverside Works IG11: Bark 38Rc 74
Riverside Yd. SW17 63Eb 133
Riverstone Cl. HA2: Harr 32Fa 66
Riverstone Ct. KT2: King T 67Pa 131
River St. EC1 3K 217 (41Qb 90)
River St. SL4: Wind 2H 103
River St. M. EC1 3K 217 (41Qb 90)
(off River St.)
River Ter. WC2 5H 223 (45Pb 90)
(off Lancaster Pl.)
River Thames Vis. Cen. 58Ma 109
Riverton Cl. W9 41Bb 89
River Twr. SW8 51Nb 112
River Vw. DA1: Dart 56Rd 119
(off Shakespeare Rd.)
River Vw. EN2: Enf 13Sb 33
River Vw. EN9: Walt A 5Dc 20
(off Powdermill La.)
River Vw. KT15: Add 78L 149
River Vw. RM16: Grays 9C 100
Riverview Ct. E14 48Bc 92
Riverview Flats RM19: Purf 50Rd 97
River Vw. Gdns. TW1: Twick 61Ha 130
Riverview Gdns. KT11: Cobh 85W 170
Riverview Gdns. SW13 51Xa 110
Riverview Gro. W4 51Ra 109
River Vw. Hgts. SE16 47Vb 91
(off Bermondsey Wall W.)
River Vw. M. CR0: Bedd 76Nb 156
RIVERVIEW PARK 3H 145
Riverview Pk. SE6 61Cc 136
Riverview Rd. DA9: Ghithe 57Wd 120
Riverview Rd. KT19: Ewe 77Sa 153
Riverview Rd. W4 52Ra 109
Riverview Ter. RM19: Purf 50Qd 97
Riverview Wlk. SE6 62Bc 136
River Wlk. E4 24Ec 52
River Wlk. KT12: Walt T 72W 150
River Wlk. KT22: Fet 93Fa 192
River Wlk. UB9: Den 36L 63
River Wlk. W6 52Ya 110
(not continuous)
Riverwalk SW1 7F 229 (50Nb 90)
Riverwalk Apts. SW6 55Eb 111
(off Central Av.)
Riverwalk Bus. Pk. EN3: Brim 14Bc 34
Riverwalk Rd. EN3: Brim 14Bc 34
River Way BR3: Beck 74Cc 158
River Way IG10: Lough 16Pc 36
River Way KT19: Ewe 78Ta 153
River Way SE10 48Hc 93
River Way TW2: Twick 61Da 129
Riverway N13 22Qb 50
Riverway TW18: Staines 67K 127
River Wharf Bus. Pk. DA17: Belv 46Fd 96
Riverwood La. BR7: Chst 67Tc 138
Rivet Ho. SE1 7K 231 (50Vb 91)
(off Cooper's Rd.)
Rivey Cl. KT14: W Byf 86H 169
Riviera Ct. E1 46Wb 91
(off St Katharine's Way)
Rivington Av. IG8: Wfd G 26Mc 53
Rivington Ct. NW10 39Wa 68
Rivington Ct. RM10: Dag 37Dd 76
Rivington Cres. NW7 24Va 48
Rivington Pl. EC2 4J 219 (41Ub 91)
Rivington St. EC2 4H 219 (41Ub 91)
Rivington Wlk. E8 39Wb 71
Rivulet Apts. N4 31Tb 71
Rivulet Rd. N17 24Sb 51
Rixon Cl. SL3: Geor G 44A 82
Rixon Ho. SE18 51Rc 116
Rixon St. N7 34Qb 70
Rixsen Rd. E12 36Nc 74
Roach RM18: E Til 9L 101
Roach Rd. E3 38Cc 72
Road Ho. Est. GU22: Wok 93C 188
Roads Pl. N19 33Nb 70
Roakes Av. KT15: Add 75J 149
Roan Gdns. CR4: Mitc 67Hb 133
Roan St. SE10 51Ec 114
Roasthill La. SL4: Dor 1B 102
Robarts Cl. HA5: Eastc 29X 45
Robbins Hall EN3: Pond E 16Zb 34
Robb Rd. HA7: Stan 23Ja 46
Robe End HP1: Hem H 1H 3
Robert Adam St. W1 2H 221 (44Jb 90)
Roberta St. E2 41Wb 91
Robert Av. AL1: St A 6P 5
Robert Bell Ho. SE16 49Wb 91
(off Rouel Rd.)
Robert Burns Ho. N17 24Xb 51
(off Northumberland Pk.)
Robert Burns M. SE24 57Rb 113
Robert Clack Leisure Cen. 32Cd 76
Robert Cl. EN6: Pot B 5Ab 16
Robert Cl. IG7: Chig 22Vc 55
Robert Cl. KT12: Hers 78X 151
Robert Cl. W9 6A 214 (42Eb 89)
Robert Cl. WD18: Wat 16V 26
Robert Ct. SE15 54Wb 113
Robert Daniels Ct. CM16: They B 9Uc 22
Robert Dashwood Way SE17 6D 230 (49Sb 91)
Robert Gentry Ho. W14 50Ab 88
(off Gledstanes Rd.)
Robert Jones Ho. SE16 49Wb 91
(off Rouel Rd.)
Robert Keen Cl. SE15 53Wb 113
Robert Lewis Ho. IG11: Bark 42Xc 95
Robert Lowe Cl. SE14 52Zb 114
Robert Morton Ho. NW8 38Eb 69
Roberton Dr. BR1: Broml 67Lc 137
Robert Owen Ho. E2 3K 219 (41Vb 91)
(off Baroness Rd.)
Robert Owen Ho. N22 25Qb 50
(off Progress Way)
Robert Owen Ho. SW6 53Za 110
Robert Peel Cl. SL9: Ger X 32D 62
Robert Rd. SL2: Hedg 3H 61
Robert Runcie Ct. SW2 56Pb 112
Roberts All. W5 47Ma 87
Robertsbridge Rd. SM5: Cars 74Eb 155
Roberts Cl. BR5: St M Cry 71Yc 161
Roberts Cl. CR7: Thor H 69Tb 135
Roberts Cl. EN8: Chesh 2Ac 20
Roberts Cl. IG11: Bark 37Sc 74
(off Tanner St.)
Roberts Cl. RM3: Rom 25Kd 57
Roberts Cl. SE16 47Zb 92
Roberts Cl. SE9 60Tc 116
Roberts Cl. SM3: Cheam 80Za 154
Roberts Cl. TW19: Stanw 58L 105
Roberts Cl. UB7: Yiew 46N 83
Roberts Ct. KT9: Chess 78Ma 153
Roberts Ct. N1 39Rb 71
(off Essex Rd.)
Roberts Ct. NW10 37Ua 68
Roberts Ct. SE20 67Yb 136
(off Maple Rd.)
Roberts La. SL9: Chal P 22C 42
Roberts M. BR6: Orp 74Wc 161
Roberts M. SW1 4H 227 (48Jb 90)
Robertson Ct. GU21: Wok 10J 167
Robertson Ct. RM17: Grays 49De 99
(off Hathaway Rd.)
Robertson Gro. SW17 64Gb 133
Robertson Rd. E16 43Jc 93
Robertson Rd. HP4: Berk 1A 2
Robertson St. SW8 55Kb 112

Roberts Pl. EC1 5A 218 (42Qb 90)
Roberts Pl. RM10: Dag 37Cd 76
Robert Sq. SE13 56Ec 114
Roberts Rd. DA17: Belv 50Cd 96
Roberts Rd. E17 25Dc 52
Roberts Rd. WD18: Wat 15Y 27
Robert St. CR0: C'don 76Sb 157
Robert St. E16 46Rc 94
Robert St. NW1 4A 216 (41Kb 90)
Robert St. SE18 50Tc 94
(not continuous)
Robert St. WC2 5G 223 (45Nb 90)
Robert Sutton Ho. E1 44Yb 92
(off Tarling St.)
Roberts Way AL10: Hat 2B 8
Roberts Way TW20: Eng G 6N 125
Roberts Wood Dr. SL9: Chal P 22B 42
Robertswood Lodge SL9: Chal P 23B 42
Robeson St. E3 43Bc 92
Robeson Way WD6: Bore 11Sa 29
Robina Cl. DA6: Bex 56Zc 117
Robina Cl. HA6: Nwood 25V 44
Robina Ct. BR8: Swan 70Jd 140
Robin Cl. KT15: Add 78M 149
Robin Cl. NW7 20Ua 30
Robin Cl. RM5: Col R 24Fd 56
Robin Cl. TW12: Hamp 64Aa 129
Robin Ct. E14 47Ec 92
Robin Ct. SE16 49Wb 91
Robin Ct. SM6: W'gton 78Lb 156
Robin Cres. E6 43Mc 93
Robin Gdns. RH1: Redh 4A 208
Robin Gro. HA3: Kenton 30Pa 47
Robin Gro. N6 33Jb 70
Robin Gro. TW8: Bford 51La 108
Robin Hill Dr. BR7: Chst 65Nc 138
ROBIN HOOD 62Ua 132
Robin Hood Cl. GU21: Wok 10K 167
Robin Hood Cl. SL1: Slou 6D 80
Robinhood Cl. CR4: Mitc 69Lb 134
Robin Hood Ct. EC4 2A 224 (44Qb 90)
(off Shoe La.)
Robin Hood Cres. GU21: Knap 9J 167
Robin Hood Dr. HA3: Hrw W 24Ha 46
Robin Hood Dr. WD23: Bush 11Ba 27
Robin Hood Gdns. E14 45Ec 92
(off Woolmore St.)
Robin Hood Grn. BR5: St M Cry 71Wc 161
Robin Hood La. DA6: Bex 57Ad 117
Robin Hood La. E14 45Ec 92
Robin Hood La. GU4: Sut G 96B 188
Robin Hood La. SM1: Sutt 78Cb 155
Robin Hood La. SW15 62Ua 132
Robinhood La. CR4: Mitc 69Lb 134
Robin Hood Rd. CM15: B'wood 17Xd 40
Robin Hood Rd. GU21: Knap 9H 167
(not continuous)
Robin Hood Rd. GU21: Wok 9H 167
(not continuous)
Robin Hood Rd. SW19 64Wa 132
Robin Hood Way SW15 62Ua 132
Robin Hood Way SW20 64Ua 132
Robin Hood Way UB6: G'frd 37Ha 66
Robin Hood Works GU21: Knap 9J 167
Robin Ho. NW8 2D 214 (40Gb 69)
(off Newcourt St.)
Robin Howard Dance Theatre (Place), The 4E 216 (41Mb 90)
(off Duke's Rd.)
Robinia Av. DA11: Nflt 59Fe 121
Robinia Cl. IG6: Ilf 22Uc 54
Robinia Cl. SE20 67Wb 135
(off Sycamore Gro.)
Robinia Cres. E10 33Dc 72
Robinia Ho. SE16 47Ac 92
(off Blondin Way)
Robin La. NW4 27Za 48
Robin Pde. SL2: Farn C 6G 60
Robin Pl. WD25: Wat 4X 13
Robins, The RM13: Rain 40Md 77
Robins Cl. AL2: Lon C 9J 7
Robins Cl. UB8: Cowl 43L 83
Robin's Ct. BR3: Beck 68Fc 137
Robins Ct. CR2: S Croy 77Ub 157
(off Birdhurst Rd.)
Robins Ct. SE12 62Lc 137
Robinscroft M. SE10 53Ec 114
Robins Dale GU21: Knap 9G 166
Robinsfield HP1: Hem H 2J 3
Robinsfield Gdns. CR3: Cat'm 95Vb 197
Robins Gro. BR4: W W'ck 76Jc 159
Robin's La. CM16: They B 8Sc 22
Robinson Av. EN7: G Oak 1Rb 19
Robinson Cl. E11 34Gc 73
Robinson Cl. EN2: Enf 13Sb 33
Robinson Cl. GU22: Wok 92E 188
Robinson Cl. RM12: Horn 38Kd 77
Robinson Ct. CR7: Thor H 72Rb 157
Robinson Ct. N1 39Rb 71
(off St Mary's Path)
Robinson Ct. TW9: Rich 56Pa 109
Robinson Cres. WD23: B Hea 18Ea 28
Robinson Ho. E14 43Cc 92
(off Selsey St.)
Robinson Ho. W10 44Za 88
(off Bramley Rd.)
Robinson Rd. E2 40Yb 72
Robinson Rd. RM10: Dag 35Cd 76
Robinson Rd. SE1 6B 224 (46Rb 91)
Robinson Rd. SW17 65Gb 133
Robinson's Cl. W13 43Ja 86
Robinson St. SW3 51Hb 111
Robinson Way DA11: Nflt 57Ce 121
Robinson Way SE14 52Zb 114
Robins Orchard SL9: Chal P 23A 42
Robins Rd. HP3: Hem H 4A 4
Robins Way AL10: Hat 3B 8
Robinsway EN9: Walt A 6Gc 21
Robinsway KT12: Hers 77Y 151
Robinswood Gdns. E17 25Bc 52
Robinswood M. N5 36Rb 71
Robin Way BR5: St P 69Xc 139
Robin Way TW18: Staines 62H 127
Robin Willis Way SL4: Old Win 8L 103
Robinwood Dr. TN15: Seal 91Pd 203
Robinwood Gro. UB8: Hil 42P 83
Robinwood Pl. SW15 63Ta 131
Roborough Wlk. RM12: Horn 37Ld 77
Robsart St. SW9 54Pb 112
Robson Av. NW10 38Wa 68
Robson Cl. E6 44Nc 94
Robson Cl. EN2: Enf 12Rb 33
Robson Cl. SL9: Chal P 22A 42
Robson Rd. SE27 62Sb 135
Robsons Cl. EN8: Chesh 1Yb 20
Roby Ho. EC1 5D 218 (42Sb 91)
(off Mitchell St.)
Robyns Cft. DA11: Nflt 2A 144
Robyns Way TN13: S'oaks 94Hd 202
Roca Ct. E11 29Jc 53
Rocastle Rd. SE4 57Ac 114
Roch Av. HA8: Edg 26Pa 47
Rochdale Rd. E17 31Cc 72
Rochdale Rd. SE2 50Xc 95
Rochdale Way SE8 52Cc 114
(not continuous)
Roche Ho. E14 45Bc 92
(off Beccles St.)
Rochelle Cl. SW11 56Fb 111
Rochelle St. E2 4K 219 (41Vb 91)
(not continuous)
Rochemont Wlk. E8 39Wb 71
(off Pownell Rd.)
Roche Rd. SW16 67Pb 134
Rochester & Cobham Pk. Golf Course 9N 145
Rochester Av. BR1: Broml 68Kc 137
Rochester Av. E13 39Lc 73
Rochester Av. TW13: Felt 61V 128
Rochester Cl. DA15: Sidc 58Xc 117
Rochester Cl. EN1: Enf 11Ub 33
Rochester Cl. SW16 66Nb 134
Rochester Ct. E2 42Xb 91
(off Wilmot St.)
Rochester Ct. NW1 38Lb 70
(off Rochester Sq.)
Rochester Dr. DA5: Bexl 58Bd 117
Rochester Dr. HA5: Pinn 29Z 45
Rochester Dr. WD25: Wat 7Y 13
Rochester Gdns. CR0: C'don 76Ub 157
Rochester Gdns. CR3: Cat'm 94Ub 197
Rochester Gdns. IG1: Ilf 31Pc 74
Rochester Ho. SE1 2F 231 (47Tb 91)
(off Manciple St.)
Rochester Ho. SE15 51Yb 114
(off Sharratt St.)
Rochester M. NW1 38Lb 70
Rochester M. W5 49La 86
Rochester Pde. TW13: Felt 61W 128
Rochester Pl. NW1 37Lb 70
Rochester Rd. DA1: Dart 59Qd 119
Rochester Rd. DA12: Grav'nd 9G 122
Rochester Rd. HA6: Nwood 27V 44
Rochester Rd. NW1 37Lb 70
Rochester Rd. RM12: Horn 38Kd 77
Rochester Rd. SM5: Cars 77Hb 155
Rochester Rd. TW18: Staines 64F 126
Rochester Row SW1 5C 228 (49Lb 90)
Rochester Sq. NW1 38Lb 70
Rochester St. SW1 4D 228 (48Mb 90)
Rochester Ter. NW1 37Lb 70
Rochester Wlk. RH2: Reig 10K 207
Rochester Wlk. SE1 6F 225 (46Tb 91)
Rochester Way DA1: Dart 59Fd 118
Rochester Way SE3 53Kc 115
Rochester Way SE9 55Pc 116
Rochester Way WD3: Crox G 14R 26
Rochester Way Relief Rd. SE3 53Kc 115
Roche Wlk. SM5: Cars 72Fb 155
Rochford N17 26Ub 51
(off Griffin Rd.)
Rochford Av. CM15: Shenf 15Ce 41
Rochford Av. EN9: Walt A 6Fc 21
Rochford Av. IG10: Lough 13Sc 36
Rochford Av. RM6: Chad H 29Yc 55
Rochford Cl. E6 40Mc 73
Rochford Cl. RM12: Horn 37Kd 77
Rochford Grn. IG10: Lough 13Sc 36
Rochfords Gdns. SL2: Slou 5N 81
Rochford Wlk. E8 38Wb 71
Rochford Way CR0: C'don 72Nb 156
Rochfort Ho. SE8 50Bc 92
Rockall Ct. SL3: L'ly 48D 82
Rock Av. SW14 55Ta 109
Rockbourne M. SE23 60Zb 114
Rockbourne Rd. SE23 60Zb 114
Rockchase Gdns. RM11: Horn 30Nd 57
Rock Cl. CR4: Mitc 68Fb 133
Rockdale TN13: S'oaks 97Kd 203
Rockdale Gdns. TN13: S'oaks 97Kd 203
Rockdale Pleasance TN13: S'oaks 98Kd 203
Rockdale Rd. TN13: S'oaks 97Ld 203
Rockell's Pl. SE22 58Xb 113
Rockfield Cl. RH8: Oxt 3K 211
Rockfield Ho. NW4 28Za 48
(off Belle Vue Est.)
Rockfield Ho. SE10 51Ec 114
(off Welland St.)
Rockfield Rd. RH8: Oxt 1K 211
Rockford Av. UB6: G'frd 40Ja 66
Rock Gdns. RM10: Dag 36Dd 76
Rock Gro. Way SE16 49Wb 91
(not continuous)
Rockhall Rd. NW2 35Za 68
Rockhall Way NW2 34Za 68
Rockhampton Cl. SE27 63Qb 134
Rockhampton Rd. CR2: S Croy 79Ub 157
Rockhampton Rd. SE27 63Qb 134
Rock Hill BR6: Well H 79Dd 162
Rock Hill SE26 63Vb 135
(not continuous)
Rockingham Av. RM11: Horn 30Kd 57
Rockingham Cl. SW15 56Va 110
Rockingham Cl. UB8: Uxb 39L 63
Rockingham Ga. WD23: Bush 16Ea 28
Rockingham Pde. UB8: Uxb 38L 63
Rockingham Rd. UB8: Uxb 39K 63
Rockingham St. SE1 4D 230 (48Sb 91)
Rockland Rd. SW15 56Ab 110
Rocklands Ct. RM17: Grays 50Ee 99
Rocklands Dr. CR2: S Croy 79Tb 157
Rocklands Dr. HA7: Stan 26Ka 46
Rockleigh Ct. CM15: Shenf 17Ce 41
Rockley Ct. W14 47Za 88
(off Rockley Rd.)
Rockley Rd. W14 47Za 88
Rockliffe Av. WD4: K Lan 2Q 12
Rockmount Rd. SE18 50Vc 95
Rockmount Rd. SE19 65Tb 135
Rock Rd. TN15: Bor G 92Be 205
Rockshaw Rd. RH1: Mers 99Lb 196
Rocks La. SW13 53Wa 110
Rock St. N4 33Qb 70
Rock Ter. Ground 69Fb 133
(off Belgrave Road)
Rockware Av. UB6: G'frd 39Fa 66
Rockware Av. Bus. Cen. UB6: G'frd 39Fa 66
Rockways EN5: Ark 16Va 30
Rockwell Ct. WD18: Wat 15U 26
Rockwell Gdns. SE19 64Ub 135
Rockwell Rd. RM10: Dag 36Dd 76
Rockwood Pl. W12 47Ya 88
Rocky La. RH1: Redh 2B 208
Rocky La. RH2: Reig 100Hb 195
Rocliffe St. N1 2C 218 (40Rb 71)
Rocombe Cres. SE23 59Yb 114
Rocque Ho. SW6 52Bb 111
(off Estcourt Rd.)
Rocque La. SE3 55Hc 115
Rodale Mans. SW18 58Db 111
Rodborough Ct. W9 42Cb 89
(off Hermes Cl.)
Rodborough Rd. NW11 32Cb 69
Rodd Est. TW17: Shep 71S 150
Roden Ct. N6 31Mb 70
Roden Gdns. CR0: C'don 72Ub 157
Rodenhurst Rd. SW4 58Lb 112
Roden St. IG1: Ilf 34Qc 74
Roden St. N7 34Pb 70
Roden Way IG1: Ilf 34Qc 74
(off Roden St.)
Rodeo Cl. DA8: Erith 53Kd 119
Roderick Ho. SE16 49Yb 92
(off Raymouth Rd.)
Roderick Rd. NW3 35Hb 69
Rodgers Cl. WD6: E'tree 16Ma 29
Rodgers Ho. SW4 59Mb 112
(off Clapham Pk. Est.)
Rodin Ct. N1 39Rb 71
(off Essex Rd.)
Roding CM14: B'wood 18Xd 40
Roding Av. IG8: Wfd G 23Nc 54
Roding Gdns. IG10: Lough 16Nc 36
Roding Ho. N1 1K 217 (39Qb 70)
(off Barnsbury Est.)
Roding La. IG7: Chig 19Qc 36
Roding La. IG9: Buck H 18Mc 35
Roding La. Nth. IG8: Wfd G 23Nc 54
Roding La. Sth. IG4: Ilf 28Mc 53
(not continuous)
Roding La. Sth. IG4: Wfd G 28Mc 53
(not continuous)
Roding La. Sth. IG8: Wfd G 27Mc 53
Roding M. E1 46Wb 91
Roding Rd. E5 35Zb 72
Roding Rd. E6 43Rc 94
Roding Rd. IG10: Lough 15Nc 36
Rodings, The IG8: Wfd G 23Lc 53
Rodings, The RM14: Upm 30Td 58
RODING SPIRE HOSPITAL 27Mc 53
Roding Trad. Est. IG11: Bark 38Rc 74
Roding Valley Meadows Nature Reserve 17Qc 36
Roding Valley Station (Underground) 21Mc 53
Roding Vw. IG9: Buck H 18Mc 35
Roding Way RM13: Rain 40Md 77
Rodmarton St. W1 1G 221 (43Hb 89)
Rodmell WC1 4G 217 (41Nb 90)
(off Regent Sq.)
Rodmell Cl. UB4: Yead 42Aa 85
Rodmell Slope N12 22Bb 49
Rodmere St. SE10 50Gc 93
Rodmill La. SW2 59Nb 112
Rodney Av. AL1: St A 4E 6
Rodney Cl. CR0: C'don 74Rb 157
Rodney Cl. HA5: Pinn 31Aa 65
Rodney Cl. KT12: Walt T 74Y 151
Rodney Cl. KT3: N Mald 71Ua 154
Rodney Ct. EN5: Barn 13Bb 31
Rodney Ct. W9 5A 214 (42Eb 89)
Rodney Gdns. BR4: W W'ck 77Jc 159
Rodney Gdns. HA5: Eastc 29X 45
Rodney Grn. KT12: Walt T 75Y 151
Rodney Ho. E14 49Dc 92
(off Cahir St.)
Rodney Ho. N1 2J 217 (40Pb 70)
(off Donegal St.)
Rodney Ho. SW1 50Lb 90
(off Dolphin Sq.)
Rodney Ho. W11 45Cb 89
(off Pembridge Cres.)
Rodney Pl. E17 26Ac 52
Rodney Pl. SE17 5E 230 (49Sb 91)
Rodney Pl. SW19 67Eb 133
Rodney Point SE16 47Bc 92
(off Rotherhithe St.)
Rodney Rd. CR4: Mitc 69Gb 133
Rodney Rd. E11 28Kc 53
Rodney Rd. KT12: Walt T 75Y 151
Rodney Rd. KT3: N Mald 71Ua 154
Rodney Rd. SE17 5E 230 (49Sb 91)
(not continuous)
Rodney Rd. TW2: Whitt 58Ca 107
Rodney St. N1 1J 217 (40Pb 70)
Rodney Way RM7: Mawney 25Dd 56
Rodney Way SL3: Poyle 53G 104
Rodona Rd. KT13: Weyb 83T 170
Rodway Rd. BR1: Broml 67Kc 137
Rodway Rd. SW15 59Wa 110
Rodwell Cl. HA4: Ruis 32Y 65
Rodwell Ct. KT12: Walt T 76X 151
Rodwell Ct. KT15: Add 77L 149
Rodwell Pl. HA8: Edg 23Qa 47
Rodwell Rd. SE22 58Vb 113
Roe NW9 24Va 48
Roebourne Way E16 46Qc 94
Roebuck Cl. KT21: Asht 92Na 193
Roebuck Cl. N17 23Vb 51
Roebuck Cl. RH2: Reig 6J 207
Roebuck Cl. TW13: Felt 63X 129
Roebuck Ct. SL1: Slou 6C 80
Roebuck Grn. SL1: Slou 6C 80
Roebuck Hgts. IG9: Buck H 17Lc 35
Roebuck La. IG9: Buck H 17Lc 35
Roebuck Rd. IG6: Ilf 22Xc 55
Roebuck Rd. KT9: Chess 78Qa 153
Roebuck Rd. Trad. Est. IG6: Ilf 23Xc 55
Roedean Av. EN3: Enf H 11Yb 34
Roedean Cl. BR6: Chels 77Xc 161
Roedean Cl. EN3: Enf H 11Yb 34
Roedean Cres. SW15 58Ua 110
Roedean Dr. RM1: Rom 28Gd 56
Roedean Ho. WD24: Wat 10Y 13
(off Exeter Cl.)
Roe End NW9 28Sa 47
Roefields Cl. HP3: Hem H 6J 3
ROE GREEN 28Sa 47
Roe Grn. NW9 29Sa 47
Roe Grn. Cen. AL10: Hat 1B 8
Roe Grn. Cl. AL10: Hat 1A 8
ROEHAMPTON 59Wa 110
Roehampton Cl. DA12: Grav'nd 9G 122
Roehampton Cl. SW15 56Wa 110
Roehampton Dr. BR7: Chst 65Sc 138
Roehampton Ga. SW15 58Ua 110
Roehampton Golf Course 56Va 110
Roehampton High St. SW15 59Wa 110
Roehampton Ho. RM8: Dag 36Wc 75
Roehampton La. SW15 56Wa 110
ROEHAMPTON LANE 60Xa 110
Roehampton Sport & Fitness Cen. 59Wa 110
Roehampton University Digby Stuart College 57Wa 110
Roehampton University Main Site 57Wa 110
Roehampton University Southlands College 57Wa 110
Roehampton University Whitelands Site 59Wa 110
Roehampton Va. SW15 62Va 132
Roe Hill Cl. AL10: Hat 1B 8
ROEHYDE 2A 8
Roehyde Way AL10: Hat 2A 8
Roe La. NW9 28Ra 47
Roesel Pl. BR5: Pet W 71Rc 160
ROESTOCK 5P 7
Roestock Gdns. AL4: Col H 4A 8
Roestock La. AL4: Col H 5P 7
Roe Way SM6: W'gton 79Nb 156
Rofant Rd. HA6: Nwood 23U 44
Roffe Gdns. RM8: Dag 32Ad 75
Roffe's La. CR3: Cat'm 96Tb 197
Roffey Cl. CR8: Purl 88Rb 177
Roffey St. E14 47Ec 92
Roffo Ct. SE17 51Tb 113
(off Boundary La.)
Roffords GU21: Wok 9M 167
Rogan Ho. SW8 53Kb 112
(off St Joseph's St.)
Rogate Ho. E5 34Wb 71
Roger Dowley Ct. E2 40Yb 72
Roger Harriss Almshouses E15 39Hc 73
(off Gift La.)
Roger Reede's Almshouses RM1: Rom 28Gd 56
Rogers Cl. CR3: Cat'm 94Xb 197
Rogers Cl. CR5: Coul 90Rb 177
Rogers Ct. BR8: Swan 70Jd 140
Rogers Ct. E14 45Cc 92
(off Premiere Pl.)
Rogers Est. E2 41Yb 92
(not continuous)
Rogers Gdns. RM10: Dag 36Cd 76
Roger's Ho. RM10: Dag 34Cd 76
Rogers Ho. SW1 5E 228 (49Mb 90)
(off Page St.)
Roger Simmons Ct. KT23: Bookh 96Ba 191
Rogers La. CR6: W'ham 90Bc 178
Rogers La. SL2: Stoke P 8K 61
Rogers Mead RH9: G'stone 4P 209
Rogers Rd. E16 44Hc 93
Rogers Rd. RM10: Dag 36Cd 76
Rogers Rd. RM17: Grays 49Ee 99
Rogers Rd. SW17 63Fb 133
Rogers Ruff HA6: Nwood 25S 44
Roger St. WC1 6J 217 (42Pb 90)
Rogers Wlk. N12 20Db 31
Rogers Wood La. DA3: Fawk 77Wd 164
Rohere Ho. EC1 3D 218 (41Sb 91)
Rojack Rd. SE23 60Zb 114
Rokeby Ct. GU21: Wok 9K 167
Rokeby Gdns. IG8: Wfd G 25Jc 53
Rokeby Ho. SW12 59Kb 112
(off Lochinvar St.)
Rokeby Ho. WC1 6H 217 (42Pb 90)
(off Lamb's Conduit St.)
Rokeby Pl. SW20 66Xa 132
Rokeby Rd. HA1: Harr 27Fa 46
Rokeby Rd. SE4 54Bc 114
Rokeby St. E15 39Fc 73
Roke Cl. CR8: Kenley 86Sb 177
Rokell Ho. BR3: Beck 64Dc 136
(off Beckenham Hill Rd.)
Roke Lodge Rd. CR8: Kenley 85Rb 177
Roke Rd. CR8: Kenley 87Sb 177
Roker Pk. Av. UB10: Ick 35N 63
Rokesby Cl. DA16: Well 54Tc 116
Rokesby Pl. HA0: Wemb 36Ma 67
Rokesby Rd. SL2: Slou 1D 80
Rokesly Av. N8 29Nb 50
Rokewood Apts. BR3: Beck 67Cc 136
Roland Gdns. SW10 7A 226 (50Eb 89)
Roland Gdns. SW7 7A 226 (50Eb 89)
Roland Ho. SW7 7A 226 (50Eb 89)
(off Old Brompton Rd.)
Roland Mans. SW7 7A 226 (50Eb 89)
(off Old Brompton Rd.)
Roland M. E1 43Zb 92
Roland Rd. E17 28Fc 53
Roland St. AL1: St A 2E 6
Roland Way KT4: Wor Pk 75Va 154
Roland Way SE17 50Tb 91
Roland Way SW7 7A 226 (50Eb 89)
Roles Gro. RM6: Chad H 28Zc 55
Rolfe Cl. EN4: E Barn 14Gb 31
Rolfe Ter. SE18 50Rc 94
Rolinsden Way BR2: Kes 78Mc 159
Rolland Ho. W7 43Ga 86
Rollason Way CM14: B'wood 20Xd 40
Rollerbowl 26Bd 55
Rollesby Rd. KT9: Chess 79Qa 153
Rollesby Way SE28 44Yc 95
Rolleston Av. BR5: Pet W 72Rc 160
Rolleston Cl. BR5: Pet W 73Rc 160
Rolleston Rd. CR2: S Croy 80Tb 157
Roll Gdns. IG2: Ilf 29Qc 54
Rolling Mills M. E14 44Ac 92
Rollins Ho. SE15 51Yb 114
(off Rollins St.)
Rollins St. SE15 51Yb 114
Rollit Cres. TW3: Houn 57Ca 107
Rollit St. N7 36Qb 70
Rollo Rd. BR8: Hext 66Hd 140
Rolls Bldgs. EC4 2K 223 (44Qb 90)
Rollscourt Av. SE24 57Sb 113
ROLLS PARK 17Tc 36
Rolls Pk. Av. E4 22Cc 52
Rolls Pk. Rd. E4 22Dc 52
Rolls Pas. EC4 2K 223 (44Qb 90)
(off Chancery La.)
Rolls Rd. SE1 7K 231 (50Vb 91)
Rolls Royce Cl. SM6: W'gton 80Nb 156
Rolt St. SE8 51Ac 114
(not continuous)
Rolvenden Gdns. BR1: Broml 66Mc 137
Rolvenden Pl. N17 25Wb 51
Roma Corte SE13 55Dc 114
(off Elmira St.)
Roma Ct. TN13: S'oaks 93Kd 203
Roma Ho. AL1: St A 3D 6
Roman RM18: E Til 9L 101
Romana Ct. CR0: C'don 75Vb 157
Romana Ct. TW18: Staines 63J 127
Roman Apts. E8 38Xb 71
(off Silesia Bldgs.)
Romanby Ct. RH1: Redh 7P 207
Roman Cl. CM15: Mount 11Fe 41
Roman Cl. RM13: Rain 40Fd 76
Roman Cl. TW14: Felt 57Y 107
Roman Cl. UB9: Hare 25K 43
Roman Cl. W3 47Ra 87
Roman Ct. N7 37Pb 70
Roman Ct. TN15: Bor G 92Be 205
Romanfield Rd. SW2 59Pb 112
Roman Gdns. WD4: K Lan 2R 12
Roman Ho. AL2: Lon C 9H 7
Roman Ho. EC2 1E 224 (43Sb 91)
(off Wood St.)
Roman Ho. RM13: Rain 40Fd 76
Romanhurst Av. BR2: Broml 70Gc 137
Romanhurst Gdns. BR2: Broml 70Gc 137
Roman Ind. Est. CR0: C'don 73Ub 157
Roman Mosaic & Hypocaust, The 3N 5
Roman Ri. SE19 65Tb 135
Roman Rd. NW2: Edgware Rd. 34Ya 68
Roman Rd. NW2: Temple Rd. 34Ya 68
Roman Rd. CM15: Mount 13Ee 41
Roman Rd. CM15: Shenf 13Ee 41
Roman Rd. DA11: Nflt 62Ee 143
Roman Rd. E2 41Yb 92
Roman Rd. E3 40Ac 72
Roman Rd. E6 42Mc 93
Roman Rd. IG1: Ilf 37Rc 74
Roman Rd. N10 24Kb 50
Roman Rd. W4 49Ua 88
Roman Rd. Mkt. 39Bc 72
(off Roman Rd.)
Romans End AL3: St A 4A 6
Roman Sq. SE28 46Wc 95
Romans Way GU22: Pyr 87J 169
Roman Temple Virginia Water 9J 125
Roman Theatre of Verulamium 2N 5
Roman Villa Rd. DA2: Daren 64Sd 142
Roman Villa Rd. DA4: S Dar 65Sd 142
Roman Villa Rd. DA4: Sut H 65Sd 142
Roman Wlk. WD7: R'lett 8Ha 14
Roman Wall of St Albans 3N 5
Roman Way CR0: C'don 75Rb 157
Roman Way DA1: Cray 57Gd 118
Roman Way EN1: Enf 15Vb 33
Roman Way EN9: Walt A 7Dc 20
Roman Way KT17: Ewe 82Wa 174
Roman Way N7 37Pb 70
Roman Way SE15 52Yb 114
Roman Way SM5: Cars 81Hb 175
Roman Way Ind. Est. N7 38Pb 70
(off Roman Way)
Romany Ct. HP2: Hem H 1C 4
Romany Gdns. E17 25Ac 52
Romany Gdns. SM3: Sutt 73Cb 155
Romany Ri. BR5: Farnb 74Sc 160
Romany Rd. GU21: Knap 7G 166
Roma Read Cl. SW15 59Xa 110
Roma Rd. E17 27Ac 52
Romayne Ho. SW4 55Mb 112
Romberg Rd. SW17 62Jb 134
Romborough Gdns. SE13 57Ec 114
Romborough Way SE13 57Ec 114
Rom Cres. RM7: Rush G 31Hd 76
Romeland AL3: St A 2A 6
Romeland EN9: Walt A 5Ec 20
Romeland WD6: E'tree 16Ma 29
Romeland Hill AL3: St A 2A 6
Romeo Bus. Cen. RM15: Avel 47Pd 97
Romero Cl. SW9 55Pb 112
Romeyn Rd. SW16 62Pb 134
ROMFORD 29Gd 56
Romford Golf Course 27Kd 57
Romford Greyhound Stadium 30Ed 56
Romford Rd. E12 36Mc 73
Romford Rd. E15 37Gc 73
Romford Rd. E7 37Hc 73
Romford Rd. IG7: Chad H 20Xc 37
Romford Rd. IG7: Chig 20Xc 37
Romford Rd. RM15: Avel 41Sd 98
Romford Rd. RM5: Col R 24Ad 55
Romford Station (Rail & Crossrail) 30Gd 56
Romford St. E1 43Wb 91
Romilly Dr. WD19: Wat 21Aa 45
Romilly Ho. W11 45Ab 88
(off Wilsham St.)
Romilly Rd. N4 33Rb 71
Romilly St. W1 4E 222 (45Mb 90)
Romily Ct. SW6 53Bb 111
Rommany Rd. SE27 63Tb 135
(not continuous)
Romney Chase RM11: Horn 30Pd 57
Romney Cl. HA2: Harr 31Ca 65
Romney Cl. KT9: Chess 77Na 153
Romney Cl. N17 25Xb 51
Romney Cl. NW11 32Eb 69
Romney Cl. SE14 52Yb 114
Romney Cl. TW15: Ashf 64S 128
Romney Ct. NW3 37Gb 69
Romney Ct. UB5: N'olt 40Z 65
(off Parkfield Dr.)
Romney Ct. W12 47Za 88
(off Shepherd's Bush Grn.)
Romney Dr. BR1: Broml 66Mc 137
Romney Dr. HA2: Harr 31Ca 65
Romney Gdns. DA7: Bex 53Bd 117
Romney Ho. SW1 4E 228 (48Mb 90)
(off Marsham St.)
Romney Lock SL4: Wind 1J 103
Romney Lock Rd. SL4: Wind 2H 103
Romney M. W1 7H 215 (43Jb 90)
Romney Pde. UB4: Hayes 40T 64
Romney Rd. DA11: Nflt 2A 144
Romney Rd. KT3: N Mald 72Ta 153
Romney Rd. SE10 51Fc 115
Romney Rd. UB4: Hayes 40T 64
Romney Row NW2 33Za 68
(off Brent Ter.)
Romney St. SW1 4F 229 (48Nb 90)
Romney St. TN15: Knat 83Qd 183
ROMNEY STREET 83Qd 183
Romney St. Cvn. Pk. TN15: Knat 84Qd 183
Romney Wlk. SL4: Wind 2H 103
Romola Rd. SE24 60Rb 113
Romsey Cl. BR6: Farnb 77Rc 160
Romsey Cl. SL3: L'ly 48B 82
Romsey Cl. SS17: Stan H 2K 101
Romsey Dr. SL2: Farn C 4H 61
Romsey Gdns. RM9: Dag 39Zc 75
Romsey Rd. RM9: Dag 39Zc 75
Romsey Rd. W13 45Ja 86
Romside Pl. RM7: Rom 28Fd 56
Romulus Ct. TW8: Bford 52Ma 109
Romulus Rd. DA12: Grav'nd 8F 122
Rom Valley Way RM7: Rush G 30Gd 56
Rom Valley Way Retail Pk. 31Hd 76
Ronald Av. E15 41Gc 93
Ronald Buckingham Ct. SE16 47Yb 92
(off Kenning St.)
Ronald Cl. BR3: Beck 70Bc 136
Ronald Cl. AL2: Brick W 1Ba 13
Ronald Ct. EN5: New Bar 13Db 31
Ronald Rd. RM3: Hrld W 25Qd 57
Ronaldsay Spur SL1: Slou 3J 81
Ronaldshay N4 31Qb 70
Ronalds Rd. BR1: Broml 67Jc 137
Ronalds Rd. N5 36Qb 70
(not continuous)
Ronaldstone Rd. DA15: Sidc 58Uc 116
Ronald St. E1 44Yb 92
Rona Maclean Cl. KT19: Eps 84Qa 173
Ronan Way UB9: Den 33H 63
Rona Rd. NW3 35Jb 70
Ronart St. HA3: W'stone 27Ha 46

Rona Wlk. N1 37Tb 71
Rondel Ct. DA5: Bexl 58Ad 117
Rondu Rd. NW2 36Ab 68
Ronelean Rd. KT6: Surb 75Pa 153
Roneo Cnr. RM12: Horn 32Hd 76
Roneo Link RM12: Horn 32Hd 76
Ronfearn Av. BR5: St M Cry 71Zc 161
Ron Grn. Ct. DA8: Erith 51Fd 118
Ron Leighton Way E6 39Nc 74
Ronley Ct. TN13: S'oaks 93Ld 203
(off Hillingdon Av.)
Ronneby Cl. KT13: Weyb 76U 150
Ronnie La. E12 35Qc 74
(not continuous)
Ronson Way KT22: Lea 93Ja 192
Ron Todd Cl. RM10: Dag 39Cd 76
Ronver Rd. SE12 59Hc 115
Rood La. EC3 4H 225 (45Ub 91)
Roof Ter. Apts., The EC1 6C 218 (42Rb 91)
(off Gt. Sutton St.)
Rookby Ct. N21 19Rb 33
Rook Cl. HA9: Wemb 34Ra 67
Rook Cl. RM12: Horn 38Jd 76
Rookdean TN13: Chips 94Ed 202
Rookeries Cl. TW13: Felt 62X 129
Rookery, The RM20: W Thur 51Wd 120
ROOKERY, THE 17X 27
Rookery Cl. KT22: Fet 96Ga 192
Rookery Cl. NW9 29Va 48
Rookery Cl. SS17: Stan H 2K 101
Rookery Ct. E10 34Dc 72
Rookery Ct. RM20: W Thur 51Wd 120
Rookery Cres. RM10: Dag 38Dd 76
Rookery Dr. BR7: Chst 67Qc 138
Rookery Gdns. BR5: St M Cry 71Yc 161
Rookery Hill KT21: Asht 90Qa 173
Rookery La. BR2: Broml 72Mc 159
Rookery La. RM17: Grays 50Fe 99
Rookery Mead CR5: Coul 94Mb 196
Rookery Rd. BR6: Downe 82Pc 180
Rookery Rd. SW4 56Lb 112
Rookery Rd. TW18: Staines 64K 127
Rookery Vw. RM17: Grays 50Fe 99
Rookery Vs. BR6: Downe 82Pc 180
Rookery Way KT20: Lwr K 99Bb 195
Rookery Way NW9 29Va 48
Rookesley Rd. BR5: St M Cry 73Zc 161
Rooke Way SE10 50Hc 93
Rookfield Av. N10 28Lb 50
Rookfield Cl. N10 28Lb 50
Rook La. CR3: Cat'm 97Pb 196
Rookley Cl. SM2: Sutt 81Db 175
Rookley Ct. RM19: Purf 50Rd 97
(off Linnet Way)
Rooks Hill WD3: Loud 14M 25
Rooksmead Rd. TW16: Sun 68V 128
ROOKS NEST 2C 210
Rooks Ter. UB7: W Dray 47N 83
Rookstone Rd. SW17 64Hb 133
Rook Wlk. E6 44Mc 93
Rookwood Av. IG10: Lough 13Sc 36
Rookwood Av. KT3: N Mald 70Wa 132
Rookwood Av. SM6: Bedd 77Mb 156
Rookwood Cl. RH1: Mers 1B 208
Rookwood Cl. RM17: Grays 49De 99
Rookwood Gdns. E4 19Hc 35
Rookwood Gdns. IG10: Lough 13Sc 36
Rookwood Ho. IG11: Bark 40Tc 74
Rookwood Pl. RH1: Mers 100Kb 196
(off London Rd. Sth.)
Rookwood Rd. N16 31Vb 71
Rookwood Way E3 38Cc 72
Roosevelt Memorial 4J 221 (45Jb 90)
Roosevelt Way RM10: Dag 37Fd 76
Rootes Dr. W10 43Za 88
Ropemaker Rd. SE16 47Ac 92
Ropemaker's Flds. E14 45Bc 92
Ropemaker St. EC2 7F 219 (43Tb 91)
Roper Cres. TW16: Sun 67W 128
Roper La. SE1 2J 231 (47Ub 91)
Ropers Av. E4 22Dc 52
Ropers Orchard SW3 51Gb 111
(off Danvers St.)
Roper St. SE9 57Pc 116
Ropers Wlk. SW2 59Qb 112
Ropers Yd. CM14: B'wood 19Yd 40
Roper Way CR4: Mitc 68Jb 134
Ropery Bus. Pk. SE7 49Lc 93
Ropery St. E3 42Bc 92
Rope St. SE16 49Ac 92
Rope Ter. E16 47Kc 93
Rope Wlk. TW16: Sun 69Y 129
Ropewalk Gdns. E1 44Wb 91
Ropewalk M. E8 38Wb 71
Ropeworks, The IG11: Bark 39Sc 74
Ropley St. E2 40Wb 71
Rosa Alba M. N5 35Sb 71
Rosa Av. TW15: Ashf 63Q 128
Rosalind Ct. IG11: Bark 38Wc 75
(off Meadow Rd.)
Rosalind Ho. N1 2J 219 (40Ub 71)
(off Arden Est.)
Rosaline Rd. SW6 52Ab 110
Rosaline Ter. SW6 52Ab 110
(off Rosaline Rd.)
Rosa M. E17 27Bc 52
Rosamond St. SE26 62Xb 135
Rosamund Cl. CR2: S Croy 77Tb 157
Rosamun St. UB2: S'hall 49Aa 85
Rosa Parks Ho. SE17 5E 230 (49Sb 91)
(off Munton Rd.)
Rosary Cl. TW3: Houn 54Aa 107
Rosary Ct. EN6: Pot B 2Db 17
Rosary Gdns. SW7 6A 226 (49Eb 89)
Rosary Gdns. TW15: Ashf 63R 128
Rosary Gdns. WD23: Bush 17Ga 28
Rosaville Rd. SW6 52Bb 111
Roscoe St. EC1 6E 218 (42Sb 91)
(not continuous)
Roscoe St. Est. EC1 6E 218 (42Sb 91)
Roscoff Cl. HA8: Edg 25Sa 47
Rosea Apts. SW11 56Gb 111
(off Danvers Av.)
Roseacre RH8: Oxt 6L 211
Roseacre Cl. RM11: Horn 32Pd 77
Roseacre Cl. SM1: Sutt 75Eb 155
Roseacre Cl. TW17: Shep 71Q 150
Roseacre Cl. W13 43Ka 86
Roseacre Rd. DA16: Well 55Xc 117
Rose Acre Vs. HP3: Hem H 7E 4
Rose All. EC2 1J 225 (43Ub 91)
(off Bishopsgate)
Rose All. SE1 6E 224 (46Sb 91)
Rose & Crown Ct. EC2 2D 224 (44Sb 91)
(off Foster La.)
Rose & Crown M. TW7: Isle 53Ja 108
Rose & Crown Yd. SW1 7C 222 (46Lb 90)
Rose Apts. SE11 51Qb 112
(off St Agnes Pl.)
Roseary Cl. UB7: W Dray 49M 83
Rose Av. CR4: Mitc 67Hb 133
Rose Av. DA12: Grav'nd 10G 122
Rose Av. E18 26Kc 53
Rose Av. SM4: Mord 71Eb 155
Rose Bank CM14: B'wood 20Zd 41
Rosebank EN9: Walt A 5Gc 21
Rosebank KT18: Eps 86Sa 173
Rosebank SE20 66Xb 135
Rosebank SW6 52Ya 110
Rosebank W3 44Ta 87
Rosebank Av. HA0: Wemb 35Ha 66
Rosebank Av. RM12: Horn 36Ld 77
Rosebank Cl. N12 22Gb 49
Rosebank Cl. TW11: Tedd 65Ja 130
Rose Bank Cotts. GU22: Wok 94A 188
Rosebank Dr. GU24: Chob 2L 167
Rosebank Gdns. DA11: Nflt 10A 122
Rosebank Gdns. E3 40Bc 72
Rosebank Gdns. W3 44Ta 87
Rosebank Gdns. Nth. E3 40Bc 72
Rosebank Gro. E17 27Bc 52
Rosebank Rd. E17 30Dc 52
Rosebank Rd. W7 47Ga 86
Rosebank Vs. E17 28Cc 52
Rosebank Wlk. NW1 38Mb 70
Rosebank Wlk. SE18 49Nc 94
Rosebank Way W3 44Ta 87
Rose Bates Dr. NW9 28Qa 47
Rosebay Av. SL3: Wex 2M 81
Rosebay Dr. N17 26Vb 51
Rosebay Ho. E3 43Cc 92
(off Hawgood St.)
Roseberry Cl. RM14: Upm 30Vd 58
Roseberry Ct. WD17: Wat 11W 26
(not continuous)
Roseberry Dr. HA8: Edg 19Ra 29
Roseberry Dr. IG7: Chig 21Tc 54
Roseberry Gdns. BR6: Orp 76Uc 160
Roseberry Gdns. DA1: Dart 59Ld 119
Roseberry Gdns. N4 30Rb 51
Roseberry Gdns. RM14: Upm 30Ud 58
Roseberry Ho. KT17: Eps 84Ua 174
Roseberry Mans. N1 39Nb 70
(off Tapper Wlk.)
Roseberry M. SW2 58Nb 112
Roseberry Pl. E8 37Vb 71
Roseberry St. SE16 49Xb 91
Rosebery Av. CR7: Thor H 68Sb 135
Rosebery Av. DA15: Sidc 59Uc 116
Rosebery Av. E12 37Nc 74
Rosebery Av. EC1 6K 217 (42Qb 90)
Rosebery Av. HA2: Harr 35Aa 65
Rosebery Av. KT17: Eps 86Ua 174
Rosebery Av. KT3: N Mald 68Va 132
Rosebery Av. N17 26Wb 51
Rosebery Cl. SM4: Mord 72Za 154
Rosebery Ct. DA11: Grav'nd 10B 122
Rosebery Ct. EC1 6K 217 (42Qb 90)
(off Rosebery Av.)
Rosebery Ct. W1 6K 221 (46Kb 90)
(off Charles St.)
Rosebery Cres. GU22: Wok 92B 188
Rosebery Gdns. N8 29Nb 50
Rosebery Gdns. SM1: Sutt 77Db 155
Rosebery Gdns. W13 44Ja 86
Rosebery Ho. E2 40Yb 72
(off Sewardstone Rd.)
Rosebery Ind. Pk. N17 26Xb 51
Rosebery M. N10 26Lb 50
Rosebery Pl. NW7 24Va 48
Rosebery Rd. KT1: King T 68Ra 131
Rosebery Rd. KT18: Eps D 91Ta 193
Rosebery Rd. N10 26Lb 50
Rosebery Rd. RM17: Grays 51Ae 121
Rosebery Rd. SM1: Sutt 79Bb 155
Rosebery Rd. SW2 58Nb 112
Rosebery Rd. TW3: Houn 57Ea 108
Rosebery Rd. WD23: Bush 17Da 27
Roseberys, The KT18: Eps 86Ua 174
Rosebery Sq. EC1 6K 217 (42Qb 90)
(off Rosebery Av.)
Rosebery Sq. KT1: King T 68Ra 131
Rosebine Av. TW2: Twick 59Fa 108
Rosebriar Cl. GU22: Pyr 88J 169
Rosebriars CR3: Cat'm 92Ub 197
Rosebriars KT10: Esh 78Ea 152
(not continuous)
Rosebriar Wlk. WD24: Wat 8V 12
Rosebury Ct. CM13: Hut 16Fe 41
Rosebury Dr. GU24: Bisl 7E 166
Rosebury Rd. SW6 54Db 111
Rosebury Sq. IG8: Wfd G 24Qc 54
Rosebury Va. HA4: Ruis 33W 64
Rose Bush Ct. NW3 36Hb 69
Rose Bushes KT17: Eps D 88Xa 174
Rose Cotts. AL2: Brick W 2Da 13
Rose Cotts. BR2: Kes 83Lc 179
Rose Ct. E1 1K 225 (43Vb 91)
(off Wentworth St.)
Rose Ct. E8 38Vb 71
(off Richmond Rd.)
Rose Ct. HA0: Wemb 40Na 67
(off Vicars Bri. Cl.)
Rose Ct. HA2: Harr 33Ea 66
Rose Ct. IG10: Lough 13Hc 35
Rose Ct. N1 39Rb 71
(off Collin's Yd.)
Rose Ct. RM1: Rom 28Hd 56
Rose Ct. SE8 49Zb 92
Rosecourt Rd. CR0: C'don 72Pb 156
Rosecrest Ct. N15 28Vb 51
Rosecroft N14 19Nb 32
Rosecroft Av. NW3 34Cb 69
Rosecroft Cl. BR5: St M Cry 72Yc 161
Rosecroft Cl. TN16: Big H 90Pc 180
Rosecroft Ct. HA6: Nwood 23S 44
Rosecroft Dr. WD17: Wat 8U 12
Rosecroft Gdns. NW2 34Wa 68
Rosecroft Gdns. TW2: Twick 60Fa 108
Rosecroft Rd. UB1: S'hall 42Ca 85
Rosecroft Wlk. HA0: Wemb 36Ma 67
Rosecroft Wlk. HA5: Pinn 29Z 45
Rose Dale BR6: Farnb 75Rc 160
Rosedale CR3: Cat'm 95Ub 197
Rosedale KT21: Asht 90La 172
Rosedale Av. EN7: Chesh 1Vb 19
Rosedale Av. UB3: Hayes 43T 84
Rosedale Cl. AL2: Brick W 2Aa 13
Rosedale Cl. DA2: Dart 59Rd 119
Rosedale Cl. HA7: Stan 23Ka 46
Rosedale Cl. SE2 48Xc 95
Rosedale Cl. W7 47Ha 86
Rosedale Ct. HA1: Harr 35Ha 66
Rosedale Ct. N5 35Rb 71
Rosedale Dr. RM9: Dag 38Xc 75
Rosedale Gdns. RM9: Dag 38Xc 75
Rosedale Ho. N16 32Tb 71
Rosedale Pl. CR0: C'don 73Zb 158
Rosedale Rd. E7 36Lc 73
Rosedale Rd. KT17: Ewe 78Wa 154
Rosedale Rd. RM1: Rom 26Ed 56
Rosedale Rd. RM17: Grays 50Fe 99
Rosedale Rd. RM9: Dag 38Xc 75
Rosedale Rd. TW9: Rich 55Na 109
Rosedale Ter. W6 48Xa 88
(off Dalling Rd.)
Rosedene NW6 39Za 68
Rosedene Av. CR0: C'don 73Nb 156
Rosedene Av. SM4: Mord 71Cb 155
Rosedene Av. SW16 62Pb 134
Rosedene Av. UB6: G'frd 41Ca 85
Rosedene Ct. DA1: Dart 59Ld 119
Rosedene Ct. HA4: Ruis 32U 64
Rosedene End AL2: Chis G 8N 5
Rosedene Gdns. IG2: Ilf 28Qc 54
Rosedene M. UB8: Uxb 41L 83
Rosedene Ter. E10 33Dc 72
Rosedew Rd. W6 51Za 110
Rose End KT4: Wor Pk 74Za 154
Rosefield TN13: S'oaks 96Jd 202
Rosefield Cl. SM5: Cars 78Gb 155
Rosefield Gdns. E14 45Cc 92
Rosefield Gdns. KT16: Ott 79F 148
Rosefield Rd. TW18: Staines 63J 127
Roseford Ct. W12 47Za 88
(off Shepherd's Bush Grn.)
Rose Gdn. Cl. HA8: Edg 23Na 47
Rose Gdn. Cl. RM3: Hrld W 25Nd 57
Rose Gdns. RM15: S Ock 42Yd 98
Rose Gdns. TW13: Felt 61W 128
Rose Gdns. TW19: Stanw 59M 105
Rose Gdns. UB1: S'hall 42Ca 85
Rose Gdns. W5 48Ma 87
Rose Gdns. WD18: Wat 15W 26
Rosegarth DA13: Ist R 7A 144
Rosegate Ho. E3 40Bc 72
(off Hereford Rd.)
Rose Glen NW9 28Ta 47
Rose Glen RM7: Rush G 32Gd 76
ROSEHALL GREEN 5G 10
Rosehart M. W11 44Cb 89
Rose Hatch Av. RM6: Chad H 27Zc 55
Roseheath HP1: Hem H 2G 2
Roseheath Rd. TW4: Houn 57Ba 107
Rose Hill SM1: Sutt 75Db 155
Rosehill KT10: Clay 79Ja 152
Rosehill RH2: Reig 5M 207
ROSEHILL 73Eb 155
Rosehill TW12: Hamp 67Ca 129
Rosehill Av. GU21: Wok 8N 167
Rosehill Av. SM1: Sutt 74Eb 155
Rosehill Ct. HP1: Hem H 4J 3
Rosehill Ct. SL1: Slou 8L 81
Rosehill Ct. SM4: Mord 73Eb 155
(off St Helier Av.)
Rosehill Ct. Pde. SM4: Mord 73Eb 155
(off St Helier Av.)
Rosehill Farm Mdw. SM7: Bans 87Db 175
Rosehill Gdns. SM1: Sutt 75Db 155
Rosehill Gdns. UB6: G'frd 36Ha 66
Rosehill Gdns. WD5: Ab L 4S 12
Rose Hill Pk. W. SM1: Sutt 74Eb 155
Rosehill Rd. SW18 58Eb 111
Rosehill Rd. TN16: Big H 89Lc 179
ROSE HILL RDBT. 73Eb 155
Rosehip Cl. RM3: Rom 24Ld 57
Rose Ho. CR8: Kenley 88Vb 177
Rose Joan M. NW6 35Cb 69
Rosekey Ct. SE8 52Bc 114
(off Baildon St.)
Roseland Cl. N17 24Tb 51
Rose La. GU23: Rip 93L 189
Rose La. HP3: Hem H 8A 4
Rose La. RM6: Chad H 27Zc 55
Rose Lawn WD23: B Hea 18Ea 28
Roseleigh Av. N5 35Rb 71
Roseleigh Cl. TW1: Twick 58Ma 109
Rose Lipman Bldg. 39Ub 71
Rosemarie Cl. WD25: Wat 5V 12
Rosemary Av. EN2: Enf 11Ub 33
Rosemary Av. KT8: W Mole 69Ca 129
Rosemary Av. N3 26Db 49
Rosemary Av. N9 18Xb 33
Rosemary Av. RM1: Rom 27Hd 56
Rosemary Av. TW4: Houn 54Z 107
Rosemary Branch Theatre 39Tb 71
(off Rosemary St.)
Rosemary Cl. CR0: C'don 72Nb 156
Rosemary Cl. RH8: Oxt 5L 211
Rosemary Cl. RM15: S Ock 41Yd 98
Rosemary Cl. UB8: Hil 43Q 84
Rosemary Ct. DA8: Erith 52Hd 118
(off Furners Cl.)
Rosemary Ct. HA5: Hat E 24Ca 45
(off The Avenue)
Rosemary Ct. SE15 52Ub 113
Rosemary Ct. SE8 51Bc 114
(off Dorking Cl.)
Rosemary Dr. AL2: Lon C 8E 6
Rosemary Dr. E14 44Fc 93
Rosemary Dr. IG4: Ilf 29Mc 53
Rosemary Gdns. KT9: Chess 77Na 153
Rosemary Gdns. RM8: Dag 32Bd 75
Rosemary Gdns. SW14 55Sa 109
Rosemary Ga. KT10: Esh 78Ea 152
Rosemary Ho. N1 39Tb 71
(off Colville Est.)
Rosemary Ho. NW10 39Xa 68
(off Uffington Rd.)
Rosemary La. SW14 55Sa 109
Rosemary La. TN15: Hod S 81Fe 185
Rosemary La. TW20: Thorpe 69D 126
Rosemary M. SS17: Stan H 1M 101
Rosemary Rd. DA16: Well 53Vc 117
Rosemary Rd. SE15 52Vb 113
Rosemary Rd. SW17 62Eb 133
Rosemary St. N1 39Tb 71
Rosemary Works N1 39Tb 71
(off Branch Pl.)
Rose Mead EN6: Pot B 2Eb 17
Rosemead KT16: Chert 73K 149
Rosemead Av. CR4: Mitc 69Lb 134
Rosemead Av. HA9: Wemb 36Na 67
Rosemead Av. TW13: Felt 61V 128
Rosemead Cl. KT6: Surb 74Qa 153
Rosemead Cl. RH1: Redh 8M 207
Rosemead Gdns. CM13: Hut 14Fe 41
Rose Mdw. GU24: W End 5E 166
Rosemere Pl. BR2: Broml 70Gc 137
Rose M. N18 21Xb 51
Rosemont Av. N12 23Eb 49
Rosemont Ct. W3 46Ra 87
(off Rosemont Rd.)
Rosemont Rd. HA0: Wemb 39Na 67
Rosemont Rd. KT3: N Mald 69Sa 131
Rosemont Rd. NW3 37Eb 69
Rosemont Rd. TW10: Rich 58Na 109
Rosemont Rd. W3 45Ra 87
Rosemoor Ho. W13 46Ja 86
(off Broadway)
Rosemoor St. SW3 6F 227 (49Hb 89)
Rosemount SM6: W'gton 79Lb 156
(off Clarendon Rd.)
Rosemount Av. KT14: W Byf 85J 169
Rosemount Cl. IG8: Wfd G 23Pc 54
Rosemount Dr. BR1: Broml 70Pc 138
Rosemount Point KT14: W Byf 85J 169
(off Rosemount Av.)
Rosemount Point SE23 62Zb 136
Rosemount Rd. W13 44Ja 86
Rosenau Cres. SW11 53Hb 111
Rosenau Rd. SW11 53Gb 111
Rosenburg Rd. W3 47Ra 87
Rosen Cres. CM13: Hut 16Fe 41
Rosendale Rd. SE21 59Sb 113
Rosendale Rd. SE24 59Sb 113
Roseneath Av. N21 18Rb 33
Roseneath Cl. BR6: Chels 80Yc 161
Roseneath Ct. CR3: Cat'm 97Wb 197
Roseneath Pl. SW16 63Pb 134
(off Curtis Fld. Rd.)
Roseneath Rd. SW11 58Jb 112
Roseneath Wlk. EN1: Enf 14Ub 33
Rosen's Wlk. HA8: Edg 20Ra 29
Rosenthal Rd. SE6 58Dc 114
Rosenthorpe Rd. SE15 57Zb 114
Rose Pk. KT15: Wdhm 81G 168
Rose Pk. Cl. UB4: Yead 43Y 85
Rosepark Ct. IG5: Ilf 26Pc 54
Rose Pl. N8 30Nb 50
Rose Playhouse 6E 224 (46Sb 91)
(off Park St.)
Roserton St. E14 47Ec 92
Rosery, The CR0: C'don 72Zb 158
Rosery, The TW20: Thorpe 68G 126
Roses, The IG8: Wfd G 24Hc 53
Roses La. SL4: Wind 4B 102
Rose Sq. SW3 7C 226 (50Fb 89)
Rose Stapleton Ter. SE1 4J 231 (48Ub 91)
(off Page's Wlk.)
Rose St. DA11: Nflt 58De 121
Rose St. EC4 2C 224 (44Rb 91)
Rose St. WC2 4F 223 (45Nb 90)
(not continuous)
Rose Theatre Kingston 68Ma 131
Rose Theatre Sidcup 60Xc 117
Rosethorn Cl. SW12 59Mb 112
Rose Tree M. IG8: Wfd G 23Nc 54
Rosetree Pl. TW12: Hamp 66Ca 129
Rosetta Cl. SW8 52Nb 112
(off Kenchester Cl.)
Rosetta Ct. SE19 66Ub 135
Rosetti Ter. RM8: Dag 35Xc 75
(off Marlborough Rd.)
Rose Valley CM14: B'wood 20Yd 40
Roseveare Rd. SE12 63Lc 137
Rose Vw. KT15: Add 78L 149
Rose Vs. DA1: Dart 59Rd 119
Roseville N21 18Qb 32
(off The Green)
Roseville Av. TW3: Houn 57Ca 107
Roseville Rd. UB3: Harl 50W 84
Rosevine Rd. SW20 67Ya 132
Rose Wlk. BR4: W W'ck 75Ec 158
Rose Wlk. CR8: Purl 83Mb 176
Rose Wlk. KT5: Surb 71Ra 153
Rose Wlk. SL2: Slou 3F 80
Rose Wlk., The WD7: R'lett 8Ka 14
Rosewarne Cl. GU21: Wok 10L 167
Rose Way HA8: Edg 21Sa 47
Rose Way SE12 57Jc 115
Roseway SE21 58Tb 113
Rosewell Cl. SE20 66Xb 135
Rosewood DA2: Wilm 63Gd 140
Rosewood GU22: Wok 91C 188
Rosewood KT7: T Ditt 75Ja 152
Rosewood SM2: Sutt 82Eb 175
Rosewood Av. RM12: Horn 36Jd 76
Rosewood Av. UB6: G'frd 36Ja 66
Rosewood Cl. DA14: Sidc 62Yc 139
Rosewood Cl. RM15: S Ock 41Ae 99
Rosewood Ct. BR1: Broml 67Lc 137
Rosewood Ct. E11 35Fc 73
Rosewood Ct. HP1: Hem H 1G 2
Rosewood Ct. KT14: Byfl 84N 169
Rosewood Ct. KT2: King T 66Qa 131
Rosewood Ct. RM6: Chad H 29Yc 55
Rosewood Dr. EN2: Crew H 7Qb 18
Rosewood Dr. TW17: Shep 71P 149
Rosewood Gdns. SE13 54Ec 114
Rosewood Gro. SM1: Sutt 75Eb 155
Rosewood Ho. SW8 51Pb 112
Rosewood M. DA12: Grav'nd 3F 144
Rosewood Sq. W12 44Wa 88
Rosewood Way GU24: W End 5C 166
Rosewood Way SL2: Farn C 6G 60
Rosher Cl. E15 38Fc 73
Rosher Ho. DA11: Nflt 8B 122
ROSHERVILLE 8B 122
Rosherville Way DA11: Nflt 9A 122
Roshni Ho. SW17 65Gb 133
Rosie's Way RM15: S Ock 44Zd 99
Rosina Gro. DA10: Swans 60Be 121
Rosina St. E9 37Zb 72
Rosing Apts. BR2: Broml 70Lc 137
(off Homesdale Rd.)
Roskeen Ct. SW20 66Ya 132
Roskell Rd. SW15 55Za 110
Rosken Gro. SL2: Farn R 10F 60
Rosler Bldg. SE1 7D 224 (46Sb 91)
(off Ewer St.)
Roslin Ho. E1 45Zb 92
(off Brodlove La.)
Roslin Rd. W3 48Ra 87
Roslin Way BR1: Broml 64Jc 137
Roslyn Cl. CR4: Mitc 68Fb 133
Roslyn Ct. GU21: Wok 10L 167
Roslyn Gdns. RM2: Rom 26Hd 56
Roslyn Rd. N15 29Tb 51
Rosmead Rd. W11 45Ab 88
Rosoman Pl. EC1 5A 218 (42Qb 90)
Rosoman St. EC1 4A 218 (41Qb 90)
Rossall Cl. RM11: Horn 30Jd 56
Rossall Cres. NW10 41Pa 87
Ross Apts. E16 45Jc 93
(off Seagull La.)
Ross Av. RM8: Dag 32Bd 75
Ross Cl. HA3: Hrw W 24Ea 46
Ross Cl. UB3: Harl 49T 84
Ross Cl. UB5: N'olt 35Fa 66
Ross Ct. E5 35Xb 71
(off Napoleon Rd.)
Ross Ct. NW9 27Ua 48
Ross Ct. SW15 59Za 110
Ross Ct. W13 43Ka 86
(off Cleveland Rd.)
Rosscourt Mans. SW1 3A 228 (48Kb 90)
(off Buckingham Pal. Rd.)
Ross Cres. WD25: Wat 7W 12
Rossdale SM1: Sutt 78Gb 155
Rossdale Dr. N9 16Yb 34
Rossdale Dr. NW9 32Sa 67
Rossdale Rd. SW15 56Ya 110
Ross Dr. WD25: Wat 3X 13
Rosse Gdns. SE13 58Fc 115
Rosse M. SE3 53Kc 115
Rossendale Cl. EN2: Enf 8Rb 19
Rossendale St. E5 33Xb 71
Rossendale Way NW1 38Lb 70
Rossetti CR0: C'don 74Sb 157
(off Saffron Central Sq.)
Rossetti Ct. WC1 7D 216 (43Mb 90)
(off Ridgmount Pl.)
Rossetti Gdn. Mans. SW3 51Hb 111
(off Flood St.)
Rossetti Gdns. CR5: Coul 90Pb 176
Rossetti Ho. SW1 6E 228 (49Mb 90)
(off Erasmus St.)
Rossetti M. NW8 1C 214 (39Fb 69)
Rossetti Rd. SE16 50Xb 91
Rossetti Studios SW3 51Gb 111
(off Flood St.)
Ross Haven Pl. HA6: Nwood 25V 44
Ross Ho. E1 46Xb 91
(off Prusom St.)
Rossignol Gdns. SM5: Cars 75Jb 156
Rossindel Rd. TW3: Houn 57Ca 107
Rossington Av. WD6: Bore 10Na 15
Rossington Cl. EN1: Enf 10Xb 19
Rossington St. E5 33Wb 71
Rossiter Cl. SE19 66Sb 135
Rossiter Cl. SL3: L'ly 49A 82
Rossiter Flds. EN5: Barn 16Ab 30
Rossiter Gro. SW9 55Qb 112
Rossiter Rd. SW12 60Kb 112
Rossland Cl. DA6: Bex 57Dd 118
Rosslare Cl. TN16: Westrm 97Tc 200
Rosslyn Av. E4 19Hc 35
Rosslyn Av. EN4: E Barn 16Gb 31
Rosslyn Av. RM3: Hrld W 26Nd 57
Rosslyn Av. RM8: Dag 31Bd 75
Rosslyn Av. SW13 55Ua 110
Rosslyn Av. TW14: Felt 58W 106
Rosslyn Cl. BR4: W W'ck 76Hc 159
Rosslyn Cl. TW16: Sun 65U 128
Rosslyn Cl. UB3: Hayes 43T 84
Rosslyn Cres. HA1: Harr 28Ha 46
Rosslyn Cres. HA9: Wemb 35Na 67
Rosslyn Gdns. HA9: Wemb 34Na 67
Rosslyn Hill NW3 35Fb 69
Rosslyn Mans. NW6 38Eb 69
(off Goldhurst Ter.)
Rosslyn M. NW3 35Fb 69
Rosslyn Pk. KT13: Weyb 77T 150
Rosslyn Pk. M. NW3 36Fb 69
Rosslyn Rd. E17 28Ec 52
Rosslyn Rd. IG11: Bark 38Tc 74
Rosslyn Rd. TW1: Twick 58La 108
Rosslyn Rd. WD18: Wat 13X 27
Rossmere M. CM14: B'wood 18Yd 40
Rossmore Cl. EN3: Pond E 14Zb 34
Rossmore Cl. NW1 6E 214 (42Gb 89)
(off Rossmore Rd.)
Rossmore Ct. NW1 5F 215 (42Hb 89)
Rossmore Rd. NW1 6E 214 (42Gb 89)
Ross Pde. SM6: W'gton 79Kb 156
Ross Rd. DA1: Dart 58Jd 118
Ross Rd. KT11: Cobh 85Y 171
Ross Rd. SE25 69Tb 135
Ross Rd. SM6: W'gton 78Lb 156
Ross Rd. TW2: Whitt 60Da 107
Ross Wlk. SE27 62Tb 135
Ross Way E14 44Ac 92
Ross Way HA6: Nwood 21V 44
Ross Way SE9 55Nc 116
Rossway Dr. WD23: Bush 15Ea 28
Rosswood Gdns. SM6: W'gton 79Lb 156
Rostella Rd. SW17 63Fb 133
Rostrevor Av. N15 30Vb 51
Rostrevor Gdns. SL0: Iver H 40F 62
Rostrevor Gdns. UB2: S'hall 50Aa 85
Rostrevor Gdns. UB3: Hayes 46U 84
Rostrevor Mans. SW6 53Bb 111
(off Rostrevor Rd.)
Rostrevor M. SW6 53Bb 111
Rostrevor Rd. SW19 64Cb 133
Rostrevor Rd. SW6 53Bb 111
Roswell Apts. E3 43Bc 92
(off Joseph St.)
Roswell Cl. EN8: Chesh 2Ac 20
Rotary Ho. RM17: Grays 48Ee 99
Rotary St. SE1 3C 230 (48Rb 91)
Rothay NW1 3A 216 (41Kb 90)
(off Albany St.)
Rothbury Av. RM13: Rain 43Kd 97
Rothbury Cotts. SE10 49Gc 93
(off Maritius Rd.)
Rothbury Gdns. TW7: Isle 52Ja 108
Rothbury Rd. E9 38Bc 72
Rothbury Wlk. N17 24Wb 51
Roth Dr. CM13: Hut 19De 41
Rotheley Ho. E9 38Yb 72
(off Balcorne St.)
Rother Cl. WD25: Wat 6Y 13
Rotherfield Ct. N1 38Tb 71
(off Rotherfield St.)
Rotherfield Rd. EN3: Enf W 9Zb 20
Rotherfield Rd. SM5: Cars 77Jb 156
Rotherfield St. N1 38Sb 71
Rotherham Wlk. SE1 7B 224 (46Rb 91)
(off Nicholson St.)
Rotherhill Av. SW16 65Mb 134
ROTHERHITHE 47Yb 92
Rotherhithe Bus. Est. SE16 49Xb 91
Rotherhithe New Rd. SE16 50Xb 91
Rotherhithe Old Rd. SE16 49Zb 92
Rotherhithe Sands Film Studios 47Yb 92
(off Tunnel Rd.)
Rotherhithe Station (Overground) 47Yb 92
Rotherhithe St. SE16 47Yb 92
Rotherhithe Tunnel 46Zb 92
Rother Ho. SE15 56Xb 113
Rotherwick Hill W5 42Pa 87
Rotherwick Ho. E1 45Wb 91
(off Thomas More St.)
Rotherwick Rd. NW11 31Cb 69
Rotherwood Cl. SW20 67Ab 132
Rotherwood Rd. SW15 55Za 110
Rothery St. N1 39Rb 71
(off St Marys Path)
Rothery Ter. SW9 52Rb 113
(off Foxley Rd.)
Rothesay Av. SW20 68Ab 132
Rothesay Av. TW10: Rich 56Ra 109

Rothesay Av. UB6: G'frd 37Ea 66
(not continuous)
Rothesay Ct. SE11 51Qb 112
(off Harleyford St.)
Rothesay Ct. SE12 62Kc 137
Rothesay Ct. SE6 61Hc 137
(off Cumberland Pl.)
Rothesay Rd. SE25 70Tb 135
Rothley Ct. NW8 5B 214 (42Fb 89)
(off St John's Wood Rd.)
Rothsay Ct. KT13: Weyb 79T 150
Rothsay Rd. E7 38Lc 73
Rothsay St. SE1 3H 231 (48Ub 91)
Rothsay Wlk. E14 49Cc 92
(off Charnwood Gdns.)
Rothschild Ho. TW8: Bford 51Pa 109
Rothschild Rd. W4 49Sa 87
Rothschild St. SE27 63Rb 135
Roth Wlk. N7 33Pb 70
Rothwell Ct. HA1: Harr 29Ha 46
Rothwell Gdns. KT15: Add 79H 149
Rothwell Gdns. RM9: Dag 38Yc 75
Rothwell Ho. TW5: Hest 51Ca 107
Rothwell Rd. RM9: Dag 39Yc 75
Rothwell St. NW1 39Hb 69
Rotten Row NW3: Hampstead Heath 32Eb 69
Rotten Row SW7: Hyde Pk. 1C 226 (47Fb 89)
Rotterdam Dr. E14 48Ec 92
Rotunda, The RM7: Rom 29Fd 56
(off Yew Tree Gdns.)
Rotunda, The SW10 52Eb 111
Rotunda Cen., The 68Na 131
Rotunda Ct. BR1: Broml 64Kc 137
(off Burnt Ash La.)
Rouel Rd. SE16 49Wb 91
Rouge La. DA12: Grav'nd 10D 122
Rougemont Av. SM4: Mord 72Cb 155
Rough, The GU22: Wok 89F 168
Roughdown Av. HP3: Hem H 5J 3
Roughdown Rd. HP3: Hem H 5K 3
Roughdown Vs. Rd. HP3: Hem H 5J 3
ROUGHETS, THE 1M 209
Roughets La. RH1: Blet 1L 209
Roughlands GU22: Pyr 87G 168
Rough Rd. GU22: Wok 4G 186
Roughs, The HA6: Nwood 20U 26
ROUGHWAY 100De 205
Roughwood Cl. WD17: Wat 10U 12
Roughwood Cft. HP8: Chal G 15A 24
Roughwood La. HP8: Chal G 17A 24
Rounce La. GU24: W End 5B 166
Roundabout Ho. HA6: Nwood 25W 44
Roundacre SW19 61Za 132
Round Ash Way DA3: Hartl 72Ae 165
Roundaway Rd. IG5: Ilf 25Pc 54
Roundburrow Cl. CR6: W'ham 89Wb 177
ROUND BUSH 10Fa 14
Roundbush La. WD25: A'ham 10Ea 14
Roundel Cl. SE4 56Bc 114
Round Gro. CR0: C'don 73Zb 158
Roundhay Cl. SE23 61Zb 136
Roundhedge Way EN2: Enf 10Pb 18
Round Hill SE26 61Yb 136
(not continuous)
Roundhill GU22: Wok 91D 188
Roundhill Dr. EN2: Enf 14Pb 32
Roundhill Dr. GU22: Wok 90D 168
Roundhills EN9: Walt A 6Gc 21
Roundhill Way KT11: Cobh 83Da 171
Roundhouse, The 38Jb 70
Round Ho. Ct. EN8: Chesh 1Zb 20
Roundhouse La. E20 37Ec 72
(off International Way)
Roundlyn Gdns. BR5: St M Cry 70Xc 139
Roundmead Av. IG10: Lough 12Qc 36
Roundmead Cl. IG10: Lough 13Qc 36
Roundmoor Dr. EN8: Chesh 1Ac 20
Round Oak Rd. KT13: Weyb 77P 149
ROUNDSHAW 80Nb 156
Roundshaw Downs Local Nature
Reserve 81Pb 176
Round St. DA13: Sole S 9E 144
ROUND STREET 10F 144
Roundtable Rd. BR1: Broml 62Hc 137
Roundthorn Way GU21: Wok 8K 167
Roundtree Rd. HA0: Wemb 36Ka 66
Roundway TN16: Big H 88Lc 179
Roundway TW20: Egh 64E 126
Roundway, The KT10: Clay 79Ha 152
Roundway, The N17 25Sb 51
Roundway, The WD18: Wat 16V 26
Roundways HA4: Ruis 34V 64
Round Wood WD4: K Lan 9N 3
Roundwood BR7: Chst 68Rc 138
Roundwood Av. CM13: Hut 18Ce 41
Roundwood Av. UB11: Stock P 46S 84
Roundwood Cl. HA4: Ruis 31T 64
Roundwood Ct. E2 41Zb 92
Roundwood Gro. CM13: Hut 17De 41
Roundwood Lake CM13: Hut 17De 41
Roundwood Rd. NW10 37Va 68
Roundwood Vw. SM7: Bans 87Za 174
Roundwood Way SM7: Bans 87Za 174
Rounton Cl. WD17: Wat 10V 12
Rounton Rd. E3 42Cc 92
Rounton Rd. EN9: Walt A 5Gc 21
Roupell Ho. KT2: King T 66Pa 131
(off Florence Rd.)
Roupell Rd. SW2 60Pb 112
Roupell St. SE1 7A 224 (46Qb 90)
Rousden St. NW1 38Lb 70
Rousebarn La. WD3: Chan C 10P 11
Rousebarn La. WD3: Crox G 10P 11
Rouse Cl. KT13: Weyb 77V 150
Rouse Ct. SL9: Ger X 29B 42
Rouse Gdns. SE21 63Ub 135
Rous Rd. IG9: Buck H 18Nc 36
Routemaster Cl. E13 41Kc 93
Routh Ct. TW14: Bedf 60T 106
Routh Rd. SW18 59Gb 111
Routh St. E6 43Pc 94
Rover Av. IG6: Ilf 23Vc 55
Rover Ho. N1 1J 219 (39Ub 71)
(off Whitmore Est.)
Row, The DA3: Nw A G 75Be 165
Rowallan Rd. SW6 52Ab 110
Rowallen Pde. RM8: Dag 32Yc 75
Rowan N10 26Kb 50
Rowan Av. E4 23Bc 52
Rowan Av. NW9 24Va 48
Rowan Av. TW20: Egh 64E 126
Rowan Cl. AL2: Brick W 3Ca 13
Rowan Cl. AL4: St A 2J 7
Rowan Cl. HA0: Wemb 34Ja 66
Rowan Cl. HA7: Stan 23Ha 46
Rowan Cl. HP2: Hem H 2P 3
Rowan Cl. IG1: Ilf 36Tc 74
Rowan Cl. KT3: N Mald 68Ua 132
Rowan Cl. RH2: Reig 8L 207
Rowan Cl. SM7: Bans 87Za 174
Rowan Cl. SW16 67Lb 134
Rowan Cl. TW15: Ashf 63M 127
Rowan Cl. W5 47Na 87
Rowan Cl. WD7: Shenl 4Na 15
Rowan Ct. E13 40Kc 73
(off High St.)
Rowan Ct. SE15 52Vb 113
(off Garnies Cl.)
Rowan Ct. SW11 58Hb 111
Rowan Ct. WD6: Bore 14Qa 29
Rowan Cres. DA1: Dart 60Ld 119
Rowan Cres. SW16 67Lb 134
Rowan Dr. NW9 27Wa 48
Rowan Gdns. CR0: C'don 76Vb 157
Rowan Gdns. SL0: Iver H 40F 62
Rowan Grn. KT13: Weyb 77T 150
Rowan Grn. E. CM13: B'wood 21Be 59
Rowan Grn. W. CM13: B'wood 20Be 41
Rowan Gro. CR5: Coul 93Kb 196
Rowan Gro. RM15: Avel 45Sd 98
Rowan Ho. BR2: Broml 68Gc 137
Rowan Ho. DA14: Sidc 62Vc 139
Rowan Ho. E3 39Bc 72
(off Hornbeam Sq.)
Rowan Ho. IG1: Ilf 36Tc 74
Rowan Ho. SE16 47Zb 92
(off Woodland Cres.)
Rowanhurst Dr. SL2: Farn C 6G 60
Rowan Lodge W8 48Db 89
(off Chantry Sq.)
Rowan Mead KT20: Tad 91Xa 194
Rowan Pl. UB3: Hayes 45V 84
Rowan Rd. BR8: Swan 69Fd 140
Rowan Rd. DA7: Bex 55Ad 117
Rowan Rd. SW16 68Lb 134
Rowan Rd. TW8: Bford 52Ka 108
Rowan Rd. UB7: W Dray 49M 83
Rowan Rd. W6 49Za 88
Rowans, The EN9: Walt A 7Lc 21
(within Woodbine Cl. Cvn. Pk.)
Rowans, The GU22: Wok 90A 168
Rowans, The HP1: Hem H 2J 3
Rowans, The N13 20Rb 33
Rowans, The RM15: Avel 46Sd 98
Rowans, The TW16: Sun 64V 128
Rowans Cl. DA3: Lfield 68Zd 143
Rowans Tenpin 33Qb 70
Rowans Way IG10: Lough 14Pc 36
Rowan Ter. SW19 66Ab 132
Rowan Ter. W6 49Za 88
Rowantree Cl. N21 18Tb 33
Rowantree Rd. EN2: Enf 12Rb 33
Rowantree Rd. N21 18Tb 33
Rowan Wlk. AL10: Hat 3C 8
Rowan Wlk. BR2: Broml 75Pc 160
Rowan Wlk. EN5: New Bar 15Db 31
Rowan Wlk. N19 33Lb 70
Rowan Wlk. N2 29Eb 49
Rowan Wlk. RM11: Horn 28Md 57
Rowan Wlk. W10 42Ab 88
Rowan Way RM15: S Ock 42Yd 98
Rowan Way RM6: Chad H 27Yc 55
Rowan Way SL2: Slou 3F 80
Rowanwood Av. DA15: Sidc 60Wc 117
Rowanwood M. EN2: Enf 12Rb 33
Rowben Cl. N20 18Db 31
Rowberry Cl. SW6 52Ya 110
Rowbourne Pl. EN6: Cuff 1Mb 18
Rowcroft HP1: Hem H 3G 2
Rowcross St. SE1 7K 231 (50Vb 91)
Rowdell Rd. UB5: N'olt 39Ca 65
Rowden Pde. E4 23Cc 52
(off Chingford Rd.)
Rowden Pk. Gdns. E4 24Cc 52
Rowden Rd. BR3: Beck 67Ac 136
Rowden Rd. E4 23Dc 52
Rowden Rd. KT19: Ewe 77Ra 153
Rowditch La. SW11 54Jb 112
Rowdon Av. NW10 38Xa 68
Row Dow TN14: Otf 88Md 183
Row Dow La. TN15: Knat 85Md 183
Rowdown Cres. CR0: New Ad 81Fc 179
Rowdowns Rd. RM9: Dag 39Bd 75
Rowe Gdns. IG11: Bark 40Vc 75
Rowe Ho. E9 37Yb 72
Rowe La. E9 36Yb 72
Rowe La. GU24: Pirb 6E 186
Rowena Cres. SW11 54Gb 111
Rowenhurst Mans. NW6 37Eb 69
(off Canfield Gdns.)
Rowe Wlk. HA2: Harr 34Ca 65
Rowfant Rd. SW17 60Jb 112
Rowhedge CM13: B'wood 20Ce 41
Rowhill KT15: Add 79H 149
ROWHILL 79H 149
Rowhill Rd. BR8: Hext 65Hd 140
Rowhill Rd. DA2: Wilm 65Hd 140
Rowhill Rd. E5 35Xb 71
Rowhurst Av. KT15: Add 79K 149
Rowhurst Av. KT22: Lea 89Ha 172
Rowington Cl. W2 43Db 89
Rowland Av. HA3: Kenton 27La 46
Rowland Cl. SL4: Wind 5B 102
Rowland Ct. E16 42Hc 93
Rowland Cres. IG7: Chig 21Uc 54
Rowland Gro. SE26 62Xb 135
(not continuous)
Rowland Hill Almshouses TW15: Ashf 64Q 128
(off Feltham Hill Rd.)
Rowland Hill Av. N17 24Sb 51
Rowland Hill Ho. SE1 1B 230 (47Rb 91)
Rowland Hill St. NW3 36Gb 69
Rowland Pl. CR8: Purl 88Qb 176
Rowland Pl. HA6: Nwood 24U 44
Rowlands Av. HA5: Hat E 22Ca 45
Rowlands Cl. EN8: Chesh 2Zb 20
Rowlands Cl. N6 30Jb 50
Rowlands Cl. NW7 24Wa 48
Rowlands Ct. EN8: Chesh 2Zb 20
(off Rowlandsfields)
Rowlandsfields EN8: Chesh 1Zb 20
Rowlands Rd. RM8: Dag 33Bd 75
Rowland Wlk. RM4: Have B 20Gd 38
Rowland Way SW19 67Db 133
Rowland Way TW15: Ashf 66T 128
Rowlatt Cl. DA2: Wilm 63Ld 141
Rowlatt Ct. AL1: St A 1C 6
(off Hillside Rd.)
Rowlatt Dr. AL3: St A 4N 5
Rowlatt Rd. DA2: Wilm 63Ld 141
Rowley Av. DA15: Sidc 59Xc 117
Rowley Cl. GU22: Pyr 88K 169
Rowley Cl. HA0: Wemb 38Pa 67
Rowley Cl. WD19: Wat 16Aa 27
Rowley Ct. CR3: Cat'm 94Tb 197
Rowley Ct. EN1: Enf 15Ub 33
(off Wellington Rd.)
Rowley Gdns. EN8: Chesh 1Zb 20
Rowley Gdns. N4 31Sb 71
ROWLEY GREEN 14Va 30
Rowley Green Common Nature Reserve 14Va 30
Rowley Grn. Rd. EN5: Ark 15Va 30
Rowley Ho. SE8 50Cc 92
(off Watergate St.)
Rowley Ind. Pk. W3 48Ra 87
Rowley La. EN5: Ark 14Ua 30
Rowley La. SL3: Wex 9N 61
Rowley La. WD6: Bore 11Ta 29
Rowley Lane Sports Ground 14Ua 30
Rowley Rd. N15 29Sb 51
Rowley Rd. RM16: Ors 3C 100
Rowley Way NW8 39Db 69
Rowlheys Pl. UB7: W Dray 48N 83
Rowlls Rd. KT1: King T 69Pa 131
Rowlock Ho. UB7: Yiew 45M 83
Rowmarsh Cl. DA11: Nflt 62Fe 143
Rowney Gdns. RM9: Dag 37Yc 75
Rowney Rd. RM9: Dag 37Xc 75
Rowntree Clifford Cl. E13 42Jc 93
Rowntree Cl. NW6 37Cb 69
Rowntree M. E17 25Bc 52
Rowntree Path SE28 46Xc 95
Rowntree Rd. TW2: Twick 60Ga 108
Rowse Cl. E15 39Ec 72
Rowsley Av. NW4 27Ya 48
Rowstock Gdns. N7 36Mb 70
Rowton Rd. SE18 52Sc 116
Row Town KT15: Add 80H 149
ROW TOWN 79J 149
Rowzill Rd. BR8: Hext 65Hd 140
Roxborough Av. HA1: Harr 31Fa 66
Roxborough Av. TW7: Isle 52Ha 108
Roxborough Hgts. HA1: Harr 30Ga 46
(off College Rd.)
Roxborough Pk. HA1: Harr 31Ga 66
Roxborough Rd. HA1: Harr 29Fa 46
Roxbourne Cl. UB5: N'olt 37Z 65
ROXBOURNE COMPLEX 33Da 65
Roxbourne Pk. Miniature Railway 33Z 65
Roxburgh Av. RM14: Upm 34Sd 78
Roxburghe Mans. W8 47Db 89
(off Kensington Ct.)
Roxburgh Pl. BR1: Broml 67Nc 138
Roxburgh Rd. SE27 64Rb 135
Roxburn Way HA4: Ruis 35W 64
Roxby Pl. SW6 51Cb 111
ROXETH 33Fa 66
Roxeth Ct. TW15: Ashf 64Q 128
Roxeth Grn. Av. HA2: Harr 34Da 65
Roxeth Grn. Av. UB5: N'olt 36Ca 65
Roxeth Gro. HA2: Harr 35Da 65
Roxeth Hill HA2: Harr 33Fa 66
Roxford Cl. TW17: Shep 71U 150
Roxford Ho. E3 42Dc 92
(off Devas St.)
Roxley Rd. SE13 58Dc 114
Roxton Gdns. CR0: Addtn 78Cc 158
Roxwell NW1 37Kb 70
(off Hartland Rd.)
Roxwell Cl. SL1: Slou 6C 80
Roxwell Gdns. CM13: Hut 15Ee 41
Roxwell Ho. IG10: Lough 17Nc 36
Roxwell Rd. IG11: Bark 40Wc 75
Roxwell Rd. W12 47Wa 88
Roxwell Trad. Pk. E10 31Ac 72
Roxwell Way IG8: Wfd G 24Lc 53
Roxy Av. RM6: Chad H 31Yc 75
Royal Academy of Arts
(Burlington House) 5B 222 (45Lb 90)
Royal Academy of Music
Mus. 6J 215 (42Jb 90)
(off Marylebone Rd.)
Royal Air Force Memorial 7G 223 (46Nb 90)
Royal Albert Hall 2B 226 (47Fb 89)
ROYAL ALBERT RDBT. 45Nc 94
(on Royal Albert Way)
Royal Albert Station (DLR) 45Nc 94
Royal Albert Way E16 45Mc 93
Royal Anglian Way RM8: Dag 32Ad 75
Royal Arc. W1 5B 222 (45Lb 90)
(off Old Bond St.)
Royal Archer SE14 52Zb 114
(off Egmont St.)
ROYAL ARSENAL WEST 48Rc 94
Royal Ascot Golf Course 7A 124
Royal Av. EN8: Walt C 5Ac 20
Royal Av. KT4: Wor Pk 75Ua 154
Royal Av. SW3 7F 227 (50Hb 89)
Royal Av. Ho. SW3 7F 227 (50Hb 89)
(off Royal Av.)
Royal Belgrave Ho. SW1 6A 228 (49Kb 90)
(off Hugh St.)
Royal Blackheath Golf Course 59Pc 116
Royal Borough of Kensington & Chelsea
Cemetery, The 46Ha 86
Royal Botanic Gdns. Kew 53Na 109
Royal Brass Foundry 48Rc 94
ROYAL BROMPTON HOSPITAL
Sydney St. 7D 226 (50Gb 89)
ROYAL BROMPTON HOSPITAL
(OUTPATIENTS) 7C 226 (50Fb 89)
Royal Carriage M. SE18 48Rc 94
Royal Cir. SE27 62Qb 134
Royal Cl. BR6: Farnb 77Rc 160
Royal Cl. IG3: Ilf 31Wc 75
Royal Cl. KT4: Wor Pk 75Ua 154
Royal Cl. N16 32Ub 71
Royal Cl. SE8 51Bc 114
Royal Cl. SW19 62Za 132
Royal Cl. UB8: Hil 44P 83
Royal College of Music 3B 226 (48Fb 89)
Royal College of Nursing 2A 222 (44Kb 90)
(off Dean's M.)
Royal College of Physicians
Mus. 5A 216 (42Kb 90)
Royal Coll. St. NW1 38Lb 70
Royal Connaught Apts. E16 46Mc 93
(off Connaught Rd.)
Royal Connaught Dr. WD23: Bush 14Ba 27
Royal Connaught Pk. 14Ba 27
Royal Ct. EC3 3G 225 (44Tb 91)
(off Cornhill)
Royal Ct. EN1: Enf 16Ub 33
Royal Ct. HA4: Ruis 30W 44
Royal Ct. HP3: Hem H 5N 3
Royal Ct. SE16 48Bc 92
Royal Ct. SE9 60Pc 116
Royal Ct. WD18: Wat 14U 26
Royal Ct. WD23: Bush 15Aa 27
Royal Courts of Justice 3J 223 (44Pb 90)
(off Strand)
Royal Court Theatre London 6H 227 (49Jb 90)
(off Sloane Sq.)
Royal Cres. HA4: Ruis 35Aa 65
Royal Cres. IG2: Ilf 30Tc 54
Royal Cres. W11 46Za 88
Royal Cres. M. W11 46Za 88
Royal Crest Av. E16 47Kc 93
Royal Docks Rd. E6 44Rc 94
Royal Docks Rd. IG11: Bark 42Rc 94
Royal Dr. KT18: Tatt C 90Xa 174
Royal Dr. N11 22Jb 50
(not continuous)
Royal Duchess M. SW12 59Kb 112
Royal Earlswood Pk. RH1: Redh 9A 208
Royale Leisure Pk. W3 42Qa 87
Royal Engineers Way NW7 23Ab 48
Royal Epping Forest Golf Course 17Gc 35
Royal Exchange 3G 225 (44Tb 91)
Royal Exchange Av. EC3 3G 225 (44Tb 91)
(off Threadneedle St.)
Royal Exchange Bldgs. EC3 3G 225 (44Tb 91)
(off Threadneedle St.)
Royal Festival Hall 7J 223 (46Pb 90)
Royal Free Ct. SL4: Wind 3H 103
(off Bachelors Acre)
ROYAL FREE HOSPITAL 36Gb 69
Royal Gdns. W7 48Ja 86
Royal George M. SE5 56Tb 113
Royal Gunpowder Mills 4Dc 20
Royal Herbert Pavilions SE18 53Pc 116
Royal Hill SE10 52Ec 114
Royal Hill Ct. SE10 52Ec 114
(off Greenwich High St.)
Royal Hill Pk. Development RH1: Redh 7B 208
Royal Holloway (University of London)
Egham Hill 5P 125
Royal Holloway (University of London)
Gower Street 7E 216 (43Mb 90)
(off Gower St.)
Royal Holloway (University of London)
Sports Cen. 66A 126
Royal Horticultural Society Cotts.
GU23: Wis 88N 169
Royal Hospital Chelsea 50Jb 90
Royal Hospital Chelsea Great Hall 50Hb 89
Royal Hospital Chelsea Mus. 7H 227 (50Jb 90)
ROYAL HOSPITAL FOR
NEURO-DISABILITY 58Ab 110
Royal Hospital Rd. SW3 51Hb 111
Royal Household Golf Course 3J 103
Royal Institution 5B 222 (45Lb 90)
Royal Jubilee Ct. RM2: Rom 27Jd 56
Royal La. UB7: Yiew 44P 83
Royal La. UB8: Hil 43P 83
Royal Langford Apts. NW6 40Db 69
(off Greville Rd.)
Royal London Bldgs. SE15 51Xb 113
(off Old Kent Rd.)
Royal London Est., The N17 23Xb 51
ROYAL LONDON HOSPITAL, THE 43Xb 91
Royal London Hospital Archives & Mus. 43Xb 91
(off Newark St.)
ROYAL LONDON HOSPITAL FOR
INTEGRATED MEDICINE 6G 217 (42Nb 90)
Royal London Ind. Est. NW10 40Ta 67
ROYAL MARSDEN HOSPITAL, THE
Fulham 7C 226 (50Fb 89)
ROYAL MARSDEN HOSPITAL, THE
Sutton 82Eb 175
Royal Mausoleum Windsor 5J 103
Royal M. KT8: E Mos 69Ga 130
Royal M. SL4: Wind 3H 103
Royal M. SW1 3A 228 (48Kb 90)
Royal Mews, The 2A 228 (47Kb 90)
Royal Mid-Surrey Golf Course 55Ma 109
Royal Mint Ct. EC3 5K 225 (45Vb 91)
Royal Mint Pl. E1 45Wb 91
Royal Mint St. E1 45Vb 91
ROYAL NAT. ORTHOPAEDIC HOSPITAL Central
London Outpatient Dept. 6A 216 (42Kb 90)
ROYAL NAT. ORTHOPAEDIC HOSPITAL
Stanmore 19La 28
ROYAL NATIONAL THROAT,
NOSE & EAR HOSPITAL
Expected Close 2020 3H 217 (41Pb 90)
ROYAL NATIONAL THROAT, NOSE & EAR
HOSPITAL New Site 6D 216 (42Mb 90)
Royal Naval Pl. SE14 52Bc 114
Royal Oak Cl. IG10: Lough 14Nc 36
Royal Oak Cotts. HP1: Hem H 1M 3
(off High St.)
Royal Oak Ct. N1 3H 219 (41Ub 91)
(off Pitfield St.)
Royal Oak Hill TN14: Knock 90Vc 181
Royal Oak M. TW11: Tedd 64Ja 130
Royal Oak Pl. SE22 58Xb 113
Royal Oak Rd. DA6: Bex 57Bd 117
(not continuous)
Royal Oak Rd. E8 37Xb 71
Royal Oak Rd. GU21: Wok 10N 167
Royal Oak Station (Underground) 43Db 89
Royal Oak Ter. DA12: Grav'nd 10E 122
(off Parrock Rd.)
Royal Oak Yd. SE1 2H 231 (47Ub 91)
Royal Observatory Greenwich 52Gc 115
Royal Opera Arc. SW1 6D 222 (46Mb 90)
Royal Opera House 3G 223 (44Nb 90)
Royal Opera House, Bob & Tamar Manoukian
Production Workshop 50Sd 98
Royal Orchard Cl. SW18 59Ab 110
Royal Pde. BR7: Chst 66Sc 138
Royal Pde. RM10: Dag 37Dd 76
Royal Pde. SE3 54Hc 115
Royal Pde. SW6 52Ab 110
Royal Pde. TW9: Kew 53Qa 109
(off Station App.)
Royal Pde. W5 41Na 87
Royal Pde. M. BR7: Chst 66Sc 138
(off Royal Pde.)
Royal Pde. M. SE3 54Hc 115
(off Royal Pde.)
Royal Pier M. DA12: Grav'nd 8E 122
Royal Pier Rd. DA12: Grav'nd 8D 122
Royal Pl. SE10 52Ec 114
Royal Quarter KT2: King T 67Na 131
Royal Quay UB9: Hare 24J 43
Royal Quay Rd. E16 45Rc 94
Royal Rd. AL1: St A 2F 6
Royal Rd. DA14: Sidc 62Zc 139
Royal Rd. DA2: Hawl 64Qd 141
Royal Rd. E16 44Mc 93
Royal Rd. SE17 51Rb 113
Royal Rd. TW11: Tedd 64Fa 130
Royal Route HA9: Wemb 35Pa 67
Royal St. SE1 3J 229 (48Pb 90)
Royal Thames Wlk. KT7: T Ditt 72Ja 152
Royal Twr. Lodge E1 45Wb 91
(off Cartwright St.)
Royalty Mans. W1 3D 222 (44Mb 90)
(off Meard St.)
Royalty M. W1 3D 222 (44Mb 90)
Royalty Studios W11 44Ab 88
(off Lancaster Rd.)
Royal Veterinary Coll. Camden Town 39Mb 70
Royal Veterinary Coll., The Hawkshead
Campus, Boltons Park Site 1Cb 17
Royal Veterinary Coll., The Hawkshead Campus,
Hawkshead La. 10F 8
Royal Victoria Dock E16 45Kc 93
Royal Victoria Gdns. SE16 49Ac 92
(off Whiting Way)
Royal Victoria Patriotic Bldg. SW18 58Fb 111
Royal Victoria Pl. E16 46Kc 93
Royal Victoria Sq. E16 45Kc 93
Royal Victoria Station (DLR) 45Jc 93
Royal Victor Pl. E3 40Zb 72
Royal Wlk. SM6: W'gton 75Kb 156
Royal Wlk. TW17: Shep 74Q 150
Royal Westminster Lodge
SW1 5D 228 (49Mb 90)
(off Elverton St.)
Royal Wharf E16 47Kc 93
Royal Wharf Wlk. E16 47Kc 93
Royal Wimbledon Golf Course 64Xa 132
Royal Windsor Racecourse 1D 102
Royal Windsor Way SL4: Wind 3F 102
Royce Av. NW9 26Wa 48
Royce Gro. WD25: Wat 6V 12
Roycraft Av. IG11: Bark 40Vc 75
Roycroft Av. IG11: Bark 40Vc 75
Roycroft Cl. E18 25Kc 53
Roycroft Cl. SW2 60Qb 112
Roydene Rd. SE18 51Uc 116
Roydon Cl. IG10: Lough 17Nc 36
Roydon Cl. SW11 54Hb 111
Roydon Ct. KT12: Hers 77W 150
Roydon Ct. TW20: Egh 65F 126
Royds La. CM14: Kel H 11Td 40
Roy Gdns. IG2: Ilf 28Uc 54
Roy Gro. TW12: Hamp 65Da 129
Royle Bldg. N1 2D 218 (40Sb 71)
(off Wenlock Rd.)
Royle Cl. RM2: Rom 29Kd 57
Royle Cl. SL9: Chal P 24B 42
Royle Cres. W13 42Ja 86
Royley Ho. EC1 5E 218 (42Sb 91)
(off Old St.)
Roymount Ct. TW2: Twick 62Ga 130
Roy Richmond Way KT19: Eps 82Ua 174
Roy Rd. HA6: Nwood 24V 44
Roy Sq. E14 45Ac 92
Royston Av. E4 22Cc 52
Royston Av. KT14: Byfl 84N 169
Royston Av. SM1: Sutt 76Fb 155
Royston Av. SM6: Bedd 77Mb 156
Royston Cl. KT12: Walt T 74W 150
Royston Cl. TW5: Cran 53X 107
Royston Ct. E13 39Jc 73
(off Stopford Rd.)
Royston Ct. KT10: Hin W 75Ha 152
Royston Ct. SE24 58Sb 113
Royston Ct. TW9: Kew 53Pa 109
Royston Ct. W8 46Cb 89
(off Kensington Chu. St.)
Royston Gdns. IG1: Ilf 30Mc 53
Royston Gro. HA5: Hat E 23Ca 45
Royston Ho. N11 21Hb 49
Royston Ho. SE15 51Xb 113
(off Friary Est.)
Royston Pde. IG1: Ilf 30Mc 53
Royston Pk. KT14: Byfl 84N 169
Royston Pk. Rd. HA5: Hat E 23Ba 45
Royston Rd. AL1: St A 3F 6
Royston Rd. DA1: Cray 58Hd 118
Royston Rd. KT14: Byfl 84N 169
Royston Rd. RM3: Hrld W 24Qd 57
Royston Rd. SE20 67Zb 136
Royston Rd. TW10: Rich 57Na 109
Roystons, The KT5: Surb 71Ra 153
Royston St. E2 40Yb 72
Royston Way SL1: Slou 3B 80
Rozel Ct. N1 39Ub 71
Rozel Rd. SW4 55Lb 112
Rozel Ter. CR0: C'don 76Sb 157
(off Church Rd.)
RQ33 SW18 56Cb 111
Rubastic Rd. UB2: S'hall 48Y 85
Rubeck Cl. RH1: Redh 4B 208
Rubens Ct. WD25: Wat 8Z 13
Rubens Gdns. SE22 59Wb 113
(off Lordship La.)
Rubens Pl. SW4 56Nb 112
Rubens Rd. UB5: N'olt 40Y 65
Rubens St. SE6 61Bc 136
Rubicon Ct. N1 39Nb 70
Rubin Pl. EN3: Enf L 9Cc 20
Rubus Cl. GU24: W End 5C 166
Ruby Cl. E5 34Zb 72
Ruby Cl. SL1: Slou 7E 80
Ruby Ct. E15 39Ec 72
(off Warton Rd.)
Ruby Ct. RM8: Dag 32Cd 76
(off Emerald Gdns.)
Ruby M. N13 22Nb 50
Ruby Rd. E17 27Cc 52
Ruby St. NW10 38Sa 67
Ruby St. SE15 51Xb 113
Ruby Triangle SE15 51Xb 113
Ruby Tuesday Dr. DA1: Dart 54Pd 119
Ruby Way NW9 25Va 48
Ruckholt Cl. E10 34Dc 72
Ruckholt Rd. E10 35Cc 72
Rucklers La. WD4: K Lan 10J 3
RUCKLERS LANE 9M 3
Rucklidge Av. NW10 40Va 68
Rucklidge Pas. NW10 40Va 68
(off Rucklidge Av.)
Rudall Cres. NW3 35Fb 69
Rudbeck Ho. SE15 52Wb 113
(off Peckham Pk. Rd.)
Ruddington Cl. E5 35Ac 72
Ruddlesway SL4: Wind 3B 102
Ruddock Cl. HA8: Edg 24Sa 47
Ruddstreet Cl. SE18 49Rc 94
Ruddy Way NW7 23Va 48
Ruden Way KT17: Eps D 88Xa 174
Rudge Ho. SE16 48Wb 91
(off Jamaica Rd.)
Rudge Ri. KT15: Add 78H 149
Rudgwick Ct. SE18 49Nc 94
(off Woodville St.)
Rudgwick Ter. NW8 39Gb 69
Rudland Rd. DA7: Bex 55Dd 118
Rudloe Rd. SW12 59Lb 112
Rudolf Pl. SW8 51Nb 112
Rudolph Rd. E13 40Hc 73
Rudolph Rd. NW6 40Db 69
Rudolph Rd. WD23: Bush 16Ca 27
Rudstone Ho. E3 41Dc 92
(off Bromley High St.)
Rudsworth Cl. SL3: Coln 53F 104
Rudyard Ct. CM14: W'ley 22Xd 58
Rudyard Ct. SE1 2G 231 (47Tb 91)
(off Long La.)

Rudyard Gro. NW7 23Sa 47
Rue de St Lawrence EN9: Walt A 6Ec 20
Ruegg Ho. SE18 51Qc 116
(off Woolwich Comn.)
Ruffets Wood DA12: Grav'nd 5E 144
Ruffetts, The CR2: Sels 80Xb 157
Ruffetts Cl. CR2: Sels 80Xb 157
Ruffetts Way KT20: Tad 90Ab 174
Ruffle Cl. UB7: W Dray 47N 83
Rufford Cl. HA3: Kenton 30Ja 46
Rufford Cl. WD17: Wat 9V 12
Rufford St. N1 39Nb 70
Rufford St. M. N1 38Nb 70
Rufford Twr. W3 46Ra 87
Rufforth Ct. NW9 25Ua 48
(off Pageant Av.)
Rufus Bus. Cen. SW18 61Db 133
Rufus Cl. HA4: Ruis 34Aa 65
Rufus Ho. SE1 3K 231 (48Vb 91)
(off St Saviour's Est.)
Rufus St. N1 4H 219 (41Ub 91)
Rugby Av. HA0: Wemb 36Ka 66
Rugby Av. N9 18Vb 33
Rugby Av. UB6: G'frd 37Fa 66
Rugby Cl. HA1: Harr 28Ga 46
Rugby Gdns. RM9: Dag 37Yc 75
Rugby La. SM2: Cheam 81Za 174
Rugby Lodge WD24: Wat 7U 12
Rugby Mans. W14 49Ab 88
(off Bishop King's Rd.)
Rugby Rd. NW9 28Ra 47
Rugby Rd. RM9: Dag 38Xc 75
Rugby Rd. TW1: Twick 57Ga 108
Rugby Rd. W4 47Ua 88
Rugby St. WC1 6H 217 (42Pb 90)
Rugby Way WD3: Crox G 15R 26
Rugged La. EN9: Walt A 5Mc 21
Ruggles-Brise Rd. TW15: Ashf 64M 127
Rugg St. E14 45Cc 92
Rugless Ho. E14 47Ec 92
(off E. Ferry Rd.)
Rugmere NW1 38Jb 70
(off Ferdinand St.)
Rugosa Rd. GU24: W End 5C 166
RUISLIP 32U 64
Ruislip Cl. UB6: G'frd 42Da 85
RUISLIP COMMON 28S 44
Ruislip Ct. HA4: Ruis 33V 64
RUISLIP GARDENS 34W 64
Ruislip Gardens Station (Underground) 35W 64
Ruislip Golf Course 33S 64
Ruislip Lido 28T 44
Ruislip Lido Railway 28T 44
Ruislip Lido Woodlands Cen. 28T 44
RUISLIP MANOR 33W 64
Ruislip Manor Station (Underground) 32W 64
Ruislip Rd. UB5: N'olt 39Y 65
Ruislip Rd. UB6: G'frd 41Ca 85
Ruislip Rd. E. UB6: G'frd 42Fa 86
Ruislip Rd. E. W13 42Ha 86
Ruislip Rd. E. W7 42Ga 86
Ruislip Social Club 33V 64
(off Cranley Dr.)
Ruislip Station (Underground) 32U 64
Ruislip St. SW17 63Hb 133
Ruislip Woods 27S 44
Rumania Wlk. DA12: Grav'nd 2H 145
Rumball Ho. SE5 52Ub 113
(off Harris St.)
Rumballs Cl. HP3: Hem H 5A 4
Rumballs Rd. HP3: Hem H 5A 4
Rumbold Rd. SW6 52Db 111
Rum Cl. E1 45Yb 92
Rumford Ho. SE1 4D 230 (48Sb 91)
(off Tiverton St.)
Rumford Shop. Hall 29Gd 56
Rumsey Cl. TW12: Hamp 65Ba 129
Rumsey M. N4 34Rb 71
Rumsey Rd. SW9 55Pb 112
Runacres Ct. SE17 7D 230 (50Sb 91)
Runbury Circ. NW9 33Ta 67
Runcie Cl. IG6: Ilf 28Tc 54
Runciman Cl. BR6: Prat B 82Yc 181
Runcorn Cl. N17 28Xb 51
Runcorn Ho. RM3: Rom 23Nd 57
(off Kingsbridge Cir.)
Runcorn Pl. W11 45Ab 88
Rundell Cres. NW4 29Xa 48
Rundell Twr. SW8 53Pb 112
Runes Cl. CR4: Mitc 70Fb 133
Runham Rd. HP3: Hem H 4N 3
Runnel Ct. IG11: Bark 40Sc 74
(off Spring Pl.)
Runnelfield HA1: Harr 34Ga 66
Runnemede Rd. TW20: Egh 63B 126
Running Horse Yd. TW8: Bford 51Na 109
Running Waters CM13: B'wood 21Ce 59
(not continuous)
Runnymede 1P 125
RUNNYMEDE 1P 125
Runnymede SW19 67Eb 133
RUNNYMEDE BMI HOSPITAL 76F 148
Runnymede Cl. TW2: Whitt 58Da 107
Runnymede Ct. DA2: Dart 60Sd 120
Runnymede Ct. SM6: W'gton 79Kb 156
Runnymede Ct. SS17: Stan H 2L 101
Runnymede Ct. SW15 60Wa 110
Runnymede Ct. TW20: Egh 63C 126
Runnymede Cres. SW16 67Mb 134
Runnymede Gdns. TW2: Whitt 58Da 107
Runnymede Gdns. UB6: G'frd 40Ga 66
Runnymede Ga. KT15: Add 78J 149
Runnymede Ho. E9 35Ac 72
Runnymede Ho. KT16: Chert 73J 149
(off London St.)
Runnymede Memorial 3P 125
Runnymede Rd. SS17: Stan H 2L 101
Runnymede Rd. TW2: Whitt 58Da 107
RUNNYMEDE RDBT. 63D 126
Runtley Wood La. GU4: Sut G 97B 188
Runway, The HA4: Ruis 36X 65
Runway Cl. NW9 26Va 48
Rupack St. SE16 47Yb 92
Rupert Av. HA9: Wemb 36Na 67
Rupert Ct. KT8: W Mole 70Ca 129
(off St Peter's Rd.)
Rupert Ct. W1 4D 222 (45Mb 90)
Rupert Gdns. SW9 54Rb 113
Rupert Ho. SE11 6A 230 (49Qb 90)
Rupert Ho. SW5 49Cb 89
(off Nevern Sq.)
Rupert Rd. N19 34Mb 70
(not continuous)
Rupert Rd. NW6 40Bb 69
Rupert Rd. W4 48Ua 88
Rupert St. W1 4D 222 (45Mb 90)
Rural Cl. RM11: Horn 32Kd 77
Rural Va. DA11: Nflt 9A 122
Rural Way RH1: Redh 6A 208
Rural Way SW16 66Kb 134
Rusbridge Cl. E8 36Wb 71
Ruscoe Dr. GU22: Wok 89C 168
Ruscoe Rd. E16 44Hc 93
Ruscombe NW1 1A 216 (39Kb 70)
(off Delancey St.)
Ruscombe Dr. AL2: Park 8A 6
Ruscombe Gdns. SL3: Dat 2L 103
Ruscombe Way TW14: Felt 59V 106
Ruscus Rd. E17 25Cc 52
Rush, The SW19 67Bb 133
(off Watery La.)
Rusham Ct. TW20: Egh 65C 126
Rusham Pk. Av. TW20: Egh 65B 126
Rusham Rd. SW12 58Hb 111
Rusham Rd. TW20: Egh 65B 126
Rushbridge Cl. CR0: C'don 72Sb 157
Rushbrook Cres. E17 25Bc 52
Rushbrook Rd. SE9 61Sc 138
Rushbury Ct. TW12: Hamp 67Ca 129
Rush Comn. M. SW2 59Pb 112
Rushcroft Rd. E4 24Dc 52
Rushcroft Rd. SW2 56Qb 112
Rushcutters Ct. SE16 49Ac 92
(off Boat Lifter Way)
Rushden Cl. SE19 66Tb 135
Rushdene SE2 48Zc 95
(not continuous)
Rushdene Av. EN4: E Barn 17Gb 31
Rushdene Cl. UB5: N'olt 40Y 65
Rushdene Cres. UB5: N'olt 40X 65
Rushdene Rd. CM15: B'wood 17Yd 40
Rushdene Rd. HA5: Eastc 30Z 45
Rushdene Wlk. TN16: Big H 89Mc 179
Rushden Gdns. IG5: Ilf 26Qc 54
Rushden Gdns. NW7 23Ya 48
Rushdon Cl. RM1: Rom 29Jd 56
Rushdon Cl. RM17: Grays 48Ce 99
Rush Dr. EN9: Walt A 8Ec 20
Rushen Wlk. SM5: Cars 74Fb 155
Rushes, The TW18: Staines 64G 126
(off Wapshott Rd.)
Rushes Mead UB8: Uxb 39L 63
Rushet Rd. BR5: St P 68Wc 139
Rushett Cl. KT7: T Ditt 74Ka 152
Rushett La. KT18: Eps 84Na 173
Rushett La. KT9: Chess 83La 172
Rushett Rd. KT7: T Ditt 73Ka 152
RUSHETTS FARM 10L 207
Rushetts Rd. RH2: Reig 10L 207
Rushetts Rd. TN15: W King 80Ud 164
Rushey Cl. KT3: N Mald 70Ta 131
Rushey Grn. SE6 59Dc 114
Rushey Hill EN2: Enf 14Pb 32
Rushey Mead SE4 57Cc 114
Rushfield EN6: Pot B 4Za 16
Rushford Rd. SE4 58Bc 114
RUSH GREEN 32Fd 76
Rush Grn. Gdns. RM7: Rush G 32Ed 76
Rush Grn. Rd. RM7: Rush G 32Dd 76
Rushgrove Av. NW9 29Ua 48
Rushgrove Ct. NW9 29Ua 48
Rushgrove Pde. NW9 29Ua 48
Rushgrove St. SE18 49Pc 94
Rush Hill M. SW11 55Jb 112
(off Rush Hill Rd.)
Rush Hill Rd. SW11 55Jb 112
Rushleigh Av. EN8: Chesh 3Zb 20
Rushley Cl. BR2: Kes 77Mc 159
Rushley Cl. RM16: Grays 46Fe 99
Rush Leys Ct. AL2: Lon C 9E 6
Rushlight M. KT9: Chess 84La 172
Rushmead E2 41Xb 91
Rushmead TW10: Ham 62Ka 130
Rushmead Cl. CR0: C'don 77Vb 157
Rushmead Cl. HA8: Edg 19Ra 29
Rushmere Av. RM14: Upm 34Sd 78
Rushmere Ct. HP3: Hem H 6N 3
Rushmere Ct. KT4: Wor Pk 75Wa 154
Rushmere Ct. TN15: Igh 92Zd 205
Rushmere Pl. SW19 64Za 132
Rushmere Pl. TW20: Eng G 64A 126
Rushmon Gdns. KT12: Walt T 76X 151
Rushmon Pl. SM3: Cheam 79Ab 154
Rushmon Vs. KT3: N Mald 70Va 132
Rushmoor Cl. HA5: Eastc 28X 45
Rushmoor Cl. WD3: Rick 19M 25
Rushmoor Ct. WD18: Wat 16S 26
Rushmore Cl. BR1: Broml 69Nc 138
Rushmore Cres. E5 35Zb 72
Rushmore Hill BR6: Prat B 81Yc 181
Rushmore Hill TN14: Knock 84Zc 181
Rushmore Hill TN14: Prat B 84Zc 181
Rushmore Ho. SW15 59Wa 110
Rushmore Ho. W14 48Ab 88
(off Russell Rd.)
Rushmore Rd. E5 35Yb 72
(not continuous)
Rusholme Av. RM10: Dag 34Cd 76
Rusholme Gro. SE19 64Ub 135
Rusholme Rd. SW15 58Za 110
Rushout Av. HA3: Kenton 30Ka 46
Rushton Av. RH9: S God 10A 210
Rushton Av. WD25: Wat 7W 12
Rushton Ct. EN8: Chesh 1Zb 20
Rushton Ho. SW8 54Mb 112
Rushton St. N1 1G 219 (40Tb 71)
Rushton Wlk. E3 42Bc 92
(off Hamlets Way)
Rushworth Rd. RH2: Reig 5J 207
Rushworth St. SE1 1C 230 (47Rb 91)
Rushymead TN15: Kems'g 90Qd 183
Rushy Mdw. La. SM5: Cars 75Gb 155
Ruskin Av. DA16: Well 54Wc 117
Ruskin Av. E12 37Nc 74
Ruskin Av. EN9: Walt A 6Gc 21
Ruskin Av. RM14: Upm 31Sd 78
Ruskin Av. TW14: Felt 58V 106
Ruskin Av. TW9: Kew 52Qa 109
Ruskin Cl. NW11 30Db 49
Ruskin Ct. N21 17Pb 32
Ruskin Ct. SE5 55Tb 113
(off Champion Hill)
Ruskin Dr. BR6: Orp 76Uc 160
Ruskin Dr. DA16: Well 55Wc 117
Ruskin Dr. KT4: Wor Pk 75Xa 154
Ruskin Gdns. HA3: Kenton 29Pa 47
Ruskin Gdns. RM3: Rom 24Kd 57
Ruskin Gdns. W5 42Ma 87
Ruskin Gro. DA1: Dart 57Qd 119
Ruskin Gro. DA16: Well 54Wc 117
Ruskin Ho. CR2: S Croy 78Tb 157
(off Selsdon Rd.)
Ruskin Ho. SW1 6E 228 (49Mb 90)
(off Herrick St.)
Ruskin Mans. W14 51Ab 110
(off Queen's Club Gdns.)
Ruskin Pde. CR2: S Croy 78Tb 157
(off Selsdon Rd.)
Ruskin Pde. HA8: Edg 21Pa 47
Ruskin Pk. Ho. SE5 55Tb 113
Ruskin Rd. CR0: C'don 75Rb 157
Ruskin Rd. DA17: Belv 49Cd 96
Ruskin Rd. N17 25Vb 51
Ruskin Rd. RM16: Grays 9C 100
Ruskin Rd. SM5: Cars 78Hb 155
Ruskin Rd. SS17: Stan H 2L 101
Ruskin Rd. TW18: Staines 65H 127
Ruskin Rd. TW7: Isle 55Ha 108
Ruskin Rd. UB1: S'hall 45Aa 85
Ruskin Sq. CR0: C'don 75Tb 157
Ruskin Wlk. BR2: Broml 72Pc 160
Ruskin Wlk. N9 19Wb 33
Ruskin Wlk. SE24 57Sb 113
Ruskin Way SW19 67Fb 133
Rusland Av. BR6: Orp 76Tc 160
Rusland Hgts. HA1: Harr 28Ga 46
Rusland Pk. Rd. HA1: Harr 28Ga 46
Rusling Ct. WD17: Wat 12X 27
Rusmon Ct. KT16: Chert 73H 149
Rusper Cl. HA7: Stan 21La 46
Rusper Cl. NW2 34Ya 68
Rusper Ct. SW9 54Nb 112
(off Clapham Rd.)
Rusper Rd. N17 27Sb 51
Rusper Rd. N22 26Rb 51
Rusper Rd. RM9: Dag 37Yc 75
Russell Av. AL3: St A 2B 6
Russell Av. N22 26Qb 50
Russell Chambers WC1 1G 223 (43Nb 90)
(off Bury Pl.)
Russell Cl. BR3: Beck 69Ec 136
Russell Cl. CM15: B'wood 17Xd 40
Russell Cl. DA1: Cray 56Jd 118
Russell Cl. DA7: Bex 56Cd 118
Russell Cl. GU21: Wok 7N 167
Russell Cl. HA4: Ruis 33Y 65
Russell Cl. HA6: Nwood 22S 44
Russell Cl. KT20: Walt H 97Wa 194
Russell Cl. NW10 38Sa 67
Russell Cl. SE7 52Lc 115
Russell Cl. W4 51Va 110
Russell Ct. AL2: Brick W 2Ca 13
Russell Ct. CR8: Purl 82Qb 176
Russell Ct. E10 31Dc 72
Russell Ct. EN5: New Bar 14Eb 31
Russell Ct. KT22: Lea 94Ka 192
Russell Ct. N14 16Mb 32
Russell Ct. SE15 54Xb 113
(off Heaton Rd.)
Russell Ct. SM6: W'gton 78Lb 156
(off Ross Rd.)
Russell Ct. SW1 7C 222 (46Lb 90)
(off Cleveland Row)
Russell Ct. SW16 64Pb 134
Russell Ct. WC1 6F 217 (42Nb 90)
(off Woburn Pl.)
Russell Cres. WD25: Wat 7V 12
Russell Dr. TW19: Stanw 58M 105
Russell Flint Ho. E16 46Kc 93
(off Pankhurst Av.)
Russell Gdns. IG2: Ilf 31Tc 74
Russell Gdns. N20 19Gb 31
Russell Gdns. NW11 30Ab 48
Russell Gdns. TW10: Ham 61La 130
Russell Gdns. UB7: Sip 50Q 84
Russell Gdns. W14 48Ab 88
Russell Gdns. M. W14 47Ab 88
Russell Grn. Cl. CR8: Purl 82Qb 176
Russell Gro. NW7 22Ua 48
Russell Gro. SW9 52Qb 112
Russell Hill CR8: Purl 82Pb 176
Russell Hill Pl. CR8: Purl 83Qb 176
Russell Hill Rd. CR8: Purl 83Qb 176
Russell Ho. BR2: Broml 72Nc 160
(off Wells Vw. Dr.)
Russell Ho. E14 44Cc 92
(off Saracen St.)
Russell Ho. SW1 7B 228 (50Lb 90)
(off Cambridge St.)
Russell Kerr Cl. W4 52Sa 109
Russell La. N20 19Gb 31
Russell La. WD17: Wat 8T 12
Russell Lodge E4 19Ec 34
Russell Lodge SE1 3F 231 (48Tb 91)
(off Spurgeon St.)
Russell Mans. WC1 7G 217 (43Nb 90)
(off Southampton Row)
Russell Mead HA3: Hrw W 25Ha 46
Russell Pde. NW11 30Ab 48
(off Golders Grn. Rd.)
Russell Pl. DA4: Sut H 67Qd 141
Russell Pl. HP3: Hem H 5K 3
Russell Pl. NW3 36Gb 69
Russell Pl. SE16 48Ac 92
Russell Pl. SM2: Sutt 80Db 155
Russell Quay DA11: Grav'nd 7C 122
Russell Rd. CR4: Mitc 69Gb 133
Russell Rd. DA12: Grav'nd 8F 122
Russell Rd. E10 30Dc 52
Russell Rd. E16 44Jc 93
Russell Rd. E17 27Bc 52
Russell Rd. E4 21Bc 52
Russell Rd. EN1: Enf 10Vb 19
Russell Rd. GU21: Wok 7N 167
Russell Rd. HA6: Nwood 22S 44
Russell Rd. IG9: Buck H 18Kc 35
Russell Rd. KT12: Walt T 72W 150
Russell Rd. N13 23Pb 50
Russell Rd. N15 29Ub 51
Russell Rd. N20 19Gb 31
Russell Rd. N8 30Mb 50
Russell Rd. NW9 30Va 48
Russell Rd. RM17: Grays 49Ce 99
Russell Rd. RM18: Tilb 3A 122
Russell Rd. SW19 66Cb 133
Russell Rd. TW17: Shep 73S 150
Russell Rd. TW2: Twick 58Ha 108
Russell Rd. UB5: N'olt 36Ea 66
Russell Rd. W14 48Ab 88
Russells KT20: Tad 94Za 194
Russell's Footpath SW16 64Nb 134
Russell Sq. DA3: Lfield 69Zd 143
Russell Sq. WC1 7F 217 (43Nb 90)
Russell Sq. Mans. WC1 7G 217 (43Nb 90)
(off Southampton Row)
Russell Square Station (Underground) 6F 217 (42Nb 90)
Russell's Ride EN8: Chesh 3Ac 20
Russell St. SL4: Wind 3H 103
Russell St. WC2 4G 223 (45Nb 90)
Russell's Way RM15: S Ock 44Zd 99
Russell's Wharf Flats W10 42Bb 89
Russell Ter. DA4: Hort K 70Sd 142
Russell Wlk. TW10: Rich 58Pa 109
Russell Way SM1: Sutt 78Db 155
Russell Way WD19: Wat 17X 27
Russell Wilson Ho. RM3: Hrld W 25Pd 57
Russell Yd. SW15 56Ab 110
Russet Av. TW17: Shep 69U 128
Russet Cl. BR6: Chels 78Xc 161
Russet Cl. KT12: Hers 76Z 151
Russet Cl. TW19: Stanw M 58H 105
Russet Cl. UB10: Hil 42S 84
Russet Cres. N7 36Pb 70
Russet Dr. AL4: St A 3G 6
Russet Dr. CR0: C'don 74Ac 158
Russet Dr. WD7: Shenl 4Na 15
Russet Ho. RM17: Grays 52Ee 121
Russets KT20: Tad 95Ya 194
Russets Cl. E4 21Fc 53
Russett Ct. CR3: Cat'm 97Wb 197
Russett Hill SL9: Chal P 27A 42
Russettings HA5: Hat E 24Ba 45
(off Westfield Pk.)
Russetts RM11: Horn 28Nd 57
Russetts Cl. GU21: Wok 87B 168
Russett Way BR8: Swan 68Fd 140
Russett Way SE13 54Dc 114
Russet Wlk. DA9: Ghithe 57Wd 120
Russia Dock Rd. SE16 46Ac 92
Russia La. E2 40Yb 72
Russia Row EC2 3E 224 (44Sb 91)
Russia Wlk. SE16 47Ac 92
Russington Rd. TW17: Shep 72T 150
Rusthall Av. W4 49Ta 87
Rusthall Cl. CR0: C'don 72Yb 158
Rustic Av. SW16 66Kb 134
Rustic Cl. RM14: Upm 32Ud 78
Rustic Pl. HA0: Wemb 35Ma 67
Rustic Wlk. E16 44Kc 93
(off Lambert Rd.)
Rustington Wlk. SM4: Mord 73Bb 155
Ruston Av. KT5: Surb 73Ra 153
Ruston Gdns. N14 16Jb 32
Ruston M. W11 44Ab 88
Ruston Rd. SE18 48Nc 94
Ruston St. E3 39Bc 72
Rust Sq. SE5 52Tb 113
Rutford Rd. SW16 64Nb 134
Ruth Cl. HA7: Stan 28Pa 47
Ruth Ct. E3 40Ac 72
Ruthen Cl. KT18: Eps 86Ra 173
Rutherford Cl. SL4: Wind 3D 102
Rutherford Cl. SM2: Sutt 79Fb 155
Rutherford Cl. UB8: Hil 42P 83
Rutherford Cl. WD6: Bore 12Sa 29
Rutherford Ho. E1 42Xb 91
(off Brady St.)
Rutherford Ho. HA9: Wemb 34Sa 67
(off Barnhill Rd.)
Rutherford Ho. SL9: Ger X 29A 42
Rutherford Ho. SW11 54Hb 111
(off Battersea Pk. Rd.)
Rutherford Pl. RM3: Rom 22Ld 57
Rutherford St. SW1 5D 228 (49Mb 90)
Rutherford Twr. UB1: S'hall 44Da 85
Rutherford Way HA9: Wemb 34Qa 67
Rutherford Way WD23: B Hea 18Fa 28
Rutherglen Rd. SE2 51Wc 117
Rutherwick Ri. CR5: Coul 89Nb 176
Rutherwyke Cl. KT17: Ewe 79Wa 154
Rutherwyk Rd. KT16: Chert 73G 148
Ruth Ho. W10 42Ab 88
(off Kensal Rd.)
Ruthin Cl. NW9 30Ua 48
Ruthin Rd. SE3 51Jc 115
Ruthven Av. EN8: Walt C 5Zb 20
Ruthven St. E9 39Zb 72
Rutland App. RM11: Horn 29Qd 57
Rutland Av. DA15: Sidc 59Wc 117
Rutland Av. SL1: Slou 3G 80
Rutland Cl. DA1: Dart 59Md 119
Rutland Cl. DA5: Bexl 61Zc 139
Rutland Cl. KT19: Ewe 82Ta 173
Rutland Cl. KT21: Asht 89Na 173
Rutland Cl. KT9: Chess 79Pa 153
Rutland Cl. RH1: Redh 5P 207
Rutland Cl. SW14 55Ra 109
Rutland Cl. SW19 66Gb 133
Rutland Ct. BR7: Chst 67Qc 138
Rutland Ct. EN3: Pond E 15Xb 33
Rutland Ct. KT1: King T 70Ma 131
(off Palace Rd.)
Rutland Ct. SE5 56Tb 113
Rutland Ct. SE9 61Sc 138
Rutland Ct. SW7 2E 226 (47Gb 89)
(off Rutland Gdns.)
Rutland Ct. W3 44Qa 87
Rutland Dr. RM11: Horn 29Qd 57
Rutland Dr. SM4: Mord 72Bb 155
Rutland Dr. TW10: Ham 60Ma 109
Rutland Gdns. CR0: C'don 77Ub 157
Rutland Gdns. HP2: Hem H 1P 3
Rutland Gdns. N4 30Rb 51
Rutland Gdns. RM8: Dag 36Yc 75
Rutland Gdns. SW7 2E 226 (47Gb 89)
Rutland Gdns. W13 43Ja 86
Rutland Gdns. M. SW7 2E 226 (47Gb 89)
Rutland Ga. BR2: Broml 70Hc 137
Rutland Ga. DA17: Belv 50Dd 96
Rutland Ga. SW7 2E 226 (47Gb 89)
Rutland Ga. M. SW7 2D 226 (47Gb 89)
(off Rutland Ga.)
Rutland Gro. W6 50Xa 88
Rutland Ho. KT18: Eps 86Ta 173
(off South St.)
Rutland Ho. NW10 41Wa 88
Rutland Ho. UB5: N'olt 37Ca 65
(off The Farmlands)
Rutland Ho. W8 48Db 89
(off Marloes Rd.)
Rutland M. NW8 39Db 69
Rutland M. E. SW7 3E 226 (48Gb 89)
(off Ennismore St.)
Rutland M. Sth. SW7 3D 226 (48Gb 89)
(off Ennismore St.)
Rutland M. W. SW7 3D 226 (48Gb 89)
(off Rutland Ga.)
Rutland Pk. NW2 37Ya 68
Rutland Pk. SE6 61Bc 136
Rutland Pk. Gdns. NW2 37Ya 68
(off Rutland Pk.)
Rutland Pk. Mans. NW2 37Ya 68
Rutland Pl. EC1 6C 218 (42Rb 91)
Rutland Pl. WD23: B Hea 18Fa 28
Rutland Rd. E11 29Kc 53
Rutland Rd. E17 30Cc 52
Rutland Rd. E7 38Mc 73
Rutland Rd. E9 39Zb 72
Rutland Rd. HA1: Harr 30Ea 46
Rutland Rd. IG1: Ilf 34Rc 74
Rutland Rd. SW19 66Gb 133
Rutland Rd. TW2: Twick 61Fa 130
Rutland Rd. UB1: S'hall 43Ca 85
Rutland Rd. UB3: Harl 49T 84
Rutland St. SW7 3E 226 (48Gb 89)
Rutland Wlk. SE6 61Bc 136
Rutland Way BR5: St M Cry 72Yc 161
Rutledge Cl. RM16: Ors 3C 100
Rutley Cl. RM3: Hrld W 26Md 57
Rutley Cl. SE17 51Rb 113
Rutlish Rd. SW19 67Cb 133
Rutson Rd. KT14: Byfl 86P 169
Rutter Gdns. CR4: Mitc 70Eb 133
Rutters Cl. UB7: W Dray 47Q 84
Rutts, The WD23: B Hea 18Fa 28
Rutt's Ter. SE14 53Zb 114
Ruvigny Gdns. SW15 55Za 110
Ruxbury Ct. TW15: Ashf 62N 127
Ruxbury Rd. KT16: Chert 72E 148
RUXLEY 66Zc 139
Ruxley Cl. DA14: Sidc 65Zc 139
Ruxley Cl. KT19: Ewe 78Ra 153
Ruxley Cl. KT20: Kgswd 93Ab 194
Ruxley Cnr. Ind. Est. DA14: Sidc 65Zc 139
Ruxley Ct. GU22: Wok 1P 187
(off West Hill Rd.)
Ruxley Ct. KT19: Ewe 78Sa 153
Ruxley Cres. KT10: Clay 79Ka 152
Ruxley Gdns. TW17: Shep 71S 150
Ruxley La. KT19: Ewe 79Ra 153
Ruxley M. KT19: Ewe 78Ra 153
Ruxley Pk. Golf Course 67Zc 139
Ruxley Ridge KT10: Clay 80Ja 152
Ruxley Towers KT10: Clay 80Ja 152
Ruxton Cl. BR8: Swan 69Gd 140
Ruxton Cl. CR5: Coul 87Lb 176
Ruxton Ct. BR8: Swan 69Gd 140
Ryall Cl. AL2: Brick W 1Aa 13
Ryalls Ct. N20 20Hb 31
Ryan Cl. HA4: Ruis 32X 65
Ryan Cl. SE3 56Lc 115
Ryan Ct. RM7: Rom 30Ed 56
Ryan Ct. SW16 66Nb 134
Ryan Ct. WD19: Wat 17Aa 27
Ryan Dr. TW8: Bford 51Ja 108
Ryan Way WD24: Wat 11Y 27
Ryarsh Cres. BR6: Orp 77Uc 160
Rybrook Dr. KT12: Walt T 75Y 151
Rycott Path SE22 59Wb 113
Rycroft SL4: Wind 5D 102
Rycroft Way N17 27Vb 51
Ryculff Sq. SE3 54Hc 115
Rydal Cl. CR8: Purl 85Tb 177
Rydal Cl. NW4 25Ab 48
Rydal Ct. HA8: Edg 22Pa 47
Rydal Ct. HA9: Wemb 31Pa 67
Rydal Ct. WD25: Wat 4X 13
Rydal Cres. UB6: G'frd 41Ka 86
Rydal Dr. BR4: W W'ck 75Gc 159
Rydal Dr. DA7: Bex 53Cd 118
Rydale Ct. TN13: Dun G 93Gd 202
Rydal Gdns. HA9: Wemb 32La 66
Rydal Gdns. NW9 29Ua 48
Rydal Gdns. SW15 64Ua 132
Rydal Gdns. TW3: Houn 58Da 107
Rydal Mt. BR2: Broml 70Hc 137
Rydal Mt. EN6: Pot B 5Ab 16
Rydal Rd. SW16 63Mb 134
Rydal Water NW1 4B 216 (41Lb 90)
Rydal Way EN3: Pond E 16Yb 34
Rydal Way HA4: Ruis 35Y 65
Rydal Way TW20: Egh 66D 126
Ryde, The TW18: Staines 67K 127
Ryde Cl. GU23: Rip 93L 189
Ryde Dr. SS17: Stan H 3L 101
Ryde Heron GU21: Knap 9J 167
Ryde Ho. NW6 39Cb 69
(off Priory Pk. Rd.)
RYDENS 76Y 151
Rydens Av. KT12: Walt T 75X 151
Rydens Cl. KT12: Walt T 75Y 151
Rydens Gro. KT12: Hers 77Z 151
Rydens Ho. SE9 62Lc 137
Rydens Pde. GU22: Wok 92D 188
Rydens Pk. KT12: Walt T 75Z 151
Rydens Rd. KT12: Walt T 76X 151
Rydens Way GU22: Wok 92C 188
Ryde Pl. TW1: Twick 58Ma 109
Ryder Av. E10 31Dc 72
Ryder Cl. BR1: Broml 64Kc 137
Ryder Cl. HP3: Bov 10C 2
Ryder Cl. WD23: Bush 16Da 27
Ryder Ct. E10 33Dc 72
Ryder Ct. SW1 6C 222 (46Lb 90)
(off Ryder St.)
Ryder Dr. SE16 50Xb 91
Ryder Gdns. RM13: Rain 37Hd 76
Ryder Ho. E1 42Yb 92
(off Colebert Av.)
Ryder M. E9 36Yb 72
Ryders Av. AL4: Col H 2A 8
Ryders Av. AL4: S'ford 2A 8
Ryder Seed M. AL1: St A 3B 6
Ryder's Ter. NW8 1A 214 (40Eb 69)
Ryder St. SW1 6C 222 (46Lb 90)
Rydes Cl. GU22: Wok 92E 188
Ryde Va. Rd. SW12 61Lb 134
Rydinghurst Ho. SL9: Chal P 22A 42
Rydings SL4: Wind 5D 102
Rydon Bus. Cen. KT22: Lea 92Ka 192
Rydon M. SW19 66Ya 132
Rydons Cl. SE9 55Nc 116
Rydon's La. CR5: Coul 92Sb 197
Rydon St. N1 39Sb 71
Rydons Way RH1: Redh 7P 207
Rydon's Wood Cl. CR5: Coul 92Sb 197
Rydston Cl. N7 38Nb 70
Rye, The N14 17Mb 32
Ryebridge Cl. KT22: Lea 90Ja 172
Ryebrook KT22: Lea 92Ja 192
Ryebrook Rd. KT22: Lea 90Ja 172
Rye Cl. DA5: Bexl 58Dd 118
Rye Cl. RM12: Horn 35Ld 77
Rye Cl. WD6: Bore 14Ta 29
Ryecotes Mead SE21 60Ub 113
Rye Ct. SL1: Slou 8L 81
Rye Cres. BR5: Orp 74Zc 161
Ryecroft AL10: Hat 2B 8
Ryecroft DA12: Grav'nd 4G 144
Ryecroft DA3: Long H 70Fe 143
Ryecroft Av. IG5: Ilf 26Rc 54
Ryecroft Av. TW2: Whitt 59Da 107
Ryecroft Cl. HP2: Hem H 2C 4
Ryecroft Ct. AL4: St A 2K 7
Ryecroft Cres. EN5: Barn 15Xa 30
Ryecroft Rd. BR5: Pet W 72Tc 160
Ryecroft Rd. SE13 57Ec 114
Ryecroft Rd. SW16 65Qb 134
Ryecroft Rd. TN14: Otf 89Jd 182
Ryecroft St. SW6 53Db 111
Ryedale SE22 58Xb 113
Ryedale Pl. RM3: Rom 24Nd 57

Rye Fld. BR5: Orp 74Zc 161
Rye Fld. KT21: Asht 88Ma 173
Ryefield Av. UB10: Hil 38R 64
Ryefield Ct. HA6: Nwood 26W 44
Ryefield Cres. HA6: Nwood 26W 44
Ryefield Pde. HA6: Nwood 26W 44
(off Joel St.)
Ryefield Path SW15 60Wa 110
Ryefield Rd. SE19 65Sb 135
Ryegates SE15 54Xb 113
(off Caulfield Rd.)
Rye Gro. GU18: Light 10D 146
Rye Hill Pk. SE15 56Yb 114
Rye Ho. SE16 47Yb 92
(off Swan Rd.)
Rye Ho. SW1 7K 227 (50Kb 90)
(off Ebury Bri. Rd.)
Ryeland Blvd. SW18 57Db 111
Ryeland Cl. UB7: View 44N 83
Ryelands Cl. CR3: Cat'm 93Ub 197
Ryelands Ct. KT22: Lea 90Ja 172
Ryelands Cres. SE12 58Lc 115
Ryelands Pl. KT13: Weyb 76U 150
Rye La. SE15 53Wb 113
Rye La. TN14: Dun G 92Hd 202
Rye La. TN14: Otf 92Hd 202
Rye Mans. E20 36Ec 72
(off Napa Cl.)
Rye Pas. SE15 55Wb 113
Rye Rd. SE15 56Zb 114
Rye Wlk. SW15 57Za 110
Rye Way HA8: Edg 23Pa 47
Ryfold Rd. SW19 62Cb 133
Ryhope Rd. N11 21Kb 50
Rykhill RM16: Grays 8D 100
Ryland Cl. TW13: Felt 63V 128
Rylandes Rd. CR2: Sels 81Xb 177
Rylandes Rd. NW2 34Wa 68
Ryland Rd. NW5 37Kb 70
Rylett Cres. W12 47Va 88
Rylett Rd. W12 47Va 88
Rylston Cl. E4 12Ec 34
Rylston Rd. N13 20Tb 33
Rylston Rd. SW6 51Bb 111
Rylton Ho. KT12: Walt T 74W 150
Ryman Ct. WD3: Chor 16E 24
Rymer Rd. CR0: C'don 73Ub 157
Rymer St. SE24 58Rb 113
Rymill Cl. HP3: Bov 10C 2
Rymill St. E16 46Qc 94
Rysbrack St. SW3 3F 227 (48Hb 89)
Rysted La. TN16: Westrm 98Sc 200
Rythe, The KT10: Esh 82Da 171
Rythe Bank Cl. KT7: T Ditt 73Ka 152
Rythe Cl. KT10: Clay 78Ga 152
Rythe Cl. KT9: Chess 80La 152
Rythe Ct. KT7: T Ditt 73Ja 152
Rythe Rd. KT10: Clay 78Fa 152
Ryvers End SL3: L'ly 48B 82
Ryvers Rd. SL3: L'ly 48B 82

S

Saatchi Gallery 7G 227 (50Hb 89)
Sabah Ct. TW15: Ashf 63Q 128
Sabella Ct. E3 40Bc 72
Sabina Rd. RM16: Grays 9E 100
Sabine Rd. SW11 55Hb 111
SABINE'S GREEN 13Nd 39
Sabine's Rd. CM14: N'side 13Nd 39
Sabine's Rd. CM14: Nave 13Nd 39
Sabine's Rd. RM4: N'side 12Md 39
Sabine's Rd. RM4: Nave 12Md 39
Sable Cl. TW4: Houn 55Y 107
Sable Ho. E20 36Dc 72
(off Scarlet Cl.)
Sable St. N1 38Rb 71
Sachfield Dr. RM16: Chaf H 48Ae 99
Sach Rd. E5 33Xb 71
Sackett Rd. IG11: Bark 42Wc 95
Sackville Av. BR2: Hayes 74Jc 159
Sackville Cl. HA2: Harr 34Fa 66
Sackville Cl. TN13: S'oaks 94Kd 203
Sackville Cotts. RH1: Blet 5K 209
Sackville Cres. RM3: Hrld W 25Nd 57
Sackville Gdns. IG1: Ilf 32Pc 74
Sackville Ho. SW16 62Nb 134
Sackville Pl. TN13: S'oaks 96Ld 203
Sackville Rd. DA2: Wilm 61Nd 141
Sackville Rd. SM2: Sutt 80Cb 155
Sackville St. W1 5B 222 (45Lb 90)
Sacombe Rd. HP1: Hem H 1H 3
Saddington St. DA12: Grav'nd 9D 122
Saddleback La. W7 48Ga 86
Saddlebrook Pk. TW16: Sun 66U 128
Saddle M. CR0: C'don 73Sb 157
Saddlers Cl. EN5: Ark 15Xa 30
Saddlers Cl. HA5: Hat E 23Ca 45
Saddlers Cl. WD6: Bore 15Ta 29
Saddlers Ct. KT18: Eps 85Sa 173
Saddlers Ho. E17 29Bc 52
(off Track St.)
Saddlers Ho. E20 36Ec 72
(off Ribbons Wlk.)
Saddlers M. HA0: Wemb 35Ha 66
Saddlers M. KT1: Hamp W 67La 130
Saddlers M. SW8 53Nb 112
Saddler's Pk. DA4: Eyns 76Md 163
Saddlers Path WD6: Bore 15Ta 29
(off Farriers Way)
Saddlers Pl. TW3: Houn 55Ea 108
Saddlers Wlk. WD4: K Lan 1Q 12
Saddlers Way KT18: Eps D 91Ta 193
Saddlery, The KT23: Bookh 98Ba 191
Saddlescombe Way N12 22Cb 49
Saddleworth Rd. RM3: Rom 23Ld 57
Saddleworth Sq. RM3: Rom 23Ld 57
Saddle Yd. W1 6K 221 (46Kb 90)
Sadleir Rd. AL1: St A 4C 6
Sadler Cl. CR4: Mitc 68Hb 133
Sadler Hgts. N1 38Pb 70
(off Caledonian Rd.)
Sadler Ho. E3 41Dc 92
(off Bromley High St.)
Sadler Ho. EC1 3B 218 (41Rb 91)
(off Spa Grn. Est.)
Sadler Pl. E9 36Ac 72
Sadlers Ct. SE1 3H 231 (48Ub 91)
(off Wilds Rents)
Sadlers Ga. M. SW15 55Ya 110
Sadlers Ride KT8: W Mole 68Ea 130
Sadler's Wells Theatre 3A 218 (41Qb 90)
Safara Ho. SE5 53Ub 113
(off Dalwood St.)
Safari Cinema 29Ha 46
Safflower La. RM3: Hrld W 25Pd 57
Saffron Av. E14 45Fc 93
Saffron Central Sq. CR0: C'don 74Sb 157
Saffron Cl. CM13: W H'dn 30Fe 59
Saffron Cl. CR0: C'don 72Nb 156
Saffron Cl. NW11 30Bb 49
Saffron Cl. SL3: Dat 3M 103
Saffron Cl. SS17: Horn H 1J 101
Saffron Ct. E15 36Gc 73
(off Maryland Pk.)
Saffron Ct. TW14: Bedf 59S 106
SAFFRON GREEN 11Va 30
Saffron Hill EC1 7A 218 (43Qb 90)
Saffron Ho. SM2: Sutt 80Db 155
Saffron Ho. TW9: Kew 53Ra 109
Saffron La. HP1: Hem H 1J 3
Saffron M. SW19 66Ab 132
Saffron Rd. RM16: Chaf H 49Yd 98
Saffron Rd. RM5: Col R 26Fd 56
Saffron St. EC1 7A 218 (43Qb 90)
Saffron Way KT6: Surb 74Ma 153
Saffron Wharf SE1 1K 231 (47Vb 91)
(off Shad Thames)
Sage Cl. E6 43Pc 94
Sage M. SE22 57Vb 113
Sage St. E1 45Yb 92
Sage Way WC1 4H 217 (41Pb 90)
(off Cubitt St.)
Sage Yd. KT6: Surb 74Pa 153
Sahara Ct. UB1: S'hall 45Aa 85
Saigasso Cl. E16 44Mc 93
Sailacre Ho. SE10 50Hc 93
Sail Ct. E14 45Fc 93
(off Newport Av.)
Sailmakers Ct. SW6 55Eb 111
Sailors Ho. E14 44Fc 93
(off Deauville Cl.)
Sail St. SE11 5J 229 (49Pb 90)
Saimet NW9 24Va 48
(off Wiggins Mead)
Sainfoin Rd. SW17 61Jb 134
Sainsbury Cen., The 73J 149
Sainsbury Rd. SE19 64Ub 135
Sainsbury Wing 5E 222 (45Mb 90)
(within National Gallery)
St Agatha's Dr. KT2: King T 65Pa 131
St Agatha's Gro. SM5: Cars 74Hb 155
St Agnes Cl. E9 39Yb 72
St Agnes Ho. E3 40Cc 72
(off Ordell Rd.)
St Agnes Pl. SE11 51Rb 113
St Agnes Quad. SE11 51Qb 112
St Agnes Well EC1 5G 219 (42Tb 91)
(off City Rd.)
St Aidans Ct. IG11: Bark 40Xc 75
St Aidan's Rd. SE22 58Xb 113
St Aidan's Rd. W13 47Ka 86
St Aidan's Way DA12: Grav'nd 2G 144
ST ALBANS 2B 6
St Albans Abbey Station (Rail) 4B 6
St Albans Av. E6 41Pc 94
St Alban's Av. RM14: Upm 33Ud 78
St Alban's Av. W4 49Ta 87
St Albans Av. KT13: Weyb 76Q 150
St Albans Av. TW13: Hanw 64Z 129
St Albans Cathedral 2A 6
St Albans City FC 1D 6
ST ALBANS CITY HOSPITAL 1A 6
St Albans Clock Tower 2B 6
(off High St.)
St Alban's Cl. DA12: Grav'nd 2F 144
St Alban's Cl. SL4: Wind 3H 103
St Albans Cl. NW11 32Cb 69
St Albans Ct. EC2 2E 224 (44Sb 91)
(off Wood St.)
St Alban's Cres. IG8: Wfd G 24Jc 53
St Albans Cres. N22 25Qb 50
St Albans Farm TW14: Houn 57Y 107
St Alban's Gdns. DA12: Grav'nd 2F 144
St Alban's Gdns. TW11: Tedd 64Ja 130
St Alban's Gro. SM5: Cars 73Gb 155
St Alban's Gro. W8 48Db 89
St Albans Hill HP3: Hem H 5N 3
St Alban's La. NW11 32Cb 69
St Albans La. WD5: Bedm 2M 11
St Albans Mans. W8 48Db 89
(off Kensington Ct. Pl.)
St Albans Museum & Gallery 2B 6
St Albans Organ Theatre 3F 6
St Albans Pl. N1 39Rb 71
St Albans Retail Pk. 4B 6
St Alban's Rd. CM16: Coop 1Zc 23
St Alban's Rd. DA1: Dart 59Pd 119
St Alban's Rd. IG8: Wfd G 24Jc 53
St Alban's Rd. KT2: King T 65Na 131
St Alban's Rd. RH2: Reig 4J 207
St Alban's Rd. SM1: Sutt 77Bb 155
St Albans Rd. EN5: Barn 7Xa 16
St Albans Rd. EN6: S Mim 3Ua 16
St Albans Rd. HP2: Hem H 4M 3
St Albans Rd. HP3: Hem H 3A 4
St Albans Rd. IG3: Ilf 32Vc 75
St Albans Rd. NW10 39Ua 68
St Albans Rd. NW5 34Jb 70
St Albans Rd. WD17: Wat 12X 27
St Albans Rd. WD24: Wat 11X 27
St Albans Rd. WD25: Wat 8Y 13
St Albans Rd. W. AL10: Hat Poplar Av. 1N 7
(not continuous)
St Albans South Signal Box 3D 6
St Albans Station (Rail) 2D 6
St Alban's St. SL4: Wind 3H 103
St Albans St. SW1 5D 222 (45Mb 90)
St Albans Studios W8 48Db 89
(off St Albans Gro.)
St Albans Sub Aqua Club 3B 6
St Albans Ter. W6 51Ab 110
St Albans Vs. NW5 34Jb 70
St Albans Visitor Info. Cen. 2B 6
St Alfege Pas. SE10 51Ec 114
St Alfege Rd. SE7 51Mc 115
St Alphage Ct. NW9 27Ta 47
St Alphage Gdn. EC2 1E 224 (43Sb 91)
St Alphage Highwalk EC2 1E 224 (43Sb 91)
(off Wood St.)
St Alphage Wlk. HA8: Edg 26Sa 47
St Alphege Rd. N9 17Yb 34
St Alphonsus Rd. SW4 56Lb 112
St Amunds Cl. SE6 63Cc 136
St Andrew's Av. HA0: Wemb 35Ja 66
St Andrew's Av. RM12: Horn 36Hd 76
St Andrew's Av. SL4: Wind 4D 102
St Andrews Chambers W1 1C 222 (43Lb 90)
(off Wells St.)
St Andrew's Cl. HA4: Ruis 33Z 65
St Andrew's Cl. HA7: Stan 26La 46
St Andrew's Cl. N12 21Eb 49
St Andrew's Cl. NW2 34Xa 68
St Andrew's Cl. SL4: Old Win 8L 103
St Andrew's Cl. TW17: Shep 70T 128
St Andrew's Cl. TW19: Wray 58A 104
St Andrew's Cl. TW7: Isle 53Ga 108
St Andrews Cl. GU21: Wok 9N 167
St Andrews Cl. KT7: T Ditt 74Ka 152
St Andrews Cl. RH2: Reig 7K 207
St Andrews Cl. SE16 50Xb 91
St Andrews Cl. SE28 44Zc 95
St Andrews Cl. SW19 65Db 133
St Andrews Cotts. SL4: Wind 4E 102
(off Cross Oak)
St Andrew's Ct. DA12: Grav'nd 8D 122
(off Queen St.)
St Andrew's Ct. SW18 61Eb 133
St Andrews Ct. BR8: Swan 69Gd 140
St Andrews Ct. E17 26Ac 52
St Andrews Ct. RM18: Tilb 4C 122
St Andrews Ct. SL1: Slou 8J 81
(off Upton Pk.)
St Andrews Ct. SM1: Sutt 76Gb 155
St Andrews Ct. WD17: Wat 11X 27
St Andrew's Cres. SL4: Wind 4D 102
St Andrew's Dr. BR5: St M Cry 72Xc 161
St Andrews Dr. AL1: St A 5F 6
St Andrews Dr. HA7: Stan 25La 46
St Andrew's Gdns. 8D 122
St Andrews Gdns. KT11: Cobh 85Y 171
St Andrew's Ga. GU22: Wok 90B 168
St Andrew's Gro. N16 32Tb 71
St Andrew's Hill EC4 3C 224 (44Rb 91)
(not continuous)
St Andrews Ho. KT17: Eps 85Ta 173
(off High St.)
St Andrews Ho. RM8: Dag 35Wc 75
St Andrews Ho. SE16 48Xb 91
(off Southwark Pk. Rd.)
St Andrews Mans. W1 1H 221 (43Jb 90)
(off Dorset St.)
St Andrews Mans. W14 51Ab 110
(off St Andrew's Rd.)
St Andrew's M. N16 32Ub 71
St Andrew's M. SE3 52Jc 115
St Andrews M. SW12 60Mb 112
St Andrew's Pl. CM15: Shenf 19Be 41
St Andrew's Pl. NW1 5K 215 (42Kb 90)
St Andrew's Rd. CR0: C'don 77Sb 157
St Andrew's Rd. CR5: Coul 88Jb 176
St Andrew's Rd. DA12: Grav'nd 9D 122
St Andrew's Rd. DA14: Sidc 62Zc 139
St Andrew's Rd. E11 30Gc 53
St Andrew's Rd. E13 41Kc 93
St Andrew's Rd. E17 26Zb 52
St Andrew's Rd. EN1: Enf 13Tb 33
St Andrew's Rd. HP3: Hem H 6M 3
St Andrew's Rd. IG1: Ilf 31Pc 74
St Andrew's Rd. KT6: Surb 72Ma 153
St Andrew's Rd. N9 17Yb 34
St Andrew's Rd. NW10 37Xa 68
St Andrew's Rd. NW11 30Bb 49
St Andrew's Rd. NW9 32Ta 67
St Andrew's Rd. RM18: Tilb 3A 122
St Andrew's Rd. RM7: Rom 30Fd 56
St Andrew's Rd. SM5: Cars 76Gb 155
St Andrew's Rd. UB10: Uxb 39N 63
St Andrew's Rd. W14 51Ab 110
St Andrew's Rd. W3 45Ua 88
St Andrew's Rd. W7 47Ga 86
St Andrews Rd. WD19: Wat 20Z 27
St Andrew's Sq. KT6: Surb 72Ma 153
St Andrews Sq. W11 44Ab 88
St Andrews Ter. WD19: Wat 22Y 45
St Andrew's Twr. UB1: S'hall 46Ea 86
(off Baird Av.)
St Andrew St. EC4 1A 224 (43Qb 90)
St Andrew's Wlk. KT11: Cobh 87X 171
St Andrews Way SL1: Slou 5B 80
St Andrews Way E3 42Dc 92
St Andrews Way SS17: Stan H 3K 101
St Andrew's Wharf SE1 1K 231 (47Vb 91)
St Anna Rd. EN5: Barn 15Za 30
St Anne's Av. TW19: Stanw 59M 105
St Annes Blvd. RH1: Redh 4B 208
St Anne's Cl. N6 34Jb 70
St Anne's Cl. WD19: Wat 21Y 45
St Annes Cl. RM16: Grays 46Dc 99
St Anne's Ct. BR4: W W'ck 77Gc 159
St Anne's Ct. NW6 39Ab 68
St Anne's Ct. W1 3D 222 (44Mb 90)
St Anne's Dr. RH1: Redh 5A 208
St Annes Dr. Nth. RH1: Redh 4A 208
St Anne's Flats NW1 3D 216 (41Mb 90)
(off Doric Way)
St Anne's Gdns. NW10 41Pa 87
St Annes M. SW20 66Za 132
St Anne's Mt. RH1: Redh 5A 208
St Anne's Pas. E14 44Bc 92
St Anne's Ri. RH1: Redh 5A 208
St Anne's Rd. AL2: Lon C 9H 7
St Anne's Rd. CM15: Mount 11Ee 41
St Anne's Rd. CM15: Shenf 11Ee 41
St Anne's Rd. E11 33Fc 73
St Anne's Rd. HA0: Wemb 36Ma 67
St Anne's Rd. UB9: Hare 27L 43
St Anne's Row E14 44Bc 92
St Annes Ter. IG6: Ilf 22Uc 54
St Anne's Trad. Est. E14 44Bc 92
(off St Anne's Row)
St Anne St. E14 44Bc 92
St Annes Way RH1: Redh 5A 208
St Ann's IG11: Bark 39Sc 74
St Anns GU22: Wok 90A 168
St Ann's Cl. KT16: Chert 72H 149
St Ann's Ct. NW4 27Xa 48
St Ann's Cres. SW18 58Db 111
St Ann's Gdns. NW5 37Jb 70
St Ann's Hill SW18 57Db 111
St Ann's Hill Rd. KT16: Chert 72E 148
ST ANN'S HOSPITAL Harringay 29Sb 51
St Ann's Ho. WC1 4K 217 (41Qb 90)
(off Margery St.)
St Ann's La. SW1 4E 228 (48Mb 90)
St Ann's M. KT16: Chert 73G 148
ST ANN'S PARK 70B 126
St Ann's Pk. Rd. SW18 58Eb 111
St Ann's Pas. SW13 55Ua 110
St Ann's Rd. HA1: Harr 30Ga 46
St Ann's Rd. IG11: Bark 39Sc 74
St Ann's Rd. KT16: Chert 72G 148
(not continuous)
St Ann's Rd. N15 29Rb 51
St Ann's Rd. N9 19Vb 33
St Ann's Rd. SW13 54Va 110
St Ann's Rd. W11 45Za 88
St Ann's Shop. Cen. 30Ga 46
St Ann's St. SW1 3E 228 (48Mb 90)
St Ann's Ter. NW8 1C 214 (40Fb 69)
St Ann's Vs. W11 46Za 88
St Anns Way CR2: S Croy 79Rb 157
St Anns Way TN16: Big H 88Rc 180
St Anselms Ct. SW16 64Nb 134
St Anselm's Pl. W1 4K 221 (45Kb 90)
St Anselm's Rd. UB3: Hayes 47V 84
St Anthony's Av. IG8: Wfd G 23Lc 53
St Anthonys Av. HP3: Hem H 4B 4
St Anthony's Cl. E1 46Wb 91
St Anthony's Cl. E9 37Bc 72
(off Wallis Rd.)
St Anthony's Cl. SW17 61Gb 133
St Anthony's Ct. BR6: Farnb 75Rc 160
St Anthony's Ct. SW17 61Jb 134
St Anthony's Flats NW1 2D 216 (40Mb 70)
(off Aldenham St.)
ST ANTHONY'S HOSPITAL 74Za 154
St Anthony's Way TW14: Felt 56V 106
St Antony's Rd. E7 38Kc 73
St Arvan's Cl. CR0: C'don 76Ub 157
St Asaph Rd. SE4 55Zb 114
St Aubins Ct. N1 39Tb 71
St Aubyn's Av. SW19 64Bb 133
St Aubyn's Av. TW3: Houn 57Ca 107
St Aubyn's Cl. BR6: Orp 76Vc 161
St Aubyns Gdns. BR6: Orp 75Vc 161
St Aubyn's Rd. SE19 65Vb 135
St Audrey Av. DA7: Bex 54Cd 118
St Audreys Cl. AL10: Hat 3D 8
St Augustine Rd. RM16: Grays 9D 100
St Augustine's Av. BR2: Broml 71Nc 160
St Augustine's Av. CR2: S Croy 79Sb 157
St Augustine's Av. HA9: Wemb 34Na 67
St Augustine's Av. W5 40Na 67
St Augustine's Ct. SE1 50Xb 91
(off Lynton Rd.)
St Augustine's Ho. NW1 3D 216 (41Mb 90)
(off Werrington St.)
St Augustine's Mans. SW1 6C 228 (49Lb 90)
(off Bloomburg St.)
St Augustine's Path N5 35Sb 71
St Augustine's Rd. DA17: Belv 49Bd 95
St Augustine's Rd. NW1 38Mb 70
St Augustine's Sports Cen. 40Cb 69
St Austell Cl. HA8: Edg 26Pa 47
St Austell Rd. SE13 54Ec 114
St Awdry's Rd. IG11: Bark 38Tc 74
St Awdry's Wlk. IG11: Bark 38Sc 74
St Barnabas Cl. BR3: Beck 68Ec 136
St Barnabas Cl. SE22 57Ub 113
St Barnabas Ct. HA3: Hrw W 25Ea 46
St Barnabas Ct. HP2: Hem H 2A 4
St Barnabas Gdns. KT8: W Mole 71Ca 151
St Barnabas M. SW1 7J 227 (50Jb 90)
(off St Barnabas St.)
St Barnabas Rd. CR4: Mitc 66Jb 134
St Barnabas Rd. E17 30Cc 52
St Barnabas Rd. IG8: Wfd G 25Kc 53
St Barnabas Rd. SM1: Sutt 78Fb 155
St Barnabas St. SW1 7J 227 (50Jb 90)
St Barnabas Ter. E9 36Zb 72
St Barnabas Vs. SW8 53Nb 112
St Bartholomew's Cl. SE26 63Xb 135
St Bartholomew's Ct. E6 40Nc 74
(off St Bartholomew's Rd.)
ST BARTHOLOMEW'S HOSPITAL 1C 224 (43Rb 91)
St Bartholomew's Hospital Mus. 1C 224 (43Rb 91)
(in St Bartholomew's Hospital)
St Bartholomew's Rd. E6 40Pc 74
St Bart's Cl. AL4: St A 3H 7
St Benedict's Av. DA12: Grav'nd 1F 144
St Benedict's Cl. SW17 64Jb 134
St Benet's Cl. SW17 61Gb 133
St Benet's Gro. SM5: Cars 73Eb 155
St Benet's Pl. EC3 4G 225 (45Tb 91)
St Benjamins Dr. BR6: Prat B 81Yc 181
St Bernards CR0: C'don 76Ub 157
St Bernard's Cl. SE27 63Tb 135
ST BERNARD'S HOSPITAL 47Fa 86
St Bernards Ho. E14 48Ec 92
(off Galbraith St.)
St Bernard's Rd. AL3: St A 1C 6
St Bernard's Rd. E6 39Mc 73
St Bernards Rd. SL3: L'ly 8N 81
St Blaise Av. BR1: Broml 68Kc 137
St Botolph Rd. DA11: Nflt 62Fe 143
St Botolph Row EC3 3K 225 (44Vb 91)
St Botolphs E1 2K 225 (44Vb 91)
(off St Botolph St.)
St Botolph's Av. TN13: S'oaks 96Jd 202
St Botolph's Rd. TN13: S'oaks 96Kd 203
St Botolph St. EC3 2K 225 (44Vb 91)
St Brelades Ct. N1 39Ub 71
St Bride's Av. EC4 3B 224 (44Rb 91)
(off Bride La.)
St Bride's Av. HA8: Edg 25Pa 47
St Brides Cl. DA18: Erith 47Zc 95
St Bride's Crypt Mus. 3B 224 (44Rb 91)
(off Fleet St.)
St Bride's Ho. E3 40Cc 72
(off Ordell Rd.)
St Bride's Pas. EC4 3B 224 (44Rb 91)
(off Salisbury Ct.)
St Bride St. EC4 2B 224 (44Rb 91)
St Catherines GU22: Wok 1N 187
St Catherines KT13: Weyb 76R 150
(off Thames St.)
St Catherine's Apts. E3 41Dc 92
(off Bow Rd.)
St Catherine's Cl. SW17 61Gb 133
St Catherine's Cl. SW20 71Ya 154
St Catherines Cl. KT9: Chess 79Ma 153
St Catherines Ct. W4 48Ua 88
St Catherines Ct. TW13: Felt 60W 106
St Catherines Ct. TW18: Staines 63J 127
St Catherine's Cross RH1: Blet 6L 209
St Catherine's Dr. SE14 54Zb 114
St Catherine's Farm Ct. HA4: Ruis 30S 44
ST CATHERINE'S HOSPICE Caterham 93Ub 197
St Catherines M. SW3 5F 227 (49Hb 89)
St Catherines Pl. TW20: Egh 64C 126
St Catherine's Rd. E4 19Cc 34
St Catherine's Rd. HA4: Ruis 30T 44
St Cecilia Pl. SE3 50Jc 93
St Cecilia Rd. RM16: Grays 9D 100
St Cecilia's Cl. SM3: Sutt 74Ab 154
St Cedd's Ct. RM16: Grays 46De 99
St Chads Cl. KT6: Surb 73La 152
St Chad's Dr. DA12: Grav'nd 2G 144
St Chad's Gdns. RM6: Chad H 31Ad 75
St Chad's Pl. WC1 3G 217 (41Nb 90)
St Chad's Rd. RM6: Chad H 31Ad 75
St Chads Rd. RM16: Grays 1C 122
St Chads Rd. RM18: Grays 4C 122
St Chads Rd. RM18: Tilb 4C 122
St Chad's St. WC1 3G 217 (41Nb 90)
(not continuous)
St Charles Ct. KT13: Weyb 78Q 150
ST CHARLES HOSPITAL 43Za 88
St Charles Pl. KT13: Weyb 78Q 150
St Charles Pl. W10 43Ab 88
St Charles Rd. CM14: B'wood 18Xd 40
St Charles Sq. W10 43Za 88
St Chloe's Ho. E3 40Cc 72
(off Ordell Rd.)
St Christopher Rd. UB8: Cowl 44M 83
St Christopher's Cl. TW7: Isle 53Ga 108
St Christophers Ct. WD3: Chor 15F 24
St Christophers Dr. UB3: Hayes 45X 85
St Christopher's Gdns. CR7: Thor H 69Qb 134
ST CHRISTOPHER'S HOSPICE (SYDENHAM) 64Yb 136
ST CHRISTOPHER'S HOSPICECARE (ORPINGTON) 77Vc 161
St Christopher's Ho. NW1 2C 216 (40Lb 70)
(off Bridgeway St.)
St Christopher's M. SM6: W'gton 78Lb 156
St Christopher's Pl. W1 2J 221 (44Jb 90)
St Clair Cl. IG5: Ilf 26Pc 54
St Clair Cl. RH2: Reig 6L 207
St Clair Cl. RH8: Oxt 2G 210
St Clair Dr. KT4: Wor Pk 76Xa 154
St Clair Ho. E3 41Bc 92
(off British St.)
St Clair Rd. E13 40Kc 73
St Clair's Rd. CR0: C'don 75Ub 157
St Clare Bus. Pk. TW12: Hamp H 65Ea 130
St Clare St. EC3 3K 225 (44Vb 91)
St Clement Cl. UB8: Cowl 44M 83
St Clement's Av. RM20: W Thur 51Xd 120
St Clements Av. E3 41Bc 92
St Clements Av. RM3: Hrld W 26Nd 57
St Clement's Cl. DA11: Nflt 2B 144
St Clement's Ct. EC4 4G 225 (45Tb 91)
(off Clements La.)
St Clement's Ct. N7 37Qb 70
St Clements Ct. EN9: Walt A 5Ec 20
St Clements Ct. RM17: Grays 51Be 121
St Clements Ct. RM19: Purf 49Qd 97
St Clements Ct. SE14 51Zb 114
(off Myers La.)
St Clements Ct. W11 45Za 88
(off Stoneleigh St.)
St Clement's Development E3 41Bc 92
St Clement's Hgts. SE26 63Wb 135
St Clements Ho. E1 1K 225 (43Vb 91)
(off Leyden St.)
St Clement's La. WC2 3J 223 (44Pb 90)
St Clements Mans. SW6 51Za 110
(off Lillie Rd.)
St Clements Rd. DA9: Ghithe 56Yd 120
St Clements Rd. RM20: Grays 52Yd 120
St Clements St. N7 37Qb 70
St Clements Way DA2: Bean 61Wd 142
St Clements Way DA2: Bluew 61Wd 142
St Clements Way DA9: Bluew 57Wd 120
St Clements Way DA9: Ghithe 57Wd 120
St Clements Way RM20: W Thur 51Vd 120
St Clements Yd. SE22 56Vb 113
St Clere TN15: Kems'g 88Vd 184
St Clere Hill Rd. TN15: W King 84Vd 184
St Clere's Hall Golf Course 3K 101
St Cloud Rd. SE27 63Sb 135
St Columba's Cl. DA12: Grav'nd 2F 144
(not continuous)
St Columba's Ct. E15 35Gc 73
(off Janson Rd.)
St Columbas Ho. E17 28Dc 52
St Columb's Ho. W10 43Ab 88
(off Blagrove Rd.)
St Crispin's Cl. NW3 35Gb 69
St Crispin's Cl. UB1: S'hall 44Ba 85
St Crispins Way KT16: Ott 81E 168
St Cross St. EC1 7A 218 (43Qb 90)
St Cuthbert La. UB8: Cowl 44M 83
St Cuthberts Cl. TW20: Eng G 5P 125
St Cuthberts Gdns. HA5: Hat E 24Ba 45
St Cuthbert's Rd. NW2 37Bb 69
St Cuthberts Rd. N13 23Qb 50
St Cyprian's St. SW17 63Hb 133
St David Cl. UB8: Cowl 43M 83
St David's CR5: Coul 89Pb 176
St David's Cl. BR4: W W'ck 73Dc 158
St David's Cl. HA9: Wemb 34Sa 67
St David's Cl. HP3: Hem H 3D 4
St David's Cl. RH2: Reig 5L 207
St David's Cl. SL0: Iver H 39F 62
St Davids Cl. SE16 50Xb 91
(off Masters Dr.)
St Davids Cl. TW19: Stanw 59M 105
St David's Ct. BR1: Broml 69Rc 138
St David's Ct. E17 27Ec 52
St Davids Ct. TW15: Ashf 61P 127
St David's Cres. DA12: Grav'nd 3F 144
St David's Dr. HA8: Edg 25Pa 47
St Davids Dr. TW20: Eng G 6N 125
St Davids M. E18 25Jc 53
St Davids M. E3 41Ac 92
(off Morgan St.)
St David's Pl. NW4 31Xa 68
St Davids Rd. BR8: Hext 65Hd 140
St Davids Sq. E14 50Dc 92
St Denis Rd. SE27 63Tb 135
St Denys Cl. CR8: Purl 82Rb 177
St Denys Cl. GU21: Knap 10H 167
St Dionis Rd. SW6 54Bb 111
St Domingo Ho. SE18 48Pc 94
(off Leda Rd.)
St Donatt's Rd. SE14 53Bc 114
ST DUNSTAN'S 79Bb 155
St Dunstan's All. EC3 5H 225 (45Ub 91)
(off St Dunstans Hill)
St Dunstans Av. W3 45Ta 87
St Dunstan's Cl. UB3: Harl 50V 84
St Dunstan's Ct. EC4 3A 224 (44Qb 90)
St Dunstan's Dr. DA12: Grav'nd 3G 144
St Dunstan's Enterprises 60Cc 114
St Dunstan's Gdns. W3 45Ta 87
St Dunstans Hill SM1: Sutt 78Ab 154
St Dunstans Hill EC3 5H 225 (45Ub 91)
St Dunstan's Ho. WC2 2K 223 (44Qb 90)
(off Chancery La.)
St Dunstan's La. BR3: Beck 72Ec 158
St Dunstan's La. EC3 5H 225 (45Ub 91)
St Dunstans M. E1 43Ac 92
(off White Horse Rd.)
St Dunstan's Rd. E7 37Kc 73
St Dunstan's Rd. SE25 70Vb 135
St Dunstan's Rd. TW13: Felt 62W 128
St Dunstan's Rd. TW4: Cran 54X 107
(not continuous)
St Dunstan's Rd. W6 50Za 88
St Dunstan's Rd. W7 47Ga 86
ST EBBA'S 81Sa 173
St Ebbas Way KT19: Eps 81Sa 173
St Edith Cl. KT18: Eps 86Sa 173
St Edith Cotts. TN15: Kems'g 88Xd 184
St Ediths Ct. TN15: Kems'g 89Qd 183
St Edith's Farm Cott. TN15: Kems'g 90Rd 183
St Edith's Rd. TN15: Kems'g 90Qd 183
St Edmund's Av. HA4: Ruis 30S 44
St Edmund's Cl. NW8 1F 215 (39Hb 69)
St Edmund's Cl. SW17 61Gb 133

St Edmunds Cl. DA18: Erith 47Zc 95
St Edmund's Ct. NW8 1F 215 (39Hb 69)
(off St Edmund's Ter.)
St Edmund's Ct. TN15: W King 81Vd 184
St Edmunds Ct. CR0: C'don 75Rb 157
St Edmunds Dr. HA7: Stan 25Ja 46
St Edmund's La. TW2: Whitt 59Da 107
St Edmund's Rd. DA1: Dart 56Qd 119
St Edmund's Rd. IG1: Ilf 30Pc 54
St Edmund's Rd. N9 17Wb 33
St Edmunds Sq. SW13 51Ya 110
St Edmund's Ter. NW8 1E 214 (39Gb 69)
St Edmunds Wlk. AL4: St A 3H 7
St Edward's Cl. CR0: New Ad 83Fc 179
St Edward's Cl. NW11 30Cb 49
St Edwards Ct. NW11 30Cb 49
St Edwards Way RM1: Rom 29Fd 56
St Egberts Way E4 18Ec 34
St Elizabeth Dr. KT18: Eps 86Sa 173
St Elmo Cl. SL2: Slou 2H 81
St Elmo Cres. SL2: Slou 2H 81
St Elmo Rd. W12 46Va 88
St Elmos Rd. SE16 47Ac 92
St Erkenwald M. IG11: Bark 39Tc 74
St Erkenwald Rd. IG11: Bark 39Tc 74
St Ermin's Hill SW1 3D 228 (48Mb 90)
(off Broadway)
St Ervan's Rd. W10 43Bb 89
St Ethelburga Ct. RM3: Hrld W 26Qd 57
St Eugene Ct. NW6 39Ab 68
(off Salusbury Rd.)
St Faith's Cl. EN2: Enf 11Sb 33
St Faith's Country Pk. 19Wd 40
St Faith's Rd. SE21 60Rb 113
St Fidelis Rd. DA8: Erith 49Fd 96
St Fillans GU22: Wok 88D 168
St Fillans Rd. SE6 60Ec 114
St Francis Av. DA12: Grav'nd 3G 144
St Francis Cl. BR5: Pet W 72Uc 160
St Francis Cl. EN6: Pot B 6Eb 17
St Francis Cl. WD19: Wat 18X 27
ST FRANCIS HOSPICE 20Gd 38
St Francis' Ho. NW1 2D 216 (40Mb 70)
(off Bridgeway St.)
St Francis Pl. KT17: Ewe 82Wa 174
St Francis Pl. SW12 58Kb 112
St Francis Rd. DA8: Erith 49Fd 96
St Francis Rd. SE22 56Ub 113
St Francis Rd. UB9: Den 30H 43
St Francis Way IG1: Ilf 35Tc 74
St Francis Way RM16: Grays 8E 100
St Frideswide's M. E14 44Ec 92
St Gabriel's Cl. E11 33Kc 73
St Gabriel's Cl. E14 43Dc 92
St Gabriels Ct. N11 24Mb 50
St Gabriels Mnr. SE5 53Rb 113
(off Cormont Rd.)
St Gabriels Rd. NW2 36Za 68
St Gabriel Wlk. SE1 5C 230 (49Rb 91)
(off Elephant & Castle)
St George Ga. KT15: Add 76M 149
St George's Antiochian Orthodox Cathedral 3A 216 (41Kb 90)
St George's Av. E7 38Kc 73
St George's Av. KT13: Weyb 79R 150
St George's Av. N7 35Mb 70
St George's Av. NW9 28Ta 47
St George's Av. RM11: Horn 31Pd 77
St George's Av. RM17: Grays 49Ee 99
St George's Av. UB1: S'hall 45Ba 85
St George's Av. W5 47Ma 87
St George's Bldgs. SE1 4B 230 (48Rb 91)
(off St George's Rd.)
St Georges Bus. Pk. KT13: Weyb 81Q 170
St George's Cen. 8D 122
St George's Cir. SE1 3B 230 (48Rb 91)
St George's Cl. HA0: Wemb 34Ja 66
St George's Cl. KT13: Weyb 78S 150
St George's Cl. NW11 30Bb 49
St George's Cl. SW8 53Lb 112
St Georges Cl. SE28 44Zc 95
St Georges Cl. SL4: Wind 3C 102
St George's Ct. CM14: B'wood 17Xd 40
St George's Ct. E6 42Pc 94
St George's Ct. KT13: Weyb 79S 150
St George's Ct. SE1 3B 230 (48Rb 91)
(off Garden Row)
St George's Ct. SW1 7B 228 (50Lb 90)
(off St George's Dr.)
St George's Ct. SW15 56Bb 111
St George's Ct. SW3 5D 226 (49Gb 89)
(off Brompton Rd.)
St George's Ct. SW7 3A 226 (48Eb 89)
St Georges Ct. AL1: St A 2C 6
(off Lemsford Rd.)
St Georges Ct. CM13: Hut 17Ee 41
St Georges Ct. E17 29Fc 53
St Georges Ct. EC4 2B 224 (44Rb 91)
St Georges Ct. HA3: Kenton 30Ja 46
(off Kenton Rd.)
St Georges Ct. KT15: Add 77L 149
St Georges Ct. TN15: Wro 88Be 185
St George's Cres. DA12: Grav'nd 3F 144
St George's Cres. SL1: Slou 5B 80
St George's Dr. SW1 6A 228 (49Kb 90)
St George's Dr. UB10: Ick 34P 63
St George's Dr. WD19: Wat 20Aa 27
ST GEORGE'S FIELD 4E 220 (45Gb 89)
St George's Flds. W2 3E 220 (44Gb 89)
St George's Gdns. KT17: Eps 86Va 174
St George's Gdns. KT6: Surb 75Ra 153
St George's Gro. SW17 62Fb 133
ST GEORGE'S HILL 82R 170
St George's Hill Golf Course 83S 170
ST GEORGE'S HOSPITAL (TOOTING) 64Fb 133
St George's Ho. NW1 2D 216 (40Mb 70)
(off Bridgeway St.)
St Georges Ho. SW11 53Jb 112
(off Charlotte Despard Av.)
St George's Ind. Est. KT2: King T 64Ma 131
St George's Ind. Est. N22 24Rb 51
St George's La. EC3 4G 225 (45Tb 91)
(off Pudding La.)
St George's La. SL5: Asc 9A 124
St George's Leisure Cen. 45Xb 91
St George's Mans. SW1 7E 228 (50Mb 90)
(off Causton St.)
St George's M. NW1 38Hb 69
St George's M. SE1 3A 230 (48Qb 90)
(off Westminster Bri. Rd.)
St George's M. SE8 49Bc 92
St Georges Pde. SE6 61Bc 136
(off Perry Hill)
St George's Path SE4 56Cc 114
(off Adelaide Av.)
St George's Pl. TW1: Twick 60Ja 108
St Georges Pl. KT10: Esh 77Ea 152
St George's Rd. BR1: Broml 68Pc 138
St George's Rd. BR3: Beck 67Dc 136
St George's Rd. BR5: Pet W 72Tc 160
St George's Rd. CR4: Mitc 69Kb 134
St George's Rd. DA14: Sidc 65Zc 139
St George's Rd. E10 34Ec 72
St George's Rd. E7 38Kc 73
St George's Rd. EN1: Enf 10Vb 19
St George's Rd. HP3: Hem H 6L 3
St George's Rd. IG1: Ilf 31Pc 74
St George's Rd. KT13: Weyb 79T 150
St George's Rd. KT15: Add 77L 149
St George's Rd. KT2: King T 66Qa 131
St George's Rd. N13 20Pb 32
St George's Rd. NW11 30Bb 49
St George's Rd. RM9: Dag 36Ad 75
St George's Rd. SE1 3A 230 (48Qb 90)
St George's Rd. SM6: W'gton 78Kb 156
St George's Rd. SW19 66Bb 133
(not continuous)
St George's Rd. TN13: S'oaks 94Kd 203
St George's Rd. TW1: Twick 57Ka 108
St George's Rd. TW13: Hanw 63Z 129
St George's Rd. W4 47Ta 87
St George's Rd. W7 46Ha 86
St George's Rd. WD24: Wat 10X 13
St Georges Rd. BR8: Swan 70Hd 140
(not continuous)
St Georges Rd. TW9: Rich 55Pa 109
St George's Rd. W. BR1: Broml 67Nc 138
St George's Shop. & Leisure Cen. 30Ga 46
St George's Sq. DA3: Lfield 69Ae 143
St George's Sq. E7 38Kc 73
St George's Sq. KT3: N Mald 69Ua 132
St George's Sq. SE8 49Bc 92
(not continuous)
St George's Sq. SW1 7D 228 (50Mb 90)
St Georges Sq. DA11: Grav'nd 8D 122
St Georges Sq. E14 45Ac 92
St George's Sq. M. SW1 7D 228 (50Mb 90)
St George's Ter. E6 41Nc 94
(off Masterman Rd.)
St George's Ter. NW1 38Hb 69
St George's Ter. SE15 52Wb 113
(off Peckham Hill St.)
St George St. W1 4A 222 (45Kb 90)
St George's University of London 64Fb 133
St George's Wlk. CR0: C'don 76Sb 157
St George's Way SE15 51Ub 113
St George's Wharf SE1 1K 231 (47Vb 91)
(off Shad Thames)
St George Wharf SW8 51Nb 112
St Gerards Cl. SW4 57Lb 112
St German's Pl. SE3 53Jc 115
St German's Rd. SE23 60Ac 114
St Giles Av. EN6: S Mim 4Xa 16
St Giles Av. RM10: Dag 38Dd 76
St Giles Av. UB10: Ick 35S 64
St Giles Churchyard EC2 1E 224 (43Sb 91)
St Giles Cir. W1 2E 222 (44Mb 90)
St Giles Cl. BR6: Farnb 78Tc 160
St Giles Cl. RM10: Dag 38Dd 76
St Giles Cl. RM16: Ors 2C 100
St Giles Cl. TW5: Hest 52Aa 107
St Giles Ct. EN1: Enf 7Yb 20
St Giles Ct. HP8: Chal G 19C 24
St Giles High St. WC2 2E 222 (44Mb 90)
St Giles Ho. EN5: New Bar 14Eb 31
St Giles Ho. SE5 53Ub 113
St Giles Quad. HP8: Chal G 19C 24
St Giles Rd. SE5 52Ub 113
St Giles Sq. WC2 2E 222 (44Mb 90)
St Giles Ter. EC2 1E 224 (43Sb 91)
(off Wood St.)
St Giles Twr. SE5 53Ub 113
(off Gables Cl.)
St Gilles Ho. E2 40Zb 72
(off Mace St.)
St Gothard Rd. SE27 63Tb 135
(not continuous)
St Gregory Cl. HA4: Ruis 35Y 65
St Gregory's Ct. DA12: Grav'nd 1G 144
St Gregory's Cres. DA12: Grav'nd 1G 144
St Helena Ho. WC1 4K 217 (41Qb 90)
(off Margery St.)
St Helena Rd. SE16 49Zb 92
St Helena St. WC1 4K 217 (41Qb 90)
St Helena Ter. TW9: Rich 57Ma 109
St Helens KT7: T Ditt 73Ha 152
St Helen's Cl. KT4: Wor Pk 74Wa 154
St Helens Cl. UB8: Cowl 43M 83
St Helen's Ct. RM13: Rain 42Jd 96
St Helen's Ct. CM16: Epp 2Wc 23
(off Hemnall St.)
St Helen's Cres. SW16 67Pb 134
St Helen's Gdns. W10 43Za 88
St Helen's M. CM14: B'wood 19Zd 41
St Helen's Pl. EC3 2H 225 (44Ub 91)
St Helens Pl. E10 31Ac 72
St Helen's Rd. DA18: Erith 47Zc 95
St Helen's Rd. IG1: Ilf 30Pc 54
St Helen's Rd. SW16 67Pb 134
St Helen's Rd. W13 46Ka 86
St Helen's Sports Cen. 23U 44
ST HELIER 73Gb 155
St Helier Av. SM4: Mord 73Eb 155
St Helier Ct. N1 39Ub 71
(off De Beauvoir Est.)
St Helier Ct. SE16 47Zb 92
(off Poolmans St.)
ST HELIER HOSPITAL 74Eb 155
St Helier's Av. TW3: Houn 57Ca 107
St Helier's Rd. E10 30Ec 52
St Helier Station (Rail) 72Cb 155
St Henera's Ct. BR1: Broml 69Qc 138
(off Brady Dr.)
St Hilary's Ct. BR1: Broml 69Rc 138
St Hildas TN15: Plax 99Be 205
St Hilda's Av. TW15: Ashf 64N 127
St Hilda's Cl. GU21: Knap 9H 167
St Hilda's Cl. NW6 38Za 68
St Hilda's Cl. SW17 61Gb 133
St Hilda's Rd. SW13 51Xa 110
St Hilda's Way DA12: Grav'nd 3F 144
St Hilda's Wharf E1 46Yb 92
(off Wapping High St.)
St Huberts Cl. SL9: Ger X 32A 62
St Hubert's Ho. E14 48Cc 92
(off Janet St.)
St Hubert's La. SL9: Ger X 33B 62
St Hughes Cl. SW17 61Gb 133
St Hugh's Rd. SE20 67Xb 135
St Ignatius College Sports Cen. 9Xb 19
St Ives Cl. RM3: Rom 24Pd 57
St Ives Pl. E14 43Ec 92
St Ivian Ct. N10 26Jb 50
St Ivian's Dr. RM2: Rom 27Jd 56
St James Apts. E17 29Ac 52
(off Pretoria Av.)
St James Av. KT17: Ewe 83Va 174
St James Av. N20 20Gb 31
St James Av. SM1: Sutt 78Cb 155
St James Av. W13 46Ja 86
St James Cl. EN4: E Barn 14Fb 31
St James Cl. GU21: Wok 10L 167
St James Cl. HA4: Ruis 33Y 65
St James Cl. KT18: Eps 86Ua 174
St James Cl. KT3: N Mald 71Va 154
St James Cl. N20 20Gb 31
St James Ct. AL1: St A 3E 6
St James Ct. CR0: C'don 73Rb 157
St James Ct. DA9: Ghithe 58Vd 120
St James Ct. E12 33Lc 73
St James Ct. E2 41Wb 91
(off Bethnal Grn. Rd.)
St James Ct. KT13: Weyb 78S 150
(off York Rd.)
St James Ct. KT21: Asht 89Ma 173
St James Ct. RM1: Rom 28Hd 56
St James Ct. SE3 53Kc 115
St James' Ct. SW1 3C 228 (48Lb 90)
St James Dr. RM3: Rom 23Nd 57
St James Gdns. RM6: Chad H 28Xc 55
St James' Gdns. HA0: Wemb 38Ma 67
St James Ga. IG9: Buck H 18Lc 35
St James Ga. SL5: S'dale 3D 146
St James Gro. SW11 54Hb 111
St James Hall N1 39Sb 71
(off Prebend St.)
St James Ho. RM1: Rom 29Hd 56
(off Eastern Rd.)
St James Ind. M. SE1 50Wb 91
St James La. DA9: Dart 60Ud 120
St James La. DA9: Ghithe 60Ud 120
St James Mans. SE1 3K 229 (48Qb 90)
(off McAuley Cl.)
St James' Mans. NW6 38Cb 69
(off West End La.)
St James M. E14 48Ec 92
St James M. E17 29Ac 52
St James M. KT13: Weyb 77R 150
St James Oaks DA11: Grav'nd 9C 122
St James Path E17 29Ac 52
St James Pl. DA1: Dart 58Md 119
St James Pl. SL1: Slou 4A 80
St James Residences W1 4D 222 (45Mb 90)
(off Brewer St.)
St James Rd. CM14: B'wood 20Yd 40
St James Rd. CR4: Mitc 66Jb 134
St James Rd. CR8: Purl 85Rb 177
St James Rd. SM1: Sutt 78Cb 155
St James Rd. SM5: Cars 76Gb 155
St James Rd. WD18: Wat 15X 27
St James' Rd. E15 36Hc 73
St James' Rd. KT6: Surb 72Ma 153
St James' Rd. N9 19Xb 33
St James's SE14 53Ac 114
ST JAMES'S 7D 222 (46Mb 90)
St James's App. EC2 6H 219 (42Ub 91)
St James's Av. BR3: Beck 69Ac 136
St James's Av. DA11: Grav'nd 9C 122
St James's Av. E2 40Yb 72
St James's Av. TW12: Hamp H 64Ea 130
St James's Chambers SW1 6C 222 (46Lb 90)
(off Jermyn St.)
St James's Cl. NW8 1F 215 (39Hb 69)
(off St James's Ter. M.)
St James's Cl. SE18 50Sc 94
St James's Cl. SW17 61Hb 133
St James's Cotts. TW9: Rich 57Ma 109
St James's Ct. HA1: Harr 30Ja 46
St James's Ct. KT1: King T 69Na 131
St James's Ct. N18 22Wb 51
(off Fore St.)
St James's Cres. SW9 55Qb 112
St James's Dr. SW12 60Hb 111
St James's Dr. SW17 60Hb 111
St James's Gdns. W11 46Ab 88
(not continuous)
St James's Ho. SE1 49Wb 91
(off Strathnairn St.)
St James's La. N10 28Kb 50
St James's Mkt. SW1 5D 222 (45Mb 90)
St James's Palace 1C 228 (47Lb 90)
St James's Pk. CR0: C'don 73Sb 157
St James's Pk. 1D 228 (47Mb 90)
St James's Park Station (Underground) 2D 228 (47Mb 90)
St James's Pas. EC3 3J 225 (44Ub 91)
(off Duke's Pl.)
St James's Pl. SW1 7B 222 (46Lb 90)
St James Sq. DA3: Lfield 69Ae 143
(off Park Dr.)
St James's Rd. CR0: C'don 73Rb 157
St James's Rd. DA11: Grav'nd 8C 122
St James's Rd. KT1: King T 68Ma 131
St James's Rd. SE1 51Wb 113
St James's Rd. SE16 48Wb 91
St James's Rd. TN13: S'oaks 94Kd 203
St James's Rd. TW12: Hamp H 64Da 129
St James's Sq. SW1 6C 222 (46Lb 90)
St James's St. DA11: Grav'nd 8C 122
St James's St. E17 29Ac 52
St James's St. SW1 6B 222 (46Lb 90)
St James's Ter. NW8 1F 215 (39Hb 69)
(off Prince Albert Rd.)
St James's Ter. M. NW8 1F 215 (39Hb 69)
St James St. W6 50Ya 88
St James Street Station (Overground) 29Ac 52
St James's Wlk. EC1 5B 218 (42Rb 91)
St James Ter. BR6: Prat B 81Yc 181
(off St Benjamins Dr.)
St James Ter. SW12 60Jb 112
St James Wlk. SL0: Rich P 47G 82
St James Way DA14: Sidc 64Ad 139
St Jeromes Gro. UB3: Hayes 44S 84
St Joan's Ho. NW1 3D 216 (41Mb 90)
(off Phoenix Rd.)
St Joan's Rd. N9 19Vb 33
St John Fisher Rd. DA18: Erith 48Zc 95
St John's RH1: Redh 8N 207
ST JOHN'S 8N 207
ST JOHN'S 94Ld 203
ST JOHNS 10K 167
ST JOHNS 54Cc 114
St John's Av. CM14: W'ley 21Zd 59
St John's Av. KT17: Eps 84Wa 174
St John's Av. KT22: Lea 93Ka 192
St John's Av. N11 22Hb 49
St John's Av. NW10 39Va 68
St John's Av. SW15 57Za 110
St Johns Cvn. Pk. EN2: Enf 9Rb 19
St Johns Chu. Rd. E9 36Yb 72
St John's Cl. EN6: Pot B 5Eb 17
St John's Cl. HA9: Wemb 36Na 67
St John's Cl. KT22: Lea 93La 192
St John's Cl. N20 20Eb 31
(off Rasper Rd.)
St John's Cl. RM13: Rain 38Jd 76
St John's Cl. SW6 52Cb 111
St Johns Cl. HP1: Hem H 4K 3
St Johns Cl. N14 16Lb 32
St Johns Cl. TN16: Big H 88Rc 180
St Johns Cl. UB8: Uxb 39K 63
St John's Cnr. RH1: Redh 8P 207
(off St John's Rd.)
St John's Cotts. SE20 66Yb 136
St John's Ct. AL1: St A 1F 6
St John's Ct. DA8: Erith 49Fd 96
St John's Ct. E1 46Xb 91
(off Scandrett St.)
St John's Ct. GU24: Brkwd 2D 186
St John's Ct. HA1: Harr 30Ha 46
St John's Ct. HA6: Nwood 24U 44
(off Murray Rd.)
St John's Ct. IG9: Buck H 18Kc 35
St John's Ct. KT1: King T 70Na 131
(off Beaufort Rd.)
St John's Ct. N4 33Rb 71
St John's Ct. RM17: Grays 49Ee 99
St John's Ct. SE13 54Ec 114
St John's Ct. TN13: S'oaks 94Ld 203
St John's Ct. TW20: Egh 64C 126
St John's Ct. TW7: Isle 54Ha 108
St John's Ct. W6 49Xa 88
(off Glenthorne Rd.)
St Johns Ct. GU21: Wok 1L 187
St Johns Ct. RH9: S God 10D 210
St Johns Ct. SW10 52Eb 111
(off Ashburnham Rd.)
St John's Cres. SW9 55Qb 112
St John's Dr. KT12: Walt T 74Y 151
St John's Dr. SL4: Wind 4E 102
St Johns Dr. SW18 60Db 111
St John's Est. N1 2G 219 (40Tb 71)
St John's Est. SE1 1K 231 (47Vb 91)
(off Fair St.)
St John's Gdns. GU21: Wok 10L 167
(off St John's Rd.)
St John's Gdns. W11 45Ab 88
St John's Gate 6B 218 (42Rb 91)
(off St John's La.)
St John's Gro. N19 33Lb 70
St John's Gro. SW13 54Va 110
St John's Gro. TW9: Rich 56Na 109
St John's Hill CR5: Coul 89Qb 176
St John's Hill SW11 57Fb 111
St John's Hill TN13: S'oaks 93Ld 203
St John's Hill Gro. SW11 56Fb 111
St John's Hill Rd. GU21: Wok 1L 187
ST JOHN'S HOSPICE London 2B 214 (40Fb 69)
(within Hospital of St John & St Elizabeth)
St John's Ho. E14 49Ec 92
(off Pier St.)
St Johns Ho. SE17 51Tb 113
(off Lytham St.)
St John's Jerusalem 66Rd 141
St John's La. DA3: Hartl 72Be 165
St John's La. EC1 6B 218 (42Rb 91)
St John's Lodge NW3 38Gb 69
(off King Henry's Rd.)
St Johns Lodge GU21: Wok 1L 187
St John's Lye GU21: Wok 1K 187
St John's Mans. EC1 3B 218 (41Rb 91)
(off St John St.)
St John's M. GU21: Wok 1L 187
St John's M. KT1: Hamp W 68La 130
St John's M. W11 44Cb 89
St John Smith Square 4F 229 (48Nb 90)
(off Smith Sq.)
St John's Pde. W13 46Ka 86
St Johns Pde. DA14: Sidc 63Wc 139
(off Sidcup High St.)
St John's Pk. SE3 52Hc 115
St John's Pk. Mans. N19 34Lb 70
St John's Pas. SW19 65Ab 132
St John's Path EC1 6B 218 (42Rb 91)
(off Britton St.)
St Johns Pathway SE23 60Yb 114
St John's Pl. EC1 6B 218 (42Rb 91)
St John's Ri. GU21: Wok 1M 187
St Johns Ri. TN16: Big H 88Rc 180
St John's Rd. BR5: Pet W 72Tc 160
St John's Rd. CM16: Epp 2Vc 23
St John's Rd. CR0: C'don 76Rb 157
St John's Rd. DA12: Grav'nd 9F 122
St John's Rd. DA14: Sidc 63Xc 139
(not continuous)
St John's Rd. DA16: Well 55Xc 117
St John's Rd. DA2: Dart 59Sd 120
St John's Rd. DA8: Erith 50Fd 96
St John's Rd. E16 44Jc 93
St John's Rd. E17 26Dc 52
St John's Rd. E4 21Dc 52
St John's Rd. E6 39Nc 74
St John's Rd. GU21: Wok 1K 187
St John's Rd. HA1: Harr 30Ha 46
St John's Rd. HA9: Wemb 35Ma 67
St John's Rd. HP1: Hem H 4J 3
St John's Rd. IG10: Lough 12Pc 36
St John's Rd. IG11: Bark 39Uc 74
St John's Rd. IG2: Ilf 31Tc 74
St John's Rd. KT1: Hamp W 68La 130
St John's Rd. KT22: Lea 93La 192
St John's Rd. KT3: N Mald 69Sa 131
St John's Rd. KT8: E Mos 70Fa 130
St John's Rd. N15 30Ub 51
St John's Rd. NW11 30Bb 49
St John's Rd. RH1: Redh 8P 207
St John's Rd. SE20 65Yb 136
St John's Rd. SL4: Wind 4E 102
St John's Rd. SM1: Sutt 75Db 155
St John's Rd. SM5: Cars 76Gb 155
St John's Rd. SW11 56Gb 111
St John's Rd. SW19 66Ab 132
St John's Rd. TN13: S'oaks 93Kd 203
St John's Rd. TW13: Hanw 63Aa 129
St John's Rd. TW7: Isle 54Ha 108
St John's Rd. TW9: Rich 56Na 109
St John's Rd. UB2: S'hall 48Aa 85
St John's Rd. UB8: Uxb 39K 63
St John's Rd. WD17: Wat 12X 27
St Johns Rd. RM16: Grays 10D 100
St Johns Rd. RM5: Col R 22Ed 56
St Johns Rd. SL2: Slou 5L 81
St John's Sq. EC1 6B 218 (42Rb 91)
St Johns Sq. SL4: Eton 1H 103
St Johns Station (Rail) 54Cc 114
St John's Ter. E7 37Kc 73
St John's Ter. EN2: Enf 9Tb 19
St John's Ter. SE18 51Sc 116
St John's Ter. SW15 62Ua 132
(off Kingston Va.)
St John's Ter. W10 42Za 88
St John's Ter. Rd. RH1: Redh 8P 207
St John St. EC1 2A 218 (40Qb 70)
St John's Va. SE8 54Cc 114
St John's Vs. N11 22Hb 49
(off Friern Barnet Rd.)
St John's Vs. N19 33Mb 70
St John's Vs. W8 48Db 89
(off St Mary's Pl.)
St Johns Waterside GU21: Wok 10K 167
(off Copse Rd.)
St John's Way KT16: Chert 74J 149
St John's Way N19 33Lb 70
ST JOHN'S WOOD 1C 214 (40Fb 69)
St John's Wood Ct. NW8 4C 214 (41Fb 89)
(off St John's Wood Rd.)
St John's Wood High St. NW8 2C 214 (40Fb 69)
St John's Wood Pk. NW8 39Fb 69
St John's Wood Rd. NW8 5B 214 (42Fb 89)
St John's Wood Station (Underground) 1B 214 (40Fb 69)
St John's Wood Ter. NW8 1C 214 (40Fb 69)
St Josephs Almshouses W6 49Za 88
(off Brook Grn.)
St Joseph's Cl. BR6: Orp 77Vc 161
St Joseph's Cl. W10 43Ab 88
St Joseph's College Sports Cen. 65Rb 135
St Joseph's Cotts. SW3 6F 227 (49Hb 89)
(off Cadagon St.)
St Joseph's Ct. E4 17Fc 35
St Josephs Ct. SE2 51Zc 117
St Josephs Ct. SE7 51Kc 115
St Joseph's Dr. UB1: S'hall 46Aa 85
St Joseph's Flats NW1 3D 216 (41Mb 90)
(off Drummond Cres.)
St Joseph's Gro. NW4 28Xa 48
ST JOSEPH'S HOSPICE London 39Xb 71
St Joseph's Ho. W6 49Za 88
(off Brook Grn.)
St Joseph's Rd. EN8: Walt C 5Ac 20
St Joseph's Rd. N9 17Xb 33
St Joseph's St. SW8 53Kb 112
St Joseph's Va. SE3 55Fc 115
St Jude's Cl. TW20: Eng G 4N 125
St Jude's Cotts. TW20: Eng G 4N 125
St Judes Ct. IG8: Wfd G 24Nc 54
St Jude's Rd. E2 40Xb 71
St Judes Rd. TW20: Eng G 2N 125
St Jude St. N16 36Ub 71
ST JULIANS 5B 6
St Julian's Cl. SW16 63Qb 134
St Julian's Farm Rd. SE27 63Qb 134
St Julian's Rd. AL1: St A 4B 6
St Julian's Rd. NW6 39Cb 69
St Justin Cl. BR5: St P 69Zc 139
St Katharine Docks 5K 225 (45Vb 91)
(off St Katharine's Way)
St Katharine's Pct. NW1 1K 215 (40Kb 70)
St Katharine's Way E1 6K 225 (46Vb 91)
(not continuous)
St Katharine's Yacht Haven 46Vb 91
(off St Katharine's Way)
St Katherine's Rd. DA18: Erith 47Zc 95
St Katherines Rd. CR3: Cat'm 97Wb 197
St Katherine's Row EC3 4J 225 (45Ub 91)
(off Fenchurch St.)
St Katherines Wlk. W11 46Za 88
St Kathryn's Pl. RM14: Upm 33Sd 78
St Keverne Rd. SE9 63Nc 138
St Kilda Rd. BR6: Orp 74Vc 161
St Kilda Rd. W13 46Ja 86
St Kilda's Rd. CM15: B'wood 17Xd 40
St Kilda's Rd. HA1: Harr 30Ga 46
St Kilda's Rd. N16 32Tb 71
St Kitts Ter. SE19 64Ub 135
St Laurence Cl. BR5: St P 69Zc 139
St Laurence Cl. NW6 39Za 68
St Laurence Cl. UB8: Cowl 43L 83
St Laurence Way SL1: Slou 8L 81
St Lawrence Bus. Cen. TW13: Felt 61X 129
St Lawrence Cl. HA8: Edg 24Pa 47
St Lawrence Cl. HP3: Bov 9C 2
St Lawrence Cl. WD5: Ab L 2U 12
St Lawrence Cotts. E14 46Ec 92
(off St Lawrence St.)
St Lawrence Ct. GU24: Chob 3J 167
St Lawrence Ct. N1 38Tb 71
St Lawrence Ct. WD5: Ab L 2U 12
St Lawrence Dr. HA5: Eastc 29X 45
St Lawrence Ho. GU24: Chob 3J 167
(off Bagshot Rd.)
St Lawrence Ho. SE1 3J 231 (48Ub 91)
(off Purbrook St.)
St Lawrence Rd. RM14: Upm 33Sd 78
St Lawrence St. E14 46Ec 92
St Lawrence's Way RH2: Reig 6J 207
St Lawrence Ter. W10 43Ab 88
St Lawrence Way AL2: Brick W 2Ba 13
St Lawrence Way CR3: Cat'm 95Sb 197
St Lawrence Way SW9 53Qb 112
St Leger Ct. NW6 38Za 68
(off Coverdale Rd.)
St Leonard M. N1 1H 219 (40Ub 71)
(off Hoxton St.)
St Leonard's Av. E4 23Fc 53
St Leonard's Av. HA3: Kenton 29La 46
St Leonard's Av. SL4: Wind 4G 102
St Leonard's Cl. DA16: Well 55Wc 117
St Leonard's Cl. WD23: Bush 14Aa 27
St Leonards Cl. RM17: Grays 51Be 121
St Leonard's Ct. N1 3G 219 (41Tb 71)
(off New North Rd.)
St Leonards Ct. SW14 55Sa 109
St Leonard's Gdns. IG1: Ilf 36Sc 74
St Leonard's Gdns. TW5: Hest 52Aa 107
ST LEONARDS HAMLET 32Kd 77
St Leonard's Hill SL4: Wind 6B 102
St Leonard's Ri. BR6: Orp 77Uc 160
St Leonard's Rd. SL4: Wind Imperial Rd. 5E 102
St Leonard's Rd. SL4: Wind Osborne M. 4G 102
St Leonard's Rd. CR0: Wadd 76Rb 157
St Leonard's Rd. E14 43Dc 92
(not continuous)
St Leonard's Rd. KT10: Clay 79Ha 152
St Leonard's Rd. KT18: Tatt C 91Ya 194
St Leonard's Rd. KT6: Surb 71Ma 153
St Leonard's Rd. KT7: T Ditt 72Ja 152
St Leonard's Rd. NW10 42Ta 87
St Leonard's Rd. SL4: Wink 8A 102
St Leonard's Rd. SW14 55Ra 109
St Leonard's Rd. W13 45La 86
(not continuous)
St Leonards Sq. KT6: Surb 71Ma 153
St Leonards Sq. NW5 37Jb 70
St Leonard's St. E3 41Dc 92
St Leonard's Studios SW3 7F 227 (50Hb 89)
(off Smith St.)
St Leonard's Ter. SW3 7F 227 (50Hb 89)
St Leonards Wlk. SW16 66Pb 134
St Leonards Wlk. SL0: Rich P 48H 83
St Leonards Way RM11: Horn 33Kd 77
St Loo Av. SW3 51Gb 111
St Loo Ct. SW3 51Gb 111
(off St Loo Av.)
St Louis Cl. EN6: Pot B 5Eb 17
St Louis Rd. SE27 63Tb 135

St Loy's Rd. N17 26Ub 51
St Lucia Dr. E15 39Hc 73
St Luke Cl. UB8: Cowl 44M 83
ST LUKE'S 5E 218 (42Sb 91)
St Luke's Av. EN2: Enf 10Tb 19
St Luke's Av. IG1: Ilf 36Rc 74
St Luke's Av. SW4 56Mb 112
St Luke's Cl. BR8: Swan 68Fd 140
St Luke's Cl. DA2: Daren 64Ud 142
St Luke's Cl. EC1 5E 218 (42Sb 91)
St Luke's Cl. SE25 72Xb 157
St Lukes Ct. E10 31Dc 72
(off Capworth St.)
St Lukes Ct. GU21: Wok 86E 168
St Lukes Ct. W11 44Bb 89
(off St Luke's Rd.)
St Luke's Est. EC1 4F 219 (41Tb 91)
ST LUKE'S HEALTHCARE FOR THE CLERGY 6B 216 (42Lb 90)
(within Fitzroy Square BMI Hosp.)
ST LUKE'S HOSPICE Harrow 29Ma 47
ST LUKE'S HOSPITAL (BOSTALL HOUSE) 50Yc 95
St Luke's M. W11 44Bb 89
St Lukes M. E14 47Cc 92
(off Strafford St.)
St Luke's Pas. KT2: King T 67Pa 131
St Luke's Path IG1: Ilf 36Rc 74
St Luke's Rd. CR3: Whyt 90Vb 177
St Luke's Rd. SL4: Old Win 8L 103
St Luke's Rd. UB10: Uxb 39N 63
St Luke's Rd. W11 43Bb 89
St Luke's Sq. E16 44Hc 93
St Luke's St. SW3 7D 226 (50Gb 89)
St Luke's Yd. W9 40Bb 69
(not continuous)
St Magnus Ct. HP3: Hem H 4B 4
St Malo Av. N9 20Yb 34
St Margaret Dr. KT18: Eps 86Sa 173
St Margaret's IG11: Bark 39Tc 74
St Margaret's KT2: King T 64Sa 131
ST MARGARETS 66Ud 142
ST MARGARETS 58Ka 108
St Margaret's Av. DA15: Sidc 62Tc 138
St Margaret's Av. HA2: Harr 34Ea 66
St Margaret's Av. N15 28Rb 51
St Margaret's Av. N20 18Eb 31
St Margaret's Av. SM3: Cheam 76Ab 154
St Margaret's Av. SS17: Stan H 3L 101
St Margaret's Av. TW15: Ashf 64R 128
St Margarets Av. TN16: Big H 88Rc 180
St Margarets Av. UB8: Hil 42Q 84
St Margarets Bus. Cen. TW1: Twick 58Ka 108
St Margaret's Cl. BR6: Chels 77Xc 161
St Margaret's Cl. DA2: Dart 61Td 142
St Margaret's Cl. HP4: Berk 2A 2
St Margarets Cl. EC2 2F 225 (44Tb 91)
St Margarets Cl. SL0: Iver H 40F 62
St Margaret's Ct. N11 21Jb 50
St Margarets Ct. SE1 7E 224 (46Sb 91)
St Margarets Ct. SL0: Iver H 40F 62
St Margarets Ct. SW15 56Xa 110
St Margaret's Cres. DA12: Grav'nd 2G 144
St Margaret's Cres. SW15 57Xa 110
St Margaret's Dr. TW1: Twick 57Ka 108
St Margarets Ga. SL0: Iver H 40F 62
St Margaret's Gro. E11 34Hc 73
St Margaret's Gro. SE18 51Sc 116
St Margaret's Gro. TW1: Twick 58Ja 108
ST MARGARET'S HOSPITAL 1Xc 23
St Margaret's Ho. NW1 2D 216 (40Mb 70)
(off Polygon Rd.)
St Margaret's La. W8 48Db 89
St Margaret's M. KT2: King T 64Sa 131
St Margaret's Pas. SE13 55Gc 115
(not continuous)
St Margarets Path SE18 50Sc 94
St Margaret's Rd. CR5: Coul 93Kb 196
St Margaret's Rd. E12 33Lc 73
St Margaret's Rd. HA4: Ruis 30T 44
St Margaret's Rd. HA8: Edg 22Ra 47
St Margaret's Rd. N17 27Ub 51
St Margaret's Rd. NW10 41Ya 88
St Margaret's Rd. W7 47Ga 86
St Margarets Rd. BR3: Beck 70Zb 136
St Margarets Rd. DA11: Nflt 10A 122
St Margarets Rd. DA2: G St G 65Ud 142
St Margarets Rd. DA4: S Dar 66Ud 142
St Margarets Rd. SE4 56Bc 114
(not continuous)
St Margarets Rd. TW1: Twick 58Ka 108
St Margarets Rd. TW7: Isle 56Ka 108
St Margarets Rd. TW7: Twick 56Ka 108
ST MARGARETS RDBT. 58Ka 108
St Margaret's Sports Cen. 18Da 27
St Margarets Station (Rail) Twickenham 58Ka 108
St Margaret's Ter. SE18 50Sc 94
St Margaret St. SW1 2F 229 (47Nb 90)
St Margarets Way HP2: Hem H 2C 4
(not continuous)
St Margaret Way SL1: Slou 7D 80
(off Mathecombe Rd.)
St Mark's Av. DA11: Nflt 9A 122
St Mark's Cl. EN5: New Bar 13Db 31
St Mark's Cl. SE10 52Ec 114
St Mark's Cl. W11 44Ab 88
St Marks Cl. AL4: Col H 4M 7
St Marks Cl. HA1: Harr 31Ka 66
St Marks Cl. SW6 53Cb 111
St Mark's Ct. GU22: Wok 91A 188
(off Brooklyn Rd.)
St Marks Ct. NW8 2A 214 (40Eb 69)
(off Abercorn Pl.)
St Marks Ct. W7 47Ga 86
(off Lwr. Boston Rd.)
St Mark's Cres. NW1 39Jb 70
St Mark's Ga. E9 38Bc 72
St Mark's Gro. SW10 52Db 111
St Mark's Hill KT6: Surb 72Na 153
ST MARK'S HOSPITAL (HARROW) 31Ka 66
St Marks Ho. SE17 51Tb 113
(off Lytham St.)
St Marks Ind. Est. E16 46Mc 93
St Mark's Pl. RM10: Dag 37Cd 76
St Mark's Pl. SW19 65Bb 133
St Mark's Pl. W11 44Ab 88
St Marks Pl. SL4: Wind 4G 102
St Mark's Ri. E8 36Vb 71
St Mark's Rd. BR2: Broml 69Jc 137
St Mark's Rd. KT18: Tatt C 90Ya 174
St Mark's Rd. SE25 70Wb 135
St Mark's Rd. TW11: Tedd 66Ka 130
St Mark's Rd. W10 44Za 88
St Mark's Rd. W11 44Ab 88
St Mark's Rd. W5 46Na 87
St Mark's Rd. W7 47Ga 86
St Marks Rd. CR4: Mitc 68Hb 133
St Marks Rd. EN1: Enf 16Vb 33
St Marks Rd. SL4: Wind 4G 102
St Mark's Sq. BR2: Broml 69Jc 137
St Mark's Sq. NW1 39Jb 70
St Mark St. E1 3K 225 (44Vb 91)
St Mark's Vs. N4 33Pb 70
(off Moray Rd.)
St Martha's Av. GU22: Wok 93B 188
St Martin Cl. UB8: Cowl 44M 83
St Martin-in-the-Fields Chu. Path WC2 5F 223 (45Nb 90)
(off St Martin's Pl.)
St Martins HA6: Nwood 22T 44
St Martin's Almshouses NW1 39Lb 70
St Martin's App. HA4: Ruis 31U 64
St Martin's Av. E6 40Mc 73
St Martin's Av. KT18: Eps 86Ua 174
St Martin's Cl. CM13: Hut 19Ee 41
St Martin's Cl. DA18: Erith 47Zc 95
St Martin's Cl. EN1: Enf 11Xb 33
St Martin's Cl. NW1 39Lb 70
St Martin's Cl. UB7: W Dray 48M 83
St Martins Cl. KT17: Eps 85Va 174
St Martins Cl. KT24: E Hor 100U 190
St Martins Cl. WD19: Wat 21Y 45
St Martin's Ct. EC4 2D 224 (44Sb 91)
(off Newgate St.)
St Martin's Ct. KT24: E Hor 100U 190
St Martin's Ct. TW15: Ashf 64L 127
St Martin's Ct. WC2 4F 223 (45Nb 90)
St Martins Ct. N1 39Ub 71
(off De Beauvoir Est.)
St Martin's Ctyd. WC2 4F 223 (45Nb 90)
(off Up. St Martin's La.)
St Martin's Dr. DA4: Eyns 77Md 163
St Martins Dr. KT12: Walt T 76Y 151
St Martins Est. SW2 60Qb 112
St Martin's Ho. NW1 3D 216 (41Mb 90)
(off Polygon Rd.)
St Martin's La. BR3: Beck 71Dc 158
St Martin's La. WC2 4F 223 (45Nb 90)
St Martin's Le-Grand EC1 2D 224 (44Sb 91)
St Martins Mdw. TN16: Bras 95Yc 201
St Martins M. GU22: Pyr 88J 169
St Martin's Pl. WC2 5F 223 (45Nb 90)
St Martins Pl. RM3: Rom 21Md 57
St Martin's Rd. DA1: Dart 58Pd 119
St Martin's Rd. N9 19Xb 33
St Martin's Rd. SW9 54Pb 112
St Martin's Rd. UB7: W Dray 48L 83
St Martin's St. WC2 5E 222 (45Mb 90)
(not continuous)
St Martin's Theatre 4F 223 (45Nb 90)
(off West St.)
St Martins Way SW17 62Eb 133
St Mary Abbot's Ct. W14 48Bb 89
(off Warwick Gdns.)
St Mary Abbot's Pl. W8 48Bb 89
St Mary Abbot's Ter. W14 48Bb 89
St Mary at Hill EC3 5H 225 (45Ub 91)
St Mary Av. SM6: W'gton 76Jb 156
St Mary Axe EC3 3H 225 (44Ub 91)
St Marychurch St. SE16 47Yb 92
ST MARY CRAY 70Yc 139
St Mary Cray Station (Rail) 70Xc 139
St Mary Graces Ct. E1 45Vb 91
St Marylebone Cl. NW10 39Ua 68
St Marylebone Crematorium 27Db 49
St Mary le-Park Ct. SW11 52Gb 111
(off Parkgate Rd.)
St Mary Magdalene Cres. SE18 48Qc 94
St Mary Magdalene Gdns. N7 37Qb 70
St Mary Newington Cl. SE17 7J 231 (50Ub 91)
(off Surrey Sq.)
St Mary Rd. E17 28Cc 52
St Mary's IG11: Bark 39Tc 74
St Marys KT13: Weyb 76T 150
St Mary's App. E12 36Pc 74
St Mary's Av. BR2: Broml 69Gc 137
St Mary's Av. CM15: Shenf 15Ce 41
St Mary's Av. E11 31Kc 73
St Mary's Av. HA6: Nwood 22U 44
St Mary's Av. N3 26Ab 48
St Mary's Av. TW11: Tedd 65Ha 130
St Mary's Av. TW19: Stanw 59M 105
St Mary's Av. Central UB2: S'hall 49Da 85
St Mary's Av. Nth. UB2: S'hall 49Da 85
St Mary's Av. Sth. UB2: S'hall 49Da 85
St Mary's Chu. Rd. TN14: Sund 97Bd 201
St Mary's Cl. BR5: St P 68Xc 139
St Mary's Cl. DA12: Grav'nd 1E 144
St Mary's Cl. IG10: Lough 14Nc 36
St Mary's Cl. KT17: Ewe 80Va 154
St Mary's Cl. KT22: Fet 95Fa 192
St Mary's Cl. KT9: Chess 80Pa 153
St Mary's Cl. N17 25Wb 51
St Mary's Cl. RH8: Oxt 1J 211
St Mary's Cl. TN15: Plat 92Ee 205
St Mary's Cl. TW16: Sun 70W 128
St Mary's Cl. TW19: Stanw 59M 105
St Mary's Cl. UB9: Hare 27K 43
St Mary's Cl. WD18: Wat 14Y 27
(off George St.)
St Marys Cl. HP1: Hem H 1L 3
St Marys Cl. RM17: Grays 51Fe 121
St Mary's Community Gdn. 1K 219 (40Vb 71)
(off Appleby St.)
St Mary's Copse KT4: Wor Pk 75Ua 154
St Mary's Ct. E3 41Dc 92
(off Bow Rd.)
St Mary's Ct. E6 42Pc 94
St Mary's Ct. EN6: Pot B 4Db 17
St Mary's Ct. HP2: Hem H 1M 3
St Mary's Ct. KT3: N Mald 69Ua 132
St Mary's Ct. SE7 52Mc 115
St Mary's Ct. SM6: W'gton 77Lb 156
St Mary's Ct. TN16: Westrm 98Tc 200
St Mary's Ct. W5 47Ma 87
St Mary's Ct. W6 48Va 88
St Marys Ct. WD3: Rick 18N 25
St Mary's Cres. NW4 27Xa 48
St Mary's Cres. TW19: Stanw 59M 105
St Mary's Cres. TW7: Isle 52Fa 108
St Mary's Cres. UB3: Hayes 45V 84
St Mary's Dr. TN13: Riv 95Gd 202
St Mary's Dr. TW14: Bedf 59S 106
St Mary's Est. SE16 47Yb 92
(off Elephant La.)
St Mary's Flats NW1 3D 216 (41Mb 90)
(off Drummond Cres.)
St Marys Gdn. GU3: Worp 9J 187
St Mary's Gdns. SE11 5A 230 (49Qb 90)
St Mary's Ga. W8 48Db 89
St Mary's Grn. N2 26Eb 49
St Mary's Grn. TN16: Big H 90Lc 179
St Mary's Gro. N1 37Rb 71
St Mary's Gro. SW13 55Xa 110
St Mary's Gro. TN16: Big H 90Lc 179
St Mary's Gro. TW9: Rich 56Pa 109
St Mary's Gro. W4 51Ra 109
St Mary's Hill SL5: S'hill 2A 146
ST MARY'S HOSPITAL London 2C 220 (44Fb 89)
St Mary's Ho. N1 39Rb 71
(off St Mary's Path)
St Mary's La. CM13: W H'dn 31De 79
St Mary's La. RM14: Upm 33Qd 77
St Mary's Mans. W2 7B 214 (43Fb 89)
St Mary's M. NW6 38Db 69
St Marys M. TW10: Ham 61La 130
St Mary's Mt. CR3: Cat'm 96Vb 197
St Marys Open Space 39Ua 68
(off Challenge Cl.)
St Mary's Path N1 39Rb 71
St Mary's Pl. SE9 58Qc 116
St Mary's Pl. W5 47Ma 87
St Mary's Pl. W8 48Db 89
ST MARYS PLATT 92Ee 205
St Mary's Rd. KT6: Surb St Chads Cl. 73La 152
St Mary's Rd. KT6: Surb Victoria Rd. 72Ma 153
St Mary's Rd. BR8: Swan 70Fd 140
St Mary's Rd. CR2: Sande 82Tb 177
St Mary's Rd. DA5: Bexl 60Ed 118
St Mary's Rd. DA9: Ghithe 57Ud 120
St Mary's Rd. E10 34Ec 72
St Mary's Rd. E13 40Kc 73
St Mary's Rd. EN4: E Barn 17Hb 31
St Mary's Rd. EN8: Chesh 1Yb 20
St Mary's Rd. GU21: Wok 9N 167
St Mary's Rd. HP2: Hem H 1M 3
St Mary's Rd. IG1: Ilf 33Sc 74
St Mary's Rd. KT13: Weyb 77T 150
St Mary's Rd. KT22: Lea 94Ka 192
St Mary's Rd. KT4: Wor Pk 75Ua 154
St Mary's Rd. KT8: E Mos 71Fa 152
St Mary's Rd. N8 28Nb 50
St Mary's Rd. N9 18Xb 33
(not continuous)
St Mary's Rd. NW10 39Ua 68
St Mary's Rd. NW11 31Ab 68
St Marys Rd. RH2: Reig 7K 207
St Mary's Rd. RM16: Grays 8D 100
St Mary's Rd. SE15 53Yb 114
St Mary's Rd. SE25 69Ub 135
St Mary's Rd. SL3: L'ly 46A 82
St Mary's Rd. SL5: Asc 3A 146
St Mary's Rd. SW19 64Ab 132
St Mary's Rd. TN15: Wro 89Ce 185
St Mary's Rd. UB3: Hayes 45V 84
St Mary's Rd. UB9: Den 30H 43
St Mary's Rd. UB9: Hare 27K 43
St Mary's Rd. WD18: Wat 14X 27
St Marys Rd. W5 47Ma 87
St Mary's Sq. W2 7B 214 (43Fb 89)
St Mary's Sq. W5 47Ma 87
St Mary's Sq. WD18: Wat 14Y 27
St Mary's Ter. W2 7B 214 (43Fb 89)
St Mary's Twr. EC1 6E 218 (42Sb 91)
(off Fortune St.)
St Mary St. SE18 49Pc 94
St Mary's University Coll. 62Ha 130
St Mary's University College Sports Cen. 63Ha 130
St Mary's Vw. HA3: Kenton 29La 46
St Mary's Vw. WD18: Wat 14Y 27
(off King St.)
St Mary's Wlk. RH1: Blet 5K 209
St Mary's Wlk. SE11 5A 230 (49Qb 90)
St Mary's Wlk. UB3: Hayes 45V 84
St Mary's Way DA3: Lfield 69Ae 143
St Mary's Way IG7: Chig 22Qc 54
St Mary's Way SL9: Chal P 26A 42
St Matthew Cl. UB8: Cowl 44M 83
St Matthew's Av. KT6: Surb 74Na 153
St Matthew's Cl. RM13: Rain 38Jd 76
St Matthews Cl. WD19: Wat 16Z 27
St Matthew's Ct. TW15: Ashf 63Q 128
(off Feltham Rd.)
St Matthews Ct. E10 31Dc 72
St Matthews Ct. N10 26Jb 50
St Matthews Ct. SE1 4D 230 (48Sb 91)
(off Meadow Row)
St Matthew's Dr. BR1: Broml 69Pc 138
St Matthews Ho. SE17 51Tb 113
(off Phelp St.)
St Matthew's Lodge NW1 1C 216 (40Lb 70)
(off Oakley Sq.)
St Matthew's Rd. RH1: Redh 5P 207
St Matthew's Rd. SW2 56Pb 112
St Matthew's Rd. W5 46Na 87
St Matthew's Row E2 41Wb 91
St Matthew St. SW1 4D 228 (48Mb 90)
St Matthias Cl. NW9 29Va 48
St Maur Rd. SW6 53Bb 111
St Mawes Cl. WD3: Crox G 14R 26
St Meddens BR7: Chst 66Tc 138
St Mellion Cl. SE28 44Zc 95
St Merryn Cl. SE18 52Tc 116
St Merryn Ct. BR3: Beck 66Cc 136
St Michael's RH8: Limp 2L 211
St Michaels AL3: St A 2P 5
St Michael's All. EC3 3G 225 (44Tb 91)
St Michael's Av. HA9: Wemb 37Qa 67
St Michael's Av. N9 17Yb 34
St Michaels Av. HP3: Hem H 4B 4
St Michael's Cl. BR1: Broml 69Nc 138
St Michael's Cl. DA18: Erith 47Zc 95
St Michael's Cl. KT12: Walt T 75Y 151
St Michael's Cl. KT4: Wor Pk 75Va 154
St Michael's Cl. N12 22Gb 49
St Michael's Cl. N3 26Bb 49
St Michael's Cl. RM15: Avel 45Sd 98
St Michaels Cl. E16 43Mc 93
St Michael's Ct. CR0: C'don 74Sb 157
(off Poplar Wlk.)
St Michael's Ct. KT13: Weyb 78S 150
(off Pine Gro.)
St Michael's Ct. SE1 2E 230 (47Sb 91)
(off Trinity St.)
St Michaels Ct. E14 43Ec 92
(off St Leonard's Rd.)
St Michaels Ct. SL2: Slou 2B 80
St Michael's Cres. HA5: Pinn 30Aa 45
St Michael's Dr. TN14: Otf 88Md 183
St Michaels Dr. WD25: Wat 5X 13
St Michael's Flats NW1 2D 216 (40Mb 70)
(off Aldenham St.)
St Michael's Gdns. W10 43Ab 88
ST MICHAEL'S HOSPITAL Enfield 11Tb 33
St Michael's M. SW1 6H 227 (49Jb 90)
St Michael's Pde. WD24: Wat 10X 13
St Michael's Ri. DA16: Well 53Xc 117
St Michael's Rd. CR0: C'don 74Sb 157
St Michael's Rd. CR3: Cat'm 94Tb 197
St Michael's Rd. DA16: Well 55Xc 117
St Michael's Rd. GU21: Wok 86F 168
St Michael's Rd. NW2 35Ya 68
St Michael's Rd. RM16: Grays 10D 100
St Michael's Rd. SM6: W'gton 79Lb 156
St Michael's Rd. SW9 54Pb 112
St Michael's Rd. TW15: Ashf 64Q 128
St Michael's St. AL3: St A 2P 5
St Michael's St. W2 2C 220 (44Fb 89)
St Michael's Ter. N22 25Nb 50
St Michaels Ter. N6 32Jb 70
(off South Gro.)
St Michaels Way EN6: Pot B 2Db 17
St Mildred's Ct. EC2 3F 225 (44Tb 91)
St Mildreds Rd. SE12 59Hc 115
St Mildreds Rd. SE6 59Gc 115
St Mirren Ct. EN5: New Bar 15Eb 31
St Monica's Rd. KT20: Kgswd 93Bb 195
St Nazaire Cl. TW20: Egh 64E 126
St Neots Cl. WD6: Bore 10Qa 15
St Neot's Rd. RM3: Rom 24Pd 57
St Nicholas Av. KT23: Bookh 97Da 191
St Nicholas Av. RM12: Horn 34Jd 76
St Nicholas Cen. 78Db 155
St Nicholas Cl. UB8: Cowl 44M 83
St Nicholas Cl. WD6: E'tree 16Ma 29
St Nicholas Ct. KT1: King T 70Na 131
(off Surbiton Rd.)
St Nicholas Ct. TN13: S'oaks 97Kd 203
(off Lime Tree Wlk.)
St Nicholas Cres. GU22: Pyr 88J 169
St Nicholas Dr. TN13: S'oaks 98Kd 203
St Nicholas Dr. TW17: Shep 73Q 150
St Nicholas' Flats NW1 2D 216 (40Mb 70)
(off Werrington St.)
St Nicholas Glebe SW17 64Jb 134
St Nicholas Gro. CM13: Ingve 22Ee 59
St Nicholas Hill KT22: Lea 94Ka 192
St Nicholas Ho. SE8 51Cc 114
(off Deptford Grn.)
St Nicholas Ho's. EN2: Enf 7Jb 18
St Nicholas M. KT7: T Ditt 72Ha 152
St Nicholas Mt. HP1: Hem H 2H 3
St Nicholas Pl. IG10: Lough 14Qc 36
St Nicholas Rd. KT7: T Ditt 72Ha 152
St Nicholas Rd. SE18 50Vc 95
St Nicholas Rd. SM1: Sutt 78Db 155
St Nicholas St. SE8 53Bc 114
St Nicholas Way SM1: Sutt 77Db 155
St Nicolas La. BR7: Chst 67Nc 138
St Ninian's Ct. N20 20Hb 31
St Norbert Grn. SE4 56Ac 114
St Norbert Rd. SE4 57Zb 114
St Normans Way KT17: Ewe 82Wa 174
St Olaf Ho. SE1 6G 225 (46Tb 91)
(off Tooley St.)
St Olaf's Rd. SW6 52Ab 110
St Olaf Stairs SE1 6G 225 (46Tb 91)
(off Tooley St.)
St Olaves Cl. TW18: Staines 66H 127
St Olave's Ct. EC2 3F 225 (44Tb 91)
St Olave's Est. SE1 1J 231 (47Ub 91)
St Olave's Gdns. SE11 5K 229 (49Qb 90)
St Olaves Ho. SE11 5K 229 (49Qb 90)
(off Walnut Tree Wlk.)
St Olave's Mans. SE11 5K 229 (49Qb 90)
(off Walnut Tree Wlk.)
St Olave's Rd. E6 39Qc 74
St Olaves Wlk. SW16 68Lb 134
St Olav's Sq. SE16 47Yb 92
St Onge Pde. EN1: Enf 13Tb 33
(off Southbury Rd.)
St Oswald's Pl. SE11 7H 229 (50Pb 90)
St Oswald's Rd. SW16 67Rb 135
St Oswalds Studios SW6 51Cb 111
(off Sedlescombe Rd.)
St Oswulf St. SW1 6E 228 (49Mb 90)
St Owen Ho. SE1 3J 231 (48Ub 91)
(off St Saviour's Est.)
ST PANCRAS 4G 217 (41Nb 90)
St Pancras Commercial Cen. NW1 39Lb 70
(off Pratt St.)
St Pancras Ct. N2 26Fb 49
St Pancras Gdns. NW1 1E 216 (40Mb 70)
(off Pancras Rd.)
St Pancras Gdns. 1D 216 (40Mb 70)
ST PANCRAS HOSPITAL 1D 216 (39Mb 70)
St Pancras International Station (Rail) 2F 217 (40Nb 70)
St Pancras Way NW1 38Lb 70
St Patrick's Ct. IG8: Wfd G 24Gc 53
St Patrick's Gdns. DA12: Grav'nd 2F 144
St Patrick's Pl. RM16: Grays 9D 100
St Patricks Pl. RM3: Rom 25Kd 57
St Paul Cl. UB8: Cowl 43M 83
St Paulinus Ct. DA1: Cray 56Gd 118
(off Manor Rd.)
St Paul's All. EC4 3C 224 (44Rb 91)
(off St Paul's Chyd.)
St Paul's Av. HA3: Kenton 28Pa 47
St Paul's Av. NW2 37Ya 68
St Paul's Av. SE16 46Zb 92
St Pauls Av. SL2: Slou 5K 81
St Paul's Bldgs. EC1 5C 218 (42Rb 91)
(off Dallington St.)
St Paul's Cathedral 3D 224 (44Sb 91)
St Paul's Churchyard EC4 3C 224 (44Rb 91)
St Paul's Cl. DA10: Swans 59Ae 121
St Paul's Cl. KT15: Add 78J 149
St Paul's Cl. KT9: Chess 77Ma 153
St Paul's Cl. RM15: Avel 45Sd 98
St Paul's Cl. SM5: Cars 74Gb 155
St Paul's Cl. TW15: Ashf 64S 128
St Paul's Cl. TW3: Houn 54Aa 107
St Paul's Cl. UB3: Harl 50T 84
St Paul's Cl. W5 47Pa 87
St Pauls Cl. SE7 50Mc 93
St Pauls Cl. WD6: Bore 15Sa 29
St Paul's Ct. TW4: Houn 55Aa 107
St Pauls Ct. SW4 57Mb 112
St Pauls Ct. WD4: Bucks 5M 11
St Pauls Ct. WD4: Chfd 3J 11
St Pauls Ctyd. SE8 52Cc 114
(off Crossfield St.)
ST PAUL'S CRAY 68Xc 139
St Paul's Cray Rd. BR7: Chst 67Tc 138
St Paul's Cres. NW1 38Mb 70
(not continuous)
St Paul's Dr. E15 36Fc 73
St Pauls Ho. SE8 52Cc 114
(off Market Yd.)
St Paul's M. NW1 38Mb 70
St Paul's Pl. N1 37Tb 71
St Pauls Pl. AL1: St A 2E 6
St Pauls Pl. RM15: Avel 45Sd 98
St Paul's Ri. N13 23Rb 51
St Paul's Rd. CR7: Thor H 69Sb 135
St Paul's Rd. DA8: Erith 52Ed 118
St Paul's Rd. GU22: Wok 89C 168
St Paul's Rd. HP2: Hem H 1M 3
St Paul's Rd. IG11: Bark 39Sc 74
St Paul's Rd. N1 37Rb 71
St Paul's Rd. N17 24Wb 51
St Paul's Rd. TW18: Staines 64F 126
St Paul's Rd. TW8: Bford 51Ma 109
St Paul's Rd. TW9: Rich 55Pa 109
St Paul's Shrubbery N1 37Tb 71
St Paul's Sq. BR2: Broml 68Hc 137
St Paul's Station (Underground) 2D 224 (44Sb 91)
St Paul's Studios W14 50Ab 88
(off Talgarth Rd.)
St Pauls Ter. SE17 51Rb 113
St Paul St. N1 1D 218 (39Sb 71)
(not continuous)
St Pauls Vw. Apts. EC1 4K 217 (41Qb 90)
(off Amwell St.)
St Paul's Wlk. KT2: King T 66Qa 131
St Paul's Way E3 43Bc 92
St Paul's Way N3 24Db 49
St Pauls Way EN9: Walt A 5Fc 21
St Pauls Way WD24: Wat 12Y 27
St Paul's Wood Hill BR5: St P 68Uc 138
St Peters HP8: Chal G 19C 24
St Peters KT16: Chert 76E 148
St Peter's All. EC3 3G 225 (44Tb 91)
(off Gracechurch St.)
St Peter's Av. E17 28Gc 53
St Peter's Av. E2 40Wb 71
St Peter's Av. N18 21Wb 51
St Peters Av. TN16: Big H 88Rc 180
St Petersburgh M. W2 45Db 89
St Petersburgh Pl. W2 45Db 89
St Peter's Cen. E1 46Xb 91
(off Reardon St.)
St Peter's Chu. Ct. N1 1C 218 (39Rb 71)
(off St Peter's St.)
St Peter's Cl. AL1: St A 1B 6
St Peter's Cl. BR7: Chst 66Tc 138
St Peter's Cl. DA10: Swans 59Be 121
St Peter's Cl. E2 40Wb 71
St Peter's Cl. EN5: Barn 15Xa 30
St Peter's Cl. GU22: Wok 92E 188
St Peter's Cl. HA4: Ruis 33Z 65
St Peter's Cl. IG2: Ilf 28Uc 54
St Peter's Cl. SL1: Burn 2A 80
St Peter's Cl. SL4: Old Win 7L 103
St Peter's Cl. SW17 61Gb 133
St Peter's Cl. TW18: Staines 65H 127
St Peters Cl. SL9: Chal P 25A 42
St Peters Cl. WD23: B Hea 18Fa 28
St Peters Cl. WD3: Rick 18K 25
St Peter's Ct. NW4 29Ya 48
St Peter's Ct. SL9: Chal P 25A 42
St Peter's Ct. WC1 4G 217 (41Nb 90)
(off Seaford St.)
St Peters Ct. E1 42Yb 92
(off Cephas St.)
St Peters Ct. KT8: W Mole 70Ca 129
St Peters Ct. SE12 57Hc 115
St Peter's Gdns. SE27 62Qb 134
St Peter's Gro. W6 49Wa 88
ST PETER'S HOSPITAL Chertsey 76F 148
St Peter's Ho. WC1 4G 217 (41Nb 90)
(off Regent Sq.)
St Peters Ho. SE17 51Tb 113
St Peter's La. BR5: St P 68Wc 139
St Peter's M. AL1: St A 1B 6
St Peters M. N4 29Rb 51
St Peters M. N8 29Qb 50
St Peter's Path E17 27Gc 53
St Peters Pl. W9 42Db 89
St Peter's Rd. AL1: St A 2C 6
St Peter's Rd. CM14: W'ley 21Xd 58
St Peter's Rd. CR0: C'don 77Tb 157
St Peter's Rd. GU22: Wok 93D 188
St Peter's Rd. KT1: King T 68Qa 131
St Peter's Rd. KT8: W Mole 70Ca 129
St Peter's Rd. N9 18Xb 33
St Peter's Rd. RM16: Grays 9D 100
St Peter's Rd. TW1: Twick 57Ka 108
St Peter's Rd. UB1: S'hall 43Ca 85
St Peter's Rd. W6 50Wa 88
St Peters Rd. SL4: Old Win 7L 103
St Peters Rd. UB8: Cowl 43M 83
St Peter's Sq. E2 40Wb 71
St Peter's Sq. W6 49Va 88
St Peter's St. AL1: St A 2B 6
(not continuous)
St Peter's St. CR2: S Croy 78Tb 157
St Peter's St. N1 1C 218 (39Rb 71)
St Peter's St. M. N1 1C 218 (40Rb 71)
(off St Peters St.)
St Peter's Ter. SW6 52Bb 111
St Peter's Vs. W6 49Wa 88
St Peter's Way KT15: Add 77G 148
St Peter's Way KT16: Chert 77F 148
St Peter's Way N1 38Ub 71
St Peter's Way W5 43Ma 87
St Peters Way UB3: Harl 50T 84
St Peters Way WD3: Chor 14D 24
St Peter's Wharf W4 50Wa 88
St Philip Ho. WC1 4K 217 (41Qb 90)
(off Lloyd Baker St.)
St Philip's Av. KT4: Wor Pk 75Xa 154
St Philip's Ga. KT4: Wor Pk 75Xa 154
St Philip Sq. SW8 54Kb 112
St Philip's Rd. E8 37Wb 71
St Philips Rd. KT6: Surb 72Ma 153
St Philip St. SW8 54Kb 112
St Philip's Way N1 39Sb 71
St Pinnock Av. TW18: Staines 67J 127
St Quentin Ho. SW18 58Fb 111
St Quentin Rd. DA16: Well 55Vc 117
St Quintin Av. W10 43Ya 88
St Quintin Gdns. W10 43Ya 88
St Quintin Ho. W10 43Za 88
(off Princess Louise Wlk.)
St Quintin Rd. E13 41Kc 93
St Quintin Vw. W10 43Ya 88
St Raphaels Ct. AL1: St A 1C 6
(off Avenue Rd.)
ST RAPHAEL'S HOSPICE 75Za 154
St Raphael's Way NW10 36Sa 67
St Regis Cl. N10 26Kb 50
St Regis Hgts. NW3 34Db 69
St Richard's Ho. NW1 3D 216 (41Mb 90)
(off Eversholt St.)
St Ronan's Cl. EN4: Had W 10Fb 17
St Ronan's Cres. IG8: Wfd G 24Jc 53
St Ronans Vw. DA1: Dart 59Pd 119
St Rule St. SW8 54Lb 112
St Saviour's Ct. CR8: Purl 85Pb 176
(off OLd Lodge La.)
St Saviours Ct. HA1: Harr 29Ga 46
St Saviours Ct. N22 26Mb 50
St Saviour's Est. SE1 2K 231 (47Vb 91)
St Saviour's Rd. CR0: C'don 72Rb 157
St Saviour's Rd. SW2 57Pb 112
St Saviours Vw. AL1: St A 1D 6
(off Lemsford Rd.)
St Saviour's Wharf SE1: Mill St. 47Vb 91
(off Mill St.)

St Saviour's Wharf SE1:
Shad Thames ... 1K 231 (47Vb 91)
(off Shad Thames)
Saints Cl. SE27 ... 63Rb 135
Saints Dr. E7 ... 36Mc 73
St Silas Pl. NW5 ... 37Jb 70
St Simon's Av. SW15 ... 57Ya 110
Saints M. CR4: Mitc ... 69Gb 133
ST STEPHENS ... 4A 6
St Stephen's Av. AL3: St A ... 4P 5
St Stephen's Av. E17 ... 29Ec 52
St Stephen's Av. KT21: Asht ... 88Na 173
St Stephen's Av. W12 ... 46Xa 88
(not continuous)
St Stephen's Av. W13 ... 44Ka 86
St Stephen's Cl. E17 ... 29Dc 52
St Stephen's Cl. NW8 ... 39Gb 69
St Stephen's Cl. UB1: S'hall ... 43Ca 85
St Stephens Cl. AL3: St A ... 5P 5
St Stephens Cl. NW5 ... 36Hb 69
(off Malden Rd.)
St Stephen's Ct. EN1: Enf ... 16Ub 33
(off Park Av.)
St Stephens Ct. N8 ... 30Pb 50
St Stephens Ct. RH9: S God ... 10C 210
(off Oaklands)
St Stephens Ct. W13 ... 44Ka 86
St Stephen's Cres. CM13: B'wood ... 21Ce 59
St Stephen's Cres. CR7: Thor H ... 69Qb 134
St Stephen's Cres. W2 ... 44Cb 89
St Stephens Cres. RM16: Grays ... 10D 100
St Stephen's Gdns. SW15 ... 57Bb 111
St Stephen's Gdns. TW1: Twick ... 58La 108
St Stephen's Gdns. W2 ... 44Cb 89
(not continuous)
St Stephens Gro. SE13 ... 55Ec 114
St Stephens Hill AL1: St A ... 4A 6
St Stephens Ho. SE17 ... 51Tb 113
(off Lytham St.)
St Stephen's M. W2 ... 43Cb 89
St Stephens Pde. E7 ... 38Lc 73
St Stephen's Pas. TW1: Twick ... 58La 108
St Stephen's Rd. E17 ... 29Dc 52
St Stephen's Rd. E3 ... 39Ac 72
St Stephen's Rd. E6 ... 38Lc 73
St Stephen's Rd. EN5: Barn ... 15Za 30
St Stephen's Rd. TW3: Houn ... 58Ca 107
St Stephen's Rd. UB7: Yiew ... 46M 83
St Stephen's Rd. W13 ... 44Ka 86
St Stephens Rd. EN3: Enf W ... 9Zb 20
St Stephen's Row EC4 ... 3F 225 (44Tb 91)
(off Walbrook)
St Stephen's Ter. SW8 ... 52Pb 112
St Stephen's Wlk. SW7 ... 5A 226 (49Eb 89)
(off Southwell Gdns.)
Saint's Wlk. RM16: Grays ... 9E 100
St Swithins La. EC4 ... 4F 225 (45Tb 91)
St Swithun's Rd. SE13 ... 58Fc 115
St Teresa Wlk. RM16: Grays ... 9D 100
St Theresa Cl. KT18: Eps ... 86Sa 173
St Theresa Ct. E4 ... 17Fc 35
St Theresa's Cl. E9 ... 35Cc 72
St Theresa's Rd. TW14: Felt ... 56V 106
St Thomas' Almshouses
DA11: Grav'nd ... 10C 122
St Thomas Cl. GU21: Wok ... 9N 167
St Thomas Cl. KT6: Surb ... 74Pa 153
St Thomas Ct. DA5: Bexl ... 59Cd 118
St Thomas Ct. E10 ... 31Dc 72
(off Lake Rd.)
St Thomas Ct. HA5: Pinn ... 25Aa 45
St Thomas Ct. NW1 ... 38Lb 70
(off Wrotham Rd.)
St Thomas Dr. BR5: Farnb ... 74Sc 160
St Thomas' Dr. HA5: Pinn ... 25Aa 45
St Thomas Gdns. IG1: Ilf ... 37Sc 74
ST THOMAS' HOSPITAL ... 3H 229 (48Pb 90)
St Thomas Ho. E1 ... 44Zb 92
(off W. Arbour St.)
St Thomas M. SW18 ... 57Cb 111
St Thomas More Cl. IG10: Lough ... 11Tc 36
St Thomas Rd. DA11: Nflt ... 1A 144
St Thomas Rd. DA17: Belv ... 47Ed 96
St Thomas Rd. E16 ... 44Jc 93
St Thomas Rd. N14 ... 17Mb 32
St Thomas Rd. W4 ... 51Sa 109
St Thomas' Rd. CM14: B'wood ... 19Zd 41
St Thomas's Av. DA11: Grav'nd ... 10D 122
St Thomas's Cl. EN9: Walt A ... 5Kc 21
St Thomas's Gdns. NW5 ... 37Jb 70
St Thomas's M. SE7 ... 49Nc 94
St Thomas's Pl. E9 ... 38Yb 72
St Thomas's Pl. RM17: Grays ... 51De 121
St Thomas's Rd. N4 ... 33Qb 70
St Thomas's Rd. NW10 ... 39Ua 68
St Thomas's Sq. E9 ... 38Yb 72
St Thomas St. SE1 ... 7G 225 (46Tb 91)
St Thomas's Way SW6 ... 52Bb 111
St Thomas Wlk. SL3: Coln ... 52F 104
St Timothys M. BR1: Broml ... 67Kc 137
St Ursula Gro. HA5: Pinn ... 29Z 45
St Ursula Rd. UB1: S'hall ... 44Ca 85
St Valery Pl. TW5: Hest ... 52Z 107
St Vincent Cl. SE27 ... 64Rb 135
St Vincent De Paul Ho. E1 ... 43Yb 92
(off Jubilee St.)
St Vincent Dr. AL1: St A ... 5E 6
St Vincent Ho. SE1 ... 3K 231 (48Vb 91)
(off St Saviour's Est.)
St Vincent Rd. KT12: Walt T ... 76X 151
St Vincent Rd. TW2: Whitt ... 58Ea 108
St Vincents Av. DA1: Dart ... 57Qd 119
St Vincent's Cotts. WD18: Wat ... 14X 27
(off Marlborough Rd.)
ST VINCENT'S HAMLET ... 18Rd 39
St Vincent's La. NW7 ... 21Ya 48
St Vincents Rd. DA1: Dart ... 58Qd 119
St Vincent St. W1 ... 1J 221 (43Jb 90)
St Vincents Vs. DA1: Dart ... 58Pd 119
St Wilfrids EN6: Pot B ... 6Eb 17
St Wilfrid's Cl. EN4: E Barn ... 15Gb 31
St Wilfrid's Rd. EN4: E Barn ... 15Fb 31
St Williams Ct. N1 ... 38Nb 70
St Winefride's Av. E12 ... 36Pc 74
St Winifreds CR8: Kenley ... 87Sb 177
St Winifred's Cl. IG7: Chig ... 22Sc 54
St Winifred's Rd. TN16: Big H ... 90Pc 180
St Winifred's Rd. TW11: Tedd ... 65Ka 130
St Yon Ct. AL4: St A ... 2J 7
Sakura Dr. N22 ... 25Mb 50
Saladin Dr. RM19: Purf ... 49Qd 97
Salamanca Pl. IG11: Bark ... 40Xc 75
Salamanca Pl. SE1 ... 6H 229 (49Pb 90)
Salamanca Sq. SE1 ... 6H 229 (49Pb 90)
(off Salamanca Pl.)
Salamanca St. SE1 ... 6H 229 (49Pb 90)
Salamanca St. SE11 ... 6H 229 (49Pb 90)
Salamander Cl. KT2: King T ... 64La 130
Salamander Quay KT1: Hamp W ... 67Ma 131
Salamander Quay UB9: Hare ... 24J 43
Salamons Way RM13: Rain ... 44Gd 96
Salcombe Ct. E14 ... 43Ec 92
(off St Ives Pl.)
Salcombe Dr. RM6: Chad H ... 30Bd 55
Salcombe Dr. SM4: Mord ... 74Za 154
Salcombe Gdns. NW7 ... 23Ya 48
Salcombe Pk. IG10: Lough ... 15Mc 35
Salcombe Rd. E17 ... 31Bc 72
Salcombe Rd. N16 ... 36Ub 71
Salcombe Rd. TW15: Ashf ... 62N 127
Salcombe Vs. TW10: Rich ... 57Na 109
Salcombe Way HA4: Ruis ... 33W 64
Salcombe Way UB4: Hayes ... 41T 84
Salcot Cres. CR0: New Ad ... 82Ec 178
Salcote Rd. DA12: Grav'nd ... 4G 144
Salcott Rd. CR0: Bedd ... 76Nb 156
Salcott Rd. SW11 ... 57Gb 111
Salehurst Cl. HA3: Kenton ... 29Na 47
Salehurst Rd. SE4 ... 58Bc 114
Salem Pl. CR0: C'don ... 76Sb 157
Salem Pl. DA11: Nflt ... 59Fe 121
Salem Rd. W2 ... 45Db 89
Salento Cl. N3 ... 24Cb 49
Sale Pl. W2 ... 1D 220 (43Gb 89)
Salesian Gdns. KT16: Chert ... 74J 149
Sale St. E2 ... 42Wb 91
Salford Ho. E14 ... 49Ec 92
(off Seyssel St.)
Salford Rd. SW2 ... 60Mb 112
Salhouse Cl. SE28 ... 44Yc 95
Salisbury Av. AL1: St A ... 1F 6
Salisbury Av. BR8: Swan ... 70Jd 140
Salisbury Av. IG11: Bark ... 38Tc 74
Salisbury Av. N3 ... 27Bb 49
Salisbury Av. SL2: Slou ... 2G 80
Salisbury Av. SM1: Sutt ... 79Bb 155
Salisbury Av. SS17: Stan H ... 2M 101
Salisbury Cl. EN6: Pot B ... 4Eb 17
Salisbury Cl. KT4: Wor Pk ... 76Va 154
Salisbury Cl. RM14: Upm ... 33Ud 78
Salisbury Cl. SE17 ... 6F 231 (49Tb 91)
Salisbury Ct. E9 ... 36Ac 72
(off Mabley St.)
Salisbury Ct. EC4 ... 3B 224 (44Rb 91)
Salisbury Ct. EN2: Enf ... 14Tb 33
(off London Rd.)
Salisbury Ct. SE16 ... 48Wb 91
(off Stork's Rd.)
Salisbury Ct. SM5: Cars ... 78Hb 155
Salisbury Ct. UB5: N'olt ... 36Da 65
(off Newmarket Av.)
Salisbury Cres. EN8: Chesh ... 4Zb 20
Salisbury Gdns. IG9: Buck H ... 19Mc 35
Salisbury Gdns. SW19 ... 66Ab 132
Salisbury Hall Gdns. E4 ... 23Cc 52
Salisbury Ho. AL3: St A ... 3A 6
Salisbury Ho. E14 ... 44Dc 92
(off Hobday St.)
Salisbury Ho. EC2 ... 1G 225 (43Tb 91)
(off London Wall)
Salisbury Ho. HA7: Stan ... 23Ja 46
Salisbury Ho. N1 ... 39Rb 71
(off St Mary's Path)
Salisbury Ho. SM6: W'gton ... 78Kb 156
Salisbury Ho. SW1 ... 7E 228 (50Mb 90)
(off Drummond Ga.)
Salisbury Ho. SW9 ... 52Qb 112
(off Cranmer Rd.)
Salisbury Mans. N15 ... 29Rb 51
Salisbury M. SW6 ... 52Bb 111
Salisbury Pas. SW6 ... 52Bb 111
(off Dawes Rd.)
Salisbury Pavement SW6 ... 52Bb 111
(off Dawes Rd.)
Salisbury Pl. KT14: W Byf ... 83L 169
Salisbury Pl. SW9 ... 52Rb 113
Salisbury Pl. W1 ... 7F 215 (43Hb 89)
Salisbury Prom. N8 ... 29Rb 51
Salisbury Rd. BR2: Broml ... 71Nc 160
Salisbury Rd. DA11: Grav'nd ... 10B 122
Salisbury Rd. DA2: Dart ... 60Sd 120
Salisbury Rd. DA5: Bexl ... 60Cd 118
Salisbury Rd. E10 ... 33Ec 72
Salisbury Rd. E12 ... 36Mc 73
Salisbury Rd. E17 ... 29Ec 52
Salisbury Rd. E4 ... 20Cc 34
Salisbury Rd. E7 ... 37Jc 73
Salisbury Rd. EN3: Enf L ... 9Bc 20
Salisbury Rd. EN5: Barn ... 13Ab 30
Salisbury Rd. GU22: Wok ... 91A 188
Salisbury Rd. HA1: Harr ... 29Fa 46
Salisbury Rd. HA5: Eastc ... 28W 44
Salisbury Rd. IG3: Ilf ... 33Uc 74
Salisbury Rd. KT3: N Mald ... 69Ta 131
Salisbury Rd. KT4: Wor Pk ... 77Ta 153
Salisbury Rd. N22 ... 25Rb 51
Salisbury Rd. N4 ... 29Rb 51
Salisbury Rd. RH9: G'stone ... 3A 210
Salisbury Rd. RM10: Dag ... 37Dd 76
Salisbury Rd. RM17: Grays ... 51Ee 121
Salisbury Rd. RM2: Rom ... 29Kd 57
Salisbury Rd. SE25 ... 72Wb 157
Salisbury Rd. SM5: Cars ... 79Hb 155
Salisbury Rd. SM7: Bans ... 86Db 175
Salisbury Rd. SW19 ... 66Ab 132
Salisbury Rd. TW13: Felt ... 60Y 107
Salisbury Rd. TW4: Houn ... 55Y 107
Salisbury Rd. TW6: H'row A ... 58S 106
(not continuous)
Salisbury Rd. TW9: Rich ... 56Na 109
Salisbury Rd. UB2: S'hall ... 49Aa 85
Salisbury Rd. UB8: Uxb ... 40K 63
Salisbury Rd. W13 ... 47Ka 86
Salisbury Rd. WD24: Wat ... 10X 13
Salisbury Sq. EC4 ... 3A 224 (44Qb 90)
Salisbury St. NW8 ... 6D 214 (42Gb 89)
Salisbury St. W3 ... 47Sa 87
Salisbury Ter. SE15 ... 55Yb 114
Salisbury Wlk. N19 ... 33Lb 70
Salix Cl. KT22: Fet ... 95Da 191
Salix Cl. TW16: Sun ... 66X 129
Salix Ct. N3 ... 23Cb 49
Salix La. IG8: Wfd G ... 25Nc 54
Salix Rd. RM17: Grays ... 51Fe 121
Salk Cl. NW9 ... 26Ua 48
Salliesfield TW2: Whitt ... 58Fa 108
Sallow Rd. RM3: Hrld W ... 26Nd 57
Sallows Shaw DA13: Sole S ... 10E 144
Sally Murray Cl. E12 ... 35Qc 74
Salmen Rd. E13 ... 40Hc 73
Salmon Cl. HA7: Stan ... 23Ja 46
Salmonds Gro. CM13: Ingve ... 22Ee 59
Salmon La. E14 ... 44Ac 92
Salmon Mdw. HP3: Hem H ... 6M 3
Salmon M. NW6 ... 36Cb 69
Salmon Rd. DA1: Dart ... 55Pd 119
Salmon Rd. DA17: Belv ... 50Cd 96
Salmons La. CR3: Cat'm ... 92Ub 197
Salmons La. CR3: Whyt ... 92Ub 197
Salmons La. W. CR3: Cat'm ... 92Ub 197
Salmons Rd. KT24: Eff ... 100X 191
Salmons Rd. KT9: Chess ... 79Na 153
Salmons Rd. N9 ... 18Wb 33
Salmon St. E14 ... 44Bc 92
Salmon St. NW9 ... 32Ra 67
Salomons Rd. E13 ... 43Lc 93
Salop Rd. E17 ... 30Zb 52
Salsabil Apts. E3 ... 42Bc 92
Saltash Cl. SM1: Sutt ... 77Bb 155
Saltash Rd. DA16: Well ... 53Yc 117
Saltash Rd. IG6: Ilf ... 24Tc 54
Salt Box Cotts. GU2: Guild ... 10L 187
Saltbox Hill TN16: Big H ... 85Kc 179
Saltbox Hill Nature Reserve ... 84Kc 179
Salt Box Rd. GU3: Guild ... 10K 187
Salt Box Rd. GU4: Guild ... 10M 187
Saltcoats Rd. W4 ... 47Ua 88
Saltcote Cl. DA1: Cray ... 58Gd 118
Saltcroft Cl. HA9: Wemb ... 32Ra 67
Saltdene N4 ... 32Pb 70
Salter Cl. CR5: Coul ... 88Lb 176
Salter Cl. HA2: Harr ... 35Ba 65
Salterford Rd. SW17 ... 65Jb 134
Saltern Ct. IG11: Bark ... 41Xc 95
(off Galleons Dr.)
Salter Rd. SE16 ... 46Zb 92
Salters Cl. WD3: Rick ... 18N 25
Salters Ct. EC4 ... 3E 224 (44Sb 91)
(off Bow La.)
Salters Gdns. WD17: Wat ... 11W 26
Salter's Hall Ct. EC4 ... 4F 225 (45Tb 91)
(off Cannon St.)
Salter's Hill SE19 ... 64Tb 135
Salters Rd. E17 ... 28Fc 53
Salters Rd. W10 ... 42Za 88
Salters Row N1 ... 37Tb 71
(off Tilney Gdns.)
Salter St. E14 ... 45Bc 92
(not continuous)
Salter St. NW10 ... 41Wa 88
Salterton Rd. N7 ... 34Pb 70
Saltford Cl. DA8: Erith ... 50Gd 96
SALT HILL ... 6G 80
Salt Hill Av. SL1: Slou ... 6G 80
Salt Hill Cl. UB8: Uxb ... 36N 63
Salt Hill Dr. SL1: Slou ... 6G 80
Salt Hill Mans. SL1: Slou ... 6G 80
Salt Hill Way SL1: Slou ... 6G 80
Saltings, The DA9: Ghithe ... 56Yd 120
Salting St. IG11: Bark ... 39Tc 74
Saltley Cl. E6 ... 44Nc 94
Salton Cl. N3 ... 26Cb 49
Salton Sq. E14 ... 44Bc 92
Saltoun Rd. SW2 ... 56Qb 112
Saltram Cl. N15 ... 28Vb 51
Saltram Cres. W9 ... 41Bb 89
Saltun Cl. DA12: Grav'nd ... 10J 123
Saltwell St. E14 ... 45Cc 92
Saltwood Cl. BR6: Chels ... 77Yc 161
Saltwood Gro. SE17 ... 7F 231 (50Tb 91)
Saltwood Ho. SE15 ... 51Yb 114
(off Lovelinch Cl.)
Salusbury Rd. NW6 ... 39Ab 68
Salus Ct. CR0: C'don ... 77Sb 157
(off Parker Rd.)
Salutation Rd. SE10 ... 49Gc 93
Salvador SW17 ... 64Hb 133
Salvation Pl. KT22: Fet ... 96Ja 192
Salvia Ct. GU24: Bisl ... 8E 166
Salvia Gdns. UB6: G'frd ... 40Ja 66
Salvin Rd. SW15 ... 55Za 110
Salway Cl. IG8: Wfd G ... 24Jc 53
Salway Pl. E15 ... 37Fc 73
Salway Rd. E15 ... 37Fc 73
Samantha Cl. E17 ... 31Bc 72
Samantha M. RM4: Have B ... 20Gd 38
Samaras Mans. E20 ... 36Ec 72
(off Liberty Bri. Rd.)
Samas Way DA1: Cray ... 57Jd 118
Sam Bartram Cl. SE7 ... 50Lc 93
Sambourne Family Home, The ... 48Cb 89
(off Stafford Ter.)
Sambroke Sq. EN4: E Barn ... 14Fb 31
Sambrooke Ct. EN1: Enf ... 16Ub 33
Sambrook Ho. E1 ... 43Yb 92
(off Jubilee St.)
Sambrook Ho. SE11 ... 6K 229 (49Qb 90)
(off Hotspur St.)
Sambruck M. SE6 ... 60Dc 114
Samels Ct. W6 ... 50Wa 88
Samford Ho. N1 ... 1K 217 (39Qb 70)
(off Barnsbury Est.)
Samford St. NW8 ... 6C 214 (42Fb 89)
Samian Ga. AL3: St A ... 4M 5
Samira Cl. E17 ... 30Cc 52
Sam King Wlk. SE5 ... 52Tb 113
(off Edmund St.)
Sam Manners Ho. SE10 ... 50Gc 93
(off Tuskar St.)
Sam March Ho. E14 ... 44Fc 93
(off Blair St.)
Sammi Ct. CR7: Thor H ... 70Sb 135
Samos Rd. SE20 ... 68Xb 135
Samphire Ct. RM17: Grays ... 1A 122
Samphire Hgts. E20 ... 36Ec 72
(off Napa Cl.)
Sampson Av. EN5: Barn ... 15Za 30
Sampson Cl. DA17: Belv ... 48Zc 95
Sampson Ct. TW17: Shep ... 71S 150
Sampson Ho. SE1 ... 6B 224 (46Rb 91)
Sampson's Grn. SL2: Slou ... 1D 80
Sampson St. E1 ... 46Wb 91
Samson Cl. DA1: Dart ... 57Nd 119
Samson St. E13 ... 40Lc 73
Samuda Est. E14 ... 48Ec 92
Samuel Cl. E8 ... 39Vb 71
Samuel Cl. HA7: Stan ... 19Ja 28
Samuel Cl. SE14 ... 51Zb 114
Samuel Cl. SE18 ... 49Nc 94
Samuel Ct. N1 ... 4H 219 (41Ub 91)
(off Pitfield St.)
Samuel Ferguson Pl. IG11: Bark ... 40Vc 75
Samuel Gray Gdns. KT2: King T ... 67Ma 131
Samuel Johnson Cl. SW16 ... 63Pb 134
Samuel Jones Ct. SE15 ... 52Ub 113
Samuel Lewis Bldgs. N1 ... 37Qb 70
Samuel Lewis Trust Dwellings E8 ... 36Wb 71
(off Amhurst Rd.)
Samuel Lewis Trust Dwellings N16 ... 30Ub 51
Samuel Lewis Trust Dwellings
SW3 ... 6D 226 (49Gb 89)
Samuel Lewis Trust Dwellings SW6 ... 52Cb 111
(off Vanston Pl.)
Samuel Lewis Trust Dwellings W14 ... 49Bb 89
(off Lisgar Ter.)
Samuel Lewis Trust Est. SE5 ... 53Sb 113
(off Warner Rd.)
Samuel Palmer Ct. BR6: Orp ... 73Wc 161
(off Chislehurst Rd.)
Samuel Richardson Ho. W14 ... 49Bb 89
(off North End Cres.)
Samuel's Cl. W6 ... 49Ya 88
Samuelson Pl. TW7: Isle ... 54Ga 108
Samuel Sq. AL1: St A ... 3B 6
(off Pageant Rd.)
Samuel St. E8 ... 39Vb 71
Samuel St. SE15 ... 52Vb 113
Samuel St. SE18 ... 49Pc 94
Samuel Wallis Lodge SE10 ... 49Gc 93
(off Banning St.)
Samuel Wallis Lodge SE3 ... 55Hc 115
(off Banning St.)
Sanchia Ct. E2 ... 41Wb 91
(off Wellington Row)
Sancroft Cl. NW2 ... 34Xa 68
Sancroft Ho. SE11 ... 7J 229 (50Pb 90)
(off Sancroft St.)
Sancroft Rd. HA3: W'stone ... 26Ha 46
Sancroft St. SE11 ... 7J 229 (50Pb 90)
Sanctuary, The DA5: Bexl ... 58Zc 117
Sanctuary, The SM4: Mord ... 72Cb 155
Sanctuary, The SW1 ... 3E 228 (48Mb 90)
(off Broad Sanctuary)
Sanctuary Cl. DA1: Dart ... 58Ld 119
Sanctuary Cl. UB9: Hare ... 24L 43
Sanctuary Gdns. SS17: Stan H ... 1N 101
Sanctuary M. E8 ... 37Vb 71
Sanctuary Rd. TW6: H'row A ... 58Q 106
Sanctuary St. SE1 ... 2E 230 (47Sb 91)
Sandale Cl. N16 ... 34Tb 71
Sandall Cl. W5 ... 42Na 87
Sandall Ho. E3 ... 40Ac 72
Sandall Rd. NW5 ... 37Lb 70
Sandall Rd. W5 ... 42Na 87
Sandal Rd. KT3: N Mald ... 71Ta 153
Sandal Rd. N18 ... 22Wb 51
Sandal St. E15 ... 39Gc 73
Sandalwood Av. KT16: Chert ... 76G 148
Sandalwood Cl. E1 ... 42Ac 92
Sandalwood Cl. EN5: Ark ... 15Va 30
Sandalwood Dr. HA4: Ruis ... 31S 64
Sandalwood Ho. DA15: Sidc ... 61Vc 139
Sandalwood Mans. W8 ... 48Db 89
(off Stone Hall Gdns.)
Sandalwood Rd. TW13: Felt ... 62X 129
Sanday Cl. HP3: Hem H ... 4B 4
Sandbach Pl. SE18 ... 49Sc 94
Sandbanks TW14: Felt ... 60U 106
Sandbanks Hill DA2: G St G ... 65Xd 142
Sandbourne NW8 ... 39Db 69
(off Abbey Rd.)
Sandbourne W11 ... 44Cb 89
(off Dartmouth Cl.)
Sandbourne Av. SW19 ... 68Db 133
Sandbourne Rd. SE4 ... 54Ac 114
Sandbrook Cl. NW7 ... 23Ta 47
Sandbrook Rd. N16 ... 34Ub 71
Sandby Ct. NW10 ... 41Xa 88
(off Plough Cl.)
Sandby Grn. SE9 ... 55Nc 116
Sandby Ho. NW6 ... 39Cb 69
Sandcliff Rd. DA8: Erith ... 49Fd 96
Sandcroft Cl. N13 ... 23Rb 51
Sandcross La. RH2: Reig ... 9H 207
Sandell's Av. TW15: Ashf ... 63S 128
Sandell St. SE1 ... 1K 229 (47Qb 90)
Sanderling Cl. RM18: E Til ... 9K 101
Sanderling Ct. SE28 ... 45Yc 95
Sanderling Ct. SE8 ... 51Bc 114
(off Abinger Gro.)
Sanderling Lodge E1 ... 45Vb 91
(off Star Pl.)
Sanderling Way DA9: Ghithe ... 58Wd 120
Sanders Cl. AL2: Lon C ... 9H 7
Sanders Cl. HP3: Hem H ... 6P 3
Sanders Cl. TW12: Hamp H ... 64Ea 130
Sanders Ct. CM14: W'ley ... 21Yd 58
Sandersfield Gdns. SM7: Bans ... 87Cb 175
Sandersfield Rd. SM7: Bans ... 87Db 175
Sanders Ho. WC1 ... 3K 217 (41Qb 90)
(off Gt. Percy St.)
Sanders La. NW7: Bittacy Ri. ... 24Ya 48
Sanders La. NW7: Grants Cl. ... 24Za 48
Sanders La. TW4: Houn ... 57Ba 107
Sanderson Bldg. NW2 ... 33Ya 68
Sanderson Cl. CM13: W H'dn ... 30Ee 59
Sanderson Cl. NW5 ... 35Kb 70
Sanderson Ho. E16 ... 43Jc 93
(off Hammersley Rd.)
Sanderson Ho. SE8 ... 50Bc 92
(off Grove St.)
Sanderson Rd. UB8: Uxb ... 37L 63
Sandersons Av. TN14: Bad M ... 82Cd 182
Sandersons La. W4 ... 50Ta 87
(off Chiswick High Rd.)
Sanderson Sq. BR1: Broml ... 69Qc 138
Sanders Pl. AL1: St A ... 3E 6
(off Camp Rd.)
Sanders Rd. HP3: Hem H ... 6A 4
SANDERSTEAD ... 84Wb 177
Sanderstead Av. NW2 ... 33Ab 68
Sanderstead Cl. SW12 ... 59Lb 112
Sanderstead Ct. Av. CR2: Sande ... 85Wb 177
Sanderstead Hill CR2: Sande ... 83Ub 177
Sanderstead Rd. BR5: St M Cry ... 72Xc 161
Sanderstead Rd. CR2: S Croy ... 80Tb 157
Sanderstead Rd. CR2: Sande ... 80Tb 157
Sanderstead Rd. E10 ... 32Ac 72
Sanderstead Station (Rail) ... 81Tb 177
Sanders Way N19 ... 32Mb 70
Sandes Ct. CR7: Thor H ... 69Sb 135
Sandes Pl. KT22: Lea ... 90Ja 172
Sandfield WC1 ... 4G 217 (41Nb 90)
(off Cromer St.)
Sandfield Gdns. CR7: Thor H ... 69Rb 135
Sandfield Pas. CR7: Thor H ... 69Sb 135
Sandfield Pl. CR7: Thor H ... 69Sb 135
Sandfield Rd. AL1: St A ... 2E 6
Sandfield Rd. CR7: Thor H ... 69Rb 135
Sandfields GU23: Send ... 96F 188
Sandford Av. IG10: Lough ... 13Sc 36
Sandford Av. N22 ... 25Sb 51
Sandford Cl. E6 ... 42Pc 94
Sandford Ct. EN5: New Bar ... 13Db 31
Sandford Ct. N16 ... 32Ub 71
Sandford Rd. BR2: Broml ... 69Jc 137
Sandford Rd. DA7: Bex ... 56Ad 117
Sandford Rd. E6 ... 41Nc 94
Sandford St. SW6 ... 52Db 111
Sandgate Cl. RM7: Rush G ... 31Ed 76
Sandgate Ho. E5 ... 36Xb 71
Sandgate Ho. W5 ... 43La 86
Sandgate La. SW18 ... 60Gb 111
Sandgate Rd. DA16: Well ... 52Yc 117
Sandgates KT16: Chert ... 75G 148
Sandgate St. SE15 ... 51Xb 113
Sandgate Trad. Est. SE15 ... 51Xb 113
(off Sandgate St.)
Sandham Ct. SW4 ... 53Nb 112
Sandham Point SE18 ... 49Rc 94
(off Vincent Rd.)
Sandhills SM6: Bedd ... 77Mb 156
Sandhills, The SW10 ... 51Eb 111
(off Limerston St.)
Sandhills Ct. GU25: Vir W ... 71A 148
Sandhills La. GU25: Vir W ... 71A 148
Sandhills Mdw. TW17: Shep ... 73S 150
Sandhills Rd. RH2: Reig ... 7J 207
Sandhurst Av. HA2: Harr ... 30Da 45
Sandhurst Av. KT5: Surb ... 73Ra 153
Sandhurst Cl. CR2: Sande ... 81Ub 177
Sandhurst Cl. NW9 ... 27Qa 47
Sandhurst Ct. SW2 ... 56Nb 112
Sandhurst Dr. IG3: Bark ... 35Vc 75
Sandhurst Dr. IG3: Ilf ... 35Vc 75
Sandhurst Ho. E1 ... 43Yb 92
(off Wolsy St.)
Sandhurst Ho. E17 ... 25Bc 52
(off Robinswood Gdns.)
Sandhurst Mkt. SE6 ... 60Ec 114
(off Sandhurst Rd.)
Sandhurst Rd. BR6: Chels ... 76Wc 161
Sandhurst Rd. DA15: Sidc ... 62Vc 139
Sandhurst Rd. DA5: Bexl ... 57Zc 117
Sandhurst Rd. N9 ... 16Yb 34
Sandhurst Rd. NW9 ... 27Qa 47
Sandhurst Rd. RM18: Tilb ... 4E 122
Sandhurst Rd. SE6 ... 60Fc 115
Sandhurst Way CR2: Sande ... 80Ub 157
Sandifer Dr. NW2 ... 34Za 68
Sandifield AL10: Hat ... 3D 8
Sandiford Rd. SM3: Sutt ... 75Bb 155
Sandiland Cres. BR2: Hayes ... 75Hc 159
Sandilands CR0: C'don ... 75Wb 157
Sandilands TN13: Chips ... 94Fd 202
Sandilands Rd. SW6 ... 53Db 111
Sandilands Stop
(London Tramlink) ... 75Vb 157
Sandison St. SE15 ... 55Wb 113
Sandlands Gro. KT20: Walt H ... 95Wa 194
Sandlands Rd. KT20: Walt H ... 95Wa 194
Sandland St. WC1 ... 1J 223 (43Pb 90)
Sandlers End SL2: Slou ... 2F 80
Sandling Ri. SE9 ... 62Qc 138
Sandlings, The N22 ... 26Qb 50
Sandlings Cl. SE15 ... 54Xb 113
Sandmartin Way SM6: W'gton ... 74Jb 156
Sandmere Cl. HP2: Hem H ... 3A 4
Sandmere Rd. SW4 ... 56Nb 112
Sandon Cl. KT10: Esh ... 73Fa 152
Sandon Rd. EN8: Chesh ... 2Yb 20
Sandover Ho. SE16 ... 48Wb 91
(off Spa Rd.)
Sandow Cres. UB3: Hayes ... 48V 84
Sandown Av. KT10: Esh ... 78Ea 152
Sandown Av. RM10: Dag ... 37Ed 76
Sandown Av. RM12: Horn ... 33Md 77
Sandown Cl. RM16: Ors ... 3G 100
Sandown Cl. TW5: Cran ... 53W 106
Sandown Ct. HA7: Stan ... 22La 46
Sandown Ct. RH1: Redh ... 5N 207
(off Station Rd.)
Sandown Ct. RM10: Dag ... 37Ed 76
(off Sandown Av.)
Sandown Ct. SE26 ... 62Xb 135
Sandown Ct. SM2: Sutt ... 80Db 155
Sandown Dr. SM5: Cars ... 81Jb 176
Sandown Ga. KT10: Esh ... 76Fa 152
Sandown Ho. TW18: Staines ... 63H 127
(off High St.)
Sandown Ind. Pk. KT10: Esh ... 75Ca 151
Sandown Lodge KT18: Eps ... 86Ta 173
Sandown Pk. Golf Course ... 76Da 151
Sandown Pk. Racecourse ... 75Ea 152
Sandown Rd. CR5: Coul ... 88Jb 176
Sandown Rd. DA12: Grav'nd ... 5E 144
Sandown Rd. KT10: Esh ... 77Ea 152
Sandown Rd. RM16: Ors ... 3G 100
Sandown Rd. SE25 ... 71Xb 157
Sandown Rd. SL2: Slou ... 3D 80
Sandown Rd. WD24: Wat ... 10Y 13
Sandown Rd. Ind. Est. WD24: Wat ... 9Y 13
Sandown Sports Club ... 76Da 151
Sandown Way UB5: N'olt ... 37Aa 65
Sandpiper Cl. AL10: Hat ... 1C 8
Sandpiper Cl. DA9: Ghithe ... 58Wd 120
Sandpiper Cl. E17 ... 24Zb 52
Sandpiper Cl. RM18: E Til ... 9K 101
Sandpiper Cl. SE16 ... 47Bc 92
Sandpiper Ct. E1 ... 45Wb 91
(off Thomas More St.)
Sandpiper Ct. E14 ... 48Ec 92
(off New Union Cl.)
Sandpiper Ct. SE8 ... 51Cc 114
(off Edward Pl.)
Sandpiper Dr. DA8: Erith ... 52Kd 119
Sandpiper Dr. HA2: Harr ... 33Da 65
Sandpiper Ho. UB7: Yiew ... 45M 83
(off Wraysbury Dr.)
Sandpiper Rd. CR2: Sels ... 83Zb 178
Sandpiper Rd. SM1: Sutt ... 78Bb 155
Sandpipers, The DA12: Grav'nd ... 1F 144
Sandpiper Ter. IG5: Ilf ... 27Rc 54
Sandpiper Way BR5: St P ... 70Zc 139
Sandpit Cotts. GU24: Pirb ... 4C 186
Sandpit Hall Rd. GU24: Chob ... 4L 167
Sandpit La. AL1: St A ... 1C 6
Sandpit La. AL4: St A ... 1F 6
Sandpit La. CM14: Pil H ... 18Vd 40
Sandpit La. CM14: S Weald ... 18Vd 40
Sandpit La. CM15: Pil H ... 15Vd 40
Sandpit La. GU21: Knap ... 6F 166
(not continuous)
Sandpit Pl. SE7 ... 50Nc 94
Sandpit Rd. BR1: Broml ... 64Gc 137
Sandpit Rd. DA1: Dart ... 56Ld 119
Sandpit Rd. RH1: Redh ... 7N 207
Sandpits Rd. CR0: C'don ... 77Zb 158
Sandpits Rd. TW10: Ham ... 61Ma 131
Sandra Cl. KT8: E Mos ... 70Fa 130
Sandra Cl. N22 ... 25Sb 51
Sandra Cl. TW3: Houn ... 57Da 107
Sandra Ct. CR4: Mitc ... 65Hb 133
Sandridge Cl. EN4: Had W ... 9Gb 17
Sandridge Cl. HA1: Harr ... 28Ga 46
Sandridge St. N19 ... 33Lb 70
Sandringham Av. SW20 ... 67Ab 132
Sandringham Bldgs. SE17 ... 5F 231 (49Tb 91)
(off Balfour St.)
Sandringham Cl. EN1: Enf ... 12Ub 33
Sandringham Cl. GU22: Pyr ... 88J 169
Sandringham Cl. IG6: Ilf ... 27Sc 54
Sandringham Cl. SW19 ... 60Za 110
Sandringham Cl. WD6: Bore ... 15Sa 29
Sandringham Ct. DA15: Sidc ... 58Vc 117

Sandringham Ct. KT2: King T 67Na 131
(off Skerne Wlk.)
Sandringham Ct. SE16 .... 46Zb 92
(off King & Queen Wharf)
Sandringham Ct. SL1: Slou 4B 80
Sandringham Ct. SM2: Sutt 81Cb 175
Sandringham Ct. UB10: Hil 42S 84
Sandringham Ct. W1 .... 3C 222 (44Lb 90)
(off Dufour's Pl.)
Sandringham Ct. W9 .... 4A 214 (41Eb 89)
(off Maida Vale)
Sandringham Cres. HA2: Harr 33Ca 65
Sandringham Dr. DA16: Well 54Uc 116
Sandringham Dr. DA2: Wilm 61Gd 140
Sandringham Dr. TW15: Ashf 63M 127
Sandringham Flats WC2 .... 4E 222 (45Mb 90)
(off Charing Cross Rd.)
Sandringham Gdns. IG6: Ilf 27Sc 54
Sandringham Gdns. KT8: W Mole 70Ca 129
Sandringham Gdns. N12 .... 23Fb 49
Sandringham Gdns. N8 .... 30Nb 50
Sandringham Gdns. TW5: Cran 53W 106
Sandringham Ho. SE1 .... 7J 225 (46Ub 91)
(off Potters Flds.)
Sandringham Ho. W14 .... 49Ab 88
(off Windsor Way)
Sandringham M. TW12: Hamp 67Ba 129
Sandringham M. W5 .... 45Ma 87
Sandringham Pk. KT11: Cobh 84Ca 171
Sandringham Rd. BR1: Broml 64Jc 137
Sandringham Rd. CM15: Pil H 16Xd 40
Sandringham Rd. CR7: Thor H 71Sb 157
Sandringham Rd. E10 .... 30Fc 53
Sandringham Rd. E7 .... 36Lc 73
Sandringham Rd. E8 .... 36Vb 71
Sandringham Rd. EN6: Pot B 2Db 17
Sandringham Rd. IG11: Bark 36Vc 75
Sandringham Rd. KT4: Wor Pk 76Wa 154
Sandringham Rd. N22 .... 27Sb 51
Sandringham Rd. NW11 .... 31Ab 68
Sandringham Rd. NW2 .... 37Xa 68
Sandringham Rd. TW6: H'row A 57N 105
Sandringham Rd. UB5: N'olt 38Ca 65
Sandringham Rd. WD24: Wat 9Y 13
Sandringham Way EN8: Walt C 6Yb 20
Sandrock Pl. CR0: C'don 77Zb 158
Sandrock Rd. SE13 .... 55Cc 114
Sandroyd Way KT11: Cobh 85Ca 171
SANDS END 53Eb 111
Sand's End La. SW6 .... 53Db 111
Sands Farm Dr. SL1: Burn 2A 80
Sandshaw Ct. DA3: Hartl 71Ae 165
Sandstone Cl. RH1: Mers 100Nb 196
Sandstone La. E16 .... 45Kc 93
Sandstone Pl. N19 .... 33Kb 70
Sandstone Rd. SE12 .... 61Kc 137
Sands Way IG8: Wfd G 23Pc 54
Sandtoft Rd. SE7 .... 51Kc 115
Sandway Path BR5: St M Cry 70Yc 139
(off Okemore Gdns.)
Sandway Rd. BR5: St M Cry 70Yc 139
Sandwell Cres. NW6 .... 37Cb 69
Sandwich Ho. SE16 .... 47Yb 92
(off Swan Rd.)
Sandwich Ho. WC1 .... 4F 217 (41Nb 90)
(off Sandwich St.)
Sandwich St. WC1 .... 4F 217 (41Nb 90)
Sandwick Cl. NW7 .... 24Wa 48
Sandy Bank Rd. DA12: Grav'nd 10D 122
Sandy Bury BR6: Orp 76Tc 160
Sandy Cl. GU22: Wok 89E 168
Sandycombe Rd. TW14: Felt 60W 106
Sandycombe Rd. TW9: Kew 55Pa 109
Sandycombe Rd. TW9: Rich 55Pa 109
Sandycoombe Rd. TW1: Twick 58La 108
Sandy Ct. KT11: Cobh 85Ba 171
Sandy Cft. KT17: Ewe 82Ya 174
Sandycroft SE2 .... 51Wc 117
Sandy Dr. KT11: Cobh 83Ca 171
Sandy Dr. TW14: Felt 60U 106
Sandy Gro. WD6: Bore 12Pa 29
Sandy Hill Av. SE18 .... 50Rc 94
Sandy Hill Rd. SE18 .... 50Rc 94
Sandy Hill Rd. SM6: W'gton 81Lb 176
Sandyhill Rd. IG1: Ilf 35Rc 74
Sandy Holt KT11: Cobh 85Ba 171
Sandy Ho. IG11: Bark 42Wc 95
Sandy La. BR5: St P 68Zc 139
Sandy La. BR6: Orp 73Wc 161
Sandy La. CR4: Mitc 67Jb 134
Sandy La. DA14: Sidc 66Zc 139
Sandy La. DA2: Bean 61Yd 142
Sandy La. DA2: Sflt 61Yd 142
Sandy La. GU22: Pyr 89J 169
(not continuous)
Sandy La. GU22: Wok 89D 168
Sandy La. GU23: Send 95E 188
Sandy La. GU24: Chob 1J 167
Sandy La. GU25: Vir W 70A 126
(not continuous)
Sandy La. HA3: Kenton 30Pa 47
Sandy La. HA6: Nwood 19V 26
Sandy La. KT1: Hamp W 67La 130
Sandy La. KT11: Cobh 84Ba 171
Sandy La. KT12: Walt T 72X 151
Sandy La. KT20: Kgswd 96Bb 195
Sandy La. KT22: Oxs 84Da 171
Sandy La. RH1: Blet 4H 209
Sandy La. RH1: S Nut 7D 208
Sandy La. RH2: Reig 7D 206
Sandy La. RH3: B'wth 7A 206
Sandy La. RH8: Limp 99Kc 199
Sandy La. RH8: Oxt 1G 210
Sandy La. RM15: Avel 45Pd 97
Sandy La. RM15: Wenn 45Pd 97
Sandy La. RM16: Grays 1D 122
Sandy La. RM20: W Thur 50Xd 98
Sandy La. SL5: S'dale 1E 146
Sandy La. SM2: Cheam 80Ab 154
Sandy La. SM6: W'gton 79Mb 156
Sandy La. TN13: S'oaks 95Ld 203
Sandy La. TN15: Igh 97Xd 204
Sandy La. TN15: Ivy H 97Xd 204
Sandy La. TN16: Westrm 97Tc 200
Sandy La. TW10: Ham 61La 130
Sandy La. TW11: Hamp W 66Ja 130
Sandy La. TW11: Tedd 66Ja 130
Sandy La. WD23: Bush 13Ea 28
Sandy La. WD25: A'ham 13Ea 28
Sandy La. Cvn. Site WD25: A'ham 13Ea 28
Sandy La. Nth. SM6: W'gton 79Mb 156
Sandy La. Sth. SM6: W'gton 81Lb 176
Sandy Lodge HA5: Hat E 23Ca 45
Sandy Lodge HA6: Nwood 19U 26
Sandy Lodge Ct. HA6: Nwood 22U 44
Sandy Lodge Golf Course 19T 26
Sandy Lodge La. HA6: Nwood 19T 26
Sandy Lodge Rd. WD3: Rick 19R 26
Sandy Lodge Way HA6: Nwood 22T 44
Sandy Mead KT19: Eps 82Qa 173
Sandymount Av. HA7: Stan 22La 46
Sandy Ride SL5: S'hill 10C 124
Sandy Ridge BR7: Chst 65Qc 138
Sandy Ridge TN15: Bor G 92Ce 205
Sandy Ri. SL9: Chal P 25A 42
Sandy Rd. DA8: Erith 50Ed 96
Sandy Rd. KT15: Add 79J 149
Sandy Rd. NW3 .... 33Db 69
Sandys Row E1 .... 1J 225 (43Ub 91)
Sandy Way CR0: C'don 76Bc 158
Sandy Way GU22: Wok 89E 168
Sandy Way KT11: Cobh 84Ca 171
Sandy Way KT12: Walt T 74V 150
Sanford La. N16 .... 34Vb 71
Sanford St. SE14 .... 51Ac 114
Sanford Ter. N16 .... 34Vb 71
Sanford Wlk. N16 .... 33Vb 71
Sanford Wlk. SE14 .... 51Ac 114
Sangam Cl. UB2: S'hall 48Aa 85
Sanger Av. KT9: Chess 78Na 153
Sanger Dr. GU23: Send 95E 188
Sangley Rd. SE25 .... 70Ub 135
Sangley Rd. SE6 .... 59Dc 114
Sangora Rd. SW11 .... 56Fb 111
San Juan Dr. RM16: Chaf H 49Yd 98
Sankey Ho. E2 .... 40Yb 72
(off St James's Av.)
San Luis Dr. RM16: Chaf H 49Yd 98
San Marcos Dr. RM16: Chaf H 49Yd 98
Sansom Cl. WD3: Crox G 15T 26
Sansom Rd. E11 .... 33Hc 73
Sansom St. SE5 .... 53Tb 113
Sans Wlk. EC1 .... 5A 218 (42Qb 90)
Santa Maria Ct. E1 .... 43Ac 92
(off Ocean Est.)
Santers La. EN6: Pot B 5Ab 16
Sant Ho. SE17 .... 6E 230 (49Sb 91)
(off Browning St.)
Santiago Ct. E1 .... 43Ac 92
(off Ocean Est.)
Santiago Way RM16: Chaf H 50Zd 99
Santina Apartment CR0: C'don 74Tb 157
(off Cherry Orchard Rd.)
Santley Ho. SE1 .... 2A 230 (47Qb 90)
Santley St. SW4 .... 56Pb 112
Santos Rd. SW18 .... 57Cb 111
Santway, The HA7: Stan 22Ga 46
SANWAY 86N 169
Sanway Cl. KT14: Byfl 86N 169
Sanway Rd. KT14: Byfl 86N 169
Sapcote Trad. Cen. NW10 .... 37Va 68
Saperton Wlk. SE11 .... 5J 229 (49Pb 90)
(off Juxon St.)
Saphire Ct. E15 .... 39Ec 72
(off Warton Rd.)
Sapho Pk. DA12: Grav'nd 3H 145
Saphora Cl. BR6: Farnb 78Tc 160
Sapperton Ct. EC1 .... 5D 218 (42Sb 91)
(off Gee St.)
Sapperton Ho. W2 .... 43Cb 89
(off Westbourne Pk. Rd.)
Sapphire Cl. E6 .... 44Qc 94
Sapphire Cl. RM8: Dag 32Yc 75
Sapphire Ct. E1 .... 45Wb 91
(off Cable St.)
Sapphire Rd. NW10 .... 38Sa 67
Sapphire Rd. SE8 .... 49Ac 92
Sappho Ct. GU21: Wok 8J 167
Saracen Cl. CR0: C'don 72Tb 157
Saracen Est. HP2: Hem H 1B 4
Saracens FC 24Ya 48
Saracens Head HP2: Hem H 1A 4
Saracens Head Yd. AL1: St A 2B 6
Saracens Head Yd. EC3 .... 3K 225 (44Vb 91)
(off Jewry St.)
Saracen St. E14 .... 44Cc 92
Sara Cres. DA9: Ghithe 56Xd 120
Sarah Ct. UB5: N'olt 39Ba 65
Sarah Ho. E1 .... 44Xb 91
(off Commercial Rd.)
Sarah Swift Ho. SE1 .... 1G 231 (47Tb 91)
(off Kipling St.)
Sara La. Ct. N1 .... 1J 219 (40Ub 71)
(off Stanway St.)
Sara Pk. DA12: Grav'nd 3G 144
Saratoga Rd. E5 .... 35Yb 72
Sara Turnbull Ho. SE18 .... 49Pc 94
Saravia Ct. SE13 .... 58Gc 115
Sardinia St. WC2 .... 3H 223 (44Pb 90)
Sargeant Cl. UB8: Cowl 41M 83
Sarita Cl. HA3: Hrw W 26Fa 46
Sarjant Path SW19 .... 61Za 132
(off Blincoe Cl.)
Sarjeant Ct. BR4: W W'ck 75Fc 159
(off Bencurtis Pk.)
Sark Cl. TW5: Hest 52Ca 107
Sark Ho. EN3: Enf W 10Zb 20
Sark Ho. WD18: Wat 16V 26
(off Scammell Way)
Sark Twr. SE28 .... 47Sc 94
Sark Wlk. E16 .... 44Kc 93
Sarnes Ct. N11 .... 21Kb 50
(off Oakleigh Rd. Sth.)
Sarnesfield Ho. SE15 .... 51Xb 113
(off Pencraig Way)
Sarnesfield Rd. EN2: Enf 14Tb 33
SARRATT 8J 11
SARRATT BOTTOM 9G 10
SARRATT HALL 7J 11
Sarratt Ho. W10 .... 43Ya 88
(off Sutton Way)
Sarratt La. WD3: Chor 12K 25
Sarratt La. WD3: Loud 12K 25
Sarratt La. WD3: Sarr 12K 25
Sarratt Rd. WD3: Chan C 9K 11
Sarratt Rd. WD3: Crox G 9K 11
Sarratt Rd. WD3: Sarr 9K 11
Sarre Av. RM12: Horn 37Ld 77
Sarre Rd. BR5: St M Cry 71Yc 161
Sarre Rd. NW2 .... 36Bb 69
Sarsby Dr. TW19: Wray 61C 126
Sarsen Av. TW3: Houn 54Ca 107
Sarsens Cl. DA12: Cobh 9H 145
Sarsfeld Rd. SW12 .... 61Hb 133
Sarsfield Rd. UB6: G'frd 40Ka 66
Sartoria Cl. RM20: W Thur 51Wd 120
Sartor Rd. SE15 .... 56Zb 114
Sarum Complex UB8: Uxb 41K 83
Sarum Grn. KT13: Weyb 76U 150
Sarum Ho. W11 .... 45Bb 89
(off Portobello Rd.)
Sarum Ter. E3 .... 42Bc 92
Saskia M. SE15 .... 54Xb 113
Sassoon NW9 .... 25Va 48
Sassoon M. DA11: Nflt 61De 143
Satanita Cl. E16 .... 44Mc 93
Satchell Mead NW9 .... 25Va 48
Satchwell Rd. E2 .... 41Wb 91
Satchwell St. E2 .... 41Wb 91
Satin Ho. E1 .... 44Wb 91
(off Piazza Wlk.)
Satinwood Ct. HP3: Hem H 4N 3
Satis Ct. KT17: Ewe 83Va 174
Satis Ho. SL3: Dat 2N 103
Sattar M. N16 .... 34Tb 71
Saturn Ho. E15 .... 39Fc 73
(off High St.)
Saturn Ho. E3 .... 39Cc 72
(off Garrison Rd.)
Sauls Grn. E11 .... 34Gc 73
Saundby La. SE3 .... 56Kc 115
Saunders Apts. E3 .... 41Cc 92
(off Marchant St.)
Saunders Cl. DA11: Nflt 1A 144
Saunders Cl. E14 .... 45Bc 92
(off Limehouse C'way.)
Saunders Cl. IG1: Ilf 32Tc 74
Saunders Copse GU22: Wok 4M 187
Saunders Ho. SE16 .... 47Zb 92
(off Quebec Way)
Saunders La. GU22: Wok 4J 187
Saunders Ness Rd. E14 .... 50Ec 92
Saunders Rd. SE18 .... 50Vc 95
Saunders Rd. UB10: Uxb 38P 63
Saunders St. SE11 .... 5K 229 (49Qb 90)
Saunders Way DA1: Dart 61Pd 141
Saunders Way SE28 .... 45Xc 95
Saunderton Rd. HA0: Wemb 36Ka 66
Saunton Av. UB3: Harl 52V 106
Saunton Ct. UB1: S'hall 45Ea 86
(off Haldane Rd.)
Saunton Rd. RM12: Horn 33Jd 76
Savage Gdns. E6 .... 44Pc 94
Savage Gdns. EC3 .... 4J 225 (45Ub 91)
(not continuous)
Savanna Ct. WD18: Wat 14V 26
(off Rickmansworth Rd.)
Savannah Cl. SE15 .... 52Vb 113
Savay Cl. UB9: Den 31J 63
Savay La. UB9: Den 30J 43
Savera Cl. UB2: S'hall 48Y 85
Savernake Ct. HA7: Stan 23Ka 46
Savernake Ho. N4 .... 31Sb 71
Savernake Rd. N9 .... 16Wb 33
Savernake Rd. NW3 .... 35Hb 69
Savery Dr. KT6: Surb 73Ka 152
Savile Cl. KT3: N Mald 71Ua 154
Savile Cl. KT7: T Ditt 74Ha 152
Savile Gdns. CR0: C'don 75Vb 157
Savile Row W1 .... 4B 222 (45Lb 90)
Savill Building, The 5K 125
Saville Ct. KT19: Eps 83Ra 173
Saville Cres. TW15: Ashf 65T 128
Saville Pl. TW20: Eng G 4P 125
Saville Rd. E16 .... 46Nc 94
Saville Rd. RM6: Chad H 30Bd 55
Saville Rd. TW1: Twick 60Ha 108
Saville Rd. W4 .... 48Ta 87
Saville Row BR2: Hayes 74Hc 159
Saville Row EN3: Enf H 12Zb 34
Savill Gdn., The 5J 125
Savill Gdns. SW20 .... 69Wa 132
Savill Ho. E16 .... 46Rc 94
(off Robert St.)
Savill Ho. SW4 .... 58Mb 112
Savill M. TW20: Eng G 5P 125
Savill Row IG8: Wfd G 23Hc 53
Savin Lodge SM2: Sutt 80Eb 155
(off Walnut M.)
Savona Cl. SW19 .... 66Za 132
Savona Ho. SW8 .... 52Lb 112
(off Savona St.)
Savona St. SW8 .... 52Lb 112
Savoy Av. UB3: Harl 50U 84
Savoy Bldgs. WC2 .... 5H 223 (45Pb 90)
(off Strand)
SAVOY CIRCUS 45Va 88
Savoy Cl. E15 .... 39Gc 73
Savoy Cl. HA8: Edg 22Qa 47
Savoy Cl. UB9: Hare 26M 43
Savoy Ct. HA2: Harr 29Da 45
Savoy Ct. NW3 .... 34Eb 69
Savoy Ct. SW5 .... 49Cb 89
(off Cromwell Rd.)
Savoy Ct. WC2 .... 5H 223 (45Pb 90)
Savoy Gro. RM11: Horn 30Kd 57
Savoy Hill WC2 .... 5H 223 (45Pb 90)
Savoy M. AL1: St A 5P 5
Savoy M. SW9 .... 55Nb 112
Savoy Pde. EN1: Enf 13Ub 33
Savoy Pl. W12 .... 46Za 88
Savoy Pl. WC2 .... 5G 223 (45Nb 90)
Savoy Rd. DA1: Dart 57Md 119
Savoy Row WC2 .... 5H 223 (45Pb 90)
(off Savoy St.)
Savoy Steps WC2 .... 5H 223 (45Pb 90)
(off Savoy St.)
Savoy St. WC2 .... 4H 223 (45Pb 90)
Savoy Theatre 5G 223 (45Nb 90)
(off Strand)
Savoy Way WC2 .... 5H 223 (45Pb 90)
(off Savoy Hill)
Sawbill Cl. UB4: Yead 43Z 85
Sawbridgeworth Ct. WD23: Bush 15Ea 28
(off Goddard Dr.)
Sawcotts Way RM16: Grays 8A 100
Sawkins Cl. SW19 .... 61Ab 132
Saw Mill Way N16 .... 30Wb 51
Sawmill Yd. E3 .... 39Ac 72
Sawston Ct. RM19: Purf 50Rd 97
(off Linnet Way)
Sawtry Cl. SM5: Cars 73Gb 155
Sawtry Way WD6: Bore 10Qa 15
Sawyer Cl. N9 .... 19Wb 33
Sawyer Ho. WD10 .... 38Ta 67
Sawyers Apts. SW11 .... 56Gb 111
(off Danvers Avenue)
Sawyers Chase RM4: Abr 13Xc 37
Sawyers Cl. RM10: Dag 37Ed 76
Sawyers Cl. SL4: Wind 2C 102
Sawyers Ct. CM15: Shenf 17Be 41
Sawyers Ct. EN8: Walt C 5Zb 20
Sawyers Gro. CM15: B'wood 18Zd 41
Sawyers Hall La. CM15: B'wood 17Yd 40
Sawyer's Hill TW10: Rich 59Pa 109
Sawyers La. EN6: Pot B 5Za 16
Sawyers Lawn W13 .... 44Ja 86
Sawyer St. SE1 .... 1D 230 (47Sb 91)
Sawyers Way HP2: Hem H 2P 3
Saxby Rd. SW2 .... 59Nb 112
Saxbys Rd. TN15: Seal 94Td 204
Saxham Rd. IG11: Bark 40Uc 74
Saxlingham Rd. E4 .... 20Fc 35
Saxon Av. TW13: Hanw 61Aa 129
Saxonbury Av. TW16: Sun 69X 129
Saxonbury Cl. CR4: Mitc 69Fb 133
Saxonbury Ct. N7 .... 36Nb 70
Saxonbury Gdns. KT6: Surb 74La 152
Saxon Bus. Cen. SW19 .... 68Eb 133
Saxon Chase N8 .... 30Pb 50
Saxon Cl. CM13: B'wood 20Ce 41
Saxon Cl. DA11: Nflt 62Ee 143
Saxon Cl. E17 .... 31Cc 72
Saxon Cl. KT6: Surb 72Ma 153
Saxon Cl. RM3: Hrld W 26Pd 57
Saxon Cl. SL3: L'ly 47B 82
Saxon Cl. TN14: Otf 89Hd 182
Saxon Cl. UB8: Hil 43P 83
Saxon Ct. HA6: Nwood 24V 44
Saxon Ct. N1 .... 39Nb 70
Saxon Ct. SL0: Iver 44G 82
Saxon Ct. WD6: Bore 11Na 29
Saxon Dr. W3 .... 44Qa 87
Saxonfield Cl. SW2 .... 59Pb 112
Saxon Gdns. UB1: S'hall 45Aa 85
Saxon Hall W2 .... 45Db 89
(off Palace Ct.)
Saxon Ho. E1 .... 1K 225 (43Vb 91)
(off Thrawl St.)
Saxon Ho. KT1: King T 70Pa 131
Saxon Ho. SM6: W'gton 74Jb 156
(off London Rd.)
Saxon Ho. TN14: S'oaks 93Md 203
Saxon Ho. TW13: Hanw 61Ba 129
Saxon Lea Ct. E3 .... 40Bc 72
(off Saxon Rd.)
Saxon Lodge CR0: C'don 74Sb 157
(off Tavistock Rd.)
Saxon Pl. DA4: Hort K 71Sd 164
Saxon Rd. BR1: Broml 66Hc 137
Saxon Rd. DA2: Hawl 63Nd 141
Saxon Rd. E3 .... 40Bc 72
Saxon Rd. E6 .... 42Pc 94
Saxon Rd. HA9: Wemb 34Sa 67
Saxon Rd. IG1: Ilf 37Rc 74
Saxon Rd. KT12: Walt T 76Z 151
Saxon Rd. KT2: King T 67Na 131
Saxon Rd. N22 .... 25Rb 51
Saxon Rd. SE25 .... 71Tb 157
Saxon Rd. TW15: Ashf 65T 128
Saxon Rd. UB1: S'hall 45Aa 85
Saxons KT20: Tad 93Za 194
Saxon Ter. SE6 .... 61Bc 136
Saxon Wlk. DA14: Sidc 65Yc 139
Saxon Way EN9: Walt A 5Ec 20
Saxon Way N14 .... 16Mb 32
Saxon Way RH2: Reig 5H 207
Saxon Way SL4: Old Win 8M 103
Saxon Way UB7: Harm 51L 105
Saxon Way Ind. Est. UB7: Harm 51L 105
Saxony Pde. UB3: Hayes 43S 84
Saxton Cl. RM17: Grays 51De 121
Saxton Cl. SE13 .... 55Fc 115
Saxton M. WD17: Wat 12W 26
Saxton Pl. KT8: W Mole 70Ba 129
Saxville Rd. BR5: St P 69Xc 139
Sayer Cl. DA9: Ghithe 57Wd 120
Sayers Cl. KT22: Fet 95Ea 192
Sayers Ct. W5 .... 47Ma 87
Sayers Ho. N2 .... 26Fb 49
(off The Grange)
Sayer St. SE17 .... 5D 230 (49Sb 91)
Sayer's Wlk. TW10: Rich 59Pa 109
Sayesbury La. N18 .... 22Wb 51
Sayes Ct. KT15: Add 78L 149
Sayes Ct. SE8 .... 51Bc 114
Sayes Ct. Farm Dr. KT15: Add 78K 149
Sayes Ct. Rd. BR5: St P 70Wc 139
Sayes Ct. St. SE8 .... 51Bc 114
Scadbury Gdns. BR5: St P 68Wc 139
Scadbury Pk. 66Vc 139
Scads Hill Cl. BR6: Pet W 72Vc 161
Scafell NW1 .... 3B 216 (41Lb 90)
(off Stanhope St.)
Scafell Rd. SL2: Slou 2D 80
Scala 3G 217 (41Nb 90)
(off Pentonville Rd.)
Scala St. W1 .... 7C 216 (43Lb 90)
Scales Rd. N17 .... 27Vb 51
Scammell Way WD18: Wat 16V 26
Scampston M. W10 .... 44Za 88
Scandrett St. E1 .... 46Xb 91
Scarab Cl. E16 .... 45Hc 93
Scarba Wlk. N1 .... 37Tb 71
(off Essex Rd.)
Scarborough Cl. SM2: Cheam 83Bb 175
Scarborough Cl. TN16: Big H 90Lc 179
Scarborough Dr. WD3: Crox G 14R 26
Scarborough Rd. E11 .... 32Fc 73
Scarborough Rd. N4 .... 31Qb 70
Scarborough Rd. N9 .... 17Yb 34
Scarborough Rd. TW6: H'row A 58S 106
Scarborough St. E1 .... 3K 225 (44Vb 91)
Scarborough Way SL1: Slou 8F 80
Scarbrook Rd. CR0: C'don 76Sb 157
Scarle Rd. HA0: Wemb 37Ma 67
Scarlet Cl. BR5: St P 70Xc 139
Scarlet Cl. E20 .... 36Dc 72
Scarlet Rd. DA8: Erith 52Jd 118
Scarlet Rd. SE6 .... 62Gc 137
Scarlett Cl. GU21: Wok 10K 167
Scarlette Mnr. Way SW2 .... 59Qb 112
Scarlet Wlk. EN3: Pond E 15Zb 34
Scarsbrook Rd. SE3 .... 55Mc 115
Scarsdale Pl. W8 .... 48Db 89
Scarsdale Rd. HA2: Harr 34Ea 66
Scarsdale Studios W8 .... 48Cb 89
(off Stratford Rd.)
Scarsdale Vs. W8 .... 48Cb 89
Scarth Rd. SW13 .... 55Va 110
Scatterdells La. WD4: Chfd 2H 11
Scatterdells Pk. WD4: Chfd 2J 11
Scawen Cl. SM5: Cars 77Jb 156
Scawen Rd. SE8 .... 50Ac 92
Scawfell St. E2 .... 40Vb 71
Scaynes Link N12 .... 21Cb 49
SCC Smallholdings Rd. KT17: Eps 86Ya 174
(not continuous)
Sceaux Gdns. SE5 .... 53Ub 113
Scena Way SE5 .... 52Sb 113
Sceptre Ct. EC3 .... 5K 225 (45Vb 91)
(off Tower Hill)
Sceptre Ho. E1 .... 42Yb 92
(off Malcolm Rd.)
Sceptre Rd. E2 .... 41Yb 92
Schafer Ho. NW1 .... 4B 216 (41Lb 90)
Schofield Wlk. SE3 .... 52Jc 115
Scholars, The WD18: Wat 15Y 27
(off Lady's Cl.)
Scholars Cl. EN5: Barn 14Ab 30
Scholar's Ct. AL1: St A 2C 6
(off Newsom Pl.)
Scholars Ct. AL4: Col H 5N 7
Scholars Ct. RM2: Rom 29Jd 56
(off Academy Flds. Rd.)
Scholars Ho. NW6 .... 39Cb 69
(off Glengall Rd.)
Scholars Pl. KT12: Walt T 74Y 151
Scholars Pl. N16 .... 34Ub 71
Scholars Rd. E4 .... 18Fc 35
Scholars Rd. SW12 .... 60Lb 112
Scholars Vw. KT7: T Ditt 73Ga 152
Scholar's Wlk. SL9: Ger X 28A 42
Scholars Wlk. AL10: Hat 3C 8
(not continuous)
Scholars Wlk. SL3: L'ly 47C 82
Scholars Wlk. SL9: Chal P 23A 42
Scholars Way RM2: Rom 29Kd 57
Scholars Way RM8: Dag 35Wc 75
Scholefield Rd. N19 .... 33Mb 70
Scholey Ho. SW11 .... 55Gb 111
Schomberg Ho. SW1 .... 5E 228 (49Mb 90)
(off Page St.)
Schonfeld Sq. N16 .... 33Tb 71
School All. TW1: Twick 60Ja 108
School Allotment Ride SL4: Wink 9A 102
School App. E2 .... 3J 219 (41Ub 91)
(off Kingsland Rd.)
School App. TN15: Bor G 92Ce 205
School App. UB4: Hayes 42V 84
Schoolbank Rd. SE10 .... 48Hc 93
Schoolbell M. E3 .... 40Ac 72
School Cl. GU24: Bisl 7D 166
School Cotts. GU22: Wok 4N 187
(off Mayford Grn.)
School Cres. DA1: Cray 56Hd 118
Schoolfield Rd. RM20: W Thur 51Wd 120
Schoolfield Way RM20: W Thur 51Wd 120
Schoolgate Dr. SM4: Mord 71Db 155
School Hill RH1: Mers 100Lb 196
School Ho. SE1 .... 5H 231 (49Ub 91)
(off Page's Wlk.)
Schoolhouse Gdns. IG10: Lough 14Rc 36
School Ho. La. NW7 .... 23Ab 48
School Ho. La. TW11: Tedd 66Ka 130
Schoolhouse La. E1 .... 45Zb 92
Schoolhouse Yd. SE18 .... 50Rc 94
School La. AL2: Brick W 6Ba 13
School La. BR8: Swan 67Kd 141
School La. CM13: Ingve 23Ee 59
School La. CR3: Cat'm 98Vb 197
School La. DA16: Well 55Xc 117
School La. DA2: Bean 63Yd 142
School La. DA3: Fawk 73Vd 164
School La. DA4: Hort K 70Sd 142
School La. GU20: W'sham 8B 146
School La. GU23: Ock 94R 190
School La. GU24: Pirb 4C 186
School La. HA5: Pinn 28Aa 45
School La. IG7: Chig 21Vc 55
School La. KT1: Hamp W 67La 130
School La. KT15: Add 78J 149
School La. KT20: Walt H 97Wa 194
School La. KT22: Fet 94Fa 192
School La. KT24: W Hor 100R 190
School La. KT6: Surb 74Pa 153
School La. RH5: Mick 99La 192
School La. RM16: Ors 3C 100
School La. SE23 .... 61Xb 135
School La. SL2: Slou 5K 81
School La. SL2: Stoke P 8M 61
School La. SL9: Chal P 26A 42
School La. TN11: Plax 100Ae 205
School La. TN11: S'brne 100Ae 205
School La. TN15: Plax 100Ae 205
School La. TN15: Seal 93Pd 203
School La. TN15: W King 84Ud 184
School La. TW17: Shep 72R 150
School La. TW20: Egh 64C 126
School La. WD23: Bush 17Da 27
School La. WD25: Wat 6Ba 13
School Mead WD5: Ab L 4U 12
School M. E1 .... 45Xb 91
(off Hawksmoor M.)
School Nook E5 .... 33Zb 72
School of Oriental & African Studies
Vernon Sq. Campus 3J 217 (41Pb 90)
School of Pharmacy, The 5G 217 (42Nb 90)
(off Brunswick Sq.)
School Pde. UB9: Hare 26L 43
School Pas. KT1: King T 68Pa 131
School Pas. UB1: S'hall 45Ba 85
School Rd. BR7: Chst 67Sc 138
School Rd. DA12: Grav'nd 2E 144
School Rd. E12 .... 35Pc 74
School Rd. EN6: Pot B 2Eb 17
School Rd. KT1: Hamp W 67La 130
School Rd. KT8: E Mos 70Fa 130
School Rd. NW10 .... 42Ta 87
School Rd. RM10: Dag 39Cd 76
School Rd. SL5: S'hill 1B 146
School Rd. TW12: Hamp H 65Ea 130
School Rd. TW15: Ashf 65R 128
School Rd. TW3: Houn 55Ea 108
School Rd. UB7: Harm 51M 105
School Rd. Av. TW12: Hamp H 65Ea 130
SCHOOL ROAD JUNC. 66R 128
School Row HP1: Hem H 3H 3
School Sq. SE10 .... 48Hc 93
School Wlk. SL2: Slou 5M 81
School Wlk. TW16: Sun 70V 128
School Way N12 .... 23Fb 49
School Way RM8: Dag 34Yc 75
Schooner Cl. E14 .... 48Fc 93
Schooner Cl. IG11: Bark 41Xc 95
Schooner Cl. SE16 .... 47Zb 92
Schooner Ct. DA2: Dart 56Sd 120
Schooner Ho. DA8: Erith 50Gd 96
Schooner Pk. DA2: Dart 57Sd 120
Schooner Rd. E16 .... 47Kc 93
Schopwick Pl. WD6: E'tree 16Ma 29
Schroder Ct. TW20: Eng G 4M 125
Schubert Rd. SW15 .... 57Bb 111
Schubert Rd. WD6: E'tree 16Ma 29
Schurlock Pl. TW2: Twick 61Ga 130
Schwartz Wharf E9 .... 38Cc 72
Science Mus. 4B 226 (48Fb 89)
Scilla Ct. RM17: Grays 51Fe 121
SCILLY ISLES 75Ga 152
Sclater St. E1 .... 5K 219 (42Vb 91)
Scoble Pl. N16 .... 35Vb 71
Scoles Cres. SW2 .... 60Rb 113
Scoop, The 7J 225 (46Ub 91)
Scope Way KT1: King T 70Na 131
Score Complex Leyton, The 34Dc 72
Scoresby St. SE1 .... 7B 224 (46Rb 91)
Scorton Av. UB6: G'frd 40Ja 66
Scorton Ho. N1 .... 1J 219 (40Ub 71)
(off Whitmore Est.)
Scotch Comn. W13 .... 43Ja 86
SCOTCH HOUSE 2G 227 (47Hb 89)
Scoter Cl. IG8: Wfd G 24Kc 53
Scoter Ct. SE8 .... 51Bc 114
(off Abinger Gro.)

Scot Gro. HA5: Pinn 24Z 45
Scotia Bldg. E1 45Zb 92
(off Jardine Rd.)
Scotia Ct. SE16 47Yb 92
(off Canada Est.)
Scotia Rd. SW2 59Qb 112
Scotland Bri. Lock KT15: New H 83J 169
Scotland Bri. Rd. KT15: New H 83J 169
Scotland Grn. N17 26Vb 51
Scotland Grn. Rd. EN3: Pond E 15Zb 34
Scotland Grn. Rd. Nth. EN3: Pond E 14Zb 34
Scotland La. DA12: Cobh 9H 145
(not continuous)
Scotland Pl. SW1 6F 223 (46Nb 90)
Scotland Rd. IG9: Buck H 18Lc 35
Scotlands Dr. SL2: Farn C 7F 60
Scotney Cl. BR6: Farnb 77Qc 160
Scotney Cl. WD3: Crox G 14R 26
Scotney Ho. E9 37Yb 72
Scotney Wlk. RM12: Horn 36Ld 77
Scots Cl. TW19: Stanw 60M 105
Scotscraig WD7: R'lett 7Ha 14
Scotsdale Cl. BR5: Pet W 70Uc 138
Scotsdale Cl. SM3: Cheam 80Ab 154
Scotsdale Rd. SE12 57Kc 115
Scotshall La. CR6: W'ham 87Ec 178
Scots Hill WD3: Crox G 16P 25
Scots Hill Cl. WD3: Crox G 16P 25
Scotsmill La. WD3: Crox G 16P 25
Scotson Ho. SE11 6K 229 (49Qb 90)
(off Marylee Way)
Scotspine La. RM3: Hrld W 26Nd 57
Scotswood Pk. GU21: Wok 86F 168
Scotswood St. EC1 5A 218 (42Qb 90)
Scotswood Wlk. N17 24Wb 51
Scott Av. SW15 58Ab 110
Scott Cl. AL3: St A 4N 5
Scott Cl. DA1: Dart 57Pd 119
Scott Cl. KT19: Ewe 78Sa 153
Scott Cl. SL2: Farn C 6G 60
Scott Cl. SW16 67Pb 134
Scott Cl. UB7: W Dray 49P 83
Scott Ct. W3 47Ta 87
Scott Cres. DA8: Erith 53Hd 118
Scott Cres. HA2: Harr 32Da 65
Scott Ellis Gdns. NW8 4B 214 (41Fb 89)
Scottes La. RM8: Dag 32Zc 75
Scott Farm Cl. KT7: T Ditt 74Ka 152
Scott Gdns. TW5: Hest 52Z 107
Scott Ho. DA17: Belv 50Bd 95
Scott Ho. E13 40Jc 73
(off Queens Rd. W.)
Scott Ho. E14 47Cc 92
(off Admirals Way)
Scott Ho. KT17: Eps 84Ua 174
(off Winter Cl.)
Scott Ho. N18 22Wb 51
(off Woolmer Rd.)
Scott Ho. N7 37Pb 70
(off Caledonian Rd.)
Scott Ho. NW10 38Ta 67
(off Stonebridge Pk.)
Scott Ho. NW8 6D 214 (42Gb 89)
(off Broadley St.)
Scott Ho. RM11: Horn 30Jd 56
(off Benjamin Cl.)
Scott Ho. SE8 50Bc 92
(off Grove St.)
Scott Lidgett Cres. SE16 47Wb 91
Scott Rd. DA12: Grav'nd 4F 144
Scott Rd. HA8: Edg 26Ra 47
Scott Rd. RM16: Grays 9C 100
Scott Russell Pl. E14 50Dc 92
Scotts Av. BR2: Broml 68Fc 137
Scotts Av. TW16: Sun 66U 128
Scotts Cl. RM12: Horn 36Ld 77
Scotts Ct. W12 47Ya 88
(off Scott's Rd.)
Scotts Dr. TW12: Hamp 66Da 129
Scotts Farm Rd. KT19: Ewe 79Sa 153
Scott's Gro. Cl. GU24: Chob 5H 167
Scott's Gro. Rd. GU24: Chob 5F 166
Scott's La. BR2: Broml 69Fc 137
Scotts La. KT12: Hers 77Z 151
Scotts Pas. SE18 49Rc 94
Scott's Rd. E10 32Ec 72
Scott's Rd. UB2: S'hall 48Y 85
Scott's Rd. W12 47Xa 88
Scotts Rd. BR1: Broml 66Jc 137
Scott's Sufferance Wharf
SE1 2K 231 (47Vb 91)
(off Mill St.)
Scotts Ter. SE9 61Nc 138
Scott St. E1 42Xb 91
Scotts Way TN13: Riv 94Gd 202
Scotts Way TW16: Sun 66U 128
Scottswood Cl. WD23: Bush 12Aa 27
Scottswood Rd. WD23: Bush 12Aa 27
Scott's Yd. EC4 4F 225 (45Tb 91)
Scott Trimmer Way TW3: Houn 54Aa 107
Scottwell Dr. NW9 29Va 48
Scoulding Ho. E14 48Cc 92
(off Mellish St.)
Scoulding Rd. E16 44Jc 93
Scouler St. E14 45Ec 92
Scout App. NW10 35Ua 68
Scout La. SW4 55Lb 112
Scout Pk. N11 24Mb 50
Scout Way NW7 21Ta 47
Scovell Cres. SE1 2D 230 (47Sb 91)
(off McCoid Way)
Scovell Rd. SE1 2D 230 (47Sb 91)
Scratchers La. DA3: Fawk 75Td 164
Scratton Rd. SS17: Stan H 1M 101
Scrattons Farm Eco-Park 40Ad 75
Scrattons Ter. IG11: Bark 40Zc 75
Screenworks N5 36Sb 71
Scrimgoeur Pl. N4 32Sb 71
Scriven Ct. E8 39Vb 71
Scriveners Cl. HP2: Hem H 2N 3
Scriven St. E8 39Vb 71
Scrooby St. SE6 58Dc 114
Scrope Ho. EC1 7K 217 (43Qb 90)
(off Bourne Est.)
Scrubbitts Pk. Rd. WD7: R'lett 7Ja 14
Scrubbitts Sq. WD7: R'lett 7Ja 14
Scrubs La. NW10 41Wa 88
Scrubs La. W10 43Xa 88
Scrutton Cl. SW12 59Mb 112
Scrutton St. EC2 6H 219 (42Ub 91)
Scudamore La. NW9 28Sa 47
Scudders Hill DA3: Fawk 72Xd 164
Sculpture Ho. E1 43Zb 92
(off Duckett St.)
Scutari Rd. SE22 57Yb 114
Scutley La. GU18: Light 1C 166
Scutley La. GU20: W'sham 10C 146
Scylla Cres. TW6: H'row A 59R 106
(not continuous)
Scylla Pl. GU21: Wok 1L 187
Scylla Rd. SE15 55Wb 113
(not continuous)
Scylla Rd. TW6: H'row A 58R 106
Seaborne Wharf E3 43Cc 92
(off Invicta Cl.)
Seaborough Rd. RM16: Grays 8E 100
Seabright St. E2 41Xb 91
Seabrook Dr. BR4: W W'ck 75Gc 159
Seabrooke Ri. RM17: Grays 51De 121
Seabrook Gdns. RM7: Rush G 31Cd 76
Seabrook Rd. RM8: Dag 34Zc 75
Seabrook Rd. WD4: K Lan 9D 4
Seaburn Cl. RM13: Rain 40Gd 76
Seacole Cl. W3 44Ta 87
Seacole Lodge N21 15Pb 32
(off Pennington Dr.)
Sea Containers Ho. SE1 5A 224 (45Qb 90)
(off Upper Ground)
Seacon Twr. E14 47Bc 92
Seacourt Rd. SE2 47Zc 95
Seacourt Rd. SL3: L'ly 49D 82
Seacroft Gdns. WD19: Wat 20Z 27
Seafarer Way SE16 49Ac 92
Seafield Rd. N11 21Mb 50
Seaford Cl. HA4: Ruis 32T 64
Seaford Ho. SE16 47Yb 92
(off Swan Rd.)
Seaford Rd. E17 27Dc 52
Seaford Rd. EN1: Enf 14Ub 33
Seaford Rd. N15 29Tb 51
Seaford Rd. TW6: H'row A 57M 105
Seaford Rd. W13 46Ka 86
Seaford St. WC1 4G 217 (41Nb 90)
Seaforth Av. KT3: N Mald 71Xa 154
Seaforth Cl. RM1: Rom 24Gd 56
Seaforth Cres. N5 36Sb 71
Seaforth Dr. EN8: Walt C 6Zb 20
Seaforth Gdns. IG8: Wfd G 22Lc 53
Seaforth Gdns. KT19: Ewe 77Va 154
Seaforth Gdns. N21 17Pb 32
Seaforth Pl. SW1 3C 228 (48Lb 90)
(off Buckingham Ga.)
Seager Pl. SE8 53Cc 114
(off Deptford Bri.)
Seagrave Cl. E1 43Zb 92
Seagrave Lodge SW6 51Cb 111
(off Seagrave Rd.)
Seagrave Rd. SW6 51Cb 111
Seagry Rd. E11 31Jc 73
Seagull Cl. IG11: Bark 41Wc 95
Seagull La. E16 45Jc 93
Seahorse Sailing Club 33Ta 67
SEAL 93Pd 203
Sealand Rd. TW6: H'row A 58Q 106
Sealand Wlk. UB5: N'olt 41Z 85
SEAL CHART 95Sd 204
Sealcroft Cotts. TN15: Seal 91Pd 203
Seal Dr. TN15: Seal 93Pd 203
Seale Hill RH2: Reig 8J 207
Seal Hollow Rd. TN13: S'oaks 96Ld 203
Seal Hollow Rd. TN15: S'oaks 96Md 203
Seal Ho. SE1 3G 231 (48Tb 91)
(off Weston St.)
Sea Life London 1H 229 (47Pb 90)
Seally Rd. RM17: Grays 50Ce 99
Seal Rd. TN14: S'oaks 93Ld 203
Seal Rd. TN15: S'oaks 93Nd 203
Seal St. E8 35Vb 71
Sealy Way HP3: Hem H 6M 3
Seaman Cl. AL2: Park 7B 6
Searches La. WD5: Bedm 9J 5
Searchwood Hgts. CR6: W'ham 90Xb 177
Searchwood Rd. CR6: W'ham 90Xb 177
Searle Ho. NW8 1E 214 (39Gb 69)
(off Cecil Gro.)
Searle Ho. SW11 53Jb 112
(off Macduff Rd.)
Searle Pl. N4 32Pb 70
Searles Cl. SW11 52Gb 111
Searles Ct. BR8: Swan 71Jd 162
Searles Dr. E6 43Rc 94
Searles Rd. SE1 5G 231 (49Tb 91)
Searlwood Ct. GU24: W End 6D 166
Searson Ho. SE17 6C 230 (49Rb 91)
(off Canterbury Pl.)
Sears St. SE5 52Tb 113
Seasalter Ho. SW9 53Qb 112
(off Gosling Way)
Seasons Cl. W7 46Ha 86
Seasons Ho. E20 37Ec 72
(off Mirabelle Gdns.)
Seasprite Cl. UB5: N'olt 41Z 85
Seaton Av. IG3: Ilf 36Vc 75
Seaton Cl. E13 42Jc 93
Seaton Cl. SE11 7A 230 (50Qb 90)
Seaton Cl. SW15 60Xa 110
Seaton Cl. TW2: Whitt 58Fa 108
Seaton Dr. TW15: Ashf 61N 127
Seaton Gdns. HA4: Ruis 34V 64
Seaton Point E5 35Wb 71
Seaton Rd. AL2: Lon C 8H 7
Seaton Rd. CR4: Mitc 68Gb 133
Seaton Rd. DA1: Dart 59Jd 118
Seaton Rd. DA16: Well 52Yc 117
Seaton Rd. HA0: Wemb 40Na 67
Seaton Rd. HP3: Hem H 5M 3
Seaton Rd. TW2: Whitt 58Ea 108
Seaton Rd. UB3: Harl 49T 84
Seaton Sq. NW7 24Za 48
Seaton St. N18 22Wb 51
Seawall Ct. IG11: Bark 40Sc 74
(off Dock Rd.)
Sebastian Av. CM15: Shenf 16Ce 41
Sebastian Ct. IG11: Bark 39Vc 75
Sebastian Ho. N1 2H 219 (40Ub 71)
(off Hoxton St.)
Sebastian St. EC1 4C 218 (41Rb 91)
Sebastopol Rd. N9 21Wb 51
Sebbon St. N1 38Rb 71
Sebergham Gro. NW7 24Wa 48
Sebert Rd. E7 36Kc 73
Sebright Ho. E2 40Wb 71
(off Coate St.)
Sebright Pas. E2 40Wb 71
Sebright Rd. EN5: Barn 12Za 30
Sebright Rd. HP1: Hem H 4J 3
Secker Cres. HA3: Hrw W 25Ea 46
Secker Ho. SW9 54Rb 113
(off Loughborough Est.)
Secker St. SE1 7K 223 (46Qb 90)
Second Av. E12 35Nc 74
Second Av. E13 41Jc 93
Second Av. E17 29Cc 52
Second Av. EN1: Enf 15Vb 33
Second Av. EN9: Walt A 2Kc 21
Second Av. HA9: Wemb 33Ma 67
Second Av. KT12: Walt T 72X 151
Second Av. KT20: Lwr K 97Ab 194
Second Av. N18 21Yb 52
Second Av. NW4 28Za 48
Second Av. RM10: Dag 40Dd 76
Second Av. RM20: W Thur 51Wd 120
Second Av. RM6: Chad H 29Yc 55
Second Av. SW14 55Ua 110
Second Av. UB3: Hayes 46V 84
Second Av. W10 42Ab 88
Second Av. W3 46Va 88
Second Av. WD25: Wat 7Z 13
Second Cl. KT8: W Mole 70Ea 130
Second Cres. SL1: Slou 3G 80
Second Cross Rd. TW2: Twick 61Ga 130
Second Way HA9: Wemb 35Ra 67
Sedan Way SE17 7H 231 (50Ub 91)
Sedcombe Cl. DA14: Sidc 63Xc 139
Sedcote Rd. EN3: Pond E 15Yb 34
Sedding St. SW1 5H 227 (49Jb 90)
Sedding Studios SW1 5H 227 (49Jb 90)
(off Sedding St.)
Seddon Highwalk EC2 7D 218 (43Sb 91)
(off Aldersgate St.)
Seddon Ho. EC2 7D 218 (43Sb 91)
(off Aldersgate St.)
Seddon Rd. SM4: Mord 71Fb 155
Seddon St. WC1 4H 217 (41Pb 90)
Sedgebrook Rd. SE3 54Mc 115
Sedgecombe Av. HA3: Kenton 29La 46
Sedge Ct. RM17: Grays 2A 122
Sedgefield Cl. RM3: Rom 21Pd 57
Sedgefield Ct. UB5: N'olt 36Da 65
(off Newmarket Av.)
Sedgefield Cres. RM3: Rom 22Pd 57
Sedgeford Rd. W12 46Va 88
Sedge Gdns. IG11: Bark 41Wc 95
Sedgehill Rd. SE6 63Cc 136
Sedgemere Av. N2 27Eb 49
Sedgemere Rd. SE2 48Yc 95
Sedgemoor Dr. RM10: Dag 35Cd 76
Sedge Rd. N17 24Yb 52
Sedgeway SE6 60Hc 115
Sedgewood Cl. BR2: Hayes 73Hc 159
Sedgmoor Pl. SE5 52Ub 113
Sedgwick Av. UB10: Hil 38R 64
Sedgwick Ho. E3 43Cc 92
(off Gale St.)
Sedgwick Pl. WD17: Wat 15Y 27
Sedgwick Rd. E10 33Ec 72
Sedgwick St. E9 36Zb 72
Sedleigh Rd. SW18 58Bb 111
Sedlescombe Rd. SW6 51Cb 111
Sedley DA13: Sflt 65Ce 143
Sedley Cl. EN1: Enf 10Xb 19
Sedley Ct. SE26 61Xb 135
Sedley Gro. UB9: Hare 28L 43
Sedley Ho. SE11 7J 229 (50Pb 90)
(off Newburn St.)
Sedley Pl. W1 3K 221 (44Kb 90)
Sedley Ri. IG10: Lough 12Pc 36
Sedona Ho. E20 37Dc 72
(off Victory Pde.)
Sedum Cl. NW9 29Ra 47
Sedum M. EN2: Enf 13Qb 32
Seedbed Cen., The RM7: Rush G 31Gd 76
Seeley Dr. SE21 63Ub 135
Seelig Av. NW9 31Wa 68
Seely Rd. SW17 65Jb 134
Seetha Ho. IG1: Ilf 33Tc 74
(off High Rd.)
Seething La. EC3 4J 225 (45Ub 91)
SEETHING WELLS 72La 152
Seething Wells La. KT6: Surb 72La 152
Sefton Av. HA3: Hrw W 26Fa 46
Sefton Av. NW7 22Ta 47
Sefton Cl. AL1: St A 1D 6
Sefton Cl. BR5: St M Cry 70Vc 139
Sefton Cl. GU24: W End 5D 166
Sefton Cl. SL2: Stoke P 9K 61
Sefton Ct. EN2: Enf 12Rb 33
Sefton Ct. TW3: Houn 53Da 107
Sefton Paddock SL2: Stoke P 8L 61
Sefton Pk. SL2: Stoke P 8L 61
Sefton Rd. BR5: St M Cry 70Vc 139
Sefton Rd. CR0: C'don 74Wb 157
Sefton Rd. KT19: Ewe 82Ta 173
Sefton St. SW15 55Ya 110
Sefton Way UB8: Cowl 44L 83
Segal Cl. SE23 59Ac 114
Segrave Cl. KT13: Weyb 80Q 150
Sejant Ho. RM17: Grays 51De 121
(off Bridge Rd.)
Sekforde St. EC1 6B 218 (42Rb 91)
Sekhon Ter. TW13: Hanw 62Ca 129
(off Church Rd.)
Selah Dr. BR8: Swan 67Ed 140
Selan Gdns. UB4: Yead 43X 85
Selbie Av. NW10 36Va 68
Selborne Av. DA5: Bexl 60Ad 117
Selborne Av. E12 35Qc 74
Selborne Gdns. NW4 28Wa 48
Selborne Gdns. UB6: G'frd 40Ja 66
Selborne Pl. KT13: Weyb 79T 150
Selborne Rd. CR0: C'don 76Ub 157
Selborne Rd. DA14: Sidc 63Xc 139
Selborne Rd. E17 29Bc 52
Selborne Rd. IG1: Ilf 33Qc 74
Selborne Rd. KT3: N Mald 68Ua 132
Selborne Rd. N14 20Nb 32
Selborne Rd. N22 25Pb 50
Selborne Rd. SE5 54Tb 113
Selborne Wlk. E17 29Bc 52
Selborne Wlk. Shop. Cen. 28Bc 52
Selbourne Av. KT15: New H 81K 169
Selbourne Av. KT6: Surb 75Pa 153
Selbourne Cl. DA3: Lfield 69Fe 143
Selbourne Cl. KT15: New H 81K 169
Selbourne Ho. SE1 2F 231 (47Tb 91)
(off Gt. Dover St.)
Selbourne Sq. RH9: G'stone 2A 210
Selby Av. AL3: St A 2B 6
Selby Chase HA4: Ruis 33X 65
Selby Cl. BR7: Chst 65Qc 138
Selby Cl. E6 43Nc 94
Selby Cl. KT9: Chess 80Na 153
Selby Gdns. UB1: S'hall 42Ca 85
Selby Grn. SM5: Cars 73Gb 155
Selby Rd. E11 34Gc 73
Selby Rd. E13 43Kc 93
Selby Rd. N17 24Ub 51
Selby Rd. SE20 68Wb 135
Selby Rd. SM5: Cars 73Gb 155
Selby Rd. TW15: Ashf 65S 128
Selby Rd. W5 42Ka 86
Selby Sq. W10 41Ab 88
(off Dowland St.)
Selby St. E1 42Wb 91
Selby Wlk. GU21: Wok 10M 167
Selcroft Ho. SE10 50Hc 93
(off Glenister Rd.)
Selcroft Rd. CR8: Purl 84Rb 177
Selden Hill HP1: Hem H 3M 3
Selden Hill HP2: Hem H 3M 3
Selden Ho. SE15 54Yb 114
(off Selden Rd.)
Selden Rd. SE15 54Yb 114
Selden Wlk. N7 33Pb 70
Seldon Ho. SW1 50Lb 90
(off Churchill Gdns.)
Seldon Ho. SW8 52Lb 112
(off Stewart's Rd.)
Selfridges 3J 221 (44Jb 90)
(off Oxford St.)
SELHURST 72Ub 157
Selhurst Cl. GU21: Wok 87B 168
Selhurst Cl. SW19 60Za 110
Selhurst New Rd. SE25 72Ub 157
Selhurst Pk. 70Ub 135
Selhurst Pl. SE25 72Ub 157
Selhurst Rd. N9 20Tb 33
Selhurst Rd. SE25 72Ub 157
Selhurst Station (Rail) 71Ub 157
Selig Ct. NW11 31Ab 68
Selim Ct. SL1: Slou 7M 81
(off Clifton Rd.)
Selina Ho. NW8 5C 214 (42Fb 89)
(off Frampton St.)
Selinas La. RM8: Dag 31Ad 75
Selkirk Dr. DA8: Erith 53Gd 118
Selkirk Ho. N1 39Pb 70
(off Bingfield St.)
Selkirk Rd. SW17 63Gb 133
Selkirk Rd. TW2: Twick 61Ea 130
Sellers Cl. WD6: Bore 11Sa 29
Sellers Hall Cl. N3 24Cb 49
Sellincourt Rd. SW17 64Gb 133
Sellindge Cl. BR3: Beck 66Bc 136
Sellons Av. NW10 39Va 68
Sellwood Dr. EN5: Barn 15Za 30
Selman Ho. E9 37Ac 72
Selman Way SL9: Chal P 2B 42
SELSDON 82Yb 178
Selsdon Av. CR2: S Croy 79Tb 157
Selsdon Cl. KT6: Surb 71Na 153
Selsdon Cl. RM5: Col R 25Ed 56
Selsdon Cres. CR2: Sels 81Yb 178
Selsdon Pk. Hotel Golf Course 83Yb 178
Selsdon Pk. Rd. CR0: Sels 80Bc 158
Selsdon Pk. Rd. CR2: Sels 81Zb 178
Selsdon Rd. CR2: S Croy 78Tb 157
Selsdon Rd. E11 31Jc 73
Selsdon Rd. E13 39Lc 73
Selsdon Rd. KT15: New H 83J 169
Selsdon Rd. NW2 33Va 68
Selsdon Rd. SE27 62Qb 134
Selsdon Way E14 48Dc 92
Selsdon Wood Nature Reserve 83Ac 178
Selsea Pl. N16 36Ub 71
Selsey WC1 4G 217 (41Nb 90)
(off Tavistock Pl.)
Selsey Cres. DA16: Well 53Zc 117
Selsey St. E14 43Cc 92
Selvage La. NW7 22Ta 47
Selway Cl. HA5: Eastc 28X 45
Selway Ho. SW8 53Nb 112
(off Sth. Lambeth Rd.)
Selwin Ct. KT12: Walt T 74Y 151
Selwood Cl. TW19: Stanw 58L 105
Selwood Gdns. TW19: Stanw 58L 105
Selwood Pl. SW7 7B 226 (50Fb 89)
Selwood Rd. CM14: B'wood 20Vd 40
Selwood Rd. CR0: C'don 75Xb 157
Selwood Rd. GU22: Wok 92D 188
Selwood Rd. KT9: Chess 77Ma 153
Selwood Rd. SM3: Sutt 74Bb 155
Selwood St. SW2 59Qb 112
Selwood Ter. SW7 7B 226 (50Fb 89)
Selworthy Cl. E11 29Jc 53
Selworthy Ho. SW11 53Fb 111
(off Battersea Church Rd.)
Selworthy Rd. SE6 62Bc 136
Selwyn Av. AL10: Hat 1P 7
Selwyn Av. E4 23Ec 52
Selwyn Av. IG3: Ilf 30Vc 55
Selwyn Av. TW9: Rich 55Na 109
Selwyn Cl. SL4: Wind 4B 102
Selwyn Cl. TW4: Houn 56Aa 107
Selwyn Ct. E17 29Cc 52
(off Yunus Khan Cl.)
Selwyn Ct. HA8: Edg 24Ra 47
Selwyn Ct. HA9: Wemb 34Sa 67
Selwyn Ct. SE3 55Hc 115
Selwyn Ct. TW10: Rich 57Pa 109
(off Church Rd.)
Selwyn Cres. AL10: Hat 1A 8
Selwyn Cres. DA16: Well 55Xc 117
Selwyn Dr. AL10: Hat 1P 7
Selwyn Pl. BR5: St P 69Xc 139
Selwyn Pl. SL1: Slou 5D 80
Selwyn Rd. E13 39Kc 73
Selwyn Rd. E3 40Bc 72
Selwyn Rd. KT3: N Mald 71Ta 153
Selwyn Rd. NW10 38Ua 68
Selwyn Rd. RM18: Tilb 4B 122
Semley Ga. E9 37Bc 72
(not continuous)
Semley Ho. SW1 6K 227 (49Kb 90)
(off Semley Pl.)
Semley Pl. SW1 6J 227 (49Jb 90)
Semley Rd. SW16 68Nb 134
Semper Cl. GU21: Knap 9J 167
Semper Rd. RM16: Grays 7E 100
Sempill Rd. HP3: Hem H 5N 3
Semples SS17: Stan H 1P 101
Senate St. SE15 54Yb 114
Senators Lodge E3 40Ac 72
(off Roman Rd.)
Senator Wlk. SE28 48Tc 94
SEND 95F 188
Sendall Ct. SW11 55Fb 111
(off Winstanley Rd.)
Send & Ripley History Society Mus. 93K 189
Send Barns La. GU23: Send 96F 188
Send Cl. GU23: Send 95E 188
Send Hill GU23: Send 97E 188
SEND MARSH 95H 189
Send Marsh Grn. GU23: Rip 95H 189
Send Marsh Rd. GU23: Rip 96F 188
Send Marsh Rd. GU23: Send 96F 188
Send Pde. Cl. GU23: Send 95E 188
Send Rd. GU23: Send 95D 188
Seneca Rd. CR7: Thor H 70Sb 135
Sener Ct. CR2: S Croy 79Sb 157
Senga Rd. SM6: W'gton 74Jb 156
Senhouse Rd. SM3: Cheam 76Za 154
Senior St. W2 43Db 89
Senlac Rd. SE12 60Kc 115
Sennen Rd. EN1: Enf 17Vb 33
Sennen Wlk. SE9 62Nc 138
Sennocke Ct. TN13: S'oaks 97Kd 203
Senrab St. E1 44Zb 92
Sentamu Cl. SE24 60Rb 113
Sentinel WD17: Wat 12X 27
Sentinel Bldg. E15 36Fc 73
(off Property Row)
Sentinel Cl. UB5: N'olt 42Aa 85
Sentinel Sq. NW4 28Ya 48
Sentis Ct. HA6: Nwood 23V 44
September Ct. UB1: S'hall 46Da 85
(off Dormer's Wells La.)
September Way HA7: Stan 23Ka 46
Septimus Pl. EN1: Enf 15Wb 33
Sequoia Cl. WD23: B Hea 18Fa 28
Sequoia Gdns. BR6: Orp 73Vc 161
Sequoia Pk. HA5: Hat E 23Da 45
Serap Ct. CR0: C'don 77Tb 157
(off Dean Rd.)
Seraph Ct. EC1 3D 218 (41Sb 91)
(off Moreland St.)
Serbin Cl. E10 31Ec 72
Serenaders Rd. SW9 54Qb 112
Serene M. RM3: Rom 23Kd 57
Serenity Apts. E17 29Dc 52
Serenity Apts. SW11 56Gb 111
(off Monarch Sq.)
Serenity Cl. HA2: Harr 33Da 65
Serenity Ct. DA9: Ghithe 56Wd 120
(off Evelyn Wlk.)
Seren Pk. Gdns. SE3 51Gc 115
Seren Pl. SL2: Farn C 8F 60
Sergeant Ind. Est. SW18 58Db 111
Sergeants Grn. La. EN9: Walt A 5Lc 21
Sergeants Pl. CR3: Cat'm 94Sb 197
Sergehill La. WD5: Bedm 9F 4
Serica Ct. SE10 52Ec 114
Serjeants Inn EC4 3A 224 (44Qb 90)
Serlby Ct. W14 48Bb 89
(off Somerset Sq.)
Serle St. WC2 2J 223 (44Pb 90)
Sermed Ct. SL2: Slou 6N 81
Sermon Dr. BR8: Swan 69Ed 140
Sermon La. EC4 3D 224 (44Sb 91)
(off Carter La.)
Serpentine, The 7D 220 (46Gb 89)
Serpentine Cl. RM6: Chad H 31Yc 75
Serpentine Ct. SE16 47Zb 92
(off Christopher Cl.)
Serpentine Ct. TN13: S'oaks 94Md 203
Serpentine Gallery 1C 226 (47Fb 89)
Serpentine Grn. RH1: Mers 1D 208
Serpentine Lido, The 7D 220 (46Gb 89)
Serpentine Rd. TN13: S'oaks 95Ld 203
Serpentine Rd. W2 7D 220 (46Gb 89)
Serpentine Sackler Gallery 6C 220 (46Fb 89)
Serra Ho. AL1: St A 3D 6
Service Rd., The EN6: Pot B 4Cb 17
Serviden Dr. BR1: Broml 67Mc 137
Servite Ho. BR3: Beck 67Bc 136
Servite Ho. GU21: Knap 9H 167
Servite Ho. KT4: Wor Pk 75Va 154
(off The Avenue)
Servite Ho. N14 15Kb 32
(off Bramley Rd.)
Servius Ct. TW8: Bford 52Ma 109
Setchell Rd. SE1 5K 231 (49Vb 91)
Setchell Way SE1 5K 231 (49Vb 91)
Seth St. SE16 47Yb 92
Seton Gdns. RM9: Dag 38Yc 75
Settle Point E13 40Jc 73
Settle Rd. RM3: Rom 21Qd 57
Settlers Ct. E14 45Fc 93
Settles St. E1 43Wb 91
Settrington Rd. SW6 54Db 111
Seven Acres BR8: Crock 72Fd 162
Seven Acres DA3: Nw A G 76Ae 165
Seven Acres HA6: Nwood 23X 45
Seven Acres SM5: Cars 75Gb 155
Seven Arches App. KT13: Weyb 80P 149
Seven Arches Rd. CM14: B'wood 20Zd 41
Seven Dials WC2 3F 223 (44Nb 90)
Seven Dials Ct. WC2 3F 223 (44Nb 90)
(off Shorts Gdns.)
Sevenex Pde. HA9: Wemb 36Na 67
Seven Hills Cl. KT12: W Vill 81U 170
Seven Hills Rd. KT11: Cobh 83U 170
Seven Hills Rd. KT12: Hers 81U 170
Seven Hills Rd. KT12: W Vill 81U 170
Seven Hills Rd. SL0: Iver H 36D 62
Seven Hills Rd. Sth. KT11: Cobh 85U 170
Seven Islands Leisure Cen. 48Yb 92
SEVEN KINGS 32Vc 75
SEVEN KINGS HEALTH CEN. 33Uc 74
(off Salisbury Rd.)
Seven Kings Rd. IG3: Ilf 32Uc 74
Seven Kings Station
(Rail & Crossrail) 32Uc 74
Seven Kings Way KT2: King T 67Na 131
SEVENOAKS 97Kd 203
Sevenoaks Bus. Cen. TN14: S'oaks 93Ld 203
Sevenoaks Bus Station 97Ld 203
Sevenoaks By-Pass TN14: Sund 95Ed 202
Sevenoaks Cl. DA7: Bex 56Dd 118
Sevenoaks Cl. RM3: Rom 21Ld 57
Sevenoaks Cl. SM2: Sutt 82Cb 175
SEVENOAKS COMMON 100Kd 203
Sevenoaks Ct. HA6: Nwood 24S 44
SEVENOAKS HOSPITAL 93Ld 203
Sevenoaks Leisure Cen. 97Ld 203
Sevenoaks Rd. BR6: Chels 78Vc 161
Sevenoaks Rd. BR6: Orp 78Vc 161
Sevenoaks Rd. BR6: Prat B 80Vc 161
Sevenoaks Rd. SE4 58Ac 114
Sevenoaks Rd. TN14: Hals 81Ad 181
Sevenoaks Rd. TN14: Off 88Kd 183
Sevenoaks Rd. TN14: S'oaks 88Kd 183
Sevenoaks Rd. TN15: Bor G 92Ae 205
Sevenoaks Rd. TN15: Igh 95Vd 204
Sevenoaks Rd. TN15: Seal 95Vd 204
Sevenoaks Station (Rail) 96Jd 202
Sevenoaks Way BR5: St P 66Yc 139
Sevenoaks Way DA14: Sidc 66Yc 139
Sevenoaks Wildlife Reserve 93Hd 202
Seven Sea Gdns. E3 43Dc 92
Sevenseas Rd. TW6: H'row A 58S 106
SEVEN SISTERS 29Vb 51
Seven Sisters Rd. N15 30Tb 51
Seven Sisters Rd. N4 33Qb 70
Seven Sisters Rd. N7 34Pb 70
Seven Sisters Station (Underground
& Overground) 29Ub 51
Seven Stars Cnr. W12 48Wa 88
Seven Stars Yd. E1 7K 219 (43Vb 91)
(off Brick La.)
Seventeen SM1: Sutt 79Eb 155
(off Sutton Ct. Rd.)
Seventh Av. E12 35Pc 74
Seventh Av. KT20: Lwr K 97Ab 194
Seventh Av. UB3: Hayes 46W 84
Seven Ways Pde. IG2: Ilf 29Qc 54

Severin Ct. SE17 7H 231 (50Ub 91)
(off East St.)
Severn RM18: E Til 8K 101
Severnake Cl. E14 49Cc 92
Severn Av. RM2: Rom 27Kd 57
Severn Av. W10 41Ab 88
Severn Ct. KT2: King T 67Ma 131
(off John Williams Cl.)
Severn Cres. SL3: L'ly 50D 82
Severn Dr. EN1: Enf 10Wb 19
Severn Dr. KT10: Hin W 75Ja 152
Severn Dr. KT12: Walt T 75Z 151
Severn Dr. RM14: Upm 30Td 58
Severn Ho. SW18 56Cb 111
(off Enterprise Way)
Severn Rd. RM15: Avel 44Sd 98
Severns Fld. CM16: Epp 1Wc 23
Severnvale AL2: Lon C 9K 7
Severn Way NW10 36Va 68
Severn Way WD25: Wat 6Y 13
Severus Ho. UB3: Hayes 44T 84
Severus Rd. SW11 56Gb 111
Seville Ho. E1 46Wb 91
(off Wapping High St.)
Seville M. N1 38Ub 71
Seville St. SW1 2G 227 (47Hb 89)
Sevington Rd. NW4 30Xa 48
Sevington St. W9 42Db 89
Seward Rd. BR3: Beck 68Zb 136
Seward Rd. W7 47Ja 86
SEWARDSTONE 11Ec 34
SEWARDSTONEBURY 15Gc 35
Sewardstone Cl. E4 11Ec 34
Sewardstone Gdns. E4 15Dc 34
Sewardstone Rd. E2 40Yb 72
Sewardstone Rd. E4 17Dc 34
Sewardstone Rd. EN9: Walt A 6Ec 20
Sewardstone St. EN9: Walt A 6Ec 20
Seward St. EC1 4C 218 (41Rb 91)
Sewdley St. E5 34Zb 72
Sewell Cl. AL4: St A 2J 7
Sewell Cl. RM16: Chaf H 50Yd 98
Sewell Rd. SE2 48Wc 95
Sewell St. E13 41Jc 93
Sextant Av. E14 49Fc 93
Sexton Cl. RM13: Rain 39Hd 76
Sexton Ct. E14 45Fc 93
(off Newport Av.)
Sexton Rd. RM18: Tilb 3B 122
Sextons Ho. SE10 51Ec 114
(off Bardsley La.)
Seymer Rd. RM1: Rom 27Fd 56
Seymore M. SE14 52Bc 114
(off New Cross Rd.)
Seymour Av. CR3: Cat'm 95Sb 197
Seymour Av. KT17: Ewe 81Xa 174
Seymour Av. N17 26Wb 51
Seymour Av. SM4: Mord 73Za 154
Seymour Chase CM16: Epp 1Xc 23
Seymour Cl. HA5: Hat E 25Ba 45
Seymour Cl. IG10: Lough 16Nc 36
Seymour Cl. KT8: E Mos 71Ea 152
Seymour Ct. E4 19Hc 35
Seymour Ct. KT1: Hamp W 67Ma 131
Seymour Ct. KT11: Cobh 85W 170
Seymour Ct. KT19: Ewe 81Ua 174
Seymour Ct. N10 26Jb 50
Seymour Ct. N21 16Pb 32
Seymour Ct. NW2 33Xa 68
Seymour Cres. HP2: Hem H 2N 3
Seymour Dr. BR2: Broml 74Pc 160
Seymour Gdns. HA4: Ruis 32Z 65
Seymour Gdns. IG1: Ilf 32Pc 74
Seymour Gdns. KT5: Surb 71Pa 153
Seymour Gdns. SE4 55Ac 114
Seymour Gdns. TW1: Twick 59Ka 108
Seymour Gdns. TW13: Hanw 63Y 129
Seymour Gro. WD19: Wat 17Y 27
Seymour Ho. E16 46Jc 93
(off De Quincey M.)
Seymour Ho. NW1 3E 216 (41Mb 90)
(off Churchway)
Seymour Ho. SL3: L'ly 47A 82
Seymour Ho. SM2: Sutt 79Db 155
(off Mulgrave Rd.)
Seymour Ho. WC1 5F 217 (42Nb 90)
(off Tavistock Pl.)
Seymour Leisure Cen. 1F 221 (43Hb 89)
Seymour M. KT17: Ewe 82Wa 174
Seymour M. W1 2H 221 (44Jb 90)
Seymour Pl. GU22: Wok 2M 187
Seymour Pl. KT13: Weyb 76R 150
Seymour Pl. RM11: Horn 31Md 77
Seymour Pl. SE25 70Xb 135
Seymour Pl. W1 1F 221 (43Hb 89)
Seymour Rd. CR4: Mitc 73Jb 156
Seymour Rd. DA11: Nflt 10B 122
Seymour Rd. E10 32Bc 72
Seymour Rd. E4 18Dc 34
Seymour Rd. E6 40Mc 73
Seymour Rd. KT1: Hamp W 67Ma 131
Seymour Rd. KT8: E Mos 71Ea 152
Seymour Rd. KT8: W Mole 71Ea 152
Seymour Rd. N3 24Db 49
Seymour Rd. N8 29Qb 50
Seymour Rd. N9 19Xb 33
Seymour Rd. RM18: Tilb 3B 122
Seymour Rd. SL1: Slou 7G 80
Seymour Rd. SM5: Cars 78Jb 156
Seymour Rd. SW18 59Bb 111
Seymour Rd. SW19 62Za 132
Seymour Rd. TW12: Hamp H 64Ea 130
Seymour Rd. W4 49Sa 87
Seymours, The IG10: Lough 11Qc 36
Seymour St. SE18 48Sc 94
Seymour St. W1 3F 221 (44Hb 89)
Seymour St. W2 3F 221 (44Hb 89)
Seymour Ter. SE20 67Xb 135
Seymour Vs. SE20 67Xb 135
Seymour Wlk. DA10: Swans 59Ae 121
Seymour Wlk. SW10 51Eb 111
Seymour Way TW16: Sun 66V 128
Seyssel St. E14 49Ec 92
Shaa Rd. W3 45Ta 87
Shabana Rd. W12 46Xa 88
Shabden Cotts. CR5: Chips 93Hb 195
Shab Hall Cotts. TN13: Dun G 90Ed 182
Shacklands Rd. TN14: Bad M 83Dd 182
Shacklands Rd. TN14: S'ham 83Dd 182
Shackleford Rd. GU22: Wok 92C 188
Shacklegate La. TW11: Tedd 63Ga 130
Shackleton Cl. SE23 61Xb 135
Shackleton Ct. E14 50Cc 92
(off Maritime Quay)
Shackleton Ct. TW19: Stanw 58N 105
(off Whitley Cl.)
Shackleton Ct. W12 47Xa 88
(off Scott's Rd.)
Shackleton Dr. DA1: Dart 57Pd 119
Shackleton Ho. CR0: Wadd 74Pb 156
Shackleton Ho. E1 46Yb 92
(off Prusom St.)
Shackleton Ho. NW10 38Ta 67
Shackleton Rd. SL1: Slou 5K 81
Shackleton Rd. UB1: S'hall 45Ba 85
Shackleton Ter. DA12: Grav'nd 3E 144
(off Christian Flds.)
Shackleton Way E16 45Sc 94
Shackleton Way WD5: Ab L 4W 12
(off Lysander Way)
SHACKLEWELL 35Vb 71
Shacklewell Grn. E8 35Vb 71
Shacklewell Ho. E8 35Vb 71
Shacklewell La. E8 36Vb 71
Shacklewell La. N16 35Vb 71
Shacklewell Rd. N16 35Vb 71
Shacklewell Row E8 35Vb 71
Shacklewell St. E2 4K 219 (41Vb 91)
Shadbolt Av. E4 22Ac 52
Shadbolt Cl. KT4: Wor Pk 75Va 154
Shade M. RM13: Rain 40Ld 77
Shad Thames SE1: Anchor Brewhouse 7K 225 (46Vb 91)
Shad Thames SE1:
Jamaica Rd. 2K 231 (47Vb 91)
SHADWELL 45Xb 91
Shadwell Ct. UB5: N'olt 40Ba 65
Shadwell Dr. UB5: N'olt 41Ba 85
Shadwell Gdns. E1 45Yb 92
Shadwell Pierhead E1 45Yb 92
Shadwell Pl. E1 45Yb 92
(off Sutton St.)
Shadwell Station (Overground & DLR) 45Xb 91
Shady Bush Cl. WD23: Bush 17Ea 28
Shady La. WD17: Wat 12X 27
Shaef Way TW11: Tedd 66Ja 130
Shafter Rd. RM10: Dag 37Ed 76
Shaftesbury IG10: Lough 13Mc 35
Shaftesbury Av. EN3: Enf H 12Zb 34
Shaftesbury Av. EN5: New Bar 14Eb 31
Shaftesbury Av. HA2: Harr 32Da 65
Shaftesbury Av. HA3: Kenton 29Ma 47
Shaftesbury Av. TW14: Felt 58W 106
Shaftesbury Av. UB2: S'hall 49Ca 85
Shaftesbury Av. W1 5D 222 (45Mb 90)
Shaftesbury Av. WC1 2F 223 (44Nb 90)
Shaftesbury Av. WC2 2F 223 (44Nb 90)
Shaftesbury Barnet Harriers 25Ya 48
Shaftesbury Cen. W10 42Za 88
(off Barlby Rd.)
Shaftesbury Circ. HA2: Harr 32Ea 66
Shaftesbury Ct. E6 44Qc 94
(off Sapphire Cl.)
Shaftesbury Ct. N1 2F 219 (40Tb 71)
(off Shaftesbury St.)
Shaftesbury Ct. SE1 3F 231 (48Tb 91)
(off Alderney M.)
Shaftesbury Ct. SE5 56Tb 113
Shaftesbury Ct. SL1: Slou 7J 81
Shaftesbury Ct. SW16 62Mb 134
Shaftesbury Ct. WD3: Crox G 14R 26
Shaftesbury Cres. TW18: Staines 66M 127
Shaftesbury Gdns. NW10 42Ua 88
Shaftesbury Ho. CR5: Coul 94Mb 196
Shaftesbury La. CR5: Coul 89Lb 176
Shaftesbury La. DA1: Dart 56Rd 119
Shaftesbury Lodge E14 44Dc 92
(off Upper Nth. St.)
Shaftesbury M. SW4 57Lb 112
Shaftesbury M. W8 48Cb 89
Shaftesbury Pde. HA2: Harr 32Ea 66
Shaftesbury Pl. EC2 1D 224 (43Sb 91)
(off London Wall)
Shaftesbury Pl. W14 49Bb 89
(off Warwick Rd.)
Shaftesbury Point E13 40Jc 73
(off High St.)
Shaftesbury Rd. BR3: Beck 68Bc 136
Shaftesbury Rd. CM16: Epp 1Vc 23
Shaftesbury Rd. E10 32Cc 72
Shaftesbury Rd. E17 30Dc 52
Shaftesbury Rd. E4 18Fc 35
Shaftesbury Rd. E7 38Lc 73
Shaftesbury Rd. GU22: Wok 89D 168
Shaftesbury Rd. GU24: Bisl 8D 166
Shaftesbury Rd. N18 23Ub 51
Shaftesbury Rd. N19 32Nb 70
Shaftesbury Rd. RM1: Rom 30Hd 56
Shaftesbury Rd. SM5: Cars 73Fb 155
Shaftesbury Rd. TW9: Rich 55Na 109
Shaftesbury Rd. WD17: Wat 13Y 27
Shaftesbury Row SE8 52Cc 114
(off Speedwell St.)
Shaftesburys, The IG11: Bark 40Sc 74
Shaftesbury St. N1 2E 218 (40Sb 71)
(not continuous)
Shaftesbury Theatre London 2F 223 (44Nb 90)
(off Shaftesbury Av.)
Shaftesbury Vs. W8 48Cb 89
(off Allen St.)
Shaftesbury Way TW2: Twick 62Fa 130
Shaftesbury Way WD4: K Lan 10C 4
Shaftesbury Waye UB4: Yead 43Y 85
Shafto M. SW1 4F 227 (48Hb 89)
Shafton M. E9 39Zb 72
Shafton Rd. E9 39Zb 72
Shaftsbury Ct. DA8: Erith 53Hd 118
(off Selkirk Dr.)
Shafts Ct. EC3 3H 225 (44Ub 91)
Shaftswood Ct. SW17 62Hb 133
(off Lynwood Rd.)
Shaggy Calf La. SL2: Slou 5L 81
Shahjalal Ho. E2 40Wb 71
(off Pritchards Rd.)
Shakespeare Av. N11 22Lb 50
Shakespeare Av. NW10 39Ta 67
Shakespeare Av. RM18: Tilb 4D 122
Shakespeare Av. TW14: Felt 58W 106
Shakespeare Av. UB4: Hayes 44W 84
(not continuous)
Shakespeare Av. UB4: Yead 42X 85
Shakespeare Av. UB4: Yead 44W 84
(not continuous)
Shakespeare Cl. AL4: St A 2H 7
Shakespeare Cl. HA3: Kenton 31Qa 67
Shakespeare Ct. EN5: New Bar 13Db 31
Shakespeare Ct. HA3: Kenton 30Pa 47
Shakespeare Ct. NW6 38Eb 69
(off Fairfax Rd.)
Shakespeare Cres. E12 37Pc 74
Shakespeare Dr. HA3: Kenton 30Pa 47
Shakespeare Dr. WD6: Bore 14Qa 29
Shakespeare Gdns. N2 28Hb 49
Shakespeare Ho. E9 38Yb 72
(off Lyme Gro.)
Shakespeare Ho. N14 19Mb 32
Shakespeare Ind. Est. WD24: Wat 10W 12
Shakespeare Rd. DA1: Dart 56Qd 119
Shakespeare Rd. DA7: Bex 53Ad 117
Shakespeare Rd. E17 26Zb 52
Shakespeare Rd. KT15: Add 77M 149
Shakespeare Rd. N3 25Cb 49
Shakespeare Rd. NW10 39Ta 67
Shakespeare Rd. NW7 21Va 48
Shakespeare Rd. RM1: Rom 30Hd 56
Shakespeare Rd. SE24 57Rb 113
Shakespeare Rd. W3 46Sa 87
Shakespeare Rd. W7 45Ha 86
Shakespeare's Globe &
Exhibition 6D 224 (46Sb 91)
Shakespeare Sq. IG6: Ilf 23Sc 54
Shakespeare St. WD24: Wat 10X 13
Shakespeare Twr. EC2 7D 218 (43Sb 91)
(off Beech St.)
Shakespeare Way TW13: Hanw 63Y 129
Shakspeare M. N16 35Ub 71
Shakspeare Wlk. N16 35Ub 71
Shalbourne Sq. E9 37Bc 72
Shalcomb St. SW10 51Eb 111
Shalcross Dr. EN8: Chesh 2Bc 20
Shalden Ho. SW15 58Va 110
Shaldon Dr. HA4: Ruis 34Y 65
Shaldon Dr. SM4: Mord 71Ab 154
Shaldon Rd. HA8: Edg 26Pa 47
Shaldon Way KT12: Walt T 76Y 151
Shale Grn. RH1: Mers 1D 208
Shalfleet Dr. W10 45Za 88
Shalford Cl. BR6: Farnb 77Sc 160
Shalford Ct. N1 1B 218 (40Rb 71)
Shalford Ho. SE1 3G 231 (48Tb 91)
Shalimar Gdns. W3 45Sa 87
Shalimar Rd. W3 45Sa 87
Shallcross Cres. AL10: Hat 3C 8
Shallons Rd. SE9 63Rc 138
Shalstone Rd. SW14 55Ra 109
Shalston Vs. KT6: Surb 72Pa 153
Shambles, The TN13: S'oaks 97Ld 203
Shamrock Cl. KT22: Fet 93Fa 192
Shamrock Cotts. GU3: Worp 10M 187
Shamrock Ho. SE26 63Wb 135
(off Talisman Sq.)
Shamrock Rd. CR0: C'don 72Pb 156
Shamrock Rd. DA12: Grav'nd 9G 122
Shamrock St. SW4 55Mb 112
Shamrock Way N14 18Kb 32
Shandon Rd. SW4 58Lb 112
Shand St. SE1 1J 231 (47Ub 91)
Shandy St. E1 43Zb 92
Shan Ho. WC1 6H 217 (42Pb 90)
(off Millman St.)
Shanklin Cl. EN7: Chesh 1Vb 19
Shanklin Gdns. WD19: Wat 21Y 45
Shanklin Ho. E17 26Bc 52
Shanklin Rd. N15 28Wb 51
Shanklin Rd. N8 29Mb 50
Shannon Cl. NW2 34Za 68
Shannon Cl. UB2: S'hall 50Z 85
Shannon Commercial Cen.
KT3: N Mald 70Wa 132
SHANNON CORNER 70Wa 132
Shannon Cnr. Retail Pk. 70Wa 132
Shannon Ct. CR0: C'don 74Sb 157
(off Tavistock Rd.)
Shannon Ct. N16 34Ub 71
Shannon Ct. SE15 52Vb 113
(off Garnies Cl.)
Shannon Gro. SW9 56Pb 112
Shannon M. SE3 56Hc 115
Shannon Pl. NW8 1E 214 (40Gb 69)
Shannon Way BR3: Beck 65Dc 136
Shannon Way RM15: Avel 45Sd 98
Shanti Ct. SW18 60Cb 111
Shantock Hall La. HP3: Bov 1A 10
Shantock La. HP3: Bov 2A 10
Shap Cres. SM5: Cars 74Hb 155
Shapland Way N13 22Pb 50
Shapwick Cl. N11 22Hb 49
Shardcroft Av. SE24 57Rb 113
Shardeloes Rd. SE14 54Bc 114
Shardeloes Rd. SE4 55Bc 114
Shard's Sq. SE15 51Wb 113
Sharland Cl. CR7: Thor H 72Qb 156
Sharland Rd. DA12: Grav'nd 1E 144
Sharman Ct. DA14: Sidc 63Wc 139
Sharman Row SL3: L'ly 50B 82
Sharman Way E3 43Dc 92
Sharnbrooke Cl. DA16: Well 55Yc 117
Sharnbrook Ho. W14 51Cb 111
Sharney Av. SL3: L'ly 48D 82
Sharon Cl. KT19: Eps 85Sa 173
Sharon Cl. KT23: Bookh 96Ca 191
Sharon Cl. KT6: Surb 74La 152
Sharon Ct. CR2: S Croy 78Sb 157
(off Warham Rd.)
Sharon Gdns. E9 39Yb 72
Sharon Rd. EN3: Enf H 12Ac 34
Sharon Rd. W4 50Ta 87
Sharpe Cl. W7 43Ha 86
Sharpes La. HP1: Hem H 4D 2
Sharp Ho. SW8 55Kb 112
Sharp Ho. TW1: Twick 58Ma 109
Sharpleshall St. NW1 38Hb 69
Sharpley Ct. SE1 1C 230 (47Rb 91)
(off Pocock St.)
Sharpness Cl. UB4: Yead 43Aa 85
Sharp's La. HA4: Ruis 31T 64
Sharp Way DA1: Dart 55Pd 119
Sharratt St. SE15 51Yb 114
Sharsted St. SE17 50Rb 91
Sharvel La. UB5: N'olt 39W 64
Sharwood WC1 2J 217 (40Pb 70)
(off Penton Ri.)
Shaver's Pl. SW1 5D 222 (45Mb 90)
(off Coventry St.)
Shaw, The BR7: Chst 66Sc 138
Shaw Av. IG11: Bark 40Ad 75
Shawbrooke Rd. SE9 57Lc 115
Shawbury Cl. NW9 25Ua 48
Shawbury Rd. SE22 57Vb 113
Shaw Cl. CR2: Sande 84Vb 177
Shaw Cl. EN8: Chesh 1Yb 20
Shaw Cl. KT16: Ott 79E 148
Shaw Cl. KT17: Ewe 83Va 174
Shaw Cl. RM11: Horn 32Kd 77
Shaw Cl. SE28 46Xc 95
Shaw Cl. TW19: Stanw 60N 105
Shaw Cl. WD23: B Hea 19Ga 28
Shaw Ct. CR3: Cat'm 93Tb 197
Shaw Ct. SL4: Old Win 7L 103
Shaw Ct. SM4: Mord 73Eb 155
Shaw Cres. CM13: Hut 14Fe 41
Shaw Cres. CR2: Sande 84Vb 177
Shaw Cres. E14 43Ac 92
Shaw Cres. RM18: Tilb 3D 122
Shaw Dr. KT12: Walt T 73Y 151
SHAW FARM 6J 103
Shawfield Ct. UB7: W Dray 48N 83
Shawfield Pk. BR1: Broml 68Mc 137
Shawfield St. SW3 7E 226 (50Gb 89)
Shawford Ct. SW15 59Wa 110
Shawford Rd. KT19: Ewe 79Ta 153
Shaw Gdns. IG11: Bark 40Ad 75
Shaw Gdns. SL3: L'ly 50B 82
Shaw Gro. CR5: Coul 92Sb 197
Shaw Ho. DA17: Belv 50Bd 95
Shaw Ho. E16 46Qc 94
(off Claremont St.)
Shaw Ho. SM7: Bans 91Eb 195
Shawley Cres. KT18: Tatt C 90Ya 174
Shawley Way KT18: Tatt C 90Xa 174
Shaw Path BR1: Broml 62Hc 137
Shaw Pl. N2 27Jb 50
Shaw Rd. BR1: Broml 62Hc 137
Shaw Rd. EN3: Enf H 11Zb 34
Shaw Rd. SE22 56Ub 113
Shaw Rd. TN16: Tats 92Lc 199
SHAW'S CNR. 6N 207
Shaws Cotts. SE23 62Ac 136
Shaw Sq. E17 25Ac 52
Shaw Theatre 3E 216 (41Mb 90)
Shaw Way SM6: W'gton 80Nb 156
Shaxton Cres. CR0: New Ad 81Ec 178
Shead Ct. E1 44Xb 91
(off James Voller Way)
Sheaf Cotts. KT7: T Ditt 74Ga 152
(off Weston Grn.)
Shearing Dr. SM5: Cars 73Eb 155
Shearling Way N7 37Nb 70
Shearman Rd. SE3 56Hc 115
SHEAR MEADOW 4E 2
SHEARS, THE 66U 128
Shears Cl. DA1: Dart 61Ld 141
Shears Ct. TW16: Sun 66U 128
Shears Grn. Ct. DA11: Nflt 1C 144
Shears La. SW16 66Nb 134
Shearsmith Ho. E1 45Wb 91
(off Hindmarsh Cl.)
Shears Way TW16: Sun 67U 128
Shearwater DA3: Lfield 69De 143
Shearwater Av. RM18: E Til 10J 101
Shearwater Cl. IG11: Bark 41Wc 95
Shearwater Ct. DA9: Ghithe 58Wd 120
(off Waterstone Way)
Shearwater Ct. E1 45Wb 91
(off Star Pl.)
Shearwater Ct. SE8 51Bc 114
(off Abinger Gro.)
Shearwater Dr. NW9 31Wa 68
Shearwater Rd. HP3: Hem H 6L 3
Shearwater Rd. SM1: Sutt 78Bb 155
Shearwater Way UB4: Yead 44Z 85
Shearwood Cres. DA1: Cray 55Hd 118
Sheath Cotts. KT7: T Ditt 72Ka 152
(off Ferry Rd.)
Sheath La. KT22: Oxs 85Da 171
Sheaveshill Av. NW9 28Ua 48
Sheaveshill Ct. NW9 28Ta 47
Sheaveshill Pde. NW9 28Ua 48
(off Sheaveshill Av.)
Sheba Ct. N17 23Wb 51
(off Altair Cl.)
Sheba Pl. E1 6K 219 (42Vb 91)
Sheehy Way SL2: Slou 5M 81
Sheen Comn. Dr. TW10: Rich 56Qa 109
Sheen Ct. TW10: Rich 56Qa 109
Sheen Ct. Rd. TW10: Rich 56Qa 109
Sheendale Rd. TW9: Rich 56Pa 109
Sheenewood SE26 63Xb 135
Sheen Ga. Gdns. SW14 56Sa 109
Sheengate Mans. SW14 56Ta 109
Sheen Gro. N1 39Qb 70
Sheen La. SW14 57Sa 109
Sheen Pk. TW9: Rich 56Pa 109
Sheen Rd. BR5: St M Cry 70Vc 139
Sheen Rd. TW10: Rich 56Qa 109
Sheen Rd. TW9: Rich 57Na 109
Sheen Way SM6: W'gton 78Pb 156
Sheen Wood SW14 57Sa 109
Sheepbarn La. CR6: W'ham 84Hc 179
Sheepcot Dr. WD25: Wat 6Y 13
Sheepcote Cl. TW5: Cran 52W 106
Sheepcote Gdns. UB9: Den 30J 43
Sheepcote La. BR5: St M Cry 71Bd 161
Sheepcote La. BR8: Swan 70Cd 140
Sheepcote La. SW11 54Hb 111
Sheepcote Rd. HA1: Harr 30Ha 46
Sheepcote Rd. HP2: Hem H 2P 3
Sheepcote Rd. SL4: Eton W 10E 80
Sheepcote Rd. SL4: Wind 4C 102
Sheepcotes Rd. RM6: Chad H 28Ad 55
Sheepcot La. WD25: Wat 5W 12
(not continuous)
Sheephouse Rd. HP3: Hem H 4P 3
Sheephouse Way KT3: N Mald 74Ta 153
Sheep La. E8 39Xb 71
Sheep Wlk. KT18: Eps D 93Ta 193
Sheep Wlk. RH2: Reig 3H 207
Sheep Wlk. TW17: Shep 73P 149
Sheep Wlk., The GU22: Wok 90F 168
Sheep Wlk. M. SW19 65Za 132
Sheerness M. E16 47Rc 94
SHEERWATER 86F 168
Sheerwater Av. KT15: Wdhm 84G 168
Sheerwater Bus. Cen. GU21: Wok 87E 168
Sheerwater Rd. E16 43Mc 93
Sheerwater Rd. GU21: Wok 84G 168
Sheerwater Rd. KT14: W Byf 85G 168
Sheerwater Rd. KT15: Wdhm 84G 168
Sheethanger La. HP3: Hem H 5J 3
Sheet Hill TN15: Plax 97Zd 205
SHEET HILL 97Be 205
SHEET'S HEATH 1D 186
Sheet's Heath La. GU24: Brkwd 1E 186
Sheet St. SL4: Wind 4H 103
Sheet St. Rd. SL4: Wind 3B 124
Sheffield Dr. RM3: Rom 22Qd 57
Sheffield Gdns. RM3: Rom 22Qd 57
Sheffield Rd. SL1: Slou 4G 80
Sheffield Rd. TW6: H'row A 58S 106
Sheffield Sq. E3 41Bc 92
Sheffield St. WC2 3H 223 (44Pb 90)
Sheffield Ter. W8 46Cb 89
Sheffield Way TW6: H'row A 57T 106
Shefton Ri. HA6: Nwood 24W 44
Sheila Cl. RM5: Col R 24Dd 56
Sheila Rd. RM5: Col R 24Dd 56
Sheilings, The RM11: Horn 29Pd 57
Sheilings, The TN15: Seal 92Pd 203
Shelbourne Cl. HA5: Pinn 27Ba 45
Shelbourne Pl. BR3: Beck 66Bc 136
Shelbourne Rd. N17 26Xb 51
Shelburne Dr. TW4: Houn 58Ca 107
Shelburne Rd. N7 35Pb 70
Shelbury Cl. DA14: Sidc 62Wc 139
Shelbury Rd. SE22 57Xb 113
Sheldon Av. IG5: Ilf 26Rc 54
Sheldon Av. N2 30Hb 49
Sheldon Av. N6 31Gb 69
Sheldon Cl. RH2: Reig 7K 207
Sheldon Cl. SE12 57Kc 115
Sheldon Cl. SE20 67Xb 135
Sheldon Ct. EN5: New Bar 14Db 31
Sheldon Ct. RM7: Rush G 30Fd 56
(off Union Rd.)
Sheldon Ct. SW8 52Nb 112
(off Lansdowne Grn.)
Sheldon Hgts. DA12: Grav'nd 6G 144
Sheldon Ho. N1 39Ub 71
(off Kingsland Rd.)
Sheldon Pl. E2 40Wb 71
(not continuous)
Sheldon Rd. DA7: Bex 53Bd 117
Sheldon Rd. N18 21Ub 51
Sheldon Rd. NW2 35Za 68
Sheldon Rd. RM9: Dag 38Ad 75
Sheldon Sq. W2 1A 220 (43Eb 89)
Sheldon St. CR0: C'don 76Sb 157
Sheldrake Cl. E16 46Pc 94
Sheldrake Ho. SE16 49Zb 92
(off Tawny Way)
Sheldrake Pl. W8 47Cb 89
Sheldrick Cl. SW19 68Fb 133
Shelduck Cl. E15 36Hc 73
Shelduck Ct. SE8 51Bc 114
(off Pilot Cl.)
Sheldwich Ter. BR2: Broml 72Nc 160
Shelford KT1: King T 68Qa 131
Shelford Cl. RM16: Ors 3C 100
Shelford Pl. N16 34Tb 71
Shelford Ri. SE19 66Vb 135
Shelford Rd. EN5: Barn 16Ya 30
Shelgate Rd. SW11 57Gb 111
Shellbank La. DA2: Bean 65Wd 142
Shellbank La. DA2: G St G 65Wd 142
Shell Cl. BR2: Broml 72Nc 160
Shellduck Cl. NW9 26Ua 48
Shelley N8 27Nb 50
(off Boyton Rd.)
Shelley Av. E12 37Nc 74
Shelley Av. RM12: Horn 33Hd 76
Shelley Av. UB6: G'frd 41Fa 86
Shelley Cl. BR6: Orp 76Uc 160
Shelley Cl. CR5: Coul 89Pb 176
Shelley Cl. HA6: Nwood 22V 44
Shelley Cl. HA8: Edg 21Ra 47
Shelley Cl. SE15 54Xb 113
Shelley Cl. SL3: L'ly 50B 82
Shelley Cl. SM7: Bans 87Za 174
Shelley Cl. UB4: Hayes 43W 84
Shelley Cl. UB6: G'frd 41Fa 86
Shelley Cl. WD6: Bore 14Qa 29
Shelley Ct. E10 31Dc 72
(off Skelton's La.)
Shelley Ct. E11 28Kc 53
(off Makepeace Rd.)
Shelley Ct. EN9: Walt A 5Hc 21
(off Ninefields)
Shelley Ct. N19 32Pb 70
Shelley Ct. SW3 51Hb 111
(off Tite St.)
Shelley Cres. TW5: Hest 53Z 107
Shelley Cres. UB1: S'hall 44Ba 85
Shelley Dr. DA16: Well 53Uc 116
Shelley Gdns. HA0: Wemb 33La 66
Shelley Gro. IG10: Lough 14Pc 36
Shelley Ho. E2 41Yb 92
(off Cornwall Av.)
Shelley Ho. N16 35Ub 71
Shelley Ho. SE17 7E 230 (50Sb 91)
(off Browning St.)
Shelley Ho. SW1 51Lb 112
(off Churchill Gdns.)
Shelley La. UB9: Hare 25J 43
Shelley Lodge EN2: Enf 11Tb 33
Shelley M. HP3: Hem H 5K 3
Shelley Pl. N1 39Rb 71
Shelley Pl. RM18: Tilb 3D 122
Shelley Rd. CM13: Hut 17Fe 41
Shelley Rd. NW10 39Ta 67
Shelleys La. TN14: Knock 88Vc 181
Shelley Way SW19 64Fb 133
Shellfield Cl. TW19: Stanw M 57J 105
Shellness Rd. E5 36Xb 71
Shell Rd. SE13 55Dc 114
Shell Twr. SE1 1J 229 (47Pb 90)
(off Milner Pl.)
Shellwood Rd. SW11 54Hb 111
Shelmerdine Cl. E3 43Cc 92
Shelson Av. TW13: Felt 62V 128
Shelson Pde. TW13: Felt 62V 128
Shelton Av. CR6: W'ham 89Yb 178
Shelton Cl. CR6: W'ham 89Yb 178
Shelton Ct. SL3: L'ly 8N 81
Shelton Rd. SW19 67Cb 133
Shelton St. WC2 3F 223 (44Nb 90)
(not continuous)
Shelvers Grn. KT20: Tad 93Ya 194
Shelvers Hill KT20: Tad 93Xa 194
Shelvers Spur KT20: Tad 93Ya 194
Shelvers Way KT20: Tad 93Ya 194
Shenden Cl. TN13: S'oaks 99Ld 203
Shenden Way TN13: S'oaks 100Ld 203
Shendish Edge HP3: Hem H 7C 3
Shendish Manor Golf Course 8M 3
Shene Ho. EC1 7K 217 (43Qb 90)
(off Bourne Est.)
Shene Sports & Fitness Cen. 56Ua 110
SHENFIELD 16Ce 41
Shenfield Cl. CR5: Coul 91Lb 196
Shenfield Cres. CM15: B'wood 19Ae 41
Shenfield Gdns. CM13: Hut 16De 41
Shenfield Grn. CM15: Shenf 17Ce 41
Shenfield Ho. SE18 53Mc 115
(off Portway Gdns.)
Shenfield Pl. CM15: Shenf 17Ae 41
Shenfield Rd. CM15: B'wood 19Zd 41
Shenfield Rd. CM15: Shenf 19Zd 41
Shenfield Rd. IG8: Wfd G 24Kc 53
Shenfield Station (Rail & Crossrail) 17Ce 41
Shenfield St. N1 2J 219 (40Ub 71)
(not continuous)
SHENLEY 6Qa 15
Shenley Av. HA4: Ruis 33V 64
Shenleybury WD7: Shenl 2Na 15
SHENLEYBURY 3Na 15
Shenleybury Cotts. WD7: Shenl 3Na 15
Shenleybury Vs. WD7: Shenl 3Na 15
Shenley Cl. CR2: Sande 82Vb 177
Shenley Hill WD7: R'lett 7Ka 14
Shenley La. AL2: Lon C 7F 6
Shenley Pk. 5Na 15
Shenley Rd. DA1: Dart 58Qd 119
Shenley Rd. SE5 53Ub 113
Shenley Rd. TW5: Hest 53Aa 107
Shenley Rd. WD6: Bore 14Qa 29

Shenley Rd. WD7: R'lett 6Ka 14
Shen Pl. Almshouses CM15: B'wood 19Zd 41
Shenston Ct. SL4: Wind 3H 103
(off James St.)
Shenstone W13 46La 86
Shenstone Cl. DA1: Cray 56Fd 118
Shenstone Dr. SL1: Burn 2B 80
Shenstone Gdns. RM3: Rom 25Ld 57
Shenstone Ho. SW16 64Lb 134
Shenstone Pk. SL5: S'hill 10C 124
Shenwood Ct. WD6: Bore 9Qa 15
Shepheard's Ho. BR7: Chst 67Tc 138
SHEPHERD CEN, THE 18J 25
Shepherd Cl. TW13: Hanw 63Aa 129
Shepherd Cl. WD5: Ab L 2V 12
Shepherdess Pl. N1 3E 218 (41Sb 91)
Shepherdess Wlk. N1 1E 218 (40Sb 71)
Shepherd Ho. E14 44Dc 92
(off Annabel Cl.)
Shepherd Ho. E16 45Rc 94
(off University Way)
Shepherd Mkt. W1 6K 221 (46Kb 90)
SHEPHERD'S BUSH 47Ya 88
Shepherd's Bush Empire Theatre 47Ya 88
Shepherd's Bush Grn. W12 47Ya 88
Shepherd's Bush Mkt. W12 47Ya 88
(not continuous)
Shepherd's Bush Market Station (Underground) 46Ya 88
Shepherd's Bush Pl. W12 47Za 88
Shepherd's Bush Rd. W6 49Ya 88
Shepherd's Bush Station (Rail, Underground & Overground) 47Za 88
Shepherd's Cl. BR6: Orp 76Vc 161
Shepherd's Cl. N6 30Kb 50
Shepherds Cl. HA7: Stan 22Ja 46
(not continuous)
Shepherds Cl. RM6: Chad H 29Zc 55
Shepherds Cl. TW17: Shep 72R 150
Shepherds Cl. UB8: Cowl 42L 83
Shepherds Cl. W1 4H 221 (45Jb 90)
(off Lees Pl.)
Shepherds Ct. SL4: Wind 4C 102
Shepherds Ct. W12 47Za 88
(off Shepherd's Bush Grn.)
Shepherds Farm WD3: Rick 18J 25
Shepherds Grn. BR7: Chst 66Tc 138
Shepherds Grn. HP1: Hem H 3G 2
(not continuous)
Shepherd's Hill N6 30Kb 50
Shepherd's Hill RH1: Mers 98Lb 196
Shepherds Hill RM14: Upm 26Sd 58
Shepherds Hill RM3: Hrld W 26Qd 57
Shepherd's La. DA1: Dart 60Jd 118
Shepherd's La. WD3: Chor 16F 24
Shepherd's La. WD3: Rick 16F 24
Shepherds La. E9 37Zb 72
Shepherds La. GU20: W'sham 8D 146
Shepherds La. SE28 46Uc 94
Shepherds Leas SE9 56Sc 116
Shepherd's Path CM14: S Weald 17Ud 40
Shepherd's Path NW3 36Fb 69
(off Lyndhurst Rd.)
Shepherds Path UB5: N'olt 37Aa 65
(off Arnold Rd.)
Shepherds Pl. W1 4H 221 (45Jb 90)
Shepherd's Rd. WD18: Wat 13V 26
Shepherd St. DA11: Nflt 59Fe 121
Shepherd St. W1 7K 221 (46Kb 90)
Shepherd's Wlk. KT18: Eps D 93Ra 193
Shepherd's Wlk. NW3 36Fb 69
Shepherds Wlk. NW2 33Wa 68
Shepherds Wlk. WD23: B Hea 19Fa 28
Shepherds Way AL9: Brk P 9L 9
Shepherds Way CR2: Sels 80Zb 158
Shepherds Way WD3: Rick 17K 25
Shepiston La. UB3: Harl 49R 84
Shepley Cl. SM5: Cars 76Jb 156
Shepley Dr. SL5: S'dale 2G 146
Shepley End SL5: S'dale 1G 146
Shepley M. EN3: Enf L 9Cc 20
Sheppard Cl. EN1: Enf 11Xb 33
Sheppard Cl. KT1: King T 70Na 131
Sheppard Dr. SE16 50Xb 91
Sheppard Ho. E2 40Wb 71
(off Warner Pl.)
Sheppard Ho. SW2 60Qb 112
Sheppards Coll. BR1: Broml 67Jc 137
(off London Rd.)
Sheppard St. E16 42Hc 93
Sheppards Yd. HP2: Hem H 1M 3
(off Figtree Hill)
SHEPPERTON 72R 150
Shepperton Bus. Pk. TW17: Shep 71S 150
Shepperton Cl. WD6: Bore 11Ta 29
Shepperton Ct. TW17: Shep 72R 150
Shepperton Ct. Dr. TW17: Shep 71R 150
Shepperton Film Studios 69P 127
SHEPPERTON GREEN 70Q 128
Shepperton Marina TW17: Shep 72U 150
Shepperton Rd. BR5: Pet W 72Sc 160
Shepperton Rd. N1 39Sb 71
Shepperton Rd. TW18: Lale 69L 127
Shepperton Rd. TW18: Shep 69L 127
Shepperton Station (Rail) 71S 150
Sheppey Cl. DA8: Erith 52Kd 119
Sheppey Gdns. RM9: Dag 38Yc 75
Sheppey Rd. RM9: Dag 38Xc 75
Sheppey's La. WD4: K Lan 1S 12
Sheppey's La. WD5: Bedm 10E 4
Sheppy Pl. DA12: Grav'nd 9D 122
Shepton Ho's. E2 41Yb 92
(off Welwyn St.)
Sherard Ct. N7 34Nb 70
Sherard Ho. E9 38Yb 72
(off Frampton Pk. Rd.)
Sherard Rd. SE9 57Nc 116
Sheraton Bus. Cen. UB6: G'frd 40Ka 66
Sheraton Cl. WD6: E'tree 15Pa 29
Sheraton Dr. KT19: Eps 85Sa 173
UB7: Harm
Sheraton Ho. SW1 51Kb 112
(off Churchill Gdns.)
Sheraton Ho. WD3: Chor 14E 24
Sheraton M. WD18: Wat 14U 26
UB3: Harl
Sheraton St. W1 3D 222 (44Mb 90)
Sherborne NW1 38Mb 70
(off Agar Gro.)
Sherborne Av. EN3: Enf H 12Yb 34
Sherborne Av. UB2: S'hall 49Ca 85
Sherborne Cl. KT18: Tatt C 89Ya 174
Sherborne Cl. SL3: Poyle 53G 104
Sherborne Cl. UB4: Yead 44Y 85
Sherborne Cotts. WD18: Wat 15Y 27
(off Muriel Av.)
Sherborne Cres. SM5: Cars 73Gb 155
Sherborne Gdns. NW9 27Qa 47
Sherborne Gdns. RM5: Col R 22Cd 56
Sherborne Gdns. W13 43Ka 86
Sherborne Gro. TN15: Kems'g 89Qd 183
Sherborne Ho. SW1 7A 228 (50Kb 90)
(part of Abbots Mnr.)
Sherborne Ho. SW8 52Pb 112
(off Bolney St.)
Sherborne La. EC4 4F 225 (45Tb 91)
Sherborne Pl. HA6: Nwood 23T 44
Sherborne Rd. BR5: St M Cry 70Vc 139
Sherborne Rd. KT9: Chess 78Na 153
Sherborne Rd. SM3: Sutt 75Cb 155
Sherborne Rd. TW14: Bedf 60T 106
(not continuous)
Sherborne St. N1 39Tb 71
Sherborne Wlk. KT22: Lea 93La 192
Sherborne Way WD3: Crox G 14R 26
Sherboro Rd. N15 30Vb 51
Sherbourne Cl. DA1: Dart 57Pd 119
Sherbourne Cl. HP2: Hem H 3N 3
Sherbourne Cl. TN15: W King 79Ud 164
Sherbourne Ct. AL1: St A 2D 6
(off Beaconsfield Rd.)
Sherbourne Ct. SM2: Sutt 79Eb 155
Sherbourne Ct. SW5 49Db 89
(off Cromwell Rd.)
Sherbourne Dr. SL4: Wind 6D 102
Sherbourne Dr. SL5: S'dale 1H 147
Sherbourne Gdns. TW17: Shep 73U 150
Sherbourne Ho. WD18: Wat 16T 26
Sherbourne Pl. HA7: Stan 23Ja 46
Sherbourne Wlk. SL2: Farn C 5G 60
Sherbrooke Cl. DA6: Bex 56Cd 118
Sherbrooke Ho. E2 40Yb 72
(off Bonner Rd.)
Sherbrooke Ho. SW1 4E 228 (48Mb 90)
(off Monck St.)
Sherbrooke Rd. SW6 52Ab 110
Sherbrooke Ter. SW6 52Ab 110
(off Sherbrooke Rd.)
Sherbrooke Way KT4: Wor Pk 73Xa 154
Sherbrook Gdns. N21 17Rb 33
Sherbrook Ho. SE16 47Yb 92
(off Albatross Way)
Shere Av. SM2: Cheam 82Ya 174
Shere Cl. KT9: Chess 78Ma 153
Sheredan Rd. E4 22Fc 53
Shere Ho. SE1 3F 231 (48Tb 91)
(off Gt. Dover St.)
Shere Rd. IG2: Ilf 29Qc 54
Sherfield Av. WD3: Rick 20M 25
Sherfield Cl. KT3: N Mald 70Ra 131
Sherfield Gdns. SW15 58Va 110
Sherfield M. UB3: Hayes 44U 84
Sherfield Rd. RM17: Grays 51De 121
Sheridan Bldgs. WC2 3G 223 (44Nb 90)
(off Martlett Ct.)
Sheridan Cl. BR8: Swan 70Hd 140
Sheridan Cl. HP1: Hem H 3K 3
Sheridan Cl. RM3: Rom 24Ld 57
Sheridan Cl. UB10: Hil 42S 84
Sheridan Ct. CR0: C'don 77Ub 157
(off Coombe Rd.)
Sheridan Ct. DA1: Dart 56Qd 119
Sheridan Ct. HA1: Harr 30Fa 46
Sheridan Ct. NW6 38Eb 69
(off Belsize Rd.)
Sheridan Ct. SL1: Slou 5C 80
Sheridan Ct. SW5 49Db 89
(off Barkston Gdns.)
Sheridan Ct. TW4: Houn 57Aa 107
Sheridan Ct. UB5: N'olt 36Da 65
Sheridan Ct. W7 45Ha 86
(off Milton Rd.)
Sheridan Cres. BR7: Chst 68Rc 138
Sheridan Dr. RH2: Reig 4K 207
Sheridan Gdns. HA3: Kenton 30Ma 47
Sheridan Grange SL5: S'dale 2E 146
Sheridan Hgts. E1 44Xb 91
(off Watney St.)
Sheridan Ho. KT22: Lea 93Ja 192
Sheridan Ho. SE11 6A 230 (49Qb 90)
(off Wincott St.)
Sheridan Lodge BR2: Broml 70Lc 137
(off Homesdale Rd.)
Sheridan M. E11 30Kc 53
Sheridan Pl. BR1: Broml 68Mc 137
Sheridan Pl. SW13 55Va 110
Sheridan Pl. TW12: Hamp 67Da 129
Sheridan Rd. DA17: Belv 49Cd 96
Sheridan Rd. DA7: Bex 55Ad 117
Sheridan Rd. E12 36Nc 74
Sheridan Rd. E7 34Hc 73
Sheridan Rd. SW19 67Bb 133
Sheridan Rd. TW10: Ham 62La 130
Sheridan Rd. WD19: Wat 17Z 27
Sheridans Rd. KT23: Bookh 98Ea 192
Sheridan Ter. UB5: N'olt 36Da 65
Sheridan Wlk. NW11 30Cb 49
Sheridan Wlk. SM5: Cars 78Hb 155
Sheridan Way BR3: Beck 67Bc 136
Sheriden Pl. HA1: Harr 31Ga 66
Sheriff Way WD25: Wat 5W 12
Sheringham NW8 39Fb 69
Sheringham Av. E12 35Pc 74
Sheringham Av. N14 15Mb 32
Sheringham Av. RM7: Rom 30Ed 56
Sheringham Av. TW2: Whitt 60Ba 107
Sheringham Ct. TW13: Felt 62W 128
(off Sheringham Av.)
Sheringham Ct. UB3: Hayes 47V 84
Sheringham Dr. IG11: Bark 36Vc 75
Sheringham Ho. NW1 7D 214 (43Gb 89)
(off Lisson St.)
Sheringham Rd. N7 37Pb 70
Sheringham Rd. SE20 69Yb 136
Sheringham Twr. UB1: S'hall 45Da 85
Sherington Av. HA5: Hat E 24Ca 45
Sherington Rd. SE7 51Kc 115
Sherland Ct. WD7: R'lett 8Ja 14
(off The Dell)
Sherland Rd. TW1: Twick 60Ha 108
Sherleys Ct. HA4: Ruis 33U 64
Sherlies Av. BR6: Orp 75Uc 160
Sherlock Cl. SW16 68Pb 134
Sherlock Ct. NW8 39Fb 69
(off Dorman Way)
Sherlock Holmes Mus. 6G 215 (42Hb 89)
(off Baker St.)
Sherlock M. W1 7H 215 (43Jb 90)
Sherman Av. KT12: Walt T 71Y 151
Shermanbury Cl. DA8: Erith 52Hd 118
Sherman Gdns. RM6: Chad H 30Yc 55
Sherman Ho. E14 44Ec 92
(off Dee St.)
Sherman Ho. UB3: Harl 48S 84
(off Nine Acres Cl.)
Sherman Rd. BR1: Broml 67Jc 137
Sherman Rd. SL1: Slou 3J 81
Shernbroke Rd. EN9: Walt A 6Hc 21
Shernhall St. E17 27Ec 52
Sherrard Rd. E12 36Mc 73
Sherrard Rd. E7 37Lc 73
Sherrards Way EN5: Barn 15Cb 31
Sherren Ho. E1 42Yb 92
Sherrick Grn. Rd. NW10 36Xa 68
Sherriff Cl. KT10: Esh 75Da 151
Sherriff Ct. NW6 37Cb 69
(off Sherriff Rd.)
Sherriff Rd. NW6 37Cb 69
Sherringham Av. N17 26Wb 51
Sherringham Av. TW13: Felt 62W 128
Sherringham Ct. E16 44Hc 93
(off Silvertown Sq.)
Sherrington Ct. E16 43Hc 93
(off Rathbone St.)
Sherrin Rd. E10 35Dc 72
Sherrock Gdns. NW4 28Xa 48
Sherry M. IG11: Bark 38Tc 74
Sherston Ct. SE1 5C 230 (49Rb 91)
(off Newington Butts)
Sherston Ct. WC1 4K 217 (41Qb 90)
(off Attneave St.)
Sherwin Ho. SE11 51Qb 112
(off Kennington Rd.)
Sherwin Rd. SE14 53Zb 114
Sherwood KT6: Surb 75Ma 153
Sherwood NW6 38Ab 68
Sherwood RM16: N Stif 46Ae 99
Sherwood Av. E18 27Kc 53
Sherwood Av. EN6: Pot B 4Ab 16
Sherwood Av. HA4: Ruis 30U 44
Sherwood Av. SW16 66Mb 134
Sherwood Av. UB4: Yead 42X 85
Sherwood Av. UB6: G'frd 37Ga 66
Sherwood Cl. DA5: Bexl 58Yc 117
Sherwood Cl. E17 26Bc 52
Sherwood Cl. KT22: Fet 95Ea 192
Sherwood Cl. SL3: L'ly 48A 82
Sherwood Cl. SW13 55Xa 110
Sherwood Cl. W13 46Ka 86
Sherwood Ct. CR2: S Croy 78Sb 157
(off Nottingham Rd.)
Sherwood Ct. HA2: Harr 33Da 65
Sherwood Ct. SL3: L'ly 50B 82
Sherwood Ct. SW11 55Eb 111
Sherwood Ct. W1 1F 221 (43Hb 89)
(off Bryanston Pl.)
Sherwood Ct. WD25: Wat 6V 12
Sherwood Cres. RH2: Reig 10K 207
Sherwood Gdns. E14 49Cc 92
Sherwood Gdns. IG11: Bark 38Tc 74
Sherwood Gdns. SE16 50Wb 91
Sherwood Ho. WD5: Ab L 3V 12
(off College Rd.)
Sherwood Pk. Av. DA15: Sidc 59Wc 117
Sherwood Pk. Rd. CR4: Mitc 70Lb 134
Sherwood Pk. Rd. SM1: Sutt 78Cb 155
Sherwood Rd. CR0: C'don 73Xb 157
Sherwood Rd. CR5: Coul 88Lb 176
Sherwood Rd. DA16: Well 54Uc 116
Sherwood Rd. GU21: Knap 9J 167
Sherwood Rd. HA2: Harr 33Ea 66
Sherwood Rd. IG6: Ilf 28Tc 54
Sherwood Rd. NW4 27Ya 48
Sherwood Rd. SW19 66Bb 133
Sherwood Rd. TW12: Hamp H 64Ea 130
Sherwoods Rd. WD19: Wat 17Aa 27
Sherwood St. N20 20Fb 31
Sherwood St. W1 4C 222 (45Lb 90)
Sherwood Ter. E16 44Lc 93
(off Bingley Rd.)
Sherwood Ter. N20 20Fb 31
Sherwood Way BR4: W W'ck 75Ec 158
Sherwood Way KT19: Eps 84Pa 173
Shetland Cl. GU21: Brkwd 1E 186
Shetland Cl. WD6: Bore 16Ta 29
Shetland Ho. DA17: Belv 47Dd 96
Shetland Ho. WD18: Wat 16V 26
(off Pioneer Way)
Shetland Rd. E3 40Bc 72
Shetland Rd. TW6: H'row A 58S 106
Shevon Way CM14: B'wood 21Vd 58
Shewens Rd. KT13: Weyb 77T 150
Shey Copse GU22: Wok 89E 168
Shield Dr. TW8: Bford 51Ja 108
Shieldhall St. SE2 49Yc 95
Shield Rd. DA1: Dart 54Sd 120
Shield Rd. TW15: Ashf 63S 128
Shield St. SE15 52Vb 113
Shiers Av. DA1: Dart 55Pd 119
Shifford Path SE23 62Zb 136
Shilburn Way GU21: Wok 10L 167
Shillaker Ct. W3 46Va 88
Shillibeer Pl. W1 1E 220 (43Gb 89)
(off Harcourt St.)
Shillibeer Wlk. IG7: Chig 20Vc 37
Shillingford Cl. NW7 24Za 48
Shillingford Ho. E3 41Dc 92
(off Talwin St.)
Shillingford St. N1 38Rb 71
Shilling Pl. W7 47Ja 86
Shillingshaw Lodge E16 44Jc 93
(off Butchers Rd.)
Shillingstone Ho. W14 48Ab 88
(off Russell Rd.)
Shillington Gro. WD17: Wat 11W 26
Shillington St. Open Space 55Gb 111
Shillitoe Av. EN6: Pot B 4Za 16
Shinecroft TN14: Otf 88Jd 182
Shinfield St. W12 44Ya 88
Shingle Ct. EN9: Walt A 5Jc 21
Shinglewell Rd. DA8: Erith 52Cd 118
Shingly Pl. E4 18Ec 34
Shinners Cl. SE25 71Wb 157
Ship All. W4 51Qa 109
Ship & Mermaid Row SE1 1G 231 (47Tb 91)
Shipbuilding Way E13 40Mc 73
Shipfield Cl. TN16: Tats 93Lc 199
Ship Hill SL1: Burn 2C 60
Ship Hill TN16: Tats 93Lc 199
Shipka Rd. SW12 60Kb 112
Shiplake Ho. E2 4K 219 (41Vb 91)
(off Arnold Cir.)
Ship La. BR8: Swan 67Md 141
Ship La. DA4: Sut H 67Md 141
Ship La. RM15: Avel 46Td 98
Ship La. RM15: Purf 46Td 98
Ship La. RM19: Purf 48Ud 98
Ship La. SW14 55Sa 109
Shipley Hills Rd. DA13: Meop 75Fe 165
Shipman Rd. E16 44Kc 93
Shipman Rd. SE23 61Zb 136
Ship St. SE8 53Cc 114
Ship Tavern Pas. EC3 4H 225 (45Ub 91)
Shipton Cl. RM8: Dag 34Zc 75
Shipton Ho. E2 40Vb 71
(off Shipton St.)
Shipton Rd. UB10: Ick 35P 63
Shipton St. E2 40Vb 71
Shipwright Rd. SE16 47Ac 92
Shipwright St. E16 47Kc 93
Shipwright Yd. SE1 7H 225 (46Ub 91)
Ship Yd. E14 50Dc 92
Ship Yd. KT13: Weyb 76R 150
Shirburn Cl. SE23 59Yb 114
Shirebrook Rd. SE3 55Mc 115
Shire Ct. DA18: Erith 48Zc 95
Shire Ct. HP2: Hem H 1A 4
Shire Ct. KT17: Ewe 80Va 154
Shirehall Cl. NW4 30Za 48
Shirehall Gdns. NW4 30Za 48
Shirehall La. NW4 30Za 48
Shirehall Pk. NW4 29Za 48
Shirehall Rd. DA2: Hawl 64Ld 141
Shirehall Rd. DA2: Wilm 64Ld 141
Shire Horse Way TW7: Isle 55Ha 108
Shire Ho. E3 41Dc 92
(off Talwin St.)
Shire Ho. EC1 6E 218 (42Sb 91)
(off Lambs Pas.)
Shire La. WD3: Chor Bullsland La. 18C 24
Shire La. WD3: Chor Chalfont La. 15D 24
Shire La. BR2: Kes 81Nc 180
Shire La. BR6: Chels 80Rc 160
(not continuous)
Shire La. BR6: Downe 80Rc 160
(not continuous)
Shire La. HP8: Chal G 19C 24
Shire La. SL9: Chal P 20D 24
(not continuous)
Shire London Golf Course, The 11Za 30
Shiremeade WD6: E'tree 15Pa 29
Shire M. TW2: Whitt 58Ea 108
Shire Pl. RH1: Redh 8P 207
Shire Pl. SW18 59Eb 111
Shire Pl. TW8: Bford 52Ma 109
Shires, The TW10: Ham 63Na 131
Shires, The TW18: Staines 64G 126
(off Wapshott Rd.)
Shires, The WD25: Wat 3X 13
Shires Cl. KT21: Asht 91Ma 193
Shires Ho. KT14: Byfl 85N 169
Shirland M. W9 41Bb 89
Shirland Rd. W9 41Bb 89
Shirlbutt St. E14 45Dc 92
SHIRLEY 75Zb 158
Shirley Av. CR0: C'don 74Yb 158
Shirley Av. CR5: Coul 91Rb 197
Shirley Av. DA5: Bexl 59Zc 117
Shirley Av. RH1: Redh 10P 207
Shirley Av. SL4: Wind 3D 102
Shirley Av. SM1: Sutt 77Fb 155
Shirley Av. SM2: Cheam 81Bb 175
Shirley Chu. Rd. CR0: C'don 76Zb 158
Shirley Cl. DA1: Dart 56Ld 119
Shirley Cl. DA12: Grav'nd 1K 145
Shirley Cl. E17 29Dc 52
Shirley Cl. EN8: Chesh 1Yb 20
Shirley Cl. TW3: Houn 57Ea 108
Shirley Ct. IG10: Lough 12Pc 36
Shirley Ct. SW16 66Nb 134
Shirley Cres. BR3: Beck 70Ac 136
Shirley Dr. TW3: Houn 57Ea 108
Shirley Gdns. IG11: Bark 37Uc 74
Shirley Gdns. RM12: Horn 33Ld 77
Shirley Gdns. W7 46Ha 86
Shirley Gro. N9 17Yb 34
Shirley Gro. SW11 55Jb 112
Shirley Hgts. SM6: W'gton 81Lb 176
Shirley Hills Rd. CR0: C'don 78Yb 158
Shirley Ho. SE5 52Tb 113
(off Picton St.)
Shirley Ho. Dr. SE7 52Lc 115
Shirleyhyrst KT13: Weyb 79T 150
SHIRLEY OAKS 74Zb 158
SHIRLEY OAKS BMI HOSPITAL 73Yb 158
Shirley Oaks Rd. CR0: C'don 74Zb 158
Shirley Pk. CR0: C'don 75Yb 158
Shirley Pk. Golf Course 75Xb 157
Shirley Pk. Rd. CR0: C'don 74Xb 157
Shirley Pl. GU21: Knap 9G 166
Shirley Rd. AL1: St A 3D 6
Shirley Rd. CR0: C'don 73Xb 157
Shirley Rd. DA15: Sidc 62Uc 138
Shirley Rd. E15 38Gc 73
Shirley Rd. EN2: Enf 13Sb 33
Shirley Rd. SM6: W'gton 81Lb 176
Shirley Rd. W4 47Ta 87
Shirley Rd. WD17: Wat 11W 26
Shirley Rd. WD5: Ab L 4V 12
Shirley Sherwood Gallery of Botanical Art, The 54Pa 109
Shirley St. E16 44Hc 93
Shirley Way CR0: C'don 76Ac 158
Shirley Windmill 76Yb 158
Shirlock Rd. NW3 35Hb 69
Shirwell Cl. NW7 24Za 48
Shobden Rd. N17 25Tb 51
Shobroke Cl. NW2 34Ya 68
Shoebury Rd. E6 38Pc 74
Shoelands Ct. NW9 27Ta 47
Shoe La. EC4 2A 224 (44Qb 90)
Sholden Gdns. BR5: St M Cry 71Yc 161
Sholto Rd. TW6: H'row A 57P 105
Shona Ho. E13 43Lc 93
Shooters Av. HA3: Kenton 28La 46
Shooters Hill DA16: Well 54Sc 116
Shooters Hill SE18 53Pc 116
SHOOTERS HILL 53Qc 116
Shooters Hill Golf Course 53Sc 116
Shooters Hill Rd. SE10 53Fc 115
Shooters Hill Rd. SE18 53Mc 115
Shooters Hill Rd. SE3 52Lc 115
Shooters Rd. EN2: Enf 11Rb 33
SHOOTING STAR HOUSE, CHILDREN'S HOSPICE 65Ba 129
Shoot Up Hill NW2 36Ab 68
Shopping Hall, The E6 39Nc 74
Shop Rd. SL4: Wind 2A 102
Shord Hill CR8: Kenley 88Tb 177
Shore, The DA11: Nflt Clifton Marine Pde. 8B 122
Shore, The DA11: Nflt Granby Rd. 57Fe 121
Shore Bus. Cen. E9 38Yb 72
Shore Cl. TW12: Hamp 65Aa 129
Shore Cl. TW14: Felt 59W 106
Shorediche Cl. UB10: Ick 34P 63
SHOREDITCH 3H 219 (41Ub 91)
Shoreditch Ct. E8 38Vb 71
(off Queensbridge Rd.)
Shoreditch High St. E1 6J 219 (42Ub 91)
Shoreditch High Street Station (Overground) 6K 219 (42Vb 91)
Shoreditch Ho. BR2: Broml 72Nc 160
Shoreditch Ho. N1 4G 219 (41Tb 91)
(off Charles Sq.)
Shore Gro. TW13: Hanw 61Ca 129
SHOREHAM 83Hd 182
Shoreham Aircraft Mus., The 83Hd 182
Shoreham Cl. CR0: C'don 72Yb 158
Shoreham Cl. DA5: Bexl 60Zc 117
Shoreham Cl. SW18 57Db 111
Shoreham La. BR6: Well H 79Cd 162
Shoreham La. TN13: Riv 94Hd 202
Shoreham La. TN13: S'oaks 94Hd 202
Shoreham La. TN14: Hals 84Bd 181
Shoreham Pl. TN14: S'ham 84Jd 182
Shoreham Ri. SL2: Slou 2B 80
Shoreham Rd. BR5: St P 67Xc 139
Shoreham Rd. DA4: Eyns 79Ld 163
Shoreham Rd. TN14: Otf 83Kd 183
Shoreham Rd. TN14: S'ham 83Kd 183
Shoreham Rd. E. TW6: H'row A 57N 105
Shoreham Rd. W. TW6: H'row A 57N 105
Shoreham Station (Rail) 83Kd 183
Shoreham Way BR2: Hayes 72Jc 159
Shorehill Ct. TN15: Kems'g 89Pd 183
Shorehill La. TN15: Kems'g 87Pd 183
Shorehill La. TN15: Knat 87Pd 183
Shore Ho. SW8 55Kb 112
Shore M. E9 38Yb 72
(off Shore Rd.)
Shore Pl. E9 38Yb 72
Shore Point IG9: Buck H 19Kc 35
Shore Rd. E9 38Yb 72
Shores Rd. GU21: Wok 86A 168
Shore Way SW9 54Qb 112
(off Crowhurst Cl.)
Shorncliffe Rd. SE1 7K 231 (50Vb 91)
Shorndean St. SE6 60Ec 114
SHORNE 4N 145
Shorne Cl. BR5: St M Cry 70Zc 139
Shorne Cl. DA15: Sidc 58Xc 117
Shornefield Cl. BR1: Broml 69Qc 138
Shorne Ifield Rd. DA12: Shorne 5J 145
Shornells Way SE2 49Yc 95
SHORNE RIDGEWAY 6N 145
Shorne Woods Country Pk. 6K 145
Shorne Woods Country Pk. Vis. Cen. 6L 145
Shorrold's Rd. SW6 52Bb 111
Shortacres RH1: Nutf 5F 208
Short Blue Pl. IG11: Bark 38Sc 74
Shortcroft Rd. KT17: Ewe 80Va 154
Shortcrofts Rd. RM9: Dag 37Bd 75
Shorter Av. CM15: Shenf 17Be 41
Shorter St. E1 4K 225 (45Vb 91)
Shortfern SL2: Slou 4N 81
Shortgate N12 21Bb 49
Short Hedges TW3: Houn 53Ca 107
Short Hill HA1: Harr 32Ga 66
SHORTLANDS 68Gc 137
Shortlands UB3: Harl 51T 106
Shortlands W6 49Za 88
Shortlands Cl. DA17: Belv 48Bd 95
Shortlands Cl. N18 20Tb 33
Shortlands Gdns. BR2: Broml 68Gc 137
Shortlands Golf Course 67Gc 137
Shortlands Gro. BR2: Broml 69Fc 137
Shortlands Rd. BR2: Broml 69Fc 137
Shortlands Rd. E10 31Dc 72
Shortlands Rd. KT2: King T 66Pa 131
Shortlands Station (Rail) 68Gc 137
Short La. AL2: Brick W 1Aa 13
Short La. RH8: Limp 4M 211
Short La. TN15: Igh 95Xd 204
Short La. TW19: Stanw 59P 105
Shortmead Dr. EN8: Chesh 3Ac 20
Short Path SE18 51Rc 116
Short Rd. E11 33Gc 73
Short Rd. TW6: H'row A 58N 105
Short Rd. W4 51Ua 110
Shorts Cft. NW9 28Ra 47
Shorts Gdns. WC2 3F 223 (44Nb 90)
Shorts Rd. SM5: Cars 77Gb 155
Short St. NW4 28Ya 48
(off Foster St.)
Short St. SE1 1A 230 (47Qb 90)
Short Wall E15 41Ec 92
Short Way SE9 55Nc 116
Short Way TW2: Whitt 59Ea 108
Shortway N12 23Gb 49
Shortwood GU22: Wok 1P 187
(off Mt. Hermon Rd.)
Shortwood Av. TW18: Staines 62K 127
Shortwood Comn. TW18: Staines 63K 127
Shorwell Ct. RM19: Purf 50Rd 97
Shotfield SM6: W'gton 79Kb 156
Shothanger Way HP3: Bov 7F 2
Shott Cl. SM1: Sutt 78Eb 155
Shottendane Rd. SW6 53Cb 111
Shottery Cl. SE9 62Nc 138
Shottfield Av. SW14 56Ua 110
Shottsford W2 44Cb 89
(off Talbot Rd.)
Shoulder of Mutton All. E14 45Ac 92
Shouldham St. W1 1E 220 (43Gb 89)
Showcase Cinema Barking 41Tc 94
Showcase Cinema Bluewater 60Vd 120
Showers Way UB3: Hayes 46W 84
Shrapnel Cl. SE18 52Nc 116
Shrapnel Rd. SE9 55Pc 116
SHREDING GREEN 44E 82
Shrek's Adventure! 1H 229 (47Pb 90)
Shrewsbury Av. HA3: Kenton 28Na 47
Shrewsbury Av. SW14 56Sa 109
Shrewsbury Cl. KT6: Surb 75Na 153
Shrewsbury Ct. EC1 6E 218 (42Sb 91)
(off Whitecross St.)
Shrewsbury Ho. SW3 51Gb 111
(off Cheyne Wlk.)
Shrewsbury Ho. SW8 51Pb 112
(off Kennington Oval)
Shrewsbury La. SE18 53Rc 116
Shrewsbury M. W2 43Cb 89
(off Chepstow Rd.)
Shrewsbury Rd. BR3: Beck 69Ac 136
Shrewsbury Rd. E7 36Mc 73
Shrewsbury Rd. N11 23Lb 50
Shrewsbury Rd. NW10 39Ta 67
Shrewsbury Rd. RH1: Redh 6N 207
Shrewsbury Rd. SM5: Cars 72Gb 155
Shrewsbury Rd. TW6: H'row A 58S 106
(not continuous)
Shrewsbury Rd. W2 44Cb 89
Shrewsbury St. W10 42Ya 88
Shrewsbury Wlk. TW7: Isle 55Ja 108
(off Magdala Rd.)
Shrewton Rd. SW17 66Hb 133
Shroffold Rd. BR1: Broml 63Gc 137
Shropshire Cl. CR4: Mitc 70Nb 134
Shropshire Ct. W7 44Ha 86
(off Copley Cl.)
Shropshire Ho. N18 22Xb 51
(off Cavendish Cl.)
Shropshire Pl. WC1 6C 216 (42Lb 90)
Shropshire Rd. N22 24Pb 50
Shroton St. NW1 7D 214 (43Gb 89)

Shrubberies, The AL1: St A 3B 6
Shrubberies, The E18 26Jc 53
Shrubberies, The IG7: Chig 22Sc 54
Shrubbery, The E11 29Kc 53
Shrubbery, The HP1: Hem H 1G 2
Shrubbery, The KT6: Surb 74Na 153
Shrubbery, The RM14: Upm 34Sd 78
Shrubbery Cl. N1 39Sb 71
Shrubbery Gdns. N21 17Rb 33
Shrubbery Rd. DA12: Grav'nd 10D 122
Shrubbery Rd. DA4: S Dar 67Td 142
Shrubbery Rd. N9 20Wb 33
Shrubbery Rd. SW16 63Nb 134
Shrubbery Rd. UB1: S'hall 46Ba 85
Shrubbs Hill GU24: Chob 1G 166
Shrubbs Hill La. SL5: S'dale 2G 146
Shrubhill Rd. HP1: Hem H 3H 3
Shrubland Ct. SM7: Bans 88Bb 175
(off Garratts La.)
Shrubland Gro. KT4: Wor Pk 76Ya 154
Shrubland Rd. E10 31Cc 72
Shrubland Rd. E17 29Cc 52
Shrubland Rd. E8 39Vb 71
Shrubland Rd. SM7: Bans 88Bb 175
Shrublands AL9: Brk P 8K 9
Shrublands, The EN6: Pot B 5Ab 16
Shrublands Av. CR0: C'don 76Cc 158
Shrublands Cl. IG7: Chig 23Sc 54
Shrublands Cl. N20 18Fb 31
Shrublands Cl. SE26 62Yb 136
Shrubsall Cl. SE9 60Nc 116
Shrubshall Mdw. TN15: Plax 99Ce 205
SHRUBS HILL 2G 146
Shrubs Rd. UB9: Rick 23P 43
Shuna Wlk. N1 37Tb 71
Shurland Av. EN4: E Barn 16Fb 31
Shurland Gdns. SE15 52Vb 113
Shurlock Av. BR8: Swan 68Fd 140
Shurlock Dr. BR6: Farnb 77Sc 160
Shushan Cl. N16 31Ub 71
Shuters Sq. W14 50Bb 89
Shuttle Cl. DA15: Sidc 59Vc 117
Shuttlemead DA5: Bexl 59Bd 117
Shuttle Rd. DA1: Cray 55Jd 118
Shuttle St. E1 42Wb 91
Shuttleworth Rd. SW11 54Gb 111
Siamese M. N3 25Cb 49
Siani M. N8 28Rb 51
Sibella Rd. SW4 54Mb 112
Sibley Cl. BR1: Broml 71Nc 160
Sibley Cl. DA6: Bex 57Ad 117
Sibley Ct. BR2: Broml 68Fc 137
Sibley Ct. UB8: Hil 43S 84
Sibley Gro. E12 38Nc 74
Sibthorpe Rd. AL9: Wel G 6F 8
Sibthorpe Rd. SE12 58Kc 115
Sibthorp Rd. CR4: Mitc 68Hb 133
Sibton Rd. SM5: Cars 73Gb 155
Sicilian Av. WC1 1G 223 (43Nb 90)
(off Vernon Pl.)
Sickle Cnr. RM9: Dag 42Dd 96
Sidbury Cl. SL5: S'dale 1E 146
Sidbury St. SW6 53Ab 110
SIDCUP 63Wc 139
Sidcup By-Pass BR7: Chst 62Tc 138
Sidcup Family Golf 62Sc 138
Sidcup Golf Course 60Xc 117
Sidcup High St. DA14: Sidc 63Wc 139
Sidcup Hill DA14: Sidc 63Xc 139
Sidcup Hill Gdns. DA14: Sidc 64Yc 139
Sidcup Leisure Cen. 61Wc 139
Sidcup Pl. DA14: Sidc 64Wc 139
Sidcup Place 64Wc 139
Sidcup Rd. SE12 58Lc 115
Sidcup Rd. SE9 60Pc 116
Sidcup Station (Rail) 61Wc 139
Sidcup Technology Cen. DA14: Sidc 64Zc 139
Siddeley Dr. TW4: Houn 55Aa 107
Siddeley Rd. DA1: Cray 57Jd 118
Siddeley Rd. E17 26Ec 52
Siddons Cl. SS17: Linf 8J 101
Siddons La. NW1 6G 215 (42Hb 89)
Siddons Rd. CR0: Wadd 76Qb 156
Siddons Rd. N17 25Wb 51
Siddons Rd. SE23 61Ac 136
Side Rd. UB9: Den 31F 62
Sidewood Rd. SE9 60Tc 116
Sidford Cl. HP1: Hem H 2H 3
Sidford Ho. SE1 4K 229 (48Qb 90)
(off Cosser St.)
Sidford Pl. SE1 4K 229 (48Qb 90)
Sidgwick Ho. SW9 54Pb 112
(off Stockwell Rd.)
Sidi Ct. N15 27Rb 51
Sidings, The AL10: Hat 1A 8
Sidings, The E11 32Ec 72
Sidings, The HP2: Hem H 2M 3
Sidings, The IG10: Lough 16Nc 36
Sidings, The TN13: Dun G 92Gd 202
Sidings, The TW18: Staines 63K 127
Sidings Apts., The E16 47Qc 94
Sidings M. N7 34Qb 70
Siding St. E20 39Dc 72
Sidlaw Av. AL2: Lon C 8E 6
Sidlaw Ho. N16 32Vb 71
Sidmouth Av. TW7: Isle 54Ga 108
Sidmouth Cl. WD19: Wat 19X 27
Sidmouth Ct. DA1: Dart 60Rd 119
(off Churchill Cl.)
Sidmouth Dr. HA4: Ruis 34W 64
Sidmouth Ho. SE15 52Wb 113
(off Lindsey Est.)
Sidmouth Ho. W1 2E 220 (44Gb 89)
(off Cato St.)
Sidmouth M. WC1 4H 217 (41Pb 90)
Sidmouth Pde. NW2 38Ya 68
Sidmouth Rd. BR5: St M Cry 71Xc 161
(not continuous)
Sidmouth Rd. DA16: Well 52Yc 117
Sidmouth Rd. E10 34Ec 72
Sidmouth Rd. NW2 38Ya 68
Sidmouth St. WC1 4G 217 (41Nb 90)
Sidney Av. N13 22Pb 50
Sidney Boyd Ct. NW6 38Cb 69
Sidney Cl. UB8: Uxb 38L 63
Sidney Elson Way E6 40Qc 74
Sidney Est. E1: Bromhead St. 44Yb 92
Sidney Est. E1: Lindley St. 43Yb 92
Sidney Gdns. TN14: Otf 88Ld 183
Sidney Gdns. TW8: Bford 51Ma 109
Sidney Godley (VC) Ho. E2 41Yb 92
(off Digby St.)
Sidney Gro. EC1 2B 218 (40Rb 71)
Sidney Ho. E2 40Zb 72
(off Old Ford Rd.)
Sidney Miller Ct. W3 46Ra 87
(off Crown St.)
Sidney Rd. BR3: Beck 68Ac 136
Sidney Rd. CM16: They B 8Tc 22
Sidney Rd. E7 34Jc 73
Sidney Rd. HA2: Harr 27Ea 46
Sidney Rd. KT12: Walt T 73W 150
Sidney Rd. N22 24Pb 50
Sidney Rd. SE25 71Wb 157
Sidney Rd. SL4: Wind 5A 102
Sidney Rd. SW9 54Pb 112
Sidney Rd. TW1: Twick 58Ja 108
Sidney Rd. TW18: Staines 63J 127
Sidney Sq. E1 43Yb 92
Sidney St. E1 43Xb 91
(not continuous)
Sidney Webb Ho. SE1 3G 231 (48Tb 91)
(off Tabard St.)
Sidonie Apts. SW11 56Gb 111
(off Danvers Av.)
Sidworth St. E8 38Xb 71
Siebel Ct. TW20: Egh 63D 126
Siebert Rd. SE3 51Jc 115
Siege Ho. E1 44Xb 91
(off Sidney St.)
Siemens Brothers Way E16 45Jc 93
Siemens Rd. SE18 48Mc 93
Sienna SE28 48Wc 95
Sienna Alto SE13 55Ec 114
(off Cornmill La.)
Sienna Cl. KT9: Chess 79Ma 153
Sienna Ho. E20 37Dc 72
(off Victory Pde.)
Sienna Ter. NW2 33Wa 68
Sigdon Pas. E8 36Wb 71
Sigdon Rd. E8 36Wb 71
Sigers, The HA5: Eastc 30X 45
Sigmund Freud Statue 37Eb 69
Signal Ho. E8 38Xb 71
(off Martello Ter.)
Signal Ho. SE1 2D 230 (47Sb 91)
(off Gt. Suffolk St.)
Signal Wlk. E4 23Ec 52
Signmakers Yd. NW1 39Kb 70
(off Delancey St.)
Sigrist Sq. KT2: King T 67Na 131
Sikorski Mus. 2C 226 (47Fb 89)
Silas Ct. WD17: Wat 11W 26
(off Lockhart Rd.)
Silbury Av. CR4: Mitc 67Gb 133
Silbury Ho. SE26 62Wb 135
Silbury St. N1 3F 219 (41Tb 91)
Silchester Ct. CR7: Thor H 70Qb 134
Silchester Ct. TW15: Ashf 61N 127
Silchester Rd. W10 44Za 88
Silecroft Rd. DA7: Bex 53Cd 118
Silesia Bldgs. E8 38Xb 71
Silex St. SE1 2C 230 (47Rb 91)
Silicon Bus. Cen. UB6: G'frd 40La 66
Silicon M. E3 39Cc 72
Silicon Way N1 4G 219 (41Tb 91)
(off Corsham St.)
Silistria Cl. GU21: Knap 10G 166
Silk Cl. SE12 57Jc 115
Silk Ct. E2 41Wb 91
(off Squirries St.)
Silkfield Rd. NW9 29Ua 48
Silkham Rd. RH8: Oxt 99Fc 199
Silk Ho. E1: Leman St. 44Wb 91
(off Leman St.)
Silk Ho. E1: Trafalgar Gdns. 43Zb 92
(off Trafalgar Gdns.)
Silk Ho. E2 1K 219 (40Vb 71)
(off How's St.)
Silk Ho. NW9 27Ta 47
Silkin M. SE15 52Wb 113
Silk M. SE11 7A 230 (50Qb 90)
(off Kennington Rd.)
Silk Mill Ct. WD19: Wat 16X 27
Silk Mill Rd. WD19: Wat 17X 27
Silk Mills Cl. TN14: S'oaks 93Ld 203
Silk Mills Pas. SE13 54Dc 114
Silk Mills Path SE13 54Ec 114
(not continuous)
Silk Mills Sq. E9 37Bc 72
Silkmore La. KT24: W Hor 97Q 190
Silk Rd. SM6: W'gton 75Jb 156
Silks Ct. E11 32Hc 73
Silkstream Pde. HA8: Edg 25Sa 47
Silkstream Rd. HA8: Edg 25Sa 47
Silk St. EC2 7E 218 (43Sb 91)
Silk Weaver Way E2 40Xb 71
Sillitoe Ho. N1 39Tb 71
(off Colville Est.)
Silsoe Ho. NW1 2A 216 (40Kb 70)
Silsoe Rd. N22 26Pb 50
Silverbeck Way TW19: Stanw M 57J 105
Silver Birch Av. E4 22Bc 52
Silver Birch Cl. DA2: Wilm 63Gd 140
Silver Birch Cl. KT15: Wdhm 84G 168
Silver Birch Cl. N11 23Jb 50
Silver Birch Cl. SE28 46Wc 95
Silver Birch Cl. SE6 62Bc 136
Silver Birch Cl. UB10: Ick 35N 63
Silver Birch Ct. EN8: Chesh 3Zb 20
Silver Birches CM13: Hut 18Ce 41
Silver Birch Gdns. E6 42Pc 94
Silver Birch M. IG6: Ilf 23Sc 54
Silver Birch M. RM14: Upm 32Ud 78
Silverbirch Wlk. NW3 37Hb 69
Silverburn Ho. SW9 53Rb 113
(off Lothian Rd.)
Silvercliffe Gdns. EN4: E Barn 14Gb 31
Silver Cl. HA3: Hrw W 24Fa 46
Silver Cl. KT20: Kgswd 96Ab 194
Silver Cl. SE14 52Ac 114
Silver Ct. KT22: Lea 92Ka 192
Silver Cres. W4 49Ra 87
Silverdale DA3: Hartl 70Be 143
Silverdale EN2: Enf 14Nb 32
Silverdale NW1 3B 216 (41Lb 90)
(off Harrington St.)
Silverdale SE26 63Yb 136
Silverdale Av. IG2: Ilf 29Uc 54
Silverdale Av. KT12: Walt T 75V 150
Silverdale Av. KT22: Oxs 86Ea 172
Silverdale Cen., The HA0: Wemb 39Pa 67
Silverdale Cl. SM1: Sutt 77Bb 155
Silverdale Cl. UB5: N'olt 36Ba 65
Silverdale Cl. W7 46Ga 86
Silverdale Ct. EC1 5C 218 (42Rb 91)
(off Goswell Rd.)
Silverdale Ct. TW18: Staines 63K 127
Silverdale Dr. RM12: Horn 36Kd 77
Silverdale Dr. SE9 61Nc 138
Silverdale Dr. TW16: Sun 68X 129
Silverdale Factory Cen. UB3: Hayes 48W 84
Silverdale Gdns. UB3: Hayes 47W 84
Silverdale Ind. Est. UB3: Hayes 47W 84
Silverdale Rd. BR5: Pet W 70Sc 138
Silverdale Rd. BR5: St P 69Wc 139
Silverdale Rd. DA7: Bex 54Dd 118
Silverdale Rd. E4 23Fc 53
Silverdale Rd. UB3: Hayes 47V 84
Silverdale Rd. WD23: Bush 15Aa 27
Silver Dell WD24: Wat 7V 12
Silverdene N12 23Db 49
(off Thyra Gro.)
Silvergate KT19: Ewe 78Sa 153
Silverglade Bus. Pk. KT9: Chess 84La 172
Silverhall St. TW7: Isle 55Ja 108
Silver Hill WD6: Bore 8Sa 15
Silverholme Cl. HA3: Kenton 31Na 67
Silver Jubilee Way TW4: Cran 54X 107
Silverlands Cl. KT16: Chert 76F 148
Silverland St. E16 46Pc 94
Silver La. BR4: W W'ck 75Fc 159
Silver La. CR8: Purl 84Mb 176
Silverleigh Rd. CR7: Thor H 70Pb 134
Silverlocke Rd. RM17: Grays 51Fe 121
Silver Mead E18 25Jc 53
Silvermere Av. RM5: Col R 23Dd 56
Silvermere Ct. CR3: Cat'm 96Vb 197
Silvermere Ct. CR8: Purl 84Qb 176
Silvermere Dr. N18 23Zb 52
Silvermere Golf Course 85S 170
Silvermere Rd. SE6 59Dc 114
Silver Pl. W1 3C 222 (44Lb 90)
Silver Pl. WD18: Wat 14U 26
Silver Rd. DA12: Grav'nd 1G 144
Silver Rd. SE13 55Dc 114
Silver Rd. W12 45Za 88
Silvers IG9: Buck H 18Lc 35
(off Palmerston Rd.)
Silversmiths Way GU21: Wok 10N 167
Silver Spring Cl. DA8: Erith 51Dd 118
Silverstead La. TN16: Westrm 93Tc 200
Silverstone Ct. RH1: Redh 4P 207
Silverston Way HA7: Stan 23La 46
Silver St. EN1: Enf 13Tb 33
Silver St. EN7: G Oak 2Rb 19
Silver St. EN7: Walt C 2Rb 19
Silver St. EN9: Walt A 6Ec 20
Silver St. N18 21Tb 51
Silver St. RM4: Abr 13Xc 37
Silver Street Station (Overground) 21Vb 51
Silverthorn NW8 39Db 69
(off Abbey Rd.)
Silverthorn Dr. HP3: Hem H 6B 4
Silverthorne Loft Apts. SE5 51Tb 113
(off Albany Rd.)
Silverthorne Rd. SW8 54Kb 112
Silverthorn Gdns. E4 19Cc 34
Silverton Rd. W6 51Za 110
SILVERTOWN 46Lc 93
Silvertown Av. SS17: Stan H 1M 101
Silvertown Quay Development E16 46Lc 93
Silvertown Sq. E16 44Hc 93
Silvertown Viaduct E16 44Hc 93
Silvertown Way E16: Clarkson Rd. 44Gc 93
Silvertown Way E16: Hanover Av. 46Jc 93
Silver Train Gdns. DA1: Dart 54Pd 119
Silver Tree Cl. KT12: Walt T 76W 150
Silvertree La. UB6: G'frd 41Fa 86
Silver Trees AL2: Brick W 2Ba 13
Silver Wlk. SE16 46Ac 92
Silver Way RM7: Mawney 27Dd 56
Silver Way UB10: Hil 40R 64
Silver Wing Ind. Est. CR0: Wadd 79Pb 156
Silverwood Cl. BR3: Beck 66Cc 136
Silverwood Cl. CR0: Sels 81Bc 178
Silverwood Cl. HA6: Nwood 25S 44
Silverwood Cl. RM16: Grays 45Ce 99
Silverwood Pl. SE10 53Ec 114
Silverworks Cl. NW9 27Ta 47
Silvester Ho. E1 44Xb 91
(off Varden St.)
Silvester Ho. E2 41Yb 92
(off Sceptre Rd.)
Silvester Ho. W11 44Bb 89
(off Basing St.)
Silvester Rd. SE22 57Vb 113
Silvester St. SE1 2F 231 (47Tb 91)
Silvocea Way E14 44Fc 93
Silwood Cl. SL5: Asc 8B 124
Silwood Est. SE16 49Yb 92
Silwood Pk. 9C 124
Silwood Rd. SL5: S'dale 10D 124
Silwood Rd. SL5: S'hill 10D 124
Silwood St. SE16 49Yb 92
(off Rotherhithe New Rd.)
Simkins Cl. SW2 56Nb 112
Simla Ct. N7 38Nb 70
(off Brewery Rd.)
Simla Ho. SE1 2G 231 (47Tb 91)
(off Kipling Est.)
Simmil Rd. KT10: Clay 78Ga 152
Simmonds Ct. SW5 49Db 89
(off Earl's Ct. Gdns.)
Simmonds Dr. DA3: Hartl 71Ce 165
Simmonds Ho. TW8: Bford 50Na 87
(off Clayponds La.)
Simmonds Ri. HP3: Hem H 4M 3
Simmons Cl. KT9: Chess 79La 152
Simmons Cl. N20 19Gb 31
Simmons Cl. SL3: L'ly 49C 82
Simmons Dr. RM8: Dag 34Ad 75
Simmons Ga. KT10: Esh 78Ea 152
Simmons La. E4 19Fc 35
Simmons Pl. RM16: Grays 46Ce 99
Simmons Pl. TW18: Staines 64G 126
Simmons Way N20 19Gb 31
Simms Cl. SM5: Cars 75Gb 155
Simms Gdns. N2 26Eb 49
Simms Rd. SE1 49Wb 91
Simnel Rd. SE12 59Kc 115
Simon Cl. W11 45Bb 89
Simon Ct. W9 41Cb 89
(off Saltram Cres.)
Simon Ct. WD23: Bush 16Ca 27
Simon Dean HP3: Bov 10A 2
Simonds Rd. E10 33Cc 72
Simone Cl. BR1: Broml 67Mc 137
Simone Ct. SE26 62Yb 136
Simone Dr. CR8: Kenley 88Sb 177
Simons Cl. KT16: Ott 79E 148
Simons Ct. N16 33Vb 71
Simons Wlk. E15 36Fc 73
Simons Wlk. TW20: Eng G 6N 125
Simplemarsh Ct. KT15: Add 77K 149
Simplemarsh Rd. KT15: Add 77J 149
Simpson Cl. CR0: C'don 71Sb 157
Simpson Cl. N21 15Nb 32
Simpson Dr. W3 44Ta 87
Simpson Ho. NW8 5D 214 (42Gb 89)
Simpson Ho. SE11 7J 229 (50Pb 90)
Simpson Rd. RM13: Rain 37Hd 76
Simpson Rd. TW10: Ham 63La 130
Simpson Rd. TW4: Houn 58Ba 107
Simpson's Rd. BR1: Broml 69Jc 137
Simpson's Rd. E14 45Dc 92
Simpson St. SW11 54Gb 111
Simpsons Way SL1: Slou 6H 81
Simpson Way KT6: Surb 72La 152
Simrose Ct. SW18 57Cb 111
Sims Cl. DA8: Erith 52Jd 118
Sims Cl. RM1: Rom 28Hd 56
Sim St. N4 33Sb 71
Sims Wlk. SE3 56Hc 115
Sinclair Ct. CR0: C'don 75Ub 157
Sinclair Dr. SM2: Sutt 81Db 175
Sinclair Gdns. W14 47Za 88
Sinclair Gro. NW11 30Za 48
Sinclair Ho. E15 37Fc 73
(off Leyton Rd.)
Sinclair Ho. WC1 4F 217 (41Nb 90)
(off Sandwich St.)
Sinclair Mans. W12 47Za 88
(off Richmond Way)
Sinclair Pl. SE4 58Cc 114
Sinclair Rd. E4 22Bc 52
Sinclair Rd. SL4: Wind 5G 102
Sinclair Rd. W14 47Za 88
Sinclairs Ho. E3 40Bc 72
(off St Stephen's Rd.)
Sinclair Way DA2: Daren 63Td 142
Sinclare Cl. EN1: Enf 11Vb 33
Sincots Rd. RH1: Redh 6P 207
Sinderby Cl. WD6: Bore 11Pa 29
Sindercombe M. W12 48Wa 88
Singapore Rd. W13 46Ja 86
Singer M. SW4 54Nb 112
Singer St. EC2 4G 219 (41Tb 91)
Singles Cross TN14: Knock 87Zc 181
Singles Cross La. TN14: Knock 86Yc 181
Single St. TN16: Big H 87Rc 180
SINGLE STREET 87Rc 180
Singleton Cl. CR0: C'don 73Sb 157
Singleton Cl. RM12: Horn 35Hd 76
Singleton Cl. SW17 66Hb 133
Singleton Rd. RM9: Dag 36Bd 75
Singleton Scarp N12 22Cb 49
SINGLEWELL 4F 144
Singlewell La. DA11: Grav'nd 2D 144
Singlewell Rd. DA11: Grav'nd 2D 144
Singret Pl. UB8: Cowl 42L 83
Sinnott Rd. E17 25Zb 52
Siobhan Davies Dance Studios 4B 230 (48Rb 91)
(off St George's Rd.)
Sion Ct. TW1: Twick 60Ka 108
Sion Rd. TW1: Twick 60Ka 108
Sippets Ct. IG1: Ilf 32Tc 74
SIPSON 51Q 106
Sipson Cl. UB7: Sip 51Q 106
Sipson La. UB3: Harl 51R 106
Sipson La. UB7: Sip 51Q 106
Sipson Rd. UB7: Sip 48P 83
(not continuous)
Sipson Rd. UB7: W Dray 48P 83
(not continuous)
Sipson Way UB7: Sip 52Q 106
Sir Abraham Dawes Cotts. SW15 56Ab 110
Sir Alexander Cl. W3 46Va 88
Sir Alexander Rd. W3 46Va 88
Sir Christopher France Ho. E1 41Ac 92
Sir Cyril Black Way SW19 66Cb 133
Sirdar Rd. CR4: Mitc 65Jb 134
Sirdar Rd. N22 27Rb 51
Sirdar Rd. W11 45Za 88
Sirdar Strand DA12: Grav'nd 4H 145
Sireen Apts. E3 42Bc 92
Sir Francis Drake Ct. SE10 50Fc 93
Sir Francis Way CM14: B'wood 19Xd 40
Sir Henry Peakes Dr. SL2: Farn C 7E 60
Sirinham Point SW8 51Pb 112
(off Meadow Rd.)
Sirius Bldg. E1 45Zb 92
(off Jardine Rd.)
Sirius Ho. SE16 49Ac 92
(off Seafarer Way)
Sirius Rd. HA6: Nwood 22V 44
Sir James Altham Pool 22Z 45
Sir James Black Ho. SE5 54Tb 113
(off Coldharbour La.)
Sir James Moody Way CR5: Coul 89Kb 176
Sir John Kirk Cl. SE5 52Sb 113
Sir John Lyon Ho. EC4 4D 224 (45Sb 91)
(off High Timber St.)
Sir John Morden Wlk. SE3 54Jc 115
Sir John Soane's Mus. 2H 223 (44Pb 90)
Sir Nicholas Garrow Ho. W10 42Ab 88
(off Kensal Rd.)
Sirocco Twr. E14 47Dc 92
Sir Oswald Stoll Mans. SW6 52Db 111
(off Fulham Rd.)
Sir Robert M. SL3: L'ly 50C 82
Sir Simon Milton Sq. SW1 4A 228 (48Kb 90)
Sir Steve Redgrave Bridge 45Rc 94
Sir Sydney Camm Ho. SL4: Wind 3F 102
Sir Walter Raleigh Ct. SE10 50Gc 93
Sir William Atkins Ho. KT18: Eps 85Ta 173
Sir William Powell's Almshouses SW6 54Ab 110
Sise La. EC4 3F 225 (44Tb 91)
Siskin Cl. WD23: Bush 14Aa 27
Siskin Cl. WD6: Bore 14Qa 29
Siskin Dr. DA9: Ghithe 58Wd 120
Siskin Dr. HP3: Hem H 7M 3
Siskin Ho. SE16 49Zb 92
(off Tawny Way)
Siskin Ho. WD18: Wat 16T 26
Siskin Pl. UB4: Yead 43Y 85
Sisley Rd. IG11: Bark 39Uc 74
Sispara Gdns. SW18 58Bb 111
Sissinghurst Cl. BR1: Broml 64Gc 137
Sissinghurst Ho. SE15 51Yb 114
(off Sharratt St.)
Sissinghurst Rd. CR0: C'don 73Wb 157
Sissulu Ct. E6 39Lc 73
Sister Mabel's Way SE15 52Wb 113
Sisters Av. SW11 55Hb 111
Sistova Rd. SW12 60Kb 112
Sisulu Pl. SW9 55Qb 112
Sitarey Ct. W12 46Xa 88
Sitka Ho. SE16 47Ac 92
Sittingbourne Av. EN1: Enf 16Tb 33
Sitwell Gro. HA7: Stan 22Ha 46
Siverst Cl. UB5: N'olt 37Da 65
Sivill Ho. E2 3K 219 (41Vb 91)
(off Columbia Rd.)
Siviter Way RM10: Dag 38Dd 76
Siward Rd. BR2: Broml 69Kc 137
Siward Rd. N17 25Tb 51
Siward Rd. SW17 62Eb 133
Six Acres HP3: Hem H 5A 4
Six Acres Est. N4 33Pb 70
Six Bells La. TN13: S'oaks 98Ld 203
Six Bridges Ind. Est. SE1 50Wb 91
(not continuous)
Sixpenny Ct. IG11: Bark 37Sc 74
Sixteenth Av. KT20: Lwr K 98Ab 194
Sixth Av. E12 35Pc 74
Sixth Av. KT20: Lwr K 97Ab 194
Sixth Av. UB3: Hayes 46V 84
Sixth Av. W10 41Ab 88
Sixth Av. WD25: Wat 7Z 13
Sixth Cross Rd. TW2: Twick 62Ea 130
Siyah Gdn. E3 41Bc 92
Skardu Rd. NW2 36Ab 68
Skarnings Ct. EN9: Walt A 5Jc 21
Skeena Hill SW18 59Ab 110
Skeet Hill La. BR5: Orp 76Cd 162
Skeet Hill La. BR6: Orp 74Ad 161
Skeffington Rd. E6 39Pc 74
Skeffington St. SE18 48Sc 94
Skeggs Ho. E14 48Ec 92
(off Glengall St.)
Skegness Ho. N7 38Pb 70
(off Sutterton St.)
Skelbrook St. SW18 61Eb 133
Skelgill Rd. SW15 56Bb 111
Skelley Rd. E15 38Hc 73
Skelton Cl. E8 37Vb 71
Skelton Lodge SE10 48Jc 93
(off Billinghurst Way)
Skelton Rd. E7 37Jc 73
Skelton's La. E10 31Dc 72
Skelwith Rd. W6 51Ya 110
Skene Cl. GU23: Send 95D 188
Skenfrith Ho. SE15 51Xb 113
(off Commercial Way)
Skerne Rd. KT2: King T 67Ma 131
Skerne Wlk. KT2: King T 67Ma 131
Skerries Ct. SL3: L'ly 49C 82
Sketch Apts. E1 43Zb 92
(off Shandy St.)
Sketchley Gdns. SE16 50Zb 92
Sketty Rd. EN1: Enf 13Vb 33
Skibbs La. BR5: Orp 75Bd 161
Skibbs La. BR5: St M Cry 75Bd 161
Skibbs La. BR6: Chels 78Ad 161
Skibbs La. BR6: Orp 78Ad 161
Skid Hill La. CR6: W'ham 84Hc 179
Skidmore Way WD3: Rick 18N 25
Skieasy 54Ta 109
Skiers St. E15 39Gc 73
Skiffington Cl. SW2 60Qb 112
Skillen Lodge HA5: Pinn 25Z 45
Skimmington Cotts. RH2: Reig 7F 206
Skimpans Cl. AL9: Wel G 6F 8
Skinner Ct. E3 42Dc 92
(off Barry Blandford Way)
Skinner Pl. SW1 6H 227 (49Jb 90)
(off Bourne St.)
Skinners Ct. N13 20Qb 32
Skinners La. EC4 4E 224 (45Sb 91)
Skinners La. KT21: Asht 90Ma 173
Skinners La. TW5: Hest 53Da 107
Skinner's Row SE10 53Dc 114
Skinner St. EC1 4A 218 (41Qb 90)
Skinney La. DA4: Hort K 69Td 142
Skinney La. DA4: S Dar 69Td 142
Skip La. UB9: Hare 32M 63
Skipper Ct. IG11: Bark 39Sc 74
Skippers Cl. DA9: Ghithe 57Xd 120
Skipsea Ho. SW18 58Gb 111
Skipsey Av. E6 41Pc 94
Skipton Cl. N11 23Jb 50
Skipton Dr. UB3: Harl 48S 84
Skipton Ho. SE4 56Ac 114
Skipwith Ho. EC1 7K 217 (43Qb 90)
(off Bourne Est.)
Skipworth Rd. E9 39Yb 72
Skua Ct. SE8 51Bc 114
(off Dorking Cl.)
Sky Bus. Pk. TW20: Thorpe 68E 126
Skydmore Path SL2: Slou 1D 80
Skye Ho. WD18: Wat 16V 26
Skye La. HA8: Edg 21Pa 47
Skye Lodge SL1: Slou 6J 81
(off Lansdowne Av.)
Sky Gdn. Wlk. EC3 4H 225 (45Ub 91)
(off Philpot La.)
Skylark Av. DA9: Ghithe 58Wd 120
Skylark Cl. KT17: Ewe 83Wa 174
Skylark Ct. RM13: Rain 40Fd 76
Skylark Ct. SE1 2E 230 (47Sb 91)
(off Swan St.)
Skylark Ct. SW15 58Ya 110
Skylark Gro. RM3: Rom 21Md 57
Skylark Rd. UB9: Den 32E 62
Skyline Apts. N4 31Tb 71
(off Devan Gro.)
Skyline Ct. CR0: C'don 76Tb 157
(off Park La.)
Skyline Ct. SE1 4K 231 (48Vb 91)
Skyline Plaza Bldg. E1 44Wb 91
(off Commercial Rd.)
Skylines E14 47Ec 92
Skylines Village E14 47Ec 92
Sky Peals Rd. IG8: Wfd G 24Fc 53
Skyport Dr. UB7: Harm 52M 105
Sky Studios E16 47Qc 94
Skyvan Cl. TW6: H'row A 57S 106
Skyview Apts. CR0: C'don 75Sb 157
(off Park St.)
Sky View Twr. E15 40Dc 72
(off High Street)
Skyway 14 SL3: Poyle 55H 105
Slade, The SE18 51Uc 116
Sladebrook Rd. SE3 55Mc 115
Slade Ct. EN5: New Bar 13Db 31
Slade Ct. KT16: Ott 79F 148
Slade Ct. WD7: R'lett 7Ja 14
Sladedale Rd. SE18 50Uc 94
Slade End CM16: They B 8Uc 22
Slade Gdns. DA8: Erith 53Hd 118
SLADE GREEN 53Jd 118
Slade Grn. Rd. DA8: Erith 52Hd 118
Slade Green Station (Rail) 53Jd 118
Slade Ho. TW4: Houn 58Ba 107
Sladen Pl. E5 35Xb 71
Slade Oak La. SL9: Ger X 27D 42
Slade Oak La. UB9: Den 29E 42
Slade Rd. GU24: Brkwd 2B 186
Slade Rd. KT16: Ott 79F 148
Slades Cl. EN2: Enf 13Qb 32
Slades Dr. BR7: Chst 62Sc 138
Slades Gdns. EN2: Enf 12Qb 32
Slades Hill EN2: Enf 13Qb 32
Slades Ri. EN2: Enf 13Qb 32
Slade Twr. E10 33Cc 72
(off Leyton Grange Est.)
Slade Wlk. SE17 51Rb 113
Slade Way CR4: Mitc 67Jb 134
Slagrove Pl. SE13 57Cc 114
Slaidburn St. SW10 51Eb 111
Slaithwaite Rd. SE13 56Ec 114
Slaney Ct. NW10 38Ya 68

Slaney Pl. N7 36Qb 70
Slaney Rd. RM1: Rom 29Gd 56
Slapleys GU22: Wok 92A 188
Slate Ho. E14 44Bc 92
(off Keymer Pl.)
Slater Cl. SE18 50Qc 94
Slater M. SW4 55Lb 112
(off Grafton Sq.)
Slatter NW9 24Va 48
Slattery Rd. TW13: Felt 60Z 107
Sleaford Grn. WD19: Wat 20Z 27
Sleaford Ho. E3 42Cc 92
(off Fern St.)
Sleaford Ind. Est. SW8 52Lb 112
Sleaford St. SW8 52Lb 112
Sleapcross Gdns. AL4: S'ford 3M 7
SLEAPSHYDE 3M 7
Sleapshyde La. AL4: S'ford 3M 7
Sleat Ho. E3 40Bc 72
(off Saxon Rd.)
Sledmere Ct. TW14: Bedf 60U 106
Sleepers Farm Rd. RM16: Grays 7D 100
Sleigh Ho. E2 41Yb 92
(off Bacton St.)
Slewins Cl. RM11: Horn 29Ld 57
Slewins La. RM11: Horn 29Ld 57
Slide, The 38Dc 72
Slievemore Cl. SW4 55Mb 112
Sligo Ho. E1 42Zb 92
(off Beaumont Gro.)
Slindon Ct. N16 34Vb 71
Slines Oak Rd. CR3: W'ham 95Cc 198
Slines Oak Rd. CR3: Wold 95Cc 198
Slines Oak Rd. CR6: W'ham 91Cc 198
Slingsby Pl. WC2 4F 223 (45Nb 90)
Slip, The TN16: Westrm 98Sc 200
Slippers Hill HP2: Hem H 1M 3
Slippers Pl. SE16 48Xb 91
Slipshatch Rd. RH2: Reig 10F 206
Slipshoe St. RH2: Reig 6H 207
Slipway Ho. E14 50Dc 92
(off Burrells Wharf Sq.)
Sloane Av. SW3 6D 226 (49Gb 89)
Sloane Av. Mans. SW3 6F 227 (49Hb 89)
SLOANE BMI HOSPITAL, THE 67Fc 137
Sloane Ct. TW7: Isle 53Ga 108
Sloane Ct. E. SW3 7H 227 (50Jb 90)
Sloane Ct. W. SW3 7H 227 (50Jb 90)
Sloane Gdns. BR6: Farnb 76Sc 160
Sloane Gdns. SW1 6H 227 (49Jb 90)
Sloane Ga. Mans. SW1 5H 227 (49Jb 90)
(off D'Oyley St.)
Sloane Ho. E9 38Yb 72
(off Loddiges Rd.)
Sloane M. N8 29Nb 50
Sloane Sq. DA3: Lfield 69Ae 143
Sloane Sq. SW1 6H 227 (49Jb 90)
Sloane Square Station (Underground) 6H 227 (49Jb 90)
Sloane St. SW1 2G 227 (47Hb 89)
Sloane Ter. SW1 5H 227 (49Jb 90)
Sloane Ter. Mans. SW1 5H 227 (49Jb 90)
(off Sloane Ter.)
Sloane Wlk. CR0: C'don 72Bc 158
Slocock Hill GU21: Wok 9N 167
Slocum Cl. SE28 45Yc 95
SLOUGH 7K 81
Slough Bus Station 6K 81
Slough Crematorium 3K 81
Slough Ice Arena 6H 81
Slough Ind. Est. SL1: Slou 3E 80
(not continuous)
Slough Interchange Ind. Est. SL2: Slou 6L 81
Slough La. KT18: Head 97Sa 193
Slough La. NW9 29Sa 47
Slough La. RH3: Bkld 4C 206
Slough Mus. 7K 81
Slough Retail Pk. 6F 80
Slough Rd. SL0: Iver H 41E 82
Slough Rd. SL1: Slou 9J 81
Slough Rd. SL3: Dat 10L 81
Slough Rd. SL4: Eton 1H 103
Slough Station (Rail & Crossrail) 6K 81
Slough Town FC 4L 81
Slough Trad. Est. SL1: Slou Ajax Av. 5F 80
Slough Trad. Est. SL1: Slou Liverpool Rd. 4F 80
Slough Trad. Est. SL1: Slou Oxford Av. 3D 80
Slowmans Cl. AL2: Park 10A 6
Sly St. E1 44Xb 91
Smaldon Cl. UB7: W Dray 48Q 84
Small Acre HP1: Hem H 2H 3
Smallberry Av. TW7: Isle 54Ha 108
Smallbrook M. W2 3B 220 (44Fb 89)
Smalley Cl. N16 34Vb 71
Smalley Rd. Est. N16 34Vb 71
(off Smalley Cl.)
SMALLFORD 2M 7
Smallford La. AL4: S'ford 3M 7
Smallford Works AL4: S'ford 3M 7
Small Grains DA3: Fawk 76Xd 164
Small Heath Av. RM3: Rom 22Ld 57
Smallwood Rd. SW17 63Fb 133
Smarden Cl. DA17: Belv 50Cd 96
Smarden Gro. SE9 63Pc 138
Smart Cl. RM3: Rom 25Kd 57
Smart's Heath La. GU22: Wok 5L 187
Smart's Heath Rd. GU22: Wok 5K 187
Smart's La. IG10: Lough 14Mc 35
Smart's Pl. N18 22Wb 51
Smart's Pl. WC2 2G 223 (44Nb 90)
Smarts Rd. DA12: Grav'nd 2D 144
Smart St. E2 41Zb 92
Smead Way SE13 55Dc 114
Smeathman Ct. HP1: Hem H 3L 3
Smeaton Cl. EN9: Walt A 4Gc 21
Smeaton Cl. KT9: Chess 79Ma 153
Smeaton Ct. SE1 4D 230 (48Sb 91)
Smeaton Dr. GU22: Wok 92D 188
Smeaton Rd. EN3: Enf L 9Cc 20
Smeaton Rd. IG8: Wfd G 22Pc 54
Smeaton Rd. SW18 59Cb 111
Smeaton St. E1 46Xb 91
Smedley St. SW4 54Mb 112
Smedley St. SW8 54Mb 112
Smeed Rd. E3 38Cc 72
Smikle Ct. SE14 53Zb 114
(off Hatcham Pk. M.)
Smiles Pl. GU22: Wok 88D 168
Smiles Pl. SE13 54Ec 114
Smitham Bottom La. CR8: Purl 83Lb 176
Smitham Downs Rd. CR8: Purl 85Mb 176
Smith Cl. SE16 46Zb 92
Smith Ct. GU21: Wok 85F 168
Smithfield HP2: Hem H 1M 3
Smithfield Ct. E1 45Wb 91
(off Cable St.)
Smithfield Market 1B 224 (43Rb 91)
(off West Smithfield)
Smithfield Sq. N8 28Nb 50
Smithfield St. EC1 1B 224 (43Rb 91)
Smith Gro. KT22: Fet 97Ea 192
Smith Hill TW8: Bford 51Na 109
Smithies Rd. SE2 49Xc 95
Smith Rd. RH2: Reig 9H 207
Smith's Ct. W1 4D 222 (45Mb 90)
(off Gt. Windmill St.)
Smiths Cres. AL4: S'ford 3M 7
Smithsland Rd. RM3: Rom 22Nd 57
Smith's La. SL4: Wind 4C 102
Smith's Lawn (Polo & Equestrian Grounds) 6J 125
Smithson Rd. N17 25Tb 51
Smiths Point E13 39Jc 73
(off Brooks Rd.)
Smith Sq. SW1 4F 229 (48Nb 90)
Smiths Sq. W6 50Za 88
Smith St. KT5: Surb 72Pa 153
Smith St. SW3 7F 227 (50Hb 89)
Smith St. WD18: Wat 14Y 27
Smiths Yd. CR0: C'don 76Sb 157
(off St George's Wlk.)
Smiths Yd. SW18 61Eb 133
Smith Ter. SW3 7F 227 (50Hb 89)
Smithwood Cl. SW19 60Ab 110
Smithy Cl. KT20: Lwr K 98Bb 195
Smithy La. KT20: Lwr K 99Bb 195
Smithy La. TW3: Houn 55Da 107
Smithy's Grn. GU20: W'sham 9B 146
Smithy St. E1 43Yb 92
Smock Wlk. CR0: C'don 72Sb 157
Smokehouse Yd. EC1 7C 218 (43Rb 91)
(off St John St.)
Smoke La. RH2: Reig 8K 207
Smoothfield Ct. TW3: Houn 56Ca 107
Smugglers Wlk. DA9: Ghithe 57Xd 120
Smugglers Way SW18 56Db 111
Smugglers Yd. W12 46Xa 88
(off Devonport Rd.)
SMUG OAK 2Da 13
Smug Oak Grn. Bus. Cen. AL2: Brick W 1Da 13
Smug Oak La. AL2: Brick W 2Da 13
Smug Oak La. AL2: Col S 2Da 13
Smyrk's Rd. SE17 7J 231 (50Ub 91)
Smyrna Mans. NW6 38Cb 69
(off Smyrna Rd.)
Smyrna Rd. NW6 38Cb 69
Smythe Cl. N9 20Wb 33
Smythe Rd. DA4: Sut H 67Qd 141
Smythe St. E14 45Dc 92
Snag La. BR6: Prat B 82Uc 180
Snag La. TN14: Cud 84Tc 180
Snakes Hill CM14: N'side 12Sd 40
Snakes Hill CM14: Pil H 12Sd 40
Snakes La. N14 13Kb 32
Snakes La. E. IG8: Buck H 23Lc 53
Snakes La. E. IG8: Wfd G 23Lc 53
Snakes La. W. IG8: Wfd G 22Jc 53
Snakey La. TW13: Felt 63W 128
Snape Spur SL1: Slou 4J 81
SNARESBROOK 29Jc 53
Snaresbrook Dr. HA7: Stan 21Ma 47
Snaresbrook Hall E18 28Jc 53
Snaresbrook Ho. E18 28Hc 53
Snaresbrook Rd. E11 28Gc 53
Snaresbrook Station (Underground) 29Jc 53
Snarsgate St. W10 43Ya 88
Snatts Hill RH8: Oxt 1K 211
Sneath Av. NW11 31Bb 69
Snelling Av. DA11: Nflt 1A 144
Snellings Rd. KT12: Hers 78Y 151
Snells Pk. N18 23Vb 51
Sneyd Rd. NW2 35Ya 68
Snipe Cl. DA8: Erith 52Kd 119
Snodland Cl. BR6: Downe 82Qc 180
Snowberry Cl. E15 35Fc 73
Snowberry Cl. EN5: Barn 13Bb 31
Snowbury Rd. SW6 54Db 111
Snow Cen., The 4P 3
Snowcrete SL3: Wex 10N 61
Snowden Av. UB10: Hil 40R 64
Snowden Cl. SL4: Wind 6B 102
Snowden Hill DA11: Nflt 57Ce 121
Snowden St. EC2 6H 219 (42Ub 91)
Snowdon Aviary 1G 215 (39Hb 69)
(in London Zoo)
Snowdon Ct. RM2: Rom 28Ld 57
Snowdon Cres. UB3: Harl 48S 84
Snowdon Dr. NW9 30Ua 48
Snowdon Rd. TW6: H'row A 58S 106
Snowdown Cl. SE20 67Zb 136
Snowdrop Cl. TW12: Hamp 65Ca 129
Snowdrop Ct. RM13: Rain 40Ed 76
Snowdrop M. HA5: Pinn 26Y 45
Snowdrop Path RM3: Rom 24Md 57
Snowdrop Way GU24: Bisl 9E 166
Snowerhill Rd. RH3: B'wth 8A 206
Snow Hill EC1 1B 224 (43Rb 91)
Snow Hill Ct. EC1 2C 224 (44Rb 91)
Snowman Ho. NW6 39Db 69
Snowsfields SE1 1G 231 (47Tb 91)
Snowshill Rd. E12 36Nc 74
Snows Paddock GU20: W'sham 6A 146
Snow's Ride GU20: W'sham 8A 146
Snowy Fielder Waye TW7: Isle 54Ka 108
Soames Pl. EN4: Had W 12Db 31
Soames St. SE15 55Vb 113
Soames Wlk. KT3: N Mald 67Ua 132
Soane Cl. W5 47Ma 87
Soane Ct. NW1 38Lb 70
(off St Pancras Way)
Soane Ho. SE17 50Tb 91
(off Roland Way)
Soane Sq. HA7: Stan 20Ga 28
Soap Ho. La. TW8: Bford 52Na 109
Sobell Leisure Cen. 34Pb 70
Sobraon Ho. KT2: King T 66Pa 131
(off Elm Rd.)
Socket La. BR2: Hayes 72Kc 159
Soda Studios E8 39Vb 71
(off Kingsland Rd.)
Soham Rd. EN3: Enf L 9Bc 20
SOHO 3C 222 (44Lb 90)
Soho Ho. W12 45Ya 88
Soho Sq. W1 2D 222 (44Mb 90)
Soho St. W1 2D 222 (44Mb 90)
Soho Theatre 3D 222 (44Mb 90)
(off Dean St.)
Sojourner Truth Cl. E8 37Xb 71
Sola Ct. CR0: C'don 74Tb 157
(off Sydenham Rd.)
Solander Gdns. E1: Cable St. 45Xb 91
(off Cable St.)
Solander Gdns. E1: The Highway 45Yb 92
Solar Ct. N3 25Db 49
Solar Ct. SE16 47Wb 91
(off Chambers St.)
Solar Ct. WD18: Wat 15V 26
Solar Ho. E15 37Gc 73
(off Romford Rd.)
Solar Ho. E6 43Qc 94
Solarium Ct. SE1 5K 231 (49Vb 91)
(off Alscot Rd.)
Solar Way EN3: Enf L 8Bc 20
Soldene Ct. N7 37Pb 70
Solebay St. E1 42Ac 92
Solecote KT23: Bookh 97Ca 191
Sole Farm Av. KT23: Bookh 97Ba 191
Sole Farm Cl. KT23: Bookh 96Ba 191
Sole Farm Rd. KT23: Bookh 97Ba 191
Solefields Rd. TN13: S'oaks 100Kd 203
Solent Ct. SW16 68Pb 134
Solent Ho. E1 43Ac 92
(off Ben Jonson Rd.)
Solent Ri. E13 41Jc 93
Solent Rd. NW6 36Cb 69
Soleoak Dr. TN13: S'oaks 99Kd 203
Solesbridge Cl. WD3: Chor 13H 25
Solesbridge La. WD3: Chor 13H 25
Solesbridge La. WD3: Sarr 13H 25
Soley M. WC1 3K 217 (41Qb 90)
Solid La. CM15: Dodd 11Vd 40
Solid La. CM15: Pil H 11Vd 40
Solna Av. SW15 57Ya 110
Solna Rd. N21 18Tb 33
Solomon Av. N9 21Wb 51
Solomons Ct. N12 24Eb 49
Solomon's Hill WD3: Rick 17M 25
Solomon's Pas. SE15 56Xb 113
Solomon Way E1 43Ac 92
Soloms Ct. Rd. SM7: Bans 89Fb 175
Solon New Rd. SW4 56Nb 112
Solon New Rd. Est. SW4 56Nb 112
Solon Rd. SW2 56Nb 112
Solway RM18: E Til 8L 101
Solway Cl. E8 37Vb 71
(off Queensbridge Rd.)
Solway Cl. TW4: Houn 55Aa 107
Solway Ho. E1 42Zb 92
(off Ernest St.)
Solway Rd. N22 25Rb 51
Solway Rd. SE22 56Wb 113
Somaford Gro. EN4: E Barn 16Fb 31
Somali Rd. NW2 36Bb 69
Sombourne Ho. SW15 59Wa 110
(off Fontley Way)
Somerby Rd. IG11: Bark 38Tc 74
Somercoates Cl. EN4: Cockf 13Gb 31
Somer Ct. SW6 51Cb 111
(off Anselm Rd.)
Somerden Rd. BR5: St M Cry 73Zc 161
Somerfield Cl. KT20: Tad 91Ab 194
Somerfield Rd. N4 33Rb 71
(not continuous)
Somerfield St. SE16 50Zb 92
Somerford Cl. HA5: Eastc 28W 44
Somerford Gro. N16 35Vb 71
Somerford Gro. N17 24Wb 51
(not continuous)
Somerford Gro. Est. N16 35Vb 71
Somerford St. E1 42Xb 91
Somerford Way SE16 47Ac 92
Somerhill Av. DA15: Sidc 59Xc 117
Somerhill Rd. DA16: Well 54Xc 117
Someries Rd. HP1: Hem H 1H 3
Somerleyton Pas. SW9 56Rb 113
Somerleyton Rd. SW9 56Qb 112
Somersby Gdns. IG4: Ilf 29Pc 54
Somers Cl. NW1 1D 216 (40Mb 70)
Somers Cl. RH2: Reig 5J 207
Somers Cres. W2 3D 220 (44Gb 89)
Somerset Av. DA16: Well 57Vc 117
Somerset Av. KT9: Chess 77Ma 153
Somerset Av. SW20 68Xa 132
Somerset Cl. IG8: Wfd G 25Jc 53
Somerset Cl. KT12: Hers 78X 151
Somerset Cl. KT19: Ewe 81Ta 173
Somerset Cl. KT3: N Mald 72Ua 154
Somerset Cl. N17 26Tb 51
Somerset Cl. SM3: Wor Pk 77Ya 154
Somerset Ct. DA11: Nflt 61Ee 143
Somerset Ct. IG9: Buck H 19Lc 35
Somerset Ct. NW1 2D 216 (40Mb 70)
Somerset Ct. TW11: Tedd 64Ga 130
Somerset Ct. W7 44Ha 86
(off Copley Cl.)
Somerset Est. SW11 53Fb 111
Somerset Gdns. HA0: Wemb 36La 66
Somerset Gdns. N17 24Ub 51
Somerset Gdns. N6 31Jb 70
Somerset Gdns. RH1: Redh 8M 207
Somerset Gdns. RM11: Horn 32Qd 77
Somerset Gdns. SE13 54Dc 114
Somerset Gdns. SW16 69Pb 134
Somerset Gdns. TW11: Tedd 64Ga 130
Somerset Hall N17 24Ub 51
Somerset Ho. GU22: Wok 89B 168
(off Oriental Rd.)
Somerset Ho. RH1: Redh 5P 207
Somerset Ho. SW19 62Za 132
Somerset House 4H 223 (45Pb 90)
Somerset Lodge TW8: Bford 51Ma 109
Somerset Rd. BR6: Orp 73Wc 161
Somerset Rd. DA1: Dart 58Kd 119
Somerset Rd. E17 29Cc 52
Somerset Rd. EN3: Enf L 10Cc 20
Somerset Rd. EN5: New Bar 15Db 31
Somerset Rd. HA1: Harr 29Ea 46
Somerset Rd. KT1: King T 68Pa 131
Somerset Rd. N17 27Vb 51
Somerset Rd. N18 22Vb 51
Somerset Rd. NW4 28Ya 48
Somerset Rd. RH1: Redh 8M 207
Somerset Rd. SS17: Linf 7J 101
Somerset Rd. SW19 62Za 132
Somerset Rd. TW11: Tedd 64Ga 130
Somerset Rd. TW8: Bford 51La 108
Somerset Rd. UB1: S'hall 43Ba 85
Somerset Rd. W13 46Ka 86
Somerset Rd. W4 48Ta 87
Somerset Sq. W14 47Ab 88
Somerset Way SL0: Rich P 47H 83
Somerset Waye TW5: Hest 51Aa 107
Somersham Rd. DA7: Bex 54Ad 117
Somers Pl. RH2: Reig 5J 207
Somers Pl. SW2 59Pb 112
Somers Rd. AL9: Wel G 6E 8
Somers Rd. E17 28Bc 52
Somers Rd. RH2: Reig 5J 207
Somers Rd. SW2 58Pb 112
Somers Sq. AL9: Wel G 5E 8
Somerston Ho. NW1 39Lb 70
(off St Pancras Way)
SOMERS TOWN 2D 216 (40Mb 70)
Somers Town Community Sports Cen. 2D 216 (40Mb 70)
Somers Way WD23: Bush 17Ea 28
Somerton Av. TW9: Rich 55Ra 109
Somerton Cl. CR8: Purl 87Qb 176
Somerton Ho. WC1 4E 216 (41Mb 90)
(off Euston Rd.)
Somerton Rd. NW2 34Za 68
Somerton Rd. SE15 56Xb 113
Somertrees Av. SE12 61Kc 137
Somervell Rd. HA2: Harr 36Ba 65
Somerville Av. SW13 51Xa 110
Somerville Cl. SW9 53Pb 112
Somerville Ct. RH1: Redh 5N 207
(off Oxford Rd.)
Somerville Point SE16 47Bc 92
Somerville Rd. DA1: Dart 58Pd 119
Somerville Rd. KT11: Cobh 86Ca 171
Somerville Rd. RM6: Chad H 30Yc 55
Somerville Rd. SE20 66Zb 136
Somerville Rd. SL4: Eton 10G 80
Somery Wlk. WD25: A'ham 8Da 13
Sommerville Ct. WD6: Bore 11Pa 29
(off Alconbury Cl.)
Sonderburg Rd. N7 33Pb 70
Sondes St. SE17 51Tb 113
Sonesta Apts. SE15 53Xb 113
Songhurst Cl. CR0: C'don 72Pb 156
Sonia Cl. WD19: Wat 17Y 27
Sonia Ct. HA1: Harr 30Ha 46
Sonia Ct. HA8: Edg 24Pa 47
Sonia Gdns. N12 21Eb 49
Sonia Gdns. NW10 35Va 68
Sonia Gdns. TW5: Hest 52Ca 107
Sonnets, The HP1: Hem H 1K 3
Sonnet Wlk. TN16: Big H 90Kc 179
Sonning Gdns. TW12: Hamp 65Aa 129
Sonning Ho. E2 4K 219 (41Vb 91)
(off Swanfield St.)
Sonning Rd. SE25 72Wb 157
Soper Cl. E4 22Bc 52
Soper Cl. SE23 60Zb 114
Soper Dr. CR3: Cat'm 95Tb 197
Soper M. EN3: Enf L 10Cc 20
Sopers Rd. EN6: Cuff 1Pb 18
Sophia Cl. N7 37Pb 70
Sophia Ho. W6 50Ya 88
(off Queen Caroline St.)
Sophia Rd. E10 32Dc 72
Sophia Rd. E16 44Kc 93
Sophia Sq. SE16 45Ac 92
(off Sovereign Cres.)
Sophie Gdns. SL3: L'ly 7P 81
Sophora Ho. SW11 52Kb 112
Soprano Ct. E15 39Hc 73
(off Plaistow Rd.)
Soprano Way KT10: Surb 76Ka 152
SOPWELL 5C 6
Sopwell La. AL1: St A 3B 6
Sopwell Nunnery 4C 6
Sopwell Nunnery Grn. Space 4C 6
(off Old Sopwell Gdns.)
Sopwith NW9 24Va 48
Sopwith Av. E17 28Zb 52
Sopwith Av. KT9: Chess 78Na 153
Sopwith Cl. KT2: King T 64Pa 131
Sopwith Cl. RM12: Horn 36Kd 77
Sopwith Cl. TN16: Big H 88Mc 179
Sopwith Dr. KT13: Weyb 83N 169
Sopwith Rd. TW5: Hest 52Y 107
Sopwith Way KT15: Add 76M 149
Sopwith Way KT2: King T 67Na 131
Sopwith Way SW11 52Kb 112
Sorbie Cl. KT13: Weyb 79T 150
Sorbus Ct. EN2: Enf 12Rb 33
Sorensen Ct. E10 33Dc 72
(off Leyton Grange Est.)
Sorrel Bank CR0: Sels 82Ac 178
(not continuous)
Sorrel Cl. AL10: Hat 1C 8
Sorrel Cl. SE28 46Wc 95
Sorrel Cl. RM17: Grays 51Fe 121
Sorrel Gdns. E6 43Nc 94
Sorrel La. E14 44Fc 93
Sorrell Cl. SE14 52Ac 114
Sorrell Cl. SW9 54Qb 112
Sorrells, The SS17: Stan H 1P 101
Sorrel Mead NW9 31Wa 68
Sorrel Wlk. RM1: Rom 27Hd 56
Sorrel Way DA11: Nflt 3A 144
Sorrento Rd. SM1: Sutt 76Db 155
Sospel Ct. SL2: Farn R 10G 60
Sotheby Rd. N5 34Rb 71
Sotheran Cl. E8 39Wb 71
Sotherby Lodge E2 40Yb 72
(off Sewardstone Rd.)
Sotheron Pl. SW6 52Db 111
Sotheron Rd. WD17: Wat 13Y 27
Soudan Rd. SW11 53Hb 111
Souldern Rd. W14 48Za 88
Souldern St. WD18: Wat 15X 27
Soul St. SE6 60Ec 114
Sounding All. E3 39Cc 72
Sounds Lodge BR8: Crock 72Ed 162
South Access Rd. E17 31Ac 72
Southacre W2 3D 220 (44Gb 89)
(off Hyde Pk. Cres.)
Southacre Way HA5: Pinn 25Y 45
SOUTH ACTON 47Sa 87
South Acton Station (Overground) 48Sa 87
South Africa Rd. W12 46Xa 88
South Albert Rd. RH2: Reig 5H 207
SOUTHALL 46Ba 85
Southall Ct. UB1: S'hall 45Ba 85
Southall Ent. Cen. UB2: S'hall 47Ca 85
SOUTHALL GREEN 48Aa 85
Southall Ho. RM3: Rom 23Nd 57
(off Kingsbridge Cir.)
Southall La. TW5: Cran 51X 107
Southall La. UB2: S'hall 49Y 85
Southall Pl. SE1 2F 231 (47Tb 91)
Southall Sports Cen. 46Aa 85
Southall Station (Rail & Crossrail) 47Ba 85
Southall Waterside UB1: S'hall 47Z 85
Southall Way CM14: B'wood 21Vd 58
Southam Ho. KT15: Add 78K 149
(off Addlestone Pk.)
Southam Ho. W10 42Ab 88
(off Southam St.)
Southam M. WD3: Crox G 16R 26
Southampton Bldgs. WC2 1K 223 (43Qb 90)
Southampton Gdns. CR4: Mitc 71Nb 156
Southampton M. E16 46Kc 93
Southampton Pl. WC1 1G 223 (43Nb 90)
Southampton Rd. NW5 36Hb 69
Southampton Rd. E. TW6: H'row A 58P 105
Southampton Rd. W. TW6: H'row A 58N 105
Southampton Row WC1 7G 217 (43Nb 90)
Southampton St. WC2 4G 223 (45Nb 90)
Southampton Way SE15 52Ub 113
Southampton Way SE5 52Tb 113
Southam St. W10 42Ab 88
South App. HA6: Nwood 20T 26
South Ash Rd. TN15: Ash 82Yd 184
South Audley St. W1 5J 221 (45Jb 90)
South Av. E4 17Dc 34
South Av. KT12: W Vill 82U 170
South Av. KT13: Weyb 82Q 170
South Av. SM5: Cars 80Jb 156
South Av. TW20: Egh 65E 126
South Av. TW9: Kew 54Qa 109
South Av. UB1: S'hall 45Ba 85
South Av. Gdns. UB1: S'hall 45Ba 85
Southbank BR8: Hext 66Hd 140
(not continuous)
Southbank KT7: T Ditt 73Ka 152
South Bank KT6: Surb 72Na 153
South Bank SE1 7H 223 (46Pb 90)
South Bank TN16: Westrm 98Tc 200
Southbank Bus. Cen. SW11 53Hb 111
Southbank Cen. 6J 223 (46Pb 90)
South Bank Ter. KT6: Surb 72Na 153
SOUTH BARNET 18Jb 32
SOUTH BEDDINGTON 79Mb 156
South Bermondsey Station (Rail) 50Yb 92
South Birkbeck Rd. E11 34Fc 73
South Black Lion La. W6 50Wa 88
South Block RM2: Rom 27Md 57
South Block SE1 2H 229 (47Pb 90)
(off Belvedere Rd.)
South Bolton Gdns. SW5 50Db 89
South Border, The CR8: Purl 83Mb 176
SOUTHBOROUGH 71Pc 160
SOUTHBOROUGH 74Na 153
Southborough Cl. KT6: Surb 74Na 153
Southborough Ho. SE17 7H 231 (50Ub 91)
(off Kinglake Est.)
Southborough La. BR2: Broml 71Nc 160
Southborough Rd. BR1: Broml 69Nc 138
Southborough Rd. E9 39Zb 72
Southborough Rd. KT6: Surb 74Na 153
Southbourne BR2: Hayes 73Jc 159
Southbourne Av. NW9 26Sa 47
Southbourne Cl. HA5: Pinn 31Aa 65
Southbourne Ct. NW9 26Sa 47
Southbourne Cres. NW4 28Ab 48
Southbourne Gdns. HA4: Ruis 32X 65
Southbourne Gdns. IG1: Ilf 36Sc 74
Southbourne Gdns. SE12 57Kc 115
Southbridge Pl. CR0: C'don 77Sb 157
Southbridge Rd. CR0: C'don 77Sb 157
Southbridge Way UB2: S'hall 47Aa 85
SOUTH BROMLEY 45Ec 92
Southbrook M. SE12 58Hc 115
Southbrook Rd. SE12 58Hc 115
Southbrook Rd. SW16 67Nb 134
South Buckinghamshire Academy, The 3K 81
South Buckinghamshire Golf Course, The 9J 61
Southbury NW8 39Eb 69
(off Loudoun Rd.)
Southbury Av. EN1: Enf 14Wb 33
Southbury Cl. RM12: Horn 36Md 77
Southbury Ho. EN8: Walt C 4Zb 20
(off High St.)
Southbury Leisure Cen. 13Wb 33
Southbury Rd. EN1: Enf 13Ub 33
Southbury Rd. EN3: Pond E 14Xb 33
Southbury Station (Overground) 14Xb 33
South Carriage Dr. SW7 2C 226 (47Fb 89)
SOUTH CHINGFORD 22Bc 52
Southchurch Ct. E6 40Pc 74
(off High St. Sth.)
Southchurch Rd. E6 40Pc 74
South City Ct. SE15 52Ub 113
Southcliffe Dr. SL9: Chal P 22A 42
South Cl. AL2: Chis G 7P 5
South Cl. DA6: Bex 56Zc 117
South Cl. EN5: Barn 13Bb 31
South Cl. GU21: Wok 8N 167
South Cl. HA5: Pinn 31Ba 65
South Cl. N6 30Kb 50
South Cl. RM10: Dag 39Cd 76
South Cl. RM15: S Ock 42Yd 98
South Cl. SL1: Slou 5B 80
South Cl. SM4: Mord 72Cb 155
South Cl. TW2: Twick 62Ca 129
South Cl. UB7: W Dray 48P 83
South Cl. Grn. RH1: Mers 1B 208
South Colonnade, The E14 46Cc 92
(not continuous)
Southcombe St. W14 49Ab 88
South Comn. Rd. UB8: Uxb 37N 63
Southcote GU21: Wok 7P 167
Southcote Av. KT5: Surb 73Ra 153
Southcote Av. TW13: Felt 61V 128
Southcote Ho. KT15: Add 75M 149
Southcote Ri. HA4: Ruis 31T 64
Southcote Rd. CR2: Sande 82Ub 177
Southcote Rd. E17 29Zb 52
Southcote Rd. N19 35Lb 70
Southcote Rd. RH1: Mers 2C 208
Southcote Rd. SE25 71Xb 157
South Cottage Dr. WD3: Chor 15H 25
South Cottage Gdns. WD3: Chor 15H 25
Southcott Ho. E3 41Dc 92
(off Devons Rd.)
Southcott Ho. W9 6A 214 (42Eb 69)
(off Clifton Gdns.)
Southcott M. NW8 2D 214 (40Gb 69)
Southcott Rd. TW11: Hamp W 67La 130
South Countess Rd. E17 27Bc 52
South Cres. E16 42Fc 93
South Cres. WC1 1D 222 (43Mb 90)
South Crescent M. WC1 5F 217 (42Nb 90)
Southcroft SL2: Slou 2F 80
Southcroft TW20: Eng G 4M 125
Southcroft Av. BR4: W W'ck 75Ec 158
Southcroft Av. DA16: Well 55Uc 116
Southcroft Rd. BR6: Orp 76Uc 160
Southcroft Rd. SW16 65Kb 134
Southcroft Rd. SW17 65Jb 134
South Cross Rd. IG6: Ilf 29Sc 54
South Croxted Rd. SE21 62Tb 135
SOUTH CROYDON 78Tb 157
South Croydon Sports Club 78Ub 157
South Croydon Station (Rail) 78Tb 157
Southdale IG7: Chig 22Uc 54
SOUTH DARENTH 67Sd 142
Southdean Gdns. SW19 61Bb 133
Southdene TN14: Hals 85Bd 181
(not continuous)
South Dene NW7 20Ta 29
Southdene Ct. N11 20Kb 32
Southdown Av. W7 48Ja 86
Southdown Ct. AL10: Hat 3C 8
Southdown Cres. HA2: Harr 32Ea 66
Southdown Cres. IG2: Ilf 29Uc 54
Southdown Dr. SW20 66Za 132
Southdown Rd. AL10: Hat 3C 8

Southdown Rd. CR3: Wold 94Bc 198
Southdown Rd. KT12: Hers 77Aa 151
Southdown Rd. RM11: Horn 31Kd 77
Southdown Rd. SM5: Cars 81Jb 176
Southdown Rd. SW20 67Za 132
Southdowns DA4: S Dar 68Td 142
South Dr. AL4: St A 2H 7
South Dr. BR6: Orp 78Uc 160
South Dr. CM14: W'ley 21Zd 59
South Dr. CR5: Coul 87Mb 176
South Dr. EN6: Cuff 2Nb 18
South Dr. GU24: Brkwd 3A 186
South Dr. GU25: Vir W 4L 147
South Dr. HA4: Ruis 32U 64
South Dr. RM2: Rom 27Ld 57
South Dr. SM2: Cheam 82Ab 174
South Dr. SM7: Bans 85Gb 175
South Ealing Rd. W5 47Ma 87
South Ealing Station (Underground) 48Ma 87
South Eastern Av. N9 20Vb 33
South Eaton Pl. SW1 5J 227 (49Jb 90)
South Eden Pk. Rd. BR3: Beck 72Dc 158
South Edwardes Sq. W8 48Bb 89
SOUTHEND 63Fc 137
South End CR0: C'don 77Sb 157
South End CR2: S Croy 78Sb 157
South End KT23: Bookh 98Da 191
South End W8 48Db 89
Southend Arterial Rd. CM13: Dun 29Ae 59
Southend Arterial Rd. CM13: Gt War 29Ae 59
Southend Arterial Rd. CM13: L War 29Ae 59
Southend Arterial Rd. CM13: W H'dn 29Ae 59
Southend Arterial Rd. RM11: Horn 26Nd 57
Southend Arterial Rd. RM11: Upm 26Nd 57
Southend Arterial Rd. RM14: Gt War 28Sd 58
Southend Arterial Rd. RM14: Upm 28Sd 58
Southend Arterial Rd. RM2: Rom 26Md 57
Southend Arterial Rd. RM3: Hrld W 26Md 57
Southend Cl. SE9 58Rc 116
South End Cl. NW3 35Gb 69
Southend Cres. SE9 58Rc 116
South End Grn. NW3 35Gb 69
Southend La. EN9: Walt A 6Kc 21
Southend La. SE26 63Bc 136
Southend La. SE6 63Bc 136
Southend Rd. BR3: Beck 67Cc 136
Southend Rd. E17 25Dc 52
Southend Rd. E18 25Jc 53
Southend Rd. E4 22Ac 52
Southend Rd. E6 38Pc 74
Southend Rd. IG8: Wfd G 26Lc 53
Southend Rd. RM17: Grays 49Ee 99
South End Rd. NW3 35Gb 69
South End Rd. RM12: Horn 37Kd 77
South End Rd. RM13: Horn 40Jd 76
South End Rd. RM13: Rain 40Jd 76
South End Row W8 48Db 89
Southerland Cl. KT13: Weyb 77S 150
Southern Av. SE25 69Vb 135
Southern Av. TW14: Felt 60W 106
Southern Cotts. TW19: Stanw M 57J 105
Southern Dr. IG10: Lough 16Pc 36
Southerngate Way SE14 52Ac 114
Southern Gro. E3 41Bc 92
Southernhay IG10: Lough 14Mc 35
Southern Perimeter Rd. TW6: H'row A 57K 105
(not continuous)
Southern Perimeter Rd. TW6: Stanw 57K 105
(not continuous)
Southern Pl. BR8: Swan 70Fd 140
Southern Pl. HA1: Harr 35Ha 66
Southern Rd. E13 40Kc 73
Southern Rd. N2 28Hb 49
Southern Row W10 42Ab 88
Southerns La. CR5: Coul 97Eb 195
Southern St. N1 1H 217 (40Pb 70)
Southern Ter. W12 46Ya 88
Southern Valley Golf Course 3J 145
Southern Way RM7: Rom 30Cd 56
Southern Way SE10 49Hc 93
Southernwood Cl. HP2: Hem H 1A 4
Southernwood Retail Pk. 7K 231 (50Vb 91)
Southerton Rd. W6 49Ya 88
Southerton Way WD7: Shenl 5Na 15
South Esk Rd. E7 37Lc 73
South Essex Crematorium 36Td 78
South Ct. KT23: Bookh 96Da 191
Southey Ho. SE17 7E 230 (50Sb 91)
(off Browning St.)
Southey M. E16 46Jc 93
Southey Rd. N15 29Ub 51
Southey Rd. SW19 66Cb 133
Southey Rd. SW9 53Qb 112
Southey St. SE20 66Zb 136
Southey Wlk. RM18: Tilb 3D 122
Southfield EN5: Barn 16Za 30
Southfield Av. WD24: Wat 10Y 13
Southfield Cl. SL4: Dor 8A 80
Southfield Cl. UB8: Hil 42Q 84
Southfield Cotts. W7 47Ha 86
Southfield Ct. E11 34Hc 73
Southfield Gdns. SL1: Burn 3A 80
Southfield Gdns. TW1: Twick 63Ha 130
Southfield Pk. HA2: Harr 28Da 45
Southfield Pl. KT13: Weyb 80R 150
Southfield Rd. BR7: Chst 69Wc 139
Southfield Rd. EN3: Pond E 16Xb 33
Southfield Rd. EN8: Walt C 4Ac 20
Southfield Rd. N17 26Ub 51
Southfield Rd. W4 47Ta 87
Southfields BR8: Hext 66Gd 140
Southfields KT8: E Mos 72Ga 152
Southfields NW4 27Xa 48
SOUTHFIELDS 4F 100
SOUTHFIELDS 60Bb 111
Southfields Av. TW15: Ashf 65R 128
Southfields Ct. SM1: Sutt 75Cb 155
Southfields Grn. DA11: Grav'nd 4D 144
Southfields Ho. DA11: Grav'nd 4D 144
(off Southfields Grn.)
Southfields M. SW18 58Cb 111
Southfields Pas. SW18 58Cb 111
Southfields Rd. CR3: Wold 94Dc 198
Southfields Rd. SW18 58Cb 111
Southfields Rd. TN15: W King 80Vd 164
Southfields Station (Underground) 60Bb 111
SOUTHFLEET 64Ce 143
Southfleet NW5 37Jb 70
Southfleet Av. DA3: Lfield 68De 143
Southfleet Rd. BR6: Orp 76Uc 160
Southfleet Rd. DA10: Swans 59Be 121
Southfleet Rd. DA11: Nflt 10B 122
Southfleet Rd. DA2: Bean 63Yd 142
South Gdns. HA9: Wemb 33Qa 67
South Gdns. SE17 6E 230 (49Sb 91)
South Gdns. SW19 66Fb 133
SOUTHGATE 18Mb 32
Southgate RM19: Purf 49Sd 98
Southgate Av. TW13: Felt 63T 128
SOUTHGATE CIR. 18Mb 32
Southgate Cotts. WD3: Rick 17M 25
Southgate Ct. N1 38Tb 71
(off Downham Rd.)
Southgate Gro. N1 38Tb 71
Southgate Hockey Cen. 13Kb 32
Southgate Ho. EN8: Chesh 2Ac 20
(off Turner's Hill)
Southgate Leisure Cen. 17Mb 32
Southgate Rd. EN6: Pot B 5Eb 17
Southgate Rd. N1 39Tb 71
Southgate Station (Underground) 18Mb 32
South Gipsy Rd. DA16: Well 55Zc 117
South Glade, The DA5: Bexl 60Bd 117
SOUTH GODSTONE 10C 210
South Grn. NW9 25Ua 48
(off Parklea Cl.)
South Grn. SL1: Slou 5J 81
South Greenford Station (Rail) 41Ga 86
South Gro. E17 29Bc 52
South Gro. KT16: Chert 72H 149
South Gro. N15 29Tb 51
South Gro. N6 32Jb 70
South Gro. Ho. N6 32Jb 70
South Guildford St. KT16: Chert 74H 149
SOUTH HACKNEY 39Zb 72
South Hall Cl. DA4: Farni 73Pd 163
South Hall Dr. RM13: Rain 43Kd 97
SOUTH HAMPSTEAD 38Eb 69
South Hampstead Station (Overground) 38Eb 69
SOUTH HAREFIELD 29L 43
SOUTH HARROW 34Ea 66
South Harrow Ind. Est. HA2: Harr 33Ea 66
South Harrow Station (Underground) 34Ea 66
SOUTH HATFIELD 2B 8
South Herts Golf Course 18Cb 31
South Hill SS17: Horn H Saffron Cl. 1J 101
South Hill BR7: Chst 65Pc 138
South Hill HA6: Nwood 25U 44
South Hill Av. HA1: Harr 34Fa 66
South Hill Av. HA2: Harr 34Fa 66
South Hill Cres. SS17: Horn H 1J 101
South Hill Gro. HA1: Harr 35Ga 66
South Hill Pk. NW3 35Gb 69
South Hill Pk. Gdns. NW3 34Gb 69
South Hill Rd. BR2: Broml 69Gc 137
South Hill Rd. DA12: Grav'nd 10E 122
South Hill Rd. HP1: Hem H 2L 3
Southholme Cl. SE19 67Ub 135
SOUTH HORNCHURCH 40Hd 76
Southill Ct. BR2: Broml 71Hc 159
Southill La. HA5: Eastc 28X 45
Southill Rd. BR7: Chst 66Nc 138
Southill St. E14 44Dc 92
South Island Pl. SW9 52Pb 112
SOUTH KENSINGTON 6C 226 (49Fb 89)
South Kensington Sta. Arc. SW7 5C 226 (49Fb 89)
(off Pelham St.)
South Kensington Station (Underground) 5C 226 (49Fb 89)
South Kent Av. DA11: Nflt 58Ee 121
South Kenton Station (Underground & Overground) 32La 66
SOUTH LAMBETH 52Nb 112
South Lambeth Pl. SW8 51Nb 112
South Lambeth Rd. SW8 51Nb 112
Southland Rd. SE18 52Vc 117
Southlands Av. BR6: Orp 77Tc 160
Southlands Cl. CR5: Coul 89Pb 176
Southlands Coll. Roehampton University 57Wa 110
Southlands Dr. SW19 61Za 132
Southlands Gro. BR1: Broml 69Nc 138
Southlands La. RH8: Oxt 6F 210
Southlands La. RH8: Tand 6F 210
Southlands Rd. BR1: Broml 70Nc 138
Southlands Rd. BR2: Broml 71Lc 159
Southlands Rd. SL0: Den 37G 62
Southlands Rd. SL0: Iver H 37G 62
Southlands Rd. UB9: Den 35H 63
Southland Ter. RM19: Purf 50Rd 97
Southland Way TW3: Houn 57Fa 108
South La. KT1: King T 69Ma 131
South La. KT3: N Mald 70Ta 131
South La. W. KT3: N Mald 70Ta 131
South Lawn SL4: Eton 1G 102
(off South Mdw. La.)
South Lawns DA11: Nflt 58Ee 121
(off Lawn Rd.)
SOUTHLEA 4M 103
Southlea Rd. SL3: Dat 3M 103
Southlea Rd. SL4: Wind 6L 103
South Lodge E16 46Kc 93
(off Audley Dr.)
South Lodge NW8 2B 214 (40Fb 69)
South Lodge SW7 2E 226 (47Gb 89)
(off Knightsbridge)
South Lodge TW2: Whitt 58Ea 108
South Lodge Av. CR4: Mitc 70Nb 134
South Lodge Cres. EN2: Enf 14Mb 32
(not continuous)
South Lodge Dr. N14 14Mb 32
South Lodge Rd. KT12: Hers 81W 170
South London Crematorium 68Lb 134
South London Gallery 53Ub 113
South London Theatre 62Rb 135
(off Norwood High St.)
South Mall N9 20Wb 33
(off Plevna Rd.)
South Mall RM17: Grays 51Ce 121
(off Grays Shop. Cen.)
South Mall SW18 58Db 111
South Mall TW18: Staines 63H 127
(within The Elmsleigh Cen.)
South Mead KT19: Ewe 80Va 154
South Mead NW9 25Va 48
South Mead RH1: Redh 3P 207
Southmead Cres. EN8: Chesh 2Ac 20
Southmead Gdns. TW11: Tedd 65Ja 130
South Meadow La. SL4: Eton 1G 102
Southmead Rd. SW19 60Ab 110
Southmere Boating Cen. 47Xc 95
Southmere Dr. SE2 47Zc 95
Southmere Ho. E15 40Dc 72
(off Highland Street)
SOUTH MERSTHAM 2C 208
South Merton Station (Rail) 69Bb 133
South Mill Apts. E2 1K 219 (39Vb 71)
(off Hebden St.)
SOUTH MIMMS 4Wa 16
SOUTH MIMMS SERVICE AREA 6Xa 16
South Molton La. W1 3K 221 (44Kb 90)
South Molton Rd. E16 44Jc 93
South Molton St. W1 3K 221 (44Kb 90)
Southmont Rd. KT10: Hin W 75Ga 152
Southmoor Way E9 37Bc 72
South Mt. N20 19Eb 31
(off High Rd.)
SOUTH NORWOOD 70Vb 135
South Norwood Country Pk. 70Yb 136
South Norwood Country Pk. Vis. Cen. 70Xb 135
South Norwood Hill SE25 67Ub 135
South Norwood Leisure Cen. 71Xb 157
SOUTH NUTFIELD 8F 208
SOUTH OCKENDON 44Wd 98
South Oak Rd. SW16 63Pb 134
Southold Ri. SE9 62Pc 138
Southolm St. SW11 53Kb 112
South Ordnance Rd. EN3: Enf L 10Cc 20
Southover BR1: Broml 64Jc 137
Southover N12 20Cb 31
SOUTH OXHEY 21Y 45
South Pde. HA8: Edg 26Qa 47
South Pde. RH1: Mers 100Lb 196
South Pde. RM15: S Ock 43Yd 98
South Pde. SM6: W'gton 79Lb 156
South Pde. SW3 7C 226 (50Fb 89)
South Pde. W4 49Ta 87
SOUTH PARK 9N 209
SOUTH PARK 9J 207
South Pk. SL9: Ger X 29B 42
South Pk. TN13: S'oaks 97Kd 203
South Pk. Av. WD3: Chor 15H 25
South Pk. Ct. BR3: Beck 66Cc 136
South Pk. Ct. SL9: Ger X 29B 42
(off South Pk.)
South Pk. Cres. IG1: Ilf 34Tc 74
South Pk. Cres. SE6 60Gc 115
South Pk. Cres. SL9: Ger X 28A 42
South Pk. Dr. IG11: Bark 36Uc 74
South Pk. Dr. IG3: Ilf 33Uc 74
South Pk. Dr. SL9: Ger X 28A 42
South Pk. Gro. KT3: N Mald 70Sa 131
South Pk. Hill Rd. CR2: S Croy 78Tb 157
South Pk. La. RH1: Blet 8N 209
South Pk. M. SW6 55Db 111
South Pk. Rd. IG1: Ilf 34Tc 74
South Pk. Rd. SW19 65Cb 133
South Pk. Ter. IG1: Ilf 34Uc 74
South Pk. Vw. SL9: Ger X 28B 42
South Pk. Vs. IG3: Ilf 35Uc 74
South Pk. Way HA4: Ruis 37Y 65
South Path SL4: Wind 3G 102
South Pl. EC2 7G 219 (43Tb 91)
South Pl. EN3: Pond E 15Yb 34
South Pl. EN9: Walt A 5Ec 20
South Pl. KT5: Surb 73Pa 153
South Pl. M. EC2 1G 225 (43Tb 91)
Southport Rd. SE18 49Tc 94
South Quay Plaza E14 47Dc 92
South Quay Sq. E14 47Dc 92
South Quay Station (DLR) 47Dc 92
South Ridge KT13: Weyb 82R 170
Southridge Pl. SW20 66Za 132
South Riding AL2: Brick W 2Ba 13
South Ri. SM5: Cars 81Gb 175
South Ri. W2 4E 220 (45Gb 89)
(off St George's Flds.)
South Ri. Way SE18 50Tc 94
South Rd. KT13: Weyb Birchwood Ct. 78S 150
South Rd. RM6: Chad H Dunmow Cl. 29Yc 55
South Rd. RM6: Chad H Mill La. 30Ad 55
South Rd. KT13: Weyb West Rd. 81R 170
South Rd. DA8: Erith 52Hd 118
South Rd. GU21: Wok 7N 167
South Rd. GU24: Bisl 8D 166
South Rd. HA1: Harr 32Ja 66
South Rd. HA8: Edg 25Ra 47
South Rd. N9 18Wb 33
South Rd. RH2: Reig 7K 207
South Rd. RM15: S Ock 43Yd 98
(not continuous)
South Rd. SE23 61Zb 136
South Rd. SW19 65Eb 133
South Rd. TW12: Hamp 65Aa 129
South Rd. TW13: Hanw 64Z 129
South Rd. TW2: Twick 62Fa 130
South Rd. TW20: Eng G 5N 125
South Rd. TW5: Hest 51Z 107
South Rd. UB1: S'hall 47Ba 85
South Rd. UB7: W Dray 48Q 84
South Rd. W5 49Ma 87
South Rd. WD3: Chor 15E 24
South Row SE3 54Hc 115
South Row SL3: Ful 35A 62
SOUTH RUISLIP 35Y 65
South Ruislip Station (Rail & Underground) 36Y 65
Southsea Av. WD18: Wat 14W 26
Southsea Ho. RM3: Rom 22Md 57
(off Darlington Gdns.)
Southsea Rd. KT1: King T 70Na 131
South Sea St. SE16 48Bc 92
Southside N7 35Mb 70
South Side EN9: Walt A 2Kc 21
South Side KT16: Chert 69J 127
South Side N15 28Vb 51
South Side SL9: Chal P 27A 42
South Side W6 48Va 88
Southside Cl. UB10: Uxb 38N 63
Southside Comn. SW19 65Ya 132
Southside Halls SW7 3C 226 (48Fb 89)
(off Prince's Gdns.)
Southside House 65Ya 132
Southside Ind. Est. SW8 53Lb 112
(off Havelock Ter.)
Southside Shop. Cen. 58Db 111
Southspring DA15: Sidc 59Tc 116
South Sq. NW11 30Db 49
South Sq. WC1 1K 223 (43Qb 90)
South Stand N5 35Rb 71
South Station App. RH1: S Nut 8E 208
SOUTH STIFFORD 51Zd 121
South St. BR1: Broml 68Jc 137
South St. CM14: B'wood 19Yd 40
South St. DA12: Grav'nd 9D 122
South St. EN3: Pond E 15Yb 34
South St. KT18: Eps 85Ta 173
South St. RM1: Rom 29Gd 56
South St. RM13: Rain 40Ed 76
South St. TW18: Staines 64H 127
South St. TW7: Isle 55Ja 108
South St. W1 6J 221 (46Jb 90)
SOUTH STREET 91Qc 200
South St. Studios BR1: Broml 68Jc 137
(off South St.)
South Tadworth Farm Cl. KT20: Tad 92Xa 194
South Tenter St. E1 4K 225 (45Vb 91)
South Ter. DA4: Farni 73Pd 163
South Ter. KT6: Surb 72Na 153
South Ter. SL4: Wind 3J 103
South Ter. SW7 5D 226 (49Gb 89)
SOUTH TOTTENHAM 29Vb 51
South Tottenham Station (Overground) 29Vb 51
South Va. HA1: Harr 35Ga 66
South Va. SE19 65Ub 135
Southvale Rd. SE3 54Gc 115
South Vw. BR1: Broml 68Lc 137
South Vw. KT19: Eps 82Qa 173
South Vw. RM16: Ors 3D 100
South Vw. SL4: Eton 10F 80
South Vw. SL4: Eton W 10F 80
South Vw. SW19 65Za 132
Southview Av. NW10 36Va 68
Southview Av. RM18: Tilb 3C 122
South Vw. Cl. DA5: Bexl 58Bd 117
Southview Cl. BR8: Swan 70Jd 140
Southview Cl. SW17 64Jb 134
South Vw. Ct. GU22: Wok 90A 168
South Vw. Ct. SE19 66Sb 135
Southview Cres. IG2: Ilf 30Rc 54
South Vw. Dr. E18 27Kc 53
South Vw. Dr. RM14: Upm 34Qd 77
Southview Gdns. SM6: W'gton 80Lb 156
South Vw. Hgts. RM20: Grays 51Yd 120
Southview Pde. RM13: Rain 41Fd 96
South Vw. Rd. DA2: Wilm 62Md 141
South Vw. Rd. HA5: Pinn 23X 45
South Vw. Rd. IG10: Lough 16Pc 36
South Vw. Rd. KT21: Asht 91Ma 193
South Vw. Rd. N8 27Mb 50
South Vw. Rd. RM20: Grays 51Yd 120
Southview Rd. BR1: Broml 63Fc 137
Southview Rd. CR3: Wold 96Dc 198
Southview Rd. CR6: W'ham 91Wb 197
Southviews CR2: Sels 82Zb 178
South Vw. Vs. HP4: Berk 2A 2
South Vs. NW1 37Mb 70
Southville SW8 53Mb 112
Southville Cl. KT19: Ewe 81Ta 173
Southville Cl. TW14: Bedf 60U 106
(not continuous)
Southville Cl. TW14: Felt 60U 106
(not continuous)
Southville Cres. TW14: Felt 60U 106
Southville Rd. KT7: T Ditt 73Ja 152
Southville Rd. TW14: Felt 60U 106
South Wlk. BR4: W W'ck 76Gc 159
South Wlk. RH2: Reig 6K 207
South Wlk. UB3: Hayes 43T 84
SOUTHWARK 7C 224 (46Rb 91)
Southwark Bri. SE1 5E 224 (45Sb 91)
Southwark Bri. Bus. Cen. SE1 7E 224 (46Sb 91)
(off Southwark Bri. Rd.)
Southwark Bri. Rd. SE1 3C 230 (48Rb 91)
Southwark Cathedral 6F 225 (46Tb 91)
Southwark Ho. WD6: Bore 12Qa 29
(off Stratfield Rd.)
Southwark Pk. 48Xb 91
Southwark Pk. Est. SE16 49Xb 91
(off Southwark Pk. Rd.)
Southwark Pk. Rd. SE16 5K 231 (49Vb 91)
Southwark Pk. Sports Cen. Track 49Yb 92
Southwark Pk. Sports Complex 49Yb 92
Southwark Pl. BR1: Broml 69Pc 138
Southwark Playhouse 3D 230 (48Sb 91)
(off Newington C'way)
Southwark St George's RC Cathedral 3A 230 (48Qb 90)
Southwark Station (Underground) 7B 224 (46Rb 91)
Southwark St. SE1 6C 224 (46Rb 91)
Southwater Cl. BR3: Beck 66Dc 136
Southwater Cl. E14 44Bc 92
Southway BR2: Hayes 73Jc 159
Southway N20 19Cb 31
Southway NW11 30Db 49
Southway SM5: Cars 82Fb 175
Southway SM6: W'gton 77Lb 156
Southway SW20 71Ya 154
South Way AL10: Hat 4B 8
South Way AL9: Hat 3D 8
South Way CR0: C'don 76Ac 158
South Way EN9: Walt A 8Ec 20
South Way HA2: Harr 28Ca 45
South Way HA9: Wemb 36Qa 67
South Way N11 23Lb 50
South Way N9 19Yb 34
South Way RM19: Purf 48Ud 98
South Way WD5: Ab L 5T 12
Southway Cl. W12 47Xa 88
SOUTH WEALD 19Ud 40
South Weald Dr. EN9: Walt A 5Fc 21
South Weald Rd. CM14: B'wood 20Wd 40
Southwell Av. UB5: N'olt 37Ca 65
Southwell Cl. RM16: Chaf H 50Yd 98
Southwell Gdns. SW7 5A 226 (49Eb 89)
Southwell Gro. Rd. E11 33Gc 73
Southwell Ho. SE16 49Xb 91
(off Anchor St.)
Southwell Rd. CR0: C'don 72Qb 156
Southwell Rd. HA3: Kenton 30Ma 47
Southwell Rd. SE5 55Sb 113
South Western Rd. TW1: Twick 58Ja 108
South W. India Dock Entrance E14 47Ec 92
SOUTH WEST LONDON ELECTIVE ORTHOPAEDIC CEN. 87Sa 173
(within Epsom General Hospital)
South W. Middlesex Crematorium 60Aa 107
Southwest Rd. E11 32Fc 73
South Wharf Rd. W2 2B 220 (44Fb 89)
Southwick M. W2 2C 220 (44Fb 89)
Southwick Pl. W2 3D 220 (44Gb 89)
Southwick St. W2 2D 220 (44Gb 89)
Southwick Yd. W2 3D 220 (44Gb 89)
(off Titchborne Row)
SOUTH WIMBLEDON 65Db 133
South Wimbledon Station (Underground) 66Db 133
Southwold Dr. IG11: Bark 36Wc 75
Southwold Mans. W9 41Cb 89
(off Widley Rd.)
Southwold Rd. DA5: Bexl 58Dd 118
Southwold Rd. E5 33Xb 71
Southwold Rd. WD24: Wat 9Y 13
Southwold Spur SL3: L'ly 47E 82
Southwood Av. CR5: Coul 87Lb 176
Southwood Av. GU21: Knap 10H 167
Southwood Av. KT16: Ott 80E 148
Southwood Av. KT2: King T 67Sa 131
Southwood Av. N6 31Kb 70
Southwood Cl. BR1: Broml 70Pc 138
Southwood Cl. KT4: Wor Pk 74Za 154
Southwood Ct. EC1 4B 218 (41Rb 91)
(off Wynyatt St.)
Southwood Ct. KT13: Weyb 78R 150
Southwood Ct. NW11 29Db 49
Southwood Dr. KT5: Surb 73Sa 153
SOUTH WOODFORD 26Jc 53
South Woodford Station (Underground) 26Kc 53
South Woodford to Barking Relief Rd. E11 29Mc 53
Southwood Gdns. IG2: Ilf 28Rc 54
Southwood Gdns. KT10: Hin W 76Ja 152
Southwood Hall N6 30Kb 50
Southwood Hgts. N6 31Kb 70
Southwood Ho. W11 45Ab 88
(off Avondale Pk. Rd.)
Southwood La. N6 31Jb 70
Southwood Lawn Rd. N6 31Jb 70
Southwood Mans. N6 30Jb 50
(off Southwood La.)
Southwood Pk. N6 31Jb 70
Southwood Rd. SE28 46Xc 95
Southwood Rd. SE9 61Rc 138
Southwood Smith Ho. E2 41Xb 91
(off Florida St.)
Southwood Smith St. N1 39Rb 71
South Worple Av. SW14 55Ua 110
South Worple Way SW14 55Ta 109
Southwyck Ho. SW9 56Rb 113
Soval Ct. HA6: Nwood 24T 44
Sovereign Beeches SL2: Farn C 7F 60
Sovereign Bus. Cen. EN3: Brim 13Bc 34
Sovereign Cl. CR8: Purl 82Pb 176
Sovereign Cl. E1 45Xb 91
Sovereign Cl. HA4: Ruis 32U 64
Sovereign Cl. W5 43La 86
Sovereign Ct. CR2: S Croy 78Sb 157
(off Warham Rd.)
Sovereign Ct. DA4: Sut H 67Rd 141
(off Ship La.)
Sovereign Ct. HA6: Nwood 25W 44
Sovereign Ct. HA7: Stan 24Ma 47
Sovereign Ct. KT22: Lea 92Ha 192
Sovereign Ct. KT8: W Mole 70Ba 129
Sovereign Ct. SL5: S'dale 3F 146
Sovereign Ct. TW3: Houn 55Ca 107
Sovereign Ct. W8 48Db 89
(off Wright's La.)
Sovereign Ct. WD18: Wat 14W 26
Sovereign Cres. SE16 45Ac 92
Sovereign Gro. HA0: Wemb 34Ma 67
Sovereign Hgts. SL3: Dat 51C 104
Sovereign Ho. E1 42Xb 91
(off Cambridge Heath Rd.)
Sovereign Ho. SE18 48Pc 94
(off Leda Rd.)
Sovereign Ho. TW15: Ashf 63N 127
Sovereign M. E2 1K 219 (40Vb 71)
Sovereign M. EN4: Cockf 13Hb 31
Sovereign M. SL5: Asc 9A 124
Sovereign Pk. AL4: St A 3H 7
Sovereign Pk. HP2: Hem H 1B 4
Sovereign Pk. NW10 42Ra 87
Sovereign Pk. Trad. Est. NW10 42Ra 87
Sovereign Pl. HA1: Harr 29Ha 46
Sovereign Pl. IG10: Lough 12Pc 36
Sovereign Rd. IG11: Bark 41Yc 95
Sovereign Way AL3: St A 2B 6
(off Chequer St.)
Sowerby Cl. SE9 57Pc 116
Sowrey Av. RM13: Rain 37Hd 76
Soyer Ct. GU21: Wok 10J 167
Spa at Beckenham, The 67Bc 136
Spa Bus. Pk. SE16 48Wb 91
Space Arts Cen., The 49Cc 92
(off Westferry Rd.)
Space Bus. Pk. NW10 41Ra 87
Spaces Bus. Cen. SW8 53Lb 112
Space Waye TW14: Felt 57W 106
Spackmans Way SL1: Slou 8G 80
Spa Cl. SE25 67Ub 135
Spa Ct. SE16 48Wb 91
Spa Ct. SW16 63Pb 134
Spa Dr. KT18: Eps 86Qa 173
Spafield St. EC1 5K 217 (42Qb 90)
Spa Grn. Est. EC1 3B 218 (41Rb 91)
Spa Hill SE19 67Tb 135
Spalding Cl. HA8: Edg 24Ua 48
Spalding Ho. SE4 56Ac 114
Spalding Rd. NW4 31Ya 68
Spalding Rd. SW17 64Kb 134
Spalt Cl. CM13: Hut 19De 41
Spanbrook IG7: Chig 20Rc 36
Spanby Rd. E3 42Cc 92
Spaniards Cl. NW11 32Fb 69
Spaniards End NW3 32Eb 69
Spaniards Rd. NW3 33Eb 69
Spanish Pl. W1 2J 221 (44Jb 90)
Spanish Rd. SW18 57Eb 111
Spanswick Lodge N15 28Rb 51
Spareleaze Hill IG10: Lough 14Pc 36
Sparepenny La. DA4: Eyns 75Md 163
Sparepenny La. DA4: Farni 75Md 163
Spare St. SE17 6D 230 (49Sb 91)
Sparkbridge Rd. HA1: Harr 28Ga 46
Sparkbrook Way EN3: Enf W 9Zb 20
Sparkes Cl. BR2: Broml 70Kc 137
Sparkes Cotts. SW1 6H 227 (49Jb 90)
(off Graham Ter.)
Sparkford Gdns. N11 22Jb 50
Sparkford Ho. SW11 53Fb 111
(off Battersea Church Rd.)
Sparks Cl. RM8: Dag 33Zc 75
Sparks Cl. TW12: Hamp 65Aa 129
Sparks Cl. W3 44Ta 87
Spa Rd. SE16 4K 231 (48Vb 91)
Sparrick's Row SE1 1G 231 (47Tb 91)
Sparrow Cl. TW12: Hamp 65Aa 129
Sparrow Dr. BR5: Farnb 74Sc 160
Sparrow Farm Dr. TW14: Felt 59Y 107
Sparrow Farm Rd. KT17: Ewe 77Wa 154
Sparrow Grn. RM10: Dag 34Dd 76
Sparrowhawk Pl. AL10: Hat 1C 8
Sparrow Ho. E1 42Yb 92
(off Cephas Av.)
Sparrow Row GU24: Chob 9F 166
SPARROW ROW 9F 146
Sparrow's Farm Leisure Cen. 59Sc 116
Sparrows Herne WD23: Bush 17Da 27
Sparrows La. SE9 59Sc 116
Sparrows Mead RH1: Redh 3A 208
Sparrows Way WD23: Bush 17Ea 28
Sparrow Wlk. WD25: Wat 7W 12
Sparsholt Cl. IG11: Bark 39Uc 74
(off St John's Rd.)
Sparsholt Rd. IG11: Bark 39Uc 74
Sparsholt Rd. N19 32Pb 70
Spartan Cl. SM6: W'gton 80Nb 156
Sparta St. SE10 53Ec 114
Sparvell Rd. GU21: Knap 1F 186
Spa SPC 1H 81
Speakers' Corner 4G 221 (45Hb 89)
Speakers Ct. CR0: C'don 74Tb 157
Speakman Ho. SE4 55Ac 114
(off Arica Rd.)
Spearman Ho. E14 44Cc 92
(off Upper Nth. St.)

Spearman St. SE18 ....51Qc **116**
Spear M. SW5 ....49Cb **89**
Spearpoint Gdns. IG2: Ilf 29Vc **55**
Spears Cl. CR6: W'ham 89Zb **178**
Spears Rd. N19 ....32Nb **70**
Speart La. TW5: Hest 52Aa **107**
Spectacle Works E13 ....41Lc **93**
Spectrum Bldg., The N1 ....3G **219** (41Tb **91**)
....*(off East Rd.)*
Spectrum Pl. SE17 ....51Tb **113**
....*(off Lytham St.)*
Spectrum Twr. IG1: Ilf 33Sc **74**
....*(off Hainault St.)*
Spectrum Way SW18 ....57Cb **111**
Spedan Cl. NW3 ....34Eb **69**
Speechly M. E8 ....36Vb **71**
Speedbird Way UB7: Harm 52K **105**
SPEEDGATE 75Vd **164**
Speedgate Hill DA3: Fawk 75Wd **164**
Speed Highwalk EC2 ....7F **219** (43Tb **91**)
....*(off Silk St.)*
Speed Ho. EC2 ....7F **219** (43Tb **91**)
....*(off Silk St.)*
Speedway Ind. Est. UB3: Hayes 47T **84**
Speedwell Cl. HP1: Hem H 3G **2**
Speedwell Ct. RM17: Grays 2A **122**
Speedwell Ho. N12 ....21Db **49**
Speedwell St. SE8 ....52Cc **114**
Speedy Pl. WC1 ....4F **217** (41Nb **90**)
....*(off Cromer St.)*
Speer Rd. KT7: T Ditt 72Ha **152**
Speirs Cl. KT3: N Mald 72Va **154**
Speirs Gdns. RM8: Dag 32Ad **75**
Spekehill SE9 ....62Pc **138**
Speke Rd. CR7: Thor H 68Tb **135**
Speke's Monument 6B **220** (46Fb **89**)
Speldhurst Cl. BR2: Broml 71Hc **159**
Speldhurst Rd. E9 ....38Zb **72**
Speldhurst Rd. W4 ....48Ta **87**
Spellbrook Wlk. N1 ....39Sb **71**
Spelman Ho. E1 ....43Wb **91**
....*(off Spelman St.)*
Spelman St. E1 ....43Wb **91**
....(not continuous)
Spelthorne Gro. TW16: Sun 66V **128**
Spelthorne La. TW15: Ashf 67S **128**
Spelthorne Leisure Cen. 64J **127**
Spelthorne Mus. 64H **127**
Spence Av. KT14: Byfl 86N **169**
Spence Cl. SE16 ....47Bc **92**
Spencer Av. N13 ....23Pb **50**
Spencer Av. UB4: Hayes 43W **84**
Spencer Cl. BR6: Orp 75Uc **160**
Spencer Cl. CM16: Epp 1Xc **23**
Spencer Cl. GU21: Wok 85E **168**
Spencer Cl. IG8: Wfd G 22Lc **53**
Spencer Cl. KT18: Eps D 91Ua **194**
Spencer Cl. N3 ....26Cb **49**
Spencer Cl. NW10 ....41Pa **87**
Spencer Cl. UB8: Cowl 41L **83**
Spencer Cl. WD7: R'lett 9Ha **14**
Spencer Ct. BR6: Farnb 78Sc **160**
Spencer Ct. DA12: Grav'nd 8F **122**
Spencer Ct. KT22: Lea 95La **192**
Spencer Ct. NW8 ....40Eb **69**
....*(off Marlborough Pl.)*
Spencer Ct. SW20 ....67Xa **132**
Spencer Ctyd. N3 ....26Bb **49**
....*(off Regents Pk. Rd.)*
Spencer Dr. N2 ....30Eb **49**
Spencer Gdns. SE9 ....57Pc **116**
Spencer Gdns. SW14 ....57Sa **109**
Spencer Gdns. TW20: Eng G 4P **125**
Spencer Ga. AL1: St A 1C **6**
Spencer Hill SW19 ....65Ab **132**
Spencer Hill Rd. SW19 ....66Ab **132**
Spencer House 7B **222** (46Lb **90**)
....*(off St James's Pl.)*
Spencer Mans. W14 ....51Ab **110**
....*(off Queen's Club Gdns.)*
Spencer M. SW8 ....53Pb **112**
....*(off Lansdowne Way)*
Spencer M. W6 ....51Ab **110**
Spencer Pk. KT8: E Mos 71Ea **152**
Spencer Pk. SW18 ....57Fb **111**
SPENCER PARK 57Fb **111**
Spencer Pl. CR0: C'don 73Tb **157**
Spencer Pl. N1 ....38Rb **71**
Spencer Ri. NW5 ....35Kb **70**
Spencer Rd. CR4: Mitc Commonside E. 69Jb **134**
Spencer Rd. CR4: Mitc Wood St. 73Jb **156**
Spencer Rd. BR1: Broml 66Hc **137**
Spencer Rd. CR2: S Croy 78Ub **157**
Spencer Rd. CR3: Cat'm 93Tb **197**
Spencer Rd. E17 ....26Ec **52**
Spencer Rd. E6 ....39Mc **73**
Spencer Rd. HA0: Wemb 33La **66**
Spencer Rd. HA3: W'stone 26Ga **46**
Spencer Rd. IG3: Ilf 32Vc **75**
Spencer Rd. KT11: Cobh 87X **171**
Spencer Rd. KT8: E Mos 70Ea **130**
Spencer Rd. N11 ....21Kb **50**
Spencer Rd. N17 ....25Wb **51**
Spencer Rd. N8 ....29Pb **50**
....(not continuous)
Spencer Rd. RM13: Rain 41Fd **96**
Spencer Rd. SL3: L'ly 48B **82**
Spencer Rd. SW18 ....56Fb **111**
Spencer Rd. SW20 ....67Xa **132**
Spencer Rd. TW2: Twick 62Ga **130**
Spencer Rd. TW7: Isle 53Ea **108**
Spencer Rd. W3 ....46Sa **87**
Spencer Rd. W4 ....52Sa **109**
Spencers Cotts. TN15: Bor G 92Ce **205**
Spencer St. AL3: St A 2B **6**
Spencer St. DA11: Grav'nd 9C **122**
Spencer St. EC1 ....4B **218** (41Rb **91**)
Spencer St. UB2: S'hall 47Z **85**
Spencer Wlk. NW3 ....35Fb **69**
Spencer Wlk. RM18: Tilb 4D **122**
Spencer Wlk. SW15 ....56Za **110**
Spencer Wlk. WD3: Rick 15L **25**
Spencer Way E1 ....44Xb **91**
Spencer Yd. SE3 ....54Hc **115**
....*(off Tranquil Va.)*
Spenlow Ho. SE16 ....48Wb **91**
....*(off Jamaica Rd.)*
Spenser Av. KT13: Weyb 81Q **170**
Spenser Cres. RM14: Upm 31Sd **78**
Spenser Gro. N16 ....36Ub **71**
Spenser M. SE21 ....61Tb **135**
Spenser Rd. SE24 ....57Rb **113**
Spenser St. SW1 ....3C **228** (48Lb **90**)
Spens Ho. WC1 ....6H **217** (42Pb **90**)
....*(off Lamb's Conduit St.)*
Spensley Wlk. N16 ....34Tb **71**
Speranza St. SE18 ....50Vc **95**
Sperling Rd. N17 ....26Ub **51**
Spert St. E14 ....45Ac **92**
Speyhawk Pl. EN6: Pot B 1Eb **17**
Speyside N14 ....16Lb **32**
Spey St. E14 ....43Ec **92**
Spey Way RM1: Rom 24Gd **56**
Spezia Rd. NW10 ....40Wa **68**
Sphere, The E16 ....44Hc **93**
....*(off Hallsville Rd.)*
Sphere Ind. Est., The AL1: St A 2E **6**
Sphinx Way EN5: Barn 15Bb **31**
Spice Ct. E1 ....45Wb **91**
Spice Quay Hgts. SE1 ....7K **225** (46Vb **91**)
Spicer Cl. KT12: Walt T 72Y **151**
Spicer Cl. SW9 ....54Rb **113**
Spicer Ct. EN1: Enf 13Ub **33**
Spicer M. SL3: L'ly 9N **81**
Spicers Fld. KT22: Oxs 85Fa **172**
Spicer St. AL3: St A 2A **6**
Spice's Yd. CR0: C'don 77Sb **157**
Spielman Rd. DA1: Dart 56Pd **119**
Spiers Gdns. RM8: Dag 32Ad **75**
....*(off Ager Av.)*
Spigurnell Rd. N17 ....25Tb **51**
Spikes Bri. Moorings UB4: Yead 45Aa **85**
Spikes Bri. Rd. UB1: S'hall 44Aa **85**
Spilsby Rd. RM3: Rom 24Md **57**
Spindle Cl. KT19: Eps 81Ta **173**
Spindle Cl. SE18 ....48Nc **94**
Spindle M. BR6: Farnb 77Rc **160**
Spindles RM18: Tilb 2C **122**
Spindlewood Gdns. CR0: C'don 77Ub **157**
Spindlewoods KT20: Tad 94Xa **194**
Spindrift Av. E14 ....49Cc **92**
Spinel Cl. SE18 ....50Vc **95**
Spingate Cl. RM12: Horn 36Ld **77**
Spinnaker Cl. IG11: Bark 41Xc **95**
Spinnaker Cl. KT11: Cobh 86Z **171**
Spinnaker Ct. KT1: Hamp W 67Ma **131**
....*(off Becketts Pl.)*
Spinnaker Ho. E14 ....47Cc **92**
....*(off Byng St.)*
Spinnaker Ho. E16 ....47Kc **93**
....*(off Waypoint Way)*
Spinnaker Ho. SW18 ....56Eb **111**
....*(off Juniper Dr.)*
Spinnells Rd. HA2: Harr 32Ba **65**
Spinner Ho. E8 ....39Vb **71**
....*(off Lovelace St.)*
Spinners Wlk. SL4: Wind 3G **102**
Spinney SL1: Slou 6F **80**
Spinney, The BR8: Swan 68Gd **140**
Spinney, The CM13: Hut 16Ee **41**
Spinney, The CR8: Purl 83Rb **177**
Spinney, The DA14: Sidc 64Ad **139**
Spinney, The EN5: New Bar 12Db **31**
Spinney, The EN6: Pot B 3Fb **17**
Spinney, The EN7: Chesh 2Xb **19**
Spinney, The GU23: Send 99L **189**
Spinney, The HA0: Wemb 34Ja **66**
Spinney, The HA6: Nwood 23W **44**
Spinney, The HA7: Stan 21Na **47**
Spinney, The IG10: Lough 14Rc **36**
Spinney, The KT18: Head 98Sa **193**
Spinney, The KT18: Tatt C 91Xa **194**
Spinney, The KT22: Oxs 84Ea **172**
Spinney, The KT23: Bookh 96Da **191**
Spinney, The N21 ....17Qb **32**
Spinney, The RM16: Ors 2C **100**
Spinney, The SL5: S'dale 1C **146**
Spinney, The SL9: Ger X 2P **61**
Spinney, The SM3: Cheam 77Ya **154**
Spinney, The SW13 ....52Xa **110**
Spinney, The SW16 ....62Lb **134**
Spinney, The TW16: Sun 67W **128**
Spinney, The WD17: Wat 11W **26**
Spinney, The WD25: A'ham 10Fa **14**
Spinney Cl. BR3: Beck 70Dc **136**
Spinney Cl. KT11: Cobh 83Ca **171**
Spinney Cl. KT3: N Mald 71Ua **154**
Spinney Cl. KT4: Wor Pk 75Va **154**
Spinney Cl. RM13: Rain 40Gd **76**
Spinney Cl. SL5: Asc 8A **124**
Spinney Cl. UB7: Yiew 45N **83**
Spinneycroft KT22: Oxs 87Fa **172**
Spinney Dr. TW14: Bedf 59S **106**
Spinney Gdns. KT10: Esh 76Ca **151**
Spinney Gdns. RM9: Dag 36Ad **75**
Spinney Gdns. SE19 ....64Vb **135**
Spinney Hill KT15: Add 78G **148**
Spinney Oak BR1: Broml 68Nc **138**
Spinney Oak KT16: Ott 79F **148**
Spinney Oaks KT15: Add 78G **148**
Spinney Row AL2: Lon C 8F **6**
Spinneys, The BR1: Broml 68Pc **138**
Spinneys Dr. AL3: St A 4P **5**
Spinney Way TN14: Cud 83Tc **180**
Spinning Wheel Way SM6: W'gton 75Jb **156**
SPIRE BUSHEY HOSPITAL 17Ha **28**
Spire Cl. DA12: Grav'nd 10D **122**
Spire Ct. BR3: Beck 68Dc **136**
....*(off Crescent Rd.)*
Spire Ho. W2 ....4A **220** (45Eb **89**)
....*(off Lancaster Ga.)*
Spire Pl. CR6: W'ham 90Ac **178**
Spires, The CM14: B'wood 19Zd **41**
Spires, The DA1: Dart 61Md **141**
Spires, The HP2: Hem H 3M **3**
Spires Shop. Cen., The 13Ab **30**
Spirit Quay E1 ....46Wb **91**
SPITAL 6F **102**
SPITALFIELDS 7K **219** (43Vb **91**)
Spitalfields City Farm 42Wb **91**
Spital La. CM14: B'wood 20Vd **40**
Spital Sq. E1 ....7J **219** (43Ub **91**)
Spital St. DA1: Dart 58Md **119**
Spital St. E1 ....43Wb **91**
Spital Yd. E1 ....7J **219** (43Ub **91**)
Spitfire Bldg. N1 ....2H **217** (40Pb **70**)
....*(off Collier St.)*
Spitfire Bus. Pk. CR0: Wadd 79Qb **156**
Spitfire Cl. SL3: L'ly 49C **82**
Spitfire Est., The TW5: Cran 50Y **85**
Spitfire Rd. SM6: W'gton 80Nb **156**
Spitfire Rd. TW6: H'row A 58S **106**
Spitfire Way TW5: Cran 50Y **85**
Splendour Wlk. SE16 ....50Yb **92**
....*(off Verney Rd.)*
Spode Ho. SE11 ....4K **229** (48Qb **90**)
....*(off Lambeth Wlk.)*
Spode Wlk. NW6 ....36Db **69**
Spondon Rd. N15 ....28Wb **51**
Spoonbill Way UB4: Yead 43Z **85**
Spooner Ho. TW5: Hest 51Ca **107**
Spooners Dr. AL2: Park 9A **6**
Spooners M. W3 ....46Ta **87**
Spooner Wlk. SM6: W'gton 78Nb **156**
Sporle Rd. Ct. SW11 ....55Fb **111**
Sports Academy (LSBU) 3C **230** (48Rb **91**)
....(off London Rd.)
Sportsbank St. SE6 ....59Ec **114**
Sports Direct Fitness Croydon 80Qb **156**
Sports Direct Fitness Epsom 77Ta **153**
SportsDock 45Rc **94**
Sportsman Pl. E2 ....39Wb **71**
Sportspace Hemel Hempstead 4L **3**
Sportspace Longdean 5B **4**
Sportspace Athletics Track 4P **3**
Spottiswood Ct. CR0: C'don 72Sb **157**
....*(off Harry Cl.)*
Spottons Gro. N17 ....25Sb **51**
Spout Hill CR0: Addtn 78Cc **158**
Spout La. TW19: Stanw M 57J **105**
Spout La. Nth. TW19: Stanw M 56K **105**
Spratt Hall Rd. E11 ....30Jc **53**
Spratts All. KT16: Ott 79G **148**
Spratts La. KT16: Ott 79G **148**
Spray La. TW2: Whitt 58Ga **108**
Spray St. SE18 ....49Rc **94**
Spread Eagle Wlk. KT19: Eps 85Ta **173**
Spreighton Rd. KT8: W Mole 70Da **129**
Spriggs Ct. CM16: Epp 1Wc **23**
....*(off Palmers Hill)*
Spriggs Ho. N1 ....38Rb **71**
....*(off Canonbury Rd.)*
Spriggs Oak CM16: Epp 1Wc **23**
....*(off Palmers Hill)*
Sprimont Pl. SW3 ....7F **227** (50Hb **89**)
Springall St. SE15 ....52Xb **113**
Springalls Wharf SE16 ....47Wb **91**
....*(off Bermondsey Wall W.)*
Spring Apts. CR0: C'don 75Ub **157**
....*(off Addiscombe Gro.)*
Spring Apts. E14 ....49Ec **92**
....*(off Stebondale St.)*
Springate Fld. SL3: L'ly 47A **82**
Spring Av. TW20: Egh 65A **126**
Springbank N21 ....16Pb **32**
Springbank Av. RM12: Horn 36Ld **77**
Springbank Rd. SE13 ....58Fc **115**
Springbank Wlk. NW1 ....38Mb **70**
Springbottom La. RH1: Blet 99Qb **196**
Springbourne Ct. BR3: Beck 67Ec **136**
Spring Bri. M. W5 ....45Ma **87**
Spring Bri. Rd. W5 ....45Ma **87**
Spring Cl. EN5: Barn 15Za **30**
Spring Cl. HP5: Lat 8A **10**
Spring Cl. RM8: Dag 32Zc **75**
Spring Cl. UB9: Hare 25M **43**
Spring Cl. WD6: Bore 11Qa **29**
Springclose La. SM3: Cheam 79Ab **154**
Springcopse Rd. RH2: Reig 8L **207**
Spring Cnr. TW13: Felt 62W **128**
Spring Cotts. KT6: Surb 71Ma **153**
Spring Ct. KT17: Ewe 81Va **174**
Spring Ct. NW6 ....37Bb **69**
Spring Ct. W7 ....45Fa **86**
Spring Ct. Rd. EN2: Enf 10Qb **18**
Springcroft DA3: Hartl 72Ce **165**
Springcroft Av. N2 ....28Hb **49**
Spring Crofts WD23: Bush 15Ca **27**
Spring Cross DA3: Nw A G 76Ce **165**
....(not continuous)
Springdale M. N16 ....35Tb **71**
Springdale Rd. N16 ....35Tb **71**
Spring Dr. HA5: Eastc 30W **44**
Springer Ct. E3 ....42Ec **92**
....*(off Navigation Rd.)*
Springett Ho. SW2 ....57Qb **112**
....*(off St Matthews Rd.)*
Springfarm Cl. RM13: Rain 41Md **97**
Springfield CM16: Epp 4Vc **23**
Springfield E5 ....32Xb **71**
Springfield GU18: Light 3B **166**
Springfield RH8: Oxt 2H **211**
Springfield SE25 ....69Wb **135**
Springfield SL1: Slou 8M **81**
Springfield WD23: B Hea 18Fa **28**
Springfield Av. BR8: Swan 70Hd **140**
Springfield Av. CM13: Hut 17Fe **41**
Springfield Av. N10 ....27Lb **50**
Springfield Av. SW20 ....69Bb **133**
Springfield Av. TW12: Hamp 65Da **129**
Springfield Cl. EN6: Pot B 3Gb **17**
Springfield Cl. GU21: Knap 10J **167**
Springfield Cl. HA7: Stan 20Ja **28**
Springfield Cl. N12 ....22Db **49**
Springfield Cl. SL4: Wind 4F **102**
Springfield Cl. WD3: Crox G 15R **26**
Springfield Ct. IG1: Ilf 36Rc **74**
Springfield Ct. KT1: King T 69Na **131**
....*(off Springfield Rd.)*
Springfield Ct. NW3 ....38Gb **69**
....*(off Eton Av.)*
Springfield Ct. RM14: Upm 34Sd **78**
Springfield Ct. SM6: W'gton 78Kb **156**
Springfield Ct. WD3: Rick 18K **25**
Springfield Dr. IG2: Ilf 29Sc **54**
Springfield Dr. KT22: Lea 91Ga **192**
Springfield Gdns. BR1: Broml 70Pc **138**
Springfield Gdns. BR4: W W'ck 75Dc **158**
Springfield Gdns. E5 ....32Xb **71**
Springfield Gdns. HA4: Ruis 32X **65**
Springfield Gdns. IG8: Wfd G 24Lc **53**
Springfield Gdns. NW9 ....29Ta **47**
Springfield Gdns. RM14: Upm 34Rd **77**
Springfield Gro. SE7 ....51Lc **115**
Springfield Gro. TW16: Sun 67V **128**
Springfield La. KT13: Weyb 77R **150**
Springfield La. NW6 ....39Db **69**
Springfield Mdws. KT13: Weyb 77R **150**
Springfield Mt. NW9 ....29Ua **48**
Springfield Pde. M. N13 ....21Qb **50**
Springfield Pk. 61Fb **133**
Springfield Pl. KT3: N Mald 70Sa **131**
Springfield Pl. SL9: Ger X 29A **42**
Springfield Ri. SE26 ....62Xb **135**
Springfield Rd. AL1: St A 3E **6**
Springfield Rd. AL4: S'ford 2M **7**
Springfield Rd. BR1: Broml 70Pc **138**
Springfield Rd. CR7: Thor H 67Sb **135**
Springfield Rd. DA16: Well 55Xc **117**
Springfield Rd. DA7: Bex 56Dd **118**
Springfield Rd. E15 ....41Gc **93**
Springfield Rd. E17 ....30Bc **52**
Springfield Rd. E4 ....18Gc **35**
Springfield Rd. E6 ....38Pc **74**
Springfield Rd. EN8: Chesh 4Ac **20**
Springfield Rd. HA1: Harr 30Ga **46**
Springfield Rd. HP2: Hem H 1P **3**
Springfield Rd. KT1: King T 69Na **131**
Springfield Rd. KT17: Ewe 82Ya **174**
Springfield Rd. KT22: Lea 91Ha **192**
Springfield Rd. N11 ....22Kb **50**
Springfield Rd. N15 ....28Wb **51**
Springfield Rd. NW8 ....1A **214** (39Eb **69**)
Springfield Rd. RM16: Grays 47Fe **99**
Springfield Rd. SE26 ....64Xb **135**
Springfield Rd. SL3: L'ly 52D **104**
Springfield Rd. SL4: Wind 4F **102**
Springfield Rd. SM6: W'gton 78Kb **156**
Springfield Rd. SW19 ....64Bb **133**
Springfield Rd. TW11: Tedd 64Ja **130**
Springfield Rd. TW15: Ashf 64P **127**
Springfield Rd. TW2: Whitt 60Ca **107**
Springfield Rd. UB4: Yead 46Y **85**
Springfield Rd. W7 ....46Ga **86**
Springfield Rd. WD25: Wat 5X **13**
Springfield Rd. Link DA10: Ebbs 59De **121**
Springfields EN5: New Bar 15Db **31**
....*(off Somerset Rd.)*
Springfields EN9: Walt A 6Gc **21**
Springfields Cl. KT16: Chert 74K **149**
SPRINGFIELD UNIVERSITY HOSPITAL 61Gb **133**
Springfield Wlk. BR6: Orp 74Tc **160**
....*(off Place Farm Av.)*
Springfield Wlk. NW6 ....39Db **69**
Spring Gdns. BR6: Chels 79Xc **161**
Spring Gdns. IG8: Wfd G 24Lc **53**
Spring Gdns. KT8: W Mole 71Da **151**
Spring Gdns. N5 ....36Sb **71**
Spring Gdns. RM12: Horn 35Kd **77**
Spring Gdns. RM7: Rom 29Ed **56**
Spring Gdns. SM6: W'gton 78Lb **156**
Spring Gdns. SW1 ....6E **222** (46Mb **90**)
....(not continuous)
Spring Gdns. TN16: Big H 90Lc **179**
Spring Gdns. WD25: Wat 7Y **13**
Spring Gdns. Bus. Pk. RM7: Rom 30Ed **56**
Spring Glen AL10: Hat 1B **8**
Spring Gro. CR4: Mitc 67Jb **134**
Spring Gro. DA12: Grav'nd 10D **122**
Spring Gro. IG10: Lough 16Mc **35**
Spring Gro. KT22: Fet 95Da **191**
Spring Gro. SE19 ....66Vb **135**
Spring Gro. TW12: Hamp 67Da **129**
Spring Gro. W4 ....50Qa **87**
Spring Gro. W7 ....45Ga **86**
SPRING GROVE 53Ga **108**
Spring Gro. Cres. TW3: Houn 53Ea **108**
Spring Gro. Rd. TW10: Rich 57Pa **109**
Spring Gro. Rd. TW3: Houn 53Da **107**
Spring Gro. Rd. TW3: Isle 53Da **107**
Spring Gro. Rd. TW7: Isle 53Da **107**
Spring Head Cl. TN15: Kems'g 89Pd **183**
Springhead Ent. Pk. DA11: Nflt 60De **121**
Springhead Parkway DA11: Nflt 60De **121**
Spring Head Rd. TN15: Kems'g 89Pd **183**
Springhead Rd. DA11: Nflt 61Ee **143**
Springhead Rd. DA8: Erith 51Hd **118**
SpringHealth Leisure Club
Richmond 56Ma **109**
....(within Pools on the Pk.)
Spring Hill E5 ....31Wb **71**
Spring Hill SE26 ....63Yb **136**
Springhill Cl. SE5 ....55Tb **113**
Springholm Cl. TN16: Big H 90Lc **179**
Spring Ho. E17 ....26Ec **52**
....*(off Fulbourne Rd.)*
Spring Ho. WC1 ....4K **217** (41Qb **90**)
....*(off Margery St.)*
Springhouse La. SS17: Corr 1P **101**
Springhurst Cl. CR0: C'don 77Bc **158**
Spring Lake HA7: Stan 21Ka **46**
Spring La. E5 ....31Xb **71**
Spring La. HP1: Hem H 1H **3**
Spring La. N10 ....27Jb **50**
Spring La. RH8: Oxt 3H **211**
Spring La. SE25 ....72Xb **157**
Spring La. SL1: Slou 6D **80**
Spring La. SL2: Farn R 8F **60**
Spring La. TN15: Igh 94Xd **204**
Spring M. KT17: Ewe 81Va **174**
Spring M. SE11 ....7H **229** (50Pb **90**)
Spring M. TW9: Rich 56Na **109**
....*(off Rosedale Rd.)*
Spring M. W1 ....7G **215** (43Hb **89**)
SPRING PARK 76Cc **158**
Spring Pk. Av. CR0: C'don 75Zb **158**
Spring Pk. Dr. N4 ....32Sb **71**
Springpark Dr. BR3: Beck 69Ec **136**
Spring Pk. Rd. CR0: C'don 75Zb **158**
Spring Pas. SW15 ....55Za **110**
Spring Path NW3 ....36Fb **69**
Spring Pl. IG11: Bark 40Sc **74**
Spring Pl. KT11: Cobh 85Aa **171**
Spring Pl. N3 ....27Cb **49**
Spring Pl. NW5 ....36Kb **70**
Springpond Rd. RM9: Dag 36Ad **75**
Spring Prom. UB7: W Dray 47P **83**
Springrice Rd. SE13 ....58Fc **115**
Spring Ri. TW20: Egh 65A **126**
Spring Rd. TW13: Felt 62V **128**
Springs Cl. TW19: Stanw 60N **105**
Springshaw Cl. TN13: Bes G 95Fd **202**
Spring Shaw Rd. BR5: St P 67Wc **139**
Spring St. KT17: Ewe 81Va **174**
Spring St. W2 ....3B **220** (44Fb **89**)
Spring Ter. TW9: Rich 57Na **109**
Spring Tide Cl. SE15 ....53Xb **113**
Spring Va. DA7: Bex 56Dd **118**
Spring Va. DA9: Ghithe 58Yd **120**
Springvale Av. TW8: Bford 50Ma **87**
Spring Va. Cl. BR8: Swan 67Hd **140**
Springvale Cl. KT23: Bookh 97Da **191**
Springvale Ct. DA11: Nflt 61Ee **143**
Spring Va. Nth. DA1: Dart 59Md **119**
Springvale Retail Pk. 69Yc **139**
....(not continuous)
Spring Va. Sth. DA1: Dart 59Md **119**
Springvale Ter. W14 ....48Za **88**
Springvale Way BR5: St P 69Yc **139**
Spring Villa Pk. HA8: Edg 24Qa **47**
Spring Villa Rd. HA8: Edg 24Qa **47**
Spring Wlk. E1 ....43Wb **91**
Springwater WC1 ....7H **217** (43Pb **90**)
....*(off New North St.)*
Springwater Cl. SE18 ....53Qc **116**
Spring Way HP2: Hem H 1B **4**
Spring Way SE5 ....53Sb **113**
Springway HA1: Harr 31Fa **66**
Springwell Av. NW10 ....39Va **68**
Springwell Av. WD3: Rick 19J **25**
Springwell Cl. SW16 ....63Pb **134**
Springwell Ct. WD3: Rick 19J **25**
Springwell Ct. TW4: Houn 54Y **107**
Springwell La. UB9: Hare 22K **43**
Springwell La. WD3: Rick 20J **25**
Springwell Rd. SW16 ....63Qb **134**
Springwell Rd. TW4: Houn 54Z **107**
Springwell Rd. TW5: Hest 54Z **107**
Springwood Cl. E3 ....40Cc **72**
Springwood Cl. UB9: Hare 25M **43**
Springwood Ct. CR2: S Croy 77Ub **157**
....*(off Birdhurst Rd.)*
Springwood Cres. HA8: Edg 19Ra **29**
Springwood Pl. KT13: Weyb 80R **150**
Spring Woods GU25: Vir W 10M **125**
Springwood Way RM1: Rom 29Jd **56**
Sproggit Ind. Est. TW19: Stanw 58P **105**
Sprowston M. E7 ....37Jc **73**
Sprowston Rd. E7 ....36Jc **73**
Spruce Cl. RH1: Redh 5P **207**
Spruce Ct. SL1: Slou 8K **81**
Spruce Ct. W5 ....48Na **87**
Sprucedale Cl. BR8: Swan 68Gd **140**
Sprucedale Gdns. CR0: C'don 77Zb **158**
Sprucedale Gdns. SM6: W'gton 81Nb **176**
Spruce Hills Rd. E17 ....26Ec **52**
Spruce Ho. SE16 ....47Zb **92**
....*(off Woodland Cres.)*
Spruce Pk. BR2: Broml 70Hc **137**
Spruce Rd. TN16: Big H 88Mc **179**
Spruce Way AL2: Park 9P **5**
Sprules Rd. SE4 ....54Ac **114**
Spur, The GU21: Knap 10F **166**
Spur, The KT12: Walt T 75Y **151**
Spur, The SL1: Slou 3B **80**
Spur Cl. RM4: Abr 13Xc **37**
Spur Cl. WD5: Ab L 5T **12**
Spurfield KT8: W Mole 69Da **129**
Spurgate CM13: Hut 19Ce **41**
Spurgeon Av. SE19 ....67Tb **135**
Spurgeon Cl. RM17: Grays 51Ee **121**
Spurgeon Rd. SE19 ....67Tb **135**
Spurgeon St. SE1 ....3F **231** (48Tb **91**)
Spurling Rd. RM9: Dag 37Bd **75**
Spurling Rd. SE22 ....56Vb **113**
Spurrell Av. DA5: Bexl 63Fd **140**
Spurrell Ct. DA8: Erith 52Hd **118**
....*(off Furners Cl.)*
Spur Rd. BR6: Orp 75Wc **161**
Spur Rd. HA8: Edg 21Na **47**
Spur Rd. N15 ....28Tb **51**
Spur Rd. SE1 ....1K **229** (47Qb **90**)
Spur Rd. SW1 ....2B **228** (47Lb **90**)
Spur Rd. TW14: Felt 56X **107**
Spur Rd. TW7: Isle 52Ka **108**
Spurstowe Rd. E8 ....37Xb **71**
Spurstowe Ter. E8 ....36Wb **71**
Spurway Pde. IG2: Ilf 29Pc **54**
....*(off Woodford Av.)*
Spynes Mere Nature Reserve 2F **208**
Squadrons App. RM12: Horn 37Ld **77**
Square, The BR8: Swan 69Fd **140**
Square, The CM14: B'wood 19Yd **40**
Square, The CR3: Cat'm 96Wb **197**
Square, The E10 ....34Ec **72**
Square, The GU18: Light 2A **166**
Square, The GU23: Wis 88N **169**
Square, The HP1: Hem H 2M **3**
Square, The HP5: Lat 9A **10**
Square, The IG1: Ilf 31Qc **74**
Square, The IG10: Lough 13Rc **36**
Square, The IG8: Wfd G 22Jc **53**
Square, The KT13: Weyb 77S **150**
Square, The KT22: Lea 92Ja **192**
Square, The RM8: Dag 32Ad **75**
Square, The SM5: Cars 78Jb **156**
Square, The TN13: Riv 94Gd **202**
Square, The TN16: Tats 92Lc **199**
Square, The TW9: Rich 57Ma **109**
Square, The UB11: Stock P 46T **84**
Square, The UB2: S'hall 49Y **85**
Square, The UB7: Lford 53K **105**
Square, The W6 ....50Ya **88**
Square, The WD24: Wat 9X **13**
Square of Fame 35Qa **67**
....(off Arena Sq.)
Square Rigger Row SW11 ....55Eb **111**
Squarey St. SW17 ....62Eb **133**
Squerryes 100Sc **200**
Squerryes, The CR3: Cat'm 93Ub **197**
Squerryes Ct. TN16: Westrm 100Sc **200**
Squerryes Mede TN16: Westrm 99Sc **200**
Squerryes Pk. Cotts. TN16:
Westrm 99Sc **200**
Squire Gdns. NW8 ....4B **214** (41Fb **89**)
....*(off Grove End Rd.)*
Squires, The RM7: Rom 30Ed **56**
Squire's Bri. Rd. TW17: Shep 70P **127**
Squires Ct. KT16: Chert 74K **149**
Squires Ct. SW19 ....63Cb **133**
Squires Ct. SW4 ....53Nb **112**
Squires Fld. BR8: Hext 67Jd **140**
Squires La. N3 ....26Db **49**
Squires Mt. NW3 ....34Fb **69**
Squire's Rd. TW17: Shep 70Q **128**
Squires Wlk. TW15: Ashf 66T **128**
....(not continuous)
Squires Way DA2: Wilm 63Fd **140**
Squires Wood Dr. BR7: Chst 66Nc **138**
Squirrel Chase HP1: Hem H 1G **2**
Squirrel Cl. TW4: Houn 55Y **107**
Squirrel Dr. SL4: Wink 10A **102**
Squirrel Keep KT14: W Byf 84K **169**
Squirrel La. SL4: Wink 10A **102**
Squirrel M. W13 ....45Ha **86**
Squirrels, The HA5: Pinn 27Ba **45**
Squirrels, The SE13 ....55Fc **115**
Squirrels, The WD23: Bush 16Fa **28**
Squirrels Chase RM16: Ors 7C **100**
Squirrels Cl. BR6: Orp 74Uc **160**
Squirrels Cl. BR8: Swan 69Hd **140**
Squirrels Cl. N12 ....21Eb **49**
Squirrels Cl. UB10: Hil 38Q **64**
Squirrels Ct. KT4: Wor Pk 75Va **154**
....*(off The Avenue)*
Squirrels Drey BR2: Broml 68Gc **137**
....*(off Park Hill Rd.)*
Squirrels Grn. KT23: Bookh 95Ca **191**
Squirrels Grn. KT4: Wor Pk 75Va **154**
Squirrels Grn. RH1: Redh 5P **207**
SQUIRREL'S HEATH 27Md **57**
Squirrels Heath Av. RM2: Rom 27Kd **57**
Squirrels Heath La. RM11: Horn 28Md **57**
Squirrels Heath La. RM11: Rom 28Md **57**
Squirrels Heath La. RM2: Horn 28Md **57**
Squirrels Heath La. RM2: Rom 28Ld **57**
Squirrels Heath Rd. RM3: Hrld W 27Nd **57**
Squirrel's La. IG9: Buck H 20Mc **35**
Squirrels Trad. Est. UB3: Hayes 48V **84**
Squirrels Way KT18: Eps 86Ta **173**
Squirrel Wood KT14: W Byf 84K **169**
Squirries St. E2 ....41Wb **91**
SSE Arena Wembley, The 35Qa **67**
SS Robin 45Kc **93**
Stable Block, The RH1: Blet 3K **209**
Stable Cl. KT18: Eps D 91Ua **194**
Stable Cl. KT2: King T 65Pa **131**
Stable Cl. UB5: N'olt 40Ca **65**
Stable Ct. AL1: St A 1C **6**
Stable Ct. CR3: Cat'm 94Wb **197**
Stable Ct. EC1 ....6C **218** (42Rb **91**)
....*(off Clerkenwell Rd.)*
Stable Ct. SM6: W'gton 76Jb **156**
Stable Ct. TN13: S'oaks 98Ld **203**
Stableford Cl. CR2: Sande 83Ub **177**
Stable La. DA5: Bexl 61Dd **140**

Stable M. AL1: St A 1C 6
(off Hillside Rd.)
Stable M. NW5 37Kb 70
Stable M. RH2: Reig 6J 207
Stable M. SE6 60Gc 115
Stable M. TW1: Twick 60Ha 108
Stable Pl. N4 33Rb 71
Stables, The DA13: Nflt G 64Fe 143
Stables, The IG9: Buck H 17Lc 35
Stables, The KT11: Cobh 86Ba 171
Stables, The TN15: Seal 93Pd 203
Stables, The WD25: A'ham 8Da 13
Stables, The WD6: E'tree 14Na 29
Stables End BR6: Farnb 76Sc 160
Stables End WD25: A'ham 11Da 27
Stables Gallery & Arts Cen. 34Wa 68
Stables Lodge E8 38Xb 71
(off Mare St.)
Stables Mkt., The NW1 38Kb 70
Stables M. AL9: Brk P 10M 9
Stables M. SE27 64Sb 135
Stables Row E11 29Jc 53
Stable St. N1 1F 217 (39Nb 70)
Stable St. SE18 50Rc 94
Stables Way SE11 7K 229 (50Qb 90)
Stables Yd. SW18 58Cb 111
Stable Vs. BR1: Broml 65Mc 137
Stable Wlk. E1 44Wb 91
(off Boulevard Walkway)
Stable Wlk. N2 25Fb 49
Stable Way W10 44Ya 88
Stable Yd. SW1 1B 228 (47Lb 90)
(off Stable Yd. Rd.)
Stable Yd. SW15 55Ya 110
Stable Yd. TN15: Kems'g 89Qd 183
Stableyard, The SW9 54Pb 112
Stableyard M. TW11: Tedd 65Ha 130
Stable Yd. Rd. SW1 1C 228 (47Lb 90)
(not continuous)
Staburn Ct. HA8: Edg 26Sa 47
Stacey Av. N18 21Yb 52
Stacey Cl. DA12: Grav'nd 4G 144
Stacey Cl. E10 29Fc 53
Stacey Ct. RH1: Mers 1C 208
Stacey St. N7 34Qb 70
Stacey St. WC2 3E 222 (44Mb 90)
Stack Ho. RH8: Oxt 2J 211
Stack Ho. SW1 6J 227 (49Jb 90)
(off Cundy St.)
Stackhouse St. SW3 3F 227 (48Hb 89)
(off Rysbrack St.)
Stacklands Cl. TN15: W King 79Ud 164
Stack La. DA3: Hartl 71Be 165
Stack Rd. DA4: Hort K 70Td 142
Stacy Path SE5 52Ub 113
Stadbury, The KT13: Weyb 75Q 150
Staddleswood Pl. TN15: Plat 92De 205
Staddon Cl. BR3: Beck 70Ac 136
Stadium Bus. Cen. HA9: Wemb 34Ra 67
Stadium M. N5 34Qb 70
Stadium Retail Pk. 34Qa 67
Stadium Rd. SE18 52Nc 116
Stadium Rd. E. NW4 31Xa 68
Stadium St. SW10 52Eb 111
Stadium Way DA1: Cray 57Gd 118
Stadium Way HA9: Wemb 35Pa 67
Stadium Way WD18: Wat 15X 27
Staffa Rd. E10 32Ac 72
Staffhurst Wood Nature Reserve 9N 211
Staffhurst Wood Rd. RH8: Eden 9M 211
Staffhurst Wood Rd. TN8: Eden 9N 211
Stafford Av. RM11: Horn 27Md 57
Stafford Av. SL2: Slou 2G 80
Stafford Cl. CR3: Cat'm 95Vb 197
Stafford Cl. DA9: Ghithe 57Vd 120
Stafford Cl. E17 30Bc 52
(not continuous)
Stafford Cl. EN8: Chesh 1Xb 19
Stafford Cl. N14 15Lb 32
Stafford Cl. NW6 41Cb 89
Stafford Cl. RM16: Chaf H 49Yd 98
Stafford Cl. SL6: Tap 4A 80
Stafford Cl. SM3: Cheam 79Ab 154
Stafford Cl. SS17: Linf 8J 101
Stafford Ct. DA5: Bexl 59Bd 117
Stafford Ct. SW8 52Nb 112
Stafford Ct. W7 44Ha 86
(off Copley Cl.)
Stafford Ct. W8 48Cb 89
Stafford Cripps Ho. E2 41Yb 92
(off Globe Rd.)
Stafford Cripps Ho. SW6 51Bb 111
(off Clem Attlee Ct.)
Stafford Cross Bus. Pk. CR0: Wadd 78Pb 156
Stafford Gdns. CR0: Wadd 78Pb 156
Stafford Ho. SE1 7K 231 (50Vb 91)
(off Cooper's Rd.)
Stafford Ind. Est. RM11: Horn 27Md 57
Stafford Lake GU21: Knap 10D 166
Stafford Lake GU24: Bisl 10D 166
STAFFORDLAKE 10D 166
Stafford Mans. SW1 3B 228 (48Lb 90)
(off Stafford Pl.)
Stafford Mans. SW11 52Hb 111
(off Albert Bri. Rd.)
Stafford Mans. SW4 56Nb 112
Stafford Mans. W14 48Za 88
(off Haarlem Rd.)
Stafford Pl. SW1 3B 228 (48Lb 90)
Stafford Pl. TW10: Rich 59Pa 109
Stafford Ri. CR3: Cat'm 94Wb 197
Stafford Rd. CR0: Wadd 77Qb 156
Stafford Rd. CR3: Cat'm 95Vb 197
Stafford Rd. DA14: Sidc 63Uc 138
Stafford Rd. E3 40Bc 72
Stafford Rd. E7 38Lc 73
Stafford Rd. HA3: Hrw W 24Ea 46
Stafford Rd. HA4: Ruis 35V 64
Stafford Rd. KT3: N Mald 69Sa 131
Stafford Rd. NW6 41Cb 89
Stafford Rd. SM6: W'gton 79Lb 156
Staffordshire St. SE15 53Wb 113
Stafford Sq. KT13: Weyb 77T 150
Stafford St. W1 6B 222 (46Lb 90)
Stafford Ter. W8 48Cb 89
Stafford Way TN13: S'oaks 99Ld 203
Staff St. EC1 4G 219 (41Tb 91)
Stagbury Av. CR5: Chips 90Gb 175
Stagbury Cl. CR5: Chips 91Gb 195
Stagbury Ho. CR5: Chips 91Gb 195
Stag Cl. HA8: Edg 26Ra 47
Stag Community Arts Cen. 97Kd 203
Stag Ct. KT2: King T 67Qa 131
(off Coombe Rd.)
Staggart Grn. IG7: Chig 22Vc 55
Staggart Grn. IG7: Ilf 22Vc 55
Stagg Hill EN4: Had W 8Gb 17
Stagg Hill EN6: Pot B 7Fb 17
Stag La. HA8: Edg 26Ra 47
Stag La. IG9: Buck H 19Kc 35
Stag La. NW9 27Sa 47
Stag La. SL5: S'dale 3E 146
Stag La. SW15 62Va 132
Stag La. WD3: Chor 16E 24
STAG LANE 61Va 132
Stag Leys KT21: Asht 92Na 193
Stag Leys Cl. SM7: Bans 87Gb 175
Stags Way TW7: Isle 51Ha 108
Stainash Cres. TW18: Staines 64K 127
Stainash Pde. TW18: Staines 64K 127
(off Kingston Rd.)
Stainbank Rd. CR4: Mitc 69Kb 134
Stainby Cl. UB7: W Dray 48N 83
Stainby Rd. N15 28Vb 51
Stainer Rd. WD6: Bore 11Ma 29
Staines Av. SM3: Cheam 75Za 154
Staines Boat Club 64G 126
Staines Bri. 64G 126
Staines By-Pass TW19: Staines 61E 126
Staines La. KT16: Chert 72H 149
Staines La. Cl. KT16: Chert 72H 149
Staines Moor Nature Reserve 60G 104
Staines Rd. IG1: Ilf 36Sc 74
Staines Rd. KT16: Chert 68H 127
Staines Rd. TW14: Bedf 60Q 106
Staines Rd. TW14: Felt 60Q 106
Staines Rd. TW18: Lale 67K 127
Staines Rd. TW18: Staines 67K 127
Staines Rd. TW19: Wray 59A 104
Staines Rd. TW2: Twick 62Ca 129
Staines Rd. TW3: Houn 55Da 107
Staines Rd. TW4: Houn 57Z 107
Staines Rd. E. TW16: Sun 66W 128
Staines Rd. W. TW15: Ashf 65R 128
Staines Rd. W. TW16: Sun 66U 128
Staines Station (Rail) 64J 127
Staines Town FC 66J 127
STAINES-UPON-THAMES 63H 127
Staines Wlk. DA14: Sidc 65Yc 139
Stainford Cl. TW15: Ashf 64T 128
Stainforth Rd. E17 28Cc 52
Stainforth Rd. IG2: Ilf 31Tc 74
Staining La. EC2 2E 224 (44Sb 91)
Stainmore Cl. BR7: Chst 67Tc 138
Stainsbury St. E2 40Yb 72
Stainsby Rd. E14 44Cc 92
Stains Cl. EN8: Chesh 1Ac 20
Stainton Ct. WD23: Bush 15Da 27
(off Farrington Av.)
Stainton Rd. EN3: Enf H 11Yb 34
Stainton Rd. SE6 58Fc 115
Stainton Wlk. GU21: Wok 10N 167
Stairfoot La. TN13: Chips 94Ed 202
Staith Ct. E3 41Ec 92
(off Bolinder Way)
Staiths Way KT20: Tad 92Xa 194
Stalbridge Flats W1 3J 221 (44Jb 90)
(off Lumley St.)
Stalbridge Ho. NW1 2B 216 (40Lb 70)
(off Hampstead Rd.)
Stalbridge St. NW1 7E 214 (43Gb 89)
Staleys Acre TN15: Bor G 92Be 205
Staleys Rd. TN15: Bor G 92Ae 205
Stalham St. SE16 48Xb 91
Stalham Way IG6: Ilf 25Rc 54
Stalisfield Pl. BR6: Downe 82Qc 180
Stambourne Way BR4: W W'ck 75Ec 158
Stambourne Way SE19 66Ub 135
Stambourne Woodland Wlk. SE19 66Ub 135
Stamford Bridge 52Db 111
Stamford Bri. Studios SW6 52Db 111
(off Wandon Rd.)
Stamford Brook Arches W6 49Wa 88
Stamford Brook Av. W6 48Wa 88
Stamford Brook Gdns. W6 48Wa 88
Stamford Brook Mans. W6 49Wa 88
(off Goldhawk Rd.)
Stamford Brook Rd. W6 48Wa 88
Stamford Brook Station (Underground) 49Wa 88
Stamford Bldgs. SW8 52Nb 112
(off Meadow Pl.)
Stamford Cl. EN6: Pot B 4Fb 17
Stamford Cl. HA3: Hrw W 24Ga 46
Stamford Cl. N15 28Wb 51
Stamford Cl. NW3 34Eb 69
(off Heath St.)
Stamford Cl. UB1: S'hall 45Ca 85
Stamford Cotts. SW10 52Db 111
(off Billing St.)
Stamford Ct. W6 49Wa 88
Stamford Dr. BR2: Broml 70Hc 137
Stamford Gdns. RM9: Dag 38Yc 75
Stamford Ga. SW6 52Db 111
Stamford Ga. Ho. SW6 52Db 111
(off Stamford Ga.)
STAMFORD GREEN 85Ra 173
Stamford Grn. Rd. KT18: Eps 85Ra 173
Stamford Gro. E. N16 32Wb 71
Stamford Gro. W. N16 32Wb 71
Stamford Hill N16 33Vb 71
STAMFORD HILL 32Vb 71
Stamford Hill Station (Overground) 31Ub 71
Stamford Ho. GU24: Chob 3J 167
(off Bagshot Rd.)
Stamford Lodge N16 31Vb 71
Stamford Rd. E6 39Nc 74
Stamford Rd. KT12: Walt T 76Z 151
Stamford Rd. N1 38Ub 71
Stamford Rd. N15 29Wb 51
Stamford Rd. RM9: Dag 39Xc 75
Stamford Rd. WD17: Wat 12X 27
Stamford Sq. SW15 57Ab 110
Stamford St. SE1 7K 223 (46Qb 90)
Stamp Pl. E2 2K 219 (40Vb 71)
Stanacre Ct. KT20: Kgswd 93Bb 195
Stanard Cl. N16 31Ub 71
Stanborough Av. WD6: Bore 9Qa 15
Stanborough Cl. TW12: Hamp 65Ba 129
Stanborough Cl. WD6: Bore 10Qa 15
Stanborough Ho. E3 42Dc 92
(off Empson St.)
Stanborough Pk. WD25: Wat 7Y 13
Stanborough Pas. E8 37Vb 71
Stanborough Rd. TW3: Houn 55Fa 108
Stanbridge Pl. N21 19Rb 33
Stanbridge Rd. SW15 55Ya 110
Stanbrook Rd. DA11: Grav'nd 10B 122
Stanbrook Rd. SE2 47Xc 95
Stanbury Av. WD17: Wat 9U 12
Stanbury Ct. NW3 37Hb 69
Stanbury Rd. SE15 53Xb 113
(not continuous)
Stancroft NW9 29Ua 48
Standale Gro. HA4: Ruis 29S 44
Standard Ind. Est. E16 47Pc 94
Standard Pl. EC2 4J 219 (41Ub 91)
(off Rivington St.)
Standard Rd. BR6: Downe 82Qc 180
Standard Rd. DA17: Belv 50Cd 96
Standard Rd. DA6: Bex 56Ad 117
Standard Rd. EN3: Enf W 10Ac 20
Standard Rd. NW10 42Sa 87
Standard Rd. TW4: Houn 55Aa 107
Standcumbe Ct. BR3: Beck 71Bc 158
Standen Av. RM12: Horn 34Nd 77
Standen Rd. SW18 59Bb 111
Standfield WD5: Ab L 3U 12
Standfield Gdns. RM10: Dag 37Cd 76
Standfield Rd. RM10: Dag 36Cd 76
Standish Ho. W6 49Wa 88
(off St Peter's Gro.)
Standish Rd. W6 49Wa 88
Standlake Point SE23 62Zb 136
Standring Ri. HP3: Hem H 5K 3
Stane Cl. SW19 66Db 133
Stane Gro. SW9 54Nb 112
Stanesgate Ho. SE15 52Wb 113
(off Friary Est.)
Stane Way KT17: Ewe 82Wa 174
Stane Way SE18 52Mc 115
Stanfield Ho. NW8 5C 214 (42Fb 89)
(off Frampton St.)
Stanfield Ho. UB5: N'olt 40Z 65
(off Academy Gdns.)
Stanfield Rd. E3 40Ac 72
Stanford Cl. HA4: Ruis 30S 44
Stanford Cl. IG8: Wfd G 22Nc 54
Stanford Cl. RM7: Rom 30Dd 56
Stanford Cl. TW12: Hamp 65Ba 129
STANFORD COMMON 8C 186
Stanford Cotts. GU24: Pirb 8C 186
Stanford Ct. EN9: Walt A 5Jc 21
Stanford Ct. SW6 53Db 111
Stanford Ct. W8 48Db 89
(off Cornwall Gdns.)
Stanford Gdns. RM15: Avel 46Ud 98
Stanford Ind. Est. SS17: Stan H 2L 101
STANFORD-LE-HOPE 2M 101
Stanford-le-Hope Station (Rail) 2L 101
Stanford M. E8 36Wb 71
Stanford Pl. SE17 6H 231 (49Ub 91)
Stanford Rd. N11 22Hb 49
Stanford Rd. RM16: Grays 48Fe 99
Stanford Rd. RM16: Ors 48Fe 99
Stanford Rd. SS17: Stan H 3H 101
Stanford Rd. SW16 68Mb 134
Stanford Rd. W8 48Db 89
Stanfords, The KT17: Eps 84Ua 174
(off East St.)
Stanford St. SW1 6D 228 (49Mb 90)
Stanford Warren Nature Reserve 4M 101
Stanford Way SW16 68Mb 134
Stangate SE1 3J 229 (48Pb 90)
(off Royal St.)
Stangate Cres. WD6: Bore 15Ta 29
Stangate Gdns. HA7: Stan 21Ka 46
Stangate Lodge N21 16Pb 32
Stanger Rd. SE25 70Wb 135
Stanham Pl. DA1: Cray 56Jd 118
Stanham Rd. DA1: Dart 57Ld 119
Stanhill Cotts. DA2: Wilm 66Fd 140
Stanhope Av. BR2: Hayes 74Hc 159
Stanhope Av. HA3: Hrw W 25Fa 46
Stanhope Av. N3 27Bb 49
Stanhope Cl. SE16 47Zb 92
Stanhope Gdns. IG1: Ilf 32Pc 74
Stanhope Gdns. N4 30Rb 51
Stanhope Gdns. N6 30Kb 50
Stanhope Gdns. NW7 22Va 48
Stanhope Gdns. RM8: Dag 34Bd 75
Stanhope Gdns. SW7 5A 226 (49Eb 89)
Stanhope Ga. W1 6J 221 (46Jb 90)
Stanhope Gro. BR3: Beck 71Bc 158
Stanhope Heath TW19: Stanw 58L 105
Stanhope Ho. N11 21Kb 50
(off Coppies Gro.)
Stanhope Ho. SE8 52Bc 114
(off Adolphus St.)
Stanhope Ind. Pk. SS17: Stan H 4P 101
Stanhope M. E. SW7 5A 226 (49Eb 89)
Stanhope M. Sth. SW7 6A 226 (49Eb 89)
Stanhope M. W. SW7 5A 226 (49Eb 89)
Stanhope Pde. NW1 3B 216 (41Lb 90)
Stanhope Pk. Rd. UB6: G'frd 42Ea 86
Stanhope Pl. W2 4F 221 (45Hb 89)
Stanhope Rd. AL1: St A 2D 6
Stanhope Rd. CR0: C'don 76Ub 157
Stanhope Rd. DA10: Swans 58Be 121
Stanhope Rd. DA15: Sidc 63Wc 139
Stanhope Rd. DA7: Bex 54Ad 117
Stanhope Rd. E17 29Dc 52
Stanhope Rd. EN5: Barn 16Ya 30
Stanhope Rd. EN8: Walt C 5Ac 20
Stanhope Rd. N12 22Eb 49
Stanhope Rd. N6 30Lb 50
Stanhope Rd. RM13: Rain 40Jd 76
Stanhope Rd. RM8: Dag 33Bd 75
Stanhope Rd. SL1: Slou 4B 80
Stanhope Rd. SM5: Cars 80Jb 156
Stanhope Rd. UB6: G'frd 42Ea 86
Stanhope Row W1 7K 221 (46Kb 90)
Stanhopes RH8: Limp 100Kc 199
Stanhope St. NW1 2B 216 (40Lb 70)
Stanhope Ter. TW2: Twick 59Ha 108
Stanhope Ter. W2 4C 220 (45Fb 89)
Stanhope Way TN13: Riv 94Fd 202
Stanhope Way TW19: Stanw 58L 105
Stanier Cl. W14 50Bb 89
Stanier Ho. SW6 53Eb 111
(off Station Ct.)
Staniland Dr. KT13: Weyb 83P 169
Stanlake M. W12 46Ya 88
Stanlake Rd. W12 46Ya 88
Stanlake Vs. W12 46Ya 88
Stanley Av. AL2: Chis G 7N 5
Stanley Av. BR3: Beck 68Ec 136
Stanley Av. HA0: Wemb 38Na 67
Stanley Av. IG11: Bark 40Vc 75
Stanley Av. KT3: N Mald 71Wa 154
Stanley Av. RM2: Rom 28Jd 56
Stanley Av. RM8: Dag 32Bd 75
Stanley Av. UB6: G'frd 39Ea 66
Stanley Bri. Studios SW6 52Db 111
(off King's Rd.)
Stanley Cl. CR5: Coul 89Pb 176
Stanley Cl. DA9: Ghithe 57Ud 120
Stanley Cl. HA0: Wemb 38Na 67
Stanley Cl. RM12: Horn 33Ld 77
Stanley Cl. RM2: Rom 28Jd 56
Stanley Cl. SE9 60Sc 116
Stanley Cl. SW8 51Pb 112
Stanley Cl. UB8: Uxb 40M 63
Stanley Cohen Ho. EC1 6E 218 (42Sb 91)
(off Golden La.)
Stanley Cotts. DA2: Daren 64Ud 142
Stanley Cotts. SL2: Slou 6K 81
Stanley Ct. SM2: Sutt 80Db 155
Stanley Ct. SM5: Cars 80Jb 156
Stanley Ct. W5 43La 86
Stanley Cres. DA12: Grav'nd 4F 144
Stanley Cres. W11 45Bb 89
Stanleycroft Cl. TW7: Isle 53Ga 108
Stanley Dr. AL10: Hat 2D 8
Stanley Gdns. CR2: Sande 84Wb 177
Stanley Gdns. CR4: Mitc 65Jb 134
Stanley Gdns. KT12: Hers 79Y 151
Stanley Gdns. NW2 36Ya 68
Stanley Gdns. SM6: W'gton 79Lb 156
Stanley Gdns. W11 45Bb 89
Stanley Gdns. W3 47Ua 88
Stanley Gdns. WD6: Bore 11Na 29
Stanley Gdns. M. W11 45Bb 89
(off Kensington Pk. Rd.)
Stanley Gdns. Rd. TW11: Tedd 64Ga 130
Stanley Grn. E. SL3: L'ly 49B 82
Stanley Grn. W. SL3: L'ly 49B 82
Stanley Gro. CR0: C'don 72Qb 156
Stanley Gro. SW8 54Jb 112
Stanley Hill GU24: Pirb 4A 186
Stanley Holloway Ct. E16 44Jc 93
(off Coolfin Rd.)
Stanley Ho. E14 44Cc 92
(off Saracen St.)
Stanley Ho. SW10 52Eb 111
(off Coleridge Gdns.)
Stanley Kubrick Rd. UB9: Den 29J 43
Stanley Mans. SW10 51Eb 111
(off Park Wlk.)
Stanley Maude Ho. SL9: Chal P 21A 42
(off Micholls Av.)
Stanley M. SW10 52Eb 111
(off Coleridge Gdns.)
Stanley Pk. Dr. HA0: Wemb 39Pa 67
Stanley Pk. Rd. SM5: Cars 80Gb 155
Stanley Pk. Rd. SM6: W'gton 79Kb 156
Stanley Picker Gallery 69Na 131
(off Springfield Rd.)
Stanley Rd. BR2: Broml 70Kc 137
Stanley Rd. BR6: Orp 74Wc 161
Stanley Rd. CR0: C'don 73Qb 156
Stanley Rd. CR4: Mitc 66Jb 134
Stanley Rd. DA10: Swans 58Be 121
Stanley Rd. DA11: Nflt 1A 144
Stanley Rd. DA14: Sidc 62Wc 139
Stanley Rd. E10 30Dc 52
Stanley Rd. E12 36Nc 74
Stanley Rd. E18 25Hc 53
Stanley Rd. E4 18Fc 35
Stanley Rd. EN1: Enf 13Ub 33
Stanley Rd. GU21: Wok 88B 168
Stanley Rd. HA2: Harr 33Ea 66
Stanley Rd. HA6: Nwood 25W 44
Stanley Rd. HA9: Wemb 37Pa 67
Stanley Rd. IG1: Ilf 33Tc 74
Stanley Rd. N10 24Kb 50
Stanley Rd. N11 23Mb 50
Stanley Rd. N15 28Rb 51
Stanley Rd. N2 27Fb 49
Stanley Rd. N9 18Vb 33
Stanley Rd. NW9 31Wa 68
Stanley Rd. RM12: Horn 33Ld 77
Stanley Rd. RM17: Grays 50De 99
Stanley Rd. SM2: Sutt 79Db 155
Stanley Rd. SM4: Mord 70Cb 133
Stanley Rd. SM5: Cars 80Jb 156
Stanley Rd. SW14 56Ra 109
Stanley Rd. SW19 65Cb 133
Stanley Rd. TW11: Tedd 63Ga 130
Stanley Rd. TW15: Ashf 64N 127
Stanley Rd. TW2: Twick 62Fa 130
Stanley Rd. TW3: Houn 56Ea 108
Stanley Rd. UB1: S'hall 45Aa 85
Stanley Rd. W3 48Sa 87
Stanley Rd. WD17: Wat 14Y 27
Stanley Rd. Nth. RM13: Rain 39Gd 76
Stanley Rd. Sth. RM13: Rain 40Hd 76
Stanley Sq. SM5: Cars 81Hb 175
Stanley St. CR3: Cat'm 94Sb 197
Stanley St. SE8 52Bc 114
Stanley Studios SW10 51Eb 111
(off Park Wlk.)
Stanley Ter. DA6: Bex 56Cd 118
Stanley Ter. N19 33Nb 70
Stanley Way BR5: St M Cry 71Xc 161
Stanliff Ho. E14 48Cc 92
Stanmer St. SW11 53Gb 111
STANMORE 22Ka 46
Stanmore & Edgware Golf Cen. 20Ma 29
Stanmore Chase AL4: St A 3H 7
Stanmore Common Local Nature Reserve 19Ha 28
Stanmore Country Pk. & Local Nature Reserve 20La 28
Stanmore Gdns. SM1: Sutt 76Eb 155
Stanmore Gdns. TW9: Rich 55Pa 109
Stanmore Golf Course 24Ka 46
Stanmore Hill HA7: Stan 20Ja 28
Stanmore Lodge HA7: Stan 21Ka 46
Stanmore Pl. NW1 39Kb 70
Stanmore Rd. DA17: Belv 49Ed 96
Stanmore Rd. E11 32Hc 73
Stanmore Rd. N15 28Rb 51
Stanmore Rd. TW9: Rich 55Pa 109
Stanmore Rd. WD24: Wat 11X 27
Stanmore Station (Underground) 21Ma 47
Stanmore St. N1 39Pb 70
Stanmore Ter. BR3: Beck 68Cc 136
Stanmore Way IG10: Lough 11Qc 36
Stanmount Rd. AL2: Chis G 7N 5
Stannard Cotts. E1 42Yb 92
(off Fox Cl.)
Stannard Ct. SE6 60Dc 114
Stannard Ho. SW19 63Eb 133
Stannard M. E8 37Wb 71
(off Stannard Rd.)
Stannard Rd. E8 37Wb 71
Stannary Pl. SE11 50Qb 90
Stannary St. SE11 51Qb 112
STANNERS HILL 10P 147
Stannet Way SM6: W'gton 77Lb 156
Stannington Path WD6: Bore 11Qa 29
Stansbury Sq. W10 41Ab 88
Stansfeld Ho. SE1 6K 231 (49Vb 91)
(off Longfield Est.)
Stansfeld Rd. E16 45Nc 94
Stansfeld Rd. E6 43Mc 93
Stansfield Rd. SW9 55Pb 112
Stansfield Rd. TW4: Cran 54X 107
Stansgate Rd. RM10: Dag 33Cd 76
Stanstead WC1 4G 217 (41Nb 90)
(off Tavistock Pl.)
Stanstead Cl. BR2: Broml 71Hc 159
Stanstead Cl. CR3: Cat'm 96Ub 197
Stanstead Gro. SE6 60Bc 114
Stanstead Ho. E3 42Ec 92
(off Devas St.)
Stanstead Mnr. SM1: Sutt 79Cb 155
Stanstead Rd. CR3: Cat'm 99Tb 197
Stanstead Rd. E11 29Kc 53
Stanstead Rd. SE23 60Zb 114
Stanstead Rd. SE6 60Bc 114
STANSTED 82Be 185
Stansted Cl. RM12: Horn 37Kd 77
Stansted Cres. DA5: Bexl 60Zc 117
Stansted Hill TN15: Stans 82Be 185
Stansted La. TN15: Ash 82Xd 184
Stansted Rd. TW6: H'row A 58P 105
Stanswood Gdns. SE5 52Ub 113
Stanthorpe Cl. SW16 64Nb 134
Stanthorpe Rd. SW16 64Nb 134
Stanton Av. TW11: Tedd 65Ga 130
Stanton Cl. BR5: Orp 73Yc 161
Stanton Cl. KT19: Ewe 78Ra 153
Stanton Cl. KT4: Wor Pk 74Za 154
Stanton Ct. CR2: S Croy 78Ub 157
(off Birdhurst Ri.)
Stanton Gro. KT20: Tad 93Za 194
Stanton Ho. SE10 51Ec 114
(off Thames St.)
Stanton Ho. SE16 47Bc 92
(off Rotherhithe St.)
Stanton Rd. CR0: C'don 73Sb 157
Stanton Rd. SE26 63Bc 136
Stanton Rd. SW13 54Ua 110
Stanton Rd. SW20 67Za 132
Stanton Sq. SE26 63Bc 136
Stanton Way SE26 63Bc 136
Stanton Way SL3: L'ly 49A 82
Stanway Cl. IG7: Chig 22Uc 54
Stanway Cotts. KT16: Chert 74J 149
Stanway Ct. N1 2J 219 (40Ub 71)
(off Shenfield St.)
Stanway Gdns. HA8: Edg 22Sa 47
Stanway Gdns. W3 46Qa 87
Stanway Rd. EN9: Walt A 5Jc 21
Stanway St. N1 1J 219 (40Ub 71)
STANWELL 58M 105
Stanwell Cl. TW19: Stanw 58M 105
Stanwell Gdns. TW19: Stanw 58M 105
TW19: Stanw
STANWELL MOOR 57J 105
Stanwell Moor Rd. TW18: Staines 62J 127
Stanwell Moor Rd. TW19: Staines 62K 127
Stanwell Moor Rd. TW19: Stanw M 62K 127
Stanwell Moor Rd. UB7: Lford 54K 105
Stanwell New Rd. TW18: Staines 62J 127
Stanwell Pl. TW19: Stanw 58L 105
Stanwell Pl. TW19: Stanw M 58L 105
Stanwell Rd. SL3: Hort 55C 104
Stanwell Rd. SL3: Poyle 55C 104
Stanwell Rd. TW14: Bedf 59R 106
Stanwell Rd. TW15: Ashf 61N 127
Stanwick Rd. W14 49Bb 89
Stanworth Ct. TW5: Hest 52Ba 107
Stanworth St. SE1 3K 231 (48Vb 91)
Stanwyck Dr. IG7: Chig 22Sc 54
Stanwyck Gdns. RM3: Rom 22Kd 57
Stanyhurst SE23 60Ac 114
Stapenhill Rd. HA0: Wemb 34Ka 66
Staplands Mnr. KT13: Weyb 76U 150
Staple Cl. DA5: Bexl 62Fd 140
Staple Cl. RM5: Col R 23Ed 56
Staplefield Cl. HA5: Pinn 24Aa 45
Staplefield Cl. SW2 60Nb 112
Stapleford N17 26Ub 51
(off Willan Rd.)
STAPLEFORD ABBOTTS 15Ed 38
Stapleford Abbotts Golf Course 17Jd 38
Stapleford Airfield 12Cd 38
Stapleford Av. IG2: Ilf 29Uc 54
Stapleford Cl. E4 20Ec 34
Stapleford Cl. KT1: King T 68Qa 131
Stapleford Cl. SW19 59Ab 110
Stapleford Ct. TN13: S'oaks 96Hd 202
Stapleford Gdns. RM5: Col R 23Cd 56
Stapleford Rd. HA0: Wemb 38Ma 67
Stapleford Rd. RM4: Stap A 14Dd 38
Stapleford Rd. RM4: Stap T 14Dd 38
Staple Hill Rd. GU24: Chob 9H 147
Staplehurst Cl. RH2: Reig 10L 207
Staplehurst Rd. RH2: Reig 10L 207
Staplehurst Rd. SE13 57Fc 115
Staplehurst Rd. SM5: Cars 80Gb 155
Staple Inn WC1 1K 223 (43Qb 90)
(off Staple Inn Bldgs.)
Staple Inn Bldgs. WC1 1K 223 (43Qb 90)
Staples, The BR8: Swan 67Kd 141
Staples Cl. SE16 46Ac 92
STAPLES CORNER 32Xa 68
Staples Cnr. Bus. Pk. NW2 32Xa 68
Staples Cnr. Retail Pk. 32Xa 68
Staples Ho. E6 44Qc 94
(off Savage Gdns.)
Staple's Rd. IG10: Lough 13Mc 35
Staple St. SE1 2G 231 (47Tb 91)
Stapleton Cl. EN6: Pot B 3Fb 17
Stapleton Cres. RM13: Rain 37Jd 76
Stapleton Gdns. CR0: Wadd 78Qb 156
Stapleton Hall Rd. N4 32Pb 70
Stapleton Ho. E2 41Xb 91
(off Ellsworth St.)
Stapleton Rd. BR6: Orp 77Vc 161
Stapleton Rd. DA7: Bex 52Bd 117
Stapleton Rd. SW17 62Jb 134
Stapleton Rd. WD6: Bore 10Qa 15
Stapleton Vs. N16 35Ub 71
(off Wordsworth Rd.)
Stapley Rd. AL3: St A 1B 6
Stapley Rd. DA17: Belv 50Cd 96
Stapylton Rd. EN5: Barn 13Ab 30
Star All. EC3 4J 225 (45Ub 91)
(off Mark La.)
Star & Garter Hill TW10: Rich 60Na 109
Star Apts. KT12: Hers 79Y 151
Starboard Av. DA9: Ghithe 58Xd 120
Starboard Way E14 48Cc 92
Starboard Way E16 46Lc 93
Starbuck Cl. SE9 59Qc 116
Star Bus. Cen. RM13: Rain 43Fd 96
Starch Ho. La. IG6: Ilf 26Tc 54
Star Cl. EN3: Pond E 16Yb 34
Starcross St. NW1 4C 216 (41Lb 90)
Starfield Rd. W12 47Wa 88
Star Hill DA1: Cray 57Gd 118
Star Hill Rd. TN14: Dun G 88Bd 181
Starkey Pl. DA8: Erith 52Gd 118
Star La. BR5: St M Cry 70Yc 139
Star La. BR5: St P 70Yc 139
Star La. CM16: Epp 2Wc 23
Star La. CR5: Coul 93Jb 196
Star La. E16 42Gc 93
Star Lane Station (DLR) 42Gc 93
Starley Cl. E17 25Fc 53

Starlight Way AL4: St A 4G 6
Starlight Way TW6: H'row A 57S 106
Starling Cl. CR0: C'don 72Ac 158
Starling Cl. DA3: Lfield 69De 143
Starling Cl. HA5: Pinn 27Y 45
Starling Cl. IG9: Buck H 18Jc 35
Starling Cres. SL3: L'ly 9N 81
Starling Ho. NW8 1D 214 (40Gb 69)
(off Charlbert St.)
Starling M. KT5: Surb 72Pa 153
Starling Pl. WD25: Wat 4Y 13
Starlings, The KT22: Oxs 85Ea 172
Starlings, The SL0: Iver H 42F 82
Starling Wlk. TW12: Hamp 64Aa 129
Starmans Cl. RM9: Dag 39Ad 75
Star Path UB5: N'olt 40Ca 65
(off Brabazon Rd.)
Star Pl. E1 45Vb 91
Star Rd. TW7: Isle 54Fa 108
Star Rd. UB10: Hil 42S 84
Star Rd. W14 51Bb 111
Starrock La. CR5: Chips 92Hb 195
Starrock Rd. CR5: Coul 91Kb 196
Star St. W2 2D 220 (44Gb 89)
Starts Cl. BR6: Farnb 76Qc 160
Starts Hill Av. BR6: Farnb 77Rc 160
Starts Hill Rd. BR6: Farnb 76Qc 160
Starveall Cl. UB7: W Dray 48P 83
Star Wharf NW1 39Lb 70
(off St Pancras Way)
Starwood Cl. KT14: W Byf 83L 169
Starwood Ct. SL3: L'ly 8N 81
Star Yd. WC2 2K 223 (44Qb 90)
State Farm Av. BR6: Farnb 77Rc 160
Staten Bldg. E3 40Cc 72
(off Fairfield Rd.)
Staten Gdns. TW1: Twick 60Ha 108
State Pde. IG6: Ilf 26Sc 54
Statham Ct. N19 34Nb 70
(off Alexander Rd.)
Statham Gro. N16 35Tb 71
Statham Gro. N18 22Ub 51
Statham Ho. SW8 53Lb 112
(off Wadhurst Rd.)
Station App. IG10: Lough Alderton Hill 15Nc 36
Station App. DA7: Bex Barnehurst Rd 54Ed 118
Station App. BR7: Chst Bennetts Copse 65Nc 138
Station App. SE9: Bercta Rd 61Sc 138
Station App. SE9: Crossmead 60Pc 116
Station App. SW16: Estreham Rd 65Mb 134
Station App. KT17: Ewe Fennells Mead 81Va 174
Station App. SW16: Gleneagle Rd 64Mb 134
Station App. SE26: Lower Sydenham 64Bc 136
Station App. HA4: Ruis Mahlon Av 36X 65
Station App. KT17: Ewe Nonsuch Ct. Av. 82Xa 174
Station App. HA4: Ruis Pembroke Rd 32U 64
Station App. DA7: Bex Percy Rd 54Ad 117
Station App. IG10: Lough Torrington Dr 14Sc 36
Station App. BR7: Chst Vale Rd 67Qc 138
Station App. BR1: Broml 69Jc 137
(off High St.)
Station App. BR2: Hayes 74Jc 159
Station App. BR3: Beck 67Cc 136
Station App. BR4: W W'ck 73Ec 158
Station App. BR5: St M Cry 70Xc 139
Station App. BR6: Chels 78Xc 161
Station App. BR6: Orp 75Vc 161
Station App. BR8: Swan 70Gd 140
Station App. CM13: W H'dn 31Ee 79
Station App. CM16: They B 8Uc 22
Station App. CR2: Sande 81Tb 177
Station App. CR3: Whyt 89Wb 177
Station App. CR5: Chips 90Hb 175
Station App. CR5: Coul 88Mb 176
Station App. CR8: Purl 83Qb 176
Station App. DA1: Cray 58Hd 118
Station App. DA1: Dart 58Nd 119
Station App. DA13: Meop 10C 144
Station App. DA16: Well 54Wc 117
Station App. DA5: Bexl 60Cd 118
Station App. E11 29Jc 53
Station App. E17 29Cc 52
Station App. E18 26Kc 53
Station App. E4 23Fc 53
Station App. E7 35Kc 73
Station App. EN5: New Bar 14Eb 31
Station App. EN8: Walt C 6Ac 20
Station App. GU22: Wok 90B 168
Station App. GU25: Vir W 10P 125
Station App. HA0: Wemb 37Ka 66
Station App. HA1: Harr 31Ga 66
Station App. HA5: Pinn 27Aa 45
Station App. HA6: Nwood 24U 44
Station App. HP3: Hem H 5J 3
Station App. IG8: Wfd G 23Kc 53
Station App. IG9: Buck H 21Mc 53
Station App. KT1: King T 67Qa 131
Station App. KT10: Hin W 76Ha 152
Station App. KT13: Weyb 79Q 150
Station App. KT14: W Byf 84J 169
Station App. KT19: Eps 85Ta 173
Station App. KT19: Ewe 78Wa 154
Station App. KT20: Tad 94Ya 194
Station App. KT22: Lea 93Ja 192
Station App. KT22: Oxs 84Ea 172
Station App. KT24: E Hor 98U 190
Station App. KT4: Wor Pk 74Wa 154
Station App. N11 22Kb 50
Station App. N12 21Db 49
Station App. N16 33Vb 71
(off Stamford Hill)
Station App. NW1 6G 215 (42Hb 89)
Station App. NW10 41Va 88
Station App. NW11 31Za 68
Station App. RH1: Redh 5A 208
(off Redstone Hill)
Station App. RH8: Oxt 1J 211
Station App. RM14: Upm 33Sd 78
Station App. RM15: S Ock 41Yd 98
Station App. RM17: Grays 51Ce 121
Station App. RM18: Tilb 6D 122
Station App. SE12 58Jc 115
(off Burnt Ash Hill)
Station App. SL3: L'ly 47C 82
Station App. SL9: Ger X 29A 42
Station App. SM2: Cheam 80Ab 154
Station App. SM2: Sutt 82Db 175
Station App. SM5: Cars 77Hb 155
Station App. SW14 55Sa 109
Station App. SW20 68Xa 132
Station App. SW6 55Ab 110
Station App. TN13: Dun G 92Gd 202
Station App. TN15: Bor G 92Be 205
Station App. TW12: Hamp 67Ca 129
Station App. TW15: Ashf 63P 127
Station App. TW16: Sun 67W 128
Station App. TW17: Shep 71S 150
Station App. TW18: Staines 64J 127
Station App. TW8: Bford 51La 108
(off Sidney Gdns.)
Station App. TW9: Kew 53Qa 109
Station App. UB3: Hayes 48V 84
Station App. UB6: G'frd 38Ea 66
Station App. UB7: Yiew 46N 83
Station App. UB9: Den 31F 62
Station App. W7 46Ga 86
Station App. WD19: Wat 20Z 27
Station App. WD3: Chor 14F 24
Station App. WD4: K Lan 2R 12
Station App. WD7: R'lett 7Ja 14
Station App. E. RH1: Redh 8P 207
Station App. Nth. DA15: Sidc 61Wc 139
Station App. Rd. CR5: Coul 87Mb 176
Station App. Rd. SE1 2K 229 (47Qb 90)
Station App. Rd. W4 52Sa 109
Station App. Sth. DA15: Sidc 61Wc 139
(off Jubilee Way)
Station App. Southside SE9 60Pc 116
Station App. W. RH1: Redh 8P 207
Station Arc. W1 6A 216 (42Kb 90)
(off Gt. Portland St.)
Station Av. CR3: Cat'm 96Wb 197
Station Av. KT12: Walt T 77W 150
Station Av. KT19: Ewe 81Ua 174
Station Av. KT3: N Mald 69Ua 132
Station Av. SW9 55Rb 113
Station Av. TW9: Kew 53Qa 109
Station Bldgs. KT1: King T 68Na 131
(off Fife Rd.)
Station Chambers E6 38Nc 74
(off High St. Nth.)
Station Cl. AL9: Brk P 8G 8
Station Cl. EN6: Pot B 3Bb 17
Station Cl. N12 21Db 49
Station Cl. N3 25Cb 49
Station Cl. TW12: Hamp 67Da 129
Station Cotts. BR6: Orp 75Vc 161
Station Ct. KT12: Hers 77W 150
Station Ct. N15 29Vb 51
Station Ct. SW6 53Eb 111
Station Ct. TN15: Bor G 91Be 205
Station Cres. HA0: Wemb 37Ka 66
Station Cres. N15 28Tb 51
Station Cres. SE3 50Jc 93
Station Cres. TW15: Ashf 62M 127
Stationer's Hall Ct. EC4 3C 224 (44Rb 91)
Stationers Pl. HP3: Hem H 7N 3
Station Est. BR3: Beck 69Zb 136
Station Est. E18 26Kc 53
Station Est. Rd. TW14: Felt 60X 107
Station Footpath WD4: K Lan 2R 12
(not continuous)
Station Forecourt WD3: Rick 17M 25
(off Homestead Rd.)
Station Garage M. SW16 65Mb 134
Station Gdns. W4 52Sa 109
Station Gro. HA0: Wemb 37Na 67
Station Hill BR2: Hayes 75Jc 159
Station Ho. SE8 52Cc 114
(off Deptford High St.)
Station Ho. M. N9 21Wb 51
Station La. RM12: Horn 34Md 77
Station M. EN6: Pot B 3Cb 17
Station M. Ter. SE3 50Jc 93
Station Pde. UB5: N'olt Accock Gro. 35Da 65
Station Pde. UB5: N'olt Court Farm Rd 38Ca 65
Station Pde. BR1: Broml 67Jc 137
(off Tweedy Rd.)
Station Pde. DA15: Sidc 61Wc 139
Station Pde. E11 29Jc 53
Station Pde. E13 39Lc 73
(off Green St.)
Station Pde. E5 34Xb 71
(off Up. Clapton Rd.)
Station Pde. E6 38Nc 74
Station Pde. EN4: Cockf 14Jb 32
Station Pde. GU25: Vir W 10P 125
Station Pde. HA2: Harr 35Da 65
Station Pde. HA3: Kenton 26Ja 46
Station Pde. HA4: Ruis 33T 64
Station Pde. HA8: Edg 24Na 47
Station Pde. IG11: Bark 38Sc 74
Station Pde. IG9: Buck H 21Mc 53
Station Pde. KT24: E Hor 98U 190
Station Pde. N14 18Mb 32
Station Pde. NW2 37Ya 68
Station Pde. RM1: Rom 30Gd 56
Station Pde. RM12: Horn 35Kd 77
Station Pde. RM9: Dag 37Cd 76
Station Pde. SL5: S'dale 3E 146
Station Pde. SM2: Sutt 79Eb 155
(off High St.)
Station Pde. SW12 60Jb 112
Station Pde. TN13: S'oaks 96Jd 202
Station Pde. TW14: Felt 60X 107
Station Pde. TW15: Ashf 63P 127
Station Pde. TW9: Kew 53Qa 109
Station Pde. UB9: Den 31J 63
Station Pde. W3 44Qa 87
Station Pde. W4 52Sa 109
Station Pde. W5 46Pa 87
Station Pas. E18 26Kc 53
Station Pas. E20 37Ec 72
Station Pas. SE15 53Yb 114
Station Path SW6: Putney Bridge 55Bb 111
Station Path TW18: Staines 63H 127
Station Pl. N4 33Qb 70
Station Ri. SE27 61Rb 135
Station Rd. AL2: Brick W 3Ca 13
Station Rd. AL4: S'ford 1M 7
Station Rd. AL9: Brk P 6E 8
Station Rd. AL9: N Mym 6E 8
Station Rd. AL9: Wel G 6E 8
Station Rd. BR1: Broml 67Jc 137
Station Rd. BR2: Broml 68Gc 137
Station Rd. BR4: W W'ck 74Ec 158
Station Rd. BR5: St P 70Yc 139
Station Rd. BR6: Orp 75Vc 161
Station Rd. BR8: Swan 70Gd 140
Station Rd. CM13: W H'dn 30Ee 59
Station Rd. CM16: Epp 3Vc 23
Station Rd. CR0: C'don 74Sb 157
Station Rd. CR3: Whyt 90Vb 177
Station Rd. CR3: Wold 94Ac 198
Station Rd. CR8: Kenley 86Sb 177
Station Rd. DA1: Cray 59Hd 118
Station Rd. DA11: Nflt 58De 121
Station Rd. DA13: Meop 10C 144
Station Rd. DA13: Sflt 63Be 143
Station Rd. DA15: Sidc 61Wc 139
Station Rd. DA17: Belv 48Cd 96
Station Rd. DA3: Lfield 69Ae 143
Station Rd. DA4: Eyns 76Md 163
Station Rd. DA4: S Dar 68Rd 141
Station Rd. DA7: Bex 55Ad 117
Station Rd. DA9: Ghithe 57Wd 120
(not continuous)
Station Rd. E12 35Nc 74
Station Rd. E17 30Ac 52
Station Rd. E4 18Fc 35
Station Rd. E7 35Jc 73
Station Rd. EN5: New Bar 15Db 31
Station Rd. EN6: Cuff 1Pb 18
Station Rd. EN9: Walt A 6Cc 20
Station Rd. EN9: Walt C 6Cc 20
Station Rd. GU24: Chob 3K 167
Station Rd. HA1: Harr 28Ha 46
Station Rd. HA2: Harr 29Da 45
Station Rd. HA8: Edg 23Qa 47
Station Rd. HP1: Hem H 4K 3
Station Rd. IG1: Ilf 34Rc 74
Station Rd. IG10: Lough 14Nc 36
Station Rd. IG6: Ilf 27Tc 54
Station Rd. IG7: Chig 20Rc 36
Station Rd. KT1: Hamp W 67La 130
Station Rd. KT10: Clay 79Ga 152
Station Rd. KT10: Esh 75Fa 152
Station Rd. KT11: Stoke D 89Aa 171
Station Rd. KT14: W Byf 84J 169
Station Rd. KT15: Add 77L 149
Station Rd. KT16: Chert 74H 149
Station Rd. KT2: King T 67Qa 131
Station Rd. KT22: Lea 93Ja 192
Station Rd. KT3: N Mald 71Xa 154
Station Rd. KT7: T Ditt 73Ha 152
Station Rd. KT9: Chess 78Na 153
Station Rd. N11 22Kb 50
Station Rd. N17 27Wb 51
Station Rd. N19 34Lb 70
Station Rd. N21 18Rb 33
Station Rd. N22 26Nb 50
(not continuous)
Station Rd. N3 25Cb 49
Station Rd. NW10 40Va 68
Station Rd. NW4 30Wa 48
Station Rd. NW7 23Ua 48
Station Rd. RH1: Mers 100Lb 196
Station Rd. RH1: Redh 5N 207
Station Rd. RH3: B'wth 3A 206
Station Rd. RH9: S God 10C 210
Station Rd. RM14: Upm 33Sd 78
Station Rd. RM18: E Til 1H 123
Station Rd. RM18: W Til 1H 123
Station Rd. RM2: Rom 28Kd 57
Station Rd. RM3: Hrld W 25Pd 57
Station Rd. RM6: Chad H 31Zc 75
Station Rd. RM6: Dag 31Zc 75
Station Rd. SE13 55Ec 114
Station Rd. SE20 65Yb 136
Station Rd. SE25 70Vb 135
Station Rd. SL1: Slou 4C 80
Station Rd. SL3: L'ly 48C 82
Station Rd. SL5: S'dale 2E 146
Station Rd. SL9: Ger X 29A 42
Station Rd. SM2: Sutt 82Cb 175
Station Rd. SM5: Cars 77Hb 155
Station Rd. SW13 54Va 110
Station Rd. SW19 67Eb 133
Station Rd. TN13: Dun G 92Gd 202
Station Rd. TN14: Hals 82Bd 181
Station Rd. TN14: Otf 88Kd 183
Station Rd. TN14: S'ham 83Jd 182
Station Rd. TN15: Bor G 92Be 205
Station Rd. TN16: Bras 95Xc 201
Station Rd. TW1: Twick 60Ha 108
Station Rd. TW11: Tedd 65Ja 130
Station Rd. TW12: Hamp 67Ca 129
Station Rd. TW15: Ashf 63P 127
Station Rd. TW16: Sun 66W 128
Station Rd. TW17: Shep 71S 150
Station Rd. TW19: Wray 58B 104
Station Rd. TW20: Egh 64C 126
Station Rd. TW3: Houn 56Da 107
Station Rd. UB3: Harl 49U 84
(not continuous)
Station Rd. UB3: Hayes 49U 84
(not continuous)
Station Rd. UB7: W Dray 47N 83
Station Rd. UB8: Cowl 42L 83
Station Rd. W5 44Pa 87
Station Rd. W7 46Ga 86
Station Rd. WD17: Wat 12X 27
Station Rd. WD3: Rick 17M 25
Station Rd. WD4: K Lan 1R 12
Station Rd. WD6: Bore 14Qa 29
Station Rd. WD7: R'lett 7Ja 14
Station Rd. E. RH8: Oxt 1J 211
Station Rd. Nth. DA17: Belv 48Ed 96
Station Rd. Nth. RH1: Mers 100Lb 196
Station Rd. Nth. TW20: Egh 64C 126
Station Rd. Sth. RH1: Mers 100Lb 196
Station Rd. W. RH8: Oxt 1J 211
Station Sq. BR5: Pet W 71Sc 160
Station Sq. RM2: Rom 28Kd 57
Station St. E15 38Fc 73
Station St. E16 46Rc 94
Station Ter. AL2: Park 8B 6
Station Ter. NW10 40Za 68
Station Ter. RM19: Purf 50Qd 97
Station Ter. SE5 53Sb 113
Station Vw. UB6: G'frd 39Fa 66
Station Wlk. IG1: Ilf 33Rc 74
(within The Exchange)
Station Wlk. W11 45Za 88
(off Bramley Rd.)
Station Way AL1: St A 2D 6
Station Way IG9: Buck H 21Lc 53
Station Way KT10: Clay 79Ga 152
Station Way KT19: Eps 85Ta 173
Station Way SE15 54Wb 113
Station Way SE18 48Rc 94
Station Way SM3: Cheam 79Ab 154
Station Yd. AL4: S'ford 2M 7
Station Yd. CR8: Purl 84Rb 177
Station Yd. HA4: Ruis 33Ac 64
Station Yd. KT20: Kgswd 93Bb 195
Station Yd. TW1: Twick 59Ja 108
Station Yd. UB9: Den 31J 63
Staton Ct. E10 31Dc 72
(off Kings Cl.)
Staunton Ho. SE17 6H 231 (49Ub 91)
(off Wansey St.)
Staunton Rd. KT2: King T 65Na 131
Staunton Rd. SL2: Slou 3H 81
Staunton St. SE8 51Bc 114
Stave Hill Ecological Pk. 47Ac 92
Staveley NW1 3B 216 (41Lb 90)
(off Varndell St.)
Staveley Cl. E9 36Yb 72
Staveley Cl. N7 35Nb 70
Staveley Cl. SE15 53Xb 113
Staveley Ct. E11 29Jc 53
Staveley Gdns. W4 53Ta 109
Staveley Rd. TW15: Ashf 65T 128
Staveley Rd. W4 51Sa 109
Staveley Way GU21: Knap 9J 167
Stavers Ho. E3 40Bc 72
(off Tredegar Rd.)
Staverton Pl. BR1: Broml 70Pc 138
Staverton Rd. NW2 38Ya 68
Staverton Rd. RM11: Horn 30Md 57
Stave Yd. Rd. SE16 46Ac 92
Stavordale Lodge W14 48Bb 89
(off Melbury Rd.)
Stavordale Rd. N5 35Rb 71
Stavordale Rd. SM5: Cars 73Eb 155
Stayne End GU25: Vir W 10L 125
Stayner's Rd. E1 42Zb 92
Stayton Rd. SM1: Sutt 76Cb 155
Stead Cl. BR7: Chst 64Qc 138
Steadfast Rd. KT1: King T 67Ma 131
Steadman Ct. EC1 5E 218 (42Sb 91)
(off Old St.)
Steadman Ho. RM10: Dag 34Cd 76
(off Uvedale Rd.)
Stead St. SE17 6F 231 (49Tb 91)
Steam Farm La. TW14: Felt 56V 106
Stean St. E8 39Vb 71
Stebbing Ho. W11 46Za 88
(off Queensdale Cres.)
Stebbing Way IG11: Bark 40Wc 75
Stebondale St. E14 49Ec 92
Stede Cl. CR2: Sande 84Wb 177
Stedham Pl. WC1 2F 223 (44Nb 90)
(off New Oxford St.)
Stedman Cl. DA5: Bexl 62Gd 140
Stedman Cl. UB10: Ick 34Q 64
Steed Cl. RM11: Horn 33Kd 77
Steedman St. SE17 6D 230 (49Sb 91)
Steeds Rd. N10 25Hb 49
Steeds Way IG10: Lough 13Nc 36
Steele Av. DA9: Ghithe 57Vd 120
Steele Ct. TW11: Tedd 66La 130
Steele Ho. E15 40Gc 73
(off Eve Rd.)
Steele Rd. E11 35Gc 73
Steele Rd. N17 27Ub 51
Steele Rd. NW10 40Sa 67
Steele Rd. TW7: Isle 56Ja 108
Steele Rd. W4 48Sa 87
Steele's M. Nth. NW3 37Hb 69
Steele's M. Sth. NW3 37Hb 69
Steele's Rd. NW3 37Hb 69
Steele's Studios NW3 37Hb 69
Steele Wlk. DA8: Erith 52Dd 118
Steel's La. E1 44Yb 92
Steel's La. KT22: Oxs 86Da 171
Steelyard Pas. EC4 5F 225 (45Tb 91)
(off Cousin La.)
Steen Way SE22 57Ub 113
Steep Cl. BR6: Chels 79Vc 161
Steep Hill CR0: C'don 77Ub 157
Steep Hill GU24: Chob 10G 146
Steep Hill SW16 62Mb 134
Steeplands WD23: Bush 17Da 27
Steeple Cl. SW19 64Ab 132
Steeple Cl. SW6 54Ab 110
Steeple Ct. E1 42Xb 91
Steeple Ct. TW20: Egh 64D 126
Steeple Gdns. KT15: Add 78K 149
Steeple Hgts. Dr. TN16: Big H 89Mc 179
Steeple Point SL5: Asc 9A 124
Steeplestone Cl. N18 22Sb 51
Steeple Wlk. N1 39Sb 71
(off New Nth. Rd.)
Steerforth St. SW18 61Eb 133
Steering Cl. N9 18Yb 34
Steers Mead CR4: Mitc 67Hb 133
Steers Way SE16 47Ac 92
Stelfox Ho. WC1 3J 217 (41Pb 90)
(off Penton Ri.)
Stella Cl. UB8: Hil 43R 84
Stellar Ho. N17 23Vb 51
Stella Rd. SW17 65Hb 133
Stelling Rd. DA8: Erith 52Fd 118
Stellman Cl. E5 34Wb 71
Stembridge Rd. SE20 68Xb 135
Stem Ct. KT16: Chert 74L 149
Sten Cl. EN3: Enf L 9Ac 20
Stenning Av. SS17: Linf 9K 101
Stents La. KT11: Stoke D 92Ba 191
Stepbridge Path GU21: Wok 9P 167
Stepgates KT16: Chert 73K 149
Stepgates Cl. KT16: Chert 73K 149
Stephan Cl. E8 39Wb 71
Stephen Av. RM13: Rain 37Jd 76
Stephen Cl. BR6: Orp 76Uc 160
Stephen Cl. TW20: Egh 65E 126
Stephendale Rd. SW6 55Db 111
Stephendale Yd. SW6 55Db 111
(off Stephendale Rd.)
Stephen Fox Ho. W4 50Ua 88
(off Chiswick La.)
Stephen Jewers Gdns. IG11: Bark 38Vc 75
Stephen M. W1 1D 222 (43Mb 90)
Stephen Pl. SW4 55Lb 112
Stephen Rd. DA7: Bex 55Ed 118
Stephens Cl. RM3: Rom 22Ld 57
Stephens Ct. E16 42Hc 93
Stephens Ct. SE4 55Ac 114
Stephens Lodge N12 20Eb 31
(off Woodside La.)
Stephenson Av. RM18: Tilb 3C 122
Stephenson Cl. DA16: Well 54Wc 117
Stephenson Cl. E3 41Dc 92
Stephenson Cl. SL1: Slou 7K 81
(off Osborne St.)
Stephenson Ct. SM2: Cheam 80Ab 154
(off Station App.)
Stephenson Dr. SL4: Wind 2F 102
Stephenson Ho. SE1 3D 230 (48Sb 91)
Stephenson Pl. RH1: Mers 100Lb 196
(off Station Rd. Nth.)
Stephenson Rd. E17 29Ac 52
Stephenson Rd. TW2: Whitt 59Ca 107
Stephenson Rd. W7 44Ha 86
Stephenson St. E16 42Gc 93
Stephenson St. NW10 41Ua 88
Stephenson Way NW1 5C 216 (42Lb 90)
Stephenson Way WD23: Bush 14Z 27
Stephenson Wharf HP3: Hem H 7P 3
Stephen's Rd. E15 39Gc 73
Stephen St. W1 1D 222 (43Mb 90)
Stephen Woods Cen., The 82Ua 174
Stephyns Chambers HP1: Hem H 3M 3
STEPNEY 43Zb 92
Stepney C'way. E1 44Zb 92
Stepney City Apts. E1 43Yb 92
Stepney City Farm 43Zb 92
Stepney Cl. CR4: Mitc 67Jb 134
Stepney Grn. E1 43Yb 92
Stepney Grn. Ct. E1 43Zb 92
(off Stepney Grn.)
Stepney Green Station (Underground) 42Zb 92
Stepney High St. E1 43Zb 92
Stepney Way E1 43Xb 91
Sterling Av. EN8: Walt C 6Zb 20
Sterling Av. HA8: Edg 21Pa 47
Sterling Cl. NW10 38Wa 68
Sterling Ct. KT16: Chert 73J 149
Sterling Gdns. SE14 51Ac 114
Sterling Ind. Est. RM10: Dag 35Dd 76
Sterling Pl. KT13: Weyb 77U 150
Sterling Pl. W5 49Na 87
Sterling Rd. DA7: Bex 56Dd 118
Sterling Rd. EN2: Enf 11Tb 33
Sterling St. SW7 3E 226 (48Gb 89)
Sterling Way N18 21Tb 51
Sterling Way N7 37Pb 70
Stern Cl. IG11: Bark 40Yc 75
Stern Ct. E3 41Ec 92
(off Culvert Dr.)
Sterndale Rd. DA1: Dart 59Pd 119
Sterndale Rd. W14 48Za 88
Sterne St. W12 47Za 88
Sternhall La. SE15 55Wb 113
Sternhold Av. SW2 61Mb 134
Sterry Cres. RM10: Dag 36Cd 76
Sterry Dr. KT19: Ewe 77Ua 154
Sterry Dr. KT7: T Ditt 72Ga 152
Sterry Gdns. RM10: Dag 37Cd 76
Sterry Rd. IG11: Bark 39Vc 75
Sterry Rd. RM10: Dag 35Cd 76
Sterry St. SE1 2F 231 (47Tb 91)
Steucers La. SE23 60Ac 114
Stevannie Ct. DA17: Belv 50Bd 95
Steve Biko Ct. W10 42Za 88
(off St John's Ter.)
Steve Biko La. SE6 63Cc 136
Steve Biko Lodge E13 40Jc 73
(off London Rd.)
Steve Biko Rd. N7 34Qb 70
Steve Biko Way TW3: Houn 55Ca 107
Stevedale Rd. DA16: Well 54Yc 117
Stevedore St. E1 46Xb 91
Stevenage Cres. WD6: Bore 11Na 29
Stevenage Rd. E6 37Qc 74
Stevenage Rd. SW6 52Za 110
Stevens Av. E9 37Yb 72
Stevens Cl. BR3: Beck 65Cc 136
Stevens Cl. DA2: Daren 64Ud 142
Stevens Cl. DA5: Bexl 63Fd 140
Stevens Cl. EN6: Pot B 5Za 16
Stevens Cl. HA5: Eastc 29Y 45
Stevens Cl. KT17: Eps 84Ua 174
Stevens Cl. TW12: Hamp 65Ba 129
Stevens Grn. WD23: B Hea 18Ea 28
Stevens La. KT10: Clay 80Ja 152
Stevenson Cl. DA8: Erith 52Kd 119
Stevenson Cl. EN5: New Bar 17Fb 31
Stevenson Ct. CR4: Mitc 69Gb 133
Stevenson Ct. SE6 61Hc 137
Stevenson Cres. SE16 50Wb 91
Stevenson Ho. NW8 39Eb 69
(off Boundary Rd.)
Stevenson Rd. SL2: Hedg 3H 61
Stevens Pl. CR8: Purl 85Rb 177
Stevens Rd. RM8: Dag 33Yc 75
Stevens St. SE1 3J 231 (48Ub 91)
Stevens Way IG7: Chig 21Uc 54
Steventon Rd. W12 45Va 88
Steward Cl. EN8: Chesh 2Ac 20
Steward Ho. E3 41Cc 92
(off Trevithick Way)
Stewards Cl. CM16: Epp 5Wc 23
STEWARD'S GREEN 4Xc 23
Stewards Grn. La. CM16: Epp 4Xc 23
Stewards Grn. Rd. CM16: Epp 5Wc 23
Stewards Grn. Rd. CM16: Fidd H 5Wc 23
Stewards Holte Wlk. N11 21Kb 50
Steward St. E1 1J 225 (43Ub 91)
Stewards Wlk. RM1: Rom 29Gd 56
Stewart KT20: Tad 93Za 194
Stewart Av. RM14: Upm 34Rd 77
Stewart Av. SL1: Slou 3K 81
Stewart Av. TW17: Shep 70Q 128
Stewart Cl. BR7: Chst 64Rc 138
Stewart Cl. GU21: Wok 9K 167
Stewart Cl. NW9 30Sa 47
Stewart Cl. TW12: Hamp 65Aa 129
Stewart Cl. WD5: Ab L 4V 12
Stewart Ct. CM16: Epp 3Vc 23
Stewart Ct. UB9: Den 30H 43
Stewart Ho. KT1: King T 69Pa 131
Stewart Quay UB3: Hayes 47U 84
Stewart Rainbird Ho. E12 36Qc 74
(off Parkhurst Rd.)
Stewart Rd. E15 35Fc 73
Stewartsby Cl. N18 22Sb 51
Stewart's Dr. SL2: Farn C 5F 60
Stewart's Gro. SW3 7C 226 (50Fb 89)
Stewarts Holte Wlk. WD5: Ab L 3V 12
Stewart's Pl. SW2 58Pb 112
Stewart's Rd. SW8 52Lb 112
Stewart St. E14 47Ec 92
Stew La. EC4 4D 224 (45Sb 91)
Steyne Horn La. W3 46Ra 87
Steyne Ho. W3 46Sa 87
(off Narrow St.)
Steyne Rd. W3 46Ra 87
Steyning Cl. CR8: Kenley 88Rb 177
Steyning Gro. SE9 63Pc 138
Steynings Way N12 22Cb 49
Steyning Way TW4: Houn 56Y 107
Steynton Av. DA5: Bexl 61Zc 139
Stibbington Ho. NW1 1C 216 (40Lb 70)
(off Cranleigh St.)
Stickland Rd. DA17: Belv 49Cd 96
Stickle Ho. SE8 52Dc 114
(off Creative Rd.)
Stickleton Cl. UB6: G'frd 41Da 85
Stifford Clays Rd. RM16: Grays 46Be 99
(not continuous)
Stifford Clays Rd. RM16: N Stif 46Be 99
(not continuous)
Stifford Clays Rd. RM16: Ors 46Be 99
(not continuous)
Stifford Hill RM15: N Stif 45Yd 98
Stifford Hill RM15: S Ock 45Yd 98
Stifford Hill RM16: N Stif 46Zd 99
Stifford Ho. E1 43Yb 92
(off Stepney Way)
Stifford Rd. RM15: Avel 46Ud 98
Stifford Rd. RM15: S Ock 46Ud 98
Stigand La. DA9: Ghithe 58Xd 120
Stilecroft Gdns. HA0: Wemb 34Ka 66
Stile Footpath RH1: Redh 6P 207
Stile Hall Gdns. W4 50Qa 87
Stile Hall Mans. W4 50Qa 87
(off Wellesley Rd.)
Stile Hall Pde. W4 50Qa 87

Stileman Ho. E3 43Bc **92**
(off Ackroyd Dr.)
Stile Path TW16: Sun 69W **128**
Stile Rd. SL3: L'ly 8P **81**
Stiles Cl. BR2: Broml 72Pc **160**
Stiles Cl. DA8: Erith 50Dd **96**
Stillingfleet Rd. SW13 51Wa **110**
Stillington St. SW1 5C **228** (49Lb **90**)
Stillness Rd. SE23 58Ac **114**
Still Wlk. SE1 7K **225** (46Vb **91**)
(off Duchess Wlk.)
Stilton Path WD6: Bore 10Qa **15**
Stilwell Cl. BR5: St P 67Xc **139**
Stilwell Dr. UB8: Hil 42P **83**
STILWELL RDBT. 45Q **84**
Stipularis Dr. UB4: Yead 42Z **85**
Stirling Av. HA5: Pinn 32Z **65**
Stirling Av. SM6: W'gton 80Nb **156**
Stirling Av. TW17: Shep 69U **128**
Stirling Bus. Pk. EN8: Walt C 6Bc **20**
Stirling Cl. DA14: Sidc 63Uc **138**
Stirling Cl. RM13: Rain 41Kd **97**
Stirling Cl. SL4: Wind 4B **102**
Stirling Cl. SM7: Bans 89Bb **175**
Stirling Cl. SW16 67Mb **134**
Stirling Cl. UB8: Cowl 41L **83**
Stirling Cnr. EN5: Ark 16Ta **29**
STIRLING CORNER 16Ta **29**
Stirling Ct. EC1 5B **218** (42Rb **91**)
(off St John St.)
Stirling Ct. W13 45Ka **86**
Stirling Dr. BR6: Chels 78Xc **161**
Stirling Dr. CR3: Cat'm 93Sb **197**
Stirling Gro. TW3: Houn 54Ea **108**
Stirling Ho. RH1: Redh 6P **207**
Stirling Ho. WD6: Bore 14Sa **29**
Stirling Retail Pk. 16Ta **29**
Stirling Rd. E13 40Kc **73**
Stirling Rd. E17 27Ac **52**
Stirling Rd. HA3: W'stone 27Ha **46**
Stirling Rd. N17 25Wb **51**
Stirling Rd. N22 25Rb **51**
Stirling Rd. SL1: Slou 3E **80**
Stirling Rd. SW9 54Nb **112**
Stirling Rd. TW2: Whitt 59Ca **107**
Stirling Rd. TW6: H'row A 58P **105**
Stirling Rd. UB3: Hayes 45X **85**
Stirling Rd. W3 48Ra **87**
Stirling Rd. Path E17 27Ac **52**
Stirling Wlk. KT5: Surb 72Ra **153**
Stirling Way CR0: Bedd 73Nb **156**
Stirling Way WD5: Ab L 4W **12**
Stirling Way WD6: Bore 16Ta **29**
Stites Hill Rd. CR5: Coul 92Rb **197**
Stiven Cres. HA2: Harr 34Ba **65**
Stoats Nest Rd. CR5: Coul 86Nb **176**
Stoats Nest Village CR5: Coul 87Nb **176**
Stockbeck NW1 2C **216** (40Lb **70**)
(off Ampthill Est.)
Stockbridge Ho. SW18 56Eb **111**
(off Eltringham St.)
Stockbury Rd. CR0: C'don 72Yb **158**
Stockdale Rd. RM8: Dag 33Bd **75**
Stockdales Rd. SL4: Eton W 9D **80**
Stockdove Way UB6: G'frd 41Ha **86**
Stocker Gdns. RM9: Dag 38Yc **75**
Stockers Farm Rd. WD3: Rick 20M **25**
Stocker's Lake 20K **25**
Stockers La. GU22: Wok 92B **188**
(not continuous)
Stockfield Rd. KT10: Clay 78Ga **152**
Stockfield Rd. SW16 62Pb **134**
Stockford Av. NW7 24Za **48**
Stockham's Cl. CR2: Sande 82Tb **177**
Stock Hill TN16: Big H 88Mc **179**
Stockholm Apts. NW1 38Jb **70**
(off Chalk Farm Rd.)
Stockholm Ho. E1 45Wb **91**
(off Swedenborg Gdns.)
Stockholm Rd. SE16 50Yb **92**
Stockholm Way E1 46Wb **91**
Stockhurst Cl. SW15 54Ya **110**
Stockingswater La. EN3: Brim 12Bc **34**
Stockland Rd. RM7: Rom 30Fd **56**
Stock La. DA2: Wilm 62Ld **141**
Stockleigh Hall NW8 1E **214** (40Gb **69**)
(off Prince Albert Rd.)
Stockley Cl. UB7: W Dray 47R **84**
Stockley Country Pk. 45Q **84**
Stockley Farm Rd. UB7: W Dray 48R **84**
STOCKLEY PARK 46R **84**
Stockley Pk. Golf Course 45S **84**
Stockley Rd. UB11: Stock P 45R **84**
Stockley Rd. UB7: W Dray 49R **84**
Stockley Rd. UB8: Hil 44Q **84**
Stock Orchard Cres. N7 36Pb **70**
Stock Orchard St. N7 36Pb **70**
Stockport Rd. SW16 67Mb **134**
Stockport Rd. WD3: Herons 17E **24**
Stocksfield Rd. E17 27Ec **52**
Stocks Ho. KT22: Lea 94Ka **192**
Stocks Mdw. HP2: Hem H 1A **4**
Stocks Pl. E14 45Bc **92**
Stocks Pl. UB10: Hil 39Q **64**
Stock St. E13 40Jc **73**
Stockton Cl. EN5: New Bar 14Eb **31**
Stockton Ct. SW1 4D **228** (48Mb **90**)
(off Greycoat St.)
Stockton Gdns. N17 24Sb **51**
Stockton Gdns. NW7 20Ua **30**
Stockton Ho. E2 41Xb **91**
(off Ellsworth St.)
Stockton Ho. HA2: Harr 32Ca **65**
Stockton Rd. N17 24Sb **51**
Stockton Rd. N18 23Wb **51**
Stockton Rd. RH2: Reig 9J **207**
STOCKWELL 54Pb **112**
Stockwell Av. SW9 55Pb **112**
Stockwell Cl. BR1: Broml 68Kc **137**
Stockwell Cl. HA8: Edg 26Sa **47**
Stockwell Gdns. SW9 53Pb **112**
Stockwell Gdns. Est. SW9 54Nb **112**
Stockwell Grn. SW9 54Pb **112**
Stockwell Grn. Ct. SW9 54Pb **112**
Stockwell La. EN7: Chesh 1Xb **19**
(not continuous)
Stockwell La. SW9 54Pb **112**
Stockwell M. SW9 54Pb **112**
Stockwell Pk. Cres. SW9 54Pb **112**
Stockwell Pk. Est. SW9 54Pb **112**
Stockwell Pk. Rd. SW9 53Pb **112**
Stockwell Pk. Wlk. SW9 55Pb **112**
Stockwell Rd. SW9 54Pb **112**
Stockwell Station (Underground) 53Nb **112**
Stockwell St. SE10 51Ec **114**
Stockwell Ter. SW9 53Pb **112**
Stodart Rd. SE20 67Yb **136**
Stoddart Ho. SW8 51Pb **112**
Stodmarsh Ho. SW9 53Qb **112**
(off Cowley Rd.)
Stofield Gdns. SE9 62Mc **137**
Stoford Cl. SW19 59Ab **110**
Stoke Av. IG6: Ilf 23Wc **55**
Stoke Cl. KT11: Stoke D 88Ba **171**
STOKE COMMON 5L **61**
Stoke Comn. Rd. SL3: Ful 5L **61**
Stoke Cotts. KT22: Fet 93Fa **192**
Stoke Ct. Dr. SL2: Stoke P 9J **61**
STOKE D'ABERNON 89Aa **171**
Stoke Gdns. SL1: Slou 6J **81**
STOKE GREEN 2L **81**
Stoke Grn. SL2: Stoke P 2L **81**
Stoke Ho. SL2: Stoke P 2L **81**
Stokenchurch St. SW6 53Db **111**
STOKE NEWINGTON 34Vb **71**
Stoke Newington Chu. St. N16 34Tb **71**
Stoke Newington Comn. N16 34Vb **71**
Stoke Newington High St. N16 34Vb **71**
Stoke Newington Rd. N16 36Vb **71**
Stoke Newington Station (Overground) 33Vb **71**
Stoke Pk. Stoke Poges 8J **61**
Stoke Pk. Av. SL2: Farn R 1G **80**
Stoke Pl. NW10 41Va **88**
STOKE POGES 8L **61**
Stoke Poges Golf Course 1H **81**
Stoke Poges La. SL1: Slou 6J **81**
Stoke Poges La. SL2: Slou 3J **81**
Stoke Poges La. SL2: Stoke P 3J **81**
Stoke Rd. KT11: Cobh 87Y **171**
Stoke Rd. KT11: Stoke D 87Y **171**
Stoke Rd. KT12: Walt T 76Y **151**
Stoke Rd. KT2: King T 66Sa **131**
Stoke Rd. RM13: Rain 40Md **77**
Stoke Rd. SL2: Slou 6K **81**
Stoke Rd. SL2: Stoke P 6K **81**
Stokesay SL2: Slou 5K **81**
Stokesay Ct. DA2: Dart 58Rd **119**
(off Osbourne Rd.)
Stokesby Rd. KT9: Chess 79Pa **153**
Stokes Cotts. IG6: Ilf 25Sc **54**
Stokes Ct. N2 28Gb **49**
Stokes Field Local Nature Reserve 75Ka **152**
Stokesheath Rd. KT22: Oxs 83Ea **172**
Stokesley St. W12 44Va **88**
Stokes M. TW11: Tedd 64Ja **130**
Stokes Ridings KT20: Tad 95Za **194**
Stokes Rd. CR0: C'don 72Zb **158**
Stokes Rd. E6 42Nc **94**
Stoke Vw. SL1: Slou 6K **81**
Stoke Wood SL2: Stoke P 5K **61**
Stokley Ct. N8 28Nb **50**
Stoll Cl. NW2 34Ya **68**
Stompond La. KT12: Walt T 75W **150**
Stoms Path SE6 64Cc **136**
(off Maroons Way)
Stonard Rd. N13 20Qb **32**
Stonard Rd. RM8: Dag 36Xc **75**
Stonards Hill CM16: Coop 1Wc **23**
Stonards Hill CM16: Epp 1Wc **23**
Stonards Hill IG10: Lough 16Pc **36**
Stondon Pk. SE23 58Ac **114**
Stondon Wlk. E6 40Mc **73**
STONE 57Ud **120**
Stonebanks KT12: Walt T 73W **150**
STONEBRIDGE 39Ta **67**
Stonebridge Cen. N15 29Vb **51**
Stonebridge Fld. SL4: Eton 10F **80**
Stonebridge Gdns. 39Vb **71**
(off Arbutus St.)
Stonebridge M. SE19 66Tb **135**
Stonebridge Pk. NW10 38Ta **67**
Stonebridge Park Station (Underground & Overground) 38Ra **67**
Stonebridge Rd. DA11: Nflt 57Ce **121**
Stonebridge Rd. N15 29Vb **51**
Stonebridge Road 57Ce **121**
Stonebridge Way HA9: Wemb 37Ra **67**
Stone Bldgs. WC2 1J **223** (43Pb **90**)
(off Chancery La.)
Stone Castle Dr. DA9: Ghithe 58Wd **120**
Stonechat M. DA9: Ghithe 58Wd **120**
Stonechat M. SW15 56Wa **110**
Stonechat Sq. E6 43Nc **94**
Stone Cl. RM8: Dag 33Bd **75**
Stone Cl. SW4 54Lb **112**
Stone Cl. UB7: Yiew 46P **83**
Stonecot Cl. SM3: Sutt 74Ab **154**
Stonecot Hill SM3: Sutt 74Ab **154**
Stone Ct. CR3: Cat'm 97Ub **197**
Stone Ct. DA8: Erith 50Hd **96**
Stone Ct. TN15: Bor G 91Ce **205**
Stone Cres. TW14: Felt 59V **106**
Stonecroft Av. SL0: Iver 44G **82**
Stonecroft Cl. EN5: Ark 14Xa **30**
Stonecroft Rd. DA8: Erith 52Ed **118**
Stonecroft Way CR0: C'don 73Nb **156**
Stonecrop Cl. NW9 27Ta **47**
Stonecross AL1: St A 1C **6**
Stonecross Cl. AL1: St A 1C **6**
Stone Crossing Station (Rail) 57Ud **120**
Stonecutter St. EC4 2B **224** (44Rb **91**)
Stonefield N4 33Pb **70**
Stonefield Cl. DA7: Bex 55Cd **118**
Stonefield Cl. HA4: Ruis 36Aa **65**
Stonefield Mans. N1 39Qb **70**
(off Cloudesley St.)
Stonefield St. N1 39Qb **70**
Stonefield Way HA4: Ruis 35Aa **65**
Stonefield Way SE7 52Mc **115**
Stonegate Cl. BR5: St P 69Yc **139**
Stone Ga. Ct. TW18: Staines 63J **127**
Stonegrove HA8: Edg 21Na **47**
STONEGROVE 21Pa **47**
Stone Gro. Ct. HA8: Edg 22Pa **47**
Stonegrove Gdns. HA8: Edg 22Na **47**
Stone Hall W8 48Db **89**
(off Stone Hall Gdns.)
Stonehall Av. IG1: Ilf 30Nc **54**
Stone Hall Gdns. W8 48Db **89**
Stone Hall Pl. W8 48Db **89**
Stone Hall Rd. N21 17Pb **32**
Stoneham Rd. N11 22Lb **50**
Stoneham Rd. SS17: Stan H 3K **101**
STONEHILL 79A **148**
Stonehill Cl. KT23: Bookh 97Ca **191**
Stonehill Cl. SW14 57Ta **109**
Stonehill Ct. E4 17Dc **34**
Stonehill Cres. KT16: Ott 79A **148**
Stonehill Ga. SL5: S'hill 2B **146**
STONEHILL GREEN 66Ed **140**
Stonehill Grn. DA2: Wilm 66Ed **140**
Stonehill Rd. GU24: Chob 2N **167**
Stonehill Rd. KT16: Ott 78B **148**
Stonehill Rd. SW14 57Sa **109**
Stonehill Rd. W4 50Qa **87**
Stonehills Ct. SE21 62Ub **135**
Stonehill Woods Pk. DA14: Sidc 65Dd **140**
Stonehorse Rd. EN3: Pond E 15Yb **34**
Stonehouse NW1 1C **216** (39Lb **70**)
(off Plender St.)
Stone Ho. Fld. TN15: Plat 92De **205**
Stone Ho. Gdns. CR3: Cat'm 97Ub **197**
Stonehouse Ho. W2 43Cb **89**
(off Westbourne Pk. Rd.)
Stone Ho. La. DA2: Dart 58Sd **120**
Stonehouse La. RM19: Purf 50Ud **98**
Stonehouse La. TN14: Hals 81Zc **181**
Stonehouse Rd. TN14: Hals 82Yc **181**
Stoneings La. TN14: Knock 90Vc **181**
Stone Lake Ind. Pk. SE7 49Lc **93**
Stone Lake Retail Pk. 49Lc **93**
Stonelea Rd. HP3: Hem H 4P **3**
STONELEIGH 78Wa **154**
Stoneleigh Av. EN1: Enf 10Xb **19**
Stoneleigh Av. KT4: Wor Pk 77Wa **154**
Stoneleigh B'way. KT17: Ewe 78Wa **154**
Stoneleigh Cl. EN8: Walt C 5Zb **20**
Stoneleigh Cl. IG5: Ilf 27Nc **54**
Stoneleigh Ct. KT11: Stoke D 89Aa **171**
Stoneleigh Cres. KT19: Ewe 78Va **154**
Stoneleigh M. E3 40Ac **72**
Stoneleigh Pk. KT13: Weyb 78S **150**
Stoneleigh Pk. Av. CR0: C'don 72Zb **158**
Stoneleigh Pk. Rd. KT19: Ewe 79Va **154**
Stoneleigh Pk. Rd. KT4: Wor Pk 77Wa **154**
Stoneleigh Pl. W11 45Za **88**
Stoneleigh Rd. BR1: Broml 69Rc **138**
Stoneleigh Rd. IG5: Ilf 27Nc **54**
Stoneleigh Rd. N17 27Vb **51**
Stoneleigh Rd. RH8: Limp 2P **211**
Stoneleigh Rd. SM5: Cars 73Gb **155**
Stoneleigh Station (Rail) 78Wa **154**
Stoneleigh St. W11 45Za **88**
Stoneleigh Ter. N19 33Kb **70**
Stonell's Rd. SW11 58Hb **111**
Stonely Cres. DA9: Ghithe 56Yd **120**
Stonemason Ct. SE1 2D **230** (47Sb **91**)
(off Borough Rd.)
Stonemason Ho. SE14 53Zb **114**
(off Fishers Ct.)
Stonemasons Cl. N15 28Tb **51**
Stonemasons Yd. SW18 59Fb **111**
Stoneness Rd. RM20: W Thur 51Xd **120**
Stonenest St. N4 32Pb **70**
Stone Pk. Av. BR3: Beck 70Cc **136**
Stone Pl. KT4: Wor Pk 75Wa **154**
Stone Pl. Rd. DA9: Ghithe 57Ud **120**
Stone Rd. BR2: Broml 71Hc **159**
Stones All. WD18: Wat 14X **27**
Stones Av. DA1: Dart 54Pd **119**
Stones Cross Rd. BR8: Crock 71Ed **162**
Stones End St. SE1 2D **230** (47Sb **91**)
Stone's Rd. KT17: Eps 84Ua **174**
Stone St. DA11: Grav'nd 9D **122**
STONE STREET 97Vd **204**
Stone St. Rd. TN15: Ivy H 95Sd **204**
Stone St. Rd. TN15: Seal 95Sd **204**
Stoneswood Rd. RH8: Limp 2M **211**
Stonewall E6 43Qc **94**
Stoneway Wlk. E3 40Zb **72**
Stone Well Rd. TW19: Stanw 60N **105**
Stonewold Ct. W5 44Ma **87**
Stonewood DA2: Bean 62Yd **142**
STONEWOOD 62Zd **143**
Stonewood Rd. DA8: Erith 50Gd **96**
Stoney All. SE18 54Qc **116**
Stoneyard La. E14 45Dc **92**
Stoney Bri. Dr. EN9: Walt A 6Jc **21**
Stoney Cft. CR5: Coul 94Lb **196**
Stoneycroft HP1: Hem H 2J **3**
Stoneycroft Cl. SE12 59Hc **115**
Stoneycroft Rd. IG8: Wfd G 23Nc **54**
Stoneydeep TW11: Tedd 63Ja **130**
Stoneydown E17 28Ac **52**
Stoneydown Av. E17 28Ac **52**
Stoneydown Ho. E17 28Ac **52**
(off Stoneydown)
Stoneyfield SL9: Ger X 2N **61**
Stoneyfield Rd. CR5: Coul 89Pb **176**
Stoneyfields Gdns. HA8: Edg 21Ta **47**
Stoneyfields La. HA8: Edg 22Sa **47**
Stoneylands Ct. TW20: Egh 64B **126**
Stoneylands Rd. TW20: Egh 64B **126**
Stoney La. E1 2K **225** (44Vb **91**)
Stoney La. HP1: Hem H 5C **2**
Stoney La. HP3: Bov 9D **2**
Stoney La. SE19 65Vb **135**
Stoney La. SL2: Farn R 9E **60**
Stoney La. WD4: Chfd 2G **10**
Stoney Meade SL1: Slou 6F **80**
Stoney St. SE1 6F **225** (46Tb **91**)
Stonhouse St. SW4 56Mb **112**
Stonny Cft. KT21: Asht 89Pa **173**
Stonor Rd. W14 49Bb **89**
Stony Cnr. DA13: Meop 10A **144**
Stonycroft Cl. EN3: Enf H 12Ac **34**
Stony Hill KT10: Esh 80Ba **151**
Stony La. HP6: L Chal 10A **10**
Stony Path IG10: Lough 11Pc **36**
Stonyshotts EN9: Walt A 6Gc **21**
Stoop Ct. KT14: W Byf 84K **169**
Stopes Av. DA10: Swans 60Ce **121**
Stopes St. SE15 52Vb **113**
Stopford Rd. E13 39Jc **73**
Stopford Rd. SE17 7C **230** (50Rb **91**)
Stopher Ho. SE1 2C **230** (47Rb **91**)
(off Webber St.)
Storas Ct. RM19: Purf 50Rd **97**
(off Linnet Way)
Storehouse M. E14 45Cc **92**
Storer Dr. DA16: Well 55Xc **117**
Store Rd. E16 47Qc **94**
Storers Quay E14 49Fc **93**
Store St. E15 36Fc **73**
Store St. WC1 1D **222** (43Mb **90**)
Storey Cl. UB10: Ick 34S **64**
Storey Ct. NW8 4B **214** (41Fb **89**)
(off St John's Wood Rd.)
Storey Ct. WD23: Bush 15Da **27**
Storey Ho. E14 45Dc **92**
(off Cottage St.)
Storey Rd. E17 28Bc **52**
Storey Rd. N6 30Hb **49**
Storey's Ga. SW1 2E **228** (47Mb **90**)
Storey St. E16 46Qc **94**
Storey St. HP3: Hem H 6M **3**
Stories M. SE5 54Ub **113**
Stories Rd. SE5 55Ub **113**
Stork Rd. E7 37Hc **73**
Storksmead Rd. HA8: Edg 24Ua **48**
Stork's Rd. SE16 48Wb **91**
Stormont Lawn Tennis & Squash Club 29Hb **49**
Stormont Rd. N6 31Hb **69**
Stormont Rd. SW11 55Jb **112**
Stormont Way KT9: Chess 78La **152**
Stormount Dr. UB3: Harl 47S **84**
Stornaway Rd. SL3: L'ly 49E **82**
Stornaway Strand DA12: Grav'nd 2H **145**
Stornoway HP3: Hem H 4B **4**
Storr Gdns. CM13: Hut 15Fe **41**
Storrington WC1 4G **217** (41Nb **90**)
(off Regent Sq.)
Storrington Rd. CR0: C'don 74Vb **157**
Storr's La. GU3: Worp 6G **186**
Storth Oaks Mead BR7: Chst 64Pc **138**
Story St. N1 38Pb **70**
Stothard Ho. E1 42Yb **92**
(off Amiel St.)
Stothard Pl. E1 7J **219** (43Ub **91**)
Stothard St. E1 42Yb **92**
Stott Cl. SW18 58Fb **111**
Stoughton Av. SM3: Cheam 78Za **154**
Stoughton Cl. SE11 6J **229** (49Pb **90**)
Stoughton Cl. SW15 60Wa **110**
Stour Av. UB2: S'hall 48Ca **85**
Stourcliffe Cl. W1 2F **221** (44Hb **89**)
Stourcliffe St. W1 3F **221** (44Hb **89**)
Stour Cl. BR2: Kes 77Lc **159**
Stour Cl. SL1: Slou 8F **80**
Stourhead Cl. SW19 59Za **110**
Stourhead Gdns. SW20 69Wa **132**
Stourhead Ho. SW1 7D **228** (50Mb **90**)
(off Tachbrook St.)
Stour Rd. DA1: Cray 55Jd **118**
Stour Rd. E3 38Cc **72**
Stour Rd. RM10: Dag 33Cd **76**
Stour Rd. RM16: Grays 10C **100**
Stourton Av. TW13: Hanw 63Ba **129**
Stour Way RM14: Upm 30Ud **58**
Stovell Rd. SL4: Wind 2F **102**
Stowage SE8 51Cc **114**
Stow Ct. DA2: Dart 59Sd **120**
Stow Cres. E17 24Ac **52**
Stowe Cres. HA4: Ruis 30R **44**
Stowe Gdns. N9 18Vb **33**
Stowe Ho. NW11 30Eb **49**
Stowell Av. CR0: New Ad 82Fc **179**
Stowell Ho. N8 28Nb **50**
(off Pembroke Rd.)
Stowe Pl. N15 27Ub **51**
Stowe Rd. BR6: Chels 77Xc **161**
Stowe Rd. SL1: Slou 5C **80**
Stowe Rd. W12 47Xa **88**
Stowting Rd. BR6: Orp 77Uc **160**
Stox Mead HA3: Hrw W 25Ea **46**
Stracey Rd. E7 35Jc **73**
Stracey Rd. NW10 39Ta **67**
Strachan Pl. SW19 65Ya **132**
Stradbroke Dr. IG7: Chig 23Qc **54**
Stradbroke Gro. IG5: Ilf 27Nc **54**
Stradbroke Gro. IG9: Buck H 18Mc **35**
Stradbroke Pk. IG7: Chig 23Rc **54**
Stradbroke Rd. N5 35Sb **71**
Stradbrook Cl. HA2: Harr 34Ba **65**
Stradella Rd. SE24 58Sb **113**
Strafford Av. IG5: Ilf 26Qc **54**
Strafford Cl. EN6: Pot B 4Cb **17**
Strafford Ga. EN6: Pot B 4Cb **17**
Strafford Ho. SE8 50Bc **92**
(off Grove St.)
Strafford Rd. EN5: Barn 13Ab **30**
Strafford Rd. TW1: Twick 59Ja **108**
Strafford Rd. TW3: Houn 55Ba **107**
Strafford Rd. W3 47Sa **87**
Strafford St. E14 47Cc **92**
Strahan Rd. E3 41Ac **92**
Straight, The UB1: S'hall 47Z **85**
Straight Mile Pl. KT18: Tatt C 90Ya **174**
Straight Rd. RM3: Rom 22Kd **57**
Straight Rd. SL4: Old Win 7L **103**
Straightsmouth SE10 52Ec **114**
Strait Rd. E6 45Nc **94**
Strakers Rd. SE15 56Xb **113**
Strale Ho. N1 1H **219** (39Ub **71**)
(off Whitmore Est.)
Strand WC2 5F **223** (45Nb **90**)
Strandburgh Pl. HP3: Hem H 4B **4**
Strand Cl. KT18: Eps D 91Ta **193**
Strand Ct. SE18 50Vc **95**
Strand Dr. TW9: Kew 52Ra **109**
Strand E. Twr. E15 40Ec **72**
Strandfield Cl. SE18 50Uc **94**
Strand Ho. BR2: Broml 72Mc **159**
(off Wells Vw. Dr.)
Strand Ho. SE28 46Tc **94**
Strand La. WC2 4J **223** (45Pb **90**)
STRAND ON THE GREEN 51Qa **109**
Strand on the Grn. W4 51Qa **109**
Strand Pl. N18 21Ub **51**
Strand School App. W4 51Qa **109**
Strangeways WD17: Wat 8U **12**
Strang Ho. N1 39Sb **71**
Strangways Ter. W14 48Bb **89**
Stranraer Gdns. SL1: Slou 6J **81**
Stranraer Way N1 38Nb **70**
Stranraer Way TW6: H'row A 58N **105**
Strasburg Rd. SW11 53Jb **112**
Strata SE1 5D **230** (49Sb **91**)
(off Walworth Rd.)
Strata Ct. KT12: Walt T 74V **150**
Strata Rd. DA8: Erith 50Ed **96**
Stratfield Pk. Cl. N21 17Rb **33**
Stratfield Rd. SL1: Slou 8N **81**
Stratfield Rd. WD6: Bore 13Qa **29**
STRATFORD 38Fc **73**
Stratford Av. UB10: Hil 40P **63**
Stratford Bus Station 38Fc **73**
Stratford Cen., The 38Fc **73**
Stratford Circus (Performing Arts Cen.) 37Fc **73**
Stratford City Bus Station 37Ec **72**
Stratford Cl. IG11: Bark 38Wc **75**
Stratford Cl. RM10: Dag 38Ed **76**
Stratford Cl. SL2: Slou 2B **80**
Stratford Ct. KT3: N Mald 70Ta **131**
Stratford Ct. WD17: Wat 11X **27**
Stratford Eye E15 37Fc **73**
Stratford Gdns. SS17: Stan H 1M **101**
Stratford Gro. SW15 56Za **110**
Stratford High Street Station (DLR) 39Fc **73**
Stratford Ho. RM3: Rom 23Md **57**
(off Dartfields)
Stratford Ho. Av. BR1: Broml 69Nc **138**
Stratford International Station (Rail & DLR) 37Ec **72**
STRATFORD NEW TOWN 37Fc **73**
Stratford Office Village, The E15 38Gc **73**
(off Romford Rd.)
Stratford One E20 37Dc **72**
Stratford Picturehouse London 37Fc **73**
Stratford Pl. E20 37Ec **72**
(within Westfield Shop. Cen.)
Stratford Pl. W1 3K **221** (44Kb **90**)
Stratford Rd. CR7: Thor H 70Qb **134**
Stratford Rd. E13 39Hc **73**
(not continuous)
Stratford Rd. NW4 28Za **48**
Stratford Rd. TW6: H'row A 58R **106**
Stratford Rd. UB2: S'hall 49Aa **85**
Stratford Rd. UB4: Yead 42X **85**
Stratford Rd. W8 48Cb **89**
Stratford Rd. WD17: Wat 12W **26**
Stratford Station (Rail, Underground, Overground, DLR & Crossrail) 38Ec **72**
Stratford Studios W8 48Cb **89**
Stratford Vs. NW1 38Lb **70**
Stratford Way AL2: Brick W 1Ba **13**
Stratford Way HP3: Hem H 5K **3**
Stratford Way WD17: Wat 12V **26**
Stratford Workshops E15 39Fc **73**
(off Burford Rd.)
Strathan Cl. SW18 58Ab **110**
Strathaven Rd. SE12 58Kc **115**
Strathblaine Rd. SW11 57Fb **111**
Strathbrook Rd. SW16 66Pb **134**
Strathcona Av. KT23: Bookh 100Aa **191**
Strathcona Gdns. GU21: Knap 10H **167**
(not continuous)
Strathcona Rd. HA9: Wemb 33Ma **67**
Strathdale SW16 64Pb **134**
Strathdon Dr. SW17 62Fb **133**
Strathearn Av. TW2: Whitt 60Da **107**
Strathearn Av. UB3: Harl 52V **106**
Strathearn Ho. W2 4D **220** (45Gb **89**)
(off Strathern Pl.)
Strathearn Pl. W2 3D **220** (44Gb **89**)
Strathearn Rd. SM1: Sutt 78Cb **155**
Strathearn Rd. SW19 64Cb **133**
Stratheden Pde. SE3 52Jc **115**
Stratheden Rd. SE3 53Jc **115**
Strathfield Gdns. IG11: Bark 37Tc **74**
Strathleven Rd. SW2 57Nb **112**
Strathmore RM18: E Til 9L **101**
Strathmore Cl. CR3: Cat'm 93Ub **197**
Strathmore Ct. NW8 3D **214** (41Gb **89**)
(off Park Rd.)
Strathmore Gdns. HA8: Edg 26Ra **47**
Strathmore Gdns. N3 25Db **49**
Strathmore Gdns. RM12: Horn 32Hd **76**
Strathmore Gdns. W8 46Cb **89**
Strathmore Rd. CR0: C'don 73Tb **157**
Strathmore Rd. SW19 62Cb **133**
Strathmore Rd. TW11: Tedd 63Ga **130**
Strathnairn St. SE1 49Wb **91**
(not continuous)
Strathray Gdns. NW3 37Gb **69**
Strath Ter. SW11 56Gb **111**
Strathville Rd. SW18 61Cb **133**
(not continuous)
Strathyre Av. SW16 69Qb **134**
Stratosphere Twr. E15 38Fc **73**
(off Gt. Eastern Rd.)
STRATTON 4A **210**
Stratton Av. EN2: Enf 9Tb **19**
Stratton Av. SM6: W'gton 81Mb **176**
Stratton Cl. DA7: Bex 55Ad **117**
Stratton Cl. HA8: Edg 23Pa **47**
Stratton Cl. KT12: Walt T 74Y **151**
Stratton Cl. SW19 68Cb **133**
Stratton Cl. TW3: Houn 53Ca **107**
Stratton Ct. HA5: Hat E 24Ba **45**
(off Devonshire Rd.)
Stratton Ct. N1 38Ub **71**
(off Hertford Rd.)
Strattondale St. E14 48Ec **92**
Stratton Dr. IG11: Bark 36Uc **74**
Stratton Gdns. UB1: S'hall 44Ba **85**
Stratton Rd. DA7: Bex 55Ad **117**
Stratton Rd. RM3: Rom 22Qd **57**
Stratton Rd. SW19 68Cb **133**
Stratton Rd. TW16: Sun 68V **128**
Stratton St. W1 6A **222** (46Kb **90**)
Stratton Ter. TN16: Westrm 99Sc **200**
Stratton Wlk. RM3: Rom 22Qd **57**
Strauss Rd. W4 47Ta **87**
Strawberry Cl. GU24: Brkwd 3B **186**
Strawberry Cres. AL2: Lon C 8F **6**
Strawberry Fld. AL10: Hat 3C **8**
Strawberry Flds. BR6: Farnb 78Rc **160**
Strawberry Flds. BR8: Swan 67Gd **140**
Strawberry Flds. GU24: Bisl 7E **166**
Strawberry Flds. KT15: Add 80H **149**
Strawberry Hill KT9: Chess 79Ma **153**
Strawberry Hill SL9: Ger X 33A **62**
Strawberry Hill TW1: Twick 62Ha **130**
STRAWBERRY HILL 62Ha **130**
Strawberry Hill 62Ha **130**
Strawberry Hill Cl. TW1: Twick 62Ha **130**
Strawberry Hill Golf Course 62Ga **130**
Strawberry Hill Rd. TW1: Twick 62Ha **130**
Strawberry Hill Station (Rail) 62Ha **130**
Strawberry La. SM5: Cars 76Jb **156**
Strawberry M. HP2: Hem H 2D **4**
Strawberry Ri. GU24: Bisl 7E **166**
Strawberry Ter. N10 25Hb **49**
Strawberry Va. N2 25Fb **49**
Strawberry Va. TW1: Twick 62Ja **130**
(not continuous)
Straw Cl. CR3: Cat'm 95Sb **197**
Strayfield Rd. EN2: Enf 8Qb **18**
Streakes Fld. Rd. NW2 33Wa **68**
Stream Cl. KT14: Byfl 84M **169**
Streamdale SE2 51Xc **117**
Stream La. HA8: Edg 22Ra **47**
Streamline Ct. SE22 60Wb **113**
(off Streamline M.)
Streamline M. SE22 60Wb **113**
Streamside SL1: Slou 6D **80**
Streamside Cl. BR2: Broml 70Jc **137**
Streamside Cl. N9 18Vb **33**
Streamway DA17: Belv 51Cd **118**
Streatfeild Av. E6 39Pc **74**
Streatfield Rd. HA3: Kenton 27La **46**
STREATHAM 62Nb **134**
Streatham Cl. SW16 61Nb **134**
STREATHAM COMMON 65Mb **134**
Streatham Comn. Nth. SW16 64Nb **134**
Streatham Comn. Sth. SW16 65Nb **134**
Streatham Common Station (Rail) 65Mb **134**
Streatham Ct. SW16 62Nb **134**
Streatham Green 63Mb **134**
(off Streatham High Road)
Streatham High Rd. SW16 63Nb **134**
Streatham Hill SW2 61Nb **134**
STREATHAM HILL 61Nb **134**
Streatham Hill Station (Rail) 61Nb **134**
STREATHAM PARK 64Lb **134**
Streatham Pl. SW2 59Nb **112**
Streatham Rd. CR4: Mitc 67Jb **134**
Streatham Rd. SW16 66Kb **134**
Streatham Station (Rail) 64Mb **134**
Streatham St. WC1 2F **223** (44Nb **90**)

Streatham Va. SW16 67Lb 134
STREATHAM VALE 66Lb 134
Streathbourne Rd. SW17 61Jb 134
Streatley Pl. NW3 35Eb 69
Streatley Rd. NW6 38Bb 69
Street, The DA12: Cobh 9H 145
Street, The DA12: Shorne 4N 145
Street, The DA4: Hort K 70Rd 141
Street, The E20 37Ec 72
(within Westfield Shop. Cen.)
Street, The GU4: W Cla 100J 189
Street, The KT21: Asht 91Na 193
Street, The KT22: Fet 94Fa 192
Street, The KT24: Eff 99Z 191
Street, The KT24: W Hor 100R 190
Street, The RH3: B'wth 7A 206
Street, The TN15: Ash 78Zd 165
Street, The TN15: Igh 93Zd 205
Street, The TN15: Plax 99Ae 205
Street, The WD4: Chfd 3J 11
Streeters La. SM6: Bedd 76Mb 156
Streetfield M. SE3 55Jc 115
Streets Heath GU24: W End 4D 166
Streimer Rd. E15 40Ec 72
Strelley Way W3 45Ua 88
Stretton Mans. SE8 50Cc 92
Stretton Rd. CR0: C'don 73Ub 157
Stretton Rd. TW10: Ham 61La 130
Stretton Way WD6: Bore 10Na 15
Strickland Av. DA1: Dart 55Pd 119
(not continuous)
Strickland Ct. SE15 55Wb 113
Strickland Ho. E2 4K 219 (41Vb 91)
(off Chambord St.)
Strickland Row SW18 59Fb 111
Strickland St. SE8 54Cc 114
Strickland Way BR6: Orp 77Vc 161
Stride Rd. E13 40Hc 73
Strides Ct. KT16: Ott 79E 148
(off Brox Rd.)
Strimon Cl. N9 19Yb 34
Stringer Ho. N1 1J 219 (39Ub 71)
(off Whitmore Est.)
Stringer's Av. GU4: Jac W 10P 187
STRINGERS COMMON 10N 187
Stringers Cotts. SL9: Chal P 25A 42
(off The Vale)
Stringhams Copse GU23: Rip 96H 189
Stripling Way WD18: Wat 16W 26
Strode Cl. N10 24Jb 50
Strode Rd. E7 35Jc 73
Strode Rd. N17 26Ub 51
Strode Rd. NW10 37Wa 68
Strode Rd. SW6 52Ab 110
Strode's Coll. La. TW20: Egh 64B 126
Strode's Cres. TW18: Staines 64L 127
Strode St. TW20: Egh 63C 126
Stroma Cl. HP3: Hem H 4C 4
Stroma Ct. SL1: Slou 5B 80
Strome Ho. NW6 40Db 69
(off Carlton Vale)
Stromness Wlk. W3 46Sa 87
Strone Rd. E12 37Mc 73
Strone Rd. E7 37Lc 73
Strone Way UB4: Yead 42Aa 85
Strongbow Cres. SE9 57Pc 116
Strongbow Rd. SE9 57Pc 116
Strongbridge Cl. HA2: Harr 32Ca 65
Stronsa Rd. W12 47Va 88
Stronsay Cl. HP3: Hem H 4C 4
Strood Av. RM7: Rush G 32Fd 76
Strood Ho. SE1 2G 231 (47Tb 91)
(off Staple St.)
Strood La. SL4: Wink 5A 124
Stroud Cl. SL4: Wind 5B 102
Stroud Cres. SW15 62Wa 132
STROUDE 68B 126
Stroude Rd. GU25: Vir W 71A 148
Stroude Rd. TW20: Egh 65C 126
Stroudes Cl. KT4: Wor Pk 73Ua 154
Stroud Fld. UB5: N'olt 37Aa 65
Stroud Ga. HA2: Harr 35Da 65
STROUD GREEN 31Pb 70
Stroud Grn. Gdns. CR0: C'don 73Yb 158
Stroud Grn. Rd. N4 32Pb 70
Stroud Grn. Way CR0: C'don 73Xb 157
Stroud Ho. RM3: Rom 22Md 57
(off Montgomery Cres.)
Stroudley Ho. SW8 53Lb 112
Stroudley Wlk. E3 41Dc 92
Stroud Rd. SE25 72Wb 157
Stroud Rd. SW19 62Cb 133
Stroud's Cl. RM6: Chad H 29Xc 55
Stroudwater Pk. KT13: Weyb 79R 150
Stroud Way TW15: Ashf 65R 128
Stroud Wood Bus. Cen. AL2: F'mre 9C 6
Strouts Pl. E2 3K 219 (41Vb 91)
Struan Gdns. GU21: Wok 87A 168
Strudwick Ct. SW4 53Nb 112
(off Binfield Rd.)
Strutton Ct. SW1 4D 228 (48Mb 90)
(off Gt. Peter St.)
Strutton Ground SW1 3D 228 (48Mb 90)
Struttons Av. DA11: Nflt 1B 144
Strype St. E1 1K 225 (43Vb 91)
Stuart Av. BR2: Hayes 74Jc 159
Stuart Av. HA2: Harr 34Ba 65
Stuart Av. KT12: Walt T 74X 151
Stuart Av. NW9 31Wa 68
Stuart Av. W5 47Pa 87
Stuart Cl. BR8: Hext 66Hd 140
Stuart Cl. CM15: Pil H 15Xd 40
Stuart Cl. SL4: Wind 4D 102
Stuart Cl. UB10: Hil 37Q 64
Stuart Ct. CR0: C'don 76Rb 157
(off St John's Rd.)
Stuart Ct. RH1: Redh 5A 208
(off St Anne's Ri.)
Stuart Ct. WD6: E'tree 16Ma 29
Stuart Cres. CR0: C'don 76Dc 158
Stuart Cres. N22 25Pb 50
Stuart Cres. RH2: Reig 9J 207
Stuart Cres. UB3: Hayes 44S 84
Stuart Evans Cl. DA16: Well 55Yc 117
Stuart Gro. TW11: Tedd 64Ga 130
Stuart Ho. E16 46Kc 93
(off Beaulieu Av.)
Stuart Ho. E9 37Zb 72
(off Queen Anne Rd.)
Stuart Ho. W14 49Ab 88
(off Windsor Way)
Stuart Lodge KT18: Eps 85Ta 173
(off Ashley Rd.)
Stuart Mantle Way DA8: Erith 52Fd 118
(not continuous)
Stuart Mill Ho. N1 2H 217 (40Pb 70)
(off Killick St.)
Stuart Pl. AL1: St A 4E 6
Stuart Pl. CR4: Mitc 67Hb 133
Stuart Rd. CR6: W'ham 92Xb 197
Stuart Rd. CR7: Thor H 70Sb 135
Stuart Rd. DA11: Grav'nd 8C 122
Stuart Rd. DA16: Well 53Xc 117
Stuart Rd. EN4: E Barn 17Gb 31
Stuart Rd. HA3: W'stone 27Ha 46
Stuart Rd. IG11: Bark 38Vc 75
Stuart Rd. NW6 41Cb 89
Stuart Rd. RH2: Reig 9J 207
Stuart Rd. RM17: Grays 50De 99
Stuart Rd. SE15 56Yb 114
Stuart Rd. SW19 62Cb 133
Stuart Rd. TW10: Ham 61Ka 130
Stuart Rd. W3 46Sa 87
Stuarts RM11: Horn 32Pd 77
(off High St.)
Stuarts Cl. HP3: Hem H 4M 3
Stuart Twr. W9 4A 214 (41Eb 89)
(off Maida Vale)
Stuart Way EN7: Chesh 3Xb 19
Stuart Way GU25: Vir W 10L 125
Stuart Way SL4: Wind 4C 102
Stuart Way TW18: Staines 65K 127
Stubbers Adventure Cen. 37Ud 78
Stubbers La. RM14: Upm 37Td 78
Stubbings Hall La. EN9: Walt A 1Ec 20
Stubbs Cl. NW9 29Sa 47
Stubbs Ct. W4 50Ra 87
(off Chaseley Dr.)
Stubbs Dr. SE16 50Xb 91
Stubbs Hill BR6: Prat B 85Yc 181
Stubbs Ho. E2 41Zb 92
(off Bonner St.)
Stubbs Ho. SW1 6E 228 (49Mb 90)
(off Erasmus St.)
Stubbs La. KT20: Lwr K 100Bb 195
Stubbs M. RM8: Dag 35Xc 75
(off Marlborough Rd.)
Stubbs Point E13 42Jc 93
Stubbs Way SW19 67Fb 133
Stucley Pl. NW1 38Kb 70
Stucley Rd. TW5: Hest 52Ea 108
Studdridge St. SW6 54Cb 111
Studd St. N1 39Rb 71
Stud Grn. WD25: Wat 4W 12
Studholme Ct. NW3 35Cb 69
Studholme St. SE15 52Xb 113
Studio Ct. N15 28Ub 51
Studio La. W5 46Ma 87
Studio M. NW4 28Ya 48
Studio Pl. SW1 2G 227 (47Hb 89)
(off Kinnerton St.)
Studio Plaza 74W 150
Studios, The DA3: Nw A G 75Be 165
(off The Row)
Studios, The SW4 56Lb 112
(off Crescent La.)
Studios, The W8 46Cb 89
(off Edge St.)
Studios, The WD17: Wat 13X 27
(off The Parade)
Studios, The WD23: Bush 16Ca 27
Studios Rd. TW17: Shep 69P 127
Studio Tour Dr. WD25: Wat 6U 12
Studio Way WD6: Bore 12Sa 29
Studland SE17 7F 231 (50Tb 91)
(off Portland St.)
Studland Cl. DA15: Sidc 62Vc 139
Studland Rd. KT14: Byfl 85P 169
Studland Rd. KT2: King T 65Na 131
Studland Rd. SE26 64Zb 136
Studland Rd. W7 44Fa 86
Studland St. W6 49Xa 88
Studley Av. E4 24Fc 53
Studley Cl. E5 36Ac 72
Studley Ct. DA14: Sidc 64Xc 139
Studley Ct. E14 45Fc 93
(off Jamestown Way)
Studley Cres. DA3: Lfield 68Ee 143
Studley Dr. IG4: Ilf 30Mc 53
Studley Est. SW4 53Nb 112
Studley Grange Rd. W7 47Ga 86
Studley Rd. E7 37Kc 73
Studley Rd. RM9: Dag 38Zc 75
Studley Rd. SW4 53Nb 112
Stukeley Rd. E7 38Kc 73
Stukeley St. WC2 2G 223 (44Nb 90)
Stumps Hill La. BR3: Beck 65Cc 136
Stumps La. CR3: Whyt 89Vb 177
(not continuous)
Stunell Ho. SE14 51Zb 114
(off John Williams Cl.)
Sturdee Ho. E2 40Wb 71
(off Horatio St.)
Sturdy Ho. E3 40Ac 72
(off Gernon Rd.)
Sturdy Rd. SE15 54Xb 113
Sturge Av. E17 26Dc 52
Sturgeon Rd. SE17 50Sb 91
Sturges Fld. BR7: Chst 65Tc 138
Sturgess Av. NW4 31Xa 68
Sturge St. SE1 1D 230 (47Sb 91)
Sturlas Way EN8: Walt C 5Zb 20
Sturmer Cl. AL4: St A 3G 6
Sturmer Way N7 36Pb 70
Sturminster NW1 38Mb 70
(off Agar Gro.)
Sturminster Cl. UB4: Yead 44Y 85
Sturminster Ho. SW8 52Pb 112
(off Dorset Rd.)
Sturrock Cl. N15 28Tb 51
Sturry St. E14 44Dc 92
Sturts Apts. N1 39Tb 71
(off Branch Pl.)
Sturt's La. KT20: Walt H 99Va 194
Sturt St. N1 2E 218 (40Sb 71)
Stutfield St. E1 44Wb 91
Stuttle Ho. E1 42Wb 91
(off Buxton St.)
STYANTS BOTTOM 94Vd 204
Styants Bottom Rd. TN15: Seal 93Vd 204
Stychens Cl. RH1: Blet 5J 209
Stychens La. RH1: Blet 3J 209
Styles End KT23: Bookh 99Da 191
Styles Gdns. SW9 55Rb 113
Styles Ho. SE1 7B 224 (46Rb 91)
(off Hatfields)
Styles Way BR3: Beck 70Ec 136
Stylus Ho. E1 44Yb 92
Styventon Pl. KT16: Chert 73H 149
Subrosa Dr. RH1: Mers 2B 208
Subrosa Pk. RH1: Mers 2B 208
Success Ho. SE1 7K 231 (50Vb 91)
(off Cooper's Rd.)
Succession Wlk. E3 38Cc 72
Succombs Hill CR3: W'ham 92Xb 197
Succombs Hill CR6: W'ham 92Xb 197
Succombs Pl. CR6: W'ham 91Xb 197
Sudbourne Rd. SW2 57Nb 112
Sudbrooke Rd. SW12 58Hb 111
Sudbrook Gdns. TW10: Ham 62Ma 131
Sudbrook La. TW10: Ham 60Na 109
Sudbury E6 43Qc 94
SUDBURY 36Ka 66
Sudbury & Harrow Road Station (Rail) 36Ka 66
Sudbury Av. HA0: Wemb 34La 66
Sudbury Cl. RM3: Rom 22Nd 57
Sudbury Ct. AL1: St A 3C 6
Sudbury Ct. SW8 53Mb 112
(off Allen Edwards Dr.)
Sudbury Ct. Dr. HA1: Harr 34Ha 66
Sudbury Ct. Rd. HA1: Harr 34Ha 66
Sudbury Cres. BR1: Broml 65Jc 137
Sudbury Cres. HA0: Wemb 36Ka 66
Sudbury Cft. HA0: Wemb 35Ha 66
Sudbury Gdns. CR0: C'don 77Ub 157
Sudbury Golf Course 38La 66
Sudbury Hgts. Av. UB6: G'frd 36Ha 66
Sudbury Hill HA1: Harr 33Ga 66
Sudbury Hill Cl. HA0: Wemb 35Ha 66
Sudbury Hill Harrow Station (Rail) 35Ga 66
Sudbury Hill Station (Underground) 35Ga 66
Sudbury Ho. SW18 57Db 111
Sudbury Rd. IG11: Bark 36Vc 75
Sudbury Town Station (Underground) 37Ka 66
Sudeley Ct. E17 25Bc 52
(off Broughton Pl.)
Sudeley St. N1 2C 218 (40Rb 71)
Sudicamps Ct. EN9: Walt A 5Jc 21
Sudlow Rd. SW18 57Cb 111
Sudrey St. SE1 2D 230 (47Sb 91)
Suez Av. UB6: G'frd 40Ha 66
Suez Rd. EN3: Brim 14Ac 34
Suffield Cl. CR2: Sels 84Zb 178
SUFFIELD HATCH 21Ec 52
Suffield Ho. SE17 7C 230 (50Rb 91)
(off Berryfield Rd.)
Suffield Rd. E4 20Dc 34
Suffield Rd. N15 29Vb 51
Suffield Rd. SE20 68Yb 136
Suffolk Cl. AL2: Lon C 7G 6
Suffolk Cl. SL1: Slou 4C 80
Suffolk Cl. WD6: Bore 15Ta 29
Suffolk Ct. E10 31Cc 72
Suffolk Ct. IG3: Ilf 30Uc 54
Suffolk Ct. RM6: Chad H 30Yc 55
Suffolk Ho. CR0: C'don 75Tb 157
(off George St.)
Suffolk Ho. KT13: Weyb 78S 150
(off Princes Rd.)
Suffolk Ho. SE20 67Zb 136
(off Croydon Rd.)
Suffolk La. EC4 4F 225 (45Tb 91)
Suffolk Pk. Rd. E17 28Ac 52
Suffolk Pl. SE2 50Yc 95
Suffolk Pl. SW1 6E 222 (46Mb 90)
Suffolk Rd. DA1: Dart 58Nd 119
Suffolk Rd. DA12: Grav'nd 8F 122
Suffolk Rd. DA14: Sidc 65Yc 139
Suffolk Rd. E13 41Jc 93
Suffolk Rd. EN3: Pond E 15Xb 33
Suffolk Rd. EN6: Pot B 4Ab 16
Suffolk Rd. HA2: Harr 30Ba 45
Suffolk Rd. IG11: Bark 38Tc 74
Suffolk Rd. IG3: Ilf 30Uc 54
Suffolk Rd. KT4: Wor Pk 75Va 154
Suffolk Rd. N15 29Tb 51
Suffolk Rd. NW10 38Ua 68
Suffolk Rd. RM10: Dag 36Ed 76
Suffolk Rd. SE25 70Vb 135
Suffolk Rd. SW13 52Va 110
Suffolk St. E7 35Jc 73
Suffolk St. SW1 5E 222 (45Mb 90)
Suffolk Way RM11: Horn 28Qd 57
Suffolk Way TN13: S'oaks 97Ld 203
Sugar Bakers Ct. EC3 3J 225 (44Ub 91)
(off Creechurch La.)
Sugar Ho. E1 44Wb 91
(off Leman St.)
Sugar Ho. Island Development E15 40Ec 72
Sugar Ho. La. E15 40Ec 72
Sugar La. HP1: Hem H 4D 2
Sugar La. SE16 47Wb 91
Sugar Loaf Wlk. E2 41Yb 92
Sugar Quay EC3 5J 225 (45Ub 91)
(off Sugar Quay Wlk.)
Sugar Quay Wlk. EC3 5J 225 (45Ub 91)
Sugden Rd. KT7: T Ditt 74Ka 152
Sugden Rd. SW11 55Jb 112
Sugden Way IG11: Bark 40Vc 75
Sulby Ho. SE4 56Ac 114
(off Turnham Rd.)
Sulgrave Gdns. W6 47Ya 88
Sulgrave Rd. W6 48Ya 88
Sulina Rd. SW2 59Nb 112
Sulivan Ct. SW6 54Cb 111
Sulivan Ent. Cen. SW6 55Db 111
Sulivan Rd. SW6 55Cb 111
Sulkin Ho. E2 41Zb 92
(off Knottisford St.)
Sullivan Av. E16 43Mc 93
Sullivan Cl. DA1: Dart 58Kd 119
Sullivan Cl. KT8: W Mole 69Da 129
Sullivan Cl. SW11 55Gb 111
Sullivan Cl. UB4: Yead 43Y 85
Sullivan Ct. E3 42Bc 92
(off Eric St.)
Sullivan Ct. N16 31Vb 71
Sullivan Ct. SW5 49Cb 89
(off Earls Ct. Rd.)
Sullivan Cres. UB9: Hare 26M 43
Sullivan Dr. SE9: Sidc 60Tc 116
Sullivan Ho. SE11 6J 229 (49Pb 90)
(off Vauxhall St.)
Sullivan Ho. SW1 51Kb 112
(off Churchill Gdns.)
Sullivan Rd. RM18: Tilb 3C 122
Sullivan Rd. SE11 5A 230 (49Qb 90)
Sullivan Row BR2: Broml 72Mc 159
Sullivans Reach KT12: Walt T 73V 150
Sullivan Way WD6: E'tree 16La 28
Sultan Ho. SE1 50Wb 91
(off St James's Rd.)
Sultan Rd. E11 28Kc 53
Sultan St. BR3: Beck 68Zb 136
Sultan St. SE5 52Sb 113
Sultan Ter. N22 26Qb 50
Sumatra Rd. NW6 36Cb 69
Sumburgh Rd. SW12 58Jb 112
Sumburgh Way SL1: Slou 3J 81
Sumeria Ct. SE16 49Yb 92
(off Rotherhithe New Rd.)
Summer Av. KT8: E Mos 71Ga 152
Summerbee Ho. SW18 56Eb 111
(off Eltringham St.)
Summer Cl. KT14: Byfl 86P 169
Summer Ct. HP2: Hem H 1M 3
Summercourt Rd. E1 44Yb 92
Summer Crossing KT7: T Ditt 71Ga 152
Summer Dr. UB7: W Dray 47P 83
Summerene Cl. SW16 66Lb 134
Summerfield AL10: Hat 3C 8
Summerfield BR1: Broml 67Kc 137
(off Freelands Rd.)
Summerfield KT21: Asht 91Ma 193
Summerfield Av. NW6 40Ab 68
Summerfield Cl. AL2: Lon C 8G 6
Summerfield Cl. KT15: Add 78H 149
Summerfield La. KT6: Surb 75Ma 153
Summerfield Pl. KT16: Ott 79F 148
Summerfield Rd. IG10: Lough 16Mc 35
Summerfield Rd. W5 42Ka 86
Summerfield Rd. WD25: Wat 7W 12
Summerfields Av. N12 23Gb 49
Summerfield St. SE12 59Hc 115
Summer Gdns. KT8: E Mos 71Ga 152
Summer Gdns. UB10: Ick 33S 64
Summer Gro. BR4: W W'ck 75Gc 159
Summer Gro. WD6: E'tree 16Ma 29
Summerhayes Cl. GU21: Wok 86A 168
Summerhays KT11: Cobh 85Z 171
Summer Hill BR7: Chst 68Qc 138
Summer Hill WD6: E'tree 15Qa 29
Summerhill Cl. BR6: Orp 76Uc 160
Summerhill Ct. AL1: St A 1D 6
(off Avenue Rd)
Summerhill Gro. EN1: Enf 16Ub 33
Summerhill Rd. DA1: Dart 59Md 119
Summerhill Rd. N15 28Tb 51
Summerhill Vs. BR7: Chst 67Qc 138
(off Susan Wood)
Summerhill Way CR4: Mitc 67Jb 134
Summerhouse Av. TW5: Hest 53Aa 107
Summerhouse Dr. DA2: Wilm 63Fd 140
Summerhouse Dr. DA5: Bexl 63Fd 140
Summerhouse Dr. DA5: Dart 63Fd 140
Summerhouse Dr. HA7: Stan 22Ha 46
Summerhouse La. UB7: Harm 51M 105
Summerhouse La. UB9: Hare 24J 43
Summerhouse La. WD25: A'ham 12Ea 28
Summerhouse Rd. N16 33Ub 71
Summerhouse Way WD5: Ab L 2V 12
Summerland Gdns. N10 27Kb 50
Summerland Grange N10 27Kb 50
Summerlands Av. W3 45Sa 87
Summerlands Lodge BR6: Farnb 77Qc 160
Summerlay Cl. KT20: Tad 92Ab 194
Summerlea SL1: Slou 6F 80
Summerleas Cl. HP2: Hem H 2N 3
Summerlee Av. N2 28Hb 49
Summerlee Gdns. N2 28Hb 49
Summerleigh KT13: Weyb 79T 150
(off Gower Rd.)
Summerley St. SW18 61Db 133
Summerly Av. RH2: Reig 5J 207
Summer Pl. WD18: Wat 16V 26
Summer Rd. KT7: T Ditt 71Ha 152
Summer Rd. KT8: E Mos 71Fa 152
(not continuous)
Summersby Ct. SL3: L'ly 10N 81
Summersby Rd. N6 30Kb 50
Summers Cl. HA9: Wemb 32Ra 67
Summers Cl. KT13: Weyb 83Q 170
Summers Cl. SM2: Sutt 80Cb 155
Summerskill Cl. SE15 55Xb 113
Summerskille Cl. N9 20Xb 33
Summers La. N12 24Fb 49
Summers Rd. SL1: Burn 1A 80
Summers Row N12 23Gb 49
Summers St. EC1 6K 217 (42Qb 90)
SUMMERSTOWN 62Eb 133
Summerstown SW17 62Eb 133
Summers Way AL2: Lon C 9J 7
Summerswood Cl. CR8: Kenley 88Tb 177
Summerswood La. WD6: Bore 6Ua 16
Summerton Ho. E16 47Lc 93
(off Starboard Way)
Summerton Way SE28 44Zc 95
Summer Trees TW16: Sun 67X 129
Summerville Gdns. SM1: Sutt 79Bb 155
Summerwood SL5: S'dale 3D 146
Summerwood Rd. TW7: Isle 57Ha 108
Summit, The IG10: Lough 11Pc 36
Summit Av. NW9 29Ta 47
Summit Bus. Pk. TW16: Sun 66W 128
Summit Cen. EN6: Pot B 2Ab 16
Summit Cl. HA8: Edg 24Qa 47
Summit Cl. N14 19Lb 32
Summit Cl. NW9 28Ta 47
Summit Ct. NW2 36Ab 68
Summit Dr. IG8: Wfd G 26Mc 53
Summit Est. N16 31Wb 71
Summit Ho. BR4: W W'ck 75Ec 158
Summit Pl. KT13: Weyb 80Q 150
Summit Rd. E17 28Dc 52
Summit Rd. EN6: Pot B 2Ab 16
Summit Rd. UB5: N'olt 38Ca 65
Summit Way N14 19Kb 32
Summit Way SE19 66Ub 135
Sumner Av. SE15 53Vb 113
Sumner Bldgs. SE1 6D 224 (46Sb 91)
(off Sumner St.)
Sumner Cl. BR6: Farnb 77Sc 160
Sumner Cl. KT22: Fet 96Fa 192
Sumner Ct. SW8 52Nb 112
Sumner Est. SE15 52Vb 113
Sumner Gdns. CR0: C'don 74Qb 156
Sumner Ho. E3 43Dc 92
(off Watts Gro.)
Sumner Pl. KT15: Add 78J 149
Sumner Pl. SW7 6C 226 (49Fb 89)
Sumner Pl. M. SW7 6C 226 (49Fb 89)
Sumner Rd. CR0: C'don 74Qb 156
Sumner Rd. HA1: Harr 31Ea 66
Sumner Rd. SE15 51Vb 113
Sumner Rd. Sth. CR0: C'don 74Qb 156
Sumner St. SE1 6C 224 (46Rb 91)
Sumpter Cl. NW3 37Eb 69
Sumpter Yd. AL1: St A 3B 6
Sun All. TW9: Rich 56Na 109
Sunbeam Cres. W10 42Ya 88
Sunbeam Rd. NW10 42Sa 87
Sunbeam Way NW10 42Ta 87
Sunbird Wlk. HA2: Harr 33Ca 65
(off Sandpiper Dr.)
SUNBURY 69Y 129
Sunbury Av. NW7 22Ta 47
Sunbury Av. SW14 56Ta 109
Sunbury Av. Pas. SW14 56Ua 110
Sunbury Bus. Cen. TW16: Sun 67V 128
Sunbury Cl. KT12: Walt T 72W 150
SUNBURY COMMON 66V 128
Sunbury Ct. EN5: Barn 14Ab 30
Sunbury Ct. SL4: Eton 1H 103
Sunbury Ct. Island TW16: Sun 69Z 129
Sunbury Ct. M. TW16: Sun 68Z 129
Sunbury Ct. Rd. TW16: Sun 68Y 129
Sunbury Cres. TW13: Felt 63V 128
SUNBURY CROSS 66W 128
Sunbury Cross Cen. TW16: Sun 66V 128
Sunbury Embroidery Gallery, The 69X 129
Sunbury Gdns. NW7 22Ta 47
Sunbury Golf Course 70T 128
Sunbury Ho. E2 4K 219 (41Vb 91)
(off Swanfield St.)
Sunbury Ho. SE14 51Zb 114
(off Myers La.)
Sunbury Ho. TW16: Sun 67U 128
(off Brooklands Cl.)
Sunbury La. KT12: Walt T 72W 150
Sunbury La. SW11 53Fb 111
(not continuous)
Sunbury Leisure Cen. 67V 128
Sunburylock Ait KT12: Walt T 70X 129
Sunbury Pk. Walled Gdn. 69X 129
Sunbury Rd. SL4: Eton 1H 103
Sunbury Rd. SM3: Cheam 76Za 154
Sunbury Rd. TW13: Felt 62V 128
Sunbury Station (Rail) 67W 128
Sunbury St. SE18 48Pc 94
Sunbury Way TW13: Hanw 64Y 129
Sunbury Workshops E2 4K 219 (41Vb 91)
(off Swanfield St.)
Sun Bus. Pk. BR5: St M Cry 70Xc 139
Sun Cl. SL4: Eton 1H 103
Sun Ct. DA8: Erith 54Hd 118
Sun Ct. E3 42Ec 92
(off Navigation Rd.)
Sun Ct. EC3 3G 225 (44Tb 91)
(off Cornhill)
Suncroft Pl. SE26 62Yb 136
Sundale AL1: St A 1D 6
(off Althorp Rd.)
Sundale Av. CR2: Sels 82Yb 178
Sundeala Cl. TW16: Sun 66W 128
(off Hanworth Rd.)
Sunderland Av. AL1: St A 1E 6
Sunderland Ct. SE22 59Wb 113
Sunderland Ct. TW19: Stanw 58N 105
(off Whitley Cl.)
Sunderland Est. WD4: K Lan 1R 12
Sunderland Gro. WD25: Wat 6V 12
Sunderland Ho. W2 43Cb 89
(off Westbourne Pk. Rd.)
Sunderland Mt. SE23 61Zb 136
Sunderland Point E16 46Sc 94
Sunderland Rd. SE23 60Zb 114
Sunderland Rd. W5 48Ma 87
Sunderland Ter. W2 44Db 89
Sunderland Way E12 33Mc 73
Sundew Av. W12 45Wa 88
Sundew Cl. GU18: Light 3B 166
Sundew Cl. W12 45Wa 88
Sundew Ct. HA0: Wemb 40Na 67
(off Elmore Cl.)
Sundew Ct. RM17: Grays 51Fe 121
Sundew Rd. HP1: Hem H 3G 2
Sundial Av. SE25 69Vb 135
Sundial Ct. EC1 7E 218 (43Sb 91)
(off Chiswell St.)
Sundon Cres. GU25: Vir W 1M 147
Sundorne Rd. SE7 50Lc 93
Sundown Av. CR2: Sande 83Vb 177
Sundown Rd. TW15: Ashf 64S 128
Sundra Wlk. E1 42Zb 92
SUNDRIDGE 65Kc 137
SUNDRIDGE 96Ad 201
Sundridge Av. BR1: Broml 67Mc 137
Sundridge Av. BR7: Chst 66Nc 138
Sundridge Av. DA16: Well 54Tc 116
Sundridge Cl. DA1: Dart 58Qd 119
Sundridge Hill TN14: Knock 90Yc 181
Sundridge Hill TN14: Sund 90Yc 181
Sundridge Ho. E9 38Zb 72
(off Church Cres.)
Sundridge La. TN14: Knock 89Xc 181
Sundridge Pde. BR1: Broml 66Kc 137
SUNDRIDGE PARK 66Kc 137
Sundridge Pk. Golf Course 65Kc 137
Sundridge Park Mans. BR1: Broml 65Mc 137
Sundridge Park Station (Rail) 66Kc 137
Sundridge Pl. CR0: C'don 74Wb 157
Sundridge Rd. CR0: C'don 73Vb 157
Sundridge Rd. GU22: Wok 91C 188
Sundridge Rd. TN14: Dun G 92Cd 202
Sunfields Pl. SE3 52Kc 115
Sunflower Cl. NW2 33Bb 69
Sunflower Ct. RM13: Rain 40Ed 76
Sunflower M. HA7: Stan 22Ha 46
Sunflower Way RM3: Hrld W 25Md 57
Sungate Cotts. RM5: Col R 25Bd 55
Sun Ga. Ho. KT3: N Mald 70Va 132
Sunguard Ct. SE1 6B 224 (46Rb 91)
Sun Hill DA3: Fawk 76Wd 164
Sun Hill GU22: Wok 3L 187
SUN-IN-THE-SANDS 52Kc 115
Sunken Rd. CR0: C'don 78Yb 158
Sunkist Way SM6: W'gton 81Nb 176
Sunland Av. DA6: Bex 56Ad 117
Sun La. DA12: Grav'nd 1E 144
Sun La. SE3 52Kc 115
Sunleigh Rd. HA0: Wemb 39Na 67
Sunley Gdns. UB6: G'frd 39Ja 66
Sun Life Trad. Est. TW14: Felt 55W 106
Sunlight Cl. SW19 65Eb 133
Sunlight M. SW6 54Db 111
Sunlight Sq. E2 41Xb 91
Sunliner Way RM15: S Ock 42Xd 98
Sun Marsh Way DA12: Grav'nd 10J 123
Sunmead Cl. KT22: Fet 94Ha 192
Sunmead Rd. HP2: Hem H 1M 3
Sunmead Rd. TW16: Sun 69W 128
Sunna Gdns. TW16: Sun 68X 129
Sunniholme Ct. CR2: S Croy 78Sb 157
(off Warham Rd.)
Sunning Av. SL5: S'dale 3C 146
Sunningdale N14 22Mb 50
SUNNINGDALE 1E 146
Sunningdale W13 43Ka 86
(off Hardwick Grn.)
Sunningdale Av. HA4: Ruis 32Y 65
Sunningdale Av. IG11: Bark 39Tc 74
Sunningdale Av. RM13: Rain 42Kd 97
Sunningdale Av. TW13: Hanw 61Aa 129
Sunningdale Av. W3 45Ua 88
Sunningdale Cl. E6 41Pc 94
Sunningdale Cl. HA7: Stan 23Ja 46
Sunningdale Cl. KT6: Surb 75Na 153
Sunningdale Cl. SE16 50Xb 91
Sunningdale Cl. SE28 44Ad 95
Sunningdale Gdns. NW9 29Sa 47
Sunningdale Gdns. W8 48Cb 89
(off Stratford Rd.)
Sunningdale Golf Course 4E 146
Sunningdale Hgts. SL5: S'dale 4E 146

Sunningdale Ladies Golf Course 4E 146
Sunningdale Lodge HA8: Edg 22Pa 47
(off Stonegrove)
Sunningdale Pk. 1D 146
Sunningdale Rd. BR1: Broml 70Nc 138
Sunningdale Rd. RM13: Rain 38Jd 76
Sunningdale Rd. SM1: Sutt 76Bb 155
Sunningdale Station (Rail) 3E 146
Sunningfields Cres. NW4 26Xa 48
Sunningfields Rd. NW4 26Xa 48
Sunninghill DA11: Nflt 1A 144
SUNNINGHILL 1B 146
Sunninghill Cl. SL5: S'hill 10B 124
Sunninghill Ct. SL5: S'hill 10B 124
Sunninghill Ct. W3 47Sa 87
SUNNINGHILL PARK 7A 124
Sunninghill Rd. SE13 54Dc 114
Sunninghill Rd. SL4: Wink 4B 124
Sunninghill Rd. SL5: Asc 5B 124
Sunninghill Rd. SL5: S'hill 1B 146
Sunnings La. RM14: Upm 36Sd 78
Sunningvale Av. TN16: Big H 87Lc 179
Sunningvale Cl. TN16: Big H 88Mc 179
Sunny Bank SE25 69Wb 135
Sunnybank CR6: W'ham 89Ac 178
Sunnybank KT18: Eps 88Sa 173
Sunnybank Rd. EN6: Pot B 5Cb 17
Sunnybank Vs. RH1: Blet 4M 209
Sunny Cres. NW10 38Sa 67
Sunnycroft Gdns. RM14: Upm 31Vd 78
Sunnycroft Rd. SE25 69Wb 135
Sunnycroft Rd. TW3: Houn 54Da 107
Sunnycroft Rd. UB1: S'hall 43Ca 85
Sunnydale BR6: Farnb 75Qc 160
Sunnydale Gdns. NW7 23Ta 47
Sunnydale Rd. SE12 57Kc 115
Sunnydell AL2: Chis G 8P 5
Sunnydene Av. E4 22Fc 53
Sunnydene Av. HA4: Ruis 32W 64
Sunnydene Cl. RM3: Hrld W 24Pd 57
Sunnydene Gdns. HA0: Wemb 37La 66
Sunnydene Rd. CR8: Purl 85Rb 177
Sunnydene St. SE26 63Ac 136
Sunnyfield NW7 21Va 48
Sunnyfield Rd. BR7: Chst 69Wc 139
Sunny Gdns. Pde. NW4 26Xa 48
Sunny Gdns. Rd. NW4 26Xa 48
Sunny Hill NW4 27Xa 48
Sunnyhill Cl. E5 35Ac 72
Sunnyhill Rd. HP1: Hem H 2K 3
Sunnyhill Rd. SW16 63Nb 134
Sunnyhill Rd. WD3: W Hyd 23F 42
Sunnyhurst Cl. SM1: Sutt 76Cb 155
Sunnymead Av. CR4: Mitc 69Mb 134
Sunnymead Rd. NW9 31Ta 67
Sunnymead Rd. SW15 57Xa 110
SUNNYMEADS 56A 104
Sunnymeads Station (Rail) 55A 104
Sunnymede IG7: Chig 20Xc 37
Sunnymede Av. KT19: Ewe 81Ua 174
Sunnymede Av. SM5: Cars 83Fb 175
Sunnymede Dr. IG2: Ilf 29Rc 54
Sunnymede Dr. IG6: Ilf 28Rc 54
Sunny M. NW1 38Jb 70
Sunny M. RM5: Col R 24Ed 56
Sunny Nook Gdns. CR2: S Croy 79Tb 157
Sunny Pl. NW4 28Ya 48
Sunny Ri. CR3: Cat'm 96Tb 197
Sunny Rd., The EN3: Enf H 11Zb 34
Sunnyside GU21: Knap 1F 186
Sunnyside KT12: Walt T 71Y 151
Sunnyside NW2 34Bb 69
Sunnyside SE6 59Bc 114
(off Blythe Hill)
Sunnyside SW19 65Ab 132
Sunnyside Dr. E4 17Ec 34
Sunnyside Gdns. RM14: Upm 34Sd 78
Sunnyside Ho's. NW2 34Bb 69
(off Sunnyside)
Sunnyside Pas. SW19 65Ab 132
Sunnyside Pl. SW19 65Ab 132
Sunnyside Rd. CM16: Epp 5Vc 23
Sunnyside Rd. E10 32Cc 72
Sunnyside Rd. IG1: Ilf 34Sc 74
Sunnyside Rd. N19 31Mb 70
Sunnyside Rd. TW11: Tedd 63Fa 130
Sunnyside Rd. W5 46Ma 87
Sunnyside Rd. E. N9 20Wb 33
Sunnyside Rd. Nth. N9 20Wb 33
Sunnyside Rd. Sth. N9 20Wb 33
Sunnyside Ter. NW9 27Ta 47
Sunny Vw. NW9 29Ta 47
Sunny Way N12 24Gb 49
Sun Pas. SE16 48Wb 91
(off Old Jamaica Rd.)
Sun Pas. SL4: Wind 3H 103
Sunray Av. BR2: Broml 72Nc 160
Sunray Av. CM13: Hut 16Fe 41
Sunray Av. KT5: Surb 75Ra 153
Sunray Av. SE24 56Tb 113
Sunray Av. UB7: W Dray 47M 83
Sunrise Av. RM12: Horn 34Ld 77
Sunrise Cl. E20 36Ec 72
Sunrise Cl. TW13: Hanw 62Ba 129
Sunrise Cotts. TN13: S'oaks 94Hd 202
Sunrise Cres. HP3: Hem H 5N 3
Sunrise Vw. NW7 23Va 48
Sun Rd. DA10: Swans 58Be 121
Sun Rd. W14 50Bb 89
Sunset Av. E4 18Dc 34
Sunset Av. IG8: Wfd G 21Hc 53
Sunset Cl. DA8: Erith 52Kd 119
Sunset Ct. IG8: Wfd G 24Lc 53
Sunset Dr. RM4: Have B 22Kd 57
Sunset Gdns. SE25 68Vb 135
Sunset Lodge NW10 38Ya 68
(off Hanover Rd.)
Sunset M. RM5: Col R 23Ed 56
Sunset Rd. SE28 46Wc 95
Sunset Rd. SE5 56Sb 113
Sunset Rd. SW19 64Xa 132
Sunset Vw. EN5: Barn 12Ab 30
Sunshine Way CR4: Mitc 68Hb 133
Sun Sq. HP1: Hem H 1M 3
(off Chapel St.)
Sunstone Gro. RH1: Mers 1D 208
Sun St. EC2: Finsbury Sq. 7G 219 (43Tb 91)
Sun St. EC2: Primrose St. 7H 219 (43Ub 91)
Sun St. EN9: Walt A 5Ec 20
Sun St. Pas. EC2 1H 225 (43Ub 91)
Sun Wlk. E1 45Vb 91
Sunwell Cl. SE15 53Xb 113
Sun Wharf SE8 52Dc 114
(off Creekside)
Superior Dr. BR6: Chels 79Vc 161
Supreme Court 2F 229 (47Nb 90)
Supreme Point E16 43Jc 93
(off Butchers Rd.)
SURBITON 72Ma 153
Surbiton Ct. KT6: Surb 72La 152
Surbiton Cres. KT1: King T 70Na 131
Surbiton Golf Course 77Ka 152
Surbiton Hall Cl. KT1: King T 70Na 131
Surbiton Hill Pk. KT5: Surb 71Pa 153
Surbiton Hill Rd. KT6: Surb 70Na 131
Surbiton Pde. KT6: Surb 72Na 153
Surbiton Plaza KT6: Surb 72Ma 153
(off St Mary's Rd.)
Surbiton Raceway 76Ta 153
Surbiton Rd. KT1: King T 70Ma 131
Surbiton Station (Rail) 72Na 153
Surlingham Cl. SE28 45Zc 95
Surly Hall Wlk. SL4: Wind 3D 102
Surma Cl. E1 42Xb 91
Surman Cres. CM13: Hut 17Ee 41
Surmans Cl. RM9: Dag 39Zc 75
Surrendale Pl. W9 42Cb 89
Surrey Av. SL2: Slou 3G 80
Surrey Canal Rd. SE14 51Zb 114
Surrey Canal Trade Pk. SE14 51Zb 114
Surrey Cl. N3 27Ab 48
Surrey Cl. TN15: W King 80Ud 164
Surrey County Cricket Club 51Pb 112
Surrey Cres. W4 50Qa 87
Surrey Docks Farm 47Bc 92
Surrey Docks Watersports Cen. 48Ac 92
Surrey Downs Golf Course 94Eb 195
Surrey Dr. RM11: Horn 28Qd 57
Surrey Gdns. KT24: Eff J 94W 190
Surrey Gdns. N4 30Sb 51
Surrey Gro. SE17 7H 231 (50Ub 91)
Surrey Gro. SM1: Sutt 76Fb 155
Surrey History Cen. 10P 167
Surrey Ho. CR0: C'don 76Sb 157
(off Surrey St.)
Surrey Ho. SE16 46Zb 92
(off Rotherhithe St.)
Surrey La. SW11 53Gb 111
Surrey La. Est. SW11 53Gb 111
Surrey Lodge KT12: Hers 78X 151
(off Queens Rd.)
Surrey M. SE27 63Ub 135
Surrey Mt. SE23 60Xb 113
Surrey Nat. Golf Course 95Rb 197
Surrey Quays Rd. SE16 48Yb 92
Surrey Quays Shop. Cen. 48Zb 92
Surrey Quays Station (Overground) 49Zb 92
Surrey Rd. BR4: W W'ck 74Dc 158
Surrey Rd. HA1: Harr 29Ea 46
Surrey Rd. IG11: Bark 38Uc 74
Surrey Rd. RM10: Dag 36Dd 76
Surrey Rd. SE15 57Zb 114
Surrey Row SE1 1B 230 (47Rb 91)
Surrey Sq. SE17 7H 231 (50Ub 91)
Surrey Steps WC2 4J 223 (45Pb 90)
(off Surrey St.)
Surrey St. CR0: C'don 75Sb 157
Surrey St. E13 41Kc 93
Surrey St. WC2 4J 223 (45Pb 90)
Surrey Ter. SE17 7J 231 (50Ub 91)
Surrey Towers KT15: Add 78L 149
(off Bush Cl.)
Surrey Water Rd. SE16 46Zb 92
Surridge Cl. RM13: Rain 41Ld 97
Surridge Ct. SW9 54Nb 112
(off Clapham Rd.)
Surridge Gdns. SE19 65Tb 135
Surr St. N7 36Nb 70
Surry Cres. UB6: G'frd 41Ca 85
Sury Basin KT2: King T 67Na 131
Susan Cl. RM7: Mawney 27Ed 56
Susan Constant Ct. E14 45Fc 93
(off Newport Av.)
Susan Edwards Ho. SL9: Chal P 21A 42
(off Micholls Av.)
Susan Lawrence Ho. E12 35Qc 74
(off Walton Rd.)
Susan Lawrence Ho. E3 40Ac 72
(off Zealand Rd.)
Susannah St. E14 44Dc 92
Susan Rd. SE3 54Kc 115
Susan Wood BR7: Chst 67Qc 138
Sussex Av. RM3: Hrld W 24Pd 57
Sussex Av. TW7: Isle 55Ga 108
Sussex Cl. GU21: Knap 10G 166
Sussex Cl. IG4: Ilf 29Pc 54
Sussex Cl. KT3: N Mald 70Ua 132
Sussex Cl. N19 33Nb 70
Sussex Cl. RH2: Reig 7M 207
Sussex Cl. SL1: Slou 7M 81
Sussex Cl. TN15: W King 80Ud 164
Sussex Cl. TW1: Twick 58Ka 108
Sussex Ct. GU21: Knap 9G 166
Sussex Ct. KT15: Add 78L 149
Sussex Ct. KT18: Eps 86Ua 174
(off Downside)
Sussex Ct. SE10 51Ec 114
(off Roan St.)
Sussex Ct. W2 3B 220 (44Fb 89)
(off Spring St.)
Sussex Cres. UB5: N'olt 37Ca 65
Sussex Gdns. KT9: Chess 79Ma 153
Sussex Gdns. N4 29Sb 51
Sussex Gdns. N6 29Hb 49
Sussex Gdns. W2 4B 220 (45Fb 89)
Sussex Ga. N6 29Hb 49
Sussex Ho. NW1 1D 216 (40Mb 70)
(off Chalton St.)
Sussex Ho. SL2: Farn C 7G 60
Sussex Keep SL1: Slou 7M 81
Sussex Lodge W2 3C 220 (44Fb 89)
(off Sussex Pl.)
Sussex Mans. SW7 6B 226 (49Fb 89)
(off Old Brompton Rd.)
Sussex Mans. WC2 4G 223 (45Nb 90)
(off Maiden La.)
Sussex M. SE6 59Cc 114
Sussex M. E. W2 3C 220 (44Fb 89)
(off Clifton Pl.)
Sussex M. W. W2 4C 220 (45Fb 89)
Sussex Pl. GU21: Knap 10G 166
Sussex Pl. KT3: N Mald 70Ua 132
Sussex Pl. NW1 5F 215 (42Hb 89)
Sussex Pl. SL1: Slou 7L 81
Sussex Pl. W2 3C 220 (44Fb 89)
Sussex Pl. W6 50Ya 88
Sussex Ring N12 22Cb 49
Sussex Rd. BR4: W W'ck 74Dc 158
Sussex Rd. BR5: Orp 72Yc 161
Sussex Rd. CM14: W'ley 21Xd 58
Sussex Rd. CR2: S Croy 79Tb 157
Sussex Rd. CR4: Mitc 71Nb 156
Sussex Rd. DA1: Dart 59Qd 119
Sussex Rd. DA14: Sidc 64Xc 139
Sussex Rd. DA8: Erith 52Dd 118
Sussex Rd. E6 39Qc 74
Sussex Rd. GU21: Knap 10G 166
Sussex Rd. HA1: Harr 29Ea 46
Sussex Rd. KT3: N Mald 70Ua 132
Sussex Rd. SM5: Cars 79Hb 155
Sussex Rd. UB10: Ick 35S 64
Sussex Rd. UB2: S'hall 48Z 85
Sussex Rd. WD24: Wat 9W 12
Sussex Sq. W2 4C 220 (45Fb 89)
Sussex St. E13 41Kc 93
Sussex St. SW1 7A 228 (50Kb 90)
Sussex Ter. RM19: Purf 50Rd 97
Sussex Ter. SE20 66Yb 136
(off Graveney Gro.)
Sussex Way EN4: Cockf 15Kb 32
Sussex Way N19 32Mb 70
(not continuous)
Sussex Way N7 33Nb 70
Sutcliffe Cl. NW11 29Db 49
Sutcliffe Cl. WD23: Bush 14Ea 28
Sutcliffe Ho. UB3: Hayes 44W 84
Sutcliffe Pk. Athletics Track 57Lc 115
Sutcliffe Rd. DA16: Well 54Yc 117
Sutcliffe Rd. SE18 51Uc 116
Sutherland Av. BR5: St M Cry 72Vc 161
Sutherland Av. DA16: Well 56Uc 116
Sutherland Av. GU4: Jac W 100A 188
Sutherland Av. TN16: Big H 89Mc 179
Sutherland Av. TW16: Sun 68V 128
Sutherland Av. UB3: Hayes 49W 84
Sutherland Av. W13 44Ka 86
Sutherland Av. W9 42Cb 89
Sutherland Cl. DA12: Grav'nd 1K 145
Sutherland Cl. DA9: Ghithe 57Vd 120
Sutherland Cl. EN5: Barn 14Ab 30
Sutherland Ct. N16 34Tb 71
Sutherland Ct. NW9 29Ra 47
Sutherland Ct. W9 42Cb 89
(off Marylands Rd.)
Sutherland Ct. WD18: Wat 17S 26
Sutherland Dr. SW19 67Fb 133
Sutherland Gdns. KT4: Wor Pk 74Xa 154
Sutherland Gdns. SW14 55Ua 110
Sutherland Gdns. TW16: Sun 68V 128
Sutherland Grange SL4: Wind 2B 102
Sutherland Gro. SW18 58Ab 110
Sutherland Gro. TW11: Tedd 64Ga 130
Sutherland Ho. IG8: Wfd G 24Qc 54
Sutherland Ho. W8 48Db 89
Sutherland Pl. W2 44Cb 89
Sutherland Rd. CR0: C'don 73Qb 156
Sutherland Rd. DA17: Belv 48Cd 96
Sutherland Rd. E17 26Zb 52
Sutherland Rd. E3 40Bc 72
Sutherland Rd. EN3: Pond E 16Zb 34
Sutherland Rd. N17 24Wb 51
Sutherland Rd. N9 18Xb 33
Sutherland Rd. UB1: S'hall 44Ba 85
Sutherland Rd. W13 44Ja 86
Sutherland Rd. W4 51Ua 110
Sutherland Rd. Path E17 27Zb 52
Sutherland Row SW1 7A 228 (50Kb 90)
Sutherland Sq. SE17 50Sb 91
Sutherland St. SW1 7K 227 (50Kb 90)
Sutherland Wlk. SE17 50Sb 91
Sutherland Way EN6: Cuff 1Mb 18
Sutlej Rd. SE7 52Lc 115
Sutterton St. N7 37Pb 70
SUTTON 50E 82
SUTTON 78Db 155
SUTTON AT HONE 66Rd 141
Sutton Av. GU21: Wok 1J 187
Sutton Av. SL3: L'ly 7N 81
Sutton Cl. BR3: Beck 67Dc 136
Sutton Cl. CR3: Cat'm 94Wb 197
Sutton Cl. HA5: Eastc 29W 44
Sutton Cl. IG10: Lough 17Nc 36
Sutton Cl. W4 51Sa 109
(off Sutton La. Sth.)
Sutton Comn. Rd. SM1: Sutt 75Cb 155
Sutton Comn. Rd. SM3: Sutt 73Bb 155
Sutton Common Station (Rail) 75Db 155
Sutton Ct. KT8: W Mole 71Ba 151
Sutton Ct. SE19 66Vb 135
Sutton Ct. SM2: Sutt 79Eb 155
Sutton Ct. W4 51Sa 109
Sutton Ct. W5 46Na 87
Sutton Ct. Rd. E13 41Lc 93
Sutton Ct. Rd. SM1: Sutt 79Eb 155
Sutton Ct. Rd. UB10: Hil 39R 64
Sutton Ct. Rd. W4 52Sa 109
Sutton Cres. EN5: Barn 15Za 30
Sutton Dene TW3: Houn 53Da 107
Sutton Est. EC1 4G 219 (41Tb 91)
(off Old St.)
Sutton Est. SW3 7D 226 (50Gb 89)
Sutton Est. W10 43Ya 88
Sutton Est., The N1 38Rb 71
Sutton Gdns. CR0: C'don 71Vb 157
Sutton Gdns. IG11: Bark 39Uc 74
Sutton Gdns. RH1: Mers 1D 208
SUTTON GREEN 98B 188
Sutton Grn. IG11: Bark 39Uc 74
(off Sutton Rd.)
Sutton Green Golf Course 95A 188
Sutton Grn. Rd. GU4: Sut G 98A 188
Sutton Gro. SM1: Sutt 77Fb 155
Sutton Hall Rd. TW5: Hest 52Ca 107
Sutton Hgts. SM2: Sutt 80Fb 155
SUTTON HOSPITAL 82Db 175
Sutton La. EC1 6C 218 (42Rb 91)
(off Gt. Sutton St.)
Sutton La. SL3: L'ly 51D 104
Sutton La. SM2: Sutt 83Db 175
Sutton La. SM7: Bans 85Db 175
Sutton La. TW3: Houn 55Ba 107
Sutton La. Nth. W4 50Sa 87
Sutton La. Sth. W4 51Sa 109
Sutton Pde. NW4 28Ya 48
(off Church Rd.)
SUTTON PARK 98B 188
Sutton Pk. Rd. SM1: Sutt 79Db 155
Sutton Path WD6: Bore 13Qa 29
Sutton Pl. E9 36Yb 72
Sutton Pl. SL3: L'ly 51D 104
Sutton Plaza SM1: Sutt 79Eb 155
Sutton Rd. AL1: St A 3F 6
Sutton Rd. E13 42Hc 93
Sutton Rd. E17 25Zb 52
Sutton Rd. IG11: Bark 40Uc 74
Sutton Rd. N10 25Jb 50
Sutton Rd. TW5: Hest 53Ca 107
Sutton Rd. WD17: Wat 13Y 27
(not continuous)
Sutton Row W1 2D 222 (44Mb 90)
Suttons Av. RM12: Horn 34Ld 77
Suttons Bus. Pk. RM13: Rain 41Gd 96
Suttons Gdns. RM12: Horn 34Md 77
Suttons La. RM12: Horn 36Md 77
SUTTONS MANOR 11Gd 38
Sutton Sports Village 74Db 155
Sutton Sq. E9 36Yb 72
Sutton Sq. TW5: Hest 53Ba 107
Sutton Station (Rail) 79Eb 155
Sutton St. E1 45Yb 92
Sutton's Way EC1 6E 218 (42Sb 91)
(off Lambs Pas.)
Suttons Wharf E2 41Zb 92
Sutton Tennis Academy 74Db 155
Sutton United FC 77Cb 155
Sutton Wlk. SE1 7J 223 (46Pb 90)
Sutton Way TW5: Hest 53Ba 107
Sutton Way W10 42Ya 88
Swabey Rd. SL3: L'ly 49C 82
Swaby Rd. SW18 60Eb 111
Swaffam Ct. RM6: Chad H 30Yc 55
Swaffham Way N22 24Rb 51
Swaffield Rd. SW18 59Db 111
Swaffield Rd. TN13: S'oaks 94Ld 203
Swail Ho. KT18: Eps 85Ta 173
Swain Cl. SW16 65Kb 134
Swain Rd. CR7: Thor H 71Sb 157
Swains Cl. UB7: W Dray 47N 83
Swain's La. N6 32Jb 70
Swainson Rd. W3 47Va 88
Swains Rd. SW17 66Hb 133
Swain St. NW8 5D 214 (42Gb 89)
Swaisland Rd. DA1: Dart 57Kd 119
Swaislands Dr. DA1: Cray 57Hd 118
Swakeleys Dr. UB10: Ick 35Q 64
Swakeleys Rd. UB10: Ick 35N 63
SWAKELEYS RDBT. 35N 63
Swalecliffe Rd. DA17: Belv 50Dd 96
Swale Cl. RM15: Avel 44Sd 98
Swaledale Cl. N11 23Jb 50
Swaledale Rd. DA2: Dart 60Sd 120
Swale Rd. DA1: Cray 56Jd 118
Swaley's Way RM15: S Ock 44Zd 99
Swallands Rd. SE6 62Cc 136
(not continuous)
Swallow Cl. DA8: Erith 53Gd 118
Swallow Cl. DA9: Ghithe 57Vd 120
Swallow Cl. RM16: Chaf H 49Yd 98
Swallow Cl. SE14 53Zb 114
Swallow Cl. TW18: Staines 63H 127
Swallow Cl. WD23: Bush 18Ea 28
Swallow Cl. WD3: Rick 17L 25
Swallow Ct. EN3: Enf W 9Yb 20
Swallow Ct. HA4: Ruis 32Y 65
Swallow Ct. IG2: Ilf 29Rc 54
Swallow Ct. SE1 2E 230 (47Sb 91)
(off Swan St.)
Swallow Ct. SE12 59Jc 115
Swallow Ct. W9 43Cb 89
(off Admiral Wlk.)
Swallowdale CR2: Sels 81Zb 178
Swallowdale SL0: Iver H 41F 82
Swallow Dr. NW10 37Ta 67
Swallow Dr. UB5: N'olt 40Ca 65
Swallowfield NW1 4A 216 (41Kb 90)
(off Munster Sq.)
Swallowfield TW20: Eng G 5M 125
Swallowfield Rd. SE7 50Kc 93
Swallow Flds. SL0: Iver 42F 82
Swallowfields DA11: Nflt 2A 144
Swallowfield Way UB3: Hayes 47T 84
Swallow Gdns. AL10: Hat 2C 8
Swallow Gdns. SW16 64Mb 134
Swallow Ho. NW8 1D 214 (40Gb 69)
(off Allitsen Rd.)
Swallow La. AL1: St A 5F 6
Swallow Pk. Cl. KT6: Surb 76Pa 153
Swallow Pas. W1 3A 222 (44Kb 90)
(off Swallow Pl.)
Swallow Pl. E14 44Bc 92
(off Newell St.)
Swallow Pl. KT17: Ewe 82Wa 174
Swallow Pl. W1 3A 222 (44Kb 90)
Swallow Ri. GU21: Knap 9G 166
Swallows, The UB9: Den 29H 43
(off Patrons Way W.)
Swallows Cl. SM1: Sutt 76Cb 155
Swallows Ct. DA1: Dart 55Pd 119
(off Vickers La.)
Swallows Oak WD5: Ab L 3V 12
Swallow St. E6 43Nc 94
Swallow St. SL0: Iver 41F 82
Swallow St. SL0: Iver H 41F 82
Swallow St. W1 5C 222 (45Lb 90)
Swallowtail Cl. BR5: St P 70Zc 139
Swallowtail Ho. E20 36Ec 72
(off Sunrise Cl.)
Swallow Wlk. HP3: Hem H 6L 3
Swallow Wlk. RM12: Horn 37Kd 77
SWAN, THE 75Ec 158
Swanage Ct. N1 38Ub 71
(off Hertford Rd.)
Swanage Ho. SW8 52Pb 112
(off Dorset Rd.)
Swanage Rd. E4 24Ec 52
Swanage Rd. SW18 58Eb 111
Swanage Waye UB4: Yead 44Y 85
Swan & Pike Rd. EN3: Enf L 10Cc 20
Swan App. E6 43Nc 94
Swan Av. RM14: Upm 32Vd 78
Swanbourne Dr. RM12: Horn 36Ld 77
Swanbourne Ho. NW8 5D 214 (42Gb 89)
(off Capland St.)
Swanbridge Rd. DA7: Bex 53Cd 118
Swan Bus. Pk. DA1: Dart 56Md 119
Swan Cen., The 93Ka 192
Swan Cen., The SW17 62Db 133
Swan Cl. BR5: St P 69Wc 139
Swan Cl. CR0: C'don 73Ub 157
Swan Cl. E17 25Ac 52
Swan Cl. KT12: Walt T 73V 150
Swan Cl. TW13: Hanw 63Aa 129
Swan Cl. WD3: Rick 17M 25
Swan Ct. E1 45Wb 91
(off Star Pl.)
Swan Ct. E14 44Bc 92
(off Agnes St.)
Swan Ct. HA4: Ruis 31T 64
Swan Ct. HP1: Hem H 3L 3
Swan Ct. KT22: Lea 94Ka 192
Swan Ct. SW3 50Gb 89
Swan Ct. SW6 52Cb 111
(off Fulham Rd.)
Swan Ct. TW7: Isle 55Ka 108
(off Swan St.)
Swandon Way SW18 56Db 111
Swandrift TW18: Staines 66H 127
Swan Dr. NW9 26Ua 48
Swan Fld. Ho. WD3: Rick 17M 25
Swanfield Rd. EN8: Walt C 5Ac 20
Swanfield St. E2 4K 219 (41Vb 91)
Swan Ho. E15 38Gc 73
(off Broadway)
Swan Ho. EN3: Pond E 15Yb 34
Swan Ho. N1 38Tb 71
(off Oakley Rd.)
Swan Island TW1: Twick 62Ja 130
Swanland Rd. AL9: N Mym 7D 8
Swanland Rd. EN6: S Mim 6Xa 16
Swan La. CM14: B'wood 19Yd 40
Swan La. DA1: Dart 59Hd 118
Swan La. EC4 5G 225 (45Tb 91)
Swan La. IG10: Lough 17Lc 35
Swan La. N20 20Eb 31
Swan La. N4 32Sb 71
SWANLEY 69Gd 140
SWANLEY BAR 10K 9
Swanley Bar La. EN6: Pot B 10K 9
Swanley By-Pass DA14: Sidc 67Dd 140
Swanley Cen. BR8: Swan 69Gd 140
Swanley Ct. WD24: Wat 9Y 13
Swanley Cres. EN6: Pot B 1Db 17
Swanley Ho. SE17 7J 231 (50Ub 91)
(off Kinglake Est.)
SWANLEY INTERCHANGE 71Kd 163
Swanley La. BR8: Swan 69Hd 140
Swanley Rd. DA16: Well 53Yc 117
Swanley Station (Rail) 70Fd 140
SWANLEY VILLAGE 67Kd 141
Swanley Village Rd. BR8: Swan 67Kd 141
Swan Mead HP3: Hem H 7P 3
Swan Mead SE1 4H 231 (48Ub 91)
Swan M. CR4: Mitc 67Hb 133
Swan M. RM7: Mawney 28Dd 56
Swan M. SW6 53Bb 111
Swan M. SW9 54Pb 112
Swann Ct. SL1: Slou 8J 81
Swann Ct. TW7: Isle 55Ja 108
(off South St.)
Swanne Ho. SE10 52Ec 114
(off Gloucester Cir.)
Swannells Wlk. WD3: Rick 17H 25
Swannell Way NW2 32Za 68
Swanns Mdw. KT23: Bookh 98Ca 191
Swann St. DA10: Swans 60Ae 121
Swan Paddock CM14: B'wood 19Yd 40
Swan Pas. E1 45Vb 91
(off Cartwright St.)
Swan Path KT1: King T 69Pa 131
Swan Pl. SW13 54Va 110
Swan Pl. TN16: Westrm 99Tc 200
Swan Rd. EN8: Walt C 6Ac 20
Swan Rd. SE16 47Yb 92
Swan Rd. SE18 48Mc 93
Swan Rd. SL0: Iver 44H 83
Swan Rd. TW13: Hanw 64Aa 129
Swan Rd. UB1: S'hall 44Da 85
Swan Rd. UB7: W Dray 47M 83
Swans, The UB9: Den 29H 43
Swan Sanctuary, The 72U 150
Swans Cl. AL4: St A 3J 7
SWANSCOMBE 57Be 121
Swanscombe Bus. Cen. DA10: Swans 56Ae 121
Swanscombe Heritage Pk. 58Zd 121
Swanscombe Ho. W11 46Za 88
(off St Ann's Rd.)
Swanscombe Leisure Cen. 58Zd 121
Swanscombe Rd. W11 46Za 88
Swanscombe Rd. W4 50Ua 88
Swanscombe Skull Site Nat. Nature Reserve 58Zd 121
Swanscombe Station (Rail) 57Be 121
Swanscombe St. DA10: Swans 59Ae 121
Swans Ct. EN8: Walt C 6Ac 20
Swansea Cl. RM5: Col R 24Fd 56
Swansea Ct. E16 46Rc 94
(off Fishguard Way)
Swansea Rd. EN3: Pond E 14Yb 34
Swansea Rd. TW14: Felt 58S 106
Swanshope IG10: Lough 12Rc 36
Swansland Gdns. E17 25Ac 52
Swansmere Cl. KT12: Walt T 74Y 151
Swanston Ho. WD18: Wat 14W 26
(off Whippendell Rd.)
Swanston Path WD19: Wat 20Y 27
Swan St. SE1 3E 230 (48Sb 91)
Swan St. TW7: Isle 55Ka 108
Swansway, The KT13: Weyb 76Q 150
Swan Ter. SL4: Wind 2F 102
Swanton Ct. SE13 55Dc 114
Swanton Gdns. SW19 60Za 110
Swanton La. TN11: Roug 100Fe 205
Swanton Rd. DA8: Erith 52Cd 118
Swan Wlk. RM1: Rom 29Gd 56
Swan Wlk. SW3 51Hb 111
Swan Wlk. TW17: Shep 73U 150
Swan Way EN3: Enf H 12Zb 34
Swan Wharf Bus. Cen. UB8: Uxb 40K 63
Swanwick Cl. SW15 59Va 110
Swanworth La. RH5: Mick 100Ja 192
Swan Yd. N1 37Rb 71
Swanzy Rd. TN14: S'oaks 92Ld 203
Sward Rd. BR5: St M Cry 72Wc 161
Swathling Ho. SW15 58Va 110
(off Tunworth Cres.)
Swaton Rd. E3 42Cc 92
Swaylands Rd. DA17: Belv 51Cd 118
Swaythling Cl. N18 21Xb 51
Swedeland Ct. E1 1J 225 (43Ub 91)
(off Bishopsgate)
Swedenborg Gdns. E1 45Xb 91
Sweden Ga. SE16 48Ac 92
Swedish Quays SE16 48Ac 92
(not continuous)
Sweeney Cres. SE1 2K 231 (47Vb 91)
Sweeps Ditch Cl. TW18: Staines 66J 127
Sweeps La. BR5: St M Cry 71Zc 161
Sweeps La. TW20: Egh 64B 126
Sweetbriar Av. SM5: Cars 74Hb 155
Sweet Briar Grn. N9 20Vb 33
Sweet Briar Gro. N9 20Vb 33
Sweet Briar La. KT18: Eps 86Ta 173
Sweet Briar Wlk. N18 21Vb 51
Sweetcroft La. UB10: Hil 38P 63
Sweetmans Av. HA5: Pinn 27Z 45
Sweets Way N20 19Fb 31
Swetenham Wlk. SE18 50Sc 94
Swete St. E13 40Jc 73
Sweyne Rd. DA10: Swans 58Ae 121
Sweyn Pl. SE3 54Jc 115
Swievelands Rd. TN16: Big H 91Kc 199
Swift Cen. CR0: Wadd 80Pb 156
Swift Cl. E17 24Ac 52
Swift Cl. HA2: Harr 33Da 65
Swift Cl. KT17: Ewe 82Wa 174
Swift Cl. RM14: Upm 32Ud 78
Swift Cl. SE28 45Xc 95
Swift Cl. SL1: Slou 5D 80
Swift Cl. UB3: Hayes 44V 84
Swift Ct. SM2: Sutt 80Db 155
Swift Ct. TN15: Seal 93Qd 203
Swift Ho. E3 39Bc 72
(off Old Ford Rd.)

Swift Ho. NW6 40Bb 69
(off Albert Rd.)
Swift La. SE13 56Gc 115
Swift Lodge W9 43Cb 89
(off Admiral Wlk.)
Swift Rd. TW13: Hanw 62Aa 129
Swift Rd. UB2: S'hall 48Ca 85
Swiftsden Way BR1: Broml 65Gc 137
Swiftstone Twr. SE10 48Jc 93
Swift St. SW6 53Bb 111
Swiftsure Rd. RM16: Chaf H 49Yd 98
Swiller's La. DA12: Shorne 4N 145
SWILLET, THE 16D 24
Swimmers La. E2 1K 219 (39Vb 71)
Swinbrook Rd. W10 43Ab 88
Swinburne Ct. SE5 56Tb 113
(off Basingdon Way)
Swinburne Cres. CR0: C'don 72Yb 158
Swinburne Gdns. RM18: Tilb 4D 122
Swinburne Ho. E2 41Yb 92
(off Roman Rd.)
Swinburne Rd. SW15 56Wa 110
Swinderby Rd. HA0: Wemb 37Na 67
Swindon Cl. IG3: Ilf 33Uc 74
Swindon Cl. RM3: Rom 22Pd 57
Swindon Gdns. RM3: Rom 22Pd 57
Swindon La. RM3: Rom 22Pd 57
Swindon Rd. TW6: H'row A 57S 106
Swindon St. W12 46Xa 88
Swinfield Cl. TW13: Hanw 62Aa 129
Swinford Gdns. SW9 55Rb 113
Swingate La. SE18 51Uc 116
Swingfield Ho. E9 39Yb 72
(off Templecombe Rd.)
Swing Ga. La. HP4: Berk 3A 2
Swinley Ho. NW1 3A 216 (41Kb 90)
(off Redhill St.)
Swinnerton St. E9 36Ac 72
Swinson Ho. N11 22Lb 50
Swinton Cl. HA9: Wemb 32Ra 67
Swinton Pl. WC1 3H 217 (41Pb 90)
Swinton St. WC1 3H 217 (41Pb 90)
Swires Shaw BR2: Kes 77Mc 159
Swiss Av. WD18: Wat 14U 26
Swiss Cl. WD18: Wat 13U 26
SWISS COTTAGE 38Fb 69
Swiss Cott. Pl. IG10: Lough 15Mc 35
Swiss Cottage Sports Cen. 38Fb 69
Swiss Cottage Station (Underground) 38Fb 69
Swiss Ct. W1 5E 222 (45Mb 90)
(off Panton St.)
Swiss Ter. NW6 38Fb 69
Switch Ho. E14 45Fc 93
Swithland Gdns. SE9 63Qc 138
Swyncombe Av. W5 49Ka 86
Swynford Gdns. NW4 28Wa 48
Sybil M. N4 30Rb 51
Sybil Phoenix Cl. SE8 50Zb 92
Sybil Thorndike Casson Ho. SW5 50Cb 89
(off Kramer M.)
Sybourn St. E17 31Bc 72
Sycamore App. WD3: Crox G 15S 26
Sycamore Av. AL10: Hat 1C 8
Sycamore Av. DA15: Sidc 58Vc 117
Sycamore Av. E3 39Bc 72
Sycamore Av. GU22: Wok 92A 188
Sycamore Av. RM14: Upm 34Qd 77
Sycamore Av. UB3: Hayes 45U 84
Sycamore Av. W5 48Ma 87
Sycamore Cl. CM13: Gt War 23Xd 58
Sycamore Cl. CR2: S Croy 78Ub 157
Sycamore Cl. DA12: Grav'nd 9F 122
Sycamore Cl. E16 42Gc 93
Sycamore Cl. EN4: E Barn 16Fb 31
Sycamore Cl. HA8: Edg 21Sa 47
Sycamore Cl. IG10: Lough 12Rc 36
Sycamore Cl. KT19: Ewe 78Ra 153
Sycamore Cl. KT22: Fet 95Ha 192
Sycamore Cl. N9 21Wb 51
Sycamore Cl. RM18: Tilb 4C 122
Sycamore Cl. SE9 61Nc 138
Sycamore Cl. SM5: Cars 77Hb 155
Sycamore Cl. TW13: Felt 62W 128
Sycamore Cl. UB5: N'olt 39Aa 65
Sycamore Cl. UB7: Yiew 45P 83
Sycamore Cl. W3 46Ua 88
Sycamore Cl. WD23: Bush 12Aa 27
Sycamore Cl. WD25: Wat 7X 13
Sycamore Ct. DA8: Erith 50Fd 96
(off Sandcliff Rd.)
Sycamore Ct. DA9: Ghithe 59Ud 120
Sycamore Ct. E7 37Jc 73
Sycamore Ct. KT13: Weyb 76V 150
Sycamore Ct. KT3: N Mald 69Ua 132
Sycamore Ct. NW6 39Cb 69
(off Bransdale Cl.)
Sycamore Ct. RH8: Oxt 1J 211
Sycamore Ct. RM12: Horn 32Hd 76
Sycamore Ct. SE1 2H 231 (47Ub 91)
(off Royal Oak Yd.)
Sycamore Ct. SL4: Wind 5G 102
Sycamore Ct. TW4: Houn 56Aa 107
Sycamore Dr. AL2: Park 9B 6
Sycamore Dr. BR8: Swan 69Gd 140
Sycamore Dr. CM14: B'wood 18Yd 40
Sycamore Gdns. CR4: Mitc 68Fb 133
Sycamore Gdns. N15 28Vb 51
Sycamore Gdns. W6 47Xa 88
Sycamore Gro. KT3: N Mald 69Ta 131
Sycamore Gro. NW9 31Sa 67
Sycamore Gro. RM2: Rom 26Jd 56
Sycamore Gro. SE20 67Wb 135
Sycamore Gro. SE6 58Ec 114
Sycamore Hill N11 23Jb 50
Sycamore Ho. BR2: Broml 68Gc 137
Sycamore Ho. CR6: W'ham 87Dc 178
Sycamore Ho. IG9: Buck H 19Mc 35
Sycamore Ho. N2 26Fb 49
(off The Grange)
Sycamore Ho. SE16 47Zb 92
(off Woodland Cres.)
Sycamore Ho. W6 47Xa 88
Sycamore Ho. WD23: Bush 15Z 27
(off Plantation Cl.)
Sycamore Lodge BR6: Orp 75Vc 161
Sycamore Lodge TW16: Sun 66V 128
Sycamore Lodge W8 48Db 89
(off Stone Hall Pl.)
Sycamore M. CR3: Cat'm 95Tb 197
Sycamore M. DA8: Erith 50Fd 96
(off St John's Rd.)
Sycamore M. SW4 55Lb 112
Sycamore Path E17 30Dc 52
(off Poplars Rd.)
Sycamore Pl. BR1: Broml 69Qc 138
Sycamore Pl. IG7: Chig 21Tc 54
Sycamore Ri. SM7: Bans 86Za 174
Sycamore Rd. DA1: Dart 60Md 119
Sycamore Rd. SW19 65Ya 132
Sycamore Rd. WD3: Crox G 15S 26
Sycamores, The AL1: St A 3B 6
Sycamores, The BR8: Hext 65Hd 140
Sycamores, The EN9: Walt A 7Lc 21
Sycamores, The HP3: Hem H 5H 3
Sycamores, The KT23: Bookh 96Ea 192
Sycamores, The RM15: Avel 46Td 98
Sycamores, The WD7: R'lett 6Ka 14
Sycamore St. EC1 6D 218 (42Sb 91)
Sycamore Wlk. IG6: Ilf 28Sc 54
Sycamore Wlk. RH2: Reig 9L 207
Sycamore Wlk. SL3: Geor G 44A 82
Sycamore Wlk. TW20: Eng G 5M 125
Sycamore Wlk. W10 42Ab 88
Sycamore Way CR7: Thor H 71Qb 156
Sycamore Way RM15: S Ock 42Zd 99
Sycamore Way TW11: Tedd 65La 130
Sydcote SE21 60Sb 113
SYDENHAM 63Yb 136
Sydenham Av. N21 15Pb 32
Sydenham Av. SE26 64Xb 135
Sydenham Cl. RM1: Rom 27Hd 56
Sydenham Cotts. SE12 61Lc 137
Sydenham Ct. CR0: C'don 74Tb 157
(off Sydenham Rd.)
Sydenham Gdns. SL1: Slou 7G 80
Sydenham Hill SE23 60Xb 113
Sydenham Hill SE26 63Vb 135
Sydenham Hill Station (Rail) 62Vb 135
Sydenham Hill Wood & Cox's Walk Nature Reserve 61Wb 135
Sydenham Pk. SE26 62Yb 136
Sydenham Pk. Mans. SE26 62Yb 136
(off Sydenham Pk.)
Sydenham Pk. Rd. SE26 62Yb 136
Sydenham Pl. SE27 62Rb 135
Sydenham Ri. SE23 61Xb 135
Sydenham Rd. CR0: C'don 74Sb 157
Sydenham Rd. SE26 64Zb 136
Sydenham Sta. App. SE26: Sydenham 63Yb 136
Sydenham Station (Rail & Overground) 63Yb 136
Sydmons Ct. SE23 59Yb 114
Sydner M. N16 35Vb 71
Sydner Rd. N16 35Vb 71
Sydney Av. CR8: Purl 84Pb 176
Sydney Chapman Way EN5: Barn 12Bb 31
Sydney Cl. SW3 6C 226 (49Fb 89)
Sydney Ct. UB4: Yead 42Y 85
Sydney Cres. TW15: Ashf 65R 128
Sydney Gro. NW4 29Ya 48
Sydney Gro. SL1: Slou 4G 80
Sydney M. SW3 6C 226 (49Fb 89)
Sydney Pl. AL3: St A 2A 6
Sydney Pl. SW7 6C 226 (49Fb 89)
Sydney Rd. DA14: Sidc 63Uc 138
Sydney Rd. DA6: Bex 56Zc 117
Sydney Rd. E11 30Kc 53
Sydney Rd. EN2: Enf 14Tb 33
Sydney Rd. IG6: Ilf 26Sc 54
Sydney Rd. IG8: Wfd G 21Jc 53
Sydney Rd. N10 25Jb 50
Sydney Rd. N8 28Qb 50
Sydney Rd. RM18: Tilb 4C 122
Sydney Rd. SE2 48Yc 95
Sydney Rd. SM1: Sutt 77Cb 155
Sydney Rd. SW20 68Za 132
Sydney Rd. TW11: Tedd 64Ha 130
Sydney Rd. TW14: Felt 60W 106
Sydney Rd. TW9: Rich 56Na 109
Sydney Rd. W13 46Ja 86
Sydney Rd. WD18: Wat 15U 26
Sydney Russell Leisure Cen. 36Ad 75
Sydney Simmons KT21: Asht 90Pa 173
Sydney St. SW3 7D 226 (50Gb 89)
Sydney Ter. KT10: Clay 79Ha 152
(off The Green)
Syke Cluan SL0: Rich P 47G 82
Syke Ings SL0: Rich P 48G 82
Sykes Dr. TW18: Staines 64K 127
Sykes Rd. SL1: Slou 4F 80
Sylva Cotts. SE8 53Cc 114
Sylvana Cl. UB10: Hil 40Q 62
Sylvan Av. N22 24Pb 50
Sylvan Av. N3 26Cb 49
Sylvan Av. NW7 23Va 48
Sylvan Av. RM11: Horn 30Nd 57
Sylvan Av. RM6: Chad H 30Bd 55
Sylvan Cl. CR2: Sels 82Xb 177
Sylvan Cl. GU22: Wok 89D 168
Sylvan Cl. HP3: Hem H 3A 4
Sylvan Cl. RH8: Limp 1M 211
Sylvan Cl. RM16: Chaf H 49Ae 99
Sylvan Ct. N12 20Db 31
Sylvan Ct. NW6 39Db 69
(off Abbey Rd.)
Sylvan Est. SE19 67Vb 135
Sylvan Gdns. KT6: Surb 73Ma 153
Sylvan Gro. NW2 35Za 68
Sylvan Gro. SE15 51Xb 113
Sylvan Hgts. AL1: St A 4E 6
Sylvan Hill SE19 67Ub 135
Sylvan M. DA9: Ghithe 56Xd 120
(off Watermans Way)
Sylvan Rd. E11 29Jc 53
Sylvan Rd. E17 29Cc 52
Sylvan Rd. E7 37Kc 73
Sylvan Rd. IG1: Ilf 33Sc 74
Sylvan Rd. SE19 67Vb 135
Sylvan Ter. SE15 51Xb 113
(off Sylvan Gro.)
Sylvan Wlk. BR1: Broml 69Pc 138
Sylvan Way BR4: W W'ck 77Gc 159
Sylvan Way IG7: Chig 20Xc 37
Sylvan Way RH1: Redh 7A 208
Sylvan Way RM8: Dag 35Xc 75
Sylverdale Rd. CR0: C'don 76Rb 157
Sylverdale Rd. CR8: Purl 85Rb 177
Sylvester Av. BR7: Chst 65Pc 138
Sylvester Gdns. IG6: Ilf 22Xc 55
Sylvester Path E8 37Xb 71
Sylvester Pl. CR3: Cat'm 95Tb 197
Sylvester Rd. E17 31Bc 72
Sylvester Rd. E8 37Xb 71
Sylvester Rd. HA0: Wemb 36La 66
Sylvester Rd. N2 26Eb 49
Sylvestres TN13: Riv 93Fd 202
Sylvestrian Leisure Cen. 28Gc 53
Sylvestrus Cl. KT1: King T 67Qa 131
Sylvia Av. CM13: Hut 19Ee 41
Sylvia Av. HA5: Hat E 23Aa 45
Sylvia Ct. HA9: Wemb 38Ra 67
Sylvia Ct. N1 2G 219 (40Tb 71)
(off Wenlock St.)
Sylvia Gdns. HA9: Wemb 38Ra 67
Sylvia Pankhurst Ho. RM10: Dag 34Cd 76
(off Wythenshawe Rd.)
Sylvia Pankhurst St. E16 44Hc 93
Symes M. NW1 1B 216 (40Lb 70)
Symington Ho. SE1 4F 231 (48Tb 91)
(off Deverell St.)
Symington M. E9 36Zb 72
Symister M. N1 4H 219 (41Ub 91)
(off Coronet St.)
Symonds Cl. TN15: W King 78Ud 164
Symonds Ct. EN8: Chesh 1Zb 20
Symons Cl. SE15 54Yb 114
Symons St. SW3 6G 227 (49Hb 89)
Sympathy Va. DA1: Dart 54Pd 119
Symphony Cl. HA8: Edg 24Ra 47
Symphony M. W10 41Ab 88
Syon Ct. AL1: St A 3E 6
Syon Ct. E11 29Kc 53
Syon Ga. Way TW8: Bford 52Ja 108
Syon House 53La 108
Syon La. TW7: Isle 51Ga 108
Syon Lane Station (Rail) 52Ja 108
Syon Lodge SE12 59Jc 115
Syon Pk. 53Ka 108
Syon Pk. Gdns. TW7: Isle 52Ha 108
Syracuse Av. RM13: Rain 41Nd 97
Syringa Ct. RM17: Grays 52Fe 121
Syringa Ho. SE4 55Bc 114
Sythwood GU21: Wok 9M 167

T

Tabard Ct. E14 44Ec 92
(off Lodore St.)
Tabard Gdn. Est. SE1 3F 231 (48Tb 91)
Tabard Ho. SE1 3G 231 (48Tb 91)
(off Manciple St.)
Tabard St. SE1 1F 231 (47Tb 91)
Tabard Theatre 49Ua 88
Tabarin Way KT17: Eps D 88Ya 174
Tabernacle, The 44Bb 89
(off Powis Sq.)
Tabernacle Av. E13 42Jc 93
Tabernacle Gdns. E2 4K 219 (41Vb 91)
(off Hackney Rd.)
Tabernacle St. EC2 6G 219 (42Tb 91)
Tableer Av. SW4 57Lb 112
Tabley Rd. N7 35Nb 70
Tabor Ct. SM3: Cheam 79Ab 154
Tabor Gdns. SM3: Cheam 79Bb 155
Tabor Gro. SW19 66Bb 133
Tabor Rd. W6 48Xa 88
Tabors Ct. CM15: Shenf 17Ce 41
Tabrums Way RM14: Upm 31Ud 78
TA Cen. 1M 3
(off Queensway)
TA Cen. 37Fb 69
(off College Cres.)
TA Cen. 55Ab 110
(off Ranelagh Gdns.)
Tachbrook Est. SW1 7E 228 (50Mb 90)
Tachbrook M. SW1 5B 228 (49Lb 90)
Tachbrook Rd. TW14: Felt 59V 106
Tachbrook Rd. UB2: S'hall 49Z 85
Tachbrook Rd. UB7: W Dray 46N 83
Tachbrook Rd. UB8: Uxb 40L 63
Tachbrook St. SW1 6C 228 (49Lb 90)
Tack M. SE4 55Cc 114
Tadema Ho. NW8 6C 214 (42Fb 89)
(off Penfold St.)
Tadema Rd. SW10 52Eb 111
Tadlow KT1: King T 69Qa 131
(off Washington Rd.)
Tadlows Cl. RM14: Upm 36Rd 77
Tadmor Cl. TW16: Sun 70V 128
Tadmor St. W12 46Za 88
Tadorne Rd. KT20: Tad 93Ya 194
TADWORTH 94Ya 194
Tadworth Av. KT3: N Mald 70Va 132
Tadworth Cl. KT20: Tad 94Za 194
Tadworth Ct. KT20: Tad 93Za 194
Tadworth Ct. RH1: Redh 4B 208
(off Reynolds Av.)
Tadworth Ho. SE1 2B 230 (47Rb 91)
(off Webber St.)
Tadworth Leisure & Community Cen. 92Ya 194
Tadworth Pde. RM12: Horn 35Kd 77
TADWORTH PARK 93Za 194
Tadworth Rd. NW2 33Wa 68
TADWORTH RDBT. 94Ab 194
Tadworth Station (Rail) 94Ya 194
Tadworth St. KT20: Tad 95Ya 194
Taeping St. E14 49Dc 92
Taffeta Ho. E20 37Ec 72
(off De Coubertin St.)
Taff Ho. KT2: King T 67Ma 131
(off Henry Macaulay Av.)
Taffrail Ho. E14 50Dc 92
(off Burrells Wharf Sq.)
Taffy's How CR4: Mitc 69Gb 133
Taft Way E3 41Dc 92
Tagalie Pl. WD7: Shenl 4Na 15
Taggs Ho. KT1: King T 68Ma 131
(off Market Sq.)
Taggs Island TW12: Hamp 68Fa 130
Tagore Cl. HA3: W'stone 27Ha 46
Tagwright Ho. N1 3F 219 (41Tb 91)
(off Westland Pl.)
Tailor Ho. WC1 6F 217 (42Nb 90)
(off Colonnade)
Tailworth St. E1 43Wb 91
(off Chicksand St.)
Tait Ct. E3 39Bc 72
(off St Stephen's Rd.)
Tait Ct. SW8 53Nb 112
(off Lansdowne Grn.)
Tait Ho. SE1 7A 224 (46Qb 90)
(off Greet St.)
Tait Pl. WD25: Wat 4X 13
Tait Rd. CR0: C'don 73Ub 157
Tait Rd. Ind. Est. CR0: C'don 73Ub 157
(off Tait Rd.)
Taits SS17: Stan H 1P 101
Tait St. E1 44Xb 91
Taj Apts. E1 7K 219 (43Vb 91)
(off Brick La.)
Takeley Cl. EN9: Walt A 5Fc 21
Takeley Cl. RM5: Col R 26Fd 56
Takhar M. SW11 54Gb 111
Tala Cl. KT6: Surb 76Pa 153
Talacre Community Sports Cen. 37Jb 70
Talacre Rd. NW5 37Jb 70
Talbot Av. N2 27Fb 49
Talbot Av. SL3: L'ly 47B 82
Talbot Av. WD19: Wat 17Aa 27
Talbot Cl. CR4: Mitc 70Lb 134
Talbot Cl. N15 28Vb 51
Talbot Cl. RH2: Reig 7K 207
Talbot Cl. TN15: Bor G 91Be 205
Talbot Ct. EC3 4G 225 (45Tb 91)
(off Gracechurch St.)
Talbot Ct. HP3: Hem H 4M 3
Talbot Ct. NW9 34Ta 67
Talbot Ct. SL4: Wind 5F 102
Talbot Cres. NW4 29Wa 48
Talbot Gdns. IG3: Ilf 33Wc 75
Talbot Gro. Ho. W11 44Ab 88
(off Lancaster Rd.)
Talbot Ho. E14 44Dc 92
(off Girauld St.)
Talbot Ho. N7 34Qb 70
Talbot Ho. SW11 55Fb 111
(off York Place)
Talbot La. DA10: Swans 61Ce 143
Talbot Lodge KT10: Esh 78Ca 151
Talbot Pl. DA8: Erith 52Hd 118
Talbot Pl. SE3 54Gc 115
Talbot Pl. SL3: Dat 3N 103
Talbot Rd. CR7: Thor H 70Tb 135
Talbot Rd. E6 40Qc 74
Talbot Rd. E7 35Jc 73
Talbot Rd. HA0: Wemb 37Ma 67
Talbot Rd. HA3: W'stone 26Ha 46
Talbot Rd. N15 28Vb 51
Talbot Rd. N22 26Lb 50
Talbot Rd. N6 30Jb 50
Talbot Rd. RM9: Dag 37Bd 75
Talbot Rd. SE22 56Ub 113
Talbot Rd. SM5: Cars 78Jb 156
Talbot Rd. TW15: Ashf 64N 127
Talbot Rd. TW2: Twick 60Ga 108
Talbot Rd. TW7: Isle 56Ja 108
Talbot Rd. UB2: S'hall 49Aa 85
Talbot Rd. W11 44Bb 89
(not continuous)
Talbot Rd. W13 46Ja 86
Talbot Rd. W2 44Cb 89
Talbot Rd. WD3: Rick 18N 25
Talbot Sq. W2 3C 220 (44Fb 89)
Talbot Wlk. NW10: Church End 37Ua 68
Talbot Wlk. W11: Notting Hill 44Ab 88
(off St Mark's Rd.)
Talbot Yd. SE1 7F 225 (46Tb 91)
Talbrook CM14: B'wood 20Vd 40
Talcott Path SW2 60Qb 112
Talehangers Cl. DA6: Bex 56Zc 117
Taleworth Cl. KT21: Asht 92Ma 193
Taleworth Pk. KT21: Asht 92Ma 193
Taleworth Rd. KT21: Asht 91Ma 193
Talfourd Pl. SE15 53Vb 113
Talfourd Rd. SE15 53Vb 113
Talfourd Way RH1: Redh 9P 207
Talgarth Ho. RM3: Rom 23Nd 57
(off Kingsbridge Cir.)
Talgarth Mans. W14 50Ab 88
(off Talgarth Rd.)
Talgarth Rd. W14 50Za 88
Talgarth Rd. W6 50Za 88
Talgarth Wlk. NW9 29Ua 48
Talia Ho. E14 48Ec 92
(off Manchester Rd.)
Talina Cen. SW6 53Eb 111
Talisman Cl. IG3: Ilf 32Xc 75
Talisman Sq. SE26 63Wb 135
Talisman Way HA9: Wemb 34Pa 67
Talisman Way KT17: Eps D 88Ya 174
Tallack Cl. HA3: Hrw W 24Ga 46
Tallack Rd. E10 32Bc 72
Tall Elms Cl. BR2: Broml 71Hc 159
Tallents Cl. DA4: Sut H 66Rd 141
Talleyrand Ho. SE5 54Sb 113
(off Lilford Rd.)
Tallis Cl. E16 44Kc 93
Tallis Ct. SS17: Stan H 1L 101
Tallis Gro. SE7 51Kc 115
Tallis St. EC4 4A 224 (45Qb 90)
Tallis Vw. NW10 37Ta 67
Tallis Way CM14: W'ley 22Xd 58
Tallis Way WD6: Bore 11Ma 29
Tallon Rd. CM13: Hut 15Fe 41
Tallow Cl. RM9: Dag 38Zc 75
Tallow Rd. TW8: Bford 51La 108
Tall Pines Cl. RM11: Horn 29Pd 57
Tall Trees SL3: Coln 53F 104
Tall Trees SW16 69Pb 134
Tall Trees Cl. RM11: Horn 30Md 57
Talma Gdns. TW2: Twick 58Ga 108
Talmage Cl. SE23 59Yb 114
Talman Gro. HA7: Stan 23Ma 47
Talma Rd. SW2 56Qb 112
Talus Cl. RM19: Purf 49Td 98
Talwin St. E3 41Dc 92
Tamar Cl. E3 39Bc 72
Tamar Cl. RM14: Upm 30Ud 58
Tamar Dr. RM15: Avel 44Sd 98
Tamar Ho. E14 47Ec 92
(off Plevna St.)
Tamar Ho. SE11 7A 230 (50Qb 90)
(off Kennington La.)
Tamarind Ct. SE1 1K 231 (47Vb 91)
(off Gainsford St.)
Tamarind Ct. TW20: Egh 64B 126
Tamarind Ct. W8 48Db 89
(off Stone Hall Gdns.)
Tamarind Gro. IG7: Chig 22Ad 55
Tamarind Ho. SE15 52Wb 113
(off Reddins Rd.)
Tamarind Yd. E1 46Wb 91
(off Kennet St.)
Tamarisk Rd. RM15: S Ock 41Yd 98
Tamarisk Sq. W12 45Va 88
Tamarisk Way SL1: Slou 7E 80
Tamarix Cres. AL2: Lon C 8E 6
Tamar Sq. IG8: Wfd G 23Kc 53
Tamar St. SE7 48Nc 94
Tamar Way N17 27Vb 51
Tamar Way SL3: L'ly 50D 82
Tamerton Sq. GU22: Wok 91A 188
Tamesis Gdns. KT4: Wor Pk 75Ua 154
Tamesis Strand DA12: Grav'nd 4G 144
Tamian Ind. Est. TW4: Houn 56Y 107
Tamian Way TW4: Houn 56Y 107
Tamworth Av. IG8: Wfd G 23Gc 53
Tamworth La. CR4: Mitc 68Kb 134
Tamworth Pk. CR4: Mitc 70Kb 134
Tamworth Pl. CR0: C'don 75Sb 157
Tamworth Rd. CR0: C'don 75Rb 157
Tamworth St. SW6 51Cb 111
Tancred Rd. N4 30Rb 51
Tandem Cen. 67Eb 133
Tandem Ho. E17 29Bc 52
(off Track St.)
Tandem Way SW19 67Eb 133
TANDRIDGE 5E 210
Tandridge Ct. CR3: Cat'm 94Wb 197
Tandridge Dr. BR6: Orp 74Tc 160
Tandridge Gdns. CR2: Sande 85Vb 177
Tandridge Golf Course 3G 210
Tandridge Hill La. RH9: G'stone 100Bc 198
Tandridge La. RH8: C'rst 4E 210
Tandridge La. RH8: Tand 4E 210
Tandridge Leisure Cen. 1J 211
Tandridge Pl. BR6: Orp 74Tc 160
Tandridge Pl. SS17: Stan H 3K 101
Tandridge Rd. CR6: W'ham 91Zb 198
Tanfield Av. NW2 35Va 68
Tanfield Rd. CR0: C'don 77Sb 157
Tangent Link RM3: Rom 25Md 57
Tangerine Ho. SE1 2G 231 (47Tb 91)
(off Long La.)
Tangier Ct. SL4: Eton 1H 103
Tangier La. SL4: Eton 1H 103
Tangier Rd. TW10: Rich 56Qa 109
Tangier Way KT20: Tad 89Ab 174
Tangier Wood KT20: Tad 90Ab 174
Tangleberry Cl. BR1: Broml 70Pc 138
Tangle Tree Cl. N3 26Db 49
Tanglewood Cl. CR0: C'don 76Yb 158
Tanglewood Cl. GU22: Pyr 88F 168
Tanglewood Cl. HA7: Stan 19Ga 28
Tanglewood Cl. KT16: Longc 6M 147
Tanglewood Cl. UB10: Hil 42Q 84
Tanglewood Ct. BR5: St P 69Yc 139
Tanglewood Ride GU24: W End 4B 166
Tanglewood Way TW13: Felt 62X 129
Tangley Gro. SW15 58Va 110
Tangley Pk. Rd. TW12: Hamp 65Ba 129
Tanglyn Av. TW17: Shep 71R 150
Tangmere N17 26Tb 51
(off Willan Rd.)
Tangmere WC1 4H 217 (41Pb 90)
(off Sidmouth St.)
Tangmere Cres. RM12: Horn 37Kd 77
Tangmere Cres. UB10: Uxb 39N 63
Tangmere Gdns. UB5: N'olt 40Y 65
Tangmere Gro. KT2: King T 64Ma 131
Tangmere Way NW9 26Ua 48
Tan Ho. E9 36Ac 72
(off Sadler Pl.)
Tanhouse Fld. NW5 36Mb 70
(off Torriano Av.)
Tan Ho. La. CM14: N'side 14Pd 39
Tanhouse Rd. RH8: Oxt 4H 211
Tanhouse Way SL3: Coln 52F 104
Tanhurst Ho. SW2 59Pb 112
(off Redlands Way)
Tanhurst Wlk. SE2 48Zc 95
(off Alsike Rd.)
Tankerfield Pl. AL3: St A 2A 6
(off Romeland Hill)
Tankerton Ho's. WC1 4G 217 (41Nb 90)
(off Tankerton St.)
Tankerton Rd. KT6: Surb 75Pa 153
Tankerton St. WC1 4G 217 (41Nb 90)
Tankerton Ter. CR0: C'don 72Pb 156
Tankerville Ct. TW3: Houn 55Ea 108
Tankerville Rd. SW16 66Mb 134
Tank Hill Rd. RM19: Avel 50Qd 97
Tank Hill Rd. RM19: Purf 50Qd 97
Tank La. RM19: Purf 49Qd 97
Tankridge Rd. NW2 33Xa 68
Tanner Cl. NW9 26Ua 48
Tanner Ho. E1 43Zb 92
(off White Horse La.)
Tanner Ho. SE1 2J 231 (47Ub 91)
(off Tanner St.)
Tanner Ho. WD19: Wat 17X 27
Tanneries, The E1 42Yb 92
(off Cephas Av.)
Tanner Point E13 39Jc 73
(off Pelly Rd.)
Tanners Cl. AL3: St A 1A 6
Tanners Cl. DA1: Cray 55Gd 118
Tanners Cl. KT12: Walt T 72X 151
Tanners Dean KT22: Lea 94La 192
Tanners End La. N18 21Ub 51
Tanner's Hill SE8 53Bc 114
Tanners Hill WD5: Ab L 3V 12
Tanners La. IG6: Ilf 27Sc 54
Tanners M. SE8 53Bc 114
(off Tanner's Hill)
Tanner St. IG11: Bark 37Sc 74
Tanner St. SE1 2J 231 (47Ub 91)
Tanners Wood Cl. WD5: Ab L 4U 12
Tanners Wood Ct. WD5: Ab L 4U 12
Tanners Wood La. WD5: Ab L 4U 12
Tanners Yd. E2 40Xb 71
(off Treadway St.)
Tannery, The SE1: Grange Rd 4J 231 (48Ub 91)
Tannery, The SE1: Whites Grounds 1J 231 (47Ub 91)
(off Black Swan Yd.)
Tannery, The RH1: Redh 6P 207
Tannery Cl. BR3: Beck 71Zb 158
Tannery Cl. RM10: Dag 34Dd 76
Tannery Ho. E1 43Wb 91
(off Deal St.)
Tannery La. GU23: Send 95F 188
Tannery Sq. SE1 4J 231 (48Ub 91)
(off Tannery Way)
Tannery Way SE1 4K 231 (48Vb 91)
Tannington Ter. N5 34Rb 71
Tannoy Sq. SE27 63Tb 135
Tannsfeld Rd. SE26 64Zb 136
Tannsfield Dr. HP2: Hem H 1P 3
Tannsmore Cl. HP2: Hem H 1P 3
Tansley Cl. N7 36Mb 70
Tanswell St. SE1 2K 229 (47Qb 90)
Tansy Cl. E6 44Qc 94
Tansy Cl. RM3: Rom 23Nd 57
Tantallon Rd. SW12 60Jb 112
Tant Av. E16 44Hc 93
Tantivy Ct. WD17: Wat 13Y 27
Tantony Gro. RM6: Chad H 27Zc 55
Tanworth Cl. HA6: Nwood 23S 44
Tanworth Gdns. HA5: Pinn 26X 45
Tanyard Cotts. DA12: Shorne 5N 145
Tanyard Hill DA12: Shorne 5N 145
Tanyard Ho. TW8: Bford 52La 108
(off High St.)
Tan Yd. La. DA5: Bexl 59Cd 118
Tany Mead RM15: Avel 45Td 98
Tanza Rd. NW3 35Hb 69
Tapestries Hall SL4: Old Win 7L 103
Tapestry Apts. N1 39Mb 70
Tapestry Bldg. EC2 1J 225 (43Ub 91)
(off New St.)
Tapestry Cl. SM2: Sutt 80Db 155
Tapley Cl. E14 44Ec 92
Tapley Ho. SE1 47Wb 91
(off Wolseley St.)
Taplow NW3 38Fb 69
Taplow SE17 7G 231 (50Tb 91)
(off Thurlow St.)
Taplow Ct. CR4: Mitc 70Gb 133
Taplow Ho. E2 4K 219 (41Vb 91)
(off Palissy St.)
Taplow Rd. N13 21Sb 51
Taplow St. N1 2E 218 (40Sb 71)
Tapper Wlk. N1 39Nb 70

Tappesfield Rd. SE15 55Yb **114**
Tapping Cl. KT2: King T 66Qa **131**
Tapp St. E1 42Xb **91**
Tapster St. EN5: Barn 13Bb **31**
Tara Arts Cen. **60Db 111**
Tara Ct. BR3: Beck 68Dc **136**
Tara Ho. E14 49Cc **92**
(off Deptford Ferry Rd.)
Tara M. N8 30Nb **50**
Taransay HP3: Hem H 4B **4**
Taransay Wlk. N1 37Tb **71**
Tarbay La. SL4: Oak G 5A **102**
Tarbert M. N15 29Ub **51**
Tarbert Rd. SE22 57Ub **113**
Tarbert Wlk. E1 45Yb **92**
Tarbuck Ho. SE10 49Gc **93**
(off Manilla Wlk.)
Target Cl. TW14: Felt 58U **106**
Target Ho. W13 46Ka **86**
(off Sherwood Cl.)
TARGET RDBT. 39Ba **65**
Tariff Cres. SE8 49Bc **92**
Tariff Rd. N17 23Wb **51**
Tarleton Ct. N22 26Qb **50**
Tarleton Gdns. SE23 61Xb **135**
Tarling Cl. DA14: Sidc 62Xc **139**
Tarling Rd. E16 44Hc **93**
Tarling Rd. N2 26Eb **49**
Tarling St. E1 44Xb **91**
Tarling St. Est. E1 44Yb **92**
Tarmac Way UB7: Harm 52K **105**
Tarnbank EN2: Enf 15Nb **32**
Tarnbrook Ct. SW1 6H **227** (49Jb **90**)
(off Holbein Pl.)
Tarns, The NW1 3B **216** (41Lb **90**)
(off Varndell St.)
Tarn St. SE1 4D **230** (48Sb **91**)
Tarnwood Pk. SE9 59Pc **116**
Tarnworth Rd. RM3: Rom 22Qd **57**
Tarplett Ho. SE14 51Zb **114**
(off John Williams Cl.)
Tarquin Ho. SE26 63Wb **135**
(off High Level Dr.)
Tarragon Cl. SE14 52Ac **114**
Tarragon Ct. IG1: Ilf 33Uc **74**
Tarragon Gro. SE26 65Zb **136**
Tarranbrae NW6 38Ab **68**
Tarrant Ho. E2 41Yb **92**
(off Roman Rd.)
Tarrant Ho. W14 48Ab **88**
(off Russell Rd.)
Tarrant Pl. W1 1F **221** (43Hb **89**)
Tarrington Cl. SW16 62Mb **134**
Tartan Ho. E14 44Ec **92**
(off Dee St.)
Tartar Rd. KT11: Cobh 85Y **171**
Tarver Rd. SE17 7C **230** (50Rb **91**)
Tarves Way SE10: Lit. Cottage Pl. 52Dc **114**
Tarves Way SE10: Norman Rd. 52Dc **114**
Taryn Gro. BR1: Broml 69Pc **138**
Tash Pl. N11 22Kb **50**
Tasker Cl. UB3: Harl 52S **106**
Tasker Ho. E14 43Bc **92**
(off Wallwood St.)
Tasker Ho. IG11: Bark 40Tc **74**
Tasker Lodge W8 47Cb **89**
(off Campden Hill)
Tasker Rd. NW3 36Hb **69**
Tasker Rd. RM16: Grays 8D **100**
Tasman Ct. E14 49Dc **92**
(off Westferry Rd.)
Tasman Ct. TW16: Sun 66U **128**
Tasman Ho. E1 46Xb **91**
(off Clegg St.)
Tasmania Ho. RM18: Tilb 3C **122**
Tasmania Ter. N18 23Sb **51**
Tasman Rd. SW9 55Nb **112**
Tasman Wlk. E16 44Mc **93**
Tasso Rd. W6 51Ab **110**
Tasso Yd. W6 51Ab **110**
(off Tasso Rd.)
Tatam Rd. NW10 38Ta **67**
Tatchbury Ho. SW15 58Va **110**
(off Tunworth Cres.)
Tate Apts. E1 44Xb **91**
(off Sly St.)
Tate Britain **6F 229 (49Nb 90)**
Tate Cl. KT22: Lea 95La **192**
Tate Gdns. WD23: Bush 17Ga **28**
Tate Ho. E2 40Zb **72**
(off Mace St.)
Tate Modern **6C 224 (46Rb 91)**
Tate Rd. E16 46Pc **94**
(not continuous)
Tate Rd. SL9: Chal P 22A **42**
(not continuous)
Tate Rd. SM1: Sutt 78Cb **155**
Tates Orchard DA3: Hartl 72Be **165**
Tatham Pl. NW8 1C **214** (40Fb **69**)
TATLING END 32D 62
Tatnell Rd. SE23 58Ac **114**
TATSFIELD 93Lc 199
Tatsfield App. Rd. TN16: Tats 95Kc **199**
TATSFIELD GREEN 93Nc 200
Tatsfield Ho. SE1 3G **231** (48Tb **91**)
(off Pardoner St.)
Tatsfield La. TN16: Tats 93Pc **200**
TATTENHAM CORNER 90Xa 174
Tattenham Cnr. Rd. KT18: Eps D 89Va **174**
Tattenham Cnr. Rd. KT18: Tatt C 89Va **174**
Tattenham Corner Station (Rail) 90Xa 174
Tattenham Cres. KT18: Tatt C 90Wa **174**
Tattenham Gro. KT18: Tatt C 90Xa **174**
Tattenham Way KT20: Tad 90Za **174**
Tattersall Cl. SE9 57Nc **116**
Tatton Cl. SM5: W'gton 74Jb **156**
Tatton Cres. N16 31Vb **71**
Tatum St. SE17 6G **231** (49Tb **91**)
Tauber Cl. WD6: E'tree 14Pa **29**
Tauheed Cl. N4 33Sb **71**
Taunton Av. CR3: Cat'm 95Vb **197**
Taunton Av. SW20 68Xa **132**
Taunton Av. TW3: Houn 54Ea **108**
Taunton Cl. DA7: Bex 54Fd **118**
Taunton Cl. IG6: Ilf 23Vc **55**
Taunton Cl. SM3: Sutt 74Cb **155**
Taunton Dr. EN2: Enf 13Qb **32**
Taunton Dr. N2 26Eb **49**
Taunton Ho. RM3: Rom 22Pd **57**
(off Redcar Rd.)
Taunton Ho. W2 44Eb **89**
(off Hallfield Est.)
Taunton La. CR5: Coul 91Qb **196**
Taunton M. NW1 6F **215** (42Hb **89**)
Taunton Pl. NW1 5F **215** (42Hb **89**)
Taunton Rd. DA11: Nflt 57Ce **121**
Taunton Rd. RM3: Rom 21Ld **57**
Taunton Rd. SE12 57Gc **115**
Taunton Rd. UB6: G'frd 39Da **65**
Taunton Va. DA12: Grav'nd 2F **144**
Taunton Way HA7: Stan 26Na **47**
Tavern Cl. SM5: Cars 73Gb **155**
Tavern Cl. TN15: Bor G 92Be **205**
Tavern Ct. SE1 4E **230** (48Sb **91**)
(off New Kent Rd.)
Taverners HP2: Hem H 1N **3**
Taverners Cl. W11 46Ab **88**
Taverners Ct. E3 41Ac **92**
(off Grove Rd.)
Taverner Sq. N5 35Sb **71**
Taverners Way E4 18Gc **35**
Tavern La. SW9 54Qb **112**
Tavern Quay SE16 49Ac **92**
Tavistock Av. AL1: St A 5A **6**
Tavistock Av. E17 27Zb **52**
Tavistock Av. NW7 24Za **48**
Tavistock Av. UB6: G'frd 40Ja **66**
Tavistock Cl. AL1: St A 6B **6**
Tavistock Cl. EN6: Pot B 3Fb **17**
Tavistock Cl. N16 36Ub **71**
Tavistock Cl. RM3: Rom 25Md **57**
Tavistock Cl. TW18: Staines 66M **127**
Tavistock Ct. CR0: C'don 74Tb **157**
(off Tavistock Rd.)
Tavistock Ct. WC1 5E **216** (42Mb **90**)
(off Tavistock Sq.)
Tavistock Ct. WC2 4G **223** (45Nb **90**)
(off Tavistock Street)
Tavistock Cres. CR4: Mitc 70Nb **134**
Tavistock Cres. W11 43Bb **89**
(not continuous)
Tavistock Gdns. IG3: Ilf 35Uc **74**
Tavistock Ga. CR0: C'don 74Tb **157**
Tavistock Gro. CR0: C'don 73Tb **157**
Tavistock Ho. IG8: Wfd G 23Qc **54**
Tavistock Ho. WC1 5E **216** (42Mb **90**)
Tavistock M. N19 34Nb **70**
(off Tavistock Ter.)
Tavistock M. W11 44Bb **89**
Tavistock Pl. N14 16Kb **32**
Tavistock Pl. WC1 5F **217** (42Nb **90**)
Tavistock Rd. BR2: Broml 70Hc **137**
Tavistock Rd. CR0: C'don 74Tb **157**
Tavistock Rd. DA16: Well 53Yc **117**
Tavistock Rd. E15 37Hc **73**
Tavistock Rd. E18 27Jc **53**
Tavistock Rd. E7 35Hc **73**
Tavistock Rd. HA8: Edg 25Qa **47**
Tavistock Rd. N4 30Tb **51**
Tavistock Rd. NW10 40Va **68**
Tavistock Rd. SM5: Cars 74Fb **155**
Tavistock Rd. UB10: Ick 36T **64**
Tavistock Rd. UB7: Yiew 46M **83**
Tavistock Rd. W11 44Bb **89**
(not continuous)
Tavistock Rd. WD24: Wat 11Z **27**
Tavistock Sq. WC1 5E **216** (42Mb **90**)
Tavistock St. WC2 4G **223** (45Nb **90**)
Tavistock Ter. N19 34Mb **70**
Tavistock Twr. SE16 48Ac **92**
Tavistock Wlk. SM5: Cars 74Fb **155**
Tavistock Youth Cen. **40Va 68**
(off Tavistock Rd.)
Taviton St. WC1 5D **216** (42Mb **90**)
Tavy Bri. SE2 47Yc **95**
Tavy Cl. SE11 7A **230** (50Qb **90**)
(off White Hart St.)
Tawney Rd. SE28 45Xc **95**
Tawny Av. RM14: Upm 36Rd **77**
Tawny Cl. TW13: Felt 62W **128**
Tawny Cl. W13 46Ka **86**
Tawny Way SE16 49Zb **92**
Tayben Av. TW2: Twick 58Ga **108**
Tayberry Ho. E20 37Dc **72**
(off Ravens Wlk.)
Taybridge Rd. SW11 55Jb **112**
Tay Bldgs. SE1 3H **231** (48Ub **91**)
(off Decima St.)
Tayburn Cl. E14 44Ec **92**
Tay Ct. E2 41Zb **92**
(off Meath Cres.)
Tay Ct. SE1 3H **231** (48Ub **91**)
(off Decima St.)
Tayfield Cl. UB10: Ick 34T **64**
Tay Ho. E3 40Bc **72**
(off St Stephen's Rd.)
Tayler Ct. NW8 39Fb **69**
Taylers Cotts. EN6: Ridge 6Va **16**
Tayles Hill Dr. KT17: Ewe 82Va **174**
Taylor Av. TW9: Kew 54Ra **109**
Taylor Cl. BR6: Orp 77Vc **161**
Taylor Cl. KT19: Eps 83Qa **173**
Taylor Cl. N17 24Wb **51**
Taylor Cl. RM5: Col R 24Cd **56**
Taylor Cl. SE8 51Bc **114**
Taylor Cl. TW12: Hamp H 64Ea **130**
Taylor Cl. TW3: Houn 53Ea **108**
Taylor Cl. UB9: Hare 25L **43**
Taylor Cl. WD25: Wat 8X **13**
Taylor Ct. SE20 68Yb **136**
(off Elmers End Rd.)
Taylor Ct. WD6: Bore 12Sa **29**
Taylor Gdns. RM3: Rom 19Ld **39**
Taylor Ho. E14 45Cc **92**
(off Storehouse M.)
Taylor Pl. E3 40Dc **72**
Taylor Rd. CR4: Mitc 66Gb **133**
Taylor Rd. KT21: Asht 89Ma **173**
Taylor Rd. SM6: W'gton 78Kb **156**
Taylor Row DA2: Wilm 62Kd **141**
Taylors IG7: Chig 17Wc **37**
Taylors Bldgs. SE18 49Rc **94**
Taylor's Bushes Ride SL4: Wind 1A **124**
Taylors Cl. DA14: Sidc 62Vc **139**
Taylors Ct. TW13: Felt 61W **128**
Taylors Grn. W3 44Ua **88**
Taylor's La. SE26 63Xb **135**
Taylors La. EN5: Barn 11Bb **31**
Taylors La. NW10 37Ua **68**
Taylors Mead NW7 22Wa **48**
Taylors Yd. E1 6K **219** (42Vb **91**)
(off Brick La.)
Taymount Grange SE23 61Yb **136**
Taymount Ri. SE23 61Yb **136**
Taynton Dr. RH1: Mers 2D **208**
Tayport Cl. N1 38Nb **70**
Tayside Ct. SE5 56Tb **113**
Tayside Dr. HA8: Edg 20Ra **29**
Tay Way RM1: Rom 25Hd **56**
Taywood Rd. UB5: N'olt 42Ba **85**
Tazzeta Ho. E20 36Ec **72**
(off Victory Pde.)
Teak Cl. SE16 46Ac **92**
Teal Av. BR5: St P 70Zc **139**
Tealby Cl. KT20: Lwr K 98Bb **195**
Tealby Ct. N7 37Pb **70**
Teal Cl. CR2: Sels 83Zb **178**
Teal Cl. E16 43Mc **93**
Teal Cl. EN3: Enf W 9Yb **20**
Teal Ct. E1 45Wb **91**
(off Star Pl.)
Teal Ct. NW10 37Ta **67**
Teal Ct. SE8 51Bc **114**
(off Taylor Cl.)
Teal Ct. SM6: W'gton 78Lb **156**
Teal Dr. HA6: Nwood 24S **44**
Teale St. E2 40Wb **71**
Teal Ho. WD25: Wat 8Aa **13**
Tealing Dr. KT19: Ewe 77Ta **153**
Teal Pl. SM1: Sutt 78Bb **155**
Teal St. SE10 48Hc **93**
Teal Way HP3: Hem H 7P **3**
TeamSport Indoor Karting Southwark **48Xb 91**
Teamsport Karting Edmonton **21Zb 52**
Teardrop Cen. BR8: Swan 71Kd **163**
Teasel Cl. CR0: C'don 74Zb **158**
Teasel Cres. SE28 46Uc **94**
Teasel Way E15 41Gc **93**
Tea Trade Wharf SE1 47Vb **91**
(off Shad Thames)
Tea Tree Cl. TW15: Ashf 64S **128**
Teazlewood Pk. KT22: Lea 89Ja **172**
Tebbs Way TN15: Igh 96Yd **204**
Tebworth Rd. N17 24Vb **51**
Technology Pk., The NW9 27Ua **48**
Teck Cl. TW7: Isle 54Ja **108**
Tedder Cl. HA4: Ruis 36W **64**
Tedder Cl. KT9: Chess 78La **152**
Tedder Cl. UB10: Uxb 38P **63**
Tedder Rd. CR2: Sels 80Yb **158**
Tedder Rd. HP2: Hem H 1A **4**
TEDDINGTON 64Ja 130
Teddington Bus. Pk. TW11: Tedd 65Ha **130**
(off Station Rd.)
Teddington Cl. KT19: Eps 82Ta **173**
Teddington Lock TW10: Ham 63Ka **130**
Teddington Lock **63Ka 130**
TEDDINGTON MEMORIAL HOSPITAL **65Ga 130**
Teddington Pk. TW11: Tedd 64Ha **130**
Teddington Pk. Rd. TW11: Tedd 63Ha **130**
Teddington Pool & Fitness Cen. **64Ja 130**
Teddington Riverside TW11: Tedd 64Ka **130**
Teddington Sports Cen. **65Ma 131**
Teddington Station (Rail) 65Ja 130
Ted Hennem Ho. RM10: Dag 34Dd **76**
Ted Roberts Ho. E2 40Xb **71**
(off Parmiter St.)
Tedworth Gdns. SW3 50Hb **89**
Tedworth Sq. SW3 50Hb **89**
Tee, The W3 44Ua **88**
Tees Av. UB6: G'frd 40Ga **66**
Tees Cl. RM14: Upm 30Td **58**
Tees Ct. W7 44Fa **86**
(off Hanway Rd.)
Teesdale Av. TW7: Isle 53Ja **108**
Teesdale Cl. E2 40Wb **71**
Teesdale Gdns. SE25 68Ub **135**
Teesdale Gdns. TW7: Isle 53Ja **108**
Teesdale Rd. DA2: Dart 60Sd **120**
Teesdale Rd. E11 30Hc **53**
Teesdale Rd. SL2: Slou 3D **80**
Teesdale St. E2 40Xb **71**
Teesdale Yd. E2 40Xb **71**
(off Teesdale St.)
Tees Dr. RM3: Rom 20Md **39**
Teeswater Ct. DA18: Erith 48Zc **95**
Teevan Cl. CR0: C'don 73Wb **157**
Teevan Rd. CR0: C'don 74Wb **157**
Tegan Cl. SM2: Sutt 80Cb **155**
Tegg's La. GU22: Pyr 88H **169**
Teign M. SE9 61Nc **138**
Teignmouth Cl. HA8: Edg 26Pa **47**
Teignmouth Cl. SW4 56Mb **112**
Teignmouth Gdns. UB6: G'frd 40Ja **66**
Teignmouth Pde. UB6: G'frd 40Ja **66**
Teignmouth Rd. DA16: Well 54Yc **117**
Teignmouth Rd. NW2 36Za **68**
Telcon Way SE10 49Gc **93**
Telcote Way HA4: Ruis 31Y **65**
Telegraph Av. NW9 27Ua **48**
Telegraph Av. SE10 49Gc **93**
Telegraph Hill NW3 34Db **69**
Telegraph La. KT10: Clay 77Ha **152**
Telegraph M. IG3: Ilf 32Wc **75**
Telegraph Pas. SW2 59Nb **112**
(off New Pk. Rd.)
Telegraph Path BR7: Chst 64Rc **138**
Telegraph Pl. E14 49Dc **92**
Telegraph Rd. SW15 59Xa **110**
Telegraph St. EC2 2F **225** (44Tb **91**)
Telegraph Track SM5: Cars 83Jb **176**
Telemann Sq. SE3 56Kc **115**
Telephone Pl. SW6 51Bb **111**
Telfer Cl. W3 47Sa **87**
Telfer Ho. EC1 4D **218** (41Sb **91**)
(off Lever St.)
Telferscot Rd. SW12 60Mb **112**
Telford Av. SW2 60Mb **112**
Telford Cl. E17 31Ac **72**
Telford Cl. SE19 65Vb **135**
Telford Cl. WD25: Wat 7Z **13**
Telford Ct. AL1: St A 3C **6**
Telford Ct. AL10: Hat 2B **8**
Telford Dr. KT12: Walt T 73Y **151**
Telford Dr. SL1: Slou 7E **80**
Telford Ho. SE1 3D **230** (48Sb **91**)
(off Tiverton St.)
Telford Rd. AL2: Lon C 9G **6**
Telford Rd. N11 22Lb **50**
Telford Rd. NW9 30Wa **48**
Telford Rd. SE9 61Tc **138**
Telford Rd. TW2: Whitt 59Ca **107**
Telford Rd. UB1: S'hall 45Da **85**
Telford Rd. W10 43Ab **88**
Telford Sq. DA1: Dart 55Pd **119**
Telfords Yd. E1 45Wb **91**
Telford Ter. SW1 51Lb **112**
Telford Way UB4: Yead 43Aa **85**
Telford Way W3 43Ua **88**
Telham Rd. E6 40Qc **74**
Tell Gro. SE22 56Vb **113**
Tellisford KT10: Esh 77Da **151**
Tellson Av. SE18 53Mc **115**
Telscombe Cl. BR6: Orp 75Uc **160**
Telston La. TN14: Otf 89Gd **182**
Temair Ho. SE10 52Dc **114**
(off Tarves Way)
Temeraire Pl. TW8: Bford 50Pa **87**
Temeraire St. SE16 47Yb **92**
Tempelhof Av. NW4 31Ya **68**
Temperance St. AL3: St A 2A **6**
Temperley Rd. SW12 59Jb **112**
Tempest Av. EN6: Pot B 4Fb **17**
Tempest Rd. TW20: Egh 65E **126**
Tempest Way RM13: Rain 37Jd **76**
Templar Ct. NW8 4B **214** (41Fb **89**)
(off St John's Wood Rd.)
Templar Ct. RM7: Mawney 28Dd **56**
Templar Dr. DA11: Grav'nd 4C **144**
Templar Dr. SE28 44Zc **95**
Templar Ho. E15 37Fc **73**
(off Leyton Rd.)
Templar Ho. HA2: Harr 33Ea **66**
Templar Ho. NW2 37Bb **69**
Templar Pl. TW12: Hamp 66Ca **129**
Templars Av. NW11 30Bb **49**
Templars Ct. DA1: Dart 57Qd **119**
Templars Cres. N3 26Cb **49**
Templars Dr. HA3: Hrw W 23Fa **46**
Templars Ho. E16 45Rc **94**
(off University Way)
Templar St. SE5 54Rb **113**
Temple Av. CR0: C'don 75Bc **158**
Temple Av. EC4 4A **224** (45Qb **90**)
Temple Av. N20 17Fb **31**
Temple Av. RM8: Dag 32Cd **76**
Temple Bar **3K 223 (44Qb 90)**
Temple Bar Gate **3C 224 (44Rb 91)**
(off Paternoster Sq.)
Temple Bar Rd. GU21: Wok 1K **187**
Temple Chambers EC4 4A **224** (45Qb **90**)
(off Temple Av.)
Temple Cl. E11 31Gc **73**
Temple Cl. EN7: Chesh 3Wb **19**
Temple Cl. KT19: Eps 84Ta **173**
Temple Cl. N3 26Bb **49**
Temple Cl. SE28 48Sc **94**
Temple Cl. WD17: Wat 12V **26**
Templecombe M. GU22: Wok 88D **168**
Templecombe Rd. E9 39Yb **72**
Templecombe Way SM4: Mord 71Ab **154**
Temple Ct. E1 43Zb **92**
(off Rectory Sq.)
Temple Ct. EN6: Pot B 3Ab **16**
Temple Ct. KT13: Weyb 77R **150**
Temple Ct. KT19: Eps 84Ta **173**
Temple Ct. SW8 52Nb **112**
(off Thorncroft St.)
Templecroft TW15: Ashf 65T **128**
Templedene Av. TW18: Staines 66K **127**
Temple Dwellings E2 40Xb **71**
(off Temple St.)
Templefield Cl. KT15: Add 79K **149**
TEMPLE FORTUNE 29Bb 49
Temple Fortune Hill NW11 29Cb **49**
Temple Fortune La. NW11 30Bb **49**
Temple Fortune Pde. NW11 29Bb **49**
Temple Gdns. EC4 4K **223** (45Qb **90**)
(off Middle Temple La.)
Temple Gdns. N21 19Rb **33**
Temple Gdns. NW11 30Bb **49**
Temple Gdns. RM8: Dag 34Zc **75**
Temple Gdns. TW18: Staines 67H **127**
Temple Gdns. WD3: Rick 21R **44**
Temple Gro. EN2: Enf 12Rb **33**
Temple Gro. NW11 30Cb **49**
Temple Hall Ct. E4 19Fc **35**
Temple Hill DA1: Dart 58Pd **119**
TEMPLE HILL 57Pd 119
Temple Hill Sq. DA1: Dart 57Pd **119**
Temple Ho. EN7: Walt C 5Wb **19**
Temple La. EC4 3A **224** (44Qb **90**)
Templeman Cl. CR8: Purl 88Rb **177**
Templeman Rd. W7 43Ha **86**
Temple Mkt. KT13: Weyb 77R **150**
Temple Mead HP2: Hem H 1M **3**
Templemead SL9: Ger X 2N **61**
Temple Mead Cl. HA7: Stan 23Ka **46**
Templemead Cl. W3 44Ua **88**
Templemead Ho. E9 35Ac **72**
Templemere KT13: Weyb 76T **150**
TEMPLE MILLS 35Dc 72
Temple Mills La. E10 35Dc **72**
Temple Mills La. E15 36Ec **72**
Temple Mills La. E20 35Ec **72**
Templepan La. WD3: Chan C 9N **11**
Temple Pde. EN5: New Bar 17Fb **31**
(off Netherlands Rd.)
Temple Pk. UB8: Hil 41Q **84**
Temple Pl. WC2 4J **223** (45Pb **90**)
Templer Av. RM16: Grays 9C **100**
Temple Rd. CR0: C'don 77Tb **157**
Temple Rd. E6 39Nc **74**
Temple Rd. KT19: Eps 84Ta **173**
Temple Rd. N8 28Pb **50**
Temple Rd. NW2 35Ya **68**
Temple Rd. SL4: Wind 4G **102**
Temple Rd. TN16: Big H 89Mc **179**
Temple Rd. TW3: Houn 56Da **107**
Temple Rd. TW9: Rich 54Pa **109**
Temple Rd. W4 48Sa **87**
Temple Rd. W5 48Ma **87**
Temple Sheen SW14 56Sa **109**
Temple Sheen Rd. SW14 56Ra **109**
Temple Station (Underground) 4J 223 (45Pb 90)
Temple St. E2 40Xb **71**
Temple Ter. N22 26Qb **50**
(off Vincent Rd.)
Templeton Av. E4 21Cc **52**
Templeton Cl. N15 30Tb **51**
Templeton Cl. N16 36Ub **71**
Templeton Cl. SE19 67Tb **135**
Templeton Ct. AL1: St A 1C **6**
(off Newsom Pl.)
Templeton Ct. EN3: Enf W 10Yb **20**
Templeton Ct. WD6: Bore 11Pa **29**
(off Lyndhurst Wlk.)
Templeton Pl. SW5 49Cb **89**
Templeton Rd. N15 30Tb **51**
Temple Vw. AL3: St A 1A **6**
Temple Way SL2: Farn C 6G **60**
Temple Way SM1: Sutt 76Fb **155**
Temple W. M. SE11 4B **230** (48Rb **91**)
Templewood W13 43Ka **86**
Templewood Av. NW3 34Db **69**
Temple Wood Dr. RH1: Redh 3P **207**
Templewood Gdns. NW3 34Db **69**
Templewood Ga. SL2: Farn C 6G **60**
Templewood La. SL2: Farn C 6G **60**
Templewood La. SL2: Stoke P 6G **60**
Templewood Point NW2 33Bb **69**
(off Granville Rd.)
Temple Yd. E2 40Xb **71**
(off Temple St.)
Tempo Ho. UB5: N'olt 41Z **85**
Tempsford Av. WD6: Bore 14Ta **29**
Tempsford Cl. EN2: Enf 14Sb **33**
Tempsford Ct. HA1: Harr 30Ha **46**
Tempus Apts. EC1 2B **218** (40Rb **71**)
(off Goswell Rd.)
Tempus Ct. E18 25Jc **53**
Tempus Wharf SE16 47Wb **91**
(off Bermondsey Wall W.)
Temsford Cl. HA2: Harr 26Ea **46**
Tenacre GU21: Wok 10L **167**
Ten Acre La. TW20: Thorpe 68E **126**
Ten Acres KT22: Fet 96Fa **192**
Ten Acres Cl. KT22: Fet 96Fa **192**
Tenbury Cl. E7 36Mc **73**
Tenbury Ct. SW2 60Mb **112**
Tenby Av. HA3: Kenton 26Ka **46**
Tenby Cl. N15 28Vb **51**
Tenby Cl. RM6: Chad H 30Ad **55**
Tenby Ct. E17 29Ac **52**
Tenby Dr. SL5: S'hill 1B **146**
Tenby Gdns. UB5: N'olt 37Ca **65**
Tenby Ho. UB3: Harl 48S **84**
Tenby Ho. W2 3A **220** (44Eb **89**)
(off Hallfield Est.)
Tenby Mans. W1 7J **215** (43Jb **90**)
(off Nottingham St.)
Tenby Rd. DA16: Well 53Zc **117**
Tenby Rd. E17 29Ac **52**
Tenby Rd. EN3: Pond E 14Yb **34**
Tenby Rd. HA8: Edg 25Pa **47**
Tenby Rd. RM6: Chad H 30Ad **55**
Tenchley's La. RH8: Limp 3N **211**
Tench St. E1 46Xb **91**
Tenda Rd. SE16 49Xb **91**
Tendring Ct. CM13: Hut 15Fe **41**
Tendring Way RM6: Chad H 29Yc **55**
Tenham Av. SW2 60Mb **112**
Tenison Ct. W1 4B **222** (45Lb **90**)
Tenison Way SE1 7K **223** (46Qb **90**)
(off York Rd.)
Tennants Row RM18: Tilb 4A **122**
Tenniel Cl. W2 45Eb **89**
Tennis Close KT12: Walt T 75W **150**
Tennis Ct. La. KT8: E Mos 69Ha **130**
Tennison Av. WD6: Bore 15Ra **29**
Tennison Cl. CR5: Coul 92Rb **197**
Tennison Rd. SE25 70Vb **135**
Tennis St. SE1 1F **231** (47Tb **91**)
Tenniswood Rd. EN1: Enf 11Ub **33**
Tenny Ho. RM17: Grays 52De **121**
Tennyson CR0: C'don 74Sb **157**
Tennyson Av. E11 31Jc **73**
Tennyson Av. E12 38Nc **74**
Tennyson Av. EN9: Walt A 6Gc **21**
Tennyson Av. KT3: N Mald 71Xa **154**
Tennyson Av. NW9 27Sa **47**
Tennyson Av. RM17: Grays 48De **99**
Tennyson Av. TW1: Twick 60Ha **108**
Tennyson Cl. DA16: Well 53Uc **116**
Tennyson Cl. EN3: Pond E 15Zb **34**
Tennyson Cl. TW14: Felt 58V **106**
Tennyson Ct. NW1 6F **215** (42Hb **89**)
(off Dorset Sq.)
Tennyson Ho. DA17: Belv 50Bd **95**
Tennyson Ho. N8 27Nb **50**
(off Boyton Cl.)
Tennyson Ho. SE17 7E **230** (50Sb **91**)
(off Browning St.)
Tennyson Mans. SW3 51Gb **111**
(off Lordship Pl.)
Tennyson Mans. W14 51Bb **111**
(off Queen's Club Gdns.)
Tennyson Rd. AL2: Chis G 8N **5**
Tennyson Rd. CM13: Hut 17Ee **41**
Tennyson Rd. DA1: Dart 57Qd **119**
Tennyson Rd. E10 32Dc **72**
Tennyson Rd. E15 38Gc **73**
Tennyson Rd. E17 30Bc **52**
Tennyson Rd. KT15: Add 77N **149**
Tennyson Rd. NW6 39Bb **69**
Tennyson Rd. NW7 22Wa **48**
Tennyson Rd. RM3: Rom 24Ld **57**
Tennyson Rd. SE20 66Zb **136**
Tennyson Rd. SW19 65Eb **133**
Tennyson Rd. TW15: Ashf 64N **127**
Tennyson Rd. TW3: Houn 54Ea **108**
Tennyson Rd. W7 45Ha **86**
Tennyson St. SW8 54Kb **112**
Tennyson Wlk. DA11: Nflt 62Fe **143**
Tennyson Wlk. RM18: Tilb 4D **122**
Tennyson Way RM12: Horn 33Hd **76**
Tennyson Way SL2: Slou 2C **80**
Tenpin Acton **42Qa 87**
Tenpin Bexleyheath **56Bd 117**
Tenpin Croydon **75Pb 156**
Tenpin Feltham **61X 129**
Tenpin Kingston upon Thames **68Na 131**
(within The Rotunda Cen.)
Tensing Av. DA11: Nflt 2A **144**
Tensing Rd. UB2: S'hall 48Ca **85**
Tentelow La. UB2: S'hall 50Ca **85**
Tenterden Cl. NW4 27Za **48**
Tenterden Cl. SE9 63Pc **138**
Tenterden Dr. NW4 27Za **48**
Tenterden Gdns. CR0: C'don 73Wb **157**
Tenterden Gdns. NW4 27Za **48**
Tenterden Gro. NW4 27Za **48**
Tenterden Ho. SE17 7H **231** (50Ub **91**)
(off Surrey Gro.)
Tenterden Rd. CR0: C'don 73Wb **157**
Tenterden Rd. N17 24Vb **51**
Tenterden Rd. RM8: Dag 33Bd **75**
Tenterden St. W1 3A **222** (44Kb **90**)
Tenter Ground E1 1K **225** (43Vb **91**)
Tenth Av. KT20: Lwr K 98Ab **194**
Tent Peg La. BR5: Pet W 71Sc **160**
Tent St. E1 42Xb **91**
Tenzing Ct. E14 43Dc **92**
(off Hillary M.)
Tenzing Rd. HP2: Hem H 2A **4**
Tequila Wharf E14 44Ac **92**
Tera 40 UB6: G'frd 38Ea **66**
Terborch Way SE22 57Ub **113**
Tercelet Ter. NW3 35Eb **69**
Tercel Path IG7: Chig 21Xc **55**
Teredo St. SE16 48Zb **92**
Terence Cl. DA12: Grav'nd 10H **123**
Terence Ct. DA17: Belv 51Bd **117**
(off Nuxley Rd.)
Terence McMillan Stadium **42Lc 93**
Terence Messenger Twr. E10 33Dc **72**
(off Alpine Rd.)
Teresa Gdns. EN8: Walt C 6Yb **20**
Teresa M. E17 28Cc **52**
Teresa Wlk. N10 29Kb **50**
Terling Cl. E11 34Hc **73**
Terling Ho. W10 43Ya **88**
(off Sutton Way)
Terling Rd. RM8: Dag 33Cd **76**
Terlings, The CM14: B'wood 20Wd **40**
Terling Wlk. N1 39Sb **71**
(off Popham St.)
TERMINAL 4 RDBT. 58S **106**
TERMINAL 5 RDBT. 54K **105**
Terminal Ho. HA7: Stan 22Ma **47**
Terminus Pl. SW1 4A **228** (48Kb **90**)
Tern Gdns. RM14: Upm 32Ud **78**
Tern Way CM14: B'wood 21Ud **58**
Terrace, The DA12: Grav'nd 8D **122**

Terrace, The E2 41Yb 92
(off Old Ford Rd.)
Terrace, The E4 20Gc 35
(off Newgate St.)
Terrace, The EC4 3A 224 (44Qb 90)
(off King's Bench Wlk.)
Terrace, The GU22: Wok 92C 188
(not continuous)
Terrace, The IG8: Wfd G 23Jc 53
Terrace, The KT15: Add 78N 149
Terrace, The N3 26Bb 49
Terrace, The NW6 39Cb 69
Terrace, The SE23 59Ac 114
Terrace, The SE8 49Bc 92
(off Longshore)
Terrace, The SL5: S'hill 1B 146
Terrace, The SW13 54Ua 110
Terrace, The TN13: Riv 94Fd 202
Terrace Apts. N5 36Qb 70
Terrace Gdns. SW13 54Va 110
Terrace Gdns. WD17: Wat 12X 27
Terrace La. TW10: Rich 58Na 109
Terrace Rd. E13 40Jc 73
Terrace Rd. E9 38Yb 72
Terrace Rd. KT12: Walt T 73W 150
Terraces, The DA2: Dart 59Sd 120
Terraces, The E2 40Wb 71
(off Garner St.)
Terraces, The NW8 1B 214 (40Fb 69)
(off Queen's Ter.)
Terrace St. DA12: Grav'nd 8D 122
(not continuous)
Terrace Wlk. RM9: Dag 36Ad 75
Terrace Wlk. SW11 52Hb 111
Terracotta Rd. RH9: S God 10A 210
Terrano Ho. TW9: Kew 52Ra 109
Terrapin Rd. SW17 62Kb 134
Terrent Ct. SL4: Wind 3E 102
Terretts Pl. N1 38Rb 71
(off Upper St.)
Terrick Rd. N22 25Nb 50
Terrick St. W12 44Xa 88
Terrilands HA5: Pinn 27Ba 45
Territorial Ho. SE11 6A 230 (49Qb 90)
(off Reedworth St.)
Terront Rd. N15 28Sb 51
Terry Pl. UB8: Cowl 43L 83
Terry's Lodge Rd. TN15: Wro 86Xd 184
Terry Spinks Pl. E16 43Hc 93
(off Barking Rd.)
Tersha St. TW9: Rich 56Pa 109
Tesla Ct. W3 47Ua 88
Tessa Sanderson Pl. SW8 55Kb 112
(off Daley Thompson Way)
Tessa Sanderson Way UB6: G'frd 36Fa 66
Tester's Cl. RH8: Oxt 3M 211
Testerton Rd. W11 45Za 88
(off Hurstway Wlk.)
Testerton Wlk. W11 45Za 88
Testwood Ct. W7 45Ga 86
Testwood Rd. SL4: Wind 3B 102
Tetbury Pl. N1 1B 218 (39Rb 71)
Tetcott Rd. SW10 52Eb 111
(not continuous)
Tetherdown N10 27Jb 50
Tetty Way BR1: Broml 68Jc 137
Tevatree Ho. SE1 50Wb 91
(off Old Kent Rd.)
Teversham La. SW8 53Nb 112
Teviot Av. RM15: Avel 44Sd 98
Teviot Cl. DA16: Well 53Xc 117
Teviot Est. E14 43Dc 92
Teviot St. E14 43Ec 92
Tewin Rd. HP2: Hem H 2C 4
Tewkesbury Av. HA5: Pinn 29Aa 45
Tewkesbury Av. SE23 60Xb 113
Tewkesbury Cl. EN4: E Barn 14Fb 31
Tewkesbury Cl. IG10: Lough 16Nc 36
Tewkesbury Cl. KT14: Byfl 83M 169
Tewkesbury Cl. N15 30Tb 51
Tewkesbury Gdns. NW9 27Ra 47
Tewkesbury Rd. N15 30Tb 51
Tewkesbury Rd. SM5: Cars 74Fb 155
Tewkesbury Rd. W13 46Ja 86
Tewkesbury Ter. N11 23Lb 50
Tewson Rd. SE18 50Uc 94
Texcel Bus. Pk. DA1: Erith 54Hd 118
Texryte Ho. N1 39Tb 71
(off Southgate Rd.)
Textile Ho. E1 43Zb 92
(off Duckett St.)
Teynham Av. EN1: Enf 16Tb 33
Teynham Ct. BR3: Beck 69Dc 136
Teynham Grn. BR2: Broml 71Jc 159
Teynham Rd. DA2: Dart 60Sd 120
Teynton Ter. N17 25Sb 51
Thackeray Av. N17 26Wb 51
Thackeray Av. RM18: Tilb 3D 122
Thackeray Cl. SW19 66Za 132
Thackeray Cl. TW7: Isle 54Ja 108
Thackeray Cl. UB8: Hil 44R 84
Thackeray Ct. NW6 38Eb 69
(off Fairfax Rd.)
Thackeray Ct. SW3 7F 227 (50Hb 89)
(off Elystan Pl.)
Thackeray Ct. W14 48Ab 88
(off Blythe Rd.)
Thackeray Ct. W5 44Pa 87
(off Hanger Va. La.)
Thackeray Dr. DA11: Nflt 61De 143
Thackeray Dr. RM6: Chad H 31Wc 75
Thackeray Ho. WC1 5F 217 (42Nb 90)
(off Herbrand St.)
Thackeray Lodge TW14: Bedf 58T 106
Thackeray M. E8 37Wb 71
Thackeray Rd. E6 40Mc 73
Thackeray Rd. SW8 54Kb 112
Thackeray St. W8 48Db 89
Thackery Ct. EC1 7B 218 (43Rb 91)
(off Turnmill St.)
Thackrah Cl. N2 26Eb 49
(off Simms Gdns.)
Thakeham Cl. SE26 63Xb 135
Thalia Cl. SE10 51Fc 115
Thalia Ct. E8 38Vb 71
(off Albion Dr.)
Thalmassing Cl. CM13: Hut 19De 41
Thame Rd. SE16 47Zb 92
Thames Av. KT16: Chert 69J 127
Thames Av. KT4: Wor Pk 74Ya 154
Thames Av. RM9: Dag 42Dd 96
Thames Av. RM9: Rain 42Dd 96
Thames Av. SL4: Wind 2H 103
Thames Av. SW10 53Eb 111
Thames Av. UB6: G'frd 40Ha 66
Thames Bank SW14 54Sa 109
Thamesbank Pl. SE28 44Yc 95
Thames Barrier 47Mc 93
Thames Barrier Ind. Area SE18 48Mc 93
(off Faraday Way)
Thames Barrier Info. Cen. 48Mc 93
Thames Barrier Pk. 47Lc 93
Thamesbrook SW3 7D 226 (50Gb 89)
(off Dovehouse St.)
Thames Chase Forest Cen. 34Wd 78
Thames Circ. E14 49Cc 92
Thames Cl. KT16: Chert 73K 149
Thames Cl. RM13: Rain 44Kd 97
Thames Cl. TW12: Hamp 68Da 129
Thames Cotts. KT7: T Ditt 72Ka 152
Thames Ct. KT8: W Mole 68Da 129
Thames Ct. NW6 40Bb 69
(off Albert Rd.)
Thames Ct. SE15 52Vb 113
(off Daniel Gdns.)
Thames Ct. W7 44Ga 86
(off Hanway Rd.)
Thames Cres. W4 52Ua 110
Thamesdale AL2: Lon C 9K 7
THAMES DITTON 72Ja 152
Thames Ditton & Esher Golf Course 75Fa 152
Thames Ditton Station (Rail) 73Ha 152
Thames Ditton Youth Cen. 73Ja 152
Thames Dr. HA4: Ruis 30S 44
Thames Dr. RM16: Grays 10C 100
Thames Edge Ct. TW18: Staines 63G 126
(off Clarence St.)
Thames Exchange Bldg. EC4 5E 224 (45Sb 91)
(off Queen St. Pl.)
Thames Eyot TW1: Twick 60Ja 108
Thamesfield Ct. TW17: Shep 73S 150
Thamesfield M. TW17: Shep 73S 150
Thames Ga. DA1: Dart 57Qd 119
Thamesgate TW18: Staines 67K 127
Thamesgate Cl. TW10: Ham 63Ka 130
Thamesgate Shop. Cen. 8D 122
Thames Gateway RM13: Rain 43Gd 96
Thames Gateway RM13: Wenn 43Gd 96
Thames Gateway RM15: Avel 46Pd 97
Thames Gateway RM15: Purf 46Pd 97
Thames Gateway RM9: Dag 40Bd 75
Thames Gateway Pk. RM9: Dag 41Bd 95
Thames Haven KT6: Surb 71Ma 153
Thames Hgts. SE1 1K 231 (47Vb 91)
(off Gainsford St.)
Thameshill Av. RM5: Col R 26Ed 56
THAMES HOSPICECARE (WINDSOR) 5E 102
Thames Ho. DA1: Cray 55Jd 118
Thames Ho. EC4 5E 224 (45Sb 91)
(off Queen St. Pl.)
Thames Ho. KT1: King T 70Ma 131
(off Surbiton Rd.)
Thames Ho. SW1 5F 229 (49Nb 90)
(off Millbank)
Thameside KT16: Chert 72L 149
Thameside KT8: W Mole 69Da 129
Thameside TW11: Tedd 66Ma 131
Thameside TW18: Lale 64H 127
(not continuous)
Thameside TW18: Staines 64H 127
(not continuous)
Thameside Aviation Mus. 3M 123
Thameside Ind. Est. DA8: Erith 51Md 119
Thameside Ind. Est. E16 47Mc 93
Thameside Pl. KT1: Hamp W 67Ma 131
Thameside Theatre 50Ce 99
Thameside Wlk. SE28 44Wc 95
Thames Ind. Pk. RM18: E Til 10K 101
Thames Innovation Cen. DA18: Erith 47Bd 95
Thames Lock 69Y 129
Thames Lock KT13: Weyb 76Q 150
Thames Lock 52Ma 109
(off Dock Rd.)
Thames Mead SL4: Wind 3C 102
Thamesmead KT12: Walt T 72W 150
THAMESMEAD 46Xc 95
THAMESMEAD CENTRAL 46Wc 95
THAMESMEAD EAST 47Bd 95
THAMESMEAD NORTH 44Zc 95
Thames Mdw. KT8: W Mole 68Ca 129
Thames Mdw. TW17: Shep 74T 150
THAMESMEAD SOUTH 47Zc 95
THAMESMEAD SOUTH WEST 47Vc 95
THAMESMEAD WEST 48Sc 94
Thamesmere Dr. SE28 45Wc 95
Thamesmere Leisure Cen. 45Wc 95
Thames Path SE10 48Jc 93
Thames Pl. SW15 55Za 110
Thames Point SW6 54Eb 111
Thamespoint TW11: Tedd 66Ma 131
Thames Police Mus. 46Xb 91
Thames Quay E14 47Dc 92
Thames Quay SW10 53Eb 111
(off Chelsea Harbour)
Thames Reach KT1: Hamp W 67Ma 131
Thames Reach SE28 47Uc 94
Thames Reach W6 51Ya 110
(off Rainville Rd.)
Thames Retreat TW18: Staines 67H 127
Thames Rd. DA1: Cray 54Hd 118
Thames Rd. E16 46Mc 93
Thames Rd. IG11: Bark 41Vc 95
Thames Rd. RM17: Grays 52De 121
Thames Rd. SL3: L'ly 49C 82
Thames Rd. SL4: Wind 2A 102
Thames Rd. W4 51Qa 109
Thames Rd. Ind. Est. E16 46Mc 93
Thames Side KT1: King T 67Ma 131
Thames Side KT7: T Ditt 72Ka 152
Thames Side SL4: Wind 2H 103
Thames St. DA9: Ghithe 56Vd 120
Thames St. KT1: King T 68Ma 131
Thames St. KT12: Walt T 73V 150
Thames St. KT13: Weyb 75R 150
Thames St. SE10 51Dc 114
Thames St. SL4: Wind 3H 103
Thames St. TW12: Hamp 67Da 129
Thames St. TW16: Sun 70W 128
Thames St. TW18: Staines 64H 127
Thames Tunnel Mills SE16 47Yb 92
Thamesvale Cl. TW3: Houn 55Ca 107
Thames Valley Athletics Cen. 10J 81
THAMES VALLEY SPIRE HOSPITAL 9N 61
Thames Vw. IG1: Ilf 33Sc 74
(off Axon Pl.)
Thames Vw. RM16: Grays 10C 100
Thamesview Bus. Cen. RM13: Rain 43Gd 96
Thamesview Ho's. KT12: Walt T 72W 150
Thames Vw. Lodge IG11: Bark 41Uc 94
Thames Village W4 53Sa 109
Thames Wlk. KT12: Walt T 73W 150
(off Manor Rd.)
Thames Wlk. SW11 52Gb 111
Thames Way DA11: Grav'nd 57Ce 121
Thames Way DA11: Nflt 57Ce 121
Thames Wharf Studios W6 51Ya 110
(off Rainville Rd.)
Thamley RM19: Purf 49Qd 97
Thanescroft Gdns. CR0: C'don 76Ub 157
Thanet Ct. W3 44Qa 87
Thanet Dr. BR2: Kes 76Mc 159
Thanet Ho. CR0: C'don 77Sb 157
(off Coombe Rd.)
Thanet Ho. DA11: Nflt 10B 122
Thanet Ho. WC1 4F 217 (41Nb 90)
(off Thanet St.)
Thanet Ho. WD18: Wat 16V 26
(off Explorer Dr.)
Thanet Lodge NW2 37Ab 68
(off Mapesbury Rd.)
Thanet Pl. CR0: C'don 77Sb 157
Thanet Rd. DA5: Bexl 59Cd 118
Thanet Rd. DA8: Erith 52Gd 118
Thanet St. WC1 4F 217 (41Nb 90)
Thanet Wharf SE8 51Dc 114
(off Copperas St.)
Thane Vs. N7 34Pb 70
Thane Works N7 34Pb 70
Thanington Ct. SE9 58Uc 116
Thant Cl. E10 34Dc 72
Tharp Rd. SM6: W'gton 78Mb 156
Thatcham Ct. N20 17Eb 31
Thatcham Gdns. N20 17Eb 31
Thatcher Cl. UB7: W Dray 47N 83
Thatcher Ct. DA1: Dart 59Md 119
Thatchers Cl. IG10: Lough 12Sc 36
Thatchers La. GU3: Worp 9H 187
Thatchers Way TW7: Isle 57Fa 108
Thatches Gro. RM6: Chad H 28Ad 55
Thavie's Inn EC4 2A 224 (44Qb 90)
Thaxted Bold CM13: Hut 15Fe 41
Thaxted Ct. N1 2G 219 (40Tb 71)
(off Murray Gro.)
Thaxted Grn. CM13: Hut 15Ee 41
Thaxted Ho. RM10: Dag 38Dd 76
Thaxted Ho. SE16 49Yb 92
(off Abbeyfield Est.)
Thaxted Pl. SW20 66Za 132
Thaxted Rd. IG9: Buck H 17Nc 36
Thaxted Rd. SE9 62Sc 138
Thaxted Wlk. RM13: Rain 38Gd 76
Thaxted Way EN9: Walt A 5Fc 21
Thaxton Pl. E4 17Fc 35
Thaxton Rd. W14 51Bb 111
Thayers Farm Rd. BR3: Beck 67Ac 136
Thayer St. W1 2J 221 (44Jb 90)
Thaynesfield EN6: Pot B 3Fb 17
Theatre Hackney Wick, The 36Cc 72
Theatre Bldg. E3 41Cc 92
(off Paton Cl.)
Theatre Ct. KT18: Eps 85Ta 173
Theatre Pl. SE8 52Cc 114
(off Speedwell Ct.)
Theatre-Rites 2B 230 (47Rb 91)
(off Blackfriars Rd.)
Theatre Royal Drury Lane 3G 223 (44Nb 90)
(off Catherine St.)
Theatre Royal Haymarket 5E 222 (45Mb 90)
(off Haymarket)
Theatre Royal Stratford East 38Fc 73
Theatre Royal Windsor 2H 103
Theatre Sq. E15 37Fc 73
Theatre St. SW11 55Hb 111
Theatre Vw. Apts. SE1 1B 230 (47Rb 91)
(off Short St.)
Theatro Technis 1C 216 (39Lb 70)
(off Crowndale Rd.)
Theatro Twr. SE8 51Cc 114
Theberton St. N1 39Qb 70
Theed St. SE1 7K 223 (46Qb 90)
Thelbridge Ho. E3 41Dc 92
(off Bruce Rd.)
Thellusson Way WD3: Rick 18H 25
Thelma Cl. DA12: Grav'nd 4H 145
Thelma Gdns. SE3 53Mc 115
Thelma Gro. TW11: Tedd 65Ja 130
Thelusson Ct. WD7: R'lett 7Ja 14
Theobald Cres. HA3: Hrw W 25Ea 46
Theobald Rd. CR0: C'don 75Rb 157
Theobald Rd. E17 31Bc 72
Theobalds Av. N12 21Eb 49
Theobalds Av. RM17: Grays 50Ee 99
Theobalds Cl. EN6: Cuff 2Pb 18
Theobalds Cl. TN15: Kems'g 90Qd 183
Theobalds Ct. EN8: Chesh 4Zb 20
(off Crossbrook St.)
Theobalds Ct. N4 34Sb 71
Theobalds Grove Station (Overground) 4Zb 20
Theobalds La. EN7: Walt C 4Wb 19
Theobalds La. EN8: Chesh 4Yb 20
Theobalds Pk. Rd. EN2: Crew H 7Rb 19
Theobalds Pk. Rd. EN2: Enf 7Rb 19
Theobald's Rd. WC1 7H 217 (43Pb 90)
Theobalds Rd. EN6: Cuff 2Nb 18
Theobald St. SE1 4F 231 (48Tb 91)
Theobald St. WD6: Bore 11Na 29
Theobald St. WD7: R'lett 8Ka 14
Theodora Way HA5: Eastc 27V 44
Theodor Ct. NW9 26Ta 47
Theodore Ct. SE13 58Fc 115
Theodore Rd. SE13 58Fc 115
Thepps Cl. RH1: S Nut 9F 208
Therapia La. CR0: Bedd 73Mb 156
Therapia La. CR0: C'don 72Nb 156
Therapia Lane Stop (London Tramlink) 73Nb 156
Therapia Rd. SE22 58Yb 114
Theresa Rd. W6 49Wa 88
Theresa's Wlk. CR2: Sande 81Tb 177
Therfield Ct. N4 33Sb 71
Thermopylae Ga. E14 49Dc 92
Theseus Wlk. N1 2C 218 (40Rb 71)
(off City Gdn. Row)
Thesiger Rd. SE20 66Zb 136
Thessaly Ho. SW8 52Lb 112
(off Thessaly Rd.)
Thessaly Rd. SW8 52Lb 112
(not continuous)
Thesus Ho. E14 44Ec 92
(off Blair St.)
Thetford Cl. N13 23Rb 51
Thetford Gdns. RM9: Dag 38Ad 75
Thetford Ho. SE1 3K 231 (48Vb 91)
(off St Saviour's Est.)
Thetford Rd. KT3: N Mald 72Ta 153
Thetford Rd. RM9: Dag 38Zc 75
Thetford Rd. TW15: Ashf 63N 127
Thetis Ter. TW9: Kew 51Qa 109
Theven St. E1 42Yb 92
THEYDON BOIS 8Uc 22
Theydon Bois Golf Course 6Tc 22
Theydon Bois Station (Underground) 8Vc 23
Theydon Bold CM13: Hut 15Fe 41
Theydon Bower CM16: Epp 3Wc 23
Theydon Ct. EN9: Walt A 5Jc 21
Theydon Gdns. RM13: Rain 38Gd 76
THEYDON GARNON 7Yc 23
Theydon Gro. CM16: Epp 2Wc 23
Theydon Gro. IG8: Wfd G 23Lc 53
Theydon Pk. Rd. CM16: They B 9Uc 22
Theydon Pl. CM16: Epp 3Vc 23
Theydon Rd. CM16: Epp 4Tc 22
Theydon Rd. E5 33Yb 72
Theydon St. E17 31Bc 72
Theydon Towers CM16: Epp 5Tc 22
Thicket, The UB7: Yiew 44N 83
Thicket Cres. SM1: Sutt 77Eb 155
Thicket Gro. RM9: Dag 37Yc 75
Thicket Gro. SE20 66Wb 135
Thicket Rd. SE20 66Wb 135
Thicket Rd. SM1: Sutt 77Eb 155
Thicketts TN13: S'oaks 95Ld 203
Thickthorne La. TW18: Staines 66L 127
Thimble Cres. SM6: W'gton 75Jb 156
Third Av. DA11: Nflt 10A 122
Third Av. E12 35Nc 74
Third Av. E13 41Jc 93
Third Av. E17 29Cc 52
Third Av. EN1: Enf 15Vb 33
Third Av. EN9: Walt A 2Kc 21
Third Av. HA9: Wemb 33Ma 67
Third Av. KT20: Lwr K 97Ab 194
Third Av. RM10: Dag 39Dd 76
Third Av. RM20: W Thur 51Wd 120
Third Av. RM6: Chad H 30Yc 55
Third Av. UB3: Hayes 46V 84
Third Av. W10 41Ab 88
Third Av. W3 46Va 88
Third Av. WD25: Wat 7Z 13
Third Cl. KT8: W Mole 70Ea 130
Third Cres. SL1: Slou 3G 80
Third Cross Rd. TW2: Twick 61Fa 130
Third Way HA9: Wemb 35Ra 67
Thirkleby Cl. SL1: Slou 6G 80
Thirleby Rd. HA8: Edg 25Ta 47
Thirleby Rd. NW7 22Ab 48
Thirleby Rd. SW1 4C 228 (48Lb 90)
Thirlestane AL1: St A 1D 6
Thirlestane Ct. N10 26Jb 50
Thirlmere NW1 3A 216 (41Kb 90)
(off Cumberland Mkt.)
Thirlmere Av. SL1: Slou 3A 80
Thirlmere Av. UB6: G'frd 41La 86
Thirlmere Cl. TW20: Egh 66D 126
Thirlmere Dr. AL1: St A 4F 6
Thirlmere Gdns. HA6: Nwood 22R 44
Thirlmere Gdns. HA9: Wemb 32La 66
Thirlmere Ho. N16 35Tb 71
(off Howard Rd.)
Thirlmere Ho. TW7: Isle 57Ha 108
Thirlmere Ri. BR1: Broml 65Hc 137
Thirlmere Rd. DA7: Bex 53Ed 118
Thirlmere Rd. N10 25Kb 50
Thirlmere Rd. SW16 63Mb 134
Thirsk Cl. UB5: N'olt 37Ca 65
Thirsk Rd. CR4: Mitc 66Jb 134
Thirsk Rd. SE25 70Tb 135
Thirsk Rd. SW11 55Jb 112
Thirsk Rd. WD6: Bore 9Qa 15
Thirston Path WD6: Bore 12Qa 29
Thirza Ho. E1 44Yb 92
(off Devonport St.)
Thirza Rd. DA1: Dart 58Pd 119
Thistlebrook SE2 48Yc 95
Thistlebrook Ind. Est. SE2 48Yc 95
Thistle Cl. HP1: Hem H 3G 2
Thistle Cl. IG7: Chig 22Ad 55
Thistle Ct. DA1: Dart 60Rd 119
(off Churchill Cl.)
Thistle Ct. SE12 61Jc 137
Thistlecroft HP1: Hem H 3K 3
Thistlecroft Gdns. HA7: Stan 25Ma 47
Thistlecroft Rd. KT12: Hers 77Y 151
Thistledene KT14: W Byf 85H 169
Thistledene KT7: T Ditt 72Ga 152
Thistledene Av. HA2: Harr 34Aa 65
Thistledene Av. RM5: Col R 22Dd 56
Thistledown DA12: Grav'nd 5F 144
Thistlefield Cl. DA5: Bexl 60Zc 117
Thistle Gro. SW10 7A 226 (50Eb 89)
Thistle Ho. E14 44Ec 92
(off Dee St.)
Thistle Mead IG10: Lough 13Qc 36
Thistlemead BR7: Chst 68Rc 138
Thistle Rd. DA12: Grav'nd 9G 122
Thistles, The HP1: Hem H 1K 3
Thistles, The KT22: Lea 94La 192
Thistleton Ho. NW9 27Va 48
Thistlewaite Rd. E5 34Xb 71
Thistlewood Cl. N7 33Pb 70
Thistlewood Cres. CR0: New Ad 84Fc 179
Thistleworth Cl. TW7: Isle 52Fa 108
Thistleworth Marine TW7: Isle 56Ka 108
(off Railshead Rd.)
Thistley Cl. CR5: Coul 94Mb 196
Thistley Cl. N12 23Gb 49
Thistley Ct. SE8 51Dc 114
Thomas a Beckett Cl. HA0: Wemb 35Ha 66
Thomas Av. CR3: Cat'm 93Sb 197
Thomas Baines Rd. SW11 55Fb 111
Thomas Barnardo Way IG6: Ilf 27Sc 54
Thomas Bata Av. RM18: E Til 9K 101
Thomas Burt Ho. E2 41Xb 91
(off Canrobert St.)
Thomas Cl. CM15: B'wood 19Ae 41
Thomas Coulter Ho. E16 47Kc 93
(off Shipwright Street)
Thomas Ct. IG6: Ilf 28Sc 54
Thomas Cribb M. E6 44Qc 94
Thomas Darby Ct. W11 44Ab 88
Thomas Dean Rd. SE26 63Bc 136
Thomas Dinwiddy Rd. SE12 61Kc 137
Thomas Doyle St. SE1 3C 230 (48Rb 91)
Thomas Dr. DA12: Grav'nd 1F 144
Thomas Dr. RM2: Rom 28Ld 57
Thomas Dr. UB8: Uxb 37N 63
Thomas Earle Ho. W14 49Bb 89
(off Warwick La.)
Thomas England Ho. RM7: Rom 30Fd 56
(off Waterloo Gdns.)
Thomas Frye Ct. E15 40Dc 72
(off High St.)
Thomas Fyre Dr. E3 40Cc 72
Thomas Hardy Ho. N22 24Pb 50
Thomas Hardy M. SW16 64Lb 134
Thomas Hewlett Ho. HA1: Harr 35Ga 66
Thomas Hollywood Ho. E2 40Yb 72
(off Approach Rd.)
Thomas Ho. SL2: Stoke P 8L 61
(off Bells Hill Grn.)
Thomas Ho. SM2: Sutt 80Db 155
Thomas Jacomb Pl. E17 28Bc 52
Thomas Joseph Ho. SE4 57Zb 114
(off St Norbert Rd.)
Thomas La. SE6 59Cc 114
Thomas Lodge E17 29Dc 52
Thomas Moore Ho. RH2: Reig 6L 207
(off Reigate Rd.)
Thomas More Gdns. KT10: Esh 76Ca 151
Thomas More Highwalk EC2 1D 224 (43Sb 91)
(off Aldersgate St.)
Thomas More Ho. EC2 1D 224 (43Sb 91)
(off Aldersgate St.)
Thomas More Ho. HA4: Ruis 32U 64
Thomas More Sq. E1 45Wb 91
Thomas More St. E1 45Wb 91
Thomas More Way N2 27Eb 49
Thomas Neal's Cen. 3F 223 (44Nb 90)
Thomas Parmiter Sports Cen., The 3Z 13
Thomas Pl. W8 48Db 89
Thomas Rd. DA1: Cray 55Jd 118
Thomas Rd. E14 44Bc 92
Thomas Rd. Ind. Est. E14 43Cc 92
(not continuous)
Thomas Sawyer Way WD18: Wat 16X 27
Thomas Sims Ct. RM12: Horn 37Kd 77
Thomas Spencer Hall of Residence SE18 49Qc 94
(off Grand Depot Rd.)
Thomas St. SE18 49Rc 94
Thomas Tallis Sports Cen. 55Kc 115
Thomas Twr. E8 37Vb 71
(off Dalston Sq.)
Thomas Turner Path CR0: C'don 75Sb 157
(off George St.)
Thomas Wall Cl. SM1: Sutt 78Db 155
Thomas Watson Cott. Homes EN5: Barn 14Ab 30
(off Leecroft Rd.)
Thomas Wyatt Way TN15: Wro 88Be 185
Thompkins La. SL2: Farn R 8D 60
Thompson Av. TW9: Rich 55Qa 109
Thompson Cl. IG1: Ilf 33Sc 74
Thompson Cl. SL3: L'ly 49B 82
Thompson Cl. SM3: Sutt 74Cb 155
Thompson Ho. SE14 51Zb 114
(off John Williams Cl.)
Thompson Ho. W10 42Ab 88
(off Wornington Rd.)
Thompson Rd. RM9: Dag 34Bd 75
Thompson Rd. SE22 58Vb 113
Thompson Rd. TW3: Houn 56Da 107
Thompson Rd. UB10: Uxb 38N 63
Thompson's Av. SE5 52Sb 113
Thompson's Cl. GU24: Pirb 5B 186
Thompsons Cl. EN7: Chesh 1Vb 19
Thompson's La. GU24: Chob 1H 167
Thompson's La. IG10: Lough 10Jc 21
Thompson Way WD3: Rick 18J 25
Thomson Cres. CR0: C'don 74Qb 156
Thomson Ho. E14 44Cc 92
(off Saracen St.)
Thomson Ho. SE17 6H 231 (49Ub 91)
(off Tatum St.)
Thomson Ho. SW1 7E 228 (50Mb 90)
(off Bessborough Pl.)
Thomson Ho. UB1: S'hall 45Aa 85
(off The Broadway)
Thomson Rd. HA3: W'stone 27Ga 46
THONG 5J 145
Thong La. DA12: Grav'nd 3H 145
Thong La. DA12: Shorne 3H 145
Thong La. TN15: Bor G 93Ae 205
Thonrey Cl. NW9 27Va 48
Thorburn Ho. SW1 2G 227 (47Hb 89)
(off Kinnerton St.)
Thorburn Sq. SE1 49Wb 91
Thorburn Way SW19 67Fb 133
Thoresby St. N1 3E 218 (41Sb 91)
Thorkhill Gdns. KT7: T Ditt 74Ja 152
Thorkhill Rd. KT7: T Ditt 74Ja 152
Thorley Cl. KT14: W Byf 86J 169
Thorley Gdns. GU22: Pyr 87J 169
Thorley Rd. RM16: Grays 46Ce 99
Thornaby Gdns. N18 23Wb 51
Thornaby Ho. E2 41Xb 91
(off Canrobert St.)
Thorn Apts. E3 43Cc 92
(off St Paul's Way)
Thornash Cl. GU21: Wok 7N 167
Thornash Rd. GU21: Wok 7N 167
Thornash Way GU21: Wok 7N 167
Thorn Av. WD23: B Hea 18Ea 28
Thornbank Cl. TW19: Stanw M 57J 105
Thornbill Ho. SE15 52Wb 113
(off Bird in Bush Rd.)
Thornbridge Rd. SL0: Iver H 39E 62
Thornbury NW4 28Xa 48
(off Prince of Wales Cl.)
Thornbury Av. TW7: Isle 52Fa 108
Thornbury Cl. N16 36Ub 71
Thornbury Cl. NW7 24Za 48
Thornbury Ct. CR2: S Croy 78Tb 157
(off Blunt Rd.)
Thornbury Ct. CR3: Whyt 92Vb 197
Thornbury Ct. TW7: Isle 52Ga 108
Thornbury Ct. W11 45Cb 89
(off Chepstow Vs.)
Thornbury Gdns. WD6: Bore 14Sa 29
Thornbury Ho. RM3: Rom 22Md 57
(off Bridgwater Wlk.)
Thornbury Lodge EN2: Enf 13Rb 33
Thornbury Rd. SW2 58Nb 112
Thornbury Rd. TW7: Isle 52Fa 108
Thornbury Sq. N6 32Lb 70
Thornbury Way E17 25Bc 52
Thornby Rd. E5 34Yb 72
Thorncliffe Rd. SW2 58Nb 112
Thorncliffe Rd. UB2: S'hall 50Ba 85
Thorn Cl. BR2: Broml 72Qc 160
Thorn Cl. UB5: N'olt 41Ba 85
Thorncombe Rd. SE22 57Ub 113
Thorncroft HP3: Hem H 4B 4
Thorncroft RM11: Horn 30Kd 57
Thorncroft TW20: Eng G 6N 125
Thorncroft Cl. CR5: Coul 91Qb 196
Thorncroft Dr. KT22: Lea 95Ka 192
Thorncroft Rd. SM1: Sutt 78Db 155
Thorncroft St. SW8 52Nb 112
Thorndales CM14: W'ley 21Zd 59
Thorndean St. SW18 61Eb 133
Thorndene Av. N11 18Jb 32
Thorndike SL2: Slou 3E 80
Thorndike Av. UB5: N'olt 39Z 65
Thorndike Cl. SW10 52Eb 111
Thorndike Ho. SW1 7D 228 (50Mb 90)
(off Vauxhall Bri. Rd.)
Thorndike Rd. N1 37Tb 71
Thorndike St. SW1 6D 228 (49Mb 90)
Thorndon App. CM13: Heron 24Ee 59
Thorndon Av. CM13: W H'dn 28Ee 59
Thorndon Cl. BR5: St P 68Vc 139

Thorndon Country Pk. 23Ce 59
Thorndon Country Pk. North 24Be 59
Thorndon Country Pk. South 26Ee 59
Thorndon Countryside Cen. 24Be 59
Thorndon Gdns. KT19: Ewe 78Ua 154
Thorndon Ga. CM13: Ingve 22Ee 59
Thorndon Hall CM13: Ingve 23De 59
Thorndon Pk. Golf Course 23De 59
Thorndon Rd. BR5: St P 68Vc 139
Thorndown La. GU20: W'sham 10B 146
Thorn Dr. SL3: Geor G 44A 82
Thorndyke Ct. HA5: Hat E 23Ba 45
Thorndyke Way TN15: Wro 88Be 185
Thorne Cl. DA8: Erith 51Dd 118
Thorne Cl. E11 35Fc 73
Thorne Cl. E16 44Jc 93
Thorne Cl. HP1: Hem H 4K 3
Thorne Cl. KT10: Clay 80Ja 152
Thorne Cl. TW15: Ashf 66S 128
Thorne Ho. AL1: St A 1C 6
Thorne Ho. E14 48Ec 92
(off Launch St.)
Thorne Ho. E2 41Yb 92
(off Roman Rd.)
Thorneloe Gdns. CR0: Wadd 78Qb 156
Thorne Pas. SW13 54Ua 110
Thorne Rd. SL3: L'ly 10N 81
Thorne Rd. SW8 52Nb 112
Thornes Cl. BR3: Beck 69Ec 136
Thornes Ho. SW11 52Mb 112
(off Ponton Rd.)
Thorne St. SW13 55Ua 110
Thornet Wood Rd. BR1: Broml 69Qc 138
Thornewill Ho. E1 45Yb 92
(off Cable St.)
THORNEY 48K 83
Thorney Bus. Pk. SL0: Iver 47F 82
Thorney Country Pk. 48K 83
Thorney Ct. W8 2A 226 (47Eb 89)
(off Palace Ga.)
Thorney Cres. SW11 52Fb 111
Thorneycroft Cl. KT12: Walt T 72Y 151
Thorneycroft Dr. EN3: Enf L 9Cc 20
Thorneycroft Rd. N20 19Fb 31
Thorney Hedge Rd. W4 49Ra 87
Thorney La. Nth. SL0: Iver 44H 83
Thorney La. Sth. SL0: Rich P 47H 83
Thorney Mill Rd. SL0: Thorn 48J 83
(not continuous)
Thorney Mill Rd. UB7: W Dray 48L 83
Thorney Pk. Golf Course 47K 83
Thorney St. SW1 5F 229 (49Nb 90)
Thornfield Av. NW7 25Ab 48
Thornfield Ct. NW7 24Ab 48
Thornfield Ho. E14 45Cc 92
(off Rosefield Gdns.)
Thornfield Pde. NW7 24Ab 48
(off Holders Hill Rd.)
Thornfield Rd. SM7: Bans 89Cb 175
Thornfield Rd. W12 47Xa 88
(not continuous)
Thornford Rd. SE13 57Ec 114
Thorngate Rd. W9 42Cb 89
Thorngrove Rd. E13 39Kc 73
Thornham Gro. E15 36Fc 73
Thornham Ind. Est. E15 36Fc 73
Thornham St. SE10 51Dc 114
Thornhaugh M. WC1 6E 216 (42Mb 90)
Thornhaugh St. WC1 6E 216 (42Mb 90)
Thornhill Av. KT6: Surb 75Na 153
Thornhill Av. SE18 52Uc 116
Thornhill Bri. 1H 217 (40Pb 70)
(off Caledonian Rd.)
Thornhill Bri. Wharf N1 1H 217 (39Pb 70)
Thornhill Cres. N1 38Pb 70
Thornhill Gdns. E10 33Dc 72
Thornhill Gdns. IG11: Bark 38Uc 74
Thornhill Gro. N1 38Pb 70
Thornhill Ho. W4 50Ua 88
(off Wood St.)
Thornhill Ho's. N1 38Qb 70
(off Thornhill Rd.)
Thornhill M. SW15 56Bb 111
Thornhill Rd. CR0: C'don 73Sb 157
Thornhill Rd. E10 33Dc 72
Thornhill Rd. HA6: Nwood 21S 44
Thornhill Rd. KT6: Surb 75Na 153
Thornhill Rd. N1 38Qb 70
Thornhill Rd. UB10: Ick 35P 63
Thornhill Sq. N1 38Pb 70
Thornhill Way TW17: Shep 71Q 150
Thorn Ho. WD6: Bore 12Sa 29
(off Elstree Way)
Thornicroft Ho. SW9 54Pb 112
(off Stockwell Rd.)
Thorn La. RM13: Rain 40Md 77
Thornlaw Rd. SE27 63Qb 134
Thornleas Pl. KT24: E Hor 98U 190
Thornley Cl. N17 24Wb 51
Thornley Dr. HA2: Harr 33Da 65
Thornley Pl. SE10 50Gc 93
Thornridge CM14: B'wood 18Xd 40
Thorns, The CM15: Kel H 11Ud 40
Thornsbeach Rd. SE6 60Ec 114
Thornsett Pl. SE20 68Xb 135
Thornsett Rd. SE20 68Xb 135
Thornsett Rd. SW18 60Db 111
Thornsett Ter. SE20 68Xb 135
(off Croydon Rd.)
Thorns Mdw. TN16: Bras 95Yc 201
Thorn Ter. CM16: Epp 4Xc 23
Thorn Ter. SE15 55Yb 114
Thornton Av. CR0: C'don 72Pb 156
Thornton Av. SW2 60Mb 112
Thornton Av. UB7: W Dray 48P 83
Thornton Av. W4 49Ua 88
Thornton Cl. KT22: Lea 92Ja 192
Thornton Cl. UB7: W Dray 48P 83
Thornton Cres. CR5: Coul 91Qb 196
Thornton Dene BR3: Beck 68Cc 136
Thornton Gdns. SW12 60Mb 112
Thornton Gro. HA5: Hat E 23Ca 45
THORNTON HEATH 70Sb 135
Thornton Heath Leisure Cen. 70Sb 135
THORNTON HEATH POND 71Qb 156
Thornton Heath Station (Rail) 70Sb 135
Thornton Hill SW19 66Ab 132
Thornton Ho. SE17 6H 231 (49Ub 91)
(off Townsend St.)
Thornton Pl. W1 7F 215 (43Hb 89)
Thornton Rd. BR1: Broml 64Jc 137
Thornton Rd. CR0: C'don 73Pb 156
Thornton Rd. CR7: Thor H 72Pb 156
Thornton Rd. DA17: Belv 49Dd 96
Thornton Rd. E11 33Fc 73
Thornton Rd. EN5: Barn 13Ab 30
Thornton Rd. EN6: Pot B 2Eb 17
Thornton Rd. IG1: Ilf 35Rc 74
Thornton Rd. N18 20Yb 34
Thornton Rd. SM5: Cars 74Fb 155
Thornton Rd. SW12 59Mb 112
Thornton Rd. SW14 56Ta 109
Thornton Rd. SW19 65Za 132
Thornton Rd. E. SW19 65Za 132
Thornton Rd. Ind. Est. CR0: C'don 72Pb 156
Thornton Row CR7: Thor H 71Qb 156
Thorntons CM13: Ingve 23Ee 59
Thornton's Farm Av. RM7: Rush G 32Ed 76
Thornton Side RH1: Mers 3B 208
Thornton St. AL3: St A 1A 6
Thornton St. E20 38Dc 72
Thornton St. SW9 54Qb 112
Thornton Way NW11 29Db 49
Thorntree Ct. W5 43Na 87
Thorntree Rd. SE7 50Mc 93
Thornville Gro. CR4: Mitc 68Fb 133
Thornville St. SE8 53Cc 114
Thornwell Ct. W7 47Ga 86
(off Lwr. Boston Rd.)
Thornwood Cl. E18 26Kc 53
Thornwood Gdns. W8 47Cb 89
Thornwood Ho. IG9: Buck H 17Nc 36
Thornwood Lodge W8 47Cb 89
(off Thornwood Gdns.)
Thornwood Rd. CM16: Epp 1Xc 23
Thornwood Rd. SE13 57Gc 115
Thornycroft Ho. W4 50Ua 88
(off Fraser St.)
Thorogood Gdns. E15 36Gc 73
Thorogood Way RM13: Rain 39Gd 76
Thorold Cl. CR2: Sels 82Zb 178
Thorold Ho. SE1 1D 230 (47Sb 91)
(off Pepper St.)
Thorold Rd. IG1: Ilf 33Rc 74
Thorold Rd. N22 24Nb 50
Thoroughfare, The KT20: Walt H 96Wa 194
Thorparch Rd. SW8 53Mb 112
THORPE 69E 126
Thorpebank Rd. W12 46Wa 88
Thorpe Bold CM13: Hut 15Fe 41
Thorpe By-Pass TW20: Thorpe 68D 126
Thorpe Cl. BR6: Orp 75Uc 160
Thorpe Cl. CR0: New Ad 83Ec 178
Thorpe Cl. SE26 63Zb 136
Thorpe Cl. W10 44Ab 88
THORPE COOMBE HOSPITAL 27Ec 52
Thorpe Ct. EN2: Enf 13Rb 33
Thorpe Cres. E17 26Bc 52
Thorpe Cres. WD19: Wat 17Y 27
Thorpedale Gdns. IG2: Ilf 28Qc 54
Thorpedale Gdns. IG6: Ilf 28Rc 54
Thorpedale Rd. N4 33Nb 70
THORPE GREEN 70C 126
Thorpe Hall Rd. E17 25Ec 52
Thorpe Hay Meadow Nature
Reserve 66F 126
Thorpe Ho. N1 1J 217 (39Pb 70)
(off Barnsbury Est.)
Thorpe Ind. Pk. TW20: Thorpe 67E 126
Thorpe Lakes Wakeboard, Waterski
& Wakesurf Cen. 70F 126
THORPE LEA 65E 126
Thorpe Lea Rd. TW20: Egh 65D 126
Thorpe Lea Rd. TW20: Thorpe 65D 126
Thorpe Lodge RM11: Horn 30Nd 57
Thorpe Pk. 70G 126
Thorpe Rd. AL1: St A 3B 6
Thorpe Rd. E17 26Ec 52
Thorpe Rd. E6 39Pc 74
Thorpe Rd. E7 35Hc 73
Thorpe Rd. IG11: Bark 38Tc 74
Thorpe Rd. KT16: Chert 71F 148
Thorpe Rd. KT2: King T 66Na 131
Thorpe Rd. N15 30Ub 51
Thorpe Rd. TW18: Staines 65F 126
Thorpeside Cl. TW18: Staines 68G 126
Thorpewood Av. SE26 61Xb 135
Thorpland Av. UB10: Ick 34S 64
Thorrington Bold CM13: Hut 15Fe 41
Thorsden Cl. GU22: Wok 91A 188
Thorsden Ct. GU22: Wok 90A 168
Thorsden Way SE19 64Ub 135
Thors Oak SS17: Stan H 1N 101
Thorverton Rd. NW2 34Ab 68
Thoydon Rd. E3 40Ac 72
Thrale Rd. SW16 63Lb 134
Thrale St. SE1 7E 224 (46Sb 91)
Thrapston Ho. RM3: Rom 22Nd 57
(off Lindfield Rd.)
Thrasher Cl. E8 39Vb 71
Thrawl St. E1 1K 225 (43Vb 91)
Thrayle Ho. SW9 55Pb 112
(off Benedict Rd.)
Threadgold Ho. N1 37Tb 71
(off Dovercourt Est.)
Threadneedle St. EC2 3G 225 (44Tb 91)
Threadneedle Wlk. EC2 3G 225 (44Tb 91)
(off Throgmorton St.)
Thread St. SM6: W'gton 75Jb 156
Three Angels Cl. SM6: W'gton 78Pb 156
Three Arch Bus. Pk. RH1: Redh 10A 208
Three Arches Pk. RH1: Redh 10P 207
Three Arch Rd. RH1: Redh 10P 207
Three Bridges Bus. Cen. UB2: S'hall 47Ea 86
Three Colts La. E2 42Xb 91
Three Colt St. E14 44Bc 92
Three Corners DA7: Bex 54Dd 118
Three Corners HP3: Hem H 4A 4
Three Cups Yd. WC1 1J 223 (43Pb 90)
(off Sandland St.)
Three Gates Rd. DA3: Fawk 76Vd 164
Three Kings Yd. W1 4K 221 (45Kb 90)
Three Meadows M. HA3: Hrw W 25Ha 46
Three Mill La. E3 41Ec 92
Three Mills Studios 41Ec 92
Three Oak La. SE1 1K 231 (47Vb 91)
Three Oaks Cl. UB10: Ick 34P 63
Three Quays EC3 5J 225 (45Ub 91)
(off Tower Hill)
Three Quays Wlk. EC3 5J 225 (45Ub 91)
Three Rivers Mus., The 17N 25
Three Valleys Way WD23: Bush 15Z 27
Threshers Pl. W11 45Ab 88
Threshold Way GU24: Chob 82A 168
Thriffwood SE26 62Yb 136
Thrift, The DA2: Bean 62Yd 142
Thrift Farm La. WD6: Bore 12Sa 29
Thrift Grn. CM13: B'wood 20Ce 41
Thrift La. TN14: Cud 89Uc 180
Thrifts Hall Farm M. CM16: They B 9Vc 23
Thrifts Mead CM16: They B 9Uc 22
Thriftwood Holiday Pk. TN15: Stans 85Ae 185
Thriftwood International Scout
Activity Cen. 19Ce 41
Thrigby Rd. KT9: Chess 79Pa 153
Thring Ho. SW9 54Pb 112
(off Stockwell Rd.)
Throckmorton Rd. E16 44Kc 93
Throgmorton Av. EC2 2G 225 (44Tb 91)
(not continuous)
Throgmorton St. EC2 2G 225 (44Tb 91)
Throstle Pl. WD25: Wat 4Y 13
Throwley Cl. SE2 48Yc 95
(not continuous)
Throwley Rd. SM1: Sutt 78Db 155
Throwley Way SM1: Sutt 77Db 155
Thrums, The WD24: Wat 9X 13
Thrupp Cl. CR4: Mitc 68Kb 134
Thrupp's Av. KT12: Hers 78Z 151
Thrupp's La. KT12: Hers 78Z 151
Thrush Av. AL10: Hat 2C 8
Thrush Grn. HA2: Harr 28Ca 45
Thrush Grn. WD3: Rick 17L 25
Thrush St. SE17 7D 230 (50Sb 91)
Thumpers HP2: Hem H 1N 3
Thunderer St. E13 40Lc 73
Thunderer Wlk. SE18 48Rc 94
Thurbarn Rd. SE6 64Dc 136
Thurland Ho. SE16 49Xb 91
(off Camilla Rd.)
Thurland Rd. SE16 48Wb 91
Thurlby Cl. HA1: Harr 30Ja 46
Thurlby Cl. IG8: Wfd G 22Pc 54
Thurlby Cft. NW4 27Ya 48
(off Mulberry Cl.)
Thurlby Rd. HA0: Wemb 37Ma 67
Thurlby Rd. SE27 63Qb 134
Thurleigh Av. SW12 58Jb 112
Thurleigh Ct. SW12 58Jb 112
Thurleigh Rd. SW12 59Hb 111
Thurleston Av. SM4: Mord 71Ab 154
Thurlestone Av. IG3: Bark 35Vc 75
Thurlestone Av. IG3: Ilf 35Vc 75
Thurlestone Av. N12 23Hb 49
Thurlestone Cl. TW17: Shep 72S 150
Thurlestone Ct. UB1: S'hall 44Da 85
(off Howard Rd.)
Thurlestone Pde. TW17: Shep 72S 150
(off High St.)
Thurlestone Rd. SE27 62Qb 134
Thurloe Cl. SW7 5D 226 (49Gb 89)
Thurloe Ct. SW3 6D 226 (49Gb 89)
(off Fulham Rd.)
Thurloe Gdns. RM1: Rom 30Hd 56
Thurloe Pl. SW7 5C 226 (49Fb 89)
Thurloe Pl. M. SW7 5C 226 (49Fb 89)
(off Thurloe Pl.)
Thurloe Sq. SW7 5D 226 (49Gb 89)
Thurloe St. SW7 5C 226 (49Fb 89)
Thurloe Wlk. RM17: Grays 48Ce 99
Thurlow Cl. E4 23Ec 52
Thurlow Gdns. HA0: Wemb 36Ma 67
Thurlow Gdns. IG6: Ilf 23Tc 54
Thurlow Hill SE21 60Sb 113
Thurlow Ho. SW16 62Nb 134
Thurlow Pk. Rd. SE21 61Rb 135
Thurlow Rd. NW3 36Fb 69
Thurlow Rd. W7 47Ja 86
Thurlow St. SE17 7G 231 (50Tb 91)
(not continuous)
Thurlow Ter. NW5 36Jb 70
Thurlow Wlk. SE17 7H 231 (50Ub 91)
(not continuous)
Thurlstone Rd. HA4: Ruis 34W 64
Thurlton Ct. GU21: Wok 88A 168
Thurnby Ct. TW2: Twick 62Ga 130
Thurnham Way KT20: Tad 92Za 194
Thurnscoe NW1 39Lb 70
(off Pratt St.)
Thurrock Athletics Stadium 47Fe 99
Thurrock Bus. Cen. RM20: W Thur 51Vd 120
Thurrock Commercial Cen.
RM15: Avel 47Pd 97
THURROCK COMMUNITY HOSPITAL 46Ee 99
Thurrock Ent. Cen. RM17: Grays 51Ce 121
THURROCK LAKESIDE 48Wd 98
Thurrock Lakeside 48Xd 98
Thurrock Mus. 50De 99
Thurrock Pk. Way RM18: Tilb 52Fe 121
THURROCK SERVICE AREA 47Vd 98
Thurrock Trade Pk. RM20: W Thur 52Wd 120
Thurrock Yacht Club 52Ce 121
Thursby Rd. GU21: Wok 10L 167
Thursland Rd. DA14: Sidc 64Ad 139
Thursley Cres. CR0: New Ad 80Ec 158
Thursley Gdns. SW19 61Za 132
Thursley Ho. SW2 59Pb 112
(off Holmewood Gdns.)
Thursley Rd. SE9 62Pc 138
Thurso Cl. RM3: Hrld W 23Rd 57
Thurso Ho. NW6 40Db 69
Thurso St. SW17 63Fb 133
Thurstan Dwellings WC2 2G 223 (44Nb 90)
(off Newton St.)
Thurstan Rd. SW20 66Xa 132
Thurstan St. SW6 53Eb 111
Thurston Ho. BR3: Beck 65Dc 136
Thurston Ho. N1 1J 217 (39Pb 70)
(off Barnsbury Est.)
Thurston Rd. SE13 54Dc 114
Thurston Rd. SL1: Slou 4J 81
Thurston Rd. UB1: S'hall 44Ba 85
Thurtle Rd. E2 1K 219 (39Vb 71)
Thwaite Cl. DA8: Erith 51Ed 118
Thyer Cl. BR6: Farnb 77Sc 160
Thyme Cl. SE3 55Lc 115
Thyme Ct. NW7 25Ab 48
Thyme Wlk. E5 35Zb 72
Thyra Gro. N12 23Db 49
Tibbatt's Rd. E3 42Dc 92
Tibbenham Pl. SE6 61Cc 136
Tibbenham Wlk. E13 40Hc 73
Tibberton Sq. N1 39Sb 71
Tibbet's Cl. SW19 60Za 110
TIBBET'S CORNER 59Za 110
Tibbet's Ride SW15 59Za 110
Tibbles Cl. WD25: Wat 7Aa 13
Tibbs Hill Rd. WD5: Ab L 2V 12
Tiber Cl. E3 39Cc 72
Tiber Gdns. N1 1G 217 (39Nb 70)
Tiberius Sq. AL3: St A 4N 5
Ticehurst Cl. BR5: St P 66Wc 139
Ticehurst Rd. SE23 61Ac 136
Tichborne WD3: Map C 22F 42
Tichmarsh KT19: Eps 82Sa 173
Tickford Cl. SE2 47Yc 95
Tickford Ho. NW8 4D 214 (41Gb 89)
Tickners Way CR5: Coul 88Lb 176
Ticknor Dr. DA10: Swans 61Be 143
Tidal Basin Rd. E16 45Hc 93
(not continuous)
Tidbury Ct. SW8 52Lb 112
(off Stewart's Rd.)
Tide Cl. CR4: Mitc 67Jb 134
Tideham Ho. SE28 46Tc 94
Tidelea Twr. SE28 47Sc 94
Tidemill Sq. SE10 49Gc 93
Tidemill Way SE8 52Cc 114
Tidenham Gdns. CR0: C'don 76Ub 157
Tideside Ct. SE18 48Nc 94
Tideslea Path SE28 46Tc 94
Tideswell Rd. CR0: C'don 76Cc 158
Tideswell Rd. SW15 56Ya 110
Tidewaiters Ho. E14 44Fc 93
(off Blair St.)
Tideway Cl. TW10: Ham 63Ka 130
Tideway Ct. SE16 46Zb 92
Tideway Ho. E14 47Cc 92
(off Strafford St.)
Tideway Ho. E17 28Zb 52
Tideway Wlk. SW11 51Lb 112
Tidey St. E3 43Cc 92
Tidford Rd. DA16: Well 54Vc 117
Tidlock Ho. SE28 47Tc 94
Tidworth Rd. E3 42Cc 92
Tidy's La. CM16: Epp 1Xc 23
Tiepigs La. BR2: Hayes 74Hc 159
Tiepigs La. BR4: W W'ck 75Gc 159
Tierney Ct. CR0: C'don 75Ub 157
Tierney La. W6 51Ya 110
Tierney Rd. SW2 60Nb 112
Tiffany Hgts. SW18 59Cb 111
Tiffin Girls Community Sports Cen. 65Na 131
Tiffin Sports Cen. 68Pa 131
Tiger Cl. IG11: Bark 40Xc 75
Tiger Ho. WC1 4E 216 (41Mb 90)
(off Burton St.)
Tiger La. BR2: Broml 70Kc 137
Tiger Way E5 35Xb 71
Tiggap Ho. SE10 49Gc 93
Tigris Cl. N9 19Yb 34
Tilbrook Rd. SE3 55Lc 115
Tilburstow Hill Rd. RH9: G'stone 4A 210
(not continuous)
Tilburstow Hill Rd. RH9: S God 4A 210
(not continuous)
TILBURY 4C 122
Tilbury Bowls & Squash Cen. 4C 122
(off Civic Sq.)
Tilbury Cl. BR5: St P 68Xc 139
Tilbury Cl. HA5: Hat E 24Ba 45
Tilbury Cl. SE15 52Vb 113
TILBURY DOCKS 54Fe 121
Tilbury Fort 6E 122
Tilbury Gdns. RM18: Tilb 3C 122
Tilbury Ho. SE14 51Zb 114
(off Myers La.)
Tilbury Lodge CR2: S Croy 79Ub 157
Tilbury Rd. E10 31Ec 72
Tilbury Rd. E6 40Pc 74
Tilbury Town Station (Rail) 4B 122
Tilbury Wlk. SL3: L'ly 48D 82
Tildesley Rd. SW15 58Ya 110
Tile Farm Rd. BR6: Orp 76Tc 160
Tile Ho. N1 39Nb 70
(off Beaconsfield St.)
Tilehouse Cl. WD6: Bore 13Pa 29
Tilehouse La. UB9: Den 27G 42
Tilehouse La. WD3: W Hyd 26G 42
(not continuous)
Tilehouse Way UB9: Den 31H 63
Tilehurst NW1 4A 216 (41Kb 90)
(off Lit. Albany St.)
Tilehurst Point SE2 47Yc 95
(not continuous)
Tilehurst Rd. SM3: Cheam 78Ab 154
Tilehurst Rd. SW18 60Fb 111
Tile Kiln Cl. HP3: Hem H 3B 4
Tilekiln Cl. EN7: Chesh 1Vb 19
Tile Kiln Cres. HP3: Hem H 3B 4
Tile Kiln La. DA5: Bexl Dartford Rd. 61Ed 140
Tile Kiln La. DA5: Bexl Faesten Way 62Gd 140
Tile Kiln La. HP3: Hem H 3A 4
Tile Kiln La. N13 22Sb 51
Tile Kiln La. N6 32Kb 70
Tile Kiln La. UB9: Hare 31R 64
Tile Kiln Studios N6 31Lb 70
Tilemakers Yd. SW18 61Eb 133
Tilers Cl. RH1: Mers 3C 208
Tiler's Wlk. RH2: Reig 10L 207
Tiler's Way RH2: Reig 10L 207
Tileyard Rd. N7 38Nb 70
Tilford Av. CR0: New Ad 81Ec 178
Tilford Gdns. SW19 60Za 110
Tilford Ho. SW2 59Pb 112
(off Holmewood Gdns.)
Tilgate Comn. RH1: Blet 5J 209
Tilgate Gdns. CR5: Coul 91Qb 196
Tilia Cl. SM1: Sutt 78Bb 155
Tilia Rd. E5 35Xb 71
Tilia Wlk. SW9 56Rb 113
Tillage Cl. AL4: St A 4H 7
Till Av. DA4: Farni 73Pd 163
Tiller Ho. N1 39Ub 71
(off Whitmore Est.)
Tiller Ho. UB2: S'hall 48Ca 85
(off Lockwood Rd.)
Tiller Leisure Cen. 48Cc 92
Tiller Rd. E14 48Cc 92
Tillers Cl. TW18: Staines 65G 126
Tillett Cl. NW10 37Sa 67
Tillett Sq. SE16 47Ac 92
Tillett Way E2 41Wb 91
Tilley La. KT18: Eps D 94Sa 193
Tilley La. KT18: Head 94Sa 193
Tilley Rd. TW13: Felt 60W 106
Tillingbourne Gdns. N3 27Bb 49
Tillingbourne Grn. BR5: St M Cry 70Vc 139
Tillingbourne Way N3 28Bb 49
Tillingdown Hill CR3: Cat'm 94Xb 197
Tillingdown Hill CR3: Wold 94Xb 197
Tillingdown La. CR3: Cat'm 96Xb 197
(not continuous)
Tillingdown La. CR3: Wold 96Xb 197
(not continuous)
Tillingham Bold CM13: Hut 15Fe 41
Tillingham Ct. EN9: Walt A 5Jc 21
Tillingham Way N12 21Cb 49
Tilling Rd. NW2 32Ya 68
Tillings Cl. SE5 53Sb 113
Tilling Way HA9: Wemb 33Ma 67
Tillmans TN15: Bor G 92Ce 205
Tillman St. E1 44Xb 91
Tilloch St. N1 38Pb 70
Tillotson Ct. SW8 52Mb 112
(off Wandsworth Rd.)
Tillotson Rd. HA3: Hrw W 24Da 45
Tillotson Rd. IG1: Ilf 31Qc 74
Tillotson Rd. N9 19Vb 33
Tillys La. TW18: Staines 63H 127
Tilmans Mead DA4: Farni 73Pd 163
Tilney Cl. KT22: Lea 92Ja 192
Tilney Ct. EC1 5E 218 (42Sb 91)
Tilney Ct. IG9: Buck H 19Jc 35
Tilney Dr. IG9: Buck H 19Jc 35
Tilney Gdns. N1 37Tb 71
Tilney Rd. RM9: Dag 37Bd 75
Tilney Rd. UB2: S'hall 49Y 85
Tilney St. W1 6J 221 (46Jb 90)
Tilson Cl. SE5 52Ub 113
Tilson Gdns. SW2 59Nb 112
Tilson Ho. SW2 59Nb 112
Tilson Rd. N17 25Wb 51
Tilston Bright Sq. SE2 48Yc 95
Tilston Cl. E11 34Hc 73
Tilstone Av. SL4: Eton W 10C 80
Tilstone Cl. SL4: Eton W 10C 80
Tilt Cl. KT11: Cobh 88Aa 171
Tiltman Pl. N7 34Pb 70
Tilt Mdw. KT11: Cobh 88Aa 171
Tilt Rd. TN15: Bor G 92Be 205
Tilton St. SW6 51Ab 110
Tilt Rd. KT11: Cobh 87Y 171
Tilt Rd. KT11: Stoke D 87Y 171
Tilt Vw. KT11: Cobh 87Y 171
Tiltwood, The W3 45Sa 87
Tilt Yd. App. SE9 58Pc 116
Timber Cl. BR7: Chst 68Qc 138
Timber Cl. GU22: Pyr 86H 169
Timber Cl. KT23: Bookh 98Ea 192
Timber Ct. RM17: Grays 51Ce 121
Timbercroft KT19: Ewe 77Ua 154
Timbercroft La. SE18 51Uc 116
TIMBERDEN BOTTOM 82Gd 182
Timberdene NW4 26Za 48
Timberdene Av. IG6: Ilf 25Sc 54
Timberhill KT21: Asht 91Na 193
Timber Hill Cl. KT16: Ott 80E 148
Timber Hill Rd. CR3: Cat'm 96Wb 197
Timberidge WD3: Loud 14M 25
Timberland Cl. SE15 52Wb 113
Timberland Rd. E1 44Xb 91
Timber La. CR3: Cat'm 96Wb 197
Timberley Ct. DA14: Sidc 64Vc 139
Timberling Gdns. CR2: Sande 81Tb 177
Timber Mill Way SW4 55Mb 112
Timber Pond Rd. SE16 46Zb 92
Timbers, The SM3: Cheam 79Ab 154
Timberslip Dr. SM6: W'gton 81Mb 176
Timber St. EC1 6D 218 (42Sb 91)
Timbertop Rd. TN16: Big H 90Lc 179
Timber Wharf E2 39Vb 71
Timberwharf Rd. N16 30Wb 51
Timber Wharves Est. E14 49Cc 92
(off Copeland Dr.)
Timberwood SL2: Farn C 4H 61
Timber Yd., The N1 4J 219 (41Ub 91)
(off Drysdale St.)
Timberyard M. KT4: Wor Pk 75Xa 154
Timbralls SL4: Eton 10H 81
(off Slough Rd.)
Timbrell Pl. SE16 46Bc 92
Time Sq. E8 36Vb 71
Times Sq. E1 44Wb 91
Times Sq. SM1: Sutt 78Db 155
Timians Way BR1: Broml 63Kc 137
Timmins Apts. E2 41Zb 92
Timms Cl. BR1: Broml 70Pc 138
Timor Ho. E1 42Ac 92
(off Duckett St.)
Timothy Cl. DA6: Bex 57Ad 117
Timothy Cl. SW4 57Lb 112
Timothy Ho. DA18: Erith 47Ad 95
(off Kale Rd.)
Timothy Pl. KT8: W Mole 71Ba 151
Timpani Hill CR6: W'ham 90Xb 177
Timperley Ct. RH1: Redh 4N 207
(off Timperley Gdns.)
Timperley Ct. SW19 60Ab 110
Timperley Gdns. RH1: Redh 4N 207
Timplings Row HP1: Hem H 1K 3
Timsbury Wlk. SW15 60Wa 110
Timsway TW18: Staines 64H 127
Tina Ct. SE6 59Bc 114
Tindale Cl. CR2: Sande 83Tb 177
Tindall Cl. RM3: Hrld W 26Pd 57
Tindall M. RM12: Horn 34Ld 77
Tindal St. SW9 53Rb 113
Tinderbox All. SW14 55Ta 109
Tinderbox Ho. SE8 52Cc 114
(off Octavius St.)
Tinefields KT20: Tad 91Ab 194
Tine Rd. IG7: Chig 22Uc 54
(not continuous)
Tingeys Top La. EN2: Crew H 8Qb 18
Tinker Pot La. TN15: W King 86Rd 183
Tinkerpot Ri. TN15: W King 86Sd 184
Tinkers La. SL4: Wind 4B 102
Tinkers La. SL6: S'dale 2F 146
Tinmans Row KT11: D'side 90Y 171
Tinniswood Cl. N5 36Qb 70
Tinsey Cl. TW20: Egh 64D 126
Tinsley Cl. SE25 69Xb 135
Tinsley Rd. E1 43Yb 92
Tintagel Cl. KT17: Eps 86Va 174
Tintagel Ct. EC1 5B 218 (42Rb 91)
(off St John St.)
Tintagel Ct. RM11: Horn 32Qd 77
Tintagel Cres. SE22 56Vb 113
Tintagel Dr. HA7: Stan 21Ma 47
Tintagel Gdns. SE22 56Vb 113
Tintagel Rd. BR5: Orp 75Yc 161
Tintagel Way GU22: Wok 88C 168
Tintells La. KT24: W Hor 100R 190
Tintern Av. NW9 27Ra 47
Tintern Cl. SL1: Slou 8G 80
Tintern Cl. SW15 57Ab 110
Tintern Cl. SW19 65Eb 133
Tintern Ct. W13 45Ja 86
Tintern Gdns. N14 17Nb 32
Tintern Ho. NW1 2A 216 (40Kb 70)
(off Augustus St.)
Tintern Ho. SW1 6K 227 (49Kb 90)
(part of Abbots Mnr.)
Tintern Path NW9 30Ua 48
(off Fryent Gro.)
Tintern Rd. N22 25Sb 51
Tintern Rd. SM5: Cars 74Fb 155
Tintern St. SW4 56Nb 112
Tintern Way HA2: Harr 32Da 65
Tinto Rd. E16 42Jc 93
Tinwell M. WD6: Bore 15Ta 29
Tinworth Ct. SE17 51Rb 113
(off Royal Rd.)
Tinworth St. SE11 7G 229 (50Nb 90)
Tippendell La. AL2: Chis G 7N 5
Tippendell La. AL2: Park 7N 5
Tippett Ct. E6 40Pc 74
Tippett La. RH8: Oxt 5L 211
Tippetts Cl. EN2: Enf 11Sb 33
Tippler Wlk. E15 36Fc 73
Tipthorpe Rd. SW11 55Jb 112
Tipton Dr. CR0: C'don 77Ub 157
Tiptree NW1 38Kb 70
(off Castlehaven Rd.)

Tiptree Cl. E4 20Ec 34
Tiptree Cl. RM11: Horn 32Qd 77
Tiptree Ct. CM13: Hut 15Fe 41
Tiptree Cres. IG5: Ilf 27Qc 54
Tiptree Dr. EN2: Enf 14Tb 33
Tiptree Rd. HA4: Ruis 35X 65
Tiree Cl. HP3: Hem H 4B 4
Tiree Ho. SL2: Slou 3F 80
Tirlemont Rd. CR2: S Croy 80Sb 157
Tirrell Rd. CR0: C'don 72Sb 157
Tisbury Ct. W1 4D 222 (45Mb 90)
(off Wardour St.)
Tisbury Rd. SW16 68Nb 134
Tisdall Ho. SE17 6G 231 (49Tb 91)
(off Barlow St.)
Tisdall Pl. SE17 6G 231 (49Tb 91)
Tissington Ct. SE16 49Yb 92
Titan Bus. Est. SE8 52Cc 114
(off Ffinch St.)
Titan Ct. TW8: Bford 50Pa 87
Titan Ho. E20 36Ec 72
(off Napa Cl.)
Titan Ind. Est. RM17: Grays 49Ce 99
Titanium Point E2 41Zb 92
(off Palmer's Rd.)
Titan Rd. RM17: Grays 50Ce 99
Titan Way RM17: Grays 50Ce 99
Titchborne Row W2 3D 220 (44Gb 89)
Titchfield Rd. EN3: Enf W 9Ac 20
Titchfield Rd. NW8 1E 214 (39Gb 69)
Titchfield Rd. SM5: Cars 74Fb 155
Titchfield Wlk. SM5: Cars 73Fb 155
Titchwell Rd. SW18 60Fb 111
Tite Hill TW20: Egh 4P 125
Tite Hill TW20: Eng G 4P 125
Tite St. SW3 50Hb 89
Tithe Barn Cl. AL1: St A 5A 6
Tithe Barn Cl. KT2: King T 67Pa 131
Tithe Barn Ct. WD5: Ab L 1V 12
(off Dairy Way)
Tithebarns La. GU23: Send 98J 189
Tithe Barn Way UB5: N'olt 40X 65
Tithe Cl. GU25: Vir W 2P 147
Tithe Cl. KT12: Walt T 72X 151
Tithe Cl. NW7 25Wa 48
Tithe Cl. UB4: Hayes 43V 84
Tithe Ct. SL3: L'ly 49C 82
Tithe Farm Av. HA2: Harr 34Ca 65
Tithe Farm Cl. HA2: Harr 34Ca 65
Tithe Farm Social Club 33Ca 65
Tithe La. GU24: W End 4E 166
Tithe La. TW19: Wray 58C 104
Tithe Mdw. WD18: Wat 17T 26
Tithe Mdws. GU25: Vir W 2N 147
Tithepit Shaw La. CR6: W'ham 89Xb 177
Tithe Wlk. NW7 25Wa 48
Titian Av. WD23: B Hea 17Ga 28
Titian Hgts. E20 36Dc 72
(off Scarlet Cl.)
Titlarks Hill SL5: S'dale 5F 146
Titley Cl. E4 22Cc 52
Titmus Cl. UB8: Hil 44S 84
Titmuss Av. SE28 45Xc 95
Titmuss St. W12 47Ya 88
Titness Pk. SL5: S'hill 9E 124
TITSEY 97Kc 199
Titsey Hill RH8: T'sey 95Jc 199
Titsey Place & Gdns. 96Kc 199
Titsey Rd. RH8: Limp 97Kc 199
Titsey Rd. RH8: T'sey 97Kc 199
Tivendale N8 27Nb 50
Tiverton Av. IG5: Ilf 27Qc 54
Tiverton Cl. CR0: C'don 73Vb 157
Tiverton Dr. SE9 60Sc 116
Tiverton Gro. RM3: Rom 22Qd 57
Tiverton Ho. EN3: Enf H 13Zb 34
Tiverton Rd. EN6: Pot B 3Fb 17
Tiverton Rd. HA0: Wemb 40Na 67
Tiverton Rd. HA4: Ruis 34W 64
Tiverton Rd. HA8: Edg 26Pa 47
Tiverton Rd. N15 30Tb 51
Tiverton Rd. N18 22Ub 51
Tiverton Rd. NW10 39Za 68
Tiverton Rd. TW3: Houn 54Ea 108
Tiverton St. SE1 3D 230 (48Sb 91)
Tiverton Way KT9: Chess 78Ma 153
Tiverton Way NW7 24Za 48
Tivoli Ct. SE16 46Bc 92
Tivoli Gdns. DA12: Grav'nd 10D 122
Tivoli Gdns. SE18 49Nc 94
(not continuous)
Tivoli M. E14 44Bc 92
Tivoli Rd. N8 29Mb 50
Tivoli Rd. SE27 64Sb 135
Tivoli Rd. TW4: Houn 56Aa 107
Tivoli Way E14 44Bc 92
Tizzard Gro. SE3 56Kc 115
Toad La. TW4: Houn 56Ba 107
Tobacco Dock 45Xb 91
Tobago St. E14 47Cc 92
Tobermory Cl. SL3: L'ly 49A 82
Tobin Cl. KT19: Eps 83Ra 173
Tobin Cl. NW3 38Gb 69
Toby Ct. N9 17Yb 34
(off Tramway Av.)
Toby La. E1 42Ac 92
Toby Way KT5: Surb 75Ra 153
Tockley Rd. SL1: Burn 1A 80
Tockwith Ct. TN13: S'oaks 95Ld 203
Todber Ho. W14 48Ab 88
(off Russell Rd.)
Todd Cl. RM13: Rain 42Md 97
Todd Cl. WD6: Bore 12Sa 29
Todd Ho. N2 26Fb 49
(off The Grange)
Todds Wlk. N7 33Pb 70
Todhunter Ter. EN5: New Bar 14Cb 31
Toft Av. RM17: Grays 49Fe 99
Token Ho. CR0: C'don 76Sb 157
(off Robert St.)
Tokenhouse Yd. EC2 2F 225 (44Tb 91)
Token Yd. SW15 56Ab 110
TOKYNGTON 37Ra 67
Tokyngton Av. HA9: Wemb 37Qa 67
Toland Sq. SW15 57Wa 110
Tolbut Ct. RM1: Rom 30Hd 56
Tolcairn Ct. DA17: Belv 50Cd 96
Tolcarne Dr. HA5: Eastc 26W 44
Tolchurch W11 44Bb 89
(off Dartmouth Cl.)
Toley Av. HA9: Wemb 31Na 67
Tolgate Ct. RH1: Redh 10P 207
Tolhurst Dr. W10 41Ab 88
Tollard Ho. W14 48Bb 89
(off Kensington High St.)
Toll Bar Ct. SM2: Sutt 81Db 175
Tollbridge Cl. W10 42Ab 88
Tolldene Cl. GU21: Knap 9J 167
Tollers La. CR5: Coul 90Pb 176
Tollesbury Ct. CM13: Hut 15Fe 41
Tollesbury Gdns. IG6: Ilf 27Tc 54
Tollet St. E1 42Zb 92
Tollgate Cl. WD3: Chor 14H 25
Tollgate Ct. CR2: S Croy 78Tb 157
Tollgate Dr. SE21 61Ub 135
Tollgate Dr. UB4: Yead 45Z 85
Tollgate Gdns. NW6 40Db 69
(off Oxford Rd.)
Tollgate Ho. NW6 40Db 69
(off Tollgate Gdns.)
Tollgate Lodge BR7: Chst 67Qc 138
(off Caveside Cl.)
Tollgate M. TN15: Bor G 92Ce 205
Tollgate Rd. AL4: Col H 5P 7
Tollgate Rd. AL9: N Mym 7C 8
Tollgate Rd. DA2: Dart 59Td 120
Tollgate Rd. E16 43Lc 93
Tollgate Rd. E6 43Mc 93
Tollgate Rd. EN8: Walt C 7Zb 20
Tollgate Sq. 43Pc 94
Tollhouse La. SM6: W'gton 81Lb 176
Tollhouse Way N19 33Lb 70
Tollington Pk. N4 33Pb 70
Tollington Pl. N4 33Pb 70
Tollington Rd. N7 35Pb 70
Tollington Way N7 34Nb 70
Tolmers Av. EN6: Cuff 1Nb 18
Tolmers Gdns. EN6: Cuff 1Nb 18
Tolmers Sq. NW1 5C 216 (42Lb 90)
(not continuous)
Tolpaide Ho. SE11 6K 229 (49Qb 90)
(off Marylee Way)
Tolpits Cl. WD18: Wat 15V 26
Tolpits La. WD18: Wat 18S 26
Tolpits La. Cvn. Site WD18: Wat 17U 26
Tolpuddle Av. E13 39Lc 73
Tolpuddle St. N1 1K 217 (40Qb 70)
Tolsey Mead TN15: Bor G 91Ce 205
Tolsford Rd. E5 36Xb 71
Tolson Rd. TW7: Isle 55Ja 108
Tolvaddon Cl. GU21: Wok 9L 167
Tolverne Rd. SW20 67Ya 132
TOLWORTH 75Ra 153
Tolworth B'way. KT6: Surb 74Ra 153
Tolworth Cl. KT6: Surb 74Ra 153
Tolworth Gdns. RM6: Chad H 29Zc 55
TOLWORTH HOSPITAL 75Qa 153
TOLWORTH JUNC. (TOBY JUG) 75Ra 153
Tolworth Pde. RM6: Chad H 29Ad 55
Tolworth Pk. Rd. KT6: Surb 75Pa 153
Tolworth Recreation Cen. 76Pa 153
Tolworth Ri. Nth. KT5: Surb 74Ra 153
Tolworth Ri. Sth. KT5: Surb 75Ra 153
Tolworth Rd. KT6: Surb 75Na 153
Tolworth Station (Rail) 75Ra 153
Tolworth Twr. KT6: Surb 75Ra 153
Tomahawk Gdns. UB5: N'olt 41Z 85
Tomblin Cl. WD6: Bore 12Ra 29
Tomblin M. SW16 68Lb 134
Tom Coombs Cl. SE9 56Nc 116
Tom Cribb Rd. SE28 48Sc 94
Tom Groves Cl. E15 36Fc 73
Tom Hood Cl. E15 36Fc 73
Tom Jenkinson Rd. E16 46Jc 93
Tomkins Cl. SS17: Stan H 1L 101
Tomkins Cl. WD6: Bore 11Na 29
Tomkyns Ho. SE11 6K 229 (49Qb 90)
(off Distin St.)
Tomkyns La. RM14: Upm 26Td 58
Tomlin Cl. KT19: Eps 83Ta 173
Tomlin Ct. KT19: Eps 83Ta 173
Tomline Ho. SE1 7D 224 (46Sb 91)
(off Union St.)
Tomlin Rd. SL2: Slou 2C 80
Tomlins All. TW1: Twick 60Ja 108
Tomlin's Gro. E3 41Cc 92
Tomlinson Cl. E2 4K 219 (41Vb 91)
Tomlinson Cl. W4 50Ra 87
Tomlins Orchard IG11: Bark 39Sc 74
Tomlins Ter. E14 44Ac 92
Tomlins Wlk. N7 33Pb 70
Tomlyns Cl. CM13: Hut 16Fe 41
Tom Mann Cl. IG11: Bark 39Uc 74
Tommy Flowers M. NW7 22Za 48
Tom Nolan Cl. E15 40Gc 73
Tomo Ind. Est. UB8: Cowl 44L 83
Tompion Ho. EC1 5C 218 (42Rb 91)
(off Percival St.)
Tompion St. EC1 4C 218 (41Rb 91)
(not continuous)
Toms Cft. HP2: Hem H 3N 3
Tomsfield AL10: Hat 1A 8
Tom's Hill WD3: Chan C 8N 11
Toms La. WD4: K Lan 1R 12
Toms La. WD5: Bedm 9E 4
Tom Smith Cl. SE10 51Gc 115
Tomson Ho. SE1 3K 231 (48Vb 91)
(off St Saviour's Est.)
Tomswood Ct. IG6: Ilf 25Sc 54
Tomswood Hill IG6: Ilf 23Rc 54
Tomswood Rd. IG7: Chig 23Qc 54
Tom Williams Ho. SW6 51Bb 111
(off Clem Attlee Ct.)
Tonbridge Cl. SM7: Bans 86Hb 175
Tonbridge Cres. HA3: Kenton 28Na 47
Tonbridge Ho's. WC1 4F 217 (41Nb 90)
(off Tonbridge St.)
Tonbridge Rd. KT8: W Mole 70Ba 129
Tonbridge Rd. RM3: Rom 24Md 57
Tonbridge Rd. TN13: S'oaks 99Ld 203
Tonbridge Rd. TN15: Igh 97Yd 204
Tonbridge Rd. TN15: Plax 97Yd 204
Tonbridge St. WC1 3F 217 (41Nb 90)
Toneborough NW8 39Db 69
(off Abbey Rd.)
Tonfield Rd. SM3: Sutt 74Bb 155
Tonge Cl. BR3: Beck 71Cc 158
Tonman Ho. AL1: St A 2C 6
Tonsley Hill SW18 57Db 111
Tonsley Pl. SW18 57Db 111
Tonsley Rd. SW18 57Db 111
Tonsley St. SW18 57Db 111
Tonstall Rd. CR4: Mitc 68Jb 134
Tonstall Rd. KT19: Eps 82Ta 173
Tony Cannell M. E3 41Bc 92
Tony Law Ho. SE20 67Xb 135
Tony Rawson Way RM10: Dag 34Ed 76
Tooke Cl. HA5: Pinn 25Aa 45
Tookey Cl. HA3: Kenton 31Pa 67
Took's Ct. EC4 2K 223 (44Qb 90)
Tooley St. DA11: Nflt 59Fe 121
Tooley St. SE1 6G 225 (46Tb 91)
Toorack Rd. HA3: Hrw W 26Fa 46
TOOTING 64Gb 133
TOOTING BEC 62Hb 133
Tooting Bec Gdns. SW16 63Mb 134
(not continuous)
Tooting Bec Lido 63Lb 134
Tooting Bec Rd. SW16 63Kb 134
Tooting Bec Rd. SW17 62Jb 134
Tooting Bec Station (Underground) 62Jb 134
Tooting B'way. SW17 64Gb 133
Tooting Broadway Station
(Underground) 64Gb 133
TOOTING GRAVENEY 65Hb 133
Tooting Gro. SW17 64Gb 133
Tooting High St. SW17 65Gb 133
Tooting Leisure Cen. 63Fb 133
Tooting Mkt. SW17 63Hb 133
Tooting Station (Rail) 65Hb 133
Tootswood Rd. BR2: Broml 71Gc 159
Tooveys Mill Cl. WD4: K Lan 1Q 12
Topaz Cl. SL1: Slou 6F 80
Topaz Ct. E11 32Gc 73
Topaz Wlk. NW2 31Za 68
Topcliffe Dr. BR6: Farnb 77Tc 160
Top Dartford Rd. BR8: Hext 66Hd 140
Top Dartford Rd. DA2: Wilm 65Kd 141
TopGolf Surrey 80L 149
TopGolf Watford 10Aa 13
Topham Ho. SE10 52Ec 114
(off Prior St.)
Topham Sq. N17 25Sb 51
Topham St. EC1 5K 217 (42Qb 90)
Topham Yd. SW19 67Db 133
Top Ho. Ri. E4 17Ec 34
Topiary, The KT21: Asht 92Na 193
Topiary Sq. TW9: Rich 55Pa 109
Toplands Av. RM15: Avel 46Sd 98
Topley St. SE9 56Lc 115
Topmast Point E14 47Cc 92
Top Meadow Golf Course 36Zd 79
Top Pk. BR3: Beck 71Gc 159
Top Pk. SL9: Ger X 1N 61
Topping La. UB8: Cowl 41M 83
Topp Wlk. NW2 33Ya 68
Topsfield Cl. N8 29Mb 50
Topsfield Pde. N8 29Nb 50
(off Tottenham La.)
Topsfield Rd. N8 29Nb 50
Topsham Rd. SW17 62Hb 133
Tora Ct. IG9: Buck H 19Lc 35
Torbay Ct. NW1 38Kb 70
Torbay Mans. NW6 39Bb 69
(off Willesden La.)
Torbay Rd. HA2: Harr 33Aa 65
Torbay Rd. NW6 38Bb 69
Torbay St. NW1 38Kb 70
Torbitt Way IG2: Ilf 29Vc 55
Torbridge Cl. HA8: Edg 24Na 47
Torbrook Cl. DA5: Bexl 58Ad 117
Tor Ct. W8 47Cb 89
Torcross Dr. SE23 61Yb 136
Torcross Rd. HA4: Ruis 34X 65
Torel Way RM16: Grays 9B 100
Toreno Mansion E20 36Dc 72
(off Olympic Pk. Av.)
Tor Gdns. W8 47Cb 89
Tor Gro. SE28 46Uc 94
Tor Ho. N6 30Kb 50
Torin Ct. TW20: Eng G 4N 125
Torino Way RM15: S Ock 42Xd 98
Torland Dr. KT22: Oxs 85Fa 172
Tor La. KT13: Weyb 83S 170
Tormead Cl. SM1: Sutt 79Cb 155
Tormount Rd. SE18 51Uc 116
Tornay Ho. E9 38Yb 72
Tornay Ho. N1 1H 217 (40Pb 70)
(off Priory Grn. Est.)
Toronto Av. E12 35Pc 74
Toronto Ho. SE16 47Zb 92
Toronto Rd. IG1: Ilf 32Rc 74
Toronto Rd. RM18: Tilb 4C 122
Torquay Ct. E14 43Ec 92
(off St Ives Pl.)
Torquay Gdns. IG4: Ilf 28Mc 53
Torquay Spur SL2: Slou 1F 80
Torquay St. W2 43Db 89
Torrance Cl. RM11: Horn 32Ld 77
Torrance Cl. SE7 51Mc 115
Torrens Ct. SE5 55Tb 113
Torrens Rd. E15 37Hc 73
Torrens Rd. SW2 57Pb 112
Torrens Sq. E15 37Hc 73
Torrens St. EC1 2A 218 (40Qb 70)
Torrens Wlk. DA12: Grav'nd 4G 144
Torrent Lodge SE10 52Dc 114
(off Merryweather Pl.)
Torres Sq. E14 50Cc 92
Torre Vista SE13 55Ec 114
(off Loampit Va.)
Torre Wlk. SM5: Cars 74Gb 155
Torrey Dr. SW9 54Qb 112
Torriano Av. NW5 36Mb 70
Torriano Cotts. NW5 36Lb 70
Torriano M. NW5 36Lb 70
Torridge RM18: E Til 9L 101
Torridge Gdns. SE15 56Yb 114
Torridge Rd. CR7: Thor H 71Rb 157
Torridge Rd. SL3: L'ly 51D 104
Torridon Cl. GU21: Wok 9M 167
Torridon Ho. NW6 40Db 69
(off Randolph Gdns.)
Torridon Rd. SE13 59Gc 115
Torridon Rd. SE6 59Fc 115
Torrington Av. N12 22Fb 49
Torrington Cl. KT10: Clay 79Ga 152
Torrington Cl. N12 21Fb 49
Torrington Ct. SE26 64Wb 135
(off Crystal Palace Pk. Rd.)
Torrington Dr. EN6: Pot B 3Fb 17
Torrington Dr. HA2: Harr 35Da 65
Torrington Dr. IG10: Lough 13Sc 36
Torrington Gdns. IG10: Lough 14Sc 36
Torrington Gdns. N11 23Lb 50
Torrington Gdns. UB6: G'frd 39La 66
Torrington Gro. N12 22Gb 49
Torrington Pk. N12 22Eb 49
Torrington Pl. E1 46Wb 91
Torrington Pl. WC1 7D 216 (43Mb 90)
Torrington Rd. E18 27Jc 53
Torrington Rd. HA4: Ruis 34V 64
Torrington Rd. KT10: Clay 79Ga 152
Torrington Rd. RM8: Dag 32Bd 75
Torrington Rd. UB6: G'frd 39La 66
Torrington Sq. CR0: C'don 73Tb 157
Torrington Sq. WC1 6E 216 (42Mb 90)
Torrington Way SM4: Mord 72Cb 155
Tor Rd. DA16: Well 53Yc 117
Torr Rd. SE20 66Zb 136
Tortington Ho. SE15 52Wb 113
(off Friary Est.)
Torver Rd. HA1: Harr 28Ga 46
Torver Way BR6: Orp 76Tc 160
Torwood La. CR3: Whyt 92Vb 197
Torwood Rd. SW15 57Wa 110
Torworth Rd. WD6: Bore 11Pa 29
To the Woods Cvn. & Camping Site
TN15: W King 79Td 164
TOT HILL 98Sa 193
Tothill Ho. SW1 5E 228 (49Mb 90)
(off Page St.)
Tot Hill La. KT18: Head 98Sa 193
Tothill St. SW1 2D 228 (47Mb 90)
Totnes Rd. DA16: Well 52Xc 117
Totnes Vs. N11 22Lb 50
(off Telford Rd.)
Totnes Wlk. N2 28Fb 49
Tottan Ter. E1 44Zb 92
Tottenhall NW1 38Jb 70
(off Ferdinand St.)
Tottenhall Rd. N13 23Qb 50
TOTTENHAM 26Vb 51
Tottenham Community Sports Cen. 25Vb 51
Tottenham Ct. Rd. W1 5C 216 (42Lb 90)
Tottenham Court Road Station
(Underground & Crossrail) 2E 222 (44Mb 90)
Tottenham Ent. Cen. N17 26Vb 51
Tottenham Grn. E. N15 28Vb 51
(not continuous)
Tottenham Green Pools & Fitness 28Vb 51
TOTTENHAM HALE 26Wb 51
Tottenham Hale Bus Station 27Wb 51
TOTTENHAM HALE GYRATORY 28Wb 51
Tottenham Hale Retail Pk. 28Wb 51
Tottenham Hale Station
(Rail & Underground) 27Xb 51
Tottenham Hotspur FC 24Vb 51
Tottenham La. N8 30Nb 50
Tottenham M. W1 7C 216 (43Lb 90)
Tottenham Rd. N1 37Ub 71
Tottenham St. W1 1C 222 (43Lb 90)
Totterdown St. SW17 63Hb 133
TOTTERIDGE 18Bb 31
Totteridge & Whetstone Station
(Underground) 19Eb 31
Totteridge Comn. N20 19Wa 30
Totteridge Fields Local Nature
Reserve 18Wa 30
Totteridge Grn. N20 19Cb 31
Totteridge Ho. SW11 54Fb 111
(off Yelverton Rd.)
Totteridge La. N20 19Cb 31
Totteridge Rd. EN3: Enf W 9Zb 20
Totteridge Village N20 18Ab 30
Totternhoe Cl. HA3: Kenton 29La 46
Totters Ct. SE17 51Tb 113
(off Westmoreland Rd.)
Totton Rd. CR7: Thor H 69Qb 134
Toucan Cl. NW10 41Qa 87
Touchard Ho. N1 4G 219 (41Ub 91)
(off Chart St.)
Toulmin St. SE1 2D 230 (47Sb 91)
Toulon St. SE5 52Sb 113
Toulouse Ct. SE16 50Xb 91
(off Rossetti Rd.)
Tounson Ct. SW1 6E 228 (49Mb 90)
(off Montaigne Cl.)
Tourers Ho. E17 29Bc 52
(off Track St.)
Tournay Rd. SW6 52Bb 111
Tours Pas. SW11 56Eb 111
Tourtel Yd. DA11: Nflt 60Ee 121
Toussaint Wlk. SE16 48Wb 91
Tovey Cl. AL2: Lon C 8H 7
Tovil Cl. SE20 68Xb 135
Tovy Ho. SE1 50Wb 91
(off Avondale Sq.)
Towcester Rd. E3 42Dc 92
Tower, The SE1 7K 225 (46Vb 91)
(off Potters Flds.)
Tower, The SW8 51Nb 112
Tower, The TW8: Bford 50Ma 87
(off Ealing Rd.)
Tower Bridge 7K 225 (46Vb 91)
Tower Bri. App. E1 6K 225 (46Vb 91)
Tower Bri. Bus. Complex SE16 48Wb 91
Tower Bri. Bus. Sq. SE16 49Xb 91
Tower Bridge Exhibition 7K 225 (46Vb 91)
(off Tower Bri.)
Tower Bri. M. HA1: Harr 35Ha 66
Tower Bri. Plaza SE1 7K 225 (46Vb 91)
Tower Bri. Rd. SE1 4H 231 (48Ub 91)
Tower Bri. Sq. SE1 1K 231 (47Vb 91)
(off Queen Elizabeth St.)
Tower Bri. Wharf E1 46Wb 91
Tower Bldgs. E1 46Xb 91
(off Brewhouse La.)
Tower Cl. BR6: Orp 75Vc 161
Tower Cl. DA12: Grav'nd 4G 144
Tower Cl. GU21: Wok 9P 167
Tower Cl. IG6: Ilf 23Rc 54
Tower Cl. NW3 36Fb 69
Tower Cl. SE20 66Xb 135
Tower Ct. CM14: B'wood 19Yd 40
Tower Ct. DA17: Belv 49Ed 96
Tower Ct. N1 38Sb 71
(off Canonbury St.)
Tower Ct. NW8 1E 214 (40Gb 69)
(off Mackennal St.)
Tower Ct. TW20: Egh 64C 126
Tower Ct. WC2 3F 223 (44Nb 90)
(off Tower St.)
Tower Cft. DA4: Eyns 75Nd 163
Tower Gdns. KT10: Clay 80Ja 152
Tower Gdns. Rd. N17 25Sb 51
Towergate SE1 5H 231 (49Ub 91)
(off Page's Wlk.)
Towergate Cl. UB8: Uxb 36N 63
Towergate Ho. E3 40Bc 72
(off Ordell Rd.)
Tower Gateway Station
(DLR) 4K 225 (45Vb 91)
Tower Gro. KT13: Weyb 75U 150
TOWER HAMLETS CENTRE FOR
MENTAL HEALTH 42Zb 92
Tower Hamlets Rd. E17 27Cc 52
Tower Hamlets Rd. E7 35Hc 73
TOWER HILL 5K 225 (45Vb 91)
TOWER HILL 2H 11
Tower Hill CM14: B'wood 19Yd 40
Tower Hill EC3 5K 225 (45Vb 91)
Tower Hill WD4: Chfd 1G 10
Tower Hill Ct. DA17: Erith 49Ed 96
(off Tower Ct.)
Tower Hill Memorial 5K 225 (45Vb 91)
Tower Hill Station
(Underground) 4K 225 (45Vb 91)
Tower Hill Ter. EC3 5J 225 (45Ub 91)
(off Tower Hill)
Tower Ho. E1 43Wb 91
(off Fieldgate St.)
Tower Ho. KT14: Byfl 84N 169
(off High Rd.)
Tower Ho. SE13 55Ec 114
Tower Ho. SL0: Iver 44G 82
Tower Ho. SL1: Slou 7J 81
Tower Ho. UB8: Uxb 38L 63
(off High St.)
Tower Ind. Est. TN15: Wro 86Zd 185
Tower La. HA9: Wemb 34Ma 67
Tower La. RH2: Reig 100Gb 195
Tower Mans. SE1 4K 231 (48Vb 91)
(off Grange Rd.)
Tower M. E17 28Cc 52
Tower M. E5 35Ac 72
Tower Mill Rd. SE15 52Ub 113
Tower of London, The 5K 225 (45Vb 91)
Tower of London Jewel
House 5K 225 (45Vb 91)
Tower of London Welcome
Cen. 5J 225 (45Ub 91)
Tower Pk. Rd. DA1: Cray 57Hd 118
Tower Pier EC3 5J 225 (45Ub 91)
Tower Pl. CR6: W'ham 86Zc 178
Tower Pl. EC3 5J 225 (45Ub 91)
Tower Pl. E. EC3 5J 225 (45Ub 91)
(off Lwr. Thames St.)
Tower Pl. W. EC3 5J 225 (45Ub 91)
(off Lwr. Thames St.)
Tower Point EN2: Enf 14Tb 33
Tower Retail Pk. 57Hd 118
Tower Ride SL4: Wind 2C 124
Tower Ri. TW9: Rich 55Na 109
Tower Rd. BR6: Orp 75Vc 161
Tower Rd. CM16: Epp 2Uc 22
Tower Rd. DA1: Dart 59Ld 119
Tower Rd. DA17: Belv 49Ed 96
Tower Rd. DA7: Bex 56Cd 118
Tower Rd. KT20: Tad 95Ya 194
Tower Rd. NW10 38Wa 68
Tower Rd. TW1: Twick 62Ha 130
Towers, The CR8: Kenley 87Sb 177
Towers Av. UB10: Hil 41S 84
Towers Bus. Pk. HA9: Wemb 35Sa 67
(off Carey Way)
Towers Ct. UB10: Hil 41S 84
Towers Pl. TW9: Rich 57Na 109
Towers Rd. HA5: Pinn 25Aa 45
Towers Rd. HP2: Hem H 1N 3
Towers Rd. RM17: Grays 50Ee 99
Towers Rd. UB1: S'hall 42Ca 85
Tower St. WC2 3F 223 (44Nb 90)
Towers Wlk. KT13: Weyb 79R 150
Towers Wood DA4: S Dar 67Td 142
Tower Ter. N22 26Pb 50
Tower Vw. CR0: C'don 73Ac 158
Tower Vw. WD23: B Hea 17Ga 28
Tower Wlk. SE1 5H 231 (49Ub 91)
(off Leroy St.)
Tower Wharf DA11: Nflt 56De 121
Tower Wharf SE1 1K 231 (47Vb 91)
(off Tooley St.)
Tower Workshops SE1 2K 231 (47Vb 91)
(off Riley Rd.)
Towfield Ct. TW13: Hanw 61Ba 129
Towfield Rd. TW13: Hanw 61Ba 129
Town, The EN2: Enf 13Tb 33
Towncourt Cres. BR5: Pet W 71Sc 160
Towncourt La. BR5: Pet W 72Tc 160
Towneley Cotts. RM4: Stap A 18Gd 38
Town End CR3: Cat'm 94Ub 197
Town End Cl. CR3: Cat'm 94Ub 197
Townend Ct. BR1: Broml 67Hc 137
Town End Pde. KT1: King T 69Ma 131
(off High St.)
Towney Mead UB5: N'olt 40Ba 65
Towney Mead Ct. UB5: N'olt 40Ba 65
Town Farm Way TW19: Stanw 59M 105
Townfield WD3: Rick 17L 25
Townfield Cnr. DA12: Grav'nd 10E 122
Townfield Rd. UB3: Hayes 46V 84
Townfield Sq. UB3: Hayes 46V 84
Town Fld. Way TW7: Isle 54Ja 108
Towngate KT11: Cobh 87Aa 171
Town Hall App. N16 35Tb 71
(off Albion Rd.)
Town Hall App. Rd. N15 28Vb 51
Town Hall Av. W4 50Ta 87
Town Hall Rd. SW11 55Hb 111
Townhall Sq. DA1: Cray 57Hd 118
Townholm Cres. W7 48Ha 86
Townhouse M. W12 47Xa 88
Town La. TW19: Stanw 58M 105
(not continuous)
Townley Ct. E15 37Hc 73
Townley Rd. DA6: Bex 57Bd 117
Townley Rd. SE22 57Ub 113
Townley St. SE17 7F 231 (50Tb 91)
(not continuous)
Town Mead RH1: Blet 5K 209
Townmead Bus. Cen. SW6 55Eb 111
Town Mdw. TW8: Bford 51Ma 109
Town Mdw. Rd. TW8: Bford 52Ma 109
Townmead Rd. EN9: Walt A 6Ec 20
Townmead Rd. SW6 55Db 111
Townmead Rd. TW9: Kew 54Ra 109
Town Pier DA11: Grav'nd 7D 122
Town Quay IG11: Bark 39Rc 74
Town Quay TW18: Lale 69K 127
(off Blacksmiths La.)
Town Quay Wharf IG11: Bark 39Rc 74
Town Rd. N9 19Xb 33
TOWNSEND 1B 6
Townsend Av. AL1: St A 1C 6
Townsend Av. N14 21Mb 50
Townsend Ga. HP4: Berk 2A 2
Townsend Ho. SE1 49Wb 91
(off Strathnairn St.)
Townsend Ind. Est. NW10 40Sa 67
Townsend La. GU22: Wok 93D 188
Townsend La. NW9 31Ta 67
Townsend M. SW18 61Eb 133
Townsend Rd. N15 29Vb 51
Townsend Rd. SE3 56Lc 115
Townsend Rd. TW15: Ashf 64N 127
Townsend Rd. UB1: S'hall 46Aa 85
Townsend St. SE17 5H 231 (49Ub 91)
Townsend Way HA6: Nwood 24V 44
Townsend Yd. N6 32Kb 70
Townshend Cl. DA14: Sidc 65Xc 139
Townshend Ct. NW8 1E 214 (40Gb 69)
(off Townshend Rd.)
Townshend Est. NW8 1D 214 (40Gb 69)
Townshend Rd. BR7: Chst 64Rc 138
Townshend Rd. NW8 1D 214 (39Gb 69)
(not continuous)
Townshend Rd. TW9: Rich 56Pa 109
Townshend Ter. TW9: Rich 56Pa 109
Townshott Cl. KT23: Bookh 97Ca 191
Towns Ho. SW4 55Mb 112
Townslow La. GU23: Wis 88L 169
Townson Av. UB5: N'olt 40W 64
Townson Way UB5: N'olt 40W 64
Town Sq. DA8: Erith 51Gd 118

Town Sq. IG11: Bark 39Sc 70
(off Clockhouse Av.)
Town Sq. SL1: Slou 7K 81
Town Tree Rd. TW15: Ashf 64Q 128
Town Wharf TW7: Isle 55Ka 108
Towpath, The SW10 53Fb 111
Towpath KT12: Walt T 71W 150
Towpath TW17: Shep 74P 149
Towpath Rd. N18 23Zb 52
Towpath Wlk. E9 36Bc 72
Towpath Way CR0: C'don 72Vb 157
Towton M. N11 23Mb 50
Towton Rd. SE27 61Sb 135
Toye Av. N20 19Fb 31
Toynbec Cl. BR7: Chst 63Rc 138
Toynbee Rd. SW20 67Ab 132
Toynbee St. E1 1K 225 (43Vb 91)
Toynbee Studios 2K 225 (44Vb 91)
(off Commercial St.)
Toyne Way N6 30Hb 49
Tozer Wlk. SL4: Wind 5B 102
Tracery, The SM7: Bans 87Db 175
Tracey Av. NW2 36Ya 68
Tracey Bellamy Ct. E14 44Ac 92
(off Repton St.)
Tracious Cl. GU21: Wok 8M 167
Tracious La. GU21: Wok 8M 167
Track St. E17 29Bc 52
Tracy Av. SL3: L'ly 50B 82
Tracy Ct. HA7: Stan 24La 46
Tracy Ho. E3 41Bc 92
(off Mile End Rd.)
Trade City KT13: Weyb 82N 169
Trade City Bus. Pk. TW16: Sun 67U 128
Trade City Bus. Pk. UB8: Uxb 40L 63
Trade City Watford WD18: Wat 16X 27
Trade Cl. N13 21Qb 50
Trader Rd. E6 44Rc 94
Tradescant Ho. E9 38Yb 72
(off Frampton Pk. Rd.)
Tradescant Rd. SW8 52Nb 112
Tradewind Hgts. SE16 46Zb 92
(off Rotherhithe St.)
Tradewinds Ct. E1 45Wb 91
Trading Est. Rd. NW10 42Sa 87
Trafalgar Av. KT4: Wor Pk 74Za 154
Trafalgar Av. N17 23Ub 51
Trafalgar Av. SE15 7K 231 (50Vb 91)
Trafalgar Bldg. KT2: King T 67Ma 131
(off Henry Macaulay Av.)
Trafalgar Bus. Cen. IG11: Bark 42Vc 95
Trafalgar Chambers SW3 7C 226 (50Fb 89)
(off South Pde.)
Trafalgar Cl. SE16 48Ac 92
Trafalgar Ct. DA8: Erith 52Hd 118
(off Frobisher Rd.)
Trafalgar Ct. E1 46Yb 92
(off Wapping Wall)
Trafalgar Ct. KT11: Cobh 85W 170
Trafalgar Dr. KT12: Walt T 76X 151
Trafalgar Gdns. E1 43Zb 92
Trafalgar Gdns. W8 48Db 89
Trafalgar Gro. SE10 51Fc 115
Trafalgar Ho. SE17 7E 230 (50Sb 91)
(off Bronti Cl.)
Trafalgar Ho. SW18 56Eb 111
Trafalgar M. E9 37Bc 72
Trafalgar Pl. E11 28Jc 53
Trafalgar Pl. N18 22Wb 51
Trafalgar Point N1 38Tb 71
(off Downham Rd.)
Trafalgar Quarters SE10 51Fc 115
(off Park Row)
Trafalgar Rd. DA1: Dart 61Nd 141
Trafalgar Rd. DA11: Grav'nd 9C 122
Trafalgar Rd. RM13: Rain 40Hd 76
Trafalgar Rd. SE10 51Fc 115
Trafalgar Rd. SW19 66Db 133
Trafalgar Rd. TW2: Twick 61Fa 130
Trafalgar Sq. SW1 6E 222 (46Mb 90)
Trafalgar Sq. WC2 6E 222 (46Mb 90)
Trafalgar Square 6E 222 (46Mb 90)
Trafalgar St. SE17 7F 231 (50Tb 91)
Trafalgar Studios 7F 223 (46Nb 90)
(off Whitehall)
Trafalgar Ter. HA1: Harr 32Ga 66
Trafalgar Trad. Est. EN3: Brim 14Ac 34
Trafalgar Way CR0: Wadd 75Qb 156
Trafalgar Way E14 46Ec 92
Trafford Cl. IG6: Ilf 23Vc 55
Trafford Cl. WD7: Shenl 4Na 15
Trafford Ho. N1 1G 219 (40Tb 71)
(off Cranston Est.)
Trafford Rd. CR7: Thor H 71Pb 156
Trafford Way BR3: Beck 65Cc 136
Traherne Lodge TW11: Tedd 64Ha 130
Trahorn Cl. E1 42Xb 91
Trail St. E17 29Bc 52
Traitors' Gate 6K 225 (46Vb 91)
(in The Tower of London)
Tralee Ct. SE16 50Xb 91
(off Masters Dr.)
Tram Cl. SE24 55Rb 113
(off Milkwood Rd.)
Tramlink, The SW19 68Eb 133
Tramsheds, The CR0: Bedd 73Mb 156
Tramway Av. E15 38Gc 73
Tramway Av. N9 17Xb 33
Tramway Cl. SE20 67Yb 136
Tramway Ct. E1 43Ac 92
Tramway Path CR4: Mitc 70Gb 133
(not continuous)
Tranby M. E9 36Zb 72
(off Brooksby's Wlk.)
Tranley M. NW3 35Gb 69
(off Fleet Rd.)
Tranmere Ct. SM2: Sutt 80Eb 155
Tranmere Rd. N9 17Vb 33
Tranmere Rd. SW18 61Eb 133
Tranmere Rd. TW2: Whitt 59Da 107
Tranquil Dale RH3: Bkld 4B 206
Tranquil La. HA2: Harr 32Da 65
Tranquil Pas. SE3 54Hc 115
(off Montpelier Va.)
Tranquil Ri. DA8: Erith 50Gd 96
Tranquil Va. SE3 54Gc 115
Transenna Works N1 37Rb 71
(off Laycock St.)
Transept St. NW1 1E 220 (43Gb 89)
Transmere Cl. BR5: Pet W 72Sc 160
Transmere Rd. BR5: Pet W 72Sc 160
Transom Cl. SE16 49Ac 92
Transom Sq. E14 50Dc 92
Transport Av. TW8: Bford 50Ja 86
Tranton Rd. SE16 48Wb 91
Trappes Ho. SE16 49Xb 91
(off Camilla Rd.)
Trap's Hill IG10: Lough 13Pc 36
Traps La. KT3: N Mald 67Ua 132
Traq Motor Racing 72Lb 156
Travellers Cl. AL9: Wel G 5E 8
Travellers La. AL10: Hat 7C 8
Travellers La. AL9: Wel G 4E 8
Travellers Path E6 39Mc 73
Travellers Way TW4: Cran 54Y 107
Travers Cl. E17 25Zb 52
Travers Ho. SE10 51Fc 115
(off Trafalgar Gro.)
Travers Rd. N7 34Qb 70
Travic Rd. SL2: Slou 1D 80
Travis Ct. SL2: Farn R 1F 80
Trayford Ct. RM19: Purf 50Rd 97
(off Wingrove Dr.)
Trays Hill Cl. N19 31Lb 70
TREACLE MINE RDBT. 47Be 99
Treacy Cl. WD23: B Hea 19Ea 28
Treadgold Ho. W11 45Za 88
(off Bomore Rd.)
Treadgold St. W11 45Za 88
Treadway St. E2 40Xb 71
Treadwell Rd. KT18: Eps 88Ua 174
Treasury Cl. SM6: W'gton 78Mb 156
Treasury M. DA5: Bexl 59Dd 118
Treasury Pas. SW1 1F 229 (47Nb 90)
(off Downing St.)
Treaty Cen. 55Da 107
Treaty St. N1 1H 217 (39Pb 70)
Trebble Rd. DA10: Swans 58Ae 121
Trebeck St. W1 6K 221 (46Kb 90)
Trebellan Dr. HP2: Hem H 1P 3
Trebovir Rd. SW5 50Cb 89
Treby St. E3 42Bc 92
Trecastle Way N7 35Mb 70
Tredegar Ho. E3 41Cc 92
(off Bow Rd.)
Tredegar M. E3 41Bc 92
Tredegar Rd. DA2: Wilm 61Jd 140
Tredegar Rd. E3 40Bc 72
Tredegar Rd. N11 24Mb 50
Tredegar Sq. E3 41Bc 92
Tredegar Ter. E3 41Bc 92
Trederwen Rd. E8 39Wb 71
Tredown Rd. SE26 64Yb 136
Tredwell Cl. BR2: Broml 70Nc 138
Tredwell Cl. SW2 61Pb 134
Tredwell Rd. SE27 63Rb 135
Treebank Gdns. W7 45Ga 86
Treebourne Rd. TN16: Big H 89Lc 179
Treebys Av. GU4: Jac W 10P 187
Tree Cl. TW10: Ham 60Ma 109
Tree La. TN15: Plax 99Ae 205
Treemount Ct. KT17: Eps 85Ua 174
Treen Av. SW13 55Va 110
Tree Rd. E16 44Lc 93
Treeside Cl. UB7: W Dray 49M 83
Treeside Pl. N10 28Kb 50
Treetop Ct. CR7: Thor H 71Sb 157
Tree Top M. RM10: Dag 37Fd 76
Treetop M. NW6 38Ab 68
Tree Tops CM15: B'wood 18Yd 40
Tree Tops SL9: Chal P 22A 42
Treetops CR3: W'ham 90Wb 177
Treetops DA12: Grav'nd 4D 144
Treetops RH9: S Gods 9C 210
Treetops TN15: Kems'g 89Qd 183
Treetops Cl. HA6: Nwood 22T 44
Treetops Cl. RM17: Grays 49De 99
Treetops Cl. SE2 50Ad 95
Treetops Vw. IG10: Lough 16Mc 35
Tree Vw. Cl. SE19 67Ub 135
Treeview Ct. RH2: Reig 6M 207
(off Wray Comn. Rd.)
Treewall Gdns. BR1: Broml 63Kc 137
Tree Way RH2: Reig 3K 207
Trefgarne Rd. RM10: Dag 33Cd 76
Trefil Wlk. N7 35Nb 70
Trefoil Ho. DA18: Erith 47Ad 95
(off Kale Rd.)
Trefoil Ho. RM17: Grays 52De 121
Trefoil Ho. SE10 49Gc 93
Trefoil Rd. SW18 57Eb 111
Trefusis Ct. TW5: Cran 53X 107
Trefusis Wlk. WD17: Wat 11U 26
Tregaron Av. N8 30Nb 50
Tregaron Gdns. KT3: N Mald 70Ua 132
Tregarthen Pl. KT22: Lea 93La 192
Tregarth Pl. GU21: Wok 9K 167
Tregarvon Rd. SW11 56Jb 112
Tregenna Av. HA2: Harr 35Ca 65
Tregenna Cl. N14 15Lb 32
Tregenna Ct. HA2: Harr 35Ca 65
Treglos Ct. KT13: Weyb 74U 150
Tregonwell Ter. SE9 63Rc 138
Tregony Rd. BR6: Chels 77Vc 161
Trego Rd. E9 38Cc 72
Tregothnan Rd. SW9 55Nb 112
Tregunter Rd. SW10 51Db 111
Trehaven Pde. RH2: Reig 9K 207
Trehearn Rd. IG6: Ilf 24Tc 54
Treherne Ct. SW17 63Jb 134
Trehern Rd. SW14 55Ta 109
Trehurst St. E5 36Ac 72
Trelawn Cl. KT16: Ott 80E 148
Trelawney Av. SL3: L'ly 8P 81
Trelawney Est. E9 37Yb 72
Trelawney Gro. KT13: Weyb 79Q 150
Trelawney Ho. SE1 1D 230 (47Sb 91)
(off Union St.)
Trelawney Pl. RM16: Chaf H 48Yd 98
Trelawney Rd. IG6: Ilf 24Tc 54
Trelawn Rd. E10 34Ec 72
Trelawn Rd. SW2 57Qb 112
Trelawny Cl. E17 28Dc 52
Trellick Twr. W10 42Bb 89
(off Golborne Rd.)
Trellis Ho. SW19 66Eb 133
Trellis Sq. E3 41Bc 92
Treloar Gdns. SE19 65Tb 135
Tremadoc Rd. SW4 56Mb 112
Tremaine Cl. SE4 54Cc 114
Tremaine Rd. SE20 68Xb 135
Trematon Bldg. N1 1G 217 (40Nb 70)
(off Trematon Wlk.)
Trematon Ho. SE11 7A 230 (50Qb 90)
(off Kennings Way)
Trematon M. N1 1G 217 (40Nb 70)
Trematon Pl. TW11: Tedd 66La 130
Trematon Wlk. N1 1G 217 (40Nb 70)
(off Trematon M.)
Tremelo Grn. RM8: Dag 32Ad 75
Tremlett Gro. N19 34Lb 70
Tremlett M. N19 34Lb 70
Trenance GU21: Wok 9L 167
Trenance Gdns. IG3: Ilf 34Wc 75
Trenchard Av. HA4: Ruis 35X 65
Trenchard Cl. HA7: Stan 23Ja 46
Trenchard Cl. KT12: Hers 78Y 151
Trenchard Cl. NW9 25Ua 48
Trenchard Ct. NW4 29Wa 48
Trenchard Ct. SM4: Mord 72Cb 155
Trenchard St. SE10 50Fc 93
Trenches La. SL3: L'ly 45C 82
Trenchold St. SW8 51Nb 112
Trendell Ho. E14 44Cc 92
(off Dod St.)
Trenear Cl. BR6: Chels 77Wc 161
Trenham Dr. CR6: W'ham 88Yb 178
Trenholme Cl. SE20 66Xb 135
Trenholme Ct. CR3: Cat'm 95Wb 197
Trenholme Rd. SE20 66Xb 135
Trenholme Ter. SE20 66Xb 135
Trenmar Gdns. NW10 41Xa 88
Trent RM18: E Til 9L 101
Trent Av. RM14: Upm 30Td 58
Trent Av. W5 48La 86
Trentbridge Cl. IG6: Ilf 23Vc 55
Trent Cl. WD7: Shenl 4Na 15
Trent Ct. CR2: S Croy 78Sb 157
(off Nottingham Rd.)
Trent Ct. E11 29Jc 53
Trent Gdns. N14 16Kb 32
Trentham Cres. GU22: Wok 93C 188
Trentham Dr. BR5: St M Cry 70Wc 139
Trentham Ho. W3 43Ta 87
Trentham Rd. RH1: Redh 8P 207
Trentham St. SW18 60Cb 111
Trent Ho. KT2: King T 67Ma 131
Trent Ho. SE15 56Yb 114
Trent Pk. EN4: Cockf 12Lb 32
Trent Pk. 11Kb 32
TRENT PARK 12Kb 32
Trent Pk. Golf Course 14Lb 32
Trent Rd. IG9: Buck H 18Kc 35
Trent Rd. SL3: L'ly 51D 104
Trent Rd. SW2 57Pb 112
Trent Vs. SL3: Dat 3M 103
(off Datchet Pl.)
Trent Way KT4: Wor Pk 76Ya 154
Trent Way UB4: Hayes 40U 64
Trentwood Side EN2: Enf 13Pb 32
Treport St. SW18 59Db 111
Tresco Cl. BR1: Broml 65Gc 137
Trescoe Gdns. HA2: Harr 31Aa 65
Trescoe Gdns. RM5: Col R 22Ed 56
Tresco Gdns. IG3: Ilf 33Wc 75
Tresco Ho. SE11 7K 229 (50Qb 90)
(off Sancroft St.)
Tresco Rd. SE15 56Xb 113
Tresham Cres. NW8 5D 214 (42Gb 89)
Tresham Ho. WC1 7H 217 (43Pb 90)
(off Red Lion Sq.)
Tresham Rd. IG11: Bark 38Vc 75
Tresham Wlk. E9 36Yb 72
Tresidder Ho. SW4 59Mb 112
Tresilian Av. N21 15Pb 32
Tresillian Way GU21: Wok 8L 167
Tressell Cl. N1 38Rb 71
Tressillian Cres. SE4 55Cc 114
Tressillian Rd. SE4 56Bc 114
Tress Pl. SE1 6B 224 (46Rb 91)
(off Blackfriars Rd.)
Tresta Wlk. GU21: Wok 8L 167
Trestis Cl. UB4: Yead 43Z 85
Trestle Theatre 3G 6
Treswell Rd. RM9: Dag 39Ad 75
Tretawn Gdns. NW7 21Ua 48
Tretawn Pk. NW7 21Ua 48
Trevanion Rd. W14 49Ab 88
Treve Av. HA1: Harr 31Fa 66
Trevellance Way WD25: Wat 5Z 13
Trevelyan Av. E12 35Pc 74
Trevelyan Cl. DA1: Dart 56Pd 119
Trevelyan Ct. KT3: N Mald 73Ua 154
Trevelyan Ct. SL4: Wind 4F 102
Trevelyan Cres. HA3: Kenton 31Ma 67
Trevelyan Gdns. IG10: Lough 11Pc 36
Trevelyan Gdns. NW10 39Ya 68
Trevelyan Ho. E2 41Zb 92
(off Morpeth St.)
Trevelyan Ho. SE5 52Rb 113
(off John Ruskin St.)
Trevelyan Pl. AL1: St A 4A 6
Trevelyan Rd. E15 35Hc 73
Trevelyan Rd. SW17 64Gb 133
Trevenna Ho. SE23 62Zb 136
(off Dacres Rd.)
Trevera Ct. EN3: Pond E 14Ac 34
Trevera Ct. EN8: Walt C 5Ac 20
(off Eleanor Rd.)
Treveris St. SE1 7C 224 (46Rb 91)
Treversh Ct. BR1: Broml 67Gc 137
Treverton St. W10 42Za 88
Treverton Towers W10 43Za 88
(off Treverton St.)
Treves Cl. N21 15Pb 32
Treves Ho. E1 42Wb 91
(off Vallance Rd.)
Treville St. SW15 59Xa 110
Treviso Rd. SE23 61Zb 136
Trevithick Cl. TW14: Felt 60V 106
Trevithick Dr. DA1: Dart 56Pd 119
Trevithick Ho. SE16 49Xb 91
(off Rennie Est.)
Trevithick St. SE8 51Cc 114
Trevithick Way E3 41Cc 92
Trevone Gdns. HA5: Pinn 30Aa 45
Trevor Cl. BR2: Hayes 73Hc 159
Trevor Cl. EN4: E Barn 16Fb 31
Trevor Cl. HA3: Hrw W 24Ha 46
Trevor Cl. TW7: Isle 57Ha 108
Trevor Cl. UB5: N'olt 40Y 65
Trevor Cres. HA4: Ruis 35V 64
Trevor Gdns. HA4: Ruis 35W 64
Trevor Gdns. HA8: Edg 25Ta 47
Trevor Gdns. UB5: N'olt 40Y 65
Trevor Pl. SW7 2E 226 (47Gb 89)
Trevor Rd. HA8: Edg 25Ta 47
Trevor Rd. IG8: Wfd G 24Jc 53
Trevor Rd. SW19 66Ab 132
Trevor Rd. UB3: Hayes 47U 84
Trevor Roper Cl. IG1: Ilf 33Pc 74
Trevor Sq. SW7 2F 227 (47Hb 89)
Trevor St. SW7 2E 226 (47Gb 89)
Trevor Wlk. SW7 2F 227 (47Hb 89)
(off Lancelot Pl.)
Trevose Av. KT14: W Byf 86H 169
Trevose Ho. SE11 7J 229 (50Pb 90)
(off Orsett St.)
Trevose Ho. SL2: Slou 2F 80
Trevose Rd. E17 25Fc 53
Trevose Way WD19: Wat 20Y 27
Trewarden Av. SL0: Iver H 40F 62
Trewarden Row SL0: Iver H 40F 62
Trewenna Dr. EN6: Pot B 4Fb 17
Trewenna Dr. KT9: Chess 78Ma 153
Trewince Rd. SW20 67Ya 132
Trewint St. SW18 61Eb 133
Trewsbury Ho. SE2 46Zc 95
Trewsbury Rd. SE26 64Zb 136
Tria Apts. E2 41Wb 91
(off Durant St.)
Triandra Way UB4: Yead 43Z 85
Triangle, The DA15: Sidc 59Wc 117
(off Burnt Oak La.)
Triangle, The E8 39Xb 71
Triangle, The EC1 5C 218 (42Rb 91)
(off Goswell Rd.)
Triangle, The GU21: Wok 10N 167
Triangle, The IG11: Bark 37Sc 74
Triangle, The KT1: King T 68Ra 131
Triangle, The N13 21Pb 50
Triangle Bus. Cen., The NW10 41Va 88
Triangle Cen. UB1: S'hall 46Fa 86
Triangle Ct. E16 43Mc 93
Triangle Ct. SE1 7E 224 (46Sb 91)
(off Redcross Way)
Triangle Est. SE11 7K 229 (50Qb 90)
(off Kennington Rd.)
Triangle Ho. SE1 1K 231 (47Vb 91)
(off Three Oak La.)
Triangle Pas. EN4: E Barn 14Eb 31
Triangle Pl. SW4 56Mb 112
Triangle Rd. E8 39Xb 71
Triangle Way W3 48Qa 87
Triangle Works N9 19Zb 34
Tribeca Apts. E1 43Vb 91
(off Heneage St.)
Trickett Ho. SM2: Sutt 81Db 175
Trico Ho. TW8: Bford 50Ma 87
(off Ealing Rd.)
Tricorn Ho. SE28 46Uc 94
Trident Bus. Cen. SW17 64Hb 133
Trident Cen., The WD24: Wat 11Y 27
Trident Gdns. UB5: N'olt 41Z 85
Trident Ho. E14 44Ec 92
(off Blair St.)
Trident Ho. SE28 46Tc 94
Trident Ho. TW19: Stanw 59N 105
(off Clare Rd.)
Trident Ind. Est. SL3: Poyle 55G 104
Trident Pl. SW3 51Fb 111
(off Old Church St.)
Trident Point HA1: Harr 30Fa 46
Trident Rd. WD25: Wat 6V 12
Trident St. SE16 49Zb 92
Trident Way UB2: S'hall 48X 85
Trigg's Cl. GU22: Wok 1P 187
Trigg's La. GU21: Wok 1N 187
Trigg's La. GU22: Wok 1N 187
Trig La. EC4 4D 224 (45Sb 91)
Trigo Ct. KT19: Eps 83Ta 173
Trigon Rd. SW8 52Pb 112
Tri Ho. KT12: Hers 77W 150
Trilby Rd. SE23 61Zb 136
Trillo Ct. IG2: Ilf 31Uc 74
Trimdon NW1 1C 216 (39Lb 70)
Trimmer Wlk. TW8: Bford 51Na 109
Trim St. SE14 51Bc 114
Trinder Gdns. N19 32Nb 70
Trinder M. SW12 60Jb 112
Trinder M. TW11: Tedd 64Ja 130
Trinder Rd. EN5: Barn 15Ya 30
Trinder Rd. N19 32Nb 70
Trindles Rd. RH1: S Nut 8F 208
Tring Av. HA9: Wemb 37Qa 67
Tring Av. UB1: S'hall 44Ba 85
Tring Av. W5 46Pa 87
Tring Cl. IG2: Ilf 29Tc 54
Tring Cl. RM3: Rom 21Nd 57
Tring Ct. TW1: Twick 63Ja 130
Tring Gdns. RM3: Rom 21Nd 57
Tring Grn. RM3: Rom 21Nd 57
Tringham Cl. GU21: Knap 10G 166
Tringham Cl. KT16: Ott 78E 148
Tringham Cotts. GU24: W End 4D 166
Tring Ho. WD18: Wat 17U 26
Tring Wlk. RM3: Rom 21Nd 57
Trinidad Gdns. RM10: Dag 38Fd 76
Trinidad Ho. E14 45Bc 92
(off Gill St.)
Trinidad St. E14 45Bc 92
Trinity Av. EN1: Enf 16Vb 33
Trinity Av. N2 27Fb 49
Trinity Buoy Wharf E14 45Gc 93
Trinity Chu. Pas. SW13 51Xa 110
Trinity Chu. Rd. SW13 51Xa 110
Trinity Chu. Sq. SE1 3E 230 (48Sb 91)
Trinity Cl. BR2: Broml 74Nc 160
Trinity Cl. CR2: Sande 81Ub 177
Trinity Cl. E11 33Gc 73
Trinity Cl. E8 37Vb 71
Trinity Cl. HA6: Nwood 23U 44
Trinity Cl. KT20: Tad 91Ya 194
Trinity Cl. NW3 35Fb 69
Trinity Cl. SE13 56Fc 115
Trinity Cl. SW4 56Lb 112
Trinity Cl. TW19: Stanw 58L 105
Trinity Cl. TW4: Houn 56Aa 107
Trinity Cotts. TW9: Rich 55Pa 109
Trinity Ct. DA1: Dart Baker Cres. 59Md 119
Trinity Ct. AL1: St A 1C 6
(off Newsom Pl.)
Trinity Ct. BR1: Broml 67Hc 137
(off Highland Rd.)
Trinity Ct. CR0: C'don 75Sb 157
Trinity Ct. DA1: Dart 60Rd 119
(off Churchill Cl.)
Trinity Ct. EN2: Enf 12Sb 33
Trinity Ct. N1 39Ub 71
(off Downham Rd.)
Trinity Ct. N18 23Vb 51
Trinity Ct. NW2 36Ya 68
Trinity Ct. SE25 72Ub 157
Trinity Ct. SE26 62Yb 136
Trinity Ct. SE7 49Mc 93
Trinity Ct. SE8 50Ac 92
(off Evelyn St.)
Trinity Ct. SL4: Wind 4G 102
(off Hawtrey Rd.)
Trinity Ct. SW9 55Pb 112
Trinity Ct. W2 2A 220 (44Eb 89)
(off Gloucester Ter.)
Trinity Ct. W9 41Bb 89
(off Croxley Rd.)
Trinity Ct. WC1 5J 217 (42Pb 90)
(off Gray's Inn Rd.)
Trinity Ct. WD17: Wat 11V 26
Trinity Ct. WD3: Rick 19N 25
Trinity Cres. SL5: S'dale 1E 146
Trinity Cres. SW17 61Hb 133
Trinity Dr. UB8: Hil 43S 84
Trinity Gdns. DA1: Dart 58Md 119
Trinity Gdns. E16 43Hc 93
(not continuous)
Trinity Gdns. SW9 56Pb 112
Trinity Grn. E1 42Yb 92
Trinity Gro. SE10 53Ec 114
Trinity Hall Cl. WD24: Wat 13Y 27
TRINITY HOSPICE 56Kb 112
Trinity Hospital SE10 50Fc 93
TRINITY HOSPITAL 50Fc 93
Trinity Ho. EN8: Walt C 4Ac 20
Trinity Ho. RM17: Grays 52De 121
(off Argent St.)
Trinity Ho. RM8: Dag 36Xc 75
Trinity Ho. SE1 3E 230 (48Sb 91)
(off Bath Ter.)
Trinity Ho. W14 48Bb 89
Trinity Ho. WD6: Bore 14Qa 29
Trinity Laban 51Ec 114
(within Old Royal Naval College)
Trinity La. EN8: Walt C 4Ac 20
Trinity M. E1 43Yb 92
(off Redman's Rd.)
Trinity M. HP2: Hem H 3D 4
Trinity M. SE20 67Xb 135
Trinity M. W10 44Za 88
Trinity Pk. E4 23Bc 52
Trinity Path SE26 62Yb 136
(not continuous)
Trinity Pl. DA6: Bex 56Bd 117
Trinity Pl. EC3 4K 225 (45Vb 91)
Trinity Pl. SL4: Wind 4G 102
Trinity Ri. SW2 60Qb 112
Trinity Rd. DA12: Grav'nd 9E 122
Trinity Rd. GU21: Knap 10F 166
Trinity Rd. IG6: Ilf 27Sc 54
Trinity Rd. N2 27Fb 49
Trinity Rd. N22 24Nb 50
(not continuous)
Trinity Rd. SW17 60Gb 111
Trinity Rd. SW18 56Eb 111
Trinity Rd. SW19 65Cb 133
Trinity Rd. TW9: Rich 55Pa 109
Trinity Rd. UB1: S'hall 46Aa 85
Trinity Sq. E14 44Gc 93
Trinity Sq. EC3 4J 225 (45Ub 91)
Trinity St. E16 43Hc 93
Trinity St. EN2: Enf 12Sb 33
Trinity St. SE1 2E 230 (47Sb 91)
(not continuous)
Trinity Ter. IG9: Lough 17Kc 35
Trinity Twr. E1 45Wb 91
(off Vaughan Way)
Trinity Twr. E14 48Dc 92
Trinity Wlk. HP2: Hem H 3D 4
Trinity Wlk. NW3 37Eb 69
Trinity Way E4 23Bc 52
Trinity Way W3 45Ua 88
Trio Pl. SE1 2E 230 (47Sb 91)
Triptych Ho. SE8 52Cc 114
(off Watson's St.)
Triscott Ho. UB3: Hayes 46W 84
Tristan Ct. SE8 51Bc 114
(off Dorking Cl.)
Tristan Lodge WD23: Bush 14Z 27
Tristan Sq. SE3 55Gc 115
Tristram Cl. E17 27Fc 53
Tristram Dr. N9 20Wb 33
Tristram Rd. BR1: Broml 63Hc 137
Triton Ct. E16 43Jc 93
(off Robertson Rd.)
Triton Ho. E14 49Dc 92
(off Cahir St.)
Triton Sq. NW1 5B 216 (42Lb 90)
Triton St. NW1 5A 216 (42Kb 90)
Tritton Av. CR0: Bedd 77Nb 156
Tritton Rd. SE21 62Tb 135
Trittons KT20: Tad 93Za 194
Triumph Cl. RM16: Chaf H 49Yd 98
Triumph Cl. UB3: Harl 53S 106
Triumph Ho. IG11: Bark 41Wc 95
Triumph Ho. TW18: Staines 65K 127
Triumph Rd. E6 44Pc 94
Triumph Trad. Est. N17 23Wb 51
Trivett Cl. DA9: Ghithe 57Wd 120
Trocette Mans. SE1 3H 231 (48Ub 91)
(off Bermondsey St.)
Trojan Ct. NW6 38Ab 68
Trojan Ind. Est. NW10 37Va 68
Trojan M. SW19 66Cb 133
Trojan Way CR0: Wadd 76Pb 156
Troon Cl. SE16 50Xb 91
Troon Cl. SE28 44Zc 95
Troon Ct. SL5: S'hill 1A 146
Troon Ho. E1 44Ac 92
(off White Horse Rd.)
Troon St. E1 44Ac 92
Troopers Dr. RM3: Rom 21Md 57
Tropical Ct. W10 41Za 88
(off Kilburn La.)
Trosley Av. DA11: Grav'nd 1D 144
Trosley Country Pk. Vis. Cen. 84Fe 185
Trosley Rd. DA17: Belv 51Cd 118
Trossachs Rd. SE22 57Ub 113
Troston Ct. TW18: Staines 64H 127
Trothy Rd. SE1 49Wb 91
Trotman Ho. SE14 53Yb 114
(off Pomeroy St.)
Trotsworth Av. GU25: Vir W 70A 126
Trotsworth Ct. GU25: Vir W 10P 125
Trotters Bottom EN5: Barn 9Wa 16
Trotters La. GU24: Chob 4M 167
Trotter Way KT19: Eps 84Qa 173
Trott Rd. N10 24Hb 49
Trotts La. TN16: Westrm 99Sc 200
Trott St. SW11 53Gb 111
Trotwood IG7: Chig 23Tc 54
Trotwood Cl. CM15: Shenf 18Ae 41
Trotwood Ho. SE16 47Xb 91
(off Wilson Gro.)
Troubridge Sq. E17 27Ec 52
Troughton Rd. SE7 50Kc 93
Troutbeck NW1 4A 216 (41Kb 90)
(off Albany St.)
Troutbeck Cl. SL2: Slou 5L 81
Troutbeck Rd. SE14 53Ac 114
Trout La. UB7: Yiew 45L 83
Trout Ri. WD3: Loud 13K 25
Trout Rd. UB7: Yiew 46M 83
Troutstream Way WD3: Loud 14J 25
Trouvere Pk. HP1: Hem H 1K 3
Trouville Rd. SW4 58Lb 112
Trowbridge Est. E9 37Bc 72
(off Osborne Rd.)
Trowbridge Ho. E9 37Bc 72
(off Felstead St.)
Trowbridge Rd. E9 37Bc 72
Trowbridge Rd. RM3: Rom 23Md 57
Trowers Way RH1: Redh 3B 208
Trowers Way Cen. RH1: Redh 3B 208
Trowley Ri. WD5: Ab L 3U 12
Trowlock Av. TW11: Tedd 65La 131
Trowlock Island TW11: Tedd 65Ma 131
Trowlock Way TW11: Tedd 65Ma 131

Troy Cl. KT20: Tad 92Xa 194
Troy Ct. SE18 49Rc 94
Troy Ct. W8 48Cb 89
(off Kensington High St.)
Troy Ind. Est. HA1: Harr 29Ha 46
Troy Rd. SE19 65Tb 135
Troy Town SE15 55Wb 113
Trubshaw Rd. UB2: S'hall 48Da 85
True Lovers Ct. HA6: Nwood 24T 44
Trueman Cl. HA8: Edg 24Ra 47
Trueman Rd. CR8: Kenley 92Tb 197
Truesdale Dr. UB9: Hare 28L 43
Truesdale Rd. E6 44Pc 94
Truesdales UB10: Ick 33S 64
truGym Bromley 67Jc 137
(off East St.)
truGym Uxbridge 38M 63
(off Vine St.)
Trulock Ct. N17 24Wb 51
Trulock Rd. N17 24Wb 51
Trumans Rd. N16 36Vb 71
Truman Wlk. E3 42Dc 92
Trumpers Way W7 48Ga 86
Trumper Way SL1: Slou 6D 80
Trumper Way UB8: Uxb 38L 63
Trumpets Hill Rd. RH2: Reig 7D 206
Trumpington Dr. AL1: St A 5B 6
Trumpington Rd. E7 35Hc 73
TRUMPS GREEN 2P 147
Trumpsgreen Av. GU25: Vir W 2P 147
Trumps Grn. Cl. GU25: Vir W 71A 148
Trumpsgreen Rd. GU25: Vir W 4N 147
Trumps Mill La. GU25: Vir W 72B 148
Trump St. EC2 3E 224 (44Sb 91)
Trundle Ho. SE1 1D 230 (47Sb 91)
(off Trundle St.)
Trundlers Way WD23: B Hea 18Ga 28
Trundle St. SE1 1D 230 (47Sb 91)
Trundleys Rd. SE8 50Zb 92
Trundley's Ter. SE8 49Zb 92
Trunks All. BR8: Swan 68Dd 140
Truro Gdns. IG1: Ilf 31Nc 74
Truro Ho. HA5: Hat E 24Ba 45
Truro Ho. W2 43Cb 89
(off Westbourne Pk. Rd.)
Truro Rd. DA12: Grav'nd 2F 144
Truro Rd. E17 28Bc 52
Truro Rd. N22 24Nb 50
Truro St. NW5 37Jb 70
Truro Wlk. RM3: Rom 23Ld 57
Truro Way UB4: Hayes 41U 84
Truslove Rd. SE27 64Qb 134
Truss Hill Rd. SL5: S'hill 1A 146
Trussley Rd. W6 48Ya 88
Trustees Cl. UB9: Den 30H 43
Truston's Gdns. RM11: Horn 31Jd 76
Trust Rd. EN8: Walt C 6Ac 20
Trust Wlk. SE21 60Rb 113
Trycewell La. TN15: Igh 93Zd 205
Tryfan Cl. IG4: Ilf 29Mc 53
Tryon Cres. E9 39Yb 72
Tryon St. SW3 7F 227 (50Hb 89)
Trystings Cl. KT10: Clay 79Ja 152
Tuam Rd. SE18 51Tc 116
Tubbenden Cl. BR6: Orp 76Uc 160
Tubbenden Dr. BR6: Orp 77Tc 160
Tubbenden La. BR6: Orp 77Tc 160
Tubbenden La. Sth. BR6: Farnb 78Tc 160
Tubbs Rd. NW10 40Va 68
Tubs Hill TN13: S'oaks 96Kd 203
Tubs Hill Ho. TN13: S'oaks 96Jd 202
Tubs Hill Pde. TN13: S'oaks 96Jd 202
Tubwell Rd. SL2: Stoke P 9M 61
Tucana Ct. E1 5K 219 (42Vb 91)
(off Cygnet St.)
Tucana Hgts. E20 36Ec 72
(off Cheering La.)
Tucker Rd. KT16: Ott 79F 148
Tucker St. WD18: Wat 15Y 27
Tuckey Gro. GU23: Rip 95H 189
Tucklow Wlk. SW15 59Va 110
Tuck Rd. RM13: Rain 37Jd 76
Tudor Av. EN7: Chesh 3Wb 19
Tudor Av. KT4: Wor Pk 76Xa 154
Tudor Av. RM2: Rom 27Jd 56
Tudor Av. TW12: Hamp 66Ca 129
Tudor Av. WD24: Wat 10Z 13
Tudor Bus. Cen. KT20: Kgswd 93Bb 195
Tudor Cl. AL10: Hat 3B 8
Tudor Cl. BR7: Chst 67Pc 138
Tudor Cl. CM15: Shenf 16Ce 41
Tudor Cl. CR2: Sande 87Xb 177
Tudor Cl. CR5: Coul 90Qb 176
Tudor Cl. DA1: Dart 58Kd 119
Tudor Cl. DA11: Nflt 10A 122
Tudor Cl. EN7: Chesh 3Xb 19
Tudor Cl. GU22: Wok 89C 168
Tudor Cl. HA5: Eastc 29W 44
Tudor Cl. IG7: Chig 21Qc 54
Tudor Cl. IG8: Wfd G 22Kc 53
Tudor Cl. KT11: Cobh 85Ba 171
Tudor Cl. KT17: Ewe 82Va 174
Tudor Cl. KT23: Bookh 96Ba 191
(not continuous)
Tudor Cl. KT9: Chess 78Na 153
Tudor Cl. N6 31Lb 70
Tudor Cl. NW3 36Gb 69
Tudor Cl. NW7 23Wa 48
Tudor Cl. NW9 33Sa 67
Tudor Cl. SM3: Cheam 78Za 154
Tudor Cl. SM6: W'gton 80Lb 156
Tudor Cl. SM7: Bans 87Ab 174
Tudor Cl. SW2 58Pb 112
Tudor Cl. TW12: Hamp H 64Ea 130
Tudor Cl. TW15: Ashf 63N 127
Tudor Ct. AL2: Lon C 8H 7
Tudor Ct. BR8: Crock 73Ed 162
Tudor Ct. CM14: W'ley 22Xd 58
Tudor Ct. DA14: Sidc 62Wc 139
Tudor Ct. E17 31Bc 72
Tudor Ct. GU21: Knap 9H 167
Tudor Ct. N1 37Ub 71
Tudor Ct. N22 24Nb 50
Tudor Ct. RH1: Redh 5A 208
(off St Anne's Ri.)
Tudor Ct. RM3: Hrld W 23Rd 57
Tudor Ct. SE16 46Zb 92
(off Princes Riverside Rd.)
Tudor Ct. SE9 56Nc 116
Tudor Ct. TN16: Big H 90Nc 180
Tudor Ct. TW11: Tedd 65Ha 130
Tudor Ct. TW13: Hanw 63Y 129
Tudor Ct. TW19: Stanw 58N 105
Tudor Ct. TW20: Egh 64C 126
Tudor Ct. W3 47Qa 87
Tudor Ct. WD3: Rick 18J 25
Tudor Ct. WD6: Bore 12Na 29
Tudor Ct. Nth. HA9: Wemb 36Qa 67
Tudor Ct. Sth. HA9: Wemb 36Qa 67
Tudor Cres. EN2: Enf 11Sb 33
Tudor Cres. IG6: Ilf 23Rc 54
Tudor Cres. TN14: Otf 88Ld 183
Tudor Dr. KT12: Walt T 74Z 151
Tudor Dr. KT2: King T 64Ma 131
Tudor Dr. RM2: Rom 28Jd 56
Tudor Dr. SM4: Mord 72Za 154
Tudor Dr. TN14: Otf 88Ld 183
Tudor Dr. WD24: Wat 10Z 13
Tudor Ent. Pk. HA1: Harr 34Ha 66
Tudor Ent. Pk. HA3: W'stone 27Fa 46
Tudor Est. NW10 40Ra 67
Tudor Gdns. BR4: W W'ck 76Ec 158
Tudor Gdns. HA3: Hrw W 26Fa 46
Tudor Gdns. NW9 33Sa 67
Tudor Gdns. RM14: Upm 33Sd 78
Tudor Gdns. RM2: Rom 28Jd 56
Tudor Gdns. SL1: Slou 4A 80
Tudor Gdns. SW13 55Ua 110
Tudor Gdns. TW1: Twick 60Ha 108
Tudor Gdns. W3 43Qa 87
Tudor Grange KT13: Weyb 75U 150
Tudor Gro. E9 38Yb 72
Tudor Gro. N20 19Gb 31
Tudor Ho. E16 46Kc 93
(off Wesley Av.)
Tudor Ho. E9 38Yb 72
Tudor Ho. HA5: Pinn 26Y 45
(off Pinner Hill Rd.)
Tudor Ho. KT13: Weyb 79Q 150
Tudor Ho. SE1 7K 225 (46Vb 91)
(off Duchess Wlk.)
Tudor Ho. W14 49Za 88
(off Windsor Way)
Tudor La. SL4: Old Win 9N 103
Tudor Lodge KT20: Kgswd 93Bb 195
Tudor Mnr. Gdns. WD25: Wat 4Z 13
Tudor M. E17 28Bc 52
Tudor M. RM1: Rom 29Hd 56
Tudor Pde. RM6: Chad H 31Zc 75
Tudor Pde. SE9 56Nc 116
Tudor Pde. WD3: Rick 17J 25
Tudor Pk. Footgolf 13Db 31
Tudor Pl. CR4: Mitc 66Gb 133
Tudor Pl. IG9: Buck H 19Nc 36
Tudor Pl. SE19 66Vb 135
Tudor Rd. BR3: Beck 69Ec 136
Tudor Rd. E4 23Dc 52
Tudor Rd. E6 39Lc 73
Tudor Rd. E9 39Xb 71
Tudor Rd. EN5: New Bar 13Cb 31
Tudor Rd. HA3: Hrw W 26Fa 46
Tudor Rd. HA3: W'stone 26Fa 46
Tudor Rd. HA5: Pinn 26Y 45
Tudor Rd. IG11: Bark 39Vc 75
Tudor Rd. KT2: King T 66Qa 131
Tudor Rd. N9 17Xb 33
Tudor Rd. SE19 66Vb 135
Tudor Rd. SE25 71Xb 157
Tudor Rd. TW12: Hamp 66Ca 129
Tudor Rd. TW15: Ashf 65T 128
Tudor Rd. TW3: Houn 56Fa 108
Tudor Rd. UB1: S'hall 45Aa 85
Tudor Rd. UB3: Hayes 44T 84
Tudors, The RH2: Reig 3L 207
Tudor Sq. UB3: Hayes 43T 84
Tudor Stacks SE24 56Sb 113
Tudor St. EC4 4A 224 (45Qb 90)
Tudor Vs. EN7: G Oak 1Ub 19
(not continuous)
Tudor Wlk. DA5: Bexl 58Ad 117
Tudor Wlk. KT13: Weyb 76R 150
Tudor Wlk. KT22: Lea 92Ha 192
Tudor Wlk. WD24: Wat 9Z 13
Tudor Way BR5: Pet W 72Tc 160
Tudor Way EN9: Walt A 5Fc 21
Tudor Way GU21: Knap 1G 186
Tudor Way N14 18Mb 32
Tudor Way SL4: Wind 3C 102
Tudor Way UB10: Hil 37Q 64
Tudor Way W3 47Qa 87
Tudor Way WD3: Rick 18J 25
Tudor Well Cl. HA7: Stan 22Ka 46
Tudor Works UB4: Yead 46Y 85
Tudway Rd. SE3 56Lc 115
Tuffnell Ct. EN8: Chesh 1Zb 20
(off Coopers Wlk.)
Tufnail Rd. DA1: Dart 58Pd 119
Tufnell Ct. E3 39Bc 72
(off Old Ford Rd.)
TUFNELL PARK 35Lb 70
Tufnell Pk. Hall N7 34Lb 70
Tufnell Pk. Rd. N19 35Lb 70
Tufnell Pk. Rd. N7 35Lb 70
Tufnell Park Station (Underground) 35Lb 70
Tufter Rd. IG7: Chig 22Vc 55
Tufton Ct. SW1 4F 229 (48Nb 90)
(off Tufton St.)
Tufton Gdns. KT8: W Mole 68Da 129
Tufton Rd. E4 21Cc 52
Tufton St. SW1 3F 229 (48Nb 90)
Tugboat St. SE28 47Uc 94
Tugela Rd. CR0: C'don 72Tb 157
Tugela St. SE6 61Bc 136
Tugwood Cl. CR5: Coul 93Mb 196
Tulett Av. N20 19Fb 31
Tulip Cl. CM15: Pil H 15Xd 40
Tulip Cl. CR0: C'don 74Zb 158
Tulip Cl. E6 43Pc 94
Tulip Cl. RM3: Rom 23Ld 57
Tulip Cl. TW12: Hamp 65Ba 129
Tulip Cl. UB2: S'hall 47Ea 86
Tulip Gdns. E4 20Fc 35
Tulip Gdns. IG1: Ilf 37Rc 74
Tulip Tree Ct. SM2: Sutt 83Cb 175
Tulip Way UB7: W Dray 49M 83
Tulk Ho. KT16: Ott 80D 148
Tullick Way CR5: Coul 88Mb 176
Tullis Ho. E9 38Yb 72
(off Frampton Pk. Rd.)
Tull St. CR4: Mitc 73Hb 155
Tulse Cl. BR3: Beck 69Ec 136
Tulse Hill SW2 58Qb 112
TULSE HILL 60Rb 113
Tulse Hill Est. SW2 58Qb 112
Tulse Hill Station (Rail) 61Rb 135
Tulse Ho. SW2 58Qb 112
Tulsemere Rd. SE27 61Sb 135
Tulyar Cl. KT20: Tad 92Xa 194
Tumber St. KT18: Head 97Sa 193
Tumblefield Rd. TN15: Stans 82Be 185
Tumblefield Rd. TN15: Wro 82Be 185
Tumblewood Rd. SM7: Bans 88Ab 174
Tumbling Bay KT12: Walt T 72W 150
Tumbling Dice M. DA1: Dart 54Pd 119
Tummons Gdns. SE25 68Ub 135
Tump Ho. SE28 46Uc 94
Tunbridge Ct. AL1: St A 1D 6
(off Manor Rd.)
Tunbridge Ho. EC1 3B 218 (41Rb 91)
(off St John St.)
Tuncombe Rd. N18 21Ub 51
Tunis Rd. W12 46Ya 88
Tunley Grn. E14 43Bc 92
Tunley Rd. NW10 39Ua 68
Tunley Rd. SW17 60Jb 112
Tunmarsh La. E13 41Kc 93
Tunnan Leys E6 44Qc 94
Tunnel App. E14 45Ac 92
Tunnel App. SE10 47Gc 93
Tunnel App. SE16 47Yb 92
Tunnel Av. SE10 47Fc 93
Tunnel Av. Trad. Est. SE10 47Fc 93
Tunnel Est. RM20: W Thur 49Vd 98
Tunnel Gdns. N11 24Lb 50
Tunnel Ind. Est. RM20: W Thur 50Vd 98
Tunnel Link Rd. TW6: H'row A 58Q 106
Tunnel Rd. RH2: Reig 6J 207
Tunnel Rd. SE16 47Yb 92
Tunnel Rd. E. TW6: H'row A 53R 106
Tunnel Rd. W. TW6: H'row A 53Q 106
Tunnel Wood Cl. WD17: Wat 9V 12
Tunnel Wood Rd. WD17: Wat 9V 12
Tuns La. SL1: Slou 8G 80
Tunstall Av. IG6: Ilf 23Wc 55
Tunstall Cl. BR6: Orp 77Uc 160
Tunstall Rd. CR0: C'don 74Ub 157
Tunstall Rd. SW9 56Pb 112
Tunstall Wlk. TW8: Bford 51Na 109
Tunstock Way DA17: Belv 48Ad 95
Tunworth Cl. NW9 30Sa 47
Tunworth Cres. SW15 58Va 110
Tun Yd. SW8 54Kb 112
(off Peardon St.)
Tupelo Rd. E10 33Dc 72
Tupman Ho. SE16 47Wb 91
(off Scott Lidgett Cres.)
Tuppy St. SE28 48Sc 94
Tupwood Ct. CR3: Cat'm 96Wb 197
Tupwood Gdns. CR3: Cat'm 97Wb 197
Tupwood La. CR3: Cat'm 98Wb 197
Tupwood Scrubbs Rd. CR3: Cat'm 100Wb 197
Turenne Cl. SW18 56Eb 111
Turfhouse La. GU24: Chob 1J 167
Turing St. E20 38Ec 72
Turin Ho. WD18: Wat 14U 26
Turin Rd. N9 17Yb 34
Turin St. E2 41Wb 91
Turkey Oak Cl. SE19 67Ub 135
Turkey St. EN1: Enf 8Wb 19
Turkey St. EN3: Enf W 9Yb 20
TURKEY STREET 9Xb 19
Turkey Street Station (Overground) 9Yb 20
Turks Boatyard KT1: King T 67Ma 131
Turks Cl. UB8: Hil 41Q 84
Turks Head Ct. SL4: Eton 2H 103
Turks Head Yd. EC1 7B 218 (43Rb 91)
Turk's Row SW3 7G 227 (50Hb 89)
Turle Rd. N4 33Pb 70
Turle Rd. SW16 68Nb 134
Turlewray Cl. N4 32Pb 70
Turley Cl. E15 39Gc 73
Turnagain La. EC4 2B 224 (44Rb 91)
(off Farringdon St.)
Turnage Rd. RM8: Dag 32Ad 75
Turnant Rd. N17 25Sb 51
Turnberry Cl. NW4 26Za 48
Turnberry Cl. SE16 50Xb 91
Turnberry Ct. WD19: Wat 20Y 27
Turnberry Dr. AL2: Brick W 2Aa 13
Turnberry Quay E14 48Dc 92
Turnberry Way BR6: Orp 74Tc 160
Turnbull Cl. DA9: Ghithe 59Ud 120
Turnbull Ho. N1 39Rb 71
Turnbury Cl. SE28 44Zc 95
Turnchapel M. SW4 55Kb 112
Turner Av. CR4: Mitc 67Hb 133
Turner Av. N15 28Ub 51
Turner Av. TN16: Big H 84Lc 179
Turner Av. TW2: Twick 62Ea 130
Turner Cl. CM13: Gt War 23Xd 58
Turner Cl. HA0: Wemb 37Ma 67
Turner Cl. NW11 30Db 49
Turner Cl. SW9 52Rb 113
Turner Cl. UB4: Hayes 40S 64
Turner Ct. DA1: Dart 57Ld 119
Turner Ct. KT22: Lea 93Ja 192
(off Highbury Dr.)
Turner Ct. SE16 47Yb 92
(off Albion St.)
Turner Cres. CR0: C'don 72Sb 157
Turner Dr. NW11 30Db 49
Turner Ho. E14 47Cc 92
(off Cassilis Rd.)
Turner Ho. NW6 36Eb 69
(off Dresden Cl.)
Turner Ho. NW8 1D 214 (40Gb 69)
(off Townshend Est.)
Turner Ho. SW1 6E 228 (49Mb 90)
(off Herrick St.)
Turner Ho. TW1: Twick 58Ma 109
(off Clevedon Rd.)
Turner M. SM2: Sutt 80Db 155
Turner Pde. N1 38Qb 70
(off Barnsbury Pk.)
Turner Pl. CR5: Coul 88Lb 176
Turner Pl. SW11 57Gb 111
Turner Rd. DA2: Bean 62Xd 142
Turner Rd. E17 27Ec 52
Turner Rd. HA8: Edg 26Na 47
Turner Rd. KT3: N Mald 73Ta 153
Turner Rd. RM12: Horn 33Hd 76
Turner Rd. SL3: L'ly 7N 81
Turner Rd. WD23: Bush 14Ea 28
Turners Cl. N20 20Hb 31
Turners Cl. TW18: Staines 64K 127
Turners Ct. E15 35Ec 72
(off Drapers Rd.)
Turners Ct. EN8: Chesh 3Zb 20
Turners Ct. N15 29Tb 51
Turners Ct. RM4: Abr 13Xc 37
Turners Ct. TN15: W King 79Ud 164
Turners Gdns. TN13: S'oaks 100Ld 203
Turner's Hill EN8: Chesh 1Zb 20
Turners Hill HP2: Hem H 3N 3
Turners La. KT12: Hers 79X 151
Turners Mdw. Way BR3: Beck 67Bc 136
Turners Oak DA3: Nw A G 76Ae 165
Turners Pl. DA4: S Dar 68Sd 142
Turners Rd. E3 43Bc 92
Turner St. E1 43Xb 91
Turner St. E16 44Hc 93
Turners Way CR0: Wadd 75Qb 156
Turners Wood NW11 31Eb 69
Turneville Rd. W14 51Bb 111
Turney Rd. SE21 59Sb 113
Turneys Orchard WD3: Chor 15F 24
TURNHAM GREEN 49Ua 88
Turnham Green Station (Underground) 49Ua 88
Turnham Grn. Ter. W4 49Ua 88
Turnham Grn. Ter. M. W4 49Ua 88
Turnham Rd. SE4 57Ac 114
Turnmill St. EC1 6B 218 (42Rb 91)
Turnoak Av. GU22: Wok 92A 188
Turnoak La. GU22: Wok 92A 188
Turnoak Pk. SL4: Wind 6C 102
Turnour Ho. E1 44Xb 91
(off Walburgh St.)
Turnpike Cl. DA16: Well 55Wc 117
Turnpike Cl. SE8 52Bc 114
Turnpike Ct. DA6: Bex 56Zc 117
Turnpike Ct. EN8: Walt C 6Ac 20
Turnpike Dr. BR6: Prat B 81Yc 181
Turnpike Ho. EC1 4C 218 (41Rb 91)
Turnpike La. N8 28Pb 50
Turnpike La. RM18: W Til 9F 100
Turnpike La. SM1: Sutt 78Eb 155
Turnpike La. UB10: Uxb 41N 83
Turnpike Lane Station (Underground) 27Rb 51
Turnpike Link CR0: C'don 75Ub 157
Turnpike M. N8 27Qb 50
(off Turnpike La.)
Turnpike Pde. N15 27Rb 51
(off Green Lanes)
Turnpike Way TW7: Isle 53Ja 108
Turnpin La. SE10 51Ec 114
Turnstone DA3: Lfield 69Ce 143
Turnstone Cl. CR2: Sels 82Ac 178
Turnstone Cl. E13 41Jc 93
Turnstone Cl. NW9 26Ua 48
Turnstone Cl. RM18: E Til 9K 101
Turnstone Cl. UB10: Ick 36R 64
Turnstone Ho. E1 45Wb 91
(off Star Pl.)
Turnstones, The DA12: Grav'nd 1F 144
Turnstones, The WD25: Wat 8Aa 13
Turp Av. RM16: Grays 47Ee 99
Turpentine La. SW1 7K 227 (50Kb 90)
Turpin Av. RM5: Col R 23Cd 56
Turpin Cl. E1 45Zb 92
Turpin Cl. EN3: Enf L 9Cc 20
Turpin Ct. WD18: Wat 15V 26
Turpington Cl. BR2: Broml 73Nc 160
Turpington La. BR2: Broml 73Nc 160
Turpin Ho. SW11 53Kb 112
Turpin La. DA8: Erith 51Jd 118
Turpin Rd. TW14: Felt 58V 106
Turpin's La. IG8: Wfd G 22Pc 54
Turpins Yd. NW2 36Za 68
Turpins Yd. SE10 52Ec 114
Turpin Way N19 33Mb 70
Turpin Way SM6: W'gton 80Kb 156
Turquand St. SE17 6E 230 (49Sb 91)
Turret Gro. SW4 55Lb 112
Turton Rd. HA0: Wemb 36Na 67
Turton Way SL1: Slou 8H 81
Turville Ct. KT23: Bookh 97Da 191
Turville Ho. NW8 5D 214 (42Gb 89)
(off Grendon St.)
Turville St. E2 5K 219 (42Vb 91)
Tuscan Ho. E2 41Yb 92
(off Knottisford St.)
Tuscan Rd. SE18 50Tc 94
Tuscany Corte SE13 55Dc 114
(off Loampit Va.)
Tuscany Ho. E17 26Bc 52
Tuscany Ho. IG3: Ilf 30Wc 55
Tuskar St. SE10 51Gc 115
Tussah Ho. E2 40Yb 72
(off Russia La.)
Tussauds Cl. WD3: Crox G 15Q 26
Tustin Est. SE15 51Yb 114
Tuttlebee La. IG9: Buck H 19Jc 35
Tuttleby Cotts. RM4: Abr 17Ad 37
Tuttle Ho. SW1 7D 228 (50Mb 90)
(off Aylesford St.)
Tutton Ho. DA12: Grav'nd 3E 144
Tuxford Cl. WD6: Bore 10Na 15
Twankhams All. CM16: Epp 2Wc 23
Tweed RM18: E Til 9L 101
Tweed Ct. W7 44Ga 86
(off Hanway Rd.)
Tweeddale Gro. UB10: Ick 34S 64
Tweeddale Rd. SM5: Cars 74Fb 155
Tweed Glen RM1: Rom 24Fd 56
Tweed Grn. RM1: Rom 24Fd 56
Tweedmouth Rd. E13 40Kc 73
Tweed Rd. SL3: L'ly 51D 104
Tweed Wlk. E14 42Ec 92
Tweed Way RM1: Rom 24Fd 56
Tweedy Cl. EN1: Enf 15Vb 33
Tweedy Rd. BR1: Broml 67Jc 137
Tweezer's All. WC2 4K 223 (45Qb 90)
(off Milford La.)
Twelfth Av. KT20: Lwr K 98Ab 194
Twelve Acre Cl. KT23: Bookh 96Ba 191
Twelve Acre Ho. E12 34Qc 74
(off Grantham Rd.)
Twelvetrees Bus. Pk. E3 42Fc 93
Twelvetrees Cres. E3 42Ec 92
(not continuous)
Twentyman Cl. IG8: Wfd G 22Jc 53
TWICKENHAM 60Ja 108
Twickenham Bri. 57La 108
Twickenham Cl. CR0: Bedd 76Pb 156
Twickenham Gdns. HA3: Hrw W 24Ga 46
Twickenham Gdns. UB6: G'frd 36Ja 66
Twickenham Ho. DA1: Dart 54Pd 119
Twickenham Mus. 60Ja 108
(off The Embankment)
Twickenham Pl. KT7: T Ditt 75Ha 152
(off Woodfield Rd.)
Twickenham Rd. E11 33Ec 72
Twickenham Rd. TW11: Tedd 63Ja 130
(not continuous)
Twickenham Rd. TW13: Hanw 62Ba 129
Twickenham Rd. TW7: Isle 57Ja 108
Twickenham Rd. TW9: Rich 56La 108
Twickenham Stadium 58Ga 108
Twickenham Station (Rail) 59Ja 108
Twickenham Stoop 59Ga 108
Twickenham Tourist Info. Cen. 60Ka 108
(off Church St.)
Twickenham Trad. Est. TW1: Twick 58Ha 108
Twig Folly Cl. E2 40Zb 72
Twigg Cl. DA8: Erith 52Gd 118
Twilley St. SW18 59Db 111
Twill Way SM6: W'gton 75Jb 156
Twin Bridges Bus. Pk. CR2: S Croy 79Tb 157
Twinches La. SL1: Slou 6F 80
Twine Cl. IG11: Bark 41Xc 95
Twine Ct. E1 45Yb 92
Twineham Grn. N12 21Cb 49
Twine Ter. E3 42Bc 92
(off Ropery St.)
Twining Av. TW2: Twick 62Ea 130
Twinners Rd. CR5: Coul 89Kb 176
Twinoaks KT11: Cobh 85Ca 171
Twin Tumps Way SE28 45Wc 95
Twisden Rd. NW5 35Kb 70
Twisted Stone Golf Course 87K 169
Twist Ho. SE1 4J 231 (48Ub 91)
Twistleton Ct. DA1: Dart 58Md 119
Twist Way SL2: Slou 2D 80
Twitten Gro. BR1: Broml 69Pc 138
TWITTON 88Gd 182
Twitton La. TN14: Otf 87Fd 182
Twitton Mdws. TN14: Otf 88Gd 182
Twitton Stream Cotts. TN14: Otf 88Gd 182
Two Mile Dr. SL1: Slou 7C 80
Two Oaks KT13: Weyb 77R 150
Two Rivers Retail Pk. 63G 126
Two Rivers Shop. Cen. 63H 127
Two Southbank Pl. SE1 1J 229 (47Pb 90)
(off York Rd.)
TWO WATERS 5M 3
Two Waters Rd. HP3: Hem H 4L 3
Two Waters Way HP3: Hem H 6L 3
Twybridge Way NW10 38Sa 67
Twycross M. SE10 50Gc 93
Twyford Abbey Rd. NW10 41Pa 87
Twyford Av. N2 27Hb 49
Twyford Av. W3 45Qa 87
Twyford Ct. HA0: Wemb 40Na 67
(off Vicars Bri. Cl.)
Twyford Ct. N10 27Jb 50
Twyford Cres. W3 46Qa 87
Twyford Ho. N15 30Ub 51
(off Chisley Rd.)
Twyford Ho. N5 34Rb 71
Twyford Pl. WC2 2H 223 (44Pb 90)
Twyford Rd. HA2: Harr 32Da 65
Twyford Rd. IG1: Ilf 36Sc 74
Twyford Rd. SM5: Cars 74Fb 155
Twyford Sports Cen. 46Ra 87
Twyford St. N1 39Pb 70
Twynersh Av. KT16: Chert 72H 149
Twynholm Mans. SW6 52Ab 110
(off Lillie Rd.)
Twysdens Ter. AL9: Wel G 6E 8
Tyas Rd. E16 42Hc 93
Tybenham Rd. SW19 69Cb 133
Tyberry Rd. EN3: Enf H 13Xb 33
Tyburn Ho. NW8 5C 214 (42Fb 89)
(off Fisherton St.)
Tyburn La. HA1: Harr 31Ha 66
Tyburns, The CM13: Hut 19Ee 41
Tyburn Tree (site of) 4F 221 (45Hb 89)
(off Marble Arch)
Tyburn Way W1 4G 221 (45Hb 89)
Tycehurst Hill IG10: Lough 14Pc 36
Tydcombe Rd. CR6: W'ham 91Yb 198
Tye La. BR6: Farnb 78Sc 160
Tye La. KT18: Head 99Ua 194
Tyers Est. SE1 1H 231 (47Ub 91)
(off Bermondsey St.)
Tyer's Ga. SE1 2H 231 (47Ub 91)
Tyers St. SE11 7H 229 (50Pb 90)
Tyers Ter. SE11 7H 229 (50Pb 90)
Tyeshurst Cl. SE2 50Ad 95
Tyfield Cl. EN8: Chesh 2Yb 20
Tygan Ho. SM3: Cheam 79Ab 154
(off The Broadway)
Tylecroft Rd. SW16 68Nb 134
Tyle Grn. RM11: Horn 28Nd 57
Tylehurst Dr. RH1: Redh 7P 207
Tylehurst Gdns. IG1: Ilf 36Sc 74
Tyle Pl. SL4: Old Win 7L 103
Tyler Cl. DA11: Nflt 60Ee 121
Tyler Cl. DA8: Erith 52Dd 118
Tyler Cl. E2 1K 219 (40Vb 71)
Tyler Ct. SE17 5F 231 (49Tb 91)
(off New Paragon Wlk.)
Tyler Gdns. KT15: Add 77L 149
Tyler Gro. DA1: Dart 56Pd 119
Tyler Rd. UB2: S'hall 48Da 85
Tylers Cl. IG10: Lough 17Nc 36
Tylers Cl. RH9: G'stone 2P 209
Tylers Cl. WD4: K Lan 10N 3
Tyler's Ct. W1 3D 222 (44Mb 90)
(off Wardour St.)
Tylers Ct. E17 28Cc 52
(off Westbury Rd.)
Tylers Ct. HA0: Wemb 40Na 67
Tylers Cres. RM12: Horn 36Ld 77
Tylersfield WD5: Ab L 3V 12
Tylers Ga. HA3: Kenton 30Na 47
TYLER'S GREEN 1P 209
Tylers Grn. Rd. BR8: Crock 72Ed 162
Tylers Path SM5: Cars 77Hb 155
Tyler St. SE10 50Gc 93
(not continuous)
Tylers Way WD25: A'ham 13Ea 28
Tyler Wlk. SL3: L'ly 50B 82
Tyler Way CM14: B'wood 18Xd 40
Tylney Av. SE19 64Vb 135
(not continuous)
Tylney Cl. IG7: Chig 21Vc 55
Tylney Ho. E1 44Xb 91
(off Nelson St.)
Tylney Rd. BR1: Broml 68Mc 137
Tylney Rd. E7 35Lc 73
Tynamara KT1: King T 70Ma 131
(off Portsmouth Rd.)
Tynan Cl. TW14: Felt 60W 106
Tyndale Ct. E14 50Dc 92
(off Transom Sq.)
Tyndale Ct. E9 37Ac 72
(off Brookfield Rd.)
Tyndale Ho. N1 38Rb 71
(off Tyndale La.)
Tyndale La. N1 38Rb 71
Tyndale Mans. N1 38Rb 71
(off Upper St.)
Tyndale M. SL1: Slou 7F 80
Tyndale Ter. N1 38Rb 71
Tyndall Gdns. E10 33Ec 72
Tyndall Rd. DA16: Well 55Vc 117
Tyndall Rd. E10 33Ec 72
Tyndal Way DA1: Dart 54Nd 119
Tyne RM18: E Til 9L 101
Tyne Cl. RM14: Upm 30Td 58
Tyne Ct. W7 44Ga 86
(off Hanway Rd.)
Tynedale AL2: Lon C 9K 7
Tynedale Cl. DA2: Dart 60Td 120
Tyne Gdns. RM15: Avel 45Sd 98
Tyneham Cl. SW11 55Jb 112
Tyneham Rd. SW11 54Jb 112
Tyne Ho. KT2: King T 67Ma 131
Tynemouth Cl. E6 44Rc 94
Tynemouth Dr. EN1: Enf 10Wb 19
Tynemouth Rd. CR4: Mitc 66Jb 134
Tynemouth Rd. N15 28Vb 51
Tynemouth Rd. SE18 50Uc 94
Tynemouth St. SW6 54Eb 111

Tyne St. E1 .... 2K **225** (44Vb **91**)
Tynley Gro. GU4: Jac W 10P **187**
Tynne Ct. E14 .... 42Ec **92**
.... *(off Teviot St.)*
Tynsdale Rd. NW10 .... 38Ua **68**
Tynte Ct. E9 .... 36Ac **72**
.... *(off Mabley St.)*
Tynwald Ho. SE26 .... 62Wb **135**
Type St. E2 .... 40Zb **72**
Typhoon Way SM6: W'gton 80Nb **156**
Tyrawley Rd. SW6 .... 53Db **111**
Tyre La. NW9 .... 28Ua **48**
Tyrell Cl. HA1: Harr 35Ga **66**
Tyrell Ct. SM5: Cars 77Hb **155**
Tyrell Gdns. SL4: Wind 5D **102**
Tyrell Ho. BR3: Beck 64Dc **136**
.... *(off Beckenham Hill Rd.)*
Tyrell Pl. CM15: Shenf 18Be **41**
Tyrell Ri. CM14: W'ley 22Yd **58**
Tyrells Cl. RM14: Upm 33Rd **77**
Tyrian Pl. E1 .... 43Wb **91**
Tyrols Rd. SE23 .... 60Zb **114**
Tyrone Rd. E6 .... 40Pc **74**
Tyron Way DA14: Sidc 63Uc **138**
Tyrrell Av. DA16: Well 57Wc **117**
Tyrrell Ho. SW1 .... 51Lb **112**
.... *(off Churchill Gdns.)*
Tyrrell Rd. SE22 .... 56Wb **113**
Tyrrells Cl. KT22: Fet 95Fa **192**
Tyrrells Hall Cl. RM17: Grays 51Fe **121**
Tyrrell Sq. CR4: Mitc 67Gb **133**
TYRRELL'S WOOD 96Qa **193**
Tyrrells Wood Golf Course **96Pa 193**
Tyrrel Way NW9 .... 31Va **68**
Tyrsal Ct. RM11: Horn 29Qd **57**
Tyrwhitt Rd. SE4 .... 55Cc **114**
Tysea Hill RM4: Noak H 18Hd **38**
Tysea Hill RM4: Stap A 18Hd **38**
TYSEA HILL 17Hd **38**
Tysoe Av. EN3: Enf L 8Bc **20**
Tysoe St. EC1 .... 4K **217** (41Qb **90**)
Tyson Gdns. SE23 .... 59Yb **114**
Tyson Rd. SE23 .... 59Yb **114**
Tyssen Pas. E8 .... 37Vb **71**
Tyssen Pl. RM15: S Ock 41Yd **98**
Tyssen Rd. N16 .... 34Vb **71**
Tyssen St. E8 .... 37Vb **71**
Tyssen St. N1 .... 1J **219** (40Ub **71**)
Tytherton E2 .... 40Yb **72**
.... *(off Cyprus St.)*
Tytherton Rd. N19 .... 34Mb **70**
TYTTENHANGER 5J **7**
Tyttenhanger Grn. AL4: St A 5H **7**

U

UAL (London College of Fashion) 1J **223** (43Pb **90**)
.... *(off High Holborn)*
Uamvar St. E14 .... 43Dc **92**
Uber E17 .... 28Bc **52**
UCH MACMILLAN CANCER CEN. .... **6C 216 (42Lb 90)**
Uckfield Gro. CR4: Mitc 66Jb **134**
Uckfield Rd. EN3: Enf W 9Zb **20**
Udall Gdns. RM5: Col R 23Cd **56**
Udall St. SW1 .... 6C **228** (49Lb **90**)
Udimore Ho. W10 .... 43Ya **88**
.... *(off Sutton Way)*
Udney Pk. Rd. TW11: Tedd 65Ja **130**
Uffington Rd. NW10 .... 39Wa **68**
Uffington Rd. SE27 .... 63Qb **134**
Ufford Cl. HA3: Hrw W 24Da **45**
Ufford Rd. HA3: Hrw W 24Da **45**
Ufford St. SE1 .... 1A **230** (47Qb **90**)
Ufton Ct. UB5: N'olt 41Z **85**
Ufton Gro. N1 .... 38Tb **71**
Ufton Rd. N1 .... 38Tb **71**
.... (not continuous)
Uhura Sq. N16 .... 34Ub **71**
Ujima Ct. SW16 .... 63Nb **134**
Ullathorne Rd. SW16 .... 63Lb **134**
Ulleswater Rd. N14 .... 20Nb **32**
Ullin St. E14 .... 43Ec **92**
Ullswater E18 .... 26Jc **53**
Ullswater Bus. Pk. CR5: Coul 88Nb **176**
Ullswater Cl. BR1: Broml 66Gc **137**
Ullswater Cl. GU18: Light 2A **166**
Ullswater Cl. SL1: Slou 3A **80**
Ullswater Cl. SW15 .... 63Ta **131**
Ullswater Cl. UB4: Hayes 40U **64**
Ullswater Ct. HA2: Harr 31Ca **65**
Ullswater Cres. CR5: Coul 88Mb **176**
Ullswater Cres. SW15 .... 63Ta **131**
Ullswater Ho. SE15 .... 51Yb **114**
.... *(off Hillbeck Cl.)*
Ullswater Rd. GU18: Light 2A **166**
Ullswater Rd. HP3: Hem H 4C **4**
Ullswater Rd. SE27 .... 61Rb **135**
Ullswater Rd. SW13 .... 52Wa **110**
Ullswater Way RM12: Horn 36Jd **76**
Ulstan Cl. CR3: Wold 95Cc **198**
Ulster Gdns. N13 .... 21Sb **51**
Ulster Pl. NW1 .... 6K **215** (42Kb **90**)
Ulster Ter. NW1 .... 5K **215** (42Kb **90**)
Ulundi Rd. SE3 .... 51Gc **115**
Ulva Rd. SW15 .... 57Za **110**
Ulverscroft Rd. SE22 .... 57Vb **113**
Ulverston RM19: Purf 50Sd **98**
Ulverston Cl. AL1: St A 2D **6**
Ulverstone Rd. SE27 .... 61Rb **135**
Ulverston Rd. RM3: Rom 23Nd **57**
.... *(off Kingsbridge Cir.)*
Ulverston Rd. E17 .... 26Fc **53**
Ulwin Av. KT14: Byfl 85N **169**
Ulysses Rd. NW6 .... 36Bb **69**
Umberstones GU25: Vir W 2P **147**
Umberston St. E1 .... 44Wb **91**
Umberville Way SL2: Slou 1D **80**
Umbria St. SW15 .... 58Wa **110**
Umbriel Pl. E13 .... 40Jc **73**
Umfreville Rd. N4 .... 30Rb **51**
Umpire Vw. HA1: Harr 29Ea **46**
Underacres Cl. HP2: Hem H 1A **4**
Undercliff Rd. SE13 .... 55Cc **114**
Underhill EN5: Barn 15Cb **31**
UNDERHILL 15Cb **31**
Underhill Ct. EN5: Barn 15Cb **31**
Underhill Gdns. W5 .... 45La **86**
Underhill Ho. E14 .... 43Cc **92**
.... *(off Burgess St.)*
Underhill Pk. Rd. RH2: Reig 3J **207**
Underhill Pas. NW1 .... 39Kb **70**
.... *(off Camden High St.)*
Underhill Rd. SE22 .... 57Wb **113**
Underhill St. NW1 .... 39Kb **70**
Underne Av. N14 .... 19Kb **32**
Undershaft EC3 .... 3H **225** (44Ub **91**)
Undershaw Rd. BR1: Broml 62Hc **137**
Underwood, The SE9 .... 61Pc **138**
Underwood CR0: New Ad 78Ec **158**
Underwood Bldg., The EC1 .... 1D **224** (43Sb **91**)
.... *(off Bartholomew Cl.)*
Underwood Ct. CR3: Cat'm 97Ub **197**
Underwood Ct. E10 .... 32Dc **72**
.... *(off Leyton Grange Est.)*
Underwood Ho. KT8: W Mole 71Ca **151**
.... *(off Approach Rd.)*
Underwood Ho. W6 .... 48Xa **88**
.... *(off Sycamore Gdns.)*
Underwood Rd. CR3: Cat'm 98Ub **197**
Underwood Rd. E1 .... 42Wb **91**
Underwood Rd. E4 .... 22Dc **52**
Underwood Rd. IG8: Wfd G 24Lc **53**
Underwood Row N1 .... 3E **218** (41Sb **91**)
Underwood St. N1 .... 3E **218** (41Sb **91**)
Undine Rd. E14 .... 49Dc **92**
Undine St. SW17 .... 64Hb **133**
Uneeda Dr. UB6: G'frd 39Fa **66**
Unex Twr. E15 .... 38Fc **73**
.... *(off Station St.)*
Unicorn Bldg. E1 .... 45Zb **92**
.... *(off Jardine Rd.)*
Unicorn Theatre London .... **7J 225 (46Ub 91)**
Unicorn Vw. EN5: Barn 16Bb **31**
Unicorn Wlk. DA9: Ghithe 57Vd **120**
Unicorn Works N17 .... 24Yb **52**
Union Bus. Pk. UB8: Uxb 38K **63**
Union Cl. E11 .... 35Fc **73**
Union Cotts. E15 .... 38Gc **73**
Union Ct. EC2 .... 2H **225** (44Ub **91**)
.... *(off Old Broad St.)*
Union Ct. SW4 .... 54Nb **112**
Union Ct. TW9: Rich 57Na **109**
Union Ct. W9 .... 43Cb **89**
.... *(off Elmfield Way)*
Union Dr. E1 .... 42Ac **92**
Union Grn. HP2: Hem H 1M **3**
Union Gro. SW8 .... 54Mb **112**
Union Ho. CR0: C'don 73Sb **157**
Union La. TW7: Isle 54Ja **108**
Union M. SW4 .... 54Nb **112**
Union Mills Apts. E8 .... 39Vb **71**
.... *(off Samuel St.)*
Union Pk. SE10 .... 50Hc **93**
Union Rd. BR2: Broml 71Mc **159**
Union Rd. CR0: C'don 73Sb **157**
Union Rd. E17 .... 30Bc **52**
Union Rd. HA0: Wemb 37Na **67**
Union Rd. N11 .... 23Mb **50**
Union Rd. RM7: Rush G 30Fd **56**
Union Rd. SW4 .... 54Mb **112**
Union Rd. SW8 .... 54Mb **112**
Union Rd. UB5: N'olt 40Ca **65**
Union Sq. N1 .... 1D **218** (39Sb **71**)
Union St. EN5: Barn 13Ab **30**
Union St. KT1: King T 68Ma **131**
Union St. SE1 .... 7B **224** (46Rb **91**)
Union Theatre .... **1C 230 (47Rb 91)**
Union Wlk. E2 .... 3J **219** (41Ub **91**)
Union Wharf N1: Arlington Av. 1E **218** (39Sb **71**)
Union Wharf N1: Wenlock Rd. 2D **218** (40Sb **71**)
.... *(off Wenlock Rd.)*
Union Wharf UB7: Yiew 46M **83**
.... *(off Bentinck Rd.)*
Union Yd. W1 .... 3A **222** (44Kb **90**)
Unitair Cen. TW14: Bedf 58S **106**
United Cen. TW14: Felt 59V **106**
United Ho. SE16 .... 47Yb **92**
.... *(off Brunel Rd.)*
Unit Workshops E1 .... 44Wb **91**
.... *(off Adler St.)*
Unity Cl. CR0: New Ad 81Dc **178**
Unity Cl. NW10 .... 37Wa **68**
Unity Cl. SE19 .... 64Sb **135**
Unity Ct. SE1 .... 50Vb **91**
.... *(off Fortune Pl.)*
Unity M. NW1 .... 1D **216** (40Mb **70**)
Unity Pl. E17 .... 27Ac **52**
Unity Rd. EN3: Enf W 9Yb **20**
Unity Ter. HA2: Harr 32Da **65**
Unity Trad. Est. IG8: Wfd G 26Mc **53**
Unity Way SE18 .... 48Mc **93**
Unity Wharf SE1 .... 1K **231** (47Vb **91**)
.... *(off Mill St.)*
Universal Ho. UB1: S'hall 45Aa **85**
University Cl. NW7 .... 24Wa **48**
University Cl. WD23: Bush 14Ca **27**
UNIVERSITY COLLEGE HOSPITAL Main Site .... **5C 216 (42Lb 90)**
UNIVERSITY COLLEGE HOSPITAL Maternity Department .... **6C 216 (42Lb 90)**
UNIVERSITY COLLEGE HOSPITAL Westmoreland Street .... **1J 221 (43Jb 90)**
University College London Art Mus. .... **5D 216 (42Mb 90)**
.... (off Gower St.)
University College London Bloomsbury Campus **5D 216 (42Mb 90)**
University College London Department of Geological Collections .... **6D 216 (42Mb 90)**
.... (off Gower St.)
University College London Institute of Neurology .... **6G 217 (42Nb 90)**
.... *(off Queen Sq.)*
University College London Slade School of Fine Art .... **5D 216 (42Mb 90)**
.... *(off Gower Ct.)*
University College London The Institute of Cancer Research .... **7A 226 (50Eb 89)**
University College London Wolfson Ho. .... **4C 216 (41Lb 90)**
.... *(off Stephenson Way)*
University College London Medical School Whittington Hospital Campus **33Lb 70**
.... *(within Whittington Hospital)*
University for the Creative Arts Epsom Campus **86Ua 174**
University Gdns. DA5: Bexl 59Bd **117**
UNIVERSITY HOSPITAL, LEWISHAM **57Dc 114**
University of East London Docklands Campus **45Qc 94**
University of East London Stratford Campus **37Gc 73**
University of Greenwich Avery Hill Campus **58Sc 116**
University of Greenwich Greenwich Campus, King William Wlk. **51Fc 115**
University of Greenwich Maritime Greenwich Campus **51Ec 114**
University of Hertfordshire Art & Design Gallery **1A 8**
University of Hertfordshire College Lane Campus **2B 8**
University of London Birkbeck College .... **7E 216 (43Mb 90)**
University of London Heythrop College **48Db 89**
.... *(off Kensington Sq.)*
University of London Institutes of Education & Advanced Legal Studies .... **6E 216 (42Mb 90)**
University of London School of Hygiene & Tropical Medicine .... **7E 216 (43Mb 90)**
.... *(off Gower St.)*
University of London School of Oriental & African Studies .... **6E 216 (42Mb 90)**
.... *(off Torrington Sq.)*
University of London Senate House .... **7E 216 (43Mb 90)**
University of London Warburg Institute .... **6E 216 (42Mb 90)**
University of London Observatory **23Va 48**
University of North London Hornsey Rd. **35Qb 70**
University of North London Ladbrooke House **36Sb 71**
University of North London North London Campus, Spring House **37Qb 70**
University of the Arts London Camberwell College of Arts, Peckham Rd. **53Ub 113**
University of the Arts London Camberwell College of Arts, Wilson Rd. **53Ub 113**
University of the Arts London Chelsea College of Art & Design .... **7E 228 (50Mb 90)**
University of the Arts London London College of Communication .... **5C 230 (49Rb 91)**
University of the Arts London London College of Fashion, Curtain Rd. .... **5J 219 (42Ub 91)**
.... *(off Curtain Rd.)*
University of the Arts London London College of Fashion, Golden La. .... **6D 218 (42Sb 91)**
.... *(off Golden La.)*
University of the Arts London London College of Fashion, Mare St. **38Yb 72**
University of the Arts London Wimbledon College of Art **67Ab 132**
University of West London Brentford Campus **50La 86**
University of West London Ealing Campus, Grove House **46Ma 87**
University of West London Ealing Campus, Spesom House **45La 86**
University of West London Ealing Campus, St Marys Road **46Ma 87**
University of West London Ealing Campus, Vestry Hall **47Ma 87**
University of West London Ealing Campus, Walpole House **45Ma 87**
University of Westminster Cavendish Campus, Hanson St. .... **7B 216 (43Lb 90)**
University of Westminster Cavendish Campus, Lit. Titchfield St. .... **1B 222 (43Lb 90)**
.... *(off Lit. Titchfield St.)*
University of Westminster Harrow Campus **31Ja 66**
University of Westminster Marylebone Campus .... **7H 215 (43Jb 90)**
University of Westminster Regent Campus, Regent St. .... **2A 222 (44Kb 90)**
.... *(off Regent St.)*
University of Westminster Regent Campus, Wells St. .... **1C 222 (43Lb 90)**
.... *(off Wells St.)*
University Pl. DA8: Erith 52Ed **118**
University Rd. SW19 .... 65Fb **133**
University St. WC1 .... 6C **216** (42Lb **90**)
University Way DA1: Dart 56Rd **119**
University Way E16 .... 45Qc **94**
Unwin Av. TW14: Felt 57T **106**
Unwin Cl. SE15 .... 51Wb **113**
Unwin Ct. N2 .... 28Gb **49**
Unwin Mans. W14 .... 51Bb **111**
.... *(off Queen's Club Gdns.)*
Unwin Rd. SW7 .... 3C **226** (48Fb **89**)
Unwin Rd. TW7: Isle 55Ga **108**
Unwin Way HA7: Stan 24Ma **47**
Up at O2, The **46Gc 93**
Upbrook M. W2 .... 3A **220** (44Eb **89**)
Upcerne Rd. SW10 .... 52Eb **111**
Upchurch Cl. SE20 .... 66Xb **135**
Upcott Ho. E3 .... 41Dc **92**
.... *(off Bruce Rd.)*
Upcott Ho. E9 .... 38Yb **72**
.... *(off Frampton Pk. Rd.)*
Upcroft SL4: Wind 5F **102**
Upcroft Av. HA8: Edg 22Sa **47**
Updale Cl. EN6: Pot B 5Ab **16**
Updale Rd. DA14: Sidc 63Vc **139**
Updown Hill GU20: W'sham 9B **146**
Upfield CR0: C'don 76Xb **157**
Upfield Rd. W7 .... 43Ha **86**
Upgrove Mnr. Way SW2 .... 59Qb **112**
Uphall Rd. IG1: Ilf 36Rc **74**
Upham Pk. Rd. W4 .... 49Ua **88**
Uphavering Ho. RM12: Horn 33Ld **77**
Uphill BR2: Broml 70Hc **137**
.... *(off Westmoreland Rd.)*
Uphill Dr. NW7 .... 22Ua **48**
Uphill Dr. NW9 .... 29Sa **47**
Uphill Gro. NW7 .... 21Ua **48**
Uphill Rd. NW7 .... 21Ua **48**
Upland Ct. HA6: Nwood 21V **44**
Upland Ct. Rd. RM3: Hrld W 26Pd **57**
Upland Dr. AL9: Brk P 7K **9**
Upland Dr. KT18: Tatt C 90Ya **174**
Upland M. SE22 .... 57Wb **113**
Upland Rd. CR2: S Croy 78Tb **157**
Upland Rd. CR3: W'ham 92Cc **198**
Upland Rd. CR3: Wold 92Cc **198**
Upland Rd. DA7: Bex 55Bd **117**
Upland Rd. E13 .... 42Jc **93**
Upland Rd. SE22 .... 57Wb **113**
.... (not continuous)
Upland Rd. SM2: Sutt 80Fb **155**
Uplands, The AL2: Brick W 2Aa **13**
Uplands, The HA4: Ruis 32W **64**
Uplands, The IG10: Lough 13Pc **36**
Uplands, The SL9: Ger X 32A **62**
Uplands BR3: Beck 68Cc **136**
Uplands CR6: W'ham 90Bc **178**
Uplands KT21: Asht 92Ma **193**
Uplands WD3: Crox G 16P **25**
Uplands Av. E17 .... 26Zb **52**
Uplands Bus. Pk. E17 .... 27Zb **52**
Uplands Cl. SE18 .... 50Rc **94**
Uplands Cl. SL9: Ger X 32A **62**
Uplands Cl. SW14 .... 57Ra **109**
Uplands Cl. TN13: Riv 95Hd **202**
Uplands Ct. N21 .... 17Qb **32**
.... *(off The Green)*
Uplands Dr. KT22: Oxs 86Fa **172**
Uplands End IG8: Wfd G 24Nc **54**
Uplands Pk. Rd. EN2: Enf 12Qb **32**
Uplands Rd. BR6: Orp 74Xc **161**
Uplands Rd. CM14: W'ley 22Ae **59**
Uplands Rd. CR8: Kenley 88Sb **177**
Uplands Rd. EN4: E Barn 18Jb **32**
Uplands Rd. IG8: Wfd G 24Nc **54**
Uplands Rd. N8 .... 29Pb **50**
Uplands Rd. RM6: Chad H 27Zc **55**
Uplands Way N21 .... 15Qb **32**
Uplands Way TN13: Riv 95Hd **202**
Upland Way KT18: Tatt C 90Ya **174**
UPMINSTER 33Sd **78**
Upminster Bridge Station (Underground) 33Qd **77**
Upminster Golf Course **31Td 78**
Upminster Rd. RM11: Horn 33Pd **77**
Upminster Rd. RM12: Horn 33Pd **77**
Upminster Rd. RM14: Upm 33Qd **77**
Upminster Rd. Nth. RM13: Rain 41Ld **97**
Upminster Rd. Sth. RM13: Rain 42Jd **96**
Upminster Station (Rail & Underground) 33Sd **78**
Upminster Tithe Barn Mus. of Nostalgia **31Td 78**
Upminster Trad. Pk. RM14: Upm 31Zd **79**
Upminster Windmill **33Rd 77**
Upnall Ho. SE15 .... 51Yb **114**
Upney Ct. RM12: Horn 36Md **77**
Upney La. IG11: Bark 37Uc **74**
Upney Station (Underground) 38Vc **75**
Upnor Way SE17 .... 7J **231** (50Ub **91**)
Uppark Dr. IG2: Ilf 30Sc **54**
Upper Abbey Rd. DA17: Belv 49Bd **95**
Upper Addison Gdns. W14 .... 47Ab **88**
Upper Austin Lodge Rd. DA4: Eyns 77Md **163**
Upper Av. DA13: Ist R 7A **144**
Upper Bank St. E14 .... 46Dc **92**
.... (not continuous)
Upper Bardsey Wlk. N1 .... 37Sb **71**
.... *(off Douglas Rd. Nth.)*
Upper Barn HP3: Hem H 5P **3**
Upper Belgrave St. SW1 .... 3J **227** (48Jb **90**)
Upper Berenger Wlk. SW10 .... 52Fb **111**
.... *(off Worlds End Est.)*
Upper Berkeley St. W1 .... 3F **221** (44Hb **89**)
Upper Beulah Hill SE19 .... 67Ub **135**
Upper Blantyre Wlk. SW10 .... 52Fb **111**
.... *(off Worlds End Est.)*
Upper Bourne End La. HP1: Hem H 6A **2**
.... (not continuous)
Upper Brentwood Rd. RM2: Rom 28Ld **57**
Upper Bri. Rd. RH1: Redh 6N **207**
Upper Brighton Rd. KT6: Surb 72Ma **153**
Upper Brockley Rd. SE4 .... 55Bc **114**
.... (not continuous)
Upper Brook St. W1 .... 5H **221** (45Jb **90**)
Upper Butts TW8: Bford 51La **108**
Upper Caldy Wlk. N1 .... 37Sb **71**
.... *(off Caldy Wlk.)*
Upper Camelford Wlk. W11 .... 44Ab **88**
Upper Cavendish Av. N3 .... 27Cb **49**
Upper Cheapside Pas. EC2 .... 3D **224** (44Sb **91**)
.... *(off Cheapside)*
Upper Cheyne Row SW3 .... 51Gb **111**
Upper Church Hill DA9: Ghithe 57Ud **120**
UPPER CLAPTON 33Xb **71**
Upper Clapton Rd. E5 .... 32Xb **71**
Upper Clarendon Wlk. W11 .... 44Ab **88**
.... *(off Clarendon Rd.)*
Upper Cornsland CM14: B'wood 20Zd **41**
Upper Ct. Rd. CR3: Wold 95Cc **198**
Upper Ct. Rd. KT19: Eps 83Sa **173**
Upper Culver Rd. AL1: St A 1C **6**
Upper Dagnall St. AL3: St A 2B **6**
Upper Dartrey Wlk. SW10 .... 52Eb **111**
.... *(off Worlds End Est.)*
Upper Dengie Wlk. N1 .... 39Sb **71**
.... *(off Baddow Wlk.)*
Upper Dock Wlk. E16 .... 45Sc **94**
.... *(off Frobisher Yd.)*
Upper Dr. TN16: Big H 90Lc **179**
Upper Dunnymans SM7: Bans 86Bb **175**
UPPER EDMONTON 22Wb **51**
UPPER ELMERS END 71Bc **158**
Upper Elmers End Rd. BR3: Beck 70Ac **136**
Upper Fairfield Rd. KT22: Lea 93Ka **192**
Upper Farm Rd. KT8: W Mole 70Ba **129**
Upper Feilde W1 .... 4H **221** (45Jb **90**)
.... *(off Park St.)*
Upper Fosters NW4 .... 28Ya **48**
.... *(off New Brent St.)*
UPPER GATTON 99Fb **195**
Upper Green E. CR4: Mitc 69Hb **133**
Upper Green W. CR4: Mitc 68Hb **133**
.... (not continuous)
Upper Grosvenor St. W1 .... 5H **221** (45Jb **90**)
Upper Grotto Rd. TW1: Twick 61Ha **130**
Upper Ground SE1 .... 6K **223** (46Qb **90**)
Upper Gro. SE25 .... 70Ub **135**
Upper Grove Rd. DA17: Belv 51Bd **117**
Upper Guild Hall DA9: Bluew 59Vd **120**
Upper Gulland Wlk. N1 .... 38Sb **71**
.... *(off Church Rd.)*
UPPER HALLIFORD 70U **128**
Upper Halliford By-Pass TW17: Shep 71U **150**
Upper Halliford Grn. TW17: Shep 70U **128**
Upper Halliford Rd. TW17: Shep 69U **128**
.... (not continuous)
Upper Halliford Station (Rail) 68U **128**
Upper Hall Pk. HP4: Berk 2A **2**
Upper Hampstead Wlk. NW3 .... 35Eb **69**
Upper Ham Rd. KT2: King T 63Ma **131**
Upper Ham Rd. TW10: Ham 63Ma **131**
Upper Handa Wlk. N1 .... 37Tb **71**
.... *(off Handa Wlk.)*
Upper Harestone CR3: Cat'm 99Wb **197**
Upper Hawkwell Wlk. N1 .... 39Sb **71**
.... *(off Maldon Cl.)*
Upper Heath Rd. AL1: St A 1D **6**
Upper High St. KT17: Eps 85Ua **174**
Upper Highway WD4: Hunt C 4S **12**
Upper Highway WD5: Ab L 4S **12**
Upper Hill Ri. WD3: Rick 16J **25**
Upper Hitch WD19: Wat 18Aa **27**
UPPER HOLLOWAY 33Lb **70**
Upper Holloway Station (Overground) 33Mb **70**
Upper Holly Hill Rd. DA17: Belv 50Dd **96**
Upper James St. W1 .... 4C **222** (45Lb **90**)
Upper John St. W1 .... 4C **222** (45Lb **90**)
Upper Lattimore Rd. AL1: St A 2C **6**
Upper Lees Rd. SL2: Slou 1F **80**
Upper Lismore Wlk. N1 .... 37Sb **71**
.... *(off Clephane Rd.)*
Upper Lodge W8 .... 46Db **89**
.... *(off Palace Grn.)*
Upper Lodge M. TW12: Hamp H 65Fa **130**
Upper Lodge Way CR5: Coul 94Mb **196**
Upper Mall W6 .... 50Wa **88**
.... (not continuous)
Upper Marlborough Rd. AL1: St A 2C **6**
Upper Marsh SE1 .... 3J **229** (48Pb **90**)
Upper Mdw. SL9: Ger X 2N **61**
Upper Montagu St. W1 .... 7F **215** (43Hb **89**)
Upper Mulgrave Rd. SM2: Cheam 80Ab **154**
Upper Nth. St. E14 .... 43Cc **92**
UPPER NORWOOD 67Ub **135**
Upper Nursery SL5: S'dale 1E **146**
Upper Paddock Rd. WD19: Wat 16Aa **27**
Upper Pk. IG10: Lough 14Mc **35**
Upper Pk. Rd. BR1: Broml 67Kc **137**
Upper Pk. Rd. DA17: Belv 49Dd **96**
Upper Pk. Rd. KT2: King T 65Qa **131**
Upper Pk. Rd. N11 .... 22Kb **50**
Upper Pk. Rd. NW3 .... 36Hb **69**
Upper Phillimore Gdns. W8 .... 47Cb **89**
Upper Pillory Down SM5: Cars 85Jb **176**
Upper Pines SM7: Bans 89Hb **175**
Upper Rainham Rd. RM12: Horn 32Hd **76**
Upper Ramsey Wlk. N1 .... 37Tb **71**
.... *(off Ramsey Wlk.)*
Upper Rawreth Wlk. N1 .... 39Sb **71**
.... *(off Basire St.)*
Upper Richmond Rd. SW15 .... 56Va **110**
Upper Richmond Rd. W. SW14 .... 56Ra **109**
Upper Richmond Rd. W. TW10: Rich 56Qa **109**
Upper Rd. E13 .... 41Jc **93**
Upper Rd. SM6: W'gton 78Mb **156**
Upper Rd. UB9: Den 31F **62**
Upper Rose Gallery DA9: Bluew 59Wd **120**
UPPER RUXLEY 66Cd **140**
Upper Ryle CM14: B'wood 17Xd **40**
Upper St Martin's La. WC2 .... 4F **223** (45Nb **90**)
Upper Sales HP1: Hem H 3H **3**
Upper Sawleywood SM7: Bans 86Bb **175**
Upper Selsdon Rd. CR2: Sande 80Vb **157**
Upper Selsdon Rd. CR2: Sels 80Vb **157**
Upper Sheridan Rd. DA17: Belv 49Cd **96**
UPPER SHIRLEY 77Zb **158**
Upper Shirley Rd. CR0: C'don 75Yb **158**
Upper Spring La. TN15: Igh 94Xd **204**
Upper Sq. TW7: Isle 55Ja **108**
Upper Stanford GU24: Pirb 7D **186**
Upper Station Rd. WD7: R'lett 7Ja **14**
Upper St. N1 .... 1A **218** (40Qb **70**)
Upper St. Nth. DA3: Nw A G 75Be **165**
.... *(off The Row)*
Upper St. Sth. DA3: Nw A G 75Be **165**
.... *(off The Row)*
Upper Sunbury Rd. TW12: Hamp 67Aa **129**
Upper Sutton La. TW5: Hest 52Ca **107**
Upper Swaines CM16: Epp 2Vc **23**
UPPER SYDENHAM 62Xb **135**
Upper Tachbrook St. SW1 .... 5B **228** (49Lb **90**)
Upper Tail WD19: Wat 20Aa **27**
Upper Talbot Wlk. W11 .... 44Ab **88**
.... *(off Talbot Wlk.)*
Upper Teddington Rd. KT1: Hamp W 66La **130**
Upper Ter. NW3 .... 34Eb **69**
Upper Thames St. EC4 .... 4C **224** (45Rb **91**)
Upper Thames Wlk. DA9: Bluew 60Vd **120**
Upper Tollington Pk. N4 .... 32Qb **70**
.... (not continuous)
Upperton Rd. DA14: Sidc 64Vc **139**
Upperton Rd. E. E13 .... 41Lc **93**
Upperton Rd. W. E13 .... 41Lc **93**
UPPER TOOTING 63Hb **133**
Upper Tooting Pk. SW17 .... 61Hb **133**
Upper Tooting Rd. SW17 .... 63Hb **133**
Upper Town Rd. UB6: G'frd 42Da **85**
Upper Tulse Hill SW2 .... 59Pb **112**
Upper Vernon Rd. SM1: Sutt 78Fb **155**
Upper Village Rd. SL5: S'hill 1A **146**
Upper Wlk. GU25: Vir W 70A **126**
UPPER WALTHAMSTOW 28Fc **53**
Upper Walthamstow Rd. E17 .... 28Ec **52**
UPPER WARLINGHAM 89Wb **177**
Upper Warlingham Station (Rail) 90Wb **177**
Upper West St. RH2: Reig 6H **207**
Upper Whistler Wlk. SW10 .... 52Eb **111**
.... *(off Worlds End Est.)*
Upper Wickham La. DA16: Well 52Xc **117**
Upper Wimpole St. W1 .... 7J **215** (43Jb **90**)
Upper Woburn Pl. WC1 .... 4E **216** (41Mb **90**)
Upper Woodcote Village CR8: Purl 84Mb **176**
Uppingham Av. HA7: Stan 25Ka **46**
Upsdell Av. N13 .... 23Qb **50**
UPSHIRE 5Mc **21**
Upshirebury Grn. EN9: Walt A 5Mc **21**
Upshire Ho. E17 .... 26Bc **52**
Upshire Rd. EN9: Walt A 4Hc **21**
Upshot La. GU22: Pyr 89H **169**
Upstairs at the Gatehouse Theatre Highgate **32Jb 70**
Upstall St. SE5 .... 53Rb **113**
UPTON 57Zc **117**
UPTON 38Jc **73**
Upton GU21: Wok 9M **167**
UPTON 8L **81**
Upton Av. AL3: St A 1B **6**
Upton Av. E7 .... 38Jc **73**
UPTON CEN. **56Ad 117**
Upton Cl. AL2: Park 7B **6**
Upton Cl. DA5: Bexl 58Bd **117**
Upton Cl. NW2 .... 34Ab **68**
Upton Cl. SL1: Slou 8K **81**
Upton Cl. SS17: Stan H 1M **101**
Upton Ct. SE20 .... 66Yb **136**
.... *(off Blean Gro.)*
Upton Ct. Rd. SL3: L'ly 8L **81**
Upton Ct. Rd. SL3: Slou 8L **81**
Upton Dene SM2: Sutt 80Db **155**
Upton Gdns. HA3: Kenton 29Ka **46**
Upton Hgts. E7 .... 38Jc **73**
UPTON HOSPITAL **7K 81**
Upton Ho. E9 .... 36Ac **72**
.... *(off Ward La.)*
Upton Ho. RM3: Rom 22Md **57**
.... *(off Barnstaple Rd.)*
Upton La. E7 .... 38Jc **73**
Upton Lea Pde. SL2: Slou 5M **81**
Upton Lodge E7 .... 37Jc **73**
Upton Lodge Cl. WD23: Bush 17Ea **28**
UPTON PARK 39Mc **73**
Upton Pk. SL1: Slou 8J **81**
UPTON PARK 8K **81**
Upton Pk. Boleyn Cinema **40Mc 73**
Upton Pk. Rd. E7 .... 38Kc **73**
Upton Park Station (Underground) 39Lc **73**
Upton Rd. CR7: Thor H 68Tb **135**
Upton Rd. DA5: Bexl 58Bd **117**
Upton Rd. DA6: Bex 56Ad **117**
Upton Rd. N18 .... 22Wb **51**
Upton Rd. SE18 .... 51Sc **116**
Upton Rd. SL1: Slou 8L **81**
Upton Rd. TW3: Houn 55Ca **107**
Upton Rd. WD18: Wat 13X **27**
Upton Rd. Sth. DA5: Bexl 58Bd **117**
Upton Vs. DA6: Bex 56Ad **117**
Upward Ct. RM1: Rom 28Hd **56**
Upway N12 .... 23Gb **49**
Upway SL9: Chal P 25B **42**
Upwey Ho. N1 .... 39Ub **71**
Upwood Rd. SE12 .... 58Jc **115**
Upwood Rd. SW16 .... 67Nb **134**

Uranus Rd. HP2: Hem H 1P 3
Urban Av. RM12: Horn 34Ld 77
Urbanest King's Cross N1 .... 39Nb 70
Urban M. N4 .... 30Rb 51
Urdang, The 4A 218 (41Qb 90)
(off Rosebery Av.)
URGENT CARE CENTRE (ANGEL MEDICAL PRACTICE) 1A 218 (40Qb 70)
URGENT CARE CENTRE (BARNET) 14Za 30
URGENT CARE CENTRE (BECKENHAM BEACON) 68Bc 136
URGENT CARE CENTRE (CARSHALTON) 74Eb 155
URGENT CARE CENTRE (CENTRAL MIDDLESEX HOSPITAL) 41Sa 87
URGENT CARE CENTRE (CHASE FARM HOSPITAL) 11Qb 32
URGENT CARE CENTRE (CHELSEA & WESTMINSTER HOSPITAL) 51Eb 111
URGENT CARE CENTRE (EALING) 47Fa 86
URGENT CARE CENTRE (ERITH & DISTRICT HOSPITAL) 51Fd 118
URGENT CARE CENTRE (FULHAM) 50Za 88
URGENT CARE CENTRE (HAMMERSMITH HOSPITAL) 44Wa 88
URGENT CARE CENTRE (HAMPSTEAD) 36Gb 69
URGENT CARE CENTRE (HEMEL HEMPSTEAD) 2M 3
URGENT CARE CENTRE (HILLINGDON HOSPITAL) 43P 83
URGENT CARE CENTRE (HOMERTON UNIVERSITY HOSPITAL) 36Zb 72
URGENT CARE CENTRE (KING GEORGE HOSPITAL) 28Wc 55
URGENT CARE CENTRE (NEWHAM) 42Lc 93
URGENT CARE CENTRE (NORTH MIDDLESEX UNIVERSITY HOSPITAL) 22Ub 51
URGENT CARE CENTRE (NORTHWICK PARK HOSPITAL) 31Ja 66
URGENT CARE CENTRE (PRINCESS ROYAL UNIVERSITY HOSPITAL) 77Qc 160
URGENT CARE CENTRE (QUEEN ELIZABETH HOSPITAL) Charlton 51Nc 116
URGENT CARE CENTRE (QUEEN'S HOSPITAL) 31Gd 76
URGENT CARE CENTRE (ST CHARLES CENTRE) 43Za 88
URGENT CARE CENTRE (ST GEORGE'S HOSPITAL) 64Gb 133
URGENT CARE CENTRE (ST MARY'S HOSPITAL) 2B 220 (44Fb 89)
URGENT CARE CENTRE (SIDCUP) 65Wc 139
URGENT CARE CENTRE (THORNTON HEATH) 72Rb 157
URGENT CARE CENTRE (UNIVERSITY COLLEGE HOSPITAL) 5C 216 (42Lb 90)
URGENT CARE CENTRE (UNIVERSITY HOSPITAL LEWISHAM) 57Dc 114
URGENT CARE CENTRE (WEST MIDDLESEX UNIVERSITY HOSPITAL) 54Ja 108
URGENT CARE CENTRE (WHIPPS CROSS UNIVERSITY HOSPITAL) 30Fc 53
URGENT CARE CENTRE (WHITTINGTON HOSPITAL) 33Lb 70
Urlwin St. SE5 .... 51Sb 113
Urlwin Wlk. SW9 .... 53Qb 112
Urmston Dr. SW19 .... 60Ab 110
Urmston Ho. E14 .... 49Ec 92
(off Seyssel St.)
Urquhart Ct. BR3: Beck 66Bc 136
Ursa Mans. E20 .... 36Ec 72
(off Cheering La.)
Ursula Gould Way E14 .... 43Cc 92
Ursula Lodges DA14: Sidc 64Xc 139
Ursula M. N4 .... 32Sb 71
Ursula St. SW11 .... 53Gb 111
Urswick Gdns. RM9: Dag 38Ad 75
Urswick Rd. E9 .... 36Yb 72
Urswick Rd. RM9: Dag 38Zc 75
Usborne M. SW8 .... 52Pb 112
Usher Hall NW4 .... 28Xa 48
(off The Burroughs)
Usher Rd. E3 .... 39Bc 72
(not continuous)
Usk Rd. RM15: Avel 43Sd 98
Usk Rd. SW11 .... 56Eb 111
Usk St. E2 .... 41Zb 92
Utah Bldg. SE13 .... 53Dc 114
(off Deal's Gateway)
Utopia Village NW1 .... 38Jb 70
Uvedale Cl. CR0: New Ad 83Fc 179
Uvedale Cres. CR0: New Ad 83Fc 179
Uvedale Rd. EN2: Enf 15Tb 33
Uvedale Rd. RH8: Oxt 2K 211
Uvedale Rd. RM10: Dag 34Cd 76
Uverdale Rd. SW10 .... 52Eb 111
UXBRIDGE 38M 63
Uxbridge Bus Station 38M 63
Uxbridge Ct. KT1: King T 71Ma 153
(off Uxbridge Rd.)
Uxbridge Golf Course 33N 63
Uxbridge Ind. Est. UB8: Uxb 40K 63
Uxbridge Lido 37N 63
UXBRIDGE MOOR 40K 63
Uxbridge Moor Nature Reserve 38K 63
Uxbridge Rd. HA3: Hrw W 24Ea 46
Uxbridge Rd. HA5: Hat E 26Y 45
Uxbridge Rd. HA5: Pinn 26Y 45
Uxbridge Rd. HA7: Stan 23Ha 46
Uxbridge Rd. KT1: King T 70Ma 131
Uxbridge Rd. SL0: Iver H 41D 82
Uxbridge Rd. SL1: Slou 7L 81
Uxbridge Rd. SL2: Slou 6M 81
Uxbridge Rd. SL3: Geor G 43A 82
Uxbridge Rd. SL3: Wex 43A 82
Uxbridge Rd. TW12: Hamp 63Ca 129
Uxbridge Rd. TW12: Hamp H 63Ca 129
Uxbridge Rd. TW13: Felt 61Y 129
Uxbridge Rd. UB1: S'hall 46Ca 85
Uxbridge Rd. UB10: Hil 41Q 84
Uxbridge Rd. UB4: Hayes 43U 84
Uxbridge Rd. UB4: Yead 43U 84
Uxbridge Rd. W12 .... 46Wa 88
Uxbridge Rd. W13 .... 46Ka 86
Uxbridge Rd. W3 .... 46Qa 87
Uxbridge Rd. W5 .... 45Na 87
Uxbridge Rd. W7 .... 46Ha 86
Uxbridge Rd. WD3: Rick 20H 25
Uxbridge Rd. Retail Pk. 45Y 85
Uxbridge Station (Underground) 38M 63
Uxbridge St. W8 .... 46Cb 89
Uxbridge Tourist Info. Cen. 38M 63
Uxendon Cres. HA9: Wemb 32Na 67
Uxendon Hill HA9: Wemb 32Pa 67

V

Vaillant Rd. KT13: Weyb 77S 150
Vaine Ho. E9 .... 37Ac 72
Vaizeys Wharf SE7 .... 48Kc 93
Valance Av. E4 .... 18Gc 35
Valan Leas BR2: Broml 69Gc 137
Vale, The CM14: B'wood 18Yd 40
Vale, The CR0: C'don 75Zb 158
Vale, The CR5: Coul 86Mb 176
Vale, The HA4: Ruis 35Y 65
Vale, The IG8: Wfd G 24Jc 53
Vale, The N10 .... 25Jb 50
Vale, The N14 .... 17Mb 32
Vale, The NW11 .... 34Za 68
Vale, The SL9: Chal P 25A 42
Vale, The SW3 .... 51Fb 111
Vale, The TW14: Felt 58X 107
Vale, The TW16: Sun 65W 128
Vale, The TW5: Hest 51Aa 107
Vale, The W3 .... 46Ta 87
Vale Av. WD6: Bore 15Ra 29
Vale Border CR2: Sels 83Zb 178
Vale Cl. BR6: Farnb 77Qc 160
Vale Cl. CM15: Pil H 15Vd 40
Vale Cl. CR5: Coul 86Nb 176
Vale Cl. GU21: Wok 88A 168
Vale Cl. KT13: Weyb 76T 150
Vale Cl. KT18: Eps D 91Ua 194
Vale Cl. N2 .... 27Hb 49
Vale Cl. SL9: Chal P 25A 42
Vale Cl. TW1: Twick 62Ja 130
Vale Cl. W9 .... 4A 214 (41Eb 89)
Vale Cotts. SW15 .... 62Ua 132
Vale Ct. EN5: New Bar 14Db 31
Vale Ct. KT13: Weyb 76T 150
Vale Ct. W3 .... 46Va 88
Vale Ct. W9 .... 4A 214 (41Eb 89)
Vale Cres. SW15 .... 63Ua 132
Vale Cft. HA5: Pinn 29Aa 45
Vale Cft. KT10: Clay 81Ha 172
Vale Dr. EN5: Barn 14Bb 31
Vale End SE22 .... 56Vb 113
Vale Est., The W3 .... 46Ua 88
Vale Farm Rd. GU21: Wok 89A 168
Vale Farm Sports Cen. 35Ka 66
Vale Gro. N4 .... 31Sb 71
Vale Gro. SL1: Slou 8J 81
Vale Gro. W3 .... 47Ta 87
Vale Ho. GU21: Wok 89A 168
Vale Ind. Est. WD18: Wat 17R 26
Vale La. W3 .... 43Qa 87
Vale Lodge SE23 .... 61Yb 136
Valence Av. RM8: Dag 32Zc 75
Valence Cir. RM8: Dag 34Zc 75
Valence House Mus. 34Ad 75
Valence Rd. DA8: Erith 52Fd 118
Valence Wood Rd. RM8: Dag 34Zc 75
Valencia Cl. E14 .... 44Fc 93
Valencia Rd. HA7: Stan 21La 46
Valencia Twr. EC1 .... 3D 218 (41Sb 91)
(off Bollinder Place)
Valency Cl. HA6: Nwood 21V 44
Valency Dr. HA6: Nwood 21V 44
Valentia Pl. SW9 .... 56Qb 112
Valentina Av. NW9 .... 26Va 48
Valentine Av. DA5: Bexl 61Ad 139
Valentine Ct. SE23 .... 61Zb 136
(not continuous)
Valentine Ho. E3 .... 39Bc 72
(off Garrison Rd.)
Valentine Pl. SE1 .... 2B 230 (47Rb 91)
Valentine Rd. E9 .... 37Zb 72
Valentine Rd. HA2: Harr 34Da 65
Valentine Row SE1 .... 2B 230 (47Rb 91)
Valentines Mansion & Gdns. 31Qc 74
Valentines Rd. IG1: Ilf 32Rc 74
Valentine's Way RM7: Rush G 33Gd 76
Valentine Vs. RM15: S Ock 43Yd 98
Valentine Way HP8: Chal G 20A 24
Valentyne Cl. CR0: New Ad 83Gc 179
Vale of Health NW3 .... 34Fb 69
VALE OF HEALTH 34Eb 69
Vale Pde. SW15 .... 62Ua 132
Valerian Wlk. N11 .... 19Jb 32
Valerian Way E15 .... 41Gc 93
Valerie Cl. AL1: St A 2F 6
Valerie Ct. SM2: Sutt 80Db 155
Valerie Ct. WD23: Bush 17Ea 28
Valerio M. N1 .... 37Tb 71
Vale Ri. NW11 .... 32Bb 69
Vale Rd. BR1: Broml 67Qc 138
Vale Rd. CR4: Mitc 69Mb 134
Vale Rd. DA1: Dart 60Kd 119
Vale Rd. DA11: Nflt 59Fe 121
Vale Rd. E7 .... 37Kc 73
Vale Rd. KT10: Clay 81Ga 172
Vale Rd. KT13: Weyb 76T 150
Vale Rd. KT19: Ewe 77Va 154
Vale Rd. KT4: Wor Pk 76Va 154
Vale Rd. N4 .... 31Sb 71
Vale Rd. SL4: Wind 2D 102
Vale Rd. SM1: Sutt 77Db 155
Vale Rd. WD23: Bush 15Aa 27
Vale Rd. Nth. KT6: Surb 75Na 153
Vale Rd. Sth. KT6: Surb 75Na 153
Vale Row N5 .... 34Rb 71
Vale Royal N7 .... 38Nb 70
Vale Royal Ho. WC2 .... 4E 222 (45Mb 90)
(off Charing Cross Rd.)
Valery Pl. TW12: Hamp 66Ca 129
Valeside Ct. EN5: New Bar 14Db 31
Vale St. SE27 .... 62Tb 135
Valeswood Rd. BR1: Broml 64Hc 137
Vale Ter. N4 .... 30Sb 51
Valetta Gro. E13 .... 40Jc 73
Valetta Ho. SW11 .... 52Kb 112
Valetta Rd. W3 .... 47Ua 88
Valette Ct. N10 .... 28Kb 50
(off St James's La.)
Valette Ho. E9 .... 37Yb 72
Valette St. E9 .... 37Yb 72
Valiant Cl. RM7: Mawney 26Dd 56
Valiant Cl. UB5: N'olt 41Z 85
Valiant Ho. E14 .... 47Ec 92
(off Plevna St.)
Valiant Ho. SE7 .... 50Lc 93
Valiant Path NW9 .... 24Ua 48
Valiant Way E6 .... 43Pc 94
Vallance Rd. E1 .... 42Wb 91
Vallance Rd. E2 .... 41Wb 91
Vallance Rd. N22 .... 26Lb 50
Vallentin Rd. E17 .... 28Ec 52
Valley, The 50Lc 93
Valley Av. N12 .... 21Fb 49
Valley Cl. DA1: Cray 58Hd 118
Valley Cl. EN9: Walt A 4Ec 20
Valley Cl. HA5: Pinn 26X 45
Valley Cl. IG10: Lough 16Pc 36
Valley Ct. CR3: Cat'm 94Wb 197
Valley Ct. RH2: Reig 5M 207
Valley Dr. DA12: Grav'nd 3F 144
Valley Dr. NW9 .... 30Qa 47
Valley Dr. TN13: S'oaks 98Kd 203
VALLEY END 9E 146
Valley End SL3: Wex 2N 81
Valley End Rd. GU24: Chob 9E 146
Valleyfield Rd. SW16 .... 64Pb 134
Valley Flds. Cres. EN2: Enf 12Qb 32
Valley Gdns., The 8J 125
Valley Gdns. DA9: Ghithe 58Xd 120
Valley Gdns. HA0: Wemb 38Pa 67
Valley Gdns. SW19 .... 66Fb 133
Valley Gro. SE7 .... 50Lc 93
Valley Hgts. DA1: Dart 56Pd 119
Valley Hill IG10: Lough 17Nc 36
Valley Ho. EN8: Chesh 4Zb 20
Valley Leisure Pk. 74Nb 156
Valleylink Est. EN3: Pond E 16Ac 34
Valley Lodge IG10: Lough 16Pc 36
Valley M. TW1: Twick 61Ha 130
Valley Pk. BR8: Hext 66Kd 141
Valley Pk. WD18: Wat 17S 26
Valley Point Ind. Est. CR0: Bedd 73Nb 156
Valley Ri. WD25: Wat 5X 13
Valley Rd. BR2: Broml 68Gc 137
Valley Rd. BR5: St P 67Xc 139
Valley Rd. CR8: Kenley 87Tb 177
Valley Rd. DA1: Cray 58Hd 118
Valley Rd. DA17: Belv 49Dd 96
Valley Rd. DA3: Fawk 74Xd 164
Valley Rd. DA8: Erith 49Ed 96
Valley Rd. SW16 .... 64Pb 134
Valley Rd. UB10: Uxb 40N 63
Valley Rd. WD3: Rick 15J 25
Valley Side E4 .... 19Cc 34
Valley Side SE7 .... 50Mc 93
Valleyside HP1: Hem H 2H 3
Valley Side Pde. E4 .... 19Cc 34
Valley Vw. DA9: Ghithe 58Xd 120
Valley Vw. EN5: Barn 16Ab 30
Valley Vw. EN7: G Oak 1Sb 19
Valley Vw. TN16: Big H 90Lc 179
Valley Vw. Gdns. CR8: Kenley 87Ub 177
Valley Vw. Ter. DA4: Farni 74Pd 163
Valley Wlk. CR0: C'don 75Yb 158
Valley Wlk. WD3: Crox G 15S 26
Valley Way SL9: Ger X 1N 61
Valliere Rd. NW10 .... 41Wa 88
Valliers Wood Rd. DA15: Sidc 60Uc 116
Vallings Pl. KT6: Surb 73Ka 152
Vallis Way KT9: Chess 77Ma 153
Vallis Way W13 .... 43Ja 86
Val McKenzie Av. N7 .... 34Qb 70
Valmar Av. SS17: Stan H 2K 101
Valmar Rd. SE5 .... 53Sb 113
Valmar Trad. Est. SE5 .... 53Sb 113
Valnay St. SW17 .... 64Hb 133
Valognes Av. E17 .... 25Ac 52
Valois Ho. SE1 .... 3K 231 (48Vb 91)
(off St Saviour's Est.)
Valonia Gdns. SW18 .... 58Bb 111
Vambery Rd. SE18 .... 51Sc 116
Vamey Ct. BR1: Broml 69Jc 137
V&A Mus. of Childhood 41Yb 92
Vanbrough Cres. UB5: N'olt 39Y 65
Van Brugh Av. CR5: Coul 89Lb 176
Vanbrugh Castle SE10 .... 51Gc 115
(off Maze Hill)
Vanbrugh Cl. E16 .... 43Mc 93
Vanbrugh Ct. SE11 .... 6A 230 (49Qb 90)
(off Wincott St.)
Vanbrugh Dr. KT12: Walt T 73Y 151
Vanbrugh Flds. SE3 .... 51Hc 115
Vanbrugh Hill SE10 .... 50Hc 93
Vanbrugh Hill SE3 .... 51Hc 115
Vanbrugh Ho. E9 .... 38Yb 72
(off Loddiges Rd.)
Vanbrugh M. E15 .... 35Gc 73
Vanbrugh M. KT12: Walt T 73Y 151
Vanbrugh Pk. SE3 .... 52Hc 115
Vanbrugh Pk. Rd. SE3 .... 52Hc 115
Vanbrugh Pk. Rd. W. SE3 .... 52Hc 115
Vanbrugh Rd. W4 .... 48Ta 87
Vanbrugh Ter. SE3 .... 53Hc 115
Vanburgh Cl. BR6: Orp 74Uc 160
Vanburgh Ho. E1 .... 7K 219 (43Vb 91)
(off Folgate St.)
Vancouver Cl. BR6: Chels 77Wc 161
Vancouver Cl. KT19: Eps 83Sa 173
Vancouver Ho. E1 .... 46Xb 91
(off Reardon Path)
Vancouver Ho. SE16 .... 47Zb 92
(off Needleman St.)
Vancouver Mans. HA8: Edg 25Ra 47
Vancouver Rd. HA8: Edg 25Ra 47
Vancouver Rd. SE23 .... 61Ac 136
Vancouver Rd. TW10: Ham 63La 130
Vancouver Rd. UB4: Yead 42X 85
Vanda Cres. AL1: St A 3D 6
Vanderbilt Rd. SW18 .... 60Db 111
Vanderbilt Vs. W12 .... 47Za 88
(off Sterne St.)
Vandervell Ct. W3 .... 47Ua 88
(off Amber Way)
Vanderville Gdns. N2 .... 26Fb 49
Vandome Cl. E16 .... 44Kc 93
Vandon Ct. SW1 .... 3C 228 (48Lb 90)
(off Petty France)
Vandon Pas. SW1 .... 3C 228 (48Lb 90)
Vandon St. SW1 .... 3C 228 (48Lb 90)
Van Dyck Av. KT3: N Mald 73Ta 153
Vandyke Cl. RH1: Redh 3P 207
Vandyke Cl. SW15 .... 59Za 110
Vandyke Cross SE9 .... 57Nc 116
Vandy St. EC2 .... 6H 219 (42Ub 91)
Vane Cl. HA3: Kenton 30Pa 47
Vane Cl. NW3 .... 36Fb 69
Vanessa Cl. DA17: Belv 50Cd 96
Vanessa Wlk. DA12: Grav'nd 4H 145
Vanessa Way DA5: Bexl 62Fd 140
Vane St. SW1 .... 5C 228 (49Lb 90)
Vange Ho. W10 .... 43Ya 88
(off Sutton Way)
Van Gogh Cl. TW7: Isle 55Ja 108
Van Gogh Ct. E14 .... 48Fc 93
Vanguard NW9 .... 24Ua 48
Vanguard Bldg. E14 .... 47Bc 92
Vanguard Cl. CR0: C'don 74Rb 157
Vanguard Cl. E16 .... 43Jc 93
Vanguard Cl. RM7: Mawney 26Cd 56
Vanguard Ct. SE5 .... 53Ub 113
Vanguard Ho. E8 .... 38Xb 71
Vanguard St. SE8 .... 53Cc 114
Vanguard Way SM6: W'gton 80Nb 156
Vanguard Way TW6: H'row A 54U 106
Vanilla & Sesame Ct. SE1 .... 1K 231 (47Vb 91)
(off Curlew St.)
Vanneck Sq. SW15 .... 57Wa 110
Vanners Pde. KT14: Byfl 85N 169
Vanoc Gdns. BR1: Broml 63Jc 137
Vanquish Cl. TW2: Whitt 59Ca 107
Vanquisher Wlk. DA12: Grav'nd 2H 145
Vanryne Ho. IG10: Lough 13Nc 36
Vansittart Est. SL4: Wind 2G 102
Vansittart Rd. E7 .... 35Hc 73
Vansittart Rd. SL4: Wind 3F 102
Vansittart St. SE14 .... 52Ac 114
Vanstone Ct. N7 .... 37Qb 70
(off Blackthorn Av.)
Vanston Pl. SW6 .... 52Cb 111
Vantage Bldg. UB3: Hayes 48V 84
(off Station App.)
Vantage Ct. GU21: Wok 9P 167
Vantage Ct. UB3: Harl 52U 106
Vantage M. E14 .... 46Ec 92
(off Coldharbour)
Vantage M. HA6: Nwood 23T 44
Vantage Pl. TW14: Felt 58W 106
Vantage Pl. W8 .... 48Cb 89
Vantage Point BR3: Beck 67Fc 137
(off Albemarle Rd.)
Vantage Point CR2: Sande 81Tb 177
Vantage Point DA9: Ghithe 59Wd 120
Vantage Point EN5: Barn 14Bb 31
(off Victors Way)
Vantage Rd. SL1: Slou 6F 80
Vantage W. TW8: Bford 49Pa 87
Vantrey Ho. SE11 .... 6K 229 (49Qb 90)
(off Marylee Way)
Vant Rd. SW17 .... 64Hb 133
Vapery La. GU24: Pirb 3B 186
Varcoe Gdns. UB3: Hayes 44T 84
Varcoe Rd. SE16 .... 50Xb 91
Vardens Rd. SW11 .... 56Fb 111
Varden St. E1 .... 44Xb 91
Vardon Cl. W3 .... 44Ta 87
Vardy Dr. TW1: Isle 56Ka 108
Varley Ho. NW6 .... 39Cb 69
Varley Ho. SE1 .... 4E 230 (48Sb 91)
(off County St.)
Varley Pde. NW9 .... 28Ua 48
Varley Rd. E16 .... 44Kc 93
Varley Way CR4: Mitc 68Fb 133
Varna Rd. SW6 .... 52Ab 110
Varna Rd. TW12: Hamp 67Da 129
Varndell St. NW1 .... 3B 216 (41Lb 90)
Varney Cl. HP1: Hem H 2H 3
Varney Rd. HP1: Hem H 2H 3
Varnishers Yd. N1 .... 2G 217 (40Nb 70)
(off York Way)
Varsity Dr. TW1: Twick 57Ga 108
Varsity Row SW14 .... 54Sa 109
Vartry Rd. N15 .... 30Tb 51
Vascroft Est. NW10 .... 42Ra 87
Vassall Ho. E3 .... 41Ac 92
(off Antill Rd.)
Vassall Rd. SW9 .... 52Qb 112
Vat Ho. SW8 .... 52Nb 112
(off Rita Rd.)
Vauban Est. SE16 .... 4K 231 (48Vb 91)
Vauban St. SE16 .... 4K 231 (48Vb 91)
Vaudeville Cl. RM12: Horn 33Jd 76
Vaudeville Theatre 5G 223 (45Nb 90)
(off Strand)
Vaughan Almshouses TW15: Ashf 64R 128
(off Feltham Hill Rd.)
Vaughan Av. DA9: Ghithe 56Yd 120
Vaughan Av. NW4 .... 29Wa 48
Vaughan Av. RM12: Horn 35Md 77
Vaughan Av. W6 .... 49Va 88
Vaughan Cl. DA1: Dart 59Md 119
Vaughan Cl. TW12: Hamp 65Aa 129
Vaughan Copse SL4: Eton 9H 81
Vaughan Est. E2 .... 3K 219 (41Vb 91)
(off Diss St.)
Vaughan Gdns. IG1: Ilf 31Pc 74
Vaughan Gdns. SL4: Eton W 9D 80
Vaughan Ho. SE1 .... 1B 230 (47Rb 91)
(off Blackfriars Rd.)
Vaughan Ho. SW4 .... 59Lb 112
Vaughan Rd. DA16: Well 54Vc 117
Vaughan Rd. E15 .... 37Hc 73
Vaughan Rd. HA1: Harr 30Ea 46
Vaughan Rd. KT7: T Ditt 73Ka 152
Vaughan Rd. SE5 .... 54Sb 113
Vaughan St. SE16 .... 47Bc 92
Vaughan Way E1 .... 45Wb 91
Vaughan Way SL2: Slou 2C 80
Vaughan Williams Cl. SE8 .... 52Cc 114
Vaughan Williams Way CM14: Gt War 23Wd 58
Vaughan Williams Way CM14: W'ley 23Wd 58
Vaughn St. RM8: Dag 32Ad 75
Vaux Cres. KT12: Hers 79X 151
VAUXHALL 51Nb 112
Vauxhall Bri. SW1 .... 7F 229 (50Nb 90)
Vauxhall Bri. Rd. SW1 .... 4B 228 (48Lb 90)
Vauxhall Bus Station 51Nb 112
Vauxhall City Farm 7H 229 (50Pb 90)
(off Tyers St.)
Vauxhall Cl. DA11: Nflt 9B 122
VAUXHALL CROSS 50Nb 90
Vauxhall Gdns. CR2: S Croy 79Sb 157
Vauxhall Gro. SW8 .... 51Pb 112
Vauxhall Pl. DA1: Dart 59Nd 119
Vauxhall Rd. HP2: Hem H 2A 4
Vauxhall Road 2A 4
Vauxhall Station (Rail & Underground) 50Nb 90
Vauxhall St. SE11 .... 7J 229 (50Pb 90)
Vauxhall Wlk. SE11 .... 7H 229 (50Pb 90)
VauxWall East Climbing Cen. 6J 229 (49Pb 90)
VauxWall West Climbing Cen. 51Nb 112
(off Sth. Lambeth Rd.)
Vawdrey Cl. E1 .... 42Yb 92
Veals Mead CR4: Mitc 67Gb 133
Vectis Gdns. SW17 .... 65Kb 134
Vectis Rd. SW17 .... 65Kb 134
Veda Rd. SE13 .... 56Cc 114
Vega Cres. HA6: Nwood 22V 44
Vega Ho. E20 .... 36Ec 72
(off Prize Wlk.)
Vegal Cres. TW20: Eng G 4M 125
Vega Rd. WD23: Bush 17Ea 28
Veitch Cl. TW14: Felt 59V 106
Veldene Way HA2: Harr 34Ba 65
Velde Way SE22 .... 57Ub 113
Vellacott Cl. RM19: Purf 51Td 120
Velletri Ho. E2 .... 40Zb 72
(off Mace St.)
Vellum Ct. E17 .... 26Ac 52
Vellum Dr. SM5: Cars 76Jb 156
Velocity Way EN3: Enf L 9Bc 20
Velodrome Queen Elizabeth Olympic Pk. 36Dc 72
Velo Ho. E17 .... 29Bc 52
(off Track St.)
Velo Pl. E20 .... 36Dc 72
Velvet Ho. E2 .... 1K 219 (40Vb 71)
(off Whiston Rd.)
Venables Cl. RM10: Dag 35Dd 76
Venables St. NW8 .... 6C 214 (42Fb 89)
Vencourt Pl. W6 .... 49Wa 88
Venden Hgts. CR8: Purl 83Pb 176
Veneer Bldg., The 47U 84
Venerable Ho. E3 .... 42Bc 92
(off Portia Way)
Venetian Ho. E20 .... 37Dc 72
(off Victory Pde.)
Venetian Rd. SE5 .... 54Sb 113
Venetia Rd. N4 .... 30Rb 51
Venetia Rd. W5 .... 47Ma 87
Venette Cl. RM13: Rain 43Kd 97
Venice Av. WD18: Wat 14U 26
Venice Corte SE13 .... 55Ec 114
(off Elmira St.)
Venice Ct. SE5 .... 52Sb 113
(off Bowyer St.)
Venice Ho. HA0: Wemb 39Na 67
Venice Wlk. W2 .... 7A 214 (43Eb 89)
Venner Cl. RH1: Redh 5A 208
Venner Rd. SE26 .... 65Yb 136
(not continuous)
Venners Cl. DA7: Bex 54Gd 118
Venn Ho. N1 .... 1J 217 (39Pb 70)
(off Barnsbury Est.)
Venn St. SW4 .... 56Lb 112
Ventnor Av. HA7: Stan 25Ka 46
Ventnor Dr. N20 .... 20Db 31
Ventnor Gdns. IG11: Bark 37Uc 74
Ventnor Rd. SE14 .... 52Zb 114
Ventnor Rd. SM2: Sutt 80Db 155
Venton Cl. GU21: Wok 9M 167
Ventura Pk. AL2: Col S 1Ha 14
Venture Cl. DA5: Bexl 59Ad 117
Venture Ct. DA12: Grav'nd 8F 122
Venture Ct. SE1 .... 3J 231 (48Ub 91)
(off Market Yd. M.)
Venture Ct. SE12 .... 59Jc 115
Venture Ho. W10 .... 44Za 88
(off Bridge Cl.)
Venue (Leisure Cen.), The 12Sa 29
Venue St. E14 .... 43Ec 92
Venus Cl. SL2: Slou 2D 80
Venus Hill HP3: Bov 3C 10
VENUS HILL 3C 10
Venus Ho. E14 .... 49Cc 92
(off Westferry Rd.)
Venus Ho. E3 .... 39Cc 72
(off Garrison Rd.)
Venus M. CR4: Mitc 69Gb 133
Venus Rd. SE18 .... 48Pc 94
Veny Cres. RM12: Horn 36Md 77
Vera Av. N21 .... 15Qb 32
Vera Ct. E3 .... 41Dc 92
(off Grace Pl.)
Vera Ct. WD19: Wat 17Z 27
Vera Lynn Cl. E7 .... 35Jc 73
Vera Rd. SW6 .... 53Ab 110
Verbena Cl. E16 .... 42Hc 93
Verbena Cl. RM15: S Ock 44Yd 98
Verbena Cl. UB7: W Dray 50M 83
Verbena Gdns. W6 .... 50Wa 88
Verdana Ct. WD18: Wat 15U 26
(off Whippendell Rd.)
Verdant Ct. SE6 .... 59Gc 115
(off Verdant La.)
Verdant La. SE6 .... 59Gc 115
Verdayne Av. CR0: C'don 75Zb 158
Verdayne Gdns. CR6: W'ham 88Yb 178
Verderers Rd. IG7: Chig 22Wc 55
Verdi Cres. W10 .... 40Ab 68
Verdon Ct. SL2: Farn R 1F 80
Verdon Roe Ct. E4 .... 20Dc 34
Verdun Rd. SE18 .... 51Wc 117
Verdun Rd. SW13 .... 51Wa 110
Verdure Cl. WD25: Wat 4Aa 13
Vere Ct. W2 .... 44Db 89
(off Westbourne Gdns.)
Vereker Dr. TW16: Sun 69W 128
Vereker Rd. W14 .... 50Ab 88
Vere Rd. IG10: Lough 14Sc 36
Vere St. W1 .... 3K 221 (44Kb 90)
Veridion Way DA18: Erith 47Bd 95
Verini Cl. WD23: Bush 15Z 27
Veritas Ho. DA14: Sidc 61Wc 139
(off Jubilee Way)
Verity Cl. W11 .... 45Ab 88
Verity Ho. E3 .... 41Bc 92
(off Merchant St.)
Veritys AL10: Hat 1C 8
Vermeer Ct. E14 .... 48Fc 93
Vermeer Gdns. SE15 .... 56Yb 114
Vermilion Apts. E3 .... 39Ac 72
(off Gunmaker's La.)
Vermont Cl. EN2: Enf 14Rb 33
Vermont Ho. E17 .... 26Bc 52
Vermont Rd. SE19 .... 65Tb 135
Vermont Rd. SL2: Slou 2D 80
Vermont Rd. SM1: Sutt 76Db 155
Vermont Rd. SW18 .... 58Db 111
Verna Ho. E20 .... 36Ec 72
(off Sunrise Cl.)
Verney Gdns. RM9: Dag 35Ad 75
Verney Ho. NW8 .... 5D 214 (42Gb 89)
(off Jerome Cres.)
Verney Rd. RM9: Dag 35Ad 75
(not continuous)
Verney Rd. SE16 .... 51Wb 113
Verney Rd. SL3: L'ly 49C 82
Verney St. NW10 .... 34Ta 67
Verney Way SE16 .... 50Xb 91
Vernham Rd. SE18 .... 51Sc 116
Vernon Av. E12 .... 35Pc 74
Vernon Av. EN3: Enf W 8Ac 20
Vernon Av. IG8: Wfd G 24Kc 53
Vernon Av. SW20 .... 68Za 132
Vernon Cl. AL1: St A 3B 6
Vernon Cl. BR5: St P 69Xc 139
Vernon Cl. KT16: Ott 79F 148
Vernon Cl. KT19: Ewe 79Sa 153
Vernon Cl. TN15: W King 81Vd 184
Vernon Cl. TW19: Stanw 60N 105
Vernon Ct. HA7: Stan 25Ka 46
Vernon Ct. NW2 .... 34Bb 69
Vernon Ct. W5 .... 45La 86
Vernon Cres. CM13: B'wood 20Ce 41
Vernon Cres. EN4: E Barn 16Jb 32
Vernon Dr. CR3: Cat'm 94Sb 197
Vernon Dr. HA7: Stan 25Ja 46
Vernon Dr. UB9: Hare 25L 43
Vernon Ho. SE11 .... 7J 229 (50Pb 90)
(off Vauxhall St.)
Vernon Ho. WC1 .... 1G 223 (43Nb 90)
(off Vernon Pl.)
Vernon Mans. W14 .... 51Bb 111
(off Queen's Club Gdns.)
Vernon M. E17 .... 29Bc 52
Vernon M. W14 .... 49Ab 88
Vernon Pl. WC1 .... 1G 223 (43Nb 90)

Vernon Ri. UB6: G'frd 36Fa 66
Vernon Ri. WC1 3J 217 (41Pb 90)
Vernon Rd. DA10: Swans 58Be 121
Vernon Rd. E11 32Gc 73
Vernon Rd. E15 38Gc 73
Vernon Rd. E17 29Bc 52
Vernon Rd. E3 40Bc 72
Vernon Rd. IG3: Ilf 32Vc 75
Vernon Rd. N8 27Qb 50
Vernon Rd. RM5: Col R 22Ed 56
Vernon Rd. SM1: Sutt 78Eb 155
Vernon Rd. SW14 55Ta 109
Vernon Rd. TW13: Felt 61V 128
Vernon Rd. WD23: Bush 15Aa 27
Vernon Sq. WC1 3J 217 (41Pb 90)
Vernon St. W14 49Ab 88
Vernon Wlk. KT20: Tad 92Za 194
Vernon Yd. W11 45Bb 89
Vern Pl. TN16: Tats 93Lc 199
Veroan Rd. DA7: Bex 54Ad 117
Verona Cl. UB8: Cowl 44L 83
Verona Ct. SE14 51Zb 114
(off Myers La.)
Verona Ct. TW15: Ashf 63R 128
Verona Ct. W4 50Ua 88
Verona Dr. KT6: Surb 75Na 153
Verona Gdns. DA12: Grav'nd 3G 144
Verona Ho. CR4: Mitc 69Kb 134
(off Aventine Av.)
Verona Rd. E7 38Jc 73
Veronica Cl. RM3: Rom 24Ld 57
Veronica Gdns. SW16 67Lb 134
Veronica Ho. E3 41Dc 92
(off Talwin St.)
Veronica Ho. SE4 55Bc 114
Veronica Rd. SW17 61Kb 134
Veronique Gdns. IG6: Ilf 29Sc 54
Verrall Cl. CM16: Coop 1Zc 23
Verralls GU22: Wok 89D 168
(not continuous)
Verran Rd. SW12 59Kb 112
Ver Rd. AL3: St A 2A 6
Versailles Rd. SE20 66Wb 135
Vert Ho. RM17: Grays 52Ee 121
Verulam Av. CR8: Purl 84Lb 176
Verulam Av. E17 30Bc 52
Verulam Bldgs. WC1 7J 217 (43Pb 90)
(off Grays Inn)
Verulam Ct. NW9 31Wa 68
Verulam Ct. UB1: S'hall 44Ea 86
(off Haldane Rd.)
Verulam Golf Course 4D 6
Verulam Ho. W6 47Ya 88
(off Hammersmith Gro.)
Verulam Ind. Est. AL1: St A 4D 6
Verulamium Mus. 2P 5
Verulamium Pk. 3P 5
Verulamium Roman Town 2N 5
Verulam Pas. WD17: Wat 12X 27
Verulam Rd. AL3: St A 1A 6
Verulam Rd. UB6: G'frd 42Ca 85
Verulam St. WC1 7K 217 (43Qb 90)
Vervian Ho. SE15 52Wb 113
(off Reddins Rd.)
Verwood Dr. EN4: Cockf 13Hb 31
Verwood Ho. SW8 52Pb 112
(off Cobbett St.)
Verwood Lodge E14 48Fc 93
(off Manchester Rd.)
Verwood Rd. HA2: Harr 26Ea 46
Veryan GU21: Wok 9L 167
Veryan Cl. BR5: St P 70Yc 139
Veryan Ct. N8 29Mb 50
Vesage Ct. EC1 1A 224 (43Qb 90)
(off Leather La.)
Vesey Path E14 44Dc 92
Vespan Rd. W12 47Wa 88
Vespucci Ct. E14 43Ac 92
(off Oman Way)
Vesta Av. AL1: St A 5A 6
Vesta Ct. SE1 2H 231 (47Ub 91)
(off City Wlk.)
Vesta Ho. E20 36Ec 72
(off Liberty Bri. Rd.)
Vesta Ho. E3 39Cc 72
(off Garrison Rd.)
Vesta Rd. SE4 54Ac 114
Vestris Rd. SE23 61Zb 136
Vestry Cotts. DA3: Lfield 70Ee 143
Vestry Cotts. TN14: S'oaks 91Ld 203
Vestry Ct. RM7: Rush G 30Gd 56
Vestry Ct. SW1 4E 228 (48Mb 90)
(off Monck St.)
Vestry House Mus. 28Dc 52
Vestry Ind. Est. TN14: S'oaks 91Ld 203
Vestry M. SE5 53Ub 113
Vestry M. SW18 57Fb 111
Vestry Rd. E17 28Dc 52
Vestry Rd. SE5 53Ub 113
Vestry Rd. TN14: S'oaks 91Kd 203
Vestry St. N1 3F 219 (41Tb 91)
Vesuvius Apts. E3 40Bc 72
(off Centurion La.)
Vevers Rd. RH2: Reig 9L 207
Vevey St. SE6 61Bc 136
Vexil Cl. RM19: Purf 49Td 98
Veysey Cl. HP1: Hem H 4K 3
Veysey Gdns. RM10: Dag 34Cd 76
Viaduct, The E18 26Jc 53
Viaduct, The N10 28Kb 50
Viaduct Bldgs. EC1 1A 224 (43Qb 90)
Viaduct Gdns. SW11: Ace Way 51Mb 112
Viaduct Gdns. SW11: Nine Elms La. 52Lb 112
Viaduct Pl. E2 41Xb 91
Viaduct Rd. N2 26Fb 49
Viaduct St. E2 41Xb 91
Viaduct Ter. DA4: S Dar 68Sd 142
Vian Av. EN3: Enf W 7Ac 20
Vian St. SE13 55Dc 114
Viant Ho. NW10 38Ta 67
(off Fawood Av.)
Via Romana DA12: Grav'nd 10K 123
Vibart Gdns. SW2 59Pb 112
Vibart Wlk. N1 39Nb 70
(off Outram Pl.)
Vibeca Apts. E1 43Vb 91
(off Chicksand St.)
Vibia Cl. TW19: Stanw 59M 105
Viburnum Ct. GU24: W End 5C 166
Viburnum Ga. WD7: R'lett 9Ha 14
Vicarage Av. SE3 52Jc 115
Vicarage Av. TW20: Egh 64D 126
Vicarage Cl. AL1: St A 5A 6
Vicarage Cl. CM14: B'wood 21Ud 58
Vicarage Cl. DA8: Erith 51Ed 118
Vicarage Cl. EN6: N'thaw 2Hb 17
Vicarage Cl. EN6: Pot B 4Ab 16
Vicarage Cl. HA4: Ruis 31T 64
Vicarage Cl. HP1: Hem H 4L 3
Vicarage Cl. KT20: Kgswd 96Ab 194
Vicarage Cl. KT23: Bookh 97Ca 191
Vicarage Cl. KT4: Wor Pk 74Ua 154
Vicarage Cl. UB5: N'olt 38Ba 65
Vicarage Ct. BR3: Beck 69Ac 136
Vicarage Ct. DA12: Grav'nd 10J 123
Vicarage Ct. EN9: Walt A 6Jc 21
(off Horseshoe Cl.)
Vicarage Ct. IG1: Ilf 36Rc 74
Vicarage Ct. TW14: Bedf 59S 106
Vicarage Ct. TW20: Egh 65D 126
Vicarage Ct. W8 47Db 89
Vicarage Cres. SW11 53Fb 111
Vicarage Cres. TW20: Egh 64D 126
Vicarage Dr. BR3: Beck 67Cc 136
Vicarage Dr. DA11: Nflt 58Ee 121
Vicarage Dr. IG11: Bark 38Sc 74
Vicarage Dr. SW14 57Ta 109
Vicarage Farm Ct. TW5: Hest 52Ba 107
Vicarage Farm Rd. TW3: Houn 54Aa 107
Vicarage Farm Rd. TW5: Hest 53Aa 107
Vicarage Flds. KT12: Walt T 72Y 151
Vicarage Fld. Shop. Cen. 38Sc 74
Vicarage Gdns. CR4: Mitc 69Gb 133
Vicarage Gdns. SW14 57Sa 109
Vicarage Gdns. W8 46Cb 89
Vicarage Ga. W8 46Db 89
Vicarage Ga. M. KT20: Kgswd 96Ab 194
Vicarage Gro. SE5 53Tb 113
Vicarage Hill TN16: Westrm 98Tc 200
Vicarage Ho. KT1: King T 68Pa 131
(off Cambridge Rd.)
Vicarage La. DA12: Grav'nd 1J 145
Vicarage La. E15 38Gc 73
Vicarage La. E6 41Pc 94
Vicarage La. GU23: Send 98E 188
Vicarage La. HP3: Bov 8D 2
Vicarage La. IG1: Ilf 32Tc 74
Vicarage La. IG7: Chig 19Sc 36
Vicarage La. KT17: Ewe 81Wa 174
(not continuous)
Vicarage La. KT20: Kgswd 95Ab 194
Vicarage La. KT22: Lea 94Ka 192
Vicarage La. TN13: Dun G 91Fd 202
Vicarage La. TW18: Lale 69L 127
Vicarage La. TW19: Wray 60A 104
Vicarage La. WD4: K Lan 1P 11
Vicarage M. KT16: Longc 6M 147
Vicarage M. NW9 33Ta 67
Vicarage M. W4 51Ua 110
(off Bennett St.)
Vicarage Pde. N15 28Sb 51
Vicarage Pk. SE18 50Sc 94
Vicarage Path N8 31Nb 70
Vicarage Rd. CM16: Coop 1Yc 23
Vicarage Rd. CR0: Wadd 76Qb 156
Vicarage Rd. DA5: Bexl 60Dd 118
Vicarage Rd. E10 31Cc 72
Vicarage Rd. E15 38Hc 73
Vicarage Rd. GU22: Wok 93B 188
Vicarage Rd. GU24: Chob 3H 167
Vicarage Rd. IG8: Wfd G 24Nc 54
Vicarage Rd. KT1: Hamp W 67La 130
Vicarage Rd. KT1: King T 68Ma 131
Vicarage Rd. N17 25Wb 51
Vicarage Rd. NW4 30Wa 48
Vicarage Rd. RM10: Dag 38Dd 76
Vicarage Rd. RM12: Horn 32Jd 76
Vicarage Rd. SE18 50Sc 94
(not continuous)
Vicarage Rd. SM1: Sutt 76Db 155
Vicarage Rd. SW14 57Sa 109
Vicarage Rd. TW11: Tedd 64Ja 130
Vicarage Rd. TW16: Sun 64Y 128
Vicarage Rd. TW18: Staines 62G 126
Vicarage Rd. TW2: Twick 61Ga 130
Vicarage Rd. TW2: Whitt 58Ea 108
Vicarage Rd. TW20: Egh 64C 126
Vicarage Rd. WD18: Wat 17V 26
Vicarage Rd. Pct. WD18: Wat 14X 27
(off Vicarage Rd.)
Vicarage Road Stadium 15X 27
Vicarage Sq. RM17: Grays 51Ce 121
Vicarage Wlk. KT12: Walt T 73W 150
Vicarage Wlk. SW11 53Fb 111
Vicarage Way HA2: Harr 31Ca 65
Vicarage Way NW10 34Ta 67
Vicarage Way SL3: Coln 52E 104
Vicarage Way SL9: Ger X 30B 42
Vicars Bri. Cl. HA0: Wemb 40Na 67
Vicar's Cl. E9 39Yb 72
Vicars Cl. E15 39Jc 73
Vicars Cl. EN1: Enf 12Ub 33
Vicar's Hill SE13 56Dc 114
Vicars Moor La. N21 17Qb 32
Vicars Oak Rd. SE19 65Ub 135
Vicar's Rd. NW5 36Jb 70
Vicars Wlk. RM8: Dag 34Xc 75
Vicary Ho. EC1 1D 224 (43Sb 91)
(off Bartholomew Cl.)
Vicentia Ct. SW11 55Eb 111
Viceroy Cl. N2 28Gb 49
(off East End Rd.)
Viceroy Ct. CR0: C'don 74Tb 157
Viceroy Ct. HA6: Nwood 23U 44
Viceroy Ct. NW8 1E 214 (40Gb 69)
(off Prince Albert Rd.)
Viceroy Pde. N2 28Gb 49
(off High Rd.)
Viceroy Rd. SW8 53Nb 112
Vicinity Ho. E14 45Cc 92
(off Storehouse M.)
Vic Johnson Ho. E3 39Bc 72
(off Armagh Rd.)
Vickers Cl. KT16: Vir W 5L 147
Vickers Cl. SM6: W'gton 80Pb 156
Vickers Ct. N17 27Xb 51
Vickers Ct. SE20 66Zb 136
Vickers Ct. TW19: Stanw 58N 105
(off Whitley Cl.)
Vickers Dr. Nth. KT13: Weyb 82N 169
Vickers Dr. Sth. KT13: Weyb 83N 169
Vickers La. DA1: Dart 55Qd 119
Vickers Rd. DA8: Erith 50Fd 96
Vickers Way TW4: Houn 57Aa 107
Vickery Ct. EC1 5E 218 (42Sb 91)
(off Mitchell St.)
Vickery's Wharf E14 44Cc 92
Victor App. RM12: Horn 32Md 77
Victor Beamish Av. CR3: Cat'm 92Ub 197
Victor Cazalet Ho. N1 39Rb 71
(off Gaskin St.)
Victor Cl. RM12: Horn 32Md 77
Victor Ct. RM12: Horn 32Md 77
(off Victor App.)
Victor Gdns. RM12: Horn 32Md 77
Victor Gro. HA0: Wemb 38Na 67
Victor Ho. SE7 51Lc 115
Victoria Almshouses RH1: Redh 3A 208
Victoria Almshouses RH2: Reig 6L 207
Victoria & Albert Mus. 4C 226 (48Fb 89)
Victoria Arc. SW1 4A 228 (48Kb 90)
(off Victoria St.)
Victoria Av. CR2: Sande 82Sb 177
Victoria Av. DA12: Grav'nd 9D 122
Victoria Av. E6 39Mc 73
Victoria Av. EC2 1J 225 (43Ub 91)
Victoria Av. EN4: E Barn 14Fb 31
Victoria Av. HA9: Wemb 37Ra 67
Victoria Av. KT6: Surb 72Ma 153
Victoria Av. KT8: W Mole 69Da 129
Victoria Av. N3 25Bb 49
Victoria Av. RM16: Grays 47Ee 99
Victoria Av. RM5: Col R 23Dd 56
Victoria Av. SM6: W'gton 76Jb 156
Victoria Av. TW3: Houn 57Ca 107
Victoria Av. UB10: Hil 37R 64
Victoria Bldgs. E8 39Xb 71
(off Mare St.)
Victoria Bus Station 4A 228 (48Kb 90)
Victoria Chambers EC2 5H 219 (42Ub 91)
(off Paul St.)
Victoria Cl. EN4: E Barn 14Fb 31
Victoria Cl. EN8: Chesh 2Zb 20
Victoria Cl. HA1: Harr 30Ha 46
Victoria Cl. KT13: Weyb 76T 150
Victoria Cl. KT8: W Mole 69Ca 129
Victoria Cl. RM16: Grays 47Ee 99
Victoria Cl. SE22 57Wb 113
Victoria Cl. UB3: Hayes 44T 84
Victoria Cl. WD3: Rick 17M 25
Victoria Coach Station 6K 227 (49Kb 90)
Victoria Colonnade WC1 1G 223 (43Nb 90)
(off Southampton Row)
Victoria Cotts. E1 43Wb 91
(off Deal St.)
Victoria Cotts. N10 26Jb 50
Victoria Cotts. TW9: Kew 53Pa 109
Victoria Ct. CM14: W'ley 21Yd 58
Victoria Ct. E18 27Kc 53
Victoria Ct. HA7: Stan 24Ma 47
(off Howard Rd.)
Victoria Ct. HA9: Wemb 37Qa 67
Victoria Ct. RH1: Redh 9A 208
Victoria Ct. RM1: Rom 29Jd 56
Victoria Ct. SE1 6J 231 (49Ub 91)
(off Hendre Rd.)
Victoria Ct. SE26 65Yb 136
Victoria Ct. SL1: Slou 6J 81
(off Blair Rd.)
Victoria Ct. SS17: Stan H 1L 101
Victoria Ct. TN14: Dun G 92Hd 202
Victoria Ct. W3 47Qa 87
Victoria Ct. WD17: Wat 13Y 27
Victoria Cres. N15 29Ub 51
Victoria Cres. SE19 65Ub 135
Victoria Cres. SL0: Iver 45H 83
Victoria Cres. SW19 66Bb 133
Victoria Dock Rd. E16 45Hc 93
Victoria Dr. DA4: S Dar 68Td 142
Victoria Dr. SL1: Burn 7C 60
Victoria Dr. SL1: Farn C 7C 60
Victoria Dr. SL2: Farn C 7C 60
Victoria Dr. SW19 59Za 110
Victoria Emb. EC4 4K 223 (45Qb 90)
Victoria Emb. SW1 2G 229 (47Nb 90)
Victoria Emb. WC2 6G 223 (46Nb 90)
Victoria Embankment Gdns.
WC2 6G 223 (46Nb 90)
(off Victoria Embankment)
Victoria Gdns. TN16: Big H 87Lc 179
Victoria Gdns. TW5: Hest 53Aa 107
Victoria Gdns. W11 46Cb 89
Victoria Gro. N12 22Fb 49
Victoria Gro. W8 48Eb 89
Victoria Gro. M. W2 45Cb 89
Victoria Hall E16 46Jc 93
(off Wesley Av.)
Victoria Hill Rd. BR8: Hext 67Hd 140
Victoria Ho. SW1:
Ebury Bri. Rd. 7K 227 (50Kb 90)
(off Ebury Bri. Rd.)
Victoria Ho. SW1: Francis St. 5C 228 (49Lb 90)
(off Francis St.)
Victoria Ho. E6 44Qc 94
Victoria Ho. HA8: Edg 23Ra 47
Victoria Ho. KT22: Lea 93La 192
Victoria Ho. RM2: Rom 28Ld 57
Victoria Ho. SE16 47Yb 92
(off Surrey Quays Rd.)
Victoria Ho. SL9: Chal P 21A 42
(off Micholls Av.)
Victoria Ho. SW8 52Nb 112
(off Sth. Lambeth Rd.)
Victoria Ind. Est. W3 43Ua 88
Victoria La. EN5: Barn 14Bb 31
Victoria La. UB3: Harl 50T 84
Victoria Mans. NW10 38Xa 68
Victoria Mans. SW8 52Nb 112
(off Sth. Lambeth Rd.)
Victoria Mans. W14 51Bb 111
(off Queen's Club Gdns.)
Victoria M. E8 37Wb 71
Victoria M. KT13: Weyb 77Q 150
(off Balfour Rd.)
Victoria M. NW6 39Cb 69
Victoria M. SW18 60Eb 111
Victoria M. SW4 56Kb 112
Victoria M. TW20: Eng G 5N 125
Victoria M. W11 46Cb 89
Victoria Mills Studios E15 39Fc 73
(off Burford Rd.)
Victorian Gro. N16 35Ub 71
Victorian Hgts. SW8 54Kb 112
(off Thackeray Rd.)
Victorian Rd. N16 34Ub 71
Victoria Palace Theatre 4B 228 (48Lb 90)
(off Victoria St.)
Victoria Pde. SE10 51Dc 114
Victoria Pde. TW9: Kew 53Qa 109
(off Sandycombe Rd.)
Victoria Pk. Hackney 39Zb 72
Victoria Pk. Ct. E9 38Yb 72
(off Well St.)
Victoria Pk. Ind. Cen. E9 38Cc 72
(off Rothbury Rd.)
Victoria Pk. Ind. Est. DA1:
Dart 57Nd 119
Victoria Pk. Rd. E9 39Yb 72
Victoria Pk. Sq. E2 41Yb 92
Victoria Pk. Studios E9 37Yb 72
(off Milborne St.)
Victoria Pas. NW8 5B 214 (42Fb 89)
(off Fisherton St.)
Victoria Pas. WD18: Wat 14X 27
Victoria Pl. GU21: Wok 88C 168
(off North Rd.)
Victoria Pl. HP2: Hem H 2M 3
Victoria Pl. KT10: Esh 77Da 151
(off Esher Pk. Av.)
Victoria Pl. KT11: Cobh 86X 171
Victoria Pl. KT17: Eps 84Ua 174
Victoria Pl. TW9: Rich 57Ma 109
Victoria Pl. Shop. Cen. 5A 228 (49Kb 90)
(off Buckingham Pal. Rd.)
Victoria Point E13 40Jc 73
(off Victoria Rd.)
Victoria Retail Pk. South Ruislip 36Z 65
Victoria Ri. NW6 38Eb 69
(off Hilgrove Rd.)
Victoria Ri. SW4 55Kb 112
Victoria Rd. BR2: Broml 71Mc 159
Victoria Rd. BR7: Chst 64Qc 138
Victoria Rd. CM14: W'ley 21Yd 58
Victoria Rd. CR4: Mitc 66Gb 133
Victoria Rd. CR5: Coul 87Mb 176
Victoria Rd. DA1: Dart 57Md 119
Victoria Rd. DA11: Nflt 10B 122
Victoria Rd. DA15: Sidc 62Vc 139
Victoria Rd. DA6: Bex 56Cd 118
Victoria Rd. DA8: Erith 51Gd 118
Victoria Rd. E11 35Gc 73
Victoria Rd. E13 40Jc 73
Victoria Rd. E17 26Ec 52
Victoria Rd. E18 26Kc 53
Victoria Rd. E4 18Gc 35
Victoria Rd. EN4: E Barn 14Fb 31
Victoria Rd. EN9: Walt A 6Ec 20
Victoria Rd. GU21: Knap 9H 167
Victoria Rd. GU22: Wok 89A 168
Victoria Rd. HA4: Ruis 32W 64
Victoria Rd. IG11: Bark 37Rc 74
Victoria Rd. IG9: Buck H 19Mc 35
Victoria Rd. KT1: King T 68Pa 131
Victoria Rd. KT13: Weyb 76T 150
Victoria Rd. KT15: Add 77M 149
Victoria Rd. KT6: Surb 72Ma 153
Victoria Rd. N15 28Wb 51
Victoria Rd. N18 21Vb 51
Victoria Rd. N22 25Lb 50
Victoria Rd. N4 31Pb 70
Victoria Rd. N9 20Vb 33
Victoria Rd. NW10 43Ta 87
Victoria Rd. NW4 28Ya 48
Victoria Rd. NW6 40Bb 69
Victoria Rd. NW7 22Va 48
Victoria Rd. RH1: Redh 7A 208
Victoria Rd. RM1: Rom 30Hd 56
Victoria Rd. RM10: Dag 36Dd 76
Victoria Rd. SL2: Farn C 7G 60
Victoria Rd. SL2: Slou 6M 81
Victoria Rd. SL4: Eton W 9C 80
Victoria Rd. SM1: Sutt 78Fb 155
Victoria Rd. SS17: Stan H 2L 101
Victoria Rd. SW14 55Ta 109
Victoria Rd. TN13: S'oaks 97Kd 203
Victoria Rd. TW1: Twick 59Ka 108
Victoria Rd. TW11: Tedd 65Ja 130
Victoria Rd. TW13: Felt 60X 107
Victoria Rd. TW18: Staines 62G 126
Victoria Rd. UB2: S'hall 48Ba 85
Victoria Rd. UB8: Uxb 38L 63
Victoria Rd. W3 43Ta 87
Victoria Rd. W5 43Ka 86
Victoria Rd. W8 48Eb 89
Victoria Rd. WD23: Bush 18Da 27
Victoria Rd. WD24: Wat 10X 13
Victoria Road 36Dd 76
Victoria Scott Ct. DA1: Cray 55Hd 118
Victoria Sq. AL1: St A 3D 6
Victoria Sq. SW1 3A 228 (48Kb 90)
Victoria Station
(Rail & Underground) 5A 228 (49Kb 90)
Victoria St. AL1: St A 2B 6
Victoria St. DA17: Belv 50Bd 95
Victoria St. E15 38Gc 73
Victoria St. SL1: Slou 7K 81
Victoria St. SL4: Wind 3H 103
Victoria St. SW1 4B 228 (48Lb 90)
Victoria St. TW20: Eng G 5N 125
Victoria's Way RM15: S Ock 44Yd 98
Victoria Ter. HA1: Harr 32Ga 66
Victoria Ter. N4 32Qb 70
Victoria Ter. NW10 42Va 88
Victoria Ter. W5 46Ma 87
Victoria Vs. TW9: Rich 55Pa 109
Victoria Way GU21: Wok 89A 168
Victoria Way HA4: Ruis 36Z 65
Victoria Way KT13: Weyb 76T 150
Victoria Way SE7 50Kc 93
Victoria Wharf E14 45Ac 92
Victoria Wharf E2 40Zb 72
(off Palmers Rd.)
Victoria Wharf SE8 50Bc 92
(off Dragoon Rd.)
Victoria Works NW2 33Xa 68
Victoria Yd. E1 44Wb 91
Victor Rd. HA2: Harr 27Ea 46
Victor Rd. NW10 41Xa 88
Victor Rd. SE20 66Zb 136
Victor Rd. SL4: Wind 5G 102
Victor Rd. TW11: Tedd 63Ga 130
Victor's Cres. CM13: Hut 19De 41
Victors Dr. TW12: Hamp 65Aa 129
Victors Way EN5: Barn 13Bb 31
Victor Vs. N9 20Tb 33
Victor Wlk. NW9 26Ua 48
Victor Wlk. RM12: Horn 32Md 77
Victor Way AL2: Col S 3Ha 14
Victor Wharf SE1 6F 225 (46Tb 91)
(off Clink St.)
Victory Av. SM4: Mord 71Eb 155
Victory Bus. Cen. TW7: Isle 56Ha 108
Victory Cl. RM16: Chaf H 49Yd 98
Victory Cl. TW19: Stanw 60N 105
Victory Cotts. KT24: Eff 100Aa 191
Victory Ct. DA8: Erith 52Hd 118
(off Frobisher Rd.)
Victory Ct. IG11: Bark 42Xc 95
Victory Ct. W9 42Cb 89
(off Hermes Cl.)
Victory Ho. KT18: Eps 85Sa 173
(off West St.)
Victory M. UB2: S'hall 48Aa 85
Victory Pde. E20 37Dc 72
Victory Pde. SE18 48Rc 94
Victory Pk. HA9: Wemb 34Ma 67
Victory Pk. M. KT15: Add 77L 149
(off Victory Pk. Rd.)
Victory Pk. Rd. KT15: Add 76L 149
Victory Pl. E14 45Ac 92
Victory Pl. SE17 5F 231 (49Tb 91)
Victory Pl. SE19 66Ub 135
Victory Rd. E11 28Jc 53
Victory Rd. KT16: Chert 74J 149
Victory Rd. RM13: Rain 40Jd 76
Victory Rd. SW19 66Eb 133
Victory Rd. M. SW19 66Eb 133
(off Victory Rd.)
Victory Wlk. SE8 53Cc 114
Victory Way DA2: Dart 56Sd 120
Victory Way RM7: Mawney 26Dd 56
Victory Way SE16 47Ac 92
Victory Way TW5: Cran 50Y 85
Vida Ho. SE8 50Zb 92
Video Ct. N4 31Pb 70
Vidler Cl. KT9: Chess 79La 152
Vienna Cl. IG5: Ilf 26Mc 53
View, The SE2 50Ad 95
View Cl. HA1: Harr 28Fa 46
View Cl. IG7: Chig 22Tc 54
View Cl. N6 31Hb 69
View Cl. TN16: Big H 88Lc 179
View Cres. N8 29Mb 50
Viewfield Cl. HA3: Kenton 31Na 67
Viewfield Rd. DA5: Bexl 60Yc 117
Viewfield Rd. SW18 58Bb 111
Viewland Rd. SE18 50Vc 95
Viewlands Av. TN16: Westrm 92Uc 200
View Rd. EN6: Pot B 4Eb 17
View Rd. N6 31Hb 69
View Tube, The 39Dc 72
(off Greenway)
Viga Rd. N21 16Qb 32
Vigerons Way RM16: Grays 9D 100
Vigers Ct. NW10 41Xa 88
(off Harrow Rd.)
Viggory La. GU21: Wok 7N 167
Vigilant Cl. SE26 63Wb 135
Vigilant Way DA12: Grav'nd 4H 145
Vignoles Rd. RM7: Rush G 31Cd 76
Vigo Hill ME19: Tros 85Fe 185
Vigo Rd. TN15: Fair 83De 185
Vigors Cft. AL10: Hat 1B 8
Vigo St. W1 5B 222 (45Lb 90)
Viking Bus. Cen. RM7: Rush G 31Ed 76
Viking Cl. E3 40Ac 72
Viking Ct. SW6 51Cb 111
Viking Gdns. E6 42Nc 94
Viking Ho. SE18 49Nc 94
(off Pett St.)
Viking Ho. SE5 54Sb 113
(off Denmark Rd.)
Viking Pl. E10 32Bc 72
Viking Rd. DA11: Nflt 62Ee 143
Viking Rd. UB1: S'hall 45Aa 85
Viking Way CM15: Pil H 16Xd 40
Viking Way DA8: Erith 48Ed 96
Viking Way RM13: Rain 42Jd 96
Viking Way TN15: W King 78Ud 164
Villa Cl. DA12: Grav'nd 1K 145
Villa Ct. DA1: Dart 61Nd 141
Villacourt Rd. SE18 52Wc 117
Village, The DA9: Bluew 59Vd 120
Village, The HP5: Lat 9A 10
Village, The NW3 33Eb 69
Village, The SE7 51Lc 115
VILLAGE, THE 2E 124
Village, The 46Za 88
Village Arc. AL3: St A 2B 6
(off High St.)
Village Arc. E4 18Fc 35
Village Cen. HP3: Hem H 3C 4
Village Cl. E4 22Ec 52
Village Cl. KT13: Weyb 76T 150
Village Cl. NW3 36Fb 69
(off Belsize La.)
Village Ct. E17 29Dc 52
(off Eden Rd.)
Village Ct. SE3 55Gc 115
(off Hurren Cl.)
Village Ctyd. SW11 51Kb 112
(off Arches La.)
Village Cres., The DA9: Bluew 59Vd 120
Village Gdns. KT17: Ewe 82Va 174
Village Ga. TW17: Shep 71R 150
Village Grn. Av. TN16: Big H 89Nc 180
Village Grn. Rd. DA1: Cray 56Jd 118
Village Grn. Way TN16: Big H 89Nc 180
Village Health Club, The 94Sb 197
Village Hgts. IG8: Wfd G 22Hc 53
Village La. SL2: Hedg 2H 61
Village M. HP3: Bov 9C 2
Village M. NW9 33Ta 67
Village M. SL5: S'hill 10A 124
Village M. SW18 60Bb 111
(off Elsenham St.)
Village Mt. NW3 35Eb 69
(off Perrins Ct.)
Village Pk. Cl. EN1: Enf 16Ub 33
Village Rd. EN1: Enf 15Ub 33
Village Rd. N3 26Ab 48
Village Rd. SL4: Dor 8A 80
Village Rd. TW20: Thorpe 69E 126
Village Rd. UB9: Den 33H 63
Village Row SM2: Sutt 80Cb 155
Village Sq., The CR5: Coul 94Mb 196
Village Way BR3: Beck 68Cc 136
Village Way CR2: Sande 85Wb 177
Village Way HA5: Pinn 31Aa 65
Village Way IG6: Ilf 27Sc 54
Village Way NW10 35Ta 67
Village Way SE21 58Tb 113
Village Way TW15: Ashf 63P 127
Village Way E. HA2: Harr 31Ca 65
Villa Rd. SW9 55Qb 112
Villas on the Heath NW3 34Eb 69
Villas Rd. SE18 50Sc 94
Villa St. SE17 7G 231 (50Tb 91)
Villa Wlk. SE17 50Tb 91
(off Villa St.)
Villiers, The KT13: Weyb 79T 150
Villiers Av. KT5: Surb 71Pa 153
Villiers Av. TW2: Whitt 60Ba 107
Villiers Cl. E10 33Cc 72
Villiers Cl. KT5: Surb 70Pa 131
Villiers Ct. SL4: Wind 2E 102
Villiers Ct. SW11 53Gb 111
(off Battersea Bri. Rd.)
Villiers Gdns. E20 36Dc 72
Villiers Gro. SM2: Cheam 81Za 174
Villiers Ho. SL4: Eton 10G 80
(off Common La.)
Villiers M. NW2 37Wa 68
Villiers Path KT6: Surb 71Na 153
Villiers Rd. BR3: Beck 68Zb 136
Villiers Rd. KT1: King T 70Pa 131
Villiers Rd. NW2 37Wa 68
Villiers Rd. SL2: Slou 3H 81
Villiers Rd. TW7: Isle 54Ga 108
Villiers Rd. UB1: S'hall 46Ba 85
Villiers Rd. WD19: Wat 16Aa 27
Villiers St. WC2 6G 223 (46Nb 90)
Villier St. UB8: Uxb 41M 83
Vimy Cl. TW4: Houn 57Ba 107
Vimy Dr. DA1: Dart 55Pd 119

Vimy Ridge Ct. E3 40Bc 72
(off Festubert Pl.)
Vimy Way DA1: Cray 57Jd 118
Vincam Cl. TW2: Whitt 59Ca 107
Vince Ct. N1 4G 219 (41Tb 91)
Vincennes Est. SE27 63Tb 135
Vincent Av. KT5: Surb 75Sa 153
Vincent Av. SM5: Cars 83Fb 175
Vincent Cl. BR2: Broml 70Kc 137
Vincent Cl. CR5: Chips 92Hb 195
Vincent Cl. DA15: Sidc 60Uc 116
Vincent Cl. EN5: New Bar 13Db 31
Vincent Cl. EN8: Chesh 1Ac 20
Vincent Cl. IG6: Ilf 23Sc 54
Vincent Cl. KT10: Esh 76Da 151
Vincent Cl. KT16: Chert 73G 148
Vincent Cl. KT22: Fet 95Da 191
Vincent Cl. SE16 47Ac 92
Vincent Cl. UB7: Sip 51Q 106
Vincent Ct. HA6: Nwood 25V 44
Vincent Ct. N4 32Nb 70
Vincent Ct. NW4 28Za 48
Vincent Ct. SW9 53Pb 112
Vincent Ct. W1 2F 221 (44Hb 89)
(off Seymour Pl.)
Vincent Dr. TW17: Shep 69U 128
Vincent Dr. UB10: Uxb 39P 63
Vincent Gdns. NW2 34Va 68
Vincent Ho. SW1: Regency St. 5E 228 (49Mb 90)
(off Regency St.)
Vincent Ho. SW1: Vincent Sq. 6D 228 (49Mb 90)
(off Vincent Sq.)
Vincent M. E3 40Cc 72
Vincent Rd. CR0: C'don 73Ub 157
Vincent Rd. CR5: Coul 88Lb 176
Vincent Rd. E4 23Fc 53
Vincent Rd. HA0: Wemb 38Pa 67
Vincent Rd. KT1: King T 69Qa 131
Vincent Rd. KT11: Stoke D 88Aa 171
Vincent Rd. KT16: Chert 73G 148
Vincent Rd. N15 28Sb 51
Vincent Rd. N22 26Qb 50
Vincent Rd. RM13: Rain 42Ld 97
Vincent Rd. RM9: Dag 38Ad 75
Vincent Rd. SE18 49Rc 94
Vincent Rd. TW4: Houn 54Z 107
Vincent Rd. TW7: Isle 53Fa 108
Vincent Row TW12: Hamp H 65Ea 130
Vincents Path UB5: N'olt 37Aa 65
(off Arnold Rd.)
Vincent Sq. N22 26Qb 50
Vincent Sq. SW1 5D 228 (49Mb 90)
Vincent Sq. TN16: Big H 85Lc 179
Vincent Sq. Mans. SW1 5C 228 (49Lb 90)
(off Walcott St.)
Vincent St. E16 43Hc 93
Vincent St. SW1 5D 228 (49Mb 90)
Vincent's Yd. SW9 53Qb 112
(off Alphabet M.)
Vincent Ter. N1 1B 218 (40Rb 71)
Vincenzo Cl. AL9: Wel G 5E 8
Vince St. EC1 4G 219 (41Tb 91)
Vine, The TN13: S'oaks 96Kd 203
Vine Av. TN13: S'oaks 96Kd 203
Vine Cl. E5 35Wb 71
Vine Cl. GU3: Worp 8H 187
Vine Cl. KT5: Surb 72Pa 153
Vine Cl. SM1: Sutt 76Eb 155
Vine Cl. TW19: Stanw M 57J 105
Vine Cl. UB7: W Dray 49Q 84
Vine Cotts. E1 44Yb 92
(off Sidney Sq.)
Vine Cotts. W7 46Ga 86
Vine Ct. E1 43Wb 91
Vine Ct. HA3: Kenton 30Na 47
Vine Ct. KT12: Hers 79Y 151
Vine Ct. Rd. TN13: S'oaks 96Ld 203
Vinegar All. E17 28Dc 52
Vine Gdns. IG1: Ilf 36Sc 74
Vinegar St. E1 46Xb 91
Vinegar Yd. SE1 1H 231 (47Ub 91)
Vine Gro. UB10: Hil 38Q 64
Vine Hill EC1 6K 217 (42Qb 90)
Vine La. SE1 7J 225 (46Ub 91)
Vine La. UB10: Hil 39P 63
Vine Lodge TN13: S'oaks 96Ld 203
Vine Lodge Ct. TN13: S'oaks 96Ld 203
Vine Pl. TW3: Houn 56Da 107
Vine Pl. W5 46Na 87
(off St Mark's Rd.)
Viner Cl. KT12: Walt T 72Y 151
Vineries, The EN1: Enf 13Ub 33
Vineries, The N14 16Lb 32
Vineries, The SE6 60Cc 114
Vineries Bank NW7 22Xa 48
Vineries Cl. RM9: Dag 37Bd 75
Vineries Cl. UB7: Sip 51Q 106
Vine Rd. BR6: Chels 79Vc 161
Vine Rd. E15 38Hc 73
Vine Rd. KT8: E Mos 70Ea 130
Vine Rd. SL2: Stoke P 7K 61
Vine Rd. SW13 55Va 110
Viner Pl. E17 28Cc 52
Vinery, The SW8 52Nb 112
(off Regent's Bri. Gdns.)
Vinery Way W6 48Xa 88
Vines Av. N3 25Db 49
Vine Sq. W14 50Bb 89
(off Star Rd.)
Vine St. E17 28Dc 52
Vine St. EC3 3K 225 (44Vb 91)
Vine St. RM7: Rom 28Ed 56
Vine St. UB8: Uxb 39M 63
Vine St. W1 5C 222 (45Lb 90)
Vine St. Bri. EC1 6A 218 (42Qb 90)
Vine Tree Ct. WD3: Rick 19K 25
Vine Way CM14: B'wood 18Yd 40
Vineyard, The TW10: Rich 57Na 109
Vine Yd. SE1 2E 230 (47Sb 91)
(off Sanctuary St.)
Vineyard Av. NW7 24Ab 48
Vineyard Cl. KT1: King T 69Pa 131
Vineyard Cl. SE6 60Cc 114
Vineyard Gro. N3 25Db 49
Vineyard Hill EN6: N'thaw 1Jb 18
Vineyard Hill Rd. SW19 63Bb 133
Vineyard M. TW10: Rich 57Na 109
Vineyard Pas. TW10: Rich 57Na 109
Vineyard Path SW14 55Ta 109
Vineyard Rd. TW13: Felt 62W 128
Vineyard Row KT1: Hamp W 67La 130
Vineyards, The TW13: Felt 62W 128
(off High St.)
Vineyards, The TW16: Sun 69W 128
Vineyards Rd. EN6: N'thaw 2Hb 17
Vineyard Wlk. EC1 5K 217 (42Qb 90)
Viney Bank CR0: Sels 81Bc 178
Viney Rd. SE13 55Dc 114
Vining St. SW9 56Qb 112
Vinlake Av. UB10: Ick 34P 63
Vinson Cl. BR6: Orp 74Wc 161
Vinson Ho. N1 2G 219 (40Tb 71)
(off Cranston Est.)
Vintage M. E4 21Cc 52
Vinter Ct. TW17: Shep 71Q 150
Vintner's Ct. EC4 4E 224 (45Sb 91)
Vintner's Pl. EC4 5E 224 (45Sb 91)
Vintry Ct. SE1 2G 231 (47Tb 91)
(off Porlock St.)
Vintry Gdn., The 2B 6
(off Holywell Hill)
Vintry M. E17 28Cc 52
Vinyl Pl. UB3: Hayes 46T 84
(off High Rd.)
Viola Av. SE2 49Xc 95
Viola Av. TW14: Felt 58Y 107
Viola Av. TW19: Stanw 60M 105
Viola Cl. RM15: S Ock 41Yd 98
Viola Sq. W12 45Va 88
Violet Av. EN2: Enf 10Tb 19
Violet Av. UB8: Hil 43P 83
Violet Cl. E16 42Gc 93
Violet Cl. SE8 51Bc 114
Violet Cl. SM3: Sutt 74Ab 154
Violet Cl. SM6: W'gton 74Jb 156
Violet Ct. E15 38Gc 73
(off Victoria St.)
Violet Ct. NW9 25Ua 48
Violet Gdns. CR0: Wadd 78Rb 157
Violet Hill NW8 2A 214 (40Eb 69)
Violet Hill Ho. NW8 2A 214 (40Eb 69)
(off Violet Hill)
Violet La. CR0: Wadd 79Rb 157
Violet Rd. E17 30Cc 52
Violet Rd. E18 26Kc 53
Violet Rd. E3 42Dc 92
Violet St. E2 42Xb 91
Violet Ter. UB8: Hil 43Q 84
Violet Way WD3: Loud 14L 25
VIP Trading Est. SE7 49Lc 93
Virgil Pl. W1 1F 221 (43Hb 89)
Virgil St. SE1 3J 229 (48Pb 90)
Virgin Active Aldersgate 1D 224 (43Sb 91)
(off Aldersgate St.)
Virgin Active Bank 2G 225 (44Tb 91)
Virgin Active Barbican 7D 218 (43Sb 91)
(off Aldersgate St.)
Virgin Active Broadgate 7H 219 (43Ub 91)
(off Exchange Pl.)
Virgin Active Bromley 70Mc 137
Virgin Active Chelsea 51Eb 111
Virgin Active Chiswick 53Ua 110
Virgin Active Chiswick Pk. 49Ra 87
Virgin Active Clearview 29Be 59
Virgin Active Cricklewood 34Ab 68
Virgin Active Crouch End 29Nb 50
(off Tottenham La.)
Virgin Active Fulham Pools 51Ab 110
Virgin Active Hammersmith 49Za 88
(off Hammersmith Rd.)
Virgin Active Islington 2B 218 (40Rb 71)
Virgin Active Kensington 47Db 89
(off Old Court Pl.)
Virgin Active Mansion Ho. 4E 224 (45Sb 91)
(off Lit. Trinity La.)
Virgin Active Mayfair 3H 221 (44Jb 90)
Virgin Active Mill Hill East 24Ab 48
Virgin Active Moorgate 6F 219 (42Tb 91)
Virgin Active Notting Hill 44Ab 88
Virgin Active Repton Park 24Qc 54
Virgin Active Smugglers Way 56Db 111
Virgin Active Strand 5G 223 (45Nb 90)
Virgin Active Streatham 62Nb 134
Virgin Active Swiss Cottage 37Eb 69
(within O2 Centre)
Virgin Active Twickenham Club, The 58Ga 108
Virgin Active Tower Bridge 4K 225 (45Vb 91)
(off Haydon St.)
Virgin Active Walbrook 4F 225 (45Tb 91)
Virgin Active West London 46Ua 88
Virgin Active Wimbledon, Worple Rd. 65Bb 133
Virginia Av. GU25: Vir W 1N 147
Virginia Beeches GU25: Vir W 9M 125
Virginia Cl. BR2: Broml 69Gc 137
Virginia Cl. KT13: Weyb 79S 150
Virginia Cl. KT21: Asht 90Ma 173
Virginia Cl. KT3: N Mald 70Sa 131
Virginia Cl. RM5: Col R 24Ed 56
Virginia Cl. TW18: Lale 69L 127
Virginia Ct. DA1: Dart 55Qd 119
Virginia Ct. GU25: Vir W 10P 125
Virginia Ct. SE16 47Zb 92
(off Eleanor Cl.)
Virginia Ct. WC1 5E 216 (42Mb 90)
(off Burton St.)
Virginia Dr. GU25: Vir W 1N 147
Virginia Gdns. IG6: Ilf 26Sc 54
Virginia Ho. E14 45Ec 92
(off Newby Pl.)
Virginia Ho. TW11: Tedd 64Ka 130
Virginia Pk. GU25: Vir W 70A 126
Virginia Pl. KT11: Cobh 86W 170
Virginia Quay Pk. 45Fc 93
(off Newport Avenue)
Virginia Rd. CR7: Thor H 67Rb 135
Virginia Rd. DA1: Cray 57Jd 118
Virginia Rd. E2 4K 219 (41Vb 91)
Virginia St. E1 45Wb 91
Virginia Wlk. DA12: Grav'nd 5F 144
Virginia Wlk. SW2 58Pb 112
VIRGINIA WATER 1P 147
Virginia Water 9J 125
Virginia Water Station (Rail) 71A 148
Viridian Apts. SW8 52Lb 112
Viridian M. BR6: Orp 73Xc 161
Visage Apts. NW3 38Fb 69
(off Winchester Rd.)
Viscount Cl. N11 23Kb 50
Viscount Ct. SL4: Wind 3G 102
Viscount Ct. W2 44Cb 89
(off Pembridge Vs.)
Viscount Dr. E6 43Pc 94
Viscount Gdns. KT14: Byfl 84N 169
Viscount Gro. UB5: N'olt 41Z 85
Viscount Ind. Est. SL3: Poyle 55G 104
Viscount M. BR7: Chst 65Rc 138
Viscount Point SW19 66Cb 133
(off The Broadway)
Viscount Rd. TW19: Stanw 60N 105
Viscount St. EC1 6D 218 (42Sb 91)
Viscount Way TW6: H'row A 56U 106
Vision Ind. Pk. W3 43Ra 87
Vista, The DA14: Sidc 64Vc 139
Vista, The SE9 58Mc 115
Vista Av. EN3: Enf H 12Zb 34
Vista Bldg. E3 41Bc 92
(off Bow Rd.)
Vista Bldg. SE18 49Qc 94
Vista Ct. E1 43Ac 92
(off Ocean Est.)
Vista Dr. IG4: Ilf 29Mc 53
Vista Ho. N4 33Qb 70
Vista Ho. SW19 67Fb 133
(off Chapter Way)
Vista Way HA3: Kenton 30Na 47
Vistec Ho. CR0: C'don 73Rb 157
Vita Apts. CR0: C'don 75Tb 157
Vitae Apts. W6 48Wa 88
Vita Ho. AL1: St A 3D 6
Vitali Cl. SW15 58Wa 110
Vittoria Ho. N1 1J 217 (39Pb 70)
(off High Rd.)
Viveash Cl. UB3: Hayes 48V 84
Vivenne Ho. TW18: Staines 64J 127
Vivian Av. HA9: Wemb 36Qa 67
Vivian Av. NW4 29Xa 48
Vivian Cl. WD19: Wat 18W 26
Vivian Comma Cl. N4 34Rb 71
Vivian Ct. N12 22Db 49
Vivian Ct. W9 40Db 69
Vivian Gdns. HA9: Wemb 36Qa 67
Vivian Gdns. WD19: Wat 18W 26
Vivian Mans. NW4 29Xa 48
(off Vivian Av.)
Vivian Rd. E3 40Ac 72
Vivian Sq. SE15 55Xb 113
Vivian Way N2 29Fb 49
Vivien Cl. KT9: Chess 80Na 153
Vivienne Cl. TW1: Twick 58Ma 109
Vixen M. E8 38Vb 71
(off Haggerston Rd.)
Voce Rd. SE18 52Tc 116
Voewood Cl. KT3: N Mald 72Va 154
Vogan Cl. RH2: Reig 9K 207
Vogans Mill SE1 47Vb 91
Vogler Ho. E1 45Yb 92
(off Cable St.)
Vogue Ct. BR1: Broml 67Kc 137
Vollasky Ho. E1 43Wb 91
(off Daplyn St.)
Volta Cl. N9 20Yb 34
Voltaire Rd. SW4 55Mb 112
Voltaire Way UB3: Hayes 45U 84
Volt Av. NW10 41Ta 87
Volta Way CR0: Wadd 74Pb 156
Voluntary Pl. E11 30Jc 53
Vorley Rd. N19 33Lb 70
Voss Ct. SW16 65Nb 134
Voss St. E2 41Wb 91
Voyager Bus. Est. SE16 48Wb 91
(off Spa Rd.)
Voyager Ct. E16 43Jc 93
(off Hammersley Rd.)
Voyagers Cl. SE28 44Yc 95
Voysey Cl. N3 27Ab 48
Voysey Sq. E3 42Dc 92
Vue Cinema Acton 42Qa 87
Vue Cinema Apollo 5D 222 (45Mb 90)
(off Regent St.)
Vue Cinema Bromley 69Jc 137
Vue Cinema Croydon, High St. 76Sb 157
Vue Cinema Croydon, Purley Way 74Pb 156
Vue Cinema Dagenham 39Ad 75
Vue Cinema Finchley Rd. 37Eb 69
(within O2 Centre)
Vue Cinema Fulham Broadway 52Cb 111
Vue Cinema Harrow 30Ga 46
(within St George's Shop. & Leisure Cen.)
Vue Cinema Islington 1A 218 (40Qb 70)
Vue Cinema Leicester Square 4E 222 (45Mb 90)
(off Cranbourn St.)
Vue Cinema North Finchley 24Fb 49
Vue Cinema Romford 30Gd 56
Vue Cinema Shepherds Bush 47Za 88
Vue Cinema Staines Upon Thames 63G 126
Vue Cinema Stratford City 37Ec 72
(in Shopping Cen.)
Vue Cinema Watford 5Y 13
Vue Cinema West Thurrock 48Wd 98
Vue Cinema Westfield 46Ya 88
Vue Cinema Wood Green 26Qb 50
Vulcan Bus. Cen. CR0: New Ad 81Gc 179
Vulcan Cl. E6 44Qc 94
Vulcan Cl. TN16: Big H 85Lc 179
Vulcan Ga. EN2: Enf 12Qb 32
Vulcan Rd. SE4 54Bc 114
Vulcan Sq. E14 49Dc 92
Vulcan Ter. SE4 54Bc 114
Vulcan Way CR0: New Ad 82Gc 179
Vulcan Way N7 37Pb 70
Vulcan Way SM6: W'gton 81Nb 176
Vulcan Wharf E15 40Dc 72
(off Cook's Rd.)
Vulliamy Cl. E4 19Fc 35
Vyne, The DA7: Bex 55Dd 118
Vyner Rd. W3 45Ta 87
Vyner St. E2 39Xb 71
Vyners Way UB10: Ick 36Q 64
Vyse Cl. EN5: Barn 14Ya 30

W

Wacky Warehouse 19Y 27
Wadard Ter. BR8: Swan 71Ld 163
Wadbrook St. KT1: King T 68Ma 131
Wadding St. SE17 6F 231 (49Tb 91)
Waddington Av. CR5: Coul 92Qb 196
Waddington Cl. CR5: Coul 91Rb 197
Waddington Cl. EN1: Enf 14Ub 33
Waddington Rd. AL3: St A 2B 6
Waddington Rd. E15 36Fc 73
Waddington St. E15 37Fc 73
Waddington Way SE19 66Sb 135
WADDON 76Qb 156
Waddon Cl. CR0: Wadd 76Qb 156
Waddon Ct. Rd. CR0: Wadd 76Qb 156
Waddon Leisure Cen. 78Qb 156
Waddon Marsh Stop (London Tramlink) 75Qb 156
Waddon Marsh Way CR0: Wadd 74Pb 156
Waddon New Rd. CR0: C'don 76Rb 157
Waddon Pk. Av. CR0: Wadd 77Qb 156
Waddon Rd. CR0: C'don 76Qb 156
Waddon Rd. CR0: Wadd 76Qb 156
Waddon Station (Rail) 77Qb 156
Waddon Way CR0: Wadd 79Qb 156
Wade Av. BR5: Orp 73Zc 161
Wade Ct. N10 24Kb 50
Wade Dr. SL1: Slou 6E 80
Wade Ho. EN1: Enf 15Tb 33
Wade Ho. SE1 47Wb 91
(off Parkers Row)
Wades, The AL10: Hat 3C 8
Wades Gro. N21 17Qb 32
Wades Hill N21 16Qb 32
Wades La. TW11: Tedd 64Ja 130
Wadeson St. E2 40Xb 71
Wade's Pl. E14 45Dc 92
Wadeville Av. RM6: Chad H 30Ad 55
Wadeville Cl. DA17: Belv 50Cd 96
Wadham Av. E17 24Dc 52
Wadham Cl. TW17: Shep 73S 150
Wadham Gdns. NW3 39Gb 69
Wadham Gdns. UB6: G'frd 37Fa 66
Wadham Ho. N18 22Vb 51
Wadham M. SW14 54Sa 109
Wadham Rd. E17 24Dc 52
Wadham Rd. SW15 56Ab 110
Wadham Rd. WD5: Ab L 3V 12
Wadhurst Cl. SE20 68Xb 135
Wadhurst Rd. SW8 53Lb 112
Wadhurst Rd. W4 48Ta 87
Wadley Cl. HP2: Hem H 3P 3
Wadley Rd. E11 31Gc 73
Wadsworth Bus. Cen. UB6: G'frd 40La 66
Wadsworth Cl. EN3: Pond E 15Zb 34
Wadsworth Cl. UB6: G'frd 40La 66
Wadsworth Rd. UB6: G'frd 40Ka 66
Wager St. E3 42Bc 92
WAGGONERS RDBT. 53X 107
Waggon La. N17 23Wb 51
Waggon M. N14 18Lb 32
Waghorn Rd. E13 39Lc 73
Waghorn Rd. HA3: Kenton 27Ma 47
Waghorn St. SE15 55Wb 113
Wagner M. KT6: Surb 71Na 153
(off Avenue Elmers)
Wagner St. SE15 52Yb 114
Wagon Rd. EN4: Barn 8Cb 17
Wagon Rd. EN4: Had W 8Cb 17
Wagon Rd. EN5: Barn 7Cb 17
Wagon Way WD3: Loud 13L 25
Wagstaff Gdns. RM9: Dag 38Yc 75
Wagtail Cl. EN1: Enf 11Xb 33
Wagtail Cl. NW9 26Ua 48
Wagtail Ct. SW15 58Ya 110
Wagtail Gdns. CR2: Sels 82Ac 178
Wagtail Rd. TW6: H'row A 54K 105
Wagtail Wlk. BR3: Beck 71Ec 158
Wagtail Way BR5: St P 70Zc 139
Waid Cl. DA1: Dart 58Pd 119
Waight's Ct. KT2: King T 67Na 131
Wain Cl. EN6: Pot B 1Db 17
Wainfleet Av. RM5: Col R 26Ed 56
Wainford Cl. SW19 60Za 110
Wainscot SL5: S'dale 2D 146
Wainwright Av. CM13: Hut 16Fe 41
Wainwright Av. DA9: Ghithe 56Yd 120
Wainwright Gro. TW7: Isle 56Fa 108
Wainwright Ho. E1 46Yb 92
(off Garnet St.)
Waite Davies Rd. SE12 59Hc 115
Waite Ho. W3 46Va 88
Waite St. SE15 51Vb 113
Waithman St. EC4 3B 224 (44Rb 91)
(off Black Friars La.)
Wakefield Cl. KT14: Byfl 84N 169
Wakefield Ct. SE26 65Yb 136
Wakefield Cres. SL2: Stoke P 7K 61
Wakefield Gdns. IG1: Ilf 30Nc 54
Wakefield Gdns. SE19 66Ub 135
Wakefield Ho. SE15 53Wb 113
Wakefield M. WC1 4G 217 (41Nb 90)
Wakefield Rd. DA9: Ghithe 57Yd 120
Wakefield Rd. N11 22Mb 50
Wakefield Rd. N15 29Vb 51
Wakefield Rd. TW10: Rich 57Ma 109
Wakefield St. DA11: Grav'nd 8D 122
Wakefield St. E6 39Mc 73
Wakefield St. N18 22Wb 51
Wakefield St. WC1 4G 217 (41Nb 90)
Wakefields Wlk. EN8: Chesh 3Ac 20
Wakeford Cl. DA5: Bexl 60Zc 117
Wakeford Cl. SW4 57Lb 112
Wakeford Ct. KT16: Chert 74H 149
Wakehams Hill HA5: Pinn 27Ba 45
Wakeham St. N1 37Tb 71
Wakehurst Path GU21: Wok 86E 168
Wakehurst Rd. SW11 57Gb 111
Wakeling Ct. CR8: Purl 82Nb 176
Wakeling La. HA0: Wemb 34Ka 66
Wakeling Rd. W7 43Ha 86
Wakeling St. E14 44Ac 92
Wakelin Ho. N1 38Rb 71
(off Sebbon St.)
Wakelin Rd. E15 40Gc 73
Wakely Cl. TN16: Big H 90Lc 179
Wakely Ct. AL1: St A 2C 6
(off Hatfield Rd.)
Wakeman Ho. NW10 41Za 88
(off Wakeman Rd.)
Wakeman Rd. NW10 41Ya 88
Wakemans Hill Av. NW9 29Ta 47
Wakerfield Cl. RM11: Horn 29Pd 57
Wakering Rd. IG11: Bark 37Sc 74
(not continuous)
Wakerley Cl. E6 44Pc 94
Wake Rd. IG10: H Beech 10Lc 21
Wakeup Docklands 45Jc 93
Wakley St. EC1 3B 218 (41Rb 91)
Walberswick St. SW8 52Nb 112
Walbrook EC4 4F 225 (45Tb 91)
(not continuous)
Walbrook Bldg., The EC4 4F 225 (45Tb 91)
(off Walbrook)
Walbrook Ct. N1 1H 219 (40Ub 71)
(off Hemsworth St.)
Walbrook Ho. N9 19Yb 34
(off Huntingdon Rd.)
Walbrook Wharf EC4 5E 224 (45Sb 91)
(off Cousin La.)
Walburgh St. E1 44Xb 91
Walburton Rd. CR8: Purl 85Lb 176
Walcorde Av. SE17 6E 230 (49Sb 91)
Walcot Gdns. SE11 5K 229 (49Qb 90)
(off Kennington Rd.)
Walcot Rd. EN3: Brim 12Bc 34
Walcot Sq. SE11 5A 230 (49Qb 90)
Walcott St. SW1 5C 228 (49Lb 90)
Waldair Ct. E16 47Rc 94
Waldeck Gro. SE27 62Rb 135
Waldeck Rd. DA1: Dart 59Pd 119
Waldeck Rd. N15 28Rb 51
Waldeck Rd. SW14 55Sa 109
Waldeck Rd. W13 44Ka 86
Waldeck Rd. W4 51Qa 109
Waldeck Ter. SW14 55Sa 109
(off Waldeck Rd.)
Waldegrave Ct. IG11: Bark 39Tc 74
Waldegrave Ct. RM14: Upm 32Rd 77
Waldegrave Gdns. RM14: Upm 32Rd 77
Waldegrave Gdns. TW1: Twick 61Ha 130
Waldegrave Pk. TW1: Twick 63Ha 130
Waldegrave Rd. BR1: Broml 70Nc 138
Waldegrave Rd. N8 27Qb 50
Waldegrave Rd. RM8: Dag 33Yc 75
Waldegrave Rd. SE19 66Vb 135
Waldegrave Rd. TW1: Twick 63Ha 130
Waldegrave Rd. TW11: Tedd 63Ha 130
Waldegrave Rd. W5 45Pa 87
Waldegrove CR0: C'don 76Vb 157
Waldemar Av. SW6 53Ab 110
Waldemar Av. W13 46La 86
Waldemar Rd. SW19 64Cb 133
Walden Av. BR7: Chst 63Pc 138
Walden Av. N13 21Sb 51
Walden Av. RM13: Rain 40Fd 76
Walden Cl. DA17: Belv 50Bd 95
Walden Ct. SW8 53Mb 112
Walden Gdns. CR7: Thor H 69Pb 134
Walden Ho. SW1 6J 227 (49Jb 90)
(off Pimlico Rd.)
Walden Ho. SW11 53Jb 112
(off Dagnall St.)
Waldenhurst Rd. BR5: St M Cry 73Zc 161
Walden Pde. BR7: Chst 65Pc 138
(not continuous)
Walden Rd. BR7: Chst 65Pc 138
Walden Rd. N17 25Tb 51
Walden Rd. RM11: Horn 30Md 57
Waldens Cl. BR5: St M Cry 73Zc 161
Waldenshaw Rd. SE23 60Yb 114
Waldens Pk. Rd. GU21: Wok 8N 167
Waldens Rd. BR5: St M Cry 73Ad 161
Waldens Rd. GU21: Wok 9P 167
Walden St. E1 44Xb 91
(not continuous)
Walden Way IG6: Ilf 24Uc 54
Walden Way NW7 23Za 48
Walden Way RM11: Horn 30Md 57
Waldo Cl. SW4 57Lb 112
Waldo Ho. NW10 41Xa 88
(off Waldo Rd.)
Waldo Ind. Est. BR1: Broml 69Mc 137
Waldon RM18: E Til 8K 101
Waldo Pl. CR4: Mitc 66Gb 133
Waldorf Cl. CR2: S Croy 81Rb 177
Waldo Rd. BR1: Broml 69Mc 137
Waldo Rd. NW10 41Wa 88
Waldram Cres. SE23 60Yb 114
Waldram Pk. Rd. SE23 60Zb 114
Waldram Pl. SE23 60Yb 114
Waldrist Way DA18: Erith 48Bd 95
Waldron Gdns. BR2: Broml 69Fc 137
Waldronhyrst CR2: S Croy 77Rb 157
Waldron M. SW3 51Fb 111
Waldron Rd. HA1: Harr 32Ga 66
Waldron Rd. HA2: Harr 32Ga 66
Waldron Rd. SW18 62Eb 133
Waldrons, The CR0: C'don 77Rb 157
Waldrons, The RH8: Oxt 3K 211
Waldron's Path CR2: S Croy 77Sb 157
Waldrons Yd. HA2: Harr 33Fa 66
Waldstock Rd. SE28 45Wc 95
Waleorde Rd. SE17 7E 230 (50Sb 91)
Waleran Cl. HA7: Stan 22Ha 46
Walerand Rd. SE13 54Ec 114
Waleran Flats SE1 5H 231 (49Ub 91)
Wales Av. SM5: Cars 78Gb 155
Wales Cl. SE15 51Xb 113
Wales Farm Rd. W3 43Ta 87
Wales Office 7F 223 (46Nb 90)
(off Whitehall)
Waleton Acres SM6: W'gton 79Lb 156
Waley St. E1 43Ac 92
Walfield Av. N20 17Db 31
Walford Ho. E1 44Xb 91
Walford Rd. N16 35Ub 71
Walford Rd. UB8: Uxb 40L 63
Walfrey Gdns. RM9: Dag 38Bd 75
WALHAM GREEN 53Cb 111
Walham Grn. Ct. SW6 52Db 111
(off Waterford Rd.)
Walham Gro. SW6 52Cb 111
Walham Ri. SW19 65Ab 132
Walham Yd. SW6 52Cb 111
Walk, The EN6: Pot B 4Cb 17
Walk, The N13 20Qb 32
(off Fox La.)
Walk, The RH8: Tand 5E 210
Walk, The RM11: Horn 33Pd 77
Walk, The SL4: Eton W 10E 80
Walk, The TW16: Sun 66V 128
Walkato Lodge IG9: Buck H 18Lc 35
Walkden Rd. BR7: Chst 64Qc 138
Walker Cl. CR0: New Ad 80Ec 158
Walker Cl. DA1: Cray 55Hd 118
Walker Cl. DA10: Swans 59Be 121
Walker Cl. N11 21Lb 50
Walker Cl. SE18 49Sc 94
Walker Cl. TW12: Hamp 65Ba 129
Walker Cl. TW14: Felt 59V 106
Walker Cl. W7 46Ga 86
Walker Cres. SL3: L'ly 5B 82
Walker Ho. NW1 2D 216 (40Mb 70)
Walker Ho. SE16 48Bc 92
(off Redriff Est.)
Walker M. SW2 57Qb 112
Walker Pl. TN15: Igh 93Zd 205
Walker's Ct. W1 4D 222 (45Mb 90)
(off Brewer St.)
Walkerscroft Mead SE21 60Sb 113
Walkers Lodge E14 47Ec 92
(off Manchester Rd.)
Walkers Pl. SW15 56Ab 110
Walkers Sq. SS17: Stan H 2M 101
Walkfield Dr. KT18: Tatt C 89Xa 174
Walkinshaw Ct. N1 38Sb 71
(off Rotherfield St.)
Walkley Rd. DA1: Dart 57Kd 119
Walks, The N2 27Fb 49
Walkynscroft SE15 54Xb 113
(off Caulfield Rd.)
Wallace Bldg. NW8 6C 214 (42Fb 89)
(off Penfold St.)
Wallace Cl. SE28 45Zc 95
Wallace Cl. TW17: Shep 70T 128
Wallace Cl. UB10: Uxb 40N 63
Wallace Collection 2H 221 (44Jb 90)
Wallace Ct. NW1 1E 220 (43Gb 89)
(off Old Marylebone Rd.)
Wallace Ct. SE3 56Kc 115
Wallace Cres. SM5: Cars 78Hb 155
Wallace Flds. KT17: Eps 84Va 174
Wallace Gdns. DA10: Swans 58Ae 121
Wallace Ho. N7 37Pb 70
(off Caledonian Rd.)
Wallace Rd. N1 37Sb 71
Wallace Rd. RM17: Grays 48Ce 99
Wallace Sq. CR5: Coul 94Mb 196
Wallace Wlk. KT15: Add 77L 149
Wallace Wlk. SL4: Eton 10K 81
Wallace Way N19 33Mb 70
(off St John's Way)

Wallace Way RM1: Rom 25Fd 56
Wallasey Cres. UB10: Ick 33Q 64
Wallbrook Bus. Cen. TW4: Houn 55X 107
Wallbutton Rd. SE4 54Ac 114
Wallcote Av. NW2 32Za 68
Wall Ct. N4 32Pb 70
(off Stroud Grn. Rd.)
Walled Gdn., The KT20: Tad 94Za 194
Walled Gdn., The RH3: B'wth 7A 206
Walled Gdn. Cl. BR3: Beck 70Dc 136
Walled Gdn. Ct. HA7: Stan 20Ga 28
Wallenberg Pl. W1 3G 221 (44Hb 89)
(off Gt. Cumberland Pl.)
WALLEND 39Qc 74
Wall End Ct. E6 39Qc 74
(off Wall End Rd.)
Wall End Rd. E6 38Qc 74
Wallenger Av. RM2: Rom 27Kd 57
Waller Dr. HA6: Nwood 26W 44
Waller La. CR3: Cat'm 95Vb 197
Waller Rd. SE14 53Zb 114
Wallers Cl. IG8: Wfd G 23Pc 54
Wallers Cl. RM9: Dag 39Ad 75
Waller's Hoppet IG10: Lough 12Pc 36
Waller Way SE10 52Dc 114
Wallfield Pk. RH2: Reig 6H 207
Wallflower St. W12 45Va 88
Wallgrave Rd. SW5 49Db 89
Wall Hall WD25: A'ham 9Da 13
Wall Hall Dr. WD25: A'ham 8Da 13
Wallhouse Rd. DA8: Erith 52Kd 119
Wallingford Av. W10 43Za 88
Wallingford Ho. RM3: Rom 23Nd 57
(off Kingsbridge Rd.)
Wallingford Rd. UB8: Uxb 40K 63
Wallingford Wlk. AL1: St A 5B 6
WALLINGTON 79Kb 156
Wallington Cl. HA4: Ruis 30S 44
Wallington Cnr. SM6: W'gton 77Kb 156
(off Manor Rd. Nth.)
Wallington Ct. SM6: W'gton 79Kb 156
(off Stanley Pk. Rd.)
WALLINGTON GREEN 77Kb 156
WALLINGTON RESOURCE CEN. 80Lb 156
Wallington Rd. IG3: Ilf 31Vc 75
Wallington Sq. SM6: W'gton 79Kb 156
Wallington Station (Rail) 79Kb 156
Wallis All. SE1 1E 230 (47Sb 91)
(off Marshalsea Rd.)
Wallis Cl. DA2: Wilm 62Hd 140
Wallis Cl. RM11: Horn 32Kd 77
Wallis Cl. SW11 55Fb 111
Wallis Ct. SL1: Slou 7L 81
Wallis Ho. HA4: Ruis 32T 64
Wallis Ho. SE14 53Ac 114
Wallis Ho. TW8: Bford 50Na 87
Wallis M. KT22: Lea 94Ja 192
Wallis M. N8 27Qb 50
(off Courcy Rd.)
Wallis Pk. DA11: Nflt 57De 121
Wallis Rd. E9 37Bc 72
Wallis Rd. TW6: H'row A 54K 105
Wallis Rd. UB1: S'hall 44Da 85
Wallis's Cotts. SW2 59Nb 112
Wallman Pl. N22 25Pb 50
Wallorton Gdns. SW14 56Ta 109
Wallpaper Apts., The N1 38Qb 70
(off Offord Rd.)
Wallside EC2 1E 224 (43Sb 91)
(off Monkwell Sq.)
Wall St. N1 37Tb 71
Wallwood Rd. E11 31Fc 73
Wallwood St. E14 43Bc 92
Walmar Cl. EN4: Had W 11Fb 31
Walmer Cl. BR6: Farnb 77Tc 160
Walmer Cl. E4 19Dc 34
Walmer Cl. RM7: Mawney 26Dd 56
Walmer Ct. KT5: Surb 71Na 153
(off Cranes Pk.)
Walmer Gdns. W13 47Ja 86
Walmer Ho. W10 44Za 88
(off Bramley Rd.)
Walmer Pl. W1 7F 215 (43Hb 89)
(off Walmer St.)
Walmer Rd. W10 44Ya 88
Walmer Rd. W11 45Ab 88
Walmer St. W1 7F 215 (43Hb 89)
Walmer Ter. SE18 49Sc 94
Walmgate Rd. UB6: G'frd 39Ka 66
Walmington Fold N12 23Cb 49
Walm La. NW2 37Ya 68
Walney Wlk. N1 37Sb 71
Walnut Av. UB7: W Dray 48Q 84
Walnut Cl. AL2: Park 9P 5
Walnut Cl. DA4: Eyns 76Md 163
Walnut Cl. IG6: Ilf 28Sc 54
Walnut Cl. KT18: Eps 87Va 174
Walnut Cl. SE8 51Bc 114
Walnut Cl. SM5: Cars 78Hb 155
Walnut Cl. UB3: Hayes 45U 84
Walnut Ct. E17 28Ec 52
Walnut Ct. W5 47Na 87
Walnut Ct. W8 48Db 89
(off St Mary's Ga.)
Walnut Dr. KT20: Kgswd 96Ab 194
Walnut Flds. KT17: Ewe 81Va 174
Walnut Gdns. E15 36Gc 73
Walnut Grn. WD23: Bush 12Ba 27
Walnut Gro. EN1: Enf 15Tb 33
Walnut Gro. HP2: Hem H 2M 3
Walnut Gro. RH2: Horn 32Md 77
Walnut Gro. SM7: Bans 86Za 174
Walnut Hill Rd. DA13: Ist R 9A 144
Walnut Hill Rd. DA13: Meop 9A 144
Walnut Ho. E3 39Bc 72
(off Barge La.)
Walnut Ho. RH2: Reig 8L 207
Walnut Lodge SL1: Slou 8H 81
Walnut M. N22 27Qb 50
(off High Rd.)
Walnut M. SM2: Sutt 80Eb 155
Walnut Rd. E10 33Cc 72
Walnuts, The BR6: Orp 75Wc 161
Walnuts Leisure Cen. 74Wc 161
Walnuts Rd. BR6: Orp 74Xc 161
Walnut Tree Av. CR4: Mitc 69Gb 133
(off De'Arn Gdns.)
Walnut Tree Av. DA1: Dart 61Nd 141
Walnut Tree Cl. BR7: Chst 67Tc 138
Walnut Tree Cl. EN8: Chesh 3Zb 20
Walnut Tree Cl. KT23: Fet 97Fa 192
Walnut Tree Cl. SM7: Bans 84Ab 174
Walnut Tree Cl. SW13 53Va 110
Walnut Tree Cl. TN16: Westrm 98Tc 200
Walnut Tree Cl. TW17: Shep 69S 128
Walnut Tree Cl. UB10: Ick 35N 63
Walnut Tree Cotts. SW19 64Ab 132
Walnut Tree Ho. SW10 51Db 111
(off Tregunter Rd.)
Walnut Tree La. KT14: Byfl 84M 169
Walnut Tree Pl. GU23: Send 95F 188
Walnut Tree Rd. DA8: Erith 50Gd 98
Walnut Tree Rd. RM8: Dag 33Ad 75
Walnut Tree Rd. SE10 50Gc 93
(not continuous)
Walnut Tree Rd. TW17: Shep 68S 128
Walnut Tree Rd. TW5: Hest 51Ba 107
Walnut Tree Rd. TW8: Bford 51Na 109
Walnut Tree Wlk. SE11 5K 229 (49Qb 90)
Walnut Way BR8: Swan 68Fd 140
Walnut Way HA4: Ruis 37Y 65
Walnut Way IG9: Buck H 20Mc 35
Walpole Av. CR5: Chips 91Hb 195
Walpole Av. TW9: Kew 54Pa 109
Walpole Cl. HA5: Hat E 23Ca 45
Walpole Cl. RM17: Grays 49Ee 99
Walpole Cl. W13 47La 86
Walpole Ct. NW6 38Db 69
(off Fairfax Rd.)
Walpole Ct. TW2: Twick 61Ga 130
Walpole Ct. W14 48Za 88
(off Blythe Rd.)
Walpole Cres. TW11: Tedd 64Ha 130
Walpole Gdns. TW2: Twick 61Ga 130
Walpole Gdns. W4 50Sa 87
Walpole Ho. KT8: W Mole 71Ca 151
(off Approach Rd.)
Walpole Ho. SE1 2K 229 (47Qb 90)
(off Westminster Bri. Rd.)
Walpole Ho. SL4: Eton 1G 102
(off Eton Wick Rd.)
Walpole Ho. SW15 57Ab 110
(off Plaza Gdns.)
Walpole Lodge W13 46La 86
Walpole M. NW8 1B 214 (39Fb 69)
Walpole M. SW19 65Fb 133
Walpole Pk. KT13: Weyb 80Q 150
Walpole Pl. SE18 49Rc 94
Walpole Pl. TW11: Tedd 64Ha 130
Walpole Rd. BR2: Broml 71Mc 159
Walpole Rd. CR0: C'don 75Tb 157
Walpole Rd. E17 28Ac 52
Walpole Rd. E18 25Hc 53
Walpole Rd. E6 38Lc 73
Walpole Rd. KT6: Surb 73Na 153
Walpole Rd. N17 26Sb 51
(not continuous)
Walpole Rd. SL1: Slou 4B 80
Walpole Rd. SL4: Old Win 9M 103
Walpole Rd. SW19 65Fb 133
Walpole Rd. TW11: Tedd 64Ha 130
Walpole Rd. TW2: Twick 61Ga 130
Walpole St. SW3 7F 227 (50Hb 89)
Walrond Av. HA9: Wemb 36Na 67
Walsham Cl. N16 32Wb 71
Walsham Cl. SE28 45Zc 95
Walsham Ent. Cen. RM17: Grays 50Ee 99
Walsham Ho. SE14 54Zb 114
Walsham Ho. SE17 7F 231 (50Tb 91)
(off Blackwood St.)
Walsham How Cl. E17 29Fc 53
Walsham M. GU23: Rip 93L 189
Walsham Rd. SE14 54Zb 114
Walsham Rd. TW14: Felt 59X 107
Walsh Cres. CR0: New Ad 84Gc 179
Walshford Way WD6: Bore 10Qa 15
Walsingham NW8 39Fb 69
Walsingham Gdns. KT19: Ewe 77Ua 154
Walsingham Ho. E4 17Fc 35
Walsingham Lodge SW13 53Wa 110
Walsingham Mans. SW6 52Db 111
(off Fulham Rd.)
Walsingham Pk. 9H 7
Walsingham Pk. BR7: Chst 68Tc 138
Walsingham Pl. SW4 58Jb 112
Walsingham Rd. BR5: St P 67Xc 139
Walsingham Rd. CR0: New Ad 82Ec 178
Walsingham Rd. CR4: Mitc 71Hb 155
Walsingham Rd. E5 34Wb 71
Walsingham Rd. EN2: Enf 14Tb 33
Walsingham Rd. W13 46Ja 86
Walsingham Wlk. DA17: Belv 51Cd 118
Walsingham Way AL2: Lon C 9G 6
Walston Ho. SW1 7D 228 (50Mb 90)
(off Aylesford St.)
Walter Besant Ho. E1 41Zb 92
(off Bancroft Rd.)
Walter Ct. W3 44Sa 87
(off Lynton Ter.)
Walter Grn. Ho. SE15 53Yb 114
(off Lausanne Rd.)
Walter Ho. SW10 52Fb 111
(off Riley St.)
Walter Hurford Ho. E12 35Qc 74
(off Grantham Rd.)
Walter Langley Ct. SE16 47Yb 92
(off Watford Rd.)
Walter Rodney Cl. E6 37Pc 74
Walter Savil Twr. E17 30Cc 52
(off Colchester Rd.)
Walters Cl. SE17 6F 231 (49Tb 91)
(off Brandon St.)
Walters Cl. UB3: Hayes 47V 84
Walters Ct. WD17: Wat 10W 12
Walters Ho. N1 39Rb 71
(off Essex Rd.)
Walters Ho. SE17 51Rb 113
(off Otto St.)
Walter Sickert Hall N1 2D 218 (40Sb 71)
(off Graham St.)
Walters Mead KT21: Asht 89Na 173
Walters Rd. EN3: Pond E 14Yb 34
Walters Rd. SE25 70Ub 135
Walter St. E2 41Zb 92
Walter St. KT2: King T 67Na 131
Walters Way SE23 58Zb 114
Walters Yd. BR1: Broml 68Jc 137
Walter Ter. E1 44Zb 92
Walterton Rd. W9 42Bb 89
Walter Wlk. HA8: Edg 23Sa 47
WALTHAM ABBEY 5Ec 20
Waltham Abbey Gatehouse 5Ec 20
(off Church St.)
Waltham Abbey Leisure Cen. 5Hc 21
Waltham Abbey Sports Cen. 5Hc 21
Waltham Abbey Swimming Pool 6Fc 21
Waltham Abbey Tourist Info. Cen. 5Ec 20
Waltham Av. NW9 30Qa 47
Waltham Av. UB3: Harl 48S 84
Waltham Cl. BR5: Orp 74Zc 161
Waltham Cl. CM13: Hut 16Ee 41
Waltham Cl. DA1: Dart 58Jd 118
Waltham Cl. RM3: Rom 23Pd 57
WALTHAM CROSS 5Ac 20
Waltham Cross Bus Station 6Ac 20
Waltham Cross Eleanor Cross 6Ac 20
Waltham Cross Station (Rail) 6Bc 20
Waltham Dr. HA8: Edg 26Qa 47
Waltham Forest Feel Good Cen. 26Dc 52
Waltham Gdns. EN3: Enf W 8Yb 20
Waltham Ho. NW8 39Eb 69
Waltham Pk. Way E17 25Cc 52
Waltham Rd. CR3: Cat'm 94Xb 197
Waltham Rd. IG8: Wfd G 23Nc 54
Waltham Rd. SM5: Cars 73Fb 155
Waltham Rd. UB2: S'hall 48Aa 85
WALTHAMSTOW 27Cc 52
Walthamstow Av. E4 23Bc 52
Walthamstow Bus. Cen. E17 26Ec 52
Walthamstow Bus Station 28Cc 52
Walthamstow Central Station
(Underground & Overground) 29Cc 52
Walthamstow Leisure Cen. 30Bc 52
Walthamstow Marsh Nature Reserve 31Xb 71
Walthamstow Pumphouse Mus. 30Ac 52
Walthamstow Queens Road Station
(Overground) 29Cc 52
Waltham Way E4 20Bc 34
Waltheof Av. N17 25Tb 51
Waltheof Gdns. N17 25Tb 51
Walton Av. HA2: Harr 36Ba 65
Walton Av. HA9: Wemb 34Ra 67
Walton Av. KT3: N Mald 70Va 132
Walton Av. SM3: Cheam 76Bb 155
Walton Bri. KT12: Walt T 73U 150
Walton Bri. 73U 150
Walton Bri. Rd. TW17: Shep 73U 150
Walton Cl. E4 22Cc 52
Walton Cl. E5 34Zb 72
(off Orient Way)
Walton Cl. HA1: Harr 28Fa 46
Walton Cl. NW2 33Xa 68
Walton Cl. SW8 52Nb 112
WALTON COMMUNITY HOSPITAL 75X 151
Walton Ct. CR2: S Croy 78Sb 157
(off Warham Rd.)
Walton Ct. EN5: New Bar 15Eb 31
Walton Ct. GU21: Wok 88C 168
Walton Ct. NW6 38Eb 69
(off Fairfax Rd.)
Walton Cft. HA1: Harr 35Ga 66
Walton Dr. HA1: Harr 28Fa 46
Walton Dr. NW10 37Ta 67
Walton Gdns. CM13: Hut 15Ee 41
Walton Gdns. EN9: Walt A 5Dc 20
Walton Gdns. HA9: Wemb 33Na 67
Walton Gdns. TW13: Felt 63V 128
Walton Gdns. W3 43Ra 87
Walton Grn. CR0: New Ad 81Dc 178
Walton Hall Campsite SS17: Stan H 6J 101
WALTON HEATH 99Va 194
Walton Heath Cl. SS17: Stan H 4K 101
Walton Heath Golf Course 97Xa 194
Walton Ho. E17 27Dc 52
Walton Ho. E2 5K 219 (42Vb 91)
(off Montclare St.)
Walton Ho. NW1 5A 216 (42Kb 90)
(off Longford St.)
Walton Ho. SW3 4F 227 (48Hb 89)
(off Walton St.)
Walton La. KT12: Walt T 75S 150
Walton La. KT13: Weyb 75R 150
Walton La. SL2: Farn R 10D 60
Walton La. TW17: Shep 73T 150
WALTON-ON-THAMES 74W 150
Walton on Thames Camping
& Cvn. Site KT12: Walt T 73Ca 151
Walton-on-Thames Station (Rail) 77W 150
WALTON ON THE HILL 96Wa 194
Walton Pk. KT12: Walt T 75Z 151
Walton Pk. La. KT12: Walt T 75Z 151
Walton Pl. SW3 3F 227 (48Hb 89)
Walton Rd. DA14: Sidc 61Yc 139
Walton Rd. E12 35Qc 74
(not continuous)
Walton Rd. E13 40Lc 73
Walton Rd. GU21: Wok 88B 168
Walton Rd. HA1: Harr 28Fa 46
Walton Rd. KT12: Walt T 71Y 151
Walton Rd. KT18: Eps D 89Va 174
Walton Rd. KT18: Eps D 93Sa 193
Walton Rd. KT18: Head 93Sa 193
Walton Rd. KT8: E Mos 70Ba 129
Walton Rd. KT8: W Mole 70Ba 129
Walton Rd. N15 28Vb 51
Walton Rd. RM5: Col R 24Bd 55
Walton Rd. WD23: Bush 14Z 27
Walton's Hall Rd. SS17: Stan H 7J 101
Walton St. AL1: St A 1D 6
Walton St. EN2: Enf 11Tb 33
Walton St. KT20: Walt H 96Wa 194
Walton St. SW3 5E 226 (49Gb 89)
Walton Ter. GU21: Wok 87D 168
Walton Ter. WD6: E'tree 16Ma 29
(off Watford Rd.)
Walton Vs. N1 38Ub 71
(off Downham Rd.)
Walton Way CR4: Mitc 70Lb 134
Walton Way W3 43Ra 87
Walt Whitman Cl. SE24 56Rb 113
Walverns Cl. WD19: Wat 16Y 27
WALWORTH 7D 230 (50Sb 91)
Walworth Pl. SE17 7E 230 (50Sb 91)
Walworth Rd. SE1 5D 230 (49Sb 91)
Walworth Rd. SE17 5D 230 (49Sb 91)
Walworth Sq. SE17 6D 230 (49Sb 91)
Walwyn Av. BR1: Broml 69Mc 137
Wambrook Cl. CM13: Hut 18Ee 41
Wanborough Dr. SW15 60Xa 110
Wanderer Dr. IG11: Bark 41Yc 95
Wander Wharf WD4: K Lan 1R 12
Wandle Ct. CR2: S Croy 78Tb 157
Wandle Bank CR0: Bedd 76Nb 156
Wandle Bank SW19 66Fb 133
Wandle Ct. CR0: Bedd 76Nb 156
Wandle Ct. KT19: Ewe 77Sa 153
Wandle Ct. Gdns. CR0: Bedd 76Nb 156
Wandle Ho. BR1: Broml 64Fc 137
Wandle Ho. NW8 7D 214 (43Gb 89)
(off Penfold St.)
Wandle Industrial Mus. 69Hb 133
Wandle Meadow Nature Pk. 64Eb 133
Wandle Pk. 75Rb 157
Wandle Park Stop
(London Tramlink) 75Qb 156
Wandle Pk. Trad. Est., The CR0: C'don 75Rb 157
Wandle Recreation Cen. 58Db 111
Wandle Rd. CR0: Bedd 76Nb 156
Wandle Rd. CR0: C'don 76Sb 157
Wandle Rd. SM4: Mord 70Eb 133
Wandle Rd. SM6: W'gton 76Kb 156
Wandle Rd. SW17 61Gb 133
Wandle Side CR0: Wadd 76Pb 156
Wandle Side SM6: W'gton 76Kb 156
Wandle Trad. Est. CR4: Mitc 73Hb 155
Wandle Way CR4: Mitc 71Hb 155
Wandle Way SW18 60Db 111
Wandon Rd. SW6 52Db 111
WANDSWORTH 57Db 111
Wandsworth Bri. 55Db 111
Wandsworth Bri. Rd. SW6 53Db 111
WANDSWORTH COMMON 60Hb 111
Wandsworth Common Station (Rail) 60Hb 111
Wandsworth Comn. W. Side SW18 57Eb 111
WANDSWORTH GYRATORY 57Db 111
Wandsworth High St. SW18 57Cb 111
Wandsworth Mus. 57Cb 111
Wandsworth Plain SW18 57Db 111
Wandsworth Rd. SW8 55Kb 112
Wandsworth Road Station
(Overground) 54Lb 112
Wandsworth Town Station (Rail) 56Db 111
Wangey Rd. RM6: Chad H 31Zc 75
Wangford Ho. SW9 56Rb 113
(off Loughborough Pk.)
Wanless Rd. SE24 55Sb 113
Wanley Rd. SE5 56Tb 113
Wanlip Rd. E13 42Kc 93
Wanmer Ct. RH2: Reig 5J 207
(off Birkheads Rd.)
Wannock Gdns. IG6: Ilf 24Rc 54
Wansbeck Ct. EN2: Enf 13Rb 33
(off Waverley Rd.)
Wansbeck Rd. E9 38Bc 72
Wansbury Way BR8: Swan 71Jd 162
Wansdown Pl. SW6 52Db 111
Wansey St. SE17 6D 230 (49Sb 91)
Wansford Cl. CM14: B'wood 20Vd 40
Wansford Grn. GU21: Wok 9K 167
Wansford Pk. WD6: Bore 14Ta 29
Wansford Rd. IG8: Wfd G 25Lc 53
WANSTEAD 30Kc 53
Wanstead Cl. BR1: Broml 68Lc 137
Wanstead Gdns. IG4: Ilf 30Mc 53
Wanstead Golf Course 31Lc 73
Wanstead La. IG1: Ilf 30Mc 53
Wanstead Leisure Cen. 30Lc 53
Wanstead Pk. Av. E12 32Mc 73
Wanstead Pk. Rd. IG1: Ilf 30Mc 53
Wanstead Park Station (Overground) 35Kc 73
Wanstead Pl. E11 30Jc 53
Wanstead Rd. BR1: Broml 68Lc 137
Wanstead Station (Underground) 30Kc 53
Wansunt Rd. DA5: Bexl 60Ed 118
Wantage Rd. SE12 57Hc 115
Wantz La. RM13: Rain 42Kd 97
Wantz Rd. RM10: Dag 35Dd 76
Waplings, The KT20: Walt H 96Xa 194
WAPPING 46Xb 91
Wapping Dock St. E1 46Xb 91
Wapping High St. E1 46Wb 91
Wapping La. E1 45Xb 91
Wapping Station (Overground) 46Yb 92
Wapping Wall E1 46Yb 92
Wapses Lodge CR3: Wold 92Xb 197
WAPSES LODGE RDBT. 92Xb 197
Wapseys La. SL2: Hedg 1J 61
Wapshott Rd. TW18: Staines 65G 126
Waratah Dr. BR7: Chst 64Pc 138
Warbank Cl. CR0: New Ad 82Gc 179
Warbank Cres. CR0: New Ad 82Gc 179
Warbank La. KT2: King T 66Ua 132
Warbeck Ho. KT13: Weyb 78T 150
(off Queens Rd.)
Warbeck Rd. W12 47Xa 88
Warberry Rd. N22 25Pb 50
Warbler Ct. HP3: Hem H 7L 3
Warbler's Grn. KT11: Cobh 86Ba 171
Warboys App. KT2: King T 65Ra 131
Warboys Cres. E4 22Ec 52
Warboys Rd. KT2: King T 65Ra 131
Warburg Ct. NW9 27Ua 48
(off Mornington Cl.)
Warburton Cl. HA3: Hrw W 23Fa 46
Warburton Cl. N1 37Ub 71
(off Culford Rd.)
Warburton Ct. HA4: Ruis 33W 64
Warburton Ho. E8 39Xb 71
(off Warburton St.)
Warburton Rd. E8 39Xb 71
Warburton Rd. TW2: Whitt 60Da 107
Warburtons SS17: Stan H 1P 101
Warburton St. E8 39Xb 71
Warburton Ter. E17 26Dc 52
Warburton La. GU21: Knap 7F 166
War Coppice Rd. CR3: Cat'm 99Tb 197
Wardalls Gro. SE14 52Yb 114
Wardalls Ho. SE8 51Bc 114
(off Staunton St.)
Ward Av. RM17: Grays 49Ce 99
Ward Cl. CR2: S Croy 79Ub 157
Ward Cl. DA8: Erith 51Fd 118
Ward Cl. SL0: Iver 45H 83
Wardell Cl. NW7 24Ua 48
Wardell Fld. NW9 25Ua 48
Wardell Ho. SE10 51Ec 114
(off Welland St.)
Wardell M. SW4 55Kb 112
Warden Av. HA2: Harr 32Ba 65
Warden Av. RM5: Col R 22Ed 56
Warden Rd. NW5 37Jb 70
Wardens Fld. Cl. BR6: Chels 79Uc 160
Wardens Gro. SE1 7D 224 (46Sb 91)
Ward Gdns. RM3: Hrld W 25Md 57
Ward Gdns. SL1: Slou 5C 80
Ward La. CR6: W'ham 88Yb 178
Ward La. E9 36Ac 72
Wardle St. E9 36Zb 72
Wardley St. SW18 59Db 111
Wardo Av. SW6 53Ab 110
Wardona Ct. DA10: Swans 58Be 121
Wardona Ct. DA11: Nflt 57De 121
Wardona Ho. DA10: Swans 58Be 121
Wardour Ct. DA2: Dart 58Rd 119
(off Bow Arrow La.)
Wardour M. W1 3C 222 (44Lb 90)
(off D'Arblay St.)
Wardour St. W1 2C 222 (44Lb 90)
Ward Point SE11 6K 229 (49Qb 90)
Ward Rd. E15 39Fc 73
Ward Rd. N19 34Lb 70
Ward Rd. SW19 67Eb 133
Ward Rd. WD24: Wat 8W 12
Wardrobe, The TW9: Rich 57Ma 109
(off Old Palace Yd.)
Wardrobe Pl. EC4 3C 224 (44Rb 91)
(off Carter La.)
Wardrobe Ter. EC4 4C 224 (45Rb 91)
(off Addle Hill)
Wardroper Ho. SE1 4C 230 (48Rb 91)
(off St George's Rd.)
Ward Royal SL4: Wind 3G 102
Ward Royal Pde. SL4: Wind 3G 102
(off Alma Rd.)
Wards Dr. WD3: Sarr 8H 11
Wards La. WD6: E'tree 12Ha 28
Ward's Pl. TW20: Egh 65E 126
Wards Rd. IG2: Ilf 31Tc 74
Wards Wharf App. E16 46Mc 93
Wardur Ho. KT12: Walt T 76W 150
Ware Ct. SM1: Sutt 77Bb 155
Wareham Cl. TW3: Houn 56Da 107
Wareham Ct. N1 38Ub 71
(off Hertford Rd.)
Wareham Ho. SW8 52Pb 112
Warehome M. E13 42Jc 93
(off Jutland Rd.)
Warehouse, The 7A 224 (46Qb 90)
(off Theed St.)
Warehouse Ct. SE18 48Rc 94
Warehouse Way E16 45Kc 93
Waremead Rd. IG2: Ilf 29Rc 54
Warenford Way WD6: Bore 11Qa 29
Warenne Hgts. RH1: Redh 8M 207
Warenne Rd. KT22: Fet 94Ea 192
Warepoint Dr. SE28 47Tc 94
Warescot Cl. CM15: B'wood 17Xd 40
Warescot Rd. CM15: B'wood 17Xd 40
Warfield Rd. NW10 41Za 88
Warfield Rd. TW12: Hamp 67Da 129
Warfield Rd. TW14: Felt 59U 106
Warfield Yd. NW10 41Za 88
(off Warfield Rd.)
Wargrave Av. N15 30Vb 51
Wargrave Ho. E2 4K 219 (41Vb 91)
(off Navarre St.)
Wargrave Rd. HA2: Harr 34Ea 66
Warham Cl. EN8: Chesh 2Xb 19
Warham Rd. CR2: S Croy 78Rb 157
Warham Rd. HA3: W'stone 26Ha 46
Warham Rd. N4 29Qb 50
Warham Rd. TN14: Otf 88Kd 183
Warham St. SE5 52Rb 113
Waring & Gillow Est. W3 42Qa 87
Waring Cl. BR6: Chels 79Vc 161
Waring Dr. BR6: Chels 79Vc 161
Waring Rd. DA14: Sidc 65Yc 139
Waring St. SE27 63Sb 135
Warkworth Gdns. TW7: Isle 52Ja 108
Warkworth Rd. N17 24Tb 51
Warland Rd. SE18 52Tc 116
Warland Rd. TN15: W King 81Vd 184
WARLEY 22Yd 58
Warley Av. RM8: Dag 31Bd 75
Warley Av. UB4: Hayes 44W 84
Warley Cl. E10 32Bc 72
Warley Country Pk. 21Wd 58
Warley Gap CM13: Gt War 24Xd 58
Warley Gap CM13: L War 24Xd 58
Warley Hall La. RM14: Upm 31Be 79
Warley Hill CM13: Gt War 23Xd 58
Warley Hill CM13: W'ley 23Xd 58
Warley Hill CM14: W'ley 22Yd 58
Warley Hill Bus. Pk., The
CM13: Gt War 23Yd 58
Warley Mt. CM14: W'ley 21Yd 58
Warley Pk. Golf Course 26Zd 59
Warley Place Nature Reserve 24Wd 58
Warley Rd. CM13: Gt War 25Wd 58
Warley Rd. IG5: Ilf 25Qc 54
Warley Rd. IG8: Wfd G 24Kc 53
Warley Rd. N9 19Yb 34
Warley Rd. RM14: Gt War 26Sd 58
Warley Rd. RM14: Upm 26Sd 58
Warley Rd. UB4: Hayes 44W 84
Warley St. CM13: Gt War 28Yd 58
Warley St. CM13: Upm 28Yd 58
Warley St. E2 41Zb 92
Warley St. RM14: Upm 30Yd 58
Warleywoods Cres. CM14: W'ley 21Xd 58
WARLINGHAM 90Zb 178
Warlingham Ct. SE13 58Ec 114
Warlingham Rd. CR7: Thor H 70Rb 135
Warlock Rd. W9 42Bb 89
Warlow Cl. EN3: Enf L 9Cc 20
Warlters Cl. N7 35Nb 70
Warlters Rd. N7 35Nb 70
Warltersville Mans. N19 31Nb 70
Warltersville Rd. N19 31Nb 70
Warmark Rd. HP1: Hem H 1G 2
War Memorial Sports Ground 77Gb 155
Warmington Cl. E5 34Zb 72
Warmington Rd. SE24 58Sb 113
Warmington St. E13 42Jc 93
Warminster Gdns. SE25 68Wb 135
Warminster Ho. RM3: Rom 22Pd 57
(off Redcar Rd.)
Warminster Rd. SE25 68Vb 135
Warminster Sq. SE25 68Wb 135
Warminster Way CR4: Mitc 67Kb 134
Warmsworth NW1 39Lb 70
(off Pratt St.)
Warmwell Av. NW9 25Ua 48
Warndon St. SE16 49Zb 92
Warneford Pl. WD19: Wat 16Aa 27
Warneford Rd. HA3: Kenton 27Ma 47
Warneford Rd. TW6: H'row A 54K 105
Warneford St. E9 39Xb 71
Warne Pl. DA15: Sidc 58Xc 117
Warner Av. SM3: Cheam 75Ab 154
Warner Bros. Studios Leavesden 6U 12
Warner Cl. E15 36Gc 73
Warner Cl. EN4: Had W 9Gb 17
Warner Cl. NW9 31Va 68
Warner Cl. SL1: Slou 6C 80
Warner Cl. TW12: Hamp 64Ba 129
Warner Cl. UB3: Harl 52T 106
Warner Dr. WD25: Wat 5V 12
Warner Ho. BR3: Beck 65Dc 136
Warner Ho. NW8 41Eb 89
Warner Ho. SE13 54Dc 114
(off Russett Way)
Warner Pl. E2 40Wb 71
Warner Rd. BR1: Broml 66Hc 137
Warner Rd. E17 28Ac 52
Warner Rd. N8 28Mb 50
Warner Rd. SE5 53Sb 113
Warners Cl. IG8: Wfd G 22Jc 53
WARNERS END 2J 3
Warners End Rd. HP1: Hem H 2J 3
Warners La. KT2: King T 63Ma 131
Warners Path IG8: Wfd G 22Jc 53
Warner St. EC1 6K 217 (42Qb 90)
Warner Ter. E14 43Dc 92
(off Broomfield St.)
Warner Yd. EC1 6K 217 (42Qb 90)
(off Warner St.)
Warnford Ct. EC2 2G 225 (44Tb 91)
(off Throgmorton Av.)
Warnford Ho. SW15 58Ua 110
(off Tunworth Cres.)
Warnford Ind. Est. UB3: Hayes 47U 84
Warnford Rd. BR6: Chels 78Vc 161
Warnham WC1 4H 217 (41Pb 90)
(off Sidmouth St.)

Warnham Ct. Rd. SM5: Cars 80Hb **155**
Warnham Gro. BR5: St P 68Yc **139**
Warnham Ho. SW2 59Pb **112**
(off Up. Tulse Hill)
Warnham Rd. N12 22Gb **49**
Warple M. W3 47Ua **88**
Warple Way W3 47Ua **88**
Warre Ho. SL4: Eton 10G **80**
(off Common La.)
Warren, The AL2: Park 10A **6**
Warren, The DA12: Grav'nd 3F **144**
Warren, The E12 35Nc **74**
Warren, The KT20: Kgswd 95Ab **194**
Warren, The KT21: Asht 91Na **193**
Warren, The KT22: Oxs 84Ea **172**
Warren, The KT4: Wor Pk 77Ta **153**
Warren, The SL9: Chal P 24B **42**
Warren, The SM5: Cars 81Fb **175**
Warren, The SS17: Stan H 4N **101**
Warren, The TW5: Hest 52Ba **107**
Warren, The UB4: Hayes 44W **84**
Warren, The 47Y **85**
Warren, The WD4: K Lan 1P **11**
Warren, The WD7: R'lett 5Ja **14**
Warren Av. BR1: Broml 66Gc **137**
Warren Av. BR6: Chels 78Vc **161**
Warren Av. CR2: Sels 80Zb **158**
Warren Av. E10 34Ec **72**
Warren Av. SM2: Cheam 82Bb **175**
Warren Av. TW10: Rich 56Ra **109**
Warren Cl. DA6: Bex 57Cd **118**
Warren Cl. HA9: Wemb 33Ma **67**
Warren Cl. KT10: Esh 77Da **151**
Warren Cl. N9 17Zb **34**
Warren Cl. SE21 59Sb **113**
Warren Cl. SL3: L'ly 48A **82**
Warren Cl. UB4: Yead 43Y **85**
Warren Ct. BR3: Beck 66Cc **136**
Warren Ct. CR0: C'don 74Ub **157**
Warren Ct. IG7: Chig 21Tc **54**
Warren Ct. KT13: Weyb 78Q **150**
Warren Ct. KT21: Asht 91Na **193**
Warren Ct. N17 27Wb **51**
(off High Cross Rd.)
Warren Ct. NW1 5B **216** (42Lb **90**)
(off Warren St.)
Warren Ct. SE7 51Lc **115**
Warren Ct. SL2: Farn C 6G **60**
Warren Ct. TN13: S'oaks 96Ld **203**
Warren Ct. W5 43La **86**
Warren Ct. WD25: Wat 2X **13**
Warren Ct. Farm TN14: Hals 85Ad **181**
Warren Cres. N9 17Vb **33**
Warren Cutting KT2: King T 66Ta **131**
Warrender Rd. N19 34Lb **70**
Warrender Way HA4: Ruis 31W **64**
Warren Dr., The E11 31Lc **73**
Warren Dr. BR6: Chels 78Xc **161**
Warren Dr. HA4: Ruis 31Z **65**
Warren Dr. KT20: Kgswd 94Bb **195**
Warren Dr. RM12: Horn 35Jd **76**
Warren Dr. UB6: G'frd 42Da **85**
Warren Dr. Nth. KT5: Surb 74Ra **153**
Warren Dr. Sth. KT5: Surb 74Sa **153**
Warrene Cl. SS17: Stan H 2M **101**
Warreners La. KT13: Weyb 80T **150**
Warren Farm Cl. KT17: Bans 87Ya **174**
Warren Farm Cotts. RM6: Chad H 28Bd **55**
Warren Farm Mobile Home Pk.
GU22: Pyr 91K **189**
Warren Fld. CM16: Epp 4Wc **23**
Warren Fld. SL0: Iver H 40E **62**
Warrenfield Cl. EN7: Chesh 3Wb **19**
Warren Flds. HA7: Stan 21La **46**
Warren Footpath TW1: Twick 60La **108**
Warren Gdns. BR6: Chels 78Wc **161**
Warren Gdns. E15 36Fc **73**
Warrengate La. EN6: S Mim 3Ya **16**
Warrengate Rd. AL9: N Mym 8D **8**
Warren Gro. WD6: Bore 14Ta **29**
Warren Hastings Ct. DA11: Nflt 8B **122**
Warren Hgts. IG10: Lough 15Lc **35**
Warren Hgts. RM16: Chaf H 49Ae **99**
Warren Hill IG10: Lough 15Lc **35**
Warren Hill KT18: Eps 88Ta **173**
Warren Ho. E3 41Dc **92**
(off Bromley High St.)
Warren Ho. N17 27Wb **51**
(off High Cross Rd.)
Warren Ho. W14 49Bb **89**
Warrenhurst Gdns. KT13: Weyb 79T **150**
Warren La. CM15: Dodd 11Vd **40**
Warren La. CM15: Kel H 11Vd **40**
Warren La. GU22: Pyr 90J **169**
Warren La. HA7: Stan 19Ja **28**
Warren La. KT22: Oxs 83Ea **172**
Warren La. RH8: Oxt 6L **211**
Warren La. RM16: Chaf H 49Zd **99**
Warren La. SE18 48Rc **94**
Warren Lodge KT20: Kgswd 96Ab **194**
Warren Lodge Dr. KT20: Kgswd 96Ab **194**
Warren Mead SM7: Bans 87Ya **174**
Warren M. KT13: Weyb 76V **150**
Warren M. W1 6B **216** (42Lb **90**)
Warrenne Way RH2: Reig 6J **207**
Warren Pde. SL2: Slou 6N **81**
Warren Pk. CR6: W'ham 90Zb **178**
Warren Pk. KT2: King T 65Sa **131**
Warren Pk. Rd. SM1: Sutt 79Fb **155**
Warren Pl. E1 44Zb **92**
(off Pitsea St.)
Warren Pl. KT12: Walt T 75Y **151**
Warren Pond Rd. E4 18Hc **35**
Warren Ri. KT3: N Mald 67Ta **131**
Warren Rd. AL1: St A 6A **6**
Warren Rd. BR2: Hayes 75Jc **159**
Warren Rd. BR6: Chels 78Vc **161**
(not continuous)
Warren Rd. CR0: C'don 74Ub **157**
Warren Rd. CR8: Purl 84Rb **177**
Warren Rd. DA1: Dart 62Nd **141**
Warren Rd. DA13: Sflt 64De **143**
Warren Rd. DA14: Sidc 62Yc **139**
Warren Rd. DA6: Bex 57Cd **118**
Warren Rd. E10 34Ec **72**
Warren Rd. E11 30Lc **53**
Warren Rd. E4 19Ec **34**
Warren Rd. IG6: Ilf 29Tc **54**
Warren Rd. KT15: New H 82J **169**
Warren Rd. KT2: King T 65Sa **131**
Warren Rd. NW2 33Va **68**
Warren Rd. RH2: Reig 5K **207**
Warren Rd. SM7: Bans 86Ya **174**
Warren Rd. SW19 65Gb **133**
Warren Rd. TW15: Ashf 66U **128**
Warren Rd. TW2: Whitt 58Ea **108**
Warren Rd. UB10: Ick 35N **63**
Warren Rd. WD23: B Hea 18Ea **28**
Warrens, The DA3: Hartl 72Be **165**
Warrens End KT15: Add 79H **149**
Warren Sports Cen. 29Bd **55**
Warrens Shawe La. HA8: Edg 19Ra **29**
Warren St. W1 6B **216** (42Lb **90**)
Warren Street Station
(Underground) 5B **216** (42Lb **90**)
Warren Ter. RM16: N Stif 47Zd **99**
Warren Ter. RM6: Chad H 28Zc **55**
Warren Vw. DA12: Shorne 4N **145**
Warren Wlk. SE7 51Lc **115**
Warren Way HA8: Edg 26Ra **47**
Warren Way KT13: Weyb 78S **150**
WARREN WOOD 4M **9**
Warren Wood Cl. BR2: Hayes 75Hc **159**
Warrenwood M. AL9: Hat 4M **9**
Warriner Av. RM12: Horn 33Md **77**
Warriner Dr. N9 20Wb **33**
Warriner Gdns. SW11 53Hb **111**
Warrington Av. SL1: Slou 4G **80**
Warrington Ct. CR0: Wadd 76Rb **157**
(off Warrington Rd.)
Warrington Cres. W9 5A **214** (42Eb **89**)
Warrington Gdns. RM11: Horn 30Ld **57**
Warrington Gdns. W9 42Eb **89**
(not continuous)
Warrington Rd. CR0: Wadd 76Rb **157**
Warrington Rd. HA1: Harr 29Ga **46**
Warrington Rd. RM8: Dag 33Zc **75**
Warrington Rd. TW10: Rich 57Ma **109**
Warrington Spur SL4: Old Win 9M **103**
Warrington Sq. RM8: Dag 33Zc **75**
Warrior Av. DA12: Grav'nd 3E **144**
Warrior Cl. SE28 46Tc **94**
Warrior Ct. SW9 55Rb **113**
(off Coldharbour Rd.)
Warrior Sq. E12 35Qc **74**
Warsaw Cl. HA4: Ruis 37X **65**
Warspite Ho. E14 49Dc **92**
(off Cahir St.)
Warspite Rd. SE18 48Nc **94**
Warton Ct. E1 45Zb **92**
(off Cable St.)
Warton Ct. W3 48Sa **87**
(off All Saints Rd.)
Warton Ho. E15 39Ec **72**
(off High St.)
Warton Rd. E15 39Ec **72**
Warwall E6 44Rc **94**
Warwark Ct. KT6: Surb 75Na **153**
Warwick W14 49Bb **89**
(off Kensington Village)
Warwick Av. HA2: Harr 35Ba **65**
Warwick Av. HA8: Edg 20Ra **29**
Warwick Av. SL2: Slou 2G **80**
Warwick Av. TW18: Staines 65L **127**
Warwick Av. TW20: Egh 67E **126**
Warwick Av. W2 7A **214** (43Eb **89**)
Warwick Av. W9 42Db **89**
Warwick Avenue Station
(Underground) 6A **214** (42Eb **89**)
Warwick Bldg. SW11 51Kb **112**
Warwick Chambers W8 48Cb **89**
(off Pater St.)
Warwick Cl. BR6: Chels 76Wc **161**
Warwick Cl. DA5: Bexl 59Bd **117**
Warwick Cl. EN4: E Barn 15Fb **31**
Warwick Cl. RM11: Horn 28Pd **57**
Warwick Cl. TW12: Hamp 66Ea **130**
Warwick Cl. W8 48Bb **89**
(off Kensington High St.)
Warwick Cl. WD23: B Hea 17Ga **28**
Warwick Ct. BR2: Broml 68Gc **137**
Warwick Ct. DA8: Erith 52Hd **118**
Warwick Ct. EC4 3C **224** (44Rb **91**)
(off Warwick La.)
Warwick Ct. EN5: New Bar 15Db **31**
(off Station Rd.)
Warwick Ct. HA1: Harr 27Ga **46**
Warwick Ct. KT13: Weyb 78Q **150**
Warwick Ct. N11 23Mb **50**
Warwick Ct. SL4: Wind 4G **102**
(off Queen's Rd.)
Warwick Ct. TN13: S'oaks 97Kd **203**
Warwick Ct. UB5: N'olt 36Ca **65**
(off Newmarket Av.)
Warwick Ct. W7 44Ha **86**
(off Copley Cl.)
Warwick Ct. WC1 1J **223** (43Pb **90**)
Warwick Ct. WD3: Chor 13H **25**
Warwick Cres. UB4: Hayes 42V **84**
Warwick Cres. W2 7A **214** (43Eb **89**)
Warwick Deeping KT16: Ott 78E **148**
Warwick Dene W5 46Na **87**
Warwick Dr. EN8: Chesh 1Zb **20**
Warwick Dr. SW15 55Xa **110**
Warwick Est. W2 43Db **89**
Warwick Gdns. CR7: Thor H 69Qb **134**
Warwick Gdns. EN5: Barn 10Bb **17**
Warwick Gdns. IG1: Ilf 32Rc **74**
Warwick Gdns. KT21: Asht 89La **172**
Warwick Gdns. KT7: T Ditt 71Ha **152**
Warwick Gdns. N4 29Sb **51**
Warwick Gdns. RM2: Rom 27Ld **57**
Warwick Gdns. W14 48Bb **89**
Warwick Gro. E5 32Xb **71**
Warwick Gro. KT5: Surb 73Pa **153**
Warwick Ho. AL1: St A 2B **6**
(off London Rd.)
Warwick Ho. BR8: Swan 70Gd **140**
Warwick Ho. E16 46Jc **93**
(off Wesley Av.)
Warwick Ho. KT2: King T 67Na **131**
(off Acre Rd.)
Warwick Ho. SW9 54Qb **112**
Warwick Ho. St. SW1 6E **222** (46Mb **90**)
Warwick La. EC4 2C **224** (44Rb **91**)
Warwick La. GU21: Wok 1L **187**
Warwick La. RM13: Rain 41Pd **97**
Warwick La. RM14: Avel 40Pd **77**
Warwick La. RM14: Upm 40Pd **77**
Warwick La. W14 49Bb **89**
Warwick Lodge TW2: Twick 62Da **129**
Warwick Mans. SW5 49Cb **89**
(off Cromwell Cres.)
Warwick M. WD3: Crox G 16Q **26**
Warwick Pde. HA3: Kenton 26Ka **46**
Warwick Pas. EC4 2C **224** (44Rb **91**)
(off Old Bailey)
Warwick Pl. CM14: Pil H 14Sd **40**
Warwick Pl. DA11: Nflt 57De **121**
Warwick Pl. KT7: T Ditt 72Ja **152**
Warwick Pl. RH2: Reig 6M **207**
Warwick Pl. UB8: Uxb 38L **63**
Warwick Pl. W5 47Ma **87**
Warwick Pl. W9 43Eb **89**
Warwick Pl. WD6: Bore 13Ta **29**
Warwick Pl. Nth. SW1 6B **228** (49Lb **90**)
Warwick Quad. RH1: Redh 5A **208**
Warwick Rd. AL1: St A 1D **6**
Warwick Rd. CR5: Coul 86Lb **176**
Warwick Rd. CR7: Thor H 69Qb **134**
Warwick Rd. DA14: Sidc 64Xc **139**
Warwick Rd. DA16: Well 55Yc **117**
Warwick Rd. E11 29Kc **53**
Warwick Rd. E12 36Nc **74**
Warwick Rd. E15 37Hc **73**
Warwick Rd. E17 25Bc **52**
Warwick Rd. E4 22Cc **52**
Warwick Rd. EN3: Enf L 9Bc **20**
Warwick Rd. EN5: New Bar 14Db **31**
Warwick Rd. KT1: Hamp W 67La **130**
Warwick Rd. KT3: N Mald 69Sa **131**
Warwick Rd. KT7: T Ditt 71Ha **152**
Warwick Rd. N11 23Mb **50**
Warwick Rd. N18 21Ub **51**
Warwick Rd. RH1: Redh 5P **207**
Warwick Rd. RM13: Rain 42Ld **97**
Warwick Rd. SE20 69Xb **135**
Warwick Rd. SM1: Sutt 77Eb **155**
Warwick Rd. SW5 49Cb **89**
Warwick Rd. TW15: Ashf 64N **127**
Warwick Rd. TW2: Twick 60Ga **108**
Warwick Rd. TW4: Houn 55X **107**
Warwick Rd. UB2: S'hall 48Ba **85**
Warwick Rd. UB7: W Dray 47N **83**
Warwick Rd. W14 49Bb **89**
Warwick Rd. W5 47Ma **87**
Warwick Rd. WD6: Bore 13Ta **29**
Warwick Row SW1 3B **228** (48Lb **90**)
Warwickshire Path SE8 52Bc **114**
Warwickshire Rd. N16 35Ub **71**
Warwick Sq. EC4 2C **224** (44Rb **91**)
Warwick Sq. SW1 7B **228** (50Lb **90**)
(not continuous)
Warwick Sq. M. SW1 6B **228** (49Lb **90**)
Warwick St. W1 4C **222** (45Lb **90**)
Warwick Ter. E17 29Fc **53**
(off Lea Bri. Rd.)
Warwick Ter. SE18 51Tc **116**
Warwick Vs. TW20: Egh 67E **126**
Warwick Way DA1: Dart 61Nd **141**
Warwick Way SW1 7K **227** (50Kb **90**)
Warwick Way WD3: Crox G 14S **26**
WARWICK WOLD 1G **208**
Warwick Wold Rd. RH1: Mers 1G **208**
Warwick Yd. EC1 6E **218** (42Sb **91**)
Wasdale NW1 3A **216** (41Kb **90**)
(off Cumberland Mkt.)
Washbourne Ct. N9 19Wb **33**
(off Acton Cl.)
Washbourne Rd. NW10 39Ta **67**
Washington Av. E12 35Pc **74**
Washington Bldg. SE13 53Dc **114**
(off Deal's Gateway)
Washington Cl. E3 41Dc **92**
Washington Cl. RH2: Reig 4J **207**
Washington Dr. SL1: Slou 5B **80**
Washington Dr. SL4: Wind 5C **102**
Washington Ho. E17 26Bc **52**
Washington Ho. SW3 2F **227** (47Hb **89**)
(off Basil St.)
Washington Ho. WD23: Bush 14Ba **27**
(off King Edward Pl.)
Washington Rd. E18 26Hc **53**
Washington Rd. E6 38Lc **73**
Washington Rd. KT1: King T 68Qa **131**
Washington Rd. KT4: Wor Pk 75Xa **154**
Washington Rd. SW13 52Wa **110**
Washington Rd. TW6: H'row A 55L **105**
Wash La. EN6: S Mim 5Xa **16**
WASHMILLS 60Yd **120**
Washneys Rd. BR6: Prat B 86Wc **181**
Washpond La. CR6: W'ham 90Ec **178**
Wash Rd. CM13: Hut 16Fe **41**
Wash Rd. CM13: Mount 16Fe **41**
Wasp Rd. TW6: H'row A 54K **105**
(off Welland Rd.)
Wastdale Rd. SE23 60Zb **114**
Watch, The N12 21Eb **49**
Watchfield Ct. W4 50Sa **87**
Watchgate DA2: Daren 64Td **142**
(not continuous)
Watcombe Cotts. TW9: Kew 51Qa **109**
Watcombe Pl. SE25 71Xb **157**
Watcombe Rd. SE25 71Xb **157**
Waterbank Ho. SW18 60Db **111**
(off Knaresborough Dr.)
Waterbank Rd. SE6 62Dc **136**
Waterbeach Cl. SL1: Slou 4H **81**
Waterbeach Rd. RM9: Dag 37Yc **75**
Waterbeach Rd. SL1: Slou 4H **81**
Waterbourne Way CR8: Kenley 86Tb **177**
Water Brook La. NW4 29Ya **48**
Watercress Cl. TN14: S'oaks 92Ld **203**
Watercress Dr. TN14: S'oaks 92Ld **203**
Watercress Pl. N1 38Ub **71**
Watercress Way DA11: Nflt 59Fe **121**
Watercress Way GU21: Wok 9M **167**
Watercroft Rd. TN14: Hals 82Bd **181**
Waterdale AL2: Brick W 2Aa **13**
WATERDALE 2Y **13**
WATERDALE 2Z **13**
Waterdale Rd. SE2 51Wc **117**
Waterdales DA11: Nflt 61Ee **143**
Waterden Ct. W11 46Ab **88**
Waterden Rd. E20 36Cc **72**
Water Dr. WD3: Rick 18N **25**
WATER END 8D **8**
Water End Cl. WD6: Bore 12Pa **29**
Waterer Gdns. KT20: Tad 90Za **174**
Waterer Ho. SE6 63Ec **136**
Waterer Ri. SM6: W'gton 79Mb **156**
Waterers Ri. GU21: Knap 9H **167**
Waterfall Cl. GU25: Vir W 9L **125**
Waterfall Cl. N14 20Lb **32**
Waterfall Cotts. SW19 65Fb **133**
Waterfall Ho. SW2 57Qb **112**
(off Brixton Water La.)
Waterfall Rd. N11 21Kb **50**
Waterfall Rd. N14 20Lb **32**
Waterfall Rd. SW19 65Fb **133**
Waterfall Ter. SW17 65Gb **133**
Waterfall Wlk. N14 18Kb **32**
Waterfield KT20: Tad 92Xa **194**
Waterfield WD3: Herons 17E **24**
Waterfield Cl. DA17: Belv 48Cd **96**
Waterfield Cl. SE28 46Xc **95**
Waterfield Dr. CR6: W'ham 91Yb **198**
Waterfield Dr. KT18: Tatt C 91Xa **194**
Waterfield Dr. KT20: Tad 91Xa **194**
Waterfield Gdns. SE25 70Tb **135**
Waterfield Grn. KT20: Tad 92Xa **194**
Waterfields KT22: Lea 91Ka **192**
Waterfields Retail Pk. 14Z **27**
Waterfields Way WD17: Wat 14Z **27**
Waterford Ho. KT11: Cobh 83Aa **171**
Waterford Ho. BR1: Broml 67Jc **137**
(off Newman Rd.)
Waterford Ho. UB7: W Dray 48L **83**
Waterford Ho. W11 45Bb **89**
(off Kensington Pk. Rd.)
Waterford Point SW8 52Nb **112**
Waterford Rd. SW6 52Db **111**
Waterford Way NW10 36Xa **68**
Waterfront, The WD6: E'tree 16Ka **28**
Waterfront W6 51Ya **110**
Waterfront Dr. SW10 53Eb **111**
Waterfront Ho. E5 33Yb **72**
(off Harry Zeital Way)
Waterfront Leisure Cen. Woolwich 48Qc **94**
Waterfront M. N1 1E **218** (40Sb **71**)
Waterfront Studios Bus. Cen. E16 46Hc **93**
(off Dock Rd.)
Water Gdns., The W2 2D **220** (44Gb **89**)
Watergardens, The KT2: King T 65Sa **131**
Water Gdns. HA7: Stan 23Ka **46**
Water Gdns. Sq. SE16 47Zb **92**
Watergate, The WD19: Wat 19Z **27**
Watergate EC4 4B **224** (45Rb **91**)
Watergate St. SE8 51Cc **114**
Watergate Wlk. E14 48Dc **92**
Watergate Wlk. WC2 6G **223** (46Nb **90**)
Waterglade Ind. Pk. RM20:
W Thur 51Vd **120**
WATERHALES 15Md **39**
Waterhall Av. E4 21Gc **53**
Waterhall Cl. E17 25Zb **52**
Waterhead NW1 3B **216** (41Lb **90**)
(off Varndell St.)
Waterhouse, The HP1: Hem H 3L **3**
Waterhouse CR0: C'don 74Sb **157**
(off Saffron Central Sq.)
Waterhouse Cl. E16 43Mc **93**
Waterhouse Cl. NW3 36Fb **69**
Waterhouse Cl. W6 49Za **88**
Waterhouse La. CR8: Kenley 91Sb **197**
Waterhouse La. KT20: Kgswd 93Ab **194**
Waterhouse La. RH1: Blet 4M **209**
Waterhouse Sq. EC1 1A **224** (43Qb **90**)
Waterhouse St. HP1: Hem H 2L **3**
Wateridge Cl. E14 48Cc **92**
Wateringbury Cl. BR5: St P 69Xc **139**
Water La. AL2: Lon C 10H **7**
Water La. DA14: Sidc 61Bd **139**
Water La. E15 37Gc **73**
Water La. EC3 5J **225** (45Ub **91**)
Water La. GU24: Bisl 1C **186**
Water La. GU24: Chob 1F **166**
Water La. HP3: Bov 1C **10**
Water La. IG3: Ilf 34Uc **74**
Water La. KT1: King T 67Ma **131**
Water La. KT11: Cobh 87Aa **171**
Water La. KT23: Bookh 97Z **191**
Water La. N9 18Xb **33**
Water La. NW1 38Kb **70**
Water La. RH1: Blet 2H **209**
Water La. RH8: Limp 98Jc **199**
Water La. RH9: S God 10B **210**
Water La. RM19: Purf 49Qd **97**
Water La. SE14 52Yb **114**
Water La. TN14: S'ham 84Hd **182**
Water La. TN16: Westrm 99Tc **200**
Water La. TW1: Twick 60Ja **108**
Water La. TW9: Rich 57Ma **109**
Water La. WD17: Wat 14Y **27**
Water La. WD4: K Lan 1R **12**
Water Lily Cl. UB2: S'hall 47Ea **86**
Waterline Ho. W2 1D **220** (43Gb **89**)
(off Harbet Rd.)
Waterloo Bri. WC2 5H **223** (45Pb **90**)
Waterloo Cl. E9 36Yb **72**
Waterloo Cl. TW14: Felt 60V **106**
Waterloo Ct. AL1: St A 3F **6**
Waterloo Ct. KT12: Hers 77W **150**
Waterloo East Station (Rail) 7A **224** (46Qb **90**)
Waterloo East Theatre 7A **224** (46Qb **90**)
(off Brad St.)
Waterloo Gdns. E2 40Yb **72**
Waterloo Gdns. N1 38Rb **71**
Waterloo Gdns. RM7: Rom 30Fd **56**
Waterloo Pas. NW6 38Bb **69**
Waterloo Pl. SM5: Cars 76Hb **155**
(off Wrythe Grn.)
Waterloo Pl. SW1 6D **222** (46Mb **90**)
Waterloo Pl. TW9: Rich 56Na **109**
Waterloo Rd. CM14: B'wood 18Yd **40**
Waterloo Rd. E10 31Cc **72**
Waterloo Rd. E6 38Lc **73**
Waterloo Rd. E7 36Hc **73**
Waterloo Rd. IG6: Ilf 26Sc **54**
Waterloo Rd. KT19: Eps 84Ta **173**
Waterloo Rd. NW2 32Wa **68**
Waterloo Rd. RM7: Rom 30Gd **56**
Waterloo Rd. RM7: Rush G 30Gd **56**
Waterloo Rd. SE1 6J **223** (46Pb **90**)
Waterloo Rd. SM1: Sutt 78Fb **155**
Waterloo Rd. UB8: Uxb 39L **63**
Waterloo Station
(Rail & Underground) 1K **229** (47Qb **90**)
Waterloo St. DA12: Grav'nd 9E **122**
Waterloo Ter. KT13: Weyb 77R **150**
(off Baker St.)
Waterloo Ter. N1 38Rb **71**
Waterlow Bldg., The E3 43Bc **92**
(off Weatherley Cl.)
Waterlow Ct. NW11 31Db **69**
Waterlow Pk. Cen. 32Kb **70**
Waterlow Rd. N19 32Lb **70**
Waterlow Rd. RH2: Reig 7L **207**
Waterman Bldg. E14 47Bc **92**
Waterman Cl. WD19: Wat 16X **27**
Waterman Ct. SL1: Slou 6C **80**
Waterman Ho. E15 37Fc **73**
(off Forrester Way)
Watermans RM1: Rom 29Hd **56**
**Watermans Art Cen., Cinema
& Theatre** 51Na **109**
Watermans Bus. Pk. TW18: Staines 63G **126**
Watermans Cl. KT2: King T 66Na **131**
Watermans Ct. TW8: Bford 51Ma **109**
(off High St.)
Watermans Ho. E14 44Fc **93**
(off New Village Av.)
Watermans M. W5 45Na **87**
Waterman's Quay SW6 54Eb **111**
Watermans St. SW15 55Za **110**
Watermans Wlk. SE16 47Ac **92**
Watermans Way DA9: Ghithe 56Xd **120**
Watermans Way E1 46Xb **91**
Watermark Ct. RM6: Chad H 30Xc **55**
(off Quarles Pk. Rd.)
Water Mead CR5: Chips 89Hb **175**
Watermead GU21: Wok 8K **167**
Watermead KT20: Tad 93Xa **194**
Watermead TW14: Felt 60U **106**
Watermead Ho. E9 36Ac **72**
Watermead La. SM5: Cars 73Hb **155**
Watermead Lodge SE16 46Zb **92**
(off Princes Riverside Rd.)
Watermeadow Cl. DA8: Erith 52Kd **119**
Watermeadow La. SW6 54Eb **111**
Water Mdws. AL2: F'mre 10B **6**
Watermead Rd. SE6 63Ec **136**
Watermead Way N17 27Xb **51**
Watermen's Sq. SE20 66Yb **136**
Water M. SE15 56Yb **114**
Watermill Bus. Cen. EN3: Brim 12Bc **34**
Watermill Cl. TN16: Bras 96Yc **201**
Watermill Cl. TW10: Ham 62La **130**
Water Mill Ho. TW13: Hanw 61Ca **129**
Watermill La. N18 22Ub **51**
Watermill Pl. KT11: Cobh 86X **171**
Water Mill Way DA4: S Dar 68Rd **141**
Watermill Way SW19 67Eb **133**
Watermill Way TW13: Hanw 61Ba **129**
Watermint Cl. BR5: St P 70Zc **139**
Watermint Quay N16 31Wb **71**
Waterperry La. GU24: Chob 2K **167**
Water Rd. HA0: Wemb 39Pa **67**
Waters Dr. TW18: Staines 62H **127**
Water's Edge SW6 53Ya **110**
(off Palemead Cl.)
Watersedge KT19: Ewe 77Sa **153**
Waters Edge Ct. DA8: Erith 50Hd **96**
Watersfield Way HA8: Edg 24Ma **47**
Waters Gdns. RM10: Dag 36Cd **76**
Waterside AL2: Lon C 9J **7**
(not continuous)
Waterside BR3: Beck 67Bc **136**
Waterside DA1: Cray 57Gd **118**
Waterside DA11: Nflt 8A **122**
Waterside E17 30Yb **52**
Waterside HP4: Berk 1A **2**
Waterside N1 2D **218** (40Sb **71**)
Waterside TW18: Staines 63G **126**
Waterside UB7: Harm 52L **105**
Waterside UB8: Cowl 43L **83**
Waterside W2 1B **220** (43Fb **89**)
(off Nth. Wharf Rd.)
Waterside WD4: K Lan 1Q **12**
Waterside WD7: R'lett 6Ka **14**
Waterside Apts. N4 32Sb **71**
(off Goodchild Rd.)
Waterside Av. BR3: Beck 71Ec **158**
(off Adamson Way)
Waterside Bus. Cen. TW7: Isle 56Ka **108**
Waterside Cl. E3 39Bc **72**
Waterside Cl. HA9: Wemb 34Sa **67**
Waterside Cl. IG11: Bark 35Wc **75**
Waterside Cl. KT6: Surb 75Na **153**
Waterside Cl. RM3: Hrld W 24Qd **57**
Waterside Cl. SE16 47Wb **91**
Waterside Cl. SE28 46Vc **95**
Waterside Cl. TW17: Shep 67S **128**
Waterside Cl. UB5: N'olt 41Ba **85**
Waterside Cl. WD4: K Lan 1R **12**
Waterside Ct. DA2: Dart 57Td **120**
Waterside Ct. HP3: Hem H 6N **3**
Waterside Ct. SM5: Cars 76Jb **156**
(off Millpond Pl.)
Waterside Dr. KT12: Walt T 71W **150**
Waterside Dr. SL3: L'ly 47B **82**
Waterside Hgts. E16 47Lc **93**
(off Booth Rd.)
Waterside M. UB9: Hare 23J **43**
Waterside Pl. NW1 39Jb **70**
Waterside Point SW11 52Gb **111**
Waterside Rd. UB2: S'hall 48Ca **85**
Waterside Twr. SW6 54Eb **111**
(off The Boulevard)
Waterside Trad. Cen. W7 48Ga **86**
Waterside Trad. Est. KT15: Add 77N **149**
Waterside Way GU21: Wok 10M **167**
Waterside Way N17 27Xb **51**
Waterside Way SW17 63Eb **133**
Waterslade RH1: Redh 6N **207**
Watersmeet Pl. N4 32Sb **71**
Watersmeet Theatre 18N **25**
Watersmeet Way SE28 44Yc **95**
Waterson Rd. RM16: Grays 9D **100**
Waterson St. E2 3J **219** (41Ub **91**)
Waters Pl. SW15 54Ya **110**
Watersplash Cl. KT1: King T 69Na **131**
Watersplash Ct. AL2: Lon C 9K **7**
Watersplash La. SL5: Asc 7B **124**
Watersplash La. TW5: Cran 50X **85**
Watersplash La. UB3: Harl 49W **84**
Watersplash Rd. TW17: Shep 71Q **150**
Watersreach Apts. N4 32Sb **71**
(off Kayani Av.)
Waters Rd. KT1: King T 68Ra **131**
Waters Rd. SE6 62Gc **137**
Waters Sq. KT1: King T 69Ra **131**
Waterstone Way DA9: Ghithe 58Wd **120**
Water St. E14 46Dc **92**
Water St. WC2 4K **223** (45Qb **90**)
(off Maltravers St.)
Waterton BR8: Swan 70Fd **140**
Waterton Av. DA12: Grav'nd 9G **122**
Water Twr. CM14: W'ley 22Xd **58**
Water Twr. Cl. UB8: Uxb 36N **63**
Water Twr. Hill CR0: C'don 77Tb **157**
Water Twr. Pl. N1 39Qb **70**
Water Twr. Rd. CM14: Gt War 22Yd **58**
Waterview Cl. DA6: Bex 57Zc **117**
Waterview Dr. SE10 46Fc **93**
Waterview Ho. E14 43Ac **92**
(off Carr St.)
Waterway Av. SE13 55Dc **114**
Waterway Pk. UB3: Hayes 47S **84**
Waterway Rd. KT22: Lea 94Ja **192**
Waterways Bus. Cen. EN3: Enf L 10Bc **20**
Waterworks Cen. 33Ac **72**
WATERWORKS CORNER 25Gc **53**
Waterworks La. E5 33Zb **72**
Waterworks Nature Reserve 33Ac **72**
Waterworks Rd. SW2 58Pb **112**
Waterworks Vs. TN13: S'oaks 98Kd **203**
Waterworks Yd. CR0: C'don 76Sb **157**
(off Charles St.)
Watery La. AL10: Hat 1A **8**
Watery La. DA14: Sidc 65Xc **139**
Watery La. GU24: Chob 2H **167**
Watery La. KT16: Lyne 73F **148**
Watery La. SW19 67Bb **133**
Watery La. SW20 68Bb **133**
Watery La. TN15: Kems'g 93Sd **204**
Watery La. TN15: Seal 93Sd **204**
Watery La. UB3: Harl 50T **84**
Watery La. UB5: N'olt 40Y **65**
Wates Way CM15: B'wood 18Zd **41**
Wates Way CR4: Mitc 72Hb **155**
Wateville Rd. N17 25Sb **51**
WATFORD 14Y **27**
Watford Arches Retail Pk. 15Z **27**
Watford By-Pass WD6: E'tree 16Ja **28**

Watford Cl. SW11 53Gb 111
Watford Colosseum 13W 26
Watford Ent. Cen. WD18: Wat 16U 26
Watford FC 15X 27
Watford Fld. Rd. WD18: Wat 15Y 27
WATFORD GENERAL HOSPITAL 15X 27
Watford Health Campus WD18: Wat 15X 27
Watford Heath WD19: Wat 17Z 27
WATFORD HEATH 17Z 27
Watford Heath Farm WD19: Wat 17Aa 27
Watford High Street Station (Overground) 14Y 27
Watford Ho. RM3: Rom 22Pd 57
(off Redruth Rd.)
Watford Ho. La. WD17: Wat 13X 27
Watford Indoor Bowls Club 5Y 13
Watford Interchange WD24: Wat 11Y 27
Watford Junction Station (Rail & Overground) 12Y 27
Watford Leisure Cen. Central 13W 26
Watford Leisure Cen. Woodside 5Y 13
Watford Mus. 14Y 27
Watford North Station (Rail) 9Y 13
Watford Palace Theatre 13X 27
Watford Rd. AL1: St A 6P 5
Watford Rd. AL2: Chis G 9N 5
Watford Rd. E16 43Jc 93
Watford Rd. HA0: Wemb 34Ka 66
Watford Rd. HA1: Harr 31Ja 66
Watford Rd. HA6: Nwood 24V 44
Watford Rd. WD3: Crox G 16Q 26
Watford Rd. WD4: Hunt C 2Q 12
Watford Rd. WD4: K Lan 2Q 12
Watford Rd. WD6: E'tree 16Ka 28
Watford Rd. WD7: R'lett 8Ga 14
Watford Station (Underground) 13V 26
Watford Way NW4 28Wa 48
Watford Way NW7 21Ua 48
Watkin M. EN3: Enf L 9Cc 20
Watkin Rd. HA9: Wemb 34Ra 67
Watkins Cl. HA6: Nwood 25V 44
Watkins Ho. E14 47Ec 92
(off Manchester Rd.)
Watkinson Rd. N7 37Pb 70
Watkins Ri. EN6: Pot B 4Db 17
Watkins Way RM8: Dag 32Ad 75
WATLING 24Ta 47
Watling Av. HA8: Edg 25Sa 47
Watling Ct. EC4 3E 224 (44Sb 91)
(off Watling St.)
Watling Ct. WD6: E'tree 16Ma 29
Watling Farm Cl. HA7: Stan 18La 28
Watling Gdns. NW2 37Ab 68
Watling Ga. NW9 28Ua 48
Watling Ho. AL3: St A 4P 5
(off King Harry La.)
Watling Ho. SE1 5E 230 (49Sb 91)
(off New Kent Rd.)
Watling Knoll WD7: R'lett 5Ha 14
Watling Mans. WD7: R'lett 7Ka 14
Watlings Cl. CR0: C'don 72Ac 158
Watling St. DA1: Dart Broomhill Rd. 58Kd 119
Watling St. WD7: R'lett Harper La. 3Ha 14
Watling St. WD7: R'lett Loom La. 9Ja 14
Watling St. DA1: Dart The Brent 59Qd 119
Watling St. AL1: St A 4A 6
Watling St. AL2: Park 7B 6
Watling St. DA1: Cray 57Fd 118
Watling St. DA11: Grav'nd 4D 144
Watling St. DA11: Nflt 61De 143
Watling St. DA12: Cobh 7J 145
Watling St. DA12: Grav'nd 7J 145
Watling St. DA2: Bean 60Td 120
Watling St. DA2: Dart 60Td 120
Watling St. DA6: Bex 56Dd 118
Watling St. DA7: Bex 56Dd 118
Watling St. EC4 3D 224 (44Sb 91)
Watling St. SE15 51Ub 113
Watling St. WD6: E'tree 12La 28
Watling St. Cvn. Pk. AL2: Park 7A 6
Watlington Gdns. CM13: Gt War 23Xd 58
Watlington Gro. SE26 64Ac 136
Watling Vw. AL1: St A 5A 6
Watney Cl. CR8: Purl 85Pb 176
Watney Cotts. SW14 55Sa 109
Watney Mkt. E1 44Xb 91
Watney Rd. SW14 55Sa 109
Watney's Rd. CR4: Mitc 71Mb 156
Watney St. E1 44Xb 91
Watson Av. E6 38Qc 74
Watson Av. SM3: Cheam 75Ab 154
Watson Cl. N16 36Tb 71
Watson Cl. RM20: W Thur 53Wd 120
Watson Cl. SW19 65Gb 133
Watson Ct. E3 41Cc 92
(off Campbell Rd.)
Watson Ct. WD18: Wat 15X 27
Watson Gdns. RM3: Hrld W 26Md 57
Watson Ho. RH2: Reig 5J 207
Watson Ho. SW11 52Mb 112
(off Ponton Rd.)
Watson Pl. SE25 71Vb 157
Watsons Ho. N1 1J 219 (39Ub 71)
(off Nuttall St.)
Watson's M. W1 1E 220 (43Gb 89)
Watsons Rd. N22 25Pb 50
Watsons St. SE8 52Cc 114
Watson St. E13 40Kc 73
Watson's Wlk. AL1: St A 3C 6
Watsons Yd. BR6: Orp 73Xc 161
Watt Ct. W3 47Ua 88
Watteau Sq. CR0: C'don 74Qb 156
Wattendon Rd. CR8: Kenley 88Rb 177
Wattisfield Rd. E5 34Yb 72
WATTON'S GREEN 16Kd 39
Watts Apts. SW8 52Nb 112
(off Cellini St.)
Watts Bri. Rd. DA8: Erith 51Hd 118
Watt's Cl. KT20: Tad 94Za 194
Watts Cl. N15 29Ub 51
Watt's Cres. RM19: Purf 49Sd 98
Wattsdown Cl. E13 39Jc 73
Watts Farm Pde. GU24: Chob 2K 167
(off Barnmead)
Watts Gro. E3 43Cc 92
Watts Ho. W10 43Ab 88
(off Wornington Rd.)
Watt's La. KT20: Tad 94Za 194
Watts La. BR7: Chst 67Rc 138
Watts La. TW11: Tedd 64Ja 130
Watts Lea GU21: Wok 7L 167
Watt's Mead KT20: Tad 94Za 194
Watts M. IG6: Ilf 27Tc 54
Watts M. SW16 65Mb 134
Watts Rd. KT7: T Ditt 73Ja 152
Watts St. E1 46Xb 91
Watts St. SE15 53Vb 113
Wat Tyler Ho. N8 27Nb 50
(off Boyton Rd.)
Wat Tyler Rd. SE10 54Ec 114
Wat Tyler Rd. SE3 54Fc 115
Wauthier Cl. N13 22Rb 51
Wave Ct. RM7: Rush G 30Gd 56
Wavel Ct. CR0: C'don 78Tb 157
(off Hurst Rd.)
Wavel Ct. E1 46Yb 92
(off Garnet St.)
Wavelengths Leisure Cen. 52Cc 114
Wavell Dr. DA15: Sidc 58Uc 116
Wavell Gdns. SL2: Slou 1D 80
Wavell Ho. AL1: St A 4F 6
Wavell Ho. N6 30Jb 50
(off Hillcrest)
Wavel M. N8 28Mb 50
Wavel M. NW6 38Db 69
Wavel Pl. SE26 63Vb 135
Wavendene Av. TW20: Egh 66D 126
Wavendon Av. W4 50Ta 87
Waveney Av. SE15 56Xb 113
Waveney Cl. E1 46Wb 91
Waveney Ho. SE15 56Xb 113
Waverley Av. CR8: Kenley 88Ub 177
Waverley Av. E17 27Fc 53
Waverley Av. E4 21Bc 52
Waverley Av. HA9: Wemb 36Pa 67
Waverley Av. KT5: Surb 72Ra 153
Waverley Av. SM1: Sutt 75Db 155
Waverley Av. TW2: Whitt 60Ba 107
Waverley Cl. BR2: Broml 71Mc 159
Waverley Cl. E18 25Lc 53
Waverley Cl. KT8: W Mole 71Ca 151
Waverley Cl. UB3: Harl 49T 84
Waverley Ct. EN2: Enf 13Rb 33
Waverley Ct. GU22: Wok 90A 168
Waverley Ct. NW3 37Hb 69
Waverley Ct. NW6 38Ab 68
Waverley Ct. SE26 64Yb 136
Waverley Cres. RM3: Rom 24Ld 57
Waverley Cres. SE18 50Tc 94
Waverley Dr. GU25: Vir W 9L 125
Waverley Dr. KT16: Chert 76F 148
Waverley Gdns. E6 43Nc 94
Waverley Gdns. HA6: Nwood 25W 44
Waverley Gdns. IG11: Bark 40Uc 74
Waverley Gdns. IG6: Ilf 26Sc 54
Waverley Gdns. NW10 40Pa 67
Waverley Gdns. RM16: Grays 47Ce 99
Waverley Gro. N3 27Za 48
Waverley Ind. Est. HA1: Harr 27Fa 46
Waverley Lodge AL3: St A 1B 6
(off Falmouth Ct.)
Waverley Lodge E15 37Gc 73
(off Litchfield Av.)
Waverley Pl. KT22: Lea 94Ka 192
Waverley Pl. N4 33Rb 71
Waverley Pl. NW8 1B 214 (40Fb 69)
Waverley Rd. E17 27Ec 52
Waverley Rd. E18 25Lc 53
Waverley Rd. EN2: Enf 13Rb 33
Waverley Rd. HA2: Harr 33Aa 65
Waverley Rd. KT11: Oxs 86Da 171
Waverley Rd. KT11: Stoke D 86Da 171
Waverley Rd. KT13: Weyb 78Q 150
Waverley Rd. KT17: Ewe 78Xa 154
Waverley Rd. KT22: Oxs 86Da 171
Waverley Rd. N17 24Xb 51
Waverley Rd. N8 30Nb 50
Waverley Rd. RM13: Rain 42Kd 97
Waverley Rd. SE18 50Sc 94
Waverley Rd. SE25 70Xb 135
Waverley Rd. SL1: Slou 3G 80
Waverley Rd. UB1: S'hall 45Ca 85
Waverley Way SM5: Cars 79Gb 155
Waverton Ho. E3 39Bc 72
Waverton Rd. SW18 59Eb 111
Waverton St. W1 6J 221 (46Jb 90)
Wavertree Ct. SW2 60Nb 112
Wavertree Rd. E18 26Jc 53
Wavertree Rd. SW2 60Pb 112
Waxham NW3 36Hb 69
Waxhouse Ga. AL3: St A 2B 6
Waxlow Cres. UB1: S'hall 44Ca 85
Waxlow Ho. UB4: Yead 43Z 85
Waxlow Rd. NW10 40Sa 67
Waxlow Way UB5: N'olt 42Ba 85
Waxwell Cl. HA5: Pinn 26Z 45
Waxwell Farm Ho. HA5: Pinn 26Z 45
Waxwell La. HA5: Pinn 26Z 45
Way, The RH2: Reig 5M 207
Wayborne Gro. HA4: Ruis 30S 44
Waycross Rd. RM14: Upm 31Ud 78
Waye Av. TW5: Cran 53W 106
Wayfarer Rd. TW6: H'row A 54K 105
Wayfarer Rd. UB5: N'olt 41Z 85
Wayfaring Grn. RM17: Grays 50Be 99
Wayfield Link SE9 58Tc 116
Wayford St. SW11 54Gb 111
Wayland Av. E8 36Wb 71
Wayland Ho. SW9 54Qb 112
(off Robsart St.)
Waylands BR8: Swan 70Hd 140
Waylands TW19: Wray 58A 104
Waylands UB3: Hayes 43T 84
Waylands Cl. TN14: Knock 87Ad 181
Waylands Mead BR3: Beck 67Dc 136
Waylen Gdns. DA1: Dart 58Pd 119
Waylett Ho. SE11 7J 229 (50Pb 90)
(off Loughborough St.)
Waylett Pl. HA0: Wemb 35Ma 67
Waylett Pl. SE27 62Rb 135
Wayman Ct. E8 37Xb 71
Wayne Cl. BR6: Orp 76Vc 161
Wayneflete M. KT12: Hers 77Aa 151
Wayneflete Pl. KT10: Esh 76Da 151
Wayneflete Twr. Av. KT10: Esh 76Ca 151
Wayne Kirkum Way NW6 36Bb 69
Waynflete SL4: Eton 1G 102
(off Eton Wick Rd.)
Waynflete Av. CR0: Wadd 76Rb 157
Waynflete Ho. KT10: Esh 77Da 151
(off High St.)
Waynflete Ho. SE1 7D 224 (46Sb 91)
(off Union St.)
Waynflete Sq. W10 45Za 88
(not continuous)
Waynflete St. SW18 61Eb 133
Waypoint Way E16 47Kc 93
Wayside, The HP3: Hem H 3C 4
Wayside CR0: New Ad 79Dc 158
Wayside EN6: Pot B 5Fb 17
Wayside NW11 32Ab 68
Wayside SW14 57Sa 109
Wayside WD4: Chfd 2K 11
Wayside WD7: Shenl 5Ma 15
Wayside Av. RM12: Horn 33Md 77
Wayside Av. WD23: Bush 16Fa 28
Wayside Cl. N14 16Lb 32
Wayside Cl. RM1: Rom 27Hd 56
Wayside Commercial Est. IG11: Bark 39Vc 75
Wayside Ct. AL2: Brick W 2Ba 13
Wayside Ct. GU21: Wok 8J 167
(off Langmans Way)
Wayside Ct. HA9: Wemb 34Qa 67
Wayside Ct. TW1: Twick 58La 108
Wayside Gdns. RM10: Dag 36Cd 76
Wayside Gdns. SL9: Ger X 1P 61
Wayside Gro. SE9 63Pc 138
Wayside M. IG2: Ilf 29Qc 54
Wayville Rd. DA1: Dart 59Rd 119
Way Volante DA12: Grav'nd 3G 144
WC1 WC1 6J 217 (42Pb 90)
Weald, The BR7: Chst 65Pc 138
Weald Cl. BR2: Broml 75Nc 160
Weald Cl. CM14: B'wood 20Wd 40
Weald Cl. DA13: Ist R 6A 144
Weald Cl. SE16 50Xb 91
Weald Country Pk. 17Td 40
Weald Country Pk. Visitor Cen. 18Td 40
Wealden Ho. E3 41Dc 92
(off Talwin St.)
Wealden Pl. TN13: S'oaks 93Ld 203
Weald La. HA3: Hrw W 26Fa 46
Weald Pk. Way CM14: S Weald 20Ud 40
Weald Ri. HA3: Hrw W 24Ha 46
Weald Rd. CM14: B'wood 18Qd 39
Weald Rd. CM14: S Weald 18Qd 39
Weald Rd. UB10: Hil 40Q 64
Weald Sq. E5 33Wb 71
WEALDSTONE 27Ga 46
WEALDSTONE CEN., THE 27Ga 46
(off High St.)
Wealdstone FC 33V 64
Wealdstone Rd. SM3: Sutt 75Bb 155
Weald Village Open Space 25Ga 46
Weald Way CR3: Cat'm 100Ub 197
Weald Way RH2: Reig 10L 207
Weald Way RM7: Rom 30Dd 56
Weald Way UB4: Hayes 41U 84
Wealdwood Gdns. HA5: Hat E 23Da 45
Weale Rd. E4 20Fc 35
Weall Cl. CR8: Purl 84Pb 176
Weall Ct. HA5: Pinn 28Aa 45
Weall Grn. WD25: Wat 4X 13
Weardale Av. DA2: Dart 61Sd 142
Weardale Gdns. EN2: Enf 11Tb 33
Weardale Rd. SE13 56Fc 115
Wearmouth Ho. E3 42Bc 92
(off Joseph St.)
Wear Pl. E2 41Xb 91
(not continuous)
Wearside Rd. SE13 56Dc 114
Weasdale Ct. GU21: Wok 8K 167
Weatherall Cl. KT15: Add 78K 149
Weatherbury W2 44Cb 89
(off Talbot Rd.)
Weatherbury Ho. N19 34Mb 70
(off Wedmore St.)
Weatherley Cl. E3 43Bc 92
Weave Ct. RM1: Rom 29Jd 56
Weaver Cl. CR0: C'don 77Vb 157
Weaver Cl. E6 45Rc 94
Weaver Ho. E1 42Wb 91
(off Pedley St.)
Weavers Almshouses E11 30Hc 53
(off Cambridge Rd.)
Weavers Cl. DA11: Grav'nd 10C 122
Weavers Cl. TW7: Isle 56Ga 108
Weavers Ho. E11 30Jc 53
(off New Wanstead)
Weavers La. SE1 7J 225 (46Ub 91)
Weavers La. TN14: S'oaks 93Ld 203
Weavers Orchard DA13: Sflt 65Ce 143
Weavers Row E20 36Dc 72
Weavers Ter. SW6 51Cb 111
(off Micklethwaite Rd.)
Weaver St. E1 42Wb 91
(not continuous)
Weavers Way NW1 39Mb 70
Weaver Wlk. HA9: Wemb 35Qa 67
Weaver Wlk. SE27 63Sb 135
Webb Cl. SL3: L'ly 9P 81
Webb Cl. W10 42Ya 88
Webb Ct. SE28 45Xc 95
(off Attlee Rd.)
Webber Cl. DA8: Erith 52Kd 119
Webber Cl. WD6: E'tree 16Ma 29
Webber Ho. IG11: Bark 38Sc 74
(off North St.)
Webber Path E14 45Ec 92
(off Bullivant St.)
Webber Row SE1 2B 230 (47Rb 91)
Webber St. SE1 1A 230 (47Qb 90)
Webb Est. E5 31Wb 71
Webb Gdns. E13 42Jc 93
Webb Ho. E3 41Cc 92
(off Trevithick Way)
Webb Ho. RM10: Dag 34Cd 76
(off Kershaw Rd.)
Webb Ho. SW8 52Mb 112
Webb Ho. TW13: Hanw 62Aa 129
Webb Pl. NW10 41Va 88
Webb Rd. SE3 51Hc 115
Webb's All. TN13: S'oaks 97Ld 203
(not continuous)
Webbscroft Rd. RM10: Dag 35Dd 76
Webbs Mdw. TN13: S'oaks 97Ld 203
Webb's Rd. SW11 56Hb 111
Webbs Rd. UB4: Yead 41X 85
Webb St. SE1 4H 231 (48Ub 91)
Webheath NW6 38Bb 69
(not continuous)
Webley Ct. EN3: Enf L 9Cc 20
(off Sten Cl.)
Webster Cl. EN9: Walt A 5Hc 21
Webster Cl. KT22: Oxs 86Da 171
Webster Cl. RM12: Horn 34Md 77
Webster Ct. WD3: Rick 18N 25
Webster Gdns. W5 46Ma 87
Webster Rd. E11 34Ec 72
Webster Rd. SE16 48Wb 91
Webster Rd. SS17: Stan H 1N 101
Websters Cl. GU22: Wok 2M 187
Weddell Ho. E1 42Zb 92
(off Duckett St.)
Wedderburn Ho. SW1 7H 227 (50Jb 90)
(off Lwr. Sloane St.)
Wedderburn Rd. IG11: Bark 39Uc 74
Wedderburn Rd. NW3 36Fb 69
Wedgewood Apts. SW8 52Nb 112
Wedgewood Cl. CM16: Epp 2Wc 23
Wedgewood Cl. HA6: Nwood 24S 44
Wedgewood Ct. BR2: Broml 69Hc 137
(off Cumberland Rd.)
Wedgewood Ct. DA5: Bexl 59Cd 118
Wedgewood Ho. SW1 50Kb 90
(off Churchill Gdns.)
Wedgewood M. SW6 54Ab 110
Wedgewood M. W1 3E 222 (44Mb 90)
Wedgwood Ct. N7 35Pb 70
Wedgwood Ho. E2 41Zb 92
(off Warley St.)
Wedgwood Ho. SE11 4K 229 (48Qb 90)
(off Lambeth Wlk.)
Wedgwood Pl. KT11: Cobh 85W 170
Wedgwoods TN16: Tats 93Lc 199
Wedgwood Wlk. NW6 36Db 69
(off Dresden Cl.)
Wedgwood Way SE19 66Sb 135
Wedlake Cl. RM11: Horn 32Nd 77
Wedlake St. W10 42Ab 88
Wedmore Av. IG5: Ilf 25Qc 54
Wedmore Gdns. N19 33Mb 70
Wedmore M. N19 34Mb 70
Wedmore Rd. UB6: G'frd 41Fa 86
Wedmore St. N19 34Mb 70
Wednesbury Gdns. RM3: Rom 24Pd 57
Wednesbury Grn. RM3: Rom 24Pd 57
Wednesbury Rd. RM3: Rom 24Pd 57
Weech Rd. NW6 35Cb 69
Weedington Rd. NW5 36Jb 70
Weedon Ho. W12 44Wa 88
Weedon M. RM13: Rain 39Jd 76
Weekes Dr. SL1: Slou 6F 80
Weekley Sq. SW11 55Fb 111
Weelkes Cl. SS17: Stan H 1L 101
Weigall Rd. SE12 57Jc 115
Weighhouse St. W1 3J 221 (44Jb 90)
Weightman Ho. SE16 48Wb 91
Weighton Rd. HA3: Hrw W 27Ea 46
Weighton Rd. SE20 68Xb 135
Weihurst Ct. SM1: Sutt 78Gb 155
Weihurst Gdns. SM1: Sutt 78Fb 155
Weimar St. SW15 55Ab 110
Weind, The CM16: They B 8Uc 22
Weint, The SL3: Coln 52E 104
Weir Ct. KT13: Weyb 75R 150
Weirdale Av. N20 19Hb 31
Weird Wood DA3: Lfield 69Ee 143
Weir Hall Av. N18 23Tb 51
Weir Hall Gdns. N18 22Tb 51
Weir Hall Rd. N17 23Tb 51
Weir Hall Rd. N18 22Tb 51
Weir Pl. TW18: Staines 67G 126
Weir Rd. DA5: Bexl 59Dd 118
Weir Rd. KT12: Walt T 72W 150
Weir Rd. KT16: Chert 73K 149
Weir Rd. SW12 59Lb 112
Weir Rd. SW19 62Db 133
Weirside Gdns. UB7: W Dray 46M 83
Weir's Pas. NW1 3E 216 (41Mb 90)
Weiss Rd. SW15 55Za 110
Welbeck Av. BR1: Broml 63Jc 137
Welbeck Av. DA15: Sidc 60Wc 117
Welbeck Av. UB4: Yead 42X 85
Welbeck Cl. KT17: Ewe 80Wa 154
Welbeck Cl. KT3: N Mald 71Va 154
Welbeck Cl. N12 22Fb 49
Welbeck Cl. WD6: Bore 13Qa 29
Welbeck Ct. UB4: Yead 42X 85
Welbeck Ct. W14 49Bb 89
(off Addison Bri. Pl.)
Welbeck Ho. W1 2K 221 (44Kb 90)
(off Welbeck St.)
Welbeck Rd. E6 41Mc 93
Welbeck Rd. EN4: E Barn 16Gb 31
Welbeck Rd. HA2: Harr 32Da 65
Welbeck Rd. SM1: Sutt 75Fb 155
Welbeck Rd. SM5: Cars 75Fb 155
Welbeck St. W1 1J 221 (43Jb 90)
Welbeck Vs. N21 19Sb 33
Welbeck Wlk. SM5: Cars 74Fb 155
Welbeck Way W1 2K 221 (44Kb 90)
Welbury Ct. E8 38Ub 71
(off Kingsland Rd.)
Welby Ho. N19 31Mb 70
Welby St. SE5 53Rb 113
Welch Pl. HA5: Pinn 25Y 45
Welclose St. AL3: St A 2A 6
Welcombes Vw. CR5: Coul 88Kb 176
Welcome Ct. E17 31Cc 72
(off Saxon Cl.)
Welcome Ct. SS17: Stan H 2L 101
SE9: Lon
Welcomes Cotts. CR3: Wold 95Cc 198
Welcomes Rd. CR8: Kenley 89Tb 177
Welcomes Ter. CR3: Whyt 88Vb 177
Welcote Dr. HA6: Nwood 23T 44
Welden SL2: Slou 4N 81
Weldin M. SW18 57Cb 111
(off Lebanon Rd.)
WELDON 60Be 121
Weldon Cl. HA4: Ruis 37X 65
Weldon Ct. N21 15Pb 32
Weldon Dr. KT8: W Mole 70Ba 129
Weldon Rd. DA10: Swans 59Be 121
Weldon Way RH1: Mers 1D 208
Weld Pl. N11 22Kb 50
(not continuous)
Weld Works M. SW2 58Pb 112
Welfare Rd. E15 38Gc 73
Welford Cl. E5 34Zb 72
Welford Ct. HA8: Edg 21Na 47
(off Lacey Dr.)
Welford Ct. NW1 38Kb 70
(off Castlehaven Rd.)
Welford Ct. SW8 54Lb 112
Welford Ct. W9 43Cb 89
(off Elmfield Way)
Welford Ho. UB5: N'olt 42Ba 85
(off Waxlow Way)
Welford Pl. SW19 63Ab 132
Welham Cl. AL9: Wel G 6E 8
Welham Cl. WD6: Bore 11Qa 29
Welham Ct. AL9: Wel G 6E 8
(off Dixons Hill Rd.)
WELHAM GREEN 5E 8
Welham Green Station (Rail) 6E 8
Welham Mnr. AL9: Wel G 6E 8
Welham Rd. SW16 65Kb 134
Welham Rd. SW17 64Jb 134
Welhouse Rd. SM5: Cars 74Gb 155
Welkin Grn. HP2: Hem H 1C 4
Wellacre Rd. HA3: Kenton 30Ka 46
Wellan Cl. DA15: Sidc 57Xc 117
Welland RM18: E Til 9L 101
Welland Cl. SL3: L'ly 51D 104
Welland Ct. SE6 61Bc 136
(off Oakham Cl.)
Welland Gdns. UB6: G'frd 40Ha 66
Welland Ho. SE15 56Yb 114
Welland M. E1 46Wb 91
Welland Rd. TW6: H'row A 54K 105
Wellands Cl. BR1: Broml 68Pc 138
Welland St. SE10 51Ec 114
Well App. EN5: Barn 15Ya 30
Wellbrook Rd. BR6: Farnb 77Qc 160
Wellbury Ter. HP2: Hem H 2C 4
Wellby Cl. N9 18Wb 33
Wellby Ct. E13 39Lc 73
Well Cl. GU21: Wok 9N 167
Well Cl. HA4: Ruis 34Aa 65
Well Cl. SW16 63Pb 134
Wellclose Sq. E1 45Wb 91
(not continuous)
Wellclose St. E1 45Wb 91
Wellcome Av. DA1: Dart 56Nd 119
Wellcome Collection 5D 216 (42Mb 90)
(off Euston Rd.)
Wellcome Mus., The 2J 223 (44Pb 90)
(within Royal College of Surgeons)
Well Cott. Cl. E11 30Lc 53
Well Ct. EC4 3E 224 (44Sb 91)
(not continuous)
Wellcroft HP1: Hem H 1K 3
Wellcroft Rd. SL1: Slou 6F 80
Wellday Ho. E9 37Ac 72
(off Hedger's Gro.)
Welldon Ct. HA1: Harr 29Ga 46
Welldon Cres. HA1: Harr 29Ga 46
WELL END 10Ta 15
Well End Rd. WD6: Bore 9Sa 15
Wellen Ri. HP3: Hem H 5N 3
Weller Ct. W11 46Bb 89
(off Ladbroke Rd.)
Weller Ho. SE16 47Wb 91
(off George Row)
Weller M. BR2: Broml 70Kc 137
Weller M. EN2: Enf 11Qb 32
Weller Pl. BR6: Downe 83Qc 180
Wellers Cl. TN16: Westrm 99Sc 200
Weller St. SE1 1D 230 (47Sb 91)
Welles Ct. E14 45Cc 92
(off Premiere Pl.)
Wellesford Cl. SM7: Bans 89Bb 175
Wellesley Av. HA6: Nwood 22V 44
Wellesley Av. SL0: Rich P 48H 83
Wellesley Av. W6 48Xa 88
Wellesley Cl. SE7 50Lc 93
Wellesley Cnr. DA11: Nflt 60De 121
Wellesley Ct. NW2 33Wa 68
Wellesley Ct. SE1 4D 230 (48Sb 91)
(off Rockingham St.)
Wellesley Ct. SL0: Rich P 47H 83
Wellesley Ct. SM3: Sutt 74Ab 154
Wellesley Ct. W9 3A 214 (41Eb 89)
Wellesley Ct. Rd. CR0: C'don 75Tb 157
Wellesley Cres. EN6: Pot B 5Ab 16
Wellesley Cres. TW2: Twick 61Ga 130
Wellesley Gro. CR0: C'don 75Tb 157
Wellesley Ho. NW1 4D 216 (41Mb 90)
(off Wellesley Pl.)
Wellesley Ho. SL4: Wind 3F 102
(off Vansittart Rd.)
Wellesley Ho. SW1 7K 227 (50Kb 90)
(off Ebury Bri. Rd.)
Wellesley Mans. W14 50Bb 89
(off Edith Vs.)
Wellesley Pde. TW2: Twick 62Ga 130
Wellesley Pk. M. EN2: Enf 12Rb 33
Wellesley Pas. CR0: C'don 75Sb 157
Wellesley Path SL1: Slou 7L 81
Wellesley Pl. NW1 4D 216 (41Mb 90)
Wellesley Pl. NW5 36Jb 70
Wellesley Rd. CM14: B'wood 18Yd 40
Wellesley Rd. CR0: C'don 74Sb 157
Wellesley Rd. E11 29Jc 53
Wellesley Rd. E17 30Cc 52
Wellesley Rd. HA1: Harr 29Ga 46
Wellesley Rd. IG1: Ilf 33Rc 74
Wellesley Rd. N22 26Qb 50
Wellesley Rd. NW5 36Jb 70
Wellesley Rd. SE18 52Qc 116
Wellesley Rd. SL1: Slou 6L 81
Wellesley Rd. SM2: Sutt 79Eb 155
(not continuous)
Wellesley Rd. TW2: Twick 62Fa 130
Wellesley Rd. W4 50Qa 87
Wellesley Road Stop (London Tramlink) 75Tb 157
Wellesley St. E1 43Zb 92
Wellesley Ter. N1 3E 218 (41Sb 91)
Welley Av. TW19: Wray 56A 104
Welley Rd. SL3: Hort 55A 104
Welley Rd. TW19: Wray 58A 104
Well Farm Hgts. CR3: Whyt 91Wb 197
Well Farm Rd. CR6: W'ham 91Wb 197
Wellfield DA3: Hartl 70Be 143
Wellfield Av. N10 27Kb 50
Wellfield Gdns. SM5: Cars 81Gb 175
Wellfield Rd. SW16 63Nb 134
Wellfields IG10: Lough 13Qc 36
Wellfield Wlk. SW16 63Pb 134
Wellfit St. SE24 55Rb 113
Wellgarth UB6: G'frd 37Ka 66
Wellgarth Rd. NW11 32Db 69
Well Gro. N20 18Eb 31
Well Hall Pde. SE9 56Pc 116
Well Hall Rd. SE9 55Nc 116
WELL HALL RDBT. 56Nc 116
Well Hill BR6: Well H 79Dd 162
WELL HILL 79Dd 162
Well Hill La. BR6: Well H 79Dd 162
Well Ho. SM7: Bans 87Db 175
Wellhouse La. EN5: Barn 14Ya 30
Wellhouse Rd. BR3: Beck 70Cc 136
Wellhurst Cl. BR6: Chels 80Vc 161
WELLING 55Xc 117
Wellingborough Ho. RM3: Rom 22Pd 57
(off Redruth Rd.)
Welling High St. DA16: Well 55Xc 117
Welling Rd. RM16: Ors 4E 100
Wellings Ho. UB3: Hayes 46X 85
Welling Station (Rail) 54Wc 117
Wellington N8 28Nb 50
(not continuous)
Wellington Arch 1J 227 (47Jb 90)
Wellington Av. DA15: Sidc 58Wc 117
Wellington Av. E4 19Cc 34
Wellington Av. GU25: Vir W 1M 147
Wellington Av. HA5: Hat E 25Ba 45
Wellington Av. KT4: Wor Pk 76Ya 154
Wellington Av. N15 30Vb 51
Wellington Av. N9 20Xb 33
Wellington Av. TW3: Houn 57Ca 107
Wellington Bldgs. E3 41Cc 92
(off Wellington Way)
Wellington Bldgs. SW1 7J 227 (50Jb 90)
Wellington Cl. KT12: Walt T 74V 150
Wellington Cl. RM10: Dag 38Ed 76
Wellington Cl. SE14 53Zb 114
Wellington Cl. W11 44Cb 89
Wellington Cl. WD19: Wat 20Ba 27
Wellington Ct. KT19: Eps 83Ta 173
Wellington Ct. NW8 2B 214 (40Fb 69)
(off Wellington Rd.)
Wellington Ct. RM16: Grays 46De 99

Wellington Ct. SW1 2F 227 (47Hb 89)
(off Knightsbridge)
Wellington Ct. SW6 53Db 111
(off Maltings Pl.)
Wellington Ct. TW12: Hamp H 64Fa 130
Wellington Ct. TW15: Ashf 64N 127
Wellington Ct. TW19: Stanw 59N 105
Wellington Cres. KT3: N Mald 69Sa 131
Wellington Dr. CR8: Purl 82Pb 176
Wellington Dr. RM10: Dag 38Ed 76
Wellington Gdns. SE7 51Lc 115
Wellington Gdns. TW2: Twick 63Fa 130
Wellington Gro. SE10 52Fc 115
Wellington Hill IG10: H Beech 9Jc 21
Wellington Hill IG10: Lough 9Jc 21
WELLINGTON HOSPITAL, THE 3C 214 (41Fb 89)
Wellington Ho. E16 46Jc 93
(off Pepys Cres.)
Wellington Ho. NW3 37Hb 69
(off Eton Rd.)
Wellington Ho. RM2: Rom 28Ld 57
Wellington Ho. SE17 51Sb 113
(off Arnside St.)
Wellington Ho. UB5: N'olt 38Ca 65
(off The Farmlands)
Wellington Ho. W5 41Na 87
Wellington Ho. WD23: Bush 14Ba 27
Wellington Ho. WD24: Wat 12Y 27
(off Exeter Cl.)
Wellingtonia Av. RM4: Have B 21Ed 56
Wellingtonia Ho. KT15: Add 78J 149
Wellingtonia Pl. RH2: Reig 5J 207
Wellington Lodge SE1 2A 230 (47Qb 90)
(off Waterloo Rd.)
Wellington Lodge SL4: Wink 1A 124
Wellington Mans. E10 32Cc 72
Wellington Mans. SE7 50Lc 93
(off Wellington Gdns.)
Wellington Mans. W14 51Bb 111
(off Queen's Club Gdns.)
Wellington M. N7 37Pb 70
(off Roman Way)
Wellington M. SE22 56Wb 113
Wellington M. SE7 51Lc 115
Wellington M. SW16 62Mb 134
Wellington Monument
London 1J 227 (47Jb 90)
Wellington Mus. London 1J 227 (47Jb 90)
Wellington Pde. DA15: Sidc 57Wc 117
Wellington Pk. Est. NW2 33Wa 68
Wellington Pas. E11 29Jc 53
Wellington Pl. CM14: W'ley 22Yd 58
Wellington Pl. KT11: Cobh 84Ca 171
Wellington Pl. N2 29Gb 49
Wellington Pl. NW8 3C 214 (41Fb 89)
Wellington Rd. AL1: St A 3F 6
Wellington Rd. AL2: Lon C 8H 7
Wellington Rd. BR2: Broml 70Lc 137
Wellington Rd. BR5: St M Cry 72Xc 161
Wellington Rd. CR0: C'don 73Rb 157
Wellington Rd. CR3: Cat'm 94Sb 197
Wellington Rd. DA1: Dart 58Ld 119
Wellington Rd. DA17: Belv 50Bd 95
Wellington Rd. DA5: Bexl 57Zc 117
Wellington Rd. E10 32Ac 72
Wellington Rd. E11 29Jc 53
Wellington Rd. E17 28Ac 52
Wellington Rd. E6 39Pc 74
Wellington Rd. E7 35Hc 73
Wellington Rd. EN1: Enf 15Ub 33
Wellington Rd. HA3: W'stone 27Ga 46
Wellington Rd. HA5: Hat E 25Ba 45
Wellington Rd. NW10 41Za 88
Wellington Rd. NW8 2C 214 (40Fb 69)
Wellington Rd. RM18: Tilb 4C 122
Wellington Rd. SW19 61Cb 133
Wellington Rd. TW12: Hamp H 64Fa 130
Wellington Rd. TW14: Felt 57U 106
Wellington Rd. TW15: Ashf 64N 127
Wellington Rd. TW2: Twick 63Fa 130
Wellington Rd. TW6: H'row A 55L 105
(off Whittle Rd.)
Wellington Rd. UB8: Uxb 39L 63
Wellington Rd. W5 48La 86
Wellington Rd. WD17: Wat 12X 27
Wellington Rd. Nth. TW4: Houn 55Ba 107
Wellington Rd. Sth. TW4: Houn 56Ba 107
Wellington Row E2 41Vb 91
Wellington Sq. N1 39Nb 70
Wellington Sq. SW3 7F 227 (50Hb 89)
Wellington St. DA12: Grav'nd 9E 122
Wellington St. SE18 49Qc 94
Wellington St. SL1: Slou 6K 81
Wellington St. WC2 4H 223 (45Pb 90)
Wellington Ter. E1 46Xb 91
Wellington Ter. GU21: Knap 10J 167
Wellington Ter. HA1: Harr 32Fa 66
Wellington Ter. N8 27Ob 50
(off Turnpike La.)
Wellington Ter. W2 45Cb 89
Wellington Way E3 41Cc 92
Wellington Way KT13: Weyb 82P 169
Welling United FC 55Yc 117
Welling Way DA16: Well 55Tc 116
Welling Way SE9 55Sc 116
Well La. CM15: Pil H 13Vd 40
Well La. GU21: Wok 9N 167
Well La. RM16: N Stif 46Ae 99
Well La. SW14 57Sa 109
Wellmeade Dr. TN13: S'oaks 99Kd 203
Wellmeadow Rd. SE13 58Gc 115
(not continuous)
Wellmeadow Rd. SE6 59Gc 115
Wellmeadow Rd. W7 49Ja 86
Wellow Wlk. SM5: Cars 74Fb 155
Well Path GU21: Wok 9N 167
Well Pl. NW3 34Fb 69
Well Rd. EN5: Barn 15Ya 30
Well Rd. EN6: N'thaw 10N 9
Well Rd. NW3 34Fb 69
Well Rd. TN14: Otf 88Ld 183
WELLS, THE 86Qa 173
Wells, The N14 17Mb 32
Wellsborough M. SW20 68Ab 132
Wells Cl. AL3: St A 1A 6
Wells Cl. CR2: S Croy 78Ub 157
Wells Cl. KT23: Bookh 96Ea 192
Wells Cl. SL4: Wind 3E 102
Wells Cl. UB5: N'olt 41Y 85
Wells Ct. BR2: Broml 68Fc 137
Wells Ct. DA11: Nflt 60De 121
Wells Ct. NW6 40Cb 69
(off Cambridge Av.)
Wells Ct. WD17: Wat 15Y 27
Wells Dr. NW9 32Ta 67
Wellsfield WD23: Bush 15Aa 27
Wells Gdns. IG1: Ilf 31Nc 74
Wells Gdns. RM10: Dag 36Dd 76
Wells Gdns. RM13: Rain 37Hd 76
Wells Ga. Cl. IG8: Wfd G 21Jc 53
Wells Ho. BR1: Broml 64Kc 137
(off Pike Cl.)
Wells Ho. EC1 3A 218 (41Qb 90)
(off Spa Grn. Est.)
Wells Ho. IG11: Bark 38Wc 75
(off Margaret Bondfield Av.)
Wells Ho. KT18: Eps 86Qa 173
Wells Ho. SE16 48Yb 92
(off Howland Est.)
Wells Ho. W10 42Ab 88
(off Wornington Rd.)
Wells Ho. W5 45Ma 87
(off Grove Rd.)
Wells Ho. Rd. NW10 43Ua 88
Wellside Cl. EN5: Barn 14Ya 30
Wellside Gdns. SW14 56Sa 109
Wells La. SL5: Asc 9A 124
Wells La. TW3: Houn 55Da 107
Wells M. N11 22Mb 50
Wells M. W1 1C 222 (43Lb 90)
Wellsmoor Gdns. BR1: Broml 69Qc 138
Wells Pk. Rd. SE26 62Wb 135
Wells Path UB4: Hayes 41U 84
Wells Pl. RH1: Mers 2B 208
Wells Pl. SW18 59Eb 111
Wells Pl. TN16: Westrm 99Sc 200
Wells Pl. Ind. Est. RH1: Mers 1B 208
Wellspring Cl. E14 43Ec 92
Wellspring Cres. HA9: Wemb 34Ra 67
Wellspring M. SE26 62Xb 135
Wellspring Way WD17: Wat 15Y 27
Wells Ri. NW8 1F 215 (39Hb 69)
Wells Rd. BR1: Broml 68Pc 138
Wells Rd. KT18: Eps 86Qa 173
Wells Rd. W12 47Ya 88
Wells Sq. WC1 4H 217 (41Pb 90)
Wells St. W1 1B 222 (43Lb 90)
Wellstead Av. N9 17Zb 34
Wellstead Rd. E6 40Qc 74
Wells Ter. N4 33Qb 70
Wellston Cres. N14 16Lb 32
Wellstones WD17: Wat 14X 27
Well St. E15 37Gc 73
Well St. E9 38Yb 72
Wells Vw. Dr. BR2: Broml 72Nc 160
Wells Way SE5 51Tb 113
Wells Way SW7 3B 226 (48Fb 89)
Wellswood SL5: Asc 9A 124
Wellswood Cl. HP2: Hem H 1B 4
Wells Yd. N7 36Qb 70
Wells Yd. WD17: Wat 13X 27
Well Wlk. NW3 35Fb 69
Well Way KT18: Eps 87Qa 173
Wellwood Cl. CR5: Coul 86Nb 176
Wellwood Rd. IG3: Ilf 32Wc 75
Welmar M. SW4 57Mb 112
(off Clapham Pk. Rd.)
Welsby Ct. W5 43La 86
Welsford St. SE1 49Wb 91
(not continuous)
Welsh Cl. E13 41Jc 93
Welsh Harp Reservoir 32Va 68
Welsh Ho. E1 46Xb 91
(off Wapping La.)
Welshpool Ho. E8 39Wb 71
(off Welshpool St.)
Welshpool St. E8 39Wb 71
(not continuous)
Welshside NW9 30Ua 48
(off Ruthin Cl.)
Welshside Wlk. NW9 30Ua 48
Welstead Ho. E1 44Xb 91
(off Cannon St. Rd.)
Welstead Way W4 49Va 88
Weltje Rd. W6 49Wa 88
Welton Ct. SE5 53Ub 113
Welton Ho. E1 43Zb 92
(off Stepney Way)
Welton Rd. SE18 52Uc 116
Welwyn Av. TW14: Felt 58V 106
Welwyn St. E2 41Yb 92
Welwyn Way UB4: Hayes 42U 84
WEMBLEY 36Na 67
Wembley & Sudbury Tennis
& Squash Club 36Ma 67
Wembley Central Station
(Rail, Underground & Overground) 36Na 67
Wembley Cl. RM5: Col R 24Ed 56
Wembley Commercial Cen.
HA9: Wemb 33Ma 67
Wembley Hill Rd. HA9:
Wemb East La. 34Pa 67
Wembley Hill Rd.
HA9: Wemb South Way 36Pa 67
Wembley Leisure Cen. 34Qa 67
WEMBLEY PARK 35Qa 67
Wembley Pk. Blvd. HA9: Wemb 36Qa 67
Wembley Pk. Bus. Cen. HA9: Wemb 35Ra 67
Wembley Pk. Dr. HA9: Wemb 35Pa 67
Wembley Pk. Ga. HA9: Wemb 35Qa 67
Wembley Park Station (Underground) 34Qa 67
Wembley Rd. TW12: Hamp 67Ca 129
Wembley Sailing Club 33Ta 67
Wembley Stadium 35Qa 67
Wembley Stadium Ind. Est.
HA9: Wemb 35Ra 67
Wembley Stadium Station (Rail) 36Pa 67
Wembley Way HA9: Wemb 37Ra 67
Wemborough Rd. HA7: Stan 25Ka 46
Wembury M. N6 31Lb 70
Wembury Rd. N6 31Kb 70
Wemyss Rd. SE3 54Hc 115
Wend, The CR5: Coul 86Mb 176
Wendela Cl. GU22: Wok 90B 168
Wendela Ct. HA1: Harr 33Ga 66
Wendell M. W12 47Va 88
Wendell Rd. W12 48Va 88
Wenderholme CR2: S Croy 78Tb 157
(off South Pk. Hill Rd.)
Wendle Ct. SW8 51Nb 112
Wendle Sq. SW11 53Gb 111
Wendley Dr. KT15: New H 82H 169
Wendling NW5 36Hb 69
Wendling Rd. SM1: Sutt 74Fb 155
Wendon St. E3 39Bc 72
Wendover SE17 7H 231 (50Ub 91)
(not continuous)
Wendover Cl. UB4: Yead 42Aa 85
Wendover Ct. BR2: Broml 69Kc 137
(off Wendover Rd.)
Wendover Ct. NW10 42Ra 87
Wendover Ct. NW2 34Cb 69
Wendover Ct. W1 1H 221 (43Jb 90)
(off Chiltern St.)
Wendover Dr. KT3: N Mald 72Va 154
Wendover Gdns. CM13: B'wood 19De 41
Wendover Ho. W1 1H 221 (43Jb 90)
(off Chiltern St.)
Wendover Ho. WD18: Wat 17U 26
(off Chenies Way)
Wendover Pl. TW18: Staines 64F 126
Wendover Rd. BR2: Broml 70Kc 137
Wendover Rd. NW10 40Va 68
Wendover Rd. SE9 55Mc 115
Wendover Rd. SL1: Burn 3A 80
Wendover Rd. TW18: Staines 64E 126
Wendover Way BR6: St M Cry 72Wc 161
Wendover Way DA16: Well 57Wc 117
Wendover Way RM12: Horn 36Ld 77
Wendover Way WD23: Bush 16Ea 28
Wendron Cl. GU21: Wok 10L 167
Wendy Cl. EN1: Enf 16Vb 33
Wendy Way HA0: Wemb 39Na 67
Wenham Gdns. CM13: Hut 16Ee 41
Wenham Ho. SW8 52Lb 112
Wenlack Cl. UB9: Den 34J 63
Wenlake Ho. EC1 5D 218 (42Sb 91)
(off Old St.)
Wenlock Barn Est. N1 2F 219 (40Tb 71)
(off Wenlock St.)
Wenlock Ct. N1 2G 219 (40Tb 71)
Wenlock Gdns. NW4 28Xa 48
Wenlock M. E10 31Cc 72
Wenlock Rd. HA8: Edg 24Ra 47
Wenlock Rd. N1 1D 218 (40Sb 71)
Wenlock St. N1 2E 218 (40Sb 71)
WENNINGTON 45Md 97
Wennington Rd. E3 40Zb 72
Wennington Rd. RM13: Rain 42Jd 96
Wennington Rd. RM13: Wenn 42Jd 96
Wensdale Ho. E5 33Wb 71
Wensley Av. IG8: Wfd G 24Hc 53
Wensley Cl. N11 23Jb 50
Wensley Cl. RM5: Col R 22Cd 56
Wensley Cl. SE9 58Pc 116
Wensleydale Av. IG5: Ilf 26Nc 54
Wensleydale Gdns. TW12: Hamp 66Da 129
Wensleydale Pas. TW12: Hamp 67Ca 129
Wensleydale Rd. TW12: Hamp 65Ca 129
Wensley Rd. N18 23Xb 51
Wensum Ct. WD3: Rick 18M 25
Wensum Pl. BR2: Hayes 73Jc 159
Wensum Way WD3: Rick 18M 25
Wenta Bus. Cen., The WD24: Wat 9Z 13
Wentbridge Path WD6: Bore 10Qa 15
Wentland Cl. SE6 61Fc 137
Wentland Rd. SE6 61Fc 137
Wentway Ct. W13 42Ha 86
(off Ruislip Rd. E.)
WENTWORTH 1K 147
Wentworth Av. N3 24Cb 49
Wentworth Av. SL2: Slou 10E 60
Wentworth Av. WD6: E'tree 15Pa 29
Wentworth Cl. BR2: Hayes 75Jc 159
Wentworth Cl. BR6: Farnb 78Uc 160
Wentworth Cl. DA11: Grav'nd 4C 144
Wentworth Cl. EN6: Pot B 3Cb 17
Wentworth Cl. GU23: Rip 93K 189
Wentworth Cl. KT6: Surb 75Ma 153
Wentworth Cl. N3 24Db 49
Wentworth Cl. SE28 44Zc 95
Wentworth Cl. SM4: Mord 73Cb 155
Wentworth Cl. TW15: Ashf 63R 128
Wentworth Cl. WD17: Wat 10V 12
Wentworth Ct. SW1 7K 227 (50Kb 90)
Wentworth Ct. SW18 58Db 111
(off Garratt La.)
Wentworth Ct. TW2: Twick 62Ga 130
Wentworth Ct. W6 51Ab 110
(off Paynes Wlk.)
Wentworth Cres. SE15 52Wb 113
Wentworth Cres. UB3: Harl 48T 84
Wentworth Dene KT13: Weyb 78R 150
Wentworth Dr. DA1: Dart 58Jd 118
Wentworth Dr. GU25: Vir W 10K 125
Wentworth Dr. HA5: Eastc 29W 44
Wentworth Dr. TW6: H'row A 54K 105
Wentworth Dr. WD19: Pinn 22Z 45
Wentworth Dr. WD19: Wat 22Z 45
Wentworth Dwellings E1 2K 225 (44Vb 91)
(off Wentworth St.)
Wentworth Flds. UB4: Hayes 40T 64
Wentworth Gdns. N13 20Rb 33
Wentworth Golf Course (East Course) 2M 147
Wentworth Golf Course (Edinburgh
Course) 4K 147
Wentworth Golf Course (West Course) 1K 147
Wentworth Hall GU25: Vir W 1L 147
Wentworth Hill HA9: Wemb 32Pa 67
Wentworth Ho. IG8: Wfd G 24Qc 54
Wentworth Ho. KT15: Add 77K 149
Wentworth M. E3 42Ac 92
Wentworth M. W3 44Ua 88
Wentworth Pk. N3 24Cb 49
Wentworth Pl. HA7: Stan 23Ka 46
Wentworth Pl. RM16: Grays 48Fe 99
Wentworth Rd. CR0: C'don 73Qb 156
Wentworth Rd. E12 35Mc 73
Wentworth Rd. EN5: Barn 13Za 30
Wentworth Rd. NW11 30Bb 49
Wentworth Rd. SS17: Stan H 3K 101
Wentworth Rd. UB2: S'hall 49Y 85
Wentworth St. E1 2K 225 (44Vb 91)
Wentworth Tennis & Health Club, The 1L 147
Wentworth Way CR2: Sande 86Wb 177
Wentworth Way HA5: Pinn 28Aa 45
Wentworth Way RM13: Rain 41Kd 97
Wenvoe Av. DA7: Bex 54Dd 118
Wepham Cl. UB4: Yead 43Z 85
Wernbrook St. SE18 51Sc 116
Werndee Rd. SE25 70Wb 135
Werneth Hall Rd. IG5: Ilf 27Qc 54
Werrington St. NW1 2C 216 (40Lb 70)
Werter Rd. SW15 56Ab 110
Wesco Ct. GU21: Wok 88C 168
Wescott Way UB8: Uxb 40L 63
Wesleyan Pl. NW5 35Kb 70
Wesley Apts. SW8 53Mb 112
Wesley Av. E16 46Jc 93
Wesley Av. NW10 41Ta 87
Wesley Av. TW3: Houn 54Aa 107
Wesley Cl. BR5: St P 69Yc 139
Wesley Cl. EN7: G Oak 1Sb 19
Wesley Cl. HA2: Harr 33Ea 66
Wesley Cl. KT19: Ewe 78Sa 153
Wesley Cl. N7 33Pb 70
Wesley Cl. RH2: Reig 7H 207
Wesley Cl. SE17 6C 230 (49Rb 91)
Wesley Ct. SE16 48Xb 91
Wesley Dr. TW20: Egh 65C 126
Wesley Ho. AL1: St A 2B 6
(off Marlborough Rd.)
Wesley Pl. KT18: Tatt C 90Za 174
Wesley Pl. SL4: Wink 1A 124
Wesley Rd. E10 31Ec 72
Wesley Rd. KT22: Lea 95La 192
Wesley Rd. NW10 39Sa 67
Wesley Rd. UB3: Hayes 45W 84
Wesley's House & Mus. of
Methodism 5G 219 (42Tb 91)
Wesley Sq. W11 44Ab 88
Wesley St. W1 1J 221 (43Jb 90)
Wessels KT20: Tad 93Za 194
Wessex Av. SW19 69Cb 133
Wessex Cl. IG3: Ilf 30Uc 54
Wessex Cl. KT1: King T 67Ra 131
Wessex Cl. KT7: T Ditt 75Ha 152
Wessex Ct. BR3: Beck 67Ac 136
Wessex Ct. EN5: Barn 14Za 30
Wessex Ct. HA9: Wemb 33Pa 67
Wessex Ct. TW19: Stanw 58N 105
Wessex Dr. DA8: Erith 54Gd 118
Wessex Dr. HA5: Hat E 24Aa 45
Wessex Gdns. NW11 32Ab 68
Wessex Ho. SE1 7K 231 (50Vb 91)
Wessex Ho. WD23: Bush 14Ba 27
Wessex La. RM3: Hrld W 25Pd 57
Wessex La. UB6: G'frd 41Fa 86
Wessex St. E2 41Yb 92
Wessex Ter. CR4: Mitc 71Gb 155
Wessex Wlk. DA2: Wilm 61Gd 140
Wessex Way NW11 32Ab 68
Wesson Mead SE5 52Sb 113
(off Camberwell Rd.)
West 12 Shop. Cen. 47Za 88
Westacott UB4: Hayes 43U 84
Westacott Cl. N19 32Mb 70
West Acre HA2: Harr 33Ga 66
West Acres KT10: Esh 80Ba 151
WEST ACTON 44Qa 87
West Acton Station (Underground) 44Qa 87
Westall Rd. IG10: Lough 13Rc 36
West App. BR5: Pet W 71Sc 160
West Arbour St. E1 44Zb 92
West Av. AL2: Chis G 7P 5
West Av. E17 28Dc 52
West Av. HA5: Pinn 30Ba 45
West Av. KT12: W Vill 82U 170
West Av. N3 23Cb 49
West Av. NW4 29Za 48
West Av. SM6: W'gton 78Nb 156
West Av. UB1: S'hall 45Ba 85
West Av. UB3: Hayes 45V 84
West Av. Rd. E17 28Cc 52
West Bank EN2: Enf 12Sb 33
West Bank IG11: Bark 39Rc 74
West Bank N16 31Ub 71
Westbank Rd. TW12: Hamp H 65Ea 130
WEST BARNES 71Xa 154
West Barnes La. KT3: N Mald 69Xa 132
West Barnes La. SW20 69Xa 132
WEST BECKTON 44Mc 93
WEST BEDFONT 58P 105
Westbeech Rd. N22 27Qb 50
Westbere Dr. HA7: Stan 22Ma 47
Westbere Rd. NW2 35Ab 68
West Block SE1 2J 229 (47Pb 90)
(off York Rd.)
Westbourne Apts. SW6 55Db 111
Westbourne Av. SM3: Cheam 75Ab 154
Westbourne Av. W3 44Ta 87
Westbourne Bri. W2 43Eb 89
Westbourne Cl. UB4: Yead 42Y 85
Westbourne Ct. W2 43Eb 89
Westbourne Cres. W2 4B 220 (45Fb 89)
Westbourne Cres. M. W2 4B 220 (45Fb 89)
(off Westbourne Cres.)
Westbourne Dr. CM14: B'wood 21Vd 58
Westbourne Dr. SE23 61Zb 136
Westbourne Gdns. W2 44Db 89
WESTBOURNE GREEN 43Cb 89
Westbourne Gro. W11 45Bb 89
Westbourne Gro. W2 44Cb 89
Westbourne Gro. M. W11 44Cb 89
Westbourne Gro. Ter. W2 44Db 89
Westbourne Ho. SW1 7K 227 (50Kb 90)
(off Ebury Bri. Rd.)
Westbourne Ho. TW5: Hest 51Ca 107
Westbourne M. AL1: St A 2B 6
Westbourne Pde. UB10: Hil 42R 84
Westbourne Pk. Pas. W2 43Cb 89
(off Harrow Rd.)
Westbourne Pk. Rd. W11 44Ab 88
Westbourne Pk. Rd. W2 43Cb 89
Westbourne Park Station
(Underground) 43Bb 89
Westbourne Pk. Vs. W2 43Cb 89
Westbourne Pl. N9 20Xb 33
Westbourne Rd. CR0: C'don 72Vb 157
Westbourne Rd. DA7: Bex 52Zc 117
Westbourne Rd. N7 37Pb 70
Westbourne Rd. SE26 65Zb 136
Westbourne Rd. TW13: Felt 62V 128
Westbourne Rd. TW18: Staines 66K 127
Westbourne Rd. UB8: Hil 42R 84
Westbourne St. W2 4B 220 (45Fb 89)
Westbourne Ter. SE23 61Zb 136
(off Westbourne Dr.)
Westbourne Ter. W2 2A 220 (44Eb 89)
Westbourne Ter. M. W2 2A 220 (44Eb 89)
Westbourne Ter. Rd. W2 43Eb 89
Westbourne Ter. Rd. Bri. 43Eb 89
(off Westbourne Ter. Rd.)
Westbridge Cl. W12 47Wa 88
Westbridge Ho. SW11 53Gb 111
(off Westbridge Rd.)
Westbridge Rd. SW11 53Fb 111
WEST BROMPTON 51Eb 111
West Brompton Station
(Underground & Overground) 51Cb 111
Westbrook Av. TW12: Hamp 66Ba 129
Westbrook Cl. EN4: Cockf 13Fb 31
Westbrook Cres. EN4: Cockf 13Fb 31
Westbrook Dr. BR5: Orp 74Zc 161
Westbrooke Cres. DA16: Well 55Yc 117
Westbrooke Rd. DA15: Sidc 61Tc 138
Westbrooke Rd. DA16: Well 55Xc 117
Westbrook Ho. E2 41Yb 92
(off Victoria Pk. Sq.)
Westbrook Rd. CR7: Thor H 67Tb 135
Westbrook Rd. SE3 53Kc 115
Westbrook Rd. TW18: Staines 64H 127
Westbrook Rd. TW5: Hest 52Ba 107
Westbrook Sq. EN4: Cockf 13Fb 31
Westbury EN8: Chesh 2Zb 20
Westbury SL4: Eton 1G 102
(off Eton Wick Rd.)
Westbury Av. HA0: Wemb 38Na 67
Westbury Av. KT10: Clay 79Ha 152
Westbury Av. N22 27Rb 51
Westbury Av. UB1: S'hall 42Ca 85
Westbury Cl. CR3: Whyt 90Vb 177
Westbury Cl. HA4: Ruis 31W 64
Westbury Cl. TW17: Shep 72R 150
Westbury Ct. IG11: Bark 39Tc 74
(off Ripple Rd.)
Westbury Dr. CM14: B'wood 19Yd 40
Westbury Gro. N12 23Cb 49
Westbury Ho. E17 28Bc 52
Westbury Ho. W11 43Cb 89
(off Aldridge Rd. Vs.)
Westbury La. IG9: Buck H 19Kc 35
Westbury Lodge Cl. HA5: Pinn 27Z 45
Westbury Pde. SW12 58Kb 112
(off Balham Hill)
Westbury Pl. TW8: Bford 51Ma 109
Westbury Rd. BR1: Broml 67Mc 137
Westbury Rd. BR3: Beck 69Ac 136
Westbury Rd. CM14: B'wood 19Yd 40
Westbury Rd. CR0: C'don 72Tb 157
Westbury Rd. E17 28Bc 52
Westbury Rd. E7 37Kc 73
Westbury Rd. HA0: Wemb 38Na 67
Westbury Rd. HA6: Nwood 21U 44
Westbury Rd. IG1: Ilf 33Qc 74
Westbury Rd. IG11: Bark 39Tc 74
Westbury Rd. IG9: Buck H 19Lc 35
Westbury Rd. KT3: N Mald 70Ta 131
Westbury Rd. N11 23Nb 50
Westbury Rd. N12 23Cb 49
Westbury Rd. SE20 67Zb 136
Westbury Rd. TW13: Felt 60Z 107
Westbury Rd. W5 44Na 87
Westbury Rd. WD18: Wat 15X 27
Westbury Ter. E7 37Kc 73
Westbury Ter. RM14: Upm 33Ud 78
Westbury Ter. TN16: Westrm 99Sc 200
Westbush Ct. W12 47Xa 88
(off Goldhawk M.)
WEST BYFLEET 85J 169
West Byfleet Golf Course 85H 169
West Byfleet Station (Rail) 84J 169
West Cadet Apts. SE18 52Qc 116
(off Langhorne St.)
Westcar La. KT12: Hers 79X 151
West Carriage Dr.
W2: Nth. Ride 5D 220 (45Gb 89)
West Carriage Dr.
W2: Rotten Row 1C 226 (47Fb 89)
West Carriage Ho. SE18 48Rc 94
(off Royal Carriage M.)
West Central St. WC1 2F 223 (44Nb 90)
West Chantry HA3: Hrw W 25Da 45
Westchester Dr. NW4 27Za 48
WEST CLANDON 100J 189
Westcliffe Apts. W2 1C 220 (43Fb 89)
West Cl. EN4: Cockf 14Jb 32
West Cl. EN5: Barn 15Xa 30
West Cl. HA9: Wemb 32Pa 67
West Cl. N9 20Vb 33
West Cl. RM13: Rain 42Kd 97
West Cl. TW12: Hamp 65Aa 129
West Cl. TW15: Ashf 63N 127
West Cl. UB6: G'frd 40Ea 66
Westcombe Av. CR0: C'don 73Nb 156
Westcombe Ct. SE3 52Hc 115
Westcombe Dr. EN5: Barn 15Cb 31
Westcombe Hill SE10 50Jc 93
Westcombe Hill SE3 52Jc 115
Westcombe Lodge Dr. UB4: Hayes 43U 84
Westcombe Pk. Rd. SE3 51Gc 115
Westcombe Park Station (Rail) 50Jc 93
West Comn. Cl. SL9: Ger X 29A 42
West Comn. Rd. BR2: Hayes 74Jc 159
West Comn. Rd. BR2: Kes 74Jc 159
West Comn. Rd. UB8: Uxb 36M 63
Westcoombe Av. SW20 67Va 132
Westcote Ri. HA4: Ruis 31S 64
Westcote Rd. KT19: Eps 83Ra 173
Westcote Rd. SW16 64Lb 134
West Cotts. NW6 36Cb 69
Westcott Av. DA11: Nflt 2B 144
Westcott Cl. BR1: Broml 71Pc 160
Westcott Cl. CR0: New Ad 81Dc 178
Westcott Cl. N15 30Vb 51
Westcott Cres. W7 44Ga 86
Westcott Ho. E14 45Cc 92
Westcott Rd. SE17 51Rb 113
Westcott Way SM2: Cheam 82Ya 174
West Ct. E17 28Cc 52
West Ct. HA0: Wemb 33La 66
West Ct. TW5: Isle 52Ea 108
WESTCOURT 1F 144
Westcourt La. DA12: Grav'nd 10H 123
Westcourt Pde. DA12: Grav'nd 2G 144
West Cres. SL4: Wind 3D 102
West Cres. Rd. DA12: Grav'nd 8D 122
Westcroft SL2: Slou 2F 80
Westcroft Cl. EN3: Enf W 10Yb 20
Westcroft Cl. NW2 35Ab 68
Westcroft Est. NW2 35Ab 68
Westcroft Gdns. SM4: Mord 69Bb 133
Westcroft Leisure Cen. 77Jb 156
Westcroft Rd. SM5: Cars 77Jb 156
Westcroft Rd. SM6: W'gton 77Jb 156
Westcroft Sq. W6 49Wa 88
Westcroft Way NW2 35Ab 68
West Cromwell Rd. SW5 49Cb 89
West Cromwell Rd. W14 50Bb 89
West Cross Cen. TW8: Bford 51Ja 108
West Cross Route W10 45Za 88
West Cross Way TW8: Bford 51Ka 108
West Croydon Bus Station 74Sb 157
West Croydon Station (Rail, Overground
& London Tramlink) 74Sb 157
Westdale Pas. SE18 51Rc 116
Westdale Rd. SE18 51Rc 116
Westdean Av. SE12 60Kc 115
Westdean Cl. SW18 58Db 111
West Dene SM3: Cheam 79Ab 154
Westdene CR0: C'don 77Tb 157
(off Chatsworth Rd.)
Westdene KT12: Hers 77X 151
Westdene Cl. CR8: Purl 85Pb 176
West Dene Dr. RM3: Rom 22Md 57
Westdene Way KT13: Weyb 76U 150
West Down KT23: Bookh 99Da 191
Westdown Rd. E15 35Ec 72
Westdown Rd. SE6 59Cc 114
WEST DRAYTON 47N 83
West Drayton Pk. Av. UB7: W Dray 48N 83
West Drayton Rd. UB8: Hil 44R 84
West Drayton Station
(Rail & Crossrail) 46N 83
West Dr. GU25: Vir W 3J 147
(not continuous)
West Dr. HA3: Hrw W 23Fa 46
West Dr. KT15: New H 82K 169
West Dr. KT20: Tad 90Za 174
West Dr. SL5: S'dale 1H 147
West Dr. SL5: Vir W 1H 147
West Dr. SM2: Cheam 81Za 174
West Dr. SM5: Cars 82Fb 175
West Dr. SW16 63Lb 134
West Dr. WD25: Wat 8X 13

West Dr. Gdns. HA3: Hrw W 23Fa 46
WEST DULWICH 61Tb 135
West Dulwich Station (Rail) 61Tb 135
WEST EALING 46Ka 86
West Ealing Bus. Cen. W13 45Ja 86
West Ealing Station (Rail & Crossrail) 45Ka 86
West Eaton Pl. SW1 5H 227 (49Jb 90)
West Eaton Pl. M. SW1 5H 227 (49Jb 90)
(off W. Eaton Pl.)
Wested La. BR8: Crock 73Jd 162
(not continuous)
Wested La. BR8: Swan 73Jd 162
(not continuous)
West Ella Rd. NW10 38Ua 68
West Elms Studios SW8 53Lb 112
West End TN15: Kems'g 89Pd 183
West End TN16: Bras 97Xc 201
WEST END 4D 166
WEST END 79Ba 151
WEST END 40Z 65
West End Av. E10 29Fc 53
West End Av. HA5: Pinn 28Z 45
West End Cl. NW10 38Sa 67
West End Common 80Aa 151
West End Ct. HA5: Pinn 28Z 45
West End Ct. NW6 38Db 69
West End Ct. SL2: Stoke P 9K 61
West End Gdns. KT10: Esh 78Ba 151
West End Gdns. UB5: N'olt 40Y 65
West End Grn. 36Cb 69
(off West End Lane)
West End La. EN5: Barn 14Za 30
West End La. HA5: Pinn 27Z 45
West End La. KT10: Esh 80Ba 151
West End La. NW6 36Cb 69
(not continuous)
West End La. SL2: Stoke P 9J 61
West End La. UB3: Harl 52S 106
West End M. WD18: Wat 14W 26
West End Quay W2 1C 220 (43Fb 89)
(off Harbet Rd.)
West End Rd. HA4: Ruis 33U 64
West End Rd. UB1: S'hall 46Aa 85
West End Rd. UB5: N'olt 38Y 65
Westerdale Ct. N5 35Rb 71
(off Hamilton Pk. W.)
Westerdale Rd. SE10 50Jc 93
Westerfield Rd. N15 29Vb 51
Westerfolds Cl. GU22: Wok 89E 168
Westergate W5 43Na 87
Westergate Ho. KT1: King T 70Ma 131
(off Portsmouth Rd.)
Westergate Rd. SE2 51Ad 117
Westerham NW1 1C 216 (39Lb 70)
(off Bayham St.)
WESTERHAM 98Tc 200
Westerham Av. N9 20Tb 33
Westerham Cl. KT15: Add 79L 149
Westerham Cl. SM2: Sutt 82Cb 175
Westerham Dr. DA15: Sidc 58Xc 117
Westerham Golf Course 98Wc 201
Westerham Hill TN16: Westrm 93Rc 200
WESTERHAM HILL 92Qc 200
Westerham Ho. SE1 3G 231 (48Tb 91)
(off Law St.)
Westerham Lodge BR3: Beck 66Cc 136
(off Park Rd.)
Westerham Rd. BR2: Kes 79Mc 159
Westerham Rd. E10 31Dc 72
Westerham Rd. RH8: Limp 1K 211
Westerham Rd. RH8: Oxt 1K 211
Westerham Rd. TN13: Bes G 95Ed 202
Westerham Rd. TN14: Sund 95Dd 202
Westerham Rd. TN16: Bras 97Vc 201
Westerham Rd. TN16: Westrm 100Pc 200
Westerham Rd. TN16: Westrm 97Vc 201
Westerham Trade Cen. TN16: Westrm 97Tc 200
Westerley Cres. SE26 64Bc 136
Westerley Ware TW9: Kew 51Qa 109
(off Kew Grn.)
Westerley Ware Recreation Ground 51Qa 109
Westermain KT15: New H 82L 169
Western Av. CM14: B'wood 18Yd 40
Western Av. CM16: Epp 4Vc 23
Western Av. HA4: Ruis 37S 64
Western Av. KT16: Chert 69J 127
Western Av. NW11 30Za 48
Western Av. RM10: Dag 37Ed 76
Western Av. RM2: Rom 26Ld 57
Western Av. TW20: Thorpe 69D 126
Western Av. UB10: Hil 36Q 64
Western Av. UB10: Uxb 36Q 64
Western Av. UB5: N'olt 38Z 65
Western Av. UB6: G'frd 40Fa 66
Western Av. UB9: Den 35L 63
Western Av. W3 42Qa 87
Western Av. W5 42Pa 87
Western Av. Bus. Pk. W3 42Ra 87
Western Beach Apts. E16 46Jc 93
Western Cl. KT16: Chert 69J 127
Western Ct. N3 23Cb 49
Western Ct. NW6 40Bb 69
Western Ct. RM1: Rom 29Gd 56
(off Chandlers Way)
Western Ct. W3 44Ta 87
Western Ctyd. EC2 1J 225 (43Ub 91)
(off Cutlers Gdns.)
WESTERN CROSS 60Xd 120
Western Cross Cl. DA9: Ghithe 58Yd 120
Western Dr. TW17: Shep 72T 150
WESTERN EYE HOSPITAL 7F 215 (43Hb 89)
Western Gdns. CM14: B'wood 19Yd 40
Western Gdns. W5 45Qa 87
Western Gateway E16 45Jc 93
Western Intl. Mkt. UB2: S'hall 49X 85
Western La. SW12 59Jb 112
Western Mans. EN5: New Bar 15Db 31
(off Great Nth. Rd.)
Western M. W9 42Bb 89
Western Pde. EN5: New Bar 15Cb 31
Western Pde. RH2: Reig 9K 207
Western Pathway RM13: Rain 38Kd 77
Western Perimeter Rd. TW6: H'row A 54K 105
Western Perimeter Rd. TW6: Lford 54K 105
Western Pl. SE16 47Yb 92
Western Rd. CM14: B'wood 19Yd 40
Western Rd. CM16: Epp 4Vc 23
Western Rd. CR4: Mitc 67Fb 133
Western Rd. E13 40Lc 73
Western Rd. E17 29Ec 52
Western Rd. N2 28Hb 49
Western Rd. N22 26Pb 50
Western Rd. NW10 42Sa 87
Western Rd. RM1: Rom 29Fd 56
Western Rd. SM1: Sutt 78Cb 155
Western Rd. SW19 67Fb 133
Western Rd. SW9 55Qb 112
Western Rd. TN15: Bor G 92Be 205
Western Rd. UB2: S'hall 49Y 85
Western Rd. W5 45Ma 87
Western Sq. N18 23Yb 52
Western Ter. DA1: Dart 58Kd 119
Western Ter. W6 50Wa 88
Western Transit Shed N1 1F 217 (39Nb 70)
Western Vw. UB3: Hayes 47V 84
Westernville Gdns. IG2: Ilf 31Sc 74
Western Wlk. KT1: King T 68Na 131
(off Eden Walk Shop. Cen.)
Western Way EN5: Barn 16Cb 31
Western Way SE28 48Tc 94
West Essex Golf Course 14Gc 35
WEST EWELL 81Ta 173
West Farm Av. KT21: Asht 90La 172
West Farm Cl. KT21: Asht 91La 192
West Farm Dr. KT21: Asht 91Ma 193
Westferry Cir. E14 46Bc 92
Westferry Rd. E14 45Bc 92
Westferry Station (DLR) 45Cc 92
Westfield AL9: Hat 5H 9
Westfield DA3: Nw A G 77Be 165
WESTFIELD 93A 188
Westfield IG10: Lough 15Mc 35
Westfield KT21: Asht 90Pa 173
Westfield RH2: Reig 3K 207
Westfield TN13: S'oaks 94Ld 203
Westfield Av. CR2: Sande 85Tb 177
Westfield Av. E20 37Dc 72
Westfield Av. GU22: Wok 93A 188
Westfield Av. WD24: Wat 9Y 13
Westfield Cl. DA12: Grav'nd 5E 144
Westfield Cl. EN3: Enf H 13Ac 34
Westfield Cl. EN8: Walt C 3Bc 20
Westfield Cl. GU22: Wok 93B 188
Westfield Cl. NW9 27Sa 47
Westfield Cl. SM1: Sutt 77Bb 155
Westfield Cl. SW10 52Eb 111
Westfield Comn. GU22: Wok 94A 188
Westfield Community Sports Cen. 15V 26
Westfield Ct. KT6: Surb 71Ma 153
(off Portsmouth Rd)
Westfield Ct. NW10 41Za 88
(off Chamberlayne Rd.)
Westfield Dr. HA3: Kenton 28Ma 47
Westfield Dr. KT23: Bookh 94Ca 191
Westfield Gdns. HA3: Kenton 28Ma 47
Westfield Gdns. RM6: Chad H 30Yc 55
Westfield Gro. GU22: Wok 92A 188
Westfield Ho. SE16 49Zb 92
(off Rotherhithe New Rd.)
Westfield Ho. SW10 52Fb 111
(off Worlds End Est.)
Westfield La. HA3: Kenton 29Ma 47
Westfield La. SL3: Geor G 4P 81
Westfield London Bus Station 46Ya 88
Westfield Pde. KT15: New H 82M 169
Westfield Pk. HA5: Hat E 24Ba 45
Westfield Pk. Dr. IG8: Wfd G 23Nc 54
Westfield Rd. BR3: Beck 68Bc 136
Westfield Rd. CR0: C'don 75Rb 157
Westfield Rd. CR4: Mitc 68Gb 133
Westfield Rd. DA7: Bex 55Ed 118
Westfield Rd. GU22: Wok 4P 187
Westfield Rd. KT12: Walt T 73Aa 151
Westfield Rd. KT6: Surb 71Ma 153
Westfield Rd. NW7 20Ta 29
Westfield Rd. RM9: Dag 35Ad 75
Westfield Rd. SL2: Slou 2F 80
Westfield Rd. SM1: Sutt 77Bb 155
Westfield Rd. W13 46Ja 86
Westfields AL3: St A 4N 5
Westfields SW13 55Va 110
Westfields Av. SW13 55Ua 110
Westfield Shop. Cen. London 46Ya 88
Westfield Shop. Cen. Stratford City 37Ec 72
Westfield Sq. GU22: Wok 94A 188
Westfields Rd. W3 43Ra 87
Westfield St. SE18 48Mc 93
Westfield Wlk. EN8: Walt C 3Bc 20
Westfield Way E1 41Ac 92
Westfield Way GU22: Wok 94A 188
Westfield Way HA4: Ruis 34U 64
West Finchley Station
(Underground) 23Db 49
West Gdn. Pl. W2 3E 220 (44Gb 89)
West Gdns. E1 45Xb 91
West Gdns. KT17: Ewe 82Ua 174
West Gdns. SW17 65Gb 133
Westgate W5 41Na 87
Westgate Apts. E16 45Jc 93
(off Western Gateway)
Westgate Cl. KT18: Eps 87Ta 173
Westgate Ct. EN8: Walt C 7Zb 20
Westgate Ct. SE12 60Jc 115
(off Burnt Ash Hill)
Westgate Ct. SW9 55Qb 112
(off Canterbury Cres.)
Westgate Cres. SL1: Slou 5D 80
Westgate Est. TW14: Bedf 60R 106
Westgate Ho. KT18: Eps 87Ta 173
(off Chalk La.)
Westgate Ho. TW7: Isle 54Fa 108
Westgate Ho. TW8: Bford 50Ma 87
(off Ealing Rd.)
West Ga. M. WD18: Wat 15U 26
Westgate M. W10 42Ab 88
(off West Row)
Westgate Retail Pk. Chippenham 5E 80
Westgate Rd. BR3: Beck 67Ec 136
Westgate Rd. DA1: Dart 58Md 119
Westgate Rd. SE25 70Xb 135
Westgate St. E8 39Xb 71
Westgate Ter. SW10 51Db 111
Westglade Ct. HA3: Kenton 29Ma 47
WEST GREEN 28Rb 51
West Grn. Pl. UB6: G'frd 39Fa 66
West Grn. Rd. N15 28Rb 51
West Gro. IG8: Wfd G 23Lc 53
West Gro. KT12: Hers 78X 151
West Gro. SE10 53Ec 114
West Gro. SE17 5D 230 (49Sb 91)
Westgrove La. SE10 53Ec 114
West Gro. Sq. SE17 5D 230 (49Sb 91)
West Halkin St. SW1 3H 227 (48Jb 90)
West Hall KT14: W Byf 85M 169
West Hallowes SE9 60Mc 115
Westhall Pk. CR6: W'ham 91Yb 198
West Hall Rd. TW9: Kew 53Ra 109
Westhall Rd. CR6: W'ham 90Wb 177
WEST HAM 39Hc 73
West Ham La. E15 38Fc 73
West Ham Pk. E7 38Jc 73
WEST HAMPSTEAD 37Db 69
West Hampstead M. NW6 37Db 69
West Hampstead Station
(Underground & Overground) 37Cb 69
West Hampstead Thameslink Station
(Rail) 37Cb 69
West Ham Station (DLR) 41Gc 93
West Ham Station
(Rail & Underground) 41Gc 93
West Ham Utd FC 39Dc 72
West Handyside Canopy N1 1F 217 (39Nb 70)
West Harding St. EC4 2A 224 (44Qb 90)
West Harold BR8: Swan 69Fd 140
WEST HARROW 31Ea 66
West Harrow Station (Underground) 30Ea 46
West Hatch Mnr. HA4: Ruis 32V 64
Westhay Gdns. SW14 57Ra 109
West Heath GU24: Pirb 5B 186
WEST HEATH 2L 211
WEST HEATH 51Zc 117
West Heath Av. NW11 32Cb 69
West Heath Cl. DA1: Cray 58Hd 118
West Heath Cl. NW3 34Cb 69
West Heath Cotts. TN13: S'oaks 100Kd 203
West Heath Ct. NW11 32Cb 69
West Heath Dr. NW11 32Cb 69
West Heath Gdns. NW3 33Cb 69
West Heath La. TN13: S'oaks 100Kd 203
West Heath Rd. DA1: Cray 58Hd 118
West Heath Rd. NW3 33Cb 69
West Heath Rd. SE2 51Yc 117
WEST HENDON 31Wa 68
West Hendon B'way. NW9 30Va 48
West Hertfordshire Crematorium 3Z 13
West Herts Bus. Cen. WD6: Bore 13Ra 29
(off Eldon Av.)
West Herts Golf Course 13T 26
West Hill BR6: Downe 84Pc 180
West Hill CR2: Sande 82Ub 177
West Hill DA1: Dart 58Md 119
West Hill HA2: Harr 33Ga 66
West Hill HA9: Wemb 32Pa 67
West Hill KT19: Eps 85Ra 173
West Hill RH8: Oxt 2H 211
West Hill SW15 59Za 110
West Hill SW18 57Bb 111
WEST HILL 58Bb 111
West Hill Av. KT19: Eps 85Ra 173
West Hill Bank RH8: Oxt 2H 211
West Hill Cl. GU24: Brkwd 2F 186
Westhill Cl. DA12: Grav'nd 10D 122
West Hill Ct. KT19: Eps 85Sa 173
(off Court La.)
West Hill Ct. N6 34Jb 70
Westhill Ct. W11 45Bb 89
(off Denbigh Rd.)
West Hill Dr. DA1: Dart 58Ld 119
West Hill Golf Course 2F 186
West Hill Pk. N6 33Hb 69
(not continuous)
West Hill Pl. RH8: Oxt 1J 211
West Hill Ri. DA1: Dart 58Md 119
West Hill Rd. GU22: Wok 1P 187
West Hill Rd. SW18 58Bb 111
West Hill Way N20 18Db 31
Westholm NW11 28Db 49
West Holme DA8: Erith 53Ed 118
Westholme BR6: Orp 73Uc 160
Westholme Gdns. HA4: Ruis 32W 64
Westhope Ho. E2 42Wb 91
(off Derbyshire St.)
WEST HORNDON 30Fe 59
West Horndon Station (Rail) 30Ee 59
Westhorne Av. SE12 59Jc 115
Westhorne Av. SE9 57Mc 115
Westhorpe Gdns. NW4 27Ya 48
Westhorpe Rd. SW15 55Ya 110
WEST HORSLEY 100R 190
West Ho. IG11: Bark 37Rc 74
West Ho. Cl. SW19 60Ab 110
West Ho. Cotts. HA5: Pinn 28Z 45
Westhurst Dr. BR7: Chst 64Rc 138
WEST HYDE 24G 42
West Hyde La. SL9: Chal P 24B 42
West India Av. E14 46Cc 92
West India Dock Rd. E14 45Bc 92
West India Ho. E14 45Cc 92
(off W. India Dock Rd.)
West India Quay Station (DLR) 45Cc 92
WEST KENSINGTON 49Ab 88
West Kensington Ct. W14 50Bb 89
(off Edith Vs.)
West Kensington Mans. W14 50Bb 89
(off Beaumont Cres.)
West Kensington Station
(Underground) 50Bb 89
West Kent Av. DA11: Nflt 58Ee 121
West Kent Golf Course 84Nc 180
WEST KILBURN 41Bb 89
Westking Pl. WC1 4H 217 (41Pb 90)
WEST KINGSDOWN 79Ud 164
West Kingsdown Ind. Est.
TN15: W King 81Ud 184
West Kingsdown Windmill 82Wd 184
Westlake SE16 49Yb 92
(off Rotherhithe New Rd.)
Westlake Cl. N13 20Qb 32
Westlake Cl. UB4: Yead 42Aa 85
Westlake Rd. HA9: Wemb 33Ma 67
Westland Av. RM11: Horn 32Nd 77
Westland Cl. TW19: Stanw 58N 105
Westland Cl. WD25: Wat 6V 12
Westland Ct. UB5: N'olt 41Z 85
(off Seasprite Cl.)
Westland Dr. AL9: Brk P 9G 8
Westland Dr. BR2: Hayes 75Hc 159
Westland Ho. E16 46Qc 94
(off Rymill St.)
Westland Pl. N1 3F 219 (41Tb 91)
Westland Rd. WD17: Wat 12X 27
Westlands Av. SL1: Slou 4A 80
Westlands Cl. SL1: Slou 4A 80
Westlands Cl. UB3: Harl 49W 84
Westlands Ct. KT18: Eps 87Sa 173
Westlands Ct. KT8: E Mos 70Fa 130
Westlands Est. UB3: Harl 48U 84
Westlands Ter. SW12 58Lb 112
Westlands Way RH8: Oxt 99Fc 199
Westland Vw. RM16: Grays 46Ce 99
West La. SE16 47Xb 91
West Lawn WD25: A'ham 8Da 13
Westlea Av. WD25: Wat 8Aa 13
Westlea Rd. W7 48Ja 86
Westleigh Av. CR5: Coul 88Jb 176
Westleigh Av. SW15 57Xa 110
Westleigh Ct. CR2: S Croy 77Ub 157
(off Birdhurst Rd.)
Westleigh Ct. E11 29Jc 53
(off Nightingale La.)
Westleigh Dr. BR1: Broml 67Nc 138
Westleigh Gdns. HA8: Edg 25Qa 47
West Links HA0: Wemb 41Ma 87
Westlinton Cl. NW7 23Ab 48
West Lodge E16 46Jc 93
(off Britannia Ga.)
West Lodge Av. W3 46Qa 87
West Lodge Ct. W3 46Qa 87
West London Crematorium 42Xa 88
West London Golf Course 39Y 65
West London Studios SW6 52Db 111
(off Fulham Rd.)
Westlyn Cl. RM13: Rain 42Ld 97
Westmacott Dr. TW14: Felt 60V 106
Westmacott Ho. NW8 6C 214 (42Fb 89)
(off Hatton St.)
West Mall N9 20Wb 33
West Mall RM17: Grays 51Ce 121
(off Grays Shop. Cen.)
West Mall W8 46Cb 89
(off Kensington Mall)
West Malling Way RM12: Horn 36Ld 77
Westmark Point SW15 60Xa 110
(off Norley Va.)
West Mead HA4: Ruis 35Y 65
West Mead KT19: Ewe 79Ua 154
Westmead GU21: Wok 9M 167
Westmead SL4: Wind 5F 102
Westmead SW15 58Xa 110
Westmead Cnr. SM5: Cars 77Gb 155
Westmead Ct. SE6 58Cc 114
Westmeade Cl. EN7: Chesh 1Xb 19
Westmead Ho. SM1: Sutt 77Fb 155
Westmead Rd. SM1: Sutt 77Fb 155
Westmede IG7: Chig 23Sc 54
Westmere Dr. NW7 20Ta 29
West Mersea Cl. E16 46Kc 93
West M. N17 24Xb 51
West M. SW1 6A 228 (49Kb 90)
(off W. Warwick Pl.)
West Middlesex Golf Course 45Ea 86
WEST MIDDLESEX UNIVERSITY
HOSPITAL 54Ja 108
West Mill DA11: Grav'nd 8B 122
WESTMINSTER 2F 229 (47Nb 90)
Westminster Abbey Chapter
House 3F 229 (48Nb 90)
(within Abbey)
Westminster Abbey Mus. 3F 229 (48Nb 90)
Westminster Abbey Pyx C
hamber 3F 229 (48Nb 90)
(within Westminster Abbey)
Westminster Av. CR7: Thor H 68Rb 135
Westminster Boating Base & Pier 51Mb 112
Westminster Bri. SW1 2G 229 (47Nb 90)
Westminster Bri. Ho. SE1 3B 230 (48Rb 91)
(off Lambeth Rd.)
Westminster Bri. Rd. SE1 2H 229 (47Pb 90)
Westminster Bus. Sq. SE11 51Pb 112
(off Durham St.)
Westminster Cl. HA6: Nwood 24S 44
Westminster Cl. IG6: Ilf 26Tc 54
Westminster Cl. TW11: Tedd 64Ja 130
Westminster Cl. TW14: Felt 60W 106
Westminster Ct. AL1: St A 4A 6
Westminster Ct. E11 30Kc 53
(off Cambridge Pk.)
Westminster Ct. EN8: Walt C 5Cc 20
Westminster Ct. GU22: Wok 93D 188
Westminster Ct. NW8 6B 214 (42Fb 89)
(off Aberdeen Pl.)
Westminster Ct. SE16 46Zb 92
(off King & Queen Wharf)
Westminster Dr. N13 22Nb 50
Westminster Gdns. E4 18Gc 35
Westminster Gdns. IG11: Bark 40Uc 74
Westminster Gdns. IG6: Ilf 26Sc 54
Westminster Gdns. SW1 5F 229 (49Nb 90)
(off Vincent St.)
Westminster Hall 2F 229 (47Nb 90)
(off St Margaret St.)
Westminster Ho. WD24: Wat 12Y 27
(off Hallam Cl.)
Westminster Ind. Est. SE18 48Mc 93
Westminster Lodge Leisure Cen. 3A 6
Westminster Mans. SW1 4E 228 (48Mb 90)
(off Gt. Smith St.)
Westminster Pal. Gdns. SW1 4D 228 (48Mb 90)
(off Artillery Row)
Westminster RC Cathedral 4B 228 (48Lb 90)
Westminster Rd. N9 18Xb 33
Westminster Rd. SM1: Sutt 75Fb 155
Westminster Rd. W7 46Ga 86
Westminster Station
(Underground) 2G 229 (47Nb 90)
Westmoat Cl. BR3: Beck 66Ec 136
WEST MOLESEY 70Ca 129
Westmont Rd. KT10: Hin W 75Ga 152
Westmoor Gdns. EN3: Enf H 12Zb 34
Westmoor Rd. EN3: Enf H 12Zb 34
Westmoor St. SE7 48Lc 93
Westmore Ct. SW15 57Ab 110
Westmore Grn. TN16: Tats 92Lc 199
Westmoreland Av. DA16: Well 55Uc 116
Westmoreland Av. RM11: Horn 29Ld 57
Westmoreland Dr. SM2: Sutt 80Db 155
Westmoreland Ho. E16 46Jc 93
(off Gatcombe Rd.)
Westmoreland Ho. NW10 41Wa 88
Westmoreland Pl. BR1: Broml 69Jc 137
Westmoreland Pl. SW1 7A 228 (50Kb 90)
Westmoreland Pl. W5 43Ma 87
Westmoreland Rd. BR1: Broml 69Jc 137
Westmoreland Rd. BR2: Broml 71Gc 159
Westmoreland Rd. NW9 27Pa 47
Westmoreland Rd. SE17 51Tb 113
(not continuous)
Westmoreland Rd. SW13 53Va 110
Westmoreland St. W1 1J 221 (43Jb 90)
Westmoreland Ter. SE20 66Xb 135
Westmoreland Ter. SW1 7A 228 (50Kb 90)
Westmoreland Wlk. SE17 51Tb 113
(not continuous)
Westmore Rd. TN16: Tats 93Lc 199
Westmorland Cl. E12 33Mc 73
Westmorland Cl. KT19: Ewe 82Ua 174
Westmorland Cl. TW1: Twick 58Ka 108
Westmorland Ct. KT6: Surb 73Ma 153
Westmorland Rd. E17 30Cc 52
Westmorland Rd. HA1: Harr 29Da 45
Westmorland Way CR4: Mitc 70Mb 134
Westmount Apts. WD18: Wat 14V 26
(off Metropolitan Sta. App.)
Westmount Cen. UB4: Yead 45Z 85
Westmount Cl. KT4: Wor Pk 74Ya 154
Westmount Ct. W5 44Pa 87
Westmount Rd. SE9 54Pc 116
WEST NORWOOD 63Sb 135
West Norwood Crematorium 62Sb 135
West Norwood Station (Rail) 63Rb 135
West Oak BR3: Beck 67Ec 137
Westoe Rd. N9 19Xb 33
West Officers Apts. SE18 52Qc 116
Weston Av. KT15: Add 77K 149
Weston Av. KT7: T Ditt 73Ga 152
Weston Av. KT8: W Mole 69Aa 129
Weston Av. RM20: W Thur 51Vd 120
Westonbirt Ct. SE15 51Vb 113
(off Ebley Cl.)
Weston Cl. CM13: Hut 17Ee 41
Weston Cl. CR5: Coul 92Pb 196
Weston Cl. EN6: Pot B 4Bb 17
Weston Cl. KT15: Add 80J 149
Weston Ct. KT1: King T 69Na 131
(off Grove Cres.)
Weston Ct. N20 17Eb 31
(off Farnham Cl.)
Weston Ct. N4 34Sb 71
Weston Dr. CR3: Cat'm 94Sb 197
Weston Dr. HA7: Stan 25Ka 46
West One Ho. W1 1B 222 (43Lb 90)
(off Wells St.)
Westone Mans. IG11: Bark 38Vc 75
(off Upney La.)
West One Shop. Cen. 3J 221 (44Jb 90)
(off Gilbert St.)
Weston Gdns. GU22: Pyr 88G 168
Weston Gdns. TW7: Isle 53Ga 108
WESTON GREEN 74Ga 152
Weston Grn. KT7: T Ditt Weston Grn. Rd. 74Ga 152
Weston Grn. KT7: T Ditt Weston Rd. 75Ga 152
Weston Grn. RM9: Dag 35Bd 75
Weston Grn. Rd. KT10: Esh 74Fa 152
Weston Grn. Rd. KT7: T Ditt 74Ga 152
Weston Gro. BR1: Broml 67Hc 137
Weston Ho. E9 39Yb 72
(off King Edward's Rd.)
Weston Ho. NW6 38Ab 68
Westonia Ct. EN3: Enf W 8Zb 20
Weston Lea KT24: W Hor 97T 190
Weston Pk. KT1: King T 68Na 131
Weston Pk. KT7: T Ditt 74Ga 152
Weston Pk. N8 30Nb 50
Weston Pk. Cl. KT7: T Ditt 74Ga 152
Weston Ri. WC1 3J 217 (41Pb 90)
Weston Rd. BR1: Broml 66Hc 137
Weston Rd. EN2: Enf 12Tb 33
Weston Rd. KT17: Eps 83Ua 174
Weston Rd. KT7: T Ditt 74Ga 152
Weston Rd. RM9: Dag 35Ad 75
Weston Rd. SL1: Slou 3D 80
Weston Rd. W4 48Sa 87
Weston St. SE1 2H 231 (47Ub 91)
(not continuous)
Westons Yd. SL4: Eton 1H 103
Weston Wlk. E8 38Xb 71
Weston Way GU22: Pyr 88G 168
Westover Cl. SM2: Sutt 81Db 175
Westover Hill NW3 33Cb 69
Westover Rd. SW18 59Eb 111
Westow Hill SE19 65Ub 135
Westow St. SE19 65Ub 135
West Pal. Gdns. KT13: Weyb 76R 150
West Pk. SE9 61Nc 138
Westpark W5 44Ma 87
West Pk. Av. TW9: Kew 53Qa 109
West Park Cl. RM6: Chad H 29Zc 55
West Park Cl. TW5: Hest 51Ba 107
West Pk. Hill CM14: B'wood 20Wd 40
West Pk. Rd. KT19: Eps 83Pa 173
West Pk. Rd. TW9: Kew 53Qa 109
West Pk. Rd. UB2: S'hall 46Ea 86
West Parkside CR6: W'ham 87Cc 178
West Parkside SE10 47Gc 93
West Pk. Wlk. E20 37Ec 72
West Pl. SW19 64Ya 132
West Plaza TW15: Ashf 61N 127
West Point E14 45Bc 92
(off Grenade St.)
West Point KT19: Ewe 81Ua 174
West Point SE1 50Wb 91
West Point SL1: Slou 6B 80
Westpoint Apts. N8 27Pb 50
West Point Cl. TW4: Houn 55Ba 107
(off Grosvenor Rd.)
Westpoint Trad. Est. W3 43Ra 87
Westpole Av. EN4: Cockf 14Jb 32
Westport Ct. UB4: Yead 42Y 85
Westport Rd. E13 42Kc 93
Westport St. E1 44Zb 92
West Poultry Av. EC1 1B 224 (43Rb 91)
West Quarters W12 44Wa 88
West Quay SW10 53Eb 111
West Quay Dr. UB4: Yead 43Aa 85
West Quay Wlk. E14 48Dc 92
West Ramp TW6: H'row A 53Q 106
Westray HP3: Hem H 4C 4
West Reservoir Cen. 32Sb 71
Westridge Cl. HP1: Hem H 2H 3
West Ridge Gdns. UB6: G'frd 40Ea 66
West Riding AL2: Brick W 2Ba 13
West Ri. W2 4E 220 (45Gb 89)
(off St George's Flds.)
West Rd. E15 39Hc 73
West Rd. EN4: E Barn 18Jb 32
West Rd. KT13: Weyb 81R 170
West Rd. KT2: King T 67Sa 131
West Rd. KT9: Chess 84La 172
West Rd. N17 23Xb 51
West Rd. RH2: Reig 7K 207
West Rd. RM15: S Ock 41Xd 98
West Rd. RM6: Chad H 30Zc 55
West Rd. RM7: Rush G 31Fd 76
West Rd. SE1 1J 229 (47Pb 90)
West Rd. SW3 50Hb 89
West Rd. SW4 57Mb 112
West Rd. TW14: Bedf 58T 106
West Rd. UB7: W Dray 48P 83
West Rd. W5 43Na 87
Westrovia Ct. SW1 7D 228 (50Mb 90)
(off Moreton St.)
West Row W10 42Ab 88
Westrow SW15 58Ya 110
Westrow Dr. IG11: Bark 36Wc 75
Westrow Gdns. IG3: Ilf 33Vc 75
WEST RUISLIP 33S 64
West Ruislip Ct. HA4: Ruis 33T 64
(off Ickenham Rd.)
West Ruislip Station
(Rail & Underground) 33S 64
West Shaw DA3: Lfield 68Zd 143
West Sheen Va. TW9: Rich 56Pa 109
Westside HP3: Hem H 7P 3
Westside N2 27Hb 49
Westside NW4 26Xa 48
Westside Apts. IG1: Ilf 34Qc 74
(off Roden St.)
West Side Comn. SW19 64Ya 132
West Side Ct. TW16: Sun 66U 128
(off Scotts Av.)
Westside Ct. GU24: W End 5C 166
Westside Ct. W9 42Cb 89
(off Elgin Av.)
West Silvertown Station (DLR) 46Jc 93
West Smithfield EC1 1B 224 (43Rb 91)

West Spur Rd. UB8: Cowl....41M 83
West Sq. SE11....4B 230 (48Rb 91)
West Sq. SL0: Iver....44H 83
West Stand N5....35Rb 71
West St. BR1: Broml....67Jc 137
West St. CR0: C'don....77Sb 157
West St. DA11: Grav'nd....8C 122
West St. DA7: Bex....55Bd 117
West St. DA8: Erith....49Fd 96
West St. E11....34Gc 73
West St. E17....29Dc 52
West St. E2....40Xb 71
West St. GU21: Wok....89B 168
West St. HA1: Harr....32Fa 66
West St. KT17: Ewe....82Ua 174
West St. KT18: Eps....85Sa 173
West St. RH2: Reig....6G 206
West St. RM17: Grays....51Ce 121
West St. SM1: Sutt....78Db 155
West St. SM5: Cars....76Hb 155
West St. TN15: Wro....88Be 185
West St. WC2....3E 222 (44Mb 90)
West St. WD17: Wat....12X 27
West St. La. SM5: Cars....77Hb 155
West St. Pl. CR0: C'don....77Sb 157
....(off West St.)
West Sutton Station (Rail)....77Cb 155
West Temple Sheen SW14....57Ra 109
West Tenter St. E1....3K 225 (44Vb 91)
West Ter. DA15: Sidc....60Uc 116
West Thamesmead Bus. Pk. SE28....48Uc 94
....(not continuous)
WEST THURROCK....51Wd 120
West Thurrock Way RM20: Chaf H....48Vd 98
West Thurrock Way RM20: Grays....48Vd 98
West Thurrock Way RM20: W Thur....48Vd 98
WEST TILBURY....1G 122
West Twr. E14....47Dc 92
....(off Pan Peninsula Sq.)
West Twr. SW10....53Fb 111
West Towers HA5: Pinn....29Z 45
Westvale M. W3....47Ua 88
West Valley Rd. HP3: Hem H....7L 3
West Vw. IG10: Lough....13Pc 36
West Vw. KT21: Asht....91La 192
West Vw. NW4....28Ya 48
West Vw. TW14: Bedf....59S 106
Westview GU22: Wok....90B 168
....(off Park Dr.)
West View Apts. N7....38Nb 70
....(off York Way)
Westview Av. CR3: Whyt....90Vb 177
Westview Cl. NW10....36Va 68
Westview Cl. RH1: Redh....8N 207
Westview Cl. RM13: Rain....41Ld 97
Westview Cl. W10....44Ya 88
Westview Cl. W7....44Ga 86
West Vw. Ct. WD6: E'tree....16Ma 29
Westview Ct. N20....18Eb 31
Westview Cres. N9....17Ub 33
Westview Dr. IG8: Wfd G....26Mc 53
West Vw. Gdns. WD6: E'tree....16Ma 29
Westview Ri. HP2: Hem H....1M 3
West Vw. Rd. AL3: St A....1B 6
West Vw. Rd. BR8: Crock....72Fd 162
West Vw. Rd. BR8: Swan....70Jd 140
West Vw. Rd. DA1: Dart....58Pd 119
Westview Rd. CR6: W'ham....91Xb 197
Westville Rd. KT7: T Ditt....74Ja 152
Westville Rd. W12....47Wa 88
West Wlk. EN4: E Barn....17Jb 32
West Wlk. UB3: Hayes....46W 84
West Wlk. W5....43Na 87
Westward Pde. E14....48Dc 92
....(off Pepper St.)
Westward Rd. E4....22Bc 52
....(not continuous)
Westward Way HA3: Kenton....30Na 47
West Warwick Pl. SW1....6B 228 (49Lb 90)
WEST WATFORD....14W 26
West Way BR4: W W'ck....72Fc 159
West Way BR5: Pet W....71Tc 160
West Way CM14: B'wood....20Wd 40
West Way CR0: C'don....75Ac 158
West Way HA4: Ruis....32V 64
West Way HA5: Pinn....28Z 45
West Way HA8: Edg....23Ra 47
West Way N18....21Tb 51
West Way NW10....34Ta 67
West Way SM5: Cars....82Fb 175
West Way TW17: Shep....72T 150
West Way TW5: Hest....53Ba 107
West Way WD3: Rick....18K 25
Westway CR3: Cat'm....94Tb 197
Westway SW20....69Xa 132
Westway W10....43Bb 89
Westway W12....45Va 88
Westway Cl. SW20....69Xa 132
Westway Ct. CR3: Cat'm....95Tb 197
Westway Ct. UB5: N'olt....39Ca 65
Westway Cross Shop. Pk....39Ga 66
Westway Est. W3....43Ua 88
West Way Gdns. CR0: C'don....75Zb 158
Westway Gdns. RH1: Redh....3A 208
Westway Lodge W9....43Cb 89
....(off Amberley Rd.)
West Ways HA6: Nwood....26W 44
Westways KT19: Ewe....77Va 154
Westways TN16: Westrm....98Sc 200
Westway Sports Cen....44Za 88
Westway Travellers Site W12....45Za 88
....(off Stable Way)
Westwell Cl. BR5: Orp....74Zc 161
Westwell M. SW16....65Nb 134
Westwell Rd. SW16....65Nb 134
Westwell Rd. App. SW16....65Nb 134
Westwick KT1: King T....68Qa 131
....(off Chesterton Ter.)
Westwick Cl. HP2: Hem H....3D 4
Westwick Gdns. TW4: Cran....54X 107
Westwick Gdns. W14....47Za 88
WEST WICKHAM....74Ec 158
West Wickham Leisure Cen....74Ec 158
West Wickham Station (Rail)....73Ec 158
Westwick Pl. WD25: Wat....6Y 13
Westwick Row HP2: Hem H....2D 4
West Wing DA2: Dart....58Sd 120
West Wintergarden....46Dc 92
....(off Bank St.)
Westwode Cl. RM5: Col R....23Cd 56
Westwood DA11: Grav'nd....4E 144
WESTWOOD....66Zd 143
Westwood Av. CM14: B'wood....21Wd 58
Westwood Av. HA2: Harr....35Da 65
Westwood Av. KT15: Wdhm....84H 169
Westwood Av. SE19....67Sb 135
Westwood Bus. Cen. NW10....42Ua 88
Westwood Cl. BR1: Broml....68Mc 137
Westwood Cl. EN6: Pot B....2Cb 17
Westwood Cl. HA4: Ruis....30R 44
Westwood Cl. HP6: L Chal....11A 24
Westwood Cl. KT10: Esh....76Fa 152
Westwood Cl. RH2: Reig....6G 206
Westwood Ct. EN1: Enf....16Ub 33
....(off Village Rd.)
Westwood Ct. HA0: Wemb....35Ka 66
Westwood Ct. UB6: G'frd....36Fa 66
Westwood Gdns. SW13....55Va 110
Westwood Hill SE26....64Wb 135
Westwood Ho. W12....46Ya 88
....(off Wood La.)
Westwood La. DA15: Sidc....57Wc 117
Westwood La. DA16: Well....55Vc 117
Westwood M. E3....41Cc 92
....(off Addington Rd.)
Westwood Pk. SE23....59Xb 113
Westwood Pk. Trad. Est. W3....43Ra 87
Westwood Pl. SE26....63Wb 135
Westwood Rd. CR5: Coul....90Mb 176
Westwood Rd. DA13: Sflt....66Ae 143
Westwood Rd. E16....46Kc 93
Westwood Rd. GU20: W'sham....5C 146
Westwood Rd. IG3: Ilf....32Vc 75
Westwood Rd. SW13....55Va 110
West Woodside DA5: Bexl....59Ad 117
Westwood Way TN13: S'oaks....94Hd 202
West Yoke TN15: Ash....75Zd 165
WEST YOKE....76Ae 165
West Yoke Rd. DA3: Nw A G....76Ae 165
Wetheral Dr. HA7: Stan....25Ka 46
Wetherall M. AL1: St A....3C 6
Wetherby Cl. UB5: N'olt....37Da 65
Wetherby Gdns. SW5....49Eb 89
Wetherby Mans. SW5....50Db 89
....(off Earls Ct. Sq.)
Wetherby M. SW5....50Db 89
Wetherby Pl. SW7....6A 226 (49Eb 89)
Wetherby Rd. EN2: Enf....11Sb 33
Wetherby Rd. WD6: Bore....11Na 29
Wetherby Way KT9: Chess....80Na 153
Wetherden St. E17....31Bc 72
Wethered Dr. SL1: Burn....3A 80
Wetherell Rd. E9....39Zb 72
Wetherill Rd. N10....25Jb 50
Wettern Cl. CR2: Sande....82Ub 177
Wetton Ct. TW20: Egh....64B 126
....(off Wetton Pl.)
Wetton Pl. TW20: Egh....64B 126
Wevco Wharf SE15....51Xb 113
Wexfenne Gdns. GU22: Pyr....88K 169
Wexford Ho. E1....43Yb 92
....(off Sidney St.)
Wexford Rd. SW12....59Hb 111
WEXHAM....2M 81
WEXHAM COURT....4N 81
Wexham Lodge SL2: Wex....3M 81
Wexham Pk. Golf Course....10N 61
WEXHAM PARK HOSPITAL....2N 81
Wexham Pk. La. SL3: Wex....2N 81
Wexham Pl. SL2: Wex....7P 61
Wexham Rd. SL1: Slou....7L 81
Wexham Rd. SL2: Slou....4M 81
Wexham Rd. SL2: Wex....4M 81
Wexham Springs SL3: Wex....8P 61
Wexham St. SL2: Stoke P....2M 81
Wexham St. SL3: Stoke P....1M 81
Wexham St. SL3: Wex....1M 81
WEXHAM STREET....10N 61
Wexham Woods SL3: Wex....3N 81
Wexner Bldg. E1....1K 225 (43Vb 91)
....(off Strype St.)
Wey Av. KT16: Chert....69J 127
Weybank GU23: Wis....88N 169
Wey Barton KT14: Byfl....85P 169
Weybourne Pl. CR2: Sande....82Tb 177
Weybourne St. SW18....61Eb 133
Weybourne Way KT15: New H....81L 169
WEYBRIDGE....77Q 150
Weybridge Bus. Pk. KT15: Add....77N 149
Weybridge Ct. SE16....50Xb 91
Weybridge Ho. KT13: Weyb....78T 150
Weybridge Lawn Tennis Club....75R 150
Weybridge Library....77Q 150
Weybridge Pk. KT13: Weyb....78Q 150
Weybridge Point SW11....54Hb 111
Weybridge Rd. CR7: Thor H....70Qb 134
Weybridge Rd. KT13: Weyb....77P 149
Weybridge Rd. KT15: Add....77N 149
Weybridge Sailing Club....75R 150
Weybridge Station (Rail)....79Q 150
Wey Cl. KT14: W Byf....85K 169
Wey Ct. GU22: Wok....91A 188
....(off Claremont Av.)
Wey Ct. KT15: New H....81M 169
Wey Ct. KT19: Ewe....77Sa 153
Weydown Cl. SW19....60Ab 110
Weyhill Rd. E1....44Wb 91
Wey Ho. NW8....6C 214 (42Fb 89)
....(off Church St. Est.)
Wey Ho. UB5: N'olt....42Ba 85
....(off Taywood Rd.)
Weylands Cl. KT12: Walt T....74Ba 151
Weylands Ct. KT15: Add....77M 149
....(off Corrie Rd.)
Weylands Pk. KT13: Weyb....79T 150
Weylond Rd. RM8: Dag....34Bd 75
Wey Mnr. Rd. KT15: New H....81M 169
Weyman Rd. SE3....53Lc 115
Weymarks, The N17....23Tb 51
Weymead Cl. KT16: Chert....74L 149
Wey Mdws. KT13: Weyb....78N 149
Weymede KT14: Byfl....84P 169
Weymouth Av. NW7....22Ua 48
Weymouth Av. W5....48La 86
Weymouth Cl. E6....44Rc 94
Weymouth Ct. E2....1K 219 (40Vb 71)
Weymouth Ct. SM2: Sutt....80Cb 155
Weymouth Dr. RM16: Chaf H....50Zd 99
Weymouth Ho. BR2: Broml....68Hc 137
....(off Hill Ho. M.)
Weymouth Ho. SW8....52Pb 112
....(off Bolney St.)
Weymouth M. W1....7K 215 (43Kb 90)
Weymouth Pl. SE2....48Wc 95
Weymouth Rd. UB4: Hayes....41U 84
Weymouth St. HP3: Hem H....6M 3
Weymouth St. W1....1J 221 (43Jb 90)
WEYMOUTH STREET HOSPITAL....7J 215 (43Jb 90)
....(off Weymouth St.)
Weymouth Ter. E2....1K 219 (40Vb 71)
Weymouth Vs. N4....33Pb 70
....(off Moray Rd.)
Weymouth Wlk. HA7: Stan....23Ja 46
Wey Retail Pk....84N 169
Wey Rd. KT13: Weyb....76P 149
Weyside Cl. KT14: Byfl....84P 169
Weystone Rd. KT13: Weyb....77P 149
Weyver Ct. AL1: St A....1C 6
....(off Avenue Rd.)
Weyview Ct. KT15: New H....80L 149
Whadcoat St. N4....33Qb 70
Whaddon Ho. SE22....55Ub 113
Whalebone Av. RM6: Chad H....30Bd 55
Whalebone Ct. EC2....2F 225 (44Tb 91)
....(off Telegraph St.)
Whalebone Gro. RM6: Chad H....30Bd 55
Whalebone La. E15....38Gc 73
Whalebone La. Nth. RM6: Chad H....24Ad 55
Whalebone La. Nth. RM6: Col R....24Ad 55
Whalebone La. Sth. RM6: Chad H....31Bd 75
Whalebone La. Sth. RM6: Dag....31Bd 75
Whalebone La. Sth. RM8: Dag....31Bd 75
Whales Yd. E15....38Gc 73
....(off West Ham La.)
Whaley Rd. EN6: Pot B....5Eb 17
Wharf, The EC3....6K 225 (46Vb 91)
Wharf, The KT13: Weyb....75Q 150
Wharf Cl. SS17: Stan H....2M 101
Wharfdale Cl. N11....23Jb 50
Wharfdale Rd. N1....1G 217 (40Nb 70)
Wharfedale Ct. E5....35Zb 72
Wharfedale Gdns. CR7: Thor H....70Pb 134
Wharfedale Ho. NW6....39Db 69
....(off Kilburn Vale)
Wharfedale Rd. DA2: Dart....60Sd 120
Wharfedale St. SW10....50Db 89
Wharfedale Yd. N1....1G 217 (40Nb 70)
....(off Wharfedale Rd.)
Wharf Ho. DA8: Erith....50Gd 96
....(off West St.)
Wharf La. E14....44Bc 92
Wharf La. GU23: Rip....90M 169
Wharf La. GU23: Send....95E 188
Wharf La. TW1: Twick....60Ja 108
Wharf La. WD3: Rick....18N 25
Wharf Mill Apts. E2....1K 219 (39Vb 71)
....(off Laburnum St.)
Wharf Pl. E2....39Xb 71
Wharf Rd. N1: City Rd....1D 218 (40Sb 71)
Wharf Rd. N1: York Way....1F 217 (39Nb 70)
Wharf Rd. CM14: B'wood....20Yd 40
Wharf Rd. DA12: Grav'nd....8G 122
Wharf Rd. E16....46Lc 93
Wharf Rd. EN3: Pond E....16Ac 34
Wharf Rd. HP1: Hem H....4K 3
Wharf Rd. RM17: Grays....51Be 121
Wharf Rd. SS17: Stan H....2M 101
Wharf Rd. TW19: Wray....9N 103
Wharf Rd. Ind. Est. EN3: Pond E....16Ac 34
Wharf Rd. Sth. RM17: Grays....51Be 121
Wharfside Cl. DA8: Erith....50Hd 96
Wharfside Point Nth. E14....45Ec 92
....(off Poplar High St.)
Wharfside Point Sth. E14....45Ec 92
Wharfside Rd. E16....43Gc 93
Wharf St. E16....43Gc 93
Wharf St. SE8....50Cc 92
Wharf Vw. Ct. E14....44Ec 92
....(off Blair St.)
Wharf Vs. HP1: Hem H....4K 3
Wharf Way WD4: Hunt C....5S 12
Wharncliffe Dr. UB1: S'hall....46Fa 86
Wharncliffe Gdns. SE25....68Ub 135
Wharncliffe M. SW4....58Mb 112
Wharncliffe Rd. SE25....68Ub 135
Wharneclifffe DA9: Ghithe....58Xd 120
Wharton Cl. NW10....37Ua 68
Wharton Cotts. WC1....4K 217 (41Qb 90)
Wharton Ho. E2....41Zb 92
Wharton Ho. SE1....3K 231 (48Vb 91)
....(off St Saviour's Est.)
Wharton Rd. BR1: Broml....67Kc 137
Wharton St. WC1....4J 217 (41Pb 90)
Whatcote Cotts. TN15: Plat....92Ee 205
Whatcott's Yd. N16....35Ub 71
Whateley Rd. SE20....66Zb 136
Whateley Rd. SE22....57Vb 113
Whatley Av. SW20....69Za 132
Whatman Ho. E14....44Bc 92
....(off Wallwood St.)
Whatman Rd. SE23....59Zb 114
Whatmore Cl. TW19: Stanw M....58J 105
Whattingdon Rd. CR5: Coul....89Lb 176
Wheatash Rd. KT15: Add....75K 149
Wheatbutts, The SL4: Eton W....9D 80
Wheatcroft EN7: Chesh....1Xb 19
Wheatcroft Ct. SM1: Sutt....74Db 155
....(off Cleeve Way)
Wheatfield Ho. NW6....41Cb 89
....(off Kilburn Pk. Rd.)
Wheatfields CM14: W'ley....21Yd 58
Wheatfields E6....44Rc 94
Wheatfields EN3: Enf H....11Ac 34
Wheatfields Ct. EN9: Walt A....6Jc 21
....(off Farthingale La.)
Wheatfield Way KT1: King T....68Na 131
Wheathill Ho. SE20....68Xb 135
....(off Penge Rd.)
Wheathill Rd. SE20....69Xb 135
Wheat Knoll CR8: Kenley....88Sb 177
Wheatland Ho. SE22....55Ub 113
Wheatlands TW5: Hest....51Ca 107
Wheatlands Rd. SL3: Slou....8M 81
Wheatlands Rd. SW17....62Jb 134
Wheatley Cl. DA9: Ghithe....57Wd 120
Wheatley Cl. NW4....26Wa 48
Wheatley Cl. RM11: Horn....29Md 57
Wheatley Ct. E3....41Dc 92
....(off Bruce Rd.)
Wheatley Cres. UB3: Hayes....45W 84
Wheatley Dr. WD25: Wat....6Y 13
Wheatley Gdns. N9....19Ub 33
Wheatley Ho. SW15....59Wa 110
....(off Ellisfield Dr)
Wheatley Mans. IG11: Bark....38Wc 75
....(off Lansbury Av.)
Wheatley M. KT8: E Mos....70Fa 130
Wheatley Rd. TW7: Isle....55Ha 108
Wheatley's Eyot TW16: Sun....71W 150
Wheatley St. W1....1J 221 (43Jb 90)
Wheatley Ter. Rd. DA8: Erith....51Hd 118
Wheatley Way SL9: Chal P....23A 42
Wheat Sheaf Cl. E14....49Dc 92
Wheatsheaf Cl. GU21: Wok....88A 168
Wheatsheaf Cl. KT16: Ott....79F 148
Wheatsheaf Cl. UB5: N'olt....36Aa 65
Wheatsheaf Hill TN14: Hals....81Bd 181
Wheatsheaf La. SW6....52Ya 110
Wheatsheaf La. SW8....52Nb 112
Wheatsheaf La. TW18: Staines....66H 127
Wheatsheaf Pde. SL4: Old Win....7L 103
....(off St Luke's Rd.)
Wheatsheaf Pk....67J 127
Wheatsheaf Rd. RM1: Rom....30Hd 56
Wheatsheaf Ter. SW6....52Bb 111
Wheatstone Cl. CR4: Mitc....67Gb 133
Wheatstone Cl. SL3: Slou....8L 81
Wheatstone Ho. SE1....4E 230 (48Sb 91)
....(off County St.)
Wheatstone Rd. DA8: Erith....50Fd 96
Wheeler Av. RH8: Oxt....1H 211
Wheeler Cl. DA1: Dart....57Pd 119
Wheeler Cl. IG8: Wfd G....23Pc 54
Wheeler Gdns. N1....39Nb 70
....(off Outram Pl.)
Wheeler Pl. BR2: Broml....70Kc 137
Wheelers CM16: Epp....1Vc 23
Wheelers Cross IG11: Bark....40Tc 74
Wheelers Dr. HA4: Ruis....30S 44
Wheelers La. CM14: N'side....14Rd 39
Wheelers La. CM14: Pil H....14Rd 39
Wheelers La. HP3: Hem H....4N 3
Wheelers La. KT18: Eps....86Ra 173
Wheelers Orchard SL9: Chal P....23A 42
Wheel Farm Dr. RM10: Dag....34Ed 76
Wheel Ho. E14....50Dc 92
....(off Burrells Wharf Sq.)
Wheel Ho. E17....27Zb 52
Wheelock Cl. DA8: Erith....52Dd 118
Wheelwright Cl. WD23: Bush....16Da 27
Wheelwrights TN15: Plax....100Ae 205
Wheelwrights Pl. SL3: Coln....52E 104
Wheelwright St. N7....38Pb 70
Whelan Rd. W3....48Ra 87
Whelan Way SM6: Bedd....76Mb 156
Wheler Ho. E1....6K 219 (42Vb 91)
....(off Quaker St.)
Wheler St. E1....6K 219 (42Vb 91)
Whellock Rd. W4....48Ua 88
WHELPLEY HILL....8A 2
Whelpley Hill Pk. HP5: Whel H....8A 2
Whenman Av. DA5: Bexl....61Ed 140
Whernside Cl. SE28....45Yc 95
WHETSTONE....19Eb 31
Whetstone Cl. N20....19Fb 31
Whetstone Pk. WC2....2H 223 (44Pb 90)
Whetstone Rd. SE3....54Lc 115
Whewell Rd. N19....33Nb 70
Whidborne Bldgs. WC1....4G 217 (41Nb 90)
....(off Whidborne St.)
Whidborne Cl. SE8....54Cc 114
Whidborne St. WC1....4G 217 (41Nb 90)
....(not continuous)
Whidbourne M. SW8....53Mb 112
Whiffins Orchard CM16: Coop....1Zc 23
Whimbrel Cl. CR2: Sande....83Tb 177
Whimbrel Cl. SE28....45Yc 95
Whimbrel Way UB4: Yead....44Z 85
Whinchat Rd. SE28....48Tc 94
Whinfell Cl. SW16....64Mb 134
Whinfell Way DA12: Grav'nd....3H 145
Whinshill Ct. SL5: S'dale....4E 146
Whinyates Rd. SE9....55Nc 116
Whippendell Cl. BR5: St P....67Xc 139
Whippendell Hill WD4: Chfd....2L 11
Whippendell Rd. WD18: Wat....15U 26
Whippendell Way BR5: St P....67Xc 139
Whippingham Ho. E3....41Bc 92
....(off Merchant St.)
Whipps Cross E17....29Fc 53
Whipps Cross Ho. E17....29Fc 53
....(off Wood St.)
Whipps Cross Rd. E11....29Fc 53
....(not continuous)
WHIPPS CROSS UNIVERSITY HOSPITAL....29Fc 53
Whiskin St. EC1....4B 218 (41Rb 91)
Whisper Wood WD3: Loud....13K 25
Whisperwood Cl. HA3: Hrw W....25Ga 46
Whistler Ct. WD25: Wat....8Z 13
Whistler Gdns. HA8: Edg....26Pa 47
Whistler M. RM8: Dag....36Xc 75
....(off Fitzstephen Rd.)
Whistler M. SE15....52Vb 113
Whistlers Av. SW11....52Fb 111
Whistlers Gro. DA15: Sidc....59Uc 116
Whistler St. N5....36Rb 71
Whistler Twr. SW10....52Eb 111
....(off Worlds End Est.)
Whistler Wlk. SW10....52Fb 111
Whiston Ho. N1....38Rb 71
....(off Richmond Gro.)
Whiston Rd. E2....1K 219 (40Vb 71)
Whitacre M. SE11....50Qb 90
Whitakers Lodge EN2: Enf....11Tb 33
Whitakers Way IG10: Lough....11Pc 36
Whitbread Cl. N17....25Wb 51
Whitbread Ho. SW11....52Mb 112
....(off Charles Clowes Wlk.)
Whitbread Pl. CM14: B'wood....20Yd 40
....(off Rollason Way)
Whitbread Rd. SE4....56Ac 114
Whitburn Rd. SE13....56Dc 114
Whitby Av. CM13: Ingve....23Fe 59
Whitby Av. NW10....41Ra 87
Whitby Cl. DA9: Ghithe....57Wd 120
Whitby Cl. TN16: Big H....91Kc 199
Whitby Ct. N7....35Nb 70
Whitby Gdns. NW9....27Qa 47
Whitby Gdns. SM1: Sutt....75Fb 155
Whitby Ho. NW8....39Eb 69
....(off Boundary Rd.)
Whitby Pde. HA4: Ruis....33Y 65
Whitby Rd. HA2: Harr....34Ea 66
Whitby Rd. HA4: Ruis....34X 65
Whitby Rd. SE18....49Pc 94
Whitby Rd. SL1: Slou....5G 80
Whitby Rd. SM1: Sutt....75Fb 155
Whitby Rd. Bus. Cen. SL1: Slou....5G 80
Whitby St. E1....5K 219 (42Vb 91)
....(not continuous)
Whitcher Cl. SE14....51Ac 114
Whitcher Pl. NW1....37Lb 70
Whitchurch Av. HA8: Edg....24Pa 47
Whitchurch Cl. HA8: Edg....23Pa 47
Whitchurch Gdns. HA8: Edg....23Pa 47
Whitchurch Ho. W10....44Za 88
....(off Kingsdown Cl.)
Whitchurch La. HA8: Edg....24Ma 47
Whitchurch Pde. HA8: Edg....24Qa 47
Whitchurch Rd. RM3: Rom....21Md 57
Whitchurch Rd. W11....45Za 88
Whitchurch Ter. SL9: Chal P....22B 42
Whitcomb Ct. WC2....5E 222 (45Mb 90)
....(off Whitcomb St.)
Whitcombe M. TW9: Kew....53Ra 109
Whitcomb St. WC2....5E 222 (45Mb 90)
Whiteadder M. TW9: Kew....53Ra 109
Whiteadder Way E14....49Dc 92
Whitear Wlk. E15....37Ec 73
White Av. DA11: Nflt....2B 144
Whitebarn La. RM10: Dag....39Cd 76
Whitebeam Av. BR2: Broml....73Oc 160
Whitebeam Cl. KT17: Eps D....87Xa 174
Whitebeam Cl. SW9....52Pb 112
Whitebeam Cl. TN15: Kems'g....89Qd 183
Whitebeam Cl. UB8: Uxb....38L 63
Whitebeam Cl. WD7: Shenl....5Pa 15
Whitebeam Dr. RH2: Reig....9K 207
Whitebeam Dr. RM15: S Ock....41Yd 98
Whitebeam Ho. E15....41Gc 93
....(off Teasel Way)
Whitebeam Pl. CR3: Whyt....89Wb 177
White Beams AL2: Park....10A 6
Whitebeams AL10: Hat....3C 8
White Beam Way KT20: Tad....93Wa 194
White Bear Pl. NW3....35Fb 69
White Bear Yd. EC1....6K 217 (42Qb 90)
....(off Clerkenwell Rd.)
White Bri. Av. CR4: Mitc....69Fb 133
Whitebridge Cl. TW14: Felt....58V 106
Whitebroom Rd. HP1: Hem H....1G 2
WHITE BUSHES....10A 208
Whitebushes RH1: Redh....10A 208
White Butts Rd. HA4: Ruis....34Z 65
WHITECHAPEL....43Wb 91
Whitechapel Art Gallery....44Vb 91
Whitechapel High St. E1....2K 225 (44Vb 91)
Whitechapel Rd. E1....43Wb 91
Whitechapel Sports Cen....43Xb 91
Whitechapel Station (Underground, Overground & Crossrail)....43Xb 91
White Church La. E1....44Wb 91
White Church Pas. E1....44Wb 91
....(off White Church La.)
WHITE CITY....44Ya 88
WHITE CITY....45Xa 88
White City Cl. W12....45Ya 88
White City Est. W12....45Xa 88
White City Rd. W12....45Ya 88
White City Station (Underground)....45Ya 88
Whitecliff DA10: Swans....59Ae 121
White Cl. SL1: Slou....6H 81
White Collar Factory EC1....5F 219 (42Tb 91)
....(off City Rd.)
White Conduit St. N1....1A 218 (40Qb 70)
Whitecote Rd. UB1: S'hall....44Ea 86
White Craig Cl. HA5: Hat E....22Ca 45
Whitecroft AL1: St A....5F 6
Whitecroft BR8: Swan....68Gd 140
Whitecroft Cl. BR3: Beck....70Fc 137
Whitecroft Way BR3: Beck....71Ec 158
Whitecross Pl. EC2....7G 219 (43Tb 91)
Whitecross St. EC1....5E 218 (42Sb 91)
Whitefield Av. CR8: Purl....88Qb 176
Whitefield Av. NW2....32Ya 68
Whitefield Cl. BR5: St P....69Yc 139
Whitefield Cl. SW15....58Ab 110
Whitefields CR3: Cat'm....93Ub 197
Whitefields Rd. EN8: Chesh....1Yb 20
Whitefoot La. BR1: Broml....63Ec 136
Whitefoot La. SE6....63Ec 136
Whitefoot Ter. BR1: Broml....62Hc 137
Whiteford Rd. SL2: Slou....3J 81
White Friars TN13: S'oaks....99Jd 202
Whitefriars Av. HA3: W'stone....26Ga 46
Whitefriars Close KT21: Asht....91Na 193
Whitefriars Ct. N12....22Fb 49
Whitefriars Dr. HA3: Hrw W....26Fa 46
Whitefriars St. EC4....3A 224 (44Qb 90)
Whitefriars Trad. Est. HA3: W'stone....27Fa 46
White Gables Ct. CR2: S Croy....78Ub 157
White Gdns. RM10: Dag....37Cd 76
White Ga. GU22: Wok....92B 188
Whitegate Gdns. HA3: Hrw W....24Ha 46
White Gates KT7: T Ditt....73Ja 152
White Gates RM12: Horn....33Ld 77
Whitegates CR3: W'ham....91Wb 197
Whitegates Av. TN15: W King....79Ud 164
Whitegates Cl. WD3: Crox G....14Q 26
Whitegate Way KT20: Tad....92Xa 194
White Hall RM4: Abr....13Xc 37
Whitehall....79Ab 154
Whitehall SW1....7F 223 (46Nb 90)
Whitehall Cl. IG7: Chig....20Wc 37
Whitehall Cl. UB8: Uxb....39L 63
Whitehall Cl. WD6: Bore....14Qa 29
Whitehall Ct. SW1....7F 223 (46Nb 90)
....(not continuous)
Whitehall Cres. KT9: Chess....78Ma 153
Whitehall Farm La. GU25: Vir W....68A 126
Whitehall Gdns. E4....18Gc 35
Whitehall Gdns. SW1....7F 223 (46Nb 90)
....(off Horseguards Av.)
Whitehall Gdns. W3....46Qa 87
Whitehall Gdns. W4....51Ra 109
Whitehall La. DA8: Erith....54Hd 118
Whitehall La. IG9: Buck H....19Jc 35
Whitehall La. RH2: Reig....10H 207
Whitehall La. RM17: Grays....50Ee 99
Whitehall La. TW19: Wray....58C 104
Whitehall La. TW20: Egh....66B 126
Whitehall Lodge N10....26Jb 50
Whitehall Pde. DA12: Grav'nd....2E 144
Whitehall Pk. N19....32Lb 70
Whitehall Pk. Rd. W4....51Ra 109
Whitehall Pl. E7....36Jc 73
Whitehall Pl. SM6: W'gton....77Kb 156
Whitehall Pl. SW1....7F 223 (46Nb 90)
Whitehall Rd. BR2: Broml....71Mc 159
Whitehall Rd. CR7: Thor H....71Qb 156
Whitehall Rd. E4....19Gc 35
Whitehall Rd. HA1: Harr....31Ga 66
Whitehall Rd. IG8: Wfd G....19Hc 35
Whitehall Rd. RM17: Grays....49Ee 99
Whitehall Rd. UB8: Uxb....39M 63
Whitehall Rd. W7....47Ja 86
Whitehall St. N17....24Vb 51
White Hart Av. SE18....49Vc 95
White Hart Av. SE28....48Vc 95
White Hart Cl. GU23: Rip....93L 189
White Hart Cl. TN13: S'oaks....100Ld 203
White Hart Cl. UB3: Harl....51T 106
White Hart Ct. EC2....1H 225 (43Ub 91)
....(off Liverpool St.)
White Hart Ct. EN8: Walt C....5Ac 20
White Hart Dr. HP2: Hem H....3P 3
White Hart La. N17....24Sb 51
White Hart La. N22....25Pb 50
White Hart La. NW10....37Va 68
White Hart La. RM7: Col R....25Cd 56
White Hart La. RM7: Mawney....25Cd 56
White Hart La. SW13....54Ua 110
White Hart Lane Station (Overground)....24Vb 51
White Hart Mdws. GU23: Rip....93L 189
White Hart Pde. TN13: Riv....94Gd 202
White Hart Rd. BR6: Orp....73Wc 161
White Hart Rd. HP2: Hem H....3A 4
White Hart Rd. SE18....49Uc 94
White Hart Rd. SL1: Slou....8H 81
WHITE HART RDBT. Greenford....40Z 65
White Hart Row KT16: Chert....73J 149
White Hart Slip BR1: Broml....68Jc 137

White Hart St. EC4 ... 2C 224 (44Rb 91)
White Hart St. SE11 ... 7A 230 (50Qb 90)
White Hart Triangle SE28 ... 47Vc 95
White Hart Triangle Bus. Pk. SE28 ... 47Wc 95
White Hart Yd. DA11: Grav'nd ... 8D 122
(off High St.)
White Hart Yd. SE1 ... 7F 225 (46Tb 91)
Whitehaven SL1: Slou ... 5K 81
Whitehaven Cl. BR2: Broml ... 70Jc 137
Whitehaven Cl. EN7: G Oak ... 1Ub 19
Whitehaven Dr. KT23: Bookh ... 98Da 191
Whitehaven St. NW8 ... 6D 214 (42Gb 89)
Whitehead Cl. DA2: Wilm ... 62Ld 141
Whitehead Cl. N18 ... 22Tb 51
Whitehead Cl. SW18 ... 59Eb 111
Whiteheads Gro. SW3 ... 6E 226 (49Gb 89)
White Heart Av. UB8: Hil ... 43S 84
Whiteheath Av. HA4: Ruis ... 31S 64
White Heather Ho. WC1 ... 4G 217 (41Nb 90)
(off Cromer St.)
White Hedge Dr. AL3: St A ... 1A 6
White Hermitage SL4: Old Win ... 7N 103
White Heron M. TW11: Tedd ... 65Ha 130
White Hill CR2: Sande ... 82Tb 177
White Hill CR5: Chips ... 95Gb 195
White Hill GU20: W'sham ... 7A 146
White Hill HP1: Hem H ... 3H 3
White Hill TN15: Wro ... 88De 185
White Hill WD3: Rick ... 23Q 44
White Hill Cl. CR3: Cat'm ... 97Ub 197
White Hill La. RH1: Blet ... 99Tb 197
Whitehill La. DA12: Grav'nd ... 2E 144
Whitehill La. GU23: Ock ... 95S 190
Whitehill Pl. GU25: Vir W ... 71A 148
Whitehill Rd. DA1: Cray ... 57Jd 118
Whitehill Rd. DA12: Grav'nd ... 1E 144
Whitehill Rd. DA13: Sflt ... 67Zd 143
Whitehill Rd. DA3: Dart ... 68Zd 143
Whitehill Rd. DA3: Lfield ... 68Zd 143
Whitehills Rd. IG10: Lough ... 13Qc 36
White Horse All. EC1 ... 7B 218 (43Rb 91)
(off Cowcross St.)
White Horse Apts. N1 ... 39Qb 70
(off Liverpool Rd.)
White Horse Dr. KT18: Eps ... 86Sa 173
White Horse Hill BR7: Chst ... 63Qc 138
White Horse La. AL2: Lon C ... 8H 7
White Horse La. E1 ... 43Zb 92
White Horse La. GU23: Rip ... 93L 189
Whitehorse La. SE25 ... 70Tb 135
Whitehorse M. EN3: Enf H ... 12Zb 34
Whitehorse M. SE1 ... 3A 230 (48Qb 90)
White Horse Rd. E1 ... 44Ac 92
(not continuous)
White Horse Rd. E6 ... 41Pc 94
White Horse Rd. SL4: Wind ... 5B 102
Whitehorse Rd. CR0: C'don ... 73Sb 157
Whitehorse Rd. CR7: Thor H ... 71Tb 157
White Horse St. W1 ... 7A 222 (46Kb 90)
White Horse Yd. EC2 ... 2F 225 (44Tb 91)
White Ho., The NW1 ... 5A 216 (42Kb 90)
(off Albany St.)
White Ho. CR0: C'don ... 77Tb 157
(off Coombe Rd.)
White Ho. SW11 ... 53Fb 111
White Ho. SW4 ... 59Mb 112
(off Clapham Pk. Est.)
Whitehouse E10 ... 30Ec 52
(off Leyton Grn. Rd.)
Whitehouse Apts. SE1 ... 7J 223 (46Pb 90)
Whitehouse Av. WD6: Bore ... 13Ra 29
White Ho. Cl. SL9: Chal P ... 24A 42
White Ho. Ct. N14 ... 19Nb 32
White Ho. Dr. HA7: Stan ... 21La 46
White Ho. Dr. IG8: Wfd G ... 23Hc 53
White Ho. La. EN2: Enf ... 11Sb 33
Whitehouse La. WD5: Bedm ... 8H 5
White Ho. M. E10 ... 30Ec 52
White Ho. Rd. EN5: Barn ... 7Cb 17
Whitehouse Way N14 ... 19Kb 32
Whitehouse Way SL0: Iver H ... 41F 82
Whitehouse Way SL3: L'ly ... 8P 81
White Kennett St. E1 ... 2K 225 (44Vb 91)
White Knights Rd. KT13: Weyb ... 80S 150
White Knobs Way CR3: Cat'm ... 97Wb 197
Whitelands Av. WD3: Chor ... 13D 24
Whitelands Cres. SW18 ... 59Ab 110
Whitelands Ho. SW3 ... 7F 227 (50Hb 89)
(off Cheltenham Ter.)
Whitelands Way RM3: Hrld W ... 25Md 57
White La. RH8: T'sey ... 95Kc 199
White La. TN16: T'sey ... 95Kc 199
White La. TN16: Tats ... 95Kc 199
Whiteleaf Rd. HP3: Hem H ... 5L 3
Whiteledges W13 ... 44La 86
Whitelegg Rd. E13 ... 40Hc 73
Whiteley SL4: Wind ... 2C 102
Whiteley Rd. SE19 ... 64Tb 135
Whiteleys Cen. ... 44Db 89
Whiteleys Pde. UB10: Hil ... 42R 84
Whiteley's Way TW13: Hanw ... 62Ca 129
WHITELEY VILLAGE ... 81U 170
White Lillies Island SL4: Wind ... 2E 102
White Lion Cen. ... 2K 217 (40Qb 70)
(off White Lion St.)
White Lion Ct. EC3 ... 3H 225 (44Ub 91)
(off Cornhill)
White Lion Ct. SE15 ... 51Yb 114
White Lion Ct. TW7: Isle ... 55Ka 108
White Lion Ga. KT11: Cobh ... 86W 170
White Lion Hill EC4 ... 4C 224 (45Rb 91)
White Lion St. HP3: Hem H ... 6M 3
White Lion St. N1 ... 2K 217 (40Qb 70)
White Lodge KT21: Asht ... 92Na 193
White Lodge SE19 ... 66Rb 135
White Lodge TN13: S'oaks ... 95Jd 203
White Lodge W5 ... 43La 86
White Lodge Cl. KT20: Tad ... 95Ya 194
White Lodge Cl. N2 ... 30Fb 49
White Lodge Cl. SM2: Sutt ... 80Eb 155
White Lodge Cl. TN13: S'oaks ... 95Kd 203
White Lodge Cl. TW7: Isle ... 54Ja 108
White Lodge Ct. TW16: Sun ... 67Y 129
White Lyon Ct. EC2 ... 6D 218 (42Sb 91)
(off Fann St.)
White Lyons Rd. CM14: B'wood ... 19Yd 40
White Oak Ct. BR8: Swan ... 69Gd 140
Whiteoak Ct. BR7: Chst ... 65Qc 138
White Oak Dr. BR3: Beck ... 68Ec 136
White Oak Gdns. DA15: Sidc ... 59Vc 117
White Oak Leisure Cen. ... 68Fd 140
Whiteoaks SM7: Bans ... 85Db 175
Whiteoaks La. UB6: G'frd ... 41Fa 86
White Oak Sq. BR8: Swan ... 69Gd 140
(off London Rd.)
White Orchards HA7: Stan ... 22Ja 46
White Orchards N20 ... 18Bb 31
White Pillars GU22: Wok ... 2M 187
WHITE POST ... 5L 209
White Post Hill DA4: Farni ... 73Qd 163
Whitepost Hill RH1: Redh ... 6N 207
(not continuous)
White Post La. DA13: Sole S ... 10D 144
White Post La. E9 ... 38Bc 72
Whitepost La. RM13: Rain ... 40Hd 76
White Post St. SE15 ... 52Yb 114
White Rd. E15 ... 38Gc 73
White Rose Ct. E1 ... 1J 225 (43Ub 91)
(off Widegate St.)
White Rose La. GU22: Wok ... 90B 168
White Rose Lane Local Nature Reserve ... 91D 188
Whiterose Trad. Est. EN4: E Barn ... 15Fb 31
(off Margaret Rd.)
Whites Av. IG2: Ilf ... 30Uc 54
Whites Cl. DA9: Ghithe ... 58Yd 120
Whites Grounds SE1 ... 2J 231 (47Ub 91)
White's Grounds Est. SE1 ... 1J 231 (47Ub 91)
(off White's Grounds)
White Shack La. WD3: Chan C ... 9P 11
Whites La. GU24: Pirb ... 6E 186
Whites La. SL3: Dat ... 1M 103
White's Mdw. BR1: Broml ... 70Qc 138
White's Row E1 ... 1K 225 (43Vb 91)
Whites Sq. SW4 ... 56Mb 112
Whitestile Rd. TW8: Bford ... 50La 86
Whitestone Cl. EN4: Had W ... 10Gb 17
Whitestone La. NW3 ... 34Eb 69
Whitestone Wlk. NW3 ... 34Eb 69
Whitestone Way CR0: Wadd ... 75Qb 156
White Swan M. W4 ... 50Ua 88
Whitethorn Av. CR5: Coul ... 87Jb 176
Whitethorn Av. UB7: Yiew ... 45N 83
Whitethorn Gdns. CR0: C'don ... 75Xb 157
Whitethorn Gdns. EN2: Enf ... 15Tb 33
Whitethorn Gdns. RM11: Horn ... 30Ld 57
Whitethorn Ho. E1 ... 46Yb 92
(off Prusom St.)
Whitethorn Pas. E3 ... 42Cc 92
(off Whitethorn St.)
Whitethorn Pl. UB7: Yiew ... 46P 83
Whitethorn St. E3 ... 43Cc 92
White Tower, The ... 5K 225 (45Vb 91)
White Twr. Way E1 ... 43Ac 92
Whiteway KT23: Bookh ... 98Da 191
Whiteways Ct. TW18: Staines ... 66K 127
Whitewebbs Golf Course ... 9Tb 19
Whitewebbs La. EN2: Enf ... 7Ub 19
Whitewebbs Mus. of Transport ... 7Rb 19
Whitewebbs Pk. ... 8Sb 19
Whitewebbs Rd. EN2: Crew H ... 7Rb 19
Whitewebbs Rd. EN2: Enf ... 7Rb 19
Whitewebbs Way BR5: St P ... 67Vc 139
Whitewood Cotts. TN16: Tats ... 92Lc 199
Whitfield Ct. IG1: Ilf ... 31Pc 74
Whitfield Ct. SW20 ... 68Za 132
Whitfield Cres. DA2: Dart ... 58Sd 120
Whitfield Ho. NW8 ... 6D 214 (42Gb 89)
(off Salisbury St.)
Whitfield Pl. W1 ... 6B 216 (42Lb 90)
(off Whitfield St.)
Whitfield Rd. DA7: Bex ... 52Bd 117
Whitfield Rd. E6 ... 38Lc 73
Whitfields SS17: Stan H ... 1P 101
Whitfield St. W1 ... 6C 216 (42Lb 90)
Whitfield Way WD3: Rick ... 18H 25
Whitford Gdns. CR4: Mitc ... 69Hb 133
Whitgift Av. CR2: S Croy ... 78Rb 157
Whitgift Cen. ... 75Sb 157
Whitgift Ct. CR2: S Croy ... 78Sb 157
(off Nottingham Rd.)
Whitgift Ho. SE11 ... 5H 229 (49Pb 90)
Whitgift Ho. SW11 ... 53Gb 111
Whitgift Sq. CR0: C'don ... 75Sb 157
Whitgift St. CR0: C'don ... 76Sb 157
Whitgift St. SE11 ... 5H 229 (49Pb 90)
Whit Hern Ct. EN8: Chesh ... 2Yb 20
Whiting Av. IG11: Bark ... 38Rc 74
Whitings IG2: Ilf ... 29Uc 54
Whitings Rd. EN5: Barn ... 15Ya 30
Whitings Way E6 ... 43Qc 94
Whiting Way SE16 ... 49Ac 92
Whitland Rd. SM5: Cars ... 74Fb 155
Whitlars Dr. WD4: K Lan ... 10P 3
Whitley Cl. TW19: Stanw ... 58N 105
Whitley Cl. WD5: Ab L ... 4W 12
Whitley Ct. AL1: St A ... 2C 6
(off Hatfield Rd.)
Whitley Ho. SW1 ... 51Lb 112
(off Churchill Gdns.)
Whitley Rd. N17 ... 26Ub 51
WHITLEY ROW ... 100Dd 202
Whitlock Dr. SW19 ... 59Ab 110
Whitman Ho. E2 ... 41Yb 92
(off Cornwall Av.)
Whitman Rd. E3 ... 42Ac 92
Whitmead Cl. CR2: S Croy ... 79Ub 157
Whitmoor & Rickford Commons Local Nature Reserve ... 9L 187
WHITMOOR COMMON ... 8L 187
Whitmoor La. GU4: Sut G ... 8P 187
Whitmore Av. RM16: Grays ... 46De 99
Whitmore Av. RM3: Hrld W ... 26Nd 57
Whitmore Bldg. SE16 ... 3K 231 (48Vb 91)
(off Arts La.)
Whitmore Cl. N11 ... 22Kb 50
Whitmore Cl. RM16: Ors ... 4G 100
Whitmore Est. N1 ... 1J 219 (39Ub 71)
Whitmore Gdns. NW10 ... 40Ya 68
Whitmore Ho. N1 ... 1J 219 (39Ub 71)
(off Whitmore Est.)
Whitmore La. SL5: S'dale ... 1E 146
Whitmore La. SL5: S'hill ... 1E 146
Whitmore Rd. BR3: Beck ... 69Bc 136
Whitmore Rd. HA1: Harr ... 31Ea 66
Whitmore Rd. N1 ... 1H 219 (39Ub 71)
Whitmores Cl. KT18: Eps ... 87Sa 173
Whitmores Wood HP2: Hem H ... 1B 4
Whitnell Way SW15 ... 57Ya 110
(not continuous)
Whitney Av. IG4: Ilf ... 28Mc 53
Whitney Rd. E10 ... 31Dc 72
Whitney Wlk. DA14: Sidc ... 65Ad 139
Whitstable Cl. BR3: Beck ... 67Bc 136
Whitstable Cl. HA4: Ruis ... 33U 64
Whitstable Ho. W10 ... 44Za 88
(off Silchester Rd.)
Whitstable Pl. CR0: C'don ... 77Sb 157
Whitstone Av. W7 ... 44Ha 86
Whitstone La. BR3: Beck ... 71Dc 158
Whittaker Av. TW9: Rich ... 57Ma 109
Whittaker Ct. KT21: Asht ... 89Ma 173
Whittaker Pl. TW9: Rich ... 57Ma 109
(off Whittaker Av.)
Whittaker Rd. E6 ... 38Lc 73
Whittaker Rd. SL2: Slou ... 2B 80
Whittaker Rd. SM3: Sutt ... 76Bb 155
Whittaker St. SW1 ... 6H 227 (49Jb 90)
Whittaker Way SE1 ... 49Wb 91
Whitta Rd. E12 ... 35Mc 73
Whittell Gdns. SE26 ... 62Yb 136
Whittenham Cl. SL2: Slou ... 6L 81
Whittets Ait KT13: Weyb ... 75Q 150
Whittingham N17 ... 24Xb 51
Whittingham Ct. W4 ... 52Ua 110
Whittingstall Rd. SW6 ... 53Bb 111
Whittington Apts. E1 ... 44Zb 92
(off E. Arbour St.)
Whittington Av. EC3 ... 3H 225 (44Ub 91)
Whittington Av. UB4: Hayes ... 43V 84
Whittington Ct. N2 ... 29Hb 49
WHITTINGTON HOSPITAL ... 33Lb 70
Whittington Ho. N19 ... 33Mb 70
(off Holloway Rd.)
Whittington M. N12 ... 21Eb 49
Whittington Rd. CM13: Hut ... 16Ee 41
Whittington Rd. N22 ... 24Nb 50
Whittington Way HA5: Pinn ... 29Aa 45
Whittlebury Cl. SM5: Cars ... 80Hb 155
Whittlebury M. E. NW1 ... 38Jb 70
Whittlebury M. W. NW1 ... 38Jb 70
Whittle Cl. E17 ... 30Ac 52
Whittle Cl. UB1: S'hall ... 44Da 85
Whittle Cl. WD25: Wat ... 6V 12
Whittle Parkway SL1: Slou ... 4B 80
Whittle Rd. TW5: Hest ... 52Y 107
Whittle Rd. TW6: H'row A ... 55K 105
Whittle Rd. UB2: S'hall ... 47Da 85
Whittlesea Cl. HA3: Hrw W ... 24Ea 46
Whittlesea Path HA3: Hrw W ... 25Ea 46
Whittlesea Rd. HA3: Hrw W ... 24Ea 46
Whittlesey St. SE1 ... 7K 223 (46Qb 90)
WHITTON ... 59Ea 108
Whitton Av. E. UB6: G'frd ... 36Ga 66
Whitton Av. W. UB5: N'olt ... 36Da 65
Whitton Av. W. UB6: G'frd ... 36Fa 66
Whitton Cl. UB6: G'frd ... 37Ka 66
Whitton Dene TW3: Houn ... 57Ea 108
Whitton Dene TW3: Isle ... 57Ea 108
Whitton Dene TW7: Isle ... 58Fa 108
Whitton Dr. UB6: G'frd ... 37Ja 66
Whitton Mnr. Rd. TW7: Isle ... 58Ea 108
Whitton Rd. TW1: Twick ... 58Ha 108
Whitton Rd. TW2: Twick ... 58Ga 108
Whitton Rd. TW3: Houn ... 56Da 107
WHITTON ROAD RDBT. ... 58Ha 108
Whitton Sports & Fitness Cen. ... 61Da 129
Whitton Station (Rail) ... 59Ea 108
Whitton Wlk. E3 ... 41Cc 92
(not continuous)
Whitton Waye TW3: Houn ... 58Ca 107
Whitwell Ho. E13 ... 41Jc 93
Whitwell Rd. WD25: Wat ... 7Z 13
Whitworth Av. RM3: Rom ... 22Ld 57
Whitworth Cl. DA11: Nflt ... 61Ee 143
Whitworth Cres. EN3: Enf L ... 9Cc 20
Whitworth Ho. SE1 ... 4E 230 (48Sb 91)
Whitworth Pl. TW4: Houn ... 59Aa 107
Whitworth Rd. SE18 ... 52Qc 116
Whitworth Rd. SE25 ... 69Ub 135
Whitworth St. SE10 ... 50Gc 93
Whopshott Av. GU21: Wok ... 8N 167
Whopshott Cl. GU21: Wok ... 8N 167
Whopshott Dr. GU21: Wok ... 8N 167
Whorlton Rd. SE15 ... 55Xb 113
Whybrews SS17: Stan H ... 1P 101
Whybridge Cl. RM13: Rain ... 39Hd 76
Whychcote Point NW2 ... 32Ya 68
(off Whitefield Av.)
Whymark Av. N22 ... 27Qb 50
Whyteacre CR3: W'ham ... 92Xb 197
Whytebeam Vw. CR3: Whyt ... 90Vb 177
Whytecliffe Rd. Nth. CR8: Purl ... 83Rb 177
Whytecliffe Rd. Sth. CR8: Purl ... 83Qb 176
Whytecroft TW5: Hest ... 52Z 107
WHYTELEAFE ... 90Vb 177
Whyteleafe Bus. Village CR3: Whyt ... 89Vb 177
Whyteleafe Hill CR3: Whyt ... 92Ub 197
Whyteleafe Rd. CR3: Cat'm ... 92Ub 197
Whyteleafe South Station (Rail) ... 91Wb 197
Whyteleafe Station (Rail) ... 89Vb 177
Whyte M. SM3: Cheam ... 80Ab 154
Whyteville Rd. E7 ... 37Kc 73
Whytlaw Ho. E3 ... 43Bc 92
(off Baythorne St.)
Wiblin M. NW5 ... 35Kb 70
Wichling Cl. BR5: Orp ... 74Zc 161
Wickenden Rd. TN13: S'oaks ... 94Ld 203
Wickens Cvn. Site TN14: Dun G ... 91Jd 202
Wickens Mdw. TN14: Dun G ... 91Hd 202
Wickersley Rd. SW11 ... 54Jb 112
Wickers Oake SE19 ... 63Vb 135
Wicker St. E1 ... 44Xb 91
Wicket, The CR0: Addtn ... 78Cc 158
Wicket Rd. UB6: G'frd ... 41Ja 86
Wickets, The TW15: Ashf ... 63N 127
Wickets Cl. BR5: St M Cry ... 71Yc 161
Wickets End WD7: Shenl ... 5Na 15
Wickets Way IG6: Ilf ... 23Vc 55
Wickfield Apts. E15 ... 37Fc 73
(off Grove Cres. Rd.)
Wickfield Ho. SE16 ... 47Xb 91
(off Wilson Gro.)
Wickfields IG7: Chig ... 23Tc 54
Wickford Cl. RM3: Rom ... 22Pd 57
Wickford Dr. RM3: Rom ... 22Pd 57
Wickford Ho. E1 ... 42Yb 92
(off Wickford St.)
Wickford St. E1 ... 42Yb 92
Wickford Way E17 ... 28Zb 52
Wickham Av. CR0: C'don ... 75Ac 158
Wickham Av. SM3: Cheam ... 78Ya 154
Wickham Chase BR4: W W'ck ... 74Fc 159
Wickham Cl. E1 ... 43Yb 92
Wickham Cl. EN3: Enf H ... 13Xb 33
Wickham Cl. KT3: N Mald ... 72Va 154
Wickham Cl. UB9: Hare ... 25M 43
Wickham Ct. KT5: Surb ... 71Pa 153
(off Cranes Pk.)
Wickham Ct. Rd. BR4: W W'ck ... 75Ec 158
Wickham Cres. BR4: W W'ck ... 75Ec 158
Wickham Fld. TN14: Otf ... 88Hd 182
Wickham Gdns. SE4 ... 55Bc 114
Wickham Ho. N1 ... 39Ub 71
(off Halcomb St.)
Wickham La. DA16: Well ... 52Xc 117
Wickham La. SE2 ... 50Wc 95
Wickham La. TW20: Egh ... 66C 126
Wickham M. SE4 ... 54Bc 114
Wickham Noakes Ct. BR3: Beck ... 67Dc 136
Wickham Rd. BR3: Beck ... 68Dc 136
Wickham Rd. CR0: C'don ... 75Zb 158
Wickham Rd. E4 ... 24Ec 52
Wickham Rd. HA3: Hrw W ... 26Fa 46
Wickham Rd. RM16: Grays ... 7E 100
Wickham Rd. SE4 ... 56Bc 114
Wickham St. DA16: Well ... 54Uc 116
Wickham St. SE11 ... 7H 229 (50Pb 90)
Wickhams Way DA3: Hartl ... 71Be 165
Wickham Theatre Cen. ... 75Fc 159
Wickham Way BR3: Beck ... 70Ec 136
Wick Ho. KT1: Hamp W ... 67Ma 131
(off Station Rd.)
Wick La. E3 ... 39Cc 72
Wick La. TW20: Eng G ... 5K 125
Wickliffe Av. N3 ... 26Ab 48
Wickliffe Gdns. HA9: Wemb ... 33Ra 67
Wicklow Ho. N16 ... 32Vb 71
Wicklow St. WC1 ... 3H 217 (41Pb 90)
Wick Rd. E9 ... 37Zb 72
Wick Rd. TW11: Tedd ... 66Ka 130
Wick Rd. TW20: Eng G ... 7L 125
Wicks Cl. SE9 ... 63Mc 137
Wicksteed Cl. DA5: Bexl ... 62Fd 140
Wicksteed Ho. SE1 ... 4E 230 (48Sb 91)
Wicksteed Ho. TW8: Bford ... 50Pa 87
Wicks Way SE19 ... 65Tb 135
Wickway Ct. SE15 ... 51Vb 113
(off Cator St.)
Wickwood Ct. AL1: St A ... 1F 6
Wickwood St. SE5 ... 54Rb 113
Wid Cl. CM13: Hut ... 15Fe 41
Widdenham Rd. N7 ... 35Pb 70
Widdicombe Av. HA2: Harr ... 33Aa 65
Widdin St. E15 ... 38Gc 73
Widecombe Cl. RM3: Rom ... 25Md 57
Widecombe Gdns. IG4: Ilf ... 28Nc 54
Widecombe Rd. SE9 ... 62Nc 138
Widecombe Way N2 ... 29Fb 49
Widecroft Rd. SL0: Iver ... 44G 82
Wideford Dr. RM7: Rush G ... 30Fd 56
Widegate St. E1 ... 1J 225 (43Ub 91)
Widenham Cl. HA5: Eastc ... 29Y 45
Widewater Pl. UB9: Hare ... 29L 43
Wide Way CR4: Mitc ... 69Mb 134
Widewing Cl. TW11: Tedd ... 66Ka 130
Widford NW1 ... 37Kb 70
(off Lewis St.)
Widford Ho. N1 ... 2B 218 (40Rb 71)
(off Colebrooke Rd.)
Widgeon Cl. E16 ... 44Kc 93
Widgeon Rd. DA8: Erith ... 52Kd 119
Widgeon Rd. TW6: H'row A ... 54K 105
Widgeon Way WD25: Wat ... 9Aa 13
Widley Rd. W9 ... 41Cb 89
Widmer Ct. TW3: Houn ... 54Aa 107
WIDMORE ... 69Lc 137
WIDMORE GREEN ... 68Mc 137
Widmore Lodge Rd. BR1: Broml ... 68Mc 137
Widmore Rd. BR1: Broml ... 68Jc 137
Widmore Rd. UB8: Hil ... 42R 84
Widvale Rd. CM15: Mount ... 13Ee 41
Widvale Rd. CM15: Mount ... 13Ee 41
Widvale Rd. CM15: Shenf ... 13Ee 41
Widworthy Hayes CM13: Hut ... 18De 41
Wieland Rd. HA6: Nwood ... 24W 44
Wigan Ho. E5 ... 32Xb 71
Wigeon Path SE28 ... 48Tc 94
Wigeon Way UB4: Yead ... 44Z 85
Wiggenhall Ind. Est. WD18: Wat ... 15Y 27
Wiggenhall Rd. WD18: Wat ... 15X 27
Wiggie La. RH1: Redh ... 4A 208
Wiggington Ho. SL4: Eton ... 2H 103
(off High St.)
Wiggins La. TW10: Ham ... 61La 130
Wiggins Mead NW9 ... 24Va 48
Wigginton Av. HA9: Wemb ... 37Ra 67
Wight Ho. KT1: King T ... 69Ma 131
(off Portsmouth Rd.)
Wight Ho. WD18: Wat ... 15V 26
Wightman Rd. N4 ... 29Qb 50
Wightman Rd. N8 ... 28Qb 50
Wighton M. TW7: Isle ... 54Ga 108
Wigley Bush La. CM14: B'wood ... 19Ud 40
Wigley Bush La. CM14: S Weald ... 19Ud 40
Wigley Rd. TW13: Felt ... 61Z 129
Wigmore Hall ... 2K 221 (44Kb 90)
(off Wigmore St.)
Wigmore Pl. E17 ... 27Ac 52
Wigmore Pl. W1 ... 2K 221 (44Kb 90)
Wigmore Rd. SM5: Cars ... 75Fb 155
Wigmore St. W1 ... 3H 221 (44Jb 90)
Wigmore Wlk. SM5: Cars ... 75Fb 155
Wigram Ho. E14 ... 45Dc 92
(off Wade's Pl.)
Wigram Rd. E11 ... 30Lc 53
Wigram Sq. E17 ... 27Ec 52
Wigston Cl. N18 ... 22Ub 51
Wigston Rd. E13 ... 42Kc 93
Wigton Gdns. HA7: Stan ... 25Na 47
Wigton Pl. SE11 ... 7A 230 (50Qb 90)
Wigton Rd. E17 ... 25Bc 52
Wigton Rd. RM3: Rom ... 21Nd 57
Wigton Way RM3: Rom ... 21Nd 57
Wilberforce Ct. BR2: Kes ... 80Mc 159
Wilberforce Ct. KT18: Eps ... 86Ta 173
(off Heathcote Rd.)
Wilberforce M. SW4 ... 56Mb 112
Wilberforce Rd. N4 ... 33Rb 71
Wilberforce Rd. NW9 ... 30Wa 48
Wilberforce Wlk. E15 ... 36Gc 73
Wilberforce Way DA12: Grav'nd ... 4F 144
Wilberforce Way SE25 ... 70Vb 135
Wilberforce Way SW19 ... 65Za 132
Wilbraham Ho. SW8 ... 52Nb 112
(off Wandsworth Rd.)
Wilbraham Mans. SW1 ... 5H 227 (49Jb 90)
(off Wilbraham Pl.)
Wilbraham Pl. SW1 ... 5G 227 (49Hb 89)
Wilbrahams Almshouses EN5: Barn ... 12Bb 31
Wilbrooke Pl. SE3
53Kc 115
Wilbury Av. SM2: Cheam ... 82Bb 175
Wilbury Rd. GU21: Wok ... 9P 167
Wilbury Way N18 ... 22Tb 51
Wilby M. W11 ... 46Bb 89
Wilcon Way WD25: Wat ... 6Z 13
Wilcot Av. WD19: Wat ... 17Aa 27
Wilcot Cl. GU24: Bisl ... 8E 166
Wilcot Cl. WD19: Wat ... 17Aa 27
Wilcot Gdns. GU24: Bisl ... 8E 166
Wilcox Cl. SW8 ... 52Nb 112
(not continuous)
Wilcox Cl. WD6: Bore ... 11Sa 29
Wilcox Gdns. TW17: Shep ... 69N 127
Wilcox Ho. E3 ... 43Bc 92
(off Ackroyd Dr.)
Wilcox Pl. SW1 ... 4C 228 (48Lb 90)
Wilcox Rd. SM1: Sutt ... 77Db 155
Wilcox Rd. SW8 ... 52Nb 112
Wilcox Rd. TW11: Tedd ... 63Fa 130
Wildacres HA6: Nwood ... 21V 44
Wildacres KT14: W Byf ... 83L 169
Wildbank Ct. GU22: Wok ... 90B 168
Wildberry Cl. W7 ... 49Ja 86
Wildbore Ho. N1 ... 38Qb 70
(off Liverpool Rd.)
Wildcary La. RM3: Hrld W ... 26Pd 57
Wildcat Rd. TW6: H'row A ... 55K 105
(off Wayfarer Rd.)
Wild Ct. WC2 ... 3H 223 (44Pb 90)
Wildcroft Gdns. HA8: Edg ... 23Ma 47
Wildcroft Mnr. SW15 ... 59Ya 110
Wildcroft Rd. SW15 ... 59Ya 110
Wilde Cl. E8 ... 39Wb 71
Wilde Cl. RM18: Tilb ... 4E 122
Wilde Ct. AL2: Lon C ... 9F 6
Wilde Ho. W2 ... 4B 220 (45Fb 89)
(off Gloucester Ter.)
Wilde Pl. N13 ... 23Rb 51
Wilde Pl. SW18 ... 59Fb 111
Wilder Cl. HA4: Ruis ... 32X 65
Wilderness, The KT8: E Mos ... 71Ea 152
Wilderness, The KT8: W Mole ... 71Ea 152
Wilderness, The TW12: Hamp H ... 63Da 129
WILDERNESSE ... 94Nd 203
Wildernesse Av. TN15: S'oaks ... 94Nd 203
Wildernesse Av. TN15: Seal ... 94Nd 203
Wilderness Cl. HA8: Edg ... 21Sa 47
Wildernesse Golf Course ... 94Qd 203
Wildernesse Mt. TN13: S'oaks ... 94Md 203
Wilderness Island Local Nature Reserve ... 75Jb 156
Wilderness Local Nature Reserve, The ... 52Kb 112
(off Queenstown Rd.)
Wilderness M. SW4 ... 56Kb 112
Wilderness Rd. BR7: Chst ... 66Rc 138
Wilderness Rd. RH8: Oxt ... 2H 211
Wilde Rd. DA8: Erith ... 52Dd 118
Wilders Cl. GU21: Wok ... 10N 167
Wilderton Rd. N16 ... 31Ub 71
Wilder Wlk. W1 ... 5C 222 (45Lb 90)
(off Glasshouse St.)
Wildfell Rd. SE6 ... 59Dc 114
Wild Goose Dr. SE14 ... 53Yb 114
Wild Grn. Nth. SL3: L'ly ... 49C 82
Wild Grn. Sth. SL3: L'ly ... 49C 82
Wild Hatch NW11 ... 30Cb 49
WILDHILL ... 3M 9
Wildhill Rd. AL9: Hat ... 5G 8
Wilding Ct. WD6: Bore ... 14Qa 29
(off Whitehall Cl.)
Wildmarsh Ct. EN3: Enf W ... 9Ac 20
(off Manly Dixon Dr.)
Wildoaks Cl. HA6: Nwood ... 23V 44
Wild's Rents SE1 ... 3H 231 (48Ub 91)
Wild St. WC2 ... 3G 223 (44Nb 90)
Wildwood HA6: Nwood ... 23T 44
Wildwood Av. AL2: Brick W ... 2Ba 13
Wildwood Cl. GU22: Pyr ... 87H 169
Wildwood Cl. KT24: E Hor ... 97V 190
Wildwood Cl. SE12 ... 59Hc 115
Wildwood Ct. CR8: Kenley ... 87Tb 177
Wildwood Ct. WD3: Chor ... 14H 25
Wildwood Gro. NW3 ... 32Eb 69
Wildwood Ri. NW11 ... 32Eb 69
Wildwood Rd. NW11 ... 30Db 49
Wildwood Ter. NW3 ... 32Eb 69
Wilford Cl. EN2: Enf ... 13Tb 33
Wilford Cl. HA6: Nwood ... 24T 44
Wilford Rd. CR0: C'don ... 72Sb 157
Wilford Rd. SL3: L'ly ... 49A 82
Wilfred Av. RM13: Rain ... 43Jd 96
Wilfred Cl. WD18: Wat ... 15V 26
Wilfred Ct. N15 ... 29Tb 51
(off South Gro.)
Wilfred Owen Cl. SW19 ... 65Eb 133
Wilfred St. DA12: Grav'nd ... 8D 122
Wilfred St. GU21: Wok ... 10P 167
Wilfred St. SW1 ... 3B 228 (48Lb 90)
Wilfred Wood Ct. W6 ... 49Ya 88
(off Samuel's Cl.)
Wilfrid Gdns. W3 ... 43Sa 87
Wilhelmina Av. CR5: Coul ... 91Lb 196
Wilkens Pl. SL9: Chal P ... 22B 42
Wilkes Cl. NW7 ... 23Za 48
Wilkes Rd. CM13: Hut ... 15Fe 41
Wilkes Rd. TW8: Bford ... 51Na 109
Wilkes St. E1 ... 7K 219 (43Vb 91)
Wilkie Ho. SW1 ... 7E 228 (50Mb 90)
(off Cureton St.)
Wilkins Cl. CR4: Mitc ... 67Gb 133
Wilkins Cl. UB3: Harl ... 50V 84
WILKINS GREEN ... 1N 7
Wilkin's Grn. La. AL10: Hat ... 1N 7
Wilkin's Grn. La. AL4: S'ford ... 1N 7
Wilkins Grn. Ter. AL4: S'ford ... 2M 7
Wilkins Ho. SW1 ... 51Kb 112
(off Churchill Gdns.)
Wilkinson Cl. DA1: Dart ... 56Pd 119
Wilkinson Cl. NW2 ... 33Ya 68
Wilkinson Cl. UB10: Hil ... 39R 64
Wilkinson Ct. CM14: B'wood ... 20Xd 40
Wilkinson Ct. SW17 ... 63Fb 133
Wilkinson Gdns. SE25 ... 67Ub 135
Wilkinson Ho. N1 ... 2G 219 (40Tb 71)
(off Cranston Est.)
Wilkinson Rd. E16 ... 44Lc 93
Wilkinson St. SW8 ... 52Pb 112
Wilkinson Way HP3: Hem H ... 6P 3
Wilkinson Way W4 ... 47Ta 87
Wilkin St. NW5 ... 37Jb 70
Wilkin St. M. NW5 ... 37Kb 70
Wilkins Way TN16: Bras ... 96Xc 201
Wilks Av. DA1: Dart ... 61Pd 141
Wilks Gdns. CR0: C'don ... 74Ac 158
Wilks Pl. N1 ... 2J 219 (40Ub 71)
Wilks Rd. DA2: Dart ... 58Rd 119
Willan Rd. N17 ... 26Tb 51
Willard St. SW8 ... 55Kb 112
Willats Cl. KT16: Chert ... 72H 149
Willcocks Cl. KT9: Chess ... 76Na 153
Willcott Rd. W3 ... 46Ra 87
Will Crooks Gdns. SE9 ... 56Lc 115
Willen Fld. Rd. NW10 ... 40Sa 67
Willenhall Av. EN5: New Bar ... 16Eb 31
Willenhall Ct. EN5: New Bar ... 16Eb 31
Willenhall Dr. UB3: Hayes ... 45U 84
Willenhall Rd. SE18 ... 50Rc 94
Willersley Av. BR6: Orp ... 76Tc 160
Willersley Av. DA15: Sidc ... 60Vc 117
Willersley Cl. DA15: Sidc ... 60Vc 117
Willerton Lodge KT13: Weyb ... 79T 150
WILLESDEN ... 37Wa 68
WILLESDEN CENTRE FOR HEALTH & CARE ... 38Wa 68
WILLESDEN GREEN ... 37Ya 68
Willesden Green Station (Underground) ... 37Ya 68
Willesden Junction Station (Underground & Overground) ... 41Va 88
Willesden La. NW2 ... 37Ya 68
Willesden La. NW6 ... 37Za 68

Willesden Section Ho. NW6 38Za 68
(off Willesden La.)
Willesden Sports Cen. 39Xa 68
Willesden Sports Stadium 39Xa 68
Willes Rd. NW5 37Kb 70
Willets Cl. RM3: Rom 24Ld 57
Willett Cl. BR5: Pet W 72Uc 160
Willett Cl. UB5: N'olt 41Y 85
Willett Ho. E13 40Kc 73
(off Queens Rd. W.)
Willett Pl. CR7: Thor H 71Qb 156
Willett Rd. CR7: Thor H 71Qb 156
Willetts La. UB9: Den 36H 63
Willett Way BR5: Pet W 71Tc 160
Willey Broom La. CR3: Cat'm 97Qb 196
Willey Farm La. CR3: Cat'm 98Sb 197
Willey La. CR3: Cat'm 97Tb 197
William IV St. WC2 5F 223 (45Nb 90)
William Allen Ho. HA8: Edg 24Pa 47
William Ash Cl. RM9: Dag 37Xc 75
William Banfield Ho. SW6 54Bb 111
(off Munster Rd.)
William Barefoot Dr. SE9 63Qc 138
William Blake Ho. SW11 53Gb 111
William Bonney Est. SW4 56Mb 112
William Booth Ho. E14 44Cc 92
(off Hind Gro.)
William Booth Rd. SE20 67Wb 135
William Carey Way HA1: Harr 30Ga 46
William Caslon Ho. E2 40Xb 71
(off Patriot Sq.)
William Channing Ho. E2 41Xb 91
(off Canrobert St.)
William Cl. N2 26Fb 49
William Cl. RM5: Col R 25Ed 56
William Cl. SE13 55Ec 114
William Cl. SE7 50Mc 93
William Cl. UB2: S'hall 47Ea 86
William Cobbett Ho. W8 48Db 89
(off Scarsdale Pl.)
William Congreve M. N1 1D 218 (39Sb 71)
William Cory Prom. DA8: Erith 50Gd 96
William Cotton Ct. E14 43Cc 92
(off Selsey St.)
William Ct. HP3: Hem H 6M 3
William Ct. NW8 3A 214 (41Eb 89)
William Ct. SE10 53Dc 114
(off Greenwich High Rd.)
William Ct. SE25 69Vb 135
(off Chalfont Rd.)
William Ct. SW16 66Pb 134
(off Streatham High Rd.)
William Ct. W5 43La 86
William Covell Cl. EN2: Enf 10Pb 18
William Crook Ho. HP1: Hem H 2H 3
William Dr. HA7: Stan 23Ja 46
William Dromey Ct. NW6 38Bb 69
William Dunbar Ho. NW6 40Bb 69
(off Albert Rd.)
William Dyce M. SW16 63Mb 134
William Ellis Cl. SL4: Old Win 7L 103
William Ellis Way SE16 48Wb 91
(off St James's Rd.)
William Evans Ho. SE8 49Zb 92
(off Haddonfield)
William Evans Rd. KT19: Eps 83Qa 173
William Farm La. SW15 55Xa 110
William Fenn Ho. E2 41Wb 91
(off Shipton Rd.)
William Foster La. DA16: Well 54Wc 117
William Fry Ho. E1 44Zb 92
(off W. Arbour St.)
William Gdns. SW15 57Xa 110
William Gibbs Ct. SW1 4D 228 (48Mb 90)
(off Old Pye St.)
William Gunn Ho. NW3 36Gb 69
William Guy Gdns. E3 41Dc 92
William Harvey Ho. SW19 60Ab 110
(off Whitlock Dr.)
William Henry Wlk. SW11 51Mb 112
William Hope Cl. IG11: Bark 40Uc 74
William Ho. BR1: Broml 69Jc 137
William Ho. DA12: Grav'nd 9D 122
William Hunter Way CM14: B'wood 19Yd 40
William Hunt Mans. SW13 51Ya 110
William Margrie Cl. SE15 54Wb 113
William Marshall Cl. E17 30Ac 52
William M. N1 38Ub 71
William M. SW1 2G 227 (47Hb 89)
William Morley Cl. E6 39Mc 73
William Morris Cl. E17 27Bc 52
William Morris Gallery 27Cc 52
William Morris Ho. W6 51Za 110
William Morris Way SW6 55Eb 111
William Mundy Way DA1: Dart 57Nd 119
William Murdoch Rd. E14 43Fc 93
William Nash Ct. BR5: St P 69Yc 139
William Owston Ct. E16 46Nc 94
(off Connaught Rd.)
William Parry Ho. E16 47Kc 93
(off Shipwright Street)
William Penn Leisure Cen. 18H 25
William Perkin Ct. UB6: G'frd 37Ga 66
William Petty Way BR5: Orp 74Yc 161
William Pike Ho. RM7: Rom 30Fd 56
(off Waterloo Gdns.)
William Pl. E3 40Bc 72
William Rathbone Ho. E2 41Xb 91
(off Florida St.)
William Rd. CR3: Cat'm 94Tb 197
William Rd. NW1 4B 216 (41Lb 90)
William Rd. SM1: Sutt 78Eb 155
William Rd. SW19 66Ab 132
William Rushbrooke Ho. SE16 49Wb 91
(off Rouel Rd.)
William Russell Ct. GU21: Wok 10J 167
Williams Av. E17 25Bc 52
William Saville Ho. NW6 40Bb 69
(off Denmark Rd.)
William's Bldgs. E2 42Yb 92
Williamsburg Plaza E14 45Ec 92
Williams Cl. KT15: Add 78K 149
Williams Cl. N8 30Mb 50
Williams Cl. SW6 52Ab 110
Williams Ct. EN8: Chesh 1Zb 20
Williams Dr. TW3: Houn 56Ca 107
William Sellars Cl. CR3: Cat'm 93Ub 197
Williams Gro. KT6: Surb 72La 152
Williams Gro. N22 25Qb 50
Williams Ho. E3 41Cc 92
(off Alfred St.)
Williams Ho. E9 39Xb 71
(off King Edward's Rd.)
Williams Ho. NW2 34Ya 68
(off Stoll Cl.)
Williams Ho. SW1 6E 228 (49Mb 90)
(off Montaigne Cl.)
William's La. SW14 55Sa 109
Williams La. SM4: Mord 71Eb 155
Williams M. SE4 58Bc 114
William Smith Ho. DA17: Belv 48Cd 96
(off Ambrooke Rd.)
William Smith Ho. E3 41Cc 92
(off Ireton St.)
Williamson Cl. SE10 50Hc 93
Williamson Ct. SE17 7D 230 (50Sb 91)
Williamson Rd. N4 30Rb 51
Williamson Rd. WD24: Wat 8W 12
Williamson St. N7 35Nb 70
Williamson Way WD3: Rick 18J 25
William Sq. SE16 45Ac 92
(off Sovereign Cres.)
Williams Rd. RH8: Oxt 5L 211
Williams Rd. UB2: S'hall 49Aa 85
Williams Rd. W13 46Ja 86
William Stanley Ho. SE25 70Wb 135
Williams Ter. CR0: Wadd 79Qb 156
William St. DA12: Grav'nd 9D 122
William St. E10 30Dc 52
William St. HP4: Berk 1A 2
William St. IG11: Bark 38Sc 74
William St. N17 24Vb 51
William St. RM17: Grays 51De 121
(not continuous)
William St. SL1: Slou 7K 81
William St. SL4: Wind 3H 103
William St. SM5: Cars 76Gb 155
William St. SW1 2G 227 (47Hb 89)
William St. WD23: Bush 13Z 27
Williams Way DA2: Wilm 61Fd 140
Williams Way HA0: Wemb 36Ka 66
Williams Way WD7: R'lett 7Ka 14
William Whiffin Sq. E3 42Bc 92
William Wood Ho. SE26 62Yb 136
(off Shrublands Cl.)
Willifield Way NW11 28Bb 49
Willingale Cl. CM13: Hut 16Fe 41
Willingale Cl. IG10: Lough 12Sc 36
Willingale Cl. IG8: Wfd G 23Lc 53
Willingale Ct. CM16: They B 8Uc 22
Willingale Rd. IG10: Lough 11Sc 36
Willingdon Rd. N22 26Rb 51
Willinghall Cl. EN9: Walt A 4Fc 21
Willingham Cl. NW5 36Lb 70
Willingham Ter. NW5 36Lb 70
Willingham Way KT1: King T 69Qa 131
Willington Ct. E5 34Ac 72
Willington Rd. SW9 55Nb 112
Willis Av. SM2: Sutt 79Gb 155
Willis Cl. KT18: Eps 86Ra 173
Willis Ct. BR4: W W'ck 75Fc 159
Willis Ct. CR7: Thor H 72Qb 156
Willis Ho. E14 45Dc 92
(off Hale St.)
Willis Pl. SL3: L'ly 10P 81
Willis Rd. CR0: C'don 73Sb 157
Willis Rd. DA8: Erith 49Ed 96
Willis Rd. E15 40Hc 73
Willis St. E14 44Dc 92
Willis Yd. N14 17Mb 32
Will Miles Ct. SW19 66Eb 133
Willmore End SW19 67Db 133
Willoners SL2: Slou 3E 80
Willoughby Av. CR0: Bedd 77Pb 156
Willoughby Av. UB10: Uxb 40N 63
Willoughby Ct. AL2: Lon C 8H 7
Willoughby Dr. RM13: Rain 38Gd 76
Willoughby Gro. N17 24Xb 51
Willoughby Highwalk EC2 1F 225 (43Tb 91)
(off Moor La.)
Willoughby Ho. E1 46Xb 91
(off Reardon Path)
Willoughby Ho. EC2 1F 225 (43Tb 91)
(off Moor La.)
Willoughby La. BR1: Broml 66Kc 137
Willoughby La. N17 23Xb 51
Willoughby M. N17 24Xb 51
Willoughby M. SW4 56Kb 112
(off Cedars M.)
Willoughby Pk. Rd. N17 24Xb 51
Willoughby Pas. E14 46Cc 92
(off W. India Av.)
Willoughby Rd. KT2: King T 67Pa 131
Willoughby Rd. N8 27Qb 50
Willoughby Rd. NW3 35Fb 69
Willoughby Rd. SL3: L'ly 48C 82
Willoughby Rd. TW1: Twick 57La 108
(not continuous)
Willoughbys, The SW14 55Ua 110
Willoughby St. WC1 1F 223 (43Nb 90)
(off Gt. Russell St.)
Willoughby Way SE7 49Kc 93
Willow Av. BR8: Swan 69Hd 140
Willow Av. DA15: Sidc 58Wc 117
Willow Av. SW13 54Va 110
Willow Av. UB7: Yiew 45P 83
Willow Av. UB9: Den 37L 63
Willow Bank GU22: Wok 94A 188
Willow Bank SW6 55Ab 110
Willow Bank TW10: Ham 62Ka 130
Willowbank CR5: Coul 86Nb 176
Willowbank KT7: T Ditt 74Ja 152
WILLOWBANK 36L 63
Willowbank Gdns. KT20: Tad 94Xa 194
Willowbank Pl. CR8: Purl 81Rb 177
Willowbay Cl. EN5: Barn 16Za 30
Willow Bri. Rd. N1 37Sb 71
(not continuous)
Willowbrook SL4: Eton 9H 81
Willowbrook TW12: Hamp H 64Da 129
Willowbrook Est. SE15 52Wb 113
Willow Brook Rd. SE15 52Vb 113
Willowbrook Rd. TW19: Stanw 61N 127
Willowbrook Rd. UB2: S'hall 48Ca 85
Willow Bus. Pk. SE26 62Yb 136
Willow Cl. BR2: Broml 71Pc 160
Willow Cl. BR5: St M Cry 73Xc 161
Willow Cl. CM13: Hut 16De 41
Willow Cl. DA5: Bexl 58Bd 117
Willow Cl. IG9: Buck H 20Mc 35
Willow Cl. KT15: Wdhm 83H 169
Willow Cl. KT16: Chert 75G 148
Willow Cl. RM12: Horn 34Kd 77
Willow Cl. SE6 60Hc 115
Willow Cl. SL3: Coln 52E 104
Willow Cl. SL9: Chal P 26A 42
Willow Cl. SM7: Bans 86Ab 174
Willow Cl. TW8: Bford 51La 108
Willow Cotts. TW13: Hanw 62Aa 129
Willow Cotts. TW9: Kew 51Qa 109
Willow Ct. E11 33Gc 73
(off Trinity Cl.)
Willow Ct. EC2 5H 219 (42Ub 91)
Willow Ct. HA3: Hrw W 25Ha 46
Willow Ct. HA8: Edg 21Na 47
Willow Ct. HP3: Hem H 6N 3
Willow Ct. N12 21Db 49
Willow Ct. NW6 38Ab 68
Willow Ct. SM6: W'gton 80Kb 156
(off Willow Rd.)
Willow Ct. TW16: Sun 66U 128
(off Staines Rd. W.)
Willow Ct. W4 52Ua 110
(off Corney Reach Way)
Willow Ct. W9 43Cb 89
(off Admiral Wlk.)
Willowcourt Av. HA3: Kenton 29Ka 46
Willow Cres. AL1: St A 2G 6
Willow Cres. E. UB9: Den 36L 63
Willow Cres. W. UB9: Den 36L 63
Willow Dene HA5: Pinn 26Z 45
Willow Dene WD23: B Hea 17Ga 28
Willowdene CM15: Pil H 15Vd 40
Willowdene N6 31Hb 69
Willowdene SE15 52Xb 113
Willowdene Cl. TW2: Whitt 59Ea 108
Willowdene Ct. CM14: W'ley 21Yd 58
Willowdene Ct. N20 17Eb 31
(off High Rd. Whetstone)
Willow Dr. EN5: Barn 14Ab 30
Willow Dr. GU23: Rip 96J 189
Willow Edge WD4: K Lan 1Q 12
Willow End HA6: Nwood 23W 44
Willow End KT6: Surb 74Na 153
Willow End N20 19Cb 31
Willowfields Cl. SE18 50Uc 94
Willow Gdns. HA4: Ruis 33V 64
Willow Gdns. TW3: Houn 53Ca 107
Willow Glade RH2: Reig 9L 207
Willow Grange DA14: Sidc 62Xc 139
Willow Grange WD17: Wat 11W 26
Willow Grn. GU24: W End 5D 166
Willow Grn. NW9 25Ua 48
Willow Grn. WD6: Bore 15Ta 29
Willow Gro. BR7: Chst 65Qc 138
Willow Gro. E13 40Jc 73
Willow Gro. HA4: Ruis 33V 64
Willowhayne Ct. KT12: Walt T 73X 151
(off Willowhayne Dr.)
Willowhayne Dr. KT12: Walt T 73X 151
Willowhayne Gdns. KT4: Wor Pk 76Ya 154
Willowherb Wlk. RM3: Rom 24Ld 57
Willow Ho. BR2: Broml 68Gc 137
Willow Ho. CR6: W'ham 87Dc 178
Willow Ho. SE1 5K 231 (49Vb 91)
(off Curtis St.)
Willow Ho. SE4 55Ac 114
(off Dragonfly Pl.)
Willow Ho. W10 42Za 88
(off Maple Wlk.)
Willow La. CR4: Mitc 71Hb 155
Willow La. SE18 49Pc 94
Willow La. WD18: Wat 15W 26
Willow La. Bus. Pk. CR4: Mitc 72Hb 155
Willow La. Ind. Est. CR4: Mitc 72Hb 155
Willow Lodge RM7: Rom 29Fd 56
Willow Lodge SW6 53Za 110
Willow Lodge TW16: Sun 66V 128
(off Grangewood Dr.)
Willowmead IG7: Chig 20Wc 37
Willowmead TW18: Staines 67K 127
Willowmead Cl. GU21: Wok 8L 167
Willowmead Cl. W5 42Ma 87
Willowmere KT10: Esh 77Ea 152
Willow M. CR3: Cat'm 95Tb 197
Willow M. TN15: Plat 92Ee 205
Willow Mt. CR0: C'don 76Ub 157
Willow Pde. RM14: Upm 32Ud 78
Willow Pde. SL3: L'ly 48C 82
Willow Pk. SL2: Stoke P 8L 61
Willow Pk. TN14: Otf 89Hd 182
Willow Path EN9: Walt A 6Gc 21
Willow Pl. SL4: Eton 1G 102
Willow Pl. SW1 5C 228 (49Lb 90)
Willow Rd. DA1: Dart 60Ld 119
Willow Rd. DA8: Erith 53Jd 118
Willow Rd. EN1: Enf 13Ub 33
Willow Rd. IG7: Chig 21Tc 54
Willow Rd. KT3: N Mald 70Sa 131
Willow Rd. NW3 35Fb 69
Willow Rd. RH1: Redh 9L 207
Willow Rd. RM6: Chad H 30Ad 55
Willow Rd. SL3: Poyle 54G 104
Willow Rd. SM6: W'gton 80Kb 156
Willow Rd. W5 47Na 87
Willows, The AL1: St A 6F 6
Willows, The BR3: Beck 67Cc 136
Willows, The E6 38Pc 74
Willows, The GU18: Light 3A 166
Willows, The HP2: Hem H 1B 4
Willows, The IG10: Lough 15Mc 35
Willows, The KT10: Clay 79Ga 152
Willows, The KT13: Weyb 76Q 150
Willows, The KT14: Byfl 85N 169
Willows, The RH1: Redh 7P 207
Willows, The RM17: Grays 51Fe 121
Willows, The SE1 5J 231 (49Ub 91)
Willows, The SL4: Wind 2B 102
Willows, The TN15: Igh 93Zd 205
Willows, The WD19: Wat 17X 27
Willows, The WD3: Rick 19J 25
Willows, The WD6: Bore 11Qa 29
Willows Activity Farm 8K 7
Willows Av. SM4: Mord 71Db 155
Willows Cl. HA5: Pinn 26Y 45
Willowside AL2: Lon C 9J 7
Willowside Ct. EN2: Enf 13Rb 33
Willows Lodge SL4: Wind 2B 102
Willows Path KT18: Eps 86Ra 173
Willows Path SL4: Wind 3A 102
Willows Riverside Pk. SL4: Wind 2A 102
Willows Ter. NW10 40Va 68
(off Rucklidge Av.)
Willow St. E4 17Fc 35
Willow St. EC2 5H 219 (42Ub 91)
Willow St. RM7: Rom 28Ed 56
Willow Ter. DA4: Eyns 75Nd 163
Willow Tree Cen. UB9: Hare 29R 44
Willow Tree Cl. E3 39Bc 72
Willow Tree Cl. RM4: Abr 13Xc 37
Willow Tree Cl. SW18 60Db 111
Willow Tree Cl. UB4: Yead 42Y 85
Willow Tree Cl. UB5: N'olt 37Ba 65
Willowtree Cl. UB10: Ick 34S 64
Willow Tree Ct. DA14: Sidc 64Wc 139
Willow Tree Ct. HA0: Wemb 36Ma 67
Willow Tree La. UB4: Yead 42Y 85
Willow Tree Lodge HA6: Nwood 22V 44
WILLOW TREE RDBT. 43Z 85
Willow Tree Wlk. BR1: Broml 67Kc 137
Willowtree Way CR7: Thor H 67Qb 134
Willow Va. BR7: Chst 65Rc 138
Willow Va. KT22: Fet 95Da 191
(not continuous)
Willow Va. W12 46Wa 88
Willow Vw. SW19 67Fb 133
Willow Wlk. BR6: Farnb 76Rc 160
Willow Wlk. DA1: Cray 56Ld 119
(not continuous)
Willow Wlk. DA1: Dart 56Ld 119
(not continuous)
Willow Wlk. E17 29Bc 52
Willow Wlk. IG1: Ilf 33Rc 74
Willow Wlk. KT16: Chert 73J 149
Willow Wlk. KT23: Bookh 97Ea 192
Willow Wlk. N1 1C 218 (40Rb 71)
Willow Wlk. N15 28Rb 51
Willow Wlk. N2 26Fb 49
Willow Wlk. N21 16Pb 32
Willow Wlk. RH1: Redh 8B 208
Willow Wlk. RM14: Upm 32Ud 78
Willow Wlk. SE1 4J 231 (48Ub 91)
Willow Wlk. SM3: Sutt 76Bb 155
Willow Wlk. TW20: Eng G 4N 125
Willow Way AL10: Hat 3B 8
Willow Way AL2: Chis G 9N 5
Willow Way EN6: Pot B 5Db 17
Willow Way GU22: Wok 93A 188
Willow Way HA0: Wemb 34Ja 66
Willow Way HP1: Hem H 1K 3
Willow Way KT14: W Byf 83L 169
Willow Way KT19: Ewe 79Ta 153
Willow Way N3 24Db 49
Willow Way RH9: G'stone 4P 209
Willow Way RM3: Hrld W 23Rd 57
Willow Way SE26 62Yb 136
Willow Way TW16: Sun 70W 128
Willow Way TW2: Twick 61Da 129
Willow Way W11 45Za 88
Willow Way WD7: R'lett 8Ga 14
Willow Wood Cres. SE25 72Ub 157
Willow Wren Wharf UB2: S'hall 49X 85
Will Perrin Ct. RM13: Rain 39Jd 76
Willrose Cres. SE2 50Xc 95
Willsbridge Ct. SE15 51Vb 113
Wills Cres. TW3: Houn 58Da 107
Wills Gro. NW7 22Wa 48
Wills Hill SS17: Stan H 1M 101
Willson Rd. TW20: Eng G 4M 125
Will Wyatt Ct. N1 2H 219 (40Ub 71)
(off Pitfield St.)
Wilman Gro. E8 38Wb 71
Wilmar Cl. UB4: Hayes 42T 84
Wilmar Cl. UB8: Uxb 38M 63
Wilmar Gdns. BR4: W W'ck 74Dc 158
Wilmar Way TN15: Seal 92Pd 203
Wilmcote Ho. W2 43Db 89
(off Woodchester Sq.)
Wilment Ct. NW2 34Ya 68
Wilmer Cl. KT2: King T 64Pa 131
Wilmer Cres. KT2: King T 64Pa 131
Wilmer Gdns. N1 1H 219 (39Ub 71)
(not continuous)
Wilmerhatch La. KT18: Eps 90Ra 173
Wilmer Ho. E3 40Ac 72
(off Daling Way)
Wilmer Lea Cl. E15 38Fc 73
Wilmer Pl. N16 33Vb 71
Wilmer Way N14 22Mb 50
WILMINGTON 62Ld 141
Wilmington Av. BR6: Orp 75Yc 161
Wilmington Av. W4 52Ta 109
Wilmington Cl. WD18: Wat 13X 27
Wilmington Ct. SW16 66Nb 134
Wilmington Ct. Rd. DA2: Wilm 62Jd 140
Wilmington Gdns. IG11: Bark 37Tc 74
Wilmington Sq. WC1 4K 217 (41Qb 90)
(not continuous)
Wilmington St. WC1 4K 217 (41Qb 90)
Wilmington Ter. SE16 49Xb 91
(off Camilla Rd.)
Wilmot Cl. N2 26Eb 49
Wilmot Cl. SE15 52Wb 113
Wilmot Cotts. SM7: Bans 87Db 175
Wilmot Grn. CM13: Gt War 23Yd 58
Wilmot Ho. SE11 5B 230 (49Rb 91)
(off George Mathers Rd.)
Wilmot Pl. NW1 38Lb 70
(not continuous)
Wilmot Pl. W7 46Ga 86
Wilmot Rd. CR8: Purl 84Qb 176
Wilmot Rd. DA1: Dart 57Jd 118
Wilmot Rd. E10 33Dc 72
Wilmot Rd. N17 27Tb 51
Wilmot Rd. SL1: Burn 1A 80
Wilmot Rd. SM5: Cars 78Hb 155
Wilmots Cl. RH2: Reig 5L 207
Wilmot St. E2 42Xb 91
Wilmot Way SM7: Bans 86Cb 175
Wilmount St. SE18 49Rc 94
Wilmslow Ho. RM3: Rom 22Nd 57
(off Chudleigh Rd.)
Wilna Rd. SW18 59Eb 111
Wilsham St. W11 46Za 88
Wilshaw Cl. NW4 27Wa 48
Wilshaw Ho. SE8 52Cc 114
Wilshaw St. SE14 53Cc 114
Wilshere Av. AL1: St A 4A 6
Wilsman Rd. RM15: S Ock 40Yd 78
Wilsmere Dr. HA3: Hrw W 24Ga 46
Wilsmere Dr. UB5: N'olt 37Aa 65
Wilson Av. CR4: Mitc 66Gb 133
Wilson Cl. CR2: S Croy 78Tb 157
Wilson Cl. DA10: Swans 60Ce 121
Wilson Cl. HA9: Wemb 31Pa 67
Wilson Cl. SS17: Stan H 3L 101
Wilson Ct. NW9 26Ta 47
Wilson Ct. RM7: Rush G 30Fd 56
(off Union Rd.)
Wilson Ct. SE28 48Sc 94
Wilson Dr. HA9: Wemb 31Pa 67
Wilson Dr. KT16: Ott 78D 148
Wilson Gdns. HA1: Harr 31Ea 66
Wilson Gro. SE16 47Xb 91
WILSON HOSPITAL, THE 70Hb 133
Wilson Ho. NW6 38Eb 69
(off Goldhurst Ter.)
Wilson La. DA4: S Dar 68Vd 142
Wilson Rd. E6 41Mc 93
Wilson Rd. IG1: Ilf 31Pc 74
Wilson Rd. KT9: Chess 79Pa 153
Wilson Rd. SE5 53Ub 113
Wilsons KT20: Tad 93Za 194
Wilson's Av. N17 26Vb 51
WILSONS CNR. 19Zd 41
(off High St.)
Wilson's Rd. W6 50Za 88
Wilson St. E17 29Ec 52
Wilson St. EC2 7G 219 (43Tb 91)
Wilson St. N21 17Qb 32
Wilson Wlk. W4 49Va 88
(off Prebend Gdns.)
Wilson Way GU21: Wok 8P 167
Wilstone Cl. UB4: Yead 42Aa 85
Wiltern Ct. NW2 37Ab 68
Wilthorne Gdns. RM10: Dag 38Dd 76
Wilton Av. W4 50Ua 88
Wilton Cl. UB7: Harm 51M 105
Wilton Ct. E1 44Xb 91
(off Cavell St.)
Wilton Ct. WD17: Wat 13Y 27
Wilton Cres. SL4: Wind 6B 102
Wilton Cres. SW1 2H 227 (47Jb 90)
Wilton Cres. SW19 67Bb 133
Wilton Dr. RM5: Col R 24Ed 56
Wilton Est. E8 37Wb 71
Wilton Gdns. KT12: Walt T 74Z 151
Wilton Gdns. KT8: W Mole 69Ca 129
Wilton Gro. KT3: N Mald 72Va 154
Wilton Gro. SW19 67Bb 133
Wilton Ho. CR2: S Croy 78Sb 157
(off Nottingham Rd.)
Wilton M. E8 37Xb 71
Wilton M. SW1 3J 227 (48Jb 90)
Wilton Pde. TW13: Felt 61W 128
Wilton Pl. E4 23Fc 53
Wilton Pl. HA1: Harr 30Ha 46
Wilton Pl. KT15: New H 81M 169
Wilton Pl. SW1 2H 227 (47Jb 90)
Wilton Plaza SW1 5B 228 (49Lb 90)
(off Wilton Rd.)
Wilton Rd. EN4: Cockf 14Hb 31
Wilton Rd. N10 26Jb 50
Wilton Rd. RH1: Redh 7P 207
Wilton Rd. SE2 49Yc 95
Wilton Rd. SW1 4A 228 (48Kb 90)
Wilton Rd. SW19 66Gb 133
Wilton Rd. TW4: Houn 55Z 107
Wilton Row SW1 2H 227 (47Jb 90)
Wilton's Music Hall 45Wb 91
(off Graces All.)
Wilton Sq. N1 39Tb 71
Wilton St. SW1 3K 227 (48Kb 90)
Wilton Ter. SW1 3H 227 (48Jb 90)
Wilton Vs. N1 39Tb 71
(off Wilton Sq.)
Wilton Way E8 37Wb 71
Wiltshire Av. RM11: Horn 28Pd 57
Wiltshire Av. SL2: Slou 2G 80
Wiltshire Cl. DA2: Dart 59Td 120
Wiltshire Cl. NW7 22Va 48
Wiltshire Cl. SW3 5F 227 (49Hb 89)
Wiltshire Ct. CR2: S Croy 78Sb 157
Wiltshire Ct. IG1: Ilf 37Sc 74
Wiltshire Ct. N4 32Pb 70
(off Marquis Rd.)
Wiltshire Gdns. N4 30Sb 51
Wiltshire Gdns. TW2: Twick 60Ea 108
Wiltshire La. HA5: Eastc 27V 44
Wiltshire Rd. BR6: Orp 73Wc 161
Wiltshire Rd. CR7: Thor H 69Qb 134
Wiltshire Rd. SW9 55Qb 112
Wiltshire Row N1 39Tb 71
Wilverley Cres. KT3: N Mald 72Ua 154
Wimbart Rd. SW2 59Pb 112
WIMBLEDON 65Ab 132
Wimbledon All England Lawn Tennis & Croquet Club 63Ab 132
Wimbledon Bri. SW19 65Bb 133
Wimbledon Chase Station (Rail) 68Ab 132
Wimbledon Cl. SW20 66Za 132
Wimbledon Common 63Wa 132
Wimbledon Common Golf Course 64Xa 132
Wimbledon Hill Rd. SW19 65Ab 132
Wimbledon Lawn Tennis Mus. 62Ab 132
Wimbledon Leisure Cen. 65Db 133
WIMBLEDON PARK 62Cb 133
Wimbledon Pk. Athletics Track 61Bb 133
Wimbledon Pk. Ct. SW19 60Bb 111
Wimbledon Pk. Golf Course 62Bb 133
Wimbledon Pk. Rd. SW18 60Bb 111
Wimbledon Pk. Rd. SW19 61Ab 132
Wimbledon Pk. Side SW19 62Za 132
Wimbledon Park Station (Underground) 62Cb 133
Wimbledon Pk. Watersports Cen. 61Bb 133
Wimbledon Rd. SW17 63Eb 133
Wimbledon Stadium Bus. Cen. SW17 62Db 133
Wimbledon Station (Rail, Underground & London Tramlink) 65Bb 133
Wimbledon Theatre 66Cb 133
Wimbledon Windmill Mus. 62Xa 132
Wimbolt St. E2 41Wb 91
Wimborne Av. BR5: St P 70Vc 139
Wimborne Av. BR7: Chst 69Vc 139
Wimborne Av. RH1: Redh 10P 207
Wimborne Av. UB2: S'hall 49Ca 85
Wimborne Av. UB4: Yead 44X 85
Wimborne Cl. IG9: Buck H 19Kc 35
Wimborne Cl. KT17: Eps 85Ua 174
Wimborne Cl. KT4: Wor Pk 74Ya 154
Wimborne Cl. SE12 57Hc 115
Wimborne Ct. SW12 62Lb 134
Wimborne Ct. UB5: N'olt 37Ca 65
Wimborne Dr. HA5: Pinn 31Z 65
Wimborne Dr. NW9 27Qa 47
Wimborne Gdns. W13 43Ka 86
Wimborne Gro. WD17: Wat 9U 12
Wimborne Ho. DA12: Grav'nd 2F 144
Wimborne Ho. E16 45Kc 93
(off Victoria Dock Rd.)
Wimborne Ho. NW1 6E 214 (42Gb 89)
(off Harewood Av.)
Wimborne Ho. SW12 62Lb 134
Wimborne Ho. SW8 52Pb 112
(off Dorset Rd.)
Wimborne Rd. N17 26Ub 51
Wimborne Rd. N9 19Wb 33
Wimborne Way BR3: Beck 69Zb 136
Wimbourne Ct. N1 1F 219 (40Tb 71)
(off Wimbourne St.)
Wimbourne St. N1 1F 219 (40Tb 71)
Wimpole Cl. BR2: Broml 70Lc 137
Wimpole Cl. KT1: King T 68Pa 131
Wimpole M. W1 7K 215 (43Kb 90)
Wimpole Rd. UB7: Yiew 46M 83
Wimpole St. W1 1K 221 (43Kb 90)
Wimshurst Cl. CR0: Wadd 74Nb 156
Winans Wlk. SW9 54Qb 112
Winant Ho. E14 45Dc 92
(off Simpson's Rd.)
Wincanton Ct. N11 23Jb 50
(off Martock Gdns.)
Wincanton Cres. UB5: N'olt 36Ca 65
Wincanton Gdns. IG6: Ilf 27Rc 54
Wincanton Rd. RM3: Rom 20Md 39
Wincanton Rd. SW18 59Bb 111
Winchcombe Rd. SM5: Cars 73Fb 155
Winchcomb Gdns. SE9 55Mc 115
Winchdells HP3: Hem H 5A 4
Winchelsea Av. DA7: Bex 52Bd 117
Winchelsea Cl. SW15 57Za 110
Winchelsea Ho. SE16 47Yb 92
(off Swan Rd.)
Winchelsea Rd. E7 34Jc 73

Winchelsea Rd. N17 27Ub 51
Winchelsea Rd. NW10 39Ta 67
Winchelsea Rd. RM18: Tilb 3A 122
Winchelsey Ri. CR2: S Croy 79Vb 157
Winchendon Rd. SW6 53Bb 111
Winchendon Rd. TW11: Tedd 63Fa 130
Winchester Av. NW6 39Ab 68
Winchester Av. NW9 27Qa 47
Winchester Av. RM14: Upm 32Vd 78
Winchester Av. TW5: Hest 51Ba 107
Winchester Bldgs. SE1 1D 230 (47Sb 91)
(off Copperfield St.)
Winchester Cl. BR2: Broml 69Hc 137
Winchester Cl. E6 44Pc 94
Winchester Cl. EN1: Enf 15Ub 33
Winchester Cl. EN9: Walt A 5Dc 20
Winchester Cl. KT10: Esh 77Ca 151
Winchester Cl. KT2: King T 66Ra 131
Winchester Cl. SE17 6C 230 (49Rb 91)
Winchester Cl. SL3: Poyle 53G 104
Winchester Ct. AL1: St A 2D 6
(off Lemsford Rd.)
Winchester Ct. W8 47Cb 89
(off Vicarage Ga.)
Winchester Cres. DA12: Grav'nd 2F 144
Winchester Dr. HA5: Pinn 29Z 45
Winchester Gro. TN13: S'oaks 95Kd 203
Winchester Ho. AL3: St A 3A 6
Winchester Ho. E14 44Cc 92
(off New Festival Av.)
Winchester Ho. E3 41Bc 92
(off Hamlets Way)
Winchester Ho. IG11: Bark 38Wc 75
(off Margaret Bondfield Av.)
Winchester Ho. KT19: Eps 84Qa 173
(off Phoenix Cl.)
Winchester Ho. SE18 52Mc 115
(off Portway Gdns.)
Winchester Ho. SW3 51Fb 111
(off Beaufort St.)
Winchester Ho. SW9 52Qb 112
Winchester Ho. W2 3A 220 (44Eb 89)
(off Hallfield Est.)
Winchester M. KT4: Wor Pk 75Za 154
Winchester M. NW3 38Fb 69
(off Winchester Rd.)
Winchester Palace 6F 225 (46Tb 91)
(off Stoney St.)
Winchester Pk. BR2: Broml 69Hc 137
Winchester Pl. E8 36Vb 71
Winchester Pl. N6 32Kb 70
Winchester Rd. BR2: Broml 69Hc 137
Winchester Rd. BR6: Chels 77Yc 161
Winchester Rd. DA7: Bex 54Zc 117
Winchester Rd. E4 24Ec 52
Winchester Rd. HA3: Kenton 28Na 47
Winchester Rd. HA6: Nwood 26V 44
Winchester Rd. IG1: Ilf 34Tc 74
Winchester Rd. KT12: Walt T 74W 150
Winchester Rd. N6 31Kb 70
Winchester Rd. N9 18Vb 33
Winchester Rd. NW3 38Fb 69
Winchester Rd. TW1: Twick 58Ka 108
Winchester Rd. TW13: Hanw 62Ba 129
Winchester Rd. UB3: Harl 52U 106
Winchester Sq. SE1 6F 225 (46Tb 91)
(off Winchester Wlk.)
Winchester Sq. SE8 49Bc 92
Winchester St. SW1 7A 228 (50Kb 90)
Winchester St. W3 46Sa 87
Winchester Wlk. SE1 6F 225 (46Tb 91)
Winchester Wlk. SW15 57Bb 111
(off Up. Richmond Rd.)
Winchester Way WD3: Crox G 15R 26
Winchester Wharf SE1 6F 225 (46Tb 91)
(off Clink St.)
Winchet Wlk. CR0: C'don 72Yb 158
Winchfield Cl. HA3: Kenton 30La 46
Winchfield Ho. SW15 58Va 110
Winchfield Rd. SE26 64Ac 136
Winchfield Way WD3: Rick 17L 25
Winch Ho. E14 48Dc 92
(off Tiller Rd.)
Winch Ho. SW10 52Eb 111
(off King's Rd.)
Winchilsea Cres. KT8: W Mole 68Ea 130
Winchilsea Ho. NW8 4B 214 (41Fb 89)
(off St John's Wood Rd.)
WINCHMORE HILL 17Qb 32
Winchmore Hill Rd. N14 18Mb 32
Winchmore Hill Rd. N21 17Nb 32
Winchmore Hill Station (Rail) 17Rb 33
Winchmore School Sports Cen. 19Sb 33
Winchmore Vs. N21 17Pb 32
(off Winchmore Hill Rd.)
Winch's Mdw. SL1: Burn 1A 80
Winchstone Cl. TW17: Shep 70P 127
Winckley Cl. HA3: Kenton 29Pa 47
Winckworth Ct. N1 4G 219 (41Tb 91)
(off Charles Sq. Est.)
Wincott Pde. SE11 5A 230 (49Qb 90)
(off Wincott St.)
Wincott St. SE11 6A 230 (49Qb 90)
Wincrofts Dr. SE9 56Tc 116
Windall Cl. SE19 67Wb 135
Windborough Rd. SM5: Cars 80Jb 156
Windermere NW1 4A 216 (41Kb 90)
(off Albany St.)
Windermere Av. AL1: St A 4F 6
(not continuous)
Windermere Av. HA4: Ruis 31Y 65
Windermere Av. HA9: Kenton 31La 66
Windermere Av. HA9: Wemb 31La 66
Windermere Av. N3 27Cb 49
Windermere Av. NW6 39Ab 68
Windermere Av. RM12: Horn 36Jd 76
Windermere Av. RM19: Purf 50Sd 98
Windermere Av. SW19 69Db 133
Windermere Cl. BR6: Farnb 76Rc 160
Windermere Cl. DA1: Dart 60Kd 119
Windermere Cl. HP3: Hem H 3C 4
Windermere Cl. TW14: Felt 60V 106
Windermere Cl. TW19: Stanw 60N 105
Windermere Cl. TW20: Egh 66D 126
Windermere Cl. WD3: Chor 15F 24
Windermere Ct. CR8: Kenley 87Rb 177
Windermere Ct. GU21: Wok 10L 167
(off St John's Rd.)
Windermere Ct. HA9: Wemb 31La 66
Windermere Ct. SM5: Cars 76Jb 156
Windermere Ct. SW13 51Va 110
Windermere Ct. WD17: Wat 12W 26
Windermere Gdns. IG4: Ilf 29Nc 54
Windermere Gro. HA9: Wemb 32La 66
Windermere Hall HA8: Edg 22Pa 47
Windermere Ho. E3 42Bc 92
Windermere Ho. EN5: New Bar 14Db 31
Windermere Ho. TW7: Isle 57Ha 108
Windermere Point SE15 52Yb 114
(off Old Kent Rd.)
Windermere Rd. BR4: W W'ck 75Gc 159
Windermere Rd. CR0: C'don 74Vb 157
Windermere Rd. CR5: Coul 87Nb 176
Windermere Rd. DA7: Bex 54Ed 118
Windermere Rd. GU18: Light 2A 166
Windermere Rd. N10 25Kb 50
Windermere Rd. N19 33Lb 70
Windermere Rd. SW15 63Ua 132
Windermere Rd. SW16 67Lb 134
Windermere Rd. UB1: S'hall 43Ba 85
Windermere Rd. W5 48La 86
Windermere Way RH2: Reig 5N 207
Windermere Way SL1: Slou 3A 80
Windermere Way UB7: Yiew 46N 83
Winders Rd. SW11 54Gb 111
(not continuous)
Windfield KT22: Lea 93Ka 192
Windfield Cl. SE26 63Zb 136
Windham Av. CR0: New Ad 82Fc 179
Windham Rd. TW9: Rich 55Pa 109
Windhover Way DA12: Grav'nd 3G 144
Windings, The CR2: Sande 83Vb 177
Winding Shot HP1: Hem H 1J 3
Winding Way RM8: Dag 34Yc 75
Windlass Ho. E16 47Kc 93
(off Schooner Rd.)
Windlass Pl. SE8 49Ac 92
Windlebrook Pk. KT16: Longc 75B 148
Windle Cl. GU20: W'sham 9B 146
WINDLESHAM 9B 146
Windlesham Ct. GU20: W'sham 6A 146
Windlesham Ct. Dr. GU20: W'sham 7A 146
Windlesham Gro. SW19 60Za 110
Windlesham Ho. SE1 7K 225 (46Vb 91)
(off Duchess Wlk.)
Windlesham M. TW12: Hamp H 65Ea 130
Windlesham Rd. GU24: Chob 10E 146
Windlesham Rd. GU24: W End 3C 166
Windley Cl. SE23 61Yb 136
Windmill WC1 7H 217 (43Pb 90)
(off New North St.)
Windmill Av. KT17: Ewe 83Va 174
Windmill Av. UB2: S'hall 46Ea 86
Windmill Bri. Ho. CR0: C'don 74Ub 157
(off Freemasons Rd.)
Windmill Bus. Village TW16: Sun 67U 128
Windmill Cl. CR3: Cat'm 93Sb 197
Windmill Cl. EN9: Walt A 6Gc 21
Windmill Cl. KT17: Eps 84Va 174
Windmill Cl. KT6: Surb 74La 152
Windmill Cl. RM14: Upm 33Qd 77
Windmill Cl. SE1 49Wb 91
(off Beatrice Rd.)
Windmill Cl. SE13 54Ec 114
Windmill Cl. SL4: Wind 4F 102
Windmill Cl. TW16: Sun 66U 128
Windmill Ct. E4 19Gc 35
Windmill Ct. HA4: Ruis 32W 64
Windmill Ct. NW2 37Ab 68
Windmill Ct. W5 49La 86
(off Windmill Rd.)
Windmill Dr. BR2: Kes 77Lc 159
Windmill Dr. KT22: Lea 95La 192
Windmill Dr. NW2 34Ab 68
Windmill Dr. RH2: Reig 4M 207
Windmill Dr. SW4 57Kb 112
Windmill Dr. WD3: Crox G 16P 25
Windmill End KT17: Eps 84Va 174
Windmill Fld. GU20: W'sham 9A 146
Windmill Gdns. EN2: Enf 13Qb 32
Windmill Grange TN15: W King 80Ud 164
Windmill Grn. TW17: Shep 73U 150
Windmill Gro. CR0: C'don 72Sb 157
Windmill Hill EN2: Enf 13Rb 33
Windmill Hill HA4: Ruis 31V 64
Windmill Hill NW3 34Eb 69
Windmill Hill TN15: Wro H 93Fe 205
Windmill Hill WD4: Chfd 4H 11
(not continuous)
Windmill Ho. E14 49Cc 92
Windmill Ho. SE1 1A 230 (47Qb 90)
(off Windmill Wlk.)
Windmill La. E15 37Fc 73
Windmill La. EN5: Ark 16Va 30
Windmill La. EN8: Chesh 2Ac 20
Windmill La. KT17: Eps 84Va 174
Windmill La. KT6: Surb 72Ka 152
Windmill La. TW7: Isle 49Ga 86
Windmill La. UB2: S'hall 46Ea 86
Windmill La. UB6: G'frd 44Ea 86
Windmill La. WD23: B Hea 18Ga 28
Windmill M. W4 49Ua 88
Windmill Pk. TN15: Wro H 92Fe 205
Windmill Pas. W4 49Ua 88
Windmill Pl. CR5: Coul 92Rb 197
Windmill Pl. UB2: S'hall 46Ea 86
Windmill Ri. CR6: W'ham 92Yb 198
Windmill Ri. KT2: King T 66Ra 131
Windmill Rd. CR0: C'don 73Sb 157
Windmill Rd. CR4: Mitc 71Lb 156
Windmill Rd. HP2: Hem H 2N 3
Windmill Rd. N18 21Tb 51
Windmill Rd. SL1: Slou 6H 81
Windmill Rd. SL3: Ful 6P 61
Windmill Rd. SW18 58Fb 111
Windmill Rd. SW19 61Xa 132
Windmill Rd. TW12: Hamp H 64Da 129
Windmill Rd. TW16: Sun 67U 128
Windmill Rd. TW8: Bford 50La 86
Windmill Rd. W4 49Ua 88
Windmill Rd. W5 49La 86
Windmill Rd. W. TW16: Sun 68U 128
Windmill Row SE11 7K 229 (50Qb 90)
Windmills, The KT22: Lea 95La 192
Windmill Shott TW20: Egh 65B 126
Windmill St. DA12: Grav'nd 8D 122
Windmill St. W1 1D 222 (43Mb 90)
(not continuous)
Windmill St. WD23: B Hea 18Ga 28
Windmill Ter. TW17: Shep 73U 150
Windmill Wlk. SE1 7A 224 (46Qb 90)
Windmill Way HA4: Ruis 32V 64
Windmill Way RH2: Reig 4M 207
Windmore Av. EN6: Pot B 3Ya 16
Windmore Cl. HA0: Wemb 36Ja 66
Windridge Cl. AL3: St A 4N 5
Windrose Cl. SE16 47Zb 92
Windrush KT3: N Mald 70Ra 131
Windrush SE28 46Xc 95
Windrush Av. SL3: L'ly 48D 82
Windrush Cl. E8 38Wb 71
Windrush Cl. N17 25Ub 51
Windrush Cl. SW11 56Fb 111
Windrush Cl. UB10: Ick 35P 63
Windrush Cl. W4 53Sa 109
Windrush Ct. DA8: Erith 50Gd 96
Windrushes CR3: Cat'm 97Wb 197
Windrush Ho. NW8 6C 214 (42Fb 89)
(off Church St. Est.)
Windrush La. SE23 62Zb 136
Windrush Rd. NW10 39Ta 67
Windrush Rd. RM18: Tilb 2A 122
Windrush Sq. SW2 56Qb 112
Winds End Cl. HP2: Hem H 1A 4
Windsock Cl. SE16 49Bc 92
Windsock Way TW6: H'row A 54K 105
WINDSOR 3H 103
Windsor & Eton Central Station (Rail) 3H 103
Windsor & Eton Riverside Station (Rail) 2H 103
Windsor & Royal Borough Mus. 3H 103
Windsor Av. E17 26Ac 52
Windsor Av. HA8: Edg 21Ra 47
Windsor Av. KT3: N Mald 71Sa 153
Windsor Av. KT8: W Mole 69Ca 129
Windsor Av. RM16: Grays 47De 99
Windsor Av. SM3: Cheam 76Ab 154
Windsor Av. SW19 67Eb 133
Windsor Av. UB10: Hil 39R 64
Windsor Boys' School Sports Cen. 3F 102
Windsor Bus. Cen. SL4: Wind 2G 102
Windsor Castle 2J 103
Windsor Cen., The N1 39Rb 71
(off Windsor St.)
Windsor Cl. BR7: Chst 64Rc 138
Windsor Cl. EN7: Chesh 2Wb 19
Windsor Cl. HA2: Harr 34Ca 65
Windsor Cl. HA6: Nwood 26W 44
Windsor Cl. HP2: Hem H 4N 3
Windsor Cl. HP3: Bov 10C 2
Windsor Cl. N3 26Ab 48
Windsor Cl. SE27 63Sb 135
Windsor Cl. SL1: Burn 2A 80
Windsor Cl. TW6: H'row A 55L 105
(off Whittle Rd.)
Windsor Cl. TW8: Bford 51Ka 108
Windsor Cl. WD6: Bore 11Qa 29
Windsor Club, The 3F 102
Windsor Cotts. SE14 52Bc 114
(off Amersham Gro.)
Windsor Ct. AL1: St A 3E 6
Windsor Ct. CR3: Whyt 90Vb 177
Windsor Ct. E3 40Cc 72
(off Mostyn Gro.)
Windsor Ct. GU24: Chob 1J 167
Windsor Ct. HA5: Pinn 27Z 45
Windsor Ct. KT1: King T 70Ma 131
(off Palace Rd.)
Windsor Ct. KT18: Eps 85Ta 173
(off Ashley Rd.)
Windsor Ct. N11 22Hb 49
Windsor Ct. N14 17Lb 32
Windsor Ct. NW11 30Ab 48
(off Golders Grn. Rd.)
Windsor Ct. NW3 35Cb 69
Windsor Ct. SE16 45Zb 92
(off King & Queen Wharf)
Windsor Ct. SW11 54Fb 111
Windsor Ct. SW3 7E 226 (50Gb 89)
(off Jubilee Pl.)
Windsor Ct. TW16: Sun 66W 128
Windsor Ct. W10 44Za 88
(off Bramley Rd.)
Windsor Ct. W2 45Db 89
(off Moscow Rd.)
Windsor Ct. WD23: Bush 17Ea 28
(off Catsey La.)
Windsor Ct. WD4: K Lan 1Q 12
Windsor Ct. WD6: Bore 12Sa 29
Windsor Ct. Rd. GU24: Chob 1J 167
Windsor Cres. HA2: Harr 34Ca 65
Windsor Cres. HA9: Wemb 34Ra 67
Windsor Dr. BR6: Chels 79Wc 161
Windsor Dr. DA1: Dart 58Jd 118
Windsor Dr. EN4: E Barn 16Hb 31
Windsor Dr. TW15: Ashf 63M 127
Windsor Gdns. CR0: Bedd 76Nb 156
Windsor Gdns. UB3: Harl 48T 84
Windsor Gdns. W9 43Cb 89
Windsor Great Pk. 2F 124
Windsor Grey Cl. SL5: Asc 9A 124
Windsor Gro. SE27 63Sb 135
Windsor Hall E16 46Kc 93
(off Wesley Av.)
Windsor Home Park Park & Ride 1J 103
Windsor Ho. E2 41Zb 92
(off Knottisford St.)
Windsor Ho. E20 36Dc 72
(off Peloton Av.)
Windsor Ho. KT16: Chert 72J 149
(off Windsor St.)
Windsor Ho. N1 1E 218 (40Sb 71)
Windsor Ho. NW1 3A 216 (41Kb 90)
(off Cumberland Mkt.)
Windsor Ho. NW2 37Ab 68
(off Chatsworth Rd.)
Windsor Ho. SW16 68Pb 134
Windsor Ho. TW18: Staines 63H 127
Windsor Ho. UB5: N'olt 37Ca 65
(off The Farmlands)
Windsor Ho. WD23: Bush 14Ba 27
Windsor Info. Cen. 3H 103
Windsor La. SL1: Burn 2A 80
Windsor Lawn Tennis Club 3F 102
Windsor Legoland Park & Ride 7C 102
Windsor Leisure Cen. 2F 102
Windsor M. SE23 60Ac 114
Windsor M. SE6 60Ec 114
Windsor Pk. Rd. UB3: Harl 52V 106
Windsor Pl. KT16: Chert 72J 149
Windsor Pl. SW1 4C 228 (48Lb 90)
Windsor Rd. CM15: Pil H 16Xd 40
Windsor Rd. CR7: Thor H 68Rb 135
Windsor Rd. DA12: Grav'nd 2D 144
Windsor Rd. DA6: Bex 56Ad 117
Windsor Rd. E10 33Dc 72
Windsor Rd. E11 32Jc 73
Windsor Rd. E4 21Dc 52
Windsor Rd. E7 36Kc 73
Windsor Rd. EN3: Enf W 8Zb 20
Windsor Rd. EN5: Barn 16Za 30
Windsor Rd. GU24: Chob 7G 146
Windsor Rd. HA3: Hrw W 25Ea 46
Windsor Rd. IG1: Ilf 35Rc 74
Windsor Rd. KT2: King T 66Na 131
Windsor Rd. KT4: Wor Pk 75Wa 154
Windsor Rd. N13 20Qb 32
Windsor Rd. N17 26Wb 51
Windsor Rd. N3 26Ab 48
Windsor Rd. N7 34Nb 70
Windsor Rd. NW2 37Xa 68
Windsor Rd. RM11: Horn 31Ld 77
Windsor Rd. RM8: Dag 34Ad 75
Windsor Rd. SL1: Slou 8J 81
Windsor Rd. SL2: Ger X 5L 61
Windsor Rd. SL2: Stoke P 5L 61
Windsor Rd. SL3: Dat 2L 103
Windsor Rd. SL4: Old Win 10P 103
Windsor Rd. SL4: Wat O 2A 102
Windsor Rd. SL4: Wind 2A 102
Windsor Rd. SL4: Wink 5A 124
Windsor Rd. SL9: Ger X 2M 61
Windsor Rd. TW11: Tedd 64Fa 130
Windsor Rd. TW16: Sun 65W 128
Windsor Rd. TW19: Wray 58A 104
Windsor Rd. TW20: Egh 62B 126
Windsor Rd. TW4: Cran 54X 107
Windsor Rd. TW9: Kew 54Pa 109
Windsor Rd. UB2: S'hall 48Ba 85
Windsor Rd. W5 45Na 87
(not continuous)
Windsor Rd. WD24: Wat 10Y 13
Windsors, The IG9: Buck H 19Nc 36
Windsor St. KT16: Chert 72J 149
Windsor St. N1 39Rb 71
Windsor St. UB8: Uxb 39L 63
Windsor Ter. N1 3E 218 (41Sb 91)
Windsor Theatre Royal 2H 103
Windsor Wlk. KT12: Walt T 74Z 151
Windsor Wlk. KT13: Weyb 78R 150
Windsor Wlk. SE5 54Tb 113
Windsor Way GU22: Wok 88E 168
Windsor Way W14 49Za 88
Windsor Way WD3: Rick 18J 25
Windsor Wharf E9 36Bc 72
Windsor Wood EN9: Walt A 5Gc 21
Windspoint Dr. SE15 51Xb 113
Winds Ridge GU23: Send 97E 188
Windus M. N16 32Vb 71
Windus Rd. N16 32Vb 71
Windus Wlk. N16 32Vb 71
Windward Cl. EN3: Enf W 7Zb 20
Windward Ct. E16 45Rc 94
(off Gallions Rd.)
Windycroft Cl. CR8: Purl 85Mb 176
Windy Hill CM13: Hut 18Ee 41
Windy Ridge BR1: Broml 67Nc 138
Windy Ridge Cl. SW19 64Za 132
Wine Cl. E1 46Yb 92
(not continuous)
Wine Office Ct. EC4: Peterborough Ct. 3A 224 (44Qb 90)
Winern Glebe KT14: Byfl 85M 169
Winery La. KT1: King T 69Pa 131
Winey Cl. KT9: Chess 80La 152
Winfield La. TN15: Bor G 97Ae 205
Winfield Mobile Home Pk. WD25: A'ham 11Da 27
Winford Ct. SE15 53Xb 113
Winford Ho. E3 38Bc 72
Winford Pde. UB1: S'hall 44Da 85
(off Marconi Way)
Winforton St. SE10 53Ec 114
Winfrith Rd. SW18 59Eb 111
Wingate & Finchley FC 24Fb 49
Wingate Bus. Cen. AL9: Wel G 5E 8
Wingate Cres. CR0: C'don 72Nb 156
Wingate Ho. E3 41Dc 92
(off Bruce Rd.)
Wingate Rd. DA14: Sidc 65Yc 139
Wingate Rd. IG1: Ilf 36Rc 74
Wingate Rd. W6 48Xa 88
Wingate Sq. SW4 55Lb 112
Wingate Way AL1: St A 3E 6
Wingfield RM17: Grays 50Be 99
Wingfield Bank DA11: Nflt 61Ee 143
Wingfield Cl. CM13: B'wood 20Ce 41
Wingfield Cl. KT15: New H 82K 169
Wingfield Ct. DA15: Sidc 61Vc 139
Wingfield Ct. E14 45Fc 93
(off Newport Av.)
Wingfield Ct. SM7: Bans 87Cb 175
Wingfield Ct. WD18: Wat 16S 26
Wingfield Dr. RM16: Ors 3C 100
Wingfield Gdns. RM14: Upm 30Ud 58
Wingfield Ho. E2 4K 219 (41Vb 91)
(off Virginia Rd.)
Wingfield M. SE15 55Wb 113
Wingfield Rd. DA12: Grav'nd 9D 122
Wingfield Rd. E15 35Gc 73
Wingfield Rd. E17 29Dc 52
Wingfield Rd. KT2: King T 65Pa 131
Wingfield St. SE15 55Wb 113
Wingfield Way HA4: Ruis 36X 65
Wingford Rd. SW2 58Nb 112
Wingletye La. RM11: Horn 28Pd 57
Wingmore Rd. SE24 55Sb 113
Wingrad Ho. E1 43Yb 92
(off Jubilee St.)
Wingrave Cres. CM14: B'wood 21Ud 58
Wingrave Rd. W6 51Ya 110
Wingreen NW8 39Db 69
(off Abbey Rd.)
Wingrove E4 17Cc 34
Wingrove Ct. RM7: Rom 29Ed 56
Wingrove Dr. RM19: Purf 50Rd 97
Wingrove Rd. SE6 61Gc 137
Wings Cl. SM1: Sutt 77Cb 155
Wings Rd. TW6: H'row A 55K 105
(off Whittle Rd.)
Wingway CM14: B'wood 18Yd 40
Wing Yip Bus. Cen. NW2 33Xa 68
Winicotte Ho. W2 7C 214 (43Fb 89)
(off Paddington Grn.)
Winifred Av. RM12: Horn 35Md 77
Winifred Cl. EN5: Ark 16Va 30
Winifred Dell Ho. CM13: Gt War 23Yd 58
Winifred Pl. N12 22Eb 49
Winifred Rd. CR5: Coul 88Jb 176
Winifred Rd. DA1: Dart 57Kd 119
Winifred Rd. DA8: Erith 50Gd 96
Winifred Rd. HP3: Hem H 6M 3
Winifred Rd. RM8: Dag 33Ad 75
Winifred Rd. SW19 67Cb 133
Winifred Rd. TW12: Hamp H 63Ca 129
Winifred St. E16 46Pc 94
Winifred Ter. EN1: Enf 17Vb 33
Winkers Cl. SL9: Chal P 25B 42
Winkers La. SL9: Chal P 25B 42
Winkfield Rd. E13 40Kc 73
Winkfield Rd. N22 25Qb 50
Winkfield Rd. SL4: Wind 10A 102
Winkfield Rd. SL4: Wink 10A 102
Winkley Ct. HA2: Harr 34Ca 65
Winkley Ct. N10 28Kb 50
(off St James's La.)
Winkley St. E2 40Xb 71
Winkwell HP1: Hem H 4F 2
WINKWELL 4F 2
Winkworth Cotts. E1 42Yb 92
(off Cephas St.)
Winkworth Pl. SM7: Bans 86Bb 175
Winkworth Rd. SM7: Bans 86Cb 175
Winlaton Rd. BR1: Broml 63Fc 137
Winmill Rd. RM8: Dag 34Bd 75
Winnards GU21: Wok 10L 167
Winn Comn. Rd. SE18 51Uc 116
Winnepeg Ho. SE16 47Yb 92
(off Province Dr.)
Winnett St. W1 4D 222 (45Mb 90)
Winningales Ct. IG5: Ilf 26Nc 54
Winnings Wlk. UB5: N'olt 37Aa 65
Winnington Cl. N2 30Fb 49
Winnington Ho. SE5 52Sb 113
(off Wyndham Est.)
Winnington Ho. W10 42Ab 88
(off Southern Row)
Winnington Rd. EN3: Enf W 10Yb 20
Winnington Rd. N2 30Fb 49
Winnington Way GU21: Wok 10M 167
Winnipeg Dr. BR6: Chels 79Vc 161
Winnock Rd. UB7: Yiew 46M 83
Winn Rd. SE12 60Jc 115
Winns Av. E17 27Bc 52
Winns M. N15 28Ub 51
Winns Ter. E17 27Cc 52
Winsbeach E17 26Fc 53
Winscombe Cres. W5 42Ma 87
Winscombe St. N19 33Kb 70
Winscombe Way HA7: Stan 22Ja 46
Winsford Rd. SE6 62Bc 136
Winsford Ter. N18 22Tb 51
Winsham Gro. SW11 57Jb 112
Winsham Ho. NW1 3E 216 (41Mb 90)
(off Churchway)
Winslade Rd. SW2 57Nb 112
Winslade Way SE6 59Dc 114
Winsland M. W2 2B 220 (44Fb 89)
Winsland St. W2 2B 220 (44Fb 89)
Winsley St. W1 2B 222 (44Lb 90)
Winslow SE17 7H 231 (50Ub 91)
Winslow Cl. HA5: Eastc 30X 45
Winslow Cl. NW10 34Ua 68
Winslow Gro. E4 19Gc 35
Winslow Rd. W6 51Ya 110
Winslow Way KT12: Walt T 76Y 151
Winslow Way TW13: Hanw 62Z 129
Winsmoor Ct. EN2: Enf 13Rb 33
WINSOR PARK 43Rc 94
Winsor Ter. E6 43Qc 94
Winstanley Cl. KT11: Cobh 86X 171
Winstanley Est. SW11 55Fb 111
Winstanley Rd. SW11 55Fb 111
(not continuous)
Winstanley Wlk. KT11: Cobh 86X 171
(off Winstanley Cl.)
Winstead Gdns. RM10: Dag 36Ed 76
Winston Av. NW9 31Ua 68
Winston Churchill School Sports Cen. 10J 167
Winston Churchill Way EN8: Walt C 5Yb 20
Winston Cl. DA9: Ghithe 58Vd 120
Winston Cl. HA3: Hrw W 23Ha 46
Winston Cl. KT8: E Mos 72Fa 152
Winston Cl. RM7: Mawney 28Dd 56
Winston Ct. BR1: Broml 67Kc 137
(off Widmore Rd.)
Winston Ct. HA3: Hrw W 24Da 45
Winston Dr. KT11: Stoke D 88Aa 171
Winston Dr. TN16: Big H 89Mc 179
Winston Ho. W13 47Ja 86
(off Balfour Rd.)
Winston Ho. WC1 5E 216 (42Mb 90)
(off Endsleigh St.)
Winston Rd. N16 35Tb 71
Winston Wlk. W4 48Ta 87
Winston Way EN6: Pot B 5Cb 17
Winston Way GU22: Wok 92D 188
Winston Way IG1: Ilf 34Rc 74
Winstre Rd. WD6: Bore 11Qa 29
Winter Av. E6 39Nc 74
Winterborne Av. BR6: Orp 76Tc 160
Winterbourne Gro. KT13: Weyb 79S 150
Winterbourne Ho. W11 45Ab 88
(off Portland Rd.)
Winterbourne M. RH8: Oxt 2G 210
Winterbourne Rd. CR7: Thor H 70Qb 134
Winterbourne Rd. RM8: Dag 33Yc 75
Winterbourne Rd. SE6 60Bc 114
Winter Box Wlk. TW10: Rich 57Pa 109
Winterbrook Rd. SE24 58Sb 113
Winterburn Cl. N11 23Jb 50
Winter Cl. KT17: Eps 84Ua 174
Winterdown Gdns. KT10: Esh 79Ba 151
Winterdown Rd. KT10: Esh 79Ba 151
Winterfold Cl. SW19 61Ab 132
Wintergarden DA9: Bluew 59Wd 120
Wintergarden Cres. DA9: Bluew 59Wd 120
Winter Gdns. TW11: Tedd 63Ja 130
Wintergreen Blvd. UB7: W Dray 47P 83
Wintergreen Cl. E6 43Nc 94
Winterleys NW6 40Bb 69
(off Denmark Rd.)
Winter Lodge SE16 50Wb 91
(off Fern Wlk.)
Winter's Ct. E4 20Dc 34
Winters Cft. DA12: Grav'nd 5F 144
Wintersells Ind. Est. KT14: Byfl 82N 169
Wintersells Rd. KT14: Byfl 82M 169
Winterslow Ho. SE5 54Sb 113
(off Flaxman Rd.)
Winterslow Rd. SW9 53Rb 113
Winters Rd. KT7: T Ditt 73Ka 152
Winterstoke Gdns. NW7 22Wa 48
Winterstoke Rd. SE6 60Bc 114
Winters Way EN9: Walt A 5Jc 21
Winterton Ct. KT1: Hamp W 67Ma 131
(off Lwr. Teddington Rd.)
Winterton Ct. SE20 68Wb 135
Winterton Ct. TN16: Westrm 99Tc 200
(off Market Sq.)
Winterton Ho. E1 44Yb 92
(off Deancross St.)
Winterton Pl. SW10 51Eb 111
Winterwell Rd. SW2 57Nb 112
Winthorpe Gdns. WD6: Bore 11Pa 29
Winthorpe Rd. SW15 56Ab 110
Winthrop Ho. W12 45Xa 88
(off White City Est.)
Winthrop St. E1 43Xb 91
Winthrop Wlk. HA9: Wemb 34Na 67
(off Everard Way)
Winton App. WD3: Crox G 15S 26
Winton Av. N11 24Lb 50
Winton Cl. N9 17Zb 34
Winton Ct. BR8: Swan 70Gd 140
Winton Ct. N1 2H 217 (40Pb 70)
(off Calshot St.)
Winton Cres. WD3: Crox G 15R 26
Winton Dr. EN8: Chesh 1Ac 20
Winton Dr. WD3: Crox G 16R 26
Winton Gdns. HA8: Edg 24Pa 47
Winton Rd. BR6: Farnb 77Rc 160
Winton Ter. AL1: St A 3C 6
(off Old London Rd.)

Winton Way SW16 64Qb 134
Wintoun Path SL2: Slou 2C 80
Winvale SL1: Slou 8J 81
Winwood SL2: Slou 4N 81
Winwood SL4: Wind 3C 102
Wireless Rd. TN16: Big H 87Mc 179
Wireworks Ct. SE1 2D 230 (47Sb 91)
(off Gt. Sufolk St.)
Wirrall Ho. SE26 62Wb 135
Wirral Wood Cl. BR7: Chst 65Qc 138
Wirra Rd. TW6: H'row A 54K 105
(off Wayfarer Rd.)
Wisbeach Rd. CR0: C'don 71Tb 157
Wisbech N4 32Pb 70
(off Lorne Rd.)
Wisborough Rd. CR2: Sande 81Vb 177
Wisden Ho. SW8 51Pb 112
Wisdom Ct. TW7: Isle 55Ja 108
(off South St.)
Wisdons Cl. RM10: Dag 32Dd 76
Wise Ct. WD18: Wat 14U 26
(off Raven Cl.)
Wise La. NW7 22Wa 48
Wise La. UB7: W Dray 48M 83
Wiseman Rd. E10 33Cc 72
Wise Rd. E15 39Fc 73
Wise's La. AL9: N Mym 9E 8
Wise's La. TN15: Ash 81Zd 185
Wise's La. TN15: Stans 81Zd 185
Wiseton Rd. SW17 60Gb 111
Wishart Rd. SE3 54Mc 115
Wishaw Wlk. N13 23Nb 50
Wishbone Way GU21: Wok 8K 167
Wishford Ct. KT21: Asht 90Pa 173
WISLEY 88N 169
WISLEY COMMON 88Q 170
Wisley Common, Ockham & Chatley Heath Nature Reserve 90T 170
Wisley Ct. CR2: Sande 82Tb 177
Wisley Ct. RH1: Redh 5P 207
(off Clarendon Rd.)
Wisley Golf Course 89M 169
Wisley Ho. SW1 7D 228 (50Mb 90)
(off Rampayne St.)
WISLEY INTERCHANGE 88S 170
Wisley La. GU23: Wis 88L 169
Wisley Rd. BR5: St P 66Wc 139
Wisley Rd. SW11 57Jb 112
Wistaria Cl. BR6: Farnb 75Rc 160
Wistaria Cl. CM15: Pil H 15Yd 40
Wistaria Dr. AL2: Lon C 8F 6
Wisteria Apts. E9 37Yb 72
(off Chatham Pl.)
Wisteria Cl. IG1: Ilf 36Rc 74
Wisteria Cl. NW7 23Va 48
Wisteria Gdns. BR8: Swan 68Fd 140
Wisteria Rd. SE13 56Fc 115
Wistlea Cres. AL4: Col H 4M 7
Wistow Ho. E2 39Wb 71
(off Whiston Rd.)
Witanhurst La. N6 32Jb 70
Witan St. E2 41Xb 91
Witches La. TN13: Riv 94Fd 202
Witchwood Ho. SW9 55Qb 112
(off Gresham Rd.)
Witham Cl. IG10: Lough 16Nc 36
Witham Ct. E10 34Dc 72
Witham Ct. SW17 62Hb 133
Witham Gdns. CM13: W H'dn 30Fe 59
Witham Rd. RM10: Dag 36Cd 76
Witham Rd. RM2: Rom 29Kd 57
Witham Rd. SE20 69Yb 136
Witham Rd. TW7: Isle 53Fa 108
Witham Rd. W13 46Ja 86
Withens Cl. BR5: St M Cry 70Yc 139
Witherby Cl. CR0: C'don 78Ub 157
Witherings, The RM11: Horn 29Nd 57
Witherington Rd. N5 36Qb 70
Withers Cl. KT9: Chess 79La 152
Withers Mead NW9 25Wa 48
Withers Pl. EC1 5E 218 (42Sb 91)
Witherston Way SE9 61Qc 138
Withey Beds Local Nature Reserve, The 19R 26
Withey Cl. SL4: Wind 3C 102
Witheygate Av. TW18: Staines 65K 127
Withies, The GU21: Knap 9J 167
Withies, The KT22: Lea 92Ka 192
Withybed Cnr. KT20: Walt H 95Xa 194
Withy Cl. GU18: Light 2A 166
Withycombe Rd. SW19 59Za 110
Withycroft SL3: Geor G 44A 82
Withy Ho. E1 42Zb 92
(off Globe Rd.)
Withy La. HA4: Ruis 29S 44
Withy Mead E4 20Fc 35
Withy Pl. AL2: Park 10A 6
Witley Ct. WC1 5F 217 (42Nb 90)
(off Coram St.)
Witley Cres. CR0: New Ad 79Ec 158
Witley Gdns. UB2: S'hall 49Ba 85
Witley Ho. SW2 59Nb 112
Witley Ind. Est. UB2: S'hall 49Ba 85
Witley Point SW15 60Xa 110
(off Wanborough Dr.)
Witley Rd. N19 33Lb 70
Witney Cl. HA5: Hat E 23Ba 45
Witney Cl. UB10: Ick 35P 63
Witney Path SE23 62Zb 136
Wittenham Way E4 20Fc 35
Wittering Cl. KT2: King T 64Ma 131
Wittering Wlk. RM12: Horn 37Ld 77
Wittersham Rd. BR1: Broml 64Hc 137
Witts Ho. KT1: King T 69Pa 131
(off Winery La.)
Wivenhoe Cl. SE15 55Xb 113
Wivenhoe Ct. TW3: Houn 56Ba 107
Wivenhoe Rd. IG11: Bark 40Wc 75
Wiverton Rd. SE26 65Yb 136
Wiverton Twr. E1 44Vb 91
(off Aldgate Pl.)
Wix Rd. RM9: Dag 39Zc 75
Wix's La. SW4 55Kb 112
WLA Community Sports Cen. 39Aa 65
Woburn W13 43Ka 86
(off Clivedon Ct.)
Woburn Av. CM16: They B 9Uc 22
Woburn Av. CR8: Purl 83Qb 176
Woburn Av. RM12: Horn 35Jd 76
Woburn Cl. SE28 44Zc 95
Woburn Cl. SW19 65Eb 133
Woburn Cl. WD23: Bush 16Ea 28
Woburn Ct. CR0: C'don 74Sb 157
Woburn Ct. E18 26Jc 53
Woburn Ct. SE16 50Xb 91
(off Masters Dr.)
Woburn Ct. WC1 6F 217 (42Nb 90)
(off Bernard St.)
Woburn Hill KT15: Add 75L 149
Woburn Hill Pk. KT15: Add 76M 149
Woburn Mans. WC1 7D 216 (43Mb 90)
(off Torrington Pl.)
Woburn M. WC1 6E 216 (42Mb 90)
WOBURN PARK 75L 149
Woburn Pl. WC1 6F 217 (42Nb 90)
Woburn Rd. CR0: C'don 74Sb 157
Woburn Rd. SM5: Cars 74Gb 155
Woburn Sq. WC1 6E 216 (42Mb 90)
(not continuous)
Woburn Twr. UB5: N'olt 41Z 85
(off Broomcroft Av.)
Woburn Wlk. WC1 4E 216 (41Mb 90)
Wodeham Gdns. E1 43Wb 91
Wodehouse Av. SE5 53Vb 113
Wodehouse Rd. DA1: Dart 56Qd 119
Woffington Cl. KT1: Hamp W 67La 130
Wogan Ho. W1 1A 222 (43Kb 90)
(off Portland Pl.)
Wokindon Rd. RM16: Grays 8D 100
WOKING 89B 168
Woking Bus. Pk. GU21: Wok 87D 168
Woking Cl. SW15 56Va 110
WOKING COMMUNITY HOSPITAL 90B 168
Woking Crematorium 1J 187
Woking FC 92B 188
Woking Golf Course 2L 187
WOKING HOSPICE 9L 167
Woking Leisure Cen. 91B 188
WOKING NUFFIELD HEALTH HOSPITAL 86A 168
WOKING PRIORY HOSPITAL 8G 166
Woking Rd. GU4: Jac W 10N 187
Woking Station (Rail) 89B 168
Wolcot Ho. NW1 2C 216 (40Lb 70)
(off Aldenham St.)
Wold, The CR3: Wold 94Cc 198
Woldham Pl. BR2: Broml 70Lc 137
Woldham Rd. BR2: Broml 70Lc 137
WOLDINGHAM 95Cc 198
WOLDINGHAM GARDEN VILLAGE 92Ac 198
Woldingham Golf Course 92Zb 198
Woldingham Rd. CR3: Wold 92Xb 197
Woldingham Station (Rail) 94Zb 198
Wolds Dr. BR6: Farnb 77Qc 160
Wolseley Ct. GU21: Knap 10G 166
(off Tudor Way)
Wolfe Cl. BR2: Hayes 72Jc 159
Wolfe Cl. UB4: Yead 41X 85
Wolfe Cotts. TN16: Westrm 99Tc 200
Wolfe Cres. SE16 47Zb 92
Wolfe Cres. SE7 50Mc 93
Wolfe Ho. W12 45Xa 88
(off White City Est.)
Wolfe Ho. W14 49Bb 89
Wolfendale Cl. RH1: Mers 2C 208
Wolferton Rd. E12 35Pc 74
Wolffe Gdns. E15 37Hc 73
Wolfington Rd. SE27 63Rb 135
Wolf La. SL4: Wind 5B 102
Wolfram Cl. SE13 57Gc 115
Wolf's Hill RH8: Oxt 3L 211
Wolf's Rd. RH8: Limp 2M 211
Wolf's Row RH8: Limp 1M 211
Wolfs Wood RH8: Oxt 4L 211
Wolftencroft Cl. SW11 55Gb 111
Wollaston Cl. SE1 5D 230 (49Sb 91)
Wollaton Ho. N1 1A 218 (40Qb 70)
(off Batchelor St.)
Wollett Ct. NW1 38Lb 70
(off St Pancras Way)
Wollstonecraft St. N1 1E 216 (39Mb 70)
Wolmer Cl. HA8: Edg 21Qa 47
Wolmer Gdns. HA8: Edg 20Qa 29
Wolseley Av. SW19 61Cb 133
Wolseley Gdns. W4 51Ra 109
Wolseley Rd. CR4: Mitc 73Jb 156
Wolseley Rd. E7 38Kc 73
Wolseley Rd. HA3: W'stone 27Ga 46
Wolseley Rd. N22 25Pb 50
Wolseley Rd. N8 30Mb 50
Wolseley Rd. RM7: Rush G 31Fd 76
Wolseley Rd. W4 49Sa 87
Wolseley St. SE1 47Wb 91
Wolsey Av. E17 27Bc 52
Wolsey Av. E6 41Qc 94
Wolsey Av. EN7: Chesh 1Vb 19
Wolsey Av. KT7: T Ditt 71Ha 152
Wolsey Bus. Pk. WD18: Wat 17T 26
Wolsey Cl. KT2: King T 67Ra 131
Wolsey Cl. KT4: Wor Pk 77Wa 154
Wolsey Cl. SE2 47Yc 95
Wolsey Cl. SW20 66Xa 132
Wolsey Cl. TW3: Houn 56Ea 108
Wolsey Cl. UB2: S'hall 48Ea 86
Wolsey Ct. NW6 38Eb 69
Wolsey Ct. SE9 58Pc 116
(off Court Rd.)
Wolsey Ct. SW11 53Gb 111
(off Westbridge Rd.)
Wolsey Cres. CR0: New Ad 81Ec 178
Wolsey Cres. DA9: Ghithe 58Wd 120
Wolsey Cres. SM4: Mord 73Ab 154
Wolsey Dr. KT12: Walt T 74Z 151
Wolsey Dr. KT2: King T 64Na 131
Wolsey Gdns. IG6: Ilf 23Rc 54
Wolsey Gro. HA8: Edg 24Ta 47
Wolsey Gro. KT10: Esh 77Da 151
Wolsey M. BR6: Chels 78Vc 161
Wolsey M. NW5 37Lb 70
Wolsey Rd. EN1: Enf 12Xb 33
Wolsey Rd. HA6: Nwood 19S 26
Wolsey Rd. HP2: Hem H 3M 3
Wolsey Rd. KT10: Esh 77Da 151
Wolsey Rd. KT8: E Mos 70Fa 130
Wolsey Rd. N1 36Tb 71
Wolsey Rd. TW12: Hamp H 65Da 129
Wolsey Rd. TW15: Ashf 63N 127
Wolsey Rd. TW16: Sun 66V 128
Wolsey St. E1 43Yb 92
Wolsey Wlk. GU21: Wok 89A 168
Wolsey Way KT9: Chess 78Qa 153
Wolsley Cl. DA1: Cray 57Gd 118
Wolstan Cl. UB9: Den 34J 63
Wolstenholme HA7: Stan 22Ka 46
Wolstonbury N12 22Cb 49
Wolvercote Rd. SE2 47Zc 95
Wolverley St. E2 41Xb 91
Wolverton SE17 7H 231 (50Ub 91)
(not continuous)
Wolverton Av. KT2: King T 67Qa 131
Wolverton Gdns. W5 45Pa 87
Wolverton Gdns. W6 49Za 88
Wolverton Ho. RM3: Rom 22Nd 57
(off Chudleigh Rd.)
Wolverton Rd. HA7: Stan 23Ka 46
Wolverton Way N14 15Lb 32
Wolves La. N13 23Qb 50
Wolves La. N22 24Qb 50
Wombwell Gdns. DA11: Nflt 1A 144
WOMBWELL PARK 61Fe 143
Womersley Rd. N8 30Pb 50
Wonersh Way SM2: Cheam 81Za 174
Wonford Cl. KT2: King T 67Ua 132
Wonford Cl. KT20: Walt H 98Wa 194
Wonham La. RH3: B'wth 7A 206
Wonham Pl. RH9: S God 7D 210
Wonnacott Pl. EN3: Enf W 8Zb 20
Wontford Rd. CR8: Purl 87Qb 176
Wontner Cl. N1 38Sb 71
Wontner Rd. SW17 61Hb 133
Wooburn Cl. UB8: Hil 42R 84
Wooburn Comn. Rd. SL1: Burn 3A 60
Wood, The WD19: Wat 19Aa 27
Woodall Av. EN3: Pond E 16Zb 34
Woodall Cl. E14 45Dc 92
Woodall Cl. KT9: Chess 80La 152
Woodall Rd. EN3: Pond E 16Zb 34
Wood Av. RM19: Purf 49Sd 98
Woodbank WD3: Rick 16L 25
Woodbank Rd. BR1: Broml 62Hc 137
Woodbastwick Rd. SE26 64Zb 136
Woodberry Av. HA2: Harr 28Da 45
Woodberry Av. N21 19Qb 32
Woodberry Cl. NW7 24Za 48
Woodberry Cl. TW16: Sun 65W 128
Woodberry Cres. N10 27Kb 50
Woodberry Down CM16: Epp 1Wc 23
Woodberry Down N4 31Sb 71
Woodberry Down Est. N4 31Sb 71
(not continuous)
Woodberry Gdns. N12 23Eb 49
Woodberry Gro. DA5: Bexl 62Fd 140
Woodberry Gro. N12 23Eb 49
Woodberry Gro. N4 31Sb 71
Woodberry Way E4 17Ec 34
Woodberry Way N12 23Eb 49
Woodbine Cl. EN9: Walt A 7Lc 21
Woodbine Cl. TW2: Twick 61Fa 130
Woodbine Cl. Cvn. Pk. EN9: Walt A 7Lc 21
Woodbine Gro. EN2: Enf 10Tb 19
Woodbine Gro. SE20 66Xb 135
Woodbine La. KT4: Wor Pk 76Xa 154
Woodbine Pl. E11 30Jc 53
Woodbine Rd. DA15: Sidc 60Uc 116
Woodbines Av. KT1: King T 69Ma 131
Woodbine Ter. E9 37Yb 72
Woodborough Rd. SW15 56Xa 110
Woodbourne Av. SW16 62Mb 134
Woodbourne Cl. SW16 62Nb 134
Woodbourne Dr. KT10: Clay 79Ha 152
Woodbourne Gdns. SM6: W'gton 80Kb 156
Woodbridge Av. KT22: Lea 90Ja 172
Woodbridge Cl. N7 33Pb 70
Woodbridge Cl. NW2 34Wa 68
Woodbridge Cl. RM3: Rom 21Md 57
Woodbridge Cnr. KT22: Lea 90Ja 172
Woodbridge Ct. IG8: Wfd G 24Nc 54
Woodbridge Gro. KT22: Lea 90Ja 172
Woodbridge Ho. E11 32Hc 73
Woodbridge La. RM3: Rom 20Md 39
Woodbridge Rd. IG11: Bark 36Vc 75
Woodbridge St. EC1 5B 218 (42Rb 91)
(not continuous)
Woodbridge Ter. RM6: Chad H 30Xc 55
Woodbridge Way SS17: Stan H 3K 101
Woodbrook Gdns. EN9: Walt A 5Gc 21
Woodbrook Rd. SE2 51Wc 117
Woodburn Cl. NW4 29Za 48
Woodbury Cl. CR0: C'don 75Vb 157
Woodbury Cl. E11 28Kc 53
Woodbury Cl. TN16: Big H 90Pc 180
Woodbury Cres. IG5: Ilf 25Pc 54
Woodbury Dr. SM2: Sutt 82Eb 175
Woodbury Gdns. SE12 62Kc 137
Woodbury Hill IG10: Lough 13Nc 36
Woodbury Hollow IG10: Lough 12Nc 36
Woodbury Ho. SE26 62Wb 135
Woodbury Pk. Rd. W13 42Ka 86
Woodbury Rd. E17 28Dc 52
Woodbury St. SW17 64Gb 133
WOODBURY UNIT 30Gc 53
Woodby Dr. SL5: S'dale 3D 146
Woodchester Ho. E14 48Dc 92
(off Selsdon Way)
Woodchester Sq. W2 43Db 89
Woodchurch Cl. DA14: Sidc 62Tc 138
Woodchurch Dr. BR1: Broml 66Mc 137
Woodchurch Rd. NW6 38Cb 69
Wood Cl. DA5: Bexl 62Gd 140
Wood Cl. E2 42Wb 91
Wood Cl. HA1: Harr 31Fa 66
Wood Cl. NW9 31Ta 67
Wood Cl. SL4: Wind 6G 102
Woodclyffe Dr. BR7: Chst 68Qc 138
Woodcock Cl. TW6: H'row A 56K 105
Woodcock Ct. HA3: Kenton 31Na 67
Woodcock Dell Av. HA3: Kenton 31Ma 67
Woodcock Dr. GU24: Chob 10G 146
Woodcock Hill HA3: Kenton 29La 46
Woodcock Hill WD3: Rick 22N 43
WOODCOCK HILL 22N 43
Woodcock Hill WD6: E'tree 16Ra 29
Woodcock Hill Est. WD3: Rick 21N 43
Woodcock Ho. E14 43Cc 92
(off Burgess St.)
Woodcock La. GU24: Chob 10F 146
Woodcocks E16 43Lc 93
Woodcombe Cres. SE23 60Yb 114
WOODCOTE 84Mb 176
WOODCOTE 87Sa 173
Woodcote Av. CR7: Thor H 70Rb 135
Woodcote Av. NW7 23Ya 48
Woodcote Av. RM12: Horn 35Jd 76
Woodcote Av. SM6: W'gton 81Kb 176
Woodcote Cl. EN3: Pond E 16Yb 34
Woodcote Cl. EN8: Chesh 2Yb 20
Woodcote Cl. KT18: Eps 86Ta 173
Woodcote Cl. KT2: King T 64Pa 131
Woodcote Cl. WD23: Bush 16Ea 28
Woodcote Ct. KT18: Eps 86Ta 173
(off Dorking Rd.)
Woodcote Dr. BR6: Orp 74Tc 160
Woodcote Dr. CR8: Purl 82Mb 176
Woodcote End KT18: Eps 87Ta 173
WOODCOTE GREEN 81Lb 176
Woodcote Grn. SM6: W'gton 81Lb 176
Woodcote Grn. Rd. KT18: Eps 87Sa 173
WOODCOTE GROVE 85Kb 176
Woodcote Gro. Rd. CR5: Coul 87Mb 176
Woodcote Hall KT18: Eps 86Ta 173
Woodcote Ho. KT18: Eps 87Sa 173
Woodcote Ho. SE8 51Bc 114
(off Prince St.)
Woodcote Ho. Ct. KT18: Eps 87Ta 173
Woodcote Hurst KT18: Eps 88Sa 173
Woodcote La. CR8: Purl 83Mb 176
Woodcote Lodge KT18: Eps 87Sa 173
Woodcote M. IG10: Lough 17Mc 35
Woodcote M. SM6: W'gton 79Kb 156
WOODCOTE PARK 89Sa 173
Woodcote Pk. Av. CR8: Purl 84Lb 176
Woodcote Pk. Golf Course 86Kb 176
Woodcote Pk. Rd. KT18: Eps 88Sa 173
Woodcote Pl. SE27 64Rb 135
Woodcote Rd. CR8: Purl 82Lb 176
Woodcote Rd. E11 31Jc 73
Woodcote Rd. KT18: Eps 86Ta 173
Woodcote Rd. SM6: W'gton 79Kb 156
Woodcote Side KT18: Eps 87Ra 173
Woodcote Valley Rd. CR8: Purl 85Mb 176
Woodcote Vs. SE27 64Sb 135
(off Woodcote Pl.)
Wood Cres. HP3: Hem H 3M 3
Wood Cres. W12 45Ya 88
Wood Crest SM2: Sutt 80Eb 155
(off Christchurch Pk.)
Woodcrest Rd. CR8: Purl 85Nb 176
Woodcrest Wlk. RH2: Reig 4N 207
Woodcroft N21 18Qb 32
Woodcroft SE9 62Pc 138
Woodcroft UB6: G'frd 37Ja 66
Woodcroft Av. HA7: Stan 25Ja 46
Woodcroft Av. NW7 23Ua 48
Woodcroft Cl. SE9 58Qc 116
Woodcroft Cres. UB10: Hil 39R 64
Woodcroft Ho. CR8: Kenley 89Sb 177
Woodcroft M. SE8 49Ac 92
Woodcroft Rd. CR7: Thor H 71Rb 157
Woodcutter Pl. AL2: Park 9A 6
Woodcutters Av. RM16: Grays 47Ee 99
Woodcutters Cl. RM11: Horn 28Md 57
Wood Dr. BR7: Chst 65Nc 138
Wood Dr. TN13: S'oaks 98Hd 202
Woodedge Cl. E4 18Hc 35
Wood End, The SM6: W'gton 81Kb 176
Wood End AL2: Park 10A 6
Wood End BR8: Swan 70Ed 140
WOOD END 5A 124
Wood End UB3: Hayes 44U 84
WOOD END 43V 84
WOOD END 36Ea 66
Wood End WD3: Crox G 17R 26
Woodend KT10: Esh 75Ea 152
Woodend KT22: Lea 97La 192
Woodend SE19 65Sb 135
Woodend SM1: Sutt 75Eb 155
Wood End Av. HA2: Harr 35Da 65
Wood End Av. UB5: N'olt 35Ea 66
Wood End Cl. HP2: Hem H 1C 4
Wood End Cl. SL2: Farn C 4H 61
Wood End Cl. UB5: N'olt 36Fa 66
Woodend Cl. GU21: Wok 1L 187
Woodend Cl. RM17: Grays 49De 99
Woodend Dr. SL5: S'hill 1A 146
Wood End Gdns. UB5: N'olt 36Ea 66
Woodend Gdns. EN2: Enf 14Nb 32
WOOD END GREEN 43T 84
Wood End Grn. Rd. UB3: Hayes 43T 84
Wood End La. UB5: N'olt 37Da 65
Woodend Pk. KT11: Cobh 87Z 171
Wood End Rd. HA1: Harr 35Fa 66
Woodend Rd. E17 26Ec 52
Wood End Way UB5: N'olt 36Ea 66
Wooder Gdns. E7 35Jc 73
Wooderson Cl. SE25 70Ub 135
Woodfall Av. EN5: Barn 15Bb 31
Woodfall Dr. DA1: Cray 56Gd 118
Woodfall Rd. N4 33Qb 70
Woodfall St. SW3 7F 227 (50Hb 89)
Wood Farm Cl. HA7: Stan 19Ka 28
Woodfarm Rd. HP2: Hem H 2N 3
Woodfarrs SE5 56Tb 113
Wood Fld. NW3 36Hb 69
Woodfield KT21: Asht 89Ma 173
Woodfield Av. DA11: Grav'nd 10D 122
Woodfield Av. HA0: Wemb 34La 66
Woodfield Av. HA6: Nwood 21U 44
Woodfield Av. NW9 28Ua 48
Woodfield Av. SM5: Cars 79Jb 156
Woodfield Av. SW16 62Mb 134
Woodfield Av. W5 42La 86
Woodfield Cl. CR5: Coul 91Lb 196
Woodfield Cl. EN1: Enf 14Ub 33
Woodfield Cl. KT21: Asht 89Ma 173
Woodfield Cl. RH1: Redh 4N 207
Woodfield Cl. SE19 66Sb 135
Woodfield Cres. W5 42La 86
Woodfield Dr. EN4: E Barn 18Jb 32
Woodfield Dr. HP3: Hem H 4D 4
Woodfield Dr. RM2: Rom 28Jd 56
Woodfield Gdns. HP3: Hem H 4D 4
Woodfield Gdns. KT3: N Mald 71Va 154
Woodfield Gro. SW16 62Mb 134
Woodfield Hill CR5: Coul 91Kb 196
Woodfield Ho. KT7: T Ditt 75Ha 152
(off Woodfield Rd.)
Woodfield Ho. SE23 62Zb 136
(off Dacres Rd.)
Woodfield La. AL9: Hat 5M 9
Woodfield La. KT21: Asht 89Na 173
Woodfield La. SG13: New S 5P 9
Woodfield La. SW16 62Mb 134
Woodfield Pl. W9 42Bb 89
Woodfield Ri. WD23: Bush 17Fa 28
Woodfield Rd. KT21: Asht 89Ma 173
Woodfield Rd. KT7: T Ditt 75Ha 152
Woodfield Rd. TW4: Cran 54X 107
Woodfield Rd. W5 42La 86
Woodfield Rd. W9 43Bb 89
Woodfield Rd. WD7: R'lett 8Ja 14
Woodfields, The CR2: Sande 83Vb 177
Woodfields TN13: Chips 95Fd 202
Woodfields WD18: Wat 14Y 27
(off George St.)
Woodfield Ter. UB9: Hare 26K 43
Woodfield Way N11 24Mb 50
Woodfield Way RH1: Redh 4N 207
Woodfield Way RM12: Horn 32Md 77
Woodfines, The RM11: Horn 30Md 57
WOODFORD 23Lc 53
Woodford Av. IG2: Ilf 29Pc 54
Woodford Av. IG4: Ilf 27Mc 53
Woodford Av. IG4: Wfd G 27Mc 53
WOODFORD BRIDGE 23Nc 54
Woodford Bri. Rd. IG4: Ilf 27Mc 53
Woodford Ct. EN9: Walt A 5Jc 21
Woodford Ct. W12 47Za 88
(off Shepherd's Bush Grn.)
Woodford Cres. HA5: Pinn 26X 45
Woodforde Ct. UB3: Harl 50T 84
Woodford Golf Course 22Jc 53
WOODFORD GREEN 23Jc 53
Woodford Green Athletics Club 23Nc 54
Woodford Hall Path E18 25Hc 53
Woodford Ho. E18 28Jc 53
Woodford New Rd. E17 28Gc 53
Woodford New Rd. E18 25Gc 53
Woodford New Rd. IG8: Wfd G 24Hc 53
Woodford Pl. HA9: Wemb 32Na 67
Woodford Rd. E18 28Jc 53
Woodford Rd. E7 34Kc 73
Woodford Rd. WD17: Wat 12Y 27
WOODFORD SIDE 22Hc 53
Woodford Station (Underground) 23Kc 53
Woodford Trad. Est. IG8: Wfd G 26Mc 53
Woodford Way SL2: Slou 1E 80
WOODFORD WELLS 20Kc 35
Woodgate WD25: Wat 5X 13
Woodgate Av. EN6: N'thaw 5Kb 18
Woodgate Av. KT9: Chess 78Ma 153
Woodgate Cl. KT11: Cobh 85X 171
Woodgate Ct. RM11: Horn 27Md 57
Woodgate Cres. HA6: Nwood 23W 44
Woodgate Dr. SW16 66Mb 134
Woodgate M. WD17: Wat 11W 26
Woodgates Ho. DA12: Grav'nd 4E 144
(off Nursery M.)
Woodgavil SM7: Bans 88Bb 175
Woodger Rd. W12 47Ya 88
Woodgers Gro. BR8: Swan 68Hd 140
Woodget Cl. E6 44Nc 94
Woodgrange Av. EN1: Enf 16Wb 33
Woodgrange Av. HA3: Kenton 29La 46
Woodgrange Av. N12 23Fb 49
Woodgrange Av. W5 46Qa 87
Woodgrange Cl. HA3: Kenton 29Ma 47
Woodgrange Gdns. EN1: Enf 16Wb 33
Woodgrange Ho. W5 46Pa 87
(off Woodgrange Av.)
Woodgrange Mans. HA3: Kenton 29Ma 47
Woodgrange Park Station (Overground) 36Mc 73
Woodgrange Rd. E7 36Kc 73
Woodgrange Ter. EN1: Enf 16Wb 33
WOOD GREEN 7Lc 21
WOOD GREEN 26Pb 50
Wood Grn. Animal Shelter 25Rb 51
(off Lordship La.)
Wood Grn. Hall N22 26Pb 50
(off Station Rd.)
Woodgreen Rd. EN9: Walt A 5Kc 21
Wood Grn. Shop. City 26Qb 50
Wood Green Station (Underground) 26Qb 50
Wood Grn. Way EN8: Chesh 3Ac 20
Woodhall NW1 4B 216 (41Lb 90)
(off Robert St.)
Woodhall Av. HA5: Pinn 25Aa 45
Woodhall Av. SE21 62Vb 135
Woodhall Cl. UB8: Uxb 36M 63
Woodhall Cres. RM11: Horn 31Pd 77
Woodhall Dr. HA5: Pinn 25Z 45
Woodhall Dr. SE21 62Vb 135
Woodhall Ga. HA5: Pinn 24Z 45
Woodhall Ho. SW18 58Fb 111
Woodhall La. HP2: Hem H 1N 3
Woodhall La. SL5: S'dale 5C 146
Woodhall La. WD19: Wat 20Z 27
Woodhall La. WD7: Shenl 7Na 15
Woodhall Rd. HA5: Pinn 24Z 45
WOODHAM 83H 169
Woodham Ct. E18 28Hc 53
Woodham Ga. GU21: Wok 85E 168
Woodham Hall Est. GU21: Wok 86D 168
Woodham La. GU21: Wok 86D 168
Woodham La. KT15: New H 84G 168
Woodham La. KT15: Wdhm 84G 168
Woodham Lock KT14: W Byf 84H 169
Woodham Pk. Rd. KT15: Wdhm 81H 169
Woodham Pk. Way KT15: Wdhm 83H 169
Woodham Pl. GU21: Wok 86B 168
Woodham Ri. GU21: Wok 86B 168
Woodham Rd. GU21: Wok 87A 168
Woodham Rd. SE6 62Ec 136
Woodham Waye GU21: Wok 85D 168
WOODHATCH 9K 207
Woodhatch Cl. E6 43Nc 94
Woodhatch Rd. RH1: Redh 10M 207
Woodhatch Rd. RH2: Reig 9K 207
Woodhatch Spinney CR5: Coul 88Nb 176
Woodhaven Gdns. IG6: Ilf 28Sc 54
Woodhaven M. KT12: Walt T 77W 150
Woodhaw TW20: Egh 63D 126
Woodhayes Rd. SW19 66Ya 132
Woodhead Dr. BR6: Orp 76Uc 160
Woodheyes Rd. NW10 36Ta 67
WOODHILL 5M 9
Woodhill GU23: Send 98F 188
Woodhill SE18 49Nc 94
Woodhill Av. SL9: Ger X 30C 42
Woodhill Ct. GU23: Send 97F 188
Woodhill Ct. SL9: Ger X 31A 62
Woodhill Cres. HA3: Kenton 30Ma 47
Wood Ho. NW6 40Bb 69
(off Albert Rd.)
Woodhouse Av. UB6: G'frd 40Ha 66
Woodhouse Cl. SE22 56Wb 113
Woodhouse Cl. UB3: Harl 48U 84
Woodhouse Cl. UB6: G'frd 39Ha 66
Woodhouse Eaves HA6: Nwood 22W 44
Woodhouse Gro. E12 37Nc 74
Woodhouse Rd. E11 34Hc 73
Woodhouse Rd. N12 23Fb 49
Woodhurst Av. BR5: Pet W 72Sc 160
Woodhurst Av. WD25: Wat 6Z 13
Woodhurst Dr. UB9: Den 29H 43
Woodhurst La. RH8: Oxt 2J 211
Woodhurst Pk. RH8: Oxt 2J 211
Woodhurst Rd. SE2 50Wc 95
Woodhurst Rd. W3 45Sa 87
Woodhyrst Gdns. CR8: Kenley 87Rb 177
Woodies La. KT3: N Mald 72Ta 153
Woodin Cl. DA1: Dart 58Md 119
Woodington Cl. SE9 58Qc 116
Woodknoll Dr. BR7: Chst 67Pc 138
Woodland App. UB6: G'frd 37Ja 66
Woodland Av. CM13: Hut 15Ee 41
Woodland Av. DA3: Hartl 71Be 165
Woodland Av. HP1: Hem H 3K 3
Woodland Av. SL1: Slou 5H 81
Woodland Av. SL4: Wind 6D 102
Woodland Chase WD3: Crox G 17R 26
Woodland Cl. CM13: Hut 15Ee 41
Woodland Cl. DA3: Lfield 69Ee 143
Woodland Cl. HP1: Hem H 3K 3
Woodland Cl. IG8: Wfd G 20Kc 35
Woodland Cl. KT13: Weyb 77T 150
Woodland Cl. KT19: Ewe 79Ua 154
Woodland Cl. KT24: E Hor 99V 190
Woodland Cl. NW9 30Sa 47
Woodland Cl. SE19 65Ub 135
Woodland Cl. UB10: Ick 33R 64
Woodland Ct. E11 30Jc 53
(off New Wanstead)
Woodland Ct. KT17: Eps 84Va 174
Woodland Ct. N7 37Nb 70
Woodland Ct. RH8: Oxt 100Fc 199

Woodland Cres. SE10 ... 51Gc **115**
Woodland Cres. SE16 ... 47Zb **92**
Woodland Dr. AL4: St A ... 1G **6**
Woodland Dr. KT11: Cobh ... 83Ba **171**
Woodland Dr. KT24: E Hor ... 99V **190**
Woodland Dr. WD17: Wat ... 11V **26**
... (not continuous)
Woodland Gdns. CR2: Sels ... 83Yb **178**
Woodland Gdns. N10 ... 29Kb **50**
Woodland Gdns. TW7: Isle ... 55Ga **108**
Woodland Glade SL2: Farn C ... 4H **61**
Woodland Grange SL0: Rich P ... 48G **82**
Woodland Gro. CM16: Epp ... 4Wc **23**
Woodland Gro. KT13: Weyb ... 77T **150**
Woodland Gro. SE10 ... 50Gc **93**
Woodland Hill SE19 ... 65Ub **135**
Woodland La. WD3: Chor ... 13F **24**
Woodland M. SE13 ... 55Dc **114**
... *(off Loampit Hill)*
Woodland M. SW16 ... 62Nb **134**
Woodland Pl. HP1: Hem H ... 3K **3**
Woodland Pl. WD3: Chor ... 14H **25**
Woodland Ri. N10 ... 28Kb **50**
Woodland Ri. RH8: Oxt ... 2J **211**
Woodland Ri. TN15: S'oaks ... 95Nd **203**
Woodland Ri. TN15: Seal ... 95Nd **203**
Woodland Ri. UB6: G'frd ... 37Ja **66**
Woodland Rd. CR7: Thor H ... 70Qb **134**
Woodland Rd. E4 ... 18Ec **34**
Woodland Rd. IG10: Lough ... 13Nc **36**
Woodland Rd. IG7: Chig ... 21Tc **54**
Woodland Rd. N11 ... 22Kb **50**
Woodland Rd. SE19 ... 64Ub **135**
Woodland Rd. TN14: Dun G ... 93Hd **202**
Woodland Rd. WD3: Map C ... 22F **42**
Woodlands, The BR6: Chels ... 79Xc **161**
Woodlands, The HA1: Harr ... 33Ga **66**
Woodlands, The HA7: Stan ... 22Ka **46**
Woodlands, The KT10: Esh ... 75Ea **152**
Woodlands, The N12 ... 23Eb **49**
Woodlands, The N14 ... 18Kb **32**
Woodlands, The N5 ... 35Sb **71**
Woodlands, The SE13 ... 59Fc **115**
Woodlands, The SE19 ... 66Sb **135**
Woodlands, The SM6: W'gton ... 81Kb **176**
Woodlands, The SW9 ... 52Rb **113**
... *(off Langton Rd.)*
Woodlands, The TW7: Isle ... 54Ha **108**
Woodlands AL2: Park ... 9A **6**
Woodlands AL9: Brk P ... 8K **9**
Woodlands BR2: Broml ... 70Hc **137**
Woodlands CM16: Epp ... 3Wc **23**
Woodlands DA6: Bex ... 57Dd **118**
Woodlands GU22: Wok ... 90B **168**
Woodlands GU23: Send ... 97H **189**
Woodlands HA2: Harr ... 28Ca **45**
Woodlands KT15: Add ... 76N **149**
Woodlands KT21: Asht ... 90Na **173**
Woodlands NW11 ... 29Ab **48**
Woodlands SL9: Ger X ... 30B **42**
Woodlands SW20 ... 70Ya **132**
WOODLANDS ... 85Sd 184
WOODLANDS ... 55Ga 108
Woodlands WD7: R'lett ... 6Ja **14**
Woodlands Av. DA15: Sidc ... 60Uc **116**
Woodlands Av. E11 ... 32Kc **73**
Woodlands Av. HA4: Ruis ... 31Y **65**
Woodlands Av. KT14: W Byf ... 85H **169**
Woodlands Av. KT3: N Mald ... 67Sa **131**
Woodlands Av. KT4: Wor Pk ... 75Va **154**
Woodlands Av. N3 ... 24Eb **49**
Woodlands Av. RH1: Redh ... 7P **207**
Woodlands Av. RM11: Horn ... 29Md **57**
Woodlands Av. RM6: Chad H ... 31Ad **75**
Woodlands Av. W3 ... 46Ra **87**
Woodlands Bridge ... 57Bb **111**
... *(off Woodlands Way)*
Woodlands Cl. BR1: Broml ... 68Pc **138**
Woodlands Cl. BR8: Swan ... 69Hd **140**
Woodlands Cl. KT10: Clay ... 80Ha **152**
Woodlands Cl. KT16: Ott ... 82D **168**
Woodlands Cl. NW11 ... 29Ab **48**
Woodlands Cl. RH1: Mers ... 1E **208**
Woodlands Cl. RM16: Grays ... 8A **100**
Woodlands Cl. SL9: Ger X ... 30C **42**
Woodlands Cl. WD6: Bore ... 14Ra **29**
Woodlands Copse KT21: Asht ... 88Ma **173**
Woodlands Cotts. SL2: Farn C ... 6G **60**
Woodlands Ct. BR1: Broml ... 67Hc **137**
Woodlands Ct. GU21: Wok ... 10L **167**
Woodlands Ct. GU22: Wok ... 91A **188**
Woodlands Ct. HA1: Harr ... 29Ha **46**
Woodlands Ct. IG10: Lough ... 15Qc **36**
Woodlands Ct. KT12: Walt T ... 74X **151**
Woodlands Ct. NW10 ... 39Za **68**
... *(off Wrentham Av.)*
Woodlands Ct. RH1: Redh ... 8P **207**
Woodlands Ct. SE23 ... 59Xb **113**
Woodlands Dr. HA7: Stan ... 23Ha **46**
Woodlands Dr. RH9: S God ... 9C **210**
Woodlands Dr. TW16: Sun ... 68Y **129**
Woodlands Dr. WD4: K Lan ... 10C **4**
Woodlands Gdns. E17 ... 28Gc **53**
Woodlands Gdns. KT18: Tatt C ... 89Ya **174**
Woodlands Ga. SW15 ... 57Bb **111**
... *(off Woodlands Way)*
Woodlands Gro. CR5: Coul ... 89Jb **176**
Woodlands Gro. TW7: Isle ... 54Ga **108**
Woodlands Hgts. SE3 ... 51Hc **115**
... *(off Vanburgh Hill)*
Woodlands Hill HP9: Beac ... 1C **60**
Woodlands Ho. GU21: Wok ... 86E **168**
Woodlands La. DA12: Shorne ... 6M **145**
Woodlands La. GU20: W'sham ... 9B **146**
Woodlands La. KT11: Lea ... 89Ca **171**
Woodlands La. KT11: Stoke D ... 89Ca **171**
Woodlands Manor Golf Course ... **85Td 184**
Woodlands Pde. TW15: Ashf ... 65S **128**
Woodlands Pk. DA5: Bexl ... 63Fd **140**
Woodlands Pk. GU21: Wok ... 86E **168**
Woodlands Pk. KT15: Add ... 78H **149**
Woodlands Pk. Rd. N15 ... 29Rb **51**
Woodlands Pk. Rd. SE10 ... 51Gc **115**
... (not continuous)
Woodlands Pl. CR3: Cat'm ... 98Xb **197**
Woodlands Ri. BR8: Swan ... 68Hd **140**
Woodlands Rd. BR1: Broml ... 68Nc **138**
Woodlands Rd. BR6: Chels ... 79Wc **161**
Woodlands Rd. DA7: Bex ... 55Ad **117**
Woodlands Rd. E11 ... 33Gc **73**
Woodlands Rd. E17 ... 27Ec **52**
Woodlands Rd. EN2: Enf ... 10Tb **19**
Woodlands Rd. GU25: Vir W ... 10N **125**
Woodlands Rd. HA1: Harr ... 29Ha **46**
Woodlands Rd. HP3: Hem H ... 9A **4**
Woodlands Rd. IG1: Ilf ... 34Sc **74**
Woodlands Rd. KT14: W Byf ... 86H **169**
Woodlands Rd. KT18: Eps ... 87Qa **173**
Woodlands Rd. KT22: Lea ... 89Fa **172**
Woodlands Rd. KT23: Bookh ... 100Aa **191**
Woodlands Rd. KT6: Surb ... 73Ma **153**
Woodlands Rd. N9 ... 18Yb **34**
Woodlands Rd. RH1: Redh ... 8P **207**
Woodlands Rd. RM1: Rom ... 27Hd **56**
Woodlands Rd. RM3: Hrld W ... 25Qd **57**
Woodlands Rd. SW13 ... 55Va **110**
Woodlands Rd. TW7: Isle ... 55Fa **108**
Woodlands Rd. UB1: S'hall ... 46Z **85**
Woodlands Rd. WD23: Bush ... 15Aa **27**
Woodlands Rd. E. GU25: Vir W ... 10N **125**
Woodlands Rd. W. GU25: Vir W ... 10N **125**
Woodlands St. SE13 ... 59Fc **115**
Woodlands Ter. BR8: Crock ... 72Dd **162**
Woodland St. E8 ... 37Vb **71**
Woodlands Vw. TN14: Bad M ... 82Dd **182**
Woodlands Way KT21: Asht ... 88Qa **173**
Woodlands Way SW15 ... 57Bb **111**
Woodland Ter. SE7 ... 49Nc **94**
Woodland Wlk. SE10: Maze Hill ... 50Gc **93**
Woodland Wlk. BR1: Broml Southend ... 63Fc **137**
... (not continuous)
Woodland Wlk. KT19: Ewe ... 79Qa **153**
Woodland Wlk. NW3 ... 36Gb **69**
Woodland Way BR4: W W'ck ... 77Dc **158**
Woodland Way BR5: Pet W ... 70Sc **138**
Woodland Way CM16: They B ... 8Tc **22**
Woodland Way CR0: C'don ... 74Ac **158**
Woodland Way CR3: Cat'm ... 100Ub **197**
Woodland Way CR4: Mitc ... 66Jb **134**
Woodland Way CR8: Purl ... 85Qb **176**
Woodland Way DA9: Ghithe ... 57Wd **120**
Woodland Way IG8: Wfd G ... 20Kc **35**
Woodland Way KT13: Weyb ... 78T **150**
Woodland Way KT20: Kgswd ... 94Ab **194**
Woodland Way KT5: Surb ... 75Ra **153**
Woodland Way N21 ... 19Qb **32**
Woodland Way NW7 ... 23Va **48**
Woodland Way SE2 ... 49Zc **95**
Woodland Way SM4: Mord ... 70Bb **133**
Woodland Way WD5: Bedm ... 9F **4**
Wood La. CR3: Cat'm ... 96Tb **197**
Wood La. DA2: Daren ... 63Td **142**
Wood La. GU21: Knap ... 10H **167**
Wood La. HA4: Ruis ... 32T **64**
Wood La. HA7: Stan ... 20Ja **28**
Wood La. HP2: Hem H ... 3M **3**
Wood La. IG8: Wfd G ... 21Hc **53**
Wood La. KT13: Weyb ... 81S **170**
Wood La. KT20: Tad ... 89Bb **175**
Wood La. N6 ... 30Kb **50**
Wood La. NW9 ... 31Ta **67**
Wood La. RM10: Dag ... 33Cd **76**
Wood La. RM12: Horn ... 36Jd **76**
Wood La. RM8: Dag ... 35Zc **75**
Wood La. RM9: Dag ... 35Zc **75**
Wood La. SL0: Iver ... 41E **82**
Wood La. SL0: Iver H ... 41E **82**
Wood La. SL1: Slou ... 8D **80**
Wood La. SL2: Hedg ... 3J **61**
Wood La. TW7: Isle ... 51Ga **108**
Wood La. W12 ... 44Ya **88**
Wood La. Cl. SL0: Iver H ... 41D **82**
Wood La. End HP2: Hem H ... 1A **4**
Wood Lane Station (Underground) ... 45Ya 88
Wood La. Studios W12 ... 44Ya **88**
Woodlark Ct. KT10: Clay ... 79Ha **152**
Woodlark Gro. RM3: Rom ... 21Nd **57**
Woodlawn Cl. SW15 ... 57Bb **111**
Woodlawn Cres. TW2: Whitt ... 61Da **129**
Woodlawn Dr. TW13: Felt ... 61Z **129**
Woodlawn Gro. GU21: Wok ... 87B **168**
Woodlawn Rd. SW6 ... 52Za **110**
Woodlawns KT19: Ewe ... 80Ta **153**
Woodlea AL2: Chis G ... 7N **5**
Woodlea DA3: Lfield ... 69Ee **143**
Woodlea Dr. BR2: Broml ... 71Gc **159**
Woodlea Gro. HA6: Nwood ... 23T **44**
Woodlea Rd. N16 ... 34Ub **71**
Woodlee Cl. GU25: Vir W ... 8N **125**
Woodleigh E18 ... 25Jc **53**
Woodleigh Av. N12 ... 23Gb **49**
Woodleigh Gdns. SW16 ... 62Nb **134**
Woodley Cl. SW17 ... 66Hb **133**
Woodley La. SM5: Cars ... 76Gb **155**
Woodley Rd. BR6: Chels ... 75Yc **161**
Woodlodge KT21: Asht ... 89Na **173**
Wood Lodge Gdns. BR1: Broml ... 66Nc **138**
Wood Lodge Grange TN13: S'oaks ... 94Ld **203**
Wood Lodge La. BR4: W W'ck ... 76Ec **158**
Woodmancote Gdns. KT14: W Byf ... 85J **169**
Woodman La. E4 ... 15Gc **35**
Woodman M. TW9: Kew ... 53Ra **109**
Woodman Pde. E16 ... 46Qc **94**
... *(off Woodman St.)*
Woodman Path IG6: Ilf ... 23Uc **54**
Woodman Rd. CM14: W'ley ... 22Yd **58**
Woodman Rd. CR5: Coul ... 87Lb **176**
Woodman Rd. HP3: Hem H ... 4N **3**
Woodman Rd. IG6: Ilf ... 22Uc **54**
Woodmans Gro. NW10 ... 36Va **68**
Woodmans Ho. WD17: Wat ... 14Y **27**
Woodman's M. W12 ... 43Xa **88**
WOODMANSTERNE ... 87Hb 175
Woodmansterne La.
SM5: Cars Croydon La. ... 84Hb **175**
Woodmansterne La. SM6: W'gton ... 83Kb **176**
Woodmansterne La. SM7: Bans ... 87Db **175**
Woodmansterne Rd. CR5: Coul ... 87Lb **176**
Woodmansterne Rd. SM5: Cars ... 84Hb **175**
Woodmansterne Rd. SW16 ... 66Lb **134**
Woodmansterne Station (Rail) ... 88Kb 176
Woodmansterne St. SM7: Bans ... 87Gb **175**
Woodman St. E16 ... 46Qc **94**
Woodman Vs. DA3: Fawk ... 76Xd **164**
Wood Martyn Ct. BR6: Orp ... 75Wc **161**
... *(off Orchard Gro.)*
Wood Mead N17 ... 23Wb **51**
Wood Meads CM16: Epp ... 1Wc **23**
Woodmere SE9 ... 60Pc **116**
Woodmere Av. CR0: C'don ... 73Yb **158**
Woodmere Av. WD24: Wat ... 10Z **13**
Woodmere Cl. CR0: C'don ... 73Zb **158**
Woodmere Cl. SW11 ... 55Jb **112**
Woodmere Ct. N14 ... 17Kb **32**
Woodmere Gdns. CR0: C'don ... 73Zb **158**
Woodmere Way BR3: Beck ... 71Fc **159**
Woodmill Cl. SW15 ... 58Wa **110**
Woodmill Rd. E5 ... 33Yb **72**
Woodmill St. SE16 ... 3K **231** (48Vb **91**)
Woodmount BR8: Crock ... 73Fd **162**
Woodnook Rd. SW16 ... 64Kb **134**
Woodpecker Cl. AL10: Hat ... 3B **8**
Woodpecker Cl. HA3: Hrw W ... 25Ha **46**
Woodpecker Cl. KT11: Cobh ... 84Aa **171**
Woodpecker Cl. N9 ... 16Xb **33**
Woodpecker Cl. WD23: Bush ... 18Ea **28**
Woodpecker Dr. DA9: Ghithe ... 58Wd **120**
Woodpecker Dr. HP3: Hem H ... 6M **3**
Woodpecker M. SE13 ... 56Fc **115**
... *(off Freshfield Cl.)*
Woodpecker Mt. CR0: Sels ... 81Ac **178**
Woodpecker Rd. SE14 ... 51Ac **114**
Woodpecker Rd. SE28 ... 45Yc **95**
Woodpeckers, The UB9: Den ... 29H **43**
... *(off Patrons Way W.)*
Woodpecker Way GU22: Wok ... 6P **187**
Woodplace Cl. CR5: Coul ... 91Lb **196**
Woodplace La. CR5: Coul ... 90Lb **176**
Woodquest Av. SE24 ... 57Sb **113**
Woodredon Farm La. EN9: Walt A ... 7Mc **21**
Woodredon Rd. EN9: Walt A ... 7Mc **21**
Wood Retreat SE18 ... 52Tc **116**
Woodridden Hill EN9: Epp ... 7Mc **21**
Woodridden Hill EN9: They B ... 7Mc **21**
Woodridden Hill EN9: Walt A ... 7Mc **21**
Wood Ride BR5: Pet W ... 70Tc **138**
Wood Ride EN4: Had W ... 11Fb **31**
Woodridge Cl. EN2: Enf ... 11Qb **32**
Woodridge Way HA6: Nwood ... 23U **44**
Wood Riding GU22: Pyr ... 87G **168**
Woodridings KT13: Weyb ... 79Q **150**
Woodridings Av. HA5: Hat E ... 25Ba **45**
Woodridings Cl. HA5: Hat E ... 24Aa **45**
Woodridings Ct. N22 ... 25Mb **50**
Woodriffe Rd. E11 ... 31Fc **73**
Wood Ri. HA5: Eastc ... 29W **44**
Wood Rd. NW10 ... 38Sa **67**
Wood Rd. TN16: Big H ... 90Lc **179**
Wood Rd. TW17: Shep ... 70Q **128**
Woodrow SE18 ... 49Pc **94**
Woodrow Av. UB4: Hayes ... 43V **84**
Woodrow Cl. UB6: G'frd ... 38Ka **66**
Woodrow Ct. N17 ... 24Xb **51**
Woodrow Ct. SE5 ... 53Sb **113**
... *(off Camberwell Sta. Rd.)*
Woodrush Cl. SE14 ... 52Ac **114**
Woodrush Way RM6: Chad H ... 28Zc **55**
Woods, The HA6: Nwood ... 22W **44**
Woods, The UB10: Ick ... 35R **64**
Woods, The WD7: R'lett ... 6Ka **14**
Wood's Bldgs. E1 ... 43Xb **91**
... *(off Winthrop St.)*
Woods Cl. TW3: Houn ... 55Da **107**
... *(off High St.)*
Woods Dr. SL2: Farn C ... 6D **60**
Woodseer St. E1 ... 43Vb **91**
Woodsford SE17 ... 7F **231** (50Tb **91**)
... *(off Portland St.)*
Woodsford Sq. W14 ... 47Ab **88**
Woodshire Rd. RM10: Dag ... 34Dd **76**
Woodshore Cl. GU25: Vir W ... 2M **147**
Woodshots Mdw. WD18: Wat ... 15T **26**
Woods Ho. SW1 ... 7K **227** (50Kb **90**)
Woods Ho. SW8 ... 53Lb **112**
... *(off Wadhurst Rd.)*
WOODSIDE ... 3J 9
Woodside BR6: Chels ... 78Wc **161**
Woodside EN7: Chesh ... 3Wb **19**
Woodside IG9: Buck H ... 19Lc **35**
Woodside KT12: Walt T ... 74W **150**
Woodside KT15: New H ... 81L **169**
Woodside KT20: Lwr K ... 100Bb **195**
Woodside KT22: Fet ... 94Da **191**
Woodside KT24: W Hor ... 98S **190**
Woodside N10 ... 27Jb **50**
Woodside NW11 ... 29Cb **49**
WOODSIDE ... 72Wb 157
WOODSIDE ... 4A 124
Woodside SW19 ... 65Bb **133**
Woodside WD24: Wat ... 8W **12**
WOODSIDE ... 4X 13
Woodside WD6: E'tree ... 14Na **29**
Woodside Av. BR7: Chst ... 64Sc **138**
Woodside Av. HA0: Wemb ... 39Na **67**
Woodside Av. KT10: Esh ... 73Ga **152**
Woodside Av. KT12: Hers ... 77X **151**
Woodside Av. N10 ... 28Jb **50**
Woodside Av. N12 ... 21Db **49**
Woodside Av. N6 ... 29Hb **49**
Woodside Av. SE25 ... 72Xb **157**
Woodside Cl. CM13: Hut ... 15Fe **41**
Woodside Cl. CR3: Cat'm ... 96Ub **197**
Woodside Cl. DA7: Bex ... 56Fd **118**
Woodside Cl. GU21: Knap ... 9H **167**
Woodside Cl. HA0: Wemb ... 39Na **67**
Woodside Cl. HA4: Ruis ... 30T **44**
Woodside Cl. HA7: Stan ... 22Ka **46**
Woodside Cl. KT5: Surb ... 73Sa **153**
Woodside Cl. RM13: Rain ... 42Ld **97**
Woodside Cl. RM16: Grays ... 8A **100**
Woodside Cl. SL9: Chal P ... 26A **42**
Woodside Cotts. CR3: Wold ... 95Cc **198**
Woodside Ct. E12 ... 32Lc **73**
Woodside Ct. N12 ... 21Db **49**
Woodside Ct. RM7: Mawney ... 28Cd **56**
Woodside Ct. W5 ... 46Na **87**
Woodside Ct. WD25: Wat ... 5Y **13**
Woodside Ct. Rd. CR0: C'don ... 73Wb **157**
Woodside Cres. DA15: Sidc ... 62Uc **138**
Woodside Dr. DA2: Wilm ... 63Gd **140**
Woodside End HA0: Wemb ... 39Na **67**
Woodside Gdns. E4 ... 23Dc **52**
Woodside Gdns. N17 ... 26Ub **51**
Woodside Grange Rd. N12 ... 21Db **49**
WOODSIDE GREEN ... 3K 9
Woodside Grn. SE25 ... 72Wb **157**
Woodside Gro. N12 ... 20Eb **31**
Woodside Hill SL9: Chal P ... 26A **42**
Woodside Ho. SW19 ... 65Bb **133**
Woodside La. AL9: Hat ... 3H **9**
Woodside La. DA5: Bexl ... 58Zc **117**
Woodside La. N12 ... 20Eb **31**
Woodside La. SL4: Wink ... 4A **124**
Woodside Leisure Pk. ... **5X 13**
Woodside M. DA12: Grav'nd ... 1E **144**
Woodside M. SE22 ... 57Vb **113**
Woodside Pde. DA15: Sidc ... 62Uc **138**
WOODSIDE PARK ... 21Cb 49
Woodside Pk. SE25 ... 72Xb **157**
Woodside Pk. Av. E17 ... 28Fc **53**
Woodside Pk. Rd. N12 ... 21Db **49**
Woodside Park Station
(Underground) ... 21Db 49
Woodside Pl. AL9: Hat ... 3J **9**
Woodside Pl. CM13: Gt War ... 22Xd **58**
Woodside Pl. HA0: Wemb ... 39Na **67**
Woodside Rd. AL2: Brick W ... 2Ba **13**
Woodside Rd. BR1: Broml ... 71Nc **160**
Woodside Rd. CR8: Purl ... 85Mb **176**
Woodside Rd. DA15: Sidc ... 62Uc **138**
Woodside Rd. DA7: Bex ... 56Fd **118**
Woodside Rd. E13 ... 42Lc **93**
Woodside Rd. HA6: Nwood ... 24V **44**
Woodside Rd. IG8: Wfd G ... 21Jc **53**
Woodside Rd. KT11: Cobh ... 85Ca **171**
Woodside Rd. KT2: King T ... 66Na **131**
Woodside Rd. KT3: N Mald ... 68Ta **131**
Woodside Rd. N22 ... 24Pb **50**
Woodside Rd. SE25 ... 72Xb **157**
Woodside Rd. SL4: Wink ... 4A **124**
Woodside Rd. SM1: Sutt ... 76Eb **155**
Woodside Rd. TN13: S'oaks ... 95Jd **202**
Woodside Rd. TN14: Sund ... 96Ad **201**
Woodside Rd. WD25: Wat ... 3X **13**
Woodside Rd. WD5: Ab L ... 3W **12**
Woodside Stadium ... **5Y 13**
Woodside Stop
(London Tramlink) ... 72Xb 157
Woodside Wlk. HA6: Nwood ... 22X **45**
Woodside Way RH1: Redh Redstone
Hollow ... 7A **208**
Woodside Way CR0: C'don ... 72Yb **158**
Woodside Way CR4: Mitc ... 67Kb **134**
Woodside Way GU25: Vir W ... 9M **125**
Woods M. W1 ... 4H **221** (45Jb **90**)
Woodsome Lodge KT13: Weyb ... 79S **150**
Woodsome Rd. NW5 ... 34Jb **70**
Woods Pl. SE1 ... 4J **231** (48Ub **91**)
Woodspring Rd. SW19 ... 61Ab **132**
Woods Rd. SE15 ... 53Xb **113**
Woodstar Ho. SE15 ... 52Wb **113**
... *(off Reddins Rd.)*
Woodstead Gro. HA8: Edg ... 23Na **47**
WOODSTOCK, THE ... 73Bb **155**
Woodstock GU4: W Cla ... 100K **189**
Woodstock Av. NW11 ... 31Ab **68**
Woodstock Av. RM3: Rom ... 22Rd **57**
Woodstock Av. SL3: L'ly ... 9P **81**
Woodstock Av. SM3: Sutt ... 73Bb **155**
Woodstock Av. TW7: Isle ... 57Ja **108**
Woodstock Av. UB1: S'hall ... 41Ba **85**
Woodstock Av. W13 ... 48Ja **86**
Woodstock Cl. DA5: Bexl ... 59Bd **117**
Woodstock Cl. GU21: Wok ... 88A **168**
Woodstock Cl. HA7: Stan ... 26Na **47**
Woodstock Ct. KT19: Eps ... 85Ta **173**
Woodstock Ct. SE11 ... 7J **229** (50Pb **90**)
Woodstock Ct. SE12 ... 58Jc **115**
Woodstock Cres. N9 ... 16Xb **33**
Woodstock Dr. UB10: Ick ... 35N **63**
Woodstock Gdns. BR3: Beck ... 67Dc **136**
Woodstock Gdns. IG3: Ilf ... 33Wc **75**
Woodstock Gdns. UB4: Hayes ... 43V **84**
Woodstock Grange W5 ... 46Na **87**
Woodstock Gro. W12 ... 47Za **88**
Woodstock La. KT9: Chess ... 77Ka **152**
Woodstock La. Nth. KT6: Surb ... 75La **152**
Woodstock La. Sth. KT10: Clay ... 78Ka **152**
Woodstock La. Sth. KT9: Chess ... 77La **152**
Woodstock M. W1 ... 1J **221** (43Jb **90**)
... *(off Westmoreland St.)*
Woodstock Ri. SM3: Sutt ... 73Bb **155**
Woodstock Rd. CR0: C'don ... 76Tb **157**
Woodstock Rd. CR5: Coul ... 88Kb **176**
Woodstock Rd. E17 ... 26Fc **53**
Woodstock Rd. E7 ... 38Lc **73**
Woodstock Rd. HA0: Wemb ... 39Pa **67**
Woodstock Rd. N4 ... 32Qb **70**
Woodstock Rd. NW11 ... 31Bb **69**
Woodstock Rd. SM5: Cars ... 78Jb **156**
Woodstock Rd. W4 ... 49Ua **88**
Woodstock Rd. WD23: B Hea ... 17Ga **28**
Woodstock Rd. Nth. AL1: St A ... 1F **6**
Woodstock Rd. Sth. AL1: St A ... 2F **6**
Woodstock St. W1 ... 3K **221** (44Kb **90**)
Woodstock Studios W12 ... 47Za **88**
... *(off Woodstock Gro.)*
Woodstock Ter. E14 ... 45Dc **92**
Woodstock Way CR4: Mitc ... 68Kb **134**
Woodstone Av. KT17: Ewe ... 78Wa **154**
Wood St. BR8: Swan ... 68Ld **141**
Wood St. CR4: Mitc ... 73Jb **156**
Wood St. E17 ... 27Ec **52**
Wood St. EC2 ... 3E **224** (44Sb **91**)
... (not continuous)
Wood St. EN5: Barn ... 14Za **30**
Wood St. KT1: King T ... 68Ma **131**
Wood St. RH1: Mers ... 1C **208**
Wood St. RM17: Grays ... 51Ee **121**
Wood St. W4 ... 50Ua **88**
WOOD STREET ... 27Ec **52**
Wood Street Walthamstow Station
(Overground) ... 28Fc 53
Woodsway KT22: Oxs ... 86Ga **172**
Woodsyre SE26 ... 63Vb **135**
Wood Ter. NW2 ... 34Xa **68**
Woodthorpe Rd. SW15 ... 56Xa **110**
Woodthorpe Rd. TW15: Ashf ... 65M **127**
Woodtree Cl. NW4 ... 26Za **48**
Wood Va. N10 ... 29Lb **50**
Wood Va. SE23 ... 60Xb **113**
Woodvale EN8: Walt C ... 7Yb **20**
Woodvale Av. SE25 ... 69Vb **135**
Woodvale Ct. BR1: Broml ... 67Kc **137**
... *(off Widmore Rd.)*
Wood Va. Est. SE23 ... 59Yb **114**
Woodvale Pk. AL1: St A ... 2F **6**
Woodvale Wlk. SE27 ... 64Sb **135**
Woodvale Way NW11 ... 34Za **68**
Wood Vw. HP1: Hem H ... 1K **3**
Wood Vw. RM16: Grays ... 48Fe **99**
Wood Vw. RM17: Grays ... 48Fe **99**
Woodview DA2: G St G ... 65Wd **142**
Woodview KT9: Chess ... 83La **172**
Woodview Av. E4 ... 21Ec **52**
Woodview Cl. BR6: Farnb ... 75Sc **160**
Woodview Cl. CR2: Sande ... 86Xb **177**
Woodview Cl. KT21: Asht ... 88Qa **173**
Woodview Cl. N4 ... 31Rb **71**
Woodview Cl. SW15 ... 63Ta **131**
Woodview Cl. TN15: W King ... 79Ud **164**
Woodview Ct. KT13: Weyb ... 78S **150**
Woodview Ct. WD17: Wat ... 11W **26**
... *(off Grandfield Av.)*
Wood Vw. M. RM1: Rom ... 25Fd **56**
Woodview M. SE19 ... 67Ub **135**
Woodview Rd. BR8: Swan ... 68Ed **140**
Woodview Way CR3: Cat'm ... 95Tb **197**
Woodville, The ... 9D **122**
Woodville, The W5 ... 44Ma **87**
... *(off Woodville Rd.)*
Woodville Cl. SE12 ... 57Jc **115**
Woodville Cl. SE3 ... 53Kc **115**
Woodville Cl. TW11: Tedd ... 63Ja **130**
Woodville Ct. KT22: Lea ... 92Ka **192**
Woodville Ct. N14 ... 15Lb **32**
Woodville Ct. SE10 ... 53Ec **114**
... *(off Blissett St.)*
Woodville Ct. SE19 ... 67Vb **135**
Woodville Ct. WD17: Wat ... 12W **26**
Woodville Gdns. HA4: Ruis ... 31S **64**
Woodville Gdns. IG6: Ilf ... 27Rc **54**
Woodville Gdns. KT6: Surb ... 73Ma **153**
Woodville Gdns. NW11 ... 31Za **68**
Woodville Gdns. W5 ... 44Na **87**
Woodville Gro. DA16: Well ... 55Wc **117**
Woodville Ho. SE1 ... 3K **231** (48Vb **91**)
... *(off St Saviour's Est.)*
Woodville Pl. DA12: Grav'nd ... 9D **122**
Woodville Rd. CR7: Thor H ... 70Sb **135**
Woodville Rd. E11 ... 32Hc **73**
Woodville Rd. E17 ... 28Bc **52**
Woodville Rd. E18 ... 26Kc **53**
Woodville Rd. EN5: New Bar ... 13Db **31**
Woodville Rd. N16 ... 36Ub **71**
Woodville Rd. NW11 ... 31Za **68**
Woodville Rd. NW6 ... 40Bb **69**
Woodville Rd. SM4: Mord ... 70Cb **133**
Woodville Rd. TW10: Ham ... 62Ka **130**
Woodville Rd. W5 ... 44Ma **87**
Woodville St. SE18 ... 49Nc **94**
Woodvill Rd. KT22: Lea ... 92Ka **192**
Wood Wlk. WD3: Chor ... 12G **24**
Woodward Av. NW4 ... 29Wa **48**
Woodward Cl. KT10: Clay ... 79Ha **152**
Woodward Cl. RM17: Grays ... 49De **99**
Woodwarde Rd. SE22 ... 58Ub **113**
Woodward Gdns. HA7: Stan ... 24Ha **46**
Woodward Gdns. RM9: Dag ... 38Yc **75**
Woodward Hgts. RM17: Grays ... 49De **99**
Woodward Rd. RM9: Dag ... 38Xc **75**
Woodward's Footpath TW2:
Whitt ... 58Ea **108**
Woodward Ter. DA9: Ghithe ... 58Ud **120**
Wood Way BR6: Farnb ... 75Qc **160**
Woodway CM13: Hut ... 18Ce **41**
Woodway CM15: Hut ... 18Ce **41**
Woodway CM15: Shenf ... 18Ce **41**
Woodway Cres. HA1: Harr ... 30Ja **46**
Woodwaye WD19: Wat ... 17Y **27**
Woodwell St. SW18 ... 57Eb **111**
Wood Wharf SE10 ... 51Dc **114**
Woodwicks WD3: Map C ... 22F **42**
Woodyard, The CM16: Epp ... 1Yc **23**
Woodyard Cl. NW5 ... 36Jb **70**
Woodyard La. SE21 ... 59Ub **113**
Woodyates Rd. SE12 ... 58Jc **115**
Woolacombe Rd. SE3 ... 53Lc **115**
Woolacombe Way UB3: Harl ... 49U **84**
Woolaton M. CM14: B'wood ... 18Yd **40**
Woolbrook Rd. DA1: Cray ... 58Gd **118**
Woolcombes Ct. SE16 ... 46Zb **92**
... *(off Princes Riverside Rd.)*
Wooldridge Cl. TW14: Bedf ... 60S **106**
Wooler St. SE17 ... 7F **231** (50Tb **91**)
Woolf Cl. SE28 ... 46Xc **95**
Woolf M. WC1 ... 5E **216** (42Mb **90**)
... *(off Burton Pl.)*
Woolford Ct. SE5 ... 54Sb **113**
... *(off Coldharbour La.)*
Woolf Wlk. RM18: Tilb ... 4E **122**
... *(off Brennan Rd.)*
Woolgar M. N16 ... 36Ub **71**
... *(off Gillett St.)*
Woolhampton Way IG7: Chig ... 20Xc **37**
Woolhams CR3: Cat'm ... 98Wb **197**
Woolhouse Pl. DA2: Dart ... 58Rd **119**
Woolings Cl. RM16: Ors ... 5B **100**
Woollard St. EN9: Walt A ... 6Ec **20**
Woollaston Rd. N4 ... 30Rb **51**
Woollett Cl. DA1: Cray ... 56Jd **118**
Woolley Ho. SW9 ... 55Rb **113**
... *(off Loughborough Rd.)*
Woollon Ho. E1 ... 44Yb **92**
... *(off Clark St.)*
Woolman Rd. WD17: Wat ... 10W **12**
Woolmead Av. NW9 ... 31Wa **68**
Woolmer Cl. WD6: Bore ... 10Qa **15**
Woolmerdine Ct. WD23: Bush ... 13Z **27**
Woolmer Dr. HP2: Hem H ... 2C **4**
Woolmer Gdns. N18 ... 23Wb **51**
Woolmer Rd. N18 ... 22Wb **51**
Woolmore St. E14 ... 45Ec **92**
Woolneigh St. SW6 ... 55Db **111**
Woolridge Way E9 ... 38Yb **72**
Wool Rd. SW20 ... 65Xa **132**
Woolstaplers Way SE16 ... 48Wb **91**
Woolston Cl. E17 ... 26Zb **52**
Woolstone Ho. E2 ... 39Wb **71**
... *(off Whiston Rd.)*
Woolstone Rd. SE23 ... 61Ac **136**
Woolston Manor Golf Course ... **15Tc 36**
WOOLWICH ... 48Pc 94
Woolwich Arsenal Station (Rail, DLR &
Crossrail) ... 49Rc 94
Woolwich Cen., The ... 49Qc **94**
... *(off Wellington St.)*
Woolwich Chu. St. SE18 ... 48Nc **94**
Woolwich Comn. SE18 ... 51Qc **116**
Woolwich Comn. Youth Club ... **51Qc 116**
Woolwich Dockyard Ind. Est. SE18 ... 48Nc **94**
Woolwich Dockyard Station (Rail) ... 49Pc 94
Woolwich High St. SE18 ... 48Qc **94**
Woolwich Ho. UB3: Harl ... 48S **84**
... *(off Nine Acres Cl.)*
Woolwich Mnr. Way E16 ... 47Rc **94**
Woolwich Mnr. Way E6 ... 42Pc **94**
Woolwich New Rd. SE18 ... 50Qc **94**
Woolwich Rd. DA17: Belv ... 50Bd **95**
Woolwich Rd. DA6: Bex ... 56Cd **118**
Woolwich Rd. DA7: Bex ... 55Cd **118**
Woolwich Rd. SE10 ... 50Hc **93**
Woolwich Rd. SE2 ... 51Zc **117**
Woolwich Rd. SE7 ... 50Jc **93**
Woolwich Trade Pk. SE28 ... 48Tc **94**
Wooster Gdns. E14 ... 44Fc **93**
Wooster M. HA2: Harr ... 27Ea **46**
Wooster Pl. SE1 ... 5G **231** (49Tb **91**)
... *(off Searles Rd.)*
Wootton Cl. KT18: Eps ... 88Va **174**
Wootton Cl. RM11: Horn ... 29Md **57**
Wootton Cl. WD7: R'lett ... 7Ja **14**
Wootton Grange GU22: Wok ... 91A **188**
... *(off Langley Wlk.)*
Wootton Gro. N3 ... 25Cb **49**
Wootton Pl. KT10: Esh ... 77Ea **152**
Wootton St. SE1 ... 7A **224** (46Qb **90**)
Worbeck Rd. SE20 ... 68Xb **135**
Worcester Av. N17 ... 24Wb **51**
Worcester Av. RM14: Upm ... 33Vd **78**
Worcester Cl. CR0: C'don ... 75Cc **158**
Worcester Cl. CR4: Mitc ... 68Jb **134**
Worcester Cl. DA13: Ist R ... 6B **144**
Worcester Cl. DA9: Ghithe ... 56Xd **120**
Worcester Cl. NW2 ... 34Xa **68**
Worcester Cl. SE20 ... 67Wb **135**
Worcester Ct. AL1: St A ... 3E **6**
Worcester Ct. HA1: Harr ... 27Ga **46**
Worcester Ct. KT12: Walt T ... 75Y **151**
Worcester Ct. KT4: Wor Pk ... 76Ua **154**
Worcester Ct. N12 ... 22Db **49**
Worcester Ct. RH1: Redh ... 4N **207**
... *(off Timperley Gdns.)*
Worcester Ct. W7 ... 44Ha **86**
... *(off Copley Cl.)*

Worcester Ct. W9 43Cb 89
(off Elmfield Way)
Worcester Cres. IG8: Wfd G 21Kc 53
Worcester Cres. NW7 20Ua 30
Worcester Dr. BR8: Swan 70Hd 140
Worcester Dr. TW15: Ashf 64R 128
Worcester Dr. W4 47Ua 88
Worcester Gdns. IG1: Ilf 31Nc 74
Worcester Gdns. KT4: Wor Pk 76Ua 154
Worcester Gdns. SL1: Slou 7H 81
Worcester Gdns. SW11 57Hb 111
(off Grandison Rd.)
Worcester Gdns. UB6: G'frd 37Fa 66
Worcester Ho. SE11 4K 229 (48Qb 90)
(off Kennington Rd.)
Worcester Ho. SW9 52Qb 112
(off Cranmer Rd.)
Worcester Ho. W2 44Eb 89
(off Hallfield Est.)
Worcester Ho. WD6: Bore 12Qa 29
(off Stratfield Rd.)
Worcester M. NW6 37Db 69
WORCESTER PARK 74Wa 154
Worcester Pk. Rd. KT4: Wor Pk 76Ta 153
Worcester Park Station (Rail) 74Wa 154
Worcester Point EC1 4D 218 (41Sb 91)
Worcester Rd. E12 35Pc 74
Worcester Rd. E17 26Zb 52
Worcester Rd. RH2: Reig 5H 207
Worcester Rd. SM2: Sutt 80Cb 155
Worcester Rd. SW19 64Bb 133
Worcester Rd. UB8: Cowl 43L 83
Worcesters Av. EN1: Enf 10Wb 19
Wordsworth Av. CR8: Kenley 87Tb 177
Wordsworth Av. E12 38Nc 74
Wordsworth Av. E18 27Hc 53
Wordsworth Av. UB6: G'frd 41Fa 86
Wordsworth Cl. AL3: St A 4P 5
Wordsworth Cl. RM18: Tilb 4D 122
Wordsworth Cl. RM3: Rom 25Ld 57
Wordsworth Ct. HA1: Harr 31Ga 66
Wordsworth Dr. SM3: Cheam 77Ya 154
Wordsworth Gdns. WD6: Bore 15Qa 29
Wordsworth Ho. NW6 41Cb 89
(off Stafford Rd.)
Wordsworth Ho. SE18 51Qc 116
(off Woolwich Comn.)
Wordsworth Mans. W14 51Bb 111
(off Queens Club Gdns.)
Wordsworth Mead RH1: Redh 4A 208
Wordsworth Pde. N8 28Rb 51
Wordsworth Pl. NW5 36Hb 69
Wordsworth Rd. DA16: Well 53Uc 116
Wordsworth Rd. KT15: Add 77M 149
Wordsworth Rd. N16 35Ub 71
Wordsworth Rd. SE1 6K 231 (49Vb 91)
Wordsworth Rd. SE20 66Zb 136
Wordsworth Rd. SL2: Slou 2B 80
Wordsworth Rd. SM6: W'gton 79Lb 156
Wordsworth Rd. TW12: Hamp 63Ba 129
Wordsworth Wlk. NW11 28Cb 49
Wordsworth Way DA1: Dart 56Qd 119
(not continuous)
Wordsworth Way UB7: W Dray 49N 83
Worfield St. SW11 52Gb 111
Worgan St. SE11 7H 229 (50Pb 90)
Worgan St. SE16 49Zb 92
Worland Rd. E15 38Gc 73
World Bus. Cen. TW6: H'row A 53S 106
World Gdn. 77Ld 163
World of Golf Cen. Croydon 71Yb 158
World of Golf Cen. New Malden 69Wa 132
World Rugby Mus. 58Ga 108
WORLD'S END 13Pb 32
World's End KT11: Cobh 86W 170
Worlds End KT18: Eps 88Ta 173
Worlds End Est. SW10 52Fb 111
World's End La. EN2: Enf 14Pb 32
World's End La. N21 15Pb 32
Worlds End La. BR6: Chels 79Vc 161
World's End Pas. SW10 52Fb 111
(off Worlds End Est.)
World's End Pl. SW10 52Fb 111
(off Worlds End Est.)
Worley Rd. AL3: St A 1B 6
Worleys Dr. BR6: Orp 77Tc 160
Worlidge St. W6 50Ya 88
Worlingham Rd. SE22 56Vb 113
Wormholt Rd. W12 45Wa 88
Wormley Ct. EN9: Walt A 5Jc 21
Wormwood Scrubs Pk. 43Va 88
Wormwood St. EC2 2H 225 (44Ub 91)
Wormyngford Ct. EN9: Walt A 5Jc 21
Wornington Rd. W10 42Ab 88
(not continuous)
Woronzow Rd. NW8 39Fb 69
Worple, The TW19: Wray 58B 104
Worple Av. SW19 66Za 132
Worple Av. TW18: Staines 65K 127
Worple Av. TW7: Isle 57Ja 108
Worple Cl. HA2: Harr 32Ba 65
Worple Rd. KT18: Eps 87Ta 173
Worple Rd. KT22: Lea 94Ka 192
(not continuous)
Worple Rd. SW19 66Za 132
Worple Rd. SW20 68Ya 132
Worple Rd. TW18: Staines 65K 127
(not continuous)
Worple Rd. TW7: Isle 56Ja 108
Worple Rd. M. SW19 65Bb 133
WORPLESDON 9J 187
Worplesdon Golf Course 4J 187
Worplesdon Hill GU22: Wok 4G 186
Worplesdon Hill Ho. GU22: Wok 4H 187
Worplesdon Rd. GU2: Guild 10K 187
Worplesdon Rd. GU3: Guild 7G 186
Worplesdon Rd. GU3: Worp 7G 186
Worplesdon Station (Rail) 6M 187
Worple St. SW14 55Ta 109
Worple Way HA2: Harr 32Ba 65
Worple Way TW10: Rich 57Na 109
Worrall La. UB8: Uxb 38N 63
Worrin Cl. CM15: Shenf 18Be 41
Worrin Pl. CM15: Shenf 19Be 41
Worrin Rd. CM15: Shenf 19Be 41
Worsfold Cl. GU23: Send 95D 188
Worships Hill TN13: Riv 95Gd 202
Worship St. EC2 6G 219 (42Tb 91)
(not continuous)
Worslade Rd. SW17 63Fb 133
Worsley Bri. Rd. BR3: Beck 66Cc 136
Worsley Bri. Rd. SE26 63Bc 136
Worsley Grange BR7: Chst 65Sc 138
Worsley Gro. E5 35Wb 71
Worsley Ho. SE23 61Yb 136
Worsley Rd. E11 35Gc 73
Worsopp Dr. SW4 57Lb 112
Worsted Grn. RH1: Mers 1C 208
Worth Cl. BR6: Orp 77Uc 160
Worthfield Cl. KT19: Ewe 80Ta 153
Worth Gro. SE17 7F 231 (50Tb 91)
Worthing Cl. E15 39Gc 73
Worthing Cl. RM17: Grays 51Ae 121
Worthing Rd. TW5: Hest 51Ba 107
Worthington Cl. CR4: Mitc 70Kb 134
Worthington Ho. EC1 3A 218 (41Qb 90)
(off Myddelton Pas.)
Worthington Rd. KT6: Surb 74Pa 153
Wortley Rd. CR0: C'don 73Qb 156
Wortley Rd. E6 38Mc 73
Worton Ct. TW7: Isle 56Ga 108
Worton Gdns. TW7: Isle 54Fa 108
Worton Hall Ind. Est. TW7: Isle 56Ga 108
Worton Rd. TW7: Isle 56Fa 108
Worton Way TW3: Houn 54Fa 108
Worton Way TW7: Houn 54Fa 108
Worton Way TW7: Isle 54Fa 108
Wotton Ct. E14 45Fc 93
(off Jamestown Way)
Wotton Grn. BR5: St M Cry 70Zc 139
Wotton Ho. SL4: Eton 10H 81
(off Common La.)
Wotton Rd. NW2 34Ya 68
Wotton Rd. SE8 51Bc 114
Wotton Way SM2: Cheam 82Ya 174
Wouldham Rd. E16 44Hc 93
Wouldham Rd. RM20: Grays 51Ae 121
Wrabness Way TW18: Staines 67K 127
Wragby Rd. E11 34Gc 73
Wrampling Pl. N9 18Wb 33
Wrangley Ct. EN9: Walt A 5Jc 21
Wrangthorn Wlk. CR0: Wadd 77Qb 156
Wraxall Rd. E3 42Bc 92
(off Hamlets Way)
Wray Av. IG5: Ilf 27Qc 54
Wrayburn Ho. SE16 47Wb 91
(off Llewellyn St.)
Wray Cl. RM11: Horn 31Ld 77
WRAY COMMON 5M 207
Wray Comn. Rd. RH2: Reig 5L 207
Wray Cres. N4 33Nb 70
Wrayfield Av. RH2: Reig 5L 207
Wrayfield Rd. SM3: Cheam 76Za 154
Wraylands Dr. RH2: Reig 4M 207
Wray La. RH2: Reig 2L 207
Wraymead Pl. RH2: Reig 5K 207
Wray Mill Pk. RH2: Reig 4M 207
Wray Pk. Rd. RH2: Reig 5K 207
Wray Rd. SM2: Cheam 81Bb 175
WRAYSBURY 58B 104
Wraysbury Cl. TW4: Houn 57Aa 107
Wraysbury Dive Cen. 58B 104
Wraysbury Dr. UB7: Yiew 45M 83
Wraysbury Gdns. TW18: Staines 63G 126
Wraysbury Rd. TW18: Staines 62E 126
Wraysbury Rd. TW19: Staines 61D 126
Wraysbury Station (Rail) 58C 104
Wrays Way UB4: Hayes 42U 84
Wrekin Rd. SE18 52Sc 116
Wren Av. NW2 35Ya 68
Wren Av. UB10: Uxb 39P 63
Wren Av. UB2: S'hall 49Ba 85
Wren Cl. BR5: St P 69Zc 139
Wren Cl. CR2: Sels 81Zb 178
Wren Cl. E16 44Hc 93
Wren Cl. KT19: Eps 82Ta 173
Wren Cl. N9 18Zb 34
Wren Cl. TW6: H'row A 55K 105
Wren Ct. CR0: C'don 77Tb 157
(off Coombe Rd.)
Wren Ct. SL3: L'ly 48C 82
Wren Cres. KT15: Add 78M 149
Wren Cres. WD23: Bush 18Ea 28
Wren Dr. EN9: Walt A 6Jc 21
Wren Dr. UB7: W Dray 48M 83
Wren Gdns. RM12: Horn 32Hd 76
Wren Gdns. RM9: Dag 36Zc 75
Wren Ho. E3 40Ac 72
(off Gernon Rd.)
Wren Ho. KT1: Hamp W 68Ma 131
(off High St.)
Wren Ho. SW1 7D 228 (50Mb 90)
(off Aylesford St.)
Wren Landing E14 46Cc 92
Wren La. HA4: Ruis 30X 45
Wren M. SE11 6J 229 (49Pb 90)
Wren M. SE13 56Gc 115
Wrenn Ho. SW13 51Ya 110
Wren Path SE28 48Tc 94
Wren Pl. CM14: B'wood 20Zd 41
Wren Rd. DA14: Sidc 63Yc 139
Wren Rd. RM9: Dag 36Zc 75
Wren Rd. SE5 53Tb 113
Wren's Av. TW15: Ashf 63S 128
Wrens Cft. DA11: Nflt 3A 144
Wrensfield HP1: Hem H 2J 3
Wrensfield Cl. WD17: Wat 9V 12
Wrens Hill KT22: Oxs 87Ea 172
Wren's Pk. Ho. E5 33Xb 71
Wren St. WC1 5J 217 (42Pb 90)
Wren Ter. IG10: Lough 11Rc 36
Wrentham Av. NW10 40Za 68
Wrenthorpe Rd. BR1: Broml 63Gc 137
Wren Vw. N6 31Lb 70
Wren Wlk. RM18: Tilb 2D 122
Wrenwood Way HA5: Eastc 28X 45
Wrexham Rd. E3 40Cc 72
Wrexham Rd. RM3: Rom 20Md 39
Wricklemarsh Rd. SE3 54Kc 115
(not continuous)
Wrigglesworth St. SE14 52Zb 114
Wright SL4: Wind 5A 102
Wright Cl. DA10: Swans 58Zd 121
Wright Cl. SE13 56Fc 115
Wright Cl. WD23: Bush 15Aa 27
Wright Gdns. TW17: Shep 71Q 150
Wright Rd. TW5: Hest 52Y 107
Wrights All. SW19 65Ya 132
Wrightsbridge Rd. CM14: S Weald 18Pd 39
Wright's Bldgs. WD17: Wat 12X 27
(off St Albans Rd.)
Wrights Cl. RM10: Dag 35Dd 76
Wright's Cotts. SL9: Chal P 25A 42
(off Church La.)
Wrights Grn. SW4 56Mb 112
Wright's La. W8 48Db 89
Wrights Pl. NW10 37Sa 67
Wright Sq. SL4: Wind 5B 102
Wright's Rd. E3 40Bc 72
(not continuous)
Wrights Rd. SE25 69Ub 135
Wrights Row SM6: W'gton 77Kb 156
Wrights Wlk. SW14 55Ta 109
Wright Way SL4: Wind 5A 102
Wright Way TW6: H'row A 54K 105
Wrigley Cl. E4 22Fc 53
Wrington Ho. RM3: Rom 22Pd 57
(off Redruth Rd.)
Wriotsley Way KT15: Add 79J 149
Writtle Ho. NW9 26Va 48
Writtle Wlk. RM13: Rain 39Gd 76
WROTHAM 88Ce 185
Wrotham Bus. Pk. EN5: Barn 9Bb 17
Wrotham By-Pass TN15: Wro 89Ce 185
WROTHAM HEATH 90Fe 185
Wrotham Hill Picnic Site 86Ae 185
Wrotham Hill Rd. TN15: Wro 85Be 185
Wrotham Ho. BR3: Beck 66Bc 136
(off Sellindge Cl.)
Wrotham Ho. SE1 3G 231 (48Tb 91)
(off Law St.)
Wrotham Pk. 8Ab 16
Wrotham Rd. DA11: Grav'nd 3C 144
Wrotham Rd. DA13: Ist R 9B 144
Wrotham Rd. DA13: Meop 9B 144
Wrotham Rd. DA16: Well 53Yc 117
Wrotham Rd. EN5: Barn 12Ab 30
Wrotham Rd. NW1 38Lb 70
Wrotham Rd. TN15: Bor G 92Be 205
Wrotham Rd. TN15: Wro 92Be 205
Wrotham Rd. W13 46La 86
Wrotham Water La. ME19: Tros 87Fe 185
Wrotham Water Rd. TN15: Wro H 86Fe 185
Wroth's Path IG10: Lough 11Pc 36
Wrott & Hill Ct. DA4: Sut H 67Rd 141
Wrottesley Rd. NW10 40Wa 68
Wrottesley Rd. SE18 51Sc 116
Wroughton Rd. SW11 57Hb 111
Wroughton Ter. NW4 28Xa 48
Wroxall Ct. RM19: Purf 50Rd 97
(off Linnet Way)
Wroxall Rd. RM9: Dag 37Yc 75
Wroxham Av. HP3: Hem H 4M 3
Wroxham Gdns. EN2: Crew H 7Rb 19
Wroxham Gdns. EN6: Pot B 3Za 16
Wroxham Gdns. N11 24Mb 50
Wroxham Rd. SE28 45Zc 95
Wroxham Way IG6: Ilf 25Rc 54
Wroxton Rd. SE15 54Yb 114
WRYTHE, THE 76Hb 155
Wrythe Grn. SM5: Cars 76Hb 155
Wrythe Grn. Rd. SM5: Cars 76Hb 155
Wrythe La. SM5: Cars 74Eb 155
Wulfred Way TN15: Kems'g 90Rd 183
Wulfstan St. W12 43Va 88
Wulstan Pk. EN6: Pot B 4Fb 17
Wyatt Cl. GU24: Bisl 9E 166
Wyatt Cl. SE16 47Bc 92
Wyatt Cl. TN15: Bor G 92Be 205
Wyatt Cl. TW13: Felt 60Z 107
Wyatt Cl. UB4: Hayes 43W 84
Wyatt Cl. WD23: Bush 17Fa 28
Wyatt Ct. HA0: Wemb 38Na 67
Wyatt Ct. W3 48Sa 87
(off All Saints Rd.)
Wyatt Dr. SW13 51Xa 110
Wyatt Ho. NW8 6B 214 (42Fb 89)
(off Frampton St.)
Wyatt Ho. SE3 54Hc 115
Wyatt Ho. TW1: Twick 58Ma 109
Wyatt Pk. Rd. SW2 61Nb 134
Wyatt Point SE28 47Sc 94
Wyatt Rd. DA1: Cray 55Hd 118
Wyatt Rd. E7 37Jc 73
Wyatt Rd. N5 34Sb 71
Wyatt Rd. SL4: Wind 5B 102
Wyatt Rd. TW18: Staines 64J 127
Wyatt's Cl. WD3: Chor 13J 25
Wyatt's Covert UB9: Den 28H 43
Wyatts La. E17 27Ec 52
Wyatt's Rd. WD3: Chor 14H 25
Wybert St. NW1 5B 216 (42Lb 90)
Wyborne Ho. NW10 38Sa 67
Wyborne Way NW10 38Sa 67
Wyburn Av. EN5: Barn 13Bb 31
Wychcombe Studios NW3 37Hb 69
Wyche Gro. CR2: S Croy 80Tb 157
Wych Elm Cl. KT2: King T 66Pa 131
Wych Elm Cl. RM11: Horn 31Qd 77
Wych Elm Lodge BR1: Broml 66Hc 137
Wych Elm Pas. KT2: King T 66Pa 131
Wych Elm Rd. RM11: Horn 30Qd 57
Wychelm Rd. GU18: Light 3A 166
Wych Elms AL2: Park 10P 5
Wycherley Cl. SE3 52Hc 115
Wycherley Cres. EN5: New Bar 16Db 31
Wych Hill GU22: Wok 1N 187
Wych Hill La. GU22: Wok 1P 187
Wych Hill Pk. GU22: Wok 1P 187
Wych Hill Ri. GU22: Wok 1N 187
Wych Hill Way GU22: Wok 2P 187
Wychwood Av. CR7: Thor H 69Sb 135
Wychwood Av. HA8: Edg 23Ma 47
Wychwood Cl. HA8: Edg 23Ma 47
Wychwood Cl. KT22: Oxs 86Fa 172
Wychwood Cl. TW16: Sun 65W 128
Wychwood End N6 31Lb 70
Wychwood Gdns. IG5: Ilf 28Pc 54
Wychwood Way HA6: Nwood 24V 44
Wychwood Way SE19 65Tb 135
Wyckham Ho. RH8: Oxt 1J 211
(off Station App.)
Wyclif Ct. EC1 4B 218 (41Rb 91)
(off Wyclif St.)
Wycliffe Av. KT15: Add 80J 149
Wycliffe Cl. DA16: Well 53Vc 117
Wycliffe Cl. EN8: Chesh 1Zb 20
Wycliffe Ct. WD5: Ab L 4U 12
Wycliffe Ho. DA11: Nflt 10B 122
(off Wycliffe Row)
Wycliffe Rd. SW11 54Jb 112
Wycliffe Rd. SW19 65Db 133
Wycliffe Row DA11: Nflt 10B 122
(not continuous)
Wyclif St. EC1 4B 218 (41Rb 91)
Wycombe Gdns. NW11 33Cb 69
Wycombe Ho. NW8 5D 214 (42Gb 89)
(off Grendon St.)
Wycombe Pl. SW18 58Eb 111
Wycombe Rd. HA0: Wemb 39Qa 67
Wycombe Rd. IG2: Ilf 29Pc 54
Wycombe Rd. N17 25Wb 51
Wycombe Sq. W8 46Bb 89
Wydehurst Rd. CR0: C'don 73Wb 157
Wydell Cl. SM4: Mord 72Za 154
Wydeville Mnr. Rd. SE12 63Kc 137
Wyecliffe Gdns. RH1: Mers 2C 208
Wye Cl. BR6: Orp 73Vc 161
Wye Cl. HA4: Ruis 30S 44
Wye Cl. TW15: Ashf 63R 128
Wye Ct. W13 43Ka 86
(off Malvern Way)
Wyedale AL2: Lon C 9K 7
Wyemead Cres. E4 19Gc 35
Wye Rd. DA12: Grav'nd 1F 144
Wye Rd. TN15: Bor G 91Ce 205
Wye St. SW11 54Fb 111
Wyeth Cl. SL6: Tap 5A 80
Wyeths M. KT17: Eps 85Va 174
Wyeths Rd. KT17: Eps 85Va 174
Wyevale Cl. HA5: Eastc 27W 44
Wyfields IG5: Ilf 25Rc 54
Wyfold Ho. SE2 47Zc 95
(off Wolvercote Rd.)
Wyfold Rd. SW6 52Ab 110
Wyhill Wlk. RM10: Dag 37Ed 76
Wyke Cl. TW7: Isle 51Ha 108
Wyke Gdns. W7 48Ja 86
Wyke Green Golf Course 51Ha 108
Wykeham Av. RM11: Horn 30Md 57
Wykeham Av. RM9: Dag 37Yc 75
Wykeham Cl. DA12: Grav'nd 5G 144
Wykeham Cl. HA3: W'stone 25Ha 46
Wykeham Cl. UB7: Sip 51Q 106
Wykeham Ct. NW4 29Ya 48
(off Wykeham Rd.)
Wykeham Grn. RM9: Dag 37Yc 75
Wykeham Hill HA9: Wemb 32Pa 67
Wykeham Ho. SE1 7D 224 (46Sb 91)
(off Union St.)
Wykeham Ri. N20 18Ab 30
Wykeham Rd. HA3: Kenton 28Ka 46
Wykeham Rd. NW4 28Ya 48
Wyke Rd. E3 38Cc 72
Wyke Rd. SW20 68Ya 132
Wylands Rd. SL3: L'ly 49C 82
Wylchin Cl. HA5: Eastc 27V 44
Wyldes Cl. NW11 32Eb 69
Wyldewoods SL5: S'hill 2B 146
Wyldfield Gdns. N9 19Vb 33
Wyld Way HA9: Wemb 37Ra 67
Wyleu St. SE23 59Ac 114
Wyllen Cl. E1 42Yb 92
Wyllyotts Cen. & Cinema 4Bb 17
Wyllyotts Cl. EN6: Pot B 4Bb 17
Wyllyotts La. EN6: Pot B 4Bb 17
Wyllyotts Pl. EN6: Pot B 4Bb 17
Wyllyotts Theatre 4Bb 17
Wylo Dr. EN5: Ark 16Wa 30
Wymark Cl. RM13: Rain 40Jd 76
Wymering Mans. W9 41Cb 89
(off Wymering Rd.)
Wymering Rd. W9 41Cb 89
Wymers Cl. SL1: Burn 10A 60
Wymondham Ct. NW8 39Fb 69
(off Queensmead)
Wymond St. SW15 55Ya 110
Wynan Rd. E14 50Dc 92
Wynash Gdns. SM5: Cars 78Gb 155
Wynaud Ct. N22 23Pb 50
Wyncham Av. DA15: Sidc 60Uc 116
Wyncham Ho. DA15: Sidc 61Wc 139
(off Longlands Rd.)
Wynches Farm Dr. AL4: St A 2H 7
Wynchgate HA3: Hrw W 24Ga 46
Wynchgate N14 18Mb 32
Wynchgate N21 17Nb 32
Wynchgate UB5: N'olt 36Aa 65
Wynchlands Cres. AL4: St A 2H 7
Wyncote Way CR2: Sels 81Zb 178
Wyncroft Cl. BR1: Broml 69Pc 138
Wyndale Av. NW9 30Qa 47
Wyndcliff Rd. SE7 51Kc 115
Wyndcroft Cl. EN2: Enf 13Rb 33
Wyndham Apts. SE10 50Gc 93
Wyndham Av. KT11: Cobh 85W 170
Wyndham Cl. BR6: Farnb 74Sc 160
Wyndham Cl. SM2: Sutt 80Cb 155
Wyndham Ct. W7 49Ja 86
Wyndham Cres. N19 34Lb 70
Wyndham Cres. SL1: Burn 10A 60
Wyndham Cres. TW4: Houn 58Ca 107
Wyndham Deedes Ho. E2 40Wb 71
(off Hackney Rd.)
Wyndham Est. SE5 52Sb 113
Wyndham Ho. E14 47Dc 92
(off Marsh Wall)
Wyndham Ho. SW1 6H 227 (49Jb 90)
(off Sloane Sq.)
Wyndham M. W1 1F 221 (43Hb 89)
Wyndham Pl. W1 1F 221 (43Hb 89)
Wyndham Rd. E6 38Mc 73
Wyndham Rd. EN4: E Barn 18Hb 31
Wyndham Rd. GU21: Wok 10M 167
Wyndham Rd. KT2: King T 66Pa 131
Wyndham Rd. SE5 52Sb 113
Wyndham Rd. W13 48Ka 86
Wyndhams Ct. E8 38Vb 71
(off Celandine Dr.)
Wyndham's Theatre 4F 223 (45Nb 90)
(off Charing Cross Rd.)
Wyndham St. W1 7F 215 (43Hb 89)
Wyndham Yd. W1 1F 221 (43Hb 89)
Wyndhurst Cl. CR2: S Croy 80Rb 157
Wyneham Rd. SE24 57Tb 113
Wynell Rd. SE23 62Zb 136
Wynford Gro. BR5: St P 69Xc 139
Wynford Pl. DA17: Belv 51Cd 118
Wynford Rd. N1 1H 217 (40Pb 70)
Wynford Way SE9 62Pc 138
Wynlie Gdns. HA5: Pinn 26X 45
Wynn Bri. Cl. IG8: Wfd G 25Mc 53
Wynndale Rd. E18 25Kc 53
Wynne Ct. WD18: Wat 15U 26
(off Raven Cl.)
Wynne Rd. SW9 54Qb 112
Wynn's Av. DA15: Sidc 57Vc 117
Wynnstay Gdns. W8 48Cb 89
Wynnstow Pk. RH8: Oxt 3K 211
Wynter St. SW11 56Eb 111
Wynton Gdns. SE25 71Vb 157
Wynton Gro. KT12: Walt T 76W 150
Wynton Pl. W3 44Ra 87
Wynyard Cl. WD3: Sarr 8J 11
Wynyard Ho. SE11 7J 229 (50Pb 90)
(off Newburn St.)
Wynyard Ter. SE11 7J 229 (50Pb 90)
Wynyatt St. EC1 4B 218 (41Rb 91)
Wyre Gro. HA8: Edg 20Ra 29
Wyre Gro. UB3: Harl 49W 84
Wyresdale Cres. UB6: G'frd 41Ha 86
Wyses M. RM1: Rom 30Jd 56
Wyteleaf Cl. HA4: Ruis 30S 44
Wytham Ho. NW8 6C 214 (42Fb 89)
(off Church St. Est.)
Wythburn Ct. W1 2F 221 (44Hb 89)
(off Wythburn Pl.)
Wythburn Pl. W1 3F 221 (44Hb 89)
Wythegate TW18: Staines 66H 127
Wythenshawe Rd. RM10: Dag 34Cd 76
Wythens Wlk. SE9 58Rc 116
Wythes Cl. BR1: Broml 68Pc 138
Wythes Rd. E16 46Nc 94
Wythfield Rd. SE9 58Pc 116
Wyvell Cl. CR0: C'don 73Zb 158
Wyvenhoe Rd. HA2: Harr 35Ea 66
Wyvern Cl. BR6: Chels 76Xc 161
Wyvern Cl. DA1: Dart 59Ld 119
Wyvern Est. KT3: N Mald 70Wa 132
Wyvern Ho. RM17: Grays 51De 121
(off Bridge Rd.)
Wyvern Pl. KT15: Add 77K 149
Wyvern Rd. CR8: Purl 82Rb 177
Wyvern Way UB8: Uxb 38K 63
Wyvil Rd. SW8 52Nb 112
Wyvis St. E14 43Dc 92

X

XC Hemel Hempstead 3P 3
Xchange, The WD18: Wat 13X 27
(off Exchange Rd.)
Xylon Ho. KT4: Wor Pk 75Xa 154

Y

Yabsley St. E14 46Ec 92
Yaffle Rd. KT13: Weyb 82S 170
Yaldham Ho. SE1 6H 231 (49Ub 91)
(off Old Kent Rd.)
Yaldham Mnr. Dr. TN15: Kems'g 89Xd 184
Yalding Gro. BR5: St M Cry 70Zc 139
Yalding Rd. SE16 48Wb 91
Yale Cl. TW4: Houn 57Ba 107
Yale Ct. NW6 36Db 69
Yale Way RM12: Horn 35Jd 76
Yarborough Rd. SW19 67Fb 133
Yarbridge Cl. SM2: Sutt 82Db 175
Yard, The N1 2G 217 (40Nb 70)
(off Caledonia Rd.)
Yardley Cl. E4 15Dc 34
Yardley Cl. RH2: Reig 4K 207
Yardley Ct. SM3: Cheam 77Ya 154
Yardley La. E4 15Dc 34
Yardleys TW20: Thorpe 69D 126
Yardley St. WC1 4K 217 (41Qb 90)
(not continuous)
Yardmaster Ho. CR0: C'don 74Tb 157
Yard Mead TW20: Egh 62C 126
Yarlington Ct. N11 22Jb 50
(off Sparkford Gdns.)
Yarm Cl. KT22: Lea 95La 192
Yarm Ct. Rd. KT22: Lea 95La 192
Yarm Holt KT22: Lea 96La 192
Yarmouth Cres. N17 29Xb 51
Yarmouth Pl. W1 7K 221 (46Kb 90)
Yarmouth Rd. SL1: Slou 5G 80
Yarmouth Rd. WD24: Wat 10Y 13
Yarm Way KT22: Lea 95Ma 193
Yarnfield Sq. SE15 53Wb 113
Yarnton Way DA18: Belv 48Bd 95
Yarnton Way DA18: Erith 48Bd 95
Yarnton Way SE2 47Yc 95
Yarra Ho. AL1: St A 2C 6
(off Beaconsfield Rd.)
Yarrell Mans. W14 51Bb 111
(off Queen's Club Gdns.)
Yarrow Ct. TN14: Dun G 92Hd 202
Yarrow Cres. E6 43Nc 94
Yarrow Ho. E14 48Ec 92
(off Stewart St.)
Yarrow Ho. W10 43Ya 88
(off Sutton Way)
Yateley Ct. CR8: Kenley 86Sb 177
Yateley St. SE18 48Mc 93
Yatesbury Ct. E5 36Ac 72
(off Studley Cl.)
Yates Ct. NW2 37Za 68
(off Willesden La.)
Yates Ct. SE1 47Vb 91
(off Abbey St.)
Yates Ho. E2 41Wb 91
(off Roberta St.)
Yatton Ho. W10 43Ya 88
(off Sutton Way)
YEADING 42Y 85
Yeading Av. HA2: Harr 33Aa 65
Yeading Brook Meadows Nature Reserve 42W 84
Yeading Ct. UB4: Yead 42Y 85
Yeading Fork UB4: Yead 43Y 85
Yeading Gdns. UB4: Yead 43X 85
Yeading Ho. UB4: Yead 43Z 85
Yeading La. UB4: Yead 44X 85
Yeading La. UB5: N'olt 41Y 85
Yeading Wlk. HA2: Harr 29Ba 45
Yeadon Ho. W10 43Ya 88
(off Sutton Way)
Yeames Cl. W13 44Ja 86
Yearby Ho. W10 42Ya 88
(off Sutton Way)
Yeate St. N1 38Tb 71
Yeatman Rd. N6 30Hb 49
Yeats Cl. NW10 36Ua 68
Yeats Cl. RH1: Redh 9L 207
Yeats Cl. SE13 54Fc 115
Yeats Ct. W7 45Ha 86
Yeats Ho. AL3: St A 4P 5
Ye Cnr. WD19: Wat 15Z 27
Yeend Cl. KT8: W Mole 70Ca 129
Yeldham Ho. W6 50Za 88
(off Yeldham Rd.)
Yeldham Rd. W6 50Za 88
Yeldham Vs. W6 50Za 88
(off Yeldham Rd.)
Yellowcress Dr. GU24: Bisl 8E 166
Yellowhammer Ct. NW9 26Ua 48
(off Eagle Dr.)
Yellowpine Way IG7: Chig 21Xc 55
Yellow Stock M. RM14: Upm 36Wd 78
Yelverton Cl. RM3: Rom 25Md 57
Yelverton Lodge TW1: Twick 59La 108
(off Richmond Rd.)
Yelverton Rd. SW11 54Fb 111
Ye Mkt. CR2: S Croy 78Tb 157
(off Selsdon Rd.)
Yenston Cl. SM4: Mord 72Cb 155
Yeoman Cl. E6 45Rc 94
Yeoman Cl. SE27 62Rb 135
Yeoman Ct. E14 42Ec 92
(off Tweed Wlk.)
Yeoman Ct. KT20: Tad 91Ab 194
Yeoman Ct. TW5: Hest 52Ba 107
Yeoman Dr. TW19: Stanw 60N 105
Yeoman Rd. UB5: N'olt 38Aa 65
Yeomanry Cl. KT17: Eps 84Va 174
Yeomans Acre HA4: Ruis 30W 44
Yeomans Cft. KT23: Bookh 97Ca 191
Yeomans Keep WD3: Chor 13H 25
Yeoman's Mdws. TN13: S'oaks 98Jd 202
Yeomans M. TW7: Isle 58Fa 108
Yeoman's Row SW3 4E 226 (48Gb 89)
Yeoman St. SE8 49Ac 92
Yeomans Way EN3: Enf H 12Yb 34

Yeoman's Yd. E1 4K 225 (45Vb 91) *(off Chamber St.)*
Yeoman Way RH1: Redh 10B 208
Yeomen Way IG6: Ilf 23Sc 54
Yeo St. E3 43Dc 92
Yeoveney Cl. TW19: Staines 61F 126
Yeovil Cl. BR6: Orp 75Uc 160
Yeovil Ent. Cen. SL2: Slou 3D 80
Yeovil Ho. W10 42Ya 88 *(off Sutton Way)*
Yeovil Rd. SL1: Slou 3C 80
Yeovilton Pl. KT2: King T 64La 130
Yerbury Rd. N19 34Mb 70 (not continuous)
Yester Dr. BR7: Chst 66Nc 138
Yester Pk. BR7: Chst 66Pc 138
Yester Rd. BR7: Chst 66Nc 138
Yetev Lev Ct. E5 32Wb 71
Yevele Way RM11: Horn 31Nd 77
Yew Av. UB7: Yiew 45N 83
Yewbank Cl. CR8: Kenley 87Tb 177
Yew Cl. IG9: Buck H 19Mc 35
Yewdale Cl. BR1: Broml 65Gc 137
Yewdells Cl. RH3: Bkld 5C 206
Yewfield Rd. NW10 37Va 68
Yew Gro. NW2 35Za 68
Yew Ho. SE16 47Zb 92 *(off Woodland Cres.)*
Yew Ho. WD23: Bush 15Z 27 *(off Plantation Cl.)*
Yewlands Cl. SM7: Bans 87Eb 175
Yew Pl. KT13: Weyb 77V 150
Yews, The DA12: Grav'nd 1F 144
Yews, The KT14: Byfl 84N 169
Yews, The TW15: Ashf 63R 128
Yews Av. EN1: Enf 8Xb 19
Yewsley Ter. SL3: L'ly 50C 82
Yewstone Ct. WD18: Wat 13W 26
Yewtree Av. RM10: Dag 36Ed 76
Yew Tree Bottom Rd. KT17: Eps D 88Xa 174
Yew Tree Bottom Rd. KT18: Tatt C 89Ya 174
Yew Tree Cl. CM13: Hut 16De 41
Yew Tree Cl. CR5: Chips 91Hb 195
Yew Tree Cl. DA16: Well 53Wc 117
Yew Tree Cl. DA3: Lfield 68Fe 143
Yew Tree Cl. KT17: Eps D 88Xa 174
Yew Tree Cl. KT4: Wor Pk 74Ua 154
Yew Tree Cl. N21 17Qb 32
Yew Tree Cl. SE13 55Ec 114
Yew Tree Cl. TN13: Bes G 95Fd 202
Yewtree Cl. HA2: Harr 28Da 45
Yewtree Cl. N22 25Lb 50
Yew Tree Cotts. TN14: Hals 84Bd 181
Yew Tree Ct. HP1: Hem H 4J 3
Yew Tree Ct. NW11 29Bb 49 *(off Bridge La.)*
Yew Tree Ct. SM2: Sutt 80Eb 155 *(off Walnut M.)*
Yew Tree Ct. WD6: E'tree 16Ma 29
Yew Tree Dr. CR3: Cat'm 97Vb 197
Yew Tree Dr. HP3: Bov 10D 2
Yewtree End AL2: Park 9A 6
Yew Tree Gdns. KT18: Eps 87Sa 173
Yew Tree Gdns. RM6: Chad H 29Ad 55
Yew Tree Gdns. RM7: Rom 29Fd 56
Yew Tree La. RH2: Reig 3K 207
Yew Tree Lodge HA4: Ruis 30W 44
Yew Tree Lodge RM7: Rom 29Fd 56 *(off Yew Tree Gdns.)*
Yew Tree Lodge SW16 63Lb 134
Yew Tree M. TN16: Westrm 99Tc 200 *(off Market Sq.)*
Yew Tree Rd. BR3: Beck 69Bc 136
Yew Tree Rd. SL1: Slou 8L 81
Yew Tree Rd. TN14: Dun G 93Hd 202
Yew Tree Rd. UB10: Uxb 39P 63
Yew Tree Rd. W12 45Va 88
Yew Trees TW17: Shep 70P 127
Yew Trees TW20: Thorpe 69E 126
Yew Tree Wlk. CR8: Purl 82Sb 177
Yew Tree Wlk. KT24: Eff 99Z 191
Yew Tree Wlk. TW4: Houn 57Ba 107
Yew Tree Way CR0: Sels 82Ac 178
Yew Wlk. HA1: Harr 32Ga 66
YIEWSLEY 46N 83
Yiewsley Ct. UB7: Yiew 46N 83
YMCA RH1: Redh East Surrey 9A 208
YMCA E16: George Williams College 43Kc 93
YMCA W2: German 4A 220 (45Eb 89) *(off Craven Ter.)*
YMCA KT2: King T Hawker 64Ma 131
YMCA RH1: Redh Hillbrook Ho., E. Surrey 7P 207 *(off Brighton Rd.)*
YMCA WD5: Ab L Leavesden Country Pk. 3W 12
YMCA UB8: Uxb W. London, Uxbridge 39M 63
Yoakley Rd. N16 33Ub 71
Yoga Way KT4: Wor Pk 75Wa 154
Yoke Cl. N7 37Nb 70
Yolande Gdns. SE9 57Nc 116
Yonge Pk. N4 34Qb 70
YOPPS GREEN 98Ae 205
Yopps Grn. TN15: Plax 98Ae 205
York Av. DA15: Sidc 61Uc 138
York Av. HA7: Stan 25Ka 46
York Av. SL1: Slou 4G 80
York Av. SL4: Wind 4F 102
York Av. SW14 57Sa 109
York Av. UB3: Hayes 43S 84
York Av. W7 46Ga 86
York Bri. NW1 5H 215 (42Jb 90)
York Bldgs. WC2 5G 223 (45Nb 90)
York Cl. CM15: Shenf 17Be 41
York Cl. E6 44Pc 94
York Cl. KT14: Byfl 84N 169
York Cl. SE5 54Sb 113 *(off Lilford Rd.)*
York Cl. SM4: Mord 70Db 133
York Cl. TW18: Staines 65N 127
York Cl. W7 46Ga 86
York Cl. WD4: K Lan 1Q 12
York Ct. KT19: Eps 84Qa 173
York Ct. N14 20Nb 32
York Cres. IG10: Lough 13Nc 36
York Cres. WD6: Bore 12Ta 29
Yorke Gdns. RH2: Reig 5J 207
Yorke Ga. CR3: Cat'm 94Tb 197
Yorke Ga. WD17: Wat 12W 26
Yorke Rd. RH2: Reig 5H 207
Yorke Rd. WD3: Crox G 16Q 26
York Gdns. KT12: Walt T 75Z 151
York Ga. N14 17Nb 32
York Ga. NW1 6H 215 (42Jb 90)
York Gro. SE15 53Yb 114
York Hall Leisure Cen. 40Xb 71
York Hill IG10: Lough 13Nc 36
York Hill SE27 62Rb 135
York Ho. AL3: St A 3A 6
York Ho. E16 46Jc 93 *(off De Quincey M.)*
York Ho. EN2: Enf 11Tb 33
York Ho. HA9: Wemb 35Qa 67
York Ho. KT2: King T 66Pa 131 *(off Elm Rd.)*
York Ho. KT20: Lwr K 99Bb 195
York Ho. SE1 4J 229 (48Pb 90)
York Ho. SW3 7G 227 (50Hb 89) *(off Turks Row)*
York Ho. W1 7F 215 (43Hb 89) *(off York St.)*
York Ho. W8 47Db 89 *(off York Ho. Pl.)*
York Ho. WD23: Bush 14Ca 27
York Ho. WD6: Bore 12Qa 29 *(off Canterbury Rd.)*
York Ho. Pl. W8 47Db 89
Yorkland Av. DA16: Well 55Vc 117
York La. CR4: Mitc 68Lb 134
Yorkley Ho. W10 42Ya 88 *(off Sutton Way)*
York Mans. SE17 7E 230 (50Sb 91) *(off Browning St.)*
York Mans. SW11 53Jb 112 *(off Prince of Wales Dr.)*
York Mans. SW5 50Db 89 *(off Earls Ct. Rd.)*
York Mans. W1 7H 215 (43Jb 90) *(off Chiltern St.)*
York M. IG1: Ilf 34Qc 74
York M. NW5 36Kb 70
York Pde. NW9 30Wa 48 *(off W. Hendon B'way.)*
York Pde. TW8: Bford 50Ma 87
York Pl. IG1: Ilf 33Qc 74
York Pl. RM10: Dag 37Ed 76
York Pl. SW11 55Fb 111
York Pl. WC2 5G 223 (45Nb 90) *(off Villiers St.)*
York Pl. Mans. W1 7G 215 (43Hb 89) *(off Baker St.)*
York Ri. BR6: Orp 75Uc 160
York Ri. NW5 34Kb 70
York Rd. AL1: St A 1D 6
York Rd. CM15: Shenf 17Be 41
York Rd. CR0: C'don 73Qb 156
York Rd. CR2: Sels 82Zb 178
York Rd. DA1: Dart 59Pd 119
York Rd. DA11: Nflt 59Fe 121
York Rd. DA12: Grav'nd 2E 144
York Rd. E10 34Ec 72
York Rd. E17 29Zb 52
York Rd. E4 21Cc 52
York Rd. E7 37Jc 73
York Rd. EN5: New Bar 15Eb 31
York Rd. EN8: Walt C 5Ac 20
York Rd. GU22: Wok 1P 187
York Rd. HA6: Nwood 26W 44
York Rd. IG1: Ilf 34Qc 74
York Rd. KT13: Weyb 78S 150
York Rd. KT14: Byfl 84N 169
York Rd. KT2: King T 66Pa 131
York Rd. N11 23Mb 50
York Rd. N18 23Xb 51
York Rd. N21 17Tb 33
York Rd. RM13: Rain 38Fd 76
York Rd. SE1 1J 229 (47Pb 90)
York Rd. SL4: Wind 4F 102
York Rd. SM1: Sutt 79Cb 155
York Rd. SM2: Sutt 79Cb 155
York Rd. SW11 56Eb 111
York Rd. SW18 56Eb 111
York Rd. SW19 65Eb 133
York Rd. TN16: Big H 91Kc 199
York Rd. TW10: Rich 57Pa 109
York Rd. TW11: Tedd 63Ga 130
York Rd. TW3: Houn 55Da 107
York Rd. TW8: Bford 50Ma 87
York Rd. UB8: Uxb 38M 63
York Rd. W3 44Sa 87
York Rd. W5 48La 86
York Rd. WD18: Wat 15Y 27
York Rd. Bus. Cen. SW11 54Fb 111
Yorkshire Cl. N16 34Ub 71
Yorkshire Gdns. N18 22Xb 51
YORKSHIRE GREY 57Mc 115
Yorkshire Grey Pl. NW3 35Eb 69
Yorkshire Grey Yd. WC1 1H 223 (43Pb 90) *(off Eagle St.)*
Yorkshire Rd. CR4: Mitc 71Nb 156
Yorkshire Rd. E14 44Ac 92
York Sq. E14 44Ac 92
York St. CR4: Mitc 73Jb 156
York St. TW1: Twick 60Ja 108
York St. W1 1F 221 (43Hb 89)
York St. Chambers W1 7F 215 (43Hb 89) *(off York St.)*
York Ter. DA8: Erith 53Ed 118
York Ter. EN2: Enf 10Sb 19
York Ter. SE7 49Lc 93 *(off Gallon Cl.)*
York Ter. E. NW1 6J 215 (42Jb 90)
York Ter. W. NW1 6H 215 (42Jb 90)
Yorkton St. E2 40Wb 71
York Way HP2: Hem H 3N 3
York Way KT9: Chess 80Na 153
York Way N1 2G 217 (40Nb 70)
York Way N20 20Hb 31
York Way N7 37Mb 70
York Way TW13: Hanw 62Ba 129 (not continuous)
York Way WD25: Wat 8Z 13
York Way WD6: Bore 12Ta 29
York Way Ct. N1 1G 217 (39Nb 70) (not continuous)
York Way Est. N7 37Nb 70
TW6: H'row A
Young Ct. NW6 38Ab 68
Youngfield Rd. HP1: Hem H 1H 3
Youngmans Cl. EN2: Enf 11Sb 33
Young Rd. E16 44Lc 93
Youngs Bldgs. EC1 5E 218 (42Sb 91) *(off Old St.)*
Youngs Ct. SW11 53Jb 112
Youngs Rd. IG2: Ilf 29Tc 54
Young St. KT22: Fet 97Ga 192
Young St. KT22: Lea 97Ga 192
Young St. W8 47Db 89
Youngstroat La. GU24: Chob 82A 168
Young Vic Theatre, The 1A 230 (47Qb 90) (off The Cut)
Yoxall Ho. W10 42Ya 88 *(off Sutton Way)*
Yoxall M. RH1: Redh 4B 208
Yoxford Ct. RM6: Chad H 29Xc 55
Yoxley App. IG2: Ilf 30Sc 54
Yoxley Dr. IG2: Ilf 30Sc 54
Ypres Pl. RM9: Dag 39Xc 75
Yukon Rd. SW12 59Kb 112
Yule Cl. AL2: Brick W 2Ba 13
Yunus Khan Cl. E17 29Cc 52
Yvon Ho. SW11 53Jb 112 *(off Alexandra Av.)*

Z

Zachary Cl. CR5: Coul 89Lb 176
Zachary Ct. SW1 6E 228 (49Mb 90) *(off Montaigne Cl.)*
Zachary Ho. SW9 53Pb 112 *(off Lett Rd.)*
Zachary M. E3 41Bc 92 *(off Tredegar Sq.)*
Zahra Ho. NW10 41Ya 88 *(off Harrow Rd.)*
Zambezie Dr. N9 20Yb 34
Zambra Way TN15: Seal 92Pd 203
Zampa Rd. SE16 50Yb 92
Zanara Ct. SE26 64Yb 136
Zander Ct. E2 41Wb 91
Zangwill Ho. NW6 40Bb 69 *(off Carlton Va.)*
Zangwill Rd. SE3 53Mc 115
Zapotec Ho. SE8 49Zb 92 *(off Chilton Gro.)*
Zealand Av. UB7: Harm 52M 105
Zealand Ho. SE18 49Nc 94
Zealand Ho. SE5 54Sb 113 *(off Denmark Rd.)*
Zealand Rd. E3 40Ac 72
Zelah Rd. BR5: Orp 73Yc 161
Zeller Ho. E20 36Dc 72 *(off Scarlet Cl.)*
Zenith Cl. NW9 27Ta 47
Zenith Lodge N3 24Db 49
Zennor Rd. SW12 60Lb 112
Zennor Rd. Ind. Est. SW12 60Lb 112
Zenobia Mans. W14 51Bb 111 *(off Queen's Club Gdns.)*
Zenoria St. SE22 56Vb 113
Zermatt Rd. CR7: Thor H 70Sb 135
Zetec Apts. RM1: Rom 28Hd 56 *(off Mercury Gdns.)*
Zetland Ho. W8 48Db 89 *(off Marloes Rd.)*
Zetland St. E14 43Dc 92
Zeus Ct. UB7: Yiew 46N 83
Zig Zag Rd. CR8: Kenley 88Sb 177
Zinc Bldg., The SE13 56Gc 115
Zinnia Dr. GU24: Bisl 8E 166
Zinnia Mans. E20 37Dc 72 *(off Olympic Pk. Av.)*
Zion Ho. E1 44Yb 92 *(off Jubilee St.)*
Zion Pl. CR7: Thor H 70Tb 135
Zion Pl. DA12: Grav'nd 9D 122
Zion Rd. CR7: Thor H 70Tb 135
Zion St. TN15: Seal 93Pd 203
Zoar St. SE1 6D 224 (46Sb 91)
Zodiac Bus. Pk. UB8: Cowl 44M 83
Zodiac Cl. HA8: Edg 24Qa 47
Zodiac Ct. CR0: C'don 74Rb 157
Zodiac Ho. E3 42Cc 92 *(off Wellington Way)*
Zoffany St. N19 33Mb 70
Zona Ct. SE1 4K 231 (48Vb 91) *(off Grange Wlk.)*
Zulu M. SW11 54Gb 111

Published by Geographers' A-Z Map Company Limited
An imprint of HarperCollins Publishers
Westerhill Road
Bishopbriggs
Glasgow
G64 2QT

HarperCollinsPublishers
1st Floor, Watermarque Building, Ringsend Road, Dublin 4, Ireland

www.az.co.uk
a-z.maps@harpercollins.co.uk

18th edition 2022

A catalogue record for this book is available from the British Library.

ISBN 978-0-00-851368-9

10 9 8 7 6 5 4 3 2 1

Printed in Bosnia and Herzegovina

M1
A1(M)
B157
A416
Bovingdon
Kings Langley
21/6a
1a
Chiswell Green
Frogmore
London Colney
22
Brookmans Park
South Mimms
Potters Bar
B4505
B485
Chesham
Chipperfield
Abbots Langley
20
6
Bricket Wood
Colney Street
Shenleybury
SOUTH MIMMS
23/1
24
Chesham Bois
Latimer
Sarratt
19
Chenies
Radlett
Shenley
Aldenham
Patchetts Green
Green Street
Bentley Heath
Hadley Wood
Botany Bay
Hadley
A404
Chandler's Cross
Mickleffeld Green
WATFORD
5
Borehamwood
BARNET
Arkley
Cockfosters
Amersham
A413
Chorleywood
18
Croxley Grn.
Bushey
The Swillett
Oxhey
Bushey Heath
LONDON GATEWAY
Totteridge
East Barnet
Whetstone
Southgate
Penn
A355
Chalfont St Giles
17
Rickmansworth
Maple Cross
Eastbury
Northwood
South Oxhey
Stanmore
4
Harrow Weald
Edgware
Mill Hill
Friern Barnet
Seer Green
Chalfont Common
West Hyde
Harefield
Hatch End
2
Beaconsfield
Chalfont St Peter
M25
BEACONSFIELD
Wooburn Green
Wooburn
Pinner
Harrow
Queensbury
Kingsbury
Hendon
Muswell
Golders Green
Denham Green
Ruislip
Eastcote
Hedgerley
Gerrards Cross
Denham
Rayners Lane
Harrow on the Hill
Ickenham
1
Cricklewood
Highgate
Hampstead
Neasden
M40
16/1a
Northolt
Wembley
Willesden
Kilburn
Camden Town
Farnham Common
Farnham Royal
Stoke Poges
Iver Heath
UXBRIDGE
North Hillingdon
Hillingdon
Greenford
Perivale
Maida Vale
Marylebone
Taplow
Burnham
Cippenham
SLOUGH
Cowley
George Green
Iver
Yiewsley
Yeading
Ealing
Paddington
Notting Hill
Hayes
Acton
Kensington
Mayfair
HEAD
7
M4
6
Eton
Wick
Upton
Langley
Richings Park
West Drayton
15/4b
4
Hayes Town
Southall
Hanwell
Shepherd's Bush
Westmin
3
HESTON
Gunnersbury
Hammersmith
Chelsea
A308
5
Eton
Windsor
Datchet
Colnbrook
Horton
Harmondsworth
Sipson
4a
Harlington
Osterley
Heston
Brentford
Kew
Chiswick
Barnes
Fulham
Battersea
Spital
Poyle
Cranford
LONDON HEATHROW
14
Hatton
Hounslow
Isleworth
RICHMOND
Putney
Wandsworth
Clapham
Old Windsor
Wraysbury
Stanwell Moor
Stanwell
East Bedfont
Feltham
Twickenham
Roehampton
Winkfield
Woodside
Englefield Green
Hythe End
Egham
13
STAINES-UPON-THAMES
Ashford
Felthamhill
Hanworth
Ham
Teddington
Wimbledon
Upper Tooting
Cheapside
1
Thorpe
Sunbury-on-Thames
Hampton
Hampton Wick
KINGSTON UPON THAMES
New Malden
Colliers Wood
Mitcham
Ascot
Virginia Water
Laleham
Littleton
Upper Halliford
West Molesey
East Molesey
Thames Ditton
Surbiton
Old Malden
Morden
Sunningdale
12/2
Chertsey
Shepperton
Walton-on-Thames
Long Ditton
Tolworth
Worcester Park
Lyne
Windlesham
Longcross
Ottershaw
11
Addlestone
Weybridge
Esher
Hersham
Claygate
Hook
Stoneleigh
Cheam
Carshalton
Sutton
Wallington
Chessington
West Ewell
Ewell
Bagshot
M3
3
Burrowhill
Chobham
Row Town
New Haw
Whiteley Village
Belmont
Lightwater
West End
Woodham
Byfleet
Fairmile
Oxshott
Malden Rushett
Epsom
B311
Bisley
Horsell
West Byfleet
Pyrford
Pyrford Green
Wisley
10
Cobham
Stoke D'Abernon
9
Ashtead
Nork
Banstead
Woodmansterne
Chipstead
Knaphill
COBHAM
9
Brookwood
WOKING
Old Woking
Ripley
Ockham
Fetcham
Leatherhead
Burgh Heath
Tadworth
Kingswood
Hooley
Pirbright
Mayford
Send
Little Bookham
Great Bookham
7
Effingham
Headley
Walton on the Hill
Lower Kingswood
West Clandon
Mickleham
Pebble Coombe
8
Merstham
East Clandon
West Horsley
East Horsley
A24
Redhill
A246
Dorking
A25
Buckland
A31
Westcott
Brockham
Betchworth
Reigate
South Park
South Nutfield
SCALE
0 1 2 3 4 Miles
0 1 2 3 4 5 6 Kilometers

LOW EMISSION ZONE
ULTRA LOW EMISSION ZONE
CONGESTION CHARGING ZONE
Visit www.tfl.gov.uk/modes/driving for more information on London's driving zones.
The Low Emission Zone (LEZ) is a specified area of Greater London that operates 24 hours a day, every day of the year. A daily charge applies to a range of the most polluting diesel-engined vehicles that fail to meet specific emissions standards (e.g. lorries, coaches, buses, minibuses, service vehicles and specialist vehicles, including motorhomes and motorised horseboxes).
The Ultra Low Emissions Zone (ULEZ) is a specified area within the North Circular (A406) and the South Circular (A205), which also operates 24 hours a day, every day of the year. A daily charge applies to a range of vehicles which fail to meet specific emissions standards.
A small number of vehicles are exempt.
CHESHUNT
Waltham Abbey
Epping
North Weald Bassett
Chipping Ongar
Blackmore
Kelvedon Hatch
Ingatestone
Doddinghurst
Theydon Bois
Loughton
Enfield
Billericay
Brentwood
Chigwell
Walthamstow
ROMFORD
ILFORD
Barking
South Ockendon
Stanford-le-Hope
Woolwich
GRAYS
Tilbury
DARTFORD
GRAVESEND
Northfleet
Swanscombe
Lewisham
BROMLEY
Orpington
Swanley
CROYDON
New Addington
Biggin Hill
Warlingham
Caterham
Oxted
Godstone
Limpsfield
Westerham
Shoreham
London City
London Biggin Hill
M11
M25
M20
M23
A414
A12
A127
A13
A129
A128
A113
B1007
B148
B194
B181
B269
B2026
CLACKET LANE
THURROCK